THE MERCK INDEX

TWELFTH EDITION

OTHER
MERCK PROFESSIONAL HANDBOOKS

THE MERCK MANUAL OF DIAGNOSIS AND THERAPY
First Edition, 1899

THE MERCK VETERINARY MANUAL
First Edition, 1955

THE MERCK MANUAL OF GERIATRICS
First Edition, 1990

Merck Professional Handbooks
are published on a nonprofit basis
as a service to the scientific community.

THE
MERCK INDEX

AN ENCYCLOPEDIA OF

CHEMICALS, DRUGS, AND BIOLOGICALS

TWELFTH EDITION

Susan Budavari, *Editor*

Maryadele J. O'Neil, *Senior Associate Editor*

Ann Smith, *Associate Editor*

Patricia E. Heckelman, *Assistant Editor*

Joanne F. Kinneary, *Assistant Editor*

Published by
Merck Research Laboratories
Division of

MERCK & CO., INC.

Whitehouse Station, NJ

1996

MERCK & CO., INC.
Whitehouse Station, N.J.
U.S.A.

1st Edition—1889
2nd Edition—1896
3rd Edition—1907
4th Edition—1930
5th Edition—1940
6th Edition—1952
7th Edition—1960
8th Edition—1968
9th Edition—1976
10th Edition—1983
11th Edition—1989

Library of Congress Catalog
Card Number 89-60001
ISBN Number 0911910-12-3

First Printing—March 1996
Second Printing—April 1997

EDITOR'S PREFACE

With the Twelfth Edition, The Merck Index marks the beginning of its second century of publication. A comparison of this one-volume, 2600-page, multidisciplinary encyclopedia with the thin catalog of Merck products that was published in 1889, underscores the vast changes that have occurred in this publication over twelve editions, and in the realm of scientific information. Although the text continues to be designed primarily to serve the needs of chemists, biochemists, pharmacists, and those in allied professions, it contains information of interest to a broad spectrum of scientists and others in a large number of disciplines. The material is assembled with today's quick reference needs in mind, but the commitment to enhanced scientific communication and to excellence in scientific scholarship remains paramount.

This edition contains a diverse collection of over 10,000 monographs of which more than 4,000 are devoted to a wide variety of drugs and pharmaceuticals, over 2,000 describe common organic chemicals and laboratory reagents, and another 2,000 cover naturally occurring substances and plants. An additional 1,000 monographs focus on the elements and on inorganic chemicals, almost 1,000 pertain to compounds of agricultural significance, and several hundred describe endogenous substances and biological agents.

A number of changes have been made since the Eleventh Edition was published in 1989. The monograph section has been extensively revised. The chemical structures have been redrawn employing current conventions for chemical depictions. Nomenclature has been reviewed and stereochemical descriptors have been added, where pertinent. In response to requests from our readership, the section on Organic Name Reactions, which last appeared in 1983 in the Tenth Edition, has been updated and reintroduced. The compilation of Chemical Abstracts Service Registry Numbers has been significantly expanded. Several new tables have been added including a glossary emphasizing some of the newer terminology employed in the fields of molecular biology and immunology.

In recognition of the growing utilization of electronic versions of traditional reference works, THE MERCK INDEX ONLINE is made available through major online database vendors. A CD-ROM version of the Twelfth Edition, which is both text and structure searchable, will also be published in 1996.

Compared with their counterparts of earlier eras, scientists of the 1990s have unparalleled access to information from a multitude of sources. In this environment, it is especially gratifying that so many readers inform us that they refer to The Merck Index on a daily basis. We hope that this edition, published by Merck & Co., Inc., as a service to the scientific community, will continue the tradition and prove to be an indispensable reference to all who consult it.

Susan Budavari, *Editor*
Merck Research Laboratories
Rahway, New Jersey 07065

ACKNOWLEDGMENTS

The creation process for each edition of The Merck Index involves the efforts of many dedicated people with special talents and varied expertise. The successful publication is a tribute to all involved.

The Merck Index Editorial staff wishes to especially thank our coworkers for their untiring efforts and cooperation: *Jo Ann Gallipeau* for assistance in researching and preparing various portions of the manuscript; *Mary Ann D'Arecca* for organizing the data entry of the tabular material and for procuring research materials; *Edward Hendrzak* for preparing tabular material and skillfully drawing the chemical structures; and *Mariann E. Lucas* for diligent editorial assistance in evaluating various topics and standardizing the structures. We also wish to thank *Dr. Norman Penix* for providing computer expertise, custom programming and technical guidance at critical times; *Debra Shelinsky Greene, Susan C. Mattson* and *Jon Filderman,* for advice and counsel on intellectual property and contractual issues; and especially, *Gary Zelko*, our Publisher, for facilitating all business aspects of this edition and for providing needed support.

We would like to express our gratitude to *Professor David A. Evans* of Harvard University and *Professor Barry M. Trost* of Stanford University who reviewed various portions of the manuscript for the Organic Name Reactions section and provided salient comments and valuable suggestions.

Finally, we wish to mention and thank the numerous Merck scientists, past and present, who contributed in varying degrees to this book, as well as the many readers who took the time to contact us with comments, suggestions, and corrections.

TABLE OF CONTENTS

The Twelfth edition of The Merck Index contains 10,330 monographs describing significant chemicals, drugs, and biological substances. The entries are not a listing of Merck & Co., Inc. products, but rather cover a wide range of compounds which have been selected on the basis of present, or historic, importance and interest. Since the publication of the Eleventh Edition in 1989, over 5000 monographs have been significantly revised and updated, while several hundred monographs have either been condensed, or deleted from the manuscript to make room for hundreds of new monographs.

Entries are generally limited to single substances and related compounds (isomers, salts, etc.). While multi-component drugs are, for the most part, excluded, there are a number of monographs devoted to families of natural products or biological substances. Monographs vary greatly in length. The length of a monograph, however, is not necessarily indicative of the importance of a compound, but rather may simply be an indication of the amount of relevant published information available for the compound.

For the purpose of illustrating the general monograph format, a typical drug monograph is depicted and the components are identified. While all the possible categories of information in a monograph are described below, it must be emphasized that not all categories are present in every monograph.

Monograph Number. Sequential accession numbers are assigned to monographs which are alphabetized by title. Entries in the indices are referenced to these numbers, not to page numbers. (*Note:* Monograph numbers in the Twelfth Edition do not correspond to Eleventh Edition monograph numbers.)

Monograph Title. Titles are usually generic (United States Adopted Names (USAN) or International Nonproprietary Names (INN), in the case of drugs), trivial, or simple chemical names. Trademarks, designated by ®, are used for a small number of entry titles, primarily when nonproprietary terms are not available.

Chemical Abstracts Name(s). Synonyms in ***boldface italic*** following the title are uninverted forms of Chemical Abstracts Service (CAS) Index names. When a specific isomer is being discussed, stereochemical descriptors have been included, when possible, to more clearly delineate that isomer. The Chemical Abstracts Names (and CAS Registry Numbers starting on page REG-1) are provided as aids for further searching of the compound of interest in *Chemical Abstracts* and elsewhere.

Alternate Name(s). Other chemical names which identify the entry, trivial names, experimental drug codes, and current (and former) trademarks are listed. Listing of trademarks is for information purposes only and it should not be assumed that the trademarks are in current use. The first letter of each trademark is capitalized; absence of capitalization, however, does not preclude that a name may either currently be a proprietary name, or may once have been the subject of proprietary rights. In those cases where it has come to our attention that a trademark is obsolete, we have indicated that fact. If known, the company associated with a particular trademark (as a manufacturer, distributor or trademark owner), is listed alongside the trademark in the Name Index. Names

Monograph number

4979. Indinavir. *[1(1S,2R),5(S)]-2,3,5-Trideoxy-N-(2,3-dihydro-2-hydroxy-1H-inden-1-yl)-5-[2-[[(1,1-dimethylethyl)-amino]carbonyl]-4-(3-pyridinylmethyl)-1-piperazinyl]-2-(phenylmethyl)-D-erythro-pentonamide;* (αR,γS,2S)-α-benzyl-2-(*tert*-butylcarbamoyl)-γ-hydroxy-*N*-[(1*S*,2*R*)-2-hydroxy-1-indanyl]-4-(3-pyridylmethyl)-1-piperazinevaleramide; *N*-(2(*R*)-hydroxy-1(*S*)-indanyl)-2(*R*)-(phenylmethyl)-4(*S*)-hydroxy-5-[1-[4-(3-pyridylmethyl)-2(*S*)-(*N*-*tert*-butylcarbamoyl) piperazinyl]] pentanamide. $C_{36}H_{47}N_5O_4$; mol wt 613.80. C 70.45%, H 7.72%, N 11.41%, O 10.43%. Member of the novel hydroxyaminopentane amide class of HIV-1 protease inhibitors. Prepn: J. P. Vacca *et al., Eur. pat. Appl.* **541,168;** *eidem,* **U.S.** pat. **5,413,999** (1993, 1995 both to Merck & Co.); B. D. Dorsey *et al., J. Med. Chem.* **37,** 3443 (1994). Diastereoselective synthesis: D. Askin *et al., Tetrahedron Letters* **35,** 673 (1994); P. E. Maligres *et al., ibid.* **36,** 2195 (1995). Crystal structure of HIV-II protease complex: Z. Chen *et al., J. Biol. Chem.* **269,** 26344 (1994). Antiviral activity and pharmacokinetics: J. P. Vacca *et al., Proc. Nat. Acad. Sci. USA* **91,** 4096 (1994). HPLC determn in plasma and urine: E. Woolf *et al., J. Chromatog. A.* **692,** 45 (1995). Metabolism in humans: S. K. Balani *et al., Drug Metab. Dispos.* **23,** 266 (1995).

Percentage composition

Molecular weight

Literature references

Alternate chemical names (lightface)

Molecular formula

Patent and chemical information

Biological, pharmacological information, etc.

Structure

Physical data for title compound

Occurs as monohydrate, crystals from wet ethyl or iso-propyl acetate, loss of water below 100° followed by recrystallization of anhydrous form with mp 153-154°; second anhydrous form, mp 167.5-168°. Soly in water (mg/ml): 0.015 (unbuffered), > 1.5 (pH 4.0). $[\alpha]_D^{22}$ + 24.1° (c = 0.0133 in chloroform).

Drug code number

Derivative of title compound

Sulfate, $C_{36}H_{47}N_5O_4.H_2SO_4$, ***MK-639, Crixivan.*** Occurs as the monoethanolate, crystals from absolute ethanol, softens at 135°, mp 150-153° (dec). Converts to hydrate upon loss of ethanol and exposure to moist air.

Therapeutic category (in humans)

THERAP CAT: Antiviral.

Physical data for derivative

Trademarks (capitalized) and/or synonyms of derivative (***boldface italic***) appearing elsewhere in the monograph in boldface italic, also appear in the Name Index.

Molecular Formula, Molecular Weight, % Composition. Elements in the molecular formula are listed according to the Hill convention (C, H, then other elements in alphabetical order). Formula and molecular weight are provided for all title compounds having a specific known structure. The 1993 IUPAC Standard Atomic Weights have been used to calculate all molecular weights.

Literature References. A concise reference history of each compound is provided. Frequently, there is a brief description or capsule statement, although in some monographs, particularly those on significant biological substances, a lengthier description is given. References to isolation, preparation or synthesis, patent information, and structural studies are cited. While reference is made to various methods of synthesis, the intent is to give a representative, but by no means, exhaustive list. Patent numbers are provided merely as a source of preparative information. However, whenever possible, the United States product patent has been cited in monographs which were newly added for the Eleventh and Twelfth Editions. References to pharmacology or biological activity, clinical trials, and toxicity studies may be included, where appropriate. Review articles, where available, are usually cited at the end of the references. Reviews pertaining to a group of closely related compounds or to a family of natural substances, are generally listed only in the monograph for the parent element or compound. Literature references are cited in the conventional manner; journal abbreviations (with notable exceptions listed in the Table of Abbreviations) generally correspond to those in Chemical Abstracts Service Source Index (CASSI). The number of the first page of the reference is given; first and last page numbers are listed for reviews.

Structure. For this edition, all structures have been drawn according to current conventions using CambridgeSoft's ChemDraw™ software package. Structural depictions, including stereochemistry, if relevant, are included in over 6500 monographs. Structures which do not correspond precisely to the monograph title, have been labeled to identify the specific form depicted. Standard conventions of heavy wedges and dotted lines to show bonds directed above, or below the plane of the paper are used where appropriate; whenever possible, double bond geometry has been defined within the structure. Amino acid residues are assumed to be L unless specified otherwise. In addition, more than 2000 monographs contain line formulae showing molecular arrangements.

Physical Data. Data are cited as found in the literature. When several alternate data values appear in the literature, the data is evaluated and representative selections are made. The values are then reported with the corresponding source. Whenever possible, the color of a substance is stated, but the absence of color (white or colorless) is often omitted. Temperatures are given in degrees Celsius (centigrade), unless otherwise noted. When solubilities are determined at room temperature (about 25°C), the temperature is generally omitted. When optical rotations are measured in water, the solvent is usually not specified. For ultraviolet absorption measurements, the solvent is given within parentheses.

An effort has been made to provide toxicity data (e.g. LD_{50}, LC_{50}, etc.) and to identify the source of this information. ***Caution and/or Note*** statements are also provided in a number of monographs. Specific statements are given for compounds on the U.S. Government's Schedules of Controlled Substances in Title 21 of the 1995 Code of Federal Regulations, for compounds listed as suspected or confirmed carcinogens in the *Seventh Annual Report on Carcinogens* issued in 1994 by the U.S. Department of Health and Human Services (USDHHS), and for chemicals considered potential occupational hazards as described in sources such as the *NIOSH Pocket Guide to Chemical Hazards (USDHHS)*. *Note:* Absence of toxicity data or specific cautions does not imply that toxic effects do not exist.

Derivatives. When derivatives (isomers, salts, etc.) of the title compound are described in a monograph, the information appears in the paragraph(s) directly following the physical data. Data for derivatives are presented in the same format as for the parent compound. Molecular formulae for parent substances appear in the Formula Index. Molecular formulae for derivatives other than acid addition salts or simple hydrates, are listed in the Formula Index if there are names associated with them.

Derivative paragraphs may also be used to describe specific members of a large family of natural substances. Literature references pertaining to these specific substances may be included here.

Use. Descriptions of specific uses, which are not medical or veterinary therapeutic applications, are summarized under this heading.

Therapeutic Category and Therapeutic Category (Veterinary). In most cases, therapeutic categories correspond to those published in the *USP Dictionary of USAN and International Drug Names*. However, in instances where there is no listing, or where the USAN Council has listed a mechanism of action, a therapeutic category has been assigned which most closely describes the indication claimed by the manufacturer, or reported in the clinical literature. When available, mode of action information is included in the literature references section of the monograph. Monographs for human drugs have been indexed by both therapeutic category and biological activity beginning on page THER-1.

Indices. More than 60,000 synonyms, including titles, CAS names, alternate names, trademarks and derivative names, are contained in the Name Index, and more than 11,500 entries appear in the Formula Index. Each entry directs the reader to the number of the monograph in which the substance of interest is described. The effort to match trademarks with associated companies has been continued for this edition. In the Name Index an abbreviated form of the company name appears in brackets following the trademark. The complete company name and location is listed in an updated and expanded Company Register in the TABLES section of the book. Company names are provided as a source of additional information and do not necessarily imply trademark ownership. Due to reorganizations or mergers, some company names may have changed since the original matching process was completed.

In the Therapeutic Category and Biological Activity Index, monographs describing human drugs have been listed by one or more therapeutic indications and/or mechanisms of action. Cross references to closely related categories and mechanisms have been included. Whenever appropriate, subclassifications have been developed by grouping compounds according to structural similarities.

Two CAS Registry Number indices are also provided: one arranged alphabetically by substance name, the second by ascending Registry Number order. Registry Numbers are provided for title substances and for selected derivatives. Where appropriate, Registry Numbers are provided for isomeric and unspecified forms.

The Merck Index is not intended as an official therapeutic guide. Inclusion of a drug, or any other compound in this book, is not an endorsement, but merely a statement of the fact that such a substance exists. THERAPEUTIC CATEGORY and THERAPEUTIC CATEGORY (VETERINARY) paragraphs are intended only as summary statements of major pharmacological properties or indications for the individual compounds. For additional information on uses, dosage, side effects and adverse reactions, readers are directed to consult pertinent scientific and professional publications, product circulars, information sheets or material safety data sheets prepared, or published, by the respective manufacturers.

Organic Name Reactions. This section, which has been significantly updated since it was last published in 1983, is comprised of 425 named reactions, each containing a concise reference history and reaction scheme, and an index.

Tables. A compilation of over 100 pages of tables including a glossary, is provided to supplement the material presented in the monographs.

Great care has been taken to assure the accuracy of the information contained in The Merck Index. However, the Editorial Staff and the Publisher cannot be responsible for errors incurred in publication or for any consequences arising from use of the information published in The Merck Index. Accordingly, reference to original sources is strongly encouraged, as is reporting of errors and omissions in order to assure that appropriate changes may be made in the next edition.

Abbreviations and Selected Definitions

Please consult the Glossary (Misc. Tables Section) for additional definitions.

A	adenine; adenosine; absorbance (extinction)
Å	Angstrom units
AAT	α_1-antitrypsin
Ab	antibody
abs	absolute; absorption
abs config	absolute configuration
abstr	abstract
Abu	α-aminobutyric acid
Ac	acetyl CH_3CO—
acc	according
ACE	angiotensin converting enzyme
Acetyl CoA	acetyl coenzyme A
Ac_2O	acetic anhydride
AcOEt	ethyl acetate
AcOH	acetic acid
ACP	acyl carrier protein
A.C.S.	American Chemical Society
ACTH	adrenocorticotropic hormone
AcylSCoA	acyl coenzyme A
ADCC	antibody-dependent cellular cytotoxicity
add(n)	adding, addition
Ade	adenine
ADH	alcohol dehydrogenase
ADP	adenosine diphosphate
AEC	(United States) Atomic Energy Commission
Ag	antigen, silver
AI	anaphylotoxin inhibitor
AIDS	acquired immunodeficiency syndrome
Ala	alanine
alc, alcoh	alcohol(ic); ethanol; ethyl alcohol
ALG	anti-lymphocyte globulin
alk	alkali(ne)
$[\alpha]_D^{25}$	specific optical rotation at 25° C for D (sodium) line; absence of brackets indicates optical rotation of a liquid in a 1 decimeter cell, neat
ALS	anti-lymphocyte serum, amyotrophic lateral sclerosis
a_M	molar absorptivity (concn in g-moles/l)
amorph	amorphous
AMP	adenosine 5'-monophosphate (adenylic acid)
cAMP	cyclic AMP
amps	ampules
amt(s)	amount(s)
ANF	anti-nuclear factor; atrial natriuretic factor
anhydr	anhydrous
anti	stereodescriptor, *see Glossary*
Ann.	*Justus Liebig's Annalen der Chemie*
APhA	American Pharmaceutical Association
approx	approximate(ly)
APT	alum precipitated toxoid
aq	aqueous
Ar	aryl
A.R.	analytical reagent
ARC	AIDS related complex
Archiv Exp. Pathol. Pharmakol.	*Naunyn Schmiedebergs Archiv für Experimentelle Pathologie und Pharmakologie*
ArCO-	aromatic acyl radical
Arg	arginine
Asa	β-carboxyaspartic acid
Asn	asparagine
Asp	aspartic acid
assoc(d)	association; associated

A.S.T.M.	American Society for Testing Materials
as-, asym-	asymmetrical, unsymmetrical
at.	atomic
ATCC	American Type Culture Collection
atm, atmos	atmosphere(s), atmospheric
at. no.	atomic number
ATP	adenosine triphosphate
ATPase	adenosine triphosphatase
at. wt.	atomic weight
BAN	British Approved Name
B cell	bone marrow or bursa of Fabricius derived cell
BCG	Bacillus Calmette-Guerin
Bé	Baumé (a specific gravity scale)
Beilstein	*Beilstein's Handbuch der Organischen Chemie*
Belg. pat.	Belgian patent
Ber.	*Chemische Berichte* (Berichte der Deutschen Chemischen Gesellschaft)
BGG	bovine gamma globulin
biol	biological
B.I.O.S.	British Intelligence Objectives Subcommittee
BOC	*tert*-butoxycarbonyl
B.O.D.	biochemical oxygen demand
boil.	boiling
bp	basepair; boiling point; boils at; boiling at Example: bp_{70} 48° means *boils at 48°C if the pressure is 70 mm Hg*
B.P.	*British Pharmacopeia*
B.P.C.	*British Pharmaceutical Codex*
Brit. pat.	British patent
BRM	biological response modifier
BSA	bovine serum albumin
Btu	British thermal units
Bu	butyl
Bz	benzoyl C_6H_5CO—
BzH	benzaldehyde
BzOH	benzoic acid
c	concentration by volume (g/100ml) after optical rotations only
C	cytosine; cytidine; complement
°C	Centigrade degrees; Celsius degrees
C_p	heat capacity (constant pressure)
ca.	(*circa*) about
C.A.	*Chemical Abstracts*
cal	calorie(s)
calc(d)	calculate, calculated
Can. pat	Canadian patent
Cbz	carbobenzoxy
cbc	complete blood count
cc	cubic centimeter(s) (milliliter)
CCK	cholecystokinin
CD	circular dichroism
CDP	cytidine diphosphate
CEA	carcinoembryonic antigen
cf.	(*confer*) compare
CFA	complete Freund's adjuvant (same as FCA)
CFT	complement fixation test
CHD	coronary heart disease
CHF	congestive heart failure
chem	chemical
Chem. Commun.	*Journal of the Chemical Society, Chemical Communications.*
Ci	curie
C.I.	*Colour Index* (British)
cis-	stereodescriptor, *see Glossary*
cm	centimeter(s)

CM-cellulose	O-(carboxymethyl) cellulose	**D.R.P.**	(Deutsches Reichs-Patent)–German patent
CMI	cell-mediated immunity	ds	double stranded
CML	cell-mediated lymphocytotoxicity	DTT	dithiothreitol
CMP	cytidine 5′-monophosphate (cytidylic acid)	dyn	dynes
CNS	central nervous system	(E)-	entgegen, stereodescriptor, see Glossary
CoA or CoASH	coenzyme A	$E_{1cm}^{1\%}$	the absorbance of a solution containing one gram per 100 ml contained in a cell having an absorption path of one cm
coll. vol.	collective volume		
compd	compound		
compn	composition		
Con A	concanavalin A		
conc(d)	concentrated	E_M	molar extinction coefficient (concn in g-moles/l)
concn	concentration		
config	configuration	EAA	excitatory amino acid
constit	constituent	EAC	erythrocyte coated by antibody and complement
contd	continued		
contg	containing	EAE	experimental allergic encephalomyelitis
CoQ	coenzyme Q (ubiquinone)		
cor(r)	corrected	EC	electron capture
corresp	corresponding, corresponds	ECF-A	eosinophil chemotactic factor of anaphylaxis
cp	centipoise		
C.P.	chemically pure	ECG	electrocardiogram
cpd	compound	E.C. No.	Enzyme Commission Number
crit press	critical pressure	ed.	edition
crit temp	critical temperature	ED	effective dose
cryst	crystalline, crystals	Ed(s).	editor(s)
crystn	crystallization	EDTA	ethylenediaminetetraacetic acid
CSF	colony stimulating factor; cerebral spinal fluid	EEG	electroencephalogram
		e.g.	(exempli gratia) for example
CTFA	Cosmetic, Toiletry and Fragrance Assoc.	EGF	epidermal growth factor
		eidem	the same (authors), plural of idem
CTP	cytidine triphosphate	EINECS	European Inventory of Existing Chemical Substances
Cys	cysteine		
Cyt	cytosine	EKG	electrocardiogram
d	density; specific gravity (d_4^{19} specific gravity at 19° referred to water at 4°)	ELISA	enzyme-linked immunosorbent assay
		emf	electromotive force
		en	ethylenediamine (in formulas)
d-	dextro(rotatory), the opposite of l	endo-	stereochemical descriptor, see Glossary
D-	dextro (in configurational sense only), the opposite of L	EPA	Environmental Protection Agency
		EPO	erythropoietin; European Patent Office
Da	daltons		
DEAE cellulose	O-(diethylaminoethyl)cellulose	ϵ (epsilon)	molar extinction coefficient (concn in g-moles/l); dielectric constant
dec, decomp	decompose(s), decomposition		
decompn	decomposition	eq	equation
deg	degree	equilib	equilibrium
deliquesc	deliquescent	equiv	equivalent
Δ (delta)	indicates the locant of the double bond	esp	especially
		esu	electrostatic units of electrical charge
deriv	derivative	Et	ethyl C_2H_5—
determn	determination	η (eta)	viscosity
DFP, DIFP, DIPF	diisopropyl fluorophosphate or diisopropyl phosphofluoridate	et al.	(et alii) and others
		etc.	et cetera
Dha	dihydroalanine	Et_2O	ether
Dhb	dehydrobutyrine, β-methyldehydroalanine	EtOH	ethyl alcohol
		Eur. pat. **Appl.**	European patent application
diff	difference	ev	electron volt
dil(d), (n)	dilute, diluted, dilution	evac	evacuated
distln	distillation	evapn	evaporation
dl-	racemic	exo-	stereochemical descriptor, see Glossary
DL-	optically inactive by external compensation as contrasted with meso-	expt(ly)	experimental(ly)
		ext(d)	extract, extracted
dm	decimeter(s)	extern	externally
DMA	dimethylacetamide	°F	Fahrenheit degrees; also Fourneau
DMARD	disease modifying antirheumatic drug	F-1-P	fructose 1-phosphate
DMF	dimethylformamide	F-6-P	fructose 6-phosphate
DMSO	dimethylsulfoxide	FA	fatty acid
DNA	deoxyribonucleic acid	FAB	fast atom bombardment
cDNA	complementary DNA	FAD (FADH₂)	flavin adenine dinucleotide (reduced form)
mtDNA	mitochondrial DNA		
DNAase	deoxyribonuclease	FCA	Freund's complete adjuvant (same as CFA)
DNFB	2,4-dinitro-1-fluorobenzene		
DNP	2,4-dinitrophenyl or 2,4-dinitrophenol	Fd	ferredoxin
		F.D.A.	Food and Drug Administration (U.S.A.)
Dopa	dihydroxyphenylalanine		
dp, DP	degree of polymerization (number of monomeric units in the polymer)	FD & C	Food, Drug and Cosmetic (U.S.A.)

FDNB	1-fluoro-2,4-dinitrobenzene
FDP	fructose 1,6-diphosphate
ff	following
FFA	free fatty acid
FFC	free from chlorine
FIA	Freund's incomplete adjuvant (same as ICFA)
FIAT	Field Information Agency, Technical (U.S. reports)
FMN (FMNH$_2$)	flavin mononucleotide (reduced form); same as riboflavin phosphate
Fmoc	9-fluorenylmethoxycarbonyl
Fortschr. Chem. Org. Naturst.	*Fortschritte der Chemie Organischer Naturstoffe* (Progress in the Chemistry of Organic Natural Products)
fp	freezing point
FP	flavoprotein
FPLC	fast protein liquid chromatography
Fr. pat.	French patent
Frdl.	*P. Friedlander Fortschritte der Teer-farbenfabrikation,* a collection of patents
Fru	fructose
FSH	follicle-stimulating hormone
FT	Fourier transform
g	gram(s)
G	guanine; guanosine
G-1-P	glucose 1-phosphate
G-3-P	glyceraldehyde 3-phosphate
G-6-P	glucose 6-phosphate
GABA	γ-aminobutyric acid
gal	gallon(s)
Gal	galactose
GalNAc	*N*-acetyl-D-galactosamine
γ (gamma)	microgram(s)
GC	gas chromatography
GDH	glutamate dehydrogenase
GDP	guanosine diphosphate
gem-	geminal, stereodescriptor *see Glossary*
geol	geological
Ger. pat.	German patent
GH	growth hormone
GI	gastrointestinal
g/l	grams per liter
Gla	γ-carboxyglutamic acid
Glc	glucose
GLC	gas-liquid chromatography
GlcA	gluconic acid
GlcN	glucosamine
GlcNAc	*N*-acetyl-D-glucosamine
GlcUA	glucuronic acid
Gln	glutamine
Glu	glutamic acid
GluA	glucuronic acid
Gly	glycine
Glycerol-3-P	glycerol 3-phosphate
Gmelin's	*Gmelin's Handbuch der Anorganischen Chemie,* a comprehensive German encyclopedia of inorganic chemistry
GMP	guanosine 5'-monophosphate (guanylic acid)
cGMP	cyclic GMP
GM-CSF	granulocyte macrophage colony stimulating factor
gov't	government
GP	glycoprotein
GRF, GHRF	growth hormone releasing factor
GSH	glutathione (reduced)
GSSG	glutathione (oxidized)
GTP	guanosine triphosphate
GU	genitourinary
Gua	guanine
GVH	graft-*versus*-host
habit.	habitat
Har	homoarginine
Hb	hemoglobin
HbCO	carbon monoxide hemoglobin; carboxyhemoglobin
HbO$_2$	oxyhemoglobin
HCG	human chorionic gonadotropin
Hcy	homocysteine
HDL	high density lipoproteins
HeLa	cells of the first continuously cultured (human cervical) carcinoma strain
HGG	human gamma globulin
HGH	human growth hormone
His	histidine
HIV	human immunodeficiency virus
Hiv	α-hydroxyisovaline
HLA	human leukocyte antigen
HMG-CoA	3-hydroxy-3-methylglutaryl coenzyme A
Houben	a German collection of medicinal patents
Houben Weyl	*Houben-Weyl Methoden der Organischen Chemie,* a German collection of preparative methods in organic chemistry
HPLC	high performance (pressure, power) liquid chromatography
hr	hour
HSA	human serum albumin
Hse	homoserine
HSV	herpes simplex virus
HT	hydroxytryptamine (serotonin)
HTLV	human T lymphotropic virus *see also* HIV
Hyl	hydroxylysine
Hyp	hydroxyproline
i-	optically inactive by internal compensation as *i*-inositol; archaic for *meso-*
I	inosine
Ia	I-region antigen
IACR	International Association of Cancer Registries
IARC	International Agency for Research on Cancer
IARC Monographs	*IARC Monographs on the Evaluation of Carcinogenic Risk of Chemicals to Man*
ibid.	(*ibidem*) at the same place
I.C.C.	Interstate Commerce Commission
ICFA	incomplete Freund's adjuvant (same as FIA)
ICSH	interstitial cell-stimulating hormone (same as LH)
IDDM	insulin dependent diabetes mellitis
idem	the same (author); plural: *eidem,* the same (authors)
IDP	inosine diphosphate
i.e.	(*id est*) that is
IEF	Isoelectric focusing
IF, IFN	interferon
i.g.	intragastric
Ig	immunoglobulin
I.G. Farben	*Interessengemeinschaft der Farbenindustrie, Aktiengesellschaft–* the German dye trust
IGF-1	insulin-like growth factor 1
IL	interleukin
Ile	isoleucine
i.m.	intramuscular
IMP	inosine 5'-monophosphate (inosinic acid)
incl	including
incompat	incompatibility
INN	International Nonproprietary Name

inorg	inorganic
insol	insoluble
intern	internal
Intl	International
i.p.	intraperitoneal
IR	infrared
Ir genes	immune response genes
ISO	Internal Organization for Standardization
isoln	isolation
ITP	inosine triphosphate; idiopathic thrombocytopenic purpura
I.U.	international unit
I.U.C.	International Union of Chemistry
I.U.P.A.C.	International Union of Pure and Applied Chemistry
i.v.	intravenous
Japan. Kokai	Japanese patent (unexamined)
Japan. pat.	Japanese patent
K	dissociation constant; equilibrium constant
°K	degrees Kelvin
Kb	kilobase
kcal	kilocalories(s)
K cell	killer cell
kg	kilogram(s)
αKG	α-ketoglutarate
KLH	keyhole limpet hemocyanin
KPA	kidney plasminogen activator
l	liter
l-	*levo*(rotatory), the opposite of *d*
L-	*levo* (in configurational sense only), the opposite of D
Lac	lactose
LAD	lymphocyte-activating determinant
λ (lambda)	wavelength
LAMMA	laser microprobe mass analysis
LATS	long acting thyroid stimulator
lb	pound(s)
LC	lethal concentration; liquid chromatography
LC$_{50}$	median lethal concentration, the concn of a chemical that is estimated to be fatal to 50% of the organisms tested
LCM	lymphocyte choriomeningitis
LD	lethal dose
LD$_{50}$	median lethal dose, the quantity of a chemical that is estimated to be fatal to 50% of the organisms tested
LDH	lactate dehydrogenase
LDL	low density lipoproteins
Leu	leucine
LH	luteinizing hormone (same as ICSH)
ln	natural logarithm
loc. cit.	(*loco citato*) in the place cited
log	logarithm (common)
l.o.i.	limit of impurities
LPS	lipopolysaccharide
Lys	lysine
m	meter; given after mass number signifies metastable isomer
m-	*meta* chemical locant for ring substituents
M	molar (concentration)
Mab, mAb	monoclonal antibody
MAC	maximum allowable concentration
MAF	macrophage activating factor
MAO	monoamine oxidase
MAOI	monoamine oxidase inhibitor
mass spec	mass spectrometry
max	maximum, maxima
Mb	myoglobin
MbO$_2$	oxymyoglobin
M.C.A.	Manufacturing Chemists Association (U.S.A.)

mcg	microgram
mCi	millicurie
MD	molecular rotation $\dfrac{[\alpha]_D \times \text{mol wt}}{100}$
MDH	malate dehydrogenase
Me	methyl CH_3—
Me$_2$CO—	acetone
MeOH	methyl alcohol
Mellor's	*Mellor's Comprehensive Treatise on Inorganic and Theoretical Chemistry*
mEq	milli-equivalent (0.001 of an equivalent)
Met	methionine
fMet	*N*-formylmethionine
MetHb	methemoglobin
meV	millielectron volts
mfg, manuf	manufacturing
mfr	manufacture
mg	milligram
MHC	major histocompatibility complex
mCi	microcurie
mg	microgram
microcryst	microcrystalline
MIF	migration inhibition factor
min	minimum; minute(s)
MIS	Mullerian inhibiting substance
misc	miscible
mixt	mixture
ml	milliliter (cubic centimeter)
MLD	minimum lethal dose
MLR	mixed lymphocyte reaction
mm	millimeter
mμ	millimicron(s), nm
mol wt	molecular weight
Monatsh.	*Monatshefte für Chemie*
mOsm	milliosmolar
mp	melting point; melts, melting at, when followed by a figure denoting temperature
M$_r$	relative molecular mass
MRI	magnetic resonance imaging
ms-	*meso-* (internally compensated)
MS	mass spectrometry
MSH	melanocyte-stimulating hormone (melanotropin)
n	index of refraction (n_D^{20} for 20° and sodium light); normal, as *n*-propyl
N	normal (equivalents per liter, as applied to concentration); nitrogen (as in *N*-methylpyridine)
NAcneu	*N*-acetylneuraminic acid
NAD$^+$ (NADH)	nicotinamide adenine dinucleotide (reduced form)
NADP$^+$ (NADPH)	nicotinamide adenine dinucleotide phosphate (reduced form)
NANSAIDS	nonaspirin nonsteroidal anti-inflammatory drugs
NBS	National Bureau of Standards
NCTC	*National Collection of Type Cultures*
NDP	nucleoside 5′-phosphate
Neth. pat. **Appl.**	Netherlands patent application
N.F.	*National Formulary*
ng	nanogram (10^{-9} grams)
NGF	nerve growth factor
NIDDM	Non-insulin dependent diabetes mellitis
NIOSH	National Institute for Occupational Safety and Health
nm	nanometer
NMN$^+$ (NMNH)	nicotinamide mononucleotide (reduced form)
NMP	nucleoside 5′-monophosphate
NMR	nuclear magnetic resonance
N.N.D.	*New and Nonofficial Drugs* (Lippincott, 1959-1964)

N.N.R.	*New and Nonofficial Remedies* (Lippincott, 1933-1958)	R	alkyl, univalent hydrocarbon radical (or hydrogen)
no.	number	*(R)-*	*rectus* (right), stereodescriptor *see Glossary*
nor-	chemical nomenclature descriptor, *see Glossary*	RAST	radioallergosorbent test
NRDC	National Research Development Corporation	rbc	red blood cell; red blood count
NSAID(s)	nonsteroidal anti-inflammatory drugs	RCO-	aliphatic acyl radical
		recryst(n)	recrystallize, recrystallization
NSC	National Service Center	ref	reference
NTP	nucleoside 5′-triphosphate; National Toxicology Program (U.S.)	rep [REP]	''roentgen equivalent physical'' means a dose of ionizing radiation capable of producing energy absorption of 93 ergs per gram of tissue
o-	*ortho* chemical locant for ring substituents		
O	denoting attachment to oxygen, as in *O*- acetylhydroxylamine	resp	respectively
OA	ovalbumin	R_f or R_F	(in paper chromatography) ratio of movement of the band to the front of the solvent
OAA	oxaloacetate		
OD	optical density	RIA	radioimmunoassay
op. cit.	*(opere citato)* in the work cited	Rib	D-ribose
org	organic	RIST	radioimmunosorbent test
OSHA	Occupational Safety and Health Act	RNA	ribonucleic acid
OsM	osmolar	hnRNA	heterogeneous nuclear RNA
oz.	ounce(s)	mRNA	messenger RNA
P, p	concentration by weight (after optical rotations only)	nRNA	nuclear RNA
		rRNA	ribosomal RNA
p, pp	page(s)	snRNA	small nuclear RNA
p-	*para* chemical locant for ring substituents	tRNA	transfer (soluble) RNA
		RNase	ribonuclease
Pi	inorganic phosphate	*RTECS*	*Registry of Toxic Effects of Chemical Substances*
Pa	pascal		
PABA	*p*-aminobenzoic acid	*s-*	symmetrical
PAF	platelet-activating factor	*S*	denoting attachment to sulfur as in *S*-methylcysteine
PAGE	polyacrylamide gel electrophoresis		
passim	here and there, scattered	*(S)-*	*sinister* (left), stereodescriptor *see Glossary*
pat.	patent		
PB report	Publication Board Report (United States Department of Commerce, Scientific and Industrial Reports)	S.A.E.	Society of Automotive Engineers
		SAM	*S*-adenosylmethionine
		sapon(if)	saponification
PCA	passive cutaneous anaphylaxis	Sar	sarcosine
PCT	Patent Co-operation Treaty	satd	saturated
PEP	phosphoenolpyruvate	s.c.	subcutaneous
PET	positron emission tomography	SDS	sodium dodecyl sulfate
petr, petrol	petroleum	sec	second(s)
PFC	plaque-forming cell	*sec-*	secondary
3PG	3-phosphoglycerate	sepn	separation
PG	prostaglandin	Ser	serine
PGA	pteroylglutamic acid (folic acid)	SI	International System of Units
PGP	3-phosphoglyceroyl phosphate	SLE	systemic lupus erythematosus
pH	acid-base scale; log of reciprocal of hydrogen ion concentration	SNRI	serotonin-noradrenaline reuptake inhibitor
PHA	phytohemagglutin	sol; soly	soluble; solubility
Phe	phenylalanine	solidif	solidifies, solidification
physiol	physiological	soln	solution
pK	log of the reciprocal of the dissociation constant, $1/\log K$	sp.	species; specific
		spec	spectroscopy
PKU	phenylketonuria	SPECT	single photon emission computed tomography
PMN	polymorphonuclear leukocyte		
PPi	inorganic pyrophosphate	sp gr	specific gravity
ppm	parts per million	*spp.*	species (plural)
ppt(g)	precipitate, precipitating	sq	square
pptd	precipitated	*sqq*	*(sequentia)* and following
PQ	plastoquinone	SRBC	sheep red blood cells
Pr	propyl (normal)	SRS-A	slow reacting substance of anaphylaxis
prepd	prepared		
prepn	preparation	ss	single stranded
press.	pressure	SSPE	subacute sclerosing panencephalitis
Pro	proline	SSRI	selective serotonin reuptake inhibitor
PRPP	5-phosphoribosyl 1-pyrophosphate	S.T.P.	standard temperature and pressure
ψ (psi)	pseudo	subl	sublimes
pt	point	suppl	supplement
PTH	parathyroid hormone	*sym-*	symmetrical
Q	coenzyme Q (ubiquinone)	*syn-*	stereochemical descriptor *see Glossary*
q.q.v.	*(quae vide)* which see, plural		
q.v.	*(quod vide)* which see	T	thymine; thymidine
r	''roentgen'' unit of radiation	$t_{1/2}$	half-life
r-	racemic	tabl	tablet(s)

TB, tb	tuberculosis		*U.S.D.*	*United States Dispensatory*
TCA	trichloroacetic acid; tricarboxylic acid		U.S.D.A.	United States Department of Agriculture
T cell	thymus derived cell		*U.S.P.*	*United States Pharmacopeia*
TDL	thoracic duct lymphocyte		**U.S.** pat.	United States patent
TDP	thymidine diphosphate		UTP	uridine triphosphate
TEAE cellulose	*O*-(triethylaminoethyl)-cellulose		uv	ultraviolet
tech	technical		v	volt(s)
temp	temperature		*v*-	vicinal (adjacent)
tert-	tertiary		Val	valine
TH	thyrotropic hormone		var	variety
THF	tetrahydrofuran		*viz.*	*(videlicet)* that is to say; namely
THFA	tetrahydrofolic acid		VLDL	very low density lipoproteins
Thr	threonine		vol	volume
Thy	thymine		vs	versus
TLC	thin-layer chromatography		v/v	percent "volume in volume" expresses the number of milliliters of an active constituent in 100 milliliters of solution
TMP	thymidine 5'-monophosphate			
TMV	tobacco mosaic virus			
TNF	tumor necrosis factor			
TNP	trinitrophenyl		wbc	white blood cell; white blood count
TPA	tissue plasminogen activator		WHO	World Health Organization
TPP, ThPP	thiamine pyrophosphate		wks	weeks
trans-	stereodescriptor *see Glossary*		wt	weight
Tris	tris(hydroxymethyl)aminomethane; 2-amino-2-hydroxymethylpropane-1,3-diol		w/v	percent "weight in volume" expresses the number of grams of an active constituent in 100 milliliters of solution, and is used regardless of whether water or another liquid is the solvent
Trp	tryptophan			
TSH	thyroid-stimulating hormone			
TSTA	tumor-specific transplantation antigens		w/w	percent "weight in weight" expresses the number of grams of an active constituent in 100 grams of solution or mixture
TTP, dTTP	thymidine triphosphate			
Tyr	tyrosine			
U	uracil or uridine			
UDP	uridine diphosphate		XMP	xanthosine monophosphate
UDP-gal	uridine diphosphogalactose		yr(s)	year(s)
UDP-glc	uridine disphosphoglucose		Z	carbobenzoxy
U.K.	United Kingdom		(Z)-	*zusammen* stereodescriptor *see Glossary*
UMP	uridine 5'-monophosphate (uridylic acid)			
			Z. Physiol.	*Hoppe-Seyler's Zeitschrift für Physiologische Chemie*
uncor(r)	uncorrected			
uns-, unsym-	unsymmetrical, asymmetrical		~	approximately
Ura	uracil		≈	approximately equal
U.S.A.E.C.	United States Atomic Energy Commission		>	greater than
			<	less than
USAN	United States Adopted Names			

MONOGRAPHS

THE MERCK INDEX

OF CHEMICALS, DRUGS, AND BIOLOGICALS

A

1. Abamectin. *Avermectin B₁;* 5-*O*-demethylavermectin A₁ₐ and 5-*O*-demethyl-25-de(1-methylpropyl)-25-(1-methylethyl)avermectin A₁ₐ (4:1); avermectin B₁ₐ/ᵦ; MK-936; Affirm; Agrimek; Avid; Avomec; Vertimec; Zephyr. Mixture of avermectins, *q.v.,* containing at least 80% of avermectin B₁ₐ (C₄₈H₇₂O₁₄) and not more than 20% of avermectin B₁ᵦ (C₄₇H₇₀O₁₄). Isoln from *Streptomyces avermitilis:* G. Albers-Schönberg *et al.,* **Ger.** pat. **2,717,040;** *eidem,* **U.S.** pat. **4,310,519** (1977, 1982 both to Merck & Co.). Separation of components: T. W. Miller *et al., Antimicrob. Ag. Chemother.* **15,** 368 (1979); by semi-preparative HPLC: C. C. Ku *et al., J. Liq. Chromatog.* **7,** 2905 (1984). Structure determn: G. Albers-Schönberg *et al., J. Am. Chem. Soc.* **103,** 4216 (1981). Absolute configuration: J. P. Springer *et al., ibid.* 4221. Partial synthesis of B₁ₐ: K. C. Nicolaou *et al., ibid.* **106,** 4189 (1984). Total synthesis: S. Hanessian *et al., ibid.* **108,** 2776 (1986). Anthelmintic activity: L. S. Blair, W. C. Campbell, *J. Parasitol.* **64,** 1032 (1978); J. R. Egerton *et al., Antimicrob. Ag. Chemother.* **15,** 372 (1979); K. S. Todd *et al., Am. J. Vet. Res.* **45,** 976 (1984). Pesticidal activity: I. Putter *et al., Experientia* **37,** 963 (1981); R. A. Dybas, A. St. J. Green, *Brit. Crop. Prot. Conf. - Pests Dis.* **1984,** 947. Control of red imported fire ants: J. A. Greenblatt *et al., J. Agr. Entomol.* **3,** 233 (1986). Interaction with GABA receptors: S. S. Pong *et al., J. Neurochem.* **34,** 351 (1980); T. N. Mellin *et al., Neuropharmacol.* **22,** 89 (1983). Receptor binding studies: S. S. Pong, C. C. Wang, *ibid.* **19,** 311 (1980); G. Drexler, W. Sieghart, *Eur. J. Pharmacol.* **99,** 269 (1984). Fate in soil and plants: D. L. Bull *et al., J. Agr. Food Chem.* **32,** 94 (1984); H. A. Moye *et al., ibid.* **35,** 859 (1987); M. S. Maynard *et al., ibid.* **37,** 178, 184 (1989). *Review:* J. R. Babu, *ACS Symposium Series* **380,** 91-108 (1988). Reviews of modes of action: C. C. Wang, S. S. Pong, *Prog. Clin. Biol. Res.* **97,** 373-395 (1982); D. J. Wright, *Biochem. Soc. Trans.* **15,** 65-67 (1987); of insecticidal activities and use: L. Strong, T. A. Brown, *Bull. Ent. Res.* **77,** 357-389 (1987). *Book: Ivermectin and Abamectin,*

W. C. Campbell, Ed. (Springer-Verlag, New York, 1989) 363 pp.

Odorless, off-white to yellow crystals from methanol, mp 150-155° (dec). [α]_D +55.7° ±2° (c = 0.87 in CHCl₃). uv max (methanol): 237, 245, 253 nm (log ε 4.48, 4.53, 4.34). Vapor pressure: 1.5×10^{-9} torr. Soly at 21° (μg/l): water 10; (mg/ml): acetone 100; *n*-butanol 10; chloroform 25; cyclohexane 6; ethanol 20; isopropanol 70; kerosene 0.5; methanol 19.5; toluene 350. Hydrolysis does not occur in aq soln at pH 3, 5, 7. LD₅₀ (technical grade) orally in sesame oil in mouse, rat: 13.5, 10.0 mg/kg; dermally in rabbit: >2000 mg/kg. LD₅₀ in mallard duck, bobwhite quail: 84.6, >2000 mg/kg. LC₅₀ (96 hr) in rainbow trout, bluegill: 3.6, 9.6 μg/l; LC₅₀ (48 hr) in *Daphnia magna:* 0.34 μg/l (Merck Technical Data Sheet).

USE: Acaricide; insecticide.

THERAP CAT (VET): Anthelmintic.

2. Abecarnil. *4-(Methoxymethyl)-6-(phenylmethoxy)-9H-pyrido[3,4-b]indole-3-carboxylic acid 1-methylethyl ester;* isopropyl 6-(benzyloxy)-4-(methoxymethyl)-9H-pyrido[3,4-b]indole-3-carboxylate; isopropyl 6-(benzyloxy)-4-(methoxymethyl)-β-carboline-3-carboxylate; ZK-112119. C₂₄H₂₄N₂O₄; mol wt 404.47. C 71.27%, H 5.98%, N 6.93%, O 15.82%. Benzodiazepine receptor agonist. Prepn: H. Biere *et al.,* **Ger.** pat. **3,609,699** (1987 to Schering AG), *C.A.* **108,** 21872z (1988). Pharmacology and receptor binding studies: D. N. Stephens *et al., J. Pharmacol. Exp. Ther.* **253,** 334 (1990). Determn in biological samples: W. Krause *et al., J. Pharm. Sci.* **78,** 622 (1989). Clinical pharmacokinetics: W. Krause *et al., Xenobiotica* **21,** 763 (1991). Clinical trial in anxiety: J. C. Ballenger *et al., Advan. Biochem. Psychopharmacol.* **47,** 431 (1992). Series of articles on pharmacology, tolerance and clinical evaluation: *Psychopharmacol. Ser.* **11,** 50-147 (1993).

Crystals from ethyl acetate, mp 150-151°.
THERAP CAT: Anxiolytic.

3. Abietic Acid. *[1R-(1α,4aβ,4bα,10aα)]-1,2,3,4,-4a,4b,5,6,10,10a-Decahydro-1,4a-dimethyl-7-(1-methylethyl)-1-phenanthrenecarboxylic acid; 13-isopropylpodocarpa-7,13-dien-15-oic acid;* sylvic acid. C₂₀H₃₀O₂; mol wt 302.46. C 79.42%, H 10.00%, O 10.58%. A widely available organic acid, prepared by isomerization of rosin: Harris, Sanderson, *Org. Syn.,* **coll. vol. IV,** 1 (1963); Fieser, Fieser, *The Chemistry of Natural Products Related to Phenanthrene* (New York, 3rd ed., 1949). Synthesis from dehydroabietic acid: A. W. Burgstahler, L. W. Worden, *J. Am. Chem. Soc.* **83,** 2587 (1961); E. Wenkert *et al., ibid.* **86,** 2038 (1964). Chromatographic study: A. G. Douglas, T. G. Powell, *J. Chromatog.* **43,** 241 (1969). Metabolism in rabbits: Y. Asakawa *et al., Xenobiotica* **16,** 753 (1986).

Consult the Name Index before using this section.

Monoclinic plates from alcohol + water, mp 172-175°. $[\alpha]_D^{24}$ —106° (c = 1 in abs alc). uv max: 235, 241.5, 250 nm (ϵ 19500, 22000, 14300). Insol in water. Sol in alc, benzene, chloroform, ether, acetone, carbon disulfide, dil NaOH soln. Commercial abietic acid made by heating rosin alone or with acids may be glassy or partly crystalline, usually of yellow color and melting as low as 85°.

Methyl ester, see Methyl Abietate.

USE: Manufacture of esters (ester gums), e.g., methyl, vinyl and glyceryl esters for use in lacquers and varnishes. Manufacture of "metal resinates", soaps, plastics and paper sizes. Assists growth of lactic and butyric acid bacteria.

4. Abikoviromycin. *7-Ethylidene-1a,2,3,7-tetrahydrocyclopent[b]oxireno[c]pyridine;* 4,4a-epoxy-5-ethylidene-2,3,4,4a-tetrahydro-5H-1-pyridine; abicoviromycin; latumcidin. $C_{10}H_{11}NO$; mol wt 161.20. C 74.51%, H 6.88%, N 8.69%, O 9.92%. Antiviral antibiotic produced by *Streptomyces abikoensis* and *Streptomyces rubescens*. Chromatographic isoln from broth cultures: Umezawa et al., *Japan. Med. J.* **4**, 331 (1951); *C.A.* **46**, 7167 (1952); Umezawa, *Japan. pat.* **6200('54)** (to Nippon). Identity with latumcidin: Sakagami et al., *J. Antibiot.* **11A**, 231 (1958). Structure: Gurevich et al., *Tetrahedron Letters* **1968**, 2209. Stereochemistry: Kono et al., *J. Antibiot.* **23**, 572 (1970); Gurevich et al., *Khim. Prir. Soedin.* **7**, 104 (1971), *C.A.* **75**, 5752e (1971). Crystal and molecular structure of the selenate: Y. Kono et al., *Acta Crystallog.* Sect. B, **27**, 2341 (1971). *In vitro* antiviral activity: V. M. Roikhel, N. A. Zeitlenok, *Antibiotiki (Moscow)* **14**, 969 (1969), *C.A.* **72**, 19394q (1969).

Highly unstable and polymerizes promptly on isolation even at —50°; however, it can be handled in dilute solutions and in the form of its salts. uv max (neutral ethanol or 0.1N KOH): 218, 244, 289 nm (log ϵ 3.83, 3.99, 3.94); (0.1N HCl) 236, 341 nm (log ϵ 3.99, 4.05).

5. Abrin. Agglutinin; toxalbumin. A toxic lectin and hemagglutinin obtained from seeds of *jequirity*, *Abrus precatorius* L., *Leguminosae*, a common vine of tropical countries, also found in central and southern Florida. Isoln and purification: J. Y. Lin et al., *J. Formosan Med. Assoc.* **68**, 518 (1969), *C.A.* **72**, 98695 (1970); eidem, *Toxicon* **9**, 97 (1971). The high toxicity of abrin was originally believed to result from its hemagglutinating activity, but subsequent studies have shown that separate proteins are responsible for the toxicity and agglutination: S. Olsnes, A. Pihl, *Eur. J. Biochem.* **35**, 179 (1973). Five glycoproteins have been purified from the seeds of *A. precatorius*: *Abrus agglutinin* and the toxic principles, abrins a-d. *Abrus* agglutinin is a tetramer of 134,900 Da, is non-toxic to animal cells and a potent hemagglutinator. Abrins a through d (mol wt 63,000-67,000 Da) are composed of two disulfide-linked polypeptide chains. The smaller A-chain inhibits protein synthesis and causes cell death; the larger B-chain binds to the cell plasma membrane. Purification of major components: C. H. Wei et al., *J. Biol. Chem.* **249**, 3061 (1974). Crystallographic study: C. H. Wei, J. R. Einstein, *ibid.* 2985. Improved purification, properties, crystallography of *Abrus* agglutinin: C. H. Wei et al., *ibid.* **250**, 4790 (1975). Physical properties of the toxic principles: M. S. Herrmann,

W. D. Behnke, *Biochim. Biophys. Acta* **621**, 43 (1980); eidem, *ibid.* **667**, 397 (1981). Isoln and purification of all five proteins: J. Y. Lin et al., *Toxicon* **19**, 41 (1981). Amino acid sequence of the A-chain of abrin-a and comparison with ricin: G. Funatsu et al., *Agr. Biol. Chem.* **52**, 1095 (1988). Antitumor effects in animals: V. V. S. Reddy, M. Sirsi, *Cancer Res.* **29**, 1447 (1969); J. Y. Lin et al., *Nature* **227**, 292 (1970); O. Fodstad et al., *Cancer Res.* **37**, 4559 (1977). Immunoelectron microscopy studies of abrin toxic action on tumor cells: C. T. Lin et al., *J. Ultrastruc. Res.* **73**, 310 (1980). Studies on toxicity and binding kinetics: M. Witten et al., *Exp. Cell Biol.* **49**, 306 (1981); C. E. Bennett et al., *ibid.* 319. Use of A-chain in cell-type-specific cytotoxic agents known as "immunotoxins": A. J. Cumber et al., *Methods in Enzymology* **112**, 207 (1985). Toxicity study: J. Y. Lin et al., *J. Formosan Med. Assoc.* **68**, 322 (1969), *C.A.* **71**, 121926 (1969). See also Ricin, Lectins.

Yellowish-white powder. Sol in solns of sodium chloride, usually with turbidity. The toxic portion is heat-stable to incubation at 60° for 30 min; at 80°, most of the toxicity is lost in 30 min. LD_{50} i.p. in mice: 0.020 mg/kg (Lin p. 322).

Caution: Seeds of *A. precatorius* are extremely toxic; one seed, if thoroughly masticated, can cause fatal poisoning, *cf.* J. M. Kingsbury, *Poisonous Plants of the United States and Canada* (Prentice-Hall, New Jersey, 1964) p 303; K. Genest et al., *Arzneimittel-Forsch.* **21**, 888 (1971).

Note: Do not confuse with abrine, *q.v.*

USE: Exptly in cancer research.

6. Abrine. *N-Methyl-L-tryptophan;* α-methylamino-β-(3-indole)propionic acid. $C_{12}H_{14}N_2O_2$; mol wt 218.26. C 66.04%, H 6.47%, N 12.84%, O 14.66%. Not to be confused with the albuminous substance abrin, *q.v.* Obtained from the seeds of *Abrus precatorius* L., *Leguminosae* (jequirity): Hoshino, *Ann.* **520**, 31 (1935). Synthesis: Miller, Robson, *J. Chem. Soc.* **1938**, 1910. Configuration: Cahill, Jackson, *J. Biol. Chem.* **126**, 29 (1938).

Prisms from water, dec 295°. $[\alpha]_D^{21}$ +44° (0.28 g in 10 ml 0.5N HCl). One gram dissolves in about 100 ml methanol, slightly sol in water, insol in ether. Sol in dil acids, alkalies.

Hydrochloride, $C_{12}H_{14}N_2O_2 \cdot$HCl, needles, mp 222°, soluble in water.

Nitrate, $C_{12}H_{14}N_2O_2 \cdot$HNO_3, needles, dec 143°.

Acetyl derivative, $C_{14}H_{16}N_2O_3$. mp 176°. $[\alpha]_D^{25}$ —148° (43 mg in 5 ml 0.1N NaOH).

7. Abscisic Acid. *[S-(Z,E)]-5-(1-Hydroxy-2,6,6-trimethyl-4-oxo-2-cyclohexen-1-yl)-3-methyl-2,4-pentadienoic acid;* abscisin II; dormin; ABA. $C_{15}H_{20}O_4$; mol wt 264.32. C 68.16%, H 7.63%, O 24.21%. An abscission-accelerating plant hormone. Presence of the naturally occurring (+)-cis,trans-form (also designated as (S)-abscisic acid) in sycamore, birch, rose leaves, cabbage, potato, lemon, avocado: Cornforth et al., *Nature* **210**, 627; **211**, 742 (1966). Identity with dormin: Cornforth et al., *Nature* **205**, 1269 (1965); idem, *Tetrahedron, Suppl.* No. 8, Part II, 603 (1967). Isoln from young cotton fruit: Ohkuma et al., *Science* **142**, 1592 (1963). Identification of (+)-abscisic acid in rat, pig brain: M.-Th. Le Page-Degivry et al., *Proc. Nat. Acad. Sci. USA* **83**, 1155 (1986). Synthesis of (±)-cis,trans-forms: Cornforth et al., *Nature* **206**, 715 (1965); of (+)-cis,trans- and (—)-cis,-trans-forms: F. Kienzle et al., *Helv. Chim. Acta* **61**, 2616 (1978); of (±)-trans,cis-form: Ohkuma, *Agr. Biol. Chem.* **29**, 962 (1965); **30**, 434 (1966); Findlay, MacKay, *Can. J. Chem.* **49**, 2369 (1971). Structure: Cornforth et al., *Nature* **206**, 715 (1965); Ohkuma et al., *Tetrahedron Letters* **1965**, 2529. Absolute configuration of (+)-cis,trans-form: Cornforth et al., *Chem. Commun.* **1967**, 114; revised stereochemistry: Ryback, *ibid.* **1972**, 1190; Koreeda et al., *J. Am. Chem. Soc.* **95**, 239 (1973). Crystal and molecular structure: H. W. Schmalle et al., *Acta Crystallogr.* **B33**, 2218 (1977). Effect on seed set in wheat: J. M. Morgan, *Nature* **285**, 655

(1980). *Review:* Addicott, Lyon, *Ann. Rev. Plant Physiol.* **20**, 139 (1969). Review of role in root growth regulation: P. E. Pilet, P. W. Barlow, *Plant Growth Regul.* **6**, 217-265 (1987); of metabolism: J. A. D. Zeevaart, R. A. Creelman, *Ann. Rev. Plant Physiol. Plant Mol. Biol.* **39**, 439-473 (1988). Book: *Abscisic Acid,* F. T. Addicott, Ed. (Praeger, New York, 1983) 608 pp.

(+)-*cis,trans*-Form, crystals from ethyl acetate + hexane, mp 161-163°. Sublimes at 120°. $[\alpha]_D^{20}$ +411.1° (c = 1 in ethanol). $[\alpha]_D^{20}$ +426.5° (c = 1 in 0.005N methanolic H_2SO_4). Sol in aq $NaHCO_3$, chloroform, acetone, ethyl acetate, ether; slightly sol in benzene, water; sparingly sol in petr ether. uv max (methanol): 252 nm (ε 25200).

(−)-*cis,trans*-Form, mp 162-163°. $[\alpha]_D^{20}$ −426.2° (c = 1 in 0.005N H_2SO_4).

(±)-*cis,trans*-Form, crystals, mp 188-190°.

Note: Abscisin I, an abscission-accelerating substance. Nomenclature: Ohkuma *et al., Science* **142**, 1592 (1963). Isoln from mature fruit walls of cotton: Liu, Carns, *Science* **134**, 384 (1961). Crystals, mp 197-198°. uv max (methanol): 250 nm. Acidic reaction. Sol in chloroform, dil NaOH; slightly sol in dimethyl ether. Practically insol in dil HCl.

8. Absinthin. *[3S-(3α,3aα,6β,6aα,6bβ,7α,7aβ,8α,-10aβ,11β,13aα,13bα,13cβ,14bβ)]-3,3a,4,5,6,6a,6b,7,7a,8,9,-10,10a,13a,13c,14b-Hexadecahydro-6,8-dihydroxy-3,6,8,11,-14,15-hexamethyl-2H-7,13b-ethenopentaleno[1″,2″:6,7;5″,4″:-6′,7′]dicyclohepta[1,2-b:1′,2′-b′]difuran-2,12(11H)-dione;* absinthiin; absynthin. $C_{30}H_{40}O_6$; mol wt 496.64. C 72.55%, H 8.12%, O 19.33%. Chief bitter principle of wormwood, *Artemisia absinthium* L., *Compositae.* Isoln by chromatography: Herout *et al., Coll. Czech. Chem. Commun.* **21**, 1485 (1956); see also *Chem. & Ind. (London)* **1955**, 569. Structural studies: Novotny *et al., ibid.* **1958**, 465; *Coll. Czech. Chem. Commun.* **25**, 1492 (1960); Vokac *et al., Tetrahedron Letters* **1968**, 3855. Structure: J. Beauhaire *et al., ibid.* **21**, 3191 (1980).

Very bitter orange needles from abs ether, mp 179-180° (dec). Solvated crystals from benzene, decomp 165°. $[\alpha]_D^{20}$ +180.0° (c = 1.9 in CHCl$_3$). Bitterness threshold 1:70,000.

9. Absinthium. Wormwood; Absinthe; Armoise. Dried leaves and flowering tops of *Artemisia absinthium* L., *Compositae. Habit.* Grows as weed or is cultivated in Europe, U.S., Canada, North and West Asia, Africa. Has been used as stomachic tonic and anthelmintic. *Constit.* Absinthin, anabsinthin, dark green or brown volatile oil (chiefly thujone). E. Guenther, *The Essential Oils,* V, 487 (Van Nostrand, New York 1952). Isolation of various constituents: Cekan, Herout, *Coll. Czech. Chem. Commun.* **21**, 79 (1956); Herout *et al., ibid.* 1485. Characterization of oil: T. Sacco, F. Chialva, *Planta Med.* **54**, 93 (1988).

Very strong odor, acrid taste.

Caution: Ingestion of the volatile oil or of the liqueur, absinthe, may cause G.I. symptoms, nervousness, stupor, convulsions, death.

USE: As flavoring in alcoholic beverages, *e.g. vermouth,* which is a blend of white wines, contg traces of absinthium and other flavors.

10. Acacetin. *5,7-Dihydroxy-2-(4-methoxyphenyl)-4H-1-benzopyran-4-one; 5,7-dihydroxy-4′-methoxyflavone;* apigenin-4′-methyl ether. $C_{16}H_{12}O_5$; mol wt 284.27. C 67.60%, H 4.25%, O 28.14%. The aglycon of linarin, *q.v.,* and of acaciin. Isoln from linarin: Zemplén, Bognar, *Ber.* **74B**, 1818 (1941). From acaciin: Hattori, *Acta Phytochim.* **2**, 105 (1925). Isoln from *Robinia pseudoacacia* L., *Leguminosae:* Nakazawa, Matsuura, *J. Pharm. Soc. Japan* **73**, 481 (1953). Structure: Baker *et al., J. Chem. Soc.* **1951**, 691. Synthesis: Robinson, Venkataraman, *ibid.* **1926**, 2348; Zemplén, Bognar, *Ber.* **76B**, 452 (1943); Narasimhachari, Seshadri, *Proc. Indian Acad. Sci.* **30A**, 151 (1949); Simpson, *Sci. Proc. Roy. Dublin Soc.* **27**, 111 (1956), *C.A.* **51**, 8082a (1957).

Yellow needles from 95% alcohol, mp 263°. Sol in hot alc, practically insol in ether. Sol in alkalies with yellow color.

Diacetate, $C_{20}H_{16}O_7$, lustrous needles from alc, mp 203°.

7-Rhamnoglucoside, $C_{28}H_{32}O_{14}$, *acaciin.* From *Robinia pseudoacacia* L., *Leguminosae:* Freudenberg, Hartmann, *Ann.* **587**, 207 (1954). Structure: Zemplén, Mester, *Magyar Kém. Folyoirat* **56**, 2 (1950), *C.A.* **45**, 7977d (1951). Needles from pyridine + water, mp 263°. $[\alpha]$ −85.3° (pyridine); −99.5° (glacial acetic acid). Sparingly soluble in cold, more sol in boiling water; slightly sol in organic solvents.

11. Acacia. Gum Arabic. Estimations of mol wt range from about 240,000: Oakley, *Trans. Faraday Soc.* **31**, 136 (1935), to 580,000: Anderson *et al., Carbohyd. Res.* **3**, 308 (1967). According to the U.S.P., acacia is the dried gummy exudation from the stems and branches of *Acacia senegal* (L.) Willd., *Leguminosae,* or other African species of Acacia. According to C. L. Mantell, *The Water-Soluble Gums* (New York, 1947), Kordofan gum (hashab geneina), the gum from *Acacia verek* Guill. & Perr. from plantations in the Kordofan province (Sudan) is considered the best commercial variety. Grades of Kordofan gum which are clear, white (sun bleached) and tasteless are preferred for food prepns and pharmaceuticals. (There is a close relationship between color and flavor due to the presence of tannins.) Acacia was originally thought to be composed only of (−)-arabinose, (+)-galactose, (−)-rhamnose, (+)-glycuronic acid. Revised composition and structural studies: Anderson *et al., J. Chem. Soc. (C)* **1966**, 1959. *See also* Swenson *et al., J. Polym. Sci. Part A-2* **6**, 1593 (1968). General review: Anderson, Dea, *J. Soc. Cosmet. Chem.* **22**, 61-76 (1971). Review of use as food additive: D. M. W. Anderson, *Food Addit. Contam.* **3**, 225-230 (1986).

Occurs in spheroidal tears up to 32 mm in diameter. Also flakes and powder. Solns of gum from *Acacia verek* are levorotatory; other acacia species are dextrorotatory: Hamy, *Bull. Sci. Pharmacol.* **35**, 421 (1928). Specific gravity: 1.35-1.49 (samples dried at 100° are heavier). Moisture content usually varies from 13-15%. U.S.P. limit 15%. Material containing less than 12% chips easily and produces dust during transportation. Insol in alcohol. Almost completely sol in twice its weight of water. 100 grams of a satd soln contains 37 g at 25°; 38 g at 50°; 40 g at 90°: Taft, Malm, *Trans. Kans. Acad. Sci.* **32**, 49 (1929). Aq soln acid to litmus. Also sol in glycerol and in propylene glycol, but prolonged heating (several days) may be necessary for complete solution (about 5%).

Incompat. Precipitates or jellies result upon addition of solns of ferric salts, borax, basic lead acetate (lead subacetate, but not neutral lead acetate), alcohol, sodium silicate, gelatin, ammoniated tincture of guaiac.

USE: As mucilage, excipient for tablets, size, emulsifier, thickener, also in candy, other foods; as colloidal stabilizer. In the manufacture of spray-dried "fixed" flavors—stable, powdered flavors used in packaged dry-mix products (puddings, desserts, cake mixes) where flavor stability and long shelf life are important.

12. Acacic Acid. *(3β,16α,21β)-Trihydroxyolean-12-en-28-oic acid.* $C_{30}H_{48}O_5$; mol wt 488.71. C 73.73%, H 9.90%, O 16.37%. Isoln from pods of *Acacia concinna* D.C., *Leguminosae:* Varshney, Shamsuddin, *Tetrahedron Letters* **1964**, 2055; from bark of *A. concinna* D.C.: R. Banerji, S. K. Nigam, *J. Ind. Chem. Soc.* **57**, 1043 (1980). Structure and stereochemistry: Varshney *et al., ibid.* **1965**, 1187. Revised structure: A. K. Barua *et al., Trans. Bose Res. Inst., Calcutta* **39**, 61 (1976), *C.A.* **87**, 53460c (1977).

Needles from methanol, mp 280-281°.

Methyl ester, $C_{31}H_{50}O_5$, needles from methanol, mp 223-224°.

Diacetyl lactone, $C_{34}H_{50}O_6$, crystals, mp 235-236°.

13. Acadesine. *5-Amino-1-β-D-ribofuranosyl-1H-imidazole-4-carboxamide;* 5-amino-4-imidazolecarboxamide ribonucleoside; AICA-riboside; Arasine; Protara. $C_9H_{14}N_4O_5$; mol wt 258.23. C 41.86%, H 5.46%, N 21.70%, O 30.98%. Purine nucleoside analog; prototype adenosine regulating agent. Biosynthesis, isoln and characterization: G. R. Greenberg, E. L. Spilman, *J. Biol. Chem.* **219**, 411 (1956). Chemical synthesis: J. Baddiley *et al., Proc. Chem. Soc.* **1957**, 149; J. P. Ferris *et al., J. Org. Chem.* **50**, 747 (1985); T. Saito *et al., Chem. Pharm. Bull.* **42**, 2263 (1994). Pharmacology: Z.-Q. Zhao *et al., Cardiovasc. Res.* **29**, 495 (1995). Antiplatelet activity: D. A. Bullough *et al., J. Clin. Invest.* **94**, 1524 (1994). Disposition and metabolism: R. Dixon *et al., J. Clin. Pharmacol.* **33**, 955 (1993). HPLC determn in human plasma: L.-S. Chen *et al., J. Liq. Chromatog.* **18**, 1451 (1995). Clinical evaluation in coronary artery bypass graft surgery: J. M. Leung *et al., Anesth. Analg.* **78**, 420 (1994). Review of pharmacology: K. Mullane, M. Young, *Drug Dev. Res.* **28**, 336-343 (1993). Historical perspective: K. Mullane *et al., Trends Cardiovasc. Med.* **3**, 227-234 (1993). Brief overview: A. M. Alkhulaifi, W. B. Pugsley, *Brit. Heart J.* **73**, 304-305 (1995).

Crystals from methanol or water, mp 213-214° (dec). Also reported as slightly brownish prisms from 90% aq ethanol, mp 206-208° (dec) (Saito). uv max (pH 7 and 1N NaOH): 265 nm (ε 12400). uv max (1N HCl): 245, 265 nm (ε 8670, 10320).

THERAP CAT: Cardioprotective.

14. Acamprosate Calcium. *3-(Acetylamino)-1-propanesulfonic acid calcium salt (2:1);* calcium acetyl homotaurinate; Ca-AOTA; calcium bisacetyl homotaurine; Aotal. $C_{10}H_{20}CaN_2O_8S_2$; mol wt 400.49. C 29.99%, Ca 10.01%, N 6.99%, O 31.96%, S 16.01%. GABA (γ-aminobutyric acid, *q.v.*) agonist. Prepn: J. P. Durlach, **Ger.** pat. **3,019,350**; *idem,* **U.S.** pat. **4,355,043** (1980, 1982 both to Lab. Meram). Physicochemical and pharmacological study: C. Chabenat *et al., Methods Find. Exp. Clin. Pharmacol.* **10**, 311 (1988). Pharmacology: J. Durlach *et al., ibid.* 437; A. Guiet-Bara *et al., Alcohol* **5**, 63 (1988). Suppression of ethanol intake in rats: F. Boismare *et al., Pharmacol. Biochem. Behav.* **21**, 787 (1984); J. Le Magnen *et al., Alcohol* **4**, 97 (1987). Evaluation of abuse potential: K. A. Grant, W. L. Woolverton, *Pharmacol. Biochem. Behav.* **32**, 607 (1989). HPLC determn in plasma: C. Chabenat *et al., J. Chromatog.* **414**, 417 (1987). Clinical evaluation in relapse prevention in weaned alcoholics: J. P. L'Huintre *et al., Lancet* **1**, 1014 (1985); J. P. L'Huintre *et al., Alcohol Alcoholism* **25**, 613 (1990). Brief review: M. Daoust *et al., Drugs of Today* **27**, 75-77 (1991).

Colorless crystalline powder, mp 270°. uv max (water): 192 nm (ε 7360). LD_{50} i.p. in male mice: 1.87 g/kg (Durlach, 1982).

THERAP CAT: In treatment of alcoholism.

15. Acarbose. *O-4,6-Dideoxy-4-[[[1S-(1α,4α,5β,6α)]-4,5,6-trihydroxy-3-(hydroxymethyl)-2-cyclohexen-1-yl]amino]-α-D-glucopyranosyl-(1→4)-O-α-D-glucopyranosyl-(1→4)-D-glucose; 4'',6''-dideoxy-4''-[[(1S)-(1,4,6/5)-4,5,6-trihydroxy-3-hydroxymethyl-2-cyclohexenylamino]maltotriose;* Bay g 5421; Glucobay. $C_{25}H_{43}NO_{18}$; mol wt 645.61. C 46.51%, H 6.71%, N 2.17%, O 44.61%. Pseudotetrasaccharide containing an unsaturated cyclitol moiety. An α-glucosidase inhibitor that reduces sugar absorption in the gastrointestinal tract. Isoln from strains of *Actinoplanes:* W. Frommer *et al.,* **Ger.** pat. **2,347,782**; *eidem,* **U.S.** pat. **4,062,950** (1975, 1977 both to Bayer). Total synthesis: S. Ogawa, Y. Shibata, *Chem. Commun.* **1988**, 605; *eidem, Carbohyd. Res.* **189**, 309 (1989). Biosynthetic studies: U. Degwert *et al., J. Antibiot.* **40**, 855 (1987). Glucosidase inhibition studies: D. D. Schmidt *et al., Naturwiss.* **64**, 535 (1977); W. Puls *et al., ibid.* 536. Use in treatment of diabetic adults: D. Sailor, G. Roder, *Arzneimittel-Forsch.* **30**, 2182 (1980); H. Laube *et al., ibid.* 1154. Long-term study in sulfonylurea-treated diabetics: H. Vierhapper *et al., Diabetologia* **20**, 586 (1981). Potential use in prophylaxis of dental caries: N. E. Fiehn, D. Moe, *Scand. J. Dent. Res.* **90**, 124 (1982). Review of pharmacodynamics, pharmacokinetics and therapeutic potential: S. P. Clissold, C. Edwards, *Drugs* **35**, 214-243 (1988).

Amorphous powder. $[\alpha]_D^{18}$ +165° (c = 0.4 in water).

THERAP CAT: Antidiabetic.

16. Acebutolol. *N-[3-Acetyl-4-[2-hydroxy-3-[(1-methylethyl)amino]propoxy]phenyl]butanamide; 3'-acetyl-4'-[2-hydroxy-3-(isopropylamino)propoxy]butyranilide;* 1-(2-acetyl-4-n-butyramidophenoxy)-2-hydroxy-3-isopropylaminopropane; 5'-butyramido-2'-(2-hydroxy-3-isopropylaminopropoxy)acetophenone; Prent. $C_{18}H_{28}N_2O_4$; mol wt 336.43. C 64.26%, H 8.39%, N 8.33%, O 19.02%. Cardioselective β-adrenergic blocker. Prepn: K. R. H. Wooldridge, B. Basil,

S. Afr. pat. **68 08,345**; *eidem,* U.S. pat. **3,857,952** (1969, 1974 both to May & Baker). Pharmacology: Cuthbert, Owusu-Ankomah, *Brit. J. Pharmacol.* **43**, 639 (1971); Basil *et al., ibid.* **48**, 198 (1973); Lewis *et al., Brit. Heart J.* **35**, 743 (1973). HPLC determn in plasma and urine: M. Piquette-Miller *et al., J. Chromatog.* **526**, 129 (1990). Crystal structure: A. Carpy *et al., Acta Crystallogr.* **B35**, 185 (1979). Review of pharmacology, therapeutic efficacy in hypertension, angina pectoris, arrhythmia: B. N. Singh *et al., Drugs* **29**, 531-569 (1985); of pharmacokinetics, clinical pharmacology and clinical efficacy: G. DeBono *et al., Am. Heart J.* **109**, 1211-1223 (1985). Comprehensive description: R. T. Foster, R. A. Carr in *Analytical Profiles of Drug Substances* vol. **19**, K. Florey Ed. (Academic Press, New York, 1990) pp 1-26.

Crystals, mp 119-123°.

Hydrochloride, $C_{18}H_{28}N_2O_4 \cdot HCl$, *IL-17803A, M & B 17803A, Acetanol, Neptall, Sectral.* Crystals from anhydr methanol-anhydr diethyl ether, mp 141-143°. Freely sol in water. Soly at room temperature (mg/ml): water 200; ethanol 70.

THERAP CAT: Antihypertensive; antianginal; antiarrhythmic (class II).

17. Acecainide. *4-(Acetylamino)-N-[2-(diethylamino)ethyl]benzamide;* 4'-[[2-(diethylamino)ethyl]carbamoyl]-acetanilide; *N*-acetylprocainamide. $C_{15}H_{23}N_3O_2$; mol wt 277.37. C 64.96%, H 8.36%, N 15.15%, O 11.54%. Metabolite of procainamide, *q.v.* Prepn: E. C. Schreiber, **Ger.** pat. **2,062,978** (1971 to Squibb), *C.A.* **75**, 76427 (1972). Pharmacology studies: R. D. Reynolds, B. L. Kamath, *Eur. J. Pharmacol.* **59**, 115 (1979); R. D. Reynolds, R. J. Gorczynski, *J. Pharmacol. Exp. Ther.* **212**, 579 (1980). Pharmacokinetics: M. Wierzchowiecki *et al., Int. J. Clin. Pharmacol. Ther. Toxicol.* **18**, 272 (1980). Clinical pharmacology and anti-arrhythmic efficacy: J. Kluger *et al., Am. J. Cardiol.* **45**, 1250 (1980); R. A. Winkle *et al., ibid.* **47**, 123 (1981). HPLC determn in plasma: D. Raphanaud *et al., Ther. Drug Monit.* **8**, 365 (1986).

Hydrochloride, $C_{15}H_{23}N_3O_2 \cdot HCl$, *ASL-601, NAPA.* Cryst, mp 190-193°.

THERAP CAT: Antiarrhythmic (cardiac depressant).

18. Acecarbromal. *N-[(Acetylamino)carbonyl]-2-bromo-2-ethylbutanamide; N*-acetyl-*N*-bromodiethylacetylurea; acetylbromodiethylacetylcarbamide; *N*-acetyl-*N'*-α-bromo-α-ethylbutyrylcarbamide; acetylcarbromal; Abasin; Carbased; Sedamyl; Acetyl Adalin. $C_9H_{15}BrN_2O_3$; mol wt 279.13. C 38.73%, H 5.42%, Br 28.63%, N 10.04%, O 17.20%. Prepd by acetylating carbromal with acetic anhydride in the presence of $ZnCl_2$: **Ger.** pat. **327,129**; *see also* H. P. Kaufmann, *Arzneimittelsynthese* (Springer, 1953).

Crystals, slightly bitter taste, mp 109°. Slightly sol in water; freely sol in alcohol, ethyl acetate.

Caution: This substance may be habit forming and is listed in the U.S. Code of Federal Regulations, Title 21 Part 329.1 (1995).

THERAP CAT: Sedative, hypnotic.

19. Aceclofenac. *2-[(2,6-Dichlorophenyl)amino]benzeneacetic acid carboxymethyl ester;* 2-[(2,6-dichlorophenyl)-amino]phenylacetoxyacetic acid; glycolic acid [o-(2,6-dichloroanilino)phenyl]acetate ester; PR-82/3; Airtal; Biofenac; Tresquim. $C_{16}H_{13}Cl_2NO_4$; mol wt 354.19. C 54.26%, H 3.70%, Cl 20.02%, N 3.95%, O 18.07%. Prepn: A. V. Casas, **Span.** pat. **8,404,783**; *idem,* U.S. pat. **4,548,952** (1984, 1985 both to Prodes). Gastrointestinal tolerance in rats in comparison with diclofenac sodium, *q.v.*: V. Rimbau *et al., Farmaco Ed. Prat.* **43**, 19 (1988). Clinical trial in comparison with acetaminophen, *q.v.*, in episiotomal pain: A. Yscla, *Drugs Exp. Clin. Res.* **14**, 491 (1988). Clinical evaluation in rheumatoid arthritis: R. Ballesteros *et al., Clin. Trials J.* **27**, 12 (1990).

White crystals from cyclohexane, mp 149-150°. uv max (ethanol): 275 nm (log ε 4.14).

THERAP CAT: Anti-inflammatory; analgesic.

20. Acedapsone. *N,N'-(Sulfonyldi-4,1-phenylene)bis-acetamide; 4',4'''-sulfonylbis[acetanilide];* bis(4-acetamidophenyl)sulfone; 4,4'-diacetyldiaminodiphenyl sulfone; 4,4'-diacetylaminodiphenyl sulfone; *N,N'*-diacetyl-4,4'-diaminodiphenyl sulfone; DADDS; diacetyldapsone; sulfadiamine; 1399F; CI-556; Hansolar; Rodilone. $C_{16}H_{16}N_2O_4S$; mol wt 332.38. C 57.82%, H 4.85%, N 8.43%, O 19.25%, S 9.65%. Prepn: Fromm, Wittmann, *Ber.* **41**, 2270 (1908); Raiziss *et al., J. Am. Chem. Soc.* **61**, 2763 (1939); Elslager *et al., J. Med. Chem.* **12**, 357 (1969). Properties: Elslager, Worth, *Nature* **206**, 630 (1965).

Crystalline solid, mp 289-292°. uv max (methanol): 256, 284 nm (ε 25500, 36200). Soly in water: 0.003 mg/ml; in 40% benzyl benzoate-60% castor oil: 0.026 mg/ml.

THERAP CAT: Antimalarial; antibacterial (leprostatic).

21. Acediasulfone. *N-[4-[(4-Aminophenyl)sulfonyl]-phenyl]glycine;* N-p-sulfanilylphenylglycine; p-amino-p'-(carboxymethylamino)diphenyl sulfone; 4-carboxymethylamino-4'-aminodiphenylsulfone; diaminodiphenylsulfone-*N*-acetic acid. $C_{14}H_{14}N_2O_4S$; mol wt 306.34. C 54.89%, H 4.61%, N 9.14%, O 20.89%, S 10.47%. Prepn: Jackson, *J. Am. Chem. Soc.* **70**, 680 (1948); **Swiss** pats. **254,803** and **278,482** (1949, 1952, to Cilag Ltd.); Rawlins, U.S. pat. **2,589,211** (1952 to Parke, Davis).

Crystals, mp 194°. Sol in methanol, dil sodium hydroxide, acetone.

Sodium salt, $C_{14}H_{13}N_2NaO_4S$, *Sulfon-Cilag.*

Morpholine salt, $C_{18}H_{23}N_3O_5S$, *Bentrofene.* Prepn: Martin, Habicht, U.S. pat. 2,751,382 (1956 to Cilag Ltd.). Glittering crystals, mp 133-135° (dec).

THERAP CAT: Antibacterial.

22. Acefylline. *1,2,3,6-Tetrahydro-1,3-dimethyl-2,6-dioxopurine-7-acetic acid;* carboxymethyltheophylline; 7-theophyllineacetic acid. $C_9H_{10}N_4O_4$; mol wt 238.20. C 45.38%, H 4.23%, N 23.52%, O 26.87%. Prepd from theophylline and chloroacetic acid in hot dil NaOH: **Ger. pat. 352,980** (1922 to E. Merck); *Frdl.* **14,** 1320. Prepn of free acid and sodium salt: Milletti, Virgili, *Chimica (Milan)* **6,** 394 (1951), *C.A.* **46,** 8615h (1952).

Crystals from water, mp 271°.
Sodium salt, $C_9H_9N_4NaO_4$, *Aminodal.* Silky needles, mp > 300°.

THERAP CAT: Diuretic; cardiotonic; bronchodilator.

23. Acefylline Piperazine. *1,2,3,6-Tetrahydro-1,3-dimethyl-2,6-dioxo-7H-purine-7-acetic acid compd with piperazine;* acepifylline; piperazine theophylline-7-acetate; 7-theophyllineacetic acid piperazine salt; piperazine theophylline ethanoate; Dynaphylline; Etaphydel; Etaphylline; Etafillina. $C_{13}H_{20}N_6O_4$; mol wt 324.34. C 48.14%, H 6.22%, N 25.91%, O 19.73%. Undefined mixture of the 1:1 and 2:1 salts which contains 75-78% theophylline acetic acid and 22-25% anhydrous piperazine. Active metabolite is acefylline, *q.v.* Prepn: Baisse, *Bull. Soc. Chim. France* **1949,** 769. Pharmacokinetics: J. Zuidema, F. W. H. Merkus, *Pharm. Weekbl.* **113,** 605 (1978); S. Sved *et al., Biopharm. Drug Dispos.* **2,** 177 (1981). HPLC determn in serum: J. Zuidema, F. W. H. Merkus, *J. Chromatog.* **145,** 489 (1978). GC determn in urine: J. Zuidema, H. Hilbers, *ibid.* **182,** 445 (1980).

THERAP CAT: Bronchodilator.

24. Aceglatone. *D-Glucaric acid di-γ-lactone diacetate;* 2,5-di-*O*-acetyl-D-glucaro-1,4:6,3-dilactone; 2,5-di-*O*-acetyl-D-glucosaccharo-1,4:6,3-dilactone; Aceglaton; Glucaron. $C_{10}H_{10}O_8$; mol wt 258.18. C 46.52%, H 3.90%, O 49.58%. Prepn and structure: Hirasaka, Umemoto, *Chem. Pharm. Bull.* **13,** 325 (1965), *C.A.* **63,** 3024h (1965); Ishidate *et al.,* **Japan. pat. 14,956('67)** (1967 to Tokyo Biochem. Res. Com.), *C.A.* **68,** 78558m (1968). Pharmacological studies: Iida *et al., Japan. J. Pharmacol.* **15,** 88 (1965), *C.A.* **63,** 5961g (1965). Review and toxicity data: *Japan. Med. Gaz.* **8**(8), 15 (1971).

White, odorless and tasteless crystalline powder, mp 185-186° (Hirasaka). Sol in dimethylformamide, sparingly sol in acetone, slightly sol in dioxane, methanol and ethanol. Practically insol in water. LD_{50} in mice, rats (g/kg): > 20, > 10 orally; > 20, > 10 s.c.; 5.80-6.35, 6.10-6.15 i.p. *(Japan. Med. Gaz.).*

THERAP CAT: Antineoplastic.

25. Aceglutamide. *N²-Acetyl-L-glutamine;* α-N-acetyl-L-glutamine; Acutil-S. $C_7H_{12}N_2O_4$; mol wt 188.18. C 44.68%, H 6.43%, N 14.89%, O 34.01%. Prepn: P. Karrer

et al., Helv. Chim. Acta **9,** 301 (1926); **Brit. pat. 792,576** (1958 to Merck & Co.), *C.A.* **53,** 2109a (1959); I. J. Maschler, N. Lichtenstein, *Biochim. Biophys. Acta* **57,** 252 (1962). Stability study: G. Sekules, G. Guadagnini, *Farmaco Ed. Prat.* **21,** 22 (1966). NMR study: W. Voelter *et al., Z. Naturforsch. B* **26,** 213 (1971). Fermentation study: T. Nakanishi, *J. Ferment. Technol.* **56,** 573 (1978). Prepn of the aluminum complex: T. Kagawa *et al.,* **Ger. pat. 2,127,-176** corresp to U.S. pat. 3,787,466 (1971, 1974 both to Kyowa). Physico-chemical properties: E. Hayakawa *et al., Yakugaku Zasshi* **97,** 731 (1977), *C.A.* **87,** 141198 (1977). Effect on exptl chronic gastric ulcer: H. Tanaka *et al., Oyo Yakuri* **7,** 1035 (1973), *C.A.* **81,** 33283v (1974). Cytoprotective effect: H. Tanaka, *Arzneimittel-Forsch.* **36,** 1485 (1986).

Crystals from ethanol, mp 197°. $[\alpha]_D^{20} -12.5°$ (c = 2.9 in water).

Aluminum complex, $C_{35}H_{59}Al_3N_{10}O_{24}$, *pentakis (N²-acetyl-L-glutaminato)tetrahydroxytrialuminum, aceglutamide aluminum, KW-110, Glumal.* White powder, mp 221° (dec). Sol in water. Practically insol in methanol, ethanol, acetone. LD_{50} in male mice, rats (g/kg): 14.3, > 14.5 orally; 5.0, 4.2 i.p.; 0.46, 0.40 i.v. (Kagawa).

THERAP CAT: Free acid as nootropic; aluminum complex as anti-ulcerative.

26. Acemannan. Carrisyn. Long-chain, polydispersed, β-(1,4)-linked acetylated mannan obtained from the mucilage of *Aloe barbadensis,* Miller, *Liliaceae* (aloe vera). Isolation, purification, and characterization: B. H. McAnalley, **PCT Int. pat. Appl. 87 00,052;** *idem,* **U.S. pat. 4,735,935** (1987, 1988 both to Carrington Labs.). Structural study: S. Manna, B. H. McAnalley, *Carbohyd. Res.* **241,** 317 (1993). Immunostimulant effect *in vitro:* D. Womble, J. H. Helderman, *Int. J. Immunopharmacol.* **10,** 967 (1988). Inhibition of HIV-1 replication *in vitro:* J. B. Kahlon *et al., Mol. Biother.* **3,** 127 (1991). Veterinary trial in feline-leukemia virus infection: M. A. Sheets *et al., ibid.* 41; in canine and feline spontaneous tumors: C. Harris *et al., ibid.* 207. Toxicological evaluation: R. W. Fogleman *et al., Vet. Hum. Toxicol.* **34,** 201 (1992).

THERAP CAT: Antiviral; immunomodulator.

THERAP CAT (VET): Antiviral.

27. Acemetacin. *1-(4-Chlorobenzoyl)-5-methoxy-2-methyl-1H-indole-3-acetic acid carboxymethyl ester;* [[1-(4-chlorobenzoyl)-5-methoxy-2-methylindol-3-yl]acetoxy]acetic acid; TV-1322; Acemix; Emflex; Rantudil; Rheumibis; Solart. $C_{21}H_{18}ClNO_6$; mol wt 415.83. C 60.66%, H 4.36%, Cl 8.53%, N 3.37%, O 23.09%. Deriv of indomethacin, *q.v.* Prepn: K. H. Boltze *et al.,* **Ger. pat. 2,234,651;** *eidem,* **U.S. pat. 3,910,952** (1972, 1975 both to Troponwerke). Series of articles on chemistry, analysis, pharmacodynamics, toxicology and clinical trials: *Arzneimittel-Forsch.* **30,** 1313-1468 (1980). Toxicity data: H. Jacobi, H.-D. Dell, *ibid.* 1398.

Very fine pale yellow crystals from petr ether, mp 150-153°. LD_{50} in male, female mice, male, female rats (mg/kg):

55.5, 18.42, 24.2, 30.1 orally; 34.1, 51.1, 38.1, 28.3 i.v. (Jacobi, Dell).

THERAP CAT: Anti-inflammatory.

28. Acenaphthene. *1,2-Dihydroacenaphthylene; peri-ethylenenaphthalene; 1,8-ethylenenaphthalene.* $C_{12}H_{10}$; mol wt 154.21. C 93.46%, H 6.54%. Occurs in coal tar. Isoln: Ges. f. Teerverwertung, **Ger.** pat. **277,110**; *Chem. Zentr.* **1914,** II, 597. In petroleum residues: Orloff *et al., C.A.* **31,** 2800⁹ (1937). By passing ethylene and benzene or naphthalene though a red hot tube: Berthelot, *Bull. Soc. Chim.* [2] **7,** 274; **8,** 226, 245 (1867). By heating tetrahydroacenaphthene with sulfur to 180°: Braun *et al., Ber.* **55,** 1694 (1922). From acenaphthenone or acenaphthenequinone by high pressure hydrogenation in decalin with nickel at 180-240°: Braun, Bayer, *Ber.* **59,** 921, 923 (1926). From acenaphthenone oxime: Morgan, Stanley, *J. Soc. Chem. Ind. (London)* **44,** 494T (1925).

Orthorhombic bipyramidal needles from alcohol. d 1.189. mp 95°. bp 279°. uv spectrum: Seshan, *Proc. Indian Acad. Sci.* **A3,** 148 (1936). Factors influencing the uv spectrum: Jones, *J. Am. Chem. Soc.* **67,** 2127 (1945). Insol in water. One gram dissolves in 31 ml alcohol, 56 ml methanol, 25 ml propanol, 2.5 ml chloroform, 5 ml benzene or toluene. 3.2 g are sol in 100 ml glacial acetic acid. Forms water-sol, cryst complexes with desoxycholic acid, containing two molecules of the bile acid as a rule. The complexes crystallize when concentrated solns of the proper amount of the components in alcohol or dioxane are allowed to cool slowly.

USE: Dye intermediate; manuf plastics; insecticide; fungicide.

29. Acenocoumarol. *4-Hydroxy-3-[1-(4-nitrophenyl)-3-oxobutyl]-2H-1-benzopyran-2-one; 3-(α-acetonyl-p-nitrobenzyl)-4-hydroxycoumarin; 3-(α-p-nitrophenyl-β-acetylethyl)-4-hydroxycoumarin; acenocoumarin; nicoumalone; G-23350; Minisintrom; Sinthrome; Sintrom.* $C_{19}H_{15}NO_6$; mol wt 353.33. C 64.59%, H 4.28%, N 3.96%, O 27.17%. Prepn: Stoll, Litvan, **U.S.** pat. **2,648,682** (1953 to Geigy). Pharmacology: Jindal, Shah, *Arzneimittel-Forsch.* **16,** 878 (1966).

Crystals, mp 196-199°. Almost insol in water or organic solvents. Forms water-sol salt with alkalies. LD_{50} in mice: 114.7 mg/kg i.p.

THERAP CAT: Anticoagulant.

30. Aceperone. *N-[[1-[4-(4-Fluorophenyl)-4-oxobutyl]-4-phenyl-4-piperidinyl]methyl]acetamide; N-[[1-[3-(p-fluorobenzoyl)propyl]-4-phenyl-4-piperidyl]methyl]acetamide; 1-(p-fluorophenyl)-4-(4-phenyl-4-acetamidomethylpiperidino)-1-butanone; 1-[γ-(4-fluorobenzoyl)propyl]-4-acetamidomethyl-4-phenylpiperidine; N-[1-[3-(p-fluorobenzoyl)propyl]-4-phenylpiperidin-4-ylmethyl]acetamide; 4'-fluoro-4-(4-acetamidomethyl-4-phenylpiperido)butyrophenone; acetabutone; acetobuton; R-3248.* $C_{24}H_{29}FN_2O_2$; mol wt 396.50. C 72.70%, H 7.37%, F 4.79%, N 7.07%, O 8.07%. Prepd by reacting γ-chloro-4-fluorobutyrophenone with 4-acetamidomethyl-4-phenylpiperidine: P. A. J. Janssen, **Belg.** pat. **606,849** (1961), corresp to **U.S.** pat. **3,083,205** (1961, 1963

both to Janssen). Crystal structure: N. Van Opdenbosch *et al., Acta Crystallogr.* **33B,** 171 (1977).

Crystals, mp 97-100°.

31. Acephate. *Acetylphosphoramidothioic acid O,S-dimethyl ester;* Ortho 12420; Orthene. $C_4H_{10}NO_3PS$; mol wt 183.17. C 26.23%, H 5.50%, N 7.65%, O 26.20%, P 16.91%, S 17.51%. Prepn and activity: Magee, **U.S.** pats. **3,716,600** and **3,845,172** (1973, 1974 both to Chevron).

White solid, mp 64-68° (impure). Very sol in water; moderately sol in acetone, alc. Low soly in aromatic solvents. LD_{50} orally in rats: 700 mg/kg (Magee, 1973).

USE: Contact and systemic insecticide.

32. Acepromazine. *1-[10-[3-(Dimethylamino)propyl]-10H-phenothiazin-2-yl]ethanone; 10-[3-(dimethylamino)propyl]phenothiazin-2-yl methyl ketone;* 2-acctyl-10-(3-dimethylaminopropyl)phenothiazine; 3-acetyl-10-(3-dimethylaminopropyl)phenothiazine; 10-(3-dimethylaminopropyl)-phenothiazine-3-ethylone; acetazine; acepromazine; acetylpromazine; 1522-CB; Vetranquil. $C_{19}H_{22}N_2OS$; mol wt 326.46. C 69.90%, H 6.79%, N 8.58%, O 4.90%, S 9.82%. Prepn: Schmitt *et al., Bull. Soc. Chim. France* **1957,** 938.

Orange-colored oil, $bp_{0.5}$ 220-240°. Maleate, $C_{19}H_{22}N_2OS.C_4H_4O_4$, *Atravet, Calmivet, Notensil, Plegicil, Sedalin, Soprontin.* Yellow crystals from ethyl acetate, mp 135-136°. Sol in water. pH of 0.1% aq soln 5.2. LD_{50} in rats (calcd as free base): 130 mg/kg orally; 70 mg/kg i.v.

THERAP CAT (VET): Tranquilizer.

33. Acerin. An extract from the dried fruit of the Norway maple *Acer plantanoides* L., *Aceraceae.* Freshly prepd solutions of acerin destroy *E. coli* and vaccinia virus in 5 min but are less effective on the Staphylococcus Twort virus. Phagicidal action also exists even in the presence of the host cells. Extraction and properties: G. Fischer, *Acta Pathol. Microbiol. Scand.* **31,** 433 (1952); **34,** 482 (1954); Fischer *et al., Experientia* **10,** 329 (1954).

Practically insol in methanol, ethanol, ether, acetone. Does not dialyze through a cellophane membrane at room temp but passes through a Seitz filter. Sodium chloride precipitates the active principle in acerin preparations as does also normal serum with 1:800 solns of acerin. Strong acerin preparations acquire reddish-brown color after treatment with alkali and oxidation in air. In general, the substance behaves similarly to vegetable tannins. It is practically nontoxic to rabbits and mice when administered s.c.

34. Acerola. The ripe fruit of *Malpighia punicifolia* L., *Malpighiaceae,* also called West Indian Cherry. *Habit.* Central America, Puerto Rico. Probably the richest natural source of ascorbic acid. Analysis of pitted fruit: Water

92.28%; ascorbic acid 1690 mg/100 g; vitamin A 11.0 mg/100 g; niacin 407 γ/100 g; vitamin B₆ 8.7 γ/100 g; thiamine-HCl 30.0 γ/100 g; fluorine 10 γ/100 g; carbohydrates 6.4%. The pits comprise 19.25% of the fruit. *Ref:* Derse, Elvehjem, *J. Am. Med. Assoc.* **156,** 1501 (1954). Commercial aspects: *J. Agr. Food Chem.* **2,** 1155 (1954). *Review:* Moscoso, *Econ. Bot.* **10,** 280 (1956). Method of preparing acerola juice concentrates: Morse, U.S. pats. 3,012,942-3 (1961 to Nutrilite Products).

THERAP CAT: Nutrient.

35. Acesulfame. *6-Methyl-1,2,3-oxathiazin-4(3H)-one 2,2-dioxide;* 6-methyl-3,4-dihydro-1,2,3-oxathiazin-4-one 2,2-dioxide; acetosulfam. $C_4H_5NO_4S$; mol wt 163.15. C 29.45%, H 3.09%, N 8.58%, O 39.23%, S 19.65%. Nonnutritive artificial sweetener. Prepn: K. Clauss, H. Jensen, **Ger.** pat. **2,001,017;** *eidem,* **U.S.** pat. **3,689,486** (1971, 1972 to Hoechst); K. Clauss *et al., Z. Lebensm.-Unters. Forsch.* **162,** 37 (1976). Crystal structure: E. F. Paulus, *Acta Crystallogr.* **B31,** 1191 (1975). HPLC analysis: H. Grosspietsch, H. Hachenberg, *Z. Lebensm.-Unters. Forsch.* **171,** 41 (1980). *Book: Acesulfame-K,* D. G. Mayer, F. H. Kemper, Eds. (Marcel Dekker, Inc., New York, 1991) 243 pp.

Needles from benzene or chloroform, mp 123-123.5°. Potassium salt, $C_4H_4KNO_4S$, *acesulfame-K, HOE-095K, Sunette.* Colorless, odorless powder. Dec pt ~225°. d (solid) 1.81 g/cm³; d (bulk) 1.1-1.3 kg/dm³. uv max: 227 nm (ε 1.0762×10⁴). Very sol in water, DMF, DMSO. Sol in alc, glycerin-water. Soly in water (g/l): ~150 (0°); ~210 (10°); ~270 (20°); ~360 (30°); ~460 (40°); ~580 (50°); ~830 (70°); ~1300 (100°). Soly at 20° (g/l): methanol ~10; ethanol (anhydrous) ~1; glycerol (anhydrous) ~30; glycerol/water (80:20, v/v) ~82; glycerol/water (50:50, v/v) ~162; acetone ~0.8; glacial acetic acid ~130. Soly at 23° (g/l): ethanol/water (80:20, v/v) ~46; ethanol/water (60:40, v/v) ~100; ethanol/water (40:60, v/v) ~155; ethanol/water (20:80, v/v) ~221. LD₅₀ in rats (mg/kg): 7431 orally, 2243 i.p. (Mayer, Kemper).

USE: Potassium salt as sweetener for foods, cosmetics.

36. Acetal. *1,1-Diethoxyethane;* diethylacetal; acetaldehyde diethyl acetal; ethylidene diethyl ether. $C_6H_{14}O_2$; mol wt 118.18. C 60.98%, H 11.94%, O 27.08%. CH₃CH-(OC₂H₅)₂. Made from acetaldehyde and alcohol in the presence of anhydrous calcium chloride or of small quantities of mineral acid: H. Adkins, B. H. Nissen, *J. Am. Chem. Soc.* **44,** 2750 (1922); *eidem, Org. Syn.* coll. vol. I, 1 (2nd ed., 1941). Toxicity study: H. F. Smyth *et al., J. Ind. Hyg. Toxicol.* **31,** 60 (1949).

Volatile liquid. d²⁰₄ 0.8254. bp₇₆₀ 102.7°; bp₂₀₀ 66.3°; bp₆₀ 39.8°; bp₄₀ 31.9°; bp₂₀ 19.6°; bp₁₀ +8.0°; bp₅ −2.3°; bp₁.₀ −23°. n²⁰_D 1.38193. Flash pt, closed cup: 97°F (36°C). uv spectrum: Purvis, *J. Chem. Soc.* **127,** 9 (1925). 100 g water dissolve 5 g acetal. Misc with alcohol, 60% alcohol, ether. Sol in heptane, methylcyclohexane, propyl-, isopropyl-, butyl-, isobutyl alcohol, and ethyl acetate. Tends to polymerize on standing. Stable to alkalies. LD₅₀ orally in rats: 4.57 g/kg (Smyth).

USE: Solvent; in synthetic perfumes such as jasmine; in organic syntheses.

THERAP CAT: Hypnotic.

37. Acetaldehyde. Ethanal; "aldehyde"; acetic aldehyde; ethylaldehyde. C_2H_4O; mol wt 44.05. C 54.53%, H 9.15%, O 36.32%. CH₃CHO. Produced by oxidation of alcohol with Na₂Cr₂O₇ and H₂SO₄; usually from acetylene, dil H₂SO₄ and mercuric oxide as catalyst; also by passing alcohol vapor over a heated metallic catalyst. Lab procedure from ethanol: Wertheim, *J. Am. Chem. Soc.* **44,** 2658 (1922); Fricke, Havestadt, *Angew. Chem.* **36,** 546 (1923); Gattermann-Wieland, *Praxis des organischen Chemikers* (de Gruyter, Berlin, 40th ed., 1961) p 180; from acetylene: Gattermann-Wieland, *op. cit.* 183; from paraldehyde: A. I.

Vogel, *Practical Organic Chemistry* (Longmans, London, 3rd ed., 1959) p 324; by catalytic oxidation of ethylene in aq soln: J. Smidt *et al., Angew Chem.* **71,** 176 (1959); by oxidation of ethylene in fuel cells in the gas phase: K. Otsuka *et al., Chem. Commun.* **1988,** 1272. Manuf: Faith, Keyes & Clark's *Industrial Chemicals,* F. A. Lowenheim, M. K. Moran, Eds. (Wiley-Interscience, New York, 4th ed., 1975) pp 1-7. Toxicity data: Smyth, *Arch. Ind. Hyg. Occup. Med.* **4,** 119 (1951). *Review:* H. J. Hagemeyer in Kirk-Othmer *Encyclopedia of Chemical Technology* vol. 1 (John Wiley & Sons, New York, 4th ed., 1991) pp 94-109.

Flammable liquid; characteristic, pungent odor. d¹⁶₄ 0.788. mp −123.5°. bp 21°. n²⁰_D 1.3316. Flash pt, closed cup: −36°F (−38°C). Miscible with water, alcohol. *Keep cold. Chill thoroughly before opening.* LD₅₀ orally in rats: 1930 mg/kg (Smyth).

Caution: Potential symptoms of overexposure are eye, nose and throat irritation; conjunctivitis; coughing; central nervous system depression; eye and skin burns; dermatitis; delayed pulmonary edema. *See NIOSH Pocket Guide to Chemical Hazards* (DHHS/NIOSH 90-117, 1990) p 30. *See also Clinical Toxicology of Commercial Products,* R. E. Gosselin *et al.,* Eds. (Williams & Wilkins, Baltimore, 5th ed., 1984) Section II, p 186. This substance may reasonably be anticipated to be a carcinogen: *Seventh Annual Report on Carcinogens* (PB95-109781, 1994) p 74.

USE: Manuf paraldehyde, acetic acid, butanol, perfumes, flavors, aniline dyes, plastics, synthetic rubber; silvering mirrors, hardening gelatin fibers. Flavoring agent in foods and beverages.

38. Acetaldehyde Ammonia. *1-Aminoethanol;* α-aminoethyl alcohol; aldehyde ammonia. C_2H_7NO; mol wt 61.08. C 39.33%, H 11.55%, N 22.93%, O 26.19%. CH₃-CH(OH)NH₂. Prepd from acetaldehyde and ammonia: Aschan, *Ber.* **48,** 874 (1915).

Crystals; gradually turns yellow to brown in air. mp 97°. bp 110°, partly decomposing. Freely sol in water, slightly in ether. *Protect from light and air.*

Caution: Irritates eyes, mucous membranes.

USE: For preparing pure acetaldehyde; in organic syntheses.

39. Acetaldehyde Sodium Bisulfite. *1-Hydroxyethanesulfonic acid sodium salt.* $C_2H_5NaO_4S$; mol wt 148.12. C 16.22%, H 3.40%, Na 15.52%, O 43.21%, S 21.65%. CH₃CH(OH)SO₃Na.

Hemihydrate, crystals. Decomposed by acids. Freely sol in water. Insol in alcohol.

Caution: Irritates skin, mucous membranes.

USE: Making pure acetaldehyde; in organic synthesis.

40. Acetaldoxime. *Acetaldehyde oxime;* aldoxime; ethylidenehydroxylamine. C_2H_5NO; mol wt 59.07. C 40.67%, H 8.53%, N 23.71%, O 27.09%. CH₃CH=NOH. Prepn: Dunstan, Dymond, *J. Chem. Soc.* **61,** 470 (1892). Manuf: Donaruma, U.S. pat. **2,763,686** (1956 to du Pont).

Two crystalline modifications, mp 12° (β-form), mp 46.5° (α-form). d 0.966. bp 114.5°. n²⁰_D 1.415. Dec by aq HCl into acetaldehyde and hydroxylamine. Very sol in water, alcohol, ether.

41. Acetamide. Acetic acid amide. C_2H_5NO; mol wt 59.07. C 40.67%, H 8.53%, N 23.71%, O 27.09%. CH₃CONH₂. Prepd by fractional distillation of ammonium acetate: Coleman, Alvarado, *Org. Syn.;* coll. vol. I, 3 (2nd ed., 1941); Gattermann-Wieland, *Praxis des organischen Chemikers* (40th ed., 1961) p 118; Vogel, *Practical Organic Chemistry* (3rd ed., 1959) p 401. Prepn from methyl acetate, W. P. Munro *et al.,* U.S. pat. **2,106,697** (1936 to Calco Chem.); from ethyl acetate, Vogel, *op. cit.,* p 403. Studies of acetamide as an ionizing solvent: Jauder, Winkler, *J. Inorg. Nucl. Chem.* **9,** 24, 32, 39 (1959). Toxicological study: Weisburger *et al., Toxicol. Appl. Pharmacol.* **14,** 163 (1969).

Deliquescent hexagonal crystals. Odorless when pure, but frequently has a mousy odor. d²⁰₄ 1.159. mp 81°. bp₇₆₀ 222°; bp₁₀₀ 158°; bp₄₀ 136°; bp₂₀ 120°; bp₁₀ 105°; bp₅ 92°. n⁷⁸_D 1.4274. Neutral reaction. pKb (25°): 14.51. One gram dissolves in 0.5 ml water, 2 ml alcohol, 6 ml pyridine. Sol in chloroform, glycerol, hot benzene.

USE: Solvent; molten acetamide is an excellent solvent for many organic and inorganic compounds. Solubilizer; renders sparingly soluble substances more soluble in water by mere addition or by fusion. Plasticizer; stabilizer. Manuf methylamine, denaturing alcohol. In organic syntheses.

42. Acetamidine Hydrochloride. *Ethanimidamide hydrochloride;* ethanamidine hydrochloride; α-amino-α-iminoethane hydrochloride; ethenylamidine hydrochloride; acediamine hydrochloride; SN-4455. $C_2H_7ClN_2$; mol wt 94.54. C 25.41%, H 7.46%, Cl 37.50%, N 29.63%. Prepd by passing HCl into a soln of acetonitrile in abs alcohol, then passing NH_3 into the reaction mixture: Pinner, *Ber.* **16**, 1654 (1883); **17**, 178 (1884); Dox, *Org. Syn.*, coll. vol. I, 5 (New York, 2nd ed., 1941). *See also* Fargher, *J. Chem. Soc.* **117**, 674 (1920). *Review:* Shriner, Neumann, *Chem. Rev.* **35**, 351-425 (1944).

Long prisms from alcohol, somewhat deliquescent, mp 174° (Fargher); mp 164-166° (Dox). Very sol in water. Sol in alcohols. Practically insol in acetone, ether. Should be stored in a closed container and in a cool place. If alkali is added to an aq soln, the free base is liberated.

Free base, $C_2H_6N_2$. uv max: 224 nm (ϵ 4000). pK_1 (25°): 12.1. Has a strong alkaline reaction and on slight warming dissociates into ammonia and acetic acid.

Caution: Irritates mucous membranes, skin. Avoid tasting, swallowing, inhalation of the dust, skin contact.

USE: In the synthesis of imidazoles, pyrimidines, triazines.

43. ϵ-Acetamidocaproic Acid. *6-(Acetylamino)hexanoic acid; 6-acetamidohexanoic acid;* acetaminocaproic acid; acexamic acid; *N*-acetyl-6-aminohexanoic acid. $C_8H_{15}NO_3$; mol wt 173.21. C 55.47%, H 8.73%, N 8.09%, O 27.71%. $CH_3CONH(CH_2)_5COOH$. Prepn: Offe, *Z. Naturforsch.* **2b**, 182 (1947); **Fr. pat.** **M2332** (1964 to Rowa), *C.A.* **61**, 577b (1964). Anti-inflammatory activity of salts: O. Guillard *et al., Pharmacology* **34**, 296 (1987). Clinical comparison with ranitidine in gastroduodenal ulcer: M. J. Varas Lorenzo, *Curr. Ther. Res.* **39**, 19 (1986).

Crystals from acetone, mp 104-105.5° (**Fr. pat.** **M2332**), mp 112° (Offe).

Sodium salt, $C_8H_{14}NNaO_3$, *Plastenan.*

Zinc salt, $C_{16}H_{28}N_2O_6Zn$, *zinc acexamate, Copinal.*

THERAP CAT: Anti-inflammatory; zinc salt as antiulcerative.

44. Acetamidoeugenol. *N,N-Diethyl-2-[2-methoxy-4-(2-propenyl)phenoxy]acetamide; 2-(4-allyl-2-methoxyphenoxy)-N,N-diethylacetamide; N,N*-diethyl-2-methoxy-4-allylphenoxyacetamide; 2-methoxy-4-allylphenoxyacetic acid *N,N*-diethylamide; 2-M-4-A; G-29505; Detrovel; Estil. $C_{16}H_{23}NO_3$; mol wt 277.36. C 69.29%, H 8.36%, N 5.05%, O 17.31%. Prepn: **Brit. pats.** **792,490; 837,995** (1958, 1960 to Geigy).

Oil. $bp_{0.001}$ 143-146°; n_D^{25} 1.5300. Soluble in water, pH of 5% aq soln 7.3.

THERAP CAT: Anesthetic (intravenous).

45. Acetaminophen. *N-(4-Hydroxyphenyl)acetamide; 4'*-hydroxyacetanilide; *p*-hydroxyacetanilide; *p*-acetamidophenol; *p*-acetaminophenol; *p*-acetylaminophenol; *N*-acetyl-*p*-aminophenol; paracetamol; Abensanil; Acamol; Acetal-

gin; Alpiny; Amadil; Anaflon; Anhiba; Apamide; APAP; Ben-u-ron; Bickie-mol; Calpol; Captin; Claratal; Cetadol; Dafalgan; Datril; Dirox; Disprol; Doliprane; Dolprone; Dymadon; Enelfa; Eneril; Eu-Med; Exdol; Febrilex; Finimal; Gelocatil; Hedex; Homoolan; Korum; Momentum; Naprinol; Nobedon; Ortensan; Pacemol; Paldesic; Panadol; Panaleve; Panasorb; Panets; Panex; Panodil; Parelan; Paraspen; Parmol; Pasolind N; Salzone; Tabalgin; Tapar; Tempra; Tralgon; Tylenol; Valadol. $C_8H_9NO_2$; mol wt 151.17. C 63.56%, H 6.00%, N 9.27%, O 21.17%. Prepn from *p*-nitrophenol: Morse, *Ber.* **11**, 232 (1878); Tingle, Williams, *Am. Chem. J.* **37**, 63 (1907); from *p*-aminophenol: Lumière *et al., Bull. Soc. Chim. France* [3] **33**, 785 (1905); Fierz-David, Kuster, *Helv. Chim. Acta* **22**, 94 (1939); Wilbert, De Angelis, U.S. pat. **2,998,450** (1961 to Warner-Lambert); Bergmann, **Ger. pat. 453,577;** *Chem. Zentr.* **1928, I,** 2663; *Frdl.* **16**, 238; from *p*-hydroxyacetophenone hydrazone: Pearson *et al., J. Am. Chem. Soc.* **75**, 5907 (1953). Toxicity data: G. A. Starmer *et al., Toxicol. Appl. Pharmacol.* **19**, 20 (1971); D. C. Dahlin, S. D. Nelson, *J. Med. Chem.* **25**, 885 (1982). Evaluation of renal effects: D. P. Sandler *et al., N. Engl. J. Med.* **320**, 1238 (1989). Comprehensive description: J. E. Fairbrother, *Anal. Profiles Drug Subs.* **3**, 1-109 (1974). Review of pharmacology: B. Ameer, D. J. Greenblatt, *Ann. Int. Med.* **87**, 202-209 (1977). Review of acetaminophen-induced hepatotoxicity: J. A. Hinson, *Rev. Biochem. Toxicol.* **2**, 103-129 (1980); *idem, Life Sci.* **29**, 107-116 (1981); and proposed protective agents: M. Davis, *Sem. Liver Dis.* **6**, 138-147 (1986). For proposed toxic metabolite *see* Acetimidoquinone.

Large monoclinic prisms from water, mp 169-170.5°. d_4^{21} 1.293. uv max (ethanol): 250 nm (ϵ 13800). Very slightly sol in cold water, considerably more sol in hot water. Sol in methanol, ethanol, dimethylformamide, ethylene dichloride, acetone, ethyl acetate. Slightly sol in ether. Practically insol in petr ether, pentane, benzene. LD_{50} in mice (mg/kg): 338 orally (Starmer), 500 i.p. (Dahlin, Nelson).

USE: Manuf azo dyes, photographic chemicals.

THERAP CAT: Analgesic; antipyretic.

46. Acetaminosalol. *2-Hydroxybenzoic acid 4-(acetylamino)phenyl ester; p*-acetamidophenyl salicylate; acetylaminophenyl salicylate; acetyl-*p*-aminosalol; *p*-acetylaminophenol salicylic acid ester; phenetsal; Salophen; Phenosal. $C_{15}H_{13}NO_4$; mol wt 271.27. C 66.41%, H 4.83%, N 5.16%, O 23.59%. Prepn: Brewster, *J. Am. Chem. Soc.* **40**, 1136 (1918).

Crystals from hot ethanol, mp 187°. Practically insol in petr ether, cold water, more sol in warm water. Sol in alcohol, ether, benzene. Incompatible with alkalies and alkaline solns which dissolve it with decompn. The alkaline soln gradually becomes blue when boiled, the blue color being discharged upon continued boiling and again produced upon cooling and exposure to air.

THERAP CAT: Analgesic; antipyretic; anti-inflammatory.

THERAP CAT (VET): Analgesic; antipyretic.

47. Acetanilide. *N-Phenylacetamide;* antifebrin; acetylaniline; acetylaminobenzene. C_8H_9NO; mol wt 135.17. C 71.07%, H 6.71%, N 10.36%, O 11.84%. Usually prepd from aniline and acetic acids: A. I. Vogel, *Practical Organic Chemistry* (London, 3rd ed., 1959) p 577. From aniline and acetyl chloride: Gattermann-Wieland, *Praxis des Organischen Chemikers* (Berlin, 40th ed., 1961) p 114. From aniline, acetone and ketene: Hurd, *Org. Syn.*, **coll. vol. I** (New York, 2nd ed., 1941) p 332. Chemistry: G. C. Derick, J. H. Bornman, *J. Am. Chem. Soc.* **35,** 1269 (1913). Toxicity data: P. K. Smith, W. E. Hambourger, *J. Pharmacol. Exp. Ther.* **54,** (1935). Monograph: Gross, *Acetanilid* (Hillhouse Press, New Haven, 1946).

Orthorhombic plates, scales from water. mp 113-115°; bp 304-305°. Slightly burning taste. Appreciably volatile at 95°. d_4^{15} 1.219. pK (28°): 13.0. One gram dissolves in 185 ml water, 20 ml boiling water, 3.4 ml alcohol, 3 ml methanol, 0.6 ml boiling alcohol, 3.7 ml chloroform, 4 ml acetone, 5 ml glycerol, 8 ml dioxane, 18 ml ether, 47 ml benzene. Very sparingly sol in petr ether. Chloral hydrate increases the soly of acetanilide in water. LD_{50} intragastric in rats: 800 mg/kg (Smith, Hambourger).
Pharmaceut. Incompat. Ethyl nitrate, amyl nitrite, acid solns of nitriles, alkalies (aniline liberated), alkali bromide and iodides in aq soln (insol compounds formed); chloral hydrate, phenol, resorcinol, thymol (these produce a liquid or a soft mass on trituration); ferric salts (red color or precipitate).
USE: Manufacture of other medicinals and of dyes; stabilizer for H_2O_2 soln; as addition to cellulose ester varnishes.
THERAP CAT: Antipyretic, analgesic.
THERAP CAT (VET): Antipyretic, analgesic.

48. *p*-Acetanisidine.

N-(4-Methoxyphenyl)acetamide; p-acetanisidide; methacetin; *p*-methoxyacetanilide. C_9H_{11}-NO_2; mol wt 165.19. C 65.44%, H 6.71%, N 8.48%, O 19.37%. $CH_3OC_6H_4NHCOCH_3$. Prepn by acetylation of *p*-anisidine: Reverdin, Bucky, *Ber.* **39,** 2679 (1906); from *p*-acetylanisole and hydroxylamine-*O*-sulfonic acid: Sanford *et al., J. Am. Chem. Soc.* **67,** 1941 (1945).
Cryst powder, feeble, bitter taste, mp 130-132°. Slightly sol in water; sol in alcohol, acetone, chloroform, dil acids, alkalies.

49. Acetarsone.

[3-(Acetylamino)-4-hydroxyphenyl]-arsonic acid; N-acetyl-4-hydroxy-m-arsanilic acid; 3-acetamido-4-hydroxyphenylarsonic acid; 3-acetamido-4-hydroxybenzenearsonic acid; acetarsol; acetphenarsine; Ehrlich 594; Fourneau 190; F-190; Stovarsol; Amarsan; Arsaphen; Dynarsan; Goyl; Kharophen; Limarsol; Malagride; Gynoplix; Oralcid; Devegan; Orarsan; Osarsal; Osarsol; Osvarsan; Paroxyl; Sanogyl; Spirocid; S.V.C.; Monargan; Ginarsol; Stovarsolan. $C_8H_{10}AsNO_5$; mol wt 275.09. C 34.93%, H 3.66%, As 27.24%, N 5.09%, O 29.08%. The commercial product contains from 27.1 to 27.4% As. Various synthetic routes: Raiziss, Fisher, *J. Am. Chem. Soc.* **48,** 1323 (1926); Hewitt, King, *J. Chem. Soc.* **1926,** 823. Toxicity data: H. H. Anderson, C. D. Leake, *Proc. Soc. Exp. Biol. Med.* **27,** 267 (1930).

Stout prisms from water, dec 240-250°; slight acid taste. Slightly sol in water; freely sol in solns of alkalies and alkali carbonates. Stable at ordinary temps. MLD in rabbits, cats (mg/kg): 125-150, 150-175 orally (Anderson, Leake).
Calcium salt (with hexosediphosphoric acid), *Realphene.*
Bismuth salt, *Bistovol.*
Diethylamine salt, $C_{12}H_{21}AsN_2O_5$, *Acetarsin, Acetilarsano, Acetylarsan, Arsaphenan, Golarsyl, Syntharsol.* Prepn: **Brit. pat. 224,764** (1954 to Rhône-Poulenc).
THERAP CAT: Antiprotozoal (Trichomonas). Diethylamine salt formerly as antisyphilitic.
THERAP CAT (VET): Antiprotozoal (Histomonas) in turkeys. Antibacterial (spirocheticide) in poultry. Has been used as a tonic.

50. Acetazolamide.

N-[5-(Aminosulfonyl)-1,3,4-thiadiazol-2-yl]acetamide; 5-acetamido-1,3,4-thiadiazole-2-sulfonamide; 2-acetylamino-1,3,4-thiadiazole-5-sulfonamide; 6063; Acetamox; Atenezol; Défiltran; Diamox; Didoc; Diuriwas; Donmox; Edemox; Fonurit; Glaupax. $C_4H_6N_4$-O_3S_2; mol wt 222.25. C 21.62%, H 2.72%, N 25.21%, O 21.60%, S 28.86%. Carbonic anhydrase inhibitor. Prepn: R. O. Roblin, J. W. Clapp, *J. Am. Chem. Soc.* **72,** 4890 (1950); J. W. Clapp, R. O. Roblin, U.S. pat. **2,554,816** (1951 to Am. Cyanamid). HPLC determn in pharmaceuticals: Z. S. Gomaa, *Biomed. Chromatog.* **7,** 134 (1993). Effect on retinal circulation: S. M. B. Rassam *et al., Eye* **7,** 697 (1993). Clinical trial in postoperative elevation of intraocular pressure: I. D. Ladas *et al., Brit. J. Ophthalmol.* **77,** 136 (1993). Review of efficacy in acute mountain sickness: L. D. Ried *et al., J. Wilderness Med.* **5,** 34-48 (1994).

Crystals from water, mp 258-259° (effervescence). Weak acid. pKa 7.2. Sparingly sol in cold water.
Sodium salt, *Vetamox.*
THERAP CAT: Antiglaucoma; diuretic; in treatment of acute mountain sickness.
THERAP CAT (VET): Diuretic.

51. Acetiamine.

Ethanethioic acid S-[1-[2-(acetyloxy)ethyl]-2-[[(4-amino-2-methyl-5-pyrimidinyl)methyl]formylamino]-1-propenyl] ester; thioacetic acid S-ester with N-[(4-amino-2-methyl-5-pyrimidinyl)methyl]-N-(4-hydroxy-2-mercapto-1-methyl-1-butenyl)formamide acetate; N-[(4-amino-2-methyl-5-pyrimidinyl)methyl]-*N*-(4-hydroxy-2-mercapto-1-methyl-1-butenyl)formamide *O,S*-diacetate; 3-acetylthio-4-[(4-amino-2-methyl-5-pyrimidinyl)methyl-*N*-formylamino]-3-pentenyl acetate; 5-acetoxy-3-acetylthio-2-[(4-amino-2-methyl-5-pyrimidinyl)methyl-*N*-formylamino]-2-pentene; diacethiamine; *O,S*-diacetylthiamine; vitamin B_1 *O,S*-diacetate; D.A.T.; Thianeuron. $C_{16}H_{22}N_4O_4S$; mol wt 366.44. C 52.44%, H 6.05%, N 15.29%, O 17.46%, S 8.75%. A fat-soluble deriv. of vitamin B_1. Prepn: Matsukawa, Kawasaki, *J. Pharm. Soc. Japan* **23,** 705, 709 (1953), *C.A.* **48,** 7017e (1954); Takamizawa *et al., Bull. Chem. Soc. Japan* **36,** 1214 (1963). Synthesis and pharmacology: Gauthier *et al., Ann. Pharm. Franc.* **21,** 655 (1963). Clinical studies: Wagner, Wagner, *Arzneimittel-Forsch.* **16,** 1643 (1966); Blum, Thomas, *Pharmacol. Clin.* **2,** 177 (1970).

cis-form

Colorless prisms from benzene-petr ether, mp 122-123° (dec); from water, mp 123-124°. Soluble in water, methanol, ethanol.
Hydrochloride, $C_{16}H_{22}N_4O_4S.HCl$, *Nevriton Comprimés.*
THERAP CAT: Vitamin (enzyme cofactor).

52. Acetic Acid Glacial.

Aci-Jel. $C_2H_4O_2$; mol wt 60.05. C 40.00%, H 6.71%, O 53.28%. CH_3COOH. Obtained in the destructive distillation of wood; from acetylene and water, via acetaldehyde by oxidation with air. Manuf processes: Bhattacharyya, Sourirajan, *J. Appl. Chem. (London)* **6,** 442 (1956); *eidem, ibid.* **9,** 126 (1959); Elce *et al.,* U.S. pat. **2,800,504** (1957 to Distillers Co.); Wirth, U.S. pat. **2,818,428** (1957 to British Petroleum); McKusick and Hoover, U.S. pats. **2,940,913** and **2,940,914** (both 1960 to Du Pont); Faith, Keyes & Clark's *Industrial Chemicals,* F. A. Lowenheim, M. K. Moran, Eds. (Wiley-Interscience, New York, 4th ed., 1975) pp 8-15. Toxicity data: H. F. Smyth *et al., Arch. Ind. Hyg. Occup. Med.* **4,** 119 (1951).

Review: F. S. Wagner in Kirk-Othmer *Encyclopedia of Chemical Technology* vol. **1** (Wiley-Interscience, New York, 3rd ed., 1978) pp 124-147.

Liquid; pungent odor. *Produces burns of the skin!* $d^{16.67}$ (liq) 1.053; $d^{16.60}$ (solid) 1.266. d_{25}^{25} 1.049. bp 118°. mp 16.7°. n_D^{20} 1.3718. Flash pt, closed cup: 103°F (39°C). *Caution: Flammable!* Contracts slightly on freezing. It is an excellent solvent for many organic compounds; also dissolves phosphorus, sulfur and halogen acids. Miscible with water, alcohol, glycerol, ether, carbon tetrachloride. Practically insol in carbon disulfide. Weakly ionized in aq solns: pKa 4.74. pH of aq solns $1.0M = 2.4$; $0.1M = 2.9$; $0.01M = 3.4$. LD_{50} in rats (g/kg): 3.53 orally (Smyth).

Incompat. Carbonates, hydroxides, many oxides, and phosphates, etc.

Caution: Ingestion may cause severe corrosion of mouth and G.I. tract, with vomiting, hematemesis, diarrhea, circulatory collapse, uremia, death. Chronic exposure may cause erosion of dental enamel, bronchitis, eye irritation, *cf.* Patty's *Industrial Hygiene and Toxicology* vol. **2C**, G. D. Clayton, F. E. Clayton, Eds. (Wiley-Interscience, New York, 3rd ed., 1982) p 4909-4911.

USE: Manuf various acetates, acetyl compounds, cellulose acetate, acetate rayon, plastics and rubber in tanning; as laundry sour; printing calico and dyeing silk; as acidulant and preservative in foods; solvent for gums, resins, volatile oils and many other substances. Widely used in commercial organic syntheses. Pharmaceutic aid (acidifier).

THERAP CAT (VET): Vesicant, caustic, destruction of warts.

53. Acetic Anhydride. Acetic oxide; acetyl oxide. $C_4H_6O_3$; mol wt 102.09. C 47.06%, H 5.92%, O 47.02%. $(CH_3CO)_2O$. Equivalent to 117.64% acetic acid. Made formerly from sod. acetate and acetyl or sulfuryl chloride; now usually obtained from acetaldehyde or acetic acid: Faith, Keyes & Clark's *Industrial Chemicals*, F. A. Lowenheim, M. K. Moran, Eds. (Wiley-Interscience, New York, 4th ed., 1975) pp 16-20. Of industrial importance is also the ketene process, starting with the thermal decomposition of acetone: Schmidlin, Bergmann, *Ber.* **43**, 2821 (1910). Toxicity study: H. F. Smyth *et al., Arch. Ind. Hyg. Occup. Med.* **4**, 119 (1951).

Very refractive liquid; strong acetic odor. *Readily combustible. Fire hazard.* Flash pt 130°F. d_4^{15} 1.080. mp −73°. bp 139°. n_D^{20} 1.3904. Slowly soluble in water, forming acetic acid; with alcohol forms ethyl acetate; sol in chloroform, ether. LD_{50} orally in rats: 1.78 g/kg (Smyth).

Caution: Potential symptoms of overexposure are conjunctivitis, lacrimation, corneal edema, opacity and photophobia; nasal, pharyngeal irritation; cough, dyspnea and bronchitis; skin burns, vesiculation and sensitization dermatitis. *See NIOSH Pocket Guide to Chemical Hazards* (DHHS/NIOSH 90-117, 1990) p 30.

USE: Manuf acetyl compounds, cellulose acetates. As acetulizer and solvent in examining wool fat, glycerol, fatty and volatile oils, resins; detection of rosin. Widely used in organic syntheses, e.g., as dehydrating agent in nitrations, sulfonations and other reactions where removal of water is necessary.

54. Acetimidoquinone. *N-(4-oxo-2,5-cyclohexadien-1-ylidene)acetamide; N*-acetyl-*p*-benzoquinonimine; *N*-acetylimidoquinone; NAPQI. $C_8H_7NO_2$; mol wt 149.15. C 64.42%, H 4.73%, N 9.39%, O 21.45%. Proposed "ultimate" toxic metabolite of acetaminophen, *q.v.* Initially prepared by oxidation of acetaminophen with lead tetraacetate and characterized as a Diels-Alder adduct: I. C. Calder *et al., J. Med. Chem.* **16**, 499 (1973). Electrochemical generation in buffer: D. J. Miner, P. T. Kissinger, *Biochem. Pharmacol.* **28**, 3285 (1979). Synthesis in stable benzene soln and reactivity: I. A. Blair *et al., Tetrahedron Letters* **21**, 4947 (1980). Prepn in cryst form, decomposition kinetics, preliminary toxicological studies: D. C. Dahlin, S. D. Nelson, *J. Med. Chem.* **25**, 885 (1982). Microsomal reactivity study: G. B.

Corcoran, *Adv. Exp. Med. Biol.* **136B**, 1085 (1982). Review of acetaminophen-induced hepatotoxicity: J. A. Hinson *et al., Life Sci.* **29**, 107-116 (1981).

Yellow cubic cryst, mp 74-75°. Sublimes at 45-50° (0.07 mm Hg). uv max (*n*-hexane): 263, 376 nm (ϵ 3.3×10^4, 1.6×10^2 M^{-1} cm^{-1}). LD_{50} i.p. in male BALB/c mice: 20 mg/kg (Dahlin, Nelson).

55. Acetoacetanilide. *3-Oxo-N-phenylbutanamide;* α-acetylacetanilide; acetoacetic anilide; β-ketobutyranilide. $C_{10}H_{11}NO_2$; mol wt 177.20. C 67.78%, H 6.26%, N 7.90%, O 18.06%. $C_6H_5NHCOCH_2COCH_3$. Prepd by the reaction of ketene dimer with aniline: Chick, Wilsmore, *J. Chem. Soc.* **1908**, 946; Boese, *Ind. Eng. Chem.* **32**, 16 (1940); Williams, Krynitsky, *Org. Syn. coll. vol. III*, 10 (1955). By the reaction of aniline with ethyl acetoacetate: Knorr, *Ann.* **26**, 69 (1886); Roos, *Ber.* **21**, 624 (1883); Knorr, Reuter, *Ber.* **27**, 1169 (1894). Enol form (unstable) prepd by pouring an alkaline soln of the keto form in cold, dilute H_2SO_4.

Keto form: Leaflets from dilute alcohol, mp 85°. Slightly soluble in water; sol in alcohol, chloroform, ether, hot benzene, hot petr ether, acids, or alkali hydroxide solns. Gives a violet color with ferric chloride.

USE: Manuf of yellow dyes, such as Hansa and benzidine yellows. In rubber compounding. In organic syntheses.

56. Acetoacetic Acid. *3-Oxobutanoic acid;* diacetic acid; acetylacetic acid; acetonecarboxylic acid; 3-ketobutyric acid; 2-ketobutyric acid. $C_4H_6O_3$; mol wt 102.09. C 47.06%, H 5.92%, O 47.02%. CH_3COCH_2COOH. Prepd by hydrolysis of ethyl acetoacetate: Krueger, *J. Am. Chem. Soc.* **74**, 5536 (1952).

Crystals from ether, mp 36-37°. A strong, but unstable acid. At 100° violently decomposes into acetone and CO_2. Miscible with water, alc.

Caution: Severely irritating to skin, mucous membranes.

USE: In organic syntheses.

57. Acetobromglucose. α-D-*Glucopyranosyl bromide 2,3,4,6-tetraacetate;* α-acetobromoglucose; 2,3,4,6-tetraacetyl-α-D-glucopyranosyl bromide. $C_{14}H_{19}BrO_9$; mol wt 411.20. C 40.89%, H 4.66%, Br 19.43%, O 35.02%. Prepd by the action of hydrogen bromide upon anhydrous glucose in acetic anhydride: Redemann, Niemann, *Org. Syn.* **coll. vol. III**, 11 (1955). *See also* Fischer, *Ber.* **49**, 584 (1916); Freudenberg, *Ber.* **60**, 241 (1927); Gattermann-Wieland, *Praxis des organischen Chemikers* (de Gruyter, Berlin, 40th ed., 1961) p 340.

Crystals from isopropyl ether, mp 88-89°. $[\alpha]_D^{19}$ +199.3° (c = 3 in chloroform); $[\alpha]_D^{15}$ +230.3° (c = 9 in benzene). Best stored in a vacuum desiccator. Dec on contact with water. One gram dissolves in 20 ml absolute ethanol, more soluble in methanol. Freely sol in ether, chloroform, acetone, ethyl acetate, benzene. Slightly sol in petr ether.

58. Acetochlor. *2-Chloro-N-(ethoxymethyl)-N-(2-ethyl-6-methylphenyl)acetamide; 2-chloro-N-(ethoxymethyl)-6'-ethyl-o-acetotoluidide;* 2'-ethyl-6'-methyl-*N*-(ethoxymethyl)-2-chloroacetanilide; MON-097; MG-02; Acenit; Harness; Partner. $C_{14}H_{20}ClNO_2$; mol wt 269.77. C 62.33%, H 7.47%, Cl 13.14%, N 5.19%, O 11.86%. Pre-emergent herbicide. Prepn: J. F. Olin, U.S. pat. **3,547,620** (1970 to Monsanto). Description of physical and biological properties: R. P. Upchurch *et al., Proc. North Central Weed Control Conf.* **25**, 56 (1970). Synthesis: B. G. Zupancic, M. Sopcic, *Synthesis* **1982**, 942. Field trials on peanuts and soybeans: C. W. Derting *et al., Proc. South Weed Sci. Soc.* **25**, 131 (1972); on peanuts: L. R. Hawf, K. R. Frost , Jr., *ibid.* **38**, 20 (1985). Soil adsorption and bioactivity: J. B. Weber, C. J. Peter, *Weed. Sci.* **30**, 14 (1982). Soil transformation: P. C. C. Feng, *Pest. Biochem. Physiol.* **40**, 136 (1991).

Pale straw-colored oil. $bp_{0.4 torr}$ 134°. n_D^{20} 1.5272. Sol in water (25°): 400 ppm. LD_{50} orally in rats: 1160 mg/kg (Upchurch).

USE: Herbicide.

59. Acetohexamide. *4-Acetyl-N-[(cyclohexylamino)-carbonyl]benzenesulfonamide; 1-[(p-acetylphenyl)sulfonyl]-3-cyclohexylurea;* 3-cyclohexyl-1-(*p*-acetylphenylsulfonyl)urea; *N*-(*p*-acetylbenzenesulfonyl)-*N'*-cyclohexylurea; cyclamide; tsiklamid; Dimelor; Dymelor; Dimelin; Ordimel. $C_{15}H_{20}N_2O_4S$; mol wt 324.40. C 55.54%, H 6.21%, N 8.64%, O 19.73%, S 9.88%. Prepn: **Brit.** pat. **912,789** (1962 to Lilly); Marshall *et al., J. Med. Chem.* **6**, 60 (1963). Comprehensive description: C. E. Shafer in *Analytical Profiles of Drug Substances* **vol. 1**, K. Florey, Ed. (Academic Press, New York, 1972) pp 1-14; A. A. Al-Badr, H. A. El-Obeid in *Analytical Profiles of Drug Substances and Excipients* **vol. 21**, H. G. Brittain, Ed. (Academic Press, San Diego, 1992) pp 1-41.

Crystals from 90% aq ethanol, mp 188-190° (**Brit.** pat. **912,789**); from dil ethanol, mp 175-177° (Marshall). uv max (methanol): 247, 283 nm. Sol in pyridine; slightly sol in alcohol, chloroform. Insol in water, ether.

THERAP CAT: Antidiabetic.

60. Acetohydroxamic Acid. *N-Hydroxyacetamide; N*-acetylhydroxylamine; acetic acid oxime; AHA; Lithostat. $C_2H_5NO_2$; mol wt 75.07. C 32.00%, H 6.71%, N 18.66%, O 42.63%. $CH_3CONHOH$. Urease inhibitor. Prepn: A. Miolati, *Ber.* **25**, 699 (1892); W. M. Wise, W. W. Brandt, *J. Am. Chem. Soc.* **77**, 1058 (1955); H. A. Staab *et al., Ber.* **95**, 1275 (1962); G. Sosnovsky, J. A. Krogh, *Synthesis* **1980**, 654. Inhibition of urease activity: K. Kobashi *et al., Biochim. Biophys. Acta* **65**, 380 (1962); W. N. Fishbein *et al., Nature* **208**, 46 (1965); W. N. Fishbein, P. P. Carbone, *J. Biol. Chem.* **240**, 2407 (1965); D. P. Griffith *et al., Invest. Urol.* **11**, 234 (1973). Metabolism: E. Wolpert *et al., Proc. Soc. Exp. Biol. Med.* **136**, 592 (1971); W. N. Fishbein *et al., J. Pharmacol. Exp. Ther.* **186**, 173 (1973). Pharmacokinetics: S. Feldman *et al., Invest. Urol.* **15**, 498 (1978). Clinical studies in treatment of kidney stones: D. B. Griffith *et al., J. Urol.* **119**, 9 (1978); A. Martelli *et al., Urology* **17**, 320 (1981).

mp 89-92°. pKa 8.70. pH (aq soln): 9.39.

THERAP CAT: Antiurolithic. Antibacterial adjunct (urinary tract infection).

61. Acetoin. *3-Hydroxy-2-butanone;* 2,3-butanolone; acetyl methyl carbinol; dimethylketol; *γ*-hydroxy-*β*-oxobutane. $C_4H_8O_2$; mol wt 88.11. C 54.53%, H 9.15%, O 36.32%. $CH_3CH(OH)COCH_3$. A product of fermentation, also in cream ripened for churning. Obtained by the action of sorbose bacterium or *Mycoderma aceti* on 2,3-butanediol or by the action of fungi, such as *Aspergillus, Penicillium, Mycoderma* on sugar cane juice: Browne, *J. Am. Chem. Soc.* **28**, 467 (1906). By action of yeast on diacetyl: Nagelschmidt, *Biochem. Z.* **186**, 317 (1927). From diacetyl by partial reduction with Zn and acid: Diels, Stephan, *Ber.* **40**, 4338 (1907). Fermentation process: Vergnaud, U.S. pat. **2,529,061** (1950 to Usines de Melle).

Liquid. Pleasant odor. mp 15°. d_4^{17} 0.9972. bp_{760} 148°. $n_D^{17.3}$ 1.4190. Miscible with water, alcohol. Sparingly sol in ether, petr ether. Reduces Fehling's soln forming acetic

acid. Forms a solid dimer $C_8H_{16}O_4$ on standing or on treatment with granulated zinc. The dimer is easily converted back to the monomer by melting, distilling or dissolving.

62. Acetol®. The name ACETOL appears in the literature covering a variety of substances.

According to earlier indices to *Chemical Abstracts* it is synonymous with 1-hydroxy-2-propanone or hydroxyacetone.

According to *Gehes Codex* (5th ed, 1929) ACETOL is registered in Canada as a name for acetylsalicylic acid by the Ontario College of Pharmacy upon stipulation by the Canadian Pharmaceutical Association.

In W. Gardner's *Chemical Synonyms and Trade Names* (5th ed 1948) ACETOL is also found as a proprietary name to designate cellulose acetate material in the form of flake, and as a synonym for acetal or diethyl aldehyde used as a hypnotic, and last as a trademark for a liquid fuel for internal combustion engines containing 80% alcohol and 20% other materials including acetylene.

63. Acetomeroctol. *(Acetato-O)[2-hydroxy-5-(1,1,3,3-tetramethylbutyl)phenyl]mercury; 2-(acetoxymercuri)-4-(1,1,-3,3-tetramethylbutyl)phenol;* Merbak. $C_{16}H_{24}HgO_3$; mol wt 464.95. C 41.33%, H 5.20%, Hg 43.14%, O 10.32%. Prepn: Niederl, Shukis, *J. Am. Chem. Soc.* **66**, 844 (1944).

Crystals, mp 158°. Practically insol in water. Sol in alc, ether, chloroform. Sparingly sol in benzene. Marketed as a 1:1000 soln contg 50% (v/v) alc and 10% (v/v) acetone.

THERAP CAT: Topical anti-infective.

64. Acetone. *2-Propanone;* dimethyl ketone; *β*-ketopropane; pyroacetic ether. C_3H_6O; mol wt 58.08. C 62.04%, H 10.41%, O 27.55%. CH_3COCH_3. Obtained by fermentation (by-product of butyl alcohol manufacture) or by chemical synthesis from isopropanol (as chief product); from cumene (by-product in phenol manufacture); from propane (by-product of oxidation-cracking); Faith, Keyes & Clark's *Industrial Chemicals*, F. A. Lowenheim, M. K. Moran, Eds. (Wiley-Interscience, New York, 4th ed., 1975) pp 21-25. Toxicity: Smyth *et al., Ind. Hyg. J.* **23**, 95 (1962). *Review:* Weiss, *Chem. Eng. News* **36**, 79 (1958); D. L. Nelson, B. P. Webb in Kirk-Othmer *Encyclopedia of Chemical Technology* **vol. 1** (Wiley-Interscience, New York, 3rd ed., 1978) pp 179-191.

Volatile, *highly flammable* liquid; characteristic odor; pungent, sweetish taste. d_D^{25} 0.788. bp 56.5°. mp −94°. n_D^{20} 1.3591. Flash pt, closed cup: 0°F (−18°C). Miscible with water, alcohol, dimethylformamide, chloroform, ether, most oils. *Keep away from fire!* Keep away from plastic eyeglass frames, jewelry, pens and pencils, rayon stockings and other rayon garments. LD_{50} in rats: 10.7 ml/kg orally (Smyth).

Caution: Potential symptoms of overexposure are irritation of eyes, nose and throat; headache, dizziness; dermatitis. *See NIOSH Pocket Guide to Chemical Hazards* (DHHS/NIOSH 90-117, 1990) p 30.

USE: Solvent for fats, oils, waxes, resins, rubber, plastics, lacquers, varnishes, rubber cements. Manuf methyl isobutyl ketone, mesityl oxide, acetic acid (ketene process), diacetone alcohol, chloroform, iodoform, bromoform, explosives, aeroplane dopes, rayon, photographic films, isoprene; storing acetylene gas (takes up about 24 times its vol of the gas); extraction of various principles from animal and plant substances; in paint and varnish removers; purifying paraffin; hardening and dehydrating tissues. Pharmaceutic aid (solvent).

65. Acetone Cyanohydrin. *2-Hydroxy-2-methylpropanenitrile; 2-methyllactonitrile; α*-hydroxyisobutyronitrile. C_4H_7NO; mol wt 85.11. C 56.45%, H 8.29%, N 16.46%, O 18.80%. $(CH_3)_2C(OH)CN$. Prepd by adding acetone to sodium or potassium cyanide in water and treating with H_2SO_4 at below 20°: Welch, Clemo, *J. Chem. Soc.* **1928**,

2629; Cox, Stormont, *Org. Syn.* **coll. vol. II,** 7 (1943). Continuous production from HCN and aq acetone: G. Barsky, **U.S. pat. 2,731,490** (1956). Toxicity study: H. F. Smyth *et al., Am. Ind. Hyg. Assoc. J.* **23,** 95 (1962).

Liquid; d_4^{25} 0.9267; d_4^{19} 0.932. mp −19°. bp_{15} 81°; bp_{20} 88-90°; bp_{23} 82°; bp_{760} 95°. n_D^{19} 1.40002; n_D^{15} 1.3980. Freely sol in water and in the usual organic solvents. Practically insol in petr ether and carbon disulfide. LD_{50} orally in rats: 0.17 g/kg (Smyth).

Caution: May be slightly irritating to skin, mucous membranes. Decomposes readily to form hydrogen cyanide, *q.v.,* which is highly toxic.

USE: In preparative organic chemistry for transcyanohydrination, such as preparing the 17-monocyanohydrin of a 3,17-diketo steroid by hydrogen cyanide exchange with the reagent, *e.g.* Ercoli, de Ruggieri, *J. Am. Chem. Soc.* **75,** 650 (1953).

66. Acetonedicarboxylic Acid. *3-Oxopentanedioic acid; 3-oxoglutaric acid; β-ketoglutaric acid.* $C_5H_6O_5$; mol wt 146.10. C 41.11%, H 4.14%, O 54.76%. $CO_2HCH_2CO-CH_2CO_2H$. Prepd by the action of fuming sulfuric acid on citric acid: Willstätter, Pfannenstiel, *Ann.* **422,** 5 (1921); Ingold, Nickolls, *J. Chem. Soc.* **121,** 1642 (1922); R. Adams *et al., Org. Syn.* **coll. vol. I,** 10 (2nd ed., 1941). By the action of *Aspergillus niger* on ammonium citrate mixed with some citric acid: Walker *et al., J. Chem. Soc.* **1927,** 3050; also by *B. pyocyaneus:* Butterworth, Walker, *Biochem. J.* **23,** 931 (1929). Decomposition studies: Wiig, *J. Phys. Chem.* **32,** 961 (1928). Intermediate in synthesis of tropinone: R. Robinson, *J. Chem. Soc.* **111,** 762 (1917); C. Schöpf, G. Lehmann, *Ann.* **518,** 1 (1921).

Needles from ethyl acetate, mp 138° (dec). The pure compound may be stored over P_2O_5 for several months. Crude material dec after a few hours. Decomposed by hot water, acids or alkalies to CO_2 and acetone. K (25°): 7.9 × 10^{-4}. Very sol in water and alcohol; slightly sol in ethyl acetate, ether. Insol in chloroform, ligroin, benzene.

USE: In organic synthesis.

67. Acetone Sodium Bisulfite. *2-Hydroxy-2-propanesulfonic acid sodium salt;* acetone sulfite. $C_3H_7NaO_4S$; mol wt 162.14. C 22.22%, H 4.35%, Na 14.18%, O 39.47%, S 19.78%. $(CH_3)_2C(OH)SO_3Na$.

Crystals, fatty feel, slight SO_2 odor. Freely sol in water, sparingly in alcohol. Decomposed by acids.

USE: In photography; manuf of pure acetone; dyeing and printing texiles; in organic syntheses.

68. Acetonitrile. Methyl cyanide; cyanomethane; ethanenitrile. C_2H_3N; mol wt 41.05. C 58.52%, H 7.37%, N 34.12%. CH_3CN. Small amounts occur in coal tar. Obtained commercially as a byproduct in manuf of acrylonitrile, *q.v.:* R. A. Smiley in Kirk-Othmer *Encyclopedia of Chemical Technology,* **vol. 15** (Wiley-Interscience, New York, 3rd ed., 1978) p 896. Prepn by dehydration of acetamide: Adkins, Nissen, *J. Am. Chem. Soc.* **46,** 143 (1924); A. I. Vogel, *Practical Organic Chemistry* (Longmans, London, 3rd ed., 1959) p 407; Gattermann-Wieland, *Praxis des organischen Chemikers* (de Gruyter, Berlin, 40th ed., 1961) p 125; or from acetylene and ammonia: **Ger. pat. 365,432;** *Chem. Zentr.* **1924 I,** 2398; **1925 II,** 1563. Review of purification methods: *Techniques of Chemistry,* **vol. II,** *Organic Solvents,* J. A. Riddick, W. B. Bunger, Eds. (Wiley, 3rd ed., 1970) pp 798-805. Toxicity: H. F. Smyth, C. P. Carpenter, *J. Ind. Hyg. Toxicol.* **30,** 63 (1968).

Liquid. Ether-like odor. *Poisonous!* Burns with a luminous flame. Flash pt 12.8°C (55°F). d_4^{15} 0.78745; d_4^{30} 0.7138. mp −45°. bp_{760} 81.6°. n_D^{15} 1.34604, n_D^{30} 1.33934. Dielectric constant at 20° = 38.8. Surface tension at 20° = 29.04 dynes/cm. Misc with water, methanol, methyl acetate, ethyl acetate, acetone, ether, acetamide solutions, chloroform, carbon tetrachloride, ethylene chloride and many unsaturated hydrocarbons. Immiscible with many saturated hydrocarbons (petroleum fractions). Dissolves some in organic salts, *e.g.,* silver nitrate, lithium nitrate, magnesium bromide. Constant boiling mixture with water contains 16% H_2O and bp 76°. LD_{50} orally in rats: 3800 mg/kg (Smyth).

Caution: Potential symptoms of overexposure are asphyxia; nausea, vomiting; chest pain; weakness; stupor, convul-

sions. Direct contact may cause skin and eye irritation. *See NIOSH Pocket Guide to Chemical Hazards* (DHHS/NIOSH 90-117, 1990) p 32.

USE: In organic synthesis as starting material for acetophenone, α-naphthaleneacetic acid, thiamine, acetamidine. To remove tars, phenols, and coloring matter from petroleum hydrocarbons which are not soluble in acetonitrile. To extract fatty acids from fish liver oils and other animal and vegetable oils. Can be used to recrystallize steroids. As an indifferent medium in physicochemical investigations. Wherever a polar solvent having a rather high dielectric constant is required. As medium for promoting reactions involving ionization. As a solvent in non-aqueous titrations. As a non-aqueous solvent for inorganic salts.

69. Acetonylacetone. *2,5-Hexanedione; α,β-diacetylethane.* $C_6H_{10}O_2$; mol wt 114.14. C 63.14%, H 8.83%, O 28.03%. $CH_3COCH_2CH_2COCH_3$. Prepn by decarboxylation of diacetosuccinic ester (from sodium acetoacetic ester and iodine): Paal, *Ber.* **18,** 58 (1885); Knorr, *Ber.* **22,** 2100 (1889); by hydrolysis of 2,5-dimethylfuran: Perkins, Toussaint, **U.S. pat. 2,052,652** (1936 to Union Carbide). Toxicity data: H. F. Smyth, C. P. Carpenter, *J. Ind. Hyg. Toxicol.* **26,** 269 (1944).

Liquid, gradually turns yellow. d_4^{20} 0.970. mp −9°. bp 188°, n_D^{20} 1.449. Misc with water, alcohol, ether. LD_{50} orally in rats: 2.7 g/kg (Smyth, Carpenter).

Caution: A mild local irritant. High concns cause narcosis.

70. Acetophenazine. *1-[10-[3-[4-(2-Hydroxyethyl)-1-piperazinyl]propyl]-10H-phenothiazin-2-yl]ethanone; 10-[3-[4-(2-hydroxyethyl)-1-piperazinyl]propyl]phenothiazin-2-yl methyl ketone; 2-acetyl-10-[3-[4-(β-hydroxyethyl)piperazinyl]propyl]phenothiazine; 1-(2-hydroxyethyl)-4-[3-(2-acetyl-10-phenothiazyl)propyl]piperazine.* $C_{23}H_{29}N_3O_2S$; mol wt 411.57. C 67.12%, H 7.10%, N 10.21%, O 7.77%, S 7.79%. Prepd from 1-(2-hydroxyethyl)piperazine and 10-(3-chloropropyl)-2-acetylphenothiazine: Sherlock, Sperber, **U.S. pat. 2,985,654** (1961 to Schering). Toxicity study: E. W. Schafer, *Toxicol. Appl. Pharmacol.* **21,** 315 (1972).

Dimaleate, $C_{23}H_{29}N_3O_2S.2C_4H_4O_4$, *Tindal.* Crystals from ethanol, mp 167-168.5°. LD_{50} orally in rats: 433 mg/kg (Schafer).

THERAP CAT: Antipsychotic.

71. Acetophenone. *1-Phenylethanone;* phenyl methyl ketone; acetylbenzene; hypnone. C_8H_8O; mol wt 120.15. C 79.97%, H 6.71%, O 13.32%. $C_6H_5COCH_3$. Made from benzene and acetylchloride in presence of aluminum chloride; catalytically from acetic and benzoic acids. Prepn from benzene and acetic anhydride: Adams, *J. Am. Chem. Soc.* **46,** 1889 (1924); A. I. Vogel, *Practical Organic Chemistry* (Longmans, London, 3rd ed., 1959) p 730; Gattermann-Wieland, *Die Praxis des Organischen Chemikers* (de Gruyter, Berlin, 40th ed., 1961) p 297. Toxicity study: H. F. Smyth, C. P. Carpenter, *J. Ind. Hyg. Toxicol.* **30,** 63 (1948).

Liquid. Forms laminar crystals at low temp. mp 20.5°. d_4^{15} 1.033. bp 202°. n_D^{20} 1.5339. Flash pt, closed cup: 221°F (105°C). Slightly sol in water; freely in alcohol, chloroform, ether, fatty oils, glycerol. Sol in concd H_2SO_4 with orange color. LD_{50} orally in rats: 0.90 g/kg (Smyth, Carpenter).

USE: In perfumery to impart an orange-blossom-like odor; catalyst for the polymerization of olefins; in organic syntheses, esp. as photosensitizer.

THERAP CAT: Hypnotic.

72. Acetorphan. (±)-*N-[2-[(Acetylthio)methyl]-1-oxo-3-phenylpropyl]glycine phenylmethyl ester; N-[(R,S)-3-acetylthio-2-benzylpropanoyl]glycine benzyl ester; Tiorfan.*

$C_{21}H_{23}NO_4S$; mol wt 385.48. C 65.43%, H 6.01%, N 3.63%, O 16.60%, S 8.32%. Antisecretory enkephalinase inhibitor. Prepn: B. Roques *et al.*, **Eur.** pat. **Appl. 38,758** (1981); *eidem*, **U.S.** pat. **4,513,009** (1985 to Bioprojet). Pharmacology: J.-M. Lecomte *et al.*, *J. Pharmacol. Exp. Ther.* **237**, 937 (1986). Effect on intestinal transit: J. F. Bergmann *et al.*, *Aliment. Pharmacol. Ther.* **6**, 305 (1992). Clinical trial in acute diarrhea: P. Baumer *et al.*, *Gut* **33**, 753 (1992).

White crystals from ether, mp 89°.
(S)-Form, *ecadotril, sinorphan.* Prepn: P. Duhamel *et al.*, **Eur. Appl. 318,377;** *eidem*, **U.S.** pat. **5,208,255** (1989, 1993 both to Bioprojet); and pharmacology: B. Giros *et al.*, *J. Pharmacol. Exp. Ther.* **243**, 666 (1987). Clinical effect on plasma ANF levels in CHF: J. C. Kahn *et al.*, *Lancet* **335**, 118 (1990). Clinical evaluation in hypertension: P. Lefrancois *et al.*, *ibid.* **336**, 307. Effect on renal function in humans: F. Schmitt *et al.*, *Am. J. Physiol.* **267**, F20 (1994). mp 71°. $[\alpha]_D^{25}$ −24.1° (c = 1.2 in methanol). LD_{50} i.v. in mice: > 100 mg/kg (Duhamel, 1993).
THERAP CAT: Racemate as antidiarrheal; *(S)*-form as antihypertensive.

73.   Acetosulfone Sodium. *N-[[5-Amino-2-[(4-aminophenyl)sulfonyl]phenyl]sulfonyl]acetamide monosodium salt; N-6-sulfanilylmetanilylacetamide N-sodium salt; N^1-acetyl-6-(sulfanilylmetanilamido)sodium; sodium 2-N-acetylsulfamyl-4,4′-diaminodiphenylsulfone; sodium 4,4′-diaminodiphenylsulfone-2-N-acetylsulfonamide; sulfadiasulfone sodium; acetosulphone; Internal Antiseptic no. 307; I.A. 307; Promacetin.* $C_{14}H_{14}N_3NaO_5S_2$; mol wt 391.40. C 42.96%, H 3.61%, N 10.74%, Na 5.87%, O 20.44%, S 16.39%. Prepn: Tullar, **U.S.** pat. **2,358,365** (1944 to Parke, Davis).

Crystals. Soly in water about 3%. mp ~285° (free sulfonamide).
THERAP CAT: Antibacterial (leprostatic).

74.   Acetotoluide. $C_9H_{11}NO$; mol wt 149.19. C 72.46%, H 7.43%, N 9.39%, O 10.72%. Prepn of *m*-, *o*-, and *p*-acetotoluidides, from *m*-, *o*-, or *p*-toluidine and acetic acid: Gasopoulos, *Ber.* **59B**, 2187 (1926).

m-Acetotoluide, N-(3-methylphenyl)acetamide, m-acetotoluidide, m-tolylacetamide, aceto-m-aminotoluene, m-methylacetanilide, N-acetyl-m-toluidine. Needles, mp 65.5°, bp 303°. Slightly sol in water; freely sol in alc, ether.
o-Acetotoluide, crystals, mp 110°, bp 296°. Slightly sol in water; sol in alc, chloroform, ether. Determn of soly at various temps: Hall *et al.*, *J. Phys. Chem.* **37**, 1087 (1933).
p-Acetotoluide, crystals, mp 153°, bp 307°. Very slightly sol in water; sol in alc, ether, ethyl acetate, glacial acetic acid.

75.   Acetoxime. *2-Propanone oxime;* acetone oxime; β-isonitrosopropane. C_3H_7NO; mol wt 73.09. C 49.30%, H 9.65%, N 19.16%, O 21.89%. $(CH_3)_2C=NOH$. Prepd by shaking an aqueous soln of hydroxylamine with acetone and extracting with ether: W. L. Semon, *Org. Syn.* **coll. vol. I,** 318 (2nd ed., 1941). Ionization constants: C. V. King, A. P. Marion, *J. Am. Chem. Soc.* **66**, 977 (1944).
Columnar prisms. Volatilizes quickly in air. Odor resembling that of chloral hydrate. d_4^{62} 0.9113. mp 60°. bp_{728} 134.8°. Neutral reaction. pK (24.9°): 12.42. uv spectrum: Hartley, Dobbie, *J. Chem. Soc.* **77**, 326 (1900). Freely soluble in water, alcohol, ether, petr ether. Can be extracted with ether from neutral water solns, but not from alkaline or acidic solns.

76.   Acetoxolone. *(3β,20β)-3-(Acetyloxy)-11-oxoolean-12-en-29-oic acid; 3β-hydroxy-11-oxoolean-12-en-30-oic acid acetate;* 3-O-acetyl-18β-glycyrrhetic acid; acetylglycyrrhetinic acid; glycyrrhetic acid acetate; $C_{32}H_{48}O_5$; mol wt 512.73. C 74.96%, H 9.44%, O 15.60%. Deriv of enoxolone, *q.v.;* homolog of carbenoxolone, *q.v.* Prepn: J. M. Beaton, F. S. Spring, *J. Chem. Soc.* **1955**, 3126. Prepn of the aluminum salt: A. Bonati, **S. Afr.** pat. **70 06901;** *idem*, **U.S.** pat. **3,764,618** (1971, 1973 both to Dott. Inverni & Della Beffa). NMR study of the acetate: M. Mousseron-Canet *et al.*, *Bull. Soc. Chim. France* **12**, 4668 (1967). Pharmacology of the aluminum salt: C. Capra, *Filoterapia* **40**, 8 (1969), *C.A.* **73**, 64754 (1970). Absorption in rats: B. D. Cameron *et al.*, *Arzneimittel-Forsch.* **26**, 1680 (1976). Analysis of gastric protective effect: P. Altmayer *et al.*, *ibid.* **31**, 853 (1981).

Crystals, mp 322-325°. $[\alpha]_D^{20}$ +141°.
Aluminium salt, $C_{96}H_{141}AlO_{15}$, *Oriens.* White powder, mp 286-290°. $[\alpha]_D^{20}$ +126°±2° (c = 1 in chloroform). Insol in water. Sol in most organic solvents. LD_{50} in male rats (mg/kg): > 3300 orally, and i.p. (Bonati).
THERAP CAT: Antiulcerative.

77.   21-Acetoxypregnenolone. *(3β)-21-(Acetyloxy)-3-hydroxypregn-5-en-20-one; 3β,21-dihydroxypregn-5-en-20-one 21-acetate;* 5-pregnene-3β,21-diol-20-one 21-acetate; 21-acetoxy-5-pregnen-3-ol-20-one; 3-hydroxy-21-acetoxy-5-pregnen-20-one; prebediolone acetate; A.O.P.; Acetoxanon; Artisone acetate. $C_{23}H_{34}O_4$; mol wt 374.52. C 73.76%, H 9.15%, O 17.09%. Intermediate product in the synthesis of desoxycorticosterone acetate: Steiger, Reichstein, *Helv. Chim. Acta* **20**, 1164 (1937). Prepd by reacting 5,6-pregnen-3-ol-20-one with bromine in chloroform and treating the obtained oily bromide with potassium acetate in alcohol: Herloff, Inhoffen, **U.S.** pat. **2,409,043** (1946 to Schering).

Needles from acetone. Becomes opaque at about 80° or on prolonged standing. mp 184-185°. Very slightly sol in ether, in pentane; sol in chloroform, toluene. The esterified keto group is relatively stable to the action of chromic acid in cold and to such reducing agents as potassium iodide or zinc. On heating with aluminum isopropoxide in isopropyl alcohol it is reduced to 5-pregnene-3,20,21-triol.

THERAP CAT: Anti-inflammatory.

78. Acetozone. *Acetyl benzoyl peroxide;* benzoyl acetyl peroxide; benzozone. $C_9H_8O_4$; mol wt 180.16. C 60.00%, H 4.48%, O 35.52%. Prepd from benzaldehyde and acetic anhydride in the presence of O-contg gas and dibenzoyl peroxide: Carruthers, U.S. pat. **1,985,886** (1935 to Carbide & Carbon Chemicals); Appell, U.S. pat. **3,397,245** (1968 to Koppers).

Crystals, mp 36-37°. bp_{19} 130°. Decomposes on contact with moisture or in a warm place. Soluble in carbon tetrachloride, chloroform, ether, oils. Marketed as a mixture with an equal part of an absorbent powder. *Incompat.* Water, alkalies, heat.

Caution: Strong oxidizing agent. Can cause severe skin burns.

USE: Germicide; in bleaching of flour and food oils.

79. Acetrizoate Sodium. *3-(Acetylamino)-2,4,6-triiodobenzoic acid sodium salt;* 3-acetamido-2,4,6-triiodobenzoic acid sodium salt; acetrizoic acid sodium salt; Acétiodone; Bronchoselectan; Cystokon; Diaginol; Iodopaque; Pyelokon-R; Salpix; Thixokon; Tri-Abrodil; Triopac; Triurol; Urokon Sodium; Vesamin; Visotrast. $C_9H_5I_3NNaO_3$; mol wt 578.85. C 18.67%, H 0.87%, I 65.77%, N 2.42%, Na 3.97%, O 8.29%. Prepn: Wallingford *et al., J. Am. Chem. Soc.* **74**, 4365 (1952); Wallingford, *J. Am. Pharm. Assoc., Sci. Ed.* **42**, 721 (1953). Toxicity data: J. O. Hoppe *et al., J. Pharmacol. Exp. Ther.* **116**, 394 (1956).

Not isolated from soln. Marketed as a sterile 30% aq soln, d about 1200, pH between 4.8 and 5.2. The free acid is a white powder, decompn 278-283°. Sol in alcohol. Solubility in water at 25°: 94.2 g/100 ml. Very slightly sol in chloroform. Practically insol in benzene. LD_{50} in rabbits, cats, dogs (mg/kg): 5200, 5600, 6300 i.v. (Hoppe).

Mixture with meglumine acetrizoate, *Opacoron.*

THERAP CAT: Diagnostic aid (radiopaque medium).

80. Acetulan®. Acetylated lanolin alcohols. Prepn: Conrad, Motiuk, *J. Soc. Cosmet. Chem.* **6**, 344 (1955); *eidem,* U.S. pat. **2,725,334** (1955 to American Cholesterol Prod.).

Pale yellow, practically odorless liquid. Sp gr at 25°: 0.867. Neutral to litmus. Acid no. 0.35. Hydroxyl no. 2.0. Saponification no. 190.0. Hydrophobic, practically insol in water with no emulsification. Miscible with mineral oil, castor oil, vegetable oils, isopropanol, 95% ethanol, isopropyl myristate, isopropyl palmitate, butyl stearate.

USE: In cosmetic formulations and shampoo.

81. Aceturic Acid. *N-Acetylglycine;* acetamidoacetic acid; acetylaminoacetic acid; acetylglycocoll; ethanoylaminoethanoic acid. $C_4H_7NO_3$; mol wt 117.10. C 41.03%, H 6.03%, N 11.96%, O 40.99%. $CH_3CONHCH_2COOH$. Prepd by warming glycine with a slight excess of acetic anhydride in benzene: Radenhausen, *J. Prakt. Chem.* [2] **52**, 437 (1895); by warming glycine in glacial acetic acid with the stoichiometric amount of acetic anhydride: Dakin, *J. Biol. Chem.* **82**, 443 (1929).

Long needles from water, mp 206-208°. K at 25° = 2.3 × 10^{-4}. Sol in water at 15° = 2.7%. Moderately soluble in

alcohol. Slightly sol in acetone, chloroform, glacial acetic acid. Practically insol in ether, benzene. Forms stable salts with organic bases.

82. Acetylacetone. *2,4-Pentanedione;* diacetylmethane. $C_5H_8O_2$; mol wt 100.12. C 59.98%, H 8.05%, O 31.96%. $CH_3COCH_2COCH_3$. Made by the action of sodium on ethyl acetate and acetone or by the action of anhydrous aluminum chloride on acetyl chloride in the presence of an inert solvent: Combes, *Compt. Rend.* **103**, 814 (1886); *Ann. Chim. (Paris)* **12**, 199 (1887); Claisen, *Ber.* **38**, 695 (1905); Hunt, U.S. pat. **2,737,528** (1956 to Shawinigan Chem.); Georgieff, U.S. pat. **2,834,811** (1958); Gattermann-Wieland, *Praxis des Organischen Chemikers* (de Gruyter, Berlin, 40th ed, 1961) p 219. Toxicity data: Carpenter, *J. Ind. Hyg. Toxicol.* **31**, 343 (1949).

Colorless or slightly yellow, flammable liquid; pleasant odor. d 0.976. bp 140.5°. mp −23°. n_D^{20} 1.4512. One part dissolves in about 8 parts of water. Miscible with alcohol, benzene, chloroform, ether, acetone, glacial acetic acid. LC_{50} (4 hrs) in rats: 1000 ppm (Carpenter).

Caution: Mild irritant to skin, mucous membranes.

USE: Forms organometallic complexes which are used as gasoline additives, lubricant additives, driers for varnishes and printer's inks, fungicides, insecticides, colors.

83. Acetyl Bromide. C_2H_3BrO; mol wt 122.95. C 19.54%, H 2.46%, Br 64.99%, O 13.01%. CH_3COBr. Prepd from phosphorus tribromide and glacial acetic acid or acetic anhydride: Burton, Degering, *J. Am. Chem. Soc.* **62**, 227 (1940).

Fuming liquid; *irritating to the eyes.* bp 76°. d^9 1.52. mp −96°. Violently decomposed by water, or by alcohol; miscible with ether, chloroform, benzene. *Protect from water.*

84. α-Acetylbutyrolactone. *3-Acetyldihydro-2(3H)-furanone;* α-(2-hydroxyethyl)acetoacetic acid γ-lactone; α-acetyl-γ-hydroxybutyric acid γ-lactone; α-acetobutyrolactone. $C_6H_8O_3$; mol wt 128.13. C 56.25%, H 6.29%, O 37.46%. Prepn from sodium acetoacetate and ethylene oxide in abs alcohol: Knunyantz *et al., Compt. Rend. Acad. Sci. (U.R.S.S.)* [N.S.] **1**, 312 (1934), *C.A.* **28**, 4383 (1934); Feofilaktov, Onishchenko, *J. Gen. Chem. USSR* **9**, 304 (1939), *C.A.* **34**, 378 (1940); Forman; Johnson, U.S. pats. **2,397,134** and **2,443,827** (1946, 1948 both to U.S. Industrial Chemicals); *see also* Matukawa *et al., Japan.* pat. **134,284**; *C.A.* **35**, 7421 (1941).

Liquid; ester-like odor. Solubility in water 20% v/v; soly of water in lactone 12% v/v. d_4^{20} 1.1846; d_{20}^{20} 1.185-1.189; bp_{30} 142-143°; bp_{18} 130-132°; bp_5 107-108°. n_D^{20} 1.4562 (Knunyantz *et al.*); n_D^{20} 1.4590 (Feofilaktov *et al.*). Acquires a blue to bluish-purple color when in contact with iron.

Caution: Irritating to skin, mucous membranes.

USE: In synthesis of 3,4-disubstituted pyridines; of 5-(β-hydroxyethyl)-4-methylthiazole.

85. Acetylcarnitine. *2-(Acetyloxy)-3-carboxy-N,N,N-trimethyl-1-propanaminium inner salt;* (3-carboxy-2-hydroxypropyl)trimethylammonium hydroxide inner salt acetate; carnitine acetyl ester; vitamin B_T acetate. $C_9H_{17}NO_4$; mol wt 203.24. C 53.19%, H 8.43%, N 6.89%, O 31.49%. $(CH_3)_3$-$N^+CH_2CH(OCOCH_3)CH_2COO^-$. O-Acetyl deriv of carnitine, *q.v.* Prepn of racemate: R. Engeland, *Ber.* **42**, 2457 (1909); of L-form: R. Krimberg, W. Wittandt, *Biochem. Z.* **251**, 229 (1932); E. Strack *et al., Z. Physiol. Chem.* **238**, 183 (1936). Resolution of isomers: K. Brendel, R. Bressler, *Biochim. Biophys. Acta* **137**, 98 (1967). Extraction from calf brain tissue: E. A. Hosein, A. Orzeck, *Int. J. Neuropharmacol.* **3**, 71 (1964). Vitamin B_T activity: P. K. Bhattacharyya *et al., Arch. Biochem. Biophys.* **54**, 424 (1955). Neuropharmacology of isomers: E. Tempesta *et al., Neuropharmacology* **24**, 43 (1985). Determn in urine by carboxylic acid analyzer and reversed-phase column chromatography: K. Kidouchi *et al., J. Chromatog.* **423**, 297 (1987). Clinical evaluation of L-form in Alzheimer's disease: G. Acierno, *Clin. Ter.* **105**, 135 (1983); in senile brain: E. Bonavita, *Int. J. Clin. Pharmacol. Ther. Toxicol.* **24**, 511 (1986); in depression: E. Tempesta *et al., Drugs Exp. Clin. Res.* **13**, 417 (1987). Review of pharmacology in the CNS: L. Janiri, E. Tempesta, *Int. J. Clin. Pharmacol. Res.* **3**, 295-306 (1983).

L-Form, $C_9H_{17}NO_4$, hygroscopic crystals, mp 145° (dec). $[\alpha]_D^{20}$ −19.52° (c = 6). Very sol in water, alcohol. Practically insol in ether.

DL-Form hydrochloride, $C_9H_{17}NO_4HCl$, crystals, mp 187-188° (dec).

L-Form hydrochloride, $C_9H_{17}NO_4 \cdot HCl$, *acetyl-L-carnitine chloride, levacecarnine hydrochloride, ST-200, Alcar, Branigen, Nicetile, Normobren, Zibren.* Stable crystals, mp 187° (dec). Also reported as small rod-shaped crystals from butanol, mp 181° (Strack). $[\alpha]_D^{20}$ −26.9° (c = 9). $[\alpha]_D^{25}$ −28° (c = 2 in water). Very sol in water, sol in alcohol. Practically insol in ether.

THERAP CAT: Nootropic.

86. Acetyl Chloride. C_2H_3ClO; mol wt 78.50. C 30.60%, H 3.85%, Cl 45.16%, O 20.38%. CH_3COCl. Prepd from acetic acid and chlorine in the presence of phosphorus; from acetic acid and salts of chlorosulfonic acid; from sodium acetate and sulfuryl chloride. Details of prepn: Gattermann-Wieland, *Praxis des Organischen Chemikers* (de Gruyter, Berlin, 40th ed., 1961) p 111; A. I. Vogel, *Practical Organic Chemistry* (Longmans, London, 3rd ed., 1959) p 367. Lab prepn from acetic anhydride and calcium chloride: Gmünder, *Helv. Chim. Acta* **36**, 2021 (1953).

Flammable liquid; pungent odor. *Extremely irritating to the eyes.* Imparts a green tinge to a colorless flame. d 1.104. mp −112°. bp 52°. n_D^{20} 1.3898. Decomposed violently by water or alc. Miscible with benzene, chloroform, ether, glacial acetic acid, petr ether. *Protect from water.*

Caution: Irritant, corrosive. Causes severe burns. Avoid contact with skin, eyes, mucous membranes.

USE: Acetylating agent, in testing for cholesterol, determination of H_2O in organic liquids.

87. Acetylcholine Bromide. *2-(Acetyloxy)-N,N,N-trimethylethanaminium bromide;* Pragmoline; Tonocholin B. $C_7H_{16}BrNO_2$; mol wt 226.11. C 37.18%, H 7.13%, Br 35.34%, N 6.19%, O 14.15%. $(CH_3)_3N(Br)CH_2CH_2OOC-CH_3$. From trimethylamine and β-bromoethyl acetate: Fourneau, Page, *Bull. Soc. Chim. France* [4] **15**, 544 (1914).

Deliquesc crystals. Very soluble in cold water; decomposed by hot water or alkalies; soluble in alcohol; practically insol in ether. *Keep tightly closed.*

THERAP CAT: Cholinergic.

88. Acetylcholine Chloride. *2-(Acetyloxy)-N,N,N-trimethylethanaminium chloride;* Acecoline; Arterocoline; Miochol; Ovisot. $C_7H_{16}ClNO_2$; mol wt 181.66. C 46.28%, H 8.88%, Cl 19.52%, N 7.71%, O 17.61%. $(CH_3)_3N(Cl)-CH_2CH_2OOCCH_3$. Prepd from trimethylamine and β-chloroethyl acetate: Fourneau, Page, *Bull. Soc. Chim. France* [4] **15**, 544 (1914).

Crystalline powder, very deliquescent, mp 149-152°. Very sol in cold water, alcohol; decomposed by hot water or alkalies. Practically insol in ether. *Keep tightly closed.* LD_{50} s.c. in rats: 250 mg/kg.

THERAP CAT: Cholinergic.

89. Acetylcysteine. *N-Acetyl-L-cysteine;* L-α-acetamido-β-mercaptopropionic acid; *N*-acetyl-3-mercaptoalanine; Airbron; Broncholysin; Brunac; Fabrol; Fluatox; Fluimucil; Fluimucetin; Fluprowit; Mucocedyl; Mucolator; Mucolyticum; Mucomyst; Muco Sanigen; Mucosolvin; Mucret; NAC; Neo-Fluimucil; Parvolex; Respaire (obsolete); Tixair. $C_5H_9NO_3S$; mol wt 163.20. C 36.80%, H 5.56%, N 8.58%, O 29.41%, S 19.65%. Prepn: Smith, Gorin, *J. Org. Chem.* **26**, 820 (1961). Prepn and use in treatment of respiratory diseases: Martin, Waller, U.S. pat. **3,184,505** (1965 to Mead Johnson). Effect in corneal healing: F. Menna *et al., Bull. Mem. Soc. Fr. Opthalmol.* **94**, 425 (1982); G. Petroutsos *et al., Ophthal. Res.* **14**, 241 (1982). Toxicity: E. I. Goldenthal, *Toxicol. Appl. Pharmacol.* **18**, 185 (1971). Multicenter clinical trial in acetaminophen

overdose: M. J. Smilkstein *et al., N. Engl. J. Med.* **319**, 1557 (1988). *Review:* G. R. McKinney, G. M. Sisson, in *Pharmacological and Biochemical Properties of Drug Substances* vol. 2, M. E. Goldberg, Ed. (Am. Pharm. Assoc., Washington, DC, 1979) pp 479-488. Review of use in clinical toxicology: R. J. Flanagan, T. J. Meredith, *Am. J. Med.* **91**, Suppl. 3C, 131S-139S (1991).

Crystals from water, mp 109-110°. LD_{50} orally in rats: 5050 mg/kg (Goldenthal).

THERAP CAT: Mucolytic; corneal vulnerary; antidote (acetaminophen poisoning).

THERAP CAT (VET): Expectorant.

90. Acetyldigitoxins. $C_{43}H_{66}O_{14}$; mol wt 806.99. C 64.00%, H 8.24%, O 27.76%. Obtained by enzymatic hydrolysis of lanatoside A *(q.v.).* Composed of the aglycone digitoxigenin and 3 mols digitoxose, to one of which an acetyl group is attached. Acetyldigitoxin-α differs from acetyldigitoxin-β in the position of the acetyl group. The β-form is obtained either by splitting off the glucose residue from lanatoside A by means of enzymes, or by extraction from the leaves of *Digitalis ferruginea* L., *Scrophulariaceae.* The α-form can be obtained from acetyldigitoxin-β by heating in an anhydrous or aq organic solvent at pH 3.5 to 8: Stoll, Kreis, U.S. pat. **2,776,963** (1957 to Sandoz).

α-Form, *Acylanid.* Platelets from methanol, mp 217-221°. $[\alpha]_D^{20}$ +5.0° (c = 0.7 in pyridine). Slightly sol in chloroform.

β-Form, stout, solvated prisms from methanol losing their solvent of crystn in desiccator. When dry, dec 225°. $[\alpha]_D^{20}$ +16.7° (pyridine). One gram dissolves in 7 to 9 ml chloroform, in 150 ml methanol, in 220 ml amyl alc; almost insol in ether, water (one gram dissolves in 200 liters H_2O at 20°).

THERAP CAT: Cardiotonic.

91. Acetylene. *Ethyne;* ethine. C_2H_2; mol wt 26.04. C 92.26%, H 7.74%. HC≡CH. Manuf from calcium carbide and water: Eastman, U.S. pat. **3,017,259** (1962 to Texaco); from methane: Anderson, U.S. pat. **3,051,639** (1962 to Union Carbide). Toxicity data: Riggs, *Proc. Soc. Exp. Biol. Med* **22**, 269 (1925). Review of manuf processes: Faith, Keyes & Clark's *Industrial Chemicals,* F. A. Lowenheim, M. K. Moran, Eds. (Wiley-Interscience, New York, 4th ed., 1975) pp 26-35. *Review:* Nieuwland, Vogt, *The Chemistry of Acetylene* (Reinhold, New York, 1945) pp 1-219. Comprehensive monograph in 2 vols: S. A. Miller, *Acetylene* (Academic Press, New York, 1965); several authors in Kirk-Othmer *Encyclopedia of Chemical Technology* vol. 1 (Wiley-Interscience, New York, 3rd ed., 1978) pp 192-243.

Gas; odor not unpleasant when pure, but disagreeable when impure (due to phosphine). *Toxic when inhaled.* mp −81° (subl). At 0° liquifies at 21.5 atm; below 37° (crit temp) liquifies at 68 atm. One liter at 0° and 760 mm weighs 1.165 g; d gas (air = 1) 0.90. Burns brilliantly in air with very sooty flame. Heat of combustion 313 cal. Not explosive at ordinary atmospheric pressure, but at 2 atms or more it is explosive by spark or decomposition. Mixture with air containing more than 3% or less than 65% gas is explosive, max being 1 vol gas and 12.5 vol air. Forms insoluble explosive compounds with copper and silver; hence copper or brass containers must be avoided. One vol dissolves in 1 vol water, in 6 vols glacial acetic acid or alcohol; soluble in ether, benzene. Acetone dissolves 25 vols acetylene at 15° and 760 mm; but 300 vols at 12 atm. LC in rats: 900000 ppm (Riggs).

Caution: A simple asphyxiant. High concentrations cause narcosis. 20% concn may cause dyspnea, headache; 40% or more concn may cause collapse.

USE: Illuminant, oxyacetylene welding, cutting, and soldering metals, signalling; pptg metals, particularly Cu; manuf acetaldehyde, acetic acid; fuel for motor boats.

92. Acetylene Dibromide. *1,2-Dibromoethene; 1,2-dibromoethylene; sym-dibromoethylene.* $C_2H_2Br_2$; mol wt 185.85. C 12.93%, H 1.08%, Br 85.99%. BrCH=CHBr. Prepd by reaction of tetrabromoethylene with Zn and alc followed by sepn of *cis-* and *trans-*forms by fractional distillation: Noyes *et al., J. Am. Chem. Soc.* **72**, 33 (1950).

Liquid, gradually decomposed by air, moisture, or light. d_4^{17} 2.21. n_D^{20} 1.5428. Practically insol in water; sol in many organic solvents. LD_{50} in rats: 117 mg/kg orally.

cis-Form, n_D^{25} 1.5370.

trans-Form, n_D^{25} 1.5440.

Caution: Narcotic in high concns.

93. Acetylene Dichloride. *1,2-Dichloroethene; 1,2-dichloroethylene; sym-dichloroethylene; Dioform.* $C_2H_2Cl_2$; mol wt 96.94. C 24.78%, H 2.08%, Cl 73.14%. ClCH=CH-Cl. Prepn: Bordner, U.S. pat. **2,504,919** (1950 to du Pont). Prepn of *trans*-form: Adler, U.S. pat. **2,440,997** (1948 to Stockholms Superfosfot Fabriks Aktiebolag). Sepn of *cis*- and *trans*-forms by fractional distillation: Wood, Dickinson, *J. Am. Chem. Soc.* **61**, 3259 (1939); Johnsen, Fitzpatrick, *Rec. Trav. Chim.* **70**, 823 (1951); Truce, Barney, *J. Org. Chem.* **27**, 128 (1962). Toxicity: D. Gradiski *et al., Eur. J. Toxicol.* **7**, 247 (1974); K. J. Freundt *et al., Toxicology* **7**, 141 (1977). *Review:* V. L. Stevens in Kirk-Othmer *Encyclopedia of Chemical Technology,* **Vol. 5** (Wiley-Interscience, New York, 3rd ed., 1979) pp 742-745.

Liquid; ethereal, slightly acrid odor; gradually decomposed by air, light and moisture, forming HCl. d ~1.28. bp ~55°. Insol in water. Sol in alc, ether and most other organic solvents. LD_{50} i.p. in mice: ~2150 mg/kg (Gradiski).

cis-Form, mp −81.5°. bp_{745} 59.6°, bp_{760} 60°. n_D^{25} 1.4435.

trans-Form, mp −49.4°. bp_{745} 47.2°. Noticeably subject to air oxidation. LD_{50} in rats (ml/kg): 1.0 orally; 60 i.p.; in mice (ml/kg): 3.2 i.p. (Freundt).

Caution: Potential symptoms of overexposure are irritation of eyes and respiratory system; CNS depression. *See NIOSH Pocket Guide to Chemical Hazards* (DHHS/NIOSH 90-117, 1990) p 88. *See also* Patty's *Industrial Hygiene and Toxicology* vol. **2B**, G. D. Clayton, F. E. Clayton, Eds. (Wiley-Interscience, New York, 3rd ed., 1981) pp 3550-3553.

USE: Solvent for fats, phenol, camphor, etc.

94. Acetyleneurea. *Tetrahydroimidazo[4,5-d]imidazole-2,5-(1H,3H)-dione; glycoluril;* acetylenediureine; glyoxaldiureine; acetylene carbamide. $C_4H_6N_4O_2$; mol wt 142.12. C 33.81%, H 4.26%, N 39.42%, O 22.52%. Prepd by the sodium amalgam reduction of allantoin: Biltz, Schiemann, *J. Prakt. Chem.* **113**, 77 (1926); from glyoxal and urea: Reibnitz, U.S. pat. **2,731,472** (1956 to BASF).

Crystals, dec at about 300°. Slightly sol in cold, more sol in hot water; sol in warm mineral acids, warm ammonia.

95. Acetyl Iodide. C_2H_3IO; mol wt 169.95. C 14.13%, H 1.78%, I 74.67%, O 9.41%. CH_3COI. Prepd from acetyl chloride and HI: Gustus, Stevens, *J. Am Chem. Soc.* **55**, 374 (1933).

Liquid; suffocating odor; fumes and turns brown in air. bp 108°, bp_{735} 104-106°; d_4^{20} 2.0674; n_D^{20} 1.5491. Dec by water or alcohol; sol in benzene, ether. *Keep tightly closed and protected from light.*

Caution: A strong irritant. Avoid contact with skin. Vapors can cause pulmonary edema.

96. Acetylleucine Monoethanolamine. *N-Acetyl-DL-leucine compd with 2-aminoethanol (1:1);* monoethanolamine DL-acetylleucinate; DL-acetylleucine monoethanolamine salt; monoethanolamine salt of α-acetamidoisocaproic acid; RP-7452; Tanganil. $C_{10}H_{22}N_2O_4$; mol wt 234.30. C 51.26%, H 9.46%, N 11.96%, O 27.31%. Prepn: Gailliot *et al.,* U.S. pat. **2,941,924** (1960 to Rhône-Poulenc).

Crystals, mp ~150°. Soly in water: >20%; slight soly in alcohol: ~1%. pH of 10% aq soln: ~6.

THERAP CAT: Antivertigo agent.

97. *N*-Acetylmethionine. $C_7H_{13}NO_3S$; mol wt 191.25. C 43.96%, H 6.85%, N 7.32%, O 25.10%, S 16.77%. $CH_3SCH_2CH_2CH(NHCOCH_3)COOH$. Prepn of racemic mixture: Wheeler, Ingersoll, *J. Am. Chem. Soc.* **73**, 4604 (1951); Callanan, Patton, U.S. pat. **2,745,873** (1956 to Distillers Products). Resolution: Wheeler, Ingersoll, *loc. cit.;* Gillingham, U.S. pat. **3,028,395** (1962 to Parke, Davis); Tullar, U.S. pat. **3,056,799** (1962 to Sterling Drug); **Brit.** pat. **1,072,876** (1967 to Tanabe Seiyaku), *C.A.* **67**, 114015x (1967).

DL-Form, *Methionamine.* Crystals from water, mp 114-115°.

D(+)-Form, crystals from water or ethyl acetate, mp 104-105°. $[\alpha]_D^{25}$ +20.3° (c = 4 in water).

L(−)-Form, crystals, mp 104°. $[\alpha]_D^{25}$ −20.3°.

THERAP CAT: DL-Form as lipotropic.

98. 5-Acetyl-2-methoxybenzaldehyde. 3-Acetyl-6-methoxybenzaldehyde. $C_{10}H_{10}O_3$; mol wt 178.19. C 67.41%, H 5.66%, O 26.94%. A natural plant growth inhibitor found in the leaves of *Encelia farinosa* A. Gray, Compositae. Isoln and synthesis from 3-acetyl-6-methoxybenzonitrile: Gray, Bonner, *J. Am. Chem. Soc.* **70**, 1249 (1948).

Needles from alcohol or ether, mp 144°. Sublimes without decomp. Emits fragrant odor on prolonged heating. Sol in hot water, warm ether, alcohol, benzene, chloroform. Practically insol in cold water, 5% HCl, 5% NaOH, petr ether, carbon tetrachloride. Sol with orange color in concd mineral acids.

99. Acetyl Nitrate. *Acetic acid anhydride with nitric acid.* $C_2H_3NO_4$; mol wt 105.05. C 22.87%, H 2.88%, N 13.33%, O 60.92%. CH_3COONO_2. Prepn from acetic anhydride and N_2O_5: Boh, *Annales de Chimie* [11] **20**, 437 (1945).

Fuming, mobile, hygroscopic liquid. Should be colorless. d_4^{15} 1.24. bp_{70} 22°. Although it may be stored in the dark over P_2O_5 for the weekend, it should be used in *statu nascendi* to avoid explosions. Always explodes when heated suddenly over 60° or when in contact with HgO. Explosions have occurred upon contact with ground glass surfaces: König, *Angew. Chem.* **67**, 157 (1955).

Caution: Irritant, corrosive.

USE: In nitrations, especially to introduce a single nitro group in an ortho position on an aromatic ring.

100. *N*-Acetylpenicillamine. *N-Acetyl-3-mercapto-D-valine.* $C_7H_{13}NO_3S$; mol wt 191.25. C 43.96%, H 6.85%, N 7.32%, O 25.10%, S 16.77%. Prepn: Crooks in *The Chemistry of Penicillin* (Princeton Univ. Press, 1949) p 470; Sheehan *et al.,* U.S. pats. **2,477,148** and **2,496,416** (1949, 1950, both to Merck & Co.).

dl-Form, crystals from hot water, mp 183°.
d-Form, crystals from water, mp 189-190°; [α]$_D^{25}$ +18° (50% ethanol).

101. Acetylpheneturide. *N-[(Acetylamino)carbonyl]-α-ethylbenzeneacetamide; 1-acetyl-3-(2-phenylbutyryl)urea; N-(α-ethylphenylacetyl)-N'-acetylurea;* Crampol. $C_{13}H_{16}N_2O_3$; mol wt 248.28. C 62.89%, H 6.50%, N 11.28%, O 19.33%. Synthesis: Umemoto *et al., Yakugaku Zasshi* **83**, 753 (1963), *C.A.* **59**, 13849d (1963); Takamatsu *et al.,* U.S. pat. **3,110,728** (1963 to Dainippon). Pharmacology and toxicology: Nakamura *et al., Arzneimittel-Forsch.* **18**, 524 (1968).

Crystals from dil alcohol, mp 100-101°. LD$_{50}$ orally in mice: 4.73 mMole/kg (1.17 g/kg) (Nakamura).
THERAP CAT: Anticonvulsant.

102. Acetylsalicylsalicylic Acid. *2-(Acetyloxy)benzoic acid 2-carboxyphenyl ester; 2-hydroxybenzoic acid acetate 2-carboxyphenyl ester;* salicylacetylsalicylic acid; Diplosal Acetate. $C_{16}H_{12}O_6$; mol wt 300.27. C 64.00%, H 4.03%, O 31.97%. Prepd by treating salicylsalicylic acid *q.v.,* with acetic anhydride or by fractionating an acetylsalicylic acid melt: Ger. pat. **236,196** (1908 to Boehringer).

Leaflets from dilute acetic acid, mp 159°; also reported as mp 152°, Rudoi, *Soviet Med.* **1941**, no. 12, 26-27. Practically insol in water, benzene. One part dissolves in 3 parts boiling alcohol, in 70 parts boiling ether, in 45 parts boiling benzene. Sol in alkaline solns, but is saponified by heating in alk soln. On acidification of the soln salicylic acid is precipitated.
THERAP CAT: Analgesic.

103. Acetyl Sulfamethoxypyrazine. *N-[(4-Aminophenyl)sulfonyl]-N-(3-methoxypyrazinyl)acetamide; N^1-acetyl-N^1-(3-methoxypyrazinyl)sulfanilamide; 2-(N^1-acetylsulfanilamido)-3-methoxypyrazine;* Acetylazide. $C_{13}H_{14}N_4O_4S$; mol wt 322.34. C 48.44%, H 4.38%, N 17.38%, O 19.85%, S 9.95%. Prepn: Camerino, Palamidessi, *Gazz. Chim. Ital.* **90**, 1815 (1960); *eidem,* U.S. pat. **3,098,069** (1963 to Soc. Farmaceutici Italia).

Crystals, mp 199°.
THERAP CAT: Antibacterial.

104. N^4-Acetylsulfanilamide. *N-[4-(Aminosulfonyl)phenyl]acetamide; p-sulfamylacetanilide.* $C_8H_{10}N_2O_3S$; mol wt 214.25. C 44.85%, H 4.70%, N 13.08%, O 22.40%, S 14.97%. Formed in the animal body where as much as 60% of ingested sulfanilamide is acetylated by the liver; some extrahepatic acetylation occurs also. Prepd in the laboratory by condensing acetanilide with chlorosulfonic acid, and treating the resulting N^4-acetylsulfanilyl chloride with ammonia. The formation of N^4-acetylsulfanilamide is a step in the manufacture of sulfanilamide: Miller *et al., J. Am. Chem. Soc.* **61**, 1198 (1939); *cf.* Smiles, Stewart, *Org. Syn.* coll. vol. I, 8 (1932).

Needles from alcohol + water, mp 216°. Slightly sol in water, alcohol. Max soly is reached in a 65% alcoholic soln.

105. N-Acetylsulfanilic Acid. *4-(Acetylamino)benzenesulfonic acid; N-p-acetylanilinosulfonic acid;* acetanilidsulfonic acid; 4-acetamidobenzenesulfonic acid. $C_8H_9NO_4S$; mol wt 215.23. C 44.64%, H 4.21%, N 6.51%, O 29.73%, S 14.90%. Prepd by the action of oleum or sulfuric acid on acetanilide in acetic anhydride: Junghahn, Neumann, *Ber.* **33**, 1366 (1900); Söll, Stutzer, *Ber.* **42**, 4539 (1909).

Dihydrate, crystals. Freely sol in water, sol in alcohol, sparingly sol in glacial acetic acid, insol in ether. Slowly hydrolyzes in aq soln.
Sodium salt, $C_8H_8NNaO_4S$, *Cosaprin.* Prisms, soluble in water, moderately sol in alcohol, insol in ether.
Ethyl ester, $C_{10}H_{13}NO_4S$, needles from benzene, mp 115°.

106. N-Acetylsulfanilyl Chloride. *4-(Acetylamino)benzenesulfonyl chloride; p-acetamidobenzenesulfonyl chloride; p-acetaminobenzenesulfonyl chloride;* acetanilide-p-sulfonyl chloride; ASC. $C_8H_8ClNO_3S$; mol wt 233.68. C 41.12%, H 3.45%, Cl 15.17%, N 5.99%, O 20.54%, S 13.72%. Prepd by sulfonation of acetanilide: Stewart, *J. Chem. Soc.* **121**, 2558 (1922); Smiles, Stewart, *Org. Syn.* coll. vol. I, 8 (2nd ed., 1941).

Thick, light tan prisms from benzene, mp 149°. Has slight odor of acetic acid. Benzene and ether are poor solvents for purification by recrystn. Much better results are obtained using chloroform or ethylene dichloride. *Cf.* Northey, *Sulfonamides* (New York, 1948) p 12.
Caution: Irritant to skin, eyes, and mucous membranes of nose and throat.
USE: Intermediate in the prepn of sulfanilamide and its derivs.

107. Acetyltannic Acid. Diacetyltannic acid; tannyl acetate; Acetannin; Tannigen. Acetylation of tannin: Ciusa, Sollazzo, *Ann. Chim. Appl.* **33**, 72 (1942); *C.A.* **38**, 5794^4 (1944).
Yellowish- or grayish-white odorless powder; darkens on exposure to light. Softens under water at about 70°. Slightly soluble in water or alcohol; sol in ethyl acetate, solns of borax or sodium phosphate, or with gradual decompn in

solns of alkali hydroxides and carbonates. *Incompat.* Alkalies, iron salts.

THERAP CAT: Astringent (intestinal).

THERAP CAT (VET): Astringent (intestinal) for foals, calves, dogs.

108. Achillea. Milfoil; yarrow; thousand-leaf. Flowering herb of *Achillea millefolium* L., *Compositae. Habit.* Europe, Asia, naturalized in U.S. Extensively used as herbal remedy by many cultures. Constituents of plant and of the volatile oil may vary depending on origin of plant. Identification of constituents: Pailer, Kump, *Monatsh.* **90,** 395 (1959); *eidem, Arch. Pharm.* **293,** 646 (1960); A. J. Falk *et al., J. Nat. Prod.* **37,** 598 (1974); R. F. Chandler *et al., J. Pharm. Sci.* **71,** 690 (1982). Isoln of anti-inflammatory constituents: A. S. Goldberg, *J. Pharm. Sci.* **58,** 938 (1969). Phytopharmacology: J. P. Tewari *et al., Ind. J. Med. Sci.* **28,** 331 (1974). Review of medicinal uses and composition: R. F. Chandler *et al., Econ. Bot.* **36,** 203-223 (1982).

109. Acid Fuchsin. *2-Amino-5-[(4-amino-3-sulfophenyl)(4-imino-3-sulfo-2,5-cyclohexadien-1-ylidene)methyl]-3-methylbenzenesulfonic acid disodium salt; C.I. Acid Violet 19;* 2-amino-α^5-(4-amino-3-sulfophenyl)-α^5-(4-imino-3-sulfo-2,5-cyclohexadien-1-ylidene)-3,5-xylenesulfonic acid disodium salt; acid magenta; acid rubin; fuchsin(e) acid; acid roseine; C.I. 42685. $C_{20}H_{17}N_3Na_2O_9S_3$; mol wt 585.55. C 41.02%, H 2.93%, N 7.18%, Na 7.85%, O 24.59%, S 16.43%. The trisulfonic acid disodium salt of magenta I, *q.v.* Prepn: *Colour Index* vol. 4 (3rd ed., 1971) p 4399. Brief review: H. J. Conn's *Biological Stains,* R. D. Lillie, Ed. (Williams & Wilkins, Baltimore, 9th ed., 1977) pp 285-286.

Olive to dark olive-green coarse powder. Absorption max: 540-545 nm (10 mg/l of 0.01% HCl). One gram dissolves in 7 ml water. Slightly sol in alc. Dil aq solns are purplish red. Very dil aq solns (1:10,000) are decolorized by drops of concd aq NaOH solns, but not returned by concd HCl. The decolorized soln neutralized with NaOH is called the *Andrade indicator.* Changes from red to colorless at pH 12-14.

USE: As pH indicator; biological stain.

110. Acid Violet 7B. *N-[4-[Bis[4-[ethyl(3-sulfophenyl)amino]phenyl]methylene]-2,5-cyclohexadien-1-ylidene]-N-methylmethanaminium inner salt, sodium salt; C.I. Acid*

Violet 25; C.I. 42745. $C_{37}H_{36}N_3NaO_6S_2$; mol wt 705.83. C 62.96%, H 5.14%, N 5.95%, Na 3.26%, O 13.60%, S 9.09%. Prepd by sulfonation of the condensation product from *p*-dimethylaminobenzoyl chloride with *N*-ethyldiphenylamine: Müller, Ger. pat. **34,463;** *Frdl.* **1,** 88; U.S. pat. **353,266;** Turnbull, Evans, U.S. pat. **1,402,195.** *See also: Colour Index* vol. 4 (3rd ed., 1971) p 4401.

Violet powder. Sol in water, alcohol with deep blue color and violet tinge. Insol in ether. Absorption max (water): 607.5 nm.

USE: As dye; in inks and stains.

111. Acifluorfen. *5-[2-Chloro-4-(trifluoromethyl)-phenoxy]-2-nitrobenzoic acid.* $C_{14}H_7ClF_3NO_5$; mol wt 361.66. C 46.49%, H 1.95%, Cl 9.80%, F 15.76%, N 3.87%, O 22.12%. Selective pre- and post-emergence herbicide. Prepn: H. O. Bayer *et al.,* Ger. pat. **2,311,638;** *eidem,* U.S. pats. **3,928,416, 4,063,929** (1973, 1975, 1977 all to Rohm & Haas). Synthesis, activity and toxicity: W. O. Johnson *et al., J. Agr. Food Chem.* **26,** 285 (1978). Carcinogenicity studies: J. A. Quest *et al., Regul. Toxicol. Pharmacol.* **10,** 149 (1989).

Off-white solid, mp 151.5-157°.

Sodium salt, $C_{14}H_6ClF_3NNaO_5$, *scifluorfen, RH-6201, Blazer.* White powder, mp 124-125°. Soly in water > 25%. LD_{50} orally in rats: 1300 mg/kg (Johnson).

Caution: Skin and eye irritant. Do not incorporate in soil. Do not mix with oils, surfactants, liquid fertilizers or other pesticides.

USE: Herbicide.

112. Acifran. *4,5-Dihydro-5-methyl-4-oxo-5-phenyl-2-furancarboxylic acid;* 2-methyl-2-phenyl-3(2*H*)-furanone-5-carboxylic acid; AY-25712; Reductol. $C_{12}H_{10}O_4$; mol wt 218.21. C 66.05%, H 4.62%, O 29.33%. Prepd (not claimed): I. L. Jirkovsky *et al.,* U.S. pat. **4,169,202;** prepn and resolution of enantiomers: *eidem,* U.S. pat. **4,244,958** (1979, 1981 both to American Home); I. L. Jirkovsky, M. N. Cayen, *J. Med. Chem.* **25,** 1154 (1982). Metabolism and pharmacokinetics in dogs and rats: M. N. Cayen *et al., Xenobiotica* **16,** 251 (1986). LC determn in serum: R. Gonzalez, D. Thibeault, *Clin. Biochem.* **16,** 244 (1983). Clinical evaluation in hyperlipoproteinemia: D. B. Hunninghake *et al., Clin. Pharmacol. Ther.* **38,** 313 (1985); J. C. LaRosa *et al., Artery* **14,** 338 (1987).

Crystals, mp 176°. uv max (methanol): 281 nm (ϵ 7960). LD_{50} orally in rats: 3 g/kg (Jirkovsky, 1982).

(−)-Form, mp 87-89°. $[\alpha]_D^{25}$ −144.7° (c = 2 in methanol).

(+)-Form, mp 87-89°. $[\alpha]_D^{25}$ +146.4° (c = 2 in methanol).

THERAP CAT: Antihyperlipoproteinemic.

113. Acipimox. *5-Methylpyrazinecarboxylic acid 4-oxide;* 2-carboxy-5-methylpyrazine 4-oxide; K-9321; Olbemox; Olbetam. $C_6H_6N_2O_3$; mol wt 154.13. C 46.76%, H 3.92%, N 18.18%, O 31.14%. Prepn: V. Ambrogi *et al.,* Ger. pat. **2,319,834;** *eidem,* U.S. pat. **4,002,750** (1973, 1977 both to Carlo Erba). Prepn and toxicology: *eidem, Eur. J. Med. Chem.* **15,** 157 (1980). Pharmacological profile: P. P. Lovisolo *et al., Pharmacol. Res. Commun.* **13,** 151, 163 (1981). Pharmacokinetics: L. M. Fuccella *et al., Clin. Pharmacol. Ther.* **28,** 790 (1980); L. Musatti *et al., J. Int. Med. Res.* **9,** 381 (1981). Mechanism of action: K. Aktories *et al., Arzneimittel-Forsch.* **33,** 1525 (1983).

Crystals from water, mp 177-180°. LD$_{50}$ orally in mice: 3500 mg/kg (Ambrogi).

THERAP CAT: Antihyperlipoproteinemic.

114. Acitretin. *(all-E)-9-(4-Methoxy-2,3,6-trimethylphenyl)-3,7-dimethyl-2,4,6,8-nonatetraenoic acid;* etretin; Ro-10-1670; Neotigason; Soriatane. C$_{21}$H$_{26}$O$_3$; mol wt 326.44. C 77.27%, H 8.03%, O 14.70%. Synthetic retinoid; free acid form and major metabolite of etretinate, *q.v.* Prepn: W. Bollag *et al.,* Ger. pat. **2,414,619;** *eidem,* U.S. pat. **4,105,681** (1974, 1978 both to Hoffmann-La Roche). Teratogenicity study: A. Kistler, H. Hummler, *Arch. Toxicol.* **58,** 50 (1985). HPLC determn in plasma: N. R. Al-Mallah *et al., Anal. Letters* **21,** 1603 (1988). Pharmacokinetics in humans: F. G. Larsen *et al., Pharmacol. Toxicol.* **62,** 159 (1988). Clinical evaluation in cutaneous lupus erythematosus: T. Ruzicka *et al., Arch. Dermatol.* **124,** 897 (1988). Review of clinical pharmacology: A. Vahlquist, O. Rollman, *Dermatologica* **175,** Suppl. 1, 20-27 (1987). Review of clinical studies in psoriatic and nonpsoriatic dermatoses: J.-M. Geiger, B. M. Czarnetzki, *ibid.* **176,** 182-190 (1988).

Crystals from hexane, mp 228-230°. LD$_{50}$ i.p. in mice (mg/kg): >4000 (1 day), 700 (10 days), 700 (20 days) (Bollag, 1978).

THERAP CAT: Antipsoriatic.

115. Aclacinomycins. Antitumor antibiotic complex of the anthracycline group, produced by *Streptomyces galilaeus.* Thirteen yellow and seven red-colored components have been identified. Isoln of the major components, aclacinomycins A and B: H. Umezawa *et al.,* Ger. pat. **3,532,-568;** *eidem,* U.S. pat. **3,988,315** (1974, 1976 both to Microbiochem. Res. Found.). See also: T. Oki *et al., J. Antibiot.* **28,** 830 (1975). Structure, taxonomy, production, properties: *eidem, ibid.* **32,** 791, 801 (1979). The aglycone portion of aclacinomycin A is known as *aklavinone.* Isoln of aklavinone: J. J. Gordon *et al., Tetrahedron Letters* **8,** 28 (1960). Synthesis of racemic aklavinone: B. A. Pearlman *et al., J. Am. Chem. Soc.* **103,** 4248 (1981); P. A. Confalone, G. Pizzolato, *ibid.* **4251;** R. K. Boeckman, F. W. Sum, *ibid.* **104,** 4604 (1982). Synthesis of optically active aklavinone: A. S. Kende, J. P. Rizzi, *ibid.* **103,** 4247 (1981); J. M. McNamara, Y. Kishi, *ibid.* **104,** 7371 (1982). *In vitro* metabolism of aclacinomycins: T. Komiyama *et al., Gann* **70,** 395 (1979), *C.A.* **92,** 15143x (1980). HPLC determn of aclacinomycin A and its metabolites: T. Ogasawara *et al., J. Antibiot.* **34,** 47, 52 (1981). Pharmacokinetics: M. J. Egorin *et al., Cancer Chemother. Pharmacol.* **8,** 41 (1982). Series of articles on absorption, excretion, distribution and general pharmacology: *Japan. J. Antibiot.* **33,** 169-213 (1980), *C.A.* **93,** 19316, 125326, 125649, 142986 (1980). Immunological study: M. Ishizuka *et al., J. Antibiot.* **34,** 331 (1981). Clinical studies: R. Maral, *Drugs Exp. Clin. Res.* **9,** 375 (1983); R. P. Warrell, Jr., S. J. Kempin, *Am. J. Clin. Oncol.* **6,** 81 (1983); A. Y. Bedikian *et al., ibid.* 187 (1983). Review of pharmacology of aclacinomycin A: T. Oki, *Anthracyclines [Proc. Workshop],* S. T. Crooke, S. D. Reich, Eds. (Academic Press, New York, 1980) pp 323-342, *C.A.* **93,** 160778h (1980).

Aclacinomycin A

Aclacinomycin A, C$_{42}$H$_{53}$NO$_{15}$, *[1R-(1α,2β,4β)]-2-ethyl-1,-2,3,4,6,11-hexahydro-2,5,7-trihydroxy-6,11-dioxo-4-[[2,3,6-trideoxy-4-O-[2,6-dideoxy-4-O-[(2R-trans)-tetrahydro-6-methyl-5-oxo-2H-pyran-2-yl]-α-L-lyxo-hexopyranosyl]-3-(dimethylamino)-α-L-lyxo-hexopyranosyl]oxy]-1-naphthacenecarboxylic acid methyl ester,* aclarubicin, antibiotic MA 144A1, NSC-208734, Jaclacin. Yellow microcryst powder from chloroform/hexane, mp 151-153° (dec). [α]$_D^{24}$ −11.5° (c = 1 in methylene chloride). uv max (methanol): 229.5, 259, 289.5, 431 nm (E$_{1cm}^{1\%}$ 550, 326, 135, 161); (0.1N HCl) 229.5, 258.5, 290, 431 nm (E$_{1cm}^{1\%}$ 571, 338, 130, 161); (0.1N NaOH) 239, 287, 523 nm (E$_{1cm}^{1\%}$ 450, 113, 127). Sol in CHCl$_3$, ethyl acetate. Insol in ethyl ether, *n*-hexane, petr ether. Addn of alkali to aq solns gives an intense reddish-purple color; in conc HCl the soln is yellow. LD$_{50}$ in mice (mg/kg): 22.6 i.p., 33.7 i.v. (Oki).

Aclacinomycin A hydrochloride, C$_{42}$H$_{53}$NO$_{15}$·HCl, *aclarubicin hydrochloride, Aclacin, Aclacinon, Aclaplastin.*

Aclacinomycin B, C$_{42}$H$_{51}$NO$_{15}$, *[1R-(1α,2β,4β)]-4-[[[2′′′,-3′′-anhydro]-O-3,6-dideoxy-α-L-erythro-hexopyranos-4-ulos-1-yl-(1→4)-O-2,6-dideoxy-α-L-lyxo-hexopyranosyl-(1→4)-2,3,6-trideoxy-3-(dimethylamino)-α-L-lyxo-hexopyranosyl]oxy]-2-ethyl-1,2,3,4,6,11-hexahydro-2,5,7-trihydroxy-6,11-dioxo-1-naphthacenecarboxylic acid methyl ester, antibiotic MA 144B1.* Yellow microcryst from chloroform/hexane, mp 163-167° (dec). [α]$_D^{24}$ +3° (c = 1 in methylene chloride). Other physical properties similar to aclacinomycin A. LD$_{50}$ in mice (mg/kg): 13.7 i.p., 16.4 i.v. (Oki).

THERAP CAT: Antineoplastic.

116. Aclatonium Napadisilate. *2-[2-(Acetyloxy)-1-oxopropoxy]-N,N,N-trimethylethanaminium 1,5-naphthalenedisulfonate (2:1); 1,5-naphthalenedisulfonic acid bis [2-[2-(acetyloxy)-1-oxopropoxy]-N,N,N-trimethylethanaminium]ion (2−);* acetyllactoylcholine 1,5-naphthalenedisulfonate; choline 1,5-naphthalenedisulfonate (2:1), dilactate, diacetate; TM-723; Abovis. C$_{30}$H$_{46}$N$_2$O$_{14}$S$_2$; mol wt 722.83. C 49.85%, H 6.41%, N 3.88%, O 30.99%, S 8.87%. Spasmolytic agent related structurally to acetylcholine chloride, *q.v.* Prepn: K. Miura *et al.,* Ger. pat. **2,425,983;** *eidem,* U.S. pat. **3,903,137** (1973, 1975 both to Toyama). Synthesis and smooth muscle activity: *eidem, Yakugaku Zasshi* **99,** 1245 (1979), *C.A.* **92,** 146197r (1980). Toxicity study: A. Takai *et al., Oyo Yakuri* **19,** 93 (1980), *C.A.* **93,** 613367 (1980). Series of articles on metabolism, pharmacology, toxicity studies: *Oyo Yakuri* **18**, 695-942 (1979), *C.A.* **92**, 191015r, 191281z, 191282a, 208995k-7n (1980).

Crystals, mp 189-191°. LD$_{50}$ in mice: 15 g/kg orally, 826 mg/kg s.c. (Miura); in dogs: > 10 g/kg orally (Takai).

THERAP CAT: Cholinergic.

117. Aconine. *(1α,3α,6α,14α,15α,16β)-20-Ethyl-1,6,-16-trimethoxy-4-(methoxymethyl)aconitane-3,8,13,14,15-pentol.* C$_{25}$H$_{41}$NO$_9$; mol wt 499.60. C 60.10%, H 8.27%, N 2.80%, O 28.82%. Obtained by hydrolysis of aconitine: Schulze, *Arch. Pharm.* **244**, 165 (1906); Schneider, *Ber.* **89**, 768 (1956). Structure: Schneider, Tausend, *Ann.* **628**, 114 (1959); McCaldin, Marion, *Can. J. Chem.* **37**, 1071 (1959).

Amorphous powder, mp 132°. Bitter taste. [α]$_D$ +23°. pK 9.52. Absorption spectrum: Nath, *J. Indian Chem. Soc.* **32**, 75 (1955). Very sol in water, alcohol; moderately sol in chloroform; slightly sol in benzene. Practically insol in ether, petr ethcr.

Hydrochloride dihydrate, C$_{25}$H$_{42}$ClNO$_9$.2H$_2$O, crystals, mp 175-176°. [α]$_D$ −8°.

Hydrobromide sesquihydrate, C$_{25}$H$_{42}$BrNO$_9$.1½H$_2$O, crystals from water, mp 225°.

118. Aconite. Monkshood; wolf's-bane; friar's cowl; mouse-bane. Dried tuberous root of *Aconitum napellus* L., *Ranunculaceae*. *Habit.* Mountainous regions of Europe, Asia, and North America. *Constit.* 0.4-0.8% aconitine; aconine; napelline (isoaconitine, pseudaconitine); picraconitine; aconitic acid; itaconic acid; succinic acid; malonic acid; fat; levulose. *Quite toxic. Ref:* Freudenberg, *Ber.* **69**, 1962 (1936); Rogers, Freudenberg, *ibid.* **70**, 349 (1937); Lascombes, *Ann. Pharm. Franc.* **16**, 429 (1958). *See also* "Aconite," A. Ph. A. Monograph No. 1 (Am. Pharm. Assoc., Washington, D.C., 1938).

THERAP CAT: Antipyretic.

THERAP CAT (VET): Formerly used as an antihypertensive.

119. Aconitic Acid. *1-Propene-1,2,3-tricarboxylic acid;* equisetic acid; citridic acid; achilleic acid. C$_6$H$_6$O$_6$; mol wt 174.11. C 41.39%, H 3.47%, O 55.14%. Found in leaves and tubers of *Aconitum napellus* L., *Ranunculaceae*, in various species of *Achillea (Compositae)* and *Equisetum (Equisetaceae)*, in beet root, and in sugar cane. Can be prepd commercially from calcium magnesium aconitate recovered from sugar cane juice: McCalip, Seibert, *Ind. Eng. Chem.* **33**, 637 (1941); from molasses: Regna, Bruins, *ibid.* **48**, 1268 (1956). Most of the commercial aconitic acid is, however, manufactured by sulfuric acid dehydration of citric acid: Bruce, *Org. Syn.* coll. vol. II, p 12 (1943); by using methanesulfonic acid instead of sulfuric: Cranston, U.S. pat. 2,727,066 (1955 to Daniel F. Kelly). Aconitic acid prepd by any of the above methods has the *trans*-configuration which is the form described here.

trans-form

Leaflets, plates from water. Decompn 198-199° (capillary inserted in oil bath at 190°); decompn 204-205° (capillary in oil bath at 195°); decompn 209° (electrically heated bar). K$_1$ at 25° = 1.58 × 10^{-3}; K$_2$ = 3.5 × 10^{-5}. One gram dissolves in 5.5 ml water at 13°, in 2 ml water at 25°. Soluble in 2 parts of 88% alcohol at 12°. Slightly sol in ether.

Triethyl ester, bp$_5$ 155°.

Tributyl ester, bp$_3$ 190°.

USE: Manuf itaconic acid. As plasticizer for buna rubber and plastics. Used in form of triethyl or tributyl ester.

120. Aconitine. *(1α,3α,6α,14α,15α,16β)-20-Ethyl-1,6,-16-trimethoxy-4-(methoxymethyl)aconitane-3,8,13,14,15-pentol 8-acetate 14-benzoate.* C$_{34}$H$_{47}$NO$_{11}$; mol wt 645.75. C 63.24%, H 7.34%, N 2.17%, O 27.25%. Several isomers from *Aconitum napellus* L., *Ranunculaceae* and other aconites. Majima *et al.*, *Ber.* **57**, 1456 (1924); *Proc. Imp. Acad. Tokyo* **5**, 415 (1929); Freudenberg, *Ber.* **69**, 1964 (1936); Swanson *et al.*, *Aconite*, A. Ph. A. Monograph no. 1 (1938); *Methods of Analysis, A.O.A.C.*, 8th ed., 598, 651 (1955). Structure: Wiesner *et al.*, *Coll. Czech. Chem. Commun.* **28**, 2462 (1963); Wiesner *et al.*, *Can. J. Chem.* **47**, 2734 (1969). Stereochemistry: Bachelor *et al.*, *Tetrahedron Letters* **1960**, no. 10, 1; Gilman, Marion, *ibid.* **1961**, 923; Tsuda, Marion, *Can. J. Chem.* **41**, 1634 (1963); Birnbaum *et al.*, *Tetrahedron Letters* **1971**, 867. Pharmacology: H. Sato *et al.*, *Tohoku J. Exp. Med.* **128**, 175 (1979). Inotropic effects: P. Honerjäger, A. Meissner, *Arch. Pharmacol.* **322**, 49 (1983). Toxicity: Dybing *et al.*, *Acta Pharmacol. Toxicol.* **7**, 337 (1951).

Hexagonal plates, mp 204°. [α]$_D$ +17.3° (chloroform). Aq soln is alkaline to litmus. pK 5.88. One gram dissolves in 2 ml chloroform, 7 ml benzene, 28 ml abs alcohol, 50 ml ether, 3300 ml water. Slightly sol in petr ether. *Poisonous!* LD$_{50}$ in mice (mg/kg): 0.166 i.v.; 0.328 i.p.; approx 1 orally (Dybing); also reported as LD$_{50}$ in mice (mg/kg): 1.8 orally, 0.270 s.c.; 0.380 i.p.; 0.12 i.v. (Sato).

Hydrobromide hemipentahydrate, C$_{34}$H$_{47}$NO$_{11}$.HBr. 2½H$_2$O, hexagonal tablets from water, mp 200-207° (sinters at 160°). Amorphous form mp 115-120°. Also crystallizes from ethanol + ether with ½H$_2$O, mp 206-207°; [α]$_D$ −30.8°. *Violent poison! Ref:* Paech, Tracey, *Modern Methods of Plant Analysis* vol. IV (Springer-Verlag, Berlin, 1955) p 375.

Hydrochloride hemipentahydrate, C$_{34}$H$_{47}$NO$_{11}$.HCl. 2½H$_2$O, crystals, mp 149-153°. mp 194-195° (dry). [α]$_D$ −30.9°. *Violent poison! Ref:* Paech, Tracey.

Nitrate, C$_{34}$H$_{47}$NO$_{11}$.HNO$_3$, crystals, mp about 200° (dec). [α]$_D^{20}$ −35° (c = 2 in water). *Violent poison!* One gram dissolves in 10 ml boiling water. Less sol in cold water. Sol in alcohol.

USE: Used in producing heart arrhythmia in experimental animals: Boyadzhiev, *C.A.* **73**, 86526e (1970).

THERAP CAT: Has been used topically in neuralgia.

121. Aconitine, Amorphous. Mild aconitine. Mixture of amorphous alkaloids from *Aconitum napellus* L., *Ranunculaceae*. Contains aconitine, mesaconitine, hypaconitine, neopelline, *l*-ephedrine, sparteine, neoline and napelline. *Ref:* Rogers, Freudenberg, *Ber.* **70**, 349 (1937); *eidem*, *J. Am. Chem. Soc.* **59**, 2572 (1937).

Yellowish-white, amorphous powder. *Violent poison!* Insol in water. Sol in alcohol, chloroform, ether, dil acids. MLD s.c. in rats: 0.175 mg/kg, *Handbook of Toxicology* Vol. 1, W. S. Spector, Ed. (Saunders, Philadelphia, 1956) pp 12-13.

Caution: A highly toxic alkaloid. Can be absorbed through skin.

122. Aconitum Ferox. Indian aconite; bish; visha; bishma; bikhroot. Tuber of *Aconitum ferox* Wall., *Ranunculaceae*. *Habit.* Nepal, Himalaya Mountains, India. *Constit.* Pseudaconitine, *q.v.* Most powerful of all the aconites. Used by natives as arrow poison in big game hunting; in neuralgia and rheumatism. *Ref: Aconite*, A. Ph. A.

Monograph no. 1, p 78 and *passim* (Am. Pharm. Assoc., Washington, D.C., 1938); N. B. Dutt, *Indian J. Pharm.* **1**, 81-84 (1939); *C.A.* **34**, 6768 (1940).

Caution: Extremely toxic. A very small dose can cause fatal cardiac depression.

123. Acranil®. *1-[(6-Chloro-2-methoxy-9-acridinyl)-amino]-3-(diethylamino)-2-propanol dihydrochloride;* 5-[(γ-diethylamino-β-hydroxypropyl)amino]-3-methoxy-8-chloro-acridine dihydrochloride; SKF-16214-A2; SN-186. $C_{21}H_{28}$-$Cl_3N_3O_2$; mol wt 460.83. C 54.73%, H 6.12%, Cl 23.08%, N 9.12%, O 6.94%. Prepn: Mietzsch, Mauss, **Ger. pat. 553,-072,** *Frdl.* **19,** 1167; **Brit. pat. 363,392;** U.S. pat. **2,113,357** (1930, 1930, 1938 all to I. G. Farben.); Maghidson, Grigo-rovski, *Ber.* **69,** 396 (1936); *J. Gen. Chem. (USSR)* **6,** 806 (1936). Antiviral activity: T. E. Gláz *et al., Antimicrob. Ag. Chemother.* **3,** 537 (1973). Interferon induction and radio-protective activity: T. E. Glaz, M. Talas, *Arch. Virol.* **48,** 375 (1975); M. Talas, E. Szolgay, *ibid.* **56,** 309 (1978).

Bitter, yellow crystals, mp 237-239° (dec). Sol in water. Free base, $C_{21}H_{26}ClN_3O_2$, yellow crystals, mp 105-107°. Sparingly sol in ether.

THERAP CAT: Antiprotozoal (Giardia).

124. Acridine. Dibenzo[*b,e*]pyridine; 10-azaanthra-cene. $C_{13}H_9N$; mol wt 179.22. C 87.12%, H 5.06%, N 7.82%. Occurs in coal tar. Isoln: Graebe, Caro, *Ann.* **158,** 265 (1871); from high boiling tar oils: Wirth, **Ger. pat. 440,771** (1926). Prepn from *N*-phenylanthranilic acid: Per-kin, Clemo, **Brit. pat. 214,756** (1923); from Ca-anthranilate: Koller, Krakauer, *Monatsh.* **50,** 51 (1928); by passing benz-ylaniline vapor over red-hot platinum wire: Meyer, Hof-mann, *ibid.* **37,** 698 (1916); *cf.* Ullmann, *Ber.* **40,** 2521 (1907). Absorption spectrum: Pinnow, *J. Prakt. Chem.* [2] **66,** 276 (1902). Toxicity: S. D. Rubbo, *Brit. J. Exp. Pathol.* **28,** 1 (1947). *Reviews:* A. Albert, *The Acridines* (St. Mar-tin's Press, New York, 2nd ed., 1966); R. M. Acheson, *Acridines* (Interscience, New York, 2nd ed., 1973).

Pale yellow orthorhombic plates, needles from diluted alcohol. There are five crystalline forms of acridine all of which melt below 111°. Purified acridine mp 110-111°; com-mercial samples mp 107° or mp 109°. Begins to sublime at 100°. bp$_{760}$ 346° (also reported as > 360°). Volatile with steam. Slightly sol in boiling water, liquid ammonia, liquid sulphur dioxide; sparingly sol in light petroleum; freely sol in alcohol, ether, hydrocarbons, carbon disulfide. 1 gm sol in < 1 ml boiling benzene or alcohol; 5 ml benzene (20°); 6 ml alcohol (20°); 16 ml of ether (20°); 1.8 ml boiling cyclo-hexane. d 1.27-1.28. Dilute solns of acridine and its salts have a violet and green fluorescence, respectively. Weak base, colors litmus paper blue, forms yellow crystalline salts with mineral acids. Forms colored quaternary ammonium compounds (acridinium compounds), by the action of alkyl and aryl halides and sulfates. LD$_{50}$ s.c. in mice: 0.40 g/kg (Rubbo).

Caution: Strongly irritating to skin, mucous membranes; *Clinical Toxicology of Commercial Products,* R. E. Gosselin *et al.,* Eds. (Williams & Wilkins, Baltimore, 5th ed., 1984) Section II, p 384.

USE: Manuf dyes and intermediates; some dyes derived from it are used as antiseptics, e.g. 9-aminoacridine, acri-flavine and proflavine. The hydrochloride has been used as reagent for cobalt, iron and zinc.

125. Acriflavine. *3,6-Diamino-10-methylacridinium chloride mixt with 3,6-acridinediamine;* neutral acriflavine; euflavine; trypaflavine; neutroflavine; gonacrine. Contains 30-40% unmethylated compd. Prepn: L. Benda, *Ber.* **45,** 1787 (1912); M. Gaillot, *Quart. J. Pharm. Pharmacol.* **7,** 63 (1934); J. Marshall, *ibid.* 514. Activity studies: C. H. Browning, W. Gilmour, *J. Path. Bact.* **18,** 144 (1913); C. H. Browning *et al., Brit. Med. J.* **1,** 73 (1917); H. Berry, *Quart. J. Pharm. Pharmacol.* **14,** 149 (1941); H. McIlwain, *Bio-chem. J.* **35,** 1311 (1941). Toxicology: J. Ungar *et al., J. Pharmacol. Exp. Ther.* **80,** 217 (1944). *Reviews:* A. Adrien, *The Acridines* (St. Martin's Press, New York, 2nd ed., 1966); R. M. Acheson, *Acridines* (Interscience, New York, 2nd ed., 1973).

Deep orange, granular powder. One gram dissolves in about 3 ml water. Incompletely sol in alcohol; nearly insol in ether, chloroform, fixed oils. Aqueous solns are reddish orange and fluoresce on dilution. pH (1% soln) ~3.5.

Dihydrochloride, *acid acriflavine, acid trypaflavine, acri-flavine hydrochloride, Panflavin.* Deep reddish-brown, crystalline powder. pH (1% soln) ~1.5.

THERAP CAT: Anti-infective.

THERAP CAT (VET): Antiseptic; locally in wounds in tricho-moniasis of bulls.

126. Acrilan®. A brand of acrylic fiber, a copolymer of acrylonitrile and of a minor constituent with mildly basic character. Prepn of such copolymers: Mowry, Craig, U.S. pat. **2,744,086** and Craig, U.S. pat. **2,749,325** (both 1956 to Chemstrand). *Review:* R. W. Moncrieff, *Man-Made Fibres* (John Wiley & Sons, New York, 4th ed., 1963) pp 471-482.

Solid, d^{25} 1.17. Fiber decomposes before it melts. Sticks at 245°. Prolonged exposure in air at elevated temps causes some yellowing. Fabrics are not readily ignited and their rate of burning is less than that of cotton, viscose rayon, or acetate rayon. Fiber is practically insoluble in, and unaffec-ted by common solvents. Has fairly good resistance to weak alkalies, very good resistance to mineral acids, sun-light, weathering. Has good dyeing properties. Its wear resistance is better than that of wool, but generally greatly inferior to that of nylon or Terylene. It is unaffected by mildew, molds, moth and carpet beetle larvae.

USE: Fabrics, sweaters, blankets, carpets.

127. Acrinathrin. *[1R-[1α(S*),3α(Z)]]-2,2-Dimethyl-3-[3-oxo-3-[2,2,2-trifluoro-1-(trifluoromethyl)ethoxy]-1-pro-penyl]cyclopropanecarboxylic acid cyano(3-phenoxyphenyl)-methyl ester;* (S)-α-cyano-3-phenoxybenzyl (1R-cis)-2,2-di-methyl-3-[(Z)-3-oxo-3-[2-(1,1,1,3,3,3-hexafluoropropoxy)]-1-propenyl]cyclopropanecarboxylate; RU-38702; Rufast. $C_{26}H_{21}F_6NO_5$; mol wt 541.45. C 57.68%, H 3.91%, F 21.05%, N 2.59%, O 14.77%. Synthetic pyrethroid acari-cide. Prepn: J. Martel *et al.,* **French pat. 2,486,073;** *eidem,* U.S. pat. **4,542,142** (1982, 1985 both to Roussel-UCLAF); J. R. Tessier *et al.,* in *Pesticide Chemistry: Human Welfare and the Environment* vol. 1, J. Miyamoto, P. C. Kearney, Eds. (Pergamon Press, Oxford, 1983) pp 95-100. Field trial vs spider mite: B. H. Labanowska, C. Tkaczuk, *Fruit Sci. Rep.* **18,** 185 (1991). *Review:* J. J. Heller *et al., Meded. Fac. Landbouwwet. Univ. Gent* **57,** 931-939 (1992).

mp 82°. $[\alpha]_D^{20}$ +23.5° (c = 0.5% in benzene). LD$_{50}$ in rats (mg/kg): > 5000 orally; > 2000 dermally (Heller).

USE: Insecticide.

128. Acrisorcin. *4-Hexyl-1,3-benzenediol compd with 9-acridinamine (1:1); 9-aminoacridine compd with 4-hexylresorcinol;* 9-aminoacridinium 4-hexylresorcinolate; Sch-7056; Akrinol. $C_{25}H_{28}N_2O_2$; mol wt 388.51. C 77.29%, H 7.26%, N 7.21%, O 8.24%. Prepd from 9-aminoacridine and 4-hexylresorcinol: Seneca, *Antibiot. & Chemother.* **11**, 587 (1961).

Yellow crystals.

THERAP CAT: Antifungal.

129. Acrivastine. *(E,E)-3-[6-[1-(4-Methylphenyl)-3-(1-pyrrolidinyl)-1-propenyl]-2-pyridinyl]-2-propenoic acid; (E)-6-[(E)-3-(1-pyrrolidinyl)-1-p-tolylpropenyl]-2-pyridineacrylic acid;* BW-270C; BW-825C; BW-A825C; Semprex. $C_{22}H_{24}N_2O_2$; mol wt 348.44. C 75.83%, H 6.94%, N 8.04%, O 9.18%. Nonsedating type histamine H$_1$-receptor antagonist; analog of triprolidine, *q.v.* Prepn: G. G. Coker, J. W. A. Findlay, *Eur. pat. Appl.* **85,959** (1983 to Wellcome); J. W. A. Findlay, G. G. Coker, U.S. pat. **4,501,893** (1985). Pharmacodynamics and pharmacokinetics in humans: A. F. Cohen *et al.*, *Eur. J. Clin. Pharmacol.* **28**, 197 (1985). Evaluation of CNS effects: A. F. Cohen *et al.*, *Clin. Pharmacol. Ther.* **38**, 381 (1985). Clinical trials in idiopathic urticaria: J. G. Gibson *et al.*, *Dermatologica* **169**, 179 (1984); H. Neittaanmaki *et al., ibid.* **177**, 98 (1988); in allergic rhinitis: T. G. Gibbs *et al., J. Int. Med. Res.* **16**, 413 (1988). Review of pharmacology and therapeutic efficacy: R. N. Brogden, D. McTavish, *Drugs* **41**, 927-940 (1991).

Crystals from isopropanol, mp 222° (dec).

Combination with pseudoephedrine, *Duact.*

THERAP CAT: Antihistaminic.

130. Acrolein. *2-Propenal;* acrylic aldehyde; acrylaldehyde; acraldehyde; Aqualin. C_3H_4O; mol wt 56.06. C 64.27%, H 7.19%, O 28.54%. CH_2=CHCHO. Prepd industrially by passing glycerol vapors over magnesium sulfate heated to 330-340°. Lab prepn by heating a mixture of anhydr glycerol, acid potassium sulfate and potassium sulfate in the presence of a small amount of hydroquinone and distilling in the dark: H. Adkins, W. H. Hartung, *Org. Syn.* **coll. vol. I**, 15 (1941). Formation from glycerol by the action of *B. amaracrylus:* Voisenet, *Compt. Rend.* **188**, 941, 1271 (1929); by *B. welchii:* Humphreys, *J. Infect. Dis.* **35**, 282; *Chem. Zentr.* **1925**, II, 309. Toxicity study: H. F. Smyth *et al., Arch. Ind. Hyg. Occup. Med.* **4**, 119 (1951). *Review:* L. G. Hess *et al.*, in Kirk-Othmer *Encyclopedia of Chemical Technology* vol. 1 (Wiley-Interscience, New York, 3rd ed., 1978) pp 277-297.

Flammable liquid with pungent odor. mp −88°. d^0 0.8621; d^{20} 0.8389; d^{50} 0.8075. bp$_{760}$ 52.5°; bp$_{200}$ 17.5°; bp$_{100}$ +2.5°; bp$_{60}$ −7.5°; bp$_{1.0}$ −64.5°. n_D^{19} 1.4022. Sol in 2 to 3 parts water; in alcohol, ether. Flash pt, open cup: < 0°F (−18°C). Vapor pressure at 20°: 210 mm Hg. Unstable, polymerizes (especially under light or in the presence of alkali or strong acid) forming disacryl, a plastic solid. Absorption spectrum: Lüthy, *Z. Physik. Chem.* **107**, 291, 298 (1923). LD$_{50}$ orally in rats: 0.046 g/kg (Smyth).

Caution: Potential symptoms of overexposure are irritation of eyes, skin and mucous membranes; abnormal pulmonary function; delayed pulmonary edema; chronic respiratory disease. *See NIOSH Pocket Guide to Chemical Hazards* (DHHS/NIOSH 90-117, 1990) p 32. *See also Clinical Toxicology of Commercial Products,* R. E. Gosselin *et al.*, Eds. (Williams & Wilkins, Baltimore, 5th ed., 1984) Section II, p 186.

USE: Manuf colloidal forms of metals; making plastics, perfumes; warning agent in methyl chloride refrigerant. Has been used in military poison gas mixtures. Used in organic syntheses. Aquatic herbicide.

131. Acrylamide. *2-Propenamide.* C_3H_5NO; mol wt 71.08. C 50.69%, H 7.09%, N 19.71%, O 22.51%. CH_2=CHCONH$_2$. Prepd from acrylonitrile by treatment with H$_2$SO$_4$ or HCl: Bayer, *Angew. Chem.* **61**, 240 (1949); Weisgerber, U.S. pat. **2,535,245** (1950 to Hercules). *Reviews:* Carpenter, Davis, *J. Appl. Chem. (London)* **7**, 671 (1957); C. E. Habermann in Kirk-Othmer, *Encyclopedia of Chemical Technology* vol. 1 (John Wiley & Sons, New York, 4th ed., 1991) pp 251-266. Toxicity: R. E. Peterson, N. K. Sheth, *Toxicol. Appl. Pharmacol.* **33**, 142 (1975). Review of carcinogenic risk: *IARC Monographs* **60**, 389-433 (1994).

Monomer, flake-like crystals from benzene. d$_4^{30}$ 1.122. mp 84.5°. bp$_2$ 87°; bp$_5$ 103°; bp$_{25}$ 125°. Solubilities in g/100 ml solvent at 30°: water 215.5; methanol 155; ethanol 86.2; acetone 63.1; ethyl acetate 12.6; chloroform 2.66; benzene 0.346; heptane 0.0068. The solid may be stored in a cool, dark place. Readily polymerizes at the mp or under uv light. Commercial solns of the monomer may be stabilized with hydroquinone, *tert*-butylpyrocatechol, *N*-phenyl-2-naphthylamine or other antioxidants. LD$_{50}$ i.p. in mice: 170 mg/kg (Peterson, Sheth).

Polymer, various forms, sol and insol in water, are obtained by heating with various polymerization catalysts: C. E. Schildknecht, *Vinyl and Related Polymers* (Wiley, New York, 1952) pp 314-322; D. Lipp, J. Kozakiewicz in Kirk-Othmer, *Encyclopedia of Chemical Technology* vol. 1 (John Wiley & Sons, New York, 4th ed., 1991) pp 266-287.

Caution: Potential symptoms of overexposure to the monomer are ataxia, numbness of limbs, paresthesia; muscle weakness; absence of deep tendon reflex; sweating of hands; fatigue, lethargy; irritation of eyes and skin. *See NIOSH Pocket Guide to Chemical Hazards* (DHHS/NIOSH 90-117, 1990) p 34. Readily absorbed through intact skin from aqueous solutions. *See Clinical Toxicology of Commercial Products,* R. E. Gosselin *et al.*, Eds. (Williams & Wilkins, Baltimore, 5th ed., 1984) Section II, p 409. This substance may reasonably be anticipated to be a carcinogen: *Seventh Annual Report on Carcinogens* (PB95-109781, 1994) p 79.

USE: Monomer as chemical intermediate in production of polyacrylamides; in synthesis of dyes; in copolymers for contact lenses; in construction of dam foundations, tunnels and sewers. Polymers as additives for water treatment, enhanced oil recovery, flocculants, papermaking aids, thickeners, soil conditioning agents, sewage and waste treatment, ore processing, permanent-press fabrics.

132. Acrylic Acid. *2-Propenoic acid;* vinylformic acid. $C_3H_4O_2$; mol wt 72.06. C 50.00%, H 5.59%, O 44.40%. CH_2=CHCO$_2$H. Prepd by hydrolysis of acrylonitrile: Kaszuba, *J. Am. Chem. Soc.* **67**, 1227 (1945); by oxidation of acrolein: U.S. pats. **1,911,219** (1933 to Rohm & Haas); **2,288,566** (1942 to Acrolein Corp.); **2,341,339** (1944 to Distillers). Various other syntheses, see *Org. Syn.* coll. vol. III, 30-34 (1955). Toxicity study: H. F. Smyth *et al., Am. Ind. Hyg. Assoc. J.* **23**, 95 (1962). *Review:* J. W. Nemec, W. Bauer in Kirk-Othmer *Encyclopedia of Chemical Technology* vol. 1 (Wiley-Interscience, New York, 3rd ed., 1978) pp 330-354.

Corrosive liquid; acrid odor and fumes. d_4^{16} 1.0621. mp 14°. bp 141.0°; bp_{400} 122.0°; bp_{200} 103.3°; bp_{100} 86.1°; bp_{40} 66.2°; bp_{10} 39.0°; bp_5 27.3°. n_D^{20} 1.4224. Flash pt, open cup: 155°F (68°C). pKa (25°): 4.25. Miscible with water, alc, ether. Polymerizes readily in the presence of oxygen. LD_{50} orally in rats: 2.59 g/kg (Smyth).

Caution: Strong irritant.
USE: In the manuf of plastics.

133. Acrylonitrile. *2-Propenenitrile;* vinyl cyanide; cyanoethylene; Acritet; Fumigrain; Ventox. C_3H_3N; mol wt 53.06. C 67.91%, H 5.70%, N 26.40%. CH_2=CHCN. Prepn by dehydration of ethylene cyanohydrin or acrylamide: Moureu, *Ann. Chem. Phys.* [7] **2**, 186 (1893). Manuf by ammoxidation of propylene: Faith, Keyes & Clark's *Industrial Chemicals,* F. A. Lowenheim, M. K. Moran, Eds. (Wiley-Interscience, New York, 4th ed., 1975) pp 46-49. Toxicity: H. F. Smyth, C. P. Carpenter, *J. Ind. Hyg. Toxicol.* **30**, 63 (1948). Causes acute and chronic adrenocortical insufficiency: S. Szabo *et al., Lab. Invest.* **42**, 533 (1980); *eidem, J. Appl. Toxicol.* **4**, 131 (1984). Review of carcinogenic risk: *IARC Monographs* **19**, 73-133 (1979); of toxicology and human exposure: *Toxicological Profile for Acrylonitrile* (PB91-180489, 1990) 136 p. Comprehensive review: *The Chemistry of Acrylonitrile* (Am. Cyanamid, New York, 2nd ed., 1959) 272 pp; J. F. Brazdil in Kirk-Othmer *Encyclopedia of Chemical Technology* vol. 1 (John Wiley & Sons, New York, 4th ed., 1991) pp 352-369.

Explosive, flammable and toxic liquid. Should be stored and used in closed systems whenever possible. Work areas should be adequately ventilated, and should be free from open lights, flames, and equipment that is not explosion-proof. *Handle in hood.* May polymerize spontaneously, particularly in the absence of oxygen or on exposure to visible light. Polymerizes violently in the presence of concentrated alkali. On standing may slowly develop a yellow color particularly after excessive exposure to light. bp_{760} 77.3°; bp_{500} 64.7°; bp_{250} 45.5°; bp_{100} 23.6°; bp_{50} 8.7°. mp −83.55°. d_4^{20} 0.8060; d_4^{25} 0.8004. n_D^{25} 1.3888. Flash pt, open cup: 32°F (0°C). Explosive mixtures in air at 25°: 3.05% low limit; 17.0% upper limit. At 20° 7.35 parts dissolve in 100 parts water and 3.1 parts water dissolve in 100 parts acrylonitrile. Miscible with most organic solvents. LD_{50} orally in rats: 0.093 g/kg (Smyth, Carpenter).

Caution: Potential symptoms of overexposure are asphyxia; irritation of eyes; headache; sneezing; nausea, vomiting; weakness, lightheadedness; skin vesiculation; scaling dermatitis. *See NIOSH Pocket Guide to Chemical Hazards* (DHHS/NIOSH 90-117, 1990) p 34. *See also Clinical Toxicology of Commercial Products,* R. E. Gosselin *et al.,* Eds. (Williams & Wilkins, Baltimore, 5th ed., 1984) Section II, p 215. This substance may reasonably be anticipated to be a carcinogen: *Seventh Annual Report on Carcinogens* (PB95-109781, 1994) p 83.

USE: Manufacture of acrylic fibers. In the plastics, surface coatings, and adhesives industries. As a chemical intermediate in the synthesis of antioxidants, pharmaceuticals, dyes, surface-active agents, etc. In organic synthesis to introduce a cyanoethyl group. As a modifier for natural polymers. As a pesticide fumigant for stored grain. Experimentally to induce adrenal hemorrhagic necrosis in rats.

134. Actaplanins. A-4696; Kamoran. Complex of glycopeptide antibiotics produced by *Actinoplanes missouriensis.* Six actaplanins (A, B_1, B_2, B_3, C_1, G) have been isolated and characterized as having a central peptide core with the amino sugar *ristosamine* and up to four neutral sugars attached. Isoln: R. L. Hamill *et al.,* **Ger.** pat. **2,-209,018** (1972 to Lilly), *C.A.* **77**, 138338n (1972); A. P. Raun, **U.S.** pat. **3,816,618**; R. L. Hamill *et al.,* **U.S.** pat. **4,115,552** (1974, 1978 to Lilly). Chemical characterization: M. Debono *et al., J. Antibiot.* **37**, 85 (1984). ^{1}H NMR studies and structures: A. H. Hunt *et al., J. Org. Chem.* **49**, 635 (1984); *eidem, ibid.* 641. Growth promotant activity: C. Tsaltos *et al., Bull. Hellenic Vet. Med. Soc.* **33**, 139 (1982). Use to increase milk production in ruminants: C. Scheifinger, **Eur.** pat. **Appl. 63,491** corresp to **U.S.** pat. **4,430,328** (1982, 1984 to Eli Lilly). Determn in milk: K. H. Hahne *et al., Milchwissenschaft* **39**, 473 (1984).

	R	R'
Actaplanin A	mannosylglucose	mannose
Actaplanin B_1	rhamnosylglucose	mannose
Actaplanin B_2	glucose	mannose
Actaplanin B_3	mannosylglucose	H
Actaplanin C_1	rhamnosylglucose	H
Actaplanin G	glucose	H

Hydrochloride, white cryst solid, mp >220°. Approx mol wt 1158. $[\alpha]_D^{25}$ −42.3° (c = 1 in water). uv max (acidic and neutral solns): 276 nm ($E_{1 cm}^{1\%}$ 65). Sol in water. Insol in most organic solvents. Stable over pH 1.0 to 10.0 up to 27°.

THERAP CAT (VET): Growth stimulant.

135. Actarit. *4-(Acetylamino)benzeneacetic acid; (p-acetamidophenyl)acetic acid;* MS-932; Moba; Orcl. $C_{10}H_{11}NO_3$; mol wt 193.20. C 62.17%, H 5.74%, N 7.25%, O 24.84%. Immunomodulator. Prepn: S. Gabriel, *Chem. Ber.* **15**, 834 (1882). *See also:* H. Munakata *et al.,* **Ger.** pat. **3,317,107**; *eidem,* **U.S.** pat. **4,720,506** (1983, 1988 both to Mitsubishi Chem. Ind. and Nippon Shinyaku). Pharmacology: H. Fujisawa *et al., Arzneimittel-Forsch.* **40**, 693 (1990). Pharmacokinetics: K. Sugihara *et al., ibid.* 800, 806. Antiarthritic effect in mice: H. Fujisawa *et al., ibid.* **44**, 64 (1994). Acute toxicity: K. Toshida *et al., Oyo Yakuri* **40**, 117 (1990), *C.A.* **114**, 17276q (1991).

Odorless, white crystals or crystalline powder. mp 173-175° (Munakata); also reported as mp 168-170° (Gabriel). Freely sol in methanol; sol in ethanol; sparingly sol in acetone; slightly sol in water; very slightly sol in ether. LD_{50} in male female mice, male, female rats (mg/kg): 1.06, 1.30, 1.95, 2.03 i.p.; 5.68, 5.48, 5.48, 6.12 s.c.; 15.3, 14.7, 14.8, 15.4 orally (Toshida).

THERAP CAT: Antiarthritic.

136. ACTH. *Corticotropin;* adrenocorticotrop(h)in; corticotrophin; adrenocorticotrop(h)ic hormone of the pituitary gland; Acethropan; Acortan; Acorto; Acthar; Acton; Actonar; Adrenomone; Alfatrofin; Cibachten; Corstiline; Cortiphyson; Cortrophin; Isactid; Reacthin; Solacthyl; Tubex. Pituitary hormone which stimulates the secretion of adrenal cortical steroids and induces growth of the adrenal cortex. Occurs also in female human urine and in serum of pregnant mares. Isoln procedure from swine pituitaries: Sayers *et al., J Biol. Chem.* **149**, 425 (1943); *Proc. Soc. Exp. Biol. Med.* **52**, 199 (1943); from sheep pituitaries: Li *et al., J. Biol. Chem.* **149**, 413 (1943); Li, *J. Am. Chem. Soc.* **74**, 2124 (1952); from human pituitaries: Pickering *et al., Biochim. Biophys. Acta* **74**, 763 (1963). Purification: Johnson, **U.S.** pat. **3,124,509** (1964 to Upjohn). Corticotropin is a single chain polypeptide containing 39 amino acids. The first 24 residues are identical in all species. *In vivo* studies show that this portion of the peptide chain is responsible for the biological activity and that the remaining residues, while not necessary for the hormonal action, are essential in the species immunological specificity of ACTH. The term α-ACTH is used to distinguish it from the pepsin- or acid-degraded product, β-ACTH. Prepn of active cleavage products by hydrolysis and simultaneous dialysis: Wettstein, Benz, U.S. pat. 2,734,015 (1956 to Ciba). Amino acid sequence of bovine ACTH: Li et al., J. Am. Chem. Soc. 80, 2587 (1958); proposed structure of human ACTH: Lee et al., J. Biol. Chem. 236, 2970 (1961). Revised amino acid sequences of porcine and human ACTH: Riniker et al., Nature New Biol. 235, 114 (1972). Revised sequences of ovine and bovine ACTH: Li, Biochem. Biophys. Res. Commun. 49, 835 (1972). Synthesis of the revised human corticotropin: Sieber et al., Helv. Chim. Acta 55, 1243 (1972); solid phase synthesis: Yamashiro, Li, J. Am. Chem. Soc. 95, 1310 (1973).

White powder. Freely sol in water. Partly precipitated at the isoelectric point (pH 4.65-4.80). Appreciably soluble in 60 to 70% alcohol or acetone. Almost completely precipitated in 2.5% trichloroacetic acid soln. Also precipitated from dilute soln by 20% sulfosalicylic acid and by 5% lead acetate soln. Solns are stable to heat. An aq soln buffered to pH 7.5 may be put in a boiling water bath for at least 120 minutes. In 0.10 molar HCl an 0.2% soln retains its biological potency when kept at 100° for 60 min, but in 0.10 molar NaOH the activity is lost within 30 min. At 60° at 0.2% soln at pH 10.8 maintains its potency for 60 min. In general the substance is more stable in acid soln. Pure ACTH has 150 to 200 potency units per mg. The biological activity is not destroyed by digestion with pepsin, and hydrolysis products have the biological activity of ACTH. One U.S.P. unit or one international unit or one Armour unit or one potency unit denotes the same activity.

Corticotrophin zinc hydroxide suspension, Cortrophin-Z. ACTH absorbed on zinc hydroxide. Used as an aqueous suspension.

THERAP CAT: Adrenocorticotropic hormone.

THERAP CAT (VET): Stimulates glucocorticoid production.

137. Actin. One of the major proteins of muscle and an important component of all eukaryotic cells. Formerly believed to be a single, highly conserved protein in all cell types, but multiple forms have been shown to exist. Muscle actin, or *α-actin*, is found in differentiated muscle cells; *β-actin* and *γ-actin* are present in all non-muscle cell types. All three forms contain equimolar amounts of N-methylhistidine, although α-actin differs from β- and γ-actins by several peptides; the two "non-muscle" actins are nearly identical. Several "minor actins" are also known. The various forms, although similar in activity, mol wt and amino acid composition, show immunological differences and are considered to be synthesized under the control of different genes. In addition to its role in muscle relaxation and contraction (together with myosin, q.v.), actin is involved in a variety of cellular events: cell movement, cytokinesis, phagocytosis, exocytosis and chromosome movement. Identification in muscle: F. B. Straub, Stud. Szeged 2, 3 (1942). Isoln: Feuer et al., Hung. Acta Physiol. 1, 150 (1948). Separation of actin from muscle material: A. Szent-Gyorgyi, U.S. pat. 2,742,456 (1956 to Armour). In soln, actin and myosin combine to give *actomyosin*: Tonomura et al., J. Biol. Chem. 237, 1074 (1962). Actin, in the absence of salts, exists in a globular form, designated as *G-actin;* in the presence of ATP and potassium, sodium, or magnesium chlorides it polymerizes to a fibrous form, *F-actin.* Removal of bound ATP results in complete loss of polymerizability: Straub, Feuer, Biochim. Biophys. Acta 4, 455 (1950). Structure of G-actin: Nagy, Jencks, Biochemistry 1, 987 (1962); of F-actin: Hanson, Lowy, J. Mol. Biol. 6, 46 (1963). Complete amino acid sequence of rabbit skeletal muscle actin: Elzinga et al., Proc. Nat. Acad. Sci. USA 70, 2687 (1973). Identification and characterization of multiple forms of actin: J. I. Garrels, W. Gibson, Cell 9, 793 (1976). Partial amino acid sequence of calf brain actin: R. C. Lu, M. Elzinga, Biochemistry 16, 5801 (1977). Comparison of actins from calf thymus, bovine brain, mouse cells and rabbit skeletal muscle: J. Vandeker, K. Weber, Eur. J. Biochem. 90, 451 (1978). Isoln, characterization of porcine brain actin: J. P. Weir, D. W. Frederik, Arch. Biochem. 203, 1 (1980). In vivo and in vitro synthesis of multiple forms of rat

brain actin: E. Palmer, J. L. Saborio, J. Biol. Chem. 253, 7482 (1978). Evidence for control of synthesis of human heart and platelet actins by different genes: M. Elzinga et al., Science 191, 94 (1976). Immunological differences between cardiac muscle, skeletal muscle and brain actins: J. L. Morgan et al., Proc. Nat. Acad. Sci. USA 77, 2069 (1980). History of discovery: Finck, Science 160, 332 (1968). Reviews: Several authors, Biochemistry of Muscle Contraction, J. Gergely, Ed. (Little, Brown, Boston, 1964); Laki, "Actin" in Contractile Proteins and Muscle, K. Laki, Ed. (Marcel Dekker, New York, 1971) pp 97-133; E. D. Korn, Proc. Nat. Acad. Sci. USA 75, 588-599 (1978); R. Kassab et al., Biochimie 63, 273-289 (1981); S. Highsmith, Biochem. Biophys. Acta 639, 31-39 (1981); E. D. Korn, Physiol. Rev. 62, 672-737 (1982).

138. Actinium. Ac; at. no. 89; valence 3. Group IIIB(3). Known isotopes (mass numbers): 209-233; naturally occurring isotopes: 227, 228. Longest lived isotope: ^{227}Ac ($T_{1/2}$ 21.773 ±0.003 years, mode of decay α, β, γ-emission, rel. at. mass 227.0278). Daughter of ^{231}Pa; parent of ^{227}Th (radioactinium) and ^{223}Fr. Originally extracted from uranium ores where it occurs naturally as a member of the ^{235}U decay series; now prepd by neutron bombardment of ^{226}Ra in nuclear reactors. ^{228}Ac or *mesothorium II* (MsTh$_2$) occurs naturally as a member of the ^{232}Th decay series. $T_{1/2}$ 6.13 ±0.03 hrs; decays by β and γ-emission. Daughter of ^{228}Ra (mesothorium I); parent of ^{228}Th (radiothorium). Discovery of ^{227}Ac: A. Debierne, C. R. Acad. Sci. 129, 593 (1899); F. O. Geisel, Ber. 55, 3608 (1902). Prepn of metal: J. G. Stites, Jr., et al., J. Am. Chem. Soc. 77, 237 (1955); J. D. Farr et al., J. Inorg. Nucl. Chem. 18, 42 (1961). Discovery of ^{228}Ac: O. Hahn, Ber. 40, 1462 (1907). Review of actinium and the actinides: Comprehensive Inorganic Chemistry vol. 5, J. C. Bailar, Jr. et al., Eds. (Pergamon Press, Oxford, 1973) passim; H. W. Kirby in The Chemistry of the Actinide Elements vol. 1, J. J. Katz et al., Eds. (Chapman and Hall, New York, 1986) pp 14-40; G. T. Seaborg in Kirk Othmer Encyclopedia of Chemical Technology vol. 1 (Wiley-Interscience, New York, 4th ed., 1991) 442-444.

Cubic crystal, d^{25} 10.07. mp 1050 ±50°. bp ~3300°. Homologous with lanthanum; more basic than lanthanum. Strong electropositive element. Forms insol salts, e.g. carbonate, oxalate, phopshate etc. just as the lanthanides. Rapidly oxidizes to AcO$_3$ in moist air.

Caution: Radiation hazard; handling requires special equipment and shieldling facilities (Katz et al., loc. cit., vol. 2, p 1128).

139. Actinobolin. *[3R-[3α,4α(S*),4aβ,5β,6α]]-2-Amino-N-(3,4,4a,5,6,7-hexahydro-5,6,8-trihydroxy-3-methyl-1-oxo-1H-2-benzopyran-4-yl)propanamide; 4-(2-aminopropionamido)-3,4,4a,5,6,7-hexahydro-5,6,8-trihydroxy-3-methylisocoumarin.* $C_{13}H_{20}N_2O_6$; mol wt 300.31. C 51.99%, H 6.71%, N 9.33%, O 31.97%. Antibiotic produced by *Streptomyces griseoviridus* var *atrofaciens;* (+)-form is naturally occurring. Isoln and characterization: T. H. Haskell, Q. R. Bartz, Antibiot. Ann. 1958-1959, 505; T. H. Haskell et al., U.S. pat. 3,043,830 (1962 to Parke, Davis). Structural studies: Struck et al., Tetrahedron Letters 1967, 1589; Munk et al., J. Am. Chem. Soc. 89, 4158 (1967). Structure: eidem, ibid. 90, 1087 (1968). Absolute configuration and chemistry: Antosz et al., ibid. 92, 4933 (1970). Crystal structure: J. B. Wetherington, J. W. Moncrief, Acta Crystallogr. B31, 501 (1975). Total synthesis: M. Yoshioka et al., J. Am. Chem. Soc. 106, 1133 (1984); eidem, Heterocycles 21, 151 (1984); R. S. Garigipati et al., J. Am. Chem. Soc. 107, 7790

(1985). Antibacterial activity and toxicity: R. F. Pittillo *et al., Antibiot. Ann.* **1958-1959,** 497. Series of articles on antineoplastic activity: *ibid.* 515-532. Inhibition of protein synthesis: D. Smithers *et al., Mol. Pharmacol.* **5,** 433 (1969). Experimental use as cariostat: D. E. Hunt *et al., J. Dent. Res.* **50,** 371 (1971). Reduction of periodontal syndrome in mice, rats: J. H. Shaw, J. K. Ivimey, *Arch. Oral Biol.* **18,** 357 (1973).

Amorphous, fluffy, very hygroscopic powder. $[\alpha]_D^{28}$ +59° (c = 0.5 in pH 7 phosphate buffer). pKa: 7.5, 8.8. Freely sol in water; mod sol in methanol, ethanol. Unstable in aqueous solns pH >7. uv max: 262 nm (0.1N HCl); 288 nm (0.1N NaOH). LD_{50} in mice, rats (mg/kg): 800 ±27, 1550 ±26 i.v. (Pittillo).

Acetate, $C_{13}H_{20}N_2O_6 \cdot C_2H_4O_2$, needles from ethanol; softens at 130°, resolidifies at 145°, mp 263-266°. $[\alpha]_D^{26}$ +58°. pKa: 4.6, 7.5, 8.8. Very sol in water; sol in warm methanol, ethanol, acetone; sparingly sol in ethyl acetate. uv max: 264 nm (pH 7 phosphate buffer).

Hydrochloride, $C_{13}H_{20}N_2O_6 \cdot HCl$, hygroscopic. $[\alpha]_D^{22}$ +59° (c = 0.41) (natural); $[\alpha]_D^{22}$ +55° (c = 0.47) (synthetic).

140. Actinodaphnine. *6,7,7a,8-Tetrahydro-11-methoxy-5H-benzo[g]-1,3-benzodioxolo[6,5,4-de]quinolin-10-ol; 10-methoxy-1,2-(methylenedioxy)-6aα-noraporphin-9-ol;* 1,2-methylenedioxy-9-hydroxy-10-methoxynoraporphine. $C_{18}H_{17}NO_4$; mol wt 311.34. C 69.44%, H 5.50%, N 4.50%, O 20.56%. In bark of *Actinodaphne hookeri* Meissn., Lauraceae. Isoln: Krishna, Ghose, *J. Indian Chem. Soc.* **9,** 429 (1932). Structure: Ghose *et al., Helv. Chim. Acta* **17,** 919 (1934). Photolytic synthesis: M. S. Premila *et al., Indian J. Chem.* **13,** 945 (1975).

Needles from alcohol, mp 211°. $[\alpha]_D^{20}$ +33° (ethanol). Sol in acetone, alcohol, benzene, chloroform. Moderately sol in ether. Practically insol in water.

Hydrochloride, $C_{18}H_{17}NO_4 \cdot HCl$, needles from alcohol-ether, mp 281° (dec). $[\alpha]_D^{20}$ +9°.

Methyl ether, $C_{19}H_{19}NO_4$, needles from ether + alc, mp 114-115°. Synthesis: Hey, Lobo, *J. Chem. Soc.* **1954,** 2246.

141. Actinomycetin. Bacteriolytic cell-free fluid of culture filtrates of actinomycetes. Produced by most species of *Streptomyces;* studied mostly in *Streptomyces albus.* Protein-like in nature. Stated to consist of a bactericidal fatty acid fraction and an enzyme: Welsch, *J. Bacteriol.* **42,** 801 (1941); **44,** 571 (1942); **53,** 101 (1947). Dissolves dead gram-negative, and, with more difficulty, dead gram-positive organisms. Also dissolves living organisms in aq suspensions, such as *Staph. aureus.* In answer to the question whether actinomycetin deserves the name "antibiotic", *see* the polemic between Hoogerheide and Welsch, *Bot. Rev.* **10,** 599 (1944) and *J. Bacteriol.* **53,** 101 (1947). Purification and prepn from *Streptomyces albus:* Ghuysen, **Belg.** pats. **517,-191** and **521,114** (both 1953 to Soc. Belge de l'Azote), *C.A.* **53,** 5600g, 12593e (1959). *Review:* Caltrider in *Antibiotics* **vol. 1,** D. Gottlieb, P. D. Shaw, Eds. (Springer Verlag, New York, 1967) pp 681-683.

Sol in water. Pptd by alcohol, acetone, ammonium sulfate and other protein precipitants. Destroyed by strong acid. More stable at pH 10 than at pH 4. Rapidly inactivated by heat, 60-70° destroys activity.

142. Actinomycin F₁. Actinomycin KS4; KS4. An antibiotic produced by cultures of actinomycin C-elaborating strains of Streptomyces, such as Streptomyces BOP 476 (NRRL 2580) and *Streptomyces chrysomallus* (NRRL 2250). Considered to be a derivative of actinomycin C₂; amino acid analysis indicates that the prolines present in actinomycin C₂ are substituted by sarcosine in actinomycin F₁. Production by controlled biosynthesis: Schmidt-

Kastner, *Naturwiss.* **43,** 131 (1956); *idem, Ann. N.Y. Acad. Sci.* **89,** 299 (1960); Schmidt-Kastner, Hackmann, U.S. pat. **3,219,544** (1965 to Bayer).

THERAP CAT: Antineoplastic.

143. Actinoquinol. *8-Ethoxy-5-quinolinesulfonic acid.* $C_{11}H_{11}NO_4S$; mol wt 253.28. C 52.16%, H 4.38%, N 5.53%, O 25.27%, S 12.66%. Prepn: Orlov, Bogdanov, *J. Appl. Chem. USSR* **5,** 803 (1932), *C.A.* **27,** 724[8] (1933); Ghosh, Roy, *J. Indian Chem. Soc.* **22,** 39 (1945); Dorier *et al., Trav. Soc. Pharm. Montpellier* **14**(3), 152 (1954), *C.A.* **50,** 1011a (1956).

Brown needles from water, mp 286-288° (dec). Sol in dil $NaHCO_3$.

Sodium salt, $C_{11}H_{10}NNaO_4S$, *sodium etoquinol, etoquinol sodium, Corodenin, Uviban.*

THERAP CAT: Sodium salt as ultraviolet screen.

144. Actinorhodine. $C_{32}H_{26}O_{14}$; mol wt 634.55. C 60.57%, H 4.13%, O 35.30%. Antibiotic pigment produced by *Streptomyces coelicolor*, found in woods near Göttingen, Germany: Brockmann, Pini, *Naturwiss.* **34,** 190 (1947); Brockmann *et al., Ber.* **83,** 161 (1950). Tentative structure: Brockmann, Hieronymus, *Ber.* **88,** 1379 (1955); Brockmann *et al., Naturwiss.* **49,** 131 (1962). Structure: Brockmann, *Angew. Chem.* **76,** 863 (1964); Brockmann *et al., Ann.* **698,** 209 (1966). Stereochemistry: Zeeck, Christiansen, *Ann.* **724,** 172 (1969). Biosynthesis and determn of point of dimerization: C. P. Gorst-Allman *et al., J. Org. Chem.* **46,** 455 (1981).

Fine red needles from dioxane, dec 270°. Absorption max (dioxane): 560, 523 nm. Sol in pyridine, piperidine, tetrahydrofuran, dioxane, phenol; slightly sol in alcohol, acetic acid, acetone. Red soln with acetone, blue soln with pyridine. Practically insol in aq acid; sol in aq alkali with bright blue color.

145. Actiphenol. *4-[2-(2-Hydroxy-3,5-dimethylphenyl)-2-oxoethyl]-2,6-piperidinedione; 3-(2-hydroxy-3,5-dimethylphenacyl)glutarimide; β-(3,5-dimethyl-2-hydroxybenzoylmethyl)glutarimide; C-73.* $C_{15}H_{17}NO_4$; mol wt 275.30. C 65.44%, H 6.22%, N 5.09%, O 23.25%. Metabolic product found in culture filtrates of cycloheximide-producing actinomycetes *(Streptomyces albulus):* Highet, Prelog, *Helv. Chim. Acta* **42,** 1523 (1959); Rao, *J. Org. Chem.* **25,** 661 (1960). Synthesis: Johnson, *ibid.* **27,** 3658 (1962). May be prepd from cycloheximide: Highet, Prelog, *loc. cit.* Structure-activity studies: Ennis, *Biochem. Pharmacol.* **17,** 1197 (1968).

Fluffy needles from methylene chloride + methanol. mp 200-202°. Pure crystals are white or colorless. uv max: 262, 345 nm (ϵ 10870, 4550). Exhibits bright yellow fluorescence under u.v. Soluble in aqueous NaOH (forming bright yellow solns), in anhydr alcohol, tetrahydrofuran. Insol in aq solns of sodium bicarbonate.

Acetate, $C_{17}H_{19}NO_5$, crystals from methylene chloride + ether, mp 155.5°.

146. Activins. Polypeptide hormone, identified in ovarian follicular fluid, which selectively stimulates secretion of FSH, q.v. Dimer composed of β subunits of inhibin, q.v. (either β_A or β_B), mol wt 24,000 daltons. Isolation and characterization of the heterodimer, *activin AB:* N. Ling *et al., Nature* **321,** 779 (1986); of the homodimer, *activin A:* W. Vale *et al., ibid.* 776; N. Ling *et al., Biochem. Biophys. Res. Commun.* **138,** 1129 (1986). Comparison of FSH stimulation with gonadotropin releasing hormone (LH-RH, q.v.): R. H. Schwall *et al., ibid.* **151,** 1099 (1988). Approach to nonradiometric assays: R. H. Schwall *et al., Prog. Clin. Biol. Res.* **285,** 205 (1988). Brief review of chemical properties and biological activity: S.-Y. Ying, *Proc. Soc. Exp. Biol. Med.* **186,** 253-264 (1987).

147. ACV. *(S)-N-[N-(5-Amino-5-carboxy-1-oxopentyl)-L-cysteinyl]-D-valine; N-[N-(L-5-amino-5-carboxyvaleryl)-L-cysteinyl]-D-valine;* δ-(L-α-aminoadipyl)-L-cysteinyl-D-valine. $C_{14}H_{25}N_3O_6S$; mol wt 363.44. C 46.27%, H 6.93%, N 11.56%, O 26.41%, S 8.82%. Biosynthetic precursor of penicillins and cephalosporins. Isoln from mycelial extracts of *Penicillium chrysogenum:* H. R. V. Arnstein *et al., Biochem. J.* **76,** 353 (1960); structure determn of tripeptide: H. R. V. Arnstein, D. Morris, *ibid.* 357. Isoln from *Cephalosporium* sp., identification and optical configuration of constituent amino acids: P. B. Loder, E. P. Abraham, *ibid.* **123,** 471 (1971). Configuration of peptide from *P. chrysogenum,* identity with peptide from *Cephalosporium:* P. Adriaens *et al., Antimicrob. Ag. Chemother.* **8,** 638 (1975). Synthesis of labelled tripeptide and role in penicillin biosynthesis: P. A. Fawcett *et al., Biochem. J.* **157,** 651 (1976). Synthesis of ACV as the disulfide: S. Wolfe, M. G. Jokinen, *Can. J. Chem.* **57,** 1388 (1979). Cellfree conversion of ACV directly into isopenicillin N: J. O'Sullivan *et al., Biochem. J.* **184,** 421 (1979); T. Konomi *et al., ibid.* 427. Confirmation by ^{13}C NMR: J. E. Baldwin *et al., Chem. Commun.* **1980,** 1271. *Reviews:* D. J. Aberhart, *Tetrahedron* **33,** 1545-1559 (1977); E. P. Abraham, *Japan. J. Antibiot.* **30,** Suppl., S1-S26 (1977); S. Wolfe *et al., Science* **226,** 1386-1392 (1984).

148. Acyclovir. *2-Amino-1,9-dihydro-9-[(2-hydroxyethoxy)methyl]-6H-purin-6-one; acycloguanosine;* 9-[(2-hydroxyethoxy)methyl]guanine; BW-248U; Wellcome 248U; Acicloftal; Aclovir; Cargosil; Poviral; Virorax; Zovirax. $C_8H_{11}N_5O_3$; mol wt 225.21. C 42.67%, H 4.92%, N 31.10%, O 21.31%. Orally active acyclic nucleoside with inhibitory activity towards several herpes viruses. Prepn: H. J. Schaeffer, *Ger. pat.* **2,539,963;** *idem, U.S. pat.* **4,199,574** (1976, 1980 to Wellcome). Convenient synthesis from guanine: H. Matsumoto *et al., Chem. Pharm. Bull.* **36,** 1153 (1988). Selectivity of action: G. B. Elion *et al., Proc. Nat.*

Acad. Sci. USA **74,** 5716 (1977). Chemistry, antiviral activity, metabolism: H. J. Schaeffer *et al., Nature* **272,** 583 (1978). *In vitro* activity: P. Collins, D. J. Bauer, *J. Antimicrob. Chemother.* **5,** 431 (1979). Effect on herpes simplex infections in mice: H. J. Field *et al., Antimicrob. Ag. Chemother.* **15,** 554 (1979); on herpes zoster in immunocompromised patients: H. H. Balfour *et al., N. Engl. J. Med.* **308,** 1448 (1983). Treatment of primary episodes of genital herpes simplex infection: Y. J. Bryson *et al., ibid.* 916; of recurrent genital herpes: S. E. Straus *et al., ibid.* **310,** 1545 (1984); J. M. Douglas *et al., ibid.* 1551. Symposia on pharmacology and clinical studies: *Am. J. Med.* **73,** Suppl. 1A, 1-392 (1982); *J. Antimicrob. Chemother.* **12,** Suppl. B, 1-202 (1983); *Scand. J. Infect. Dis.* **Suppl. 47,** 1-176 (1985). *Review:* R. J. Whitley, J. W. Gnann, Jr., *N. Engl. J. Med.* **327,** 782-789 (1992).

Crystals from methanol, mp 256.5-257°. LD_{50} in mice (mg/kg): > 10,000 orally; 1000 i.p. (Schaeffer).

THERAP CAT: Antiviral.

149. Adamantane. *Tricyclo[3.3.1.1^{3,7}]decane;* diamantane (obsolete). $C_{10}H_{16}$; mol wt 136.24. C 88.16%, H 11.84%. Isoln from petroleum from Moravia province in Czechoslovakia: Landa, Machacek, *Coll. Czech. Chem. Commun.* **3,** 1 (1933); from American petroleum: Mair *et al., Anal. Chem.* **31,** 2082 (1959). Synthesis: Prelog, Seiwerth, *Ber.* **74,** 1769 (1941); Stetter *et al., Ber.* **89,** 1922 (1956); by aluminum chloride-catalyzed isomerization of tetrahydrodicyclopentadiene: Yan, Shreh, *Bull. Inst. Chem. Acad. Sinica,* **Dec. 1965,** no. 11, pp 79-81. Two step synthesis starting with dicyclopentadiene: Schleyer, *J. Am. Chem. Soc.* **79,** 3292 (1957); Schleyer, Donaldson, *ibid.* **82,** 4645 (1960); Ludwig *U.S. pat.* **2,937,211** (1960 to du Pont). *Reviews:* Stetter, *Angew. Chem.* **66,** 217 (1954); Fort, Schleyer, *Chem. Revs.* **64,** 277 (1964).

Crystallizes at − 30°; can be purified by recrystallization from acetone or by sublimation. mp 269.6-270.8°.

Note: The name diamantane has been abandoned as a synonym for adamantane and proposed as the name for the second member of the adamantane series, congressane: Vogl, Anderson, *Tetrahedron Letters* **1966,** 415.

USE: The diamine as curing agent for epoxy resins [U.S. pat. **3,053,907** (1962 to du Pont)].

150. Adenine. *1H-Purin-6-amine;* 6-aminopurine; 6-amino-1H-purine; 6-amino-3H-purine; 6-amino-9H-purine; 1,6-dihydro-6-iminopurine; 3,6-dihydro-6-iminopurine; Leuco-4. $C_5H_5N_5$; mol wt 135.13. C 44.44%, H 3.73%, N 51.83%. Also referred to as *vitamin B_4:* Lecoq, *Int. Z. Vitaminforsch.* **27,** 291 (1957). Widespread throughout animal and plant tissues combined with niacinamide, D-ribose, and phosphoric acids; a constituent of nucleic acids and coenzymes, such as codehydrase I and II, adenylic acid, coalaninedehydrase. Isoln from bovine pancreas: Kossel, *Ber.* **18,** 79, 1928 (1885). Syntheses: Fischer, *ibid.* **30,** 2226 (1897); Traube, *Ann.* **331,** 64 (1904); Hoffer, *Jubilee Vol. Emil Barell* **1946,** 428-434; Taylor *et al., Ciba Foundation Symposium, Chem. and Biol. Purines* **1957,** 20, *C.A.* **53,** 6238b (1959); Bredereck *et al., Angew. Chem.* **71,** 524 (1959); Morita *et al., Chem. & Ind. (London)* **1968,** 1117; Sekiya, Suzuki, *Chem. Pharm. Bull.* **20,** 209 (1972); N. J. Kos *et al., J. Org. Chem.* **44,** 3140 (1979). Toxicity study: Philips *et al., J. Pharmacol. Exp. Ther.* **104,** 20 (1952). *Review:* Ts'o, "Bases, Nucleosides and Nucleotides" in *Basic*

Principles in Nucleic Acid Chemistry **vol. 1,** P. O. P. Ts'o, Ed. (Academic Press, New York, 1974) pp 453-584. *See also* Nucleic Acids.

Trihydrate, orthorhombic needles. Anhydr at 110°, dec 360-365°, subl 220°. uv max (pH 7.0): 207, 260.5 nm ($\epsilon \times 10^{-3}$ 23.2, 13.4). One gram of anhydr compd dissolves in 2000 ml water, 40 ml boiling water; slightly sol in alc. Practically insol in ether, $CHCl_3$. Aq solns are neutral. Combines with acids and bases. LD_{50} orally in rats: 745 mg/kg (Philips).

Hydrochloride hemihydrate, monoclinic prisms. One gram dissolves in 42 ml water.

Sulfate dihydrate, crystals. One gram dissolves in 150 ml water; slightly sol in alc.

USE: In microbial determination of niacin; in research on heredity, virus diseases, and cancer.

151. A-Denopterin. *N-[4-[[1-(2,4-Diamino-6-pteridinyl)ethyl]methylamino]benzoyl]-L-glutamic acid;* 4-amino-N^{10},9-dimethylpteroylglutamic acid. $C_{21}H_{24}N_8O_5$; mol wt 468.47. C 53.84%, H 5.16%, N 23.92%, O 17.08%. Prepn: Hultquist *et al., J. Am. Chem. Soc.* **71,** 619 (1949); Hultquist, Smith, **Brit.** pat. **667,098** (1952 to Am. Cyanamid).

Dihydrate, yellow-orange microcrystals. uv max (0.1N NaOH): 255, 306, 369 nm (ϵ 25,000, 25,000, 9000); (0.1N HCl): 244, 316 nm (ϵ 22,000, 14,000).

152. Adenosine. *9-β-D-Ribofuranosyl-9H-purin-6-amine;* 6-amino-9-β-D-ribofuranosyl-9H-purine; 9-β-D-ribofuranosidoadenine; adenine riboside; Adenocard; Adenocor; Adenoscan. $C_{10}H_{13}N_5O_4$; mol wt 267.24. C 44.94%, H 4.90%, N 26.21%, O 23.95%. Nucleoside; widely distributed in nature. From yeast nucleic acid: Levene and Bass, *Nucleic Acids* (New York, 1931) p 163. Structure: Levene, Tipson, *J. Biol. Chem.* **94,** 809 (1932); Bredereck, *Ber.* **66,** 198 (1933); *Z. Physiol. Chem.* **223,** 61 (1934); Gulland, Holiday, *J. Chem. Soc.* **1936,** 765. *Cf.* Szent-Györgyi, *J. Physiol.* **68,** 213 (1930); Lythgoe *et al., J. Chem. Soc.* **1947,** 355; **1948,** 965. Synthesis: Davoll *et al., ibid.* **1948,** 967; H. Vorbrueggen, K. Krolikiewicz, *Angew. Chem. Intl. Ed.* **14,** 421 (1975). Crystal structure: T. F. Lai, R. E. Marsh, *Acta Crystallogr.* **B28,** 1982 (1972). Conformational properties: D. B. Davies, A. Rabczenko, *J. Chem. Soc. Perkin Trans. 2* **1975,** 1703. Symposium on cardiac electrophysiology, pharmacology and clinical efficacy in supraventricular tachycardia: *Prog. Clin. Biol. Res.* **230,** 1-395 (1987). *Reviews: see* Adenine; Nucleic Acids.

Crystals from water, mp 234-235°. $[\alpha]_D^{11}$ −61.7° (c = 0.706 in water); $[\alpha]_D^9$ −58.2° (c = 0.658 in water). uv max: 260 nm (ϵ 15100). Practically insol in alcohol.

THERAP CAT: Antiarrhythmic.

153. Adenosine Diphosphate. *Adenosine 5'-(trihydrogen diphosphate);* ADP; adenosine 5'-pyrophosphoric acid; 5'-adenylphosphoric acid; adenosinediphosphoric acid. $C_{10}H_{15}N_5O_{10}P_2$; mol wt 427.20. C 28.12%, H 3.54%, N 16.39%, O 37.45%, P 14.50%. Formed from ATP in the muscle by the enzyme adenosinetriphosphatase upon stimulation of the muscle unless hydrolysis is prevented by injection of magnesium sulfate. Prepd by hydrolysis of ATP by means of adenosinetriphosphatase from lobster or rabbit muscle: LePage, *Biochem. Prepns.* **I,** 1 (1949). Synthesis: Chambers *et al., J. Am. Chem. Soc.* **82,** 970 (1960). *See also* the ref and data *under* Adenosine Triphosphate.

Barium salt, $Ba_3(C_{10}H_{12}N_5O_{10}P_2)_2$. The purity of the prepns can be checked by analyses for nitrogen, ribose (orcinol reaction), total phosphorus, easily hydrolyzable phosphorus, and inorganic phosphorus. ADP should give a ratio of 2:1 for total organic phosphorus to easily hydrolyzable phosphorus, see LePage, *loc. cit.* a_M (molar absorbancy of pure ADP): 15.4 × 10^3 at 259 nm and pH 7.0.

154. Adenosine Triphosphate. *Adenosine 5'-(tetrahydrogen triphosphate);* ATP; adenosine 5'-triphosphoric acid; Adephos; Adetol; Atipi; Atriphos; Striadyne; Triadenyl; Triphosaden. $C_{10}H_{16}N_5O_{13}P_3$; mol wt 507.18. C 23.68%, H 3.18%, N 13.81%, O 41.01%, P 18.32%. Coenzyme valuable in the transfer of phosphate bond energy. Mammalian skeletal muscle at rest contains 350-400 mg ATP per 100 g. Upon stimulation of the muscle the ATP is hydrolyzed to ADP by the myosin-actin complex unless hydrolysis is prevented by injection of magnesium sulfate. Isoln from rabbit muscle: LePage, *Biochem. Prepn.* **1,** 5 (1949); *cf.* Fiske, Subbarow, *Science* **70,** 381 (1929); Lohmann, *Biochem. Z.* **223,** 460 (1931); **254,** 381 (1932); Barrenscheen, Filz, *ibid* **250,** 281 (1932); Kerr, *J. Biol. Chem.* **139,** 121 (1941); Needham, *Biochem. J.* **36,** 113 (1942). Synthetic routes: Tanaka, Honjo, *U.S.* pat. **3,079,379** (1963 to Takeda). Reviews of biosynthesis: Racker, *Advan. Enzymol.* **23,** 323-399 (1961); Deamer, *J. Chem. Ed.* **46,** 198-206 (1971). Reviews of nucleotide coenzymes: Bock in *The Enzymes* vol. 2A, P. D. Boyer *et al.,* Eds. (Academic Press, New York, 2nd ed., 1960) pp 3-38; A. M. Michelson, *The Chemistry of Nucleosides and Nucleotides* (Academic Press, New York, 1963) pp 153-250; D. W. Hutchinson, *Nucleotides and Coenzymes* (John Wiley, New York, 1964) pp 36-82.

Free ATP, isolated as a glass (by treatment of the Ba salt with H_2SO_4 and treating the concd aq soln with acetone). $[\alpha]_D^{22}$ −26.7° (c = 3.095). a_M(molar absorbancy): 15.4 × 10^3 at 259 nm and pH 7.0. Freely sol in water. A 1% aq soln has a pH of about 2 and is stable at 0° for several hrs. ATP is a tetrabasic acid and after hydrolytic cleavage it is hexabasic. It is usually pptd as the dibarium salt with 4 or 6 mols of water of crystn which can be removed by prolonged drying at 100° over P_2O_5 and in a vacuum. The anhydr salt is stable, but the hydrated salt slowly decomp forming 5'-adenylic acid and barium pyrophosphate. For use the barium salt is converted to the sodium or potassium salt by treatment with sodium or potassium sulfate in HCl soln. At pH 6.8-7.4 the Na salt is stable in aq soln provided the product is pure. Ba^{2+} catalyzes breakdown.

Injectable solns of the disodium salt are marketed as *Trinosin, Adetphos, Circulen "Kyorin", Atenen.*

USE: In biochemical research. To inhibit enzymatic browning of raw edible plant materials, such as sliced apples, potatoes, etc.

THERAP CAT: Coenzyme.

155. S-Adenosylmethionine. *(3S)-5'-[(3-Amino-3-carboxypropyl)methylsulfonio]-5'-deoxyadenosine inner salt; S-(5'-desoxyadenosin-5'-yl)-L-methionine; "active methionine"; ademetionine; AdoMet; SAMe; Donamet; S. Amet.* $C_{15}H_{22}$-N_6O_5S; mol wt 398.44. C 45.22%, H 5.57%, N 21.09%, O 20.08%, S 8.05%. Physiological methyl donor involved in enzymatic transmethylation reactions; present in all living organisms. Enzymatic prepn: G. L. Cantoni, *J. Am. Chem. Soc.* **74,** 2942 (1952). Isoln from cultures of *Torulopsis utilis:* F. Schlenk, R. E. DePalma, *J. Biol. Chem.* **229,** 1037 (1957). Synthesis of *S*-adenosyl-D,L-methionine: J. Baddiley, G. A. Jamieson, *J. Chem. Soc.* **1954,** 4280. Synthesis of optically active chloride: C. H. Shunk, J. W. Richter, U.S. pat. **2,969,353** (1961 to Merck & Co.). Fermentative production by *Aspergillus tamarii* IAM 2137: H. Kusakabe *et al.,* Japan. pat. **72 37,037** (1972 to Yamaza Shoyu), *C.A.* **78,** 2745f (1973); production by yeasts: T. Tsuchida *et al.,* Japan. Kokai **75 82,288;** *eidem,* U.S. pat. **3,962,034** (1975, 1976 both to Ajinomoto). Pharmacology: G. Stramentinoli *et al., Minerva Med.* **66,** 1541 (1975). Clinical efficacy in hypertriglyceridemia: F. Consolo *et al., Curr. Ther. Res.* **22,** 75 (1977); in hepatic steatosis: R. Mazzanti *et al., ibid.* **25,** 25 (1979); G. Mascio *et al., Gazz. Med. Ital.* **140,** 37 (1981). Clinical antidepressant effects: A. Agnoli *et al., J. Psychiat. Res.* **13,** 43 (1976); J. F. Lipinski *et al., Am. J. Psychiat.* **141,** 448 (1984). Anti-inflammatory activity in rats: M. Gualano *et al., Int. J. Tissue React.* **7,** 41 (1985). Clinical comparison with ibuprofen in osteoarthritis: S. Glorioso *et al., Int. J. Clin. Pharm. Res.* **5,** 39 (1985). Review of preparative methods and physiological activity: S. Shimizu, H. Yamada, *Trends Biotechnol.* **2,** 137 (1984). Books: *The Biochemistry of Adenosylmethionine,* F. Salvatore *et al.,* Eds. (Columbia Univ. Press, New York, 1977) 588 pp; *Biochemistry of S-Adenosylmethionine and Related Compounds,* E. Usdin *et al.,* Eds. (Macmillan Press, London, 1982) 760 pp.

Chloride, $C_{15}H_{23}ClN_6O_5S$, uv max (water): 260 nm. $[\alpha]_D^{25}$ +32° (c = 3.3 in water).

Disulfate ditosylate, $C_{29}H_{42}N_6O_{19}S_5$, *Gumbaral, Samyr.* LD_{50} in mice (mg/kg): 560 i.v., 2500 i.p., > 6000 orally (Stramentinoli).

THERAP CAT: Anti-inflammatory. In treatment of chronic liver disease.

156. 3'-Adenylic Acid. Adenosine 3'-monophosphate; adenosine-3'-phosphoric acid; adenosine-3'-monophosphoric acid; yeast adenylic acid; syn-adenylic acid; h-adenylic acid. $C_{10}H_{14}N_5O_7P$; mol wt 347.22. C 34.59%, H 4.06%, N 20.17%, O 32.25%, P 8.92%.

Early prepns from yeast nucleic acid: Levene, Bass, *Nucleic Acids,* (Chemical Catalogue Co., New York, 1931). Early work probably done on mixtures of 2'- and 3'-adenylic acids; both compds isomerize readily to form an equilibrium mixture under acid conditions: Carter, Cohn, *Fed. Proc.* **8,** 190 (1949); Baddily, in *The Nucleic Acids* vol. 1, E. Chargaff, J. N. Davidson, Eds. (Academic Press, New York, 1955) pp 165-168; A. M. Michelson, *The Chemistry of Nucleosides and Nucleotides* (Academic Press, New York, 1963) pp 100-106. Synthesis: Brown, Todd, *J. Chem. Soc.* **1952,** 44. Structure of dihydrate: Brown *et al., Nature* **172,** 1184 (1953); Sundaralingam, *Acta Cryst.* **21,** 495 (1966). *Reviews: see* Adenosine; Nucleic Acids.

Long, fine, colorless, needles. mp 197° (dec). Dihydrate, d (calc) 1.698. Absorption spectrum: Voet *et al., Biopolymers* **1,** 193 (1963). Difficultly sol in boiling water. Gives quantitative yield of furfural when distilled with 20% HCl for 3 hrs (different from 5'-adenylic acid which yields only traces). *cf.* Hoffman, *J. Biol. Chem.* **73,** 15 (1927); Embden, Schmidt, *Z. Physiol. Chem.* **181,** 130 (1929).

157. 5'-Adenylic Acid. Muscle adenylic acid; erga-denylic acid; t-adenylic acid; adenosine 5'-monophosphate; adenosine phosphate; adenosine-5'-phosphoric acid; adeno-sine-5'-monophosphoric acid; A-5MP; AMP; NSC-20264; Adényl; Lycedan; Myoston. $C_{10}H_{14}N_5O_7P$; mol wt 347.22. C 34.59%, H 4.06%, N 20.17%, O 32.25%, P 8.92%. Nucleotide; widely distributed in nature. Prepn from tissues: Embden, Zimmerman, *Z. Physiol. Chem.* **167,** 137 (1927); Embden, Schmidt, *ibid.* **181,** 130 (1929); *cf.* Kalckar, *J. Biol. Chem.* **167,** 445 (1947). Prepn by hydrolysis of ATP with barium hydroxide: Kerr, *J. Biol. Chem.* **139,** 131 (1941). Synthesis: Baddiley, Todd, *J. Chem. Soc.* **1947,** 648. Commercial prepn by enzymatic phosphorylation of adenosine. Monograph on synthesis of nucleotides: G. R. Pettit, *Synthetic Nucleotides* vol. 1 (Van Nostrand-Reinhold, New York, 1972) 252 pp. Crystal structure: Kraut, Jensen, *Acta Cryst* **16,** 79 (1963). *Reviews: see* Adenosine; Nucleic Acids.

Crystals from water + acetone, mp 196-200°. $[\alpha]_D^{20}$ −47.5° (c = 2, 2% NaOH); $[\alpha]_D^{20}$ −26.0° (c = 2, 10% HCl). pK_1 = 3.8; pK_2 = 6.2. a_M: 15.4 × 10^3 at 259 nm (pH 7.0). Readily sol in boiling water. The compound is readily deaminated by nitrous acid to form inosinic acid; less rapidly hydrolyzed than 3'-adenylic acid by sulfuric acid. Furfural is formed only in traces on distillation with 20%

HCl, *cf.* Levene, Bass, *Nucleic Acids* (New York, 1931) pp 230-232.

THERAP CAT: Nutrient.

158. Adhatoda. Malabar nut; arusa; adulsa; vasaca; adhatodai; bakash. Leaves of *Adhatoda vasica* Nees, *Acanthacae. Habit.* East India. *Constit.* Adhatodic acid, vasicine. *Ref:* Krishnaswami, David, *Indian J. Pharm.* **2**, 141 (1940).

THERAP CAT: Expectorant.

159. Adinazolam. *8-Chloro-N,N-dimethyl-6-phenyl-4H-[1,2,4]triazolo[4,3-a][1,4]benzodiazepine-1-methanamine;* 8-chloro-1-[(dimethylamino)methyl]-6-phenyl-4*H*-*s*-triazolo-[4,3-a][1,4]benzodiazepine; U-41123. $C_{19}H_{18}ClN_5$; mol wt 351.84. C 64.86%, H 5.16%, Cl 10.08%, N 19.91%. Triazolobenzodiazepine with antidepressant and anxiolytic properties. Dimethylamino derivative of alprazolam, *q.v.* Prepn: H. Allgeier, A. Gagneux, Ger. pat. 2,201,210 corresp to **Brit.** pat. **1,393,256** (1972, 1975 both to Ciba-Geigy); J. B. Hester, Jr., U.S. pat. 4,250,094 (1984 to Upjohn); *idem, J. Heterocyclic Chem.* **17**, 575 (1980). Chromatographic determn in human plasma: G. W. Peng, *J. Pharm. Sci.* **73**, 1173 (1984). Pharmacological profile: R. A. Lahti *et al., Neuropharmacology* **22**, 1277 (1983). Biological activity and metabolism: V. H. Sethy *et al., J. Pharm. Pharmacol.* **36**, 546 (1984). Use as antipsychotic: R. Lahti, R. E. Pyke, **Ger.** pat. **3,401,290** corresp to U.S. pat. 4,472,397 (both 1984 to Upjohn). Preliminary clinical study in severe depression: R. E. Pyke *et al., Psychopharmacol. Bull.* **19**, 96 (1983).

Crystals from ethyl acetate, mp 171-172.5° (Hester, U.S. pat.). Also reported as mp 165-166° (Allgeier, Gagneaux).
Monomethanesulfonate, $C_{19}H_{18}ClN_5\cdot CH_3SO_3H$, *adinazolam mesylate, Deracyn.* Crystals from methanol-ether, mp 230-244°.

THERAP CAT: Antidepressant.

160. Adiphenine Hydrochloride. α-*Phenylbenzeneacetic acid 2-(diethylamino)ethyl ester hydrochloride;* 2-diethylaminoethyl diphenylacetate hydrochloride; diphenylacetyl-diethylaminoethanol hydrochloride; Trasentine hydrochloride; Diphacil hydrochloride; Spasmolytin; Difacil hydrochloride; Patrovina. $C_{20}H_{26}ClNO_2$; mol wt 347.88. C 69.05%, H 7.53%, Cl 10.19%, N 4.03%, O 9.20%. Antimuscarinic antispasmodic. Prepd by the action of diphenylacetyl chloride or diphenyl ketene on diethylaminoethanol: **Swiss** pat. **190,541** (1937 to Ciba); **Ger.** pats. **626,539; 653,778** (1937). Crystal and molecular structure: Guy, Hamor, *J. Chem. Soc. Perkin Trans. II* **1973**, 942.

Crystals, mp 113-114°. Freely sol in water; very sparingly sol in alcohol, ether. A 5% aq soln is neutral to litmus.
Methyl bromide, $C_{21}H_{28}BrNO_2$, *2-diethylaminoethyl diphenylacetate methyl bromide, Lunal.* LD$_{50}$ i.v. in rabbits: 22.5-27.5 mg/kg.

THERAP CAT: Anticholinergic.

161. Adipic Acid. *Hexanedioic acid;* 1,4-butanedicarboxylic acid. $C_6H_{10}O_4$; mol wt 146.14. C 49.31%, H 6.90%,

O 43.79%. HOOC$(CH_2)_4$COOH. Found in beet juice. Prepn from cyclohexanol: Bouveault, Locquin, *Bull. Soc. Chim.* [4] **3**, 438 (1908); Ellis, *Org. Syn.* coll. vol. I, 18 (2nd ed., 1941); Feagen, Copenhaver, *J. Am. Chem. Soc.* **62**, 869 (1940); U.S. pats. **2,191,786** (1940); **2,196,357** (1940); Zilberman *et al., J. Appl. Chem. (USSR)* **29**, 621 (1956). Convenient lab prepn from cyclohexanone: L. F. Fieser, *Organic Experiments* (Heath, Boston, 1964) pp 106-108. Prepn by one-step oxidation of cyclohexane: Onopchenko, Schulz, *J. Org. Chem.* **38**, 3729 (1973); Tanaka, 167th Am. Chem. Soc. Meet. (Los Angeles, March-April, 1974) *Abstracts of Papers,* p 29. Manuf: Faith, Keyes & Clark's *Industrial Chemicals,* F. A. Lowenheim, M. K. Moran, Eds. (Wiley-Interscience, New York, 4th ed., 1975) pp 50-54. *Review:* D. E. Danly, C. R. Campbell in Kirk-Othmer *Encyclopedia of Chemical Technology* vol. 1 (Wiley-Interscience, New York, 3rd ed., 1978) pp 510-531.
Monoclinic prisms from ethyl acetate, from water or from acetone + petr ether. d_4^{25} 1.360. mp 152°. bp$_{760}$ 337.5°; bp$_{100}$ 265°; bp$_{40}$ 240.5°; bp$_{20}$ 222°; bp$_{10}$ 205.5°; bp$_5$ 191°; bp$_{1.0}$ 159.5°. K_1 (25°) = 3.90 × 10^{-5}; K_2 = 5.29 × 10^{-6}. Absorption spectrum: Ramart-Lucas, Salmon-Legagneur, *Compt. Rend.* **189**, 916 (1929). Freely soluble in methanol, ethanol; sol in acetone. 100 ml of a satd aq soln contains 1.44 g; 100 ml of boiling water dissolves 160 g; 100 parts of ether dissolve 0.633 parts (w/w) at 19°. Slightly sol in cyclohexane. Practically insol in benzene, petr ether. pH of satd aq soln at 25° = 2.7; pH of 0.1% soln = 3.2.
Dimethyl ester, $C_8H_{14}O_4$, liquid, solidifies at 0°, bp$_{10}$ 112°.
Diethyl ester, $C_{10}H_{18}O_4$, *ethyl adipate.* Liquid, bp 240-245°. d^{20} 1.009. Insol in water; sol in alcohol and many other organic solvents.

USE: Manuf artificial resins, plastics (nylon), urethan foams. Used as acidulant in baking powders instead of tartaric acid, cream of tartar, and phosphates because adipic acid is not hygroscopic. As an intermediate in lubricating oil additives.

162. Adipsin. Serine protease homolog synthesized by mammalian adipocytes and secreted into the bloodstream; also produced by sciatic nerve. Developmentally regulated glycoprotein; mol wt ranges from 37-44 kilodaltons depending on the degree of glycosylation. Circulating levels of adipsin protein are reduced in certain genetic and experimental models of obesity. Originally identified as a 28 kDa protein encoded by a cDNA clone corresponding to messenger RNA specifically induced during adipocyte differentiation: B. M. Spiegelman *et al., J. Biol. Chem.* **258**, 10083 (1983). Nucleotide analysis and predicted amino acid sequence: K. S. Cook *et al., Proc. Nat. Acad. Sci. USA* **82**, 6480 (1985). Structure of adipsin gene: H. Y. Min, B. M. Spiegelman, *Nucleic Acids Res.* **14**, 8879 (1986). Isoln from adipose tissue, reportedly the primary site of synthesis, and from sciatic nerve: K. S. Cook *et al., Science* **237**, 402 (1987). Studies in murine obesity models and potential biological role: J. S. Flier *et al., ibid.* 405.

163. Adlumidine. *[S-(R*,R*)]-6-(5,6,7,8-Tetrahydro-6-methyl-1,3-dioxolo[4,5-g]isoquinolin-5-yl)furo[3,4-e]-1,3-benzodioxol-8(6H)-one.* $C_{20}H_{17}NO_6$; mol wt 367.36. C 65.39%, H 4.66%, N 3.81%, O 26.13%. An alkaloid present in the d-form in the entire plant of *Adlumia fungosa* (Ait.) Greene (*A. cirrhosa* Raf.), *Corydalis thalictrifolia* Franch. and *C. incisa* (Thumb.) Pers., *Fumariaceae.* Schlotterbeck, Watkins, *J. Am. Chem. Soc.* **25**, 596 (1903); R. H. F. Man-

d-Adlumidine

ske, *Can. J. Res.* **21B,** 111 (1943). *l*-Adlumidine was isolated from *Corydalis sempervirens* (L.) Pers., *C. scouleri* Hook., and *C. crystallina* Engelm., *Fumariaceae:* R. H. F. Manske, *ibid.* **8,** 407 (1932); **14B,** 347 (1936); **17B,** 57 (1939). Structure: *idem, J. Am. Chem. Soc.* **72,** 3207 (1950). Stereochemistry: Blaha *et al., Coll. Czech. Chem. Commun.* **29,** 2328 (1964); Snatzke *et al., Tetrahedron* **25,** 5059 (1969). *Review:* Stanek, Manske in *The Alkaloids* vol. **IV,** R. H. F. Manske, H. L. Holmes, Eds. (Academic Press, New York, 1954) pp 167-198. Stereoisomer of bicuculline, *q.v.*

Rhombic plates from chloroform + methanol, mp 236-237°. $[\alpha]_D^{25}$ +116.2° (c = 22 in chloroform). pKa 4.27. Practically insol in water. Very sparingly sol in alcohol, ether, hexane.

l-Adlumidine, *capnoidine.* Stout prisms from chloroform + methanol, mp 238°. $[\alpha]_D^{22}$ −113.2° (in chloroform). pKa 4.24. Practically insol in water. Sparingly sol in methanol; sol in chloroform.

164. Adlumine. *6-(1,2,3,4-Tetrahydro-6,7-dimethoxy-2-methyl-1-isoquinolinyl)furo[3,4-e]-1,3-benzodioxol-8(6H)-one.* $C_{21}H_{21}NO_6$; mol wt 383.40. C 65.79%, H 5.52%, N 3.65%, O 25.04%. An alkaloid present in the entire plant of some *Fumariaceae* species. *d*-Adlumine was isolated from *Adlumina fungosa* (Ait.) Greene (*A. cirrhosa* Raf.); *l*-adlumine was isolated from *Corydalis scouleri* Hook. and *C. sempervirens* (L.) Pers., *C. ophiocarpa* Hook. f. et Thoms *Fumariaceae:* Manske, *Can. J. Res.* **8,** 210 (1933); **14B,** 347 (1936); **16B,** 81 (1938); **17B,** 51 (1939). Preliminary stereochemical studies of *d*-form: Safe, Moir, *Can. J. Chem.* **42,** 160 (1964). Revised stereochemistry: Blaha *et al., Coll. Czech. Chem. Commun.* **29,** 2328 (1964); Snatzke *et al., Tetrahedron* **25,** 5059 (1969). *Review:* Stanek, Manske in *The Alkaloids* Vol. **IV,** R. H. F. Manske, H. L. Holmes, Eds. (Academic Press, New York, 1954) pp 167-198.

d-Adlumine

l-Form, hexagonal plates from chloroform + methanol, mp 180°. $[\alpha]_D^{22}$ −42.5° (c = 0.8 in chloroform).
dl-Form, stout short prisms, mp 190°.

165. Adonis vernalis. False hellebore; vernal pheasant's eye; bird's eye. Herb of *Adonis vernalis* L., *Ranunculaceae. Habit.* Lime-rich soil of Eastern and Southern Europe, especially Hungary and Serbia. Also found in Labrador, Asian U.S.S.R. *Constit.* Adonidoside, adonitoxin *q.v.;* acetyladonitoxin *see* adonitoxin; adonivernoside; adonitol *q.v.;* adonivernith; vernadin; strophanthidin *q.v.;* 16-hydroxystrophanthidin *see* strophanthidin; vernadigin; aconitic acid (?) *q.v.;* 2,6-dimethoxyquinone *q.v.;* phytosterols *q.v.;* choline *q.v.;* resin. Physiologically active extracts are known as *Adovern; Adonilen; Adonisid; Adonigen; Adonival.* Monograph: Jean Sassard, *Essai sur les propriétés et les modes d'application de l'Adonis vernalis* (Lyon 1930); Heyl *et al., J. Am. Chem. Soc.* **40,** 436 (1918); Karrer, *Festschrift E. C. Barell* **1936,** 240; Meyer, *Pharmazie* **4,** 432 (1949); Reinhard, Hufschmidt, *Deut. Apoth. Ztg.* **98,** 1197 (1958).
THERAP CAT: Cardiotonic.

166. Adonitol. *Ribitol;* adonite. $C_5H_{12}O_5$; mol wt 152.15. C 39.47%, H 7.95%, O 52.58%. A pentitol from *Adonis vernalis* L., *Ranunculaceae;* also from *A. amurensis* L., *Ranunculaceae:* Santavy, Reichstein, *Pharm. Acta Helv.* **23,** 153 (1948).

Large, optically inactive crystals. mp 102°. Does not reduce Fehling's soln. Freely soluble in water, hot alcohol; insol in ether.

167. Adonitoxin. *(3β,5β,16β)-3-[(6-Deoxy-α-L-mannopyranosyl)oxy]-14,16-dihydroxy-19-oxocard-20(22)-enolide.* $C_{29}H_{42}O_{10}$; mol wt 550.65. C 63.26%, H 7.69%, O 29.06%. From *Adonis vernalis* L., *Ranunculaceae:* Katz, Reichstein, *Pharm. Acta Helv.* **22,** 437 (1947); *C.A.* **43,** 1790i (1949); Pitra, Cekan, *Coll. Czech. Chem. Commun.* **26,** 1551 (1961). Structure: Poláková, Cekan, *Chem. & Ind. (London)* **1963,** 1766.

Crystals from chloroform + methanol + ether, decomp 251-258°.
3-*O*-Acetyladonitoxin, $C_{31}H_{44}O_{11}$, needles from dil methanol, prisms from methanol+ ether, decomp 213-219°. $[\alpha]_D^{22}$ −19.9° (c = 0.33 in methanol).

168. Adrafinil. *2-[(Diphenylmethyl)sulfinyl]-N-hydroxyacetamide;* 2-(benzhydrylsulfinyl)acetohydroxamic acid; CRL-40028; Olmifon. $C_{15}H_{15}NO_3S$; mol wt 289.36. C 62.26%, H 5.23%, N 4.84%, O 16.59%, S 11.08%. α-Adrenergic agonist. Prepn, pharmacology: L. Lafon, **Belg.** pat. **846,880;** U.S. pat. **4,066,686** (1977, 1978 both to Lafon). Mode of action: J. Duteil *et al., Eur. J. Pharmacol.* **59,** 121 (1979); C. Rozé *et al., Arch. Int. Pharmacodyn. Ther.* **265,** 119 (1983). Psychopharmacology in mice: F. A. Rambert *et al., J. Pharmacol.* **17,** 37 (1986).

Crystals from ethyl acetate-isopropyl alcohol, mp 159-160°. Soly in water < 1 g/l. LD_{50} in mice (mg/kg): < 2048 i.p.; 1950 gastric admin (Lafon).
THERAP CAT: Treatment of depression.

169. Adrenal Cortical Extract. ACE; Corticopan; Cortidin; Cortidyn; Cortine Naturelle; Cortisural; Escatin; Eschatin; Eucortone; Maxicortex; Mencortex; Pancorton; Supracort. Isoln and purification: Swingle, Pfiffner, U.S. pats. **2,074,492** and **2,074,493** (both 1937 to Parke, Davis).

THERAP CAT: Corticosteroid.

THERAP CAT (VET): Has been used for bovine ketosis.

170. Adrenalone. *1-(3,4-Dihydroxyphenyl)-2-(methyl-amino)ethanone; 3′,4′-dihydroxy-2-(methylamino)acetophe-none;* 3,4-dihydroxy-α-methylaminoacetophenone; adre-none; 4-methylaminocetopyrocatechol; Kephrine; Stryph-none; Stypnone. $C_9H_{11}NO_3$; mol wt 181.19. C 59.66%, H 6.12%, N 7.73%, O 26.49%. Intermediate in some epineph-rine manuf processes. Prepd from 4-chloroacetylpyrocate-chol and methylamine: **Ger. pat. 152,814** (Hoechst); Stolz, *Ber.* **37**, 4152 (1904); by heating α-(p-toluene sulfonylmeth-ylamino)-3,4-dimethoxyacetophenone with HCl under pres-sure: **Ger. pat. 277,540** (Bayer); *Frdl.* **12**, 764; *Chem. Zentr.* **1914**, II, 740.

Needles, dec 235-236°. Sparingly sol in water, alc, ether. Hydrochloride, $C_9H_{11}NO_3.HCl$, *Stryphnonasal.* Crystals, mp 243°. Freely sol in water; sol in alcohol. Insol in ether. Aq solns are neutral to litmus.

THERAP CAT: Hemostatic.

171. Adrenochrome. *2,3-Dihydro-3-hydroxy-1-methyl-1H-indole-5,6-dione; 3-hydroxy-1-methyl-5,6-indolinedione.* $C_9H_9NO_3$; mol wt 179.18. C 60.33%, H 5.06%, N 7.82%, O 26.79%. Pigment obtained by the oxidation of epinephrine. Prepn using Ag_2O as oxidizing agent: MacCarthy, *Chim. Ind. Paris* **55**, 435 (1946). Structure: D. E. Green, D. Richter, *Biochem. J.* **31**, 596 (1937); J. Harley-Mason, *Ex-perientia* **4**, 307 (1948). Probably has the zwitterionic quin-onimine structure shown here. Spectral data: A. Lund, *Acta Pharmacol.* **5**, 121 (1949). *Review:* R. A. Heacock, *Chem. Rev.* **59**, 181-237 (1959). Review, as a psychoto-mimetic agent: R. A. Heacock, *Chim. Ther.* **6**, 300 (1971).

Brilliant red crystals may exist as hemihydrate from methanol-formic acid, dec 115-120°. Absorption max (1 mg % aq sol): 220, 300, 485 nm (log ε 4.33, 4.01, 3.64). Well-formed and well-dried crystals can be kept in a vacuum desiccator for several weeks. Easily oxidized to melanin. Freely sol in water; fairly sol in alcohol. Almost insol in benzene and ether. Solns are unstable. Optimum pH of water soln 4.0.

Oxime sesquihydrate, $C_9H_{10}N_2O_3.1\frac{1}{2}H_2O$, orange needles from water, mp 278°. Much more stable than adreno-chrome.

Monosemicarbazone, $C_{10}H_{12}N_4O_3$, *carbazochrome, Adren-oxyl, Cromadrenal, Cromosil.* Prepn: Dechamps *et al.*, U.S. pat. **2,506,294** (1950 to Société Belge de l'Azote). Orange-red crystals from dil alcohol, mp 203° (dec). *See also* Carb-azochrome Salicylate.

Thiosemicarbazone, $C_{10}H_{12}N_4O_2S$, mp 215-220°. Descrip-tion: Fleischhacker, Barsel, U.S. pat. **2,712,024** (1955 to International Hormones).

THERAP CAT: Monosemicarbazone as hemostatic.

172. Adrenoglomerulotropin. *2,3,4,9-Tetrahydro-6-methoxy-1-methyl-1H-pyrido[3,4-b]indole;* aldosterone-stimulating hormone; ASH; 1-methyl-6-methoxy-1,2,3,4-tetrahydro-2-carboline. $C_{13}H_{16}N_2O$; mol wt 216.28. C 72.19%, H 7.46%, N 12.95%, O 7.40%. Found in extracts of pineal gland tissue: Farrell, *Circulation* **21**, 1009 (1960); Farrell, McIsaac, *Arch. Biochem. Biophys.* **94**, 543 (1961). Synthesis from 5-methoxytryptamine and acetaldehyde:

McIsaac, *Biochim. Biophys. Acta* **52**, 607 (1961); Meek *et al.*, *Chem. & Ind. (London)* **1964**, 622.

Crystals, mp 150-151°. uv max (ethanol): 225, 280 nm (log ε 4.34, 3.86); (0.1N HCl): 220, 273 nm (log ε 4.40, 3.86).

173. Adrenolutin. *1-Methyl-1H-indole-3,5,6-triol;* N-methyl-5,6-dihydroxyindoxyl; 3,5,6-trihydroxy-N-meth-ylindole. $C_9H_9NO_3$; mol wt 179.18. C 60.33%, H 5.06%, N 7.82%, O 26.79%. Alkaline rearrangement product of adrenochrome. Isoln and identification: Lund, *Acta Phar-macol. Toxicol.* **5**, 75 (1949). Structure and synthesis: Bal-singer *et al.*, *Helv. Chim. Acta* **36**, 708 (1953); Heacock, Mahon, *Can. J. Chem.* **36**, 1550 (1958). NMR studies and its existence in the ketonic form in solution: Powell, Hea-cock, *Chim. Ther.* **7**, 133 (1972).

Monohydrate, bright yellow prisms from water, mp 236° (dec); mp 195° (Balsinger). Anhydrous, bright yellow prisms, mp 245°.

174. Adrenosterone. *Androst-4-ene-3,11,17-trione;* Reichstein's substance G. $C_{19}H_{24}O_3$; mol wt 300.40. C 75.97%, H 8.05%, O 15.98%. Prepn: Reichstein, *Helv. Chim. Acta* **19**, 29, 1107 (1936); Kendall *et al.*, *J. Biol. Chem.* **116**, 267 (1936); Reichstein, *Helv. Chim. Acta* **20**, 817, 953 (1937). Total synthesis: Velluz *et al.*, *Compt. Rend.* **250**, 1293 (1960). Crystal structure: Ohrt *et al.*, *Acta Cryst.* **19**, 479 (1965). Metabolism: Bradlow *et al.*, *Steroids* **10**, 233 (1967).

Needles from alcohol, mp 220-224°. Sublimes in high vacuum. Soly in water: 9.85 (23°), 15.2 (37°) mg/100 ml. Sol in alcohol, acetone, ether. $[\alpha]_D^{20}$ +262° (abs alcohol); $[\alpha]_{5461}^{25}$ +364° ± 5° (c = 0.18 in abs alc). uv max: 235 nm.

175. Aequorin. Photoprotein from the luminescent jellyfish *Aequorea*, mol wt 20,000 Da. Contains a single chromophore, *coelentrazine,* which emits a blue light via an intramolecular reaction when traces of Ca^{2+} are present. The protein is converted to *apoaequorin, coelenteramide,* and CO_2 with the emission of light. Isolation and purifica-tion: O. Shimomura *et al., J. Cell. Comp. Physiol.* **59**, 223 (1962). Mechanism of luminescence: O. Shimomura *et al., Biochemistry* **13**, 3278 (1974). Verification of coelentrazine as chromophore: S. Inoue *et al., Chem. Letters* **1975**, 141. Characterization and properties of isoforms: O. Shimo-mura, *Biochem. J.* **234**, 271 (1986); of semisynthetic aequor-in: O. Shimomura *et al., ibid.* **251**, 405 (1988); and recombi-nant forms: *eidem, ibid.* **270**, 309 (1990). Use in measure-ment of cellular calcium: A. Azzi, B. Chance, *Biochim. Biophys. Acta* **189**, 141 (1969); J. M. Kendall *et al., Anal. Biochem.* **221**, 173 (1994); M. N. Badminton *et al., Exptl. Cell Res.* **216**, 236 (1995). Review: O. Shimomura in *Natu-*

ral Products and Biological Activities, H. Imura et al., Eds (Elsevier Science Publishers, Amsterdam, 1986) p 33-44.

Sol in aq buffers > 30 mg/ml. $E_{1cm}^{1\%}$ at 280 nm: 27.0; $E_{1cm}^{1\%}$ at 460 nm: 0.81. pI 4.2-4.9.

USE: For quantitation of free calcium in biological systems.

176. AET. *Carbamimidothioic acid 2-aminoethyl ester dihydrobromide; 2-(2-aminoethyl)-2-thiopseudourea dihydrobromide; S-(2-aminoethyl)isothiuronium bromide hydrobromide; β-aminoethylisothiuronium bromide hydrobromide; Antirad; Surrectan; Antiradon.* $C_3H_{11}Br_2N_3S$; mol wt 281.01. C 12.82%, H 3.95%, Br 56.87%, N 14.95%, S 11.41%. Prepd by refluxing thiourea with 2-bromoethylamine HBr in isopropanol: Clinton et al., J. Am. Chem. Soc. **70**, 950 (1948); Funahashi, Miyano, J. Agr. Chem. Soc. Japan **27**, 775 (1953), C.A. **49**, 15737b (1955); Doherty et al., J. Am. Chem. Soc. **79**, 5667 (1957). Toxicity study: J. Pospisil et al., Cas. Lekaru Cesk. **105**, 1165 (1966), C.A. **67**, 72261s (1967).

Crystals from abs ethanol + ethyl acetate, mp 194-195°. Hygroscopic, cakes together, and is converted in significant amounts of 2-aminothiazoline, a transformation that can be detected by a drop in melting point by as much as 30°. The animal organism appears to convert it to 2-mercaptoethylguanidine hydrobromide. LD_{50} in mice (mg/kg): 100 i.v., 280 s.c., 480 i.p., 1600 orally (Pospisil).

THERAP CAT: Radioprotective agent.

177. Affinin. *(E,E,Z)-N-(2-Methylpropyl)-2,6,8-decatrienamide; (E,E,Z)-N-isobutyl-2,6,8-decatrienamide; N-isobutyldeca-trans-2-cis-6-trans-8-trienamide; spilanthol.* $C_{14}H_{23}NO$; mol wt 221.34. C 75.97%, H 10.47%, N 6.33%, O 7.23%. Insecticidal lipid amide isolated from Heliopsis longipes (A. Gray) Blake, Compositae. Isoln and structure: Acree et al., J. Org. Chem. **10**, 236, 449 (1945). Identity with spilanthol: Jacobson, Chem. & Ind. (London) **1957**, 50. Stereochemistry and synthesis: Crombie, Krasinski, ibid. **1962**, 983; Crombie et al., J. Chem. Soc. **1963**, 4970.

Pale yellow viscous oil, $bp_{0.2}$ 141°; $bp_{0.3-0.5}$ 160-165°. mp 23°. n_D^{25} 1.5134. Soluble in organic solvents. Practically insoluble in aq alkali and acid. uv max (95% ethanol): 228.5 nm (ε 33700).

USE: Insecticide synergist.

178. Aflatoxins B. Aflatoxins are a closely related group of secondary fungal metabolites shown to be mycotoxins. They are produced by Aspergillus flavus Link ex Fries, the causative principle of turkey "X" disease; and by Aspergillus parasiticus: Sargeant et al., Nature **192**, 1096 (1961); Hesseltine et al., Proc. 1st U.S. Japan Conf. Toxic Microorganisms, Honolulu, Hawaii 1968, p 202. Aflatoxins have been reported to naturally occur in peanuts, peanut meal, cottonseed meal, corn, dried chili peppers etc. However, the growth of the mold itself does not always indicate the presence of toxin since the yield of aflatoxin depends on growth conditions such as moisture, temperature, substrates, and aeration as well as genetic requirements. These heterocycles are now characterized as aflatoxins B_1, B_2, G_1, G_2, M_1 and M_2 (milk toxins) and B_{2a}, G_{2a}: Büchi, Rae in Aflatoxin, L. Goldblatt, Ed. (Academic Press, New York, 1969) pp 55-75. Toxic material is separated chromatographically into 4 distinct compounds based on fluorescent color (blue = B, green = G with subscripts relating to relative mobility): Nesbitt et al., Nature **195**, 1062 (1962); Sargeant et al., Chem. & Ind. (London) **1963**, 53. B_1 is one of the most potent environmental mutagens and carcinogens

known. B_2 and G_2 are the less toxic dihydro derivs of B_1 and G_1: Asao et al., J. Am. Chem. Soc. **85**, 1706 (1963); **87**, 882 (1965), and B_{2a} and G_{2a} are the 2-hydroxy derivs of B_2 and G_2: Dutton, Heathcote, Biochem. J. **101**, 21P (1967); Chem. & Ind. (London) **1968**, 418. Isoln of B_3 (parasiticol), a possible metabolite of G_1, from A. flavus: J. G. Heathcote, M. F. Dutton, Tetrahedron **25**, 1497 (1969). Aflatoxins R_0 (aflatoxicol), P_1, Q_1, RB_1, RB_2, and D_1 are also known: P. F. Schuda, Topics in Current Chemistry **91**, 77-106 (1980). Total syntheses of aflatoxins B_1 and G_1: Buchi et al., J. Am. Chem. Soc. **88**, 4534 (1966); Knight et al., Chem. Commun. **1966**, 706; Büchi, Weinreb, J. Am. Chem. Soc. **93**, 746 (1971). The extreme toxicity and carcinogenicity of aflatoxins may be due to their inhibition of nucleic acid synthesis by either direct interaction with enzymes involved or by a toxin-DNA template: Clifford, Rees, Nature **209**, 312 (1966); Sporn et al., Science **151**, 1539 (1966). See also Wogan et al., Food Cosmet. Toxicol. **12**, 681 (1974). Inhibition of salt-induced conversion of B-DNA to Z-DNA by aflatoxin B_1: A. Nordheim et al., Science **219**, 1434 (1983). Prepn of the exo-8,9-epoxide of B_1, the metabolic deriv thought to be responsible for B_1's carcinogenicity: S. W. Baertschi et al., J. Am. Chem. Soc. **110**, 7929 (1988). uv spectrum of B_2: Hartley et al., Nature **198**, 1056 (1963). Physicochemical data for B_1: A. E. Pohland et al., Pure Appl. Chem. **54**, 2219 (1982). Toxicity data: Carnaghan et al., ibid. **200**, 1101 (1963); G. Büchi et al., Life Sci. **13**, 1143 (1973). Symposium on toxicology and synthesis: J. Toxicol. Toxin Rev. **8**, 1-416 (1991). Review and evaluation of studies of carcinogenic action in laboratory animals and humans: IARC Monographs **10**, 51-72 (1976). Comprehensive reviews: Goldblatt, Econ. Bot. **22**, 51-62 (1968); Detroy et al., "Aflatoxin and Related Compounds" in Microbial Toxins, Vol. VI, A. Ciegler et al., Eds. (Academic Press, New York, 1971) pp 3-178; W. F. Busby, Jr., G. N. Wogan, "Aflatoxins" in Chemical Carcinogens, C. E. Searle, Ed., A.C.S. Monograph Series, no. **182** (American Chemical Society, Washington DC, 1984) pp 945-1136. Review of chemistry and syntheses: P. F. Schuda, loc. cit.; of biosynthesis: M. W. Sinz, W. T. Shier, J. Toxicol. Toxin Rev. **10**, 87-121 (1991).

Aflatoxin B₁

Aflatoxin B_1, $C_{17}H_{12}O_6$, (6aR-cis)-2,3,6a,9a-tetrahydro-4-methoxycyclopenta[c]furo[3',2':4,5]furo[2,3-h][1]benzopyran-1,11-dione. Crystals, mp 268-269°. Exhibits blue fluorescence. $[\alpha]_D$ −558° (c = 0.1 in $CHCl_3$). $[\alpha]_D$ −480° (c = 0.1 in DMF). uv max (ethanol): 223, 265, 362 nm (ε 25600, 13400, 21800). LD_{50} orally in day old duckling: 18.2 μg/50 gm body wt (Carnaghan); i.p. in newborn mice: 9.50 mg/kg body wt (Büchi).

Aflatoxin B_2, $C_{17}H_{14}O_6$, (6aR-cis)-2,3,6a,8,9,9a-hexahydro-4-methoxycyclopenta[c]furo[3',2':4,5]furo[2,3-h][1]benzopyran-1,11-dione. The 8,9-dihydro deriv of aflatoxin B_1. Crystals, mp 286-289°. Exhibits blue fluorescence. $[\alpha]_D$ −492° (c = 0.1 in $CHCl_3$). uv max (ethanol): 265, 363 nm (ε 11700, 23400). LD_{50} orally in day old duckling: 84.8 μg/50 gm body wt (Carnaghan).

Note: The aflatoxins have been listed as known carcinogens: Seventh Annual Report on Carcinogens (PB95-109781, 1994) p 17.

179. Aflatoxins G. Toxic metabolites of fungi Aspergillus flavus Link ex Fries and Aspergillus parasiticus. Total synthesis of G_1: Büchi, Weinreb, J. Am. Chem. Soc. **93**, 746 (1971). Physical and chemical data: Hartley et al., Nature **198**, 1056 (1963); Asao, J. Am. Chem. Soc. **87**, 882 (1965). Toxicity data: R. B. A. Carnaghan et al., Nature **200**, 1101 (1963). For general refs see Aflatoxins B.

Aflatoxin G₁

Aflatoxin G₁. $C_{17}H_{12}O_7$. *3,4,7aα,10aα-Tetrahydro-5-methoxy-1H,12H-furo[3',2':4,5]furo[2,3-h]pyrano[3,4-c][1]benzopyran-1,12-dione.* Crystals, mp 244-246°. Exhibits green fluorescence. $[\alpha]_D$ −556° (chloroform). uv max (ethanol): 243, 257, 264, 362 nm (ε 11500, 9900, 10000, 16100). LD_{50} orally in day old duckling: 39.2 μg/50 gm body wt (Carnaghan).

Aflatoxin G₂. $C_{17}H_{14}O_7$. *3,4,7aα,9,10,10aα-Hexahydro-5-methoxy-1H,12H-furo[3',2':4,5]furo[2,3-h]benzopyran-1,12-dione.* The 9,10-dihydro deriv of aflatoxin G₁. Crystals, mp 237-240°. Exhibits green-blue fluorescence. $[\alpha]_D$ −473° (c = 0.084 in chloroform). uv max (ethanol): 265, 363 nm (ε 9700, 21000). LD_{50} orally in day old duckling: 172.5 μg/50 gm body wt (Carnaghan).

Note: The aflatoxins have been listed as known carcinogens: *Seventh Annual Report on Carcinogens* (PB95-109781, 1994) p 17.

180. Aflatoxins M. Highly toxic 4-hydroxylated aflatoxin B derivs found in the milk of cows fed toxic meal. Initial reports: Allcroft, Carnaghan, *Vet. Rec.* **75**, 259 (1963). Isoln: De Iongh *et al., Nature* **202**, 466 (1964); Allcroft *et al., ibid.* **209**, 154 (1966). Isoln and structure of M₁ and M₂: C. W. Holzapfel, P. S. Steyn, *Tetrahedron Letters* **1966**, 2799. Total synthesis of M₁: Büchi, Weinreb, *J. Am. Chem. Soc.* **91**, 5408 (1969); **93**, 746 (1971); G. Büchi *et al., ibid.* **103**, 3497 (1981). Carcinogenicity studies: Wogan, Paglialunga, *Food Cosmet. Toxicol.* **12**, 381 (1974). For general refs *see* Aflatoxins B.

Aflatoxin M₁

Aflatoxin M₁. $C_{17}H_{12}O_7$. *2,3,6a,9a-Tetrahydro-9a-hydroxy-4-methoxycyclopenta[c]furo[3',2':4,5]furo[2,3-h][1]benzopyran-1,11-dione; 4-hydroxyaflatoxin B₁.* Crystals from methanol, mp 299° (dec). Exhibits blue-violet fluorescence. $[\alpha]_D$ −280° (c = 0.1 in DMF). uv max (ethanol): 226, 265, 357 nm (ε 23100, 11600, 19000). LD_{50} orally in day old Pekin ducklings: 16.6 μg/duckling (Holzapfel, Steyn).

Aflatoxin M₂. $C_{17}H_{14}O_7$. *2,3,6a,8,9,9a-Hexahydro-9a-hydroxy-4-methoxycyclopenta[c]furo[3',2':4,5]furo[2,3-h][1]benzopyran-1,11-dione; 4-hydroxyaflatoxin B₂.* The 8,9-dihydro deriv of aflatoxin M₁. Exhibits violet fluorescence. Crystals from methanol-chloroform, mp 293° (dec). uv max (ethanol): 221, 264, 357 nm (ε 20000, 10900, 21000). LD_{50} orally in day old Pekin ducklings: 62 μg/duckling (Holzapfel, Steyn).

Note: The aflatoxins have been listed as known carcinogens: *Seventh Annual Report on Carcinogens* (PB95-109781, 1994) p 17.

181. Afloqualone. *6-Amino-2-(fluoromethyl)-3-(2-methylphenyl)-4(3H)-quinazolinone;* 6-amino-2-fluoromethyl-3-(o-tolyl)-4(3H)-quinazolinone; HQ-495; Aroft; Arofuto. $C_{16}H_{14}FN_3O$; mol wt 283.31. C 67.83%, H 4.98%, F 6.71%, N 14.83%, O 5.65%. A centrally acting muscle relaxant. Prepn: I. Inoue *et al., Ger.* pat. **2,449,113**; *eidem,* U.S. pat. **3,966,731** (1975, 1976 to Tanabe); J. Tani *et al., J. Med. Chem.* **22**, 95 (1979). Pharmacology: T. Ochiai, R. Ishida,

Japan. J. Pharmacol. **31**, 491 (1981); **32**, 427 (1982). Metabolism: N. Otsuka *et al., J. Pharmacobio-Dyn.* **5**, S-59 (1982); S. Furuuchi *et al., Drug Metab. Dispos.* **11**, 371 (1983).

Pale yellow prisms from 2-propanol, mp 195-196°. LD_{50} in mice (mg/kg): 315.1 i.p. (Tani).

THERAP CAT: Muscle relaxant (skeletal).

182. Agar. Agar-agar; gelose; Japan agar; Bengal isinglass; Ceylon isinglass; Chinese isinglass; Japan isinglass; Layor Carang. A polysaccharide complex extracted from the agarocytes of algae of the *Rhodophyceae.* Predominant agar-producing genera are *Gelidium, Gracilaria, Acanthopeltis, Ceramium, Pterocladia* found in the Pacific and Indian Oceans and Japan Sea. Can be separated into a neutral gelling fraction, *agarose,* and a sulfated non-gelling fraction, *agaropectin:* Araki, *J. Chem. Soc. Japan* **58**, 1338 (1937). Structure believed to be a complex range of polysaccharide chains having alternating α-(1→3) and β-(1→4) linkages and varying in total charge content; three extremes of structure noted, namely neutral agarose, pyruvated agarose having little sulfation, and a sulfated galactan: Duckworth *et al., Carbohyd. Res.* **16**, 189, 435, 446 (1971). *Reviews:* V. J. Chapman, *Seaweeds and Their Uses* (Pitman Publ., New York, 1952) pp 89-123; Humm, *Econ. Bot.* **1**, 317 (1947); Mori, *Advan. Carbohyd. Chem.* **8**, 317 (1953); Selby, Wynne, in *Industrial Gums,* R. L. Whistler, Ed. (Academic Press, N.Y., 2nd ed., 1973) pp 29-48.

Transparent, odorless, tasteless strips or coarse or fine powder. Insol in cold water, alc; slowly sol in hot water to a viscid soln. A 1% soln forms a stiff jelly on cooling.

USE: Substitute for gelatin, isinglass, etc. in making emulsions including photographic, gels in cosmetics, and as thickening agent in foods esp. confectionaries and dairy products; in meat canning; in production of medicinal encapsulations and ointments; as dental impression mold base; as corrosion inhibitor; sizing for silks and paper; in the dyeing and printing of fabrics and textiles; in adhesives. In nutrient media for bacterial cultures.

THERAP CAT: Cathartic.

THERAP CAT (VET): Laxative in dogs, cats. Demulcent.

183. Agaric. Larch agaric; touch wood; white agaric; purging agaric; amadou; German fungus. The dried fruit body of *Fomes laricis* (Jacq.) Murrill *(Polyporus officinalis* Fries), *Polyporaceae,* deprived of its outer rind. *Habit.* European and Asiatic Russia. Grows upon various species of *Pinus, Larix,* and *Picea. Constit.* 14-16% Agaric acid; agaricoresin, agaricol, phytosterin, ricinoleic acid, cetyl alcohol, glucose, malic acid, carbohydrates. Isoln and separation of some constituents: Valentin, Knütter, *Pharm. Zentralhalle* **96**, 478 (1957). Impregnated with potassium nitrate and dried, it constitutes punk or tinder.

Light, fibrous, grayish-white to pale brown, spongy, friable pieces of irregular shape; feeble odor and bitter, acrid, yet somewhat sweetish taste.

USE: Anhidrotic.

184. Agaric Acid. *2-Hydroxy-1,2,3-nonadecanetricarboxylic acid;* agaric acid; agaricin; agaricinic acid; laricic acid; n-hexadecylcitric acid; α-cetylcitric acid. $C_{22}H_{40}O_7$; mol wt 416.56. C 63.44%, H 9.68%, O 26.89%. Active principle of *Fomes laricis* (Jacq.) Murrill *(Polyporus officinalis* Fries), *Polyporaceae:* Thoms, Vogelsang, *Ann.* **357**, 145 (1907). Attempted syntheses: Evans, *J. Chem. Soc.* **1959**, 1313; Graf, Liu, *Arch. Pharm.* **306**, 366 (1973). Activity studies: Bacchi *et al., J. Bacteriol.* **98**, 23 (1969).

Sesquihydrate, odorless, almost tasteless, crystalline powder. Anhydrous, mp 142° (dec). $[\alpha]_D^{19}$ $-9°$ (NaOH). Slightly sol in cold water, chloroform or ether; freely sol in boiling water, alkalies, hot glacial acetic acid. One gram dissolves in 180 ml cold, 10 ml boiling alcohol.

Note: The name agaric acid has also been used for a lanostane-like mixture [*C.A.* **68**, 29894j (1968)].

USE: Has been used as antiperspirant.

185. Agaritine. L-*Glutamic acid 5-[2-[4-(hydroxymethyl)phenyl]hydrazide]; β-N-[γ-L(+)-glutamyl]-4-hydroxymethylphenylhydrazine.* $C_{12}H_{17}N_3O_4$; mol wt 267.28. C 53.92%, H 6.41%, N 15.72%, O 23.94%. Constituent of the commercial, edible mushroom *Agaricus bisporus* (Lange) Sing. [*Psalliota hortensis* Cooke var. *bispora*], Agaricaceae (Agaricales). Isoln and structure: Levenberg, *Fed. Proc.* **19**, 6 (1960); *J. Am. Chem. Soc.* **83**, 503 (1961); Daniels *et al.*, *ibid.* 3333; *see also* Levenberg in *Methods Enzymol.* **17** (Part A), 877 (1970). Synthesis: Kelly *et al.*, *J. Org. Chem.* **27**, 3229 (1962); Hinman, Kelly, U.S. pats. **3,274,232; 3,288,-848** (both 1966 to Upjohn).

Glistening crystals from dil alcohol, dec 205-209°. $[\alpha]_D^{25}$ 17° (c = 0.8). uv max (water): 237.5, 280 nm (ε 12,000, 1,400). Very freely sol in water, practically insol in the usual anhydr organic solvents. pKa in water: 3.4 and 8.86.

186. Agmatine. 4-*(Aminobutyl)guanidine;* 1-amino-4-guanidobutane. $C_5H_{14}N_4$; mol wt 130.19. C 46.13%, H 10.84%, N 43.03%. Decarboxylated arginine. Found in pollen of *Ambrosia artemisifolia* L., *Compositae*, in ergot, in sponges, in herring sperm, in octopus muscle: F. W. Heyl, *J. Am. Chem. Soc.* **41**, 670 (1919); A. Kossel, *Z. Physiol. Chem.* **66**, 257 (1910); J. L. Irvin, D. W. Wilson, *J. Biol. Chem.* **127**, 565 (1939). Prepn of salts: **Ger. pat. 463,576** (1928 to Schering-Kahlbaum AG). Synthesis: A. Kossel, *Z. Physiol. Chem.* **68**, 170 (1910); K. Odo, *J. Chem. Soc. Japan* **67**, 132 (1946); K. Dose, *Ber.* **90**, 1251 (1957). Use as marker for tumor cells: F. S. Steven *et al.*, *Anticancer Res.* **9**, 247 (1989); F. S. Steven *et al.*, *J. Enzyme Inhibition* **4**, 63 (1990).

Sulfate, $C_5H_{14}N_4 \cdot H_2SO_4$, needles from dil methanol, mp 231°. Fairly sol in water. Nearly insol in alcohol.

Gold chloride, $C_5H_{14}N_4 \cdot 2HCl \cdot 2AuCl_3$, yellow needles from water, dec 223°.

USE: Biochemical probe

187. Agroclavine. 8,9-*Didehydro-6,8-dimethylergoline.* $C_{16}H_{18}N_2$; mol wt 238.33. C 80.63%, H 7.61%, N 11.75%. A non-peptide ergot alkaloid obtained from cultures of fungi parasitic on *Elymus mollis* Trin.: Abe *et al.*, **Japan. pat. 178,336('49)** (to Takeda), *C.A.* **45**, 6352c (1951); *Ann. Rep. Takeda Res. Lab.* **10**, 145, 167, 171 (1951); **Japan. pat. 7498('54);** *C.A.* **50**, 6000[b] (1956); **U.S. pat. 2,835,675** (1958). Found in fungi parasitic on *Pennisetum typhoideum* Rich.: Stoll *et al.*, *Helv. Chim. Acta* **37**, 1815 (1954). Structure and stereochemistry: Schreier, *ibid.* **41**, 1984 (1958). Biosynthesis: Floss *et al.*, *J. Am. Chem. Soc.* **90**, 6500 (1968). Synthesis: Plieninger *et al.*, *Ann.* **743**, 95 (1971). Metabolism: Ramstad, *Lloydia* **31**, 327 (1968).

Rods from ether, dec 198-203°; needles from acetone, dec 205-206°. $[\alpha]_D^{20}$ $-155°$ (c = 0.9 in chloroform); $[\alpha]_D^{20}$ $-182°$ (c = 0.5 in pyridine). uv max: 225, 284, 293 nm (ε 4.47, 3.88, 3.81). Freely sol in alc, chloroform, pyridine; sol in benzene, ether; very slightly sol in water.

188. Agrocybin. 8-*Hydroxy-2,4,6-octatriynamide.* $C_8H_5NO_2$; mol wt 147.13. C 65.31%, H 3.43%, N 9.52%, O 21.75%. $CH_2OHC\equiv C-C\equiv C-C\equiv C-CONH_2$. Antibiotic substance produced by the basidiomycete *Agrocybe dura*: Kavanagh *et al.*, *Proc. Nat. Acad. Sci. USA* **36**, 102 (1950). Structure and synthesis: Bu'Lock *et al.*, *Chem. & Ind. (London)* **1954**, 990; Ashworth *et al.*, *J. Chem. Soc.* **1958**, 950. Origin of the carbon skeleton: E. R. H. Jones *et al.*, *J. Chem. Res. (M)* **1977**, 744.

Crystals from 20% alc or ether, mp 140° (conflagrates). Stable in air for about one day after which the crystals turn black and become insol in ether. Sol in alc, acetone, ether, chloroform, methyl isobutyl ketone; slightly sol in water; practically insol in hexane. Absorption spectrum: Ashworth *et al.*, *loc. cit.* Has antifungal activity. Causes dermatitis. Because of high toxicity and inactivation by blood no therapeutic tests have been attempted.

189. Ahistan. 10-*(Dimethylamino)acetyl-10H-phenothiazine; 10-(N,N-dimethylglycyl)phenothiazine.* $C_{16}H_{16}N_2$-OS; mol wt 284.38. C 67.58%, H 5.67%, N 9.85%, O 5.63%, S 11.28%. Prepd from phenothiazine and chloroacetyl chloride, followed by treatment with dimethylamine: Cusic, U.S. pat. **2,694,705** (1954 to Searle).

Crystals, mp 144-145°.

Hydrochloride, $C_{16}H_{16}N_2OS \cdot HCl$. Crystals, mp 230-231°.

THERAP CAT: Antihistaminic.

190. AICAR. 5-*Amino-1-(5-O-phosphono-β-D-ribofuranosyl)-1H-imidazole-4-carboxamide; 5-amino-1-ribofuranosylimidazole-4-carboxamide 5'-phosphate;* 5-amino-4-imidazolecarboxamide ribonucleotide; 5-amino-4-imidazolecarboxamide ribotide; 5-amino-1-(5'-phosphofuranoribosyl)-4-imidazolecarboxamide. $C_9H_{15}N_4O_8P$; mol wt 338.21. C 31.96%, H 4.47%, N 16.57%, O 37.84%, P 9.16%. Synthesis from AIR: Lukens, Buchanan, *J. Am. Chem. Soc.* **79**, 1511 (1957); *J. Biol. Chem.* **234**, 1791 (1959).

Barium salt, $C_9H_{13}BaN_4O_8P$. uv max (pH 7.0): 269 nm (ε 12600); *see* Flaks *et al.*, *J. Biol. Chem.* **228**, 201 (1957). In the presence of Bratton-Marshall reagents [*ibid.* **128**, 537

(1939)] AICAR is converted to a purple dye; absorption max: 540 nm (ϵ 26400).

191. AIR. *1-(5-O-Phosphono-β-D-ribofuranosyl)-1H-imidazol-5-amine; 5-amino-1-ribofuranosylimidazole 5'-phosphate;* 5-aminoimidazole ribonucleotide; 5-aminoimidazole ribotide; 5-amino-1-(5'-phosphofuranoribosyl)imidazole. $C_8H_{14}N_3O_7P$; mol wt 295.19. C 32.55%, H 4.78%, N 14.24%, O 37.94%, P 10.49%. An intermediate in the biosynthesis of purines. Synthesis from formylglycinamide ribonucleotide: Levenberg, Buchanan, *J. Biol. Chem.* **224**, 1005 (1957); Lukens, Buchanan, *ibid.* **234**, 1791 (1959).

Isolated in soln only or as the barium salt. The ultraviolet absorption spectrum at pH 7 shows no selective absorption between 215 and 300 nm, although a moderately strong general absorption is evident below 250 nm. In the presence of Bratton-Marshall reagents [*J. Biol. Chem.* **128**, 537 (1939)] AIR is converted into an orange chromophor (absorption max 500 nm).

192. Ajacine. *4-[[[2-(Acetylamino)benzoyl]oxy]methyl]-20-ethyl-1,6,14,16-tetramethoxyaconitane-7,8-diol;* N-acetylanthranilic acid ester of lycoctonine. $C_{34}H_{48}N_2O_9$; mol wt 628.76. C 64.95%, H 7.69%, N 4.46%, O 22.90%. From seeds of larkspur *(Delphinium ajacis L., Ranunculaceae):* Keller, Völker, *Arch. Pharm.* **251**, 207 (1913); Hunter, *Quart. J. Pharm.* **17**, 302 (1944); Goodson, *J. Chem. Soc.* **1944**, 108; **1945**, 245; from *D. barbeyi:* Cook, Beath, *J. Am. Chem. Soc.* **74**, 1411 (1952). Structure: Kuzovkov, Platonova, *Zh. Obshch. Khim.* **29**, 2782 (1959). For structure *see* Lycoctonine.

Needles from 70% alcohol, mp 154°. $[\alpha]_D^{22}$ +49.5° (c = 2 in abs alcohol); $[\alpha]_D^{16}$ +53° (c = 0.66 in chloroform). uv max: 223, 252, 310 nm (ϵ 28400, 16600, 5400). Slightly sol in water; sol in ether, alcohol with blue fluorescence.

193. Ajaconine. $C_{22}H_{33}NO_3$; mol wt 359.51. C 73.50%, H 9.25%, N 3.90%, O 13.35%. From seeds of larkspur, *Delphinium ajacis L., Ranunculaceae:* Keller, Volker, *Arch. Pharm.* **251**, 207 (1913); Goodson, *J. Chem. Soc.* **1945**, 245. Structure: Dvornik, Edwards, *Tetrahedron* **14**, 54 (1961); Nabors *et al., Tetrahedron Letters* **1969**, 2445. Stereochemistry: Solo, Pelletier, *Chem. & Ind. (London)* **1960**, 1108; Whalley, *Tetrahedron* **18**, 43 (1962). Synthetic studies: Nabors *et al., ibid.* **27**, 2385 (1971). Rearrangement: S. W. Pelletier, N. V. Mody, *J. Am. Chem. Soc.* **101**, 492 (1979).

Prisms from dil alcohol, mp 172°. $[\alpha]_D^{18}$ −119° (c = 2 in anhydr alcohol).
Sulfate heptahydrate, hexagonal plates from dil acetone, mp 113°. $[\alpha]_D^{20}$ +5.5° (c = 2).

194. Ajmaline. *Ajmalan-17,21-diol;* rauwolfine; Gilurytmal; Cardiorythmine; Ritmos; Tachmalin. $C_{20}H_{26}N_2O_2$; mol wt 326.44. C 73.59%, H 8.03%, N 8.58%, O 9.80%. From roots of *Rauwolfia serpentina* (L.) Benth. *(Ophioxylon serpentinum* L.), *Apocynaceae.* Isolation: S. Siddiqui, R. H. Siddiqui, *J. Indian Chem. Soc.* **8**, 667 (1931); **9**, 539 (1932); **12**, 37 (1935); L. van Itallie, A. J. Steenhauer,

Arch. Pharm. **270**, 313 (1932). Structure: A. Chatterjee, S. Bose, *J. Indian Chem. Soc.* **31**, 17 (1954); F. A. L. Anet *et al., J. Chem. Soc.* **1954**, 1242. Stereochemistry: M. F. Bartlett *et al., J. Am. Chem. Soc.* **84**, 622 (1962). Synthesis: S. Masamune *et al., ibid.* **89**, 2506 (1967); E. E. Van Tamelen, L. K. Oliver, *ibid.* **92**, 2136 (1970); K. Mashimo, Y. Sato, *Chem. Pharm. Bull.* **18**, 353 (1970). Physico-chemical properties: A. Petter, *Arzneimittel-Forsch.* **24**, 874 (1974). Antiarrhythmic activity: A. Petter, K. Engelmann, *ibid.* 876. *Reviews:* R. Robinson in *Festschrift Arthur Stoll* (Birkhäuser-Verlag, Basel, 1957) pp 457-467; A. Koskinen, M. Lounasmaa in *Progress in the Chemistry of Natural Products,* **vol. 43**, W. Herz *et al.,* Eds. (Springer-Verlag, New York, 1983) pp 268-346.

Pale amber, solvated, tetragonal prisms from methanol, $C_{20}H_{26}N_2O_2 \cdot CH_3OH$, mp 158-160°. $[\alpha]_D^{18}$ +131° (c = 0.4 in chloroform). Anhydr mp 205-207°. $[\alpha]_D^{20}$ +144° (c = 0.8 in chloroform). uv max (ethanol): 247, 295 nm (log ϵ 3.94, 3.49). Sol in methanol, ethanol, ether, chloroform; slightly sol in water.
Hydrochloride dihydrate, $C_{20}H_{26}N_2O_2 \cdot 2HCl \cdot 2H_2O$, hexagonal bipyramidal crystals from water, mp 140°. $[\alpha]_D^{18}$ +96.6°. One gram dissolves in 40 ml water.

THERAP CAT: Antihypertensive; antiarrhythmic.

195. Ajoene. *2-Propenyl 3-(2-propenylsulfinyl)-1-propenyl disulfide;* 4,5,9-trithiadodeca-1,6,11-triene 9-oxide. $C_9H_{14}OS_3$; mol wt 234.41. C 46.12%, H 6.02%, O 6.83%, S 41.04%. Antithrombotic principle found in garlic *(Allium sativum* L., *Liliaceae).* Formed from allicin, *q.v.,* in the approximate ratio of 4:1 (*E:Z*). The (*Z*)-isomer appears to be more bioactive. Isoln and effect on human platelet aggregation *in vitro:* R. Apitz-Castro *et al., Thromb. Res.* **32**, 155 (1983). [*Note:* Structure initially assigned as *allyl 1,5-hexadienyltrisulfide.*] Structural study, isoln of isomers and synthesis of (*E,Z*)-form: E. Block *et al., J. Am. Chem. Soc.* **106**, 8295 (1984). Structure determn and structure-activity study: *eidem, ibid.* **108**, 7045 (1986). Synthesis, not claimed: E. Block, S. Ahmad, U.S. pat. **4,643,994** (1987 to Res. Found. SUNY). Mechanism of antiplatelet activity: R. Apitz-Castro *et al., Biochem. Biophys. Res. Commun.* **141**, 145 (1986). Synergism with other antiplatelet compounds: *eidem, Thromb. Res.* **42**, 303 (1986). *In vitro* antifungal activity: S. Yoshida *et al., Appl. Environ. Microbiol.* **53**, 615 (1987).

(Z)–Ajoene

(E)–Ajoene

(*E*)-Form, colorless, odorless oil. uv max: 240 nm.

196. Ajowan Oil. Ptychotis oil. Obtained by distillation of seeds of *Carum copticum* (DC.) Benth. & Hook. *(Ptychotis ajowan* DC.), *Umbelliferae.* Contains thymol, α-pinene, *p*-cymene, dipentene, γ-terpinene; fixed oil content includes petroselenic acid, oleic acid, linoleic acid, resin acids, and palmitic acid. *Ref:* Farook *et al. J. Sci. Food Agr.* **4**, 132 (1953); E. Guenther, *The Essential Oils* **IV**, 551-2 (Van Nostrand, New York, 1950).

Almost colorless or brownish liquid; thyme odor; sharp burning taste. d 0.910-0.930. Rotation 0° to +5°. n_D^{20} 1.498-1.504. Sol in 1-2.5 vol and more of 80% alcohol. The oil is toxic to earthworms, while alcoholic extracts are toxic to staphylococci and *Escherichia coli*. Dil extracts are useful as expectorants: Umanskii, Krutik, *Farmatsiya* **8**, no. 6, 19 (1945), *C.A.* **41**, 2209 (1947).

197. Ajugarins. Diterpenes isolated from the leaves of *Ajuga remota (Labiatae).* Five different ajugarins have been isolated and identified; of these ajugarins I-III possess antifeedant activity against the African army worm, ajugarin-IV has insecticidal activity, ajugarin V is inactive. Isoln and structure of ajugarins I-III: I. Kubo *et al., Chem. Commun.* **1976**, 949; of ajugarin IV: *eidem, ibid.* **1982**, 618; of ajugarin V: *eidem, Chem. Letters* **1983**, 223. *See also:* **Japan. Kokai 82 48,979** (1982 to Otsuka). X-ray crystal structure of 12-bromoajugarin-I and absolute configuration: I. Kubo *et al., Chem. Commun.* **1980**, 897; ^{13}C-NMR shift data: J. M. Luteijn *et al., Org. Magn. Reson.* **19**, 95 (1982). Synthetic approaches: D. J. Goldsmith *et al., J. Org. Chem.* **43**, 3182 (1978); J. M. Luteijn, Ae. de Groot, *Tetrahedron Letters* **1981**, 789. Total synthesis of ajugarin-I: S. V. Ley *et al., Chem. Commun.* **1983**, 503.

Ajugarin-I   R = CH$_2$OCOCH$_3$

Ajugarin-I, C$_{24}$H$_{34}$O$_7$, *[1R-(1α,4aβ,5β,6α,8α,8aα)]-4-[2-[8-(acetyloxy)-8a-[(acetyloxy)methyl]octahydro-5,6-dimethylspiro[naphthalene-1(2H),2'-oxiran]-5-yl]ethyl]-2(5H)furanone.* mp 155-157°. uv max (methanol): 212 nm (ε 10000).
Ajugarin-II, C$_{22}$H$_{32}$O$_6$, 8-hydroxyanalog of ajugarin I. mp 188-189°.
Ajugarin-III, C$_{24}$H$_{36}$O$_8$, 1,1-dihydroxy analog of ajugarin I. mp 243-245°.
Ajugarin-IV, C$_{23}$H$_{34}$O$_6$, mp 119-120.5°. [α]$_D$ −57.5° (c = 0.06 in CHCl$_3$). uv max (ethanol): 215 nm (ε 17000).
Ajugarin-V, C$_{22}$H$_{32}$O$_5$, mp 217-218°. [α]$_D$ −13.5° (c = 0.18 in CHCl$_3$). uv max (ethanol): 210 nm (ε 11000).

198. Aklomide. *2-Chloro-4-nitrobenzamide;* Aklomix. C$_7$H$_5$ClN$_2$O$_3$; mol wt 200.58. C 41.92%, H 2.51%, Cl 17.67%, N 13.97%, O 23.93%. Prepn from 2-chloro-4-nitrobenzoylchloride and ammonium carbonate: Grohmann, *Ber.* **24**, 3813 (1891); Cosulich *et al., J. Am. Chem. Soc.* **75**, 4675 (1953). Used vs. infections caused by protozoa and bacteria: **Belg. pat. 608,763** (1962 to Salsbury's Labs).

Gray scales from alcohol, mp 172°.
THERAP CAT (VET): Coccidiostat.

199. Akuammicine. *2,16,19-20-Tetradehydrocuran-17-oic acid methyl ester.* C$_{20}$H$_{22}$N$_2$O$_2$; mol wt 322.41. C 74.51%, H 6.88%, N 8.69%, O 9.92%. From the seeds of *Picralima klaineana*, Pierre, *Apocynaceae* found in the Gold Coast: Henry, Sharp, *J. Chem. Soc.* **1927**, 1950; Henry, *ibid.* **1932**, 2759. Structure: Aghoramurthy, Robinson, *Tetrahedron* **1**, 172 (1957); Bernauer *et al., Helv. Chim. Acta* **43**, 717 (1960); Edwards, Smith, *J. Chem. Soc.* **1961** 152.

Derivs: Robinson, Thomas, *ibid.* **1955**, 2049. Total synthesis: Kutney, Fuller, *Heterocycles* **3**, 197 (1975).

Plates from ethanol+ water, mp 182°. [α]$_D^{16}$ −745° (c = 0.994 in ethanol). pKa 7.45. uv max (ethanol): 227, 300, 330 nm (log ε 4.09, 4.07, 4.24).
Hydrochloride dihydrate, C$_{20}$H$_{22}$N$_2$O$_2$.HCl.2H$_2$O, leaflets from alcohol or water, mp 171°. [α]$_D^{21}$ −610° (c = 1.430 in ethanol).
Perchlorate monohydrate, C$_{20}$H$_{22}$N$_2$O$_2$.HClO$_4$.H$_2$O, needles from ethanol + water, mp 134-136°.
Hydriodide monohydrate, C$_{20}$H$_{22}$N$_2$O$_2$.HI.H$_2$O, square plates from water, mp 128°.
Methiodide, crystals from water, mp 252°.
Nitrate, needles from hot water, mp 182.5°.

200. Akuammine. Vincamajoridine. C$_{22}$H$_{26}$N$_2$O$_4$; mol wt 382.46. C 69.09%, H 6.85%, N 7.32%, O 16.73%. From seeds of *Picralima nitida* (Stapf.) Th. & H. Dur. (*P. klaineana* Pierre), *Apocynaceae*: T. A. Henry, T. M. Sharp, *J. Chem. Soc.* **131**, 1950 (1927); T. A. Henry, *ibid.* **1932**, 2759. Identity with vincamajoridine from *Vinca major* L., *Apocynaceae*: M. M. Janot *et al., Experientia* **11**, 343 (1955). Structure and chemistry: M. F. Millson *et al., ibid.* **9**, 89 (1953); Saxton, *Quart. Rev.* **10**, 141 (1956); J. A. Joule, G. F. Smith, *J. Chem. Soc.* **1962**, 312. Revised structure and absolute configuration: L. Olivier *et al., Bull. Soc. Chim. France* **1965**, 868.

Crystals, decomp 255°. [α]$_D^{16}$ −105.3° (c = 1.31 in pyridine). pKa 7.5. uv max (ethanol): 245, 312 nm (log ε 3.96, 3.66). Slightly sol in water, cold alcohol; sol in boiling alcohol, chloroform, acetone.
Hydrochloride ethanolate, C$_{22}$H$_{26}$N$_2$O$_4$.HCl.C$_2$H$_5$OH, crystals, mp 232-233°.
Methyl ether, C$_{23}$H$_{28}$N$_2$O$_4$, needles from acetone, mp 242-243°. uv max (ethanol): 244, 308 nm (ε 11500, 4200).

201. AL 721. Active lipid; Altrigen. Membrane fluidizer composed of neutral glycerides, phosphatidylcholine, and phosphatidylethanolamine in a ratio of 7:2:1. Prepn from hen egg yolk: M. Shinitzky *et al., Eur. pat. Appl.* **74,251;** *eidem,* **U.S. pat. 4,474,773** (1983, 1984 both to Yeda Res. Dev.). Effect on drug withdrawal symptoms in morphine-addicted mice: D. S. Heron *et al., Eur. J. Pharmacol.* **83**, 253 (1982). Characterization and mode of action study: M. Lyte, M. Shinitzky, *Biochim. Biophys. Acta* **812**, 133 (1985). Effect on HIV-1 infectivity *in vitro:* P. S. Sarin *et al., N. Engl. J. Med.* **313**, 1289 (1985). Effect on lymphocyte responsiveness in the elderly: H. Rabinowich *et al., Mech. Ageing Devel.* **40**, 131 (1987).
USE: Nutritional supplement.

202. Alacepril. *(S)-N-[1-[3-(Acetylthio)-2-methyl-1-oxopropyl]-L-prolyl]-L-phenylalanine;* 1-(D-3-acetylthio-2-methylpropanoyl)-L-prolyl-L-phenylalanine; *N-[1-[(S)-3-mercapto-2-methylpropionyl]-L-prolyl]-3-phenyl-L-alanine* acetate (ester); DU-1219; Cetapril. C$_{20}$H$_{26}$N$_2$O$_5$S; mol wt 406.50. C 59.09%, H 6.45%, N 6.89%, O 19.68%, S 7.89%. Angiotensin-converting enzyme (ACE) inhibitor. Prepn: T.

Sawayama *et al.,* **Japan. Kokai 80 9058;** *eidem,* U.S. pat. **4,248,883** (1980, 1981 both to Dainippon Pharm.). Pharmacology in animals: K. Takeyama *et al., Arzneimittel-Forsch.* **35,** 1502 (1985); *eidem, ibid.* 1507. Series of articles on pharmacology, metabolism, enzyme inhibiting activity: *ibid.* **36,** 47-83 (1986). Metabolism to captopril, *q.v.:* K. Matsumoto *et al., ibid.* 40. HPLC determn of metabolites in plasma and urine: K. Hayashi *et al., J. Chromatog.* **338,** 161 (1985). Pharmacokinetics in humans: K. Onoyama *et al., Clin. Pharmacol.* **38,** 462 (1985). Preclinical evaluation in essential hypertension: K. Mizuno *et al., Res. Commun. Chem. Pathol. Pharmacol.* **49,** 175 (1985); H. Shionoiri *et al., Curr. Ther. Res.* **38,** 537 (1985). Series of articles on toxicology: M. Iida *et al., Yakuri to Chiryo* **13,** 7033-7121 (1985), *C.A.* **104,** 21888h-21890c; **105,** 396v (1986).

Crystals from ethanol/*n*-hexane, mp 155-156°. $[\alpha]_D^{25}$ −81.3° (c = 1.02 in ethanol). LD_{50} in rats, mice (mg/kg): > 5000, > 5000 orally; > 3000, > 3000 s.c.; ~2000, ~3000 i.p. (Iida, pp 7033-40).

THERAP CAT: Antihypertensive.

203. Alachlor. *2-Chloro-N-(2,6-diethylphenyl)-N-(methoxymethyl)acetamide; 2-chloro-2',6'-diethyl-N-(methoxymethyl)acetanilide;* metachlor; CP-50144; Lasso; Alanex. $C_{14}H_{20}ClNO_2$; mol wt 269.77. C 62.33%, H 7.47%, Cl 13.14%, N 5.19%, O 11.86%. Pre-emergence herbicide. Prepn: Neth. pat. **Appl. 6,602,564** (1967 to Monsanto), *C.A.* **67,** 99832r (1967). Activity: D. M. Evans, *Chem. & Ind. (London)* **1969,** 615. Soil degradn: R. S. Hargrove, M. G. Merkle, *Weed Sci.* **19,** 652 (1971). Mechanism of action: L. M. Deal, F. D. Hess, *ibid.* **28,** 168 (1980).

Cryst solid, mp 40-41°. $d_{15.6}^{25}$ 1.133. Soly in water at 23°: 140 mg/l. Sol in ether, acetone, benzene, ethanol, ethyl acetate. Hydrolyzed under strong acid or alkaline conditions. LD_{50} orally in rats: 1200 mg/kg (Evans).

Note: The EPA has determined that alachlor is oncogenic in rats and mice: *Fed. Reg.* **50,** 1115 (1985).

USE: Herbicide.

204. Alafosfalin. *[R-(R*,S*)]-[1-[(2-Amino-1-oxopropyl)amino]ethyl]phosphonic acid;* 1*R*-1-(L-alanylamino)ethylphosphonic acid; alaphosphin; Ro-3-7008. $C_5H_{13}N_2O_4P$; mol wt 196.14. C 30.62%, H 6.68%, N 14.28%, O 32.63%, P 15.79%. Synthetic phosphonodipeptide with antibacterial activity. Prepn: F. R. Atherton *et al.,* **Ger.** pat. **2,602,193;** *eidem,* U.S. pat. **4,016,148** (1976, 1977 both to Hoffmann-La Roche); J. G. Allen *et al., Nature* **272,** 56 (1978); F. R. Atherton *et al., Antimicrob. Ag. Chemother.* **15,** 677 (1979). Improved process: E. K. Baylis, **Eur.** pat. **Appl. 10,872;** *eidem,* U.S. pat. **4,331,591** (1980, 1982 both to Ciba-Geigy). Separation of diastereoisomers: J. Szewczyk *et al., Experientia* **38,** 983 (1982). Antibacterial spectrum: F. R. Atherton *et al., Antimicrob. Ag. Chemother.* **15,** 684 (1979); W. H. Traub, *Chemotherapy* **26,** 103 (1980). Synergism with β-lactams: H. B. Maruyama *et al., Antimicrob. Ag. Chemother.* **16,** 444 (1979); F. R. Atherton *et al., ibid.* **20,** 470 (1981); M. Arisawa *et al., ibid.* **21,** 706 (1982). Pharmacokinetics: J. D. Allen, L. J. Lees, *ibid.* **17,** 973 (1980). Comprehensive review: C. H. Hassall in *Anti-*

biotics **VI,** F. E. Hahn, Ed. (Springer-Verlag, New York, 1983) pp 1-11.

Crystals from ethanol-water, mp 295-296° (dec). $[\alpha]_D^{20}$ −44.0° (c = 1 in H_2O).

205. Alanine. *L-Alanine;* Ala; A; L-α-alanine; L-α-aminopropionic acid; *(S)*-2-aminopropanoic acid. $C_3H_7NO_2$; mol wt 89.09. C 40.44%, H 7.92%, N 15.72%, O 35.92%. Non-essential amino acid for human development. First syllable of the name denotes its origin from aldehyde. First synthesized and named by: A. Strecker, *Ann.* **75,** 27 (1850); prior to identification in silk hydrolysate: P. Schützenberger, A. Bourgeous, *Compt. Rend.* **81,** 191 (1875); Th. Weyl, *Ber.* **21,** 1407 (1888). Early chemistry and biochemistry: *Amino Acids and Proteins,* D. M. Greenberg, Ed. (Charles C. Thomas, Springfield, IL, 1951) 950 pp., *passim;* J. P. Greenstein, M. Winitz, *Chemistry of the Amino Acids* **vols 1-3** (John Wiley and Sons, Inc., New York, 1961) pp. 1819-1840, *passim.* Enantiomeric selective synthesis: R. Amoroso *et al., J. Org. Chem.* **57,** 1082 (1992). Review of metabolism in man: J. E. Liljenquist *et al., Clin. Nutr. Update,* H. L. Greene *et al.,* Eds. (American Medical Association, Chicago, IL, 1977) pp 22-34; T. N. Palmer *et al., Biosci. Reports* **5,** 1015-1033 (1985). Review of industrial production by microorganisms: S. Takamatsu, T. Tosa, *Bioproc. Technol.* **16,** 25-35 (1993).

Orthorhombic crystals from water, dec 297°. d 1.401. $[\alpha]_D^{25}$ +2.42° (c = 10 in H_2O); +13.7° (c = 2.06 in 6*N* HCl). Soly (g/l) in water at 0°: 127.3; at 25°: 166.5; at 50°: 217.9; at 75°: 285.1; at 100°: 373.0. Soly in cold 80% ethanol: 0.2%. Insol in ether.

D-Form, Occurs in bacterial cell walls and in some peptide antibiotics. *Review:* M. Bodanszky, D. Perlman, *Science* **163,** 352-358 (1969); C. T. Walsh, *J. Biol. Chem.* **264,** 2393-2396 (1989). Crystals, mp 289°-291°. $[\alpha]_D$ −14.1° (c = 0.9 in 1*N* HCl).

DL-Form, orthorhombic bipyramidal needles or rods from water. d 1.424. Sublimes above 200°. Decomp 264-296° depending on rate of heating. pK_1 2.35; pK_2 9.87. Soly in water at 0°: 121 g/l, at 25°: 167 g/l, at 50°: 231 g/l, at 75°: 319 g/l, at 100°: 440 g/l; in ethanol 0.0087 g/100 g at 25°. Insol in ether.

206. β-Alanine. Beta-alanine; β-aminopropionic acid; 3-aminopropanoic acid; 3-aminopropionic acid; Abufène. $C_3H_7NO_2$; mol wt 89.09. C 40.44%, H 7.92%, N 15.72%, O 35.92%. $NH_2CH_2CH_2CO_2H$. Prepd by the action of KOBr and KOH upon succinimide: Clarke, Behr, *Org. Syn.* **16,** 1 (1936). By the action of liq ammonia upon methyl acrylate: Morsch, *Monatsh.* **63,** 220 (1933), *cf. C.A.* **41,** 4104 (1947); by the addition of NH_4OH to acrylonitrile: Ford *et al., J. Am. Chem. Soc.* **69,** 844 (1947). By electrolytic oxidation of 3-amino-1-propanol in H_2SO_4 using Pb electrodes without diaphragm: *Jubilee Vol. Emil Barell* **1946,** 85-91. For industrial methods of prepn *see* several pats. by T. L. Gresham to B. F. Goodrich. Prepn from ethylene cyanohydrin (β-hydroxypropionitrile): Boatright, U.S. pat. **2,734,081** (1956 to Am. Cyanamid); from β-aminopropionitrile: Ford, *Org. Syn.* **coll. vol. III,** 34 (1955). Improved process: Beutel, Klemchuk, U.S. pat. **2,956,080** (1960 to Merck & Co.).

Orthorhombic bipyramidal crystals from water, decomp 207° (very rapid heating). Decomp 197-198° (Ford, *Org. Syn. loc. cit.*). Slightly sweet taste. pK_1 3.60; pK_2 10.19. pH of 5% aq soln: 6.0 to 7.3. Freely sol in water, slightly in alcohol. Practically insol in ether, acetone.

Hydrochloride, $C_3H_7NO_2.HCl$, plates, leaflets, mp 122.5°. Freely sol in water, less sol in alcohol, insol in ether.

Platinichloride, $2C_3H_7NO_2.2HCl.PtCl_4$, yellow leaflets from alcohol + HCl, decomp 210°. Freely sol in water, sparingly in abs alcohol.

USE: In the synthesis of pantothenic acid and derivatives; as buffer in electroplating.

207. L-Alanosine. *3-(Hydroxynitrosoamino)-L-alanine; L-2-amino-3-(hydroxynitrosoamino)propanoic acid;* L-2-amino-3-[(N-nitroso)hydroxylamino]propionic acid. $C_3H_7N_3O_4$; mol wt 149.11. C 24.17%, H 4.73%, N 28.18%, O 42.92%. Antibiotic substance from the fermentation of *Streptomyces alanosinicus* n. sp. Prepn: **Neth. pat. Appl. 6,509,543** corresp to J. Thiemann, Y. S. K. Murthy, **U.S. pat. 3,676,-490** (1966, 1972 to Lepetit). The first natural product found to have a *N*-nitrosohydroxylamino group on an aliphatic chain. Isoln and structure: Coronelli *et al., Farmaco Ed. Sci.* **21,** 269 (1966). Characterization: Thiemann, Beretta, *J. Antibiot.* **19A,** 155 (1966). Structure and synthesis of L-, D- and DL-forms: Lancini *et al., Tetrahedron Letters* **1966,** 1769; *eidem, Farmaco Ed. Sci.* **24,** 169 (1969). Synthesis of L-form: Isowa *et al., Bull. Chem. Soc. Japan* **46,** 1847 (1973). Improved synthesis of DL-form: Eaton *et al., J. Med. Chem.* **16,** 289 (1973). Pharmacology: Murthy *et al., Nature* **211,** 1198 (1966). Mode of action studies: Gale *et al., Biochem. Pharmacol.* **17,** 363, 1823 (1968). Effect on insect growth: E. E. Kenaga, *J. Econ. Entomol.* **62,** 1006 (1969); S. Matsumoto *et al., Agric. Biol. Chem.* **48,** 827 (1984).

Crystalline powder from slightly acidic water, dec 190°. $[\alpha]_D$ +8°, −46°, −37.8° (in $1N$ HCl, $0.1N$ NaOH, water). uv max ($0.1N$ HCl): 228 nm ($E_{1cm}^{1\%}$ 505); in $0.1N$ NaOH: 250 nm ($E_{1cm}^{1\%}$ 630). pKa 4.8. Slightly sol in water; practically insol in the common organic solvents. Sol in alkaline and acidic solns, from which it ppts by adjusting the pH between 4 and 6. LD_{50} in mice (mg/kg): 600 i.p.; 300 i.v. (Thiemann, Murthy).

D-Form, mp 183°. $[\alpha]_D$ +45° (c = $0.5N$ in $1N$ HCl).

USE: Experimental insect reproduction inhibitor.

208. Alantolactone. *[3aR-(3aα,5β,8aβ,9aα)]-3a,5,6,7,-8,8a,9,9a-Octahydro-5,8a-dimethyl-3-methylenenaphtho[2,3-b]furan-2(3H)-one; 8β-hydroxy-4αH-eudesm-5-en-12-oic acid γ-lactone;* helenin; alant camphor; elecampane camphor; inula camphor; Eupatal. $C_{15}H_{20}O_2$; mol wt 232.32. C 77.55%, H 8.68%, O 13.77%. A terpene from roots of *Inula helenium* L., *Compositae:* Kallen, *Ber.* **6,** 1506 (1873); **9,** 154 (1876); Ruzicka *et al., Helv. Chim. Acta* **14,** 397, 1090 (1931); **16,** 268 (1933). Structure: Marshall, Cohen, *J. Org. Chem.* **29,** 3727 (1964). Stereoselective synthesis: Marshall *et al., J. Am. Chem. Soc.* **88,** 3408 (1966).

Crystals from alcohol, mp 78-79°. bp 275°. $[\alpha]_D$ +175° (chloroform). uv max (ethanol): 212 nm (ε 9500). Volatile with steam. Freely sol in alcohol, chloroform, benzene, ether, oils. Practically insol in water.

THERAP CAT: Anthelmintic (Nematodes).

209. Alazopeptin. L-Alanyl-(6-diazo-5-oxo)-L-nor-leucyl-(6-diazo-5-oxo)-L-norleucine. $C_{15}H_{20}N_6O_5$; mol wt 364.36. C 49.45%, H 5.53%, N 23.07%, O 21.96%. Tumor-inhibiting antibiotic produced by *Streptomyces griseoplanus* from soil near Williamsburg, Iowa: De Voe *et al., Antibiot. Ann.* **1956-7,** p 730. Peptide consisting of one mole α-ala-

nine and two moles of a C_6 diazo keto amino acid oxidizable to glutamic acid: E. L. Patterson *et al., Antimicrob. Ag. Chemother.* **1965,** 115-118. Anti-tumor activity: T. Hata *et al., J. Antibiot.* **26,** 181 (1973). Toxicity: J. B. Thiersch, *Proc. Soc. Exp. Biol. Med.* **97,** 888 (1958).

Monohydrate, crystals from dilute acetone. Poor stability. Has no definite melting point. $[\alpha]_D^{25}$ +9.5° (c = 4.7 in H_2O). uv max (pH 7.0 phosphate buffer): 242, 274 nm ($E_{1cm}^{1\%}$ 321, 549). Freely sol in water. Somewhat sol in acetic acid, formamide, DMSO, aq solns of methanol, ethanol, acetone. Practically insol in anhyd alcohols, acetone, ethyl acetate, ether. LD_{50} i.p. in rats: 150 mg/kg (Thiersch).

210. Albaspidin. *2,2′-Methylenebis[6-butyryl-3,5-dihy-droxy-4,4-dimethyl-2,5-cyclohexadien-1-one];* polystichalbin; methylenebis(butyrylfilicinic acid); albaspidin-BB. $C_{25}H_{32}-O_8$; mol wt 460.52. C 65.20%, H 7.00%, O 27.79%. Found in the rhizomes of the male fern, *Aspidium filix mas* (L.) Schott., *Dryopteris filix mas* (L.) Schott., *A. spinulosum, Polypodiaceae* and other ferns. Isoln from *Aspidium* extract: Boehm, *Ann.* **318,** 305 (1901); McGookin *et al., J. Chem. Soc.* **1953,** 1828; *see also* Tryon *et al., Phytochemistry* **12,** 683 (1973). Isoln as one component of a homologous mixture: Penttila, Sundman, *Acta Chem. Scand.* **18,** 344 (1964). Synthesis: Riedl, Mitteldorf, *Ber.* **89,** 2595 (1956); Inagaki *et al., J. Pharm. Soc. Japan* **76,** 1258 (1956). Biosynthetic studies: Penttila, Fales, *J. Am. Chem. Soc.* **88,** 2327 (1966).

Crystals from ethanol or methanol, mp 149°. Freely sol in chloroform; moderately sol in ether, benzene; sparingly sol in alc, acetone, glacial acetic acid, sol in KOH solns; very sparingly sol in Na_2CO_3 solns. Practically insol in methanol.

211. Albendazole. *[5-(Propylthio)-1H-benzimidazol-2-yl]carbamic acid methyl ester;* methyl 5-(propylthio)-2-benz-imidazolecarbamate; 5-(propylthio)-2-carbomethoxyamino-benzimidazole; SKF-62979; Bilutac; Eskazole; Proftril; Val-bazen; Zentel. $C_{12}H_{15}N_3O_2S$; mol wt 265.34. C 54.32%, H 5.70%, N 15.84%, O 12.06%, S 12.09%. Prepn: R. J. Gyu-rik, V. J. Theodorides, U.S. pat. **3,915,986** (1975 to Smith-Kline). Anthelmintic activity: V. J. Theodorides, *Experientia* **32,** 702 (1976); D. A. Denham *et al., J. Helminthol.* **54,** 199 (1980). Comparative effect vs cestodes: W. S. Evans *et al., J. Parasitol.* **66,** 935 (1980). Efficacy studies: J. C. Williams *et al., Am. J. Vet. Res.* **42,** 318 (1981); P. Stevenson *et al., Vet. Rec.* **109,** 82 (1981). Clinical trial in humans: P. Pene *et al., Am. J. Trop. Med. Hyg.* **31,** 263 (1982).

Colorless crystals, mp 208-210°.
THERAP CAT (VET): Anthelmintic.

212. Albizziin. *3-[(Aminocarbonyl)amino]-L-alanine; 2-amino-3-ureidopropionic acid.* $C_4H_9N_3O_3$; mol wt 147.13. C 32.65%, H 6.17%, N 28.56%, O 32.62%. $NH_2CONH-CH_2CH(NH_2)COOH$. Constituent of several plants belonging to the family *Mimosaceae.* First obtained from the seeds of *Albizzia julibrissin* Durazz., *Mimosaceae:* Gmelin *et al., Z. Naturforsch.* **13b,** 252 (1958); *Z. Physiol. Chem.* **314,** 28 (1959). Structure: Kjaer *et al., Experientia* **15,** 253 (1959). Synthesis: Kjaer, Larsen, *Acta Chem. Scand.* **13,** 1565 (1959); Rudinger *et al., Coll. Czech. Chem. Commun.* **25,** 2022 (1960).

Needles from dil alcohol, decomp 218-220°. $[\alpha]_D^{25}$ −66.2° (c = 4 in H_2O); $[\alpha]_D^{24}$ −22.2° (c = 4.24 in 1.0N HCl); $[\alpha]_D^{24}$ +3.2° (c = 4.7 in 1.0N NaOH). ORD data: Kjaer *et al., Acta Chem. Scand.* **18,** 2412 (1964).

213. Albofungin. *[1S-(1α,4α,8aα)]-13-Amino-3,4,8a,-13-tetrahydro-1,15,16-trihydroxy-4-methoxy-12-methyl-1H-xantheno[4′,3′,2′:4,5][1,3]benzodioxino[7,6-g]isoquinoline-14,17(2H,9H)-dione.* $C_{27}H_{24}N_2O_9$; mol wt 520.50. C 62.31%, H 4.65%, N 5.38%, O 27.66%. Antifungal antibiotic produced by *Streptomyces albus* var *fungistaticus (fungatus)* Solovyeva and Rudaya. Isoln procedure: Khokhlov, Liberman, *Proc. Symp. Antibiotics Prague* (May 1959) p 81. Structure: A. I. Gurevich *et al., Tetrahedron Letters* **1972,** 1751. Stereochemistry: A. I. Gurevich *et al., ibid.* **1974,** 2801. Thought to be identical with *Ba-180265* described by Liu *et al.,* in *Antimicrob. Ag. Chemother.* **1962,** 767. Chemical and biological properties: A.I. Gurevich *et al., Antibiotiki* **17,** 771 (1972).

Crystals from nitromethane, mp 304-307°. Also reported as dec 190°. $[\alpha]_D^{20}$ −670° (DMF). uv max (ethanol): 228, 254, 303, 376 nm (log ε 4.58, 4.58, 4.19, 4.42). Practically insol in water, petr ether. Sparingly sol in alcohol; quite sol in chloroform, dichloroethane, acetone, chlorobenzene, formamide, dimethylformamide, glacial acetic acid.

214. Albomycin. Iron-containing antibiotic produced by *Actinomyces subtropicus:* Gauze, Braznikova, *Novosti Med.* **23,** 3 (1951). Used as the sulfate. Consists of six components, α, β, γ, δ_1, δ_2, ε, of which δ_1 and δ_2 are the main components. δ_2 is the unstable, highly active albomycin A_1; the other components are its degradation products: Braznikova *et al., Biokhimiya* **22,** 111 (1959); Turková *et al., Antibiotiki* **7,** 878 (1962). Early studies proposed cyclic hexapeptide structures containing three serines and three ornithine derivatives for δ_1-, δ_2- and ε-albomycin (*cf.* Ferrichromes). Structural studies: *eidem., Coll. Czech. Chem. Commun.* **30,** 118 (1965); Poddubnaya, el′Naggar, *Zh. Obshch. Khim.* **38,** 450 (1968). Conflicting structural analysis shows a 3:1 ornithine : serine ratio: Maehr, Pitcher, *J. Antibiot.* **24,** 830 (1971). Similarity to or identity with grisein, *q.v.:* Stapley, Ormond, *Science* **125,** 587 (1957); Turková *et al., Coll. Czech. Chem. Commun.* **31,** 2444 (1966). Review and antibacterial spectrum of albomycin and other hydroxamic acids: Gauze, *Brit. Med. J.* **2,** 1177 (1955); Bhuyan in *Antibiotics* **1,** D. Gottlieb, P. D. Shaw, Eds. (Springer-Verlag, New York, 1967) pp 153-155; Maehr, *Pure Appl. Chem.* **28,** 603-636 (1971); Emery, *Advan. Enzymol. Relat. Areas Mol. Biol.* **35,** 135-185 (1971).

Sulfate, amorphous red powder. uv max: 283 nm ($E_{1cm}^{1\%}$ 880). Freely sol in water. Slightly sol in methanol. Practically insol in most other organic solvents. Aq solns have a bright orange color. Effective against penicillin-resistant pneumococci and staphylococci. The toxicity is comparable to that of penicillin.

Note: Not to be confused with albamycin.

215. Alborixin. Antibiotic S 14750A. $C_{48}H_{84}O_{14}$; mol wt 885.19. C 65.13%, H 9.56%, O 25.30%. Polycyclic polyether antibiotic ionophore, isolated from a strain of

Streptomyces albus. Initial description and x-ray structure: M. Allèaume *et al., Chem. Commun.* **1975,** 411. Production from *Streptomyces hygroscopicus* and use: M. Kuhn, H. D. King, **Ger.** pat. **2,608,337** (1976 to Sandoz), *C.A.* **86,** 28491k (1977). Isoln from *S. albus,* structure, properties: P. Gachon *et al., J. Antibiot.* **29,** 603 (1976); C. Delhomme *et al., ibid.* 692. NMR spectrum and solution conformation: N. A. Rodios, M. J. O. Anteunis, *Bull. Soc. Chim. Belg.* **88,** 279 (1979). Revised structure: H. Seto *et al., J. Antibiot.* **32,** 970 (1979). *In vitro* study: M. Chapel *et al., ibid.* 740. Effects on cardiovascular function and plasma cation concentration in the dog: N. Moins *et al., J. Cardiovasc. Pharmacol.* **1,** 659 (1979).

White amorphous powder, mp 100-105°. $[\alpha]^{20}$ −7° (c = 4 in acetone). pKa 10.02 (25° in methanol). LD_{50} in mice: 15 mg/kg s.c. (Delhomme). LD_{50} originally reported as 150 mg/kg, corrected in *J. Antibiot.* **30,** Suppl. (1977), facing p 77-12.

216. Albumen. Egg white; dried egg white. By convention the pure protein is spelled "albumin" and commercial egg white is called "albumen." The word "albumen" goes back to Roman times and is classical Latin for egg white: *Ref:* Plinius (Major), *Historia Naturalis* **28,** 6, 18 paragraph 66. Constitutes about six parts by weight of a hen's egg (wet basis). Average weight of raw egg white 32.9 g. Albumen contains 75% ovalbumin *(q.v.),* ovoconalbumin, ovomucoid, ovomucin, ovoglobulin, lysozyme *(q.v.)* and avidin *(q.v.).* Monograph: A. L. Romanoff, A. J. Romanoff, *The Avian Egg* (Wiley, New York, 1949) 918 pp.

Raw egg albumen, clear, colloidal, flowing, limpid mass. White rubbery solid when denatured. d_{25}^{25} 1.035. n_D^{25} 1.356. Coagulating temp 61°. mp −0.42°. pH 7.6.

Dried albumen, yellow, transparent, amorphous lumps or scales, or yellow powder. Decomposes in moist air. In water it swells at first, then dissolves gradually. The soln coagulates at 61° (because of denaturation of the proteins). Albumen solns are also denatured on contact with the following chemicals: Salts of copper, iron, mercury and silver; H_2O_2, phenol, picric acid, alum, tannic acid, formaldehyde, ether, alcohol.

USE: For clarifying and refining wines and vinegars. As textile dye mordant, in lithography as vehicle for substances that sensitize plates, in gilding leather (Venetian decorations), stamping with gold and bronze powder (especially in book binding). In adhesives and veneers. Ingredient of compositions used in sizing and in making papers. In fine color preparations for artist's paints. As activating agent for certain enzymes. Formerly in sugar refining. Ingredient of bakery products, confectionery, food preparations. In pharmaceutical compounding, i.e. to make various albuminates. As analytical reagent in testing for ionic mercury, foreign oil in olive oil, candy or wine colorings.

THERAP CAT: Antidote to mercury poisoning.

217. Albuterol. α^1-*[[(1,1-Dimethylethyl)amino]methyl]-4-hydroxy-1,3-benzenedimethanol;* α^1-*[(tert-butylamino)methyl]-4-hydroxy-m-xylene-α,α′-diol;* 2-(*tert*-butylamino)-1-(4-hydroxy-3-hydroxymethylphenyl)ethanol; 4-hydroxy-3-hydroxymethyl-α-[(*tert*-butylamino)methyl]benzyl alcohol; salbutamol; AH-3365. $C_{13}H_{21}NO_3$; mol wt 239.31. C 65.25%, H 8.84%, N 5.85%, O 20.06%. Prepn: L. H. C. Lunts *et al.,* **S. Afr.** pat. **67 05,591;** *eidem,* **U.S.** pat. **3,644,-353** (1968, 1972, to Allen and Hanburys). Synthesis and structure-activity studies: Collin *et al., J. Med. Chem.* **13,** 674 (1970). Absolute configuration of optical isomers: Hartley, Middlemiss, *ibid.* **14,** 895 (1971). Pharmacology: Brittain *et al., Nature* **219,** 862 (1968); Callum *et al., Brit. J. Pharmacol.* **35,** 141 (1969). Metabolism: Martin *et al., Eur. J. Pharmacol.* **14,** 183 (1971). Clinical studies in asthma: S. Chodosh, *Arch. Intern. Med.* **138,** 1394 (1978); R. W. Light *et al., ibid.* **139,** 639 (1979). Use in prevention of premature labor: G. J. Addis, *Lancet* **1,** 42 (1981). *Review: Drugs* **1,** 274-302 (1971); R. T. Brittain, D. M. Harris, in *Pharmaco-*

logical and Biochemical Properties of Drug Substances **vol. 1**, M. E. Goldberg, Ed. (Am. Pharm. Assoc., Washington, DC, 1977) pp 257-276. Comprehensive description: H. Y. Aboul-Enein et al., Anal. Profiles Drug Subs. **10**, 665-689 (1981). Review of clinical toxicology: J. D. Truwit, Crit. Care Clinics **7**, 639-657 (1991).

Crystalline powder from ethanol-ethyl acetate or ethyl acetate-cyclohexane, mp 151° (Lunts); 157-158° (Collin). Sol in most organic solvents.

Sulfate, $(C_{13}H_{21}NO_3)_2.H_2SO_4$, *Aerolin, Asmaven, Broncovaleas, Cetsim, Cobutolin, Ecovent, Loftan, Proventil, Salbulin, Salbumol, Salbutard, Salbutine, Salbuvent, Sultanol, Venetlin, Ventodisks, Ventolin, Volma.*

THERAP CAT: Bronchodilator; tocolytic.

218. Albutoin. *5-(2-Methylpropyl)-3-(2-propenyl)-2-thioxo-4-imidazolidinone; 3-allyl-5-isobutyl-2-thiohydantoin; 3-allyl-5-sec-butyl-2-thiohydantoin; CO-ORD; Euprax.* $C_{10}H_{16}N_2OS$; mol wt 212.32. C 56.57%, H 7.60%, N 13.19%, O 7.54%, S 15.10%. Prepn: Oba et al., C.A. **46**, 3885f (1952).

Crystals, mp 210-211°.
USE: Fog inhibitor in photography.
THERAP CAT: Anticonvulsant.

219. Alcian Blue. *C.I. Ingrain Blue 1;* C.I. 74240; alcian blue 8GX; michrome no. 24. When R = toluyl, the mol wt is about 1300. Discovered by Haddock and Wood in 1944: Haddock, Research **1**, 685 (1948); Brit. pats. **586,-340** and **587,636** (both to ICI). During dyeing the X groups split off. Prepn: Colour Index vol. 4 (3rd ed.) p 4620.

X = an onium group, e.g. —CH₂S—C(NR₂)(NR₂)⁺ Cl⁻
(R = alkyl or aryl)

Greenish-black crystals with metallic sheen. When R = toluyl the solubility in water at 20° is about 9.5% w/w giving bright greenish-blue solutions. Soly in absolute ethanol about 6.0, Cellosolve 6.0, ethylene glycol 3.25. Practically insoluble in xylene.

USE: Gelling agent for lubricating fluids. As bacterial stain. To dye histiocytes and fibroblasts in vivo: McManus, Bailie, Fed. Proc. **22**, no. 2, part I, p 190 (March-April 1963).

220. Alclofenac. *3-Chloro-4-(2-propenyloxy)benzeneacetic acid; [4-(allyloxy)-3-chlorophenyl]acetic acid; W-7320;* Allopydin; Epinal; Medifenac; Mervan; Neoston; Prinalgin; Reufenac; Zumaril. $C_{11}H_{11}ClO_3$; mol wt 226.66. C 58.29%, H 4.89%, Cl 15.64%, O 21.18%. Prepn: N. P. Buu-Hoi et al., Belg. pat. **704,368** (1968 to Madan). Pharmacology: eidem, Naturwiss. **56**, 330 (1969). Series of articles on pharmacology, toxicology and metabolism: Arzneimittel-Forsch. **20**, 610-636 (1970). Toxicity data: G. Lambelin et al., ibid. 618. Clinical studies: F. Lambotte, ibid. 569; Van Hoek, Curr. Ther. Res. **12**, 551 (1970).

Prisms from cyclohexane, mp 92-93°. LD_{50} in mice, rats (mg/kg): 1100, 1050 orally; 600, 630 s.c.; 555, 530 i.p. (Lambelin).

Monoethanolamine complex, $C_{13}H_{18}ClNO_4$, *Mirvan A.*
THERAP CAT: Analgesic; antipyretic; anti-inflammatory.

221. Alclometasone. *(7α,11β,16α)-7-Chloro-11,17,21-trihydroxy-16-methylpregna-1,4-diene-3,20-dione;* 7α-chloro-16α-methylprednisolone. $C_{22}H_{29}ClO_5$; mol wt 408.92. C 64.62%, H 7.15%, Cl 8.67%, O 19.56%. Non-fluorinated corticosteroid with few systemic effects. Prepn: M. J. Green et al., U.S. pat. **4,076,708**; M. J. Green, H. J. Shue, U.S. pat. **4,124,707** (both 1978 to Schering); M. J. Green et al., J. Steroid Biochem. **11**, 61 (1979); H. J. Shue et al., J. Med. Chem. **23**, 430 (1980). Topical anti-inflammatory activity of the 17,21-dipropionate: B. Lutsky et al., Arzneimittel-Forsch. **29**, 992 (1979); M. J. Green et al., ibid. **30**, 1618 (1980).

Crystals from acetone/hexane, mp 176-179°. $[\alpha]_D^{26}$ +47.5° (c = 0.3 in DMF). uv max (methanol): 242 nm (ϵ 15500).

17,21-Dipropionate, $C_{28}H_{37}ClO_7$, Sch-22219, *Aclosone, Aclovate, Almeta, Delonal, Legederm, Modrasone, Perderm, Vaderm.* Crystals from acetone/methanol/isopropyl ether, mp 212-216°. $[\alpha]_D^{26}$ +42.6° (c = 0.3 in DMF). uv max (methanol): 242 nm (ϵ 15600).

THERAP CAT: Anti-inflammatory (topical).

222. Alcuronium. *4,4'-Didemethyl-4,4'-di-2-propenyltoxiferine I; N,N'-diallylnortoxiferinium;* diallylbis(nortoxiferine); diallylnortoxiferine; diallyltoxiferine. $[C_{44}H_{50}N_4O_2]^{2+}$. Prepn of dichloride and diiodide: Boller et al., U.S. pat. **3,080,373** (1963 to Hoffmann-La Roche).

Dichloride, $C_{44}H_{50}Cl_2N_4O_2$, *N,N'-diallylnortoxiferinium dichloride, Ro-4-3816, Alloferin, Toxiferene.* Crystals from methanol or ethanol. Compd contains 5 moles of water of crystn after equilibration in air; $[\alpha]_D^{22}$ −348° (methanol); uv max (methanol): 292 nm (ϵ 43000).

Consult the Name Index before using this section.

Diiodide, $C_{44}H_{50}I_2N_4O_2$, solid. uv max (methanol): 291 nm (ϵ 39900).

THERAP CAT: Muscle relaxant (skeletal).

223. Aldicarb. *2-Methyl-2-(methylthio)propanal O-[(methylamino)carbonyl]oxime; 2-methyl-2-(methylthio)propionaldehyde O-(methylcarbamoyl)oxime;* UC-21149; Temik. $C_7H_{14}N_2O_2S$; mol wt 190.27. C 44.19%, H 7.42%, N 14.72%, O 16.82%, S 16.85%. $CH_3SC(CH_3)_2CH=NOCONHCH_3$. Synthesis: Payne *et al., J. Agr. Food Chem.* **14**, 356 (1966). C^{14}-labeled synthesis: Bartley *et al., ibid.* 604. Metabolism: Bartley *et al., ibid.* **18**, 446 (1970). Mass spectrum: Benson, Damico, *J. Assoc. Offic. Anal. Chem.* **1968**, 347. Crystal and molecular structure: F. Takusagawa, R. A. Jacobson, *J. Agr. Food Chem.* **25**, 333 (1977). Toxicity study: Weiden *et al., J. Econ. Entomol.* **58**, 154 (1965). *Review:* Romine, *Anal. Methods Pestic. Plant Growth Regul.* **7**, 147-162 (1973).

Crystals from isopropyl ether, mp 99-100°. Soly at 25° (w/w) in water: 0.6%; acetone: 35%; benzene: 15%; xylene: 5%; methylene chloride: 30%. LD_{50} orally in female rats: ~1 mg/kg (Weiden).

USE: Insecticide, acaricide, nematocide.

224. Aldioxa. *[(2,5-Dioxo-4-imidazolidinyl)ureato]-dihydroxyaluminum; (allantoinato)dihydroxyaluminum;* dihydroxy[(4,5-dihydro-2-hydroxy-5-oxo-1H-imidazol-4-yl)ureato]aluminum; aluminum dihydroxy allantoinate; RC-172; Alanetorin; Alusa; Arlanto; Ascomp; Chlokale; Isalon; Nische; Peptilate. $C_4H_7AlN_4O_5$; mol wt 218.11. C 22.03%, H 3.23%, Al 12.37%, N 25.69%, O 36.68%. Prepn: S. B. Mecca, U.S. pat. **2,761,867** (1956 to Schuylkill Chem.). Pharmacology: R. Cahen, J. F. Clement, *Ann. Pharm. Franc.* **20**, 693 (1962); R Cahen, A. Pessonnier, *ibid.* 704. Toxicology: R. Cahen, A. Pessonnier, *ibid.* 623. Metabolism studies: K. Fukawa *et al., Oyo Yakuri* **11**, 421 (1976), *C.A.* **88**, 130701g (1978). HPLC determn in plasma: K. Hirota *et al., J. Chromatog.* **277**, 165 (1983). Clinical studies of combination therapy for peptic ulcer: M. Sakai *et al., Japan. Arch. Intern. Med.* **31**, 275 (1984); *idem et al., ibid.* **33**, 61 (1986); *idem et al., ibid.* **36**, 357 (1989). Review of cosmetic use and safety assessment: S. N. J. Pang, *J. Am. Coll. Toxicol.* **12**, 237-242 (1993).

White powder. mp 230°. Insol in polar and nonpolar solvents.

USE: Astringent and skin conditioning agent in cosmetics.
THERAP CAT: Antiulcerative.

225. Aldol. *3-Hydroxybutanal; 3-hydroxybutyraldehyde;* acetaldol. $C_4H_8O_2$; mol wt 88.11. C 54.53%, H 9.15%, O 36.32%. $CH_3CH(OH)CH_2CHO$. Manuf by condensation of acetaldehyde in aq NaOH: Alhéritière, Gobron, U.S. pat. **2,713,598** (1955 to Usines de Melle). Toxicity study: H. F. Smyth *et al., J. Ind. Hyg. Toxicol.* **31**, 60 (1949).

Colorless, thick liquid. d^{16} 1.109. bp_{20} 83°; dec about 85°. Miscible with water, alcohol, ether. LD_{50} orally in rats: 2.18 g/kg (Smyth).

USE: Manuf rubber vulcanizers, accelerators and age resisters; in perfumes; ore flotation.
THERAP CAT: Sedative; hypnotic.

226. Aldosterone. *(11β)-11,21-Dihydroxy-3,20-dioxopregn-4-en-18-al;* 3,20-diketo-11β,18-oxido-4-pregnene-18,21-diol; Aldocorten; Aldocortin; Electrocortin. $C_{21}H_{28}O_5$; mol wt 360.45. C 69.99%, H 7.83%, O 22.19%. Adrenocortical steroid which exerts regulatory influence on metabolism of electrolytes and water. Isoln: Simpson *et al., Experientia* **9**, 333 (1953); *Helv. Chim. Acta* **37**, 1163 (1954); Mattox *et al., J. Am. Chem. Soc.* **75**, 4869 (1953); Harman *et al., ibid.* **76**, 5035 (1954). Solutions contain an equilibrium mixture of the aldehyde and the hemiacetal, the equilibrium favoring the latter. Structure: Tait *et al., Experientia* **10**, 132 (1954); *Helv. Chim. Acta* **37**, 1200 (1954). Crystal

structure and molecular conformation: Duax, Hauptmann, *J. Am. Chem. Soc.* **94**, 5467 (1972). ^{13}C-NMR spectrum: P. Gerard, *Org. Magn. Resonance* **11**, 478 (1978). Total synthesis: Schmidlin *et al., Helv. Chim. Acta* **40**, 1438 (1957); Johnson *et al., J. Am. Chem. Soc.* **80**, 2585 (1958); **85**, 1409 (1963). Three-step synthesis from corticosterone: Barton, Beaton, *ibid.* **82**, 2640 (1960); **83**, 4083 (1961). Alternate synthesis: D. H. R. Barton *et al., J. Chem. Soc., Perkin Trans. I* **1975**, 2243; M. Miyano, *J. Org. Chem.* **46**, 1846 (1981). Biosynthesized in the zona glomerulosa and transported chiefly by albumin. In man, 400 μg secreted normally in one day. Secretion influenced by ACTH, growth hormone, plasma sodium and potassium, and the renin-angiotensin system. Causes reabsorption of Na^+, Cl^-, and HCO_3^- and diuresis of K^+. *Review:* L. F. Fieser, M. Fieser, *Steroids* (Reinhold, New York, 1959) pp 701-720.

Hydrated crystals from dilute acetone, mp 108-112° (when anhydr mp 164°). $[\alpha]_D^{23}$ +152.2° (anhydr; c = 2 in acetone). $[\alpha]_D^{25}$ +161° (c = 0.1 in chloroform). uv max: 240 nm (log ϵ 4.20 for the monohydrate; ϵ_{mol} 15,000 for the anhydr).

21-Acetate, $C_{23}H_{30}O_6$. Synthesis: Wettstein *et al.;* Jeger, U.S. pats. **3,002,972** and **3,014,029** (both 1958 to Ciba). Flat needles from acetone + ether, mp 198-199°. $[\alpha]_D^{24}$ +121.7° (c = 0.71 in chloroform).

THERAP CAT: Mineralocorticoid.
THERAP CAT (VET): Mineralocorticoid.

227. Aldrin. *1,2,3,4,10,10-Hexachloro-1,4,4a,5,8,8a-hexahydro-1,4:5,8-dimethanonaphthalene;* HHDN; compd 118; Octalene. $C_{12}H_8Cl_6$; mol wt 364.91. C 39.50%, H 2.21%, Cl 58.29%. Activity: C. W. Kearns *et al., J. Econ. Entomol.* **42**, 127 (1949). Prepn of aldrin and endo,endo-isomer: Lidov, U.S. pat. **2,635,977** (1953 to Shell). Alternate syntheses: Schmerling, U.S. pat. **2,911,447** (1959 to Universal Oil Prod.); Korte, Rechmeier, *Ann.* **656**, 131 (1962). Toxicity study: T. B. Gaines, *Toxicol. Appl. Pharmacol.* **14**, 515 (1969).

Crystals, mp 104°. Vapor pressure at 20°: 7.5×10^{-5} mm Hg. Very sol in most organic solvents. Practically insol in water. Stable in presence of organic and inorganic alkalies; stable to the action of hydrated metal chlorides. LD_{50} in male, female rats (mg/kg): 39, 60 orally (Gaines).

endo,endo-Isomer, *isodrin,* compd 711. Crystals, mp 240-242°. LD_{50} in male, female rats (mg/kg): 15, 7.0 orally (Gaines).

Caution: Poisoning may occur by ingestion, inhalation, skin absorption. Severe symptoms may result from ingestion or percutaneous absorption of 1 to 3 g, especially in presence of liver disease. *Acute toxicity:* Renal damage, tremors, ataxia, convulsions followed by CNS depression, respiratory failure, death. *Chronic toxicity:* Prolonged exposure may cause hepatic damage, cf. Patty's *Industrial Hygiene and Toxicology* vol. **2B**, G. D. Clayton, F. E. Clayton, Eds. (Wiley-Interscience, New York, 3rd ed., 1981) pp 3702-3707.

USE: Formerly as insecticide; manuf and use has been discontinued in the U.S.

228. Alendronic Acid. *(4-Amino-1-hydroxybutylidene)bisphosphonic acid;* α-hydroxy-δ-aminobutylidenediphosphonic acid; ABDP. $C_4H_{13}NO_7P_2$; mol wt 249.10. C 19.29%, H 5.26%, N 5.62%, O 44.96%, P 24.87%. Bisphosphonate inhibitor of bone resorption; structurally related to etidronic and pamidronic acids, *q.q.v.* Prepn: M. I. Kabachnik *et al., Bull. Acad. Sci. USSR,* **27,** 374 (1978). *See also:* **Belg.** pat. **903,519;** G. Staibano, U.S. pat. **4,705,651** (1986, 1987 both to Gentili). Pharmacokinetics: J. H. Lin *et al., Drug Metab. Dispos.* **19,** 926 (1991). Mechanism of action: M. Sato *et al., J. Clin. Invest.* **88,** 2095 (1991). Determn by HPLC in urine: W. F. Kline *et al., J. Chromatog.* **534,** 139 (1990); by IC in pharmaceutical formulations: E. W. Tsai *et al., ibid.* **596,** 217 (1992). Clinical evaluation in malignant hypercalcaemia: R. Rizzoli *et al., Int. J. Cancer* **50,** 706 (1992). Clinical trials in osteoporosis: C. H. Chesnut *et al., Am. J. Med.* **99,** 144 (1995); U. A. Liberman *et al., N. Engl. J. Med.* **333,** 1437 (1995).

Fine white powder, mp 233-235° (dec). pK_1 2.72 ±0.05; pK_2 8.73 ±0.05; pK_3 10.5 ±0.1; pK_4 11.6 ±0.1 at 25°, (0.1*M* KCl).

Monosodium salt trihydrate, $C_4H_{12}NNaO_7P_2 \cdot 3H_2O$, *alendronate sodium, MK-217, G-704650, Adronat, Alendros, Fosamax, Onclast.*

THERAP CAT: Suppressant (bone resorption).

229. Aletris. Star grass; starwort; true unicorn root; blazing star; colic root. Rhizome of *Aletris farinosa* L., *Liliaceae. Habit.* Eastern U.S., Ontario. *Constit.* Starch, diosgenin. Isoln of sapogenin: Marker *et al., J. Am. Chem. Soc.* **62,** 2620 (1940). Pharmacological studies: Butler, Costello, *J. Am. Pharm. Assoc., Sci. Ed.* **33,** 177 (1944).

THERAP CAT: Antiflatulent.

230. Aleuritic Acid. *(R*,S*)-(±)-9,10,16-Trihydroxyhexadecanoic acid;* 9,10,16-trihydroxypalmitic acid; 8,9,15-trihydroxypentadecane-1-carboxylic acid. $C_{16}H_{32}O_5$; mol wt 304.43. C 63.13%, H 10.60%, O 26.23%. One of the constituent acids of shellac. Obtained in 43% yield from dewaxed shellac: Schaeffer, Gardner, *Ind. Eng. Chem.* **30,** 333 (1938); Gidvani, *J. Chem. Soc.* **1944,** 306; Sengupta, Bose, *J. Sci. Ind. Res. (India)* **11B,** 458 (1952). The acid obtained from shellac is optically inactive, although it contains two asymmetric carbon atoms. It has been shown to be the DL-*erythro* or *dl-cis* form, and is the only form described here. Synthesis of diastereoisomers: Mitter *et al., Sci. Cult. (Calcutta)* **8,** 273 (1942); Hunsdiecker, *Ber.* **76,** 142 (1943); **77,** 185 (1944); Baudart, *Compt. Rend.* **221,** 205 (1945).

Crystals from dilute ethanol, mp 100-101°. Sol in methanol. Forms a crystalline sodium salt.

Methyl ester, $C_{17}H_{34}O_5$, fine feathery needles, mp 72-73°; $bp_{0.1}$ 235°. Sol in methanol, ethanol, chloroform, acetone. Less sol in benzene. Insol in petr ether.

Ethyl ester, $C_{18}H_{36}O_5$, needles from dil ethanol, mp 59°. Hydrazide, $C_{16}H_{34}N_2O_4$, crystals from abs ethanol, mp 139-140°.

231. Alexidine. *N,N''-Bis(2-ethylhexyl)-3,12-diimino-2,4,11,13-tetraazatetradecanediimidamide; 1,1'-hexamethylenebis[5-(2-ethylhexyl)biguanide];* Win-21904; Sterwin 904; Bisguadine. $C_{26}H_{56}N_{10}$; mol wt 508.80. C 61.38%, H 11.09%, N 27.53%. Prepn from 1,1'-hexamethylenebis(3-cyanoguanide) and 2-ethylhexylamine hydrochloride: **Fr.** pat. **1,463,818** (1965 to Sterling Drug). Evaluation as antimicrobial agent: McNamara *et al., J. Soc. Cosmet. Chem.* **16,** 499 (1965).

Dihydrochloride, crystals from methanol + ether, mp 220.6-223.4°.

THERAP CAT: Antibacterial.

232. Alexitol Sodium. Sodium polyhydroxyaluminum monocarbonate hexitol complex; aluminum sodium carbonate hexitol complex; Actal. Description and antacid properties: J. R. Gwilt *et al., J. Pharm. Pharmacol.* **10,** 770 (1958). Stabilization with a hexitol (sorbitol or mannitol): Alford, U.S. pat. **2,999,790** (1962 to Sterling Drug). Neutralizing capacity: R. E. Barry, J. Ford, *Br. Med. J.* **1,** 413 (1978).

Tasteless, odorless powder. Dec without melting when strongly heated. Practically insol in water. Readily sol in dil acids. At normal temps can be stored indefinitely without apparent change in its physical or chemical properties.

THERAP CAT: Antacid.

233. Alfadolone Acetate. *(3α,5α)-21-(Acetyloxy)-3-hydroxypregnane-11,20-dione; 3α,21-dihydroxy-5α-pregnane-11,20-dione 21-acetate;* 21-acetoxy-3α-hydroxy-5α-pregnane-11,20-dione; alphadolone acetate; GR-2/1574. $C_{23}H_{34}O_5$; mol wt 390.52. C 70.74%, H 8.78%, O 20.48%. Prepn: Brown, Kirk, *J. Chem. Soc. (C)* **1969,** 1653; Davis *et al.,* **Ger.** pat. **2,030,402** and **S. Afr.** pat. **70 03,861** (both 1971 to Glaxo), *C.A.* **75,** 20793n, 64114w (1971).

Crystals from acetone-hexane, mp 175-177°. $[α]_D^{26}$ +97° (c = 1.02 in chloroform).

Mixture with alfaxalone, *see* Alfaxalone.

THERAP CAT: Combination with alfaxalone as anesthetic (intravenous).

THERAP CAT (VET): Combination with alfaxalone as anesthetic (intravenous).

234. Alfaprostol. *[1R-[1α(Z),2β(S*),3α,5α]]-7-[2-(5-Cyclohexyl-3-hydroxy-1-pentynyl)-3,5-dihydroxycyclopentyl]-5-heptenoic acid methyl ester;* 18,19,20-trinor-17-cyclohexyl-13,14-didehydro-PGF$_{2α}$ methyl ester; Ro-22-9000; K-11941; Alfavet; Alphacept. $C_{24}H_{38}O_5$; mol wt 406.56. C 70.90%, H 9.42%, O 19.68%. Synthetic analog of prostaglandin F$_{2α}$, *q.v.* Synthesis: C. Gandolfi *et al.,* **Ger.** pat. **2,539,116;** *eidem,* U.S. pat. **4,035,415** (1976, 1977 both to Carlo Erba). Efficacy in bovine estrus synchronization: E. Schilling *et al., Theriogenology* **18,** 413 (1982). Treatment of infertility in cows caused by persistent corpus luteum: G. Maffeo *et al., Prostaglandins* **25**, 541 (1983).

THERAP CAT (VET): Estrus control.

235. Alfaxalone. *(3α,5α)-3-Hydroxypregnane-11,20-dione;* alphaxalone; GR-2/234. $C_{21}H_{32}O_3$; mol wt 332.48. C 75.86%, H 9.70%, O 14.44%. Prepn: Nagata *et al., Helv. Chim. Acta* **42**, 1399 (1959); Browne, Kirk, *J. Chem. Soc. (C)* **1969**, 1653; Davis *et al.,* **Ger. pat. 2,030,402** and **S. Afr. pat. 70 03,861** (both 1971 to Glaxo), *C.A.* **75**, 20793n, 64114w (1971). Mass spectral data: Ende, Spiteller, *Monatsh.* **102**, 929 (1971). Review of pharmacology and clinical efficacy of mixture with alfadolone acetate, *q.v.: Postgrad. Med. J.* **48**, Suppl. 2, 1-139 (1972). *See also* Child *et al.;* Campbell *et al.; Brit. J. Anaesth.* **43**, 2-24 (1971); *Lancet* **1**, 888 (1972).

Colorless prisms from ether, mp 172-174°. $[\alpha]_D^{26}$ +113.4° (c = 1.2 in chloroform).

Mixture with alfadolone acetate, *alphadione, alfadione, CT-1341, Alfatésin, Althesin, Saffan.* LD_{50} i.v. in mice: 54.7 mg/kg.

THERAP CAT: Combination with alfadolone acetate as anesthetic (intravenous).

THERAP CAT (VET): Combination with alfadolone acetate as anesthetic (intravenous).

236. Alfentanil. *N-[1-[2-(4-Ethyl-4,5-dihydro-5-oxo-1H-tetrazol-1-yl)ethyl]-4-(methoxymethyl)-4-piperidinyl]-N-phenylpropanamide; N-[1-[2-(4-ethyl-5-oxo-2-tetrazolin-1-yl)ethyl]-4-(methoxymethyl)-4-piperidyl]propionanilide.* $C_{21}H_{32}N_6O_3$; mol wt 416.52. C 60.56%, H 7.74%, N 20.18%, O 11.52%. Tetrazole derivative of fentanyl, *q.v.* Prepn: F. Janssens, **Ger. pat. 2,819,873;** *idem,* **U.S. pat. 4,167,574** (1978, 1979 both to Janssen); F. Janssens *et al., J. Med. Chem.* **29**, 2290 (1986). Determn in human plasma by GC: T. J. Gillespie *et al., J. Anal. Toxicol.* **5**, 133 (1981); by radioimmunoassay: M. Michaels *et al., J. Pharm. Pharmacol.* **35**, 86 (1983). Pharmacology in animals: C. J. E. Niemegeers, P. A. J. Janssen, *Drug. Dev. Res.* **1**, 83 (1981); in humans: B. Kay, B. Pleuvry, *Anaesthesia* **35**, 952 (1980). Cardiovascular effects during surgery: P. S. Sebel *et al., Brit. J. Anaesth.* **54**, 1185 (1982). Clinical evaluation as analgesic component of anesthesia: B. Kay *et al., Ann. Roy. Coll. Surg. Engl.* **65**, 316 (1983); as anesthetic induction agent: J. Nauta *et al., Anesth. Analg. (Cleveland)* **61**, 267 (1982); M. E. Sinclair, G. M. Cooper, *Anaesthesia* **38**, 435 (1983). Review of clinical pharmacokinetics: L. E. Mather, *Clin. Pharmacokinet.* **8**, 422 (1983); of clinical trials of alfentanil and sufentanil: S. De Lange, *Mt. Sinai J. Med.* **50**, 312 (1983).

Hydrochloride monohydrate, $C_{21}H_{32}N_6O_3 \cdot HCl \cdot H_2O$, *alfentanil hydrochloride, R-39209, Limifen, Rapifen, Alfenta.* Crystals from 2-propanone, mp 140.8° (Janssens, 1979); also reported as crystals from acetone, mp 138.4° (Janssens, 1986). Sol in water. LD_{50} in rats, dogs (mg/kg): 47.5, 20 i.v. (Niemegeers, Janssen).

Note: This is a controlled substance (opiate) listed in the U.S. Code of Federal Regulations, Title 21 Part 1308.12 (1995).

THERAP CAT: Analgesic (narcotic).

THERAP CAT (VET): Analgesic (narcotic).

237. Alfuzosin. *N-[3-[(4-Amino-6,7-dimethoxy-2-quinazolinyl)methylamino]propyl]tetrahydro-2-furancarboxamide; N₁-(4-amino-6,7-dimethoxyquinazol-2-yl)-N₁-methyl-N₂-(tetrahydrofuroyl-2)-propylenediamine;* SL-77.499. $C_{19}H_{27}N_5O_4$; mol wt 389.45. C 58.60%, H 6.99%, N 17.98%, O 16.43%. α_1-Adrenoceptor antagonist structurally similar to prazosin, *q.v.* Prepn: P. M. J. Manoury, **Ger. pat. 2,904,445;** *idem,* **U.S. pat. 4,315,007** (1979, 1982 both to Synthelabo); and antihypertensive activity in rats: P. M. Manoury *et al., J. Med. Chem.* **29**, 19 (1986). Pharmacology: A. G. Ramage, *Eur. J. Pharmacol.* **129**, 307 (1986). HPLC determn in biological fluids: P. Guinebault *et al., J. Chromatog.* **353**, 361 (1986). Pharmacology in humans: A. H. Deering, *Brit. J. Clin. Pharmacol.* **25**, 417 (1988). Clinical evaluation in essential hypertension: S. Leto Di Priolo *et al., Eur. J. Clin. Pharmacol.* **35**, 25 (1988); A. K. Ghosh, S. Ghosh, *Ger. Cardiovasc. Med.* **1**, 81 (1988).

Hydrochloride, $C_{19}H_{27}N_5O_4 \cdot HCl$, *SL-77.499-10, Alfoten, Urion, Xatral.* Crystals from ethanol + ether, mp 225° (Manoury, 1986), also reported earlier as mp 235° (dec) (Manoury, 1982). pKa 8.13.

THERAP CAT: Antihypertensive.

238. Algestone. *(16α)-16,17-Dihydroxypregn-4-ene-3,-20-dione;* 16α,17-dihydroxyprogesterone; 4-pregnene-16α,-17α-diol-3,20-dione; alphasone. $C_{21}H_{30}O_4$; mol wt 346.47. C 72.80%, H 8.73%, O 18.47%. Prepn from 16-dehydroprogesterone: Inhoffen *et al., Ber.* **87**, 593 (1954); Cooley *et al., J. Chem. Soc.* **1955**, 4373; Allen, Bernstein, *J. Am. Chem. Soc.* **78**, 1909 (1956); Hydorn *et al., Steroids* **3**, 493 (1964). Manuf from 16-dehydroprogesterone: Colton, **U.S. pat. 2,-727,909** (1955 to Searle); Hydorn *et al.,* **U.S. pat. 3,165,541** (1965 to Olin Mathieson); from its 16α-acetate: Diassi, **U.S. pat. 3,027,384** (1962 to Olin Mathieson).

Needles from ethanol + dichloromethane, mp 225°. $[\alpha]_D^{22}$ +95° (c = 0.81 in $CHCl_3$). uv max: 240 nm (ε 16600).

Cyclic acetal with acetone, $C_{24}H_{34}O_4$, *16α,17-[(1-methyl-ethylidene)bis(oxy)]pregn-4-ene-3,20-dione, algestone acetonide, alphasone acetonide, 16α,17α-isopropylidenedioxyprogesterone.* Prepn: Cooley *et al., loc. cit;* Fried *et al., Chem. & Ind. (London)* **1961**, 465. Needles from aq ethanol, mp 210°. $[\alpha]_D^{20} +137°$ (c = 0.7 in CHCl$_3$).

16α-Methyl ether, $C_{22}H_{32}O_4$, *17-hydroxy-16α-methoxy-pregn-4-ene-3,20-dione.* Prepd from the free diol via 16α,17-dihydroxypregn-4-ene-3,20-dione cyclic borate: Fried, U.S. pat. **3,006,930** (1961 to Olin Mathieson). Crystals from 95% ethanol, mp 142-143°. $[\alpha]_D^{23} +60°$ (c = 0.15 in CHCl$_3$). uv max: 234 nm (ε 15400).

THERAP CAT: Acetonide as anti-inflammatory (topical).

239. Algestone Acetophenide. *[16α(R)]-16,17-[(1-Phenylethylidene)bis(oxy)]pregn-4-ene-3,20-dione; 16α,17-dihydroxypregn-4-ene-3,20-dione cyclic acetal with acetophenone;* 16α,17α-dihydroxyprogesterone acetophenide; alphasone acetophenide; P-DHP; SQ-15101; Deladroxone (obsolete); Droxone (obsolete); Neolutin Depositum. $C_{29}H_{36}O_4$; mol wt 448.60. C 77.65%, H 8.09%, O 14.27%. Progestogen. Prepn: J. Fried, U.S. pat. **2,941,997** (1960 to Olin Mathieson); J. Fried *et al., Chem. & Ind. (London)* **1961**, 465. Improved prepn: **Irish** pat. **6,800,457**; S. J. Brancato *et al.,* U.S. pat. **3,488,347** (1968, 1970 both to Smith Kline & French). In estrus synchronization: J. N. Wiltbank *et al., J. Anim. Sci.* **26**, 764 (1967). Clinical evaluations of combination with estradiol enanthate as injectable contraceptive: R. Plesner, *Acta Endocrinol.* **61**, 494 (1969); idem, ibid. **65**, 683 (1970); R. Recio *et al., Contraception* **33**, 579 (1986).

Crystals from 95% ethanol, mp 150-151°. $[\alpha]_D^{23} +51°$ (CHCl$_3$). Stable to boiling mineral acids; readily cleaved by warming with formic acid, with subsequent deformylation.

THERAP CAT: Antiacne.

240. Algin. *Alginic acid sodium salt;* sodium alginate; sodium polymannuronate; Alto; Alman; Alloid; Allose; Kelgin; Protanal. A gelling polysaccharide extracted from giant brown seaweed (giant kelp, *Macrocystis pyrifera* (L.) Ag., *Lessoniaceae*) or from horsetail kelp (*Laminaria digitata* (L.) Lamour, *Laminariaceae*) or from sugar kelp (*Laminaria saccharina* (L.) Lamour). Process of manuf: Tseng, *Chem. Met. Eng.* **52**, 97 (1945); Mantell, *The Water-Soluble Gums* (New York, 1947); Green, **U.S.** pat. **2,036,934** (1936 to Kelco); Gloahec, Herter, U.S. pat. **2,128,551** (1938 to Algin Corp. of America). Wound healing properties and use in hemostatic dressings: J. H. M. Miller, **Brit.** pat. **1,328,088** (1973 to Wallace, Cameron & Co.), *C.A.* **80**, 6974u (1974). Series of articles on hemostatic effects: *Yakugaku Zasshi* **101**, 452-469 (1981), *C.A.* **95**, 35654e, 35655f, 35405z (1981). Immunoadjuvant effect: G. H. Scherr, A. S. Markowitz, U.S. pat. **3,075,883** (1963 to Consolidated Labs.). Clinical comparison with alum immunoadjuvant: G. Bruno *et al., Ann. Allergy* **56**, 384 (1986). Review of structural studies: D. A. Rees, E. J. Welsh, *Angew. Chem. Int. Ed.* **16**, 214 (1977). Review of production, properties and use in the food industry: A. Askar, *Alimenta* **21**, 165-169 (1982). *Reviews:* McNeely, Pettitt, in *Industrial Gums,* R. L. Whistler, Ed. (Academic Press, New York, 2nd ed., 1973) pp 49-81; I. W. Cottrell, J. K. Baird, "Gums" in *Kirk-Othmer Encyclopedia of Chemical Technology* **vol. 12** (Wiley-Interscience, New York, 3rd ed., 1980) pp 48-51.

Cream-colored powder. Sol in water, forming a viscous, colloidal soln. Insol in alcohol and in hydro-alcoholic solns in which the alcohol content is > 30% w/w. Insol in chloroform, ether, in aq acid solns when the pH is below 3.

USE: In the manufacture of ice cream where it serves as a stabilizing colloid, insuring creamy texture and preventing the growth of ice crystals. In drilling muds; in coatings; in the flocculation of solids in water treatment; as sizing agent; thickener; emulsion stabilizer; suspending agent in soft drinks; in dental impression preparations. Pharmaceutic aid (suspending agent).

THERAP CAT: Hemostatic.

241. Alginic Acid. Norgine; polymannuronic acid; Sazio. Mol wt about 240,000. A hydrophilic, colloidal polysaccharide obtained from seaweeds which, in the form of mixed salts of calcium, magnesium, and other bases, occurs as a structural component of the cell wall. Isoln from fronds of *Laminaria digitata* (L.) Edmonson, *Laminariaceae:* Stanford, *J. Chem. Soc.* **44**, 943 (1883); Bird, Haas, *Biochem. J.* **25**, 403 (1931); from *Macrocystis pyrifera* (L.) C. Ag., *Lessoniaceae:* Nelson, Cretcher, *J. Am. Chem. Soc.* **51**, 1914 (1929). Polysaccharides similar to alginic acid are also secreted by the bacteria *Pseudomonas aeruginosa* and *Azotobacter vinelandii:* G. H. Cohen, D. B. Johnstone, *J. Bacteriol.* **88**, 329 (1964); L. R. Evans, A. Linker, *ibid.* **116**, 915 (1973). Alginic acid is a linear polymer of β-(1→4)-D-mannosyluronic acid and α-(1→4)-L-gulosyluronic acid residues, the relative proportions of which vary with the botanical source and state of maturation of the plant. Clinical use of combination with antacid in gastric reflux esophagitis: D. Y. Graham *et al., Curr. Ther. Res.* **22**, 653 (1977). Clinical evaluation of calcium salt as hemostatic dressing: T. Gilchrist, A. M. Martin, *Biomaterials* **4**, 317 (1983); A. R. Groves, J. C. Lawrence, *Ann. Roy. Coll. Surg. Engl.* **68**, 27 (1986). Review of structure studies and of secondary and tertiary structure in solutions and gels: D. A. Rees, E. J. Welsh, *Angew. Chem. Int. Ed.* **16**, 214 (1977). Review of production, properties and use in the food industry: A. Askar, *Alimenta* **21**, 165-169 (1982).

Very slightly sol in water. Tasteless. Capable of absorbing 200-300 times its weight of water and salts to the extent of 60%. Resists hydrolysis. Sol in alkaline solns. pH of a 3 in 100 suspension in water is between 2.0 and 3.4.

Calcium salt, *Sorbsan.* Forms gelatinous precipitate in water.

Potassium salt, *Stercofuge.*

Sodium salt, *See* Algin.

USE: Sizing paper and textiles; as binder for briquettes; manuf artificial horn, ivory, celluloid; emulsionizing mineral oils; mucilage. Additional uses are described under Algin.

THERAP CAT: Calcium salt as hemostatic.

242. Alglucerase. *Glucosylceramidase (human placental isoenzyme protein moiety reduced);* mannose-terminated human placental glucocerebrosidase; macrophage-targeted β-glucocerebrosidase; Ceredase. Modified form of β-glucocerebrosidase, an enzyme involved in glycolipid metabolism, which is deficient in patients with Gaucher's disease. Designed to target the tissue macrophages in which the lipid accumulates. Monomeric glycoprotein of 497 amino acids. Prepd by sequential deglycosolation of the native placental-derived enzyme to expose terminal mannose residues that can be recognized by macrophage cell surface receptors. Prepn: F. S Furbish *et al., Biochim. Biophys. Acta* **673**, 425 (1981); G. J. Murray, *Meth. Enzymol.* **149**, 25 (1987). Clinical evaluation in Gaucher's disease: N. W. Barton *et al., N. Engl. J. Med.* **324**, 1464 (1991); M. L. Figueroa *et al., ibid.* **327**, 1632 (1992). Review of pharmacology and therapeutic efficacy: R. Whittington, K. L. Goa, *Drugs* **44**, 72-93 (1992).

Imiglucerase, rGCR, Cerezyme. Mannose-terminated human glucocerebrosidase produced by recombinant DNA technology. Prepn: J. Rasmussen *et al.,* **PCT Int.** pat. **Appl. 90 07,573;** eidem, U.S. pat. **5,236,838** (1990, 1993 both to Genzyme).

THERAP CAT: Enzyme replacement therapy for Gaucher's disease.

243. Alibendol. *2-Hydroxy-N-(2-hydroxyethyl)-3-methoxy-5-(2-propenyl)benzamide; 5-allyl-2-hydroxy-N-(2-hydroxyethyl)-m-anisamide;* 2-hydroxy-3-methoxy-5-allyl-N-(β-hydroxyethyl)benzamide; EB-1856; FC-54; H-3774; Cebera. $C_{13}H_{17}NO_4$; mol wt 251.28. C 62.14%, H 6.82%, N 5.57%, O 25.47%. Amide analog of eugenol, *q.v.* Prepn: **Fr.** pat.

1,584,715; F. Clémence, O. Le Martret, U.S. pat. **3,668,238** (1967, 1972 both to Roussel). Synthesis and pharmacologic activity: F. Clémence *et al., Chim. Ther.* **5,** 188 (1970).

Crystals from benzene, mp 95°. uv max (ethanol): 316, 218 nm. LD_{50} in Swiss male mice (mg/kg): > 3000 orally; > 2000 s.c.; 209 i.p.; 217 i.v. (Clémence).
THERAP CAT: Choleretic; antispasmodic.

244. Alinidine. *N-(2,6-Dichlorophenyl)-4,5-dihydro-N-2-propenyl-1H-imidazol-2-amine; 2-(N-allyl-2,6-dichloro-anilino)-2-imidazoline; 2-[N-allyl-N-(2,6-dichlorophenyl)-amino]-2-imidazoline;* ST-567. $C_{12}H_{13}Cl_2N_3$; mol wt 270.16. C 53.35%, H 4.85%, Cl 26.25%, N 15.55%. Specific brady-cardic agent; analog of clonidine, *q.v.* Prepn: H. Stähle *et al.,* Ger. pat. **1,958,201;** *eidem,* U.S. pat. **3,708,485** (1971, 1973 both to Boehringer, Ing.); *eidem, J. Med. Chem.* **23,** 1217 (1980). Clinical pharmacology: D. W. G. Harron *et al., J. Cardiovasc. Pharmacol.* **4,** 213 (1982). Mode of action study: J. S. Millar, E. M. Vaughan Williams, *Lancet* **1,** 1291 (1981). HPLC determn in human plasma: U.-W. Wiegand *et al., J. Chromatog.* **223,** 238 (1981). Clinical pharmacokinetics: *eidem, J. Cardiovasc. Pharmacol.* **4,** 59 (1982). Hemodynamic effects in angina or infarction: M. L. Simoons, P. G. Hugenholtz, *Eur. Heart J.* **5,** 227 (1984). Review of pharmacology and potential therapeutic uses: D. W. G. Harron, R. G. Shanks, *ibid.* **6,** 722-729 (1985).

Crystals, mp 130-131° (Stähle, 1973); also reported as mp 127-129° (Stähle, 1980). pka 10.42.
Hydrobromide, $C_{12}H_{14}BrCl_2N_3$, crystals from methanol + water, mp 193-194°.

245. Alitame. *L-α-Aspartyl-N-(2,2,4,4-tetramethyl-3-thietanyl)-D-alaninamide;* L-aspartyl-D-alanine-*N*-(2,2,4,4-tetramethylthietan-3-yl)amide; 3-(L-aspartyl-D-alaninami-do)-2,2,4,4-tetramethylthietane; CP-54802. $C_{14}H_{25}N_3O_4S$; mol wt 331.44. C 50.73%, H 7.60%, N 12.68%, O 19.31%, S 9.67%. Dipeptide amide reported to be approx 2000 times sweeter than sucrose. Prepn: T. M. Brennan, M. E. Hendrick, **Eur. pat. Appl. 34,876;** *eidem,* U.S. pat. **4,411,925** (1981, 1983 both to Pfizer). Prepn of aromatic sulfonic acid salts: C. Sklavounos, U.S. pat. **4,375,430** (1983 to Pfizer).

USE: Non-nutritive sweetener.

246. Alizapride. *6-Methoxy-N-[[1-(2-propenyl)-2-pyr-rolidinyl]methyl]-1H-benzotriazole-5-carboxamide; N-[(1-allyl-2-pyrrolidinyl)methyl]-6-methoxy-1H-benzotriazole-5-carboxamide.* $C_{16}H_{21}N_5O_2$; mol wt 315.38. C 60.94%, H 6.71%, N 22.21%, O 10.15%. Prepn: Belg. pat. **825,605** corresp to G. Bulteau *et al.,* U.S. pat. **4,039,672** (1975, 1977 both to Soc. Etudes Sci. Ind. de Ile de France). Series of articles on pharmacokinetics, bioavailability, pharmacody-namics, clinical studies: *Sem. Hop. Paris* **58,** 323-374 (1982).

Cryst from acetone, mp 139°. LD_{50} i.v. in mice (5 days): 92.7 mg/kg (Bulteau).
Hydrochloride, $C_{16}H_{21}N_5O_2$.HCl, *Nausilen, Plitican, Vergentan.* Cryst from methanol/methyl ethyl ketone, mp 206-208°.
THERAP CAT: Antipsychotic; antiemetic.

247. Alizarin. *1,2-Dihydroxy-9,10-anthracenedione; 1,2-dihydroxyanthraquinone;* C.I. Mordant Red 11; C.I. Pigment Red 83; C.I. 58000. $C_{14}H_8O_4$; mol wt 240.22. C 70.00%, H 3.36%, O 26.64%. Occurs in the root of the madder plant (*Rubia tinctorum* L., *Rubiaceae;* Krappwur-zel) in combination with 2 mols glucose, called ruberythric acid. Was known and used in ancient Egypt, Persia, and India. Synthesized from 2-anthraquinonesulfonic acid sodium salt : Caro *et al., Ber.* **3,** 359 (1870); Perkin, *Ber.* **9,** 281 (1876). Historical *review:* Fieser, *J. Chem. Ed.* **7,** 2609 (1930). Laboratory prepn: Gattermann-Wieland, *Laboratory Methods of Organic Chemistry* (New York, 1937). Modern methods of manufacture: Pohl, *Ullmann's Enzy-klopädie der technischen Chemie* vol. I, p 200; Fierz-David and Blangey, *Grundlegende Operationen der Farbenchemie* (Vienna, 5th ed., 1943). *See also Colour Index* vol. **4,** (3rd ed., 1971) p 4513.

Orthorhombic, orange needles by sublimation or from abs alc. Solvated scales from dil alc or by evaporation from ether. Sublimes at 110° (2 mm Hg). mp 290°. bp 430°. Absorption spectrum: Moir, *J. Chem. Soc.* **1927,** 1810. Soly in water at 18°: 2.1×10^{-6} mols/l; at 25°: 2.5×10^{-6} mols/l. Sol in 300 parts boiling water; moderately sol in alcohol, freely in hot methanol and in ether at 25°. Also sol in benzene, toluene, xylene, pyridine, carbon disulfide, gla-cial acetic acid. Sol in water solns of alkalies with blue color, but without fluorescence. Fluorescent solns indicate unchanged 2-anthraquinone sodium sulfonate.
1-Methyl ether, $C_{15}H_{10}O_4$, orange needles with $1H_2O$ from dil methanol. When dried at 100° mp 179°.
2-Methyl ether, $C_{15}H_{10}O_4$, orange needles from alcohol, mp 231°.
Dimethyl ether, $C_{16}H_{12}O_4$, golden-yellow needles from alcohol, mp 215°.
USE: In the manufacture of acid and chrome dyes for wool; acid-base indicator (in 0.5% alcoholic soln; pH: yel-low 5.5, red 6.8); in spot tests as reagent for aluminum, indium, mercury, zinc, and zirconium; biological stain.

248. Alizarin Cyanine Green F. *2,2-[(9,10-Dihydro-9,10-dioxo-1,4-anthracenediyl)diimino]bis(5-methylbenzene-sulfonic acid) disodium salt; 6,6'-(1,4-anthraquinonylenedi-imino)di-m-toluenesulfonic acid disodium salt;* D & C Green No. 5; C.I. Acid Green 25; C.I. 61570. $C_{28}H_{20}N_2Na_2O_8S_2$; mol wt 622.59. C 54.02%, H 3.24%, N 4.50%, Na 7.39%, O 20.56%, S 10.30%. Discovered by R. E. Schmidt in 1894: *Colour Index* vol. **4** (3rd ed., 1971) p 4541.

Green powder. Slightly sol in acetone, alc, pyridine. Insol in chloroform, toluene. Dull blue soln in conc H_2SO_4, turning turquoise on dilution. Absorption spectra: C. F. H. Allen et al., J. Org. Chem. **7**, 63, 169 (1942).

USE: For nylon sutures: Fed. Regist. **42**, 52395 (1977). Permitted for use in drugs and cosmetics, excluding use in eye area: ibid. **47**, 49628 (1982).

249. Alizarine Blue. 5,6-Dihydroxynaphtho[2,3-f]-quinoline-7,12-dione; 7,8-dihydroxy-5,6-phthalylquinoline; Alizarin Blue R; C.I. 67410. $C_{17}H_9NO_4$; mol wt 291.26. C 70.10%, H 3.11%, N 4.81%, O 21.97%. Prepn from 3-nitroalizarin, glycerol, and concd sulfuric acid: Auerbach, J. Chem. Soc. **35**, 799 (1879); Colour Index vol. **4** (3rd ed., 1971) p 4567.

Lustrous brownish violet needles from benzene, mp 268-270°. Practically insol in water; sparingly sol in alc, ether; slightly sol in cold benzene; sol in amyl alcohol, glacial acetic acid, hot benzene.

USE: As indicator in saturated alcoholic soln. pH: pink 0.0 to yellow 1.6; yellow 6.0 to green 7.6.

250. Alizarine Orange. 1,2-Dihydroxy-3-nitro-9,10-anthracenedione; 1,2-dihydroxy-3-nitroanthraquinone; 3-nitroalizarin; C.I. Mordant Orange 14; C.I. 58015. $C_{14}H_7$-NO_6; mol wt 285.21. C 58.96%, H 2.47%, N 4.91%, O 33.66%. From 3-bromoalizarin in acetic acid with nitric acid: Barnett, Cook, J. Chem. Soc. **121**, 1376 (1922). Additional prepns: Colour Index vol. **4** (3rd ed., 1971) p 4514.

Orange needles or plates from acetic acid, dec 244°. Sublimes with partial decompn. Yellow in organic solvents, purple-red in dil aq alkali, orange in H_2SO_4.

USE: Dyes cloth mordanted with Al orange, with Fe red to violet. As indicator in satd alc soln. pH 2.0-4.0, color change from golden orange to flat yellow (in water). pH 5.0-6.5 from yellow to purplish red.

251. Alizarine Yellow R. 2-Hydroxy-5-[(4-nitrophenyl)azo]benzoic acid; 5-(p-nitrophenylazo)salicylic acid; p-nitrobenzeneazosalicylic acid; Alizarine Yellow RW; C.I. Mordant Orange 1; C.I. 14030. $C_{13}H_9N_3O_5$; mol wt 287.23. C 54.36%, H 3.16%, N 14.63%, O 27.85%. Prepd by coupling diazotized p-nitroaniline with salicylic acid: Armento,

U.S. pat. **2,746,955** (1956 to General Aniline and Film); Colour Index vol. **4** (3rd ed., 1971) p 4058.

Orange-brown needles from dil glacial acetic acid, dec 253-254°. Sol in water, alc; slightly sol in acetone, Cellosolve; practically insol in other org solvents.

Sodium salt, $C_{13}H_8N_3NaO_5$, Mordant Yellow 3R. Brownish yellow powder. Sol in water.

USE: Sodium salt used as indicator in 0.1% aq soln pH: yellow to red, range 10.2 to 12.0.

252. Alkanet. Alkanna; orcanette; dyer's alkanet; anchusa; orkanet. The root of Alkanna tinctoria Tausch, Boraginaceae. Habit. Asia Minor, Hungary, Greece, Mediterranean region. Constit. Alkannin (the coloring principle) and tannin. Extraction of roots: Betrabet, Chakravarti, J. Indian Inst. Sci. **16A**, 51 (1933); C.A. **28**, 1038 (1934); Majamdar, Chakravarti, J. Indian Chem. Soc. **17**, 272 (1940); C.A. **34**, 6262 (1940).

Alkannin paper, anchusin paper, Boettger's paper. White paper impregnated with a 1% alc tincture of alkanet root and dried. A blue paper can be made from the red paper by treating the latter with a 1% Na_2CO_3 soln. This paper acts nearly like litmus paper. Alkalies = green or blue; acids = red. pH: 8.0 red; 10.0 blue.

USE: For coloring wines, cosmetics, confectionery. Alkannin paper as indicator.

THERAP CAT: Astringent.

253. Alkannin. (S)-5,8-Dihydroxy-2-(1-hydroxy-4-methyl-3-pentenyl)-1,4-naphthalenedione; (−)-5,8-dihydroxy-2-(1-hydroxy-4-methyl-3-pentenyl)-1,4-naphthoquinone; anchusa acid; anchusin; alkanna red; alkanet extract; (1-hydroxy-3-isohexenyl)naphthazarine; 2-(1-hydroxy-4-methyl-3-pentenyl)-5,8-dihydroxy-1,4-naphthoquinone; C.I. Natural Red 20; C.I. 75530. $C_{16}H_{16}O_5$; mol wt 288.30. C 66.66%, H 5.59%, O 27.75%. Isoln from the root of Alkanna tinctoria Tausch, Boraginaceae: Brockmann, Ann. **521**, 1 (1936); Toribara, Underwood, Anal. Chem. **21**, 1352 (1949). Absolute configuration: Arakawa, Nakagaki, Chem. & Ind. (London) **1961**, 947. Toxicity study: L. Majlathova, Nahrung **15**, 505 (1971), C.A. **76**, 122513j (1972).

Brownish-red prisms with a metallic sheen from benzene, mp 149°. Can be sublimed in high vac at 140-150°. $[\alpha]_{Cd}^{20}$ −165° (benzene); −226° (chloroform). Also reported as −254° ± 7° (chloroform) (Toribara). Sol in organic solvents, sparingly sol in water. Buffered aq solns are red at pH 6.1; purple at pH 8.8; blue at pH 10.0. LD_{50} in male, female mice, rats (g/kg): 3.0 ± 1.0; 3.1 ± 0.1; >1.0 orally (Majlathova).

(+)-Form, shikonin.
(±)-Form, shikalkin.

USE: Red dye for cosmetics and food; spectrophotometric microdetermination of beryllium.

THERAP CAT: Astringent.

254. Alkofanone. 3-[(4-Aminophenyl)sulfonyl]-1,3-diphenyl-1-propanone; 3-phenyl-3-sulfanilylpropiophenone; p-aminophenyl(2-benzoyl-1-phenylethyl)sulfone; Nu-404; Ro-2-0404; Alfone. $C_{21}H_{19}NO_3S$; mol wt 365.45. C 69.02%, H 5.24%, N 3.83%, O 13.13%, S 8.77%. Prepn: Goldberg et al., Jubilee Volume Emil Barell (Basle, 1946) p 341; Goldberg, U.S. pat. **2,421,836** (1947 to Hoffmann-La Roche).

Decomp 221-223°. Slightly soluble in acetone, dioxane. Practically insol in most of the common organic solvents.

THERAP CAT: Antidiarrheal.

255. Allantoin. *(2,5-Dioxo-4-imidazolidinyl)urea; 5-ureidohydantoin;* glyoxyldiureide; cordianine; Psoralon; Septalan. $C_4H_6N_4O_3$; mol wt 158.12. C 30.39%, H 3.82%, N 35.43%, O 30.36%. Product of purine metabolism. Prepd synthetically by the oxidation of uric acid with alkaline potassium permanganate: *Org. Syn. coll. vol. II*, 23 (1943). By heating urea with dichloroacetic acid: C. N. Zellner, J. R. Stevens, U.S. pat. 2,158,098 (1939 to Merck & Co.) Acetyl derivs: Biltz, Loewe, *J. Prakt. Chem.* 141, 291 (1934). Optically active forms have been obtained by extraction procedures.

Racemic form, monoclinic plates or prisms from water, mp 238°. One gram dissolves in 190 ml water, 500 ml alcohol; more sol in hot water and hot alcohol. Almost insol in ether. pH of satd water soln: 5.5.

THERAP CAT: Vulnerary (topical); skin ulcer therapy.

THERAP CAT (VET): Has been used topically to stimulate healing of suppurating wounds, resistant ulcers.

256. Allenolic Acid. *6-Hydroxy-2-naphthalenepropanoic acid;* 2-hydroxy-6-naphthalenepropionic acid; amphihydroxynaphthyl-β-propionic acid. $C_{13}H_{12}O_3$; mol wt 216.24. C 72.21%, H 5.59%, O 22.20%. Prepn from the corresponding methoxy compd: Jacques, Horeau, *Bull. Soc. Chim. France* **1948**, 714.

Crystals from aq methanol, mp 180-181°. Sol in alcohol, methanol, pyridine.

USE: In biochemical research, in the prepn of estrogenic compds.

257. Allethrins. Allyl cinerins. Synthetic analogs of the naturally occurring insecticides cinerin, jasmolin, and pyrethrin *(q.v.).* Review: Barthel, *World Rev. Pest Contr.* 6, 59 (1967); World Health Org., *Environ. Health Criteria* 87, 1-75 (1989).

Allethrin I   R = CH₃
Allethrin II   R = COOCH₃

Allethrin I, $C_{19}H_{26}O_3$, *2,2-dimethyl-3-(2-methyl-1-propenyl)cyclopropanecarboxylic acid 2-methyl-4-oxo-3-(2-propenyl)-2-cyclopenten-1-yl ester, allethrolone ester of chrysanthemum-monocarboxylic acid, Pynamin, Pyresyn.* Synthesis: M. S. Schechter *et al., J. Am. Chem. Soc.* 71, 1517, 3165 (1949);

H. A. Stansbury, H. R. Guest, U.S. pat. 2,768,965 (1956 to UCC). Commercial process: H. J. Sanders, A. W. Taff, *Ind. Eng. Chem.* 46, 414 (1954). The commercial product is a mixture of 8 optically active isomers. Clear, pale yellow oil. bp $_{0.1}$ 140°. Decomp by rapid pyrolysis at > 400°. Vaporizes without dec when heated at 150°. Unstable in light, air, alkaline conditions. d_{20}^{20} 1.010; n_D^{20} 1.5040; n_D^{30} 1.5023. Practically insol in water. Sol in alcohol, petr ether, kerosene, carbon tetrachloride, ethylene dichloride, nitromethane. Incompatible with alkalies.

Allethrin II, $C_{20}H_{26}O_5$, *3-(3-methoxy-2-methyl-3-oxo-1-propenyl)-2,2-dimethylcyclopropanecarboxylic acid 2-methyl-4-oxo-3-(2-propenyl)-2-cyclopenten-1-yl ester, allethrolone ester of chrysanthemumdicarboxylic acid monomethyl ester, ENT-17510.* Synthesis: M. Matsui, Y. Yamada, *Agr. Biol. Chem.* 27, 373 (1963); M. Matsui, H. Meguro, *ibid.* 28, 27 (1964); M. Elliott, N. F. Janes, *Chem. & Ind. (London)* **1969**, 270; A. Kobayashi *et al., Agr. Biol. Chem.* 35, 1961 (1971); T. Sugiyama *et al., ibid.* 36, 565 (1972). Oily pale yellow liq. $n_D^{20.6}$ 1.5156. uv max (ethanol): 232 nm (ε 23000). Unstable in light, air, alkaline conditions. Sol in organic solvents. Practically insol in water.

Caution: Toxic symptoms similar to those of the pyrethrins, *q.v.*

USE: Insecticide.

258. Allicin. *2-Propene-1-sulfinothioic acid S-2-propenyl ester;* thio-2-propene-1-sulfinic acid *S*-allyl ester; diallyl disulfide-oxide. $C_6H_{10}OS_2$; mol wt 162.28. C 44.41%, H 6.21%, O 9.86%, S 39.52%. A biologically active constit of freshly crushed garlic *(Allium sativum* L., *Liliaceae).* Naturally formed by the action of the enzyme allicinase on alliin, *q.v.,* when the tissue of the garlic bulb is disrupted. Isoln and antibacterial activity: C. J. Cavallito, J. H. Bailey, *J. Am. Chem. Soc.* 66, 1950 (1944). Structure: Cavallito *et al., ibid.* 1952. Synthesis: Stoll, Seebeck, *Experientia* 6, 330 (1950); Cavallito, Small, U.S. pat. 2,508,745 (1950 to Sterling Drug). Antifungal activity: Y. Yamada, K. Azuma, *Antimicrob. Ag. Chemother.* 11, 743 (1977). Pharmacological effects and prepn: P. R. Mayeux *et al., Agents Actions* 25, 182 (1988). Stability in blood, solvents, and simulated physiological fluids: F. Freeman, Y. Kodera, *J. Agric. Food Chem.* 43, 2332 (1995).

Yellow liquid. True odor of garlic. Decomp on distilling. d_4^{20} 1.112. n_D^{20} 1.561. Soly in water at 10° about 2.5% w/w. pH about 6.5. Upon standing an oily precipitate forms from aq solns. Miscible with alcohol, ether, benzene; fairly insol in the Skellysolves; unstable to hot alkali; stable to acids. LD_{50} in mice (mg/kg): 60 i.v.; 120 s.c. (Cavallito, Bailey).

259. Allidochlor. *2-Chloro-N,N-di-2-propenylacetamide;* α-chloro-*N,N*-diallylacetamide; *N,N*-diallyl-2-chloroacetamide; CDAA; CP-6343; Randox. $C_8H_{12}ClNO$; mol wt 173.64. C 55.34%, H 6.97%, Cl 20.42%, N 8.07%, O 9.21%. Selective pre-planting or pre-emergence herbicide. Prepd from chloroacetyl chloride and diallylamine: Speziale, Hamm, *J. Am. Chem. Soc.* 78, 2556 (1956); *eidem., J. Agr. Food Chem.* 4, 518 (1956); Bruce, Hanslick, U.S. pat. 2,844,629 (1958 to Am. Home Prods.); Hamm, Speziale, U.S. pat. 2,864,683 (1958 to Monsanto). Toxicity: G. W. Bailey, J. L. White, *Residue Rev.* 10, 97 (1965).

Liquid. $bp_{0.3}$ 74°; bp_1 115-117°; bp_2 92°. n_D^{25} 1.4932. Slightly sol in water (2%), sol in alcohol, hexane, xylene. LD_{50} orally in rats: 700 mg/kg (Bailey, White).

Caution: A strong irritant. Can be absorbed through skin.

USE: Herbicide.

260. Alliin. *(S)-3-(2-Propenylsulfinyl)-L-alanine; 3-((S)-allylsulfinyl)-L-alanine;* S-allyl-L-cysteine sulfoxide. $C_6H_{11}NO_3S$; mol wt 177.22. C 40.66%, H 6.26%, N 7.90%, O 27.08%, S 18.09%. Constituent of garlic, *Allium sativum* L., *Liliaceae;* also found in other *Allium* spp. Isoln: Stoll, Seebeck, *Helv. Chim. Acta* **31,** 189 (1948). Synthesis: *eidem, Experientia* **6,** 330 (1950); *Helv. Chim. Acta* **34,** 481 (1951).

Hemihydrate, odorless, bunched needles from dil acetone, mp 164-166° (effervescence). $[\alpha]_D^{20}$ +63.5° (c = 2). Freely sol in water. Practically insol in abs ethanol, chloroform, acetone, ether, benzene. Upon cleavage by the specific enzyme alliinase, an odor of garlic develops, and the fission products show antibacterial action similar to allicin, *q.v.*

261. Allobarbital. *5,5-Di-2-propenyl-2,4,6(1H,3H,5H)-pyrimidinetrione; 5,5-diallylbarbituric acid;* allobarbitone; Malilum; Diadol; Dial. $C_{10}H_{12}N_2O_3$; mol wt 208.22. C 57.69%, H 5.81%, N 13.45%, O 23.05%. Prepn: U.S. pat. **1,042,265** (1912). Acute toxicity: F. Sandberg, *Acta Physiol. Scand.* **24,** 7 (1951).

Crystals, leaflets. Slightly bitter taste, mp 171-173°. One part dissolves in about 300 parts water, 50 parts boiling water, 20 parts cold alcohol, 20 parts ether; very sol in hot alcohol and in acetone; sol in ethyl acetate. Insol in aliphatic hydrocarbons. A satd aq soln is acid to litmus. LD_{50} i.p. in rats: 127.3 mg/kg (Sandberg).

Caution: May be habit forming. This is a controlled substance (depressant) listed in the U.S. Code of Federal Regulations, Title 21 Parts 329.1 and 1308.13 (1995).

THERAP CAT: Sedative, hypnotic.

262. Allocholesterol. *(3β)-Cholest-4-en-3-ol;* coprostenol; 4:5-coprosten-3-ol. $C_{27}H_{46}O$; mol wt 386.66. C 83.87%, H 11.99%, O 4.14%. Prepn: Windaus, *Ann.* **453,** 101 (1927); Schoenheimer, Evans, *J. Biol. Chem.* **114,** 567 (1936). Sepn from cholesterol: Stoll, *Z. Physiol. Chem.* **246,** 10 (1937).

Needles from ether-methanol, mp 132°. $[\alpha]_D^{23}$ +43.7° (c = 1 in benzene). Freely sol in benzene, acetone, ether, chloroform, dioxane, pyridine; less sol in methanol, alcohol. Pptd by digitonin. Deep-red color with 90% trichloroacetic acid.

Acetate, $C_{29}H_{48}O_2$, long needles from dil methanol, mp 85°.

263. Alloclamide. *4-Chloro-N-[2-(diethylamino)ethyl]-2-(2-propenyloxy)benzamide; 2-(allyloxy)-4-chloro-N-[2-(diethylamino)ethyl]benzamide.* $C_{16}H_{23}ClN_2O_2$; mol wt 310.82. C 61.83%, H 7.46%, Cl 11.41%, N 9.01%, O

10.29%. Prepn: Mauvernay, U.S. pat. **3,160,557** (1964 to C.E.R.M.).

Hydrochloride, an ingredient of *Dégryp.* Crystals from abs alcohol + ether, mp 125-127°. Sol in abs alcohol. LD_{50} orally in mice: 740 mg/kg.

THERAP CAT: Antitussive, antihistaminic.

264. Allocryptopine. *5,7,8,15-Tetrahydro-3,4-dimethoxy-6-methylbenzo[e][1,3]dioxolo[4,5-k][3]benzazecin-14-(6H)-one.* $C_{21}H_{23}NO_5$; mol wt 369.42. C 68.28%, H 6.28%, N 3.79%, O 21.65%. Isomeric with cryptopine, *q.v.* Obtained from *Chelidonium majus* L., *Bocconia cordata* Willd., *Sanguinaria canadensis* L. and allied *Papaveraceae:* Manske in Manske, Holmes, *The Alkaloids* vol. **IV** (Academic Press, New York, 1954) p 159. Structure: Gadamer, *Arch. Pharm.* **257,** 298 (1919). Exists in two isomeric modifications designated as α- and β-forms. Identity of α-form with α-fagarine: Redemann *et al., J. Am. Chem. Soc.* **71,** 1030 (1949). Synthesis: Haworth, Perkin, *J. Chem. Soc.* **1926,** 445; Bentley, Murray, *ibid.* **1963,** 2497.

α-Allocryptopine, α-*fagarine.* Crystals from ethanol, mp 160-161°. Soluble in alcohol, chloroform, ether, ethyl acetate and dil acids.

β-Allocryptopine, crystallizes with ½ mol of alcohol or ethyl acetate. Melts after expelling the solvent, at 169-171°.

265. Allocupreide Sodium. *3-[[(2-Propenylamino)thioxomethyl]amino]benzoic acid monocopper (1+) monosodium salt; m-[(N-allyl-1-mercaptoformimidoyl)amino]benzoic acid copper derivative sodium salt;* 3-[1-(3-allyl-S-cupropseudothioureido)]benzoic acid sodium salt; *m-[[(allylimino)mercaptomethyl]amino]benzoic acid S-copper derivative sodium salt;* sodium 3-(3-allyl-S-cuproisothioureido)-benzoate; sodium 3-(3-allyl-S-cupropseudothioureido)-benzoate; cuprothiosinamine *m-*benzoate sodium; cupralylnatrium; cupralylsodium; Cuprelon; Cupralene; Cuprion; Ebesal. $C_{11}H_{11}CuN_2NaO_2S$; mol wt 321.82. C 41.05%, H 3.45%, Cu 19.75%, N 8.70%, Na 7.14%, O 9.94%, S 9.96%. Prepd by heating *m-*aminobenzoic acid with allyl isothiocyanate and treating the resulting *m,ω-*allylthioureidobenzoic acid with cuprous chloride: Bockmühl, Fritzsche, **Ger.** pat. **551,421** (1932 to I. G. Farben).

Dark brown powder. Freely sol in water. Insol in alcohol, ether.

THERAP CAT: Antirheumatic.

266. Allopregnane. *(5α)-Pregnane.* $C_{21}H_{36}$; mol wt 288.52. C 87.42%, H 12.58%. Prepn from corticosterone: Steiger, Reichstein, *Helv. Chim. Acta* **21,** 161 (1938); from allopregnan-3-one: Ruzicka *et al., ibid.* **22,** 1294 (1939); from 21-hydroxyallopregnane: Plattner *et al., ibid.* **27,** 1177

(1944); Casanova, Reichstein, *ibid.* **32**, 647 (1949); by degradation of conessine: Haworth *et al., J. Chem. Soc.* **1949**, 831; by Hofmann decomposition of 3β- and 3α-dimethylaminopregnanes: Haworth *et al., ibid.* **1953**, 1110.

Crystals from acetone + methanol, mp 84-85°. [α]$_D^{19}$ +18.4° (c = 1.69 in chloroform).

267.   Allopregnane-3α,20α-diol.   *5α-Pregnane-3α,20α-diol;* 3α,20α-dihydroxy-5α-pregnane. $C_{21}H_{36}O_2$; mol wt 320.52.   C 78.70%, H 11.32%, O 9.98%.   Progesterone metabolite. Isoln from human pregnancy urine: Beall, *Biochem. J.* **31**, 35 (1937); from bulls' urine: Marker *et al., J. Am. Chem. Soc.* **60**, 2931 (1938). Prepn by reduction of 5α-pregn-1-ene-3,20-dione with lithium aluminum hydride: Schütt, Tamm, *Helv. Chim. Acta* **41**, 1751 (1958).
Crystals from methanol, mp 243-245°. [α]$_D^{20}$ +17° (c = 0.148 in ethanol).
Diacetate, $C_{25}H_{40}O_4$, crystals, mp 139.5-140.5°. [α]$_D^{20}$ +18° (c = 0.408 in benzene).

268.   Allopregnane-3α,20β-diol.   *(3α,5α,20R)-Pregnane-3,20-diol;* 3α,20β-dihydroxy-5α-pregnane. $C_{21}H_{36}O_2$; mol wt 320.52. C 78.70%, H 11.32%, O 9.98%. Progesterone metabolite. Prepn by hydrogenation of progesterone: Marker, Lawson, *J. Am. Chem. Soc.* **61**, 588 (1939); by catalytic hydrogenation of epiallopregnane-3-ol-20-one: Marker, U.S. pat. **2,231,019** (1941 to Parke, Davis); by reduction of 5α-pregn-1-ene-3,20-dione with lithium aluminum hydride: Schütt, Tamm, *Helv. Chim. Acta* **41**, 1751 (1958).
Needles from acetone, mp 207-209°. [α]$_D^{26}$ +12° (c = 1.132 in chloroform).

269.   Allopregnane-3β,20α-diol.   *5α-Pregnane-3β,20α-diol;* 3β,20α-dihydroxy-5α-pregnane. $C_{21}H_{36}O_2$; mol wt 320.52.   C 78.70%, H 11.32%, O 9.98%.   Progesterone metabolite. Isoln from pregnant mares' urine: Brooks *et al., Biochem. J.* **51**, 694 (1952). Prepn by heating 3α,20α-dihydroxy-5α-pregnane with sodium, or by hydrogenation of 5α-pregnane-20-ol-3-one in acid: Marker, U.S. pats. **2,196,220** and **2,250,962** (1940, 1941, both to Parke, Davis); by reduction of allopregnane-3β-ol-20-one acetate with sodium: Klyne, Barton, *J. Am. Chem. Soc.* **71**, 1500 (1949); by reduction of 5,16-pregnadien-3β-ol-20-one with Zn + acetic acid: Ercoli, De Ruggieri, *Farm. Sci. Tec. Pavia* **7**, 11 (1952), *C.A.* **46**, 10186b (1952).
Crystals from acetone, mp 218-219°. [α]$_D^{22}$ +23° (c = 0.93 in chloroform).
Diacetate, $C_{25}H_{40}O_4$, crystals from petr ether, mp 164-165°. [α]$_D^{19}$ +0.8° (c = 1.2 in chloroform).
Dibenzoate, $C_{35}H_{44}O_4$, crystals, mp 170.5-172°. [α]$_D^{20}$ +27.7° (c = 0.84 in chloroform).

270.   Allopregnane-3β,20β-diol.   *(3β,5α,20R)-Pregnane-3,20-diol;* 3β,20β-dihydroxy-5α-pregnane. $C_{21}H_{36}O_2$; mol wt 320.52. C 78.70%, H 11.32%, O 9.98%. Progesterone metabolite. Isoln from pregnant mares' urine: Brooks *et al., Biochem. J.* **51**, 694 (1952). Prepn by catalytic reduction of allopregnane-3,20-dione: Marker *et al.,* **Brit.** pat. **512,940** (1939 to Parke, Davis); by hydrogenation of pregn-5-ene-3β-ol-20-one: Klyne, Barton, *J. Am. Chem. Soc.* **71**, 1500 (1949); by hydrogenation of pregn-5-ene-3β,20β-diol: Klyne, Miller, *J. Chem. Soc.* **1950**, 1972.
Leaflets from ethyl acetate + petr ether, mp 194.5-195.5°. [α]$_D$ +4.4° (c = 1.04 in chloroform).
Diacetate, $C_{25}H_{40}O_4$, needles from ethyl acetate + petr ether, mp 141-142.5°. [α]$_D^{20}$ +21° (c = 1 in chloroform).
Dibenzoate, $C_{35}H_{44}O_4$, crystals, mp 237.5-239°. [α]$_D^{20}$ −10.1° (c = 2.08 in chloroform).

271.   Allopregnane-3β,21-diol-11,20-dione.   *3β,21-Dihydroxy-5α-pregnane-11,20-dione;* 3β,21-dihydroxy-11,20-dioxo-5α-pregnane; Kendall's compound H; Reichstein's substance N. $C_{21}H_{32}O_4$; mol wt 348.48. C 72.38%, H 9.26%, O 18.36%.   Isoln from adrenal glands: Steiger, Reichstein, *Helv. Chim. Acta* **21**, 546 (1938). Prepn by reduction of allopregnan-21-ol-3,11,20-trione acetate: Mancera *et al., J. Am. Chem. Soc.* **77**, 5669 (1955).
Hydrated crystals from dil acetone; anhyd shiny spears from benzene or acetone, mp 189-191°. [α]$_D^{19}$ +93.8° (c = 1.4 in abs ethanol).
3,21-Diacetate, $C_{25}H_{36}O_6$, clusters of needles from ether-pentane, mp 148-149.5°. [α]$_D^{18}$ +77.5° ± 2.5° (acetone); [α]$_D^{19}$ +85.6 ± 2° (dioxane).

272.   Allopregnane-3β,17α-diol-20-one.   *3β,17-Dihydroxy-5α-pregnan-20-one;* 3β,17-dihydroxy-20-oxo-5α-pregnane; 17-(1-ketoethyl)androstane-3,17-diol; Reichstein's substance L; Wintersteiner's compound G. $C_{21}H_{34}O_3$; mol wt 334.50.   C 75.41%, H 10.25%, O 14.35%.   Isoln from adrenal cortex: Wintersteiner, Pfiffner, *J. Biol. Chem.* **116**, 291 (1936). Structure: Reichstein, Gätzi, *Helv. Chim. Acta* **21**, 1497 (1938). Prepn from allopregnan-3β-ol-20-one 3-acetate: Rosenkranz *et al., J. Am. Chem. Soc.* **72**, 4081 (1950); from pregn-5-ene-3β-ol-20-one: Ramirez, Stafie, *ibid.* **77**, 134 (1955).
Crystals from abs alcohol, mp 264-266°. [α]$_D^{21}$ +30.6° (c = 0.54 in abs alcohol).
3-Acetate, $C_{23}H_{36}O_4$, crystals from acetone, mp 187-189°. [α]$_D^{20}$ +18° (acetone).

273.   3,20-Allopregnanedione.   *5α-Pregnane-3,20-dione;* 3,20-dioxo-5α-pregnane. $C_{21}H_{32}O_2$; mol wt 316.48. C 79.70%, H 10.19%, O 10.11%. From pregnancy urine: Hartmann, Locher, *Helv. Chim. Acta* **18**, 160 (1935); Lieberman *et al., J. Biol. Chem.* **172**, 263 (1948). Prepn from pregna-4,11-diene-3,20-dione: Shoppee, Reichstein, *Helv. Chim. Acta* **24**, 356 (1941); by oxidation of allopregnan-3β-ol-20-one with chromium trioxide: Billeter, Miescher, *ibid.* **30**, 1409 (1947); by hydrogenation of pregnenolone: Pappas, Nace, *J. Am. Chem. Soc.* **81**, 4556 (1959); from funtumine: Janot *et al., Bull. Soc. Chim. France* **1960**, 1669.
Crystals from methylene chloride + hexane, mp 200°. [α]$_D$ +125° (c = 1.2 in chloroform).
USE: In the synthesis of progesterone, Fr. pat. **845,034**, *C.A.* **34**, 8184² (1940).

274.   Allopregnane-3β,11β,17α,20β,21-pentol.   *5α-Pregnane-3β,11β,17,20β,21-pentol;* 3β,11β,17,20β,21-pentahydroxy-5α-pregnane; 17-(1,2-dihydroxyethyl)androstane-3,11,17-triol; Kendall's compound D; Reichstein's substance A; Wintersteiner's compound A. $C_{21}H_{36}O_5$; mol wt 368.51.   C 68.45%, H 9.85%, O 21.71%.   Occurs in adrenal cortex: Mason *et al., J. Biol. Chem.* **114**, 613 (1936). Prepn from 21-acetoxy-3β,17α-dihydroxy-5α-pregnane-11,20-dione + lithium aluminum hydride: Klyne, Ridley, *J. Chem. Soc.* **1956**, 4825; by hydrogenation of cortisol with platinum oxide: Caspi, *J. Org. Chem.* **24**, 669 (1959).
Crystals from methanol, mp 159-163°, resolidifies and mp 215-216°. [α]$_D^{22}$ +9.8° (c = 0.56 in methanol).

275.   Allopregnane-3β,17α,20β,21-tetrol.   *Pregnane-3,17,20,21-tetrol;* 3β,17,20β,21-tetrahydroxy-5α-pregnane; 17-(1,2-dihydroxyethyl)androstane-3,17-diol; Reichstein's substance K. $C_{21}H_{36}O_4$; mol wt 352.51. C 71.55%, H 10.29%, O 18.15%.   Isoln from adrenal cortex: Steiger, Reichstein, *Helv. Chim. Acta* **21**, 546 (1938). Prepn by reduction of allopregn-17-ene-3,17-diol 3,21-diacetate with osmium tetroxide: Serini *et al., Ber.* **72**, 391 (1939).
Leaflets from dilute methanol, mp 198-200°. [α]$_D^{21}$ −1.0° (c = 1 in abs ethanol).
3,20,21-Triacetate, $C_{27}H_{42}O_7$, rods from ether + pentane, mp 179°. [α]$_D^{19}$ +53.2° (c = 1.9 in acetone).

276.   Allopregnane-3α,11β,17α,21-tetrol-20-one.   *3α,11β,17,21-Tetrahydroxy-5α-pregnan-20-one;* 3α,11β,17,21-tetrahydroxy-20-oxo-5α-pregnane; 17-(1-keto-2-hydroxyethyl)androstane-3,11,17-triol; 3α-allotetrahydrocortisol; Kendall's compound C; Reichstein's substance C; Wintersteiner's compound D. $C_{21}H_{34}O_5$; mol wt 366.50. C 68.82%, H 9.35%, O 21.83%.   Isoln from adrenal cortex: Mason *et al.,*

J. Biol. Chem. **124**, 459 (1938); Kuizenga, Cartland, *Endocrinology* **24**, 526 (1939); v. Euw, Reichstein, *Helv. Chim. Acta* **25**, 988 (1942); v. Euw *et al., ibid.* **41**, 1516 (1958). Prepn by hydrogenation of cortisol with rhodium: Caspi, *J. Org. Chem.* **24**, 669 (1959); from bismethylenedioxyhydrocortisone: Fukushima, Daum, *ibid.* **26**, 520 (1961).

Crystals from methanol, mp 244-245°. $[\alpha]_D^{21}$ +59.7° (c = 0.34 in methanol).

3,21-Diacetate, $C_{25}H_{38}O_7$, crystals, dec 204-205°. $[\alpha]_D^{20}$ +73.8°; $[\alpha]_{542}^{20}$ +90.5° (dioxane).

277. Allopregnane-3β,11β,17α,21-tetrol-20-one. *3β,- 11β,17,21-Tetrahydroxy-5α-pregnan-20-one;* 3β,11β,17,21-tetrahydroxy-20-oxo-5α-pregnane; Reichstein's substance V. $C_{21}H_{34}O_5$; mol wt 366.50. C 68.82%, H 9.35%, O 21.83%. Isoln from adrenal cortex: von Euw, Reichstein, *Helv. Chim. Acta* **25**, 988 (1942). Prepn from 5α-pregnan-3β-ol-11,20-dione: Chamberlin, Chemerda, *J. Am. Chem. Soc.* **77**, 1221 (1955); by hydrogenation of cortisol with rhodium: Caspi, *J. Org. Chem.* **24**, 669 (1959); from bismethylenedioxyhydrocortisone: Fukushima, Daum, *ibid.* **26**, 520 (1961).

Needles from dilute methanol, decomp 220-225°. $[\alpha]_D^{13}$ +50.7° (c = 0.8672 in dioxane).

3,21-Diacetate, $C_{25}H_{38}O_7$, flat prisms from acetone-ether, decomp 225-227°. $[\alpha]_D^{18}$ +62.6° (c = 1.101 in dioxane).

278. Allopregnane-3β,17α,20α-triol. *5α-Pregnane-3β,17,20α-triol;* 3β,17,20α-trihydroxy-5α-pregnane; Reichstein's substance O. $C_{21}H_{36}O_3$; mol wt 336.52. C 74.95%, H 10.78%, O 14.26%. Isoln from adrenal glands: Steiger, Reichstein, *Helv. Chim. Acta* **21**, 546 (1938). Prepn from allopregn-16-en-3β-ol-20-one: Plattner *et al., ibid.* **31**, 2210 (1948); Julian *et al.,* U.S. pat. **2,662,904** (1953 to Glidden).

Leaflets from dil methanol, mp 222-223°. $[\alpha]_D^{20}$ −12.55° (c = 1.195 in methanol).

3,20-Diacetate, $C_{25}H_{40}O_5$, crystals by sublimation, mp 252°. $[\alpha]_D^{21}$ −30.1° (c = 1.097 in acetone).

279. Allopregnane-3β,17α,20β-triol. *5α-Pregnane-3β,17,20β-triol;* 3β,17,20β-trihydroxy-5α-pregnane; 17-(1-hydroxyethyl)androstane-3,17-diol; Reichstein's substance J. $C_{21}H_{36}O_3$; mol wt 336.52. C 74.95%, H 10.78%, O 14.26%. Isoln from adrenal cortex: Reichstein, *Helv. Chim. Acta* **19**, 1126 (1936); Steiger, Reichstein, *ibid.* **21**, 546 (1938). Prepn from 3β-acetoxy-16α,17α-oxido-20-oxo-5-allopregnane: Plattner *et al., ibid.* **31**, 2210 (1948); *Brit. pat.* **665,254** (1952 to Ciba); from 3β-acetoxy-17α-hydroxyallopregnan-20-one: Fukushima, Meyer, *J. Org. Chem.* **23**, 174 (1958).

Rhomboid needles from dil acetone, mp 216-217°. Sublimes in high vacuum. $[\alpha]_D^{20}$ −7.9° (c = 1.77 in abs alc).

3,20-Diacetate, $C_{25}H_{40}O_5$, crystals from pentane, mp 161-162°. $[\alpha]_D^{20}$ +24.6° (c = 2.11 in acetone).

280. Allopregnane-3β,17α,21-triol-11,20-dione. *3β,-17,21-Trihydroxy-5α-pregnane-11,20-dione;* 3β,17,21-trihydroxy-11,20-dioxo-5α-pregnane; 17-(1-keto-2-hydroxyethyl)-androstane-3,17-diol-11-one; Kendall's compound G; Reichstein's substance D; Wintersteiner's compound B. $C_{21}H_{32}O_5$; mol wt 364.48. C 69.20%, H 8.85%, O 21.95%. Occurs in adrenal cortex: Kuizenga, Cartland, *Endocrinology* **24**, 526 (1939); von Euw, Reichstein, *Helv. Chim. Acta* **25**, 988 (1942); **41**, 1516 (1958). Prepn by reduction of allopregnan-17α,21-diol-3,11,20-trione: Kaufmann, Pataki, *Experientia* **7**, 260 (1951); from 5α-pregnan-3β-ol-11,20-dione: Chamberlin, Chemerda, *J. Am. Chem. Soc.* **77**, 1221 (1955).

Crystals from abs alc, mp 238-242°. $[\alpha]_D^{20}$ +61.8° (c = 1.07 in dioxane).

3,21-Diacetate, $C_{25}H_{36}O_7$, crystals, mp 223-224°. $[\alpha]_D^{20}$ +72.3° (dioxane).

281. Allopregnane-3β,11β,21-triol-20-one. *3β,11β,-21-Trihydroxy-5α-pregnan-20-one;* 3β,11β,21-trihydroxy-20-oxo-5α-pregnane; Reichstein's substance R. $C_{21}H_{34}O_4$; mol wt 350.50. C 71.96%, H 9.78% O 18.26%. Isolation from adrenal glands: Reichstein, von Euw, *Helv. Chim. Acta* **21**, 1197 (1938); Reichstein, *ibid.* 1490. Partial synthesis by hydrogenation of corticosterone acetate: Pataki *et al., J. Biol. Chem.* **195**, 751 (1952). Prepn from 3β,11β-dihydroxy-alloetiocholanic acid: Lardon, Reichstein, *Helv. Chim. Acta*

37, 443 (1954); from hydrocortisone: Mancera *et al, J. Am. Chem. Soc.* **77**, 5669 (1955); from 3β,21-diacetoxy-17(20)-allopregnene + osmium tetroxide and triethylamine oxide peroxide: Schneider, Hanze, U.S. pat. **2,769,823** (1956 to Upjohn).

Needles from alc, mp 202-204°. $[\alpha]_D$ +110° (ethanol).

3,21-Diacetate, $C_{25}H_{38}O_6$, crystals from acetone + ether, mp 170-172°. $[\alpha]_D^{22}$ +82.5° (c = 1.38 in dioxane); $[\alpha]_D^{20}$ +101° (acetone).

282. Allopregnane-3β,17α,21-triol-20-one. *3β,17,21-Trihydroxy-5α-pregnan-20-one;* 3α,17,21-trihydroxy-20-oxo-5α-pregnane; Reichstein's substance P. $C_{21}H_{34}O_4$; mol wt 350.50. C 71.96%, H 9.78%, O 18.26%. Isoln from adrenal glands: Steiger, Reichstein, *Helv. Chim. Acta* **21**, 546 (1938); Reichstein, Gatzi, *ibid.* 1185. Partial synthesis from cholesterol: Reichstein, von Euw, *ibid.* **24**, 401 (1941). Prepn from allopregnan-3β-ol-20-one: Rosenkranz *et al., J. Am. Chem. Soc.* **72**, 4081 (1950); from 3β-acetoxy-21-bromo-17α-hydroxy-2-oxoallopregnane: Kaufmann *et al.,* U.S. pat. **2,596,562** (1952 to Syntex).

Pointed needles from abs ethanol, decomp 230-239°. $[\alpha]_D^{20}$ +48.0° (c = 0.938 in abs ethanol). Freely sol in alcohol, acetone. Sparingly sol in ether, water.

3,21-Diacetate, $C_{25}H_{38}O_6$, crystals from benzene, mp 208-210°. $[\alpha]_D^{20}$ +41.5° (chloroform).

283. Allopregnan-3α-ol-20-one. *3α-Hydroxy-5α-pregnan-20-one;* 3α-hydroxy-20-oxo-5α-pregnane; epiallopregnan-3-ol-20-one. $C_{21}H_{34}O_2$; mol wt 318.50. C 79.19%, H 10.76%, O 10.05%. Isoln from pregnancy urine of women: Marker *et al., J. Am. Chem. Soc.* **59**, 616 (1937); Lieberman *et al., J. Biol. Chem.* **172**, 263 (1948); Davis, Plotz, *Acta Endocrinol.* **21**, 245 (1956). Prepn from pregnenolone: Fleischer *et al., J. Am. Chem. Soc.* **60**, 79 (1938); Schwenk *et al.,* U.S. pat. **2,180,614** (1940 to Schering); by reduction of allopregnane-3,20-dione: Soloway *et al., J. Am. Chem. Soc.* **75**, 2356 (1953).

Crystals from abs alc, mp 176-178°. $[\alpha]_D$ +87.7° (abs alc).

Acetate, $C_{23}H_{36}O_3$, crystals from aq ethanol, mp 141-142°. $[\alpha]_D^{22}$ +94.5° (abs ethanol).

284. Allopregnan-3β-ol-20-one. *3β-Hydroxy-5α-pregnan-20-one;* 3β-hydroxy-20-oxo-5α-pregnane. $C_{21}H_{34}O_2$; mol wt 318.50. C 79.19%, H 10.76%, O 10.05%. Isoln from adrenal cortex: von Euw, Reichstein, *Helv. Chim. Acta* **24**, 885 (1941); from corpus luteum: Butenandt, Mamoli, *Ber.* **68**, 1847 (1935); Prelog, Meister, *Helv. Chim. Acta* **32**, 2435 (1949); from human pregnancy urine: Lieberman *et al., J. Biol. Chem.* **172**, 263 (1948); from human placenta: Pearlman, Cerceo, *ibid.* **194**, 807 (1952). Prepn from 20-methyl-Δ^{20}-allopregnen-3β-ol: Koechlin, Reichstein, *Helv. Chim. Acta* **27**, 549 (1944); from pregnenolone: Mancera *et al., J. Org. Chem.* **16**, 192 (1951); Pappas, Nace, *J. Am. Chem. Soc.* **81**, 4556 (1959); by redn of allopregnane-3,20-dione with sodium borohydride: Mancera *et al., ibid.* **75**, 1286 (1953); from progesterone 20-cycloethylene ketal: Sondheimer, Klibansky, *Tetrahedron* **5**, 15 (1959).

Plates from dil methanol, mp 194-195°. $[\alpha]_D^{27}$ +91.2° (c = 0.4 in ethanol).

Acetate, $C_{23}H_{36}O_3$, plates from methanol, mp 144-146°. $[\alpha]_D^{25}$ +69° (chloroform).

285. Allopregnan-20α-ol-3-one. *20α-Hydroxy-5α-pregnan-3-one;* 20α-hydroxy-3-oxo-5α-pregnane. $C_{21}H_{34}O_2$; mol wt 318.50. C 79.19%, H 10.76%, O 10.05%. Prepn from funtumidine: Janot *et al., Bull. Soc. Chim. France* **1960**, 1669.

Crystals, mp 179°. $[\alpha]_D$ +36.6° (c = 0.8 in chloroform).

Acetate, $C_{23}H_{36}O_3$, crystals from ethanol, mp 155°. $[\alpha]_D$ +24° (c = 1.5 in chloroform).

286. Allopregnan-20β-ol-3-one. *20β-Hydroxy-5α-pregnan-3-one;* 20β-hydroxy-3-oxo-5α-pregnane. $C_{21}H_{34}O_2$; mol wt 318.50. C 79.19%, H 10.76%, O 10.05%. Prepn from allopregnane-3β,20β-diol-3-acetate: Rubin *et al., J. Am. Chem. Soc.* **73**, 2338 (1951); from isofuntumidine: Janot *et al., Bull. Soc. Chim. France* **1960**, 1669.

Crystals from heptane, mp 185°. $[\alpha]_D$ +20° (c = 1.2 in chloroform).

Acetate, $C_{23}H_{36}O_3$, crystals from ethanol, mp 148°. $[\alpha]_D$ +57° (c = 1.9 in chloroform).

287. Allopurinol. *1,5-Dihydro-4H-pyrazolo[3,4-d]pyrimidin-4-one; 1H-pyrazolo[3,4-d]pyrimidin-4-ol;* 4-hydroxypyrazolo[3,4-*d*]pyrimidine; HPP; BW-56158; Adenock; Aloral; Alositol; Allo-Puren; Allozym; Allural; Anoprolin; Anzief; Apulonga; Apurol; Apurin; Bleminol; Bloxanth; Caplenal; Cellidrin; Cosuric; Dabroson; Embarin; Epidropal; Foligan; Geapur; Gichtex; Hamarin; Hexanurat; Ketanrift; Ketobun-A; Ledopur; Lopurin; Lysuron; Miniplanor; Monarch; Nektrohan; Remid; Riball; Sigapurol; Suspendol; Takanarumin; Urbol; Uricemil; Uripurinol; Urobenyl; Urosin; Urtias; Xanturat; Zyloprim; Zyloric. $C_5H_4N_4O$; mol wt 136.11. C 44.12%, H 2.96%, N 41.16%, O 11.75%. Xanthine oxidase inhibitor which decreases uric acid production. Prepn: Robins, *J. Am. Chem. Soc.* **78**, 784 (1956); Schmidt, Druey, *Helv. Chim. Acta* **39**, 986 (1956); Druey, Schmidt, U.S. pat. **2,868,803** (1959 to Ciba); **Brit.** pat. **798,-646** (1958 to Wellcome Found.); Hitchings, Falco, U.S. pat. **3,474,098** (1969 to Burroughs Wellcome). Physiological and biochemical studies: Hitchings, in *Biochem. Aspects Antimetab. Drug Hydroxylation,* D. Shugar, Ed. (Academic Press, London, 1969) pp 11-22, *C.A.* **75**, 3531h (1971). Clinical trial in treatment of renal calculi: M. J. V. Smith, *J. Urol.* **117**, 690 (1977); B. Ettinger *et al., N. Engl. J. Med.* **315**, 1386 (1986). Use in hyperuricemia and gout: G. R. Boss, J. E. Seegmiller, *ibid.* **300**, 1459 (1977). Effect on renal function in treatment of gout: T. Gibson, *Ann. Rheum. Dis.* **41**, 59 (1982). Comprehensive description: S. A. Benezra, T. R. Bennett, *Anal. Profile Drug Subs.* **7**, 1-17 (1978).

Crystals, mp above 350°. uv max (0.1*N* NaOH): 257 nm (ϵ 7200); (0.1*N* HCl): 250 nm (ϵ 7600); (methanol): 252 nm (ϵ 7600). Soly in mg/ml at 25°: water 0.48; *n*-octanol <0.01; chloroform 0.60; ethanol 0.30; DMSO 4.6. pKa 10.2.

THERAP CAT: Treatment of hyperuricemia and chronic gout. Antiurolithic.

288. D-Allose. *β-D-Allopyranose.* $C_6H_{12}O_6$; mol wt 180.16. C 40.00%, H 6.71%, O 53.28%. An aldohexose sugar. Obtained from leaves of *Protea rubropilosa:* Beylis, Perold, *Chem. Commun.* **1971**, 597. Prepd by reduction of D-allonic lactone with sodium amalgam: Levene, Jacobs, *Ber.* **43**, 3147 (1910). Synthesis: Hughes, Speakman, *J. Chem. Soc.* **1965**, 2236; Baker *et al., Carbohyd. Res.* **24**, 192 (1972).

Crystals from dilute methanol, mp 128-128.5°. Shows mutarotation: $[\alpha]_D^{20}$ +0.58° (4 min) → +3.26° (10 min) → +14.41° (final value, 20 hr; c = 5). Soluble in water. Practically insol in alcohol.
Phenylosazone, $C_{18}H_{22}N_4O_4$, mp 178°.

289. Allotetrahydrocortisone. *(3α,5α)-3,17,21-Trihydroxy-pregnane-11,20-dione;* Reichstein's substance Dehydro-C; 11-dehydro C; allopregnane-3α,17α,21-triol-11,20-dione. $C_{21}H_{32}O_5$; mol wt 364.48. C 69.20%, H 8.85%, O 21.95%. Isoln from beef adrenals: v. Euw *et al., Helv. Chim. Acta* **41**, 1516 (1958). Synthesis of the 3α,21-diacetate: Nagata *et al., ibid.* **42**, 1399 (1959).

Rhombohedric leaflets from acetone + water, mp 212-214°.
3α,21-Diacetate, $C_{25}H_{36}O_7$, rhombohedric platelets, mp 225-227°; $[\alpha]_D^{25}$ +94.2° ±1.5° (c = 1.45 in dioxane).

290. Alloxan. *2,4,5,6(1H,3H)-pyrimidinetetrone;* 2,4,-5,6-tetraoxohexahydropyrimidine; mesoxalylurea; mesoxalylcarbamide. $C_4H_2N_2O_4$; mol wt 142.07. C 33.82%, H 1.42%, N 19.72%, O 45.05%. Found by Liebig in mucus excreted during dysentery. Prepn by direct oxidation of uric acid: G. Brugnatelli, *Ann. Chim. Phys. [2]* **8**, 201 (1818); J. Liebig, F. Wöhler, *Ann.* **26**, 241 (1838); H. Biltz, M. Hehn, *ibid.* **413**, 60 (1917). Prepd from alloxantin: J. Liebig, *ibid.* **147**, 366 (1868); W. W. Hartman, O. E. Sheppard, *Org. Syn.* **coll. vol. III**, 37 (1955). *See also* A. V. Holmgren, W. W. Wenner, *ibid.* **coll. vol. IV**, 23 (1963). Produces diabetes in animals by selective necrosis of pancreatic islet β-cells: J. S. Dunn, N. G. B. McLetchie, *Lancet* **2**, 384 (1943); W. B. Kennedy, F. D. W. Lukens, *Proc. Soc. Exp. Biol. Med.* **57**, 143 (1944). Mechanism of action study: L. Boquist, *Acta Pathol. Microbiol. Scand. Sect. A* **88**, 201 (1980). *In vitro* antineoplastic activity: P. Grobon, *Compt. Rend. Ser. D.* **280**, 2413 (1975). Antibacterial and antifungal activity: J. D. Douros, A. F. Kerst, **Japan. Kokai 72 4900**; *eidem,* **U.S.** pat. **3,728,454** (1972, 1973 both to Gates Rubber Co.). Toxicological study in mice: B. A. Waisbren, *Proc. Soc. Exp. Biol. Med.* **67**, 154 (1948).

Anhydrous, orthorhombic crystals from anhydr acetone or glacial acetic acid or by sublimation in vacuo. Turns pink at 230° and dec at 256°. Acid to litmus. K at 25° = 2.32×10^{-7}. Absorption spectrum: Hartley, *J. Chem. Soc.* **87**, 1802, 1808 (1905). Freely sol in water. Hot aqueous solns are yellow and become colorless on cooling. Aqueous solns, after being in contact with the human skin for some time, give it a red color and a disagreeable odor. Sol in acetone, alcohol, methanol, glacial acetic acid; slightly sol in chloroform, petr ether, toluene, ethyl acetate and acetic anhydride. Insol in ether.
Tetrahydrate, large triclinic prisms or oblique monoclinic rhombs from water.
Monohydrate, by heating the tetrahydrate at 100° or by exposing it to the air. Forms triclinic pinacoidal crystals.
Compound with urea, $CH_4N_2O \cdot C_4H_2N_2O_4 \cdot H_2O$, minute, yellow needles; red at 170°, decomp at 185-186°.
USE: In production of diabetes in experimental animals; in nutrition experiments; in organic syntheses.

291. Alloxantin. *5,5'-Dihydroxy-[5,5'-bipyrimidine]-2,2',4,4',6,6'(1H,1'H,3H,3'H,5H,5'H)-hexone; 5,5'-dihydroxy-5,5'-bibarbituric acid;* uroxin. $C_8H_6N_4O_8$; mol wt 286.16. C 33.58%, H 2.11%, N 19.58%, O 44.73%. Prepd from uric acid: Nightingale, *Org. Syn.* **coll. vol. III**, 42 (1955); from alloxan monohydrate: Tipson, *ibid.*, **coll. vol. IV**, 25 (1963). Structure: Singh, *Acta Cryst.* **19**, 767 (1965).

Dihydrate, cryst powder. On exposure to air becomes red. Becomes yellow at 225°; dec at 253-255°. Sparingly sol in cold water, alcohol or ether. Its aq soln is acid; reduces Ag salts and gives a blue ppt with $Ba(OH)_2$ soln. *Keep tightly closed.* Emits toxic fumes upon decompn.

Caution: Ingestion causes disturbed carbohydrate metabolism leading to diabetes, *cf. Dangerous Properties of Industrial Materials,* N. I. Sax, Ed. (Van Nostrand Reinhold, New York, 4th ed., 1975) p 366.

292. Allura® Red AC. *6-Hydroxy-5-[(2-methoxy-5-methyl-4-sulfophenyl)azo]-2-naphthalenesulfonic acid disodium salt; 6-hydroxy-5-[(6-methoxy-4-sulfo-m-tolyl)azo]-2-naphthalenesulfonic acid disodium salt;* 1-[(6-methoxy-4-sulfo-*m*-tolyl)azo]-2-naphthol-6-sulfonic acid disodium salt; FD & C Red No. 40; C.I. 16035; C.I. Food Red 17. $C_{18}H_{14}N_2Na_2O_8S_2$; mol wt 496.43. C 43.55%, H 2.84%, N 5.64%, Na 9.26%, O 25.78%, S 12.92%. Prepn: **Brit. 1,164,249,** *C.A.* **72,** 68195j (1970) corresp to Rast, Steiner, **U.S. pat. 3,519,617** (1969, 1970, both to Allied Chem.).

Dark red powder. Soly at 25°: water, 22.5%; 50% alcohol, 1.3%.

USE: Color additive. Approved by FDA for use in foods, drugs, cosmetics.

293. Allyl acetate. *Acetic acid 2-propenyl ester;* 3-acetoxy-1-propene. $C_5H_8O_2$; mol wt 100.12. C 59.98%, H 8.05%, O 31.96%. $CH_2{=}CHCH_2OCOCH_3$. Prepn: N. Zinin, *Ann.* **1855,** 361; K. N. Gurudutt *et al., Tetrahedron* **38,** 1843 (1982); and physical properties: G. H. Jeffery, A. I. Vogel, *J. Chem. Soc.* **1948,** 658. Vibrational spectra and thermodynamic functions: B. Singh *et al., Proc. Indian Acad. Sci.* **89,** 201 (1980). Acute toxicity: H. F. Smyth, Jr. *et al., J. Ind. Hygiene Toxicol.* **31,** 60 (1949); P. M. Jenner *et al., Food Cosmet. Toxicol.* **2,** 327 (1964). Effect on respiratory function in mice: G. D. Nielsen *et al., Acta Pharmacol. Toxicol.* **54,** 292 (1984).

bp_{760} 104° (Gurudutt); also reported as bp_{773} 104° (Jeffery, Vogel), bp 105° (Zinin). n_D 1.40396. d_4^{20} 0.9277. LD_{50}: 0.13 g/kg orally in rats; 1.1 ml/kg dermally in rabbits (Smyth). LD_{50} in rats, mice (mg/kg): 142, 170 orally (Jenner).

294. Allyl Alcohol. *2-Propen-1-ol;* 1-propenol-3; vinyl carbinol. C_3H_6O; mol wt 58.08. C 62.04%, H 10.41%, O 27.55%. $CH_2{=}CHCH_2OH$. Prepd by heating glycerol with formic acid: O. Kamm, C. S. Marvel, *Org. Syn.* coll. vol. I, 42 (2nd ed., 1941); *cf.* Delaby, Dubois, *Compt. Rend.* **188,** 710 (1929). Toxicity study: Smyth, Carpenter, *J. Ind. Hyg. Toxicol.* **30,** 63 (1948).

Colorless liquid; pungent, mustard-like odor. d_4^{20} 0.8540. bp 96-97°. mp −50°. n_D^{20} 1.41345. Absorption spectrum: Lüthy, *Z. Phys. Chem.* **107,** 289 (1923). Flash pt 70°F (open cup); 75°F (closed cup); autoignition temp 713°F. 72.3% of the alcohol with 27.7% water forms a constant boiling mixture, bp at 87.5°. Miscible with water, alcohol, chloroform, ether, petr ether. *Keep tightly closed.* Upon storage for several years allyl alcohol polymerizes and a thick syrup is formed (insol in water, sol in chloroform) which on treat-

ment with ether yields a brittle resinoid mass: Blicke, *J. Am. Chem. Soc.* **45,** 1563 (1923). LD_{50} orally in rats: 64 mg/kg (Smyth, Carpenter).

Caution: Potential symptoms of overexposure are eye irritation, tissue damage; irritation of upper respiratory system and skin; pulmonary edema. *See NIOSH Pocket Guide to Chemical Hazards* (DHHS/NIOSH 90-117, 1990) p 34.

USE: Manuf allyl compds, war gas, resins, plasticizers.

295. Allylamine. *2-Propen-1-amine;* 3-aminopropylene. C_3H_7N; mol wt 57.10. C 63.11%, H 12.36%, N 24.53%. $CH_2{=}CHCH_2NH_2$. Manuf from allyl chloride and ammonia: Ploetz, **U.S. pat. 2,915,385** (1959 to Feldmühle Papier- and Zellstoffwerke). Toxicity study: C. H. Hine *et al., Arch. Environ. Health* **1,** 34 (1960).

Liquid; burning taste; strong ammonia odor causing sneezing and tears. d_{20}^{20} 0.760. bp 55-58°. n_D^{20} 1.4186. Flash pt, closed cup: 10°F (−12°C). Misc with water, alc, chloroform, ether. *Keep tightly closed.* LD_{50} i.p. in mice: 49 mg/kg (Hine).

Caution: A strong eye and respiratory tract irritant; intolerable at 14 ppm, *cf.* Patty's *Industrial Hygiene and Toxicology,* vol. **2B,** G. D. Clayton, F. E. Clayton, Eds. (Wiley-Interscience, New York, 3rd ed., 1981) pp 3157-3158.

USE: In the manuf of mercurial diuretics.

296. Allyl Bromide. *3-Bromo-1-propene;* 3-bromopropylene. C_3H_5Br; mol wt 120.98. C 29.79%, H 4.17%, Br 66.05%. $CH_2{=}CHCH_2Br$. Prepd from hydrobromic acid and allyl alcohol: O. Kamm, C. S. Marvel, *Org. Syn.* **coll. vol. 1,** 27 (2nd ed., 1941); from triphenylphosphite, allyl alcohol and benzyl bromide: Landauer, Rydon, *J. Chem. Soc.* **1953,** 2224.

Colorless liquid; unpleasant, pungent odor. d_4^{20} 1.398. bp_{760} 71.3°. mp −119°. n_D^{20} 1.46545. Slightly sol in water; miscible with alcohol, chloroform, ether, carbon disulfide, carbon tetrachloride. *Keep tightly closed.*

Caution: Produces irritation of eyes, respiratory passages.

USE: Manuf synthetic perfumes, other allyl compounds.

297. Allyl Chloride. *3-Chloro-1-propene;* 3-chloropropylene; Chlorallylene. C_3H_5Cl; mol wt 76.53. C 47.09%, H 6.59%, Cl 46.33%. $CH_2{=}CHCH_2Cl$. Manuf by chlorination of propylene: Samples, Hilbert, **U.S. pat. 3,054,831** (1962 to Union Carbide). Prepd from diphenylphosphite, allyl alcohol and benzyl chloride: Landauer, Rydon, *J. Chem. Soc.* **1953,** 2224. Toxicity data: H. F. Smyth, C. P. Carpenter, *J. Ind. Hyg. Toxicol.* **30,** 63 (1948).

Liquid; unpleasant, pungent odor. d_4^{20} 0.938. bp 44-45°. mp −134.5°. n_D^{20} 1.4154. Flash pt, closed cup: −25°F (−31°C). Slightly sol in water; misc with alcohol, chloroform, ether, petrol ether. *Keep tightly closed.* LD_{50} orally in rats: 0.7 g/kg (Smyth, Carpenter).

Caution: Potential symptoms of overexposure are irritation of eyes, nose and skin; pulmonary edema. *See NIOSH Pocket Guide to Chemical Hazards* (DHHS/NIOSH 90-117, 1990) p 36.

USE: In the synthesis of allyl compds.

298. Allyl Cyanide. *3-Butenenitrile;* vinylacetonitrile; β-butenonitrile. C_4H_5N; mol wt 67.09. C 71.61%, H 7.51%, N 20.88%. $CH_2{=}CHCH_2CN$. Found in some mustard oils. Prepd by treating dry CuCN with allyl bromide: Bruylants, *Bull. Soc. Chim. Belg.* **31,** 175 (1922); Supniewski, Salzberg, *Org. Syn.* **coll. vol. I,** 46 (2nd ed., 1941). Toxicity study: H. F. Smyth *et al., Am. Ind. Hyg. Assoc. J.* **30,** 470 (1969).

Liquid. Agreeable onion-like odor. Stable to heat. d_4^{20} 0.8341. mp −87°. bp_{760} 119°; bp_{400} 98°; bp_{200} 78°; bp_{100} 60.2°; bp_{60} 48.8°; bp_{40} 40.0°; bp_{20} 26.6°; bp_{10} 14.1°; bp_5 +2.9°; $bp_{1.0}$ −19.6°. n_D^{20} 1.4060. Absorption spectrum in hexane, water and dil alkali: Bruylants, Castille, *op. cit.* **34,** 265; *Chem. Zentr.* **1926, I,** 1962. LD_{50} orally in rats: 0.115 g/kg (Smyth).

299. Allylestrenol. *(17β)-17-(2-Propenyl)estr-4-en-17-ol; 17-allylestr-4-en-17β-ol;* 17α-allyl-17-hydroxy-19-nor-4-androstene; 17-hydroxy-17α-allyl-4-estrene; Gestanin; Gestanol; Orageston; Gestanyn; Gestanon; Turinal. $C_{21}H_{32}O$; mol wt 300.48. C 83.94%, H 10.73%, O 5.32%. Prepn: **Brit. pat. 841,411** (1960 to Organon).

Crystals, mp 79.5-80°. Practically insol in water. Sol in alcohol, ether, acetone, chloroform. Sensitive to oxidizing agents.

THERAP CAT: Progestogen.

300. Allyl Ether. *3,3′-Oxybis-1-propene;* diallyl ether. $C_6H_{10}O$; mol wt 98.14. C 73.43%, H 10.27%, O 16.30%. $(CH_2=CHCH_2)_2O$. Prepn from allyl alcohol in the presence of $CuCl$-H_2SO_4: Stephenson, **Brit.** pat. **913,919** (1962 to Monsanto); from allyl alcohol with K_2CO_3 and allyl bromide: Riemschneider, Kötzsch, *Monatsh.* **90,** 787 (1959). Toxicity: H. F. Smyth *et al., J. Ind. Hyg. Toxicol.* **31,** 60 (1949).

Liquid; radish-like odor. d_0^{18} 0.805. bp 94°. n_D^{20} 1.4240. Practically insol in water. Miscible with alc, ether. LD_{50} orally in rats: 0.32 g/kg (Smyth).

Caution: An irritant. Can be absorbed through skin.

301. Allyl Ethyl Ether. *3-Ethoxy-1-propene;* ethyl allyl ether. $C_5H_{10}O$; mol wt 86.13. C 69.72%, H 11.70%, O 18.58%. $CH_2=CHCH_2OCH_2CH_3$. Prepd from allyl bromide and sodium alcoholate: Brühl, *Ann.* **200,** 139 (1880).

Liquid. d 0.765. $bp_{742.9}$ 66-67°. n_D^{20} 1.3881. Practically insol in water. Miscible with alc, ether.

Caution: Irritating to eyes, skin, mucous membranes.

302. Allyl Iodide. *3-Iodo-1-propene;* 3-iodopropylene. C_3H_5I; mol wt 167.98. C 21.45%, H 3.00%, I 75.55%. $CH_2=CHCH_2I$. Prepd from allyl alcohol, methyl iodide, and triphenyl phosphite: Landauer, Rydon, *J. Chem. Soc.* **1953,** 2224; Rydon, Landauer, **Brit.** pat. **695,468** (1953 to Nat. Res. Dev. Corp.).

Yellowish liquid; darkens on exposure to light and air, liberating iodine. Unpleasant, pungent odor. d^{12} 1.848. bp 103°. n_D^{11} 1.5540. Practically insol in water. Miscible with alc, chloroform, ether. *Keep tightly closed and protected from light.*

Caution: See Allyl Chloride.

303. Allyl Isothiocyanate. *3-Isothiocyanato-1-propene; isothiocyanic acid allyl ester;* allyl isosulfocyanate; "mustard oil"; allyl mustard oil; Redskin. C_4H_5NS; mol wt 99.16. C 48.45%, H 5.08%, N 14.13%, S 32.34%. $CH_2=CHCH_2NCS$. Principal constituent of *volatile oil of mustard.* Occurs in seeds as the glucoside, sinigrin, *q.v.* Isolated from *Brassica nigra* (L.) Koch, *Cruciferae* (black mustard seed), or prepd from allyl iodide and potassium thiocyanate: Dulière, *J. Pharm. Belg.* **2,** 981 (1920), *C.A.* **15,** 571⁴ (1921). Description: E. Guenther, D. Althausen, *The Essential Oils* vol. II (Van Nostrand, New York, 1949) pp. 734-7. Review of isolation: E. Guenther, *ibid.* vol VI, pp. 55-60 (1952); and synthesis: G. M. Dyson, *Perfum. Essent. Oil Rec.* **20,** 42 (1929). Evaluation in relieving muscle pain: J. G. Macarthur, S. Alstead, *Lancet* **2,** 1060 (1953). Toxicity study: P. M. Jenner *et al., Food Cosmet. Toxicol.* **2,** 327 (1964). HPLC determn in brown mustard: K. Kanemaru *et al., J. Japan. Soc. Food Sci. Technol.* **37,** 565 (1990).

Colorless or pale yellow, very refractive liquid; very pungent, irritating odor; acrid taste. d_4^{15} 1.024. bp 152°. n_D^{25} 1.5248. Sparingly sol in water; misc with alcohol and most organic solvents. One ml dissolves in 8 ml 70% alcohol. LD_{50} orally in rats: 339 mg/kg (Jenner).

USE: Manuf flavors; war gas.

THERAP CAT: Counterirritant.

THERAP CAT (VET): Counterirritant.

304. Allylprodine. *1-Methyl-4-phenyl-3-(2-propenyl)-4-piperidinol propanoate;* α-3-allyl-1-methyl-4-phenyl-4-propionoxypiperidine; α-3-allyl-1-methyl-4-phenyl-4-piperidinol propionate; NIH-7440; Ro-2-7113; Alperidine. $C_{18}H_{25}NO_2$; mol wt 287.40. C 75.22%, H 8.77%, N 4.87%, O

11.13%. Prepn: Ziering *et al., J. Org. Chem.* **22,** 1521 (1957); Lee, Ziering, **U.S.** pat. **2,798,073** (1957 to Hoffmann-La Roche).

Hydrochloride, $C_{18}H_{25}NO_2 \cdot HCl$, crystals from acetone + methanol, mp 186-187°.

Note: This is a controlled substance (opiate) listed in the U.S. Code of Federal Regulations, Title 21 Part 1308.11 (1995).

THERAP CAT: Analgesic (narcotic).

305. Allyl Sulfide. *3,3′-Thiobis-1-propene;* diallyl sulfide; thioallyl ether; "oil garlic". $C_6H_{10}S$; mol wt 114.21. C 63.10%, H 8.83%, S 28.08%. $(CH_2=CHCH_2)_2S$. Prepared from allyl bromide and sulfur in soln of Na in liquid NH_3: Brandsma, Wijers, *Rec. Trav. Chim.* **82,** 68 (1963).

Liquid, garlic odor. bp 139°. d_4^{27} 0.888. n_D^{27} 1.4877. Practically insoluble in water. Miscible with alc, chloroform, ether, carbon tetrachloride.

USE: Manuf flavors.

306. Allylurea. *N-2-Propenylurea;* allylcarbamide. $C_4H_8N_2O$; mol wt 100.12. C 47.99%, H 8.05%, N 27.98%, O 15.98%. $CH_2=CHCH_2NHCONH_2$. Prepn: Neville, McGee, *Can. J. Chem.* **41,** 2123 (1963).

Crystals, mp 85°. Freely sol in water, alc. Practically insol in chloroform, ether, toluene, carbon disulfide.

USE: Manuf allylthiourea; corrosion inhibitors.

307. Almagate. *[Carbonato(2)]heptahydroxy(aluminum)trimagnesium dihydrate; carbonic acid, aluminummagnesium complex;* Almax. $C_2H_{14}Al_2Mg_6O_{20}\cdot4H_2O$; mol wt 629.98. C 3.81%, H 3.52%, Al 8.57%, Mg 23.15%, O 60.95%. $Al_2Mg_6(OH)_{14}(CO_3)_2\cdot4H_2O$. A crystalline hydrated aluminum-magnesium hydroxycarbonate. Prepn: R. G. W. Spickett *et al.,* **Belg.** pat. **873,843;** *eidem,* **U.S.** pat. **4,447,417** (1979, 1984 both to Anphar). Crystal structure: J. Moragues *et al., Arzneimittel-Forsch.* **34,** 1346 (1984). pH curves, acid consuming capacity: J. E. Beneyto *et al., ibid.* 1350. Effect of proteolytic enzymes, polypeptides: J. E. Beneyto, J. L. Fábregas, *ibid.* 1357. Serum levels of magnesium, aluminum not increased by high doses of almagate: J. Jauregui, J. Segura, *ibid.* 1364. Pharmacology: P. R. Beckett *et al., ibid.* 1367. Clinical trials: A. Suau *et al., ibid.* 1380.

Dehydrates at 510°K. USP acid neutralization capacity 28.3 mEq HCl/g.

THERAP CAT: Antacid.

308. Alminoprofen. *α-Methyl-4-[(2-methyl-2-propenyl)amino]benzeneacetic acid;* 2-(p-methylallylaminophenyl)propionic acid. EB-382; Minalfene. $C_{13}H_{17}NO_2$; mol wt 219.28. C 71.21%, H 7.81%, N 6.39%, O 14.59%. Prepn: Fr. pat. **2,137,211;** E. Bouchera, **U.S.** pat. **3,957,850** (1971, 1976 both to Bouchera). Synthesis, pharmacology and toxicity data: B. Dumaitre *et al., Eur. J. Med. Chem.* **14,** 207 (1979). Pharmacokinetic study: A. Premel-Cabic *et al., Eur. J. Clin. Pharmacol.* **18,** 419 (1980).

Crystals from cyclohexane, mp 107°. LD$_{50}$ orally in mice: 2400 mg/kg (Dumaitre).

THERAP CAT: Anti-inflammatory; analgesic.

309. Almitrine. 6-[4-[Bis(4-fluorophenyl)methyl]-1-piperazinyl]-N,N'-di-2-propenyl-1,3,5-triazine-2,4-diamine; 2,4-bis[allylamino]-6-[4-[bis(p-fluorophenyl)methyl]-1-piperazinyl]-s-triazine; S-2620. C$_{26}$H$_{29}$F$_2$N$_7$; mol wt 477.56. C 65.39%, H 6.12%, F 7.96%, N 20.53%. Prepn: G. Regnier, R. Canevari, Ger. pat. 1,947,332; eidem, U.S. pat. 3,647,794 (1970, 1972 both to Sci. Union et Cie.-Soc. Franç. Recher. Méd.). Physical properties and toxicity: G. Regnier et al., Experientia 28, 814 (1972). Pharmacology: M. Laubie, F. Diot, J. Pharmacol. (Paris) 3, 363 (1972); M. Laubie, H. Schmitt, Eur. J. Pharmacol. 61, 125 (1980). Evaluation of combination with raubasine, q.v., in cerebral ischemia in rats: M. G. Borzeix, J. Cahn, Arzneimittel-Forsch. 37, 491 (1987). Clinical effect on hypoxemia in chronic obstructive pulmonary disease: R. C. Bell et al., Ann. Int. Med. 105, 342 (1986); B. Gothe et al., Am. J. Med. 84, 436 (1988).

Crystals, mp 181°.

Dimethanesulfonate, C$_{26}$H$_{29}$F$_2$N$_7$.2CH$_3$SO$_3$H, almitrine dimesylate, Vectarion. mp 243° (dec). uv max (ethanol): 227, 246 nm (log ε 4.52, 4.53). LD$_{50}$ in mice: 210 mg/kg i.v., 390 mg/kg i.p., >2 g/kg orally (Regnier).

THERAP CAT: Respiratory stimulant.

310. Almond, Bitter. Ripe seed of Prunus amygdalus Stokes var amara Focke (P. communis Arcang. var amara (Focke) Schneid.), Rosaceae. Habit. Italy, Spain, and Southern France. Constit. 35-50% fixed oil, about 3% amygdalin, proteins, emulsin (synaptase), sugar. Ref: E. W. Eckey, Vegetable Fats and Oils (Reinhold, New York, 1954) p 455.

Caution: Cyanide poisoning from ingestion of burnt bitter almonds has been reported: C.A. 50, 6666a (1956).

USE: Preparing amygdalin, essential and expressed oils, almond and bitter-almond water; in perfumery and in manuf of liqueurs.

311. Almond, Sweet. Jordan almond. Ripe seed of Prunus amygdalus Stokes var dulcis (D.C.) Baill. (P. communis Arcang. var dulcis (Focke) Schneid.); P. amygdalus var sativa (Focke), Rosaceae. Habit. Italy, Spain and Southern France. Constit. About 50% fixed oil, proteins, emulsin, sugar. Ref: E. W. Eckey, Vegetable Fats and Oils (Reinhold, New York, 1954) p 455; Subrahmanyam, Achaya, J. Sci. Food Agr. 8, 657 (1957).

USE: In perfumery and in confectionery; preparing expressed oil of almond, almond milk, almond meal, etc.

312. Aloe. Genus of succulent plants (family Liliaceae) having triangular, spear-like leaves with thorny ridges. Habit. Africa, southern Arabia, Madagascar. Naturalized in the West Indies, southern U.S., Central America, Asia. Several species have been used medicinally, primarily Aloe barbadensis Miller (also known as Aloe vera Linné, Curaçao aloe, or Barbados aloe) and Aloe ferox Miller (also known as Cape aloe). "Drug aloe", used as a cathartic, is the dried latex or bitter, yellow exudate from the leaf bundle sheath cells. The latex contains varying amounts of aloin, aloe-emodin, chrysophanic acid, q.q.v., volatile oil, and resins. Aloe vera gel is the fresh mucilaginous gel obtained from the parenchymatous tissue in the leaf center, used for its emollient and wound healing activity. Physical and chemical properties of the gel: Y.-T. Wang, K. J. Strong, Phytother. Res. 7, S1 (1993). Effects in wound healing: J. P.

Heggers et al., ibid. S48. Reviews: D. Grindlay, T. Reynolds, J. Ethnopharmacol. 16, 117-151 (1986); A. D. Klein et al., J. Am. Acad. Dermatol. 18, 714-720 (1988); R. M. Shelton, Int. J. Dermatol. 30, 679-683 (1991).

Aloeresin A, C$_{28}$H$_{28}$O$_{11}$, (E)-3-(4-hydroxyphenyl)-2-propenoic acid 2'-ester with 8-β-D-glucopyranosyl-7-hydroxy-5-methyl-2-(2-oxopropyl)-4H-1-benzopyran-4-one. A major constituent of Cape aloe latex. Structure: P. Gramatica et al., Tetrahedron Letters 23, 2423 (1982). mp 148-150°. uv max in methanol: 228, 252, 300 nm (ε 34250, 25000, 37960).

THERAP CAT: Latex as laxative; gel as emollient, vulnerary.

313. Aloe-Emodin. 1,8-Dihydroxy-3-(hydroxymethyl)-9,10-anthracenedione; 1,8-dihydroxy-3-(hydroxymethyl)anthraquinone; 3-hydroxymethylchrysazin; rhabarberone. C$_{15}$H$_{10}$O$_5$; mol wt 270.24. C 66.67%, H 3.73%, O 29.60%. Occurs in the free state and as a glycoside in Rheum (rhubarb), in senna leaves and in various species of Aloe (Liliaceae). Isolation: Condo-Vissicchio, Arch. Pharm. 247, 81 (1909); Mary et al., J. Am. Pharm. Assoc. 45, 229 (1956). Prepn: Cahn, Simonsen, J. Chem. Soc. 1932, 2573; Hay, Haynes, ibid. 1956, 3141; Bapat et al., Tetrahedron Letters no. 5, 15 (1960).

Orange needles from toluene, mp 223-224°. Sublimes in CO$_2$ stream. Absorption spectrum: Stone, Furman, J. Am. Chem. Soc. 68, 2742 (1946). Freely sol in hot alcohol, in ether, in benzene with yellow color, in ammonia water and in sulfuric acid with crimson color.

Trimethyl ether, C$_{18}$H$_{16}$O$_5$, orange needles from acetic acid, mp 163°.

Triacetate, C$_{21}$H$_{16}$O$_8$, yellow needles from benzene, mp 175-177°.

Note: See also Emodin.

THERAP CAT: Cathartic.

314. Aloin. 10-Glucopyranosyl-1,8-dihydroxy-3-(hydroxymethyl)-9(10H)-anthracenone; 1,8-dihydroxy-3-hydroxymethyl-10-(6-hydroxymethyl-3,4,5-trihydroxy-2-pyranyl)anthrone; 10-(1',5'-anhydroglucosyl)-aloe-emodin-9-anthrone; barbaloin. C$_{21}$H$_{22}$O$_9$; mol wt 418.40. C 60.28%, H 5.30%, O 34.42%. Isoln from various species of aloe (Cape, Uganda and Socotrine): Léger, Ann. Chim. 6, 318 (1916); 8, 265 (1918); Cahn, Simonsen, J. Chem. Soc. 1932, 2573; Rosenthaler, Arch. Pharm. 270, 214 (1932); Mühlemann, Pharm. Acta Helv. 27, 17 (1952); Böhme, Bertram, Arch. Pharm. 288, 510 (1955). The molecule is built from aloe-emodin, q.v., and from glucose. Structure: Hay, Haynes, J. Chem. Soc. 1956, 3141.

Lemon-yellow crystals, mp 148-149°. Quickly forms a monohydrate, mp 70-80°. Slight odor of aloe; bitter taste. Solubility at 18°: 57% in pyridine, 7.3% in glacial acetic acid, 5.4% in methanol, 3.2% in acetone, 2.8% in methyl acetate, 1.9% in ethanol, 1.8% in water, 1.6% in propanol, 0.78% in ethyl acetate, 0.27% in isopropanol. Very slightly soluble in isobutanol, chloroform, carbon disulfide, ether. Incompatible with alkali hydroxides, tannin, ferric chloride.

THERAP CAT: Cathartic.

THERAP CAT (VET): Purgative.

315. Aloxidone. *5-Methyl-3-(2-propenyl)-2,4-oxazolidinedione; 3-allyl-5-methyl-2,4-oxazolidinedione;* allomethadione; Malidone; Malazol. $C_7H_9NO_3$; mol wt 155.15. C 54.19%, H 5.85%, N 9.03%, O 30.94%. Prepn: Davies, Hook, **Brit.** pats. **627,971; 632,423** (1949 to British Schering).

Oily liquid. $bp_{0.5}$ 86-87°; $bp_{1.8}$ 88.90°. n_D^{20} 1.4710. Slightly sol in water; sol in alcohol. Not miscible with vegetable oils. Forms a sodium salt.

THERAP CAT: Anticonvulsant.

316. Aloxiprin. *2-(Acetyloxy)benzoic acid polymer with aluminum oxide;* polyoxyaluminum acetylsalicylate; Alaprin; Lyman; Palaprin; Rumatral; Superpyrin. A polymeric condensation product of aluminum oxide and aspirin. Approx formula, $Al_3O_2[C_6H_4(OOCCH_3)COO]_5$. Prepd from aluminum isopropoxide and aspirin: Cummings *et al., J. Pharm. Pharmacol.* **15,** 56 (1963).

Powder with greater bulk density and relatively low free salicylic acid content as compared with other aluminum salicylates. Rates of soln of aspirin from aloxiprin in buffer solns of pH 2-8 are lower than those of aspirin, particularly in more acid buffers.

THERAP CAT: Analgesic.

317. Alphaprodine. *1,3-Dimethyl-4-phenyl-4-piperidinol propanoate;* α-1,3-dimethyl-4-phenyl-4-piperidinyl propionate; α-1,3-dimethyl-4-phenyl-4-propionoxypiperidine; α-prodine. $C_{16}H_{23}NO_2$; mol wt 261.36. C 73.53%, H 8.87%, N 5.36%, O 12.24%. Prepn of alphaprodine and its stereoisomer *betaprodine* as hydrochlorides: Lee, Ziering, **U.S.** pat. **2,498,433** (1950 to Hoffmann-La Roche); Beckett *et al., J. Pharm. Pharmacol.* **9,** 939 (1957); Ziering *et al., J. Org. Chem.* **22,** 1521 (1957). Abs config of alphaprodine: F. R. Ahmed *et al., Chem. & Ind. (London)* **1959,** 485; of betaprodine: *eidem, ibid.* **1962,** 97. Metabolism of both isomers: M. M. Abdel-Monem *et al., J. Med. Chem.* **15,** 494 (1972). Pharmacology and toxicology of alphaprodine: G. M. Gruber *et al., J. Pharmacol. Exp. Ther.* **99,** 312 (1950).

dl-Form hydrochloride, $C_{16}H_{23}NO_2$·HCl, *Nu-1196, Nisentil, Nisintel, Prisilidene.* Crystals from acetone, mp 220-221°. Slightly saline taste. Freely sol in water, alc, chloroform. Practically insol in ether. pH of 1% aq soln 4.5-5.2. LD_{50} in mice: 54 mg/kg i.v., 73 mg/kg i.p.; in rats: 22 mg/kg i.p. (Gruber).

Betaprodine *dl*-form hydrochloride, $C_{16}H_{23}NO_2$·HCl, *β-prodine, Nu-1779.* Crystals from methyl ethyl ketone, mp 195-196°; from acetone + methanol, mp 199-200°.

Note: This is a controlled substance (opiate) listed in the U.S. Code of Federal Regulations, Title 21 Parts 1308.11 and 1308.12 (1995).

THERAP CAT: Analgesic (narcotic).

318. Alpidem. *6-Chloro-2-(4-chlorophenyl)-N,N-dipropylimidazo[1,2-a]pyridine-3-acetamide;* SL-80.0342; Ananxyl. $C_{21}H_{23}Cl_2N_3O$; mol wt 404.34. C 62.38%, H 5.73%, Cl 17.54%, N 10.39%, O 3.96%. Non-benzodiazepine tran-

quilizer. Prepn: J. P. Kaplan, P. George, **Eur.** pat. **Appl. 50,563;** *eidem,* **U.S.** pat. **4,382,938** (1982, 1983 both to Synthelabo). Pharmacology and pharmacokinetics in humans: B. Saletu *et al., Int. Clin. Psychopharmacol.* **1,** 145 (1986). HPLC determn in human plasma: V. Ascalone *et al., J. Chromatog.* **414,** 101 (1987). Clinical evaluation of tranquilizing vs hypnotic effects: B. Saletu *et al., Curr. Ther. Res.* **40,** 769 (1986). Pilot study in anxiety: S. F. Langer *et al., Psychopharmacol. Bull.* **24,** 161 (1988).

mp 140-141°.

THERAP CAT: Anxiolytic.

319. Alpiropride. *4-Amino-2-methoxy-5-[(methylamino)sulfonyl]-N-[[1-(2-propenyl)-2-pyrrolidinyl]methyl]benzamide;* (±)-*N*-[(1-allyl-2-pyrrolidinyl)methyl]-4-amino-5-(methylsulfamoyl)-*o*-anisamide; *N*-(1-allyl-2-pyrrolidinylmethyl)-2-methoxy-4-amino-5-methylsulfamoylbenzamide; RIV-2093; Revistel. $C_{17}H_{26}N_4O_4S$; mol wt 382.48. C 53.38%, H 6.85%, N 14.65%, O 16.73%, S 8.38%. Dopamine D_2-receptor antagonist structurally similar to sulpiride, *q.v.* Prepn: J. Perrot, M. Thominet, **Eur.** pat. **Appl. 47,207;** *eidem,* **U.S.** pat. **4,550,179** (1982, 1985 both to Soc. d'Etudes Sci. Ind. l'Ile de France). Dopamine receptor binding study: P. Sokoloff *et al., Arch. Pharmacol.* **327,** 221 (1984). HPLC determn in plasma and urine: F. Bressolle *et al., J. Chromatog.* **343,** 443 (1985). Pharmacokinetics and metabolism in humans: F. Bressolle *et al., J. Pharm. Clin.* **4,** 261 (1985). Preliminary study in experimentally induced migraine: A. Bès *et al., Int. J. Clin. Pharm. Res.* **6,** 189 (1986).

Crystals from absolute ethanol, mp 168.5-169°. LD_{50} in male mice (mg/kg): 44 i.v.; 184 i.p.; 204 s.c.; 3600 orally (Perrot, Thominet, 1985).

THERAP CAT: Antimigraine.

320. Alprazolam. *8-Chloro-1-methyl-6-phenyl-4H-[1,2,4]triazolo[4,3-a][1,4]benzodiazepine; 8-chloro-1-methyl-6-phenyl-4H-s-triazolo[4,3-a][1,4]benzodiazepine;* D-65MT; U-31889; Alplax; Tafil; Tranquinal; Trankimazin; Xanax; Xanor. $C_{17}H_{13}ClN_4$; mol wt 308.77. C 66.13%, H 4.24%, Cl 11.48%, N 18.15%. Prepn: J. B. Hester, **Ger.** pat. **2,012,190;** *idem,* **U.S.** pat. **3,987,052** (1970, 1976 both to Upjohn); J. B. Hester *et al., Tetrahedron Letters* **1971,** 1609; A. Walser, G. Zenchoff, *J. Med. Chem.* **20,** 1694 (1977). Central depressant activity: R. Nakajima *et al., Japan. J. Pharmacol.* **21,** 497 (1971). Pharmacology: V. H. Sethy, *Arch. Pharmacol.* **301,** 157 (1978). Clinical studies: L. F. Fabre, *Curr. Ther. Res.* **19,** 661 (1976); J. B. Cohn, *J. Clin. Psychiat.* **42,** 347 (1981). Pharmacokinetics: D. R. Abernethy *et al., ibid.* **44,** 45 (1983). Review of pharmacokinetics, clinical efficacy, and mechanism of action: J. A. Fawcett, H. M. Kravitz, *Pharmacotherapy* **2,** 242-254 (1982); of pharmacology and efficacy in anxiety and depression: *Drugs* **27,** 132 (1984).

Crystals from ethyl acetate, mp 228-228.5°. Sol in alc. Insol in water. uv max (ethanol): 222 nm (ε 40250). LD$_{50}$ in mice, rats (mg/kg): 1020, > 2000 orally; 540, 610 i.p. (Nakajima).

Note: This is a controlled substance (depressant) listed in the U.S. Code of Federal Regulations, Title 21 part 1308.14 (1995).

THERAP CAT: Anxiolytic.

321. Alprenolol. *1-[(1-Methylethyl)amino]-3-[2-(2-propenyl)phenoxy]-2-propanol; 1-(o-allylphenoxy)-3-(isopropylamino)-2-propanol;* H-56/28. C$_{15}$H$_{23}$NO$_2$; mol wt 249.35. C 72.25%, H 9.30%, N 5.62%, O 12.83%. β-Adrenergic blocker. Prepn: Brandstrom *et al., Acta Pharm. Suecica* **3**, 303 (1966); **Neth.** pat. **Appl.** 6,605,692 (1966 to AB Hassle), *C.A.* **66**, 46214p (1967); **Neth.** pat. **Appl.** 6,612,958 (1967 to ICI), *C.A.* **67**, 99851w (1967). Pharmacology: Marmo, *Clin. Ter.* **56**, 121-176 (1971). Series of articles on clinical effect in myocardial infarction: *Acta Med. Scand.* **1984**, Suppl. 680, 1-64.

Hydrochloride, C$_{15}$H$_{23}$NO$_2$.HCl, *Apllobal, Aprobal, Aptine, Aptol Duriles, Gubernal, Regletin, Yobir.* Crystals from ethyl acetate, mp 107-109°. LD$_{50}$ in mice, rats, rabbits (mg/kg): 278.0, 597.0, 337.3 orally (Marmo).

THERAP CAT: Antihypertensive; antianginal; antiarrhythmic.

322. Alsactide. *1-β-Alanine-17-[N-(4-aminobutyl)-L-lysinamide]-α$^{1-17}$-corticotropin;* 1-β-alanine-17-[L-2,6-diamino-N-(4-aminobutyl)hexanamide]-α$^{1-17}$-corticotropin; [β-ala^1,lys^{17}]corticotropin-(1-17)-heptadecapeptide-4-amino-N-butylamide; [β-ala^1,lys^{17}]ACTH$^{1-17}$-4-amino-N-butylamide; alisactide; HOE-433; Synchrodyn. C$_{99}$H$_{155}$N$_{29}$O$_{21}$S; mol wt 2119.57. C 56.10%, H 7.37%, N 19.16%, O 15.85%, S 1.51%. Short-chain ACTH analog. Synthesis: R. Geiger, H. G. Schroeder, **Ger.** pat. **1,954,794**; *eidem,* **U.S.** pat. **3,749,704** (1971, 1973, both to Hoechst); R. Geiger, *Ann.* **750**, 165 (1971). Pharmacology: J. Sandow *et al., Arch. Pharmacol.* **297**, Suppl. 2, R41 (1977). Dose dependent effects on cortisol and aldosterone plasma levels in man: A. Angeli *et al., Horm. Metab. Res.* **13**, 24 (1981).

β-Ala–Tyr–Ser–Met–Glu–His–Phe–Arg–Trp–Gly–Lys–Pro–Val–Gly–Lys–Lys–LysNH(CH$_2$)$_4$NH$_2$

Acetate, C$_{99}$H$_{155}$N$_{29}$O$_{21}$S.C$_2$H$_3$O$_2$, [α]$_D^{20}$ −68.6° (c = 0.5 in 1% acetic acid).

THERAP CAT: Diagnostic aid (adrenal function).

323. Alstonidine. *3β-(Hydroxymethyl)-2α-methyl-4β-[(9-methyl-9H-pyrido[3,4-b]indol-1-yl)methyl]-2H-pyran-5-carboxylic acid methyl ester.* C$_{22}$H$_{24}$N$_2$O$_4$; mol wt 380.44. C 69.46%, H 6.36%, N 7.36%, O 16.82%. From the bark or root of *Alstonia constricta* F. Muell., *Apocynaceae:* Hesse, *Ann.* **205**, 360 (1880); Svoboda, *J. Am. Pharm. Assoc.* **46**, 508 (1957). Structure: Boaz *et al., ibid.* 510. Stereochemistry: Crow *et al., Aust. J. Chem.* **23**, 2489 (1970).

Fine crystals from ether, mp 188-190°. Practically insol in water. Sol in alcohol, chloroform, ether, acetone, 5% HCl. uv max (methanol): 238, 291, 360 nm (log ε 4.66, 4.25, 3.74). The acid solutions are fluorescent.

Acetyl deriv trihydrate, C$_{24}$H$_{26}$N$_2$O$_5$.3H$_2$O, fine needles from dil acetone, mp 92-96°.

324. Alstonine. *3,4,5,6,16,17-Hexadehydro-16-(methoxycarbonyl)-19α-methyl-20α-oxayohimbanium.* C$_{21}$H$_{20}$N$_2$-O$_3$; mol wt 348.40. C 72.40%, H 5.79%, N 8.04%, O 13.78%. From *Alstonia constricta* F. Mull., *Rauwolfia vomitoria* Afzel., *Rauwolfia obscura* K. Schum., *Vinca rosea* L., and other *Apocynaceae:* Sharp, *J. Chem. Soc.* **1934**, 287; Schlittler *et al., Helv. Chim. Acta* **35**, 271 (1952); Pillay, Kumari, *J. Sci. Ind. Res.* **20B**, 458 (1961); *C.A.* **56**, 7425b (1962). Structure: Bader, *Helv. Chim. Acta* **36**, 215 (1953). Stereoisomer of serpentine (alkaloid) *q.v.* Stereochemistry: Wenkert, Roychaudhuri, *J. Am. Chem. Soc.* **79**, 1519 (1957); **80**, 1613 (1958); Shamma, Richey, *ibid.* **85**, 2507 (1963). Biosynthesis: Woodward, *Angew. Chem.* **68**, 13 (1956).

Fine, yellow needles from acetone, dec above 300° (brown at 200°, black at 260°). uv max (methanol): 252, 289, 309, 336, 369 nm (log ε 4.54, 4.08, 4.36, 3.39, 3.60). Quickly decomposes on standing in organic solvents, yielding red solns with blue fluorescence.

Hydrochloride, C$_{21}$H$_{20}$N$_2$O$_3$.HCl, yellow plates from abs alcohol, dec 286°. [α]$_D$ +132° (c = 1.06). Sol in water.

325. Althea. Marshmallow. Dried root of *Althaea officinalis* L., *Malvaceae* (marshmallow), deprived of brown, corky layer. *Habit.* Europe, Western and Northern Asia, naturalized in Eastern U.S. *Constit.* Asparagine, 25-35% mucilage, sugar, pectin.

USE: Demulcent.

326. Althiazide. *6-Chloro-3,4-dihydro-3-[(2-propenylthio)methyl]-2H-1,2,4-benzothiadiazine-7-sulfonamide 1,1-dioxide; 3-[(allylthio)methyl]-6-chloro-3,4-dihydro-2H-1,2,4-benzothiadiazine-7-sulfonamide 1,1-dioxide;* 3-allylthiomethyl-6-chloro-7-sulfamoyl-3,4-dihydrobenzothiadiazine 1,1-dioxide; 6-chloro-3,4-dihydro-7-sulfamoyl-3-(2-thiapent-4-enyl)-2H-1,2,4-benzothiadiazine 1,1-dioxide; altizide; P-1779. C$_{11}$H$_{14}$ClN$_3$O$_4$S$_3$; mol wt 383.90. C 34.42%, H 3.68%, Cl 9.23%, N 10.95%, O 16.67%, S 25.06%. Prepn: **Brit.** pat. **902,658** (1962 to Pfizer).

Solid, mp 206-207°.

Combination with spironolactone, *Aldactazine.*
THERAP CAT: Diuretic, antihypertensive.

327. Altrenogest. *(17β)-17-Hydroxy-17-(2-propenyl)-*
estra-4,9,11-trien-3-one; 17α-allyl-17-hydroxyestra-4,9,11-
trien-3-one; 13β-methyl-17α-allyl-$\Delta^{4,9,11}$-gonatriene-17β-ol-3-
one; allytrenbolone; A-35957; RU-2267; Regumate. $C_{21}H_{26}$-
O_2; mol wt 310.44. C 81.25%, H 8.44%, O 10.31%. Synthetic oral progestogen. Prepn: **Neth. pat. Appl. 6,401,555;**
G. Nomine *et al.,* **U.S. pat. 3,257,278** (1964, 1966 both to
Roussel). Effect on estrous cycle and fertility of mares: E.
L. Squires *et al., J. Anim. Sci.* **49,** 729 (1979). Synchronization of estrus in swine: R. R. Kraeling *et al., ibid.* **52,** 831
(1981); in gilts: V. G. Pursel *et al., ibid.* 130.

Cryst, mp 120°. $[\alpha]_D^{20}$ −72° (c = 0.5 in ethanol).
THERAP CAT (VET): Progestogen.

328. Altretamine. *N,N,N',N',N'',N''-Hexamethyl-1,3,5-*
triazine-2,4,6-triamine; 2,4,6-tris(dimethylamino)-*s*-triazine;
hemel; hexamethylmelamine; HMM; ENT-50852; NSC-
13875; Hexalen; Hexastat. $C_9H_{18}N_6$; mol wt 210.28. C
51.41%, H 8.63%, N 39.97%. An antitumor agent which
also acts as a chemosterilant for male houseflies and other
insects. Synthesis: A. W. Hofmann, *Ber.* **18,** 2755 (1885);
D. W. Kaiser *et al., J. Am. Chem. Soc.* **73,** 2984 (1951); Y.
Bessiere-Chretien, H. Serne, *Bull. Soc. Chim. France* **6,** 2039
(1973). Production process: H. von Brachel, H. Kindler,
Ger. pat. 1,240,870 corresp to **U.S. pat. 3,424,752** (1967,
1969 both to Casella Farbwerke). Chemosterilant effect: S.
C. Chang *et al., Science* **144,** 57 (1964); A. De Milo *et al., J.
Econ. Entomol.* **65,** 1548 (1972). Chromatographic detn in
plasma or serum: A. Hulshoff *et al., J. Chromatog.* **181,** 363
(1980). Disposition and metabolism in rabbits and humans:
M. M. Ames *et al., Cancer Res.* **39,** 5016 (1979). Evaluation
in breast cancer: C. J. Fabian *et al., Cancer Treat. Rep.* **63,**
1359 (1979); in ovarian cancer: J. T. Wharton *et al., Am. J.
Obstet. Gynecol.* **133,** 833 (1979). Mammalian toxicity
study: R. L. Jasper *et al., Fed. Proc.* **24,** 641 (1965). Toxicological studies in dogs and mice: D. C. Thake *et al.,
Gov. Rep. Announce. Index (USA)* **79,** 92 (1979), *C.A.* **91,**
168372a (1979). *In vitro* cytotoxicity study: M. D'Incalci *et
al., Brit. J. Cancer* **41,** 630 (1980).

Needles from abs ethanol, mp 172-174°. uv max (etha-
nol): 226 nm (ε 49400), H. J. Anderson *et al., Can. J.
Chem.* **49,** 2315 (1971). LD$_{50}$ in rats, guinea pigs (mg/kg):
350, 255 orally (Jasper).
USE: As exptl insect chemosterilant.
THERAP CAT: Antineoplastic.

329. D-Altrose. D-Altropyranose. $C_6H_{12}O_6$; mol wt
180.16. C 40.00%, H 6.71%, O 53.28%. Prepd by the
reduction of D-altronic lactone with sodium amalgam: Levene, Jacobs, *Ber.* **43,** 3143 (1910).

Prisms from dilute alcohol, mp 103-105°. $[\alpha]_D^{20}$ +32.6° (c
= 7.6). Soluble in water. Practically insol in alcohol.
Phenylosazone, $C_{18}H_{22}N_4O_4$, mp 178°.

330. Aluminon. *5-[(3-Carboxy-4-hydroxyphenyl)(3-*
carboxy-4-oxo-2,5-cylohexadien-1-ylidene)methyl]-2-hydroxy-
benzoic acid triammonium salt; 3-[bis(3-carboxy-4-hydroxy-
phenyl)methylene]-6-oxo-1,4-cyclohexadiene-1-carboxylic acid
triammonium salt; ammonium salt of aurintricarboxylic
acid; Lysofon. $C_{22}H_{23}N_3O_9$; mol wt 473.44. C 55.81%, H
4.90%, N 8.88%, O 30.41%. Prepd by reacting sodium
nitrite with salicylic acid and adding formaldehyde, then
treating with ammonia: G. B. Heisig, W. M. Lauer, *Org.
Syn.* **coll. vol. I,** 54 (2nd ed., 1941).

Yellowish-brown, glassy powder. Freely sol in water.
USE: Forms brilliantly colored lakes with aluminum,
chromium, iron, beryllium. Generally used for the detection
and colorimetric estimation of aluminum in water, foods,
tissues. *Cf.* Scherrer, Mogerman, *J. Res. Nat. Bur. Stand.*
21, 105 (1938).
THERAP CAT: Pharyngeal aerosol spray.

331. Aluminum. Aluminium. Al; at. wt 26.981539;
at. no. 13; valence 3. Group IIIa(13). Naturally occurring
isotope (mass number): 27 (100%); known artificial radioactive isotopes: 22-25, 26 (T$_{1/2}$ 7.2×10^5 years, longest-lived
known isotope, β^+ and γ emitter, found in meteors), 28-32.
One of the most abundant metals in earth's crust: 8.3 % by
wt (83,000 ppm); occurs in nature primarily in combination
with silica, also as oxide (*see* Aluminum Silicate; Aluminum
Oxide). First obtained in impure form by Oersted in 1825;
prepd as metal powder by Wöhler in 1827. Commercially
important source is bauxite. Reviews of aluminum, its
alloys and compds: Brandt, "Aluminum and Aluminum
Alloys" in *Proc. Met. Soc. Conf.* **vol. 40,** E. D. Verink, Ed.
(Gordon & Breach, New York, 1966); *Aluminum,* **3 vols.** K.
R. Van Horn, Ed. (American Society for Metals, Metal
Park, Ohio, 1967); Wade, Bannister, "Aluminum, Gallium,
Indium and Thallium" in *Comprehensive Inorganic Chemis-
try* **vol. 1,** J. C. Bailar, Jr. *et al.,* Eds. (Pergamon Press,
Oxford, 1973) pp 993-1064; *Chemistry of the Elements,* N.
N. Greenwood, A. Earnshaw, Eds. (Pergamon Press, New
York, 1984) pp 243-295; J. T. Stanley, W. Haupin in Kirk
Othmer *Encyclopedia of Chemical Technology* **vol. 2** (John
Wiley & Sons, 4th ed., 1992) pp 184-251; W. C. Sleppy *et
al., ibid.* 252-345. Review of clinical toxicology: C. D.
Hewitt *et al., Clin. Lab. Med.* **10,** 403-422 (1990); of toxicology and human exposure: *Toxicological Profile for Aluminum and Compounds* (PB93-110633, 1992) 158 pp. Book:
Chemistry of Aluminum, Gallium, Indium and Thallium, A.
J. Downs, Ed. (Blackie Academic & Professional, London,
1993) 515 pp.
Tin-white, malleable, ductile metal, with somewhat bluish
tint; capable of taking brilliant polish which is retained in
dry air. In moist air, oxide film forms which protects metal
from corrosion. Available in bars, leaf, powder, sheets, or
wire. d 2.70. mp 660°. bp 2327°. Does not vaporize even
at high temps, but finely divided aluminum dust is easily

Aluminum Bis(acetylsalicylate) 338

ignited, and may cause explosions. Reacts with dil HCl, H_2SO_4, KOH and NaOH with evolution or hydrogen. Reduces the cations of many heavy metals to the metallic state $E°(aq)$ Al^{3+}/Al -1.66 V. Solns of Al^{3+} in dil HCl or neutral or slightly acid solns of most aluminum salts, yield with Na_2S, a white ppt soluble in excess of Na_2S. Dil neutral soln of aluminum salts yields white gelatinous ppt on boiling with sodium acetate.

Caution: Potential symptoms of overexposure are irritation of eyes, skin and respiratory system. *See NIOSH Pocket Guide to Chemical Hazards* (DHHS/NIOSH 94-116, 1994) p 12. *See also Patty's Industrial Hygiene and Toxicology* vol. 2C, G. D. Clayton, F. E. Clayton, Eds. (John Wiley & Sons, New York, 4th ed., 1994) pp 1881-1902.

USE: As pure metal or alloys (magnalium, aluminum bronze, etc.) for structural material in construction, automotive, electrical and aircraft industries. In cooking utensils, highway signs, fencing, containers and packaging, foil, machinery, corrosion resistant chemical equipment, dental alloys. The coarse powder in aluminothermics (thermite process); the fine powder as flashlight in photography; in explosives, fireworks, paints; for absorbing occluded gases in manuf of steel. In testing for Au, As, Hg; coagulating colloidal solns of As or Sb; pptg Cu; reducer for determining nitrates and nitrites; instead of Zn for generating hydrogen in testing for As. Forms complex hydrides with lithium and boron, such as $LiAlH_4$, which are used in preparative organic chemistry.

332. Aluminum Acetate Solution. *Burow's Solution;* Domeboro. Contains about 5% neutral aluminum acetate. $Al(C_2H_3O_2)_3 = 0.66\%$ Al, 4.4% acetic acid (1.25% Al_2O_3). Prepn from a dry mix consisting of an alkali metal acetate, acetic acid and a dibasic aluminum acetate stabilized with boric acid: Gibbons *et al.,* U.S. pat. **2,824,042** (1958).

Colorless liquid; slight odor of acetic acid. d $\sim$1.002. pH of 1:20 aq soln 4.2.

THERAP CAT: Astringent, antiseptic.

THERAP CAT (VET): External astringent, antiseptic.

333. Aluminum Acetotartrate. Essitol. Consists of approx 70% basic aluminum acetate and 30% tartaric acid. Prepn: *Gmelin's Aluminum* (8th ed.) **35B,** 304 (1934).

Crystals or powder; slight acetic odor; astringent, acidulous taste. On long keeping loses acetic acid and becomes incompletely sol. Slowly sol in cold water. Practically insol in alcohol.

Aq solns, *liquor aluminii aceticotartarici, Alsol.*

THERAP CAT: Astringent, antiseptic.

334. Aluminum Alkyls. Highly reactive compounds of the form R_3Al, *trialkylaluminum;* $[R_nAlX_{3-n}]_2$, $n = 1$, 2, *alkylaluminum halide;* or $R_3Al_2X_3$, *alkylaluminum sesquihalide.* (The sesquihalides are actually equilibrium mixtures of $R_3Al_2X_3$, $[R_2AlX]_2$ and $[RAlX_2]_2$.) First prepn of aluminum alkyls: ethylaluminum sesquiiodide $[(C_2H_5)_3$-$Al_2I_3]$ prepd by Hallwachs, Schaferik, *Ann.* **109,** 207 (1859); trimethylaluminum, $[(CH_3)_3Al]_2$, and triethylaluminum, $[(C_2H_5)_3Al]_2$, prepd by Buckton, Odling, *Ann.* suppl. **4,** 109 (1865). The alkylaluminum halides are prepd from alkyl halides and aluminum, or by halogenation of trialkylaluminums. The trialkylaluminums are prepd from aluminum, hydrogen and olefins; this improved "direct synthesis" was developed by Ziegler and co-workers. Comprehensive reviews of prepn, properties and chemistry: Schultz, *Advan. Chem. Ser.* **23,** 163-171 (1959); Ziegler, "Organo-Aluminum Compounds" in *Organometallic Chemistry,* H. Zeiss, Ed., A.C.S. Monograph Series no. **147** (Reinhold, New York, 1960) p 194-269; Köster, Binger, *Advan. Inorg. Chem. Radiochem.* **7,** 263-348 (1965); T. Mole, E. A. Jeffery, *Organoaluminum Compounds* (Elsevier, New York, 1972) 465 pp; Wade, Banister in *Comprehensive Inorganic Chemistry* vol **1,** J. C. Bailar, Jr. *et al.,* Eds. (Pergamon Press, Oxford, 1973) pp 1058-1064.

Trialkylaluminum compounds, colorless liquids at room temperature. Must be stored in an inert atm; sensitive to oxidation and hydrolysis in air; the lighter trialkylaluminums ignite spontaneously in air. The low mol wt, linear-chain alkyl compounds exist as dimers; the branched-chain alkyl compounds exist primarily as monomers. Among the industrially important trialkyl aluminums are *triethylalumi-*

num, dimeric liq, d^{25} 0.832, bp_{760} 194°, bp_{13} 100° and *triisobutylaluminum,* primarily monomeric liq, d^{25} 0.781; bp_{10} 86°, mp 6°.

Alkylaluminum halides, low melting solids or colorless, volatile liquids. Strongly associated forming dimers; held together by bridging halogen bonds. Less sensitive than trialkylaluminums to oxidation upon exposure to air. Halogen aluminum bonds cleaved by water, alcohol. Industrially important halides include *chlorodiethylaluminum,* $[(C_2H_5)_2AlCl]_2$, liq, d^{25} 0.961, bp_{50} 127°, bp_{17} 100°, $bp_{1.9}$ 60° and *dichloroethylaluminum,* $[C_2H_5AlCl_2]_2$, solid, d^{50} 1.207, mp 32°, bp_{50} 113°, bp_{30} 100°, bp_5 60°.

USE: Catalyst; with compds of early transition metals as Ziegler-Natta polymerization catalysts; intermediates in organic syntheses.

335. Aluminum Ammonium Sulfate. Burnt ammonium alum; exsiccated ammonium alum. $AlH_4NO_8S_2$; mol wt 237.15. Al 11.38%, H 1.70%, N 5.91%, O 53.97%, S 27.04%. $AlNH_4(SO_4)_2$. Prepn: *Gmelin's, Aluminum* (8th ed) **35B,** pp 508-515 (1934). Crystal structure of dodecahydrate: Larson, Cromer, *Acta Cryst.* **22,** 793 (1967).

About 97-98% pure, the balance is chiefly excess Al_2O_3. White powder. One gram dissolves in about 20 ml cold, 1.5 ml boiling water, usually incompletely; practically insol in alcohol. *Keep well closed.*

Dodecahydrate, *ammonium alum.* It is about 99.5% pure. Colorless crystals, white granules or powder; styptic taste. d 1.65. mp 94.5°. At about 250° becomes anhydr; decomposes above 280°. One gram dissolves in 7 ml water, 0.5 ml boil. water; freely soluble in glycerol; practically insol in alc. The aq soln is acid to litmus; pH of 0.05 molar soln 4.6.

USE: Purifying drinking water; in baking powders; dyeing and printing fabrics; manuf pigments, lakes, artificial gems, paper, vegetable glue, marble and porcelain cements; fireproofing; tanning; electrolytic copperplating. *See also* Aluminum Potassium Sulfate.

THERAP CAT: Astringent, styptic.

THERAP CAT (VET): Astringent, styptic.

336. Aluminum Antimonide. AlSb; mol wt 148.74. Al 18.14%, Sb 81.86%. Prepn by fusion of the component elements: Kover, *Compt. Rend.* **243,** 648 (1956). Prepn of high purity crystals, resistance measurements, activation energies: Justi, Lautz, *Abhandl. Braunschweig. Wiss. Ges.* **5,** 36 (1953); *C.A.* **48,** 1756i (1954). Zone melting to remove impurities and to improve electr properties: Scheel, *Z. Metallkunde* **46,** 58 (1955), *C.A.* **49,** 4486f (1955). Oxidation by water: Rudorff, Kohlmeyer, *Z. Metallkunde* **45,** 608 (1954), *C.A.* **49,** 7432c (1955).

Lattice constant: 6.1361×10^{-8} cm. mp 1050°. Electrical properties: Kover, *loc. cit.* Absorption const, n, width of forbidden zone (opt & electr) = 0.78; 3.4; 16; 1.65 e.v.: Oswald, Schade, *Z. Naturforsch.* **9a,** 611 (1954); *C.A.* **49,** 52a (1955). Width of forbidden band = 1.6 e.v. at 0°K: Blunt *et al., Phys. Rev.* **96,** 578 (1954); *C.A.* **49,** 2861b (1955). Other electrical properties: Willardson *et al., J. Electrochem. Soc.* **101,** 354 (1954).

USE: In semiconductor research.

337. Aluminum Benzoate. $C_{21}H_{15}AlO_6$; mol wt 390.33. C 64.62%, H 3.87%, Al 6.91%, O 24.59%. $Al(C_6H_5COO)_3$. Prepn: Kränzlein *et al.,* **Ger. pat. 569,946** (1933 to I. G. Farbenind.); *Frdl.* **19,** 597.

Cryst powder. Very slightly sol in water.

338. Aluminum Bis(acetylsalicylate). *Bis[2-(acetyloxy)benzoato-O']hydroxyaluminum;* hydroxybis(acetylsalicylato)aluminum; aluminum diacetylsalicylate; aluminum acetylsalicylate N.F.; acetylsalicylate aluminum; aluminum diaspirin; aluminum aspirin; aspirin aluminum; monohydroxyaluminum diacetylsalicylate; monohydroxyaluminum bis(acetylsalicylate); Rumasal. $C_{18}H_{15}AlO_9$; mol wt 402.29. C 53.74%, H 3.76%, Al 6.71%, O 35.79%. Prepn: Wolf, U.S. pat. **1,967,649** (1934 to Chinoin); **Brit. pat. 826,299** (1959 to Lewis-Howe). *See also* Dihydroxyaluminum Acetylsalicylate.

Amorphous white powder or granules. Practically insol in water, alcohol, ether. Decomposes in dil acids or alkalies and alkali carbonates.

THERAP CAT: Analgesic, antipyretic.

339. Aluminum Borate. Occurs in nature as the mineral *eremeyevite* or *jeremejevite*, $Al_2O_3.B_2O_3$. Prepd by heating Al_2O_3 with B_2O_3; when heated to $1000°$ $2Al_2O_3.B_2O_3$ is formed; when heated to $1100°$ $9Al_2O_3.2B_2O_3$ is formed: Scholze, *Z. Anorg. Allgem. Chem.* **284,** 272 (1956).
$2Al_2O_3.B_2O_3$, mol wt 273.56. Al 39.45%, B 7.91%, O 52.64%. Needles, mp ~$1050°$. Practically insol in water.
$9Al_2O_3.2B_2O_3$, mol wt 1056.92. Al 45.95%, B 4.09%, O 49.96%. Needles, mp ~$1440°$. Practically insol in water.

USE: Catalyst for polymerizations; in glass manuf.

340. Aluminum Boroformate. Aluminum boro-formicum. Contains about 33% Al_2O_3, 20% H_3BO_3, 15% formic acid, 32% H_2O. Prepn: *Hager's Handb. Pharm. Praxis* **1,** (Aluminum) p 368 (1930).
Lustrous scales. Freely sol in water, alc. The aq soln is acid, has a sweetish astringent taste and is not precipitated by alkalies.

341. Aluminum Borohydride. *Aluminum tetrahydroborate.* AlB_3H_{12}; mol wt 71.51. Al 37.73%, B 45.35%, H 16.91%. $Al(BH_4)_3$. Prepn from alkali metal hydride and aluminum halide: Schlesinger *et al.*, *J. Am. Chem. Soc.* **75,** 209 (1953); Hinkamp, Hnizda, *Ind. Eng. Chem.* **47,** 1560 (1955); Kollonitsch, Fuchs, *Nature* **176,** 1081 (1955); Hinkamp, U.S. pat. **2,854,312** (1958 to Ethyl Corp.); Schechter, U.S. pat. **2,913,306** (1959 to Callery Chem.); from trimethylaluminum and diborane: Schlesinger *et al.*, *J. Am. Chem. Soc.* **61,** 536 (1939).
Liquid. mp $-64.5°$; bp $44.5°$; bp_{119} $0°$. Reacts vigorously with water and hydrogen chloride to liberate hydrogen; ignites in air; decomposes slowly even at room temp evolving hydrogen. Forms addn products with dimethyl ether, trimethylamine and ammonia.

USE: Reducing agent; prepn of borohydrides of heavy metals; fuel for jet engines and rockets.

342. Aluminum Bromide. $AlBr_3$; mol wt 266.69. Al 10.12%, Br 89.88%. Prepd from aluminum and bromine: Nicholson *et al.*, *Inorg. Syn.* **3,** 30 (1950).
White to yellowish-red very hygroscopic lumps, mp $97°$; bp reported within the range $250-270°$; d_4^{18} 3.205. Fumes strongly in air; combines with water *with violence. Keep tightly closed and protect from moisture.* Sol in many organic solvents such as benzene, nitrobenzene, toluene, xylene, simple hydrocarbons.
Hexahydrate, colorless to slightly yellow deliquesc crystals. mp $93°$; d 2.5. Sol in water, alcohol, ether, carbon disulfide. *Keep well closed.*

USE: *Anhydrous* form as acid catalyst in organic syntheses. It is similar to anhydr $AlCl_3$ but is more reactive and more sol in organic media.

343. Aluminum *tert*-Butoxide. *2-Methyl-2-propanol aluminum salt.* $C_{12}H_{27}AlO_3$; mol wt 246.33. C 58.51%, H 11.05%, Al 10.95%, O 19.49%. $Al[OC(CH_3)_3]_3$. Prepd from aluminum *tert*-butyl alcohol and mercuric chloride: Wayne, Adkins, *Org. Syn.* coll. vol. III, 48 (1955).
Powder; can be recryst from benzene. Sublimes at $180°$. Pure compd does not melt or decomp upon heating up to $300°$ in a sealed tube, but traces of moisture or *tert*-butyl alc cause it to melt at $160-200°$. Very sol in organic solvents; approx 9 g dissolves in 5 g ethyl propionate at $120°$.

USE: Reagent for oxidation of alcohols to ketones; in dealcoholation of orthoesters.

344. Aluminum Calcium Hydride. *Calcium tetrahydroaluminate;* calcium aluminum hydride. Al_2CaH_8; mol wt 102.10. Al 52.85%, Ca 39.25%, H 7.90%. $Ca(AlH_4)_2$.

Prepd by the interaction of aluminum chloride and calcium hydride in tetrahydrofuran: Schwab, Wintersberger, *Z. Naturforsch.* **8b,** 690 (1953); Conn, Taylor, U.S. pat. **2,999,005** (1961 to Merck & Co.).
Slate-gray mass. The dry pulverized material can ignite spontaneously in moist air and is best handled under dry nitrogen. Reacts violently with water, the ensuing conflagration resembles a display of fireworks. Slightly less violent reaction with alcohols. Sol in dry tetrahydrofuran. Practically insol in dry ether, dioxane, benzene.

USE: Reducing agent for aldehydes, ketones, acid chlorides. Also in the reduction of esters to alcohols, nitriles to amines, aromatic nitro compounds to azo compounds.

345. Aluminum Carbide. C_3Al_4; mol wt 143.96. C 25.03%, Al 74.97%. Al_4C_3. Prepd by heating aluminum powder with carbon: Becher in *Handbook of Preparative Inorganic Chemistry* vol. 1, G. Brauer, Ed. (Academic Press, New York, 2nd ed., 1963) p 832.
Yellow hexagonal crystals or powder. mp $2100°$; decomposes above $2200°$. d 2.36. Decomposed by water with evolution of methane (fire hazard).

USE: In generating methane; reducing metal oxides; in manuf of aluminum nitride.

346. Aluminum Cesium Sulfate. $AlCsO_8S_2$; mol wt 352.01. Al 7.66%, Cs 37.76%, O 36.36%, S 18.22%. $CsAl(SO_4)_2$. Prepn: *Gmelin's, Aluminum* (8th ed.) **35B,** pp 529-531 (1934).
Dodecahydrate, *cesium alum.* Crystals; mp reported from $105-122°$. Slightly sol in cold, freely in hot water. Practically insol in alcohol.

347. Aluminum Chlorate. *Chloric acid, aluminum salt.* $AlCl_3O_9$; mol wt 277.33. Al 9.73%, Cl 38.35%, O 51.92%. $Al(ClO_3)_3$. Occurs as hexahydrate and nonahydrate. Prepn: *Gmelin's Aluminum* (8th ed.) **35B,** 216-217 (1934).
Nonahydrate, *Mallebrin.* Deliquesc crystals. Freely sol in water; sol in alc. *Keep well closed.*

USE: Disinfectant; ClO_2 manuf; prevention of yellowing of acrylic fibers.

THERAP CAT: Antiseptic, astringent.

348. Aluminum Chloride. $AlCl_3$; mol wt 133.34. Al 20.24%, Cl 79.76%. Prepd from aluminum metal in a heated stream of HCl gas: Gattermann-Wieland, *Praxis des Organischen Chemikers* (de Gruyter, Berlin, 40th ed., 1961) p 295; H. J. Becher in *Handbook of Preparative Inorganic Chemistry*, **vol. 1,** G. Brauer, Ed. (Academic Press, New York, 2nd ed., 1963) p 812. Manufacture: Faith, Keyes & Clark's *Industrial Chemicals,* F. A. Lowenheim, M. K. Moran, Eds. (Wiley-Interscience, New York, 4th ed., 1975) pp 72-75. Monograph: C. A. Thomas, *Anhydrous Aluminum Chloride in Organic Chemistry,* A.C.S. Monograph Series no. **87** (Reinhold, New York, 1941).
White when pure; ordinarily gray or yellow to greenish. Fumes in air; strong odor of HCl; when heated in small quantities volatilizes without melting. Combines with water *with explosive violence* and liberation of much heat. Freely sol in many organic solvents, such as benzophenone, benzene nitrobenzene, carbon tetrachloride, chloroform. *Keep tightly closed and protected from moisture.* For physical properties see C. A. Thomas, *loc. cit.*
Hexahydrate, *Aluwets, Anhydrol, Driclor.* Colorless crystals, or white or slightly yellow deliquesc, cryst powder; odorless or slight HCl odor. One gram dissolves in 0.9 ml water, 4 ml alc; sol in ether, glycerol, propylene glycol. *Keep well closed.*
Caution: Anhydrous form is a strong irritant.

USE: The *anhydrous* form suitable as an acid catalyst, esp in Friedel-Crafts type reactions; in cracking of petroleum; in manuf rubbers, lubricants. The *hexahydrate* form used in preserving wood; disinfecting stables, slaughterhouses, etc.; in deodorants and antiperspirant preparations; refining crude oil; dyeing fabrics; manuf parchment paper.

THERAP CAT: The *hexahydrate* as a topical astringent.

349. Aluminum Diacetate. *Bis(aceto-O)hydroxyaluminum; hydroxybis(acetato)aluminum;* basic aluminum acetate; aluminum subacetate; aluminum hydroxyacetate; Lenicet; Casil. $C_4H_7AlO_5$; mol wt 162.08. C 29.64%, H

4.35%, Al 16.65%, O 49.36%. $Al(OH)(CH_3CO_2)_2$. Prepd from aluminum hydroxide and acetic acid or from sodium acetate and aluminum chloride hexahydrate: Hood, Ihde, *J. Am. Chem. Soc.* **72**, 2094 (1950). Other methods of prepn: *Gmelin's Aluminum* (8th ed.) **35B**, p 296 (1934). Also prepd in aq solution, *see* Aluminum Subacetate Solution.

White curdy precipitate or white amorphous powder. Material that has been oven-dried at 110° is practically insoluble in water. Freshly prepared material forms numerous hydrates and is quite sol in water. Greatest soly is obtained by formation in solution. The pharmacist's stock soln (*see* Aluminum Subacetate Solution) contains about 8% $Al(OH)(CH_3CO_2)_2$, while commercial solns used for waterproofing, contain 22 to 25% $Al(OH)(CH_3CO_2)_2$. When heated, the salt loses acetic acid, and aluminum to oxygen bonding occurs, resulting in a very insol compd of disputed nature, *see Gmelin's, loc. cit.* and Thomas, *Paper Trade J.* **100**, 36 (1935). Aq solns are generally acid to litmus; gradually become turbid and gelatinous. Sometimes a more basic salt precipitates out and settles to the bottom of the container. Increasing the pH to a marked degree will clear up an old soln due to formation of sodium acetate and sodium aluminate. Urea and thiourea have been suggested as stabilizers for aq solns.

USE: Manuf color lakes; mordant in dyeing; in waterproofing and fireproofing fabrics (solns for these purposes are known as *red liquor* or *mordant rouge* because they were originally used for preparing red color lakes); in antiperspirant formulations; as disinfectant by embalmers. Dusting powder.

350. Aluminum Ethoxide. *Ethanol aluminum salt; aluminum ethylate.* $C_6H_{15}AlO_3$; mol wt 162.16. C 44.44%, H 9.32%, Al 16.64%, O 29.60%. $Al(OC_2H_5)_3$. Prepd by reacting aluminum powder with absolute ethanol in xylene using small amounts of mercuric chloride and iodine as catalysts: Meerwein, Schmidt, *Ann.* **444**, 232 (1925); *Newer Methods of Preparative Organic Chemistry* (Interscience, New York, 1948) p 132; *see also* Farbwerke vorm. Meister, Lucius und Brüning, **Ger. pat. 286,596;** *J. Soc. Chem. Ind. (London)* **34**, 1168 (1915); Adkins, *J. Am. Chem. Soc.* **44**, 2178 (1922). Laboratory procedure: Gattermann-Wieland, *Praxis des Organischen Chemikers* (de Gruyter, Berlin, 40th ed., 1961) p 333.

Liquid. bp_{6-8} 200°; bp_3 175-180°. Slowly solidifies to a whole white solid, mp 140°. May crystallize with alcohol of crystallization. Decomposed by water. Slightly sol in hot xylene, chlorobenzene, other high boiling solvents.

USE: In the reduction of aldehydes and ketones; as catalyst for polymerizations.

351. Aluminum Fluoride. Aluminum trifluoride. AlF_3; mol wt 83.98. Al 32.13%, F 67.87%. Prepd by heating $(NH_4)_3AlF_6$ to red heat in a stream of nitrogen: Witt, Barrow, *Trans. Faraday Soc.* **55**, 730 (1959); Kwasnik in *Handbook of Preparative Inorganic Chemistry*, **Vol. 1**, G. Brauer, Ed. (Academic Press, New York, 2nd ed., 1963) p 225. *Review:* Kemmitt, Sharp, *Advan. Fluorine Chem.* **4**, 154-155 (1965).

Hexagonal crystals. Sublimes (760 mm) 1272°. Soly in water (25°) 0.559 g/100 ml. Sparingly sol in acids and alkalies, even hot concd H_2SO_4 has little effect. Hydrolyzed by superheated steam at 300-400°.

Monohydrate, *fluellite*, orthorhombic crystals, d 2.17.

Trihydrate, usually $AlF_3.3.1H_2O$. Prepn: Ehret, Frere, *J. Am. Chem. Soc.* **67**, 64 (1945). Loses water at 100°, more at 200°. It does not seem possible to obtain the anhydrous compd free from oxides by dehydration of the hydrates.

Caution: Less toxic on ingestion than other fluorides because of slight solubility.

USE: In ceramics, as flux in metallurgy, in aluminum manufacture, as inhibitor of fermentation, as catalyst in organic reactions.

352. Aluminum Hexafluorosilicate. Aluminum fluosilicate; aluminum silicofluoride. $Al_2F_{18}Si_3$; mol wt 480.19. Al 11.24%, F 71.22%, Si 17.55%. $Al_2(SiF_6)_3$. Occurs in nature as *topaz*, $Al_2SiO_4(OH,F)_2$. Prepn: Sanfourche, Krapivine, *Compt. Rend.* **208**, 2080 (1939).

Nonahydrate, hexagonal prisms. Easily sol in water; aq soln decomposes on heating or neutralization. Solid loses water on heating to temperatures below 500°, leaving a hexahydrate form; decomposes completely on heating to 1000°.

USE: Protection and preservation of construction materials; manuf of glass.

353. Aluminum Hexaurea Sulfate Triiodide. *Hexakis(urea-O)aluminum sulfate (triiodide)(1:1:1);* aluminum hexacarbamide sulfate triiodide; Hexadine-S; Alure. C_6H_{24}-$AlI_3N_{12}O_{10}S$; mol wt 864.09. C 8.34%, H 2.80%, Al 3.12%, I 44.06%, N 19.45%, O 18.52%, S 3.71%. $Al[CO(NH_2)_2]_6$-SO_4I_3. Prepn: Barbieri, *Atti. Reale. Acad. Lincei.* **24**, 919 (1915); Morris *et al., Ind. Eng. Chem.* **45**, 1013 (1953).

Crystals. Soly in water at 25°: 590 g/l.

USE: Decontamination of drinking water in emergencies. Used in amounts sufficient to yield 8 ppm of active iodine. A tablet contg 30 mg plus 82 mg $Na_2H_2PO_7$ plus 4 mg talc will decontaminate one quart of water. Such tablets after 7 days of storage at 60° retained 97% of their original active iodine. More stable than tetraglycine hydroperiodide.

354. Aluminum Hydride. AlH_3; mol wt 30.01. Al 89.92%, H 10.08%. Prepd by treating lithium hydride with an ether solution of aluminum chloride: Finholt *et al., J. Am. Chem. Soc.* **69**, 1199-1203 (1947).

Colorless solid, nonvolatile, probably highly polymerized and containing residual ether which cannot be completely removed.

USE: As catalyst for polymerizations; reducing agent. Lithium aluminum hydride is a more useful reagent because of its greater solubility.

355. Aluminum Hydroxide. Aluminum hydrate; aluminum trihydrate; hydrated alumina. AlH_3O_3; mol wt 78.00. Al 34.59%, H 3.88%, O 61.53%. $Al(OH)_3$. Prepn and properties: Dominé-Berges, *Ann. Chim.* [12], **5**, 106 (1950); Hennig, *Chem. Tech.* **1**, 66 (1949), *C.A.* **44**, 965f (1950); *Gmelin's Aluminum* (8th ed.) **35B**, pp 98-132 (1934); Becher in *Handbook of Preparative Inorganic Chemistry*, vol. **1**, G. Brauer, Ed. (Academic Press, New York, 2nd ed., 1963) pp 820-821; Wagner, *ibid.,* vol. 2 (1965) pp 1652-1654. Clinical comparison with calcium carbonate as phosphate binder in chronic renal failure: R. H. K. Mak *et al., Brit. Med. J.* **291**, 623 (1985).

Usually obtained as a white, bulky, amorphous powder. Practically insol in water, but sol in alkaline aq solns or in HCl, H_2SO_4 and other strong acids in the presence of some water. Forms gels on prolonged contact with water. Absorbs acids, CO_2.

Note: Used in pharmacy as the gel or dried gel, *algeldrate.* Some commercial names for such prepns are: *Aldrox, Alkagel I.S., ALternaGEL, Alucol, Al-U-Creme, Aludrox, Aludyal, Amphojel, Antidiar, Creamalin, Cremorin, Gelumina, Merlium, Pepsamar, Uracid.*

USE: Adsorbent; emulsifier; ion-exchanger; in chromatography; mordant in dyeing; filtering medium; manuf glass, fire clay, paper, pottery, printing inks, lubricating compositions, detergents; waterproofing fabrics; in antiperspirants, dentifrices.

THERAP CAT: Antacid; antihyperphosphatemic.

THERAP CAT (VET): Vaccine adjuvant. Gastric antacid, and gastrointestinal protective in dogs.

356. Aluminum Hydroxychloride. *Basic aluminum chloride;* aluminum chlorohydroxide; aluminum chlorohydrate; Astringen; Chlorhydrol; Hyperdrol; Locron; Phosphonorm. The generally accepted empirical formula is $Al_2(OH)_5Cl.2H_2O$. Prepd by electrolyzing solns of suitable Al salts: **Fr. pat. 837,862** and **Brit. pat. 509,815** (both 1939 to I. G. Farben.); H. Huehn, W. Haufe, **U.S. pat. 2,392,531** (1946); Andersen, **U.S. pat. 2,492,085** (1949 to Elizabeth Arden). Structural studies and physicochemical properties: D. L. Teagarden *et al., J. Pharm. Sci.* **70**, 758, 762 (1981). Phosphate binding activity and use in hyperphosphatemia: **Ger. pat. 3,147,869** (1983 to Nefro-Pharma), *C.A.* **99**, 110747a (1983).

Glassy solid. Dissolves in water, forming slightly turbid colloidal solns (up to 55% w/w). pH of 15% aq soln ∼4.3.

USE: In antiperspirants.

THERAP CAT: Astringent; antihyperphosphatemic.

357. Aluminum Hypophosphite. $AlH_6O_6P_3$; mol wt 221.95. Al 12.16%, H 2.72%, O 43.25%, P 41.87%. Al-

(PO$_2$H$_2$)$_3$. Prepd by heating Al(OH)$_3$ or a solution of an aluminum salt with hypophosphorous acid or sodium hypophosphite: Everest, *J. Chem. Soc.* **1952**, 2945.

Cryst powder; decomposes without melting at ~220° with evolution of phosphine. Practically insol in water. Sol in warm NaOH soln, dilute sulfuric and dilute or concd hydrochloric acid.

USE: In acrylonitrile polymer fiber finishes.

358. Aluminum Iodide. AlI$_3$; mol wt 407.69. Al 6.62%, I 93.38%. Prepn from aluminum and iodine: Watt, Hall, *Inorg. Syn.* **4**, 117 (1953); H. J. Becher in *Handbook of Preparative Inorganic Chemistry* vol. 1, G. Brauer, Ed. (Academic Press, New York, 2nd ed., 1963) p 814; Wilson, Worrall, *J. Chem. Soc. (A)* **1968**, 316.

White leaflets if pure; commercial grade yellowish- to blackish-brown lumps. mp 191°; bp 382°; d^{17} 3.948. Fumes in moist air; strong exothermic reaction with water. *Keep tightly closed and protected from light.* Sol in carbon disulfide, alcohol, ether, liquid ammonia.

Hexahydrate, yellowish, deliquesc cryst powder. Sol in water, alcohol, ether. *Keep tightly closed.*

USE: Catalyst in organic reactions.

359. Aluminum Isopropoxide. *2-Propanol aluminum salt;* aluminum isopropylate. C$_9$H$_{21}$AlO$_3$; mol wt 204.25. C 52.93%, H 10.36%, Al 13.21%, O 23.50%. Al[OCH(CH$_3$)$_2$]$_3$. Prepd from aluminum and isopropyl alcohol in the presence of mercuric chloride: Young *et al., J. Am. Chem. Soc.* **58**, 100 (1936); by adding excess isopropyl alcohol to a benzene soln of AlCl$_3$ at 6°: Teichner, *Compt. Rend.* **237**, 810 (1953). Forms trimers and tetramers: Shiner *et al., J. Am. Chem. Soc.* **85**, 2318 (1963); Oliver *et al., J. Inorg. Nucl. Chem.* **31**, 1609 (1969); Worrall, *J. Chem. Ed.* **46**, 510 (1969). Toxicity: Smyth *et al., Am. Ind. Hyg. Assoc. J.* **30**, 470 (1969). *Review:* Whitaker, *Advan. Chem. Series* **23**, 184-189 (1959).

Hygroscopic white solid, mp 119°. Solidifies rather slowly after distillation. bp$_{10}$ 135°; bp$_{7.5}$ 131°; bp$_{5.5}$ 125.5°; bp$_{2.5}$ 113°; bp$_{1.5}$ 106°; bp$_{0.5}$ 94°. Sol in ethanol, isopropanol, benzene, toluene, chloroform, carbon tetrachloride, petroleum hydrocarbons. Decomposed by water. LD$_{50}$ orally in rats: 11.3 g/kg (Smyth).

USE: Meerwein-Ponndorf reactions; alcoholysis and ester exchange; synthesis of higher alkoxides, chelates, and acylates; formation of aluminum soaps, formulation of paints; waterproofing finishes for textiles.

360. Aluminum Lactate. Aluctyl. C$_9$H$_{15}$AlO$_9$; mol wt 294.19. C 36.74%, H 5.14%, Al 9.17%, O 48.95%. Prepn from lactic acid and aluminum isopropoxide or aluminum chloride: Rai *et al., J. Prakt. Chem.* **20**, 105 (1963); from lactic acid and aluminum foil: Jones, Cluskey, *Cereal Chem.* **40**, 589 (1963).

Powder. Freely sol in water.

USE: In foam fire extinguishers; in dental-impression materials.

361. Aluminum Lithium Hydride. *Lithium tetrahydroaluminate;* lithium aluminum hydride; lithium aluminohydride; lithium alanate. AlH$_4$Li; mol wt 37.95. Al 71.09%, H 10.62%, Li 18.29%. LiAlH$_4$. Prepd by treating lithium hydride with an ether soln of AlCl$_3$: Finholt, *et al., J. Am. Chem. Soc.* **69**, 1199 (1947). Crystal structure: Sklar, Post, *Inorg. Chem.* **6**, 669 (1967). Review of chemistry: J. S. Pizey, *Synthetic Reagents,* **Vol. 1** (John Wiley, New York, 1974) pp 101-294.

Microcrystalline white powder when pure; gray when aluminum impurity present. Monoclinic crystals. d 0.92. Stable in dry air at room temperature, decomp above 125°, slowly loses hydrogen at 120°, decomp in moist air, may ignite on grinding in air. Soly (parts/100 parts solvent): 30 (ether); 13 (tetrahydrofuran); 10 (dimethylcellosolve); 2 (dibutyl ether); 0.1 (dioxane). Reacts rapidly with water and alcohols; reduces aldehydes, ketones, acid chlorides and

esters to alcohols; nitriles to amines; aromatic nitro compounds to azo compounds. Does not attack olefinic double bonds unless they are conjugated with a phenyl group and a carbonyl or nitrile group.

USE: Reducing agent; in preparation of other hydrides.

362. Aluminum Magnesium Silicate. Magnesium aluminum silicate. Al$_2$MgO$_8$Si$_2$; mol wt 262.43. Al 20.56%, Mg 9.26%, O 48.77%, Si 21.40%. MgAl$_2$(SiO$_4$)$_2$. Occurs in nature in the minerals: *colerainite, leuchtenbergite, pyrope, saponite, sapphirine, sheridanite, zebedassite.* Prepn: **Brit. pat. 834,517** (1960 to Fuji Chem.)

Hydrates, *Ervasil, Gelusil, Ultin.*

USE: As suspending agent, thickening agent.

THERAP CAT: Antacid.

363. Aluminum β-Naphtholdisulfonate. *2-Hydroxynaphthalenedisulfonic acid aluminum salt;* Alumnol. C$_{30}$H$_{18}$Al$_2$O$_{21}$S$_6$; mol wt 960.82. C 37.50%, H 1.89%, Al 5.62%, O 34.97%, S 20.02%. Al$_2$[C$_{10}$H$_5$(OH)(SO$_3$)$_2$]$_3$. Prepd from barium β-naphtholdisulfonate and aluminum sulfate: *Hagers' Handb. Pharm. Praxis* **Band 1**, 204 (Berlin, 1930).

Fine, almost white, powder. Sol in 1.5 parts water; sol in glycerol, slightly in alcohol. Practically insol in ether.

THERAP CAT: Astringent, antiseptic.

THERAP CAT (VET): Has been used as an internal astringent, antiseptic.

364. Aluminum Nicotinate. *3-Pyridinecarboxylic acid aluminum salt;* tris(nicotinato)aluminum; nicotinic acid aluminum salt; Alunitine; Nicalex. C$_{18}$H$_{12}$AlN$_3$O$_6$; mol wt 393.29. C 54.97%, H 3.08%, Al 6.86%, N 10.68%, O 24.41%. Each tablet of 625 mg is a complex consisting of aluminum nicotinate, nicotinic acid and aluminum hydroxide equivalent in activity to 500 mg nicotinic acid. Manuf of suitable compositions: Miale, U.S. pat. **2,970,082** (1961 to Walker Labs.).

THERAP CAT: Vasodilator (peripheral); antihyperlipoproteinemic.

365. Aluminum Nitrate. AlN$_3$O$_9$; mol wt 213.00. Al 12.67%, N 19.73%, O 67.60%. Al(NO$_3$)$_3$. Occurs in several states of hydration of which the nonahydrate is the most stable. Prepn: *Gmelin's, Aluminum* (8th ed.) **35B**, p 149-152 (1934). Toxicity data: Smyth *et al., Am. Ind. Hyg. Assoc. J.* **30**, 470 (1969).

Nonahydrate, deliquesc crystals; mp 73°; dec at 135°. Very sol in water, alc; very slightly sol in acetone. Almost insol in ethyl acetate and pyridine. The aq soln is acid. *Keep well closed.* LD$_{50}$ orally in rats: 4.28 g/kg (Smyth).

USE: Tanning leather; antiperspirant; corrosion inhibitor; extraction of uranium; nitrating agent.

366. Aluminum Nitride. AlN; mol wt 40.99. Al 65.83%, N 34.17%. Prepd commercially by heating bauxite and coal in a stream of nitrogen. Laboratory prepn from powdered aluminum metal: Becher in *Handbook of Preparative Inorganic Chemistry* Vol. 1, G. Brauer, Ed. (Academic Press, New York, 2nd ed., 1963) p 827.

Orthorhombic or hexagonal, bluish-white crystals. In moist air, odor of ammonia. d$_4^{25}$ 3.05. Hardness no. 9 to 10 on Mohs' scale. mp 2150-2200° at 4.3 atm. Spec heat at 0°: 0.180 cal/g/°C; at 100°: 0.207 cal/g/°C; at 500°: 0.313 cal/g/°C. Heat of formation: −74 kcal/mol. Decomposed by water into Al(OH)$_3$ and NH$_3$.

USE: In semiconductor electronics; in steel manuf.

367. Aluminum Oleate. *9-Octadecenoic acid aluminum salt; oleic acid aluminum salt.* C$_{54}$H$_{99}$AlO$_6$; mol wt 871.36. C 74.43%, H 11.45%, Al 3.10%, O 11.02%. [CH$_3$-(CH$_2$)$_7$CH=CH(CH$_2$)$_7$COO]$_3$Al. Prepd from freshly pptd Al$_2$(OH)$_6$ and oleic acid: Stich, *Pharm. Zentralhalle* **63**, 261 (1922), *C.A.* **16**, 2755 (1922).

Yellowish, viscid mass. Practically insol in water. Sol in alcohol, benzene, ether, oil turpentine.

USE: In oil or turpentine soln as lacquer for metals, as size, waterproofing agent, drier, for paints, high-pressure and high-temp greases for thickening lubricating oils.

368. Aluminum Oxalate. $C_6Al_2O_{12}$; mol wt 318.02. C 22.66%, Al 16.97%, O 60.37%. $Al_2(C_2O_4)_3$. Prepn: **Brit. pats. 348,789** and **348,790** (both 1930 to I.G. Farben).

Hydrate, powder. Practically insol in water, alc. Sol in mineral acids.

USE: Mordant in printing textiles, dyeing cotton.

369. Aluminum Oxide. Alumina. Al_2O_3; mol wt 101.96. Al 52.93%, O 47.07%. Occurs in nature as the minerals: *bauxite, bayerite, boehmite, corundum, diaspore, gibbsite.* Prepn and properties: *Mellor's* vol. V, 263-273 (1929); *Gmelin's, Aluminum* (8th ed.) **35B**, pp 7-98 (1934); Becher in *Handbook of Preparative Inorganic Chemistry* vol. 1, G. Brauer, Ed. (Academic Press, New York, 2nd ed., 1963) pp 822-823; Wagner, *ibid.* vol. 2 (1965) pp 1660-1663. Use as column matrix in ion chromatography: W. Buchberger, K. Winsauer, *J. Chromatog.* **482**, 401 (1989); in HPLC: M. T. Kelly, M. R. Smyth, *J. Pharm. Biomed. Anal.* **7**, 1757 (1989). Clinical evaluation in hip replacement: L. Sedel *et al., J. Bone Joint Surg. [Brit].* **72-B**, 658 (1990); of wear in hip replacement: L. P. Zichner, H.-G. Willert, *Clin. Ortho. Rel. Res.* **282**, 86 (1992). Review of properties, biocompatibility and clinical use: P. Boutin *et al., J. Biomed. Mat. Res.* **22**, 1203-1232 (1988); of biocompatibility: P. S. Christel, *Clin. Ortho. Rel. Res.* **282**, 10-18 (1992).

Approximate characteristics of native aluminum oxide: White cryst powder. Very hard, about 8.8 on Moh's scale. An electrical insulator; electrical resistivity at 300° about 1.2 $\times$ 10^{13} ohms-cm. When heated above 800° it becomes insol in acid and specific gravity increases from 2.8 to 4.0. Insol in water. Very hygroscopic.

USE: As adsorbent, desiccant, abrasive; as filler for paints and varnishes; in manuf of alloys, ceramic materials, electrical insulators and resistors, dental cements, glass, steel, artificial gems; in coatings for metals, etc.; as catalyst for organic reactions. As a chromatographic matrix, originally called *Brockmann aluminum oxide* when used for this purpose. The minerals *corundum* (hardness = 9) and *Alundum* (obtained by fusing bauxite in an electric furnace) are used as abrasives and polishes; in manuf of refractories.

370. Aluminum Palmitate. *Hexadecanoic acid aluminum salt;* palmitic acid aluminum salt. $C_{48}H_{93}AlO_6$; mol wt 793.24. C 72.68%, H 11.82%, Al 3.40%, O 12.10%. $[CH_3-(CH_2)_{14}COO]_3Al$. Prepd from $AlCl_3$ and palmitic acid: Mehrotra, Rai, *J. Indian Chem. Soc.* **39**, 1 (1962).

White to yellow mass or powder. Practically insol in water or alcohol; when fresh, dissolves in petrol ether or oil turpentine.

USE: Thickening petroleum and lubricants; water-proofing fabrics; sizing and glazing paper and leather.

371. Aluminum Phosphate. Aluminum orthophosphate. AlO_4P; mol wt 121.95. Al 22.12%, O 52.48%, P 25.40%. $AlPO_4$. Occurs in nature as the minerals *angelite; coeruleolactite; evansite; lucinite; metavariscite; sterretite; variscite; vashegyite; wavellite; zepharovich.* Prepn from $NaAlO_2$ and H_3PO_4: Becher in *Handbook of Preparative Inorganic Chemistry* vol. 1, G. Brauer, Ed. (Academic Press, New York, 2nd ed., 1963) p 831.

White, infusible powder. mp >1460°; d^{23} 2.56. Practically insol in water or acetic acid. Very slightly sol in concentrated hydrochloric acid and nitric acid. Isomorphous with quartz.

USE: As cement in admixture with calcium sulfate and sodium silicate; as flux for ceramics; dental cements; for special glasses. Used in pharmacy as the gel or dried gel, such as *Aluphos, Fosfalugel, Ulcocid, Phosphaljel, Phosphalugel, Phosphalutab.*

THERAP CAT: Antacid.

372. Aluminum Phosphide. Celphos; Detia; Phostoxin. AlP; mol wt 57.96. Al 46.56%, P 53.44%. Prepd from red phosphorus and aluminum powder: White, Bushey, *J. Am. Chem. Soc.* **66**, 1666 (1944); *Inorg. Syn.* **4**, 23 (1953); Montignie, *Bull. Soc. Chim. France* **1946**, 276; from Al and Zn_2P_3: Wang *et al., J. Inorg. Nucl. Chem.* **25**, 326

(1963). Use as insecticidal fumigant: W. Freyberg, W. Haupt, U.S. pat. **2,117,158** (1938 to Freyberg).

Dark gray or dark yellow crystals. Cubic zinc blende structure. Must be protected from moist air since it reacts readily to produce phosphine which is highly toxic. d_4^{15} 2.85 (Montignie); d 2.40 (Wang *et al.*). Does not melt or decompose thermally at temps up to 1000°. Treatment with water and acid produces phosphine in quantitative yields.

USE: Source of phosphine; in semiconductor research; as fumigant.

373. Aluminum Potassium Sulfate. $AlKO_8S_2$; mol wt 258.21. Al 10.45%, K 15.14%, O 49.57%, S 24.84%. $KAl(SO_4)_2$. Prepn: *Gmelin's, Aluminum* (8th ed.) **35B**, 453-477 (1934). Crystal structure: Manoli *et al., Bull. Soc. Chim. France* **1970**, 98; of dodecahydrate: Larson, Cromer, *Acta Cryst.* **22**, 793 (1967).

Anhydrous, *burnt alum, exsiccated alum.* Usually about 97-98% pure. White powder; styptic taste; attracts moisture from the air. One gram dissolves in about 20 ml cold or 1 ml boiling water, usually incompletely; practically insol in alcohol. *Keep well closed.*

Dodecahydrate, *alum, potassium alum, kalinite.* The technical product is also known as *alum flour, alum meal, cube alum.* It is about 99.5% pure. Colorless, odorless, hard, large, transparent crystals or cryst fragments or white, cryst powder; sweetish, astringent taste. Stable at ordinary temp; when kept for a long time at 60-65° (or over H_2SO_4) loses 9 H_2O which is reabsorbed on exposure to air. Becomes anhyd at about 200°; at higher temp loses SO_3 and becomes basic and incompletely sol in water. d 1.725. mp 92.5°. One gram dissolves in 7.2 ml water, in 0.3 ml boiling water; freely sol in glycerol. Insol in alcohol. The aq soln is acid; pH of 0.2 molar aq soln 3.3.

USE: In dyeing, printing fabrics; manuf dyes, lakes, paper, vegetable glue, marble cement, porcelain cement, explosives; in tanning, hardening gelatin, baking powders, purifying water, clarifying sugar, hardening plaster casts; electrolytic copperplating; as catalyst in synthesis of ammonia; as mordant in staining with carmine, eosine, hematoxylin; clarifier (as alumina cream); identifying coloring matters; hardening agent in microscopy.

THERAP CAT: Astringent.

THERAP CAT (VET): Astringent.

374. Aluminum Rubidium Sulfate. AlO_8RbS_2; mol wt 304.58. Al 8.86%, O 42.02%, Rb 28.06%, S 21.06%. $RbAl(SO_4)_2$. Prepn: *Gmelin's Aluminum* (8th ed) **35B**, p 525-527 (1934). Crystal structure of dodecahydrate: Larson, Cromer, *Acta Cryst.* **22**, 793 (1967).

Dodecahydrate, *rubidium alum.* Crystals, mp reported from 99° to 109°. Sol in water. Practically insol in alc.

375. Aluminum Salicylates, Basic. Basic aluminum salicylates. $(C_7H_5O_3)_nAl(OH)_{3-n}.xH_2O$.

Monosalicylate, $C_7H_7AlO_5$, *Alunozal.* Prepn: **Ger. pat. 354,698** and **Brit. pat. 182,446** (both 1922 to Soc. Chim. des Usines du Rhône), *Frdl.* **14**, 1234 and *C.A.* **16**, 4301⁴ (1922). Fine, colorless powder, occasionally pink.

Disalicylate, $C_{14}H_{11}AlO_7.H_2O$, *Baluvet.* $[(C_7H_5O_3)_2Al-(OH)(H_2O)]$, Prepn: Burrows, Wark, *J. Chem. Soc.* **1928**, 222. Pink needles. Sparingly sol in water. Readily loses 1 mol H_2O on drying at 100°, a second mol is driven off with difficulty.

Note: The neutral salicylate, $(C_7H_5O_3)_6Al_2.3H_2O$, is called *Saluminium insoluble* or *insoluble Salumin.*

THERAP CAT: Antidiarrheal.

THERAP CAT (VET): Antidiarrheal.

376. Aluminum Selenide. Al_2Se_3; mol wt 290.84. Al 18.55%, Se 81.45%. Prepn from aluminum and selenium: Waitkins, Shutt, *Inorg. Syn.* **2**, 184 (1946); Becher in *Handbook of Preparative Inorganic Chemistry* vol. 1, G. Brauer, Ed. (Academic Press, New York, 2nd ed., 1963) p 825.

Yellowish to light-brown powder. d_4^{15} 3.437. Unstable in air. Decomp in water and acid.

USE: In prepn of hydrogen selenide; in semiconductor research.

377. Aluminum Silicate. Al_2O_5Si; mol wt 162.05. Al 33.30%, O 49.37%, Si 17.33%. Al_2SiO_5. Usually contains some water. Polymorphous; the three naturally occurring forms are *andalusite, cyanite, sillimanite.* Other aluminum silicate minerals are *anauxite, dickite, kaolinite, kochite, mullite, newtonite, pyrophyllite, takizolite, termierite, ton.* For prepn and properties see *Gmelin's Aluminum* (8th ed.) **35B**, p 313-317 (1934).

USE: In dental cements, glass industry; manuf of semiprecious stones, enamels, ceramics, and colored lakes; paint filler; in washing compounds.

378. Aluminum Sodium Sulfate. $AlNaO_8S_2$; mol wt 242.10. Al 11.14%, Na 9.50%, O 52.87%, S 26.49%. NaAl$(SO_4)_2$. Prepn: *Gmelin's Aluminum* (8th ed) **35B**, p 378-384 (1934).

Dodecahydrate, *sodium alum, soda alum.* Colorless crystals or white granules or powder. d 1.61. mp about 60°. Sol in 1 part water. Practically insol in alc.

USE: Industrially, like aluminum potassium sulfate.

THERAP CAT: Astringent.

THERAP CAT (VET): Astringent.

379. Aluminum Stearate. *Octadecanoic acid aluminum salt; stearic acid aluminum salt;* aluminum tristearate. $C_{54}H_{105}AlO_6$; mol wt 877.41. C 73.92%, H 12.06%, Al 3.08%, O 10.94%. $[CH_3(CH_2)_{16}COO]_3Al$. Prepn: Gilmour *et al., J. Chem. Soc.* **1956**, 1972.

Hard material, passes into plastic form on heating. mp 117-120°. Practically insol in water. When freshly made, sol in alcohol, benzene, oil turpentine, mineral oils. Forms cryst pyridine complex.

USE: Waterproofing fabrics, ropes; in paint and varnish driers; thickening lubricating oils; in cements; in light-sensitive photographic compositions.

380. Aluminum Subacetate Solution. Essigsäure Tonerde. Contains about 8% aluminum diacetate, Al(OH)(CH$_3$CO$_2$)$_2$. Prepd from aluminum sulfate, acetic acid, calcium carbonate and water. Details of prepn: *USP XVIII*, pp 29-30. Several other routes of prepn, *see* Aluminum Diacetate.

Clear, colorless liquid. Odor of acetic acid. d 1.045. Acid to litmus. Gradually becomes turbid and colloidal. Sometimes a more basic salt precipitates out and settles to the bottom of the container. Heat promotes the decomposition. *Keep in a cool place and discard when cloudy.*

Caution: Ingestion of large doses may cause severe nausea, vomiting, diarrhea, melena, hematemesis.

USE: As mordant in cotton dyeing, in printing fabrics, in water- and fireproofing fabrics, in waterproofing tissues, in formulations for antiperspirants and embalming fluids. *See also* Aluminum Diacetate.

THERAP CAT: Astringent, antiseptic.

THERAP CAT (VET): Astringent.

381. Aluminum Sulfate. $Al_2O_{12}S_3$; mol wt 342.15. Al 15.77%, O 56.11%, S 28.12%. $Al_2(SO_4)_3$. Occurs in nature as the mineral *alunogenite.* Prepn: *Gmelin's Aluminum* (8th ed.) **35B**, pp 267-269 (1934).

White, lustrous crystals, pieces, granules, or powder. d 1.61. Melts when gradually heated. At 250° loses its water; decomposes at red heat. Sol in 1 part water. Practically insol in alcohol. On long boiling of the aq soln, insol basic salt ppts. The aq soln is acid.

Octadecahydrate. The commercial product is also known as *cake alum* or *patent alum.* It is about 99.5% pure. The article of commerce usually contains 5 to 10% less water than theory.

USE: Tanning leather, sizing paper, mordant in dyeing; purifying water; manuf lakes, aluminum resinate; fireproofing and waterproofing cloth; clarifying oils and fats; treating sewage; waterproofing concrete; deodorizing and decolorizing petroleum; antiperspirants; agricultural pesticides; manuf aluminum salts.

THERAP CAT: Anti-infective.

382. Aluminum Sulfide. Al_2S_3; mol wt 150.16. Al 35.94%, S 64.06%. Prepd by heating aluminum with sulfur: Flahaut, *Compt. Rend.* **232**, 334 (1951); Becher in *Handbook of Preparative Inorganic Chemistry* vol. 1, G. Brauer, Ed. (Academic Press, New York, 2nd ed., 1963) p 823.

Yellowish-gray, compact lumps; H_2S odor; decomposes in moist air to a gray powder. d 2.02. mp 1100°. Hydrolyzed by water to Al(OH)$_3$ and H_2S. *Keep tightly closed.*

383. Aluminum Tartrate. *2,3-Dihydroxybutanedioic acid aluminum salt.* $C_{12}H_{12}Al_2O_{18}$; mol wt 498.18. C 28.93%, H 2.43%, Al 10.83%, O 57.81%. $Al_2(C_4H_4O_6)_3$. Usually contains some water. Prepn: Goldman, *Biochem. Z.* **133**, 459 (1922); *Brit.* pat. **348,790** (1930 to I. G. Farben.).

Odorless granules. Slowly sol in cold, readily in hot water; sol in ammonia.

USE: In textile dyeing.

384. Aluminum Thiocyanate. Aluminum sulfocyanate. $C_3AlN_3S_3$; mol wt 201.23. C 17.91%, Al 13.41%, N 20.88%, S 47.80%. Al(CNS)$_3$. Prepn: *Gmelin's, Aluminum* (8th ed.) **35B**, p 306 (1934).

USE: Aq solution used as mordant in dye industry.

385. Aluminum Zinc Sulfate. $Al_2O_{16}S_4Zn$; mol wt 503.61. Al 10.72%, O 50.83%, S 25.47%, Zn 12.98%. $Al_2Zn(SO_4)_4$. Prepn: *Gmelin's Alumimum* (8th ed.) **35B**, p 610 (1934).

Tetracosahydrate, *zinc alum.* Crystals; sol in water.

386. Alverine. *N-Ethyl-N-(3-phenylpropyl)benzenepropanamine; N-ethyl-3,3'-diphenyldipropylamine; bis*(γ-phenylpropyl)ethylamine; di(phenylpropyl)ethylamine. $C_{20}H_{27}N$; mol wt 281.44. C 85.35%, H 9.67%, N 4.98%. Prepd from phenylpropyl chloride and ethylamine in alkaline medium: Külz *et al., Ber.* **72**, 2165 (1939); alternate route: Stühner, Elbrächter, *Arch. Pharm.* **287**, 139 (1954).

Liquid, bp$_{0.3}$ 165-168°; bp$_{13}$ 210-215°.
Hydrochloride, sirup, sol in water.
Citrate, $C_{20}H_{27}N.C_6H_8O_7$, *Antispasmin, Calmabel, Gamatran, Profenine, Prophelan, Proverine, Spacolin, Spasmaverine.* Crystals, freely sol in water. A 10% aq soln is neutral to litmus.

THERAP CAT: Anticholinergic.

387. Amanitin. Group 1 mushroom toxin from the poisonous mushroom *Amanita phalloides* (Fr.) Secr., *Agaricaceae.* Inhibits protein synthesis of mammalian cells. Comprised of α-, β-, γ-amanitin and *amanin.* α-Amanitin is the major poisonous constituent of *A. phalloides;* it is 10-20 times more toxic than phalloidin, *q.v.* Isoln of α- and β-amanitin: Wieland, *Ann.* **564**, 152 (1949). Prepn of α-amanitin from β-amanitin: Wieland, Boehringer, *ibid.* **635**, 178 (1960). Structure: Wieland, *Pure Appl. Chem.* **9**, 145 (1964); Wieland, Gebert, *Ann.* **700**, 157 (1967). Review of chemistry and toxicology of the toxins of *Amanita phalloides:* Wieland, Wieland, *Pharmacol. Rev.* **11**, 87-107 (1959); T. Wieland, *Fortschr. Chem. Org. Naturst.* **25**, 214-250 (1967); T. Wieland, H. Faulstich, *Crit. Rev. Biochem.* **5**, 185-260 (1978). Book: H. Faulstich *et al., Amanita Toxins and Poisoning: International Amanita Symposium* (Lubrecht & Cramer, Heidelberg, 1980) 246 pp.

α-Amanitin, $C_{39}H_{54}N_{10}O_{14}S$. Needles from methanol, mp 254-255°. $[α]_D^{20}$ +191°. uv max: 302 nm. LD$_{50}$ i.p. in albino mice: 0.1 mg/kg (Wieland, Wieland).

Consult the Name Index before using this section.

β-*Amanitin*, $C_{39}H_{53}N_9O_{15}S$. Needles from methanol, mp 300°. uv max: 302 nm. Sol in water, methanol, ethanol, aqueous butanol. LD_{50} i.p. in albino mice: 0.4 mg/kg (Wieland, Wieland).

Caution: Highly toxic. Following a characteristic asymptomatic period of 6-15 hrs, potential symptoms of intoxication due to ingestion include violent gastroenteritis; fever, tachycardia, hyperglycemia, dehydration, electrolyte imbalance; liver dysfunction and necrosis; renal failure; may be fatal. *See Clinical Toxicology of Commercial Products,* R. E. Gosselin *et al.,* Eds. (Williams & Wilkins, Baltimore, 5th ed., 1984) Section II, p 246; M. J. Ellenhorn, D. G. Barceloux, *Medical Toxicology: Diagnosis and Treatment of Human Poisoning* (Elsevier, New York, 1988) pp 1331-1338. USE: As a tool in molecular biology.

388. Amanozine. *N-Phenyl-1,3,5-triazine-2,4-diamine; 2-amino-4-anilino-s-triazine;* 4-amino-2-anilino-1,3,5-triazine; 2-amino-6-anilino-s-triazine; *N-phenylformoguanamine;* W-1191-2; Urofort. $C_9H_9N_5$; mol wt 187.20. C 57.74%, H 4.85%, N 37.41%. Prepd by the action of formic acid on phenylbiguanide: Papini, Folena, *Gazz. Chim. Ital.* **80,** 837 (1950); **Austrian pat. 168,063** (1951 to Gedeon Richter), *C.A.* **47,** 8097h (1953); from phenylbiguanide and ethyl formate in methanol in 67% yield: Overberger, Shapiro, *J. Am. Chem. Soc.* **76,** 93 (1954).

Crystals from dioxane or 50% ethanol, mp 235-236°. Precipitated by alkalies.

Hydrochloride, $C_9H_9N_5 \cdot HCl$, long needles, dec 258-260°.

THERAP CAT: Diuretic.

389. Amantadine. *Tricyclo[3.3.1.1³,⁷]decan-1-amine; 1-adamantanamine;* 1-aminoadamantane; 1-aminodiamantane (obsolete); 1-aminotricyclo[3.3.1.1³,⁷]decane. $C_{10}H_{17}N$; mol wt 151.25. C 79.41%, H 11.33%, N 9.26%. Synthesis (two different routes): Stetter *et al., Ber.* **93,** 226 (1960); *cf.* *Angew. Chem.* **71,** 429-430 (1959); from adamantyl chloride: Gerzon *et al., J. Med. Chem.* **6,** 760 (1963). Prepn: **Brit. pat. 925,728** corresp to **U.S. pat. 3,152,180** (1963, 1964 both to Studiengesellschaft Kohle); **Belg. pat. 646,581** (1964 to Du Pont), *C.A.* **63,** 14726h (1965). Pharmacology and toxicology: Vernier *et al., Toxicol. Appl. Pharmacol.* **15,** 642 (1969). Metabolism: Bleidner *et al., J. Pharmacol. Exp. Ther.* **150,** 484 (1965). Clinical trial in prophylaxis of influenza A infection: R. Dolin *et al., N. Engl. J. Med.* **307,** 580 (1982). *See also* Adamantane, Rimantadine. Comprehensive description: J. Kirschbaum, *Anal. Profiles Drug Subs.* **12,** 1-36 (1983).

Hexakistetrahedral crystals by sublimation, mp 160-190° (closed tube). Also reported as mp 180-192°. Sparingly sol in water.

Hydrochloride, $C_{10}H_{17}N \cdot HCl$, *EXP-105-1, NSC-83653, Amazolon, Mantadix, Mantadan, Mantadine, Midantan, Mydantane, Symmetrel, Virofral.* Crystals from abs ethanol + anhydr ether, dec 360°. Sol in water (at least 1:20), alc. Practically insol in ether. LD_{50} orally in mice, rats: 700, 1275 mg/kg (Vernier).

Sulfate, *PK-Merz, Trivaline.*

THERAP CAT: Antiviral (Influenza A); antiparkinsonian; treatment of drug-induced extrapyramidal reactions.

390. Amantanium Bromide. *N,N-Dimethyl-N-[2-[(tricyclo[3.3.1.1³,⁷]dec-1-ylcarbonyl)oxy]ethyl]-1-decanaminium bromide;* 2-(1'-adamantanecarbonyloxy)ethyldimethyldecylammonium bromide; decyl(2-hydroxyethyl)dimethylammonium bromide 1-adamantanecarboxylate; Amantol. $C_{25}H_{46}BrNO_2$; mol wt 472.55. C 63.54%, H 9.81%, Br 16.91%, N 2.96%, O 6.77%. Prepn: R. A. Bauman, **U.S. pat. 3,928,411** (1975 to Colgate-Palmolive); and activity: L. Rovati *et al.,* **U.S. pat. 4,288,609** (1981 to Rotta).

Crystals, mp 182-184°. LD_{50} orally in mice: 910 mg/kg (Rovati).

THERAP CAT: Antiseptic.

391. Amaranth (Dye). *3-Hydroxy-4-[(4-sulfo-1-naphthalenyl)azo]-2,7-naphthalenedisulfonic acid trisodium salt; C.I. Acid Red 27;* C.I. 16185; FD & C Red No. 2; Red no. 2; trisodium salt of 1-(4-sulfo-1-naphthylazo)-2-naphthol-3,6-disulfonic acid. $C_{20}H_{11}N_2Na_3O_{10}S_3$; mol wt 604.48. C 39.74%, H 1.83%, N 4.63%, Na 11.41%, O 26.47%, S 15.91%. Prepd by coupling diazotized α-naphthylamine-4-sulfonic acid with β-naphthol-3,6-disulfonic acid: Knecht, *J. Soc. Dyers Colour.* **2,** 24 (1886); Farbw. Hoechst, **Ger. pat. 3229; Brit. pat. 1715;** BASF, **Ger. pat. 5411; U.S. pat. 204,799.** Metabolism: J. L. Radomski, T. J. Mellinger, *J. Pharmacol. Exp. Ther.* **136,** 259 (1962). *See also Colour Index* vol. 4 (3rd ed., 1971) p 4093.

Dark, reddish-brown powder. Absorption max (water): 522.5 nm. One gram dissolves in about 15 ml water. Also reported as 7.20 g/100 ml H_2O at 26°. Very slightly soluble in alcohol. The aq soln is vivid red (1 cm layer). Discharge white by hydrosulfite on wool and silk. HCl does not change the color intensity of the soln, NaOH increases it. The aq soln is stable toward light.

Caution: Banned by the FDA in 1976 for use in foods, drugs and cosmetics.

USE: Dyeing wool and silk bright bluish-red from an acid bath. As indicator in hydrazine titrations. In color photography.

392. Amaranth (Plant). Genus of the *Amaranthaceae* L. family which contains approx 60 species having worldwide distribution. Many species are considered weeds but *Amaranthus caudatus* L. (*love-lies-bleeding*), *A. hybridus* variety *hypochondriacus* L. (*prince's feather*), *A. tricolor* have been cultivated as ornamentals. *A. retroflexus* L. and some of the other weedy species are known as *pigweed, redroot* and *water hemp. A. spinosus* L. has been used in the treatment of gonorrhea: W. H. Brown, *Useful Plants of the Philippines* **1** (Philippines Dept. Agr. and Natl. Resources, Manila, 1951) pp 510-515; as a poultice in the treatment of inflammation, bruises and eczema: T. H. P. de Tavera, *The Medicinal Plants of the Philippines* (P. Blakiston's Son, Philadelphia, 1901) pp 200-202. Most species are hardy, herbaceous and fast-growing cereal-like plants. Leaves and grain are used for food in parts of South America, Africa and Asia. Plants are high in protein; the amino acid composition is complementary to that of wheat. The grain was a basic food in pre-Columbian South and Central America and was important in Aztec ritual. Grain amaranths (*A. hypochondriacus, A. cruentas, A. caudatus*) produce large seedheads containing many edible seeds. The seed can be made into low gluten flour, cooked into gruel or popped like corn. Compositional study of grain: R. Becker *et al., J. Food Sci.* **46**, 1175 (1981); K. C. Pant, *Nutr. Rep. Int.* **28**, 1445 (1983). Properties of seed starches: P. V. S. Rao, J. K. Goering, *Cereal Chem.* **47**, 655 (1970); Y. Tomita *et al., J. Nutr. Sci. Vitaminol.* **27**, 471 (1981); saccharides: K. Lorenz, M. Gross, *Nutr. Rep. Int.* **29**, 721 (1984); lipids: F. I. Opute, *J. Exp. Botany* **30**, 601 (1979). Baking potential of flour: K. Lorenz, *Starch* **33**, 149 (1981). Most amaranth species have edible leaves with mild spinach-like flavor. *A. cruentas, A. dubius, A. hybridus, A. lividus* and *A. tricolor* are some of the species grown as vegetables. Magnesium and copper content of leaf: N. M. Guttiker *et al., J. Nutr. Diet.* **3**, 4 (1966); vitamin A content: C. N. Rao, ibid. **4**, 10 (1967); amino acids: I. G. Vasi, V. P. Kalintha, *J. Inst. Chemists (India)* **52**, 13 (1980). Nutritive value of leaf protein concentrate: P. R. Cheeke *et al., Can. J. Animal Sci.* **61**, 199 (1981). Toxicology: R. M. Hill, P. D. Rawate: *J. Agr. Food Chem.* **30**, 465 (1982). Book: J. N. Cole, *Amaranth, From the Past for the Future* (Rodale Press, Emmaus, Pennsylvania, 1979) 311 pp; Nat'l. Academy of Sciences Report: *Amaranth, Modern Prospects for an Ancient Crop* (Nat'l. Academy Press, Washington, D.C., 1984) 80 pp.

393. Amarogentin. *[4aS-(4aα,5β,6α)]-3,3′,5-Trihydroxy[1,1′-biphenyl]-2-carboxylic acid 2-ester with 5-ethenyl-6-(β-D-glucopyranosyloxy)-4,4a,5,6-tetrahydro-1H,3H-pyrano-[3,4-c]pyran-1-one;* sweroside-2′-(3″,5″,3‴-trihydroxydiphenyl)-2″-carboxylic acid ester. $C_{29}H_{30}O_{13}$; mol wt 586.55. C 59.38%, H 5.16%, O 35.46%. Strongly bitter glucoside. Isoln from *Swertia chirata* Buch-Ham., *Gentianaceae:* Korte, *Ber.* **88**, 704 (1955); **89**, 2404 (1956); from *Swertia japonica* Makino: Inouye *et al., Chem. Pharm. Bull.* **18**, 1856 (1970). Occurrence in gentianaceous plants: Inouye, Nakamura, *J. Pharm. Soc. Japan* **91**, 755 (1971), *C.A.* **75**, 95431b (1971). Structure: *eidem, Tetrahedron Letters* **1968**, 4919; *eidem, Tetrahedron* **27**, 1951 (1971).

Colorless needles, mp 229-230° (monohydrate). $[α]_D^{20}$ −116.6° (methanol). uv max: 230, 266, 306 nm (log ε 4.46, 4.07, 3.68). Slightly sol in benzene, water; freely sol in acetone, anhydr dioxane, tetrahydrofuran, methanol, ethanol. Practically insol in petr ether, ether, cyclohexane, $CHCl_3$.

394. Amarolide. *(2α,11α)-2,11-Dihydroxypicrasane-1,12,16-trione.* $C_{20}H_{28}O_6$; mol wt 364.44. C 65.92%, H 7.74%, O 26.34%. Isoln from *Ailanthus altissima* (Mill.) Swingle (*A. glandulosa* Desf.), *Simaroubaceae,* and proposed structure: Casinovi *et al., Tetrahedron Letters* **1965**, 2273.

Isoln from *Castela nicholsoni* Hook, *Simaroubaceae,* and revised structure: Stöcklin *et al., ibid.* **1970**, 2399.
mp 253-255°. Practically insol in weak alkaline solns. Slightly sol in sodium carbonate, easily dissolved by sodium hydroxide.
Monoacetate, $C_{22}H_{30}O_7$, mp 264-265° for 11-acetate; mp 225° for 2-acetate.
Diacetate, $C_{24}H_{32}O_8$, mp 269-270° (from alcohol). Also reported as mp 260-262°.

395. Ambazone. *2-[4-[(Aminoiminomethyl)hydrazono]-2,5-cyclohexadien-1-ylidene]hydrazinecarbothioamide; [(4-oxo-2,5-cyclohexadien-1-ylidene)amino]guanidine thiosemicarbazone;* 1-amidinohydrazono-4-thiosemicarbazono-2,5-cyclohexadiene; *p*-benzoquinone amidinohydrazone thiosemicarbazone; benzoquinone guanylhydrazone thiosemicarbazone; Iversal; Primal; Promassol. $C_8H_{11}N_7S$; mol wt 237.29. C 40.49%, H 4.67%, N 41.32%, S 13.51%. Prepn from *p*-benzoquinone and aminoguanidine followed by treatment of the reaction product with thiosemicarbazide in acid soln: Petersen, Domagk, *Naturwiss.* **41**, 10 (1954); *eidem,* Ger. pat. **965,723** and Brit. pat. **774,794** (both 1957 to Bayer).

Monohydrate, brown or copper-colored crystals from water, dec 195°. Sparingly soluble in water. Moderately sol in alcohol, acetone. Freely sol in DMF, dilute acids.
THERAP CAT: Antibacterial.

396. Ambenonium Chloride. *N,N′-[(1,2-Dioxo-1,2-ethanediyl)bis(imino-2,1-ethanediyl)]bis[2-chloro-N,N-diethylbenzenemethanaminium] dichloride;* [oxalylbis(iminoethylene)]bis[(o-chlorobenzyl)diethylammonium chloride]; *N,N′*-bis[2-diethylaminoethyl]oxamide bis[2-chlorobenzyl chloride]; Win-8077; Mytelase; Mysuran. $C_{28}H_{42}Cl_4N_4O_2$; mol wt 608.48. C 55.27%, H 6.96%, Cl 23.31%, N 9.21%, O 5.26%. Prepn: F. K. Kirchner, Ger. pat. **1,024,517** (1958 to Sterling Drug), *C.A.* **54**, 18366g (1960). Determn in serum: C. Tharasse-Bloch *et al., J. Chromatog.* **421**, 407 (1987). Pharmacokinetics in dogs: *eidem, Arzneimittel-Forsch.* **39**, 257 (1989).

Crystals, mp 196-199°. Freely soluble in water.
THERAP CAT: Cholinesterase inhibitor.

397. Amber. Baltic amber; bernstein; succinite. A fossil resin from the extinct pine tree *Pinites succinifera* (Goepp.) Conway, *Pinaceae.* Found along the Baltic coast, also mined in Samland (East Prussia). Baltic amber contains: C 79%, H 10.5%, O 10.5%; succinic acid 3-8%; *α*-amyrin 20-30%. *Refs:* Plonait, *Angew. Chem.* **48**, 184 (1935); Schmid and co-workers: *Ann.* **503**, 269 (1933); *Monatsh.* **63**, 210 (1933); **65**, 348 (1935); **72**, 290, 311 (1939). *Review:* Berthelot, *Chim. & Ind.* **50**, 78-9 (1943); Frondel, *Econ. Bot.* **22**, 371 (1968). Infrared spectroscopy of different varieties of powdered amber: C. W. Beck *et al., Nature* **201**, 256 (1964). Chemical constitution: J. B. Lambert, J. S. Frye, *Science* **217**, 55 (1982).
Pale-yellow to reddish-brown resin. Transparent or cloudy (due to enclosed air bubbles and free succinic acid). Brittle; conchoidal fracture. d 1.05 to 1.10. Harder than most other resins. n_D 1.539-1.545. Softens at 150°, mp 350-375° giving off a choking, aromatic odor. When rubbed it is a good generator of static electricity.
USE: The best quality is machined into beads and other personal ornaments. For teething strings. Also used for making mouthpieces of tobacco pipes and cigarette holders. Small pieces are pressed into "ambroid" and then used for

the same purpose. Impure material goes into the manufacture of "amber" varnishes. *See also* Oil of Amber.

398. Ambergris. Concretion from intestinal tract of the sperm whale, *Physeter catodon* L., *Physeteridae.* Found in tropical seas or seashores. Perfumers have used ambergris for centuries for its desirable odoriferous and fixative properties. Three major components isolated are the triterpene alcohol ambrein, epicoprostanol and coprostanone: Ruzicka, Lardon, *Helv. Chim. Acta* **29,** 912 (1946); Lederer *et al., ibid.,* 1354; Hardwick, Laws, *The Analyst* **76,** 662 (1951). Ambergris falls under the Marine Mammal Protection Act of 1972 and is illegal to import in the U.S.A. Analytical method for the detection and identification of ambergris: T. F. Governo *et al., J. Assoc. Offic. Anal. Chem.* **60,** 160 (1977). Hypothesis on the biological origin of ambergris: P. A. Dubois, *Parfums, Cosmet., Aromes* **19,** 35 (1978).

Gray to black, waxy mass; characteristic odor. d 0.8-0.92. mp ~60°. Flammable; almost completely volatile by heat. Insoluble in water or in alkali hydroxides. Sol in hot alcohol, chloroform, ether, fats, volatile oils.

USE: Chiefly in perfumery as tincture and essence for fixing delicate odors.

399. Amberlite®. Synthetic, high capacity cation and anion exchange resins and ion exchange-impregnated papers. *Reviews:* Nachod, Schubert, *Ion Exchange Technology* (Academic Press, New York, 1956); Kunin, *Ion Exchange Resins* (John Wiley, New York, 1958); Kunin, *Elements of Ion Exchange* (Reinhold, New York, 1960); Rieman, Walton, *Ion Exchange in Analytical Chemistry* (Pergamon Press, 1970); Dorfner, *Ion Exchangers Properties and Applications* (Ann Arbor Science, 1972).

400. Ambrosin. *3,3a,4,5,6,6a,9a,9b-Octahydro-6,9a-dimethyl-3-methyleneazuleno[4,5-b]furan-2,9-dione; 6β-hydroxy-4-oxo-10αH-ambrosa-2,11(13)-dien-12-oic acid γ-lactone.* $C_{15}H_{18}O_3$; mol wt 246.31. C 73.15%, H 7.37%, O 19.49%. From herb of *Ambrosia maritima* L., *Compositae:* Abu-Shady, Soine, *J. Am. Pharm. Assoc.* **42,** 387 (1953); **43,** 365 (1954). Structure: Bernardi, Büchi, *Experientia* **13,** 466 (1957); Sorm *et al., Coll. Czech. Chem. Commun.* **24,** 1548 (1959); Nerz *et al., J. Am. Chem. Soc.* **84,** 2601 (1962). Absolute configuration: Emerson *et al., Tetrahedron Letters* **1966,** 6151. Stereospecific total synthesis of (±)-form: P. A. Grieco *et al., J. Am. Chem. Soc.* **99,** 7393 (1977).

Crystals from alc, mp 146°. $[\alpha]_D^{22}$ −154.50° (c = 2 in alcohol). uv max (95% ethanol): 217, 324 nm (ε 13,465; 36). Very sol in chloroform; sol in alc, benzene, sparingly sol in ether, petr ether. Practically insol in water, cold dil NaOH, dil acids.

(±)-Form, mp 188-190°.

401. Ambroxol. *4-[[(2-Amino-3,5-dibromophenyl)-methyl]amino]cyclohexanol; N-(trans-p-hydroxycyclohexyl)-(2-amino-3,5-dibromobenzyl)amine; NA-872.* $C_{13}H_{18}Br_2N_2$-O; mol wt 378.11. C 41.30%, H 4.80%, Br 42.27%, N 7.41%, O 4.23%. Metabolite of bromhexine, *q.v.* Structure: E. Schraven *et al., Eur. J. Pharmacol.* **7,** 445 (1967). Synthesis: J. Keck, *Ann.* **707,** 107 (1967); Fr. pat. **1,522,709;** J. Keck *et al.,* U.S. pat. **3,536,713** (1968, 1970 both to Thomae). Toxicity: S. Püschmann, R. Engelhorn, *Arzneimittel-Forsch.* **28,** 889 (1978). Series of articles on pharmacology, metabolism, and clinical studies: *ibid.* 889-935; on pharmacology and clinical efficacy of combination with amoxicillin, *q.v.,* in bronchopulmonary disease: *ibid.* **37,** 965-971 (1987). Symposium on pharmacology and efficacy in multicenter studies: *Respiration* **51,** Suppl. 1, 1-68 (1987).

Hydrochloride, $C_{13}H_{18}Br_2N_2O \cdot HCl$, *Abramen, Ambril, Bronchopront, Duramucal, Fluibron, Fluixol, Frenopect, Lindoxyl, Motosol, Muco-Burg, Mucofar, Mucosan, Mucosolvan, Mucoclear, Mucovent, Pect, Solvolan, Stas-Hustenlöser, Surbronc, Surfactal.* Crystals from ethanol, mp 233-234.5° (dec). LD_{50} in mice, rats (mg/kg): 268, 380 i.p.; 2720, 13400 orally (Püschmann, Engelhorn).

THERAP CAT: Expectorant.

402. Ambucaine. *4-Amino-2-butoxybenzoic acid 2-diethylaminoethyl ester;* 2-butoxy-4-aminobenzoic acid β-diethylaminoethyl ester; β-diethylaminoethyl 2-butoxy-4-aminobenzoate; 2-diethylaminoethyl 4-amino-2-n-butoxybenzoate; ambutoxate. $C_{17}H_{28}N_2O_3$; mol wt 308.42. C 66.20%, H 9.15%, N 9.08%, O 15.56%. Prepn: Büchi *et al., Helv. Chim. Acta* **34,** 1002 (1951); Clinton *et al., J. Am. Chem. Soc.* **74,** 592 (1952); Clinton, Laskowski, U.S. pat. **2,689,248** (1954 to Sterling Drug); Grimme, Schmitz, *Ber.* **87,** 179 (1954).

Hydrochloride, $C_{17}H_{28}N_2O_3 \cdot HCl$, *Win-3706, Sympocaine.* Crystals, mp 127°.

Dihydrochloride, crystals, decomp 156.8-159°.

THERAP CAT: Anesthetic (local).

403. Ambucetamide. *α-(Dibutylamino)-4-methoxybenzeneacetamide; N-(trans-p-hydroxycyclohexyl)acetamide; α-dibutylamino-α-(p-methoxyphenyl)acetamide; α-p-methoxyphenyl-α-di-n-butylaminoacetamide; A-16; Dibutamide; Bersen; Meritin.* $C_{17}H_{28}N_2O_2$; mol wt 292.42. C 69.83%, H 9.65%, N 9.58%, O 10.94%. Prepn starting with dibutylamine, anisaldehyde and potassium cyanide: Janssen, *J. Am. Chem. Soc.* **76,** 6192 (1954).

Rods from ethanol + 10% ether. mp 134° (first reported as mp 125-127°). Practically insol in water. Sol in ethanol, isopropanol, glacial acetic acid.

THERAP CAT: Antispasmodic.

404. Ambuphylline. *3,7-Dihydro-1,3-dimethyl-1H-purine-2,6-dione compd with 2-amino-2-methyl-1-propanol (1:1);* theophylline aminoisobutanol; theophylline compd with 2-amino-2-methyl-1-propanol; bufylline (rescinded USAN); Butaphyllamine; Buthoid. $C_{11}H_{19}N_5O_3$; mol wt 269.30. C 49.06%, H 7.11%, N 26.01%, O 17.82%. Prepn: Shelton, U.S. pat. **2,404,319** (1946 to Wm. S. Merrell). Toxicity studies: Thompson, Warren, *J. Clin. Med.* **31,** 1337 (1946).

Crystals, mp 254-256°. Soly in water about 55%. LD_{50} i.v. in rabbits: 163 mg/kg; orally in mice: 600 mg/kg, (Thompson, Warren).

THERAP CAT: Bronchodilator.

405. Ambuside. *4-Chloro-6-[(3-hydroxy-2-butenylidene)amino]-N¹-2-propenyl-1,3-benzenedisulfonamide; N'-allyl-4-chloro-6-[(3-hydroxy-2-butenylidene)amino]-m-benzenedisulfonamide;* 2-(allylsulfamoyl)-5-chloro-4-sulfamoyl-*N*-(3-hydroxy-2-butenylidene)aniline; 2-allylsulfamyl-4-sulfamyl-5-chloro-*N*-(3-hydroxy-2-butenylidene)aniline; 5-allylsulfamoyl-2-chloro-4-(3-hydroxybut-2-enylideneamino)benzenesulfonamide; EX-4810; Hydrion; Novohydrin. $C_{13}H_{16}ClN_3O_5S_2$; mol wt 393.87. C 39.64%, H 4.09%, Cl 9.00%, N 10.67%, O 20.31%, S 16.28%. Prepn: Robertson, Fr. pat. **1,331,680** (1963 to Lakeside), corresp to U.S. pat. **3,188,329** (1965 to Colgate-Palmolive); Robertson *et al., J. Med. Chem.* **8**, 90 (1965).

Crystals, mp 205-207°. uv max: 343 nm (ε 32900).
THERAP CAT: Diuretic; antihypertensive.

406. Ambutonium Bromide. *γ-(Aminocarbonyl)-N-ethyl-N,N-dimethyl-γ-phenylbenzenepropanaminium bromide;* 3-(carbamoyl-3,3-diphenylpropyl)ethyldimethylammonium bromide; 4-dimethylamino-2,2-diphenylbutyramide ethyl bromide; R-100. $C_{20}H_{27}BrN_2O$; mol wt 391.35. C 61.38%, H 6.95%, Br 20.42%, N 7.16%, O 4.09%. Prepn: Janssen *et al., Arch. Int. Pharmacodyn.* **103**, 82 (1955).

Crystals, mp 228-229° (dec).
Combination with oxazepam, *q.v., Praxiten SP.*
THERAP CAT: Anticholinergic.

407. Amcinonide. *(11β,16α)-21-(Acetyloxy)-16,17-[cyclopentylidenebis(oxy)]-9-fluoro-11-hydroxypregna-1,4-diene-3,20-dione;* 9-fluoro-11β,16α,17,21-tetrahydroxypregna-1,4-diene-3,20-dione cyclic 16,17-acetal with cyclopentanone 21-acetate; CL-34699; Amciderm; Cyclocort; Penticort. $C_{28}H_{35}FO_7$; mol wt 502.58. C 66.92%, H 7.02%, F 3.78%, O 22.28%. Prepn: W. Shultz *et al.,* **Ger.** pat. **2,437,847** (1975 to Am. Cyanamid), *C.A.* **83**, 10608g (1975). Bioavailability study: R. Woodford, J. M. Haigh, *Curr. Ther. Res.* **26**, 301 (1979). Therapeutic use in eczematoid con-

ditions: G. L. Rocha *et al., ibid.* **19**, 538 (1976); E. W. Rosenberg, *Cutis* **24**, 642 (1979).

THERAP CAT: Glucocorticoid.

408. Amdinocillin. *[2S-(2α,5α,6β)]-6-[[(Hexahydro-1H-azepin-1-yl)methylene]amino]-3,3-dimethyl-7-oxo-4-thia-1-azabicyclo[3.2.0]heptane-2-carboxylic acid;* 6-[(hexahydro-1H-azepin-1-yl)methyleneamino]penicillanic acid; mecillinam; FL-1060; Ro-10-9070; Selexidin. $C_{15}H_{23}N_3O_3S$; mol wt 325.43. C 55.36%, H 7.12%, N 12.91%, O 14.75%, S 9.85%. Semi-synthetic antibiotic related to penicillin. Prepn: F. J. Lund, **Ger.** pat. **2,055,531;** *idem,* U.S. pat. **3,957,764** (1971, 1976 both to Lövens Kemiske Fabrik). Synthesis and chemical properties: H. B. König *et al., Arzneimittel-Forsch.* **33**, 88 (1983). X-ray structural study: J. W. Krajewski *et al., J. Antibiot.* **34**, 282 (1981). *In vitro* study: B. Chattopadhyay, I. Hall, *J. Antimicrob. Chemother.* **5**, 549 (1979). Activity against gram-negative bacteria: D. S. Reeves, *J. Antimicrob. Chemother.* **3**, Suppl. B, 5 (1977). Studies on mechanism of action: B. G. Spratt, *ibid.* 13. Metabolism: A. P. Ball *et al., ibid.* **4**, 241 (1978). Use in urinary tract infections and septicemia: N. Frimodt-Miller, T. J. Ravn, *Infection* **7**, 35 (1979). Determn in plasma and urine by HPLC: T. L. Lee, M. A. Brooks, *J. Chromatog.* **227**, 137 (1982). Pharmacokinetics in man: B. R. Meyers *et al., Antimicrob. Ag. Chemother.* **23**, 827 (1983). Symposium on pharmacology, pharmacokinetics, clinical studies: *Am. J. Med.* **75**, no. 2A, 1-138 (1983). Review of pharmacology and clinical efficacy: H. C. Neu, *Pharmacotherapy* **5**, 1-10 (1985).

Crystals from methanol-acetone, mp 156° (dec). $[\alpha]_D^{20}$ +285° (c = 1 in 0.1N HCl). Sol in water.
Pivaloyloxymethyl ester, *see* Amdinocillin Pivoxil.
THERAP CAT: Antibacterial.

409. Amdinocillin Pivoxil. *[2S-(2α,5α,6β)]-6-[[(Hexahydro-1H-azepin-1-yl)methylene]amino]-3,3-dimethyl-7-oxo-4-thia-1-azabicyclo[3.2.0]heptane-2-carboxylic acid (2,2-dimethyl-1-oxopropoxy)methyl ester;* pivaloyloxymethyl 6-[(hexahydro-1H-azepin-1-yl)methyleneamino]penicillanate; pivamdinocillin; pivmecillinam; FL-1039; Selexid (susp.). $C_{21}H_{33}N_3O_5S$; mol wt 439.58. C 57.38%, H 7.57%, N 9.56%, O 18.20%, S 7.29%. Semi-synthetic antibiotic related to penicillin. Pivaloyloxymethyl ester of amdinocillin, *q.v.* Prepn: F. J. Lund, **Ger.** pat. **2,055,531;** *idem,* U.S. pat. **3,957,764** (1971, 1976 both to Lövens Kemiske Fabrik). HPLC stability analysis: R. B. Hagel, E. H. Waysek, *J. Chromatog.* **178**, 97 (1979). Bacteriological and pharmacokinetic study: T. Damsgaard *et al., J. Antimicrob. Chemother.* **5**, 267 (1979). Metabolism: J. D. Anderson, M. A. Adams, *Chemotherapy* **25**, 1 (1979). Clinical study: B. T. Andersen *et al., Infection* **8**, 27 (1980). Toxicity study: S. Sato *et al., Takeda Kenkysho Ho* **35**, 179 (1976), *C.A.* **86**, 115154w (1977).

Crystals from cyclohexane, mp 118.5-119.5°. $[\alpha]_D^{20}$ +231° (c = 1 in 96% ethanol). LD_{50} in mice, rats (mg/kg): 475-480, 465 i.v.; 1736-1930, 1935-2100 s.c.; 3020, 9500-10000 orally (Sato).

Hydrochloride, $C_{21}H_{33}N_3O_5S \cdot HCl$, *Melysin, Negaxid, Selexid (tabl.).* Crystals from methanol-diisopropyl ether, mp 172-173°. $[\alpha]_D^{20}$ +219° (c = 1 in 0.1N HCl).

THERAP CAT: Antibacterial.

410. Americium. Am; at. no. 95; valences 3, 4, 5, 6. Man-made radioactive element. No stable nuclides; known isotopes (mass numbers): 234, 237-247. Longest-lived known isotope 243 ($T_{1/2}$ 7.38 × 10^3 years, rel. at. mass. 243.0614, α-emitter). First isotope prepared: ^{241}Am ($T_{1/2}$ 432.7 years, α and γ-emitter); prepd in 1944 by G. T. Seaborg *et al.* in *The Transuranium Elements*, G.T. Seaborg *et al.*, Eds., (McGraw-Hill, New York, 1949) p 1525-1553; *eidem, Phys. Rev.* **78**, 472 (1950). Isoln: Armstrong *et al., A. I. Ch. E. J.* **3**, 286 (1957); Coleman, *J. Inorg. Nucl. Chem.* **3**, 327 (1957). Prepn of metal: E. F. Westrum, Jr., L. Eyring, *J. Am. Chem. Soc.* **73**, 3396 (1951); Cunningham, Lohr, *ibid.* 2026. Superconducting properties: J. L. Smith, R. G. Haire, *Science* **200**, 535 (1978). Clinical application in bone mineral determn: E. G. De Puey *et al., J. Nucl. Med.* **16**, 891 (1975); in cancer radiotherapy: R. Nath *et al., Int. J. Rad. Oncol. Biol. Phys.* **14**, 969 (1988). *Reviews:* C. Keller, *The Chemistry of the Transuranium Elements* (Verlag Chemie, Weinheim, English Ed., 1971) pp 485-527; *Comprehensive Inorganic Chemistry* vol. **5**, J. C. Bailar, Jr. *et al.*, Eds. (Pergamon Press, Oxford, 1973) *passim; Handbk. Exp. Pharmakol.* **36**, 689-940 (1973); W. W. Schulz, R. A. Penneman in *The Chemistry of the Actinide Elements* vol. **2**, J. J. Katz *et al.*, Eds. (Chapman and Hall, New York, 1986) pp 887-961. *See also* metabolism study of internal contamination of ^{241}Am in man: N. Cohen *et al., Science* **206**, 64 (1979).

Silvery, ductile, very maleable, non-magnetic metal. Allotropic forms: double hexagonal close-packed α-form transforms to β-form at 658°; face-centered cubic β-form exists from 793-1004°; γ-form exists from ~1050° to mp. mp 1173°. bp (calc) 2067°. d 13.671. Dissolves readily in aq HCl; insol in liquid NH_3.

Trivalent americium, the most common in aq soln. Color light pink changing to yellow with increasing concn. Sharp absorption peak at 5027 Å. Americium dioxide, AmO_2, is obtained by ignition of most trivalent Am compounds.

Tetravalent americium, known only in the solid state. When AmO_2 or AmF_3 is treated with fluoride, solid AmF_4 results.

Pentavalent and hexavalent americium compds, when obtained in soln are doubly oxygenated and have the general formula AmO_2^+n, where n = +1 for Am(V) and n = +2 for Am(VI). Hexavalent americium is yellow or light brown in dilute perchloric or nitric acid, green in fluoride solns and dark brown in sulfuric acid. A deep red ion complex is formed in bicarbonate-carbonate solns.

Caution: Radiation hazard; handling requires special equipment and sheilding facilities (Katz *et al., loc. cit.,* vol. 2, p 1128).

USE: ^{241}Am as radiation source for thickness gauging, density and radiographic measurements; parent for production of ^{242}Cf and ^{242}Pu; in ionization smoke detectors; for dissipation of static electrical charges. ^{241}Am-Be mixture as neutron source. ^{243}Am as target material for production of transcurium elements in high neutron-flux reactors.

THERAP CAT: ^{241}Am as diagnostic aid (bone mineral analyzer); as antineoplastic (radiation source).

411. Ametryn. *N-Ethyl-N'-(1-methylethyl)-6-(methylthio)-1,3,5-triazine-2,4-diamine; 2-(ethylamino)-4-(isopropylamino)-6-(methylthio)-s-triazine;* 2-ethylamino-4-isopropyl-

amino-6-methylmercapto-*s*-triazine; 2-methylthio-4-ethyl-amino-6-isopropylamino-*s*-triazine; ametryne; G-34162; Ametrex; Evik; Gesapax. $C_9H_{17}N_5S$; mol wt 227.33. C 47.55%, H 7.54%, N 30.81%, S 14.11%. Prepn: Gysin, Knuesli, **Swiss** pat. 337,019, *C.A.* **57**, 14226c (1962); Rufener *et al.,* **U.S.** pat. 3,558,622 (1959, 1971 both to Geigy). Acute toxicity: T. B. Gaines, R. E. Linder, *Fundam. Appl. Toxicol.* **7**, 299 (1986).

Crystals, mp 88-89°. Aqueous soly data: Ward, Weber, *J. Agr. Food Chem.* **16**, 959 (1968). LD_{50} in adult male, female rats (mg/kg): 508, 590 orally (Gaines, Linder).

USE: Herbicide.

412. Amezinium Methyl Sulfate. *4-Amino-6-methoxy-1-phenylpyridazinium methyl sulfate;* LU-1631; Regulton; Risumic; Supratonin. $C_{12}H_{15}N_3O_5S$; mol wt 313.33. C 46.00%, H 4.83%, N 13.41%, O 25.53%, S 10.23%. Sympathomimetic agent with vascular and cardiac activity. Prepn: F. Richenender, R. Kropp, **Ger.** pat. 1,912,941; *eidem,* **U.S.** pat. 3,631,038 (1970, 1971 both to BASF). Series of articles on synthesis, pharmacology, mechanism of action, metabolism, pharmacokinetics, bioavailability, clinical trials: *Arzneimittel-Forsch.* **31**, 1527-1671 (1981). Acute toxicity data: H. J. Teschendorf, *ibid.* 1580. HPLC determn in human plasma: D. Hotz, E. Brode, *J. Chromatog.* **277**, 217 (1983). Disposition and identification of major metabolites in rats: K. Nambu *et al., Arzneimittel-Forsch.* **38**, 909 (1988).

Crystals from water, mp 176° (dec). LD_{50} in mice, rats (mg/kg): 1630, 1410 orally; 40.4, 45.5 i.v. (Teschendorf).

THERAP CAT: Antihypotensive.

413. Amfenac. *2-Amino-3-benzoylbenzeneacetic acid;* 2-amino-3-benzoylphenylacetic acid. $C_{15}H_{13}NO_3$; mol wt 255.27. C 70.58%, H 5.13%, N 5.49%, O 18.80%. Prepn: W. J. Welstead, H. W. Moran, **Ger.** pat. 2,324,768, *C.A.* **80**, 59708s (1974) and **U.S.** pat. 4,045,576 (1973, 1977 both to A. H. Robbins). Synthesis and anti-inflammatory activity: W. J. Welstead *et al., J. Med. Chem.* **22**, 1074 (1979). Anti-inflammatory, analgesic, antipyretic activities: H. Fujimura *et al., Oyo Yakuri* **22**, 381 (1981), *C.A.* **97**, 49427m (1982). Platelet aggregation inhibition: *eidem, ibid.* 399, *C.A.* **97**, 49502g (1982). Effect on polymorphonuclear leukocytes: T. Matsumoto *et al., Pharmacol. Res. Commun.* **14**, 523 (1982). Toxicity data: L. F. Sancilio *et al.,* 170th Am. Chem. Soc. Meet. (Chicago, Aug. 1975), Abstracts of Papers, MEDI 17.

Consult the Name Index before using this section.

mp 121-123° (dec). LD$_{50}$ in mice, rats (mg/kg): 615, 311 orally (Sancilio).

Sodium salt monohydrate, C$_{15}$H$_{12}$NNaO$_3$.H$_2$O, *AHR-5850D, Fenamate, Fenazox.* Yellow solid from ethanol/isopropyl ether, mp 254-255.5°.

THERAP CAT: Anti-inflammatory.

414. Amicarbalide. *3,3'-(Carbonyldiimino)bisbenzenecarboximidamide; N,N'-bis[3-(aminoiminomethyl)phenyl]benzenecarboximidamide; 3,3'-diamidinocarbanilide; N,N'-bis(m-amidinophenyl)urea; N,N'-di(m-amidinophenyl)urea.* C$_{15}$H$_{16}$N$_6$O; mol wt 296.33. C 60.80%, H 5.44%, N 28.36%, O 5.40%.

Diisethionate, C$_{19}$H$_{28}$N$_6$O$_9$S$_2$, *M & B 5062A, Diampron.* Prepn: Ashley *et al., Nature* **185,** 461 (1960); Berg, **Belg.** pat. **585,595** (1960 to May & Baker); *idem, J. Chem. Soc.* **1961,** 5097. Crystals, mp 209° (decomp 256°). Soly in water: ~100 g/100 ml. LD$_{50}$ s.c. in mice: 120 mg/kg.

THERAP CAT (VET): Babesiacide in cattle.

415. Amicetin. *[2R-[2α(S*),5β,6α]]-4-[(2-Amino-3-hydroxy-2-methyl-1-oxopropyl)amino]-N-[1-[5-[[4,6-dideoxy-4-(dimethylamino)-α-D-glucopyranosyl]oxy]tetrahydro-6-methyl-2H-pyran-2-yl]-1,2-dihydro-2-oxo-4-pyrimidinyl]benzamide.* C$_{29}$H$_{42}$N$_6$O$_9$; mol wt 618.69. C 56.30%, H 6.84%, N 13.58%, O 23.27%. Antibiotic substance produced by *Streptomyces vinaceus-drappus* isolated from soil near Kalamazoo, Mich.: De Boer *et al., J. Am. Chem. Soc.* **75,** 499, 5864 (1953); McKormich, Hoehn, *Antibiot. & Chemother.* **3,** 718 (1953); De Boer, Hinman, U.S. pats. **2,909,463, 2,909,464** and **2,909,517** (all 1959 to Upjohn). Structure: Haskell, *J. Am. Chem. Soc.* **80,** 747 (1958); Stevens *et al., J. Org. Chem.* **27,** 2991 (1962). Structure of *amosamine:* Stevens *et al., J. Am. Chem. Soc.* **85,** 1552 (1963). Configuration: Hanessian, Haskell, *Tetrahedron Letters* **1964,** 2451. Synthesis of *amosamine:* Stevens *et al., J. Org. Chem.* **31,** 2822 (1966).

Needles from cold water, mp 165-169°. Crystals from warm water, dec 243-244°. pKa = 10.4, 7.0. [α]$_D^{25}$ +116.5° (c = 0.5 in 0.1N HCl). uv max in water: 305 nm (E$_{1cm}^{1\%}$ 465); in 0.1N HCl: 316 nm (E$_{1cm}^{1\%}$ 433); in 0.1N NaOH: 322 nm (E$_{1cm}^{1\%}$ 470). Soly in water at 25° = 2 mg/ml. Slightly sol in common organic solvents.

Dihydrochloride, C$_{29}$H$_{44}$Cl$_2$N$_6$O$_9$, crystals from methanol + acetone, mp 190-192°. [α]$_D^{24.5}$ +117° (c = 0.4 in alc).

416. Amicibone. *1-[2-(Hexahydro-1H-azepin-1-yl)ethyl]-2-oxocyclohexanecarboxylic acid phenylmethyl ester; 2-(β-hexamethyleniminoethyl)cyclohexanone-2-carboxylic acid benzyl ester; Biotussal; Pectipront.* C$_{22}$H$_{31}$NO$_3$; mol wt 357.49. C 73.92%, H 8.74%, N 3.92%, O 13.43%. Prepn:

Belg. pats. **639,472; 639,473;** Frank *et al.,* **Austrian** pat. **237,593** (all 1964 to Biochemie GmbH), *C.A.* **62,** 5207e (1965).

Hydrochloride, C$_{22}$H$_{31}$NO$_3$.HCl. Crystalline powder, mp 136-137°. Sol in water, ethanol and other common organic solvents.

THERAP CAT: Antitussive.

417. Amicoumacin A. *3-Amino-2,3-dideoxy-N^6-[1-(3,4-dihydro-8-hydroxy-1-oxo-1H-2-benzopyran-3-yl)-3-methylbutyl]hexaramide.* C$_{20}$H$_{29}$N$_3$O$_7$; mol wt 423.47. C 56.73%, H 6.90%, N 9.92%, O 26.45%. Major component of a complex of antibiotics produced by *Bacillus pumilus* BN-103. Also exhibits anti-inflammatory activity *in vivo.* Isoln and biol activity: *Japan. Kokai* **83 18,379** (1983 to Meiji Seika), *C.A.* **99,** 20906x (1983); J. Itoh *et al., J. Antibiot.* **34,** 611 (1981); *eidem, Agr. Biol. Chem.* **46,** 1255 (1982). Structure: *eidem, ibid.* 2659. Use as acaricide: *Japan. Kokai* **83 216,107** (1983 to Meiji Seika), *C.A.* **100,** 116494b (1984).

Hydrochloride, colorless powder, mp 132-135° (dec). uv max (methanol): 208, 247, 315 nm (ε 27300, 6400, 4380). [α]$_D^{23}$ −97.2° (c = 1.0 in methanol). LD$_{50}$ orally in mice: 132 mg/kg (Itoh).

418. Amidephrine. *N-[3-[1-Hydroxy-2-(methylamino)ethyl]phenyl]methanesulfonamide; 3'-[1-hydroxy-2-(methylamino)ethyl]methanesulfonanilide; MJ-1996.* C$_{10}$H$_{16}$N$_2$O$_3$S; mol wt 244.31. C 49.16%, H 6.60%, N 11.47%, O 19.65%, S 13.12%. Prepn: Larsen, Uloth, **Fr.** pat. **M3027** (1965 to Mead Johnson), *C.A.* **62,** 13091a (1965); Uloth *et al., J. Med. Chem.* **9,** 88 (1966). Stimulates adrenergic α-receptors. Pharmacology studies of the racemate: Dungan *et al., Int. J. Neuropharmacol.* **4,** 219 (1965); Stanton *et al., ibid.* 235; of the isomers: Larsen, Lish, *Nature* **203,** 1283 (1964); Buchthal, Jenkinson, *Eur. J. Pharmacol.* **10,** 293 (1970). Toxicology: Weikel, Harper, *Toxicol. Appl. Pharmacol.* **23,** 589 (1972).

Crystals, mp 159-161°. pKa 9.1.

Monomethanesulfonate, C$_{10}$H$_{16}$N$_2$O$_3$S.CH$_3$SO$_3$H, *amidephrine mesylate, MJ-5190, Dricol, Fentrinol, Nalde.* Crystals from ethanol, mp 207-209°. LD$_{50}$ in female rats: 13-36 mg/kg orally; 5-25 mg/kg i.p.

THERAP CAT: Vasoconstrictor; decongestant (nasal).

419. Amidinomycin. *3-Amino-N-(3-amino-3-iminopropyl)cyclopentanecarboxamide; N-(2'-amidinoethyl)-3-aminocyclopentanecarboxamide; myxoviromycin.* C$_9$H$_{18}$N$_4$O; mol wt 198.27. C 54.52%, H 9.15%, N 28.26%, O 8.07%.

Antibiotic substance produced by *Streptomyces flavochromogenes* isolated from Japanese soil (Shiuoka Prefecture). Isoln and structure: S. Nakamura *et al., J. Antibiot.* **14A**, 103 (1961); S. Nakamura, *Chem. Pharm. Bull.* **9**, 641 (1961). Identity with myxoviromycin: S. Nakamura *et al., J. Antibiot.* **14A**, 163 (1961). Prepn: Katsube, Saito, **Japan.** pat. **21,418('68)** (to Sumitomo), *C.A.* **70**, 87135q (1969). Synthesis of amidinomycin and *trans* isomer: H. Paul *et al., Arch. Pharm.* **301**, 512 (1968). Crystal and molecular structure: M. Kaneda *et al., J. Antibiot.* **33**, 778 (1980).

Sulfate, $C_9H_{18}N_4O.H_2SO_4$, plates or needles from water + methanol, dec 285-288°. $[\alpha]_D^{21}$ −3.9° (c = 3). Absorption spectra: S. Nakamura, *loc. cit.* Soluble in water. Practically insol in ether, benzene, ethyl acetate, methanol, ethanol, butanol, acetone.

THERAP CAT: Antiviral.

420. Amidochlor. *N-[(Acetylamino)methyl]-2-chloro-N-(2,6-diethylphenyl)acetamide;* N-acetamidomethyl-2-chloro-2',6'-diethylacetanilide; MON-4621; Limit. $C_{15}H_{21}$-ClN_2O_2; mol wt 296.80. C 60.70%, H 7.13%, Cl 11.95%, N 9.44%, O 10.78%. Plant growth regulator for cool season grasses. Prepn: **Neth. pat. Appl. 7,207,261;** K. W. Ratts, U.S. pat. **3,830,841** (1972, 1974 both to Monsanto); K. W. Ratts, J. P. Chupp, *J. Org. Chem.* **39**, 3745 (1974). Use as plant growth regulator: K. W. Ratts *et al.*, U.S. pat. **3,829,306** (1974 to Monsanto). Effect on growth and seedhead suppression of annual bluegrass: A. M. Petrovic *et al., Agron. J.* **77**, 670 (1985); of wild and cultivated proso millet: J. L. Carpenter, H. J. Hopen, *HortScience* **20**, 942 (1985); on established turfgrass: P. C. Bhowmik, *Proc. 5th Int. Turfgrass Res. Conf.* 735 (1985).

Crystals from methanol, mp 148-149°.
USE: Turf growth regulator.

421. Amido-G-Acid. *7-Amino-1,3-naphthalenedisulfonic acid;* 2-naphthylamine-6,8-disulfonic acid; amino-G-acid. $C_{10}H_9NO_6S_2$; mol wt 303.32. C 39.60%, H 2.99%, N 4.62%, O 31.65%, S 21.14%. Prepd by sulfonation of β-naphthylamine: Fierz-David, Braunschweig, *Helv. Chim. Acta* **6**, 1146 (1923).

Tetrahydrate, fine monoclinic needles. Sol in water, less sol in alc. Soly in water at 20°: 9.24 g in 100 g of satd soln.
USE: Manufacture of dyes.

422. Amidomycin. $C_{40}H_{68}N_4O_{12}$; mol wt 797.00. C 60.28%, H 8.60%, N 7.03%, O 24.09%. Antibiotic substance produced by an unidentified *Streptomyces* culture (PRL 1642). Composed of 4 moles each of D-(−)-valine and D-(−)-α-hydroxyisovaleric acid, linked alternately by ester and amide bonds to form a 24-membered ring: Vining, Taber, *Can. J. Chem.* **35**, 1109 (1957). Structure studies: Shemyakin *et al., Tetrahedron Letters* **1963**, 351, *Tetrahedron* **19**, 995 (1963).

D-amino acids only

Needles from dilute ethanol or petr ether, mp 192°. $[\alpha]_D^{26}$ +19.2° (c = 1.2 in ethanol). Neutral reaction. Practically insol in water. Readily sol in most organic solvents. Primarily active against yeasts.

423. Amido-R-Acid. *3-Amino-2,7-naphthalenedisulfonic acid;* 2-naphthylamine-3,6-disulfonic acid. $C_{10}H_9$-NO_6S_2; mol wt 303.32. C 39.60%, H 2.99%, N 4.62%, O 31.65%, S 21.14%. Prepd by treating 2-hydroxy-3,6-naphthalenedisulfonic acid with ammonium sulfite and ammonium hydroxide: Petitcolas, Josué, *Bull. Soc. Chim. France* **1952**, 89.

Crystals or powder. Soluble in water. Solutions show a violet-blue fluorescence.
USE: Manufacture of dyes.

424. Amifostine. *2-[(3-Aminopropyl)amino]ethanethiol dihydrogen phosphate ester; phosphorothioic acid S-[2-[(3-aminopropyl)amino]ethyl] ester;* aminopropylaminoethyl thiophosphate; ethiofos; gammaphos; SAPEP; NSC-296961; WR-2721; YM-08310; Ethyol. $C_5H_{15}N_2O_3PS$; mol wt 214.23. C 28.03%, H 7.06%, N 13.08%, O 22.41%, P 14.46%, S 14.97%. Thiophosphate derivative of cysteamine, *q.v.*; provides normal cells with selective protection against the toxic effects of cancer chemotherapy and radiation treatment. Prepn of monohydrate: J. R. Piper *et al., J. Med. Chem.* **12**, 236 (1969); J. R. Piper, T. P. Johnston, U.S. pat. **3,892,824** (1975 to Southern Res. Inst.). Differential radioprotective activity: J. M. Yuhas, J. B. Storer, *J. Nat. Cancer Inst.* **42**, 331 (1969). Mechanism of action study: G. D. Smoluk *et al., Cancer Res* **48**, 3641 (1988). Bioavailability: L. Fleckenstein *et al., Pharmacol. Ther.* **39**, 203 (1988). Clinical pharmacokinetics: L. M. Shaw *et al., ibid.* 195. HPLC determn in plasma: N. F. Swynnerton *et al., Int. J. Radiat. Oncol. Biol. Phys.* **12**, 1495 (1986). Review of development as radioprotector: D. Q. Brown *et al., Pharmacol. Ther.* **39**, 157-168 (1988); of role in chemotherapy: R. L. Capizzi *et al., Cancer* **72**, 3495-3501 (1993); M. Treskes, W. J. M. van der Vijgh, *Cancer Chemother. Pharmacol.* **33**, 93-106 (1993).

Monohydrate, white solid from methanol/ether, mp 160-161° (dec). LD_{50} in mice (mg/kg): 700 i.p. (Piper, Johnston).

Trihydrate, soly in water: > 9 g/100 ml. pKa_1 < 2.0; pKa_2 4.2; pKa_3 9.0; pKa_4 11.7.

THERAP CAT: Radioprotective agent.

425. Amikacin. *(S)-O-3-Amino-3-deoxy-α-D-glucopyranosyl-(1→6)-O-[6-amino-6-deoxy-α-D-glucopyranosyl-(1→4)]-N¹-(4-amino-2-hydroxy-1-oxobutyl)-2-deoxy-D-streptamine;* 1-*N*-[L(−)-4-amino-2-hydroxybutyryl]kanamycin A; Lukadin. $C_{22}H_{43}N_5O_{13}$; mol wt 585.61. C 45.12%, H 7.40%, N 11.96%, O 35.52%. Semisynthetic aminoglycoside antibiotic derived from kanamycin A. Prepn: Kawaguchi *et al., J. Antibiot.* **25,** 695 (1972); H. Kawaguchi, T. Naito, **Ger.** pat. **2,234,315;** H. Kawaguchi *et al.,* U.S. pat. **3,781,268** (both 1973 to Bristol-Myers). Biological formation from kanamycin A: L. M. Cappelletti, R. Spagnoli, *J. Antibiot.* **36,** 328 (1983). Microbiological evaluation: Price *et al., ibid.* **25,** 709 (1972). Pharmacokinetics: Cabana, Taggart, *Antimicrob. Ag. Chemother.* **3,** 478 (1973). *In vitro* studies: Yu, Washington, *ibid.* **4,** 133 (1973); Bodey, Stewart, *ibid.* 186. Pharmacology in humans: Bodey *et al., ibid.* **5,** 508 (1974). Toxicity studies: Fujisawa *et al., J. Antibiot.* **27,** 677 (1974). Review: K. A. Kerridge in *Pharmacological and Biochemical Properties of Drug Substances* vol. 1, M. E. Goldberg, Ed. (Am. Pharm. Assoc., Washington, DC, 1977) pp 125-153. Comprehensive description: P. M. Monteleone *et al.* in *Analytical Profiles of Drug Substances* vol. 12, K. Florey, Ed. (Academic Press, New York, 1983) pp 37-71.

White crystalline powder from methanol-isopropanol, mp 203-204° (sesquihydrate). $[\alpha]_D^{23}$ +99° (c = 1.0 in water). LD_{50} in mice of solns pH 6.6, pH 7.4 (mg/kg): 340, 560 i.v. (Kawaguchi).

Sulfate, $C_{22}H_{43}N_5O_{13} \cdot 2H_2SO_4$, *Amiglyde-V, Amikavet, Amikin, Amiklin, BB-K8, Biklin, Fabianol, Kaminax, Mikavir, Novamin, Pierami.* Amorphous form, dec 220-230°. $[\alpha]_D^{22}$ +74.75° (water).

THERAP CAT: Antibacterial.

THERAP CAT (VET): Antibacterial.

426. Amiloride. *3,5-Diamino-N-(aminoiminomethyl)-6-chloropyrazinecarboxamide; N-amidino-3,5-diamino-6-chloropyrazinamide;* N-amidino-3,5-diamino-6-chloropyrazinamide; 1-(3,5-diamino-6-chloropyrazinecarboxyl)guanidine; 1-(3,5-diamino-6-chloropyrazinoyl)guanidine; guanamprazine; amipramidin; amipramizide. $C_6H_8ClN_7O$; mol wt 229.63. C 31.38%, H 3.51%, Cl 15.44%, N 42.70%, O 6.97%. Potassium-sparing diuretic. Prepn: Cragoe, **Belg.** pat. **639,386** (1964 to Merck & Co.), *C.A.* **62,** 14698f (1965). NMR study on tautomerism and conformation: R. L. Smith *et al., J. Am. Chem. Soc.* **101,** 191 (1979). Pharmacology: Baer *et al., J. Pharmacol. Exp. Ther.* **157,** 472 (1967); Baba *et al., Clin. Pharmacol. Ther.* **9,** 318 (1968); Lant *et al., ibid.* **10,** 50 (1969). Metabolism: Weiss *et al., ibid.* 401. Reviews: H. L. Macfie *et al., Drug Intell. Clin. Pharmacol.* **15,** 94-98 (1981); D. E. Hyams, *Int. Congr. Symp. Ser.-R. Soc. Med.* **44,** 65-73 (1981). Comprehensive description of hydrochloride: D. J. Mazzo, *Anal. Profiles Drug Subs.* **15,** 1-34 (1986).

Solid, mp 240.5-241.5°.

Hydrochloride dihydrate, $C_6H_8ClN_7O \cdot HCl \cdot 2H_2O$, *MK-870, Amikal, Colectril (obsolete), Midamor, Modamide.*

Crystalline solid, dec 285-288°. Anhydr, mp 293.5°. uv max (water): 212, 285, 362 nm ($E_{1cm}^{1\%}$ 642, 555, 617). pKa 8.7. Freely sol in DMSO, slightly sol in water, isopropanol, ethanol. Practically insol in acetone, chloroform, diethyl ether, ethyl acetate.

Hydrochloride dihydrate mixture with hydrochlorothiazide, *co-amilozide, Amilco, Aquaretic, Ecodurex, Grodurex, Hexarese, Hydrocomp, Moduretic, Moduretik, Normetic, Normorix.*

THERAP CAT: Diuretic.

427. Aminacrine. *9-Acridinamine;* 5-aminoacridine; 9-aminoacridine. $C_{13}H_{10}N_2$; mol wt 194.24. C 80.39%, H 5.19%, N 14.42%. Prepd from 9-chloroacridine which is obtained by heating *N*-phenylanthranilic acid with PCl_5: A. Albert, B. Ritchie, *Org. Syn.* coll. vol. III, 53 (1955). Toxicity study: D. C. Brodie, E. Lowenhaupt, *J. Am. Pharm. Assoc.* **38,** 498 (1949).

Sulfur-yellow needles from alcohol or acetone, mp 241°. Moderately strong base. pka (25°): 4.53. Freely sol in alcohol; slightly sol in chloroform, toluene, pyridine; sol in acetone.

Hydrochloride, $C_{13}H_{10}N_2 \cdot HCl$, *Monacrin, Acramine Yellow.* Pale yellow crystals. Neutral reaction. One of the most highly fluorescent substances. One gram dissolves in 300 ml water giving a faintly yellow soln showing bluish-violet fluorescence. LD_{50} orally in mice: 78 mg/kg (Brodie, Lowenhaupt).

THERAP CAT: Anti-infective.

THERAP CAT (VET): Antiseptic.

428. Amine 220®. *2-(8-Heptadecenyl)-4,5-dihydro-1H-imidazole-1-ethanol; 2-(8-heptadecenyl)-2-imidazoline-1-ethanol;* 1-(2-hydroxyethyl)-2-(8-heptadecenyl)-2-imidazoline. $C_{22}H_{42}N_2O$; mol wt 350.59. C 75.37%, H 12.07%, N 7.99%, O 4.56%. Prepared from oleic acid and 2-[(2-aminoethyl)amino]ethanol: Stromberg, Hughes, and Hughes, Stromberg, U.S. pats. **2,987,515** and **3,020,276** (1961, 1962 both to Petrolite Corp.).

USE: Fungicide, emulsifier, soil stabilizer.

429. Amineptine. *7-[(10,11-Dihydro-5H-dibenzo[a,d]cyclohepten-5-yl)amino]heptanoic acid.* $C_{22}H_{27}NO_2$; mol wt 337.46. C 78.30%, H 8.06%, N 4.15%, O 9.48%. Prepn: C. Malen *et al.,* **Ger.** pat. **2,011,806;** *eidem,* U.S. pats. **3,758,-528** and **3,821,249** (1970, 1973, 1974 all to Sci. Union et Cie Soc. Franc. Recher. Med.). Kinetics: C. Sbarra *et al., J. Chromatog.* **162,** 31 (1970). Biochemical and pharmacological study: R. Samanin *et al., J. Pharm. Pharmacol.* **29,** 555 (1977). Pharmacodynamics: J. Dankova *et al., Eur. J. Pharmacol.* **42,** 113 (1977); J. C. Poignant, A. Avril, *Arzneimittel-Forsch.* **28,** 267 (1978). Clinical study: P. Van Amerongden, *Encephale* **4,** 131 (1978).

Hydrochloride, $C_{22}H_{27}NO_2.HCl$, *S-1694, Maneon, Survector.* Crystals from distilled water, mp 226-230°.

THERAP CAT: CNS stimulant.

430. Aminitrozole. *N-(5-Nitro-2-thiazolyl)acetamide;* 2-acetamido-5-nitrothiazole; 2-acetylamino-5-nitrothiazole; acinitrazole; Tritheon; Trichorad; Trichoral; Gynofon; Enheptin-A; Pleocide. $C_5H_5N_3O_3S$; mol wt 187.18. C 32.08%, H 2.69%, N 22.45%, O 25.64%, S 17.13%. Prepd by nitration of 2-acetamidothiazole: Ganapathi, Venkataraman, *Proc. Indian Acad. Sci.* **22A**, 343 (1945); Bellavita, *Ann. Chim. Applicata* **38**, 449 (1948); Hurd, Wehrmeister, *J. Am. Chem. Soc.* **71**, 4007 (1949); Yamamoto, *J. Pharm. Soc. Japan* **72**, 1017 (1952). As an antihistomonad: Waletzky, Marson, U.S. pat. **2,531,756** (1950 to Am. Cyanamid).

Needles from alc, elongated plates from acetic acid. mp 264-265°. The commercial product may be yellow. Sol in aq solns of NaOH and NH_3 with deep orange color.

THERAP CAT: Antiprotozoal (Trichomonas).

THERAP CAT (VET): Antihistomonad for turkeys.

431. p-Aminoacetanilide. *N-(4-Aminophenyl)acetamide; 4'-aminoacetanilide;* acetyl-*p*-phenylenediamine. $C_8H_{10}N_2O$; mol wt 150.18. C 63.98%, H 6.71%, N 18.65%, O 10.65%. Prepd by catalytic hydrogenation of 4'-nitroacetanilide: Atkinson *et al., J. Chem. Soc.* **1954**, 2023. Crystal and molecular structure: M. Haisa *et al., Acta Crystallogr.* **33B**, 2449 (1977).

White or slightly reddish crystals, mp 163.5-166.0°; darkens in air. Slightly soluble in cold water; freely soluble in hot water, alc, ether.

USE: Intermediate in the manufacture of azo dyes, pharmaceuticals.

432. Aminoacetonitrile. *Glycinonitrile;* cyanomethylamine; glycine nitrile. $C_2H_4N_2$; mol wt 56.07. C 42.84%, H 7.19%, N 49.96%. H_2NCH_2CN. Prepd by the action of alcoholic HCl or H_2SO_4 on dimolecular or trimolecular methyleneaminoacetonitrile: Anslow, King, *J. Chem. Soc.* **1929**, 2465.

Oily liquid. bp_{15} 58° (partial decompn).

Hydrochloride, $C_2H_4N_2.HCl$, hygroscopic crystals from alcohol, dec 165°.

Sulfate (1:1), $C_2H_4N_2.H_2SO_4$, Prepn: H. Stephen, *J. Chem. Soc.* **1931**, 871. Hygroscopic leaflets from ethanol + ether, mp 121°.

Sulfate (2:1), $(C_2H_4N_2)_2.H_2SO_4$, long flat prisms from water/ethanol, mp 166° (dec).

433. Aminoacetophenone. C_8H_9NO; mol wt 135.17. C 71.09%, H 6.71%, N 10.36%, O 11.84%. Prepn of *m-, o-,* and *p*-isomers: Grammaticakis, *Compt. Rend.* **235**, 546 (1952); Braude *et al., J. Chem. Soc.* **1954**, 3586. Prepn of *m*-isomer: Tinsley, U.S. pat. **2,797,244** (1957 to Union Carbide); of *p*-isomer: Norman *et al., Can. J. Chem.* **40**, 1547 (1962). Toxicity studies: J. M. Vandenbelt *et al., J. Pharmacol. Exp. Ther.* **80**, 31 (1944); H. F. Smyth *et al., Arch. Ind. Hyg. Occup. Med.* **10**, 61 (1954).

m-Aminoacetophenone, 1-(3-aminophenyl)ethanone, 3'-aminoacetophenone, m-aminoacetylbenzene. Yellow leaflets,

mp 98-99°. Partly volatile in steam. LD_{50} orally in rats: 1.87 g/kg (Smyth).

o-Aminoacetophenone, yellow oily liquid. bp_{760} 250-252° (some dec); bp_{17} 135°. Volatile with steam. Practically insol in water. Sol in alc.

p-Aminoacetophenone, yellow needles, pleasant, characteristic odor. mp 106°. bp 293-295°. Sparingly sol in cold, freely in hot water; sol in alc, ether, HCl; sparingly sol in benzene. LD_{50} i.p. in rats: 260 mg/kg (Vandenbelt).

434. D-Amino Acid Oxidase. DAAO. Flavoprotein that catalyzes the oxidative deamination of D-amino acids to the corresponding α-keto acids. Found in the kidney and liver of nearly all mammals studied. Isoln from pig kidney: Krebs, *Enzymologia* **7**, 53 (1939); Straub, *Nature* **141**, 603 (1938); Negelein, Brömel, *Biochem. Z.* **300**, 225 (1939); Massey *et al., Biochim. Biophys. Acta* **48**, 1 (1961); from sheep kidney: Burton, *Methods Enzymol.* **2**, 199 (1955). Physical properties: Yagi *et al., J. Biochem.* **61**, 580 (1967). Partial structure: Kotaki *et al., ibid.* 598. Studies on specificity and inhibition: Dixon, Kleppe, *Biochim. Biophys. Acta* **96**, 368 (1965). Reaction intermediates: K. Yagi, *Front. Physiochem. Biol., Proc. Int. Symp.* **1977**, B. Pullman, Ed. (Academic Press, New York, 1978) pp 299-308. *Review:* Meister, Wellner in *The Enzymes* 7, P. Boyer *et al.,* Eds. (Academic Press, New York, 1963) pp 634-648.

435. L-Amino Acid Oxidase. LAAO. Flavoprotein that catalyzes the oxidative deamination of L-amino acids to the corresponding α-keto acids. Found in microorganisms and in animal tissue, esp in kidney and liver. Occurs also in many snake venoms. Isoln from rat kidney: Blanchard *et al., J. Biol. Chem.* **161**, 583 (1945). Isoln of crystalline enzyme from the venom of the eastern diamondback rattlesnake (*Crotalus adamanteus):* Wellner, Meister, *J. Biol. Chem.* **235**, 2013 (1960). Structural studies: DeKok, Rawitch, *Biochem.* **8**, 1405 (1969). Studies on inhibitors: DeKok, Veeger, *Biochim. Biophys. Acta* **167**, 35 (1968). Mechanism of reversible activation-deactivation: C. J. Coles *et al., Flavins Flavoproteins, Proc. Int. Symp. 6th* **1978**, K. Yagi, T. Yamano, Eds. (Japan Sci. Soc. Press, Tokyo, 1980) pp 101-105. Multiple conformational states: D. Wellner, L. A. Lichtenberg, *Dev. Biochem.* **21**, 78 (1982). *Review:* Meister, Wellner in *The Enzymes* 7, P. Boyer *et al.,* Eds. (Academic Press, New York, 1963) pp 609-634.

436. α-Aminoadipic Acid. *2-Aminohexanedioic acid.* $C_6H_{11}NO_4$; mol wt 161.16. C 44.72%, H 6.88%, N 8.69%, O 39.71%. $HO_2CCH(NH_2)CH_2CH_2CH_2CO_2H$. An amino acid isolated from *Cholera vibrio:* Blass, Macheboeuf, *Helv. Chim. Acta* **29**, 1315 (1946). Occurrence in proteins or protein containing material: Windsor, *J. Biol. Chem.* **192**, 595 (1951). Synthesis: Dieckmann, *Ber.* **38**, 1656 (1905); T. P. Waalkes *et al., J. Am. Chem. Soc.* **72**, 5760 (1952); A. I. Scott, T. J. Wilkinson, *Synth. Commun.* **10**, 127 (1980).

Forms a monohydrate if crystallized from water below 20°. Anhydrous platelets from water above 20°. mp 206° (effervescence). One gram dissolves in 450 ml water. Sparingly sol in alcohol, ether.

Diethyl ester, $C_{10}H_{19}NO_4$, bp_{13} 155-156°.

437. 1-Aminoanthraquinone. *1-Amino-9,10-anthracenedione.* $C_{14}H_9NO_2$; mol wt 223.23. C 75.33%, H 4.06%, N 6.27%, O 14.33%. Prepd by reduction of 1-nitroanthraquinone: Graham, Hort, U.S. pat. **2,874,168** (1959 to General Aniline & Film).

Ruby-red crystals. mp ~250°; also stated as mp 243°. Practically insoluble in water; freely soluble in alcohol, benzene, chloroform, ether, glacial acetic acid, HCl.
USE: In the mfg of dyes and pharmaceuticals.

438.   1-Aminoanthraquinone-2-carboxylic Acid. *1-Amino-9,10-dihydro-9,10-dioxo-2-anthracenecarboxylic acid.* $C_{15}H_9NO_4$; mol wt 267.24. C 67.42%, H 3.39%, N 5.24%, O 23.95%. Prepd by reduction of 1-nitroanthraquinone-2-carboxylic acid with sodium sulfide: Terres, *Ber.* **46**, 1639 (1913); by boiling 1-nitro-2-methylanthraquinone with alcoholic potassium hydroxide: Scholl, *Monatsh.* **34**, 1011 (1913); by heating 1-nitro-2-bromomethylanthraquinone with sodium acetate and *o*-dichlorobenzene: Locher, Fierz, *Helv. Chim. Acta* **10**, 667 (1927).

Red needles from nitrobenzene, mp 295-296°. Soluble in aniline, boiling nitrobenzene, in sodium hydroxide and aq pyridine with formation of a deep red soln, in concd sulfuric acid with formation of a yellowish-brown soln; slightly sol in ether, benzene, alcohol. Insol in water, ligroin.
Methyl ester, $C_{16}H_{11}NO_4$, red needles from glacial acetic acid, mp 228°.
Ethyl ester, $C_{17}H_{13}NO_4$, mp 198°.
Phenyl ester, $C_{21}H_{13}NO_4$, reddish-golden crystals from glacial acetic acid, mp 198°.
USE: For the detection of traces of aluminum, magnesium, or zinc.

439.   *p*-Aminoazobenzene. *4-(Phenylazo)benzenamine; C.I. Solvent Yellow 1; p-(phenylazo)aniline; p-aminodiphenylimide; aniline yellow; C.I. 11000.* $C_{12}H_{11}N_3$; mol wt 197.24. C 73.07%, H 5.62%, N 21.30%. Prepd from aniline, $NaNO_2$, and HCl: A. I. Vogel, *A Text-Book of Practical Organic Chemistry* (Longmans, Green & Co., New York, 3rd ed., 1956) p 627; W. R. Hydro, T. L. Willard, **U.S. pat. 2,894,942** (1959 to Goodrich); *Colour Index* vol. 4 (3rd ed., 1971) p 4014.

Brownish-yellow needles with a bluish cast. mp 128°. bp above 360°. Slightly sol in water; freely sol in alc, benzene, chloroform, ether. Reduces alcoholic ammoniacal $AgNO_3$.
USE: In form of its salts in dyeing; intermediate in manuf of Acid Yellow, diazo dyes and indulines.

440.   *o*-Aminoazotoluene. *2-Methyl-4-[(2-methylphenyl)azo]benzenamine; C.I. Solvent Yellow 3;* 4-(*o*-tolylazo)-*o*-toluidine; 4'-amino-2,3'-dimethylazobenzene; 5-(*o*-tolylazo)-2-aminotoluene; toluazotoluidine; C.I. 11160. $C_{14}H_{15}N_3$; mol wt 225.29. C 74.64%, H 6.71%, N 18.65%. Prepd from *o*-toluidine, $NaNO_2$, and HCl: Shulman, **U.S. pat. 2,538,431** (1951 to Pfister Chem. Works); *Colour Index* vol. 4 (3rd ed., 1971) p 4017. Crystal and molecular structure: S. Kurosaki *et al.*, *Acta Crystallogr.* **32B**, 3160 (1976). Carcinogenic activity: Maini, Stich, *J. Nat. Cancer Inst.* **26**, 1413 (1961). *See also: H. J. Conn's Biological Stains,* R. D. Lillie, Ed. (Williams & Wilkins, Baltimore, 9th ed., 1977) p 87.

Golden crystals, mp 101-102°. Practically insoluble in water. Sol in alc, ether, chloroform, acetone, ethyl acetate, benzene, ligroin, linseed oil, oleic acid, stearic acid. Very sol in cellosolve. Absorption max (50% alcoholic 1*N* HCl):

326, 490 nm (ε 19000, 2500), Sawicki, *J. Org. Chem.* **21**, 605 (1956).
Note: This substance may reasonably be anticipated to be a carcinogen: *Seventh Annual Report on Carcinogens* (PB95-109781, 1994) p 89.
USE: Coloring oils, fats and waxes; manuf of pigments. Chemical intermediate for the production of dyes.

441.   *m*-Aminobenzoic Acid. *3-Aminobenzoic acid.* $C_7H_7NO_2$; mol wt 137.14. C 61.31%, H 5.14%, N 10.21%, O 23.33%. Prepn: Toland, Heaton, **U.S. pat. 2,878,281** (1959 to California Res. Corp.); Neilson *et al.*, *J. Chem. Soc.* **1962**, 371.

Cryst, mp 174°. d 1.51. Aq soln turns brown on standing in air. Forms soluble salts with mineral acids; slightly sol in cold, freely in boiling water or alcohol; sol in ether.

442.   *o*-Aminobenzoic Acid. *2-Aminobenzoic acid; anthranilic acid.* $C_7H_7NO_2$; mol wt 137.14. C 61.31%, H 5.14%, N 10.21%, O 23.33%. Found to be the same as *vitamin L_1*: Nakahara *et al.*, *Sci. Papers Inst. Phys. Chem. Res.* (Tokyo) **42**, 39 (1945), *C.A.* **41**, 6317d (1947). *See also* Vitamins L. Prepd by reduction of *o*-nitrobenzoic acid: Neilson *et al.*, *J. Chem. Soc.* **1962**, 371. Purification: Sugihara, Newman, *J. Org. Chem.* **21**, 1445 (1956).
White to pale yellow, cryst powder; sweetish taste. mp 144-146°. Sparingly sol in cold, freely in hot water, alcohol, ether. The solns in alcohol or ether and particularly in glycerol exhibit an amethyst fluorescence.
THERAP CAT (VET): The cadmium salt has been used as an ascaricide in swine.

443.   *p*-Aminobenzoic Acid. *4-Aminobenzoic acid;* vitamin B_x; bacterial vitamin H^1; chromotrichia factor; trichochromogenic factor; anticanitic vitamin; PABA; Amben; Paraminol; Sunbrella. $C_7H_7NO_2$; mol wt 137.14. C 61.31%, H 5.14%, N 10.21%, O 23.33%. Widely distributed in nature as a B complex factor. Baker's yeast contains 5 to 6 ppm, brewer's yeast from 10 to 100 ppm. Occurs free and in ester form. Prepn: Toland, Heaton, **U.S. pat. 2,878,281** (1959 to Calif. Research Corp.); Spiegler, **U.S. pat. 2,947,781** (1960 to Du Pont); Nielson *et al.*, *J. Chem. Soc.* **1962**, 371. Purification: Lyding, **U.S. pat. 2,735,865** (1956 to Heyden Chem.). Antirickettsial activity: M. L. Robbins *et al.*, *J. Immunol.* **64**, 431 (1950). Toxicity studies: C. C. Scott, E. B. Robbins, *Proc. Soc. Exp. Biol. Med.* **49**, 184 (1942); R. K. Richards, *Fed. Proc.* **1**, 71 (1942); G. Cronheim, *ibid.* **10**, 289 (1952).

Monoclinic prisms from dil alcohol. May turn slightly yellow on prolonged exposure to light and air. mp 187.0°-187.5°. pKa: 4.65, 4.80; pH (0.5% soln): 3.5. uv max (water): 266 nm ($E_{1cm}^{1\%}$ 1070); (isopropanol): 288 nm ($E_{1cm}^{1\%}$ 137). One gram dissolves in 170 ml water at 25°, in 90 ml boiling water; in 8 ml alcohol, in 50 ml ether. Sol in ethyl acetate, glacial acetic acid; slightly sol in benzene. Practically insol in petr ether. Incompatible with ferric salts and oxidizing agents. LD_{50} in mice, rats (g/kg): 2.85, > 6.0 orally (Scott, Robbins). LD_{50} in rabbits (g/kg): 2.0 i.v. (Richards); 1.83 orally (Cronheim).
Ingredient in *Pabanol, PreSun.*
Diethylamine salt, $C_{11}H_{18}N_2O_2$, *Navanide.* Crystals from acetone, mp 170-173°. Very soluble in water.
Potassium salt, *see* Potassium *p*-Aminobenzoate.

Note: *p*-Aminobenzoic acid is also known as *anti-gray-hair factor* (in rats only).

USE: Manuf various esters (local anesthetics), folic acid, and azo dyes; in sunburn preventives. Used in laboratories as sulfonamide antagonist.

THERAP CAT: Ultraviolet screen. Formerly as antirickettsial.

THERAP CAT (VET): Has been used in *eczema nasi* ("collie nose") in dogs.

444. 2-Aminobenzothiazole. *2-Benzothiazolamine.* $C_7H_6N_2S$; mol wt 150.20. C 55.98%, H 4.03%, N 18.65%, S 21.35%. Prepd from 2-chlorobenzothiazole by treatment with alcoholic ammonia at 150-160°: Hofmann, *Ber.* **12**, 1129 (1880); **13**, 11 (1881); from benzothiazole by boiling with hydroxylamine in water or in 2*N* NaOH: Skraup, *Ann.* **419**, 65 (1919). Toxicity study: E. F. Domino *et al., J. Pharmacol. Exp. Ther.* **105**, 486 (1952).

Leaflets from water, mp 132°. Distills without decompn. Very sparingly sol in water; freely in alcohol, ether, chloroform; sol in concd acids. LD_{50} i.v. in mice: 126 mg/kg (Domino).

USE: In the prepn of azo dyes.

445. 6-Aminobenzothiazole. *6-Benzothiazolamine.* $C_7H_6N_2S$; mol wt 150.20. C 55.98%, H 4.03%, N 18.65%, S 21.35%. Prepd from 2,5-diaminothiophenol by boiling with concd formic acid or from 6-nitrobenzothiazole by reduction with tin and HCl: Mylius, *Thesis* (Berlin, 1883).

Prisms from water, mp 87°. Sol in alcohol. Practically insol in water and ether.

USE: In the prepn of azo dyes.

446. N-(*p*-Aminobenzoyl)glutamic Acid. *N-(4-Aminobenzoyl)-L-glutamic acid;* $C_{12}H_{14}N_2O_5$; mol wt 266.25. C 54.13%, H 5.30%, N 10.52%, O 30.05%. Prepd by catalytic hydrogenation (Pd charcoal in AcOH) of the corresponding *p*-nitrobenzylglutamic acid: Auhagen, *Z. Physiol. Chem.* **277**, 203 (1943). *See also* Landsteiner, Scheer, *J. Exp. Med.* **55**, 781 (1932); *J. Immunol.* **29**, 371 (1935); Carter, Stevens, *J. Biol. Chem.* **138**, 627 (1941). Inhibition of dihydrofolate reductase: G. C. K. Roberts *et al., Biochemistry* **13**, 5351 (1974).

L-Form, crystals from water, mp 173°. $[\alpha]_D^{20}$ −16° (c = 2 in 0.1*N* HCl).

D-Form, crystals from water, mp 172°. $[\alpha]_D^{20}$ +16.7° (c = 2 in 0.1*N* HCl).

DL-Form, crystals from water, mp 197°.

447. 2-Amino-1-butanol. 2-Amino-*n*-butyl alcohol. $C_4H_{11}NO$; mol wt 89.14. C 53.90%, H 12.44%, N 15.71%, O 17.95%. $CH_3CH_2CH(NH_2)CH_2OH$. Prepd from 2-nitro-1-butanol by reduction or catalytic hydrogenation: Stiénon, *Chem. Zentr.* **1902**, I, 717; Vanderbilt, Haas, U.S. pat. **2,174,242** (1940); Johnson, Degering, *J. Org. Chem.* **8**, 7 (1943).

Liquid. d_{20}^{20} 0.944. mp −2°. bp_{760} 178°; bp_{10} 79-80°. n_D^{20} 1.453. Miscible with water. Sol in alcohols. pH of 0.1 molar aq soln 11.1.

Hydrochloride, $C_4H_{11}NO.HCl$, deliquesc needles.

Platinichloride monohydrate, $(C_4H_{11}NO)_2.2HCl.PtCl_4$- .$H_2O$, yellow leaflets, dec 189-190°. Moderately sol in water, freely sol in alcohol.

Oxalate, $(C_4H_{11}NO)_2.C_2H_2O_4$, mp 176°.

USE: In the synthesis of surface-active agents, vulcanization accelerators, pharmaceuticals. As emulsifying agent for cosmetic creams and lotions, mineral oil and paraffin wax emulsions, leather dressings, textile specialties, polishes, cleaning compounds, so-called soluble oils.

448. α-Aminobutyric Acid. *2-Aminobutanoic acid;* α-amino-*n*-butyric acid. $C_4H_9NO_2$; mol wt 103.12. C 46.59%, H 8.80%, N 13.58%, O 31.03%. $CH_3CH_2CH(NH_2)COOH$. Prepn of the DL-form from α-bromobutyric acid and ammonia: Fisher, Mounegrat, *Ber.* **33**, 2388 (1900); from potassium cyanide, ammonium chloride and propionaldehyde: Zelinsky, Stadnikow, *Ber.* **41**, 2062 (1908); by reduction of α-oxobutyric acid: Knoop, Oesterlin, *Z. Physiol. Chem.* **148**, 305 (1925); from α-ethylacetoacetic ester and hydrazoic acid: Schmidt, *Ber.* **57**, 706 (1924). The L(+)-form has been isolated from proteins: Oikawa, *Chem. Zentr.* **1926**, 1, 148. Configuration: Clough, *J. Chem. Soc.* **113**, 544, 551; Levene, *Chem. Rev.* **2**, 203 (1926); Vogler, *Helv. Chim. Acta* **30**, 1766 (1947).

DL-Form, crystals, mp 304° (begins to sublime when heated above 300°). Sol in water. One liter of water will dissolve 210.5 g at 25°. Sparingly sol in alcohol. One liter of boiling ethanol dissolves about 1.8 g. Insol in ether.

DL-Form ethyl ester, viscous liquid. bp_{11} 61°. Sol in water and in organic solvents.

L-Form, leaflets from dil alc. Sweet taste. mp 270-280° (depending on speed of heating, *see* Vogler, *loc. cit.).* $[M]_D$ +21.2° (5*N* HCl); $[M]_D$ +43.3° (glacial acetic acid). $[\alpha]_D^{16}$ +8.40° (c = 4); $[\alpha]_D^{16}$ +18.65° (c = 4.8 in 6*N* HCl).

L-Form hydrochloride, needles; $[\alpha]_D^{19}$ +12.90° (c = 3.64). Readily sol in water.

449. β-Aminobutyric Acid. *3-Aminobutanoic acid;* β-amino-*n*-butyric acid. $C_4H_9NO_2$; mol wt 103.12. C 46.59%, H 8.80%, N 13.58%, O 31.03%. $CH_3CH(NH_2)CH_2COOH$. Prepd from pyrotartaric acid diamide: Weidel, Roithmer, *Monatsh.* **17**, 185 (1896); from β-chlorobutyric acid ethyl ester and alcoholic NH_3: Balbiano, *Ber.* **13**, 312 (1880); from crotonic acid and concd NH_3: Engel, *Bull. Soc. Chim.* [2] **50**, 102 (1888); Curtius, Gumlich, *J. Prakt. Chem.* [2] **70**, 204 (1904); Stadnikow, *Chem. Zentr.* **1909**, II, 1988; *see* Fischer, Roeder, *Ber.* **34**, 3755 (1901) footnote; Stoermer, Robert, *Ber.* **55**, 1038 (1922). Prepn of HCl salt from β-aminobutyronitrile: Bruylants, *Bull. Soc. Chim. Belg.* **32**, 259 (1923); *Chem. Zentr.* **1924**, I, 1668. By reducing acetoacetic ester phenylhydrazone and saponifying the resulting ester: Fischer, Groh, *Ann.* **383**, 338 (1911). The D(−)-form has been obtained by hydrolysis of its ester: Fischer, Scheibler, *ibid.* 346.

DL-Form, crystals from alcohol, mp 193-194°. Practically tasteless. Sol in water. One liter of water dissolves 1250 g. Insol in cold absolute alcohol and ether.

Methyl ester, $C_5H_{11}NO_2$, odoriferous liquid; d^{20} 0.993; bp_{13} 54-55°. Sol in water, alcohol, ether, petr ether.

D-Form, prisms from methanol. Dec near 220° without melting. $[\alpha]_D^{20}$ −35.20° (p = 10).

450. γ-Aminobutyric Acid. *4-Aminobutanoic acid;* γ-amino-*n*-butyric acid; piperidic acid; GABA; Gammalon. $C_4H_9NO_2$; mol wt 103.12. C 46.59%, H 8.80%, N 13.58%, O 31.03%. $H_2NCH_2CH_2CH_2COOH$. Nonprotein amino acid that functions as a neurotransmitter. Prepn from succinimide: Tafel, Stern, *Ber.* **33**, 2224 (1900); from piperylurethan and fuming nitric acid: Schotten, *Ber.* **16**, 643 (1883); Abderhalden, *Chem. Zentr.* **1926**, II, 779; from *N*-(β-bromoethyl)phthalimide and sodiomalonic ester: Aschan, *Ber.* **24**, 2450 (1891); from γ-chlorobutyronitrile and potassium phthalimide: Gabriel, *Ber.* **22**, 3335 (1889); **23**, 1771 (1890); DeWitt, *Org. Syn.* coll. vol. II, 25 (1943). Review of biochemical pharmacology: R. Tapia in *Handbook of Psychopharmacology,* L. L. Iversen *et al.,* Eds. (Plenum, New York, 1975) pp 1-58; L. L. Iversen in *Psychopharmacology: A Generation of Progress,* M. A. Lipton *et al.,* Eds. (Raven, New York, 1978) pp 25-38; C. C. Mao, E. Costa, *ibid.* pp 307-318.

Leaflets from methanol + ether, needles from water + alcohol, mp 202° (dec on rapid heating). Ka 3.7 × 10⁻¹¹; Kb 1.7 × 10⁻¹⁰ at 25°. Freely sol in water. Insol or poorly sol in other solvents. On melting it dec forming pyrrolidone and water.

Hydrochloride, $C_4H_9NO_2.HCl$, crystals, mp 135-136°.
Ethyl ester, $C_6H_{13}NO_2$, liquid, bp_{12} 76°.
THERAP CAT: Antihypertensive.

451. ε-Aminocaproic Acid. *6-Aminohexanoic acid;* epsilon-aminocaproic acid; epsilcapramin; EACA; CY-116; Ipsilon; Hemocaprol; Caprocid; Capramol; Amicar; Afibrin; Epsikapron; Hepin. $C_6H_{13}NO_2$; mol wt 131.17. C 54.94%, H 9.99%, N 10.68%, O 24.39%. $H_2N(CH_2)_5COOH$. Prepn from ε-benzoylaminocapronitrile: Ruzicka, Hugoson, *Helv. Chim. Acta* **4**, 472 (1921); from ε-caprolactam: Galat, Mallin, *J. Am. Chem. Soc.* **68**, 2729 (1946); Koch, U.S. pat. **2,453,234** (1948 to Am. Enka); J. C. Eck, *Org. Syn.* **coll. vol. II**, 28 (1943); C. Y. Meyers, L. E. Miller, *ibid.* **coll. vol. IV**, 39 (1963); Garmaise *et al., Can. J. Chem.* **34**, 743 (1956); Sulzbacher, *Mfg. Chemist* **33**, 463 (1963); from 1-hydroxycyclohexylhydroperoxide: Minisci, Portolani, *Gazz. Chim. Ital.* **89**, 1941 (1959). Metabolic studies in man: McNicol *et al., J. Lab. Clin. Med.* **59**, 15 (1962). Toxicity study: D. W. Hallesy *et al., Pharmacologist* **3**, 62 (1961).

Crystals from alcohol, mp 204-206°. pK_1 4.43; pK_2 10.75. Freely sol in water; sparingly in methanol. Practically insol in ethanol. LD_{50} in rats (g/kg): 7.0 i.p.; ∼3.3 i.v. (Hallesy).

Hydrobromide, $C_6H_{13}NO_2.HBr$, crystalline powder from alcohol + ether, mp 105°.

Hydrochloride, $C_6H_{13}NO_2.HCl$, mp 128-129°.

THERAP CAT: Hemostatic.

452. Aminocarb. *4-(Dimethylamino)-3-methylphenol methylcarbamate (ester); methylcarbamic acid 4-(dimethylamino)-m-tolyl ester;* 4-dimethylamino-m-tolyl methylcarbamate; A-363; Bay 44646; ENT-25784; Matacil. $C_{11}H_{16}N_2O_2$; mol wt 208.26. C 63.44%, H 7.74%, N 13.45%, O 15.36%. Prepn: **Brit.** pat. **913,439,** *C.A.* **58**, 10122e (1963); R. Heiss *et al.,* **Ger.** pat. **1,145,162,** *C.A.* **59**, 9885b (1963) (1962, 1963 both to Bayer). Metabolism: A. Strother, *Toxicol. Appl. Pharmacol.* **21**, 112 (1972). Photochemistry: J. B. Addison *et al., Bull. Environ. Contam. Toxicol.* **11**, 250 (1974); J. B. Addison, *ibid.* **27**, 250 (1981). Toxicity study: T. B. Gaines, *Toxicol. Appl. Pharmacol.* **14**, 515 (1969).

Crystals, mp 93-94°. uv max (ethanol): 248.5 nm (ε 6.67 × 10⁴). Slightly sol in water; moderately sol in aromatic solvents; sol in most polar organic solvents. LD_{50} orally in male, female rats: 40, 38 mg/kg (Gaines).

USE: Insecticide.

453. 7-Aminocephalosporanic Acid. *(6R-trans)-3-[(Acetyloxy)methyl]-7-amino-8-oxo-5-thia-1-azabicyclo[4.2.0]oct-2-ene-2-carboxylic acid; 7-amino-3-(hydroxymethyl)-8-oxo-5-thia-1-azabicyclo[4.2.0]oct-2-ene-2-carboxylic acid acetate ester;* 3-acetoxymethyl-7-aminoceph-3-em-4-oic acid; 7-ACA. $C_{10}H_{12}N_2O_5S$; mol wt 272.28. C 44.11%, H 4.44%, N 10.29%, O 29.38%, S 11.78%. Starting material for semisynthetic cephalosporins. Obtained by mild acid hydrolysis of cephalosporin C *(q.v.):* Loder *et al., Biochem. J.* **79**, 408 (1961); Morin *et al., J. Am. Chem. Soc.* **84**, 3400 (1962); Morin *et al.,* **Belg.** pat. **615,955** (1962 to Lilly), *C.A.* **58**, 11373c (1963); by enzymatic hydrolysis of cephalosporin C: Walton, U.S. pat. **3,239,394** (1966 to Merck & Co.). Improved prepn: Fechtig *et al., Helv. Chim. Acta* **51**, 1108 (1968). Review of preparative methods: Huber *et al.,* in *Cephalosporins and Penicillins,* E. H. Flynn, Ed. (Academic Press, New York, 1972) pp 27-73.

Crystals. pI 3.5. R_f 0.14 in 1-butanol-ethanol-water (4:1:5 by vol).

454. 4-Amino-4'-chlorodiphenyl. *4'-Chloro[1,1'-biphenyl]-4-amine;* 4'-chloroxenylamine; 4'-chloro-4-aminodiphenyl; p'-chloro-p-phenylaniline; p-amino-p'-chlorobiphenyl. $C_{12}H_{10}ClN$; mol wt 203.67. C 70.77%, H 4.95%, Cl 17.41%, N 6.88%. Prepd from 4'-amino-4-biphenyldiazonium chloride by Sandmeyer's reaction: Gelmo, *Ber.* **39**, 4176 (1906); by chlorination of 4-nitrodiphenyl and reduction of 4-chloro-4'-nitrodiphenyl with iron and HCl in ethanol: Belcher *et al., J. Chem. Soc.* **1953,** 1334.

Crystals from light petroleum, mp 134° (Gelmo); mp 128° (Belcher). uv max (0.1N HCl): 254 nm (ε 22090). Sol in warm alcohol, ether, benzene, acetone, glacial acetic acid. Practically insol in water and alkalies.

USE: As a reagent in the determination of sulfur, in coal, rubber, etc.

455. Aminochlorthenoxazin. *6-Amino-2-(2-chloroethyl)-2,3-dihydro-4H-1,3-benzoxazin-4-one;* 2-(β-chloroethyl)-2,3-dihydro-4-oxo-6-amino-1,3-benzoxazine. $C_{10}H_{11}ClN_2O_2$; mol wt 226.66. C 52.99%, H 4.89%, Cl 15.64%, N 12.36%, O 14.12%. Prepn and pharmacology: F. Baroli *et al., Arzneimittel-Forsch.* **13**, 884 (1963).

Crystalline powder, mp 164°. Practically insol in water. LD_{50} in mice (mg/kg): 1941 ±108 i.p., 10,000 ±615 orally; in rats (mg/kg): 1500 ±21 i.p. (Baroli).

Hydrochloride, $C_{10}H_{11}ClN_2O_2.HCl$, *A-350, ICI-350.* Crystals, mp 209-210°. Sol in water. LD_{50} in mice (mg/kg): 924 ±13 i.p., 2250 ±312 orally, 293 ±16 i.v., 1188 ±44 s.c.; in rats (mg/kg): 607 ±14 i.p., 1958 ±126 orally (Baroli).

Mixture with aminopyrine, **Dereuma.**

THERAP CAT: Antipyretic; analgesic.

456. Aminochromes. Family of highly colored 2,3-dihydroindole-5,6-quinones obtained by oxidative cyclization of catecholamines: Sobotka, Austin, *J. Am. Chem. Soc.* **73**, 3077 (1951). Best represented by a zwitterionic structure with the substituents determined by the catecholamine used, e.g. for epinephrine as starting material, R = CH_3, R' = H, and R'' = OH. Numerous physiological activities, such as hallucinogenic, hemostatic, radioprotective, have been ascribed to aminochromes. Reviews: Sobotka *et al., Fortschr. Chem. Org. Naturst.* **14**, 217 (1957); Heacock, *Advan. Heterocycl. Chem.* **5**, 205 (1965); Heacock, Powell, *Progr. Med. Chem.* **9**, 275 (1972). Review of toxicity: A. Bindoli *et al., Toxicol. Lett.* **48**, 3-20 (1989).

Consult the Name Index before using this section.

See also adrenochrome, adrenolutin, carbazochrome salicylate.

THERAP CAT: Hemostatic.

457. 2-Amino-4,6-dichlorophenol. 2,4-Dichloro-6-aminophenol; 4,6-dichloro-*o*-aminophenol. $C_6H_5Cl_2NO$; mol wt 178.02. C 40.48%, H 2.83%, Cl 39.83%, N 7.87%, O 8.99%. Prepd by reduction of the corresponding nitrophenol: F. Fischer, *Z. Chem.* [N.F.] **4**, 386 (1868); *Ann.,* suppl **vol. 7**, 189 (1870); Katz, Cohen, *J. Org. Chem.* **19**, 758 (1954). Purification: J. Meyer, *Helv. Chim. Acta* **41**, 1890 (1958).

Long needles from carbon disulfide, warts from benzene, mp 95-96°. Sublimes (0.06 torr) 70-80° (bath temp). Freely sol in benzene, somewhat less in carbon disulfide, much less in petr ether. The stability of the free base (snow-white when pure) seems to be impaired by impurities.

Hydrochloride, $C_6H_5Cl_2NO.HCl$, crystals, dec 280-285°. Very stable when pure. The commercial product may be dark brown. Sol in water, alcohol. Precipitated from aq soln by the addition of concd HCl.

USE: Important azo-dye intermediate.

458. 1-[(2-Aminoethyl)amino]-2-propanol. *N*-(2-Hydroxypropyl)ethylenediamine; Monolene. $C_5H_{14}N_2O$; mol wt 118.18. C 50.82%, H 11.94%, N 23.70%, O 13.54%. $CH_3CH(OH)CH_2NHCH_2CH_2NH_2$. Prepn: Kitchen, Pollard, *J. Org. Chem.* **8**, 342 (1943).

Viscous liquid; mild ammoniacal odor. $bp_{3.0}$ 94°; $bp_{10.0}$ 112°. d_4^{25} 0.9837. n_D^{25} 1.4738.

Dihydrochloride, $C_5H_{16}Cl_2N_2O$, mp 184.7-185.0°.

USE: Rapid curing agent in the manuf of epoxy resins.

459. 2-Amino-2-ethyl-1,3-propanediol. $C_5H_{13}NO_2$; mol wt 119.16. C 50.40%, H 11.00%, N 11.75%, O 26.85%. An amino glycol prepd by reduction of cataytic hydrogenation of the corresp nitro compound: Vanderbilt, Hass, U.S. pat. **2,174,242** (1940); Johnson, Degering, *J. Org. Chem.* **8**, 7 (1943). Manuf: McMillan, U.S. pat. **2,485,982** (1949 to Comm. Solvents).

Crystalline mass, mp 37.5-38.5°. (The commercial product may be a viscous liquid.) d_{20}^{20} 1.099. n_D^{20} 1.490. bp_{10} 152-153°. Miscible with water. Sol in alcohols. pH of 0.1 molar aq soln 10.8.

USE: In the synthesis of surface-active agents, vulcanization accelerators, pharmaceuticals. As emulsifying agent for cosmetic creams and lotions, mineral oil and paraffin wax emulsions, leather dressings, textile specialties, polishes, cleaning compounds, so-called soluble oils. For absorbing CO_2 and H_2S from industrial gases.

460. Aminoglutethimide. *3-(4-Aminophenyl)-3-ethyl-2,6-piperidinedione; 2-(p-aminophenyl)-2-ethylglutarimide;* 3-ethyl-3-(*p*-aminophenyl)-2,6-dioxopiperidine; *Cytadren; Elipten; Orimeten.* $C_{13}H_{16}N_2O_2$; mol wt 232.28. C 67.22%, H 6.94%, N 12.06%, O 13.78%. Adrenocortical suppressant that also inhibits conversion of androgens to estrogens by the aromatase enzyme system. Prepn: Hoffmann, Urech, U.S. pat. **2,848,455** (1958 to Ciba). Metabolism: Douglas, Nicholls, *J. Pharm. Pharmacol. Suppl.* **17**, 115S (1965). Mass spectrum: Ruecker, Bohn, *Arch. Pharm. (Weinheim)* **302**, 204 (1969). Resolution and abs config of antipodes: Finch *et al., Experientia* **31**, 1002 (1975). Review of its role as an inhibitor of adrenocortical steroidogenesis: Touitou *et al., Biomedicine* **18**, 185-191, 272-278 (1973). Clinical stud-

ies: Küchel, *Pharmacol. Clin.* **2**, 138 (1970). Use in treatment of Cushing's syndrome: R. I. Misbin *et al., J. Clin. Pharmacol.* **16**, 645 (1976); of metastatic breast cancer: R. F. Asbury *et al., Cancer* **47**, 1954 (1981). Comparative clinical trial with tamoxifen: I. E. Smith *et al., Brit. Med. J.* **283**, 1432 (1981). Hematologic toxicity study: A. A. Messeih *et al., Cancer Treat. Rep.* **69**, 1003 (1985). Clinical trial in advanced prostatic carcinoma: R. Murray, P. Pitt, *Eur. J. Cancer Clin. Oncol.* **21**, 453 (1985). Comprehensive guide to therapeutic use: *Pharmanual (Basel)* **vol. 2**, R. J. Santen, I. C. Henderson, Eds. (Karger, Basel, 1981) 160 pp. Reviews of mechanisms of action, endocrinological effects and clinical experience in breast cancer: R. C. Stuart-Harris, I. E. Smith, *Cancer Treat. Rev.* **11**, pp 189-204 (1984); A. L. Harris, *Exp. Cell. Biol.* **53**, pp 1-8 (1985). Comprehensive description: H. Y. Aboul-Enein in *Analytical Profiles of Drug Substances* **vol. 15**, K. Florey, Ed. (Academic Press, New York, 1986) pp 35-69.

Crystals from methanol or ethyl acetate, mp 149-150°. Freely sol in most organic solvents; poorly sol in ethyl acetate, 0.1*N* HCl and absolute ethanol; readily sol in acetone and 100% acetic acid. Practically insol in water.

Hydrochloride, $C_{13}H_{17}ClN_2O_2$, mp 223-225°. Freely sol in water.

THERAP CAT: In Cushing's syndrome and other adrenal hormone disorders. Palliative treatment of breast and prostatic cancer. Formerly as anticonvulsant.

461. Aminoguanidine. *Hydrazinecarboximidamide;* guanylhydrazine; pimagedine. CH_6N_4; mol wt 74.09. C 16.21%, H 8.16%, N 75.62%. Nucleophilic hydrazine; inhibits the formation of advanced glycosylation end products (AGEs) that have been implicated in the etiology of diabetic complications. Prepn: J. Thiele, *Ann.* **270**, 1 (1892); G. B. L. Smith, E. Anzelmi, *J. Am. Chem. Soc.* **57**, 2730 (1935). Review of chemistry of aminoguanidine and related compounds: E. Lieber, G. B. L. Smith, *Chem. Rev.* **25**, 213-271 (1939); of preparative methods: F. Kurzer, L. E. A. Godfrey, *Chem. & Ind. (London)* **1962**, 1584-1595. Prevention of glucose-derived aortic collagen cross-linking in diabetic rats: M. Brownlee *et al., Science* **232**, 1629 (1986). Mechanism of action study: D. Edelstein, M. Brownlee, *Diabetes* **41**, 26 (1992). Review of therapeutic potential: B. H. R. Wolffenbuttel, M. S. P. Huijberts, *Neth. J. Med.* **42**, 205-208 (1993).

Crystals. Sol in water, alc. Practically insol in ether. Aq soln is strongly alkaline and reddens on standing in air; ammonia is evolved on heating.

Hydrochloride, $CH_6N_4.HCl$, *GER-11.* Large prisms from dil alc, mp 163°. Very sol in water; sol in alc. Practically insol in ether.

THERAP CAT: In prevention and treatment of chronic diabetic complications.

462. *p*-Aminohippuric Acid. *N-(4-Aminobenzoyl)glycine; N-(p-aminobenzoyl)aminoacetic acid;* PAH. $C_9H_{10}N_2O_3$; mol wt 194.19. C 55.67%, H 5.19%, N 14.43%, O 24.72%. Prepd by reduction of *p*-nitrohippuric acid: Shimizu, Okano, *J. Pharm. Soc. Japan* **73**, 523 (1953).

Needles from hot water, mp 198-199°. Soluble in alc, chloroform, benzene, acetone. Practically insol in water, ether, carbon tetrachloride.

Sodium salt, $C_9H_9N_2NaO_3$, *Nephrotest.* Soluble in water. Aq solns are alkaline and must be buffered with citric acid to pH 7.0 for i.v. injection.

THERAP CAT: Diagnostic aid (renal function).

463. 3-Amino-4-hydroxybutyric Acid. *3-Amino-4-hydroxybutanoic acid;* γ-hydroxy-β-aminobutyric acid; GOBAB. $C_4H_9NO_3$; mol wt 119.12. C 40.33%, H 7.62%, N 11.76%, O 40.29%. HOCH$_2$CH(NH$_2$)CH$_2$COOH. Prepn: Jollès, Fromageot, *Bull. Soc. Chim. France* **1951**, 862; Chibnall *et al., Biochem. J.* **68**, 122 (1958); Piskov, *Zh. Obshch. Khim.* **32**, 3407 (1962); Nagai *et al., Arzneimittel-Forsch.* **17**, 1575 (1967); Kondo, Tanaka, **Japan.** pat. **12127**('68) (1968 to Kaken Kagaku), *C.A.* **70**, 77328r (1969).

Prismatic crystals, mp 216°. Also reported as mp 228° (Nagai); mp 232-233.5° (Chibnall).

Hydrochloride, $C_4H_9NO_3$.HCl, mp 156°.

THERAP CAT: Anti-inflammatory, antifungal.

464. 4-Amino-3-hydroxybutyric Acid. *4-Amino-3-hydroxybutanoic acid;* γ-amino-β-hydroxybutyric acid; buksamin; GABOB; Gabomade; Gamibetal. $C_4H_9NO_3$; mol wt 119.12. C 40.33%, H 7.62%, N 11.76%, O 40.29%. H$_2$NCH$_2$CH(OH)CH$_2$COOH. Prepn of DL-form: Tomita, *Z. Physiol. Chem.* **124**, 255 (1923); Balenovic *et al., J. Org. Chem.* **19**, 1589 (1964); Hayashi *et al.*, **Japan.** pat. **58 772,** *C.A.* **53**, 1172d (1959); Sakai *et al.*, **Japan.** pat. **62 12,264** (to Kaken), *C.A.* **59**, 9805e (1963); **Spanish** pat. **278,780** (1963 to Antonio Gallardo, S.A.), *C.A.* **60**, 2779a (1964); Hayashi, **French** pat. **1,348,105** (1964 to Kaken), *C.A.* **60**, 11956b (1964); D'Alo, Masserini, *Farmaco Ed. Sci.* **19**, 30 (1964); M. Pinza, G. Pifferi, *J. Pharm. Sci.* **67**, 120 (1978). Prepn of L-form: Tomita, Sendju, *Z. Physiol. Chem.* **169**, 214 (1927); Kaneko, Yoshida, *Bull. Chem. Soc. Japan* **35**, 1153 (1962); M. E. Jung, T. J. Shaw, *J. Am. Chem. Soc.* **102**, 6304 (1980); of D-form: Tomita, Sendju, *loc. cit.* Improved synthesis of L-form: S. Takano *et al., Tetrahedron Letters* **28**, 1783 (1987). Purification: Yamagiwa, Tanaka, **Japan.** pat. **63 24,365** (to Kaken), *C.A.* **60**, 10515a (1964).

DL-Form, crystals from dil alc, dec 218°. Fairly readily sol in water; very sparingly sol in methanol, alc, ether, chloroform, ethyl acetate.

D(+)-Form, crystals from water, dec 214°. $[\alpha]_D^{20}$ +18.3° (p = 2 in H$_2$O).

L(—)-Form, crystals from water or water + ethanol, dec 212°, also reported as dec 216-217°. $[\alpha]_D^{20}$ —21.06° (p = 2 in H$_2$O); —20.7° (c = 1.83 in H$_2$O).

THERAP CAT: Anticonvulsant.

465. α-Aminoisobutyric Acid. *2-Methylalanine;* 2-aminoisobutyric acid; 2-amino-2-methylpropanoic acid. $C_4H_9NO_2$; mol wt 103.12. C 46.59%, H 8.80%, N 13.58%, O 31.03%. (CH$_3$)$_2$C(NH$_2$)COOH. Prepd by the treatment of acetone with hydrocyanic acid and then with alcoholic ammonia (Strecker synthesis): Tiemann, Friedländer, *Ber.* **14**, 1970 (1881), *see also* p 1965; Marckwald *et al., Ber.* **24**, 3283 (1891); Bailey, Randolph, *Ber.* **41**, 2507 (1908); Clarke, Bean, *Org. Syn.* coll. vol. II, 29 (1943); or directly with ammonium cyanide: Gulewitsch, *Ber.* **33**, 1900 (1900) or with a mixture of KCN and NH$_4$Cl: Zelinsky, Stadnikow, *Ber.* **39**, 1726 (1906); *cf.* Hellsing, *Ber.* **37**, 1921 (1904); and subsequent hydrolysis of the nitrile formed. By heating dimethylhydantoin (obtained from acetone, hydrocyanic and cyanic acids) with concd HCl: Urech, *Ann.* **164**, 268 (1872), *cf.* Heilpern, *Monatsh.* **17**, 241 (1896).

Monoclinic prisms, tables, mp 335° (sealed capillary). Begins to sublime at 280°. Sweetish taste. Absorption spectrum: Ley, Arends, *Ber.* **61**, 219 (1928); Abderhalden, Rossner, *Z. Physiol. Chem.* **176**, 253 (1928). Freely sol in water. Difficultly sol in alcohol; insol in ether.

Hydrochloride, $C_4H_9NO_2$.HCl, platelets from water, dec 236-237°. Readily sol in water, methanol, alcohol.

466. δ-Aminolevulinic Acid. *5-Amino-4-oxopentanoic acid.* $C_5H_9NO_3$; mol wt 131.13. C 45.80%, H 6.92%, N 10.68%, O 36.60%. H$_2$NCH$_2$COCH$_2$CH$_2$COOH. Tetrahydropyrrole precursor; intermediate in heme, chlorophyll biosynthesis. Prepn as hydrochloride by the hydrogenation of oxaloacetone: R. W. Wynn, A. H. Corwin, *J. Org. Chem.* **15**, 203 (1950); from furfurylamine: A. A. Marei, R. H. Raphael, *J. Chem. Soc.* **1958**, 2624; by a malonic ester synthesis: D. P. Tschudy, A. Collins, *J. Org. Chem.* **24**, 556 (1959); from 5-hydroxy-2-pyridone: C. Herdeis, A. Dimmerlung, *Arch. Pharm.* **317**, 304 (1984). Effect on dark-grown barley seedlings: E. C. Sisler, W. H. Klein, *Physiol. Plant.* **16**, 315 (1963); A. K. Stobart, I. Ameen-Bukhari, *Biochem. J.* **222**, 419 (1984). Accumulation in light-treated barley seedlings: S. Gough, *Carlsberg Res. Commun.* **43**, 497 (1978). Biosynthesis and metabolism: S. I. Beale, *Plant Physiol.* **48**, 316 (1971). Use of labelled acid in porphyrin biosynthesis studies: C. A. Rebeiz *et al., ibid.* **46**, 543 (1970). Enhancement of chlorophyll formation: S. Ochiai, E. Hase, *Plant Cell Physiol.* **11**, 663 (1970). Proposed use as photodynamic herbicide: C. A. Rebeiz *et al., Enzyme Microb. Technol.* **6**, 390 (1984); *Chem. & Eng. News* **62**, 8 (Sept. 24, 1984).

Hydrochloride, $C_5H_{10}ClNO_3$, needles from methanol-ether, mp 144-147° (decomp) (Marei, Raphael). uv max (water): 266.5 nm (ε 23.0); after addition of 2 equivalents of NaOH, uv max: 276 nm (ε 2000).

USE: Experimental herbicide.

467. β-Amino-α-methylphenethyl Alcohol. *β-Amino-α-methylbenzeneethanol;* 1-amino-1-phenyl-2-propanol; 1-phenyl-1-amino-2-propanol. $C_9H_{13}NO$; mol wt 151.21. C 71.49%, H 8.67%, N 9.26%, O 10.58%. Prepn from α-isonitroso-α-phenylacetone: Sichner, Pankova, *Coll. Czech. Chem. Commun.* **20**, 1419 (1955); from an oxazole: Viscontini, *Helv. Chim. Acta* **44**, 636 (1961).

DL-*erythro*-Form, so called *dl-norisoephedrine.* Crystals, mp 85°.

DL-*erythro*-Hydrochloride, $C_9H_{13}NO$.HCl, crystals from ethanol + ether, mp 170-171°.

468. 2-Amino-2-methyl-1,3-propanediol. 1,1-Di(hydroxymethyl)ethylamine; 1,3-dihydroxy-2-methyl-2-propylamine; AMPD. $C_4H_{11}NO_2$; mol wt 105.14. C 45.70%, H 10.55%, N 13.32%, O 30.44%. An amino glycol prepd by reduction of the corresponding nitro compd: O. Piloty, O. Ruff, *Ber.* **30**, 2057 (1897); B. M. Vanderbilt, H. B. Hass, *Ind. Eng. Chem.* **32**, 34 (1940); Johnson, Degering, *J. Org. Chem.* **8**, 7 (1943). Review of uses: G. A. Bergy, *J. Am. Pharm. Assoc., Pract. Pharm. Ed.*, **3**, 358-364 (1942). Toxicological evaluation and review: Cosmetic, Toiletry and Fragrance Assoc., *J. Am. Coll. Toxicol.* **9**, 203-228 (1990).

Crystalline mass, mp 109-111°. bp$_{10}$ 151-152°. 250 grams dissolve in 100 ml water at 20°. Sol in alcohols. pH of 0.1M aq soln 10.8.

USE: In the synthesis of surface-active agents, vulcanization accelerators, pharmaceuticals. As emulsifying agent for cosmetic creams and lotions, mineral oil and paraffin wax emulsions, leather dressings, textile specialties, polishes, cleaning compounds, so-called soil oils. In hair sprays, wave sets and other noncoloring hair preparations. Absorbent for acidic gases. Biological buffer.

469. 2-Amino-2-methyl-1-propanol. 2,2-Diethylethanolamine; 2-hydroxymethyl-2-propylamine; AMP. $C_4H_{11}NO$; mol wt 89.14. C 53.90%, H 12.44%, N 15.71%, O 17.95%. An amino alcohol prepd by reduction of the corresp nitro compound: H. B. Hass, B. M. Vanderbilt, U.S. pat. **2,139,122** (1938 to Purdue Res. Found.); Johnson, Degering, *J. Org. Chem.* **8**, 7 (1943). Toxicological evaluation and review: Cosmetic, Toiletry and Fragrance Assoc. (CTFA), *J. Am. Coll. Toxicol.* **9**, 203-228 (1990).

Crystalline mass, mp 30-31°. (The commercial product may be a viscous liquid.) d_{20}^{20} 0.934. n_D^{20} 1.449. bp_{760} 165°; bp_{10} 67.4°. Miscible with water. Sol in alcohols. pH of 0.1M aq soln 11.3. LD_{50} in mice, rats (g/kg): 2.15 ± 0.2, 2.90 ± 0.14 orally (CTFA).

USE: In the synthesis of surface-active agents, vulcanization accelerators, pharmaceuticals. As emulsifying agent for cosmetic creams and lotions, mineral oil and paraffin wax emulsions, leather dressings, textile specialties, polishes, cleaning compounds, so-called soluble oils. In hair sprays, wave sets and hair dyes. Absorbent for acidic gases. For use as a drug *see* Pamabrom.

470. 2-Amino-4-methylthiazole. *4-Methyl-2-thiazolamine;* 4-methyl-2-thiazolylamine; Normotiroide. $C_4H_6N_2S$; mol wt 114.17. C 42.08%, H 5.30%, N 24.54%, S 28.09%. Prepn: Sprague *et al., J. Am. Chem. Soc.* **68**, 2155 (1946); Rossi, *Gazz. Chim. Ital.* **85**, 898 (1955).

Crystals, mp 45-46°. bp_{20} 124-126°; $bp_{0.4}$ 70°. Very sol in water, alcohol, ether.
THERAP CAT: Antihyperthyroid.

471. Aminometradine. *6-Amino-3-ethyl-1-(2-propenyl)-2,4(1H,3H)-pyrimidinedione; 1-allyl-6-amino-3-ethyl-2,4(1H,3H)-pyrimidinedione;* 1-allyl-6-amino-3-ethyl-2,4(1H,3H)-pyrimidinedione; 1-allyl-3-ethyl-6-amino-1,2,3,4-tetrahydro-2,4-pyrimidinedione; aminometramide; Mictine; Katapyrin; Mincard; Catapyrin. $C_9H_{13}N_3O_2$; mol wt 195.22. C 55.37%, H 6.71%, N 21.52%, O 16.39%. Prepn: Papesch, Schroeder, U.S. pat. **2,650,922** (1953 to Searle).

Monohydrate, crystals from water, mp 75-115°. When anhydrous, mp 143-144°.
THERAP CAT: Diuretic.

472. 3-Amino-2-naphthoic Acid. *3-Amino-2-naphthalenecarboxylic acid;* 3-aminoisonaphthoic acid. $C_{11}H_9NO_2$; mol wt 187.20. C 70.58%, H 4.85%, N 7.48%, O 17.09%. Prepd from 3-hydroxy-2-naphthoic acid: C. F. H. Allen, A. Bell, *Org. Syn.* **coll. vol. III,** 78 (1955).

Yellow scales from dil alcohol, mp 214°. Sol in alcohol, ether. Solns are yellow with greenish fluorescence.

Sodium salt, $C_{11}H_8NNaO_2$, leaflets, very sparingly sol in water, alcohol.
Ethyl ester, $C_{13}H_{13}NO_2$, yellow needles from dil alcohol. Sol in the usual organic solvents, mp 115-115.5°.
USE: In the determination of copper, nickel, cobalt.

473. 4-Amino-1-naphthol. *4-Amino-1-naphthalenol;* 4-hydroxy-α-naphthylamine. $C_{10}H_9NO$; mol wt 159.19. C 75.45%, H 5.70%, N 8.80%, O 10.05%. Prepd by treating α-naphthol with benzenediazonium chloride and reducing the benzeneazo-α-naphthol with sodium hydrosulfite: Conant *et al., Org. Syn.* **3**, 7 (1923); *cf.* Fieser, Fieser, *J. Am. Chem. Soc.* **57**, 493 (1935). From α-naphthylhydroxylamine by rearrangement in acetone: Neunhoffer, Liebich, *Ber.* **71B**, 2247 (1938).

Needles. Unless kept absolutely dry, it acquires a violet discoloration on storage and oxidizes to 1,4-naphthoquinone. Usually isolated as the hydrochloride, $C_{10}H_9NO$-.HCl, needles, very sol in water.
N-Acetyl deriv, $C_{12}H_{11}NO_2$, *4-acetamido-1-naphthol, naphthacetol,* needles from alcohol, mp 188°, sparingly sol in water, sol in alcohol.
USE: Polymerization inhibitor; prepn of 2-allyl-4-amino-1-naphthol hydrochloride (an antihemorrhagic cpd); prepn of 1,4-naphthoquinone.

474. 1-Amino-2-naphthol-4-sulfonic Acid. *4-Amino-3-hydroxy-1-naphthalenesulfonic acid;* 1,2,4-acid. $C_{10}H_9NO_4S$; mol wt 239.25. C 50.20%, H 3.79%, N 5.85%, O 26.75%, S 13.40%. Prepd by treatment of nitroso-β-naphthol with sodium bisulfite and sulfuric acid: L. F. Fieser, *Org. Syn.* **coll. vol. II,** 42 (1943); *cf. idem, J. Am. Chem. Soc.* **57**, 494 (1935).

White or gray needles, usually contg 0.5 H_2O. May turn pink on exposure to light, especially when moist. Insol in water, alcohol, ether, benzene; sol in hot sodium bisulfite soln; sol in alkaline soln but such solns oxidize quickly on exposure to air yielding a brown substance which is sol in hot water giving a green soln. Controlled oxidation with HNO_3 yields ammonium 1,2-naphthoquinone-4-sulfonate, *see* Fieser, *loc. cit.*
Sodium salt, $C_{10}H_8NNaO_4S$, needles, sol in hot water with blue fluorescence.
USE: Manuf of azo dyes, 1,2-naphthoquinone-4-sulfonic acid.

475. 1-Amino-2-naphthol-6-sulfonic Acid. *5-Amino-6-hydroxy-2-naphthalenesulfonic acid.* $C_{10}H_9NO_4S$; mol wt 239.25. C 50.20%, H 3.79%, N 5.85%, O 26.75%, S 13.40%. Prepd by reduction of sodium 1-nitroso-2-naphthol-6-sulfonate with zinc and acetic acid: Fierz-David *et al., Helv. Chim. Acta* **29**, 1765 (1946).

Needles or prisms. Slightly sol in boiling water, less sol in alc. Practically insol in ether.

Sodium salt hemipentahydrate, $C_{10}H_8NNaO_4S.2\frac{1}{2}H_2O$, *Eikonogen.* Powder. Sol in water. Practically insol in alc, ether. Strong reducing action on silver salts.

USE: Free acid formerly used in manuf of azo dyes. Sodium salt used as photographic developer and for detection of potassium.

476. 6-Aminonicotinic Acid. *6-Amino-3-pyridinecarboxylic acid;* 6-amino-3-carboxypyridine. $C_6H_6N_2O_2$; mol wt 138.13. C 52.17%, H 4.38%, N 20.28%, O 23.17%. Prepd by heating 6-chloronicotinic acid with ammonia: Marckwald, *Ber.* **26**, 2188 (1893); **27**, 1319 (1894); Räth, Prauge, *Ann.* **467**, 4 (1928); Johnson *et al., J. Biol. Chem.* **153**, 37 (1944).

Dihydrate, crystals from dil acetic acid. Dec above 300° yielding 2-aminopyridine and CO_2. Sparingly sol in most solvents.

Potassium salt, $C_6H_5KN_2O_2$, crystals, freely sol in water, insol in alcohol.

Hydrochloride, $C_6H_6N_2O_2.HCl$, needles, freely sol in water, slightly sol in alcohol.

477. 2-Amino-5-nitrothiazole. *5-Nitro-2-thiazolamine;* Enheptin; Entramin. $C_3H_3N_3O_2S$; mol wt 145.14. C 24.83%, H 2.08%, N 28.95%, O 22.05%, S 22.09%. Prepd by deacetylation of 2-acetamido-5-nitrothiazole: Hubbard; Steahly, U.S. pats. **2,573,641; 2,573,656; 2,573,657** (all 1951 to Monsanto).

Greenish-yellow to orange-yellow fluffy powder, dec 202°. Slightly bitter taste. uv max (0.0005% in water): 386 nm (ε 0.540); min: 295 nm. Very sparingly soluble in water. (To make a water soln, a 15% stock soln in propylene glycol should first be prepd.) One gram dissolves in 150 g of 95% alcohol; in 250 g ether. Almost insol in chloroform. Sol in dilute mineral acids.

THERAP CAT (VET): Antihistomonad in turkeys, chickens. For trichomoniasis in pigeons.

478. 6-Aminopenicillanic Acid. *[2S-(2α,5α,6β)]-6-Amino-3,3-dimethyl-7-oxo-4-thia-1-azabicyclo[3.2.0]heptane-2-carboxylic acid;* 6-APA; penicin; penin. $C_8H_{12}N_2O_3S$; mol wt 216.26. C 44.43%, H 5.59%, N 12.95%, O 22.19%, S 14.83%. Obtained from cultures of *Penicillium chrysogenum* in the absence of side chain precursors: F. R. Batchelor *et al., Nature* **183**, 257 (1959). Synthesis: J. C. Sheehan, K. R. Henery-Logan, *J Am. Chem. Soc.* **81**, 5838 (1959); **84**, 2983 (1962). Intermediate in the manuf of synthetic penicillins: J. C. Sheehan, U.S. pat. **3,159,617** (1964 to Arthur D. Little). HPLC determn: J. Haginaka, J. Wakai, *Anal. Biochem.* **158**, 146 (1986). Review of prepn: F. M. Huber *et al.* in *Cephalosporins Penicillins: Chem. Biol.*, E. H. Flynn, Ed. (Academic, New York, 1972) pp 27-73. Review of chemistry: E. J. Vandamme, J. P. Voets, *Adv. Appl. Microbiol.* **17**, 311-369 (1974). General review: G. N. Rolinson, *J. Antimicrob. Chemother.* **22**, 5-14 (1988).

Crystals from water + HCl, dec 208-209°, also reported as 207-208° (dried) (Sheehan). $[\alpha]_D^{31}$ +273° (c = 1.2 in 0.1*N* HCl).

USE: In manuf of synthetic penicillins.

479. Aminopentamide. α-*[2-(Dimethylamino)propyl]-α-phenylbenzeneacetamide; 4-dimethylamino-2,2-diphenylvaleramide;* α,α-diphenyl-γ-dimethylaminovaleramide; BL-139; Centrine; Valeramide-OM. $C_{19}H_{24}N_2O$; mol wt 296.41. C 76.99%, H 8.16%, N 9.45%, O 5.40%. Prepn similar to that of methadone, starts by condensing diphenylmethyl cyanide with chlorodimethylaminopropane: Walton, *et al., J. Chem. Soc.* **1949**, 648; Cheney *et al., J. Org. Chem.* **17**, 770 (1952); Specter, U.S. pat. **2,647,926** (1953 to Bristol). Prepn of optical isomers: Wheatley *et al., J. Org. Chem.* **19**, 794 (1954). *See also* Moffett, Aspergren, *J. Am. Chem. Soc.* **79**, 4451 (1957). Toxicity study: Cazort, *J. Pharmacol. Exp. Ther.* **100**, 325 (1950).

dl-Form, long prisms from dilute alcohol, mp 183-184°. Practically insol in water.

d-Form, crystals from petr ether, mp 136.5-137.5° (sintering). $[\alpha]_D^{23}$ +98.9° (methanol).

l-Form, crystals from petr ether, mp 136.5-137.5° (sintering). $[\alpha]_D^{23}$ −101.9° (methanol).

dl-Form hydrochloride, $C_{19}H_{24}N_2O.HCl$, slightly deliquescent leaflets from alcohol + ether, dec 190-191°. Bitter taste. Soluble in water, alcohol. A 1% aq soln has a pH of 6.8. LD_{50} in mice (mg/kg): 34.7 i.v.; 396 orally (Cazort).

dl-Form acid sulfate; $C_{19}H_{24}N_2O.H_2SO_4$, deliquescent crystals from isopropanol + ethyl acetate, mp 185-187°. The commercial medicinal grade, mp 178-181°. Bitter taste. uv max (1% H_2SO_4): 258.5 nm ($A_{1cm}^{1\%}$ 10.3); min: 249 nm. Freely sol in water, alcohol. Very slightly sol in chloroform. Practically insol in ether. pH of 2.5% aq soln 1.3-2.2.

THERAP CAT: Anticholinergic.

THERAP CAT (VET): Anticholinergic; antiemetic; anticonvulsant.

480. m-Aminophenol. 3-Amino-1-hydroxybenzene; 3-hydroxyaniline. C_6H_7NO; mol wt 109.13. C 66.04%, H 6.47%, N 12.84%, O 14.66%. Manuf by reduction of *m*-nitrophenol: Freifelder, Robinson, U.S. pat. **3,079,435** (1963 to Abbott). Toxicity study: Koelzer, Giesen, *Z. Naturforsch.* **6b**, 183 (1951).

Crystals, mp 122-123°. Sol in 40 parts cold water, freely in hot water, alcohol, ether, amyl alcohol; slightly in benzene, very slightly in petr ether. LD_{50} i.p. in mice: 4.5 mg/20g (Koelzer, Giesen).

USE: Dye intermediate, manuf *p*-aminosalicylic acid.

481. o-Aminophenol. *2-Aminophenol;* 2-amino-1-hydroxybenzene; 2-hydroxyaniline. C_6H_7NO; mol wt 109.13. C 66.04%, H 6.47%, N 12.84%, O 14.66%. Manuf by reduction of *o*-nitrophenol: Freifelder, Robinson, U.S. pat. **3,079,435** (1963 to Abbott).

Crystals, rapidly becoming brown, mp 170-174°; sublimes. One gram dissolves in 50 ml cold water, 23 ml alcohol; freely soluble in ether, very slightly in benzene. *Keep tightly closed and protected from light.*

Hydrochloride, $C_6H_7NO.HCl$, crystals readily becoming gray on exposure to light. Freely sol in water or alcohol.

USE: Manuf azo and sulfur dyes; dyeing furs and hair. Hydrochloride used in dyeing fur, hair, leather, etc.

482. p-Aminophenol. p-Hydroxyaniline; 4-amino-1-hydroxybenzene; Activol; Azol; Certinal; Citol; Paranol; Rodinal; Unal; Ursol P. C_6H_7NO; mol wt 109.13. C 66.04%, H 6.47%, N 12.84%, O 14.66%. Usually prepd by the reduction of p-nitrophenol: *BIOS Final Report* **986**; Freifelder, Robinson, U.S. pat. **3,079,435** (1963 to Abbott). Orthorhombic plates from water. *Deteriorates under the influence of air and light.* mp 189.6-190.2°. The commercial product is usually pink, mp 186°. Can be sublimed at 0.3 mm and 110° without decompn. bp$_{760}$ 284° (decompn); bp$_{8.0}$ 167°; bp$_{3.0}$ 150°; bp$_{0.3}$ 130.2°. Kb at 15° = 6.6 × 10⁻⁹. Forms salts with acids and bases. Soly in water: 0.39% at 13°; 0.65% at 24°; 0.80% at 30°; 1.5% at 50°; 4.7% at 80°; 8.5% at 96°. Soly in ethyl methyl ketone: 9.3% at 58.5°; in abs ethanol: 4.5% at 0°. Practically insol in benzene, chloroform.
Hydrochloride, $C_6H_7NO.HCl$, cryst powder; gradually becomes darker. Decomp about 306°. Very sol in water; sol in alc.
Caution: May cause skin sensitization, dermatitis. Inhalation can cause asthma, methemoglobin formation.
USE: Photographic developer; intermediate in the manufacture of sulfur and azo dyes; in dyeing furs and feathers.

483. p-Aminophenylacetic Acid. 4-Aminobenzeneacetic acid; p-amino-α-toluic acid. $C_8H_9NO_2$; mol wt 151.17. C 63.56%, H 6.00%, N 9.27%, O 21.17%. Prepd by the reduction of p-nitrophenylacetic acid with hydrogen sulfide in the presence of ammonia: G. R. Robertson, *Org. Syn.* coll. vol. I, 52 (2nd ed., 1941).

Plates, leaflets from water, mp 199-200° (dec). Moderately sol in hot water; sol in alcohol, in alkalies.
Hydrochloride, $C_8H_9NO_2.HCl$, rods from HCl, freely sol in water, sol in alcohol (about 3% w/w).
Ethyl ester, $C_{10}H_{13}NO_2$, platelets from water, mp 51°.
N-Benzoyl deriv, needles from alcohol, mp 205-206°.

484. 4-Amino-3-phenylbutyric Acid. β-(Aminomethyl)benzenepropanoic acid; β-(aminomethyl)hydrocinnamic acid; 4-amino-3-phenylbutanoic acid; β-phenyl-γ-aminobutyric acid; phenygam; PhGABA; Phenigam; Fenigam; Phenigama; Fenigama. $C_{10}H_{13}NO_2$; mol wt 179.22. C 67.02%, H 7.31%, N 7.82%, O 17.85%. Prepn starting with α,γ-diamino-β-phenylpropane: Jackson, Kenner, *J. Chem. Soc.* **1928**, 1657. Alternate prepn: Cologne, Pouchol, *Bull. Soc. Chim. France* **1962**, 598. Structure-activity studies: Khaunina, *Farmakol. Toksikol.* **31**, 202 (1968). Activity studies of the isomers: *eidem, Byull. Eksp. Biol. Med.* **72**, 49 (1971), *C.A.* **76**, 81208t (1972).

mp 250-253° (dec).
Hydrobromide, $C_{10}H_{13}NO_2.HBr$, irregular platelets from benzene, mp 114°. LD$_{50}$ i.p. in mice: 900 mg/kg.
THERAP CAT: Mood elevator, tranquilizer.

485. Aminophylline. 3,7-Dihydro-1,3-dimethyl-1H-purine-2,6-dione compd with 1,2-ethanediamine (2:1); theophylline compd with ethylenediamine; theophylline ethylenediamine; theophyllamine; Carena; Inophylline; Metaphyllin; Theophyldine; Aminocardol; Aminodur; Ammophyllin; Cardiofilina; Cardophylin; Phylcardin; Tefamin; Cardiomin; Grifomin; Minaphil; Pecram; Peterphyllin; Phyllocontin; Somophyllin; Stenovasan; Theodrox; Cardophyllin; Diophyllin; Euphyllin CR; Genophyllin; Phyllindon; Theolamine; Theomin. $C_{16}H_{24}N_{10}O_4$; mol wt 420.43.

C 45.71%, H 5.75%, N 33.32%, O 15.22%. $(C_7H_8N_4O_2)_2$.$C_2H_4(NH_2)_2$.2H₂O. Prepn: Grüter, U.S. pat. **919,161** (1909 to Byk). Toxicity data: C. R. Thompson, M. R. Warren, *J. Lab. Clin. Med.* **31**, 1337 (1946). Comprehensive description: K. D. Thakker, L. T. Grady, *Anal. Profiles Drug Subs.* **11**, 1-44 (1982). Review of clinical toxicology: J. D. Truwit, *Crit. Care Clinics* **7**, 639-657 (1991).
Dihydrate, white or slightly yellowish granules or powder. Slight ammoniacal odor. Bitter taste. Gradually absorbs carbon dioxide from the air and becomes incompletely sol due to liberation of theophylline. One gram dissolves in about 5 ml water, but the soln may become turbid on standing. Insol in alcohol, ether. *Keep tightly closed.* LD$_{50}$ orally in mice: 540 mg/kg (Thompson, Warren).
Caution: Acute poisoning: Causes restlessness, anorexia, nausea, fever, vomiting, dehydration. Eventual bloody, syrupy or "coffee-ground" vomitus, tremors, delirium and coma can occur. May result in cardiovascular and respiratory collapse, shock, cyanosis and death, *Clinical Toxicology of Commercial Products*, R. E. Gosselin *et al.*, Eds. (Williams & Wilkins, Baltimore, 4th ed., 1976) Section III, pp 16-20.
THERAP CAT: Bronchodilator.
THERAP CAT (VET): Smooth muscle relaxant. For heaves in horses. Diuretic in dogs with congestive heart failure.

486. 2-Amino-4-picoline. 4-Methyl-2-pyridinamine; 2-amino-4-methylpyridine; α-amino-γ-picoline; 4-methyl-2-aminopyridine; W-45. $C_6H_8N_2$; mol wt 108.14. C 66.64%, H 7.46%, N 25.90%. Prepd by heating 4-picoline with sodamide in xylene: Seide, *Ber.* **57**, 791 (1924); **58**, 1733 (1925); Räth, *ibid.* 347; Kakimoto, Nishie, *C.A.* **50**, 14744g (1956). Formulations: von Haxthausen *et al.*, U.S. pat. **2,937,118** (1960 to Raschig). Pharmacology: von Haxthausen, *Arch. Exp. Pathol. Pharmakol.* **226**, 163; **227**, 234 (1955); Marchetti *et al.*, *Arch. Int. Pharmacodyn. Ther.* **143**, 385 (1963).

Leaflets from petr ether, mp 100-100.5°. bp$_{11}$ 115-117°. Sublimes on slow heating. Freely sol in water, lower alcohols, dimethylformamide, coal tar bases. Slightly sol in petr ether, aliphatic hydrocarbons. Component of *Ascensil*, *Askensil*.
Hydrochloride, $C_6H_8N_2.HCl$, prisms from alcohol, mp 176-177°. Freely sol in water, alcohol.
Camphorsulfonate, $C_{16}H_{24}N_2O_4S$, *Piricardio, Varunax.*
THERAP CAT: Analgesic; cardiac stimulant.

487. Aminopromazine. N,N,N',N'-Tetramethyl-3-(10H-phenothiazin-10-yl)-1,2-propanediamine; 10-[2,3-bis-(dimethylamino)propyl]phenothiazine; proquamezine; RP-3828; Tetrameprozine. $C_{19}H_{25}N_3S$; mol wt 327.49. C 69.68%, H 7.69%, N 12.83%, S 9.79%. Prepn: Jacob *et al.*, *Compt. Rend.* **243**, 1637 (1956); Horclois, **Brit.** pat. **800,635** (1958 to Rhône-Poulenc).

Fumarate, $(C_{19}H_{25}N_3S)_2.C_4H_4O_4$, *Lispamol, Lorusil, Spamol.* Crystals, sensitive to light, dec 166-170°. Soly at 20° (g/100 ml): water 9.0; methanol 5.0; ethanol 0.5. Very slightly sol in isopropanol, acetone. Practically insol in benzene, ether. pH of a 2% aq soln 5.0 to 7.0.

THERAP CAT: Antispasmodic.
THERAP CAT (VET): Antispasmodic.

488. 2-Aminopropanol. *2-Amino-1-propanol;* 2-aminopropyl alcohol; β-propanolamine; 2-hydroxyisopropylamine. C_3H_9NO; mol wt 75.11. C 47.97%, H 12.08%, N 18.65%, O 21.30%. $CH_3CH(NH_2)CH_2OH$. Obtained together with isopropylamine on reduction of acetylcarbinoloxime with sodium amalgam in dilute acetic acid: Gabriel, *Ber.* **49,** 2121 (1916); from DL-alanylglycine by reduction with sodium and abs alcohol: Abderhalden, Schwab, *Z. Physiol. Chem.* **143,** 292 (1925); by boiling the ethyl ester of DL-acetylalanine with sodium in abs alcohol: Karrer, *Helv. Chim. Acta* **4,** 98 (1921); as hydrolytic cleavage product of ergonovine and ergometrinine: Stoll, Hofmann, *ibid.* **26,** 956 (1943); by catalytic hydrogenation of the ethyl ester of alanine: Adkins, Pavlic, *J. Am. Chem. Soc.* **69,** 3039 (1947).

dl-Form, liquid, fishy odor. bp 173-176°. Freely sol in water, alcohol, ether.

Hydrochloride, $C_3H_9NO.HCl$, leaflets from abs alcohol + acetone, dec 86-87.5°.

489. 3-Aminopropionitrile. *3-Aminopropanenitrile;* β-aminopropionitrile. $C_3H_6N_2$; mol wt 70.09. C 51.41%, H 8.63%, N 39.97%. $H_2NCH_2CH_2CN$. Prepd by the reaction of acrylonitrile with ammonia: Buc *et al., J. Am. Chem. Soc.* **67,** 93 (1945); *Org. Syn. coll. vol. III,* 93 (1955); Weijlard, Sullivan, U.S. pat. **2,742,491** (1956 to Merck & Co.).

Liquid. Amine odor. bp_{760} 185°; bp_{20} 87-89°; bp_5 66-69°; bp_3 50-55°. n_D^{20} 1.4396. The pure, anhydrous material may be stored in tightly stoppered bottles for several months. Storing under refrigeration is recommended to avoid development of pressure in the bottles. Polymerization takes place slowly during storage in an open container, or very rapidly in the presence of acid or acidic compds.

USE: Intermediate in the manufacture of β-alanine and pantothenic acid.

490. p-Aminopropiophenone. *1-(4-Aminophenyl)-1-propanone;* ethyl p-aminophenyl ketone; PAPP. $C_9H_{11}NO$; mol wt 149.19. C 72.46%, H 7.43%, N 9.39%, O 10.72%. Prepd by the action of propionyl chloride on aniline in carbon bisulfide in the presence of aluminum chloride: Kunckell, *Ber.* **33,** 2641 (1900); Derrick, Bornemann, *J. Am. Chem. Soc.* **35,** 1283 (1913); Hartung, Foster, *J. Am. Pharm. Assoc.* **35,** 15 (1946).

Yellow needles from water, mp 140°. Sol in water, alc.

Hydrochloride, $C_9H_{11}NO.HCl$, needles, yellowish cast, mp 198-199°. Freely sol in water. LD_{50} i.v. in dogs: 7 mg/kg.

Sulfate, $(C_9H_{11}NO)_2.H_2SO_4$, yellow plates from alcohol, dec 25°.

THERAP CAT: Antidote (cyanide).

491. α-(α-Aminopropyl)benzyl Alcohol. *α-(1-Aminopropyl)benzenemethanol;* 2-amino-1-phenyl-1-butanol; β-amino-α-phenylbutyl alcohol; 1-phenyl-1-hydroxy-2-amino-*n*-butane. $C_{10}H_{15}NO$; mol wt 165.24. C 72.69%, H 9.15%, N 8.48%, O 9.68%. Prepn of the *dl-threo*-form: Abrams, Kipping, *J. Chem. Soc.* **1936,** 1480; Rebstock *et al., J. Am. Chem. Soc.* **73,** 3666 (1951). Prepn of the *dl-erythro*-form: Hartung *et al., ibid.* **52,** 3317 (1930); Rebstock *et al., loc. cit.* See also Beilstein **13,** suppl. 3, 1791-1792 (1973).

dl-threo-Form, thick, shiny plates from benzene + petr ether, mp 79-80° (Abrams, Kipping); 78-79° (Rebstock *et*

al.). Freely sol in alcohol; moderately sol in chloroform, benzene; sparingly sol in petr ether.

dl-threo-Form hydrochloride, $C_{10}H_{15}NO.HCl$, lustrous prisms, mp 195-196° (Abrams, Kipping); 204-205° (Rebstock *et al.).* Freely sol in water; sol in alcohol; slightly sol in chloroform.

dl-erythro-Form, mp 80.5-81°.

dl-erythro-Form hydrochloride, mp 242°.

492. Aminopropylon. *N-(2,3-Dihydro-1,5-dimethyl-3-oxo-2-phenyl-1H-pyrazol-4-yl)-2-(dimethylamino)propanamide; N-(antipyrinyl)-2-(dimethylamino)propionamide;* 4-[2-(dimethylamino)propionamido]antipyrine; 4-[α-(dimethylamino)propionamido]-2,3-dimethyl-1-phenyl-3-pyrazolin-5-one; Amipylo. $C_{16}H_{22}N_4O_2$; mol wt 302.38. C 63.56%, H 7.33%, N 18.53%, O 10.58%. Prepn: Takahashi *et al., J. Pharm. Soc. Japan* **75,** 1431 (1955), *C.A.* **50,** 10086d (1956); Ogiu, Takahashi, **Japan.** pat. **8770('58)** (to Res. Found. for Practical Life), *C.A.* **54,** 4621f (1960).

Prisms from benzene, mp 181°. Very soluble in water.
THERAP CAT: Analgesic.

493. Aminopterin. *N-[4-[[(2,4-Diamino-6-pteridinyl)methyl]amino]benzoyl]-L-glutamic acid;* 4-aminofolic acid; 4-aminopteroylglutamic acid; 4-amino-PGA. $C_{19}H_{20}N_8O_5$; mol wt 440.42. C 51.82%, H 4.58%, N 25.44%, O 18.16%. Prepd from 2,4,5,6-tetraminopyrimidine sulfate, 2,3-dibromopropionaldehyde and p-aminobenzoylglutamic acid: Seeger *et al., J. Am. Chem. Soc.* **69,** 2567 (1947); from 6-(bromomethyl)-2,4-diaminopteridine HBr: Piper, Montgomery, *J. Heterocycl. Chem.* **11,** 279 (1974). Purification: Loo, *J. Med. Chem.* **8,** 139 (1965). Inhibition of dihydrofolate reductase: J. S. Erickson *et al., J. Biol. Chem.* **247,** 5661 (1972). Use as rodenticide: U.S. pat. **2,575,168** (1951 to Am. Cyanamid).

Dihydrate, clusters of yellow needles. uv max of the 0.75 hydrate (0.1*N* NaOH): 261, 282, 373 (log ε 4.41, 4.39, 3.91).

USE: Rodenticide.

494. α-Aminopyridine. *2-Pyridinamine;* 2-aminopyridine. $C_5H_6N_2$; mol wt 94.12. C 63.81%, H 6.43%, N 29.76%. Prepd from pyridine and sodamide: Tschitschibabin, *Chem. Zentr.* **1915, I,** 1065; Wibaut, *Rec. Trav. Chim.* **42,** 240 (1923); A. I. Vogel, *Practical Organic Chemistry* (Longmans, London, 3rd ed., 1959) p 1007; Gattermann-Wieland, *Praxis des Organischen Chemikers* (de Gruyter, Berlin, 40th ed., 1961) p 316.

Leaflets, or large crystals. mp 58.1°. bp 210.6°. Soluble in water, alcohol, benzene, ether, hot petrol ether.

β-Aminopyridine, 3-pyridinamine. Prepn: Allen, Wolf, *Org. Syn.* coll. vol. IV, 45 (1963). Crystals, mp 64°. bp 250-252°. Sol in water, alcohol, benzene, ether. Insol in petr ether.

Caution: Potential symptoms of overexposure to α-aminopyridine are headache, dizziness; excitement; nausea; high blood pressure; respiratory distress; weakness; convulsions; stupor. *See NIOSH Pocket Guide to Chemical Hazards* (DHHS/NIOSH 90-117, 1990) p 36.

USE: Manuf pharmaceuticals and dyes.

495. Aminopyrine. *4-(Dimethylamino)-1,2-dihydro-1,5-dimethyl-2-phenyl-3H-pyrazol-3-one; 4-(dimethylamino)-antipyrine;* dimethylaminophenyldimethylpyrazolone; 4-dimethylamino-2,3-dimethyl-1-phenyl-3-pyrazolin-5-one; aminophenazone; amidazophen; dipyrine; dimethylaminophenazone; amidopyrine; Dipirin; Brufaneuxol; Mamallet-A; Pyramidon; Amidopyrazoline; Polinalin; Amidofebrin; Anafebrina; Netsusarin; Novamidon; Pyradone; Piridol; Itamidone; Febrinina; Dimapyrin. $C_{13}H_{17}N_3O$; mol wt 231.30. C 67.51%, H 7.41%, N 18.17%, O 6.92%. Prepn: *Beilstein* **25**, 452 (1936); T. Takahashi *et al., J. Pharm. Soc. Japan* **76**, 1180 (1956); T. Takahashi, K. Kenematsu, *Chem. Pharm. Bull.* **6**, 98 (1958). Metabolism: G. F. Lockwood, J. B. Houston, *J. Pharm. Pharmacol.* **31**, 787 (1979). Hepatotoxicity: E. Bien *et al., Pharmazie* **36**, 492 (1981). Toxicity studies: Hart, *J. Pharmacol. Exp. Ther.* **89**, 205 (1947); Tubaro *et al., Arzneimittel-Forsch.* **20**, 1024 (1970).

Leaflets from ligroin, mp 107-109°. The aq soln is slightly alkaline to litmus. Soly in water is increased by the addition of sodium benzoate. One gram dissolves in 1.5 ml alcohol, 12 ml benzene, 1 ml chloroform, 13 ml ether, 18 ml water. Stable in air, but affected by light. Readily attacked by mild oxidizing agents in the presence of water. LD_{50} orally in rats: 1.7 g/kg (Hart).

Hydrochloride, $C_{13}H_{17}N_3O \cdot HCl$, deliquescent prisms, mp 143-144°. Freely sol in water with acid reaction.

Bicamphorate, $C_{13}H_{17}N_3O \cdot 2C_{10}H_{16}O_4$, *Pyramidon bicamphorate.* Crystalline powder, mp 94°. Sol in water with gradual decompn; sol in alcohol.

Salicylate, $C_{13}H_{17}N_3O \cdot C_7H_6O_3$, *Pyramidon salicylate.* Crystalline powder, mp 70°. One gram dissolves in 16 ml water, about 6 ml alcohol.

Caution: Agranulocytosis may occur: E. Urbach, H. L. Goldburgh, *J. Am. Med. Assoc.* **131**, 893 (1946); G. Discombe, *Brit. Med. J.* **1952**, 1270.

THERAP CAT: Antipyretic; analgesic.

THERAP CAT (VET): Analgesic; antipyretic.

496. Aminoquinuride. *N,N'-Bis(4-amino-2-methyl-6-quinolinyl)urea;* bis(2-methyl-4-amino-6-quinolyl)urea; di(4-aminoquinald-6-yl)urea; aminoquincarbamide; aminochinuride; aminokinuride; Surfen; $C_{21}H_{20}N_6O$; mol wt 372.43. C 67.73%, H 5.41%, N 22.57%, O 4.30%. Prepn: Jensch, **Ger.** pat. 591,480 (1934 to I. G. Farbenind). Oncogenic and heparin-neutralizing properties: D. T. Hunter, J. M. Hill, *Nature* **191**, 1378 (1961).

Crystals from butanol, dec 255° (effervescence).

THERAP CAT: Antiseptic.

497. Aminorex. *4,5-Dihydro-5-phenyl-2-oxazolamine; 2-amino-5-phenyl-2-oxazoline;* aminoxafen; aminoxaphen. McN-742. $C_9H_{10}N_2O$; mol wt 162.19. C 66.65%, H 6.21%, N 17.27%, O 9.86%. Prepn: Poos *et al., J. Med. Chem.* **6**, 266 (1963); Poos, **U.S.** pat. 3,161,650 (1964 to McNeil).

Crystals from benzene, mp 136-138°.

Fumarate, $C_9H_{10}N_2O \cdot C_4H_4O_4$, *Menocil, Apiquel.*

THERAP CAT: Anorexic.

498. *p*-Aminosalicylic Acid. *4-Amino-2-hydroxybenzoic acid;* 4-aminosalicylic acid; PAS; Apacil; Apas; Deapasil; Enteropas; Lepasen; Pamacyl; Paramycin; Para-Pas; Parasal; Parasalicil; Parasalindon; Pasalon; PASER; Pasolac; Propasa; Rezipas. $C_7H_7NO_3$; mol wt 153.14. C 54.90%, H 4.61%, N 9.15%, O 31.34%. Prepn: **Ger.** pat. 50,835 (1889); **U.S.** pat. 427,564 (1890); Sheehan, *J. Am. Chem. Soc.* **70**, 1665 (1948); Erlenmeyer *et al., Helv. Chim. Acta* **31**, 988 (1948); Wenis, Gardner, *J. Am. Pharm. Assoc.* **38**, 9 (1949); Centolella, **U.S.** pat. 2,844,625 (1958 to Miles). Toxicity study: E. M. Bavin *et al., J. Pharm. Pharmacol.* **2**, 764 (1950). Comprehensive description: M. M. A. Hassan *et al., Anal. Profiles Drug Subs.* **10**, 1-27 (1981).

Minute crystals from alcohol, mp 150-151° with effervescence. uv max (0.1*N* HCl): 265, 300 nm. pKa 3.25. pH of 0.1% aq soln: 3.5. One gram dissolves in about 500 ml water, in 21 ml alcohol. Slightly sol in ether. Practically insol in benzene. Sol in dilute nitric acid, dil sodium hydroxide. At temps above 40°, aq solns of PAS and its hydrochloride are readily decarboxylated to give brown solns consisting mainly of *m*-aminophenol. LD_{50} orally in mice: 4 g/kg (Bavin).

Hydrochloride, $C_7H_7NO_3 \cdot HCl$, crystals, dec 224°.

Sodium salt dihydrate, $C_7H_6NNaO_3 \cdot 2H_2O$, *Aminacyl, Aminopar, Bactylan, Entepas, Gabbropas, Osacyl, Pamisyl Sodium, Paracipan, Paramisan Sodium, Pasara Sodium, Pasid, Pasmed Sodium, Salvis.* pH of 1% soln ~7. One gram dissolves in 2 ml water. Very sparingly sol in acetone. Practically insol in ether, chloroform, benzene. Solns of the sodium salt are more resistant to heat than PAS, but sterilization by bacterial filtration is recommended.

Calcium salt, $C_{14}H_{12}CaN_2O_6$, *Nippas Calcium, Pasmicina.* Bittersweet crystals. One gram dissolves in about 7 ml water, slightly sol in alc. Aq solns dec slowly and darken in color. Also marketed as the hemihydrate and the trihydrate.

Potassium salt, $C_7H_6NO_3K$, *Paskalium, Paskate.* Crystals, freely sol in water. pH of 1% soln about 7. Reported to cause less gastric irritation than the free acid or the sodium salt.

Ethyl ester, $C_9H_{11}NO_3$, needles from alc, mp 115°. Prepd by reduction of the ethyl ester of 4-nitrosalicylic acid.

Phenyl ester, *see* Phenyl Aminosalicylate.

THERAP CAT: Antibacterial (tuberculostatic).

499. *p*-Aminosalicylic Acid Hydrazide. *4-Amino-2-hydroxybenzoic acid hydrazide;* Apacizin; Apacizina. $C_7H_9N_3O_2$; mol wt 167.17. C 50.30%, H 5.43%, N 25.14%, O 19.14%. Prepn: Drain *et al., J. Chem. Soc.* **1949**, 1498; Magrane, **Span.** pat. 206,645 (1952), *C.A.* **49**, 5529e (1955); **Japan.** pat. 7472('54), *C.A.* **50**, 9947c (1956).

Needles from alcohol, mp 190-200°. Slightly soluble in water, somewhat more in ethanol.

THERAP CAT: Antibacterial (tuberculostatic).

500.   4-Amino-2-sulfobenzoic Acid. *p-Amino-o-sulfobenzoic acid.* $C_7H_7NO_5S$; mol wt 217.20. C 38.71%, H 3.25%, N 6.45%, O 36.83%, S 14.76%. Prepd by treating the potassium salt of 4-nitro-2-sulfobenzoic acid with ammonium sulfide: Hart, *Am. Chem. J.* **1**, 351 (1879/80); Kastle, *ibid.* **44**, 490 (1910).

Needles from water. Soly in water: 3 g/l at 25°. Almost insol in alcohol, ether.

USE: Proposed as alkalimetric standard.

501.   2-Aminothiazole. *2-Thiazolamine;* Abadol; Basedol. $C_3H_4N_2S$; mol wt 100.14. C 35.98%, H 4.03%, N 27.97%, S 32.02%. Prepn from vinyl acetate: Christiansen, U.S. pat. 2,242,237 (1941 to Squibb); Kyrides, U.S. pat. 2,330,223 (1943 to Monsanto); *cf.* Skrimshire, **Brit.** pat. 540,032 (1941 to B.D.H.). Prepd also by condensing tribromoparaldehyde with thiourea: Leitch, Brickman, U.S. pats. 2,230,962 (1941); 2,339,083 (1944 to Mallinckrodt). Prepn from paraldehyde and thiourea: Erlenmeyer *et al., Helv. Chim. Acta* **38**, 1293 (1955). Toxicology: W. B. Deichmann *et al., J. Ind. Hyg. Toxicol.* **30**, 71 (1948).

Crystals from benzene + petr ether, mp 93°. Distills at 3 mm without decompn. Sol in hot water. Slightly sol in cold water, alc, ether. Freely sol in dil HCl and in 20% H_2SO_4. LD_{50} orally in rats: 0.48 g/kg (Deichmann).

Hydrochloride monohydrate, $C_3H_4N_2S \cdot HCl \cdot H_2O$, needles, freely sol in water. Also used as the acid tartrate. Crystals, sol in water.

THERAP CAT: Thyroid inhibitor.

502.   2-Amino-1,1,3-tricyanopropene. *2-Amino-1-propene-1,1,3-tricarbonitrile;* 2-amino-1,3,3-tricyano-2-propene; malononitrile dimer. $C_6H_4N_4$; mol wt 132.12. C 54.54%, H 3.05%, N 42.40%. $(CN)_2C{=}C(NH_2)CH_2CN$. Prepn: Carboni *et al., J. Am. Chem. Soc.* **80**, 2838 (1958); Carboni, U.S. pat. 2,719,861 (1955 to du Pont); Coenen, **Ger.** pat. 922,531 (1955 to Bayer).

Rod-like crystals from water, mp 170-173°.

503.   Amioca. *Amylopectin.* A non-linear polymer of glucose, obtained from waxy corn: Caldwell, *Converter* **17**, No. 11, 12, 14 (1943); *Mfg. Confectioner* **33**, No. 12, 15, 17 (1943), *C.A.* **38**, 1137[1] (1944); Schopmeyer, *Food Ind.* **17**, 1476 (1945), *C.A.* **40**, 2331[3] (1946). *Review:* Powell in *Industrial Gums,* R. L. Whistler, Ed. (Academic Press, New York, 2nd ed., 1973) pp 567-576.

504.   Amiodarone. *(2-Butyl-3-benzofuranyl)[4-[2-(diethylamino)ethoxy]-3,5-diiodophenyl]methanone; 2-butyl-3-benzofuranyl-4-[2-(diethylamino)ethoxy]-3,5-diiodophenyl ketone;* 2-butyl-3-[3,5-diiodo-4-(β-diethylaminoethoxy)benz-

oyl]benzofuran. $C_{25}H_{29}I_2NO_3$; mol wt 645.32. C 46.53%, H 4.53%, I 39.33%, N 2.17%, O 7.44%. Benzofuran derivative capable of blocking both α- and β-adrenoceptors. Prepn: **Fr.** pat. **1,339,389;** R. Tondeur, F. Binon, U.S. pat. **3,248,-401** (1963, 1966 to Soc. Belge l'Azote Prod. Chim. Marly). Physicochemical properties: M. Bonati *et al., J. Pharm. Sci.* **73**, 829 (1984). Pharmacology: R. Charlier *et al., Arch. Int. Pharmacodyn.* **139**, 234 (1962); A. Baudine *et al., Arch. Int. Pharmacodyn. Ther.* **169**, 469 (1967); R. Charlier *et al., Arzneimittel-Forsch.* **18**, 1408 (1968). Metabolism: J. Broekhuysen *et al., Arch. Int. Pharmacodyn. Ther.* **177**, 340 (1969). Comparative antidysrhythmic profile: C. Labrid *et al., ibid.* **249**, 87 (1981). HPLC determn in plasma: M. De Smet, D. L. Massart, *J. Pharm. Biomed. Anal.* **6**, 277 (1988). Symposia on pharmacology, pharmacokinetics, clinical studies: *Am. Heart J.* **106**, 787-964 (1983); *Drugs* **29**, Suppl. 3, 1-56 (1985). Review of pharmacology, clinical efficacy and safety: J. W. Mason, *N. Engl. J. Med.* **316**, 455-466 (1987); of pulmonary toxicity: M. J. Reasor, S. Kacew, *Proc. Soc. Exp. Biol. Med.* **196**, 1-7 (1991). Comprehensive description: T. A. Plomp in *Analytical Profiles of Drug Substances* **vol. 20**, K. Florey, Ed. (Academic Press, New York, 1991) pp 1-120.

Hydrochloride, $C_{25}H_{29}I_2NO_3 \cdot HCl$, *L-3428, Amiodar, Ancoron, Angiodarona, Atlansil, Cordarex, Cordarone, Cordarone X, Miocard, Miodaron, Ortacrone, Ritmocardyl, Rythmarone, Trangorex.* Crystalline powder, mp 156°. Also reported as crystals from acetone, mp 159 ±2° (Bonati). Freely sol in chloroform and methanol; sol in ethanol; sparingly sol in isopropanol; slightly sol in acetone, dioxane, ether and carbon tetrachloride; very slightly sol in water. Soly at 25°C (g/100ml): chloroform 44.51; methylene chloride 19.20; methanol 9.98; ethanol 1.28; benzene 0.65; tetrahydrofuran 0.60; acetonitrile 0.32; 1-octanol 0.30; ether 0.17; 1-propanol 0.13; water 0.07; hexane 0.03 petroleum ether 0.001. pH (5% soln) 3.4-3.9. pK_a (25°C) 6.56 ±0.06. uv max (methanol): 208, 242 nm ($E_{1cm}^{1\%}$ 662 ±8, 623 ±10).

THERAP CAT: Antiarrhythmic (class III); antianginal.

505.   Amiphenazole. *5-Phenyl-2,4-thiazolediamine; 2,4-diamino-5-phenylthiazole;* DAPT; phenamizole; Dizol; Daptazole; Daptazile; Fenamizol. $C_9H_9N_3S$; mol wt 191.26. C 56.52%, H 4.74%, N 21.97%, S 16.77%. Prepd by the interaction of thiourea and α-bromobenzyl cyanide in alcohol: Davies *et al., J. Chem. Soc.* **1950**, 3491; from thiourea and α-cyanobenzyl benzenesulfonate in acetone: Dodson, Turner, *J. Am. Chem. Soc.* **73**, 4517 (1951).

Flakes from water or dilute alcohol, dec 163-164°. Turns brown on exposure to light and air.

Hydrobromide, $C_9H_9N_3S \cdot HBr$, prisms from alc, dec >250°. Freely sol in hot water, moderately in cold water.

Benzenesulfonate, $C_{15}H_{15}N_3O_3S_2$, crystals from alcohol + ether, dec 261-262°. Practically insol in water, most organic solvents. Slightly sol in alcohol.

THERAP CAT: Narcotic antagonist.

THERAP CAT (VET): Barbiturate and morphine antagonist.

506.   Amiprilose. *3-O-[3-(Dimethylamino)propyl]-1,2-O-(1-methylethylidene)-α-D-glucofuranose;* 1,2-O-isopropylidene-3-O-[3'-(N,N-dimethylamino)propyl]-α-D-glucofuranose. $C_{14}H_{27}NO_6$; mol wt 305.37. C 55.07%, H 8.91%, N 4.59%, O 31.44%. Synthetic substituted monosaccharide with immunomodulatory activity. Prepn: **Belg. pat. 823,-313;** P. Gordon, **U.S. pat. 3,939,146** (1975, 1976 both to Strategic Med. Res.). Comparative *in vitro* effects on lymphocytes and macrophages: J. W. Hadden *et al., Int. J. Immunopharmacol.* **1,** 17 (1979). Prepn, properties and pharmacokinetics in dogs: E. R. Garrett *et al., J. Pharm. Sci.* **71,** 387 (1982). Human pharmacokinetics: *eidem, J. Pharmacokinet. Biopharm.* **10,** 247 (1982). *In vitro* effect on phagocytes: C. J. Morrison *et al., Antimicrob. Ag. Chemother.* **26,** 74 (1984).

Colorless, viscous oil. n^{25} 1.4687.

Hydrochloride, $C_{14}H_{27}NO_6$.HCl, *SM-1213, Therafectin.* Crystals from methanol, mp 181-183°. Sol in water, methanol, hot ethanol.

THERAP CAT: Immunomodulator.

507. Amisometradine. *6-Amino-3-methyl-1-(2-methyl-2-propenyl)-2,4(1H,3H)-pyrimidinedione; 6-amino-3-methyl-1-(2-methylallyl)uracil;* 1-methallyl-3-methyl-6-aminotetrahydro-2,4-pyrimidinedione; 6-amino-1-methallyl-3-methylpyrimidine-2,4-dione; aminoisometradine; Rolicton. $C_9H_{13}N_3O_2$; mol wt 195.22. C 55.37%, H 6.71%, N 21.52%, O 16.39%. Prepn: Papesch, Schroeder, **U.S. pat. 2,729,669** (1956 to Searle).

Crystals, mp 175°. Soly in water (25°): 2.0 g/100 ml. Freely sol in alcohol, acetone; insol in ether. LD_{50} in mice: 610 mg/kg orally; 415 mg/kg i.p.

THERAP CAT: Diuretic.

508. Amisulpride. *4-Amino-N-[(1-ethyl-2-pyrrolidinyl)methyl]-5-(ethylsulfonyl)-2-methoxybenzamide;* 4-amino-*N*-[(1-ethyl-2-pyrrolidinyl)methyl]-5-(ethylsulfonyl)-*o*-anisamide; aminosultopride; DAN-2163; Socian; Solian. $C_{17}H_{27}N_3O_4S$; mol wt 369.49. C 55.26%, H 7.37%, N 11.37%, O 17.32%, S 8.68%. Neuroleptic agent, analog of sulpiride, sultopride, *q.q.v.* Prepn and pharmacology: M. Thominet *et al.,* **Belg. pat. 872,585;** *eidem,* **U.S. pat. 4,401,822** (1979, 1983 both to Soc. Etudes Sci. Ind. l'Ile-de-France). Dopamine antagonist activity in rats: P. Protais *et al., Neuropharmacol.* **24,** 861 (1985); in mice: M. Vasse *et al., Arch. Pharmacol.* **329,** 108 (1985). Clinical study in schizophrenics: K. Mann *et al., Pharmacopsychiat.* **17,** 111 (1984).

Crystals from acetone, mp 126-127°. LD_{50} in male mice (mg/kg): 56-60 i.v.; 175-180 i.p.; 224-250 s.c.; 1024-1054 orally (Thominet).

THERAP CAT: Antipsychotic.

509. Amiton. *S-[2-(Diethylamino)ethyl]phosphorothioic acid O,O-diethyl ester; O,O-diethyl S-(β-diethylamino)-ethyl phosphorothiolate;* Inferno; Metramac; Tetram. $C_{10}H_{24}NO_3PS$; mol wt 269.35. C 44.59%, H 8.98%, N 5.20%, O 17.82%, P 11.50%, S 11.91%. Prepn: Ghosh, Newman, *Chem. & Ind. (London)* **1955,** 118; Fukuto, Stafford, *J. Am. Chem. Soc.* **79,** 6083 (1957); Lorenz, Schrader, **U.S. pat. 3,082,240** (1963 to Bayer). Toxicity studies: Shaffer, West, *Toxicol. Appl. Pharmacol.* **2,** 1 (1960); Frawley *et al., ibid.* **5,** 605 (1963).

Liquid. $bp_{0.01}$ 76°; $bp_{0.2}$ 110°. n_D^{27} 1.4655. LD_{50} orally in rats: 5.4 mg/kg (Frawley).

Acid oxalate (hydrogen oxalate), $C_{10}H_{24}NO_3PS.C_2H_2O_4$, crystals from isopropanol + ether, mp 98-99°. LD_{50} orally in male albino rats: 9 mg/kg (Shaffer, West).

Caution: Cholinesterase inhibitor.

USE: Contact insecticide, miticide.

510. Amitraz. *N'-(2,4-Dimethylphenyl)-N-[[(2,4-dimethylphenyl)imino]methyl]-N-methylmethanimidamide; N-methyl-N'-2,4-xylyl-N-(N-2,4-xylylformimidoyl)formamidine; N,N-*di-(2,4-xyliminomethyl)methylamine; 2-methyl-1,3-di-(2,4-xylylimino)-2-azapropane; 1,5-di-(2,4-dimethylphenyl)-3-methyl-1,3,5-triazapenta-1,4-diene; *N*-methylbis(2,4-xylyliminomethyl)amine; BTS-27419; U-36059; ENT-27967; BAAM; Mitaban; Mitac; Taktic. $C_{19}H_{23}N_3$; mol wt 293.41. C 77.78%, H 7.90%, N 14.32%. Prepn: I. R. Harrison *et al.,* **Ger. pat. 2,061,132;** *eidem,* **U.S. pats. 3,781,355, 3,864,-497** (1971, 1973, 1975 all to Boots). Activity: *eidem, Pestic. Sci.* **3,** 679 (1972); **4,** 901 (1973). Bacterial degradation: P. B. Baker, D. R. Woods, *J. Appl. Bacteriol.* **42,** 187 (1977); E. R. Allcock, D. R. Woods, *ibid.* **44,** 383 (1978). Metabolism: C. O. Knowles, H. J. Benezet, *J. Environ. Sci. Health* **16,** 547 (1981). Use in treatment of demodecosis in dogs: H. Farmer, A. A. Seawright, *Aust. Vet. J.* **56,** 537 (1980).

White monoclinic needles, mp 86-87°. Soly in water: 1 ppm. Sol in most org solvents. Unstable to acidic pH. LD_{50} in mice, rats (mg/kg): 1600, 400 orally (Harrison, 1975).

USE: Acaricide; insecticide.

THERAP CAT: Ectoparasiticide.

THERAP CAT (VET): In treatment of demodectic mange in dogs.

511. Amitriptyline. *3-(10,11-Dihydro-5H-dibenzo-[a,d]cyclohepten-5-ylidene)-N,N-dimethyl-1-propanamine; 10,11-dihydro-N,N-dimethyl-5H-dibenzo[a,d]cycloheptene-$\Delta^{5,\gamma}$-propylamine;* 5-(γ-dimethylaminopropylidene)-5H-dibenzo[a,d]-10,11-dihydrocycloheptene; 10,11-dihydro-5-(γ-dimethylaminopropylidene)-5H-dibenzo[a,d]cycloheptene; 5-(3-dimethylaminopropylidene)dibenzo[a,d][1,4]-cycloheptadiene. $C_{20}H_{23}N$; mol wt 277.41. C 86.59%, H 8.36%, N 5.05%. Prepn: Hoffsommer *et al., J. Org. Chem.* **27,** 4134 (1962); **28,** 1751 (1963); *J. Med. Chem.* **8,** 555 (1965); Engelhardt *et al.,* **Belg. pat. 584,061** (1960 to Merck & Co.); **Brit. pats. 858,187; 858,188** (both 1961 to Hoffmann-La Roche); Tristram, Tull, **U.S. pat. 3,205,264** (1965 to Merck & Co.). Toxicity data: A. Tobe *et al., Arzneimittel-Forsch.* **31,** 1278 (1981). Comprehensive description: K. W. Blessel *et al.* in *Analytical Profiles of Drug Substances*

Vol. 3, K. Florey, Ed. (Academic Press, New York, 1974) pp 127-148.

Hydrochloride, $C_{20}H_{23}N.HCl$, *Ro-4-1575, Adepril, Amineurin, Amitid, Amitril, Deprex, Domical, Elavil, Endep, Euplit, Laroxyl, Lentizol, Miketorin, Redomex, Saroten, Sarotex, Sylvemid, Triptizol, Tryptanol, Tryptizol.* Minute crystals, mp 196-197°. Freely sol in water, chloroform, alc. uv max (methanol): 240 nm (ϵ 13800). pKa 9.4. LD_{50} in mice, rats (mg/kg): 350, 380 orally; 65, 75 i.p. (Tobe).
THERAP CAT: Antidepressant.

512. Amitriptylinoxide. *3-(10,11-Dihydro-5H-dibenzo[a,d]cyclohepten-5-ylidene)-N,N-dimethyl-1-propanamine N-oxide; 10,11-dihydro-N,N-dimethyl-5H-dibenzo[a,d]cycloheptene-Δ$^{5,\gamma}$-propylamine N-oxide;* amitriptyline N-oxide; Ambivalon; Equilibrin. $C_{20}H_{23}NO$; mol wt 293.41. C 81.87%, H 7.90%, N 4.77%, O 5.45%. Centrally acting metabolite of amitriptyline, *q.v.* Prepn: J. B. Pedersen, **Brit. pat. 991,651;** *idem,* U.S. pat. **3,299,139** (1965, 1967 both to Dumex). Series of articles on pharmacology, pharmacokinetics, metabolism, clinical studies, toxicity studies, teratological studies: *Arzneimittel-Forsch.* **28,** 1873-1926 (1978). HPLC determn: K. M. Jensen, *J. Chromatog.* **183,** 321 (1980). Neuropharmacology: J. Hyttel *et al., Acta Pharmacol. Toxicol.* **47,** 53 (1980). Toxicity study: H. Friehe, R. Fontaine, *Arzneimittel-Forsch.* **28,** 1898 (1978).

Crystals, mp 228-230°. (Dihydrate, mp 102-103°.) LD_{50} in mice, guinea pigs, rabbits, dogs (mg/kg): between 330-460 orally; in mice, rats: 87, 25 i.v.; 320, 110 i.p. (Friehe, Fontaine).
THERAP CAT: Antidepressant.

513. Amitrole. *1H-1,2,4-Triazol-3-amine;* 3-amino-1H-1,2,4-triazole; aminotriazole; ATA; ENT-25445; Amizol; Cytrol; Weedazol. $C_2H_4N_4$; mol wt 84.08. C 28.57%, H 4.80%, N 66.63%. Non-selective post-emergence, translocated herbicide. Prepn: G. Sjostedt, L. Gringas, *Org. Syn. coll. vol. III,* 95 (1955). Use as herbicide: Allen, U.S. pat. **2,670,282** (1954 to Am. Chemical Paint Co.). Antithyroid activity: T. H. Jukes, C. B. Shaffer, *Science* **132,** 296 (1960). *Review:* E. Kröller, *Residue Rev.* **12,** 163-192 (1966). Review of carcinogenicity studies: *IARC Monographs* **7,** 31-43 (1974).

Crystals from abs ethanol, mp 159°. Soluble in water, methanol, ethanol, chloroform. Sparingly sol in ethyl acetate. Insol in ether, acetone. Aq solns are neutral. LD_{50} in mice, rats (g/kg): 14.7, 25.0 orally (Kröller).
Hydrochloride, $C_2H_4N_4.HCl$, crystals from alcohol, mp 153°.

Note: Amitrole may reasonably be anticipated to be a carcinogen: *Seventh Annual Report on Carcinogens* (PB95-109781, 1994) p 92.
USE: Herbicide.

514. Amixetrine. *1-[2-(3-Methylbutoxy)-2-phenylethyl]pyrrolidine; 1-[β-(isopentyloxy)phenethyl]pyrrolidine; N-(2-phenyl-2-isoamyloxy)ethylpyrrolidine.* $C_{17}H_{27}NO$; mol wt 261.41. C 78.11%, H 10.41%, N 5.36%, O 6.12%. Prepn by reacting styrene and isoamyl alcohol with *t*-butyl hypobromite and then condensing the product with pyrrolidine: Mauvernay, Busch, **Ger. pat. 1,811,767** (1969 to C.E.R.M.), *C.A.* **72,** 21607e (1970). Identification of urinary metabolites in humans and dogs: M. Constantin *et al., Arzneimittel-Forsch.* **26,** 80 (1976). Pharmcokinetics: M. Constantin, J. F. Pognat, *ibid.* **28,** 646 (1978).

bp_2 121°. $n_D^{21.6}$ 1.4978.
Hydrochloride, $C_{17}H_{27}NO.HCl$, *Somagest.* Crystals, mp 150°.
THERAP CAT: Anti-inflammatory; anticholinergic.

515. Amlexanox. *2-Amino-7-(1-methylethyl)-5-oxo-5H-[1]benzopyrano[2,3-b]pyridine-3-carboxylic acid;* 2-amino-7-isopropyl-5-oxo-5H-[1]benzopyrano[2,3-b]pyridine-3-carboxylic acid; 2-amino-7-isopropyl-1-azaxanthone-3-carboxylic acid; amoxanox; AA-673; CHX-3673; Elics; Solfa. $C_{16}H_{14}N_2O_4$; mol wt 298.30. C 64.42%, H 4.73%, N 9.39%, O 21.45%. Orally active lipoxygenase inhibitor. Prepn: A. Nohara *et al.,* **Belg. pat. 864,647;** U.S. pat. **4,143,042** (1978, 1979 both to Takeda); *eidem, J. Med. Chem.* **28,** 559 (1985). Series of articles on toxicity, teratogenicity, mutagenicity, reproductive and developmental effects: *Yakuri to Chiryo* **13,** 4873-4931 (1985); *C.A.* **104,** 61684p-61691p (1986). Mode of action: T. Saijo *et al., Int. Arch. Allergy Appl. Immunol.* **79,** 231 (1986).

Crystals from dimethylformamide, mp > 300°.
THERAP CAT: Antiallergic; antiasthmatic.

516. Amlodipine. *2-[(2-Aminoethoxy)methyl]-4-(2-chlorophenyl)-1,4-dihydro-6-methyl-3,5-pyridinedicarboxylic acid 3-ethyl 5-methyl ester;* (±)-2-[(2-aminoethoxy)methyl]-4-(2-chlorophenyl)-3-ethoxycarbonyl-5-methoxycarbonyl-6-methyl-1,4-dihydropyridine; UK-48340. $C_{20}H_{25}ClN_2O_5$; mol wt 408.88. C 58.75%, H 6.16%, Cl 8.67%, N 6.85%, O 19.56%. Dihydropyridine calcium channel blocker with activity residing mainly in the (−)-isomer. Prepn: S. F. Campbell *et al.,* **Eur. pat. Appl. 89,167;** *eidem,* U.S. pat. **4,572,909** (1983, 1986 both to Pfizer). Synthesis of racemate and enantiomers; preliminary pharmacology: J. E. Arrowsmith *et al., J. Med. Chem.* **29,** 1696 (1986). Antagonist activity: R. A. Burges *et al., J. Cardiovasc. Pharmacol.* **9,** 110 (1987). Pharmacokinetics: J. K. Faulkner *et al., Brit. J. Clin. Pharmacol.* **22,** 21 (1986). Metabolism in animals: A. P. Beresford *et al., Xenobiotica* **18,** 169 (1988); in man: A. P. Beresford *et al., ibid.* 245. Effects on renal system in hypertensive men: G. P. Reams *et al., Am. J. Kidney Dis.* **10,** 446 (1987). GC determn in plasma: A. P. Beresford *et al., J. Chromatog.* **420,** 178 (1987). Clinical evaluation in hypertension: J. Webster *et al., Brit. J. Clin. Pharmacol.* **24,** 713 (1987).

(±)-Form maleate, $C_{20}H_{25}ClN_2O_5 \cdot C_4H_4O_4$, *UK-48340-11*. White crystals from ethyl acetate, mp 178-179°.

Benzenesulfonate, $C_{20}H_{25}ClN_2O_5 \cdot C_6H_5SO_3H$, *amlodipine besylate, UK-48340-26, Antacal, Istin, Monopina.*

THERAP CAT: Antianginal; antihypertensive.

517. Ammonia. H_3N; mol wt 17.03. H 17.76%, N 82.24%. NH_3. Manufactured from water gas (obtained by blowing steam through incandescent coke) as source of hydrogen, and from producer gas (obtained from steam and air through incandescent coke), as source of nitrogen by the Haber-Bosch process. Manuf from natural gas: Faith, Keyes & Clark's *Industrial Chemicals*, F. A. Lowenheim, M. K. Moran, Eds. (Wiley-Interscience, New York, 4th ed., 1975) pp 83-92. Historical monograph: A. Mittasch, *Geschichte der Ammoniaksynthese* (Verlag Chemie, 1951). Reviews of prepn, properties and chemistry: Several authors in *Mellor's* Vol. VIII, supplement I, *Nitrogen* part 1 (1964) pp 240-369; Jones in *Comprehensive Inorganic Chemistry* Vol. 2, J. C. Bailar, Jr. *et al.*, Eds. (Pergamon Press, Oxford, 1973) pp 199-227; J. R. LeBlanc *et al.*, in Kirk-Othmer *Encyclopedia of Chemical Technology* vol. 2 (Wiley-Interscience, New York, 3rd ed., 1978) pp 470-516.

Colorless gas; very pungent odor (characteristic of drying urine). Lower limit of human perception: 0.04 g/cubic meter or 53 ppm. One liter of the gas weighs 0.7714 g. d 0.5967 (air = 1). mp −77.7°. bp$_{760}$ −33.35°. Densities of liq NH_3 (temp; press.): 0.6818 (−33.35°; 1 atm); 0.6585 (−15°; 2.332 atm); 0.6386 (0°; 4.238 atm); 0.6175 (15°; 7.188 atm); 0.5875 (35°; 13.321 atm). Critical temp 132.4°; critical press. 111.5 atm. Heat capacity (25°) 8.38 cal/mole/deg. Mixtures of ammonia and air will explode when ignited under favorable conditions: *Angew. Chem.* **43**, 302 (1930), but ammonia is generally regarded as nonflammable. Corrosive, alkaline gas. pH of 1.0N aq soln 11.6; of 0.1N aq soln 11.1; of 0.01N aq soln 10.6. Water at 0° holds 47%, at 15° 38%, at 20° 34%, at 25° 31%, at 30° 28%, at 50° 18%. d$_4^{20}$ (aq solns): 0.9939 (1%); 0.9811 (4%); 0.9651 (8%); 0.9362 (16%); 0.9229 (20%); 0.9101 (24%); 0.8980 (28%). fp (aq solns): −2.9° (4%); −8.1° (8%); −23.1° (16%); −34.9° (20%); −44.5° (24%); −69.2° (28%). Solution of NH_3 in water is exothermic. 95% alcohol at 20° holds 15%, at 30° 11%. Abs ethanol at 0° 20%, at 25° 10%. Methanol at 25° 16%. It is also sol in chloroform and ether. Liquid ammonia produces low temps by its own evaporation. Heat of vaporization: 5.581 kcal/mole. It is a good solvent for many elements and compds. Usually marketed in liquefied form in steel cylinders or as ammonia water (aqua ammonia, ammonium hydroxide) in drums and bottles.

Caution: Potential symptoms of overexposure are eye, nose and throat irritation; dyspnea, bronchospasm and chest pain; pulmonary edema; pink frothy sputum; skin burns, vesiculation. *See NIOSH Pocket Guide to Chemical Hazards* (DHHS/NIOSH 90-117, 1990) p 38. *See also* Patty's *Industrial Hygiene and Toxicology* vol. **2B**, G. D. Clayton, F. E. Clayton, Eds. (Wiley-Interscience, New York, 3rd ed., 1981) pp 3045-3052.

USE: Manuf nitric acid, explosives, synthetic fibers, fertilizers. In refrigeration. In the chemical industry.

518. Ammoniacum. *Gum ammoniac.* A gum-resin exuded from the flowering and fruiting stem of *Dorema ammoniacum*, D. Don, and probably other *Umbelliferae*. *Habit*. Persia, Northern India, Southern Siberia, Africa. *Constit.* 1.3-6.7% volatile oil; 50-70% resin; 18-26% gum; ash content about 2%, may be as high as 10%; salicylic acid; aminoresinol. *Ref:* W. Sandermann, *Naturharze, Terpentinöl, Tallöl* (Springer, 1960) pp 82-83.

Irregular rounded tears, yellowish or brownish outside and whitish within; brittle when cold, but soft when warm; also masses, darker in color and less homogeneous; peculiar odor; slightly sweetish, bitter, somewhat acrid taste. mp 45-55°. Acid no. 60-80. Sapon no. 97-114. d 1.207. Partly soluble in water, alcohol, ether, vinegar or alkali soln; forms emulsions with water.

USE: Ingredient of porcelain cements.

THERAP CAT: Diaphoretic; emmenagogue.

519. Ammonia Water. Ammonium hydroxide; aqua ammonia; "Spirit of Hartshorn". A soln of 28-29% NH_3 in water.

Colorless liquid; intense, pungent, suffocating odor; acrid taste; strong alkaline reaction. d$_{25}^{25}$ ~0.90, 26°Bé. It dissolves copper, zinc. Fumes are formed when ammonia water is brought near volatile acids. *Keep cool in strong glass, plastic or rubber-stoppered bottles not completely filled.* Reaction with H_2SO_4 or other strong mineral acids is exothermic; mixture becomes boiling hot.

Ammonia Water—10%, colorless liquid. Very pungent odor. d$_{25}^{25}$ 0.957. *Irritating to skin and mucous membranes.*

USE: Detergent, removing stains, bleaching, calico printing, extracting plant colors (cochineal, archil, etc.) and alkaloids; manuf ammonium salts, aniline dyes, and a wide variety of other uses.

THERAP CAT: 10% Solution as reflex respiratory stimulant.

520. Ammonium Acetate. Acetic acid ammonium salt. $C_2H_7NO_2$; mol wt 77.08. C 31.16%, H 9.15%, N 18.17%, O 41.51%. CH_3COONH_4. Commercial product contains 95-97% salt with acetic acid and some water. Prepd from acetic acid and NH_3: Zuffanti, *J. Am. Chem. Soc.* **63**, 3123 (1941). Toxicity data: Welch *et al.*, *J. Lab. Clin. Med.* **29**, 809 (1944).

Deliquesc crystals or crystalline masses. Slight acetous odor. d 1.07. mp 114° (Zuffanti, *loc. cit.*). Tends to lose NH_3. Sol in less than 1 part water; freely sol in alc; slightly sol in acetone. Very concd aq soln is slightly acid; a 0.5 molar aq soln has pH 7.0. *Keep cool and tightly closed.* LD i.v. in mice: 1.8 mg (NH_4^+)/20g (Welch).

Ammonium acetate solution, Spirit of Mindererus. Colorless, clear liquid; acid reaction to litmus. Prepd from 1 g ammonium carbonate, 20 ml dil acetic acid (6%). Contains 6.5-7.5% CH_3COONH_4.

USE: Preserving meats, dyeing, stripping; as a reagent in anal. chemistry, *e.g.*, for determining Pb, Fe; separating $PbSO_4$ from other sulfates.

THERAP CAT: Diuretic.

THERAP CAT (VET): Formerly as diuretic, antipyretic.

521. Ammonium Benzoate. *Benzoic acid, ammonium salt.* $C_7H_9NO_2$; mol wt 139.15. C 60.42%, H 6.52%, N 10.07%, O 23.00%. $C_6H_5COONH_4$. It is about 99% pure. Manuf from benzoic acid and NH_3: Spina, U.S. pat. **1,704,-636** (1929 to Hooker Electrochem.).

Lamellar crystals or crystalline powder; odorless or faint benzoic acid odor; gradually loses NH_3 on exposure to air. d 1.26. mp 198°. One gram dissolves in 4.7 ml water, 1.2 ml boil. water, 36 ml alcohol, 8 ml boil. alcohol, 8 ml glycerol. The aq soln is slightly acid. *Keep well closed. Incompat.* Ferric salts, acids, alkali hydroxides or carbonates.

USE: To preserve glue and latex.

THERAP CAT: Urinary anti-infective.

522. Ammonium Bicarbonate. Acid ammonium carbonate; ammonium hydrogen carbonate. CH_5NO_3; mol wt 79.06. C 15.19%, H 6.37%, N 17.72%, O 60.71%. NH_4HCO_3. Occurs in the urine of alligators: Coulson, Hernandez, *Proc. Soc. Exp. Biol. Med.* **88**, 682 (1955). Usually prepd by passing an excess of carbon dioxide through concd ammonia water. Manuf: Brit. pat. **304,872** (1927 to I.G. Farben.); Brooks, Brit. pat. **742,386** (1955 to I.C.I.). *See also* Ammonium Carbonate.

Shiny, hard, colorless or white prisms or crystalline mass. Faint odor of ammonia. Comparatively stable at room temp. Volatile with decompn at about 60°. The white fumes given off consist of NH_3 21.5%, CO_2 55.7%, H_2O vapor 22.8%. Rate of decompn increases as temp rises. mp 107.5° (very rapid heating). Soly in water: 14% (10°); 17.4% (20°); 21.3% (30°). Decomposed by hot water. One gram dissolves in 10 ml glycerol (pharmaceutical grade). pH of 0.1N soln in water at 25° = 7.8. Insol in alcohol and acetone. Negative heat of solution.

Pharmaceutical Incompat: Acids, caustic alkalies.

USE: In baking powder formulations; in cooling baths (one kg dissolved in 5 liters H_2O at 17° lowers it to 7°); in fire extinguishers; manuf porous plastics, ceramics; manuf dyes, pigments; in compost heaps to accelerate decompn; as fertilizer; for defatting textiles; in cold wave solns; in chrome leather tanning; to remove gypsum from heat exchanges and other processing equipment.

THERAP CAT: Expectorant.

THERAP CAT (VET): Expectorant. Used in bloat, colic.

523. Ammonium Bifluoride. *Acid ammonium fluoride; ammonium hydrogen fluoride.* F_2H_5N; mol wt 57.04. F 66.61%, H 8.83%, N 24.55%. NH_4HF_2. Prepn from hydrofluoric acid and NH_3: Hassel, Luzanski, *Z. Kristallogr.* **A83**, 449 (1932); *Gmelin's, Ammonium* (8th ed.) **23**, 148 (1936).

Orthorhombic crystals which readily etch glass. d 1.5. mp 124.6°. Freely sol in water. *Keep in plastic, rubber, wool or paraffined containers and well closed.* See also Ammonium Fluoride.

USE: In manuf of Mg and Mg alloys; in brightening of Al; for purifying and cleansing various parts of beer-dispensing apparatus, tubes, etc., sterilizing dairy and other food equipment; in glass and porcelain industries; as mordant for aluminum; as a "sour" in laundering cloth. In lab production of HF.

524. Ammonium Bimalate. *Hydroxybutanedioic acid monoammonium salt; l-malic acid monoammonium salt;* ammonium acid malate. $C_4H_9NO_5$; mol wt 151.12. C 31.79%, H 6.00%, N 9.27%, O 52.94%. $NH_4OOCCH_2CH-(OH)COOH$ or $HOOCCH_2CH(OH)COONH_4$. Prepn: Kendrick, *Ber.* **30**, 1749 (1897); Holmberg, *ibid.* **61**, 1885 (1928).

Orthorhombic disphenoidal crystals. mp 160-161°. d 1.15. Sol in 3 parts water; slightly sol in alcohol. $[\alpha]_D^{18}$ −6.4° (c = 6).

525. Ammonium Binoxalate. Ammonium acid oxalate; ammonium hydrogen oxalate. $C_2H_5NO_4$; mol wt 107.07. C 22.44%, H 4.71%, N 13.08%, O 59.77%. $NH_4-OOCCOOH$. Prepn of monohydrate: Dehn, Heuse, *J. Am. Chem. Soc.* **29**, 1137 (1907). *Review: Gmelin's, Ammonium* (8th ed.) **23**, pp 405-406 (1936).

Monohydrate, rhombic crystals. *Poisonous.* d 1.56. Sol in 25 parts water; slightly sol in alcohol.

USE: To remove ink stains.

526. Ammonium Bisulfate. Acid ammonium sulfate; ammonium hydrogen sulfate. H_5NO_4S; mol wt 115.11. H 4.38%, N 12.17%, O 55.60%, S 27.86%. NH_4HSO_4. Prepn: *Gmelin's, Ammonium* (8th ed.) **23**, 293-298 (1936).

Deliquesc crystals. mp about 147°. d 1.787. Freely sol in water. Practically insol in alcohol, acetone, pyridine. *Keep well closed.*

USE: In hair-waving preparations; as catalyst for organic reactions.

527. Ammonium Bisulfide. Ammonium hydrogen sulfide; ammonium hydrosulfide; ammonium sulfhydrate. H_5NS; mol wt 51.11. H 9.86%, N 27.40%, S 62.74%. NH_4HS. Prepd by mixing stoichiometric amts of NH_3 and H_2S gases at 0°: Thomas, Riding, *J. Chem. Soc.* **123**, 1181 (1923).

White, tetragonal or orthorhombic crystals. d 1.17. Sublimes *in vacuo.* Decomposes into H_2S and NH_3 rather easily at room temp when in crystal form. The commercial product is furnished in porcelain-like lumps which are more stable and can be stored in a closed bottle at a cool place. Dissociation press. at room temp about 350 mm Hg. Freely sol in water or alcohol giving colorless solns which turn yellow rapidly. Decomposed by boiling water. Soly (0°): 128.1 g/100 g H_2O. Slightly sol in acetone; almost insol in ether, benzene.

Caution: Very irritating to skin; penetrates more rapidly than hydrogen sulfide and may be fatal.

USE: In lubricants.

528. Ammonium Bisulfite. H_5NO_3S; mol wt 99.11. H 5.08%, N 14.13%, O 48.43%, S 32.35%. NH_4HSO_3. Prepn: *Gmelin's, Ammonium* (8th ed.) **23**, 259-260 (1936).

Crystals. Soly in water (g/100 ml H_2O): 267 (10°); 620 (60°). *Keep well closed.* Sold as solution only.

USE: Preservative.

529. Ammonium Bitartrate. *(R)-2,3-Dihydroxybutanedioic acid monoammonium salt;* L-tartaric acid mono-ammonium salt; ammonium acid tartrate; ammonium hydrogen tartrate. $C_4H_9NO_6$; mol wt 167.12. C 28.75%, H 5.43%, N 8.38%, O 57.44%. $NH_4OOCCH(OH)CH(OH)-COOH$. Prepd from ammonium tartrate and tartaric acid: Dulk, *Ann.* **2**, 39 (1832).

Odorless crystals. d 1.68. Sol in 45.6 parts water at 15°; freely sol in hot water, alkalies, alkali carbonates. Practically insol in alc. $[\alpha]_D^{20}$ +26.0° (c = 1.5 in water): Long, *J. Am. Chem. Soc.* **23**, 813 (1901).

530. Ammonium Borate. *Ammonium tetraborate;* ammonium biborate; ammonium metaborate. $B_4H_8N_2O_7$; mol wt 191.32. B 22.60%, H 4.21%, N 14.64%, O 58.54%. $(NH_4)_2B_4O_7$. Prepn: *Gmelin's, Ammonium* (8th ed.) **23**, 323 (1936).

Tetrahydrate, tetragonal crystals. Sol in water. Practically insol in alcohol.

USE: Fireproofing wood and textiles; in electrolytic condensers.

531. Ammonium Bromide. BrH_4N; mol wt 97.94. Br 81.58%, H 4.12%, N 14.30%. NH_4Br. Contains 99-99.5% NH_4Br. Prepn and properties: *Gmelin's, Ammonium* (8th ed.) **23**, 203-218 (1936); Richards, "Ammonium Bromide" in *Mellor's*, **Vol. VIII**, supplement I, *Nitrogen* (part 1) 433-447 (1964).

White, odorless, slightly hygroscopic crystals or granules; pungent, saline taste; slowly becomes yellowish in air; sublimes at high temp without melting. d^{25} 2.429. Freely sol in water, methanol, ethanol, acetone; slightly sol in ether. Practically insol in ethyl acetate. *Keep well closed.*

Incompat. Acids, acid salts, spirit nitrous ether, alkaloids; salts of lead, mercury, silver.

USE: Manuf of photographic films, plates, and papers; in process engraving and lithography; fireproofing of wood; in corrosion inhibitors.

THERAP CAT: Sedative.

THERAP CAT (VET): Sedative.

532. Ammonium Caprylate. *Octanoic acid ammonium salt;* caprylic acid ammonium salt. $C_8H_{19}NO_2$; mol wt 161.24. C 59.59%, H 11.88%, N 8.69%, O 19.84%. $C_7H_{15}-COONH_4$. Prepn from the acid and ammonia: McMaster, Magill, *J. Am. Chem. Soc.* **38**, 1793 (1916); Stumpf, *Am. Paint J.* **38** (45), 60 (1954).

Monoclinic crystals from ether + alcohol. Somewhat hygroscopic. Decomposes on standing and develops odor of caprylic acid. mp 70-85°. Easily hydrolyzed by water. Freely sol in glacial acetic acid and ethanol; less sol in methanol; slightly sol in acetone and ethyl acetate. Practically insol in chloroform, benzene. *Keep well closed.*

USE: In photographic emulsions; as insecticide and nematocide; in manuf of zinc caprylate.

533. Ammonium Carbamate. "Anhydride" of ammonium carbonate; ammonium aminoformate. $CH_6N_2O_2$; mol wt 78.07. C 15.38%, H 7.75%, N 35.88%, O 40.99%. $NH_2-COONH_4$. Prepd from dry ice and liq ammonia: Brooks, Audrieth, *Inorg. Syn.* **2**, 85 (1946).

Cryst powder; ammonia odor; gradually loses ammonia in the air changing to ammonium carbonate. Volatilizes at about 60°. Freely sol in water; sol in alcohol.

USE: Ammoniating agent, less vigorous than NH_3.

534. Ammonium Carbonate. Hartshorn. A mixture of ammonium bicarbonate and ammonium carbamate, obtained by subliming a mixture of ammonium sulfate and calcium carbonate. Contains 30-34% NH_3, about 45% CO_2. See USP **XVII**, p 941. *Review:* Allen, "Ammonium Carbonate" in *Mellor's* Vol. VIII, supplement 1, *Nitrogen* (part 1) 459-468 (1964).

Colorless, hard, translucent, crystalline masses, white cubes or powder; strong odor of ammonia; sharp taste and alkaline reactions. Decomposes on exposure to air with loss of NH_3 and CO_2, becoming white and powdery and converting into ammonium bicarbonate. Volatilizes at about 60°. Slowly sol in 4 parts water; decomposed by hot water; the carbamate portion dissolves in alcohol. *Keep tightly closed in a cool place.*

Incompat. Acids and acid salts; salts of iron, zinc; alkaloids, alum, calomel, tartar emetic.

USE: In baking powders, for washing and defatting woolens, tanning; as mordant in dyeing; manuf rubber articles, casein glue, casein colors; in fire extinguishers, smelling salts; for separating cacao constituents; as a reagent in analytical chemistry. Pharmaceutic aid (source of ammonia).

THERAP CAT: Expectorant.

THERAP CAT (VET): Expectorant. Has been used as a carminative and stomachic.

535. Ammonium Ceric Nitrate. *Ammonium hexanitratocerate(IV);* ammonium nitratocerate(IV); ceric ammonium nitrate. $CeH_8N_8O_{18}$; mol wt 548.22. Ce 25.56%, H 1.47%, N 20.44%, O 52.53%. $(NH_4)_2Ce(NO_3)_6$. Prepd by dissolving $CeO.H_2O$ in hot concd HNO_3: Smith *et al., Ind. Eng. Chem., Anal. Ed.* **8**, 449 (1936); Smith, Fly, *Anal. Chem.* **21**, 1233 (1949); Smith, *Talanta* **10**, 709 (1963).

Small, orange-red, monoclinic crystals. Very sol in water.

USE: As a standard in oxidimetry; catalyst for polymerization of olefins.

536. Ammonium Cerous Sulfate. Ammonium disulfatocerate(III); cerous ammonium sulfate. $CeH_4NO_8S_2$; mol wt 350.28. Ce 40.00%, H 1.15%, N 4.00%, O 36.54%, S 18.31%. $NH_4Ce(SO_4)_2$. Prepd by slow evaporation of a soln contg equimolar amounts of $(NH_4)_2SO_4$ and $CeSO_4$: Blandin, Rerat, *Compt. Rend.* **242**, 1740 (1956).

Tetrahydrate, monoclinic crystals. Sol in water.

537. Ammonium Chloride. Ammonium muriate; sal ammoniac; salmiac; Amchlor; Darammon. ClH_4N; mol wt 53.49. Cl 66.28%, H 7.54%, N 26.19%. NH_4Cl. Contains 99.5-99.8% NH_4Cl; principal impurity is NaCl. Prepn and properties: *Gmelin's, Ammonium* (8th ed.) **23**, pp 150-184 (1936); Kane, "Ammonium Chloride" in *Mellor's* Vol. VIII, supplement 1, *Nitrogen* (part 1), 378-432 (1964). Manuf: Faith, Keyes & Clark's *Industrial Chemicals*, F. A. Lowenheim, M. K. Moran, Eds. (Wiley-Interscience, New York, 4th ed., 1975) pp 93-96.

Colorless, odorless crystals or cryst masses; or white, granular powder; cooling, saline taste; somewhat hygroscopic. Tendency to cake. d^{25} 1.5274. Sublimes without melting. Soly in water (w/w): 22.9% (0°); 26.0% (15°); 28.3% (25°); 39.6% (80°). Strongly endothermic. HCl and NaCl decrease soly in water. pH of aq solns (25°): 1% 5.5; 3% 5.1; 10% 5.0. Sol in methanol, ethanol; almost insol in acetone, ether, ethyl acetate. LD_{50} i.m. in rats: 30 mg/kg, *RTECS* Vol. I, R. J. Lewis, R. L. Tatken, Eds. (1979) pp 99.

Incompat. Alkalies and their carbonates; lead and silver salts.

USE: As a flux for coating sheet iron with zinc; tinning; in dry and Leclanché batteries; dyeing, freezing mixtures, electroplating, to clean soldering irons, safety explosives, lustering cotton, tanning; in washing powders; manuf dyes; in cement for iron pipes; for snow treatment (slows melting on ski slopes).

THERAP CAT: Systemic acidifier.

THERAP CAT (VET): Expectorant; diaphoretic; acidifying diuretic.

538. Ammonium Chromate(VI). Neutral ammonium chromate. $CrH_8N_2O_4$; mol wt 152.07. Cr 34.19%, H 5.30%, N 18.42%, O 42.08%. $(NH_4)_2CrO_4$. Prepn: *Gmelin's, Chromium* (8th ed.) **52**, part B, pp 707-712 (1962).

Yellow acicular crystals; loses some NH_3 in air; decomposes at 185°. d 1.8. Soly in water: 19.78% (0°); 41.20% (75°). Sparingly sol in liquid ammonia, acetone. Slightly sol in methanol. Practically insol in ethanol. The aq soln is alkaline. *Keep well closed.*

USE: Sensitizing gelatin in photography, in textile printing pastes, in fixing chromate dyes on wool, and as a reagent in analytical chemistry.

539. Ammonium Chromic Sulfate. Ammonium disulfatochromate(III); chromic ammonium sulfate. $CrH_4N-O_8S_2$; mol wt 262.16. Cr 19.83%, H 1.54%, N 5.34%, O 48.82%, S 24.46%. $NH_4Cr(SO_4)_2$. Prepd by crystallization from a soln contg equimolar amounts of $Cr_2(SO_4)_3$ and $(NH_4)_2SO_4$: Howarth in *Chromium*, vol. **I**, M. J. Udy, Ed., A.C.S. Monograph Series, no. **132** (Reinhold, New York, 1956) p 287; electrolytic prepn: Nishihara *et al., Japan.* pat. **2164**('60); *C.A.* **55**, 5200e (1961).

Dodecahydrate, *chrome alum ammonium.* Small dark-violet or violet-blue, octahedral, cubic crystals; ruby-red by transmitted light. mp 94°; loses $9H_2O$ on melting and the remaining H_2O by 300°. d^{25} 1.72. Readily sol in water; slightly sol in alcohol. Aq soln is violet when cold, green when hot.

USE: Mordant in textile industry; in manuf of electrolytic Cr metal.

540. Ammonium Citrate, Dibasic. Diammonium citrate; citric acid diammonium salt. $C_6H_{14}N_2O_7$; mol wt 226.19. C 31.86%, H 6.24%, N 12.39%, O 49.51%. Prepn: Heldt, *Ann.* **47**, 157 (1843).

$$\left[\begin{array}{c} COO^- \\ | \\ CH_2 \\ | \\ HO-C-COOH \\ | \\ CH_2 \\ | \\ COO^- \end{array} \right] \cdot 2\,NH_4^+$$

Granules or crystals; acid reaction. d 1.48. Sol in about 1 part water, slightly in alc. pH of $0.1M$ soln in H_2O = 4.3.

USE: For the determination of phosphate, esp in fertilizers.

541. Ammonium Cobaltous Phosphate. Cobaltous ammonium phosphate. CoH_4NO_4P; mol wt 171.94. Co 34.27%, H 2.34%, N 8.15%, O 37.22%, P 18.01%. NH_4Co-PO_4. Prepd by reaction of a cobaltous salt with $(NH_4)_3PO_4$, $(NH_4)_2HPO_4$, or H_3PO_4 and NH_3: Grat-Cabanac, *Bull. Microscop. Appl.* **8**, 97 (1958); Salutsky *et al.;* McCullogh, Salutsky, U.S. pats. **3,126,254; 3,141,732** (both 1964 to W. R. Grace).

Hydrate, red to violet powder, or monoclinic rectangular lamellae. Practically insol in water. Sol in acids.

USE: As pigment in ceramic glazes and vitreous enamels; as temp indicator in textile industry; in fertilizers for plant nutrition; in Co analysis.

542. Ammonium Cobaltous Sulfate. Ammonium disulfatocobaltate(II); cobaltous ammonium sulfate. $CoH_8-N_2O_8S_2$; mol wt 287.14. Co 20.52%, H 2.81%, N 9.76%, O 44.58%, S 22.33%. $(NH_4)_2Co(SO_4)_2$. Prepd by mixing equimolar amounts of solns of $(NH_4)_2SO_4$ and $CoSO_4$: Malard, *Bull. Soc. Chim. France* **1961**, 2296; *Gmelin's, Cobalt* (8th ed.) **58**, part A, p 435 (1932); suppl, pp 811-815 (1961).

Hexahydrate, red monoclinic prismatic crystals. d 1.90. Sol in water; almost insol in alcohol.

543. Ammonium Cupric Chloride. *Ammonium tetrachlorocuprate(II);* ammonium chlorocuprate; cupric ammonium chloride. $Cl_4CuH_8N_2$; mol wt 241.43. Cl 58.74%, Cu 26.32%, H 3.34%, N 11.60%. $(NH_4)_2CuCl_4$. Prepd by evaporation of a 2:1 soln of NH_4Cl and $CuCl_2$: Chrobak, *Bull. Int. Acad. Polonaise* **1929A**, 361; *C.A.* **24**, 3688[7] (1930); Willet, *J. Chem. Phys.* **41**, 2243 (1964).

Yellow, hygroscopic, orthorhombic crystals. Sol in water. *Keep well closed.*

Dihydrate, *ammonium tetrachlorodiaquocuprate(II).* Blue to bluish-green tetragonal, rhombododecahedral crystals. d 2.0. Becomes anhydr at 110-120°; decomposes on stronger heating. Sol in water, alcohol, liquid NH_3; the aq soln is acid to litmus.

USE: Analytical reagent; formerly for determining carbon in iron and steel.

544. Ammonium Dichromate(VI). Ammonium bichromate. $Cr_2H_8N_2O_7$; mol wt 252.07. Cr 41.26%, H 3.20%, N 11.11%, O 44.43%. $(NH_4)_2Cr_2O_7$. Prepd from ammonium sulfate and sodium dichromate or by the interaction of ammonia gas and chromic acid in soln: Hartford, *Ind. Eng. Chem.* **41**, 1993 (1949); *Chromium Chemicals,* Publication **52** (Mutual Chem. Div., Allied Chem.) pp 34-39. *Review:* M. J. Udy, *Chromium,* Vol. 1, ACS Monograph Series no. **132** (Reinhold, New York, 1956) *passim.*

Bright orange-red crystals. *Flammable.* Odorless and non-hygroscopic. Crystal system: monoclinic. Crystal habit: prismatic. d_4^{25} 2.155. Bulk density: 82 lbs/cu ft. Dec at about 180°. Decomposition becomes self-sustaining at about 225° with spectacular swelling and evolution of heat and nitrogen, leaving Cr_2O_3. Heat of soln −23.0 cal/g. Very sol in water. Soly in water (w/w): 15.16% (0°); 26.67% (20°); 36.99% (40°); 46.14% (60°); 54.20% (80°); 60.89% (100°). Acid reaction. A 1% soln has a pH of 3.95 and a 10% soln has a pH of 3.45.

Caution: Causes skin irritation, ulceration, "chrome sores", perforation of nasal septum, pulmonary irritation.

USE: Source of pure nitrogen (esp in the laboratory); in pyrotechnics (Vesuvius fire); in lithography and photo engraving; in special mordant, catalysts, and porcelain finishes; intermediate in the manufacture of pigments; of magnetic recording materials.

545. Ammonium Dithiocarbamate. *Dithiocarbamic acid monoammonium salt;* ammonium sulfocarbamate. $CH_6N_2S_2$; mol wt 110.20. C 10.90%, H 5.49%, N 25.42%, S 58.19%. NH_2CSSNH_4. Prepd from CS_2 and NH_3: Mathes, *Inorg. Syn.* **3**, 48 (1950); Redemann *et al., Org. Syn.* coll. vol. **III**, 763 (1955); Gatlow, Hahnkamm, *Z. Anorg. Allgem. Chem.* **364**, 161 (1969). Crystal structure: Cappuchi *et al., Chem. Commun.* **1966**, 441.

Yellow, lustrous almost odorless, orthorhombic crystals when fresh. Undergoes a reversible, exothermic transition at 63°; mp 99° (dec). d_4^{20} 1.451. Sol in water. Decomposes in air, and is then no longer clearly sol; acquires an odor of H_2S; the decomposition products contain ammonium thiocyanate, ammonium sulfide, etc. *Keep in tightly closed bottles.*

USE: Instead of H_2S or $(NH_4)_2S$ for pptg metals in chemical analysis; synthesis of heterocyclic compounds.

546. Ammonium Ferric Chromate. Ferric ammonium chromate. $Cr_2FeH_4NO_8$; mol wt 305.87. Cr 34.00%, Fe 18.26%, H 1.32%, N 4.58%, O 41.85%. $NH_4Fe(CrO_4)_2$. Prepd by addn of NH_3 to an aq soln of $Fe(NO_3)_3.6H_2O$ and CrO_3: Weinland, Mergenthaler, *Ber.* **57B**, 776 (1924).

Carmine-red, microcryst powder. Practically insol in water.

547. Ammonium Ferric Citrate. Ferric ammonium citrate. Compounds of NH_3, iron, and citric acid of undetermined structure. Prepd by addn of $Fe(OH)_3$ to an aq soln of citric acid and NH_3: Kruse, Mounce, U.S. pat. **2,644,828** (1953 to Mallinckrodt); prepn of isotonic solns: Hammarlund *et al., Pharm. Acta Helv.* **35**, 593 (1960).

Ammonium ferric citrate, brown, *soluble ferric citrate.* Contains about 9% NH_3, 16.5-18.5% Fe, and about 65% hydrated citric acid. Reddish-brown granules, garnet-red transparent scales, or brownish-yellow powder. Odorless or slight NH_3 odor; saline, ferruginous taste. Very deliquesc. Reduced to ferrous salt by light. Extremely sol in water. Practically insol in alcohol. *Keep well closed and protected from light.*

Ammonium ferric citrate, green. Contains about 7.5% NH_3, 14.5-16% Fe, and about 75% hydrated citric acid. Green transparent, deliquesc scales, pearls, granules, or powder. Odorless; mild ferruginous taste. More readily reduced to the ferrous salt by light than the brown form. Very sol in water. Practically insol in alc. *Keep well closed and protected from light.*

USE: For blueprints; in photography.

THERAP CAT: Hematinic.

THERAP CAT (VET): In iron deficiency anemia.

548. Ammonium Ferric Oxalate. *Triammonium tris-[ethanedioato(2—)-O,O']ferrate(3—);* ammonium trioxalatoferrate(III); ferric ammonium oxalate. $C_6H_{12}FeN_3O_{12}$; mol

wt 374.02. C 19.27%, H 3.23%, Fe 14.93%, N 11.23%, O 51.33%. $(NH_4)_3Fe(C_2O_4)_3$. Prepn: *Gmelin's, Iron* (8th ed.) **59**, part B, pp 1020-1021 (1932).

Hydrate, bright-green, monoclinic, prismatic crystals. $d^{17.5}$ 1.78. Affected by light. Loses 3 H_2O by 100°, decomp at 160-170°. Very sol in water; practically insol in alcohol. *Protect from light.*

USE: In photography, blueprints; in coloring of Al and Al alloys.

549. Ammonium Ferric Sulfate. Ferric ammonium sulfate. $FeH_4NO_8S_2$; mol wt 266.01. Fe 20.99%, H 1.52%, N 5.27%, O 48.12%, S 24.11%. $NH_4Fe(SO_4)_2$. Prepn: *Gmelin's, Iron* (8th ed.) **59**, part B, pp 1010-1018 (1932).

Dodecahydrate, *ferric alum, iron alum.* Colorless to pale-violet, transparent, efflorescent, octahedral crystals. Odorless; acid styptic taste. mp ~37°; d 1.71. Very sol in water; practically insol in alcohol; pH of $0.1M$ aq soln 2.5.

USE: As analytical reagent; mordant in dyeing and printing textiles.

THERAP CAT: Astringent, styptic.

550. Ammonium Ferricyanide. *Triammonium hexakis(cyano-C)ferrate(3—); ammonium hexacyanoferrate(III).* $C_6H_{12}FeN_9$; mol wt 266.07. C 27.09%, H 4.55%, Fe 20.99%, N 47.38%. $(NH_4)_3Fe(CN)_6$. Prepn: *Gmelin's, Iron* (8th ed.) **59**, part B, p 1027 (1932).

Trihydrate, red crystals. Freely sol in water; practically insol in alcohol. *Protect from light.*

551. Ammonium Ferrocyanide. *Triammonium hexakis(cyano-C)ferrate(4—);* ammonium hexacyanoferrate (II). $C_6H_{16}FeN_{10}$; mol wt 284.11. C 25.37%, H 5.68%, Fe 19.66%, N 49.30%. $(NH_4)_4Fe(CN)_6$. Prepn: *Gmelin's, Iron* (8th ed.) **59**, part B, p 1024 (1932); Lux in *Handbook of Preparative Inorganic Chemistry,* Vol. 2, G. Brauer, Ed. (Academic Press, New York, 2nd ed., 1965) pp 1509-1510.

Trihydrate, yellow cryst powder. Loses NH_3 on exposure to air and light. Decomposes on heating. Freely sol in water. Practically insol in alcohol. *Protect from light.*

552. Ammonium Ferrous Sulfate. Ferrous ammonium sulfate; Mohr's salt. $FeH_8N_2O_8S_2$; mol wt 284.05. Fe 19.66%, H 2.84%, N 9.86%, O 45.06%, S 22.58%. $(NH_4)_2Fe(SO_4)_2$. Manuf from Fe, H_2SO_4 and NH_3: Demmerle *et al., Ind. Eng. Chem.* **42**, 9 (1950); from pickling waste: Brundin, U.S. pat. **2,694,657** (1954 to Ekstrand and Tholand). Toxicity study: H. F. Smyth *et al., Am. Ind. Hyg. Assoc. J.* **30**, 470 (1969).

Hexahydrate, pale blue-green crystals or crystalline powder. Slowly oxidizes and effloresces in air. d_4^{20} 1.86. Soluble in water. Practically insol in alcohol. *Keep well closed and protected from light.* LD_{50} orally in rats: 3.25 g/kg (Smyth).

USE: In photography; as analytical standard; as polymerization catalyst; in dosimeters.

553. Ammonium Fluoride. Neutral ammonium fluoride. FH_4N; mol wt 37.04. F 51.30%, H 10.89%, N 37.82%. NH_4F. Prepd by passing ammonia gas into ice-cooled 40% hydrofluoric acid or by heating 1 part NH_4Cl with 2.25 parts NaF and separatng the ammonium fluoride by sublimation: Kwasnik in *Handbook of Preparative Inorganic Chemistry* vol. 1, G. Brauer, Ed. (Academic Press, New York, 2nd ed., 1963) p 183. *Review:* Steele, "Ammonium Fluoride" in *Mellor's* vol. **VIII**, supplement I, *Nitrogen* (part 1), 370-377 (1964).

Deliquescent leaflets or needles. Hexagonal prisms by sublimation. Occurs commercially as a granular powder. d 1.015. On heating dec into NH_3 and HF. *Corrodes glass.* Soly in water (0°): 100 g/100 ml. Decomposed by hot water into NH_3 and ammonium bifluoride. $(NH_4F.HF)$. Cannot be obtained by evapn of its aq soln. Slightly sol in alcohol. The aq soln is acid. May be stored in iron vessels. *Incompat.* Quinine salts; soluble calcium salts.

Caution: Ingestion produces nausea, salivation, vomiting, abdominal pain, diarrhea, hemorrhagic gastroenteritis, muscular weakness, tremors, convulsions, vascular collapse. Increased respiration is followed by depression, death. *Chronic toxicity:* mottling of enamel, generalized osteosclerosis, calcification in tendons and ligaments; synostoses.

USE: Etching and frosting glass; as antiseptic in brewing beer; preserving wood; in printing and dyeing textiles; as mothproofing agent.

554. Ammonium Formate. *Formic acid ammonium salt.* CH_5NO_2; mol wt 63.06. C 19.05%, H 7.99%, N 22.21%, O 50.75%. HCOONH$_4$. Prepd from formic acid and NH$_3$: Zuffanti, *J. Am. Chem. Soc.* **63**, 3123 (1941); from methylformate and NH$_3$: Kelly, Cuthbert, U.S. pat. **3,122,584** (1964 to Allied Chem.).

Deliquesc crystals or granules. d 1.27. mp 116°. Sol in less than its own wt of water; sol in alc. *Keep tightly closed.*

USE: In chemical analysis, especially to ppt base metals from salts of the "noble" metals.

555. Ammonium Hexafluoroaluminate. *Triammonium hexafluoroaluminate(3—);* ammonium cryolite; ammonium aluminum fluoride; ammonium fluoaluminate. AlF$_6$-H$_{12}$N$_3$; mol wt 195.09. Al 13.83%, F 58.43%, H 6.20%, N 21.54%. (NH$_4$)$_3$AlF$_6$. Prepn from ammonium fluoride and aluminum hydroxide: v. Helmolt, *Z. Anorg. Allgem. Chem.* **3**, 127 (1893); Petersen, *J. Prakt. Chem.* [2] **40**, 55 (1889); Kwasnik in *Handbook of Preparative Inorganic Chemistry*, vol. **1**, G. Brauer, Ed. (Academic Press, New York, 2nd ed. 1963) p 236.

Cubic crystals, d 1.78. Thermally stable to above 100°. Freely sol in water. Does not attack glass.

USE: Prepn of pure aluminum fluoride.

556. Ammonium Hexafluorogallate. *Triammonium hexafluorogallate(3—);* ammonium hexafluogallate. F$_6$Ga-H$_{12}$N$_3$; mol wt 237.83. F 47.93%, Ga 29.32%, H 5.09%, N 17.67%. (NH$_4$)$_3$GaF$_6$. Prepd from Ga(OH)$_3$, HF and NH$_4$F: Hannebohn, Klemm, *Z. Anorg. Allgem. Chem.* **229**, 341 (1936); Kwasnik in *Handbook of Preparative Inorganic Chemistry*, vol. **1**, G. Brauer, Ed., (Academic Press, New York, 2nd ed., 1963) p 228.

Octahedra. On heating in air changes to Ga$_2$O$_3$; on heating *in vacuo* at 200° for several hours forms GaN.

USE: In the prepn of GaF$_3$.

557. Ammonium Hexafluorophosphate. Ammonium hexafluophosphate; ammonium phosphorus hexafluoride. F$_6$H$_4$NP; mol wt 163.00. F 69.93%, H 2.47%, N 8.59%, P 19.00%. NH$_4$PF$_6$. Prepn: Lange, Müller, *Ber.* **63**, 1063 (1930); v. Krueger, *Ber.* **65**, 1265 (1932); Woyski, *Inorg. Syn.* **3**, 111 (1950); Kwasnik in *Handbook of Preparative Inorganic Chemistry*, Vol. **1**, G. Brauer, Ed. (Academic Press, New York, 2nd ed., 1963) p 195.

Square leaflets or tables, seldom rectangular plates; cubic system. d$_4^{18}$ 2.180. Decomposes on heating to a relatively high temp without prior melting. Soly in water (20°): 74.8 g/100 ml. Sol in acetone, methanol, ethanol, methyl acetate. Does not etch glass at room temp. Slowly hydrolyzed by boiling with strong acids.

558. Ammonium Hexafluorosilicate. *Diammonium hexafluorosilicate(2—);* ammonium fluosilicate; ammonium silicofluoride. F$_6$H$_8$N$_2$Si; mol wt 178.15. F 63.98%, H 4.53%, N 15.72%, Si 15.76%. (NH$_4$)$_2$SiF$_6$. Occurs in nature as the mineral *cryptohalite*. Prepn: *Gmelin's, Ammonium* (8th ed.) **23**, pp 414-415 (1936). Crystal structure: Schlemper *et al., J. Chem. Phys.* **44**, 2499 (1966); **45**, 408 (1966). Toxicity study: Simonin, Pierron, *C.R. Soc. Biol.* **124**, 133 (1937).

Odorless crystalline powder. Two modifications at room temp: stable, cubic phase; metastable, trigonal phase. Freely sol in water. Practically insol in alcohol, acetone. LD orally in guinea pigs: 150 mg/kg (Simonin, Pierron).

Caution: Toxic symptoms similar to sodium fluoride.

USE: In pesticides; in soldering flux; etching glass.

559. Ammonium Hypophosphite. H$_6$NO$_2$P; mol wt 83.03. H 7.28%, N 16.87%, O 38.54%, P 37.31%. NH$_4$-H$_2$PO$_2$. Prepn: *Gmelin's, Ammonium* (8th ed.) **23**, 416 (1936).

Hygroscopic and deliquesc crystals or white granules. Decomposes when heated with evolution of phosphine which ignites spontaneously. One gram dissolves in about 1 ml water, 0.2 ml boiling water, 20 ml alcohol; freely sol in boiling alcohol, practically insol in acetone. The aq soln is practically neutral. *Keep well closed.*

USE: As catalyst in polyamide manuf.

560. Ammonium Iodide. H$_4$IN; mol wt 144.94. H 2.78%, I 87.55%, N 9.66%. NH$_4$I. Contains about 98% NH$_4$I; about 1% ammonium hypophosphite is usually added as a preservative. Prepd from ammonia, iodine and hydrogen peroxide; from ammonia and hydrogen iodide; or from ammonium carbonate and hydrogen iodide: Wulff, Cameron, *Z. Physik. Chem.* **B10**, 350 (1930); Schmeisser in *Handbook of Preparative Inorganic Chemistry,* vol. **1**, G. Brauer, Ed. (Academic Press, New York, 2nd ed., 1963) p 289. *Review:* Richards "Ammonium Iodide" in *Mellor's* vol. **VIII**, Supplement I, *Nitrogen* (part 1) 448-458 (1964).

White, odorless, very hygroscopic, tetragonal crystals or granular powder, sharp saline taste. Becomes yellow to brown on exposure to air and light because of liberation of iodine. When heated it partly decomposes and partly sublimes. d^{25} 2.5142. One gram dissolves in 0.6 ml water, 0.5 ml boiling water, 3.7 ml alc, 1.5 ml glycerol, 2.5 ml methanol. Except in the presence of a stabilizer, such as ammonium hypophosphite, the aq soln will quickly become yellow. The aq soln is nearly neutral to litmus. pH of 0.1M soln about 4.6. *Keep tightly closed and protected from light.*

USE: In photographic chemicals.

561. Ammonium Lactate. DL-Lactic acid ammonium salt. C$_3$H$_9$NO$_3$; mol wt 107.11. C 33.64%, H 8.47%, N 13.08%, O 44.81%. CH$_3$CH(OH)COONH$_4$. Prepd by neutralizing DL-lactic acid with NH$_4$OH: Costello, Filachione, *J. Am. Chem. Soc.* **75**, 1242 (1953).

Crystals from propanol, mp 91-94°. Sol in water, glycerol, 95% alc; slightly sol in methanol; practically insol in ethyl, *n*-propyl, isopropyl, and *n*-butyl alcohols, ether, acetone, ethyl acetate. For a 78.8% by wt soln: n_D^{20} 1.4543, n_D^{40} 1.4536, n_D^{40} 1.4503; d$_4^{20}$ 1.2006, d$_4^{25}$ 1.1984, d$_4^{40}$ 1.1904.

THERAP CAT (VET): Has been used for bovine ketosis.

562. Ammonium Magnesium Chloride. Magnesium ammonium chloride. Cl$_3$H$_4$MgN; mol wt 148.70. Cl 71.52%, H 2.71%, Mg 16.34%, N 9.42%. MgNH$_4$Cl$_3$.

Hexahydrate, deliquesc crystals. Sol in 6 parts water. *Keep well closed.*

USE: Prepn of magnesia mixture and anhydrous magnesium chloride.

563. Ammonium Mandelate. *α-Hydroxybenzeneacetic acid monoammonium salt;* mandelic acid ammonium salt. C$_8$H$_{11}$NO$_3$; mol wt 169.18. C 56.80%, H 6.55%, N 8.28%, O 28.37%. C$_6$H$_5$CH(OH)COONH$_4$. Prepd from mandelic acid and NH$_3$ or excess of strong NH$_3$-water: Tabern *et al.,* U.S. pat. **2,220,692** (1941 to Abbott); Baker, U.S. pat. **2,209,314** (1941 to Squibb).

Very deliquescent, cryst powder; odorless or with slight odor; discolors in light. Very sol in water, sparingly in alc. The aq soln is slightly acid to litmus. *Keep tightly closed and protected from light.*

THERAP CAT: Anti-infective (urinary).

THERAP CAT (VET): Has been used as a urinary antiseptic.

564. Ammonium Mercuric Chloride. Ammonium tetrachloromercurate(II); mercuric ammonium chloride. Cl$_4$-H$_8$HgN$_2$; mol wt 378.48. Cl 37.47%, H 2.13%, Hg 53.00%, N 7.40%. (NH$_4$)$_2$HgCl$_4$.

Dihydrate, powder. *Poison!* Sol in water; partly sol in alcohol.

THERAP CAT (VET): As ointment in chronic eczema. Antifungal.

565. Ammonium Molybdate(VI). Ammonium paramolybdate. H$_{24}$Mo$_7$N$_6$O$_{24}$; mol wt 1163.80. H 2.08%, Mo 57.71%, N 7.22%, O 32.99%. (NH$_4$)$_6$Mo$_7$O$_{24}$. Prepn and structure: Guiter, *Compt. Rend.* **220**, 146 (1945); Lindqvist, *Acta Chem. Scand.* **2**, 88 (1948). Toxicity studies: L. T. Fairhall *et al.,* "The Toxicity of Molybdenum," *U.S. Public Health Service Bulletin No. 293*, Washington D.C. (1945) 39 pp.

Tetrahydrate. Colorless or slightly greenish or yellowish crystals. Sol in 2.3 parts water; practically insol in alcohol. pH of 5% aq soln 5.0 to 5.5.

USE: In photography and for decorating ceramics; for detecting and determining phosphates, arsenates, lead; also as reagent for alkaloids and many other substances.

566. Ammonium Nickel Sulfate. Ammonium disulfatonickelate(II); nickel ammonium sulfate. $H_8N_2NiO_8S_2$; mol wt 286.90. H 2.81%, N 9.76%, Ni 20.46%, O 44.61%, S 22.35%. $Ni(NH_4)_2(SO_4)_2$.

Hexahydrate, bluish-green crystals; somewhat efflorescent. d 1.923. Sol in 10.3 parts water, 5 parts water at 60°. Practically insol in alcohol. pH of 0.1M aq soln 4.6.

USE: Electroplating metals.

567. Ammonium Nitrate. $H_4N_2O_3$; mol wt 80.04. H 5.04%, N 35.00%, O 59.97%. NH_4NO_3. Manuf: Faith, Keyes & Clark's *Industrial Chemicals,* F. A. Lowenheim, M. K. Moran, Eds. (Wiley-Interscience, New York, 4th ed., 1975) pp 97-102. Comprehensive description of manuf processes and physical data: B. T. Federoff *et al., Encyclopedia of Explosives and Related Items,* vol. I (Picatinny Arsenal, Dover, N.J., 1960) pp A311-A379. *Review:* Thatcher, "Ammonium Nitrate" in *Mellor's* vol. VIII, supplement I, *Nitrogen* (part 1) 506-562 (1964).

Odorless, transparent, hygroscopic, deliquesc crystals or white granules. Five solid phases exist at normal pressure. Orthorhombic at room temp. d 1.72. Dec at about 210°, mostly into H_2O and N_2O. One gram dissolves in 0.5 ml water, 0.1 ml boil. water, about 20 ml alc, about 8 ml methanol, pH of 0.1M soln in water: 5.43. *Keep well closed.*

Note: Disastrous explosions ascribed to ammonium nitrate occurred in 1947 at Texas City and at Brest.

USE: Making nitrous oxide (laughing gas); in freezing mixtures, safety explosives, matches; pyrotechnics; in fertilizers.

THERAP CAT (VET): Has been used as an expectorant, urinary acidifier.

568. Ammonium Nitroferricyanide. *Diammonium pentakis(cyano-C)nitrosylferrate(2—);* ammonium pentacyanonitrosylferrate(III); ammonium nitroprusside. C_5H_8-FeN_8O; mol wt 252.02. C 23.83%, H 3.20%, Fe 22.16%, N 44.46%, O 6.35%. $(NH_4)_2Fe(CN)_5NO$. Prepn and structure: Burrows, Turner, *J. Chem. Soc.* 115, 1429 (1919).

Red to brownish-red crystals. Freely sol in water, alc.

569. Ammonium Oleate. *(Z)-9-Octadecenoic acid ammonium salt; oleic acid ammonium salt;* ammonia soap. $C_{18}H_{37}NO_2$; mol wt 299.50. C 72.19%, H 12.45%, N 4.68%, O 10.68%. $CH_3(CH_2)_7CH=CH(CH_2)_7COONH_4$. Prepd from oleic acid and excess 28-30% NH_3 soln: Stumpf, *Am. Paint J.* 38, no. 45, 60, 64, 68 (1954), *C.A.* 48, 12428b (1954).

Yellowish-brown paste, softens at 50-55°F; mp 70-72°F. At 80°F: sol in water; slightly sol in acetone, ethanol, methanol, benzene, CCl_4, xylene, naphtha. Sol in water at 212°F, acetone at 134°F, ethanol at 172°F, methanol at 148°F, benzene at 176°F, CCl_4 at 170°F, xylene at 180°F, naphtha at 160°F.

USE: Detergent, solidifying alcohol, demonstrating liquid crystals according to Lehmann.

570. Ammonium Osmium Chloride. Ammonium hexachloroosmate(IV); ammonium chloroosmate; osmium ammonium chloride. $Cl_6H_8N_2Os$, mol wt 439.02. Cl 48.45%, H 1.84%, N 6.38%, Os 43.33%. $(NH_4)_2OsCl_6$. Prepn: Dwyer, Hogarth, *Inorg. Syn.* 5, 206 (1957).

Red powder, or dark red, octahedral crystals. Soluble in water or alcohol.

571. Ammonium Oxalate. *Ethanedioic acid diammonium salt.* $C_2H_8N_2O_4$; mol wt 124.10. C 19.36%, H 6.50%, N 22.57%, O 51.57%. $NH_4OOCCOONH_4$. Commercial product is 98-99% pure. Prepn of monohydrate from aq oxalic acid and NH_3 or ammonium carbonate: Bérard, *Ann. Chim.* 73, 277 (1810). *Review: Gmelin's, Ammonium* (8th ed.) 23, pp 400-407 (1936). Crystal structure of monohydrate: Robertson, *Acta Cryst.* 18, 410, 417 (1965).

Monohydrate, orthorhombic, odorless crystals or granules. *Poisonous!* d 1.50. One gram dissolves in 20 ml water, 2.6 ml boil water; slightly sol in alcohol. The aq soln is practically neutral. pH of 0.1 M soln 6.4.

USE: Manuf explosives; electrolytic detinning of iron; in dyeing, metal polishes; for detection and determination of Ca, Pb, and rare earth metals.

572. Ammonium Palmitate. *Hexadecanoic acid ammonium salt; palmitic acid ammonium salt.* $C_{16}H_{35}NO_2$; mol wt 273.46. C 70.28%, H 12.90%, N 5.12%, O 11.70%. $CH_3(CH_2)_{14}COONH_4$. Prepn from palmitic acid and excess 28-30% NH_3 soln: Stumpf, *Am. Paint J.* 38, No. 45, 60, 64, 68 (1954), *C.A.* 48, 12428b (1954); from palmitic acid and ammonium carbonate: Reiling, U.S. pat. 3,053,867 (1962 to Boston Chem. Prods.).

Yellow-white powder, softens at 38-40°F, mp 70-73°F. At 80°F: sol in water; slightly sol in benzene, xylene; practically insol in acetone, ethanol, methanol, CCl_4, naphtha. Sol in water at 212°F, acetone at 134°F, ethanol at 172°F, methanol at 148°F, benzene at 176°F, CCl_4 at 170°F, xylene at 180°F, naphtha at 160°F.

USE: Waterproofing fabrics, thickening lubricants.

573. Ammonium Pentachlorozincate. Zinc ammonium chloride. $Cl_5H_{12}N_3Zn$; mol wt 296.77. Cl 59.73%, H 4.08%, N 14.16%, Zn 22.03%. $ZnCl_2.3NH_4Cl$. Prepn: Klug, Alexander, *J. Am. Chem. Soc.* 66, 1056 (1944).

Hygroscopic, orthorhombic, bipyramidal crystals. d 1.81. Sublimes at 340° without melting if absolutely dry. Very sol in water.

USE: In manuf of dry cells; as flux for welding, soldering, galvanizing.

574. Ammonium Perchlorate. ClH_4NO_4; mol wt 117.49. Cl 30.18%, H 3.43%, N 11.92%, O 54.47%. NH_4-ClO_4. Prepn: *Gmelin's, Ammonium* (8th ed.) 23, pp 196-200 (1936). Review of decompn and combustion: Jacobs, Whitehead, *Chem. Rev.* 69, 551-590 (1969).

Orthorhombic crystals. Dec on heating. d 1.95. Freely sol in water; sol in methanol; slightly sol in ethanol, acetone; almost insol in ethyl acetate, ether.

USE: In explosives, pyrotechnic compositions, jet and rocket propellants.

575. Ammonium Peroxydisulfate. Ammonium persulfate. $H_8N_2O_8S_2$; mol wt 228.20. H 3.53%, N 12.28%, O 56.09%, S 28.10%. $(NH_4)_2S_2O_8$. Available oxygen 7.01%. Contains, when recently made, 95 to 98% $(NH_4)_2S_2O_8$. Prepd by anodic oxidation of a satd $(NH_4)_2SO_4$ soln: Feher in *Handbook of Preparative Inorganic Chemistry,* Vol. 1, G. Brauer, Ed. (Academic Press, New York, 2nd ed., 1963) p 390. Toxicity study: Smyth *et al., Am. Ind. Hyg. Assoc. J.* 30, 470 (1969).

Odorless platelike or prismatic (monoclinic) crystals, or white granular powder. Stable for months when pure and dry; dec in presence of moisture, gradually evolving ozonecontg oxygen; dec on heating, evolving O_2 and forming $(NH_4)_2S_2O_7$. d 1.98. Strong oxidizing agent. Freely sol in water; aq soln is acid and dec slowly at room temp and rapidly at higher temp evolving O_2 and forming NH_4HSO_4. *Keep dry, in a cool place and protected from organic matter.* LD_{50} orally in rats: 820 mg/kg (Smyth).

USE: As oxidizer and bleacher; to remove hypo; reducer and retarder in photography; in dyeing, manuf aniline dyes; oxidizer for copper; etching zinc; decolorizing and deodorizing oils; electroplating; washing infected yeast; removing pyrogallol stains; making soluble starch; depolarizer in electric batteries; in anal. chemistry chiefly for detection and determination of manganese.

576. Ammonium Phosphate, Dibasic. Secondary ammonium phosphate; diammonium hydrogen phosphate; Fyrex. $H_9N_2O_4P$; mol wt 132.06. H 6.87%, N 21.21%, O 48.46%, P 23.45%. $(NH_4)_2HPO_4$. The grade used medicinally is 98-99% pure. Prepn: *Gmelin's, Ammonium* (8th ed.) 23, pp 422-426 (1936).

Odorless crystals or cryst powder; saline, cooling taste; gradually loses about 8% NH_3 on exposure to air. One gram dissolves in 1.7 ml water, 0.5 ml boil water; practically insol in alcohol, acetone. pH about 8. *Keep well closed.*

USE: Fireproofing textiles, paper, wood, and vegetable fibers; impregnating lamp wicks; preventing afterglow in matches; flux for soldering tin, copper, brass, and zinc; purifying sugar; in yeast cultures; in dentifrices; in corrosion inhibitors; in fertilizers.

577. Ammonium Phosphate, Monobasic. Ammonium biphosphate; ammonium dihydrogen phosphate; primary ammonium phosphate. H_6NO_4P; mol wt 115.03. H 5.26%, N 12.18%, O 55.64%, P 26.93%. $(NH_4)H_2PO_4$. Prepn: *Gmelin's, Ammonium* (8th ed.) **23**, pp 426-429 (1936).

Odorless crystals or white, cryst powder; stable in air. d 1.80. One gram dissolves in about 2.5 ml water; slightly sol in alcohol; practically insol in acetone. pH of 0.2 molar aq soln 4.2.

USE: As baking powder with sodium bicarbonate; in fermentations (yeast cultures, etc.); fireproofing of paper, wood, fiberboard, etc.

578. Ammonium Phosphite. $H_9N_2O_3P$; mol wt 116.06. H 7.82%, N 24.14%, O 41.36%, P 26.69%. $(NH_4)_2$-HPO_3. Prepn: *Gmelin's, Ammonium* (8th ed.) **23**, pp 416-417 (1936).

Monohydrate, deliquesc crystals. Sol in water. *Keep tightly closed.*

USE: As reducing agent; corrosion inhibitor for lubricating grease.

579. Ammonium Phosphomolybdate. *Ammonium molybdophosphate.* $H_{12}Mo_{12}N_3O_{40}P$; mol wt 1876.35. H 0.64%, Mo 61.36%, N 2.24%, O 34.11%, P 1.65%. $(NH_4)_3$-$PO_4 \cdot 12MoO_3$. Prepn and constitution: Illingworth, Keggin, *J. Chem. Soc.* **1935**, 575; Thistlewaite, *Analyst* **72**, 531 (1947); Healy, *Radiochim. Acta* **3**, 100 (1964).

Yellow, heavy, cryst powder. Solubility in water (20°): 0.2 ±0.1 g/l. Practically insol in nitric acid; sol in fixed alkali hydroxides.

USE: In phosphorus analysis; as cation-exchanger.

580. Ammonium Phosphotungstate. *Ammonium tungstophosphate;* ammonium phosphowolframate. $H_{12}N_3$-$O_{40}PW_{12}$; mol wt 2931.15. H 0.41%, N 1.43%, O 21.83%, P 1.06%, W 75.26%. $(NH_4)_3PO_4 \cdot 12WO_3$. Prepn: Wu, *J. Biol. Chem.* **43**, 189 (1920); Healy. *Radiochim. Acta* **2**, 146 (1964).

Dihydrate, microcryst powder. Soly in water (20°): 0.15 g/l. Freely sol in fixed alkali hydroxide solns.

USE: As ion-exchanger.

581. Ammonium Picrate. *2,4,6,-Trinitrophenol ammonium salt; picric acid ammonium salt;* ammonium picronitrate; ammonium carbazoate. $C_6H_6N_4O_7$; mol wt 246.14. C 29.28%, H 2.46%, N 22.76%, O 45.50%. Prepn: Berl, Berl, U.S. pat. **2,350,322** (1944). Crystal structure: Maartmann-Moe, *Acta. Cryst.* **25B**, 1452 (1969).

Bright yellow, bitter scales or orthorhombic crystals. *Explodes easily from heat or shock.* d 1.72. Soly in water at 20°: about 1 g/100 ml. Slightly sol in alc. "Red modification" is not a distinct polymorph, but a slightly contaminated form of the yellow salt: Mitchell, Bryant, *J. Am. Chem. Soc.* **65**, 128 (1943). Physiologic effects and protective measures in cases of exposure to ammonium picrate: Foulger, *U.S. Armed Forces Med. J.* **4**, 1425 (1953).

USE: In explosives, fireworks, rocket propellants.

582. Ammonium Platinic Chloride. Ammonium hexachloroplatinate(IV); ammonium chloroplatinate; platinic ammonium chloride. $Cl_6H_8N_2Pt$; mol wt 443.87. Cl 47.92%, H 1.82%, N 6.31%, Pt 43.95%. $(NH_4)_2PtCl_6$.

Orange-red crystals or yellow powder. d 3.06. Slightly sol in water. Practically insol in alcohol.

USE: Platinum plating; manuf spongy platinum.

583. Ammonium Platinous Chloride. Ammonium tetrachloroplatinate(II); ammonium chloroplatinite; ammonium platinochloride; platinous ammonium chloride. Cl_4-H_8N_2Pt; mol wt 372.97. Cl 38.02%, H 2.16%, N 7.51%, Pt 52.30%. $(NH_4)_2PtCl_4$.

Dark ruby-red crystals. Sol in water.

USE: In photography.

584. Ammonium Salicylate. *2-Hydroxybenzoic acid monoammonium salt; salicylic acid monoammonium salt;* Salicyl-Vasogen. $C_7H_9NO_3$; mol wt 155.15. C 54.19%, H 5.85%, N 9.03%, O 30.94%. Prepd from salicylic acid and aq NH_3: Cahours, *Ann.* **52**, 336 (1844).

Odorless, lustrous crystals or white, cryst powder. Discolors on exposure to light; loses some NH_3 on long exposure to air. Readily discolored by iron compounds. One gram dissolves in 1 ml water, 3 ml alcohol. The aq soln is slightly acid. *Keep in the dark and protected from contamination by iron.*

THERAP CAT: Analgesic; topically to loosen psoriatic scales.

585. Ammonium Selenate. $H_8N_2O_4Se$; mol wt 179.03. H 4.50%, N 15.65%, O 35.75%, Se 44.10%. $(NH_4)_2SeO_4$. Prepd by treating a soln of selenic acid with ammonia: Retgers, *Z. Phys. Chem.* **8**, 36 (1891); King, *J. Phys. Chem.* **41**, 797 (1937). Crystal structure: Gatlow, *Acta Cryst.* **15**, 419 (1962).

Monoclinic crystals. d_4^{20} 2.194. Decomposed by heat. 117 parts dissolve in 100 parts water at 7°, 197 parts in 100 parts at 160°; sol in glacial acetic acid. Insol in alcohol, acetone, ammonia.

586. Ammonium Selenite. $H_8N_2O_3Se$; mol wt 163.04. H 4.95%, N 17.18%, O 29.44%, Se 48.43%. $(NH_4)_2SeO_3$. Prepd by dissolving selenious acid in a slight excess of concd aqueous ammonia and evaporating: Berzelius, *Acad. Handl. Stockholm* **39**, 13 (1818); Nilson, *Bull. Soc. Chim.* [2] **21**, 253 (1874); *ibid.* [2] **23**, 262 (1875).

White or slightly reddish crystals. Sol in water. Dec by heat. Deliquescent.

USE: In manuf red glass; reagent for alkaloids.

587. Ammonium Sodium Phosphate. *Monoammonium, monosodium phosphoric acid salt;* sodium ammonium phosphate; microcosmic salt; salt of phosphorus; phosphorsalz. H_5NNaO_4P; mol wt 137.01. H 3.68%, N 10.22%, Na 16.78%, O 46.71%, P 22.61%. $NaNH_4HPO_4$.

Tetrahydrate, odorless, monoclinic crystals or granules. Efflorescent in air with loss of some NH_3. d 1.544. mp ~80° when rapidly heated. Prolonged heating produces $NaPO_3$. Sol in about 5 parts cold, 1 part boiling water. Practically insol in alcohol. pH of 5% aq soln 7.8 to 8.2.

USE: As reagent for determination of Mg, Zn, and in blowpipe analysis. For standardizing uranium solns. In the prepn of ammonium phosphate and ammonium molybdate stock solns.

588. Ammonium Stearate. *Octadecanoic acid ammonium salt; stearic acid ammonium salt.* $C_{18}H_{39}NO_2$; mol wt 301.51. C 71.70%, H 13.04%, N 4.65%, O 10.61%. CH_3-$(CH_2)_{16}COONH_4$. Prepn from stearic acid and excess 28-30% NH_3 soln: Stumpf, *Am. Paint J.* **38**, No. 45, 60, 64, 68 (1954), *C.A.* **48**, 12428b (1954); from stearic acid and ammonium carbonate: Reiling, U.S. pat. **3,053,867** (1962 to Boston Chem. Prod.).

Yellow-white powder, softens at 35-40°F; mp 70-75°F. At 80°F: sol in methanol, ethanol; slightly sol in water, benzene, xylene, naphtha; practically insol in acetone, CCl_4. Sol in water at 212°F, acetone at 134°F, ethanol at 172°F, methanol at 148°F, benzene at 176°F, CCl_4 at 170°F, xylene at 180°F, naphtha at 160°F.

USE: In vanishing creams, in waterproofing cements.

589. Ammonium Sulfamate. *Sulfamic acid monoammonium salt;* AMS; Amcide; Ammate. H_6NO_3S; mol wt 114.13. H 5.30%, N 24.55%, O 42.06%, S 28.10%. NH_4-SO_3NH_2. Weed-killing prepn: Cupery, Tanberg, U.S. pat. **2,277,744** (1941 to du Pont). Prepn from ammonia and sulfamic acid: Sisler, Audrieth, *Inorg. Syn.* **2**, 180 (1946).

Toxicity study: Ambrose, *J. Ind. Hyg. Toxicol.* **25**, 26 (1943).

Hygroscopic crystals (large plates). mp 131°, dec 160°. Extremely soluble in water, liquid NH_3; slightly soluble in ethanol; moderately soluble in glycerol, glycol, formamide. pH of $0.27M$ soln in H_2O = 4.9. Aq solns are stable to boiling. "Ammate" brand weed killer is a brownish-gray crystalline, hygroscopic material contg a minimum of 80% ammonium sulfamate. LD_{50} orally in rats: 3.0 g/kg (Ambrose).

Caution: Potential symptoms of overexposure are irritation of eyes, nose and throat; coughing, difficult breathing. See *NIOSH Pocket Guide to Chemical Hazards* (DHHS/NIOSH 90-117, 1990) p 38.

USE: In the manuf of fire-retardant compositions, for flameproofing textiles and paper products; in the manuf of weed killing compositions; in electroplating solns; for the generation of nitrous oxide gas.

590. Ammonium Sulfate. *Sulfuric acid diammonium salt;* mascagnite. $H_8N_2O_4S$; mol wt 132.14. H 6.10%, N 21.20%, O 48.43%, S 24.27%. $(NH_4)_2SO_4$. The "pure" grade of commerce contains at least 99% $(NH_4)_2SO_4$. Prepn: *Gmelin's, Ammonium* (8th ed.) **23**, 261-280 (1936). Manuf: Faith, Keyes & Clark's *Industrial Chemicals,* F. A. Lowenheim, M. K. Moran, Eds. (Wiley-Interscience, New York, 4th ed., 1975) pp 103-108. *Review:* Call, "Ammonium Sulfate" in *Mellor's* **Vol. VIII,** Supplement I, *Nitrogen* (part 1) 473-505 (1964). Solubility studies: A. C. D. Rivett, *J. Chem. Soc.* **121**, 379 (1922); R. M. Caven, T. C. Mitchell, *ibid.* **125**, 1428 (1924).

Odorless, orthorhombic crystals or white granules. d 1.77. Dec above 280°. Soly in water (g $(NH_4)_2SO_4$ per 100 g of satd soln): 41.22 (0°); 43.47 (25°); 50.42 (100°) (Rivett). Soly in water (g/100 g H_2O): 70.6 (0°); 76.7 (25°); 103.8 (100°) (Caven, Mitchell) . Insol in alcohol, acetone. pH of 0.1 molar aq soln 5.5.

USE: Manuf ammonia alum; in the manuf of H_2SO_4 to free it from nitrogen oxides; analytical uses; freezing mixtures; flameproofing fabrics and paper; manuf viscose silk; tanning, galvanizing iron; in fractionation of proteins. The commercial grade is used as fertilizer.

591. Ammonium Sulfide. True ammonium sulfide. H_8N_2S; mol wt 68.14. H 11.83%, N 41.11%, S 47.06%. $(NH_4)_2S$. Prepn: *Gmelin's, Ammonium* (8th ed.) **23**, p 246 (1936). Prepn of high purity soln: Johnson *et al., Chemist-Analyst* **53**, 46 (1964).

Forms crystals below $-18°$; at high temps it dec into NH_4HS, NH_3, polysulfides, etc.

USE: To apply patina to bronze; in photographic developers, in textile manufacture, in trace metal analysis.

592. Ammonium Sulfide Solution, Red. Ammonium polysulfide soln. Contains about 8% NH_3 and 22% sulfur, corresponding to about 30% $(NH_4)_2S_3$.

Clear, red liquid; odor of NH_3 and H_2S. d $\sim$1.10. Miscible with water. Strong alkaline reaction. *Keep tightly closed and in a cool place.*

593. Ammonium Sulfide Solution, Yellow. Ammonium bisulfide soln; ammonium sulfhydrate soln; ammonium hydrosulfide soln. Contains 8-9% NH_3 and about the same percentage of sulfur, corresponding to 16-20% as $(NH_4)_2S$.

When freshly made it is an almost colorless liquid, but soon acquires a yellow color which intensifies with age, due to formation of some polysulfide. Odor of NH_3 and H_2S; strong alkaline reaction. *Keep tightly closed.*

594. Ammonium Sulfite. $H_8N_2O_3S$; mol wt 116.14. H 6.94%, N 24.12%, O 41.33%, S 27.61%. $(NH_4)_2SO_3$. Prepn: *Gmelin's, Ammonium* (8th ed.) **23**, pp 256-258 (1936).

Monohydrate, efflorescent crystals. Under the influence of air and heat loses all its water of crystallization and is gradually oxidized to $(NH_4)_2SO_4$. Sol in water; almost insol in alcohol, acetone. Its aq soln is alkaline to litmus. *Keep tightly closed and in a cool place.*

USE: In photography; as reducing agent; in bricks for blast-furnace linings; in lubricants for metal cold-working.

595. Ammonium Tetrachloroaluminate. Aluminum ammonium chloride; ammonium chloroaluminate. $AlCl_4$-H_4N; mol wt 186.83. Al 14.44%, Cl 75.90%, H 2.16%, N 7.50%. NH_4AlCl_4. Prepn from $AlCl_3$ and NH_4Cl: Friedman, Taube, *J. Am. Chem. Soc.* **72**, 2236 (1950).

mp 304°. Sol in water and ether.

USE: In the processing of furs.

596. Ammonium Tetrachlorozincate. $Cl_4H_8N_2Zn$; mol wt 243.28. Cl 58.29%, H 3.31%, N 11.52%, Zn 26.88%. $(NH_4)_2ZnCl_4$. Prepd by dissolving 70 g $ZnCl_2$ and 30 g NH_4Cl in 29 ml hot water and crystallizing: Meerburg, *Z. Anorg. Allgem. Chem.* **37**, 199 (1903); Wagenknecht, Juza in *Handbook of Preparative Inorganic Chemistry,* **Vol. 1,** G. Brauer, Ed. (Academic Press, New York, 2nd ed., 1963) p 1072.

White, thin, shiny platelets. Orthorhombic bipyramidal. Hygroscopic. d 1.879. mp $\sim$150° (dec). Sublimes at 341° without melting, if absolutely dry. Very sol in water with absorption of heat.

597. Ammonium Thiocyanate. *Thiocyanic acid ammonium salt;* ammonium rhodanide; ammonium sulfocyanate; ammonium sulfocyanide. CH_4N_2S; mol wt 76.12. C 15.78%, H 5.30%, N 36.80%, S 42.12%. NH_4SCN. The usual grade is 98-99% pure. Prepn: *Gmelin's, Ammonium* (8th ed.) **23**, pp 372-383 (1936).

Deliquesc crystals. mp $\sim$149°. Freely sol in water, ethanol; sol in methanol, acetone. Practially insol in $CHCl_3$, ethyl acetate. *Keep tightly closed.*

USE: In matches; double-dyeing fabrics; photography; improving and increasing strength of silks weighted with tin salts; producing grayish-black coating on Zn; manuf transparent artificial resins, thiourea; in pesticides. Detection and determn of small quantities of Fe; determn of Ag, Hg, etc.

598. Ammonium Thiosulfate. Ammonium hyposulfite. $H_8N_2O_3S_2$; mol wt 148.21. H 5.44%, N 18.90%, O 32.39%, S 43.27%. $(NH_4)_2S_2O_3$. Prepn: *Gmelin's, Ammonium* (8th ed.) **23**, pp 304-306 (1936). Crystal structure: S. T. Teng *et al., Acta Crystallogr.* **B35**, 1682 (1979).

Crystals, dec at 150°. Sol in water; insol in alc, ether.

USE: To clean "white" metal; in photography; in lubricants for metal cold-working.

599. Ammonium Titanium Oxalate. Ammonium oxodioxalatotitanate(IV); titanium ammonium oxalate; titanyl ammonium oxalate. $C_4H_8N_2O_9Ti$; mol wt 275.98. C 17.41%, H 2.92%, N 10.15%, O 52.18%, Ti 17.34%. $(NH_4)_2TiO(C_2O_4)_2$.

Monohydrate, crystals or crystalline powder. Very sol in water.

USE: As mordant in dyeing cotton and leather.

600. Ammonium Tungstate(VI). Ammonium paratungstate. $H_{40}N_{10}O_{41}W_{12}$; mol wt 3042.44. H 1.33%, N 4.60%, O 21.56%, W 72.51%. $(NH_4)_{10}W_{12}O_{41}$. Prepn: Pilloton, U.S. pat. **3,077,379** (1963 to Union Carbide).

Pentahydrate, plates. Freely sol in water; practically insol in alcohol.

USE: Manuf of tungsten alloys.

601. Ammonium Uranate(VI). Ammonium diuranate. $H_8N_2O_7U_2$; mol wt 624.13. H 1.29%, N 4.49%, O 17.94%, U 76.28%. $(NH_4)_2U_2O_7$. Incorrectly called *"Uranium Yellow";* the true uranium yellow is sodium uranate. Prepn: Miller, Armstrong, U.S. pat. **2,466,118** (1949 to USAEC); Tridot, *Ann. Chim.* [12] **5**, 358 (1950).

Reddish-yellow, amorphous powder. Practically insol in water, alkalies; sol in acids or in ammonium carbonate soln.

USE: Painting black on porcelain; as reagent for alkaloids; prepn of UO_2 and UF_4.

602. Ammonium Uranium Carbonate. Ammonium dioxotricarbonatouranate(VI); uranium ammonium carbonate; uranyl ammonium carbonate. $C_3H_{16}N_4O_{11}U$; mol wt 522.21. C 6.90%, H 3.09%, N 10.73%, O 33.70%, U 45.58%. May contain $2H_2O$.

Yellow crystals; decomposed on exposure to air. Sol in water.

USE: In uranium yellow glazes.

603. Ammonium Uranium Fluoride. *Triammonium pentafluorodioxouranate;* ammonium dioxopentafluorouranate(VI); uranium ammonium fluoride; uranyl ammonium fluoride. $F_5H_{12}N_3O_2U$; mol wt 419.14. F 22.66%, H 2.89%, N 10.03%, O 7.63%, U 56.79%. $UO_2(NH_4)_3F_5$. Usually contains some water of crystallization. Prepn and crystal structure: Nguyen-Quy-Dao, *Bull. Soc. Chim. France* **1968**, 3543; *idem, Acta. Cryst.* **25B**, 67 (1969).

Greenish-yellow, monoclinic, cryst powder. Freely sol in water; practically insol in alcohol.

USE: Has been used in x-ray work because of its fluorescence under these rays.

604. Ammonium Valerate. *Pentanoic acid ammonium salt; valeric acid ammonium salt;* ammonium valerianate. $C_5H_{13}NO_2$; mol wt 119.16. C 50.40%, H 11.00%, N 11.75%, O 26.85%. $CH_3(CH_2)_3COONH_4$. Prepn: *Hager's Handb. Pharm. Praxis* Vol. II (Springer-Verlag, Berlin, 1969) p 1060; Zuffanti, *J. Am. Chem. Soc.* **63**, 3123 (1941).

Very hygroscopic crystals, mp 108°. Very readily sol in water, alc; sol in ether. *Keep tightly closed.*

THERAP CAT: Formerly as sedative.

605. Ammonium Vanadate(V). Ammonium metavanadate. H_4NO_3V; mol wt 116.98. H 3.45%, N 11.97%, O 41.03%, V 43.55%. NH_4VO_3. Prepn: Baker *et al., Inorg. Syn.* **3**, 117 (1950). Toxicity study: H. F. Smyth *et al., Am. Ind. Hyg. Assoc. J.* **30**, 470 (1969).

White or slightly yellow, crystalline powder. Sol in 165 parts water; more sol in hot water, in dil ammonia. Loses water and ammonia on heating. LD_{50} orally in rats: 0.16 g/kg (Smyth).

USE: In dyeing and printing on woolens; staining wood black; manuf vanadium black and "indelible ink"; producing vanadium luster on pottery; as photographic developer; in hematoxylin staining in microscopy; as reagent in analytical chemistry.

606. Ammonium Zirconyl Carbonate. Ammonium tricarbonatozirconate. $C_3H_{13}N_3O_{10}Zr$; mol wt 342.37. C 10.52%, H 3.83%, N 12.27%, O 46.73%, Zr 26.64%. $(NH_4)_3$-$ZrOH(CO_3)_3$. Prepd from $ZrO_2CO_2.8H_2O$, NH_4HCO_3 and $(NH_4)_2CO_3$: Blumenthal, *J. Chem. Ed.* **39**, 604 (1962).

Dihydrate, large prisms from water. Unstable in air, gradually evolving carbon dioxide and ammonia. Sol in water. Marketed as aq soln: d_{24}^{24} 1.238. The aq soln is moderately stable at room temp, although, after standing a month or so, it is likely to deposit hydrous zirconia. At temps above 60° aq solns dec rapidly.

USE: Water-repellent. Leaves a zirconium residue after the evaporation of ammonia and carbon dioxide, without leaving other cations or anions.

607. Amobarbital. *5-Ethyl-5-(3-methylbutyl)-2,4,6-(1H,3H,5H)-pyrimidinetrione; 5-ethyl-5-isopentylbarbituric acid;* 5-ethyl-5-isoamylbarbituric acid; barbamil; amylobarbitone; 5-isoamyl-5-ethylbarbituric acid; pentymal; Somnal; Dormytal; Isomytal; Eunoctal; Amal; Mylodorm; Sednotic; Amasust; Stadadorm; Amytal. $C_{11}H_{18}N_2O_3$; mol wt 226.28. C 58.39%, H 8.02%, N 12.38%, O 21.21%. Prepn: U.S. pat. **1,514,573** (1924). Metabolism: Frey, Magnussen, *Arzneimittel-Forsch.* **16**, 612 (1966). Toxicity data: K. Irrgang, *ibid.* **15**, 688 (1965). Comprehensive description: N. A. A. Mian *et al., Anal. Profiles Drug Subs.* **19**, 27-58 (1990).

Slightly bitter crystals, mp 156-158°. One gram dissolves in 1300 ml water, in 5 ml alc, in 17 ml chloroform, in 6 ml ether. Freely sol in benzene; sol in alkaline solns. Insol in

petr ether, aliphatic hydrocarbons. A satd aq soln is acid to litmus paper. Dissolves in aq solns of alkali hydroxides and carbonates. pH of a saturated soln in water about 5.6. pKa (25°C) 8.0. LD_{50} in mice (mg/kg): 212 s.c. (Irrgang).

Sodium salt, $C_{11}H_{17}N_2NaO_3$, *Amytal Sodium, Amsebarb, Barbamyl, Dorminal, Inmetal.* Hygroscopic, friable granules of powder. Slightly bitter taste. Very soluble in water; sol in alcohol (1:1). Practically insol in ether.

Caution: May be habit forming. This is a controlled substance (depressant) listed in the U.S. Code of Federal Regulations, Title 21 Parts 329.1, 1308.12 and 1308.13 (1995).

THERAP CAT: Sedative, hypnotic.

THERAP CAT (VET): Sedative, hypnotic.

608. Amocarzine. *4-Methyl-N-[4-[(4-nitrophenyl)-amino]phenyl]-1-piperazinecarbothioamide;* 4-nitro-4'-[(N-methylpiperazinyl)thiocarbonylamino]diphenylamine; CGP-6140. $C_{18}H_{21}N_5O_2S$; mol wt 371.46. C 58.20%, H 5.70%, N 18.85%, O 8.61%, S 8.63%. Orally active macrofilaricide; derivative of amoscanate, *q.v.* Prepn: **Belg. pat. 772,053** (1971 to Agripat); R. Spaun *et al., U.S. pat.* **3,781,290** (1973 to Ciba-Geigy). Mechanism of action study: K. P. Davies *et al., Exp. Parasitol.* **68**, 382 (1989). HPLC determn in biological fluids: S. C. Bhatia *et al., J. Chromatog.* **434**, 288 (1988). Pharmacokinetics: J. B. Lecaillon *et al., Brit. J. Clin. Pharmacol.* **30**, 625, 629 (1990). Clinical trial in onchocerciasis: A. A. Poltera *et al., Lancet* **337**, 583 (1991).

mp 191-196°. Sol in acetonitrile.

THERAP CAT: Anthelmintic (Nematodes).

609. Amodiaquin. *4-[(7-Chloro-4-quinolinyl)amino]-2-[(diethylamino)methyl]phenol; 4-[(7-chloro-4-quinolyl)-amino]-α-(diethylamino)-o-cresol;* 7-chloro-4-(3-diethylaminomethyl-4-hydroxyanilino)quinoline; 7-chloro-4-(3-diethylaminomethyl-4-hydroxyphenylamino)quinoline; 4-(3'-diethylaminomethyl-4'-hydroxyanilino)-7-chloroquinoline; SN-10751. $C_{20}H_{22}ClN_3O$; mol wt 355.87. C 67.50%, H 6.23%, Cl 9.96%, N 11.81%, O 4.50%. Prepd from 4,7-dichloroquinoline and 4-acetamido-α-diethylamino-o-cresol: Burckhalter *et al., J. Am. Chem. Soc.* **70**, 1363 (1948). U.S. pats. **2,474,819; 2,474,821** (1949 to Parke, Davis). Alternate synthesis from 2-aminomethyl-p-aminophenol and 4,7-dichloroquinoline: Natarajan, Lan, *Arzneimittel-Forsch.* **22**, 1230 (1972). Spectral study: J. Kracmar *et al, Pharmazie* **29**, 773 (1974). HPLC determn in plasma, blood and urine: E. Pussard *et al., J. Chromatog.* **374**, 111 (1986). Comprehensive description: I. Ahmad *et al.* in *Analytical Profiles of Drug Substances and Excipients* vol. 21, H. G. Brittain, Ed. (Academic Press, San Diego, 1992) pp 43-73.

Crystals from abs. ethanol, mp 208° (dec).

Dihydrochloride dihydrate, $C_{20}H_{22}ClN_3O.2HCl.2H_2O$, *CAM-AQ1, Camoquin, Flavoquine, Miaquin.* Yellow, bitter crystals, dec 150-160°. uv max (methanol): 342 nm ($E_{1cm}^{1\%}$ 349); (water): 341.5 nm ($E_{1cm}^{1\%}$ 389); (0.1N HCl): 342 nm ($E_{1cm}^{1\%}$ 396). Sol in water; sparingly sol in alc; very slightly sol in benzene, chloroform, ether. pH of 1% aq soln 4.0 to 4.8.

Dihydrochloride hemihydrate, yellow crystals from methanol, mp 243°. Slightly sol in water, alc.
THERAP CAT: Antimalarial.

610. Amolanone. *3-[2-(Diethylamino)ethyl]-3-phenyl-2(3H)-benzofuranone;* 3-(2-diethylaminoethyl)-2-oxo-3-phenyl-2,3-dihydrobenzofuran; γ-diethylamino-α-(o-hydroxyphenyl)-α-phenylbutyric acid lactone; amocaine; AP-43; Amethone. $C_{20}H_{23}NO_2$; mol wt 309.41. C 77.64%, H 7.49%, N 4.53%, O 10.34%. Prepn: Weston, Brownell, *J. Am. Chem. Soc.* **74**, 653 (1952).

Crystals from petr ether, mp 43-44°. bp$_{2.0}$ 192-194°. $n_D^{24.5}$ 1.5614.
Hydrochloride, $C_{20}H_{23}NO_2$·HCl, crystals from isopropanol, mp 152-153°. Sol in water.
THERAP CAT: Anesthetic (local).

611. Amoproxan. *3,4,5-Trimethoxybenzoic acid 1-[(3-methylbutoxy)methyl]-2-(4-morpholinyl)ethyl ester; 3,4,5-trimethoxybenzoic acid 1-[(isopentyloxy)methyl]-2-morpholinoethyl ester;* α-[(isopentyloxy)methyl]-4-morpholineethanol 3,4,5-trimethoxybenzoate; 4-[3-(isoamyloxy)-2-(3,4,5-trimethoxybenzoyloxy)propyl]tetrahydro-1,4-oxazine; 3,4,5-trimethoxybenzoic acid 2-(isopentyloxy)-1-(morpholinomethyl)ethyl ester. $C_{22}H_{35}NO_7$; mol wt 425.52. C 62.10%, H 8.29%, N 3.29%, O 26.32%. Prepn: Mauvernay *et al.*, **Brit. pat. 1,211,307** corresp to **U.S. pat. 3,781,432** (1970, 1973 to C.E.R.M.) and **U.S. pat. 3,790,569** (1974). Pharmacology: Duchene-Marullaz *et al.*, *Therapie* **24**, 665 (1969); Pernod *et al.*, *ibid.* 675; Roulet *et al.*, *ibid.* **27**, 93 (1972).

Hydrochloride, $C_{22}H_{35}NO_7$·HCl, *CERM-730, Mederel.* Crystals from anhydr isopropyl alcohol, mp 145°. Sol in water, alcohol; very slightly sol in ethyl acetate.
THERAP CAT: Antiarrhythmic.

612. Amorolfine. *cis-4-[3-[4-(1,1-Dimethylpropyl)-phenyl]-2-methylpropyl]-2,6-dimethylmorpholine;* 4-[3-(4-tert-amylphenyl)-2-methylpropyl]-2,6-dimethylmorpholine; (±)-cis-2,6-dimethyl-4-[2-methyl-3-(p-tert-pentylphenyl)propyl]morpholine; Ro-14-4767/000; Loceryl. $C_{21}H_{35}NO$; mol wt 317.52. C 79.44%, H 11.11%, N 4.41%, O 5.04%. Broad spectrum topical antimycotic, morpholine deriv. Prepn: A. Pfiffner, K. Bohnen, **Ger. pat. 2,752,096**; A. Pfiffner, **U.S. pat. 4,202,894** (1978, 1980 both to Hoffmann-La Roche). *In vitro* comparative antifungal spectrum: S. Shadomy *et al.*, *Sabouraudia* **22**, 7 (1984). Mechanism of action: A. Polak-Wyss *et al.*, *ibid.* **23**, 433 (1985); A. Polak, *Ann. N.Y. Acad. Sci.* **515**, 221 (1988). Inhibitory effect on germ tube formation in *Candida albicans:* M. Schaude *et al.*, *Mykosen* **30**, 281 (1987); on *Candida* adherence: V. Vuddhakul *et al.*, *J. Antimicrob. Chemother.* **21**, 755 (1988).

bp$_{0.036}$ 134°.
Hydrochloride, $C_{21}H_{35}NO$·HCl, *Ro-14-4767/002.*
THERAP CAT: Antifungal (topical).

613. Amoscanate. *4-Isothiocyanato-N-(4-nitrophenyl)-benzeneamine; isothiocyanic acid p-(p-nitroanilino)phenyl ester;* 4-isothiocyanato-4′-nitrodiphenylamine; nithiocyamine; C-9333-Go; CGP-4540. $C_{13}H_9N_3O_2S$; mol wt 271.30. C 57.55%, H 3.34%, N 15.49%, O 11.79%, S 11.82%. Analog of nitroscanate, *q.v.* Prepn: K. Antos *et al.*, **Ger. pat. 1,932,690** (1970 to Cesk. Akad. Ved), *C.A.* **72**, 100265 (1970); S. Rajappa *et al.*, *J. Chem. Soc. Perkins Trans I* **1979**, 2001; N. Viswanathan, R. C. Desai, *Indian J. Chem.* **20B**, 308 (1981). Anthelmintic activity: H. P. Striebel, *Experientia* **32**, 457 (1976); K. R. Middleton *et al.*, *ibid.* **35**, 243 (1979); H. G. Sen, B. N. Deb, *Am. J. Trop. Med. Hyg.* **30**, 992 (1981). HPLC determn in human plasma: W. M. Kofi-Tsekpo, C. W. Karekezi, *Drugs Exp. Clin. Res.* **14**, 31 (1988). Clinical pharmacology: A. B. Vaidya *et al.*, *Brit. J. Clin. Pharmacol.* **4**, 463 (1977). Mutagenicity study: B. S. Reddy *et al.*, *Antimicrob. Ag. Chemother.* **22**, 707 (1982). Brief review: J. I. Bruce, *Int. J. Parisitol.* **17**, 131-140 (1987).

Crystals from acetone, mp 196-198°.
THERAP CAT: Anthelmintic (Schistosoma).

614. Amosulalol. *(±)-5-[1-Hydroxy-2-[[2-(2-methoxyphenoxy)ethyl]amino]ethyl]-2-methylbenzenesulfonamide.* $C_{18}H_{24}N_2O_5S$; mol wt 380.47. C 56.82%, H 6.36%, N 7.36%, O 21.03%, S 8.43%. Sulfonamide-substituted phenylethanolamine with α- and β-adrenergic blocking activity. Prepn: K. Imai *et al.*, **Ger. pat. 2,843,016**; *eidem*, **U.S. pat. 4,217,305** (1979, 1980 both to Yamanouchi). Cardiovascular pharmacology and receptor blocking activity in dogs: T. Takenaka *et al.*, *Eur. J. Pharmacol.* **85**, 35 (1982). GC determn in urine: H. Kamimura *et al.*, *J. Chromatog.* **275**, 81 (1983). Pharmacokinetics in humans: M. Nakashima *et al.*, *Clin. Pharmacol. Ther.* **36**, 436 (1984). Metabolism in humans: H. Kamimura *et al.*, *Xenobiotica* **15**, 413 (1985). Comparison of receptor blocking activity of isomers: K. Honda *et al.*, *J. Pharmacol. Exp. Ther.* **236**, 776 (1986).

Monohydrochloride, $C_{18}H_{24}N_2O_5S$·HCl, *YM-09538, Lowgan.* Colorless crystals, mp 158-160°. pK$_1$ 7.4; pK$_2$ 10.2.
R-(−)-Form hydrochloride, mp 158°. $[\alpha]_D^{20}$ −30.4° (c = 1 in methanol).
S-(+)-Form hydrochloride, mp 158°. $[\alpha]_D^{20}$ +30.7° (c = 1 in methanol).
THERAP CAT: Antihypertensive.

615. Amotriphene. *β-[Bis(4-methoxyphenyl)methylene]-4-methoxy-N,N-dimethylbenzeneethanamine; 2,3,3-tris(p-methoxyphenyl)-N,N-dimethylallylamine;* 3-dimethylamino-1,1,2-tris(4-methoxyphenyl)-1-propene; aminoxytriphene; Win-5494; Myordil. $C_{26}H_{29}NO_3$; mol wt 403.52. C 77.39%, H 7.24%, N 3.47%, O 11.89%. Prepn: Elpern, **U.S. pat. 3,010,965** (1961 to Sterling Drug).

Hydrochloride, $C_{26}H_{29}NO_3 \cdot HCl$, crystals from absolute ethanol, mp 182-184°. LD_{50} in mice (mg/kg): 30 i.v.; 385 orally.

THERAP CAT: Vasodilator (coronary).

616. Amoxapine. *2-Chloro-11-(1-piperazinyl)dibenz-[b,f][1,4]oxazepine;* CL-67772; Asendin; Asendis; Demolox; Moxadil. $C_{17}H_{16}ClN_3O$; mol wt 313.79. C 65.07%, H 5.14%, Cl 11.30%, N 13.39%, O 5.10%. Prepn: J. Schmutz *et al., Helv. Chim. Acta* **50**, 245 (1967); *eidem, Chim. Ther.* **2**, 424 (1967); C. F. Howell *et al.,* **Fr.** pat. **1,508,536**; *eidem,* **U.S.** pat. **3,663,696** (1968, 1972 both to Am. Cyanamid). Analysis of amoxapine and its metabolites by GLC: T. B. Cooper, R. G. Kelly, *J. Pharm. Sci.* **68**, 216 (1979). X-ray crystallography: D. B. Cosulich, F. M. Lovell, *Acta Crystallogr.* **B33**, 1147 (1977). Pharmacology: E. N. Greenblatt *et al., Arch. Int. Pharmacodyn. Ther.* **233**, 107 (1978). Clinical study: R. Takahashi *et al., J. Int. Med. Res.* **7**, 7 (1979). *Review:* T. A. Ban, *Psychopharmacol. Bull.* **15**, 22-25 (1979); E. N. Greenblatt *et al.,* in *Pharmacological and Biochemical Properties of Drug Substances* **vol. 2**, M. E. Goldberg, Ed. (Am. Pharm. Assoc., Washington, DC, 1979) pp 1-18. Review of pharmacology and therapeutic efficacy: S. G. Jue *et al., Drugs* **24**, 1-23 (1982).

Crystals from benzene/petr ether, mp 175-176°. LD_{50} in mice (mg/kg): 122 i.p.; 112 orally (Howell, 1972).

THERAP CAT: Antidepressant.

617. Amoxicillin. *[2S-[2α,5α,6β(S*)]]-6-[[Amino-(4-hydroxyphenyl)acetyl]amino]-3,3-dimethyl-7-oxo-4-thia-1-azabicyclo[3.2.0]heptane-2-carboxylic acid;* (−)-6-[2-amino-2-(p-hydroxyphenyl)acetamido]-3,3-dimethyl-7-oxo-4-thia-1-azabicyclo[3.2.0]heptane-2-carboxylic acid; 6-[D(−)-α-amino-p-hydroxyphenylacetamido]penicillanic acid; α-amino-p-hydroxybenzylpenicillin; 6-(p-hydroxy-α-aminophenylacetamido)penicillanic acid; p-hydroxyampicillin; amoxycillin; AMPC; Amocilline; Amolin; Amopenixin; Amoram; Amoxipen; Anemolin; Aspenil; Betamox; Bristamox; Cabermox; Delacillin; Efpenix; Grinsil; Helvamox; Optium; Ospamox; Pasetocin; Penamox; Penimox; Piramox; Sawacillin; Simoxil; Sumox; Widecillin; Wymox. $C_{16}H_{19}N_3O_5S$; mol wt 365.41. C 52.59%, H 5.24%, N 11.50%, O 21.89%, S 8.78%. Semi-synthetic antibiotic related to penicillin. Prepn: Nayler, Smith, **Brit.** pat. **978,178** (1964 to Beecham); *eidem,* **U.S.** pat. **3,192,198** (1965); Long, Nayler, **Ger.** pat. **1,942,693** and **Brit.** pat. **1,241,844** (1970 and 1971 to Beecham), *C.A.* **72**, 90447q (1970). Resolution of isomers: Long *et al., J. Chem. Soc. (C)* **1971**, 1920. Series of articles on activity, pharmacology, absorption and excretion: *Antimicrob. Ag. Chemother.* **1970**, 407-430. Comprehensive description: P. K. Bhattacharyya, W. M. Cort, *Anal. Profiles Drug Subs.* **7**, 19-41 (1978). Review of antibacterial activity, pharmacokinetics and therapeutic use: R. N. Brogden *et al., Drugs* **18**, 169-184 (1979).

Trihydrate, *BRL-2333, Agram, Alfamox, Almodan, Amodex, Amoxi, Amoxidal, Amoxidin, Amoxil, Amoxillat, Amoxi-Wolff, Amoxypen, Ardine, AX 250, Clamoxyl, Cuxacillin, Dura AX, Flemoxin, Hiconcil, Ibiamox, Larocin (obsolete), Larotid, Moxal, Moxaline, Neamoxyl, Polymox, Raylina, Robamox, Sigamopen, Silamox, Trimox, Uro-Clamoxyl, Utimox, Zamocillin.* Off-white crystalline powder. $[\alpha]_D^{20}$ +246° (c = 0.1). uv max (ethanol): 230, 274 nm (ε 10850, 1400); (0.1N HCl): 229, 272 nm (ε 9500, 1080); (0.1N KOH): 248, 291 (ε 2200, 3000). Soly (mg/ml): water 4.0; methanol 7.5; abs ethanol 3.4. Insol in hexane, benzene, ethyl acetate, acetonitrile.

Hydrochloride trihydrate, browns at 90°, then dehydrates, dec 216-218°.

β-Naphthalenesulfonate trihydrate, mp 194° (dec).

THERAP CAT: Antibacterial.

THERAP CAT (VET): Antibacterial.

618. Ampelopsin. *(2R-trans)-2,3-Dihydro-3,5,7-trihydroxy-2-(3,4,5-trihydroxyphenyl)-4H-1-benzopyran-4-one; 3,3',4',5,5',7-hexahydroxyflavanone;* ampeloptin; dihydromyricetin. $C_{15}H_{12}O_8$; mol wt 320.26. C 56.26%, H 3.78%, O 39.97%. From leaves of *Ampelopsis meliaefolia* Kudo, *Vitaceae:* Kotake, Kubota, *Ann.* **544**, 253 (1940); from bark of *Pinus contoria* Dougl., *Pinaceae:* Hergert, *J. Org. Chem.* **21**, 534 (1956); U.S. pat. **2,870,165** (1959 to Rayonier); from *Erythrophleum africanum* (Welw.) Harms, *Caesulpiniaceae:* Hansel, Klaffenbach, *Arch. Pharm.* **294**, 158 (1961). Synthesis from myricetin, *q.v.:* Kotake, Kubota, *J. Inst. Polytech., Osaka City Univ.* **1**, no. 2, 47 (1950), *C.A.* **46**, 2052e (1952).

Hemipentahydrate, needles from water, mp 245-246°. Hexaacetate, $C_{27}H_{24}O_{14}$, needles from alc, mp 174-175°. Hexabenzoate, $C_{57}H_{42}O_{14}$, needles from alc, mp 174°.

619. Amperozide. *4-[4,4-Bis(4-fluorophenyl)butyl]-N-ethyl-1-piperazinecarboxamide;* FG-5606; Hogpax. $C_{23}H_{29}F_2N_3O$; mol wt 401.50. C 68.81%, H 7.28%, F 9.46%, N 10.47%, O 3.98%. Piperazine deriv with effects on stress-induced disorders. Prepn: A. K. K. Björk *et al.,* **Ger.** pat. **2,941,880**; **U.S.** pat. **4,308,387** (1980, 1981 both to Ferrosan). Effects on psychosomatic disorders in man: E. Christensson *et al., 3rd World Congr. Biol. Psychiatr.,* Abstr. F 78 (Stockholm, 1981). Clinical effects in pig prodn: A. Björk *et al., Proc. Int. Pig Vet. Soc. Congr.,* pp 315, 316, 317 (Mexico, 1982).

Hydrochloride, $C_{23}H_{30}ClF_2N_3O$, mp 177-178°.

THERAP CAT (VET): Anti-aggressive.

620. Amphecloral. *α-Methyl-N-(2,2,2-trichloroethylidene)benzeneethanamine;* α-methyl-*N*-(2,2,2-trichloroethylidene)phenethylamine; *N*-[2-(1-phenylpropyl)]-2,2,2-trichloroethylidenimine; amfecloral; Acutran. $C_{11}H_{12}Cl_3N$; mol wt 264.58. C 49.94%, H 4.57%, Cl 40.20%, N 5.29%. Prepn: Cavallito, U.S. pat. **2,923,661** (1960 to Irwin, Neisler).

dl-Form, bp$_{0.5}$ 95°. n_D^{25} 1.530.
d-Form, $[\alpha]_D$ +49.9°±0.3° (c = 5 in dioxane).
THERAP CAT: Anorexic.

621. Amphenidone. *1-(3-Aminophenyl)-2(1H)-pyridinone;* 1-(m-aminophenyl)-2(1H)-pyridone; Dornwal. $C_{11}H_{10}N_2O$; mol wt 186.21. C 70.95%, H 5.41%, N 15.04%, O 8.59%. Prepn: Scudi *et al.,* U.S. pat. **2,947,754** (1960 to Wallace & Tiernan). Toxicity data: Plekss *et al., Fed. Proc.* **19**, 390 (1960).

Crystals, mp 182.5-184.5°. LD$_{50}$ in mice, rats (mg/kg): 1300, 3200 orally (Plekss).
THERAP CAT: Sedative; hypnotic.

622. Amphenone B. *3,3-Bis[4-aminophenyl]-2-butanone;* amphenone; 2-oxo-3,3-bis[p-aminophenyl]butane. $C_{16}H_{18}N_2O$; mol wt 254.33. C 75.56%, H 7.13%, N 11.01%, O 6.29%. Prepn: Allen, Corwin, *J. Am. Chem. Soc.* **72**, 117 (1950); U.S. pat. **2,539,388** (1951). Structure: Bencze, Allen, *J. Org. Chem.* **22**, 352 (1957). Shows antiestrogenic activity in the chick oviduct test: Hertz *et al., Recent Progr. Horm. Res.* **11**, 119-147 (1955). Decreases adrenal action. Review: *Subsidia Medica* **10**, 99-102 (1958).

Crystals, mp 137.5-138°.
Dihydrochloride, $C_{16}H_{20}Cl_2N_2O$, crystals from ethanol, dec 272-275°. Soluble in water.
Note: Formerly a pinacolone structure was assigned to amphenone B: *1,2-Bis[p-aminophenyl]-2-methyl-1-propanone.*
USE: In biological research.

623. Amphetamine. (±)-*α-Methylbenzeneethanamine; dl-α-methylphenethylamine;* 1-phenyl-2-aminopropane; (phenylisopropyl)amine; β-aminopropylbenzene; racemic desoxy-nor-ephedrine; Actedron; Allodene; Adipan; Sympatedrine; Psychedrine; Isomyn; Isoamyne; Mecodrin; Norephedrane; Novydrine; Elastonon; Ortédrine; Phenedrine; Profamina; Propisamine; Sympamine; Simpatedrin. $C_9H_{13}N$; mol wt 135.21. C 79.95%, H 9.69%, N 10.36%. Prepn: U.S. pats. **1,879,003** (1932); **1,921,424** (1933); **2,015,408** (1935); Hartung, Munch, *J. Am. Chem. Soc.* **53**, 1875 (1931). Demonstration of stereospecific binding sites for (+)-^{3}H-amphetamine in hypothalamic membranes and correlations with anorexic potency of phenylethylamines: S. M. Paul *et al., Science* **218**, 487 (1982). Toxicity data: M. R. Warren, H. W. Werner, *J. Pharmacol. Exp. Ther.* **85**, 119 (1945); W. A. Behrendt, R. Deininger, *Arzneimittel-Forsch.* **13**, 711 (1963). Series of articles on the biochemical and

behavioral effects of amphetamines in man and animals: *Handb. Exp. Pharmakol.* **45**, 3-304 (1977); *Handb. Psychopharmacol.* **11**, 1-98 (1978). Review of use and abuse: J. P. Morgan, *Substance Abuse: Clinical Problems and Perspectives,* J. H. Lowinson, P. Ruiz, Eds. (Williams & Wilkins, Baltimore, 1981) pp 167-184. Books: C. D. Leake, *The Amphetamines: Their Actions and Uses* (Thomas, Springfield, 1958) 167 pp; O. J. Kalant, *The Amphetamines: Toxicity and Addiction* (Thomas, Springfield, 1966) 151 pp.

Mobile liquid. Amine odor. Acrid, burning taste. Volatilizes slowly at room temp. d$_4^{25}$ 0.913. bp$_{760}$ 200-203°; bp$_{13}$ 82-85°. Slightly soluble in water; sol in alc, ether; readily sol in acids. Aq solns are alkaline to litmus. LD$_{50}$ in rats (mg/kg): 180 s.c. (Warren, Werner).

Sulfate, $(C_9H_{13}N)_2.H_2SO_4$, *Alentol, Benzedrine, Psychoton, Simpamina.* Crystals. Slightly bitter taste followed by a sensation of numbness. mp above 300° (dec). One part dissolves in 8.8 parts water, 515 parts 95% alc. A soln of 1 g/10 ml water has a pH 5-6. LD$_{50}$ in mice, rats (mg/kg): 24.2, 55 orally (Behrendt, Deininger).

Phosphate, $C_9H_{13}N.H_3PO_4$, *Actemin, Aktedron, Monophos, Profetamine Phosphate, Racephen, Raphetamine Phosphate.* Prepn: Goggin, U.S. pat. **2,507,468** (1950 to Clark & Clark). Crystals, bitter taste. Sinters at about 150°. Dec around 300°. More sol in water than amphetamine sulfate. Slightly sol in alcohol. Practically insol in benzene, chloroform, ether. The pH of a 10% soln is about 4.6.

d-Form tannate, *tanphetamin, Synatan.* Prepn: Cavallito, U.S. pat. **2,950,309** (1960 to Irwin, Neisler and Co.).

d-Form sulfate, *see* Dextroamphetamine Sulfate.

l-Form, *levamphetamine, levamfetamine.*

l-Form succinate, *Cydril.*

Note: This is a controlled substance (stimulant) listed in the U.S. Code of Federal Regulations, Title 21 Part 1308.12 (1995).

THERAP CAT: CNS stimulant; anorexic.

THERAP CAT (VET): CNS stimulant, in narcotic poisoning, anesthetic collapse, in depression from encephalitis.

624. Amphetaminil. *α-[(1-Methyl-2-phenylethyl)amino]benzeneacetonitrile;* N-(α-methylphenethyl)-2-phenylglycinonitrile; α-phenyl-α-(β-phenylisopropylamino)acetonitrile; α-phenyl-α-(1-methyl-2-phenyl)ethylaminoacetonitrile; α-phenyl-α-*N*-(1-phenylisopropyl)aminoacetonitrile; AN 1; Aponeuron. $C_{17}H_{18}N_2$; mol wt 250.34. C 81.56%, H 7.25%, N 11.19%. Prepd by reaction of DL-β-phenylisopropylamine with sodium cyanide and benzaldehyde or with α-phenyl-α-bromoacetonitrile: Klosa, Ger. pat. **1,112,987** (1959), *C.A.* **56**, 3409d (1962); *idem, J. Prakt. Chem.* **20**, 275 (1963). Pharmacology: Dominok, Oelssner, *Acta Biol. Med. Ger.* **20**, 625 (1968); Beyer *et al., Deut. Apoth-Ztg.* **111**, 677, 680 (1971). Metabolic studies: Remberg *et al., Arch. Toxicol.* **29**, 153 (1972). Chemistry: Beyrich *et al., Pharmazie* **27**, 28 (1972); Gloeckl, Beyrich, *ibid.* 95.

Crystals from ethanol-water, mp 85-87°.
Hydrochloride, $C_{17}H_{19}ClN_2$, sinters at 100-104°, mp 134-136°.
THERAP CAT: Psychotropic.

625. Amphomycin. Amfomycin; glumamycin. $C_{58}H_{91}N_{13}O_{20}$; mol wt 1290.44. C 53.98%, H 7.11%, N 14.11%, O 24.80%. Polypeptide antibiotic modified at the amino terminus with a fatty acid moiety; active against gram positive bacteria. Produced by *Streptomyces canus* from soil collected near Syracuse, N.Y.: B. Heinemann *et al., Antibiot. & Chemother.* **3**, 1239 (1953). Production: *eidem,* U.S.

pat. **3,126,317** (1964 to Bristol-Myers). Structure and identity with glumamycin: M. Bodanszky *et al., J. Am. Chem. Soc.* **95**, 2352 (1973). Pharmacology and toxicity: D. E. Tisch *et al., Antibiot. Ann.* **1954-1955**, 1011. Mechanism of action: H. Tanaka *et al., Biochem. Biophys. Res. Commun.* **86**, 902 (1979). Use to improve feed efficiency in ruminant animals: M. Gordon, G. J. Christie, **Ger.** pat. **3,027,370**; *eidem,* **U.S.** pat. **4,414,206** (1981, 1983 both to Bristol-Myers).

R–Asp–MeAsp–Asp–Gly–Asp–Gly–Dab⁰–Val–Pro

Pip—Dabᵗ

Dab⁰ = D-erythro-α,β–diaminobutyric acid
Dabᵗ = L-threo-α,β–diaminobutyric acid
Pip = D-pipecolic acid
R = (+)-3-anteisotridecenoic acid or (+)-3-isododecenoic acid

Acidic, surface-active polypeptide. [α]$_D^{25}$ +7.5° (c = 1 at pH 6). Isoelectric point 3.5-3.6. Soluble in water and the lower alcohols. Insol in nonpolar solvents. Aq solns at neutral pH are stable at room temp for at least one month. Forms sodium and calcium salts. Induces hemolysis. Suggested as a topical agent for animal and plant infections.
Sodium salt, amorphous solid, sol in water. LD$_{50}$ in mice (mg/kg): 177.8 i.v. (Tisch).
Calcium salt, crystalline solid, sol in water, methyl alcohol. Used in combination with other antibiotics as antibacterial ingredient in topical anti-inflammatory preparations. LD$_{50}$ in mice (mg/kg): 120.2 i.v. (Tisch).
THERAP CAT: Antibacterial.

626. Amphotalide. *2-[5-(4-Aminophenoxy)pentyl]-1H-isoindole-1,3(2H)-dione; N-[5-(p-aminophenoxy)pentyl]-phthalimide;* 1-p-aminophenoxy-5-phthalimidopentane; amphothalide; RP-6171; Schistomide. C$_{19}$H$_{20}$N$_2$O$_3$; mol wt 324.38. C 70.35%, H 6.21%, N 8.64%, O 14.80%. Prepn: Ashley *et al., J. Chem. Soc.* **1959**, 3880; Barber *et al.,* **Brit.** pat. **769,706** (1957 to May & Baker).

Crystals from ethanol, mp 113-114°.
THERAP CAT: Anthelmintic (Schistosoma).

627. Amphotericin B. Ambisome; Amphozone; Fungizone; Fungilin; Ampho-Moronal. C$_{47}$H$_{73}$NO$_{17}$; mol wt 924.09. C 61.09%, H 7.96%, N 1.52%, O 29.43%. Polyene antibiotic produced by *Streptomycetes nodosus* M4575 obtained from soil of the Orinoco river region of Venezuela: Gold *et al., Antibiot. Ann.* **1955-1956**, 579; Vandeputte *et al., ibid.* 587; Dutcher *et al., ibid.* **1956-1957**, 866; Walters *et al., J. Am. Chem. Soc.* **79**, 5076 (1957); Dutcher *et al.,* **U.S.** pat. **2,908,611** (1959 to Olin Mathieson). Structure studies: Borowski *et al., Tetrahedron Letters* **1965**, 473. Carbon skeleton, ring size, and partial structure: Cope *et al., J. Am. Chem. Soc.* **88**, 4228 (1966). Complete structure: Mechlinski *et al., Tetrahedron Letters* **1970**, 3873; Borowski *et al., ibid.* 3909; R. C. Pandey, K. L. Rinehart, *J. Antibiot.* **29**, 1035 (1976). Synthetic approaches: K. C. Nicolaou *et al., Chem. Commun.* **1986**, 413. Total synthesis: K. C. Nicolaou *et al., J. Am. Chem. Soc.* **109**, 2821 (1987). Mechanism of action: R. W. Holz in *Antibiotics* vol. 5(pt. 2), F. E. Hahn, Ed. (Springer-Verlag, New York, 1979) pp 313-340. Toxicity: G. R. Keim *et al., Science* **179**, 584 (1973). Comprehensive description: I. M. Asher *et al.,* in *Analytical Profiles of Drug Substances* vol. 6, K. Florey, Ed. (Academic Press, New York, 1977) pp 1-42.

Deep yellow prisms or needles from DMF. Dec gradually above 170°. uv max (methanol): 406, 382, 363, 345 nm. [α]$_D^{24}$ +333° (acidic DMF); −33.6° (0.1N methanolic HCl). Insol in water at pH 6 to 7. Soly at pH 2 or pH 11 in water: about 0.1 mg/ml. Water soly increased by sodium desoxycholate. Soly in DMF 2 to 4 mg/ml; in DMF + HCl: 60 to 80 mg/ml; in DMSO: 30 to 40 mg/ml. Solids and solns appear stable for long periods between pH 4 and 10 when stored at moderate temps out of contact with light and air. LD$_{50}$ in mice (mg/kg): 88 i.p., 4 i.v. (Keim).
THERAP CAT: Antifungal.
THERAP CAT (VET): Antifungal, used for systemic mycoses.

628. Ampicillin. *[2S-[2α,5α,6β(S*)]]-6-[(Aminophenylacetyl)amino]-3,3-dimethyl-7-oxo-4-thia-1-azabicyclo-[3.2.0]heptane-2-carboxylic acid;* 6-[D(−)-α-aminophenylacetamido]penicillanic acid; D(−)-α-aminobenzylpenicillin; AY-6108; BRL-1341; P-50; Albipen; Amfipen; Amipenix; Ampipenin; Ampitab; Bonapicillin; Britacil; Doktacillin; Domicillin; Dumopen; Grampenil; Nuvapen; Omnipen; Pénicline; Tokiocillin. C$_{16}$H$_{19}$N$_3$O$_4$S; mol wt 349.41. C 55.00%, H 5.48%, N 12.03%, O 18.32%, S 9.18%. Orally active, semi-synthetic antibiotic; structurally related to penicillin. Prepn: Doyle *et al.,* **U.S.** pat. **2,985,648** (1961); *eidem,* **Brit.** pat. **902,703** (1962 to Beecham); Doyle *et al., J. Chem. Soc.* **1962**, 1440. Prepn of the trihydrate: D. A. Johnson, G. A. Hardcastle, **U.S.** pat. **3,157,640** (1964 to Bristol-Myers); of the anhydrous crystalline form: N. H. Grant, H. E. Alburn, **U.S.** pat. **3,144,445** (1964 to Am. Home Prods.). Alternate syntheses: Dane, Dockner, *Ber.* **98**, 789 (1965); F. Kajfez *et al., J. Heterocycl. Chem.* **13**, 561 (1976). LC determn: M. Margosis, *J. Assoc. Off. Anal. Chem.* **70**, 206 (1987). Series of articles on pharmacology and antibacterial activity: *Brit. Med. J.* **2**, 193-206 (1961). Comprehensive description: E. Ivashkiv in *Analytical Profiles of Drug Substances* vol. 2, K. Florey, Ed. (Academic Press, New York, 1973) pp 1-61. Review of pharmacology and clinical efficacy in combination with sulbactam, *q.v.:* D. M. Campoli-Richards, R. N. Brogden, *Drugs* **33**, 577 609 (1987).

Crystals, dec 199-202°. [α]$_D^{23}$ +287.9° (c = 1 in water). Sparingly sol in water.
Monohydrate, crystals from water, dec 202°. [α]$_D^{21}$ +281° (water). Sparingly sol in water at room temp.
Sesquihydrate, dec 199-202°. [α]$_D^{20}$ +283.1° (water).
Trihydrate, *Alpen, Amblosin, Amcill, Ampilag, Ampilar, Ampi-Tablinen, Amplital, Austrapen, Binotal, Cetampin, Cymbi, Pen A, Penbristol, Penbritin, Penbrock, Pensyn, Pentrexyl, Polycillin, Princillin, Principen, Rosampline, Totacillin, Totalciclina, Totapen, Ukapen, Ultrabion, Vidopen.*
Potassium salt, C$_{16}$H$_{18}$KN$_3$O$_4$S, *Suractin.*
Sodium salt, C$_{16}$H$_{18}$NaN$_3$O$_4$S, *Alpen-N, Amcill-S, Ampicin, Cilleral, Omnipen-N, Pen A/N, Penbritin-S, Pentrex, Polycillin-N, Principen/N, Synpenin, Viccillin.*
L(+)-Form, crystals, dec at about 205°. [α]$_D^{20}$ +209° (c = 0.2 in water). Less active as an antibiotic than D(−)-form.

THERAP CAT: Antibacterial.
THERAP CAT (VET): Antibacterial.

629. Ampiroxicam. *Carbonic acid ethyl 1-[[2-methyl-3-[(2-pyridinylamino)carbonyl]-2H-1,2-benzothiazin-4-yl]oxy]ethyl ester S,S-dioxide;* (±)-4-(1-hydroxyethoxy)-2-methyl-*N*-2-pyridyl-2*H*-1,2-benzothiazine-3-carboxamide ethyl carbonate (ester) 1,1-dioxide; 4-[1-(ethoxycarbonyloxy)ethoxy]-2-methyl-*N*-(2-pyridyl)-2*H*-1,2-benzothiazine-3-carboxamide 1,1-dioxide; CP-65703; Flucam; Nasil. $C_{20}H_{21}N_3O_7S$; mol wt 447.47. C 53.68%, H 4.73%, N 9.39%, O 25.03%, S 7.17%. Prodrug of piroxicam, *q.v.* Prepn: A. Marfat, **Eur. pat. Appl.** **147,177;** *idem,* U.S. pat. **4,551,452** (both 1985 to Pfizer). Clinical pharmacokinetics: F. C. Falkner *et al., Xenobiotica* **20,** 645 (1990). Toxicity study: M. Iijima *et al., Oyo Yakuri* **43,** 1 (1992); *C.A.* **116,** 187689k (1992). Pharmacology: T. J. Carty *et al., Agents Actions* **39,** 157 (1993). Clinical trial in oral postoperative pain: K. Kurita *et al., Oral. Ther. Pharmacol.* **10,** 138 (1991); in rheumatoid arthritis: S. Irimajiri *et al., Japan. J. Inflamm.* **12,** 81 (1992).

White crystals from toluene, mp 159°-161°. LD_{50} orally in male, female rats: 1798, 747 mg/kg (Iijima).
THERAP CAT: Anti-inflammatory.

630. Ampligen. Poly I . poly (C_{12},U); poly(rI) . poly-$[r(C_{12}U)_n]$; rI_n . $r(C_{12}U)_n$. Synthetic, mismatched double-stranded RNA capable of inducing interferons, *q.v.* Consists of a strand of 5'-inosinic acid (I) units hydrogen bonded to a second strand of 5'-cytidylic (C)- 5'uridylic (U) acid units in a 12C:U ratio. Prepn and pharmacology: P. O. P. Ts'o *et al., Mol. Pharmacol.* **12,** 299 (1976). Antiviral activity: W. A. Carter *et al., ibid.* 440; against HIV: D. C. Montefiori, W. M. Mitchell, *Proc. Nat. Acad. Sci. USA* **84,** 2985 (1987); D. C. Montefiori *et al., Antiviral Res.* **9,** 47 (1988). Induction of interferon: D. A. Stringfellow, S. D. Weed, *Antimicrob. Ag. Chemother.* **17,** 988 (1980). Mechanism of action study: W. G. Hearl, M. I. Johnston, *Biochem. Biophys. Res. Commun.* **138,** 40 (1986). Antitumor effects in comparison with interferon: H. R. Hubbell *et al., Cancer Res.* **44,** 3252 (1984); and immunomodulatory effects: H. R. Hubbell *et al., ibid.* **45,** 2481 (1985); W. A. Carter *et al., J. Biol. Response Modif.* **4,** 613 (1985). HPLC determn in plasma: M. G. Rosenblum, L. Cheung, *J. Liq. Chromatog.* **9,** 2869 (1986). Clinical evaluation in metastatic cancers: I. Brodsky *et al., J. Biol. Response Modif.* **4,** 669 (1985); in AIDS or AIDS related complex: W. A. Carter *et al., Lancet* **1,** 1286 (1987).

631. Amprolium. *1-[(4-Amino-2-propyl-5-pyrimidinyl)methyl]-2-methylpyridinium chloride;* 1-[(4-amino-2-propyl-5-pyrimidinyl)methyl]-2-picolinium chloride; Corid. $C_{14}H_{19}ClN_4$; mol wt 278.78. C 60.32%, H 6.87%, Cl 12.72%, N 20.10%. Prepn: Rogers *et al., J. Am. Chem. Soc.* **82,** 2974 (1960); Rogers, Sarett, U.S. pat. **3,020,277** (1962 to Merck & Co.), *see also* U.S. pat. **3,020,200.** Formulations as poultry feed: Rogers, Sarett, U.S. pat. **3,065,132** (1962 to Merck & Co.).

Hydrochloride, $C_{14}H_{19}ClN_4 \cdot HCl$, *Amprol.* Crystals from methanol + ethanol, dec 248-249°. Freely sol in water, methanol, 95% ethanol, dimethylformamide. Sparingly sol in abs ethanol. Practically insol in isopropanol, butanol, dioxane, acetone, ethyl acetate, acetonitrile, isooctane. pH of 10% aq soln 2.5-3.0.
THERAP CAT (VET): Coccidiostat.

632. Amprotropine Phosphate. *α-(Hydroxymethyl)-benzeneacetic acid 3-(diethylamino)-2,2-dimethylpropyl ester phosphate (1:1) salt;* 3-diethylamino-2,2-dimethylpropyl tropate phosphate; Syntropan. $C_{18}H_{32}NO_7P$; mol wt 405.43. C 53.33%, H 7.96%, N 3.45%, O 27.62%, P 7.64%. Prepn: U.S. pats. **1,932,341** (1933); **1,987,546** (1935).

Bitter crystals, mp 142-145°. Freely sol in water; slightly sol in abs alcohol. Insol in chloroform, ether. The aq soln is acid to litmus.
THERAP CAT: Anticholinergic.

633. Ampyrone. *4-Amino-1,2-dihydro-1,5-dimethyl-2-phenyl-3H-pyrazol-3-one; 4-aminoantipyrine;* 4-amino-2,3-dimethyl-1-phenyl-3-pyrazolin-5-one; 1,5-dimethyl-2-phenyl-4-aminopyrazolone. $C_{11}H_{13}N_3O$; mol wt 203.24. C 65.01%, H 6.45%, N 20.67%, O 7.87%. Prepn: Waser, *Helv. Chim. Acta* **8,** 117 (1925); Freedman, Sherndal, U.S. pat. **1,877,166** (1933 to H. A. Metz Labs.); Turpechen, Kallro, *C.A.* **36,** 426 (1942); Takahashi, Ogyu, **Japan.** pat. **3663('54)** (to Nippon New Drug), *C.A.* **50,** 2687c (1956). Metabolite of aminopyrine, *q.v.:* E. S. Vesell *et al., Clin. Pharmacol. Ther.* **20,** 661 (1976); G. F. Lockwood, J. B. Houston, *J. Pharm. Pharmacol.* **31,** 787 (1979).

Pale yellow crystals from benzene, mp 109°. Sol in water, ethanol, benzene; sparingly sol in ether.

634. Amrinone. *5-Amino-(3,4'-bipyridin)-6(1H)-one;* 3-amino-5-(4-pyridinyl)-2(1*H*)-pyridinone; Win-40680; Cartonic; Inocor; Vesistol; Wincoram. $C_{10}H_9N_3O$; mol wt 187.20. C 64.16%, H 4.85%, N 22.45%, O 8.55%. Non-glycosidic cardioactive agent with selective inotropic properties and vasodilatory activity. Prepn: G. Y. Lesher, C. J. Opalka, U.S. pats. **4,004,012; 4,072,746** (1977, 1978 both to Sterling). Cardiotonic activity: A. A. Alousi *et al., Fed. Proc.* **37,** 3692 (1978). Hemodynamic assessment: J. R. Benotti *et al., N. Engl. J. Med.* **21,** 1373 (1978). Clinical pharmacology: D. Davolos *et al., Circulation* **58,** 183 (1978). Effect on intractable myocardial failure in man: T. H. Le Jemtel *et al., ibid.* **59,** 1098 (1979). Review of pharmacology: A. E. Farah, A. A. Alousi, *Life Sci.* **22,** 1139-1148 (1978). Review of pharmacology and clinical efficacy: M. B. Bottorff *et al., Pharmacotherapy* **5,** 227-237 (1985). Reviews: A. A. Alousi, J. Edelson in *Pharmacological and Biochemical Properties of Drug Substances* vol. 3, M. E. Goldberg, Ed. (Am. Pharm. Assoc., Washington, DC, 1981) pp 120-147; A. Ward *et al., Drugs* **26,** 468 (1983).

Crystals from DMF, mp 294-297° (dec).
THERAP CAT: Cardiotonic.

635. Amsacrine. *N-[4-(9-Acridinylamino)-3-methoxy-phenyl]methanesulfonamide;* 4'-(9-acridinylamino)methanesulfon-*m*-anisidide; *m*-AMSA; CI-880; NSC-249992; SN-11841; Amekrin; Amsidine; Amsidyl; Lamasine. $C_{21}H_{19}N_3$-O_3S; mol wt 393.47. C 64.10%, H 4.87%, N 10.68%, O 12.20%, S 8.15%. Cytostatic agent with antiviral and immunosuppressive properties. Prepn: B. F. Cain *et al., J. Med. Chem.* **18,** 1110 (1975). Synthesis and biological properties of spin-labeled amsacrine: B. K. Sinha *et al., ibid.* **19,** 994 (1976). Mechanism of action: W. R. Wilson, *Chem. New Zeal.* **37,** 148 (1973). Pharmacologic disposition: R. L. Cysyk *et al., Drug Metab. Disp.* **5,** 579 (1977). Antiviral activity: D. M. Byrd, *Ann. N.Y. Acad. Sci.* **284,** 463 (1977). Immunosuppressive properties: B. C. Baguley *et al., Eur. J. Cancer* **10,** 169 (1974). Exptl antitumor properties: B. F. Cain, G. J. Atwell, *ibid.* 539. Efficacy in adult acute leukemia: S. S. Legha *et al., Ann. Int. Med.* **93,** 17 (1980). Toxicologic studies: K. L. Pavkov *et al., U.S. NTIS Rep.* **PB-298106** (1979) 284 pp, *C.A.* **92,** 34335c (1980). Review of pharmacology and clinical efficacy: J. Hornedo, D. A. Van Echo, *Pharmacother.* **5,** 78-90 (1985).

LD$_{50}$ in male, female CDF$_1$ mice: 810 mg/m²; 729 mg/m² orally (Pavkov).
Hydrochloride, $C_{21}H_{19}N_3O_3S.HCl$, *NSC-141549*. Crystals, mp 197-199°. LD$_{50}$ i.p. in mice: ~60 mg/kg (Byrd).
Methanesulfonate, $C_{21}H_{19}N_3O_3S.CH_3SO_3H$, *NSC-156303*. Crystals, mp 292-293°. LD$_{50}$ i.p. in mice: ~24 mg/kg (Byrd).
THERAP CAT: Antineoplastic.

636. Amsonic Acid. *2,2'-(1,2-Ethenediyl)bis[5-aminobenzenesulfonic acid];* 4,4'-diamino-2,2'-stilbenedisulfonic acid. $C_{14}H_{14}N_2O_6S_2$; mol wt 370.41. C 45.40%, H 3.81%, N 7.56%, O 25.92%, S 17.31%. Prepn: Bender, Schultz, *Ber.* **19,** 3234 (1886); H. E. Fierz-David, L. Blangey, *Grundlegende Operationen der Farbenchemie* (Springer-Verlag, Vienna, 7th ed., 1947) p 161; Spiegler, U.S. pat. **2,784,220** (1957 to du Pont).

Yellow needles; very slightly sol in water. Forms sparingly water-sol, crystalline salts with many bisquaternary ammonium bases.
USE: In manuf of dyes, bleaching agents.

637. Amtolmetin Guacil. *N-[[1-Methyl-5-(4-methylbenzoyl)-1H-pyrrol-2-yl]acetyl]glycine 2-methoxyphenyl ester;* N-[(1-methyl-5-p-toluoylpyrrol-2-yl)acetyl]glycine o-methoxyphenyl ester; 1-methyl-5-p-toluoylpyrrole-2-acetamidoacetic acid guaicil ester; ST-679; MED-15. $C_{24}H_{24}N_2O_5$; mol wt 420.47. C 68.56%, H 5.75%, N 6.66%, O 19.03%. Ester prodrug of tolmetin, *q.v.* Prepn: A. Baglioni, **Belg.** pat. **896,018;** *idem,* U.S. pat. **4,578,481** (1983, 1986 both to

Sigma-Tau). Pharmacology: E. Arrigoni-Martelli, *Drugs Exp. Clin. Res.* **16,** 63 (1990); A. Caruso *et al., ibid.* **18,** 481 (1992). HPLC determn in plasma: A. Mancinelli *et al., J. Chromatog.* **553,** 81 (1991). Series of articles on pharmacokinetics and clinical trials: *Clin. Ter.* **142** (1 pt 2) 3-59 (1993).

Crystals from cyclohexane-benzene, mp 117-120°. Sol in common organic solvents. LD$_{50}$ in male mice, rats (mg/kg): 1370, 1100 i.p.; >1500, 1450 orally (Baglioni).
THERAP CAT: Analgesic; anti-inflammatory.

638. Amygdalin. *[(6-O-β-D-Glucopyranosyl-β-D-glucopyranosyl)oxy]benzeneacetonitrile;* amygdaloside; mandelonitrile-β-gentiobioside; D-mandelonitrile-β-D-glucosido-6-β-D-glucoside; NSC-15780. $C_{20}H_{27}NO_{11}$; mol wt 457.43. C 52.51%, H 5.95%, N 3.06%, O 38.47%. The name amygdalin is currently used interchangeably with *laetrile.* Cyanogenic glycoside which occurs in seeds of *Rosaceae;* principally in bitter almonds; also in peaches and apricots. Most common constituent of *Laetrile*® preparations. Structure and synthesis: W. N. Haworth, B. Wylam, *J. Chem. Soc.* **123,** 3120 (1923); Kuhn, *Ber.* **56,** 857 (1923); R. Campbell, W. N. Haworth, *J. Chem. Soc.* **125,** 1337 (1924); Hudson, *J. Am. Chem. Soc.* **46,** 483 (1924); Zemplén, Kunz, *Ber.* **57,** 1357 (1924); Kuhn, Sobotka, *ibid.* 1767; Baumann, Pigman, *The Carbohydrates,* W. Pigman, Ed. (Academic Press, New York, 1957) p 550. Enzymic hydrolysis studies: Haisman, Knight, *Biochem. J.* **103,** 528 (1967). The term Laetrile® has also been applied to *mandelonitrile β-glucuronide.* Purported prepn: E. T. Krebs, E. T. Krebs, Jr., **Brit.** pat. **788,855** (1958) and U.S. pat. **2,985,664** (1961). Synthesis, characterization and comparison of mandelonitrile β-glucuronide with amygdalin: C. Fenselau *et al., Science* **198,** 625 (1977). Pharmacology and cyanide toxicity studies of amygdalin (laetrile): C. G. Moertel *et al., J. Am. Med. Assoc.* **245,** 591 (1981); M. M. Ames *et al., Cancer Chemother. Pharmacol.* **6,** 51 (1981). Pharmacokinetics: A. G. Rauws *et al., Arch. Toxicol.* **49,** 311 (1982). Amygdalin (laetrile) is a toxic drug that is not effective as a cancer treatment: C. G. Moertel *et al., N. Engl. J. Med.* **306,** 201 (1982). Review of the controversial use of amygdalin (laetrile): V. Herbert, *Am. J. Clin. Nutr.* **32,** 1121-1158 (1979).

Trihydrate, orthorhombic columns from water, mp 200°; mp about 220° when anhyd. The once melted and solidif substance remelts at 125-130°. $[\alpha]_D^{20}$ −42° (anhyd basis). One gram dissolves in 12 ml water, in 900 ml alcohol, in 11 ml boiling alcohol. Very sol in boiling water; almost insol in ether. pH of satd aq soln ~7.
Note: The misleading term *vitamin B$_{17}$* has sometimes been applied to amygdalin.

639. *n*-Amylamine. *1-Pentanamine; pentylamine;* 1-aminopentane. $C_5H_{13}N$; mol wt 87.16. C 68.90%, H 15.03%, N 16.07%. $CH_3(CH_2)_3CH_2NH_2$. Prepd by reduction of valeronitrile with LiAlH$_4$ and with LiAlH$_4$-AlCl$_3$: Nystrom, *J. Am. Chem. Soc.* **77,** 2544 (1955).

Liquid. d^{19} 0.766. bp 104°. mp −55°. Very sol in water; sol in alcohol; miscible with ether.

Caution: A strong irritant.

640. Amylase. Enzymes catalyzing the hydrolysis of α-1→4 glucosidic linkages of polysaccharides such as glycogen, starch, or their degradation products. *Endoamylases* attack the α-1→4 linkage at random. A single type of endoamylase is known, *i.e.,* α-amylases (α-*1,4-glucan 4-glucanohydrolases*), so named, because the reducing hemiacetal group liberated by the hydrolysis has α optical configuration and mutarotates downward. The more common α-amylases include those isolated from human saliva, human, hog and rat pancreas, *Bacillus subtilis, B. coagulans, Aspergillus oryzae, A. candidus, Pseudomonas saccharophila,* and barley malt. *Exoamylases* attack the α-1→4 linkages only from the non-reducing outer polysaccharide chain ends. Those breaking every glucosidic bond to produce solely α-glucose are known as *glucoamylases* (γ-*amylases*). Those breaking every alternate bond to produce maltose are known as β-amylases (α-*1,4-glucan maltohydrolases*). Exoamylases are exclusively of vegetable or microbial origin. *Reviews:* Fischer, Stein, "α-Amylases" and French, "β-Amylases", in *The Enzymes,* **Vol. 4,** P. D. Boyer *et al.,* Eds., (Academic Press, New York, 2nd ed., 1960) pp 313-343, 345-368; J. A. Thoma *et al., ibid.* Vol. V (3rd ed., 1971) pp 115-189; W. M. Fogarty, C. T. Kelly, *Microbial Enzymes and Bioconversions,* A. H. Rose, Ed. (Academic Press, New York, 1980) pp 115-170.

α-Amylase (porcine), *Buclamase, Maxilase.* Enzyme derived from swine pancrease. mol wt ~45,000. Prepn: Caldwell *et al., J. Am. Chem. Soc.* **74,** 4033 (1952).

α-Amylase (bacterial), usually derived from *Bacillus subtilis.* Purification: Stein, Fischer, *Helv. Chim. Acta* **40,** 529 (1957).

β-Amylase (sweet potato), mol wt ~152,000. Prepn: Balls *et al., J. Biol. Chem.* **173,** 9 (1948).

USE: In starch processing, brewing, distilling, baking, animal feed, sewage treatment.

THERAP CAT: Enzyme (digestive aid).

641. Amylbenzene. *Pentylbenzene; n-*amylbenzene; 1-phenylpentane. $C_{11}H_{16}$; mol wt 148.25. C 89.12%, H 10.88%. $C_6H_5(CH_2)_4CH_3$. Prepd by the action of benzylmagnesium chloride on *n-*butyl *p-*toluenesulfonate: Rossander, Marvel, *J. Am. Chem. Soc.* **50,** 1491 (1928); Gilman, Heck, *ibid.* **50,** 2223 (1928); Gilman, Robinson, *Org. Syn.* **coll. Vol. II,** 47, (1943).

Liquid. mp −78.25°. bp_{760} 202.2°; bp_{10} 81°. d_4^{20} 0.8594. n_D^{20} 1.48849: Vogel, *J. Chem. Soc.* **1948,** 607. Insol in water. Sol in alcohol; miscible with ether, benzene.

642. d-Amyl Bromide. *1-Bromo-2-methylbutane; d-pri-act-*amyl bromide; L-1-bromo-2-methylbutane. $C_5H_{11}Br$; mol wt 151.05. C 39.76%, H 7.34%, Br 52.90%. CH_3CH_2-$CH(CH_3)CH_2Br$. Prepn: W. Marckwald, *Ber.* **37,** 1038 (1904); H. O. Jones, *J. Chem. Soc.* **87,** 135 (1905); R. P. Frohardt *et al., J. Am. Chem. Soc.* **81,** 5500 (1959).

Liquid. d_4^{20} 1.221. bp 120-121°. n_D^{25} 1.4425; $[\alpha]_D^{28}$ +4.69° (neat). Practically insol in water. Sol in alcohol, ether.

643. n-Amyl Bromide. *1-Bromopentane.* $C_5H_{11}Br$; mol wt 151.05. C 39.76%, H 7.34%, Br 52.90%. CH_3-$(CH_2)_3CH_2Br$. Prepn: Fournier, *Bull. Soc. Chim. France* **35,** 623 (1906); Lindstone, Morris, *Chem. & Ind. (London)* **1958,** 560.

Liquid. d_4^{15} 1.2237. bp_{740} 129.7°. mp −95°. n_D^{20} 1.4444. Practically insol in water. Sol in alcohol; miscible with ether.

644. tert-Amyl Bromide. *2-Bromo-2-methylbutane;* 2-bromoisopentane. $C_5H_{11}Br$; mol wt 151.05. C 39.76%, H 7.34%, Br 52.90%. $(CH_3)_2CBrCH_2CH_3$. Prepd from trimethylethylene and dry HBr at −78°: Michael, Weiner, *J. Org. Chem.* **4,** 531 (1939).

Liquid, bp 107.4°. d_0^0 1.2439. n_D^{20} 1.4430.

645. n-Amyl Butyrate. *Butanoic acid pentyl ester.* $C_9H_{18}O_2$; mol wt 158.24. C 68.31%, H 11.47%, O 20.22%. $CH_3CH_2CH_2COOCH_2(CH_2)_3CH_3$. Prepn: Gartenmeister, *Ann.* **233,** 269 (1886). Toxicity study: P. M. Jenner *et al., Food Cosmet. Toxicol.* **2,** 327 (1964).

Liquid. Apricot-like odor. mp −73.2°. d_0^0 0.8832; d_4^{15} 0.8713. bp_{760} 185°. n_D^{20} 1.4110. Soly in water (50°): 0.54 g/l; very sol in alc, ether. LD_{50} orally in rats: 12210 mg/kg (Jenner).

USE: Has been used in such flavors as apricot, pineapple, pear, plum, and sparingly in some perfume compositions.

646. n-Amyl Caproate. *Hexanoic acid pentyl ester;* pentyl hexanoate; *n-*caproic acid *n-*amyl ester. $C_{11}H_{22}O_2$; mol wt 186.29. C 70.92%, H 11.90%, O 17.18%. CH_3-$(CH_2)_4COOCH_2(CH_2)_3CH_3$. Prepn: Simonini, *Monatsh.* **13,** 320 (1892).

Liquid, bp 222-227°.

647. Amyl Chloride. *1-Chloropentane; n-*amyl chloride; *n-*butylcarbonyl chloride. $C_5H_{11}Cl$; mol wt 106.60. C 56.34%, H 10.40%, Cl 33.26%. $CH_3(CH_2)_3CH_2Cl$. Prepd from 1-pentanol and concd HCl in sealed tube at 120°: Conant, Kirner, *J. Am. Chem. Soc.* **46,** 245 (1924); with HCl and zinc chloride: Clark, Streight, *Trans. Roy. Soc. Can.* [3] **23, III** 77 (1929); Vogel, *J. Chem. Soc.* **1943,** 638, 640; for prepn without admixed 2- and 3-chloropentanes *see* Whitmore's procedure from 1-pentanol with $SOCl_2$ and pyridine: Whitmore *et al., J. Am. Chem. Soc.* **60,** 2540 (1938); Mixer, Young, *ibid.* **78,** 3382 (1956).

Liquid. d_4^{20} 0.8828. mp −99°. bp_{760} 107.8°. n_D^{20} 1.41280. Flash pt, closed cup: 55°F (13°C). Miscible with alc, ether. Insol in water. Forms a constant boiling mixture with water, bp 82°, with ethanol bp 72.5°.

648. 6-n-Amyl-m-cresol. *5-Methyl-2-pentylphenol;* 6-*pentyl-m-cresol.* $C_{12}H_{18}O$; mol wt 178.27. C 80.85%, H 10.18%, O 8.97%. Prepn: Coulthard, *J. Chem. Soc.* **1930,** 280.

Solid or liquid, mp 24°, bp_{15} 137-139°. Practically insol in water; sol in ethanol, acetone, ether, alkali: Baker, *Drug & Cosmet. Ind.* **34,** 464 (1934).

USE: Antiseptic, germicide and mold preventive.

649. Amylene. *2-Methyl-2-butene;* β-isoamylene; trimethylethylene. C_5H_{10}; mol wt 70.13. C 85.63%, H 14.37%. $(CH_3)_2C=CHCH_3$. Prepd by dehydration of *tert*-amyl alcohol in the presence of *p-*toluenesulfonic acid: Applequist, Babad, *J. Org. Chem.* **27,** 288 (1962); by disproportionation of isobutene with propylene or 2-butene: Banks, Regier, *Ind. Eng. Chem., Prod. Res. Dev.* **10,** 46 (1971).

Liquid, bp 37.5-38.5°. Highly flammable, flash pt 0°F. d_4^{15} 0.66. Disagreeable odor. Polymerizes on long standing. Practically insol in water. Miscible with alc, ether.

Caution: A simple asphyxiant.

650. Amylene Dichloride. *2,3-Dichloro-2-methylbutane; tert-*amylene dichloride; trimethylethylene dichloride. $C_5H_{10}Cl_2$; mol wt 141.04. C 42.58%, H 7.15%, Cl 50.27%. Prepd from trimethylethylene (2-methyl-2-butene) by treatment with Cl_2 or SO_2Cl_2: Ger. pat. **251,100** (1911); **258,555** (1912); from neopentyl iodide and chlorine: Beringer, Schultz, *J. Am. Chem. Soc.* **77,** 5533 (1955).

*dl-*Form, liquid. d_4^{15} 1.0696. bp_{760} 138°; bp_{20} 37°. n_D^{18} 1.4450. Insol in water. Sol in alcohol, ether.

651. n-Amyl Ether. *1,1'-Oxybispentane; pentyl ether;* amyl oxide; diamyl ether. $C_{10}H_{22}O$; mol wt 158.28. C 75.88%, H 14.01%, O 10.11%. $[CH_3(CH_2)_3CH_2]_2O$. Prepn from amyl alcohol with concd H_2SO_4: Hinton, Nieuwland, *Proc. Indiana Acad. Sci.* **42,** 109 (1933), *C.A.* **27,** 5716 (1933).

Liquid, bp 186.75°. mp −69.43°. Flash pt, closed cup: 57°C. d_4^{20} 0.78326, d_4^{25} 0.77924. n_D^{20} 1.41195, n_D^{25} 1.40985:

Dreisbach, Martin, *Ind. Eng. Chem.* **41**, 2875 (1949). Practically insol in water. Miscible with alcohol, ether.

Caution: Vapors narcotic in high concns.

USE: Industrial solvent.

652. tert-Amyl Isovalerate. *3-Methylbutanoic acid 1,1-dimethylpropyl ester; isovaleric acid tert-pentyl ester; tert-*pentyl isovalerate; isovaleric acid *tert-*amyl ester. $C_{10}H_{20}O_2$; mol wt 172.27. C 69.72%, H 11.70%, O 18.58%. $(CH_3)_2$-CHCH$_2$COOC(CH$_3$)$_2$C$_2$H$_5$. Prepd from trimethylethylene, isovaleric acid and zinc chloride: *Beilstein* **vol. 2**, 312.

Liquid, valerian odor and taste, bp 173-174°. d_4^0 0.8729, d_0^{14} 0.8608. Slightly sol in water; freely sol in alc; miscible with oils.

653. n-Amyl Mercaptan. *1-Pentanethiol;* amyl thioalcohol. $C_5H_{12}S$; mol wt 104.22. C 57.63%, H 11.61%, S 30.77%. CH$_3$(CH$_2$)$_3$CH$_2$SH. Prepn: Cossar *et al., J. Org. Chem.* **27**, 93 (1962).

Liquid, bp 123-124°, n_D^{25} 1.4439. Penetrating, unpleasant odor. d_4^{20} 0.857. Practically insol in water. Sol in alc.

Caution: Mild irritant to skin, mucous membranes. May cause skin sensitization.

USE: In organic syntheses.

654. tert-Amyl Methyl Ether. *2-Methoxy-2-methylbutane; methyl tert-pentyl ether;* 1,1-dimethylpropyl methyl ether; methyl *tert-*amyl ether; TAME. $C_6H_{14}O$; mol wt 102.18. C 70.53%, H 13.81%, O 15.66%. Gasoline octane booster. Prepn: A. Reychler, *Bull. Soc. Chim. Belg.* **21**, 71 (1907); and properties: T. W. Evans, K. R. Edlund, *Ind. Eng. Chem.* **28**, 1186 (1936). Industrial synthesis and processing: H. Short, *Chem. Eng.* **93**, 34 (1986); S. Randriamahefa *et al., J. Mol. Catal.* **49**, 85 (1988); W. J. Reagan, *Prepr. Pap. - A.C.S.,* Div. Fuel Chem. **39**, 337 (1994). Reaction equilibria in synthesis: L. K. Rihko *et al., J. Chem. Eng. Data* **39**, 700 (1994). Chromatographic and spectroscopic analyses: J. S. Hardman *et al., Fuel* **72**, 1563 (1993). Comparison with MTBE, *q.v.,* of effects on emissions: W. J. Koehl *et al., Soc. Automot. Eng.* SP-1000, vol 2, 289 (1993). Reaction with nitrate radical (environmental implications): S. Langer, E. Ljungström, *J. Phys. Chem.* **98**, 5906 (1994).

Liquid, bp$_{760\ mm}$ 86.3°. d_4^{15} 0.7750; d_4^{20} 0.7703; d_4^{25} 0.7656; d_4^{30} 0.7607. n_D^{20} 1.3885. Vapor pressure (25°): 75 mm Hg. Soly in water (20°): 1.15g/100g soln.

USE: Fuel additive for gasoline.

655. Amyl Nitrite. Mixture of isomers containing not less than 97.0% and not more than 100.0% of $C_5H_{11}NO_2$. Consists chiefly of isoamyl nitrite [(CH$_3$)$_2$CHCH$_2$CH$_2$-ONO], but other isomers are also present. The N.F. grade has d_{25}^{25} 0.870-0.876. *See* Isoamyl Nitrite.

656. Amylocaine Hydrochloride. *1-(Dimethylamino)-2-methyl-2-butanol benzoate hydrochloride;* 1-(dimethylaminomethyl)-1-methylpropyl benzoate hydrochloride; methylethyldimethylaminomethylcarbinol benzoyl ester hydrochloride; amyleine hydrochloride; Stovaine. $C_{14}H_{22}ClNO_2$; mol wt 271.79. C 61.87%, H 8.16%, Cl 13.04%, N 5.15%, O 11.77%. Prepd by benzoylation of 1-(dimethylamino)-2-methyl-2-butanol: Fourneau, Ribas, *Bull. Sci. Pharmacol.* **35**, 273 (1928), *C.A.* **22**, 2919[1] (1928).

Crystals, bitter taste, followed by temporary numbness of the tongue. Dec 177-179°. One gram dissolves in 2 ml water, in 3.3 ml abs ethanol. Practically insol in ether. A

5% aq soln is faintly acid to litmus and neutral to Congo red.

THERAP CAT: Anesthetic (local).

657. Amyloid. Amyloid substance; AS. Tissue deposits occurring in animals and humans, primarily composed of unique proteinaceous fibrils. The term "amyloid" was coined by Virchow in 1853 to reflect the starch-like darkening of the deposits after iodine staining. Deposition of the relatively inert fibrils by different pathogenic mechanisms leads to excessive accumulation, with pressure atrophy and death resulting from interference with physiological processes of affected organs. This disease complex is referred to as amyloidosis or the β-fibrilloses; it occurs in association with inflammatory conditions such as rheumatoid arthritis, with immunocyte-derived dyscrasias, and with tumors of various types. Aging is thought to occur in association with amyloid deposition in the brain, heart and pancreas; systemic amyloidosis has been implicated in "senile" cardiomyopathy as well as in presenile dementia (Alzheimer's syndrome) and senile dementia, *cf.* T. I. Mandybur, *Neurology* **25**, 120 (1975). Human amyloid fibrils are characterized by green polarization birefringence after staining with Congo red, by constant tryptophan content and by an x-ray diffraction crystallographic pattern indicating a β-pleated sheet conformation: E. D. Eanes, G. G. Glenner, *J. Histochem. Cytochem.* **16**, 673 (1968). Of the two chemically distinguishable classes of amyloid fibrils, one has as its major component a fragment of immunoglobulin (*q.v.*) light chain and is designated *AL-protein*: G. G. Glenner *et al., Science* **172**, 1150 (1971); **174**, 712 (1971); and the other a non-immunoglobulin protein designated *AA-protein*: E. P. Benditt, N. Eriksen, *Am. J. Pathol.* **65**, 231 (1971); M. Levin *et al., J. Clin. Invest.* **51**, 2773 (1972); D. Ein *et al., J. Biol. Chem.* **247**, 5653 (1972); M. Levin *et al., J. Exp. Med.* **138**, 373 (1973). In addition to the fibrils, a constant constituent of amyloid deposits has been found to be a normal serum globular glycoprotein called *P-component* or *SAP;* the staining of amyloid deposits with periodic acid-Schiff has been attributed to this substance, *cf.* H. Haupt *et al., Z. Physiol. Chem.* **353**, 1841 (1972). For a discussion of isoln and structure of P-component and its relationship to C-reactive protein, *q.v., see* several authors in *Ann. N.Y. Acad. Sci.* **389** (1982) 482 pp. Comprehensive review of amyloid deposits and amyloidosis: G. G. Glenner, *N. Engl. J. Med.* **302**, 1283-1292, 1333-1343 (1980). Review of amyloidosis: P. Lovisetto *et al., Minerva Med.* **71**, 1793-1813 (1980). *See also* Bence-Jones Proteins.

658. Amylpenicillin Sodium. *[2S-(2α,5α,6β)]-3,3-Dimethyl-7-oxo-6-[(1-oxohexyl)amino]-4-thia-1-azabicyclo-[3.2.0]heptane-2-carboxylic acid monosodium salt;* sodium *n-*amylpenicillinate; penicillin dihydro F sodium; flavacidin; flavicin. $C_{14}H_{21}N_2NaO_4S$; mol wt 336.39. C 49.99%, H 6.29%, N 8.33%, Na 6.83%, O 19.02%, S 9.53%. Antibiotic produced by the mold *Aspergillus flavus* from a Czapek-Dox medium supplemented with corn steep liquor or by *Penicillium chrysogenum* Q176 or by *P. notatum* strains: McKee, MacPhillamy, *Proc. Soc. Exp. Biol. Med.* **53**, 247 (1943); Bush, Goth, *J. Pharmacol. Exp. Ther.* **78**, 164 (1943); McKee *et al., J. Bacteriol.* **47**, 187 (1944); Bush *et al., J. Pharmacol. Exp. Ther.* **84**, 264 (1945); Fried *et al., J. Biol. Chem.* **163**, 341 (1946); *see also* Wintersteiner under "Flavacidin" in *Chemistry of Penicillin* (Princeton, 1949). Prepn by hydrogenation of 2-pentenylpenicillin: Catch *et al., Brit. pat.* 584,852; Cook, Heilbron in *Chemistry of Penicillin* (Princeton, 1949). Characterization and antibacterial activity: Leigh, *Nature* **163**, 95 (1949).

Monohydrate, flat, blunt-ended needles from moist acetone or moist ethyl acetate. When anhydrous, mp 188° (dec). $[\alpha]_D^{23}$ +319°. Very soluble in water.

659. α-Amyrin. *(3β)-Urs-12-en-3-ol; α-amyrenol;* viminalol. $C_{30}H_{50}O$; mol wt 426.73. C 84.44%, H 11.81%, O 3.75%. Occurs mostly as acetate in latex of rubber trees, in latex from *Ficus variegata* Blume, *Moraceae,* also in *Balanophora elongata* Blume, *Balanophoraceae,* and in *Erythroxylum coca* Lam. var. *novogranatense* Morris, and var. *spruceanum* Burck, *Erythroxylaceae.* Isoln from *Manila elemi:* Vesterberg, Westerlind, *Ann.* **428,** 247 (1922). Structural studies: Spring, Vickerstaff, *J. Chem. Soc.* **1937,** 249; Beynon *et al., ibid.* **1938,** 1233; Meisels *et al., Helv. Chim. Acta* **32,** 1075 (1949), **38,** 1298 (1955); Melera *et al., ibid.* **39,** 441 (1956). Identity with viminalol: Soldin, Marais, *J. Pharm. Soc.* **55,** 452 (1966). Formation from ursolic acid: Goodson, *J. Chem. Soc.* **1938,** 999; from boswellic acid: Ruzicka, Wirz, *Helv. Chim. Acta* **22,** 948 (1939). Partial synthesis from glycyrrhetic acid and stereochemistry: Corey, Cantrall, *J. Am. Chem. Soc.* **81,** 1745 (1959). *Review:* J. Simonsen, W. C. J. Ross, *The Terpenes,* vol. **IV** (University Press, Cambridge, 1957) pp 116-148.

Needles from alcohol, mp 186°. $bp_{0.7}$ 243°. $[\alpha]_D^{17}$ +91.6° (c = 1.3 in benzene). Sol in 22 parts 98% alc. Sol in ether, benzene, chloroform, glacial acetic acid. Slightly sol in petr ether.

Acetate, $C_{32}H_{52}O_2$, leaflets from petr ether, mp 227°. $[\alpha]_D^{20}$ +76.35° (c = 0.572 in $CHCl_3$).

Benzoate, $C_{37}H_{54}O_2$, prisms from benzene + acetone, mp 195-196°. $[\alpha]_D^{10}$ +94.6° (c = 1.9 in $CHCl_3$).

660. β-Amyrin. *(3β)-Olean-12-en-3-ol; β-amyrenol.* $C_{30}H_{50}O$; mol wt 426.73. C 84.44%, H 11.81%, O 3.75%. Occurs together with α-amyrin. Isoln and structural studies: *see* α-amyrin. *See also* Vesterberg, *Bull. Soc. Chim.* **37,** 742 (1925); Horrmann, Firzlaff, *Arch. Pharm.* **268,** 64 (1930); Ruzicka, Marxer, *Helv. Chim. Acta* **22,** 195 (1939); Jeger, Ruzicka, *ibid.* **28,** 209 (1945); Prelog *et al., ibid.* **29,** 360 (1946). Conversion of δ-amyrene to β-amyrin: Barton *et al., J. Chem. Soc. (C)* **1968,** 1031. Biogenetic-type total synthesis: van Tamelen *et al., J. Am. Chem. Soc.* **94,** 8229 (1972). Biosynthesis from squalene: Suga *et al., Chem. Letters* **1972,** 129, 313.

Needles from petr ether or alc, mp 197-197.5°. $bp_{0.8}$ 260°. $[\alpha]_D^{19}$ +99.8° (c = 1.3 in benzene). Somewhat less soluble than the α-form. Soluble in 37 parts of 98% alc.

Acetate, $C_{32}H_{52}O_2$, prisms from petr ether, mp 241°. $[\alpha]_D^{17}$ +79° (c = 0.9 in benzene).

Palmitate, $C_{46}H_{80}O_2$, *balanophorin.* mp 77°. $[\alpha]_D^{15}$ +54.5° (c = 1.1 in benzene). Occurs in *Balanophora elongata* Blume, *Balanophoraceae,* in *Erythroxylum coca* Lam. var *novogranatense* Morris, and var *spruceanum* Burck, *Erythroxylaceae,* in latex from *Ficus variegata* Blume, *Moraceae.*

Di-β-amyrin ether, $C_{60}H_{98}O$, mp 135-136°: Rollett, *Monatsh.* **47,** 437 (1926).

661. Anabasine. *3-(2-Piperidinyl)pyridine; 2-(3-pyridyl)piperidine;* neonicotine. $C_{10}H_{14}N_2$; mol wt 162.23. C 74.03%, H 8.70%, N 17.27%. In *Anabasis aphylla* L., *Chenopodiaceae:* Orechoff, Menschikoff, *Ber.* **64,** 266 (1931); in *Nicotiana glauca* Graham, *Solanaceae:* Smith, *J. Am. Chem. Soc.* **57,** 959 (1935); Pyriki, Oehler, *Pharmazie* **9,** 685 (1954). Synthesis: Späth, *Ber.* **70B,** 70 (1937). Extracted on a large scale in Russia. Industrial extraction processes: Sadykov, Timbekov, *J. Appl. Chem. USSR* **29,** 148 (1956). Abs config: Lukes *et al., Coll. Czech. Chem. Commun.* **27,** 751 (1962).

Liquid, bp 270-272°; bp_{14} 145-147°; bp_2 105°. Freezes at 9°. d_4^{20} 1.0455. n_D^{20} 1.5430. $[\alpha]_D^{20}$ −83.1°. Sol in water and in most organic solvents.

Hydrochloride, $[\alpha]_D$ +16.5° (c = 10 in water).

Caution: Acute and subacute toxicity: increased salivation, vertigo, confusion, disturbed vision and hearing, photophobia, cold extremities, nausea, vomiting, diarrhea, syncope, clonic spasms.

USE: Insecticide.

662. Anabsinthin. *3,3a,4,5,6,6a,6b,7,7a,8,9,10,10a,-13a,13c,14b-Hexadecahydro-6-hydroxy-3,6,8,11,14,15-hexamethyl-2H-8,15-epoxy-7,13b-ethanopentaleno(1″,2″:6,7;5″,-4″:6′,7′)dicyclohepta(1,2-b:1′,2′-b′)difuran-2,12(11H)-dione;* anabsynthin. $C_{30}H_{40}O_6$; mol wt 496.64. C 72.55%, H 8.12%, O 19.33%. A bitter principle isolated from *Artemisia absinthium* L., *Compositae* (wormwood). Isoln: F. Sorm *et al., Chem. & Ind. (London)* **1955,** 569; V. Herout *et al., Coll. Czech. Chem. Commun.* **21,** 1485 (1956). Structure: L. Novotny *et al., Chem. & Ind. (London)* **1958,** 465; V. Herout *et al., Coll. Czech. Chem. Commun.* **25,** 1492 (1960); J. Beauhaire *et al., Tetrahedron Letters* **21,** 3191 (1980).

Anhydrous form, crystals from benzene or isopropanol + diisopropyl ether, or by drying over phosphorus pentoxide in vacuo, mp 267°. $[\alpha]_D^{20}$ +113° (c = 1.85 in chloroform).

Monohydrate, crystals from methanol, mp 210°.

663. Anacardic Acid. Principal constituent of cashew nut-shell liquid, *Anacardium occidentalis* L., *Anacardiaceae,* member of the family of non-isoprenoid long-chain phenols. (*See also* Urushiol.) Anacardic acid is a mixture of *2-hydroxy-6-alkylbenzoic acids* in which the alkyl chain (C_{11} or higher) is fully saturated (I) or is a monoene (II), a diene (III) or a triene (IV). The name anacardic acid is also used in the literature to designate one component of the mixture: *6-pentadecyl-2-hydroxybenzoic acid* (I) (*6-pentadecylsalicylic acid*). Another component of the mixture *6-(8-pentadecenyl)-2-hydroxybenzoic acid* (II), is also known as *ginkgoic acid.* Isoln and structure: Städeler, *Ann.* **63,** 137 (1847); Ruhemann, Skinner, *Ber.* **20,** 1861 (1887); Haagen, Smit, *K. Akad. Wetensch.* **34,** 165 (1931); Backer, Haack, *Rec. Trav. Chim.* **660,** 61 (1941); Kremers, U.S. pat. **2,431,127** (1947). Structure of side chain: Sletzinger, Dawson, *J. Org. Chem.* **14,** 670, 849 (1949). Structure of unsaturated components: J. H. P. Tyman, N. Jacobs, *J. Chromatog.* **54,** 83 (1971). GLC analysis of components: J. H. P. Tyman, *ibid.* **111,**

285 (1975). Synthesis and prostaglandin synthetase inhibiting activity of I and II: Y. Yamagiwa *et al., Tetrahedron* **43,** 3387 (1987); of I: I. Kubo *et al., Chem. Letters* **1987,** 1101. Antitumor activity of I: H. Itokawa *et al., Chem. Pharm. Bull.* **35,** 3016 (1987). *Review:* J. H. P. Tyman, *Chem. Soc. Rev.* **8,** 499-537 (1979).

I R = (CH₂)₇CH₃
II R = CH=CH(CH₂)₅CH₃
III R = CH=CHCH₂CH=CH(CH₂)₂CH₃
IV R = CH=CHCH₂CH=CHCH₂CH=CH₂

Mixture, crystals from acetone, mp 34-37°. Sparingly sol in water; freely sol in alc, ether, petr ether. Forms a water-soluble sodium salt.

6-Pentadecyl component (I), needles from hexane, mp 90.2-91.5°.

6-(8-Pentadecenyl) component (II), needles, mp 45.3-48° (subl).

664. Anagestone. *(6α)-17-Hydroxy-6-methylpregn-4-en-20-one;* 6α-methyl-4-pregnen-17α-ol-20-one. C₂₂H₃₄O₂; mol wt 330.51. C 79.95%, H 10.37%, O 9.68%. Prepn of free alc, acetate, propionate, hexanoate, and cyclopentylpropionate: **Belg. pat. 624,370** (1963 to Ortho), *C.A.* **60,** 10764g (1964).

Crystals from methanol + dichloromethane, mp 190-193°. [α]$_D^{20}$ +51° (in chloroform).

Acetate, C₂₄H₃₆O₃, *Anatropin.* Crystals from methanol + dichloromethane, mp 173-175°. [α]$_D^{20}$ +24° (in chloroform).

Propionate, C₂₅H₃₈O₃, crystals from methanol + dichloromethane, mp 143-144°. [α]$_D^{20}$ +35° (in chloroform).

Hexanoate, C₂₈H₄₄O₃, crystals from methanol + dichloromethane, mp 78-79°. [α]$_D^{20}$ +20°.

Cyclopentylpropionate, C₃₀H₄₆O₃, crystals, mp 93-95°.

THERAP CAT: Progestogen.

665. Anagrelide. *6,7-Dichloro-1,5-dihydroimidazo[2,1-b]quinazolin-2(3H)-one;* 6,7-dichloro-1,2,3,5-tetrahydroimidazo[2,1-b]quinazolin-2-one. C₁₀H₇Cl₂N₃O; mol wt 256.09. C 46.90%, H 2.76%, Cl 27.69%, N 16.41%, O 6.25%. Phosphodiesterase inhibitor with antiplatelet activity. Prepn: W. N. Beverung, A. Partyka, U.S. pats. **3,932,-407; RE 31,617;** T. A. Jenks *et al.,* U.S. pat. **4,146,718** (1976, 1984, 1979 all to Bristol-Myers); H. Yamaguchi, F. Ishikawa, *J. Heterocycl. Chem.* **18,** 67 (1981). Antithrombotic and platelet aggregation inhibiting properties: J. S. Fleming, J. P. Buyniski, *Thromb. Res.* **15,** 373 (1979). Mode of action studies: S. S. Tang, M. M. Frojmovic, *J. Lab. Clin. Med.* **95,** 241 (1980); S. Seiler *et al., J. Pharmacol. Exp. Ther.* **243,** 767 (1987). GC-MS determn in human plasma: E. H. Kerns *et al., J. Chromatog.* **416,** 357 (1987). Clinical reduction of platelet counts: W. A. Andes *et al., Thromb. Haemostas.* **52,** 325 (1984). Clinical trials to control thrombocytosis in chronic myeloproliferative diseases: M. N. Silverstein *et al., N. Engl. J. Med.* **318,** 1292 (1988); Anagrelide Study Group, *Am. J. Med.* **92,** 69 (1992). Review of pharmacology and therapeutic potential: C. M. Spencer, R. N. Brogden, *Drugs* **47,** 809-822 (1994).

Hydrochloride, C₁₀H₇Cl₂N₃O.HCl, *BL-4162A, BMY-26538-01, Agrelin.* Prepd as the hemihydrate; crystals from ethanolic HCl, mp > 280°.

THERAP CAT: Antithrombocythemic.

666. Anagyrine. *[7R-(7α,7aβ,14α)]-7,7a,8,9,10,11,13,-14-Octahydro-7,14-methano-4H,6H-dipyrido[1,2-a:1',2'-e][1,-5]diazocin-4-one;* monolupine; rhombinin. C₁₅H₂₀N₂O; mol wt 244.34. C 73.74%, H 8.25%, N 11.47%, O 6.55%. Found in seeds of *Anagyris foetida* L., *Leguminosae* and in gorse (*Ulex europaeus* L., *Leguminosae*). Isoln: Ing, *J. Chem. Soc.* **1933,** 504; Orekhov *et al., Ber.* **67,** 1394 (1934); Couch, *J. Am. Chem. Soc.* **61,** 3327 (1939); Briggs, Russell, *J. Chem. Soc.* **1942,** 507; Galinsky, Stern, *Ber.* **77,** 132 (1944); Faugeras, *Ann. Pharm. Fr.* **29,** 241 (1971). Absolute configuration: Okuda *et al., Chem. & Ind. (London)* **1961,** 1116; Okuda *et al., Chem. Pharm. Bull.* **13,** 491 (1965). Synthesis: van Tamelen, Baran, *J. Am. Chem. Soc.* **80,** 4659 (1958); Goldberg, Lipkin, *J. Org. Chem.* **37,** 1823 (1972).

Pale yellow glass. bp₄ 210-215°; bp₁₂ 260-270°. [α]$_D^{25}$ −168° (c = 4.8 in ethanol). Sol in water, alcohol, chloroform; slightly sol in ether, benzene.

Hydrochloride trihydrate, C₁₅H₂₁ClN₂O.3H₂O, crystals, mp 235-236° (mp 296° when dry). [α]$_D^{25}$ −142.5° (c = 5). Freely sol in water.

667. Anastrozole. *α,α,α',α'-Tetramethyl-5-(1H-1,2,4-triazol-1-ylmethyl)-1,3-benzenediacetonitrile;* 2,2'-[5-(1H-1,-2,4-triazol-1-ylmethyl)-1,3-phenylene]di(2-methylpropionitrile); ZD-1033; ICI-D-1033; Arimidex. C₁₇H₁₉N₅; mol wt 293.37. C 69.60%, H 6.53%, N 23.87%. Aromatase inhibitor. Prepn: P. N. Edwards, M. S. Large, **Eur. pat. Appl. 296,749;** *eidem,* U.S. pat. **4,935,437** (1989, 1990 both to ICI). Review of pharmacology, clinical pharmodynamics: P. V. Plourde *et al., Breast Cancer Res. Treat.* **30,** 103-111 (1994). Review of clinical pharmacology and early clinical results: P. V. Plourde *et al., J. Steroid Biochem. Molec. Biol.* **53,** 175-179 (1995).

Crystals from ethyl acetate/cyclohexane, mp 81°-82°.

THERAP CAT: Antineoplastic; in treatment of breast cancer.

668. Anatabine. *1,2,3,6-Tetrahydro-2,3'-bipyridine;* 2-(3-pyridyl)-1,2,3,6-tetrahydropyridine; 1,2,3,6-tetrahydro-2-(3-pyridyl)pyridine. C₁₀H₁₂N₂; mol wt 160.22. C 74.97%, H 7.55%, N 17.48%. The most abundant of the minor alkaloids of tobacco: Späth, Kesztler, *Ber.* **70,** 239, 704, 2450 (1937). Fresh *Nicotiana tabacum,* the species most commonly used for the production of cigarette tobacco, contains 3.9% anatabine. Configuration: Lukes *et al., Coll. Czech. Chem. Commun.* **27,** 751 (1962). Total synthesis of *dl*-form: Quan *et al., J. Org. Chem.* **30**, 2769 (1965). Biosynthesis: E. Leete, S. Slattery, *J. Am. Chem. Soc.* **98**, 6326 (1976). Biomimetic synthesis: E. Leete, M. E. Mueller, *ibid.* **104**, 6440 (1982).

l-Form, liquid. d_4^{19} 1.091. bp_{10} 145-146°. $[\alpha]_D^{17}$ −177.8°. n_D^{20} 1.5676. Misc with water. Sol in alc, ether, benzene.
dl-Form, liquid. $bp_{6.5}$ 136°.

669. Anazolene Sodium. *4-Hydroxy-5-[[4-(phenylamino)-5-sulfo-1-naphthalenyl]azo]-2,7-naphthalenedisulfonic acid trisodium salt;* 4′-anilino-8-hydroxy-1,1′-azonaphthalene-3,5′,6-trisulfonic acid trisodium salt; trisodium 4′-anilino-8-hydroxy-1,1′-azonaphthalene-3,6,5′-trisulfonate; 1-naphthol-3,6-disulfonic acid-8-azo-4′-[*N*-phenyl-1′-naphthylamine]-8′-sulfonic acid trisodium salt; C.I. Acid Blue 92; C.I. 13390; Coomassie Blue; Coomassie Blue Medicinal; Coomassie Blue RL; Sulfon Acid Blue R; Wool Blue RL; Pontacyl Fast Blue R. $C_{26}H_{16}N_3Na_3O_{10}S_3$; mol wt 695.59. C 44.89%, H 2.32%, N 6.04%, Na 9.92%, O 23.00%, S 13.83%. Prepn: Ulrich, U.S. pat. **611,664** (1897 to Bayer); Ger. pat. **108,546** (1899 to Farbwerke Mülheim); *Frdl.* **5,** 497; *Beilstein* **16,** EII, 226; *Colour Index* vol. 4 (3rd ed., 1971) p 4053. Properties and biological behavior: S. H. Taylor, J. M. Thorp, *Brit. Heart J.* **21,** 492 (1959). Clinical applications: S. H. Taylor, J. P. Shillingford, *ibid.* 497; I. S. Menzies, *J. Clin. Pathol.* **19,** 179 (1966).

Reddish-black powder. Soluble in water, acetone, Cellosolve, giving a reddish-blue soln. Slightly sol in alc. Absorption max (water): 565-570 nm; in acetone 585 nm; in human plasma 580-590 nm ($E_{1cm}^{1\%}$ about 600). Aq solns are stable and are not affected by light. Solns up to 10% do not stain the skin appreciably. LD_{50} i.v. in mice: 450 mg/kg (Taylor, Thorp).
THERAP CAT: Diagnostic aid (cardiac output, blood volume determination).

670. Ancitabine. *[2R-(2α,3β,3aβ,9aβ)]-2,3,3a,9a-Tetrahydro-3-hydroxy-6-imino-6H-furo[2′,3′:4,5]oxazolo[3,2-a]pyrimidine-2-methanol;* 2,2′-anhydro-(1β-D-arabinofuranosyl)cytosine; 2,2′-O-cyclocytidine; $O^{2,2'}$-cyclocytidine; ancytabine; anhydroara C. $C_9H_{11}N_3O_4$, mol wt 225.20. C 48.00%, H 4.92%, N 18.66%, O 28.42%. A cytostatic agent and intermediate in the synthesis of cytarabine, *q.v.* Prepn of the hydrochloride: E. R. Walwick *et al., Proc. Chem. Soc.* **1959,** 84; T. Y. Shen, W. V. Ruyle, U.S. pat. **3,463,850** (1969 to Merck & Co.); E. K. Hamamura *et al., J. Med. Chem.* **19,** 654 (1976). General pharmacological properties: H. Hirayama *et al., Oyo Yakuri* **6,** 1259 (1972), *C.A.* **79,** 49175f (1973). Metabolism: D. H. W. Ho, *Drug Metab. Dispos.* **1,** 752 (1973). Biochemical study: *idem, Biochem. Pharmacol.* **23,** 1235 (1974). Pharmacokinetic study: H. S. Chen, J. F. Gross, *Cancer Chemother. Pharmacol.* **2,** 85 (1979). Clinical studies: J. Z. Finklestein *et al., Cancer Treat. Rep.* **63,** 1331 (1979); T. Miale *et al., ibid.* 1913. Toxicity studies: K. Sugihara *et al., Oyo Yakuri* **8,** 1469 (1974); H. Hirayama *et al., ibid.* 1693, *C.A.* **83,** 71766, 37747 (1975). HPLC study: V. Reichelova *et al., J. Chromatog.* **588,** 147 (1991).

Hydrochloride, $C_9H_{11}N_3O_4 \cdot HCl$, *NSC-145668, Cyclo-C.* Cryst, mp 248-250° (dec). $[\alpha]_D^{23}$ −21.8° (c = 2.0 in water). uv max (pH 1-7): 262, 231 nm (ϵ 10600, 9400).
THERAP CAT: Antineoplastic.

671. Ancrod. *Agkistrodon serine proteinase; Agkistrodon rhodostoma venom proteinase;* A-38414; Abbott 38414; Arvin; Arwin; Venacil (formerly). Defibrinating enzyme isolated from the venom of the Malaysian pit-viper, *Agkistrodon rhodostoma* Boie (*Calloselasma rhodostoma* Boie). Composed of approx 64% protein and 36% carbohydrate; mol wt ~35,400. Isoln: **Neth.** pat. **Appl. 6,502,120**; H. A. Reid *et al.,* U.S. pat. **3,657,416** (1965, 1972 to Nat. Res. Dev. Corp.); K. E. Chan *et al., Brit. J. Haematol.* **11,** 646 (1965). Initial purification: M. P. Esnouf, G. W. Tunnah, *Brit. J. Haematol.* **13,** 581 (1967). Improved purification, chemical composition: C. Nolan *et al., Methods Enzymol.* **45,** 205 (1976). Effects on prothrombin and fibrinogen metabolism in man: W. R. Bell *et al., J. Lab. Clin. Med.* **91,** 592 (1978). Series of articles on hematology, defibrinogenation effects on peripheral circulation and on exptl glomerulonephritis: *J. Med. Enzymol.* **3,** 510-538 (1980), *C.A.* **93,** 215548-50, 231022 (1980). Clinical studies: G. Trübestein, *Angiology* **32,** 699 (1981); J. A. Dormandy, *ibid.* 710; G. D. Lowe *et al., ibid.* **33,** 46 (1982). *Review:* W. R. Bell, *CRC Handbook Ser. Clin. Lab. Sci. Sect. I* **3,** 301-324 (1980).
Colorless substance when pure, having a light powdery texture when in the freeze-dried state. Soluble in physiological saline. Absorbable on weakly basic anion exchange materials.
THERAP CAT: Anticoagulant.

672. Ancymidol. *α-Cyclopropyl-α-(4-methoxyphenyl)-5-pyrimidinemethanol;* α-cyclopropyl-4-methoxy-α-(pyrimidin-5-yl)benzyl alcohol; EL-531; A-Rest; Reducymol. $C_{15}H_{16}N_2O_2$; mol wt 256.30. C 70.29%, H 6.29%, N 10.93%, O 12.48%. Prepn: J. D. Davenport *et al., Fr.* pat. **1,569,940** corresp to H. M. Taylor *et al.,* U.S. pat. **3,818,009** (1969, 1974 to Lilly). GLC determn: R. Frank, E. W. Day, *Anal. Methods Pestic. Plant Growth Regul.* **8,** 475 (1976); S. D. West, E. W. Day, *J. Assoc. Off. Anal. Chem.* **60,** 904 (1977). Studies on the specificity and site of action: R. C. Coolbaugh *et al., Plant Physiol.* **62,** 571 (1978).

Cryst. solid, mp 110-111°. Vapor pressure at 50°: <1 × 10^{-6} mm Hg. Soly in water at 25°: ~650 mg/l. Freely sol in acetone, methanol, ethyl acetate, chloroform, acetonitrile. Moderately sol in aromatic hydrocarbons; slightly sol in saturated hydrocarbons. LD_{50} orally in rats: 4500 mg/kg, *RTECS Vol. II,* R. J. Lewis, R. L. Tatken, Eds. (1980) p 532.
USE: Plant growth regulator.

673. Andrographolide. *[1R-[1α[E(S*)],4aβ,5α,6α,-8aα]]-3-[2-[Decahydro-6-hydroxy-5-(hydroxymethyl)-5,8a-dimethyl-2-methylene-1-naphthalenyl]ethylidene]dihydro-4-hydroxy-2(3H)-furanone; 3α,14,15,18-tetrahydroxy-5β,9βH,-10α-labda-8(20),12-dien-16-oic acid γ-lactone.* $C_{20}H_{30}O_5$; mol wt 350.46. C 68.55%, H 8.63%, O 22.83%. Main bitter constituent and active principle isolated from the leaves of *Andrographis paniculata* Nees, *Acanthaceae*, also known as **Chuanxinlian** or **Kalmegh**, which has been used in Asian traditional medicine as an antipyretic, anti-inflammatory, and hepatoprotectant. Isoln: M. K. Gorter, *Rec. Trav. Chim.* **30**, 151 (1911). Characterization: D. Chakravarti, R. N. Chakravarti, *J. Chem. Soc.* **1952**, 1697; R. Schwyzer *et al., Helv. Chim. Acta* **34**, 652 (1951). Structure: M. P. Cava, B. Weinstein, *Chem. & Ind. (London)* **1959**, 851. Revised structure: M. P. Cava *et al., Chem. & Ind. (London)* **1963**, 167; M. P. Cava *et al., Tetrahedron* **21**, 2617 (1965). X-ray crystallographic analysis: T. Fujita *et al., Chem. Pharm. Bull.* **32**, 2117 (1984). HPLC determn: A. Sharma *et al., Phytochem. Anal.* **3**, 129 (1992). Review of andrographolide and related compds: W. Tang, G. Eisenbrand, *Chinese Drugs of Plant Origin* (Springer-Verlag, Berlin, 1992) pp 97-103.

Colorless plates from ethanol or methanol, mp 218° (dec) (Gorter). Also reported as mp 218-221° (Fujita). $[\alpha]_D^{25}$ $-96.2°$ (c = 1.00 in C_5H_5N). d_4^{21} 1.2317. uv max (alcohol): 223 nm (log ε 4.09). Sparingly sol in water; sol in acetone, methanol, chloroform, ether.

Triacetyl derivative, $C_{26}H_{36}O_8$, fine needles from alcohol + ether, mp 126-126.5°.

674. Androisoxazole. *(5α,17β)-17-Methylandrostano-[3,2-c]isoxazol-17-ol; 2,3,3a,3b,4,5,5a,6,10,10a,10b,11,12,12a-tetradecahydro-1,10a,12a-trimethyl-1H-cyclopenta[7,8]phenanthro[2,3-c]isoxazol-1-ol; 17β-hydroxy-17α-methylandrostano[3,2-c]isoxazole; 17α-methylandrostan[3,2-c]isoxazol-17β-ol; Neo-Ponden.* $C_{21}H_{31}NO_2$; mol wt 329.48. C 76.55%, H 9.48%, N 4.25%, O 9.71%. Prepn: Zderic *et al., Chem. & Ind. (London)* **1960**, 1625.

Crystals, mp 169-170°. $[\alpha]_D$ +19°. uv max (ethanol): 226 nm (log ε 3.71).

THERAP CAT: Anabolic.

675. Androstane. *(5α)-Androstane; etioallocholane.* $C_{19}H_{32}$; mol wt 260.46. C 87.62%, H 12.38%. From androstane-3,17-dione: Butenandt, Tscherning, *Z. Physiol. Chem.* **229**, 185 (1934). From androstane-3,17-diol: Steiger, Reichstein, *Helv. Chim. Acta* **20**, 817 (1937). From Δ^{16}-androstene: Prelog *et al., ibid.* **27**, 66 (1944).

Leaflets from acetone-methanol, mp 50-50.5°. Sublimes at 60° and 0.003 mm Hg. $[\alpha]_D^{16}$ +2° (c = 1.2 in chloroform). Sol in acetone, alc, methanol, ether, petr ether, chloroform.

17-Amino-HCl, dec 345°: Marker, *J. Am. Chem. Soc.* **58**, 480 (1936).

676. Androstane-3β,11β-diol-17-one. *(3β,5α,11β)-3,11-Dihydroxyandrostan-17-one.* $C_{19}H_{30}O_3$; mol wt 306.45. C 74.47%, H 9.87%, O 15.66%. First obtained by degradation of allopregnane-3β,11β,17α,20β,21-pentol (Reichstein's Substance A): Reichstein, *Helv. Chim. Acta* **19**, 402 (1936), but later isolated in small quantities directly from extracts of the adrenal cortex: Reichstein, von Euw, *ibid.* **21**, 1197 (1938); *ibid.* **24**, 879 (1941). It is uncertain whether this substance occurs in the fresh adrenal gland; it may possibly occur by oxidation or decompn during the isolation procedure: Reichstein, Shoppee, *Vitamins & Hormones* **I**, 368 (1943). Structure-activity study: S. Sassa *et al., J. Biol. Chem.* **254**, 10011 (1979). Chromatographic studies: A. Kerebel *et al., J. Chromatog.* **140**, 229 (1977); J. T. Lin, E. Heftmann, *ibid.* **237**, 215 (1982).

Needles from acetone + ether, mp 235-238°. $[\alpha]_D^{20}$ +84.5° (ethanol); $[\alpha]_D^{19}$ +81.3° (dioxane); $[\alpha]_{545}^{19}$ +105° (dioxane). Precipitated by digitonin.

3-Acetate, $C_{21}H_{32}O_4$, needles from actone + ether, mp 230-231°. $[\alpha]_D^{19}$ +70.5° (dioxane); $[\alpha]_{546}^{19}$ +87.1° (dioxane).

Diacetate, $C_{23}H_{34}O_5$, crystals, mp 154-156°.

677. Androstenediol. *(3β,17β)-Androst-5-ene-3,17-diol; Δ^5-androstene-3β,17β-diol.* $C_{19}H_{30}O_2$; mol wt 290.45. C 78.57%, H 10.41%, O 11.02%. Obtained from dehydroandrosterone: Butenandt, Hanisch, *Ber.* **68**, 1859 (1935); Z. *Physiol. Chem.* **237**, 89 (1935); Ruzicka, Wettstein, *Helv. Chim. Acta* **18**, 1264 (1935); *FIAT Final Report* 996, 45 (1947); Levy, Kapp, U.S. pat. **2,521,586** (1950 to Nopco Chem.).

Leaflets from acetone + petr ether, or from methanol or ethyl acetate. Sublimes in high vacuum. mp 184°. $[\alpha]_D^{18}$ $-55.5°$ (c = 0.4 in isopropanol). Insol in water.

3-Acetate, $C_{21}H_{32}O_3$, crystals from hexane, mp 147-148°.

17-Acetate, $C_{21}H_{32}O_3$, crystals from hexane, mp 146.5-148.5°, $[\alpha]_D^{18}$ $-62.4°$ (alc).

Diacetate, $C_{23}H_{34}O_4$, leaflets from hexane, mp 165-166°, $[\alpha]_D^{18}$ $-56.5°$ (alc).

17-Benzoate, $C_{26}H_{34}O_3$, crystals from methanol, mp 220-222°.

3-Acetate-17-benzoate, $C_{28}H_{36}O_4$, crystals, mp 180-182°.

Dipropionate, $C_{25}H_{38}O_4$, *Bisexovis, Stenandiol.*

THERAP CAT: Anabolic.

678. 4-Androstene-3,17-dione. *Androst-4-ene-3,17-dione.* Androtex. $C_{19}H_{26}O_2$; mol wt 286.41. C 79.68%, H 9.15%, O 11.17%. Prepd from 5-dehydroandrosterone: Butenandt, Kudszus, *Z. Physiol. Chem.* **237**, 75 (1935); Ruzicka, Wettstein, *Helv. Chim. Acta* **18**, 986 (1935). *See also* U.S. pats. **2,384,335; 2,175,220; 2,194,235,** and U. Westphal, H. Hellmann, *Ber.* **70**, 2136 (1937). From 5-androsten-3-ol-17-one: F. Galinovsky, *Ber.* **74**, 1624 (1941). From 4-androsten-17-ol: R. E. Marker *et al., J. Am. Chem.*

Soc. **62**, 223 (1940); U.S. pat. **2,397,424.** From 5,6-dibromoandrostane-3,17-dione: P. L. Julian *et al., J. Am. Chem. Soc.* **67**, 1728 (1945); U.S. pat. **2,374,683.** From 17-hydroxyprogesterone: D. A. Prins, T. Reichstein, *Helv. Chim. Acta* **24**, 951 (1941). Isolated in small amounts from adrenal cortex: von Euw, T. Reichstein, *ibid.* 879; it is possible that it does not occur in the adrenal gland, but originates by oxidation of Reichstein's Substance S during work-up.

Dimorphous. Needles from acetone, mp 142-144°; crystals from hexane, mp 173-174°. $[\alpha]_D^{30}$ +191° (alc). uv max: 235 nm.

Dioxime, $C_{19}H_{28}N_2O_2$, crystals, mp 143°.
3-Semicarbazone, $C_{20}H_{29}N_3O_2$, crystals, dec 245°.

679. (3α,5α)-Androst-16-en-3-ol. 3α-Hydroxy-5α-androst-16-ene; Δ^{16}-androsten-3-ol. $C_{19}H_{30}O$; mol wt 274.45. C 83.15%, H 11.02%, O 5.83%. A major constituent of boar pheromone, having a pronounced musk-like odor. Isoln from swine testes: V. Prelog, L. Ruzicka, *Helv. Chim. Acta* **27**, 61 (1944). Prepn: V. Prelog *et al., ibid.* 66; J. Fishman *et al., J. Org. Chem.* **28**, 1443 (1963). Physiological role as a sex attractant for pigs: D. B. Gower, *J. Steroid Biochem.* **3**, 45 (1972). Use in pig artificial insemination: D. R. Melrose *et al.,* **Ger.** pat. **1,937,264** corresp to U.S. pat. **3,681,490** (1970, 1972 both to Nat. Res. Dev. Corp.). *In vivo* metabolism in boar testes: Y. A. Saat *et al., Biochem. J.* **144**, 347 (1974). It has also been detected in human male axillary sweat, but has no androgenic activity: B. W. L. Brooksbank *et al., Experientia* **30**, 864 (1974). Radioimmunoassay: D. C. Bickell, D. B. Gower, *J. Steroid Biochem.* **7**, 451 (1976). Receptor studies: J. N. Gennings *et al., Biochim. Biophys. Acta* **496**, 547 (1977). Biosynthetic studies: E. L. Hurden *et al., J. Endocrinol.* **81**, 161P (1979); G. M. Cook, D. B. Gower, *ibid.* **88**, 409 (1981). Discovery of the presence of androst-16-en-3-ol in truffles *(Tuber melanosporum)* has been offered as an explanation for the ability of pigs to detect truffles growing as deep as 1 meter underground: R. Claus *et al., Experientia* **37**, 1178 (1981).

Crystals, mp 142.5-143°. Purified by sublimation in high vacuum and recryst from acetone. $[\alpha]_D^{20}$ +13.1° (c = 0.957 in chloroform). Gives a blue color in the Kägi-Miescher test, *cf. Helv. Chim. Acta* **22**, 683 (1939).

USE: As an aid to estrus determn in pig artificial insemination.

680. Androsterone. *(3α,5α)-3-Hydroxyandrostan-17-one;* *cis*-androsterone; 3α-hydroxy-17-androstanone; androstan-3α-ol-17-one; 3α-hydroxyetioallocholan-17-one; 3-epihydroxyetioallocholan-17-one. $C_{19}H_{30}O_2$; mol wt 290.45. C 78.57%, H 10.41%, O 11.02%. Isolation from male urine after removal of the phenolic estrogen fraction: Butenandt, Tscherning, *Z. Physiol. Chem.* **229**, 167 (1934); v. Euw, Reichstein, *Helv. Chim. Acta* **25**, 988 (1942). Prepn from cholesterol: Ruzicka, *ibid.* **17**, 1389 (1934); Marker, *J. Am. Chem. Soc.* **57**, 1755 (1935); Schoeller *et al.,* U.S. pat. **2,-232,735** (1941 to Schering).

Crystals from acetone-ether, mp 185-185.5°. Sublimes in high vacuum. $[\alpha]_D^{20}$ +94.6° (c = 0.7 in abs alc). $[\alpha]_D^{15}$ +87.8° (c = 1.5 in dioxane). Not precipitated by digitonin. Barely soluble in water. Sol in most organic solvents.

Acetate, $C_{21}H_{32}O_3$, crystals from ether, sublimes in high vac, mp 165°, $[\alpha]_D^{14}$ +76.7° (c = 2.04 in dioxane); $[\alpha]_D^{25}$ +86° (c = 2 in ethanol).

Propionate, $C_{22}H_{34}O_3$, mp 151-152°.
Benzoate, $C_{26}H_{34}O_3$, mp 178°.

681. Anemonin. *trans-1,7-Dioxadispiro[4.0.4.2]dodeca-3,9-diene-2,8-dione;* 1,2-dihydroxy-1,2-cyclobutanediacrylic acid di-γ-lactone; Anemone camphor; Pulsatilla camphor. $C_{10}H_8O_4$; mol wt 192.17. C 62.50%, H 4.20%, O 33.30%. Found in *Anemone pulsatilla* L. and other *Ranunculaceae.* Its precursor in plants is protoanemonin. Isoln from *Ranunculus acer:* Zecher, Wohlmuth, *Sci. Pharm.* **22**, 95 (1954); *C.A.* **48**, 13169b (1954). Structure: Moriarty *et al., J. Am. Chem. Soc.* **87**, 3251 (1965); Romain, *Diss. Abstr. B* **27**, 3867 (1967). Synthesis: Sugiyama *et al., C.A.* **67**, 116604n (1967). Toxicity study: R. Brodersen, A. Kjaer, *Acta Pharmacol.* **2**, 109 (1946).

Crystals from petr ether, mp 157-158°. Volatile with steam. Slightly sol in cold, more in hot water; sol in hot alcohol, chloroform, alkalies with yellow color. Practically insol in ether. LD_{50} i.p. in mice: 150 mg/kg (Brodersen, Kjaer).

Note: Not to be confused with anemonine which is 5-(carboxymethyl)-1,1-dimethylimidazolium hydroxide inner salt.

682. Anethole. *1-Methoxy-4-(1-propenyl)benzene; p-propenylanisole;* anise camphor; Monasirup. $C_{10}H_{12}O$; mol wt 148.20. C 81.04%, H 8.16%, O 10.80%. Chief constituent of anise, star anise and fennel oils: Monad, de Dortan, *Ind. Parfum.* **5**, 401 (1950); Naves, Tucakov, *Compt. Rend.* **248**, 843 (1959). Separation of *cis* and *trans* isomers: Naves *et al., ibid.* **246**, 1734 (1958); *Bull. Soc. Chim. France* **1958**, 566; Naves, *Helv. Chim. Acta* **43**, 230 (1960); Ferroni *et al., Gazz. Chim. Ital.* **92**, 1198 (1962). Synthesis: Mueller, Röscheisen, *Ber.* **90**, 543 (1957); R. J. DePasquale, *Synth. Commun.* **10**, 225 (1980). Toxicity: J.-R. Boissier *et al., Therapie* **22**, 309 (1967). *Review:* Wagner, *Mfg. Chemist* **23**, 56 (1952).

trans - isomer

trans-Isomer, crystalline mass at 20-21°; mp 21.4°. Liquid above 23°. d_4^{20} 0.9883. $bp_{2.3}$ 81-81.5°. n_D^{20} 1.56145. uv max (ethanol): 259 nm (ϵ 22300). Practically insol in water. Misc with ether, chloroform; sol in benzene, ethyl acetate, acetone, carbon disulfide, petr ether; 1 ml dissolves in 2 ml alc. LD_{50} i.p. in rats: 900 mg/kg (Boissier).

cis-Isomer. d_4^{20} 0.9878. $bp_{2.3}$ 79-79.5°. n_D^{20} 1.55455. uv max (ethanol): 253.5 nm (ϵ 18500). LD_{50} i.p. in rats: 93 mg/kg (Boissier).

USE: Manuf anisaldehyde; flavoring agent; in perfumery, particularly for soap and dentifrices; sensitizer in bleaching colors in color photography; as an imbedding material in microscopy. Pharmaceutic aid (flavor).

THERAP CAT (VET): Has been used as a carminative.

683. Anethole Trithione. *5-(p-Methoxyphenyl)-3H-1,-2-dithiole-3-thione;* 3-(*p*-anisyl)trithione; 3-(*p*-methoxyphenyl)-4,5-dithiacyclopent-2-ene-1-thione; 3-(*p*-anisyl)-4,5-dithiacyclopent-2-ene-1-thione; (*p*-methoxyphenyl)trithiopropene; trithio-*p*-methoxyphenylpropene; 5-(*p*-methoxyphenyl)-1,2-dithiacyclopent-4-ene-3-thione; 3-(*p*-methoxyphenyl)trithione; trithioanethole; Heporal; Mucinol; Trithio; Sufralem; Tiotrifar; Felviten; Sulfogal; Sulfarlem. $C_{10}H_8O$-S_3; mol wt 240.37. C 49.97%, H 3.35%, O 6.66%, S 40.02%. Prepn: Böttcher, Lüttringhaus, *Ann.* **557**, 89 (1947); Gaudin, Lozac'h, *Compt. Rend.* **224**, 557 (1947); Lüttringhaus *et al., Ann.* **560**, 201 (1948); Gaudin, U.S. pats. **2,556,963, 2,688,620** (1951, 1954); Böttcher, Ger. pats. **855,865, 869,-799** and **874,447** (1952 and 1953); Thuiller, Vialle, *Bull. Soc. Chim. France* **1959**, 1398.

Orange-colored prisms from butyl acetate. Very bitter taste. mp 111°. Practically insol in water. Sol in pyridine, chloroform, benzene, dioxane, carbon disulfide. Slightly sol in ether, acetone, ethyl acetate, acetic acid, alc, cyclohexane, petr ether.

Oxime, $C_{10}H_9NO_2S_2$, yellow needles, mp 170°. Soluble in dioxane.

Methiodide, yellow crystals, mp 189°.

THERAP CAT: Choleretic.

684. Angelica. Fruit or root of *Angelica archangelica* L. (*A. officinalis* Moench.), Umbelliferae. Habit. Europe, Asia. Constit. Root: Volatile oil (0.3-1%), angelic acid, 6% resin, angelicol, angelicin, xanthotoxol, starch, osthole, osthenol, archangelicin, archangin, sitosterol, and acids such as aconitic, malic, quinic, chlorogenic, caffeic, fumaric, citric, angelic, and oxalic. Fruit: About 1% volatile oil, bitter substance, coumarins, resin. *Refs:* Späth, Pesta, *Ber.* **67**, 853 (1934); Späth, Vierhapper, *ibid.* **70**, 248 (1937); Svendsen, *C.A.* **52**, 2173g (1958); Sroka, *Apoth.-Ztg.* **61**, 37 (1949).

THERAP CAT: Carminative, diaphoretic, diuretic.

685. Angelic Acid. *(Z)-2-Methyl-2-butenoic acid;* cis-2-dimethylcrotonic acid; 2-methylisocrotonic acid; cis-2,3-dimethylacrylic acid. $C_5H_8O_2$; mol wt 100.12. C 59.98%, H 8.05%, O 31.96%. Stereoisomer of tiglic acid. Found in ester form in sumbul root, *Angelica archangelica* L., Umbelliferae and together with tiglic acid esters in the oil of the Roman camomile, *Anthemis nobilis* L., Compositae. Isoln from seeds of *Schoenocaulon officinale* (Lindl.) A. Gray, Liliaceae (cevadilla seeds) by alkaline hydrolysis of cevadine: Stoll, Seebeck, *Helv. Chim. Acta* **35**, 1275 (1952). Synthesis by *trans* addition of bromine to tiglic acid: Buckles, Mock, *J. Org. Chem.* **15**, 680 (1950). Review and bibliography: Buckles *et al., Chem. Rev.* **55**, 659 (1955).

Monoclinic rods, needles, plates; mp 45°. Spicy odor. Vesicant. d_4^{47} 0.983. bp_{760} 185°; bp_{12} 86°. Sublimes. Volatile with steam. n_D^{47} 1.4434. K at 25° = 5.0 × 10^{-5}. uv max (H_2O): 217 nm (ϵ 5.15 × 10^3). Molar heat of combustion 626.6 kcal. Sparingly soluble in cold water, freely sol in hot water. Sol in alcohol, ether. Prolonged boiling of aq soln causes isomerization to tiglic acid; the process is speeded up by traces of bromine and sunlight, also by strong mineral acids or alks. Dry crystals of angelic acid have been stored in bottles for years without evidence of isomerization.

Calcium salt dihydrate, $Ca(C_5H_7O_2)_2.2H_2O$, leaflets. Much more soluble in water than calcium tiglate: 100 parts of aq soln satd at 17.5° contains 23 parts of anhydr calcium angelate.

Amide, C_5H_9NO, crystals, mp 127-128°.

Methyl ester, $C_6H_{10}O_2$, liquid; d_4^{20} 0.9413; bp_{764} 128°; n_D^{20} 1.4321.

Ethyl ester, $C_7H_{12}O_2$, liquid; $d_4^{19.5}$ 0.9178; bp_{760} 141.5°, bp_{11} 49°. n_D^{20} 1.4304. Heat of formn at constant vol 963.1 kcal, at constant press. 964.2 kcal.

686. Angelica Lactone. *5-Methyl-2-furanone.* C_5H_6-O_2; mol wt 98.10. C 61.22%, H 6.16%, O 32.62%. Exists in three forms. Prepn of α and β-forms: Wolff, *Ann.* **229**, 250 (1885); Thiele, *Ann.* **319**, 184 (1901); v. Auwers, *Ber.* **56**, 1672 (1923); J. H. Helberger *et al., Ann.* **561**, 215 (1949). Prepn of α'-form: J. P. Wineburg *et al., J. Heterocycl. Chem.* **12**, 749 (1975); V. Jäger, H. J. Günther, *Tetrahedron Letters* **1977**, 2543; R.A. Amos, J. A. Katzenellenbogen, *J. Org. Chem.* **43**, 560 (1978). Toxicity data for α-form: E. J. Moran *et al., Drug Chem. Toxicol.* **3**, 249 (1980).

α-form   β-form   α'-form

α-Form, *5-methyl-2(3H)-furanone, Δ²-angelica lactone, γ-methyl-β,γ-crotonolactone, 4-hydroxy-3-pentenoic acid γ-lactone.* Volatile needles, mp 18°. d_4^{20} 1.084. bp_{12} 56°. n_{He}^{20} 1.4476. One gram dissolves in 20 ml water at 15°. Heating with triethylamine soln converts it to the β-form. LD_{50} orally in mice: 2800 mg/kg (Moran).

β-Form, *5-methyl-2(5H)-furanone, Δ¹-angelica lactone, γ-methyl-α,β-crotonolactone, 4-hydroxy-2-pentenoic acid γ-lactone.* Liquid. Not solidified at −17°. d_4^{20} 1.076. bp_{751} 208-209°. bp_{10} 87°. n_{He}^{20} 1.4603. Sol in water. Forms a dimer. More stable than α-form.

α'-Form, *dihydro-5-methylene-2(3H)-furanone, γ-methylene-γ-butyrolactone.* bp_{17} 80°.

687. Angiogenin. Single-chain, basic protein of 123 amino acids that induces the *in vivo* formation of blood vessels. Mol wt ~14,000 Da. First isolated from human adenocarcinoma cells; subsequently found in normal human plasma and shown to be produced by the liver. Angiogenin exhibits a characteristic ribonucleolytic activity toward 28S and 18S ribosomal RNA. Its amino acid sequence is 35% identical with that of human pancreatic ribonuclease. Isoln, characterization, and angiogenic activity: J. W. Fett *et al., Biochemistry* **24**, 5480 (1985). Amino acid sequence: D. J. Strydom *et al., ibid.* 5486. Cloning and DNA sequence of human angiogenin gene: K. Kurachi *et al., ibid.* 5494. Structural study: K. A. Palmer *et al., Proc. Nat. Acad. Sci. USA* **83**, 1965 (1986). Ribonucleolytic activity: R. Shapiro *et al., Biochemistry* **25**, 3527 (1986). Isoln from normal human plasma: R. Shapiro *et al., ibid.* **26**, 5141 (1987). Tissue distribution in neonatal and adult rats: H. L. Weiner *et al., Science* **237**, 280 (1987); in human tumor and normal cells: S. M. Rybak *et al., Biochem. Biophys. Res. Commun.* **146**, 1240 (1987). Inhibition of protein synthesis: D. K. St. Clair *et al., Proc. Nat. Acad. Sci. USA* **84**, 8330 (1987). Inhibition of angiogenic and ribonucleolytic activities of angiogenin by placental ribonuclease inhibitor: R. Shapiro, B. L. Vallee, *ibid.* 2238; F. S. Lee, B. L. Vallee, *Biochemistry* **28**, 3556 (1989). *Reviews:* J. F. Riordan, B. L. Vallee, *Brit. J. Cancer* **57**, 587-590 (1988); B. L. Vallee, J. F. Riordan, *Adv. Exp. Med. Biol.* **234**, 41-53 (1988).

688. Angiostatin. Circulating factor produced by tumor cells that inhibits the growth of metastases. 38 kDa protein; amino acid sequence corresponds to the kringle domains of plasminogen, *q.v.* Exhibits antiproliferative effect on endothelial cells and inhibits angiogenesis. Found in serum and urine of tumor bearing mice; corresponding fragment prepd from human plasminogen exhibits similar activity. Isoln, amino acid sequence and bioactivity: M. S. O'Reilly *et al., Cell* **79,** 315 (1994). Review and potential role in cancer treatment: R. Vile, *Curr. Biol.* **5,** 10-13 (1995).

689. Angiotensin. Formerly *hypertensin* and *angiotonin.* Pressor substance formed by the action of renin, *q.v.,* on a plasma substrate, *angiotensinogen* (*renin substrate, hypertensinogen*). The substance so formed is a decapeptide called *angiotensin I* which is converted to the active pressor agent, *angiotensin II,* by the splitting off of the *C*-terminal His-Leu residues by the "converting enzyme" or *angiotensinase.* The octapeptide angiotensin II differs among species only in the amino acid residue in position 5 being either Val or Ile. *See* reviews for refs to isoln and synthesis. Angiotensin II acts directly on the adrenal gland to stimulate the release of aldosterone, *q.v.* Rapid liquid phase synthesis of a protected angiotensin II: S. Nozaki, I. Muramatsu, *Bull. Chem. Soc. Japan* **55,** 2165 (1982). Extraction and characterization of angiotensins I and II from rat brain: D. Ganten *et al., Science* **221,** 869 (1983). *Reviews:* M. Bodanszky, M. A. Ondetti, *Peptide Synthesis* (John Wiley, New York, 1966) pp 215-223; E. Schröder, K. Lübe, *The Peptides* vol. II (English ed., New York, 1966) pp 4-62; Bumpus, Smeby, "Angiotensin" in *Renal Hypertension,* I. Page, J. McCubbin, Eds. (Year Book Medical Publishers, Chicago, 1968) pp 62-98; Lee, "Angiotensin" in *Renin and Hypertension* (Williams & Wilkins, Baltimore, 1969) pp 32-94; G. M. Molinatti, P. Limone, *Minerva Med.* **72,** 715-732 (1981); J. A. Oliver *et al., Ann. N.Y. Acad. Sci.* **394,** 275-277 (1982); M. Marin-Grez, *Biochem. Pharmacol.* **31,** 3941-3947 (1982). *See also* Tonin.

Asp–Arg–Val–Tyr–Ile–His–Pro–Phe

Angiotensin II (horse)

Angiotensins are very stable. Hydrolyzed by strong acids and bases and above pH 9.5. Sol in organic solvents, in aq solns pH 5-8. In high dilution, lost by absorption on walls of glass vessels.

Amide, $C_{49}H_{70}N_{14}O_{11}$, *5-valine-angiotensin II amide, angiotensin II aspartic-β-amide 5-valine, 1-asparagine-5-valine-angiotensin II, val⁵-angiotensin II-asp¹-β-amide, Hypertensin, Ipertensina.*

THERAP CAT: Amide as vasoconstrictor.

690. Angostura Bark. Cusparia bark; Carony bark; *Cortex Angosturae; Cortex Cuspariae.* The bark of the tree *Galipea officinalis* Hancock, habitat Venezuela, or the bark of the tree *Cusparia febrifuga* Humb., or *C. trifoliata* Engl., *Rutaceae,* habitat Brazil. *Ref:* Meyer, *Pharm. Ztg.* **80,** 120 (1935).

Unpleasant, musty odor. Bitter, slightly aromatic taste with pungent after-taste. The bitterness is due mainly to the bitter principle angosturin, $C_9H_{12}O_5$. The bark contains also the quinoline alkaloids cusparine, cuspareine; galipine, galipoidine, and galipoline.

THERAP CAT: Antipyretic.

691. Anhalamine. *1,2,3,4-Tetrahydro-6,7-dimethoxy-8-isoquinolinol;* 6,7-dimethoxy-8-hydroxy-1,2,3,4-tetrahydroisoquinoline. $C_{11}H_{15}NO_3$; mol wt 209.25. C 63.14%, H 7.23%, N 6.69%, O 22.94%. From *Lophophora williamsii* (Lemaire) Coutl. (*Anhalonium lewinii* Henn.), *Cactaceae:* Kauder, *Arch. Pharm.* **237,** 190 (1899); Späth, Becke, *Monatsh.* **66,** 327 (1935). Structure: *eidem, Ber.* **67,** 2100 (1934). Synthesis: Späth, Roder, *Monatsh.* **43,** 93 (1922); Brossi *et al., Helv. Chim. Acta* **47,** 2089 (1964); **49,** 403 (1966). Biosynthetic studies: Kapadia *et al., J. Am. Chem. Soc.* **92,** 6943 (1970). *Review:* Manske in R. H. F. Manske, H. L. Holmes, *The Alkaloids* vol. IV (Academic Press, New York, 1954) pp 8-14.

Crystals, mp 189-191°. uv max (ethanol): 274 nm (log ε 2.90). Almost insol in cold water, cold alcohol, ether. Sol in hot water, alcohol, acetone, dil acids.

Hydrochloride dihydrate, $C_{11}H_{15}NO_3.HCl.2H_2O$, crystals from water, mp 258°.

692. Anhalonidine. *1,2,3,4-Tetrahydro-6,7-dimethoxy-1-methyl-8-isoquinolinol;* 6,7-dimethoxy-8-hydroxy-1-methyl-1,2,3,4-tetrahydroisoquinoline. $C_{12}H_{17}NO_3$; mol wt 223.27. C 64.55%, H 7.67%, N 6.27%, O 21.50%. From mescal buttons, the buds of *Lophophora williamsii* (Lemaire) Coult. (*Anhalonium lewinii* Henn.) *Cactaceae:* Heffter, *Ber.* **27,** 2975 (1894); **29,** 221 (1896); Kauder, *Arch. Pharm.* **237,** 190 (1899). Structure and synthesis from 3-acetoxy-4,5-dimethoxy-*N*-acetylphenethylamine: Späth, Passl, *Ber.* **65,** 1778 (1932); from mescaline: Brossi *et al., Helv. Chim. Acta* **47,** 2089 (1964). Biosynthesis: Kapadia *et al., J. Am. Chem. Soc.* **92,** 6943 (1970).

Small octahedra from benzene, mp 160-161°. uv max (ethanol): 270 nm (log ε 2.81). Strong base. Freely sol in water, alcohol, chlorofom, hot benzene. Sparingly sol in ether. Insol in petr ether. Solns of anhalonidine acquire a reddish color on standing.

693. Anhalonine. *6,7,8,9-Tetrahydro-4-methoxy-9-methyl-1,3-dioxolo[4,5-h]isoquinoline.* $C_{12}H_{15}NO_3$; mol wt 221.26. C 65.14%, H 6.83%, N 6.33%, O 21.69%. From mescal buttons [*Lophophora williamsii* (Lemaire) Coult. (*Anhalonium lewinii* Henn.), *Cactaceae]* also in *Ariocarpus,* in *Gymnocalycium gibbosum.* Synthesis of *dl*-form and resolution: Späth, Kesztler, *Ber.* **68,** 1663 (1935); Brossi *et al., J. Am. Chem. Soc.* **93,** 6248 (1971). Configuration: Battersby, Edwards, *J. Chem. Soc.* **1960,** 1214.

Rhombic needles from petr ether, mp 86°, $bp_{0.02}$ 140°. $[\alpha]_D^{25}$ −63.8° (methanol); −56.3° (chloroform). Very sol in alcohol, ether, chloroform, benzene, petr ether.

Hydrochloride, $C_{12}H_{15}NO_3.HCl$, orthorhombic prisms, dec 255°; freely sol in hot water. Aq soln is neutral.

694. Anilazine. *4,6-Dichloro-N-(2-chlorophenyl)-1,3,-5-triazin-2-amine; 2,4-dichloro-6-(o-chloroanilino)-s-triazine;* (*o*-chloroanilino)dichlorotriazine; Dyrene. $C_9H_5Cl_3N_4$; mol wt 275.52. C 39.23%, H 1.83%, Cl 38.60%, N 20.33%. Prepn: C. N. Wolf, U.S. pat. **2,720,480** (1955 to Ethyl Corp.); E. G. Hill, E. Clinton, U.S. pat. **2,820,032** (1958); K. H. Rattenburg *et al.,* U.S. pat. **3,074,946** (1963 to Chemagro). Toxicity study: S. D. Cohen, S. D. Murphy, *J. Agr. Food Chem.* **21,** 140 (1973).

White to tan crystals, mp 159-160°. Insol in water. Soly at 30° (g/100 ml): toluene, 5; xylene, 4; acetone, 10. Subject to hydrolysis; not compatible with oils and alkaline materials. LD_{50} orally in rats: > 5000 mg/kg, Mobay Technical Information Sheet, Jan. 1979.

USE: Fungicide.

695. Anileridine. *1-[2-(4-Aminophenyl)ethyl]-4-phenyl-4-piperidinecarboxylic acid ethyl ester; 1-(p-aminophenethyl)-4-phenylisonipecotic acid ethyl ester;* ethyl 1-(4-aminophenethyl)-4-phenylisonipecotate; N-[β-(p-aminophenyl)ethyl]-4-phenyl-4-carbethoxypiperidine; N-β-(p-aminophenyl)-ethylnormeperidine; Leritine; Nipecotan; Alidine; Apodol. $C_{22}H_{28}N_2O_2$; mol wt 352.48. C 74.97%, H 8.01%, N 7.95%, O 9.08%. Synthesis: Weijlard *et al., J. Am. Chem. Soc.* **78,** 2342 (1956); U.S. pat. **2,966,490** (1960 to Merck & Co.).

mp 83°.

Dihydrochloride, $C_{22}H_{28}N_2O_2.2HCl$, crystals from methanol + ether, mp 280-287° (dec). Freely soluble in water, methanol. Solubility in ethanol: 8 mg/g. pH of aq solns 2.0 to 2.5. Solns are stable at pH 3.5 and below. At pH 4 and higher the insol free base is precipitated. uv max (pH 7 in 90% methanol contg phosphate buffer): 235, 289 nm ($A_{1cm}^{1\%}$ 293, 34.5). Distribution coefficient (water, pH 3.6/*n*-butanol): 0.9.

Note: This is a controlled substance (opiate) listed in the U.S. Code of Federal Regulations, Title 21 Part 1308.12 (1995).

THERAP CAT: Analgesic (narcotic).

696. Aniline. *Benzenamine;* aniline oil; phenylamine; aminobenzene; aminophen; kyanol. C_6H_7N; mol wt 93.13. C 77.38%, H 7.58%, N 15.04%. First obtained in 1826 by Unverdorben from dry distillation of indigo. Runge found it in coal tar in 1834. Fritzsche, in 1841, prepared it from indigo and potash and gave it the name aniline. Manuf from nitrobenzene or chlorobenzene: Faith, Keyes & Clark's *Industrial Chemicals,* F. A. Lowenheim, M. K. Moran, Eds. (Wiley-Interscience, New York, 4th ed., 1975) pp 109-116. Procedures: A. I. Vogel, *Practical Organic Chemistry* (Longmans, London, 3rd ed., 1959) p 564; Gattermann-Wieland, *Praxis des organischen Chemikers* (de Gruyter, Berlin, 40th ed., 1961) p 148. Brochure *"Aniline"* by Allied Chemical's National Aniline Division (New York, 1964) 109 pp, gives reactions and uses of aniline (877 references). Toxicity study: K. H. Jacobson, *Toxicol. Appl. Pharmacol.* **22,** 153 (1972).

Oily liquid; colorless when freshly distilled, darkens on exposure to air and light. *Poisonous!* Characteristic odor and burning taste; combustible; volatile with steam. d_{20}^{20} 1.022. bp 184-186°. Solidif −6°. Flash pt, closed cup: 169°F (76°C). n_D^{20} 1.5863. pKb 9.30. pH of 0.2 molar aq

soln 8.1. One gram dissolves in 28.6 ml water, 15.7 ml boil. water; misc with alcohol, benzene, chloroform, and most other organic solvents. Combines with acids to form salts. It dissolves alkali or alkaline earth metals with evolution of hydrogen and formation of anilides, e.g., C_6H_5NHNa. *Keep well closed and protected from light. Incompat.* Oxidizers, albumin, solns of Fe, Zn, Al, acids, and alkalies. LD_{50} orally in rats: 0.44 g/kg (Jacobson).

Hydrobromide, $C_6H_7N.HBr$, white to slightly reddish, crystalline powder, mp 286°. Darkens in air and light. Sol in water, alc. *Protect from light.*

Hydrochloride, $C_6H_7N.HCl$, crystals, mp 198°. d 1.222. Darkens in air and light. Sol in about 1 part water; freely sol in alc. *Protect from light.*

Hydrofluoride, $C_6H_7N.HF$, crystalline powder. Turns gray on standing. Freely sol in water; slightly sol in cold, freely in hot alc.

Nitrate, $C_6H_7N.HNO_3$, crystals, dec about 190°. d 1.36. Discolors in air and light. Sol in water, alc. *Protect from light.*

Hemisulfate, $C_5H_7N.\frac{1}{2}H_2SO_4$, crystalline powder. d 1.38. Darkens on exposure to air and light. One gram dissolves in about 15 ml water; slightly sol in alc. Practically insol in ether. *Protect from light.*

Acetate, $C_6H_5NH_2.HOOCCH_3$. Prepd from aniline and acetic acid: Vignon, Evieux, *Bull. Soc. Chim. France* [4] **3,** 1012 (1908). Colorless liquid. d 1.070-1.072. Darkens with age; gradually converted to acetanilide on standing. Misc with water, alc.

Oxalate, $C_6H_5NH_2.HOOCCOOH.H_2NC_6H_5$. Prepd from aniline and oxalic acid in alc soln: Hofmann, *Ann.* **47,** 37 (1843). Triclinic rods from water, mp 174-175°. Readily sol in water; sparingly sol in abs alc. Practically insol in ether.

Caution: Intoxication may occur from inhalation, ingestion, or cutaneous absorption. *Acute Toxicity:* cyanosis, methemoglobinemia, vertigo, headache, mental confusion. *Chronic Toxicity:* anemia, anorexia, wt loss, cutaneous lesions, *Clinical Toxicology of Commercial Products,* R. E. Gosselin *et al.,* Eds. (Williams & Wilkins, Baltimore, 4th ed., 1976) Section III, pp 29-35.

USE: Manuf dyes, medicinals, resins, varnishes, perfumes, shoe blacks; vulcanizing rubber; as solvent. Hydrochloride used in manuf of intermediates, aniline black and other dyes, in dyeing fabrics or wood black.

697. Aniline Mustard. *N,N-Bis(2-chloroethyl)benzenamine; N,N-bis(2-chloroethyl)aniline;* phenylbis[2-chloroethylamine]; β,β′-dichlorodiethylaniline; Lymphochin; Lymphocin; Lymphoquin. $C_{10}H_{13}Cl_2N$; mol wt 218.13. C 55.06%, H 6.01%, Cl 32.51%, N 6.42%. Prepd by the action of phosphorus pentachloride on N,N-bis-[2-hydroxyethyl]-aniline (phenyldiethanolamine): Robinson, Watt, *J. Chem. Soc.* **1934,** 1538; Korshak, Strepikheev, *J. Gen. Chem. USSR* **14,** 312 (1944).

Stout prisms from methanol, mp 45°. bp_{14} 164°; $bp_{0.5}$ 110°. Sol in hot methanol, ethanol. Very slightly sol in ether.

Hydrochloride, $C_{10}H_{14}Cl_3N$, crystals. *Vesicant.* Freely sol in water. Sol in alcohol.

USE: In cancer research.

698. Anilinephthalein. *3,3-Bis(4-aminophenyl)-1(3H)-isobenzofuranone; 3,3-bis(p-aminophenyl)phthalide.* $C_{20}H_{16}N_2O_2$; mol wt 316.36. C 75.93%, H 5.10%, N 8.85%, O 10.11%. Prepn: Hubacher, *J. Am. Chem. Soc.* **73,** 5885 (1951).

Crystals from methanol, mp 203°. Sol in acetone. Moderately sol in methanol, ethyl acetate. Insol in water, ether, benzene.

699. 1-Anilino-8-naphthalenesulfonate. *8-(Phenylamino)-1-naphthalenesulfonic acid; 8-anilino-1-naphthalenesulfonic acid;* ANS; phenylperi acid. $C_{16}H_{13}NO_3S$; mol wt 299.35. C 64.20%, H 4.38%, N 4.68%, O 16.03%, S 10.71%. Hydrophobic fluorescent probe for protein studies; originally used in prepn of aniline dyes. Prepd from 8-aminonaphthalenesulfonic acid, aniline and anilinium hydrochloride. Fluorescence spectrum: L. Stryer, *Science* **162**, 526 (1968). Binding to bovine serum albumin: G. Weber, L. B. Young, *J. Biol. Chem.* **239**, 1415 (1964); to lipoproteins: R. A. Muesing, T. Nishida, *Biochemistry* **10**, 2952 (1971); to chymotrypsin: J. D. Johnson *et al., ibid.* **18**, 1292 (1979); L. D. Weber *et al., ibid.* 1297. Interaction with rod outer segment membrane: U. P. Andley, B. Chakrabarti, *ibid.* **20**, 1687 (1981).

Magnesium salt, $C_{32}H_{24}MgN_2O_6S_2$, green crystals from water. uv max: 350 nm (ϵ 4.95 × 10^3).
Ammonium salt, $C_{16}H_{16}N_2O_3S$, mp 242-244°.
USE: Fluorescent probe. In protein conformation studies. Magnesium salt as visualization reagent for proteins.

700. Aniracetam. *1-(4-Methoxybenzoyl)-2-pyrrolidinone;* 1-*p*-anisoyl-2-pyrrolidinone; Ro-13-5057; Draganon; Sarpul. $C_{12}H_{13}NO_3$; mol wt 219.24. C 65.74%, H 5.98%, N 6.39%, O 21.89%. Cognition enhancer related to piracetam, *q.v.* Prepn: **Japan. Kokai 79 117,468;** E. Kyburz, W. Aschwanden, **U.S. pat. 4,369,139** (1979, 1983 both to Hoffmann-La Roche). Effects on learning and memory in rats: R. Cumin *et al., Psychopharmacology (Berlin)* **78**, 104 (1982); K. Yamada *et al., Pharmacol. Biochem. Behav.* **22**, 645 (1985); in monkeys and pigeons: M. J. Pontecorvo, H. L. Evans, *ibid.* 745. Clinical evaluation in geriatric patients: P. Foltyn *et al., Arzneimittel-Forsch.* **33**, 865 (1983); in Alzheimer's disease: L. B. Sourander *et al., Psychopharmacology (Berlin)* **91**, 90 (1987). Review of toxicology: B. Schläppi *et al., Drug Invest.* **5**, Suppl. 1, 50-67 (1993).

Crystals from ethanol, mp 121-122°. LD_{50} in rats, mice (mg/kg): ~4500, > 5000 orally (Cumin).
THERAP CAT: Nootropic.

701. *p*-Anisaldehyde. *4-Methoxybenzaldehyde;* anisic aldehyde. $C_8H_8O_2$; mol wt 136.15. C 70.57%, H 5.92%, O 23.50%. Metabolic product of the odoriferous fungus *Lentinus lepidus* Fr.: Birkinshaw *et al., Biochem. J.* **38**, 131 (1944); of wood-rotting fungus *Polyporus benzoinus* (Wahl.) Fr.: Birkinshaw *et al., ibid.* **50**, 509 (1952); of *Daldalea juniperina* Murr.: Birkinshaw, Chaplen, *ibid.* **60**, 255 (1955). Prepn: Niedzielski, Nord, *J. Am. Chem. Soc.* **63**, 1462

(1941); Sisti *et al., J. Org. Chem.* **27**, 279 (1962). Toxicity study: P. M. Jenner *et al., Food Cosmet. Toxicol.* **2**, 327 (1964).

Oily liquid, bp 248°, $bp_{1.5}$ 89-90°. mp 0°. d_4^{15} 1.119. n_D^{13} 1.5764. Volatile in steam. Very slightly sol in water; misc with alc, ether. LD_{50} orally in rats: 1510 mg/kg (Jenner).
USE: Perfumery and toilet soaps; odor resembles that of coumarin, but the aldehyde must be mixed with other odorous substances to yield an agreeable odor. Also used in organic syntheses.

702. Anise. Anise seed. Dried ripe fruit of *Pimpinella anisum* L., *Umbelliferae*. Habit. Western Asia, Egypt; cultivated in Southern Europe, India and U.S. Constit. 1.5-3.5% volatile oil, starch, protein, fixed oil.
USE: Manuf oil of anise; as condiment and flavor in foods or beverages.
THERAP CAT: Carminative.
THERAP CAT (VET): Carminative, flavoring agent.

703. Anise Alcohol. *4-Methoxybenzenemethanol; p-methoxybenzyl alcohol;* anisyl alcohol. $C_8H_{10}O_2$; mol wt 138.17. C 69.55%, H 7.30%, O 23.16%. Prepd by reduction of anisaldehyde with triisobutyl aluminum and with amine boranes: Ziegler *et al., Ann.* **623**, 9 (1959); Chamberlain, Schechter, **U.S. pat. 2,898,379** (1959 to Callery Chemical). Toxicity study: Woodart *et al., J. Pharmacol. Exp. Ther.* **93**, 26 (1948).

Liquid, bp 259°. mp 24-25°, solidif 17°. d_{15}^{15} 1.113. Practically insoluble in water. Freely soluble in alc, ether. LD_{50} orally in rats: 1.2 ml/kg (Woodart).

704. *p*-Anisic Acid. *4-Methoxybenzoic acid.* $C_8H_8O_3$; mol wt 152.15. C 63.15%, H 5.30%, O 31.55%. Prepd from methoxybenzene: Gross *et al., Ber.* **96**, 1382 (1963). Manuf by catalytic oxidation of *p*-methoxytoluene: **Brit. pat. 798,-619** (1958 to General Electric); Cotterill *et al.,* **Brit. pat. 842,998** (1960 to I.C.I.).

Needles, mp 184° (subl). bp 275-280°. d 1.385. Soluble in 2500 parts water; more sol in boiling water; freely sol in alc, chloroform, ether, ethyl acetate.

705. Anisidine. C_7H_9NO; mol wt 123.15. C 68.27%, H 7.37%, N 11.37%, O 12.99%. Prepn of *p*-isomer: Cahours, *Ann.* **74**, 298 (1850); of *o*-isomer: Mühlhäuser, *Ann.* **207**, 235 (1881); of *m*-isomer: Kadaba, Massie, *J. Org. Chem.* **22**, 333 (1957).

***m*-Anisidine**, *3-methoxybenzenamine, 3-methoxyaniline, 3-aminoanisole.* Pale yellow, oily liquid. Remains fluid even at −10°. bp 251°, bp$_2$ 81-86°. Sparingly sol in water; sol in alc, acids.

***o*-Anisidine**, *2-methoxybenzenamine.* Yellowish liquid; becomes brownish on exposure to air. Volatile with steam. bp 225°. mp +5°. d$_{15}^{15}$ 1.098. Practically insol in water. Miscible with alc, ether. *Keep well closed and protected from light.*

***p*-Anisidine**, *4-methoxybenzenamine.* Crystals, mp 57°. bp 246°. Sparingly sol in water; freely sol in methanol, ethanol.

Note: *o*-Anisidine hydrochloride may reasonably be anticipated to be a carcinogen: *Seventh Annual Report on Carcinogens* (PB95-109781, 1994) p 93.

USE: In the manuf of azo dyes.

706. Anisindione. *2-(4-Methoxyphenyl)-1H-indene-1,3(2H)-dione; 2-p-anisyl-1,3-indandione; 2-(p-methoxyphenyl)-1,3-indandione;* SPE-2792; Miradon; Unidone. C$_{16}$H$_{12}$O$_3$; mol wt 252.27. C 76.18%, H 4.79%, O 19.03%. Prepn: Koelsch, *J. Am. Chem. Soc.* **58**, 1331 (1936); Horeau, Jacques, *Bull. Soc. Chim. France* **1948**, 53; Sperber, U.S. pat. **2,899,358** (1959 to Schering).

Pale yellow crystals from acetic acid or ethanol, mp 156-157°.

THERAP CAT: Anticoagulant.

707. Anisole. *Methoxybenzene.* C$_7$H$_8$O; mol wt 108.14. C 77.75%, H 7.46%, O 14.80%. C$_6$H$_5$OCH$_3$. Prepn from phenol and dimethyl sulfate: Ullmann, *Ann.* **327**, 114 (1903); Graebe, *Ann.* **340**, 204 (1905); G. S. Hiers, F. D. Hager, *Org. Syn.* **coll. vol. I**, 58 (2nd ed., 1941); from bromobenzene: Agfa, Ger. pat. **411,052**; *Chem. Zentr.* **1925**, I, 2411; *Frdl.* **15**, 193; by passing methyl chloride into a suspension of sodium phenolate in liquid ammonia: White *et al., J. Am. Chem. Soc.* **46**, 965 (1924); from phenol, methyl iodide and potassium carbonate in dimethylformamide: Brieger *et al., J. Chem. Eng. Data* **13**, 581 (1968). Forms oils or resins by condensation with formaldehyde: **Ger. pats. 403,264; 406,152;** *Chem. Zentr.* **1925**, I, 307, 1816; *Frdl.* **14**, 626, 627. Absorption spectrum: Scheibe, *Ber.* **59**, 2625 (1926). Soly in glycerol, *see* McEwen, *J. Chem. Soc.* **123**, 2285 (1923). Toxicity studies: J. M. Taylor *et al., Toxicol. Appl. Pharmacol.* **6**, 378 (1964).

Liquid. Agreeable aromatic odor. d$_4^{18}$ 0.9956; d$_4^{45}$ 0.9701. mp −37.3°. bp$_{760}$ 155.5°; bp$_{100}$ 93.0°; bp$_{40}$ 70.7°; bp$_{20}$ 55.8°; bp$_{10}$ 42.2°; bp$_5$ 30.0°; bp$_{1.0}$ 5.4°. n$_D^{20}$ 1.51791. Sol in alcohol and ether. Insol in water. LD$_{50}$ orally in rats: 3700 mg/kg (Taylor).

USE: In perfumery, in organic syntheses.

708. Anisomycin. *1,4,5-Trideoxy-1,4-imino-5-(4-methoxyphenyl)-D-xylo-pentitol 3-acetate; [2R-(2α,3α,4β)]-2-[(4-methoxyphenyl)methyl]-3,4-pyrrolidinediol 3-acetate; 2-p-methoxyphenylmethyl-3-acetoxy-4-hydroxypyrrolidine;* Flagecidin. C$_{14}$H$_{19}$NO$_4$; mol wt 265.31. C 63.38%, H 7.22%, N 5.28%, O 24.12%. Protein synthesis inhibiting antibiotic isolated from *Streptomyces griseolus* and *S. roseochromogenes:* Sobin, Tanner, Jr., *J. Am. Chem. Soc.* **76**, 4053 (1954); Tanner *et al.,* U.S. pat. **2,691,618** (1954 to Pfizer). Activity: J. E. Lynch *et al., Antibiot. & Chemother.* **4**, 844, 899 (1954). Structure and stereochemistry: Beereboom *et al., J. Org. Chem.* **30**, 2334 (1965); Schaefer, Wheatley, *ibid.* **33**, 166 (1968); Butler, *ibid.* 2136. Biosynthesis: Butler,

ibid. **31**, 317 (1966). Total synthesis: Oida, Ohki, *Chem. Pharm. Bull.* **16**, 2086 (1968); *ibid.* **17**, 1405 (1969); Felner, Schenker, *Helv. Chim. Acta* **53**, 754 (1970). Chiral synthesis: J. P. H. Verheyden *et al., Pure Appl. Chem.* **50**, 1363 (1978). Stereospecific total synthesis: D. P. Schumacher, S. S. Hall, *J. Am. Chem. Soc.* **104**, 6076 (1982). Mechanism of action: A. Jiménez, D. Vázquez in *Antibiotics* vol. 5(pt. 2), F. E. Hahn, Ed. (Springer-Verlag, New York, 1979) pp 1-19. Solubility and stability data: *Antibiot. Ann.* **1954-55**, pp 809-810. Prepn of deacetylanisomycin from anisomycin: Nickell *et al.,* U.S. pat. **2,935,444** (1960 to Pfizer).

Long needles from ethyl acetate or water, mp 140-141°. [α]$_D^{23}$ −30° (methanol). uv max: 224, 277, 283 nm (ε 10800, 1800, 1600). Base is moderately sol in water; sol in lower alcohols, esters, ketones, chloroform; slightly sol in benzene, toluene and hexane. Aq solns are stable over a wide pH range at room temp.

Hydrochloride, C$_{14}$H$_{19}$NO$_4$.HCl, crystals from ethyl acetate + ethanol, mp 187-188°. Very sol in water.

Deacetylanisomycin, C$_{12}$H$_{17}$NO$_3$, mp 176-179°. [α]$_D^{25}$ −20.0° (methanol), pK 9.2.

USE: Anisomycin and deacetylanisomycin in the eradication of bean mildew; to inhibit other pathogenic fungi in plants.

THERAP CAT: Antiprotozoal (Trichomonas).

709. Anisotropine Methylbromide. *endo-8,8-Dimethyl-3-[(1-oxo-2-propylpentyl)oxy]-8-azoniabicyclo[3.2.1]octane bromide; 3α-hydroxy-8-methyl-1αH,5αH-tropanium bromide 2-propylvalerate;* 8-methyltropinium bromide 2-propylvalerate; 8-methyl-3-(2-propylpentanoyloxy)tropinium bromide; octatropine methylbromide; Lytispasm; Valpin. C$_{17}$H$_{32}$BrNO$_2$; mol wt 362.35. C 56.35%, H 8.90%, Br 22.05%, N 3.87%, O 8.83%. Prepn: Weiner, Gordon, U.S. pat. **2,962,-499** (1960 to Endo Labs.). Metabolism: Shindo *et al., Chem. Pharm. Bull.* **19**, 513 (1971).

Crystals from acetone, mp 329°.

Methyl chloride, C$_{17}$H$_{32}$ClNO$_2$, crystals from acetone, mp 289°.

THERAP CAT: Anticholinergic.

710. *o*-(*p*-Anisoyl)benzoic Acid. *2-(4-Methoxybenzoyl)benzoic acid;* S-23/46. C$_{15}$H$_{12}$O$_4$; mol wt 256.26. C 70.31%, H 4.72%, O 24.97%. Prepn from phthalic anhydride and anisole: Meyer, Turnau, *Monatsh.* **30**, 486 (1909). Alternate route: Arcus, Marks, *J. Chem. Soc.* **1956**, 1627.

p-Anisoyl Chloride

Leaflets from water. Stout crystals from alcohol or toluene, mp 146°. Very sparingly sol in water. Freely sol in alc, ether, toluene, chloroform, glacial acetic acid.

Sodium salt, $C_{15}H_{11}NaO_4$, needles. Freely sol in water. Sol in alcohol. Rated approximately 150 times as sweet as cane sugar. Bitter taste if used in concns exceeding 0.2 g/liter. *Review:* Möhler, *Z. Lebensm.-Untersuchung u. Forschung* **90**, 431 (1950), *C.A.* **44**, 8558d (1950).

USE: The sodium salt has been proposed as a sweetening agent.

711. *p*-Anisoyl Chloride. *4-Methoxybenzoyl chloride.* $C_8H_7ClO_2$; mol wt 170.60. C 56.33%, H 4.14%, Cl 20.78%, O 18.76%. $CH_3OC_6H_4COCl$. Prepd from *p*-anisic acid and thionyl chloride: Vanderhaeghe *et al., J. Pharm. Pharmacol.* **6**, 119 (1954).

Crystals or liquid. mp 22°. bp ~262-263° with slight decompn. Dec by water or alcohol. Sol in acetone, benzene.

Caution: Vapors can cause serious eye burns. Sealed containers may explode, because slow decompn at room temp may build up pressure. *Ref:* Carroll, *Chem. & Eng. News* **38**, 40 (Aug. 22, 1960).

712. Anistreplase. Anisoylated plasminogen streptokinase activator complex; APSAC; BRL-26921; Eminase; Fibrit; Iminase; Multilase. Binary complex of streptokinase and human plasminogen in which the catalytic site is reversibly blocked by a *p*-methoxybenzoyl group without affecting the fibrin-binding site. Prepn and fibrinolytic activity: R. A. G. Smith, J. G. Winchester, **Eur. pat. Appl. 28,489**; *eidem,* **U.S. pat. 4,808,405** (1981, 1989 both to Beecham). Deacylation of the fibrin-bound complex initiates thrombolysis: R. A. G. Smith *et al., Nature* **290**, 505 (1981); D. Matsuo *et al., Thromb. Res.* **24**, 347 (1981). Effect on hemostasis in humans: D. H. Staniforth *et al., Eur. J. Clin. Pharmacol.* **24**, 751 (1983). Pharmacokinetics: M. Been *et al., Int. J. Cardiol.* **11**, 53 (1986). Evaluation of thrombolytic efficacy in myocardial infarction: W. Kasper *et al., Am. J. Cardiol.* **58**, 418 (1986); in major pulmonary embolism: J. H. Brett *et al., Aust. N.Z. J. Med.* **17**, 77 (1987). Controlled clinical trial in acute myocardial infarction: S. Ikram *et al., Brit. Med. J.* **293**, 786 (1986). Review of mechanism of action, pharmacology and therapeutic use: J. P. Monk, R. C. Heel, *Drugs* **34**, 25-49 (1987).

THERAP CAT: Thrombolytic.

713. Annatto. Arnotta; annotta. Coloring matter from seeds of *Bixa orellana* L., *Bixaceae*. Extraction from seed: Barnett, Espoy, **U.S. pat. 2,815,287** (1957); Kocher, **U.S. pat. 2,831,775** (1958). Contains bixin and several yellow to orange-red pigments which give carotene reactions: Diemair *et al., Naturwiss.* **39**, 211 (1933).

Sol in alcohol, ether, oils.

USE: Dyeing silk orange in hot soap bath; coloring butter, margarine, cheese and oils; manuf wood stains, varnishes.

714. Annotinine. $C_{16}H_{21}NO_3$; mol wt 275.35. C 69.79%, H 7.69%, N 5.09%, O 17.43%. Isoln from *Lycopodium annotinum* L., *Lycopodiaceae*: Manske, Marion, *Can. J. Res.* **21B**, 92 (1943). Structure: Przybylska, Marion, *Can. J. Chem.* **35**, 1075 (1957); Wiesner *et al., Tetrahedron* **4**, 87 (1958); Przybylska, Ahmed, *Acta Cryst.* **11**, 718 (1958). Stereochemistry: Wiesner *et al., Tetrahedron Letters* **1961**, 187; Ho, *ibid.* **1969**, 1307. Synthesis: Wiesner *et al., Can. J. Chem.* **47**, 433 (1969). Biogenesis: Leete, *Tetrahedron* **3**, 313 (1958).

Brilliant prisms from chloroform + methanol, mp 232°. Sol in chloroform, dil HCl. Sparingly sol in methanol.

715. *p*-Anol. *4-(1-Propenyl)phenol; p-propenylphenol; 4-hydroxy-1-propenylbenzene.* $C_9H_{10}O$; mol wt 134.18. C 80.56%, H 7.51%, O 11.92%. Prepn from anethole: Stoermer, Kahlert, *Ber.* **34**, 1812 (1901); from 4-propenylphenylmagnesium bromide: Quelet, *Bull. Chim. Soc.* [4] **45**, 268 (1929).

Leaflets from boiling water. Weak odor of cloves, resembling that of eugenol. Spicy, pungent taste. mp 93-94°. bp_{760} 250° (dec); bp_{14} 138-140°. Slightly sol in hot water. Freely sol in dimethylformamide, other organic solvents. Sol in aq solns of KOH and NaOH.

USE: Intermediate in the synthesis of estrogens.

716. Anot. *3-Amino-2-methyl-5-nitrobenzamide; 3-amino-5-nitro-o-toluamide.* $C_8H_9N_3O_3$; mol wt 195.18. C 49.23%, H 4.65%, N 21.53%, O 24.59%. Major metabolite of chickens on a zoalene diet. Isoln procedure: Thiegs *et al., J. Agr. Food Chem.* **9**, 201 (1961).

Golden-yellow crystals, mp 198.5-199.5°. Slightly sol in water (less than 1%).

717. Anserine. *N-β-Alanyl-3-methyl-L-histidine; α-(β-aminopropionylamino)-β-(1-methyl-5-imidazolyl)propionic acid; 3-methyl-N,α-(β-alanyl)-L-histidine.* $C_{10}H_{16}N_4O_3$; mol wt 240.26. C 49.99%, H 6.71%, N 23.32%, O 19.98%. First found in muscles of geese, later in many more animals, but not in man, cow, horse, or dog. Isoln: Ackermann, *Z. Physiol. Chem.* **183**, 1 (1929); Wolff, Wilson, *J. Biol. Chem.* **109**, 565 (1935). Synthesis: Behrens, du Vigneaud, *J. Biol. Chem.* **120**, 517 (1937); Rinderknecht *et al., J. Org. Chem.* **29**, 1968 (1964). *Review:* M. Guggenheim, *Die biogenen Amine* (Karger, Basel—New York, 4th ed., 1951) pp 439-446, 502.

L-Form, needles from dil alc, dec 240-242°. $[\alpha]_D^{30}$ +12.3° (c = 5). pK_1 2.64; pK_2 7.04; pK_3 9.49. The dried material is very hygroscopic. Freely soluble in water. Slightly sol in methanol, even less in ethanol.

Nitrate, $C_{10}H_{16}N_4O_3 \cdot HNO_3$, needles from dil methanol, mp 226-228°.

718. Antazoline. *4,5-Dihydro-N-phenyl-N-(phenylmethyl)-1H-imidazole-2-methanamine; 2-(N-benzylanilinomethyl)-2-imidazoline;* phenazoline; 2-(N-phenyl-N-benzylaminomethyl)imidazoline; imidamine; 5512-M; Antistine; Antistin; Histostab; Antastan; Antasten; Antihistal; Azalone; Ben-a-hist. $C_{17}H_{19}N_3$; mol wt 265.36. C 76.95%, H 7.22%, N 15.84%. Prepd by the condensation of benzylaniline with 2-(chloromethyl)imidazoline: Miescher, Klarer, **U.S. pat. 2,449,241** (1948 to Ciba).

Consult the Name Index before using this section.

Crystals, mp 120-122°.

Hydrochloride, $C_{17}H_{19}N_3 \cdot HCl$, *Fenazolina, Histazine*. Bitter crystals producing temporary numbness of the tongue. mp 237-241°. uv max: 242 nm ($E_{1cm}^{1\%}$ 495 to 515); min 222 nm. One gram dissolves in 40 ml water, in 25 ml alc. Practically insol in ether, benzene, chloroform. pH (1% aq soln): 6.3.

Phosphate, $C_{17}H_{19}N_3 \cdot H_3PO_4$, crystals. Bitter taste. mp 194-198°. Sol in water. Sparingly sol in methanol. Practically insol in benzene, ether. pH (2% aq soln): 4.5.

THERAP CAT: Antihistaminic.

THERAP CAT (VET): Antihistaminic.

719. Antheridiol. *(3β,22S,23R)-3,22,23-Trihydroxy-7-oxostigmasta-5,24(28)-dien-29-oic acid γ-lactone;* 3β,22,23-trihydroxy-24-(carboxymethylene)cholest-5-en-7-one 23,24-lactone; Hormone A. $C_{29}H_{42}O_5$; mol wt 470.65. C 74.01%, H 8.99%, O 17.00%. The first specific functioning sex hormone to be identified in the plant kingdom. A diffusible substance secreted by the female mycelium of the filamentous water molds *Achlya bisexualis* and *A. ambisexualis*, which induces the growth of antheridial hyphae in the male plant, thereby initiating sexual reproduction in the species. Isoln and preliminary chemical data: Raper, Haagen-Smit, *J. Biol. Chem.* **143**, 311 (1942). Isoln of crystalline material: McMorris, Barksdale, *Nature* **215**, 320 (1967). Structure: Arsenault *et al., J. Am. Chem. Soc.* **90**, 5635 (1968). Synthesis: J. A. Edwards *et al., ibid.* **91**, 1248 (1969); J. H. Fried, J. A. Edwards, U.S. pat. **3,547,911** (1970 to Syntex). Stereochemical proposal: Green *et al., Tetrahedron* **27**, 1199 (1971). Absolute stereochemistry and alternate syntheses: Edwards *et al., Tetrahedron Letters* **1972**, 791; T. C. McMorris *et al., ibid.* **2673**; eidem, *J. Org. Chem.* **39**, 669 (1974). *Reviews:* Barksdale, *Ann. N.Y. Acad. Sci.* **144**, 313-319 (1967); *idem, Science* **166**, 831-837 (1969); T. C. McMorris, *Lipids* **13**, 716-722 (1978).

Colorless crystals from methanol, mp 250-255°. uv max (ethanol): 220 nm (ε 17,000). Sol in hot methanol. Slightly sol in chloroform; very slightly sol in water.

USE: Control and regulation of plant fertility.

720. Anthiolimine. *Mercaptobutanedioic acid antimony(3+) lithium salt (3:1:6); 2,2',2''-[stibilidynetris(thio)]tris-butanedioic acid hexalithium salt; mercaptosuccinic acid antimonate (III) hexalithium salt;* mercaptosuccinic acid S-antimony derivative lithium salt; lithium antimoniothiomalate; lithium antimony thiomalate; Anthiomaline. $C_{12}H_9 \cdot Li_6O_{12}S_3Sb$; mol wt 604.80. C 23.83%, H 1.50%, Li 6.89%, O 31.74%, S 15.91%, Sb 20.13%. $Sb \cdot (C_4H_6O_4S)_3 \cdot 6Li$. Prepn: M. Delépine, P. Gailliot, U.S. pat. **2,060,181** (1936 to Rhône-Poulenc). Antitrypanosomal activity: N. Ercoli *et al., Chemotherapy* (Basel) **26**, 254 (1980). Therapeutic use in buffaloes: R. K. Gupta, S. P. Pachauri, *Agric. Sci. Digest* **2**, 65 (1982); V. Kumar *et al., J. Vet. Parasitol.* **3**, 125 (1989). HPLC determn in biological fluids: N. A. El-Rabbat *et al., Talanta* **37**, 951 (1990); titrimetric determn: F. Belal *et al., Microchem. J.* **44**, 296 (1991).

Enneahydrate, White hygroscopic powder. Very soluble in water; very slightly sol in alcohol, ether.

THERAP CAT: Anthelmintic (Trematodes).

THERAP CAT (VET): Anthelmintic.

721. Anthracene. $C_{14}H_{10}$; mol wt 178.23. C 94.34%, H 5.66%. Obtained from coal tar: Dumas, Laurent, *Ann.* **5**, 10 (1833); Laurent, *Ann.* **34**, 287 (1840); Anderson, *Ann.* **122**, 294 (1862); *J. Chem. Soc.* **15**, 44 (1862); Auerbach, *Das Anthracen und seine Derivate* (Braunschweig, 1880); Perkin, *J. Soc. Arts* **27**, 572 (1879); Lunge, *Coal Tar and Ammonia* (1916); Barnett, *Anthracene and Anthraquinone* (London, 1921); Nanson, *Textile Colorist* **48**, 605, 678, 751 (1926); **49**, 19, 246, 557, 593 (1927); Houben, Fischer, *Das Anthracen und die Anthrachinone* (Leipzig, 1929); Borrmann, *Der Teer* (Leipzig, 1940); Schumann, *Kokereiteer* (Stuttgart, 1940). Extensive patent literature on purification. Prepn of very pure anthracene from synthetic anthraquinone: Clar, *Ber.* **72**, 1645 (1939). *Review:* E. Clar, *Polycyclic Hydrocarbons* 2 vols. (Academic Press, New York, 1964).

Monoclinic plates from alc. Sublimes. When pure, colorless with violet fluoresence; when impure (due to tetracene, naphthacene), yellow with green fluorescence. Strongly triboluminescent and triboelectric. d_4^{27} 1.25. mp 218°. bp$_{760}$ 342°. Absorption spectrum: Clar, *Ber.* **65**, 506 (1932). Less soluble than the isomeric phenanthrene. Insol in water; one gram dissolves in 67 ml abs alcohol, 70 ml methanol, 62 ml benzene, 85 ml chloroform, 200 ml ether, 31 ml carbon disulfide, 86 ml carbon tetrachloride, 125 ml toluene. Anthracene darkens in sunlight. According to Downs, U.S. pat. **1,303,639** (1919), when solns of crude anthracene in coal tar naphtha are exposed to ultraviolet irradiation, the anthracene is precipitated as dianthracene (*para*-anthracene) which is reconverted to anthracene by sublimation. Forms molecular addn products with nitro compounds.

Picric acid complex, mp 139°.

sym-Trinitrobenzene complex, mp 164°.

Trinitrotoluene complex, mp 162°.

USE: Important source of dyestuffs (manuf anthraquinone, alizarin dyes).

722. Anthragallol. *1,2,3-Trihydroxy-9,10-anthracenedione;* 1,2,3-trihydroxyanthraquinone; anthragallic acid; anthracene brown. $C_{14}H_8O_5$; mol wt 256.21. C 65.63%, H 3.15%, O 31.22%. From gallic and benzoic acid with sulfuric acid at 125° or from phthalic anhydride and pyrogallol with sulfuric acid at 160°: Seuberlich, *Ber.* **10**, 39 (1877). Other methods: Kubota, Perkin, *J. Chem. Soc.* **127**, 1889 (1925); Perkin, Story, *ibid.* **1929**, 1399; Cross, Perkin, *ibid.* **1930**, 292.

Brown crystals, mp 312-313°. Sublimes 290°. Absorption spectrum: Meyer, Fischer, *Ber.* **46**, 85 (1913). Slightly sol in water, chloroform; sol in alcohol, ether, glacial acetic acid. Its soln in concd H_2SO_4 is reddish-brown. Greenish-brown soln in ammonia water changes to blue on heating. Forms salts with Na, K, Ba, Tl, Pb, La, Ce, Nd, Co.

Trimethyl ether, $C_{17}H_{14}O_5$, yellow needles from benzene + petr ether, mp 168°. Insol in water solns of alkalies.

723. Anthralin. *1,8-Dihydroxy-9(10H)-anthracenone; 1,8-dihydroxyanthrone;* dithranol; Anthra-Derm; Antra-derm; Batidrol; Cignolin; Cigthranol; Dithrocream; Psoradrate; Psoriacide. $C_{14}H_{10}O_3$; mol wt 226.23. C 74.33%, H 4.46%, O 21.22%. Prepd from chrysazin by reduction with hydrogen and nickel catalyst at high pressure: Zahn, Koch, *Ber.* **71B**, 172 (1938). Revised structure: H. M. Avdovich,

G. A. Neville, *Can. J. Spectrosc.* **25**, 110 (1980); F. R. Ahmed, *Acta Crystallogr.* **B36**, 3184 (1980).

Lemon yellow leaflets or needles from ligroin, mp 176-181°. Practically insol in water. Freely sol in chloroform. Sol in acetone, benzene, pyridine, oils. Slightly sol in alcohol, ether, glacial acetic acid. Sol in dil NaOH with yellow color and green fluorescence. Under the influence of air, alkaline solns turn red and lose their fluorescence.

Triacetate, $C_{20}H_{16}O_6$, *1,8,9-anthracenetriol triacetate, 1,8,-9-triacetoxyanthracene, Exolan.*

THERAP CAT: Antipsoriatic.

THERAP CAT (VET): Antifungal.

724. Anthramycin. *[11R-[2(E),11α,11aβ]]-3-(5,10,11,-11a-Tetrahydro-9,11-dihydroxy-8-methyl-5-oxo-1H-pyrrolo-[2,1-c][1,4]benzodiazepin-2-yl)-2-propenamide; 5,10,11,11a-tetrahydro-9,11-dihydroxy-8-methyl-5-oxo-1H-pyrrolo[2,1-c]-[1,4-]benzodiazepin-2-acrylamide.* $C_{16}H_{17}N_3O_4$; mol wt 315.33. C 60.94%, H 5.43%, N 13.33%, O 20.30%. From *Streptomyces refuineus* var *thermotolerans,* NRRL 3143: Leimgruber *et al., J. Am. Chem. Soc.* **87**, 5791 (1965); Berger *et al.,* U.S. pat. **3,361,742** (1968 to Hoffmann-La Roche). Also from *Streptomyces spadicogriseus:* N. Komatsu *et al., J. Antibiot.* **33**, 54 (1980). Structure: Leimgruber *et al., J. Am. Chem. Soc.* **87**, 5793 (1965). Synthesis: *eidem, ibid.* **90**, 5641 (1968); Batcho, Leimgruber, U.S. pat. **3,524,849** (1970 to Hoffmann-La Roche). Activity studies: Horwitz, Grollman, *Antimicrob. Ag. Chemother.* **1968**, 21. Biosynthesis: L. H. Hurley *et al., J. Am. Chem. Soc.* **97**, 4372 (1975). *Review:* Kohn in *Antibiotics* vol. 3, J. W. Corcoran, F. E. Hahn, Eds. (Springer-Verlag, New York, 1975) pp 3-11.

Yellow prisms from acetone + water, dec 188-194°. uv max (acetonitrile): 235, 333 nm (ε 18,200, 31,800). $[\alpha]_D^{25}$ +930° (DMF).

Methyl ether hydrate, $C_{17}H_{19}N_3O_4 \cdot H_2O$, pale yellow needles from methanol + water, dec above 120°.

THERAP CAT: Antineoplastic.

725. Anthranol. *9-Anthracenol; 9-anthrol;* 9-hydroxy-anthracene. $C_{14}H_{10}O$; mol wt 194.23. C 86.57%, H 5.19%, O 8.24%. Prepd by dissolving anthrone in 5 to 10% boiling NaOH soln, cooling to −5° and pouring in cooled 5% H_2SO_4: Meyer, *Ann.* **379**, 56 (1911). Also prepared from anthraquinone.

Needles from glacial acetic acid, plates from dil alcohol. Orthorhombic, mp 120° (if bath is heated to 110°). Sinters at 120°, mp 152° (if bath is cold at start). Absorption spectrum (of acetate): Barnett *et al., J. Chem. Soc.* **1928**, 885. Sol in most solvents with blue fluorescence. The solid is fairly stable when kept dry. In solution it changes quickly to anthrone. Recrystallizations of anthranol may lead to a less pure compound.

Acetate, $C_{16}H_{12}O_2$, crystals from petr ether, mp 134°.

Benzoate, $C_{21}H_{14}O_2$, crystals from pyridine + alcohol, mp 170-172°.

Methyl ether, $C_{15}H_{12}O$, crystals from alcohol, mp 97-98°.

Ethyl ether, $C_{16}H_{14}O$, crystals from methanol, mp 73°.

USE: In the manuf of dyes.

726. Anthraquinone. *9,10-Anthracenedione;* 9,10-anthraquinone; 9,10-dioxoanthracene; Morkit. $C_{14}H_8O_2$; mol wt 208.22. C 80.76%, H 3.87%, O 15.37%. Produced industrially from phthalic anydride and benzene in the presence of aluminum chloride by a Friedel-Crafts reaction: Klipstein, *Ind. Eng. Chem.* **18**, 1327 (1926). From anthracene with vanadium pentoxide, sodium chlorate, glacial acetic and sulfuric acids: *Org. Syn.* coll. vol. II, 554 (1943). Convenient lab procedure: L. F. Fieser, *Organic Experiments* (Heath & Co., Boston, 1964) pp 195-200. *Reviews:* de Barry, Barnett, *Anthracene and Anthraquinone* (London, 1921); Phillips, *Chem. Rev.* **6**, 157 (1929); Houben, Fischer, *Das Anthracen und die Anthrachinone* (Leipzig, 1929); R. H. Chung in Kirk-Othmer *Encyclopedia of Chemical Technology,* **Vol.** 2 (Wiley-Interscience, New York, 3rd ed., 1978) pp 700-707.

Light yellow, slender monoclinic prisms by sublimation *in vacuo.* Almost colorless, orthorhombic, bipyramidal crystals from H_2SO_4 + H_2O. d_4^{20} 1.42-1.44. mp 286°. bp_{760} 377°. Absorption spectrum: Flexser *et al., J. Am. Chem. Soc.* **57**, 2103 (1935). Insol in water. Solubility (g/100 g) in alc at 18° 0.05; at 25° 0.44; in boiling alc 2.25; in ether at 25° 0.11; in chloroform at 20° 0.61; at 40° 1.00; at 60° 1.60; in benzene at 20° 0.26; at 40° 0.50; at 60° 1.00; at 80° 1.80; in toluene at 25° 0.30.

Caution: Low systemic toxicity, but may cause skin irritation, sensitization.

USE: Important starting material for the manufacture of vat dyes. Usually marketed as 1-anthraquinonesulfonic acid sodium salt, or as 2-anthraquinonesulfonic acid sodium salt (called "silver salt"), or as 1,5-dihydroxyanthraquinone (called "anthrarufin"). Anthraquinone itself is used to make seeds distasteful to birds: *Chem. & Eng. News* **33**, 4998 (1955).

727. Anthrarobin. *1,2,10-Anthracenetriol;* 1,2,10-anthratriol; 3,4-dihydroxyanthranol; desoxyalizarin; leucoalizarin. $C_{14}H_{10}O_3$; mol wt 226.23. C 74.33%, H 4.46%, O 21.22%. Prepd from alizarin, ammonia, and zinc dust: *Hagers Handb. Pharm. Praxis* Vol. **1**, 466 (1930).

Yellow-brown to dark brown crystals; easily oxidized to alizarin. mp 208°; not sublimable. Slightly sol in water or benzene; freely sol in alcohol, ether, acetone, glacial acetic acid. Dissolves in alkalies with greenish-yellow color; in concd H_2SO_4 with golden-yellow color.

THERAP CAT: Parasiticide.

728. Anthrarufin. *1,5-Dihydroxy-9,10-anthracenedione;* 1,5-dihydroxyanthraquinone. $C_{14}H_8O_4$; mol wt 240.22. C 70.00%, H 3.36%, O 26.64%. Prepd from 1,5-anthraquinone potassium disulfonate: Fierz-David, Blangey, *Farbenchemie,* (Vienna, 5th ed., 1943), pp 224-225.

Green to yellow crystals from acetic acid, mp 280°. Sublimes at 120°. Absorption spectrum in H_2SO_4: Meyer, Fischer, *Ber.* **46**, 85 (1913). Sol in concd H_2SO_4, in aq KOH soln (reddish color); insol in aq Na_2CO_3, NH_3, $Ba(OH)_2$. Moderately sol in alcohol, slightly in water.

USE: Important intermediate in the manuf of alizarin and indanthrene dyestuffs. Forms insol Ba and Ca lakes; has been proposed as analytical reagent for the detection of Ca.

729. Anthrimide. *1,1'-Iminobis-9,10-anthracenedione; 1,1'-iminodianthraquinone;* dianthraquinonylamine; dianthrimide. $C_{28}H_{15}NO_4$; mol wt 429.43. C 78.31%, H 3.52%, N 3.26%, O 14.90%. Prepd by heating 1-aminoanthraquinone with 1-chloroanthraquinone in nitrobenzene in the presence of copper: Bayer & Co., **Ger. pat. 162,823**; *Chem. Zentr.* **1905**, II, 1206; Eckert, Steiner, *Monatsh.* **35**, 1129 (1914); *FIAT Final Rept.* no. 1313, **vol. II** (1948), p 137.

Deep red needles from chlorobenzene, rhombs from nitrobenzene. Also described as coppery red needles with metallic sheen. Practically insol in low boiling organic solvents. Sparingly sol in aniline, nitrobenzene, chlorobenzene, quinoline. Sol in concd sulfuric acid first giving a scarlet-red color which turns olive-green on standing.

USE: Determination of boron. Used as a 0.02% soln in concd H_2SO_4. The olive-green soln turns blue in the presence of boron. The blue soln has an absorption max at 620 nm. Also used in the determination of silicon.

730. Anthrone. *9(10H)-Anthracenone; 9,10-dihydro-9-oxoanthracene;* Carbothrone. $C_{14}H_{10}O$; mol wt 194.23. C 86.57%, H 5.19%, O 8.24%. Prepn from anthraquinone by reduction with tin, glacial acetic and hydrochloric acid: *Org. Syn.* coll. vol. I, 60 (1941). By cyclization of o-benzylbenzoic acid with liquid HF: Fieser, Hershberg, *J. Am. Chem. Soc.* **61**, 1278 (1939).

Orthorhombic needles from benzene + petr ether, mp 155°. Absorption spectrum: Martin, *Ann. Combustibles Liq.* **12**, 967 (1937). Sol in most organic solvents without fluorescence. Any fluorescence present is due to anthranol. Tendency to change to anthraquinone. Equilibrium in abs alc: 89% anthrone; 11% anthranol.

Addition compound with 4 mols desoxycholic acid, $C_{110}H_{170}O_{17}$; mp 179°.

USE: In organic syntheses; in the colorimetric determination of sugar and animal starch in body fluids.

731. α-Antiarin. *(3β,5β,12β)-3-[(6-Deoxy-β-D-gulopyranosyl)oxy]-5,12,14-trihydroxy-19-oxocard-20(22)-enolide.* $C_{29}H_{42}O_{11}$; mol wt 566.65. C 61.47%, H 7.47%, O 31.06%. From latex (arrow poison) of the upas tree, *Antiaris toxicaria* Lesch., *Moraceae*, found in Indonesia: Taylor, *Brit. J. Pharmacol.* **8**, 237 (1953). Older literature and isoln: Doebel *et al., Helv. Chim. Acta* **31**, 688 (1948); Dolder *et al., ibid.* **38**, 1364 (1955). Structure of aglycone: Martin,

Tamm, *ibid.* **42**, 696 (1959). The sugar portion of α-antiarin is antiarose (D-gulomethylose), while that of the isomeric β-antiarin is L-rhamnose. Structure: Juslén *et al., ibid.* **45**, 2285 (1962). Toxicity: Chen, Henderson, *J. Pharmacol. Exp. Ther.* **150**, 53 (1965).

Tetrahydrate, crystals (six-sided leaflets) from water. Anhydrous after 3 hrs at 110° in high vac, dec 238-240°. $[\alpha]_D^{17}$ −3.9° (c = 0.905 in methanol). Sol in water, alcohol, slightly in ether. uv max: 305, 217 nm (log ε 1.8, 4.08). MLD (anhydr compd) in cats (mg/kg): 0.116 i.v. (Chen, Henderson).

Tribenzoate, $C_{50}H_{54}O_{14}$, needles from methanol, mp 240°. $[\alpha]_D^{21}$ +3.0° (acetone). Practically insol in water, very sparingly sol in methanol, ethanol, ether. Freely sol in acetone, chloroform.

732. Antigens. Substances which have the ability of inducing the formation of antibodies and of reacting with antibodies in animals and man. In order to induce antibody formation an antigen must (a) be foreign to the animal's own circulation or contain some foreign group, (b) have a molecular weight of at least 10,000-15,000 and (c) be metabolized or at least partially susceptible to the hydrolytic enzymes of the test animal: Campbell, Bulman, *Fortschr. Chem. Org. Naturst.* **9**, 443-484 (1952). *Review:* several authors in *Immunogenicity,* F. Borek, Ed. (Elsevier, New York, 1972).

733. Antimony. Stibium; regulus of antimony. Sb; at. wt 121.75; at. no. 51; valence 3, 5. Two naturally occurring isotopes: 121 (57.25%); 123 (42.75%); artificial radioactive isotopes: 112-120; 122; 124-135; isotopes 122 and 124 are useful radioactive tracers. First accurate description of antimony by Thölde (Basil Valentine) in 1604. Antimony ore is mined in China, Mexico, and Bolivia. The antimony of commerce is about 99% pure. Prepn in the laboratory by reduction of Sb_2O_5 with KCN: Schenk in *Handbook of Preparative Inorganic Chemistry* vol. 1, G. Brauer, Ed. (Academic Press, New York, 2nd ed., 1963) p 606. Toxicity study: W. R. Bradley, W. G. Fredrick, *Ind. Med.* **10**, *Ind. Hyg. Sect.* **2**, 15 (1941). *Review:* Smith, "Arsenic, Antimony and Bismuth" in *Comprehensive Inorganic Chemistry* vol. 2, J. C. Bailar, Jr. *et al.,* Eds. (Pergamon Press, Oxford, 1973) pp 547-683; S. C. Carapella in Kirk-Othmer *Encyclopedia of Chemical Technology* vol. 3 (Wiley-Interscience, New York, 3rd ed., 1978) pp 96-105.

Silver-white, lustrous, hard, brittle metal; scale-like crystalline structure; or dark gray, lustrous powder. Is not tarnished in dry air and only slowly in moist air. d 6.68. mp 630°. bp 1635°; also reported to be 1440°: *Gmelin's, Antimony* (8th ed.) **18B**, p 69 (1949). sp heat 0.049; electrical resistivity 39 μ-ohm-cm at 0°. Not affected by cold dil acids; attacked by hot concd H_2SO_4; readily by aqua regia. Nitric acid, depending on the concn, converts it to antimonous or antimonic oxide. When finely divided it reacts with hot concd HCl. Qualitative analysis for antimony: react with slight excess of HCl with aid of HNO_3; pour soln in large vol of water; white ppt forms which becomes orange-red on addn of H_2S and is sol in ammonium sulfide. LD_{50} in rats, guinea pigs (mg Sb/100 g): 10.0, 15.0 i.p. (Bradley, Fredrick).

Caution: Antimony and its compds have been reported to cause dermatitis, keratitis, conjunctivitis and nasal septal ulceration by contact, fumes or dust. Avoid conditions in which nascent hydrogen will react with antimony to form stibine (SbH_3), which is extremely toxic (nausea, vomiting, headache, hemolysis, hematuria, abdominal pain, death). Stibine can be liberated from storage batteries when nascent hydrogen reacts, in an acid medium, with antimony present in the battery plates: E. Browning, *Toxicity of Industrial Metals* (Appleton-Century Crofts, New York, 2nd ed., 1969) pp 23-38.

USE: In manufacture of alloys, such as Britannia or Babbitt metal, hard lead, white metal, type, bullets and bearing metal; in fireworks; for thermoelectric piles, blackening iron, coating metals, etc.

734. Antimony Chloride Oxide. Antimony oxychloride; basic antimony chloride; powder of Algaroth; Mercurius vitae. ClOSb; mol wt 173.21. Cl 20.47%, O 9.24%, Sb 70.30%. SbOCl. Prepd from $SbCl_3$ and water: Schenk in *Handbook of Preparative Inorganic Chemistry,* vol 1, G. Brauer, Ed. (Academic Press, N.Y., 2nd ed., 1963) p 611.

Monoclinic crystals or crystalline powder. Hydrolyzed by water to Sb_2O_3. Sol in HCl, tartaric acid, CS_2; practically insol in alcohol, ether. Heating to 250° results in formation of $Sb_2O_5Cl_2$; above 320° Sb_2O_3 is formed.

735. Antimony Dichlorotrifluoride. *Antimony chloride fluoride;* antimony dichlorofluoride; antimony trifluorodichloride. Cl_2F_3Sb; mol wt 249.66. Cl 28.40%, F 22.83%, Sb 48.77%. $SbCl_2F_3$. Prepd according to the exothermic reaction $SbF_3 + Cl_2 \rightarrow SbCl_2F_3$: Henne, *Org. Reactions* 2, 61 (1944); Kwasnik in *Handbook of Preparative Inorganic Chemistry* Vol. 1, G. Brauer, Ed. (Academic Press, New York, 2nd ed., 1963) pp 200-201. Prepd from $SbCl_5$ and ClF_3: Dehnicke, Weidlein, *Z. Anorg. Allgem. Chem.* **323**, 267 (1963).

Viscous liquid. Can be stored in iron vessels.
Caution: Highly toxic.

USE: As catalyst in the manufacture of organic fluorine compds: Slesser, Schram, *Preparation, Properties, and Technology of Fluorine and Organic Fluoro Compounds* (New York, 1951) *passim.*

736. Antimony Pentachloride. Cl_5Sb; mol wt 299.02. Cl 59.28%, Sb 40.72%. $SbCl_5$. Usually prepd by passing chlorine into molten antimony trichloride. Laboratory procedure: Schenk in *Handbook of Preparative Inorganic Chemistry* Vol. 1, G. Brauer, Ed. (Academic Press, New York, 2nd ed., 1963) p 610.

Colorless to yellow, oily liquid. Fumes in air. Cannot be distilled at atmospheric pressures without complete decompn. d_4^{16} 2.358; d_4^{36} 2.319; d_4^{52} 2.289; d_4^{78} 2.231. mp 3.5°. bp_{14} 68°; bp_{22} 79°; bp_{30} 92°; bp_{55} 85°; bp_{68} 102.5°. Dipole moment: 1.14 at 16.3° in CCl_4. Mono- and tetrahydrates are formed in the presence of small amounts of water. Large amounts of water cause hydrolysis to Sb_2O_5. Sol in hydrochloric acid, chloroform, carbon tetrachloride.

USE: As catalyst when replacing a fluorine substituent with chlorine in organic compounds.

737. Antimony Pentafluoride. F_5Sb; mol wt 216.75. F 43.83%, Sb 56.17%. SbF_5. Lewis acid. Prepd industrially (in aluminum apparatus) according to the equation $SbCl_5 + 5HF \rightarrow 5HCl + SbF_5$: Ruff, Plato, *Ber.* **37**, 673 (1904); Perkins, Irwin, U.S. pat. **2,410,358** (1946). Laboratory procedure using SbF_3 and F_2: Woolf, Greenwood, *J. Chem. Soc.* **1950**, 2200; Kwasnik in *Handbook of Preparative Inorganic Chemistry,* Vol. 1, G. Brauer, Ed. (Academic Press, New York, 2nd ed., 1963) p 200. *Reviews:* Burg in *Fluorine Chemistry,* Vol. 1, J. Simons, Ed. (Academic Press, New York, 1950) pp 104-106; Kemmitt, Sharp, *Advan. Fluorine Chem.* **4**, 210-211 (1965).

Hygroscopic, corrosive, moderately viscous liquid. *Poisonous, attacks skin.* mp 8.3°; bp 141°. $d^{25.8}$ 3.097; density data: Hoffman, Jolly, *J. Phys. Chem.* **61**, 1574 (1957). Reacts violently with water. Also forms a solid dihydrate, which reacts violently with more water to form a clear soln. Slowly hydrolyzed in NaOH solns forming $Sb(OH)_6^-$. Forms solids with sulfur chloride, carbon disulfide, benzene, toluene, petr ether (resin formation), ether, alc, acetone,

ethyl acetate. Glacial acetic acid gives a clear soln. Slowly corrodes glass, copper, lead. May be stored in aluminum vessels.

USE: In the fluorination of organic compds, *see* the monograph *Preparation, Properties and Technology of Fluorine and Organic Fluoro Compounds,* C. Slesser, S. R. Schram, Eds. (McGraw-Hill, New York, 1951) 868 pp.

738. Antimony Pentasulfide. Golden antimony sulfide; antimonic sulfide; antimonial saffron; antimony red. S_5Sb_2; mol wt 403.85. S 39.70%, Sb 60.30%. Sb_2S_5. Prepn: *Gmelin's, Antimony* (8th ed.) **18B**, pp 534-539 (1949). Existence of Sb^{5+} sulfide doubted: Birchall, Della Valle, *Chem. Commun.* **1970**, 675. Toxicity study: W. R. Bradley, W. G. Fredrick, *Ind. Med.* **10**, *Ind. Hyg. Sect.* **2**, 15 (1941).

Orange-yellow, odorless powder. Insol in water. Sol in concd HCl with evolution of H_2S; sol in solns of alkali hydroxides or sulfides, forming sulfantimoniates. *Incompat.* Acids, metal salts. LD_{50} i.p. in rats: 150.0 mg Sb/100 g (Bradley, Fredrick).

USE: As pigment; vulcanizing and coloring rubber; manuf matches and fireworks.

739. Antimony Pentoxide. Antimonic oxide; "stibic" anhydride; antimonic "acid". O_5Sb_2; mol wt 323.52. O 24.73%, Sb 75.27%. Sb_2O_5. Prepn: Schenk in *Handbook of Preparative Inorganic Chemistry,* **vol. 1**, G. Brauer, Ed. (Academic Press, New York, 2nd ed., 1963) p 616. Toxicity study: W. R. Bradley, W. G. Fredrick, *Ind. Med.* **10**, *Ind. Hyg. Sect.* **2**, 15 (1941).

Yellowish powder. Cubic. d 3.78. Loses oxygen at 300° or higher. Studies have shown that the compound does not correspond fully to Sb_2O_5, but that it is always somewhat hydrated. Slightly soluble in water. Practically insol in HNO_3. Slowly dissolves in warm HCl or in warm KOH soln. LD_{50} i.p. in rats: 400.0 mg Sb/100 g (Bradley, Fredrick).

USE: Fire retardant in clothing.

740. Antimony Potassium Oxalate. *Ethanedioic acid anhydride with antimonic acid (3:1) tripotassium salt; tripotassium tris(oxalato)antimonate(3—);* potassium oxalatoantimonate(III); "antimony salt". $C_6K_3O_{12}Sb$; mol wt 503.11. C 14.32%, K 23.31%, O 38.16%, Sb 24.20%. $K_3[Sb(OOCCOO)_3]$. Prepn of trihydrate: Graddon, *J. Inorg. Nuclear Chem.* **3**, 308 (1956-1957).

Trihydrate, cryst powder. Sol in water. *Poisonous!*

USE: A mordant in dyeing and printing fabrics instead of tartar emetic.

741. Antimony Potassium Tartrate. *Dipotassium bis-[μ-[2,3-dihydroxybutanedioato(4—)-01,02:03,04]]diantimonate(2—) trihydrate stereoisomer;* tartar emetic; tartrated antimony; tartarized antimony; potassium antimonyltartrate. $C_8H_4K_2O_{12}Sb_2 \cdot 3H_2O$; mol wt 667.88. C 14.39%, H 1.51%, K 11.71%, O 35.93%, Sb 36.46%. Manuf from potassium bitartrate and metallic antimony in the presence of HNO_3 or solid antimony oxide: Davies, U.S. pats. **2,335,585** (1943 to Am. Cream Tartar), **2,391,297** (1945 to Stauffer Chem.). Structural studies: P. Pfeiffer, E. Schmitz, *Pharmazie* **4**, 451 (1949); E. Chinoporos, N. Papathanasopoulos, *J. Phys. Chem.* **65**, 1643 (1961); D. Grdenic, B. Kamenar, *Acta Crystallogr.* **19**, 197 (1965); R. Iyer *et al., J. Inorg. Nucl. Chem.* **34**, 3351 (1972). Toxicity: N. Ercoli, *Proc. Soc. Exp. Biol. Med.* **129**, 284 (1968). Anthelmintic activity: Z. Farid *et al., Trans. Roy. Soc. Trop. Med. Hyg.* **66**, 119 (1972); R. R. C. New *et al., Nature* **272**, 56 (1978). Oxidimetric determn: Y. A. Gawargious *et al., Pharmazie* **41**, 59 (1980).

$$2K^+ \cdot 3H_2O$$

Transparent crystals (effloresce on exposure to air) or powder. Sweetish, metallic taste. *Poisonous!* d 2.6. $[\alpha]_D^{20}$

+140.69° (c = 2 in water), +139.25° (c = 2 in glycerol). One gram dissolves in 12 ml water, 3 ml boiling water, 15 ml glycerol. Insol in alcohol. The aq soln is slightly acid. LD$_{50}$ in mice (mg/kg): 55 s.c.; 65 i.v. (Ercoli).

Incompat. Mineral acids, tannic acid, gallic acids, alkali hydroxides and carbonates, lead and silver salts, mercury bichloride, lime water, albumin, soap.

USE: As mordant in the textile and leather industry.

THERAP CAT: Anthelmintic (Schistosoma).

THERAP CAT (VET): Has been used as a parasiticide (gastrointestinal and blood parasites); as an expectorant, and as a ruminatoric.

742. Antimony Sodium Gluconate. General term referring to either a trivalent or pentavalent antimony complex with sodium gluconate. Pentavalent compound may require reduction *in vivo* to the trivalent form to be bioactive. Determn in biological materials and pharmacodynamics: R. J. Ward *et al., Clin. Chim. Acta* **99**, 143 (1979). Mechanism of action study: J. C. Mottram, G. H. Coombs, *Exp. Parasitol.* **59**, 151 (1985).

Sodium antimonylgluconate, C$_6$H$_8$NaO$_7$Sb, *Triostam.* Trivalent antimony compound. Prepn: Das Gupta, *Indian J. Pharm.* **15**, 84 (1953); A. Axon *et al., U.S. pat.* **3,306,921** (1967 to Burroughs Wellcome). Antiparasitic activity: N. Ercoli *et al., Chemotherapy* **26**, 254 (1980). White, amorphous powder. Sol in water. Insol in organic solvents. pH of 2% aq soln 9-10; the soln at this pH is unstable.

Sodium stibogluconate, C$_{12}$H$_{17}$Na$_3$O$_{17}$Sb$_2$.9H$_2$O, *2,4:2',4'-O-(oxydistibylidyne)bis[D-gluconic acid] Sb,Sb'-dioxide trisodium salt nonahydrate, Pentostam, Solustibosan.* Pentavalent antimony compound. Prepn: Datta, Ghosh, *Sci. Cult.* **11**, 699 (1945-6); Bose, Ghosh, *Indian J. Pharm.* **11**, 155 (1949). Pharmacokinetics: J. D. Chulay *et al., Trans. Royal Soc. Trop. Med. Hyg.* **82**, 69 (1988). Review of clinical safety and efficacy in leishmaniasis: B. L. Herwaldt, J. D. Berman, *Am. J. Trop. Med. Hyg.* **46**, 296 (1992). Crystals. Freely sol in water. pH of 10% aq soln 5.4-5.6.

THERAP CAT: Trivalent compd as an anthelmintic (Schistosoma); pentavalent compd as an antiprotozoal (Leishmania).

743. Antimony Sodium Tartrate. Antimony sodium oxide L(+)-tartrate; sodium antimonyl tartrate; Emeto-Na; Stibunal. C$_4$H$_4$NaO$_7$Sb; mol wt 308.82. C 15.56%, H 1.31%, Na 7.44%, O 36.27%, Sb 39.43%. (SbO)NaC$_4$H$_4$O$_6$. Prepn: Fargher, Gray, *J. Pharmacol. Exp. Ther.* **18**, 341 (1921).

Hygroscopic, transparent or whitish scales or powder. Sweetish taste. Sol in 1.5 parts water. Practically insol in 90% alc. Aq solns are slightly acid to litmus. LD i.v. in mice: 25 mg/kg (Fargher, Gray).

THERAP CAT: Anthelmintic (Schistosoma).

THERAP CAT (VET): Anthelmintic (Schistosoma); antiprotozoal (Leishmania).

744. Antimony Sodium Thioglycollate. *[(5-Oxo-1,3,2-oxathiostibolan-2-yl)thio]acetic acid sodium salt;* mercaptoacetic acid antimony derivative sodium salt; antimony sodium thioacetate. C$_4$H$_4$NaO$_4$S$_2$Sb; mol wt 324.96. C 14.78%, H 1.24%, Na 7.07%, O 19.69%, S 19.74%, Sb 37.47%. Prepn of the free acid: Klason, Carlson, *Ber.* **39**, 732 (1906); of the salt: Myers, *J. Lab. Clin. Med.* **6**, 359 (1921).

Hygroscopic prismatic crystals turning pink under the influence of light. May develop slight odor of mercaptan on standing. Freely sol in water. Practically insol in alc. Decomposed by excess alkali; incompatible with alkalies.

THERAP CAT: Anthelmintic (Schistosoma).

745. Antimony Sulfate. Antimonous sulfate; antimony trisulfate. O$_{12}$S$_3$Sb$_2$; mol wt 531.71. O 36.11%, S 18.09%, Sb 45.80%. Sb$_2$(SO$_4$)$_3$. Prepn: Schenk in *Handbook of Preparative Inorganic Chemistry* **vol. 1**, G. Brauer, Ed. (Academic Press, New York, 2nd ed., 1963) pp 618-619.

Crystalline powder or lumps; deliquesc in air. *Poisonous!* d 3.62. With a little water forms a solid mass which dissolves in more water, but excess of water changes it into an insol basic salt. Sol in dil acids. Keep well closed.

746. Antimony Thioglycollamide. *Thioantimonic acid tris(2-amino-2-oxoethyl) ester;* mercaptoacetamide antimony derivative; antimony thioglycollic acid triamide. C$_6$H$_{12}$N$_3$O$_3$S$_3$Sb; mol wt 392.14. C 18.38%, H 3.08%, N 10.72%, O 12.24%, S 24.53%, Sb 31.05%. Sb(SCH$_2$CONH$_2$)$_3$. Prepd by the action of ammonia on ethyl antimony thioglycollate in absolute alc. The ethyl antimony glycollate is prepd from ethyl thioglycollate and antimony trioxide: Rowntree, Abel, *J. Pharmacol.* **2**, 109 (1910). *cf.* Christiansen, *Organic Derivatives of Antimony*, A.C.S. Monograph Series No. 24, (New York, 1925).

Crystals, mp 140°. One gram dissolves in about 200 ml water. Slightly sol in alc. Insol in ether. Incompatible with alkalies.

THERAP CAT: Anthelmintic (Schistosoma).

747. Antimony Tribromide. Br$_3$Sb; mol wt 361.47. Br 66.32%, Sb 33.68%. SbBr$_3$. Best prepd from the elements: Schenk in *Handbook of Preparative Inorganic Chemistry* **vol. 1**, G. Brauer, Ed. (Academic Press, New York, 2nd ed., 1963) p 613.

Orthorhombic bipyramidal needles. Less hygroscopic than SbCl$_3$. d$_{23}^{22}$ 4.148. mp 96°. bp$_{749}$ 288°. Cryoscopic constant: 26.7. Specific heat: 0.0709 at 33°. Critical temp: 904.5°; crit press. 56 atm. Dipole moment: 2.47. Dec by light, water, alc. Sol in dil HCl, HBr, carbon disulfide, acetone, benzene, chloroform.

748. Antimony Trichloride. *Trichlorostibine;* Butter of antimony. Cl$_3$Sb; mol wt 228.12. Cl 46.62%, Sb 53.38%. SbCl$_3$. Prepd from the elements: Hensgen, *Rec. Trav. Chim.* **9**, 301 (1890); Kendall *et al., J. Am. Chem. Soc.* **45**, 967 (1923); Schenk in *Handbook of Preparative Inorganic Chemistry* **vol. 1**, G. Brauer, Ed. (Academic Press, New York, 2nd ed., 1963) p 608. Purification: Werner, *Z. Anorg. Allgem. Chem.* **181**, 154 (1929); Schaafsma, *U.S. pat.* **2,324,240** (1943 to Shell).

Orthorhombic, deliquesc needles from carbon disulfide or by sublimation at 100°. Fumes in air. *Poisonous!* d$_4^{20}$ 3.14. mp 73°. bp 223.5°; bp$_{70}$ 143.5°; bp$_{11}$ 102°. Cryoscopic constant: 18.4. Heat of fusion at 73.2°: 13.29 cal/g. Specific heat at 33°: 0.110 cal/g/°C. Dipole moment: 3.12 at 25° in CS$_2$. One gram dissolves in 10.1 ml H$_2$O at 25°. Gradual hydrolysis to SbOCl. Considerably more sol in dil HCl. Sol in alcohol, benzene, carbon disulfide, dioxane, chloroform (about 22%), ether, acetone, carbon tetrachloride. A satd soln in carbon tetrachloride is about 1.1 molar. Insol in pyridine, quinoline, other organic bases.

Caution: Irritant, corrosive to skin.

USE: As of antimony trichloride soln; also as reagent for chloral, aromatic hydrocarbons, and vitamin A. For mol wt determinations. In chemical microscopy for the identification of drugs (forms adducts and addition compds). Commercially as a mordant, in making other antimony salts, in organic syntheses, as catalyst.

THERAP CAT (VET): Escharotic; dehorning agent (calves, goats).

749. Antimony Trichloride Solution. Antimony chloride solution; liquid butter of antimony. Contains 36-38% SbCl$_3$ and 10-12% free HCl.

Yellow to reddish-brown, clear, somewhat oily, strongly caustic liquid. d 1.5.

USE: Bronzing iron, especially gun barrels; mordant for patent leather and in dyeing; coloring zinc black; manuf lakes, particularly from dye woods; furniture polishes.

750. Antimony Trifluoride. Antimony fluoride; antimonous fluoride. F$_3$Sb; mol wt 178.76. F 31.88%, Sb 68.12%. SbF$_3$. Prepd by dissolving Sb$_2$O$_3$ in aq HF and evaporating the water: Berzelius, *Pogg. Ann.* **1**, 34 (1824); Söll, *FIAT-Review* **23**, 276 (1947); Andersen *et al., Acta Chem. Scand.* **7**, 236 (1953); Kwasnik in *Handbook of Preparative Inorganic Chemistry*, **Vol. 1**, G. Brauer, Ed. (Academic Press, New York, 2nd ed., 1963) p 199. Industrial prepn from SbCl$_3$ and HF: Midgeley *et al., U.S. pat.* **2,024,654** (1935). *Reviews:* Burg in *Fluorine Chemistry*, **Vol. 1**, J.

H. Simons, Ed. (Academic Press, New York, 1950) p 106; Kemmitt, Sharp, *Advan. Fluorine Chem.* **4**, 211 (1965).

Orthorhombic, deliquesc crystals. *Poisonous! Irritates the skin.* d_{20}^{20} 4.379. mp 292°. bp 376°. Also reported as bp 319° (Andersen). Soly in water (g/100 ml): 443 (20°); 562 (30°). Dissolves in water with limited hydrolysis, readily forms complexes such as $[SbF_4]^-$ and many sol salts. For the purpose of halogen exchange fluorination, the compd must be dry: Henne, U.S. pat. 2,082,161 (1937). May be stored in glass vessels or steel drums. *Keep well closed.*

USE: To catalyze fluorinations by HF, manuf chlorofluorides, in dyeing, usually in form of double salts, e.g., antimony sodium fluoride or antimony fluoride and ammonium sulfate double salt, manuf pottery and porcelains.

751. Antimony Triiodide. I_3Sb; mol wt 502.47. I 75.77%, Sb 24.23%. SbI_3. Prepd by the interaction of antimony and iodine in boiling benzene or tetrachloroethane: Bailar, Cundy, *Inorg. Syn.* **1**, 104 (1939); Scattergood, *J. Chem. Ed.* **20**, 40 (1943).

Ruby-red, trigonal crystals (yellowish-green modifications have been observed). d_4^{17} 4.921. The tendency to sublime becomes noticeable at 100°. mp 168°. bp 420°. Critical temp 1101°; crit press. 55 atm. Dipole moment: 1.58. Dec by water and air to SbOI (antimony oxyiodide). Sol in alcohol, acetone, carbon disulfide, HCl, soln of KI. Insol in carbon tetrachloride.

752. Antimony Trioxide. Diantimony trioxide; flowers of antimony; exitelite; senarmontite; valentinite; weisspiessglanz. O_3Sb_2; mol wt 291.52. O 16.46%, Sb 83.54%. Sb_2O_3. Laboratory prepn from $SbCl_3$ and water: Schenk in *Handbook of Preparative Inorganic Chemistry,* vol. 1, G. Brauer, Ed. (Academic Press, New York, 2nd ed., 1963) p 615. Obtained from antimony ore minerals by a volatilization (roasting) process: L. D. Freedman in Kirk-Othmer *Encyclopedia of Chemical Technology,* vol. 3 (Wiley-Interscience, New York, 3rd ed., 1978) pp 107-108. Toxicity study: H. F. Smyth *et al., J. Ind. Hyg. Toxicol.* **30**, 63 (1948).

Crystals, polymorphic. mp 655°. bp 1425°; bp_{210} 870°. Sublimes in high vacuum at 400°. Exists in the vapor phase as Sb_4O_6. Heat capacity at 21° (294.4°K): 24.11 cal/g-atom/°C, Anderson, *J. Am. Chem. Soc.* **52**, 2712 (1930). Heat of vaporization: 17.82 kcal/mol. Slightly sol in water, dilute H_2SO_4, or dilute HNO_3. Soly in dil HCl (0.1 moles HCl/kg H_2O): $\sim 1 \times 10^{-4}$ g-atoms Sb/kg H_2O, Gayer, Garrett, *ibid.* **74**, 2353 (1952). Soly increases with increasing HCl concn: Lea, Wood, *J. Chem. Soc.* **125**, 137 (1924). Sol in solns of alkali hydroxides or sulfides, in warm soln of tartaric acid, or of bitartrates. LD_{50} orally in rats: >20 g/kg (Smyth).

USE: Manuf tartar emetic; as paint pigment; in enamels and glasses; as mordant; in flame-proofing canvas.

753. Antimony Triselenide. Sb_2Se_3; mol wt 480.40. Sb 50.69%, Se 49.31%. Prepd by the action of hydrogen selenide on a soln of potassium antimonyl tartrate: Moser, Atynsky, *Monatsh.* **45**, 235 (1924); Berzelius cited in *Mellor's* Vol. 10, 793 (1930); by direct union of the elements: Chretien, *Compt. Rend.* **142**, 1339, 1412 (1906); Parravano, *Gazz. Chim. Ital.* **43**, I, 210 (1913); Berzelius, *loc. cit. See also:* Chikashige, Fujita, *Mem. Coll. Science Kyoto* **2**, 233 (1917).

Gray powder. mp 605° (Berzelius); mp 611° (Chretien); mp 617° (Parravano); mp 572° (Chikashige, Fujite). Very slightly sol in water. Forms a brown soln with hot potash lye.

754. Antimony Trisulfide. Antimonous sulfide; antimony sulfide; needle antimony; antimony glance. S_3Sb_2; mol wt 339.72. S 28.32%, Sb 71.68%. Sb_2S_3. Occurs in nature as the mineral *stibnite.* Prepn: Donges, Fricke, *Z. Anorg. Chem.* **253**, 2 (1945); Gagliardi, Pilz, *Z. Anal. Chem.* **136**, 344 (1952); *Gmelin's, Antimony* (8th ed.) **18B**, pp 503-524 (1949). Toxicity study: W. R. Bradley, W. G. Fredrick, *Ind. Med.* **10**, *Ind. Hyg. Sect.* **2**, 15 (1941).

Gray, lustrous, crystalline masses or grayish-black powder. Also exists in a red modification. mp 550°. Practically insol in water. Sol in concd HCl with evolution of H_2S; sol

in solutions of the fixed alkali hydroxides. LD i.p. in rats: 100.0 mg Sb/100 g (Bradley, Fredrick).

USE: In pyrotechnics, Bengal fires; manuf ruby glass, matches, explosives; as a pigment in paints.

755. Antimycin A₁. 3-Methylbutanoic acid 3-[[3-(formylamino)-2-hydroxybenzoyl]amino]-8-hexyl-2,6-dimethyl-4,9-dioxo-1,5-dioxonan-7-yl ester; isovaleric acid 8-ester with 3-formamido-*N*-(7-hexyl-8-hydroxy-4,9-dimethyl-2,6-dioxo-1,5-dioxonan-3-yl)salicylamide. $C_{28}H_{40}N_2O_9$; mol wt 548.63. C 61.30%, H 7.35%, N 5.11%, O 26.25%. Antibiotic substance produced by *Streptomyces* spp. Isoln: Dunshee *et al., J. Am. Chem. Soc.* **71**, 2436 (1949); Lockwood *et al., Phytopathology* **44**, 438 (1954); U.S. pat. 2,657,-170 (1953 to Wisconsin Alumni Res. Found.). Structure: van Tamelen *et al., J. Am. Chem. Soc.* **83**, 1639 (1961); Birch *et al., J. Chem. Soc.* **1961**, 889. Absolute config.: Kinoshita *et al., J. Antibiot.* **25**, 373 (1972).

Crystals from ethyl acetate + Skellysolve B, mp 149-150°. $[\alpha]_D^{26}$ +76° (c = 1 in chloroform). uv max (alc): 226, 320 nm (log ε 4.54, 3.68). Freely sol in alc, ether, acetone, chloroform. Very slightly sol in petr ether, benzene, carbon tetrachloride. Practically insol in water and in 5% aq solns of hydrochloric acid, sodium carbonate and sodium bicarbonate. In aq sodium hydroxide a milky suspension is formed. This clears on warming, but all potency is lost.

USE: Experimentally as fungicide, insecticide, miticide.

756. Antimycin A₃. [2R-(2R*,3S*,6S*,7R*,8R*)]-3-Methylbutanoic acid 8-butyl-3-[[3-(formylamino)-2-hydroxybenzoyl]amino]-2,6-dimethyl-4,9-dioxo-1,5-dioxonan-7-yl ester; isovaleric acid 8-ester with *N*-(7-butyl-8-hydroxy-4,9-dimethyl-2,6-dioxo-1,5-dioxonan-3-yl)-3-formamidosalicylamide; blastmycin. $C_{26}H_{36}N_2O_9$; mol wt 520.58. C 59.99%, H 6.97%, N 5.38%, O 27.66%. Antifungal antibiotic substance produced by *Streptomyces blastmyceticus* from Japanese soil: Watanabe *et al., J. Antibiot.* **10A**, 39 (1957). Sepn from complex: Harada *et al., ibid.* **11**, 32 (1958). Identity of blastmycin with antimycin A₃: Liu, Strong, *J. Am. Chem. Soc.* **81**, 4387 (1959). Structural studies: Yonehara, Takeuchi, *J. Antibiot.* **11**, 254 (1958). Structure: van Tamelen *et al., J. Am. Chem. Soc.* **83**, 1639 (1961). Abs config: Kinoshita *et al., J. Antibiot.* **25**, 373 (1972). Total synthesis of diastereomeric mixture: *eidem, ibid.* **22**, 580 (1969); of natural form: *eidem. ibid.* **24**, 724 (1971); *eidem, Bull. Chem. Soc. Japan* **46**, 1279 (1973). Improved synthesis: S. Aburaki, M. Kinoshita, *ibid.* **52**, 198 (1979). Chemical studies: Endo, Yonehara, *J. Antibiot.* **23**, 91 (1970).

Needles from benzene + petr ether, mp 170.5-171.5° (Liu); 174-174.5° (Kinoshita). $[\alpha]_D^{26}$ +64.3° (chloroform) (Liu); $[\alpha]_D^{24}$ +80° (Kinoshita). uv max (methanol): 225, 320 nm (log ε 4.52, 3.86). Moderately sol in methanol, ethanol, ether; freely sol in acetone, ethyl acetate, benzene, chloroform, carbon tetrachloride. Slightly sol in hexane, cyclohexane; very sparingly sol in petr ether. Practically insol in water. LD_{50} in mice (mg/kg): 1.8 i.p.; 1.6 s.c. (Watanabe).

757. Antipyrine. *1,2-Dihydro-1,5-dimethyl-2-phenyl-3H-pyrazol-3-one; 2,3-dimethyl-1-phenyl-3-pyrazolin-5-one;*

phenazone; 1,5-dimethyl-2-phenyl-3-pyrazolone; phenyldimethylpyrazolon(e); dimethyloxychinizin; dimethyloxyquinazine; oxydimethylquinizine; Analgesine; Anodynine; Parodyne; Phenylone; Sedatine. $C_{11}H_{12}N_2O$; mol wt 188.23. C 70.19%, H 6.43%, N 14.88%, O 8.50%. Prepn: Müller *et al.*, *Monatsh.* **89**, 23 (1958); *Hagers Handb. Pharm. Praxis*, Vol. 6 (Springer Verlag, Berlin, 1977) p 571. Toxicity: Hart, *J. Pharmacol. Exp. Ther.* **89**, 205 (1947). Metabolism: H. Uchino *et al.*, *Xenobiotica* **13**, 155 (1983). Clinical comparison with paracetamol: H. Quiding *et al.*, *Int. J Oral Surg.* **11**, 304 (1982). Use as indicator of hepatic drug metabolism: E. S. Vessel, *Clin. Pharm. Ther.* **26**, 275 (1979); G. C. Farrell, L. Zaluzny, *Brit. J. Clin. Pharmacol.* **18**, 559 (1984). HPLC determn in urine: E. Brendel *et al.*, *J. Pharmaceut. Biomed. Anal.* **7**, 1783 (1989). Clinical pharmacokinetics: H. A. Ali *et al.*, *Human Exp. Toxicol.* **13**, 658 (1994).

Tabular crystals or white powder; slightly bitter taste. mp 111-113°. One gram dissolves in less than 1 ml water, 1.3 ml alcohol, 1 ml chloroform, 43 ml ether. The aq soln is neutral to litmus. LD_{50} orally in rats: 1.8 g/kg (Hart).

Acetylsalicylate, $C_{20}H_{20}N_2O_5$, *Acetopyrine, Acopyrine, Acetasol.* Crystals, mp 63-65°. One part dissolves in 400 parts of water, in 20 parts of 2% aq sod. bicarbonate soln. Freely sol in hot water, cold alc, chloroform; sparingly sol in ether. *Note:* Not to be confused with antipyrine salicylacetate.

Mandelate, $C_{11}H_{12}N_2O.C_8H_8O_3$, *antipyrine amygdalate.* Cryst powder, mp 52-55°. One gram dissolves in 15 ml water, 4 ml alc, 26 ml ether.

Methylethylglycolate, $C_{16}H_{22}N_2O_4$, *antipyrine 2-hydroxy-2-methylbutyrate, Astrolin.* Crystalline powder, mp 64-65.5°. Soluble in water, alc.

Salicylacetate, $C_{20}H_{20}N_2O_6$, *α-carboxy-o-anisic acid compd with antipyrine, Pyrosal.* Crystals, mp 149-150°. Bitter, acid taste. Sparingly soluble in water; soluble in alc. *Note:* Not to be confused with antipyrine acetylsalicylate.

THERAP CAT: Analgesic.

THERAP CAT (VET): Has been used as an antipyretic, analgesic and in laminitis of horses.

758. Antipyrine Salicylate. *2-Hydroxybenzoic acid compd with 1,2-dihydro-1,5-dimethyl-2-phenyl-3H-pyrazol-3-one (1:1);* Phenazone salicylate; Saliphenazon; Salazolon; Salipyrazolon; Salipyrine. $C_{18}H_{18}N_2O_4$; mol wt 326.35. C 66.25%, H 5.56%, N 8.58%, O 19.61%. $C_{11}H_{12}N_2O.C_6H_4$-(OH)COOH. Prepd by fusing antipyrine and salicylic acid: *Hagers Handb. Pharm. Praxis* Vol. 6, (Springer Verlag, Berlin, 1977) p 574.

Slightly sweet, cryst powder. mp 91-92°. One gram dissolves in 200 ml water, 40 ml boil. water; freely sol in alc, chloroform, sparingly in ether.

THERAP CAT: Analgesic.

THERAP CAT (VET): Has been used as an antipyretic, analgesic in dogs.

759. Antireticular Cytotoxic Serum. ACS; cytoxin-anticytotoxin serum; Bogomolets' serum; Bogoserum; Sarvinal. Biologic synthesis: Loiseleur, *Ann. Inst. Pasteur* **91**, 445 (1956). Purification: *C.A.* **52**, 6574e (1958). Stimulating effect on reticuloendothelial system: I. P. Miagkaia, *Zh. Mikrobiol. Epidemiol. Immunobiol.* **1978**, 82.

760. α₁-Antitrypsin. *α₁-Trypsin inhibitor; α₁-proteinase inhibitor; α₁-protease inhibitor; alpha₁-antitrypsin;* AAT; A1AT; A1PI; Prolastin. Serum glycoprotein synthesized by the liver. Major serine protease inhibitor (serpin) in mammalian plasma; primarily inhibits neutrophil elastase. Consists of a single polypeptide chain of 394 amino acid residues and 3 carbohydrate side chains linked to

asparagine residues. Mol wt ~52,000 daltons. Highly pleomorphic protein with a number of genetic variants. Characterization of trypsin inhibitor in alpha₁-fraction of human serum: K. Jacobsson, *Scand. J. Clin. Invest.* **7**, Suppl. 14, 57-102 (1955). Isoln from human serum: F. C. Moll *et al.*, *J. Biol. Chem.* **233**, 121 (1958); H. F. Bundy, J. W. Mehl, *ibid.* **234**, 1124 (1959). Purification: I. P. Crawford, *Arch. Biochem. Biophys.* **156**, 215 (1973); M. H. Coan, W. J. Brockway, U.S. pat. 4,379,087 (1983 to Cutter). Structural study of active site: D. Johnson, J. Travis, *J. Biol. Chem.* **253**, 7142 (1978). Complete amino acid sequence and review of genetic variants: R. W. Carrell *et al.*, *Nature* **298**, 329 (1982). Crystal structure: H. Loebermann *et al.*, *J. Mol. Biol.* **177**, 531 (1984). Cloning and expression of human AAT gene: M. Courtney *et al.*, *Proc. Nat. Acad. Sci. USA* **81**, 669 (1984). AAT deficiency has been associated with degenerative lung disease: C.-B. Laurell, S. Eriksson, *Scand. J. Clin. Lab. Invest.* **15**, 132 (1963); with hepatic cirrhosis: H. L. Sharp, *Gastroenterology* **70**, 611 (1976). Inhibition of leukocyte elastase: K. Ohlsson, *Scand. J. Clin. Lab. Invest.* **28**, 251 (1971); K. Beatty *et al.*, *J. Biol. Chem.* **255**, 3931 (1980). Review of human phenotypes and relationship to disease: J. O. Morse, *N. Engl. J. Med.* **299**, 1045-1048; 1099-1105 (1978). Clinical evaluation in AAT deficiency: M. D. Werwers *et al.*, *ibid.* **316**, 1055 (1987); N. Konietzko *et al.*, *Deutsch. Med. Wochenschr.* **113**, 369 (1988). Symposium on structure, function and clinical applications: *Am. J. Med.* **84**, Suppl. 6A, 1-90 (1988). Review of role in emphysema· P. J. Stone, *Clin. Chest Med.* **4**, 405-412 (1983); J. A. Pierce, *J. Am. Med. Assoc.* **259**, 2890-2895 (1988).

THERAP CAT: In treatment of emphysema associated with inherited α₁-antitrypsin deficiency.

761. Antivenin (Crotalidae) Polyvalent. North and South American antisnakebite serum. An antitoxin obtained from serum of horses that have been immunized with venom of four crotaline snakes *(Crotalus atrox, C. adamanteus, C. durissus terrificus, Bothrops atrox). Ref:* Gingrich, Hohenadel, in *Venoms,* E. E. Buckley, N. Porges, Eds. (Publ. 44 of Am. Assoc. Adv. of Sci., Washington, 1956) p 381; Keegan, *ibid.*, p 413; *Behringwerk-Mitteilungen* (Sonderband: *Die Gift-Schlangen der Erde,* Marburg, 1963) 464 pp.

THERAP CAT: Antidote to snake venom.

THERAP CAT (VET): To neutralize snake venom.

762. Antivirin. AV. A virus inhibitory factor produced by various cells including HeLa-S₃, Hep. no. 2, KB, FL, L, G₂V, Chang's liver cells, human embryo lung, mouse embryo fibroblast, chick embryo fibroblast: Y. Seto, S. Toyoshima, *Progr. Antimicrob. Anticancer Chemother., Proc. Int. Congr. Chemother., 6th,* vol. I (Univ. Park Press, Baltimore, 1970) pp 108-113. The antiviral spectrum of this inhibitor is wide, and unlike interferon, its activity is stable after proteolytic enzyme treatment and heating. AV is produced in the absence of interferon inducers and does not possess such strict host specificity as interferon. It exerts its effects at any early stage of virus replication. The active part of AV is of smaller molecular size than interferon. Nature and mechanism of action: Toyoshima *et al.*, *Interferon, Proc. Symp.* **1969**, Y. Nagano, Ed. (Igaku Shoin, Ltd., Tokyo, 1970) pp 185-193.

763. Antrafenine. *2-[[7-(Trifluoromethyl)-4-quinolinyl]amino]benzoic acid 2-[4-[3-(trifluoromethyl)phenyl]-1-piperazinyl]ethyl ester; 2-[4-(α,α,α-trifluoro-m-tolyl)-1-piperazinyl]ethyl-N-(7-trifluoromethyl-4-quinolyl)anthranilate;* SL-73.033; Stakane. $C_{30}H_{26}F_6N_4O_2$; mol wt 588.55. C 61.22%, H 4.45%, F 19.37%, N 9.52%, O 5.44%. Analgesic with minimal anti-inflammatory and antipyretic activity, related structurally to floctafenine, *q.v.* Prepn: P. R. L. Giudicelli *et al.*, Ger. pat. 2,415,982; *eidem*, U.S. pat. 3,935,229 (1974, 1976 both to Synthelabo); P. M. Manoury *et al.*, *J. Med. Chem.* **22**, 554 (1979). Pharmacokinetic study: L. G. Dring *et al.*, *Brit. J. Pharmacol.* **63**, 368P (1978). Metabolism: V. Rovei *et al.*, *Ann. Chim.* **67**, 733 (1977). Pharmacologic study: R. D. Sofia *et al.*, *Pharmacol. Res. Commun.* **11**, 179 (1979).

Crystals from isopropyl alc, mp 88°. LD_{50} orally in mice: 4000 mg/kg (Manoury).

THERAP CAT: Analgesic.

764. ANTU. *1-Naphthalenylthiourea; 1-(1-naphthyl)-2-thiourea; α-naphthylthiourea; N-1-naphthylthiourea; α-naphthylthiocarbamide; krysid; chemical 109; Anturat; Bantu; Rattrack.* $C_{11}H_{10}N_2S$; mol wt 202.28. C 65.32%, H 4.98%, N 13.85%, S 15.85%. Prepd from α-naphthylamine and ammonium or potassium or sodium thiocyanate: de Clermont, Wehrlin, *Bull. Soc. Chim.* [2] **26**, 125 (1876); Alvarez, *C.A.* **42**, 4560 (1948); from α-naphthylisothiocyanate by treatment with alcoholic ammonia: Dyson, Hunter, *Chem. News* **134**, 4 (1927). Rodenticide activity: Richter, *J. Am. Med. Assoc.* **129**, 927 (1945). Review of carcinogenic risk: *IARC Monographs* **30**, 347-357 (1983).

Prisms from alc. Bitter taste. mp 198°. Solubility in water at 25°: 0.06 g/100 ml; in acetone: 2.43 g/100 ml; in triethylene glycol: 8.6 g/100 ml. Fairly sol in hot alc.

Caution: Potential symptoms of toxicity following ingestion of large doses are vomiting, dyspnea, cyanosis and coarse pulmonary rales; mild liver damage. *See NIOSH Pocket Guide to Chemical Hazards* (DHHS/NIOSH 94-116, 1994) p 20. May cause glomerular injury in nonlethal poisonings. *See Clinical Toxicology of Commercial Products,* R. E. Gosselin *et al.,* Eds. (Williams & Wilkins, Baltimore, 5th ed., 1984) Section III, p 40-42.

USE: Rodenticide. Specific control for the adult Norway rat; less toxic to other rat species.

765. Apafant. *4-[3-[4-(2-Chlorophenyl)-9-methyl-6H-thieno[3,2-f][1,2,4]triazolo[4,3-a][1,4]diazepin-2-yl]-1-oxopropyl]morpholine;* 3-[4-(2-chlorophenyl)-9-methyl-6H-thieno[3,2-f][1,2,4]triazolo[4,3-a][1,4]diazepin-2-yl]-1-(4-morpholinyl)-1-propanone; WEB-2086. $C_{22}H_{22}ClN_5O_2S$; mol wt 455.97. C 57.95%, H 4.86%, Cl 7.78%, N 15.36%, O 7.02%, S 7.03%. Platelet activating factor (PAF) antagonist. Prepn: K. H. Weber *et al.,* **Ger. pat. 3,502,392;** *eidem,* U.S. pat. **5,082,839** (1986, 1992 both to Boehringer, Ing.). Pharmacology: J. Casals-Stenzel *et al., J. Pharmacol. Exp. Ther.* **241**, 974 (1987). Clinical pharmacology, pharmacokinetics: H. M. Brecht *et al., Arzneimittel-Forsch.* **41**, 51 (1991). Clinical effect vs PAF-induced bronchoconstriction: W. S. Adamus *et al., Clin. Pharmacol. Ther.* **47**, 456 (1990).

Viscous, almost colorless oil. LD_{50} in mice (mg/kg): 540 i.v.; 4600 orally (Casals-Stenzel).

USE: Tool to evaluate the role of PAF in exptl models of human disease.

766. Apalcillin. *[2S-[2α,5α,6β(S*)]]-6-[[[[(4-Hydroxy-1,5-naphthyridin-3-yl)carbonyl]amino]phenylacetyl]amino]-3,3-dimethyl-7-oxo-4-thia-1-azabicyclo[3.2.0]heptane-2-carboxylic acid.* $C_{25}H_{23}N_5O_6S$; mol wt 521.55. C 57.57%, H 4.44%, N 13.43%, O 18.41%, S 6.15%. Semi-synthetic antibiotic related to penicillin. Prepn of the potassium salt: H. Tobiki *et al.,* **S. Afr. pat. 72 05,865** corresp to U.S. pat. **3,864,329** (1973, 1975 both to Sumitomo); of the sodium salt: Y. Hirotada *et al.,* U.S. pat. **4,005,075** (1977 to Sumitomo); H. Tobiki *et al., Yakugaku Zasshi* **100**, 49 (1980), *C.A.* **93**, 95179x (1980). Microbiological evaluation: H. Noguchi *et al., Antimicrob. Ag. Chemother.* **9**, 262 (1976). Analytical studies: I. Umeda *et al., Yakugaku Zasshi* **99**, 717 (1979), *C.A.* **92**, 11287z (1980). Series of articles on pharmacology, exptl and clinical effects: *Chemotherapy (Tokyo)* **26**, Suppl. 2, 111-223 (1978). Pharmacokinetics in man: U. Busch *et al., Arzneimittel-Forsch* **32**, 1131 (1982).

Sodium salt, $C_{25}H_{22}N_5NaO_6S$, *PC-904, Lumota, Palcin.* White cryst, sol in water.

THERAP CAT: Antibacterial.

767. Apamin. $C_{79}H_{131}N_{31}O_{24}S_4$; mol wt 2027.37. C 46.80%, H 6.51%, N 21.42%, O 18.94%, S 6.33%. Neurotoxic polypeptide consisting of 18 amino acid residues and 2 disulfide bridges. Comprises about 2% by wt of the dried venom of *Apis mellifica (mellifera),* the honey bee. Isoln: E. Habermann, K. G. Reiz, *Naturwiss.* **51**, 61 (1964); *eidem, Biochem. Z.* **341**, 451 (1965). Structure: P. Haux *et al., Z. Physiol. Chem.* **348**, 737 (1967); R. Shipolini *et al., Chem. Commun.* **1967**, 679. Conformation: R. C. Hider, U. Ragnarsson, *FEBS Letters* **111**, 189 (1980); B. Busetta, *ibid.* **112**, 138 (1980). Solution structure: J. H. B. Pease, D. E. Wemmer, *Biochemistry* **27**, 8491 (1988). Solid-phase synthesis: J. van Rietschoten *et al., Eur. J. Biochem.* **56**, 35 (1975); B. E. B. Sandberg, U. Ragnarsson, *Int. J. Pept. Protein Res.* **11**, 238 (1978). Pharmacology: Wellheoner, *Arch. Pharmakol. Exp. Pathol.* **262**, 29 (1969). Biochemistry: E. Habermann, K. G. Reiz, *Biochem. Z.* **343**, 192 (1965). Action on CNS: E. Habermann, D. Cheng-Raude, *Toxicon* **13**, 465 (1975). Binding and toxicity studies: C. Labbé-Jullié *et al., Eur. J. Biochem.* **196**, 639 (1991). *Review:* E. Habermann, *Science* **177**, 314-322 (1972); C. Granier, J. van Rietschoten, in *Natural Toxins,* D. Eaker, T. Wadström, Eds. (Pergamon, New York, 1980) pp 481-486; E. Habermann, *Pharmacol. Ther.* **25**, 255-270 (1984).

Cys-Asn-Cys-Lys-Ala-Pro-Glu-Thr-Ala-Leu-Cys-Ala-Arg-Arg-Cys-Gln-Gln-HisNH₂

Highly basic compd. Pharmacologic activity destroyed by oxidn with performic acid. LD_{50} i.v. in mice: 4 mg/kg (Habermann, Reiz); LD_{50} intracerebroventricularly in mice: 12 ng/animal (Labbé-Jullié).

768. Apazone. *5-(Dimethylamino)-9-methyl-2-propyl-1H-pyrazolo[1,2-a][1,2,4]benzotriazine-1,3(2H)-dione;* 3-dimethylamino-7-methyl-1,2-(n-propylmalonyl)-1,2-dihydro-1,2,4-benzotriazine; azapropazone; AHR-3018; MI-85; Cinnamin; Sinnamin; $C_{16}H_{20}N_4O_2$; mol wt 300.36. C 63.98%, H 6.71%, N 18.65%, O 10.65%. Prepn: **Fr. pat. 1,440,629** corresp to I. Molnar *et al.,* U.S. pats. **3,349,088** and **3,482,024** (1966, 1967, 1969, all to Siegfried); Mixich, *Helv. Chim. Acta* **51**, 532 (1968). Pharmacological and toxicological studies: Jahn, Adrian, *Arzneimittel-Forsch.* **19**, 36 (1969). Metabolism: Mixich, *Helv. Chim. Acta* **55**, 1031 (1972). HPLC determn in plasma and urine: B. J. Kline *et al., Arzneimittel-Forsch.* **33**, 504 (1983).

      Consult the Name Index before using this section.

mp 228°.

Dihydrate, *MI-85Di, Azapren, Prolixan, Rheumox, Tolyprin.* Almost colorless crystals, mp 247-248°.

THERAP CAT: Anti-inflammatory; analgesic.

769. Aphidicolin. *(3α,4α,5α,17α)-3,17-Dihydroxy-4-methyl-9,15-cyclo-C,18-dinor-14,15-secoandrostane-4,17-dimethanol;* tetradecahydro-3,9-dihydroxy-4,11b-dimethyl-8,-11a-methano-11aH-cyclohepta[a]naphthalene-4,9-dimethanol; ICI-69653; NSC-234714. $C_{20}H_{34}O_4$; mol wt 338.49. C 70.97%, H 10.12%, O 18.91%. Novel tetracyclic diterpene antibiotic with antiviral and antimitotic properties, isolated as (+)-form from *Cephalosporium aphidicola* Petch. Specific inhibitor of DNA α-polymerase. Description and x-ray crystallographic determn of structure: K. M. Brundret *et al., Chem. Commun.* **1972**, 1027. Prepn and antiviral activity: A. Borrow *et al.,* U.S. pat. **3,761,512** (1973 to ICI). Structure and abs config: W. Dalziel *et al., J. Chem. Soc. Perkin Trans. 1* **1973**, 2841. Total synthesis of (±)-aphidicolin: B. M. Trost *et al., J. Am. Chem. Soc.* **101**, 1328 (1979); J. E. McMurry *et al., ibid.* 1331; E. J. Corey *et al., ibid.* **102**, 1743 (1980); J. E. McMurry *et al., Tetrahedron* **37**, Suppl. 1, 319 (1981). Antiviral effects *in vitro* and *in vivo:* R. A. Bucknall *et al., Antimicrob. Ag. Chemother.* **4**, 294 (1973). Antimitotic activity: S. Ikegami *et al., Nature* **275**, 458 (1978). Effects on DNA synthesis in mouse sarcoma: S. Seki *et al., Biochim. Biophys. Acta* **610**, 413 (1980). Enzymatic determn method: G. Pedrali-Noy *et al., J. Biochem. Biophys. Methods* **4**, 113 (1981). Induction of differentiation of human myeloid leukemia cells: J. Griffin *et al., Exp. Hematol.* **10**, 774 (1982). Effects on cell proliferation: G. Iliakis *et al., Int. J. Radiat. Biol.* **42**, 417 (1982); on γ-irradiated human fibroblasts: P. J. Smith, M. C. Paterson, *Biochim. Biophys. Acta* **739**, 17 (1983).

Needles from ethyl acetate, mp 227-232°. $[\alpha]_D^{27}$ +12° (c = 1 in methanol).

(±)-Form, white cryst, mp 218-220°.

USE: As a biological tool in studies of cell proliferation and differentiation.

770. Apholate. *2,2,4,4,6,6-Hexakis(1-aziridinyl)-2,2,-4,4,6,6-hexahydro-1,3,5,2,4,6-triazatriphosphorine;* hexakis(1-aziridinyl)phosphonitrile; 1-aziridinylphosphonitrile trimer. $C_{12}H_{24}N_9P_3$; mol wt 387.30. C 37.21%, H 6.25%, N 32.55%, P 23.99%. Prepn: Rätz, Grundmann, U.S. pat. **2,858,306** (1958 to Olin Mathieson). Toxicity study: T. B. Gaines, *Toxicol. Appl. Pharmacol.* **14**, 515 (1969).

Crystals from heptane, mp 147.5°. LD_{50} in male, female rats (mg/kg): 98, 113 orally (Gaines).

USE: Insect chemosterilant.

771. Aphylline. *[7R-(7α,7aβ,14α,14aα)]-Dodecahydro-7,14-methano-2H,6H-dipyrido[1,2-a:1',2'-e][1,5]diazocin-6-one.* $C_{15}H_{24}N_2O$; mol wt 248.37. C 72.54%, H 9.74%, N 11.28%, O 6.44%. From *Anabasis aphylla* L., *Chenopodiaceae.* Extraction procedure and structure: Orechoff, Menschikoff, *Ber.* **64**, 266 (1931); **65**, 234 (1932). Structure: Galinovsky, Jarish, *Monatsh.* **84**, 199 (1953). Stereochemistry: Edwards *et al., Can. J. Chem.* **32**, 235 (1954); Galinovsky *et al., Monatsh.* **86**, 1014 (1955). Synthesis: Bohlmann *et al., Ber.* **90**, 653 (1957).

Crystals, mp 52-57°. bp_4 200°. $[\alpha]_D^{20}$ +10.3° (c = 20 in methanol). Soluble in the usual organic solvents.

Hydrochloride, $C_{15}H_{24}N_2O·HCl$, rhombohedra from alc, mp 209°. $[\alpha]_D^{20}$ +14° (c = 25). Freely soluble in water.

772. Apicycline. *α-[4-(Dimethylamino)-1,4,4a,5,5a,6,-11,12a-octahydro-3,6,10,12,12a-pentahydroxy-6-methyl-1,11-dioxo-2-naphthacenecarboxamido]-4-(2-hydroxyethyl)-1-piperazineacetic acid;* N-[(2-hydroxyethyl)piperazinocarboxymethyl]tetracycline; RIT-1140; Traserit. $C_{30}H_{38}N_4O_{11}$; mol wt 630.65. C 57.14%, H 6.07%, N 8.88%, O 27.91%. Semisynthetic antibiotic similar to tetracycline, *q.v.* Prepn: Rondelet, Neth. pat. **6,515,688**, *C.A.* **66**, 46260a (1967) and Belg. pat. **673,130**, *C.A.* **68**, 104840u (1968) (both 1966 to Recherche & Ind. Ther.); Valcavi *et al., Farmaco Ed. Sci.* **21**, 775 (1966); Rondelet, Froment, *Ann. Pharm. Franc.* **26**, 63 (1968).

Yellow, amorphous powder with slight amine odor, mp 144.5° (dec). Freely sol in water. $[\alpha]_D$ −123° (c = 0.5 in methanol); $[\alpha]_D$ −133° (c = 0.5 in water).

THERAP CAT: Antibacterial.

773. Apigenin. *5,7-Dihydroxy-2-(4-hydroxyphenyl)-4H-1-benzopyran-4-one; 4',5,7-trihydroxyflavone;* 2-(p-hydroxyphenyl)-5,7-dihydroxychromone; pelargidenon 1449; Versulin. $C_{15}H_{10}O_5$; mol wt 270.24. C 66.67%, H 3.73%, O 29.60%. The aglucon of apiin and of apigenin-7-glucoside. From apiin by boiling with acids, from apigenin-7-glucoside by enzymatic hydrolysis with emulsin or by boiling with 15% H_2SO_4. Isoln and structure: Czajkowski *et al., Ber.* **33**, 1992 (1900); Schmid, Waschkau, *Monatsh.* **49**, 83 (1928); Baker *et al., J. Chem. Soc.* **1963**, 1477. Synthesis: Hutchins, Wheeler, *ibid.* **1939**, 91; Farooq *et al., Arch. Pharm.* **292**, 792 (1959).

Yellow needles from aqueous pyridine, mp 345-350°. uv max (ethanol): 269, 340 nm (ε 18,800; 20,900). Practically insoluble in water; moderately sol in hot alcohol. Sol in dil KOH with intense yellow color.

USE: Has been used to dye Cr mordanted wool yellow. The color is fast to soap.

774. Apigetrin. *7-(β-D-Glucopyranosyloxy)-5-hydroxy-2-(4-hydroxyphenyl)-4H-1-benzopyran-4-one;* apigenin-7-D-glucoside; 7-D-glycosylapigenin; cossmetin. $C_{21}H_{20}O_{10}$; mol wt 432.38. C 58.33%, H 4.66%, O 37.00%. Isoln from flowers of *Anthemis nobilis* L., *Compositae:* Power, Browning, *J. Chem. Soc.* **105**, 1833 (1914); from parsley: Nordström, Swain, *Chem. & Ind. (London)* **1953**, 85; *J. Chem. Soc.* **1953**, 2764.

Pale yellow, cryst powder, dec 178-180°. Astringent taste. uv max (ethanol): 335, 268 nm. Sol in water, dil alcohol.

775. Apiin. *7-[(2-O-D-Apio-β-D-furanosyl-β-D-glucopyranosyl)oxy]-5-hydroxy-2-(4-hydroxyphenyl)-4H-1-benzopyran-4-one;* 4',5,7-trihydroxyflavone-7-apiosylglucoside; apigenin-7-apiosylglucoside; apioside. $C_{26}H_{28}O_{14}$; mol wt 564.50. C 55.32%, H 5.00%, O 39.68%. Isoln from parsley and from celery: Vongerichten, *Ber.* **33**, 2334, 2904 (1900); Gupta, Seshadri, *Proc. Indian Acad. Sci.* **35A**, 242 (1952), *C.A.* **47**, 3306c (1953); Rahman, *Z. Naturforsch.* **13b**, 201 (1958). Structure: Vongerichten, *Ann.* **318**, 121 (1901); Marchlewski, Skarzynski, *Biochem. Z.* **297**, 56 (1938); Hemming, Ollis, *Chem. & Ind. (London)* **1953**, 85.

Crystals from ethanol, mp 230-232°. uv max (96% ethanol): 267.5, 342.5 nm (log ε 4.18, 4.30). Sol in hot water, hot alcohol; practically insol in ether. Sol in Na_2CO_3 or NH_3 solns with intense yellow color, in NaOH with pale yellow color.

776. Apiole (Dill). *4,5-Dimethoxy-6-(2-propenyl)-1,3-benzodioxole; 1-allyl-2,3-dimethoxy-4,5-(methylenedioxy)-benzene;* dill apiole. $C_{12}H_{14}O_4$; mol wt 222.24. C 64.85%, H 6.35%, O 28.80%. Occurs in dill oil, *Anethum graveolus* L., *Umbelliferae.* Isoln: G. Ciamician, P. Silber, *Ber.* **29**, 1799 (1896); D. B. Spoelstra, *Rec. Trav. Chim.* **48**, 372 (1929). Structure: H. Thoms, *Arch. Pharm.* **242**, 344 (1904). Synthesis: W. Baker *et al., J. Chem. Soc.* **1934**, 1681; F. Dallacker, *Ber.* **102**, 2663 (1969); J. R. Cannon *et al., J. Sci. Soc. Thailand* **6**, 59 (1980). Synergistic activity with insecticides:

E. P. Lichtenstein *et al., J. Agr. Food Chem.* **22**, 658 (1974); S.S. Tomar *et al., Agr. Biol. Chem.* **43**, 1479 (1979).

Oil, mp 29.5°. bp 285°. n_D^{17} 1.5305; d_{15}^{15} 1.1598.

777. Apiole (Parsley). *4,7-Dimethoxy-5-(2-propenyl)-1,3-benzodioxole; 1-allyl-2,5-dimethoxy-3,4-methylenedioxybenzene;* parsley apiole; apiol; apioline; parsley camphor. $C_{12}H_{14}O_4$; mol wt 222.24. C 64.85%, H 6.35%, O 28.80%. Occurs in parsley oil, *Petroselinum sativum:* Blanchet, Sell, *Ann.* **6**, 259 (1833); Vongerichten, *Ber.* **9**, 1477 (1876); Kolesnikov *et al., Aptechnoe Delo* **7**(4), 27 (1958), *C.A.* **54**, 12491e (1960). Structure: Ciamician, Silber, *Ber.* **23**, 2283 (1890); Thoms, *ibid.* **36**, 1714 (1903). Synthesis: Baker, Savage, *J. Chem. Soc.* **1938**, 1602; F. Dallacker, *Ber.* **102**, 2663 (1969). Synergistic activity with insecticides: E. P. Lichtenstein *et al., J. Agr. Food Chem.* **22**, 658 (1974).

Crystals; faint parsley odor. mp 29.5°. bp 294°. n_D^{20} 1.536-1.538. Insol in water; sol in alcohol, benzene, chloroform, ether, acetone, oils. *Keep in a cool place.*

778. Apiose. *D-Apiose;* tetrahydroxyisovaleraldehyde; 3-*C*-(hydroxymethyl)-D-glyceroaldotetrose. $C_5H_{10}O_5$; mol wt 150.13. C 40.00%, H 6.71%, O 53.28%. First found in parsley in which it occurs as the flavinoid glycoside apiin, *q.v.* Isoln from apiin: Vongerichten, *Ann.* **318**, 126 (1901); **321**, 74 (1902); Hemming, Ollis, *Chem. & Ind. (London)* **1953**, 85. From the rubber plant, *Hevea brasiliensis, Euphorbiaceae:* Patrick, *Nature* **178**, 216 (1956). Discussion of structure and isoln from the Australian marine plant *Posidonia australis* Kon., *Potamogetonaceae:* Bell, *Methods in Carbohydrate Chemistry* vol. I (Academic Press, New York, 1962) pp 260-263. Synthesis: Gorin, Perlin, *Can. J. Chem.* **36**, 480 (1958); Khalique, *J. Chem. Soc.* **1962**, 2515; Ezekiel *et al., Tetrahedron Letters* **1969**, 1635. Synthesis of L-form: Weygand, Schmiechen, *Ber.* **92**, 535 (1959); of DL-form: Kinoshita, Miwa, *Carbohyd. Res.* **28**, 175 (1973); Y. Araki *et al., ibid.* **58**, C4 (1977); of D- and L-forms: P. Ho, *Can. J. Chem.* **57**, 381 (1979). Chemistry, configuration and synthesis studies: Williams, Jones, *ibid.* **42**, 69 (1964); Hulyalker *et al., ibid.* **43**, 2085 (1965). Review: Watson, Orenstein, *Advan. Carbohyd. Chem. Biochem.* **31**, 135-184 (1975).

Syrup. $[\alpha]_D^{15}$ +5.6°; $[\alpha]_D^{19}$ +9.1°. Soluble in water.
D-Apiose di-*O*-isopropylidene, $C_{11}H_{18}O_5$, plates from water containing a trace of NH_3, mp 81-83°. $[\alpha]_D^{20}$ +55.5° (c = 1.1 in ethanol).

779. Aplasmomycin. ICI-122378. $C_{40}H_{60}BNaO_{14}$; mol wt 798.71. C 60.15%, H 7.57%, B 1.35%, Na 2.88%, O 28.04%. Antibiotic produced by *Streptomyces griseus* strain SS-20, obtained from shallow sea mud: Y. Okami *et al., J. Antibiot.* **29**, 1019 (1976); *eidem,* **Japan. Kokai 77 108901** (1977 to Microbiochem. Res. Found.), *C.A.* **88**, 35843 (1978). It inhibits the growth of gram-positive bacteria *in vitro* and is active vs *Plasmodium berghei in vivo.* Aplasmomycin is a symmetrical dimer related to boromycin, *q.v.*, the only other known natural product that contains boron. Structure, x-ray crystallography: H. Nakamura *et al., J. Antibiot.* **30,** 714 (1977). Total synthesis of (+)-form: E. J. Corey *et al., J. Am. Chem. Soc.* **104,** 6816, 6818 (1982); T. Nakata *et al., Tetrahedron Letters* **27,** 6341, 6345 (1986); J. D. White *et al., J. Am. Chem. Soc.* **108,** 8105 (1986). Use as growth promotant in ruminants: D. H. Davies *et al., Eur. pat. Appl.* **2893;** *eidem,* U.S. pat. **4,225,593** (1979, 1980 both to ICI). Two minor components, aplasmomycins B and C, have also been isolated from the fermentation; both aplasmomycin and aplasmomycin B show ionophoric properties, mediating net K$^+$ transport across a bulk phase: K. Sato *et al., J. Antibiot.* **31,** 632 (1978). NMR analysis of aplasmomycin and deboroaplasmomycin: T. S. S. Chen *et al., ibid.* **33,** 1316 (1980). Comparative anti-anaerobic activity of aplasmomycin: K. Watanabe *et al., Antimicrob. Ag. Chemother.* **19,** 519 (1981). Structural studies: T. J. Stout *et al., Tetrahedron* **47,** 3511 (1991). Review of biosynthetic studies: H. G. Floss, C. Chang in *Antibiotics IV,* J. W. Corcoran, Ed. (Springer-Verlag, New York, 1981) pp 203-210.

(+)-Aplasmomycin

Colorless needles, mp 283-285° (dec). $[\alpha]_D^{22}$ +225° (c = 1.24 in chloroform). Very lipophilic. Practically insol in water. LD$_{50}$ i.p. in mice: 125 mg/kg (Okami).

Silver salt, C$_{40}$H$_{60}$AgBO$_{14}$, colorless needles, mp 218-220°. $[\alpha]_D^{23}$ +194° (c = 0.34 in chloroform).

USE: Growth promotant in ruminants.

780. Apoatropine. *endo-α-Methylenebenzeneacetic acid 8-methyl-8-azabicyclo[3.2.1]oct-3-yl ester; 1αH,5αH-tropan-3α-ol atropate;* atropamine; atropyltropeine. C$_{17}$H$_{21}$-NO$_2$; mol wt 271.36. C 75.25%, H 7.80%, N 5.16%, O 11.79%. Occurs in root of *Atropa belladonna* L., Solanaceae. Also obtained from atropine by splitting off water or by total synthesis: Ladenburg, *Ann.* **217,** 102 (1883); Merck, *Arch. Pharm.* **230,** 134 (1892); **231,** 110 (1893); Hesse, *Ann.* **261,** 87 (1891); **271,** 124 (1892); **277,** 290 (1893). Isoln by chromatography: Steinegger, Phokas, *Pharm. Acta Helv.* **31,** 284 (1956).

Prisms from chloroform, mp 62°. Absorption spectrum: Gompel, Henri, *Compt. Rend.* **156,** 1543 (1913). Freely sol in alcohol, ether, chloroform, benzene, carbon disulfide; slightly in petr ether, isoamyl alcohol. Almost insol in water. LD$_{50}$ in mice (mg/kg): 160 orally, 14.1 i.p.

Hydrochloride, C$_{17}$H$_{21}$NO$_2$.HCl, scales, mp 239°. Sol in hot water; sparingly sol in alc, acetone. Nearly insol in ether.

Sulfate pentahydrate, (C$_{17}$H$_{21}$NO$_2$)$_2$.H$_2$SO$_4$.5H$_2$O, crystals, sparingly sol in water.

Caution: Death by respiratory arrest occurs in animals with relatively small doses. It is more active by mouth than by injection.

THERAP CAT: Antispasmodic.

781. Apocodeine. *(R)-5,6,6a,7-Tetrahydro-10-methoxy-6-methyl-4H-dibenzo[de,g]quinolin-11-ol; 10-methoxy-6aβ-aporphin-11-ol.* C$_{18}$H$_{19}$NO$_2$; mol wt 281.35. C 76.84%, H 6.81%, N 4.98%, O 11.37%. A monomethyl ether of apomorphine (*q.v.*). Prepd by heating codeine with oxalic acid: Folkers, *J. Am. Chem. Soc.* **58,** 1814 (1936); by heating codeine with phosphoric acid: Small *et al., J. Org. Chem.* **5,** 344 (1940). Configuration: Corrodi, Hardegger, *Helv. Chim. Acta* **38,** 2038 (1955). Total synthesis and pharmacology: Neumeyer *et al., J. Med. Chem.* **16,** 1223 (1973).

Small prisms from methanol (lose solvent at 80°/2 mm), mp 124° (dry). $[\alpha]_D^{24}$ −97° (c = 0.45). Slightly soluble in water; sol in alcohol, ether, dilute acids.

Hydrochloride, C$_{18}$H$_{20}$ClNO$_2$, crystals from alcohol-ether, softening 140°, dec 260-263°. $[\alpha]_D^{22}$ −43° (c = 0.51). Very sol in water; sol in alcohol.

THERAP CAT: Emetic.

782. Apocynin. *1-(4-Hydroxy-3-methoxyphenyl)ethanone;* acetovanillon; 4-hydroxy-3-methoxyacetophenone. C$_9$H$_{10}$O$_3$; mol wt 166.18. C 65.05%, H 6.07%, O 28.88%. From rhizome of Canadian hemp, *Apocynum cannabinum* L., Apocynaceae and from *A. androsaemifolium:* Finnemore, *J. Chem. Soc.* **93,** 1513, 1520 (1908). Also from the essential oil (butter) of the rhizomes of *Iris* spp., Iridaceae: Naves, *Helv. Chim. Acta* **32,** 1351 (1949).

Fine needles from water, mp 115°. bp 295-300°. Faint vanilla odor. Slightly sol in cold, freely in hot water, alc, benzene, chloroform, ether. Practically insol in petr ether.

783. Apocynum androsaemifolium. Dogbane; bitter root; spreading dogbane; milk ipecac; wild ipecac; rheumatism weed. Root of *Apocynum androsaemifolium* L., Apocynaceae. *Habit.* North America. *Constit.* Apocynin, apocynein, apocynamarin, volatile oil.

THERAP CAT: Cardiotonic.

784. Apocynum cannabinum. Canadian hemp; American Indian hemp; black Indian hemp; Indian physic; Indian dogbane. Dried rhizome and roots of *Apocynum cannabinum* L., Apocynaceae. *Habit.* U.S. *Constit.* Cynotoxin, apocyncein, apocynin, cymarin, resin, tannin, bitter extractive, starch.

Note: Not to be confused with *Apocynum androsaemifolium* which has few of its properties.

THERAP CAT: Cardiotonic.

785. Apo-β-erythroidine. *4,5,7,8,9,12-Hexahydro-11H-pyrano[3,4-d]pyrrolo[3,2,1-jk][1]benzazepin-11-one.* C$_{15}$H$_{15}$NO$_2$; mol wt 241.29. C 74.67%, H 6.27%, N 5.80%, O 13.26%. A degradation product of β-erythroidine. Prepd by heating β-erythroidine to 120° with phosphoric or sulfuric acid: Sauvage *et al.*, *Science* **109**, 627 (1947); by reacting β-erythroidine with concd hydrobromic acid at 100°: Koniuszy, Folkers, *J. Am. Chem. Soc.* **73**, 333 (1951). Synthesis and structure: Blake *et al.*, *J. Am. Chem. Soc.* **87**, 1397 (1965).

Crystals, mp 128-129°, also reported as mp 132-132.5°. uv max (ethanol): 345, 240 nm (ε 3500, 24500).

786. Apolipoprotein E. ApoE. Protein component of several classes of lipoproteins; plays a central role in plasma lipoprotein metabolism and cholesterol transport. Also involved in lipid-transport in the development and repair of nervous tissue. Polymorphic glycoprotein containing 299 amino acid residues; mol wt 34.2 kDa. Three major isoforms have been identified in humans (apo-E2, E3, E4) which are the product of 3 independent alleles (ε2, ε3, ε4) at a single gene locus. The isoforms differ from each other only in cysteine-arginine interchanges at 2 substitution sites. Allele ε3 is the most common; inheritance of ε4 has been associated with increased risk of Alzheimer's disease. Primarily synthesized in the liver by parenchymal cells and in the brain by astrocytes. Also produced in significant amounts by mature macrophages. Discovery as component of VLDL: V. G. Shore, B. Shore, *Biochemistry* **12**, 502 (1973). Purification and characterization: V. I. Zannis, J. L. Breslow, *Mol. Cell. Biochem.* **42**, 3 (1982). Amino acid sequence: S. C. Rall, Jr. *et al.*, *J. Biol. Chem.* **257**, 4171 (1982). Biosynthesis and activity: R. W. Mahley, *Science* **240**, 622 (1988). Association with Alzheimer's disease: A. M. Saunders *et al.*, *Neurology* **43**, 1467 (1993). Review of role in lipoprotein metabolism: A. D. Dergunov, M. Rosseneu, *Biol. Chem. Hoppe-Seyler* **375**, 485-495 (1994); P. de Knijff *et al.*, *Hum. Mutat.* **4**, 178-194 (1994). Review of role in the nervous system: J. Poirier, *Trends Neurosci.* **17**, 525-530 (1994); K. H. Weisgraber *et al.*, *Curr. Opin. Lipidol.* **5**, 110-116 (1994); R. W. Mahley *et al.*, *ibid.* **6**, 86-91 (1995).

787. Apomorphine. *(R)-5,6,6a,7-Tetrahydro-6-methyl-4H-dibenzo[de,g]quinoline-10,11-diol;* 6aβ-aporphine-10,11-diol. $C_{17}H_{17}NO_2$; mol wt 267.33. C 76.38%, H 6.41%, N 5.24%, O 11.97%. Synthetic opiate obtained by treating morphine with concd HCl: L. Small *et al.*, *J. Org. Chem.* **5**, 334 (1940); by heating morphine with $ZnCl_2$: Mayer, *Ber.* **4**, 121 (1871). Prepn of methylbromide: R. Pschorr, *Ger. pat.* **158,620** (1905 to Riedel), *Frdl.* **7**, 699. Structure: R. Pschorr, *Ber.* **40**, 1984 (1907). Configuration: H. Corrodi, E. Hardegger, *Helv. Chim. Acta* **38**, 2038 (1955). Total synthesis of (±)-form: J. L. Neumeyer *et al.*, *J. Pharm. Sci.* **59**, 1850 (1970); J. L. Neumeyer, U.S. pat. **3,717,639** (1973 to A. D. Little); J. L. Neumeyer *et al.*, *J. Med. Chem.* **16**, 1223 (1973). Synthesis of (+)- and (−)-forms: V. J. Ram, J. L. Neumeyer, *J. Org. Chem.* **46**, 2830 (1981). Toxicity data: J. G. Cannon *et al.*, *J. Med. Chem.* **15**, 348 (1972); J. Z. Ginos *et al.*, *ibid.* **18**, 1194 (1975). Review of pharmacology and neurochemistry: G. DiChiara, G. L. Gessa, *Adv. Pharmacol. Chemother.* **15**, 87-160 (1978). Reviews: F. C. Colpaert *et al.*, *Int. Rev. Neurobiol.* **19**, 225-268 (1976); J. L. Neumeyer *et al.*, in *Apomorphine and Other Dopaminomimetics vol. 1*, G. L. Gessa, G. U. Corsini, Eds. (Raven, New York, 1981) p 1-17. Comprehensive description: F. J.

Muhtadi, M. S. Hifnawy in *Analytical Profiles of Drug Substances* **vol 20**, K. Florey, Ed. (Academic Press, New York, 1991) pp 121-166.

Hexagonal plates from chloroform and petr ether, dec 195°; subl in high vacuum. Oxidizes rapidly in air and becomes green. Sol in alcohol, acetone, chloroform. Slightly sol in water, benzene, ether, petr ether. Solns darken rapidly. pKb 7.0; pKa 8.92. uv max (98% alc): 336, 399 nm. LD_{50} i.p. in mice: 600 μmoles/kg (Cannon).

Hydrochloride, $C_{17}H_{17}NO_2 \cdot HCl$, small crystals (usually hemihydrate). Dec and turn green on exposure to light and air. $[\alpha]_D^{25}$ −48° (c = 1.2). uv spectrum: Csokan, *Z. Anal. Chem.* **124**, 344 (1942). pH of aq soln (1 in 300) = 4.8. One gram dissolves in 50 ml water, 17 ml water at 80°, 50 ml alcohol. Very slightly sol in chloroform and ether. LD_{50} i.p. in mice: 145 μg/g (Ginos).

Diacetylapomorphine, $C_{21}H_{21}NO_4$, mp 127-128°, $[\alpha]_D^{24}$ −88° (c = 1.12 in 0.1N HCl).

Methylbromide, $C_{17}H_{17}NO_2 \cdot CH_3Br$, *N-methylapomorphinium bromide, Euporphin, Bromophin.* Crystals, mp 180°. Freely sol in water, alc. Practically insol in ether.

THERAP CAT: Emetic.

THERAP CAT (VET): Emetic, expectorant in dogs.

788. Apopain. Cysteine protease thought to be the initiator of programmed cell death (apoptosis) in mammalian cells. Degrades *poly(ADP-ribose) polymerase* (PARP), an enzyme involved in DNA repair, genome surveillance and integrity. The breakdown of PARP is known to occur during the onset of apoptosis. Apopain is composed of two subunits of relative molecular mass 17K and 12K that are derived from the precursor protein known as *CPP32.* Characterization of a protease (*CED-3*) required for apoptosis in the nematode, *Caenorhabditis elegans*: J. Yuan *et al.*, *Cell*, **75**, 641 (1993). Identification of the mammalian homolog, CPP32: T. Fernandes-Alnemri *et al.*, *J. Biol. Chem.* **269**, 30761 (1994). Identification of PARP-cleaving activity: M. Tewari *et al.*, *Cell* **81**, 801 (1995); and characterization of apopain: D. W. Nicholson *et al.*, *Nature* **376**, 37 (1995).

789. Aporeine. *(S)-6,7,7a,8-Tetrahydro-7-methyl-5H-benzo[g]-1,3-benzodioxolo[6,5,4-de]quinoline;* aporheine; 1,2-methylenedioxyaporphine; (+)-roemerine. $C_{18}H_{17}NO_2$; mol wt 279.34. C 77.40%, H 6.13%, N 5.01%, O 11.46%. From *Papaver dubium* L., Papaveraceae: Pavesi, *Gazz. Chim. Ital.* **37 I**, 629 (1907); **44 I**, 398 (1914). Structure and identity with (+)-roemerine: Slavik, *Coll. Czech. Chem. Commun.* **28**, 1738 (1963). Synthesis of dl-roemerine: Marion, Grassie, *J. Am. Chem. Soc.* **66**, 1290 (1944).

Needles from ether + petroleum ether, mp 102°. $[\alpha]_D^{22}$ +80° (c = 0.50 in ethanol). uv max: 262, 315 nm (log ε 4.3, 3.7). pK 6.1. Sol in ether, methanol, ethanol, chloroform; slightly sol in petr ether. Practically insol in water, alkali.

Hydrochloride, $C_{18}H_{17}NO_2 \cdot HCl$, leaflets from ethanol or water, mp 266-267°. Slightly sol in ethanol, water.

Methiodide, $C_{18}H_{17}NO_2 \cdot CH_3I$, crystals from boiling methanol, mp 232-233°.

790. Appetize®. Blended cholesterol-reduced animal fat and vegetable oil free of *trans* fatty acids, wherein the final weight ratio of linoleic acid and myristic acid is between about 2 and 9: D. Perlman, K. C. Hayes, **PCT Int. pat. Appl. 93 22,933**; *eidem*, U.S. pat. **5,382,442** (1993, 1995 both to Brandeis University). Historical perspective: K. Kevin, *Food Proc.* **55**, 81-85 (1994).

Contains approx 8 mg cholesterol per 100 g and as many calories as natural fats.

USE: Low cholesterol fat alternative.

791. Apraclonidine. *2,6-Dichloro-N¹-(4,5-dihydro-1H-imidazol-2-yl)-1,4-benzenediamine; 2,6-dichloro-N′-2-imidazolidinylidene-1,4-benzenediamine; 2-[(4-amino-2,6-dichlorophenyl)imino]imidazolidine; p-aminoclonidine; aplonidine; NC-14.* $C_9H_{10}Cl_2N_4$; mol wt 245.11. C 44.10%, H 4.11%, Cl 28.93%, N 22.86%. α_2-Adrenergic agonist; structural analog of clonidine, *q.v.* Synthesis: B. Rouot, G. LeClerc, *Bull. Soc. Chim. France* **Pt. 2**, 520 (1979). Prepn (not claimed) and use in treatment of intraocular pressure: B. M. York, Jr., U.S. pat. **4,517,199** (1985 to Alcon). Pharmacology: B. Rouot *et al., C. R. Acad. Sci.* **286**, 909 (1978). Receptor binding studies: D. C. U'Prichard, *Prog. Clin. Biol. Res.* **71**, 53 (1981). D. C. Stump, D. E. MacFarlane, *J. Lab. Clin. Med.* **102**, 779 (1983). Clinical pharmacology: D. A. Abrams *et al., Arch. Ophthalmol.* **105**, 1205 (1987). Clinical evaluation in treatment of intraocular pressure: A. L. Robin *et al., ibid.* 1208; as additive to maximally tolerated medical therapy in glaucoma: A. L. Robin *et al., Am. J. Ophthalmol.* **120**, 423 (1995).

Solid, mp > 230°.
Hydrochloride, $C_9H_{10}Cl_2N_4$·HCl, *ALO-2145, Iopidine.*
Dihydrochloride, $C_9H_{10}Cl_2N_4$·2HCl, uv max (ethanol): 254, 304 nm (ϵ 1800, 2500).

THERAP CAT: Treatment of post-surgical elevated intraocular pressure.

792. Apramycin. *O-4-Amino-4-deoxy-α-D-glucopyranosyl-(1→8)-O-(8R)-2-amino-2,3,7-trideoxy-7-(methylamino)-D-glycero-α-D-allo-octodialdo-1,5:8,4-dipyranosyl-(1→4)-2-deoxy-D-streptamine; 4-O-[3α-amino-6α-[(4-amino-4-deoxy-α-D-glucopyranosyl)oxy]-2,3,4,4aβ,6,7,8aα-octahydro-8β-hydroxy-7β-(methylamino)pyranopyrano[3,2-b]pyran-2α-yl]-2-deoxy-D-streptamine; nebramycin factor 2; EL-857; EL-857/820; 47657; Ambylan; Apralan.* $C_{21}H_{41}N_5O_{11}$; mol wt 539.58. C 46.75%, H 7.66%, N 12.98%, O 32.62%. Broad spectrum aminocyclitol antibiotic and component of the nebramycin complex, produced by a strain of *Streptomyces tenebrarius:* R. Q. Thompson, E. A. Presti, *Antimicrob. Ag. Chemother.* **1967**, 332; W. M. Stark, U.S. pat. **3,691,279** (1972 to Lilly). Structure, abs config and properties: S. O'Connor *et al., J. Org. Chem.* **41**, 2087 (1976). ¹³C-NMR: E. Wenkert, E. W. Hagaman, *ibid.* 701. *In vitro* activity: R. Ryden *et al., J. Antimicrob. Chemother.* **3**, 609 (1977). Synthetic studies: H. C. Jarrell, W. A. Szarek, *Carbohyd. Res.* **67**, 43 (1978); *eidem, Can. J. Chem.* **56**, 144 (1978); **57**, 924 (1979). *See also* tobramycin.

Monohydrate from aqueous ethanol, mp 245-247°. pKa (H₂O): 8.5, 7.8, 7.2, 6.2, 5.4. Very soluble in water; slightly sol in lower alcohols.

THERAP CAT: Antibacterial.
THERAP CAT (VET): Antibacterial.

793. Aprindine. *N-(2,3-Dihydro-1H-inden-2-yl)-N′,N′-diethyl-N-phenyl-1,3-propanediamine; N,N-diethyl-N′-2-indanyl-N′-phenyl-1,3-propanediamine; N-[3-(diethylamino)propyl]-N-phenyl-2-indanamine; compd 99170; AC-1802; Lilly 99170.* $C_{22}H_{30}N_2$; mol wt 322.49. C 81.94%, H 9.38%, N 8.69%. Membrane-stabilizing agent. Prepn: P. M. Vanhoof, P. M. Clarebout, **Belg. pat. 760,018;** *eidem,* U.S. pat. **3,923,813** (1971, 1975 both to Christiaens). Toxicological study: A. Georges *et al., Arzneimittel-Forsch.* **23**, 519 (1973). Pharmacokinetics: T. Kobari *et al., Eur. J. Clin. Pharmacol.* **26**, 129 (1984). HPLC determn in plasma: T. Kobari *et al., J. Chromatog.* **278**, 220 (1983). Clinical studies: D. P. Zipes *et al., Am. J. Cardiol.* **40**, 586 (1977); P. J. Troup, D. P. Zipes, *Am. Heart J.* **97**, 322 (1979). *Reviews:* P. Danilo, *ibid.* 119; I. Stoel, S. F. Hagemeijer, *Eur. Heart J.* **1**, 147 (1980).

Hydrochloride, $C_{22}H_{30}N_2$·HCl, *compd 83846, Amidonal, Aspenon, Fibocil, Fiboran, Ritmusin.* Crystals from benzene, mp 120-121°.

THERAP CAT: Antiarrhythmic (class I).

794. Aprobarbital. *5-(1-Methylethyl)-5-(2-propenyl)-2,4,6(1H,3H,5H)-pyrimidinetrione; 5-allyl-5-isopropylbarbituric acid; Alurate; Numal; Allypropymal; Aprozal.* $C_{10}H_{14}N_2O_3$; mol wt 210.23. C 57.13%, H 6.71%, N 13.32%, O 22.83%. Prepn: U.S. pat. **1,444,802** (1923). Toxicity data: H. H. Frey, *Arzneimittel-Forsch.* **12**, 389 (1962).

Slightly bitter crystals, mp 140-141.5°. Almost insoluble in water, petr ether, aliphatic hydrocarbons. Sol in alcohol, chloroform, ether, acetone, benzene, glacial acetic acid, also in solns of fixed alkali hydroxides. A satd aq soln is acid to litmus. LD_{50} i.p. in mice: 200 mg/kg (Frey).
Sodium salt, $C_{10}H_{13}N_2NaO_3$, *aprobarbital sodium, sodium 5-allyl-5-isopropylbarbiturate, Alurate Sodium, Somnipron.* Hygroscopic powder. Slightly bitter taste. Very soluble in water; slightly sol in alc. Practically insol in ether. Aq solns are alkaline to litmus.

Caution: May be habit forming. This is a controlled substance (depressant) listed in the U.S. Code of Federal Regulations, Title 21 Parts 329.1 and 1308.13 (1995).

THERAP CAT: Sedative, hypnotic.

795. Apronalide. *N-(Aminocarbonyl)-2-(1-methylethyl)-4-pentenamide; (2-isopropyl-4-pentenoyl)urea; allylisopropylacetylurea; Apronal; Isodormid; Sedormid.* $C_9H_{16}N_2O_2$; mol wt 184.24. C 58.67%, H 8.75%, N 15.21%, O 17.37%. Prepn: Slotta, *Grundriss der modernen Arzneistoff-Synthese* (Stuttgart, 1931) p 35.

Crystals. Practically tasteless. mp 194°. One gram dissolves in 3000 ml cold water, in 210 ml boiling water. One part dissolves in 10 parts alc, 75 parts ether. Decolorizes $KMnO_4$ solns.

Caution: This substance may be habit forming and is listed in the U.S. Code of Federal Regulations, Title 21 Part 329.1 (1995).

THERAP CAT: Sedative, hypnotic.

796. Aprotinin. *Pancreatic basic trypsin inhibitor;* pancreatic trypsin inhibitor (Kunitz); Bayer A 128; Riker 52G; RP-9921; Antagosan; Antikrein; Fosten; Iniprol; Kir Richter; Onquinin; Repulson; Trasylol; Trazinin; Zymofren. $C_{284}H_{432}N_{84}O_{79}S_7$; mol wt 6511.53. C 52.39%, H 6.69%, N 18.07%, O 19.41%, S 3.45%. A kallikrein inhibitor which also inhibits plasmin, trypsin, chymotrypsin and various intracellular proteases. Polypeptide found in tissues and blood, but in highest concentration in bovine parotid gland, pancreas and lung. Initial description: H. Kraut et al., Z. Physiol. Chem. **192**, 1 (1930). Isoln from bovine pancreas: M. Kunitz, J. H. Northrup, J. Gen. Physiol. **19**, 991 (1936). Isoln from bovine parotid glands: H. Kraut, R. Körbel, U.S. pat. 2,890,986 (1959 to Bayer). Aprotinin is a single chain polypeptide containing 58 amino acids of known sequence: B. Kassell et al., Biochem. Biophys. Res. Commun. **18**, 255 (1965); F. A. Anderer, S. Hörnle, Z. Naturforsch. **20b**, 457, 462 (1965); B. Kassell, M. Laskowski, Biochem. Biophys. Res. Commun. **20**, 463 (1966). Two-dimensional ^{1}H-NMR studies: K. Nagayama, K. Wüthrich, Eur. J. Biochem. **114**, 365 (1981); G. Wagner et al., ibid. 375. General review: I. Trautschold et al., Biochem. Pharmacol. **16**, 59-72 (1957). Review of therapeutic action: G. Haberland, R. McConn, Fed. Proc. **38**, 2760-2767 (1979). Review of use as a proteolytic inhibitor in radioimmunoassays of polypeptide hormones: E. S. Zyzner, Life Sci. **28**, 1861-1866 (1981). Review of biochemistry and applications: H. Fritz, G. Wunderer, Arzneimittel-Forsch. **33**, 479-494 (1983).

uv max (pH 5.9): 280 nm. Isoelec pt pH 10.5. Stable in neutral or acid media at high temp. Irreversible changes in molecular structure occur in strongly alkaline media (pH > 12). Partially and reversibly denatured on treatment with 8M urea. May be kept at room temp in physiological saline soln for > 1 yr without detrimental effects. LD_{50} i.v. in mice: 2.5 million kallikrein inhibitor units/kg (Trautschold).

USE: Proteolytic inhibitor in radioimmunoassays of polypeptide hormones.

THERAP CAT: Protease inhibitor.

797. Apyrase. An enzyme found in plants, especially potatoes. Cofactor: Ca^{2+}. Hydrolyzes acid anhydride links. Substrate: Adenosine triphosphate, adenosine diphosphate (giving adenylate and phosphate). Ref: Krishnan, Arch. Biochem. **20**, 272 (1949). Partial purification and properties of potato apyrase: Liebecq et al., Bull. Soc. Chim. Biol. **45**, 573 (1963).

798. Aquaporins. Group of integral membrane proteins that function as molecular water channels to facilitate osmotic movement of water across the plasma membrane. Found in plant and animal cells characterized as having high osmotic water permeability. Part of the *MIP* family of channel proteins, typified by the major intrinsic protein of the eye, that are responsible for the permeability of ions and small molecules. Homologs exhibit ~40% sequence identity. Structure consists of six transmembrane domains with 5 connecting loops; the water pore is thought to be formed by the juxtaposition of 2 highly conserved Asn-Pro-Ala (NPA) sequences within the lipid bilayer. Isoln of aquaporin-1 from human erythrocytes and renal tubules: B. M. Denker et al., J. Biol. Chem. **263**, 15634 (1988). Cloning

and sequence homology with MIP family: G. M. Preston, P. Agre, Proc. Nat. Acad. Sci. USA **88**, 11110 (1991). Identification of function as membrane water channel: G. M. Preston et al., Science **256**, 385 (1992). Nomenclature: P. Agre et al., Am. J. Physiol. **265**, F461 (1993). Molecular structure: J. S. Jung et al., J. Biol. Chem. **269**, 14648 (1994). Review of structure and tissue distribution: C. H. van Os et al., Biochim. Biophys. Acta **1197**, 291-309 (1994); M. J. Chrispeels, P. Agre, Trends Biochem. Sci. **19**, 421-425 (1994). Review of aquaporins in plants: M. J. Chrispeels, C. Maurel, Plant Physiol. **105**, 9-13 (1994). Review of role in renal water transport: A. F. van Lieburg et al., Pediatr. Nephrol. **9**, 228-234 (1995).

Aquaporin-1, AQP1, CHIP28, AQP-CHIP, aquaporin-CHIP, channel forming integral protein. 269 amino acid residues; mol wt 28 kDa. Constitutes 2.4% of the plasma membrane protein in red blood cells; also found in renal proximal tubules and in other water-permeable epithelia.

Aquaporin-2, AQP2, AQP-CD, WCH-CD, water channel of collecting duct. 271 amino acids; mol wt 29 kDa. Vasopressin-regulated water channel in the kidney; has a primary role in concentrating the urine.

γ-TIP, γ-tonoplast intrinsic protein, AQP.At1. Found in plant vacuolar membrane (tonoplast); facilitates rapid exchange of water between the vacuole and the cytoplasm.

799. Arabinose. L-Arabinose; pectin sugar. $C_5H_{10}O_5$; mol wt 150.13. C 40.00%, H 6.71%, O 53.28%. Widely distributed in plants, usually in the form of complex polysaccharides. Also in mycobacteria. Isoln from mesquite gum: Anderson, Sands, J. Am. Chem. Soc. **48**, 3172 (1926); Org. Syn. coll. vol. I, 16 (2nd ed., 1941); White, J. Am. Chem. Soc. **69**, 715 (1947); from western red cedar *(Thuja plicata* Don., *Cupressaceae):* Anderson, Erdtman, ibid. **71**, 2927 (1949); from sapote gum: White, ibid. **75**, 257 (1953); from heartwood of port orford cedar *(Chamaecyparis lawsoniana* (Murr.) Parl, *Cupressaceae):* Kritchevsky, Anderson, ibid. **77**, 3391 (1955). Structure: Wolfrom, Christman, ibid. **58**, 39 (1936). Synthesis: Hough, Jones, J. Chem. Soc. **1951**, 1122.

β-L-Arabinose

Orthorhombic bisphenoidal crystals, mp 157-160°. Shows mutarotation. $[\alpha]_D^{12}$ +173° (6 min) → $[\alpha]_D^{20}$ +105.1° (22½ hrs c = 3). One gram dissolves in about 1 ml water, about 250 ml 90% alc. Ka at 17°, -3.7×10^{-13}. Reduces Fehling's soln. Forms furfurol on heating to 200° in closed tube also contg water.

USE: As culture medium for certain bacteria.

800. Arabitol. *Arabinitol;* 1,2,3,4,5-pentanepentol; arabite. $C_5H_{12}O_5$; mol wt 152.15. C 39.47%, H 7.95%, O 52.58%. D-Form obtained by reduction of D-arabinose or D-lyxose with sodium amalgam: Ruff, Ber. **32**, 555 (1899); Ruff, Ollendorff, Ber. **33**, 1802 (1900); Bertrand, Bull. Soc. Chim. [3] **15**, 593 (1896). Production by fermentation from molasses using *Saccharomyces rouxii* and *Saccharomyces mellis:* Lavin, Holloway, U.S. pat. 2,934,474 (1960 to Comm. Solvents). L-Form obtained by reduction of L-arabinose with sodium amalgam: Kiliani, Ber. **20**, 1234, 1571 (1887). DL-Form obtained from equal parts of D- and L-

D-Arabitol   L-Arabitol

arabitol: Ruff, *loc. cit.* Synthesis: Lespian, *Compt. Rend.* **206**, 1773 (1938).

D-Form, big prismatic crystals. Sweet taste. mp 103°. $[\alpha]_D^{20}$ +7.7° (c = 9.26 in satd borax soln). Freely sol in water. One part dissolves in 48 parts of 90% alc at 12°.

Pentanitrate, $C_5H_7N_5O_{15}$, sirup, sol in alcohol, ether, acetone. Strongly reduces Fehling's soln.

L-Form, wart-like crystals, sweet taste. mp 102°. Weakly levorotatory in satd borax soln. Freely sol in water and in boiling 90% alc. One part dissolves in 46 parts of 90% alc at 12°. Does not reduce Fehling's soln.

DL-Form, prisms from 90% alc. mp 105-106°. One part dissolves in 66 parts of 90% alc at 12°.

801. D-Araboflavin. *1-Deoxy-1-(3,4-dihydro-7,8-dimethyl-2,4-dioxobenzo[g]pteridin-10(2H)-yl)-D-arabinitol; 7,8-dimethyl-10-(arabino-2,3,4,5-tetrahydroxypentyl)benzo[g]-pteridine-2,4(3H,10H)-dione;* 6,7-dimethyl-9-D-araboflavin. $C_{17}H_{20}N_4O_6$; mol wt 376.37. C 54.25%, H 5.36%, N 14.89%, O 25.51%. An antagonist of riboflavine. Prepd from D-(2-amino-4,5-dimethylphenyl)arabinamine and alloxan: Kuhn, Weygand, *Ber.* **68**, 1286 (1935); Sahashi *et al., Bull. Inst. Phys. Chem. Res.* (Tokyo) **24**, 72 (1948), *C.A.* **42**, 5458 (1948).

Orange-yellow needles from dil acetic acid. Bitter taste, mp 302-303°. $[\alpha]_D^{20}$ +78.6° (c = 0.509 in 0.1N NaOH); $[\alpha]_D^{20}$ −441° (c = 0.253 in 0.2N NaOH satd with borax).

Tetraacetate, $C_{25}H_{28}N_4O_{10}$, flat yellow prisms from ethyl acetate, mp 221-222°.

802. Arachidic Acid. *Eicosanoic acid;* arachic acid. $C_{20}H_{40}O_2$; mol wt 312.54. C 76.86%, H 12.90%, O 10.24%. $CH_3(CH_2)_{18}COOH$. Fatty acid found in peanut oil, vegetable and fish oils, etc. Also obtained by hydrogenation of arachidonic acid: Ege *et al., J. Am. Chem. Soc.* **83**, 3080 (1961). Synthesis: Linstead *et al., J. Chem. Soc.* **1955**, 1097. *Reviews:* A .W. Ralston, *Fatty Acids and Their Derivatives* (Wiley & Sons, New York, 1948) pp 43-46; K. S. Markley, *Fatty Acids,* Part I (Interscience, New York, 2nd ed., 1960) pp 164-167, 398-400.

Crystals from alc, mp 75.5°. bp_{760} ~328° with some decompn; $bp_{1.0}$ 205°. d_4^{100} 0.8240. n_D^{100} 1.4250. Practically insoluble in water; sparingly sol in cold water; freely in hot abs alcohol, benzene, chloroform, ether, petr ether.

Methyl ester, $C_{21}H_{42}O_2$, mp 47°; $bp_{1.0}$ 180°.

Ethyl ester, $C_{22}H_{44}O_2$, mp 41°; $bp_{0.3}$ 177°.

803. Arachidonic Acid. *(all-Z)-5,8,11,14-Eicosatetra-enoic acid.* $C_{20}H_{32}O_2$; mol wt 304.47. C 78.90%, H 10.59%, O 10.51%. An essential fatty acid, *q.v.,* and a precursor in the biosynthesis of prostaglandins, thromboxanes, and leukotrienes, *q.q.v.* Structure: Mowry *et al., J. Biol. Chem.* **142**, 679 (1942); Arcus, Smedley-Maclean, *Biochem. J.* **37**, 1 (1943). Occurs in liver, brain, glandular organs, and depot fats of animals, in small amounts in human depot fats, and is a constituent of animal phosphatides. Isolation from liver lipids: Brown, *J. Biol. Chem.* **80**, 455 (1928); from beef suprarenal phosphatides: Ault, Brown, *ibid.* **107**, 615 (1934); Shinowara, Brown, *ibid.* **134**, 331 (1940). *See also* Dolby *et al., Biochem. J.* **34**, 1422 (1940). Synthesis from 2-propargyloxytetrahydropyran and 1-heptyne: Goldberg, Rachlin, *U.S. pat.* **2,934,570** (1960 to Hoffmann-La Roche); Rachlin *et al., J. Org. Chem.* **26**, 2688 (1961). Alternate route: Osbond, Wickens, *Chem. & Ind. (London)* **1959**,

1288; Ege *et al., J. Am. Chem. Soc.* **83**, 3080 (1961). *Reviews:* K. S. Markley, *Fatty Acids,* Part I (Interscience, New York, 2nd ed., 1960) pp 164-167, 398-400; T. K. Schaaf, *Ann. Rep. Med. Chem.* **12**, 182-190 (1977); B. B. Weksler, *N. Engl. Soc. Allergy Proc.* **2**, 56-61 (1981); N. A. Nelson *et al., Chem. & Eng. News* **60**, 30-44 (Aug. 16, 1982). Series of articles on metabolism, role in inflammation and therapeutic implications: *Drugs* **33**, Suppl. 1, 2-66 (1987).

Liquid. mp −49.5°. Neutralization value 184.20. Iodine value 333.50. n_D^{20} 1.4824.

Methyl ester, $C_{21}H_{34}O_2$, bp_2 200-205°, $bp_{0.001}$ 127-128°. n_D^{20} 1.4797 to 1.4810.

THERAP CAT: Nutrient (essential fatty acid).

THERAP CAT (VET): With linoleic and linolenic acids, in eczema and dermatitis in dogs and swine.

804. Aralia. Spikenard; American spikenard; spignet; petty-morrel; spice berry. Dried rhizome and roots of *Aralia racemosa* L., Araliaceae. Habit. Northeastern U.S. Constit. Starch, pectin, sugar, resin. The active constituent appears to be a volatile oil of unknown constitution. Ingredient of Compound White Pine Syrup: *N. F. XI.*

805. Aramite®. *Sulfurous acid 2-chloroethyl 2-[4-(1,1-dimethylethyl)phenoxy]-1-methylethyl ester; sulfurous acid 2-(p-tert-butylphenoxy)-1-methylethyl 2-chloroethyl ester;* 2-chloroethyl 1-methyl-2-(p-tert-butylphenoxy)ethyl sulfite; 2-(p-tert-butylphenoxy)isopropyl 2-chloroethyl sulfite; β-chloroethyl β-(p-tert-butylphenoxy)-α-methylethyl sulfite; compd 88R; ENT-16519; Aratron; Niagaramite; Ortho-Mite. $C_{15}H_{23}ClO_4S$; mol wt 334.86. C 53.80%, H 6.92%, Cl 10.59%, O 19.11%, S 9.58%. Prepn: Harris *et al., U.S. pat.* **2,529,494** (1950 to U.S. Rubber). Toxicology: B. L. Oser, M. Oser, *Toxicol. Appl. Pharmacol.* **2**, 441 (1960). Carcinogenicity studies: *eidem, ibid.* **4**, 70 (1962); Truhaut *et al., C.R. Acad. Sci., Ser. D* **281**, 599 (1975).

Liquid. d 1.1450 to 1.1620. mp −31.7°. $bp_{0.1}$ 175°; $bp_{7.0}$ 200-210°. Vapor tension at 25°: <10 mm. n_D^{20} 1.5100 to 1.5118; n_D^{27} 1.5075. Practically insol in water. Misc with many organic solvents. Soly in petroleum oils decreases rapidly with decreasing temperatures. LD_{50} orally in rats: 3.90±0.28 g/kg (Oser, Oser).

Caution: Direct contact with undiluted form or concentrated solutions may cause irritation of skin and conjunctiva. *See Clinical Toxicology of Commercial Products,* R. E. Gosselin *et al.,* Eds. (Williams & Wilkins, Baltimore, 5th ed., 1984) Section II, p 289. This substance has been listed as reasonably anticipated to be a carcinogen: *Fourth Annual Report on Carcinogens* (NTP 85-002, 1985) p 26; delisted because no U.S. residents exposed: *Fifth Annual Report on Carcinogens* (NTP 89-239, 1989) p 340.

USE: Miticide.

806. Aranidipine. *1,4-Dihydro-2,6-dimethyl-4-(2-nitrophenyl)-3,5-pyridinedicarboxylic acid methyl 2-oxopropyl ester;* (±)-methyl 2-oxopropyl 1,4-dihydro-2,6-dimethyl-4-(2-nitrophenyl)-3,5-pyridinedicarboxylate; (±)-acetonyl methyl 1,4-dihydro-2,6-dimethyl-4-(o-nitrophenyl)-3,5-pyridinedicarboxylate; MPC-1304. $C_{19}H_{20}N_2O_7$; mol wt 388.38. C 58.76%, H 5.19%, N 7.21%, O 28.84%. Calcium channel blocker. Prepn: S. Ohno *et al., Fr. pat.* **2,514,761**; *eidem, U.S. pat.* **4,446,325** (1983, 1984 both to Maruko Seiyaku); *idem et al., Chem. Pharm. Bull.* **34**, 1589 (1986). Pharmacology and receptor binding studies: K. Miyoshi *et al., Eur. J. Pharmacol.* **238**, 139 (1993). Determn by GC-MS in plasma and urine: Y. Umeno *et al., J. Chromatog.* **434**, 123 (1988); by HPLC in plasma: Y. Iida *et al., ibid* **571**, 277 (1991).

Cardiovascular profile: A. Kanda *et al., J. Cardiovasc. Pharmacol.* **22**, 167 (1993). Clinical pharmacology: S. Suzuki *et al., Arzneimittel-Forsch.* **43**, 1152 (1993). Toxicity studies: S. Nakano *et al., Yakuri to Chiryo* **21**, S931 (1993), *C.A.* **119**, 131216 (1993).

Yellow prisms from ethyl acetate/hexane, mp 155°. LD_{50} in male, female mice, rats (mg/kg): 143, 193, 1982, 1459 orally; LD_{50} in male, female mice (mg/kg): 7.3, 9.1 i.p. (Nakano).

THERAP CAT: Antihypertensive.

807. Araroba. Goa powder; crude chrysarobin; Brazil powder; ringworm powder; "Arariba". Found in cavities in the trunk of *Andira araroba*, Aguiar *(Vouacapoua araroba* [Aguiar] Lyons), *Leguminosae* and freed as much as possible from fragments of wood. *Habit.* In damp forests of Brazil and Bahia. *Constit.* About 50% chrysarobin; chrysophanic acid, gum, about 2% resin, ararobinol—said to be the methyl ether of emodin.

Brown-yellow, slightly cryst, rough, odorless powder; becomes brownish on exposure to air.

THERAP CAT: Dermatologic.

808. Arbaprostil. *(5Z,11α,13E,15R)-11,15-Dihydroxy-15-methyl-9-oxoprosta-5,13-dien-1-oic acid;* (15R)-15-methylprostaglandin E_2; (15R)-15-methyl-PGE$_2$; *(E,Z)-(1R,2R,-3R)-7-[3-hydroxy-2-[(3R)-(3-hydroxy-3-methyl-1-octenyl)]-5-oxocyclopentyl]-5-heptenoic acid;* U-42842; Arbacet. $C_{21}H_{34}O_5$; mol wt 366.50. C 68.82%, H 9.35%, O 21.83%. Antisecretory, cytoprotective derivative of prostaglandin E_2, *q.v.* Prepn: G. L. Bundy *et al., Ger. pat.* **2,121,980;** G. L. Bundy, U.S. pat. **3,804,889** (1971, 1974 both to Upjohn). Synthetic studies: G. Bundy *et al., Ann. N.Y. Acad. Sci.* **180**, 76 (1971); E. W. Yankee, G. Bundy, *J. Am. Chem. Soc.* **94**, 3651 (1972). Synthesis of *R-* and *S-*methyl esters: E. W. Yankee *et al., ibid.* **96**, 5865 (1974). HPLC separation of epimers: R. K. Lustgarten, *J. Pharm. Sci.* **65**, 1533 (1976); G. E. Peng, V. K. Sood, *J. Liquid Chromatog.* **6**, 1499 (1983). Pharmacology: J. R. Weeks *et al., J. Pharmacol. Exp. Ther.* **186**, 67 (1973). Comparative activity of epimers: A. Robert, E. W. Yankee, *Proc. Soc. Exp. Biol. Med.* **148**, 1155 (1975). Kinetics of epimerization: M. V. Merritt, G. E. Bronson, *J. Am. Chem. Soc.* **100**, 1891 (1978). Clinical study in duodenal ulcers: G. Vantrappen *et al., Gastroenterology* **83**, 357 (1982); in prevention of aspirin-induced gastric mucosal injury: D. A. Gilbert *et al., ibid.* **86**, 339 (1984).

(15S)-Methyl ester, $C_{22}H_{36}O_5$, colorless oil, $[α]_D$ −79° (c = 1.3 in chloroform). uv max (basic ethanol): 278 nm (ε = 25250).

(15R)-Methyl ester, $C_{22}H_{36}O_5$, colorless oil, $[α]_D$ −74° (c = 1.0 in chloroform). uv max (basic ethanol): 278 nm (ε = 25200).

THERAP CAT: Antiulcerative.

809. Arbekacin. *(S)-O-3-Amino-3-deoxy-α-D-glucopyranosyl-(1→6)-O-[2,6-diamino-2,3,4,6-tetradeoxy-α-D-erythro-hexopyranosyl-(1→4)]-N¹-(4-amino-2-hydroxy-1-oxobutyl)-2-deoxy-D-streptamine;* 1-N-[(S)-4-amino-2-hydr-

oxybutyryl]dibekacin; 1-N-[(S)-4-amino-2-hydroxybutyryl]-3',4'-dideoxykanamycin B; AHB-DKB; HABA-DKB; HBK; 1665-RB; Habekacin. $C_{22}H_{44}N_6O_{10}$; mol wt 552.63. C 47.82%, H 8.03%, N 15.21%, O 28.95%. Derivative of kanamycin B, *q.v.,* closely related to dibekacin, *q.v.* Prepn and antibacterial activity: S. Kondo *et al., J. Antibiot.* **26**, 412 (1973); H. Umezawa *et al., Ger. pat.* **2,350,169;** *eidem,* U.S. pat. **4,107,424** (1974, 1978 both to Microbiochem. Res. Found.). Mechanism of action: N. Tanaka *et al., Antimicrob. Ag. Chemother.* **24**, 797 (1983); K. Matsunaga *et al., J. Antibiot.* **37**, 596 (1984). Toxicology: J. Stewens *et al., Arzneimittel-Forsch.* **35**, 1440 (1985); M. Kurebe *et al., ibid.* **36**, 1511 (1986). HPLC determn in biological fluids: I. Komiya *et al., Chemotherapy* **34**, Suppl. 1, 82 (1986). Series of articles on antibacterial activity, pharmacodynamics, pharmacokinetics, and clinical evaluations: *ibid.,* pp 1-670.

Dicarbonate, $C_{24}H_{48}N_6O_{16}$, colorless, crystalline powder, mp 178° (dec). $[α]_D^{24}$ +86.8° (c = 0.77 in water). LD_{50} i.v. in mice: > 150 mg/kg (Umezawa, 1978).

THERAP CAT: Antibacterial.

810. Arborescin. *[4aS-(3aS*,4aα,6aα,7α,9aβ,9bα)]-5,6,6a,7,9a,9b-Hexahydro-1,4a,7-trimethyl-3H-oxireno[8,8a]azuleno[4,5-b]furan-8(4aH)-one; 1,10-epoxy-6β-hydroxy-1β,-5β,7α-guiai-3-en-12-oic acid γ-lactone.* $C_{15}H_{20}O_3$; mol wt 248.32. C 72.55%, H 8.12%, O 19.33%. Ancient Greek contraceptive. Isoln from *Artemisia arborescens* L. and *Matricaria globifera* (Thunb.) Druce, *Compositae:* Meisels, Weizmann, *J. Am. Chem. Soc.* **75**, 3865 (1953); Cekan *et al., Coll. Czech. Chem. Commun.* **25**, 2553 (1960). Structure: Bates *et al., Tetrahedron Letters* **1963**, 1127; M. Suchy *et al., Coll. Czech. Chem. Commun.* **29**, 1829 (1964). Synthesis and revised structure: M. Ando *et al., Chem. Letters* **1978**, 727. Stereochemistry: M. Ando *et al., J. Org. Chem.* **47**, 3909 (1982).

Crystals from ethanol, mp 140-142°. $[α]_D^{20}$ +64° (c = 0.72 in chloroform).

811. Arbutamine. *(R)-4-[1-Hydroxy-2-[[4-(4-hydroxyphenyl)butyl]amino]ethyl]-1,2-benzenediol;* (R)-3,4-dihydroxy-α-[[[4-(p-hydroxyphenyl)butyl]amino]methyl]benzyl alcohol; 1-(3,4-dihydroxyphenyl)-2-(4-(4-hydroxyphenyl)-butylamino)ethanol. $C_{18}H_{23}NO_4$; mol wt 317.38. C 68.12%, H 7.30%, N 4.41%, O 20.16%. Catecholamine cardiac stimulant. Prepn: R. R. Tuttle, C. E. Browne, III, **Eur. pat. Appl. 329,464;** *eidem,* U.S. pat. **5,395,970** (1989, 1995 both to Gensia). Pharmacology and pharmacodynamics: M. Young *et al., Drug. Dev. Res.* **32**, 19 (1994). Efficacy as exercise stimulating agent (ESA) in comparison with dobutamine, *q.v.*: H. K. Hammond, M. D. McKirnan, *J. Am. Coll. Cardiol.* **23**, 475 (1994).

Hydrochloride, *GP-2-121-3*. mp 55°-58°. $[\alpha]_D^{23}$ −18.5 (c = in ethanol).

THERAP CAT: Stress agent for diagnosis of coronary artery disease.

812. Arbutin. *4-Hydroxyphenyl-β-D-glucopyranoside;* hydroquinone-β-D-glucopyranoside; hydroquinone glucose; arbutoside; ursin; Uvasol. $C_{12}H_{16}O_7$; mol wt 272.25. C 52.94%, H 5.92%, O 41.14%. From dried leaves of *Bergenia crassifolia* (L.) Fritsch, *Saxifragaceae:* Tschitschibabin *et al., Ann.* **479**, 303 (1930); from leaves of blueberry, cranberry, and pear trees *(Pyrus communis* L., *Rosaceae):* Urban, Rogowski, *Arch. Exp. Pathol. Pharmakol.* **211**, 194 (1950); Friedrich, *Pharmazie* **15**, 650 (1960); from leaves of cowberry *(Vaccinium vitis-idaea* L., *Ericaceae)* and bearberry *(Arctostaphylos uva-ursi* Spreng., *Ericaceae):* Friedrich, *Naturwiss.* **48**, 304 (1961); Kraus, *Pharmazie* **19**, 41 (1964). Synthesis from acetobromglucose and hydroquinone: Mannich, *Arch. Pharm.* **250**, 547 (1912); from β-D-glucose pentaacetate + hydroquinone monobenzyl ether in the presence of $POCl_3$: Jarrett, **U.S.** pat. **3,201,385** (1965 to Polaroid). Frequently occurs together with methylarbutin in plants, particularly those of the family *Ericaceae.*

Forms an unstable form which appears to be converted to the stable form by melting. Needles from hot ethyl acetate, unstable form mp 165°; stable form mp 199.5-200°: Lindpainter, *Arch. Pharm.* **277**, 398 (1940). $[\alpha]_D^{25}$ −64° (c = 3). Sol in water and in alc. Very hygroscopic. Easily hydrolyzed by dil acids, or by emulsin yielding 1 mol D-glucose and 1 mol hydroquinone. Technique of fermentative degradation: Helferich, Reischel, *Ann.* **533**, 278 (1938). Gallotannin prevents enzymes such as β-glucosidase from splitting arbutin which explains why crude plant extracts are more effective medicinally than pure arbutin: Friedrich, *Arch. Pharm.* **288**, 583 (1955). Forms a complex with hexamethylenetetramine which may be used to separate it from methylarbutin.

USE: Stabilizer for color photographic images.

THERAP CAT: Diuretic, anti-infective (urinary).

813. Areca. Betel nuts; pinang. Nuts (seeds) of *Areca catechu* L., *Palmaceae. Habit.* East Indies. *Constit.* Arecoline, arecaidine, guvacine, guvacoline, arecolidine, choline, about 15% red tannin, about 14% fat. *Reviews:* L. Marion "The Alkaloids of Areca Nut" in Manske-Holmes, *The Alkaloids* **Vol. I** (Academic Press, New York, 1950) pp 171-175; Raghavan, Baruah, *Econ. Bot.* **12**, 315 (1958).

Hard and heavy; round-conical and depressed at base. Extern. brown, mottled with fawn color; intern. brownish-red with whitish veins; astringent taste; the fresh nuts have a faint, cheese-like odor.

THERAP CAT: Anthelmintic.

THERAP CAT (VET): Has been used for tapeworms, ascarids.

814. Arecaidine. *1,2,5,6-Tetrahydro-1-methyl-3-pyridinecarboxylic acid; 1,2,5,6-tetrahydro-1-methylnicotinic acid;* arecaine; methylguvacine. $C_7H_{11}NO_2$; mol wt 141.17. C 59.56%, H 7.85%, N 9.92%, O 22.67%. From seeds of *Areca catechu* L., *Palmaceae* (betel nuts). Synthesis: Wohl, Johnson, *Ber.* **40**, 4712 (1907); Hess, Leibbrandt, *ibid.* **51**, 806 (1918); Freudenberg, *ibid.* 976; Maurit, Preobrazhenskii, *Zh. Obshch. Khim.* **28**, 968 (1958), *C.A.* **52**, 17263 (1958); Merck, *Ger.* pat. **485,139**, *C.A.* **24**, 919 (1930).

Plates from dil alc, dec 232° (after drying at 102°). pH of 0.1 molar soln 5.6. Freely soluble in water, and dilute alc; almost insol in abs alc, chloroform, ether, benzene.

Hydrochloride, $C_7H_{11}NO_2.HCl$, mp 251°; needles, dec 263° (rapid heating). Freely sol in water.

Hydrobromide, $C_7H_{11}NO_2.HBr$, crystals from methanol, dec 249°.

815. Arecoline. *1,2,5,6-Tetrahydro-1-methyl-3-pyridinecarboxylic acid methyl ester; methyl 1,2,5,6-tetrahydro-1-methylnicotinate;* methyl 1-methyl-$\Delta^{3,4}$-tetrahydro-3-pyridinedinecarboxylate; methyl *N*-methyltetrahydronicotinate; arecaline; arecholine; methylarecaidin. $C_8H_{13}NO_2$; mol wt 155.20. C 61.91%, H 8.44%, N 9.03%, O 20.62%. Cholinergic alkaloid from seeds of the betel nut palm *Areca catechu* L., *Palmaceae* (catechu). Isoln: E. Johns, *Arch. Pharm.* **229**, 673 (1891). Synthesis: F. Chemnitius, *J. Prakt. Chem.* **117**, 147 (1926); Mannich, *Ber.* **75B**, 1480 (1942); Knox, **U.S.** pat. **2,506,458** (1950 to Nopco). Improved synthesis from nicotinic acid: I. A. Kozello *et al., Khim.-Farm. Zh.* **10**, 90 (1976), *C.A.* **86**, 171205a (1977). Effect on human serial learning: N. Sitaram *et al., Science* **201**, 274 (1978). Clinical evaluation in Alzheimer's disease: T. T. Soncrant *et al., Psychopharmacology* **112**, 421 (1993). *Review:* R. B. Burrows, *Progress in Drug Research* **vol. 17**, E. Jucker, Ed. (Birkhäuser Verlag, Basel, 1973) pp 108-210.

Oily liquid, bp 209°, bp_7 92-93°, bp_{12} 105°. Volatile with steam. Strong base. pK 6.84. n_D^{20} 1.4302. d^{20} 1.0495. Miscible with water, alc, ether. Soluble in chloroform. LD_{50} in mice, dogs (mg/kg): 100, 5 s.c. (Burrows).

Hydrobromide, $C_8H_{13}NO_2.HBr$, bitter, optically inactive crystals, mp 169-171°. One gram dissolves in about 1 ml water, 10 ml alc, 2 ml boiling alc; slightly soluble in chloroform, ether. The aq soln is practically neutral.

Compd with acetarsone, *q.v.,* $C_8H_{13}NO_2.C_8H_{10}AsNO_5$, *arecoline-acetarsone, arecoline-acetarsol, drocarbil, Nemural.* White to pale-yellow powder. Stable when kept dry. Freely sol in water.

Compd with *p*-stibonobenzoic acid, $C_8H_{13}NO_2.C_7H_7O_5Sb$, *arecoline p-stibonobenzoic acid, Anthelin.* Prepn: Kartsonis, Austin, **U.S.** pat. **2,557,353** (1951 to Jensen-Salsbery Labs.).

THERAP CAT (VET): Anthelmintic (Cestodes); cathartic.

816. Argatroban. *1-[5-[(Aminoiminomethyl)amino]-1-oxo-2-[[(1,2,3,4-tetrahydro-3-methyl-8-quinolinyl)sulfonyl]amino]pentyl]-4-methyl-2-piperidinecarboxylic acid; (2R,4R)-4-methyl-1-[N²-(3-methyl-1,2,3,4-tetrahydro-8-quinolinesulfonyl)-L-arginyl]-2-piperidinecarboxylic acid; (2R,4R)-4-methyl-1-[(S)-N²-[[(R,S)-1,2,3,4-tetrahydro-3-methyl-8-quinolinyl]sulfonyl]arginyl]pipecolic acid;* MQPA. $C_{23}H_{36}N_6O_5S$; mol wt 508.64. C 54.31%, H 7.13%, N 16.52%, O 15.73%, S 6.30%. Synthetic thrombin inhibitor. Prepn of argatroban and stereoisomers: R. Kikumoto *et al.,* **Eur.** pat. **Appl. 8,746;** S. Okamoto *et al.,* **U.S.** pat. **4,258,192** (1980, 1981 both to Mitsubishi Chem. Ind.); S. Okamoto *et al., Biochem. Biophys. Res. Commun.* **101**, 440 (1981). Comparison with heparin, *q.v.,* of antithrombotic effect in animals: T. Kumada, Y. Abiko, *Thromb. Res.* **24**, 285 (1981). Stereoselective inhibition of thrombin: R. Kikumoto *et al., Biochemistry* **23**, 85 (1984). Clinical evaluation in hemodialysis: K. Ota *et al., Proc. Eur. Dial. Transplant. Assoc.* **20**, 144 (1983); in disseminated intravascular coagulation: K. Kumon *et al.*, *Crit. Care Med.* **12**, 1039 (1984).

Crystals from ethanol, mp 188-191°.

Monohydrate, *DK-7419, MCI-9038, MD-805, OM-805, Novastan, Slonnon.* Crystals from aq ethanol, mp 176-180°. $[\alpha]_D^{27}$ +76.1° (c = 1 in 0.2N HCl).

THERAP CAT: Antithrombotic.

817. Arginine. *L-Arginine;* Arg; R; 2-amino-5-guanidinovaleric acid; (*S*)-2-amino-5-[(aminoiminomethyl)amino]pentanoic acid. $C_6H_{14}N_4O_2$; mol wt 174.20. C 41.37%, H 8.10%, N 32.16%, O 18.37%. An essential amino acid for human development. Precursor for nitric oxide, *q.v.* Isoln from etiolated lupine seedlings: E. Schulze, E. Steiger, *Ber.* **19**, 1177 (1886). Early chemistry and biochemistry: *Amino Acids and Proteins,* D. M. Greenberg, Ed. (Charles C. Thomas, Springfield, IL, 1951) 950 pp., *passim*; J. P. Greenstein, M. Winitz, *Chemistry of the Amino Acids* vols **1-3** (John Wiley and Sons, Inc., New York, 1961) pp. 1841-1855, *passim*. Quick isoln from gelatin: R. M. Kothari, *J. Chromatog.* **90**, 359 (1974). Fluorometric determn in serum: T. Miura *et al.*, *Anal. Biochem.* **139**, 432 (1984). Review of biosynthesis: M. E. Jones, *Adv. Enzyme Regul.* **9**, 19-49 (1970). Review of biochemistry, physiology and therapeutic implication: A. Barbul, *J. Parent. Enteral. Nutr.* **10**, 227-238 (1986). Review of immune function: idem, *Nutrition* **6**, 53-62 (1990). Review of arginine:nitric oxide pathway: S. Moncada, A. Higgs, *N. Engl. J. Med.* **329**, 2002-2012 (1993); R. M. J. Palmer, *Curr. Opin. Nephrol. Hypertens.* **2**, 122-128 (1993). Review of structural role in proteins: C. L. Borders *et al.*, *Protein Sci.* **3**, 541-548 (1994). Review of role in health and renal disease: A. A. Reyes *et al.*, *Am. J. Physiol.* **267**, F331-F346 (1994).

Prisms containing 2 mol H_2O from water; anhydrous monoclinic plates from 96% alcohol. The dihydrate becomes anhyd at 105°, browns at 230°. Dec 244°. $[\alpha]_D^{20}$ +26.9° (c = 1.65 in 6.0N HCl); $[\alpha]_D^{20}$ +12.5° (c = 3.5 in water); $[\alpha]_D^{20}$ 11.8° (c = 0.87 in 0.5N NaOH). pK$_1$ 2.18; pK$_2$ 9.09; pK$_3$ 13.2. Absorption spectrum: Castille, Ruppol, *Bull. Soc. Chim. Biol.* **10**, 643 (1928); *Chem. Zentr.* **1928, II**, 622. The satd aq soln contains 15% (w/w) at 21°. Sparingly sol in alc. Insol in ether. Strongly alkaline, its aq solns absorb CO_2 from the air.

Hydrochloride, $C_6H_{14}N_4O_2$.HCl, *R-gene.* Plates, prisms from alc. Sinters at 218°, solidifies again at 225°. Dec 235°. $[\alpha]_D^{20}$ +12.0° (c = 4). $[\alpha]_D^{21}$ +21.9° (c = 12 in dil HCl). Sol in water, slightly sol in hot alc.

THERAP CAT: Ammonia detoxicant (hepatic failure); diagnostic aid (pituitary function).

818. Arginine Glutamate. *Glutamic acid compd with L-arginine;* Ginamate; Glutargin; Modumate. $C_{11}H_{23}N_5O_6$; mol wt 321.33. C 41.12%, H 7.21%, N 21.79%, O 29.87%. [$H_2NC(NH)]HNCH_2CH_2CH_2CH(NH_2)COOH.HOOC-CH_2CH_2CH(NH_2)COOH$. Prepd from L-arginine and L-glutamic acid: Barker, Chang, U.S. pat. **2,851,482** (1958 to General Mills). Activity studies: Weisburger *et al.*, *Toxicol. Appl. Pharmacol.* **14**, 163 (1969); Yamamoto, Weisburger, *Life Sci.* **9(II)**, 285 (1970). Crystal and molecular structure: T. N. Bhat, M. Vijayan, *Acta Crystallogr.* **B33**, 1754 (1977).

Crystals, dec 193-194.5°. Supplied as a 25% (w/v) soln in water for injection; each 100 ml represents 13.5 g of arginine and 11.5 g of glutamic acid.

THERAP CAT: Ammonia detoxicant (hepatic failure).

819. Argol. Argilla vini; argil; arcilla; Weinstein; wine lees; crude cream of tartar; crude potassium bitartrate. Formed in the secondary fermentation of grapes for wine. Contains over 40% tartaric acid, potassium bitartrate, calcium. *Ref:* R. Pasternack in Kirk-Othmer *Encyclopedia of Chemical Technology* (Interscience, New York, 1954) vol. **13**, p 649.

Gray or reddish-brown crystalline crusts.

USE: Mordant; manufacturing tartaric acid; vinegar from malt; in fertilizers.

820. Argon. Ar; at. wt 39.948; at. no. 18. Group VIIIA(18), also known as Group 0. A noble gas characterized by an electronic structure in which the outer *p* subshell is entirely filled: $1s^22s^22p^63s^23p^6$. Stable naturally occurring isotopes (mass numbers): 36 (0.337%); 38 (0.063%); 40 (99.600%); known artificial radioactive isotopes: 32-35; 37; 39; 41-46. Longest-lived known isotope: 39 (T$_{1/2}$ 269 years, β-emitter). Abundance in igneous rock of the earth's crust: 4×10^{-2} ppm by wt; concentration in the atmosphere: 93.40 ppm by vol. Discovered in 1894 by Rayleigh and Ramsay. Obtained commercially from the atmosphere by distillation-liquefaction process. Monograph: *Argon, Helium and the Rare Gases,* **Vols. 1, 2**, G. A. Cook, Ed. (Interscience, New York, 1961) 818 pp. *Reviews:* A. H. Cockett, K. C. Smith, "The Monatomic Gases" in *Comprehensive Inorganic Chemistry,* vol. 1, J. C. Bailar, Jr. *et al.*, Eds. (Pergamon Press, Oxford, 1973) pp 139-211; S.-C. Hwang, W. R. Weltmer, Jr. in Kirk Othmer *Encyclopedia of Chemical Technology* vol. 13 (John Wiley & Sons, 4th ed., 1995) pp 1-38; *Chemistry of the Elements,* N. N. Greenwood, A. Earnshaw, Eds. (Pergamon Press, New York, 1984) pp 1042-1059. Review of use in inductively coupled plasma-mass spectrometry: S. F. Durrant, *Fres. J. Anal. Chem.* **347**, 389-392 (1993).

Colorless, odorless, tasteless, monatomic, inert gas; will form compds with highly electronegative elements such as O, F, Cl. Atomic radius 1.92×10^{-8} cm. Triple pt temp 83.80°K, press 68.90 kPa. Critical temp 150.86°K, critical press 4898 kPa, critical d 535.7 kg/m³. Gas: d⁰ (101.3 kPa) 1.7838 kg/m³, d (normal bp) 5.767 kg/m³. Liquid: normal bp 87.28°K, d (normal bp) 1393.9 kg/m³, d (triple pt) 1415 kg/m³, heat of vaporization (normal bp) 6469 J/mol. Solid: d (triple pt) 1623 kg/m³, heat of vaporization (triple pt) 7.785 kJ/mol, heat of fusion (triple pt) 1.191 kJ/mol, exists as face-centered cubic crystals at normal pressure. Soly of gas in water (20°): 33.6 cc/kg water. Sol in organic liquids.

Caution: Can act as a simple asphyxiant by displacing air. *See: Matheson Gas Data Book* (Matheson Co., Inc., 4th ed., East Rutherford, NJ, 1966) pp 23-28.

USE: Gas as shield in gas metal-arc welding, in metal processing; carrier in gas-liquid and gas-solid chromatography; gas filler for incandescent light bulbs. Gas in fluorescent tubes analogous to neon lights, but produces a blue-purplish light; in rectifier tubes; in thermometers above mercury; in lasers; wherever an inert atmosphere is desired and the much cheaper nitrogen cannot be used; in ionization chambers and particle counters; in mixtures with He and Ne in Geiger counters; in argon-oxygen-decarburizing process for stainless steel; in manuf of semiconducting devices; in gas mixtures as the working fluid in plasma arc devices. Liquid as cryogen to produce low temps. The isotope ^{40}Ar is always found in minerals contg potassium, since it is a product of ^{40}K decay; measuring the amount of ^{40}Ar and ^{40}K can be used for determining the geologic age of minerals and meteors.

821. Aricine. *(19α,20α)-16,17-Didehydro-10-methoxy-19-methyloxayohimban-16-carboxylic acid methyl ester;* quinovatine; cinchovatine; heterophylline. $C_{22}H_{26}N_2O_4$; mol wt 382.46. C 69.09%, H 6.85%, N 7.32%, O 16.73%. Originally found in Cusco bark, the bark of *Cinchona pelletierana* Wedd., *Rubiaceae:* Pelletier, Coriol, *J. Pharm. Chim.* [2] **15**, 565 (1829). Also from *Rauwolfia canescens* L., *R. heterophylla* Roem. & Schult., *R. sellowii* Muell. Argov. and *Aspidosperma marcgravianum* Woodson., *Apocynaceae:*

Stoll *et al., Helv. Chim. Acta* **38**, 270 (1955); Hochstein *et al., J. Am. Chem. Soc.* **77**, 3551 (1955); Hochstein, *ibid.* **77**, 5744 (1955); Gilbert *et al., J. Org. Chem.* **27**, 4702 (1962). Structure: Stoll *et al., loc. cit.* Stereochemistry: Neuss, Boaz, *J. Org. Chem.* **22**, 1001 (1957); Shamma, Richey, *J. Am. Chem. Soc.* **85**, 2507 (1963).

Orthorhombic, elongated prisms from methanol, dec 188°. Sublimes at 0.01 mm and 180°. $[\alpha]_D^{20}$ −91° (c = 1.4 in chloroform); $[\alpha]_D^{20}$ −63° (c = 1.5 in pyridine); $[\alpha]_D^{20}$ −57° (ethanol). uv max (ethanol): 229, 281 nm (log ε 4.54, 3.97). pKa in 80% methyl cellosolve 5.80; in 1:1 DMF-water 6.8. Practically insol in water. Sol in 100 parts of 90% alcohol, 33 parts ether. Very sol in chloroform.

Hydrochloride, $C_{22}H_{26}N_2O_4 \cdot HCl$, square plates from methanol + acetone, dec 241-254°. $[\alpha]_D^{20}$ −5° (c = 0.9 in 50% ethanol).

Hydrobromide, $C_{22}H_{26}N_2O_4 \cdot HBr$, needles from methanol, dec 262-263°.

Note: Aricine does not seem to share the hypotensive and sedative properties of reserpine: Hochstein *et al., J. Am. Chem. Soc.* **77**, 3551 (1955).

822. Aristolochic Acid. *8-Methoxy-6-nitrophenan-thro[3,4-d]-1,3-dioxole-5-carboxylic acid;* 3,4-methylenedi-oxy-8-methoxy-10-nitro-1-phenanthrenecarboxylic acid; aristolochic acid-I; aristolochine. $C_{17}H_{11}NO_7$; mol wt 341.28. C 59.83%, H 3.25%, N 4.10%, O 32.82%. One of a group of fourteen known, substituted 1-phenanthrenecarboxylic acids, aristolochic acids, that occur in *Aristolochiaceae* and in butterflies feeding on these plants. They are often accompanied by *aristololactams*, twelve of which have been characterized. Isoln: Gänshirt, *Pharmazie* **8**, 584 (1953); Pailer *et al., Monatsh.* **86**, 676 (1955); Coutts *et al., J. Pharm. Pharmacol.* **11**, 607 (1959); Kupchan, Doskotch, *J. Med. Pharm. Chem.* **5**, 657 (1962). Structure: Pailer *et al., Monatsh.* **87**, 249 (1956). Synthesis: Kupchan, Wormser, *J. Org. Chem.* **30**, 3792 (1965). Biosynthesis: Spenser, Tiwari, *Chem. Commun.* **1966**, 55. Acute toxicity: U. Mengs, *Arch. Toxicol.* **59**, 328 (1987). *Review:* D. B. Mix *et al., J. Nat. Prod.* **45**, 657-666 (1982).

Shiny brown leaflets from DMF + hot water, dec 281-286°. uv max (ethanol): 390, 318, 250 nm (ε 6500; 12000; 27000). Slightly sol in water; sol in alcohol, chloroform, ether, acetone, acetic acid, aniline, alkalies. Practically insol in benzene, carbon disulfide. LD_{50} in male, female mice, male, female rats (mg/kg): 38.4, 70.1, 82.5, 74.0 i.v.; 55.9, 106.1, 203.4, 183.9 orally (Mengs).

Methyl ester, $C_{18}H_{13}NO_7$, orange-yellow rods from hot methanol, mp 285-286°.

823. Armepavine. *(R)-4-[(1,2,3,4-Tetrahydro-6,7-di-methoxy-2-methyl-1-isoquinolinyl)methyl]phenol; 6,7-dimeth-oxy-1-p-hydroxybenzyl-2-methyl-1,2,3,4-tetrahydroisoquino-line.* $C_{19}H_{23}NO_3$; mol wt 313.40. C 72.82%, H 7.40%, N 4.47%, O 15.32%. Isoln from *Papaver armeniacum* (L.) DC., *Papaveraceae:* Konowalowa *et al., Ber.* **68**, 2161

(1935); from *Euonymus europaea* L., *Celastraceae:* Bishay *et al., J. Pharm. Pharmacol.* **23** (Suppl), 233S (1971); from *Rhamnus frangula* L., *Rhamnaceae:* Pailer, Haslinger, *Monatsh.* **103**, 1399 (1972). Structure: *eidem, J. Gen. Chem. USSR* **10**, 641 (1940). Synthesis: Marion *et al., J. Org. Chem.* **15**, 216 (1950); Gibson *et al., J. Hetercyclic Chem.* **3**, 99 (1966). Resolution: Knabe, Horn, *Arch. Pharm.* **300**, 547 (1967); Farber, Giacomazi, *Chem. & Ind. (London)* **1968** (2), 57.

dl-Form hydrate, needles from acetone + ether; loses its water around 100°, then melts at 140°. Freely sol in alc, acetone, chloroform; slightly sol in ether. Almost insol in water.

dl-Form hydrochloride, $C_{19}H_{23}NO_3 \cdot HCl$, needles from alcohol + ether, mp 152°. Freely soluble in water and alc.

$R(-)$-Form, mp 149-150°. $[\alpha]_D^{20}$ −117.6° $(CHCl_3)$ (Knabe); $[\alpha]_D^{27}$ −103.2° (Farber).

$R(-)$-Form 10-camphorsulfonate hydrate, $C_{29}H_{39}NO_7S \cdot H_2O$, mp 130°. $[\alpha]_D^{20}$ −65° (MeOH).

824. Armstrong's Acid. *1,5-Naphthalenedisulfonic acid.* $C_{10}H_8O_6S_2$; mol wt 288.30. C 41.66%, H 2.80%, O 33.30%, S 22.24%. Prepd by sulfonation of naphthalene: Lynch, Scanlan, *Ind. Eng. Chem.* **19**, 1010 (1927).

Crystals. Soluble in water, alc; practically insol in ether.

825. Arnica. Arnica flowers; leopard's bane; wolf's bane; mountain tobacco. Dried flowerheads of *Arnica montana* L., *Compositae.* **Habit.** Northern Europe. **Constit.** 0.5-1% volatile oil; arnicin, arnisterol (arnidiol), anthoxanthine, tannin, resin. *Review:* Faber, *Pharmazie* **8**, 179, 286, 340 (1953).

THERAP CAT: Counterirritant (topical).

THERAP CAT (VET): Counterirritant.

826. Arogenic Acid. *[1(S)-cis]-α-Amino-1-carboxy-4-hydroxy-2,5-cyclohexadiene-1-propanoic acid;* L-(8S)-β-(1-car-boxy-4-hydroxy-2,5-cyclohexadien-1-yl)alanine; arogenate; pretyrosine. $C_{10}H_{13}NO_5$; mol wt 227.22. C 52.86%, H 5.77%, N 6.16%, O 35.21%. Precursor in the biosynthesis of L-phenylalanine and L-tyrosine, distributed widely in nature. Pseudomonad bacteria and plants use both the prephenate and arogenate pathways for L-tyrosine biosynthesis. In cyanobacteria, coryniform bacteria and at least one yeast organism the arogenate pathway has been identified as the sole route of L-tyrosine biosynthesis. Identification of arogenate (pretyrosine) in blue-green algae biosynthesis of L-tyrosine: S. L. Stenmark *et al., Nature* **247**, 290 (1974). Dual enzymic routes to L-tyrosine and L-phenylalanine via arogenate (pretyrosine) in *P. aeruginosa:* N. Patel *et al., J. Biol. Chem.* **252**, 5839 (1977). Isoln and prepn: R. A. Jensen *et al., J. Bacteriol.* **132**, 896 (1977). Confirmation that arogenate is an obligatory intermediate of L-tyrosine biosynthesis: A. M. Fazel *et al., Proc. Nat. Acad. Sci. USA* **77**, 1270 (1980). Structure and config: L. O. Zamir *et al., J. Am. Chem. Soc.* **102**, 4499 (1980). Synthesis via immobilized microbial proteins: *eidem, Bioorg. Chem.* **11**, 32 (1982). Arogenate pathway of tyrosine and phenylalanine biosynthesis in *P. aureofaciens:* B. Keller *et al., J. Gen. Microbiol.* **128**, 1199 (1982).

Unstable. Quantitatively converted to phenylalanine at acidic pH. Forms a disodium salt that is stable in the lyophilized state at basic pH (7.5) and room temperature, but is thermally unstable.

827. Arotinolol. *(±)-5-[2-[[3-[(1,1-Dimethylethyl)-amino]-2-hydroxypropyl]thio]-4-thiazolyl]-2-thiophenecarbox-amide.* 2-(3'-*tert*-butylamino-2'-hydroxypropylthio)-4-(5'-carbamoyl-2'-thienyl)thiazole. $C_{15}H_{21}N_3O_2S_3$; mol wt 371.55. C 48.49%, H 5.70%, N 11.31%, O 8.61%, S 25.89%. Propanolamine deriv with α- and β-adrenergic blocking activity. Prepn: T. Hibino *et al.,* Ger. pat. 2,341,753; *eidem,* U.S. pat. 3,932,400 (1974, 1976 both to Sumitomo); Y. Hara *et al., J. Pharm. Sci.* **67,** 1334 (1978). Characterization of adrenoceptor blocking effects *in vivo:* A. Miyagishi *et al., Arch. Int. Pharmacodyn. Ther.* **271,** 249 (1984). Possible antianginal effects: M. Sakanashi *et al., Oyo Yakuri* **28,** 709 (1984); *C.A.* **102,** 39669z (1985). Long-term antihypertensive effect in rats: K. Kishi *et al., J. Pharmacobiodyn.* **8,** 50 (1985). Comparison with timolol, *q.v.,* of effect on intraocular pressure and hemodynamics: M. Nakashima *et al., Eur. J. Clin. Pharmacol.* **28,** 391 (1985).

Crystals from chloroform/petroleum ether, mp 148-149°. Hydrochloride, $C_{15}H_{21}N_3O_2S_3 \cdot HCl$, *S-596, ARL, Almarl.* Crystals from methanol/water, mp 234-235.5° (dec). LD_{50} in mice (mg/kg): 86 i.v., > 360 i.p., > 5000 orally (Hara).

THERAP CAT: Antihypertensive; antianginal; antiarrhythmic.

828. Arprinocid. *9-[(2-Chloro-6-fluorophenyl)methyl]-9H-purin-6-amine;* 9-(2-chloro-6-fluorobenzyl)adenine; MK-302; Arpocox. $C_{12}H_9ClFN_5$; mol wt 277.69. C 51.90%, H 3.27%, Cl 12.77%, F 6.84%, N 25.22%. Prepn: E. P. Lira *et al.,* **Fr.** pat. **2,128,600** corresp to U.S. pats. **3,846,426** and **3,953,597** (1972, 1974, 1976, all to Int. Minerals and Chem.); G. D. Hartman *et al., J. Org. Chem.* **43,** 960 (1978). Determination in feed: D. W. Fink *et al., J. Assoc. Off. Anal. Chem.* **61,** 1078 (1978); **62,** 1 (1979). Evaluation of anticoccidial efficacy: P. Schindler *et al., Poult. Sci.* **58,** 23 (1979); L. R. McDougald, J. K. Johnson, *ibid.* **72.**

Cryst from methanol/water, mp 245-246°.
THERAP CAT (VET): Coccidiostat.

829. Arsacetin. *[4-(Acetylamino)phenyl]arsonic acid;* p-acetamidobenzenearsonic acid; *N*-acetyl-*p*-aminobenzene-arsonic acid; *N*-acetylarsanilic acid. $C_8H_{10}AsNO_4$; mol wt 259.09. C 37.09%, H 3.89%, As 28.92%, N 5.41%, O 24.70%. Prepn of free acid and sodium salt: Bart, *Ann.* **429,** 55 (1922). Toxicity: A. Gros, *Biochem. Z.* **184,** 360 (1927).

Sodium salt tetrahydrate, $C_8H_9AsNNaO_4 \cdot 4H_2O$, cryst powder. *Poisonous!* One gram dissolves in 10 ml cold water, 3 ml hot water. LD_{50} i.v. in rabbits: 550 mg/kg (Gros).

Note: Sodium salt was formerly referred to as arsacetin.
THERAP CAT: Sodium salt formerly as antisyphilitic.

830. Arsanilic Acid. *(4-Aminophenyl)arsonic acid;* p-aminobenzenearsonic acid; atoxylic acid; AS-101. C_6H_8-$AsNO_3$; mol wt 217.06. C 33.20%, H 3.71%, As 34.52%, N 6.45%, O 22.11%. Prepd by heating aniline and arsenic acid: Lewis, Cheetham, *Org. Syn. coll. vol.* **I** (2nd ed., 1941) p 70; Hoffmann, Green, U.S. pat. **3,763,201** (1973); from *p*-nitroaniline: Bart, *Ann.* **429,** 96 (1922); Gattermann-Wieland, *Praxis des Organischen Chemikers* (de Gruyter, Berlin, 40th ed., 1961) p 254. Toxicology in swine: Ledet *et al., Clin. Toxicol.* **6,** 439 (1973). Toxicity study: E. I. Goldenthal, *Toxicol. Appl. Pharmacol.* **18,** 185 (1971).

Needles from water or alcohol. Slightly sol in cold water, alcohol or acetic acid; sol in hot water, pentanol, in solns of alkali carbonates; moderately sol in concd mineral acids. Practically insol in acetone, benzene, chloroform, ether, and in moderately dil mineral acids. LD_{50} orally in male rats: > 1000 mg/kg (Goldenthal).

USE: Manuf medicinal arsenicals.
THERAP CAT (VET): Growth promotant; to improve feed efficiency. To control swine dysentery.

831. Arsenamide. *2,2'-[[[4-(Aminocarbonyl)phenyl]-arsinidene]bis(thio)]bisacetic acid; [[(p-carbamoylphenyl)-arsylene]dithio]diacetic acid;* bis[carboxymethylmercapto](p-carbamylphenyl)-arsine; *p*-[bis(carboxymethylmercapto)-arsino]benzamide; dithioglycolyl *p*-arsenobenzamide; 4-carbamylphenyl bis[carboxymethylthio]arsenite; thioarsenite; thiacetarsamide; Caparsolate; Caparside. $C_{11}H_{12}AsNO_5S_2$; mol wt 377.27. C 35.02%, H 3.21%, As 19.86%, N 3.71%, O 21.20%, S 17.00%. Prepd by the condensation of *p*-arsenosobenzamide and thioglycolic acid: Gough, King *J. Chem. Soc.* **1930,** 669; Maren, *J. Am. Chem. Soc.* **68,** 1864 (1946).

White crystalline powder. Sparingly sol in cold water; appreciably sol in water above 90°. Sparingly sol in cold methanol, ethanol; very sol in these solvents when they are warm. Insol in warm isopropyl ether. pKa = 4. The disodium salt is stoichiometrically formed in aq soln at pH 7-8.

THERAP CAT (VET): Anthelmintic for heartworm infection in dogs, especially the adult worm.

832. Arsenic. Grey arsenic; metallic arsenic; arsen (German). As; at. wt 74.9216; at. no. 33; valences 3, 5. Group VA(15) element, classified as a metalloid. Naturally occurring isotope (mass number): 75 (100%); known artificial, radioactive isotopes: 66-74; 76-87. Arsenic compds were described and used in antiquity, especially as poisons; their reduction to the element was known to medieval alchemists. Albertus Magnus credited with isolation of the element from the mineral orpiment in ∼1250 A.D. First precise directions for the prepn of As found in Paracelsus' writings (ca. 1520). Arsenic probably occurs throughout the universe. Meteorites reported to contain from 0.0005 to 0.1% As. Occurrence in the earth's crust: 1.8 ppm. Found in nature to a small extent as the element; occurs mostly in minerals such as realgar (As_4S_4), orpiment (As_2S_3), arsenolite (As_2O_3). Commercial sources: as by-product in flue dusts from smelting copper, lead, cobalt and gold ores; by melting $FeAs_2$ or FeAsS ores. Prepn of pure As by reduction with carbon (sugar charcoal) and sublimation in N_2: Krepelka, *Coll. Czech. Chem. Commun.* **2**, 255 (1930); E. H. Archibald, *The Preparation of Pure Inorganic Substances* (Wiley, New York, 1932) p 269. Other methods: Schenk in *Handbook of Preparative Inorganic Chemistry* vol. 1, G. Brauer, Ed. (Academic Press, New York, 2nd ed., 1963) pp 591-593. *Reviews: Gmelin's, Arsenic* (8th ed.) **17**, 475 pp (1952); Smith, "Arsenic, Antimony and Bismuth" in *Comprehensive Inorganic Chemistry* vol. 2, J. C. Bailar, Jr. *et al.*, Eds. (Pergamon Press, Oxford, 1973) pp 547-683; *Chemistry of the Elements*, N. N. Greenwood, A. Earnshaw, Eds. (Pergamon Press, New York, 1984) pp 637-697; S. C. Carapella, Jr. in Kirk Othmer *Encyclopedia of Chemical Technology* vol. 3 (John Wiley & Sons, 4th ed., 1992) pp 624-633; G. O. Doak *et al., ibid.* pp 633-659. Review of carcinogenicity studies: *IARC Monographs* **23**, 39-141 (1980); of toxicology and human exposure: *Toxicological Profile for Arsenic* (PB93-182376, 1993) 198 pp. Book: "The Chemistry of Organic Arsenic, Antimony and Bismuth Compounds," S. Patai, Ed. (John Wiley & Sons, New York, 1994) 962 pp.

Allotropic forms: α-form, metallic, steel-gray, shiny, brittle, rhombohedral crystal structure; β-form, dark gray, amorphous sold, d 4.700, transforms to metallic form at 280°. Can be heated to burn in air with bluish flame, giving off an odor of garlic and dense white fumes of As_2O_3. Stable in dry air; loses its luster on exposure to humid air as surface oxidizes, forming a black modification + As_2O_3. Brinell hardness: 147; Mohs' scale: 3.5. d_4^{25} 5.778. Sublimes$_{760}$ 615° without melting. mp 818° at 36 atm. Heat of vaporization 11.2 kcal/g-atom. Heat of sublimation 30.5 kcal/g-atom. Heat of fusion: 22.4 kcal/g-atom (*Gmelin's, loc. cit.* pp 135-136). Also reported as: heat of fusion: 6.620 kcal/g-atom; heat of sublimation 7.63 kcal/g-atom: D. R. Stull, G. C. Sinke, *Thermodynamic Properties of the Elements*, Advances in Chemistry Series 18 (A.C.S., Washington, 1956) pp 11, 44. Latent heat of fusion: 27,740 J/mol K. Latent heat of sublimation: 31,974 J/mol K (Carapella). Specific heat (25°) 24.6 J/mol K. Dielectric constant = 10.23 at 20° and 60 cycles. Electrical and magnetic properties of crystalline As: Taylor *et al., J. Phys. Chem. Solids* **26**, 69 (1965). Insol in water. Not attacked by cold H_2SO_4 or HCl; converted by HNO_3 or hot H_2SO_4 into arsenous or arsenic acid.

A yellow modification which has no metallic properties has been reported from sudden cooling of As-vapor. This yellow arsenic is converted back to the gray modification upon very short exposure to ultraviolet light.

Note: In German and other languages *Arsenik* means arsenic trioxide.

Caution: Overexposure to arsenic and arsenic compounds has been associated with acute and chronic toxicity due to inhalation or ingestion. Organic forms are usually less harmful than inorganic forms. Direct contact can cause local irritation and dermatitis. Overexposure has been associated with an increased risk of skin, liver, bladder, kidney and lung cancer. *See Toxicological Profile, loc. cit.* This substance and certain arsenic compounds have been listed as known carcinogens: *Seventh Annual Report on Carcinogens* (PB95-109781, 1994) p 21.

USE: In metallurgy for hardening copper, lead, nonferrous alloys; automotive body solder. In semiconductor materials. In the manufacture of low-melting glass. As wood preservative, herbicide, pesticide.

833. Arsenic Acid. Orthoarsenic acid. AsH_3O_4; mol wt 141.94. As 52.78%, H 2.13%, O 45.09%. H_3AsO_4. Exists only as the hemihydrate. Excessive drying produces $As_2O_5 \cdot 5/3H_2O$. Conveniently prepd from As_2O_3 and HNO_3: Simon, Thaler, *Z. Anorg. Allgem. Chem.* **161**, 143 (1927); **246**, 19 (1941). Toxicity study: Joachimoglu, *Biochem. Z.* **70**, 144 (1915).

Hemihydrate, hygroscopic crystals. *Poisonous!* Converted to As_2O_5 by heating above 300°. Freely sol in water, alcohol, glycerol. LD_{50} i.v. in rabbits: 6 mg/kg (Joachimoglu).

USE: In the manuf of arsenates.

834. Arsenic Disulfide. *Arsenic sulfide;* red arsenic sulfide; realgar; red orpiment; ruby arsenic; red arsenic glass; C.I. Pigment Yellow 39; C.I. 77085. As_4S_4; mol wt 427.95. As 70.03%, S 29.97%. Prepn: *Gmelin's, Arsenic* (8th ed.) **17**, pp 417-422 (1952); Schenk in *Handbook of Preparative Inorganic Chemistry* vol 1, G. Brauer, Ed. (Academic Press, New York, 2nd ed., 1963) p 603.

Deep red, lustrous monoclinic crystals. mp 320°; bp 565°; d 3.5. Ignites at high temp. Practically insol in water; sol in alkali hydroxides; decomposed by HNO_3; very slightly sol in hot CS_2 and benzene.

USE: As pigment in painting; in fireworks as blue fire and to give an intense white flame; manuf shot; calico printing and dyeing; tanning and depilating hides.

835. Arsenic Hemiselenide. As_2Se; mol wt 228.80. As 65.49%, Se 34.51%. Prepd by melting arsenic and selenium in the correct proportions in a sealed tube filled with nitrogen: Szarvasy, *Ber.* **28**, 2654 (1895); **30**, 1244 (1897); *Gmelin's, Arsenic* (8th ed.) **17**, p 462 (1952).

Black crystals with a metallic luster. Insol in the usual organic and inorganic solvents. Dec in boil. alkali hydroxide solns; slowly decompd by concd hydrochloric and sulfuric acids.

USE: Manuf of glass.

836. Arsenic Pentafluoride. AsF_5; mol wt 169.91. As 44.09%, F 55.91%. Prepd from As + F_2: Ruff *et al., Z. Anorg. Allgem. Chem.* **206**, 59 (1932); Seel, Detmer, *ibid.* **301**, 113 (1959); Kwasnik in *Handbook of Preparative Inorganic Chemistry* vol. 1, G. Brauer, Ed. (Academic Press, New York, 2nd ed., 1963) p 198. *Reviews:* Burg in *Fluorine Chemistry* vol. I, J. Simons, Ed. (Academic Press, New York, 1950) p 102; Kemmitt, Sharp, *Advan. Fluorine Chem.* **4**, 208 (1965).

Colorless gas. Forms white clouds in moist air. $d_{liq}^{25.8}$ 2.33. mp −79.8°. bp −53.2°. Instantly hydrolyzed by water. Sol in alcohol, ether, benzene. Dry AsF_5 does not attack glass, but a minute trace of moisture or HF catalyzes the etching reaction to the point of total destruction.

837. Arsenic Pentaselenide. As_2Se_5; mol wt 544.64. As 27.51%, Se 72.49%. Prepd by melting the correct proportions of arsenic and selenium in a sealed tube containing nitrogen: Szarvasy, *Ber.* **28**, 2654 (1895); **30**, 1244 (1897); by the action of an arsenic salt on a solution of hydrogen selenide: Moser, Atynsky, *Monatsh.* **45**, 235 (1925).

Black, brittle solid with a metallic luster. *Poisonous!* Dec when heated in air. Sol in alkali hydroxides and sulfides; insol in water, dilute acids, concd hydrochloric acid, alcohol, ether, carbon disulfide; dec in nitric acid.

838. Arsenic Pentasulfide. Diarsenic pentasulfide. As_2S_5; mol wt 310.17. As 48.31%, S 51.69%. Prepd from H_3AsO_4 and H_2S: Schenk in *Handbook of Preparative Inorganic Chemistry* vol 1, G. Brauer, Ed. (Academic Press, New York, 2nd ed., 1963) p 603.

Brownish-yellow, glassy, amorphous, highly refractive mass. Insol in water; sol in alkalies and alkali sulfides. Dec into As_2O_3, S, and As_2S_3 when boiled with water.

USE: In thin sheets as a light filter; in pigments.

839. Arsenic Pentoxide. Arsenic acid anhydride. As_2O_5; mol wt 229.84. As 65.19%, O 34.81%. Prepn: *Gmelin's, Arsenic* (8th ed.) **17**, pp 273-277 (1952).

Amorphous lumps or powder. *Poisonous!* Gradually deliquesces on exposure to air. Freely sol in water or alc. Combines very slowly with H_2O to form H_3AsO_4. *Keep well closed.*

USE: Manuf of colored glass; in adhesives for metals; in wood preservatives; in weed control; as fungicide.

840. Arsenic Tribromide. $AsBr_3$; mol wt 314.63. As 23.81%, Br 76.19%. Prepn: Oddo, Giachery, *Gazz. Chim. Ital.* **53**, 56 (1923); Schenk in *Handbook of Preparative Inorganic Chemistry* vol. 1, G. Brauer, Ed. (Academic Press, New York, 2nd ed., 1963) p 597.

Deliquescent, orthorhombic prisms. *Intensely poisonous!* d_4^{25} 3.397; d_4^{50} (liq) 3.3282; d_4^{75} 3.2623; d_4^{100} 3.1995. mp 31.1°. bp_{760} 221°; bp_{11} 89°. Dipole moment 1.66. Heat of fusion 8.93 cal/g. Fumes in moist air. Dec by water with the formation of As_2O_3 and HBr. Miscible with ether, benzene. Sol in hydrocarbons, chlorinated hydrocarbons, carbon disulfide, oils and fats.

841. Arsenic Trichloride. Butter of arsenic; fuming liquid arsenic. $AsCl_3$; mol wt 181.28. As 41.33%, Cl 58.67%. Prepn: Smith, *Ind. Eng. Chem.* **11**, 109 (1919); Reisener in *Ullman's Encyklopädie der Technischen Chemie* **vol. 3**, (Urban & Schwarzenberg, Munich, 1953) p 850; Schenk in *Handbook of Preparative Inorganic Chemistry* vol. **1**, G. Brauer, Ed. (Academic Press, New York, 2nd ed., 1963) p 596.

Oily liquid. *Intensely poisonous!* Fumes in air. d_4^{25} 2.1497. mp −16°. bp 130.21°; bp_{11} 25°. Dipole moment: 2.17. Heat of vaporization 8.9 kcal/mol. Specific heat: 3.19 cal/mol/°C. n_D^{20} 1.6006. Dec by water to form As-(OH)$_3$ and HCl. One mol $AsCl_3$ can be dissolved in 9 mols H_2O, this soln (d 1.53) may be diluted again with another 9 mols H_2O giving a soln with d 1.346. Further dilution results in the precipitation of As_2O_3. Also dec by ultraviolet light. Miscible with, or solvent for, chloroform, carbon tetrachloride, ether, iodine, phosphorus, sulfur, alkali iodides, oils and fats.

USE: In the ceramic industry; in syntheses of chlorine-contg arsenicals, e.g., chloro derivs of arsine.

842. Arsenic Trifluoride. AsF_3; mol wt 131.92. As 56.79%, F 43.21%. Prepn: Russell *et al.*, *J. Am. Chem. Soc.* **63**, 2825 (1941); Woolf, Greenwood, *J. Chem. Soc.* **1950**, 2200; Hoffman, *Inorg. Syn.* **4**, 151 (1953); Kwasnik in *Handbook of Preparative Inorganic Chemistry* vol. **1**, G. Brauer, Ed. (Academic Press, New York, 2nd ed., 1963) pp 197-198. Review: Kemmitt, Sharp, *Advan. Fluorine Chem.* **4**, 208-209 (1965).

Mobile liquid. *Very poisonous!* Fumes in air. Etches glass. d_{15}^{15} 2.73. mp −5.95°. bp 57.8°. Also reported as mp −8.5°; bp 63° (Kwasnik). Hydrolyzed by water. Sol in alcohol, ether, benzene. May be stored in iron vessels.

Caution: Extremely toxic.

843. Arsenic Triiodide. AsI_3; mol wt 455.64. As 16.44%, I 83.56%. Prepd from the elements or from $AsCl_3$ and KI: Bailar, *Inorg. Syn.* **I**, 103 (1939).

Orange-red, trigonal rhombohedra from acetone. Reacts slowly with O_2 from air, liberating iodine. d_4^{25} 4.688. Some tendency to sublime below 100°. mp 140.9° forming a red liquid. bp_{760} ~400°. One gram dissolves in 12 ml water forming a yellow soln. Does not hydrolyze rapidly and may be recovered from the water soln unchanged within 5 hrs. Aqueous solns are strongly acid (pH of 0.1*N* soln about 1.1) and ultimately form HI and As_2O_3, although an equilibrium $AsI_3 + 3H_2O = H_3AsO_3 + 3HI$ has been observed. Freely sol in carbon disulfide, chloroform, benzene, toluene, xylene. Less sol in alc, ether.

THERAP CAT: Formerly in dermatitides.

844. Arsenic Trioxide. Arsenous acid; arsenous acid anhydride; arsenous oxide; arsenic sesquioxide; white arsenic. As_2O_3; mol wt 197.84. As 75.74%, O 24.26%. Prepn: Schenk in *Handbook of Preparative Inorganic Chemistry* vol. **1**, G. Brauer, Ed. (Academic Press, New York, 2nd ed., 1963) p 600. Toxicity study: J. Harrison *et al.*, *Arch. Ind. Health* **17**, 118 (1958).

White or transparent, glassy, amorphous lumps or cryst powder. Two cryst modifications, *claudetite* (monoclinic, mp 313°) and *arsenolite* (cubic, mp 275°). bp 465°. *Intensely poisonous!* When slowly heated sublimes unchanged; when rapidly heated the cryst sublimes without fusion, while the amorphous first fuses, then sublimes. Sparingly and extremely slowly sol in cold water; sol in 15 parts boiling water, in dil HCl, in alkali hydroxide or carbonate solns. Practically insol in alc, chloroform, ether. LD$_{50}$ in mice, rats (mg/kg): 39.4, 15.1 orally (Harrison).

Incompat. Tannic acid, infusion cinchona and other vegetable astringent infusions and decoctions; iron in soln.

USE: It is the primary material for all arsenic compounds. Used in manuf of glass, Paris green, enamels, weed killers, metallic arsenic; for preserving hides; killing rodents, insects; in sheep dips and weed killers; textile mordant.

THERAP CAT (VET): Formerly as parasiticide, also for non-parasitic skin and blood diseases; in rheumatism, asthma and heaves, and as an alterative.

845. Arsenic Triselenide. Arsenious selenide; arsenous selenide. As_2Se_3; mol wt 386.72. As 38.75%, Se 61.25%. Prepd by melting arsenic and selenium in the correct proportions: Uelsmann, *Ueber Selenverbindungen*, Göttingen (1860); *Ann.* **116**, 122 (1860); by the action of an arsenic salt on a soln of hydrogen selenide: Moser, Atynsky, *Monatsh.* **45**, 235 (1925).

Dark brown solid. d 4.75. mp 260°. Sol in nitric acid, alkali-lye, alkali sulfide solutions; insol in water.

846. Arsenic Trisulfide. Yellow arsenic sulfide; orpiment; auripigment; arsenic yellow; king's yellow; king's gold. As_2S_3; mol wt 246.04. As 60.90%, S 39.10%. The article of commerce contains much less sulfur than theory. Prepn: *Gmelin's, Arsenic* (8th ed.) **17**, pp 422-433 (1952).

Yellow or orange powder. mp reported from 300° to 325°. d 3.46. Practically insol in water. Sol in alkalies, alkali sulfides or carbonates; slowly sol in hot HCl; dec by HNO_3.

USE: Manuf of glass, particularly infrared-transmitting glass; manuf of oil cloth, linoleum; in electrical semiconductors, photoconductors; as pigment; for depilating hides; in pyrotechnics.

847. Arsenious Acid Solution. Arsenic chloride solution. Prepared with 1 g As_2O_3, 5 ml dil HCl and water to 100 ml. *Poisonous!*

THERAP CAT (VET): Has been used in skin and blood disorders.

848. Arsenoacetic Acid. *Arsenodiacetic acid;* diarsenoacetic acid. $C_4H_6As_2O_4$; mol wt 267.93. C 17.93%, H 2.26%, As 55.93%, O 23.89%. HOOCCH$_2$As=AsCH$_2$-COOH. Prepd by dissolving sodium arsonoacetate and sodium hypophosphite in 15% H_2SO_4: Palmer, *J. Am. Chem. Soc.* **45**, 3023 (1923); *Org. Syn. coll. vol.* I, 73 (1941).

Minute yellow needles, begin to dec about 205°, but do not melt below 260°. Practically insol in water and common organic solvents. Readily sol in pyridine, dil sodium hydroxide and dil sodium carbonate solns.

Disodium salt, $C_4H_4As_2Na_2O_4$, *disodium arsenoacetate, Tonarsan.* Yellow powder, freely sol in water. *Compare* Arsonoacetic Acid (disodium salt).

THERAP CAT (VET): Disodium salt has been used as an injectable tonic for horses.

849. Arsine. Arsenic trihydride; hydrogen arsenide. AsH_3; mol wt 77.95. As 96.12%, H 3.88%. Prepn: Wendt, Landauer, *J. Am. Chem. Soc.* **42**, 930 (1920); Jolly, Drake, *Inorg. Syn.* **7**, 41 (1963); Schenk in *Handbook of Preparative Inorganic Chemistry* vol **1**, G. Brauer, Ed. (Academic Press, New York, 2nd ed., 1963) pp 593-595. Toxicity: Doig, *Lancet* **2**, 88 (1958); Vallee *et al.*, *Arch. Ind. Health* **21**, 132 (1960).

Colorless, neutral gas. Disagreeable garlic odor. *Very poisonous!* mp −117°. bp −62.5°. Dissociation pressure at 0° = 0.806 atm. Decomposes when heated at 300°, depositing arsenic which volatilizes at 400°. On exposure to light, moist arsine decomposes quickly depositing shiny black arsenic. Slightly sol in water. Aq solns are neutral. Traces are best removed by absorption in potassium permanganate soln or in bromine water.

Caution: Potential symptoms of overexposure are headache, malaise, weakness and dizziness; dyspnea; abdominal and back pain; nausea, vomiting; bronze skin; hematuria; jaundice; peripheral neuropathy. Potential occupational carcinogen. *See NIOSH Pocket Guide to Chemical Hazards* (DHHS/NIOSH 90-117, 1990) p 42.

850. Arsonoacetic Acid. $C_2H_5AsO_5$; mol wt 183.98. C 13.06%, H 2.74%, As 40.72%, O 43.48%. $H_2O_3AsCH_2$-COOH. Prepn from sodium arsenite and sodium chloroacetate: Palmer, *J. Am. Chem. Soc.* **45**, 3023 (1923); *Org. Syn. coll. vol.* I, 73 (1941).

Shiny plates, mp 152°. Very sol in water, alc; sparingly sol in hot glacial acetic acid. Practically insol in petr ether, benzene, acetone, chloroform, ethyl acetate.

Barium salt, $C_4H_4As_2Ba_3O_{10}$, feathery needles. Slightly sol in boiling water.

Sodium salt monohydrate, $C_2H_3AsNa_2O_5 \cdot H_2O$, *disodium arsonoacetate, disodium acetarsenate, acetarsonic acid disodium salt, sodium acetoarsinate, Aricyl.* Crystals. Freely sol in water.

THERAP CAT (VET): Disodium salt has been used to treat anaplasmosis, and as a stimulant in nervous diseases.

851. Arsphenamine. *4,4'-(1,2-Diarsenediyl)bis[2-aminophenol] dihydrochloride; 4,4'-arsenobis(2-aminophenol) dihydrochloride;* arsenphenolamine hydrochloride; 3,3'-diamino-4,4'-dihydroxyarsenobenzene dihydrochloride; Ehrlich 606; Arsaminol; Kharsivan; Salvarsan; Sanluol. $C_{12}H_{14}As_2Cl_2N_2O_2$; mol wt 439.00. C 32.83%, H 3.21%, As 34.13%, Cl 16.15%, N 6.38%, O 7.29%. Prepd by Ehrlich in 1909. Prepn: Ehrlich, Bertheim, U.S. pat. **986,148** (1911 to Hoechst); Christiansen, *J. Am. Chem. Soc.* **42**, 2402 (1920). Toxicity data: J. F. Schamberg *et al., Am. J. Syph. Neurol.* **18**, 37 (1934).

Light yellow, somewhat hygroscopic powder. It is odorless, or has a slight odor of the precipitant used. Oxidizes on exposure to air, becoming darker and more toxic. *It is poisonous!* Soluble in water, alc or glycerol, very slightly sol in chloroform or ether. Its aq soln is acid; pH about 3. LD_{100} i.v. in rats: 140 mg/kg (Schamberg).

THERAP CAT: Formerly as antisyphilitic.

THERAP CAT (VET): Formerly in contagious bovine pleuropneumonia.

852. Arsthinol. *N-[2-Hydroxy-5-[4-(hydroxymethyl)-1,3,2-dithiarsolan-2-yl]phenyl]acetamide; 3-acetamido-4-hydroxydithiobenzenearsonous acid, cyclic (hydroxymethyl)ethylene ester;* 2-(3'-acetamido-4'-hydroxyphenyl)-1,3-dithia-2-arsacyclopentane-4-methanol; 3-acetamido-4-hydroxydithiobenzenearsonous acid cyclic 3-hydroxypropylene ester; 2-acetylamino-4-(methylolcycloethylenedimercaptoarsine)-phenol; Mercaptoarsenol; Balarsen. $C_{11}H_{14}AsNO_3S_2$; mol wt 347.29. C 38.04%, H 4.06%, As 21.57%, N 4.03%, O 13.82%, S 18.47%. Prepn: Friedheim, U.S. pats. **2,593,434** (1952) and **2,772,303** (1956).

Minute crystals, mp 163-166°. Soly in 95% ethanol about 2.7% w/v. Very slightly sol in water, ether.

THERAP CAT: Antiamebic.

853. Arteether. *[3R-(3α,5aβ,6β,8aβ,9α,10α,12β,-12aR*)]-10-Ethoxydecahydro-3,6,9-trimethyl-3,12-epoxy-12H-pyrano[4,3-j]-1,2-benzodioxepin;* dihydroartemisinin ethyl ether; dihydroqinghaosu ethyl ether; SM-227. $C_{17}H_{28}O_5$; mol wt 312.41. C 65.36%, H 9.03%, O 25.61%. Derivative of artemisinin, *q.v.* Prepn: Y. Li *et al., Acta Pharm. Sinica* **16**, 429 (1981). Synthesis and antimalarial activity: China Cooperative Research Group on Qinghaosu, *J. Tradit. Chinese Med.* **2**, 9 (1982); A. Brossi *et al., J. Med. Chem.* **31**, 645 (1988).

White crystalline solid, mp 80-82°. $[α]_D^{21}$ 154.5° (c = 1.0 in CHCl$_3$).

THERAP CAT: Antimalarial.

854. Artemether. *[3R-(3α,5aβ,6β,8aβ,9α,10α,12β,-12aR*)]-Decahydro-10-methoxy-3,6,9-trimethyl-3,12-epoxy-12H-pyrano[4,3-j]-1,2-benzodioxepin;* dihydroartemisinin methyl ether; dihydroqinghaosu methyl ether; o-methyldihydroartemisinin; SM-224. $C_{16}H_{26}O_5$; mol wt 298.38. C 64.41%, H 8.78%, O 26.81%. Derivative of artemisinin, *q.v.* Prepn: Y. Li *et al., K'o Hseuh T'ung Pao* **24**, 667 (1979), *C.A.* **91**, 211376u (1979); *eidem, Acta Pharm. Sinica* **16**, 429 (1981). Absolute configuration: X.-D. Luo *et al., Helv. Chim Acta* **67**, 1515 (1984). NMR spectral study: F. S. El-Feraly *et al., Spectrosc. Lett.* **18**, 843 (1985). Inhibition of protein synthesis: H. M. Gu *et al., Biochem. Pharmacol.* **32**, 2463 (1983). Antimalarial activity: S. Thaithong, G. H. Beale, *Bull. WHO* **63**, 617 (1985); and pharmacology: W. Peters *et al., Ann. Trop. Med. Parasitol.* **80**, 483 (1986). Clinical trials in falciparum malaria: W. Tongyin, X. Ruchang, *J. Tradit. Chinese Med.* **5**, 240 (1985). Toxicity data: China Cooperative Research Group on Qinghaosu, *ibid.* **2**, 31 (1982). Series of articles on chemistry, pharmacology and antimalarial efficacy: ibid. 3-50.

Crystals, mp 86-88°. $[α]_D^{19.5}$ 171° (c = 2.59 in CHCl$_3$). LD_{50} i.m. in mice: 263 mg/kg (China Cooperative Research Group on Qinghaosu).

THERAP CAT: Antimalarial.

855. Artemisin. *[3S-(3α,3aα,4α,5aβ,9bβ)]-3a,5,5a,9b-Tetrahydro-4-hydroxy-3,5a,9-trimethylnaphtho[1,2-b]furan-2,8(3H,4H)-dione; 6α,8α-dihydroxy-3-oxoeudesma-1,4-dien-12-oic acid 12,6-lactone;* 8-hydroxysantonin. $C_{15}H_{18}O_4$; mol wt 262.31. C 68.69%, H 6.92%, O 24.40%. From the closed, unexpanded flower heads of several *Artemisia* spp, especially *Artemisia maritima* L., and *A. cina* Berg., Compositae ("wormseed"). Found in the mother liquors from the extraction of santonin. Isoln: E. Merck, *Merck's Jahresber.* **1894**, 3; *Chem. Zentr.* **1895**, I, 436. Structure: Sumi, *J. Am. Chem. Soc.* **80**, 4869 (1958); Cocker, McMurry, *Tetrahedron* **8**, 181 (1960). Stereochemistry: Bolt *et al., J. Chem. Soc.* **1963**, 5235. Synthesis of (+)-artemisin: Nakazaki, Naemura, *Tetrahedron Letters* **1966**, 2615.

Crystals from abs ethanol or ethyl acetate. Bitter taste. Turns yellow on exposure to light. mp 203°. $bp_{0.1}$ 260°. Sublimes$_{760}$ 170-175°. $[α]_D^{23}$ −84.9° (c = 3 in 95% ethanol).

One gram dissolves in 60 ml boiling water, in 3 ml boiling alc. Sol in ethyl acetate. Practically insol in petr ether. Somewhat sol in chloroform.

856. Artemisinin. *[3R-(3α,5aβ,6β,8aβ,9α,12β,12a-R*)]-Octahydro-3,6,9-trimethyl-3,12-epoxy-12H-pyrano[4,3-j]-1,2-benzodioxepin-10(3H)-one;* artemisine; arteannuin; huanghuahaosu; QHS; qinghaosu; qing hau sau. $C_{15}H_{22}O_5$; mol wt 282.34. C 63.81%, H 7.85%, O 28.33%. Active antimalarial constituent of the traditional Chinese medicinal herb *Artemisia annua* L., *Compositae*, which has been known for almost 2000 years as *Qinghao*. Isolated in 1972 and shown to be a sesquiterpene lactone with a peroxide moiety: *K'o Hsueh T'ung Pao* **22**, 142 (1977), *C.A.* **87**, 98788g (1977); L. Jing-Ming *et al., Acta Chim. Sinica* **37**, 129 (1979), *C.A.* **92**, 94594w (1980). Total synthesis and absolute configuration: G. Schmid, W. Hofheinz, *J. Am. Chem. Soc.* **105**, 624 (1983). NMR spectral study: F. S. El-Feraly *et al., Spectrosc. Letters* **18**, 843 (1985). Antimalarial activity: Qinghaosu Antimalaria Coordinating Research Group, *Chinese Med. J.* **92**, 811 (1979); L. J. Bruce-Chwatt, *Brit. Med. J.* **284**, 767 (1982). J.-B. Jiang *et al., Lancet* **2**, 285 (1982). Mechanism of action: G. H. Posner *et al., J. Am. Chem. Soc.* **117**, 5885 (1995). Clinical trial in comparison with artemether, *q.v.:* W. Tongyin, X. Ruchang, *J. Tradit. Chinese Med.* **5**, 240 (1985). Toxicity data: China Cooperative Research Group on Qinghaosu, *ibid.* **2**, 31 (1982). Series of articles on chemistry, pharmacology, and antimalarial efficacy: *ibid.* 3-50. Brief reviews: H. Koch, *Pharm. Int.* **2**, 184-185 (1981); D. L. Klayman, *Science* **228**, 1049-1054 (1985). Review of chemistry, pharmacology and clinical applications: X. D. Luo, C. C. Shen, *Med. Res. Rev.* **7**, 29-52 (1987).

Needles, mp 156-157°. $[\alpha]_D^{17}$ +66.3°. (c = 1.64 in CHCl₃). Sol in most aprotic solvents. Slightly sol in oil. LD₅₀ in mice (mg/kg): 5105 orally; 2800 i.m.; 1558 i.p. (Koch). LD₅₀ in mice, rats (mg/kg): 4228, 5576 orally; 3840, 2571 i.m. (China Cooperative Research Group on Qinghaosu).

Dihydro derivative, $C_{15}H_{24}O_5$, *dihydroartemisinin, dihydroqinghaosu.* Main metabolite of artemisinin, arteether, artemether, artesunate, *qqv.* HPLC determn: X.-D. Luo *et al., Chromatographia* **23**, 112 (1987).

THERAP CAT: Antimalarial.

857. Artesunate. *[3R-(3α,5aβ,6β,8aβ,9α,10α,12β,-12aR*)]-Butanedioic acid mono(decahydro-3,6,9-trimethyl-3,12-epoxy-12H-pyrano[4,3-j]-1,2-benzodioxepin-10-yl) ester;* artesunic acid; dihydroqinghaosu hemisuccinate; $C_{19}H_{28}O_8$; mol wt 384.43. C 59.36%, H 7.34%, O 33.30%. Derivative of artemisinin, *q.v.* Prepn: China Cooperative Research Group on Qinghaosu, *J. Tradit. Chinese Med.* **2**, 9 (1982). Absolute configuration: X.-D. Luo *et al., Helv. Chim. Acta* **67**, 1515 (1984). GC/mass spec. determn.: A. D. Theoharides *et al., Anal. Chem.* **60**, 115 (1988). Pharmacology: Y. Zhao, *J. Trop. Med. Hyg.* **88**, 391 (1985). Antimalarial activity: W. Peters *et al., Ann. Trop. Med. Parasitol.* **80**, 483 (1986); A. J. Lin *et al., J. Med. Chem.* **30**, 2147 (1987). Inhibition of cytochrome oxidase: Y. Zhao *et al., J. Nat. Prod.* **49**, 139 (1986). Toxicology: China Cooperative Research Group on Qinghaosu, *J. Tradit. Chinese Med.* **2**, 31 (1982). Series of articles on chemistry, pharmacology, and antimalarial efficacy: *ibid.* 3-50.

Fine white crystalline powder.
Sodium salt, $C_{19}H_{27}NaO_8$, *SM-804.* Poor stability in aqueous solutions. LD₅₀ in mice (mg/kg): 520 i.v.; 475 i.m. (China Cooperative Research Group); also reported as 699 ± 58.5 i.v. (Zhao, 1985).

THERAP CAT: Antimalarial.

858. Asafetida. Devil's dung; food of the gods; asafoetida; asant. Gum-resin obtained as an exudation of the decapitated rhizome and roots of *Ferula assafoetida* L., *F. foetida* Regel and some other species of *Ferula, Umbelliferae.* Habit. Iran, Turkestan, Afghanistan. *Constit.* 6-17% ethereal oil; 40-60% resin, consisting of ester of asaresinotannol and ferulic acid; pinene, vanillin, about 25% gum. Three sulfur-containing compounds isolated from asafetida resin are: *1-methylpropyl 1-propenyl disulfide, 1-(methylthio)propyl 1-propenyl disulfide* and *1-methylpropyl 3-(methylthio)-2-propenyl disulfide;* the latter two have pesticidal properties. Studies of various constituents: Bézanger-Beauquesne, Chosson, *Ann. Pharm. Franc.* **16**, 665 (1958); Caglioti *et al., Helv. Chim. Acta* **41**, 2278 (1958); **42**, 2557 (1959). Isoln and structure elucidation of three sulfur-containing components: H. Naimie *et al., Coll. Czech. Chem. Commun.* **37**, 1166 (1972). Synthesis and configuration of sulfur-containing constituents: J. Meijer, P. Vermeer, *Rec. Trav. Chim.* **93**, 242 (1974); A. Kjac *et al., Acta Chem. Scand. B* **30**, 137 (1976). *Review:* Howes, *Econ. Bot.* **4**, 313 (1950); Subrahmanyam *et al., J. Sci. Ind. Res.* **13A**, 382-386 (1954).

Soft mass or irregular mass or "tears"; garlic-like odor; bitter acrid taste. When triturated with water, it makes a milky emulsion.

USE: In India, Iran, etc., as condiment and flavoring for foods; an ingredient in Worcestershire sauce. A 2% suspension as repellent against dogs, cats, rabbits, deer.

THERAP CAT: Carminative; expectorant.

THERAP CAT (VET): Has been used as a carminative, and externally to prevent bandage chewing by dogs.

859. Asaprol. *2-Hydroxy-1-naphthalenesulfonic acid calcium salt;* calcium 2-hydroxy-1-naphthalenesulfonate; calcium β-naphthol-α-monosulfonate; calcium 2-naphthol-1-sulfonate; 2-naphthol-1-sulfonic acid calcium salt; Abrastol; Calcinaphthol. $C_{20}H_{14}CaO_8S_2$; mol wt 486.54. C 49.37%, H 2.90%, Ca 8.24%, O 26.31%, S 13.18%. Prepn: *Hagers Handb. Pharm. Praxis* **Band 2**, 204 (Berlin, 1930).

Trihydrate, reddish-white, odorless powder. Dec at about 50°. One gram dissolves in 1.5 ml water, 3 ml alc.

USE: Has been used instead of gypsum to plaster wines.

860. Asarinin. *[1R-(1α,3aα,4β,6aα)]-5,5'-(Tetrahydro-1H,3H-furo[3,4-c]furan-1,4-diyl)bis-1,3-benzodioxole; l-asarinin; (−)-episesamin; xanthoxylin S.* $C_{20}H_{18}O_6$; mol wt 354.36. C 67.79%, H 5.12%, O 27.09%. Naturally occurring *l*-form isolated from *Xanthoxylum clava-herculis* L. (X. carolinianum* Lam.), *Rutaceae; Asarum sieboldi* Miguel var. *seulensis* Nakai, *A. blumei* Duch., *Aristolochiaceae:* Colton, *Am. J. Pharm.* **52**, 191 (1880); Eberhardt, *ibid.* **62**, 231

(1890); Gordin, *J. Am. Chem. Soc.* **28**, 1649 (1906); H. Dieterle *et al., Arch. Pharm.* **269**, 384 (1931); Huang-Minlon, *Ber.* **70**, 951 (1937); T. Kaku *et al., Keijo J. Med.* **9**, 1 (1934); *C.A.* **32**, 9090[1] (1938). Structure: H. Dieterle, K. Schwenger, *Arch. Pharm.* **277**, 33 (1939). Synthesis of *dl*-form: M. Beroza, M. S. Schechter, *J. Am. Chem. Soc.* **78**, 1242 (1956); K. Freudenberg, E. Fischer, *Ber.* **89**, 1230 (1956); D. R. Stevens, D. A. Whiting, *Tetrahedron Letters* **27**, 4629 (1986). Diastereoisomeric with sesamin, *q.v.* Stereochemistry of the *d*-form: K. Freudenberg, G. S. Sidhu, *ibid.* **94**, 851 (1961). Antitubercular activity: Ramaswamy, *Naturwiss.* **44**, 380 (1957).

Crystals from alc, mp 121°. $[\alpha]_D^{20}$ −118.6°; $[\alpha]_D^{23}$ −122° (chloroform). Practically insol in water. Freely sol in boiling methanol, alcohol, chloroform, acetone, benzene.

d-Form, *episesamin, (+)-episesamin.* Crystals from ethanol, mp 121.5°. $[\alpha]_D^{25}$ +124° (chloroform).

dl-Form, crystals, mp 134-135°.

861. Asarones. *1,2,4-Trimethoxy-5-(1-propenyl)benzene; 2,4,5-trimethoxy-1-propenylbenzene; asarin; asarum camphor; asarabacca camphor.* $C_{12}H_{16}O_3$; mol wt 208.26. C 69.21%, H 7.74%, O 23.05%. From root of *Asarum europaeum* L., *Aristolochiaceae* by distillation with water. Also found in the ethereal oils of *A. europaeum* and *A. arifolium* L., *Aristolochiaceae* and in *Acorus calamus* L., *Araceae.* Occurs in nature as a mixture of two isomeric forms, *α-asarone* being the *(E)*- or *trans*-isomer, *β-asarone*, the *(Z)*- or *cis*-isomer. The unqualified term asarone is often used synonymously with α-asarone. Isoln: Gattermann, Eggers, *Ber.* **32**, 289 (1899). Early syntheses: Seshadri, Thiruvengadam, *Proc. Indian Acad. Sci.* **32A**, 110 (1950); Sharma, Dandiya, *Indian J. Appl. Chem.* **32**, 236 (1969). Stereochemistry of isomers: Baxter *et al., Can. J. Chem.* **40**, 154 (1962). Insect chemosterilant activity of β-asarone: B. P. Saxena *et al., Nature* **270**, 512 (1977); see also G. Motolesy *et al., Z. Naturforsch.* **35B**, 1449 (1980). Stereospecific synthesis of β-asarone: M. T. S. Hsia *et al.,* 177th Am. Chem. Soc. Meet. (Honolulu, April 1979), *Abstracts of Papers,* PEST 98. Synthetic and HPLC study of α- and β-asarone: L. Gracza, *Arch. Pharm.* **314**, 972 (1981).

α-Asarone   β-Asarone

α-Asarone, needles from light petroleum, mp 62-63°. bp 296°. n_D^{11} 1.5719. Practically insol in water; sol in alcohol, ether, glacial acetic acid, carbon tetrachloride, chloroform, petr ether.

862. Asarum. Wild ginger; Canada snakeroot; Indian ginger. Dried rhizome and roots of *Asarum canadense* L., *Aristolochiaceae. Habit.* Canada to N. Carolina and Kansas. *Constit.* Acrid resin, arom. volatile oil, methyl eugenol.

863. Asbestos. Amianthus. A family of naturally occurring, flexible, fibrous mineral silicates. Divided into two groups: *serpentine* and *amphibole.* Most common form is *chrysotile* $[Mg_6(Si_4O_{10})(OH)_8]$, the fibrous form of serpen-

tine (*see also* magnesium silicates). Subdivisions of amphibole are *anthophyllite* $[(Mg,Fe)_7(Si_8O_{22})(OH)_2]$ (low iron content); *amosite* $[Fe_5Mg_2(Si_8O_{22})(OH)_2]$; *actinolite* $[Ca_2(Mg,Fe)_5(Si_8O_{22})(OH)_2]$; *tremolite* $[Ca_2Mg_5(Si_8O_{22})(OH)_2]$; *crocidolite* or *blue asbestos* $[Na_2Fe_3^{2+}Fe_2^{3+}(Si_8O_{22})(OH)_2]$. Review of occupational use and carcinogenic risk: *IARC Monographs* **14**, 1-106 (1977); of toxicology and human exposure: *Toxicological Profile for Asbestos* (PB91-180497, 1990) 155 p. Review of properties and industrial applications: C. R. Jolicoeur *et al.* in Kirk-Othmer *Encyclopedia of Chemical Technology* vol. **3** (John Wiley & Sons, New York, 4th ed., 1992) pp 659-688; of electron microscopic analysis of airborne fibers: P. N. Breysse, *Crit. Rev. Anal. Chem.* **22**, 201-277 (1991). Symposium on asbestos-related disease: P. J. Landrigan, H. Kazemi, Eds., *Ann. N.Y. Acad. Sci.* **643**, 1-628 (1991).

Fire resistant fibers. Chrysotile attacked by acid; amphiboles, acid resistant.

Caution: The effects of respiratory exposure to asbestos are subacute or chronic and exhibit a latent period. Nonmalignant respiratory diseases attributable to asbestos exposure include chronic pulmonary fibrosis (asbestosis), fibrotic pleural plaques, pleuritis, and diffuse pleural thickening. Neoplastic diseases associated with occupational exposure to airborne asbestos include lung cancer and mesothelioma. *See Patty's Industrial Hygiene and Toxicology* vol. **2A**, G. D. Clayton, F. E. Clayton, Eds. (Wiley-Interscience, New York, 4th ed., 1993) pp 849-864. Potential symptoms of overexposure are dyspnea, interstitial fibrosis, restricted pulmonary function and finger clubbing. *See NIOSH Pocket Guide to Chemical Hazards* (DHHS/NIOSH 90-117, 1990) p 42. This substance has been listed as a known carcinogen: *Seventh Annual Report on Carcinogens* (PB95-109781, 1994) p 26.

USE: Heat-resistant insulators, cements, furnace and hot pipe coverings, inert filler medium (laboratory & commercial), fireproof gloves, clothing, brake linings. NaOH treated asbestos, *Ascarite,* has been used to absorb CO_2 in combustion analysis.

864. Ascaridole. *1-Methyl-4-(1-methylethyl)-2,3-dioxabicyclo[2.2.2]oct-5-ene; Ascarisin.* $C_{10}H_{16}O_2$; mol wt 168.24. C 71.39%, H 9.59%, O 19.02%. An organic peroxide which constitutes 60-80% of oil of chenopodium. Synthesis from α-terpinene by treatment with oxygen, chlorophyll, and light: Schenck, Ziegler, *Naturwiss.* **1944**, 157. Purification: Beckett *et al., J. Pharm. Pharmacol.* **7**, 55 (1955).

Liquid; unstable; prone to explode when heated or when treated with organic acids. d_{20}^{20} 1.0103; d_{20}^{20} 1.0113. mp +3.3°. $bp_{0.2}$ 39-40°. $[\alpha]_D^{20}$ ±0.00. Sol in hexane, pentane, ethanol, toluene, benzene, castor oil.

THERAP CAT: Has been used as anthelmintic (Nematodes).
THERAP CAT (VET): Anthelmintic.

865. Asclepias. Pleurisy root; butterfly weed. Dried root of *Asclepias tuberosa* L., *Asclepiadaceae. Habit.* Ontario to Minnesota. *Constit.* Asclepiadin, resins, volatile oil.

866. Asclepias syriaca. Milkweed; silkweed; wild cotton. Root of *Asclepias syriaca* L. (*A. cornuti* Decaisne), *Asclepiadaceae. Habit.* Canada to North Carolina and Kansas. *Constit.* Asclepiadin, asclepion—a bitter principle; tannin, volatile oil.

867. Ascorbic Acid. L-*Ascorbic acid;* vitamin C; L-xyloascorbic acid; 3-oxo-L-gulofuranolactone; L-3-ketothreohexuronic acid lactone; antiscorbutic vitamin; cevitamic acid; Adenex; Allercorb; Ascorin; Ascorteal; Ascorvit; Cantan; Cantaxin; Catavin C; Cebicure; Cebion; Cecon; Cegiolan; Celaskon; Celin; Cenetone; Cereon; Cergona; Cescorbat; Cetamid; Cetebe; Cetemican; Cevalin; Cevatine; Cevex; Cevimin; Ce-Vi-Sol; Cevitan; Cevitex; Cewin; Ciamin; Cipca; Concemin; C-Vimin; Davitamon C; Duoscorb; Hybrin; Laroscorbine; Lemascorb; Planavit C; Proscorbin; Redoxon; Ribena; Scorbacid; Scorbu-C; Testascorbic; Vicelat; Vitacee; Vitacimin; Vitacin; Vitascorbol; Xitix. $C_6H_8O_6$; mol wt 176.13. C 40.92%, H 4.58%, O 54.50%. Widely distributed in the plant and animal kingdom. Good sources are citrus fruits, hip berries, acerola, fresh tea leaves. Isolated from the adrenal cortex of ox and later from lemons and paprika (originally called hexuronic acid): Szent-Györgyi, *Biochem. J.* **22**, 1387 (1928); Haworth, Szent-Györgyi, *Nature* **131**, 24 (1933). Structure studies: Herbert *et al., J. Chem. Soc.* **1933**, 1270. Crystal structure: Hvoself, *Acta Chem. Scand.* **18**, 841 (1964). Synthesis: Ault *et al., J. Chem. Soc.* **1933**, 1419; Reichstein *et al., Helv. Chim. Acta* **16**, 561, 1019 (1933); **17**, 311, 510 (1934); Bakke, Theander, *Chem. Commun.* **1971**, 175; R. J. Ferrier, R. H. Furneaux, *ibid.* **1977**, 332; T. C. Crawford, R. Breitenbach, *ibid.* **1979**, 388; G. Andrews *et al., ibid.* 740. Review of syntheses: T. C. Crawford, S. A. Crawford, *Advan. Carbohyd. Chem.* **37**, 79-155 (1980). *Reviews:* E. L. Hirst, *Fortschr. Chem. Org. Naturst.* **2**, 132-159 (1939); H. R. Rosenberg, *Chemistry and Physiology of the Vitamins* (Interscience, New York, 1945); *Ann. N.Y. Acad. Sci.* **92**, entitled "Vitamin C", J. J. Burns, Ed. (1961) pp 1-332; A. K. Sim, *Chem. & Ind. (London)* **1972**, 160-165; *Ann. N.Y. Acad. Sci.* **258**, entitled "Second Conference on Vitamin C", C. G. King, J. J. Burns, Eds. (1975) pp 1-552. Discussion of the use of ascorbic acid in the treatment of the common cold: Karlowski *et al., J. Am. Med. Assoc.* **231**, 1038 (1975); Dykes, Meier, *ibid.* 1073; Pauling, *Med. Trib.* **17**, 1, 18 (March 24, 1976); 37 (April 7, 1976). Clinical applications in immunology, lipid metabolism and cancer: *Int. J. Vit. Nutr. Res.* **1982**, Suppl. 23, 294 pp. Comprehensive description: I. A. Al-Meshal, M. M. A. Hassan, *Anal. Profiles Drug Subs.* **11**, 45-78 (1982). Review of biochemistry, physiology and clinical uses: M. Levine, *N. Engl. J. Med.* **314**, 892-902 (1986).

Crystals (usually plates, sometimes needles, monoclinic system). mp 190-192° (some dec). Pleasant, sharp acidic taste. Stable to air when dry. In impure prepns and in many natural products the vitamin oxidizes on exposure to air and light. d 1.65. $[\alpha]_D^{25}$ +20.5° to +21.5° (c = 1); $[\alpha]_D^{23}$ +48° (c = 1 in methanol). pH = 3 (5 mg/ml); pH = 2 (50 mg/ml); pK_1 = 4.17; pK_2 = 11.57. uv max: 245 nm (acid soln); 265 nm (neutral soln). Redox potential of first stage at pH 5.0 is ϵ_0' = +0.127 v. One gram dissolves in about 3 ml water, 30 ml alc, 50 ml abs alc, 100 ml glycerol U.S.P., 20 ml propylene glycol. Soly in water: 80.0% at 100°; 40.0% at 45°. Insol in ether, chloroform, benzene, petr ether, oils, fats, fat solvents. Possesses relatively strong reducing power, decolorizes many dyes. Aq solns are rapidly oxidized by air. The reaction is accelerated by alkalies, iron, copper. Forms stable metal salts, *see* Sodium Ascorbate.

Calcium hypophosphite, $C_6H_9CaO_8P$, *asphocalcium, Calscorbat.*

Note: One unit (U.S.P. or international) is the vitamin C activity of 0.05 mg of the U.S.P. ascorbic acid reference standard.

Pharmaceutical Incompat. Vitamin C should not be formulated with sodium salicylate, sodium nitrite, theobromine sodium salicylate, methenamine: R. K. Aliev, M. A. Etinger, *Aptechnoe Delo* **2**, 7 (1952), *C.A.* **46**, 7707b (1952).

USE: As antimicrobial and antioxidant in foodstuffs.

THERAP CAT: Treatment of vitamin C deficiency.

THERAP CAT (VET): Treatment of vitamin C deficiency in primates, guinea pigs, fish.

868. Ascorbigen. *2-C-(1H-Indol-3-ylmethyl)-β-L-lyxo-3-hexulofuranosonic acid γ-lactone mixt with 2-C-(1H-indol-3-ylmethyl)-β-L-xylo-3-hexulofuranosonic acid γ-lactone.* $C_{15}H_{15}NO_6$; mol wt 305.29. C 59.01%, H 4.95%, N 4.59%, O 31.44%. Isoln from savoy cabbage juice: Procházka, Sanda, *Coll. Czech. Chem. Commun.* **25**, 270 (1960). Prepn from L-ascorbic acid + glucobrassicin: Gmelin, Virtanan, *Suomen Kemistilehti* **34B**, 15 (1961), *C.A.* **55**, 17774f (1961); from L-ascorbic acid + gramine methiodide: Procházka, *Coll. Czech. Chem. Commun.* **28**, 544 (1963). Prepn from L-ascorbic acid + 3-hydroxymethylindole; forms two diastereoisomers: *ascorbigen A* as the main product and *ascorbigen B* as a minor product: Kiss, Neukom, *Helv. Chim. Acta* **49**, 989 (1966); Kiss, *Angew. Chem.* **78**, 1066 (1966).

Ascorbigen A　　　　　　　　　　Ascorbigen B

Ascorbigen A, amorphous powder, sinters about 65°. $[\alpha]_D^{25}$ +11.0° (c = 2 in ethanol). uv max (ethanol): 220, 273-274, 280, 290 nm.

Ascorbigen B, yellow amorphous powder, sinters about 70°. $[\alpha]_D^{25}$ +12.5° (methanol).

869. Asiaticoside. *(2α,3β,4α)-2,3,23-Trihydroxyurs-12-en-28-oic acid O-6-deoxy-α-L-mannopyranosyl-(1→4)-O-β-D-glucopyranosyl-(1→6)-O-β-D-glucopyranosyl ester;* Centelase Dermatologico; Madecassol. $C_{48}H_{78}O_{19}$; mol wt 959.14. C 60.11%, H 8.20%, O 31.69%. Active principle of *Centella asiatica* (L.) Urban, *Umbelliferae.* Trisaccharide moiety linked to the aglycone: *asiatic acid.* Isoln: J.-E. Bontems, *Bull. Sci. Pharmacol.* **49**, 186 (1941), *C.A.* **38**, 4094 (1944); *idem, Gaz. Med. Madagascar* **5**, 29 (1942). Structure of asiatic acid: J. Polonsky, *Bull. Soc. Chim. France* **1953**, 173. Structure of asiaticoside: P. Boiteau *et al., Nature* **163**, 258 (1949); J. Polonsky *et al., Bull. Soc. Chim. France* **1959**, 880. Metabolism: L. F. Chasseaud *et al., Arzneimittel-Forsch.* **21**, 1379 (1971). Wound healing properties: H. Rosen *et al., Proc. Soc. Exp. Biol. Med.* **125**, 279 (1967). Clinical study: J.-P. Bosse *et al., Ann. Plast. Surg.* **3**, 13 (1979). Review on asiatic acid: J. L. Simonsen, W. C. J. Ross, *The Terpenes* vol. **5** (University Press, Cambridge, 1957) pp 58-67.

Minute needles from 60% methanol, mp 230-233°. Insol in water. Sol in alcohol, pyridine. $[\alpha]_D^{20}$ − 14° (alc).

THERAP CAT: Vulnerary.

870. Asoxime Chloride. *1-[[[4-(Aminocarbonyl)pyridinio]methoxy]methyl]-2-[(hydroxyimino)methyl]pyridinium dichloride; 4′-carbamoyl-2-formyl-1,1′-(oxydimethylene)-dipyridinium chloride 2-oxime; 1-(2-hydroxyiminomethyl-1-pyridinio)-3-(4-carbamoyl-1-pyridinio)-2-oxapropane dichloride; HI-6.* $C_{14}H_{16}Cl_2N_4O_3$; mol wt 359.21. C 46.81%, H 4.49%, Cl 19.74%, N 15.60%, O 13.36%. Cholinesterase reactivator. Prepn: I. Hagedorn, **U.S. pat. 3,773,775** (1973 to E. Merck); L. Y. Y. Hsiao, H. A. Musallam, U. S. pat. **5,130,438** (1992 to U.S.A. Secy. of Army). Chemical stability study: P. Eyer *et al., Arch. Toxicol.* **59**, 266 (1986). HPLC determn in plasma: M. P. McCluskey *et al., J. Anal. Toxicol.* **14**, 239 (1990). Metabolism study: D. J. Ecobichon *et al., Can. J. Physiol. Pharmacol.* **68**, 614 (1990). Antidotal activity vs soman, *q.v.*: T.-M. Shih *et al., Toxicol. Lett.* **55**, 131 (1991). Clinical pharmacokinetics: R. Kusic *et al., Fund. Appl. Toxicol.* **5**, S89 (1985). Therapeutic use: R. Kusic *et al., Hum. Exp. Toxicol.* **10**, 113 (1991). Review of pharmacology: C. G. Rousseaux, A. K. Dua, *Can. J. Physiol. Pharmacol.* **67**, 1183-1189 (1989); of physical properties and acute toxicity: J. G. Clement *et al., Arch. Toxicol.* **62**, 220-223 (1988).

Monohydrate, crystals from water/ethanol (20:75), mp 145-147°. uv max (H_2O): 350, 300, 270, 218 nm (log ε 3.093, 4.018, 3.966, 4.181).

THERAP CAT: Antidote (organophosphate poisoning).

871. L-Asparaginase. L-Asparagine amidohydrolase; colaspase; L-asnase; E.C. 3.5.1.1; MK-965; NSC-109229; Crasnitin; Elspar; Kidrolase; Leunase. Enzyme which catalyzes the hydrolysis of L-asparagine to L-aspartic acid and ammonia. Inhibits the growth of tumor cells requiring exogenous asparagine. Widely distributed; occurs in bacteria, fungi, yeasts, plants, animal tissues and guinea pig serum. Commonly obtained from *E. coli* and *Erwinia* sp. Asparaginase from *E. coli* has a mol wt of ~136,000 and is composed of four identical subunits. Identification of antitumor activity: J. D. Broome, *Nature* **191**, 1114 (1961); *idem, J. Exp. Med.* **118**, 99 (1963). Purif, chemical properties of asparaginase from *E. coli*: D. H. Ho *et al., J. Biol. Chem.* **245**, 3708 (1970). Structural studies: A. C. Greenquist, J. C. Wriston, *Arch. Biochem. Biophys.* **152**, 280 (1972). Amino acid sequence: T. Maita *et al., J. Biochem.* **76**, 1351 (1974). Pharmacology and clinical effects: R. H. Adamson, S. Fabro, *Cancer Chemother. Rep.* **52**, part I, 617 (1968); R. L. Capizzi *et al., Ann. Rev. Med.* **21**, 433 (1970). *Reviews:* R. L. Capizzi, *Leuk. Lymphoma* **10**, Suppl., 147-150 (1993); M. J. Keating *et al., ibid.* 153-157.

White crystalline powder. Freely sol in water where it appears to be global in shape. $[\alpha]_D^{20}$ − 30° to − 32°. uv max (0.03M sodium phosphate, pH 7.3): 278 nm ($A_{1cm}^{1\%}$ 7.1 ±0.15). Active at pH 5-9. Practically insol in methanol, acetone, chloroform.

Compd with polyethylene glycol, *pegaspargase, PEG-asparaginase, (monomethoxypolyethylene glycol succinimidyl)$_{74}$-L-asparaginase, Oncaspar.* L-Asparaginase from *E. coli* covalently coupled to polyethylene glycol moieties of mol wt ~5000 Da. Prepn: F. F. Davis *et al.,* **Ger. pat. 2,433,883** (1976 to Research Corp.); *eidem,* **U.S. pat. 4,179,337** (1979). Prepn, pharmacology and antitumor efficacy: A. Abuchowski *et al., Cancer Biochem. Biophys.* **7**, 175 (1984). Toxicology: A. T. Viau *et al., Am. J. Vet. Res.* **47**, 1398 (1986). Review of clinical experience: F. Fuertges, A. Abuchowski, *J. Controlled Release* **11**, 139 (1990).

THERAP CAT: Antineoplastic (acute leukemia).

872. Asparagine. *L-Asparagine; Asn; N; L-β-asparagine; α-aminosuccinamic acid; aspartic acid β-amide; al-*theine; asparamide; agedoite; (S)-2,4-diamino-4-oxobutanoic acid. $C_4H_8N_2O_3$; mol wt 132.12. C 36.36%, H 6.10%, N 21.20%, O 36.33%. Non-essential amino acid for human development. First amino acid noted in natural sources, it was named for the asparagus juice from which it was isolated: L. N. Vauquelin, P. J. Robiquet, *Ann. Chim.* **57**, 88 (1806). Due to ease of conversion to aspartic acid, *q.v.,* it was not isolated from protein until 1932: M. Damodoran, *Biochem. J.* **26**, 235 (1932). Early chemistry and biochemistry: *Amino Acids and Proteins,* D. M. Greenberg, Ed. (Charles C. Thomas, Springfield, IL, 1951) 950 pp, *passim;* J. P. Greenstein, M. Winitz, *Chemistry of the Amino Acids* vols. **1-3** (John Wiley & Sons, Inc., New York, 1961) pp. 1856-1878, *passim.* As attachment point for saccharides: A. J. Mencke *et al., Methods Enzymol.* **138**, 409 (1987). Colorimetric assay: S. Sheng *et al., Anal. Biochem.* **211**, 242 (1993). Review of metabolism: D. A. Cooney, R. E. Handschumacher, *Ann. Rev. Pharmacol.* **10**, 421-440 (1970); in plants: K. A. Sieciechowicz *et al., Phytochemistry* **27**, 663-671 (1988). Review of deamidation in proteins: R. Bischoff, H. V. J. Kolbe, *J. Chromatog. B* **662**, 261-278 (1994).

Orthorhombic bisphenoidal crystals, mp 234-235° (bath preheated to 226°). d_4^{15} 1.543. Acid to litmus. pK_1 2.02; pK_2 8.80. $[\alpha]_D^{20}$ − 5.30° (c = 1.41); $[\alpha]_D^{20}$ +34.26° (c = 2.24 in 3.4N HCl); $[\alpha]_D^{20}$ − 6.35° (c = 11.23 in 2.5N NaOH). Practically insol in methanol, ethanol, ether, benzene. Sol in acids, alkalies.

D-Form. In peptidoglycans of bacterial cell walls: D. A. Reaveley, R. E. Burge, *Adv. Microbial Physiol.* **7**, 1-81 (1972).

D-Form monohydrate, crystals, mp 215°. $[\alpha]_D^{20}$ +5.41° (c = 1.3).

873. Asparagus. Shoot of *Asparagus officinalis* L., *Liliaceae. Habit.* Europe, cultivated everywhere. *Constit.* Asparagine, tyrosine, succinic acid, arginine, α-aminodimethyl-γ-butyrothetin (a methylsulfonium deriv of methionine), fat, sugar. In some humans, ingestion of asparagus is followed by excretion of a substance that produces a characteristic strong odor in the urine: M. Nencki, *Arch. Exp. Pathol. Pharmacol.* **28**, 206 (1891); C. Gautier, *C.R. Soc. Biol.* **89**, 239 (1923); A. C. Allison, K. G. McWhirter, *Nature* **178**, 748 (1956). The odor-causing substance was originally thought to be methyl mercaptan, but subsequent investigation using GC-mass spectrometry has suggested that S-methylthioacrylate and S-methyl 3-(methylthio)thiopropionate are the malodorous agents, *cf* R. H. White, *Science* **189**, 810 (1975).

874. Aspartame. *N-L-α-Aspartyl-L-phenylalanine 1-methyl ester; 3-amino-N-(α-carboxyphenethyl)succinamic acid N-methyl ester; APM; SC-18862; Canderel; Equal; NutraSweet; Sanecta; Tri-Sweet.* $C_{14}H_{18}N_2O_5$; mol wt 294.31. C 57.14%, H 6.16%, N 9.52%, O 27.18%. Dipeptide ester about 160 times sweeter than sucrose in aqueous solution. Prepn: Davey *et al., J. Chem. Soc. (C)* **1966**, 555; J. M. Schlatter, S. Afr. pat. **67 02,190**; *idem,* **U.S. pat. 3,492,131** (1968, 1970 both to Searle); H. Pietsch, *Tetrahedron Letters* **1976**, 4053; K. J. Vinick, S. Jung, *ibid.* **23**, 1315 (1982); utilizing immobilized enzyme technology: C. Fuganti, P. Grasselli, *ibid.* **27**, 3191 (1986). Structure-taste relationship: Mazur *et al., J. Am. Chem. Soc.* **91**, 2684 (1969). Potential as a low-calorie sweetener: Cloninger, Baldwin, *Science* **170**, 81 (1970). Metabolism: Oppermann *et al., J. Nutr.* **103**, 1454, 1460 (1973).

Colorless needles from water, mp 246-247°. $[\alpha]_D^{22}$ −2.3° (1N HCl).

USE: Sweetener.

875. Aspartic Acid. *L-Aspartic acid;* Asp; D; aminosuccinic acid; asparagic acid; asparaginic acid; *(S)*-aminobutanedioic acid; 1-amino-1,2-carboxyethane; Asparaginsäure (German). $C_4H_7NO_4$; mol wt 133.10. C 36.10%, H 5.30%, N 10.52%, O 48.08%. Non-essential amino acid for human development. First identified as the acid hydrolysis product of asparagine, *q.v.:* A. Plisson, *J. Pharm.* **13**, 477 (1827); later isolated from protein: H. Ritthausen, *J. Prakt. Chem.* **103**, 233, 239 (1868). Early chemistry and biochemistry: *Amino Acids and Proteins,* D. M. Greenberg, Ed. (Charles C. Thomas, Springfield, IL, 1951) 950 pp., *passim*; J. P. Greenstein, M. Winitz, *Chemistry of the Amino Acids* vol 1-3 (John Wiley and Sons, Inc., New York, 1961) pp. 1856-1878, *passim*. Synthesis of optically active forms: K. Harada, K. Matsumoto, *J. Org. Chem.* **31**, 2985 (1966); of labelled form: A. Ivanof *et al., Anal. Biochem.* **110**, 267 (1981). Crystal structure: B. Dawson, *Acta Crystallogr.* **33B**, 882 (1977). Excitatory neurotransmitter: H. McLennan, H. V. Wheal, *Can. J. Physiol. Pharmacol.* **54**, 70 (1976). Role in protein activity: D. W. Urry *et al., Biopolymers* **34**, 889 (1994); J. M. Denu *et al., Biochemistry* **34**, 3396 (1995). Racemization with age in dentin: N. Saleh *et al., Calcif. Tissue Int.* **53**, 103 (1993); in crystallins, *q.v.:* N. Fujii *et al., Biochim. Biophys. Acta* **1204**, 157 (1994). Review of metabolism and pathology: L. D. Stegink, *J. Toxicol. Environ. Health* **2**, 215-242 (1976). Brief review of role of excitatory amino acid: P. J. Roberts, S. W. Davies, *Biochem. Soc. Trans.* **15**, 218-219 (1986).

Orthorhombic bisphenoidal leaflets or rods, mp 270-271° (sealed capillary, preheated bath). $[\alpha]_D^{20}$ +25.0° (c = 1.97 in 6N HCl). $d^{12.5}$ 1.661. pK_1' 1.88; pK_2' 3.65; pK_3' 9.60. Soly in water at 20°, 30°: 1 g/222.2 ml, 1 g/149.9 ml. Forms supersatd solns easily. More soluble in salt solns; sol in acids, alkalies. Insol in alcohol.

D-Form, *D-Asp*. Occurs naturally although to a lesser extent than the L-form. Purification and characterization from bovine lens: N. Fujii *et al., Biochim. Biophys. Acta* **999**, 239 (1989); from tissue of the mollusc, *Aplysia kurodai:* M. Sato *et al., Biochem. J.* **263**, 617 (1989). Review of industrial production by microorganisms: S. Takamatsu, T. Tosa, *Bioproc. Technol.* **16**, 25-35 (1993). $[\alpha]_D^{25}$ −2.0° (c = 3.93 in 5N HCl).

Potassium salt (L-form), *Aspara K.*

Mixture of the potassium salt and magnesium salt, *Spartase, Trophicard, Aspara, Aspartat.*

Ferrous salt tetrahydrate, $C_8H_{12}FeN_2O_8.4H_2O$, *ferrous aspartate, Sideryl, Spartocine.* Prepn: Gaudin, **Brit.** pat. 910,321 (1962).

Calcium salt, *calcium aspartate, asparaginate calcium, Calciretard.*

Compound with L-arginine, $C_{10}H_{21}N_5O_6$, *arginine L-aspartate, Dynamisan, Sargenor.*

Compound with L-ornithine, *see* Ornithine.

N-Acetyl-L-aspartic acid, $C_6H_9NO_5$, *NAA.* Found in nervous system; second most abundant free amino acid in mammals. Review: D. L. Birken, W. H. Oldendorf, *Neurosci. Biobehav. Rev.* **13**, 23-31 (1989).

THERAP CAT: Roborant.

876. Aspergillic Acid. *1-Hydroxy-6-(1-methylpropyl)-3-(2-methylpropyl)-2(1H)-pyrazinone;* 6-*sec*-butyl-1-hydroxy-3-isobutyl-2(1*H*)-pyrazinone; 6-*sec*-butyl-3-isobutylpyrazinol 1-oxide; 2-hydroxy-3-isobutyl-6-(1-methylpropyl)pyrazine 1-oxide; 3-isobutyl-6-*sec*-butyl-2-hydroxypyrazine 1-oxide. $C_{12}H_{20}N_2O_2$; mol wt 224.30. C 64.26%, H 8.99%, N 12.49%, O 14.27%. Antibiotic substance produced by *Aspergillus flavus:* White, Hill, *J. Bacteriol.* **45**, 433 (1943); Dunn *et al., J. Chem. Soc.* **1949**, Suppl. S126. Extraction and purification: Dutcher, *J. Biol. Chem.* **171**, 321 (1947).

Structure: *idem, ibid.* **232**. 785 (1958). Biosynthesis: MacDonald, *ibid.* **236**, 512 (1961), **237**, 1977 (1962). Synthesis of racemate: Chigira *et al., Bull. Chem. Soc. Japan* **39**, 632 (1966); Masaki *et al., J. Org. Chem.* **31**, 4143 (1966); Ohta, Fujii, *Chem. Pharm. Bull.* **17**, 851 (1969). Industrial prepn: Omata, Ueno, **Japan.** pat. **13,794**('65) (to Dainippon), *C.A.* **63**, 11589b (1965). Proposed as a hypotensive: Jones, Martin, **U.S.** pat. **3,720,768** (1973 to Abbott). *Reviews:* MacDonald, "Aspergillic Acid and Related Compounds" in *Antibiotics,* D. Gottlieb, P. Shaw, Eds. (Springer-Verlag, New York, 1967) vol. II, pp 43-51; Wilson, "Miscellaneous Aspergillus Toxins", in *Microbial Toxins,* A. Ciegler *et al.,* Eds. (Academic Press, New York, 1971) vol. VI, pp 207-235.

Pale yellow needles having odor similar to black walnuts, mp 97-99° (methanol). pK'a 5.5. $[\alpha]_D^{18}$ +13.3° (c = 3.9 in ethanol). uv max (water pH 8): 328, 235 nm (ε 8500, 10500). Slightly sol in cold water; sol in dil acids and alkalies, alcohol, ether, acetone, benzene, chloroform, pyridine.

877. Aspergillin. The inappropriate designation of a number of different antibacterial agents produced by *Aspergilli* as "aspergillin" has resulted in confusion. The legitimate contender for this name by prior use is an allomelanin produced by mature spores of *A. niger:* Hugouneng, Florence, *Bull. Soc. Chim. Biol.* **2**, 133 (1920). This pigment is a mixture of chemically similar macromolecules of which the fundamental monomer is probably the perylene unit substituted with several oxygenated functions. *Review:* R. A. Nicolaus, *Melanins* (Hermann, Paris, 1968) pp 130-142.

Dried pigment, as shiny black blocks, does not melt but decomposes above 200°. Slightly sol in polar solvents. uv max (2% aqueous soln): 295, 450 nm.

878. Asperlicin. *[2S-[2α,9β,9(R*),9aβ]]-6,7-Dihydro-7-[[2,3,9,9a-tetrahydro-9-hydroxy-2-(2-methylpropyl)-3-oxo-1H-imidazo[1,2-a]indol-9-yl]methyl]quinazolino[3,2-a][1,4]-benzodiazepine-5,13-dione.* $C_{31}H_{29}N_5O_4$; mol wt 535.60. C 69.52%, H 5.46%, N 13.08%, O 11.95%. Naturally occurring, nonpeptide cholecystokinin (CCK) antagonist. Produced by several strains of *Aspergillus alliaceus* along with minor related compounds known as asperlicin B, C, D and E. Isoln from fermentation cultures of *A. alliaceus* Thom and Church, ATCC 20655 and ATCC 20656: R. L. Monaghan *et al.,* **Eur.** pat. **Appl. 116,150**; *eidem,* **U.S.** pat. **4,530,790** (1984, 1985 both to Merck & Co.). Fermentation, isoln and bioactivity: M. A. Goetz *et al., J. Antibiot.* **38**, 1633 (1985); of asperlicins B, C, D and E: *eidem, ibid.* **41**, 875 (1988). Structure elucidation: J. M. Liesch *et al., ibid.* **38**, 1638 (1985); of asperlicins B, C, D and E: *eidem, ibid.* **41**, 878 (1988). Biosynthetic study: D. R. Houck *et al., ibid.* 882. Total synthesis of asperlicins C and E: M. G. Bock *et al., J. Org. Chem.* **52**, 1644 (1987). Pharmacology of asperlicin: R. S. L. Chang *et al., Science* **230**, 177 (1985). Effect on pancreatic enzyme secretion *in vitro:* K. A. Zucker *et al., Surgery* **102**, 163 (1987). Effect in exptl pancreatitis in rats: J. R. Wisner, Jr., I. G. Renner, *Pancreas* **3**, 174 (1988).

White crystals, mp 211-213°. $[\alpha]_D^{26.5}$ −185.3° (c = 1.10 in methanol). uv max (methanol): 310.5 nm (ε 4075). Sol in methylene chloride, acetone and lower alcohols. Insol in water.

Asperlicin B, $C_{31}H_{29}N_5O_5$, colorless powder. uv max (methanol): 310 nm (ε 4000).

Asperlicin C, $C_{25}H_{18}N_4O_2$, off-white powder. uv max (methanol): 222, 268, 278, 310 nm (ε 56300, 15100, 14000, 4650).

Asperlicin D, $C_{25}H_{18}N_4O_2$, colorless powder. uv max (methanol): 222, 290, 310 nm (ε 61060, 14600, 5920).

Asperlicin E, $C_{25}H_{18}N_4O_3$, off-white powder. uv max (methanol): 227, 268, 324 nm (ε 38450, 8680, 3200).

879. Asperuloside. *[2aS-(2aα,4aα,5α,7bα)]-4-[(Acetyloxy)methyl]-5-(β-D-glucopyranosyloxy)-2a,4a,5,7b-tetrahydro-1H-2,6-dioxacyclopent[cd]inden-1-one;* rubichloric acid. $C_{18}H_{22}O_{11}$; mol wt 414.37. C 52.18%, H 5.35%, O 42.47%. From herb of *Asperula odorata* L., *Galium aparine* L., *Rubiaceae.* Occurs also in *Coprosma* spp. Isoln: Hérissey, *Compt. Rend.* **180**, 1695 (1925); Briggs, Nicholls, *J. Chem. Soc.* **1954**, 3940. Structure: Grimshaw, *Chem. & Ind. (London)* **1961**, 403; Briggs *et al., J. Chem. Soc.* **1965**, 2595.

Needles from alcohol or acetone, mp 131-132°. $[\alpha]_D^{25}$ −198.6° (c = 1.44 in water). Absorption spectra: Briggs, Cain, *J. Chem. Soc.* **1954**, 4182. Sol in water, methanol, ethanol, acetone, ethylacetate, dioxane, pyridine, acetic acid. Practically insol in ether, benzene, chloroform, ligroin.

880. Asphalt. Asphaltum; mineral pitch; Judean pitch; bitumen. Bituminous substance resulting from petroleum by evaporation of lighter hydrocarbons and partial oxidation of the residue. Occurs in West Indies (chiefly Trinidad), Venezuela, Dead Sea, Switzerland, etc.

The "Syriac" asphalt (from the Dead Sea) forms deep black, shining, brittle masses of conchoidal fracture; faint, pitch-like odor and luster. Burns with a bright flame. d 1.00-1.18. Insol in water, alc, acids, alkalies; sol in oil turpentine, petroleum, CS_2, chloroform, ether, acetone.

USE: Making roads, roofs; making tanks watertight.

881. Aspidin. *2-[[2,6-Dihydroxy-4-methoxy-3-methyl-5-(1-oxobutyl)phenyl]methyl]-3,5-dihydroxy-4,4-dimethyl-6-(1-oxobutyl)-2,5-cyclohexadien-1-one; 3'-[(5-butyryl-2,4-dihydroxy-3,3-dimethyl-6-oxo-1,4-cyclohexadien-1-yl)methyl]-2',4'-dihydroxy-6'-methoxy-5'-methylbutyrophenone;* polystichin. $C_{25}H_{32}O_8$; mol wt 460.52. C 65.20%, H 7.00%, O 27.79%. Active principle of fern root: Boehm, *Ann.* **302**, 171 (1898); **329**, 321 (1903); from *Dryopteris austriaca* (Jacq.) Woynar, *Polypodiaceae:* Aebi *et al., Helv. Chim. Acta* **40**, 266 (1957). Structure: *eidem, ibid.* 569. Synthesis: Riedl, Mitteldorf, *Ber.* **89**, 2595 (1956).

Crystals from ethanol, mp 124-125°. uv max (cyclohexane): 230, 290 nm (ε 25500, 21300). Soluble in ether,

benzene, chloroform; less sol in petr ether; sparingly sol in methanol, ethanol, acetone.

THERAP CAT: Anthelmintic (Cestodes).

882. Aspidinol. *1-(2,6-Dihydroxy-4-methoxy-3-methylphenyl)-1-butanone; 2',6'-dihydroxy-4'-methoxy-3'-methyl-1-butyrophenone;* 4-butyryl-2-methylphloroglucinol 1-methyl ether; 4-butyryl-3,5-dihydroxy-1-methoxy-2-methylbenzene. $C_{12}H_{16}O_4$; mol wt 224.26. C 64.27%, H 7.19%, O 28.54%. Occurs in extracts of male fern: Boehm, *Ann.* **318**, 247 (1901); Hausmann, *Arch. Pharm.* **237**, 559 (1899). Isoln from *Dryopteris austriaca* (Jacq.) Woynar, *Polypodiaceae:* Aebi *et al., Helv. Chim. Acta* **40**, 266 (1957). Synthesis from 2-methylphloroglucinol 1-methyl ether: Karrer, Widmer, *Helv. Chim. Acta* **3**, 392 (1920); Riedl, Mittledorf, *Ber.* **89**, 2589 (1956).

Needles or prisms from benzene, mp 156-161°. Freely sol in alc, ether, chloroform, acetone; sparingly sol in water, benzene. Less soluble in ligroin than pseudoaspidinol, mp 116.5°. Sol in NaOH solns. Practically insol in Na_2CO_3 solns.

THERAP CAT: Anthelmintic (Cestodes).

883. Aspidium. Male fern; male shield-fern; filix mas (B.P.). Rhizome and stipes of *Dryopteris filix-mas* (L.) Schott., *Polypodiaceae.* *Habit.* North America, Northern Asia, Europe, Northern Africa. *Constit.* Filicic and flavaspidic acids, volatile oil, asbaspidin, filicin, filmaron, filix red, resin. It yields not less than 6.5% oleoresin (U.S.P.). *See also:* Oleoresin of Aspidium.

THERAP CAT: Anthelmintic (Cestodes).

THERAP CAT (VET): Anthelmintic (Cestodes).

884. Aspidosperma. Quebracho. Dried bark of *Aspidosperma quebracho-blanco* Schlecht., *Apocynaceae.* *Habit.* Argentina. *Constit.* 0.3-1.4% alkaloids—aspidospermine, aspidospermatine, aspidosamine, quebrachine, quebrachamine, hypoquebrachine, quebrachol, quebrachit; tannin.

USE: Chlorinated quebracho can be used to control nematodes and other parasitic worms in soils: Santmyer, U.S. pat. **2,799,612** (1957 to Monsanto).

THERAP CAT: Respiratory stimulant.

885. Aspidospermine. *1-Acetyl-17-methoxyaspidospermidine.* $C_{22}H_{30}N_2O_2$; mol wt 354.49. C 74.54%, H 8.53%, N 7.90%, O 9.03%. In *Aspidosperma quebracho-blanco* Schlect., *Vallesia dichotoma* Ruiz & Pav., and *Vallesia glabra* (Cav.) Link, *Apocynaceae:* G. Fraude, *Ber.* **11**, 2189 (1878); Hesse, *Ann.* **211**, 249 (1882); Ewins, *J. Chem. Soc.* **105**, 2738 (1914); Deulofeu *et al., ibid.* **1940**, 1051; Holker *et al., J. Org. Chem.* **24**, 314 (1959). Structure: Conroy *et al., Tetrahedron Letters* no. 11, 4 (1959). Crystal structure: Mills, Nyburg, *J. Chem. Soc.* **1960**, 1458. Stereochemistry: Smith, Wrobel, *ibid.* 1463; Craven, Zacharias, *Experientia* **24**, 770 (1968). Biogenesis: Robinson, *Tetrahedron Letters* no. 18, 1F4 (1959). Synthesis of *dl*-form: Stork, Dolfini, *J. Am. Chem. Soc.* **85**, 2872 (1963); Ban *et al., Tetrahedron Letters* **1965**, 2261; Stevens *et al., Chem. Commun.* **1971**, 857; S. F. Martin *et al., J. Am. Chem. Soc.* **102**, 3294 (1980). Aspidospermine has been reported as having a wide variety of pharmacological properties, including diuretic and respiratory stimulant activity. Biological and phytochemical evaluation: R. L. Lyon *et al., J. Pharm. Sci.* **62**, 218 (1973). Microbial transformation: S. K. Lin *et al., ibid.* **64**, 2021 (1975).

Needles or prisms from alc, needles from petr ether. mp 208°. Sublimes 180°. bp_2 220°. $[\alpha]_D^{15}$ −100.2° (alc); $[\alpha]_D$ −93° (chloroform). uv max (methanol): 218, 255, 280-290 nm (log ϵ 4.52, 4.04, 3.53-3.40). One gram dissolves in 60 ml water, 50 ml alc, 100 ml ether. Also sol in benzene, chloroform, petr ether. LD_{50} in mice: 40 mg/kg i.p., *RTECS* Vol. 1, R. J. Lewis, R. L. Tatken, Eds. (1979) p 156.

N-Formyl-*N*-deacetylaspidospermine, $C_{21}H_{28}N_2O_2$, *valle-sine*. Structure: Taylor *et al.*, *Helv. Chim. Acta* 42, 2750 (1959). Long, fine needles from acetone, mp 154-156°. $[\alpha]_D^{24}$ −91 ± 2° (c = 1.814 in abs alc). uv max: 211, 250 nm (log ϵ 4.47, 3.94).

886. Aspirin. *2-(Acetyloxy)benzoic acid; salicylic acid acetate;* 2-acetoxybenzoic acid; acetylsalicylic acid; Acenterine; Aceticyl; Acetophen; Acetosal; Acetosalic Acid; Aceto-salin; Acetylin; Acetyl-SAL; Acimetten; Acylpyrin; Arthri-sin; A.S.A.; Asatard; Aspro; Asteric; Caprin; Claradin; Colfarit; Contrheuma retard; Duramax; ECM; Ecotrin; Em-pirin; Encaprin; Endydol; Entrophen; Enterosarine; Heli-con; Levius; Longasa; Measurin; Neuronika; Platet; Rho-dine; Salacetin; Salcetogen; Saletin; Solprin; Solpyron; Xaxa. $C_9H_8O_4$; mol wt 180.16. C 60.00%, H 4.48%, O 35.52%. Prepn: C. Gerhardt, *Ann.* 87, 149 (1853). Manuf from salicylic acid and acetic anhydride: Faith, Keyes & Clark's *Industrial Chemicals,* F. A. Lowenheim, M. K. Moran, Eds. (Wiley-Interscience, New York, 4th ed., 1975) pp 117-120. Crystallization from acetone: Hamer, Phillips, U.S. pat. **2,890,240** (1959 to Monsanto). Novel process involving distillation: Edmunds, U.S. pat. **3,235,583** (1966 to Norwich Pharm.). Crystal structure: P. J. Wheatley, *J. Chem. Soc. (Suppl.)* **1964**, 6036. Toxicity data: E. R. Hart, *J. Pharmacol. Exp. Ther.* 89, 205 (1947). Evaluation as a risk factor in Reye's syndrome: P. J. Waldman *et al., J. Am. Med. Assoc.* 247, 3089 (1982). Review of clinical trials in prevention of myocardial infarction and stroke: P. C. El-wood, *Drugs* 28, 1-5 (1984). Symposium on aspirin therapy: *Am. J. Med.* 74, no. 6A, 1-109 (1983). Comprehensive de-scription: K. Florey, *Anal. Profiles Drug Subs.* 8 1-46 (1979). Monograph: M. J. H. Smith, P. K. Smith, *The Salicylates* (Interscience, New York, 1966) 313 pp. Book: *Acetylsalicylic Acid,* H. J. M. Barnett *et al.,* Eds. (Raven, New York, 1982) 278 pp.

Monoclinic tablets or needle-like crystals. d 1.40. mp 135° (rapid heating); the melt solidifies at 118°. uv max $(0.1N\ H_2SO_4)$: 229 nm ($E_{1cm}^{1\%}$ 484); $(CHCl_3)$: 277 nm ($E_{1cm}^{1\%}$ 68). Is odorless, but in moist air it is gradually hydrolyzed into salicylic and acetic acids and acquires the odor of acetic acid. Stable in dry air. pK (25°) 3.49. One gram dissolves in 300 ml water at 25°, in 100 ml water at 37°, in 5 ml alcohol, 17 ml chloroform, 10-15 ml ether. Less soluble in anhydr ether. Decomp by boiling water or when dissolved in solns of alkali hydroxides and carbonates. LD_{50} orally in mice: 1.1, 1.5 g/kg (Hart).

Guaiacol ester, $C_{16}H_{14}O_5$, *guacetisal, Broncaspin, Guaia-spir.*

Methyl ester, *see* Methyl Acetylsalicylate.

Phenyl ester, *see* Phenyl Acetylsalicylate.

Inorganic salts of acetylsalicylic acid are soluble in water (esp the Ca salt, *q. v.*), but are decomposed quickly.

Pharmaceutical Incompat. (from *Remington's Pharmaceu-tical Sciences):* Aspirin forms a damp to pasty mass when triturated with acetanilide, phenacetin, antipyrine, aminopy-rine, methenamine, phenol or phenyl salicylate. Powders containing aspirin with an alkali salt such as sodium bicar-bonate become gummy on contact with atmospheric mois-ture. Hydrolysis occurs in admixture with salts contg water of crystallization. Solns of the alkaline acetates and citrates, as well as alkalies themselves, dissolve aspirin but the result-ing solns hydrolyze rapidly to form salts of acetic and sali-cylic acids. Sugar and glycerol have been shown to hinder this decompn. Aspirin very slowly liberates hydriodic acid from potassium or sodium iodide. Subsequent oxidation by air produces free iodine.

THERAP CAT: Analgesic; antipyretic; anti-inflammatory.

THERAP CAT (VET): Analgesic; antipyretic; anti-inflamma-tory; anticoagulant.

887. Aspoxicillin. *[2S-(2α,5α,6β)]-N-Methyl-D-aspara-ginyl-N-(2-carboxy-3,3-dimethyl-7-oxo-4-thia-1-azabicyclo-[3.2.0]hept-6-yl)-D-2-(4-hydroxyphenyl)glycinamide;* 6-[D-2-(D-2-amino-3-N-methylcarbamoylpropionamido)-2-p-hydr-oxyphenylacetamido]penicillanic acid; (2S,5R,6R)-6-[(2R)-2-[(2R)-2-amino-3-(methylcarbamoyl)propionamido]-2-(p-hydroxyphenyl)acetamido]-3,3-dimethyl-7-oxo-4-thia-1-aza-bicyclo[3.2.0]heptane-2-carboxylic acid; N^4-methyl-D-aspar-aginylamoxicillin; ASPC; TA-058; Doyle. $C_{21}H_{27}N_5O_7S$; mol wt 493.54. C 51.11%, H 5.51%, N 14.19%, O 22.69%, S 6.50%. Semisynthetic penicillin. Prepn: M. Kawazu *et al.,* **Ger.** pat. **2,638,067;** *eidem,* **U.S.** pat. **4,053,609** (both 1977 to Tanabe); prepn and antibacterial activity: M. Wagat-suma *et al., J. Antibiot.* 36, 147 (1985). Mechanism of action study: T. Nishino *et al., Chemotherapy (Tokyo)* 33, 132 (1985), *C.A.* 103, 34792v (1985). Toxicological study: M. Takeshita *et al., Oyo Yakuri* 30, 687 (1985), *C.A.* 104, 101990u (1986). HPLC determn in serum: J. Knöller *et al., Zentralblatt Bakteriol. Mikrobiol. Hyg.* 265, 176 (1987). Clinical evaluation in ocular infections: M. Ooishi *et al., Acta Med. Biol.* 34, 1 (1986). Series of articles on anti-bacterial activity, pharmacology and clinical efficacy: *Chemotherapy (Tokyo)* 32, Suppl. 2, 1-791 (1984).

Colorless crystalline powder, mp 195-198° (dec).

THERAP CAT: Antibacterial.

888. Astacin. *β,β-Carotene-3,3',4,4'-tetrone;* 3,4,3',4'-tetraketo-β-carotene; 3,3'-dihydroxy-2,3,2',3'-tetradehydro-β,β-carotene-4,4'-dione; astacene. $C_{40}H_{48}O_4$; mol wt 592.82. C 81.04%, H 8.16%, O 10.80%. Red carotenoid pigment isolated from biological material originating from crustacea, algae, sponges, protozoa, fish and reptiles. Small amounts were isolated from the fat of mammals (whales, *Balaenop-tera musculus).* Occurs together with astaxanthin from which it is formed by autoxidation. Appears to be an arti-fact rather than a natural product. Isoln from lobster shells: Kuhn, Lederer, *Ber.* 66, 488 (1933). Structure: Kar-rer *et al., Helv. Chim. Acta* 17, 412, 745 (1934); 18, 96 (1935); 19, 479 (1936). Total synthesis: J. B. Davis, B. C. L. Weedon, *Proc. Chem. Soc.* **1960,** 182; E. Widmer *et al., Helv. Chim. Acta* 65, 671 (1982). Prepn by autoxidation of canthaxanthin: R. D. G. Cooper *et al., J. Chem. Soc. Perkin Trans. I* **1975,** 2195.

Purple needles or leaflets with metallic luster, sometimes bent into sickle shape, esp when crystallized from pyridine + water, mp 240-243° (slow heating in evac tube, Karrer), mp 228° (Kuhn). Absorption max (pyridine): 500 nm. Practically insol in water; freely sol in chloroform, pyridine, dioxane, carbon disulfide and dil aq alkali; slightly sol in benzene, ethyl acetate, glacial acetic acid; nearly insol in ether, petr ether, methanol.

Diacetate, $C_{44}H_{52}O_6$, black to violet needles from pyridine + water, dec 235°.

Dipalmitate (astacein), $C_{72}H_{108}O_6$, almost square red leaflets from petr ether, mp 121°.

889. Astatine. At; at. no. 85. Radioactive halogen; one of the rarest elements in nature. Radioisotopes range in mass number from 200 to 219; naturally occurring isotopes found in uranium ores: 215, 217, 218, 219 (longest-lived natural isotope, $T_{1/2}$ 0.9 min); most stable artificial isotopes: 209 ($T_{1/2}$ 5.5 hrs); 210 ($T_{1/2}$ 8.3 hrs); 211 ($T_{1/2}$ 7.21 hrs). First convincingly identified by Corson et al., *Phys. Rev.* **58**, 672 (1940). [209]At, [210]At, [211]At are obtained artificially by alpha bombardment of bismuth: Kelley, Segre, *Phys. Rev.* **75**, 999 (1949); Johnson et al., *J. Chem. Phys.* **17**, 1, (1949); Neumann, *J. Inorg. & Nucl. Chem.* **4**, 349 (1957). *Reviews:* Hyde, *J. Chem. Ed.* **36**, 15 (1959); Haissinsky, Adloff, *Radiochemical Survey of the Elements* (Elsevier, 1965) pp 11-12; Appelman, "Astatine" in *MTP Int. Rev. Sci.: Inorg. Chem., Ser. One,* **vol. 3**, V. Gutmann, Ed. (Butterworths, London, 1972) 181-198; Downs, Adams, "Chlorine, Bromine, Iodine and Astatine" in *Comprehensive Inorganic Chemistry* **vol. 2**, J. C. Bailar, Jr. et al., Eds. (Pergamon Press, Oxford, 1973) pp 1107-1594. More metallic than iodine. Soluble in organic solvents. Due to short half-life of isotopes, only a few physical properties can be measured. Chemical studies are carried out on trace amounts ($< 10^{-8}M$ solns). HAt, CH_3At, AtCl, AtBr, AtI have been identified by time-of-flight mass spectrometry: Appleman et al., *Inorg. Chem.* **5**, 766 (1966). Oxidation states: -1, 0, $+5$; existence of $+1$, $+3$, $+7$ oxidation states uncertain; E^0(aq) At/At$^-$ 0.3 V: Appleman, *J. Am. Chem. Soc.* **83**, 805 (1961). When fed to guinea pigs, is found in the thyroid gland. Mammary and pituitary tumors can be induced with a single injection.

890. Astaxanthin. *3,3'-Dihydroxy-β,β-carotene-4,4'-dione;* 3,3'-dihydroxy-4,4'-diketo-β-carotene; ovoester; $C_{40}H_{52}O_4$; mol wt 596.85. C 80.50%, H 8.78%, O 10.72%. Carotenoid pigment found mostly in animal organisms, but also occurring in plants; thought to be the precursor of astacin, *q.v.* Structure and isoln from lobster eggs: Kuhn, Sörensen, *Angew. Chem.* **51**, 465 (1938); *Ber.* **71**, 1879 (1938). Isoln from other animal organisms: Kuhn et al., *Ber.* **72**, 1688 (1939). Occurrence in plants: Tischer, *Z. Physiol. Chem.* **267**, 281 (1941); in the flower petals of *Adonis annua* L., *Ranunculaceae:* Seybold, Goodwin, *Nature* **184**, 1714 (1959). Occurs also in the red feathers of birds of the *Laniarius* spp: *Z. Physiol. Chem.* **288**, 20 (1951). Stereochemistry and spectra: Grangaud, *Compt. Rend.* **242**, 1767 (1956). Abs config: Andrewes et al., *Acta Chem. Scand. B* **28**, 730 (1974). *See also* T. W. Goodwin, *Carotenoids* (Chemical Publ. Co., New York, 1954) pp 167-172. Synthesis: R. D. G. Cooper et al., *J. Chem. Soc. Perkin Trans. I* **1975**, 2195; F. Kienzle, H. Mayer, *Helv. Chim. Acta* **61**, 2609 (1978); E. Widmer et al., *ibid.* **64**, 2405 (1981); H. Mayer et al., *ibid.* 2419.

Needles from acetone/light petroleum, mp 182-183°. uv max: (CS_2) 503 nm; (methanol) 472 nm; (hexane) 466-467 nm; (chloroform) 485 nm (Cooper et al.). Also reported as shiny purple platelets with gold luster from pyridine, mp 216° (some decomp). Readily sol in pyridine, from which it can be cryst by the addn of water (Kuhn, Sörensen).

Diacetate, $C_{44}H_{56}O_6$, stout, blue-black needles, mp 203-205° (vac).

891. Astemizole. *1-[(4-Fluorophenyl)methyl]-N-[1-[2-(4-methoxyphenyl)ethyl]-4-piperidinyl]-1H-benzimidazol-2-amine;* 1-(p-fluorobenzyl)-2-[[1-(p-methoxyphenethyl)-4-piperidyl]amino]benzimidazole; R-43512; Astemisan; Hismanal; Histamen; Histaminos; Histazol; Kelp; Laridal; Metodik; Novo-Nastizol A; Paralergin; Retolen; Waruzol. $C_{28}H_{31}FN_4O$; mol wt 458.58. C 73.34%, H 6.81%, F 4.14%, N 12.22%, O 3.49%. Nonsedating-type histamine H_1-receptor antagonist. Prepn: F. Janssens et al., **Eur. pat. Appl.** 5318; *eidem,* **U.S. pat. 4,219,559** (1979, 1980 both to Janssen). Pharmacology: J. Van Wauwe et al., *Arch. Int. Pharmacodyn. Ther.* **251**, 39 (1981); A. Wauquier, C. J. E. Niemegeers, *Eur. J. Pharmacol.* **72**, 245 (1981). *In vitro* and *in vivo* binding characteristics: P. M. Laduron et al., *Mol. Pharmacol.* **21**, 294 (1982). Effect on human psychomotor performance: T. Seppala, K. Savolainen, *Curr. Ther. Res.* **31**, 638 (1982). Clinical study in treatment of hay fever: J. Callier et al., *ibid.* **29**, 24 (1981). Mutagenicity study: P. Vanparys et al., *Arch. Toxicol.* **50**, 167 (1982). Review of pharmacology and clinical trials: D. M. Richards et al., *Drugs* **28**, 38-61 (1984). Comprehensive description: A. M. Al-Obaid, M. S. Mian in *Analytical Profiles of Drug Substances* vol. 20, K. Florey Ed. (Academic Press, New York, 1991) pp 173-208.

White, odorless crystals, mp 149.1°. Freely sol in organic solvents. Practically insol in water. uv max (ethanol): 219, 249, 286 nm (ε 27250.229, 6480.293, 8634.280); (0.1N HCl): 209, 277 nm (ε 57889.908, 18073.394).

THERAP CAT: Antiallergic; antihistaminic.

892. Atenolol. *4-[2-Hydroxy-3-[(1-methylethyl)amino]propoxy]benzeneacetamide;* 2-[p-[2-hydroxy-3-(isopropylamino)propoxy]phenyl]acetamide; 1-p-carbamoylmethylphenoxy-3-isopropylamino-2-propanol; ICI-66082; AteHexal; Atenol; Cuxanorm; Ibinolo; Myocord; Prenormine; Seles Beta; Selobloc; Teno-basan; Tenoblock; Tenormin; Uniloc. $C_{14}H_{22}N_2O_3$; mol wt 266.34. C 63.14%, H 8.33%, N 10.52%, O 18.02%. Cardioselective β-adrenergic blocker. Prepn: Barrett et al., **Ger. pat. 2,007,751**; *eidem,* **U.S. pats. 3,663,607** and **3,836,671** (1970, 1972, 1974 all to I.C.I.). Pharmacology and clinical studies: Giudicelli et al., *Compt. Rend. Soc. Biol.* **167**, 232 (1973); Hansson et al., *Acta Med. Scand.* **194**, 549 (1973); Amery et al., *N. Engl. J. Med.* **290**, 284 (1974). Clinical trial in treatment of alcohol withdrawal syndrome: M. L. Kraus et al., *ibid.* **313**, 905 (1985). HPLC determn of enantiomers in plasma and urine: S. K. Chin et al., *J. Chromatog.* **489**, 438 (1989). Comprehensive description: V. Caplar et al., *Anal. Profiles Drug Subs.* **13**, 1-25 (1984). *Review:* J. D. Fitzgerald in *Pharmacological and Biochemical Properties of Drug Substances* vol. 2, M. E. Goldberg, Ed. (Am. Pharm. Assoc., Washington, DC, 1979) pp 98-147; E. Marmo, *Drugs Exp. Clin. Res.* **6**, 639-663 (1980). Symposium on clinical studies: *Drugs* **25**, Suppl. 2, 1-346 (1983).

Crystals from ethyl acetate, mp 146-148°. Also reported as mp 150-152° (Caplar). Freely sol in methanol; sol in acetic acid, DMSO; sparingly sol in 96% ethanol; slightly sol in water, isopropanol; very slightly sol in acetone, dioxane. Practically insol in acetonitrile, ethylacetate, chloroform. pKa 9.6. Partition coefficient (n-octanol/phosphate buffer): 0.008 (pH 7.0); 0.052 (pH 8.0). Dipole moment: 5.71 ±0.20 D at 20° in propionic acid. uv max (methanol):

225, 275, 283 nm. LD_{50} in mice, rats (mg/kg): 2000, 3000 orally; 98.7, 59.24 i.v. (Fitzgerald).

THERAP CAT: Antihypertensive, antianginal, antiarrhythmic (class II).

893. Athamantin. *(8S-cis)-3-Methylbutanoic acid 8,9-dihydro-8-[1-methyl-1-(3-methyl-1-oxobutoxy)ethyl]-2-oxo-2H-furo[2,3-h]-1-benzopyran-9-yl ester; isovaleric acid diester with 8,9-dihydro-9-hydroxy-8-(1-hydroxy-1-methylethyl)-2H-furo[2,3-h]-1-benzopyran-2-one;* 2,3-dihydro-3,4-dihydroxy-2-(1-hydroxy-1-methylethyl)-5-benzofuranacrylic acid δ-lactone diisovalerate. $C_{24}H_{30}O_7$; mol wt 430.50. C 66.96%, H 7.02%, O 26.02%. From *Peucedanum oreoselinum* (L.) Moench (*Athamanta oreoselinum* L.), *Ammi visnaga* Lam., *Umbelliferae.* Isoln: Schnedermann, Winckler, *Ann.* **51**, 315 (1844). Structure: Halpern *et al., Helv. Chim. Acta* **40**, 758 (1957). Absolute configuration: Nakazaki *et al., Tetrahedron Letters* **1966**, 4735.

Fine needles from petr ether, mp 58-60°; sublimes in high vacuum at 180-200°. $[\alpha]_D^{20}$ +88° (c = 1.145 in glacial acetic acid); $[\alpha]_{546}^{22}$ +129° (c = 0.5575 in methanol); $[\alpha]_{546}^{21}$ +73.9° (c = 1.024 in chloroform). $[M]_D$ +440° (methanol); +258° (chloroform). uv max (96% ethanol): 217, 322 nm (log ε 4.18, 4.17). Practically insol in water; sol in alc, ether, chloroform.

894. Atipamezole. *4-(2-Ethyl-2,3-dihydro-1H-inden-2-yl)-1H-imidazole;* 4(5)-(2-ethyl-2-indanyl)imidazole. $C_{14}H_{16}$-N_2; mol wt 212.29. C 79.21%, H 7.60%, N 13.20%. Specific α_2-adrenoceptor antagonist. Prepn: A. J. Karjalainen *et al.,* **Eur. pat. Appl. 183,492**; *eidem,* **U.S. pat. 4,689,339** (1986, 1987 both to Farmos). Behavioral and neurochemical effects: H. Scheinin *et al., Eur. J. Pharmacol.* **151**, 35 (1988). Pharmacodynamics and pharmacokinetics: X. Fargetton, T. Vähä-Vahe, *Tijdschr. Diergeneeskd* **114**, Suppl. 1, 91S (1989). Binding study in comparison with yohimbine, *q.v.:* R. Virtanen *et al., Arch. Int. Pharmacodyn.* **297**, 190 (1989). Attenuation of ethanol effects in mice: M. J. Durcan *et al., Alcohol* **6**, 189 (1989). Preliminary report as detomidine, *q.v.,* antagonist in horses: L. Nilsfors, C. Kvart, *Acta Vet. Scand.* **82**, 121 (1986). Use as medetomidine, *q.v.,* antagonist in zoo animals: H. Jalanka, *ibid.* **85**, 193 (1989).

Hydrochloride, $C_{14}H_{16}N_2 \cdot HCl$, *MPV-1248, Antisedan.* Crystals from dry HCl in ethyl acetate, mp 211°-215°. LD_{50} in male, female rats (mg/kg): > 50, 44 s.c. (Scheinin).

THERAP CAT (VET): Antidote to α_2-adrenoceptor induced sedation.

895. Atisine. Anthorine. $C_{22}H_{33}NO_2$; mol wt 343.51. C 76.92%, H 9.68%, N 4.08%, O 9.32%. From roots of the "atis" plant, *Aconitum heterophyllum* Wall., and from *A. anthora* L., *Ranunculaceae:* Broughton, *Blue Book of East India Cinchona Cultivation* **1877**, 133; Goris, Metin, *Compt. Rend.* **180**, 968 (1925); Lawson, Topps, *J. Chem. Soc.* **1937**, 1640. Structure: Wiesner *et al., Chem. & Ind. (London)* **1954**, 132; Pelletier, Jacobs, *J. Am. Chem. Soc.* **76**, 4496 (1954). Structure and stereochemistry: Dvornik, Edwards, *Can. J. Chem.* **42**, 137 (1964). Partial synthesis: Pelletier, Jacobs, *J. Am. Chem. Soc.* **78**, 4144 (1956); Pelletier, Parthasarathy, *Tetrahedron Letters* **1963**, 205. Racemic synthesis and resolution: Nagata *et al., J. Am. Chem. Soc.* **85**, 2342

(1963); **89**, 1499 (1967); Masamune, *ibid.* **86**, 291 (1964); Guthrie *et al., Tetrahedron Letters* **1966**, 4645. ^{13}C-NMR study of C-20 epimers: N. V. Mody, S. W. Pelletier, *Tetrahedron* **34**, 2421 (1978).

Solid, mp 57-60°. Distills in high vacuum at a bath temp of 140°. Strong base: pK 12.2.

Hydrochloride, $C_{22}H_{33}NO_2 \cdot HCl$, flat needles from dil alc, dec 311-312°. $[\alpha]_D^{25}$ +28°.

896. Atmosphere. Composition of Earth's atmosphere at surface for midlatitudes (in ppm): N_2 780,840, O_2 209,460, Ar 9340, CO_2 325, Ne 18.18, He 5.24, CH_4 1.4, Kr 1.14, H_2 0.5, N_2O 0.25, Xe 0.087, O_3 0.025, H_2O variable. Monograph on chemistry of atmosphere: J. Heicklen, *Atmospheric Chemistry* (Academic Press, New York, 1976) 406 pp. Other planets of our solar system have different atmospheres. The atmosphere of Jupiter consists largely of ammonia and methane, and those of Saturn, Uranus, and Neptune are practically all methane. Exact data in *Landolt-Börnstein,* 6th ed., **vol. III** (Springer, 1952) p 59; G. P. Kuiper, *The Atmospheres of the Earth and the Planets* (University of Chicago Press, 1949).

897. Atorvastatin. *[R-(R*,R*)]-2-(4-Fluorophenyl)-β,δ-dihydroxy-5-(1-methylethyl)-3-phenyl-4-[(phenylamino)-carbonyl]-1H-pyrrole-1-heptanoic acid;* (βR,δR)-2-(p-fluorophenyl)-β,δ-dihydroxy-5-isopropyl-3-phenyl-4-(phenylcarbamoyl)pyrrole-1-heptanoic acid. $C_{33}H_{35}FN_2O_5$; mol wt 558.65. C 70.95%, H 6.31%, F 3.40%, N 5.01%, O 14.32%. HMG-CoA reductase inhibitor. Prepn: B. D. Roth, **Eur. pat. Appl. 409,281**; *idem,* **U.S. pat. 5,273,995** (1991, 1993 both to Warner-Lambert); K. L. Baumann *et al., Tetrahedron Letters* **33**, 2283 (1992). Solubility and kinetic studies: A. S. Kearney *et al., Pharm. Res.* **10**, 1461 (1993). Comparative study of tissue selectivity: T. M. A. Bocan *et al., Biochim. Biophys. Acta* **1123**, 133 (1992); of antiatherosclerotic potential: *idem et al., Atherosclerosis* **111**, 127 (1994).

Calcium salt, $C_{66}H_{68}CaF_2N_4O_{10}$, *CI-981.* $[\alpha]_D$ −7.4° (c = 1 in DMSO).

Sodium salt, $C_{33}H_{34}FN_2NaO_5$. pKa 4.46. Soly in water (30°): 20.4 μg/ml (pH 2.1); 1.23 mg/ml (pH 6.0).

Lactone, $C_{33}H_{33}FN_2O_4$, *(2R-trans)-5-(4-fluorophenyl)-2-(1-methylethyl)-N,4-diphenyl-1-[2-(tetrahydro-4-hydroxy-6-oxo-2H-pyran-2-yl)ethyl]-1H-pyrrole-3-carboxamide.* mp 159.2-160.7°. $[\alpha]_D$ +26.05° (c = 1 in chloroform).

THERAP CAT: Antihyperlipoproteinemic.

898. Atovaquone. *trans-2-[4-(4-Chlorophenyl)cyclohexyl]-3-hydroxy-1,4-naphthalenedione; trans-2-[4-(4-chlorophenyl)cyclohexyl]-3-hydroxy-1,4-naphthoquinone;* 566C80; BW-566C; BW-566C-80; Acuvel; Mepron; Wellvone. C_{22}-$H_{19}ClO_3$; mol wt 366.84. C 72.03%, H 5.22%, Cl 9.66%, O 13.08%. Hydroxynaphthoquinone derivative which inhibits mitochondrial electron transport. Prepn: A. T. Hudson, A. W. Randall, **Eur. pat. Appl. 123,238**; *eidem,* **U.S. pat. 5,-053,432** (1984, 1991 both to Burroughs Wellcome); *see also:* V. S. Latter, W. E. Gutteridge, **U.S. pat. 4,981,874** (1991). Antiprotozoal activity, metabolism, and preliminary pharmacokinetics: A. T. Hudson *et al., Drugs Exptl. Clin. Res.*

17, 427 (1991). Mechanism of action study: M. Fry, M. Pudney, *Biochem. Pharmacol.* **43**, 1545 (1992). Clinical evaluation in AIDS patients with pneumocystis pneumonia: J. Falloon *et al.*, *N. Engl. J. Med.* **325**, 1534 (1991); with toxoplasmosis: J. A. Kovacs *et al.*, *Lancet* **340**, 637 (1992).

Crystals from acetonitrile, mp 216-219°.

THERAP CAT: Antiprotozoal (Toxoplasma); antipneumocystis; antimalarial.

899. Atractyloside. *(2β,4α,15α)-15-Hydroxy-2-[[2-O-(3-methyl-1-oxobutyl)-3,4-di-O-sulfo-β-D-glucopyranosyl]oxy]-19-norkaur-16-en-18-oic acid dipotassium salt;* potassium atractylate; atractylin (C_{30} glucoside). $C_{30}H_{44}K_2O_{16}S_2$; mol wt 803.00. C 44.87%, H 5.52%, K 9.74%, O 31.88%, S 7.99%. Toxic principle isolated from the thistle *Atractylis gummifera* L., *Compositae*: M. Lefranc, *Compt. Rend.* **67**, 954 (1868). Structure and stereochemistry of the aglycone, *atractyligenin:* F. Piozzi *et al.*, *Tetrahedron Suppl.* **8 II**, 515 (1966); total synthesis of (±)-form: A. K. Singh *et al.*, *J. Am. Chem. Soc.* **109**, 6187 (1987). Structure and stereochemistry of atractyloside: *eidem, Gazz. Chim. Ital.* **97**, 935 (1967). Toxicity studies: G. Cascio *et al.*, *Boll. Soc. Ital. Biol. Sper.* **44**, 253 (1968). *Review: Atractyloside: Chemistry, Biochemistry, and Toxicology,* R. Santi, S. Luciani, Eds. (Piccin Medical Books, Padova, Italy, 1978) 136 pp.

Crystals, dec 174°. $[\alpha]_D^{20}$ −53° (water). *Highly toxic!* Has strychnine-like action; produces convulsion of a hypoglycemic nature: R. Santi, G. Cascio, *C.A.* **50**, 7320i (1956); R. Santi, *C.A.* **52**, 15733c (1958). LD_{50} i.m. in rats: 431 mg/kg (Cascio).

900. Atracurium Besylate. *2,2'-[1,5-Pentanediylbis-[oxy(3-oxo-3,1-propanediyl)]]bis[1-[(3,4-dimethoxyphenyl)-methyl]-1,2,3,4-tetrahydro-6,7-dimethoxy-2-methylisoquinolinium] dibenzenesulfonate;* 2-(2-carboxyethyl)-1,2,3,4-tetrahydro-6,7-dimethoxy-2-methyl-1-veratrylisoquinolinium benzenesulfonate pentamethylene ester; *N,N'-dimethyl-N,-N'-(4,10-dioxa-3,11-dioxotridecylene)-1,13-bis-tetrahydro-papaverinium dibesylate;* BW-33A; Wellcome 33-A-74; Tracrium. $C_{65}H_{82}N_2O_{18}S_2$; mol wt 1243.50. C 62.78%, H 6.65%, N 2.25%, O 23.16%, S 5.16%. Non-depolarizing neuromuscular blocking agent. Prepn: J. B. Stenlake *et al.*, **Ger. pat. 2,655,883;** *eidem,* U.S. pat. **4,179,507** (1977, 1979 both to Burroughs Wellcome). Pharmacology: R. Hughes, D. J. Chapple, *Brit. J. Anaesth.* **53**, 31 (1981). Clinical pharmacology: S. J. Basta *et al.*, *Anesth. Analg.* **61**, 723 (1982). Neuromuscular effects in man: R. L. Katz *et al.*, *ibid.* 730. Metabolic studies: E. A. Neill, D. J. Chapple, *Xenobiotica* **12**, 203 (1982). Pharmacokinetics: S. Ward *et al.*, *Brit. J. Anaesth.* **55**, 113 (1983). Use during halothane anesthesia in humans: J. A. Stirt *et al.*, *Anesth. Analg.* **62**, 207 (1983). Symposium on pharmacology, metabolism, and clinical studies: *Brit. J. Anaesth.* **55**, Suppl. 1, 1S-139S

(1983); on pharmacokinetics and clinical experience: *ibid.* **58**, Suppl. 1, 1S-113S (1986).

Off-white powder, mp 85-90°. Softens at 60°.

1*R-cis*,1*R'-cis* form, **cisatracurium besylate, 51W89,** Nimbex. Prepn: D. A. Hill, G. L. Turner, **PCT Int. pat. Appl. 92 00,965** (1992 to Wellcome Found.). Clinical pharmacology: C. A. Lien *et al.*, *Anesthesiology* **82**, 1131 (1995); M. R. Belmont *et al.*, *ibid.* 1139. White solid.

THERAP CAT: Muscle relaxant (skeletal).

901. Atranorin. *3-Formyl-2,4-dihydroxy-6-methylbenzoic acid 3-hydroxy-4-(methoxycarbonyl)-2,5-dimethylphenyl ester;* atranoric acid. $C_{19}H_{18}O_8$; mol wt 374.35. C 60.96%, H 4.85%, O 34.19%. From a number of lichens. Belongs to the group of lichen acids. Isoln: Paterno, Ogliarori, *Gazz. Chim. Ital.* **1877**, 189; Hesse, *Ber.* **30**, 357, 1983 (1897); St. Pfau, *Helv. Chim. Acta* **9**, 650 (1926). Synthesis: Neelakantan *et al.*, *Tetrahedron Letters* **1962**, 287. TLC analysis: R. Klee, L. Steubing, *J. Chromatogr.* **129**, 478 (1976); J. L. Ramaut *et al.*, *ibid.* **155**, 450 (1978).

Bitter crystals or cryst powder from chloroform, mp 195°. Practically insol in water; slightly sol in alc; sol in boiling benzene or in chloroform; also sol in alkalies giving a yellow soln.

902. Atrazine. *6-Chloro-N-ethyl-N'-(1-methylethyl)-1,3,5-triazine-2,4-diamine;* 2-chloro-4-ethylamino-6-isopropylamine-s-triazine; G-30027; AAtrex; Atranex; Gesaprim; Primatol A. $C_8H_{14}ClN_5$; mol wt 215.69. C 44.55%, H 6.54%, Cl 16.44%, N 32.47%. Prepn: Gysin, Knüsli, **Swiss** pats. **342,784, 342,785** (both from 1960 to Geigy), *C.A.* **55**, 5552d (1961); Mel'nikov *et al.*, *Khim. Prom.* **1961**, 703, *C.A.* **58**, 526c (1963); Andriska *et al.*, **Hung. pat. 149,189** (1962 to Nehézvegyipari Kutato Intézet), *C.A.* **58**, 13972c (1963); Mildner, **Fr. pat. 1,317,812** (1963 to Radonja Kemijska Ind.), *C.A.* **59**, 8765h (1963). Toxicity study: S. Dalgaard-Mikkelsen, E. Poulsen, *Pharmacol. Rev.* **14**, 225 (1962).

Crystals, mp 171-174°. Soly at 25° in water 70 ppm; ether 12,000 ppm; chloroform 52,000 ppm; methanol 18,000 ppm. Stable in slightly acidic or basic media; hydrolyzed to inactive hydroxy deriv by alkali or mineral acids. LD_{50} orally in mice: 1750 mg/kg (Dalgaard-Mikkelsen, Poulsen).

Caution: Acutely poisoned sheep and cattle show muscular spasms, fasciculations, stiff gait, increased respiratory rates. Adrenal degeneration, congestion of lungs, liver, kidneys observed. Inhalation hazard is low in humans. No apparent skin irritation or other toxic injuries in humans: *Clinical Toxicology of Commercial Products,* R. E. Gosselin *et al.*, Eds. (Williams & Wilkins, Baltimore, 5th ed., 1984) Section II, p. 333.

USE: Selective herbicide.

903. Atrial Natriuretic Factor. *Atriopeptin;* ANF; ANP; auriculin; cardionatrin. Potent natriuretic and vasodilatory peptide or mixture of homologous peptides derived from a common precursor and isolated from the atrium of the mammalian heart. Involved in hormonal regulation of extracellular fluid volume and blood pressure homeostasis. Stored in specific granules of atrial cardiocytes, ANF produces rapid diuresis when injected into laboratory animals: A. J. de Bold *et al.*, *Life Sci.* **28**, 89 (1981); R. Garcia *et al.*, *Experientia* **38**, 1071 (1982); A. J. de Bold, *Can. J. Physiol. Pharmacol.* **60**, 324 (1982). Isoln of rat ANF (rANF): *idem*, *Proc. Soc. Exp. Biol. Med.* **170**, 133 (1982); G. Thibault *et al.*, *Hypertension* **5**, Suppl. I, I-75 (1983); of human ANF (hANF): N. C. Trippodo *et al.*, *ibid.*, I-81. Preliminary characterization of high and low molecular weight components: *eidem*, *Proc. Soc. Exp. Biol. Med.* **170**, 502 (1982). Amino acid comp of low mol wt component: A. J. de Bold, T. G. Flynn, *Life Sci.* **33**, 297 (1983); G. Thibault *et al.*, *FEBS Letters* **164**, 286 (1983). A number of biologically active ANF peptides containing 21-33 amino acids have been identified (mol wt ∼4000). Amino acid sequence of rANF peptides: T. G. Flynn *et al.*, *Biochem. Biophys. Res. Commun.* **117**, 859 (1983); D. M. Geller *et al.*, *ibid.* **120**, 333 (1984); M. G. Currie *et al.*, *Science* **223**, 67 (1984). Synthesis of bioactive rANF: N. G. Seidah *et al.*, *Proc. Nat. Acad. Sci. USA* **81**, 2640 (1984); S. A. Atlas *et al.*, *Nature* **309**, 727 (1984). hANF differs from rANF by one amino acid substitution. Identification of 3 peptide components of hANF and sequence of 28 amino acid α-hANP: K. Kangawa, H. Matsuo, *Biochem. Biophys. Res. Commun.* **118**, 131 (1984). The high mol wt component (*atriopeptigen* or *preproANF*) has been identified as the precursor for the active, low mol wt forms. Specific proteolytic conversion is required for biological activation: D. M. Geller *et al.*, *Biochem. Biophys. Res. Commun.* **121**, 802 (1984); T. Oshima *et al.*, *Circ. Res.* **54**, 612 (1984). Structure of rat preproANF: M. Maki *et al.*, *Nature* **309**, 722 (1984). Cloning and nucleotide sequence of cDNA for rat preproANF: M. Yamanaka *et al.*, *ibid.* 719; for human preproANF: S. Oikawa *et al.*, *ibid.* 724. ANF is stored in the rat atrium as a 126 amino acid peptide (proANF or γ-rANP): K. Kangawa *et al.*, *Nature* **312**, 1523 (1984). Biosynthesis, processing and secretion of rat proANF: K. D. Bloch *et al.*, *Science* **230**, 1168 (1985). Identification of 28 amino acid α-rANP as major plasma component and comparison with human: A. Miyata *et al.*, *Biochem. Biophys. Res. Commun.* **129**, 248 (1985).

Comparison of diuretic effect of ANF with furosemide: H. Sonnenberg *et al.*, *Can. J. Physiol. Pharmacol.* **59**, 1278 (1981). Studies on the mechanism of natriuresis: *eidem*, *ibid.* **60**, 1149 (1982); R. Keeler, *ibid.* 1078; M. J. F. Camargo *et al.*, *Am. J. Physiol.* **246**, F447 (1984). Inhibition of aldosterone production: K. Atarashi *et al.*, *Science* **224**, 992 (1984). Vasorelaxant activity: M. G. Currie *et al.*, *ibid.* **221**, 71 (1983); R. T. Grammer *et al.*, *Biochem. Biophys. Res. Commun.* **116**, 696 (1983); H. D. Kleinert *et al.*, *Hypertension* **6**, Suppl. I, I-143 (1984). Vasodilator profile of synthetic peptide: R. Garcia *et al.*, *Biochem. Biophys. Res. Commun.* **119**, 685 (1984); R. J. Winquist *et al.*, *Eur. J. Pharmacol.* **102**, 169 (1984). Direct renal effect of synthetic ANF: A. A. Seymour *et al.*, *Life Sci.* **36**, 33 (1985). Determn of ANF in rat atria by radioimmunoassay: J. Gutkowska *et al.*, *Proc. Soc. Exp. Biol. Med.* **176**, 105 (1984). Evidence of cardiac secretion of ANF: E. A. Espiner *et al.*, *Lancet* **2**, 398 (1985). Clinical pharmacology of synthetic peptide: A. M. Richards *et al.*, *ibid.* **1**, 545 (1985). Symposium on biochemistry, activity and clinical pharmacology: *J. Hypertension* **4**, Suppl. 2, S1-S157 (1986). *Reviews:* M. A. Napier, E. H. Blaine in *Ann. Rep. Med. Chem.* **19**, D. M. Bailey, Ed. (Academic Press, New York, 1984) pp 256-262; P. Needleman *et al.*, *Trends Pharmacol. Sci.* **5**, 506-509 (1984); P. Needleman, J. E. Greenwald, *N. Engl. J. Med.* **314**, 828-834 (1986); U. Ackermann, *Clin. Chem.* **32**, 241-247 (1986); J. D. Baxter *et al.*, *Biotechnology* **6**, 529-546 (1988).

α-hANP

904. Atrolactamide. *α-Hydroxy-α-methylbenzeneacetamide;* α-hydroxy-α-phenylpropionamide; 2-hydroxy-2-phenylpropionamide; 2-phenyl-2-hydroxypropionamide; M-144; Themisone. $C_9H_{11}NO_2$; mol wt 165.19. C 65.44%, H 6.71%, N 8.48%, O 19.37%. Prepd from atrolactic acid by treatment with ammonia water: McKenzie, Smith, *J. Chem. Soc.* **121**, 1356 (1922); from the nitrile by treatment with fuming HCl: Staudinger, Ruzicka, *Ann.* **380**, 291 (1911).

DL-Form, leaflets from dichloroethylene, mp 102°. Freely sol in water.

L-Form, leaflets from benzene, mp 62.5-63.5°. $[\alpha]_D^{15}$ +12.8° (c = 2.2 in acetone). Freely sol in water; sol in alcohol, acetone.

THERAP CAT: Anticonvulsant.

905. Atrolactic Acid. *α-Hydroxy-α-methylbenzeneacetic acid;* α-*methylmandelic acid;* 2-phenyllactic acid; α-hydroxy-α-phenylpropionic acid; 2-hydroxy-2-phenylpropionic acid; 2-phenyl-2-hydroxypropionic acid. $C_9H_{10}O_3$; mol wt 166.18. C 65.05%, H 6.07%, O 28.88%. Prepd from acetophenone cyanohydrin: McKenzie, Clough, *J. Chem. Soc.* **101**, 393 (1912); Freudenberg *et al.*, *Ann.* **501**, 213 (1932); Eliel, Freeman, *Org. Syn.* **coll. vol. IV**, 58 (1963).

DL-Form, hemihydrate, orthorhombic crystals from water, mp 88-90° (softening at 75°). Anhydr, mp 94.5° (obtained by drying the hemihydrate at 55° and 0.5 mm Hg). pKa (25°): 3.467. Soly of the anhydr acid in one liter of water at 18° = 17.04 g, at 25° = 21.17 g, at 30° = 25.65 g. Much more sol in boiling water. Slightly sol in petr ether.

906. Atropic Acid. *α-Methylenebenzeneacetic acid;* α-phenylacrylic acid. $C_9H_8O_2$; mol wt 148.16. C 72.96%, H 5.44%, O 21.60%. Prepn: Normant, Maitte, *Bull. Soc. Chim. France* **1956**, 1439.

Tabular or acicular crystals; volatilizes in steam. mp 106-107°. bp ∼267° with partial decompn. Sol in 790 parts water; sol in alc, benzene, chloroform, ether, CS_2.

Caution: Irritating to skin, mucous membranes.

907. Atropine. *endo-(±)-α-(Hydroxymethyl)benzeneacetic acid 8-methyl-8-azabicyclo[3.2.1]oct-3-yl ester; 1αH,-5αH-tropan-3α-ol (±)-tropate (ester);* dl-hyoscyamine; tropic acid ester with tropine; dl-tropyl tropate; tropine tropate. $C_{17}H_{23}NO_3$; mol wt 289.37. C 70.56%, H 8.01%, N 4.84%, O 16.59%. Parasympatholytic alkaloid isolated from *Atropa belladonna* L., and *Datura stramonium* L., and other *Solanaceae*. Extraction procedure: Chemnitius, *J. Prakt. Chem.* **116**, 276 (1927). During extraction, partial racemization of the *l*-hyoscyamine takes place which is completed by treatment with dil alkali on heating in chloroform soln: Schneider,

Arch. Pharm. **284,** 306 (1951). Structure and synthesis: Ladenburg, *Ann.* **217,** 75 (1883); Willstätter, *Ber.* **31,** 1537 (1898); *idem.*, *Ann.* **326,** 23 (1903); Schwenker *et al.*, *Ber.* **99,** 2407 (1966). Prepn of the sulfate: **Ger. pat.** 247,455 (1912 to Hoffmann-La Roche), *Frdl.* **11,** 1022. Clinical use of the mucate: J. S. Heron, *Can. Med. Assoc. J.* **72,** 302 (1955). Use as antidote to cholinesterase inhibitors: R. V. Brown, *Brit. J. Pharmacol.* **15,** 170 (1960). Effect on cardiac arrhythmias: P. Schweitzer, H. Mark, *Am. Heart J.* **100,** 119, 255 (1980). Pharmacokinetics and pharmacodynamics: P. H. Hinderling *et al.*, *J. Pharm. Sci.* **74,** 703, 711 (1985). Toxicity: R. L. Cahen, K. Tvede, *J. Pharmacol. Exp. Ther.* **105,** 166 (1952); Goldenthal, *Toxicol. Appl. Pharmacol.* **18,** 185 (1971). Review of clinical use in anesthesia: L. E. Shutt, J. B. Bowes, *Anaesthesia* **34,** 476-490 (1979). Comprehensive description: A. A. Al-Badr, F. J. Muhtadi, *Anal. Profiles Drug Subs.* **14,** 325-389 (1985). Review of clinical toxicology: J. D. Truwit, *Crit. Care Clinics* **7,** 639-657 (1991).

Long, orthorhombic prisms from acetone, mp 114-116°. Sublimes in high vacuum at 93-110°. pK 4.35; pH of 0.0015 molar soln 10.0. Absorption spectra: Dobbie, Fox, *J. Chem. Soc.* **103,** 1194 (1913); Fischer, *Arch. Exp. Pathol. Pharmakol.* **170,** 623 (1933). One gram dissolves in 455 ml water, 90 ml water at 80°, 2 ml alc, 1.2 ml alc at 60°, 27 ml glycerol, 25 ml ether, 1 ml chloroform; also sol in benzene, dil acids. LD$_{50}$ orally in rats: 750 mg/kg (Cahen, Tvede).

Hydrochloride, $C_{17}H_{23}NO_3 \cdot HCl$, granular crystals, mp 165°. Sol in water, alc. pH of 0.05 molar soln 5.8.

Methylbromide, $C_{17}H_{23}NO_3 \cdot CH_3Br$, *Tropin*. Crystals, mp 222-223°. Sol in 1 part water; slightly sol in alc. Almost insol in chloroform, ether.

Methylnitrate, $C_{17}H_{23}NO_3 \cdot CH_3NO_3$, *methylatropine nitrate, Eumydrin, Metropine, Harvatrate, Metanite, Ekomine.* Crystals, mp 163°. Freely sol in water or alc, very slightly in chloroform, ether.

Mucate, *atropine hyperduric.* Mucic acid salt of atropine.

Sulfate monohydrate, $(C_{17}H_{23}NO_3)_2 \cdot H_2SO_4 \cdot H_2O$, *Atropisol.* Granules or powder, mp 190-194°. Almost inactive optically. Very bitter. pH ~5.4. One gram dissolves in 0.4 ml water; 5 ml cold, 2.5 ml boil. alc; in 2.5 ml glycerol, 420 ml chloroform, 3000 ml ether. Bitterness threshold 1:10,-000. Incompatible with alkalies, tannin, salts of mercury or gold, vegetable decoctions or infusions, borax, bromides, iodides, benzoates. LD$_{50}$ orally in rats: 622 mg/kg (Goldenthal).

Caution: Causes blurred vision, suppressed salivation, vasodilation, hyperpyrexia, excitement, agitation, and delirium, *Clinical Toxicology of Commercial Products,* R. E. Gosselin *et al.*, Eds. (Williams & Wilkins, Baltimore, 5th Ed., 1984) Section III, pp 47-50.

THERAP CAT: Anticholinergic; mydriatic. In preanesthetic medication.

THERAP CAT (VET): Anticholinergic; mydriatic; antispasmodic. Antidote to organophosphous insecticides.

908. Atropine *N*-Oxide. *endo-(±)-α-(Hydroxymethyl)-benzeneacetic acid 8-methyl-8-azabicyclo[3.2.1]oct-3-yl ester N-oxide; 1αH,5αH-tropan-3α-ol (±)-tropate (ester), 8-oxide; atropine aminoxide; genatropine; aminoxytropine tropate.* $C_{17}H_{23}NO_4$; mol wt 305.37. C 66.86%, H 7.59%, N 4.59%, O 20.96%. Prepd by hydrogen peroxide oxidation of atropine or racemization of hyoscyamine N-oxide: Polonovski, Polonovski, *Bull. Soc. Chim. France* **39,** 1147 (1926).

Crystalline powder, mp 127-128°, dec 135°. Very hygroscopic. Soluble in alc, chloroform. Practically insol in ether.

Hydrochloride, $C_{17}H_{23}NO_4 \cdot HCl$, *Tropinox, Xtro.* Prisms from abs alc, mp 192-193°.

THERAP CAT: Anticholinergic.

909. Attacins. Immune protein P5. A major class of non-specific antibacterial proteins induced in the hemolymph of the pupae of silk moths, particularly *Hyalophora cecropia,* as part of the immune response. (*See also* Cecropins.) Originally designated as immune *protein P5,* attacins are a heterogeneous family consisting of four basic (A,B,C,-D) and two acidic (E,F) forms with similar amino acid sequences and mol wts of 20,000-23,000 daltons. Initial identification and isolation of immune response proteins: I. Faye *et al.*, *Infect. Immun.* **12,** 1426 (1975). Purification and preliminary characterization: A. F. Pye, H. G. Boman, *ibid.* **17,** 408 (1977). Amino acid sequences and antibacterial activity: D. Hultmark *et al.*, *EMBO J.* **2,** 571 (1983). Structural study on attacin F: Å. Engström *et al.*, *ibid.* **3,** 2065 (1984). Mode of action: P. Engström *et al.*, *ibid.* 3347.

910. Aucubin. *[1S-(1α,4aα,5α,7aα)]-1,4a,5,7a-Tetra-hydro-5-hydroxy-7-(hydroxymethyl)cyclopenta[c]pyran-1-yl-β-D-glucopyranoside; rhinanthin; aucuboside.* $C_{15}H_{22}O_9$; mol wt 346.33. C 52.02%, H 6.40%, O 41.58%. From leaves, roots, stalks and seeds of *Aucuba japonica* Thunb., *Cornaceae;* also occurs in 75 different plants: Paris, Chaslot, *Ann. Pharm. Franc.* **13,** 648 (1955). Isoln: Trim, Hill, *Biochem. J.* **50,** 310 (1952); Rombouts, Links, *Experientia* **12,** 78 (1956). Structure: Haegele *et al.*, *Tetrahedron Letters* **1961,** 110; Birch *et al.*, *J. Chem. Soc.* **1961,** 5194. Abs config: A. Bianco *et al.*, *Gazz. Chim. Ital.* **106,** 725 (1976). Brief review of structural studies: J. M. Bobbitt, K.-P. Segebarth, "The Iridoid Glycosides and Similar Substances" in *Cyclopentanoid Terpene Derivatives,* W. I. Taylor, A. R. Battersby, Eds. (Marcel Dekker, New York, 1969) pp 25-31.

Crystals from ethanol + ether, mp 181°. $[\alpha]_D^{21}$ −163.1° (c = 1.6). Soluble in water, alc, methanol; practically insol in chloroform, ether, petr ether. Absorption spectrum: Trim, Hill, *loc. cit.*

911. Auranofin. *(1-Thio-β-D-glucopyranose-2,3,4,6-tetraacetato-S)(triethylphosphine)gold; (2,3,4,6-tetra-O-acetyl-1-thio-β-D-glucopyranosato-S)(triethylphosphine)gold; (1-thio-β-D-glucopyranosato)(triethylphosphine)gold 2,3,4,6-tetra-acetate;* SKF-39162; Aktil; Crisinor; Crisofin; Ridaura; Ridauran. $C_{20}H_{34}AuO_9PS$; mol wt 678.49. C 35.41%, H 5.05%, Au 29.03%, O 21.22%, P 4.57%, S 4.73%. Orally active gold coordination complex. Prepn: E. R. McGusty, B. M. Sutton, **Ger. pat.** 2,051,495; *eidem,* **U.S. pat.** 3,635,-945 (1971, 1972 both to SKF). Series of articles on metabolism, distribution, pharmacokinetics, pharmacology, and clinical studies: *J. Rheumatol.* **6,** Suppl. 5, 1-164 (1979). Use in rheumatoid arthritis: A. E. Finkelstein *et al.*, *ibid.* **7,** 160 (1980). Inhibitory effect on HeLa cells: T. M. Simon *et* *al., Cancer* **44**, 1965 (1979). Screening trial with mouse lymphocyte leukemia: *eidem, Cancer Res.* **41**, 94 (1981). Toxicity study: B. H. Payne, D. T. Walz, *Vet. Pathol.* **15**, Suppl. 5, 1 (1978). *Review:* D. T. Walz in *Pharmacological and Biochemical Properties of Drug Substances* vol. 2, M. E. Goldberg, Ed. (Am. Pharm. Assoc., Washington, DC, 1979) pp 400-406. Symposium: *J. Rheumatol.* **9**, Suppl. 8 (1982) pp 1-209. Review of pharmacology and clinical studies: M. Chaffman *et al., Drugs* **27**, 378 (1984).

R = COCH₃

Colorless crystals, mp 110-111°. LD$_{50}$ in rats, mice (mg/kg): 265, 310 orally (Payne, Walz).

THERAP CAT: Antirheumatic.

912. Aurantiogliocladin. *2,3-Dimethoxy-5,6-dimethyl-2,5-cyclohexadiene-1,4-dione; 2,3-dimethoxy-5,6-dimethyl-p-benzoquinone.* C$_{10}$H$_{12}$O$_4$; mol wt 196.20. C 61.22%, H 6.16%, O 32.62%. Antibiotic substance produced by a *Gliocladium* sp similar to *Gliocladium roseum* (Link.) Bainier: Brian *et al., Experientia* **7**, 266 (1951). Structure: Vischer, *J. Chem. Soc.* **1953**, 815. Synthesis: Baker *et al., ibid.* 820; Seshadri, Venkatasubramanian, *ibid.* **1959**, 1660. Biosynthesis: Pettersson, *Acta Chem. Scand.* **19**, 1827 (1965).

Bright orange plates from petr ether, mp 62.5°. Absorption max: 407, 275 nm (ε 1200, 38400). Sol in chloroform, ethanol, pyridine. Gives a violet color with concd H$_2$SO$_4$.

Dihydro deriv, C$_{10}$H$_{14}$O$_4$, *(5S-trans)-2,3-dimethoxy-5,6-dimethylcyclohex-2-ene-1,4-dione, gliorosein.* Structure: Grave, *J. Chem. Soc. (C)* **1966**, 985. Needles, mp 48°. [α]$_D^{22}$ +125° (c = 0.2 in CHCl$_3$). uv max: 289 nm (ε 24320).

Quinhydrone deriv, *rubrogliocladin.* Dark red needles from petr ether, mp 74°. Absorption max (alcohol): 407, 275 nm (ε 650, 22000). Sol in most organic solvents, water.

913. Aureothin. *(Z,E)-2-Methoxy-3,5-dimethyl-6-[tetrahydro-4-[2-methyl-3-(4-nitrophenyl)-2-propenylidene]-2-furanyl]-4H-pyran-4-one;* mycolutein. C$_{22}$H$_{23}$NO$_6$; mol wt 397.43. C 66.49%, H 5.83%, N 3.52%, O 24.15%. Antibiotic by-product of aureothricin, *q.v.,* isolated from the culture of *Streptomyces thioluteus.* Isoln: K. Maeda, *J. Antibiot.* **6A**, 137 (1953); (as mycolutein): H. Schmitz, R. Woodside, *Antibiot. & Chemother.* **5**, 652 (1955). Structure: Y. Hirata *et al., Tetrahedron* **14**, 252 (1961). Identity with mycolutein: J. L. Schwartz *et al., J. Antibiot.* **29**, 236 (1976). Biosynthesis: R. Cardillo *et al., Tetrahedron* **30**, 459 (1974); M. Yamazaki *et al., Chem. Pharm. Bull.* **23**, 569 (1975). Synthesis: Y. Shizuri *et al., Chem. Letters* **1987**, 1381.

Yellow prisms, mp 158°. [α]$_D^{18}$ +51° (chloroform). Sol in methanol, ethanol, acetone, chloroform, tetrahydrofuran. Practically insol in water, nonpolar solvents. uv max (ethanol): 257, 346 nm (log ε 4.39, 4.27).

914. Aureothricin. *N-(4,5-Dihydro-4-methyl-5-oxo-1,2-dithiolo[4,3-b]pyrrol-6-yl)propanamide;* 4-methyl-6-propionamido-1,2-dithiolo[4,3-b]pyrrol-5(4H)-one; 5-methyl-3-propionamidopyrrolin-4-one[4,3-d]-1,2-dithiole; propionopyrrothine. C$_9$H$_{10}$N$_2$O$_2$S$_2$; mol wt 242.32. C 44.61%, H 4.16%, N 11.56%, O 13.21%, S 26.47%. Antibiotic substance produced by *Streptomyces sp* 26A (Mitaka, Tokyo, June 1947) which resembles *Nocardia farcinicus* or *Streptomyces lipmanii:* Umezawa *et al., Japan. Med. J.* **1**, 512 (1948); Umezawa *et al., J. Antibiot.* **2**, Suppl. A, 105 (1949); Maeda, *ibid.* **2**, 793 (1949); Maeda, *Japan. Med. J.* **2**, 85 (1949); Celmer *et al., J. Am. Chem. Soc.* **74**, 6304 (1952). Structure: Celmer, Solomons, *ibid.* **77**, 2861 (1955). Prepn: Celmer, U.S. pat. **2,752,359** (1956 to Pfizer). Total synthesis: Schmidt, Geiger, *Angew. Chem.* **74**, 328 (1962).

Crystals, usually solvated. Stable in air. When anhydr dec 260-270° (sublimes near 200°). uv max: 248, 312, 388 nm (ε 6100, 3900, 11000). Practically insol in water; slightly sol in ethyl acetate, butyl acetate, acetone, benzene, ether, alc.

915. Aurin. *4-[Bis(4-hydroxyphenyl)methylene]-2,5-cyclohexadien-1-one;* 4-(p,p'-dihydroxybenzhydrylidene)-2,5-cyclohexadien-1-one; *p*-rosolic acid; corallin; C.I. 43800. C$_{19}$H$_{14}$O$_3$; mol wt 290.32. C 78.61%, H 4.86%, O 16.53%. Prepd by heating a mixture of phenol, oxalic and sulfuric acids: Zulkowsky, *Ann.* **194**, 109, 122 (1878); **202**, 179 (1880); *Monatsh.* **16**, 358 (1895); *Colour Index* vol. 4 (3rd ed., 1971) pp 4407. By heating phenol with formic acid and stannous chloride: Nencki, Schmid, *J. Prakt. Chem.* [2] **23**, 549 (1881). By heating phenol and carbon tetrachloride to 140-160° in presence of AlCl$_3$ or ZnCl$_2$ and treating the reaction product with water: Heumann, Ger. pat. **68,976** (*Frdl.* **3**, 103).

Deep red (garnet-like) crystals with metallic luster. Orthorhombic bipyramidal. Technical product may occur as yellowish-brown pieces with dark green metallic fracture. Dec 308-310°. Not volatile in vacuo at 180°. Absorption max (KOH): 534.6, 479.5 nm. Practically insol in water (0.12%) and in benzene. Freely sol in alc, giving a golden yellow solution, in aq or alc NaOH and KOH giving a carmine-red solution, in concd or 70% H$_2$SO$_4$, in HCl, in HClO$_4$ giving a yellow to orange soln. Moderately soluble in glacial acetic acid; slightly sol in ether, chloroform.

Sodium salt (yellow corallin), yellowish pieces with greenish, metallic luster; sol in water, giving a carmine-red soln.

USE: Dye intermediate.

916. Aurodox. *1-Methylmocimycin;* antibiotic X-5108; goldinodox; goldinomycin; X-5108. C$_{44}$H$_{62}$N$_2$O$_{12}$; mol wt 810.98. C 65.17%, H 7.71%, N 3.45%, O 23.67%. Antibiotic produced by *Streptomyces goldiniensis* var. *goldiniensis:* J. Berger, Ger. pat. **2,140,322;** *idem,* U.S. pat. **3,708,577** (1972, 1973 both to Hoffmann-La Roche). Isoln, production, properties and toxicity data: J. Berger *et al., J.*

Antibiot. **26**, 15 (1973). Improved production: J. Unowsky, D. C. Hoppe, *ibid.* **31**, 662 (1978). Structure: H. Maehr *et al., J. Am. Chem. Soc.* **95**, 8449 (1973). Absolute stereochemistry: *eidem, ibid.* **96**, 4034 (1974). Biosynthesis: C.-M. Liu *et al., J. Antibiot.* **30**, 416 (1977); **32**, 414 (1979). Stereospecific total synthesis: R. E. Dolle, K. C. Nicolaou, *J. Am. Chem. Soc.* **107**, 1691, 1695 (1985). Growth promoting activity: W. L. Marusich *et al., Poultry Sci.* **53**, 636 (1974). Review of chemistry: H. Maehr *et al., Can. J. Chem.* **58**, 501 (1980).

Yellow amorphous solid. Weakly acidic. Sol in methanol, ethyl acetate, CHCl$_3$, acetone, methylene chloride. Insol in water. pKa = 6.1. Unstable in acidic or basic soln, but can be kept in aq soln at pH 7-9 for 4 hrs at 25° without significant loss of activity. Solns kept at 25° or heated to 56° are 2-10 times more stable in methanol than in water or pH 9 borate buffer. LD$_{50}$ in mice (mg/kg): > 1000 s.c., > 4000 orally (Berger *et al.*).

Sodium salt, C$_{44}$H$_{61}$N$_2$NaO$_{12}$, yellow solid. $[\alpha]_D^{25}$ −82.8° (c = 0.52 in ethanol). uv max (0.1N HCl): 334, 233, 206 nm (E$_{1cm}^{1\%}$ 403, 610, 500); (0.1N KOH): 327, 231 nm (E$_{1cm}^{1\%}$ 416, 647). Sol in water, methanol, ethanol, isopropanol, butanol, DMF. Slightly sol in amyl alcohol, THF, dioxane. Insol in benzene, chloroform, ethyl ether, petr ether.

THERAP CAT (VET): Poultry growth promotant.

917. Aurothioglucose. *(1-Thio-D-glucopyranosato-O2,-S1)gold;* (1-D-glucosylthio)gold; gold thioglucose; Aureotan; Solganal; Solganal B; Aurumine; Oronol. C$_6$H$_{11}$AuO$_5$S; mol wt 392.18. C 18.38%, H 2.83%, Au 50.22%, O 20.40%, S 8.18%. Prepd by adding a soln of gold bromide to an aq soln of thioglucose containing sulfur dioxide; after refluxing, the gold thioglucose is pptd with alcohol. The product is purified by dissolving in water and reprecipitating with alc: Lebeau-Janot, *Traité de Pharmacie Chimique* **II**, 661 (1956). Clinical comparison with auranofin, *q.v.*, in rheumatoid arthritis: P. L. C. M. Van Riel *et al., Clin. Rheumatol.* **5**, 359 (1986).

Yellow crystals. Slight mercaptan-like odor. Sol in water with decompn; slightly sol in propylene glycol. Insol in alc and in most other organic solvents and in vegetable oils.

USE: Exptly to produce obesity in animals: *Chem. & Eng. News* **32**, 22 (1954).

THERAP CAT: Antirheumatic.

918. Aurothioglycanide. *(2-Mercapto-N-phenylacetamidato-O,S)gold;* *[[(phenylcarbamoyl)methyl]thio]gold;* α-auromercaptoacetanilide; 2-mercaptoacetanilide S-gold derivative; aurothioglycolic acid anilide; aurothioglycolanilide; Lauron. C$_8$H$_8$AuNOS; mol wt 363.19. C 26.46%, H 2.22%, Au 54.23%, N 3.86%, O 4.41%, S 8.83%. Prepd by the action of aurous bromide on thioglycolic acid anilide: Lewenstein, U.S. pat. 2,451,841 (1948).

Grayish-yellow powder, mp 238-241°. Insol in water, benzene, ether, chloroform, acids, bases. Marketed as a suspension in sesame oil.

THERAP CAT: Antirheumatic.

919. Avermectins. AVM. A group of broad-spectrum antiparasitic antibiotics which are derivs of pentacyclic 16-membered lactones related to the milbemycins, *q.v.* Isoln from a novel actinomycete, *Streptomyces avermitilis,* and separation of major components A$_{1a}$, A$_{2a}$, B$_{1a}$, B$_{2a}$ and minor components A$_{1b}$, A$_{2b}$, B$_{1b}$, B$_{2b}$: G. Albers-Schönberg *et al.,* Ger. pat. **2,717,040;** *eidem,* U.S. pat. **4,310,519** (1977, 1982 both to Merck & Co.); R. W. Burg *et al., Antimicrob. Ag. Chemother.* **15**, 361 (1979); T. W. Miller *et al., ibid.* 368. Antiparasitic activity: S. R. Egerton *et al., ibid.* 372. Mechanism of action: L. C. Fritz *et al., Proc. Nat. Acad. Sci. USA* **76**, 2062 (1979). Pesticidal activity: I. Putter *et al., Experientia* **37**, 963 (1981). Structure determn: G. Albers-Schönberg *et al., J. Am. Chem. Soc.* **103**, 4216 (1981). Absolute configuration of avermectin B$_{1a}$ and aglycon B$_{2a}$: J. P. Springer *et al., ibid.* 4221. Prepn of aglycons: H. Mrozik *et al., J. Org. Chem.* **47**, 489 (1982). Total synthesis of A$_{1a}$ (C$_{49}$H$_{74}$O$_{14}$): S. J. Danishefsky *et al., J. Am. Chem. Soc.* **109**, 8119 (1987). Biosynthetic studies: D. E. Cane *et al., ibid.* **105**, 4110 (1983); M. D. Schulman *et al., J. Antibiot.* **39**, 541 (1986); T. S. Chen *et al., Arch. Biochem. Biophys.* **269**, 544 (1989). *Reviews:* W. C. Campbell *et al., ACS Symposium Series* **255**, 1-20 (1984); M. H. Fisher, H. Mrozik, "The Avermectin Family of Macrolide Antibiotics" in *Macrolide Antibiotics,* S. Omura, Ed. (Academic Press, New York, 1984) pp 553-606; *Southwest Entomol.* **1985**, Suppl. 7, 1-51; J. R. Babu, *ACS Symposium Series* **380**, 91-108 (1988). Review of syntheses: T. Blizzard *et al.,* in *Recent Prog. Synth. Antibiotics,* G. Lukacs, M. Ohno, Eds (Springer-Verlag, Berlin, 1990) pp 66-102. Review of structure-activity relationships: W. L. Shoop *et al., Vet. Parasitol.* **59**, 139-156 (1995). For related structures *see* abamectin, ivermectin.

Avermectin A$_{1a}$

Avermectin A$_{1a/b}$, $[\alpha]_D^{27}$ +68.5 ± 2° (c = 0.77 in chloroform). uv max (methanol): 237, 243, 252 nm (ε 28700, 31275, 20290).

Avermectin A$_{2a/b}$, $[\alpha]_D^{27}$ +48.8 ± 2° (c = 1.64 in chloroform). uv max (methanol): 237, 243, 245 nm (ε 28800, 31740, 20425).

Avermectin B$_{1a/b}$, see Abamectin.

Avermectin B$_{2a/b}$, $[\alpha]_D^{27}$ +38.3 ± 2° (c = 0.87 in chloroform). uv max (methanol): 237, 243, 252 nm (ε 27580, 30590, 20060).

920. Avidin. Basic glycoprotein isolated from raw egg white. Exhibits high binding affinity for biotin and is capable of producing biotin deficiency in rats and chicks. Occurs in the white portion of eggs and the oviducts of birds and amphibia. Destroyed by cooking or irradiation.

Isoln: Eakin *et al., J. Biol. Chem.* **136,** 801 (1940); Pennington *et al., J. Am. Chem. Soc.* **64,** 469 (1942); Fraenkel-Conrat *et al., Arch. Biochem. Biophys.* **39,** 80, 97 (1952). Improved purification and crystallization: Green *et al., Biochem. J.* **118,** 67, 71 (1970). Structure is a glycoprotein containing four essentially identical subunits: Green, *ibid.* **92,** 16c (1964). The combined mol wt of the subunits is about 66,000. Each subunit is a single polypeptide chain containing 128 amino acid residues with alanine at the *N*-terminal, glutamic acid at the *C*-terminal, and a carbohydrate moiety attached at the asparaginyl residue, position 17. Complete amino acid sequence of the protein subunit: DeLange, Huang, *J. Biol. Chem.* **246,** 698 (1971). Studies on biotin inactivation by avidin: Becker, Wilchek, *Biochim. Biophys. Acta* **264,** 165 (1972). Review of chemistry and binding properties: N. M. Green, *Advan. Protein Chem.* **29,** 85-133 (1975). Review of use of avidin-biotin complex in molecular biology: E. A. Bayer, M. Wilchek, *Methods Biochem. Anal.* **26,** 1-45 (1980); in immunoassays: M. Wilchek, E. A. Bayer, *Immunol. Today* **5,** 39-43 (1984).

USE: Biochemical tool for affinity chromatography, affinity cytochemistry and immunoassay.

921. Avilamycin. LY-048740; Maxus; Surmax. Polyether antibiotic complex produced by *Streptomyces viridochromogenes,* ETH 23575 (NRRL 2860); belongs to the orthosomycin family of antibiotics, members of which contain one or more ortho ester linkages associated with carbohydrate residues. Isoln: E. Gaeumann *et al., Ger.* pat. **1,116,-864;** *eidem,* U.S. pat. **3,131,126** (1961, 1964 both to Ciba-Geigy); and characterization: F. Buzzetti *et al., Experientia* **24,** 320 (1968). Mechanism of action studies: H. Wolf, *FEBS Letters* **36,** 181 (1973). Avilamycin A is the main component; avilamycins B through N are also known and are derivatives of A differing at the C-45 linkage and/or the C-56 ketone adduct. Isoln of avilamycin C: W. Heilman *et al., Helv. Chim. Acta* **62,** 1 (1979). Structural study of A and C: *eidem, ibid.;* W. Keller-Schierlein *et al., ibid.* 7; E. Kupfer *et al., ibid.* **65,** 3 (1982). Isoln and structure of avilamycins F through N: J. L. Mertz *et al., J. Antibiot.* **39,** 877 (1986). GC determn of avilamycin residues in swine tissue and fluids: G. Formica, C. Giannone, *J. Assoc. Off. Anal. Chem.* **69,** 763 (1986). Synthesis of the A-B fragment of avilamycins A and C: P. Jütten *et al., Tetrahedron* **43,** 4133 (1987); of the disaccharide C-D fragment: J.-M. Beau *et al., Tetrahedron Letters* **28,** 1105 (1987). Use as a feed additive: F. Knusel *et al.,* U.S. pat. **4,185,091** (1980 to Ciba-Geigy); in prevention of swine dysentery: E. E. Ose, U.S. pat. **4,436,734** (1984 to Lilly). Effect on feed conversion efficiency in swine: D. J. Jones *et al., J. Anim. Sci.* **65,** 881 (1987). Metabolism in swine: J. D. Magnussen *et al., J. Agric. Food Chem.* **39,** 306 (1991).

Avilamycin A

Colorless, needle-shaped crystals from acetone/ether, mp 188-189.5°. $[\alpha]_D^{20}$ +0.8° (c = 1.165 in abs ethanol) and −7.7° (c = 1.083 in chloroform).

Avilamycin A, $C_{61}H_{88}Cl_2O_{32}$, colorless needles from chloroform/petr ether, mp 181-182° (1-2.H$_2$O). uv max (methanol): 227, 286 br nm (log ε 4.15, 3.33).

Avilamycin C, $C_{61}H_{90}Cl_2O_{32}$. Dihydrate, colorless fine plates from acetone/ether, mp 188-189°. $[\alpha]_D^{20}$ −4.8° (c = 1.44 in chloroform). uv max (methanol): 228, 284 br nm (log ε 4.12, 3.33).

THERAP CAT (VET): Growth promotant.

922. Avoparcin. AV-290; C-254; CL-81588; LL-AV290; Avotan. Animal feed glycopeptide antibiotic complex produced by *Streptomyces candidus.* Prodn: M. P. Kunstmann, J. N. Porter, U.S. pat. **3,338,786** (1967 to Am. Cyanamid). Fermentation, isoln, characterization: M. P. Kunstmann *et al., Antimicrob. Ag. Chemother.* **8,** 242 (1968). Isoln, purif of major components: W. J. McGahren *et al., J. Antibiotic.* **36,** 1671 (1983). HPLC separation of α-, β-avoparcin: F. Sztaricskai *et al., ibid.* 1691. Structural studies: J. J. Hlavka *et al., Tetrahedron Letters* **1974,** 175; W. J. McGahren *et al., J. Am. Chem. Soc.* **101,** 2237 (1979). Structure: *eidem, ibid.* **102,** 1671 (1980). Stereochemistry and epimerization: G. A. Ellestad *et al., J. Antibiot.* **36,** 1683 (1983). Relationship between structure, antibacterial activity: S.W. Fesik *et al., Mol. Pharmacol.* **25,** 275 (1984). Effect on resistance patterns in chickens: J. R. Walton, *Zentralbl. Veterinaermed. (B)* **25,** 290 (1978), *C.A.* **89,** 173507 (1978). Prevention of exptly induced necrotic enteritis in chickens: J. F. Prescott, *Avian Dis.* **23,** 1072 (1979). Effect on feedlot performance: R. J. Johnson *et al., J. Anim. Sci.* **48,** 1338 (1979).

α-Avoparcin　　R = H
β-Avoparcin　　R = Cl

White, hygroscopic, amorphous solid; no definitive mp. uv max: 280 nm in neutral or acidic solns; 300 nm in basic solns. Sol in water, DMF, DMSO. Max stability of aq solns is at pH 4-8. Moderately sol in methanol. LD$_{50}$ in mice, rats, and chickens: >10000 mg/kg orally. (Am. Cyanamid, company literature).

α-Avoparcin, $C_{89}H_{102}ClN_9O_{36}$, $[\alpha]_D^{25}$ −96° ±2° (c = 0.62 in 0.1*N* HCl). uv max (0.1*N* HCl): 280 nm ($E_{1cm}^{1\%}$ 42.0).

β-Avoparcin, $C_{89}H_{101}Cl_2N_9O_{36}$, $[\alpha]_D^{25}$ −102° ±2° (c = 0.65 in 0.1*N* HCl). uv max (0.1*N* HCl): 280 nm ($E_{1cm}^{1\%}$ 44.0).

THERAP CAT (VET): Antibacterial; growth promotant.

923. Azacitidine. *4-Amino-1-β-D-ribofuranosyl-1,3,5-triazin-2(1H)-one;* 5-azacytidine; ladakamycin; U-18496; NSC-102816; Mylosar. $C_8H_{12}N_4O_5$; mol wt 244.21. C 39.35%, H 4.95%, N 22.94%, O 32.76%. Glycosyl deriv of *5-azocytosine* with antibiotic and antitumor activity. Chemical synthesis: A. Piskala, F. Sorm, *Coll. Czech. Chem. Commun.* **29,** 2060 (1964); *Neth.* pat. **Appl. 6,403,587;** F. Sorm, A. Piskala, U.S. pat. **3,350,388** (1965, 1967 both to Cesk. Acad. Ved.). Production by *Streptoverticillium ladakanus* and biological activity: L. J. Hanka *et al., Antimicrob. Ag. Chemother.* **1966,** 619. Isoln and structure: M. E. Bergy,

R. R. Herr, *ibid.* 625. Pharmacokinetics: C. J. Kelly *et al.,* *Cancer Treat. Rep.* **62,** 1025 (1978). Metabolism: P. G. Plagemann *et al., Cancer Res.* **38,** 2458 (1978). Use in acute leukemia: J. H. Saiki *et al., Cancer* **42,** 2111 (1979); P. L. Lomen *et al., Neoplasma* **27,** 101 (1980). Azacitidine has been shown to selectively increase γ-globulin synthesis in a patient with β-thalassemia: T. J. Ley, *N. Engl. J. Med.* **307,** 1469 (1982). Toxicology study: P. E. Palm, C. J. Kensler, *U.S. Clearinghouse Fed. Sci. Tech. Inform.,* **PB-194791** (1970) 191 pp., *C.A.* **75,** 33704j (1971). Carcinogenicity study: G. D. Stoner *et al., Cancer Res.* **33,** 3069 (1973). *Reviews:* C. C. Cheng, *Ann. Rep. Med. Chem.* **7,** 129-144 (1972); A. Cibak, *Oncology* **30,** 405-422 (1974).

Crystals from methanol, mp 228-230°. $[\alpha]_D^{25}$ +39° (c = 1 in water). uv max (water): 241 nm (ε 8767); (0.01*N* HCl): 249 nm (ε 3077); (0.01*N* KOH): 223 nm (ε 24200). LD_{50} in mice (mg/kg): 115.9 i.p.; 572.3 orally (Palm, Kensler).

THERAP CAT: Antineoplastic.

924. Azacosterol. *(3β,17β)-17-[[3-(Dimethylamino)-propyl]methylamino]androst-5-en-3-ol; N*-methyl-*N*-[3-(dimethylamino)propyl]-17β-aminoandrost-5-en-3β-ol; 20,25-diazacholesterol; diazasterol. $C_{25}H_{44}N_2O$; mol wt 388.64. C 77.26%, H 11.41%, N 7.21%, O 4.12%. Prepn: Counsell *et al., J. Med. Pharm. Chem.* **5,** 1224 (1962); Counsell, Klimstra, **U.S. pat. 3,084,156** (1963 to Searle).

Crystals from acetone + methanol, mp 146-148°. $[\alpha]_D$ −54.5°.

Dihydrochloride, $C_{25}H_{44}N_2O.2HCl$, *SC-12937, Azasterol, Ornitrol.* $[\alpha]_D$ −32°.

THERAP CAT: Hypocholesteremic.

THERAP CAT (VET): Avian chemosterilant.

925. Azacyclonol. *α,α-Diphenyl-4-piperidinemethanol; α*-(4-piperidyl)benzhydrol; diphenyl (γ-pyridyl)carbinol; gamma-pipradrol; MER-17; Psychosan; Frenoton; Ataractan; Calmeran. $C_{18}H_{21}NO$; mol wt 267.37. C 80.86%, H 7.92%, N 5.24%, O 5.98%. Prepd by hydrogenation of α,α-diphenyl-4-pyridinemethanol: Schumann *et al.,* U.S. pat. **2,804,422** (1957 to Merrell).

Crystals, mp 160-161°.

Hydrochloride, $C_{18}H_{21}NO.HCl$, *Frenquel.* Crystals from butanone, mp 283-285°. Slightly bitter taste. Moderately sol in water.

THERAP CAT: Anxiolytic.

926. Azadirachtin. $C_{35}H_{44}O_{16}$; mol wt 720.72. C 58.33%, H 6.15%, O 35.52%. A tetranortriterpinoid isolated from the seeds of the neem tree, *Azadirachta indica* A. Juss. *(Melia azadirachta* L.), *Meliaceae,* and the chinaberry tree, *M. azedarach* L. Highly active insect feeding deterrent and growth regulator. Isoln from *A. indica* and identification as feeding inhibitor in locusts: J. H. Butterworth, E. D. Morgan, *Chem. Commun.* **1968,** 23; from *M. azedarach:* E. D. Morgan, M. D. Thornton, *Phytochemistry* **12,** 391 (1973). Partial synthesis: D. Pflieger *et al., Tetrahedron Letters* **28,** 1519 (1987). Structural studies: J. H. Butterworth *et al., J. Chem. Soc. Perkin Trans. II* **1972,** 2445. ¹H- and ¹³C-NMR data and structure: P. R. Zanno *et al., J. Am. Chem. Soc.* **97,** 1975 (1975); K. Nakanishi in *Recent Advances in Phytochemistry,* **vol. 9,** V. C. Runeckles, Ed. (Plenum Press, New York, 1975) pp 283-298. Revised structure: W. Kraus *et al., Tetrahedron Letters* **26,** 6435 (1985); H. B. Broughton *et al., Chem. Commun.* **1986,** 46. Isoln by HPLC: E. C. Uebel *et al., J. Liq. Chromatog.* **2,** 875 (1979); J. D. Warthen, Jr. *et al., ibid.* **7,** 591 (1984). Antifeedant activity in locusts: J. S. Gill, C. T. Lewis, *Nature* **232,** 402 (1971); in fall army worms, cotton bollworms: J. A. Klocke, I. Kubo, *Entomol. Exp. Appl.* **32,** 299 (1982). Insect ecdysis and growth inhibition: H. Rembold, K. P. Sieber, *Z. Naturforsch.* **36C,** 466 (1981); I. Kubo, J. A. Klocke, *Agr. Biol. Chem.* **46,** 1951 (1982); K. P. Sieber, H. Rembold, *J. Insect. Physiol.* **29,** 523 (1983). Series of articles on chemistry and activity: *Natural Pesticides from the Neem Tree,* Proc. 1st Int. Neem Conf., 1980, H. Schmutterer *et al.,* Eds. (German Agency for Technical Cooperation, Eschborn, 1981) 291 pp.

Microcrystalline powder from carbon tetrachloride, mp 154-158°. $[\alpha]_D$ −53° (c = 0.5 in $CHCl_3$). uv max (methanol): 217 nm (ε 9100).

USE: Experimentally as insect control agent.

927. Azafrin. *(5R,6R)-5,6-Dihydro-5,6-dihydroxy-10'-apo-β,ψ-carotenoic acid;* escobedin. $C_{27}H_{38}O_4$; mol wt 426.60. C 76.02%, H 8.98%, O 15.00%. Carotenoid-carboxylic acid from roots of the South American plant "Azafranillo," *Escobedia scabrifolia* Ruiz & Pav., and *Escobedia laevis* Cham. & Schlecht., *Scrophulariaceae.* Isoln: R. Kuhn *et al., Ber.* **64,** 333 (1931); *ibid.* **65,** 1873 (1932). Structure: R. Kuhn, A. Deutsch, *ibid.* **66,** 883 (1933); R. Kuhn, H. Brockmann, *ibid.* **67,** 885 (1934); *Ann.* **516,** 104 (1935). Absolute configuration (5*R*:6*R*): W. Eschenmoser, C. H. Eugster, *Helv. Chim. Acta* **58,** 1722 (1975).

Orange-colored prisms from toluene, mp 213°. $[\alpha]_{6438}^{20}$ −75° (c = 0.28 in alcohol). Absorption max (chloroform): 458, 428 nm. Practically insol in water. Sol in dil NaOH or Na_2CO_3 solns, in chloroform, alc, acetic acid and benzene; sparingly sol in ether.

Methyl ester, $C_{28}H_{40}O_4$, reddish-yellow leaflets from ether, mp 191°. $[\alpha]_{6438}^{22}$ −32° (chloroform).

928. 8-Azaguanine. *5-Amino-1,4-dihydro-7H-1,2,3-triazolo[4,5-d]pyrimidin-7-one; 5-amino-1,6-dihydro-7H-v-triazolo[4,5-d]pyrimidine-7-one; 5-amino-1H-v-triazolo[d]-pyrimidin-7-ol; 5-amino-7-hydroxy-1H-v-triazolo[d]pyrim-idine; pathocidin; guanazolo.* $C_4H_4N_6O$; mol wt 152.12. C 31.58%, H 2.65%, N 55.25%, O 10.52%. Triazolo analog of guanine. Prepd from 2,4,5-triamino-6-hydroxypyrimidine: Roblin, Jr. *et al., J. Am. Chem. Soc.* **67,** 290 (1945); from 2-amino-5-nitro-4-oxopyrimidine: H. U. Blank *et al., J. Org. Chem.* **35,** 1131 (1970). Formation from guanine by *Streptomyces albus* and identity with pathocidin: K. Hirasawa, K. Isono, *J. Antibiot.* **31A,** 628 (1978). The first purine analog to show carcinostatic effects in murine malignancies; is readily incorporated into ribonucleic acids. *Review:* R. E. Parks, Jr., K. C. Agarwal, *Handb. Exp. Pharmacol.* **38,** pt. 2, 458 (1975); D. Grunberger, G. Grunberger in *Antibiotics* **vol. 5**(pt. 2), F. E. Hahn, Ed. (Springer-Verlag, New York, 1979) pp 110-123.

Crystals from dil aq NaOH. Dec above 300° without melting. Absorption spectrum: L. F. Cavalieri *et al., J. Am. Chem. Soc.* **70,** 3875 (1948). Soluble in dil caustic, in dil acids. Practically insol in water, alcohol, ether.

USE: Purine antimetabolite.

929. Azamethonium Bromide. *2,2'-(Methylimino)bis-[N-ethyl-N,N-dimethylethanaminium]dibromide; [(methylimino)diethylene]bis[ethyldimethylammonium bromide];* 3-methyl-3-azapentane-1,5-bis(ethyldimethylammonium) dibromide; *N,N,N',N'-3-pentamethyl-N,N'-diethyl-3-azapentylene-1,5-diammonium dibromide;* pentamethazene dibromide; Präparat 9295; Ciba 9295; Pendiomid; Azameton; Azamethone; Ganlion; Pentaméthazène. $C_{13}H_{33}Br_2N_3$; mol wt 391.23. C 39.91%, H 8.50%, Br 40.85%, N 10.74%. Ganglion blocking agent. Prepd from pentamethyl diethylenetriamine and ethyl bromide: Marxer, Miescher, *Helv. Chim. Acta* **34,** 924 (1951); *Swiss* pat. **284,212,** *C.A.* **48,** 7625 (1954); *U.S.* pat. **2,654,785** (1952, 1953 both to Ciba). Pharmacological and toxicity studies: H. J. Bein, R. Meier, *Experientia* **6,** 351 (1950); *eidem, Schweiz. Med. Wochenschr.* **81,** 446 (1951); I. V. Uranov, *Farmakol. Toksikol.* **21,** 110 (1958).

Prisms or rhombohedra from alcohol + ethyl acetate, mp 212-215°. Freely sol in water. pH of satd aq soln about 6.5. LD_{50} in mice (mg/kg): 2500 orally; 60 i.v. (Uranov). LD_{50} in rabbits (mg/kg): 3000 orally; 160 s.c.; 75 i.v. (Bein, Meier, 1951).

THERAP CAT: Antihypertensive.

930. Azanidazole. *(E)-4-[2-(1-Methyl-5-nitro-1H-imidazol-2-yl)ethenyl]-2-pyrimidinamine; (E)-2-amino-4-[2-(1-methyl-5-nitroimidazol-2-yl)vinyl]pyrimidine;* nitromidine; F-4; Triclose. $C_{10}H_{10}N_6O_2$; mol wt 246.23. C 48.78%, H 4.09%, N 34.13%, O 13.00%. Prepn and use: A. Garzia, **Ger.** pat. **2,358,483;** *eidem,* **U.S.** pats. **3,882,105** and **3,969,-520** (1974, 1975, 1976 all to Chemoterapico). Toxicological and teratological studies: R. Tammiso *et al., Arzneimittel-Forsch.* **28,** 2251 (1978).

Bright yellow odorless powder, mp 232-235°. Sol in DMF, DMSO, mineral oils and acids. Slightly sol in dioxane and acetone. LD_{50} in mice, rats (mg/kg): 5100, 7600 orally; 590, 860 i.p. (Tammiso).

THERAP CAT: Antiprotozoal (Trichomonas).

931. Azaperone. *1-(4-Fluorophenyl)-4-[4-(2-pyridinyl)-1-piperazinyl]-1-butanone; 4'-fluoro-4-[4-(2-pyridyl)-1-piperazinyl]butyrophenone;* 1-[3-(4-fluorobenzoyl)propyl]-4-(2-pyridyl)piperazine; R-1929; Stresnil; Suicalm. $C_{19}H_{22}$-FN_3O; mol wt 327.40. C 69.70%, H 6.77%, F 5.80%, N 12.83%, O 4.89%. Prepn: Janssen, **U.S.** pat. **2,979,508** (1961). Synthesis of labeled compound: Soudijn, van Wijngaarden, *J. Lab. Comp.* **4,** 159 (1968). Distribution and metabolism studies in rat and pig: Heykants *et al., Arzneimittel-Forsch.* **21,** 982, 1263, 1357 (1971). Veterinary clinical studies: Symoens, van den Brande, *Vet. Rec.* **85,** 64 (1969). Crystal structure: M. H. J. Koch *et al., Acta Crystallogr.* **33B,** 1975 (1977).

Crystals, mp 73-75°.

THERAP CAT (VET): Sedative; tranquilizer.

932. Azaserine. *L-Serine diazoacetate(ester); O-diazoacetyl-L-serine;* CL-337; CN-15757; P-165. $C_5H_7N_3O_4$; mol wt 173.12. C 34.69%, H 4.08%, N 24.27%, O 36.97%. $N_2CHCOOCH_2CH(NH_2)COOH$. Antibiotic substance produced by a *Streptomyces* sp. or by synthesis: Bartz *et al., Nature* **173,** 72 (1954); Fusari *et al., J. Am. Chem. Soc.* **76,** 2878, 2881 (1954); Moore *et al., ibid.,* 2884; Nicolaides *et al., ibid.,* 2887; T. J. Curphey, D. S. Daniel, *J. Org. Chem.* **43,** 4666 (1978); J. Ehrlich *et al.,* **U.S.** pat. **2,996,435;** J. A. Moore *et al.,* **U.S.** pat. **3,030,388** (1961, 1962 both to Parke, Davis). Retards growth of transplantable neoplasms in animals: Stock *et al., Nature* **173,** 71 (1954). Pathology and pharmacology studies: S. S. Sternberg, F. S. Philips, *Cancer* **10,** 889 (1957). Crystal and molecular structure: A. Fitzgerald, L. H. Jensen, *Acta Crystallogr.* **B34,** 828 (1978). Review of carcinogenicity studies: *IARC Monographs* **10,** 73-77 (1976). *Review:* Pettillo, Hunt, *Antibiotics,* **vol. 1,** D. Gottlieb, P. Shaw, Eds. (Springer-Verlag, New York, 1967) pp 481-493.

Orthorhombic, pale yellow to green crystals from 90% ethanol, dec 146-162°. $[\alpha]_D^{27.5} -0.5°$ (c = 8.46 in H_2O at pH 5.18). uv max (pH 7): 250.5 nm ($E_{1cm}^{1\%}$ 1140); in $0.1N$ NaOH: 252 nm ($E_{1cm}^{1\%}$ 1230). Very sol in water; slightly sol in methanol, abs ethanol acetone, but sol in warm aq solns of these solvents. pKa 8.55. Aqueous solutions are most stable at pH 8. Incompatible with acids. LD_{50} in mice, rats (mg/kg/day): 150, 170 orally (Sternberg, Philips).

THERAP CAT: Antifungal; antineoplastic.

933. Azasetron. *(±)-N-1-Azabicyclo[2.2.2]oct-3-yl-6-chloro-3,4-dihydro-4-methyl-3-oxo-2H-1,4-benzoxazine-8-carboxamide;* (±)-6-chloro-3,4-dihydro-4-methyl-3-oxo-N-(3-quinuclidinyl)-2H-1,4-benzoxazine-8-carboxamide; nazasetron. $C_{17}H_{20}ClN_3O_3$; mol wt 349.82. C 58.37%, H 5.76%, Cl 10.13%, N 12.01%, O 13.72%. 5-HT_3 receptor antagonist. Prepn: T. Tahara *et al.,* **Eur.** pat. **Appl. 313,393;** *eidem,* **U.S.** pat. **4,892,872** (1989, 1990 both to Yoshitomi); T. Kawakita *et al., Chem. Pharm. Bull.* **40,** 624 (1992). Receptor binding study: N. Sato *et al., Japan. J. Pharmacol.* **59,** 443 (1992). Pharmacology: K. Haga *et al., ibid.* **63,** 377 (1993). Toxicity study: S. Takagi *et al., Oyo Yakuri* **42,** 255 (1991), *C.A.* **117,** 40208q (1991).

Hydrochloride, $C_{17}H_{20}ClN_3O_3 \cdot HCl$, *Y-25130, Serotone.* Crystals from ethanolic HCl, mp 281° (dec). Also reported as crystals from ethanol, mp 305° (dec) (Kawakita). LD_{50} in male, female rats (mg/kg): 135, 132 i.v. (Takagi).

THERAP CAT: Antiemetic.

934. Azatadine. *6,11-Dihydro-11-(1-methyl-4-piperidinylidene)-5H-benzo[5,6]cyclohepta[1,2-b]pyridine;* 4-aza-5-(N-methyl-4-piperidinylidene)-10,11-dihydro-5H-dibenzo-[a,d]cycloheptene. $C_{20}H_{22}N_2$; mol wt 290.41. C 82.72%, H 7.64%, N 9.65%. Prepn: F. J. Villani, **Belg. pat. 647,043** (1964 to Scherico) corresp to U.S. pats. **3,326,924** and **3,419,565** (1967, 1968 both to Schering); *see also* U.S. pat. **3,357,986** (1967 to Schering); F. J. Villani *et al., J. Med. Chem.* **15,** 750 (1972). Pharmacology: S. Tozzi *et al., Agents Actions* **4,** 264 (1974); T. Yanagita *et al., Jitchuken, Zenrinsho Kenkyuho* **1,** 79 (1976), *C.A.* **84,** 38611x (1976). Toxicity studies: H. Tanaka *et al., ibid.* 173 (1975), *C.A.* **84,** 25951h (1976).

Crystals from isopropyl ether, mp 124-126°. Dimaleate, $C_{20}H_{22}N_2 \cdot 2C_4H_4O_4$, *Sch-10649, Atiramin, Bonamid, Idulian, Optimine, Zadine.* Crystals from ethyl acetate-methanol, mp 152-154°.

THERAP CAT: Antihistaminic.

935. Azathioprine. *6-[(1-Methyl-4-nitro-1H-imidazol-5-yl)thio]-1H-purine;* 6-(1-methyl-4-nitro-5-imidazolyl)mercaptopurine; azothioprine; BW-57-322; NSC-39084; Azamune; Azanin; Azoran; Imuran; Imurel. $C_9H_7N_7O_2S$; mol wt 277.27. C 38.99%, H 2.54%, N 35.36%, O 11.54%, S 11.57%. Prepn: Hitchings, Elion, U.S. pat. **3,056,785** (1962 to Burroughs Wellcome). Pharmacokinetics: T. L. Ding, L. Z. Benet, *Drug. Metab. Dispos.* **7,** 373 (1979). Use in combination with cyclophosphamide and hydroxychloroquine, *q.q.v.* for the treatment of rheumatoid arthritis: D. J. McCarty, G. F. Carrera, *J. Am. Med. Assoc.* **248,** 1718 (1982). Carcinogenicity studies: P. S. Mitrou *et al., Arzneimittel-Forsch.* **29,** 483, 662 (1979). Comprehensive description: W. P. Wilson, S. A. Benezra, *Anal. Profiles Drug Subs.* 10,

29-53 (1981). Review of pharmacology and clinical efficacy in renal transplantation: G. L. C. Chan *et al., Pharmacotherapy* **7,** 165-177 (1987).

Pale yellow crystals from 50% aq acetone, dec 243-244°. uv max (methanol): 276 nm (ϵ 1.82 × 10^4); (0.1N HCl): 280 nm (ϵ 1.73 × 10^4); (0.1N NaOH): 285 nm (ϵ 1.55 × 10^4). pKa_2 8.2. Slightly sol in water, chloroform, ethanol. Sodium salt, $C_9H_6N_7NaO_2S$, *Imurek.*

Note: Azathioprine has been listed as known carcinogen: *Seventh Annual Report on Carcinogens* (PB95-109781, 1994) p 32.

THERAP CAT: Immunosuppressant; antirheumatic.

936. 6-Azathymine. *6-Methyl-1,2,4-triazine-3,5(2H,4H)-dione;* 5-methyl-6-azauracil; 3,5-dihydroxy-6-methyl-1,2,4-triazine. $C_4H_5N_3O_2$; mol wt 127.10. C 37.80%, H 3.97%, N 33.06%, O 25.18%. Prepn: Thiele, Bailey, *Ann.* **303,** 82 (1898); Bailey, *Am. Chem. J.* **28,** 386 (1902); Bougault, Daniel, *Compt. Rend.* **186,** 1216 (1928); Chang, *J. Org. Chem.* **23,** 1951 (1958).

Crystals from water. mp 210-212°. pKa 7.6. uv max (0.1N HCl): 261 nm (ϵ 5200); in 0.1N NaOH: 246 nm (ϵ 4770).

937. 6-Azauridine. *2-β-D-Ribofuranosyl-1,2,4-triazine-3,5(2H,4H)-dione;* 6-azauracil riboside; 3,5-dioxo-2,3,4,5-tetrahydro-1,2,4-triazine riboside; AzUR; Ribo-Azauracil. $C_8H_{11}N_3O_6$; mol wt 245.19. C 39.19%, H 4.52%, N 17.14%, O 39.15%. Biosynthesis by *E. coli* in the presence of 6-azauracil: Skoda *et al., Experientia* **13,** 150 (1957). Chemical synthesis: Prystas *et al., Chem. & Ind. (London)* **1961,** 947; Sorm *et al., Czech. pat.* **92,697** (1959), *C.A.* **54,** 15411 (1960); **Brit. pat. 827,441** (1960 to Spofa); Cristescu, *Rev. Roum. Chem.* **13,** 365 (1968). Crystal structure and conformation: Schwalbe *et al., Biochem. Biophys. Res. Commun.* **44,** 57 (1971). Pharmacology: Smahel *et al., Neoplasma* **18,** 435 (1971), *C.A.* **76,** 20997c (1972). Toxicity study: Plevova *et al., Toxicol. Appl. Pharmacol.* **17,** 511 (1970).

Crystals from ethanol, ether, mp 160-161°. $[\alpha]_D^{24}$ −132° (pyridine). uv max (water): 262 nm (ϵ 6100). pK 6.70.

2′,3′,5′-Triacetate, $C_{14}H_{17}N_3O_9$, *azaribine, 2-(2,3,5-tri-O-acetyl-β-D-ribofuranosyl)-as-triazine-3,5(2H,4H)-dione, CB-304, Triazure.* LD_{50} in mice, rats (g/kg): 7.8, 12.0 orally (Plevova).

THERAP CAT: Base as antineoplastic; triacetate as an antipsoriatic.

938. Azelaic Acid. *Nonanedioic acid;* 1,7-heptanedicarboxylic acid; lepargylic acid; anchoic acid; Skinoren. $C_9H_{16}O_4$; mol wt 188.22. C 57.43%, H 8.57%, O 34.00%. HOOC(CH$_2$)$_7$COOH. Prepd by disruptive oxidation of ricinoleic acid: Hill, McEwen, *Org. Syn. coll. vol. II,* 53 (1943). Occurs in rancid oleic acid. Clinical comparison with tetracycline in acne: P. T. Bladon *et al., Brit. J. Dermatol.* **114,** 493 (1986). Review of clinical studies in hyperpigmentary disorders: A. S. Breathnach, M. Nazzaro-Porro, *ibid.* **111,** 115-120 (1984); of pharmacology and clinical uses: M. Nazzaro-Porro, *J. Am. Acad. Dermatol.* **17,** 1033-1041 (1987).

Monoclinic prismatic needles, mp 106.5°. Distills above 360° with partial anhydride formation. bp_{100} 286.5°; bp_{50}

265°; bp$_{15}$ 237°; bp$_{10}$ 225°. d$_4^{10.6}$ 1.0291. One liter of water dissolves 1.0 g at 1.0°; 2.4 g at 20°; 8.2 g at 50°; 22 g at 65°. Freely sol in boiling water, in alcohol. 1000 g of ether dissolves 18.8 g at 11° and 26.8 g at 15°. pK$_1$ (25°) 4.53; pK$_2$ 5.33.

Dimethyl ester, C$_{11}$H$_{20}$O$_4$, liquid, d$_4^{20}$ 1.0026. mp -3.9°. bp$_8$ 140°.

THERAP CAT: Antiacne.

939. Azelastine. *4-[(4-Chlorophenyl)methyl]-2-(hexa-hydro-1-methyl-1H-azepin-4-yl)-1(2H)-phthalazinone; 4-(p-chlorobenzyl)-2-(hexahydro-1-methyl-1H-azepin-4-yl)-1(2H)-phthalazinone; 4-(p-chlorobenzyl)-2-(N-methylper-hydroazepin-4-yl)-1(2H)-phthalazinone.* C$_{22}$H$_{24}$ClN$_3$O; mol wt 381.90. C 69.19%, H 6.33%, Cl 9.28%, N 11.00%, O 4.19%. Orally active H$_1$-histamine receptor antagonist. Prepn: **Belg.** pat. **778,269;** D. Vogelsang *et al.,* **U.S.** pat. **3,813,384** (1972, 1974 both to Asta-Werke AG). Synthesis and x-ray structure determn: G. Scheffler *et al., Arch. Pharm.* **321,** 205 (1988). Pharmacology: K. Tasaka, M. Akagi, *Arzneimittel-Forsch.* **29,** 488 (1979). Pharmacology and toxicology: H. J. Zechel *et al., ibid.* **31,** 1184 (1981). Series of articles on pharmacokinetics, pharmacology and toxicology: *ibid.* 1184-1238. Mechanism of action studies: N. Chand *et al., Int. J. Immunopharmacol.* **7,** 833 (1985); *eidem, Br. J. Pharmacol.* **87,** 443 (1986). HPLC determn in plasma: J. Pivonka *et al., J. Chromatog.* **420,** 89 (1987). Clinical evaluations in asthma: H. Magnussen, *Chest* **91,** 855 (1987); M. K. Albazzaz, K. R. Patel, *Thorax* **43,** 306 (1988).

Oil. Sol in methylene chloride.

Monohydrate, C$_{22}$H$_{24}$ClN$_3$O.H$_2$O, crystals from alcohol/water. Two distinct crystal forms have been identified (Scheffler).

Monohydrochloride, C$_{22}$H$_{24}$ClN$_3$O.HCl, *A-5610, E-0659, W-2979M, Allergodil, Azeptin, Rhinolast.* Crystals from alcohol, mp 225-229°. LD$_{50}$ in male, female mice, male, female rats (mg/kg): 36.5, 35.5, 26.9, 30.3 i.v.; 56.4, 42.8, 43.2, 46.6 i.p.; 63.0, 54.2, 66.5, 59.6 s.c.; 124, 139, 310, 417 orally (Zechel).

THERAP CAT: Antiallergic; antiasthmatic; antihistaminic.

940. 2-Azetidinecarboxylic Acid. C$_4$H$_7$NO$_2$; mol wt 101.11. C 47.52%, H 6.98%, N 13.85%, O 31.65%. From *Convallaria majalis* L. (lily-of-the-valley) and *Polygonatum officinale* Moensh., *Liliaceae:* Fowden, *Biochem. J.* **64,** 323 (1956); Fowden, Bryant, *ibid.* **70,** 626 (1958). Structure: Virtanen, *Angew. Chem.* **67,** 619 (1955); Fowden, *loc. cit.* Biosynthesis: Linko, *Acta Chem. Scand.* **12,** 101 (1958); Fowden, Bryant, *Biochem. J.* **71,** 210 (1959); Fowden, *ibid.* **71,** 643 (1959); Leete, *J. Am. Chem. Soc.* **86,** 3162 (1964). Shows growth-inhibitory activity on cultures of *Escherichia coli* and on germinating seeds of different species in which it does not normally occur; Fowden, Richmond, *Biochim. Biophys. Acta* **71,** 459 (1963). Acts as a proline analog where a stoichiometric replacement of proline occurred with production of abnormal proteins having impaired biological activity: Peterson, Fowden, *Nature* **200,** 148 (1963).

Crystals from 95% hot methanol, discolors at 200° and darkens until 310° when heating stopped. [α]$_D^{20}$ -108° (c =

3.6). Unstable to mineral acids. Soluble in cold and hot water. Practically insol in abs ethanol.

941. Azidamfenicol. *2-Azido-N-[2-hydroxy-1-(hydr-oxymethyl)-2-(4-nitrophenyl)ethyl]acetamide; D-(−)-threo-1-p-nitrophenyl-2-azidoacetylaminopropane-1,3-diol; D-(−)-threo-2-azidoacetamido-1-p-nitrophenyl-1,3-propanediol; azidoamphenicol; Leukomycin N.* C$_{11}$H$_{13}$N$_5$O$_5$; mol wt 295.25. C 44.75%, H 4.44%, N 23.72%, O 27.09%. Semi-synthetic antibiotic related to chloramphenicol, *q.v.* Prepn: Meiser, Domagk, U.S. pat. **2,882,275** (1959 to Bayer).

Crystals from ethylene chloride, mp 107°. [α]$_D^{20}$ -20° (c = 1.6 in ethyl acetate). pH 7.1. Sol in water up to 2%.

THERAP CAT: Antibacterial.

942. Azidocillin. *[2S-[2α,5α,6β(S*)]]-6-[(Azidophenyl-acetyl)amino]-3,3-dimethyl-7-oxo-4-thia-1-azabicyclo[3.2.0]-heptane-2-carboxylic acid; α-azidobenzylpenicillin; 6-[D-α-azidophenylacetamido]penicillanic acid; SPC-97D; BRL-2351.* C$_{16}$H$_{17}$N$_5$O$_4$S; mol wt 375.41. C 51.19%, H 4.56%, N 18.66%, O 17.05%, S 8.54%. Semi-synthetic antibiotic related to penicillin. Prepn: Sjöberg, Ekström *et al.,* **Brit.** pat. **940,488** corresp. to U.S. pat. **3,293,242** (1963 and 1966 to Beecham); Ekström *et al., Acta Chem. Scand.* **19,** 281 (1965). Pharmacology: Hanssen *et al., Antimicrob. Ag. Chemother.* **1967,** 568; Tunevall, Frisk, *ibid.* 573. Chemistry: Sjöberg *et al., ibid.* 560. Metabolic studies: Ramsey *et al., Arzneimittel-Forsch.* **22,** 1962 (1972).

Sodium salt, C$_{16}$H$_{16}$N$_5$NaO$_4$S, *Globacillin, Longatren.* Potassium salt, C$_{16}$H$_{16}$KN$_5$O$_4$S, *Nalpen.* mp 194° (dec).

THERAP CAT: Antibacterial.

THERAP CAT (VET): Antimicrobial. (In mastitis, as feed supplement).

943. Azimilide. *1-[[[5-(4-Chlorophenyl)-2-furanyl]-methylene]amino]-3-[4-(4-methyl-1-piperazinyl)butyl]-2,4-imidazolidinedione.* C$_{23}$H$_{28}$ClN$_5$O$_3$; mol wt 457.96. C 60.32%, H 6.16%, Cl 7.74%, N 15.29%, O 10.48%. Potassium channel blocker. Prepn: S. S. Pelosi *et al.,* **PCT Int.** pat. **Appl. 93 04,061** (1993 to Procter & Gamble). Effect on cardiac potassium channels: A. E. Busch *et al., Eur. J. Pharmacol.* **264,** 33 (1994); B. Fermini *et al., J. Cardiovasc. Pharmacol.* **26,** 259 (1995). Pharmacology: S. C. Black *et al., ibid.* **22,** 810 (1993); M. A. McIntosh *et al., ibid.* **23,** 653 (1994).

Dihydrochloride, C$_{23}$H$_{28}$ClN$_5$O$_3$.2HCl, *NE-10064.* Crystals from ethanol/water.

THERAP CAT: Antiarrhythmic (class III).

944. Azinphos-methyl. *Phosphorodithioic acid O,O-dimethyl S-[(4-oxo-1,2,3-benzotriazin-3(4H)-yl)methyl] ester; phosphorodithioic acid O,O-dimethyl ester, S-ester with 3-mercaptomethyl-1,2,3-benzotriazin-4(3H)-one;* Bayer 17147; ENT-23233; R-1582; Cotnion-methyl; Gusathion M; Guthion. C$_{10}$H$_{12}$N$_3$O$_3$PS$_2$; mol wt 317.33. C 37.85%, H 3.81%, N 13.24%, O 15.13%, P 9.76%, S 20.21%. Cholinesterase inhibitor. Prepn: Lorenz, U.S. pat. **2,758,115** (1956 to Bayer). Activity: E. E. Ivy *et al., J. Econ. Entomol.* **48,** 293 (1955). Toxicity study: T. B. Gaines, *Toxicol. Appl. Pharmacol.* **14,** 515 (1969).

Crystals from methanol, mp 73-74°. d_4^{20} 1.44; n_D^{76} 1.6115. Soly in water at 25°: 33 mg/l. Sol in methanol, ethanol, propylene glycol, xylene, other organic solvents. Solns in ethanol and propylene glycol are stable for at least 3 weeks. Unstable at temperatures > 200°. Hydrolyzes in acid or cold alkali. LD_{50} in female rats (mg/kg): 11 orally; 220 dermally (Gaines).

O,O-Diethyl analog, $C_{12}H_{16}N_3O_3PS_2$, *azinphos-ethyl, Bayer 16259, ENT-22014, R-1513, Gusathion A, Ethyl Guthion.* Colorless needles, mp 53°, $bp_{0.001}$ 111°. d_4^{20} 1.284; n_D^{53} 1.5928. Vapor pressure at 20°: 2.2×10^{-7} mm Hg.

Caution: Potential symptoms of overexposure to azinphos-methyl are miosis; aching eyes; blurred vision, lacrimation and rhinorrhea; headache; chest tightness, wheezing, laryngeal spasms; salivation; cyanosis; anorexia; nausea, vomiting and diarrhea; sweating; twitching, paralysis and convulsions; low blood pressure, cardiac irregularities. *See NIOSH Pocket Guide to Chemical Hazards* (DHHS/NIOSH 90-117, 1990) p 42.

USE: Insecticide; acaricide.

945. Azintamide. *2-[(6-Chloro-3-pyridazinyl)thio]-N,-N-diethylacetamide; N,N*-diethyl-2-[6-(3-chloropyridazinyl)-thio]acetamide; *N,N*-diethyl-2-[6-(3-chloropyridazinyl)mercapto]acetamide; (3-chloro-6-pyridazinylthio)acetic acid diethylamide; ST-9067; Oragallin. $C_{10}H_{14}ClN_3OS$; mol wt 259.76. C 46.24%, H 5.43%, Cl 13.65%, N 16.18%, O 6.16%, S 12.34%. Prepn: E. Kloimstein *et al., Arzneimittel-Forsch.* **14,** 261 (1964); *eidem,* **Ger.** pat. **1,188,604** (1965 to Lentia). Pharmacology and acute toxicity: H. Stormann, *Arzneimittel-Forsch.* **14,** 266 (1964). Toxicology: I. Lindner *et al., ibid,* 271. Clinical pharmacology: G. Hitzenberger, *ibid,* 279. Comprehensive description: E. M. Abdel-Moety, H. A. Al-Khamees in *Analytical Profiles of Drug Substances* vol. **18,** K. Florey, Ed. (Academic Press, New York, 1989) pp 1-32.

Colorless, bitter tasting microcrystals, mp 97-98°. Freely sol in benzene, chloroform, ethyl acetate, acetone. Sol in water 0.5%. uv max (ethanol): 316, 306, 258 nm (ϵ 1.404 $\times 10^3$, 1.492 $\times 10^3$, 1.421 $\times 10^4$). LD_{50} in mice, rats (g/kg): 2.34, 1.55 orally (Stormann).

THERAP CAT: Choleretic.

946. Azithromycin. *[2R-(2R*,3S*,4R*,5R*,8R*,-10R*,11R*,12S*,13S*,14R*)]*-13-[(2,6-Dideoxy-3-C-methyl-3-O-methyl-α-L-ribo-hexopyranosyl)oxy]-2-ethyl-3,4,10-trihydroxy-3,5,6,8,10,12,14-heptamethyl-11-[[3,4,6-trideoxy-3-(dimethylamino)-β-D-xylo-hexopyranosyl]oxy]-1-oxa-6-azacyclopentadecan-15-one; N*-methyl-11-aza-10-deoxo-10-dihydroerythromycin A; 9-deoxo-9a-methyl-9a-aza-9a-homoerythromycin A; CP-62993; XZ-450; Azitrocin; Sumamed; Trozocina; Zithromax; Zitromax. $C_{38}H_{72}N_2O_{12}$; mol wt 749.00. C 60.94%, H 9.69%, N 3.74%, O 25.63%. Semi-synthetic macrolide antibiotic; related to erythromycin A, *q.v.* Prepn: **Belg.** pat. **892,357;** G. Kobrehel, S. Djokic, U.S. pat. **4,517,-359** (1982, 1985 both to Sour Pliva). Antibacterial spectrum: S. C. Aronoff *et al., J. Antimicrob. Chemother.* **19,** 275 (1987); and mode of action: J. Retsema *et al., Antimicrob. Ag. Chemother.* **31,** 1939 (1987). Series of articles on

pharmacology, pharmacokinetics, and clinical experience: *J. Antimicrob. Chemother.* **31,** Suppl. E, 1-198 (1993).

Crystals, mp 113-115°. $[\alpha]_D^{20}$ −37° (c = 1 in $CHCl_3$). THERAP CAT: Antibacterial.

947. Azlocillin. *[2S-[2α,5α,6β(S*)]]-3,3-Dimethyl-7-oxo-6-[[[[(2-oxo-1-imidazolidinyl)carbonyl]amino]phenylacetyl]amino]-4-thia-1-azabicyclo[3.2.0]heptane-2-carboxylic acid;* D-α-[(imidazolidin-2-on-1-yl)carbonylamino]benzylpenicillin; Bay e 6905. $C_{20}H_{23}N_5O_6S$; mol wt 461.50. C 52.05%, H 5.02%, N 15.18%, O 20.80%, S 6.95%. Semisynthetic, broad-spectrum acylureido penicillin. Prepn: H. Disselnkotter, K. G. Metzger, **Fr.** pat. **2,100,682,** corresp to U.S. pat. **3,933,795** (1971, 1976 both to Bayer); H. B. Konig *et al., Eur. J. Med.-Chim. Ther.* **17,** 59 (1982). *In vitro* studies: D. Stewart *et al., Antimicrob. Ag. Chemother.* **11,** 865 (1977). *In vitro* and *in vivo* activity: G. K. Daikos *et al., Curr. Chemother., Proc. 10th Int. Congr. Chemother.,* 1977 (Amer. Soc. Microbiol., Washington, D.C., 1978) **1,** pp 626-8. Pharmacokinetics: P. Fiegel, K. Becker, *Antimicrob. Ag. Chemother.* **14,** 288 (1978). Comparison with other penicillins: J. M. Andrews, K. A. Bedford, *ibid.* 559. Clinical studies: H. Lode *et al., Infection* **5,** 163 (1977); E. B. Helm *et al., Deutsch. Med. Wochenschr.* **102,** 1211 (1977). Series of articles on antibacterial activity, pharmacology, and clinical trials: *Arzneimittel-Forsch.* **29,** 1915-2032 (1979); *Infection* **10,** Suppl. 3, S121-S266 (1982); *J. Antimicrob. Chemother.* **11,** Suppl. B, 1-239 (1983).

Sodium salt, $C_{20}H_{22}N_5NaO_6S$, *Azlin, Securopen.* Pale yellow crystals, sol in water, methanol, DMF. Slightly sol in ethanol, isopropanol.

THERAP CAT: Antibacterial.

948. Azobenzene. *Diphenyldiazene;* diphenyl diimide; azobenzide; azobenzol; benzeneazobenzene. $C_{12}H_{10}N_2$; mol wt 182.22. C 79.10%, H 5.53%, N 15.37%. Made by reducing nitrobenzene with Fe in NaOH soln or by electrolytic reduction of nitrobenzene in dil alkali. Lab procedure: A. I. Vogel, *Practical Organic Chemistry* (Longmans, London, 3rd ed., 1959) p 631. Review of carcinogenicity studies: *IARC Monographs* **8,** 75 (1975).

Orange-red leaflets. d 1.20. mp 68°. bp 293°. Insol in water; sol in alcohol, ether, glacial acetic acid. LD_{50} orally in rats: 1000 mg/kg, *RTECS* Vol. **I**, R. J. Lewis, R. L. Tatken, Eds. (1980) p 209.

USE: Acaricide.

949.　2,2'-Azobisisobutyronitrile. *2,2'-Azobis[2-methylpropanenitrile];* AIBN; *α,α'-*azodiisobutyronitrile; 2,2'-dicyano-2,2'-azopropane; Porofor-57. $C_8H_{12}N_4$; mol wt 164.21. C 58.52%, H 7.37%, N 34.12%. Prepn by oxidation of *α,α'-*hydrazobutyric acid dinitrile: Overberger *et al., J. Am. Chem. Soc.* **71**, 2661 (1949); Horner, Schwenk, *Ann.* **566**, 69 (1950); from 2-aminoisobutyronitrile + NaOCl: **Brit. pat. 672,106** (1952 to Rohm & Haas); Anderson, U.S. pat. **2,711,405** (1955 to du Pont); from 2,2'-dichloro-2,2'-azopropane: Benzing, *Ann.* **631**, 1 (1960); **Brit. pat. 929,182** (1963 to Monsanto).

Crystals from ethanol + water, dec 107°. uv max (ethanol): 345 nm. Soly in methanol at 0°, 20°, 40°: 1.8, 4.96, 16.06 g/100 ml. Soly in ethanol at 0°, 20°, 40°: 0.58, 2.04, 7.15 g/100 ml. LD_{50} orally in mice: 0.7 g/kg. Can explode when dissolved in acetone: Carlisle, *Chem. & Eng. News* **27**, 150 (1949).

Caution: In the organism, forms HCN which is found in blood, liver and brain: Rusin, *C.A.* **56**, 2682f (1962).

USE: Blowing agent for elastomers and plastics. Initiator for free radical reactions: Griesbaum *et al., J. Org. Chem.* **30**, 261 (1965).

950.　Azodicarbonamide. *Diazenedicarboxamide; 1,1'-azobisformamide;* azodicarboxamide; azobiscarbonamide; azobiscarboxamide; 1,1'-azobiscarbamide. $C_2H_4N_4O_2$; mol wt 116.08. C 20.69%, H 3.47%, N 48.27%, O 27.57%. NH_2-CON=NCONH$_2$. Prepn: J. Thiele, *Ann.* **270**, 1 (1892); **271**, 127 (1892); T. Curtis, K. Heindenreich, *J. Prakt. Chem.* [2] **52**, 454 (1895); J. P. Picard, J. L. Boivin, *Can. J. Chem.* **29**, 223 (1951). Crystal structure: D. T. Cromer, A. C. Larson, *J. Chem. Phys.* **60**, 176 (1974). Safety assessment as a flour-maturing agent: B. L. Oser *et al., Toxicol. Appl. Pharmacol.* **7**, 445 (1965).

Orange-red crystals, mp 225° (dec). *Flammable!* Sol in hot water; insol in cold water, alc.

USE: As blowing and foaming agent for plastics; as maturing and bleaching agent in cereal flour.

951.　Azolitmin. Purified coloring matter from litmus: I. M. Kolthoff, C. Rosenblum, *Acid-Base Indicators* (Mac-Millan Co., New York, 1937) pp 160-162, 174, 208, 355, 361, 365-366, 368-369, 373, 377, 387.

Dark violet scales or dark red powder. Sparingly soluble in water; insol in alc; freely sol in dil alkali hydroxides or carbonates. The indicator solution is prepared by dissolving 0.5 g in 80 ml of warm H_2O, then adding 20 ml alcohol.

USE: Indicator instead of litmus. pH: 4.5 red, 8.3 blue. Usable with most mineral, some organic acids (not hydroxy acids) and some alkaloids; also used for preparing litmus media for bacteriologic purposes.

952.　Azomycin. *2-Nitro-1H-imidazole.* $C_3H_3N_3O_2$; mol wt 113.08. C 31.87%, H 2.67%, N 37.16%, O 28.30%. Antibiotic substance produced by an unidentified *Streptomyces* sp. (resembling *Nocardia mesenterica* in some aspects) from soil collected at Shiba Shirokane Daimachi (Japan): Maeda *et al., J. Antibiot.* **6A**, 182 (1953); Okami *et al., ibid.* **7A**, 53 (1954); Nakamura, Umezawa, *ibid.* **8A**, 66 (1955). Structure: Nakamura, *Pharm. Bull.* **3**, 379 (1955). Synthesis: Beaman *et al., J. Am. Chem. Soc.* **87**, 389 (1965); Lancini, Lazzari, *Experientia* **21**, 83 (1965).

Crystals from methanol, dec 283°. uv max (ethanol): 313 nm (E$_{1cm}^{1\%}$ 915); in 0.1N NaOH: 374 nm (ε 12750). Soluble in methanol, ethanol, acetone, ethyl acetate, butyl acetate,

alkaline water. Practically insol in ether, petr ether, chloroform, acidic water. LD_{50} i.v. in mice: 80 mg/kg (Maeda).

Caution: Irritating to skin.

953.　Azosemide. *2-Chloro-5-(1H-tetrazol-5-yl)-4-[(2-thienylmethyl)amino]benzenesulfonamide; 2-chloro-5-(2H-tetrazol-5-yl)-N⁴-2-thenylsulfanilamide;* 5-(4'-chloro-2'-thenylamino-5'-sulfamoylphenyl)tetrazole; Ple-1053; Diart; Diurapid; Luret. $C_{12}H_{11}ClN_6O_2S_2$; mol wt 370.84. C 38.87%, H 2.99%, Cl 9.56%, N 22.66%, O 8.63%, S 17.29%. Prepn: A. Popelak *et al.,* **Ger. pat. 1,815,922;** *eidem,* U.S. pat. **3,665,002** (1968, 1972 both to Boehringer, Mann.). Separation and analysis in blood and urine by HPLC: R. Seiwell, C. Brater, *J. Chromatog.* **182**, 257 (1980). Sites of action: C. Brater, *Clin. Pharmacol. Ther.* **25**, 428 (1979). Clinical and pharmacological studies: F. Krueck *et al., Eur. J. Clin. Pharmacol.* **14**, 153 (1978). Diuretic effect in animals: J. Greven, O. Heidenreich, *Arzneimittel-Forsch.* **31**, 346, 350 (1981).

Crystals, mp 218-221°.

THERAP CAT: Diuretic.

954.　Azoxybenzene. *Diphenyldiazene 1-oxide;* azoxybenzide. $C_{12}H_{10}N_2O$; mol wt 198.22. C 72.71%, H 5.08%, N 14.13%, O 8.07%. Prepd by reduction of nitrobenzene with sodium arsenite: Bigelow, Palmer, *Org. Syn.* coll. vol. **II**, 57 (1943); also by treatment of nitrobenzene with glucose in alkaline medium: Opolonick, *Ind. Eng. Chem.* **27**, 1045 (1935). Industrial prepn by heating nitrobenzene with molasses and NaOH in high flash naptha: **Ger. pat. 228,-722.** Also prepd by catalytic reduction of nitrobenzene: Busch, Schulz, *Ber.* **62**, 1458 (1929). By peracetic acid oxidation of azobenzene: D'Ans, Kneip, *Ber.* **48**, 1145 (1915); *cf.* Greenspan, *Ind. Eng. Chem.* **39**, 847 (1947). Prepn of cis- and trans-forms by perbenzoic acid oxidation of azobenzene: G. M. Badger *et al., J. Chem. Soc.* **1953**, 2143.

trans-Azoxybenzene

*trans-*Form, pale yellow orthorhombic needles. d_4^{26} 1.1590. d_4^{50} 1.1373. mp 36°. Slightly volatile with steam; easily volatile in superheated steam at 140-150°. Absorption spectrum: Müller, *Ann.* **493**, 166 (1932). Insol in water. Sol in alcohol, ether. 100 parts of abs alc satd at 16° contain 17.5 parts (w/w) azoxybenzene. At 15° 100 g ligroin dissolves 43.5 g azoxybenzene.

Note: The stereoisomer, the so-called *cis-*azoxybenzene, melts at 87°.

USE: In organic syntheses.

955.　Aztreonam. *[2S-[2α,3β(Z)]]-2-[[[1-(2-Amino-4-thiazolyl)-2-[(2-methyl-4-oxo-1-sulfo-3-azetidinyl)amino]-2-oxoethylidene]amino]oxy]-2-methylpropanoic acid;* azthreonam; SQ-26776; Azactam; Azonam; Aztreon; Nebactam; Primbactam. $C_{13}H_{17}N_5O_8S_2$; mol wt 435.44. C 35.86%, H 3.94%, N 16.08%, O 29.39%, S 14.73%. The first totally synthetic monocyclic β-lactam (monobactam) antibiotic. It has a high degree of resistance to β-lactamases and shows specific activity vs aerobic gram-negative rods. Prepn: R. B. Sykes *et al.,* **Neth. pat. Appl. 8,100,571** (1981 to Squibb), *C.A.* **96**, 181062x (1982). Fast-atom-bombardment mass spectra: A. I. Cohen *et al., J. Pharm. Sci.* **71**, 1065 (1982). Activity vs gram-negative bacteria: R. B. Sykes *et al., Antimicrob. Ag. Chemother.* **21**, 85 (1982). Series of articles on structure-activity, *in vitro* and *in vivo* properties, pharmacokinetics: *J. Antimicrob. Chemother.* **8**, Suppl. E, 1-148 (1981). Toxicology: G. R. Keim *et al., ibid.* 141. Mechanism of action study: A. D. Russell, J. R. Furr, *ibid.* **9**, 329 (1982). Comparative stability to renal dipeptidase: H. Mikami *et al., Antimicrob. Ag. Chemother.* **22**, 693 (1982). Human pharmacokinetics: E. A. Swabb *et al., ibid.* **21**, 944 (1982). Clinical evaluation in urinary tract infection: C. Donadio *et al., Drugs Exp. Clin. Res.* **13**, 167 (1987). Clinical efficacy in neonatal sepsis: S. Sklavunu-Tsurutsoglu *et al., Rev. Infect. Dis.* **13**, Suppl. 7, S591 (1991). Comprehensive description: K. Florey in *Analytical Profiles of Drug Substances* **vol. 17**, K. Florey, Ed. (Academic Press, New York, 1988) pp 1-39.

White crystalline, odorless powder, dec 227°. Very slightly sol in ethanol, slightly sol in methanol, sol in DMF, DMSO. Practically insol in toluene, chloroform, ethyl acetate. Disodium salt, $C_{13}H_{15}N_5Na_2O_8S_2$. LD_{50} (mg/kg): 3300 i.v. in mice; 6600 i.p. in rats (Keim).

THERAP CAT: Antibacterial.

956. Azulene. *Cyclopentacycloheptene*; bicyclo-[5.3.0]-deca-2,4,6,8,10-pentaene; bicyclo-[0.3.5]-deca-1,3,5,7,9-pentaene. $C_{10}H_8$; mol wt 128.17. C 93.71%, H 6.29%. Prepn from octahydronaphthalene: Plattner, Pfau, *Helv. Chim. Acta* **20**, 224 (1937); from indan: Plattner, Magyar, *ibid.* **25**, 581 (1942).

Intensely blue leaflets or monoclinic plates from alc. Odor of naphthalene. mp 98.5-99°. Absorption spectrum: Susz *et al., Helv. Chim. Acta* **20**, 469 (1937); Plattner, *ibid.* **24**, 283 (1941). Insol in water; sol in the usual organic solvents, in concd mineral acids with decompn.

Sodium sulfonate, *Azusalen*.

THERAP CAT: Sodium sulfonate as antacid.

957. Azure A. *3-Amino-7-(dimethylamino)phenothiazin-5-ium chloride*; 7-(dimethylamino)-3-imino-3H-phenothiazine hydrochloride; 3-amino-7-dimethylaminophenazathionium chloride; asym-dimethyl-3,7-diaminophenazathionium chloride; asym-dimethylthionine chloride; methylene azure A; C.I. 52005. $C_{14}H_{14}ClN_3S$; mol wt 291.80. C 57.63%, H 4.84%, Cl 12.15%, N 14.40%, S 10.99%. One of the main constituents of methylene azure. Prepn: Bernthsen, *Ann.* **230**, 169 (1885); Kehrmann, *Ber.* **39**, 1804 (1906); MacNeal, Killian, *J. Am. Chem. Soc.* **48**, 740 (1926). Review: H. J. Conn's *Biological Stains*, R. D. Lillie, Ed. (Williams & Wilkins, Baltimore, 9th ed., 1977) pp 420-421, 603. See also: *Colour Index* **vol. 4** (3rd ed., 1971) p 4469.

Green glistening crystals or dark green powder. Soluble in water (blue soln); sparingly sol in alc. Absorption max: 620-634 nm (3 ml of a soln of 50 mg in 250 ml H_2O diluted to 200 ml; read in a Beckman spectrophotometer 1 cm cell).

USE: Biological stain.

958. Azure B. *3-(Dimethylamino)-7-(methylamino)-phenothiazin-5-ium chloride*; 7-(dimethylamino)-3-(methylimino)-3H-phenothiazine hydrochloride; 3-methylamino-7-dimethylaminophenazathonium chloride; trimethyldiaminophenazathonium chloride; trimethylthionine chloride; methylene azure B; C.I. 52010. $C_{15}H_{16}ClN_3S$; mol wt 305.83. C 58.91%, H 5.27%, Cl 11.59%, N 13.74%, S 10.48%. One of the main constituents of methylene azure. Prepn: Bernthsen, *Ann.* **230**, 169 (1885); Kehrmann, *Ber.* **39**, 1804 (1906); Kehrmann, Duttenhöfer, *ibid.* 925, 1403; Kehrmann *et al., ibid.* **46**, 2137 (1913); MacNeal, Killian, *J. Am. Chem. Soc.* **48**, 740 (1926). Review: H. J. Conn's *Biological Stains*, R. D. Lillie, Ed. (Williams & Wilkins, Baltimore, 9th ed., 1977) pp 421-423, 603-604. See also: *Colour Index* **vol. 4** (3rd ed., 1971) p 4470.

Green glistening crystals or dark green powder. Sol in water (blue soln); sparingly sol in alc. Absorption max: 648-655 nm (3 ml of a soln of 50 mg in 250 ml H_2O dil to 200 ml; read in a Beckman spectrophotometer 1 cm cell).

USE: Biological stain.

959. Azure C. *3-Amino-7-(methylamino)phenothiazin-5-ium chloride*; 3-imino-7-(methylamino)-3H-phenothiazine hydrochloride; 3-amino-7-methylaminophenazathonium chloride; monomethyldiaminodiphenazothionium chloride; monomethylthionine chloride; C.I. 52002. $C_{13}H_{12}ClN_3S$; mol wt 277.78. C 56.21%, H 4.35%, Cl 12.76%, N 15.13%, S 11.54%. Prepn: Holmes, French, *Stain Techn.* **1**, 17 (1926). Review: H. J. Conn's *Biological Stains*, R. D. Lillie, Ed. (Williams & Wilkins, Baltimore, 9th ed., 1977) pp 419-420, 604. See also: *Colour Index* **vol. 4** (3rd ed., 1971) p 4469.

Green glistening crystals or dark green powder. Sol in water (blue soln); sparingly sol in alc. Absorption max: 608-622 nm (3 ml of a soln of 50 mg in 250 ml H_2O dil to 200 ml; read in a Beckman spectrophotometer 1 cm cell).

USE: Biological stain.

B

960. Babbitt Metal. Originally an alloy of 69% Zn, 19% Sn, 4% Cu, 3% Sb, and 5% Pb. This name now covers a variety of formulations, such as lead base babbitt, silver base babbitt, tin base babbitt, cadmium base babbitt, arsenic base babbitt. Generally and better described as "white metal bearing alloys".
USE: For machinery bearings.

961. Bacampicillin. *[2S-[2α,5α,6β(S*)]]-6-[(Amino-phenylacetyl)amino]-3,3-dimethyl-7-oxo-4-thia-1-azabicyclo-[3.2.0]heptane-2-carboxylic acid 1-[(ethoxycarbonyl)oxy]ethyl ester;* (2S,5R,6R)-6-[(R)-(2-amino-2-phenylacetamido)]-3,3-dimethyl-7-oxo-4-thia-1-azabicyclo[3.2.0]heptane-2-carboxylic acid ester with ethyl 1-hydroxyethyl carbonate; 6-[(R)-2-amino-2-phenylacetamido]penicillanic acid [1-(ethoxy-carbonyloxy)ethyl] ester; 1'-ethoxycarbonyloxyethyl 6-(D-α-aminophenylacetamido)penicillanate. $C_{21}H_{27}N_3O_7S$; mol wt 465.53. C 54.18%, H 5.85%, N 9.03%, O 24.06%, S 6.89%. Semi-synthetic antibiotic related to penicillin. Prepn: B. A. Ekstrom, B. O. H. Sjoberg, **Ger.** pat. **2,144,457;** *eidem,* **U.S.** pats. **3,873,521** and **3,939,270** (1972, 1975 and 1976 all to Astra). *In vitro* and *in vivo* study: N. O. Bodin *et al., Antimicrob. Ag. Chemother.* **8,** 518 (1975). Pharmacokinetics: M. Rozencweig *et al., Clin. Pharmacol. Ther.* **19,** 592 (1976); T. Bergan, *Antimicrob. Ag. Chemother.* **13,** 971 (1978). Animal studies: C. Carbon *et al., Scand. J. Infect. Dis. (Suppl.)* **14,** 127 (1978); T. Bergan, I. Versland, *ibid.* 135; U. Forsgren *et al., ibid.* 207. Clinical studies: P. V. Maesen *et al., J. Antimicrob. Chemother.* **2,** 279 (1976); J. Sjövall, *J. Int. Med. Res.* **5,** 313 (1977); C. Ekedahl *et al., Scand. J. Infect. Dis. (Suppl.)* **14,** 279 (1978). Toxicology: M. Edanaga *et al., Chemotherapy (Tokyo)* **27,** Suppl. 4, 17 (1979).

Hydrochloride, $C_{21}H_{27}N_3O_7S.HCl$, **Ambacamp, Ambaxin, Bacacil, Bacampicine, Penglobe, Spectrobid.** White crystals from acetone-petr ether, mp 171-176° (dec). $[α]_D^{20}$ +161.5°, also reported as +173° (Bodin). Sol in water. LD_{50} in mice (mg/kg): 8529 orally; 176 i.p.; 9475 s.c.; 184 i.v. (Edanaga).
THERAP CAT: Antibacterial.

962. Bacillus thuringiensis. Agritol; Bactospeine; Bakthane; BT; Dipel; Larvatrol; Skeetal; Sporeine; Teknar; Thuricide; Tribactur. A gram-positive spore-forming bacterium which causes the death of lepidopterous larvae. First isolated by Ishiwata in 1902 from dying silkworm larvae. Later isolated as *B. thuringiensis,* Berliner, from the larvae of the flour moth, *Ephestia kuehniella* Zell: Berliner, *Z. Angew. Entomol.* **2,** 29 (1915). The insecticidal action is ascribed to protein crystals produced by the bacillus: Hannay, *Nature* **172,** 1004 (1953). Structure and function of the protein crystal: K. W. Nickerson, *Biotechnol. Bioeng.* **22,** 1305 (1950). In addition to the crystalline parasporal body, called the endotoxin, there is another toxic entity, called the exotoxin, which is released in the culture medium during bacterial growth. Isoln and stucture of the exotoxin: Sebesta *et al., Coll. Czech. Chem. Commun.* **34,** 891 (1969); Farkas *et al., ibid.* 1118. Partial synthesis: Sorm, *Pure Appl. Chem.* **25,** 253 (1971). *Reviews:* Heimpel, *Ann. Rev. Entomol.* **12,** 287 (1967); Rogoff, Yousten, *Ann. Rev. Microbiol.* **23,** 357 (1969); Lecadet in *Microbial Toxins* vol. 3, T. C. Montie *et al.,* Eds. (Academic Press, New York, 1970) pp 437-471; L. A. Bulla *et al., Crit. Rev. Microbiol.* **8,** 147-204 (1980).

The commercial product is composed of live spores of *B. thuringiensis,* Berliner. Non-toxic to humans and animals: Fisher *et al., J. Agr. Food Chem.* **7,** 687 (1959).
USE: Microbial insecticide.

963. Bacilysin. *[1R-(1α,2β,6α)]-N-L-Alanyl-3-(5-oxo-7-oxabicyclo[4.1.0]hept-2-yl)-L-alanine;* α-[(2-amino-1-oxo-propyl)amino]-5-oxo-7-oxabicyclo[4.1.0]heptane-2-propanoic acid; α-(2-aminopropionamido)-5-oxo-7-oxabicyclo[4.1.0]hep-tane-2-propionic acid, stereoisomer; bacillin; tetaine. $C_{12}H_{18}N_2O_5$; mol wt 270.29. C 53.33%, H 6.71%, N 10.36%, O 29.60%. Antibiotic produced by the soil bacillus NCTC 7197: Gilliver *et al.,* in *Antibiotics* vol. I, Florey *et al.,* Eds. (Oxford, 1949) p 458. Production by *Bacillus subtilis* and purification: Rogers *et al., Biochem. J.* **97,** 573 (1965). Identity with tetaine: K. Kaminski, T. Sololowska, *J. Antibiot.* **26,** 184 (1973). Identity with bacillin: K. Atsumi *et al., ibid.* **28,** 77 (1975). Improved isoln: Walker, Abraham, *Biochem. J.* **118,** 557 (1970). Structural study: Rogers *et al., ibid.* **97,** 579 (1965). Final structure: Walker, Abraham, *ibid.* **118,** 563 (1970).

White amorphous powder. Freely sol in water; sol in 80% alc; sparingly sol in abs alc. Stable in aq soln at 100° for 5 min at pH 7; becomes inactive at pH 2 or pH 9.

964. Bacimethrin. *4-Amino-2-methoxy-5-pyrimidine-methanol;* 4-amino-5-hydroxymethyl-2-methoxypyrimidine. $C_6H_9N_3O_2$; mol wt 155.16. C 46.45%, H 5.85%, N 27.08%, O 20.62%. Antibiotic substance produced by *Bacillus mega-therium* from Japanese soil. Isoln and structure: Tanaka *et al., J. Antibiot.* **14A,** 161 (1961); **15A,** 191, 197 (1962). Prepn: **Brit.** pat. **884,772** (1961 to Merck & Co.); Koppel *et al., J. Org. Chem.* **27,** 1492, 3614 (1962). Biological studies: Nishimura, Tanaka, *J. Antibiot.* **16A,** 179 (1963).

Needles from methanol, or ethanol + ether, mp 174°. uv max (water): 227, 271 nm (ε 7600, 7300); (0.1N HCl): 229, 261 nm (ε 8400, 9500). Sol in water, methanol, acetic acid. Mod sol in ethanol, pyridine. Practically insol in other organic solvents. LD_{50} in mice: 300 mg/kg i.v. and i.p.

965. Bacitracin. Altracin; Ayfivin; Fortracin; Penitracin; Topitracin; Zutracin. Antibiotic polypeptide complex produced by *Bacillus subtilis* and *licheniformis:* Johnson *et al., Science* **102,** 376 (1945); Anker *et al., J. Bacteriol.* **55,** 249 (1948). Commercial bacitracin is a mixture of at least nine bacitracins. Purification of bacitracin by carrier displacement method: Porath, *Acta Chem. Scand.* **6,** 1237 (1952). Purification with ion exchange resin: Chaiet, Cochrane, **U.S.** pat. **2,915,432** (1959 to Merck & Co.). Production: Johnson, Meleney, **U.S.** pat. **2,498,165** (1950 to the U.S. Secy. of War); Freaney, Allen, **U.S.** pat. **2,828,246** (1958 to Commercial Solvents). Solubilities: Weiss *et al., Antibiot. Chemother.* **7,** 374 (1957). Preliminary structure studies: Hausmann *et al., J. Am. Chem. Soc.* **77,** 723 (1955); Lockhart *et al., Biochem. J.* **61,** 534 (1955); Stoffel, Craig, *J. Am. Chem. Soc.* **145** (1961). Structure of bacitracin A: Ressler, Kashelikar, *ibid.* **88,** 2025 (1966); Galardy *et al., Biochemistry* **10,** 2429 (1971). Synthetic studies: Munekata *et al., Bull. Chem. Soc. Japan* **46,** 3187, 3835 (1973). Mechanism of action: Storm, *Ann. N.Y. Acad. Sci.* **235,** 387 (1974). Comprehensive description: G. A. Brewer in *Ana-*

lytical Profiles of Drug Substances **vol. 9**, K. Florey, Ed. (Academic Press, New York, 1980) pp 1-69. *Reviews:* Craig *et al.*, "Bacitracin" in G. E. W. Wolstenholme, C. M. O'Connor, *Ciba Foundation Symposium on Amino Acids and Peptides with Antimetabolic Activity* (Little, Brown, Boston, 1958) pp 226-246; E. D. Weinberg in *Antibiotics*, D. Gottlieb, P. D. Shaw, Eds. (Springer-Verlag, New York, 1967) **I**, pp 90-99; **II**, pp 240-245; D. R. Storm, W. A. Toscano, Jr. in *Antibiotics* **vol. 5**, pt. 1, F. E. Hahn, Ed. (Springer-Verlag, New York, 1979) pp 1-17.

Bacitracin A

Grayish-white powder. Very bitter taste. Sol in water, alcohol. Practically insol in ether, chloroform, acetone. Stable in acid soln; unstable in alkaline solns. Potency loss probably due to transformation of bacitracin A to bacitracin F, latter having little antimicrobial activity.

Bacitracin A, $C_{66}H_{103}N_{17}O_{16}S$, major component of commercial bacitracin.

THERAP CAT: Antibacterial.

THERAP CAT (VET): Antibacterial, usually used locally. Growth promotant. Enteric infections.

966. Bacitracin Methylenedisalicylic Acid. *Bacitracin methylenebis[2-hydroxybenzoate];* bacitracin methylenedisalicylate. Prepd by the reaction of two moles methylenedisalicylic acid and one mole bacitracin: Baron, U.S. pat. **2,774,712** (1956 to S. B. Penick & Co.). Contains about 18 bacitracin units/mg (minimum 14 units).

White to grayish-brown powder. Slightly disagreeable odor. Somewhat less bitter than bacitracin. uv max (dil acetic acid): 318 nm, min: 280 nm. Solubility in water about 50 mg/ml. pH of satd aq soln 3.5 to 5.0. Sol in pyridine, ethanol, less sol to insol in acetone, ether, chloroform, pentane, benzene. Sol in dil aq alkali at pH 6 and higher, increasingly insol at pH 6 to 3.

Sodium salt, creamy-white powder. Potency about 10.8 bacitracin units/mg. Sol in water. pH 2% aq soln 9.5.

THERAP CAT (VET): Feed additive.

967. Baclofen. *β-(Aminomethyl)-4-chlorobenzenepropanoic acid; β-(aminomethyl)-p-chlorohydrocinnamic acid; γ-amino-β-(p-chlorophenyl)butyric acid; β-(4-chlorophenyl)-GABA; Ba-34647; Baclon; Lioresal.* $C_{10}H_{12}ClNO_2$; mol wt 213.66. C 56.21%, H 5.66%, Cl 16.59%, N 6.56%, O 14.98%. Prepn: **Neth.** pat. **Appl. 6,407,755**; H. Keberle *et al.*, **U.S.** pat. **3,471,548** (1965, 1969 both to Ciba); Uchimaru *et al.*, **Japan** pat. **16,692** ('70) (to Daiichi), *C.A.* **73**, 77617w (1970). Clinical studies: Hudgson, Weightman, *Brit. Med. J.* **IV**, 15 (1971). Metabolism: H. Yamamoto *et al.*, *Oyo Yakuri* **14**, 97, 109, 115 (1977). Toxicity study: T. Tadokoro *et al.*, *Osaka Daigaku Igaku Zasshi* **28**, 265 (1976), *C.A.* **88**, 183016u (1978). Comprehensive description: S. Ahuja in *Analytical Profiles of Drug Substances* **vol. 14**, K. Florey, Ed. (Academic Press, New York, 1985) pp 527-548. Book: *Baclofen: A Broader Spectrum of Activity,* I. T. Boyle *et al.*, Eds. (Longman, Harlow, England, 1980) 52 pp.

Crystals from water, mp 206-208° (Keberle); 189-191°, (Uchimaru). LD_{50} in male mice, rats (mg/kg): 45, 78 i.v.; 103, 115 s.c.; 200, 145 orally (Tadokoro).

Hydrochloride, $C_{10}H_{13}Cl_2NO_2$, mp 179-181°.

THERAP CAT: Muscle relaxant (skeletal).

968. Bacteriorhodopsin. BR. The only protein of the purple membrane of *Halobacterium halobium* and other halophilic bacteria. It is a rhodopsin-like pigment containing retinal linked to lysine through a Schiff's base and has a mol wt of approx 26,000. Bacteriorhodopsin functions as an energy transducer or "proton pump"; unlike animal rhodopsins, it uses light energy to generate an electrochemical gradient and this stored energy is used by the cell for ATP synthesis and other important energy-requiring functions. Discovery in the purple membrane of *Halobacterium halobium:* D. Oesterhelt, W. Stoekenius, *Nature New Biol.* **233**, 149 (1971); A. E. Blaurock, W. Stoekenius, *ibid.* 152. Isoln and identification in *H. cutirubrum:* S. C. Kushwaha, M. Kates, *Biochim. Biophys. Acta* **316**, 235 (1973). Description of functions: D. Oesterhelt, W. Stoekenius, *Proc. Nat. Acad. Sci. USA* **70**, 2853 (1973). Proposed mechanism of the "proton pump": K. Schulten, P. Tavan, *Nature* **272**, 85 (1978). Structural elucidation: Y. A. Ovchinnikov *et al.*, *FEBS Letters* **100**, 219 (1979); H. G. Khorana *et al.*, *Proc. Nat. Acad. Sci. USA* **76**, 5046 (1979). Three-dimensional crystallographic study: H. Michel, D. Oesterhelt, *ibid.* **77**, 1283 (1980). Series of articles on structure, biosynthesis, and energy transduction: *Photochem. Photobiol.* **33**, 417-608 (1981). Review of energy transduction: H. V. Westerhoff, Z. Dancshazy, *Trends Biochem. Sci.* **9**, 112 (1984). Comprehensive reviews: W. Stoekenius *et al.*, *Biochim. Biophys. Acta* **505**, 215-278 (1979); W. Stoekenius, R. A. Bogomolni, *Ann. Rev. Biochem.* **51**, 587-616 (1982). Review of crystallization, structure and function: J. Deisenhofer, H. Michel, *Science* **245**, 1463-1473 (1989).

Bacteriorhodopsin is not bleached by light in the same fashion as animal rhodopsins; only small spectral changes are caused by transfer from light to dark. When exposed to light, BR has a broad light absorption maximum at 568 nm that slowly shifts back to 558 nm in the dark. The 558 and 568 nm states are known as the dark-adapted and light-adapted states, respectively. Light-adapted BR, upon absorption of a light quantum, undergoes a series of reversible changes, having intermediate forms with absorption maxima of 590, 550, and 412 nm. uv max (2M HCl): 565 nm. Slow bleaching occurs when membrane suspensions are illuminated in the presence of hydroxylamine or sodium borohydride. When hydroxylamine is used, the uv max is shifted to 360 nm. The bleached membrane is called the "apomembrane"; the protein after loss of the chromophore is termed *bacterio-opsin*.

USE: As a tool in biological energy transduction research.

969. Badische Acid. *7-Amino-1-naphthalenesulfonic acid;* 2-naphthylamine-8-sulfonic acid. $C_{10}H_9NO_3S$; mol wt 223.25. C 53.80%, H 4.06%, N 6.27%, O 21.50%, S 14.36%. Prepd by sulfonation of 2-naphthylamine: Green, Vakil, *J. Chem. Soc.* **113**, 35 (1918); Hennion, Schmidle, *J. Am. Chem. Soc.* **65**, 2468 (1943).

Needles or prisms. Sol in 1680 parts water, in alkalies; slightly sol in alc.

970. Bagasse. Fibrous residue from cane sugar mfg operations, which contains 50% cellulose, 25% pentosans, and 25% lignin: Wiggins, *Sugar J.* **16**, No. 8, pp 18, 22, 24 (1954), *C.A.* **48**, 9729h (1954). Review of properties and uses: Vo Tong Xuan, Samaniego, *Sugar News* **46**, 276, 284 (1970); M. A. Clarke in Kirk-Othmer *Encyclopedia of Chemical Technology,* **vol. 3** (Wiley-Interscience, New York, 3rd ed., 1978) pp 434-438.

Caution: Inhalation of bagasse dust may cause pneumonitis, asthma.

USE: Raw material for pulp, paper and board.

971. Baicalein. *5,6,7-Trihydroxy-2-phenyl-4H-1-benzopyran-4-one; 5,6,7-trihydroxyflavone;* noroxylin. $C_{15}H_{10}O_5$; mol wt 270.24. C 66.67%, H 3.73%, O 29.60%. From roots of *Scutellaria baicalensis.* Isoln and structure: Bargellini, *Gazz. Chim. Ital.* **49,** II, 47 (1919); Shibata *et al., Acta Phytochim.* **1,** 109 (1923). Synthesis: Sastri, Seshadri, *Proc. Indian Acad. Sci.* **23A,** 262 (1946), *C.A.* **41,** 449 (1947); Schönberg *et al., J. Am. Chem. Soc.* **77,** 5390 (1955); Jouanne, Mentzer, *Compt. Rend.* **254,** 727 (1962); Agasimundin, Siddappa, *J. Chem. Soc. Perkin Trans. I* **1973,** 503. Pharmacology: Koda *et al., C.A.* **75,** 47200d (1971).

Yellow prisms from alc, dec 264-265°. uv max (ethanol): 324, 276 nm (log ε 4.18, 4.42). Sol in alcohol, methanol, ether, acetone, ethyl acetate, hot glacial acetic acid. Sparingly sol in chloroform, nitrobenzene. Practically insol in water. Sol in dil NaOH with greenish-brown color. Concd H_2SO_4 gives yellow color, green fluorescence.

THERAP CAT: Astringent.

972. Bakankosin. *[3S-(3α,4β,4aα)]-4-Ethenyl-3-(β-D-glucopyranosyloxy)-3,4,4a,5,6,7-hexahydro-8H-pyrano[3,4-c]-pyridin-8-one;* bacancosin; Bakankoside. $C_{16}H_{23}NO_8$; mol wt 357.36. C 53.78%, H 6.49%, N 3.92%, O 35.82%. Nitrogenous glucoside from seed of *Strychnos vacacoua* Baill., *Loganiaceae. Ref:* E. Bourquelot, H. Hérissey, *Compt. Rend.* **144,** 575 (1907); **147,** 750 (1908); *Arch. Pharm.* **247,** 56 (1909); K. Balenovic *et al., Helv. Chim. Acta* **35,** 2519 (1952). Proposed structure: G. Büchi, R. E. Manning, *Tetrahedron Letters* no. 26, 5 (1960); H. C. Beyerman *et al., Bull. Soc. Chim. France* **1961,** 1812; H. Inouye *et al., Chem. Pharm. Bull.* **24,** 1406 (1976). Synthesis and structure: L. F. Tietze, *Tetrahedron Letters* **29,** 2535 (1976).

Hydrate, bitter crystals, mp 157°; when anhyd mp 200°, also reported as mp 211-212°. $[\alpha]_D$ −197°. uv max (alcohol): 235 nm (log ε 4.2). Freely sol in water, alc; slightly sol in ethyl acetate.

973. Balsam Canada. Canada turpentine; balsam of fir. Improperly *"Balm of Gilead".* Liquid oleoresin from *Abies balsamea* (L.), Mill., *Pinaceae. Habit.* Canada and Northern U.S. to Va., west to Minnesota. *Constit.* 27.5% Volatiles (pinene, nopinene, β-phellandrene), 44.5% resin acid (13% abietic, 8% neoabietic), 27% neutral resinous compounds. *Ref:* Lombard *et al., Peintures, Pigments, Vernis* **34,** 106 (1958), *C.A.* **52,** 12420 (1958).

Yellowish to greenish, viscid, transparent, slightly fluorescent liquid; agreeable, aromatic pine-like odor; bitter taste; on exposure to air gradually solidifies to a solid, noncryst mass. d 0.987-0.994. n_4^{20} 1.52-1.54. $[\alpha]_D^{20}$ +1° to +4°. Acid no. 84-87. Sapon no. 89.4-95.7 (2 g in xylene for 1 hr). Insoluble in water; miscible with benzene, chloroform, xylene, ethyl acetate, oil of cedar; completely sol or almost sol

in ether, oil turpentine; about 90% dissolves in alcohol or petr ether.

USE: Cement for lenses; manuf fine lacquers; for mounting in microscopy.

974. Balsam Gurjun. Wood oil; "East Indian Copaiba". Oleoresin from various species of *Dipterocarpus, Dipterocarpaceae. Habit.* Eastern India, Burma. *Constit.* About 75% volatile oil, boiling at about 255°; gurjunic acid, resin, bitter substance sol in water. Isoln of an optically inactive sesquiterpene δ-elemene, from gurjun balsam: Gough, Powell, *Tetrahedron Letters* **1961,** 763.

Clear liquid, light brown by transmitted light with greenish fluorescence; somewhat bitter, but not very acrid taste. Acid no. 5-15. Sapon no. 10-20 (2 g in 20 g xylene for 1 hr). d 0.95-0.97; $[\alpha]_D$ −23° to −70°; n_D^{28} 1.510-1.516. Insol in water. Completely sol in benzene, chloroform; incompletely in alcohol, ether, carbon disulfide, petr ether.

USE: As of copaiba; adapted for varnishes and lacquers, particularly for articles to be exposed to a temp of 80° or so.

975. Balsam Mecca. Balm of Gilead; balsam of Gilead; balsán-Katél; Duhnul-balsan. Balsam obtained from twigs of *Commiphora opobalsamum* (Kunth.) Engl. (*Balsamodendrum gilaedense* Kunth.), *Burseraceae.*

Light-colored, mobile to viscid, turbid, brownish-red liquid; aromatic odor. Insoluble in water; sol in alcohol, benzene, chloroform, acetone, glacial acetic acid, carbon disulfide, oil turpentine, ether.

USE: In perfumery.

976. Balsam Peru. Peruvian balsam; Indian balsam; China oil; Black balsam; Honduras balsam; Surinam balsam. From *Toluifera pereirae* (Klotzsch) Baill. (*Myroxylon pereirae* Klotzsch), *Leguminosae. Habit.* Central America (San Salvador) in forests near Pacific coast. *Constit.* 50-60% cinnamein—esters of cinnamic and benzoic acid; about 28% resin, styracine, vanillin. *Ref:* Bergemann, *Pharmazie* **5,** 494 (1950); Cortesi, *Bull. Soc. Pharm. Bordeaux* **89,** 141 (1951), *C.A.* **46,** 1718b (1952).

Dark brown, viscid liquid; pleasant aromatic odor; warm, bitter taste and persistent aftertaste. d 1.150-1.170. Insol in water, olive oil; sol in alcohol, chloroform and glacial acetic acid, usually with a slight opalescence. Partly sol in ether, petr ether.

USE: In perfumery and some chocolate flavorings; also in masking of odors.

THERAP CAT: Vulnerary; ectoparasiticide (scabies).

THERAP CAT (VET): Vulnerary; miticide.

977. Balsam Tolu. Thomas balsam; opobalsam; resin Tolu. From *Toluifera balsamum* L. (*Myroxylon toluiferum* H. B. K.), *Leguminosae. Habit.* South America (Venezuela, Colombia, Peru) on elevated plains and mountains. *Constit.* 12-15% free cinnamic and benzoic acids; about 40% benzyl, etc., esters of these acids (5.2-13.4% cinnamein); 1.5-3% volatile oil. *Ref:* Rosenthaler, *Pharm. Ztg.-Nachr.* **88,** 716 (1952), *C.A.* **47,** 5076 (1953).

Yellowish-brown or brown, semifluid or nearly solid, resinous mass; aromatic odor and taste; brittle when cold. Insol in water; almost insol in petr ether. Sol in alcohol, benzene, chloroform, ether, glacial acetic acid, partially in carbon disulfide or NaOH.

USE: In perfumery, confectionery and chewing gums. In pharmacy, as ingredient and vehicle for expectorants.

THERAP CAT: Expectorant.

THERAP CAT (VET): Has been used as an expectorant.

978. Balsam Traumatic. Friar's balsam; Turlington's balsam. Composed of 100 parts benzoin, 35 storax, 35 balsam Tolu, 16 balsam Peru, 8 aloe, 8 myrrh, 4 angelica, and alcohol to make 1000.

THERAP CAT: Topical protectant; expectorant.

THERAP CAT (VET): Has been used for chronic bronchitis. Wound antiseptic and styptic.

979. Bambermycins. Moenomycin; flavophospholipol; Flavomycin. Antibiotic complex comprised of at least 4 active components, moenomycins A, B_1, B_2 and C, with *moenomycin A* being the major component. Obtained from cultures of *Streptomyces bambergiensis, S. ghanaensis, S. ederensis, S. geysiriensis* and related strains: Wallhäuser *et* al., *Antimicrob. Ag. Chemother.* **1965**, 734 *sqq.;* **Neth.** pat. Appl. 6,602,132, *C.A.* **66**, 74946x (1967) and Lindner, Wallhäuser, U.S. pat. 3,674,866 (1966, 1972 both to Hoechst). Structural studies: Tschesche *et al., Ann.* **720**, 58 (1968); *Tetrahedron Letters* **1968**, 2905; **1969**, 141; Huber, *J. Antibiot.* **25**, 1226 (1972); P. Welzel *et al., Tetrahedron Letters* **1973**, 227; F. J. Witteler *et al., ibid.* **1979**, 3493; P. Welzel *et al., Tetrahedron* **39**, 1583 (1983). Structure of moenomycin A: P. Welzel *et al., Angew. Chem. Int. Ed.* **20**, 121 (1981). Synthesis of *moenocinol,* the lipid moiety: Tschesche, Reden, *Ann.* **1974**, 853; P. J. Kocienski, *J. Org. Chem.* **45**, 2037 (1980); R. M. Coates, M. W. Johnson, *ibid.* 2685. *Review:* G. Huber in *Antibiotics* vol. 5, pt. 1, F. E. Hahn, Ed. (Springer-Verlag, New York, 1979) pp 135-153.

Moenomycin A

Complex is a colorless, amorphous solid, no definite mp, decompn starting at 200°C. uv max (water pH 7): 258 nm ($E_{1cm}^{1\%}$ 60). Sol in water, methanol, DMF; less sol in ethanol, propanol; slightly sol in ether, ethyl acetate. Insol in benzene, chloroform. Stable in neutral aq and methanolic solns; slowly decomp in acid and alkaline solns. LD_{50} in mice (mg/kg): > 2000 orally, s.c. and i.p.; 1400 i.v. (Lindner, Wallhäuser).

THERAP CAT: Antibacterial.

THERAP CAT (VET): Antibacterial; feed additive (poultry, swine and calves).

980. Bambuterol. *Dimethylcarbamic acid 5-[2-[(1,1-dimethylethyl)amino]-1-hydroxyethyl]-1,3-phenylene ester;* 1-[bis(3',5'-N,N-dimethylcarbamoyloxy)phenyl]-2-N-tert-butylaminoethanol; (±)-5-[2-(tert-butylamino)-1-hydroxyethyl]-m-phenylene bis(dimethylcarbamate); terbutaline bisdimethylcarbamate. $C_{18}H_{29}N_3O_5$; mol wt 367.45. C 58.84%, H 7.95%, N 11.44%, O 21.77%. Ester prodrug of the β_2-adrenergic agonist terbutaline, *q.v.* Prepn: O. A. T. Olsson *et al.,* **Eur.** pat. **Appl.** 43,807; *eidem,* U.S. pat. 4,-419,364 (1982, 1983 both to AB Draco). Pharmacology: O. A. T. Olsson, L.-A. Svensson, *Pharm. Res.* **1984**, 19. Review of pharmacology and metabolism: L.-A. Svensson,

Acta Pharm. Suec. **24**, 333 (1987). LC determn: O. Wannerberg, B. Persson, *J. Chromatog.* **435**, 199 (1988). Clinical evaluation in asthma: T. Sandström *et al., Respiration* **53**, 31 (1988). Clinical comparison with terbutaline: G. Persson *et al., Eur. Respir. J.* **1**, 223 (1988).

Hydrochloride, $C_{18}H_{29}N_3O_5$.HCl, *KWD-2183, Bambec.*

THERAP CAT: Bronchodilator.

981. Bamethan. *α-[(Butylamino)methyl]-4-hydroxybenzenemethanol; α-[(butylamino)methyl]-p-hydroxybenzyl alcohol;* 1-(p-hydroxyphenyl)-2-butylaminoethanol; 1-(4-hydroxyphenyl)-1-hydroxy-2-butylaminoethane; 2-butylamino-1-p-hydroxyphenylethanol; Butyl-Nor-Sympatol. $C_{12}H_{19}NO_2$; mol wt 209.29. C 68.87%, H 9.15%, N 6.69%, O 15.29%. Prepn: Corrigan *et al., J. Am. Chem. Soc.* **67**, 1894 (1945); Kovács, *Pharm. Zentralhalle* **92**, 193 (1953), *C.A.* **49**, 936 (1955). Metabolism of sulfate: Hajos, Szporny, *Arzneimittel-Forsch.* **18**, 1212 (1968).

Crystals, mp 123.5-125°.

Hydrochloride, $C_{12}H_{19}NO_2$.HCl, mp 109-110°.

Sulfate, $(C_{12}H_{19}NO_2)_2.H_2SO_4$, *Vasculat, Vasculit, Vascunicol, Rotesar, Butedrin, Vaskulat, Bupatol, Garmian.*

THERAP CAT: Vasodilator (peripheral).

982. Bamifylline. *7-[2-[Ethyl(2-hydroxyethyl)amino]-ethyl]-3,7-dihydro-1,3-dimethyl-8-(phenylmethyl)-1H-purine-2,6-dione; 8-benzyl-7-[2-[ethyl(2-hydroxyethyl)amino]ethyl]-theophylline; 8-benzyl-7-[N-ethyl-N-(β-hydroxyethyl)amino-ethyl]theophylline;* benzetamophylline; 8102 CB; Bamiphylline. $C_{20}H_{27}N_5O_3$; mol wt 385.47. C 62.32%, H 7.06%, N 18.17%, O 12.45%. Prepn: **Belg.** pat. 602,888 (1961 to Christiaens), *C.A.* **56**, 5981c (1962). Pharmacology: Georges *et al., Therapie* **17**, 211 (1962). Metabolism: Dodion *et al., Arzneimittel-Forsch.* **19**, 785 (1969). Toxicological studies: Georges *et al., ibid.* **18**, 460 (1968).

Crystals, mp 80-80.5°.

Hydrochloride, $C_{20}H_{27}N_5O_3$.HCl, *AC-3810, BAX-2739Z, Briofil, Trentadil.* Crystals, mp 185-186°. LD_{50} in mice, rats (mg/kg): 246, 1139 orally; 89, 131 i.p.; 67, 65 i.v. (Georges).

THERAP CAT: Bronchodilator.

983. Bamipine. *1-Methyl-N-phenyl-N-(phenylmethyl)-4-piperidinamine; 4-N-benzylanilino-1-methylpiperidine;* N-phenyl-N-benzyl-4-amino-1-methylpiperidine; Soventol. $C_{19}H_{24}N_2$; mol wt 280.41. C 81.38%, H 8.63%, N 9.99%. Prepn: Kallischnigg, U.S. pat. 2,683,714 (1954 to Knoll). Toxicity studies: Hanna, *Toxicol. Appl. Pharmacol.* **3**, 393 (1961).

Crystals from butanone or methanol, mp 115°.

Dihydrochloride, $C_{19}H_{24}N_2 \cdot 2HCl$, *Taumidrine.* Crystals from alc, mp 189° (dec).

THERAP CAT: Antihistaminic.

984. Baptigenin. *7-Hydroxy-3-(3,4,5-trihydroxyphenyl)-4H-1-benzopyran-4-one; 7,3',4',5'-tetrahydroxyisoflavone.* $C_{15}H_{10}O_6$; mol wt 286.24. C 62.94%, H 3.52%, O 33.54%. The aglucon of baptisin. From baptisin by acid hydrolysis or vacuum sublimation. From *radix Baptisiae:* Fischer, Ehrlich, *C.A.* **31**, 4449³ (1937). Structure: Böhm, *Arzneimittel-Forsch.* **10**, 472 (1960). Synthesis: Farkas *et al.*, *Ber.* **96**, 1865 (1963). *See also* Pseudobaptigenin.

Needles from dil ethanol, mp 284-285°. Begins to sublime at 240°. Sublimes in oil pump vacuum at 180-200°. uv max (ethanol): 270.2, 247 nm. Practically insol in water, ammonia water; slightly sol in dil alc, hot glacial acetic acid; sol in acetone, in NaOH solns.

Tetraacetylbaptigenin, $C_{23}H_{18}O_{10}$, needles from methanol, mp 214°.

Tetrabenzoylbaptigenin, $C_{43}H_{26}O_{10}$, prisms from methanol, mp 191-192°.

Tetramethylbaptigenin, $C_{19}H_{18}O_6$, crystals from methanol, mp 144-145°.

985. Baptisia. Wild indigo; indigo weed; false indigo; yellow indigo. Perennial herb. *Habit.* North America. Medicinal baptisia, from root of *Baptisia tinctoria* R. Br., *Leguminosae. Constit.* Baptin—a purgative glucoside, baptisin—a bitter glucoside, baptitoxine (identical with cytisine), an alkaloid. Also contains a blue dye which has been used as a substitute for indigo.

THERAP CAT: Anti-infective.

986. Barban. *(3-Chlorophenyl)carbamic acid 4-chloro-2-butynyl ester; m-chlorocarbanilic acid 4-chloro-2-butynyl ester;* chloro-2-butynyl *m*-chlorocarbanilate; 4-chloro-2-butynyl *N*-(3-chlorophenyl)carbamate; barbamate; barbane; chlorinat; CS-847; Carbyne. $C_{11}H_9Cl_2NO_2$; mol wt 258.10. C 51.19%, H 3.51%, Cl 27.47%, N 5.43%, O 12.40%. Prepn: Hopkins *et al.*, *J. Org. Chem.* **24**, 2040 (1959); U.S. pat. **2,906,614** (1959 to Spencer Chem.); Baskakov *et al.*, *Zh. Obshch. Khim.* **33**, 46 (1963). Activity as an herbicide: Hoffmann *et al.*, *Weeds* **8**, 198 (1960). Metabolism: Grunow *et al.*, *Food Cosmet. Toxicol.* **8**, 277 (1970). Toxicity study: Bailey, White, *Residue Rev.* **10**, 97 (1965). *Review:* Abel, *Rep. Progr. Appl. Chem.* **47**, 552 (1962).

Crystals from *n*-hexane + benzene, mp 75-76°. Practically insol in water (soly at 25° = 11 ppm). Slightly sol in hexane; readily sol in benzene, ethylene dichloride. Hydrolyzed by alkali with liberation of the terminal chlorine. Hydrolysis under acidic conditions gives 3-chloroacrylic acid. LD_{50} orally in rats: 600 mg/kg (Bailey, White).

Caution: May cause skin irritation.

USE: Selective herbicide for wild oats.

987. Barbasco. Name applied in the Spanish-speaking countries of the New World to many unrelated plants used to poison or stun fish. In Mexico it usually means roots of *Dioscorea composita* Hemsl., or of *Dioscorea tepinapensis* Uline, *Dioscoreaceae* which yield up to 5% of their dry weight in diosgenin: *Chem. Week* **79**, no. 2, p 20 (July 14, 1956). Isoln procedure: Julian, U.S. pat. **3,019,220** (1962 to Julian Labs.). Book: D. G. Coursey, *Yams* (Longmans, London, 1967) 230 pp, an account of the nature, origins, cultivation and utilization of the useful members of the *Dioscoreaceae. Compare* Yam, Mexican.

988. Barberry Bark. Berberis bark; jaundice berry; woodsour; sowberry; pepperidge bush; sour-spine. Root bark of *Berberis vulgaris* L., *Berberidaceae. Habit.* Europe and Western Asia; also U.S. (New England States, Pennsylvania and Virginia). *Constit.* Berberine, berbamine, oxyacanthine, tannin, wax, fat, resin. *Ref:* Neugebauer, Brunner, *Pharm. Zent.* **80**, 113 (1939).

989. Barbital. *5,5-Diethyl-2,4,6(1H,3H,5H)-pyrimidinetrione; 5,5-diethylbarbituric acid;* barbitone; diethylmalonylurea; Veronal; Malonal; Veroletten; Sédeval; Dormonal; Hypnogène; Deba; Vespéral; Uronal. $C_8H_{12}N_2O_3$; mol wt 184.19. C 52.17%, H 6.57%, N 15.21%, O 26.06%. Prepd by the condensation of the diethyl ester of diethylmalonic acid with urea in sodium ethoxide soln: Fischer, Dilthey, *Ann.* **335**, 334 (1904); Ger. pat. **146,496** (1903), *Frdl.* **7**, 651, *Chem. Zentr.* **1903, II**, 1483; Fischer, v. Mering, *Therapie der Gegenwart*, March 1903; *Therapeutische Monatsh.* **17**, 208 (1903), *Chem. Zentr.* **1903, I**, 1155.

Faintly bitter needles (trigonal in the stable phase) from water, mp 188-190°. Can be sublimed *in vacuo.* Acid to litmus. K at 25° = 3.7 × 10⁻⁸. One gram dissolves in about 130 ml water, 13 ml boiling water, 14 ml alcohol, 75 ml chloroform, 35 ml ether. Sol in acetone, ethyl acetate, alkalies, petr ether, acetic acid, amyl alcohol, pyridine, aniline, nitrobenzene.

Sodium salt, $C_8H_{11}N_2NaO_3$, *barbital sodium, sodium 5,5-diethylbarbiturate, barbitone sodium, soluble barbital, sodium diethylmalonylurea,Veronal sodium, Medinal, Embinal.* Bitter crystals or powder. One gram dissolves in 5 ml water, 2.5 ml boiling water, 400 ml alc. Aq soln is alkaline to litmus and phenolphthalein. pH of 0.1 molar aq soln, 9.4.

Caution: May be habit forming. This is a controlled substance (depressant) listed in the U.S. Code of Federal Regulations, Title 21 Parts 329.1 and 1308.14 (1995).

THERAP CAT: Sedative, hypnotic.

THERAP CAT (VET): Sedative, hypnotic.

990. Barbituric Acid. *2,4,6(1H,3H,5H)-pyrimidinetrione;* malonylurea; 2,4,6-trioxohexahydropyrimidine. $C_4H_4N_2O_3$; mol wt 128.09. C 37.51%, H 3.15%, N 21.87%, O 37.47%. Prepn from hydurilic acid + nitric acid: Baeyer, *Ann.* **127**, 199 (1863); *ibid.* **130**, 129 (1864). Structure: Mulder, *Ber.* **6**, 1233 (1873). Prepd from ethyl malonate and urea using sodium ethoxide as a condensing agent: Dickey, Gray, *Org. Syn.* coll. vol. II, 60 (1943). Crystal structure: Bolton, *Nature* **201**, 987 (1964). Toxicity study: E. I. Goldenthal, *Toxicol. Appl. Pharmacol.* **18**, 185 (1971). Unsubstituted barbituric acid has no hypnotic properties. *Review:* Carter, *J. Chem. Ed.* **28**, 524 (1951).

Dihydrate, rhombs from water. mp ~248° when anhydrous, with some decompn. Strong acid. K at 25° = 9.9 × 10⁻⁵. uv spectrum: Hartley, *J. Chem. Soc.* **87**, 1808 (1905). Difficultly sol in cold water; freely sol in hot water, in dil acids. Forms salts with metals. LD_{50} orally in male rats: > 5000 mg/kg (Goldenthal).

USE: Manuf plastics, pharmaceuticals.

991. Barium. Ba; at. wt 137.33; at. no. 56; valence 2. An alkaline earth metal. Abundance in earth's crust 0.05% by wt. Isotopes: 138 (71.66%); 137 (11.32%); 136 (7.81%); 135 (6.59%); 134 (2.42%); 132 (0.097%); 130 (0.101%). Occurs in barite and witherite. First prepared as a mercury amalgam by Davy in 1808. Toxicity studies of barium compds: Syed, Hosain, *Toxicol. Appl. Pharmacol.* **22**, 150 (1972). *Reviews: Gmelin's Handb. Anorg. Chem., Barium* (8th ed.) **30**, (1960); Goodenough, Stenger, "Magnesium, Calcium, Strontium, Barium and Radium" in *Comprehensive Inorganic Chemistry* Vol. **1**, J. C. Bailar, Jr. *et al.,* Eds. (Pergamon Press, Oxford, 1973) pp 591-664; C. J. Kunesh in Kirk-Othmer *Encyclopedia of Chemical Technology* vol. **3** (Wiley-Interscience, New York, 3rd ed., 1978) pp 457-463.

Yellowish-white, slightly lustrous lumps; body-centered cubic structure; somewhat malleable; very easily oxidizable; must be kept under petroleum or other oxygen-free liquid to exclude air. d 3.6. mp ~710°. bp ~1600°. E^0 (aq) Ba^{2+}/Ba −2.91 V. Description of reactions which are characteristic of alkaline earth metals *see* Calcium. Solns of sol barium salts give a white ppt with H_2SO_4 or sol sulfates; they also color nonluminous flame green.

Caution: All water or acid soluble barium compounds are poisonous!

USE: Carrier for radium. The β- and γ-radiation emitted by ^{140}Ba + ^{140}La makes a large contribution to the activity of the fission products of uranium rods during the first few weeks after their withdrawal from the reactor. Alloys of Ba with Al or Mg are used as getters in electronic tubes. The emissions from ^{133}Ba and ^{137m}Ba are used as standards in γ-spectrometry: Haissinsky, Adloff, *Radiochemical Survey of the Elements* (Elsevier, 1965) pp 12-14.

992. Barium Acetate. $C_4H_6BaO_4$; mol wt 255.42. C 18.81%, H 2.37%, Ba 53.77%, O 25.06%. $Ba(C_2H_3O_2)_2$. Prepn: *Gmelin's, Barium* (8th ed.) **30**, 315 (1932) and supplement, 478 (1960). Toxicity study: Syed, Hosain, *Toxicol. Appl. Pharmacol.* **22**, 150 (1972).

Monohydrate. *Poisonous!* d 2.19. Loses its H_2O of hydration at 110°. One gram dissolves in 1.5 ml cold or boiling water, in 700 ml alc. The aq soln is neutral or slightly acid to litmus. LD_{50} i.v. in ICR mice: 23.31 mg Ba^{2+}/kg (Syed, Hosain).

USE: Mordant for printing fabrics; in lubricating oil and grease; as catalyst for organic reactions.

993. Barium Benzenesulfonate. *Benzenesulfonic acid barium salt.* $C_{12}H_{10}BaO_6S_2$; mol wt 451.67. C 31.91%, H 2.23%, Ba 30.40%, O 21.25%, S 14.20%. Prepn: Freund, *Ann.* **120**, 76 (1861).

Monohydrate, white, nacreous leaflets. *Poisonous!* Freely sol in water; slightly sol in alc.

USE: Lubricating oil additives.

994. Barium Bromate. $BaBr_2O_6$; mol wt 393.13. Ba 34.93%, Br 40.65%, O 24.42%. $Ba(BrO_3)_2$. Prepd from potassium bromate and barium chloride: Pearce, Russell, *Inorg. Syn.* **2**, 20 (1946).

Monohydrate, monoclinic crystals from hot water. *Poisonous!* May develop slight odor of bromine on long standing. d 3.99. Dec. at 260°. Sol in water (g/100 ml): 0.44 (10°); 0.96 (30°); 5.39 (100°). Sol in acetone. Practically insol in alc, most other organic solvents.

USE: In the prepn of rare earth bromates; as corrosion inhibitor for low-C steel.

995. Barium Bromide. $BaBr_2$; mol wt 297.14. Ba 46.22%, Br 53.78%. Prepn: *Gmelin's, Barium* (8th ed.) **30**, 223 (1932) and supplement 380-381 (1960).

Dihydrate, crystals or granules; loses $1H_2O$ at 75° and all the H_2O at 120°. *Poisonous!* mp ~850° when anhydr. Very sol in water; sol in methanol. Almost insol in ethanol, ethyl acetate, acetone, dioxane.

USE: In the manuf of other bromides; in the prepn of phosphors.

996. Barium Carbonate. $CBaO_3$; mol wt 197.34. C 6.09%, Ba 69.59%, O 24.32%. $BaCO_3$. Occurs in nature as the mineral *witherite.* Prepn: *Gmelin's, Barium* (8th ed) **30**, 301-303 (1932) and supplement, 186-188, 461-466 (1960). Manuf: Faith, Keyes & Clark's *Industrial Chemicals,* F. A. Lowenheim, M. K. Moran, Eds. (Wiley-Interscience, New York, 4th ed., 1975) pp 121-125.

White, heavy powder. *Poisonous!* d (witherite) 4.2865. At about 1300° dec into BaO and CO_2. Almost insol in water, 0.024 g in a liter. Slightly sol (1:1000) in CO_2-water; sol in dil HCl, HNO_3 or acetic acid; also sol in soln NH_4Cl or NH_4NO_3.

Caution: Potential symptoms of overexposure are excessive salivation, vomiting, severe abdominal pain, violent diarrhea; increased blood pressure; tinnitis, giddiness, vertigo; muscle twitching, convulsions, paralysis; dilated pupils; confusion, somnolence; cardiac arrest; death due to respiratory failure. *See Clinical Toxicology of Commercial Products,* R. E. Gosselin *et al.,* Eds. (Williams & Wilkins, Baltimore, 5th ed., 1984) Section III, pp 61-63.

USE: Rat poison; in ceramics, paints, enamels, marble substitutes, rubber; manuf of paper, barium salts, electrodes, optical glasses; as an analytical reagent.

997. Barium Chlorate. $BaCl_2O_6$; mol wt 304.23. Ba 45.14%, Cl 23.31%, O 31.55%. $Ba(ClO_3)_2$. Prepn: Vanino, *Handb. Präp. Chem., Anorgan. Teil* (2.Aufl., Stuttgart, 1925) p 297; Schmeisser in *Handbook of Preparative Inorganic Chemistry,* Vol. **1**, G. Brauer, Ed. (Academic Press, New York, 2nd ed., 1963) p 314. Large-scale process: Munroe, *Chem. Met. Eng.* **23**, 188 (1920). Also prepd by electrolysis of barium chloride.

Monohydrate, monoclinic prismatic crystals. *Poisonous!* d 3.179. Loses its water of hydration at 120°, begins to give off oxygen at 250°, mp 414°. Freely sol in water; sol in hydrochloric acid; moderately sol in ethylamine; very sparingly sol in alc, somewhat more in acetone. Practically insol in ethyl acetate, pyridine. Fire hazard when in contact with combustible material.

USE: In pyrotechnics (green fire); manuf of explosives and matches; mordant in dyeing.

998. Barium Chloride. $BaCl_2$; mol wt 208.23. Ba 65.95%, Cl 34.05%. Prepn: *Gmelin's, Barium* (8th ed) **30**, 171-175 (1932) and supplement, 179-181, 324-325 (1960). Toxicity studies: I. B. Syed, F. Hosain, *Toxicol. Appl. Pharmacol.* **22**, 150 (1972). Induction of arrhythmia in exptl animals: F. W. Eichbaum, *Basic Res. Cardiol.* **68**, 73 (1973).

Dihydrate, crystals or granules or powder; bitter salty taste. d 3.86; mp 963. *Poisonous!* Very sol in water; sol in methanol. Practically insol in ethanol, acetone, ethyl acetate. LD_{50} i.v. in ICR mice: 19.2 mg Ba^{2+}/kg (Syed, Hosain).

Caution: Potential symptoms of overexposure are upper respiratory irritation; gastroenteritis; muscle spasm; slow pulse, extrasystoles; hypokalemia; irritation of eyes and skin; skin burns. *See NIOSH Pocket Guide to Chemical Hazards* (DHHS/NIOSH 90-117, 1990) p 44.

USE: Manuf pigments, color lakes, glass, mordant for acid dyes; weighting and dyeing textile fabrics; in Al refining; as pesticide; boiler compds for softening water; tanning and finishing leather.

THERAP CAT (VET): Formerly used as purgative in horses, ruminatoric in cattle.

999. Barium Chromate(VI). C.I. 77103; C.I. Pigment Yellow 31; Baryta yellow; lemon yellow; permanent yellow; Steinbühl yellow; ultramarine yellow. $BaCrO_4$; mol wt 253.32. Ba 54.21%, Cr 20.53%, O 25.26%. Prepn: Beyer, Rieman, *J. Am. Chem. Soc.* **65**, 971 (1943); *Colour Index* vol. **4** (3rd ed., 1971) p 4656.

Yellow, heavy, monoclinic, orthorhombic crystals. *Poisonous!* d 4.50. Practically insol in water, dil acetic or chromic acids. Dissolved or dec by mineral acids.

USE: As a pigment almost entirely in anticorrosion jointing pastes to prevent electro-chemical corrosion at junctions of dissimilar metals; some use in artists' colors and in coloring glass, ceramics, porcelain. Also used in metal primers, pyrotechnic compositions.

1000. Barium Cyanide. C_2BaN_2; mol wt 189.36. C 12.69%, Ba 72.52%, N 14.79%. $Ba(CN)_2$. Prepn: **Brit.** pat.

602,393 (1948 to I.C.I.); *Gmelin's, Barium* (8th ed.) **30,** 327 (1932) and supplement 483 (1960).

Crystals; slowly dec in air. *Very poisonous!* Very sol in water; sol in alcohol.

USE: In electroplating processes; in metallurgy.

1001. Barium Dithionate. Barium "hyposulfate". BaO_6S_2; mol wt 297.46. Ba 46.17%, O 32.27%, S 21.56%. BaS_2O_6. Prepn: Pfanstiel, *Inorg. Syn.* **2,** 170 (1946).

Dihydrate, crystals. *Poisonous!* d 4.54. Loses SO_2 on heating above 150° forming $BaSO_4$. Sol in 4 parts water; more sol in hot water; slightly sol in alcohol.

1002. Barium Ferrocyanide. *Barium hexacyanoferrate-(II).* $C_6Ba_2FeN_6$; mol wt 486.61. C 14.81%, Ba 56.44%, Fe 11.48%, N 17.27%. $Ba_2Fe(CN)_6$. Prepn: Grat-Cabanac, *Bull. Soc. Chim. France* **1956,** 1743.

Hexahydrate, yellowish rectangular monoclinic crystals; loses most of water at 40° becoming colorless. Dec at 80°, evolving HCN. Almost insol in water, alcohol.

1003. Barium Fluoride. BaF_2; mol wt 175.32. Ba 78.33%, F 21.67%. Prepd by dissolving $BaCO_3$ in excess HF, evaporating to dryness, and heating to red heat: W. Olbrich, *Thesis* (Technische Hochschule, Breslau, 1929), p 2; Kwasnik in *Handbook of Preparative Inorganic Chemistry* vol. **1,** G. Brauer, Ed., (Academic Press, New York, 2nd ed., 1963) p 234.

Transparent cubic crystals (fluorite lattice). *Poisonous!* d 4.83. mp 1353. bp 2260. Soly in water (g/l): 1.586 (10°); 1.607 (20°); 1.620 (30°). Also sol in hydrochloric, nitric, hydrofluoric acids and in aq solns of ammonium chloride. May be stored in glass bottles.

USE: As a flux and opacifier in vitreous enamels; in the manuf of carbon brushes for DC motors and generators; in heat-treating metals; in embalming; in glass manuf.

1004. Barium Formate. $C_2H_2BaO_4$; mol wt 227.36. C 10.57%, H 0.89%, Ba 60.40%, O 28.15%. $Ba(HCOO)_2$. Prepn: *Gmelin's, Barium* (8th ed) **30,** 311 (1932), and supplement, 477 (1960).

Crystals. *Poisonous!* d 3.21. Soluble in 4 parts cold, 3 parts boiling water. Practically insol in alcohol.

1005. Barium Hexafluorosilicate. Barium fluosilicate; barium silicofluoride. BaF_6Si; mol wt 279.40. Ba 49.15%, F 40.80%, Si 10.05%. $BaSiF_6$. Prepd from $BaCl_2$ and H_2SiF_6: Truchot, *Compt. Rend.* **98,** 821 (1884); Hoffmann, Gutowsky, *Inorg. Syn.* **4,** 145 (1953).

Orthorhombic needles. d_4^{21} 4.29. Dec at 300°. Heat of formation (solid) −677.42 kcal/mol. Soly in water (g/100 ml H_2O): 0.015 (0°); 0.0235 (25°); 0.091 (100°). Prolonged contact with water produces hydrolysis which is much accelerated by the presence of alkali. Slightly sol in dil acids; sol in ammonium chloride soln; practically insol in alc.

Caution: Highly toxic, especially when brought into soln by alkali.

USE: Prepn of silicon tetrafluoride; as pesticide.

1006. Barium Hydroxide. Barium hydrate; caustic baryta. BaH_2O_2; mol wt 171.34. Ba 80.15%, H 1.18%, O 18.68%. $Ba(OH)_2$. Prepn: *Gmelin's, Barium* (8th ed) **30,** 106-111 and supplement 175-177, 289 (1960).

Monohydrate, *dried barium hydroxide.* Usually contains 92-95% $Ba(OH)_2.H_2O$. White powder. *Poisonous!* d 3.743. Slightly sol in water; sol in dil acids.

Octahydrate, transparent crystals or white masses. Very alkaline; rapidly absorbs CO_2 from air, becoming incompletely sol in water. *Poisonous!* mp 78°. Freely sol in water, methanol; slightly sol in ethanol; practically insol in acetone. *Keep tightly closed.*

USE: In manuf of alkali, glass; in synthetic rubber vulcanization, in corrosion inhibitors, drilling fluids, lubricants, pesticides, sugar industry; boiler scale remedy; refining animal and vegetable oils; softening water; fresco painting.

1007. Barium Hypophosphite. $BaH_4O_4P_2$; mol wt 267.30. Ba 51.37%, H 1.51%, O 23.94%, P 23.17%. $Ba(H_2PO_2)_2$. Prepd by treating white phosphorus with barium hydroxide: $8P + 3Ba(OH)_2.8H_2O + H_2O \rightarrow 3Ba(H_2PO_2)_2.H_2O + 2PH_3$: Rose, *Pogg. Ann.* **9,** 370 (1827); Klement in *Handbook of Preparative Inorganic Chemistry*, Vol. 1, G.

Brauer, Ed. (Academic Press, New York, 2nd ed., 1963) p 557.

Monohydrate, monoclinic platelets with nacreous sheen from hot water. *Poisonous!* d_4^{17} 2.90. Soly (g/100 ml H_2O): 28.6 (17°); 33.3 (100°). Practically insol in alcohol.

USE: In nickel plating.

1008. Barium Iodate. BaI_2O_6; mol wt 487.13. Ba 28.19%, I 52.10%, O 19.71%. $Ba(IO_3)_2$. Prepn: Lambert, Yasada, *Inorg. Syn.* **7,** 13 (1963).

Monohydrate, crystals; becomes anhydr at 130°. d 5.00. *Poisonous!* Sol in 3350 parts water at 25°, 625 parts boiling water; sol in HCl or HNO_3; practically insol in alcohol.

1009. Barium Iodide. BaI_2; mol wt 391.14. Ba 35.11%, I 64.89%. Prepn: *Gmelin's, Barium* (8th ed) **30,** 238-239 (1932) and supplement, 394 (1960).

Dihydrate, colorless, odorless, transparent crystals or white granules; rapidly becomes reddish in the air due to liberation of iodine. *Poisonous!* d 5.15. Freely sol in water; the aq soln is neutral or slightly alkaline. Sol in alcohol, acetone. *Keep well closed and protected from light.*

USE: In the manuf of other iodides.

1010. Barium Manganate(VI). Manganese green; Cassel's green; Rosenstiehl's green. $BaMnO_4$; mol wt 256.26. Ba 53.59%, Mn 21.44%, O 24.97%. Prepn: Schlesinger, Stems, *J. Am. Chem. Soc.* **46,** 1965 (1924); Jellinek, *J. Inorg. Nucl. Chem.* **13,** 329 (1960); R. S. Nyholm, P. R. Woolliams, *Inorg. Syn.* **11,** 56 (1968); Bayan, Aymonino, *Ber.* **101,** 3337 (1968); H. Firouzabadi, Z. Mostafavipoor, *Bull. Chem. Soc. Japan* **56,** 914 (1983).

Dark blue-green crystals. d 4.85. Disproportionates in water or dil acid to form $Ba(MnO_4)_2$ and MnO_2.

USE: As pigment in fresco painting instead of Scheele's green because not so poisonous as latter. As mild oxidizing agent in organic synthesis.

1011. Barium Mercuric Iodide. Barium tetraiodomercurate(II); mercuric barium iodide. $BaHgI_4$; mol wt 845.53. Ba 16.24%, Hg 23.72%, I 60.04%. Yellow or reddish, deliquesc crystals; said to become red even when kept in sealed tubes. *Poison!* Very sol in water or alc. *Keep well closed!*

USE: As an aq soln known as *Rohrbach's Soln,* for separating minerals of different densities—the concd soln has a d of 3.5; also for microchemical detection of alkaloids.

1012. Barium Nitrate. BaN_2O_6; mol wt 261.34. Ba 52.55%, N 10.72%, O 36.73%. $Ba(NO_3)_2$. Prepn: *Gmelin's, Barium* (8th ed) **30,** 149-151 (1932) and supplement, 178-179, 305 (1960). Toxicity study: Syed, Hosain, *Toxicol. Appl. Pharmacol.* **22,** 150 (1972).

Crystals or crystalline powder. *Poisonous!* d 3.24. mp ~590°; dec at higher temp. Freely sol in water; very slightly sol in alcohol, acetone. LD_{50} i.v. in ICR mice: 20.10 mg Ba^{2+}/kg (Syed, Hosain).

Caution: Potential symptoms of overexposure are upper respiratory irritation; gastroenteritis; muscle spasm; slow pulse, extrasystoles; hypokalemia; irritation of eyes and skin; skin burns. *See NIOSH Pocket Guide to Chemical Hazards* (DHHS/NIOSH 90-117, 1990) p 44.

USE: Manuf BaO_2; pyrotechnics for green fire; green signal lights; in the vacuum-tube industry.

1013. Barium Nitrite. BaN_2O_4; mol wt 229.34. Ba 59.88%, N 12.21%, O 27.91%. $Ba(NO_2)_2$. Prepn: *Gmelin's, Barium* (8th ed) **30,** 144-147 (1932) and suppl, 303 (1960).

Monohydrate, crystals. *Poisonous!* d 3.187. Sol in water; practically insol in alc.

USE: In diazotization reactions; prevention of corrosion of steel bars; in explosives.

1014. Barium Oxalate. *Ethanedioic acid barium salt.* C_2BaO_4; mol wt 225.35. C 10.66%, Ba 60.94%, O 28.40%. BaC_2O_4. Prepn: *Gmelin's, Barium* (8th ed) **30,** 320-323 (1932) and supplement, 480-482 (1960).

Monohydrate, cryst powder. *Poisonous!* d 2.66. Sol in 10,000 parts cold, 5000 parts boiling water; sol in dil HNO_3 or HCl.

1015. Barium Oxide. Barium monoxide; barium protoxide; calcined baryta. BaO; mol wt 153.33. Ba 89.57%, O

10.43%. Prepn: Ehrlich in *Handbook of Preparative Inorganic Chemistry*, **vol. 1**, G. Brauer, Ed. (Academic Press, New York, 2nd ed., 1963) p 933.

White to yellowish-white powder or lumps. Very alkaline; absorbs moisture and CO_2 on exposure to air. *Poisonous!* On contact with water it forms $Ba(OH)_2$ with evolution of much heat. At 450° combines with oxygen to form BaO_2, which is reduced to BaO above 600°. d 5.7. mp ~1920°. Sol in water, dil acids. Slowly, but considerably sol in methanol, ethanol forming barium alcoholate. *Keep tightly closed.*

USE: Porous grades are marketed especially for drying gases and solvents (particularly alcohols, aldehydes and petroleum solvents). Swells, but does not become sticky upon absorption of moisture. Used in manuf of lubricating oil detergents. Also used for making barium methoxide.

1016. Barium Perchlorate. $BaCl_2O_8$; mol wt 336.23. Ba 40.84%, Cl 21.09%, O 38.07%. $Ba(ClO_4)_2$. Prepn: *Gmelin's, Barium* (8th ed) **30**, 218 (1932) and supplement, 373 (1960).

Trihydrate, crystals. *Poisonous!* Sol in water, methanol; slightly sol in ethanol, ethyl acetate, acetone; practically insol in ether.

USE: In the determination of ribonuclease; as absorbent for water in C and H analysis.

1017. Barium Permanganate. Barium manganate-(VII). $BaMn_2O_8$; mol wt 375.20. Ba 36.60%, Mn 29.28%, O 34.11%. $Ba(MnO_4)_2$. Prepn: Lux in *Handbook of Preparative Inorganic Chemistry*, **vol. 1**, G. Brauer, Ed. (Academic Press, New York, 2nd ed., 1963) p 1462.

Brownish-violet to black crystals. *Poisonous!* d 3.77. Sparingly sol in water; dec by alcohol.

USE: As dry cell depolarizer.

1018. Barium Peroxide. Barium dioxide; barium superoxide. BaO_2; mol wt 169.33. Ba 81.10%, O 18.90%. Prepn: *Gmelin's, Barium* (8th ed) **30**, 92-98 (1932) and supplement, 177, 296 (1960).

White or grayish-white, heavy powder. *Poisonous!* Dec slowly in air. Insol in water, but slowly dec by contact with it; dec by dil acid or CO_2 in presence of H_2O forming H_2O_2. Combines with water to form octahydrate. *Keep well closed.* The article of commerce contains about 85% BaO_2; the remainder is chiefly BaO.

USE: Bleaching animal substances, vegetable fibers and straw; glass decolorizer; manuf H_2O_2 and oxygen; dyeing and printing textiles; with powdered aluminum in welding; in cathodes; in igniter compositions. Oxidizing agent in organic synthesis.

1019. Barium Phosphate, Dibasic. Secondary barium phosphate. $BaHO_4P$; mol wt 233.31. Ba 58.86%, H 0.43%, O 27.43%, P 13.28%. $BaHPO_4$. Prepn: *Gmelin's, Barium* (8th ed) **30**, 340 (1932) and supplement, 492 (1960).

Crystals. *Poisonous!* d 4.16. Practically insol in water; sol in dil HCl or HNO_3.

USE: In fireproofing compositions; in prepn of phosphors.

1020. Barium Platinous Cyanide. Barium tetracyanoplatinate(II); barium cyanoplatinate(II); barium platinocyanide; platinous barium cyanide. C_4BaN_4Pt; mol wt 436.48. C 11.01%, Ba 31.46%, N 12.84%, Pt 44.69%. $BaPt(CN)_4$.

Tetrahydrate, large dichroic crystals; yellowish-green by transmitted, bluish-violet by reflected light. *Poison!* d 3.05. Sol in about 35 parts water; more sol in hot water.

USE: An aq soln mixed with some adhesive and painted on paper or wood exhibits luminescence when exposed to the invisible ultraviolet rays of the spectrum or to Roentgen, radium, or cathode rays; hence used in radiography for making x-ray screens.

1021. Barium Selenide. BaSe; mol wt 216.29. Ba 63.49%, Se 36.51%. Prepn: Henglein, *Z. Anorg. Chem.* **120**, 77 (1921); *Z. Electrochem.* **30**, 11 (1924); *Gmelin's, Barium* (8th ed) **30**, 290 (1932) and supplement, 453 (1960).

Cubic microcrystalline powder. d 5.02. Turns red on exposure to air. Dec in water.

USE: In photocells, semiconductors.

1022. Barium Silicide. $BaSi_2$; mol wt 193.50. Ba 70.97%, Si 29.03%. Prepn: *Gmelin's, Barium* (8th ed) **30**, 330 (1932) and supplement, 486 (1960).

Metal-like, gray lumps; quite permanent in dry air, but dec by moisture with evolution of H_2. Melts at white heat.

USE: Deoxidizing and desulfurizing steel and for other metallurgical purposes.

1023. Barium Sulfate. Blanc fixe; Actybaryte; Bakontal; Baridol; Baritop; Barosperse; Citobaryum; Esophotrast; E-Z-Paque; Intestibar; Microbar; Micropaque; Microtrast; Mixobar; Neobar; Oratrast; Polybar; Prontobario; Radiopaque; Telebar; Unibaryt. BaO_4S; mol wt 233.39. Ba 58.84%, O 27.42%, S 13.74%. $BaSO_4$. Occurs in nature as the mineral *barite;* also as *barytes, heavy spar.* Prepn: *Gmelin's, Barium* (8th ed) **30**, 262-267 (1932) and supplement, 182-186, 412-414 (1960). Review of clinical use: R. F. Theoni, A. R. Margulis, *Radiology* **167**, 7 (1988).

Fine, heavy, odorless powder or polymorphous crystals, d 4.25-4.5. Dec above 1600°. Practically insol in water (one gram dissolves in 400,000 parts), dil acids, alc. Sol in hot concd H_2SO_4.

USE: Manuf photographic papers, artificial ivory, cellophane; filler for rubber, linoleum, oil cloth, polymeric fibers and resins, paper, lithographic inks; as a water-color pigment for colored paper, in wallpaper; as a size for modifying the colors of other pigments; in heavy concrete for radiation shield.

THERAP CAT: Diagnostic aid (radiopaque medium).
THERAP CAT (VET): X-ray contrast medium.

1024. Barium Sulfide. BaS; mol wt 169.39. Ba 81.07%, S 18.93%. The commercial article contains 80-90% BaS. Prepn: Ehrlich in *Handbook of Preparative Inorganic Chemistry*, **vol. 1**, G. Brauer, Ed. (Academic Press, New York, 2nd ed., 1963) p 938.

Heavy, grayish-white or pale yellow powder. d 4.36. mp >2000°. *Poisonous!* Oxidizes in dry air; slowly dec in damp air into carbonate, etc., with evolution of H_2S. Slightly sol in cold, more in hot water; sol in NH_4Cl soln; decomposed by acids, even CO_2, forming H_2S. *Keep in well-closed containers.*

USE: As depilatory; in luminous paints; manuf lithopone; vulcanizing rubber, generating H_2S.

1025. Barium Sulfide, Black. Black ash. This is the crude barium sulfide obtained by strong heating of a mixture of barium sulfate mineral (barite) and charcoal. Contains 60-70% BaS; the remainder is chiefly barium sulfate and carbon.

USE: Prepn of barium salts.

1026. Barium Sulfite. BaO_3S; mol wt 217.39. Ba 63.17%, O 22.08%, S 14.75%. $BaSO_3$. Prepn: *Gmelin's, Barium* (8th ed.) **30**, 261 (1932) and supplement, 411 (1960).

Odorless crystals or powder; gradually oxidizes in air to $BaSO_4$. *Poisonous!* Slightly sol in water. Practically insol in alcohol; dec by acids with liberation of SO_2.

USE: In paper manufacture.

1027. Barium Thiocyanate. Barium sulfocyanate; barium sulfocyanide; barium rhodanide. $C_2BaN_2S_2$; mol wt 253.49. C 9.48%, Ba 54.17%, N 11.05%, S 25.30%. $Ba(SCN)_2$. Prepn: Herstein, *Inorg. Syn.* **3**, 24 (1950).

Deliquesc crystals. *Poisonous!* Very sol in water, sol in acetone, methanol, ethanol. *Keep well closed.*

Trihydrate, needle-shaped crystals (from water).

USE: In dyeing; in photography; as dispersing agent for cellulose; in prepn of thiocyanates of other metals.

1028. Barium Thiosulfate. Barium hyposulfite. BaO_3S_2; mol wt 249.46. Ba 55.05%, O 19.24%, S 25.71%. BaS_2O_3. Prepn: *Gmelin's, Barium* (8th ed.) **30**, 283 (1932) and supplement, 449 (1960).

Monohydrate, cryst powder. *Poisonous!* Very slightly sol in water. Practically insol in alc, ether, acetone, CCl_4, CS_2.

USE: Manuf explosives, matches; as an iodometry standard; in photographic diffusion-transfer process.

1029. Barium Titanate(IV). Barium metatitanate. BaO$_3$Ti; mol wt 233.19. Ba 58.89%, O 20.58%, Ti 20.53%. BaTiO$_3$. Usually prepd by calcining (at about 1300°) an intimate mixture of titanium dioxide and barium carbonate. Prepn from Ti oxalate and a barium compd: Lynd, Merker, **U.S. pat. 2,758,911** (1956 to National Lead). Wet process by the addn of a titanium ester, such as tetrapropyl titanate, to an aq soln of barium hydroxide: Flaschen, *J. Am. Chem. Soc.* **77**, 6194 (1955). Prepn of high purity material by ignition of barium titanyloxalate: Clabaugh *et al., J. Res. NBS* **56**, 289 (1956). Prepn by ignition of Ba and Ti alcoholates in an organic solvent: DiVita, Fischer, **U.S. pat. 2,985,504** (1961 to U.S. Dept. of the Army).

Exists in five cryst modifications. The tetragonal form (obtained by the wet process) appears to have the most desirable electric properties and is described here: d 6.08. mp 1625°. Curie point 120°. Has ferroelectric and piezoelectric properties. Becomes permanently polarized when exposed to high voltage direct current, provided the temperature is never allowed to rise above Curie pt. Has high dielectric properties which can be influenced by temp, voltage, and frequency.

USE: In electronic devices, e.g., as voltage-sensitive dielectric in so-called dielectric amplifiers, in computer elements, magnetic amplifiers, memory devices.

1030. Barium Uranium Oxide. Barium uranate(VI); barium diuranate; uranium barium oxide. BaO$_7$U$_2$; mol wt 725.38. Ba 18.93%, O 15.44%, U 65.63%. BaU$_2$O$_7$. Prepn: Allpress, *J. Inorg. Nucl. Chem.* **26**, 1847 (1964); Klima *et al., ibid.* **28**, 1861 (1966).

Orange or yellow powder. Practically insol in water. Sol in acids.

USE: In painting on porcelain.

1031. Barnidipine. *[S-(R*,R*)]-1,4-Dihydro-2,6-dimethyl-4-(3-nitrophenyl)-3,5-pyridinedicarboxylic acid methyl 1-(phenylmethyl)-3-pyrrolidinyl ester;* (+)-(3'S,4S)-1-benzyl-3-pyrrolidinyl methyl 1,4-dihydro-2,6-dimethyl-4-(*m*-nitrophenyl)-3,5-pyridinedicarboxylate; (3S)-1-benzyl-3-pyrrolidinyl methyl (4S)-2,6-dimethyl-4-(*m*-nitrophenyl)-1,4-dihydropyridine-3,5-dicarboxylate; mepirodipine. C$_{27}$H$_{29}$-N$_3$O$_6$; mol wt 491.54. C 65.98%, H 5.95%, N 8.55%, O 19.53%. Dihydropyridine calcium channel blocker. Prepn (stereochemistry unspecified): T. Kojima, T. Takenaka, **Ger. pat. 2,904,552**; *eidem,* **U.S. pat. 4,220,649** (1979, 1980 both to Yamanouchi). Prepn and bioactivity of enantiomers: K. Tamazawa *et al., J. Med. Chem.* **29**, 2504 (1986). GC-MS determn in plasma: T. Teramura *et al., J. Chromatog.* **528**, 191 (1990). Review of pharmacology and therapeutic efficacy: H. Satoh, *Cardiovasc. Drug Rev.* **9**, 340-356 (1991).

mp 137-139°. [α]$_D^{20}$ +64.8° (c = 1 in methanol).

Hydrochloride, C$_{27}$H$_{29}$N$_3$O$_6$·HCl, *YM-09730-5, Hypoca.* mp 226-228°. [α]$_D^{20}$ +116.4° (c = 1 in methanol). Insol in water. LD$_{50}$ orally in male, female rats: 105, 113 mg/kg (Satoh).

THERAP CAT: Antihypertensive; antianginal.

1032. Barthrin. *2,2-Dimethyl-3-(2-methyl-1-propenyl)-cyclopropanecarboxylic acid (6-chloro-1,3-benzodioxol-5-yl) methyl ester; 2,2-dimethyl-3-(2-methylpropenyl)cyclopropanecarboxylic acid 6-chloropiperonyl ester;* 6-chloropiperonyl 2,2-dimethyl-3-(2-methylpropenyl)cyclopropanecarboxylate; 6-chloropiperonyl chrysanthemumate; chrysanthemum-monocarboxylic acid 6-chloropiperonyl ester. C$_{18}$H$_{21}$ClO$_4$;

mol wt 336.82. C 64.19%, H 6.28%, Cl 10.53%, O 19.00%. Prepn: Barthel, Alexander, *J. Org. Chem.* **23**, 1012 (1958); Barthel *et al.,* **U.S. pat. 2,886,485** (1959). Shows low mammalian toxicity.

trans-form

Oily liquid, bp$_{0.2}$ 155-171°, bp$_{0.7}$ 184-206°. n$_D^{25}$ 1.5383. Sol in kerosine.

USE: Insecticide.

1033. Basic Aluminum Carbonate Gel. Basaljel. An aluminum hydroxide—aluminum carbonate gel.

THERAP CAT: Phosphorus-binding agent in prevention of recurrent renal phosphatic calculi; antacid.

1034. Basic Lead Carbonate. *C.I. Pigment White 1;* C.I. 77597; lead subcarbonate; white lead; flake lead; ceruse; cerussa; bleiweiss (German). Approx (PbCO$_3$)$_2$·Pb(OH)$_2$. Occurs in nature as the mineral *hydrocerussite. See: Mellor's* **vol. VII**, 837 (1927); *Colour Index* **vol. 4** (3rd ed., 1971) p 4676.

White, heavy powder. *Poisonous!* Dec at 400°, leaving a residue of PbO. Insol in water or alcohol. Sol in acetic acid or dil HNO$_3$ with effervescence.

USE: Pigment in oil paints and water colors; in cements; for making putty and lead carbonate paper; in the processing of parchment.

1035. Basswood. Linden tree. Also called *Tilia americana* L., *Tiliaceae.* Grows in mountainous woods from Canada to Georgia and west to Texas. Various decoctions of flowers, trees, bark and wood are used in American folk medicine for disorders of the bile and liver. Preparation of active extracts from the obscure *"Tilia alburnum":* Lafon, **U.S. pat. 3,030,271** (1962 to S. A. Orsymonde).

1036. Batimastat. *[2R-[1(S*),2R*,3S*]]-N^4-Hydroxy-N^1-[2-(methylamino)-2-oxo-1-(phenylmethyl)ethyl]-2-(2-methylpropyl)-3-[(2-thienylthio)methyl]butanediamide;* (2S,-3R)-5-methyl-3-[[(αS)-α-(methylcarbamoyl)phenethyl]carbamoyl]-2-[(2-thienylthio)methyl]hexanohydroxamic acid; [4-(N-hydroxyamino)-2R-isobutyl-3S-(2-thienylthiomethyl)-succinyl]-L-phenylalanine-N-methylamide; BB-94. C$_{23}$H$_{31}$-N$_3$O$_4$S$_2$; mol wt 477.65. C 57.84%, H 6.54%, N 8.80%, O 13.40%, S 13.43%. Synthetic matrix metalloproteinase inhibitor. Prepn: C. Campion *et al.,* **PCT Int. pat. Appl. 90 5,719**; *eidem,* **U.S. pat. 5,240,958** (1990, 1993 both to British Biotech.). Effect on transplanted human ovarian carcinoma: B. Davies *et al., Cancer Res.* **53**, 2087 (1993). Inhibition of metastasis of transplanted human colorectal carcinoma: X. Wang *et al., ibid.* **54**. 4726 (1994).

Fine white powder. mp 236-238°.

THERAP CAT: Antineoplastic adjunct (antimetastatic agent).

1037. Batrachotoxin. *Batrachotoxinin A 20-(2,4-dimethyl-1H-pyrrole-3-carboxylate);* 3α,9α-epoxy-14β,18β-(epoxyethano-N-methylimino)-5β-pregna-7,16-diene-3β,-11α,20α-triol 20α-ester with 2,4-dimethylpyrrole-3-carboxylic acid. C$_{31}$H$_{42}$N$_2$O$_6$; mol wt 538.68. C 69.12%, H 7.86%,

N 5.20%, O 17.82%. One of a group of toxic steroidal alkaloids consisting of batrachotoxin, *pseudobatrachotoxin*, homobatrachotoxin, and batrachotoxinin A. Extracted originally from the skin of South American poison-dart frogs, genus *Phyllobates*. Its name is derived from the Greek word "batrachos" meaning frog. Isoln: F. Märki, B. Witkop, *Experientia* **19**, 329 (1963); and preliminary characterization: J. W. Daly *et al., J. Am. Chem. Soc.* **87**, 124 (1965). Structure: T. Tokuyama *et al., ibid.* **91**, 3931 (1969). Synthetic studies: J. F. W. Keana, R. R. Schumaker, *J. Org. Chem.* **41**, 3840 (1976); P. Magnus *et al., Chem. Commun.* **1985**, 1185; P. Hudson, *Tetrahedron Letters* **34**, 7295 (1993). Mode of action: E. X. Albuquerque, *Fed. Proc.* **31**, 1133 (1972). Kinetics and gating action on Na⁺ channels: J. A. Wasserstrom *et al., Biophys. J.* **65**, 386 (1993). Review: J. Daly, B. Witkop, *Clin. Toxicol.* **4**, 331-342 (1971); of chemistry and pharmacology: E. X. Albuquerque *et al., Science* **72**, 995-1002 (1971); J. W. Daly, *Fortschr. Chem. Org. Naturst.* **41**, 206-227 (1982). Review of sodium channel activation: G. B. Brown, *Int. Rev. Neurobiol.* **29**, 77-116 (1988).

$[\alpha]_{584}^{24}$ −5° to −10°; $[\alpha]_{300}^{24}$ −260° (c = 0.23 in methanol). uv max (0.1N HCl-MeOH): 234, 262 nm (log ε 3.99, 3.70). pKa 7.45. LD₅₀ s.c. in mice: 2 μg/kg (Tokuyama).

Batrachotoxinin A, $C_{24}H_{35}NO_5$. Abs config: R. D. Gilardi, *Acta Crystallogr.* **B 26**, 440 (1970). LD₅₀ s.c. in mice: 1000 μg/kg (Tokuyama).

Homobatrachotoxin, $C_{32}H_{44}N_2O_6$, isobatrachotoxin. Also isolated from the feathers and skin of hooded pitohui bird of New Guinea. First demonstration of chemical defense in birds: J. P. Dumbacher *et al., Science* **258**, 799 (1992). uv max (0.1N HCl-MeOH): 233, 264 nm (log ε 3.95, 3.70). LD₅₀ s.c. in mice: 3 μg/kg (Tokuyama).

USE: Biochemical tool for study of Na⁺ channels.

1038. Batroxobin. *Bothrops atrox serine proteinase; Bothrops venom proteinase;* Botropase; Defibrase. Thrombin-like enzyme from the venom of *Bothrops atrox* (Linn.), a pit viper found in several varieties in Southern and Central America. Prepn of conc extract from *B. jararaca* venom and coagulative properties: D. von Klobusitzky, *Arch. Exp. Pathol. Pharmakol.* **179**, 204 (1935). The enzyme from *B. atrox* venom is hemostatic at low doses and acts as a blood anti-coagulant at higher doses. Composition and manuf: E. E. Percs *et al.,* **Ger. pat. 2,201,993;** *eidem,* **U.S. pat. 3,849,-252** (1972, 1974 both to Pentapharm). Studies on the subunit structure of fibrin produced by batroxobin: P. Mattock, M. P. Esnouf, *Nature New Biol.* **233**, 277 (1971). *In vitro* blood-clotting activity: K. O. Wik *et al., Brit. J. Haematol.* **23**, 37 (1972). Electron microscopic study: S. Ishimaru *et al., Thromb. Haemost.* **45**, 276 (1981). Clinical pharmacology: K. Fukutake *et al., Nippon Ketsueki Gakkai Zasshi* **44**, 1178 (1981), *C.A.* **96**, 62775x (1982). Review of characterization, properties: K. G. Stocker, G. H. Barlow, *Aktuel. Probl. Angiol.* **26**, 45-62 (1975).

Rectangular crystals. Sol in physiological saline. Practically insol in distilled water. Forms complexes with phenol and phenol derivs that are practically insol in water.

Mixture with factor-X activator, *hemocoagulase, Reptilase.*

THERAP CAT: Hemostatic.

1039. Batyl Alcohol. *3-(Octadecyloxy)-1,2-propanediol;* monooctadecyl ether of glycerol. $C_{21}H_{44}O_3$; mol wt 344.58. C 73.20%, H 12.87%, O 13.93%. CH₂OHCHOHCH₂O-

$(CH_2)_{17}CH_3$. Isoln from shark liver oils: Heilbron, Owens, *J. Chem. Soc.* **1928**, 942; Davies *et al., ibid.* **1933**, 165; Nakamiya, *C.A.* **33**, 8175 (1939); from yellow bone marrow: Holmes *et al., J. Am. Chem. Soc.* **63**, 2607 (1941); from coral-reef-building animals: Kind, Bergmann, *J. Org. Chem.* **7**, 424 (1942). Synthesis from allyl octadecyl ether: Davies *et al., J. Chem. Soc.* **1930**, 2542; Kornblum, Holmes, *J. Am. Chem. Soc.* **64**, 3045 (1942). Synthesis of optically active batyl alc: Baer, Fischer, *J. Biol. Chem.* **140**, 397 (1941).

Glistening plates from dilute acetone, mp 70.5-71°. $[\alpha]_D^{20}$ +1.14° (c = 6.6 in CHCl₃). Sol in the usual fat solvents.

Bis(*p*-nitrobenzoate), $C_{35}H_{50}N_2O_9$, pale yellow needles from methanol, soft at 63°, mp 65-66°.

THERAP CAT (VET): Has been recommended for bracken fern poisoning in cattle.

1040. Bayberry Bark. Candleberry bark; myrica; myrtle wax; berry wax; tallow shrub. Dried root bark of *Myrica cerifera* L., *Myricaceae. Habit.* Maryland to Florida, west to Texas and Arkansas. *Constit.* Acrid and astringent resin, myricic acid, tannin, red coloring matter, gum, starch. Bayberry wax consists of palmitic, myristic and lauric acid esters.

THERAP CAT: Astringent; emetic.

1041. BCG. An abbreviation for *Bacillus Calmette-Guérin,* a strain of *Mycobacterium tuberculosis.* Prepn of BCG labeled with C¹⁴: Pasquier, Kurylowicz, *Rev. Immunol.* **20**, 245 (1956). Has the potential to act as a nonspecific immunopotentiating agent that stimulates the whole range of immune responses. Role as an immunotherapeutic agent: Bartlett and Zbar *et al., J. Nat. Cancer Inst.* **48**, 245, 1441, 1709 (1972); **49**, 119 (1972). Comparison of potentiation of specific tumor immunity in mice: M. T. Scott, R. Bomford, *ibid.* **57**, 555 (1976). Review of use in Hodgkin's disease: H. Zywicka-Lopaciuk *et al., Arch. Immunol. Ther. Exp.* **29**, 739-755 (1981); in operable lung cancer: B. H. Stack *et al., Thorax* **37**, 588-593 (1982). Review of use as anti-tumor agent and of the interaction of BCG with cells of the mammalian immune system: M. Davies, *Biochim. Biophys. Acta* **651**, 143-174 (1982). Review of pathogenesis of tuberculosis and the effectiveness of BCG vaccination: H. G. ten Dam, A. Pio, *Tubercle* **63**, 225-233 (1982).

THERAP CAT: BCG vaccine as active immunizing agent.

THERAP CAT (VET): Vaccination vs. tuberculosis (cattle). Use prohibited where bovine tuberculosis eradication programs are in progress.

1042. Bebeerine. *1′α 6,6′-Dimethoxy-2,2′-dimethyltubocuraran-7′,12′-diol; d*-bebeerine; chondodendrine; pelosine. $C_{36}H_{38}N_2O_6$; mol wt 594.71. C 72.71%, H 6.44%, N 4.71%, O 16.14%. From bark of *Nectandra rodioei* Hook., *Lauraceae:* Maclagan, *Ann.* **48**, 106 (1843). From root of *Chondodendron microphyllum* (Eichl.) Moldenke and stem of *Ch. candicans* (Rick ex. D.C.) Sandwith, *Menispermaceae* and *radix pareirae bravae:* King, *J. Chem. Soc.* **1940**, 737. Structure: Faltis *et al., Ber.* **69**, 1269 (1936); King, *J. Chem. Soc.* **1939**, 1157. Stereochemistry: King, *ibid.* **1948**, 265.

Needles from methanol, mp 215°. $[\alpha]_D^{20}$ +345.7° (c = 0.4 in 1N HCl). Sol in benzene, chloroform, pyridine.

Hydrochloride, $C_{36}H_{38}N_2O_6$·HCl. mp 260°. $[\alpha]_D^{20}$ +294° (c = 0.7). Sol in water and alcohol.

THERAP CAT: Antimalarial.

1043. Bebeeru Bark. Greenheart; bibiru bark; sipiri bark. From *Nectandra rodioei* Hook., *Lauraceae. Habit.* British Guiana. *Constit.* Bebeerine, sipirine, bebeeric acid, tannin, resin.

1044. Becanthone. *1-[[2-[Ethyl(2-hydroxy-2-methyl-propyl)amino]ethyl]amino]-4-methyl-9H-thioxanthen-9-one;* becantone. $C_{22}H_{28}N_2O_2S$; mol wt 384.54. C 68.72%, H 7.34%, N 7.28%, O 8.32%, S 8.34%. Prepn: Blanz, French, *J. Med. Chem.* **6**, 185 (1963).

Hydrochloride, $C_{22}H_{28}N_2O_2S \cdot HCl$, *Win-13820, Loranil.* mp 157.6-160.4°.

THERAP CAT: Anthelmintic (Schistosoma).

1045. Beclamide. *3-Chloro-N-(phenylmethyl)propanamide; N-benzyl-3-chloropropionamide; N-benzyl-β-chloropropanamide; benzchlorpropamide; chloroethylphenamide; Chloracon; Hibicon; Neuracen; Nidrane; Posedrine; Seclar.* $C_{10}H_{12}ClNO$; mol wt 197.66. C 60.76%, H 6.12%, Cl 17.94%, N 7.09%, O 8.09%. $ClCH_2CH_2CONHCH_2C_6H_5$. Prepd by the action of β-chloropropionyl chloride on benzylamine in cooled water at pH 8: Cassell, Kushner, U.S. pat. **2,569,288** (1951 to Am. Cyanamid); Kushner *et al., J. Org. Chem.* **16**, 1283 (1951).
Large crystals from methanol, mp 94°. Slightly sol in water (0.005 to 0.01%). Moderately sol in the lower alcs. Dec in hot aq acid soln and in hot aq alkaline soln.

THERAP CAT: Anticonvulsant.

1046. Beclobrate. *(±)-2-[4-[(4-Chlorophenyl)methyl]-phenoxy]-2-methylbutanoic acid ethyl ester; (±)-2-methyl-2-[p-(p'-chlorobenzyl)phenoxy]butyric acid ethyl ester; ethyl (±)-2-[[α-(p-chlorophenyl)-p-tolyl]oxy]-2-methylbutyrate;* Sgd-24774; Beclipur; Turec. $C_{20}H_{23}ClO_3$; mol wt 346.85. C 69.26%, H 6.68%, Cl 10.22%, O 13.84%. Diphenylmethane derivative. Prepn: K. Thiele *et al.,* **Ger.** pat **2,461,069;** *eidem,* **U.S.** pat. **4,483,999** (1975, 1984 both to Siegfried); *eidem, Arzneimittel-Forsch.* **29**, 711 (1979). Preliminary clinical trial in hyperlipidemia: C. Najemnik *et al., ibid.* **31**, 2168 (1981). Series of articles on pharmacology: *ibid.* **33**, 1464-1472 (1983). Metabolism: W. Roth *et al., ibid.* **35**, 244 (1985). Pharmacokinetics: I. Gikalov, U. Ifflaender, *ibid.* **37**, 1065 (1987).

Oil, $bp_{0.01-0.1}$ 200-204°. LD_{50} orally in mice: 8000 mg/kg (Thiele, 1979).

THERAP CAT: Antihyperlipoproteinemic.

1047. Beclomethasone. *(11β,16β)-9-Chloro-11,17,21-trihydroxy-16-methylpregna-1,4-diene-3,20-dione;* 9α-chloro-16β-methyl-1,4-pregnadiene-11β,17α,21-triol-3,20-dione; 9α-chloro-16β-methylprednisolone. $C_{22}H_{29}ClO_5$; mol wt 408.92. C 64.62%, H 7.15%, Cl 8.67%, O 19.56%. Glucocorticoid. Prepn of free alcohol and 21-acetate: **Brit.** pat. **912,378** (1962 to Merck & Co.); of 21-acetate: **Brit.** pat. **901,093** (1962 to Schevico); of mono and diesters: J. Elks *et al.,* **Belg.** pat. **649,170;** *eidem,* **U.S.** pat. **3,312,590** (1964, 1967 both to Glaxo). Symposium on clinical studies: *Brit. J. Clin. Pharmacol.* **4**, Suppl. 3, 249S-312S (1977). Use in chronic asthma in children: M. Rao *et al., J. Asthma* **19**, 21 (1982); in treatment of asthma in steroid-independent adults: V. A. Malfitan, *Clin. Ther.* **4**, 472 (1982). Review of use in rhinitis: P. Small *et al., Ann. Allergy* **49**, 127 (1982); of pharmacology, side effects and use in asthma and allergic rhinitis: R. N. Brogden *et al., Drugs* **28**, 99-126 (1984).

Dipropionate, $C_{28}H_{37}ClO_7$, *Sch-18020W, Aerobec, Aldecin, Anceron, Andion, Beclacin, Becloforte, Beclomet, Beclorhinol, Becloval, Beclovent, Becodisks, Beconase, Beconasol, Becotide, Clenil-A, Entyderma, Inalone O, Inalone R, Korbutone, Propaderm, Rino-Clenil, Sanasthmax, Sanasthmyl, Vancenase, Vanceril, Viarex, Viarox.* Crystals from acetone + ether, mp 117-120° (dec). $[\alpha]_D$ +98.0° (c = 1.0 in dioxane). uv max (ethanol): 238 nm (ε 15990).

THERAP CAT: Antiallergic, antiasthmatic (inhalant). Anti-inflammatory (topical).

1048. Beclotiamine. *3-[(4-Amino-2-methyl-5-pyrimidinyl)methyl]-5-(2-chloroethyl)-4-methylthiazolium chloride;* 5-chloroethylthiamine; chlorothiamine; clotiamine; CT; Cocciden. $C_{12}H_{16}Cl_2N_4S$; mol wt 319.26. C 45.15%, H 5.05%, Cl 22.21%, N 17.55%, S 10.04%. Antithiamine compound derived from thiamine; developed as an anticoccidial for chickens: Inone *et al., J. Japan. Vet. Med. Assoc.* **20**, 293 (1967); **Brit.** pat. **1,141,055** and Matsui *et al.,* **Japan.** pat. **14,276('71)** (1969 and 1971 to Sankyo), *C.A.* **71**, 111875z (1969); **75**, 36098a (1971); Matsuzawa, *Japan. J. Vet. Sci.* **34**, 157 (1972), *C.A.* **78**, 23930c (1973). Metabolism and mechanism of action studies: Shindo, Komai, *J. Vitaminol.* **18**, 41-62, 102, 172, 218 (1972). Safety studies: Matsuzawa *et al., Sankyo Kenkyusho Nempo* **23**, 192 (1971), *C.A.* **77**, 28983m (1972).

Hydrochloride, $C_{12}H_{17}Cl_3N_4S$. LD_{50} i.v. in mice, rats: 33.6, 60.5 mg/kg.
Naphthalene-1,5-disulfonate salt, $C_{22}H_{24}Cl_2N_4O_6S_3$, mp 276° (dec). Soly in water: 0.06 g/100 ml. LD_{50} i.p. in mice, rats: 294, 585 mg/kg.

THERAP CAT (VET): Coccidiostat.

1049. Beeswax. Yellow beeswax. A substance obtained from bee honeycombs. Consists of esters of straight-chain monohydric alcohols with even-numbered carbon chains from C_{24} to C_{36} esterified with straight-chain acids also having even numbers of C atoms up to C_{36} (some C_{18} hydroxy acids). Examples of such esters are triacontanol hexadecanoate and hexacosanol hexacosanoate. These esters are mixed with about 20% (w/w) of hydrocarbons having odd-numbered straight carbon chains from C_{21} to C_{33}. Propolis, pigments and unidentified substances amount to about 6%. Composition: D. T. Downing *et al., Aust. J. Chem.* **14**, 253 (1961); Callow, *Bee World* **44**, 95 (1963). Brief review: C. S. Letcher in Kirk-Othmer *Encyclopedia of Chemical Technology* Vol. 24 (Wiley-Interscience, New York, 3rd ed., 1984) pp 466-467.
Yellowish to brownish-yellow, soft to brittle; honey-like odor; slight balsamic taste. d 0.95-0.960. mp 62-65°. Saponification number 84. Acid number 20. Practically insol in water. Slightly sol in cold alc; sol in hot alc, chloroform, benzene, ether, carbon disulfide.
White beeswax, white wax, bleached yellow wax, bleached beeswax. Prepd by oxidizing yellow beeswax cakes with peroxide or in sunlight. Yellowish-white. Properties similar to those of yellow beeswax, except for a slightly different taste. Preferred to yellow beeswax in cosmetics.

USE: Manuf of wax paper, candles, cosmetics; modeling artificial fruits and flowers; in process engraving; shoe polish. Pharmaceutic aid (in ointments, plasters).

1050. Befunolol. *1-[7-[2-Hydroxy-3-[(1-methylethyl)-amino]propoxy]-2-benzofuranyl]ethanone;* 7-[2-hydroxy-3-(isopropylamino)propoxy]-2-benzofuranyl methyl ketone; 2-acetyl-7-(2-hydroxy-3-isopropylaminopropoxy)benzofuran. $C_{16}H_{21}NO_4$; mol wt 291.35. C 65.96%, H 7.27%, N 4.81%, O 21.97%. β-Adrenergic blocker. Prepn: K. Ito *et al.*, **Ger. pat.** 2,223,184; *eidem*, U.S. pat. 3,853,923 (1972, 1974 both to Kakenyaku Kako). Prepn and activity of optical isomers: J. Nakano *et al.*, *Chem. Pharm. Bull.* **36**, 1399 (1988). Determn of befunolol and metabolites in human plasma: K. Kawahara, T. Ofuji, *J. Chromatog.* **168**, 266 (1979). Absorption, distribution, excretion in rats: H. Kitagawa *et al.*, *Oyo Yakuri* **17**, 383, 393 (1979), *C.A.* **91**, 13374w-5x (1979). General pharmacology: S. Masumoto *et al.*, *Iyakuhin Kenkyu* **10**, 741 (1979), *C.A.* **92**, 88015s (1980). Pharmacodynamics: S. Harada *et al.*, *Arch. Int. Pharmacodyn. Ther.* **252**, 262 (1981).

Crystals from cyclohexane/acetone, mp 115°. LD_{50} in mice: 100-105 mg/kg i.v. (Ito).

Hydrochloride, $C_{16}H_{21}NO_4 \cdot HCl$, *BFE-60, Benfuran, Bentos, Bentox, Glauconex.* Crystals from ethyl acetate, mp 163°.

(S)-(−)-Form hydrochloride, pale yellow prisms from isopropyl alcohol, mp 151-152°. $[\alpha]_D$ −15.5° (c = 1 in methanol).

(R)-(+)-Form hydrochloride, pale yellow prisms from isopropyl alcohol, mp 151°. $[\alpha]_D$ +15.3° (c = 1 in methanol).

THERAP CAT: Antiglaucoma.

1051. Behenic Acid. *Docosanoic acid.* $C_{22}H_{44}O_2$; mol wt 340.59. C 77.58%, H 13.02%, O 9.40%. $CH_3(CH_2)_{20}-CO_2H$. Minor constituent of most seed fats, animal milk fats and marine animal oils. Large amts (∼50%) are found in (hydrogenated?) jamba oil, mustard seed oil and rape oil: Sudborough *et al.*, *J. Indian Inst. Sci.* **9A**, 25 (1926). Prepn from erucic acid by catalytic reduction: Morgan, Holmes, *J. Soc. Chem. Ind. (London) Trans. Commun.* **44**, 491 (1925). Brief review: E. S. Lower, *Mfg. Chem.* **56**(7), 44-46 (1985).

Waxy solid, mp 79.95°. Neutralization value 164.73. bp_{60} 306°. d_4^{100} 0.8221; n_D^{100} 1.4270. One hundred grams of 90% ethanol dissolve 0.102 g of behenic acid at 17°; at 25° 0.218 g of the acid dissolves in 100 ml of 91.5% ethanol, 0.116 g in 100 ml of 86.2% ethanol, 0.011 g in 100 ml of 63.07% ethanol. One hundred grams of ether dissolves 0.1922 g of behenic acid at 16°.

Methyl ester, $C_{23}H_{46}O_2$, mp 54°.

Ethyl ester, $C_{24}H_{48}O_2$, mp 50°. $bp_{0.2}$ 185°.

Amide, $C_{22}H_{45}NO$, mp 111-112°.

USE: In lubricating oils, as solvent evaporation retarder in paint removers. Amide as anti-foam in the manuf of detergents, in floor polishes, in dripless candles.

1052. Belladonna. Deadly nightshade; banewort; death's herb; dwale; poison black cherry. Dried leaves and root of *Atropa belladonna* L., *Solanaceae*. *Habit.* Southern and Central Europe, Asia Minor, Algeria; cultivated in North America. On extraction it yields atropine, hyoscyamine, scopolamine, asparagine, choline, chrysatropic acid, atroscine, leucatropic acid, phytosterol. Leaves usually contain 0.3-0.5% total alkaloids; dried root contains 0.4-0.7% total alkaloids. *See:* Romeike, *Pharmazie* **8**, 729 (1953); Phokas, Steinegger, *ibid.* **11**, 652 (1956); *Pharm. Acta Helv.* **31**, 284 (1956).

Caution: Toxicity based on content of atropine, *q.v.*, and related alkaloids.

THERAP CAT: Anticholinergic.

THERAP CAT (VET): Anticholinergic; antispasmodic.

1053. Belladonnine. *1,2,3,4-Tetrahydro-1-phenyl-1,4-naphthalenedicarboxylic acid bis(8-methyl-8-azabicyclo-[3.2.1]oct-3-yl) ester;* isatropylditropeine; tropyl isatropate; ditropyl isatropate. $C_{34}H_{42}N_2O_4$; mol wt 542.72. C 75.25%, H 7.80%, N 5.16%, O 11.79%. From *Atropa belladonna* L. and allied *Solanaceae*. Probably formed during the process of extraction. Upon hydrolysis the substance extracted from plants yields β-isatropic acid and tropine: Küssner, *Arch. Pharm.* **276**, 617 (1938); Hotovy *et al.*, U.S. pat. 2,734,062 (1956 to E. Merck).

Crystals from ethyl acetate, mp 129°. Sparingly sol in water, petr ether. Sol in alcohol, benzene, chloroform, ethyl acetate.

Bisulfate monohydrate, $C_{34}H_{42}N_2O_4 \cdot H_2SO_4 \cdot H_2O$, *Bellacristin.* Crystals, sol in water.

Bis[ethobromide] tetrahydrate, $C_{38}H_{52}Br_2N_2O_4 \cdot 4H_2O$, *belladonnine bis[bromoethylate] tetrahydrate, C-100.* A quaternary tetrahydrate salt, crystals from water + acetone, mp 98-101°. uv max: 258, 261 nm. $E_{1cm}^{1\%}$ 4.8. Sol in water, methanol, ethanol. Practically insol in ether, petr ether, benzene, ethyl acetate.

1054. Bemegride. *4-Ethyl-4-methyl-2,6-piperidinedione; 3-ethyl-3-methylglutarimide;* 4-ethyl-4-methyl-2,6-dioxopiperidine; 2,6-dioxo-4-methyl-4-ethylpiperidine; methetharimide; β,β-methylethylglutarimide; NP-13; Mikedimide; Eukraton; Malysol; Megimide. $C_8H_{13}NO_2$; mol wt 155.20. C 61.91%, H 8.44%, N 9.03%, O 20.62%. Prepn: Thole, Thorpe, *J. Chem. Soc.* **99**, 439 (1911); Sircar, *ibid.* **1937**, 602, 604; Benica, Wilson, *J. Am. Pharm. Assoc. [Sci. Ed.]* **39**, 451 (1950); Lukes, Ferles, *Chem. Listy* **49**, 510 (1955), *C.A.* **49**, 10290 (1955). Pharmacology: Delay *et al.*, *Presse Med.* **64**, 1525 (1956); Oberdorf, Meyer, *Arch. Exp. Pathol. Pharmakol.* **238**, 128 (1960); Kretzschmar *et al.*, *Arch. Int. Pharmacodyn. Ther.* **174**, 318 (1968). Metabolism: Nicholls, *Nature* **185**, 927 (1960). Reduction of barbiturate anesthesia in mice: V. K. Patel *et al.*, *Psychopharmacol.* **71**, 21 (1980). Postsynaptic effects: P. W. Gage, P. Sah, *Brit. J. Pharmacol.* **75**, 493 (1982).

Platelets from water or from acetone + ether, mp 127°. Sublimes at 100° and 2 mm press. Sol in water, acetone. LD_{50} in mice, rats: 18.8, 17.0 mg/kg i.v.

THERAP CAT: CNS stimulant. Antidote (barbiturate poisoning).

THERAP CAT (VET): CNS stimulant. Antidote (barbiturate poisoning).

1055. Benactyzine. *α-Hydroxy-α-phenylbenzeneacetic acid 2-(diethylamino)ethyl ester;* benzilic acid β-diethylaminoethyl ester; β-diethylaminoethyl benzilate; 2-diethylaminoethyl diphenylglycolate. $C_{20}H_{25}NO_3$; mol wt 327.42. C

73.37%, H 7.70%, N 4.28%, O 14.66%. An antagonist of acetylcholine in the central and peripheral nervous systems. Prepn: Horenstein, Pahlicke, *Ber.* **71**, 1654 (1938); Blicke, Maxwell, *J. Am. Chem. Soc.* **64**, 428 (1942); Hill, Holmes, U.S. pat. **2,394,770** (1946 to Am. Cyanamid). Toxicity and pharmacodynamics: Fournier, Petit, *Therapie* **17**, 1245 (1962). Metabolism: Eldeson *et al., Arch. Int. Pharmacodyn. Ther.* **187**, 139 (1970). Crystal and molecular structure determn by x-ray diffraction: T. J. Petcher, *J. Chem. Soc., Perkin Trans. II* **1974**, 1151.

Crystals, mp 51°.

Hydrochloride, $C_{20}H_{25}NO_3$·HCl, *AY-5406-1, Actozine, Amizil, Arcadine, Cafron, Cedad, Cevanol, Fobex, Ibiotyzil, Lucidil, Nervacton, Neuroleptone, Nutinal, Parasan, Parpon, Suavitil, Tranquillin.* Crystals from acetone, mp 177-178°. Soly in water (25°): 14.9/100 ml. Practically insol in ether.

Methobromide, $C_{20}H_{25}NO_3$·CH_3Br, *Finalin, Spatomac.* Crystals from alcohol + ether, mp 169-170°.

THERAP CAT: Antidepressant; anticholinergic.

1056. Benalaxyl. *N-(2,6-Dimethylphenyl)-N-(phenylacetyl)-DL-alanine methyl ester;* methyl *N*-phenylacetyl-*N*-2,6-xylyl-DL-alaninate; *N-(2,6-dimethylphenyl)-N-(1-carbomethoxyethyl)phenylacetamide;* M-9834; Galben. $C_{20}H_{23}NO_3$; mol wt 325.41. C 73.82%, H 7.12%, N 4.30%, O 14.75%. Acylalanine fungicide systemically active against phytopathogens of the order *Peronosporales.* Prepn: E. Boson *et al.,* **Ger.** pat. **2,903,612**; *eidem,* U.S. pat. **4,425,357** (1978, 1984 both to Montedison). Activity and toxicology: P. Bergamaschi *et al., Proc. Brit. Crop Prot. Conf.-Pests Dis.* **1981**, 19. Residue accumulation in edible crops: P. Cabras *et al., J. Agr. Food Chem.* **33**, 86 (1985). GC-MS study: B. D. Ripley, *ibid.* 560.

Crystals from ligroin, mp 78-80°. LD_{50} in rats (mg/kg): 4200 orally; 1100 i.p. LC_{50} (96 hr) in rainbow trout: 3.75 mg/l (Bergamaschi).

USE: Agricultural fungicide.

1057. Benapryzine. *α-Hydroxy-α-phenylbenzeneacetic acid 2-(ethylpropylamino)ethyl ester;* benzilic acid 2-(ethylpropylamino)ethyl ester; 2-(ethyl-*n*-propylamino)ethyl α,α-diphenylglycolate. $C_{21}H_{27}NO_3$; mol wt 341.45. C 73.87%, H 7.97%, N 4.10%, O 14.06%. Prepn: **Neth.** pat. **Appl.** **6,409,696** corresp to Mehta, Bainbridge, U.S. pat. **3,746,743** (1963, 1973 to Beecham). Pharmacology: Brown *et al., Nature* **223**, 416 (1969); Barthelemy *et al., Therapie* **27**, 369 (1972); Brown *et al., Brit. J. Pharmacol.* **47**, 476 (1973). Metabolism: Jeffery *et al., Xenobiotica* **1**, 169 (1971).

Hydrochloride, $C_{21}H_{27}NO_3$·HCl, *AP-1288, BRL-1288.* Rectangular plates from butanone, mp 164-166°. LD_{50} in mice (mg/kg): 400 s.c.; 500 orally.

THERAP CAT: Anticholinergic.

1058. Benazepril. *[S-(R*,R*)]-3-[[1-(Ethoxycarbonyl)-3-phenylpropyl]amino]-2,3,4,5-tetrahydro-2-oxo-1H-1-benzazepine-1-acetic acid;* (3*S*)-1-(carboxymethyl)-[[(1*S*)-1-(ethoxycarbonyl)-3-phenylpropyl]amino]-2,3,4,5-tetrahydro-1*H*-[1]benzazepin-2-one; (3*S*)-3-[[(1*S*)-1-carboxy-3-phenylpropyl]amino]-2,3,4,5-tetrahydro-2-oxo-1*H*-1-benzazepine-1-acetic acid 3-ethyl ester. $C_{24}H_{28}N_2O_5$; mol wt 424.50. C 67.91%, H 6.65%, N 6.60%, O 18.85%. Angiotensin converting enzyme (ACE) inhibitor. Hydrolyzed *in vivo* to the active diacid metabolite. Prepn: J. W. H. Watthey, **Eur.** pat. **Appl.** **72,352**; *idem,* U.S. pat. **4,410,520** (both 1983 to Ciba-Geigy); J. W. H. Watthey *et al., J. Med. Chem.* **28**, 1511 (1985). Clinical pharmacology: M. D. Schaller *et al., Eur. J. Clin. Pharmacol.* **28**, 267 (1985). GC-MS determn in plasma and urine: A. Sioufi *et al., J. Chromatog.* **434**, 239 (1988). Pharmacokinetics of benazepril and benazeprilat: G. Kaiser *et al., Biopharm. Drug Dispos.* **10**, 365 (1989). Clinical evaluation in congestive heart failure: J. Insel *et al., Clin. Pharmacol. Ther.* **45**, 312 (1989). Multicenter clinical trial in hypertension: M. Moser *et al., ibid.* **49**, 322 (1991).

mp 148-149°. $[\alpha]_D$ −159° (c = 1.2 in ethanol).

Hydrochloride, $C_{24}H_{28}N_2O_5$·HCl, *CGS-14824A, Briem, Cibacen, Cibacène, Lotensin.* Crystals from 3-pentanone + methanol (10:1), mp 188-190°. $[\alpha]_D$ −141.0° (c = 0.9 in ethanol).

Diacid, $C_{22}H_{24}N_2O_5$, *benazeprilat, CGS-14831.* mp 270-272°. $[\alpha]_D$ −200.5° (c = 1 in 3% aqueous NH_4OH).

THERAP CAT: Antihypertensive.

1059. Bence-Jones Proteins. Low molecular weight, heat-sensitive proteins isolated from urine of multiple myeloma patients: Caputo, *Aminoacidosi* **7**, 163-77 (1957), *C.A.* **54**, 6841 (1960). Used as a model in structural studies of immunoglobulins, they are structurally homogeneous within each patient and antigenically related. Crystallographic structural study: E. E. Aeola *et al., Biochemistry* **19**, 432 (1980). They constitute the light-chain component of myeloma globulin and have been implicated in the pathogenesis of amyloidosis, cf. G. G. Glenner, *N. Engl. J. Med.* **302**, 1283 (1980); G. G. Glenner *et al., Amyloid and Amyloidosis* (Excerpta Medica, Amsterdam, 1980) p 351. For structural studies and reviews *see* Immunoglobulins.

Coagulates on heating at 45-55°. Redissolves partially or wholly on boiling. Amino acid composition in g-%: Arginine 5.6, histidine 3.3, lysine 9.5, tyrosine 6.4, tryptophan 3.4, phenylalanine 4.2, cystine 2.6, methionine 1.9, serine 8.4, threonine 8.1, leucine 7.3, isolecine 2.2, valine 9.8, glycine 8.6, alanine 3.1, proline 6.9, glutamic acid 7.5, aspartic acid 9.2. The molecule also shows a carbohydrate content of 9.4% (mostly glucose, mannose, and glucosamine).

1060. Bencyclane. *N,N-Dimethyl-3-[[1-(phenylmethyl)-cycloheptyl]oxy]-1-propanamine;* 3-[(1-benzylcycloheptyl)oxy]-*N,N*-dimethylpropylamine; 1-benzyl-1-(3-dimethylaminopropoxy)cycloheptane; *N*-[3-(1-benzylcycloheptyloxy)propyl]-*N,N*-dimethylamine; benzcyclan. $C_{19}H_{31}NO$; mol wt 289.46. C 78.84%, H 10.79%, N 4.84%, O 5.53%. Prepn: Pallos *et al.,* **Hung.** pat. **151,865** (1965 to EGYT), *C.A.* **62**, 16125b (1965). Series of articles on pharmacology: *Arzneimittel-Forsch.* **20**, 1337-1460 (1970). Toxicity: E. Komlos, L. E. Petöcz, *ibid.* 1338.

bp$_3$ 146-156°.

Fumarate, $C_{19}H_{31}NO.C_4H_4O_4$, *EGYT-201, Angiociclan, Dantrium, Dilangio, Fludilat, Fluxema, Halidor, Vasorelax.* Crystals from 25% ethanol, mp 131-133°. Soly in water at 25°: 1 g/100 ml; in hot water: 2 g/100 ml. Poorly sol in acetone, easily in alc. uv spectrum: Simonyi *et al., Acta Pharm. Hung.* **36**, 257 (1966). At pH 3.4-6.6, uv max (ethanol): 207 nm. LD$_{50}$ in mice, rats (mg/kg): 445.6, 414 orally; 49.9, 41.3 i.v.; 132, 86.3 i.p.; 203, 257 s.c. (Komlos, Petöcz).

THERAP CAT: Vasodilator (peripheral, cerebral).

1061. Bendazac. *[[1-(Phenylmethyl)-1H-indazol-3-yl]-oxy]acetic acid; [(1-benzyl-1H-indazol-3-yl)oxy]acetic acid;* bendazolic acid; bindazac; AF-983; Dogalina; Versus; Zildasac. $C_{16}H_{14}N_2O_3$; mol wt 282.30. C 68.08%, H 5.00%, N 9.92%, O 17.00%. Prepn from 1-benzylindazol-3-ol sodium and chloroacetonitrile and subsequent hydrolysis: Palazzo, Silvestrini, **S. Afr. pat. 67 03,206;** Palazzo, **U.S. pat. 3,470,-194** (1967, 1969, both to Francesco Angelini). Pharmacology: Silvestrini *et al., Inflammation Biochem. Drug Interaction, Proc. Int. Symp.* **1968**, A. Bertelli, Ed. (Excerpta Med., Amsterdam, 1970) p 283. Mechanism of action: Silvestrini *et al., Arzneimittel-Forsch.* **20**, 250 (1970). Photoprotective capacity: Fuga *et al., Ann. Ital. Dermatol. Clin. Sper.* **24**, 205 (1970). Use in treatment of cataracts: M. Testa *et al., Lancet* **1**, 849 (1982). Toxicity: *Rx Bulletin* **3**, 147 (1972).

Crystals from ethanol, mp 160°. Practically insol in water. Sol in chloroform, acetone. uv max: 306 nm (E$_{1cm}^{1\%}$ 191). LD$_{50}$ in mice, rats (mg/kg): 380, 304 i.v.; 355, 388 i.p.; 440, 910 s.c.; 1105, ~1200 orally (*Rx Bulletin*).

THERAP CAT: Anti-inflammatory.

THERAP CAT (VET): In treatment of cataracts.

1062. Bendazol. *2-(Phenylmethyl)-1H-benzimidazole; 2-benzylbenzimidazole;* 2-benzylbenziminazole; bendazole. $C_{14}H_{12}N_2$; mol wt 208.26. C 80.74%, H 5.81%, N 13.45%. Prepn: Walther, v. Pulawski, *J. Prakt. Chem.* [2] **59**, 253 (1899); Feitelson, Rothstein, *J. Chem. Soc.* **1958**, 2426; Partridge, Turner, *ibid.* **1958**, 2086.

Needles from benzene, mp 187°. Practically insol in water. Freely sol in glacial acetic acid. Sol in alcohol, hot benzene, propylene glycol.

Hydrochloride, $C_{14}H_{12}N_2.HCl$, *Dibasol, Dibasole, Tromasedan.* Needles from dil HCl, mp 175° (Feitelson, Rothstein).

THERAP CAT: Vasodilator (coronary).

1063. Bendiocarb. *2,2-Dimethyl-1,3-benzodioxol-4-ol methylcarbamate; methylcarbamic acid 2,3-(isopropylidenedioxy)phenyl ester;* NC-6897; Ficam. $C_{11}H_{13}NO_4$; mol wt 223.23. C 59.19%, H 5.87%, N 6.27%, O 28.67%. Prepn: Gates, Gillon, **S. Afr. pat. 68 00736,** *C.A.* **71**, 38941m

(1969), corresp to **U.S. pat. 3,736,338** (1968, 1973 both to Fisons). Insecticidal activity: Story, *Int. Pest Contr.* **14**, 6 (1972).

White solid, mp 129-130°. Soly in water: 40 ppm; in hexane: 350 ppm.

USE: Contact insecticide.

1064. Bendroflumethiazide. *3,4-Dihydro-3-(phenyl-methyl)-6-(trifluoromethyl)-2H-1,2,4-benzothiadiazine-7-sulfonamide 1,1-dioxide;* 3-benzyl-6-trifluoromethyl-3,4-dihydro-7-sulfamoyl-2H-1,2,4-benzothiadiazine 1,1-dioxide; 3-benzyl-3,4-dihydro-7-sulfamyl-6-trifluoromethyl-1,2,4-benzothiadiazine 1,1-dioxide; bendrofluazide; benzydroflumethiazide; benzylhydroflumethiazide; Aprinox; Benzyl-Rodiuran; Berkozide; Bristuric; Bristuron; Centyl; Flumesil; Naturetin; Naturine; Neo-Naclex; Niagaril; Nikion; Orsile; Pluryle; Plusuril; Poliuron; Relan Beta; Salures; Sinesalin; Sodiuretic; Urlea. $C_{15}H_{14}F_3N_3O_4S_2$; mol wt 421.42. C 42.75%, H 3.35%, F 13.52%, N 9.97%, O 15.19%, S 15.22%. Prepn: C. T. Holdrege *et al., J. Am. Chem. Soc.* **81**, 4807 (1959); F. Lund, W. O. Godtfredsen, **Brit. pat. 863,474;** *eidem,* **U.S. pat. 3,392,168** (1961, 1968 both to Lövens Kemiske Fabrik). Comprehensive description: K. Florey, F. M. Russo-Alesi, *Anal. Profiles Drug Subs.* **5**, 1-19 (1976).

Crystals from dioxane, mp 224.5-225.5° (U.S. patent). Also reported as mp 221-223° (Holdrege). uv max (methanol): 208, 273, 326 nm (E$_{1cm}^{1\%}$ 745, 565, 96). Insol in water, chloroform, benzene, ether. Sol in acetone, alcohol.

THERAP CAT: Diuretic, antihypertensive.

1065. Benexate Hydrochloride. *trans-2-[[[4-[[(Amino-iminomethyl)amino]methyl]cyclohexyl]carbonyl]oxy]benzoic acid phenylmethyl ester monohydrochloride;* benzyl salicylate *trans-4-(guanidinomethyl)cyclohexanecarboxylate* hydrochloride; (2'-benzyloxycarbonyl)phenyl *trans-4-(guanidinomethyl)cyclohexanecarboxylate* hydrochloride. $C_{23}H_{28}ClN_3O_4$; mol wt 445.95. C 61.95%, H 6.33%, Cl 7.95%, N 9.42%, O 14.35%. Synthetic protease inhibitor with antiulcer activity. Prepn: **Belg. pat. 885,263;** M. Muramatsu *et al.,* **U.S. pat. 4,348,410** (1981, 1982 to Nippon Chemiphar; Teikoku Chem. Ind.). Prepn and protease inhibition: T. Satoh *et al., Chem. Pharm. Bull.* **33**, 647 (1985). Prepn of the clathrate compound with β-cyclodextrin: **Japan. Kokai 83 38,250;** M. Shinoda *et al.,* **U.S. pat. 4,478,995** (1983, 1984 both to Teikoku Chem. Ind.). Effect on gastric secretion and exptl ulcers in rats: S. Okabe *et al., Oyo Yakuri* **27**, 829 (1984), *C.A.* **101**, 143881c (1984); I. Tanaka, H. Tagami, *Nippon Yakurigaku Zasshi* **85**, 167 (1985), *C.A.* **103**, 64660t (1985); F. Hirose *et al., Yakuri to Chiryo* **15**, 4749 (1987), *C.A.* **108**, 137769a (1988).

Crystals from methanol + ether, mp 83°.

Compd with β-cyclodextrin (1:1), $C_{65}H_{98}ClN_3O_{39}$, *benexate-CD*, *TA-903*, *Lonmiel*, *Ulgut*.
THERAP CAT: Antiulcerative.

1066. Benfluorex. *2-[[1-Methyl-2-[3-(trifluoromethyl)-phenyl]ethyl]amino]ethanol benzoate (ester); 2-[[α-methyl-m-(trifluoromethyl)phenethyl]amino]ethanol benzoate (ester);* 1-(*m*-trifluoromethylphenyl)-2-(β-benzoyloxyethyl)aminopropane; *N*-(2-benzoyloxyethyl)norfenfluramine; benfluramate; *S-780; SE-780; Minolip.* $C_{19}H_{20}F_3NO_2$; mol wt 351.37. C 64.95%, H 5.74%, F 16.22%, N 3.99%, O 9.11%. Prepn from 1-(*m*-trifluoromethyl)-2-(β-hydroxyethyl)aminopropane and benzoyl chloride: L. Beregi *et al.*, Fr. pat. **1,517,587** corresp to U.S. pat. **3,607,909** (1968, 1971 both to Sci. Union et Cie, Soc. Franc. Rech. Med.). Metabolism studies: A. H. Beckett *et al.*, *J. Pharm. Pharmacol.* **23**, 950 (1971); **24**, 281 (1972). Pharmacology: D. N. Brindley *et al.*, *ibid.* **28**, 670 (1976); P. Pritchard *et al.*, *ibid.* **29**, 343 (1977); *eidem, Biochem. J.* **166**, 639 (1977).

Colorless oil.
Hydrochloride, $C_{19}H_{20}F_3NO_2 \cdot HCl$, *S-992, JP-992, Mediator, Mediaxal.* Crystals from ethyl acetate, mp 161-162°.
THERAP CAT: Antihyperlipoproteinemic.

1067. Benfluralin. *N-Butyl-N-ethyl-2,6-dinitro-4-(trifluoromethyl)benzenamine; N-butyl-N-ethyl-α,α,α-trifluoro-2,6-dinitro-p-toluidine;* N-butyl-N-ethyl-2,6-dinitro-4-trifluoromethylaniline; benfin; bethrodine; *EL-110; Balan; Balfin; Benefex; Quilan.* $C_{13}H_{16}F_3N_3O_4$; mol wt 335.28. C 46.57%, H 4.81%, F 17.00%, N 12.53%, O 19.09%. Selective pre-emergence herbicide. Prepn: Q. F. Soper, U.S. pat. **3,257,190** (1966 to Lilly). Activity: E. F. Alder *et al.*, *Proc. Northeast Weed Control Conf.* **15**, 298 (1961). Environmental fate: T. Golab *et al.*, *J. Agr. Food Chem.* **18**, 838 (1970). Soil degradn: J. H. Miller *et al.*, *Weed Sci.* **23**, 211 (1975); R. L. Zimdahl, S. M. Gwynn, *ibid.* **25**, 247 (1977). Toxicity study: E. I. Goldenthal, *Toxicol. Appl. Pharmacol.* **18**, 185 (1971).

Yellow-orange crystalline solid, mp 65-66.5°. Vapor pressure at 30°: 3.89×10^{-4} mm Hg. Soly in water at 25°: <1 mg/l. Sol in most organic solvents; less sol in ethanol. Decomp in uv light. LD_{50} orally in female rats: >10,000 mg/kg (Goldenthal).
USE: Herbicide.

1068. Benfotiamine. *Benzenecarbothioic acid S-[2-[[(4-amino-2-methyl-5-pyrimidinyl)methyl]formylamino]-1-[2-(phosphonooxy)ethyl]-1-propenyl] ester; thiobenzoic acid S-ester with N-[(4-amino-2-methyl-5-pyrimidinyl)methyl]-N-(4-hydroxy-2-mercapto-1-methyl-1-butenyl)formamide O-phosphate;* S-benzoylthiamine monophosphate; BTMP; Biotamin; Vitanevril. $C_{19}H_{23}N_4O_6PS$; mol wt 466.45. C 48.92%, H 4.97%, N 12.01%, O 20.58%, P 6.64%, S 6.87%. Vitamin B_1 source. Prepn: A. Ito *et al.*, Ger. pat. **1,130,811** (1962), *C.A.* **57**, 13764h (1962). Exists as 3 temperature dependent crystalline forms, α, γ, δ: A. Ito *et al.*, *Takamine Kenkyusho Nempo* **14**, 64 (1962); *C.A.* **59**, 3920a (1963). Pharmacokinetics: H. Nogami *et al.*, *Chem. Pharm. Bull. (Tokyo)* **18**, 1937 (1970). Clinical bioequivalence to thiamine: R. Bitsch *et al.*, *Ann. Nutr. Metab.* **35**, 292 (1991).

(Z)-form

Crystals, dec 165° (δ-form).
THERAP CAT: Vitamin (enzyme cofactor).

1069. Benfuracarb. *N-[[[[(2,3-Dihydro-2,2-dimethyl-7-benzofuranyl)oxy]carbonyl]methylamino]thio]-N-(1-methylethyl)-β-alanine ethyl ester; 2-methyl-4-(1-methylethyl)-7-oxo-8-oxa-3-thia-2,4-diazadecanoic acid 2,3-dihydro-2,2-dimethyl-7-benzofuranyl ester;* 2,3-dihydro-2,2-dimethylbenzofuran-7-yl N-(N-isopropyl-N-ethoxycarbonylethylaminosulfenyl)-N-methylcarbamate; 2,3-dihydro-2,2-dimethyl-7-benzofuranyl N-[N-[2-(ethoxycarbonyl)ethyl]-N-isopropylsulfenamoyl]-N-methylcarbamate; ethyl N-[2,3-dihydro-2,2-dimethylbenzofuran-7-yloxycarbonyl(methyl)aminothio]-N-isopropyl-β-alaninate; *OK-174; Oncol.* $C_{20}H_{30}N_2O_5S$; mol wt 410.53. C 58.51%, H 7.37%, N 6.82%, O 19.49%, S 7.81%. Broad spectrum soil and foliar insecticide; cholinesterase inhibitor. Prepn and insecticidal properties: A. Tanaka *et al.*, Belg. pat. **890,162**; T. Goto *et al.*, U.S. pat. **4,413,005** (1982, 1983 both to Otsuka). Biological activity and field trials: T. Goto *et al.*, *Proc. 10th Conf. Int. Congr. Plant Prot.* **1**, 360 (1983). Absorption and metabolism in plants: A. Tanaka *et al.*, *J. Agr. Food Chem.* **33**, 1049 (1985). Brief review, field trials: P. Cagnieul, *Def. Veg.* **38**, 3-10 (1984).

Viscous brown-red liquid. d^{20} 1.17. Almost insol in water (8 mg/l at 20°). Sol in organic solvents. LD_{50} in male rats, mice, dogs (mg/kg): 138, 175, 300 orally (Goto).
USE: Insecticide.

1070. Benfurodil Hemisuccinate. *Butanedioic acid mono[1-[5-(2,5-dihydro-5-oxo-2-furanyl)-3-methyl-2-benzofuranyl]ethyl] ester; succinic acid monoester with 4-[2-(1-hydroxyethyl)-3-methyl-5-benzofuranyl]-2(5H)-furanone;* 2-[1-(succinoyloxy)ethyl]-3-methyl-5-(2-oxo-2,5-dihydro-4-furyl)benzo[b]furan; 2-(1-hydroxyethyl)-β-(hydroxymethyl)-3-methyl-5-benzofuranacrylic acid γ-lactone hydrogen succinate; benzofurodil; *Eucilat; Eudilat.* $C_{19}H_{18}O_7$; mol wt 358.35. C 63.68%, H 5.06%, O 31.25%. Prepn: J. Schmitt, Fr. pat. **1,408,721** corresp to U.S. pat. **3,355,463** (1965, 1967 to Clin-Byla). Prepn and toxicity: J. Schmitt *et al.*, *Bull. Soc. Chim. France* **1967**, 74. Synthesis and pharmacology: *eidem, Chim. Ther.* **1**, 305 (1966).

White to pale yellow crystals from ethyl acetate, mp 144°. Sol in alkaline solns. LD$_{50}$ in mice: 550 mg/kg orally (Schmitt).

THERAP CAT: Cardiotonic, vasodilator.

1071. Benidipine. (R*,R*)-(±)-1,4-Dihydro-2,6-dimethyl-4-(3-nitrophenyl)-3,5-pyridinedicarboxylic acid methyl 1-(phenylmethyl)-3-piperidinyl ester; (±)-(R*)-3-[(R*)-1-benzyl-3-piperidyl] methyl 1,4-dihydro-2,6-dimethyl-4-(m-nitrophenyl)-3,5-pyridinedicarboxylate; (±)-2,6-dimethyl-4-(3-nitrophenyl)-1,4-dihydropyridine-3,5-dicarboxylic acid-3-(1-benzyl-3-piperidyl) ester-5-methyl ester. C$_{28}$H$_{31}$N$_3$O$_6$; mol wt 505.57. C 66.52%, H 6.18%, N 8.31%, O 18.99%. Dihydropyridine calcium channel blocker. Prepn: K. Muto et al., Eur. pat. Appl. 106,275 (1984 to Kyowa), C.A. 101, 110742e (1984); and toxicity data: eidem, Arzneimittel-Forsch. 38, 1662 (1988). Resolution of enantiomers: Y. Okamoto et al., J. Chromatog. 513, 375 (1990). Structural studies: N. Hirayama, E. Shimizu, Acta Cryst. C47, 458 (1991). Series of articles on properties, pharmacology, determn and clinical evaluation: Arzneimittel-Forsch. 38, 1666-1763 (1988). Review: H. Suzuki, T. Saruta, Cardiovasc. Drug Rev. 7, 25-38 (1989).

(±)-α-Form hydrochloride, C$_{28}$H$_{31}$N$_3$O$_6$.HCl, KW-3049, Coniel. Yellow crystalline powder, mp 199.4-200.4°. uv max (ethanol): 238, 359 nm (ε 2.80 × 10^4, 6.68 × 10^3). Soly at 25° (%): methanol 6.9; ethanol 2.2; water 0.19; chloroform 0.16; acetone 0.13; ethyl acetate 0.0056; toluene 0.0019; n-heptane 0.00009. pKa 7.34. Partition coefficient (n-octanol/water): 1.23 × 10^3 (pH 6.4, 22°). LD$_{50}$ orally in male mice: 218 mg/kg (Muto, 1988).

THERAP CAT: Antihypertensive.

1072. Benmoxine. Benzoic acid 2-(1-phenylethyl)hydrazide; 1-(benzoyl)-2-(α-methylbenzyl)hydrazine; 1-(α-methylbenzyl)-2-benzoylhydrazine; Neuralex. C$_{15}$H$_{16}$N$_2$O; mol wt 240.30. C 74.97%, H 6.71%, N 11.66%, O 6.66%. Prepn: Bettinetti, Farmaco Ed. Sci. 16, 823 (1961); Fr. pat. 1,314,362 (1963 to I.C.I.), C.A. 59, 2716a (1963), corresp to Brit. pat. 919,491.

Crystals from isopropyl ether or petr ether + benzene, mp 93-94°.

THERAP CAT: Antidepressant.

1073. Benomyl. [1-[(Butylamino)carbonyl]-1H-benzimidazol-2-yl]carbamic acid methyl ester; 1-(butylcarbamoyl)-2-benzimidazolecarbamic acid methyl ester; methyl 1-(butylcarbamoyl)-2-benzimidazolecarbamate; F-1991; Benlate. C$_{14}$H$_{18}$N$_4$O$_3$; mol wt 290.32. C 57.92%, H 6.25%, N 19.30%, O 16.53%. Prepn: H. L. Klopping, Fr. pat. 1,523,-597; eidem, U.S. pat. 3,631,176 (1968, 1971 both to du Pont). The degradation product, methyl 2-benzimidazolecarbamate, is thought to be the active component; see carbendazim. Antifungal activity and mode of action studies: Kilgore, White, Bull. Environ. Contam. Toxicol. 5, 67 (1970); Maxwell, Brody, Appl. Microbiol. 21, 944 (1971); Bartels-Schooley, MacNeill, Phytopathology 61, 816 (1971). Activity as catalyst in biological oxidation of sewage and fertilizers: Kouba, U.S. pat. 3,649,530 (1972 to du Pont). Toxicity: E. W. Schafer, Toxicol. Appl. Pharmacol. 21, 315

(1972). Metabolism: J. A. Gardiner et al., J. Agr. Food Chem. 22, 419 (1974). Mutagenicity study: J. P. Seiler, Mutat. Res. 32, 151 (1975); H. Sherman et al., Toxicol. Appl. Pharmacol. 32, 305 (1975). Review: W. E. Bleidner et al., Anal. Methods Pestic. Plant Growth Regul. 10, 157-171 (1978).

White crystalline solid. Sol in chloroform. Insol in water or oil. LD$_{50}$ orally in rats: > 9590 mg/kg (Schafer).

USE: Fungicide; ascaricide.

THERAP CAT (VET): Anthelmintic.

1074. Benorylate. 2-(Acetyloxy)benzoic acid 4-(acetylamino)phenyl ester; salicylic acid acetate, ester with 4-hydroxyacetanilide; 2-acetoxy-4'-(acetamino)phenylbenzoate; p-acetamidophenyl acetylsalicylate; 4'-(acetamido)phenyl-2-acetoxybenzoate; fenasprate; Win-11450; Benoral; Benortan; Quinexin; Salipran. C$_{17}$H$_{15}$NO$_5$; mol wt 313.31. C 65.17%, H 4.83%, N 4.47%, O 25.53%. Prepn: Neth. pat. Appl. 6,504,517 (1965 to Sterwin), C.A. 64, 8097c (1966), corresp to Brit. pat. 1,101,747 (1968); Robertson, U.S. pat. 3,431,-293 and Miller, Brit. pat. 1,168,289 (both 1969 to Sterling Drug). Pharmacological investigations: Rosner et al., Therapie 23, 525 (1968); Raab, Arzneimittel-Forsch. 21, 1662 (1971). Pharmacokinetic studies in animals: Liss, Palme, ibid. 19, 1177 (1969). Clinical evaluation: Bain, Burt, Clin. Trials J. 7, 307 (1970); Cardoe, ibid. 313.

Crystals from methanol or ethanol, mp 175-176°. LD$_{50}$ in mice, rats (mg/ml): 2000, ~10000 orally; 1255, 1830 i.p. (Robertson).

THERAP CAT: Analgesic; anti-inflammatory; antipyretic.

1075. Benoxaprofen. 2-(4-Chlorophenyl)-α-methyl-5-benzoxazoleacetic acid; Compd 90459; Coxigon; Opren; Oraflex; Uniprofen. C$_{16}$H$_{12}$ClNO$_3$; mol wt 301.73. C 63.69%, H 4.01%, Cl 11.75%, N 4.64%, O 15.91%. Prepn: D. Evans et al., Ger. pat. 2,324,443; eidem, U.S. pat. 3,912,-748 (1973, 1975 both to Lilly); D. W. Dunwell et al., J. Med. Chem. 18, 53 (1975). Analysis by HPLC: S. W. McKay et al., J. Chromatog. 170, 482 (1979). Pharmacokinetics: D. H. Chatfield et al., Brit. J. Clin. Pharmacol. 4, 579 (1977). Metabolism: D. H. Chatfield, J. N. Green, Xenobiotica 8, 133 (1978). Effect on lysosomal enzyme secretion: R. J. Smith, J. Pharmacol. Exp. Ther. 207, 618 (1978). Dissolution and absorption in humans: A. S. Ridolfo et al., J. Pharm. Sci. 68, 850 (1979). Series of clinical studies: Rheumatol. Rehabil. 17, 254-264 (1978).

Cream solid from ethanol, mp 189-190°. LD$_{50}$ orally in mice: 800 mg/kg (Evans, 1973).

THERAP CAT: Anti-inflammatory; analgesic.

1076. Benoxinate. 4-Amino-3-butoxybenzoic acid 2-(diethylamino)ethyl ester; 3-butoxy-4-aminobenzoic acid 2-(diethylamino)ethyl ester; 2-(diethylamino)ethyl 4-amino-3-*n*-butoxybenzoate; oxibuprokain; oxybuprocaine. $C_{17}H_{28}N_2O_3$; mol wt 308.42. C 66.20%, H 9.15%, N 9.08%, O 15.56%. Prepn: **Brit. pat. 654,484** (1951 to Wander).

bp$_2$ 215-218°.

Hydrochloride, $C_{17}H_{28}N_2O_3$.HCl, *Conjuncain, Cebesine, Dorsacaine, Novesine, Benoxil, Lacrimin.* Crystals, mp ~155° (*U.S.P.* XVI, 1960), also reported as mp 157-160° (*N.N.R.* 1955). Very sol in water, chloroform; sol in alc. Practically insol in ether. pH of aq solns, 4.5-5.2.

THERAP CAT: Anesthetic (local).

1077. Benperidol. *1-[1-[4-(4-Fluorophenyl)-4-oxobutyl]-4-piperidinyl]-1,3-dihydro-2H-benzimidazol-2-one; 1-[1-[3-(p-fluorobenzoyl)propyl]-4-piperidyl]-2-benzimidazolinone;* 1-[1-[4-(*p*-fluorophenyl)-4-oxobutyl]piperidin-4-yl]-2-benzimidazolinone; benzperidol; McN-JR-4584; R-4584; Anquil; Frénactil; Frenactyl; Glianimon. $C_{22}H_{24}FN_3O_2$; mol wt 381.45. C 69.27%, H 6.34%, F 4.98%, N 11.02%, O 8.39%. Prepn: **Belg. pat. 626,307** (1963 to Janssen), *C.A.* **60**, 10690c (1964), corresp. to **Brit. pat. 989,755.**

Solid, mp 170-171.8°.

Hydrochloride hydrate, solid, mp 134-142°.

THERAP CAT: Antipsychotic.

1078. Benproperine. *1-[1-Methyl-2-[2-(phenylmethyl)-phenoxy]ethyl]piperidine; 1-[1-methyl-2-[(α-phenyl-o-tolyl)-oxy]ethyl]piperidine;* 1-[2-(2-benzylphenoxy)-1-methylethyl]-piperidine. $C_{21}H_{27}NO$; mol wt 309.45. C 81.51%, H 8.79%, N 4.53%, O 5.17%. Prepn: **Brit. pat. 914,008** corresp to Rubinstein, **U.S. pat. 3,117,059** (1962, 1964 to Aktieselskabet Pharmacia). Pharmacology: Yamatsu *et al., Japan. J. Pharmacol.* **17**, 538 (1967); Tellini, De Fina, *Boll. Chim. Farm.* **109**, 476 (1970).

Liquid. bp$_{0.2}$ 159-161°.

Trihydrogen phosphate, $C_{21}H_{27}NO.H_3PO_4$, *Blascorid, Pirexyl.* Crystals from abs ethanol, mp 150-152°. LD$_{50}$ in albino mice (mg/kg): 192 i.p.; 1365 orally.

Pamoate, $(C_{21}H_{27}NO)_2.C_{23}H_{16}O_6$, *benzproperine ambonate, Tussafug.*

THERAP CAT: Antitussive.

1079. Benserazide. *DL-Serine 2-[(2,3,4-trihydroxyphenyl)methyl] hydrazide; N-(DL-seryl)-N'-(2,3,4-trihydroxybenzyl) hydrazine.* $C_{10}H_{15}N_3O_5$; mol wt 257.25. C 46.69%, H 5.88%, N 16.33%, O 31.10%. Peripheral decarboxylase inhibitor. Prepn: **Belg. pat. 619,015;** Hegedüs, Zeller, **U.S. pat. 3,178,476** (1962, 1965 both to Hoffmann-La Roche). Inhibition of dopa decarboxylase *in vitro* and *in vivo:* W. P.

Burkard *et al., Experientia* **18**, 411 (1962); *eidem, Arch. Biochem. Biophys.* **107**, 187 (1964). Synergy with dopa *in vivo:* G. Bartholini *et al., Nature* **215**, 852 (1967); G. Bartholini, A. Pletscher, *J. Pharmacol. Exp. Ther.* **161**, 14 (1968). Clinical comparison with carbidopa, *q.v.:* I. Kuruma *et al., J. Pharm. Pharmacol.* **24**, 289 (1972). Clinical and pharmacokinetic profile: M.-H. Marion *et al., Adv. Neurol.* **45**, 493 (1987). Review of pharmacology and clinical efficacy of decarboxylase inhibitors: R. M. Pinder *et al., Drugs* **11**, 329-377 (1976).

Hydrochloride, $C_{10}H_{15}N_3O_5$.HCl, *Ro-4-4602.* White crystalline powder, mp 146-148°. Sol in water.

Combination with levodopa, *co-beneldopa, Madopar.*

THERAP CAT: In combination with levodopa as antiparkinsonian.

1080. Bentazon. *3-(1-Methylethyl)-1H-2,1,3-benzothiadiazin-4(3H)-one 2,2-dioxide; 3-isopropyl-1H-2,1,3-benzothiadiazin-4(3H)-one 2,2-dioxide;* bentazone; bendioxide; Basagran. $C_{10}H_{12}N_2O_3S$; mol wt 240.28. C 49.99%, H 5.03%, N 11.66%, O 19.98%, S 13.35%. Selective post-emergence herbicide. Prepn: A. Zeidler *et al.,* **S. Afr.** pat. **67 05,164** corresp to **U.S. pat. 3,708,277** (1968, 1973, both to BASF); G. Hamprecht *et al.,* **U.S. pat. 3,822,257** and **Ger. pat. 2,357,063,** *C.A.* **83**, 131644z (1974, 1975, both to BASF). Fate, metabolism and toxicity in a model ecosystem: G. M. Booth *et al., J. Environ. Qual.* **2**, 408 (1973).

White crystalline powder, mp 137-139°. Soly (w/w) at 20°: water 0.05%; acetone 150.7%; benzene 3.3%; chloroform 18%; ethanol 86.1%. LD$_{50}$ orally in rats: 1100 mg/kg, *RTECS vol. II,* E. J. Fairchild, Ed. (1977) p 206.

USE: Herbicide.

1081. Bentiromide. *(S)-4-[[2-(Benzoylamino)-3-(4-hydroxyphenyl)-1-oxopropyl]amino]benzoic acid; (S)-p-(α-benzamido-p-hydroxyhydrocinnamamido)benzoic acid; N-benzoyl-L-tyrosyl-p-aminobenzoic acid; BTPABA; E-2663; Chymex; PFT Roche.* $C_{23}H_{20}N_2O_5$; mol wt 404.42. C 68.31%, H 4.98%, N 6.93%, O 19.78%. Synthetic chymotrypsin-labile peptide, used in diagnosis of exocrine pancreatic disease. Prepn: P. L. De Benneville, N. J. Greenberger, **Ger. pat. 2,156,835;** *eidem,* **U.S. pat. 3,801,562** (1972, 1974 both to Rohm & Haas). Synthesis, *in vitro* and *in vivo* data: P. L. De Benneville *et al., J. Med. Chem.* **15**, 1098 (1972). Early clinical studies: K. Gyr *et al., Schweiz. Med. Wochenschr.* **105**, 1717 (1975); W. Bornschein *et al., Clin. Chim. Acta* **67**, 21 (1976). Pediatric study: G. Dockter *et al., Eur. J. Pediatrics* **135**, 277 (1981). Human toxicity study, effects of renal insufficiency on pancreatic function test: C. Lang *et al., J. Clin. Chem. Clin. Biochem.* **18**, 551 (1980).

Cryst from methanol/water, mp 240-242°. $[\alpha]_D^{25}$ +72.3° (c = 1 in DMF) (U.S. pat.); also reported as $[\alpha]_D^{25}$ +87° (c = 1 in DMF) (De Benneville).

THERAP CAT: Diagnostic aid (pancreatic function).

1082. Bentonite. Wilkinite. A colloidal native hydrated aluminum silicate (clay) found in the midwest of the U.S.A. and in Canada. Consists principally of montmorillonite, $Al_2O_3.4SiO_2.H_2O$. Usually contains some magnesium, iron, and calcium carbonate. *Review:* J. Alexander, *Ind. Eng. Chem.* **16**, 1140 (1924).

The color in the massive condition varies from yellowish-white to almost black. The powder is cream colored to pale brown. It has the property of forming highly viscous suspensions or gels with not less than ten times its weight of water. The property of forming gels is very much increased by the addition of small amounts of alkaline substances such as magnesium oxide.

USE: As of Fuller's earth; as emulsifier for oils; as a base for plasters. Pharmaceutic aid (suspending agent).

1083. β-Benzalbutyramide. *3-Methyl-4-phenyl-3-butenamide;* 3-methyl-4-phenyl-3-butenoic acid amide; β-benzylidenebutyramide; Kata-Lipid. $C_{11}H_{13}NO$; mol wt 175.23. C 75.40%, H 7.48%, N 7.99%, O 9.13%. Prepn of the acid: Canonica *et al., Farmaco Ed. Sci.* **14**, 112 (1959); Datta, Bagchi, *J. Org. Chem.* **25**, 932 (1960). Clinical studies in atherosclerotic patients: F. Del Regno, G. Capobianco, *Minerva Med.* **54**, 2917 (1963); A. Iacobelli *et al., G. Arterioscler.* **4**, 277 (1966); M. G. Piccardo *et al., ibid.* 294. Metabolism: L. Canonica *et al., J. Biol. Chem.* **243**, 1645 (1968). Inhibition of cholesterol biosynthesis: D. Giorgini, G. Porcellati, *Farmaco Ed. Sci.* **24**, 392 (1969). Clinical evaluation in hyperlipidemia: U. Marini *et al., Farmaco Ed. Prat.* **36**, 532 (1981).

Crystals, mp 132-134°. uv max: 249 nm (log ϵ 4.1).

THERAP CAT: Antihyperlipoproteinemic.

1084. Benzal Chloride. *(Dichloromethyl)benzene; benzylidene chloride;* benzyl dichloride; α,α-dichlorotoluene; benzylene chloride. $C_7H_6Cl_2$; mol wt 161.03. C 52.21%, H 3.76%, Cl 44.03%. $C_6H_5CHCl_2$. Obtained by chlorination of toluene. Lab procedure: A. I. Vogel, *Practical Organic Chemistry* (Longmans, London, 3rd ed., 1959) p 539; Gattermann-Wieland, *Praxis des organischen Chemikers* (de Gruyter, Berlin, 40th ed., 1961) p 184.

Very refractive liquid; fumes in air; vapors irritate the eyes; pungent odor. d 1.26. bp 205°. mp −17°. Insol in water; freely sol in alcohol, ether.

USE: Manuf benzaldehyde, cinnamic acid.

1085. Benzaldehyde. Benzoic aldehyde; artificial essential oil of almond. C_7H_6O; mol wt 106.12. C 79.23%, H 5.70%, O 15.08%. Occurs in kernels of bitter almonds; made synthetically from benzal chloride and lime or by oxidation of toluene. Laboratory prepn from benzal chloride: A. I. Vogel, *Practical Organic Chemistry* (Longmans, London, 3rd ed., 1959) p 693; Gattermann-Wieland, *Praxis des organischen Chemikers* (de Gruyter, Berlin, 40th ed., 1961) p 184. Toxicity data: P. M. Jenner *et al., Food Cosmet. Toxicol.* **2**, 327 (1964). *Review:* A. E. Williams in Kirk-Othmer *Encyclopedia of Chemical Technology* vol. 3 (Wiley-Interscience, New York, 3rd ed., 1978) pp 736-743.

Strongly refractive liquid, becoming yellowish on keeping; characteristic odor of volatile oil of almond; burning aromatic taste. Oxidizes in air to benzoic acid; volatile with steam. d_4^{15} 1.050; 1.043 at 25°. bp 179°. mp −56.5°. Flash

pt 62°. n_D^{20} 1.5456. Sol in 350 parts water; miscible with alcohol, ether, oils. It reduces ammoniacal $AgNO_3$, but not Fehling's soln. *Keep tightly closed and protected from light.* LD_{50} in rats, guinea pigs (mg/kg): 1300, 1000 orally (Jenner).

Note: Benzaldehyde FFC designates a grade of benzaldehyde free from chlorine.

Caution: Narcotic in high concns. May cause contact dermatitis.

USE: Manufacture of dyes, perfumery, cinnamic and mandelic acids, as solvent; in flavors.

1086. Benzalkonium Chloride. Benirol; BTC; Capitol; Cequartyl; Drapolene; Enuclen; Germinol; Germitol; Osvan; Paralkan; Roccal; Rodalon; Zephiran Chloride; Zephirol. A mixture of alkyldimethylbenzylammonium chlorides of the general formula in which R represents a mixture of the alkyls from C_8H_{17} to $C_{18}H_{37}$. Acute toxicity data: L. M. Cummins, E. T. Kimura, *Toxicol. Appl. Pharmacol.* **20**, 89 (1971). Review of antimicrobial activity and use: W. Gump, "Disinfectants and Antiseptics" in Kirk-Othmer *Encyclopedia of Chemical Technology* vol. 7 (Wiley-Interscience, New York, 3rd ed., 1979) pp 815-818.

White or yellowish-white, amorphous powder or gelatinous pieces. Aromatic odor, very bitter taste. Very sol in water, alcohol, acetone; slightly sol in benzene. Almost insol in ether. The aq soln is slightly alkaline to litmus and foams strongly when shaken. LD_{50} orally in rats: 400 mg/kg (Cummins, Kimura).

Incompatible with anionic detergents, such as soap, and with nitrates. A white precipitate is formed in a 1:3000 aq soln of benzalkonium chloride when nitrates are present in concns greater than the equiv. of 0.5% ammonium nitrate.

Dibromide, *Callusolve.* Adduct of benzalkonium chloride with bromine which is insol in water and non-ionized.

USE: Cationic surface active agent and germicide. Pharmaceutic aid (preservative).

THERAP CAT: Antiseptic (topical).

THERAP CAT (VET): Antiseptic for skin preoperatively or for wounds, burns, etc. Udder wash.

1087. Benzamide. Benzoylamide. C_7H_7NO; mol wt 121.14. C 69.41%, H 5.82%, N 11.56%, O 13.21%. C_6H_5-$CONH_2$. Prepd from benzoyl chloride and ammonium carbonate. Lab prepn: Gattermann-Wieland, *Praxis des organischen Chemikers* (de Gruyter, Berlin, 40th ed., 1961) p 119. Alternate procedure using concd ammonia soln: A. I. Vogel, *Practical Organic Chemistry* (Longmans, London, 3rd ed., 1959) p 797.

Crystals. d^4 1.341. mp 130°. bp 288°. One gram dissolves in 74 ml water, more sol in boiling water, in 6 ml alc, 3.3 ml pyridine. Sol in hot benzene, slightly in ether; sol in ammonia with formation of a small quantity of benzonitrile.

N-chloro deriv, mp 116°. Sol in water; insol in alcohol or benzene.

1088. Benzanilide. *N-Phenylbenzamide;* N-benzoylaniline. $C_{13}H_{11}NO$; mol wt 197.24. C 79.17%, H 5.62%, N 7.10%, O 8.11%. $C_6H_5CONHC_6H_5$. Prepd by the treatment of aniline with benzoic acid: Hübner, *Ann.* **208**, 291 (1881); Nägeli, *Bull. Soc. Chim.* [3] **11**, 892 (1894); Webb, *Org. Syn.* **7**, 6 (1927); **coll. vol. I** (2nd ed., 1941) p 82.

Leaflets from alc. d 1.315. mp 163°. Sublimes. Distills without decompn. bp_{10} 117-119°. Absorption spectrum: Crymble, *J. Chem. Soc.* **99**, 459 (1911). Insoluble in water. One gram dissolves in 60 ml alc, 7 ml boiling alc. Slightly sol in ether.

USE: Manuf dyes and perfumes.

1089. 1,2-Benzanthracene. *Benz[a]anthracene;* 2,3-benzphenanthrene; tetraphene; benzanthrene; naphthanthracene. $C_{18}H_{12}$; mol wt 228.29. C 94.70%, H 5.30%. Occurs in coal tar: Cook *et al., J. Chem. Soc.* **1933**, 395. Synthesis from naphthalene and phthalic anhydride: Elbs, *Ber.* **19**,

2209 (1886). From *o*-toluylnaphthalene: Fieser, Dietz, *Ber.* **62**, 1827 (1929). From phenanthrene and succinic anhydride: Haworth, Mavin, *J. Chem. Soc.* **1933**, 1012. Absorption spectrum: Capper, Marsh, *J. Chem. Soc.* **1926**, 726; Clar, *Ber.* **65**, 507 (1932); Mayneord, Roe, *Proc. Roy. Soc. London* **A152**, 299 (1935). *Review:* E. Clar, *Polycyclic Hydrocarbons* (Academic Press, New York, 1964) 2 vols.

Plates from glacial acetic acid or alc. Greenish-yellow fluorescence. Sublimes. mp 155-157° (Fieser, Dietz); mp 160° [Fieser, Hershberg, *J. Am. Chem. Soc.* **59**, 2502 (1937)]; mp 167° (I. G. Farbenind., **Ger.** pats. **481,819; 486,766**). Difficultly sol in boiling alc; sol in most other organic solvents. Insol in water.

Note: This substance may reasonably be anticipated to be a carcinogen: *Seventh Annual Report on Carcinogens* (PB95-109781, 1994) p 328.

1090. Benzanthrone. *7H-Benz[de]anthracen-7-one.* $C_{17}H_{10}O$; mol wt 230.27. C 88.67%, H 4.38%, O 6.95%. Prepd by heating a reduction product of anthraquinone with sulfuric acid and glycerol: Macleod, Allen, *Org. Syn.* **14**, 4 (1934); *cf.* U.S. pat. **1,626,392** (1927). Review of toxicology: G. B. Singh *et al., J. Sci. Ind. Res.* **49**, 288-296 (1990).

Pale yellow needles from alcohol or xylene. mp 170°. Absorption spectrum: Clar, *Ber.* **65**, 846 (1932). Solubility at 20° = 0.52 g in 100 g glacial acetic acid; 1.61 g in 100 g benzene; 2.05 g in 100 g chlorobenzene. Solution in H_2SO_4 is orange with green fluorescence. LD_{50} in albino rabbits: > 3 g/kg dermally; LD_{50} in rats, mice (g/kg): 1.5, 0.29 i.p. (Singh).

USE: Manufacture of dyes.

1091. Benzarone. *(2-Ethyl-3-benzofuranyl)(4-hydroxyphenyl)methanone; 2-ethyl-3-benzofuranyl p-hydroxyphenyl ketone;* 2-ethyl-3-(4-hydroxybenzoyl)benzofuran; 2-ethyl-3-(*p*-hydroxybenzoyl)coumarone; L-2197; Fragivix; Fragivil; Vasoc. $C_{17}H_{14}O_3$; mol wt 266.30. C 76.68%, H 5.30%, O 18.02%. Prepn: Buu-Hoi, Beaudet, U.S. pat. **3,012,042** (1961 to Soc. Belge l'Azote Prod. Chim. Marly). Pharmacological studies: Barchewitz *et al., Arzneimittel-Forsch.* **22**, 553 (1972). Site of action studies: I. Filipovic, E. Buddecke, *ibid.* **29**, 1578 (1979); L. O. Zwillenberg *et al., ibid.* **32**, 1114 (1982). HPLC determination in serum: H. Vergin, G. Bishop, *J. Chromatog.* **183**, 383 (1980).

Crystals, mp 124.3°.
THERAP CAT: Capillary protectant.

1092. Benzathine. *N,N'-Bis(phenylmethyl)-1,2-ethanediamine; N,N'-dibenzylethylenediamine;* 1,2-bis(benzylamino)ethane; DBED. $C_{16}H_{20}N_2$; mol wt 240.35. C 79.96%, H 8.39%, N 11.66%. Prepd from *N,N'*-dibenzenesulfonyl-*N,N'*-dibenzylethylenediamine by heating with concd HCl in

bomb tube at 170-180°: Bleier, *Ber.* **32**, 1829 (1899); from ethylene chloride and benzylamine: Frost *et al., J. Am. Chem. Soc.* **71**, 3842 (1949); by reduction of *N,N'*-dibenzylideneethylenediamine: **Ger.** pat. **98,031** (1898 to Schering); *Chem. Zentr.* **1898**, II, 743; Van Alphen, *Rec. Trav. Chim.* **54**, 93 (1935); Lob, *ibid.* **55**, 859 (1936); Szabo *et al., Antibiot. & Chemother.* **1**, 499 (1951); Rebenstorf, U.S. pat. **2,-773,098** (1956 to Upjohn).

Oily liquid, mp 26°. d_4^{20} 1.024. bp_{12} 212-213°; bp_4 195°. n_D^{20} 1.5624. Insol in water. Freely sol in the usual organic solvents except CS_2, with which it gives a solid addition product.

Diacetate, $C_{20}H_{28}N_2O_4$, needles, mp 111°. Soly in water: 253 mg/ml.
Dihydrobromide, $C_{16}H_{22}Br_2N_2$, shiny leaflets, dec 300°. Soly in water: 30 mg/ml.
Dihydrochloride, $C_{16}H_{22}Cl_2N_2$, shiny leaflets, dec 298°. Soly in water: 23.9 mg/ml.
Sulfate, $C_{16}H_{22}N_2O_4S$, crystals, dec 247-250°. Soly in water: 15.8 mg/ml.
Disalicylate, $C_{30}H_{34}N_2O_6$, crystals, mp 85°. Soly in water: 2.43 mg/ml.
Compounds with penicillins, *see* separate entries.
USE: Manuf of a repository form of penicillin.

1093. Benzbromarone. *(3,5-Dibromo-4-hydroxyphenyl)(2-ethyl-3-benzofuranyl)methanone; 3,5-dibromo-4-hydroxyphenyl 2-ethyl-3-benzofuranyl ketone;* 3-(3,5-dibromo-4-hydroxybenzoyl)-2-ethylbenzofuran; 2-ethyl-3-benzofuranyl 4-hydroxy-3,5-dibromophenyl ketone; 2-ethyl-3-(3,5-dibromo-4-hydroxybenzoyl)benzofuran; 2-ethyl-3-(3,5-dibromo-4-hydroxybenzoyl)oxaindene; 2-ethyl-3-benzofuryl 3,5-dibromo-4-hydroxyphenyl ketone; L-2214; MJ-10061; Azubromaron; Besuric; Desuric; Max-Uric; Minuric; Narcaricin; Normurat; Uricovac; Urinorm. $C_{17}H_{12}Br_2O_3$; mol wt 424.09. C 48.15%, H 2.85%, Br 37.68%, O 11.32%. Prepn: **Belg.** pat. **553,621** (1957 to Labaz); Buu-Hoi *et al., J. Chem. Soc.* **1957**, 625; Buu-Hoi, Beaudet, U.S. pat. **3,012,042** (1961 to Soc. Belge l'Azote Prod. Chim. Marly). Structure-activity study: Delbarre *et al., Chim. Ther.* **3**, 470 (1968). Pharmacology of hypouricemic effect: D. S. Sinclair, I. H. Fox, *J. Rheumatol.* **2**, 437 (1975). Pharmacokinetic and clinical studies: T. F. Yu, *ibid.* **3**, 305 (1976). Pharmacokinetics and biotransformation in man: H. Ferber *et al., Eur. J. Clin. Pharmacol.* **19**, 431 (1981). HPLC determn in serum: H. Vergin *et al., J. Chromatog.* **183**, 383 (1980).

Yellowish prisms, mp 151°.
THERAP CAT: Uricosuric.

1094. Benzene. Benzol; cyclohexatriene. C_6H_6; mol wt 78.11. C 92.26%, H 7.74%. Natural component of petroleum, usually < 1.0% by weight. Discovered by Faraday in compressed oil gas in 1825. Obtained in the coking of coal and in the production of illuminating gas from coal. Manuf by catalytic reforming and separation of aromatic compounds, thermal or catalytic dealkylation of toluene, toluene disproportionation; from pyrolysis gasoline. Purification by washing with water: **Brit.** pat. **863,711** (1961 to Schloven-Chemie and H. Koppers GmbH), *C.A.* **55**, 16971f (1961). Lab prepn from aniline: Gattermann-Wieland, *Praxis des organischen Chemikers* (de Gruyter, Berlin, 40th

Consult the Name Index before using this section.

ed., 1961) p 247. Production of pure benzene: French, *Ind. Chemist* **39**, 9-12 (1963). Manuf: Faith, Keyes & Clark's *Industrial Chemicals*, F. A. Lowenheim, M. K. Moran, Eds. (Wiley-Interscience, New York, 4th ed., 1975) pp 126-137. Physical properties: Thorne *et al.*, *Ind. Eng. Chem. Anal. Ed.* **17**, 481 (1945). Solubility studies: F. P. Schwarz, *Anal. Chem.* **52**, 10 (1980). Toxicity data: Kimura *et al.*, *Toxicol. Appl. Pharmacol.* **19**, 699 (1971). *Review:* W. Fruscella in Kirk-Othmer *Encyclopedia of Chemical Technology* vol. **4** (Wiley-Interscience, New York, 4th ed., 1992) pp 73-103. Reviews of toxicology: R. Snyder *et al.*, *Rev. Biochem. Toxicol.* **3**, 123-154 (1981); D. J. Paustenbach *et al.*, *Environ. Health Perspec.* **101**, Suppl. 6, 177-200 (1993); and human exposure: *Toxicological Profile for Benzene* (PB93-182384, 1993) 205 pp. Symposium on metabolism, toxicity and carcinogenesis: *Environ. Health Perspec.* **82**, 3-310 (1989).

Clear, colorless, volatile, highly flammable liquid; characteristic odor. d_4^{15} 0.8787. bp 80.1°. mp +5.5°. n_D^{20} 1.50108. Flash pt, closed cup: 12°F (−11°C). Soly in water at 23.5°C (w/w): 0.188%. Miscible with alcohol, chloroform, ether, carbon disulfide, carbon tetrachloride, glacial acetic acid, acetone, oils. *Keep in well-closed containers in a cool place and away from fire.* LD_{50} orally in young adult rats: 3.8 ml/kg (Kimura).

Sodium deriv, C_6H_5Na, *phenyl sodium.* Prepn: Schlosser, *Angew. Chem.* **76**, 267 (1964). Solid mass, dec by water, acids, alkalies. Sol in liquid ammonia, tetrahydrofuran.

Caution: Potential symptoms of overexposure by inhalation or ingestion are dizziness, headache, vomiting, visual disturbances, staggering gait, hilarity, fatigue, CNS depression, loss of consciousness, respiratory arrest. Chronic exposure has been associated with bone marrow depression and leukemia. Direct contact may cause irritation of eyes, nose, respiratory system and skin; dermitits may develop due to defatting action. Aspiration into the lung may lead to chemical pneumonitis. *See Patty's Industrial Hygiene and Toxicology* vol. **2B**, G. D. Clayton, F. E. Clayton, Eds. (Wiley-Interscience, New York, 4th ed., 1994) pp 1306-1326; *NIOSH Pocket Guide to Chemical Hazards* (DHHS/NIOSH 90-117, 1990) p 44. Benzene has been listed as a known carcinogen: *Seventh Annual Report on Carcinogens* (PB95- 109781, 1994) p 34.

USE: Manuf of industrial chemicals such as polymers, detergents, pesticides pharmaceuticals, dyes, plastics, resins. Solvent for waxes, resins, oils, natural rubber, etc. Gasoline additive.

THERAP CAT (VET): Has been used as a disinfectant.

1095. Benzenearsonic Acid. Phenylarsonic acid. $C_6H_7AsO_3$; mol wt 202.04. C 35.67%, H 3.49%, As 37.08%, O 23.76%. $C_6H_5AsO(OH)_2$.

Cryst powder. mp 158-162° with dec. Sol in 40 parts water, 50 parts alcohol; insol in chloroform.

USE: Reagent for tin.

1096. Benzeneboronic Acid. *Phenylboronic acid;* phenylboric acid; phenylboron dihydroxide. $C_6H_7BO_2$; mol wt 121.93. C 59.10%, H 5.79%, B 8.87%, O 26.24%. Prepd by the reaction of phenylmagnesium bromide with methyl borate: Washburn *et al.*, *Advances in Chemistry Series* **23**, 102-128 (1959). Crystal and molecular structure: S. J. Rettig, J. Trotter, *Can. J. Chem.* **55**, 3071 (1977).

Crystals from water. Spontaneous conversion to C_6H_5-BO, *benzene boronic anhydride* or *phenylboroxide* on standing in dry air, although the best method is by azeotropic dehydration with toluene. mp 215-216° (anhydride). Ka 13.7. Dipole moment 1.72 (dioxane). Soly in water at 25°: 2.5%, benzene: 1.75%, xylene: 1.2%, ether: 30.2%, methanol: 178%.

1097. Benzenestibonic Acid. *Dihydroxyphenylstibine oxide;* phenylstibonic acid; phenylstibinic acid. $C_6H_7O_3Sb$; mol wt 248.88. C 28.96%, H 2.83%, O 19.29%, Sb 48.92%. $C_6H_5SbO(OH)_2$. Arylstibonic acids exist in the solid state as polymers of high mol wt, associated by H-bonds. The ions exist probably in the form $[ArSb(OH)_5]^-$. Benzenestibonic acid is prepd from benzenediazonium chloride and antimony trioxide in the presence of alkali: Schmidt, *Ann,* **421**, 176, 188 (1920); Ger. pat. 254,421; *Chem. Zentr.* **1913**, I, 345; *Frdl.* **11**, 1084. From the double compd of benzenediazonium chloride and antimony trichloride by the action of alkali: May, *J. Chem. Soc.* **101**, 1033 (1912); U.S. pat. 1,260,707.

Minute crystals from acetic acid, larger diamond-shaped, glistening crystals from alcohol-chloroform mixture on the addition of water. Not melted at 250°, dec at higher temps. Insol in water; slightly sol in alc; more sol in acetone, in alcohol + benzene, alcohol + ether, alcohol + chloroform; sol in warm chloroform or amyl acetate; freely sol in aq soln of sodium hydroxide, sodium carbonate, and ammonia.

1098. Benzenesulfonic Acid. $C_6H_6O_3S$; mol wt 158.18. C 45.56%, H 3.82%, O 30.34%, S 20.27%. $C_6H_5SO_3H$. Made by treating benzene with fuming H_2SO_4: Gattermann-Wieland, *Praxis des organischen Chemikers* (de Gruyter, Berlin, 40th ed., 1961) p 168.

Sesquihydrate, deliquescent plates, tablets, mp 43-44°. When anhydrous mp 50-51°, also reported as mp 65-66°. pKa (25°): 0.699. Freely sol in water, alc; slightly sol in benzene; insol in ether, carbon disulfide. *Keep well closed.*

Ethyl ester, $C_8H_{10}O_3S$, *ethyl benzenesulfonate.* Colorless to slightly yellow, almost odorless liq. bp_{15} 156°. d_4^{17} 1.219. Slightly sol in water; misc with alc, benzene, chloroform, ether.

Sodium salt, $C_6H_5NaO_3S$, *sodium benzenesulfonate, sodium benzosulfonate.* Crystals. Sol in water.

Caution: Highly irritant to skin, eyes, mucous membranes.

USE: Manuf phenol by fusion with NaOH.

1099. Benzenesulfonic Anhydride. $C_{12}H_{10}O_5S_2$; mol wt 298.34. C 48.31%, H 3.38%, O 26.81%, S 21.50%. $(C_6H_5SO_2)_2O$. Prepd by heating benzenesulfonic acid with excess phosphorus pentoxide mixed with inert support: Field, *J. Am. Chem. Soc.* **74**, 394 (1952).

Light tan solid, mp 60-85° (softens about 55°); may be recrystallized from ether, mp 88-91° (softens about 75°). Liquefies upon exposure to air for 2.5 hrs. Explosion occurs when mixed with 90-95% H_2O_2.

USE: In Friedel-Crafts sulfone synthesis; in other sulfonylation reactions.

1100. Benzenesulfonyl Chloride. Benzene sulfonechloride; benzenesulfonic (acid) chloride. $C_6H_5ClO_2S$; mol wt 176.62. C 40.80%, H 2.85%, Cl 20.07%, O 18.12%, S 18.16%. Prepd from benzene and chlorosulfonic acid or from the sodium salt of benzenesulfonic acid and PCl_5 or $POCl_3$: R. Adams, C. S. Marvel, *Org. Syn.* coll. vol. I, (2nd ed., 1964) p 84; H. T. Clarke *et al.*, *ibid.* p 85.

Colorless, oily liquid. d_{15}^{15} 1.3842. Solidifies 0°, mp 14.5°. bp_{760} 251-252° (decompn); bp_{100} 177°; bp_{10} 120°. Insol in and stable toward cold water; sol in ether, alc.

1101. 1,2,4-Benzenetriol. Hydroxyhydroquinone; hydroxyquinol. $C_6H_6O_3$; mol wt 126.11. C 57.14%, H 4.80%, O 38.06%. Prepd by hydrolysis of the triacetate with H_2SO_4 in methanol, the triacetate being formed by the action of acetic anhydride on quinone: Vliet, *Org. Syn.* coll. vol. I (2nd ed, 1941) p 317.

Monoclinic prismatic leaflets from ether, mp 141°. Freely sol in water, alcohol, ether, ethyl acetate; almost insol in chloroform, carbon disulfide, ligroin, benzene.

Trimethyl ether, $C_9H_{12}O_3$, bp 247°.

Triacetate, $C_{12}H_{12}O_6$, needles from abs alc, mp 96.5-97.0°. Readily hydrolyzed by acids or alkalies.

USE: In gas analysis; 1,2,4-benzenetriol in alkaline soln is just as good an absorbent for oxygen as is pyrogallol.

1102. Benzestrol. *4,4'-(1,2-Diethyl-3-methyl-1,3-propanediyl)bisphenol;* 3-ethyl-2,4-bis(*p*-hydroxyphenyl)hexane; Octofollin; Ocestrol; Chemestrogen. $C_{20}H_{26}O_2$; mol wt 298.43. C 80.50%, H 8.78%, O 10.72%. Eight isomers are possible. The compound accepted for medical use and designated benzestrol is the racemic mixture "B-2" of Stuart. Synthesis: Stuart *et al., J. Am. Chem. Soc.* **68**, 729 (1946). Improved method: Kiprianov, Kutsenko, *Zh. Prikl. Khim. (Leningrad)* **31**, 665 (1958).

Crystals, mp 162-166°. Freely sol in acetone, ether, ethanol, methanol and dil sodium hydroxide soln; sol in vegetable oils; moderately sol in glacial acetic acid; slightly sol in benzene, chloroform, petr ether, and dil ethanol; practically insol in water and in dil mineral acids. Neutral to litmus in 75% alc soln.

Dibenzoate, crystals from ethanol, mp 118-120°.

Dimethyl ether, crystals from ethanol, mp 56°.

THERAP CAT: Estrogen.

1103. Benzethonium Chloride. *N,N-Dimethyl-N-[2-[2-[4-(1,1,3,3-tetramethylbutyl)phenoxy]ethoxy]ethyl]benzenemethanaminium chloride; benzyldimethyl[2-[2-(p-1,1,3,3-tetramethylbutylphenoxy)ethoxy]ethyl]ammonium chloride;* diisobutylphenoxyethoxyethyl dimethyl benzyl ammonium chloride; Hyamine 1622; Phemerol chloride; Phemeride; Phemithyn; Quatrachlor; Solamin. $C_{27}H_{42}ClNO_2$; mol wt 448.09. C 72.37%, H 9.45%, Cl 7.91%, N 3.13%, O 7.14%. Prepn: U.S. pats. **2,115,250** (1938); **2,170,111** (1939); **2,229,024** (1941).

Monohydrate, thin, hexagonal plates from chloroform + ether. Sinters slightly at 120°; mp 164-166° (hot stage). Very sol in water giving a foamy, soapy soln. Sol in alcohol, acetone, chloroform. The pH of a 1% aq soln is between 4.8 and 5.5. Mineral acids and many salt solns precipitate benzethonium chloride from solns more concd than 2% as an oil which crystallizes on drying. Incompatible with soap, anionic detergents. LD_{50} in rats: 420 mg/kg.

Caution: Ingestion may cause vomiting, collapse, convulsions, coma.

THERAP CAT: Anti-infective (topical).

THERAP CAT (VET): Antiseptic (topical).

1104. Benzetimide. *3-Phenyl-1'-(phenylmethyl)-[3,4'-bipiperidine]-2,6-dione; 2-(1-benzyl-4-piperidyl)-2-phenylglutarimide;* 1-benzyl-4-(2,6-dioxo-3-phenyl-3-piperidyl)piperidine. $C_{23}H_{26}N_2O_2$; mol wt 362.47. C 76.21%, H 7.23%, N 7.73%, O 8.83%. Prepn: Janssen, U.S. pat. **3,125,578** (1964

to Janssen); Hermans *et al., J. Med. Chem.* **11**, 797 (1968). Synthesis of labelled compound: van Wijngaarden, Soudijn, *J. Label. Compounds* **1**, 207 (1965). Resolution of isomers and pharmacology: van Wijngaarden, *Life Sci.* **8**, 517 (1969); *ibid.* **9**, part 1, 489 (1970); Janssen *et al., Arzneimittel-Forsch.* **21**, 1365 (1971). Metabolism: van Wijngaarden, Soudijn, *Life Sci.* **7**, 225 (1968).

Crystals, mp 156-159°.

Hydrochloride, $C_{23}H_{26}N_2O_2$.HCl, *R-4929, Dioxatrine, Spasmentral.* Crystals, mp 299-301.5°. LD_{50} in rats, mice (mg/kg): 37.6, 46.0 i.v. (Janssen).

l-Form, mp 180.5-182° (from toluene). $[\alpha]_D^{20}$ —124° (chloroform). LD_{50} i.v. in mice: 38.5 mg/kg (Janssen).

d-Form, *see* Dexetimide.

THERAP CAT: Anticholinergic.

THERAP CAT (VET): Antidiarrheal.

1105. Benzhydrylamine. *α-Phenylbenzenemethanamine; 1,1-diphenylmethylamine;* α-aminodiphenylmethane. $C_{13}H_{13}N$; mol wt 183.25. C 85.21%, H 7.15%, N 7.64%. Prepd by boiling benzophenone oxime with zinc dust and ammonia: Scholl, *Ber.* **60**, 1247 (1927).

Hexagonal plates from water, mp 34°. d_4^{22} 1.0635 (supercooled liq). bp_{763} 304.1°; bp_{12} 166°. n_D^{99} 1.59631 (supercooled liq). Strong base. Absorbs CO_2 from the air forming a substance that melts at 91°. Slightly sol in water.

Hydrochloride, $C_{13}H_{13}N$.HCl, needles, mp 293°. Sparingly sol in cold water.

1106. Benzidine. *[1,1'-Biphenyl]-4,4'-diamine; p*-diaminodiphenyl. $C_{12}H_{12}N_2$; mol wt 184.24. C 78.23%, H 6.56%, N 15.20%. Discovered by Zinin in 1845. Produced by reduction of nitrobenzene with Zn and NaOH; the resultant hydrazobenzene is heated with acid. Lab proc: A. I. Vogel, *Practical Organic Chemistry* (Longmans, London, 3rd ed., 1959) p 633; Gattermann-Wieland, *Praxis des organischen Chemikers* (de Gruyter, Berlin, 40th ed., 1961) p 165. Review of carcinogenic risk: *IARC Monographs* **1**, 80-86 (1972); *ibid.* **29**, 149-183 (1982); M. Boeniger, *Carcinogenicity and Metabolism of Azo Dyes, Especially Those Derived from Benzidine* (DHHS/NIOSH 80-119, 1980) 141 pp.

White or slightly-reddish, cryst powder; darkens on exposure to air and light. *Poisonous!* mp 115-120° when slowly heated; 128° when anhydr and rapidly heated. bp ~400°. Soly: one gram dissolves in 2500 ml cold, 107 ml boiling water, 5 ml boiling alcohol, 50 ml ether. *Keep well closed and protected from light.*

Dihydrochloride, $C_{12}H_{12}N_2$.2HCl, crystals. Sol in water, alcohol.

Caution: Potential symptoms of overexposure are hematuria; secondary anemia from hemolysis; acute cystitis; acute liver disorders; dermatitis; painful and irregular urination. *See NIOSH Pocket Guide to Chemical Hazards* (DHHS/NIOSH 90-117, 1990) p 44. This substance has been listed as a known carcinogen: *Seventh Annual Report on Carcinogens* (PB95-109781, 1994) p 37.

USE: Manuf direct azo dyes; as a reagent for H_2O_2 in milk; for detection of blood. Dihydrochloride used for quantitative determn of sulfates, and as reagent for metals.

1107. Benzil. *Diphenylethanedione; bibenzoyl;* dibenzoyl; diphenylglyoxal; diphenyl-α,β-diketone. $C_{14}H_{10}O_2$; mol wt 210.23. C 79.98%, H 4.79%, O 15.22%. $C_6H_5CO\text{-}COC_6H_5$. Prepd by the oxidation of benzoin, $C_6H_5CH\text{-}(OH)COC_6H_5$, with HNO_3 or with a copper sulfate-pyridine mixture: Adams, Marvel, *Org. Syn.* **vol. I,** p 25 (1921); Clarke, Dreger, *ibid.* **coll. vol. I,** 80 (87, 2nd ed); Hatt *et al., J. Chem. Soc.* **1936,** 93; L. F. Fieser, *Experiments in Organic Chemistry* (Boston, 3rd ed. 1955) p 173; *Organic Experiments* (Boston, 1964) p 214.

Yellow prisms (trigonal trapezohedral) from alcohol. d_4^{15} 1.23. mp 95°. bp_{760} 346-348°; bp_{12} 188°. Absorption spectrum: Hantzsch, Schwiete, *Ber.* **49,** 216 (1916). uv max (ethanol): 260 nm (ε 22000). Infrared in chloroform: 5.93; 6.22; 6.85 μ. Insol in water; sol in alcohol, ether, chloroform, ethyl acetate, benzene, toluene, nitrobenzene.

α-Benzilmonoxime, $C_{14}H_{11}NO_2$, leaflets from 30% alc, mp 137-138°.

β-Benzilmonoxime, $C_{14}H_{11}NO_2$, solvated needles from benzene, when dry mp 113-114°.

Disemicarbazone, $C_{16}H_{16}N_6O_2$, leaflets from alc, dec 243-244°.

USE: In organic syntheses.

1108. Benzil Dioxime. *Diphenylethanedione dioxime; diphenylglyoxime.* $C_{14}H_{12}N_2O_2$; mol wt 240.26. C 69.99%, H 5.03%, N 11.66%, O 13.32%. Three isomers occur: α or *anti,* β or *syn,* γ or *amphi.* Prepn of α-form from benzil and hydroxylamine hydrochloride: Brady, Perry, *J. Chem. Soc.* **127,** 2874 (1925); F. J. Welcher, *Organic Analytical Reagents* **vol. III** (Van Nostrand, New York, 1947) pp 224-227; Boyer *et al., J. Am. Chem. Soc.* **77,** 5688 (1955). Prepn of β-form: Brady, Perry, *loc. cit.;* Boyer *et al., loc. cit.* Prepn of γ-form: Welcher, *loc. cit.,* pp 227-228; *see also* Boyer *et al., loc. cit.*

α-form   β-form

α-Form, crystals from methanol, mp 238-240°, Boyer *et al., loc. cit.,* also reported as mp 243-244°, Meisenheimer, Lamparter, *Ber.* **57,** 276 (1924). Practically insol in water, ether, glacial acetic acid; slightly sol in alc; readily sol in NaOH solns.

β-Form, crystals, mp 212-214°, Meisenheimer, Lamparter, *loc. cit.*

1109. Benzilic Acid. *α-Hydroxy-α-phenylbenzeneacetic acid; diphenylglycolic acid.* $C_{14}H_{12}O_3$; mol wt 228.25. C 73.67%, H 5.30%, O 21.03%. Prepd from benzil by the action of concd aq or alc KOH: Adams, Marvel, *Org. Syn.* **vol. 1,** p 29 (1921); Ballard, Dehn, *ibid.* **coll. vol. I,** 82; Kao, Ma, *J. Chem. Soc.* **1931,** 443.

Monoclinic needles from water. Bitter taste. mp 150°. Melt is deep red at higher temp. pKa (25°): 3.036. Slightly sol in cold water; freely sol in hot water, alc, ether.

Potassium salt, $C_{14}H_{11}O_3K$, crystals, very sol in water, alc.

Lead salt, $(C_{14}H_{11}O_3)_2Pb$, amorphous precipitate. Upon heating it becomes a red liquid.

Methyl ester, $C_{15}H_{14}O_3$, mp 74-75°. bp_{13} 187°.

1110. Benzilonium Bromide. *1,1-Diethyl-3-[(hydroxy-diphenylacetyl)oxy]pyrrolidinium bromide;* 1-ethyl-3-pyrrol-idinyl benzilate ethyl bromide; 3-benziloyloxy-1,1-diethyl-pyrrolidinium bromide; benzilic acid 1-ethyl-3-pyrrolidinyl ester ethyl bromide; CI-379; PU-239; Minelsin; Ortyn retard; Minelcin; Portyn; Ulcoban. $C_{22}H_{28}BrNO_3$; mol wt 434.37. C 60.83%, H 6.50%, Br 18.40%, N 3.22%, O 11.05%. Prepd from 1-ethyl-3-hydroxypyrrolidine and ethyl benzilate followed by reaction with ethyl bromide: Bowman *et al.,* **Brit. pat.** 821,436 (1956 to Parke, Davis); Ryan, Ainsworth, *J. Org. Chem.* **27,** 2901 (1962). Toxicity study: Reichertz, Schliva, *Arzneimittel-Forsch.* **12,** 414 (1962).

Crystals, mp 203-204°. LD_{50} orally in rats: 1.86 g/kg (Reichertz, Schliva).

THERAP CAT: Anticholinergic.

1111. Benzimidazole. Benziminazole; 1,3-benzodiazole; azindole; benzoglyoxaline; N,N'-methenyl-o-phenylene-diamine. $C_7H_6N_2$; mol wt 118.14. C 71.17%, H 5.12%, N 23.71%. Prepd by the reaction of o-phenylenediamine with formic acid: Wundt, *Ber.* **11,** 826 (1878); Wagner, Millett, *Org. Syn.* **coll. vol. II,** 65 (1943). Reviews: J. B. Wright, *Chem. Rev.* **48,** 397-541 (1951); K. Hofmann, *Imidazole and its Derivatives* (Interscience, New York, 1953) p 247 sqq.

Tabular crystals. Orthorhombic and monoclinic modifications. mp 170.5°. $bp_{760} > 360°$. Weak base. pKa = 5.48 at 25°. Absorption spectrum: Steck *et al., J. Am. Chem. Soc.* **70,** 3406 (1948). Dipole moment: 3.93 (dioxane). Sparingly sol in cold water, more sol in hot water. Freely sol in alcohol, sparingly sol in ether. Practically insol in benzene, petr ether. One gram dissolves in 2 g of boiling xylene. Sol in aq solns of acids and strong alkalies. High degree of chemical stability.

1112. 2-Benzimidazolethiol. *1,3-Dihydro-2H-benzimid-azole-2-thione;* 2-mercaptobenzimidazole. $C_7H_6N_2S$; mol wt 150.20. C 55.98%, H 4.03%, N 18.65%, S 21.35%. Prepn from o-phenylenediamine and potassium ethyl xanthate: Van Allan, Deacon, *Org. Syn.* **coll. vol. IV,** 569 (1963).

Glistening platelets from 95% ethanol, mp 303-304°. Slightly sol in water; sol in methanol, ethanol.

1113. Benziodarone. *2-Ethyl-3-benzofuranyl 4-hydr-oxy-3,5-diiodophenyl methanone;* 2-ethyl-3-(3',5'-diiodo-4'-hydroxybenzoyl)benzofuran; 2-ethyl-3-(3',5'-diiodo-4'-hydroxybenzoyl)oxaindene; 2-ethyl-3-benzofuryl 3',5'-diiodo-4'-hydroxyphenyl ketone; 2-ethyl-3-(3,5-diiodo-4-hydroxy-benzoyl)coumarone; 2-ethyl-3-(4-hydroxy-3,5-diiodobenzo-yl)benzofuran; 2329 Labaz; L-2329; Amplivix; Cardivix; Dilafurane; Dila-Vasal; Retrangor. $C_{17}H_{12}I_2O_3$; mol wt 518.09. C 39.41%, H 2.33%, I 48.99%, O 9.26%. Description: Charlier, *Acta Cardiol.,* Suppl. 7, 1-60 (1959). Synthesis: Beaudet, Henaux, Belg. pat. 553,621; Brit. pat. 836,272 (1957, 1960, both to Labaz); Buu-Hoi, Beaudet, U.S. pat. 3,012,042 (1961 to Soc. Belge Azote Prod. Chim. Marly).

Yellowish powder, mp 167°. Soly in water at 25° about
0.2%, at 45° about 1.0%. Sol in chloroform, acetone.

THERAP CAT: Vasodilator (coronary).

1114. Benznidazole. *2-Nitro-N-(phenylmethyl)-1H-im-*
idazole-1-acetamide; N-benzyl-2-nitroimidazole-1-acetamide;
Ro-7-1051; Radanil. $C_{12}H_{12}N_4O_3$; mol wt 260.25. C
55.38%, H 4.65%, N 21.53%, O 18.44%. Prepn: **Brit**. pat.
1,138,529 (1966 to Hoffmann-La Roche), *C.A.* **71**, 3383d
(1969). *In vitro* effect on *T. cruzi:* S. Yoneda *et al., Experi-*
entia **33**, 1201 (1977). Pharmacokinetics: J. Raaflaub, W.
H. Ziegler, *Arzneimittel-Forsch.* **29**, 1611 (1979). Antibacterial activity and electron affinity: A. V. Reynolds, *J.*
Pharm. Pharmacol. **31**, Suppl., 29P (1979). Mutagenicity
study: C. E. Voogd *et al., Mutat. Res.* **66**, 207 (1979). *In*
vitro toxicity study: G. E. Adams *et al., J. Nat. Cancer Inst.*
64, 555 (1980).

Crystals from ethanol, mp 188.5-190°. uv max (ethanol):
313 nm (ε 7,600). Soly in water at 37°: 40 mg/100 ml, J.
Raaflaub, W. H. Ziegler, *loc. cit.*

THERAP CAT: Antiprotozoal (Trypanosoma).

1115. Benzo Azurine G. *3,3'-[(3,3'-Dimethoxy[1,1'-bi-*
phenyl]-4,4'-diyl)bis(azo)]bis[4-hydroxy-1-naphthalenesul-
fonic acid] disodium salt; disodium *o*-dianisidinediazobis(1-
naphthol-4-sulfonate); C.I. Direct Blue 8; C.I. 24140. C_{34}-
$H_{24}N_4Na_2O_{10}S_2$; mol wt 758.70. C 53.83%, H 3.19%, N
7.38%, Na 6.06%, O 21.09%, S 8.45%. Prepared from diazotized *o*-dianisidine and sodium 4-hydroxy-1-naphthalenesul-
fonate: F. J. Welcher, *Organic Analytical Reagents* **vol. 4**
(D. Van Nostrand Co., New York, 1948) pp 337-338. *See*
also Colour Index **vol. 4** (3rd ed., 1971) p 4202.

Bluish-black powder. Sol in water, aq sodium hydroxide
(red soln), sulfuric acid (blue soln), Cellosolve; very slightly
sol in ethanol. Practically insol in other organic solvents.

USE: As a dye: *Colour Index* **vol. 2** (3rd ed., 1971) p 2223.
As analytical reagent in detection of Mg.

1116. Benzocaine. *4-Aminobenzoic acid ethyl ester;*
ethyl aminobenzoate; Aethoform; Americaine; Anesthesin;
Anesthone; Orthesin; Parathesin. $C_9H_{11}NO_2$; mol wt
165.19. C 65.44%, H 6.71%, N 8.48%, O 19.37%. Prepd by
the esterification of *p*-aminobenzoic acid: Salkowski, *Ber.*
28, 1921 (1895); Vorländer, Meyer, *Ann.* **320**, 135 (1902); by
the reduction of ethyl *p*-nitrobenzoate: Limpricht, *ibid.* **303**,
278 (1898); R. Adams, F. L. Cohen, *Org. Syn.* **coll. vol. I**,
240 (2nd ed., 1941). In industrial practice the reducing
agent is usually iron and water in the presence of a little
acid. Comprehensive description: S. L. Ali, *Anal. Profiles*
Drug Subs. **12**, 73-104 (1983).

Rhombohedra from ether, mp 88-90°. Stable in air. One
gram dissolves in about 2500 ml water, 5 ml alcohol, 2 ml
chloroform, in about 4 ml ether, and in 30 to 50 ml of
expressed almond oil or olive oil. Also sol in dil acids. pKa
2.5.

THERAP CAT: Anesthetic (local).

THERAP CAT (VET): Local (usually surface) anesthetic.

1117. Benzoctamine. *N-Methyl-9,10-ethanoanthracene-*
9(10H)-methanamine; 1-(methylaminomethyl)dibenzo[*b,e*]-
bicyclo[2.2.2]octadiene; 9-(methylaminomethyl)-9,10-dihy-
dro-9,10-ethanoanthracene. $C_{18}H_{19}N$; mol wt 249.36. C
86.70%, H 7.68%, N 5.62%. Prepn: **Belg**. pat. **610,863**
corresp to P. Schmidt *et al., U.S.* pat. **3,399,201** (1962, 1968
to Ciba); Wilhelm, Schmidt, *Helv. Chim. Acta* **52**, 1385
(1969). Pharmacology: Keberle *et al., Present Status Psy-*
chotropic Drugs (Excerpta Medica, New York, 1969) p 123;
Baltzer, Bein, *Arch. Int. Pharmacodyn Ther.* **201**, 25 (1973).
Toxicity: E. I. Goldenthal, *Toxicol. Appl. Pharmacol.* **18**,
185 (1971).

Hydrochloride, $C_{18}H_{19}N.HCl$, *Ba-30803, Tacitin.* Crystals, mp 320-322°. pKa: 7.6. LD$_{50}$ orally in rats: 700 ±170
mg/kg (Goldenthal).

THERAP CAT: Anxiolytic; muscle relaxant (skeletal).

1118. Benzodepa. *[Bis(1-aziridinyl)phosphinyl]car-*
bamic acid phenylmethyl ester; benzyl [bis(1-aziridinyl)phos-
phinyl]carbamate; benzyl bis(ethylenimido)phosphoroure-
than; benzcarbimine; AB-103; NSC-37096; Dualar. $C_{12}H_{16}$-
N_3O_3P; mol wt 281.25. C 51.25%, H 5.73%, N 14.94%, O
17.07%, P 11.01%. Prepn: Bardos *et al., Nature* **183**, 399
(1959); Papanastassiou, Bardos, *J. Med. Pharm. Chem.* **5**,
1000 (1962).

Crystals from benzene + cyclohexane, mp 134-135°.
Practically insol in water. Soluble in fat solvents, peanut
oil, other oils.

THERAP CAT: Antineoplastic.

1119. Benzofuran. Coumarone; cumarone. C_8H_6O;
mol wt 118.14. C 81.34%, H 5.12%, O 13.54%. Constituent
of coal tar. Isoln from coal tar oils: Kraemer, Spilker, *Ber.*
23, 78 (1890); **33**, 2261 (1900); Breston, Gauger, *Am. Gas*
Assoc. Proc. **28**, 492 (1946). Synthesis by heating phenoxy-
acetaldehyde with zinc chloride and acetic acid: Stoermer,
Ber. **30**, 1703 (1897); *Ann.* **312**, 261 (1900). The benzofuran
used in the manuf of coumarone-indene resins is derived
from the crude heavy solvent naphtha fraction of coal tar
light oil, obtained as a byproduct in the coking of bituminous coal.

Oil. Aromatic odor. Not solid at $-18°$. bp_{760} 173-175°; bp_{15} 62-63°. Volatile with steam. $d_4^{22.7}$ 1.0913. $n_D^{16.3}$ 1.56897; $n_D^{22.7}$ 1.565. Insol in water, aq alkaline solns; miscible with benzene, petr ether, abs alcohol, ether. Slowly polymerizes on standing. For additional physical data see Breston, Gauger, loc. cit.

USE: Manuf of coumarone-indene resins.

1120. Benzoguanamine. 6-Phenyl-1,3,5-triazine-2,4-diamine; 2,4-diamino-6-phenyl-s-triazine; 4,6-diamino-2-phenyl-s-triazine. $C_9H_9N_5$; mol wt 187.20. C 57.74%, H 4.85%, N 37.41%. Prepd from benzonitrile and dicyandiamide in the presence of Na and liquid ammonia: Jones, U.S. pat. 2,735,850 (1956 to British Oxygen Co.); from benzonitrile and dicyandiamide in the presence of KOH and methyl Cellosolve: Simons, Saxton, Org. Syn. 33, 13 (1953).

Crystals. d_4^{25} 1.40. mp 227-228°. uv max (ethanol): 249 nm (ϵ 25,000). Sol in alcohol, ether, dil HCl; partially sol in dimethylformamide. Practically insol in acetone, chloroform, ethyl acetate. Soly in water at 22° 0.06%; at 100° 0.6%.

USE: In the manuf of thermosetting resins, pesticides, pharmaceuticals and dyestuffs.

1121. Benzohydrol. α-Phenylbenzenemethanol; benzhydrol; diphenylcarbinol. $C_{13}H_{12}O$; mol wt 184.24. C 84.75%, H 6.57%, O 8.68%. $(C_6H_5)_2CHOH$. Prepd by reducing benzophenone with zinc dust in strongly alkaline soln: Wiselogle, Sonneborn, Org. Syn. coll. vol. I (2nd ed, 1941) p 90. By reducing benzophenone with magnesium in methanol: Zechmeister, Rom, Ann. 468, 123 (1929).

Needles from ligroin, mp 69°. bp_{748} 298°; bp_{20} 180°; bp_{13} 176°. Absorption spectrum: Orndorff et al., J. Am. Chem. Soc. 49, 1542 (1927). One gram dissolves in 2 liters of water at 20°. Freely sol in alcohol, ether, chloroform and carbon bisulfide; almost insol in cold ligroin.

Diphenylmethyl ether, $(C_6H_5)_2CHOCH(C_6H_5)_2$, monoclinic crystals from benzene, mp 110°; bp_{15} 267°.

USE: In organic syntheses.

1122. Benzoic Acid. Benzenecarboxylic acid; phenylformic acid; dracylic acid. $C_7H_6O_2$; mol wt 122.12. C 68.85%, H 4.95%, O 26.20%. Occurs in nature in free and combined forms. Gum benzoin may contain as much as 20%. Most berries contain appreciable amounts (around 0.05%). Excreted mainly as hippuric acid by almost all vertebrates, except fowl. Mfg processes include the air oxidation of toluene, the hydrolysis of benzotrichloride, and the decarboxylation of phthalic anhydride: Faith, Keyes & Clark's Industrial Chemicals, F. A. Lowenheim, M. K. Moran, Eds. (Wiley-Interscience, New York, 4th ed., 1975) pp 138-144. Lab prepn from benzyl chloride: A. I. Vogel, Practical Organic Chemistry (Longmans, London, 3rd ed, 1959) p 755; from benzaldehyde: Gattermann-Wieland, Praxis des organischen Chemikers (de Gruyter, Berlin, 40th ed, 1961) p 193. Prepn of ultra-pure benzoic acid for use as titrimetric and calorimetric standard: Schwab, Wicher, J. Res. Nat. Bur. Standards 25, 747 (1940). Review: A. E. Williams in Kirk-Othmer Encyclopedia of Chemical Technology vol. 3 (Wiley-Interscience, New York, 3rd ed., 1978) pp 778-792.

Monoclinic tablets, plates, leaflets. d 1.321 (also reported as 1.266). mp 122.4°. Begins to sublime at $\sim$100°. bp_{760}

249.2°; bp_{400} 227°; bp_{200} 205.8°; bp_{100} 186.2°; bp_{60} 172.8°; bp_{40} 162.6°; bp_{20} 146.7°; bp_{10} 132.1°. Volatile with steam. Flash pt 121-131°. K at 25°: 6.40×10^{-5}; pH of satd soln at 25°: 2.8. Soly in water (g/l) at 0° = 1.7; at 10° = 2.1; at 20° = 2.9; at 25° = 3.4; at 30° = 4.2; at 40° = 6.0; at 50° = 9.5; at 60° = 12.0; at 70° = 17.7; at 80° = 27.5; at 90° = 45.5; at 95° = 68.0. Mixtures of excess benzoic acid and water form two liquid phases beginning at 89.7°. The two liquid phases unite at the critical soln temp of 117.2°. Composition of critical mixture: 32.34% benzoic acid, 67.66% water: see Ward, Cooper, J. Phys. Chem. 34, 1484 (1930). One gram dissolves in 2.3 ml cold alc, 1.5 ml boiling alc, 4.5 ml chloroform, 3 ml ether, 3 ml acetone, 30 ml carbon tetrachloride, 10 ml benzene, 30 ml carbon disulfide, 23 ml oil of turpentine; also sol in volatile and fixed oils, slightly in petr ether. The soly in water is increased by alkaline substances, such as borax or trisodium phosphate, see also Sodium Benzoate.

Barium salt dihydrate, $C_{14}H_{10}BaO_4.2H_2O$, barium benzoate. Nacreous leaflets. Poisonous! Soluble in about 20 parts water; slightly sol in alc.

Calcium salt trihydrate, $C_{14}H_{10}CaO_4.3H_2O$, calcium benzoate. Orthorhombic crystals or powder. d 1.44. Soluble in 25 parts water; very sol in boiling water.

Cerium salt trihydrate, $C_{21}H_{15}CeO_6.3H_2O$, cerous benzoate. White to reddish-white powder. Sol in hot water or hot alc.

Copper salt dihydrate, $C_{14}H_{10}CuO_4.2H_2O$, cupric benzoate. Light blue, cryst powder. Slightly soluble in cold water, more in hot water; sol in alc or in dil acids with separation of benzoic acid.

Lead salt dihydrate, $C_{14}H_{10}O_4Pb.2H_2O$, lead benzoate. Cryst powder. Poisonous! Slightly sol in water.

Manganese salt tetrahydrate, $C_{14}H_{10}MnO_4.4H_2O$, manganese benzoate. Pale-red powder. Sol in water, alc. Also occurs with $3H_2O$.

Nickel salt trihydrate, $C_{14}H_{10}NiO_4.3H_2O$, nickel benzoate. Light-green odorless powder. Slightly sol in water; sol in ammonia; dec by acids.

Potassium salt trihydrate, $C_7H_5KO_2.3H_2O$, potassium benzoate. Crystalline powder. Sol in water, alc.

Silver salt, $C_7H_5AgO_2$, silver benzoate. Light-sensitive powder. Sol in 385 parts cold water, more sol in hot water; very slightly sol in alc.

Uranium salt, $C_{14}H_{10}O_6U$, uranium benzoate, uranyl benzoate. Yellow powder. Slightly sol in water, alc.

Caution: Mild irritant to skin, eyes, mucous membranes. USE: Preserving foods, fats, fruit juices, alkaloidal solns, etc; manuf benzoates and benzoyl compds, dyes; as a mordant in calico printing; for curing tobacco. As standard in volumetric and calorimetric analysis. Pharmaceutic aid (antifungal).

THERAP CAT (VET): Has been used with salicylic acid as a topical antifungal.

1123. Benzoic Anhydride. Benzoic acid anhydride. $C_{14}H_{10}O_3$; mol wt 226.23. C 74.33%, H 4.46%, O 21.22%. $(C_6H_5CO)_2O$. Prepd by heating benzoic acid with acetic anhydride in the presence of a trace of phosphoric acid: Clarke, Rahrs, Org. Syn. 3, 2 (1923); coll. vol. I (Wiley, New York, 2nd ed., 1941) p 91.

Orthorhombic bipyramidal prisms from benzene + petr ether. d_4^{15} 1.1989. mp 42°. bp_{760} 360°; bp_{200} 299.1°; bp_{60} 252.7°; bp_{40} 239.8°; bp_{20} 218°; bp_{10} 198°; bp_5 180°; $bp_{1.0}$ 143.8°. n_D^{15} 1.57665. Almost insol in water (0.01 g/l). Sol in alcohol, chloroform, acetone, ethyl acetate, benzene, toluene, xylene, ether, glacial acetic acid, acetic anhydride; moderately sol in petr ether; stable in water and in cold alkaline solns.

USE: In organic syntheses, as benzoylating agent in the manufacture of pharmaceuticals, dyes, and intermediates.

1124. Benzoin. 2-Hydroxy-1,2-diphenylethanone; benzoylphenylcarbinol; α-hydroxy-α-phenylacetophenone; bitter-almond-oil camphor. $C_{14}H_{12}O_2$; mol wt 212.25. C 79.23%, H 5.70%, O 15.08%. Prepd by treating an alcoholic soln of benzaldehyde with an alkali cyanide: Adams, Marvel, Org. Syn. vol. 1, p 33 (1921); coll. vol. I, 88; Arnold, Fuson, J. Am. Chem. Soc. 58, 1295 (1936); L. F. Fieser,

Organic Experiments (D. C. Heath & Co., Boston, 1964) pp 211-214.

dl-Form, six-sided monoclinic prisms from alcohol, mp 137°. bp_{768} 344°. bp_{12} 194°. Reduces Fehling's soln. uv max (ethanol): 247 nm (ε 14500); infrared in chloroform: 2.88; 5.93; 6.21; 6.28, 6.85 μ. Soluble in 3335 parts water, more in hot water, in 5 parts pyridine; sol in acetone, in boiling alc; slightly in ether.

Methyl ether, $C_{15}H_{14}O_2$, needles, mp 49°.
Ethyl ether, $C_{16}H_{16}O_2$, needles, mp 62°.
l-Form, needles, mp 132°. $[α]_D^{12}$ −118° (c = 1.2 in acetone).
d-Form, needles, mp 132°. $[α]_D^{12}$ +120.5° (c = 1.2 in acetone).

Note: Not to be confused with Gum Benzoin.

USE: In organic syntheses.

1125. Benzoin Oxime. *2-Hydroxy-1,2-diphenyletha-none oxime;* 2-hydroxy-2-phenylacetophenone oxime. $C_{14}H_{13}NO_2$; mol wt 227.26. C 73.99%, H 5.77%, N 6.16%, O 14.08%. Two isomers occur: α or *anti* and β or *syn*. Both have been prepd from benzoin and hydroxylamine hydrochloride. Prepn of α-form: Werner, Detscheff, *Ber.* **38**, 69 (1905); F. J. Welcher, *Organic Analytical Reagents* vol. III (Van Nostrand, New York, 1947) pp 239-251. Prepn of β-form: Werner, Detscheff, *loc. cit.* Configuration of α- and β-forms: Meisenheimer, Meis, *Ber.* **57**, 289 (1924).

α-form   β-form

α-Form, *Cupron.* Prisms from benzene, mp 151-152°. Darkens on exposure to light. Slightly sol in water; sol in alc, aq ammonium hydroxide soln. *Protect from light.*

β-Form, prismatic crystals from ether. Ether of crystn is lost on standing in air. Ether-free compd, mp 99°.

USE: The α-form is used in the detection and determination of Cu, Mo, and W: Welcher, *loc. cit.*

1126. 6,7-Benzomorphan. *1,2,3,4,5,6-Hexahydro-2,6-methano-3-benzazocine;* 6,7-benzmorphan. $C_{12}H_{15}N$; mol wt 173.26. C 83.19%, H 8.73%, N 8.08%. Parent compound of a series of potent analgetics. Synthesis: Kanematsu *et al.*, *J. Am. Chem. Soc.* **90**, 1064 (1968); Kanematsu *et al.*, *J. Med. Chem.* **12**, 405 (1969). Review of derivative syntheses and pharmacology: May, Sargent, "Morphine and its Modifications", in *Analgetics*, G. de Stevens, Ed. (Academic Press, New York, 1965) pp 145-171.

Hydrochloride, $C_{12}H_{15}N \cdot HCl$, needles from methanol-acetone, mp 261-262°.

1127. Benzonatate. *4-(Butylamino)benzoic acid 3,6,9,-12,15,18,21,24,27-nonaoxaoctacos-1-yl ester;* nonaethylene-glycol monomethyl ether *p-n*-butylaminobenzoate; *p*-butyl-aminobenzoic acid ω-*O*-methylnonaethyleneglycol ester; benzononatine; Exangit; Tessalon; Ventussin. $C_{30}H_{53}NO_{11}$;

mol wt 603.75. C 59.68%, H 8.85%, N 2.32%, O 29.15%. Prepn: Matter, U.S. pat. **2,714,608** (1955 to Ciba).

Colorless to faintly yellow oil. Soluble in most organic solvents except aliphatic hydrocarbons.

THERAP CAT: Antitussive.

1128. Benzonitrile. Phenyl cyanide; cyanobenzene. C_7H_5N; mol wt 103.12. C 81.53%, H 4.89%, N 13.58%. C_6H_5CN. Prepd by heating Na benzenesulfonate with NaCN or by adding benzenediazonium chloride soln to a hot aq NaCN soln contg $CuSO_4$ and distilling. Lab prepn: A. I. Vogel, *Practical Organic Chemistry* (Longmans, London, 3rd ed, 1959) p 608.

Liquid, odor of volatile oil of almond. bp_{760} 190.7°, bp_{100} 123.5°, bp_{10} 69.2°, bp_1 28.2°. mp −12.75°. d_4^{15} 1.010. n_D^{20} 1.5289. Fire pt 167°F. Slightly sol in cold water; sol to the extent of 1% in water at 100°; miscible with common organic solvents.

USE: Solvent.

1129. Benzophenone. *Diphenylmethanone;* diphenyl ketone; benzoylbenzene. $C_{13}H_{10}O$; mol wt 182.22. C 85.69%, H 5.53%, O 8.78%. $C_6H_5COC_6H_5$. Prepd by the Friedel-Crafts ketone synthesis from benzene and benzoyl chloride in the presence of $AlCl_3$: Marvel, Sperry, *Org. Syn.* **coll. vol. I** (Wiley, New York, 2nd ed., 1941) p 95. By decarboxylation of *o*-benzoylbenzoic acid in the presence of copper catalyst: L. F. Fieser, *Organic Experiments* (D. C. Heath & Co., Boston, 1964) pp 201-203.

Stable form (there are two other labile forms), ortho-rhombic bisphenoidal prisms from alcohol or ether. Geranium-like odor. d_4^{18} 1.1108. d_4^{50} 1.0869. $n_D^{45.2}$ 1.5975. Absorption spectrum: Purvis, McCleland, *J. Chem. Soc.* **101**, 1516 (1912). mp 48.5°. bp_{760} 305.4°; bp_{400} 276.8°; bp_{200} 249.8°; bp_{100} 224.4°; bp_{60} 208.2°; bp_{40} 195.7°; bp_{20} 175.8°; bp_{10} 157.6°; bp_5 141.7°; $bp_{1.0}$ 108.2°. Insoluble in water. One gram dissolves in 7.5 ml alcohol, 6 ml ether; sol in chloroform.

Oxime, $C_{13}H_{11}NO$, for prepn see *Org. Syn.* **10**, 10. Crystals from ligroin, mp 143-144°. Freely sol in ether, acetone.

USE: Fixative for heavy perfumes, such as geranium, new-mown hay, especially when used in soaps. In the manuf of antihistamines, hypnotics, insecticides.

1130. Benzophenone-6. *Bis(2-hydroxy-4-methoxyphenyl)methanone; 2,2'-dihydroxy-4,4'-dimethoxybenzophenone;* Uvinul D49. $C_{15}H_{14}O_5$; mol wt 274.27. C 65.69%, H 5.14%, O 29.17%. Preparation: Grover *et al.*, *J. Chem. Soc.* **1955**, 3982; Hardy, Forster, U.S. pat. **2,773,903** (1956 to Am. Cyanamid); Dayan, Roberts; Hahn, Stanley, U.S. pats. **2,853,522** and **2,853,523** (both 1958 to Gen. Aniline); Hosler, Storfer, U.S. pat. **2,928,878** (1960 to Am. Cyanamid).

Crystals, mp 139-140°. uv max: 284, 340 nm (log ε 4.12, 4.12).

USE: As ultraviolet light absorber, esp in paints, plastics.

1131. Benzopinacol. *1,1,2,2-Tetraphenyl-1,2-ethane-diol;* benzpinacone; tetraphenylethylene glycol. $C_{26}H_{22}O_2$; mol wt 366.46. C 85.22%, H 6.05%, O 8.73%. $(C_6H_5)_2C(OH)C(OH)(C_6H_5)_2$. Prepd by photochemical reduction of benzophenone: Bachmann, *Org. Syn.* **coll. vol. II**, 71 (1943); L. F. Fieser, *Organic Experiments* (D. C. Heath & Co., Boston, 1964) p 203. By the action of phenylmagnesium bromide on benzil: Acree, *Ber.* **37**, 2761 (1904).

Monoclinic prisms, may contain 1 mol C_6H_6 when crystallized from benzene. mp 197° (open capillary, rapid heating); mp 222° (copper block). On heating to mp it dec to benzophenone and benzohydrol. Soluble in 11.5 parts boiling glacial acetic acid; in 26 parts boiling benzene; in 39 parts boiling 95% alcohol. Freely sol in ether, carbon bisulfide, chloroform.

1132. Benzopurpurine 4B. *3,3'-[(3,3'-Dimethyl[1,1'-biphenyl]-4,4'-diyl)bis(azo)]bis[4-amino-1-naphthalenesulfonic acid] disodium salt;* disodium *o*-tolidinediazobis(1-naphthylamine-4-sulfonate); C.I. Direct Red 2; C.I. 23500; azamin 4B; eclipse red; fast scarlet; Paper Red 4B; Sultan Red 4B; Cotton Red 4B. $C_{34}H_{26}N_6Na_2O_6S_2$; mol wt 724.73. C 56.35%, H 3.62%, N 11.60%, Na 6.34%, O 13.25%, S 8.85%. Prepn from diazotized *o*-tolidine and sodium 4-amino-1-naphthalenesulfonate: **Ger. pat.** 35,615 (1885 to A. G. für Anilinfabrikation Berlin), *Frdl.* **1**, 473; F. J. Welcher, *Organic Analytical Reagents* **vol. 4** (Van Nostrand Co., New York, 1948) pp 338-339; *Colour Index* vol. 4 (3rd ed., 1971) p 4189. *See also:* H. J. Conn's *Biological Stains*, R. D. Lillie, Ed. (Williams & Wilkins, Baltimore, 9th ed., 1977) p 154.

Brown powder. Soluble in water, sodium hydroxide, sulfuric acid, ethanol, acetone, Cellosolve. Practically insol in other organic solvents.

USE: For dyeing primarily cotton and viscose rayon: *Colour Index* **vol. 2** (3rd ed., 1971) p 2101. As an analytical reagent in detection of Al, Mg, Hg, Ag, U; as a biological stain; as pH indicator, violet 1.2 to red 4.0.

1133. 1,2-Benzopyran. *2H-1-Benzopyran;* α-5:6-benzopyran; 1,2-chromene; 3-chromene. C_9H_8O; mol wt 132.16. C 81.79%, H 6.10%, O 12.11%. Prepn from *cis-o*-hydroxycinnamyl alcohol: Chatterjea, *J. Indian Chem. Soc.* **36**, 76 (1959), *C.A.* **54**, 519b (1960); from phenyl propargyl ether in diethylphenylamine: Iwai, Ide, *Chem. Pharm. Bull.* **11**, 1042 (1963); Iwai, Iwade, **Japan.** pat. **22,587**('63) (to Sankyo), *C.A.* **60**, 2901e (1964). Earlier prepns: Maitre, *Ann. Chim. (Paris)* **9**, 431 (1954); Normant, Maitre, *Compt. Rend.* **234**, 1787 (1952).

Liquid; bp_{102} 132°, bp_{15} 92-92.5°, bp_{13} 91°, bp_9 77°. d^{16} 1.0993. n_D^{24} 1.5869; n_D^{16} 1.5923 (Maitre, *loc. cit.*); also reported as 1.5837 (**Japan.** pat. *see above*).

1134. Benzo[a]pyrene. 3,4-Benzpyrene. $C_{20}H_{12}$; mol wt 252.32. C 95.21%, H 4.79%. Formerly called *1,2-benzpyrene.* Occurs in coal tar. Isoln by fractionation: Cook *et al., J. Chem. Soc.* **1933**, 395; by adsorption and fluorimetric determn: Hieger, *Am. J. Cancer* **29**, 705 (1937); Winterstein *et al., Z. Physiol. Chem.* **230**, 158, 169 (1934); *Naturwiss.* **22**, 237 (1934). Synthesis from pyrene and succinic anhydride: Cook, Hewett, *J. Chem. Soc.* **1933**, 398; Fieser, Fieser, *J. Am. Chem. Soc.* **57**, 782 (1935); Winterstein *et al., Ber.* **68**, 1079 (1935); Fieser *et al., J. Am. Chem. Soc.* **57**, 1509 (1935). Absorption spectrum: Mayneord, Roe, *Proc. Roy. Soc. London* **A152**, 299 (1935). *Review:* Clar, *Polycyclic Hydrocarbons* (Academic Press, New York, 1964), 2 vols. Review of carcinogenicity studies: *IARC Monographs* **3**, 91-136 (1973). Inhibition of the mutagenicity of the ultimate carcinogenic metabolite of benzo[a]pyrene by ellagic acid: A. W. Wood *et al., Proc. Nat. Acad. Sci. USA* **79**, 5513 (1982). Study of the reaction between this metabolite (*benzo[a]pyrene-7,8-diol 9,10-epoxide*) and ellagic acid: J. M. Sayer *et al., J. Am. Chem. Soc.* **104**, 5562 (1982). *Review:* D. H. Phillips, *Nature* **303**, 468-472 (1983).

Yellowish plates, needles from benzene + methanol, mp 179-179.3°. Crystals may be monoclinic or orthorhombic. bp_{10} 310-312°. Violet fluorescence of dil benzene solns: Hieger, *loc. cit.* Soluble in benzene, toluene, xylene; sparingly sol in alc, methanol. Practically insol in water.

Note: This substance may reasonably be anticipated to be a carcinogen: *Seventh Annual Report on Carcinogens* (PB95-109781, 1994) p 328.

1135. Benzo[e]pyrene. 1,2-Benzpyrene; 4,5-benzpyrene. $C_{20}H_{12}$; mol wt 252.32. C 95.21%, H 4.79%. Constituent of coal tar. Isoln: Cook *et al., J. Chem. Soc.* **1933**, 396. Synthesis: Cook, Hewett, *ibid.* 398; Clar, *Ber.* **76**, 609 (1943); Buchta, Kröger, *Ann.* **705**, 190 (1967). NMR spectra: Cobb, Memory, *J. Chem. Phys.* **47**, 2020 (1967). *Review:* Clar, *Polycyclic Hydrocarbons* **vol. 2**, (Academic Press, New York, 1964) p 127.

Prisms or plates from benzene, mp 178-179°.

1136. Benzo[f]quinoline. β-Naphthoquinoline; 5,6-benzoquinoline; naphthopyridine. $C_{13}H_9N$; mol wt 179.22. C 87.12%, H 5.06%, N 7.82%. Prepn from from β-naphthylamine by the Skraup reaction: Knueppel, *Ber.* **29**, 703 (1896); Clem, Hamilton, *J. Am. Chem. Soc.* **62**, 2349 (1940); Uhle, Jacobs, *J. Org. Chem.* **10**, 76 (1945).

Crystals from alcohol + water, mp 93°. bp_{721} 349-350°. uv max (ethanol): 347, 331, 316, 266, nm (log ϵ 3.54, 3.41, 3.18, 4.06). Practically insol in water. Sol in dil acids; very sol in alcohol, ether, benzene.

USE: As a reagent for the determination of cadmium which is pptd as $(C_{13}H_9N)_2H_2(CdI_4)$ from dil nitric or sulfuric acid soln in the presence of potassium iodide.

1137. Benzoquinonium Chloride. *N,N'-[(3,6-Dioxo-1,4-cyclohexadiene-1,4-diyl)bis(imino-3,1-propanediyl)]bis[N,N-diethylbenzenemethanaminium]dichloride; [2,5-p-benzoquinonylenebis(iminotrimethylene)]bis[benzyldiethylammonium chloride];* 2,5-bis(3-diethylaminopropylamino)benzoquinone bis(benzyl chloride); Win-2747; Mytolon; Amilyt. $C_{34}H_{50}Cl_2N_4O_2$; mol wt 617.70. C 66.11%, H 8.16%, Cl 11.48%, N 9.07%, O 5.18%. Prepn: C. J. Cavallito *et al., J. Am. Chem. Soc.* **72**, 2661 (1950); C. J. Cavallito, **U.S.** pat. **2,844,585** (1958 to Sterling).

Red crystals from ethanol, mp 191-195°. Freely sol in water.

THERAP CAT: Muscle relaxant (skeletal).

1138. Benzoresorcinol. *(2,4-Dihydroxyphenyl)phenyl-methanone; 2,4-dihydroxybenzophenone;* 4-benzoylresorcinol; resbenzophenone; benzophenone-1; Uvinul 400. $C_{13}H_{10}O_3$; mol wt 214.22. C 72.89%, H 4.71%, O 22.41%. Prepd from benzoyl chloride and resorcinol: Dobner, *Ann.* **210**, 246 (1881); by Fries rearrangement: Amin, Shah, *J. Indian Chem. Soc.* **29**, 351 (1952).

Needles from hot water, mp 144-145°. Practically insol in cold water. Easily sol in alcohol, ether, glacial acetic acid; scarcely sol in cold benzene.

USE: Ultraviolet light absorber, esp in paints and plastics.

1139. Benzothiazole. C_7H_5NS; mol wt 135.19. C 62.19%, H 3.73%, N 10.36%, S 23.72%. Prepd from *N,N*-dimethylaniline and sulfur: Knowles, Watt, *J. Org. Chem.* **7**, 56 (1942). Chemistry of benzothiazoles, their use as carbonyl equivalents and in carbon-carbon bond formation: E. J. Corey, D. L. Boger, *Tetrahedron Letters* **1978**, 5, 9, 13. Toxicity study: E. F. Domino *et al., J. Pharmacol. Exp. Ther.* **105**, 486 (1952).

Liquid. Odor similar to that of quinoline. Volatile with steam. d_4^{20} 1.246. bp_{765} 227-228°; bp_{34} 131°. n_D^{20} 1.6379. Slightly sol in water. Freely sol in alcohol, carbon disulfide. LD_{50} i.v. in mice: 95±3 mg/kg (Domino).

USE: In organic synthesis.

1140. 1*H*-Benzotriazole. 1,2,3-Benzotriazole; benztriazole; azimidobenzene; benzisotriazole. $C_6H_5N_3$; mol wt 119.13. C 60.50%, H 4.23%, N 35.27%. Prepd by the action of nitrous acid on *o*-phenylenediamine: Ladenburg, *Ber.* **9**, 219 (1876); Damschroder, Peterson, *Org. Syn. coll.* **vol. III**, 106 (1955).

Needles from benzene, mp 98.5°. bp_{15} 204°; $bp_{2.0}$ 159°. *Caution:* May explode during vacuum distillation, *see Chem. & Eng. News* **34**, 2450 (1956). Sparingly sol in water. Sol in alc, benzene, toluene, chloroform, DMF.

1141. Benzotrichloride. *(Trichloromethyl)benzene; α,-α,α-trichlorotoluene;* phenylchloroform; ω,ω,ω-trichlorotoluene; benzenyl trichloride; toluene trichloride. $C_7H_5Cl_3$; mol wt 195.47. C 43.01%, H 2.58%, Cl 54.41%. Produced by chlorination of boiling toluene in the presence of light and of 2% phosphorus trichloride: Swarts, *Bull. Soc. Chim. Belg.* **31**, 375 (1922); Conklin, U.S. pats. **1,828,858**; **1,828,-859** (both 1931 to Solvay Process Co.). Commercial grades may contain hydrochloric acid, benzylidene chloride, and

benzyl chloride. Purification procedures: Holleman, deMooy, *Rec. Trav. Chim.* **33**, 25, 33 (1914); Britton, U.S. pat. **1,804,458** (1931 to Dow); *Chem. Zentr.* **1931**, II, 497. Toxicity data: H. F. Smyth *et al., Arch. Ind. Hyg. Occup. Med.* **4**, 119 (1951). *Review:* H. C. Lin in Kirk-Othmer *Encyclopedia of Chemical Technology* vol. 6 (John Wiley & Sons, New York, 4th ed., 1993) pp 113-126.

Colorless, oily liquid with pungent odor. Fumes in air. mp −5.0°. d_4^{20} 1.3756. bp_{760} 220.8°. bp_{60} 129°; bp_{25} 105°; bp_{10} 89°. n_D^{20} 1.55789. Flash pt, open cup: 127°. Autoignition temp: 211°. Unstable. Hydrolyzes in the presence of moisture, forming benzoic and hydrochloric acids. Insol in water. Sol in alcohol, benzene, ether, many other organic solvents. LD_{50} orally in rats: 6.0 g/kg (Smyth).

Caution: Vapors may be stongly irritating and lacrimatory (Lin). This substance may reasonably be anticipated to be a carcinogen: *Seventh Annual Report on Carcinogens* (PB95-109781, 1994) p 94.

USE: In dye chemistry. In organic syntheses (source of benzenyl group).

1142. Benzotrifluoride. *(Trifluoromethyl)benzene; α,α,α-trifluorotoluene;* phenylfluoroform. $C_7H_5F_3$; mol wt 146.11. C 57.54%, H 3.45%, F 39.01%. Prepd by the action of hydrogen fluoride on benzotrichloride: Simons, Lewis, *J. Am. Chem. Soc.* **60**, 492 (1938); Simons, *Ind. Eng. Chem.* **32**, 178 (1940); by the action of antimony trifluoride on benzotrichloride: Swarts, *Bull. Sci. Acad. Roy. Belg.,* [4] **35**, 375 (1898); [5] **6**, 389 (1920); [5] **13**, 175 (1927); Holt *et al.,* U.S. pat. **2,058,453** (1937 to Kinetic Chemicals).

Liquid. mp −29.05°. d^{20} 1.1886. bp 103.46°. $n_D^{13.3}$ 1.41486. Soluble in many organic solvents.

USE: In dye chemistry; in the manuf of substituted benzotrifluorides contg an ethylenic group, used in high polymer chemistry; in dielectric fluids, such as transformer oils.

1143. Benzoxiquine. *8-Quinolinol benzoate (ester);* 8-benzoyloxyquinoline; 8-hydroxyquinoline benzoate (ester); benzoxyline; Dioxyline. $C_{16}H_{11}NO_2$; mol wt 249.27. C 77.10%, H 4.45%, N 5.62%, O 12.84%. Prepn from 8-quinolinol and benzoyl chloride: Bedall, Fischer, *Ber.* **14**, 1366 (1881).

Crystals from abs alc, mp 118-120°. Practically insol in water. Sol in alcohol, ether.

THERAP CAT: Disinfectant.

1144. Benzoxonium Chloride. *N-Dodecyl-N,N-bis(2-hy-droxyethyl)benzenemethanaminium chloride; benzyldodecyl-bis(2-hydroxyethyl)ammonium chloride;* dodecyldi(β-hydr-oxyethyl)benzylammonium chloride; D-301; ZY-15021; Absonal V; Bialcol; Bradophen; Orofar. $C_{23}H_{42}ClNO_2$; mol wt 400.04. C 69.06%, H 10.58%, Cl 8.86%, N 3.50%, O 8.00%. Quaternary ammonium salt with antimicrobial activity. Prepn: **Swiss** pat. **306,648** (1955 to Ciba), *C.A.* **51**, 2023g (1957). Antimicrobial activity and toxicology: H. Goeth *et al., Arzneimittel-Forsch.* **9**, 622 (1959). Potentiometric determn: S. Pinzauti, E. La Porta, *Analyst* **102**, 938 (1977). Clinical evaluation in dental plaque control: U. P. Saxer *et al., J. Clin. Periodontol.* **9**, 162 (1982). *In vitro* antibacterial spectrum: M. Cortat, P. Fels, *Arzneimittel-Forsch.* **37**, 463

(1987). Clinical evaluation in pharyngeal infections: M. A. Weibel *et al.*, *ibid.* 467.

Colorless powder from ethyl ether, mp 107-109°. Sol in water, alcohol, benzene, toluene, chlorobenzene. LD$_{50}$ orally in rats: 750 mg/kg (Goeth).

THERAP CAT: Antiseptic.
THERAP CAT (VET): Antiseptic.

1145. Benzoyl Chloride. Benzenecarbonyl chloride. C_7H_5ClO; mol wt 140.57. C 59.81%, H 3.59%, Cl 25.22%, O 11.38%. Prepd by partial hydrolysis of benzotrichloride: Davies, Dick, *J. Chem. Soc.* **1932**, 2808; by chlorination of benzaldehyde: Wöhler, *Ann.* **3**, 262 (1832); from benzoic acid and PCl$_5$ or from benzoic acid and phosgene: Uwarow, Stepanow, *U.S.S.R.* pat. **56,693** (1936). Lab prepn from benzoic acid and thionyl chloride: A. I. Vogel, *Practical Organic Chemistry* (Longmans, London, 3rd ed., 1959) p 792; Gattermann-Wieland, *Praxis des organischen Chemikers,* (de Gruyter, Berlin, 40th ed., 1961) p 112.

Liquid. Penetrating odor. d$_4^{25}$ 1.2070. mp −1.0°. bp$_{760}$ 197.2°; bp$_{35}$ 100°; bp$_{15}$ 82.3°; bp$_9$ 71°; bp$_3$ 49°. n$_D^{20}$ 1.55369. Dipole moment 3.28. Flash pt 88° (190.4°F). Parachor 289.8. Dec by water and alc. Miscible with ether, benzene, carbon disulfide, oils.

Caution: Lacrimator. Irritating to skin, eyes, mucous membranes.

USE: For acylation, i.e., introduction of the benzoyl group into alcohols, phenols, and amines (Schotten-Baumann reaction); in the manuf of benzoyl peroxide and of dye intermediates. In organic analysis for making benzoyl derivatives for identification purposes.

1146. Benzoylecgonine. *3-(Benzoyloxy)-8-methyl-8-azabicyclo[3.2.1]octane-2-carboxylic acid; 3β-hydroxy-1αH,-5αH-tropane-2β-carboxylic acid benzoate;* ecgonine benzoate. $C_{16}H_{19}NO_4$; mol wt 289.33. C 66.42%, H 6.62%, N 4.84%, O 22.12%. Cocaine metabolite. Isoln from coca leaves: De Jong, *Rec. Trav. Chim.* **42**, 980 (1923). Prepn from *l*-ecgonine and benzoic anhydride: *idem, ibid.* **66**, 544 (1947).

Tetrahydrate, orthorhombic prisms or needles from water, mp 86-92° (dec 195°, dry). [α]$_D^{15}$ −45° (c = 3 in abs alc). Soluble in alc, hot H$_2$O.

1147. Benzoyl Isothiocyanate. Benzoylthiocarbimide. C_8H_5NOS; mol wt 163.20. C 58.88%, H 3.09%, O 9.80%, S 19.65%. C_6H_5CONCS. Prepd by refluxing a mixture of KSCN and benzoyl chloride in benzene at 110-120°: Ambelang, Johnson, *J. Am. Chem. Soc.* **61**, 632 (1939). From NH$_3$SCN, acetone and benzoyl chloride: Frank, Smith, *Org. Syn.* **28**, 89 (1948).

Liquid; bp$_{18}$ 133-137°; bp$_{10}$ 119°. d$_4^{18.3}$ 1.2142; d$_4^{16}$ 1.197. n$_D^{18.3}$ 1.6382. Reacts with a primary or secondary amine to form a benzoylthiourea, which is readily hydrolyzed by alkalies to the free thiourea.

1148. Benzoylpas. *4-(Benzoylamino)-2-hydroxybenzoic acid;* N-benzoyl-p-aminosalicylic acid; 4-benzamidosalicylic acid. $C_{14}H_{11}NO_4$; mol wt 257.25. C 65.37%, H 4.31%, N 5.44%, O 24.88%. Prepn: Drain *et al., J. Chem. Soc.* **1949**, 1498; **Brit.** pat. **676,363** (1952 to Wander). Prepn of sodium and calcium salts: Suddaby, Sumpter, **Brit.** pat. **711,-163** (1954 to Herst).

Crystals, mp 260-261°.
Sodium salt, $C_{14}H_{10}NNaO_4$, *BPAS*. White powder.
Calcium salt pentahydrate, $C_{28}H_{20}CaN_2O_8.5H_2O$, *benzoylpas calcium, Benzapas, Benzacyl, Therapas.* White powder or crystals.
Mixture of calcium salt with isoniazid, *Iso-Benzacyl.*

THERAP CAT: Antibacterial (tuberculostatic).

1149. Benzoyl Peroxide. *Dibenzoyl peroxide;* benzoyl superoxide; Acetoxyl; Acnegel; Akneroxid L; Benoxyl; Benzagel 10; Benzaknen; Debroxide; Desanden; Lucidol; Nericur; Oxy-5; Oxy-L; PanOxyl; Peroxyderm; Peroxydex; Persadox; Persa-gel; Sanoxit; Theraderm; Xerac BP 5; Xerac BP 10. $C_{14}H_{10}O_4$; mol wt 242.23. C 69.42%, H 4.16%, O 26.42%. $(C_6H_5CO)_2O_2$. Prepd by interaction of benzoyl chloride and a cooled soln of sodium peroxide. Laboratory procedure: A. I. Vogel, *Practical Organic Chemistry* (Longmans, London, 3rd ed., 1954) p 807; Gattermann-Wieland, *Praxis des organischen Chemikers* (de Gruyter, Berlin, 40th ed., 1961) p 115. Comparative clinical study with clindamycin, *q.v.*, in acne vulgaris: L. J. Swinyer *et al., Br. J. Dermatol.* **119**, 615 (1988). Review of carcinogenic and allergenic potential: D. J. Hogan, *Int. J. Dermatol.* **30**, 467-470 (1991).

Crystals. mp 103-106°. *May explode when heated.* Sparingly sol in water or alcohol; sol in benzene, chloroform, ether. One gram dissolves in 40 ml carbon disulfide, in about 50 ml olive oil.

Caution: Potential symptoms of overexposure are irritation of skin, eyes and mucous membranes; sensitization dermatitis. *See NIOSH Pocket Guide to Chemical Hazards* (DHHS/NIOSH 90-117, 1990) p 44.

USE: Source of free radicals for industrial processes. Oxidizing agent in bleaching oils, flour, etc.; catalyst in plastics industry; initiator in polymerization.

THERAP CAT: Keratolytic.
THERAP CAT (VET): Keratolytic.

1150. 3,4-Benzphenanthrene. *Benzo[c]phenanthrene.* $C_{18}H_{12}$; mol wt 228.29. C 94.70%, H 5.30%. Synthesis by a Pschorr reaction from diazotized α-(2-naphthyl)-2-aminocinnamic acid: Cook, *J. Chem. Soc.* **1931**, 2524. From diphenylmethylsuccinic anhydride: Hewett, *J. Chem. Soc.* **1936**, 599. By double ring closure of β-benzohydrylglutaric acid: Newman, Joshel, *J. Am. Chem. Soc.* **60**, 487 (1938). From 4-keto-1,2,3,4-tetrahydrophenanthrene: Bachmann, Edgerton, *ibid.* **62**, 2970 (1940).

Needles from alcohol, mp 68° (needles or leaflets from petr ether, fine needles from alcohol + acetone). Absorption maxima: Mayneord, Roe, *Proc. Roy. Soc. London* **A158**, 63 (1937).

1151. Benzphetamine. *(S)-N,α-Dimethyl-N-(phenyl-methyl)benzeneethanamine; N-benzyl-N,α-dimethylphenethyl-*

amine; d-N-methyl-*N*-benzyl-β-phenylisopropylamine. C_{17}-$H_{21}N$; mol wt 239.36. C 85.31%, H 8.84%, N 5.85%. Prepd from *d*-desoxyephedrine: Heinzelman, Aspergren, U.S. pat. **2,789,138** (1957 to Upjohn).

Liquid. $bp_{0.02}$ 127°. n_D^{19} 1.5515. Practically insol in water. Sol in methanol, ethanol, ether, chloroform, acetone, benzene.

Hydrochloride, $C_{17}H_{21}N.HCl$, *Didrex, Inapetyl.* Crystals from ethyl acetate, mp 129-130°. Dextrorotatory. Sol in water, 95% ethanol.

Note: This is a controlled substance (stimulant) listed in the U.S. Code of Federal Regulations, Title 21 Part 1308.13 (1995).

THERAP CAT: Anorexic.

1152. Benzpiperylon. *1,2-Dihydro-2-(1-methyl-4-piperidinyl)-5-phenyl-4-(phenylmethyl)-3H-pyrazol-3-one; 4-benzyl-1-(1-methyl-4-piperidyl)-3-phenyl-3-pyrazolin-5-one; 1-(N-methyl-4-piperidyl)-3-phenyl-4-benzyl-5-pyrazolone;* benzpiperilone; KB-95; Benzometan; Humedil; Reublonil; Telon. $C_{22}H_{25}N_3O$; mol wt 347.46. C 76.05%, H 7.25%, N 12.09%, O 4.60%. Prepd from (*N*-methylpiperidyl)hydrazine and ethyl 2-benzyl-3-oxo-3-phenylpropionate: A. Ebnöther *et al., Helv. Chim. Acta* **42**, 1201 (1959); E. Jucker *et al.,* **Swiss.** pats. **346,885, 346,886** (1960 both to Sandoz), *C.A.* **58**, 5696a,b (1963). *See also: eidem,* U.S. pat. **2,903,460** (1959 to Sandoz). Pharmacological properties, toxicity: A. Cerletti *et al., Boll. Chim. Farm.* **102**, 602 (1963).

Crystals from ethanol, dec 181-183°. pK_1 6.73, pK_2 9.13. LD_{50} in mice, rats, rabbits (mg/kg): 160, 160, 83 i.v.; 1880, 2700, 1700 orally (Cerletti).

THERAP CAT: Analgesic; anti-inflammatory.

1153. Benzpyrinium Bromide. *3-[[(Dimethylamino)-carbonyl]oxy]-1-(phenylmethyl)pyridinium bromide; 1-benzyl-3-hydroxypyridinium bromide dimethylcarbamate; 1-benzyl-3-(dimethylcarbamyloxy)pyridinium bromide;* benzstigminum bromidum; Stigmenene bromide; Stigmonene bromide. $C_{15}H_{17}BrN_2O_2$; mol wt 337.22. C 53.43%, H 5.08%, Br 23.70%, N 8.31%, O 9.49%. Prepd from 3-hydroxypyridine dimethylcarbamate and benzyl bromide: Wuest, U.S. pat. **2,489,247** (1949 to Wm. R. Warner).

Crystals, mp 114-115°. Freely sol in water, alc. Practically insol in ether. A 1% aq soln has a pH of 4.5 to 5.5. uv max: 269 nm ($E_{1cm}^{1\%}$ 136).

THERAP CAT: Cholinergic.

1154. Benzquinamide. *2-(Acetyloxy)-N,N-diethyl-1,3,4,-6,7,11b-hexahydro-9,10-dimethoxy-2H-benzo[a]quinolizine-3-carboxamide; N,N-diethyl-1,3,4,6,7,11b-hexahydro-2-hydroxy-9,10-dimethoxy-2H-benzo[a]quinolizine-3-carboxamide acetate;* 2-acetoxy-3-(*N,N*-diethylcarboxamido)-9,10-dimethoxy-1,2,3,4,6,7-hexahydro-11b*H*-benzopyridocoline; 2-hydroxy-3-diethylcarbamyl-9,10-dimethoxy-1,2,3,4,6,7-hexahydro-11b*H*-benzoquinolizine acetate; BZQ; NSC-64375; P-2647; Emete-Con; Emeticon; Promecon; Quantril. $C_{22}H_{32}$-N_2O_5; mol wt 404.51. C 65.32%, H 7.97%, N 6.93%, O 19.78%. Prepn: Trettner, U.S. pat. **3,053,845** (1962 to Pfizer). Pharmacology: Scriabine *et al., J. Am. Med. Assoc.* **184**, 276 (1963); Kadzielawa, Gumulka, *Arch. Int. Pharmacodyn. Ther.* **163**, 139 (1966). Metabolic studies: Koe, Pinson, *J. Med. Chem.* **7**, 635 (1964); Wiseman *et al., Biochem. Pharmacol.* **13**, 1421 (1964). Toxicity: E. I. Goldenthal, *Toxicol. Appl. Pharmacol.* **18**, 185 (1971).

Crystals from diisopropyl ether, mp 130-131.5°. LD_{50} orally in rats: 990 mg/kg; i.p. in mice: 376 mg/kg (Goldenthal).

THERAP CAT: Antipsychotic, antiemetic.

1155. Benzthiazide. *6-Chloro-3-[[(phenylmethyl)thio]-methyl]-2H-1,2,4-benzothiadiazine-7-sulfonamide 1,1-dioxide; 3-[(benzylthio)methyl]-6-chloro-2H-1,2,4-benzothiadiazine-7-sulfonamide 1,1-dioxide;* 3-[(benzylthio)methyl]-6-chloro-7-sulfamoyl-2H-benzo-1,2,4-thiadiazine 1,1-dioxide; 6-chloro-7-sulfamoyl-3-benzylthiomethyl-2H-1,2,4-benzothiadiazine 1,1-dioxide; benzothiazide; Aquatag; Dihydrex; Diucen; Edemex; ExNa; Exosalt; Fovane; Freeuril; Hy-Drine; Lemazide; Proaqua; Urese. $C_{15}H_{14}ClN_3O_4S_3$; mol wt 431.94. C 41.71%, H 3.27%, Cl 8.21%, N 9.73%, O 14.82%, S 22.27%. Prepn: J. M. McManus *et al.,* *136th Am. Chem. Soc. Meet.* (Atlantic City, Sept. 1959) *Abstr. of Papers*, pp 13-O; W. M. McLamore, G. D. Laubach, Ger. pat. **1,137,-740;** *eidem,* U.S. pat. **3,440,244** (1962, 1969 both to Pfizer). Pharmacology, toxicity: S. Y. P'an *et al., J. Pharmacol. Exp. Ther.* **128**, 122 (1960).

Crystals from acetone, mp 231-232° (U.S. patent); also reported as mp 238-239° (P'an). Bitter taste. Practically insol in water. Sol in alkaline solns. LD_{50} in mice, rats (mg/kg): > 5000, > 10000 orally; 410, 422 i.v. (P'an).

THERAP CAT: Diuretic, antihypertensive.

1156. Benztropine Mesylate. *3-(Diphenylmethoxy)-8-methyl-8-azabicyclo[3.2.1]octane methanesulfonate; 3α-(di-phenylmethoxy)-1αH,5αH-tropane methanesulfonate;* benztropine methanesulfonate; tropine benzohydryl ether methanesulfonate; Cogentin; Cogentinol; Cobrentin methanesulfonate. $C_{22}H_{29}NO_4S$; mol wt 403.54. C 65.48%, H 7.24%, N 3.47%, O 15.86%, S 7.95%. Prepd by the action of diphenyldiazomethane on tropine: Phillips, U.S. pat. **2,595,-405** (1952 to Merck & Co.). Crystal structure: P. G. Jones *et al., Acta Crystallogr.* **B34**, 3125 (1978).

Crystals from acetone + ether, mp 143°. uv max: 259 nm (E_M = 437). Soluble in water. pH about 6.

THERAP CAT: Anticholinergic.

1157. Benzydamine. *N,N-Dimethyl-3-[[1-(phenylmethyl)-1H-indazol-3-yl]oxy]-1-propanamine; 1-benzyl-3-[3-(dimethylamino)propoxy]-1H-indazole; 1-benzyl-1H-indazol-3-yl 3-(dimethylamino)propyl ether; benzindamine.* $C_{19}H_{23}N_3O$; mol wt 309.41. C 73.76%, H 7.49%, N 13.58%, O 5.17%. Prepn: Fr. pat. **1,382,855**; Palazzo, U.S. pat. **3,318,-905** (1964, 1967 both to Angelini Francesco); Palazzo *et al.*, *J. Med. Chem.* **9**, 38 (1966). Pharmacology: Lisciani *et al.*, *Eur. J. Pharmacol.* **3**, 157 (1968). Metabolism: Catanese *et al.*, *Arzneimittel-Forsch.* **16**, 1354 (1966); Kataoka *et al.*, *Chem. Pharm. Bull.* **19**, 1511 (1971). Toxicology: B. Silvestrini *et al.*, *Toxicol. Appl. Pharmacol.* **10**, 148 (1967). Series of articles on pharmacology: *Arzneimittel-Forsch.* **37**, 587-646 (1987).

$bp_{0.05}$ 160°.
Hydrochloride, $C_{19}H_{23}N_3O$.HCl, *Afloben, Andolex, Benalgin, Benzyrin, Difflam, Dorinamin, Enzamin, Imotryl, Ririlim, Riripen, Salyzoron, Saniflor, Tamas, Tantum, Verax*. Crystals, mp 160°. uv max: 306 nm ($E_{1cm}^{1\%}$ 160). Very sol in water; rather sol in ethanol, chloroform, n-butanol. LD_{50} in mice, rats (mg/kg): 110, 100 i.p.; 515, 1050 orally (Silvestrini).

THERAP CAT: Analgesic; anti-inflammatory; antipyretic.

THERAP CAT (VET): Anti-inflammatory.

1158. Benzyl Acetate. *Acetic acid phenylmethyl ester; acetic acid benzyl ester.* $C_9H_{10}O_2$; mol wt 150.18. C 71.98%, H 6.71%, O 21.31%. $C_6H_5CH_2OOCCH_3$. Occurs in a number of plants, particularly jasmine: S. Arctander, *Perfume and Flavor Materials of Natural Origin* (Elizabeth, N.J., 1960) pp 313-314. Prepd from benzyl chloride, acetic acid or sodium acetate and triethylamine: Merker, Scott, *J. Org. Chem.* **26**, 5180 (1961); Hennis *et al.*, *Ind. Eng. Chem., Prod. Res. Develop.* **6**, 193 (1967). Toxicity study: P. M. Jenner *et al.*, *Food Cosmet. Toxicol.* **2**, 327 (1964).

Liquid; pear-like odor. bp 213°, bp_{102} 134°. mp −51°. d_4^{25} 1.050. n_D^{20} 1.5232, n_D^{25} 1.4998. Flash pt, closed cup: 216° F (102° C). Practically insol in water. Misc with alcohol, ether. LD_{50} orally in rats: 2490 mg/kg (Jenner).

Caution: If ingested can cause G.I. irritation with vomiting and diarrhea. Also irritating to skin, eyes, respiratory tract.

USE: In perfumery, solvent for cellulose acetate and nitrate.

1159. Benzyl Alcohol. *Benzenemethanol;* phenylcarbinol; phenylmethanol; α-hydroxytoluene. C_7H_8O; mol wt 108.14. C 77.75%, H 7.46%, O 14.80%. $C_6H_5CH_2OH$. Constituent of jasmine, hyacinth, ylang-ylang oils, Peru and Tolu balsams, storax, where it occurs in ester form also.

Originally prepd by the Cannizzaro reaction from benzaldehyde + KOH: Cannizzaro, *Ann.* **88**, 129 (1853); *cf.* Hickinbottom, *Reactions of Organic Compds.* (Longmans, London, 3rd ed., 1957) p 251; A. I. Vogel, *Practical Organic Chemistry* (Longmans, London, 3rd ed., 1959) p 711; Gattermann-Wieland, *Praxis des organischen Chemikers* (de Gruyter, Berlin, 40th ed., 1961) p 193. Produced on a large scale by the action of sodium or potassium carbonate on benzyl chloride: **Ger. pat. 484,662**; *Chem. Zentr.* **1930**, I, 1052; *Frdl.* **16**, 426; *cf.* Kirk-Othmer *Encyclopedia of Chemical Technology* vol. 3 (Interscience, New York, 1964) pp 442-449. Toxicity: Smyth *et al.*, *Arch. Ind. Hyg. Occup. Med.* **4**, 119 (1951).

Liquid. Faint aromatic odor. Sharp burning taste. d_4^{20} 1.04535; d_4^{25} 1.04156. mp −15.19°. bp_{760} 204.7°; bp_{400} 183.0°; bp_{200} 160.0°; bp_{100} 141.7°; bp_{60} 129.3°; bp_{40} 119.8°; bp_{20} 105.8°; bp_{10} 92.6°; bp_5 80.8°; $bp_{1.0}$ 58.0°. n_D^{20} 1.54035; n_D^{25} 1.53837. Dreisbach, Martin, *Ind. Eng. Chem.* **41**, 2875 (1941). Absorption spectrum: Brode, *J. Phys. Chem.* **30**, 61 (1926). Vapor density 3.72 (air = 1.00). Flash pt, closed cup 213°F, open cup 220°F. Autoignition temp 817°F. One gram dissolves in about 25 ml water. One volume dissolves in 1.5 vols of 50% ethyl alcohol. Misc with abs and 94% alcohol, ether, chloroform. LD_{50} orally in rats: 3.1 g/kg (Smyth).

USE: Manuf other benzyl compds. Pharmaceutic aid (antimicrobial). Solvent for gelatin, casein (when hot), solvent for cellulose acetate, shellac. Used in perfumery and in flavoring (mostly in form of its aliphatic esters). In microscopy as embedding material.

THERAP CAT (VET): Has been used for relief from pruritis.

1160. Benzylamine. *Benzenemethanamine;* aminotoluene; phenylmethylamine; moringine. C_7H_9N; mol wt 107.16. C 78.46%, H 8.47%, N 13.07%. Prepn from benzylchloride and ammonia: Mason, *J. Chem. Soc.* **63**, 1311 (1893); by redn of benzonitrile: Carothers, Jones, *J. Am. Chem. Soc.* **47**, 3051 (1925); from benzyl bromide + acetamide: Erikson, *Ber.* **59**, 2665 (1926); from *N*-benzylphthalimide + hydrazine hydrate: Ing, Manske, *J. Chem. Soc.* **129**, 2348 (1926). Identity with moringine: Chakravarti, *Bull. Calcutta School Trop. Med.* **3**, 162 (1955); *C.A.* **50**, 16891e (1956).

Liquid; strongly alkaline reaction. bp 185°; bp_{12} 90°. d_4^{19} 0.983. n_D^{20} 1.5401. Miscible with water, alcohol, ether.
Hydrochloride, C_7H_9N.HCl, crystals, mp 253°.
Hydroiodide, C_7H_9N.HI, leaflets, mp 162°.
Caution: Highly irritating to skin, mucous membranes.
USE: In organic synthesis.

1161. Benzylaniline. *N-Phenylbenzenemethanamine; N-phenylbenzylamine;* benzylphenylamine. $C_{13}H_{13}N$; mol wt 183.25. C 85.21%, H 7.15%, N 7.64%. $C_6H_5CH_2NHC_6H_5$. Prepn from benzyl alc and aniline in the presence of KOH: Sprinzak, *J. Am. Chem. Soc.* **78**, 3207 (1956); from benzaldehyde and aniline in the presence of NaBH$_4$: Schellenberg, *J. Org. Chem.* **28**, 3259 (1963).

Prisms, mp 37-38°. bp 306-307°. Practically insol in water; sol in alcohol, chloroform, ether.

1162. Benzyl Benzoate. *Benzoic acid phenylmethyl ester;* benzoic acid benzyl ester; benzylbenzenecarboxylate; Ascabin; Venzonate; Ascabiol. $C_{14}H_{12}O_2$; mol wt 212.25. C 79.23%, H 5.70%, O 15.08%. $C_6H_5COOCH_2C_6H_5$. Contained in Peru and Tolu balsams. Prepd by the action of sodium benzylate on benzaldehyde: Kamm, Kamm, *Org. Syn.* **coll. vol. I**, 104 (2nd ed., 1941); by the dry esterification of sodium benzoate and benzyl chloride in the presence of triethylamine: Thorp, Nottorf, *Ind. Eng. Chem.* **39**, 1300 (1947). Toxicity studies: Graham, Kuizenga, *J. Pharmacol. Exp. Ther.* **84**, 358 (1945); Draize *et al.*, *J. Pharmacol. Exp. Ther.* **93**, 26 (1948). Comprehensive description: M. M. A.

Hassan, J. S. Mossa, *Anal. Profiles Drug Subs.* **10**, 55-74 (1981).

Leaflets or oily liq; faint, pleasant, aromatic odor; sharp burning taste. mp 21°. d_4^{25} 1.118. bp 323-324°. bp_{16} 189-191°. $bp_{4.5}$ 156°. Sparingly volatile with steam. n_D^{21} 1.5681. Insol in water or glycerol. Miscible with alc, chloroform, ether, oils. LD_{50} in rats, mice, rabbits, guinea pigs (g/kg): 1.7, 1.4, 1.8, 1.0 orally (Draize).

Caution: In exptl animals, ingestion causes progressive incoordination, excitation, convulsions, death. May cause skin irritation in humans. Avoid contact with eyes, *Clinical Toxicology of Commercial Products,* R. E. Gosselin *et al.,* Eds. (Williams & Wilkins, Baltimore, 4th ed., 1976) Section II, p 137.

USE: As solvent of cellulose acetate, nitrocellulose and artificial musk; substitute for camphor in celluloid and plastic pyroxylin compds; perfume fixative; in confectionery and chewing gum flavors.

THERAP CAT: Scabicide, pediculicide.

THERAP CAT (VET): Acaricide, pediculicide. *Contraindicated in cats.*

1163. Benzyl Bromide. *(Bromomethyl)benzene;* α-bromotoluene; ω-bromotoluene. C_7H_7Br; mol wt 171.04. C 49.16%, H 4.13%, Br 46.72%. $C_6H_5CH_2Br$. Prepd by the action of bromine on toluene in ultraviolet light: v. Konek, Loczka, *Ber.* **57**, 679 (1924); Zelinsky, **Ger. pat. 478,084;** *Chem. Zentr.* **1929** II, 1216; *Frdl.* **16**, 335; by the action of bromine on dibenzyl ether: Lachman, *J. Am. Chem. Soc.* **45**, 2359 (1923).

Lacrimatory liquid. mp −3.9°. bp 198-199°. bp_{80} 127°. d_0^{22} 1.4380; d^{17} 1.443; d_4^{64} 1.3886. Slowly decomp by water.

Caution: Intensely irritating to skin, eyes, mucous membranes. Large doses cause CNS depression.

1164. Benzyl Chloride. *(Chloromethyl)benzene;* α-chlorotoluene. C_7H_7Cl; mol wt 126.59. C 66.42%, H 5.57%, Cl 28.01%. $C_6H_5CH_2Cl$. Made by cautious chlorination of toluene: A. I. Vogel, *Practical Organic Chemistry* (Longmans, London, 3rd ed., 1959) p 538; Gattermann-Wieland, *Praxis des organischen Chemikers* (de Gruyter, Berlin, 40th ed., 1961) p 92. Manuf: Faith, Keyes & Clark's *Industrial Chemicals,* F. A. Lowenheim, M. K. Moran, Eds. (Wiley-Interscience, New York, 4th ed., 1975) pp 145-148.

Very refractive liquid; rather unpleasant, irritating odor. d_{20}^{20} 1.100. bp 179°. mp −48° to −43°. n_D^{15} 1.5415. Insol in water. Miscible with alcohol, chloroform, ether. Rapidly dec when heated in the presence of iron.

Caution: Potential symptoms of overexposure are irritation of eyes and nose; weakness; irritability; headache; skin eruption; pulmonary edema. *See NIOSH Pocket Guide to Chemical Hazards* (DHHS/NIOSH 90-117, 1990) p 46.

USE: Manuf benzyl compds, perfumes, pharmaceutical products, dyes, synthetic tannins, artificial resins.

1165. Benzyl Cinnamate. *3-Phenyl-2-propenoic acid phenylmethyl ester; trans-cinnamic acid benzyl ester;* cinnamein. $C_{16}H_{14}O_2$; mol wt 238.29. C 80.65%, H 5.92%, O 13.43%. $C_6H_5CH=CHCOOCH_2C_6H_5$. Constituent of storax, Peru and Tolu balsams: Tschirch, Trog, *Arch. Pharm.* **232**, 70 (1894); Tschirch, Oberländer, *ibid.* 559. Prepn: Volwiler, Vliet, *J. Am. Chem. Soc.* **43**, 1672 (1921); Eliel, Anderson, *ibid.* **74**, 547 (1952); Bender, Zerner, *ibid.* **84**, 2550 (1962). Toxicity study: P. M. Jenner *et al.,* *Food Cosmet. Toxicol.* **2**, 327 (1964).

Crystals from 95% ethanol; sweet odor of balsam. mp 39°; also reported as mp 33-34° (Volwiler, Vliet). Dec on distillation at ordinary pressure; $bp_{0.5}$ 154-157°, bp_5 195-200°, bp_{22} 228-230°. Practically insol in water, propylene glycol and glycerin. Sol in alc, ether, oils. LD_{50} orally in rats: 5530 mg/kg (Jenner).

USE: In artificial flavors, in perfumes, mainly as a fixative.

1166. Benzyl Cyanide. *Benzeneacetonitrile;* phenylacetonitrile; α-tolunitrile; ω-cyanotoluene. C_8H_7N; mol wt 117.15. C 82.02%, H 6.02%, N 11.96%. $C_6H_5CH_2CN$. Occurs in garden cress and other plants; made from benzyl chloride, and NaCN: Adams, Thal, *Org. Syn.* vol. 2, 9 (1922); coll. vol. I, 101 (107 in 2nd ed.).

Oily liquid, aromatic odor. d_{15}^{15} 1.0214. mp −23.8°. bp_{760} 233.5°; bp_{100} 161.8°; bp_{20} 119.4°; $bp_{1.0}$ 60°. n_D^{25} 1.52105. Insoluble in water, miscible with alc, ether.

1167. Benzyl Ether. *1,1'-[Oxybis(methylene)]bis[benzene];* dibenzyl ether. $C_{14}H_{14}O$; mol wt 198.26. C 84.81%, H 7.12%, O 8.07%. $(C_6H_5CH_2)_2O$. Prepd: Lachman, *J. Am. Chem. Soc.* **45**, 2356 (1923); Staab, Wendel, *Ber.* **93**, 2902 (1960); Lichtenberger, Tritsch, *Bull. Soc. Chim. France* **1961**, 363. Manuf by reduction of benzaldehyde in the presence of $[Co(CO)_4]_2$: Wender, Orchin, U.S. pat. **2,614,-107** (1952 to U.S.A. as represented by the Secy. of Agr.). Physical properties: Svirbely *et al.,* *J. Am. Chem. Soc.* **71**, 507 (1949); Dreisbach, Martin, *Ind. Eng. Chem.* **41**, 2875 (1949). Miscibility: Jackson, Drury, *ibid.* **51**, 1491 (1959).

Unstable liquid, bp 295-298° (with dec), bp_{21} 173-174°; bp_2 125.5-126.5°. Appears to dec slowly at ordinary temps. d^{35} 1.0341; d_4^{25} 0.99735; d_4^{20} 1.00142; d^{15} 1.0482. n_D^{25} 1.5601 (Svirbely *et al.*), 1.53851 (Dreisbach, Martin); n_D^{20} 1.54057 (Dreisbach, Martin), 1.566 (Lichtenberger, Tritsch). Practically insol in water; miscible with ethanol, ether, chloroform, acetone.

USE: Plasticizer for nitrocellulose; solvent in perfumery.

1168. Benzyl Ethyl Ether. *(Ethoxymethyl)benzene.* $C_9H_{12}O$; mol wt 136.19. C 79.37%, H 8.88%, O 11.75%. $C_6H_5CH_2OC_2H_5$. Preparation from sodium ethoxide and benzyl bromide: Letsinger, Pollart, *J. Am. Chem. Soc.* **78**, 6079 (1956); by reduction of benzaldehyde diethyl acetal with $LiAlH_4$-$AlCl_3$: Eliel, Rerick, *J. Org. Chem.* **23**, 1088 (1958).

Oily liquid, aromatic odor. bp 186°; bp_{10} 65°. d 0.949, n_D^{20} 1.4955. Volatile with steam. Practically insol in water; miscible with alcohol, ether.

1169. Benzyl Formate. *Formic acid phenylmethyl ester; formic acid benzyl ester.* $C_8H_8O_2$; mol wt 136.15. C 70.57%, H 5.92%, O 23.50%. $HCOOCH_2C_6H_5$. Prepn from formic acid and benzyl alcohol: Mailhe, *Chem. Ztg.* **35**, 508 (1911).

Liquid; pleasant fruity odor. d 1.081. bp 203°. Practically insol in water. Sol in alcohol.

USE: Solvent for cellulose esters; in perfumery.

1170. Benzyl Fumarate. *(E)-2-Butenedioic acid bis-(phenylmethyl) ester; fumaric acid dibenzyl ester;* dibenzyl fumarate. $C_{18}H_{16}O_4$; mol wt 296.32. C 72.96%, H 5.44%, O 21.60%. $C_6H_5CH_2OOCCH=CHCOOCH_2C_6H_5$. Prepd from fumaric acid and benzyl alcohol: Volwiler, Vliet, *J. Am. Chem. Soc.* **43**, 1672 (1921).

Cryst powder, mp 58.5-59.5°. bp_5 210-211°. Practically insol in water. Sol in alcohol, chloroform, ether, oils.

USE: In room spray deodorant: Kulka, U.S. pat. **3,077,-457** (1963 to Fritzsche Bros.).

1171. Benzylhydrochlorothiazide. *6-Chloro-3,4-dihydro-3-(phenylmethyl)-2H-1,2,4-benzothiadiazine-7-sulfonamide 1,1-dioxide; 3-benzyl-6-chloro-3,4-dihydro-2H-1,2,4-benzothiadiazine-7-sulfonamide 1,1-dioxide; 6-chloro-7-sulfamoyl-3-benzyl-3,4-dihydro-1,2,4-benzothiadiazine 1,1-dioxide; 3-benzyl-6-chloro-3,4-dihydro-7-sulfamoyl-1,2,4-benzothiadiazine 1,1-dioxide;* Behyd. $C_{14}H_{14}ClN_3O_4S_2$; mol wt 387.87. C 43.35%, H 3.64%, Cl 9.14%, N 10.83%, O 16.50%, S 16.53%. Prepn: Werner *et al.,* *J. Am. Chem. Soc.* **82**, 1161 (1960); Novello *et al.,* *J. Org. Chem.* **25**, 970 (1960); Ugi, U.S. pat. **3,108,097** (1963).

Crystals from acetic acid + water, mp 260-262°. Also reported as crystals from water, mp 269°.

THERAP CAT: Antihypertensive; diuretic.

1172. Benzylideneacetone. *4-Phenyl-3-buten-2-one;* benzalacetone; methyl styryl ketone; cinnamyl methyl ketone; acetocinnamone. $C_{10}H_{10}O$; mol wt 146.19. C 82.16%, H 6.89%, O 10.94%. $C_6H_5CH=CHCOCH_3$. Prepd by condensing acetone and benzaldehyde by means of aq alkali: Drake, Allen, *Org. Syn.* **3**, 17 (1923).

Lustrous plates on vacuum distillation. Coumarin type odor. Purification by steam distillation: Fromm, Haas, *Ann.* **394**, 291 (1912). mp 41.5°. d_{15}^{15} 1.0377; $d_4^{45.2}$ 1.0097. bp_{760} 261°; bp_{200} 211°; bp_{100} 187.8°; bp_{40} 161.3°; bp_{20} 143.8°; bp_{10} 127.4°; bp_5 112.2°; $bp_{1.0}$ 81.7°. $n_D^{45.9}$ 1.5836. Absorption spectrum: Baker, *J. Chem. Soc.* **91**, 1492 (1907); Baly, Schafer, *ibid.* **93**, 1813 (1908). Freely sol in alcohol, benzene, chloroform, ether; sparingly sol in water, petr ether.

USE: In perfumery, organic syntheses.

1173. Benzylideneaniline. *N-(Phenylmethylene)benzenamine;* benzalaniline. $C_{13}H_{11}N$; mol wt 181.24. C 86.15%, H 6.12%, N 7.73%. $C_6H_5N=CHC_6H_5$. Prepd by the action of aniline on benzaldehyde: Bigelow, Eatough, *Org. Syn. coll. vol.* I (2nd ed., 1941) p 80. Molecular structure: M. Traetteberg *et al., J. Mol. Struct.* **48**, 395 (1978).

Plates from 85% alc. Has been obtained as the yellow modification (needles from CS_2). The colorless form melts at 48°, after solidifying melts again at 56°. d_4^{50} 1.045. bp_{760} 300°. Absorption spectrum: Baly *et al., J. Chem. Soc.* **97**, 590 (1910). Sol in alcohol, chloroform, acetic anhydride, carbon disulfide.

1174. Benzylimidobis(p-methoxyphenyl)methane. *N-[Bis(4-methoxyphenyl)methylene]benzenemethanamine.* $C_{22}H_{21}NO_2$; mol wt 331.41. C 79.73%, H 6.39%, N 4.23%, O 9.66%. Prepd by heating *p,p'*-dimethoxybenzophenone, thionyl chloride and benzylamine: Schönberg, Urban, *Ber.* **67**, 1999 (1934).

Pale-yellow crystals, mp 89-91°. Soluble in ether, chloroform. Slightly sol in petr ether.

USE: Detection of elementary sulfur.

1175. Benzyl Methyl Ether. *(Methoxymethyl)benzene;* methyl benzyl ether. $C_8H_{10}O$; mol wt 122.17. C 78.65%, H 8.25%, O 13.10%. $C_6H_5CH_2OCH_3$. Prepd from benzyl chloride, methanol, and NaOH: Olson *et al., J. Am. Chem. Soc.* **69**, 2451 (1947).

Liquid. d 0.987. bp 174°. Practically insol in water. Sol in alcohol, ether.

1176. Benzylmorphine. *7,8-Didehydro-4,5-epoxy-17-methyl-3-(phenylmethoxy)morphinan-6-ol.* $C_{24}H_{25}NO_3$; mol wt 375.47. C 76.77%, H 6.71%, N 3.73%, O 12.78%. Prepn from morphine and benzyl chloride in the presence of sodium ethylate: Ger. pat. **91,813** (1897 to E. Merck), *Frdl.* **4**, 1245; from morphine and benzyldimethylphenylammonium hydroxide: Rodionow, *Bull. Soc. Chim. France* [4] **45**, 109 (1929). Properties of hydrochloride: Dott, *Pharm. J.* **115**, 757 (1925); *see also* Ger. pat. above.

Needles from water, prisms or platelets from ether, chloroform. mp 132°. Soluble in 2500 parts cold water, more

sol in boiling water; 1.6 parts dissolves in 100 parts ether; readily sol in 50% alcohol, benzene.

Hydrochloride, $C_{24}H_{25}NO_3.HCl$, *Peronine.* Monohydrate, needles from water. Sol at 15° in 133 parts water, 218 parts 95% alcohol, 390 parts chloroform, 100 parts methanol; slightly sol in abs alcohol; very slightly sol in ether, acetone, isoamyl alcohol.

Note: This is a controlled substance (opium derivative) listed in the U.S. Code of Federal Regulations, Title 21 Part 1308.11 (1995).

THERAP CAT: Analgesic (narcotic).

1177. Benzylpenicillinic Acid. *[2S-(2α,5α,6β)]-3,3-Dimethyl-7-oxo-6-[(phenylacetyl)amino]-4-thia-1-azabicyclo-[3.2.0]heptane-2-carboxylic acid;* free benzylpenicillin; penicillin G; free penicillin II. $C_{16}H_{18}N_2O_4S$; mol wt 334.40. C 57.47%, H 5.43%, N 8.38%, O 19.14%, S 9.59%. Obtained by extraction at ice temp of the acidified (pH 2) aq soln of sodium benzylpenicillin with ether or chloroform. Shows correct analytical composition only if moisture is excluded completely during its isoln from the ether extract (by lyophilizing from benzene). When kept dry the acid retains its antibiotic potency for limited periods, but it is rapidly inactivated by small amounts of water. The product formed on slow inactivation by moisture seems to be benzylpenicilloic acid, *cf.* Clarke *et al., The Chemistry of Penicillin* (Princeton, 1949). Prepn of methyl ester: **Dan. pat.** **85,976** (1958 to Lövens Kemiske Fabrik), *C.A.* **53**, 11409e (1959). Pharmacokinetics in humans: M. Barza, L. Weinstein, *Clin. Pharmacokinet.* **1**, 297 (1976).

Amorphous white powder. $[\alpha]_D^{20}$ +269° (methanol soln 50 ml, prepd from 350 mg benzylpenicillin sodium). pK (water) at 5°: 2.74; (water) at 25°: 2.76; (80% EtOH) at 5°: 4.84. Sparingly sol in water. Sol in methanol, ethanol, ether, ethyl acetate, benzene, chloroform, acetone. Insol in petr ether.

Crystalline benzylpenicillinic acid. Contains one mol of diisopropyl ether of crystn. Prepn: Trenner, Buhs, *J. Am. Chem. Soc.* **70**, 2897 (1948). The crystals were stable to air. mp 80-87°. $[\alpha]_D^{23}$ +241° (phophate buffer at pH 7).

Methyl ester, crystals from carbon tetrachloride, mp 97°. $[\alpha]_D^{20}$ +328° (methanol); $[\alpha]_D^{20}$ +286° (chloroform). uv max: 252.5, 258, 264.5, 268.5 nm (E_M 290, 250, 180, 120). Practically insol in water. Freely sol in methanol, benzene, chloroform, amyl acetate; sol in peanut oil. Hydrolyzes in aqueous bicarbonate soln to an active principle, presumably benzylpenicillin sodium.

THERAP CAT: Antibacterial.

1178. Benzylpenicillin Sodium. *[2S-(2α,5α,6β)]-3,3-Dimethyl-7-oxo-6-[(phenylacetyl)amino]-4-thia-1-azabicyclo-[3.2.0]heptane-2-carboxylic acid monosodium salt;* sodium penicillin G; penicillin G sodium; sodium penicillin II; sodium benzylpenicillinate; benzylpenicillinic acid sodium salt; penicillin; American penicillin; Monocillin; Nalpen G; Novocillin; Penilaryn; Pen-A-Brasive; Veticillin. $C_{16}H_{17}N_2NaO_4S$; mol wt 356.38. C 53.92%, H 4.81%, N 7.86%, Na 6.45%, O 17.96%, S 9.00%. Industrial production of antibiotic from *Penicillium chrysogenum:* Fleming, *Brit. J. Exp. Pathol.* **10**, 226 (1929); Clutterbuck *et al., Biochem. J.* **26**, 1907 (1932); Chain *et al., Lancet* **2**, 226 (1940); Abraham *et al., ibid.* **2**, 177 (1941); Clarke *et al., The Chemistry of Penicillin* (Princeton, 1949). Biosynthesis from amino acids: Arnstein, Grant, *Biochem. J.* **57**, 353, 360 (1954). Production process: Demain, Somerson, U.S. pat. **3,024,169** (1962 to Merck & Co.). Synthesis: Sheehan, U.S. pat. **3,159,617** (1964 to Arthur D. Little); M. D. Bachi, R. Breiman, *J. Chem. Soc. Perkin Trans I* **1980**, 11. Solubilities: Weiss *et al., Antibiot. & Chemother.* **7**, 374 (1957). HPLC determn in body fluids: C. van Gulpen *et al., J. Chromatog.* **381**, 365

(1986). Review of early literature: E. Chain, *Ann. Rev. Biochem.* **17**, 657 (1948).

Crystals from methanol + ethyl acetate. d 1.41. $[\alpha]_D^{24.8}$ +301° (c = 2.0). uv max (water soln): 252, 258.6, 264.4 nm (E_M about 300, 240, 180). pK at 25° = 2.76; pH 5.5-6.5. One I.U. or U.S.P. penicillin unit is equiv to 0.6 μg (0.0005988 mg) of benzylpenicillin sodium. One mg of benzylpenicillin sodium is equiv to 1670 units. Very sol in water, isotonic NaCl soln, and glucose solns. Also sol in alcohol, glycerol, and other primary alcohols. Practically insol in acetone, ether, $CHCl_3$, ethyl and amyl acetate. Insol in fixed oils and in liq petrolatum. Pure cryst benzylpenicillin sodium is stable. At pH 5.5 to 6.0 refrigerated solns may be kept for a few days. Benzylpenicillin is also precipitated by many metal ions. *See also* Procaine Penicillin G.
Pharmaceutical Incompatibilities: Must not be formulated with acids, salts of heavy metals, glycerol, naphthalene oils, resorcinol, ZnO, vitamin B_1, procaine, ephedrine, iodine and iodides, alcohol and oxidizing agents. *Ref:* R. K. Aliev, M. A. Etinger, *Aptechnoe Delo* **2**, 7 (1952), *C.A.* **46**, 7707 (1952).
THERAP CAT: Antibacterial.
THERAP CAT (VET): Antibacterial.

1179. *o*-Benzylphenol. *2-(Phenylmethyl)phenol; α-phenyl-o-cresol;* (2-hydroxydiphenyl)methane. $C_{13}H_{12}O$; mol wt 184.24. C 84.75%, H 6.57%, O 8.68%. Prepn from benzyl bromide and sodium phenoxide: Kornblum, Lurie, *J. Am. Chem. Soc.* **81**, 2711 (1959).

Crystals or liquid. mp 20.2-20.9°. bp_{10} 154-156°; $bp_{1.0}$ 121-123°. n_D^{20} 1.59945. Practically insol in water. Sol in organic solvents, in fixed alkali hydroxide solns.

1180. *p*-Benzylphenol. *4-(Phenylmethyl)phenol; α-phenyl-p-cresol;* (4-hydroxydiphenyl)methane. $C_{13}H_{12}O$; mol wt 184.24. C 84.75%, H 6.57%, O 8.68%. Prepd from phenol and benzyl chloride in the presence of zinc chloride: Ziegenbein *et al., Ber.* **88**, 1906 (1955).
Crystals, mp 84°. bp 322°; bp_4 154-157°. Slightly sol in cold water; moderately sol in hot water; sol in organic solvents, glacial acetic acid, alkali hydroxide solns.
USE: Germicide, antiseptic, preservative; also in organic syntheses.

1181. Benzyl Salicylate. *2-Hydroxybenzoic acid phenylmethyl ester; salicylic acid benzyl ester.* $C_{14}H_{12}O_3$; mol wt 228.25. C 73.67%, H 5.30%, O 21.03%. Prepd from sodium salicylate and benzyl chloride: Volwiler, Vliet, *J. Am. Chem. Soc.* **43**, 1672 (1921).

Thick liquid, slight, pleasant odor. d^{20} 1.175. bp_{25} 208°. Slightly sol in water; miscible with alcohol or ether.
USE: As fixer in perfumery; in sunscreen preparations.

1182. Benzylsulfamide. *4-[(Phenylmethyl)amino]benzenesulfonamide; N^4-benzylsulfanilamide;* RP-46; M & B 125;

Septazine; Setazine; Chemodyn; Proseptazine. $C_{13}H_{14}N_2$-O_2S; mol wt 262.33. C 59.52%, H 5.38%, N 10.68%, O 12.20%, S 12.22%. Prepd by reduction of the anil (Schiff base) resulting from the reaction of benzaldehyde with sulfanilamide: Goissedet *et al., Compt. Rend. Soc. Biol.* **121**, 1082 (1936).

Crystals from dioxane + water, mp 175°. Very slightly sol in water. Somewhat more sol in alc. Sol in acetone and dioxane. Neutral reaction. Breaks down in the body with the liberation of free sulfanilamide.
THERAP CAT: Antibacterial.

1183. Benzyl Sulfide. *1,1'-[Thiobis(methylene)]bisbenzene; dibenzylsulfide.* $C_{14}H_{14}S$; mol wt 214.33. C 78.46%, H 6.58%, S 14.96%. $(C_6H_5CH_2)_2S$. Prepn: Runge *et al., J. Prakt. Chem.* **11**, 284 (1960). Manuf from benzyl chloride and Na_2S: Stucker, Brennan, U.S. pat. **2,755,305** (1956 to Pure Oil Co.).
Plates, mp 49°. Practically insol in water. Sol in alc, ether.

1184. *p*-(Benzylsulfonamido)benzoic Acid. *4-[[(Phenylmethyl)sulfonyl]amino]benzoic acid; p-α-toluenesulfonamidobenzoic acid;* 4'-carboxyphenylmethanesulfonanilide; Carinamide; Caronamide. $C_{14}H_{13}NO_4S$; mol wt 291.33. C 57.72%, H 4.50%, N 4.81%, O 21.97%, S 11.01%. Prepn: **Brit.** pat. **666,546** (1952 to Sharp & Dohme). Toxicity study: Beyer *et al., J. Pharmacol. Exp. Ther.* **91**, 263 (1947).

Crystals, mp 229-230°. uv max (0.1*N* NaOH): 280-281 nm. Moderately sol in alcohol, acetone, propylene glycol. Very slightly sol in water, carbon tetrachloride, chloroform, benzene. Both the mono- and disodium salts are very sol in water. Aq solns may be prepared by dissolving the acid in an equivalent amount of sodium carbonate or sodium hydroxide soln. It is relatively stable to hydrolysis with acid or base. LD_{50} in mice (g/kg): 2.45 orally; in mice, dogs, rabbits (g/kg): 1.405, 1.575, 1.32 i.v. (Beyer).
THERAP CAT: Adjunct in penicillin therapy.

1185. *S*-Benzylthiuronium Chloride. *Carbamimidothioic acid phenylmethyl ester monohydrochloride.* $C_8H_{11}ClN_2$-S; mol wt 202.71. C 47.40%, H 5.47%, Cl 17.49%, N 13.82%, S 15.82%. Prepd by refluxing a mixture of 126 g benzyl chloride, 76 g thiourea, and 200 ml alcohol for 0.5 hr: Donleavy, *J. Am. Chem. Soc.* **58**, 1004 (1936).

Crystals from alc or dil HCl, mp 172-174°. Metastable form, mp 146-148°. Can be converted to the higher melting form by dissolving in alcohol and seeding with crystals, mp 172-174°.
USE: Reagent for cobalt and nickel; identification and separation of carboxylic, sulfinic, and sulfonic acid.

1186. Benzylurea. *(Phenylmethyl)urea; benzylcarbamide.* $C_8H_{10}N_2O$; mol wt 150.18. C 63.98%, H 6.71%, N 18.65%, O 10.65%. $C_6H_5CH_2NHCONH_2$. Prepn: Neville, McGee, *Can. J. Chem.* **41**, 2123 (1963).

Crystals, mp 147-148°; dec at 200°. One gram dissolves in 60 ml warm water, 33 ml acetone; slightly sol in benzene, ether.

1187. Bephenium. *N,N-Dimethyl-N-(2-phenoxyethyl)-benzenemethanaminium.* Prepn of salts: Copp, **U.S. pat. 2,918,401** (1959 to Burroughs Wellcome); Copp, **Ger. pat. 1,117,600** (1961 to Wellcome Foundation), *C.A.* **56**, 14165i (1962).

Hydroxynaphthoate, $C_{28}H_{29}NO_4$, *benzyldimethyl(2-phenoxyethyl)ammonium 3-hydroxy-2-naphthoate, Alcopar(a), Befeniol, Lecibis, Nemex.* Crystals, mp 170-171°.

Pamoate, $C_{57}H_{60}N_2O_8$, *benzyldimethyl(2-phenoxyethyl)ammonium 4,4'-methylenebis[3-hydroxy-2-naphthoate], bephenium embonate, Frantin.* Dihydrate, pale yellow solid from water, mp 144-146°.

Chloride, $C_{17}H_{22}ClNO$, crystals from acetone, mp 135-136°.

Bromide, $C_{17}H_{22}BrNO$, crystals from isopropanol + ethyl acetate, mp 144.5-146°.

Iodide, $C_{17}H_{22}INO$, crystals from methanol+ ether, mp 146-147°.

THERAP CAT: Anthelmintic (Nematodes).

THERAP CAT (VET): Anthelmintic (Nematodes).

1188. Bepridil. *β-[(2-Methylpropoxy)methyl]-N-phenyl-N-(phenylmethyl)-1-pyrrolidineethanamine;* 1-[2-(*N*-benzyl-anilino)-1-(isobutoxymethyl)ethyl]pyrrolidine; 1-isobutoxy-2-pyrrolidino-3-*N*-benzylanilinopropane; 3-isobutoxy-2-pyrrolidino-*N*-phenyl-*N*-benzylpropylamine. $C_{24}H_{34}N_2O$; mol wt 366.55. C 78.64%, H 9.35%, N 7.64%, O 4.36%. Calcium channel blocker with antianginal and antiarrhythmic properties. Prepn: R. Y. Mauvernay *et al.*, **Ger. pat. 2,310,918**; *eidem,* **U.S. pat. 3,962,238**; reissued with structure correction: N. Busch *et al.*, **U.S. pat. Re. 30,577** (1973, 1976, 1981 all to CERM). *See also:* R. Y. Mauvernay *et al.,* **Ger. pat. 2,802,864** (1978 to CERM), *C.A.* **89**, 179852s (1978). Cardiovascular pharmacology in dogs: M.-T. Michelin *et al., Therapie* **32**, 485 (1977). Mechanism of action studies: S. Vogel *et al., J. Pharmacol. Exp. Ther.* **210**, 378 (1979); C. Labrid *et al., ibid.* **211**, 546 (1979). Clinical comparison of hemodynamic and coronary changes: J. P. Merillon *et al., Therapie* **36**, 123 (1981). Clinical evaluation in angina: M. K. Sharma *et al., Am. J. Cardiol.* **61**, 1210 (1988).

Viscous liquid, bp$_{0.1}$ 184°, bp$_{0.5}$ 192°. n_D^{20} 1.5538.

Hydrochloride monohydrate, $C_{24}H_{34}N_2O \cdot HCl \cdot H_2O$, *CERM-1978, Angopril, Bepadin, Cordium, Vascor.* Crystals, mp 91° ±2°. LD$_{50}$ in mice (mg/kg): 1955 orally, 23.5 i.v. (Mauvernay, 1978).

THERAP CAT: Antianginal.

1189. Beractant. *Surfactant TA; A-60386X; Surfacten; Survanta.* Organic solvent extract of bovine lung, consisting mostly of phospholipids, modified by the addition of colfosceril palmitate, palmitic acid and tripalmitin. Prepn: **Japan. pat. 83 45299**; Y. Tanaka *et al.,* **U.S. pat. 4,397,839** (both 1983 to Tokyo Tanabe). Effect of protein component on surface activity: S. B. Hall *et al., Am. Rev. Respir. Dis.* **145**, 24 (1992). Pharmacology: J. J. Cummings *et al., ibid.* 999. Metabolism in rabbits: J. F. Lewis *et al., ibid.* 19. Evaluation of establishment of normal surfactant function in comparison with Exosurf: E. M. Scarpelli *et al., Am. J.*

Perinatol. **9**, 414 (1992). Clinical trial in neonatal respiratory distress syndrome: M. D. Reller *et al., Am. J. Dis. Child.* **145**, 1017 (1991); E. M. Zola *et al., Pediatrics* **91**, 546 (1993).

THERAP CAT: Pulmonary surfactant.

1190. Beraprost. *2,3,3a,8b-Tetrahydro-2-hydroxy-1-(3-hydroxy-4-methyl-1-octen-6-ynyl)-1H-cyclopenta(b)benzofuran-5-butanoic acid;* dl-16-methyl-18,19-tetradehydro-5,-6,7-trinor-4,8-inter-*m*-phenyleneprostaglandin I$_2$; dl-4-[(1*R*,-2*R*,3a*S*,8b*S*)-1,2,3a,8b-tetrahydro-2-hydroxy-[(3*S*,4*RS*)-3-hydroxy-4-methyl-oct-6-yne-(*E*)-1-enyl]-5-cyclopenta(b)-benzofuranyl]butanoic acid; befaprost; ML-1229. $C_{24}H_{30}O_5$; mol wt 398.50. C 72.34%, H 7.59%, O 20.07%. Stable analog of PGI$_2$. Prepn: K. Ohno *et al.,* **Eur. pat. Appl. 84,856;** *eidem,* **U.S. pat. 4,474,802** (1983, 1984 both to Toray); *eidem, Adv. Prostaglandin, Thromboxane, Leukotriene Res.* **15**, 279 (1985). PGI$_2$ receptor binding: N. Kajikawa *et al., Arzneimittel-Forsch.* **39**, 495 (1989). Series of articles on pharmacology, antithrombotic activity and peripheral circulation effects: *ibid.* 856-876. Cerebroprotective effects in dog: J. Kurihara *et al., Brit. J. Pharmacol.* **99**, 91 (1990).

Sodium salt, $C_{24}H_{29}NaO_5$, *ML-1129, TRK-100, Dorner, Procylin.*

THERAP CAT: Improves ischemic syndromes.

1191. Berbamine. *6,6',7-Trimethoxy-2,2'-dimethylberbaman-12-ol.* $C_{37}H_{40}N_2O_6$; mol wt 608.73. C 73.01%, H 6.62%, N 4.60%, O 15.77%. From *Berberis vulgaris* L. and *Berberis aquifolium* Pursh, *Berberidaceae:* Rüdel, *Arch. Pharm.* **229**, 631 (1891); Pommerehne, *ibid.* **233**, 127 (1895). From *Atherosperma moschatum* Labill., *Monimiaceae:* Bick *et al., Aust. J. Chem.* **9**, 111 (1956). Isomer of oxyacanthine: Santos, *C.A.* **24**, 1647 (1930). Structure: v. Bruchhausen *et al., Ann.* **507**, 144 (1933); Inubushi, *J. Pharm. Soc. Japan.* **72**, 220 (1952), *C.A.* **47**, 6429e (1953).

Crystals from petr ether, mp 197-210°. $[\alpha]_D^{20}$ +114.6° (chloroform); pKa 7.33 at 20° in methanol. Slightly sol in water; sol in alcohol, ether, chloroform, petr ether.

Hydrochloride, $C_{37}H_{40}N_2O_6 \cdot HCl \cdot 4H_2O$. $[\alpha]_D^{20}$ +63.2°.

Dihydrochloride, $C_{37}H_{40}N_2O_6 \cdot 2HCl$, dec 270°.

Hydrobromide, $C_{37}H_{40}N_2O_6 \cdot HBr$, dec 283°.

Methyl ether, $C_{38}H_{42}N_2O_6$, mp 182°. $[\alpha]_D$ +132°. Occurs naturally as the alkaloid *isotetrandrine.*

1192. Berberine. *5,6-Dihydro-9,10-dimethoxybenzo[g]-1,3-benzodioxolo[5,6-a]quinolizinium; 7,8,13,13a-tetrahydro-9,10-dimethoxy-2,3-(methylenedioxy)berbinium;* umbellatine. $[C_{20}H_{18}NO_4]^+$. Alkaloid isolated from *Hydrastis canadensis* L., *Berberidaceae;* found in many other plants: Wasicky, Joachimowitz, *Arch. Pharm.* **255**, 497 (1917); Stermitz, *J. Pharm. Sci.* **56,** 760 (1967); Kawashima, *J. Pharm. Soc. Japan* **89,** 1386 (1969), *C.A.* **72,** 15706v (1970). Structure: Perkin, Robinson, *J. Chem. Soc.* **97,** 305 (1910). Identity with umbellatine: Govindachari *et al., Proc. Indian Acad. Sci.* **47A,** 41 (1958), *C.A.* **52,** 14630f (1958). Tautomeric pseudobase and imino aldehyde forms: Pictet, Gams, *Compt. Rend.* **152,** 102 (1911); **153,** 386 (1911); Perkin *et al., J. Chem. Soc.* **127,** 740 (1925); Skinner, *ibid.* **1950,** 823. Biosynthesis: Gear, Spenser, *Can. J. Chem.* **41,** 783 (1963). Pharmacology: Fukuda *et al., Chem. Pharm. Bull.* **18,** 1299 (1970); Shanbhag *et al., Japan. J. Pharmacol.* **20,** 482 (1970), *C.A.* **74,** 97717c (1971). Synthetic studies: Sainsbury *et al., Tetrahedron* **25,** 1881 (1969). Total synthesis of berberine iodide: Kametani *et al., J. Chem. Soc. (C)* **1969,** 2036. *Review:* Hahn, Ciak, in *Antibiotics* vol. 3, J. W. Corcoran, F. E. Hahn, Eds. (Springer-Verlag, New York, 1975) pp 577-584.

Yellow needles from ether, mp 145°: Gadamer, *Arch. Pharm.* **243,** 33 (1905). Dissolves slowly in water with alkaline reaction; behaves as a quaternary base, forming salts by replacement of the OH group: Perkin, Jr., *J. Chem. Soc.* **113,** 503 (1918). uv max: 265, 343 nm. pK 2.47. K 3.35 × 10⁻³.

Hydroxide hemihendecahydrate, $C_{20}H_{19}NO_5 \cdot 5\frac{1}{2} H_2O$, long, silky, yellow needles. Can be dried to the hemipentahydrate at 100°, begins to dec at 110°. One gram dissolves in 20 ml water, 100 ml alcohol; much more sol in hot liquids; slightly sol in acetone, ether, benzene, chloroform.

Acid sulfate, $C_{20}H_{19}NO_8S$, **berberine bisulfate.** Yellow needles. Sol in about 100 parts water; slightly sol in alc.

Chloride dihydrate, $C_{20}H_{18}NO_4 \cdot Cl \cdot 2H_2O$, yellow crystals. Slightly sol in cold, freely in boiling water. Practically insol in cold alcohol, chloroform, ether.

Sulfate trihydrate, $(C_{20}H_{18}NO_4)_2 \cdot SO_4 \cdot 3H_2O$, yellow needles. Sol in about 30 parts water; sol in alcohol. LD_{50} in mice: 24.3 mg/kg i.p. Activity studies: Amin *et al., Can. J. Microbiol.* **15,** 1067 (1969). Pharmacology: Sabir, Bhide, *Indian J. Physiol. Pharmacol.* **15,** 111 (1971).

THERAP CAT: Bitter stomachic. Antibacterial. Antimalarial, antipyretic.

1193. Berberis. Holly-leaved barberry; Oregon grape root; mountain grape. Dried rhizome and roots of species of section of *Mahonia* DC. of the genus *Berberis* L., *Berberidaceae. Habit.* U.S. and British Columbia. *Constit.* Berberine, berbamine, oxyacanthine, phytosterol, sugar.

THERAP CAT: Bitter. Antipyretic.

1194. Berberis Aristata. Indian barberry; ruswut; rusat. Dried stem of *Berberis aristata* DC., *Berberidaceae. Habit.* India, Ceylon. *Constit.* Berberine and other alkaloidal substances, tannin, resin, starch.

THERAP CAT: Bitter. Antipyretic.

1195. Berbine. *5,8,13,13a-Tetrahydro-6H-dibenzo[a,g]-quinolizine;* tetrahydroprotoberberine. $C_{17}H_{17}N$; mol wt 235.33. C 86.77%, H 7.28%, N 5.95%. Proposed as parent substance for naming berberine and similar alkaloids: Awe, *Arch. Pharm.* **270,** 156 (1932). Synthesis: Chakravarti *et al., J. Chem. Soc.* **1927,** 2278.

Clusters of needles from ether, mp 85°. Strong base, dissolves readily in alcohol, ether. Insol in petr ether.

1196. Bergapten(e). *4-Methoxy-7H-furo[3,2-g][1]benzopyran-7-one;* 5-methoxypsoralen; bergaptan; heraclin; majudin; 5-MOP; Psoraderm. $C_{12}H_8O_4$; mol wt 216.19. C 66.67%, H 3.73%, O 29.60%. Naturally occurring analog of psoralen and isomer of methoxsalen, *q.q.v.,* found in a wide variety of plants. It was first isolated from oil of bergamot from *Citrus bergamia* Risso, *Aurantiodiae:* Pomeranz, *Monatsh.* **12,** 379 (1891), **14,** 28 (1893). Isoln from *Fagara xanthoxyloides* Lam., *Rutaceae:* H. Thoms, E. Baetcke, *Ber.* **44,** 3326 (1911); **45,** 3705 (1912). Synthesis: E. Späth *et al., Ber.* **70,** 478 (1937); W. N. Howell, R. Robertson, *J. Chem. Soc.* **1937,** 293; G. Caporale, *Farmaco Ed. Sci.* **13,** 784 (1958); V. K. Ahluwalia *et al., Indian J. Chem.* **7,** 831 (1969). Use in photochemotherapy of psoriasis: H. Hönigsmann *et al., Brit. J. Dermatol.* **101,** 369 (1979). Mutagenicity studies: B. R. Scott *et al., Mutat. Res.* **39,** 29 (1976); M. J. Ashwood-Smith *et al., Nature* **285,** 407 (1980). Phototoxicity study: A. Kornhauser *et al., Science* **217,** 733 (1982).

Needles from alcohol, mp 188° (sublimes). Practically insol in boiling water. Slightly sol in glacial acetic acid, chloroform, benzene, warm phenol. Sol in abs alcohol: 1 part in 60. Its soln in sulfuric acid is yellow-gold.

USE: Has been used to promote tanning in suntan preparations.

THERAP CAT: Antipsoriatic.

1197. Bergenin. *3,4,4a,10b-Tetrahydro-3,4,8,10-tetrahydroxy-2-(hydroxymethyl)-9-methoxypyrano[3,2-c][2]benzopyran-6(2H)-one; 4-methoxy-2-[tetrahydro-3,4,5-trihydroxy-6-(hydroxymethyl)pyran-2-yl]-α-resorcylic acid δ-lactone;* bergenit; vakerin; ardisic acid B; cuscutin; peltophorin. $C_{14}H_{16}O_9$; mol wt 328.28. C 51.22%, H 4.91%, O 43.86%. From root of *Saxifraga (Bergenia) crassifolia* L., and from rhizome of *S. sibirica* L. *Saxifragaceae:* Morelle, *Compt. Rend.* **93,** 646 (1881); *Ber.* **14,** 2694 (1881); Ssadikow, Guthner, *Biochem. Z.* **190,** 340 (1927); Tschitschibabin *et al., Ann.* **469,** 93 (1929). Identity with vakerin: Carruthers *et al., Chem. & Ind. (London)* **1957,** 76. Identity with ardisic acid B: Hung, Chu, *C.A.* **52,** 15827h (1958). Identity with cuscutin: Jain, Mishra, *Indian J. Chem.* **1,** 499 (1963). Identity with peltophorin: Joshi, Kamat, *Naturwiss.* **56,** 89 (1969). Structure: Posternak, Dürr, *Helv. Chim. Acta* **41,** 1159 (1958); Fujise *et al., Bull. Chem. Soc. Japan* **32,** 97 (1959). Synthesis and structure: Hay, Haynes, *J. Chem. Soc.* **1958,** 2231. Biosynthesis: Wenkert, *Chem. & Ind. (London)* **1959,** 906.

Crystals from methanol, mp 238°. uv max: 275, 220 nm (log ε 3.92, 4.42). $[\alpha]_D^{18}$ −37.7° (c = 1.96 in ethanol); $[\alpha]_D^{24}$ −45.3° (c = 0.51 for anhydr in water). Freely sol in water; sol in alcohol.

Monohydrate, crystals from water, mp 140°. Slightly sol in water, freely in alcohol.

1198. Berkelium. Bk; at. no. 97; valence 3, 4. Manmade radioactive element; second element in the curide series. No stable nuclides, known isotopes (mass numbers): 240, 242-251. Longest-lived known isotope: 247 (T₁/₂ 1.38

$\times 10^3$ years, α-decay, rel. at. mass 247.0703). Prepn of first isotope, ^{243}Bk ($T_{1/2}$ 4.5 hrs, electron capture and α-decay) by helium ion bombardment of ^{241}Am: S. G. Thompson *et al.*, *Phys. Rev.* **77**, 838 (1950); *eidem, ibid.* **80**, 781. Prepn of macroscopic quantities as ^{249}Bk ($T_{1/2}$ 320 days, β-decay): B. B. Cunningham, *J. Chem. Ed.* **36**, 32-37 (1959). Prepn of metal and determn of crystal structure: J. A. Fahey, *U.S. At. Energy Comm.* **TID-25741** (1971) 119 pp, *C.A.* **76**, 9882r (1972); J. R. Peterson *et al.*, *J. Inorg. Nucl. Chem.* **33**, 3345 (1971). *Reviews:* M. Haissinsky, J.-P. Adloff, *Radiochemical Survey of the Elements* (Elsevier, New York, 1965) pp 14-15; C. Keller, *The Chemistry of the Transuranium Elements* (Verlag Chemie, Weinheim, English Ed., 1971) pp 553-566; Silva, "Trans-Curium Elements" in *MTP Int. Rev. Sci.: Inorg. Chem., Ser. One* vol. 8, A. G. Maddock, Ed. (University Park Press, Baltimore, 1972) pp 71-105; *Comprehensive Inorganic Chemistry* vol. 5, J. C. Bailar, Jr. *et al.*, Eds. (Pergamon Press, Oxford, 1973) *passim; Handb. Exp. Pharmakol.* **36**, 689-928 (1973); D. E. Hobart, J. R. Peterson in *The Chemistry of the Actinide Elements* vol. 1, J. J. Katz *et al.*, Eds. (Chapman and Hall, New York, 1986) pp 989-1024.

Metal. Two allotropic forms: double hexagonal close-packed α-form, d (calc) 14.78, transforms to β-form at 930°; face-centered cubic β-form, d (calc) 13.25, exists from 930-986°. mp 986 ± 25° (Fahey). Also reported as mp 1050° (Katz *et al., loc. cit.* vol. 2, p 1150). Changes from the trivalent to the tetravalent state under the influence of oxidizing agents. In the trivalent state, its chemical properties are very close to those of curium. Can be separated from other transuranium elements by ion-exchange or by extraction of Bk(IV) with dioctylphosphoric acid in heptane (Haissinsky, Adloff). Metal dissolves rapidly in aqueous mineral acids.

Caution: Radiation hazard; handling requires special equipment and sheilding facilities (Katz *et al., loc. cit.* vol. 2, p 1128).

1199. Bermoprofen. (±)-10,11-Dihydro-α,8-dimethyl-11-oxodibenz[b,f]oxepin-2-acetic acid; 2-(8-methyl-10,11-dihydro-11-oxodibenz[b,f]oxepin-2-yl)propionic acid; AD-1590; AJ-1590; Dibenon. $C_{18}H_{16}O_4$; mol wt 296.32. C 72.96%, H 5.44%, O 21.60%. Prostaglandin synthetase inhibitor. Prepn: H. Uno *et al.*, *Eur. pat. Appl.* **3,893**; *eidem*, U.S. pat. **4,238,620** (1979, 1980 both to Dainippon); Y. Nagai *et al.*, *J. Med. Chem.* **25**, 1065 (1982). Pharmacology: H. Nakamura *et al.*, *Arzneimittel-Forsch.* **33**, 1555 (1983). Mechanism of action: K. Ishii *et al.*, *Yakugaku Zasshi* **108**, 1001 (1988), *C.A.* **110**, 33462v (1989).

Crystals from toluene, mp 128-129°. LD_{50} in mice, rats (mg/kg): 500, 147 orally (Nakamura).

THERAP CAT: Anti-inflammatory; analgesic; antipyretic.

1200. Berninamycin. *Berninamycin A.* $C_{51}H_{51}N_{15}O_{15}$S; mol wt 1146.12. C 53.45%, H 4.49%, N 18.33%, O 20.94%, S 2.80%. Major component of cyclic peptide antibiotic complex also containing minor component *Berninamycin B* ($C_{51}H_{51}N_{15}O_{14}$S). Isoln from *Streptomyces bernesis* Dietz: M. E. Bergy *et al.*, U.S. pat. **3,689,639** (1972). Inhibition of protein synthesis: F. Reusser, *Biochemistry* **8**, 3303 (1969); J. Thompson *et al.*, *J. Gen. Microbiol.* **128**, 875 (1982). Structural studies and characterization of degradation product berninamycinic acid, a novel sulfur containing moiety: J. M. Liesch *et al.*, *J. Am. Chem. Soc.* **98**, 299 (1976); J. M. Liesch *et al.*, *ibid.* **8237**; J. M. Liesch, K. L. Rinehart, Jr., *ibid.* **99**, 1645 (1977). Revised structure: H. Abe *et al.*, *Tetrahedron Letters* **29**, 1401 (1988). Confirmation of structures: R. C. M. Lau, K. L. Rinehart, *J. Antibiot.* **47**, 1466 (1994). Biosynthetic study: C. J. Pearce, K. L. Rinehart, Jr., *J. Am. Chem. Soc.* **101**, 5069 (1979). Use as growth permittant: R. M. Pellegrino, *Eur. pat. Appl.* **112,233** (1984 to Merck & Co.). Synthesis of berninamy-

cinic acid: T. R. Kelly *et al.*, *Tetrahedron Letters* **25**, 2127 (1984).

Berninamycin A

White crystals, mp > 290°. Sol in DMF, methanol, ethanol, propanol, butanol. Relatively insol in water, ether, cyclohexane, benzene, acetone, ethyl acetate. uv max (methanol): 208, 236 nm (A = 62.4, 64.4).

Berninamycinic acid, $C_{12}H_6N_2O_5S$, golden needles, mp 210° (dec). pKa 5.8. Insol in water. uv max (0.01N HCl): 228, 272 nm (ϵ 13500, 13500); uv max (0.01N NaOH): 232, 294 nm (ϵ 9500, 13000).

Note: The characteristics of berninamycin are similar to those previously reported in the literature for *theiomycetin*: M. Shibata, *Takeda Kenkyusho Nempo* **18**, 44 (1959), *C.A.* **54**, 19840c (1960).

1201. Beryllium. Glucinium. Be; at. wt 9.01218; at. no. 4; valence 2. Group IIA(2). Alkaline earth metal. Estimated abundance in earth's crust 2 to 6 ppm. Natural isotopes (mass number): 9 (100%); known artificial radioactive isotopes: 6-8; 10-12. Oxide discovered by Vauquelin in 1797; free metal isolated by Wöhler and Bussy in 1828. Produced industrially from *beryl* ($3BeO.Al_2O_3.6SiO_2$); also found in *phenacite* (Be_2SiO_4), *chrysoberyl* ($BeO.Al_2O_3$). Precious forms of beryl: *emerald, aquamarine*. Reviews of beryllium and its compounds: Kjellgren, "Beryllium" in *Rare Metals Handbook*, C. A. Hampel, Ed. (Reinhold, New York, 1954) pp 31-55; D. A. Everest, *Chemistry of Beryllium* (Elsevier, New York, 1964) 151 pp. Review of properties and use: N. P. Pinto, J. Greenspan, "Beryllium" in *Modern Materials*, vol. 6, B. W. Gonser, Ed. (Academic Press, New York, 1968) pp 319-372; D. A. Everest, "Beryllium" in *Comprehensive Inorganic Chemistry*, J. C. Bailar, Jr. *et al.*, Eds. (Pergamon Press, Oxford, 1973) pp 531-590; A. J. Stonehouse *et al.*, in Kirk-Othmer *Encyclopedia of Chemical Technology* vol. 4 (Wiley-Interscience, New York, 4th ed., 1992) pp 126-146; A. J. Stonehouse, M. N. Emly, *ibid* 147-153. Review of carcinogenic risk of beryllium and its compds: *IARC Monographs* **1**, 17-28 (1972); *ibid.* **23**, 143-204 (1980); of toxicology and human exposure: *Toxicological Profile for Beryllium* (PB93-182392, 1993) 153 pp.; of immunotoxicology: A. L. Reeves, O. P. Preuss in *Immunotoxicology and Immunopharmacology*, J. H. Dean *et al.*, Eds. (Raven Press, New York, 1985) pp 441-455; of human and environmental toxicology: WHO, *Environ. Health Criteria* **106**, 1-210 (1990).

Gray, brittle metal; close-packed hexagonal structure; anisotropic; high permeability to X-rays; can serve as a neutron source through either (α, n) or $(n, 2n)$ reactions. Lightest of all solid and chemically-stable substances. Displays excellent electrical and thermal conductivities. mp 1287°. bp 2500° (extrapolated). d 1.8477. Heat capacity at constant pressure (30°) 0.437 cal/g/°C: Walker *et al.*, *J. Chem. Eng. Data* **7**, 595 (1962). Latent heat of fusion: 3.5 kcal/

mole. Brinell hardness: 60-125. Chemical properties similar to aluminum; metal resistant to attack by acid due to the formation of a thin oxide film. E° (aq) Be/Be^{2+} 1.85 V (calc.). Finely divided or amalgamated metal reacts with HCl, dil H_2SO_4 and dil HNO_3; attacked by strong bases with evolution of H_2.

Caution: Overexposure to beryllium and beryllium compounds has been associated with acute and chronic toxicity. Toxicity due to chronic inhalation is called "berylliosis" and appears after a latent period as an immunologically mediated, progressive, systemic disease leading to the formation of a characteristic granulomatous lesion. Dermal exposure may cause acute contact dermatitis; chronic dermal exposure may cause granulomatous skin ulceration. *See Patty's Industrial Hygiene and Toxicology* vol. **2C**, G. D. Clayton, F. E. Clayton, Eds. (John Wiley & Sons, Inc., New York, 4th ed., 1994) pp 1930-1948; *NIOSH Pocket Guide to Chemical Hazards* (DHHS/NIOSH 94-116, 1994) p. 28. Beryllium and certain beryllium compounds may reasonably be anticipated to be carcinogens: *Seventh Annual Report on Carcinogens* (PB95-109781, 1994) p. 96.

USE: Source of neutrons when bombarded with alpha particles according to the equation $^9_4Be + ^4_2He \rightarrow ^{12}_6C + ^1_0n$. This yields about 30 neutrons per million alpha particles. Also as neutron reflector and neutron moderator in nuclear reactors. In beryllium copper and beryllium aluminum alloys (by direct reduction of beryllium oxide with carbon in the presence of Cu or Al). In aerospace, aircraft and satellite structures; x-ray transmission windows; missile parts; nuclear weapons; fuel containers; precision instruments; rocket propellants; navigational systems; heat shields; and mirrors.

1202. Beryllium Acetate. *Acetic acid beryllium salt.* $C_4H_6BeO_4$; mol wt 127.10. C 37.80%, H 4.76%, Be 7.09%, O 50.35%. $(CH_3COO^-)_2Be^{2+}$. Prepn: Besson, Hardt, *Compt. Rend.* **237**, 1525 (1953).

Crystals, dec at 60-100° when heated slowly, 150-180° when heated rapidly. Dissolves slowly with hydrolysis in boiling water. Practically insol in abs alc and other common organic solvents.

1203. Beryllium Acetate, Basic. *Hexakis(acetato)oxotetraberyllium;* beryllium oxide acetate. $C_{12}H_{18}Be_4O_{13}$; mol wt 406.32. C 35.47%, H 4.47%, Be 8.87%, O 51.19%. Be$_4$O-(OCOCH$_3$)$_6$. Prepn: Urbain, Lacombe, *Compt. Rend.* **133**, 874 (1901); **134**, 772 (1902); Haber, van Oordt, *Z. Anorg. Chem.* **40**, 465 (1904); Moeller *et al., Inorg. Syn.* **3**, 9 (1950).

Tetrahedra from chloroform, d 1.25. mp 285-286°. bp 330-331°. Practically insol in water. Hydrolyzed by hot water and by dil acids. Sol in hot glacial acetic acid; in the usual organic solvents except alcohol, ether.

1204. Beryllium Acetylacetonate. *(T-4)-Bis(2,4-pentanedionato-O,O')beryllium.* $C_{10}H_{14}BeO_4$; mol wt 207.23. C 57.96%, H 6.81%, Be 4.35%, O 30.88%. Prepd by the action of 2,4-pentanedione on an soln of beryllium hydroxide in dil acetic acid or on a soln of beryllium chloride in presence of ammonia: Biltz, *Ann.* **331**, 336 (1904); Parsons, *J. Am. Chem. Soc.* **26**, 732 (1904); Arch, Young, *Inorg. Syn.* **2**, 17 (1946); from 2,4-pentanedione and beryllium sulfate in NaOH soln: Jones, *J. Am. Chem. Soc.* **81**, 3188 (1959).

Monoclinic crystals. mp 108°. bp 270°. d$_4^{20}$ 1.168. Practically insol in water; hydrolyzed by boiling water. Freely sol in alc, acetone, ether, benzene, CS$_2$, other organic solvents.

1205. Beryllium Borohydride. *Beryllium tetrahydroborate(1—).* B_2BeH_8; mol wt 38.70. B 55.87%, Be 23.29%, H 20.84%. Be(BH$_4$)$_2$. Prepn from diborane and dimethylberyllium: Burg, Schlesinger, *J. Am. Chem. Soc.* **62**, 3425

(1940); from LiBH$_4$ and BeCl$_2$: Schlesinger *et al., ibid.* **75**, 212 (1953).

Similar to aluminum borohydride, but less volatile. Spontaneously flammable. Sublimes at 91.3°. Dec above 123° without melting. Reacts vigorously with water, HCl to liberate hydrogen.

1206. Beryllium Bromide. BeBr$_2$; mol wt 168.82. Be 5.34%, Br 94.66%. Prepn: Ehrlich in *Handbook of Preparative Inorganic Chemistry,* vol. 1, G. Brauer, Ed. (Academic Press, New York, 2nd ed., 1963) p 891. Review of beryllium halides: Bell, *Advan. Inorg. Chem. Radiochem.* **14**, 255-332 (1972).

Orthorhombic crystals, d 3.465. mp 506-509°; also reported as 488°: Bell, *loc. cit.* Sublimes at 473°. bp 520°. Very hygroscopic. Freely sol in water. By saturating the concd viscous soln with HBr, the tetrahydrate is formed. Sol in ethanol, in pyridine (185.6 g/l), in ethyl bromide (1.0 g/l). Forms addition compounds with amines, alcohols. *Keep tightly closed.*

1207. Beryllium Carbide. CBe$_2$; mol wt 30.04. C 39.99%, Be 60.01%. Be$_2$C. Prepn: Coobs, Koshuba, *J. Electrochem. Soc.* **99**, 115 (1952); Mallett *et al., ibid.* **101**, 298 (1954); Ehrlich in *Handbook of Preparative Inorganic Chemistry,* vol. 1, G. Brauer, Ed. (Academic Press, New York, 2nd ed., 1963) p 899.

Brick-red or yellow-red octahedra, d 1.90, dec above 2100°. Very slowly dec by water, somewhat faster by mineral acids and quickly by alkalies with the evolution of methane.

USE: Nuclear reactor core material: Schwartz, U.S. pat. **3,170,812** (1965 to USAEC).

1208. Beryllium Chloride. BeCl$_2$; mol wt 79.92. Be 11.28%, Cl 88.72%. Prepn from the elements: Tannenbaum, *Inorg. Syn.* **5**, 22 (1957); from BeO, Cl$_2$ and C: Ehrlich in *Handbook of Preparative Inorganic Chemistry,* vol. 1, G. Brauer, Ed. (Academic Press, New York, 2nd ed., 1963) p 889. Toxicity data: K. W. Cochran *et al., Fed. Proc.* **9**, 264 (1950). Review of beryllium halides: Bell, *Advan. Inorg. Chem. Radiochem.* **14**, 255-332 (1972).

White to faintly yellow, very deliquesc, orthorhombic crystals or cryst mass. Reported mp ranges from 399.2° to 440°. 399.2° is considered to be the most reliable (Bell). bp 482.3°. Sublimes *in vacuo* at 300°. d 1.90. Very sol in water with evolution of heat; the aq soln is strongly acid. Sol in alcohol, ether, pyridine, CS$_2$. Insol in benzene, toluene. *Keep tightly closed.*

Tetrahydrate, monoclinic deliquesc platelets. Has been reported to have 4½H$_2$O: Semenenko, Turova, *Russ. J. Inorg. Chem.* **10**, 42 (1965). LD$_{50}$ in guinea pigs, rats (mg Be/kg): 63, 0.6 i.p. (Cochran).

USE: Manuf of beryllium. *Anhydrous* form used as acid catalyst in organic reactions, similar to AlCl$_3$.

1209. Beryllium Fluoride. BeF$_2$; mol wt 47.01. Be 19.17%, F 80.83%. Prepd by heating ammonium fluoroberyllate (NH$_4$)$_2$BeF$_4$: Lebeau, *Compt. Rend.* **126**, 1418 (1898); Kwasnik in *Handbook of Preparative Inorganic Chemistry,* vol. 1, G. Brauer, Ed. (Academic Press, New York, 2nd ed., 1963) p 231. Review of prepn and properties of beryllium halides: Bell, *Advan. Inorg. Chem. Radiochem.* **14**, 255-332 (1972).

Glassy hygroscopic mass (tetragonal system). True mp 555°; becomes free-flowing about 800°. Sublimes at 1036° under 1 mm pres. in the presence of beryllium. d$_4^{25}$ 1.986. Very freely sol in water; sparingly sol in alc; more sol in a mixture of alc and ether; insol in anhydr HF.

USE: Manuf of Be and Be alloys; manuf of glass; in nuclear reactors.

1210. Beryllium Formate. *Formic acid beryllium salt.* $C_2H_2BeO_4$; mol wt 99.05. C 24.25%, H 2.04%, Be 9.10%, O 64.61%. Be(OOCH)$_2$. Prepn: Besson, Hardt, *Compt. Rend.* **238**, 355 (1954).

Powder, decomp above 250° to the basic formate, Be$_4$O-(HCOO)$_6$, which sublimes without melting at about 320°. Very slowly hydrolyzed by water. Practically insol in the usual organic solvents. Sol in hot pyridine, but on cooling a pyridine complex crystallizes from the soln.

1211. Beryllium Hydride. BeH$_2$; mol wt 11.03. Be 81.72%, H 18.28%. Lower purity material prepd by treating dimethylberyllium with LiAlH$_4$ in ether: Barbaras *et al., J. Am. Chem. Soc.* **73**, 4585 (1951); higher purity by pyrolysis of di-*tert*-butylberyllium: Coates, Glocking, *J. Chem. Soc.* **1954**, 2526; Head *et al., J. Am. Chem. Soc.* **79**, 3687 (1957); from triphenyl phosphine and beryllium borohydride: Banford, Coates, *J. Chem. Soc.* **1964**, 5591.

White solid. Higher purity material is inert to laboratory air. Loss of hydrogen at 190-200° negligible, rapid at 220°. Reacts slowly with water, rapidly with dil acids. Insol in ether, toluene, isopentane. Reacts with diborane to form beryllium borohydride.

1212. Beryllium Hydroxide. BeH$_2$O$_2$; mol wt 43.03. Be 20.95%, H 4.69%, O 74.37%. Be(OH)$_2$. Prepn: Ehrlich in *Handbook of Preparative Inorganic Chemistry,* **vol. 1,** G. Brauer, Ed. (Academic Press, New York, 2nd ed., 1963) p 894.

Amorphous powder or crystals. d 1.92. Amphoteric. Very slightly sol in water and dil alkali. Sol in hot concd NaOH soln and acids.

USE: Manuf of beryllium and beryllium oxide.

1213. Beryllium Iodide. BeI$_2$; mol wt 262.82. Be 3.43%, I 96.57%. Prepn: Messerknecht, Biltz, *Z. Anorg. Chem.* **148**, 152 (1925); Ehrlich in *Handbook of Preparative Inorganic Chemistry,* **vol. 1,** G. Brauer, Ed. (Academic Press, New York, 2nd ed., 1963) p 892. Review of beryllium halides: Bell, *Advan. Inorg. Chem. Radiochem.* **14,** 255-332 (1972).

Needles, mp 480°, bp 488°. Very hygroscopic. Sublimes *in vacuo.* Reacts violently with water, giving off HI. Absorbs ammonia. Dissolves in alcohols, amines, with the formation of addition compds. *Keep tightly closed.*

1214. Beryllium Nitrate. BeN$_2$O$_6$; mol wt 133.02. Be 6.77%, N 21.06%, O 72.17%. Be(NO$_3$)$_2$. Prepn: *Gmelin's, Beryllium* (8th ed.) **26,** 102-104 (1930).

Trihydrate, white to slightly yellow, deliquesc cryst mass. mp ~60°. Very sol in water, alcohol. *Keep well closed in a cool place.* LD$_{50}$ i.p. in guinea pigs: 50 mg/kg, *Handbook of Toxicology* **vol. 1, W.** S. Spector, Ed. (Saunders, Philadelphia, 1956) pp 46-47.

USE: Stiffening mantles in gas and acetylene lamps.

1215. Beryllium Nitride. Be$_3$N$_2$; mol wt 55.05. Be 49.11%, N 50.89%. Prepn: Ehrlich in *Handbook of Preparative Inorganic Chemistry,* **vol. 1,** G. Brauer, Ed. (Academic Press, New York, 2nd ed., 1963) p 898; Langsdorf, Jr., **U.S.** pat. **2,567,518** (1951 to USAEC).

White crystals to grayish white powder; mp 2200 ±40°. Volatile at bp, on further heating it dissociates into Be and N$_2$. Oxidized in air at 600°. Dec slowly by water, quickly by acids and alkalies with the evolution of ammonia.

1216. Beryllium Oxide. Beryllia. BeO; mol wt 25.01. Be 36.03%, O 63.97%. Prepn: *Gmelin's, Beryllium* (8th ed.) **26,** 82-91 (1930); Ehrlich in *Handbook of Preparative Inorganic Chemistry,* **vol. 1,** G. Brauer, Ed. (Academic Press, New York, 2nd ed., 1963) p 893. *Review:* Lillie, *USAEC* **UCRL 6457,** 23 pp (1961).

Light, amorphous powder. mp 2530°. Very sparingly sol in water; slowly sol in concd acids or solns of fixed alkali hydroxides. After ignition it is almost insol in these solvents. Pure (100%) BeO insulates electrically like a ceramic, but conducts heat like a metal. Electrical resistivity in ohm-cm: >10^{16}. Dielectric const at 8.5 gigacycles: 6.57.

USE: Manuf of beryllium oxide ceramics, glass; in nuclear reactor fuels and moderators; catalyst for organic reactions.

1217. Beryllium Perchlorate. BeCl$_2$O$_8$; mol wt 207.91. Be 4.33%, Cl 34.10%, O 61.56%. Be(ClO$_4$)$_2$. Prepn: *Gmelin's, Beryllium* (8th ed.) **26,** 121 (1930).

Tetrahydrate, very hygroscopic crystals. Holds its water of crystn tenaciously. Soly in water: 148.6 g/100 ml.

1218. Beryllium Potassium Fluoride. *Potassium tetrafluoroberyllate.* BeF$_4$K$_2$; mol wt 163.20. Be 5.52%, F 46.56%, K 47.91%. K$_2$BeF$_4$. Prepn: *Gmelin's, Beryllium* (8th ed.) **26,** 172 (1930). Review of prepn and properties of

beryllium halides: Bell, *Advan. Inorg. Chem. Radiochem.* **14,** 255-332 (1972).

Hard masses. Sol in water, practically insol in alc.

1219. Beryllium Potassium Sulfate. BeK$_2$O$_8$S$_2$; mol wt 279.34. Be 3.23%, K 27.99%, O 45.82%, S 22.96%. BeSO$_4$.-K$_2$SO$_4$. Prepn: *Gmelin's, Beryllium* (8th ed) **26,** 174 (1930).

Dihydrate, brilliant crystals. Sol in water, concd K$_2$SO$_4$ solns; practically insol in alc.

USE: In chromium- and silver-plating.

1220. Beryllium Selenate. BeO$_4$Se; mol wt 151.97. Be 5.93%, O 42.11%, Se 51.96%. BeSeO$_4$. Prepn: *Gmelin's, Beryllium* (8th ed.) **26,** 144 (1930).

Tetrahydrate, orthorhombic crystals, d 2.03. Changes to the dihydrate at 100° and becomes anhydr at 300°. Freely sol in water; aq solns of beryllium selenate are good solvents for beryllium oxide.

1221. Beryllium Sodium Fluoride. *Sodium tetrafluoroberyllate.* BeF$_4$Na$_2$; mol wt 130.99. Be 6.88%, F 58.02%, Na 35.10%. Na$_2$BeF$_4$. Prepn: *Gmelin's, Beryllium* (8th ed.) **26,** 169 (1930). Review of prepn and properties of beryllium halides: Bell, *Advan. Inorg. Chem. Radiochem.* **14,** 255-332 (1972).

Orthorhombic or monoclinic crystals. mp ~350°. Sol in water.

1222. Beryllium Sulfate. BeO$_4$S; mol wt 105.08. Be 8.58%, O 60.91%, S 30.52%. BeSO$_4$. Prepn: *Gmelin's, Beryllium* (8th ed.) **26,** 130-141 (1930). Toxicity study: White *et al., J. Pharmacol. Exp. Ther.* **102**, 88 (1951).

Tetrahydrate, crystals. d 1.71. At about 100° loses 2H$_2$O. Very sol in water. Practically insol in alc. LD$_{50}$ i.v. in mice: 0.5 mg Be/kg (White).

1223. Besipirdine. *N-Propyl-N-4-pyridinyl-1H-indol-1-amine;* 1-(propyl-4-pyridylamino)indole. C$_{16}$H$_{17}$N$_3$; mol wt 251.33. C 76.46%, H 6.82%, N 16.72%. Cholinomimetic agent with noradrenergic activity. Prepn: R. C. Effland, J. T. Klein, Eur. pat. **Appl. 287,982** (1988 to Hoechst); *idem et al.,* U.S. pat. **4,970,218** (1990 to Hoechst-Roussel); of hydrochloride: S. Kongsamut *et al.,* **U.S.** pat. **5,356,910** (1994 to Hoechst-Roussel). HPLC determn in plasma: R. S. Hsu *et al., J. Chromatog.* **572**, 352 (1991). Mechanism of action study: C. P. Smith *et al., Drug Dev. Res.* **32**, 13 (1994). Pharmacokinetics: J. W. Hubbard *et al., J. Clin. Pharmacol.* **35**, 688 (1995).

Hydrochloride, C$_{16}$H$_{17}$N$_3$.HCl, *HP-749.* Crystals from methanol, mp 212-214°.

Maleate, C$_{16}$H$_{17}$N$_3$.C$_4$H$_4$O$_4$, crystals from methanol/ether, mp 115-116°.

THERAP CAT: Nootropic.

1224. Betahistine. *N-Methyl-2-pyridineethanamine; 2-[2-(methylamino)ethyl]pyridine;* [2-(2-pyridyl)ethyl]methylamine. C$_8$H$_{12}$N$_2$; mol wt 136.20. C 70.55%, H 8.88%, N 20.57%. Prepn: Löffler, *Ber.* **37**, 161 (1904); Walter *et al., J. Am. Chem. Soc.* **63**, 2771 (1941).

Liquid. bp$_{30}$ 113-114°. Soluble in water, alcohol, ether, chloroform.

Dihydrochloride, C$_8$H$_{12}$N$_2$.2HCl, *Betaserc, Serc, Vasomotal.* Crystals from alc, mp 148-149°.

Maleate, C$_8$H$_{12}$N$_2$.C$_4$H$_4$O$_4$, *Suzutolon.*

Mesylate, $C_8H_{12}N_2 \cdot CH_3SO_3H$, *Medan, Menitazine, Merislon, Remark, Ribrain, Tenyl.*

Dimesylate, $C_8H_{12}N_2 \cdot (CH_3SO_3H)_2$, *Aequamen, Melopat.*

THERAP CAT: Vasodilator.

1225. Betaine. *1-Carboxy-N,N,N-trimethylmethanaminium inner salt;* (carboxymethyl)trimethylammonium hydroxide inner salt; glycine betaine; glycocoll betaine; lycine; oxyneurine; trimethylglycine hydroxide inner salt; trimethylglycocoll anhydride. $C_5H_{11}NO_2$; mol wt 117.15. C 51.26%, H 9.46%, N 11.96%, O 27.31%. $(CH_3)_3N^+CH_2-COO^-$. Widely distributed in plants and animals: M. Guggenheim, *Die biogenen Amine* (S. Karger, Basel, 4th ed., 1951) pp 240-242. Prepn via or from hydrochloride: Stoltzenberg, *Z. Physiol. Chem.* **92,** 445 (1914); Edsall, *J. Am. Chem. Soc.* **65,** 1767 (1943). Prepn of the hydrochloride: Kuhn, Ruelius, *Ber.* **83,** 420 (1950); of the hydrate: Vassel, U.S. pat. **2,800,502** (1957 to International Minerals and Chemical Corp.). Structure of hydrate: Leifer, Lippincott, *J. Am. Chem. Soc.* **79,** 5098 (1957). Prepn of the sodium aspartate: Thuiller, **Fr.** pat. **M2462,** *C.A.* **61,** 8405F (1964); **Belg.** pat. **638,361** (both 1964 to Labs. Rolland); Cote, **Fr.** pat. **1,356,945** (1964), *C.A.* **61,** 7098f (1964).

Deliquescent scales or prisms dec around 310°. (Isomerizes at the mp to methyl ester of dimethylaminoacetic acid.) Sweet taste. Solubility (g/100 g solvent): water, 160; methanol, 55; ethanol, 8.7. Sparingly sol in ether. Betaine yields trimethylamine with concd KOH.

Monohydrate, *(carboxymethyl)trimethylammonium hydroxide, trimethylglycine hydroxide.* Formed by crystn of betaine from aq solvents. pH of satd soln about 8.0. Loses water at 100° forming inner salt again. Mixture with glycocyamine, *Betasyamine, Betacyamine.*

Hydrochloride, $C_5H_{11}NO_2 \cdot HCl$, *1-carboxy-N,N,N-trimethylmethanaminium chloride, (carboxymethyl)trimethylammonium chloride, pluchine, Acidol.* Monoclinic crystals from alc, dec 227-228° (Stoltzenberg); also reported as dec 232° (Kuhn, Ruelius). Soly in water at 25° = 64.7 g/100 ml, in ethanol 5.0 g/100 ml. pH of 5% aq soln 1.0. Practically insol in chloroform.

Sodium aspartate, $C_9H_{17}N_2NaO_6$, *Somatyl.* Crystals, pptd from aq reaction mixture with acetone, hygroscopic, contg about 15% water (Thuiller, **Belg.** pat.). Also obtained as the trihydrate (Cote). Dec 160-170°. Soluble in water. Practically insol in acetone.

USE: In soldering, resin curing fluxes, organic synthesis.

THERAP CAT: Hepatoprotectant.

1226. Betamethasone. *(11β,16β)-9-Fluoro-11,17,21-trihydroxy-16-methylpregna-1,4-diene-3,20-dione;* 9α-fluoro-16β-methylprednisolone; 16β-methyl-9α-fluoro-Δ¹-hydrocortisone; 16β-methyl-9α-fluoroprednisolone; betadexamethasone; flubenisolone; β-methasone; Sch-4831; NSC-39470; beta-Corlan; Becort; Betasolon; Betnelan; Celestene; Celestone; Dermabet; Diprolene; Visubeta. $C_{22}H_{29}FO_5$; mol wt 392.47. C 67.33%, H 7.45%, F 4.84%, O 20.38%. Prepn: Taub et al., *J. Am. Chem. Soc.* **80,** 4435 (1958); Oliveto et al., *ibid.* 6688; Taub et al., *ibid.* **82,** 4012 (1960); U.S. pat. **3,053,865** (1962 to Merck & Co.); Amiard et al., U.S. pat. **3,104,246** (1963 to Roussel-UCLAF). Also prepared from hecogenin. Comprehensive description of the dipropionate ester: M. G. Ferrante, B. C. Rudy, *Anal. Profiles Drug Subs.* **6,** 43-60 (1977).

Crystals from ethyl acetate, mp 231-234° (dec). $[\alpha]_D$ +108° (acetone). uv max (methanol): 238 nm (ε 15200).

21-Acetate, $C_{24}H_{31}FO_6$, *Betafluorene, Celestovet.* Hexagonal prisms from acetone + ether, mp 205-208° (Taub); also reported as mp 196-201° (Oliveto). $[\alpha]_D$ +140° (chloroform). uv max (methanol): 238 nm (ε 14800).

21-Adamantoate, $C_{33}H_{43}FO_6$, *Betsovet.* Prepn: Philips, English, **Ger.** pat. **2,232,827** (1973 to Glaxo), *C.A.* **78,** 97893q (1973).

17-Benzoate, $C_{29}H_{33}FO_6$, *W-5975, Bebate, Beben, Benisone, Euvaderm, Flurobate, Parbetan, Uticort.* Crystals from acetone-ether, mp 225-228°. $[\alpha]_D^{24}$ +63.5° (dioxane). Synthesis and activity: Ercoli et al., *J. Med. Chem.* **15,** 783 (1972). *See also* Cullen, *Curr. Ther. Res.* **15,** 243 (1973).

17,21-Dipropionate, $C_{28}H_{37}FO_7$, *Sch-11460, Diproderm, Diprophos, Diprosis, Diprosone, Maxivate, Rinderon-DP.* Powder, mp 170-179° (dec). $[\alpha]_D^{26}$ +65.7° (dioxane). uv max (methanol): 238 nm (ε 15700).

17-Valerate, $C_{27}H_{37}FO_6$, *Bedermin, Betnesol-V, Betneval, Betnovate, Bextasol, Celestan-V, Celestoderm-V, Dermosol, Dermovaleas, Ecoval 70, Hormezon, Tokuderm, Valisone.* Needles from acetone + petr ether, mp 183-184°. $[\alpha]_D$ +77° (dioxane). uv max: 239 nm (ε 15920). **Neth.** pat. **Appl. 6,406,615** (1964 to Glaxo).

21-Phosphate disodium salt, $C_{22}H_{28}FNa_2O_8P$, *betamethasone 21-(dihydrogen phosphate) disodium salt, Bentelan, Betnesol, Celestan, Durabetason, Vista-Methasone.*

THERAP CAT: Glucocorticoid.

THERAP CAT (VET): Glucocorticoid.

1227. Betamipron. *N-Benzoyl-β-alanine;* 3-(benzoylamino)propionic acid; β-benzamidopropionic acid; CS-443. $C_{10}H_{11}NO_3$; mol wt 193.20. C 62.17%, H 5.74%, N 7.25%, O 24.84%. Acyl amino acid that inhibits active transport of carbapenem antibiotics in the renal cortex. *See also* panipenem. Prepn: F. H. Holm, *Arch. Pharm.* **242,** 590 (1904); C. C. Barker, *J. Chem. Soc.* **1954,** 317; J. Barluenga et al., *Tetrahedron* **45,** 2183 (1989). Use as renal protectant: T. Shiokari et al., **Eur.** pat. **Appl. 226,304;** *eidem,* U.S. pat. **4,757,066** (1987, 1988 both to Sankyo). Crystal structure: N. Feeder, W. Jones, *Acta Crystallog.* **C50,** 813 (1994). Nephroprotective effect: Y. Hirouchi et al., *Japan. J. Pharmacol.* **63,** 487 (1993). Mode of action: *eidem, ibid.* **66,** 1 (1994). Pharmacokinetics: A. Kurihara et al., *Antimicrob. Ag. Chemother.* **36,** 1810 (1992).

Colorless prisms from hot water, mp 120° (Holm); also reported as crystals from water, mp 133° (Barker). Readily sol in warm water, chloroform. Very easily sol in alcohol, ether, acetone. LD_{50} in rats (mg/kg): > 3000 i.v. (Hirouchi, 1994).

THERAP CAT: Antibacterial adjunct (renal protectant).

1228. Betasine. *β-Amino-4-hydroxy-3,5-diiodobenzenepropanoic acid;* β-amino-4-hydroxy-3,5-diiodohydrocinnamic acid; β-(4-hydroxy-3,5-diiodophenyl)-β-alanine; β-(4-hydroxy-3,5-diiodophenyl)-β-aminopropionic acid; β-diiodotyrosine; betasinum; betazine. $C_9H_9I_2NO_3$; mol wt 432.98. C 24.97%, H 2.10%, I 58.62%, N 3.23%, O 11.09%. Prepn: Rodionov et al., *Zh. Obshch. Khim.* **27,** 2234 (1957); Suvorov et al., **U.S.S.R.** pat. **104,779** (1957), *C.A.* **51,** 8794b (1957).

Crystals, mp 178-179° (dec). Sparingly soluble in water; sol in aqueous solutions of hydrochloric acid, caustic alkalies and ammonia. Practically insol in organic solvents.

Ammonium salt, crystals, mp 151-152° (dec). Difficultly soluble in cold water.

THERAP CAT: Iodine source.

1229. Betaxolol. *1-[4-[2-(Cyclopropylmethoxy)ethyl]phenoxy]-3-[(1-methylethyl)amino]-2-propanol;* (±)-1-(isopropylamino)-3-[*p*-(cyclopropylmethoxyethyl)phenoxy]-2-propanol. $C_{18}H_{29}NO_3$; mol wt 307.43. C 70.32%, H 9.51%, N 4.56%, O 15.61%. Cardioselective β_1-adrenergic blocker. Prepn: P. M. J. Manoury *et al.*, *Ger.* pat. **2,649,605** corresp to U.S. pat. **4,252,984** (1977, 1981 both to Synthelabo). Blood concn and pharmacodynamic effects: S. J. Warrington *et al.*, *Brit. J. Clin. Pharmacol.* **10**, 449 (1980). Pharmacokinetics: G. Bianchetti *et al.*, *Arzneimittel-Forsch.* **30**, 1912 (1980). Cardiovascular effects in normal volunteers: P. J. Cadigan *et al.*, *Brit. J. Clin. Pharmacol.* **9**, 569 (1980). Efficacy and pharmacokinetics: K. Balnave *et al.*, *ibid.* **11**, 171 (1981). Use in treatment of glaucoma: A. R. Berrospi, H. M. Leibowitz, *Arch. Ophthalmol.* **100**, 943 (1982). Antihypertensive effect: M. Pathe *et al.*, *Therapie* **37**, 75 (1982). *Book: Betaxolol and Other β_1-Adrenoceptor Antagonists*, P. L. Morselli *et al.*, Eds. (Raven Press, New York, 1983) 385 pp.

Crystals from petr ether, mp 70-72°.
Hydrochloride, $C_{18}H_{29}NO_3 \cdot HCl$, *SLD-212, SL-75.212, Betoptic, Betoptima, Kerlone.* Crystals from acetone, mp 116°. LD_{50} in mice (mg/kg): 944 orally; 37 i.v. (Manoury).
THERAP CAT: Antihypertensive; antiglaucoma.

1230. Betazole. *1H-Pyrazole-3-ethanamine; 3-(2-aminoethyl)pyrazole;* 3-(β-aminoethyl)pyrazole; ametazole; gastramine. $C_5H_9N_3$; mol wt 111.15. C 54.03%, H 8.16%, N 37.81%. Prepd by the catalytic reduction of 3-pyrazoleacetaldehyde hydrazone: Jones, Mann, *J. Am. Chem. Soc.* **75**, 4048 (1953); Jones, U.S. pat. **2,785,177** (1957 to Lilly).

Viscous liquid. $bp_{0.5}$ 118-123°.
Dihydrochloride, $C_5H_9N_3 \cdot 2HCl$, *Histimin, Histalog.* Crystals from ethanol, mp 224-226°. Sol in water. Practically insol in chloroform. Aq solns are acid to litmus.
THERAP CAT: Diagnostic aid (gastric secretion stimulant).

1231. Betel. Dried leaves of *Piper betle* L., *Piperaceae.* *Habit.* India, Ceylon, Malay Archipelago. *Constit.* 0.2-1% volatile oil, chavibetol, chavicol, cadinene, allylpyrocatechol. *Ref:* Ueda, Sasaki, *J. Pharm. Soc. Japan* **71**, 559 (1951), *C.A.* **45**, 9137g (1951).
THERAP CAT: Counterirritant.

1232. Bethanechol Chloride. *2-[(Aminocarbonyl)oxy]-N,N,N-trimethyl-1-propanaminium chloride; carbamate of (2-hydroxypropyl)trimethylammonium chloride;* (2-hydroxypropyl)trimethylammonium chloride carbamate; 2-carbamoyloxypropyltrimethylammonium chloride; carbamylmethylcholine chloride; urethan of β-methylcholine chloride; Duvoid; Urecholine chloride; Mechothane; Myocholine; Mictone; Myotonine chloride; Uro-Carb. $C_7H_{17}ClN_2O_2$; mol wt 196.68. C 42.75%, H 8.71%, Cl 18.03%, N 14.24%, O 16.27%. Prepn: Dalmer, Diehl, U.S. pat. **1,894,162** (1933); Major, Bonnett, U.S. pat. **2,322,375** (1943 to Merck & Co.).

Hygroscopic crystals, slight amine odor, dec 218-219°. One gram dissolves in 0.6 ml of water, in 12.5 ml of 95% alc. pH of an 0.5% aq soln 5.5-6.0. Aq solns may be sterilized by autoclaving at 120° for 20 minutes.

THERAP CAT: Cholinergic.
THERAP CAT (VET): Cholinergic. Has been used in urolithiasis of cats and in atonic conditions of the gut.

1233. Bethanidine. *N,N'-Dimethyl-N''-(phenylmethyl)-guanidine; 1-benzyl-2,3-dimethylguanidine; N-benzyl-N',N''-dimethylguanidine.* $C_{10}H_{15}N_3$; mol wt 177.25. C 67.76%, H 8.53%, N 23.71%. Adrenergic neuron blocking agent. Prepn: Walton, Ruffell, U.S. pat. **3,168,562** (1965 to Wellcome Found.); Brit. pats. **1,084,461** and **1,111,564** (1967 and 1968 to Wellcome Found.). Pharmacology: A. L. A. Boura, A. F. Green, *Brit. J. Pharmacol.* **20**, 36 (1963); J. A. Oates *et al.*, *Ann. N.Y. Acad. Sci.* **179**, 302 (1971). Pharmacokinetics: C. N. Corder, *J. Clin. Pharmacol.* **19**, 428 (1979). HPLC determn in plasma: J. R. Shipe *et al.*, *Clin. Chem.* **29**, 1793 (1983). Anti-arrhythmic activity: M. B. Bacaner, D. G. Benditt, *Am. J. Cardiol.* **50**, 728 (1982); J. C. Somberg *et al.*, *ibid.* **54**, 343 (1984). *Review:* A. F. Green, *Brit. J. Clin. Pharmacol.* **13**, 25-34 (1982).

Crystals from methanol + ether, mp 195-197°.
Hydriodide, $C_{10}H_{15}N_3 \cdot HI$, crystals from ethanol + ether, mp 141-146°.
Sulfate, $(C_{10}H_{15}N_3)_2 \cdot H_2SO_4$, *benzaidin, BW-467C60, Bendogen, Benzoxine, Betaling, Betanidol, Esbatal, Eusmanid, Hypersin, Tenathan.* LD_{50} in mice (mg/kg): 12 i.v.; 150 i.p.; 260 s.c.; 520 by stomach intubation (Boura, Green).
THERAP CAT: Antihypertensive.

1234. Betonicine. *(2S-trans)-2-Carboxy-4-hydroxy-1,1-dimethylpyrrolidinium inner salt; l-N,N-dimethyl-4-hydroxypyrrolidine-2-carboxylic acid betaine; l-1-methyl-4-hydroxypyrrolidine-2-carboxylic acid methylbetaine; l-4-hydroxystachydrine; l-4-hydroxyproline betaine.* $C_7H_{13}NO_3$; mol wt 159.19. C 52.82%, H 8.23%, N 8.80%, O 30.15%. Occurs in *Croton gubouga* S. Moore; in *Stachys officinalis* (L.) Trev. (*Betonica officinalis* L.), *Labiatae*; in *Achillea moschata* Jacq. and *A. millefolium* L., *Compositae*: Goodson, Clewer, *J. Chem. Soc.* **115**, 923 (1919); Guggenheim, *Die biogenen Amine* (S. Karger, New York, 1951) p 246; Miller, Chow, *J. Am. Chem. Soc.* **76**, 1353 (1954); Pailer, Kump, *Monatsh.* **90**, 396 (1959); *Arch. Pharm.* **293**, 646 (1960). Stereoisomeric with turicine (*cis*-form). Synthesis: Patchett, Witkop, *J. Am. Chem. Soc.* **79**, 185 (1957).

Blunt prisms from ethanol, dec 254-256°. Sweet taste. $[\alpha]_D^{20}$ −34.2° (c = 1.0). Readily sol in water or hot alcohol, slightly in cold alcohol. Practically insol in benzene, ether, chloroform, carbon tetrachloride.
Hydrochloride, $C_7H_{13}NO_3 \cdot HCl$, crystals from ethanol, dec 216-217° (after drying). $[\alpha]_D^{20}$ −24.2° (c = 0.0892 in water).
Aurichloride, $C_7H_{13}NO_3 \cdot HAuCl_4$, scaly clusters from water, mp 242°.

1235. Betoxycaine. *3-Amino-4-butoxybenzoic acid 2-[2-(diethylamino)ethoxy]ethyl ester;* 2-diethylaminoethoxyethyl 3-amino-4-butoxybenzoate. $C_{19}H_{32}N_2O_4$; mol wt 352.47. C 64.74%, H 9.15%, N 7.95%, O 18.16%. Prepn and properties: E. Cuingnet, U.S. pat. **3,209,022** (1965 to Corbiere). Ionization potential: A. Cier *et al.*, *Thérapie* **26**, 941 (1971). Investigations of possible antifibrillatory properties: P. Amaud *et al.*, *Compt. Rend. Soc. Biol.* **159**, 2427 (1965); G. Faucon *et al.*, *Thérapie* **21**, 1253 (1966).

Monohydrochloride, $C_{19}H_{32}N_2O_4$.HCl, *Millicaine.* White crystals from acetone, mp 117°. Soluble in methanol.

THERAP CAT: Anesthetic (local).

1236. Betula. European white birch. Bark and leaves of *Betula alba* L., *Betulaceae. Habit.* Europe and Northern Asia, also America, north of Pennsylvania. *Constit.* 10-15% Betulin (betula camphor), betuloresinic acid, essential oil, saponins, betulol (sesquiterpene alcohol), apigenin dimethyl ether, betuloside, gaultherin, methyl salicylate, ascorbic acid. *Ref:* Kreitmair, *Pharmazie* **8**, 534 (1953).

USE: Pharmaceutic aid (flavor).

1237. Betulin. *Lup-20(29)-ene-3,28-diol;* lup-20(30)-ene-3β,28-diol; trochol; betulinol; betulol. $C_{30}H_{50}O_2$; mol wt 442.73. C 81.39%, H 11.38%, O 7.23%. In the outer portion of the bark of white birch (up to 24%), in other barks, and in lignite. Botanical distribution: Steiner, *Molisch Festschrift* (1936). Isoln: Lowitz, *Crell's Chem. Ann.* **1**, 312 (1788); Steiner, *loc. cit;* Ruzicka, Isler, *Helv. Chim. Acta* **19**, 506 (1936); from *Lemaireocereus griseus* Britton et Rose, *Cactaceae:* Djerassi *et al., J. Am. Chem. Soc.* **78**, 2312 (1956). Structure: Ames *et al., J. Chem. Soc.* **1951**, 450; Davy *et al., ibid.* **1951**, 2696, 2702. Stereochemistry: Guider *et al., ibid.* **1953**, 3024; Das, *Chem. & Ind. (London)* **1971**, 1331. *Review:* J. Simonsen, W. C. J. Ross, *The Terpenes* **vol. IV** (Cambridge Univ. Press, 1957) pp 187-328.

Crystals from methanol-chloroform, mp 248-251°; sublimes at 240° at 0.01 mm. Solvated needles from alc contg one mol EtOH. After drying sublimes at 170-180° (bath temp) at 0.08 mm. uv max (H_2SO_4): 316 nm. $[\alpha]_D^{15} +20°$ (c = 2 in pyridine). Sparingly sol in cold water, petr ether, carbon disulfide. One part is sol in 149 parts alc, 251 ether, 113 chloroform, 417 benzene. Sol in acetic acid.

Diacetate, $C_{34}H_{54}O_4$, mp 223-224°. $[\alpha]_D^{20} +22°$ (c = 1.2 in $CHCl_3$). $d_4^{228.5}$ 0.9635; $n_D^{28.5}$ 1.4661.

1238. Bevantolol. *1-[[2-(3,4-Dimethoxyphenyl)ethyl]-amino]-3-(3-methylphenoxy)-2-propanol;* 1-[(3,4-dimethoxyphenethyl)amino]-3-(m-tolyloxy)-2-propanol. $C_{20}H_{27}NO_4$; mol wt 345.44. C 69.54%, H 7.88%, N 4.05%, O 18.53%. Cardioselective β₁-adrenergic blocker. Prepn: **Belg.** pat. **790,165;** A. Holmes, R. F. Meyer, **U.S.** pat. **3,857,891** (1973, 1974 both to Parke, Davis & Co.); M. L. Hoefle *et al., J. Med. Chem.* **18**, 148 (1975). Pharmacology in animals: S. G. Hastings *et al., Arch. Int. Pharmacodyn. Ther.* **226**, 81 (1977). Cardiovascular effects in animals: I. D. Dukes, E. M. Vaughan Williams, *Brit. J. Pharmacol.* **84**, 365 (1985). Cardioselectivity in asthmatic patients: C.-G. Löfdahl *et al., Pharmacotherapy* **4**, 205 (1984). GC determn in plasma: E. J. Randinitis *et al., J. Chromatog.* **308**, 345 (1984). Pharmacokinetics in humans: P. Vermeij, P. van Brummelen, *Eur. J. Clin. Pharmacol.* **30**, 375 (1986). Comparative clinical trial with hydrochlorothiazide in hypertension: C. P. Lucas *et al., Clin. Ther.* **8**, 49 (1985). Symposium on pharmacology and clinical efficacy: *Am. J. Cardiol.* **58**, 1E-44E (1986).

Hydrochloride, $C_{20}H_{27}NO_4$.HCl, *CI-775, Ranestol, Sentiloc, Vantol.* Crystals from acetonitrile, mp 137-138°.

THERAP CAT: Antianginal; antihypertensive; antiarrhythmic.

1239. Bevonium Methyl Sulfate. *2-[[(Hydroxydiphenylacetyl)oxy]methyl]-1,1-dimethylpiperidinium methyl sulfate (salt); 2-(hydroxymethyl)-1,1-dimethylpiperidinium methyl sulfate benzilate;* piribenzil methyl sulfate; benzilic acid ester with 2-(hydroxymethyl)-1,1-dimethylpiperidinium methyl sulfate; CG-201; Acabel. $C_{23}H_{31}NO_7S$; mol wt 465.57. C 59.34%, H 6.71%, N 3.01%, O 24.06%, S 6.89%. Prepn: **Belg.** pat. **616,951** (1962 to Grünenthal), *C.A.* **58**, 7914d (1963); Beckmann, *Arzneimittel-Forsch.* **16**, 910 (1966). Series of publications on pharmacology, toxicology, clinical trials, *see ibid.* 901-988.

Crystals from petr ether, mp 134-135°.

THERAP CAT: Anticholinergic, antispasmodic, bronchodilator.

1240. Bezafibrate. *2-[4-[2-[(4-Chlorobenzoyl)amino]-ethyl]phenoxy]-2-methylpropanoic acid;* 2-[p-[2-[(p-chlorobenzamido)ethyl]phenoxy]-2-methylpropionic acid; α-[4-(4-chlorobenzoylaminoethyl)phenoxy]isobutyric acid; BM-15075; Befizal; Bezalip; Bezatol; Cedur; Difaterol. $C_{19}H_{20}$-$ClNO_4$; mol wt 361.82. C 63.07%, H 5.57%, Cl 9.80%, N 3.87%, O 17.69%. Prepn: E. Witte *et al., Ger.* pat. **2,149,-070;** *eidem,* **U.S.** pat. **3,781,328** (both 1973 to Boehringer, Mann.). Pharmacology: R. Zimmerman *et al., Atherosclerosis* **29**, 477 (1978). Clinical studies: A. G. Olsson, P. D. Lang, *ibid.* **31**, 421, 429 (1978); P. Wahl *et al., Deut. Med. Wochenschr.* **103**, 1233 (1978). Review of pharmacodynamics and therapeutic use: J. P. Monk, P. A. Todd, *Drugs* **33**, 539-576 (1987).

Crystals from acetone, mp 186°.

THERAP CAT: Antihyperlipoproteinemic.

1241. Bezitramide. *1-[1-(3-Cyano-3,3-diphenylpropyl)-4-piperidinyl]-1,3-dihydro-3-(1-oxopropyl)-2H-benzimidazol-2-one; 1-[1-(3-cyano-3,3-diphenylpropyl)-4-piperidyl]-3-propionyl-2-benzimidazolinone;* 1-(3-cyano-3,3-diphenylpropyl)-4-(2-oxo-3-propionyl-1-benzimidazolinyl)piperidine; benzitramide; R-4845; Burgodin. $C_{31}H_{32}N_4O_2$; mol wt 492.62. C 75.58%, H 6.55%, N 11.37%, O 6.50%. Prepn: **Belg.** pat. **633,495;** P. A. J. Janssen, **U.S.** pat. **3,196,157** (1963, 1965 both to Janssen). Pharmacology and clinical data: P. A. J. Janssen *et al., Arzneimittel-Forsch.* **21**, 862 (1971); W. K. P. Amery *et al., ibid.* 868; H. Knape, *Anaesthesist* **21**, 251 (1972).

White crystalline powder, mp 145-149°. Also reported as pale yellow amorphous powder, mp 124.5-126°. Solubility of >1 g/100 ml in ethyl acetate, acetone, benzene, chloroform. Almost insol in water, dilute acids. LD_{50} orally in mice, rats: 2101, 141 mg/kg (Janssen).

Note: This is a controlled substance (opiate) listed in the U.S. Code of Federal Regulations, Title 21 Part 1308.12 (1995).

THERAP CAT: Analgesic (narcotic).

1242. Bialamicol. *3,3'-Bis[(diethylamino)methyl]-5,5'-di-2-propenyl-[1,1'-biphenyl]-4,4'-diol; 5,5'-diallyl-α,α'-bis(diethylamino)-m,m'-bitolyl-4,4'-diol; 6,6'-diallyl-α,α'-bis(diethylamino)-4,4'-bi-o-cresol; biallylamicol; SN-6771; PAA-701; Camoform.* $C_{28}H_{40}N_2O_2$; mol wt 436.64. C 77.02%, H 9.23%, N 6.42%, O 7.33%. Prepd by treating 2,2'-diallyl-p,p'-biphenol with diethylamine and formaldehyde: Burckhalter *et al., J. Am. Chem. Soc.* **68**, 1894 (1946); Rawlins *et al.,* U.S. pat. **2,459,338** (1949 to Parke, Davis).

Dihydrochloride, $C_{28}H_{40}N_2O_2.2HCl$, minute crystals, mp 209-210°. Sol in water.

THERAP CAT: Antiamebic.

1243. Biapenem. *[4R-[4α,5β,6β(R*)]]-6-[[2-Carboxy-6-(1-hydroxyethyl)-4-methyl-7-oxo-1-azabicyclo[3.2.0]hept-2-en-3-yl]thio]-6,7-dihydro-5H-pyrazolo[1,2-a][1,2,4]triazol-4-ium inner salt; (1R,5S,6S)-2-[(6,7-dihydro-5H-pyrazolo[1,2-a][1,2,4]-triazolium-6-yl)thio]-6-[(R)-1-hydroxyethyl]-1-methylcarbapen-2-em-3-carboxylate; LJC-10627; L-627; CL-186-815.* $C_{15}H_{18}N_4O_4S$; mol wt 350.40. C 51.42%, H 5.18%, N 15.99%, O 18.26%, S 9.15%. 1-β-methyl-carbapenem antibiotic. Prepn: T. Kumagai *et al.,* Eur. pat. Appl. **289,801**; *eidem,* U.S. pat. **4,990,613** (1988, 1991 both to Lederle). Total synthesis: Y. Nagao *et al., J. Org. Chem.* **57**, 4243 (1992). Renal dehydropeptidase stability: M. Hikida *et al., Antimicrob. Ag. Chemother.* **36**, 481 (1992). *In-vitro* antimicrobial spectrum: H. M. Wexler *et al., J. Antimicrob. Chemother.* **33**, 629 (1994); H. Y. Chen, D. M. Livermore, *ibid.* 949. Clinical pharmacokinetics: M. Nakashima *et al., Int. J. Clin. Pharmacol. Ther. Toxicol.* **31**, 70 (1993). Clinical evaluation: H. Meguro *et al., Japan. J. Antibiot.* **47**, 903 (1994).

Amorphous, pale yellow powder from acetone/water, occurs as hemihydrate. $[\alpha]_D^{20}$ −32.9° (c = 0.5).

THERAP CAT: Antibacterial.

1244. Bibenzonium Bromide. *2-(1,2-Diphenylethoxy)-N,N,N-trimethylethanaminium bromide; [2-(1,2-diphenylethoxy)ethyl]trimethylammonium bromide;* trimethyl-(1,2-diphenylethoxy)ethylammonium bromide; 2-trimethylamino-1,2-diphenyldiethyl ether bromide; 2-(dimethylamino)-1,2-diphenyldiethyl ether methyl bromide; ES-132; Sedobex; Lysobex; Lysibex; Thoragol; Lysbex; Medipectol. $C_{19}H_{26}BrNO$; mol wt 364.33. C 62.64%, H 7.19%, Br 21.93%, N 3.84%, O 4.39%. Prepn: Brit. pat. **796,364**; H. Suter, W. Kündig, U.S. pat. **2,913,459** (1958, 1959 both to Eprova).

Crystals from ethyl acetate, mp 144-147°. Sol in cold water, methanol, and ethanol. Practically insol in hot ethyl acetate, benzene, ether and ligroine.

THERAP CAT: Antitussive.

1245. Bibenzyl. *1,1'-(1,2-Ethanediyl)bisbenzene;* dibenzyl; *sym*-diphenylethane; 1,2-diphenylethane. $C_{14}H_{14}$; mol wt 182.27. C 92.26%, H 7.74%. Prepn by the reduction of benzil or benzoin. Clemmensen, *Ber.* **47**, 688 (1914); Gattermann-Wieland, *Praxis des organischen Chemikers* (de Gruyter, Berlin, 40th ed., 1961) p 333; from stilbene, Kleiderer, Kornfeld, *J. Org. Chem.* **13**, 455 (1948).

Monoclinic prisms from methanol, mp 52.0-52.5°. bp_{760} 284°. d_4^0 1.104; d_4^{25} 0.9782; d_4^{58} 0.958. Heat of combustion (25°): 9909.9 cal (15°)/g. Moderately sol in alcohol; freely sol in carbon disulfide, ether, chloroform, amyl acetate; sol in liquid sulfur dioxide. Practically insol in water, liquid ammonia.

1246. Bibrocathol. *4,5,6,7-Tetrabromo-2-hydroxy-1,3,2-benzodioxabismole;* tetrabromopyrocatechol bismuth derivative; bismuth tetrabromopyrocatechol; bibrocathin; Noviform; Posiformin. $C_6HBiBr_4O_3$; mol wt 649.67. C 11.09%, H 0.16%, Bi 32.17%, Br 49.20%, O 7.39%. Prepn: Hundrup, *Arch. Pharm. Chemi* **54**, 537 (1947), *C.A.* **42**, 2727a (1948). Antibacterial activity: Frank, Stark, *Pharm. Acta Helv.* **29**, 283 (1954). Identity tests: Hakkesteegt, *Pharm. Weekbl.* **99**, 922 (1964).

Yellow, odorless, tasteless powder. Practically insol in water. Slightly sol in alcohol, ether. Dec in acid, alkalies.

THERAP CAT: Antiseptic.

1247. Bicalutamide. *(±)-N-[4-Cyano-3-(trifluoromethyl)phenyl]-3-[(4-fluorophenyl)sulfonyl]-2-hydroxy-2-methylpropanamide;* 4-cyano-3-trifluoromethyl-N-(3-p-fluorophenylsulfonyl-2-hydroxy-2-methylpropionyl)aniline; (±)-4'-cyano-α,α,α-trifluoro-3-[(p-fluorophenyl)sulfonyl]-2-methyl-m-lactotoluidide; ICI-176334; Casodex. $C_{18}H_{14}F_4N_2O_4S$; mol wt 430.38. C 50.23%, H 3.28%, F 17.66%, N 6.51%, O 14.87%, S 7.45%. Non-steroidal peripherally active antiandrogen. Prepn: H. Tucker, **Eur.** pat. **Appl. 100,172;** *idem,* U.S. pat. **4,636,505** (1984, 1987 both to ICI); and structure-activity: H. Tucker *et al., J. Med. Chem.* **31,** 954 (1988). Resolution and activity of enantiomers: H. Tucker, G. J. Chesterson, *ibid.* 885. Pharmacokinetics: I. D. Cockshott *et al., Eur. Urol.* **18,** Suppl. 3, 10 (1990). Clinical trials in prostate cancer: G. Lunglmayr, *Horm. Res.* **32,** Suppl. 1, 77 (1989); D. W. W. Newling, *Eur. Urol.* **18** Suppl. 3, 18 (1990). Review of pharmacology: B. J. A. Furr, in *Hormonal Therapy of Prostatic Diseases: Basic and Clinical Aspects,* M. Motta, M. Serio, Eds. (Mediocom Europe, Milan, 1987) 148-161.

Crystals from a 1:1 (v/v) mix of ethyl acetate and petroleum ether, mp 191-193°.

THERAP CAT: Antiandrogen; antineoplastic (hormonal).

1248. Bicine. *N,N-Bis(2-hydroxyethyl)glycine;* di(hydroxyethyl)glycine; *N,N*-bis(hydroxyethyl)aminoacetic acid; *N,N*-di(hydroxyethyl)aminoacetic acid; diethylolglycine; 2-HxG. $C_6H_{13}NO_4$; mol wt 163.17. C 44.17%, H 8.03%, N 8.58%, O 39.22%. (HOCH$_2$CH$_2$)$_2$NCH$_2$COOH. One of the zwitterionic amino acids known as "Good" buffers, active in the pH range 6-8.5. Prepn by hydrolysis of its lactone (obtained from glycine and ethylene oxide): Pascal, *Compt. Rend.* **245,** 1318 (1957); from diethanolamine and bromoacetic acid: N. E. Good *et al., Biochemistry* **5,** 467 (1966). Crystal structure: V. Cody *et al., Acta Crystallogr.* **33B,** 905 (1977). Use as sequestering agent: A. Grawitz, *Rev. Tech. Ind. Cuir* **65,** 187, 190 (1973), *C.A.* **80,** 49281h (1974). Temperature effects on pKa: M. L. Soni, R. C. Kapoor, *Int. J. Quant. Chem.* **20,** 385 (1981). Use as a buffer: A. L. Remisov, *Biokhimia* **25,** 323 (1960); S. Ito *et al., Histochem. J.* **16,** 489 (1984).

Crystals from dil ethanol, mp 193-195° (slight decompn). pKa (20°): 8.35. pKa$_2$ (0.1M): 0°, 8.7; 20°, 8.35; 37°, 8.2. ΔpKa/°C −0.018. Saturated aq soln is 1.1M at 0°. Slightly sol in water. Forms a water-soluble sodium salt.

USE: Biological buffer and chelating agent.

1249. Bicozamycin. Bicyclomycin; aizumycin; 8,10-diaza-6-hydroxy-5-methylene-1-(2-methyl-1,2,3-trihydroxypropyl)-2-oxabicyclo[4.2.2]decan-7,9-dione; antibiotic 5879; WS-4545 antibiotic; Bacteron; Bacfeed. $C_{12}H_{18}N_2O_7$; mol wt 302.28. C 47.68%, H 6.00%, N 9.27%, O 37.05%. Cyclic peptide antibiotic substance. Prepn by fermentation of *Streptomyces sapporensis:* T. Miyoshi *et al.,* **Ger.** pat. **2,150,-593;** H. Imanaka *et al.,* **U.S.** pat. **3,923,790** (1972, 1975 both to Fujisawa). Isoln and characterization: T. Miyoshi *et al., J. Antibiot.* **25,** 569 (1972). Isoln from *S. aizunensis* and identity with antibiotic 5879: S. Miyamura *et al., ibid.* **26,** 479 (1973). Bicyclomycin has a unique chemical structure which bears no relation to any group of known antibiotics. Structural elucidation: T. Kamiya *et al., ibid.* **25,** 576 (1972). Crystal and molecular structure: Y. Tokuma *et al., Bull. Chem. Soc. Japan* **47,** 18 (1974). *In vitro* and *in vivo* activity studies: M. Nishida *et al., J. Antibiot.* **25,** 582 (1972). Metabolism: *eidem, ibid.* 594. Mechanism of action: N. Tanaka *et al., ibid.* **29,** 155 (1976); N. Tanaka in *Antibiotics* **vol. 5**(pt. 1), E. Hahn, Ed. (Springer, New York, 1979) p 18. Synthetic approaches: L. V. Dunkerton, R. M. Ahmed, *Tetrahedron Letters* **21,** 1803 (1980); R. M. Williams, *ibid.* **22,** 2341 (1981); S. Nakatsuka *et al., ibid.* 4973; J. H. Hoare, P. Yates, *Chem. Commun.* **1981,** 1126. Total synthesis of (±)-bicyclomycin: S. Nakatsuka *et al., Tetrahedron Letters* **24,** 5627 (1983); of (+)-bicyclomycin: R. M. Williams *et al., J. Am. Chem. Soc.* **106,** 5749 (1984); *eidem, ibid.* **107,** 3253 (1985). Review of synthetic, mechanistic and biological studies: R. M. Williams, C. A. Durham, *Chem. Rev.* **88,** 511-540 (1988).

Monoclinic crystals from ethanol, mp 188-191° (dec); rhombic crystals from methanol + acetone, mp 187-189° (dec) (Imanaka), also reported as mp 166-170° (Nakatsuka *et al.*). [α]$_D^{23}$ +63.5° (methanol). Weakly basic substance. Soly in water: 192 mg/ml. Sol in methanol; sparingly sol in ethanol. Slightly sol in acetone. Practically insol in chloroform, ethyl acetate, benzene, *n*-hexane. Unstable in alkaline soln. LD$_{50}$ in mice: > 4 g/kg (Williams, Durham).

THERAP CAT (VET): Antibacterial; feed additive (livestock).

1250. Bicuculline. *[R-(R*,S*)]-6-(5,6,7,8-Tetrahydro-6-methyl-1,3-dioxolo[4,5-g]isoquinolin-5-yl)furo[3,4-e]-1,3-benzodioxol-8(6H)-one.* $C_{20}H_{17}NO_6$; mol wt 367.36. C 65.39%, H 4.66%, N 3.81%, O 26.13%. Alkaloid naturally occurring in the *d*-form; found in *Dicentra cucullaria* (L.) Bernh., *Adlumia fungosa* (Ait.) Greene, *Fumariaceae,* and several *Corydalis* species: Manske, *Can. J. Res.* **7,** 265 (1932); **8,** 210, 407 (1933); **9,** 436 (1933); Edwards, Handa, *Can. J. Chem.* **39,** 1801 (1961). Synthesis of *dl*-form: Groenewoud, Robinson, *J. Chem. Soc.* **1936,** 199. Resolution of isomers: Haworth *et al., Nature* **165,** 529 (1950). Stereoisomer of adlumidine, *q.v.,* and of capnoidine: Manske, *J. Am. Chem. Soc.* **72,** 3207 (1950). Preliminary stereochemical studies: Safe, Moir, *Can. J. Chem.* **42,** 160 (1964). Revised stereochemistry: Blaha *et al., Coll. Czech. Chem. Commun.* **29,** 2328 (1964); Snatzke *et al., Tetrahedron* **25,** 5059 (1969). Crystal and molecular structure: Gorinsky, Moss, *J. Cryst. Mol. Struct.* **3,** 299 (1973). Shows GABA *(q.v.)* antagonist activity: Curtis *et al., Nature* **226,** 1222 (1970).

Elongated plates from chloroform-methanol, mp 215°; also reported as mp 177°, solidifies and remelts 193-195°: Manske, *Can. J. Res.* **21B,** 13 (1943). uv max (acidified ethanol): 225, 296, 324 nm (ε 36700, 6390, 5870). [α]$_D^{25}$ +130.5° (CHCl$_3$). pKa 4.84. Sol in benzene, chloroform, ethyl acetate. Sparingly sol in alc and ether.

1251. Bidisomide. *(±)-α-[2-[Acetyl(1-methylethyl)-amino]ethyl]-α-(2-chlorophenyl)-1-piperidinebutanamide;* (±)-α-(o-chlorophenyl)-α-[2-(N-isopropylacetamido)ethyl]-1-piperidinebutyramide; SC-40230. $C_{22}H_{34}ClN_3O_2$; mol wt 407.98. C 64.77%, H 8.40%, Cl 8.69%, N 10.30%, O 7.84%. Sodium channel blocker. Prepn: B. N. Desai *et al.,* **Eur.** pat. **Appl. 170,901;** *eidem,* **U.S.** pat. **4,639,524** (1986, 1987 both to Searle); *eidem, J. Med. Chem.* **31,** 2158 (1988). Cardiovascular pharmacology: L. G. Frederick *et al., Cardiovasc. Res.* **23,** 897 (1989). Electrophysiologic study: S. M. Garthwaite *et al., Drug Dev. Res.* **27,** 329 (1992). Bioavailability and tolerability in humans: R. L. Page *et al., Clin. Pharmacol. Ther.* **51**, 371 (1992). Clinical pharmacokinetics: C. S. Cook *et al., Pharm. Res.* **10**, 1675 (1993). Crystals from ethyl acetate, mp 140-141°.

THERAP CAT: Antiarrhythmic (class I).

1252. Bietamiverine. α-*Phenyl-1-piperidineacetic acid 2-(diethylamino)ethyl ester;* β-diethylaminoethyl phenylpiperidinoacetate; β-diethylaminoethyl α-(1-piperidyl)phenylacetate. $C_{19}H_{30}N_2O_2$; mol wt 318.46. C 71.66%, H 9.50%, N 8.80%, O 10.05%. Prepd by reacting piperidine with phenylchloroacetic acid ethyl ester in chloroform: Reetz, **Ger. pat. 859,892** (1952 to Nordmark-Werke); from β-diethylaminoethyl phenylbromoacetate and piperidine in chloroform: Blicke *et al., J. Am. Chem. Soc.* **76**, 3161 (1954); from ethyl α-(1-piperidyl)phenylacetate and β-diethylaminoethyl chloride: Moffett, Hart, *J. Am. Pharm. Assoc. Sci. Ed.* **42**, 717 (1953). Clinical tests: Ciravegna *et al., Clin. Ter.* **57**, 227 (1971).

Liquid, d_4^{25} 1.0184. bp_1 65°. n_D^{25} 1.5070.
Hydrochloride, $C_{19}H_{30}N_2O_2$.HCl, *Novosparol.* Crystals from methyl isobutyl ketone, mp 187-189°.
Dihydrochloride, $C_{19}H_{30}N_2O_2$.2HCl, crystals from ethanol + ether, mp 194-195°.

THERAP CAT: Antispasmodic.

1253. Bietanautine. *1,2,3,6-Tetrahydro-1,3-dimethyl-2,-6-dioxo-7H-purine-7-acetic acid compd with 2-(diphenylmethoxy)-N,N-dimethylethanamine (2:1);* 2-(benzhydryloxy)-*N,N*-dimethylethylamine bis(theophylline 7-acetate); *O*-benzhydryldimethylaminoethanol bis(theophylline 7-acetate); etanautine (formerly); Nautamine. $C_{35}H_{41}N_9O_9$; mol wt 731.77. C 57.45%, H 5.65%, N 17.23%, O 19.68%. Prepn: Mizier, **U.S. pat. 2,942,000** (1960 to Delagrange).

Crystals, mp 168-170°. Sol in alc; sparingly sol in water.
THERAP CAT: Antihistaminic. Antiemetic. Antiparkinsonian.

1254. Bietaserpine. *(3β,16β,17α,18β,20α)-1-[2-(Diethylamino)ethyl]-11,17-dimethoxy-18-[(3,4,5-trimethoxybenzoyl)oxy]yohimban-16-carboxylic acid methyl ester; 1-[2-(diethylamino)ethyl]-18β-hydroxy-11,17α-dimethoxy-3β,20α-yohimban-16β-carboxylic acid methyl ester, 3,4,5-trimethoxybenzoate ester;* 1-[2-(diethylamino)ethyl]reserpine. $C_{39}H_{53}$-N_3O_9; mol wt 707.86. C 66.17%, H 7.55%, N 5.94%, O 20.34%. Prepn: Buzas, Régnier, *C. R. Acad. Sci.* **250**, 1340 (1960); Buzas *et al.,* **Fr. pat. 1,256,524** (1961 to Dautreville & Lebas and Andre Buzas), *C.A.* **57**, 2273e (1962); Buzas *et al.,* **Fr. pat. M102,** *C.A.* **58**, 9159a (1963); **Brit. pat. 894,866** (1961, 1962 both to Soc. Nogentaise de Prods. Chim. and Andre Buzas). Pharmacology: Garattini *et al., J. Pharm. Pharmacol.* **13**, 548 (1961); Quevauviller *et al., Therapie* **18**, 1429 (1963); Guerrin *et al., C. R. Soc. Biol.* **158**, 1096 (1964); P. Berthaux, M. Neuman, *Arzneimittel-Forsch.* **14**, 1040 (1964).

Free base, $[\alpha]_D^{17}$ −121° (c = 2 in CHCl₃). Sol in dil acids and the usual organic solvents except cyclohexane and petr ether.
Bitartrate, $C_{39}H_{53}N_3O_9.C_4H_6O_6$, *DL-152, Tensibar.* Dec 145-150°. LD_{50} in mice (mg/kg): 620 orally; 430 i.p.; 215 i.v. (Berthaux, Neuman).

THERAP CAT: Antihypertensive.

1255. Bifemelane. *N-Methyl-4-[2-(phenylmethyl)phenoxy]-1-butanamine;* 4-(*o*-benzylphenoxy)-*N*-methylbutylamine; 2-(4-methylaminobutoxy)diphenylmethane; 2-benzyl-1-[4-(methylamino)butoxy]benzene. $C_{18}H_{23}NO$; mol wt 269.39. C 80.26%, H 8.61%, N 5.20%, O 5.94%. Monoamine oxidase inhibitor. Prepn: R. Kikumoto *et al., Ger. pat. 2,627,227; eidem, U.S. pat. 4,091,114* (1976, 1978 both to Mitsubishi); and antidepressant activity: R. Kikumoto *et al., J. Med. Chem.* **24**, 145 (1981). Pharmacology: A. Tobe *et al., Arzneimittel-Forsch.* **31**, 1278 (1981). Effects on experimental amnesia in rats: A. Tobe *et al., Japan. J. Pharmacol.* **39**, 153 (1985). Effects on neuronal activity in cats: M. Egawa *et al., Neuropharmacol.* **26**, 379 (1987). Inhibition of MAO: M. Naoi *et al., J. Neurochem.* **50**, 243 (1988).

Hydrochloride, $C_{18}H_{23}NO$.HCl, *E-0687, MCI-2016, Alnert, Celeport.* Crystals from acetone, mp 117-121°. LD_{50} in mice, rats (mg/kg): 1000, 1080 orally; 173, 130 i.p. (Tobe, 1981).

THERAP CAT: Nootropic.

1256. Bifenox. *5-(2,4-Dichlorophenoxy)-2-nitrobenzoic acid methyl ester;* methyl 5-(2,4-dichlorophenoxy)-2-nitrobenzoate; 2,4-dichlorophenyl 3-(methoxycarbonyl)-4-nitrophenyl ether; MC-4379; Modown. $C_{14}H_9Cl_2NO_5$; mol wt 342.13. C 49.15%, H 2.65%, Cl 20.72%, N 4.09%, O 23.38%. Pre-emergence herbicide. Prepn and herbicidal activity: **Belg. pat. 749,444;** R. J. Thiessen, **U.S. pat. 3,-652,645** (1970, 1972 both to Mobil). Comparison with other herbicides in corn: W. M. Dest *et al., Proc. Northeast. Weed Sci. Soc.* **27**, 31 (1973). Metabolism in soil, plants: G. R. Leather, C. L. Foy, *Pestic. Biochem. Physiol.* **7**, 437 (1977). Brief description: R. H. Dreger, *Weeds Today* **8**(2), 18 (1977). *Review:* P. J. Kruger *et al., Proc. 12th Brit. Weed Control Conf.* **2**, 839-845 (1974).

Yellow tan crystals, mp 84-86°. Practically insol in water (0.35 ppm at 25°). Soly in xylene at 25°: 30%. Vapor pressure at 30°: 2.4 × 10⁻⁶ mm Hg. LD_{50} orally in rats, mice: > 6400, 4556 mg/kg; LC_{50} in pheasants, wild ducks: > 5000 ppm (Kruger).

USE: Herbicide.

1257. Bifenthrin. *[1α,3α(Z)]-(±)-3-(2-Chloro-3,3,3-trifluoro-1-propenyl)-2,2-dimethylcyclopropanecarboxylic acid (2-methyl[1,1'-biphenyl]-3-yl)methyl ester;* 2-methylbiphenyl-3-ylmethyl-(Z)-(1RS)-cis-3-(2-chloro-3,3,3-trifluoroprop-1-enyl)-2,2-dimethylcyclopropanecarboxylate; biphenate; biphenthrin; biphentrin; FMC-54800; Brigade; Talstar; Capture. $C_{23}H_{22}ClF_3O_2$; mol wt 422.87. C 65.33%, H 5.24%, Cl 8.38%, F 13.48%, O 7.57%. Third generation synthetic pyrethroid. Prepn: J. F. Engel, *Eur. pat. Appl.* **3,336**; *idem,* U.S. pat. **4,238,505** (1979, 1980 both to FMC); E. L. Plummer *et al., Pestic. Sci.* **14,** 560 (1983). Physical properties, toxicology and review of field studies: H. J. H. Doel *et al., Med. Fac. Landouww. Rijksuniv. Gent* **49,** 929 (1984). Insecticidal activity: M. S. Mulla, H. A. Darwazeh, *Bull. Soc. Vector Ecol.* **10,** 1 (1985); M. S. Hamed, C. O. Knowles, *J. Econ. Entomol.* **81,** 1295 (1988). Field studies: Y. Antignus *et al., Ann. Appl. Biol.* **110,** 557 (1987); J. T. Trumble *et al., J. Econ. Entomol.* **81,** 608 (1988).

Light brown viscous oil, mp 51-66°. d^{25} 1.212 g/ml. Vapor pressure at 25°: 1.81×10^{-7} Torr. Sol in methylene chloride, chloroform, acetone, ether, toluene. Slightly sol in heptane, methanol. Soly in water: < 0.1 ppb. LD_{50} orally in rats: 54.5 mg/kg; LD_{50} dermally in rabbits: > 2000 mg/kg (Doel).

USE: Insecticide, acaricide.

1258. Bifidus Factor. *Lactobacillus bifidus* factor; *Lactobacillus bifidus* growth factor. A factor found in human milk and causing a predominant occurrence of *L. bifidus* in the intestinal tract of breast-fed infants: Petuely, Kristen, *Oesterr. Z. Kinderheilk. u. Kinderfürsorge* **6,** 173 (1951); Petuely, *Naturwiss.* **40,** 349 (1953). Essential growth factor for *L. bifidus* var *Penn:* György in Ciba Found. Symp., *Chemistry and Biology of Mucopolysaccharides* (Little, Brown, Boston, 1958) pp 140-156. Isoln from human milk: György *et al., Arch. Biochem. Biophys.* **48,** 193, 202, 209, 214 (1954); György *et al.,* U.S. pat. **2,786,051** (1957 to Am. Home Prod.); from *L. bifidus* cultured together with *Escherichia coli:* Kludas, U.S. pat. **2,962,424** (1960 to J. Carl Pflüger). Isoln from carrots and identification of the major and minor bifidus factor: Samejima *et al., Chem. Pharm. Bull.* **19,** 166, 178, 186 (1971); Z. Tamura *et al., Proc. Japan. Acad.* **48,** 138, 144 (1972). Prepn from porcine gastric mucosa and use as dietetic adjuvant in infant food: *Fr. pat.* **2,101,032,** *C.A.* **78,** 28221g (1970); P. C. Wirth, *Ger. pat.* **2,040,268,** *C.A.* **78,** 41859r (1970) (both 1972 to Sogeras).

USE: As adjuvant in powdered milk formulas for infants.

1259. Bifluranol. *(R*,S*)-4,4'-(1-Ethyl-2-methyl-1,2-ethanediyl)bis[2-fluorophenol];* erythro-3,3'-difluoro-4,4'-dihydroxy-α-ethyl-α'-methyldibenzyl; BX-341; Prostarex. $C_{17}H_{18}F_2O_2$; mol wt 292.33. C 69.85%, H 6.21%, F 13.00%, O 10.95%. Orally active antiprostatic agent. Prepn: J. C. Turner, R. P. Chan, *Ger. pat.* **2,110,428** corresp to U.S. pat. **4,051,263** (1971, 1977 both to Biorex). Antiprostatic activity: J. B. Dekanski, *Brit. J. Pharmacol.* **71,** 11 (1980). Chemistry, disposition in animals: D. J. Pope *et al., J. Pharm. Pharmacol.* **33,** 297 (1981). Metabolism: *eidem, ibid.* 302.

Cryst from toluene, mp 158-159°.

THERAP CAT: Antiandrogen.

1260. Bifonazole. *1-([1,1'-Biphenyl]-4-ylphenylmethyl)-1H-imidazole;* (±)-1-(p,α-diphenylbenzyl)imidazole; Bay h 4502; Amycor; Azolmen; Bedriol; Mycospor; Mycosporan. $C_{22}H_{18}N_2$; mol wt 310.40. C 85.13%, H 5.85%, N 9.03%. Antimycotic deriv of imidazole. Prepn: E. Regel *et al., Ger. pat.* **2,461,406;** *eidem,* U.S. pat. **4,118,487** (1976, 1978 both to Bayer). Series of articles on *in vitro* and *in vivo* antimycotic efficacy, microscopic studies, pharmacokinetics, efficacy in dermatomycoses and comparison with clotrimazole and miconazole, *q.q.v.: Arzneimittel-Forsch.* **33,** 517-551, 745-754 (1983). Toxicology: G. Schlüter, *ibid.* 739.

Crystals from acetonitrile, mp 142°. Very lipophilic. Sol in alcohols, DMF, DMSO. Soly in water at pH 6: < 0.1 mg/100 ml. Stable in aq soln at pH 1-12. LD_{50} in male mice, rats (mg/kg): 2629, 2854 orally (Schlüter).

THERAP CAT: Antifungal.

1261. Biguanide. *Imidodicarbonimidic diamide;* guanylguanidine; amidinoguanidine; diguanide. $C_2H_7N_5$; mol wt 101.11. C 23.76%, H 6.98%, N 69.26%. Preparation from dicyanodiamide: Rackmann, *Ann.* **376,** 169 (1910); Karipides, Fernelius, *Inorg. Syn.* **7,** 56, 58 (1963). Crystal structure: S. R. Ernst, F. W. Cagle, *Acta Crystallogr.* **33B,** 235 (1977); S. R. Ernst, *ibid.* 237.

Crystals from alc, mp 130°; dec rapidly at about 142°. Sol in water, alcohol. Insol in ether, benzene, chloroform. The aq soln on standing or heating. Aq solns are alkaline. Biguanide is a stronger base ($pK_1 = 11.52$; $pK_2 = 2.93$) than ammonia ($pK = 9.61$).

Hydrochloride, $C_2H_7N_5 \cdot HCl$, needles, mp 235°, very sol in water, slightly sol in alc.

Neutral sulfate, $(C_2H_7N_5)_2 \cdot H_2SO_4 \cdot 2H_2O$, large crystals, sol in water.

Acid sulfate, $C_2H_7N_5 \cdot H_2SO_4 \cdot H_2O$, rhombic prisms, sol in water, the soln giving an acid reaction.

USE: Sulfates in the determination of copper and nickel.

1262. Bikhaconitine. *(1α,6α,14α,16β)-20-Ethyl-1,6,16-trimethoxy-4-(methoxymethyl)aconitane-8,13,14-triol 8-acetate 14-(3,4-dimethoxybenzoate);* acetylveratroylbikhaconine. $C_{36}H_{51}NO_{11}$; mol wt 673.80. C 64.17%, H 7.63%, N 2.08%, O 26.12%. From roots of *Aconitum spicatum* Stapf., *Ranunculaceae:* Dunstan, Andrews, *J. Chem. Soc.* **87,** 1636 (1905). Structure: Tsuda, Marion, *Can. J. Chem.* **41,** 3055 (1963).

Prisms from *n*-hexane, mp 163.5-164°.

Monohydrate, needles from ether or dil methanol, mp 105-110°. $[\alpha]_D$ +16° (c = 1.6 in ethanol). Sol in alcohol, chloroform, ether, dil acids. Practically insol in water, petr ether.

1263. Bilirubin. *2,17-Diethenyl-1,10,19,22,23,24-hexa-hydro-3,7,13,18-tetramethyl-1,19-dioxo-21H-biline-8,12-di-propanoic acid; 1,10,19,22,23,24-hexahydro-2,7,13,17-tetra-methyl-1,19-dioxo-3,18-divinylbiline-8,12-dipropionic acid; 1,3,6,7-tetramethyl-4,5-dicarboxyethyl-2,8-divinyl-(b-13)-di-hydrobilenone; bilirubin IXα.* $C_{33}H_{36}N_4O_6$; mol wt 584.67. C 67.79%, H 6.21%, N 9.58%, O 16.42%. Principal pigment of bile and constituent of many biliary calculi. Major end-product of the biological breakdown of heme, *q.v.* Bilirubin is the chromophore responsible for coloration in various forms of jaundice. Also found in blood serum, where it exists in four major forms: unconjugated bilirubin, the monoglucuronide, the diglucuronide, and albumin-bound bilirubin. Most easily obtained from ox gallstones which are largely calcium bilirubinate: Städeler, *Ann.* **132**, 323 (1864); Küster, *Z. Physiol. Chem.* **94**, 136 (1915); **99**, 86 (1917); **121**, 80 (1922); Küster, Haas, *ibid.* **141**, 279 (1924); Fischer, *ibid.* **3**, 204 (1911); Fischer, Hess, *ibid.* **194**, 193 (1931). Isoln from pig bile: Gibson, Lowe, *J. Biol. Chem.* **123**, XLI (1938); Gray *et al., J. Chem. Soc.* **1961**, 2264, 2268; from ox bile: Libowitzky, *Z. Physiol. Chem.* **263**, 267 (1940). Industrial isoln from ox bile using chlorobenzene as extractant: U.S. pats. **2,166,073; 2,331,574; 2,363,471; 2,386,716.** Structure and synthesis: Fischer, Plieninger, *Naturwiss.* **30**, 382 (1942); *Z. Physiol. Chem.* **274**, 231 (1942); *cf.* Fischer-Orth, *Die Chemie des Pyrrols* II, 1, 621 (Leipzig, 1937); Gray *et al., Nature* **181**, 183 (1958); *eidem, J. Chem. Soc.* **1961**, 2276. Structure: Fog, Jellum, *Nature* **198**, 88 (1963). Configuration: Kuenzle *et al., Biochem. J.* **133**, 364 (1973). X-ray analysis and structure: Bonnett *et al., J. Chem. Soc. Perkin Trans. II*, **1972**, 902, 1335; *Nature* **262**, 326 (1976). NMR conformation studies: D. Kaplan, G. Navon, *J. Chem. Soc. Perkin Trans. II* **1981**, 1374. Separation of bilirubin species in serum and bile by reversed-phase HPLC: J. J. Lauff *et al., J. Chromatog.* **226**, 391 (1981); *eidem, Clin. Chem.* **29**, 800 (1983). Clinical importance of albumin-bound bilirubin: J. S. Weiss *et al., N. Engl. J. Med.* **309**, 148 (1983). Comprehensive reviews: Lemberg, Legge, *Hematin Compounds and Bile Pigments* (New York, 1949); With, *Bile Pigments* (Academic Press, New York, 1968). Reviews of toxicity: D. Bratlid, *N.Y. State J. Med.* **91**, 489-493 (1991); R. P. Wennberg, *ibid.* 493-496.

Light orange to deep reddish-brown monoclinic rhomboid, prisms, plates from chloroform. Gradually blackens on heating and does not melt. The greenish solns show a red fluorescence in ultraviolet light. A 0.001% soln in chloroform shows selective abs from 490 to 400 nm with a max at 453 nm; ε mM 60.7 ± 0.8. Practically insol in water. Sol in benzene, chloroform, chlorobenzene, carbon disulfide, acids, alkalies; slightly sol in alcohol, ether. Spreads on water. Penetrates into cholesterol, octadecylamine, and protein monolayers: Stenhagen, Rideal, *Biochem. J.* **33**, 1591 (1939).

1264. Biliverdine. *3,18-Diethenyl-1,19,22,24-tetrahy-dro-2,7,13,17-tetramethyl-1,19-dioxo-21H-biline-8,12-dipro-panoic acid; 1,19,22,24-tetrahydro-2,7,13,17-tetramethyl-1,19-dioxo-3,18-divinylbiline-8,12-dipropionic acid; 1,3,6,7-tetra-methyl-4,5-dicarboxyethyl-2,8-divinylbilenone; 4,5-di(2-car-*boxyethyl)-1,3,6,7-tetramethyl-2,8-divinylbilatriene; dehy-drobilirubin; uteroverdine; oöcyan.* $C_{33}H_{34}N_4O_6$; mol wt 582.66. C 68.03%, H 5.88%, N 9.62%, O 16.48%. Precursor of bilirubin. Formed in the body from hemoglobin. The bile of amphibia and of birds contains biliverdine only. Does not occur in normal human bile or normal human serum, but regularly accompanies bilirubin in the serum of patients with carcinomatous obstruction of the bile duct, and frequently in that of patients with liver cirrhosis, catar-rhal jaundice, and bile duct occlusion by gallstones. Can be obtained by autooxidation of bilirubin in alkaline soln but the yield is poor: Lemberg, *Biochem. J.* **28**, 978 (1934); better yields by oxidation of bilirubin with ferric chloride in glacial acetic acid: Lemberg, *Ann.* **499**, 25 (1932); by coupled oxidation of hemoglobin and ascorbic acid: Lemberg *et al., Biochem. J.* **35**, 363 (1941); by oxidation of bilirubin with ferric chloride in methanol: Gray *et al., J. Chem. Soc.* **1961**, 2264. Crystal and molecular structure: W. S. Sheldrick, *J. Chem. Soc., Perkin Trans. 2* **1976**, 1457. Comprehensive reviews: Lemberg, Legge, *Hematin Compounds and Blue Pigments* (New York, 1949); With, *Bile Pigments* (Academic Press, New York, 1968).

Dark green plates or prisms with violet surface color from methanol. Does not melt, blackens and dec above 300°. Absorption spectrum: Gray *et al., loc. cit.* Soluble in methanol, ether, chloroform, carbon disulfide, benzene, soln of alkali hydroxides. Gives the later color changes of the Gmelin test starting with green.

Dimethyl ester, $C_{35}H_{38}N_4O_6$, green crystals from chloroform + peter ether. When produced from bilirubin, mp 215-223°; synthetic 206-209°; from hemin 208°; from hemoglobin 216°.

Dimethyl ester ferrichloride, $C_{35}H_{39}Cl_4FeN_4O_6$ (green hemin ester), pleochroitic elongated platelets, no definite mp. Ferric chloride and HCl can be removed by washing the chloroform soln with water.

1265. Binapacryl. *3-Methyl-2-butenoic acid 2-(1-meth-ylpropyl)-4,6-dinitrophenyl ester; 3-methylcrotonic acid 2-sec-butyl-4,6-dinitrophenyl ester; 2-sec-butyl-4,6-dinitrophenyl 3,3-dimethylacrylate; 2-sec-butyl-4,6-dinitrophenyl 3-meth-yl-2-butenoate; 2-sec-butyl-4,6-dinitrophenyl 3-methylcroto-nate; 2-sec-butyl-4,6-dinitrophenyl senecioate; 3,3-dimethyl-acrylic acid 2-sec-butyl-4,6-dinitrophenyl ester; 4,6-dinitro-2-sec-butylphenyl β,β-dimethylacrylate; 3-methyl-2-butenoic acid 2-sec-butyl-4,6-dinitrophenyl ester; dinoseb methacry-late; senecioic acid 2-sec-butyl-4,6-dinitrophenyl ester; ENT-25793; HOE-2784; Niagara 9044; Acricid; Ambox; Endo-san; Morocide.* $C_{15}H_{18}N_2O_6$; mol wt 322.32. C 55.90%, H 5.63%, N 8.69%, O 29.78%. Prepn: Belg. pat. **630,947;** Scherer *et al.,* U.S. pat. **3,370,085** (1963, 1968 both to Hoechst). Toxicity study: T. B. Gaines, *Toxicol. Appl. Pharmacol.* **14**, 515 (1969).

Prisms, mp 70°. d_4^{20} 1.25-1.28. Vapor pressure at 60°: 1 × 10⁴ mm Hg. Very sol in acetone, xylene; sol in ethanol, kerosene. Practically insol in water. LD_{50} in female, male rats (mg/kg): 58, 63 orally (Gaines).

USE: Fungicide, miticide.

1266. Binedaline. *N,N,N'-Trimethyl-N'-(3-phenyl-1H-indol-1-yl)-1,2-ethanediamine;* 1-[[2-(dimethylamino)ethyl]methylamino]-3-phenylindole; binodaline. $C_{19}H_{23}N_3$; mol wt 293.41. C 77.78%, H 7.90%, N 14.32%. Substituted hydrazine with anti-depressant activity. Prepn: F. Schatz *et al., Ger.* pat. 2,512,702 corresp to U.S. pat. 4,204,998 (1975, 1980 both to Siegfried); F. Schatz *et al., Arzneimittel-Forsch.* 30, 919 (1980). Inhibits monoamine uptake in rat brain: D. P. Benfield, D. K. Luscombe, *J. Pharm. Pharmacol.* 33, Suppl., 42P (1981); *eidem, Arzneimittel-Forsch.* 33, 847 (1983). Comparison with amitriptyline in volunteers: P. H. Joubert *et al., Eur. J. Clin. Pharmacol.* 27, 667 (1985); J. Longmore *et al., Brit. J. Clin. Pharmacol.* 19, 295 (1985).

Crystals from petroleum, mp 52-53°.

Hydrochloride, $C_{19}H_{23}N_3 \cdot HCl$, *RU-39780, Sgd-Scha-1059, Ixprim.* Colorless crystals from isopropanol, mp 195-196°. uv max (0.1*N* HCl): 222, 263 nm (ε 28000, 14500). LD_{50} in male, female mice, male, female rats (mg/kg): 760, 770, 1380, 1160 orally; 54.0, 54.0, 27.2, 26.0 i.v. (Jahn *et al.).*

THERAP CAT: Antidepressant.

1267. Binifibrate. *3-Pyridinecarboxylic acid 2-[2-(4-chlorophenoxy)-2-methyl-1-oxopropoxy]-1,3-propanediyl ester;* trihydroxypropane 2-*p*-chlorophenoxyisobutyrate-1,3-dinicotinate; 2-(*p*-chlorophenoxy)-2-methylpropionic acid ester with 1,3-dinicotinoyloxy-2-propanol; glyceryl 2-*p*-chlorophenoxyisobutyrate-1,3-dinicotinate; WAC-104; Binewas. $C_{25}H_{23}ClN_2O_7$; mol wt 498.92. C 60.19%, H 4.65%, Cl 7.11%, N 5.61%, O 22.45%. Structurally related to clofibrate, *q.v.* Prepn: R. Andreoli, X. Cicera, **Span.** pat. 463,-218 (1978 to Soc. Espan. Espec. Farm.-Ter.), *C.A.* 90, 72067h (1979). Prepn and pharmacology: *eidem,* **Span.** pat. 488,665 corresp to **Belg.** pat. 884,722 (both 1980 to Soc. Espan. Espec. Farm.-Ter.). Microhemorheological properties: L. Bruseghini *et al., Arzneimittel-Forsch.* 33, 854 (1983). Preliminary clinical study: M. Dalmau *et al., ibid.* 858.

Yellowish-white crystals from alcohol-isopropyl ether, mp 100°. Saponification number 166. LD_{50} orally in mice and rats: >4000 mg/kg (Andreoli).

THERAP CAT: Antihyperlipoproteinemic.

1268. Biocytin. *N6-[5-(Hexahydro-2-oxo-1H-thieno[3,4-d]imidazol-4-yl)-1-oxopentyl]-L-lysine;* ε-N-biotinyl-L-lysine; biotin complex of yeast. $C_{16}H_{28}N_4O_4S$; mol wt 372.49. C 51.59%, H 7.58%, N 15.04%, O 17.18%, S 8.61%. A naturally occurring complex of biotin. Contains 65.6% biotin. Isoln: Wright *et al., J. Am. Chem. Soc.* 72, 1048 (1950); 74, 1996 (1952). Structure: Peck *et al., ibid.* 72, 1048 (1950);

74, 1999 (1952). Synthesis: Wolf *et al., ibid.* 74, 2002 (1952); Weijlard *et al., ibid.* 76, 2505 (1954); Wolf, Folkers, U.S. pat. 2,710,298 (1955 to Merck & Co.). Purification: McCormick, Föry, *Methods Enzymol.* 18 (pt A), 413 (1970). *Review:* A. F. Wagner, K. Folkers, *Vitamins and Coenzymes* (Wiley, New York, 1964) pp 138-159.

Crystals, mp 241-243°. Upon rapid crystn from dil methanol or dil acetone, mp 228-230° (dec); upon slow crystn sinters at 227°. mp 245-252° (dec, microblock). Crystals from water, mp 228.5°. $[\alpha]_D^{25}$ +53° (c = 1.05 in 0.1*N* NaOH). Infrared absorption spectrum: *J. Am. Chem. Soc.* 74, 2001 (1952). Freely sol in water, glacial acetic acid. Less sol in alc. Practically insol in acetone and most other organic solvents. When subjected to strong acid hydrolysis (at least 3*N* at 120° for one hour) biocytin yields biotin and L-lysine. Forms a crystalline hydrochloride. Biocytin is characterized microbiologically by its availability as a source of biotin to *Lactobacillus casei, L. delbrückii* LD 5, *L. acidophilus, Streptococcus fecalis* R, *Neurospora crassa,* and *Saccharomyces carlsbergensis* and by its unavailability as a source of biotin to *Lactobacillus arabinosus, L. pentosus,* and *Leuconostoc mesenteroides* P-60.

1269. Bioflavonoids. Vitamin P complex; citrus flavonoid compounds; Arliflav; Pecitrol Veinogène; C.V.P. A group of compounds which contribute to the maintenance of normal blood vessel conditions by decreasing capillary permeability and fragility. Widely distributed among plants: J. B. Harborne, *Comparative Biochemistry of the Flavonoids* (Academic Press, New York, 1967). Biosynthesis: Grisebach, Barz, *Naturwiss.* 56, 538 (1969). High concentrates can be obtained from all citrus fruits, rose hips, and black currants. Commercial methods extract the rinds of oranges, tangerines, lemons, limes, kumquats, and grapefruits. Solvents used in the extraction processes are aqueous alkalies, hot water, or water-miscible organic solvents, such as isopropanol: Freedman *et al., U.S.* pat. 2,888,381 (1959 to U.S. Vitamin). Early interest in the compounds developed because of their synergistic effect with ascorbic acid. Other pharmacologic effects such as inhibition of adrenaline autoxidation and of enzyme action are also under study: several authors in *The Pharmacology of Plant Phenolics,* J. W. Fairbairn, Ed. (Academic Press, New York, 1959); several authors in *Angiologica* 9, 133-446 (1972). Metabolism: DeEds, "Flavonoid Metabolism" in *Comprehensive Biology,* Vol. 20, M. Florkin, E. H. Stotz, Eds. (Elsevier, New York, 1968). *Reviews:* Scarborough, Bacharach, *Vitam. Horm.* (*New York*) 7, 1-55 (1949); Baier *et al., Ann. N.Y. Acad. Sci.* 61, (Art. 3), 637-736 (1955); T. Robinson, *The Organic Constituents of Higher Plants* (Burgess, Minneapolis, 1967) pp 178-209; H. Geiger, C. Quinn in *Flavonoids,* J. B. Harborne *et al.,* Eds. (Academic Press, New York, 1975) pp 692-742.

THERAP CAT: Capillary protectant.

1270. Biopterin. *[S-(R*,S*)]-2-Amino-6-(1,2-dihydroxypropyl)-4(1H)-pteridinone;* 1-(2-amino-4-hydroxy-6-pteridinyl)-1,2-propanediol; 2-amino-4-hydroxy-6-(1,2-dihydroxypropyl)pteridine; pterin HB_2. $C_9H_{11}N_5O_3$; mol wt 237.22. C 45.57%, H 4.67%, N 29.52%, O 20.23%. A pteridine widely distributed in nature; naturally occurring as the L-*erythro*-form. Considered as a growth factor for some insects; *see also* Neopterin. Isoln from human urine: Patterson *et al., J. Am. Chem. Soc.* 77, 3167 (1955); 78, 5871 (1956); **Brit.** pat. 814,462 (1959 to Am. Cyanamid); from drosophila: H. S. Forrest, H. K. Mitchell, *J. Am. Chem. Soc.* 77, 4865 (1955); from queen bee jelly: Butenandt, Rembold, *Z. Physiol. Chem.* 311, 79 (1958). Absolute configuration: E. L. Patterson *et al., J. Am. Chem. Soc.* 78, 5871 (1956). Synthesis: E. L. Patterson *et al., ibid.* 5868; Tschesche *et al. Ann.* 658, 193 (1962); Rembold, Metzger, *Ber.* 96, 1395 (1963); Viscontini *et al., Helv. Chim. Acta* 55, 570, 574 (1972); B. Schircks *et al., ibid.* 60, 211 (1977); T.

Sugimoto *et al.*, *Bull. Chem. Soc. Japan* **53**, 2344 (1980). Synthesis from neopterin: A. Kaiser, H. P. Wessel, *Helv. Chim. Acta* **70**, 766 (1987). Biosynthesis: G. Kapatos *et al.*, *Science* **213**, 1129 (1981).

Minute, yellow, spherical crystals from water. Chars without melting at 250-280°. $[\alpha]_D^{24}$ −50° (c = 0.4 in 0.1N HCl); $[\alpha]_D^{24}$ −26° (c = 0.92 in 0.1N NaOH). uv max (0.08N HCl): 247 nm (ε 11000). Soly in water: 0.7 mg/ml (20°); 4 mg/ml (90°). Soly in alc, ether, acetone, benzene: <0.1 mg/ml; in 1N NaOH, 1N HCl: >25 mg/ml. Fluoresces with a blue color in alkaline soln.

1271. Bioresmethrin. *(1R-trans)-2,2-Dimethyl-3-(2-methyl-1-propenyl)cyclopropanecarboxylic acid [5-(phenylmethyl)-3-furanyl]methyl ester; trans-(+)-2,2-dimethyl-3-(2-methylpropenyl)cyclopropanecarboxylic acid (5-benzyl-3-furyl)methyl ester;* 5-benzyl-3-furylmethyl-(+)-*trans*-chrysanthemate; *(5-benzyl-3-furyl)methyl]chrysanthemumate;* (+)-*trans*-resmethrin; NRDC-107; NIA-18739; SBP-1390; Resbuthrin; Biobenzyfuroline. $C_{22}H_{26}O_3$; mol wt 338.45. C 78.07%, H 7.74%, O 14.18%. Potent synthetic pyrethroid insecticide: M. Elliott *et al.*, *Nature* **213**, 493 (1967); M. Elliott *et al.*, *Pestic. Sci.* **2**, 243 (1971). Prepn: M. Elliott, N. F. Janes, Fr. pat. **1,503,260**; *idem*, U.S. pats. **3,465,007** and **3,542,928** (1967, 1969, 1970, all to Nat. Res. Dev. Corp.). Mammalian metabolism: C. O. Abernathy, J. E. Casida, *Science* **179**, 1235 (1973). Control of cockroaches: P. R. Chadwick *et al.*, *J. Med. Entomol.* **13**, 625 (1977). Acute toxicity: T. B. Gaines, R. E. Linder, *Fundam. Appl. Toxicol.* **7**, 299 (1986). Brief review: D. S. Gunew, *Anal. Methods Pestic. Plant Growth Regul.* **10**, 19-29 (1978).

$bp_{0.0008}$ 174°. n_D^{20} 1.5346. $[\alpha]_D^{20}$ −7.8° (c = 5 in acetone). Sol in most organic solvents. Insol in water. LD_{50} in adult male, female rats (mg/kg): 1244, 1721 orally (Gaines, Linder).

USE: Insecticide.

1272. Biotin. *Hexahydro-2-oxo-1H-thieno[3,4-d]imidazole-4-pentanoic acid; cis-tetrahydro-2-oxothieno[3,4-d]imidazoline-4-valeric acid; cis-hexahydro-2-oxo-1H-thieno[3,4]-imidazole-4-valeric acid;* vitamin H; coenzyme R; bios II; Biodermatin. $C_{10}H_{16}N_2O_3S$; mol wt 244.31. C 49.16%, H 6.60%, N 11.47%, O 19.65%, S 13.12%. Growth factor present in minute amounts in every living cell. Plays an indispensable role in numerous naturally occurring carboxylation reactions. Occurs mainly bound to proteins or polypeptides. The richest sources are liver, kidney, pancreas, yeast, and milk. The biotin content of cancerous tumors is higher than that of normal tissue. Biotin combines with the proteinaceous substance, avidin, in raw egg-white and becomes inactive. When on diets contg large amounts of raw egg-white the rat or chick develops characteristic skin lesions and growth is retarded; these symptoms can be prevented by feeding additional biotin. Isoln from egg yolk: Kögl, Tönnis, *Z. Physiol. Chem.* **242**, 43 (1936); from liver: György *et al.*, *J. Biol. Chem.* **131**, 745 (1939); du Vigneaud *et al.*, *ibid.* **140**, 643 (1941). Identity of biotin with vitamin H: du Vigneaud *et al.*, *Science* **92**, 62 (1940). Structure: du Vigneaud *et al.*, *J. Biol. Chem.* **146**, 475 (1942). First synthesis: Harris *et al.*, *J. Am. Chem. Soc.* **67**, 2096 (1945). Configuration: Traub, *Nature* **178**, 649 (1956). Stereospecific total synthesis of the naturally occurring d-form: P. N. Confalone *et al.*, *J. Am. Chem. Soc.* **97**, 5936 (1975); *eidem*, *J. Org. Chem.* **42**, 1630 (1977); T. Ogawa *et al.*, *Car-*

bohyd. Res. **57**, C31 (1977); F. G. M. Vogel *et al.*, *Ann.* **1980**, 1972; R. R. Schmidt, M. Maier, *Synthesis* **1982**, 747. Synthesis of the *dl*-form: P. N. Confalone *et al.*, *Helv. Chim. Acta* **59**, 1005 (1976); M. Marx *et al.*, *J. Am. Chem. Soc.* **99**, 6794 (1977); A. Fliri, K. Hohenlohe-Oehringen, *Ber.* **113**, 607 (1980); Ph. Rossy *et al.*, *Tetrahedron Letters* **22**, 3493 (1981). Biosynthetic studies: R. J. Parry, M. G. Kunitani, *J. Am. Chem. Soc.* **98**, 4024 (1976); R. J. Parry, M. V. Naidu, *Tetrahedron Letters* **1980**, 4783. Coordination properties: H. Siegel, *Experientia* **37**, 789 (1981). Mechanism of action: M. J. Cravey, H. Kohn, *J. Am. Chem. Soc.* **102**, 3928 (1980); D. L. Vesely, *Science* **216**, 1329 (1982). *Reviews:* A. F. Wagner, K. Folkers, *Vitamins and Coenzymes* (Wiley, New York, 1964) pp 138-159; Harris in *The Vitamins*, vol. II, W. H. Sebrell, R. S. Harris, Eds. (Academic Press, New York, 2nd ed., 1968) pp 261-359; D. B. McCormick, *Nutr. Rev.* **33**, 97-102 (1975). Review of assay methods: P. György, *The Vitamins* vol. VII, P. Gyorgy, W. N. Pearson, Eds. (1967) pp 303-313.

Characteristics of biotin isolated from liver or milk (Kögl's β-biotin): fine long needles, mp 232-233°. $[\alpha]_D^{11}$ +91° (c = 1 in 0.1N NaOH). Isoelec pt pH 3.5. pH of 0.01% aq soln 4.5. Soly at 25° (mg/100 ml): water ~22; 95% alc ~80. More sol in hot water and in dil alkali. Insol in other common organic solvents. The pure compd is stable to air and temp. Moderately acid and neutral solns are stable several months; alkaline solns are less stable, but appear reasonably stable up to a pH of about 9. Aq solns are very susceptible to mold growth. Acidic solns can be heat sterilized.

THERAP CAT: Vitamin (enzyme cofactor).

THERAP CAT (VET): Vitamin (enzyme cofactor).

1273. Biotin *l*-Sulfoxide. *[3aS-(3aα,4β,5β,6aα)]-Hexahydro-2-oxo-1H-thieno[3,4-d]imidazole-4-pentanoic acid 5-oxide;* AN factor. $C_{10}H_{16}N_2O_4S$; mol wt 260.31. C 46.14%, H 6.20%, N 10.76%, O 24.58%, S 12.32%. Isolated from *Aspergillus niger* culture filtrate where growth had taken place in the presence of pimelic acid; also from milk residue concentrates or by synthesis: Melville, *J. Biol. Chem.* **208**, 495 (1954); Wright *et al.*, *J. Am. Chem. Soc.* **76**, 4163 (1954).

Polymorphic plates from water, mp 238° (some dec). $[\alpha]_D^{20}$ −39.5° (c = 1.01 in 0.1N NaOH). Ineffective in curing the biotin deficiency syndrome when administered to rats at a level 100 times the effective dose of biotin.

1274. Biperiden. *α-Bicyclo[2.2.1]hept-5-en-2-yl-α-phenyl-1-piperidinepropanol; α-5-norbornen-2-yl-α-phenyl-1-piperidinepropanol;* 3-piperidino-1-phenyl-1-bicycloheptenyl-1-propanol; 1-bicycloheptenyl-1-phenyl-3-piperidinopropanol; 3-piperidino-1-phenyl-1-(Δ^5-bicyclo[2.2.1]hepten-2-yl)-1-propanol; KL-373. $C_{21}H_{29}NO$; mol wt 311.47. C 80.98%, H 9.38%, N 4.50%, O 5.14%. Prepn: Klavehn, U.S. pat. **2,789,110** (1957 to Knoll).

Crystals, mp 112-116°; also reported as 101° (Klavehn). Sparingly sol in water; slightly sol in ethanol; readily sol in methanol.

Hydrochloride, $C_{21}H_{29}NO \cdot HCl$, *Akineton, Akinophyl.* Crystals, dec ~275°; also reported as mp 238° (Klavehn). LD_{50} in mice: 545 mg/kg orally, 56 mg/kg i.v.

THERAP CAT: Anticholinergic, antiparkinsonian.

1275. Biphenamine. β-Diethylaminoethyl 2-hydroxy-3-phenylbenzoate; 2-diethylaminoethyl 3-phenylsalicylate; 2-diethylaminoethyl 2-hydroxy-3-biphenylcarboxylate; 3-phenylsalicylic acid 2-diethylaminoethyl ester; xenysalate. $C_{19}H_{23}NO_3$; mol wt 313.40. C 72.82%, H 7.40%, N 4.47%, O 15.32%. Prepn: Sahyun, U.S. pat. 2,594,350 (1952).

Oily liquid. Sol in water.

Hydrochloride, $C_{19}H_{23}NO_3 \cdot HCl$, *Sebaclen, Sebaklen.*

THERAP CAT: Anesthetic (local); antibacterial; antifungal.

1276. p-Biphenylamine. *[1,1'-Biphenyl]-4-amine;* 4-aminobiphenyl; *p*-aminodiphenyl; anilinobenzene; xenylamine. $C_{12}H_{11}N$; mol wt 169.23. C 85.17%, H 6.55%, N 8.28%. Prepn from diazoaminobenzene: Heusler, *Ann.* 260, 232 (1890). Review of carcinogenicity studies: *IARC Monographs* 1, 74-79 (1972).

Leaflets from alc or water. mp 53°. Volatile with steam. Slightly sol in cold water, readily sol in hot water, alcohol, chloroform.

Caution: Potential symptoms of overexposure are headache, dizziness; lethargy, dyspnea; ataxia, weakness; methemoglobinemia; urinary burning; acute hemorrhagic cystitis. See *NIOSH Pocket Guide to Chemical Hazards* (DHHS/NIOSH 90-117, 1990) p 36. This substance has been listed as a known carcinogen: *Seventh Annual Report on Carcinogens* (PB95-109781, 1994) p 18.

USE: In the detection of sulfates. Formerly as rubber antioxidant. As carcinogen in cancer research.

1277. 2,4'-Biphenyldiamine. Diphenyline; 2,4'-diphenyldiamine; *o,p'*-dianiline; 2,4'-diaminodiphenyl. $C_{12}H_{12}N_2$; mol wt 184.24. C 78.23%, H 6.56%, N 15.20%. Preparation by reducing azobenzene with tin and hydrochloric acid: Fischer, *Monatsh.* 6, 547 (1885).

Needles from dil alc, mp 45°. bp 363°. Very slightly sol in alc or ether.

USE: Detection of tungsten; manuf azo dyes.

1278. Bipiperidyl Mustard. *1,1'-Bis(2-chloroethyl)-4,4'-bipiperidine;* BPM. $C_{14}H_{26}Cl_2N_2$; mol wt 293.28. C 57.34%, H 8.94%, Cl 24.18%, N 9.55%. Obesifying agent causing rapid lipid deposition in mice: Rutman *et al., Sci-*

ence 153, 1000 (1966). Activity shown only by the active form, prepd by neutralizing the conjugate acid form and generating a bicyclic ammonium deriv at pH 10. Metabolic studies: *eidem, Trans. N.Y. Acad. Sci.* 30, 244 (1967).

LD_{50} i.p. in mice: 30-35 mg/kg.

1279. Birch Tar Oil, Empyreumatic. Oil white birch; oleum rusci. Obtained by destructive distillation of the bark and wood of *Betula alba* L., Betulaceae (white birch). *Constit.* Oil turpentine, other isomeric hydrocarbons, various empyreumatic resins, guaiacol, cresol, pyrocatechol, betulin.

Black, viscid liquid; characteristic odor. d 0.926-0.955. Partly sol in alcohol; sol in chloroform, fats, oils.

USE: For preserving leather and wood.

1280. Birch Tar Oil, Rectified. "Essential oil of birch wood". Obtained from the empyreumatic oil by steam distillation. *Constit.* Phenol, cresol, xylenol, guaiacol, creosol, pyrocatechol.

Dark brown liquid. d 0.886-0.950. One ml dissolves in about 3 ml abs alcohol; sol in benzene, chloroform, ether, glacial acetic acid, carbon disulfide, oil turpentine.

THERAP CAT: Dermatologic.

THERAP CAT (VET): Has been used extern. in skin diseases.

1281. α-Bisabolol. *(R*,R*)-α,4-Dimethyl-α-(4-methyl-3-pentenyl)-3-cyclohexene-1-methanol;* 6-methyl-2-(4-methyl-3-cyclohexen-1-yl)-5-hepten-2-ol; 1-methyl-4-(1,5-dimethyl-1-hydroxyhex-4(5)-enyl)cyclohexen-1; Camilol; Dragosantol; Hydagen B. $C_{15}H_{26}O$; mol wt 222.37. C 81.02%, H 11.79%, O 7.19%. Sesquiterpene isolated from the essential oils of a variety of plants, shrubs and trees; naturally occurring as the (−), (+), (−)-epi, (+)-epi stereoisomers. Isoln of (−)-form from *Matricaria chamomilla:* F. Sorm *et al., Coll. Czech. Chem. Commun.* 16, 626 (1951); of (+)-form from *Populus balsamifera: idem et al., Chem. Listy* 46, 364 (1952); of (−)-epi-form from *Myoporum crassifolium:* K. G. O'Brien *et al., Austral. J. Chem.* 6, 166 (1953); of (+)-epi-form from *Salvia stenophylla:* E. J. Brunke, F. J. Hammerschmidt, *Proc. 15th Intl. Symp. Essent. Oils Aromat. Plants* 1985, 145. Synthesis of racemic mixture: L. Ruzicka, M. Liguori, *Helv. Chim. Acta* 15, 3 (1932); C. D. Gutsche *et al., Tetrahedron* 24, 859 (1968). Separation of four stereoisomers: K. Günther *et al., Fresenius J. Anal. Chem.* 345, 787 (1993). Pharmacology and toxicology of (−)-form: V. Jakovlev, A. von Schlichtegroll, *Arzneimittel-Forsch.* 19, 615 (1969); S. Habersang *et al., Planta Med.* 37, 115 (1979). Comparative pharmacology of racemate and enantiomers: V. Jakovlev *et al., ibid.* 35, 125 (1979). Clinical use in chemotherapy-induced mucositis: W. Carl, L. S. Emrich, *J. Prosthet. Dent.* 66, 361 (1991). Determn by HPLC in chamomile extracts: R. Herrmann, *Deut. Apoth. Ztg.* 36, 1797 (1982); by GC in cosmetic products: D. Andre *et al., Int. J. Cosmet. Sci.* 13, 137 (1991). Review of anti-inflammatory activity and use in cosmetics: N. Jellinek, *Parfums Cosmet. Aromes* 57, 55-57 (1984); of absolute configuration studies: G. W. O'Donnell, M. D. Sutherland, *Austral. J. Chem.* 42, 2021-2034 (1989).

(−)-form

bp_{12} 155-157°. d_4^{23} 0.9223. n_D^{23} 1.4917. Also reported as colorless oil, $bp_{1.5}$ 128-130° (Gutsche). Miscible with alcohols, oils and lipophilic substances.

(−)-Form, *[S-(R*,R*)]-α-bisabolol, levomenol, Kamillosan.* Active principle in natural extracts. bp_{12} 153°. d^{20}

0.9211. n_D^{20} 1.4936. $[\alpha]_D$ —55.7°. Also reported as $[\alpha]_D^{20}$ —51.02° (Günther). LD_{50} in mice, rats (mg/kg): 11350, 14850 orally (Jakovlev, von Schlichtegroll). LD_{50} in male, female mice, rats (ml/kg): 15.2, 15.0, 14.9, 15.6 orally (Habersang).

(+)-Form, *(+)-α-bisabolol*, bp$_{1.0}$ 120-122°. d^{20} 0.9213. n_D^{20} 1.4919. $[\alpha]_D^{20}$ +51.7°. Also reported as $[\alpha]_D^{20}$ +57.04° (Günther).

(—)-Epi-form, *anymol*, $[\alpha]_D^{20}$ —67.6°. Also reported as $[\alpha]_D^{20}$ —67.48° (Günther).

(+)-Epi-form, $[\alpha]_D^{20}$ +67.4°. Also reported as $[\alpha]_D^{20}$ +70.11° (Günther).

USE: In cosmetics.

THERAP CAT: Anti-inflammatory.

1282. Bisacodyl. *4,4'-(2-Pyridylmethylene)bisphenol diacetate;* bis(p-acetoxyphenyl)-2-pyridylmethane; (4,4'-diacetoxydiphenyl)(2-pyridyl)methane; 2-(4,4'-diacetoxydiphenylmethyl)pyridine; Bicol; Broxalax; Contalax; DAMP; Dulcolan; Dulcolax; Durolax; Endokolat; Eulaxan; Godalax; Laxadin; Laxanin N; Laxorex; Nigalax; Perilax; Pyrilax; Stadalax; Telemin; Theralax; Ulcolax. $C_{22}H_{19}NO_4$; mol wt 361.40. C 73.12%, H 5.30%, N 3.88%, O 17.71%. Prepn: **Brit.** pat. 730,243; A. Kottler, E. Seeger, **U.S.** pat. **2,764,-590** (1955, 1956 both to Thomae). Pharmacology and toxicology: Schmidt, *Arzneimittel-Forsch.* **3**, 19 (1953).

Tasteless crystals, mp 138°. Practically insol in water and alkaline solns. Sol in acids, alc, acetone, propylene glycol, and other organic solvents. LD_{50} orally in rats: > 3 g/kg (Schmidt).

Complex with tannin, *bisacodyl tannex, Clysodrast.*

THERAP CAT: Cathartic.

1283. Bis(4-amino-1-anthraquinonyl)amine. *1,1'-Iminobis[4-amino-9,10-anthracenedione];* 4,4'-diamino-1,1'-dianthraquinonylamine; 4,4'-diamino-1,1'-anthrimide; 4,4'-diamino-1,1'-iminobisanthraquinone. $C_{28}H_{17}N_3O_4$; mol wt 459.46. C 73.20%, H 3.73%, N 9.15%, O 13.93%. Prepn: **Ger.** pat. 255,822 (1913); *Frdl.* **11**, 615; Tinker, Stallmann, **U.S.** pat. 2,420,022 (1947), *C.A.* **41**, 5730 (1947).

On heating with fuming sulfuric acid (30% SO$_3$) at 100° yields a dark powder, sol in water or alc with bluish-black coloration, used as dyestuff for wool. On heating with molten potassium hydroxide at 190° and oxidizing the reaction product with sodium hypochlorite, an olive-green vat dye is obtained.

USE: In the textile industry; in the determination of boron: Beckett, Webster, *Analyst* **68**, 306 (1943), *C.A.* **38**, 38 (1944).

1284. Bisantrene. *9,10-Anthracenedicarboxaldehyde bis[(4,5-dihydro-1H-imidazol-2-yl)hydrazone];* 9,10-anthracenedicarboxaldehyde bis(2-imidazolin-2-ylhydrazone). $C_{22}H_{22}N_8$; mol wt 398.47. C 66.31%, H 5.56%, N 28.12%. Prepn: K. C. Murdock *et al.*, **Ger.** pat. 2,850,822; K. C. Murdock, F. E. Durr, **U.S.** pat. **4,258,181** (1979, 1981 both to Am. Cyanamid); and antitumor activity: K. C. Murdock *et al.*, *J. Med. Chem.* **25**, 505 (1982). Crystallographic characterization of hydrochloride: R. B. Bates *et al.*, *Acta Crystallogr.* **C42**, 186 (1986); C. G. Pierpont, S. A. Lang, Jr.,

ibid. 1085. HPLC determn in plasma: Y.-M. Peng *et al.*, *J. Chromatog.* **233**, 235 (1982). Antitumor activity *in vitro*: J. D. Cowan *et al.*, *Invest. New Drugs* **1**, 139 (1983); *in vivo*: R. V. Citarella *et al.*, *Cancer Res.* **42**, 440 (1982). Clinical pharmacokinetics: K. Lu *et al.*, *Cancer Chemother. Pharmacol.* **16**, 156 (1986). Clinical trials in breast cancer: H.-Y. Yap *et al.*, *Cancer Res.* **43**, 1402 (1983); C. K. Osborne *et al.*, *Cancer Treat Rep.* **68**, 357 (1984).

Dihydrochloride, $C_{22}H_{22}N_8$.2HCl, *ADAH, ADCA, NSC-337766, CL-216942, Orange Crush, Zantrene.* Crystalline orange solid from ethanol. Hemihydrate, mp 288-289° (dec). uv max (H$_2$O) 260, 415 nm (ε 72700, 16300).

THERAP CAT: Antineoplastic.

1285. 2,5-Bis(1-aziridinyl)-3,6-bis(2-methoxyethoxy)-1,4-benzoquinone. *2,5-Bis(1-aziridinyl)-3,6-bis(2-methoxyethoxy)-2,5-cyclohexadiene-1,4-dione;* Bayer E 39 soluble. $C_{16}H_{22}N_2O_6$; mol wt 338.36. C 56.80%, H 6.55%, N 8.28%, O 28.37%. Prepn: Gauss, Petersen, *Angew. Chem.* **69**, 252 (1957); **Brit.** pat. 793,796 (1958 to Bayer). Antineoplastic activity *in vitro*: M. Akhtar *et al.*, *Can. J. Chem.* **53**, 2891 (1975); J. S. Driscoll *et al.*, *J. Pharm. Sci.* **68**, 185 (1979).

Gray needles from petr ether, mp 79-80.5°: Gauss, Petersen, *Med. Chem.* (Bayer) **7**, 649 (1963), *C.A.* **60**, 2875i (1964).

1286. Bisbentiamine. *N,N'-[Dithiobis[2-[2-(benzoyloxy)ethyl]-1-methyl-2,1-ethenediyl]]bis[N-[(4-amino-2-methyl-5-pyrimidinyl)methyl]formamide]; N,N'-[dithiobis[2-(2-hydroxyethyl)-1-methylvinylene]]bis[N-[(4-amino-2-methyl-5-pyrimidinyl)methyl]formamide] dibenzoate; O-benzoylthiamine disulfide;* Béprocin; Beston. $C_{38}H_{42}N_8O_6S_2$; mol wt 770.93. C 59.20%, H 5.49%, N 14.53%, O 12.45%, S 8.32%. Prepn: **Brit.** pat. 922,444 (1963 to Tanabe Seiyaku) corresp to **U.S.** pat. 3,109,000 (1963); Yurugi *et al.*, *Ann. Rep. Takeda Res. Lab.* **27**, 25 (1968), *C.A.* **70**, 96741m (1968); Imai, Miyaji, **Ger.** pat. 1,954,519 (1971 to Hitachi), *C.A.* **75**, 63821n (1971).

cis-form

Prisms from alc, mp 146-147°.

THERAP CAT: Vitamin B_1 source.

1287. Bis(p-chlorophenoxy)methane. *1,1'-[Methylene-bis(oxy)]bis[4-chloro]benzene;* di-(p-chlorophenoxy)methane; DCPM; K-1875; Neotran. $C_{13}H_{10}Cl_2O_2$; mol wt 269.13. C 58.02%, H 3.75%, Cl 26.35%, O 11.89%. Prepd from p-chlorophenol and dichloromethane: Moyl, **U.S. pat. 2,503,-207** (1950 to Dow); from sodium phenolate and dichloromethane: Miron, Lowry, *J. Am. Chem. Soc.* **73**, 1872 (1951).

Crystals from petr ether. mp 69.7-70.2°. bp_6 189-194°. Solubilities at 25° in g/100 ml: acetone 189; benzene 40; carbon tetrachloride 28; methanol 0.5; ether 87. Practically insol in water, petr oils. LD_{50} orally in rats: 5.8 g/kg; LC for rats in air: 18,000 ppm.

Caution: No record of human poisoning. Animal expts show low toxicity, but large doses have caused liver injury. Effects may be similar to, but less severe than, DDT, *q.v.,* Spencer *et al., Arch. Ind. Hyg. Occup. Med.* **1**, 341 (1950).

USE: Miticide.

1288. Bisdequalinium Chloride. *6,7,8,9,10,11,12,13,-14,15,16,17,24,25,26,27,28,29,30,31,32,33-Docosahydro-35,-37-dimethyl-5,34:18,23-diethenodibenzo[b,r][1,5,16,20]tetra-azacyclotriacontine-23,34-diium dichloride; N^1,N^1-deca-methylene-N^4,N^4-decamethylenebis[4-aminoquinaldinium chloride];* 1,1'-decamethylene-4,4'-(1,10-decamethylene-diimino)bis[quinaldinium chloride]; Salvizol. $C_{40}H_{58}Cl_2N_4$; mol wt 665.83. C 72.16%, H 8.78%, Cl 10.65%, N 8.41%. Prepn: Stark, **Brit. pat. 895,090** (1960).

THERAP CAT: Antiseptic, disinfectant.

1289. Bis(p-dimethylaminobenzylidene)benzidine. *N,-N'-Bis[[4-(dimethylamino)phenyl]methylene][1,1'-biphenyl]-4,4'-diamine;* 4,4'-di(4-dimethylaminobenzylideneamino)-biphenyl; bis(4-dimethylaminobenzylidene)-p,p'-diamino-diphenyl; *N,N'-bis(p-dimethylaminobenzal)benzidine.* $C_{30}H_{30}N_4$; mol wt 446.60. C 80.68%, H 6.77%, N 12.55%. Prepn from hydrazobenzene and p-dimethylaminobenz-aldehyde: Sachs, Whittaker, *Ber.* **35**, 1435 (1902).

Yellow crystals from nitrobenzene; mp 318°. Reacts with solutions of the alkali tungstates to give a cinnabar-red precipitate.

USE: Determination of tungsten: Hovorka, *Coll. Czech. Chem. Commun.* **10**, 518 (1938), *C.A.* **33**, 1624 (1939).

1290. Bis(1,2-dimethylpropyl)borane. Disiamylborane; di-sec-isoamylborine; bis(3-methyl-2-butyl)borane. $C_{10}H_{23}$-B; mol wt 154.10. C 77.94%, H 15.04%, B 7.02%. Prepn: Brown, Zweifel, *J. Am. Chem. Soc.* **83**, 1241 (1961). Use in selective reductions: G. W. Kabalka, C. F. Lane, *Chem. Tech.* **6**, 324 (1976).

Crystals, mp 35-40°. Unstable to air.

USE: Selective reagent for steric control of hydroboration of olefins.

1291. Bis(2-ethylhexyl) Phthalate. *1,2-Benzenedicar-boxylic acid bis(2-ethylhexyl) ester;* di(2-ethylhexyl) phthal-ate; dioctyl phthalate; Octoil. $C_{24}H_{38}O_4$; mol wt 390.56. C 73.81%, H 9.81%, O 16.39%. Prepn: Garner, Watson, **U.S. pat. 2,508,911** (1950 to Shell); **Brit. pat. 747,260** (1956 to Chemische Werke Hüls).

Caution: Potential symptoms of overexposure are irrita-tion of eyes and mucous membranes. *See NIOSH Pocket Guide to Chemical Hazards* (DHHS/NIOSH 90-117, 1990) p 100. This substance may reasonably be anticipated to be a carcinogen: *Seventh Annual Report on Carcinogens* (PB95-109781, 1994) p 168.

USE: As plasticizer in flexible vinyl products. As replace-ment for PCBs in dielectric fluids for electric capacitors.

1292. Bis(2-ethylhexyl) Sebacate. *Decanedioic acid bis(2-ethylhexyl) ester;* di(2-ethylhexyl) sebacate; Octoil S; Plexol 201. $C_{26}H_{50}O_4$; mol wt 426.68. C 73.19%, H 11.81%, O 15.00%. $CH_3(CH_2)_3CH(C_2H_5)CH_2OOC(CH_2)_8COOCH_2$-$CH(C_2H_5)(CH_2)_3CH_3$. Prepn: Bruno, **U.S. pat. 2,628,249** (1953 to Pittsburgh Coke & Chemical); **Brit. pat. 747,260** (1956 to Chemische Werke Hüls).

Liquid, d_{25}^{25} 0.9119, n_D^{25} 1.4496.

USE: In vacuum pumps.

1293. Bismark Brown R. *4,4'-[(4-Methyl-1,3-phenyl-ene)bis(azo)]bis[6-methyl-1,3-benzenediamine] dihydrochlo-ride; C.I. Basic Brown 4;* 5,5'-[(4-methyl-m-phenylene)bis-(azo)]bis[toluene-2,4-diamine] dihydrochloride; C.I. 21010; Bismark Brown 53; Vesuvine. $C_{21}H_{26}Cl_2N_8$; mol wt 461.40. C 54.67%, H 5.68%, Cl 15.37%, N 24.29%. Prepd by reac-tion of toluene-2,4-diamine HCl with nitrous acid: *Colour Index* vol. 4 (3rd ed., 1971) p 4154.

Dark brown solid. Very sol in water (yellowish-brown soln); sol in ethanol, Cellosolve; slightly sol in acetone; practically insol in benzene. In concd H_2SO_4 gives brown soln which on dilution turns reddish-brown; in concd HNO_3 gives violet soln which turns brown.

Free base, $C_{21}H_{24}N_8$, *C.I. Solvent Brown 12, C.I. 21010:1.* mp 130-135°. Very slightly sol in water; sol in ethanol, acetone.

USE: As textile dye, leather dye, biological stain.

1294. Bismark Brown Y. *4,4'-[1,3-Phenylenebis(azo)]-bis[1,3-benzenediamine] dihydrochloride; C.I. Basic Brown 1;* 4,4'-[m-phenylenebis(azo)]bis[m-phenylenediamine] dihy-drochloride; C.I. 21000; phenylene brown. $C_{18}H_{20}Cl_2N_8$; mol wt 419.32. C 51.56%, H 4.81%, Cl 16.91%, N 26.72%. Prepd by reaction of m-phenylenediamine HCl and nitrous acid: F. J. Welcher, *Organic Analytical Reagents* (D. Van

Nostrand Co., New York, 1948) p 339; *Colour Index* vol. 4 (3rd ed., 1971) p 4154.

Blackish-brown powder. Very sol in water; slightly sol in ethanol, Cellosolve. Practically insol in acetone, benzene carbon tetrachloride. In concd H_2SO_4 gives brown soln; in concd nitric acid gives orange soln which turns yellow.

USE: As a textile dye, biological stain.

1295. Bis(1-methylamyl) Sodium Sulfosuccinate. *Sulfobutanedioic acid 1,4-bis(1-methylpentyl) ester sodium salt;* dihexyl sodium sulfosuccinate; Aerosol MA; Alphasol MA. $C_{16}H_{29}NaO_7S$; mol wt 388.46. C 49.47%, H 7.52%, Na 5.92%, O 28.83%, S 8.25%. The bis(1-methylamyl) ester of sulfosuccinic acid monosodium salt, perhaps in admixture with the dihexyl ester. Prepd by the action of the appropriate alcohols on maleic anhydride followed by addition of sodium bisulfite: A. O. Jaeger, U.S. pats. **2,028,091; 2,176,-423** (1936, 1939 both to Am. Cyanamid); Fr. pat. **776,495** (1935 to Selden Co.).

Available as white, slightly hygroscopic, wax-like pellets. Must be soaked to dissolve in cold water. Dissolves rapidly in hot water. Solubility in water at 25° = 343 g/l; at 70° = 447 g/l. Maximum concn of electrolyte in which 1% of the wetting agent is sol: 2% NaCl; 2% NH_4Cl; 14% $(NH_4)_2$-HPO_4; 3% $NaNO_3$ (slightly turbid); 3% Na_2SO_4. Also sol in pine oil, oleic acid, acetone, kerosene, carbon tetrachloride, 2B ethanol, benzene, hot olive oil, glycerol. Insol in liquid petrolatum. Stable in acid and neutral solns, hydrolyzes in alkaline solns.

USE: Wetting agent.

1296. Bis[methylthio]methane. Bis[methylmercapto]methane; methylenebis[methyl sulfide]. $C_3H_8S_2$; mol wt 108.23. C 33.29%, H 7.45%, S 59.26%. $(CH_3S)_2CH_2$. Odorous principle of the white truffle *Tuber magnatum* (Pico) Vitt., *Tuberaceae*. Isoln: Fiecchi *et al.*, *Tetrahedron Letters* **1967**, 1681. Synthesis from methyl mercaptan: Böhme, Marx, *Ber.* **74**, 1672 (1941).

Oily liquid. Odor in high dilutions reminiscent of white truffles. The odor of the neat liquid resembles that of freshly prepd mustard without its acrid and irritating qualities. bp 148-149°. Mass spectra: Fiecchi, *loc. cit.*

1297. Bismuth. Bi; at. wt 208.9804; at. no. 83; valence 3, 5. One naturally occurring isotope: 209; artificial radioactive isotopes: 199-208; 210-215. Confused with tin until 1450. First isolated by Hillot in 1737. It was, however, Geoffrey the Younger who clearly proved its individuality in 1753. Pott and Bergmann are named as the scientific discoverers. Occurrence in the earth's crust: approx 0.2 ppm. Obtained as a byproduct from the processing of lead, copper, and tin ores. *Reviews: Nouveau Traité de Chimie Minérale*, tome **11**, P. Pascal, Ed. (Masson, Paris, 1958); *Gmelin's, Bismuth* (8th ed.) **19**, pp 1-104 (1927); supplement, pp 1-621 (1964); Smith, "Arsenic, Antimony and Bismuth" in *Comprehensive Inorganic Chemistry* vol. **2**, J. C. Bailar, Jr. *et al.*, Eds. (Pergamon Press, Oxford, 1973) pp 547-683; S. C. Carapella, H. E. Howe in Kirk-Othmer *Encyclopedia of Chemical Technology* vol. 3 (Wiley-Interscience, New York, 3rd ed., 1978) pp 912-921.

Grayish-white with reddish tinge and bright metallic luster; soft and brittle; superficially oxidized by air, frequently becoming iridescent. mp 271°; contracts when melted. bp 1420°; 1490°: *Gmelin's*, p. 43. d_4^{20} 9.78; d_4^{271} 10.07. Poor conductor of electricity. Has the greatest Hall effect of any metal, *i.e.*, its resistance increases when placed in a magnetic field. Attacked by dil HNO_3, hot H_2SO_4, concd HCl. Cold solns of Bi give a white ppt with NaOH, turning yellow on boiling; with HCl a white ppt sol in excess of acid. The solns in HCl or HNO_3 yield with much water a white ppt blackened by H_2S (different from Sb).

Precipitated bismuth, prepd by treating a soln of bismuth chloride in HCl with hypophosphorous acid, washing and drying. Contains not less than 98.5% metallic bismuth. Dull gray powder. The particles are of no greater diameter than 15 microns (0.015 mm). Easily dispersed in water.

D'Arcet metal-fusible, an alloy of 49.2% Bi, 32.2% Pb and 18.4% Sn. Whitish-gray metal, mp 96-97°.

USE: Manuf Bi salts, fusible alloys, stereotype metal, fusible boiler plugs, electric fuses, low-melting solders; tempering baths for steel; "silvering" mirrors; in dental technique.

THERAP CAT (VET): Has been used externally in dusting powders for indolent, moist or suppurating lesions; internally as a protectant of the gastrointestinal lining, and as an x-ray contrast medium. Has also been recommended to treat buccal warts in dogs.

1298. Bismuth Aluminate. *Aluminum bismuth oxide;* almuth; Bisminat. $Al_6Bi_2O_{12}$; mol wt 771.84. Al 20.97%, Bi 54.15%, O 24.87%. $Bi_2(Al_2O_4)_3$. Prepn: Roques, U.S. pat. **2,901,316** (1959 to Etablissements Roques).

Decahydrate, extremely light powder. Practically insol in water.

THERAP CAT: Antacid.

1299. Bismuth Bromide. Bismuth tribromide. $BiBr_3$; mol wt 448.69. Bi 46.58%, Br 53.42%. Prepn from the elements: Schenk in *Handbook of Preparative Inorganic Chemistry* vol. 1, G. Brauer, Ed. (Academic Press, New York, 2nd ed., 1963) p 623.

Yellowish crystals; odor of HBr. d 5.7. mp 218°. bp 441°. Dec by water, forming BiOBr; sol in solns of KI, KBr, KCl, and in dil HCl. Practically insol in alc. *Keep tightly closed.*

1300. Bismuth Bromide Oxide. Basic bismuth bromide; bismuth oxybromide; bismuthyl bromide; bismuth "subbromide". BiBrO; mol wt 304.88. Bi 68.54%, Br 26.21%, O 5.25%. BiOBr. Prepn: Schenk in *Handbook of Preparative Inorganic Chemistry* vol. 1, G. Brauer, Ed. (Academic Press, New York, 2nd ed., 1963) p 624.

Crystals or amorphous powder. Very stable, melts at red heat. Practically insol in water, alc; sol in HCl, HBr, HNO_3.

USE: Manuf of dry cell cathodes.

1301. Bismuth Butylthiolaurate. *2-(Butylthio)dodecanoic acid bismuth basic salt;* Bispecia; Neocardyl. $C_{16}H_{33}$-BiO_4S; mol wt 530.48. C 36.23%, H 6.27%, Bi 39.39%, O 12.06%, S 6.04%. $CH_3(CH_2)_9CH(SC_4H_9)COOBi(OH)_2$. Prepn: P. Lebeau, M. M. Janot, *Traité Pharm. Chim.* Vol. II (Masson, Paris, 4th ed., 1955-1956) p 435.

Yellowish powder. Practically insol in water; sol in oils. Dec in acids, liberating hydrogen sulfide.

THERAP CAT: Formerly as antisyphilitic.

1302. Bismuth Chloride. Bismuth trichloride. $BiCl_3$; mol wt 315.34. Bi 66.27%, Cl 33.73%. Prepn from the elements: Schenk in *Handbook of Preparative Inorganic Chemistry* vol. 1, G. Brauer, Ed. (Academic Press, New York, 2nd ed., 1963) p 621.

White to yellowish, deliquesc crystals; HCl odor. d 4.75. mp ~230°. Sublimes at about 430°. bp 447°. Dec by water or aq alc into BiOCl; sol in HCl, HNO_3, abs alcohol, acetone, ethyl acetate. *Keep tightly closed.*

USE: In the manuf of other bismuth salts; as catalyst for organic reactions.

1303. Bismuth Chloride Oxide. Basic bismuth chloride; bismuth oxychloride; bismuthyl chloride; bismuth "subchloride"; pearl white; blanc d'Espagne; blanc de perle; Chlorbismol. BiClO; mol wt 260.43. Bi 80.24%, Cl 13.61%, O 6.14%. BiOCl. Prepn: Schenk in *Handbook of Preparative Inorganic Chemistry* vol. 1, G. Brauer, Ed. (Academic Press, New York, 2nd ed., 1963) p 622.

Fine powder or tetragonal crystals. Melts at low red heat; d 7.72. Practically insol in water, alc; sol in HCl, HNO_3.

USE: In face powders; as pigment; manuf artificial pearls, dry-cell cathodes.

THERAP CAT: Formerly as antisyphilitic.

1304. Bismuth Ethyl Camphorate. *d-Camphoric acid ethyl ester bismuth salt;* bismuth(III) salt of *d*-camphoric acid monoethyl ester. $C_{36}H_{57}BiO_{12}$; mol wt 890.82. C 48.54%, H 6.45%, Bi 23.46%, O 21.55%. Prepd from sodium ethyl camphorate and bismuth nitrate: W. M. Lauter, H. A. Braun, *J. Am. Pharm. Assoc.* **25**, 394 (1936).

Amorphous solid, faint aromatic odor. Softens at 55°. mp 61-67° (clear melt). Practically insol in water. Sol in chloroform, ether, ethylene dichloride, oils. The soly in vegetable oils is increased by the addition of camphor. MLD i.m. in rats: 250 mg Bi/kg (Lauter, Braun).

THERAP CAT: Formerly as antisyphilitic.

1305. Bismuth Fluoride. Bismuth trifluoride. BiF_3; mol wt 265.98. Bi 78.57%, F 21.43%. Prepd by dissolving bismuth oxide or hydroxide in aqueous HF. May also be prepd by reduction of BiF_5 with CO_2 in very dil H_2 at 80-150° in a Pt tube: Muir *et al., J. Chem. Soc.* **39**, 33 (1881); v. Wartenberg, *Z. Anorg. Allgem. Chem.* **244**, 344 (1940); Aurivillius, *Acta Chem. Scand.* **9**, 1206 (1955). *Review:* Kemmitt, Sharp, *Advan. Fluorine Chem.* **4**, 213 (1965).

White to gray dimorphic crystals. d 8.3. mp 725-730°. Volatilizes at higher temps slowly and without decompn. Practically insol in water. Sol in concd HF with the formation of complexes.

USE: In the prepn of bismuth pentafluoride, *q. v.*

1306. Bismuth Hydroxide. Bismuth hydrate. BiH_3O_3; mol wt 260.00. Bi 80.38%, H 1.16%, O 18.46%. $Bi(OH)_3$. Prepn: *Gmelin's, Bismuth* (8th ed.) **19**, pp 119-122 (1927).

White to yellowish-white, amorphous powder. d^{15} 4.962. Practically insol in water. When freshly pptd it is soluble in glycerol in presence of NaOH; sol in acids. Readily loses one mol H_2O forming the metahydroxide which is yellow.

USE: As absorbent for rutin and quercetin; in hydrolysis of ribonucleic acid; in separation of plutonium from irradiated uranium.

1307. Bismuthine. Bismuth trihydride; Bismutan. BiH_3; mol wt 212.00. Bi 98.57%, H 1.43%. Prepn: Wiberg, Mödritzer, *Z. Naturforsch.* **12b**, 123 (1957); Amberger, *Ber.* **94**, 1447 (1961).

Thermally unstable liquid, bp 16.8° (est).

USE: In manuf of Ge or Si semiconductors, Scott *et al., U.S.* pat. **2,910,394** (1959 to Int'l Standard Electric).

1308. Bismuth Iodide. Bismuth triiodide. BiI_3; mol wt 589.69. Bi 35.44%, I 64.56%. Best prepd from the elements: Watt *et al., Inorg. Syn.* **4**, 114 (1953).

Black, minute, hexagonal crystals with metallic sheen. d_4^{17} 5.778, also reported as d 5.64. Sublimes at 439° (760 mm), dec at 500°. Practically insol in cold water, dec slowly in hot water. Sol in liquid ammonia, abs ethanol (about 3.5% at 20°), in aq solns of KI, HI, and HCl. The solid is slowly converted to $Bi(IO_3)_3$ upon prolonged exposure to air.

1309. Bismuth Iodide Oxide. Basic bismuth iodide; bismuth oxyiodide; bismuthyl iodide; bismuth "subiodide". BiIO; mol wt 351.88. Bi 59.39%, I 36.06%, O 4.55%. BiOI. Prepn: *Gmelin's, Bismuth* (8th ed) **19**, p 159 (1927); Schenk in *Handbook of Preparative Inorganic Chemistry* vol. **1**, G. Brauer, Ed. (Academic Press, New York, 2nd ed., 1963) p 625.

Brick-red, heavy, odorless powder or copper-colored crystals. d 7.92. Fuses at red heat with partial decompn. Prac-

tically insol in water, alcohol, $CHCl_3$; sol in HCl; dec by HNO_3 or alkali.

USE: Manuf of dry cell cathodes.

THERAP CAT: Anti-infective.

1310. Bismuth Iodosubgallate. *(Gallato)hydroxyiodobismuth;* (galloyloxy)hydroxyiodobismuthine; bismuth oxyiodogallate; Airoform; Airogen. $C_7H_6BiIO_6$; mol wt 522.01. C 16.11%, H 1.16%, Bi 40.03%, I 24.31%, O 18.39%. C_6H_2-$(OH)_3COOBi(OH)I$. Prepn: *Hagers Handb. Pharm. Praxis* **Band I**, 1565 (Berlin, 1930).

Dark gray, odorless, tasteless, bulky powder. Dec by water, acids. Practically insol in alc, ether, chloroform; sol in solns of alkali hydroxides. *Protect from light.*

THERAP CAT: Anti-infective.

THERAP CAT (VET): *See* Bismuth. For external use only.

1311. Bismuth Nitrate. Bismuth trinitrate. BiN_3O_9; mol wt 395.00. Bi 52.91%, N 10.64%, O 36.45%. $Bi(NO_3)_3$. Prepn: *Gmelin's, Bismuth* (8th ed.) **19**, pp 126-129 (1927).

Pentahydrate, lustrous, hygroscopic crystals; acid reaction; odor of nitric acid. d 2.83. Sol in water containing nitric acid; dec by water alone into subnitrate; sol in glycerol, dil acids including acetic acid, acetone. Practically insol in alcohol, ethyl acetate.

USE: Prepn of other bismuth salts, luminous paints; pptn of alkaloids.

1312. Bismuth Oleate. *Oleic acid bismuth salt;* bismuth trioleate. $C_{54}H_{99}BiO_6$; mol wt 1053.36. C 61.57%, H 9.47%, Bi 19.84%, O 9.11%. $[CH_3(CH_2)_7CH=CH(CH_2)_7$-$COO]_3Bi$; Prepd from Bi_2O_3, acetic anhydride, and oleic acid: Considine, **Brit.** pat. **947,749** (1964 to Metal & Thermit).

Solid. Sol in about 1500 parts benzene.

USE: In catalysts for manufacture of aldehydes and alcohols by the oxo process: **Brit.** pat. **907,027** (1962 to Esso); Aldridge, Cull, **Ger.** pat. **1,153,006** (1963 to Esso), *C.A.* **60**, 2763 (1964).

1313. Bismuth Oxalate. Oxalic acid bismuth salt. C_6-Bi_2O_{12}; mol wt 682.02. C 10.57%, Bi 61.28%, O 28.15%. $Bi_2(C_2O_4)_3$. Prepn: Vanino, Zumbusch, *Ber.* **41**, 3994 (1908); Soerbye, Kruse, *Acta Chem. Scand.* **16**, 1662 (1962).

Powder. *Poisonous!* Practically insol in water or alcohol; sol in moderately dil HCl or HNO_3.

1314. Bismuth Oxide. Bismuth trioxide; bismuthous oxide; bismuth yellow. Bi_2O_3; mol wt 465.96. Bi 89.70%, O 10.30%. Occurs in nature as the mineral *bismite.* Prepn: *Gmelin's, Bismuth* (8th ed.) **19**, p 109-113 (1927).

Yellow, heavy, odorless powder or monoclinic crystals; stable in air. Practically insol in water; sol in HCl or HNO_3.

USE: In disinfectants, magnets, glass, rubber vulcanization; in fireproofing of papers and polymers; in catalysts.

THERAP CAT: Astringent.

1315. Bismuth Pentafluoride. BiF_5; mol wt 303.97. Bi 68.75%, F 31.25%. Prepd according to the equation $BiF_3 + F_2 \rightarrow BiF_5$: v. Wartenberg, *Z. Anorg. Allgem. Chem.* **244**, 344 (1940); Kwasnik in *Handbook of Preparative Inorganic Chemistry*, vol. **1**, G. Brauer Ed. (Academic Press, New York, 2nd ed., 1963) pp 202-203. Prepn from Bi and F_2: Fischer, Rodzitis, *J. Am. Chem. Soc.* **81**, 6375 (1959); Kemmitt, Sharp, *Advan. Fluorine Chem.* **4**, 212 (1965); Hebecker, *Z. Anorg. Allgem. Chem.* **384**, 111 (1971).

Body-centered tetragonal crystals, sublimes at 120°. d 5.55. Very sensitive to moisture, discolors quickly in moist air. Violent reaction with water forming BiF_3 and ozone. Reacts with liquid petrolatum above 50°.

Caution: Highly toxic and irritating to mucous membranes, skin, eyes, respiratory tract.

USE: Fluorinating agent.

1316. Bismuth Phosphate. Bismugel. BiO_4P; mol wt 303.95. Bi 68.75%, O 21.06%, P 10.19%. $BiPO_4$. Prepn: Schenk in *Handbook of Preparative Inorganic Chemistry*, vol. **1**, G. Brauer, Ed. (Academic Press, New York, 2nd ed., 1963) p 626.

Odorless powder or monoclinic crystals. Does not melt on heating. d^{15} 6.323. Slightly sol in water and dil acids;

not hydrolyzed by boiling water. Practically insol in alcohol, acetic acid; sol in concd HNO_3 and HCl.

USE: In the separation of plutonium from fission products; manuf of optical flint glass.

THERAP CAT: Antacid; protectant.

1317. Bismuth Potassium Iodide. *Potassium heptaiodobismuthate(4—).* BiI_7K_4; mol wt 1253.70. Bi 16.67%, I 70.86%, K 12.47%. K_4BiI_7. Prepn: *Gmelin's, Potassium* (8th ed.) **22,** p 1068 (1936).

Red crystals. Partly dec by water; completely sol in alkali iodide solns.

USE: Pptn of vitamins, particularly thiamine HCl, and antibiotics from soln.

1318. Bismuth Potassium Tartrate. Potassium bismuthotartrate; potassium bismuth tartrate; potassium bismuthyl tartrate; tartaric acid bismuth complex potassium salt. The bismuth content of the complex salts of bismuth and alkali tartrate may be made to range within wide limits (35-75% Bi). The medicinal grade contains 60-64% bismuth. Prepn: Kober, U.S. pat. **1,663,201** (1928 to Searle); *Hagers Handb. Pharm. Praxis* **1st Suppl.,** 316 (Berlin, 1949).

Odorless powder; sweetish taste; darkens on exposure to light. Sol in 2 parts water. Practically insol in alc or other organic solvents. The aq soln is slightly alkaline. Dec by mineral acids.

THERAP CAT: Formerly as antisyphilitic.

1319. Bismuth Selenide. Bismuth triselenide. Bi_2Se_3; mol wt 654.84. Bi 63.83%, Se 36.17%. Prepn by heating a stoichiometric mixture at 475° in evacuated tube: Dönges, *Z. Anorg. Allgem. Chem.* **265,** 56 (1951); Konorov, *Zh. Tekh. Fiz.* **26,** 1394 (1956); from BiO, Se and sodium oxalate at high press.: Cambi, Elli, *Chim. Ind. (Milan)* **50,** 94 (1968).

Black crystals. Rhombohedral and hexagonal crystal structure. d_4^{20} 7.70 (also reported as 6.82). mp 710°. Heat of formation: —13.9 kcal/mol. Heat of combustion: 296.4 kcal/mol. Dec when heated in air. Dec by concd nitric acid and aqua regia. Insol in water.

USE: In semiconductor research.

1320. Bismuth Sodium Iodide. *Disodium pentaiodobismuthate(2—);* sodium pentaiodobismuthate(III); sodium bismuth iodide; sodium iodobismuthite; Aniobi; Bismjol. BiI_5Na_2; mol wt 889.48. Bi 23.49%, I 71.34%, Na 5.17%. Na_2BiI_5. Prepn: Gurchot *et al., J. Pharmacol.* **45,** 427 (1932).

Tetrahydrate, odorless, red crystals; astringent taste. mp 93° (dec). Soluble in water, alcohol, acetone, ethyl acetate; slightly sol in ether; hydrolyzes in soln.

THERAP CAT: Formerly as antisyphilitic.

1321. Bismuth Sodium Tartrate. Sodium bismuth tartrate; sodium bismuthyl tartrate; tartaric acid bismuth complex sodium salt; Natrol; Tartrol. The bismuth content of the complex salts of bismuth and alkali tartrate may be made to range within wide limits (35-75% Bi). The article generally used contains 70-74% bismuth. Prepn: Kober, U.S. pat. **1,663,201** (1928 to Searle); *Hagers Handb. Pharm. Praxis* **1st Suppl.,** 316 (Berlin, 1949).

Odorless, tasteless powder; discolors in light. Sol in about 3 parts water. Insol in alc or other organic solvents. The aq soln is slightly alkaline.

THERAP CAT: Formerly as antisyphilitic.

THERAP CAT (VET): See Bismuth.

1322. Bismuth Sodium Triglycollamate. *Nitrilotriacetic acid bismuth complex sodium salt;* Bistrimate. $C_{24}H_{28}BiN_4Na_2O_{25}$; mol wt 1142.41. C 25.23%, H 2.47%, Bi 18.29%, N 4.90%, Na 14.09%, O 35.01%. A double salt of sodium bismuthyl triglycollamate and disodium triglycollamate. Prepn: Lehman, Sproull, U.S. pat. **2,348,984** (1944).

Crystals. Slightly salty taste. Very sol in water. Insol in acetone, ether, benzene. The pH of a 2% aq soln is between 7 and 8.

THERAP CAT: Lupus erythematosus suppressant.

1323. Bismuth Subacetate. *Acetic acid basic bismuth salt;* bismuth acetate basic; bismuth acetate oxide; bismuth oxide acetate; bismuthyl acetate. $C_2H_3BiO_3$; mol wt 284.02. C 8.46%, H 1.06%, Bi 73.58%, O 16.90%. $CH_3COOBiO$. Prepn: Hofmann, *Ann.* **223,** 110 (1884); Aurivillius, *Acta Chem. Scand.* **9,** 1213 (1955). Crystal structure: Aurivillius, *loc. cit.*

Thin crystal plates. Slight acetic odor. Practically insol in water. Sol in glacial acetic acid, dil HCl or HNO_3.

1324. Bismuth Subcarbonate. *Bismuth carbonate basic;* bismuth oxycarbonate. CBi_2O_5; mol wt 509.97. C 2.36%, Bi 81.96%, O 15.69%. $(BiO)_2CO_3$. Prepn: *Gmelin's, Bismuth* (8th ed.) **19,** p 178 (1927).

Hemihydrate, odorless, tasteless powder. Practically insol in water or alc. Sol in mineral acids, in concd acetic acid.

USE: In admixture with other substances in glazes on ceramics; for artificial horn products; for pearly surfaces for plastics.

THERAP CAT: Topical protectant.

THERAP CAT (VET): See Bismuth.

1325. Bismuth Subgallate. *Gallic acid bismuth basic salt;* bismuth gallate, basic; B.S.G.; Dermatol. $C_7H_5BiO_6$; mol wt 394.09. C 21.33%, H 1.28%, Bi 53.03%, O 24.36%. Prepd from bismuth nitrate and gallic acid in an acetic acid medium. The acetic acid may be replaced with a glycol or mannitol: Pfeiffer, Schmitz, *Pharmazie* **5,** 517 (1950).

Hydrate, bright yellow, odorless, tasteless powder. Practically insol in water, alc, chloroform, ether; sol in dil alkali hydroxide solns, in hot mineral acids with decompn. Forms a water-sol sodium salt which has an alkaline reaction.

Note: Major active ingredient of *Bongast.*

THERAP CAT: Astringent; antacid; protective.

THERAP CAT (VET): See Bismuth.

1326. Bismuth Subnitrate. *Bismuth hydroxide nitrate oxide;* bismuth nitrate basic; bismuth oxynitrate; bismuth subnitricum; bismuthyl nitrate; bismuth white; magistery of bismuth; novismuth; paint white; Spanish white. A basic salt, the compn of which varies with the conditions of preparation. Contains 70 to 74% Bi or 79 to 82% Bi_2O_3. Prepd by partial hydrolysis of $Bi(NO_3)_3$: *Gmelin's, Bismuth* (8th ed.) **19,** pp 132-135 (1927); *Traité Pharm. Chim.* **vol. 1,** P. Lebeau, M. M. Janot, Eds. (Masson, Paris, 1956) p 371; *Handbuch der Pharmazie,* **vol. 4**(1), H. Thoms, Ed. (Urban & Schwarzenberg, Berlin, 1927) p 325.

Odorless, tasteless, heavy, slightly hygroscopic, microcrystalline powder. Dec to Bi_2O_3 and nitrogen oxides when heated to red heat. Practically insol in water, alc. Sol in dil HCl and HNO_3. *Keep well closed and protect from light. Incompat.* Alkaline bicarbonates, soluble iodides, gallic acid, calomel, salicylic acid, tannin, sulfur.

USE: Manuf bismuth fluxes for enamels; in cosmetics.

THERAP CAT: Antacid.

THERAP CAT (VET): See Bismuth.

1327. Bismuth Subsalicylate. *(2-Hydroxybenzoato-O¹)-oxobismuth; 2-hydroxybenzoic acid bismuth (3+) salt basic;* bismuth salicylate, basic; basic bismuth salicylate; oxo(salicylato)bismuth; salicylic acid basic bismuth salt; Bismogenol "Tosse" Inj; Stabisol. $C_7H_5BiO_4$; mol wt 362.09. C 23.22%, H 1.39%, Bi 57.71%, O 17.67%. $HOC_6H_4COOBiO$. Prepn: Fischer, Grützner, *Arch. Pharm.* **231,** 680 (1893).

Microscopic prisms. Almost insol in water or alc; dec by boiling water, alkalies into a more basic salt.

USE: To impart pearly surface to cellulose-base, polystyrene, and phenol-formaldehyde resins, Deutsch *et al.,* U.S.

Consult the Name Index before using this section.

CH₂COO section:
$$CH_2COO^- (BiO)^+$$
$$\overset{+}{HN}-CH_2COO^- Na^+$$
$$CH_2COO^-$$
·
$$\left[\begin{array}{c} CH_2COO^- Na^+ \\ \overset{+}{HN}-CH_2COO^- Na^+ \\ CH_2COO^- \end{array}\right]_3$$

pat. **2,816,044** (1957 to Brit. Resin Prods.). In heat-sensitive coating for copying paper, Van Dam, U.S. pat. **2,897,-090** (1959 to Anken Chem. & Film). To stabilize tolylenediamine mixtures against deterioration and discoloration, Powers, U.S. pat. 3,138,641 (1964 to Mobay).

THERAP CAT: Lupus erythematosus suppressant.

THERAP CAT (VET): *See* Bismuth.

1328. Bismuth Sulfate. $Bi_2O_{12}S_3$; mol wt 706.15. Bi 59.19%, O 27.19%, S 13.62%. $Bi_2(SO_4)_3$. Prepn: *Gmelin's, Bismuth* (8th ed.) **19**, p 170 (1927).

White crystals. Dec by water or alc into a basic salt; sol in acids.

USE: In analysis of other metallic sulfates.

1329. Bismuth Sulfide. Bi_2S_3; mol wt 514.16. Bi 81.29%, S 18.71%. Occurs in nature as the mineral *bismuthinite*. Prepn: *Gmelin's, Bismuth* (8th ed.) **19**, pp 162-167 (1927).

Blackish-brown, orthorhombic, bipyramidal crystals. Practically insol in water, ethyl acetate. Sol in HNO_3, HCl.

1330. Bismuth Tannate. Tanbismuth; tannic acid bismuth derivative. Contains about 36% bismuth equivalent to 40% Bi_2O_3. Prepd from freshly pptd bismuth hydroxide and tannin: *Hagers Handb. Pharm. Praxis* **Band 1**, 685 (Berlin, 1930).

Yellow or brownish-yellow powder. Practically insol in water, alcohol, ether.

THERAP CAT: Astringent; protective.

1331. Bismuth Telluride. Tellurobismuthite. Bi_2Te_3; mol wt 800.76. Bi 52.20%, Te 47.80%. Prepd by heating stoichiometric amounts of the elements to 475° for several days in an evacuated glass or quartz tube: Dönges, *Z. Anorg. Allgem. Chem.* **265**, 56 (1951). Prepn of single crystals: Ainsworth, *Proc. Phys. Soc. (London)* **B69**, 606 (1956); in zone-melting apparatus: Harmon *et al.*, *J. Phys. Chem. Solids* **2**, 181 (1957). Review of different methods: Minden, *Sylvania Technologist* **11** (no. 1), 13-25 (1958).

Gray hexagonal platelets. d 7.642. mp 585°. Single crystals have been grown by the Czochralski technique in which a hydrogen atmosphere was used to minimize the evaporation of tellurium. Since the crystals cleave readily along the (0001) basal hexagonal plane, it is mechanically easier to orient the seed so that the growth direction is in this plane rather than normal to it. The resulting crystals grow more readily along the basal plane, so that they have an oval cross section, often with a characteristic notch. All crystals so pulled are the *P* type. Heat of formation: −8 kcal/mol. Resistivity: 0.00033 ohm-cm. Thermal conductivities at room temp: $\lambda_0 = 0.015$ watt/cm-deg; $\lambda_e = 1.4 \times 10^{-3}$ watt/cm-deg. Energy gap: 0.15 ev. Electron mobility: 800 cm²/volt-sec. Hole mobility: 400 cm²/volt-sec.

USE: In electronics as semiconductor.

1332. Bismuth Tetroxide. "Bismuth peroxide". Bi_2O_4; mol wt 481.96. Bi 86.72%, O 13.28%. It may contain 1 or 2 mols of H_2O. Prepn: *Gmelin's, Bismuth* (8th ed.) **19**, 115 (1927); Schenk in *Handbook of Preparative Inorganic Chemistry* vol. 1, G. Brauer, Ed. (Academic Press, New York, 2nd ed., 1963) p 629.

Orange-red to yellowish-brown heavy powder. Slowly dec by water without dissolving. Dec and dissolved by hot mineral acids.

USE: In lubricants for metal extrusion.

1333. Bismuth Tribromophenate. *Tris(2,4,6-tribromophenoxy)bismuthine;* bismuth tribromophenol; tribromophenolbismuth; Sigmaform; Xeroform. $C_{18}H_6BiBr_9O_3$; mol wt 1198.36. C 18.04%, H 0.50%, Bi 17.44%, Br 60.01%, O 4.01%. $(Br_3C_6H_2O)_3Bi$. Prepd from sodium tribromophenolate and bismuth oxychloride: Kollo, *Pharm. Post.* **45**, 1013-1014, *C.A.* **7**, 2831[1] (1913); *Hagers Handb. Pharm. Praxis* **Band 1**, 686 (Berlin, 1930); *N.N.R.* **1951**, pp 66, 505. [In early publications bismuth tribromophenate was described as a basic bismuth salt $(Br_3C_6H_2O)_2BiOH.Bi_2O_3$: *Hagers Handb., loc. cit.*]

Amorphous yellow powder. Stable below 120°. Neutral to moistened litmus paper. Slightly sol in water, alcohol, chloroform, vegetable oils. Dec by alkalies and strong acid.

Ref: N.N.R. **1951**, p 505.

THERAP CAT: Anti-infective.

THERAP CAT (VET): *See* Bismuth.

1334. Bismuth Valerate, Basic. *Valeric acid bismuth basic salt;* basic bismuth valerate; bismuth oxide *n*-valerate; bismuth valerate oxide; bismuth valerianate; bismuthyl valerate. $C_5H_9BiO_3$; mol wt 326.11. C 18.42%, H 2.78%, Bi 64.08%, O 14.72%. $CH_3(CH_2)_3COOBiO$. Prepn: *Hagers Handb. Pharm. Praxis* **Band I**, 687 (Berlin, 1930); Aurivillius, *Acta Chem. Scand.* **9**, 1213 (1955). Crystal structure: Aurivillius, *loc. cit.*

Monohydrate, powder. Valeric acid odor. Practically insol in water, alc. Sol in dil HCl or HNO_3. *Keep well closed.*

1335. Bis(1-naphthylmethyl)amine. *N-(1-Naphthalenylmethyl)-1-naphthalenemethanamine;* di(α-naphthylmethyl)amine; α-dinaphthomethylamine. $C_{22}H_{19}N$; mol wt 297.40. C 88.85%, H 6.44%, N 4.71%. Prepd by catalytic hydrogenation of α-naphthonitrile: Rupe, Becherer, *Helv. Chim. Acta* **6**, 880 (1923).

Pale yellow crystals from petr ether, mp 62°.

USE: In determination of nitrates in fertilizers: Konek, *Z. Anal. Chem.* **97**, 416 (1934), *C.A.* **28**, 5779 (1934). The acetate has been proposed as a sensitive reagent for nitric acid.

1336. Bisoprolol. *1-[4-[[2-(1-Methylethoxy)ethoxy]methyl]phenoxy]-3-[(1-methylethyl)amino]-2-propanol;* (±)-1-[[α-(2-isopropoxyethoxy)-*p*-tolyl]oxy]-3-(isopropylamino)-2-propanol; (±)-1-[*p*-(2-isopropoxyethoxymethyl)phenoxy]-3-(isopropylamino)-2-propanol; EMD-33512. $C_{18}H_{31}NO_4$; mol wt 325.45. C 66.43%, H 9.60%, N 4.30%, O 19.66%. Cardioselective β_1-adrenergic blocker. Prepn: **Belg.** pat. **859,425;** R. Jonas *et al.*, U.S. pat. **4,258,062** (1978, 1981 both to E. Merck). Clinical pharmacology: A. E. Tattersfield *et al.*, *Brit. J. Clin. Pharmacol.* **18**, 343 (1984); P. Dorow, U. Tönnesmann, *Eur. J. Clin. Pharmacol.* **27**, 135 (1984). Binding study of (−)-form: A. J. Kaumann, H. Lemoine, *Arch. Pharmacol.* **331**, 27 (1985). HPLC determn of enantiomers in plasma and urine: T. Suzuki *et al.*, *J. Chromatog.* **619**, 267 (1993). Clinical evaluation in angina: R. S. Kohli *et al.*, *Eur. Heart J.* **6**, 845 (1985); in hypertension: F. R. Bühler *et al.*, *J. Cardiovasc. Pharmacol.* **8**, Suppl. 11, S122 (1986). Review of pharmacokinetics: J. Grevel *et al.*, *Clin. Pharmacokinet.* **17**, 53-63 (1989); of clinical efficacy in combination with hydrochlorothiazide: P. K. Zachariah *et al.*, *Clin. Ther.* **15**, 779-787 (1993).

Hemifumarate, $C_{18}H_{31}NO_4 \cdot \frac{1}{2}C_4H_4O_4$, *Concor, Detensiel, Emconcor, Emcor, Euradal, Isoten, Monocor, Soprol, Zebeta.* Crystals, mp 100°. Sol in ethanol.

Mixture with hydrochlorothiazide, *Ziac.*

THERAP CAT: Antihypertensive.

1337. Bisoxatin Acetate. *2,2-Bis[4-(acetyloxy)phenyl]-2H-1,4-benzoxazin-3(4H)-one;* 2,2-bis(4-acetoxyphenyl)-3-oxo-2,3-dihydrobenz-1,4-oxazine; Wy-8138; Laxonalin; Maratan; Talsis; Tasis. $C_{24}H_{19}NO_6$; mol wt 417.42. C 69.06%, H 4.59%, N 3.36%, O 23.00%. Prepd from 2,2-dichloro-3-oxo-2,3-dihydrobenz-1,4-oxazine and phenylacetate or by acetylation of the diphenol: Seeger, U.S. pat. **3,006,917** (1961 to Thomae). Toxicity: E. I. Goldenthal, *Toxicol. Appl. Pharmacol.* **18**, 185 (1971).

Crystals from ethanol, mp 190°. LD$_{50}$ orally in rats, mice: 8000, > 10,000 mg/kg (Goldenthal).

THERAP CAT: Cathartic.

1338. Bisphenol A. *4,4'-(1-Methylethylidene)bisphenol; 4,4'-isopropylidenediphenol;* 2,2-bis(4-hydroxyphenyl)propane. $C_{15}H_{16}O_2$; mol wt 228.29. C 78.92%, H 7.06%, O 14.02%. Manuf from phenol and acetone: Jansen, **U.S. pat. 2,468,982** (1949); Faith, Keyes & Clark's *Industrial Chemicals,* F. A. Lowenheim, M. K. Moran, Eds. (Wiley-Interscience, New York, 4th ed., 1975) pp 149-152. *Review: Chem. & Eng. News* **41,** 35 (June 3, 1963); *ibid.* **51,** 5 (July 16, 1973).

Crystals or flakes. Mild phenolic odor. mp 150-155° (solidification range). bp$_4$ 220°. Dec above 8 mm pressure when heated above 220°. Practically insol in water. Sol in aq alkaline solns, alcohol, acetone. Slightly sol in carbon tetrachloride.

USE: In the manuf of epoxy resins and polycarbonates. As fungicide.

1339. Bisphenol B. *4,4'-(1-Methylpropylidene)bisphenol; p,p'-sec-butylidenediphenol;* 2,2-bis(4-hydroxyphenyl)butane. $C_{16}H_{18}O_2$; mol wt 242.32. C 79.31%, H 7.49%, O 13.21%. May be prepd from phenol and ethyl methyl ketone: Jansen, **U.S. pat. 2,468,982** (1949 to Goodrich).

Crystals or tan granules, mp 118.9-121.7°. Approximate soly per 100 g: acetone 266 g; benzene 2.3 g; carbon tetrachloride < 0.1 g; ether 133 g; methanol 166 g; V.M.P. naphtha < 0.1 g; water < 0.1 g.

USE: In the manufacture of phenolic resins.

1340. 1,4-Bis(trichloromethyl)benzene. *α,α,α,α',α',α'-Hexachloro-p-xylene; p-bis(perchloromethyl)benzene; Bitriben; Hetol.* $C_8H_4Cl_6$; mol wt 312.84. C 30.72%, H 1.29%, Cl 68.00%. Usually obtained by photochlorination of *p*-xylene: McBee *et al., Ind. Eng. Chem.* **39,** 298 (1947); *eidem* in Slesser, Schram, *Preparation, Properties, and Technology of Fluorine and Organic Fluoro Compounds* (McGraw-Hill, New York, 1951) pp 207-221; Ross *et al., J. Am. Chem. Soc.* **75,** 4697 (1953); Rabjohn, *ibid.* **76,** 5479 (1954); Harvey *et al., J. Appl. Chem.* **4,** 319 (1954); Ligett, **U.S. pat. 2,654,789** (1953 to Ethyl Corp.).

Crystals from hexane or ether, mp 108-110°.

Note: The name Hetol was previously used to designate sodium cinnamate.

USE: Insecticide.

THERAP CAT (VET): Anthelmintic (Trematodes).

1341. Bis(triphenylphosphine)dicarbonylnickel. *Dicarbonylbis(triphenylphosphine)nickel;* dicarbonyldi(triphenylphosphino)nickel. $C_{38}H_{30}NiO_2P_2$; mol wt 639.30. C 71.39%, H 4.73%, Ni 9.18%, O 5.01%, P 9.69%. [(C$_6$H$_5$)$_3$P]$_2$-Ni(CO)$_2$. Obtained by reacting one mole of nickel carbonyl with two moles of triphenylphosphine: Reppe, Schweckendiek, *Ann.* **560,** 104 (1948); Rose, Statham, *J. Chem. Soc.* **1950,** 69. Also by reacting CO with a methanol soln of triphenylphosphine and reduced Ni at 170° and 60 atm: Yamamoto, Oku, *Bull. Chem. Soc. Japan* **27,** 382 (1954), *C.A.* **49,** 6856 (1955).

Crystals from benzene, mp 210-215° (Rose); mp 206-209° (Reppe). Catalytic activity decreases with storage.

USE: Catalyst in polymerization of acetylene to benzene and styrene, trimerization of ethynyl compds, cyclization of butadiene.

1342. Bitertanol. *β-[(1,1'-Biphenyl)-4-yloxy]-α-(1,1-dimethylethyl)-1H-1,2,4-triazole-1-ethanol;* 1-(biphenyl-4-yloxy)-3,3-dimethyl-1-(1H-1,2,4-triazol-1-yl)butan-2-ol; biloxazol; BAY KWG 0599; Baycor; Sibutol. $C_{20}H_{23}N_3O_2$; mol wt 337.42. C 71.19%, H 6.87%, N 12.45%, O 9.48%. Prepn: **Belg. pat. 814,831;** W. Kramer *et al.,* **U.S. pat. 3,952,002** (1974, 1976 both to Bayer AG). Physical properties and fungicidal activity: W. Brandes *et al., Pflanzenschutz-Nachr.* **32,** 1 (1979). Mechanism of action: P. Kraus, *ibid.* 17; S. V. Overton *et al., J. Horticult. Sci.* **63,** 183 (1988). In vitro activity: T. B. Sutton *et al., Plant Dis.* **69,** 700 (1985). Field use in control of apple diseases: K. S. Yoder, *ibid.* **66,** 580 (1982); W. F. S. Schwabe, *Pflanzenschutz-Nachr.* **35,** 125 (1982); in wheat seed pretreatment: J. A. Hoffmann, D. V. Sisson, *Plant Dis.* **71,** 839 (1987).

The marketed product is a mixture of diastereomers. Colorless crystals, mp 125-129°. Vapor pressure at 20°: 10^{-5} mbar. Soly at 20° (g/100 g solvent): water 0.0005, ligroin (80-110°) 0-1, propan-2-ol 1-5, toluene 1-5, cyclohexane 5-10, methylene chloride 10-20. Stable in aqueous acid and alkaline solns. LD$_{50}$ in rats, male and female mice (mg/kg): > 5000, 4488, 4202 orally (Brandes).

USE: Agricultural fungicide.

1343. Bithionol. *2,2'-Thiobis[4,6-dichlorophenol];* TBP; bis(2-hydroxy-3,5-dichlorophenyl)sulfide; XL-7; Actamer; Bithin; Lorothidol. $C_{12}H_6Cl_4O_2S$; mol wt 356.06. C 40.48%, H 1.70%, Cl 39.83%, O 8.99%, S 9.01%. Prepn from 2,4-halo substituted phenol with S halides: Muth, **Ger. pat. 583,055** (1933 to I. G. Farbenind.), *C.A.* **28,** 179 (1934); Copper, Godfrey, **U.S. pat. 2,849,494** (1958 to Monsanto). Comprehensive review and bibliography: Shumard *et al., Soap, Sanit. Chemicals* **29,** no. 1, 34-37, 90 (1953). Anthelmintic activity of the sulfoxide: J. Guilhon, M. Graber, *Bull. Acad. Vet. Fr.* **52,** 225 (1979). Metabolism of the sulfoxide in humans: M. Sakamoto *et al., J. Toxicol. Sci.* **6,** 307 (1981).

Crystals, mp 188°. d_4^{25} 1.73. pK_1 4.82; pK_2 10.50. Vapor press. at 37°: 1.1×10^{-9} mm Hg. Practically insol in water (0.0004% at 25°). Sol in dil caustic solns. A 4% NaOH soln will dissolve 16.2% bithionol. Soly (g/100 ml): acetone 15.0; polysorbate 80 19.0; dimethylacetamide 72.5; lanolin at 42° 5.0; pine oil 4.0; corn oil 1.0; propylene glycol 0.5; 70% ethanol 0.3.

Sodium salt, $C_{12}H_4Cl_4Na_2O_2S$, *bithionolate sodium, Vancide BN,* Harvey *et al.,* U.S. pat. 3,024,163 (1962 to Vanderbilt Co.).

Sulfoxide, $C_{12}H_6Cl_4O_3S$, *BTS, Bitin-S, Disto-5.*

USE: Surfactant-formulated antimicrobial against bacteria, molds and yeast. Banned by FDA from use in cosmetics. Proposed as agricultural fungicide.

THERAP CAT: Anti-infective (topical).

THERAP CAT (VET): Anthelmintic; antiseptic.

1344. Bitolterol. *4-Methylbenzoic acid 4-[2-[(1,1-dimethylethyl)amino]-1-hydroxyethyl]-1,2-phenylene ester; p-toluic acid 4-[2-(t-butylamino)-1-hydroxyethyl]-o-phenylene ester;* α-[(*tert*-butylamino)methyl]-3,4-dihydroxybenzyl alcohol 3,4-di-*p*-toluate. $C_{28}H_{31}NO_5$; mol wt 461.56. C 72.86%, H 6.77%, N 3.03%, O 17.33%. β_2-Adrenergic agonist; a diester of *N-tert*-butylnorepinephrine. Prepn: H. Minatoya *et al.,* Ger. pat. 2,015,573; *eidem,* U.S. pat. 4,138,581 (1970, 1979 both to Sterling); B. F. Tullar *et al., J. Med. Chem.* 19, 834 (1976). Pharmacology: H. Minatoya, *J. Pharmacol. Exp. Ther.* 206, 515 (1978). Metabolism and excretion: T. Aimoto *et al., Xenobiotica* 9, 173 (1979). Comparative clinical studies with isoproterenol, *q.v.,* in asthma: J. L. Pinnas *et al., J. Allergy Clin. Immunol.* 79, 768 (1987); R. A. Nathan *et al., ibid.* 822. *Review:* S. B. Walker *et al., Pharmacotherapy* 5, 127-137 (1985).

Methanesulfonate, $C_{28}H_{31}NO_5 \cdot CH_3SO_3H$, *bitolterol mesylate, Win-32784, Biterol, Effectin, Tornalate.* White crystalline powder, mp 170-172°. Sol in DMSO.

THERAP CAT: Bronchodilator.

1345. Bitoscanate. *1,4-Diisothiocyanatobenzene;* phenylene-1,4-diisothiocyanate; isothiocyanic acid *p*-phenylene ester; Jonit. $C_8H_4N_2S_2$; mol wt 192.27. C 49.98%, H 2.10%, N 14.57%, S 33.36%. Prepn: Lieber, Slutkin, *J. Org. Chem.* 27, 2214 (1962). Prepn and use as anthelmintic: *Fr.* pat. M1652 and *Brit.* pat. 1,001,314 (1963, 1965 to Hoechst). Purification: Soeder, Laemmer, *Ger.* pat. 1,172,664 (1964 to Hoechst). Series of articles on clinical studies: *Progress in Drug Research* vol. 19, E. Jucker, Ed. (Birkhäuser Verlag, Basel and Stuttgart, 1975) pp 2-107.

Tasteless, odorless, colorless needles, from acetic acid or acetone, mp 132°.

THERAP CAT: Anthelmintic (Nematodes).

1346. Biuret. *Imidodicarbonic diamide;* carbamylurea. $C_2H_5N_3O_2$; mol wt 103.08. C 23.30%, H 4.89%, N 40.76%, O 31.04%. $NH_2CONHCONH_2$. Prepd on a large scale by the action of heat on urea. Other syntheses are based on the treatment of urea with inorganic halides, such as thionyl chloride, on the interaction of urea and cyanic acid, and on the ammonolysis of allophanic esters and related compds. Has weak bacteriostatic and diuretic properties in rats, also causes a fall in blood pressure, but produces strong irritation of the urinary tract. Increases the activity of pepsin.

Comprehensive review and bibliography: Kurzer, *Chem Rev.* 56, 95-197 (1956).

Hygroscopic, elongated plates from ethanol. d_4^{-5} 1.467. When crystallized from water, $5C_2H_5N_3O_2 \cdot 4H_2O$ is formed, which becomes anhyd at 110°, then dec at about 193°. Pyrolysis at higher temps yields melamine. Soly (g/100 g soln) in water at 25° = 2.01; at 50° = 7; at 75° = 20; at 105.5° = 53.5. Freely sol in alc, very slightly sol in ether. Aq solns treated with cupric sulfate and NaOH give a reddish-violet color (biuret reaction, details: Kurzer, *loc. cit.,* p 181).

1347. Bixin. *6,6'-Diapo-ψ,ψ-carotenedioic acid monomethyl ester.* $C_{25}H_{30}O_4$; mol wt 394.51. C 76.11%, H 7.66%, O 16.22%. Carotenoid carboxylic acid isolated from seeds of *Bixa orellana* L., Bixaceae: Kuhn, Ehmann, *Helv. Chim. Acta* 12, 904 (1929). The *cis* form occurring in the plant is unstable and changes to the stable *trans* form (complete conversion on standing in chloroform soln in the presence of iodine). Isoln of *trans* form from *Aristolochia cymbifera* Mart., *Aristolochiaceae:* Green *et al., Helv. Chim. Acta* 37, 1717 (1954). Structure: Kuhn *et al., ibid.* 11, 427 (1928); 12, 64, 904 (1929); *Ber.* 64, 1732 (1931); *Ber.* 65, 646, 1873 (1932). Stereochemistry: Barber *et al., Proc. Chem. Soc.* 1960, 23. Synthesis of methyl *cis*-bixin: Pattenden *et al., J. Chem. Soc. (C)* 1970, 235. *Reviews:* Karrer, Jucker, *Carotenoids* (Elsevier, New York, 1950) pp 256-271; Zechmeister, *Fortsch. Chem. Org. Naturst.* 18, 320-333 (1960). The following data are for the stable *trans* form; also called *isobixin.*

Orange to purple plates from acetone, dec 217°. Absorption max (chloroform): 509.5; 475; 443 nm. Considerably less sol in organic solvents than the labile form.

Methyl ester, $C_{26}H_{32}O_4$, *methyl bixin.* Broad, blue to purple needles from benzene, mp 203°. Absorption max (benzene): 475 nm (ε 125,000). Synthesis: Isler *et al., Helv. Chim. Acta* 40, 1242 (1957); Buchta, Andree, *Ber.* 92, 3111 (1959).

Ethyl ester, $C_{27}H_{34}O_4$, *ethyl bixin.* Red rhombic crystals from ethanol, mp 138°. Freely sol in chloroform, ethyl acetate, acetone.

USE: The ethyl ester as a suspension in vegetable oil for coloring foods and drugs. Imparts a golden yellow color.

1348. Blackstrap Molasses. The final, unpurified mother liquor in the cane sugar industry from which no more sugar can be crystallized by factory methods. Rich in inorganic constituents, may analyze as much as 10% ash upon ignition. Sucrose content about 30%.

1349. Blancophor® R. *2,2'-(1,2-Ethenediyl)bis[5-[[(phenylamino)carbonyl]amino]benzenesulfonic acid] disodium salt;* 4,4'-bis(3-phenylureido)-2,2'-stilbenedisulfonic acid disodium salt; [stilbene-(4,4')]bis[ω-phenylurea]-2,2'-disulfonic acid disodium salt; C.I. Fluorescent Brightener 30; C.I. 40600; Blancol C; Blankophor R; Leucophor R; Lumisol RV; Phorwite RN; Photine R; Pontamine White BR; Tintophen X. $C_{28}H_{22}N_4Na_2O_8S_2$; mol wt 652.62. C 51.53%, H 3.40%, N 8.58%, Na 7.05%, O 19.61%, S 9.83%. Prepd by treating 4,4'-diamino-2,2'-stilbenedisulfonic acid with phenyl isocyanate in aq soln at 40°: *Ger.* pat. 746,569 and *Fr.* pat. 878,155 (to I. G. Farbenind.); *Brit.* pat. 683,-895 (1952 to Bayer); *FIAT Final Rept.* No. 1302 (Sept. 15, 1947); *Colour Index* Vol. 4 (3rd ed., 1971) p 4371.

trans-form

USE: Fluorescent dye for cellulose, protein fibers, nylon, wool or paper.

1350. Blasticidin S. *4-[[3-Amino-5-[(aminoiminomethyl)methylamino]-1-oxopentyl]amino]-1-(4-amino-2-oxo-1(2H)-pyrimidinyl)-1,2,3,4-tetradeoxy-β-D-erythro-hex-2-enopyranuronic acid;* 4-[3-amino-5-(1-methylguanidino)valeramido]-1-(4-amino-2-oxo-1(2H)-pyrimidinyl)-1,2,3,4-tetradeoxy-β-D-erythro-hex-2-enopyranuronic acid; 1-(1'-cytosinyl)-4-[L-3'-amino-5'-(1''-N-methylguanidino)-valerylamino]-1,2,3,4-tetradeoxy-β-D-erythro-hex-2-enuronic acid. $C_{17}H_{26}N_8O_5$; mol wt 422.44. C 48.33%, H 6.20%, N 26.53%, O 18.94%. Nucleoside antibiotic produced by *Streptomyces griseochromogenes:* Isoln and antimicrobial activity: Takeuchi *et al., J. Antibiot.* **11A**, 1 (1958); Sumiki, Umezawa, **Japan.** pat. **60 16,449** (1961 to Japan Antibiot. Res. Assoc.), *C.A.* **55**, 21474e (1961). Structure: Otake *et al., Agr. Biol. Chem.* **30**, 132 (1966); Fox, Watanabe, *Tetrahedron Letters* **1966**, 897. Abs config: Yonehara *et al., ibid.* **1966**, 3785. Synthesis of the unsaturated carbohydrate: Goody *et al., ibid.* **1970**, 293; of the cytosinine moiety: Fox, Watanabe, *Pure Appl. Chem.* **28**, 475 (1971); Kondo *et al., Tetrahedron Letters* **1972**, 1881; *eidem, Tetrahedron* **29**, 1801 (1973). Crystal and molecular structure: V. Swaminathan *et al., Biochim. Biophys. Acta* **655**, 335 (1981).

Blastidic Acid   Cytosinine

Needles from water, dec 235-236°. $[\alpha]_D^{11}$ +108.4° (water). uv max (0.1N HCl): 275 nm ($E_{1cm}^{1\%}$ 349); (0.1N NaOH): 266-270 nm ($E_{1cm}^{1\%}$ 266). Sol in water, acetic acid. Practically insol in methanol, ethanol, acetone, benzene, ether, ethyl acetate, chloroform, carbon tetrachloride, cyclohexane, xylene, pyridine, dioxane. LD_{50} i.v. in mice: 2.82 mg/kg (Takeuchi).

Hydrochloride, crystals, dec 224-225°.

USE: Antifungal against rice blast disease in Japan.

1351. Bleomycins. NSC-125066; Bleo. A group of related glycopeptide antibiotics isolated from *Streptomyces verticillus:* Umezawa, *Antimicrob. Ag. Chemother.* **1965**, 1079. Purification and separation into bleomycins A and B and their components: Umezawa *et al., J. Antibiot.* **19**, 200, 210 (1966); T. Takita *et al., ibid.* **21**, 79 (1968); **22**, 237 (1969). Bleomycin A_2 is the main component of the bleomycin employed clinically. Total structure elucidation: T. Takita *et al., ibid.* **25**, 755 (1972). Revised structure: *eidem, ibid.* **31**, 801 (1978). Bleomycins differ from one another in the terminal amine and show varying biological activity. Structures of the terminal amines: Fujii *et al., ibid.* **26**, 398 (1973). Synthesis of new bleomycins: T. Takita *et al., ibid.* 254. Total synthesis of bleomycin A_2: *eidem, Tetrahedron Letters* **23**, 521 (1982); Y. Aoyagi *et al., J. Am. Chem. Soc.* **104**, 5537 (1982). Improved total synthesis: S. Saito *et al., J. Antibiot.* **36**, 92 (1983). Biosynthesis: Fujii *et al., ibid.* **27**, 73 (1974). Bleomycins are believed to react with DNA and cause strand scission; they have also been shown to have a type of oxygen transferase activity. Mechanism of action studies: R. M. Burger *et al., Life Sci.* **28**, 715 (1981); N. Marugesan *et al., J. Biol. Chem.* **257**, 8600 (1982). Coordination chemistry: J. C. Dabrowiak, *J. Inorg. Biochem.* **13**, 317 (1980). Clinical pharmacology: S. T. Crooke, *Cancer Chemother.* **3**, 343 (1981). Characterization of analogs: N. J. Oppenheimer *et al., J. Biol. Chem.* **257**, 1606 (1982). Total synthesis of *deglycobleomycin*, a biologically active deriv lacking sugars: Y. Aoyagi *et al., J. Am. Chem. Soc.* **104**,

5237 (1982). *Reviews:* H. Umezawa, *Pure Appl. Chem.* **28**, 665-680 (1971); C. W. Haidle, R. S. Lloyd, *Antibiotics* **vol. 5**(pt. 2), F. E. Hahn, Ed. (Springer-Verlag, New York, 1979) pp 124-154; H. Umezawa, *Anticancer Agents Based on Natural Product Models,* J. M. Cassady, J. D. Douros, Eds. (Academic Press, New York, 1980) pp 147-166.

R = terminal amine

Colorless or yellowish powder which becomes bluish depending on Cu content. Very sol in water, methanol; slightly sol in ethanol. Practically insol in acetone, ethyl acetate, butyl acetate, ether. uv max: 244-248, 289-294 nm ($E_{1cm}^{1\%}$ 121-148, 102-121.5).

Sulfate, *Blenoxane.*

Bleomycin A_2, $C_{55}H_{84}N_{17}O_{21}S_3$, N^1-[3-(dimethylsulfonio)-propyl]bleomycinamide. R = $(CH_3)_2S^+CH_2CH_2CH_2NH-$.

THERAP CAT: Antineoplastic.

1352. Bolandiol. (3β,17β)-*Estr-4-ene-3,17-diol;* 3β,17β-dihydroxyestr-4-ene. $C_{18}H_{28}O_2$; mol wt 276.42. C 78.21%, H 10.21%, O 11.58%. Prepn: Colton, U.S. pat. **2,843,608** (1958 to Searle).

Crystals from dil acetone and from ethyl acetate + petr ether, mp 169-172°.

Dipropionate, $C_{24}H_{36}O_4$, *norpropandrolate, 3β,17β-dipropionyloxy-4-estrene, SC-7525, Anabiol, Storinal.*

THERAP CAT: Anabolic.

1353. Bolasterone. (7α,17β)-*17-Hydroxy-7,17-dimethylandrost-4-en-3-one;* 7α,17α-dimethyltestosterone; Myagen. $C_{21}H_{32}O_2$; mol wt 316.48. C 79.70%, H 10.19%, O 10.11%. Epimeric with calusterone, *q.v.* Prepn: **Belg.** pat. **610,385** corresp to Babcock, Campbell, U.S. pat. **3,341,557** (1962, 1967 both to Upjohn); Campbell, Babcock, *Hormonal Steroids,* Proc. 1st Int. Congr., **2**, L. Martini, A. Pecile, Eds. (Academic Press, New York, 1965) pp 59-67. Activity studies: Stucki *et al., ibid.* 119-132.

Crystals, mp 163-165°.

THERAP CAT: Anabolic.

1354. Boldenone. (17β)-*17-Hydroxyandrosta-1,4-dien-3-one;* 1,4-androstadien-17β-ol-3-one; 3-oxo-17β-hydroxy-1,4-androstadiene; dehydrotestosterone. $C_{19}H_{26}O_2$; mol wt

286.41. C 79.68%, H 9.15%, O 11.17%. Anabolic steroid. Prepn: Meystre *et al., Helv. Chim. Acta* **39**, 734 (1956); Florey, U.S. pat. **2,875,196** (1949 to Olin Mathieson); Nobile, U.S. pat. **2,837,464** (1958 to Schering). Metabolism: Galletti, Gardi, *Steroids* **18**, 39 (1971).

Crystals, mp 164-166°. $[\alpha]_D^{25}$ +25° (in chloroform).
Acetate, $C_{21}H_{28}O_3$, crystals, mp 151-153°.
10-Undecenoate, $C_{30}H_{44}O_3$, *boldenone undecylenate, Ba-29038, Parenabol.*
Note: This is a controlled substance (anabolic steroid) in the U.S. Code of Federal Regulations, Title 21 Part 1308.13, as defined in Part 1308.02 (1995).
THERAP CAT: Androgen.
THERAP CAT (VET): Anabolic.

1355. Boldine. *5,6,6a,7-Tetrahydro-1,10-dimethoxy-6-methyl-4H-dibenzo[de,g]quinoline-2,9-diol; 1,10-dimethoxy-6aα-aporphine-2,9-diol;* 1,10-dimethoxy-2,9-dihydroxyaporphine; 2,6-dihydroxy-3,5-dimethoxyaporphine. $C_{19}H_{21}NO_4$; mol wt 327.38. C 69.71%, H 6.47%, N 4.28%, O 19.55%. Isoln from boldo *(Peumus boldus Molina, Monimiaceae):* Bourgoin, Verne, *J. Pharm. Chim.* **16**, 191 (1872); from *Laurelia novaezelandiae* A. Cunn.: Bernauer, *Helv. Chim. Acta* **50**, 1583 (1967). Structure: Warnat, *Ber.* **58**, 2768; **59**, 85 (1926); Späth, Tharrer, *ibid.* **66**, 904 (1933); Schlittler, *ibid.* 988. Synthesis of *dl*-form: S. M. Kupchan *et al., Chem. Commun.* **1976**, 91. Biosynthetic study: D. S. Bhakuni *et al., J. Chem. Soc. Perkin Trans. I* **1977**, 706. Detailed description of the boldo brush, its ingredients and pharmacological properties: H. Schindler, *Arzneimittel-Forsch.* **7**, 747 (1957).

d-Form, crystals from ether. mp 162-164°. $[\alpha]_D^{25}$ +127° (c = 0.1 in alcohol). Very slightly sol in water or ether; sol in alcohol, chloroform, dil acids.
d-Form hydrochloride, crystals from methanol-ether and methanol, mp 212-220°.
dl-Form, mp 159-162°.
Dimethyl ether, *see* Glaucine.
Note: Boldine is used as an ingredient in choleretics and laxatives. The total alkaloids from boldo are also marketed as a grayish-white to grayish-yellow, bitter powder; almost insol in water. Sol in alc, chloroform, slightly in ether.

1356. Boldo. Boldu; boldea; boldus; boldoa. Leaves of *Peumus boldus* Molina *(Boldu boldus* [Molina] Lyons, *Boldea fragrans* Gay), *Monimiaceae. Habit.* Peru, Chile. *Constit.* About 2% volatile oil, about 0.1% boldine. *Ref:* Schindler, *Arzneimittel-Forsch.* **7**, 747 (1957).
THERAP CAT: In hepatic dysfunction, cholelithiasis.

1357. Bole, Armenian. Bolus armena; bolus rubra; red bole. A red variety of clay (aluminum silicate) contg naturally occurring ferric oxide (hematite). Originally found in Armenia.
Reddish, soft, unctuous pieces; adheres to the tongue; easily reduced to powder. Insol in usual solvents. d 1.9-2.0.

USE: For coloring powders, cements, and as a pigment.
THERAP CAT: Adsorbant; protectant.

1358. Boleko Oil. Isano oil. Oil extracted from the nuts of the tree *Ongokea gore* (Hua) Pierre, *Olacaceae,* growing in equatorial Africa. Contains fatty acids (as glycerides): isanic acid 46%, isanolic acid 44%. *Refs:* Scher, *Arch. Pharm.* **287**, 548 (1954); Dupont *et al., Bull. Soc. Chim. France* **1957**, 1495; De Vries, *Oléagineux* **12**, 427 (1957); Kneeland *et al., J. Am. Oil Chem. Soc.* **35**, 361 (1958). Extraction procedure: Lambert, U.S. pat. **2,800,492** (1957 to UCB). Structure of acids: Gunstone, Sealy, *J. Chem. Soc.* **1963**, 5772; Morris, *ibid.* 5779.
Typical characteristics: Acid no. 1-10; ester no. 183.5; sapon no. 185-200; iodine no. 254-259; d_4^{20} 0.973-0.983; n_D^{20} 1.505-1.509; viscosity at 25° = 7-10 poise. Soluble in acetone, benzene, ethyl ether, carbon tetrachloride, chloroform. Slightly sol in petr ether, ethanol, hexane. The viscosity increases upon heating and the oil is transformed into a rubber-like mass. Does not polymerize when heated to a moderate temp; when heated rapidly to temp above 200° polymerization becomes so rapid as to become explosive. Polymerization is explained by the acetylenic structure of isanic acid.
USE: In fire retardant paints.

1359. Bombesin. *2-L-Glutamine-6-L-asparaginealytesin.* $C_{71}H_{110}N_{24}O_{18}S$; mol wt 1619.87. C 52.64%, H 6.84%, N 20.75%, O 17.78%, S 1.98%. Pharmacologically active tetradecapeptide found in skins of European amphibians of the family Discoglossidae, principally *Bombina bombina* and *Bombina variegata variegata.* Isoln and structure: A. Anastasi *et al., Experientia* **27**, 166 (1971). Pharmacological activity: V. Erspamer *et al., J. Pharm. Pharmacol.* **22**, 875 (1970). Synthesis: L. Bernardi *et al., Experientia* **27**, 873 (1971). It is a potent stimulant of gastric and pancreatic secretions in mammals; a bombesin-like immunoreactive peptide is found in both brain and gut, *cf.* J. M. Polak *et al., Lancet* **1**, 1109 (1976). Other actions include hypertensive reactions, antidiuresis, and hyperglycemic activity. Bombesin has been shown to have a strong effect on core temperature lowering in rats: M. Brown *et al., Science* **196**, 996 (1977). High levels of intracellular bombesin have also been found in human small-cell lung carcinoma: T. W. Moody *et al., ibid.* **214**, 1246 (1981). Brief review of chemistry, isoln, purification: V. Mutt, *Biochem. Soc. Trans.* **8**, 11 (1980). Effect on suppression of food intake: J. Gibbs *et al., Nature* **282**, 208 (1979). Pharmacological reviews: G. Bertaccini, *Pharmacol. Rev.* **28**, 127 (1976); S. R. Bloom, J. M. Polak, *Advan. Clin. Chem.* **21**, 177 (1980).

5-oxoPro-Gln-Arg-Leu-Gly-Asn-Gln-Trp-Ala-Val-Gly-His-Leu-MetNH₂

Hydrochloride, $C_{71}H_{112}Cl_2N_{24}O_{18}S$, cryst from 99% ethanol, mp 185° (dec). $[\alpha]_D^{24}$ −20.6° (c = 0.65 in DMF-HMPT 8:2).

1360. Bomyl®. *3-[(Dimethoxyphosphinyl)oxy]-2-pentenedioic acid dimethyl ester; 3-hydroxy-2-pentenedioic acid dimethyl ester dimethyl phosphate; 3-hydroxyglutaconic acid dimethyl ester dimethyl phosphate;* dimethyl 1,3-bis(carbomethoxy)-1-propen-2-yl phosphate; dimethyl 3-hydroxyglutaconate dimethyl phosphate; ENT-24833; GC-3707; Swat. $C_9H_{15}O_8P$; mol wt 282.19. C 38.31%, H 5.36%, O 45.36%, P 10.98%. Prepn: Gilbert, U.S. pat. **2,891,887** (1959 to Allied). Activity: P. E. Newallis *et al., J. Agr. Food Chem.* **15**, 940 (1967). Toxicity: E. W. Schafer, *Toxicol. Appl. Pharmacol.* **21**, 315 (1972).

Liquid; bp₁₇ 155-164°; bp₂ 155-165° (tech grade). Miscible with methanol, ethanol, acetone, xylene. Practically insol in water, petr ether, kerosene. LD_{50} orally in rats: 32 mg/kg (Schafer).

USE: Insecticide.

1361. Bongkrekic Acid. *[R-[R*,S*-(E,Z,Z,E,E,Z,E)]]-20-(Carboxymethyl)-6-methoxy-2,5,17-trimethyl-2,4,8,10,14,-18,20-docosaheptaenedioic acid;* 3-carboxymethyl-17-methoxy-6,18,21-trimethyldocosa-2,4,8,12,14,18,20-heptaenedioic acid; BA. $C_{28}H_{38}O_7$; mol wt 486.61. C 69.11%, H 7.87%, O 23.02%. One of the two toxic antibiotic principles produced by *Pseudomonas cocovenenans* on partially defatted coconut; the other being toxoflavin, *q.v.* Name derived from "bongkrek", a molded coconut product from Indonesia which becomes highly poisonous when *P. cocovenenans* outgrows the mold. Isoln: van Veen, Mertens, *Rec. Trav. Chim.* **53**, 257 (1934); **54**, 373 (1935); Nugteren, Berends, *ibid.* **76**, 13 (1957). Purification and properties: Lijmbach *et al., Tetrahedron* **26**, 5993 (1970). Structural studies: *eidem, ibid.* **27**, 1839 (1971). Revised structure: De Bruijn *et al., ibid.* **29**, 1541 (1973). Absolute configuration: Zylber *et al., Experientia* **29**, 387, 648 (1973). Influence on carbohydrate metabolism: van Veen, Mertens, *Arch. Neer. Physiol.* **21**, 73 (1936), *C.A.* **30**, 3880⁹ (1936); inhibition of adenine nucleotide translocation: Henderson, Lardy, *J. Biol. Chem.* **245**, 1319 (1970); Klingenberg *et al., Biochem. Biophys. Res. Commun.* **39**, 363 (1970).

White, amorphous solid, mp 50-60°. uv max (methanol): 237, 267 nm (ϵ 32000, 36700). $[\alpha]_D^{25}$ +162.5°. LD_{50} i.v. in mice: 1.41 mg/kg (Lijmbach).

1362. Bopindolol. $(\pm)$-*1-[(1,1-Dimethylethyl)amino]-3-[(2-methyl-1H-indol-4-yl)oxy]-2-propanol benzoate ester;* $(\pm)$-1-(*tert*-butylamino)-3-[(2-methyl-4-yl)oxy]-2-propanol benzoate (ester). $C_{23}H_{28}N_2O_3$; mol wt 380.49. C 72.61%, H 7.42%, N 7.36%, O 12.61%. Nonselective β-adrenergic blocker. Prepn: F. Troxler, F. Seemann, *Ger. pat.* **2,635,-209**; *eidem,* U.S. pat. **4,340,541** (1977, 1982 both to Sandoz). Clinical pharmacology: P. van Brummelen *et al., Eur. J. Clin. Pharmacol.* **22**, 491 (1982). HPLC determn in plasma: C. J. Oddie *et al., J. Chromatog.* **273**, 469 (1983). Effect on plasma lipid fractions: P. van Brummelen *et al., Brit. J. Clin. Pharmacol.* **17**, 86 (1984). Pharmacokinetics: R. Platzer *et al., Clin. Pharmacol. Ther.* **36**, 5 (1984). Clinical trials in hypertension: U. L. Hulthen *et al., J. Cardiovascular Pharmacol.* **5**, 426 (1983); W. J. Schiess *et al., Eur. J. Clin. Pharmacol.* **27**, 529 (1984).

Sol in ether, methylene chloride. LD_{50} i.v. in mice: 17 mg/kg (Troxler, Seemann, 1977).
Maleate, $C_{23}H_{28}N_2O_3 \cdot C_4H_4O_4$, *Sandonorm.*
Malonate, $C_{23}H_{28}N_2O_3 \cdot C_3H_4O_4$, *LT-31-200, Wandonorm.*
THERAP CAT: Antihypertensive.

1363. Borazine. Inorganic benzene; hexahydro-*s*-triazaborine; borazane; borazole; triborine triamine; triboron nitride; *s*-triazaborane. $B_3H_6N_3$; mol wt 80.50. B 40.29%, H 7.51%, N 52.20%. Isoelectronic with benzene. Prepn: Stock, Pohland, *Ber.* **59**, 2215 (1926); Wiberg, Bolz, *Ber.* **73**, 209 (1940). Structural study: H. Watanabe, W. Kubo, *J. Am. Chem. Soc.* **82**, 2428 (1960); W. Harshbarger *et al., Inorg. Chem.* **8**, 1683 (1969). Electronic structure: A. Lötz

et al., Chem. Phys. **103**, 317 (1986). Raman spectroscopy: J. V. Kainnady, A. Weber, *J. Raman. Spectrosc.* **5**, 35 (1976). Electric and magnetic properties: G. R. Dennis, G. L. D. Ritchie, *J. Phys. Chem.* **97**, 8403 (1993). Use in synthesis of ceramic composites: K. Su *et al., Chem. Mater.* **5**, 547 (1993). *Review:* E. M. Smolin, L. Rapoport, *s-Triazines and Derivatives* (Interscience, New York, 1959) pp 597-626.

Mobile liquid; mp −58°; bp +53°. Stable up to 500° when pure and totally anhydrous. d_4^0 0.824; d_4^{57} 0.898. n_D^{20} 1.3821. Critical temp 252°. Heat of vaporization 7.0 kcal. Mol vol at bp = 99.7 ml. Surface tension at mp = 31.09 dynes/cm. Parachor 207.9. Trouton's constant 21.4. Eötvös' constant 2.0. Dipole moment in benzene (25°) = 0.50. Dissolves in water, giving a soln with strong reducing properties. Hydrolyzes slowly in aq soln yielding hydrogen, boric acid, and ammonia.

1364. Boric Acid. Boracic acid; orthoboric acid; Borofax. BH_3O_3; mol wt 61.83. B 17.48%, H 4.89%, O 77.63%. H_3BO_3. Occurs in nature as the mineral *sassolite.* Manuf: Faith, Keyes & Clark's *Industrial Chemicals,* F. A. Lowenhcim, M. K. Moran, Eds. (Wiley-Interscience, New York, 4th ed., 1975) pp 153-158. Toxicity study: Smyth *et al., Am. Ind. Hyg. Assoc. J.* **30**, 470 (1969).

Colorless, odorless, transparent crystals, or white granules or powder; slightly unctuous to the touch. mp ~171°. Phase diagram for the $B_2O_3.H_2O$ system: Kracek *et al., Am. J. Sci.* **35A**, 143 (1938). Volatile with steam. pH: 5.1 (0.1 molar). One gram dissolves in 18 ml cold, 4 ml boiling water, in 18 ml cold, 6 ml boiling alcohol, in 4 ml glycerol; soly in water is increased by HCl, citric or tartaric acids. Soly of boric acid in glycerol solns of various concns: Sciarra, Elliott, *J. Am. Pharm. Assoc. (Sci. Ed.)* **49**, 116 (1960). LD_{50} orally in rats: 5.14 g/kg (Smyth).

Incompat. Alkali carbonates and hydroxides.

Caution: Ingestion or absorption may cause nausea, vomiting, diarrhea, abdominal cramps, erythematous lesions on skin and mucous membranes, circulatory collapse, tachycardia, cyanosis, delirium, convulsions, coma. Death has occurred from < 5 g in infants and from 5 to 20 g in adults. Chronic use may cause borism (dry skin, eruptions, gastric disturbances): E. Browning, *Toxicity of Industrial Metals* (Appleton-Century-Crofts, New York, 2nd ed., 1969) pp 90-97.

USE: For weatherproofing wood and fireproofing fabrics; as a preservative; manuf cements, crockery, porcelain, enamels, glass, borates, leather, carpets, hats, soaps, artificial gems; in nickeling baths; cosmetics; printing and dyeing, painting; photography; for impregnating wicks; electric condensers; hardening steel. Also used as insecticide for cockroaches and black carpet beetles.

THERAP CAT: Astringent, antiseptic.

THERAP CAT (VET): Antibacterial and antifungal. Used chiefly in aqueous solution or powders for external use.

1365. Boric Anhydride. *Boron oxide;* boron trioxide; boric oxide; boron sesquioxide. B_2O_3; mol wt 69.62. B 31.06%, O 68.94%. Improperly called *anhydrous boric acid* or *fused boric acid.* Prepn of crystalline form: McCulloch, *J. Am. Chem. Soc.* **59**, 2650 (1937).

Colorless, brittle, vitreous, semitransparent, hygroscopic lumps or hard, white crystals. d (amorph) 1.8; d (cryst) 2.46. mp (cryst) 450°. Slowly sol in 30 parts cold, or 5 parts boiling water; sol in alcohol, glycerol. *Keep dry.*

Caution: Potential symptoms of overexposure are nasal irritation, conjunctivitis and erythema. *See NIOSH Pocket Guide to Chemical Hazards* (DHHS/NIOSH 90-117, 1990) p 46.

USE: In metallurgy; in analysis of silicates to determine SiO_2 and alkalies; in blowpipe analysis.

1366. Borneol. *endo-1,7,7-Trimethylbicyclo[2.2.1]heptan-2-ol;* *endo*-2-bornanol; *endo*-2-camphanol; *endo*-2-hydroxycamphane; bornyl alcohol; Baros camphor; Sumatra camphor; Borneo camphor; Dryobalanops camphor; Bhimsaim camphor; Malayan camphor; camphol. $C_{10}H_{18}O$; mol wt 154.25. C 77.87%, H 11.76%, O 10.37%. The dextrorotatory form which predominates occurs in the oil from *Dryobalanops aromatica* Gaertn., *Dipterocarpaceae,* and in many other plants; the levorotatory form comes from *Blumea balsamifera* (L.) DC., *Compositae:* E. Gildemeister, F. Hoffman, *Die Atherischen Ole* (Schimmel, Leipzig, 3rd ed., 1928) pp 475-481. Racemic borneol is prepd synthetically by reduction of camphor: Truett, Moulton, *J. Am. Chem. Soc.* **73,** 5913 (1951); Ziegler *et al., Ann.* **623,** 9 (1959); Ziegler, **Brit.** pat. **803,178** (1958); from pinene: Schwyzer, *Pharm. Ztg.* **75,** 1275 (1930). Configuration (isoborneol = exo-form; borneol = endo-form): Toivonen *et al., Acta Chem. Scand.* **3,** 991 (1949). *Review:* J. L. Simonsen, *The Terpenes* **vol.** II (University Press, Cambridge, 2nd ed., 1949) pp 349-365.

d-Borneol

d-Form, hexagonal plates from petr ether, mp 208°. Peculiar peppery odor and burning taste somewhat resembling that of mint. Sublimes, but is less volatile than camphor. d_4^{20} 1.011. bp 212°. $[\alpha]_D^{20}$ +37.7° (c = 5 in alc); $[\alpha]_{546}^{22}$ +44.4° (c = 0.5 in toluene). Almost insol in water. Sol in alc (176 parts dissolve in 100 parts w/w of abs alc), ether, petr ether (about 1:6), benzene (about 1:5), toluene, acetone, decalin, tetralin. LD orally in rabbits: 2 g/kg.

l-Form, hexagonal plates, mp 204°. bp_{779} 210°. $[\alpha]_D^{20}$ −37.7° (c = 5 in alc); $[\alpha]_{546}^{22}$ −44.4° (c = 0.5 in toluene).

dl-Form, mp 206-207°.

Caution: May cause nausea, vomiting, mental confusion, dizziness, convulsions.

USE: Primarily in the manuf of its esters. Some free borneol and isoborneol is used in perfumery and in incense making.

1367. Bornyl Acetate. *1,7,7-Trimethylbicyclo[2.2.1]heptan-2-ol acetate; borneol acetate.* $C_{12}H_{20}O_2$; mol wt 196.29. C 73.43%, H 10.27%, O 16.30%. $CH_3COOC_{10}H_{17}$. Prepn of the *d*-form by acylation of *d*-borneol: Shishido *et al., J. Am. Chem. Soc.* **82,** 125 (1960).

d-Form, crystals, mp 29°, also reported as mp 26.5°, Considine, *J. Org. Chem.* **25,** 671 (1960). bp 225-226°, bp_8 92-93°. d 0.99. n_D^{20} 1.4623. $[\alpha]_D$ +44.38°, Haller, *Compt. Rend.* **109,** 29 (1889); $[\alpha]_D$ +41.2°, Considine, *loc. cit.;* α_D^{14} +44.72° (neat), Shishido, *loc. cit.* Very slightly sol in water; sol in alcohol, ether.

l-Form, crystals, mp 27°. bp_{14} 103°. $[\alpha]_D$ −44.45°, Haller, *loc. cit;* $[\alpha]_D$ −42.0°, Considine, *loc. cit.*

dl-Form, crystals, mp 7.0°, Considine, *loc. cit.*

1368. *d*-Bornyl α-Bromoisovalerate. *2-Bromo-3-methylbutanoic acid 1,7,7-trimethylbicyclo[2.2.1]hept-2-yl ester; 2-bromo-3-methylbutyric acid ester with d-borneol;* bornyl α-bromoisovalerate; *d*-bornyl 2-bromo-3-methylbutyrate; Brovalol; Eubornyl; Valisan. $C_{15}H_{25}BrO_2$; mol wt 317.27. C 56.79%, H 7.94%, Br 25.19%, O 10.09%. $(CH_3)_2CHCHBrCOOC_{10}H_{17}$. Prepd from *d*-borneol and α-bromoisovaleryl chloride or α-bromovaleric acid with mineral acid: *Beilstein* **vol. 6,** p 79.

Oily liquid, bp 163°. d 1.18. Practically insol in water; sol in alc, chloroform, ether.

THERAP CAT: Sedative, hypnotic.

1369. Bornyl Chloride. *endo-2-Chloro-1,7,7-trimethylbicyclo[2.2.1]heptane;* pinene hydrochloride; 2-chlorobornane; 2-chlorocamphane; "terpene" hydrochloride; "turpentine camphor". $C_{10}H_{17}Cl$; mol wt 172.70. C 69.55%, H 9.92%, Cl 20.53%. Prepd from α-pinene: Zeiss, Zwanzig, *J. Am. Chem. Soc.* **79,** 1733 (1957); Hückel, Gelchsheimer,

Ann. **625,** 12 (1959); by chlorination of camphane: Gandini, *Gazz. Chim. Ital.* **66,** 357 (1936). Configuration: Kwart, *J. Am. Chem. Soc.* **75,** 5942 (1953); Kwart, Null, *ibid.* **78,** 5943 (1956). *Review:* J. L. Simonsen, *The Terpenes* vol. II (University Press, Cambridge, 2nd ed., 1949) pp 340-349.

Crystals from alc, mp 132°. Odor resembling camphor. bp 207-208°. When prepd from optically active sources it has the same sign as the hydrocarbon from which it is prepd: Thurber, Thielke, *J. Am. Chem. Soc.* **53,** 1032 (1931). Practically insol in water; sol in alc, ether.

THERAP CAT: Antiseptic.

1370. *d*-Bornyl Isovalerate. *(1R-endo)-3-Methylbutanoic acid 1,7,7-trimethylbicyclo[2.2.1]hept-2-yl ester; isovaleric acid d-bornyl ester;* borneol isovalerate; Bornyval. $C_{15}H_{26}O_2$; mol wt 238.37. C 75.58%, H 10.99%, O 13.42%. $(CH_3)_2$-$CHCH_2COOC_{10}H_{17}$. Prepd from *d*-borneol and isovalic acid: Siedler, *Pharm. Ztg.* **48,** 772 (1903); Vavon, Peignier, *Bull. Soc. Chim. France* **39,** 924 (1926).

Liquid; odor and taste of valerian and camphor. d 0.955. bp 255-260°. $[\alpha]_D^{20}$ +27-28°. Practically insol in water. Sol in alcohol or ether.

THERAP CAT: Sedative.

1371. Bornyl Salicylate. *2-Hydroxybenzoic acid 1,7,7-trimethylbicyclo[2.2.1]hept-2-yl ester; salicylic acid bornyl ester;* borneol salicylate; Salit. $C_{17}H_{22}O_3$; mol wt 274.36. C 74.42%, H 8.08%, O 17.49%. $HOC_6H_4COOC_{10}H_{17}$. Prepd from borneol and salicylic acid: *Hager's Handb. Pharm. Praxis* **Band 1,** 210 (Berlin, 1930).

Brown, oily liquid. Slight odor and taste. Practically insol in water; miscible with alcohol, ether, chloroform, oils; slightly sol in glycerol; dec by alkalies and the intestinal fluids into its constituents.

THERAP CAT: Counterirritant.

1372. Boromycin. $C_{45}H_{74}BNO_{15}$; mol wt 879.89. C 61.43%, H 8.48%, B 1.23%, N 1.59%, O 27.28%. Antibiotic produced by *Streptomyces antibioticus* ETH 28829: Hütter *et al., Helv. Chim. Acta* **50,** 1533 (1967). The first known natural product in which boron has been found. It is a complex of boric acid with a tetradentate organic complexing agent and yields D-valine, boric acid and a polyhydroxy macrolide-type compound upon hydrolysis. Structure: Dunitz *et al., ibid.* **54,** 1709 (1971). Mechanism of action studies: Pache, Zähner, *Arch. Mikrobiol.* **67,** 156 (1969). Activity as a coccidiostat: Miller, Burg, U.S. pat. **3,864,479** (1975 to Merck & Co.). Biosynthesis: T. S. S. Chen *et al., J. Org. Chem.* **46,** 2661 (1981). Partial synthesis: M. A. Avery *et al., Tetrahedron Letters* **22,** 3123 (1981). Synthesis of C-3′ to C-17′: S. Hanessian *et al., J. Am. Chem. Soc.* **103,** 6243 (1981); of C-3 to C-17: eidem, *Can. J. Chem.* **61,** 634

(1983); of C-1 to C-17: J. D. White *et al., J. Am. Chem. Soc.* **105**, 6517 (1983). Total synthesis: *eidem, ibid.* **111**, 790 (1989). *Review:* Pache in *Antibiotics* vol. 3, J. W. Corcoran, F. E. Hahn, Eds. (Springer-Verlag, New York, 1975) pp 585-587.

Colorless crystals from methanol, mp 223-228° (dec). $[\alpha]_D$ 63.5° (c = 0.55 in $CHCl_3$). No uv absorption between 210 and 400 nm. LD_{50} orally in mice: 180 mg/kg (Hütter).

1373. Boron. B; at. no. 5; at. wt 10.81; valence 3. Two naturally-occurring isotopes: 10; 11; three short-lived, artificial isotopes: 8, 12, 13. Occurrence in the earth's crust about 0.001% in the form of its compounds, never as the element. First obtained by Moissan in 1895 by reduction of boric anhydride (B_2O_3) with magnesium in a thermite-type reaction: Moissan, *Ann. Chim. Phys.* [7] **6**, 296 (1895), still a good method for large quantities of relatively impure boron. Prepn of high purity crystalline boron by vapor phase reduction of boron trichloride with hydrogen on electrically heated filaments in a flow system: Stern, Lynds, *J. Electrochem. Soc.* **105**, 676 (1958). Reviews of prepn and properties of boron and its compds: *Boron, Metallo-Boron Compounds and Boranes*, R. M. Adams, Ed. (Interscience, New York, 1964) 765 pp; *The Chemistry of Boron and Its Compounds*, E. L. Muetterties, Ed. (John Wiley, New York, 1967) 699 pp; Greenwood, "Boron" in *Comprehensive Inorganic Chemistry* vol. 1, J. C. Bailar, Jr. *et al.*, Eds. (Pergamon Press, Oxford, 1973) pp 655-991; J. G. Bower in Kirk-Othmer *Encyclopedia of Chemical Technology* vol. 4 (Wiley-Interscience, New York, 3rd ed., 1978) pp 62-66. Review of synthesis and applications of vinylic organoboranes: H. C. Brown, J. B. Campbell, *Aldrichim. Acta* **14**, 3-11 (1981).

Polymorphic; α-rhombohedral form, clear red crystals, 12 atoms/unit cell, d 2.46; β-rhombohedral form, black, 105 atoms/unit cell, d 2.35; α-tetragonal form, black, opaque crystals with metallic luster, 50 atoms/unit cell, d 2.31; other crystal forms known but not entirely characterized. Amorphous form, black or dark brown powder, d 2.350. Crystals are almost as hard as diamond. mp ~2200°. Vapor pressure at 2413°K (2140°): 1.56 × 10^{-5} atm, Searcy, Myers, *J. Phys. Chem.* **61**, 957 (1957). Heat capacity at 25° of amorphous form: 2.858 cal/g-atom/°C; of β-rhombohedral form: 2.650 cal/g-atom/°C. Feeble conductor of electricity at room temp, good conductor at high temps. Admixture of traces of carbon improves conductivity. Self-limiting reaction with oxygen due to formation of B_2O_3 film; oxide coating evaporates above 1000°. Reacts with fluorine at room temp. Insol in water. Unaffected by aq hydrochloric and hydrofluoric acids. When finely divided, it is sol in boiling nitric and sulfuric acids and in most molten metals, such as copper, iron, magnesium, aluminum, calcium. Reacts vigorously with fused sodium peroxide, or with a fusion mixture of sodium carbonate and potassium nitrate.

USE: In nuclear chemistry as neutron absorber, in Ignitron rectifiers, in alloys, usually to harden other metals.

1374. Boron Carbide. Norbide. CB_4; mol wt 55.26. C 21.74%, B 78.26%. B_4C. Usually prepd in an electric furnace at 2500° according to the equation $2B_2O_3 + 7C \rightarrow B_4C + 6CO$: Ridgway, *Trans. Electrochem. Soc.* **66**, 117-133 (1934); also formed by reducing boric anhydride with magnesium in the presence of carbon: Dawihl, *Ger. pat.* **752,-324** (1942 to Krupp); *BIOS* ref. no. 925, p 22 (1947). Lab prepn by the reduction of boron trichloride with hydrogen in the presence of carbon or hydrocarbons: *Bell Labs. Record* **28**, 477 (1950). Comprehensive monograph: P. W. Gilles, *High Temperature Chemistry of the Binary Compounds of Boron*, Advances in Chemistry Series **no. 32**, (ACS, Washington, D.C., 1961).

Black shiny rhombohedra or octahedra. d_4^{25} 2.508-2.512. mp 2350° (no decompn); bp > 3500°. Its hardness is less than that of industrial diamonds, but higher than the hardness of silicon carbide: ca 5,000 kg/mm², on Mohs' hardness scale = 9.3. Less brittle than most ceramics. Remarkably resistant to chemical action. Not attacked by hot HF, HNO_3 or $HCrO_4$. Decomposed by molten alkalis at red heat. Does not burn in oxygen flame.

USE: Abrasive. In the manuf of hard and chemicals-resistant ceramics or wear-resistant tools. Finely pulverized B_4C can be molded under (considerable) pressure and heat.

1375. Boron Monoxide. $(BO)_x$; B 40.34%, O 59.66%. Prepd in quantitative yield by dehydration of tetrahydroxyboron at 250° at reduced pressure: Wartik, Apple, *J. Am. Chem. Soc.* **77**, 6400 (1955); prepn and proposed structure: McCloskey *et al., ibid.* **83**, 4750 (1961).

Fluffy white solid. Hygroscopic, converts back to tetrahydroxyboron on reaction with water. One gram dissolves in 100 ml methanol, in 100 ml warm ethanol or 100 ml warm isopropyl alcohol. Practically insol in dimethylamine and methyl borate. Vaporizes at 1300-1500° into gaseous B_2O_2.

1376. Boron Nitride. BN; mol wt 24.82. B 43.56%, N 56.44%. Prepd by igniting compds of boron with compds of nitrogen: Taylor, **U.S. pat. 2,855,316** (1958 to Carborundum Co.). *Reviews:* Giardini, *U.S. Bur. Mines, Inform. Circ.* **No. 7664**, 13 pp (1953); K. Niedenzu, J. W. Dawson, *Boron-Nitrogen Compounds* (Academic Press, New York, 1965) pp 147-153.

Crystals with hexagonal, graphite lattice is most common form. *Borazon*, cubic crystalline modification, is probably the hardest substance known. There exists also an amorphous modification. mp 3000°. Begins to sublime at a temp slightly below 3000°. Begins to dissociate *in vacuo* at about 2700°. The chemical behavior of BN is dependent on the method of prepn. Not attacked by mineral acids, water; in general resistant to chemical attack. Hot concd alkali cleaves boron-nitrogen bond. Oxidation in air begins above 1200°. See Niedenzu, Dawson, *loc. cit.*

USE: Manuf of alloys; in semiconductors, nuclear reactors, lubricants.

1377. Boron Tribromide. BBr_3; mol wt 250.52. B 4.32%, Br 95.68%. Prepn: Gamble, *Inorg. Syn.* **3**, 27 (1950); Becher in *Handbook of Preparative Inorganic Chemistry*, **Vol. 1**, G. Brauer, Ed. (Academic Press, New York, 2nd ed., 1963) p 781. Review of boron halides: Massey, *Advan. Inorg. Chem. Radiochem.* **10**, 1-152 (1967).

Colorless, fuming liquid; dec by water or alcohol. mp −46.0°; bp 90°. d^0 2.698. Vapor pressure data: Barber *et al., J. Chem. Eng. Data* **9**, 137 (1964).

USE: Manuf of diborane; ultra high purity boron.

1378. Boron Trichloride. BCl_3; mol wt 117.17. B 9.23%, Cl 90.77%. Prepn: Gamble, *Inorg. Syn.* **3**, 27 (1950). *Reviews:* Gerrard, Lappert, *Chem. Rev.* **58**, 1081-1111 (1958); Massey, *Advan. Inorg. Chem. Radiochem.* **10**, 1-152 (1967).

Colorless, fuming liquid at low temp; dec by water or alcohol. bp 12.5°. mp −107°. d_4^{12} 1.35. d^0 1.3728: Ward, *J. Chem. Eng. Data* **14**, 167 (1969).

USE: Manuf and purification of boron; as catalyst for organic reactions; in semiconductors; in bonding of iron, steels; in purification of metal alloys to remove oxides, nitrides and carbides.

1379. Boron Trifluoride. BF_3; mol wt 67.81. B 15.94%, F 84.06%. A strong Lewis acid. Prepn: Swinehart, **U.S. pats. 2,148,514, 2,196,907** (1939, 1940 both to Harshaw Chemical); Booth, Wilson, *Inorg. Syn.* **1**, 21 (1939); Kwasnik in *Handbook of Preparative Inorganic Chemistry*, **Vol. 1**, G. Brauer, Ed. (Academic Press, New York, 2nd ed., 1963) pp 219-222; Wiesboeck, **U.S. pat. 3,690,821** (1972 to U.S. Steel). Dihydrate: McGrath *et al., J. Am. Chem. Soc.* **66**, 1263 (1944). *Reviews:* Booth, Martin, *Boron Trifluoride and Its Derivatives* (John Wiley & Sons, 1949), 296 pp; Booth in *Fluorine Chemistry*, **Vol. 1**, J. Simons, Ed. (Academic Press, New York, 1950) pp 201-224; Topchiev *et al., Boron Fluoride and Its Compounds as Catalysts in Or-*

ganic Chemistry (Pergamon Press, 1959) 326 pp; Martin in Kirk-Othmer, *Encyclopedia of Chemical Technology*, **Vol. 9** (Interscience, New York, 2nd ed., 1966) pp 554-562; Massey, *Advan. Inorg. Chem. Radiochem.* **10**, 1-152 (1967).

Colorless gas. Pungent, suffocating odor. Forms dense white fumes in moist air. bp $-127.1°$. bp $-100.4°$. d_4 ($-100.4°$; liq) 1.57. d (gas at STP) 3.07666 g/l. Soly in water (0°): 332 g/100 g; some hydrolysis occurs to form fluoboric and boric acids. Soly in anhydrous H_2SO_4: 1.94 g/100 g acid. Forms solid complex with nitric acid ($HNO_3.2BF_3$). Sol in most saturated and halogenated hydrocarbons and in aromatic compds. Polymerizes unsaturated molecules. Easily forms coordination complexes with molecules having at least one unshared pair of electrons. Reacts with incandescence when heated with alkali metals or alkaline earth metals except magnesium.

Caution: Potential symptoms of overexposure are nasal irritation, epistaxis; burns to eyes and skin. *See NIOSH Pocket Guide to Chemical Hazards* (DHHS/NIOSH 90-117, 1990) p 46. *See also Patty's Industrial Hygiene and Toxicology* vol. **2B**, G. D. Clayton, F. E. Clayton, Eds. (Wiley-Interscience, New York, 3rd ed., 1981) pp 2996-2999.

USE: To protect molten magnesium and its alloys from oxidation; as a flux for soldering magnesium; as a fumigant; in ionization chambers for the detection of weak neutrons. By far the largest application of boron trifluoride is in catalysis with and without promoting agents.

1380. Boron Trifluoride Etherate. *Boron fluoride ethyl ether;* boron fluoride etherate; ethyl ether-boron trifluoride complex. $C_4H_{10}BF_3O$; mol wt 141.93. C 33.85%, H 7.10%, B 7.62%, F 40.16%, O 11.27%. $(CH_3CH_2)_2O.BF_3$. Prepd by vapor-phase reaction of anhydr ether with BF_3: Laubengayer, Finlay, *J. Am. Chem. Soc.* **65**, 884 (1943).

Fuming liquid, immediately hydrolyzed by moisture in air. d_4^{25} 1.125. bp 125.7°. mp $-60.4°$. n_D^{20} 1.348. Heat of formation: 12.5 kcal. Heat of soln at 0° in ether: 2.7 kcal.

Caution: On decomposition forms highly toxic fumes of fluorides.

USE: Catalyst in acetylation, alkylation, polymerization, dehydration, and condensation reactions.

1381. Bostrycoidin. *6,9-Dihydroxy-7-methoxy-3-methylbenz[g]isoquinoline-5,10-dione;* 5,8-dihydroxy-6-methoxy-3-methyl-2-aza-9,10-anthraquinone. $C_{15}H_{11}NO_5$; mol wt 285.26. C 63.16%, H 3.89%, N 4.91%, O 28.04%. Antibiotic substance produced by *Fusarium bostrycoides:* Hamilton *et al., Antibiot. & Chemother.* **3**, 853 (1953); Cajori *et al., J. Biol. Chem.* **208**, 107 (1954). Structure: Arsenault, *Tetrahedron Letters* **1965**, 4033. Synthesis: D. W. Cameron *et al., ibid.* **21**, 5089 (1980).

Dark red crystals; mp 243-244°, changes to purple in alkaline medium, yellow in acid medium. Stable at room temp and withstands autoclaving. Insol in water; sol in abs ethanol and in 60% ethanol; moderately sol in dioxane, benzene, acetone, chloroform, carbon tetrachloride. Soluble in aq sodium carbonate solns. Partly sol in corn oil. Active *in vitro* against *Mycobacterium tuberculosis.* Serum seems to interfere.

1382. β-Boswellic Acid. *(3α,4β)-3-Hydroxyurs-12-en-23-oic acid.* $C_{30}H_{48}O_3$; mol wt 456.71. C 78.90%, H 10.59%, O 10.51%. Occurs as the acetate in frankincense *(olibanum)* from *Boswellia carterii, Burseraceae.* The β-form is predominant and is accompanied by small amounts of α- and γ-boswellic acid. Isoln from olibanum tears: Winterstein, Stein, *Z. Physiol. Chem.* **208**, 9 (1932). Early structural studies: Simpson, Williams, *ibid.* **1938**, 686, 1712; Ruzicka, Wirz, *Helv. Chim. Acta* **22**, 948 (1939); **23**, 132 (1940); Ruzicka *et al., ibid.* **27**, 1859 (1944). Revised structure and stereochemistry: Beton *et al., J. Chem. Soc.* **1956**, 2904;

Allan, *Chimia* **17**, 382 (1963); *idem, Phytochemistry* **7**, 963 (1968). *Review:* J. Simonsen, W. C. J. Ross, *The Terpenes,* vol. **5** (University Press, Cambridge, 1957) pp 68-74.

Long prisms from methanol, mp 228-232° with preliminary sintering. $[\alpha]_D$ +107° (c = 0.75 in $CHCl_3$). 100 ml of boiling methanol will dissolve 8 grams of β-boswellic acid. Sol in chloroform, ether, acetone, alc.

Acetate, $C_{32}H_{50}O_4$, prisms, mp 275-278°, $[\alpha]_D$ +63° (c = 1.88 in $CHCl_3$).

Methyl ester, $C_{31}H_{50}O_3$, mp 195-196°, $[\alpha]_D$ +111° (c = 1.6 in $CHCl_3$).

1383. Bottromycin. B-mycin. A complex of five antibiotics of which the main active component is bottromycin A_2. Produced by *Streptomyces bottropensis* and *S. canadensis:* Waisvisz *et al., J. Am. Chem. Soc.* **79**, 4520, 4522, 4524 (1957); Miller *et al., Antimicrob. Ag. Chemother.* **1967**, 407. Production: **Brit.** pat. **762,736** (1956 to Koninklijke Nederlandsche Gist en Spiritus-Fabriek); Umesawa *et al., Japan.* pat. **68 10,998** (1968 to Microbiochem. Res. Found.), *C.A.* **69**, 85439x (1968); Hata *et al.,* U.S. pat. **3,650,904** (1972). Partial structure: Waisvisz, Hoeven, *J. Am. Chem. Soc.* **80**, 38 (1958). Sepn and structures of bottromycins A_1, A_2, and B: Nakamura *et al., J. Antibiot.* **18A**, 47, 60 (1965); **19A**, 10 (1966). Isoln and characterization of bottromycins A_2, B_2, C_2: Nakamura *et al., ibid.* **20A**, 1 (1967). Revised structure of bottromycin A_2: Takahashi *et al., ibid.* **29A**, 1120 (1976); D. Schipper, *ibid.* **36A**, 1076 (1983). Mode of action of bottromycin A_2: T. Otaka, A. Kaji, *J. Biol. Chem.* **251**, 2299 (1976); *eidem, FEBS Letters* **123**, 173 (1981).

Bottromycin A_2

Bottromycin A_2, $C_{42}H_{62}N_8O_7S$, *bottromycic A_2 acid methyl ester.*

1384. Botulin Toxins. One of the most potent bacterial neurotoxins known; produced by *Clostridium botulinum.* Exists in seven distinct types, A-G, with average mol wt of 150,000 Da. Zinc-proteases synthesized as nontoxic precursors which become toxic upon proteolytic cleavage. Active form consists of 2 chains, the heavy and the light, linked by a disulfide bond and acts primarily on the parasympathetic nervous system. Identified in 1897 by Van Ermengem as the agent of food poisoning. Primary structure: B. R. DasGupta in *Microbial Toxins in Food and Feeds,* A. E. Pohland *et al.,* Eds. (Plenum Press, New York, 1990) pp 75-89. Clinical trial of botulin A in blepharospasm: D. B. Sanders *et al., Neurology* **36**, 545 (1986); B.

Arthurs *et al., Can. J. Ophthalmol.* **22**, 24 (1987); in achalasia: P. J. Pasricha *et al., New Engl. J. Med.* **332**, 774 (1995). *Review:* D. A. Boroff, *Proc. 2nd. Int. Symp. Anim. Plant. Toxins* **2**, (Gordon & Breach, New York, 1972) pp 631-640; J. L. Middlebrook, *J. Toxicol. Toxin Rev.* **5**, 177-190 (1986). Review of isoln and purification: L. L. Simpson *et al., Methods Enzymol.* **165**, 76-85 (1988). Review of mechanism of action: C. Montecucco, G. Schiavo, *Mol. Microbiol.* **13**, 1-8 (1994).

MLD in mouse: 0.0003 μg/kg (Middlebrook).

Botulin A, botulinium toxin type A, oculinum, AGN-191622, BoTox.

THERAP CAT: Botulin A in treatment of blepharospasm and achlasia.

1385. Bradycor. CP-0127. $C_{128}H_{194}N_{40}O_{28}S_2$; mol wt 2805.33. C 54.80%, H 6.97%, N 19.97%, O 15.97%, S 2.29%. Bradykinin antagonist. Dimer of D-Arg⁰-[Hyp³, Cys⁶, D-Phe⁷, Leu⁸]-bradykinin joined by a bissuccinimidohexane linker via the sulfhydryl moieties of the cysteine residues. Prepn: J. C. Cheronis *et al.*, **PCT Int. pat. Appl. 92 17,201** (1992 to Cortech); *eidem, J. Med. Chem.* **35**, 1563 (1992). Effect on survival in endotoxin shock in animals: E. T. Whalley *et al., Agents Actions Suppl.* **38** (3), 413 (1992).

D-Arg–Arg–Pro–Hyp–Gly–Phe–Cys–D-Phe–Leu–Arg

(CH₂)₆

D-Arg–Arg–Pro–Hyp–Gly–Phe–Cys–D-Phe–Leu–Arg

THERAP CAT: In treatment of systemic inflammatory response syndrome.

1386. Bradykinin. Kallidin I; kallidin-9; callidin I; BRS-640. $C_{50}H_{73}N_{15}O_{11}$; mol wt 1060.22. C 56.64%, H 6.94%, N 19.82%, O 16.60%. A tissue hormone belonging to a group of hypotensive peptides known as plasma kinins. First obtained by incubation with the venom of *Bothrops jararaca* or with crystalline trypsin: Rocha e Silva *et al., Cien Cult. (Sao Paulo)* **1**, 32 (1949); *eidem, Am. J. Physiol.* **156**, 261 (1949); Prado *et al., Arch. Biochem.* **27**, 410 (1950); Werle *et al., Biochem. Z.* **320**, 372 (1950). Large scale prepn from whole bovine plasma: Hamberg, Deutsch, *Arch. Biochem. Biophys.* **76**, 262 (1958). Formed by proteolysis of a precursor in the globulin fraction of plasma referred to as kininogen by the action of enzymes such as trypsin, plasmin, and plasma kallikrein, *q.q.v.* Acts on smooth muscle, dilates peripheral vessels, increases capillary permeability. Also is a potent pain-producing agent. Structure: Elliott *et al., Biochem. Biophys. Res. Commun.* **3**, 87 (1960); Werle *et al., Z. Physiol. Chem.* **326**, 174 (1961). Synthesis: Boissonnas *et al., Helv. Chim. Acta* **43**, 1349 (1960); **45**, 170 (1962); Merrifield, *Biochemistry* **3**, 1385 (1964); Young *et al., J. Chem. Soc. (C)* **1971**, 46; Bajusz *et al.,* **Hung. pat. 3840** (1972 to Gyogyszerkutato Intézet), *C.A.* **77**, 20027g (1972); Sipos, **Ger. pat. 2,212,787** corresp to **U.S. pat 3,714,140** (1972, 1973 both to Squibb); Corley *et al., Biochem. Biophys. Res. Commun.* **47**, 1353 (1972); N. S. S. Kumari *et al., Indian J. Chem.* **17B**, 152 (1979). *Review:* Schröder, Hempel, *Experientia* **20**, 529 (1964).

Arg–Pro–Pro–Gly–Phe–Ser–Pro–Phe–Arg

Amorphous precipitate. $[\alpha]_D^{25}$ −76.5° (c = 1.37 in 1N acetic acid). Sol in glacial acetic acid, in a 10% soln of trichloroacetic acid, 70% ethanol, in hot methanol; less sol in 90% ethanol or cold methanol. Almost insol in acetone, chloroform, ethyl ether, ethyl methyl ketone, petr ether, butanol, amyl alcohol, ethyl acetate.

THERAP CAT: Vasodilator.

1387. Brallobarbital. *5-(2-Bromo-2-propenyl)-5-(2-propenyl)-2,4,6(1H,3H,5H)-pyrimidinetrione; 5-allyl-5-(2-bromoallyl)barbituric acid;* Vesperone. $C_{10}H_{11}BrN_2O_3$; mol wt 287.11. C 41.83%, H 3.86%, Br 27.83%, N 9.76%, O 16.72%. Prepn: Morren, **Belg. pat. 497,501** (1950), *C.A.* **49**, 1100e (1955). Metabolism: Keding, Schmidt, *Arzneimittel-Forsch.* **19**, 342 (1969).

Solid, mp 168-169°.

Note: This is a controlled substance (depressant) listed in the U.S. Code of Federal Regulations, Title 21 Part 1308.13 (1995).

THERAP CAT: Sedative, hypnotic.

1388. Brandy. A potable alcoholic liquid distilled from wine or from the fermented juices of peaches, cherries, apples, or other fruit. To meet the specifications of the National Formulary, brandy must have been obtained by distillation of fermented juice from sound ripe grapes and contain between 48 and 54% ethanol (v/v), d_4^{15} 0.921-0.933. It must have been stored in wooden containers for a period of not less than 2 years. Unlike whisky, brandy is never made from cereal mash or from potatoes (as some brands of vodka).

THERAP CAT: Sedative, vasodilator (peripheral).

1389. Brassidic Acid. *(E)-13-Docosenoic acid.* $C_{22}H_{42}O_2$; mol wt 338.57. C 78.05%, H 12.50%, O 9.45%. $CH_3-(CH_2)_7CH=CH(CH_2)_{11}COOH$. Prepn by isomerization of its *cis*-isomer (erucic acid): Reimer, Will, *Ber.* **19**, 3320 (1886); Skellon, *Mfg. Chemist* **33**, 405 (1962). Synthesis and separation from erucic acid by crystn: Bowman, *Nature* **163**, 95 (1949).

Thin, cryst plates from alc, mp 61-62°. bp₃₀ 282°. n_D^{57} 1.448. Practically insol in water; sparingly sol in cold alcohol; sol in ether. Solubility at 10°, 0°, −10°, −20°, and −30° in methanol, ethyl acetate, ether, acetone, toluene, *n*-heptane: Kobl, *Diss. Abstr.* **20**, 82 (1959).

1390. Brassinolide. *(2α,3α,5α,22R,23R,24S)-2,3,22,23-Tetrahydroxy-B-homo-7-oxaergostan-6-one;* 2α,3α,22,23-tetrahydroxy-24-methyl-B-homo-7-oxa-5α-cholestan-6-one. $C_{28}H_{48}O_6$; mol wt 480.69. C 69.96%, H 10.07%, O 19.97%. Plant hormone; natural steroid containing a seven-membered B-ring lactone, that promotes both cell elongation and cell division. Over ten brassinosteroids have been isolated and characterized from sources such as pollen, seedling, leaf. Isoln, structure and activity of brassinolide from rape pollen, *Brassica napus* L.: M. D. Grove *et al., Nature* **281**, 216 (1979). Stereoselective synthesis: S. Fung, J. B. Siddall, *J. Am. Chem. Soc.* **102**, 6580 (1980). Synthesis of two stereoisomers: M. J. Thompson *et al., J. Org. Chem.* **44**, 5002 (1979). Improved synthesis: T. Kametani *et al., J. Org. Chem.* **53**, 1982 (1988). Structure-activity relationship of brassinosteroids: S. Takatsuto *et al., Phytochem.* **22**, 2437 (1983); interaction with cytokinin: C. Schlagnhaufer *et al., Physiol. Plant.* **60**, 347 (1984); bioassay: K. Wada *et al., Agr. Biol. Chem.* **48**, 719 (1984).

Crystals from methanol, mp 274-275°. [α]$_D^{27}$ +16°.
USE: Plant growth regulator.

1391. Brayera. Kousso; kosso; cusso; koso; cousso; kouso; kusso. Dried panicles of pistillate flowers of *Hagenia abyssinica* J. F. Gmel. (*Brayera anthelmintica* Kunth), *Rosaceae. Habit.* Abyssinia. *Constit.* Kosin, kossein, kosidin, protokosin, kosotoxin, volatile oil, tannin. *Ref:* Leisenring, *Arch. Pharm.* **232,** 50 (1894); Kondakow, *ibid.* **237,** 481 (1899); Hess, Todd, *J. Chem. Soc.* **1937,** 562; Birch, Todd, *ibid.* **1952,** 3102.

THERAP CAT: Anthelmintic (Cestodes).

THERAP CAT (VET): Has been used as a teniacide.

1392. Brazilin. *(6aS-cis)-7,11b-Dihydrobenz[b]indeno-[1,2-d]pyran-3,6a,9,10(6H)-tetrol;* brasilin; C.I. Natural Red 24; C.I. 75280. $C_{16}H_{14}O_5$; mol wt 286.28. C 67.13%, H 4.93%, O 27.94%. (May crystallize as the mono- or hemihydrate). From *Caesalpinia echinata* Lam. (Brazil-wood), or *C. sappan* L. (sappan-wood), *Leguminosae.* Isoln and structure: Perkin *et al., J. Chem. Soc.* **1928,** 1504; Pfeiffer *et al., Ber.* **63,** 1301 (1930). Synthesis of (±)-form: Dann, Hofmann, *Ann.* **667,** 116 (1963); Kirkiacharian, Billet, *Bull. Soc. Chim. France* **1972,** 3292. Synthesis, resolution: Morsingh, Robinson, *Tetrahedron* **26,** 281 (1970). Stereochemistry: Craig *et al., J. Org. Chem.* **30,** 1573 (1965); *Colour Index* vol. 4 (3rd ed., 1971) p 4628. *Review:* Robinson, *Bull. Soc. Chim. France* **1958,** 125-134.

Amber-yellow crystals; turn orange in air and light. Dec above 130°. Sol in water, freely in alcohol, ether, also in alkali hydroxide solns with carmine-red color. *Protect from air and light.*

USE: Chiefly as a dye. Has also been recommended as indicator in acid-base titrations; acids = yellow, alkalies = carmine-red.

1393. Brefeldin A. *1,6,7,8,9,11a,12,13,14,14a-Decahydro-1,13-dihydroxy-6-methyl-4H-cyclopent[f]oxacyclotridecin-4-one;* γ,4-dihydroxy-2-(6-hydroxy-1-heptenyl)-4-cyclopentanecrotonic acid λ-lactone; ascotoxin; cyanein; decumbin. $C_{16}H_{24}O_4$; mol wt 280.36. C 68.55%, H 8.63%, O 22.83%. A fungal metabolite which is a macrocyclic lactone exhibiting a wide range of antibiotic activity. Produced by *Penicillium brefeldianum* Dodge: E. Haerri *et al., Helv. Chim. Acta* **46,** 1235 (1963). Also produced by *P. decumbens:* V. L. Singleton *et al., Nature* **181,** 1072 (1958); *P. cyaneum:* V. Betina *et al., Folia Microbiol.* **7,** 353 (1962). Structure: H. P. Sigg, *Helv. Chim. Acta* **47,** 1401 (1964). Abs configuration: H. P. Weber *et al., ibid.* **54,** 2763 (1971). Synthesis of (±)-form: E. J. Corey, R. H. Wollenberg, *Tetrahedron Letters* **1976,** 4705; E. J. Corey *et al., ibid.* **1977,** 2243; R. Baudouy *et al., ibid.* 2973; P. A. Bartlett, F. R. Green, *J. Am. Chem. Soc.* **100,** 4548 (1978); A. E. Greene *et al., ibid.* **102,** 7583 (1980); M. Honda *et al., Tetrahedron Letters* **1981,** 2679. Total synthesis of (+)-form: T. Kitahara *et al., ibid.* **1979,** 3021. Biosynthesis: B. E. Cross, P. Hendley,

Chem. Commun. **1975,** 124; C. R. Hutchinson *et al., J. Am. Chem. Soc.* **103,** 2474, 2477 (1981); M. Sunagawa *et al., J. Antibiot.* **36,** 25 (1983). Antifungal activity: V. Betina *et al., ibid.* **17A,** 93 (1964); anti-HeLa cell effect: *eidem, Naturwiss.* **49,** 241 (1962). *See also* W. Keller-Schierlein, "Chemistry of Macrolide Antibiotics" in *Fortschr. Chem. Org. Naturst.* **30,** 313-445 (1973).

Colorless prisms from methanol/ether, mp 204-205°. uv max (ethanol): 215 nm (log ε 4.05). [α]$_D^{22}$ +96 ±2° (c = 1.08 in methanol). LD$_{50}$ i.p. in mice: > 200 mg/kg (Haerri).

1394. Brequinar. *6-Fluoro-2-(2'-fluoro[1,1'-biphenyl]-4-yl)-3-methyl-4-quinolinecarboxylic acid.* $C_{23}H_{15}F_2NO_2$; mol wt 375.37. C 73.59%, H 4.03%, F 10.12%, N 3.73%, O 8.52%. Inhibitor of dihydroorotate dehydrogenase, the fourth enzyme in the *de novo* pyrimidine biosynthetic pathway; depresses lymphocyte proliferation. Prepn: D. P. Hesson, *Eur. pat.* **Appl. 133,244;** *idem,* U.S. pat. **4,680,299** (1985, 1987 both to DuPont). Mechanisms of action: T. L. Forrest *et al., Transplantation* **58,** 920 (1994). Clinical pharmacokinetics: M. de Forni *et al., Eur. J. Cancer* **29A,** 983 (1993). Symposium on pharmacology, toxicology: *Transpl. Proc.* **25,** Suppl. 2, 1-83 (1993). Review of development and clinical experience: L. Makowka *et al., Immunol. Rev.* **136,** 51-70 (1993).

White crystals from DMF + water, mp 315-317°. Sodium salt, $C_{23}H_{14}F_2NNaO_2$, *DUP-785, NSC-368390.* White solid, mp > 360°. Sol in water.

THERAP CAT: Immunosuppressant.

1395. Bretylium Tosylate. *2-Bromo-N-ethyl-N,N-dimethylbenzenemethanaminium salt with 4-methylbenzenesulfonic acid (1:1); (o-bromobenzyl)ethyldimethylammonium p-toluenesulfonate;* N-ethyl-N-o-bromobenzyl-N,N-dimethylammonium tosylate; Bretylan; Bretylate; Bretylol; Darenthin; Ornid. $C_{18}H_{24}BrNO_3S$; mol wt 414.36. C 52.18%, H 5.84%, Br 19.28%, N 3.38%, O 11.58%, S 7.74%. Prepn: Copp, Stephenson, U.S. pat. **3,038,004** (1962 to Burroughs Wellcome). Metabolism: R. Kuntzman *et al., Clin. Pharmacol. Ther.* **11,** 829 (1970). Toxicity study: E. I. Goldenthal, *Toxicol. Appl. Pharmacol.* **18,** 185 (1971). Review of pharmacology: R. H. Heissenbuttel, J. T. Bigger, *Ann. Intern. Med.* **91,** 229-238 (1979); R. J. Lee *et al.,* in *Pharmacological and Biochemical Properties of Drug Substances* vol. 2, M. E. Goldberg, Ed. (Am. Pharm. Assoc., Washington, DC, 1979) pp 148-164. Comprehensive description: J. E. Carter *et al., Anal. Profiles Drug Subs.* **9,** 71-86 (1980).

Crystalline powder, mp 97-99°. Extremely bitter taste. uv max: 278, 271, 264 nm. Freely sol in water, methanol,

ethanol. Practically insol in ether, ethyl acetate, hexane. LD_{50} in mice (mg/kg): 400 orally; 250 i.m. (Goldenthal).

THERAP CAT: Antiadrenergic; antiarrhythmic (class III).

1396. Brevetoxins. BTX. Structurally unique neurotoxins produced by the "red-tide" dinoflagellate *Ptychodicus brevis* Davis (*Gymnodinium breve* Davis). Dense growths of these algae have been responsible for massive fish kills, mollusk poisoning and human food poisoning in the Gulf of Mexico and along the Florida coast, *cf. Marine Natural Products*, P. J. Scheuer, Ed. (Academic Press, New York, 1978). Unlike previously isolated dinoflagellate toxins, such as saxitoxin, *q.v.*, which are water-soluble sodium channel blockers, the brevetoxins are lipid-soluble sodium channel activators. Isoln of brevetoxins A, B, and C and structure of B, the major component: Y. Y. Lin *et al., J. Am. Chem. Soc.* **103**, 6773 (1981). Structure of C: J. Golik *et al., Tetrahedron Letters* **23**, 2535 (1982). Structure of A, the most potent toxin: Y. Shimizu *et al., J. Am. Chem. Soc.* **108**, 514 (1986). Synthetic approaches to brevetoxin B: K. C. Nicolau *et al., Chem. Commun.* **1985**, 1359. Absolute configuration: Y. Shimizu *et al., Chem. Commun.* **22**, 1656 (1987). Biosynthetic study: H. N. Chou, Y. Shimizu, *J. Am. Chem. Soc.* **109**, 2184 (1987). Series of articles on pharmacology of brevetoxins: *Toxicon* **23**, 469-524 (1985). Review of chemistry: K. Nakanishi, *ibid.* 473. For additional information on red tide algae, *see Toxic Dinoflagellate Blooms*, D. L. Taylor, H. H. Seliger, Eds. (Elsevier, New York, 1979) pp 327-354.

Brevetoxin B : R =

Brevetoxin C : R =

Brevetoxin A, $C_{49}H_{70}O_{13}$, *GB-1.* Fine prisms from acetonitrile, mp 197-199°; 218-220° (double melting point). LC_{100} in guppies: 4 ng/ml (Shimizu).

Brevetoxin B, $C_{50}H_{70}O_{14}$, *BTX-B, GB-2, T-34, T-47.* Needles from acetonitrile, mp 270° (dec). uv max (methanol): 208 nm (ϵ 16000, enal). LC_{50} (1 hr) in fresh water "zebra" fish, *Brachydanio rerio:* 16 ng/ml (Lin).

Brevetoxin C, $C_{49}H_{69}ClO_{14}$, *BTX-C.* uv max (methanol): 208 nm (ϵ 11300, ene-lactone). LC_{50} (1 hr) in fresh water "zebra" fish: 30 ng/ml (Lin).

USE: As tools in neurochemical research.

1397. Brilliant Blue FCF. *N-Ethyl-N-[4-[[4-[ethyl[(3-sulfophenyl)methyl]amino]phenyl](2-sulfophenyl)methylene]-2,5-cyclohexadien-1-ylidene]-3-sulfobenzenemethanaminium inner salt, disodium salt;* FD & C Blue No. 1; C.I. Acid Blue 9; C.I. Food Blue 2; C.I. 42090. $C_{37}H_{34}N_2Na_2O_9S_3$; mol wt 792.86. C 56.05%, H 4.32%, N 3.53%, Na 5.80%, O 18.16%, S 12.13%. Discovered by Sandmeyer in 1896: *Colour Index* vol. 4 (3rd ed., 1971) p 4385. Also prepared as the diammonium salt. Metabolism: S. M. Hess, O. G. Fitzhugh, *J. Pharmacol. Exp. Ther.* **114**, 38 (1955); J. P. Brown *et al., Food Cosmet. Toxicol.* **18**, 1 (1980). Toxicology: E. Gross, *Z. Krebsforsch.* **64**, 287 (1961); W. A. Mannell, H. C. Grice, *J. Pharm. Pharmacol.* **16**, 56 (1964); W. H. Hansen *et al., Toxicol. Appl. Pharmacol.* **8**, 29 (1966). Chronic toxicity and carcinogenicity: J. F. Borzelleca *et al., Food Chem. Toxicol.* **28**, 221 (1990). Review of carcinogenicity studies: *IARC Monographs* **16**, 171-186 (1978).

Reddish-violet powder or granules with a metallic lustre. Absorption max: 630 nm. Sol in water, ethanol. Practically insol in vegetable oils. Pale amber soln in conc H_2SO_4, changing to yellow then greenish blue on dilution. LD_{50} s.c. in mice: 4.6 g/kg (Gross).

USE: Approved by FDA for use in food, drugs and cosmetics excluding use in eye area: *Fed. Reg.* **47**, 42563 (1982). Biological stain; textile dye; wood stain; indicator.

1398. Brilliant Green. *N-[4-[[4-(Diethylamino)phenyl]phenylmethylene]-2,5-cyclohexadien-1-ylidene]-N-ethylethanaminium sulfate (1:1);* C.I. Basic Green 1; C.I. 42040; Malachite Green G; Emerald Green; Diamond Green G; Fast Green J; Solid Green. $C_{27}H_{34}N_2O_4S$; mol wt 482.64. C 67.19%, H 7.10%, N 5.80%, O 13.26%, S 6.64%. Prepn: Doebner, *Ber.* **13**, 2222 (1880); Fischer, *ibid.* **14**, 2521 (1881); *Colour Index*, vol. 4 (3rd ed., 1971) p 4382. Toxicity: Anderson, *Proc. Soc. Exp. Biol. Med.* **31**, 825 (1934). Antibacterial activity: O. H. Paetzold, *Arch. Klin. Exp. Dermatol.* **224**, 90 (1966); idem, *Arch. Dermatol. Forsch.* **243**, 1 (1972). *Review:* H. J. Conn's *Biological Stains*, R. D. Lillie, Ed. (Williams & Wilkins, Baltimore, 9th ed., 1977) pp 251-252, 580.

Minute, glistening, golden crystals. Soluble in water or alcohol with green color. Absorption max: 623 nm. Soln changes color from yellow to green at pH 0.0 to 2.6. LD_{100} i.v. in mice: 3 mg/kg (Anderson).

USE: Dyeing silk, wool, leather, jute and cotton yellowish-green; manuf green ink; biological stain; indicator.

THERAP CAT: Antiseptic.

THERAP CAT (VET): Antiseptic for external and internal (oral) use. In wounds and scours.

1399. Brimonidine. *5-Bromo-N-(4,5-dihydro-1H-imidazol-2-yl)-6-quinoxalinamine;* 5-bromo-6-(2-imidazolin-2-ylamino)quinoxaline; UK-14304; AGN-190342. $C_{11}H_{10}BrN_5$; mol wt 292.14. C 45.23%, H 3.45%, Br 27.35%, N 23.97%. α_2-Adrenoceptor agonist. Prepn: M. B. Thomas, M. T. Williams, Ger. pat. **2,538,620** (1976 to Pfizer), *C.A.* **85**, 46752 (1976). Pharmacology: J. A. Burke, D. E. Potter, *Curr. Eye Res.* **5**, 665 (1986). Mechanism of action: J. B. Serle *et al., Arch. Ophthalmol.* **109**, 1158 (1991). Receptor binding studies and behavioral analysis in monkeys: A. F. T. Arnsten, F. M. Leslie, *Neuropharmacology* **30**, 1279 (1991). HPLC detemn: R. Gami-Yilinkou, R. Kaliszan, *J. Chromatog.* **550**, 573 (1991). Clinical pharmacology: H. Ashton, M. D. Rawlins, *Br. J. Clin. Pharmacol.* **5**, 135 (1978). Clinical study in control of intraocular pressure: R. David *et al., Arch. Ophthalmol.* **111**, 1387 (1993).

Yellow crystals from ethanol, mp 252°.

D-Tartrate, $C_{11}H_{10}BrN_5 \cdot C_4H_6O_6$, *UK-14304-18*, *AGN-190342-LF*. Crystals from acetone, mp 207.5°.

THERAP CAT: Antiglaucoma.

1400. Brodifacoum. *3-[3-(4'-Bromo[1,1'-biphenyl]-4-yl)-1,2,3,4-tetrahydro-1-naphthalenyl]-4-hydroxy-2H-1-benzopyran-2-one*; PP-581; WBA-8119; Talon; Ratak+. $C_{31}H_{23}BrO_3$; mol wt 523.43. C 71.14%, H 4.43%, Br 15.27%, O 9.17%. Prepn: M. R. Hadler, R. S. Shadbolt, *Ger. pat.* **2,424,806** corresp to U.S. pat. **3,957,824** (1975, 1976 to Ward, Blenkinsop & Co.); R. S. Shadbolt *et al., J. Chem. Soc. Perkin Trans. I* **1976**, 1190. Anticoagulant activity: M. R. Hadler, R. S. Shadbolt, *Nature* **253**, 275 (1975). Field trials: B. D. Rennison, A. C. Dubock, *J. Hyg.* **80**, 77 (1978); F. P. Rowe *et al., ibid.* **81**, 197 (1978).

Off-white powder, mp 228-230°. Insol in water. Slightly sol in alc, benzene; sol in acetone, chloroform. LD_{50} orally in rats: 270 μg/kg, *RTECS Vol. I*, R. J. Lewis, R. L. Tatken, Eds. (1980) p 541.

USE: Rodenticide.

1401. Brodimoprim. *5-[(4-Bromo-3,5-dimethoxyphenyl)methyl]-2,4-pyrimidinediamine*; 2,4-diamino-5-(4-bromo-3,5-dimethoxybenzyl)pyrimidine; Ro-10-5970; Hyprim; Unitrim. $C_{13}H_{15}BrN_4O_2$; mol wt 339.19. C 46.03%, H 4.46%, Br 23.56%, N 16.52%, O 9.43%. Dihydrofolate reductase inhibitor; structural analog of trimethoprim, *q.v.* Prepn: M. Hoffer, I. Kompis, *Ger. pat.* **2,452,889** (1975 to Hoffmann-La Roche), *C.A.* **83**, 97361 (1975); I. Kompis, U.S. pat. **4,024,145** (1977 to Hoffmann-La Roche); I. Kompis, A. Wick, *Helv. Chim. Acta* **60**, 3025 (1977). Antibacterial activity: G. Giammanco *et al., Drugs. Exptl. Clin. Res.* **9**, 721 (1983). Comparison with trimethoprim as inhibitor of dihydrofolate reductase: R. L. Then *et al., Rev. Infect. Dis.* **4**, 372 (1982); R. L. Then, F. Hermann, *Chemotherapy (Basel)* **30**, 18 (1984). Antimycobacterial activity *in vitro*: J. K. Seydel *et al., ibid.* **29**, 249 (1983). Penetration into canine bone tissue: P. Iverson, P. O. Madsen, *Acta Pharmacol. Toxicol.* **51**, 446 (1982). Pharmacokinetics in human serum, skin blister fluid: T. Kalager *et al., Chemotherapy (Basel)* **31**, 405 (1985). Clinical evaluation in respiratory infections: H. A. Salmi *et al., Drugs Exptl. Clin. Res.* **12**, 349 (1986).

Crystals from methanol, mp 225-228°. pKa 7.15.

THERAP CAT: Antibacterial.

1402. Bromacil. *5-Bromo-6-methyl-3-(1-methylpropyl)-2,4(1H,3H)-pyrimidinedione*; 5-bromo-3-sec-butyl-6-methyluracil; 5-bromo-6-methyl-3-(1-methylpropyl)uracil; Du

Pont Herbicide 976; Hyvar; Uragon; Urox B. $C_9H_{13}BrN_2O_2$; mol wt 261.12. C 41.40%, H 5.02%, Br 30.60%, N 10.73%, O 12.25%. Prepn: H. M. Loux, U.S. pat **3,235,357** (1966 to du Pont). Mode of action: C. E. Hoffman, *Pestic. Chem., Proc. 2nd Int. Congr. Pestic.* **5**, A. S. Tahori, Ed, (Gordon and Breach Science Publishers, New York, 1972) pp 65-85. Toxicology: H. Sherman, A. M. Kaplan, *Toxicol. Appl. Pharmacol.* **34**, 189 (1975). *Review:* Pease, Deye, *Anal. Methods Pestic., Plant Growth Regul., Food Additives* **5**, 335 (1967).

White crystalline solid, mp 157.5-160°. Vapor pressure at 100°: 8×10^{-4} mm Hg. Soly in water at 20°: 815 mg/l. Moderately sol in strong aq bases, acetone, acetonitrile, ethanol. LD_{50} orally in rats: 5200 mg/kg (Sherman, Kaplan).

USE: Herbicide.

1403. Bromadiolone. *3-[3-(4'-Bromo[1,1'-biphenyl]-4-yl)-3-hydroxy-1-phenylpropyl]-4-hydroxy-2H-1-benzopyran-2-one; 3-[α-[p-(p-bromophenyl)-β-hydroxyphenethyl]benzyl]-4-hydroxycoumarin*; LM-637; Maki; Bromone; Super-Caid; Super-Rozol. $C_{30}H_{23}BrO_4$; mol wt 527.41. C 68.32%, H 4.40%, Br 15.15%, O 12.13%. Anticoagulant rodenticide. Prepn: E. Boschetti *et al., Ger. pat.* **1,959,317**; *eidem,* U.S. pat. **3,764,693** (1970, 1973 both to Lipha). Activity studies: *eidem, Chim. Ther.* **7**, 20 (1972); M. Grand, *Phytiat.-Phytopharm.* **25**, 69 (1976); R. E. Marsh, *Bull. OEPP* **7**, 495 (1977); R. Redfern, J. E. Gill, *J. Hyg.* **84**, 263 (1980). Pharmacokinetics: K. Nahas, *Pharmacol. Res. Commun.* **19**, 767 (1987). HPLC determn of diastereoisomers in animal tissues: K. Hunter *et al., J. Chromatog.* **435**, 83 (1988).

White to offwhite powder, mp 200-210°. uv max (ethanol): 260 nm ($E^{1\%}_{1cm}$ 538-582). pKa (21°) 4.04. Soly at 20-25° (g/l): dimethylformamide 730.0; ethyl acetate 25.0; acetone 22.3; chloroform 10.1; ethanol 8.2; methanol 5.6; ethyl ether 3.7; hexane 0.2; water 0.019. LD_{50} in rats, mice (mg/kg): 1.125, 1.75 orally (Grand).

USE: Rodenticide.

1404. Bromal. *Tribromoacetaldehyde.* C_2HBr_3O; mol wt 280.74. C 8.56%, H 0.36%, Br 85.39%, O 5.70%. Br_3C-CHO. Preparation from ethanol and bromine: Löwig, *Ann.* **3**, 288 (1832); from chloral and a bromide: Müller, U.S. pat. **2,053,964** (1936 to Winthrop); from paraldehyde and bromine: Long, Howard, *Org. Syn.* **17**, 18 (1937).

Yellowish, oily liquid; forms with water bromal hydrate which is solid at temps below 50°. d 2.66. bp ∼174° with decompn. Sol in water, alcohol or ether.

1405. Bromal Hydrate. *2,2,2-Tribromo-1,1-ethanediol; tribromoacetaldehyde hydrate.* $C_2H_3Br_3O_2$; mol wt 298.76. C 8.04%, H 1.01%, Br 80.24%, O 10.71%. $Br_3CCH(OH)_2$. Prepd from bromal and water: Löwig, *Ann.* **3**, 288 (1832). Structure: Jain, Soundararajan, *Tetrahedron* **20**, 1589 (1964).

Deliquescent crystals, mp 53.5°. Odor of chloral and pungent taste. Dipole moment in benzene, 2.56D. Sol in water, alcohol, chloroform, ether, glycerol. *Keep tightly closed in a cool place.*

Caution: This substance may be habit forming and is listed in the U.S. Code of Federal Regulations, Title 21 Part 329.1 (1995).

1406. Bromazepam. *7-Bromo-1,3-dihydro-5-(2-pyridinyl)-2H-1,4-benzodiazepin-2-one;* 7-bromo-5-(2-pyridyl)-3H-1,4-benzodiazepin-2(1H)-one; Ro-5-3350; Compendium; Creosedin; Durazanil; Lectopam; Lexomil; Lexotan; Lexotanil; Normoc. $C_{14}H_{10}BrN_3O$; mol wt 316.16. C 53.19%, H 3.19%, Br 25.27%, N 13.29%, O 5.06%. Prepn: Berger *et al.*, **Belg. pat. 619,101;** *eidem,* Fryer *et al.,* **U.S. pat. 3,100,-770** (1962, 1963 to Hoffmann-La Roche); *eidem, J. Pharm. Sci.* **53,** 264 (1964); **U.S. pats. 3,182,065; 3,182,067** (both 1965 to Hoffmann-La Roche). Pharmacology: Korol, Brown, *Pharmacology* **1,** 115 (1968). Metabolism: M. A. Schwartz *et al., J. Pharm. Sci.* **62,** 1776 (1973); *Drug Metab. Dispos.* **2,** 31 (1974). Evaluation as pre-anesthesia medication: P. Chalmers, J. N. Horton, *Anesthesia* **39,** 370 (1984). Multicenter clinical comparison with lorazepam, *q.v.:* G. J. Cordingley *et al., Curr. Med. Res. Opin.* **9,** 505 (1985). Evaluation of adverse effects and withdrawal reactions: R. Fontaine *et al., Psychopharmacol. Bull.* **21,** 91 (1985). Toxicity: E. I. Goldenthal, *Toxicol. Appl. Pharmacol.* **18,** 185 (1971). Comprehensive description: M. M. Hassan, M. A. Abounassif, *Anal. Profiles Drug Subs.* **16,** 1-51 (1987).

Colorless prisms from acetone, mp 237-238.5° (dec). LD_{50} orally in rats: 3050 ±405 mg/kg (Goldenthal).
Note: This is a controlled substance (depressant) listed in the U.S. Code of Federal Regulations, Title 21 Part 1308.14 (1995).

THERAP CAT: Anxiolytic.

1407. Bromcresol Green. *4,4'-(3H-2,1-Benzoxathiol-3-ylidene)bis[2,6-dibromo-3-methylphenol] S,S-dioxide;* α,α-bis-(3,5-dibromo-4-hydroxy-o-tolyl)-α-hydroxytoluenesulfonic acid, γ-sultone; 3,3',5,5'-tetrabromo-m-cresolsulfonphthalein. $C_{21}H_{14}Br_4O_5S$; mol wt 698.02. C 36.14%, H 2.02%, Br 45.79%, O 11.46%, S 4.59%. Prepd by adding bromine to a suspension of m-cresolsulfonphthalein in glacial acetic acid: Clark, Lubs, *J. Wash. Acad. Sci.* **5,** 610 (1915); **6,** 481 (1916); *J. Bact.* **2,** 110 (1917); Cohen, *Biochem. J.* **16,** 31 (1922); **17,** 535 (1923); Cohen, *Public Health Repts.* **38,** 814 (1923); **41,** 3051 (1926); *Proc. Soc. Exp. Biol. Med.* **20,** 124 (1922); Orndorff, Purdy, *J. Am. Chem. Soc.* **48,** 2216 (1926).

Minute, slightly yellow crystals from acetic acid, mp 218-219°. Sparingly sol in water. Readily sol in alcohol, ether, ethyl acetate. Fairly sol in benzene. Very sensitive to alkalies, tap water being sufficiently alkaline to give the characteristic blue-green color. pK = 4.7. To prepare a soln for use as pH indicator, dissolve 0.10 g in 7.15 ml N/50 NaOH

and dil with water to 250 ml. To prepare a soln for use as indicator in volumetric work, dissolve 0.1 g in 250 ml alc.
USE: As indicator, pH 3.8 yellow; pH 5.4 blue-green.

1408. Bromcresol Purple. *4,4'-(3H-2,1-Benzoxathiol-3-ylidene)bis[2-bromo-6-methylphenol] S,S-dioxide;* α-(5-bromo-4-hydroxy-m-tolyl)-α-(3-bromo-5-methyl-4-oxo-2,5-cyclohexadien-1-ylidene)-o-toluenesulfonic acid; 5,5'-dibromo-o-cresolsulfonphthalein. $C_{21}H_{16}Br_2O_5S$; mol wt 540.23. C 46.69%, H 2.99%, Br 29.58%, O 14.81%, S 5.94%. Prepd by treating o-cresol red with bromine in glacial acetic acid: *See refs under* Bromcresol Green.

Minute, slightly yellow crystals, mp 241-242°. Practically insol in water. Sol in alcohol, dil alkalies. pK 6.3. To prepare a soln for use as pH indicator, dissolve 0.10 g in 9.25 ml N/50 NaOH and dil with water to 250 ml. To prepare a soln for use as indicator in volumetric work, dissolve 0.05 g in 250 ml alc.
USE: As indicator, pH 5.2 yellow; pH 6.8 purple.

1409. Bromelain. Bromelin; Ananase; Extranase; Inflamen; Traumanase. Protein-digesting and milk-clotting enzymes found in pineapple fruit juice and stem tissue. Enzymes from the two sources are distinguished as fruit bromelain and stem bromelain. First isolns of fruit bromelain: Marcano, *Bull. Pharm.* **5,** 77 (1891); Chittenden, *Trans. Conn. Acad. Sci.* **8,** 281 (1892). From pineapple juice by precipitation with acetone and also with ammonium sulfide: Heinicke, **U.S. pat. 3,002,891** (1961 to Pineapple Res. Inst.). Discovery in stem tissue: *idem, Science* **118,** 753 (1953). Stem bromelain has mol wt of about 33,000 and is probably the first proteolytic enzyme of plant origin to be established as a glycoprotein: Murachi *et al., Biochemistry* **3,** 48 (1964); Ota *et al., ibid.* 180. Purification of crude prepns: Gibian, Bratfisch, **U.S. pat. 2,950,227** (1960 to Schering AG). *Reviews:* Balls *et al., Ind. Eng. Chem.* **33,** 950 (1941); of stem bromelain: Murachi, "Structure and Function of Stem Bromelain," in *Proteins, Structure and Function,* vol. 2, M. Funatsu *et al.,* Eds. (Kodansha, Tokyo, Wiley, New York, 1972) pp 47-101.
Unlike papain, fruit bromelain does not disappear as the fruit ripens. Fruit and stem bromelains are acidic and basic proteins, resp. uv max (stem): 280 nm ($A_{1cm}^{1\%}$ 20.1).
USE: Tenderizing meat, chill-proofing beer, production of protein hydrolyzates.
THERAP CAT: Anti-inflammatory.

1410. Bromethalin. *N-Methyl-2,4-dinitro-N-(2,4,6-tribromophenyl)-6-(trifluoromethyl)benzenamine;* EL-614; Vengeance. $C_{14}H_7Br_3F_3N_3O_4$; mol wt 577.93. C 29.10%, H 1.22%, Br 41.48%, F 9.86%, N 7.27%, O 11.07%. Rodenticide which acts by uncoupling oxidative phosphorylation. Single-feed toxicant, effective against warfarin-resistant strains of mice, rats. Prepn and identification as rodenticide: B. A. Dreikorn, **U.S. pat. 4,187,318** (1980 to Eli Lilly). Toxicology and activity: B. A. Dreikorn *et al., Proc. Brit. Crop Prot. Conf. - Pests Dis.* **1979,** 491. *Review:* B. A. Dreikorn, G. O. P. O'Doherty, *ACS Symposium Series* **255,** 45-63 (1984).

Pale yellow odorless crystals from ethanol, mp 150-151°. Sol in chloroform, acetone; moderately sol in aromatic hydrocarbons. Insol in water. LD_{50} in mice, rats, cats, dogs (mg/kg): 2, 5, 2, 5 orally (Dreikorn).

USE: Rodenticide.

1411. Bromfenac. *2-Amino-3-(4-bromobenzoyl)benzeneacetic acid;* AHR-10282. $C_{15}H_{12}BrNO_3$; mol wt 334.17. C 53.91%, H 3.62%, Br 23.91%, N 4.19%, O 14.36%. Prostaglandin synthetase inhibitor; analog of amfenac, *q.v.* Prepn: D. A. Walsh *et al., J. Med. Chem.* **27**, 1379 (1984). Prepd, not claimed: R. G. Poser, *Eur.* pat. **Appl. 221,753;** *idem,* U.S. pat. **4,683,242** (both 1987 to A. H. Robins). Pharmacology: L. F. Sancilio *et al., Arzneimittel-Forsch.* **37**, 513 (1987). HPLC determn in plasma: M. A. Osman *et al., J. Chromatog.* **489**, 452 (1989). Clinical pharmacokinetics: P. Högger, P. Rohdewald, *Arzneimittel-Forsch.* **43**, 1114 (1993). Clinical study in postoperative pain: J. A. Forbes *et al., Pharmacotherapy* **11**, 64 (1991); Å. A. S. Boström *et al., ibid.* **14**, 305 (1994). *Review:* J. S. Wajdula *et al.* in *Nonsteroidal Anti-Inflammatory Drugs,* A. J. Lewis, D. E. Furst, Eds. (Marcel Dekker, Inc., New York, 1994) 267-284.

Monosodium salt sesquihydrate, $C_{15}H_{11}BrNNaO_3 \cdot 1\frac{1}{2}H_2O$, *AHR-10282B.* Bright orange-yellow crystalline powder, mp 284-286° (dec). pKa 4.29. Sol in water, methanol, dilute base. Insol in chloroform, dilute acid.

THERAP CAT: Analgesic; anti-inflammatory.

1412. Bromhexine. *2-Amino-3,5-dibromo-N-cyclohexyl-N-methylbenzenemethanamine; 3,5-dibromo-N$^\alpha$-cyclohexyl-N$^\alpha$-methyltoluene-α,2-diamine;* N-cyclohexyl-N-methyl-2-(2-amino-3,5-dibromo)benzylammonium; N-(2-amino-3,5-dibromobenzyl)-N-methylcyclohexylammonium. $C_{14}H_{20}Br_2N_2$; mol wt 376.13. C 44.71%, H 5.36%, Br 42.49%, N 7.45%. Prepn: Keck, *Ann.* **662**, 171 (1963); K. Thomae, **Belg.** pat. **625,022** (1963), *C.A.* **61**, 5564 (1964). Pharmacology: R. Engelhorn, S. Püschmann, *Arzneimittel-Forsch.* **13**, 474 (1963). Metabolism: R. Jauch *et al., ibid.* **25**, 1954 (1975). Toxicity: Boyd, Sheppard, *Arch. Int. Pharmacodyn. Ther.* **163**, 284 (1966).

Crystals from abs ethanol, dec 237.5-238°. One gram dissolves in 250 ml of water or 10% ethanol. Approx LD_{50} orally in rabbits: > 10 g/kg (Boyd, Sheppard).

Hydrochloride, $C_{14}H_{20}Br_2N_2 \cdot HCl$, *NA-274, Auxit, Bisolvon, Ophtosol, Quentan.*

THERAP CAT: Expectorant; mucolytic.

THERAP CAT (VET): Expectorant.

1413. Bromic Acid. $BrHO_3$; mol wt 128.91. Br 61.98%, H 0.78%, O 37.23%. Prepd from Ba(BrO$_3$)$_2$ and H_2SO_4: Burchard, *Z. Physik. Chem.* **2**, 814 (1888); Schmeisser in *Handbook of Preparative Inorganic Chemistry,*

vol. **1**, G. Brauer, Ed. (Academic Press, 2nd ed, 1963) p 315.

Known in aq soln only. *Keep refrigerated!* Colorless soln, turns yellow on standing at room temp, even in the dark. Careful evapn *in vacuo* at −12° yields about a 50% soln. Dec on heating to 100°. Can be dil with cold water.

Caution: Highly irritating to skin, eyes, mucous membranes.

USE: Oxidizing agent.

1414. Bromindione. *2-(4-Bromophenyl)-1H-indene-1,3-(2H)-dione;* 2-(4-bromophenyl)-1,3-dioxohydrindene; HL-255; MG-2555; Fluidane; Halinone. $C_{15}H_9BrO_2$; mol wt 301.14. C 59.83%, H 3.01%, Br 26.53%, O 10.63%. Prepn: Cavallini *et al., Farmaco Ed. Sci.* **10**, 710 (1955); Freedman *et al.,* U.S. pat. **2,847,474** (1958 to U.S. Vitamin).

Crystals from ligroin, mp 137-139°.

THERAP CAT: Anticoagulant.

1415. Bromine. Br; at. wt 79.904; at. no. 35; valences 1 to 7; elemental state: Br_2. A halogen. Abundance in igneous rock: $1.6 \times 10^{-4}\%$ by wt; in seawater: 0.0065% by weight. Natural isotopes: 79 (50.54%); 81 (49.46%); known isotopes range in mass no. from 74-90; radioactive tracer elements: 77, 80, 80m (metastable), 82. Discovered by Balard in 1826. Isolated from natural brines (salt lakes) and ocean water: Seaton, *Chem. & Met. Eng.* **38**(11), 638 (1931); Robertson, *Ind. Eng. Chem.* **34**, 133 (1942); *Chem. & Met. Eng.* **52**(10), 134 (1945); Gale, Pearson, U.S. pat. **2,251,353** (1941 to Am. Potash and Chem.); Faith, Keyes & Clark's *Industrial Chemicals,* F. A. Lowenheim, M. K. Moran, Eds. (Wiley-Interscience, New York, 4th ed., 1975) pp 159-163. Prepn of ultra-pure bromine for research purposes: Baxter *et al., J. Am. Chem. Soc.* **34**, 260 (1912); Noyes, *ibid.* **45**, 1194 (1923); Hönigschmid, Zintl, *Ann.* **433**, 216 (1923). Book: Z. E. Jolles, Ed., *Bromine and its Compounds* (E. Benn, London, 1966) 940 pp. *Reviews: MTP Int. Rev. Sci.: Inorg. Chem., Ser. One,* vol. **3**, V. Gutmann, Ed. (Butterworths, London, 1972); Downs, Adams, "Chlorine, Bromine, Iodine and Astatine" in *Comprehensive Inorganic Chemistry,* vol. **2**, J. C. Bailar, Jr. *et al.,* Eds. (Pergamon Press, Oxford, 1973) pp 1107-1594; C. E. Reineke in Kirk-Othmer *Encyclopedia of Chemical Technology* vol. **4** (Wiley-Interscience, New York, 3rd ed., 1978) pp 226-243.

Dark reddish-brown, volatile, diatomic liquid; suffocating odor; vaporizes rapidly at room temperature. mp −7.25°C (265.90°K); bp 59.47° *(JANAF Thermochemical Tables);* 58.78° (Mellor's Suppl. II, Part I, "The Halogens"); d_4^{25} 3.1023; crit temp: 315°; crit pressure: 102 atm. Heat capacity at constant pressure (liq, 25°) 18.089 cal/mole deg: Hildenbrand *et al., J. Am. Chem. Soc.* **80**, 4129 (1958). Vapor pressure data: A. N. Nesmeyanov, *Vapor Pressure of the Chemical Elements,* R. Gary, Ed. (Elsevier, New York, 1963) pp 354-58. Total soly in water (25°): 0.2141 moles/l with formation of 0.00115 moles/l of HOBr; freely sol in alc, ether, $CHCl_3$, CCl_4, CS_2, concd HCl, aq solns of bromides. Less reactive than chlorine; E^0 (aq) $\frac{1}{2}Br_2/Br^-$ 1.065 V; dissociation energy (25°): 46.072 kcal. For reactions, complexes and formation of monatomic bromine *see* Chlorine. Incompatible with alkali hydroxides; arsenites; ferrous, mercurous salts; hypophosphites and other oxidizable substances. *Keep sealed or glass-stoppered. When handling bromine always keep ammonia water within reach.*

Caution: Potential symptoms of overexposure are dizziness, headache; lacrimation, epistaxis; coughing, feeling of oppression, pulmonary edema and pneumonia; abdominal pain, diarrhea; measle-like eruptions; severe burns of eyes and skin. *See NIOSH Pocket Guide to Chemical Hazards* (DHHS/NIOSH 90-117, 1990) p 48. *See also* Patty's *Industrial Hygiene and Toxicology* vol. **2B**, G. D. Clayton, F. E. Clayton, Eds. (Wiley-Interscience, New York, 3rd ed., 1981) pp 2965-2972.

USE: In water disinfection; bleaching fibers and silk; manuf medicinal bromine compds, dyestuffs.

1416. Bromine Pentafluoride. BrF_5; mol wt 174.90. Br 45.69%, F 54.31%. Prepd by fluorination of bromine at 200° in iron or Monel metal apparatus: Ruff, Menzel, *Z. Anorg. Allgem. Chem.* **202**, 49 (1931); Kwasnik in *Handbook of Preparative Inorganic Chemistry*, **Vol. 1**, G. Brauer, Ed. (Academic Press, New York, 2nd ed., 1963) pp 158-159; from F_2 and KBr: Hyde, Boudakian, *Inorg. Chem.* **7**, 2648 (1968). *Reviews:* Kemmitt, Sharp, *Advan. Fluorine Chem.* **4**, 243-244 (1965); Stein, "Physical and Chemical Properties of Halogen Fluorides" in *Halogen Chemistry*, **Vol. 1**, V. Gutmann, Ed. (Academic Press, New York, 1967) pp 133-224; Meinert, *Z. Chem.* **7**, 41-57 (1967).

Liquid. Fumes in air. mp −60.5°. bp 40.76°. d^{25} 2.4604. Trouton constant 23.7. Thermostable up to 460°. Does not attack quartz when dry. *Produces an explosion on contact with water.* Very reactive, usually with conflagration.

Caution: Corrosive and irritating to eyes, skin, mucous membranes.

1417. Bromine Trifluoride. BrF_3; mol wt 136.90. Br 58.37%, F 41.63%. Prepd by fluorination of bromine at +80°: Lebeau, *Compt. Rend.* **141**, 1018 (1905); Prideaux, *J. Chem. Soc.* **89**, 316 (1906); Ruff, Braida, *Z. Anorg. Allgem. Chem.* **206**, 59 (1932); **214**, 91 (1933); Simons, *Inorg. Syn.* **3**, 184 (1950); Kwasnik in *Handbook of Preparative Inorganic Chemistry*, **Vol. 1**, G. Brauer, Ed. (Academic Press, 2nd ed., 1963) pp 156-157. *Reviews:* Kemmitt, Sharp, *Advan. Fluorine Chem.* **4**, 244-245 (1965); Stein, "Physical and Chemical Properties of Halogen Fluorides" in *Halogen Chemistry*, **Vol. 1**, V. Gutmann, Ed. (Academic Press, New York, 1967) pp 133-224; Meinert, *Z. Chem.* **7**, 41-57 (1967).

Colorless liquid; also reported to be pale yellow. Long prisms when solid. mp 8.77°. bp 125.75°. d^{25} 2.8030. *Smokes in air. Attacks skin. Very reactive.*

Caution: Corrosive and irritating to skin, eyes, mucous membranes, respiratory tract.

USE: Solvent for fluorides.

1418. Bromisovalum. *N-(Aminocarbonyl)-2-bromo-3-methylbutanamide; (α-bromoisovaleryl)urea;* 2-monobromoisovalerylurea; α-bromo-β-dimethylpropanoylurea; bromvaletone; B.V.U.; Bromural; Bromisoval; Uvaleral; Bromuvan; Somnurol; Brovalurea; Dormigene; Isobromyl; Alluval; Pivadorm; $C_6H_{11}BrN_2O_2$; mol wt 223.07. C 32.31%, H 4.97%, Br 35.82%, N 12.56%, O 14.34%. $(CH_3)_2CHCHBr$-$CONHCONH_2$. Prepn from urea and α-bromoisovaleryl bromide: U.S. pat. 914,518 (1909). Toxicity: R. I. Mrongovius *et al., Clin. Exp. Pharmacol. Physiol.* **3**, 443 (1976).

Practically tasteless needles, mp 147-149°. Sublimes. Slightly sol in cold water, but freely sol in hot water. Readily sol in alcohol, ether and in alkaline solns. LD_{50} in male mice (mmoles/kg): 3.25 i.p. (Mrongovius).

Caution: This substance may be habit forming and is listed in the U.S. Code of Federal Regulations, Title 21 Part 329.1 (1995).

THERAP CAT: Sedative, hypnotic.

1419. *N*-Bromoacetamide. Acetobromamide. C_2H_4Br-NO; mol wt 137.96. C 17.41%, H 2.92%, Br 57.92%, N 10.15%, O 11.60%. $CH_3CONHBr$. Prepd by treating a cooled (0-5°) soln of acetamide dissolved in bromine with ice cold aq 50% KOH, allowing to stand for 2-3 hr at 0-5°, treating with NaCl, and extracting with chloroform: Oliveto, Gerold, *Org Syn.* **31**, 17 (1951).

Needles from chloroform + hexane, mp 102-105°. Sol in warm water, freely sol in cold ether. Unstable to light and heat.

Monohydrate, rectangular plates from hydr ether, mp 70-80°. Freely sol in cold water, alcohol, ether, less freely in chloroform.

USE: Brominating agent; in oxidation of primary and secondary alcohols.

1420. *p*-Bromoacetanilide. *N-(4-Bromophenyl)acetamide;* 4′-bromoacetanilide; monobromoacetanilide; bromoanilide; Antisepsin; Asepsin; Bromoantifebrin. C_8H_8BrNO; mol wt 214.06. C 44.89%, H 3.77%, Br 37.33%, N 6.54%, O 7.47%. Prepd by bromination of acetanilide: Remmers, *Ber.* **7**, 346 (1874); Merker, Vona, *J. Chem. Ed.* **26**, 613

(1949); Knowles, Alt, U.S. pat. **3,012,035** (1959 to Monsanto).

Crystals, from 95% alc, mp 168° (with previous softening). d 1.72. Practically insol in cold water. Sparingly sol in hot water; sol in benzene, chloroform, ethyl acetate; moderately sol in alcohol.

THERAP CAT: Analgesic, antipyretic.

1421. Bromoacetic Acid. $C_2H_3BrO_2$; mol wt 138.95. C 17.29%, H 2.18%, Br 57.51%, O 23.03%. $BrCH_2COOH$. Prepn by bromination of acetic acid: Perkin, Duppa, *Ann.* **108**, 106 (1858); from chloroacetic acid and HBr: Lake, Asadorian; Asadorian, Burk, U.S. pats. **2,553,518**; **3,130,-222** (1951, 1964, both to Dow); from glycolic acid and HBr: Johnston, U.S. pat. **2,876,255** (1959 to Ethyl Corp.).

Hygroscopic crystals, mp 50°. bp 208°. d 1.93. Very sol in water, alcohol. *Protect from air and moisture.*

Caution: Irritant and corrosive to skin, mucous membranes.

1422. Bromoacetone. *1-Bromo-2-propanone.* C_3H_5-BrO; mol wt 136.98. C 26.31%, H 3.68%, Br 58.33%, O 11.68%. CH_3COCH_2Br. Prepn by bromination of acetone: Emmerling, Wagner, *Ann.* **204**, 29 (1880); Catch *et al., J. Chem. Soc.* **1948**, 272; Ross U.S. pat. **2,452,154** (1948 to Colgate-Palmolive-Peet); by bromination acetone enol acetate: Magerlein, U.S. pat. **2,752,341** (1956 to Upjohn).

Liquid, mp −36.5°. bp 137°. bp_{50} 63.5-64°. d^{23} 1.634. n_D^{15} 1.4697. Turns violet rapidly even in absence of air. Sparingly sol in water; sol in alcohol, acetone. *Keep tightly closed and protected from light.*

Caution: Violent lacrimator.

USE: Chemical war gas.

1423. *p*-Bromoacetophenone. *1-(4-Bromophenyl)ethanone;* methyl *p*-bromophenyl ketone. C_8H_7BrO; mol wt 199.05. C 48.27%, H 3.54%, Br 40.14%, O 8.04%. BrC_6H_4-$COCH_3$. Prepd from bromobenzene and acetic anhydride in CS_2 in the presence of anhydr $AlCl_3$: Adams, Noller, *Org. Syn.* **5**, 17 (1925).

Leaflets from alc. mp 54°. bp_{736} 255.5°; bp_{15} 130°; bp_7 117°. Easily volatile with steam. Sol in alcohol, ether, glacial acetic acid, benzene, petr ether, carbon disulfide.

Oxime, C_8H_8BrNO, needles from dil alc, mp 128.5°.

1424. ω-Bromoacetophenone. *2-Bromo-1-phenylethanone;* phenacyl bromide. C_8H_7BrO; mol wt 199.05. C 48.27%, H 3.54%, Br 40.14%, O 8.04%. $C_6H_5COCH_2Br$. Prepn from acetophenone and bromine: Rother, Reid, *J. Am. Chem. Soc.* **41**, 77 (1919); Cowper, Davidson, *Org. Syn.* **19**, 24 (1939); Shevchuk, Dombrovskii, *Zh. Obshch. Khim.* **33**(4), 1135 (1963).

Crystals, mp 50°. d 1.65. bp_{20} 133-135°. Irritating vapors. *Lacrimator!* Practically insol in water. Freely sol in alcohol, benzene, chloroform, ether.

Caution: Highly irritating to skin, eyes, mucous membranes.

1425. *p*-Bromoaniline. *4-Bromobenzeneamine;* 4-bromoaniline. C_6H_6BrN; mol wt 172.02. C 41.89%, H 3.52%, Br 46.45%, N 8.14%. Prepd by steam distilling sodium hydroxide and *p*-bromoacetanilide: Scott, *J. Chem. Soc.* **123**, 3199 (1923); by direct bromination of aniline: Kosolapoff, *J. Am. Chem. Soc.* **75**, 3596 (1953).

Rhombic crystals from dil alc; mp 66-66.5°. $d_4^{99.6}$ 1.4970 (liq). Very sol in alcohol and ether. Insol in cold water. pKb (25°): 10.28 also reported as 9.98.

USE: In prepn of azo dyes; condensed with formaldehyde in prepn of dihydroquinazolines.

1426.   5-Bromoanthranilic Acid. *2-Amino-5-bromobenzoic acid;* 5-bromo-2-aminobenzoic acid. $C_7H_6BrNO_2$; mol wt 216.03. C 38.92%, H 2.80%, Br 36.99%, N 6.48%, O 14.81%. Prepd by bromination of anthranilic acid: Wheeler, *J. Am. Chem. Soc.* **31**, 565 (1909); Wheeler, Oates, *ibid.* **32**, 771 (1910).

Crystals, mp 218-219°. Very slightly sol in water, moderately sol in alcohol, ether, chloroform, benzene, acetic acid, freely sol in acetone.

USE: Determination of cobalt, copper, nickel, and zinc.

1427.   Bromobenzene. Monobromobenzene; phenyl bromide. C_6H_5Br; mol wt 157.01. C 45.90%, H 3.21%, Br 50.89%. Prepd industrially by the action of bromide on benzene in the presence of iron powder: Gattermann-Wieland, *Praxis des organischen Chemikers* (de Gruyter, Berlin, 40th ed., 1961) p 95; alternate procedure using pyridine as halogen carrier: A. I. Vogel, *Practical Organic Chemistry* (Longmans, London, 3rd ed., 1959) p 535.

Mobile liquid. Aromatic odor. d_4^0 1.5220; d_4^{10} 1.5083; d_4^{15} 1.5017; d_4^{20} 1.4952; d_4^{30} 1.4815; d_4^{71} 1.426. mp −30.6°. bp_{760} 156.2°; bp_{400} 132.3°; bp_{200} 110.1°; bp_{40} 68.6°; bp_{20} 53.8°; bp_{10} 40.0°; bp_5 27.8°; $bp_{1.0}$ +2.9°. n_D^{15} 1.5625; n_D^{20} 1.5602. Fire pt 155°. Critical temp 397°; crit press. 33,912 mm (44.6 atm). Viscosity at 20° = 1.124 cp. Vapor density (air = 1): 5.41. Specific heat at 26.84° = 0.2368. Heat of melting 16.186 cal/g at 15°. Practically insol in water (0.045 g/100 g at 30°). Miscible with chloroform, benzene, petr hydrocarbons. Sol in alc (10.4 g/100 g at 25°), in ether (71.3 g/100 g at 25°).

Caution: Irritating to skin.

USE: In organic synthesis, especially to make phenyl magnesium bromide; as solvent, especially for crystns on a large scale and where a heavy liquid is desirable; as additive to motor oils.

1428.   *p*-Bromobenzenesulfonyl Chloride. $C_6H_4BrClO_2S$; mol wt 255.52. C 28.20%, H 1.58%, Br 31.27%, Cl 13.87%, O 12.52%, S 12.55%. Prepd from sodium *p*-bromobenzenesulfonate and phosphorus pentachloride: Marvel, Smith, *J. Am. Chem. Soc.* **45**, 2696 (1923).

Needles from ligroin, mp 74.5°.

USE: Identification of amines.

1429.   *p*-Bromobenzoic Acid. $C_7H_5BrO_2$; mol wt 201.02. C 41.83%, H 2.51%, Br 39.75%, O 15.92%. Prepd by oxidation of *p*-bromotoluene with potassium permanganate: Hale, Thorp, *J. Am. Chem. Soc.* **35**, 269 (1913).

Needles from ether, leaflets from water or 90% alcohol. mp 251-253°. Slightly sol in hot water. Sol in alc, ether. Methyl ester, $C_6H_4BrCO_2CH_3$, needles from ether or leaflets from dil alcohol, mp 81°. d 1.689. Sol in alc, ether. Ethyl ester, $C_6H_4BrCO_2C_2H_5$, liquid; $bp_{737.4}$ 262°. Phenyl ester, $C_{13}H_9BrO_2$, scales, mp 117°. Sol in alc, ether, chloroform, CS_2, benzene; less sol in petr ether.

USE: For the detection of strontium; in org syntheses.

1430.   *p*-Bromobenzyl Bromide. *1-Bromo-4-(bromomethyl)benzene;* *p*,α-dibromotoluene. $C_7H_6Br_2$; mol wt 249.93. C 33.64%, H 2.42%, Br 63.94%. Prepd by photobromination of *p*-bromotoluene: Weizmann, Patai, *J. Am. Chem. Soc.* **68**, 150 (1946). Alternate procedure: Goerner, Nametz, *ibid.* **73**, 2940 (1951).

Crystals from alc. Agreeable aromatic odor. mp 61°. bp_{12} 115-124°. Sol in water, cold alc, more sol in hot alc, ether, carbon disulfide, benzene, glacial acetic acid.

Caution: Irritating to eyes, mucous membranes of nose and throat. Action similar to, but less intense than, benzylbromide, *q. v.*

USE: Identification of aromatic carboxylic acids.

1431.   *p*-Bromobenzyl Chloride. *1-Bromo-4-(chloromethyl)benzene;* *p*-bromo-α-chlorotoluene; α-chloro-4-bromotoluene. C_7H_6BrCl; mol wt 205.48. C 40.92%, H 2.94%, Br 38.89%, Cl 17.25%. Prepd from *p*-bromobenzene, paraformaldehyde and $SnCl_4$: Quelet, *Bull. Soc. Chim.* **41**, 329 (1927); by benzoyl peroxide catalyzed chlorination of *p*-bromotoluene with sulfhydryl chloride: Goerner, Nametz, *J. Am. Chem. Soc.* **73**, 2940 (1951).

Needles from alc, mp 40-41°. bp_{12} 105-115°; bp_{27} 136-139°. Freely sol in hot alc.

1432.   *p*-Bromobenzyl Chloroformate. *Carbonochloridic acid (4-bromophenyl)methyl ester;* chloroformic acid *p*-bromobenzyl ester. $C_8H_6BrClO_2$; mol wt 249.49. C 38.51%, H 2.42%, Br 32.03%, Cl 14.21%, O 12.83%. A stable, low-melting cryst product obtained from *p*-bromobenzyl alcohol. In the carbobenzyloxy method of prepn of amino acids and peptides, gives higher melting and better cryst products than the corresp carbobenzyloxy compds: Channing *et al.*, *Nature* **167**, 487 (1951).

USE: In the prepn of amino acids and peptides.

1433.   α-Bromobenzyl Cyanide. *α-Bromobenzeneacetonitrile;* α-bromophenylacetonitrile; α-bromo-α-tolunitrile; B.B.C.; C.A.; Camite. C_8H_6BrN; mol wt 196.05. C 49.01%, H 3.08%, Br 40.76%, N 7.14%. $C_6H_5CHBrCN$. The practical industrial prepn consists of three steps: (1) Chlorination of toluene to form benzyl chloride, (2) conversion of benzyl chloride to benzyl cyanide by the action of sodium cyanide in alcoholic soln, (3) bromination of the benzyl cyanide with bromine vapor in the presence of sunlight: Steinkopf *et al.*, *Ber.* **53**, 1146 (1920); Nekrassov, *J. Prakt. Chem.* **119**, 108 (1928).

Crystalline mass, mp 29°. Odor of soured fruit. bp_{760} 242° (dec); bp_{12} 132-134°. d_4^{29} 1.539. Vapor density 6.8 (air = 1). Vapor pressure at 20° = 0.012 mm; at 30° = 0.028 mm. Slightly sol in water. Freely sol in alcohol, ether, chloroform, acetone, and other common organic solvents; also sol in phosgene, chloropicrin, benzyl cyanide. Lethal concn: 0.90 mg/l (30 min.), A. M. Prentiss, *Chemicals in War* (McGraw-Hill, New York, 1937) p 141.

USE: War gas.

1434.   α-Bromobutyric Acid. *2-Bromobutanoic acid;* *dl-2-bromobutyric acid.* $C_4H_7BrO_2$; mol wt 167.00. C 28.77%, H 4.22%, Br 47.85%, O 19.16%. $CH_3CH_2CHBrCOOH$. Prepd by bromination of butyric acid: Naumann, *Ann.* **119**, 115 (1861); Stevens, Holland, *J. Org. Chem.* **18**, 1112 (1953); Smissman, *J. Am. Chem. Soc.* **76**, 5805 (1954). Toxicity study: J. L. Morrison, *J. Pharmacol. Exp. Ther.* **86**, 336 (1946).

Oily liquid, mp −4°. bp_{250} 181-182°, bp_{25} 127-128°. d_4^4 1.5855, d_{15}^{15} 1.5735, d_{20}^{20} 1.5669, d_{25}^{25} 1.5620: Perkin, *J. Chem. Soc.* **65**, 402 (1894). Boils with decompn at ordinary pressure. Sol in 15 parts water; sol in alcohol, ether. *Corrosive!* LD_{50} orally in mice: 310 mg/kg (Morrison).

Ethyl ester, $C_6H_{11}BrO_2$, *ethyl α-bromobutyrate.* Liquid, bp 177-178° with slight dec. d_{20}^{20} 1.329. Insol in water. Misc with alc, ether. *Protect from light. Lacrimator!*

1435. 3-Bromo-d-camphor. *3-Bromo-1,7,7-trimethylbicyclo[2.2.1]heptan-2-one; 3-bromo-d-2-bornanone.* $C_{10}H_{15}$-BrO; mol wt 231.13. C 51.97%, H 6.54%, Br 34.57%, O 6.92%. Of the two configurations found, the *endo*-form is more stable than the *exo*-form: Lowry *et al., J. Chem. Soc.* **121,** 633 (1922); Cookson, *ibid.* **1954,** 282. Prepn of the *endo*-form by bromination of *d*-camphor: Kipping, Pope, *ibid.* **63,** 548 (1893); *cf.* Woods, Roberts, *J. Org. Chem.* **22,** 1124 (1957). Prepn of *exo*-form by isomerization of *endo*-form: Lowry *et al., loc. cit.* Configuration: Wiebenga, Krom, *Rec. Trav. Chim.* **65,** 663 (1946); Cookson, *loc. cit.* *Review:* J. L. Simonsen, Ed., *The Terpenes* vol. II (University Press, Cambridge, 2nd ed., 1949), pp 401-404.

endo-form   exo-form

endo-Form, *α-bromo-d-camphor, 3α-bromo-d-camphor, bromated camphor, camphor monobromated.* Crystals from benzene, mp 76°. Camphor-like odor and taste. Discolors on prolonged exposure to light. d 1.449. $[\alpha]_D^{20}$ +122.7° (14.5 g/100 g benzene soln), Cutter *et al., J. Chem. Soc.* **127,** 1260 (1925). uv max (cyclohexane): 307.5 nm (log ε 1.98), Cookson, *loc. cit.* Sublimes, bp 274°. Almost insol in water: 1 g dissolves in 6.5 ml alcohol, 0.5 ml chloroform, 1.6 ml ether; sol in olive oil, slightly in glycerol.

When phenol, chloral hydrate, salol, menthol, or thymol is triturated with bromocamphor the mixture melts; these compds, however, are not incompatible.

exo-Form, *α′-bromo-d-camphor, 3β-bromo-d-camphor.* Needles from methanol or ethanol, mp 78.5°. d 1.484. $[\alpha]_D^{20}$ −42.1° (14.5 g/100 g benzene soln), Cutter *et al., loc. cit.* uv max (cyclohexane): 312 nm (log ε 1.95), Cookson, *loc. cit.*

THERAP CAT: Topical counterirritant.

1436. α-Bromo-*n*-caproic Acid. *2-Bromohexanoic acid.* $C_6H_{11}BrO_2$; mol wt 195.06. C 36.95%, H 5.68%, Br 40.96%, O 16.40%. $CH_3(CH_2)_3CHBrCOOH$. Prepd by heating bromine with *n*-caproic acid in the presence of phosphorus trichloride as catalyst: Clarke, Taylor, *Org. Syn.* **4,** 9 (1924). Resoln into optically active forms by means of the strychnine salt: Levene *et al., J. Biol. Chem.* **75,** 352 (1927); Levene, Mardashew, *ibid.* **117,** 707 (1937).

Liquid. bp_{760} 240°; bp_{30} 148-153°; bp_{17} 128-136°; bp_8 116-125°. Soluble in alcohol, ether.

Ethyl ester, $C_8H_{15}BrO_2$, liquid, odor of anise oil. bp_{760} 205-210°; bp_{32} 99-102°; bp_{11} 103°; bp_9 95-96°.

1437. Bromocriptine. *(5′α)-2-Bromo-12′-hydroxy-2′-(1-methylethyl)-5′-(2-methylpropyl)ergotaman-3′,6′,18-trione; 2-bromoergocryptine; 2-bromo-α-ergokryptin; CB-154.* C_{32}-$H_{40}BrN_5O_5$; mol wt 654.60. C 58.72%, H 6.16%, Br 12.21%, N 10.70%, O 12.22%. Dopamine receptor agonist; derivative of the ergotoxin group of ergot alkaloids. Prepn: E. Flückiger *et al.,* Ger. pat. **1,926,045;** *eidem,* U.S. pats. **3,752,814** and **3,752,888** (1969, 1973, 1973 all to Sandoz). Activity studies: Flückiger, Wagner, *Experientia* **24,** 1130 (1968); Yanai, Nagasawa, *J. Nat. Cancer Inst.* **45,** 1105 (1970); Stähelin *et al., Experientia* **27,** 915 (1971). Relationship of stereochemistry and biological activity: H. P. Weber, *Adv. Biochem. Psychopharmacol.* **23,** 25 (1980); N. Camerman, A. Camerman, *Mol. Pharmacol.* **19,** 517 (1981). Endocrine profile: Del Pozo *et al., Schweiz. Med. Wochenschr.* **103,** 847 (1973). As an immunomodulator: E. Nagy *et al., Immunopharmacology* **6,** 231 (1983). Use in obstetrics and gynecology: R. A. Kinch, *Fertil. Steril.* **33,** 463 (1980); in parkinsonism: R. C. Duvoisin *et al., Adv. Biochem. Psy-*

chopharmacol. **23,** 271 (1980); M. M. Hoen, *J. Am. Geriatric Soc.* **29,** 251 (1981); in treatment of pituitary tumors: R. F. Spark *et al., J. Am. Med. Assoc.* **247,** 311 (1982). Clinical studies in Parkinson's disease: M. Gawel *et al., Advan. Neurol.* **45,** 535 (1986); E. N. H. Jansen, J. D. Meerwaldt, *ibid.* 539. Review of pharmacology: Flückiger, *Triangle (Engl. Ed.)* **14,** 153-157 (1975); D. Parkes, *Advan. Drug Res.* **12,** 247-344 (1977); M. O. Thorner *et al., Bromocriptine: A Clinical and Pharmacological Review* (Raven Press, New York, 1980) 181 pp. Comprehensive description: D. A. Giron-Forest, W. D. Schönleber in *Analytical Profiles of Drug Substances* Vol. 8, K. Florey, Ed. (Academic Press, New York, 1979) pp 47-81. Review of therapeutic applications in endocrine and neurological diseases: K. Y. Ho, M. O. Thorner, *Drugs* **36,** 67-82 (1988).

Crystals from methyl ethyl ketone-isopropyl ether, mp 215-218° (dec). $[\alpha]_D^{20}$ −195° (c − 1 in methylene chloride). LD_{50} in rabbits (mg/kg): >1000 orally, 12.0 i.v. (Flückiger), U.S. pat. **3,752,814).**

Methanesulfonate, $C_{32}H_{40}BrN_5O_5 \cdot CH_3SO_3H$, *CB-154 mesylate, Bagren, Parlodel, Pravidel.* Crystals from methyl ethyl ketone, mp 192-196° (dec). $[\alpha]_D^{20}$ +95° (c − 1 in methanol-methylene chloride). Soly in mg/ml at 25°: methanol 910; ethanol 23.0; water 0.8; chloroform 0.45; benzene <0.1; hexane <0.1. pKa 4.90.

THERAP CAT: Prolactin inhibitor; antiparkinsonian.

1438. Bromodichloromethane. Dichlorobromomethane; BDCM; NCI-C55243. $CHBrCl_2$; mol wt 163.83. C 7.33%, H 0.62%, Br 48.77%, Cl 43.28%. $CHBrCl_2$. Trihalomethane contaminant found in chlorinated water. Prepn: O. Jacobsen, R. Neumeister, *Ber.* **15,** 599 (1882); M. Fedorynski *et al., Syn. Commun.* **7,** 287 (1977). Disposition and metabolism: J. M. Mathews *et al., J. Toxicol. Environ. Health* **30,** 15 (1990). Toxicity studies: F. J. Bowman *et al., Toxicol. Appl. Pharmacol.* **44,** 213 (1978); Y. Aida *et al., J. Toxicol. Sci.* **17,** 51 (1992). Review of quantitative analyses, toxicity and carcinogenicity: *IARC Monographs* **52,** 179-212 (1991).

Colorless liquid. bp 88.4-88.6° (Fedorynski); also reported as bp 91-92° (Jacobsen). d^{15} 1.9254. n_D^{20} 1.4967. LD_{50} in male, female mice (mg/kg): 450, 900 orally (Bowman).

Caution: Potential symptoms of acute oral overexposure in animals are liver and kidney damage; CNS depression, incoordination, sleepiness, lethargy, sedation. Chronic exposure has been associated with cancer of liver, kidney and intestines. *See: Toxicological Profile for Bromodichloromethane* (PB90-167461, 1989) 77 pp. This substance may reasonably be anticipated to be a carcinogen: *Seventh Annual Report on Carcinogens* (PB95-109781, 1994) p 103.

USE: Chemical reagent/intermediate in organic synthesis.

1439. Bromodiphenhydramine. *2-[(4-Bromophenyl)-phenylmethoxy]-N,N-dimethylethanamine; 2-(p-bromo-α-phenylbenzyloxy)-N,N-dimethylethylamine; β-dimethylaminoethyl p-bromobenzhydryl ether; bromdiphenhydramine; bromazine; bromanautine; Bromo-Benadryl; Deserol; Histabromamine.* $C_{17}H_{20}BrNO$; mol wt 334.26. C 61.09%, H 6.03%, Br 23.91%, N 4.19%, O 4.79%. Prepn: Rieveschl, U.S. pat. **2,527,963** (1950 to Parke, Davis).

Hydrochloride, $C_{17}H_{20}BrNO \cdot HCl$, *Ambodryl*. Crystals from isopropanol, mp 144-145°.

THERAP CAT: Antihistaminic.

1440. Bromofenofos. *3,3',5,5'-Tetrabromo-(1,1'-biphenyl)-2,2'-diol mono(dihydrogen phosphate); 3,3',5,5'-tetrabromo-2,2'-biphenyldiol mono(dihydrogen phosphate); 4,4',6,6'-tetrabromobiphenyl-2,2'-diol mono(dihydrogen phosphate);* bromophenophos; bromphenphos; PH-1882; Acedist. $C_{12}H_7Br_4O_5P$; mol wt 581.77. C 24.77%, H 1.21%, Br 54.94%, O 13.75%, P 5.32%. Prepn: **Neth. pat. Appl.** 6,505,635 corresp to S. Van der Meer *et al.*, U.S. pat. 3,482,016 and 3,662,035 (1966, 1969, 1972 to Chemiefarma); S. Van der Meer, H. Pouwels, *J. Med. Chem.* **12**, 534 (1969). Activity: J. Guilhon *et al.*, *Bull. Acad. Vet. Fr.* **43**, 67 (1970). Toxicology: J. M. Poul, M. Dagorn, *Recl. Med. Vet.* **158**, 363 (1982).

Cryst, no mp, dec > 350°.

THERAP CAT (VET): Anthelmintic (Trematodes).

1441. Bromoform. *Tribromomethane.* $CHBr_3$; mol wt 252.73. C 4.75%, H 0.40%, Br 94.85%. Prepd from acetone and sodium hypobromite: Günther, *Jahresber. Fortschr. Chem.* **1887**, 741 (*Beilstein,* **vol. 1**, 68); Kergomard, *Bull. Soc. Chim. France* **1961**, 2360. Toxicity data: Kutob, Plaa, *Toxicol. Appl. Pharmacol.* **4**, 354 (1962).

Heavy liquid; chloroform odor; sweetish taste. bp 149-150°. mp +7.5°. d_4^{15} 2.9035. n_D^{15} 1.6005. Sol in about 800 parts water; miscible with alc, benzene, chloroform, ether, petr ether, acetone, oils. Gradually dec, acquiring a yellow color; air and light accelerate the decompn. May be preserved by the addition of 3-4% alc. d 2.6-2.7. *Keep in well-closed containers, protected from light. Incompat.* Caustic alkalies. LD_{50} s.c. in mice: 7.2 mmol/kg (Kutob, Plaa).

Caution: Potential symptoms of overexposure are irritation of skin, eyes and respiratory system; central nervous system depression; liver damage. *See NIOSH Pocket Guide to Chemical Hazards* (DHHS/NIOSH 90-117, 1990) p 48. This substance may be habit forming and is listed in the U.S. Code of Federal Regulations, Title 21 Part 329.1 (1995).

USE: In separating mixtures of minerals.

THERAP CAT: Has been used as sedative, hypnotic; antitussive.

1442. α-Bromoisobutyric Acid. *2-Bromo-2-methylpropanoic acid.* $C_4H_7BrO_2$; mol wt 167.00. C 28.77%, H 4.22%, Br 47.85%, O 19.16%. $(CH_3)_2CBrCOOH$. Prepd by bromination of isobutyric acid: Markownikow, *Ann.* **153**, 228 (1870); Smissman, *J. Am. Chem. Soc.* **76**, 5805 (1954).

Crystals, mp 48-49°. bp 198-200°, bp_{20} 110-116°. d 1.52. Sparingly sol in cold water; sol in alcohol, ether. Dec by hot water into the hydroxy acid.

1443. α-Bromoisovaleric Acid. *2-Bromo-3-methylbutanoic acid.* $C_5H_9BrO_2$; mol wt 181.03. C 33.17%, H 5.01%, Br 44.14%, O 17.68%. $(CH_3)_2CHCHBrCOOH$. Prepn of *dl*-form: Ley, Popow, *Ann.* **174**, 61 (1874); Smissman, *J. Am. Chem. Soc.* **76**, 5805 (1954); Marvel, *Org. Syn.*

coll. vol. III, 848 (1955). Prepn of *d*- and *l*-forms: Berlingozzi, Furia, *Gazz. Chim. Ital.* **56**, 82 (1926); Berlingozzi, Lenoci, *ibid.* **68**, 721 (1938). Configuration of *d*-form: Brewster *et al., Nature* **166**, 179 (1950).

dl-Form, prisms from ether or chloroform, mp 44°. bp ~230° with some decompn. Sparingly sol in water; sol in alcohol, ether.

d-Form, crystals from petroleum ether, mp 43-44°. $[\alpha]_D^{20}$ +22.6° (c = 4 in benzene).

l-Form, crystals from petroleum ether, mp 43-44°. $[\alpha]_D^{20}$ −22.4° (c = 4 in benzene).

1444. β-Bromoisovaleric Acid. *3-Bromo-3-methylbutanoic acid.* $C_5H_9BrO_2$; mol wt 181.03. C 33.17%, H 5.01%, Br 44.14%, O 17.68%. $(CH_3)_2CBrCH_2COOH$. Prepn from β,β-dimethylacrylic acid and HBr: Auwers, *Ber.* **28**, 1130 (1895); from β-isovalerolactone and HBr: Gresham *et al., J. Am. Chem. Soc.* **76**, 486 (1954).

Needles from ligroin, mp 73-74°. Slightly sol in water; sol in alcohol, ether, benzene.

1445. Bromolysergide. *(8β)-2-Bromo-9,10-didehydro-N,N-diethyl-6-methylergoline-8-carboxamide;* D-2-bromolysergic acid diethylamide; 2-bromo-*N,N*-diethyl-D-lysergamide; bromo-LSD; BOL-148. $C_{20}H_{24}BrN_3O$; mol wt 402.33. C 59.71%, H 6.01%, Br 19.86%, N 10.44%, O 3.98%. Prepd by bromination of lysergide, *q.v.:* Troxler, Hofmann, *Helv. Chim. Acta* **40**, 2160 (1957). Serotonin antagonist without the hallucinogenic activity of LSD: J. P. Bennett, S. H. Snyder, *Brain Res.* **94**, 523 (1975).

Needles from ether, mp 120-127°. $[\alpha]_D^{20}$ +15° (c = 0.5 in pyridine); $[\alpha]_D^{20}$ +53° (c = 0.5 in chloroform). uv max: 240, 301 nm (log ε 4.28, 3.95).

1446. p-Bromomandelic Acid. *4-Bromo-α-hydroxybenzeneacetic acid;* p-bromophenylglycolic acid. $C_8H_7BrO_3$; mol wt 231.05. C 41.59%, H 3.05%, Br 34.58%, O 20.77%. Prepd by condensing bromobenzene with ethyl oxomalonate in presence of boron trifluoride: Riebsomer *et al., J. Am. Chem. Soc.* **60**, 2974 (1938).

Crystals, mp 117-118°. Slightly sol in water. Zirconium salt. Soly in water at 25°: 0.0446 g/l. USE: Analytical reagent for zirconium.

1447. 1-Bromonaphthalene. α-Bromonaphthalene. $C_{10}H_7Br$; mol wt 207.07. C 58.00%, H 3.41%, Br 38.59%. Prepd by dropping bromine into a mixture of naphthalene and carbon tetrachloride, distilling the CCl_4, heating the residue with NaOH, and fractionating the liquid under reduced pressure: Clarke, Schram, *Org. Syn.* **1**, 35 (1921); Clarke, Brethen, *ibid.* **coll. vol. I**, 121 (2nd ed., 1941). Also prepd by heating naphthalene (liquid or gaseous) with bromine; 2-bromonaphthalene also being produced, esp with increasing temps: Wibaut, *Chem. Weekbl.* **39**, 326, 328 (1942); Suyver, Wibaut, *Rec. Trav. Chim.* **64**, 65 (1945).

Oily liquid at room temp. More pungent odor than naphthalene. d_4^{20} 1.4834; d_4^{25} 1.4785; d_4^{30} 1.4732. When solid it exists in two forms: mp 0.2-0.7° and mp 6.2°. Darkens on standing when distilled at 760 mm, but remains colorless when distilled at 16 mm. Volatile with steam. bp_{760} 281.1°; bp_{400} 252.0°; bp_{200} 224.2°; bp_{100} 198.8°; bp_{60} 183.5°; bp_{20} 170.2°; bp_{20} 150.2°; bp_{10} 133.6°; bp_5 117.5°; $bp_{1.0}$ 84.2°. $n_D^{16.5}$ 1.66011. Slightly sol in water; miscible with alcohol, ether, benzene, chloroform. Absorption spectrum: DeLaszlo, *Proc. Roy. Soc. London* [A] **111**, 356 (1926).

Compd with 1,3,5-trinitrobenzene, $C_{10}H_7Br.C_6H_3N_3O_6$, lemon-yellow needles, mp 137°.

USE: Immersion fluid in the determination of the refractive index of crystals. For the determination of water in alc by the cloud point method. For refractometric fat determinations. Mixed with polymerized castor oil as a general immersion oil in microscopy.

1448. 2-Bromonaphthalene. *β-Bromonaphthalene.* $C_{10}H_7Br$; mol wt 207.07. C 58.00%, H 3.41%, Br 38.59%. Prepn: Liebermann, Palm, *Ann.* **183**, 267 (1876); Vingiello *et al., J. Chem. Ed.* **40**, 544 (1963).

Crystals from alcohol, mp 59°; also reported as mp 54-56° (Vingiello *et al., loc. cit.*). bp 281-282°, $bp_{4.5}$ 122-127°. d 1.60. Slightly sol in water; sol in 8 parts alcohol; very sol in ether, chloroform, benzene.

1449. p-Bromophenacyl Bromide. *2-Bromo-1-(4-bromophenyl)ethanone;* 2,4'-dibromoacetophenone; p,α-dibromoacetophenone. $C_8H_6Br_2O$; mol wt 277.94. C 34.57%, H 2.18%, Br 57.50%, O 5.76%. Prepd by adding bromine to p-bromoacetophenone in glacial acetic acid at below 20°: Langley, *Org. Syn. coll. vol. I,* 127 (2nd ed., 1941).

Crystals from ethanol, mp 109-110°. Very sol in warm alcohol.

USE: Identification of carboxylic acids.

1450. Bromophenol. C_6H_5BrO; mol wt 173.01. C 41.65%, H 2.91%, Br 46.18%, O 9.25%. Prepn of *m*-isomer: Wurster, Nölting, *Ber.* **7**, 904 (1874); Carpenter *et al., J. Org. Chem.* **16**, 586 (1951); of *o*- and *p*-isomers: Medola, Streatfield, *J. Chem. Soc.* **73**, 681 (1898); Kaeding, Lindstrom, U.S. pat. **2,805,263** (1957 to Dow).

m-Bromophenol. Crystals, mp 33°, also reported as mp 31°. bp 235-236°, bp_3 88-89°. Sol in alcohol, ether, alkalies.

o-Bromophenol. Yellow to red oily liquid; unpleasant odor. d ~1.5. bp 194°. mp 6°. Sol in water; miscible in chloroform, ether. Fusion with NaOH gives resorcinol.

p-Bromophenol. Tetragonal bipyramidal crystals from chloroform or ether. mp 64°. bp 238°. d^{15} 1.840; d^{80} 1.5875. Small amounts of water depress the mp considerably and may prevent crystallization. Absorption spectrum: Ley, *Z. Physik. Chem.* **94**, 412 (1920). Soluble in about 7 parts water; freely sol in alc, chloroform, ether, glacial acetic acid.

p-Form methyl ether, C_7H_7BrO, *p-bromoanisole.* Crystals, mp 9-10°. bp 223°.

p-Form ethyl ether, C_8H_9BrO, *p-bromophenetole.* Crystals, mp 4°. bp 233°.

USE: *p*-Bromophenol as disinfectant.

1451. p-Bromophenylhydrazine. $C_6H_7BrN_2$; mol wt 187.04. C 38.53%, H 3.77%, Br 42.72%, N 14.98%. Prepn from 4-bromophenyldiazonium chloride by modified Fischer phenylhydrazine synthesis: Chattaway, Humphrey, *J. Chem. Soc.* **1927**, 1323.

Needles from water, mp 108-109°. Sol in alcohol, ether, chloroform, benzene; moderately sol in ligroin; slightly sol in water.

USE: In prepn of indoleacetic acid derivatives; in the study of transosazonation of sugar phenylosazones.

1452. p-Bromophenyl Isocyanate. *1-Bromo-4-isocyanatobenzene;* 4-bromophenylcarbimide; p-bromophenylcarbonimide; p-bromocarbanil. C_7H_4BrNO; mol wt 198.02. C 42.46%, H 2.04%, Br 40.35%, N 7.07%, O 8.08%. Prepd by heating phenyl isocyanate dibromide: Curtius, *J. Prakt. Chem.* **87**, 517 (1913); by distilling p-bromophenylurethan and P_2O_5: Dennstedt, *Ber.* **13**, 228 (1880).

Needles, pungent odor, mp 42°. bp_{14} 158°. Very sol in ether.

USE: Prepn of bromophenylurea and urethan derivatives.

1453. Bromophos. *Phosphorothioic acid O-(4-bromo-2,5-dichlorophenyl) O,O-dimethyl ester;* O-(4-bromo-2,5-dichlorophenyl)-O,O-dimethylphosphorothioate; bromophosmethyl; ENT-27162; OMS-658; S-1942; Nexion. $C_8H_8BrCl_2O_3PS$; mol wt 366.00. C 26.25%, H 2.20%, Br 21.83%, Cl 19.37%, O 13.11%, P 8.46%, S 8.76%. Prepn: **Belg.** pat. **625,198**, *C.A.* **60**, 13187a (1964); R. Sehring, K. Zeile, U.S. pats. **3,227,610, 3,275,718** (1963, 1966, 1966 all to Boehringer, Ing.). Metabolism in rats: M. Stiasni *et al., J. Agr. Food Chem.* **15**, 474 (1967). Toxicity study: T. B. Gaines, *Toxicol. Appl. Pharmacol.* **14**, 515 (1969). Review: D. Eichler, *Residue Rev.* **41**, 65-112 (1972).

Yellow crystals, mp 53-54°. $bp_{0.01}$ 140-142°. Vapor pressure at 20°C: 1.3×10^{-4} mm Hg. Soly in water at 25°: 40 mg/l. Sol in CCl_4, ether, toluene. Stable in soln up to pH 9. Non-corrosive. LD_{50} in male, female rats (mg/kg): 1600, 1730 orally (Gaines).

O,O-Diethyl analog, $C_{10}H_{12}BrCl_2O_3PS$, *bromophos-ethyl, ENT-27258, OMS-659, S-2225, Nexagan.* Colorless to pale yellow oil, $bp_{0.001}$ 122-133°. Vapor pressure at 30°: 4.6×10^{-5} mm Hg. Soly in water at 25°: 2 mg/l.

USE: Insecticide, acaricide.

1454. Bromopride. *4-Amino-5-bromo-N-[2-(diethylamino)ethyl]-2-methoxybenzamide;* 4-amino-5-bromo-N-[2-(diethylamino)ethyl]-o-anisamide; Emepride; Emoril; Viadil. $C_{14}H_{22}BrN_3O_2$; mol wt 344.25. C 48.85%, H 6.44%, Br 23.21%, N 12.21%, O 9.30%. Prepn: M. L. Thominet, **Belg.** pat. **620,543;** *idem,* U.S. pat. **3,177,252** (1962, 1965 both to Soc. Etudes Sci. Ind. l'Ile de France); H. Mori, K. Shibata,

Ger. pat. **2,119,724** (1971 to Teikoku Hormone Manuf. Co.), *C.A.* **76**, 99375e (1972). Action on guinea-pig ileum: J. Fontaine, J. J. Reuse, *Arch. Int. Pharmacodyn. Ther.* **213**, 322 (1975). Effects on intestinal peristalsis in dogs: P. Sava *et al., Arzneimittel-Forsch.* **29**, 799 (1979). Clinical study: M. Fischer *et al., Fortschr. Med.* **97**, 883 (1979). Pharmacokinetics: P. W. Lücker *et al., Arzneimittel-Forsch.* **33**, 453 (1983).

Hydrochloride, $C_{14}H_{22}BrN_3O_2$.HCl, *Cascapride, Plesium, Praiden, Valopride, Viaben.*

Dihydrochloride monohydrate, $C_{14}H_{22}BrN_3O_2$.2HCl.H_2O, *Opridan.*

THERAP CAT: Antiemetic.

1455. β-Bromopropionic Acid. *3-Bromopropanoic acid.* $C_3H_5BrO_2$; mol wt 152.98. C 23.55%, H 3.29%, Br 52.23%, O 20.92%. $BrCH_2CH_2COOH$. Prepd by the action of hydrobromic acid on acrylic acid, on hydracrylic acid, and on ethylene cyanohydrin: Kendall, McKenzie, *Org. Syn.* **3**, 25 (1923). Prepn of the ethyl ester: *eidem, ibid.* 51; of the methyl ester: Mozingo, Patterson, *ibid.* **20**, 64 (1940). Plates from CCl_4. mp 62.5°. pK (25°): 4.00. Sol in water, alcohol, ether, chloroform, benzene. Aq alkalies hydrolyze β-bromopropionic acid to hydracrylic acid.

Ethyl ester, $C_5H_9BrO_2$, liquid. Pungent odor. Becomes yellow on exposure to light. d_4^{18} 1.4123. bp_{44} 112°; bp_{12} 70°. n_D^{18} 1.4569. Insol in water; miscible with alchol, ether. *Protect from light.*

Methyl ester, $C_4H_7BrO_2$, liquid. d^{15} 1.4897. bp_{27} 80°. bp_{18} 65°.

1456. Bromopropylate. *4-Bromo-α-(4-bromophenyl)-α-hydroxybenzeneacetic acid 1-methyl ethyl ester; 4,4'-dibromobenzilic acid isopropyl ester;* isopropyl 4,4'-dibromobenzilate; phenisobromolate; ENT-27552; GS-19851; Acarol; Folbex VA; Neoron. $C_{17}H_{16}Br_2O_3$; mol wt 428.12. C 47.69%, H 3.77%, Br 37.33%, O 11.21%. Contact acaricide with residual activity. Prepn: **Fr. pat. 1,504,969** corresp to K. Gubler, **U.S. pat. 3,639,446** (1967, 1972 to Geigy). Persistence in soil: W. B. Wheeler *et al., J. Environ. Qual.* **2**, 115 (1973).

Cryst solid, mp 77°. Vapor press at 20°: 5.1×10^{-8} mm Hg. Soly in water at 20°: < 5 mg/l. Readily sol in most organic solvents. LD_{50} orally in rats: 5000 mg/kg, *RTECS* vol. I, R. J.. Lewis, R. L. Tatken, Eds. (1980) p 276.

USE: Acaricide.

1457. Bromosalicylchloranilide. *5-Bromo-N-(4-chlorophenyl)-2-hydroxybenzamide; 5-bromo-4'-chlorosalicylanilide;* N-5-bromosalicyloyl-p-chloroaniline; Multifungin. $C_{13}H_9BrClNO_2$; mol wt 326.58. C 47.81%, H 2.78%, Br 24.47%, Cl 10.86%, N 4.29%, O 9.80%. Prepn: Schuler, **U.S. pat. 2,802,029** (1957 to Knoll); Lemaire *et al., J. Pharm. Soc.* **50**, 831 (1961).

Crystals, mp 238-243°.
THERAP CAT: Antifungal.

1458. 5-Bromosalicylhydroxamic Acid. *5-Bromo-N,2-dihydroxybenzamide;* Brosalamid; Bromocyl. $C_7H_6BrNO_3$; mol wt 232.03. C 36.23%, H 2.61%, Br 34.44%, N 6.04%, O 20.69%. Prepd by direct bromination of salicylhydroxamic acid in acetic acid: Urbanski *et al., Nature* **170**, 753 (1952); *Bull. Acad. Polon. Sci.,* cl. III, **1**, 319 (1953); Hornung, Krakowska, *Gruzlica* **20**, 469 (1952).

Crystals from alcohol, dec 232°. Very sparingly sol in water. Forms a water-soluble sodium salt.
THERAP CAT: Antibacterial (tuberculostatic).

1459. 5-Bromosalicylic Acid Acetate. *2-(Acetyloxy)-5-bromobenzoic acid;* acetyl-5-bromosalicylic acid; bromoaspirin. $C_9H_7BrO_4$; mol wt 259.06. C 41.73%, H 2.72%, Br 30.84%, O 24.70%. Prepn: Robertson, *J. Chem. Soc.* **81**, 1482 (1902).

Minute crystals from alcohol, mp 168-169°. Tastes slightly sour with sweet aftertaste. One gram dissolves in about 1500 ml water, 9 ml alcohol, 21 ml ether.
THERAP CAT: Analgesic.

1460. Bromosaligenin. *5-Bromo-2-hydroxybenzenemethanol; 5-bromo-2-hydroxybenzyl alcohol;* 5-bromosaligenin; Bromsalizol. $C_7H_7BrO_2$; mol wt 203.04. C 41.41%, H 3.48%, Br 39.35%, O 15.76%. Prepd by splitting bromosalicin with emulsin: Visser, *Arch. Pharm.* **235**, 551 (1897); by the action of bromine on saligenin in water: Auwers, Büttner, *Ann.* **302**, 138 (1898); Adams *et al., J. Am. Chem. Soc.* **45**, 2419 (1923); Dunning *et al., ibid.* **58**, 1567 (1936); by the electrolytic reduction of 5-bromo-2-hydroxybenzoic acid: Mettler, *Ber.* **39**, 2939 (1906).

Shiny, pearly leaflets, mp 109°. Solubility in water at 25°: 0.70% w/w; considerably more soluble in hot water. Soly in olive oil: 4.8% w/w; freely sol in alc, ether, ethyl acetate; moderately sol in chloroform, benzene.
THERAP CAT: Anti-inflammatory.

1461. Bromosuccinic Acid. *Bromobutanedioic acid;* monobromosuccinic acid. $C_4H_5BrO_4$; mol wt 196.99. C 24.39%, H 2.56%, Br 40.56%, O 32.49%. $HOOCCH_2CHBr$-COOH. Prepn of *dl*-form by bromination of succinic acid: Kekulé, *Ann.* **117**, 120 (1861); from fumaric acid and HBr: Fittig, *Ann.* **188**, 42 (1877); by bromination of succinyl bromide or succinic anhydride: Hughes, Watson, *J. Chem. Soc.* **1930**, 1733. Prepn of *l*-form from *l*-aspartic acid: Walden,

Ber. **29**, 133 (1896); Karrer *et al., Helv. Chim. Acta* **30**, 271 (1947).

dl-Form, crystals, mp 161°. d 2.07. Soluble in 5.5 parts water; sol in alc.

d-Form, crystals, mp 172°. $[\alpha]_D^{15}$ +41.9° (c = 5); $[\alpha]_D^{20}$ +67.9° (ether). *Ref:* Levene, Mikeska, *J. Biol. Chem.* **55**, 795 (1923); Clough, *J. Chem. Soc.* **129**, 1674 (1926).

l-Form, crystals, dec 177-178°. $[\alpha]_D$ −43.8° (c = 6), −65.0° (c = 6 in abs alc), −73.5° (c = 6 in acetone).

1462. N-Bromosuccinimide. *1-Bromo-2,5-pyrrolidinedione;* succinbromimide; NBS. $C_4H_4BrNO_2$; mol wt 177.99. C 26.99%, H 2.27%, Br 44.89%, N 7.87%, O 17.98%. Prepn: K. Ziegler *et al., Ann.* **551**, 109 (1942). In bromination of olefins: *eidem, ibid.* 80. The method has been extended to other classes of compds by the use of catalysts: H. Schmid, P. Karrer, *Helv. Chim. Acta* **29**, 573 (1946); H. Schmid, *ibid.* 1144. In oxidation of aldehydes: Y. F. Cheung, *Tetrahedron Letters* **1979**, 3809. Review of uses: C. Djerassi, *Chem. Rev.* **43**, 271-317 (1948); R. Filler, *ibid.* **63**, 21-43 (1963); J. S. Pizey, *Synthetic Reagents,* **vol. 2** (John Wiley, New York, 1974) pp 1-63.

Orthorhombic bisphenoidal crystals, mp 173-175° (slight decompn). Faint odor of bromine. d 2.098. Soly (g/100 g of solvent at 25°): water 1.47; *tert*-butanol 0.73; acetone 14.40; carbon tetrachloride 0.02; hexane 0.006; glacial acetic acid 3.10.

Caution: Highly irritating to eyes, skin, mucous membranes.

USE: In bromination of olefins; in oxidation of alcohols to aldehydes and ketones and of aldehydes to acid bromides.

1463. Bromotoluene. C_7H_7Br; mol wt 171.04. C 49.16%, H 4.13%, Br 46.72%. Prepn of *o*-isomer: Bourgeois, *Ber.* **28**, 2312 (1895). Prepn of *m*-isomer: Acree, *Ber.* **37**, 994 (1904). Kohn, Bum, *Monatsh.* **33**, 924 (1912); Feitler, *Z. Physik. Chem.* **4**, 77 (1889). Prepn of *m-, o-* and *p*-isomers: Bigelow *et al., Org. Syn.* **coll. vol. I** (2nd ed., 1941) pp 133, 135, 136 resp. Isomerization of *p*-isomer yields mixture of monobromotoluenes containing 48.2% of *o*-isomer: Crump, *U.S.* pat. **3,077,503** (1963 to Dow). Separation of *o*- and *p*-isomers by clathration: Coscia, *U.S.* pat. **3,114,784** (1963 to Am. Cyanamid).

m-Bromotoluene, 1-bromo-3-methylbenzene, 3-bromotoluene, m-tolylbromide. Liquid. d_4^{20} 1.4099; d_4^{58} 1.309; d_4^{184} 1.201. mp −39.8°. bp_{760} 183.7°; bp_{400} 160°; bp_{200} 138°; bp_{100} 117.8°; bp_{60} 104.1°; bp_{40} 93.9°; bp_{20} 78.1°; bp_{10} 64.0°; bp_5 50.8°; $bp_{1.0}$ 14.8°. n_D^{20} 1.551. Absorption spectrum: Purvis, *J. Chem. Soc.* **99**, 1706, 1710 (1911). Sol in alc, ether, benzene.

o-Bromotoluene, colorless liq. d_{15}^{15} 1.431. bp 181°. mp −26°. n_D^{20} 1.555. Practically insol in water; misc with alc, benzene.

p-Bromotoluene, crystals from abs alc. d_{35}^{35} 1.3959; d_{50}^{50} 1.3856; d_{100}^{100} 1.3637; d^{184} 1.1931. mp 28.5°. bp_{760} 184.5°; bp_{100} 116.4°; bp_{60} 102.3°; bp_{40} 91.8°; bp_{20} 75.2°; bp_{10} 61.1°; bp_5 47.5°; $bp_{1.0}$ 10.3°. n_D^{20} 1.5490. Sol in alc, ether, benzene.

Caution: Irritants!

1464. 5-Bromouracil. *5-Bromo-2,4(1H,3H)-pyrimidinedione.* $C_4H_3BrN_2O_2$; mol wt 190.98. C 25.16%, H 1.58%, Br 41.84%, N 14.67%, O 16.75%. Major chemical mutagen; incorporates into DNA, altering base-pair sequencing by replacing thymine. Prepn: H. L. Wheeler, H. F. Merriam, *Am. Chem. J.* **29**, 478 (1903); P. A. Levene, F. B. LaForge, *Ber.* **45**, 608 (1912). UV effects on DNA containing 5-bromouracil: A. G. Skavronskaya *et al., Mutat. Res.* **6**, 319 (1968); F. Hutchinson, *Quart. Rev. Biophys.* **6**, 201 (1973).

Mechanisms of mutagenesis: I. Pietrzykowska, *Mutat. Res.* **19**, 1 (1973); E. M. Witkin, E. C. Parisi, *ibid.* **25**, 407 (1974); B. Rydberg, *Molec. Gen. Genet.* **152**, 19 (1977). Crystal structure and base stacking properties: H. Sternglanz, C. E. Bugg, *Biochim. Biophys. Acta* **378**, 1 (1975).

Prisms from water, mp 293°.

USE: Exptly as mutagen.

1465. Bromoxynil. *3,5-Dibromo-4-hydroxybenzonitrile;* 3,5-dibromo-4-hydroxyphenyl cyanide; 2,6-dibromo-4-cyanophenol; broxynil; ENT-20852; MB-10064; Brominal. $C_7H_3Br_2NO$; mol wt 276.91. C 30.36%, H 1.09%, Br 57.71%, N 5.06%, O 5.78%. Selective contact herbicide. Prepn: K. Auwers, J. Reis, *Ber.* **29**, 2359 (1896); E. Müller *et al., Ber.* **92**, 2278 (1959); K. Carpenter *et al., Weed Res.* **4**, 175 (1964); *Fr.* pat. **1,375,311** (1964 to May & Baker), *C.A.* **62**, 3982h (1965). Herbicidal activity: K. Carpenter, B. J. Heywood, *Nature* **200**, 28 (1963). HPLC analysis: J. C. Van Damme, M. Galoux, *J. Chromatogr.* **190**, 401 (1980). Persistence in soil: A. E. Smith, *Pestic. Sci.* **11**, 341 (1980).

Colorless solid, mp 194-195°. pKa 4.06. Subl at 135° (0.15 mm Hg). Slightly volatile in steam. Soly in (g/l) at 25°: water 0.13; methanol 90; acetone 170; tetrahydrofuran 410. LD_{50} orally in mice: 110 mg/kg (Carpenter, 1964).

Octanoate ester, $C_{15}H_{17}Br_2NO_2$, *MB-10731, NPH-1320, RP-16272, Buctril.* Waxy solid, mp 45-46°. Low volatility; subl at 90° (0.1 mm Hg). Insol in water.

USE: Herbicide.

1466. Bromperidol. *4-[4-(4-Bromophenyl)-4-hydroxy-1-piperidinyl]-1-(4-fluorophenyl)-1-butanone; 4-[4-(p-bromophenyl)-4-hydroxypiperidino]-4'-fluorobutyrophenone;* R-11333; Azurene; Impromen; Tesoprel. $C_{21}H_{23}BrFNO_2$; mol wt 420.32. C 60.01%, H 5.52%, Br 19.01%, F 4.52%, N 3.33%, O 7.61%. Bromine analog of haloperidol, *q.v.* Prepn (no data): P. A. J. Janssen, **Brit.** pat. **895,309** corresp to **U.S.** pat. **3,438,991** (1962, 1969 both to Janssen); C. J. E. Niemegeers, P. A. J. Janssen, *Arzneimittel-Forsch.* **24**, 45 (1974). Prepn of [82]Br bromperidol and preliminary tissue distribution studies: S. H. Vincent *et al., J. Med. Chem.* **23**, 75 (1980). Pharmacological study: C. J. E. Niemegeers, P. A. J. Janssen, *Life Sci.* **24**, 2201 (1979). Preliminary clinical study: B. Woggon *et al., Int. Pharmacopsychiatry* **14**, 213 (1979). Radioimmunoassay: E. Van Den Eeckhout *et al., Eur. J. Drug Metab. Pharmacokinet.* **5**, 45 (1980).

Off-white amorphous or microcrystalline powder, mp 155-158°. uv max: 245 nm. pKa 8.6-8.7. Soly in water: 0.09 mg/ml; in 0.1*M* tartaric, lactic, citric and acetic acids: greater than or equal to 10 mg/ml.

THERAP CAT: Antipsychotic.

1467. Brompheniramine. *γ-(4-Bromophenyl)-N,N-dimethyl-2-pyridinepropanamine; 2-[p-bromo-α-(2-dimethyl-*

aminoethyl)benzyl]pyridine; 1-(*p*-bromophenyl)-1-(2-pyridyl)-3-dimethylaminopropane; 3-(*p*-bromophenyl)-3-(2-pyridyl)-*N,N*-dimethylpropylamine; parabromdylamine. $C_{16}H_{19}BrN_2$; mol wt 319.24. C 60.20%, H 6.00%, Br 25.03%, N 8.77%. Prepn from α-(*p*-bromophenyl)-α-(β-dimethylaminoethyl)-2-pyridylacetonitrile: Sperber *et al.,* U.S. pats. **2,567,245** (1951) and **2,676,964** (1954 to Schering). Prepn of *d*-form: L. A. Walter, U.S. pat. **3,061,517** (1962 to Schering).

Oily liquid. Slightly yellow color. Characteristic amine-like odor. $bp_{0.5}$ 147-152°. Soluble in dilute acids.

Maleate, $C_{16}H_{19}BrN_2 \cdot C_4H_4O_4$, *Dimegan, Dimetane, Dimotane, Ilvin, Nagemid, Symptom 3, Veltane.* Crystals, mp 132-134°. Sol in water, somewhat less sol in alcohol. pH of a 2% aq soln about 5.

d-Form, *dexbrompheniramine, (+)-parabromdylamine.* Oily liquid. $[\alpha]_D^{25}$ +42.7° (c = 1 in DMF).

d-Form maleate, $C_{16}H_{19}BrN_2 \cdot C_4H_4O_4$, *Disomer, Ebalin.* Crystals, mp 103-113°.

THERAP CAT: Antihistaminic.

1468. Bromphenol Blue. *4,4′-(3H-2,1-Benzoxathiol-3-ylidene)bis[2,6-dibromophenol] S,S-dioxide; α,α-bis(3,5-dibromo-4-hydroxyphenyl)-α-hydroxy-o-toluenesulfonic acid γ-sultone;* 3,3′,5,5′-tetrabromophenolsulfonphthalein; Albutest. $C_{19}H_{10}Br_4O_5S$; mol wt 669.97. C 34.06%, H 1.50%, Br 47.71%, O 11.94%, S 4.79%. Prepd by slow addition of excess bromine to a hot soln of phenolsulfonphthalein in glacial acetic acid: White, Acree, *J. Am. Chem. Soc.* **41**, 1205 (1919); Orndorff, Sherwood, *ibid.* **45**, 495 (1923). See *also* ref given *under* Bromcresol Green.

Elongated hexagonal prisms from acetic acid + acetone. Dec 279°. (An orange discoloration with formation of a green sublimate sets in at 210°). Soluble in water (about 0.4 g/100 ml); more sol in methyl and ethyl alcohol, and in benzene. Freely sol in NaOH solns with the formation of a water-sol sodium salt. pK = 4.0. To prepare a soln for use as pH indicator, dissolve 0.10 g in 7.45 ml *N*/50 NaOH and dil with water to 250 ml.

USE: As indicator, pH 3.0 yellow; pH 4.6 purple.

1469. Bromthymol Blue. *4,4′-(3H-2,1-Benzoxathiol-3-ylidene)bis[2-bromo-3-methyl-6-(1-methylethyl)phenol] S,S-dioxide; α,α-bis(6-bromo-5-hydroxycarvacryl)-α-hydroxy-o-toluenesulfonic acid γ-sultone;* 3,3′-dibromothymolsulfonphthalein. $C_{27}H_{28}Br_2O_5S$; mol wt 624.39. C 51.94%, H 4.52%, Br 25.59%, O 12.81%, S 5.14%. Prepd by the action of bromine on thymol blue in glacial acetic acid: Clark, Lubs, *J. Wash. Acad. Sci.* **5**, 609 (1915); **6**, 482 (1916); *Chem. Zentr.* **1916, I**, 175; **II**, 1068. See *also* ref *under* Bromcresol Green.

Cream-colored crystals. Sparingly sol in water; sol in alcohol and in aq solns of alkalies. Also sol in ether. Less sol in benzene, toluene, xylene. Practically insol in petr ether. pK = 7.0. To prepare a soln for use as pH indicator, dissolve 0.10 g in 8.0 ml *N*/50 NaOH and dil with water to 250 ml. To prepare a soln for use as indicator in volumetric work, dissolve 0.1 g in 100 ml of 50% alc.

USE: As indicator, pH 6.0 yellow; pH 7.6 blue.

1470. Bronopol. *2-Bromo-2-nitro-1,3-propanediol;* β-bromo-β-nitrotrimethyleneglycol; Bronosol. $C_3H_6BrNO_4$; mol wt 199.99. C 18.02%, H 3.02%, Br 39.95%, N 7.00%, O 32.00%. $HOCH_2CBr(NO_2)CH_2OH$. Prepn: E. Schmidt, R. Wilkendorf, *Ber.* **52**, 389 (1919); R. Wessendorf, **Ger.** pats. **1,804,068, 1,954,173** corresp to U.S. pats. **3,658,921, 3,711,-561** (1970, 1971, 1972, 1973 all to Henkel & Cie). Antibacterial, antifungal and toxic properties: Z. Eckstein *et al., Bull. Acad. Polon. Sci. Ser. Sci. Chim.* **11**, 687 (1963); B. Croshaw *et al., J. Pharm. Pharmacol.* **16** (Suppl), 127T (1964); N. G. Clark *et al.,* **Brit.** pat. **1,057,131** corresp to U.S. pat. **3,558,788** (1967, 1971 both to Boots); R. J. Stretton, T. W. Mason, *J. Appl. Bacteriol.* **36**, 61 (1973). Biotransformation and distribution in rats: H. S. Buttar, R. H. Downie, *Toxicol. Lett.* **6**, 101 (1980). As a nitrosating agent for diethanolamine: I. Schmeltz, A. Wenger, *Food Cosmet. Toxicol.* **17**, 105 (1979); R. L. Elder, *J. Environ. Pathol. Toxicol.* **4**, 47 (1980). Review of bronopol and other preservatives: B. Croshaw, *J. Soc. Cosmet. Chem.* **28**, 3-16 (1977).

Odorless crystals from ethyl acetate-chloroform, mp 120-122°. Sol in water, alcohol, ethyl acetate. Slightly sol in chloroform, acetone, ether, benzene. Insol in ligroin. LD_{50} in mice, rats (mg/kg): 350, 400 orally (Croshaw).

USE: Preservative in cosmetics and toiletries. Antiseptic.

1471. Broparoestrol. *1-(2-Bromo-1,2-diphenylethenyl)-4-ethylbenzene; 1-bromo-2-(p-ethylphenyl)-1,2-diphenylethylene;* 1-bromo-2-(4-ethylphenyl)-1,2-diphenylethene; α-bromo-α,β-diphenyl-β-(*p*-ethylphenyl)ethylene; 1-(*p*-ethylphenyl)-1,2-diphenyl-2-bromoethylene; B.D.P.E.; LN-107; Acnestrol; Longestrol. $C_{22}H_{19}Br$; mol wt 363.30. C 72.73%, H 5.27%, Br 21.99%. Prepn: Dvolaitsky, Jacques, *Bull. Soc. Chim. Biol.* **40**, 939 (1958).

cis - form

cis-trans Mixture, crystals, mp 89°. (Isomer A, needles from ethanol, mp 112-113.5°; isomer B, mp 111.5-112°). The commercial medicinal product is the *cis-trans* mixture. Soluble in ether, benzene, chloroform; less sol in ethanol.

THERAP CAT: Estrogen. Used in dermatology.

1472. Brotizolam. *2-Bromo-4-(2-chlorophenyl)-9-methyl-6H-thieno[3,2-f][1,2,4]triazolo[4,3-a][1,4]diazepine;* 2-bromo-4-(*o*-chlorophenyl)-9-methyl-6*H*-thieno[3,2-f]-*s*-triazolo-[4,3-*a*][1,4]diazepine; 8-bromo-6-(*o*-chlorophenyl)-1-methyl-4*H-s*-triazolo[3,4-*c*]thieno[2,3-*e*]-1,4-diazepine; WE-941-BS; Lendorm; Lendormin; Mederantil; Sintonal. $C_{15}H_{10}BrCl$-

N_4S; mol wt 393.69. C 45.76%, H 2.56%, Br 20.30%, Cl 9.01%, N 14.23%, S 8.14%. One of a class of triazolo-1,4-thienodiazepines having psychotropic activity. Prepn: K. H. Weber *et al.*, **Ger. pat. 2,410,030**; *eidem*, **U.S. pat. 4,094,-984** (1975, 1978 to Boehringer, Ing.); *eidem*, *Ann.* **8**, 1257 (1978). Pharmacodynamics: J. Gruenberger *et al.*, *Curr. Ther. Res.* **24**, 427 (1978). Bioavailability: B. Saletu *et al.*, *Arzneimittel-Forsch.* **29**, 700 (1979). Pharmaco-EEG study: M. Fink, P. Irwin, *Clin. Pharmacol. Ther.* **30**, 336 (1981). Toxicity studies: C. Hewett *et al.*, *Arzneimittel-Forsch.* **36**, 592 (1986). Series of articles on pharmacology, pharmacokinetics, toxicology: *ibid.* 517-620. Veterinary trial as appetite stimulant: A. van Miert *et al.*, *J. Vet. Pharmacol. Ther.* **12**, 147 (1989). Review of pharmacology and clinical efficacy: M. S. Langley, S. P. Clissold, *Drugs* **35**, 104-122 (1988).

Colorless crystals from ethanol, mp 212-214°. LD_{50} in mice, rats (mg/kg): > 10000, > 10000 orally; 920, 1000 i.p. (Hewett).

THERAP CAT: Sedative, hypnotic.

THERAP CAT (VET): Appetite stimulant.

1473. Brovincamine. *(3α,14β,16α)-11-Bromo-14,15-dihydro-14-hydroxyeburnamenine-14-carboxylic acid methyl ester; cis-*11-bromovincamine. $C_{21}H_{25}BrN_2O_3$; mol wt 433.35. C 58.21%, H 5.81%, Br 18.44%, N 6.46%, O 11.08%. Cerebrovascular agent, vincamine deriv. Prepn: P. Pfaeffli, **Ger. pat. 2,458,164** corresp to **U.S. pat. 4,146,643** (1975, 1979 both to Sandoz). HPLC determn in plasma: R. R. Brodie, L. F. Chasseaud, *J. Chromatog.* **228**, 413 (1982). Effect on autonomic nervous system in laboratory animals: K. Kushiku *et al.*, *J. Pharmacobio.-Dyn.* **7**, 177 (1984). Cardiovascular effects, mechanistic studies: *eidem*, *Clin. Exp. Pharmacol. Physiol.* **12**, 121 (1985). Mode of action studies: T. Katsuragi *et al.*, *Gen. Pharmacol.* **15**, 43 (1984). Clinical trials in patients with multi-infarct dementia: S. Hagstadius *et al.*, *Psychopharmacology (Berlin)* **83**, 321 (1984).

Crystals from isopropanol, mp 214° (dec). $[\alpha]_D^{20}$ +8.7° (1% in $CHCl_3$).

Hydrogen fumarate, $C_{21}H_{25}BrN_2O_3 \cdot C_4H_4O_4$, *BV-26-723*, *Sabromin, Zabromin.* mp 144°. $[\alpha]_D^{20}$ +4.7 (0.388% in H_2O).

THERAP CAT: Vasodilator (peripheral).

1474. Broxyquinoline. *5,7-Dibromo-8-quinolinol;* 5,7-dibromo-8-hydroxyquinoline; Brodiar; Broxykinolin; Colepur; Fenilor; Intensopan. $C_9H_5Br_2NO$; mol wt 302.95. C 35.68%, H 1.66%, Br 52.75%, N 4.62%, O 5.28%. Prepn by bromination of 5-formyl-8-quinolinol: Matsumura, Ito, *J. Am. Chem. Soc.* **77**, 6671 (1955); by bromination of 8-hydroxyquinaldine: Irving, Pinnington, *J. Chem. Soc.* **1957**, 285; by bromination of 8-quinolinol: Luis, Palomo, *Afinidad* **28**, 163 (1951), *C.A.* **47**, 10533d (1953); Zinnei, Fiedler, *Arch. Pharm.* **291**, 493 (1958); Aristov, Kostina, *Zh. Obshch. Khim.* **34**, 3421 (1964). Crystal structure: Kashino, Haisa, *Bull. Chem. Soc. Japan* **46**, 1094 (1973). Metabo-

lism: Rodriguez, Close, *Biochem. Pharmacol.* **17**, 1647 (1968).

Monoclinic needles from alc. d 2.189. mp 196°. Freely sol in chloroform, alcohol, benzene, acetic acid; slightly sol in ether. Practically insol in water.

USE: In testing for Cu, Fe, Ti.

THERAP CAT: Antiseptic; disinfectant.

1475. Bruceantin. *[11β,12α,15β(E)]-15-[(3,4-Dimethyl-1-oxo-2-pentenyl)oxy]-13,20-epoxy-3,11,12-trihydroxy-2,16-dioxopicras-3-en-21-oic acid methyl ester;* NSC-165563. $C_{28}H_{36}O_{11}$; mol wt 548.59. C 61.30%, H 6.61%, O 32.08%. Antileukemic quassinoid from the simaroubaceous tree *Brucea antidysenterica* J. F. Mill. Isoln and structure: S. M. Kupchan, **Ger. pat. 2,347,576**; S. M. Kupchan, R. W. Britton, **U.S. pat. 3,969,369** (1975, 1976 both to Research Corp.); S. M. Kupchan *et al.*, *J. Org. Chem.* **40**, 648 (1975). Mode of action: L. L. Liao *et al.*, *Mol. Pharmacol.* **12**, 167 (1976). *In vivo* study: R. K. Johnson *et al.*, *Cancer Treat. Rep.* **62**, 1535 (1978). Pharmacology: S. M. Sieber *et al.*, *ibid.* **60**, 1127 (1976); M. Fresno *et al.*, *Biochim. Biophys. Acta* **518**, 104 (1978). Clinical study: A. Y. Bedikian *et al.*, *Proc. Am. Assoc. Cancer Res.* **20**, 193 (1979). Toxicologic evaluation: T. R. Castles *et al.*, *U.S. NTIS Rep.* **PB-257175** (1976) 348 pp. Synthetic studies: O. D. Dailey, P. L. Fuchs, *J. Org. Chem.* **45**, 216 (1980); R. J. Pariza, P. L. Fuchs, *ibid.* **48**, 2306 (1983). Total synthesis: M. Sasaki *et al.*, *ibid.* **55**, 528 (1990).

Crystals from ether, mp 225-226°. $[\alpha]_D^{25}$ −43° (c = 0.31 in pyridine). uv max (ethanol): 280, 221 nm (ε 8680, 18000); (ethanol, NaOH): 328, 221 nm (ε 7290, 28600). LD_{50} in male, female mice (mg/kg): 1.95, 2.58 i.v. (Castles).

1476. Brucine. *2,3-Dimethoxystrychnidin-10-one;* 10,-11-dimethoxystrychnine. $C_{23}H_{26}N_2O_4$; mol wt 394.47. C 70.03%, H 6.64%, N 7.10%, O 16.22%. From Strychnos seeds *(Strychnos nux-vomica* L. and *S. ignatii* Berg., *Loganiaceae)*: Hartwick, Geiger, *Arch. Pharm.* **239**, 491 (1901). Structure: Findlay, *J. Am. Chem. Soc.* **73**, 3008 (1951). Prepn from strychnine, *q.v.*: E.Tedeschi *et al.*, *Tetrahedron* **24**, 4573 (1968); P. Rosenmund *et al.*, *Ber.* **101**, 2754 (1968). ^{13}C-NMR study: Wenkert *et al.*, *J. Org. Chem.* **43**, 1099 (1978). *Review:* H. L. Holmes in R. H. F. Manske, H. L. Holmes, *The Alkaloids* vol. **I** (Academic Press, New York, 1950) pp 377-420.

Needles from acetone + water, mp 178°. $[\alpha]_D$ −127° (chloroform), −85° (in abs alcohol). uv max (ethanol): 263, 301 nm (log ε 4.09, 3.93).

Tetrahydrate, monoclinic prisms. Also forms a dihydrate. Very bitter taste. Bitterness threshold 1:220,000. *Very poisonous!* Becomes anhydr at 100°. One gram dissolves in 0.8 ml methanol, 1.3 ml alcohol, 5 ml chloroform, 25 ml ethyl acetate, 36 ml glycerol, about 100 ml benzene, 187 ml ether, 1320 ml water, 750 ml boiling water. pH of satd water soln 9.5. pK_1 6.04, pK_2 11.7. *Caution:* Handle dry powder in hood only. LD_{50} orally in rats: 1 mg/kg, *RTECS* Vol. 1, R. J. Lewis, R. L. Tatken, Eds. (1979) p 324.

Hydrochloride, $C_{23}H_{26}N_2O_4 \cdot HCl$, crystals. Sol in water or alcohol. The solns are neutral or slightly acid.

Nitrate dihydrate, $C_{23}H_{26}N_2O_4 \cdot HNO_3 \cdot 2H_2O$, prisms, dec 230°. Sol in water or alcohol.

Sulfate heptahydrate, $(C_{23}H_{26}N_2O_4)_2 \cdot H_2SO_4 \cdot 7H_2O$, small, bitter crystals. One gram dissolves in 75 ml cold water, about 10 ml boiling water, 105 ml alcohol, 170 ml chloroform. Soly in water is decreased by H_2SO_4.

Caution: A highly toxic alkaloid resembling strychnine.

USE: Denaturing alcohol and oils; in analytical chemistry; for separating racemic mixtures. Has been patented as addition agent to lubricants.

THERAP CAT: CNS stimulant.

1477. Bryonia. Bryony. Dried root of *Bryonia alba* L., or of *B. dioica* Jacq., *Cucurbitaceae*. *Habit.* Europe. *Constit.* Bryonin, bryonidin, bryonicine, bryoamarid glycoside, Δ^7-stigmastenol, volatile oil, resin. Isoln of constituents: Tunmann, Wienecke, *Arch. Pharm.* **293**, 195 (1960); Biglino, *Farmaco Ed. Sci.* **14**, 673 (1959). Bryogenine structure studies: Biglino *et al., Tetrahedron Letters* **1963**, 1651. *Poisonous!*

THERAP CAT: Cathartic.

THERAP CAT (VET): Formerly as a purgative.

1478. Bryostatins. Family of seventeen biologically active macrolides, bryostatins 1 through 15, A and B. Isolated from the marine bryozoans, *Bugula neritina*, L. and related organisms. Protein kinase C activators. Isoln and structure of bryostatin 1, one of the most abundant of the series: G. R. Pettit *et al., J. Am. Chem. Soc.* **104**, 6846 (1982). Large-scale isoln of 1: D. E. Schaufelberger *et al., J. Nat. Prod.* **54**, 1265 (1991). Abs config of 1, 2 and 3: G. R. Pettit *et al., J. Org. Chem.* **56**, 1337 (1991). ^{1}H- and ^{13}C-NMR analyses of 1: D. E. Schaufelberger *et al., Magn. Reson. Chem.* **29**, 366 (1991). Approaches to synthesis: R. Roy *et al., Chem. Commun.,* **1989**, 1308; K. Ohmori *et al., Tetrahedron Letters* **34**, 4981 (1993). Total synthesis of 7: M. Kageyama *et al., J. Am. Chem. Soc.* **112**, 7407 (1990). Conversion of 2 to 1: G. R. Pettit *et al., Can. J. Chem.* **69**, 856 (1991). Pharmacodynamics: R. L. Berkow *et al., Cancer Res.* **53**, 2810 (1993); C. Stanwell *et al., Int. J. Cancer* **56**, 585 (1994). Clinical evaluation of 1 in advanced cancer: J. Prendiville *et al., Brit. J. Cancer* **68**, 418 (1993). Review of pharmacology: P. M. Blumberg, G. R. Pettit, "The Bryostatins, a Family of Protein Kinase C Activators with Therapeutic Potential", in *New Leads and Targets in Drug Research*, P. Krogsgaard-Larsen *et al.,* Eds. (Munksgaard, Copenhagen, 1992) pp 273-285. Review of isolation, struc-

Bryostatin 1

ture elucidation and biological activity: G. R. Pettit, *Fortschr. Chem. Org. Naturst.* **57**, 153-195 (1991).

Bryostatin 1, $C_{47}H_{68}O_{17}$. Colorless crystals from methylene chloride/methanol, mp 230-235°. $[\alpha]_D^{25}$ +34.1° (c = 0.044 in methanol). uv max (methanol): 233, 263 nm (ϵ 25700, 28700). Relatively insol in water. Soly in ethanol >4000 μg/ml. LD_{50} in mice, rats (mg/kg): 0.038, 0.068 i.v. (Prendiville).

1479. Bucetin. *N-(4-Ethoxyphenyl)-3-hydroxybutanamide; 3-hydroxy-p-butyrophenetidide;* β*-hydroxybutyric acid p-phenetidide; p-ethoxy-N-(*β*-hydroxybutyryl)aniline;* Betadid. $C_{12}H_{17}NO_3$; mol wt 223.27. C 64.55%, H 7.67%, N 6.27%, O 21.50%. Phenetidine derivative with analgesic and antipyretic activity. Prepn: G. Ehrhart *et al.,* U.S. pat. **2,830,087** (1958 to Hoechst); K. Kummerle *et al.,* Belg. pat. **852,340**; *eidem,* U.S. pat. **4,128,578** (1977, 1978 both to Hoechst). Analgesic activity and toxicity data: G. Doll, E. Hackenthal, *Arzneimittel-Forsch.* **13**, 68 (1963); H. Fujimura, *Folia Pharmacol. Japan.* **62**, 123 (1966). Prepn and pharmacology: G. Ehrhart *et al., ibid.* **15**, 727 (1965). Metabolism in rabbits: J. Shibasaki *et al., Chem. Pharm. Bull.* **16**, 1726, 2269 (1968). Synergistic analgesic and antipyretic effect with ethenzamide, *q.v.:* H. Fujimura, K. Shinozaki, U.S. pat. **3,284,298** (1966); H. Kojima *et al., Oyo Yakuri* **16**, 549 (1978), *C.A.* **90**, 81035y (1979); H. Kojimi, T. Sakurai, **Japan.** pat. **23,132** (1979 to Daiichi), *C.A.* **90**, 210147n (1979).

Crystals from isopropanol, mp 160°. Sparingly sol in water. LD_{50} in mice (mg/kg): 790 i.p., 2800 orally (Fujimura).

Note: Component in combination analgesics containing ethenzamide, caffeine and vitamin B_1, e.g. *Butylon, Bucetalon.*

THERAP CAT: Analgesic.

1480. Buchu. Bucco; bucku; buku. Dried leaves of *Barosma betulina* Bartl. & Wendl. (short buchu) or of *B. crenulata* (Linné) Hooker (oval buchu), or of *B. seratifolia* Willd. (long buchu), *Rutaceae.* *Habit.* Southern Africa (Cape of Good Hope). *Constit.* Diosphenol (barosma camphor), diosmin, bitter extractive, resin, *l*-menthone, mucilage, hesperidin, 1-2% volatile oil. *Review:* Feldman, Youngken, *J. Am. Pharm. Assoc.* **33**, 277 (1944).

THERAP CAT: Antiseptic (urinary).

THERAP CAT (VET): Has been used as a urinary antiseptic, diuretic.

1481. Bucillamine. *N-(2-Mercapto-2-methyl-1-oxopropyl)-L-cysteine; N-(2-mercapto-2-methylpropanoyl)-L-cysteine; N-(2-mercaptoisobutyryl)-L-cysteine; tiobutarit; thiobutarit;* DE-019; SA-96; Rimatil. $C_7H_{13}NO_3S_2$; mol wt 223.32. C 37.65%, H 5.87%, N 6.27%, O 21.49%, S 28.72%. Slow acting antirheumatic drug which modulates the immune response; structurally related to tiopronin, *q.v.* Prepn: T. Fujita *et al.,* Ger. pat. **2,709,820**; *eidem,* U.S. pat. **4,305,958** (1977, 1981 both to Santen); M. Oya *et al., Chem. Pharm. Bull.* **29**, 940 (1981). General pharmacology: T. Iso *et al., Oyo Yakuri* **25**, 123 (1983), *C.A.* **98**, 119191b (1983). Stimulation of lymphocyte proliferation *in vitro:* I. Yamamoto *et al., Int. J. Tissue React.* **4**, 1 (1982). Effect on macrophage function *in vitro* and *ex vivo:* M. Hayashi *et al., Int. J. Immunopharmacol.* **8**, 299 (1986). Clinical evaluations in rheumatoid arthritis: K. Ishikawa, M. Sakaguchi, *Scand. J. Rheumatol.* **15**, 85 (1986); M. Goto *et al., Int. J. Immunopharmacol.* **11**, 327 (1989).

Crystals from ethyl acetate, mp 139-140°. $[\alpha]_D^{25}$ +32.3° (c = 1.0 in ethanol). LD_{50} in mice (mg/kg): 2285 i.p.; 989.6 i.v. (Fujita, 1981).

THERAP CAT: Antirheumatic.

1482. Buckminsterfullerene. *[5,6]Fullerene-C₆₀-Ih;* buckyball; soccerballene. C_{60}; mol wt 720.66. Named after R. Buckminster Fuller, the American engineer, whose geodesic domes possess the same symmetry as the C_{60} truncated icosohedron typified by a soccerball. The most abundant of a class of carbon molecules characterized by a hollow cage structure and known as the *fullerenes*. One of the allotropic forms of carbon, *q.v.*, and the only molecular form, these molecules, described by C_{20} + 2m, form stable closed nets composed of 12 pentagons and 'm' hexagons. Prepn: H. W. Kroto *et al., Nature* **318,** 162 (1985); in macroscopic quantities: W. Krätschmer *et al., ibid.* **347,** 354 (1990). Identification in shungite, a carbonaceous rock: P. R. Buseck *et al., Science* **257,** 215 (1992). Review of early work: R. F. Curl, R. E. Smalley, *Sci. Am.* **265,** 54-63 (1991). Review of synthesis, properties, chemistry including metal-doping: *Acc. Chem. Res.* **25,** 98-175 (1992); *Fullerenes,* G. S. Hammond, V. J. Kuck, Eds. *ACS Symposium Series* **481,** 1-195 (1992).

1483. Bucladesine. *N-(1-Oxobutyl)adenosine cyclic 3′,-5′-(hydrogen phosphate) 2′-butanoate; N-(9-β-D-ribofuranosyl-9H-purin-6-yl)butyramide cyclic 3′,5′-(hydrogen phosphate) 2′-butyrate; N⁶,2′-O-dibutyryladenosine 3′,5′-cyclic mono-phosphate; N⁶,2′-O-dibutyryl cAMP; DBcAMP;* $C_{18}H_{24}N_5$-O_8P; mol wt 469.39. C 46.06%, H 5.15%, N 14.92%, O 27.27%, P 6.60%. Vasodilating cyclic nucleotide derivative which can permeate the cell membrane. Mimics the action of endogenous cyclic AMP, *q.v.* Prepn: Th. Posternak *et al., Biochim. Biophys. Acta* **65,** 558 (1962); O. Nagase *et al., Japan. Kokai* 76 113,896, *C.A.* **86,** 140409r (1977); T. Hirayama *et al.,* **Japan. Kokai 77 39,698, 77 39,699,** *C.A.* **87,** 136330m-136331n (1977) (1976, 1977, 1977 all to Daiichi Seiyaku). Pharmacology: H. S. White *et al., Eur. J. Pharmacol.* **57,** 107 (1979); J. D. Johnson *et al., ibid.* **91,** 343 (1983); N. Bondar *et al., J. Physiol.* **355,** 33 (1984). Clinical trial in congestive heart failure: S. Matsui *et al., Am. J. Cardiol.* **51,** 1364 (1983); after cardiopulmonary bypass surgery: T. Yoshitake *et al., Prog. Clin. Biol. Res.* **111,** 211 (1983).

Sodium salt, $C_{18}H_{23}N_5NaO_8P$, *DC-2797, Actosin.*

Barium salt, $C_{36}H_{46}BaN_{10}O_{16}P_2$. uv max (ethanol-0.5M ammonium acetate, 5:2) 270 nm.

THERAP CAT: Cardiotonic.

1484. Buclizine. *1-[(4-Chlorophenyl)phenylmethyl]-4-[[4-(1,1-dimethylethyl)phenyl]methyl]piperazine; 1-(p-tert-butylbenzyl)-4-(p-chloro-α-phenylbenzyl)piperazine; 1-(p-*chlorobenzhydryl)-4-(p-tert-butylbenzyl)diethylenediamine; 1-(p-tert-butylbenzyl)-4-(p-chlorodiphenylmethyl)piperazine; histabutyzine; histabutizine; UCB-4445; Buclifen; Longifene; Posdel; Postafen; Vibazine. $C_{28}H_{33}ClN_2$; mol wt 433.04. C 77.66%, H 7.68%, Cl 8.19%, N 6.47%. Prepn: Morren, U.S. pat. **2,709,169** (1955 to U.C.B.), *see also* **Ger. pat. 964,040,** Lui *et al., C.A.* **62,** 2776a (1965). Clinical evaluation as antiemetic in migraine: P. B. Jorgensen *et al., Curr. Ther. Res.* **16,** 1276 (1974); E. I. Adam, *J. Int. Med. Res.* **15,** 71 (1987).

$bp_{0.001}$ 217-220°.

Dihydrochloride, $C_{28}H_{33}ClN_2·2HCl$, *Buclina, Softran.* mp 230-240°.

THERAP CAT: Antiemetic.

1485. Buclosamide. *N-Butyl-4-chloro-2-hydroxybenzamide; N-butyl-4-chlorosalicylamide;* 4-chloro-2-hydroxy-benzoic acid *n*-butylamide. $C_{11}H_{14}ClNO_2$; mol wt 227.69. C 58.03%, H 6.20%, Cl 15.57%, N 6.15%, O 14.05%. Prepn: Ruschig *et al.,* U.S. pat. **2,923,737** (1960 to Hoechst).

Crystals from chloroform, mp 90-92°. Combination with salicylic acid, *Jadit.*

THERAP CAT: Antifungal.

1486. Bucloxic Acid. *3-Chloro-4-cyclohexyl-α-oxobenzenebutanoic acid; 3-(3-chloro-4-cyclohexylbenzoyl)propionic acid;* 4-(4-cyclohexyl-3-chlorophenyl)-4-oxobutyric acid; bucloxonic acid. $C_{16}H_{19}ClO_3$; mol wt 294.78. C 65.19%, H 6.50%, Cl 12.03%, O 16.28%. Prepn: Krausz, Brunaud, **Ger. pat. 2,021,445** (1970 to Clin-Byla), *C.A.* **75,** 5521d (1971) corresp to **Brit. pat. 1,315,542;** Krausz *et al., Arznei-mittel-Forsch* **24,** 1360, 1364 (1974). Review of pharmacology, metabolism, toxicity: *ibid.* 1368-1444.

Crystals from acetone, mp 163°. uv max (ethanol): 255 nm (ε 15,500). LD_{50} in mice, rats: 900, 120 mg/kg orally; 1100, 210 mg/kg i.p.

Calcium salt, $C_{32}H_{36}CaCl_2O_6$, *CB-804, Esfar.* LD_{50} in mice, rats: 1700, 175 mg/kg orally; 1700, 200 mg/kg i.p.

THERAP CAT: Anti-inflammatory.

1487. Bucolome. *5-Butyl-1-cyclohexyl-2,4,6(1H,3H,-5H)pyrimidinetrione; 5-butyl-1-cyclohexylbarbituric acid;* 5-*n*-butyl-1-cyclohexyl-2,4,6-trioxoperhydropyrimidine; BCP; Paramidin. $C_{14}H_{22}N_2O_3$; mol wt 266.34. C 63.14%, H 8.33%, N 10.52%, O 18.02%. Prepn: Senda *et al.,* U.S. pats.

3,243,344 and 3,274,195 (both 1966 to Takeda). Metabolic studies in man: Yashiki *et al., Chem. Pharm. Bull.* **19**, 468, 869 (1971).

Needles from methanol, mp 84°. $bp_{0.8}$ 185-187°.
THERAP CAT: Anti-inflammatory.

1488. Bucrylate. *2-Cyano-2-propenoic acid 2-methylpropyl ester; 2-cyanoacrylic acid isobutyl ester;* isobutyl cyanoacrylate; bucrilate; IBC; IBCA. $C_8H_{11}NO_2$; mol wt 153.18. C 62.73%, H 7.24%, N 9.14%, O 20.89%. Prepn: A. E. Ardis, U.S. pat. 2,467,926 (1949 to B. F. Goodrich). Use as tissue adhesive in surgery: H. W. Coover, U.S. pat. 2,794,788 (1957 to Eastman-Kodak); Belg. pat. 636,286 (1963 to Eastman-Kodak), *C.A.* **62**, 1811b (1965). Exptl use in aneurysms: P. H. Zanetti, F. E. Sherman, *J. Neurosurg.* **36**, 72 (1972); R. P. Leather *et al., Arch. Surg.* **114**, 1402 (1979); in cerebral arteriovenous malformations: L. D. Cromwell, A. B. Harris, *J. Neurosurg.* **52**, 705 (1980).

Liquid, $bp_{2.0}$ 170°. Polymerizes in <1 second on contact with ionic solns, e.g. saline or blood.
USE: Surgical aid (tissue adhesive).

1489. Bucumolol. *8-[3-[(1,1-Dimethylethyl)amino]-2-hydroxypropoxy]-5-methyl-2H-1-benzopyran-2-one; (±)-8-[3-(t-butylamino)-2-hydroxypropoxy]-5-methylcoumarin; dl-8-(2-hydroxy-3-tert-butylaminopropoxy)coumarin. $C_{17}H_{23}-NO_4$; mol wt 305.37. C 66.86%, H 7.59%, N 4.59%, O 20.96%. β-Adrenergic blocker. Prepn: Y. Sato *et al., Ger. pat.* 2,021,958; *eidem,* U.S. pat. 3,663,570 (1970, 1972 both to Sankyo); *eidem, Chem. Pharm. Bull.* **20**, 905 (1972). Pharmacology: T. Oshima *et al., Japan. J. Pharmacol.* **23**, 497 (1973). Metabolic studies: R. Hayashi *et al., Chem. Pharm. Bull.* **23**, 1173 (1975). Antiarrhythmic activity in dogs: K. Nakayama *et al., Japan. J. Pharmacol.* **29**, 935 (1979). Early clinical study: H. Kishida *et al., Rinsho Yakuri* **11**, 99 (1980), *C.A.* **93**, 107380w (1980).

Hydrochloride, $C_{17}H_{23}NO_4 \cdot HCl$, *CS-359, Bucumarol.* Colorless crystals from isopropanol, mp 226-228° (dec). LD_{50} in male, female mice (mg/kg): 676, 692 orally; 33.1, 31.6 i.v. (Oshima).
THERAP CAT: Antianginal, antiarrhythmic.

1490. Budesonide. *(11β,16α)-16,17-[Butylidenebis-(oxy)]-11,21-dihydroxypregna-1,4-diene-3,20-dione; (R,S)-11β,16α,17,21-tetrahydroxypregna-1,4-diene-3,20-dione cyclic 16,17-acetal with butyraldehyde;* S-1320; Bidien; Budeson; Cortivent; Micronyl; Preferid; Pulmicort; Rhinocort; Spirocort. $C_{25}H_{34}O_6$; mol wt 430.54. C 69.74%, H 7.96%, O 22.30%. Non-halogenated glucocorticoid related to triamcinolone hexacetonide, *q.v.* with a high ratio of topical to systemic activity. Prepn: R. L. Brattsand *et al., Ger. pat.* 2,323,215; *eidem,* U.S. pat. 3,929,768 (1973, 1975 both to Bofors). Synthesis and anti-inflammatory properties: A. Thalén, R. L. Brattsand, *Arzneimittel-Forsch.* **29**, 1787

(1979). Pharmacokinetic study: A. Ryrfeldt *et al., J. Steroid Biochem.* **10**, 317 (1979). HPLC determn of epimers, impurities, content: G. Roth *et al., J. Pharm. Sci.* **69**, 766 (1980). Review of pharmacodynamics and efficacy in asthma and rhinitis: S. P. Clissold, R. C. Heel, *Drugs* **28**, 485-518 (1984).

Crystals, mp 221-232° (dec). Mixture of two isomers; the content of the *S*-isomer in the mixture varies between 40-51%. $[\alpha]_D^{25}$ +98.9° (c = 0.28 in methylene chloride).
THERAP CAT: Anti-inflammatory.

1491. Budipine. *1-(1,1-Dimethylethyl)-4,4-diphenylpiperidine;* 1-tert-butyl-4,4-diphenylpiperidine. $C_{21}H_{27}N$; mol wt 293.45. C 85.95%, H 9.27%, N 4.77%. Centrally acting deriv of diphenylpiperidine. Prepn: H. G. Menge, J. Klosa, Ger. pat. 1,936,452; *eidem,* U.S. pat. 4,016,280 (1969, 1977 both to Byk Gulden). Pharmacological studies: U. Brand, H. G. Menge, *Arzneimittel-Forsch.* **30**, 1242 (1980); M. Eltze, ibid. 1129; H. G. Menge, U. Brand, *ibid.* **32**, 85 (1982). Comparison to other drugs in treatment of dyskinesias: J. Siegfried, *Can. J. Neurol. Sci.* **6**, 89 (1979). Clinical evaluation with levodopa in Parkinson's disease: K. Jellinger, H. Bliesath, *J. Neurol.* **234**, 280 (1987).

Hydrochloride, $C_{21}H_{27}N \cdot HCl$, *Parkinsan.* LD_{50} in male mice, rats (mg/kg): 120, 165 orally; 33, 28 i.v. (Menge, Brand).
THERAP CAT: Antiparkinsonian.

1492. Budralazine. *1(2H)-Phthalazinone (1,3-dimethyl-2-butenylidene)hydrazone;* 4-methyl-3-penten-2-one (1-phthalazinyl)hydrazone; mesityl oxide (1-phthalazinyl)hydrazone; DJ-1461; Buterazine. $C_{14}H_{16}N_4$; mol wt 240.31. C 69.97%, H 6.71%, N 23.31%. Deriv of hydralazine, *q.v.* Prepn: K. Ueno *et al., Ger. pat.* 2,145,359; *eidem,* U.S. pat. 3,840,539 (1972, 1974 both to Daiichi Seiyaku); *eidem, Chem. Pharm. Bull.* **24**, 1068 (1976). Pharmacology: A. Akashi *et al., Eur. J. Pharmacol.* **29**, 161 (1974); *eidem, Arch. Int. Pharmacodyn. Ther.* **235**, 134 (1978). Metabolism: R. Moroi *et al., Chem. Pharm. Bull.* **24**, 2850 (1976); K. Ono *et al., Xenobiotica* **9**, 227 (1979). Toxicity study: T. Onodera *et al., Toxicol. Appl. Pharmacol.* **44**, 431 (1978). Reproduction studies: *Oyo Yakuri* **21**, 313-350 (1981), *C.A.* **95**, 73764y-7b (1981). Series of articles on cardiovascular action, α-adrenergic effects, antihypertensive and general pharmacological properties: *Arzneimittel-Forsch.* **31**, 1072-1087 (1981). Characterization of vasodilator action: T. Chiba *et al., ibid.* **33**, 112 (1983).

Crystals from isopropyl ether, mp 132-133°. uv max (methanol): 208, 240, 289, 357 nm (ϵ 27000, 89000, 20000,

15000). LD$_{50}$ in mice, rats (mg/kg): 1820, 620 orally; 4020, 3570 i.p. (Onodera).

THERAP CAT: Antihypertensive.

1493. Bufalin. *(3β,5β)-3,14-Dihydroxybufa-20,22-dienolide.* C$_{24}$H$_{34}$O$_4$; mol wt 386.53. C 74.58%, H 8.87%, O 16.56%. Cardiotonic steroid constituent of Ch'an Su (also called Senso), a galenical preparation of the dried venom of the Chinese toad, *Bufo bufo gargarizans.* Isoln and characterization: M. Kotake, K. Kuwada, *Sci. Papers Inst. Phys. Chem. Res. (Tokyo)* **36**, 106 (1939), *C.A.* **33**, 7304^9 (1939); K. Meyer, *Helv. Chim. Acta* **32**, 1238 (1949). Structure: K. Kuwada, *J. Chem. Soc. Japan* **60**, 335 (1939), *C.A.* **35**, 51123^9 (1941). Prepn from resibufogenin *(q.v.):* H. Kondo, S. Ohno, U.S. pat. 3,134,772 (1964); from digitoxigenin *(q.v.):* G. R. Pettit *et al.,* U.S. pat. 3,687,944 (1972). Total synthesis: F. Sondheimer *et al., J. Am. Chem. Soc.* **91**, 1228 (1969). Additional syntheses: G. R. Pettit *et al., J. Org. Chem.* **35**, 2895 (1970); Y. Kamano, G. R. Pettit, *ibid.* **38**, 222 (1973); E. Yoshii *et al., Chem. Pharm. Bull.* **25**, 2249 (1977); K. Wiesner *et al., Helv. Chim. Acta* **66**, 2632 (1983). Pharmacology: S. Yoshida, T. Sakai, *Japan. J. Pharmacol.* **23**, 859 (1973); **24**, 97 (1974).

Needles from methanol/chloroform, mp 242-243°. [α]$_D$ —20°. uv max (ethanol): 298 nm (log ε 3.77).

Sodium sulfate, C$_{24}$H$_{33}$NaO$_7$S, *sodium bufalin-3-sulfate.* Isoln from Japanese toad *Bufo vulgaris formosus:* K. Shimada *et al., Tetrahedron Letters* **1974**, 2767. Colorless amorphous solid, mp 165.5-166.5°. [α]$_D^{23}$ —33.1° (c = 0.05 in methanol).

1494. Bufencarb. *3-(1-Ethylpropyl)phenol methylcarbamate mixture with 3-(1-methylbutyl)phenyl methylcarbamate (1:3); methylcarbamic acid m-(1-ethylpropyl)phenyl ester mixt. with m-(1-methylbutyl)phenyl ester;* metalkamate; Ortho 5353; ENT-27127; Bux. C$_{13}$H$_{19}$NO$_2$; mol wt 221.30. C 70.56%, H 8.65%, N 6.33%, O 14.46%. Prepn: Ospenson *et al.,* U.S. pats. 3,062,864 and 3,062,867 (1962 to Calif. Res. Corp.); Brit. pat. 1,106,326 (1968 to Chevron). Acute toxicity: T. B. Gaines, R. E. Linder, *Fundam. Appl. Toxicol.* **7**, 299 (1986).

R = CH(CH$_3$)CH$_2$CH$_2$CH$_3$ and CH(CH$_2$CH$_3$)$_2$

Yellow-amber solid, mp 26-39°. d^{26} 1.024. bp$_{0.04}$ 125°. Very sol in xylene, methanol. Soly in water <50 ppm. Stable in neutral or acidic solns; increases in pH or temp increases rate of hydrolysis. Fairly rapidly degraded in the soil. LD$_{50}$ in male, female rats (mg/kg): 97, 61 orally (Gaines, Linder).

Caution: Cholinesterase inhibitor. Absorbed through the skin: *Clinical Toxicology of Commercial Products,* R. E. Gosselin *et al.,* Eds. (Williams & Wilkins, Baltimore, 5th ed., 1984) Section II, p 304.

USE: Insecticide.

1495. Bufeniode. *4-Hydroxy-3,5-diiodo-α-[1-[(1-methyl-3-phenylpropyl)amino]ethyl]benzenemethanol;* 1-(4-hydroxy-3,5-diiodophenyl)-2-(1-methyl-3-phenylpropylamino)propanol; diiodobuphenine; Diastal; Proclival. C$_{19}$H$_{23}$I$_2$NO$_2$; mol wt 551.21. C 41.40%, H 4.21%, I 46.05%, N 2.54%, O 5.81%. Prepn: J. P. Fourneau, J. M. Delourme, S. Afr. pat. 68 00,046 corresp to U.S. pat. 3,542,870 (1968, 1970 to Lab. Houdé). Metabolism: Veyre *et al., C.R. Soc. Biol.* **163**, 136 (1969).

Crystals, mp (slow heating) 185° (dec), mp (fast heating) 212°. LD$_{50}$ in mice: >600 mg/kg i.p., >2 g/kg orally (Fourneau, Delourme).

THERAP CAT: Antihypertensive; vasodilator (peripheral).

1496. Bufetolol. *1-[(1,1-Dimethylethyl)amino]-3-[2-[(tetrahydro-2-furanyl)methoxy]phenoxy]-2-propanol;* 1-(tert-butylamino)-3-[o-[(tetrahydrofurfuryl)oxy]phenoxy]-2-propanol. C$_{18}$H$_{29}$NO$_4$; mol wt 323.43. C 66.84%, H 9.04%, N 4.33%, O 19.79%. β-Adrenergic blocker. Prepn: M. Nakanishi *et al.,* Ger. pat. 2,024,001; *eidem,* U.S. pat. 3,723,476 (1970, 1973, both to Yoshitomi); *eidem, J. Med. Chem.* **15**, 45 (1972). Pharmacology: M. Nakanishi *et al., Yakugaku Zasshi* **91**, 1037 (1971); *ibid.* **92**, 375 (1972); Imamura *et al., ibid.* 1039. Metabolism: M. Nakanishi *et al., ibid.* 299; *Oyo Yakuri* **6**, 479 (1972), *C.A.* **78**, 66785b (1973). Toxicology: *eidem, ibid.* 485, 1267, *C.A.* **78**, 66945d; **79**, 337w (1973).

bp$_{0.07}$ 180-186°.

Hydrochloride, C$_{18}$H$_{29}$NO$_4$·HCl, *Y-6124, Adobiol.* White crystals from methanol-isopropanol, mp 153.5°-157°; also reported as mp 151-154°. (Diastereoisomer, mp 118°). Freely sol in water, methanol, glacial acetic acid. Very slightly sol in benzene. Practically insol in ether. Unstable on exposure to light. Average LD$_{50}$ in mice, rats (mg/kg): 409, 1142 orally; 507, 1904 s.c., *Japan. Med. Gaz.* **11**(7), 4 (1974).

THERAP CAT: Antiarrhythmic; antianginal.

1497. Bufexamac. *4-Butoxy-N-hydroxybenzeneacetamide;* 2-(p-butoxyphenyl)acetohydroxamic acid; 2-[p-(butyloxy)phenyl]acetohydroxamic acid; CP-1044-J3; Droxarol; Droxaryl; Feximac; Malipuran; Mofenar; Norfemac; Parfenac; Parfenal. C$_{12}$H$_{17}$NO$_3$; mol wt 223.27. C 64.55%, H 7.67%, N 6.27%, O 21.50%. Synthesis: Buu-Hoï *et al., Compt. Rend.* **261**, 2259 (1965); *eidem,* Belg. pat. 661,226 corresp to U.S. pat. 3,479,396 (1965, 1969 to Madan). Pharmacology and toxicology: Lambelin *et al., Med. Pharmacol. Exp.* **15**, 545 (1966).

Needles from acetone, mp 153-155°. Practically insol in water. LD$_{50}$ orally in mice, rats: >8, >4 g/kg (Lambelin).

THERAP CAT: Anti-inflammatory, analgesic, antipyretic.

1498. Buflomedil. *4-(1-Pyrrolidinyl)-1-(2,4,6-trimethoxyphenyl)-1-butanone;* 2′,4′,6′-trimethoxy-4-(1-pyrrolidinyl)butyrophenone; (2,4,6-trimethoxyphenyl)-(3-pyrrolidinopropyl) ketone. C$_{17}$H$_{25}$NO$_4$; mol wt 307.39. C 66.43%, H 8.20%, N 4.56%, O 20.82%. Prepn: L. Lafon, Ger. pat. 2,122,144; *idem,* U.S. pat. 3,895,030 (both 1971 to Orsymonde). Pharmacology: J. Duteil *et al., Therapie* **30**, 207 (1975); C. Debray *et al., ibid.* 259. HPLC determn: J. A. Badmin *et al., J. Chromatog.* **172**, 319 (1979). Clinical pharmacokinetics: U. Gundert-Remy *et al., Eur. J. Clin. Pharmacol.* **20**, 459 (1981). Comparative safety and efficacy study: G. Rosas *et al., Angiology* **32**, 291 (1981). Review of pharmacology and therapeutic effects: S. P. Clissold *et al., Drugs* **33**, 430-460 (1987).

Hydrochloride, $C_{17}H_{25}NO_4 \cdot HCl$, *LL-1656, Bufedil, Buflan, Buflocit, Buflonat, Fonzylane, Irrodan, Lofton, Loftyl, Provas.* White crystals from isopropanol, mp 192-193°. LD_{50} in mice (mg/kg): 80 ±4.6 i.v. (Lafon).
THERAP CAT: Vasodilator (peripheral).

1499. Bufogenin B. *(3β,5β,16β)-3,14,16-Trihydroxybufa-20,22-dienolide;* desacetylbufotalin. $C_{24}H_{34}O_5$; mol wt 402.53. C 71.61%, H 8.51%, O 19.87%. Isolated from the Chinese drug Ch'an Su which is prepd from Chinese toads *(Bufo asiaticus = Bufo gargarizans* Cantor); also obtained from bufotalin, *q.v.:* Ruckstuhl, Meyer, *Helv. Chim. Acta* **40**, 1270 (1957).

Elongated prisms from methanol, begins to sinter at 195°, dec 210-223°. $[\alpha]_D^{19}$ +30° (c = 1.039 in dioxane). Very sparingly sol in chloroform, methanol, acetone.

1500. Buformin. *N-Butylimidodicarbonimidic diamide; 1-butylbiguanide; n*-butylbiguanide; butyldiguanide; butformin; W-37. $C_6H_{15}N_5$; mol wt 157.22. C 45.84%, H 9.62%, N 44.55%. Prepn: Shapiro *et al., J. Am. Chem. Soc.* **81**, 3728 (1959); Shapiro, Freedman, U.S. pat. **2,961,377** (1960 to USV). Toxicity study: Rikimaru *et al., J. Antibiot.* **18A**, 196 (1965).

Strong base. Very sol in water.
Hydrochloride, $C_6H_{15}N_5 \cdot HCl$, *Andere, Biforon, Bigunal, Bufonamin, Bulbonin, Diabrin, Dibetos, Gliporal, Insulamin, Krebon, Panformin, Silubin, Sindiatil, Tidemol, Ziavetine.* Crystals, mp 174-177°. Freely sol in water, alcohol. LD_{50} i.p. in mice: 380 mg/kg (Rikimaru).
Nitrate, $C_6H_{15}N_5 \cdot HNO_3$, crystals from ethanol, dec 125-126°.
THERAP CAT: Antidiabetic.

1501. Bufotalin. *(3β,5β,16β)-16-(Acetyloxy)-3,14-dihydroxybufa-20,22-dienolide; 3β,14,16β-trihydroxy-5β-bufa-20,22-dienolide 16-acetate.* $C_{26}H_{36}O_6$; mol wt 444.57. C 70.24%, H 8.16%, O 21.59%. One of the genins found in the venom of the common European toad *(Bufo vulgaris).* Isoln and structure: Faust, *Arch. Exp. Pathol. Pharmakol.* **47**, 279 (1902); **49**, 1 (1902); Wieland, Weil, *Ber.* **46**, 3315 (1913); Wieland, Weyland, *Sitzungsber. Bayr. Akad. Wiss.* (math.-

physikal. Klasse) **1920**, 329; Wieland *et al., Ann.* **524**, 203 (1936); Meyer, *Helv. Chim. Acta* **32**, 1993 (1949); Pettit *et al., Chem. Commun.* **1970**, 1566. Synthesis: G. R. Pettit *et al., J. Org. Chem.* **52**, 3573 (1987).

Solvated crystals from alc. Sinters at 154°, dec 223°. Sublimes in high vacuum at 225-230°. uv max: 300 nm. $[\alpha]_D^{20}$ +5.4° (c = 0.5 in chloroform). Sol in alcohol, chloroform.
3-Acetate, mp 269-272°.

1502. Bufotenine. *3-[2-(Dimethylamino)ethyl]-1H-indol-5-ol;* 3-(2-dimethylaminoethyl)-5-indolol; 5-hydroxy-*N,N*-dimethyltryptamine; *N,N*-dimethylserotonin; 3-(β-dimethylaminoethyl)-5-hydroxyindole; mappine. $C_{12}H_{16}N_2O$; mol wt 204.27. C 70.56%, H 7.89%, N 13.71%, O 7.83%. Isoln from toads: Wieland *et al., Ann.* **513**, 1 (1934); Wieland, Wieland, *ibid.* **528**, 239 (1937); from toadstools: Wieland, Motzel, *ibid.* **581**, 10 (1953). Isoln from *Piptadenia peregrina* Benth., *Leguminosae:* Stromberg, *J. Am. Chem. Soc.* **76**, 1707 (1954). Synthesis: Hoshino, Shimodaira, *Ann.* **520**, 19 (1935); Harley-Mason, Jackson, *Chem. & Ind. (London)* **1952**, 954; *see also* Serotonin; and Speetor, U.S. pat. **2,708,197** (1955 to Upjohn); Stoll *et al., Helv. Chim. Acta* **38**, 1452 (1955). Activity: Bhattacharya, Sanyal, *Indian J. Physiol. Pharmacol.* **15**, 133 (1971). Crystal and molecular structure: G. Falkenberg, *Acta Crystallogr.* **28B**, 3219 (1972).

Stout prisms from ethyl acetate, mp 146-147°. $bp_{0.1}$ 320°. uv max: 220, 265 nm (log ε 4.0, 3.7). Almost insol in water. Freely sol in alcohol, less sol in ether. Sol in dil acids and alkalies.
Methyliodide, $C_{13}H_{19}IN_2O$, stout prisms from methanol, dec 214-215°.
Note: This is a controlled substance (hallucinogen) listed in the U.S. Code of Federal Regulations, Title 21 Part 1308.11 (1995).

1503. Bufotoxin. *[3β(S),5β,16β]-16-(Acetyloxy)-3-[[8-[[4-[(aminoiminomethyl)amino]-1-carboxybutyl]amino]-1,8-dioxooctyl]oxy]-14-hydroxybufa-20,22-dienolide;* vulgarobufotoxin; bufotalin 3-suberoylarginine ester. $C_{40}H_{60}N_4O_{10}$; mol wt 756.94. C 63.47%, H 7.99%, N 7.40%, O 21.14%. Principal toxin of the venom of the common European toad, *Bufo vulgaris.* Isoln and structure: Wieland, Alles, *Ber.* **55**, 1789 (1922); Wieland *et al., Ann.* **524**, 203 (1936); Meyer, *Helv. Chim. Acta* **32**, 1993 (1949); Urscheler *et al., ibid.* **38**, 883 (1955). Alternate structure: Kamano *et al., Tetrahedron Letters* **1968**, 5673; Linde-Tempel, *Helv. Chim. Acta* **53**, 2188 (1970). Partial synthesis: Pettit, Kamano, *Chem. Commun.* **1972**, 45.

Monohydrate, bitter needles from alc, dec 205°. uv max: 295 nm (log ϵ 3.74). Freely sol in methanol, pyridine, sparingly sol in abs alc. Practically insol in water, ether, acetone, chloroform, petr ether.

1504. Bufuralol. α-[[(1,1-Dimethylethyl)amino]methyl]-7-ethyl-2-benzofuranmethanol; α-[(tert-butylamino)methyl]-7-ethyl-2-benzofuranmethanol; 2-(2-tert-butylamino-1-hydroxyethyl)-7-ethylbenzofuran; 1-(7-ethylbenzofuran-2-yl)-2-tert-butylamino-1-hydroxyethane. $C_{16}H_{23}NO_2$; mol wt 261.36. C 73.53%, H 8.87%, N 5.36%, O 12.24%. β-Adrenergic blocker with peripheral vasodilating activity. Prepn: **Neth. pat. Appl. 6,606,441;** G. A. Fothergill et al., **U.S. pat. 3,929,836** (1966, 1975 both to Hoffmann-La Roche). Prepn and resolution of isomers: eidem, Experientia **31**, 1322 (1975); eidem, Arzneimittel-Forsch. **27**, 981 (1977). Pharmacology: T. C. Hamilton, M. W. Parkes, ibid. 1410. HPLC determn in plasma: P. Haefelfinger, J. Chromatog. **221**, 327 (1980). Metabolism of isomers and racemate: R. J. Francis et al., Eur. J. Clin. Pharmacol. **23**, 529 (1982). Pharmacokinetics in hypertensive patients: M. Eckert et al., ibid. **24**, 479 (1983). Hemodynamic effects in patients with angina: M. Pfisterer et al., J. Cardiovasc. Pharmacol. **6**, 417 (1984).

Hydrochloride, $C_{16}H_{23}NO_2$.HCl, Ro-3-4787, Angium. Fine, white powder from acetone, mp 146°. LD_{50} in mice (mg/kg): 29.7 i.v.; 88.0 i.p.; 177 orally; in rats (mg/kg): 1400 s.c.; 750 orally (Hamilton, Parkes).

(+)-Hydrochloride, crystals from ethyl acetate-ether, mp 122-123°, $[\alpha]_{365}^{20}$ +135.0° (c = 1.0 in ethanol).

(−)-Hydrochloride, crystals from ethyl acetate-ether, mp 122-123°, $[\alpha]_{365}^{20}$ −136.0° (c = 1.0 in ethanol).

THERAP CAT: Antianginal; antihypertensive.

1505. Bulan®. 1,1'-(2-Nitrobutylidene)bis[4-chlorobenzene]; 1,1-bis(p-chlorophenyl)-2-nitrobutane; 2-nitro-1,1-bis-(p-chlorophenyl)butane. $C_{16}H_{15}Cl_2NO_2$; mol wt 324.21. C 59.28%, H 4.66%, Cl 21.87%, N 4.32%, O 9.87%. Prepn: Hass, Blickenstaff, **U.S. pat. 2,516,186** (1950 to Purdue Res. Found.). Toxicity study: Lehman, Assoc. Food Drug Officials U.S. Quart. Bull. **14**, 82 (1951), C.A. **45**, 3517h (1951).

LD_{50} orally in rats: 375 mg/kg (Lehman).
Combination with 2-nitro-1,1-bis(p-chlorophenyl)propane, **Dilan.**
USE: Insecticide.

1506. Bulbocapnine. (S)-6,7,7a,8-Tetrahydro-11-methoxy-7-methyl-5H-benzo[g]-1,3-benzodioxolo[6,5,4-de]quinolin-12-ol; 10-methoxy-1,2-(methylenedioxy)-6aα-aporphin-11-ol. $C_{19}H_{19}NO_4$; mol wt 325.36. C 70.14%, H 5.89%, N 4.30%, O 19.67%. From root of Corydalis cava (L.) Schweigg. & Körte (C. tuberosa DC), Fumariaceae and

Dicentra canadensis Walp., Papaveraceae. Isoln: Freund, Josephi, Ber. **25**, 2411 (1892); Manske, Can. J. Res. **7**, 258 (1932). Structure: Gadamer, Chem. Ztg. **34**, 1004 (1910). Configuration: Ayer, Taylor, J. Chem. Soc. **1956**, 472; Corrodi, Hardegger, Helv. Chim. Acta **39**, 889 (1956). Synthesis: Kikkawa, C.A. **54**, 4649b (1960). Peripheral-dopamine-receptor blocking activity: R. G. Pendleton et al., Arch. Pharmacol. **289**, 171 (1975).

dl-Form, columns, mp 213-214°. Absorption max: Brustier, C.A. **49**, 12127b (1955).

l-Form, columns, mp 202-203°. $[\alpha]_D^{22}$ −239°.

d-Form, columns, mp 201-203°. $[\alpha]_D^{22}$ +231°. Practically insol in water, sol in alcohol, chloroform. LD_{50} s.c. in mice: 195 mg/kg.

Methyl ether, $C_{20}H_{21}NO_4$, crystals from petr ether, mp 129°. $[\alpha]_D^{20}$ +260°. Absorption max: Girardet, J. Chem. Soc. **1931**, 2630.

1507. Bumadizon. Butylpropanedioic acid mono(1,2-diphenylhydrazide); butylmalonic acid mono(1,2-diphenylhydrazide); N-(2-carboxycaproyl)hydrazobenzene; α-carboxycaproyl-N,N'-diphenylhydrazine. $C_{19}H_{22}N_2O_3$; mol wt 326.40. C 69.92%, H 6.79%, N 8.58%, O 14.71%. Prepn: **Neth. pat. Appl. 6,406,412;** Pfister et al., **U.S. pat. 3,455,999** (1964, 1969 to Geigy); **Neth. pat. Appl. 6,600,685** (1966 to Byk-Gulden). Main product of the hydrolysis of phenylbutazone, q.v.: Schmid, Helv. Chim. Acta **53**, 2239 (1970). Chemical characterization: Pawelczyk, Wachowiak, Acta Pol. Pharm. **26**, 433 (1969), C.A. **72**, 136355p (1970). Series of articles on pharmacology: Arzneimittel-Forsch. **23**, 1226-1251, 1813-1822 (1973). Toxicity: R. Riedel, W. Schoetensack, ibid. 1215.

Crystals from ether-petr ether, mp 116-117°. Also reported as mp 77-79° (dependent on speed of crystallization). uv max (0.1N NaOH): 234, 264 nm (e 16200, 3700).

Calcium salt hemihydrate, $C_{38}H_{42}CaN_4O_6$.½H_2O, Bumaflex, Eumotol, Rheumatol. dec 154°. Sol in chloroform, alcohol, ether. Slightly sol in water. LD_{50} in mice, rats (mg/kg): 2500, 1250 orally; 258, 263 i.v. (Riedel, Schoetensack).

THERAP CAT: Analgesic; antipyretic; anti-inflammatory.

1508. Bumetanide. 3-(Aminosulfonyl)-5-(butylamino)-4-phenoxybenzoic acid; 3-(butylamino)-4-phenoxy-5-sulfamoylbenzoic acid; PF-1593; Ro-10-6338; Bumex; Burinex; Fontego; Fordiuran; Lixil; Lunetoron. $C_{17}H_{20}N_2O_5S$; mol wt 364.42. C 56.03%, H 5.53%, N 7.69%, O 21.95%, S 8.80%. Prepn: P. W. Feit, **Ger. pats. 1,964,503** and **1,964,-504;** idem, **U.S. pat. 3,806,534** (1970, 1970, 1974 all to Lövens Kemiske Fabrik); idem, J. Med. Chem. **14**, 432 (1971). Pharmacology: E. H. Oestergaard et al., Arzneimittel-Forsch. **22**, 66 (1972). HPLC and GC/MS determn in urine: C. Y. Gradeen et al., J. Anal. Toxicol. **14**, 123 (1990). Review of pharmacology and therapeutic use: A. Ward, R. C. Heel, Drugs **28**, 426-464 (1984). Series of articles on pharmacokinetics and comparative clinical studies in combination therapy: Curr. Ther. Res. (Clin. Exp.) **50**, Suppl. A, 1-82 (1991).

Crystals from aq ethanol, mp 230-231°. LD_{50} i.v. in mice: 330 mg/kg (Oestergaard).

THERAP CAT: Diuretic.

1509. Bunaftine. *N-Butyl-N-[2-(diethylamino)ethyl]-1-naphthalenecarboxamide; N-butyl-N-[2-(diethylamino)ethyl]-1-naphthamide;* bunaphtide; bunaphtine. $C_{21}H_{30}N_2O$; mol wt 326.48. C 77.26%, H 9.26%, N 8.58%, O 4.90%. Prepn: Giannini, **Ger.** pat. **2,099,894** (1971 to Malesci), *C.A.* **75**, 20222a (1971); Giannini *et al., Farmaco Ed. Sci.* **28**, 429 (1973). Pharmacology: Ferroni, Monticelli, *Pharmacol. Res. Commun.* **5**, 151 (1973).

$bp_{0.1}$ 178°. LD_{50} i.p. in mice: 122 mg/kg.
Citrate, $C_{21}H_{30}N_2O.C_6H_8O_7$, *Meregon.*
THERAP CAT: Antiarrhythmic (class III).

1510. Bunamidine. *N,N-Dibutyl-4-(hexyloxy)-1-naphthalenecarboximidamide; N,N-dibutyl-4-hexyloxy-1-naphthamidine.* $C_{25}H_{38}N_2O$; mol wt 382.59. C 78.48%, H 10.01%, N 7.32%, O 4.18%. Prepn: **Neth.** pat. **Appl. 6,403,-016** (1964 to Wellcome Found.), *C.A.* **62**, 9083a (1965); **Neth.** pat. **Appl. 6,603,570** and **Fr.** pat. **M3766** (both 1966 to Wellcome Found.), *C.A.* **66**, 37701t, 68911r (1967); Harfenist, Baltzly, **U.S.** pat. **3,290,375** (1966 to Burroughs-Wellcome). Review of synthesis, chemistry and activity: Harfenist *et al., J. Med. Chem.* **14**, 97 (1971). Activity studies of salts: Hatton, *Vet. Rec.* **81**, 104 (1967).

Hydrochloride, $C_{25}H_{38}N_2O.HCl$, *Scolaban.* Crystals from ethanol + ethyl acetate + ether, mp 214-214.8°. LD_{50} orally in mice: 540 mg/kg (calcd as base).
3-Hydroxy-2-naphthoate, *Buban.* mp 169-170°.
THERAP CAT (VET): Anthelmintic (Cestodes).

1511. Bunamiodyl Sodium. *2-[[2,4,6-Triiodo-3-[(1-oxobutyl)amino]phenyl]methylene]butanoic acid monosodium salt; 3-butyramido-α-ethyl-2,4,6-triiodocinnamic acid sodium salt;* 3-(3-butyrylamino-2,4,6-triiodophenyl)-2-ethylacrylic acid sodium salt; α-ethyl-β-(2,4,6-triiodo-3-butyramidophenyl)acrylic acid sodium salt; α-(2,4,6-triiodo-3-butyrylaminobenzylidene)butyric acid sodium salt; sodium 3-butyramido-α-ethyl-2,4,6-triiodocinnamate; buniodyl; Bunaiod; Orabilex; Orabilix. $C_{15}H_{15}I_3NNaO_3$; mol wt 660.99. C 27.26%, H 2.29%, I 57.60%, N 2.12%, Na 3.48%, O 7.26%. Prepn: Cassebaum, Dierbach, *Pharmazie* **16**, 389 (1961).

Acute toxicity: J. O. Hoppe *et al., J. Med. Chem.* **13**, 997 (1970).

Crystals from water (free acid, mp 105-120°). Slightly sol in water. LD_{50} in mice: 418 mg/kg i.v.; 2.78 g/kg orally (Hoppe).

THERAP CAT: Diagnostic aid (radiopaque medium).

1512. Bunazosin. *1-(4-Amino-6,7-dimethoxy-2-quinazolinyl)hexahydro-4-(1-oxobutyl)-1H-1,4-diazepine;* 1-(4-amino-6,7-dimethoxy-2-quinazolinyl)-4-butyrylhexahydro-1H-1,4-diazepine; 4-amino-6,7-dimethoxy-2-[4-(n-butyryl)-homopiperazin-1-yl]quinazoline; DDQ. $C_{19}H_{27}N_5O_3$; mol wt 373.46. C 61.11%, H 7.29%, N 18.75%, O 12.85%. Aminoquinazoline derivative with potent hypotensive properties. Prepn: T. Takahashi, H. Sugimoto, **Belg.** pat. **806,626;** *eidem,* **U.S.** pat. **3,920,636** (1974, 1975 to Eisai); alternative prepn: *eidem,* **Japan. Kokai 75 140,474** (1975 to Eisai), *C.A.* **85**, 46769r (1976). *In vitro* and *in vivo* studies of adrenergic blocking action and antihypertensive properties: T. Shoji *et al., Japan. J. Pharmacol.* **30**, 763 (1980); K. Hoshi, S. Fujino, *ibid.* 427. Antihypertensive effect in volunteers: T. Kawasaki *et al., Eur. J. Clin. Pharmacol.* **20**, 399 (1981). Pharmacokinetics: C. Yamato *et al., Xenobiotica* **12**, 549 (1982).

Hydrochloride, $C_{19}H_{27}N_5O_3.HCl$, *E-643, Detantol.* Crystals from methanol/ethanol, mp 280-282°.
THERAP CAT: Antihypertensive.

1513. Bungarotoxins. Constituent proteins of the venom of the *Elapidae* snake, *Bungarus multicinctus* (Southeast Asian banded krait). Separation of the crude venom yields several fractions, the most important being *α-bungarotoxin* (*α-Bgt*) and *β-bungarotoxin* (*β-BuTX*): C. C. Chang, C. Y. Lee, *Arch. Int. Pharmacodyn. Ther.* **144**, 241 (1963). α-Bgt is a single polypeptide chain of mol wt about 8,000 containing 74 amino acid residues with 5 disulfide bridges: D. Mebs *et al., Biochem. Biophys. Res. Commun.* **44**, 711 (1971); *eidem, Z. Physiol. Chem.* **353**, 243 (1972). Purification, characterization, immunochemical studies: D. G. Clark *et al., Biochemistry* **11**, 1663 (1972). It is a postsynaptic neurotoxin with curare-like action that binds irreversibly to acetylcholine receptor sites, producing neuromuscular blockade and skeletal muscle paralysis. Activity and use as a probe for acetylcholine receptors: C. C. Chang, *Nature* **215**, 1177 (1967); J. P. Changeux *et al., Proc. Nat. Acad. Sci. USA* **67**, 1241 (1970); R. E. Oswald, J. A. Freeman, *Neuroscience* **6**, 1 (1981). β-Bungarotoxin has been shown to contain several components, the major protein being designated as $β_1$- or β-bungarotoxin. It is composed of 2 subunits of mol wts of about 13,000 and 7,000, linked by disulfide bonds. The larger chain contains 120 amino acid residues including 13 half-cystine; the smaller chain contains 60 residues including 7 half cystine. Purification: T. Abe *et al., Eur. J. Biochem.* **80**, 1 (1977); *eidem, Proc. Roy. Soc. London B* **207**, 487 (1980). Chemical properties, and amino acid sequence of the two polypeptide chains: K. Kondo *et al., J. Biochem.* **83**, 91, 101 (1978). Complete purification and characterization of its action on synaptosomal accumulation and release of acetylcholine: J.

W. Spokes, J. O. Dolly, *Biochim. Biophys. Acta* **596**, 81 (1980). β-BuTX is a pre-synaptic neurotoxin that prevents acetylcholine release at skeletal neuromuscular junctions without affecting the sensitivity of the post-synaptic membrane. Its proposed phospholipase A_2 activity is believed to be responsible for some of its effects at motor nerve terminals. Activity studies: P. N. Strong *et al.*, *Proc. Nat. Acad. Sci. USA* **75**, 1029 (1976); T. Abe, R. Miledi, *Proc. Roy. Soc. London B* **200**, 225 (1978); M. T. Alderdice, R. L. Volle, *Arch. Pharmacol.* **316**, 126 (1981). *Review:* A. T. Tu, *Venoms: Chemistry and Molecular Biology* (Wiley, New York, 1977) pp 185-187, 240-251.

The crude venom of *Bungarus multicinctus* is quite toxic, having an LD_{50} in mice (μg/g): 0.16 s.c. (Chang, Lee); also reported as 0.33 s.c. (Mebs, 1972). LD_{50} of α-Bgt in mice (μg/g): 0.21 s.c.; 0.15 i.p. (Mebs, 1972). LD_{50} of β-BuTX in mice (μg/g): 0.019 i.p. (Kondo).

USE: As exptl tools in investigating neuromuscular processes.

1514. Bunitrolol. *2-[3-[(1,1-Dimethylethyl)amino]-2-hydroxypropoxy]benzonitrile; o-[3-(tert-butylamino)-2-hydroxypropoxy]benzonitrile;* 1-(2-cyanophenoxy)-2-hydroxy-3-*tert*-butylaminopropane; Ko-1366. $C_{14}H_{20}N_2O_2$; mol wt 248.33. C 67.72%, H 8.12%, N 11.28%, O 12.89%. β-Adrenergic blocker. Prepn: H. Koeppe *et al.*, **S. Afr.** pat. **68 03,783**; *eidem*, U.S. pat. **3,541,130** (1968, 1970 both to Boehringer, Ing.). Pharmacology: T. Kimura, *Experientia* **28**, 813 (1972). Studies on absorption, distribution, metabolism, excretion in rats and mice: T. Suzuki, T. Rikihisa, *Arzneimittel-Forsch.* **29**, 1707 (1979). Clinical studies: A. Reale *et al.*, *J. Int. Med. Res.* **4**, 338 (1976); J. J. C. Jonker *et al.*, *Arzneimittel-Forsch.* **31**, 1140 (1981). Toxicology study: M. Kanda *et al.*, *Oyo Yakuri* **9**, 457, 465, 499 (1975), *C.A.* **83**, 172616 (1975).

Hydrochloride, $C_{14}H_{20}N_2O_2$.HCl, *Betriol, Stresson.* Crystals from ethanol, mp 163-165°. LD_{50} in mice, rats (mg/kg): 1344-1440, 639-649 orally; 264-265, 222-225 i.p. (Kanda).

THERAP CAT: Antihypertensive, antiarrhythmic, antianginal.

1515. Bunte Salts. Water-sol salts of certain alkyl or aralkyl thiosulfuric acids of the general formula: RSSO$_2$ONa. Prepn: Lecher, Hardy, U.S. pat. **2,712,547** (1955 to Am. Cyanamid); El-Heweki, Taeger, *J. Prakt. Chem.* **7**, 191 (1958); Tabushi *et al.*, *Kogyo Kagaku Zasshi* **67**, 478 (1964).

THERAP CAT (VET): Have been used as coccidiostats.

1516. Buparvaquone. *2-[[4-(1,1-Dimethylethyl)cyclohexyl]methyl]-3-hydroxy-1,4-naphthalenedione;* 2-[(4-*tert*-butylcyclohexyl)methyl]-3-hydroxy-1,4-naphthoquinone; 3-(4-*t*-butylcyclohexyl)methyl-2-hydroxy-1,4-naphthoquinone; BW-720C; Butalex. $C_{21}H_{26}O_3$; mol wt 326.44. C 77.27%, H 8.03%, O 14.70%. Analog of antimalarial hydroxynaphthoquinones; *t*-butyl deriv of parvaquone, *q.v.* Prepn: A. T. Hudson, A. W. Randall, **Eur.** pat. **Appl. 77,550**; *eidem*, U.S. pat. **4,485,117** (1983, 1984 both to Wellcome). *In vivo* and *in vitro* antiprotozoal activity: A. T. Hudson *et al.*, *Parasitology* **90**, 45 (1985); N. McHardy *et al.*, *Res. Vet. Sci.* **39**, 29 (1985). Treatment of bovine theileriosis: S. Dhar *et al.*, *Vet. Rec.* **119**, 635 (1986). Efficacy as a chemoimmunoprophylactic of theileriosis: *eidem*, *ibid.* **120**, 375 (1987).

Crystals, mp 124-125°. LD_{50} orally in rats: > 2 g/kg (Hudson).

THERAP CAT (VET): Antiprotozoal (Theileria).

1517. Buphanamine. *(1α)-2,3-Didehydro-7-methoxycrinan-1-ol.* $C_{17}H_{19}NO_3$; mol wt 301.34. C 67.76%, H 6.36%, N 4.65%, O 21.24%. Isoln from bulb of *Buphane (Boöphane) disticha* Herb., Appleg. *(Haemanthus toxicarius* Herb.) *Amaryllidaceae:* Humber, Taylor, *Can. J. Chem.* **33**, 1268 (1955); Hauth, Stauffacher, *Helv. Chim. Acta* **44**, 491 (1961). From *B. fischeri* Baker: Renz *et al.*, *ibid.* **38**, 1209 (1955). Identity with "oily" haemanthine: Goosen, Warren, *J. Chem. Soc.* **1960**, 1094. Structure and stereochemistry: Fales, Wildman, *J. Org. Chem.* **26**, 881 (1961). Revised structure: Wildman in *The Alkaloids*, vol. XI, R. H. F. Manske, Ed. (Academic Press, New York, 1968) p 361. NMR of revised structure: Crain *et al.*, *J. Am. Chem. Soc.* **93**, 990 (1971). High-resolution mass spectrum: P. Longevialle *et al.*, *Org. Mass Spectrom.* **7**, 401 (1973).

Prisms from ethyl acetate, mp 183-185° (Fales); from acetone, mp 192-194° (Goosen). $[\alpha]_{589}^{24}$ −195°; $[\alpha]_{436}^{24}$ −408° (c = 0.97); $[\alpha]_D^{20}$ −205° (c = 0.69, 95% ethanol); $[\alpha]_D^{21}$ −194° (c = 0.247, chloroform). uv max: 287 nm (ε 1495).

Hydrochloride, $C_{17}H_{19}NO_4$.HCl, prisms from ethanol + ether, mp 180°.

1518. Buphanitine. *(1α,3α)-7-Methoxycrinan-1,3-diol;* hemanthine; nerbowdine. $C_{17}H_{21}NO_5$; mol wt 319.36. C 63.94%, H 6.63%, N 4.39%, O 25.05%. Major alkaloid from *Buphane (Boöphane) disticha* Herb., Appleg. *(Haemanthus toxicarius* Herb.), *Amaryllidaceae.* Isoln: Lewin, *Arch. Exp. Path. Pharmakol.* **68**, 333 (1912); Tutin, *ibid.* **69**, 314 (1912). Identity with hemanthine: Goosen, Warren, *Chem. Ind.* **1957**, 267; with nerbowdine: Lyle *et al.*, *J. Am. Chem. Soc.* **82**, 2620 (1960). Structure studies: Goosen, Warren, *J. Chem. Soc.* **1960**, 1097; Goosen *et al.*, *ibid.* **1961**, 4038. High-resolution mass spectrum: P. Longevialle *et al.*, *Org. Mass Spectrom.* **7**, 401 (1973).

Needles from chloroform + ether (changes crystalline form at 210°), mp 232°. Prisms from acetone. $[\alpha]_D^{20}$ −102° (c = 1 in chloroform).

Hydrochloride, $C_{17}H_{22}ClNO_5$, mp 265°.
Nitrate, $C_{17}H_{22}N_2O_8$, mp 222-224°.

1519. Bupirimate. *Dimethylsulfamic acid 5-butyl-2-(ethylamino)-6-methyl-4-pyrimidinyl ester;* 5-butyl-2-(ethylamino)-6-methyl-4-pyrimidinyl dimethylsulfamate; PP-588; Nimrod. $C_{13}H_{24}N_4O_3S$; mol wt 316.42. C 49.35%, H 7.64%, N 17.71%, O 15.17%, S 10.13%. Prepn: A. M. Cole *et al.*, **Ger.** pat. **2,246,645**; *eidem*, U.S. pat. **3,880,852** (1973, 1975 both to ICI). Activity: J. R. Finney *et al.*, *Proc. Brit. Insect. Fung. Conf., 8th* **2**, 667 (1975).

Pale tan waxy solid, mp 50-51°. Vapor pressure at 20°: 5 × 10⁻⁷ mm Hg. Soly in water at 25°: 22 mg/l. Sol in most organic solvents except paraffins. Easily hydrolyzed by dil acids. LD_{50} orally in rats: 4000 mg/kg (Finney). USE: Fungicide.

1520. Bupivacaine. *1-Butyl-N-(2,6-dimethylphenyl)-2-piperidinecarboxamide; dl-1-butyl-2',6'-pipecoloxylidide; 1-n-butyl-2',6'-dimethyl-2-piperidinecarboxanilide; dl-N-n-butylpipecolic acid 2,6-xylidide; 1-butyl-2-(2,6-xylylcarbamoyl)piperidine; dl-1-n-butylpiperidine-2-carboxylic acid 2,6-dimethylanilide; Anekain; Marcaine.* $C_{18}H_{28}N_2O$; mol wt 288.43. C 74.96%, H 9.78%, N 9.71%, O 5.55%. Prepn: **Brit. pat. 869,978**; B. Thuresson, B. G. Pettersson, **U.S. pat. 2,955,111** (1959, 1960 both to Bofors); B. F. Tullar, Bolen, **Brit. pat. 1,166,802** (1969 to Sterling Drug). Prepn, resolution of isomers and report on comparative activities: F. P. Luduena, B. F. Tullar, **S. Afr. pat. 68 02,611** and **Brit. pat. 1,180,712** (1968 to Sterling Drug), *C.A.* **73**, 25314a (1970); B. F. Tullar, *J. Med. Chem.* **14**, 891 (1971). Pharmacological and toxicological studies: F. Henn, R. Brattsand, *Acta Anaesthesiol. Scand. Suppl.* **21**, 9 (1966), *C.A.* **66**, 17863u (1967). Comprehensive description: T. D. Wilson in *Analytical Profiles of Drug Substances* vol. **19**, K. Florey, Ed. (Academic Press, New York, 1990) pp 59-94.

mp 107-108°. pKa 8.09; also reported as 8.17. Partition coefficient: (oleyl alcohol/water) 1565; (n-heptane/pH 7.4 buffer) 27.5.
Hydrochloride monohydrate, $C_{18}H_{28}N_2O \cdot HCl \cdot H_2O$, *AH-2250, LAC-43, Marcain, Marcaina, Carbostesin.* mp 255-256°. Slightly sol in acetone, chloroform, ether. Soly (mg/ml): water 40; alcohol 125. LD_{50} in mice (mg/kg): 7.8 i.v., 82 s.c. (Henn, Brattsand).
THERAP CAT: Anesthetic (local).

1521. Bupranolol. *1-(2-Chloro-5-methylphenoxy)-3-[(1,1-dimethylethyl)amino]-2-propanol; 1-(tert-butylamino)-3-[(6-chloro-m-tolyl)oxy]-2-propanol; 1-(6-chloro-3-methylphenoxy)-3-tert-butylaminopropan-2-ol; 1-tert-butylamino-3-(2-chloro-5-methylphenoxy)-2-propanol; bupranol; Ophtorenin.* $C_{14}H_{22}ClNO_2$; mol wt 271.79. C 61.87%, H 8.16%, Cl 13.04%, N 5.15%, O 11.77%. β-Adrenergic blocker. Prepn: Kunz *et al.,* **Ger. pat. 1,236,523** (1967 to Sanol-Arzneimittel Dr. Schwarz), *C.A.* **67**, 64046k (1967) and **U.S. pat. 3,309,406** (1967). Pharmacology: Waterloh *et al., Arzneimittel-Forsch.* **19**, 153, 330, 1710 (1969); Pendleton *et al., Arch. Int. Pharmacodyn. Ther.* **187**, 75 (1970); P. Montastruc *et al., Arch. Farmacol. Toxicol.* **3**, 93 (1977).

Hydrochloride, $C_{14}H_{22}ClNO_2 \cdot HCl$, *KL-255, Betadran, Betadrenol, Looser, Panimit.* Crystals, mp 220-222°.
THERAP CAT: Antihypertensive; antianginal; antiarrhythmic; antiglaucoma.

1522. Buprenorphine. *[5α,7α(S)]-17-(Cyclopropylmethyl)-α-(1,1-dimethylethyl)-4,5-epoxy-18,19-dihydro-3-hydroxy-6-methoxy-α-methyl-6,14-ethenomorphinan-7-methanol; 21-cyclopropyl-7α-[(S)-1-hydroxy-1,2,2-trimethylpropyl]-6,14-endo-ethano-6,7,8,14-tetrahydrooripavine; 21-cyclopropyl-7α-(2-hydroxy-3,3-dimethyl-2-butyl)-6,14-endo-ethano-6,7,8,-14-tetrahydrooripavine; RX-6029-M.* $C_{29}H_{41}NO_4$; mol wt 467.65. C 74.48%, H 8.84%, N 3.00%, O 13.68%. Analgesic that demonstrates narcotic agonist-antagonist properties. Prepn: K. W. Bentley, **Brit. pat. 1,136,214**; *idem,* **U.S. pat. 3,433,791** (1968, 1969 to Reckitt & Sons). *See also:* K. W. Bentley, "The Morphine Alkaloids" in *The Alkaloids* vol. **13**, R. F. Manske, Ed. (Academic Press, New York, 1971) pp 75-120. Pharmacology: A. Cowan in *Advan. Biochem. Psychopharmacol.* vol. **8**, M. C. Braude *et al.,* Eds. (Raven Press, New York, 1974) pp 427-438; *idem, J. Pharm. Pharmacol.* **28**, 177 (1976); W. R. Martin *et al., J. Pharmacol. Exp. Ther.* **197**, 517 (1976). Metabolism: M. J. Rance, J. S. Shillingford, *Biochem. Pharmacol.* **25**, 735 (1976). Clinical study in chronic pain: M. Kjaer *et al., Brit. J. Clin. Pharmacol.* **13**, 487 (1982). Clinical evaluation in treatment of opioid addiction: P. J. Fudala *et al., Clin. Pharmacol. Ther.* **47**, 525 (1990). *Review:* J. W. Lewis, "Ring C-Bridged Derivatives of Thebaine and Oripavine" in *Advan. Biochem. Psychopharmacol.* vol. **8**, M. C. Braude *et al.,* Eds. (Raven Press, New York, 1974) pp 123-137. Review of pharmacology and therapeutic efficacy: R. C. Heel *et al., Drugs* **17**, 81-110 (1979).

Crystals, mp 209°.
Hydrochloride, $C_{29}H_{41}NO_4 \cdot HCl$, *CL-112302, NIH-8805, UM-952, Buprenex, Lepetan, Temgesic.*
Note: This is a controlled substance (narcotic) listed in the U.S. Code of Federal Regulations, Title 21 Part 1308.15 (1995).
THERAP CAT: Analgesic (narcotic).

1523. Bupropion. *1-(3-Chlorophenyl)-2-[(1,1-dimethylethyl)amino]-1-propanone; (±)-2-(tert-butylamino)-3'-chloropropiophenone; m-chloro-α-(tert-butylamino)propiophenone; amfebutamon(e).* $C_{13}H_{18}ClNO$; mol wt 239.74. C 65.13%, H 7.57%, Cl 14.79%, N 5.84%, O 6.67%. Pharmacologically similar to the tricyclic antidepressants. Prepn: N. B. Mehta, D. A. Yeowell, **Ger. pat. 2,059,618** (1971 to Wellcome Found.) corresp to N. B. Mehta, **U.S. pats. 3,819,706** and **3,885,046** (1974, 1975 to Burroughs Wellcome). Pharmacology: F. Soroko *et al., J. Pharm. Pharmacol.* **29**, 767 (1977). Clinical evaluation in attention deficit hyperactivity disorder: T. H. Clay *et al., Psychopharmacol. Bull.* **24**, 143 (1988). Review of toxicology: W. E. Tucker, *J. Clin. Psychiatry* **44**, 60-62 (1983). *Review:* R. A. Maxwell *et al.,* in *Pharmacological and Biochemical Properties of Drug Substances* vol. **3**, M. E. Goldberg, Ed. (Am. Pharm. Assoc., Washington, DC, 1981) pp 1-55; S. G. Bryant *et al., Clin. Pharm.* **2**, 525-537 (1983).

Pale yellow oil, $bp_{.005}$ 52°. Sol in methanol, ethanol, acetone, ether, benzene. Very hygroscopic and susceptible to decompn. LD_{50} in male, female mice, rats (mg/kg): 544

±29, 636 ±23, 607 ±51, 482 ±91 orally; in male mice, rats (mg/kg): 273 ±4, 263 ±18 i.p. (Tucker).

Hydrochloride, $C_{13}H_{18}ClNO.HCl$, *Wellbatrin, Wellbutrin*. Crystals from isopropanol and abs ethanol, mp 233-234°. Soly in mg/ml: water 312; alcohol 193; 0.1N HCl 333. LD_{50} in mice, rats (mg/kg): 230, 210 i.p.; 575, 600 orally (Soroko).

THERAP CAT: Antidepressant.

1524. Buquinolate. *4-Hydroxy-6,7-bis(2-methylpropoxy)-3-quinolinecarboxylic acid ethyl ester; 4-hydroxy-6,7-diisobutoxy-3-quinolinecarboxylic acid ethyl ester; ethyl 6,7-diisobutoxy-4-hydroxyquinoline-3-carboxylate;* Bonaid. $C_{20}H_{27}NO_5$; mol wt 361.44. C 66.46%, H 7.53%, N 3.88%, O 22.13%. Prepn: Watson, **Belg.** pat. **659,237** (1965 to Norwich), *C.A.* **64**, 2071c (1966); corresp to **U.S.** pat. **3,-267,106** (1964).

Crystals, mp 288-291°.
THERAP CAT (VET): Coccidiostat.

1525. Buramate. *(Phenylmethyl)carbamic acid 2-hydroxyethyl ester; benzylcarbamic acid 2-hydroxyethyl ester; 2-hydroxyethyl benzylcarbamate; glycol benzylcarbamate;* AC-601; Hyamate. $C_{10}H_{13}NO_3$; mol wt 195.22. C 61.53%, H 6.71%, N 7.17%, O 24.59%. Prepn from glycol cyclic carbonate and benzylamine: Viard, **Brit.** pat. **689,705** (1953 to Chauny & Cirey).

Crystals, mp 40°.
THERAP CAT: Anticonvulsant; antipsychotic.

1526. Burgundy Mixture. Soda Bordeaux. Variable compn, considered to be basic cupric carbonate and/or sulfate. Prepd by reaction of aq solns of $CuSO_4$ and Na_2CO_3: Frear, *Chemistry of the Pesticides* (Van Nostrand, 3rd ed., 1955) p 322; Mond, Haberlein, *J. Chem. Soc.* **115**, 908 (1919). *See also* Cupric Sulfate, Basic.

Blue to green colloidal ppt. Practically insol in water.
USE: Plant fungicide, more phytotoxic than Bordeaux mixture.

1527. Buserelin. *6-[O-(1,1-dimethylethyl)-D-serine]-9-(N-ethyl-L-prolinamide)-10-deglycinamideluteinizing hormone-releasing factor (pig).* $C_{60}H_{86}N_{16}O_{13}$; mol wt 1239.44. C 58.14%, H 6.99%, N 18.08%, O 16.78%. Synthetic nonapeptide agonist analog of LH-RH, *q.v.* Synthesis: W. Konig *et al.*, **Ger.** pat. **2,438,350**; *eidem*, **U.S.** pat. **4,024,248** (1976, 1977 both to Hoechst); A. S. Dutta *et al.*, *J. Med. Chem.* **21**, 1018 (1978). Clinical pharmacology: A. Lemay *et al.*, *Fertil. Steril.* **37**, 193 (1982). Radioimmunoassay in plasma and urine: S. Saito *et al.*, *J. Immunol. Methods* **79**, 173 (1985). Veterinary use to increase conception rate: K. Moller, E. D. Fielden, *N. Zealand Vet. J.* **29**, 214 (1981). Clinical evaluation in prostatic carcinoma: J. H. Waxman, *Brit. J. Urol.* **55**, 737 (1983); as ovulatory stimulant for *in vitro* fertilization: V. MacLachlan *et al.*, *N. Engl. J. Med.* **320**, 1233 (1989). Review of pharmacokinetics and clinical profile: R. N. Brogden *et al.*, *Drugs* **39**, 399-437 (1990); of efficacy in prostatic carcinoma: H. J. de Voogt *et al.*, *Scand. J. Urol. Nephrol.* **Suppl 138**, 131-136 (1991).

5-oxoPro–His–Trp–Ser–Tyr–D-Ser(t-Bu)–Leu–Arg–ProNHCH₂CH₃

$[\alpha]_D^{20}$ −40.4° (c = 1 in dimethylacetamide).

Monoacetate, $C_{60}H_{86}N_{16}O_{13}.C_2H_4O_2$, *HOE-766, Receptal, Suprecur, Suprefact.*

THERAP CAT: Antineoplastic (hormonal). Gonad-stimulating principle.
THERAP CAT (VET): Gonad-stimulating principle.

1528. Buspirone. *8-[4-[4-(2-Pyrimidinyl)-1-piperazinyl]butyl]-8-azaspiro[4.5]decane-7,9-dione.* $C_{21}H_{31}N_5O_2$; mol wt 385.51. C 65.43%, H 8.11%, N 18.17%, O 8.30%. Nonbenzodiazepine anxiolytic; 5-hydroxytryptamine (5-HT$_1$) receptor agonist. Prepn: Y. H. Wu *et al.*, *J. Med. Chem.* **15**, 477 (1972); Y. H. Wu, J. W. Rayburn, **Ger.** pat. **2,057,845** (1971 to Bristol-Myers); *eidem*, **U.S.** pat. **3,717,634** (1973 to Mead-Johnson). Pharmacology: L. E. Allen *et al.*, *Arzneimittel-Forsch.* **24**, 917 (1974). Comparison with diazepam in treatment of anxiety: H. L. Goldberg, R. J. Finnerty, *Am. J. Psychiat.* **136**, 1184 (1979); A. F. Jacobson *et al.*, *Pharmacotherapy* **5**, 290 (1985). Nonsynergistic effect with alcohol: T. Seppala *et al.*, *Clin. Pharmacol. Ther.* **32**, 201 (1982). Disposition and metabolism: S. Caccia *et al.*, *Xenobiotica* **13**, 147 (1983). Series of articles on chemistry, pharmacology, addictive potential, and clinical trials: *J. Clin. Psychiat.* **43**, pp 1-116 (1982); on pharmacology, safety and clinical comparison with clorazepate: *Am. J. Med.* **80**, Suppl. 3B, 1-51 (1986). Review of pharmacology and therapeutic efficacy: K. L. Goa, A. Ward, *Drugs* **32**, 114-129 (1986). *Review:* M. W. Jann, *Pharmacotherapy* **8**, 100-116 (1988); D. P. Taylor, *FASEB J.* **2**, 2445-2452 (1988).

Hydrochloride, $C_{21}H_{31}N_5O_2.HCl$, *Ansial, Ansiced, Axoren, Bespar, Buspar, Buspimen, Buspinol, Buspisal, Censpar, Lucelan, Narol, Travin.* Crystals from abs ethanol, mp 201.5-202.5°. LD_{50} i.p. in rats: 136 mg/kg (Allen).

THERAP CAT: Anxiolytic.

1529. Busulfan. *1,4-Butanediol dimethanesulfonate esters;* 1,4-bis(methanesulfonoxy)butane; 1,4-di(methanesulfonyloxy)butane; 1,4-di(methylsulfonoxy)butane; methanesulfonic acid tetramethylene ester; tetramethylene bis(methanesulfonate); busulphan; CB-2041; GT-41; Mielucin; Misulban; Mitosan; Myelosan; Myeleukon; Myeloleukon; Myleran; Sulfabutin; $C_6H_{14}O_6S_2$; mol wt 246.31. C 29.26%, H 5.73%, O 38.97%, S 26.04%. $CH_3SO_2O(CH_2)_4OSO_2CH_3$. Alkylating agent with antileukemic activity. Discovery: A. Haddow, G. M. Timmis, *Lancet* **1**, 207 (1953). Prepn: G. M. Timmis, **U.S.** pat. **2,917,432** (1959 to Burroughs Wellcome). Chemosterilant effect in boll weevils: J. W. Haynes *et al.*, *J. Econ. Entomol.* **66**, 619 (1973); J. W. Haynes, J. E. Wright, *Southwest Entomol.* **7**, 56 (1982). Pharmacokinetics: H. Ehrsson *et al.*, *Clin. Pharmacol. Ther.* **34**, 86 (1983). HPLC determn in human plasma: W. D. Henner *et al.*, *J. Chromatog.* **416**, 426 (1987). Clinical pretreatment with cyclophosphamide, *q.v.*, for bone marrow transplants: G. W. Santos *et al.*, *N. Engl. J. Med.* **309**, 1347 (1983). Comparative clinical trial with mitobronitol, *q.v.*, in chronic myeloid leukemia: R. T. Silver *et al.*, *Cancer* **60**, 1342 (1987). Toxicity data: H. R. Scherf *et al.*, *Arzneimittel-Forsch.* **20**, 1467 (1970). Review of pharmacology: C. D. R. Dunn, *Exp. Hematol. (Copenhagen)* **2**, 101-117 (1974); of toxicology: J. B. Bishop, J. S. Wassom, *Mutat. Res.* **168**, 15-45 (1986). Comprehensive description: M. Tariq, A. A. Al Badr, *Anal. Profiles Drug Subs.* **16**, 53-83 (1987).

Crystals, mp 114-118°. Soly in acetone at 25°: 2.4 g/100 ml; in alcohol: 0.1 g/100 ml. Practically insol in water, but will dissolve slowly as hydrolysis takes place. LD_{50} i.v. in rats: 1.8 mg/kg (Scherf).

Note: This substance has been listed as a known carcinogen: *Seventh Annual Report on Carcinogens* (PB95-109781, 1994) p 42.

USE: Insect sterilant.
THERAP CAT: Antineoplastic.

1530. Butabarbital Sodium. *5-Ethyl-5-(1-methylpropyl)-2,4,6(1H,3H,5H)-pyrimidinetrione sodium salt; 5-sec-butyl-5-ethylbarbituric acid sodium salt;* 5-ethyl-5-(1-methylpropyl)barbituric acid sodium salt; sodium 5-*sec*-butyl-5-ethylbarbiturate; sodium 5-ethyl-5-(1-methylpropyl)barbiturate; secbutobarbitone sodium; Asturidon; Bubarbital Sodium; Butabarbitone Sodium; Busodium; Busotran; Butabon; Butabar; Butak; Buta-Kay; Butrate; Butte; Buticaps; Butalan; Butanotic; Butex; Butatran; Butazem; Carrbutabarb; Butased; Butisol Sodium; Loubarb; Neravan; Prelital; Sarisol. $C_{10}H_{15}N_2NaO_3$; mol wt 234.23. C 51.28%, H 6.45%, N 11.96%, Na 9.82%, O 20.49%. Prepn: Shonle, **U.S. pat. 1,856,792** (1932 to Lilly).

Bitter powder. (The free acid mp 165-168°.) One gram dissolves in 2 ml water, in about 7 ml alcohol. Practically insol in ether, benzene. The pH of 10% aq soln 10.0-11.2 (USP XXI, 1985).

Caution: May be habit forming. This is a controlled substance (depressant) listed in the U.S. Code of Federal Regulations, Title 21, Parts 329.1 and 1308.13 (1995).

THERAP CAT: Sedative, hypnotic.

1531. Butacaine. *3-(Dibutylamino)-1-propanol 4-aminobenzoate;* 3-(p-aminobenzoxy)-1-di-*n*-butylaminopropane; dibutylaminopropyl-*p*-aminobenzoate; *p*-aminobenzoyldibutylaminopropanol; Butelline. $C_{18}H_{30}N_2O_2$; mol wt 306.45. C 70.55%, H 9.87%, N 9.14%, O 10.44%. Prepd from *p*-nitrobenzoyl chloride and γ-di-*n*-butylaminopropanol followed by reduction of the NO_2 group to NH_2: O. Kamm *et al.*, **U.S. pat. 1,358,751;** Adams, Volwiler, **U.S. pat. 1,676,470;** A. Weston, **U.S. pat. 2,437,984** (1920, 1928, 1948 all to Abbott); Burnett *et al., J. Am. Chem. Soc.* **59,** 2248 (1937); Kaye, Roberts, *ibid.* **73,** 4762 (1951). Toxicity study: Schmidt *et al., Toxicol. Appl. Pharmacol.* **1,** 454 (1956).

Liquid, $bp_{0.11}$ 178-182°.
Sulfate, $(C_{18}H_{30}N_2O_2)_2.H_2SO_4$, *Butyn Sulfate.* Crystals from *n*-propanol, mp 138.5-139.5°. Also reported as mp 100-103°. Produces numbness of tongue on tasting. Affected by light. One gram dissolves slowly in somewhat less than 1 ml water, more rapidly on heating. Quite sol in warm alc, in acetone; slightly sol in chloroform. Practically insol in ether. Aq soln is practically neutral to litmus and may be boiled for sterilization without dec. LD_{50} i.v. in mice: 12.4 mg/kg (Schmidt). *Pharmaceutical Incompat.* Alkalies and alkaline-reacting substances liberate the free base as an oily liquid from solns. Bicarbonates produce a precipitate of butacaine carbonate. Iodine gives a brown precipitate. Chlorides form the almost insol butacaine chloride which may precipitate.
Hydrochloride, $C_{18}H_{30}N_2O_2.HCl$, crystals from ethanol, mp 157-158.5°.

THERAP CAT: Anesthetic (local).
THERAP CAT (VET): Anesthetic (local).

1532. Butacetin. *N-[4-(1,1-Dimethylethoxy)phenyl]acetamide; 4'-tert-butoxyacetanilide;* Tromal. $C_{12}H_{17}NO_2$; mol wt 207.27. C 69.54%, H 8.27%, N 6.76%, O 15.44%. Prepn: Bowden, Green, *J. Chem. Soc.* **1954,** 1795.

Plates from 66% aq alcohol, mp 130°.
THERAP CAT: Analgesic; antidepressant.

1533. Butachlor. *N-(Butoxymethyl)-2-chloro-N-(2,6-diethylphenyl)acetamide; N-(butoxymethyl)-2-chloro-2',6'-diethylacetanilide;* 2-chloro-2,6-diethyl-*N*-(butoxymethyl)acetanilide; CP-53619; Machete; Butanex. $C_{17}H_{26}ClNO_2$; mol wt 311.85. C 65.48%, H 8.40%, Cl 11.37%, N 4.49%, O 10.26%. Selective pre-emergence and pre-plant herbicide. Prepn: J. F. Olin, **U.S. pat. 3,442,945** (1969 to Monsanto). Photodecomposition: Y.-L. Chen, C.-C. Chen, *Nippon Noyaku Gakkaishi* **3,** 143 (1978), *C.A.* **89,** 174926j (1978). Degradation: Y.-L. Chen, T.-C. Wu, *ibid.* 411, *C.A.* **91,** 1258f (1979); Y.-L. Chen, *FFTC Book Ser.* **20,** 121 (1981).

Light yellow oil, $bp_{0.5}$ 196°. d_4^{30} 1.0695. Soly in water at 20°: 20 mg/l. Sol in most organic solvents. LD_{50} orally in rats: 1740 mg/kg, *RTECS vol. I,* R. J. Lewis, R. L. Tatken, Eds. (1980) p 25.
USE: Herbicide.

1534. 1,3-Butadiene. α,γ-Butadiene; bivinyl; divinyl; erythrene; vinylethylene; biethylene; pyrrolylene. C_4H_6; mol wt 54.09. C 88.82%, H 11.18%. $CH_2=CHCH=CH_2$. Manuf as a coproduct of hydrocarbon steam cracking to produce ethylene. Also produced from ethanol; from petroleum gases, i.e., by catalytic dehydrogenation of butene or butene-butane mixtures; by cracking naphtha and light oil. Manuf: Faith, Keyes & Clark's *Industrial Chemicals,* F. A. Lowenheim, M. K. Moran, Eds. (Wiley-Interscience, New York, 4th ed., 1975) pp 164-172. Toxicity: Carpenter *et al., J. Ind. Hyg. Toxicol.* **26,** 69 (1944). *Reviews:* Norton, *Chem. Rev.* **31,** 319 (1942); Alder, "The Diene Synthesis" in *Newer Methods of Preparative Organic Chemistry* (Interscience, New York, 1948); Konrad, *Angew. Chem.* **62,** 491 (1950); Hillyer, Stallings, *Petrol. Refiner* **35**(12), 157 (1956); A. S. Onishchenko, *Diene Synthesis* (New York, 1964); Bailey, "Butadiene" in *Vinyl and Diene Monomers* (part 2), E. C. Leonard, Ed. (Interscience, New York, 1971) pp 757-995; H. N. Sun, J. P. Wristers in Kirk-Othmer *Encyclopedia of Chemical Technology* **vol. 4** (John Wiley & Sons, New York, 4th ed., 1992) pp 663-690. Review of toxicity: L. S. Birnbaum, *Env. Health Persp.* **101,** Suppl. 6, 161-167 (1993); and health effects: *Toxicological Profile* (PB93-110690, TP-91/07, 1992) 125 p. Review of toxicology and carcinogenic risk: *IARC Scientific Publ.* **127,** 1-412 (1993).

Colorless gas. Mild aromatic odor. Flammable at room temp and atmospheric pressure. mp —108.966°: Glasgow *et al., Anal. Chem.* **22,** 1521 (1950). bp_{760} —4.5°. d_4^{-6} 0.650. Freezing pt —108.902°. Flash pt —85°. Densities from —17.8° to 60°: *Ind. Eng. Chem., Anal. Ed.* **16,** 7 (1944). n_D^{-6} 1.4223. bp (at pressures greater than 1 atm): 2 atm: 15.3°; 5 atm: 47.0°; 10 atm: 76.0°; 20 atm: 114.0°; 30 atm: 139.8°; 40 atm: 160.0°. Critical temperature 161.8°; critical pressure 42.6 atm. Infrared absorption spectrum: *ibid.* 422. Stability: *Ind. Eng. Chem.* **36,** 3 (1944). Stabilization with *o*-dihydroxybenzene: **Brit. pat. 569,412;** with aliphatic mercaptans: **U.S. pat. 2,373,754.** Explosion limits in air, vol%: lower 2.0, upper 11.5. Polymerizes and copolymerizes easily, e.g. under the influence of sodium, thereby forming synthetic rubbers. Sparingly soluble in water; slightly soluble in methanol, ethanol; soluble in organic solvents such as diethyl ether, benzene, carbon tetrachloride. Alcohol dissolves about 40 vols at room temp. LC_{50} in rats (4 hr inhalation): 129,000 ppm; in mice (2 hr inhalation): 122,000 ppm (Birnbaum).

Caution: Potential symptoms of overexposure are irritation of eyes, nose and throat; blurred vision, nausea, fatigue, headache, vertigo, drowsiness, lightheadedness, decreased blood pressure, unconsciousness, respiratory paralysis; direct dermal contact may cause burns and frostbite. *See NIOSH Pocket Guide to Chemical Hazards* (DHHS/NIOSH 90-117, 1990) p 48; *Patty's Industrial Hygiene and Toxicology* vol. 2B, G. D. Clayton, F. E. Clayton, Eds. (John Wiley & Sons, Inc., New York, 4th ed., 1994) p 1242, 1250-1252. This substance may reasonably be anticipated to be a carcinogen: *Seventh Annual Report on Carcinogens* (PB95-109781, 1994) p 105.

USE: As component in the manuf of polymers such as synthetic rubbers, plastics, resins. As a chemical intermediate for the production of many industrial materials; in the manuf of adiponitrile.

1535. Butalamine. *N,N-Dibutyl-N'-(3-phenyl-1,2,4-oxadiazol-5-yl)-1,2-ethanediamine; 5-[[2-(dibutylamino)ethyl]amino]-3-phenyl-1,2,4-oxadiazole;* 3-phenyl-5-(dibutylaminoethylamino)-1,2,4-oxadiazole. $C_{18}H_{28}N_4O$; mol wt 316.45. C 68.32%, H 8.92%, N 17.71%, O 5.06%. Prepn: **Fr.** pat. **M3334;** Aron-Samuel, Sterne, **U.S.** pat. **3,338,899** (1965, 1967 to Aron-Samuel). Pharmacology and toxicity studies: J. Sterne *et al., Therapie* 24, 735 (1969); J. Sterne, *ibid.* 745; J. L. Fontaine, R. Fontaine, *ibid.* 25, 961 (1970).

Hydrochloride, $C_{18}H_{28}N_4O \cdot HCl$, *LA-1221, Adrevil, Hemotrope, Surem, Surheme.* mp 145°. LD_{50} in mice (mg/kg): 43 i.v., 2500 s.c., 625 orally; LD_{50} in rats (mg/kg): >4000 s.c., 1600 orally (Sterne).

THERAP CAT: Vasodilator (peripheral).

1536. Butalbital. *5-(2-Methylpropyl)-5-(2-propenyl)-2,4,6(1H,3H,5H)-pyrimidinetrione; 5-allyl-5-isobutylbarbituric acid;* 5-allyl-5-(2-methylpropyl)barbituric acid; 5-isobutyl-5-allylbarbituric acid; alisobumal; allylbarbital; itobarbital; tetrallobarbital; Sandoptal. $C_{11}H_{16}N_2O_3$; mol wt 224.26. C 58.91%, H 7.19%, N 12.49%, O 21.40%. Prepn: Volwiler, *J. Am. Chem. Soc.* 47, 2236 (1925).

Prisms, mp 138-139°. Slightly bitter taste. Practically insol in water and petr ether. Sol in alcohol, chloroform, ether, acetone, glacial acetic acid, also in solns of fixed alkali hydroxides. A satd aq soln is acid to litmus.

Caution: May be habit forming. This is a controlled substance (depressant) listed in the U.S. Code of Federal Regulations, Title 21, Parts 329.1 and 1308.13 (1995).

THERAP CAT: Sedative, hypnotic.

1537. Butallylonal. *5-(2-Bromo-2-propenyl)-5-(1-methylpropyl)-2,4,6(1H,3H,5H)pyrimidinetrione; 5-(2-bromoallyl)-5-sec-butylbarbituric acid;* sonbutal. $C_{11}H_{15}BrN_2O_3$; mol wt 303.16. C 43.58%, H 4.99%, Br 26.36%, N 9.24%, O 15.83%. Prepn: **U.S.** pat. **1,739,662** (1929).

Crystals. Slightly bitter taste. mp 130-133°. Freely sol in alcohol, ether, alkaline solns. Practically insol in water (acid to litmus), petr ether, aliphatic hydrocarbons.

Sodium salt, $C_{11}H_{14}BrN_2NaO_3$, *sodium 5-(2-bromoallyl)-5-sec-butylbarbiturate.* Powder or crystals, bitter taste. Soluble in water, alcohol; slightly sol in ether, chloroform. A 10% aq soln (pH of about 9.5) is alkaline to litmus and phenolphthalein.

Caution: May be habit forming. This is a controlled substance (depressant) listed in the U.S. Code of Federal Regulations, Title 21 Parts 329.1 and 1308.13 (1995).

THERAP CAT: Sedative, hypnotic.

1538. Butamben. *4-Aminobenzoic acid butyl ester;* butyl aminobenzoate; *n*-butyl *p*-aminobenzoate; Butesin; Butoform; Planoform; Scuroforme. $C_{11}H_{15}NO_2$; mol wt 193.25. C 68.37%, H 7.82%, N 7.25%, O 16.56%. Prepn: Brill, *J. Am. Chem. Soc.* 43, 1322 (1921); Adams, Volwiler, **U.S.** pat. **1,440,652** (1923 to Abbott); **Brit.** pat. **252,870;** *C.A.* 17, 1243 (1923); *C.A.* 21, 2478 (1927).

Crystals from alc, mp 57-59°. bp_8 174°. One gram dissolves in about 7 liters of water. Sol in dil acids, alcohol, chloroform, ether, and fatty oils. It is hydrolyzed slowly when boiled with water.

THERAP CAT: Anesthetic (local).

THERAP CAT (VET): Anesthetic (local).

1539. Butamirate. *α-Ethylbenzeneacetic acid 2-[2-(diethylamino)ethoxy]ethyl ester; 2-phenylbutyric acid 2-[2-(diethylamino)ethoxy]ethyl ester;* 2-[2-(diethylamino)ethoxy]ethyl 2-phenylbutyrate; butamyrate. $C_{18}H_{29}NO_3$; mol wt 307.43. C 70.32%, H 9.51%, N 4.56%, O 15.61%. Prepd by the esterification of phenylethylacetyl chloride with diethylaminoethoxyethanol: **Ger.** pat. **1,151,515;** Heusser, **U.S.** pat. **3,349,114** (1963, 1967, both to Hommel A.G.).

Practically colorless liquid with peculiar odor. bp_1 140-155°. Practically insol in water. Very sol in alcohol, acetone, ether.

Citrate, $C_{18}H_{29}NO_3 \cdot C_6H_8O_7$, *Abbott 36581, HH-197, Acodeen, Panatus, Sincodix, Sinecod.* White, hygroscopic crystals from acetone, mp 75°.

THERAP CAT: Antitussive.

1540. Butamisole. *2-Methyl-N-[3-(2,3,5,6-tetrahydroimidazo[2,1-b]thiazol-6-yl)phenyl]propanamide; (−)-2-methyl-3'-(2,3,5,6-tetrahydroimidazo[2,1-b]thiazol-6-yl)propionanilide.* $C_{15}H_{19}N_3OS$; mol wt 289.40. C 62.25%, H 6.62%, N 14.52%, O 5.53%, S 11.08%. Prepn: L. D. Spicer, J. J. Hand, S. Afr. pat. **71 05,803** (1972 to Am. Cyanamid), *C.A.* **78,** 72151h (1973). Prepn and resolution of isomers: *eidem,* **Fr.** pat. **2,199,979** corresp to **U.S.** pat. **3,899,583** (1974, 1975 both to Am. Cyanamid). Efficacy and safety in dogs: B. T. Alford *et al., Vet. Med. Small Anim. Clin.* 74, 487

(1979). Evaluation in horses and ponies: R. B. Grieve *et al., Am. J. Vet. Res.* **40**, 139 (1979).

Hydrochloride, $C_{15}H_{19}N_3OS.HCl$, *CL-206214, Styquin.*
THERAP CAT (VET): Anthelmintic.

1541. Butane. *n*-Butane. C_4H_{10}; mol wt 58.12. C 82.66%, H 17.34%. $CH_3CH_2CH_2CH_3$. Occurrence: in petroleum, Mabery, *J. Am. Chem. Soc.* **30**, 143 (1908); in natural gas and in refinery cracking products. Prepd from C_2H_5I and sodium amalgam: Löwig, *Jahresber. Fortschr. Chem.* **1860**, 397, *Beilstein*, vol. **1**, 118. Recovery of butanes from natural and refinery gases: Kirkbride, Bertelli, *Ind. Eng. Chem.* **35**, 1242 (1943); Walters, *ibid.* **47**, 2544 (1955); Gilmore, Bauer, *Oil Gas J.* **50**, 84, 90, 94, 119 (1951), *C.A.* **46**, 1743d (1952). Separation of butane and isobutane: Stone, *Petrol. Refiner* **25**(4), 164 (1946), *C.A.* **43**, 2414 (1949). Handbook: *Butane-Propane Gases,* L. C. Denny *et al.,* Eds. (Chilton Co., Los Angeles, 4th ed., 1962) 383 pp.
Flammable gas, flash pt $-138°$. bp $-0.50°$. d(gas) 2.046 (air = 1). One vol of water dissolves 0.15 vol and 1 vol of alcohol 18 vols of the gas at 17° and 770 mm; 1 vol of ether or chloroform at 17° dissolves 25 or 30 vols of the gas, resp.
Caution: May be narcotic in high concns. A simple asphyxiant.
USE: As producer gas; raw material for motor fuels, in the manuf of synthetic rubbers.

1542. Butanilicaine. *2-(Butylamino)-N-(2-chloro-6-methylphenyl)acetamide; 2-(butylamino)-6'-chloro-o-acetotoluidide;* ω-*n*-butylaminoacetic acid 2-methyl-6-chloroanilide; 1-(butylaminoacetylamino)-2-chloro-6-methylbenzene; *N*-(butylaminoacetyl)-6-chloro-*o*-toluidine. $C_{13}H_{19}ClN_2O$; mol wt 254.76. C 61.29%, H 7.52%, Cl 13.92%, N 11.00%, O 6.28%. Prepn: Häussler, Ther, *Arzneimittel-Forsch.* **3**, 609 (1953); **Brit.** pats. **726,080; 759,744** (1955, 1956, both to Cilag); **Brit.** pat. **782,971** and Ehrhart *et al.,* U.S. pat. **2,912,460** (1957, 1959, both to Hoechst); Epstein, Kaminsky, *J. Am. Chem. Soc.* **80**, 1892 (1958).

Crystals, mp 45-46°. $bp_{0.001}$ 145°. $bp_{0.5}$ 166-167°.
Hydrochloride, $C_{13}H_{19}ClN_2O.HCl$, crystals from ethanol, mp 232°, also reported as mp 236-239° (Epstein, Kaminsky). Soluble in water. Aq soln is neutral. LD_{50} s.c. in mice: 700 mg/kg.
Phosphate, present in *Hostacaine.* Crystals, mp 126-127°.
THERAP CAT: Anesthetic (local).

1543. Butaperazine. *1-[10-[3-(4-Methyl-1-piperazinyl)-propyl]-10H-phenothiazin-2-yl]-1-butanone;* 2-butyryl-10-[3-(4-methyl-1-piperazinyl)propyl]phenothiazine; *N*-[γ-(4'-methyl-1'-piperazinyl)propyl]-3-butyrylphenothiazine; butyrylperazine; Bayer 1362; Repoise; Tyrylen. $C_{24}H_{31}N_3OS$; mol wt 409.60. C 70.38%, H 7.63%, N 10.26%, O 3.91%, S 7.83%. Prepn: Hoerlein *et al.,* Ger. pat. **1,120,451** (1961 to Bayer), *C.A.* **57**, 4677c (1962).

$bp_{0.05}$ 270-280°.
Dimaleate, $C_{24}H_{31}N_3OS.2C_4H_4O_4$, *Randolectil.*
Maleate, $C_{24}H_{31}N_3OS.C_4H_4O_4$, crystals from carbon tetrachloride, mp 180-182°.
THERAP CAT: Antipsychotic.

1544. Butaverine. β-*Phenyl-1-piperidinepropanoic acid butyl ester;* β-*phenyl-1-piperidinepropionic acid butyl ester;* butyl β-phenyl-1-piperidinepropionate; butyl β-piperidinohydrocinnamate; *n*-butyl β-(*N*-piperidyl)-β-phenylpropionate; butamiverine. $C_{18}H_{27}NO_2$; mol wt 289.42. C 74.70%, H 9.40%, N 4.84%, O 11.06%. Prepn: Pollard, Mattson, *J. Am. Chem. Soc.* **78**, 4089 (1956); Pacheco *et al., Bull. Soc. Chim. France* **1962**, 1379.

Hydrochloride, $C_{18}H_{27}NO_2.HCl$, *Espasmo-Gemora, Gemora.* Crystals, mp 170°.
THERAP CAT: Antispasmodic.

1545. Butazolamide. *N-[5-(Aminosulfonyl)-1,3,4-thiadiazol-2-yl]butanamide; N-(5-sulfamoyl-1,3,4-thiadiazol-2-yl)butyramide;* 5-butyramido-1,3,4-thiadiazole-2-sulfonamide; SKF-4965; Butamide. $C_6H_{10}N_4O_3S_2$; mol wt 250.30. C 28.79%, H 4.03%, N 22.38%, O 19.18%, S 25.62%. Carbonic anhydrase inhibitor. Prepn: Vaughan *et al., J. Org. Chem.* **21**, 700 (1956); **Brit.** pat. **769,757** (1957 to Am. Cyanamid).

Crystals, mp 260-262° (dec).
THERAP CAT: Diuretic.

1546. Butedronic Acid. *(Diphosphonomethyl)butanedioic acid; (diphosphonomethyl)succinic acid;* 2,3-dicarboxypropane-1,1-diphosphonic acid; DPD. $C_5H_{10}O_{10}P_2$; mol wt 292.08. C 20.56%, H 3.45%, O 54.78%, P 21.21%. Prepn: A. Heins *et al.,* Ger. pat. **2,217,692** (1973 to Henkel); *eidem,* U.S. pat. **3,923,876** (1975 to Bayer A.G.); K. H. Worms, H. Blum, *Z. Anorg. Allgem. Chem.* **457**, 219 (1979); and physical-chemical properties: N. Vanlic-Razumenic *et al., J. Serb. Chem. Soc.* **51**, 63 (1986). Comparative biokinetics of the ^{99m}Tc complex in volunteers: C. Schümichen, H. Schmidt, *Nuklearmedizin,* Suppl. **19**, 930 (1982). Comparative evaluation in rabbits of ^{99m}Tc complex as skeletal imaging agent: G. Subramanian et al., Radiology **149**, 823 (1983); in humans: G. Godart et al., Clin. Nucl. Med. **11**, 92 (1986).

Monohydrate, white crystalline powder from glacial acetic acid and water (15:1), mp 150°. uv max (water): 208 nm (ϵ 274).

Tetrasodium salt, $C_5H_6Na_4O_{10}P_2$, Tc-924 (DPD), Teceos (non-radioactive kit).

THERAP CAT: [99m]Tc complex as diagnostic aid (radioactive imaging agent).

1547. Butenafine. N-[[4-(1,1-Dimethylethyl)phenyl]-methyl]-N-methyl-1-naphthalenemethanamine; N-(p-tert-butylbenzyl)-N-methyl-1-naphthalenemethylamine. $C_{23}H_{27}$-N; mol wt 317.47. C 87.02%, H 8.57%, N 4.41%. Benzylamine antifungal; squalene epoxidase inhibitor. Prepn: K. Arai et al., Eur. pat. Appl. **221,781**; M. Arita et al., U.S. pat. **4,822,822** (1987, 1989 both to Mitsui Toatsu). Antifungal activity in vivo: T. Arika et al., Antimicrob. Ag. Chemother. **34**, 2250 (1990). Review of antifungal activity, mechanism of action, and clinical trials: R. Fukushiro et al., in Recent Progress in Antifungal Chemotherapy, H. Yamaguchi et al., Eds. (Dekker, New York, 1992) pp. 147-157.

Hydrochloride, $C_{23}H_{27}N \cdot HCl$, KP-363, Mentax. Crystals from methanol + acetic acid, mp 200-202°. Easily sol in methanol, ethanol, dichloromethane, chloroform. Slightly sol in water.

THERAP CAT: Antifungal.

1548. 1-Butene. α-Butylene; ethylethylene. C_4H_8; mol wt 56.11. C 85.63%, H 14.37%. $CH_3CH_2CH{=}CH_2$. Occurs in oil and coal gas. Obtained by cracking of petr oils and by thermal decompn of butane or pentane or isopentane: Egloff et al., Ind. Eng. Chem. **28**, 1283 (1936); Calingaert, J. Am. Chem. Soc. **45**, 130 (1923); from butyl alcohol by treatment with conc H_2SO_4: Compt. Rend. **176**, 813 (1923).

Gas. Not solid at $-190°$. $bp_{760} -6.47°$. $d_4^{-6.47}$ 0.6255. Explodes in mixtures with oxygen. Description of properties: Oil, Gas J. **44**, 119 (1945), C.A. **39**, 3783; J. Am. Chem. Soc. **50**, 1427 (1928).

Caution: May be narcotic in high concns. A simple asphyxiant.

1549. 2-Butene. Pseudo-butylene; sym-dimethylethylene; β-butylene. C_4H_8; mol wt 56.11. C 85.63%, H 14.37%. $CH_3CH{=}CHCH_3$. Occurs in coal gas. Obtained by cracking of petroleum oils. From isobutanol by the action of hot $ZnCl_2$: LeBel, Greene, Am. Chem. J. **2**, 24 (1880). Configuration of the stereoisomeric forms: Kistiakowsky, J. Am. Chem. Soc. **57**, 879 (1935).

Flammable gas.

cis-Form, mp $-139.3°$; $bp_{760} +3.73°$.

trans-Form, mp $-105.8°$; $bp_{744} +0.3$-$0.4°$.

1550. Butethal. 5-Butyl-5-ethyl-2,4,6(1H,3H,5H)-pyrimidinetrione; 5-butyl-5-ethylbarbituric acid; butobarbitone; Soneryl; Neonal; Butobarbital; Etoval. $C_{10}H_{16}N_2O_3$; mol wt 212.25. C 56.59%, H 7.60%, N 13.20%, O 22.61%. Prepn:

U.S. pat. **1,609,520** (1926). Pharmacology and toxicity: G. A. Alles et al., J. Pharmacol. Exp. Ther. **89**, 356 (1947).

Crystals. Slightly bitter taste. mp 124-127°. One gram dissolves in about 5 ml alcohol, 10 ml ether. Practically insol in water; insol in petr ether, aliphatic hydrocarbons. LD_{50} i.p. in mice: 1.506 mM/kg (Alles).

Caution: May be habit forming. This is a controlled substance (depressant) listed in the U.S. Code of Federal Regulations, Title 21 Parts 329.1 and 1308.13 (1995).

THERAP CAT: Sedative, hypnotic.

1551. Butethamate. Benzeneacetic acid α-ethyl-2-(diethylamino)ethyl ester; 2-phenylbutyric acid 2-(diethylamino)ethyl ester; β-diethylaminoethyl ethylphenylacetate; 2-diethylaminoethyl 2-phenylbutyrate. $C_{16}H_{25}NO_2$; mol wt 263.38. C 72.97%, H 9.57%, N 5.32%, O 12.15%. Prepn: Di Paco, Tauro, Farmaco Ed. Sci. **11**, 540 (1956); Swiss pat. **291,375** (1953 to AG Hommel's Haematogen) and Swiss pat. **292,596** (1953 to Chem. Fabrik "PARA"), C.A. **49**, 2505a,c (1955). Pharmacology: Jordan, Arzneimittel-Forsch. **8**, 716 (1958); Fleisch et al., ibid. **11**, 1119 (1961).

bp_{11} 167-169°. n_D^{20} 1.4909.

Citrate, $C_{16}H_{25}NO_2 \cdot C_6H_8O_7$, Abuphenine, Convenil, Hicoseen, Phenesin, Phenetin. Crystals from abs alcohol, mp 109-110°. Freely sol in water, alcohol.

THERAP CAT: Antitussive.

1552. Butethamine. 2-[(2-Methylpropyl)amino]ethanol 4-aminobenzoate (ester); 2-(isobutylamino)ethanol p-aminobenzoate (ester); 2-(isobutylamino)ethyl p-aminobenzoate. $C_{13}H_{20}N_2O_2$; mol wt 236.31. C 66.07%, H 8.53%, N 11.85%, O 13.54%. Prepn of the hydrochloride: Goldberg, U.S. pat. **2,139,818** (1938 to Novocol Chem.); J. Büchi et al., Arzneimittel-Forsch. **14**, 161 (1964); **16**, 1657 (1966).

Formate, $C_{13}H_{20}N_2O_2 \cdot CH_2O_2$, mp 136-139°. Freely sol in water and alcohol. Slightly sol in chloroform, ether; very slightly sol in benzene. pH (1% aq soln): about 6.1.

Hydrochloride, $C_{13}H_{20}N_2O_2 \cdot HCl$, Ibylcaine, Monocaine. mp 192-196°. Sol in water, slightly sol in alcohol, chloroform, benzene. Practically insol in ether. pH (1% aq soln): about 4.7.

meta-Isomer hydrochloride, $C_{13}H_{20}N_2O_2 \cdot HCl$, metabutethamine hydrochloride. Bitter crystals, mp 181-184°. Sol in water. pH of 1:50 aq soln about 6.2. Slightly sol in alcohol, acetone, chloroform.

THERAP CAT: Anesthetic (local).

THERAP CAT (VET): Local anesthetic for nerve block.

1553. Buthalital Sodium. Dihydro-5-(2-methylpropyl)-5-(2-propenyl)-2-thioxo-4,6(1H,5H)-pyrimidinedione monosodium salt; 5-allyl-5-isobutyl-2-thiobarbituric acid sodium salt; sodium 5-allyl-5-isobutyl-2-thiobarbiturate; 5-allyl-5-(2-methylpropyl)-2-thiobarbituric acid sodium salt; thialbutone sodium; Baytinal; Buthalitone Sodium; Thialisobumalnatrium; Transithal; Ulbreval. $C_{11}H_{15}N_2NaO_2S$; mol wt 262.31. C 50.37%, H 5.76%, N 10.68%, Na 8.76%, O 12.20%, S 12.22%. Prepn: Miller *et al., J. Am. Chem. Soc.* **58**, 1090 (1936). Pharmacology: Schildt, Schildt, *Acta Pharmacol. Toxicol.* **19**, 377 (1962).

Pale yellow, hygroscopic powder. Freely sol in water, yielding alkaline solns. Partly sol in ethanol. Insol in ether, benzene.

Free acid, $C_{11}H_{16}N_2O_2S$, crystals, mp 147°.

Note: This is a controlled substance (depressant) listed in the U.S. Code of Federal Regulations, Title 21 Part 1308.13 (1995).

THERAP CAT: Anesthetic (intravenous).

1554. Buthiazide. *6-Chloro-3,4-dihydro-3-(2-methylpropyl)-2H-1,2,4-benzothiadiazine-7-sulfonamide 1,1-dioxide;* 6-chloro-3,4-dihydro-3-isobutyl-7-sulfamoyl-1,2,4-benzothiadiazine 1,1-dioxide; thiabutazide; butizide; isobutylhydrochlorothiazide; Su-6187; S-3500; Eunephran; Saltucin. $C_{11}H_{16}ClN_3O_4S_2$; mol wt 353.85. C 37.34%, H 4.56%, Cl 10.02%, N 11.88%, O 18.09%, S 18.12%. Prepd from 5-chloro-2,4-disulfamoylaniline and isovaleraldehyde: Werner *et al., J. Am. Chem. Soc.* **82**, 1161 (1960); **Brit. pats.** 861,-367; 885,078; G. de Stevens, L. H. Werner, **U.S. pat. 3,163,-645** (1961, 1961, 1964 all to Ciba); Topliss *et al., J. Org. Chem.* **26**, 3842 (1961).

Crystals, mp 241-245° (Werner, 1960); from methanol + chloroform, mp 228° (Topliss).

An ingredient of *Modenol.*

THERAP CAT: Diuretic, antihypertensive.

1555. Buthiobate. *3-Pyridinylcarbonimidodithioic acid butyl [4-(1,1-dimethylethyl)phenyl]methyl ester;* butyl 4-tert-butylbenzyl *N*-(3-pyridyl)dithiocarbonimidate; S-1358; Denmert. $C_{21}H_{28}N_2S_2$; mol wt 372.60. C 67.70%, H 7.57%, N 7.52%, S 17.21%. Prepn: S. Tanaka *et al.,* **Ger. pat. 2,119,-174;** *eidem,* **U.S. pat. 3,832,351** (1971, 1974 both to Sumitomo). Mode of action: T. Kato *et al., Agr. Biol. Chem.* **38**, 2377 (1974); **39**, 169 (1975); T. Kato, Y. Kawase, *ibid.* **40**, 2379 (1976). Metabolism: H. Ohkawa *et al., ibid.* **39**, 1605 (1975). Degradation: *eidem, ibid.* **40**, 943 (1976).

Yellowish, oily liquid, mp 31-33°. d_{25}^{25} 1.0865; $n_D^{26.5}$ 1.596. Vapor pressure at 20°: 4.52×10^{-7} mm Hg. Practically insol in water. Sol in most organic solvents. LD_{50} orally in male, female rats: 4.9, 2.7 g/kg (Ohkawa, 1975).

USE: Fungicide.

1556. Buthionine Sulfoximine. *2-Amino-4-(S-butylsulfonimidoyl)butanoic acid;* S-(3-amino-3-carboxypropyl)-S-butylsulfoximine; S-(n-butyl)homocysteine sulfoximine; BSO. $C_8H_{18}N_2O_3S$; mol wt 222.31. C 43.22%, H 8.16%, N 12.60%, O 21.59%, S 14.42%. Selective inhibitor of γ-gluta-

myl cysteine synthetase, an enzyme in the glutathione, *q.v.,* biosynthetic pathway. BSO-mediated depletion of intracellular glutathione has been associated with increased sensitivity of tumor cells to antineoplastic agents. Prepn: K. Hayashi, *Chem. Pharm. Bull.* **8**, 177 (1960). Synthesis and *in vitro* and *in vivo* enzyme inhibition: O. W. Griffith, A. Meister, *J. Biol. Chem.* **254**, 7758 (1979). Mechanism of action and metabolism in animals O. W. Griffith, *ibid.* **257**, 13704 (1982). Chemosensitization by BSO of melphalanresistant murine leukemia cells *in vivo:* S. Somfai-Relle *et al., Biochem. Pharmacol.* **33**, 485 (1984); R. A. Kramer *et al., Cancer Res.* **47**, 1593 (1987). Enhanced melphalan, *q.v.,* cytotoxicity in human cancer cells *in vitro:* J. A. Green *et al., ibid.* **44**, 5427 (1984); and in mice bearing human tumor xenografts: H. S. Friedman *et al., J. Nat. Cancer Inst.* **81**, 524 (1989).

Crystals from aq ethanol, mp 214-215.5° (dec). Sol in water. Partition coefficient (octanol/water): $4.34 \pm 0.4 \times 10^{-3}$.

USE: Biological tool for depletion of glutathione.

1557. Butibufen. *α-Ethyl-4-(2-methylpropyl)benzeneacetic acid;* 2-(4-isobutylphenyl)butyric acid; Butilopan. $C_{14}H_{20}O_2$; mol wt 220.31. C 76.33%, H 9.15%, O 14.52%. Prepn: L. Aparicio *et al.,* **Ger. pat. 2,505,813;** *eidem,* **U.S. pat. 4,031,243** (1976, 1977 both to Juste); J. M. Carretero *et al., Eur. J. Med. Chem.* **13**, 77 (1978). Pharmacology: L. Aparicio, *Arch. Int. Pharmacodyn. Ther.* **227**, 130 (1977). Pharmacokinetics: R. Revilla De Granda *et al., An. R. Acad. Farm.* **43**, 419 (1977), *C.A.* **88**, 83349 (1978). HPLC determination: L. González Tavares *et al., Arzneimittel-Forsch.* **42**, 818 (1992).

Solid, mp 51-53°. LD_{50} orally in mice: 810 mg/kg (Carretero).

THERAP CAT: Anti-inflammatory.

1558. Butidrine Hydrochloride. *5,6,7,8-Tetrahydro-α-[[(1-methylpropyl)amino]methyl]-2-naphthalenemethanol hydrochloride;* α-[(sec-butylamino)methyl]-5,6,7,8-tetrahydro-2-naphthalenemethanol hydrochloride; butydrine hydrochloride; Betabloc; Recetan. $C_{16}H_{26}ClNO$; mol wt 283.84. C 67.71%, H 9.23%, Cl 12.49%, N 4.93%, O 5.64%. β-Adrenergic blocker. Prepn: Ferrari *et al., Boll. Chim. Farm.* **103**, 32 (1964), *C.A.* **61**, 5580d (1964); **Fr. pat. 1,390,056** (1965 to Holding Ceresia), *C.A.* **62**, 16162d (1965). Pharmacology: R. Ferrini, *Arzneimittel-Forsch.* **18**, 48 (1968). Brief review of butidrine and other β-blockers: A. M. Karow *et al., Prog. Drug Res.* **15**, 103-122 (1971).

Crystals, mp 129-130°. LD_{50} in mice (mg/kg): 20.2 i.v.; 235 orally (Ferrini).

THERAP CAT: Antiarrhythmic.

1559. Butirosin. Ambutyrosin. $C_{21}H_{41}N_5O_{12}$; mol wt 555.58. C 45.40%, H 7.44%, N 12.61%, O 34.56%. Aminoglycosidic antibiotic complex obtained from fermentation filtrates of mucoid strains of *Bacillus circulans* (NRRL B-3312 and B-3313). Consists of two components, butirosin A (80-85%) and butirosin B (15-20%), isomers which differ only in the configuration at one carbon atom in the pentose moiety. Prepn: P. W. K. Woo *et al.*, **Ger. pat. 1,914,527;** *eidem,* **U.S. pat. 3,541,078** (1969, 1970 both to Parke, Davis). Isoln and characterization: Dion *et al., Antimicrob. Ag. Chemother.* **2,** 84 (1972). Structures: Woo *et al., Tetrahedron Letters* **1971,** 2617, 2621, 2625. Activity: Howells *et al., Antimicrob. Ag. Chemother.* **2,** 79 (1972); Heifetz *et al., ibid.* 89. Synthesis of butirosin B: Ikeda *et al., J. Antibiot.* **25,** 741 (1972); Akita *et al., ibid.* **26,** 365 (1973). Proposed biosynthetic pathway to the butirosins: K. Takeda *et al., ibid.* **31,** 250 (1978).

Butirosin A : R = OH   R' = H
Butirosin B : R = H   R' = OH

Butirosin sulfate dihydrate, $C_{21}H_{41}N_5O_{12}.2H_2SO_4.2H_2O$. No sharp mp, dec ~225°. $[\alpha]_D^{25}$ +29° (c = 2 in water). pKa' (water): 5.5, 7.2, 8.5, 9.4. Very sol in water; moderately sol in methanol; slightly sol in ethanol. LD_{50} i.v. in mice: 450-500 mg/kg (Howells).

Butirosin A. *(S)-O-2,6-Diamino-2,6-dideoxy-α-D-glucopyranosyl-(1→4)-O-[β-D-xylofuranosyl-(1→5)]-N¹-(4-amino-2-hydroxy-1-oxobutyl)-2-deoxystreptamine.* White, amorphous solid, melts with dec over wide range, beginning at ~149°. $[\alpha]_D^{25}$ +26° (c = 1.46 in water). pKa' (water): 5.6, 7.3, 8.7, 9.8.

Butirosin B. *1-N-[(S)-4-Amino-2-(hydroxybutyryl)]ribostamycin.* Occurs as a dihydrate, melts over wide range beginning at 146°. $[\alpha]_D^{25}$ +33° (c = 1.5 in water). pKa' (water): 5.3, 7.1, 8.6, 9.8.

THERAP CAT: Antibacterial.

1560. Butobendine. *[S-(R*,R*)]-3,4,5-Trimethoxybenzoic acid 1,2-ethanediylbis[(methylimino)(2-ethyl-2,1-ethanediyl)] ester;* (2S,2'S)-N,N'-dimethyl-N,N'-bis-1-(3',4',5'-trimethoxybenzoyloxy)butyl-2-ethylenediamine; (+)-(S,S)-ethylenebis[(methylimino)(2-ethylethylene)]bis(3,4,5-trimethoxybenzoate). $C_{32}H_{48}N_2O_{10}$; mol wt 620.74. C 61.92%, H 7.79%, N 4.51%, O 25.77%. Antiarrhythmic agent which increases cardiac blood flow. Prepn: M. Eckstein *et al.*, **Ger. pat. 2,435,380;** *eidem,* **U.S. pat. 4,021,473** (1975, 1977 both to Polfa). Series of articles on pharmacology, pharmacokinetics, toxicology: *Pol. J. Pharmacol. Pharm.* **32,** 817-953 (1980). Physical properties: L. Krowczynski *et al., ibid.* 909. Toxicity data: J. Maj *et al., ibid.* 823.

Crystals, mp 60-62°. Also reported as mp 64-65° (Krowczynski). $[\alpha]_D^{20}$ +2.4° (c = 5 in ethanol). Partition coefficient (*n*-octanol/water) at 20°: 31.49.

Dihydrochloride, $C_{32}H_{48}N_2O_{10}.2HCl$, *M-71, Craviten.* White crystalline powder. Forms mono-, di- and trihydrates, melting range 83-113°. Also reported as crystals from isopropanol, mp 81-83° (Krowczynski). $[\alpha]_D^{20}$ −6.4° (c = 2.5 in ethanol). $[\alpha]_D^{20}$ −7.5° (c = 5 in H_2O). $[\alpha]_D^{20}$ −5.5° (c = 5 in pyridine). pH (8% aq soln): 2.20. Partition coefficient (*n*-octanol/water) at 20°: 4.49. Sol in water, chloroform, ethanol. Slightly sol in methanol. Practically insol in benzene, CCl_4, ether. LD_{50} in rats (mg/kg): 142.0

i.p., 15.8 i.v.; in mice (mg/kg): 550.0 i.p.; in rabbits (mg/kg): 5.1 i.v. (Maj).

THERAP CAT: Antiarrhythmic.

1561. Butoconazole. *(±)-1-[4-(4-Chlorophenyl)-2-[(2,6-dichlorophenyl)thio]butyl]-1H-imidazole.* $C_{19}H_{17}Cl_3N_2S$; mol wt 411.78. C 55.42%, H 4.16%, Cl 25.83%, N 6.80%, S 7.79%. Imidazole derivative with antifungal properties. Prepn: K. A. M. Walker, **U.S. pat. 4,078,071** (1978 to Syntex). Prepn, toxicity, activity vs *Candida albicans* in mice: K. A. M. Walker *et al., J. Med. Chem.* **21,** 840 (1978). *In vitro* comparison with other antifungal agents: F. C. Odds *et al., J. Antimicrob. Chemother.* **14,** 105 (1984). Clinical trials in treatment of vulvovaginal candidiasis: W. Droegemueller *et al., Obstet. Gynecol. (N.Y.)* **64,** 530 (1984); J. B. Jacobson *et al., Acta Obstet. Gynecol. Scand.* **64,** 241 (1985). Comparison with miconazole, *q.v.*: C. S. Bradbeer *et al., Genitourin. Med.* **61,** 270 (1985).

Crystals from cyclohexane, mp 68-70.5°.

Nitrate, $C_{19}H_{17}Cl_3N_2S.HNO_3$, *RS-35887, Femstat, Gynomyk.* Colorless blades from acetone/ethyl acetate, mp 162-163°. LD_{50} in mice, male, female rats (mg/kg): > 3200, > 3200, 1720 orally; > 1600, 940, 940 i.p. (Walker).

THERAP CAT: Antifungal (topical).

1562. Butoctamide. *N-(2-Ethylhexyl)-3-hydroxybutanamide; N-(2-ethylhexyl)-3-hydroxybutyramide;* hexobutyramide. $C_{12}H_{25}NO_2$; mol wt 215.34. C 66.93%, H 11.70%, N 6.50%, O 14.86%. Prepn: A. Sakuma *et al.*, **Fr. pat. 8393;** *eidem,* **U.S. pat. 3,639,457;** *eidem,* **Japan. Kokai 79 24822,** *C.A.* **91,** 34947f (1979) (1971, 1972, 1979 all to Lion Dentifrice). Toxicology study: M. Kato *et al., Oyo Yakuri* **17,** 287 (1979), *C.A.* **91,** 68665e (1979). Acute toxicity: R. Goshima *et al., Toho Igakka Zasshi* **17,** 579 (1970), *C.A.* **74,** 138765n (1971). Clinical effect on rapid eye movement (REM) sleep: Y. Hayashi *et al., Psychopharmacology* **77,** 367 (1982); in Down's syndrome: J. C. Grubar *et al., ibid.* **90,** 119 (1986).

Colorless viscous oil, $bp_{0.30\ mm}$ 149-150°.

Hydrogen succinate, $C_{16}H_{29}NO_5$, *butanedioic acid mono-[3-[(2-ethylhexyl)amino]-1-methyl-3-oxopropyl] ester,* succinic acid monoester with N-(2-ethylhexyl)-3-hydroxybutyramide, N-2-ethylhexyl-β-oxybutyramide semisuccinate, *BAHS, M-2H, Listomin S.* Colorless viscous oil; mp 46.5°.

Hydrogen succinate calcium salt, $C_{32}H_{58}CaN_2O_{10}$, white powder, slightly hygroscopic. uv max: 192 ($E_{1cm}^{1\%}$ 290). LD_{50} in mice (mg/kg): 476 i.p.; 2000 orally (Sakuma, 1971).

THERAP CAT: Sedative, hypnotic.

1563. Butofilolol. *(±)-1-[2-[3-[[(1,1-Dimethylethyl)amino]-2-hydroxypropoxy]-5-fluorophenyl]-1-butanone;* (±)-5-fluoro-2-(2-hydroxy-3-*t*-butylaminopropoxy)butyrophenone; (±)-2'-[3-(*tert*-butylamino)-2-hydroxypropoxy]-5'-fluorobutyrophenone; CM-6805. $C_{17}H_{26}FNO_3$; mol wt 311.40. C 65.57%, H 8.42%, F 6.10%, N 4.50%, O 15.41%. β-Adrenergic blocker. Prepn: H. Demarne le Florence, **Ger. pat. 2,528,147;** *idem,* **U.S. pat. 4,302,601** (1976, 1981 both to C. M. Industries). Pharmacological study: G. G. Re *et al., Boll. Soc. Ital. Biol. Sper.* **56,** 1264 (1980), *C.A.* **93,** 197668y (1980). GC and HPLC determn in plasma and urine: J. P. Jeanniot *et al., J. Chromatog.* **278,** 301 (1983). Pharmacokinetics in humans: G. Houin *et al.*, *Int. J. Clin. Pharm. Res.* **4**, 175 (1984).

Crystals from diisopropyl ether, mp 88-89°.
Maleate, $C_{17}H_{26}FNO_3.C_4H_4O_4$, *Cafide.*
THERAP CAT: Antihypertensive.

1564. Butopyronoxyl. *3,4-Dihydro-2,2-dimethyl-4-oxo-2H-pyran-6-carboxylic acid butyl ester;* butyl 3,4-dihydro-2,-2-dimethyl-4-oxo-2H-pyran-6-carboxylate; butyl mesityl oxide oxalate; α,α-dimethyl-α'-carboxydihydro-γ-pyrone butyl ester; Indalone. $C_{12}H_{18}O_4$; mol wt 226.27. C 63.70%, H 8.02%, O 28.28%. Prepd by condensation of mesityl oxide with dibutyl oxalate in the presence of sodium or sodium butoxide: Ford, U.S. pat. **2,138,540** (1938 to Kilgore Dev.). Exists as the dihydropyrone in equilibrium with the open chain enol form: Hall *et al.*, *J. Am. Chem. Soc.* **67**, 1224 (1945). Toxicity study: J. H. Draize *et al.*, *J. Pharmacol. Exp. Ther.* **93**, 26 (1948).

Yellow to pale reddish-brown liquid. Aromatic odor. d_{25}^{25} 1.052-1.060. bp_{760} 256-270°. n_D^{25} 1.4745-1.4755. Practically insol in water. Misc with alcohol, chloroform, ether, glacial acetic acid. LD_{50} in mice, rats (ml/kg): 11.6, 7.4 orally (Draize).
Caution: Direct contact may be mildly irritating to skin on repeated application. *See Clinical Toxicology of Commercial Products*, R. E. Gosselin *et al.*, Eds. (Williams & Wilkins, Baltimore, 5th ed., 1984) Section II, p 345.
USE: Insect repellent.

1565. Butorphanol. *17-(Cyclobutylmethyl)morphinan-3,14-diol;* (−)-N-cyclobutylmethyl-3,14-dihydroxymorphinan; *levo*-BC-2627. $C_{21}H_{29}NO_2$; mol wt 327.47. C 77.02%, H 8.93%, N 4.28%, O 9.77%. Mixed opioid agonist-antagonist. Prepn: I. Pachter *et al.*, **Ger.** pat. **2,243,961** corresp to U.S. pat. **3,819,635** (1973, 1974 to Bristol-Myers); I. Monkovic, T. Thomas, U.S. pat. **3,775,414** (1973 to Bristol-Myers). Total synthesis and pharmacology: I. Monkovic *et al.*, *J. Am. Chem. Soc.* **95**, 7910 (1973); *eidem, Can. J. Chem.* **53**, 3094 (1975). Clinical study: F. Vargas-Arreola *et al.*, *Curr. Ther. Res., Clin. Exp.* **22**, 186 (1977). Antitussive effect: R. L. Cavanaugh *et al.*, *Arch. Int. Pharmacodyn. Ther.* **220**, 258 (1976). *Review:* R. C. Heel *et al.*, *Drugs* **16**, 473-505 (1978); F. S. Caruso *et al.*, in *Pharmacological and Biochemical Properties of Drug Substances* vol. 2, M. E. Goldberg, Ed. (Am. Pharm. Assoc., Washington, DC, 1979) pp 19-57.

Solid, mp 215-217°. $[\alpha]_D$ −70.0° (c = 0.1 in methanol). Tartrate, $C_{21}H_{29}NO_2.C_4H_6O_6$, *Stadol, Torate, Torbugesic, Torbutrol.* Solid. mp 217-219°. $[\alpha]_D^{22}$ −64.0° (c = 0.4 in methanol). LD_{50} in mice, rats (mg/kg): 40-57, 17-20 i.v.; 395-527, 570-756 orally (Heel).

THERAP CAT: Analgesic (narcotic).
THERAP CAT (VET): Analgesic (narcotic); antitussive.

1566. Butoxycaine. *4-Butoxybenzoic acid 2-(diethylamino)ethyl ester;* 2-diethylaminoethyl *p*-butoxybenzoate. $C_{17}H_{27}NO_3$; mol wt 293.41. C 69.59%, H 9.28%, N 4.77%, O 16.36%. Prepd from *p*-butoxybenzoyl chloride and β-diethylaminoethanol: Christiansen, Harris, U.S. pat. **2,-412,966** (1946 to Squibb). Alternate prepn and activity: Reynaud *et al.*, *Chim. Ther.* **2**, 25 (1967). *See also* Büchi *et al.*, *Arzneimittel-Forsch.* **18**, 610 (1968).

Hydrochloride, $C_{17}H_{27}NO_3.HCl$, *Stadacain.* Heavy crystals, mp 146°.
THERAP CAT: Anesthetic (local).

1567. Butralin. *4-(1,1-Dimethylethyl)-N-(1-methylpropyl)-2,6-dinitrobenzenamine; N-sec-butyl-4-tert-butyl-2,6-dinitroaniline;* dibutalin; Amchem 70-25; A-820; Amexine; Tamex. $C_{14}H_{21}N_3O_4$; mol wt 295.34. C 56.94%, H 7.17%, N 14.23%, O 21.67%. Pre-emergence herbicide. Prepn: J. J. Damiano, **Ger.** pat. **2,058,201**; *idem,* **U.S.** pat. **3,672,866** (1971, 1972 both to Amchem). Activity: S. R. McLane *et al., Proc. S. Weed Sci. Soc.* **24**, 58 (1971). Soil persistence and metabolism: P. C. Kearney *et al., J. Agr. Food Chem.* **22**, 856 (1974). Photochemistry: J. R. Plimmer, U. I. Klingebiel, *ibid.* 689.

Yellow-orange crystals, mp 60-61°, $bp_{0.5}$ 134-136°. Vapor pressure at 25°: 1.3×10^{-5} mm Hg. Flash pt, open cup: 97°F (36°C). Soly in water at 25°: 1 mg/l. Soly at 25° (kg/kg): methanol 0.125; acetone 4.48; benzene 2.7; xylene 3.88; butanone 9.55; carbon tetrachloride 1.46. LD_{50} orally in rats: 2500 mg/kg (McLane).
USE: Herbicide.

1568. Butriptyline. *10,11-Dihydro-N,N,β-trimethyl-5H-dibenzo[a,d]cycloheptene-5-propanamine;* 5-(3-dimethylamino-2-methylpropyl)-10,11-dihydro-5H-dibenzo[a,d]cycloheptene; 5-(2-methyl-3-dimethylaminopropyl)dibenzo[a,d]-[1,4]cycloheptadiene; butriptyline. $C_{21}H_{27}N$; mol wt 293.45. C 85.95%, H 9.27%, N 4.77%. Prepn: Winthrop, Davis, **Belg.** pat. **613,750** (1962 to Ayerst, McKenna & Harrison), *C.A.* **57**, 15036c (1962); *eidem, J. Org. Chem.* **27**, 230 (1962); Villani, **U.S.** pat. **3,409,640** (1968 to Schering). Review of pharmacology and clinical data: *see J. Med. (Basel)* **2**, 249-343 (1971). Metabolic studies: Cameron *et al., Arzneimittel-Forsch* **24**, 93 (1974). Toxicity: Voith, Herr, *Arch. Int. Pharmacodyn. Ther.* **182**, 318 (1969).

Oil, bp_1 180-185°.
Hydrochloride, $C_{21}H_{27}N.HCl$, *AY-62014, Evadene, Evadyne.* Crystals from isopropyl alcohol-ether, mp 188-190° (dec). uv max (methanol): 273, 270, 266 nm (ϵ 460, 441, 552). Freely sol in water; moderately sol in aliphatic alc,

chloroform. Insol in ether, paraffinic hydrocarbons. LD_{50} in mice (mg/kg): 120 i.p.; 345 orally (Voith, Herr).

THERAP CAT: Antidepressant.

1569. Butropium Bromide. *[3(S)-endo]-8-[(4-Butoxyphenyl)methyl]-3-(3-hydroxy-1-oxo-2-phenylpropoxy)-8-methyl-8-azoniabicyclo[3.2.1]octane bromide; 8-(p-butoxybenzyl)-3α-hydroxy-1αH,5αH-tropanium bromide (−)-tropate; l-[1-(p-n-butoxybenzyl)hyoscyaminium] bromide; BHB; Colio-pan.* $C_{28}H_{38}BrNO_4$; mol wt 532.52. C 63.15%, H 7.19%, Br 15.00%, N 2.63%, O 12.02%. Prepn: S. Tanaka, K. Hashimoto, **Ger. pat. 1,950,378;** *eidem,* **U.S. pat. 3,696,110;** Japan. pat. **72 22,715,** *C.A.* **77,** 79549g (1972) (1970, 1972, 1972 all to Eisai); Tanaka *et al., J. Pharm. Soc. Japan* **92,** 510 (1972). Prepn of the labelled compound: Fujita *et al., J. Label. Compounds* **9,** 149, 555 (1972). Activity: Akutsu, Ichikawa, *Showa Igakkai Zasshi* **32,** 494 (1972), *C.A.* **78,** 119243g (1973). *Review: Japan. Med. Gaz.* **11**(9), 10 (1974).

Crystals from ethanol-acetone, mp 166-168°; also reported as white needles from isopropanol, mp 158-160° (U.S. pat. **3,696,110).** $[α]_D^{20}$ −21.7° (c = 0.5 in water). Freely sol in glacial acetic acid; sol in chloroform, DMF. Sparingly sol in ethanol; slightly sol in water, 0.1N HCl, 0.1N NaOH. Practically insol in acetone, ether, benzene. LD_{50} in male mice (mg/kg): 1500 orally; 660 s.c.; 12.0 i.v. (U.S. pat. **3,696,110).**

THERAP CAT: Antispasmodic; anticholinergic.

1570. n-Butyl Acetate. *Acetic acid butyl ester.* $C_6H_{12}O_2$; mol wt 116.16. C 62.04%, H 10.41%, O 27.55%. $CH_3COO(CH_2)_3CH_3$. Prepd from acetic acid and butyl alcohol: Leyes, Othmer, *Ind. Eng. Chem.* **37,** 968 (1945); Vogel, *J. Chem. Soc.* **1948,** 624; Zettlemoyer *et al.,* U.S. pat. **2,644,839** (1953 to FMC); Faith, Keyes & Clark's *Industrial Chemicals,* F. A. Lowenheim, M. K. Moran, Eds. (Wiley-Interscience, New York, 4th ed., 1975) pp 171-177. Toxicity data: H. F. Smyth *et al., Arch. Ind. Hyg. Occup. Med.* **10,** 61 (1954).

Liquid. d_{20}^{20} 0.8826. bp 125-126°. mp −77°. n_D^{20} 1.3951. Flash pt, closed cup: 72°F (22°C). Sol in about 120 parts water at 25°; misc with alcohol, ether; sol in most hydrocarbons. LD_{50} orally in rats: 14.13 g/kg (Smyth).

Caution: Potential symptoms of overexposure are headache; drowsiness; dryness and irritation of eyes, upper respiratory system and skin. *See NIOSH Pocket Guide to Chemical Hazards* (DHHS/NIOSH 90-117, 1990) p 50.

USE: Manuf lacquer, artificial leather, photographic films, plastics, safety glass.

1571. sec-Butyl Acetate. *Acetic acid 1-methylpropyl ester; acetic acid sec-butyl ester.* $C_6H_{12}O_2$; mol wt 116.16. C 62.04%, H 10.41%, O 27.55%. $CH_3COOCH(CH_3)CH_2CH_3$. Prepd from sec-butanol and acetic anhydride: Altschul, *J. Am. Chem. Soc.* **68,** 2605 (1946). Prepn of d-form: Kenyon *et al., J. Chem. Soc.* **1935,** 1072; Bird, *Tetrahedron* **18,** 1 (1962). Prepn of l- form: Kenyon *et al., loc. cit.* Manuf: Faith, Keyes & Clark's *Industrial Chemicals,* F. A. Lowenheim, M. K. Moran, Eds. (Wiley-Interscience, New York, 4th ed., 1975) pp 171-177.

dl-Form, liquid. d_4^{25} 0.865. bp 112-113°. n_D^{25} 1.3866. Flash pt, open cup: 88° F (31° C). Slightly sol in water; sol in alcohol, ether.

d-Form, liquid, bp 116-117°. d_4^{19} 0.873. n_D^{18} 1.3899.

Caution: Potential symptoms of overexposure are irritation of eyes; headache; drowsiness; dryness of upper respiratory system and skin. *See NIOSH Pocket Guide to Chemical Hazards* (DHHS/NIOSH 90-117, 1990) p 50.

1572. tert-Butyl Acetate. *Acetic acid 1,1-dimethylethyl ester; acetic acid tert-butyl ester.* $C_6H_{12}O_2$; mol wt 116.16. C

62.04%, H 10.41%, O 27.55%. $CH_3COOC(CH_3)_3$. Prepn: Baker, Bordwell; Hauser *et al., Org. Syn.* **coll. vol. III,** 141, 142 (1955). Manuf from acetic acid and isobutylene: Young, Pare, U.S. pat. **3,031,495** (1962 to Sinclair); Wheeler *et al.,* U.S. pat. **3,102,905** (1963 to Celanese); Heisler *et al.,* U.S. pat. **3,096,365** (1963 to Texaco); Faith, Keyes & Clark's *Industrial Chemicals,* F. A. Lowenheim, M. K. Moran, Eds. (Wiley-Interscience, New York, 4th ed., 1975) pp 171-177.

Liquid, bp 97.8°. d_4^{20} 0.8665, d_4^{25} 0.8593. n_D^{20} 1.3870. Practically insol in water. Miscible with alcohol, ether.

Caution: Potential symptoms of overexposure are itching and inflammation of eyes; irritation of upper respiratory tract; headache; narcosis; dermatitis. *See NIOSH Pocket Guide to Chemical Hazards* (DHHS/NIOSH 90-117, 1990) p 50.

USE: As gasoline additive (Wheeler *et al.*).

1573. tert-Butylacetic Acid. *3,3-Dimethylbutanoic acid; 3,3-dimethylbutyric acid.* $C_6H_{12}O_2$; mol wt 116.16. C 62.04%, H 10.41%, O 27.55%. $(CH_3)_3CCH_2COOH$. Prepn: Homeyer *et al., J. Am. Chem. Soc.* **55,** 4209 (1933); Botterson, Shulman, *J. Org. Chem.* **27,** 1059 (1962); A. Nilsson, R. Carlson, *Acta Chem. Scand.* **B34,** 621 (1980).

Liquid, bp_{26} 96°, bp_{739} 183.0-183.3°, bp 190°. mp 6-7°. n_D^{20} 1.4115 (Botterson, Shulman), also reported as 1.4096 (Homeyer *et al.*). d_4^{20} 0.9124.

1574. n-Butyl Acrylate. *2-Propenoic acid butyl ester; acrylic acid n-butyl ester.* $C_7H_{12}O_2$; mol wt 128.17. C 65.60%, H 9.44%, O 24.97%. $CH_2=CHCOOC_4H_9$. Prepn from n-butanol and methyl acrylate: Rehberg, *Org. Syn.* **coll. vol. III,** 146 (1955). Toxicity study: H. F. Smyth *et al., Arch. Ind. Hyg. Occup. Med.* **4,** 119 (1951).

Liquid. d_4^{20} 0.8986; d_4^{15} 0.9110; d_4^{12} 0.9117; d_4^{0} 0.9202. bp_{760} 145° (also reported 138°); bp_{101} 84-86°; bp_{25} 59°; bp_{10} 39°; bp_8 35°. n_D^{20} 1.4190; n_D^{12} 1.4254. Sp heat (−60°): 0.467 cal/g/°C; heat of vaporization 8.11 kcal/mol; heat of combustion 974.46 kcal/mol. Soly in water at 20°: 0.14 g/100 ml; at 40°: 0.12 g/100 ml. Soly of water in n-butyl acrylate at 20°: 0.8 ml/100 g. LD_{50} orally in rats: 3.73 g/kg (Smyth).

Polymer, elastic, tacky substance. Brittle temp −45°.

USE: The monomer in the manuf of polymers and resins for textile and leather finishes, paint formulations, etc.

1575. n-Butyl Alcohol. *1-Butanol; butyl alcohol; propyl carbinol.* $C_4H_{10}O$; mol wt 74.12. C 64.82%, H 13.60%, O 21.58%. $CH_3CH_2CH_2CH_2OH$. Prepn by reduction of butyraldehyde with sodium borohydride: Chaikin, Brown, *J. Am. Chem. Soc.* **71,** 122 (1949). Manuf from ethylene oxide and triethylaluminum: Rudner, U.S. pat. **3,091,627** (1963 to Koppers); by oxidation of tributylborane: Mirviss, U.S. pat. **3,067,235** (1962 to Esso). Manuf by carbohydrate fermentation, by hydrogenation of butyraldehyde, from crotonaldehyde: Faith, Keyes & Clark's *Industrial Chemicals,* F. A. Lowenheim, M. K. Moran, Eds. (Wiley-Interscience, New York, 4th ed., 1975) pp 178-185. Purification and vapor pressure: Biddiscombe *et al., J. Chem. Soc.* **1963,** 1954. Toxicity study: H. F. Smyth *et al. Arch. Ind. Hyg. Occup. Med.* **4,** 119 (1951).

Highly refractive liquid; burns with a strongly luminous flame; leaves a transitory greasy spot on paper. Odor similar to that of fusel oil, but weaker. d_4^{20} 0.810. bp 117-118°. mp −90°. Flash pt 36-38°. n_D^{20} 1.3993. A mixture of 63% of the alcohol and 37% water forms a constant boiling mixture, boiling at 92°. Soly at 25°, 9.1 ml/100 ml H_2O: Booth, Everson, *Ind. Eng. Chem.* **40,** 1491 (1948). Miscible with alc, ether and many other organic solvents. LD_{50} orally in rats: 4.36 g/kg (Smyth).

Caution: Potential symptoms of overexposure are irritation of eyes, nose and throat; headache, vertigo and drowsiness; corneal inflammation, blurred vision, lacrimation and photophobia; dry cracked skin. *See NIOSH Pocket Guide to Chemical Hazards* (DHHS/NIOSH 90-117, 1990) p 52.

USE: As solvent for fats, waxes, resins, shellac, varnish, gums etc.; manuf lacquers, rayon, detergents, other butyl compds; in microscopy for preparing paraffin imbedding materials.

1576.   sec-Butyl Alcohol.   *2-Butanol;* butylene hydrate; 2-hydroxybutane; methyl ethyl carbinol. $C_4H_{10}O$; mol wt 74.12. C 64.82%, H 13.60%, O 21.58%. $CH_3CH_2CH(OH)$-CH_3. Prepn by reduction of 2-butanone: Chaikin, Brown, *J. Am. Chem. Soc.* **71**, 122 (1949); Nystrom *et al., ibid.* 3245. Manuf by hydration of 2-butene or hydrocarbons contg butene: Dale *et al., Ind. Engl. Chem.* **48**, 913 (1956); Archibald, Mottern, U.S. pat. **2,543,820** (1951 to Standard Oil); Limerick, Wylie, U.S. pat. **2,776,324** (1957 to Shell). Purification and vapor pressure: Biddiscombe *et al., J. Chem. Soc.* **1963**, 1954. Prepn of *l*-form or *d*-form by hydroboration of *cis*-2-butene: Brown, U.S. pat. **3,078,313** (1963); Brown *et al., J. Am. Chem. Soc.* **86**, 397 (1964). Resolution of *dl*-form: Kantor, Hauser, *ibid.* **75**, 1744 (1953). Absolute configuration of *l*-form: Brown *et al., ibid.* **86**, 1071 (1964). Toxicity: H. F. Smyth *et al., Arch. Ind. Hyg. Occup. Med.* **10**, 61 (1954).

dl-Form, liquid, bp 99.5°. mp $-114.7°$. d_4^{20} 0.808. n_D^{25} 1.3949. Flash pt, open cup: 88°F (31°C). Sol in 12 parts water; misc with alcohol, ether. LD_{50} orally in rats: 6.48 g/kg (Smyth).

d-Form, liquid, d_4^{27} 0.8025. n_D^{20} 1.3954. $[\alpha]_D^{27}$ +13.52°. *Ref:* Leroux, Lucas, *J. Am. Chem. Soc.* **73**, 41 (1951).

l-Form, liquid, bp_{744} 98°. d_4^{25} 0.8042. n_D^{20} 1.3970, n_D^{25} 1.3949. $[\alpha]_D^{25}$ $-13.51°$. *Ref:* Leroux, Lucas, *loc. cit.*

Caution: Potential symptoms of overexposure are eye irritation and narcosis. *See NIOSH Pocket Guide to Chemical Hazards* (DHHS/NIOSH 90-117, 1990) p 52.

USE: In the synthesis of flotation agents, flavors, perfumes, dyestuffs, wetting agents. In industrial cleaners, paint removers. Solvent for many natural resins, linseed and castor oils.

1577.   tert-Butyl Alcohol.   *2-Methyl-2-propanol;* trimethyl carbinol. $C_4H_{10}O$; mol wt 74.12. C 64.82%, H 13.60%, O 21.58%. $(CH_3)_3COH$. Prepd from acetyl chloride and dimethylzinc: Butlerow, *Ann.* **144**, 1 (1867). Manuf by catalytic hydration of isobutylene: Kreps, Nachod, U.S. pat. **2,477,380** (1949 to Atlantic Refining); Serniuk, Vanderbilt, U.S. pat. **2,534,304** (1950 to Standard Oil); by reduction of *tert*-butyl hydroperoxide: Lorand, U.S. pat. **2,484,841** (1949 to Hercules Powder); De Jong, U.S. pat. **2,853,532** (1958 to Shell). Purification: Biddiscombe *et al., J. Chem. Soc.* **1963**, 1954. Physical properties: *eidem, ibid.;* Dreisbach, Martin, *Ind. Eng. Chem.* **41**, 2875 (1949). Toxicity: Schaffarzick, Brown, *Science* **116**, 663 (1952).

Crystals, camphor-like odor, mp 25.6°. At 99.69 mol-% purity, bp 82.41°. mp 25.7°. d_4^{20} 0.78581, d_4^{25} 0.78086, d_4^{25} (calcd) 0.78080; n_D^{20} 1.38468, n_D^{25} 1.38231. Flash pt, closed cup: 52° F (11.1° C). Sol in water. Miscible with alcohol, ether. LD_{50} orally in rats: 3.5 g/kg (Schaffarzick, Brown).

Caution: Potential symptoms of overexposure are drowsiness; irritation of skin and eyes. *See NIOSH Pocket Guide to Chemical Hazards* (DHHS/NIOSH 90-117, 1990) p 52.

USE: Denaturant for ethanol, mfg flotation agents, flavors, perfumes; as solvent; in paint removers. Octane booster in gasoline.

1578.   n-Butylamine.   *1-Butanamine;* 1-aminobutane. $C_4H_{11}N$; mol wt 73.14. C 65.69%, H 15.16%, N 19.15%. $CH_3CH_2CH_2CH_2NH_2$. Prepn by reduction of butyraldoxime: Lycan *et al., Org. Syn. coll. vol.* **II**, 319 (1943). Usually manuf by catalytic alkylation of ammonia with butyl alcohol: Davies *et al.,* U.S. pat. **2,609,394** (1952 to I.C.I.); Hindley, Fisher, U.S. pat. **2,782,237** (1957 to British Celanese); Lemon, Myerly, U.S. pat. **3,022,349** (1962 to Union Carbide); Shirley, Speranza, U.S. pat. **3,128,311** (1964 to Jefferson Chem.). Manuf from butyraldehyde and ammonia in the presence of Raney nickel: Brimer *et al.,* U.S. pat. **2,518,659** (1950 to Eastman Kodak). Toxicity study: C. H. Hine *et al., Arch. Environ. Health* **1**, 343 (1960).

Liquid, ammoniacal odor, bp 78°. mp $-50°$. d_4^{25} 0.7327. n_D^{20} 1.4010. Flash pt, open cup: 30°F $(-1°C)$. Misc with water, alcohol, ether. LD_{50} orally in rats: 500 mg/kg (Hine).

Caution: Potential symptoms of overexposure are irritation of eyes, nose and throat; headache; skin flushing and burns. *See NIOSH Pocket Guide to Chemical Hazards* (DHHS/NIOSH 90-117, 1990) p 52.

USE: Intermediate for pharmaceuticals, dyestuffs, rubber chemicals, emulsifying agents, insecticides, synthetic tanning agents.

1579.   sec-Butylamine.   *2-Butanamine;* 2-aminobutane; Frucote; Deccotane; Tutane. $C_4H_{11}N$; mol wt 73.14. C 65.69%, H 15.16%, N 19.15%. $CH_3CH_2CH(NH_2)CH_3$. Prepn by reduction of ethyl methyl ketoxime: Lycan *et al., Org. Syn. coll. vol.* **II**, 319 (1943). Manuf: Taylor *et al.,* U.S. pat. **2,636,902** (1953 to I.C.I.); Thurston, U.S. pat. **2,689,868** (1954 to Am. Cyanamid). Resolution of *dl*-form: Leithe, *Ber.* **63**, 800 (1930); Bruck *et al., J. Chem. Soc.* **1956**, 921. Toxicity: Goldenthal, *Toxicol. Appl. Pharmacol.* **18**, 204 (1971).

dl-Form, liquid, bp 63°. mp $-104°$. d_4^{20} 0.724. n_D^{20} 1.394. Miscible with water, alc. LD_{50} orally in rats: 380 mg/kg (Goldenthal).

d(S)-Form, liquid, bp 63°. d_4^{15} 0.7308. n_D^{15} 1.3963. $[\alpha]_D^{15}$ +7.80° (neat).

l-Form, liquid, bp 63°. d_4^{19} 0.728. $[\alpha]_D^{19}$ $-7.64°$.

Caution: Irritating to skin and mucous membranes.

USE: Agricultural fungistat.

1580.   tert-Butylamine.   *2-Methyl-2-propanamine;* 2-aminoisobutane; 2-amino-2-methylpropane. $C_4H_{11}N$; mol wt 73.14. C 65.69%, H 15.16%, N 19.15%. $(CH_3)_3CNH_2$. Prepn: Campbell *et al., Org. Syn. coll. vol.* **III**, 148 (1955). Manuf: Gresham *et al.,* U.S. pat. **2,501,509** (1950 to du Pont); Albert, Kibler, U.S. pat. **2,773,097** (1956 to Firestone Tire & Rubber).

Liquid, bp 44-46°. mp $-72.65°$. d_4^{20} 0.6951, d_4^{25} 0.6867. n_D^{20} 1.37. Miscible with alcohol.

Hydrochloride, mp 310°. Sol in cold methanol, in boiling isopropyl alcohol.

1581.   Butylate.   *Bis(2-methylpropyl)carbamothioic acid S-ethyl ester; diisobutylthiocarbamic acid S-ethyl ester; S-*ethyl *N,N*-diisobutylthiocarbamate; butilate; diisocarb; R-1910; Sutan. $C_{11}H_{23}NOS$; mol wt 217.38. C 60.78%, H 10.66%, N 6.44%, O 7.36%, S 14.75%. Selective pre-planting herbicide. Prepn: H. Tilles, *J. Am. Chem. Soc.* **81**, 714 (1959); H. Tilles, J. Antognini, U.S. pat. **2,913,327** (1959 to Stauffer). Metabolism: J. P. Hubbell, J. E. Casida, *J. Agr. Food Chem.* **25**, 404 (1977).

Clear liquid, bp_{21} 138°. n_D^{30} 1.4701; d 0.9417. Vapor press at 25°: 1.3×10^{-3} mm Hg. Soly in water at 25°: 45 mg/l. LD_{50} orally in rats: 4000 mg/kg, *RTECS* **Vol. I**, R. J. Lewis, Sr., R. L. Tatken, Eds. (1979) p 365.

USE: Herbicide.

1582.   Butylated Hydroxyanisole.   *(1,1Dimethylethyl)-4-methoxyphenol;* 2(3)-*tert*-butyl-4-hydroxyanisole; BHA; Antrancine 12; Embanox; Nipantiox 1-F; Sustane 1-F; Tenox BHA. $C_{11}H_{16}O_2$; mol wt 180.25. C 73.30%, H 8.95%, O 17.75%. A mixture of *2-tert-butyl-4-methoxyphenol* (also called 3-*tert*-butyl-4-hydroxyanisole) and *3-tert-butyl-4-methoxyphenol* (also called 2-*tert*-butyl-4-hydroxyanisole). Prepn from *p*-methoxyphenol and isobutene: R. H. Rosenwald, U.S. pats. **2,459,540; 2,470,902** (both 1949 to Universal Oil Products). Toxicity study: A. J. Lehman *et al., Adv. Food Res.* **3**, 197 (1951). Review of carcinogenic risk: *IARC Monographs* **40**, 123-159 (1986); antimicrobial activity: M. Riccach, *J. Food Safety* **6**, 141-170 (1984); of physicochemical properties, antioxidant activity and toxicology: H. Verhagen *et al., Chem.-Biol. Interactions* **80**, 109-134 (1991).

Waxy solid, mp 48-55°. bp$_{733}$ 264-270°. Practically insol in water. Sol in petr ether (Skellysolve H), in 50% alcohol (or higher), in propylene glycol; alcohols. Sol in fats and oils. Exhibits antioxidant properties and synergism with acids, BHT, propyl gallate, hydroquinone, methionine, lecithin, thiodipropionic acid, etc. LD$_{50}$ in mice, rats (mg/kg): 2000, 2200 orally (Lehman).

Note: This substance may reasonably be anticipated to be a carcinogen: *Seventh Annual Report on Carcinogens* (PB95-109781, 1994) p 108.

USE: Antioxidant and preservative, esp in foods, cosmetics, pharmaceuticals; also rubber and petroleum products.

1583. Butylated Hydroxytoluene. *2,6-Bis(1,1-dimethylethyl)-4-methylphenol; 2,6-di-tert-butyl-p-cresol;* 2,6-di-*tert*-butyl-4-methylphenol; BHT; Antrancine 8; Tenox BHT; Ionol CP; Sustane; Dalpac; Impruvol; Vianol. C$_{15}$H$_{24}$O; mol wt 220.35. C 81.76%, H 10.98%, O 7.26%. Prepn from *p*-cresol and isobutylene: Stillson, U.S. pat. **2,428,745** (1947 to Gulf); McConnell, Davis, U.S. pat. **3,082,258** (1963 to Eastman Kodak). Toxicity study: W. A. McOmie *et al., J. Am. Pharm. Assoc.* **38,** 366 (1949). Review of physicochemical properties, antioxidant activity and toxicology: H. Babich, *Environ. Res.* **29,** 1-29 (1982); of antimicrobial activity: M. Raccach, *J. Food Safety* **6,** 141-170 (1984).

Crystals, mp 70°. d$_4^{20}$ 1.048. bp 265°. Flash pt (open cup): 260°F (127°C). Insol in water. Freely sol in toluene, sol in methanol, ethanol, isopropanol, methyl ethyl ketone, acetone, Cellosolve, petr ether, benzene, most other hydrocarbon solvents. Soly in liquid petrolatum (white oil): 0.5% w/w. More sol in food oils and fats than butylated hydroxyanisole. Good soly in linseed oil. LD$_{50}$ orally in mice: 1040 mg/kg (McOmie).

USE: Antioxidant for food, animal feed, petroleum products, synthetic rubbers, plastics, animal and vegetable oils, soaps. Antiskinning agent in paints and inks.

1584. n-Butylbenzene. 1-Phenylbutane. C$_{10}$H$_{14}$; mol wt 134.22. C 89.49%, H 10.51%. C$_6$H$_5$(CH$_2$)$_3$CH$_3$. Prepn: Radziszewski, *Ber.* **9,** 261 (1876); Balbiano, *Ber.* **10,** 296 (1877); Read, Foster, *J. Am. Chem. Soc.* **48,** 1606 (1926).

Liquid, mp −88.5°. d$_4^{20}$ 0.8604. bp$_{760}$ 183.1°; bp$_{400}$ 159.2°; bp$_{200}$ 136.9°; bp$_{100}$ 116.2°; bp$_{60}$ 102.6°; bp$_{40}$ 92.4°; bp$_{20}$ 76.3°; bp$_{10}$ 62.0°; bp$_5$ 48.8°; bp$_{1.0}$ 22.7°. n$_D^{20}$ 1.49040. Flash pt, open cup: 160° F (71° C). Insol in water. Miscible with alcohol, ether, benzene.

1585. sec-Butylbenzene. *(1-Methylpropyl)benzene;* 2-phenylbutane. C$_{10}$H$_{14}$; mol wt 134.22. C 89.49%, H 10.51%. C$_6$H$_5$CH(CH$_3$)CH$_2$CH$_3$. Prepd from benzene and *n*-butyl chloride in presence of AlCl$_3$: Schramm, *Monatsh.* **9,** 621 (1888); by the action of sodium on γ-chloro-*sec*-butylbenzene: Braun *et al., Ber.* **46,** 1277 (1913); with other products by heating *n*- or *sec*-butyl alcohol with 80% H$_2$SO$_4$: Meyer, Bernhauer, *Monatsh.* **53,** 727 (1929).

Liquid, mp −82.7°. d$_4^{20}$ 0.8608. bp$_{760}$ 173.5°; bp$_{400}$ 150.3°; bp$_{200}$ 128.8°; bp$_{100}$ 109.5°; bp$_{60}$ 96.0°; bp$_{40}$ 86.2°; bp$_{20}$ 70.6°; bp$_{10}$ 57.0°; bp$_5$ 44.2°; bp$_{1.0}$ 18.6°. n$_D^{20}$ 1.48980. Flash pt, closed cup: 126° F (52° C). Insol in water. Misc with alcohol, ether, benzene.

d-Form, [α]$_D^{25}$ +26.6°: Bonner, Greenlee, *J. Am. Chem. Soc.* **81,** 3336 (1959).

l-Form, [α]$_D^{25}$ −27.3°.

USE: Solvent; in organic syntheses.

1586. tert-Butylbenzene. *(1,1-Dimethylethyl)benzene;* 2-methyl-2-phenylpropane; trimethylphenylmethane; pseudobutylbenzene. C$_{10}$H$_{14}$; mol wt 134.22. C 89.49%, H 10.51%. Prepn: Konowalow, *Bull. Soc. Chim.* [3] **16,** 865 (1896); Shoesmith, Mackie, *J. Chem. Soc.* **1928,** 2336; Meyer, Bernhauer, *Monatsh.* **53,** 727 (1929); Wilt, Abegg, *J. Org. Chem.* **33,** 923 (1968). *See also* Groose, Ipatieff, *J. Am. Chem. Soc.* **57,** 2415 (1935); Ipatieff, Pines, *ibid.* **58,** 1056 (1936).

Liquid, mp −58.1°. d$_4^{20}$ 0.8669. bp$_{760}$ 168.5°; bp$_{400}$ 145.8°; bp$_{200}$ 123.7°; bp$_{100}$ 103.8°; bp$_{60}$ 90.6°; bp$_{40}$ 80.8°; bp$_{20}$ 65.6°; bp$_{10}$ 51.7°; bp$_5$ 39.0°; bp$_{1.0}$ 13.0°. n$_D^{20}$ 1.49235. Flash pt, open cup: 140°F (60°C). Insol in water. Misc with alcohol, ether, benzene.

1587. n-Butyl Benzoate. *Benzoic acid butyl ester.* C$_{11}$H$_{14}$O$_2$; mol wt 178.23. C 74.13%, H 7.92%, O 17.95%. C$_6$H$_5$COO(CH$_2$)$_3$CH$_3$. Prepn: Newman, Fones, *J. Am. Chem. Soc.* **69,** 1046 (1947); Justoni, **Brit.** pat. **719,891** (1954 to Vismara). Toxicity study: H. F. Smyth *et al., Arch. Ind. Hyg. Occup. Med.* **10,** 61 (1954).

Thick, oily liquid. d 1.00. mp −22°. bp 250°. Practically insoluble in water. Sol in alcohol or ether. LD$_{50}$ orally in rats: 5.14 g/kg (Smyth).

1588. n-Butyl Bromide. *1-Bromobutane.* C$_4$H$_9$Br; mol wt 137.02. C 35.06%, H 6.62%, Br 58.32%. CH$_3$(CH$_2$)$_3$Br. Prepd from *n*-butyl alc and a hydrobromic-sulfuric acid mixture: Kamm, Marvel, *Org. Syn.* vol. 1, 5 (1921); Skau, McCullough, *J. Am. Chem. Soc.* **57,** 2440 (1935).

Colorless liquid. d$_4^{25}$ 1.2686. bp$_{760}$ 101.3°. mp −112°. n$_D^{20}$ 1.4398. Insol in water. Sol in alcohol, ether.

1589. sec-Butyl Bromide. *2-Bromobutane;* methylethylbromomethane. C$_4$H$_9$Br; mol wt 137.02. C 35.06%, H 6.67%, Br 58.32%. CH$_3$CH$_2$CHBrCH$_3$. Prepn: Levene, Marker, *J. Biol. Chem.* **91,** 405 (1931); Kenyon *et al., J. Chem. Soc.* **1935,** 1080; Skau, McCullough, *J. Am. Chem. Soc.* **57,** 2440 (1935); Colson *et al., J. Chem. Soc.* **1965,** 2364. Prepn of optically pure isomers: Goodwin, Hudson, *J. Chem. Soc. (B)* **1968,** 1333.

dl-Form, colorless liquid, pleasant odor. d$_4^{25}$ 1.2530. bp 91.2°. mp −112°. n$_D^{20}$ 1.4344. Insol in water. Freely sol in alcohol, ether.

d-Form, n$_D^{20}$ 1.4359-1.4362. α$_D^{20}$ +42.64°.

l-Form, n$_D^{20}$ 1.4368. α$_D^{20}$ −43.7°.

Caution: Narcotic in high concns.

1590. tert-Butyl Bromide. *2-Bromo-2-methylpropane;* 2-bromoisobutane; trimethylbromomethane. C$_4$H$_9$Br; mol wt 137.02. C 35.06%, H 6.62%, Br 58.32%. (CH$_3$)$_3$CBr. Prepn: Brunel, *J. Am. Chem. Soc.* **39,** 1978 (1917); Bryce-Smith, Howlett, *J. Chem. Soc.* **1951,** 1141; Coe *et al., ibid.* **1954,** 2281.

Colorless liquid. d$_4^{25}$ 1.2125. bp 73.3°. mp −16.3°. At 210° changes to isobutyl bromide. n$_D^{25}$ 1.4249. Insol in water. Miscible with organic solvents.

1591. n-Butyl n-Butyrate. *Butanoic acid butyl ester; butyric acid butyl ester.* C$_8$H$_{16}$O$_2$; mol wt 144.21. C 66.63%, H 11.18%, O 22.19%. CH$_3$(CH$_2$)$_2$COO(CH$_2$)$_3$CH$_3$. Prepn from butyl alcohol: Robertson, *Org. Syn.* coll. vol. I, 138 (1941); Horton, U.S. pat. **2,522,676** (1950 to Socony-Vacuum Oil).

Liquid, bp 165°. d$_4^{20}$ 0.8692. n$_D^{20}$ 1.4064. Practically insol in water. Miscible with alcohol, ether.

1592. Butyl Carbitol®. *2-(2-Butoxyethoxy)ethanol;* diethylene glycol monobutyl ether; butyl digol; Butyl Diicinol. C$_8$H$_{18}$O$_3$; mol wt 162.23. C 59.23%, H 11.18%, O 29.59%. HOCH$_2$CH$_2$OCH$_2$CH$_2$OC$_4$H$_9$. Prepn: Riemschneider, Gross, *Monatsh.* **90**, 783 (1959). Purification: Miller, Yonan, *J. Am. Chem. Soc.* **79**, 5931 (1957); Ridley, Ridley, **Brit.** pat. **795,866** (1958 to Esso). Toxicity data: Smyth *et al., J. Ind. Hyg. Toxicol.* **23**, 259 (1941).

Practically odorless liquid, bp 230.4°. mp −68.1°. d_{20}^{20} 0.9536. n_D^{27} 1.4258. Miscible in water, oils. Miscibility in other organic solvents: Jackson, Drury, *Ind. Eng. Chem.* **51**, 1491 (1959). Flash pt 110°. LD$_{50}$ orally in rats, guinea pigs: 6.56, 2.00 g/kg (Smyth).

USE: Solvent.

1593. *n*-Butyl Carbonate. *Carbonic acid dibutyl ester;* dibutyl carbonate. $C_9H_{18}O_3$; mol wt 174.24. C 62.04%, H 10.41%, O 27.55%. $(C_4H_9O)_2CO$. Prepn from ethyl carbonate, butyl alcohol and ethylmagnesium bromide: Frank *et al., J. Am. Chem. Soc.* **66**, 1509 (1944); from butyl alcohol and CO in the presence of Pd and CuCl$_2$: Mador, Blackham, U.S. pat. **3,114,762** (1963 to National Distillers).

Liquid, bp 206.6°. d_4^{20} 0.9251, d_4^{25} 0.9388. n_D^{20} 1.4117. Practically insol in water. Miscible with ethanol, benzene, chloroform, acetone, ether and other organic solvents, *see:* Jackson, Drury, *Ind. Eng. Chem.* **51**, 1491 (1959).

1594. Butyl Cellosolve®. *2-Butoxyethanol;* ethylene glycol monobutyl ether. $C_6H_{14}O_2$; mol wt 118.18. C 60.98%, H 11.94%, O 27.08%. $HOCH_2CH_2OC_4H_9$. Prepn: L. H. Cretcher, W. H. Pittenger, *J. Am. Chem. Soc.* **46**, 1503 (1924); W. W. Carlson, U.S. pat. **2,448,767** (1948 to Mellon Inst. Ind. Res.); R. Riemschneider, P. Gross, *Monatsh.* **90**, 783 (1959). Toxicity: H. F. Smyth *et al., J. Ind. Hyg. Toxicol.* **23**, 259 (1941); C. P. Carpenter *et al., Arch. Ind. Health* **14**, 114 (1956).

Liquid, bp 171-172°. d_4^{20} 0.9012, d_{20}^{20} 0.9019. n_D^{20} 1.4196. Flash pt, closed cup: 141°F (60°C). Sol in water, mineral oil, most organic solvents. LD$_{50}$ orally in rats: 1.48 g/kg (Smyth).

Caution: Potential symptoms of overexposure are irritation of skin, eyes, nose and throat; hemolysis, hemoglobinuria. *See NIOSH Pocket Guide to Chemical Hazards* (DHHS/NIOSH 90-117, 1990) p 50.

USE: Solvent for nitrocellulose, resins, grease, oil, albumin; dry cleaning.

1595. *n*-Butyl Chloride. *1-Chlorobutane; n*-propylcarbinyl chloride; butyl chloride. C_4H_9Cl; mol wt 92.57. C 51.90%, H 9.80%, Cl 38.30%. $CH_3CH_2CH_2CH_2Cl$. Prepd from *n*-butyl alcohol by heating with HCl and anhydr ZnCl$_2$: Whaley, Copenhaver, *J. Am. Chem. Soc.* **60**, 2497 (1938); *Org. Syn.* **coll. vol. I,** 142 (2nd ed., 1941). Toxicity study: H. F. Smyth *et al., Arch. Ind. Hyg. Occup. Med.* **10**, 61 (1954).

Liquid. *Highly flammable.* d_4^{15} 0.89197; d_4^{20} 0.88648; d_4^{25} 0.88098. One gallon weighs 7.35 pounds. mp −123.1°. bp$_{760}$ 78.5°. n_D^{20} 1.40223. Flash pt −6.7° (+20°F). Dipole moment: 1.95. Practically insol in water (0.066% at 12°). Misc with alcohol, ether. LD$_{50}$ orally in rats: 2.67 g/kg (Smyth).

USE: As butylating agent in organic synthesis, e.g., in the manuf of butyl cellulose.

THERAP CAT (VET): Anthelmintic.

1596. *sec*-Butyl Chloride. *2-Chlorobutane;* 2-chloro-3-methylpropane. C_4H_9Cl; mol wt 92.57. C 51.90%, H 9.80%, Cl 38.30%. $CH_3CH_2CHClCH_3$. Prepn from *sec*-butyl alcohol, hydrochloric acid and ZnCl$_2$: Norris, Taylor, *J. Am. Chem. Soc.* **46**, 756 (1924); Copenhaver, Whaley, *Org. Syn.* **coll. vol. I,** 143 (1941); from *sec*-butyl alcohol and PCl$_3$: Coulson *et al., J. Chem. Soc.* **1965**, 2364. Prepn of optically pure isomers: Goodwin, Hudson, *J. Chem Soc. (B)* **1968**, 1333. Toxicity data: H. F. Smyth *et al., Am. Ind. Hyg. Assoc. J.* **30**, 470 (1969).

dl-Form, liquid; pleasant, ethereal odor. d_4^{20} 0.871. bp 68°. n_D^{20} 1.3960; n_D^{25} 1.3953. One gram dissolves in 1000 ml water at 25°; misc with alcohol, ether. LD$_{50}$ orally in rats: 20.0 ml/kg (Smyth).

d-Form, n_D^{20} 1.3963. α_D^{20} +30.8°.
l-Form, n_D^{20} 1.3968. α_D^{20} −31.0°.

1597. *tert*-Butyl Chloride. *2-Chloro-2-methylpropane;* 2-chloroisobutane; trimethylchloromethane. C_4H_9Cl; mol wt 92.57. C 51.90%, H 9.80%, Cl 38.30%. $(CH_3)_3CCl$. Prepd by shaking *tert*-butyl alcohol with concd HCl and

distilling: Norris, Olmsted, *Org. Syn.* **8**, 50 (1928). Prepn from *tert*-butyl alcohol and PCl$_3$: Gerrard *et al., J. Chem. Soc.* **1953**, 1920.

Liquid. d_4^{15} 0.847. n_D^{18} 1.38686. mp −26.5°. bp$_{760}$ 51.0°; bp$_{400}$ 32.6°; bp$_{200}$ +14.6°; bp$_{100}$ −1.0°; bp$_{60}$ −11.4°; bp$_{40}$ −19.0°. Sparingly sol in water, miscible with alcohol and ether. Boiling with water yields *tert*-butyl alcohol.

1598. *tert*-Butyl Chloroacetate. *Chloroacetic acid 1,1-dimethyl ester; t*-butyl chloroacetate; chloroacetic acid *tert*-butyl ester. $C_6H_{11}ClO_2$; mol wt 150.60. C 47.85%, H 7.36%, Cl 23.54%, O 21.25%. ClCH$_2$COOC(CH$_3$)$_3$. Prepd by reacting monochloroacetic acid and isobutylene in dioxane in the presence of sulfuric acid: Johnson *et al., J. Am. Chem. Soc.* **75**, 4995 (1953); from *tert*-butyl alcohol, chloroacetyl chloride and dimethylaniline: Baker, *Org. Syn.* **24**, 21 (1944).

Liquid, bp 155° (dec); bp$_{11}$ 48-49°; bp$_{16-17}$ 56-57°. n_D^{25} 1.4204-1.4210; n_D^{20} 1.4259-1.4260. Hydrolyzes to *tert*-butyl alcohol and chloroacetic acid.

USE: In the glycidic ester condensation.

1599. Butyl Citrate. *2-Hydroxy-1,2,3-propanetricarboxylic acid tributyl ester; citric acid tributyl ester; n*-butyl citrate; tributyl citrate. $C_{18}H_{32}O_7$; mol wt 360.45. C 59.98%, H 8.95%, O 31.07%. Prepn: Fodor, *Arch. Biochem.* **28**, 274 (1950). Synthesis from *n*-butyl alcohol and citric acid: Benedict, *Chemistry* **47**, 27 (1974).

Colorless or pale yellow, odorless liquid. d_{20}^{20} 1.045. bp$_{22}$ about 233°. mp −20°. Flash pt 185°. n_D^{20} 1.4460 (Benedict). Insol in water; miscible with most organic liquids.

USE: Plasticizer and solvent for nitrocellulose lacquers; in polishes, inks and similar prepns; also as anti-foam agent.

1600. α-Butylene Dibromide. *1,2-Dibromobutane.* $C_4H_8Br_2$; mol wt 215.92. C 22.25%, H 3.73%, Br 74.01%. CH$_3$CH$_2$CHBrCH$_2$Br. Prepn by bromination of butene: Wurtz, *Ann.* **152**, 21 (1869); of bromobutane: Kharasch *et al., J. Org. Chem.* **20**, 1430 (1955).

Yellowish liquid, bp 166°. mp −65°. d_4^{20} 1.7946. n_D^{20} 1.5144. Practically insol in water; miscible with alcohol.

1601. 1,3-Butylene Glycol. *1,3-Butanediol;* 1,3-dihydroxybutane; β-butyleneglycol; methyltrimethylene glycol; butane-1,3-diol. $C_4H_{10}O_2$; mol wt 90.12. C 53.31%, H 11.18%, O 35.51%. CH$_3$CHOHCH$_2$CH$_2$OH. Usually prepd by catalytic hydrogenation of aldol using Raney nickel: Hancock, Henson, *Ind. Eng. Chem.* **45**, 629 (1953); **Brit.** pat. **853,266** (1960 to Celanese); F. S. Wagner in Kirk-Othmer, *Encyclopedia of Chemical Technology* vol. **11** (Wiley-Interscience, New York, 3rd ed., 1980) pp 956-962. Toxicity: Smyth *et al., Arch. Ind. Hyg. Occup. Med.* **4**, 119 (1951).

Viscous liquid. The pure compd is colorless. d_{20}^{20} 1.004-1.006. One gallon weighs 8.398 lbs at room temp. Viscosity in centistokes: 24.6 at 50°C, 96 at 25°, 590 at 0°, 3253 at −17.7°, 6059 at −23°. melts below 50°. Because of the high viscosity at low temps heating is necessary for pumping. Very hygroscopic: Will absorb 38.5 wt % of water within 144 hrs. at 81% relative humidity. bp 207.5°. n_D^{20} 1.4401. Flash pt (tag open cup): 250°F (121°C). Dielectric constant 28.8 at 25°. Surface tension: 37.8 dynes/cm at 25°. Sol in water, acetone, methyl ethyl ketone, ethanol, dibutyl phthalate, castor oil. Practically insol in aliphatic hydrocarbons, benzene, toluene, carbon tetrachloride, ethanolamines, mineral oil, linseed oil. LD$_{50}$ orally in rats: 22.8 g/kg (Smyth).

USE: Intermediate in the manufacture of polyester plasticizers, humectant for cellophane, tobacco. Has some mold inhibiting action.

1602. 2,3-Butylene Glycol. *2,3-Butanediol;* 2,3-dihydroxybutane; dimethylethylene glycol. $C_4H_{10}O_2$; mol wt 90.12. C 53.31%, H 11.18%, O 35.51%. CH$_3$CH(OH)CH-(OH)CH$_3$. Occurs in 3 isomeric forms: *meso-* or *erythro-*,

D($-$)-*threo*-, and L($+$)-*threo*-forms. The commercial product is usually either the *meso*- or the D($-$)-form. Prepn of *meso*-form from *trans*-2,3-epoxybutane and of DL-form from *cis*-2,3-epoxybutane: Wilson, Lucas, *J. Am. Chem. Soc.* **58**, 2396 (1936). Prepn of D($-$)- and L($+$)-forms from corresponding D- and L-mannitols: Rubin *et al., ibid.* **74**, 425 (1952). Manuf of D($-$)-form by fermentation of carbohydrate solns with organisms of the *Bacillus subtilis* group: Vergnaud, U.S. pat. **2,529,061** (1950 to Usines de Melle). Manuf from 2-butene: Cosby *et al.*, U.S. pat. **2,808,429** (1957 to Allied Chem.); Keith *et al.*, U.S. pat. **2,974,161** (1961 to Sinclair); from 2-butyne: Saegebarth, U.S. pat. **3,157,704** (1964 to du Pont). Configuration: Morell, Auernheimer, *J. Am. Chem. Soc.* **66**, 792 (1944); Leroux, Lucas, *ibid.* **73**, 41 (1951); Rubin *et al., loc. cit.*

meso-Form (*erythro*-form), hygroscopic crystals from dry diisopropyl ether, mp 34.4°. bp$_{742}$ 181.7°, bp$_{16}$ 89°. d$_4^{25}$ 0.9939. n$_D^{35}$ 1.4324. Moderately sol in diisopropyl ether.

DL-*threo*-Form, hygroscopic crystals from diisopropyl ether, mp 7.6°. bp$_{742}$ 172.7°, bp$_{16}$ 86°. n$_D^{25}$ 1.4310. Very sol in diisopropyl ether.

D($-$)-*threo*-Form, mp 19.7°. bp$_{745}$ 179-180°, bp$_{10}$ 77.5-78°. d$_4^{25}$ 0.9869. n$_D^{25}$ 1.4315. [α]$_D^{25}$ -13.0° (neat).

L($+$)-*threo*-Form, bp 179-182°. d^{25} 0.9872. n$_D^{25}$ 1.4306.

1603. *n*-**Butyl Ether.** *1,1'-Oxybisbutane; butyl ether; n*-dibutyl ether. C$_8$H$_{18}$O; mol wt 130.23. C 73.78%, H 13.93%, O 12.29%. (CH$_3$CH$_2$CH$_2$CH$_2$)$_2$O. Prepn: Smyth, *J. Chem. Ed.* **39**, 212 (1962); d'Engenieres *et al., Bull. Soc. Chim. France* **1964**(10), 2471. Toxicity study: H. F. Smyth *et al., Arch. Ind. Hyg. Occup. Med.* **10**, 61 (1954).

Liquid. d$_4^{20}$ 0.769. mp -98°. bp 142-143°. Flash pt, closed cup: 100° F (37° C). Almost insol in water. Misc with alcohol or ether. Tends to form explosive peroxides, especially when anhydr. LD$_{50}$ orally in rats: 7.4 g/kg (Smyth).

1604. *tert*-**Butyl Hydroperoxide.** 1,1-Dimethylethylhydroperoxide. C$_4$H$_{10}$O$_2$; mol wt 90.12. C 53.31%, H 11.18%, O 35.51%. (CH$_3$)$_3$C—O—OH. Prepn from *tert*-butyl alcohol and 30% H$_2$O$_2$: N. A. Milas, S. A. Harris, *J. Am. Chem. Soc.* **60**, 2434 (1938); N. A. Milas, Surgenor, *ibid.* **68**, 205 (1946); U.S. pat. **2,573,947** (1951 to Shell). By oxidation of *tert*-butylmagnesium chloride: Walling, Buckler, *J. Am. Chem. Soc.* **75**, 4372 (1953); **77**, 6032 (1955).

Liquid. Stable to 75°. d$_4^{20}$ 0.896. mp -8°. bp$_{20}$ 35°. n$_D^{20}$ 1.4007. Sol in organic solvents. *Flammable.* Slow first-order decompn can be accelerated by the presence of 1 mol-% of Cu, Co and Mn salts.

USE: Catalyst in polymerization reactions. To introduce peroxy group into org molecules, in radical substitution reactions: Kharasch, Fono, *J. Org. Chem.* **23**, 325 (1948); *see also* Kharasch, Sosnovsky, *Tetrahedron* **3**, 97, 105 (1958).

1605. *tert*-**Butyl Hypochlorite.** *Hypochlorous acid 1,1-dimethylethyl ester; t*-butyl hypochlorite. C$_4$H$_9$ClO; mol wt 108.57. C 44.25%, H 8.36%, Cl 32.65%, O 14.74%. (CH$_3$)$_3$COCl. Prepd by adding aq *tert*-butyl alcohol to a cooled aq soln of NaOH and passing chlorine into the mixture: Teeter, Bell, *Org. Syn.* **32**, 20 (1952). By adding CO$_2$ to neutral aq *tert*-butyl alcohol and NaOCl satd with NaCl: Katz, U.S. pat. **2,694,722** (1954 to Bjorksten Res. Labs.).

Pale yellow liquid, irritating odor; attacks mucous membranes and eyes. d$_{20}^{20}$ 0.910; d^{18} 0.9583. bp 77-78°. n$_D^{20}$ 1.403. Hydrolyzes in presence of aq alkali but not aq acid. Dec upon exposure to bright light to methyl chloride and acetone with considerable evolution of heat. Should be stored in an inert atmosphere in a dark, refrigerated place. Violent reaction occurs if exposed to rubber, strong light or overheating. Has far greater stability than the corresponding *n*-butyl and *sec*-butyl hypochlorites.

USE: Dehydration of alcohols; *N*- and *C*-chlorinations.

1606. **Butylidene Chloride.** *1,1-Dichlorobutane.* C$_4$H$_8$Cl$_2$; mol wt 127.01. C 37.83%, H 6.35%, Cl 55.83%. CH$_3$CH$_2$CH$_2$CHCl$_2$. Prepn from butyraldehyde and PCl$_5$: Henne *et al., J. Am. Chem. Soc.* **61**, 938 (1939).

Oily liquid. bp$_{752}$ 114.8-115.1°. d^{25} 1.0797; n$_D^{25}$ 1.4305: Brown, Ash, *J. Am. Chem. Soc.* **77**, 4019 (1955). Practically insol in water. Sol in alcohol, chloroform, ether.

1607. *n*-**Butyl Iodide.** *1-Iodobutane.* C$_4$H$_9$I; mol wt 184.02. C 26.11%, H 4.93%, I 68.96%. CH$_3$CH$_2$CH$_2$CH$_2$I. Prepn: Franzen, *Ber.* **87**, 1148 (1954); Landauer, Rydon, *J. Chem. Soc.* **1954**, 2281; Stone, Shechter, *Org. Syn.* **coll. vol. IV**, 321 (1963). Manuf: Huber, Schenck, U.S. pats. **2,899,-471** and **3,053,910** (1959 and 1962, both to GAF).

Liquid, bp 130.4°. mp -103.0°. d$_4^{20}$ 1.616. n$_D^{20}$ 1.4998: Vogel, *J. Chem. Soc.* **1943**, 636. Practically insol in water. Sol in alcohol, ether. *Protect from light.*

1608. *sec*-**Butyl Iodide.** *2-Iodobutane.* C$_4$H$_9$I; mol wt 184.02. C 26.11%, H 4.93%, I 68.96%. CH$_3$CH$_2$CHICH$_3$. Prepn of *dl*-form from 2-butanol: Kornblum *et al., J. Am. Chem. Soc.* **77**, 5528 (1955); from dibutyl ether: Long, Freeguard, *Chem. & Ind. (London)* **1965**, 223. Prepn of *d*-form from *l*-butanol: Kenyon *et al., J. Chem. Soc.* **1935**, 1072. Prepn of *l*-form from *d*-2-butanol: Levene, Rothen, *J. Biol. Chem.* **115**, 415 (1936); Kornblum *et al., J. Am. Chem. Soc.* **70**, 746 (1948).

dl-Form, liquid, rapidly turns brown on exposure to light. bp 120°. mp -104°. d$_4^{20}$ 1.592, n$_D^{20}$ 1.4991: Vogel, *J. Chem. Soc.* **1943**, 636. Practically insol in water. Sol in alc, ether. *Protect from light.*

l-Form, liquid, bp 118°. d$_4^{20}$ 1.596, n$_D^{20}$ 1.495, [α]$_D^{24}$ -15.9° (neat); also reported as [α]$_D^{17}$ -31.98° (Kornblum *et al.*).

1609. **Butyl Isocyanate.** *1-Isocyanatobutane;* isocyanic acid butyl ester; *n*-butyl isocyanate; BIC. C$_5$H$_9$NO; mol wt 99.13. C 60.58%, H 9.15%, N 14.13%, O 16.14%. CH$_3$(CH$_2$)$_3$N=C=O. Prepn: R. J. Slocombe *et al., J. Am. Chem. Soc.* **72**, 1888 (1950); T. I. Bieber, *ibid.* **74**, 4700 (1952); E. T. Shawl *et al.,* Eur. pat. Appl. **408,277** (1991 to Arco). Toxicity study: J. Pauluhn, A. Eben, *Arch. Toxicol.* **66**, 118 (1992). Analytical methods development: R. J. Rando *et al., J. Liq. Chromatog.* **16**, 3977 (1993).

Flammable liquid. bp 113-116°. n$_D^{20}$ 1.4060. d$_4^{20}$ 0.880. LC$_{50}$ (4hr) in male, female rats (mg/m^3): 60, 55 by inhalation (Pauluhn, Eben).

USE: Reagent in organic synthesis.

1610. *n*-**Butylmalonic Acid.** *Butylpropanedioic acid;* pentane-1,1-dicarboxylic acid. C$_7$H$_{12}$O$_4$; mol wt 160.17. C 52.49%, H 7.55%, O 39.96%. CH$_3$(CH$_2$)$_3$CH(COOH)$_2$. Prepd by the interaction of sodium *n*-amyl chloride and carbon dioxide under pressure with ligroin as solvent: Morton *et al., J. Am. Chem. Soc.* **58**, 754 (1936).

Prisms from water, mp 102°. pK (5°): 2.96. Sol in water: 100 g of a satd aq soln contain 11.6 g at 0°; 79.3 g at 50°. Sol in alc, ether. Heating to 150° yields caproic acid.

Diethyl ester, C$_{11}$H$_{20}$O$_4$. Prepared by the interaction of *n*-butyl bromide, diethyl malonate, and sodium alcoholate: Adams, Kamm, *Org. Syn.* **4**, 11 (1925). Liquid, bp$_{760}$ 235-240°; bp$_{40}$ 140-145°; bp$_{20}$ 130-135°. n$_D^{20}$ 1.425. Very sol in alcohol, ether.

1611. *n*-**Butyl Mercaptan.** *1-Butanethiol;* normal butyl thioalcohol; thiobutyl alcohol. C$_4$H$_{10}$S; mol wt 90.19. C 53.27%, H 11.18%, S 35.55%. CH$_3$(CH$_2$)$_2$CH$_2$SH. Formed by the action of yeast on a product obtained when hydrogen sulfide is made to react with butyraldehyde in alcoholic ammonia: Nord, *Ber.* **52**, 1209 (1919); by passing vapors of butanol and hydrogen sulfide over thorium oxide catalyst: Kramer, Reid, *J. Am. Chem. Soc.* **43**, 880 (1921); together with dibutyl sulfide in distilling an aq soln of sodium butyl sulfate and sodium sulfide: Gray, Gutekunst, *J. Am. Chem. Soc.* **42**, 858 (1920); by slightly warming dithiocarbamic butyl ester with aq KOH: v. Braun, Engelbertz, *Ber.* **56**, 1574 (1923); the sodium salt is formed when dibutyl disulfide is treated with sodium in ether or alcohol: Moses, Reid, *J. Am. Chem. Soc.* **48**, 777 (1926). Physical properties: Mathias, *ibid.* **72**, 1897 (1950); W. E. Haines *et al., J. Phys. Chem.* **60**, 549 (1956). Once cited to occur in "skunk" fluid, *see* Beckmann, *Pharm. Zentralhalle* **37**, 557 (1896); T. B. Aldrich, *J. Exp. Med.* **1**, 323 (1896); however, recent literature rebuts this: K. K. Anderson, D. T. Bernstein, *J. Chem. Ecol.* **1**, 493 (1975); *eidem, J. Chem. Ed.* **55**, 159 (1978).

Mobile liquid. Heavy skunk odor. mp -115.9°. bp$_{766}$ 98.2°; bp$_{760}$ 98.4°. Flammable. d$_4^{25}$ 0.83679. n$_D^{25}$ 1.44014. Slightly sol in water. Very sol in alcohol, ether, liquid hydrogen sulfide. Forms azeotropic mixtures with butyl alcohol (bp 97.8°; 85.16% butanethiol) and with butyl alcohol and water.

Caution: Potential symptoms of overexposure in animals are narcosis, incoordination and weakness; cyanosis, pulmonary irritation; liver and kidney damage. *See NIOSH Pocket Guide to Chemical Hazards* (DHHS/NIOSH 90-117, 1990) p 54.

1612. *sec*-Butyl Mercaptan.   *2-Butanethiol; sec*-butyl thioalcohol. $C_4H_{10}S$; mol wt 90.19. C 53.27%, H 11.18%, S 35.55%. $CH_3CH_2CH(SH)CH_3$. Prepd from 2-iodobutane by means of alcoholic KSH soln: Reymann, *Ber.* **7**, 1287 (1874); from 2-bromobutane in the same manner: Ellis, Reid, *J. Am. Chem. Soc.* **54**, 1674 (1932).

Mobile liq. Heavy skunk odor. mp −165°. bp 84-85°. Flammable. d^{17} 0.8299; d_4^{25} 0.8246. n_D^{25} 1.4338. Slightly sol in water. Very sol in alc, ether, liquid H_2S.

1613. *tert*-Butyl Mercaptan.   *2-Methyl-2-propanethiol.* $C_4H_{10}S$; mol wt 90.19. C 53.27%, H 11.18%, S 35.55%. $(CH_3)_3CSH$. Prepd from *tert*-butyl iodide, ZnS and alcohol: Dobbin, *J. Chem. Soc.* **57**, 641 (1890). *See also* U.S. pats. **2,020,421**, *C.A.* **30**, 489 (1936) and **2,051,806**, *C.A.* **30**, 6760 (1936).

Mobile liquid. Heavy skunk odor. mp −0.5°. bp_{760} 63.7-64.2°. d_4^{25} 0.79426; n_D^{25} 1.41984: Mathias, *J. Am. Chem. Soc.* **72**, 1897 (1950). Flammable. Remarkably stable to oxidizing agents. Slightly sol in water. Very sol in alcohol, ether, liquid H_2S.

1614. *n*-Butylmercuric Chloride.   *Butylchloromercury.* C_4H_9ClHg; mol wt 293.16. C 16.39%, H 3.09%, Cl 12.09%, Hg 68.42%. $CH_3CH_2CH_2CH_2HgCl$. Prepn: Slotta, Jacobi, *J. Prakt. Chem.* [2] **120**, 249 (1929); R. C. Larock, H. C. Brown, *J. Am. Chem. Soc.* **92**, 2467 (1970). Distribution in mice: M. Yonaha *et al., Chem. Pharm. Bull.* **23**, 1718 (1975). For the prepn of the bromide *see* Slotta, Jacobi, *loc. cit.*; Marvel *et al., J. Am. Chem. Soc.* **47**, 3009 (1925). Review and bibliography: Krause, von Grosse, *Die Chemie der Metallorganischen Verbindungen* (Berlin, 1937).

White needles from ethanol, sometimes leaflets, mp 130°. Soly (g/100 ml) in water at 18°: 1.4×10^{-4}; at 100°: 3.3×10^{-4}; soly (g/100 g) in ethanol at 18°: 1.5; at 78°: 9.0; in chloroform at 18°: 5.6.

1615. 1-Butyl-3-metanilylurea.   *N-[(3-Aminophenyl)-sulfonyl]-N′-butylurea; N^1-metanilyl-N^2-n-butylurea; N^1-butyl-N^2-metanilylcarbamide; N^1-(3-aminobenzenesulfonyl)-N^2-n-butylurea; SB-1; Sucrida Berna.* $C_{11}H_{17}N_3O_3S$; mol wt 271.34. C 48.69%, H 6.31%, N 15.49%, O 17.69%, S 11.82%. Prepn: Haack, *Arzneimittel-Forsch.* **8**, 444 (1958); Zahler, U.S. pat. **3,015,673** (1962 to Swiss Serum and Vaccine Inst.).

Crystals, mp 117°.
THERAP CAT: Oral hypoglycemic.

1616. Butyl Methoxydibenzoylmethane.   *1-[4-(1,1-Dimethylethyl)phenyl]-3-(4-methoxyphenyl)-1,3-propanedione; 4-tert-butyl-4′-methoxydibenzoylmethane; Parsol 1789; Parsol A.* $C_{20}H_{22}O_3$; mol wt 310.39. C 77.39%, H 7.14%, O 15.46%. UV-A blocker. Prepn: K.-F. De Polo, Ger. pat. **2,945,125**; *idem*, U.S. pat. **4,387,089** (1980, 1983 both to Givaudan). Clinical efficacy as sunscreen: R. W. Gange *et al., J. Am. Acad. Dermatol.* **15**, 494 (1986); K. Kaidbey, R. W. Gange, *ibid.* **16**, 346 (1987); N. J. Lowe *et al., ibid.* **17**, 224 (1987). Assessment of photostability: A. Deflandre, G. Lang, *Int. J. Cosmet. Sci.* **10**, 53 (1988). HPLC determn in cosmetic products: L. Gagliardi *et al., J. Chromatog.* **408**, 409 (1987).

Crystals from methanol, mp 83.5°.
THERAP CAT: Ultraviolet screen.

1617. *n*-Butyl Nitrite.   *Nitrous acid butyl ester.* $C_4H_9NO_2$; mol wt 103.12. C 46.59%, H 8.80%, N 13.58%, O 31.03%. $CH_3CH_2CH_2CH_2ONO$. Prepd by the action of nitrous acid on butyl alcohol: Noyes, *Org. Syn.* **coll. vol. II**, 108 (1943); Miller, Audrieth, *Inorg. Syn.* **2**, 139 (1946).

Oily liquid. Characteristic odor. Breathing of vapor causes headache and vasodilation. d_4^0 0.9114. bp_{760} 78.2° (some decompn). Miscible with alcohol, ether. Dec on storage. Polymerization products of butyraldehyde have been found in five-month-old samples.

Caution: Lowers blood pressure through vasodilation, causing headache, throbbing, weakness. Effects resemble those produced by amyl nitrate.

USE: In the manuf of rare earth azides.

1618. *tert*-Butyl Nitrite.   *Nitrous acid 1,1-dimethyl ethyl ester; α,α*-dimethylethyl nitrite; nitrous acid *tert*-butyl ester. $C_4H_9NO_2$; mol wt 103.12. C 46.59%, H 8.80%, N 13.58%, O 31.03%. $(CH_3)_3CONO$. Prepd by adding *t*-butyl chloride to a cooled slurry of silver nitrite and anhydr ether: Kornblum *et al., J. Am. Chem. Soc.* **77**, 5528 (1955); by passing NO_2 or NO_2 + NO into *t*-butyl alcohol at 25-30°: Treacy, U.S. pat. **2,739,166** (1956); from *t*-butanol, sodium nitrite and sulfuric acid: Coe, Doumani, *J. Am. Chem. Soc.* **70**, 1516 (1948).

Yellow liquid, agreeable odor. d_4^0 0.8941; d_4^{20} 0.8671. bp_{760} 63°; bp_{250} 34°; n_D^{20} 1.3687. Very sol in alcohol, ether, chloroform, carbon disulfide; slightly sol in water; practically insol in glycerol.

USE: Jet propellant.

1619. Butylparaben.   *4-Hydroxybenzoic acid butyl ester; n-butyl p-hydroxybenzoate; Butoben; Butyl Chemosept; Butyl Parasept; Tegosept B.* $C_{11}H_{14}O_3$; mol wt 194.23. C 68.02%, H 7.27%, O 24.71%. Prepn of calcium and magnesium salts: Engels, Weijlard, U.S. pats. **2,046,324** and **2,056,176** (both 1936 to Merck & Co.). Prepn of analogous derivatives: *See* ethyl-, methyl- and propylparaben.

Crystalline powder, mp 68-69°. Very slightly soluble in water (1:6500), glycerin; freely sol in acetone, alc, ether, chloroform, propylene glycol. Preserve in well-closed containers.

Calcium salt, $C_{22}H_{26}CaO_6$, cryst powder. Soly in water: approx 1:125.

Magnesium salt, $C_{22}H_{26}MgO_6$, cryst powder. Soly in water: approx 1:110.

USE: Pharmaceutic aid (antifungal). Preservative in foods.

1620. *p-tert*-Butylphenol.   *4-(1,1-Dimethylethyl)phenol;* butylphen. $C_{10}H_{14}O$; mol wt 150.22. C 79.96%, H 9.39%, O 10.65%. Prepd by heating phenol with isobutanol in the presence of zinc chloride: Ger. pat. **17,311**; *Frdl.* **1**, 22; also from phenol, *tert*-butyl chloride and excess alkali in alcohol: Lewis, *J. Chem. Soc.* **83**, 329 (1903). *See also* Smith, *J. Am. Chem. Soc.* **55**, 3718 (1933); Natelson, *ibid.* **56**, 1583 (1934); Huston, Hsieh, *ibid.* **58**, 439 (1936); U.S. pat. **2,039,344** (1936 to Dow); Isagulyants, Bagryantseva, *C.A.* **33**, 8183 (1939). Toxicity study: H. F. Smyth *et al., Am. Ind. Hyg. Assoc. J.* **30**, 470 (1969).

Needles from water, mp 98°. d_4^{14} 0.9081. bp 237°. Volatile with steam. Practically insol in cold water. Sol in alcohol, ether. LD_{50} orally in rats: 3.25 ml/kg (Smyth).

Sodium salt, $C_{10}H_{13}NaO$, deliquescent leaflets. Sol in water.

USE: Intermediate in the manuf of varnish and lacquer resins; as a soap antioxidant; ingredient in de-emulsifiers for oil field use; in motor oil additives.

1621. 4-*tert*-Butylphenyl Salicylate. *2-Hydroxybenzoic acid 4-(1,1-dimethylethyl)phenyl ester; salicylic acid p-tert-butylphenyl ester;* TBS. $C_{17}H_{18}O_3$; mol wt 270.33. C 75.53%, H 6.71%, O 17.76%. Prepd from salicylic acid and *p-tert*-butylphenol in the presence of $POCl_3$: Stoesser, Sommerfield, U.S. pat. 2,606,920 (1952 to Dow).

Crystals, slight odor resembling that of salol, mp 62-64°. Maximum light absorption at 290-330 nm. Soly (w/w): in water <0.1%; abs ethanol 79%; ethyl acetate 153%; methyl ethyl ketone 197%; toluene 158%; Stoddard solvent 39%.

USE: Light absorber in plastic food wrappings.

1622. *n*-Butyl Phthalate. *1,2-Benzenedicarboxylic acid dibutyl ester; phthalic acid dibutyl ester;* dibutyl phthalate; DBP. $C_{16}H_{22}O_4$; mol wt 278.35. C 69.04%, H 7.97%, O 22.99%. $1,2-(C_4H_9OOC)_2C_6H_4$. Prepn: Farrar, Wienkauff, *Chem. & Ind. (London)* **1962**, 2144. Manuf: Bruno, U.S. pat. 2,628,249 (1953 to Pittsburgh Coke & Chemical). Toxicity study: Smith, *Arch. Ind. Hyg. Occup. Med.* **7**, 310 (1953), *C.A.* **47**, 12619h (1953).

Oily liquid. bp 340°. d^{20} 1.0459 and 1.0465, Kemppinen, Gokeen, *J. Phys. Chem.* **60**, 126 (1956). n_D^{20} 1.4900. Flash pt, open cup: 340°F (171°C). Sol in about 2500 parts water; very sol in alcohol, ether, acetone, benzene. Max single oral dose tolerated by rats: >8 g/kg body wt (Smith).

Caution: Potential symptoms of overexposure are irritation of upper respiratory tract and stomach. *See NIOSH Pocket Guide to Chemical Hazards* (DHHS/NIOSH 90-117, 1990) p 84.

USE: Insect repellant for the impregnation of clothing.

1623. *n*-Butyl Propionate. *Propanoic acid butyl ester; propionic acid butyl ester.* $C_7H_{14}O_2$; mol wt 130.19. C 64.58%, H 10.84%, O 24.58%. $CH_3CH_2COOC_4H_9$. Prepn from 1-butanol and propionic acid: Vogel, *J. Chem. Soc.* **1948**, 616; Spindt, Stevens, U.S. pat. 2,470,876 (1949 to Gulf).

Liquid, bp 146.8°. mp −89°. d_4^{20} 0.8754. n_D^{20} 1.401. Very slightly sol in water; very sol in alcohol, ether.

1624. *N*-Butylscopolammonium Bromide. *[7(S)-(1α, 2β,4β,5α,7β)]-9-Butyl-7-(3-hydroxy-1-oxo-2-phenylpropoxy)-9-methyl-3-oxa-9-azoniatricyclo[3.3.1.0^{2,4}]nonane bromide; α-(hydroxymethyl)benzeneacetic acid 9-butyl-9-methyl-3-oxa-9-azoniatricyclo[3.3.1.0^{2,4}]non-7-yl ester bromide; 8-butyl-6β,7β-epoxy-3α-hydroxy-1αH,5αH-tropanium bromide (−)-tropate;* butylscopolamine bromide; hyoscine-*N*-butyl bromide; scopolamine-*N*-butyl bromide; scopolamine bromobutylate; Amisepan; Buscapina; Buscol; Buscolysin; Buscopan; Butylmin; Donopon; Monospan; Scobro; Scobron; Scobutil; Sparicon; Sporamin; Stibron; Tirantil. $C_{21}H_{30}BrNO_4$; mol wt 440.38. C 57.28%, H 6.87%, Br 18.14%, N 3.18%, O 14.53%. Ganglion blocking agent. Prepn: F. Adickes *et al.*, Ger. pat. 856,890; *eidem*, U.S. pat. 2,872,452 (1952, 1959 both to Boehringer Ing.). Biological activity studies: Bauer *et al.*, *Arzneimittel-Forsch.* **18**, 1132 (1968). Toxicological studies: K. Stockhaus, H. Wick, *Arch. Int. Pharma-*

codyn. Ther. **180**, 155 (1969). Use in endoscopic examinations: J. R. Lee, *Clin. Radiol.* **33**, 273 (1982); D. N. Hupscher, O. Dommerholt, *Diagn. Imag. Clin. Med.* **53**, 77 (1984); A. C. Steger *et al.*, *Am. J. Gastroenterol.* **81**, 615 (1986).

Crystals from methanol, mp 142-144°. $[\alpha]_D^{20}$ −20.8° (c = 3 in water). LD_{50} in mice (mg/kg): 15.6 i.v.; 74 i.p.; 570 s.c.; 3000 orally (Stockhaus, Wick).

THERAP CAT: Antispasmodic.

1625. Butyl Stearate. *Octadecanoic acid butyl ester.* $C_{22}H_{44}O_2$; mol wt 340.59. C 77.58%, H 13.02%, O 9.40%. $CH_3(CH_2)_{16}COOC_4H_9$. Prepn from silver stearate and *n*-butyl iodide: Whitby, *J. Chem. Soc.* **1926**, 1458; from stearic acid and *n*-butanol: Smith, *ibid.* **1931**, 802. Physical properties: *Beilstein* vol. 2, Suppl. 3, 1016; Smith, *loc. cit.; Ind. Eng. Chem.* **32**, 880 (1940).

Crystals from alcohol, propanol, or ether, mp 27°. Also reported at 19°. bp 343°. Closed-cup flash pt 160° (320°F). Open-cup flash pt 196° (385°F). d_{25}^{25} 0.855-0.875. Slightly sol in water; sol in alcohol, ether.

USE: Solvent, spreading and softening agent in plastics, textiles, cosmetics, rubber industries.

1626. *n*-Butyl Sulfide. *1,1'-Thiobisbutane;* butylthiobutane; dibutyl sulfide. $C_8H_{18}S$; mol wt 146.30. C 65.68%, H 12.40%, S 21.92%. $(CH_3CH_2CH_2CH_2)_2S$. Prepd by refluxing a soln contg sodium sulfide and sodium *n*-butyl sulfate for several hrs: Gray, Gutekunst, *J. Am. Chem. Soc.* **42**, 856 (1920).

Liquid. mp −79.7°. d_4^{16} 0.839. d_4^0 0.852. bp 182°. Insol in water. Very sol in alcohol, ether.

1627. Butyraldehyde. *Butanal.* C_4H_8O; mol wt 72.11. C 66.63%, H 11.18%, O 22.19%. $CH_3CH_2CH_2CHO$. Prepn from butyryl chloride: Brown, Tsukamoto, *J. Am. Chem. Soc.* **83**, 2016 (1961); by reduction of corresponding nitrile: Gaiffe, Pallaud, *Compt. Rend.* **254**, 496 (1962); by alkali aluminum hydride reduction of methyl butyrate: Zakharkin *et al.*, *Tetrahedron Letters* **1963**, 208; by oxidation of butanol: Harrison, *Proc. Chem. Soc.* **1964**, 110. Usually manuf by catalytic dehydrogenation of butanol, catalytic hydrogenation of crotonaldehyde, or by the oxo process from propene, *e.g., see:* Dunbar, Arnold, *J. Org. Chem.* **10**, 501 (1945); Horn, *Ind. Eng. Chem.* **51**, 655 (1959); W. L. Faith *et al.*, *Industrial Chemicals* (John Wiley, New York, 3rd ed., 1965) pp 183, 304-305. Toxicity study: Smyth *et al.*, *Arch. Ind. Hyg. Occup. Med.* **4**, 119 (1951). *Review:* P. D. Sherman in Kirk-Othmer *Encyclopedia of Chemical Technology* vol. 4 (Wiley-Interscience, New York, 3rd ed., 1978) pp 376-386.

Flammable liquid. bp 74.8°. mp −99°. Closed-cup flash pt −6.67° (20°F). d_4^{20} 0.8016. n_D^{20} 1.379. Soly in water at 25°, 7.1 (wt-%): Smith, Bonner, *Ind. Eng. Chem.* **43**, 1169 (1951). Miscible with ethanol, ether, ethyl acetate, acetone, toluene, many other organic solvents and oils. Single-dose LD_{50} orally in rats: 5.89 g/kg (Smyth).

Caution: May act as irritant, narcotic.

USE: Chiefly in the manuf of rubber accelerators, synthetic resins, solvents, plasticizers.

1628. *n*-Butyramide. *Butanamide.* C_4H_9NO; mol wt 87.12. C 55.15%, H 10.41%, N 16.08%, O 18.36%. $CH_3CH_2CH_2CONH_2$. Prepn by a modified Willgerodt reaction: Jelinek, U.S. pats. 2,572,809; 2,572,810 (both 1951 to GAF); from butyryl chloride: Philbrook, *J. Org. Chem.* **19**, 623 (1954); from butyraldoxime: Huber, U.S. pat. 2,721,199 (1955 to du Pont); from butyric acid: Rahman, *Rec.*

Trav. Chim. **79,** 188 (1960); from butyronitrile: Gilbert, Rumanowski, U.S. pat. **3,062,883** (1962 to Allied Chem.).

Crystals, mp 115-116°. bp 216°. Sol in water, alcohol; slightly sol in ether.

1629. Butyric Acid. *Butanoic acid; n-*butyric acid; ethylacetic acid. $C_4H_8O_2$; mol wt 88.11. C 54.53%, H 9.15%, O 36.32%. $CH_3CH_2CH_2COOH$. Discovered by Lieben and Rossi in 1869. Present in butter as an ester to the extent of 4-5%. Obtained by suitable fermentation of carbohydrates; prepn from *n-*propanol + CO at 200 atm in the presence of $Ni(CO)_4$ and NiI_2: Reppe *et al., Ann.* **582,** 83 (1953); lab prepn from ethylmalonic acid: Gattermann-Wieland, *Praxis des Organischen Chemikers* (de Gruyter, Berlin, 40th ed., 1961) p 221. Toxicity study: Smyth *et al., Arch. Ind. Hyg. Occup. Med.* **10,** 61 (1954).

Oily liq; unpleasant, rancid odor. d_4^{20} 0.959. bp 163.5°. mp $-7.9°$. n_D^{20} 1.3991. Flash pt, closed cup: 170° F (77° C). Neutralization value 636.79. Miscible with water, alcohol, ether. The calcium salt of this acid is less soluble in hot than in cold water (difference from isobutyric acid). LD_{50} orally in rats: 8.79 g/kg (Smyth).

Magnesium salt, $C_8H_{14}MgO_4$, *magnesium butyrate.* Deliquesc leaflets. Sol in water. *Keep well closed.*

USE: Manuf of esters, some of which serve as bases of artificial flavoring ingredients of certain liqueurs, soda-water syrups, candies; also for varnishes; as decalcifier of hides.

1630. Butyric Anhydride. *Butanoic acid anhydride;* butyryl oxide. $C_8H_{14}O_3$; mol wt 158.20. C 60.74%, H 8.92%, O 30.34%. $CH_3CH_2CH_2COOCOCH_2CH_2CH_3$. Prepn from butyric acid: Williams, Krynitsky, *Org. Syn.* **coll. vol. III,** 165 (1955); Kuwajima, Mukaiyama, *J. Org. Chem.* **29,** 1385 (1964). Manuf by catalytic hydrogenation of crotonic acid: Smith, Hunter, U.S. pat. **2,492,403** (1949 to Celanese); by catalytic carbonylation of the corresponding acid ester: Reppe, Friederich, U.S. pat. **2,730,546** (1956 to BASF).

Liq, bp 199.4-201.4°. mp $-75°$. d_4^{20} 0.9668. n_D^{20} 1.4070. Flash pt, open cup: 190° F (88° C). Sol in water and in alc with decompn; sol in ether.

1631. Butyroin. *5-Hydroxy-4-octanone;* 5-octanol-4-one. $C_8H_{16}O_2$; mol wt 144.21. C 66.63%, H 11.18%, O 22.19%. $CH_3(CH_2)_2CH(OH)CO(CH_2)_2CH_3$. Prepd by the reaction between an ether soln of ethyl butyrate and sodium or potassium: Bouveault, Locquin, *Bull. Soc. Chim.* [3] **35,** 629 (1906); Corson *et al., J. Am. Chem. Soc.* **52,** 3988 (1930); Scheibler, Emden, *Ann.* **434,** 265 (1923); Snell, McElvain, *Org. Syn.* **coll. vol. II,** 114 (1943).

Liquid. d_4^0 0.9367; $d_4^{16.7}$ 0.91075. bp_{760} 180-190°; bp_{155} 150-154°; bp_{20} 95°; bp_{10} 85°. $n_D^{16.7}$ 1.43455. Reduces Fehling's soln.

Oxime, $C_8H_{17}NO_2$, viscous liquid, bp_{10} 143°.

1632. Butyrolactone. *Dihydro-2(3H)-furanone; γ-*butyrolactone; 1,2-butanolide; 1,4-butanolide; γ-hydroxybutyric acid lactone; 3-hydroxybutyric acid lactone; 4-hydroxybutanoic acid lactone. $C_4H_6O_2$; mol wt 86.09. C 55.81%, H 7.02%, O 37.17%. Prepd from acetylene and formaldehyde: Reppe, *Chem. Ing. Technik* **1950,** 365; *Chem. Eng.* **58,** no. 6, 176 (1951); also prepd from ethylene chlorohydrin, glutaric acid, *ν-*hydroxybutyric acid solns, tetrahydrofuran, or vinylacetic acid: F. C. Whitmore, *Organic Chemistry* (Van Nostrand, New York, 2nd ed., 1951). Alternate synthesis: Y. Ogata *et al., J. Org. Chem.* **45,** 1320 (1980). Physical properties: McKinley, Copes *J. Am. Chem. Soc.* **72,** 5331 (1950). Toxicity: H. F. Smyth *et al., Am. Ind. Hyg. Assoc. J.* **30,** 470 (1969).

Oily liquid. d_0^0 1.1441; d_0^{15} 1.1286. mp $-43.53°$. bp_{760} 204°. bp_{12} 89°. n_D^{25} 1.4348. Flash pt, open cup: 209°F (98°C). Volatile with steam. Misc with water. Sol in methanol, ethanol, acetone, ether, benzene. Hydrolyzed by hot alkaline solns. LD_{50} orally in rats: 17.2 ml/kg (Smyth).

USE: Intermediate in the synthesis of polyvinylpyrrolidone, DL-methionine, piperidine, phenylbutyric acid, thiobutyric acids. Solvent for polyacrylonitrile, cellulose acetate, methyl methacrylate polymers, polystyrene. Constituent of paint removers, textile aids, drilling oils.

1633. Butyronitrile. *Butanenitrile;* propyl cyanide; butyric acid nitrile. C_4H_7N; mol wt 69.11. C 69.52%, H 10.21%, N 20.27%. $CH_3CH_2CH_2CN$. Prepd from 1-butanol by controlled cyanation with NH_3 at 300° in the presence of $Ni—Al_2O_3$ catalysts: Popov, Shuikin, *Izvest. Akad. Nauk SSSR, Otdel. Khim. Nauk* **1958,** 713-718, *C.A.* **52,** 19924 (1958). Toxicity: H. F. Smyth *et al., Am. Ind. Hyg. Assoc. J.* **23,** 95 (1962).

Liquid. d_4^0 0.8091; d_4^{15} 0.7954; d_4^{30} 0.7817. mp $-112°$. bp_{760} 117.5°; bp_{400} 96.8°; bp_{200} 76.7°; bp_{100} 59.0°; bp_{60} 47.3°; bp_{40} 38.4°; bp_{20} 25.7°; bp_{10} 13.4°; bp_5 +2.1°; $bp_{1.0}$ $-20.0°$. n_D^{20} 1.38385. Flash pt, open cup: 85°F (29°C). Viscosity ($\eta \times 10^5$) at 15° = 624; at 30° = 515. Dipole moment 3.5. Sparingly sol in water. Misc with alcohol, ether, dimethylformamide. LD_{50} orally in rats: 0.14 g/kg (Smyth).

Caution: Highly toxic.

1634. *n-*Butyryl Chloride. *Butanoyl chloride;* butyryl chloride. C_4H_7ClO; mol wt 106.55. C 45.09%, H 6.62%, Cl 33.27%, O 15.02%. $CH_3CH_2CH_2COCl$. Prepn from butyric acid: Helferich, Schaefer, *Org. Syn.* **coll. vol. I,** 147 (1941).

Liquid, bp 101-102°. mp $-89°$. $d_4^{20.6}$ 1.0263. n_D^{20} 1.412. Dissolves slowly with decompn in water, alcohol. Miscible with ether.

1635. Buzepide. *Hexahydro-α,α-diphenyl-1H-azepine-1-butanamide;* 2,2-diphenyl-4-hexamethyleneiminobutyramide; R-658. $C_{22}H_{28}N_2O$; mol wt 336.48. C 78.53%, H 8.39%, N 8.33%, O 4.75%. Prepn: Janssen *et al., J. Med. Pharm. Chem.* **1,** 187 (1959); of methiodide as well as free base: Janssen, de Jongh, U.S. pat. **2,881,165** (1959 to N.V. Nederlandsche Combinatie Chem. Ind.). Clinical trial in irritable bowel syndrome: J. Ph. Barbier *et al., Ann. Gastroenterol. Hepatol.* **25,** 123 (1989).

Crystals from isopropanol, mp 141.5-143.5°.

Methiodide, $C_{22}H_{28}N_2O.CH_3I$, *metazepium iodide, 1-(3-carbamoyl-3,3-diphenylpropyl)hexahydro-1-methylazepinium iodide, N-(3,3-diphenyl-3-carbamoylpropyl)-N-methylperhydroazepinium iodide, diphexamide methiodide.* Crystals, dec 212-213°.

Mixture with haloperidol, *Vésadol.*

THERAP CAT: Anticholinergic.

C

1636. Cabenegrins. Orally active antidotes against snake venoms; isolated from the root of a South American plant called "Cabeca de Negra" and structurally related to pterocarpin, *q.v.* Isoln: L. L. Darko *et al.*, **Eur. pat. Appl. 89,229**; *eidem*, **U.S. pat. 4,429,141** (1983, 1984 both to Richter, Budapest). Structure determn: M. Nakagawa *et al.*, *Tetrahedron Letters* **1982**, 3855. Synthesis of (±)-cabenegrins A-I and A-II: M. Ishiguro *et al.*, *ibid.* 3859.

Cabenegrin A-I   Cabenegrin A-II

Cabenegrin A-I, $C_{21}H_{20}O_6$, *[6aR-[4(E),6aα,12aα]]-6a,12a-dihydro-4-(4-hydroxy-3-methyl-2-butenyl)-6H-[1,3]dioxolo[5,-6]benzofuro[3,2-c][1]benzopyran-3-ol.* White crystalline solid, mp 167-168°. uv max (ethanol): 309 nm (ε 13000); uv max (methanol): 209, 233, 309 nm (ε 75000, 24000, 13000).
Cabenegrin A-II, $C_{21}H_{22}O_6$, *6a,12a-dihydro-3-hydroxy-β-methyl-6H-[1,3]dioxolo[5,6]benzofuro[3,2-c][1]benzopyran-2-butanol.* uv max (methanol): 204, 230, 292, 308 nm (ε 116000, 8000, 9400, 11800).

1637. Cabergoline. *(8β)-N-[3-(Dimethylamino)propyl]-N-[(ethylamino)carbonyl]-6-(2-propenyl)ergoline-8-carboxamide;* 1-ethyl-3-(3'-dimethylaminopropyl)-3-(6'-allylergoline-8'β-carbonyl)urea; 1-[(6-allylergoline-8β-yl)carbonyl]-1-[3-(dimethylamino)propyl]-3-ethylurea; FCE-21336; Dostinex. $C_{26}H_{37}N_5O_2$; mol wt 451.61. C 69.15%, H 8.26%, N 15.51%, O 7.09%. Dopamine receptor agonist. Prepn: P. Salvati *et al.*, **Belg. pat. 888,243**; *eidem*, **U.S. pat. 4,526,892** (1981, 1985 both to Farmitalia Carlo Erba). Prepn and bioactivity: E. Brambilla *et al.*, *Eur. J. Med. Chem.* **24**, 421 (1989). Clinical pharmacology: C. Ferrari *et al.*, *J. Clin. Endocrinol. Metab.* **63**, 941 (1986). Clinical evaluation to prevent puerperal lactation: G. B. Melis *et al.*, *Obstet. Gynecol.* **71**, 311 (1988); in hyperprolactinemic disorders: C. Ferrari *et al.*, *J. Clin. Endocrinol. Metab.* **68**, 1201 (1989). Veterinary trial as abortifacient in dogs: K. Post *et al.*, *Theriogenology* **29**, 1233 (1988).

White crystals from diethyl ether, mp 102-104°. LD_{50} orally in male mice: > 400 mg/kg (Brambilla).
Diphosphate, $C_{26}H_{37}N_5O_2.2H_3PO_4$, mp 153-155°.
THERAP CAT: Prolactin inhibitor.
THERAP CAT (VET): Prolactin inhibitor.

1638. Cacao Shell. Cocoa shells. Shells of the seed of *Theobroma cacao* L., *Sterculiaceae*. *Habit.* Brazil, Central America, Mexico, West Indies, and most tropical countries. *Constit.* Theobromine, caffeine, cacao red, protein, pento-

sans, pectic acid, starch. *Ref:* Dittmar, *Engenharia e quim* **5**, no. 1, 1 (1953), *C.A.* **48**, 2949e (1954). *Monographs:* E. M. Chatt, *Cocoa* (Interscience, New York, 1953); D. H. Urquhart, *Cocoa* (Longmans, London, 1961); Powell, Harris, *Chocolate and Cocoa* in Kirk-Othmer *Encyclopedia of Chemical Tehnology,* **vol. 5** (Interscience, New York, 2nd ed., 1964) pp 363-402.
Thin, papery, reddish-brown, concavo-convex shells; weak chocolate-like odor and taste.
Caution: Occasionally causes allergic dermatitis from handling.
USE: In the manuf of caffeine, theobromine.

1639. Cacodyl. *Tetramethyldiarsine;* dicacodyl. $C_4H_{12}As_2$; mol wt 209.98. C 22.88%, H 5.76%, As 71.36%. $(CH_3)_2AsAs(CH_3)_2$. Prepn from cacodyl chloride by heating with zinc in an atm of carbon dioxide: Bunsen, *Ann.* **42**, 14 (1842); from dimethylarsine by the action of oxides of nitrogen, aqueous chromic acid, lead peroxide, cacodyl chloride or potassium ferricyanide: Dehn, Wilcox, *Am. Chem. J.* **35**, 1 (1906); by the action of free methyl on arsenic: Paneth, Loleit, *J. Chem. Soc.* **1935**, 366; by reduction of cacodyl oxide: Witten, **U.S. pat. 2,531,487** (1950 to U.S.A.); Fuson, Shive, **U.S. pat. 2,756,245** (1956 to U.S.A.).
Oily liquid. Solidifies to large quadratic plates, mp −6°. Almost intolerable garlicky odor. Inflames spontaneously in dry air. bp_{760} 165°. Slightly sol in water. Controlled oxidation with moist air yields cacodyl oxide and cacodylic acid. Reduction with tin and HCl yields *Erytrarsin* $(CH_3As)_4As_2O_3$.

1640. Cacodylic Acid. *Dimethylarsinic acid; hydroxydimethylarsine oxide;* Phytar. $C_2H_7AsO_2$; mol wt 138.00. C 17.41%, H 5.11%, As 54.29%, O 23.19%. $(CH_3)_2As(O)OH$. Prepn: Guinot, *J. Pharm. Chim.* **27**, 55 (1923), *C.A.* **17**, 2103[8] (1923); Inverni, *Boll. Chim. Farm.* **62**, 129 (1923), *C.A.* **17**, 2413[7] (1923); Challenger, Ellis, *J. Chem. Soc.* **1935**, 396; Fioretti, Portelli, *Ann. Chim. (Rome)* **53**, 1869 (1963), *C.A.* **60**, 10712c (1964). As herbicide: Sprague, **U.S. pat. 3,056,668** (1962 to Ansul). Toxicity: G. W. Bailey, J. L. White, *Residue Rev.* **10**, 97 (1965).
Crystals from alcohol + ether. mp 195-196°. Hygroscopic. *Poisonous!* Sol in 0.5 part water; very sol in alcohol; sol in acetic acid; practically insol in ether. *Keep well closed.* LD_{50} orally in rats: 1350 mg/kg (Bailey, White).
Mercury salt, $C_4H_{12}As_2HgO_4$, *mercuric cacodylate.* Hygroscopic, somewhat unstable cryst powder. Soluble in water, alcohol; practically insol in ether. *Keep well closed.*
USE: Herbicide.
THERAP CAT: Dermatologic.
THERAP CAT (VET): Has been used in chronic eczema, anemia and as a tonic.

1641. Cacotheline. *2,3-Dihydro-4-nitro-2,3-dioxo-9,10-secostrychnidin-10-oic acid.* $C_{21}H_{21}N_3O_7$; mol wt 427.41. C 59.01%, H 4.95%, N 9.83%, O 26.20%. A nitro derivative of brucine made by treating brucine with 10% nitric acid at 60-70°: Leuchs *et al.*, *Ber.* **43**, 1042 (1910). Structure: Teuber, *ibid.* **86**, 232 (1953). Used as reversible redox indicator for Sn^{+2} titrations: P. Szarvas, J. Lantos, *Talanta* **10**, 477 (1963).

Yellow crystals; sparingly sol in water.
USE: Indicator.

1642. Cactinomycin. *Actinomycin C;* HBF-386; Sanamycin. Antibiotic complex produced by *Streptomyces chrysomallus:* Brockmann, Grubhofer, *Naturwiss.* **36**, 376 (1949); **37**, 494 (1950); *Ber.* **84**, 260 (1951); **U.S. pat. 2,953,495** (1960 to Shenley Inds.); Lindenbein, *Arch. Mikrobiol.* **17**, 361 (1952). Mixture of actinomycins C_1 (dactinomycin *q. v.)*, C_2 and C_3, 10%, 45% and 45%, resp: Brockmann, Pfennig, *Naturwiss.* **39**, 429 (1952); Brockmann, Gröne, *ibid.* **40**, 222 (1953). Description of other actinomycins: Waksman *et al., Proc. Nat. Acad. Sci. U.S.A.* **44**, 602 (1958). Structures: Brockmann *et al., Angew. Chem.* **68**, 70 (1956); Brockmann, Boldt, *Naturwiss.* **50**, 19 (1963). Synthesis of actinomycin C_3: Brockmann, Lackner, *ibid.* **47**, 230 (1960); **48**, 555 (1961); **51**, 407 (1964); Brockmann *et al., Ger. pat.* **1,172,680** (1964 to Bayer); Brockmann, Lackner, *Ber.* **100**, 353 (1967); **101**, 1312 (1968). Synthesis of actinomycin C_2: Brockmann, Lackner, *Tetrahedron Letters* **1964**, 3517. Comprehensive review: H. Brockmann, in *Fortschr. Chem. Org. Naturst.* **18**, 1-54 (1960).

Actinomycin C_2 R = D-Valine   R' = D-Alloisoleucine
Actinomycin C_3 R = D-Alloisoleucine   R' = D-Alloisoleucine

Alizarin-red hexagonal bipyramids from ethyl acetate, mp 252°. $[\alpha]_D^{25}$ −325 to −349° (c = 0.25 in ethanol). Sparingly sol in water; moderately sol in ethanol; sol in chloroform, ethyl acetate, benzene, acetone. *Protect from light.*

Actinomycin C_2. $C_{63}H_{88}N_{12}O_{16}$. Red bipyramids, prisms or needles from ethyl acetate, mp 237-239°. $[\alpha]_D^{21}$ −325° ±10° (c = 0.23 in methanol). Abs max (methanol): 443 nm (ϵ 25400).

Actinomycin C_3. $C_{64}H_{90}N_{12}O_{16}$. Red hexagonal bipyramids from ethyl acetate or methanol, dec 235°. $[\alpha]_D^{17}$ −328° (c = 0.5 in ethanol). Absorption max (methanol): 443 nm (ϵ 24100). Weak base.

THERAP CAT: Antineoplastic.

1643. Cactus Grandiflorus. Night-blooming cereus; large-flowered cereus. Fresh, succulent stems of *Selenicereus grandiflorus* (L.) Britt. and Rose, Cactaceae. *Habit.* Tropical America. *Constit.* Cactine(?), acrid resinous glucoside, fat, wax. Extract containing combined principles from freshly tinctured stems and petals is known as *cactoid.*

THERAP CAT: Cardiotonic.

THERAP CAT (VET): Has been used as circulatory stimulant.

1644. Cadalene. *1,6-Dimethyl-4-(1-methylethyl)naphthalene; 4-isopropyl-1,6-dimethylnaphthalene;* cadalin. $C_{15}H_{18}$; mol wt 198.31. C 90.85%, H 9.15%. Obtained from cadinene and other sesquiterpenes or sesquiterpene alcohols by dehydrogenation: Ruzicka, Meyer, *Helv. Chim. Acta* **4**, 505 (1921); Ruzicka, Stoll, *ibid.* **7**, 84 (1924). Synthesis: Ruzicka, Seidel, *ibid.* **5**, 369 (1922); Barnett, Cook, *J. Chem. Soc.* **1933**, 22; Johnson, Jones, *J. Am. Chem. Soc.* **69**, 792 (1947); Kohli *et al., Experientia* **28**, 131 (1972).

Liquid. d_4^{25} 0.9667. bp_{720} 291-292°; bp_{10} 149°. n_D^{25} 1.5785. uv max: 228, 232, 280, 284, 295, 310, 317, 325 nm. Insol in water. Sol in fat solvents, oils.

1645. Cadaverine. *1,5-Pentanediamine;* pentamethylenediamine; *animal coniine.* $C_5H_{14}N_2$; mol wt 102.18. C 58.77%, H 13.81%, N 27.42%. $NH_2(CH_2)_5NH_2$. Biogenic polyamine and homolog of putrescine, *q.v.*, produced by

decarboxylation of lysine. Formed through action of comma bacillus on meat, fish, albumin, etc. Found in cholera discharge. Isoln: Bocklisch, *Ber.* **18**, 1922 (1885); Ackermann, *Z. Physiol. Chem.* **54**, 16 (1907). Prepn: Ladenburg, *Ber.* **18**, 2956 (1885). GC determn in foods: W. F. Staruszkiewicz, J. F. Bond, *J. Assoc. Offic. Anal. Chem.* **64**, 584 (1981). Metabolism study of ^{14}C-cadaverine in rat brain: S. K. Salzman, M. Stepita-Klauco, *J. Neurochem.* **37**, 1308 (1981). Biosynthetic study in *S. ruminantium:* Y. Kamio *et al., J. Biol. Chem.* **257**, 3326 (1982).

Colorless, syrupy liquid; characteristic odor. Strong base, fumes and attracts CO_2 on exposure to air. pKa_1 10.25; pKa_2 9.13. *Poisonous!* d_4^{25} 0.873. mp 9°. bp 178-180°. n_D^{20} 1.463. Sol in water, alcohol; slightly sol in ether. *Keep well closed.*

Dihydrochloride, $C_5H_{14}N_2.2HCl$, needles from water, mp 225-230°. Sol in water. Practically insol in abs alc.

Caution: Skin irritant and possible sensitizer.

1646. Cadexomer Iodine. Iodosorb. A hydrophilic modified starch polymer containing 0.9% (w/w) iodine within a helical matrix. Produced by the reaction of dextrin with epichlorohydrin coupled with ion exchange groups and iodine. Clinical use in venous ulcers: E. Skog *et al., Brit. J. Dermatol.* **109**, 77 (1983); M. C. Ormiston *et al., Brit. Med. J.* **291**, 308 (1985); L. Hillström, *Acta Chir. Scand. Suppl.* **544**, 53 (1988).

THERAP CAT: Vulnerary.

1647. Cadherins. Family of calcium-dependent cell adhesion molecules thought to be key regulators of morphogenesis in vertebrates. Transmembrane glycoproteins; mol wt ∼120-140 kDa. Mediate tissue development by connecting cells bearing identical cadherin types via homophilic interactions. Loss of cadherin function has been associated with tumor metastasis. Two subfamilies have been identified: classic cadherins and the desmosomal cadherins, known as *desmocollins* and *desmogleins*, which differ from the classic cadherins primarily in the structure of the cytoplasmic domains. A number of classic cadherins have been identified; nomenclature is based on the cell from which they were first identified, *e.g.* E-, N- and P-cadherins from epithelial, neural, and placental cells, resp. Among subclasses, ∼50% of amino acids are conserved within a particular species. Classic cadherins are complexed with cytoplasmic proteins known as *catenins* which anchor them to the actin-based cytoskeleton and are required for adhesive function. Review of discovery and characterization of cadherins: M. Takeichi, *Development* **102**, 639-655 (1988); idem, *Ann. Rev. Biochem.* **59**, 237-252 (1990); of properties and role in morphogenesis: idem, *Science* **251**, 1451-1455 (1991); J. Behrens, *Acta Anat.* **149**, 165-169 (1994). Review of interaction with catenins: R. Kemler, *Trends Genet.* **9**, 317-321 (1993); L. Hinck *et al., Trends Biochem. Sci.* **19**, 538-542 (1994). Review of role in tumor metastasis: M. Takeichi, *Curr. Opin. Cell Biol.* **5**, 806-811 (1993); W. Birchmeier, J. Behrens, *Biochim. Biophys. Acta* **1198**, 11-26 (1994); O. W. Blaschuk *et al., Endocrine* **3**, 83-89 (1995). Review of desmosomal cadherins: P. J. Koch, W. W. Franke, *Curr. Opin. Cell Biol.* **6**, 682-687 (1994).

1648. Cadinenes. $C_{15}H_{24}$; mol wt 204.36. C 88.16%, H 11.84%. Sesquiterpenes occurring in essential oils from Juniper species and cedars (oil of cade). Nine possible isomers differing in stereochemistry and position of the double bonds, the principal isomer being β-cadinene: Sykora *et al., Chem. Listy* **52**, 1314 (1958). Prepn and structure: Campbell, Soffer, *J. Am. Chem. Soc.* **64**, 417 (1942); Campbell *et al., ibid.* 425; Rao *et al., Tetrahedron Letters* **1960**, 27; Herout *et al., Coll. Czech. Chem. Commun.* **31**, 3012 (1966). Synthesis: Soffer, Günay, *Tetrahedron Letters* **1965**, 1355. Of the nine possible isomers, all able to yield (−)-cadinene dihydrochloride, six are known: Herout, Sykora, *Tetrahedron* **4**, 246 (1958); Kartha *et al., ibid.* **19**, 241 (1963). Structure and configuration of isomers: Sykora *et al., Coll. Czech. Chem. Commun.* **23**, 2181 (1958). Isoln of α-cadinene from Japanese hop: Y. Naya, M. Kotake, *Bull. Chem. Soc. Japan* **42**, 1468 (1969). Synthesis: O. P. Vig, *Indian J. Chem.* **21B**, 145 (1982). Brief review in *Rodd's Chemistry of Carbon Compounds,* **vol. 2**, pt. C, S. Coffey, Ed. (Elsevier, New York, 1969) pp 268-270.

β-form

β-Cadinene, oil. Slight pleasant odor. bp$_9$ 124°. d$_4^{20}$ 0.9239. n$_D^{20}$ 1.5059. [α]$_D^{20}$ −251°.

1649. Cadmium. Cd; at. wt 112.411; at. no. 48; valence 2. Group IIB(12) element. Abundance in earth's crust: 0.1 to 0.2 ppm. Naturally occurring isotopes (mass numbers): 114 (28.73%); 112 (24.13%); 111 (12.80%); 110 (12.49%); 113 (12.22%), T$_{1/2}$ 9.3 × 10^{15} yrs, β-emitter; 116 (7.49%); 106 (1.25%); 108 (0.89%); known artificial radioactive isotopes: 97-105, 017, 109, 115, 117-122, 124, 126. Found in zinc ores; also as CdS, greenockite; CdCO$_3$, otavite. Obtained in vapor form when roasting zinc ores, as sludge from zinc sulfate purification. Lab prepns from CdSO$_4$: Treadwell, *Helv. Chim. Acta* **4**, 551 (1921). Implicated as causative agent in Itai-Itai ("ouch-ouch") disease in Japan: Flick *et al.*, *Environ. Res.* **4**, 71-85 (1971); Fassett, *Ann. Rev. Pharmacol.* **15**, 425-435 (1975). *Review:* Aylett "Group IIB" in *Comprehensive Inorganic Chemistry*, vol. 3, J. C. Bailar Jr. *et al.*, Eds. (Pergamon Press, Oxford, 1973) pp 187-328; D. S. Carr in Kirk-Othmer *Encyclopedia of Chemical Technology* vol. 4 (John Wiley & Sons, New York, 4th ed., 1992) pp 748-760; N. Herron, *ibid* 760-776. Review of carcinogenic risk: *IARC Monographs* **11**, 39-74 (1976); and toxicity: *Cadmium in the Human Environment: Toxicity and Carcinogenicity*, G. F. Nordberg *et al.*, Eds., *IARC Scientific Publ. no. 118* (1992) 469 pp.

Silver-white, blue-tinged, lustrous metal; distorted hexagonal close-packed structure; easily cut with a knife; available in the form of bars, sheets or wire or a gray, granular powder. mp 321°. bp 765°. d^{25} 8.65. Specific heat at constant pressure (25°) 6.22 cal/mole deg. Slowly oxidized by moist air to form CdO. E° (aq) Cd/Cd^{2+} 0.4025 V. Insol in water. Reacts readily with dil HNO$_3$; reacts slowly with hot HCl; does not react with alkalies. Other reactions similar to those of zinc. Solns of cadmium salts and H$_2$S or Na$_2$S yield a yellow ppt insol in excess Na$_2$S.

Caution: Overexposure to cadmium and cadmium compounds has been associated with acute and chronic toxicity. Potential symptoms of acute poisoning from the inhalation of cadmium dusts or fumes include headache, chest pains, cough, metal fume fever, weakness. Potential symptoms of acute poisoning from ingestion of cadmium salts are GI disturbances, headache, muscular cramps, vertigo, convulsions. Chronic inhalation may cause pulmonary emphysema and chronic bronchitis. *See Clinical Toxicology of Commercial Products*, R. E. Gosselin *et al.*, Eds. (Williams & Wilkins, Baltimore, 5th ed., 1984) Section II, p 99; Section III, p 77-84. Potential toxic effects due to chronic overexposure by inhalation or ingestion are anemia; kidney damage; osteomalacia and osteoporosis. Itai-itai disease is a skeletal disease associated with a cadmium-induced renal disorder and is attributed to high oral intake of cadmium in food and water; characterized by progressive bone demineralization with painful joints and bones. *See Patty's Industrial Hygiene and Toxicology* vol. 2C, G. D. Clayton, F. E. Clayton, Eds. (John Wiley & Sons, New York, 4th ed., 1994) p 1954-1967. *See also NIOSH Pocket Guide to Chemical Hazards* (DHHS/NIOSH 90-117, 1990) p 56. Cadmium and certain cadmium compounds may reasonably be anticipated to be carcinogens: *Seventh Annual Report on Carcinogens* (PB 95-109781, 1994) p 111.

USE: Batteries, including Ni-Cd storage batteries; coating and electroplating steel and cast iron; pigments; plastic stabilizers; constituent of low melting or easily fusible alloys, e.g., Lichtenberg's, Abel's, Lipowitz', Newton's, and Wood's metal; electronics and optics; soft solder and solder for aluminum; reactor control rods; hardener for copper; catalytsts.

THERAP CAT (VET): Many cadmium salts, especially the oxide and anthranilate, are used or have been suggested as anthelmintics in swine and poultry.

1650. Cadmium Acetate. C$_4$H$_6$CdO$_4$; mol wt 230.50. C 20.84%, H 2.62%, Cd 48.77%, O 27.76%. Cd(CH$_3$COO)$_2$. Prepd by treating cadmium nitrate with acetic anhydride: Späth, *Monatsh.* **33**, 241 (1912); Wagenknecht, Juza, in *Handbook of Preparative Inorganic Chemistry*, vol. 2, G. Brauer, Ed. (Academic Press, New York, 2nd ed., 1965) p 1105.

Dihydrate, crystals. Slight acetic acid odor; becomes anhydr at about 130°. d 2.01. d (anhydr) 2.341. mp (anhydr) 255°. Freely sol in water; sol in alcohol. pH of 0.2 molar aq soln 7.1.

USE: Producing iridescent effects on porcelains and pottery; as a reagent for determination of S, Se, and Te; in cadmium electroplating.

1651. Cadmium Bromide. Br$_2$Cd; mol wt 272.22. Br 58.71%, Cd 41.29%. CdBr$_2$. Usually prepd from the elements: Honigschmid, Schlee, *Z. Anorg. Allgem. Chem.* **227**, 184 (1936); Wagenknecht, Juza, in *Handbook of Inorganic Chemistry*, vol. 2, G. Brauer, Ed. (Academic Press, New York, 2nd ed., 1965) p 1096.

Hexagonal, pearly flakes; highly hygroscopic. mp 566°; bp 963°; d 5.192. Freely sol in water, alc; moderately sol in acetone; slightly sol in ether. Crystallizes as the monohydrate below 36°, as the tetrahydrate above 36°.

USE: In photography, process engraving, and lithography.

1652. Cadmium Carbonate. CCdO$_3$; mol wt 172.42. C 6.97%, Cd 65.20%, O 27.84%. CdCO$_3$. Occurs in nature as the mineral, *otavite*. Prepn: de Schulten, *Bull. Soc. Chim. France* [3] **19**, 34 (1898); Biltz, *Z. Anorg. Allgem. Chem.* **220**, 312 (1934).

Powder or rhombohedral leaflets. d 4.26. Practically insol in water. Sol in concd solns of ammonium salts, dil acids.

1653. Cadmium Chloride. Caddy; Vi-Cad. CdCl$_2$; mol wt 183.32. Cd 61.32%, Cl 38.68%. Prepn: *Gmelin's, Cadmium* (8th ed.) **33**, pp 82-83 (1925); supplement, pp 464-465 (1959); Pray, *Inorg. Syn.* **5**, 153 (1957).

Hygroscopic, rhombohedral crystals. mp 568°; bp 960°; d 4.05. Freely sol in water; sol in acetone; slightly sol in methanol, ethanol. Practically insol in ether. LD$_{50}$ orally in rats: 88 mg/kg, *Toxic Substances List*, H. E. Christensen *et al.*, Eds. (1974).

Hemipentahydrate, efflorescent granules or rhombohedral leaflets. d^{25} 3.33. Freely sol in water.

USE: Photography; dyeing and calico printing; in the vacuum tube industry; manuf of cadmium yellow; galvanoplasty; manuf of special mirrors; as ice-nucleating agent; as lubricant; in analysis of sulfides to absorb the H$_2$S; in testing for pyridine bases; as fungicide.

1654. Cadmium Cyanide. C$_2$CdN$_2$; mol wt 164.45. C 14.61%, Cd 68.36%, N 17.04%. Cd(CN)$_2$. Prepn from Cd-(OH)$_2$ and HCN: Biltz, *Z. Anorg. Allgem. Chem.* **170**, 161 (1928); Wagenknecht, Juza in *Handbook of Preparative Inorganic Chemistry*, vol. 2, G. Brauer, Ed. (Academic Press, New York, 2nd ed., 1965) p 1105.

Crystals or white powder. Turns brown on heating in air. *Very poisonous!* d 2.226. Soly in water (15°) 1.71 g/100 ml. Slightly sol in alc; sol in solns of alkali cyanides or hydroxides; not appreciably attacked by organic acids, but readily dec by dil mineral acids with evolution of hydrogen cyanide. Readily forms complex cyanides.

USE: In copper bright electroplating.

1655. Cadmium Fluoride. CdF$_2$; mol wt 150.41. Cd 74.74%, F 25.26%. Prepn from CdCl$_2$ and NH$_4$F: Kurtenacker *et al.*, *Z. Anorg. Chem.* **211**, 89 (1933); from CdCO$_3$ and HF: Kwasnik in *Handbook of Preparative Inorganic Chemistry*, vol. 1, G. Brauer, Ed. (Academic Press, New York, 2nd ed., 1963) p 243.

Cubic crystals. mp 1049°; bp 1748°; d 6.33. *Poisonous!* Soly in water (25°) 4.3 g/100 ml. Sol in HF and other mineral acids; practically insol in alcohol and liquid ammonia.

USE: Manufacture of phosphors, glass; in nuclear reactor controls.

1656. Cadmium Hydroxide. Cadmium hydrate. CdH_2O_2; mol wt 146.43. Cd 76.77%, H 1.38%, O 21.85%. $Cd(OH)_2$. Usually prepd from a Cd salt by treatment with KOH: de Schulten, *Compt. Rend.* **101**, 72 (1885); Fricke, Blaschke, *Z. Elektrochem.* **46**, 46 (1940); Scholder, Staufenbiel, *Z. Anorg. Allgem. Chem.* **247**, 271 (1941).

Powder or trigonal and hexagonal crystals. d 4.79. Dehydration starts at 130° and is complete at 200°. Absorbs CO_2 from the air. Practically insol in water; slightly sol in NaOH soln; sol in dil acids, in NH_4OH or NH_4Cl solns.

USE: In storage battery electrodes.

1657. Cadmium Iodide. CdI_2; mol wt 366.22. Cd 30.69%, I 69.31%. Prepd from the elements or from the sulfate and KI: Wagenknecht, Juza in *Handbook of Preparative Inorganic Chemistry*, **vol. 2**, G. Brauer, Ed. (Academic Press, New York, 2nd ed., 1965) p 1096.

Hexagonal, lustrous, flake-like crystals; becomes yellow on long exposure to air and light. mp 388°; bp 787°; d 5.67. Sol in water, alc, ether, acetone.

USE: In electrodeposition of Cd; as nematocide; in manuf of phosphors; as lubricant; in photoconductors; in photography, process engraving; lithography; in analytical chemistry.

1658. Cadmium Nitrate. CdN_2O_6; mol wt 236.42. Cd 47.55%, N 11.85%, O 40.60%. $Cd(NO_3)_2$. Prepn: *Gmelin's Cadmium* (8th ed.) **33**, pp 76-78 (1925); suppl. p 446 (1959).

Tetrahydrate, hygroscopic, orthorhombic crystals. mp 59.5°. Sol in 0.6 part water; sol in alc, acetone, ethyl acetate; practically insol in concd HNO_3. *Keep well closed in a cool place.*

USE: In making other Cd salts; in photographic emulsions.

1659. Cadmium Oxide. Aska-Rid. CdO; mol wt 128.41. Cd 87.54%, O 12.46%. Prepn: *Gmelin's, Cadmium* (8th ed.) **33**, pp 69-70 (1925); supplement, pp 419-420 (1959). Toxicity data: Barrett *et al., J. Ind. Hyg. Toxicol.* **29**, 279 (1947).

Dark-brown, infusible powder or cubic crystals. d 8.15. Practically insol in water. Sol in dil acids; slowly sol in ammonium salts. LC_{50} in rats, monkeys (mg/m³): 500, ~15000 (Barrett).

Caution: Toxic effects due to overexposure have been observed in the respiratory system, GI tract, kidneys, skeletal system and hematopoetic system, see cadmium, *q.v.* Potential occupational carcinogen. *See NIOSH Pocket Guide to Chemical Hazards* (DHHS/NIOSH 90-117, 1990) p 56.

USE: In phosphors, semiconductors; manuf of silver alloys, glass; in storage battery electrodes; as nematocide; as catalyst for organic reactions, in cadmium electroplating; in ceramic glazes.

THERAP CAT (VET): Has been used as an ascaricide in swine.

1660. Cadmium Potassium Cyanide. *Potassium tetracyanocadmate.* $C_4CdK_2N_4$; mol wt 294.68. C 16.30%, Cd 38.15%, K 26.54%, N 19.01%. $K_2Cd(CN)_4$. Prepn: Dickinson, *J. Am. Chem. Soc.* **44**, 774 (1922); Biltz, *Z. Anorg. Allgem. Chem.* **170**, 161 (1928).

Highly refractive, cubic crystals. d 1.846. mp ~450°. When heated, melts to a colorless liquid solidifying to a gray, cryst mass on cooling. Sol in 3 parts cold, 1 part boiling water; slightly sol in alc.

1661. Cadmium Salicylate. $C_{14}H_{10}CdO_6$; mol wt 386.64. C 43.49%, H 2.61%, Cd 29.07%, O 24.83%. $Cd(C_7H_5O_3)_2$. Prepn: Prasad *et al., J. Indian Chem. Soc.* **35**, 267 (1958).

Monohydrate, small needles or plates. mp 242° (dec). Slightly sol in cold water, freely sol in boiling water; very slightly sol in methanol, ethanol.

THERAP CAT: Antiseptic.

1662. Cadmium Selenate. CdO_4Se; mol wt 255.37. Cd 44.02%, O 25.06%, Se 30.92%. $CdSeO_4$. Prepn: *Gmelin's, Cadmium* (8th ed.) **33**, p 133 (1925); suppl. p 637 (1959).

Dihydrate, orthorhombic crystals. Dec at 100°. d 3.632. Very sol in water.

1663. Cadmium Selenide. CdSe; mol wt 191.37. Cd 58.74%, Se 41.26%. Prepd by heating cadmium in a current of hydrogen selenide and subliming the product in hydrogen at a dull red heat: Margottet, *Recherches sur les Sulfures, les Séléniures, et les Tellures Métalliques,* Paris (1879); by passing hydrogen selenide over heated cadmium chloride and washing with warm water: Grzenkowsky, *Ueber Selenide und Erdalkaliferrocyanide,* (Danzig, 1925); from cadmium sulfate and hydrogen selenide: Wagenknecht, Juza in *Handbook of Preparative Inorganic Chemistry,* **vol. 2**, G. Brauer, Ed. (Academic Press, New York, 2nd ed., 1965) p 1099. Prepn of high purity CdSe: Taylor, Conn, U.S. pat. **3,540,-859** (1970 to Merck & Co.).

White to brown cubic or hexagonal crystals; turns red in sunlight. d 5.8; mp 1350°. Dec in air or acids. Practically insol in water.

USE: In photoconductors, semiconductors, photoelectric cells, and rectifiers; in phosphors.

1664. Cadmium Succinate. *Succinic acid cadmium salt;* Cadminate. $C_4H_4CdO_4$; mol wt 228.48. C 21.03%, H 1.76%, Cd 49.20%, O 28.01%. $CdC_4H_4O_4$. Prepd from cadmium carbonate and succinic acid: Schiff, *Ann.* **104**, 325 (1857).

Needles or plates. Soly in water at 40°, 0.367 g/100 ml. Practically insol in alc. LD_{50} orally in rats, mice: 660, 312 mg/kg, *RTECS*, H. E. Christensen, T. T. Laginbyhl, Eds. (1975) p 1085.

USE: Plant fungicide.

1665. Cadmium Sulfate. CdO_4S; mol wt 208.47. Cd 53.92%, O 30.70%, S 15.38%. $CdSO_4$. Prepn: *Gmelin's, Cadmium* (8th ed.) **33**, p 121 (1925); suppl. pp 609-610.

Hydrate, 8/3 moles water per mole cadmium sulfate, odorless, monoclinic crystals. On heating loses water above 40°, forming monohydrate by 80°. Does not become anhydrous on further heating. d 3.08. Freely sol in water. Almost insol in alcohol, ethyl acetate. LD s.c. in dogs: 27 mg/kg, *Toxic Substances List,* H. E. Chistensen *et al.,* Eds. (1974) p 740.

USE: In electrodeposition of Cd, Cu, and Ni; in phosphors; manuf of standard cadmium elements; catalyst in the Marsh test for As, determining H_2S and detecting fumaric acid; as nematocide.

1666. Cadmium Sulfide. C.I. 77199; Capsebon. CdS; mol wt 144.48. Cd 77.81%, S 22.19%. Occurs in nature as the mineral *greenockite*. Prepd from $CdSO_4 + H_2S$: Milligan, *J. Phys. Chem.* **38**, 797 (1934); Frerichs, *Naturwiss.* **33**, 281 (1946); prepn of single crystals: Grillot, *Compt. Rend.* **230**, 1280 (1950); Czyzak *et al., J. Appl. Phys.* **23**, 932 (1952); Wagenknecht, Juza in *Handbook of Preparative Inorganic Chemistry,* **vol. II**, G. Brauer, Ed. (Academic Press, New York, 2nd ed., 1965) pp 1098-1099. *See also Colour Index* vol. 4 (3rd ed., 1971) № 4659.

Light-yellow or orange-colored cubic or hexagonal crystals. Cubic structure: d 4.50; hexagonal structure: d 4.82. The light-yellow variety is also known as *Cadmium Yellow* or *Jaune Brilliant.* d 4.82. Sublimes at 980°. Soly in water (18°): 0.13 mg/100 g. Sol in concd or warm dil mineral acids with evolution of H_2S; readily dec and dissolved by moderately dil HNO_3.

USE: As a pigment being fast to light and not affected by H_2S; color for soaps; coloring glass yellow; coloring textiles, paper, rubber; in printing inks, ceramic glazes, fireworks; in phosphors and fluorescent screens; in scintillation counters, semiconductors, photoconductors.

THERAP CAT: Dermatologic.

1667. Cadmium Telluride. CdTe; mol wt 240.01. Cd 46.84%, Te 53.16%. Prepd by fusion of the elements or by the action of H_2Te on $CdCl_2$ solns: Dennis, Anderson, *J. Am. Chem. Soc.* **36**, 887 (1914); Kroger, de Nobel, *J. Electronics* **1**, 190 (1955); Kretschmar, Schilberg, *J. Appl. Phys.* **28**, 865 (1957); from aluminum telluride and a cadmium salt: Nitsche, U.S. pat. **2,767,049** (1956 to du Pont). Prepn of high purity CdTe: Taylor, Conn, U.S. pat. **3,540,859** (1970 to Merck & Co.); prepn of single crystals: Kyle, U.S. pat. **3,519,399** (1970 to Hughes Aircraft).

Brownish-black, cubic crystals by sublimation in hydrogen. d_4^{15} 6.2. mp 1041°. Oxidizes upon prolonged exposure to moist air. Practically insol in water and acids, except nitric, in which it is sol with decompn.

USE: In semiconductor research, in phosphors.

1668. Cadmium Tungstate(VI). CdO_4W; mol wt 360.25. Cd 31.20%, O 17.76%, W 51.03%. $CdWO_4$. Prepn: Karl, *Compt. Rend.* **196**, 1403 (1933); prepn of single crystals: Uitert, Soden, *J. Appl. Phys.* **31**, 328 (1960).

White or yellowish monoclinic crystals or powder. Practically insol in water or dil acids. Sol in solns of alkali cyanides.

USE: In x-ray screens; in scintillation counters; in phosphors; as catalyst for organic reactions.

1669. Cadralazine. *2-[6-[Ethyl(2-hydroxypropyl)amino]-3-pyridazinyl]hydrazinecarboxylic acid ethyl ester;* ethyl 6-[ethyl(2-hydroxypropyl)amino]-3-pyridazinecarbazate; 3-(2-carbethoxyhydrazino)-6-[N-(2-hydroxypropyl)ethylamino]pyridazine; DC-826; ISF-2469; Cadral; Cadraten; Cadrilan; Presmode. $C_{12}H_{21}N_5O_3$; mol wt 283.33. C 50.87%, H 7.47%, N 24.72%, O 16.94%. Peripheral vasodilator similar to hydralazine, *q.v.* Prepn: C. Carpi *et al.,* **Belg. pat. 811,847;** *eidem,* **U.S. pats. 3,925,381; 4,002,753** (1974, 1975, 1977 all to ISF); F. Parravicini *et al., Farmaco Ed. Sci.* **34**, 299 (1979). Analytical profile: L. Citerio *et al., Boll. Chim. Farm.* **120**, 222 (1981). Pharmacology: C. Semeraro *et al., J. Cardiovasc. Pharmacol.* **3**, 455 (1981). HPLC determn in plasma and urine: T. Crolla *et al., J. Chromatog.* **310**, 139 (1984). Hemodynamic effects in dogs: L. Dorigotti *et al., Arzneimittel-Forsch.* **34**, 984 (1984); in humans: B. Persson *et al., Eur. J. Clin. Pharmacol.* **31**, 513 (1987). Pharmacokinetics and metabolism in humans: H. Schütz *et al., Eur. J. Drug Metab. Pharmacokinet.* **10**, 147 (1985); S. A. Hauffe *et al., ibid.* 217. Preliminary clinical evaluations: R. Buoninconti, M. Motolese, *Int. J. Clin. Pharmacol. Ther. Toxicol.* **23**, 613 (1985); A. Salvadeo *et al., Arzneimittel-Forsch.* **35**, 623 (1985).

Crystals from acetone, mp 160-162°. pKa 6.0. uv max: 248, 340 nm (ε 22100, 2250). Soly (mg/ml): water 1.3; HCl 235.0; DMSO 323.0; methanol 21.0; dioxane 18.6; chloroform 8.5; diethyl ether, benzene, cyclohexane <0.1. LD_{50} in rats, dogs (mg/kg): 269, approx 400 i.v.; 2060, >2000 orally (Semeraro); in mice (mg/kg): 700 i.p. (Parravicini).

THERAP CAT: Antihypertensive.

1670. Cadusafos. *Phosphorodithioic acid O-ethyl S,S-bis(1-methylpropyl) ester;* S,S-di-sec-butyl O-ethyl phosphorodithioate; ebufos; FMC-67825; Apache; Rugby; Taredan. $C_{10}H_{23}O_2PS_2$; mol wt 270.40. C 44.42%, H 8.57%, O 11.83%, P 11.45%, S 23.72%. Organophosphate insecticide structurally similar to ethoprop, *q.v.* Manufacturing process: J. M. Brochard *et al.,* **Eur. pat. Appl. 235,056** (1987 to Rhone-Poulenc Agrochimie). Field trial as nematocide for bananas: P. Quénéhervé *et al., Rev. Nematol.* **14**, 251 (1991). Behavior in soils: S. Q. Zheng *et al., Sci. Total Environ.* **156**, 1 (1994).

Colorless to yellow liquid. Vapor press (25°): 120 mPa. Soly in water: 248 mg/l.

USE: Insecticide; nematocide.

1671. Cafaminol. *3,7-Dihydro-8-[(2-hydroxyethyl)methylamino]-1,3,7-trimethyl-1H-purine-2,6-dione;* 8-[(2-hydroxy-

ethyl)methylamino]caffeine; 8-(β-oxyethyl)methylaminocaffeine; methylcoffanolamine; Rhinoptil. $C_{11}H_{17}N_5O_3$; mol wt 267.29. C 49.43%, H 6.41%, N 26.20%, O 17.96%. Alkanolamine deriv of caffeine, *q.v.* Prepn: J. Klosa, **Ger. pat. 1,085,530;** *eidem,* **U.S. pat. 3,094,531** (1958, 1963 both to Delmar Chemicals Ltd.). Efficacy studies: E. Szirmai, *Praxis* **13**, 412 (1969); R. Leypoldt, *Therapiewoche* **26**, 3381 (1976). Bioavailability and absorption kinetics in humans: H. Walther, K. Kochler, *Pharmazie* **34**, 375 (1979).

Colorless crystals from ethanol, mp 162-164°. Soly in water is about 6%; pH of aq solns is 6.9. LD_{50} s.c. in male mice: 700 mg/kg (Klosa).

THERAP CAT: Decongestant (nasal).

1672. Cafestol. *[3bS-(3bα,5aβ,7β,8β,10aα,10bβ)]-3b,4,-5,6,7,8,9,10,10a,10b,11,12-Dodecahydro-7-hydroxy-10b-methyl-5a,8-methano-5aH-cyclohepta[5,6]naphtho[2,1-b]furan-7-methanol;* cafesterol. $C_{20}H_{28}O_3$; mol wt 316.44. C 75.91%, H 8.92%, O 15.17%. Diterpenoid constituent of coffee. Isoln from green coffee oil: Slotta, Neisser, *Ber.* **71**, 1991, 2342 (1938); C. Djerassi *et al., J. Org. Chem.* **18**, 1449 (1953). Prepn and purification: R. Bertholet, **U.S. pat. 4,692,534** (1987 to Nestec). Structure: C. Djerassi *et al., J. Am. Chem. Soc.* **81**, 2386 (1959); R. A. Finnegan, C. Djerassi, *ibid.* **82**, 4342 (1960). Stereochemical studies: R. A. Finnegan, *J. Org. Chem.* **26**, 3057 (1961); A. I. Scott *et al., J. Am. Chem. Soc.* **84**, 3197 (1962); A. I. Scott *et al., Tetrahedron* **20**, 1339 (1964). Stereospecific total synthesis of (±)-form: E. J. Corey *et al., J. Am. Chem. Soc.* **109**, 4717 (1987).

Crystals from hexane, mp 158-160°. $[\alpha]_D$ −101°. uv max: 222 nm (log ε 3.78).

Acetate, $C_{22}H_{30}O_4$, needles from petr ether, mp 167-168°. $[\alpha]_D$ −89°. uv max: 222 nm (log ε 3.80).

Tetrahydrocafestol, $C_{20}H_{32}O_3$, crystals from dil methanol, mp 154.5-157°.

1673. Caffeic Acid. *3-(3,4-Dihydroxyphenyl)-2-propenoic acid; 3,4-dihydroxycinnamic acid.* $C_9H_8O_4$; mol wt 180.16. C 60.00%, H 4.48%, O 35.52%. Constituent of plants, probably occurs in plants only in conjugated forms, e.g., chlorogenic acid. Isoln from green coffee: Wolfrom *et al., J. Agr. Food Chem.* **8**, 58 (1960); from roasted coffee: Krasemann, *Arch. Pharm.* **293**, 721 (1960). Formation by acid hydrolysis of chlorogenic acid: Fiedler, *Arzneimittel-Forsch.* **4**, 41 (1954); Whiting, Carr, *Nature* **180**, 1479 (1957); Guern, *C.A.* **61**, 9965h (1964). Synthesis: Hayduck, *Ber.* **36**, 2935 (1903); Posner, *J. Prakt. Chem.* **82**, 432 (1910); Mauthner, *ibid.* **142**, 33 (1935); Pandya *et al., Proc. Indian Acad. Sci.* **9A**, 511 (1939); Neish, *Can. J. Biochem. Physiol.* **37**, 1431 (1959). *Review:* Herrmann, *Pharmazie* **11**, 433 (1956).

Yellow crystals from concd aq solns. Monohydrate from dil solns. Dec 223-225° (softens at 194°). R_f values: Fiedler, *loc. cit.* Sparingly sol in cold water. Freely sol in hot water, cold alc. Alkaline solns turn from yellow to orange.

Methyl ester, $C_{10}H_{10}O_4$, colorless needles from water, mp 152-153°.

1674. Caffeine. *3,7-Dihydro-1,3,7-trimethyl-1H-purine-2,6-dione;* 1,3,7-trimethylxanthine; 1,3,7-trimethyl-2,6-dioxopurine; coffeine; thein; guaranine; methyltheobromine; No-Doz. $C_8H_{10}N_4O_2$; mol wt 194.19. C 49.48%, H 5.19%, N 28.85%, O 16.48%. Occurs in tea, coffee, maté leaves; also in guarana paste and cola nuts: Shuman, U.S. pat. **2,508,545** (1950 to General Foods). Obtained as a by-product from the manuf of caffeine-free coffee: Barch, U.S. pat. **2,817,588** (1957 to Standard Brands); Nutting, U.S. pat. **2,802,739** (1957 to Hill Bros. Coffee); Adler, Earle, U.S. pat. **2,933,395** (1960 to General Foods). Crystal structure: Sutor, *Acta Cryst.* **11**, 453 (1958). Synthesis: Fischer, Ach, *Ber.* **28**, 2473, 3135 (1895); Gepner, Kreps, *J. Gen. Chem. USSR* **16**, 179 (1946); Bredereck *et al., Ber.* **83**, 201 (1950); Crippa, Crippa, *Farmaco Ed. Sci.* **10**, 616 (1955); Swidinsky, Baizer, U.S. pats. **2,785,162** and **2,785,163** (1957 to Quinine Chem. Works); Bredereck, Gotsmann, *Ber.* **95**, 1902 (1962). Reversed-phase HPLC study: J. W. Weyland *et al., J. Chromatog.* **247**, 221 (1982). Effect of pregnancy on the pharmacokinetics of caffeine: R. Knutti *et al., Arch. Toxicol.* **5**, Suppl., 187 (1982). Binding of caffeine on benzodiazepine receptors: V. Saano, M. M. Airaksinen, *Acta Pharmacol. Toxicol.* **51**, 300 (1982). Disposition of caffeine and its metabolites in man: D. D. Tang-Liu *et al., J. Pharmacol. Exp. Ther.* **224**, 180 (1983). Arrhythmogenic effects in humans: D. J. Dobmeyer *et al., N. Engl. J. Med.* **308**, 814 (1983). Teratogenicity study: P. E. Palm *et al., Toxicol. Appl. Pharmacol.* **44**, 1 (1978). Comprehensive description: M. U. Zubair *et al.* in *Analytical Profiles of Drug Substances* **vol. 15**, K. Florey, Ed. (Academic Press, New York, 1986) pp 71-150.

Hexagonal prisms by sublimation, mp 238°. Sublimes 178°. Fast sublimation is obtained at 160-165° under 1 mm press. at 5 mm distance. d_4^{18} 1.23. pH of 1% soln 6.9. Aq solns of caffeine salts dissociate quickly. Absorption spectrum: Hartley, *J. Chem. Soc.* **87**, 1802 (1905). One gram dissolves in 46 ml water, 5.5 ml water at 80°, 1.5 ml boiling water, 66 ml alcohol, 22 ml alcohol at 60°, 50 ml acetone, 5.5 ml chloroform, 530 ml ether, 100 ml benzene, 22 ml boiling benzene. Freely sol in pyrrole; in tetrahydrofuran contg about 4% water; also sol in ethyl acetate; slightly in petr ether. Soly in water is increased by alkali benzoates, cinnamates, citrates or salicylates. LD_{50} orally in mice, hamsters, rats, rabbits (mg/kg): 127, 230, 355, 246 (males); 137, 249, 247, 224 (females) (Palm).

Monohydrate, felted needles, contg 8.5% H_2O. Efflorescent in air; complete dehydration takes place at 80°.

Hydrochloride dihydrate, $C_8H_{11}ClN_4O_2.2H_2O$, crystals, dec 80-100° with loss of water and HCl. Sol in water and in alcohol with dec.

Mixture with citric acid, *citrated caffeine*, *"caffeine citrate".* White, crystalline powder; acid reaction. Sol in about 4 parts warm water.

THERAP CAT: CNS stimulant.

THERAP CAT (VET): Has been used as a cardiac and respiratory stimulant and as a diuretic.

1675. Calamine. Eczederm. Prepd calamine. Zinc oxide with about 0.5% ferric oxide.

Pink powder. Insol in water. Almost completely sol in mineral acids.

THERAP CAT: Topical protectant.

THERAP CAT (VET): Astringent. Skin protectant.

1676. Calamus. Sweet flag; calmus; sweet cane; sweet grass. Dried rhizome of *Acorus calamus* L., *Araceae.* Habit.

Europe, North America, Western Asia; cultivated in Burma and Ceylon. *Constit.* Acorin, acoretin (choline), 1.5% volatile oil, 2.5% resins, 1.5% tannins; also reducing sugars and sterol bodies. *Ref:* Bose *et al., J. Am. Pharm. Assoc.* **49**, 32 (1960).

THERAP CAT: Carminative, anthelmintic.

1677. Calcifediol. *(3β,5Z,7E)-9,10-Secocholesta-5,7,10-(19)-triene-3,25-diol;* 25-hydroxyvitamin D_3; 25-hydroxycholecalciferol; 25-HCC; U-32070E; Calderol; Dedrogyl; Didrogyl; Hidroferol. $C_{27}H_{44}O_2$; mol wt 400.65. C 80.94%, H 11.07%, O 7.99%. The principal circulating form of vitamin D_3, formed in the liver by hydroxylation at C-25: Ponchon, DeLuca, *J. Clin. Invest.* **48**, 1273 (1969). It is the intermediate in the formation of 1α,25-dihydroxycholecalciferol, *q.v.,* the biologically active form of vitamin D_3 in the intestine. Identification in rat as an active metabolite of vitamin D_3: Lund, DeLuca, *J. Lipid Res.* **7**, 739 (1966); Morii *et al., Arch. Biochem. Biophys.* **120**, 513 (1967). Evaluation of biological activity in comparison with vitamin D_3: Blunt *et al., Proc. Nat. Acad. Sci. USA* **61**, 717 (1968); *ibid.* 1503. Isoln from porcine plasma and establishment of structure: Blunt *et al., Biochemistry* **7**, 3317 (1968). Synthesis: Blunt, DeLuca, *ibid.* **8**, 671 (1969). Review of isoln, identification and synthesis: DeLuca, *Am. J. Clin. Nutr.* **22**, 412 (1969). Review of bioassays: J. G. Haddad Jr., *Basic Clin. Nutr.* **2**, 579-597 (1980).

uv max (ethanol): 265 nm (ε 18000) (Blunt, DeLuca).

THERAP CAT: Calcium regulator.

1678. Calcimycin. *6S-[6α(2S*,3S*),8β(R*),9β,11α]-5-(Methylamino)-2-[[3,9,11-trimethyl-8-[1-methyl-2-oxo-2-(1H-pyrrol-2-yl)ethyl]-1,7-dioxaspiro[5.5]undec-2-yl]methyl]-4-benzoxazolecarboxylic acid;* antibiotic A-23187; A-23187. $C_{29}H_{37}N_3O_6$; mol wt 523.63. C 66.52%, H 7.12%, N 8.02%, O 18.33%. Polyether antibiotic produced by a strain of *Streptomyces chartreusensis* Calhoun and Johnson NRRL 3882. Activity as a divalent cation ionophore in isolated mitochondria: P. W. Reed, H. A. Lardy, *J. Biol. Chem.* **247**, 6970 (1972). Prepn and antimicrobial activity: R. M. Gale *et al.,* U.S. pat. **3,923,823** (1975 to Lilly). Elucidation of structure: M. O. Chaney *et al., J. Am. Chem. Soc.* **96**, 1932 (1974). Spectral studies of ionophore and metal ion complexes: D. R. Pfeiffer *et al., Biochemistry* **13**, 4007 (1974). Total synthesis and absolute configuration: D. A. Evans *et al., J. Am. Chem. Soc.* **101**, 6789 (1979); P. A. Grieco *et al., J. Org. Chem.* **45**, 3537 (1980). Stereospecific synthesis: G. R. Martinez *et al., J. Am. Chem. Soc.* **104**, 1436 (1982); D. P. Negri, Y. Kishi, *Tetrahedron Letters* **28**, 1063 (1987). Review of cation binding and transport properties: D. R. Pfeiffer *et al., Ann. N.Y. Acad. Sci.* **307**, 402-423 (1978). Use in model systems of calcium transport: M. Takamori *et al., J. Neurol. Sci.* **50**, 89 (1981); M. Takamori *et al., ibid.* **51**, 207 (1981); M. H. Freedman *et al., Cell. Immunol.* **58**, 134 (1981); G. Thomas, *Eur. J. Pharmacol.* **81**, 35 (1982); V. L. Lew, J. Garcia-Sancho, *Cell Calcium* **6**, 15 (1985).

Crystalline solid, mp 181-182°. $[\alpha]_D^{25}$ −56° (c = 1 in chloroform). uv max (ethanol): 204, 225, 278, 378 nm (E 28200, 26200, 18200, 8200). pKa_1 6.9 in 90% DMSO. Slightly sol in water, readily sol in ethyl acetate, chloroform, methanol, DMSO. Also reported as mp 184.5-186° (Evans, 1979). LD_{50} i.p. in mice: 10 mg/kg (Gale).

Mixed calcium-magnesium salt. Colorless crystalline solid, mp 230-250° (dec). uv max (ethanol, neutral): 202, 228, 303, 370 nm ($E_{1cm}^{1\%}$ 425, 490, 278, 109). Insol in water, pentane, hexane, heptane. Very slightly sol in methanol, DMSO. Very sol in methylene chloride, chloroform, acetone, methyl ethyl ketone, diethyl ketone, ethyl acetate.

USE: Biochemical tool used to study the role of divalent cations in various biological systems.

1679. Calcipotriene. (1α,3β,5Z,7E,22E,24S)-24-Cyclopropyl-9,10-secochola-5,7,10(19),22-tetraene-1,3,24-triol; (1S,-1'E,3R,5Z,7E,20R)-9,10-seco-20-(3'-cyclopropyl-3'-hydroxyprop-1'-enyl)-1,3-dihydroxypregna-5,7,10(19)-triene; calcipotriol; MC-903; Daivonex; Dovonex. $C_{27}H_{40}O_3$; mol wt 412.61. C 78.60%, H 9.77%, O 11.63%. Vitamin D_3 analog with low calcemic activity. Prepn: M. J. Calverley, E. T. Binderup, **PCT Int.** pat. **Appl.** 87 00,834; eidem, U.S. pat. **4,866,048** (1987, 1989 both to Leo Pharm.); M. J. Calverley, Tetrahedron **43**, 4609 (1987). Pharmacology: L. Binderup, E. Bramm, Biochem. Pharmacol. **37**, 889 (1988); P. J. Marie et al., Bone **11**, 171 (1990). Receptor binding: T. Valaja et al., Biochem. Pharmacol. **40**, 1827 (1990). Clinical evaluation in psoriasis: K. Kragballe et al., Brit. J. Dermatol. **119**, 223 (1988).

Crystals from methyl formate, mp 166-168°. uv max (96% ethanol): 264 nm (ε 17200).

THERAP CAT: Antipsoriatic.

1680. Calcitonin. Thyrocalcitonin; TCA; TCT. Mol wt about 4,500. Calcium regulating hormone secreted from the mammalian thyroid gland and in non-mammalian species from the ultimobranchial gland. Postulation of a plasma-calcium lowering substance: Copp et al., Endocrinology **70**, 638 (1962). Recognition as a hormone: Hirsch et al., ibid. **73**, 244 (1963); of thyroid origin: Foster et al., Nature **202**, 1303 (1964). Over-all action is to oppose the bone and renal effects of parathyroid hormone, q.v.; inhibits bone resorption of Ca^{2+}, with accompanying hypocalcemia and hypophosphatemia and decreased urinary Ca^{2+} concentrations. Also abolishes the osteolytic effect of toxic doses of vitamins A and D. Calcitonin is highly active biologically, e.g. 50 mγ/min infused into a 100 g rat leads to a significant (1 mg/100 ml) decrease in the concn of the plasma calcium within 60 min (together with a corresponding fall in plasma phosphate). Activity is destroyed by trypsin, chymotrypsin,

pepsin, polyphenol oxidase; also by hydrogen peroxide oxidation, photooxidation, and treatment with N-bromosuccinimide.

Calcitonin structures are single polypeptide chains containing 32 amino acid residues. Structure of porcine: Neher et al., Helv. Chim. Acta **51**, 917 (1968); Potts et al., Proc. Nat. Acad. Sci. USA **59**, 1321 (1968); Bell et al., J. Am. Chem. Soc. **90**, 2704 (1968); eidem, Biochemistry **9**, 1665 (1970). Synthesis of porcine: Rittel et al., Helv. Chim. Acta **51**, 924 (1968); Guttmann et al., ibid. 1155. Isoln of human calcitonin from non-pathological thyroid glands: Haymovits, Rosen, Endocrinology **81**, 993 (1967); from medullary carcinoma of the thyroid: Neher et al., Nature **220**, 984 (1968); Helv. Chim. Acta **51**, 1738 (1968); Neher, Riniker, **Ger.** pat. **1,929,957** (1970 to Ciba), C.A. **73**, 28902b (1970). Structure of human: Neher et al., Helv. Chim. Acta **51**, 1900 (1968). Synthesis of human: Sieber et al., ibid. 2057; J. Hirt et al., Rec. Trav. Chim. **98**, 143 (1979). Biosynthetic studies: J. W. Jacobs et al., J. Biol. Chem. **254**, 10600 (1979); S. G. Amara et al., ibid. **255**, 2645 (1980). Amino acid sequence differs among mammalian species, salmon calcitonin showing a marked difference from that of the higher vertebrae as well as a more potent biological activity. Mechanism of action: E. M. Brown, G. D. Aurbach, Vitam. Horm. (New York) **38**, 236 (1980). Anorectic activity in rats: W. J. Freed et al., Science **206**, 850 (1979). Growth inhibition of human breast cancer cells in vitro: Y. Iwasaki et al., Biochem. Biophys. Res. Commun. **110**, 235 (1983). Review of early literature: Munson, Hirsch, Clin. Orthop. **49**, 209 (1966). Review of isoln, structure, synthesis: Behrens, Grinnan, Ann. Rev. Biochem. **38**, 83 (1969); Potts et al., Vitam. Horm. (New York) **29**, 41 (1971). Comprehensive review: Calcitonin, Proc. Symp. on Thyrocalcitonin and the C Cells, S. Taylor, Ed. (Springer-Verlag, New York, 1968); Foster et al., "Calcitonin" in Clinics in Endocrinology and Metabolism, I. MacIntyre, Ed. (W. B. Saunders, Philadelphia, 1972) pp 93-124. Review of pharmacology and therapeutic use: J. C. Stevenson, I. M. A. Evans, Drugs **21**, 257-272 (1981).

Calcitonin, porcine, Calcitar(e), Staporos.

Calcitonin, human synthetic, Cibacalcin.

Calcitonin, salmon synthetic, salcatonin, Calciben, Calcimar, Calsyn, Calsynar, Catonin, Karil, Miacalcic, Miacalcin, Miadenil, Osteocalcin, Prontocalcin, Rulicalcin, Salmotonin, Stalcin, Tonocalcin. See also Elcatonin.

THERAP CAT: Calcium regulator.

1681. Calcitriol. (1α,3β,5Z,7E)-9,10-Secocholesta-5,7,-10(19)-triene-1,3,25-triol; 1α,25-dihydroxycholecalciferol; 1α,25-dihydroxyvitamin D_3; 1,25-DHCC; Ro-21-5535; Calcijex; Rocaltrol. $C_{27}H_{44}O_3$; mol wt 416.64. C 77.84%, H 10.64%, O 11.52%. The biologically active form of vitamin D_3 in intestinal calcium transport and bone calcium resorption: Haussler et al., Proc. Nat. Acad. Sci. USA **68**, 177 (1971); Raisz et al., Science **175**, 768 (1972). Formed by the sequential hydroxylation of vitamin D_3 at C-25 in the liver and at C-1 in the kidney: Blunt et al., Biochemistry **7**, 3317 (1968); Fraser, Kodicek, Nature **228**, 764 (1970); Norman et al., Biochem. Biophys. Res. Commun. **42**, 1082 (1971). Classification as a steroid hormone: Emtage et al., Nature **246**, 100 (1973). First obtained from chick intestine and designated as metabolite 4B: Myrtle et al., J. Biol. Chem. **245**, 1190 (1970); Myrtle, Norman, Science **171**, 79 (1971); Haussler et al., Proc. Nat. Acad. Sci. USA **68**, 177 (1971). Isoln and identification: Holick et al., ibid. 803; Lawson et al., Nature **230**, 228 (1971); Norman et al., Science **173**, 51 (1971). Synthesis: Semmler et al., Tetrahedron Letters **1972**, 4147; Barton et al., Chem. Commun. **1974**, 203; H. E. Paaren et al., ibid. **1977**, 890; T. Sato et al., Chem. Pharm. Bull. **26**, 2933 (1978). Review: Suda, Vitamins **45**, 175-188 (1972), C.A. **77**, 69866u (1972); E. Kodicek, Lancet **1**, 325 (1974). Another polar metabolite of vitamin D_3, 21,25-dihydroxycholecalciferol, $C_{27}H_{44}O_3$, shows marked action on mobilization of bone mineral and a small, but significant effect on intestinal calcium transport and has been isolated from pig plasma: Suda et al., Biochemistry **9**, 2917 (1970). See also DeLuca et al., Proc. Nat. Acad. Sci. USA **68**, 2131 (1971), which details the effect of calcium on in vivo synthesis of 1,25- and 21,25-dihydroxycholecalciferol. Comprehensive description: E. Debesis in Analytical Profiles of

Drug Substances **Vol. 8**, K. Florey, Ed. (Academic Press, New York, 1979) pp 83-100.

White crystalline powder, mp 111-115°. uv max (abs ethanol): 264 nm (ϵ 19000). $[\alpha]_D^{25}$ +48° (methanol). Slightly sol in methanol, ethanol, ethyl acetate, THF. Air and light sensitive.

THERAP CAT: Calcium regulator; vitamin (antirachitic).

1682. Calcium. Ca; at. wt 40.08; at. no. 20; valence 2. An alkaline earth metal. Occurrence in the earth's crust 3.64% (fifth element in order of abundance). Sea water contains about 400 g/ton. Isotopes: 40 (97%), 44 (2.06%), 42 (0.64%), 48 (0.18%), 43 (0.145%), 46 (0.003%). Found naturally only in the form of its compds, never uncombined. Principal commercial source is limestone, *q.v.* Essential constituent of bones, shells, teeth. First isolated by Davy in 1808. Produced by electrolysis of calcium chloride: Rathenau, Suter, Ger. pat. **155,433** (1903); *Z. Elektrochem.* **10**, 502 (1904); Goodwin, *J. Am. Chem. Soc.* **27**, 1403 (1905); also by thermal reduction of lime with silicon, or with aluminum. Prepn of the pure metal for laboratory use: Whaley, *Inorg. Syn.* **6**, 18 (1960). Purification of commercial material: Marshall, Whaley, *ibid.* 24. *Reviews:* Schaufler in *Ullmanns Encyklopädie der Tech. Chemie,* **vol. 4** (Munich, 3rd ed., 1953) pp 830-836; Mantell in C. A. Hampel, *Rare Metals Handbook* (Reinhold, New York, 1954) p 17-29. Review of calcium and its compounds: Goodenough, Stenger, "Magnesium, Calcium, Strontium, Barium and Radium" in *Comprehensive Inorganic Chemistry,* **Vol. 1**, J. C. Bailar Jr. *et al.*, Eds. (Pergamon Press, Oxford, 1973) pp 591-664; C. J. Kunesh in Kirk-Othmer *Encyclopedia of Chemical Technology* **vol. 4** (Wiley-Interscience, New York, 3rd ed., 1978) pp 412-421. Book: R. P. Rubin, *Calcium and Cellular Secretion* (Plenum, New York, 1982) 276 pp.

Lustrous, silver-white surface (when freshly cut); face-centered cubic structure below 300°C. Ignites in air when finely divided, then burns with crimson flame. Much harder than sodium, but softer than aluminum or magnesium. Acquires bluish-gray tarnish on exposure to moist air. d_4^{20} 1.54. mp 850°. bp 1440°. Electrical resistivity at 20°: 3.5 μohm cm. Brinell hardness: 17. Heat of combustion 151.9 cal/g. sp ht (0-100°) 0.149 cal/g. Considerably less reactive than sodium. E° (aq) Ca^{2+}/Ca -2.87 V. Reacts with water, alcohols, and dil acids with evolution of hydrogen. Reacts with halogens. Dissolves in liquid ammonia to form a blue soln. Contact with alkali hydroxides or carbonates may cause detonation. Burns in air. Calcium salts impart brick red color to a flame. Insol in and inert towards benzene, kerosene.

USE: In metallurgy as deoxidizer for copper, beryllium, steel (together with silicon). To harden lead for bearings. Alloyed with cerium to make flints for cigarette and gas lighters. In manuf of electronic vacuum tubes as "getter" to fix residual gases as oxides, nitrides, hydrides of calcium.

1683. Calcium Acetate. $C_4H_6CaO_4$; mol wt 158.17. C 30.38%, H 3.82%, Ca 25.34%, O 40.46%. $Ca(CH_3COO)_2$. The technical product is known as *"brown acetate of lime"* or *"gray acetate of lime"*. Prepn: *Gmelin's, Calcium* (8th ed.) **28B**, 162-164, 982-991 (1958); electrolytic prepn: Schmidt, *Z. Anorg. Allgem. Chem.* **270**, 188 (1952). Toxicity study: Smyth *et al.*, *Am. Ind. Hyg. Assoc. J.* **30**, 470 (1969). Very hygroscopic, rod-shaped crystals. On heating above 160° dec to acetone and $CaCO_3$. d 1.50. Sol in water;

slightly sol in methanol. Practically insol in ethanol, acetone, benzene. *Keep well closed.*

Monohydrate, needles, granules or powder. Does not lose all its water below 150°. Sol in water; slightly sol in alcohol. pH of 0.2 molar aq soln 7.6. LD_{50} orally in rats: 4.28 g/kg (Smyth).

Dihydrate, long, transparent needles. Loses water on standing in air, forming the monohydrate.

USE: Manuf of acetic acid, acetone; in dyeing, tanning, and curing skins; in lubricants; as food stabilizer; as corrosion inhibitor.

1684. Calcium Acetylsalicylate. *2-(Acetyloxy)benzoic acid calcium salt; salicylic acid acetate calcium salt;* acetylsalicylic acid calcium salt; calcium aspirin; soluble aspirin; Ascal; Cal-Aspirin; Dispril; Disprin; Kalmopyrin; Kalsetal; Solaspin; Tylcalsin. $C_{18}H_{14}CaO_8$; mol wt 398.38. C 54.27%, H 3.54%, Ca 10.06%, O 32.13%. $Ca(OOCC_6H_4OCOCH_3)_2$. Prepd from calcium carbonate and acetylsalicylic acid: Lawrence, U.S. pat. **2,003,374** (1935 to Lee Labs.).

Dihydrate, amorphous, nonhygroscopic powder. One gram dissolves in 6 ml water, 80 ml alcohol.

Complex with urea, $C_{19}H_{18}CaN_2O_9$, *carbasalate calcium, acetylsalicylic acid calcium salt complex with urea, calcium acetylsalicylate carbamide, urea calcium acetylsalicylate, carbaspirin calcium, Alcacyl, Calurin, Solupsan.* Amorphous powder, dec 243-245°. Soly in water at 37°: 231 mg/ml, pH 4.8. Description and hydrolytic stability: Parrott, *J. Pharm. Sci.* **51**, 897 (1962).

THERAP CAT: Analgesic; antipyretic; anti-inflammatory.

1685. Calcium Aluminosilicate. Aluminum calcium silicate. Many different forms of calcium aluminosilicate are known, the most common of which are $CaAl_2Si_2O_8$ and $Ca_2Al_2SiO_7$. It occurs in nature as the minerals: *anorthite, bavenite, clinozoisite, didymolite, epistilbite, gehlenite, gismondite, grossularite, heulandite, hibschite, laubanite, laumontite, lawsonite, levynite, margarite, meionite, plazolite, pumpellyite, scolecite, stellerite, vesuvianite, zoisite.* Prepn and properties: *Gmelin's, Aluminum* (8th ed.) **35B**, 576-586 (1934). Summary and references for minerals: Hey, *An Index of Mineral Species and Varieties* (British Museum, London, 2nd ed., 1962) pp 159-162.

USE: Constituent of cement; in refractories.

1686. Calcium Arsenate. Tricalcium arsenate; Pencal. $As_2Ca_3O_8$; mol wt 398.07. As 37.64%, Ca 30.20%, O 32.15%. $Ca_3(AsO_4)_2$. Prepn: Les Veaux, U.S. pat. **2,715,-562** (1955 to FMC). Toxicity study: T. B. Gaines, *Toxicol. Appl. Pharmacol.* **14**, 515 (1969). *Review:* Guerin, *Chim. & Ind. (Paris)* **77**, 1288 (1957).

Powder. *Poisonous!* Slightly sol in water; sol in dil acids. LD_{50} orally in female rats: 298 mg/kg (Gaines).

Caution: Potential symptoms of overexposure are weakness; gastrointestinal disturbance; peripheral neuropathy; skin hyperpigmentation, palmar planter hyperkeratoses; dermatitis. Direct contact may cause skin irritation. Potential occupational carcinogen. *See NIOSH Pocket Guide to Chemical Hazards* (DHHS/NIOSH 90-117, 1990) p 56.

USE: Insecticide; molluscicide.

1687. Calcium Arsenite. Variable composition. Prepd by passing steam over a dry mixture of CaO and As_2O_3: Altwegg, Dutel, U.S. pat. **1,700,756** (1929).

White, granular powder. *Poisonous!* Slightly sol in water; sol in acids.

USE: As insecticide, germicide, molluscicide.

1688. Calcium Ascorbate. *Ascorbic acid calcium salt.* $C_{12}H_{14}CaO_{12}$; mol wt 390.31. C 36.93%, H 3.62%, Ca 10.27%, O 49.19%. $(C_6H_7O_6)_2Ca$. Prepd from ascorbic acid and calcium carbonate by controlled precipitation in dilute acetone or alcohol: Ruskin, Merrill, *Science* **105**, 504 (1947); Ruskin, U.S. pat. **2,596,103** (1952); *idem*, U.S. pat. **2,631,155** (1953 to Physiological Chemicals).

Dihydrate, triclinic crystals, $[\alpha]_D^{20}$ +95.6° (c = 2.4). Freely sol in water. Practically insol in methanol, ethanol. Aq solns are neutral and oxidize quickly. After prolonged storage calcium oxalate may precipitate. The addition of various stabilizers such as cysteine, has been patented: Karrer, U.S. pat. **2,442,461** (1948 to Hoffmann-La Roche).

THERAP CAT: Vitamin (antiscorbutic).

1689. Calcium 3-Aurothio-2-propanol-1-sulfonate. *[2-Hydroxy-3-mercapto-1-propanesulfonato(2—)-O^2,S^3]aurate(1—) calcium (2:1); [(2-hydroxy-3-sulfopropyl)thio]gold calcium salt; 2-hydroxy-3-mercapto-1-propanesulfonic acid S-gold deriv calcium salt; Chrisanol; Chrysanol; Krizanol.* $C_6H_{12}Au_2CaO_8S_4$; mol wt 774.43. C 9.31%, H 1.56%, Au 50.87%, Ca 5.18%, O 16.53%, S 16.56%. (AuSCH$_2$CHOH-CH$_2$SO$_3$)$_2$Ca. Prepd by treating sodium thiopropanolsulfonate with gold chloride in the presence of SO$_2$ and converting the sodium salt to the calcium salt by means of calcium chloride: Lipovich, Mones, *J. Appl. Chem. (USSR)* **18**, 20 (1945); *cf.* **Brit.** pat. **265,777**; Lumière, Perrin, *Compt. Rend.* **184**, 289 (1927).

Pale yellow powder. Has no definite mp. Soly in water about 2%. Practically insol in organic solvents. Marketed as *Oleochrysine* which is a suspension plus some Ca gluconate in peach kernel oil.

THERAP CAT: Antiarthritic.

1690. Calcium Bisulfite, Solution. Ca(HSO$_3$)$_2$ is known only in soln. Prepn from sulfite liquor: Arend, *Chem. Products* **10**, 53 (1947); Lougheed, *Pulp Paper Mag. Can.* **49**(3), 215 (1948); Schoeffel, U.S. pat. **2,696,424** (1954 to Sterling Drug). The product here described is substantially a soln of calcium sulfite in an aq sulfur dioxide soln.

Colorless or slightly yellow liquid; strong SO$_2$ odor. On standing in the air, crystals of CaSO$_3$.2H$_2$O form. d about 1.06. This soln corrodes metals.

USE: As germicide, preservative, and disinfectant; for washing (1:1000) casks in brewing to prevent souring and cloudiness of beer and to prevent secondary fermentation; as antichlor in bleaching fabrics; largely in manuf sulfite cellulose from wood for paper-making.

1691. Calcium Borate. Calcium pyroborate; calcium tetraborate. B$_4$CaO$_7$; mol wt 195.32. B 22.14%, Ca 20.52%, O 57.34%. CaB$_4$O$_7$. Prepn by direct fusion of B$_2$O$_3$ with CaCO$_3$: Griveau, *Compt. Rend.* **166**, 993 (1918). Various calcium borate minerals occur in nature. These include: *colemanite, ginorite, inyoite, meyerhofferite, pandermite, priceite.*

Hexahydrate, powder. Almost insol in cold, moderately sol in hot water; sol in dil acids.

USE: As flux in heavy-metal metallurgy; mfr of forsterite porcelain insulators; in glycol antifreeze; in fire-retardant paint.

1692. Calcium Borogluconate. *D-Gluconic acid cyclic 4,5-ester with boric acid calcium salt (2:1);* calcium diborogluconate. $C_{12}H_{20}B_2CaO_{16}$; mol wt 481.98. C 29.90%, H 4.18%, B 4.49%, Ca 8.32%, O 53.11%. Prepn and use in milk fever: H. Dryerre, J. R. Greig, *Vet. Rec.* **15**, 456 (1935); H. T. Macpherson, J. Stewart, *Biochem. J.* **32**, 76 (1938). Physical properties: L. Seekles, E. Havinga, *Norsk. Vet.-Tids.* **58**, No. 11, 433-445 (1946), *C.A.* **43**, 1907 (1949). Effect on plasma calcium concentration in sheep: D. A. H. Farningham, *Res. Vet. Sci.* **39**, 70 (1985). Spectrophotometric determn in plasma: D. J. Lyons, K. P. Spann, *J. Assoc. Off. Anal. Chem.* **68**, 160 (1985). Evaluation in bovine milk fever: P. A. Mullen, *Vet. Rec.* **97**, 87 (1975).

Crystals, freely sol in water. Soly in water 1:1 at 15°: 2.8:1 at 100°. A 20% aq soln has a pH of 3.5. If desired, the pH may be adjusted to 7.0 by the addition of CaO which is very sol in Ca borogluconate solns.

THERAP CAT (VET): In hypocalcemic states including bovine milk fever.

1693. Calcium Bromide. Br$_2$Ca; mol wt 199.89. Br 79.95%, Ca 20.05%. CaBr$_2$. Prepn: *Gmelin's, Calcium* (8th ed.) **28B**, 100-102, 584-599 (1958). The N.F. article is a hydrated salt, contg not less than 84% and not more than 94% CaBr$_2$.

Odorless, deliquesc granules or rhombic crystals; sharp, saline taste. Becomes yellow on long exposure to air. When anhydr mp 730°; d$_4^{25}$ 3.353. When strongly heated in air, becomes alkaline due to loss of bromine and formation of lime. Very sol in water, methanol, ethanol; sol in acetone; practically insol in dioxane, chloroform, ether. The aq soln is neutral or only slightly alkaline to litmus. *Keep well closed.*

USE: In photography for making dry plates and light-sensitive papers; manuf mineral waters, NH$_4$Br, fire-extinguishing compositions.

THERAP CAT: Sedative, anticonvulsant.

THERAP CAT (VET): Has been used in hypocalcemic states such as canine eclampsia.

1694. Calcium Bromolactobionate. *D-Gluconic acid 4-O-β-D-galactopyranosyl calcium salt (2:1) compd with calcium bromide (CaBr$_2$) (1:1); lactobionic acid calcium salt compd with calcium bromide;* calcium galactogluconate bromide; Brocalcin; Calabron; Calcibromin; Calcibronat; Calciobrom. $C_{24}H_{42}Br_2Ca_2O_{24}$; mol wt 954.55. C 30.20%, H 4.43%, Br 16.74%, Ca 8.40%, O 40.23%. (C$_{12}$H$_{21}$O$_{12}$)$_2$Ca.-CaBr$_2$. Prepn: Isbell, *J. Res. Nat. Bur. Stand.* **17**, 331 (1936).

Hexahydrate, hexagonal prisms. Mild taste. Nonhygroscopic. Freely sol in water.

THERAP CAT: Sedative.

1695. Calcium N-Carbamoylaspartate. *N-(Aminocarbonyl)-L-aspartic acid calcium salt (1:1); N-carbamoylaspartic acid calcium salt;* calcium ureidosuccinate. $C_5H_6CaN_2O_5$; mol wt 214.19. C 28.04%, H 2.82%, Ca 18.71%, N 13.08%, O 37.35%. Prepn of optically active forms: Miyazaki *et al., Chem. Pharm. Bull.* **15**, 1604 (1967); Yoshimura *et al.,* **Japan.** pat. **68 14,205** (1968 to Dainippon Pharm.), *C.A.* **71**, 112456a (1969); of DL-form: Clemence, **Fr.** pat. **M6376** (1968 to Roussel-UCLAF), *C.A.* **74**, 76655r (1971).

$[\alpha]_D^{20}$ +32.2° (c = 2.53).
DL-Form, *Pacilan.* mp > 350°.
D-Form, $[\alpha]_D^{20}$ −31.5° (c = 1.54).

THERAP CAT: Anxiolytic.

1696. Calcium Carbide. Acetylenogen. C$_2$Ca; mol wt 64.10. C 37.48%, Ca 62.52%. CaC$_2$. Prepn: Ehrlich in *Handbook of Preparative Inorganic Chemistry*, **Vol. 1**, G. Brauer, Ed. (Academic Press, New York, 2nd ed., 1963) p 943. *Review:* Brennan, *J. Electrochem. Soc.* **99**, 61c (1952).

Grayish-black, irregular lumps or orthorhombic crystals; dec by water with evolution of acetylene leaving a residue of lime. d 2.22; mp 2300°.

USE: Generating acetylene gas for lighting purposes (1 kg of carbide yields ~300 liters acetylene); as reducing agent, e.g., for direct reduction of copper sulfide to metallic copper; signal fires for marine service; manuf of calcium, iron, alloys, lampblack, cyanamide; welding and cutting metals.

1697. Calcium Carbonate. *Carbonic acid calcium salt (1:1);* Cacit; Calcichew; Calcidia; Citrical. CCaO$_3$; mol wt 100.09. C 12.00%, Ca 40.04%, O 47.96%. CaCO$_3$. Exists in nature as the minerals *aragonite, calcite* and *vaterite.*

Odorless, tasteless powder or crystals. Two crystal forms are of commercial importance: Aragonite, orthorhombic, mp 825° (dec), d 2.83, formed at temps above 30°; Calcite, hexagonal-rhombohedral, mp 1339° (102.5 atm), d$^{25.2}$ 2.711, formed at temps below 30°. At about 825° is dec into CaO and CO_2. Practically insol in water. Sol in dil acids. *Incompat:* Acids, alum, ammonium salts.

Precipitated calcium carbonate, precipitated chalk, Aeromatt, Albacar, Purecal. Commercial $CaCO_3$ produced by chemical means. It is 98-99% pure. The byproduct process, the carbonation process, and the calcium chloride process of manuf from limestone are outlined in Kirk-Othmer *Encyclopedia of Chemical Technology,* vol. **4** (Interscience, New York, 2nd ed., 1964) pp 7-11. *Review:* Woerner in *Pigment Handbook* vol. **1**, T. C. Patton, Ed. (John Wiley, New York, 1973) pp 119-128.

Prepared calcium carbonate, drop chalk, prepared chalk, whiting, English white, Paris white. Native $CaCO_3$ purified by elutriation.

USE: Manuf of paint, rubber, plastics, paper, dentifrices, ceramics, putty, polishes, insecticides, inks, shoe dressings; as a filler in production of adhesives, matches, pencils; crayons, linoleum, insulating compds, welding rods. In foods, cosmetics, pharmaceuticals, antibiotics; removing acidity of wines. In anal. chem for detecting and determining halogens in organic combinations; with NH_4Cl for decomposing silicates; preparing $CaCl_2$ soln for standardizing soap solns; for water analyses.

THERAP CAT: Antacid. Calcium supplement.

THERAP CAT (VET): Antacid, calcium supplement, antidiarrheal agent.

1698. Calcium Chlorate. $CaCl_2O_6$; mol wt 206.98. Ca 19.36%, Cl 34.26%, O 46.38%. $Ca(ClO_3)_2$. Prepn: Wilderman, **Brit.** pat. **183,671** (1921); Duveau, *Bull. Soc. Chim.* **10**, 374 (1943).

Dihydrate, monoclinic, hygroscopic crystals. d 2.711. mp 100° when rapidly heated. Sol in 0.6 part water; sol in alcohol. *Keep well closed.* LD_{50} orally in rats: 4.5 g/kg, *Handbook of Toxicology* vol. **1**, W. S. Spector, Ed. (Saunders, Philadelphia, 1956) pp 58-59.

USE: Herbicide, insecticide, seed disinfectant.

1699. Calcium Chloride. Intergravin-orales. $CaCl_2$; mol wt 110.98. Ca 36.11%, Cl 63.89%. Forms mono-, di-, tetra- and hexahydrates. Obtained as a byproduct of the ammonia-soda (Solvay) process and as a joint product from natural salt brines: Faith, Keyes & Clark's *Industrial Chemicals*, F. A. Lowenheim, M. K. Moran, Eds. (Wiley-Interscience, New York, 4th ed., 1975) pp 186-190. Acute toxicity: I. B. Syed, F. Hosain, *Toxicol. Appl. Pharmacol.* **22**, 150 (1972).

Cubic crystals, granules or fused masses. Very hygroscopic. mp 772°. bp > 1600°; d_4^{15} 2.152. Freely sol in water (with liberation of much heat), alcohol. The commercial product is about 94-97% $CaCl_2$, the chief impurity being $Ca(OH)_2$. *Keep well closed.* LD_{50} i.v. in mice: 42.2 mg/kg (Syed, Hosain).

Dihydrate, hygroscopic granules, flakes or powder. Apparent (bulk) density: 0.835. Freely sol in water, alcohol. Commercial grades contain 73-80% $CaCl_2$. *Keep well closed.*

Hexahydrate, deliquesc trigonal crystals. mp 30°. d^{17} 1.68. Loses all H_2O at 200°. Extremely sol in water, alcohol. *Keep well closed.*

USE: The *anhydrous* form used as a drying and dehydrating agent for organic liquids and gases, and in desiccators. The *dihydrate* and *hexahydrate* forms are used for antifreeze and refrigerating solns, in fire extinguishers, etc. (a 40% soln freezes at −41°); to preserve wood, stone; manuf ice, glues, cements; fireproofing fabrics; automobile antifreeze mixtures; to melt ice and snow; as coagulant in rubber manuf, as size in admixture with starch paste; in concrete mixes to give quicker initial set and greater strength; freezeproofing of coal and ores; dust control on unpaved roads; sizing and finishing cotton fabrics; as brine for filling inflatable tires on tractors to increase traction.

THERAP CAT: Electrolyte replacement. Has been used as diuretic, urinary acidifier, antiallergic.

THERAP CAT (VET): May be used intravenously in hypocalcemic states such as milk fever.

1700. Calcium Chromate(VI). Calcium chrome yellow; gelbin; yellow ultramarine; C.I. 77223; C.I. Pigment Yellow 33. $CaCrO_4$; mol wt 156.07. Ca 25.68%, Cr 33.32%, O 41.01%. Prepn: Mylius, Wrochem, *Ber.* **33**, 3689 (1900); Udy, **U.S.** pats. **2,493,789; 2,494,215** (both 1950); Dunn,

O'Brien, **U.S.** pats. **2,745,764/5** (both 1956 to Vanadium Corp. of America). Also occurs as hemihydrate, monohydrate, and dihydrate.

Yellow monoclinic or rhombic crystals. Sparingly sol in water; sol in dil acids; practically insol in alcohol.

USE: As a pigment, corrosion inhibitor; manuf of chromium; in oxidizing reactions; in battery depolarization.

1701. Calcium Citrate. *2-Hydroxy-1,2,3-propanetricarboxylic acid calcium salt (2:3).* $C_{12}H_{10}Ca_3O_{14}$; mol wt 498.44. C 28.92%, H 2.02%, Ca 24.12%, O 44.94%. $Ca_3(C_6H_5O_7)_2$. Prepn from citrus fruit: Cole, **U.S.** pat. **2,389,-766** (1945 to California Fruit Growers Exchange). *Review:* Rudy, *Pharmazie* **4**, 393 (1949).

Tetrahydrate, odorless powder. Loses most of its water at 100° and all at 120°. Sol in 1050 parts cold water; somewhat more sol in hot water; insol in alcohol.

USE: In the production of citric acid and other citrates; improvement of baking properties of flour.

1702. Calcium Cyanamide. Calcium carbimide; "cyanamide"; nitrolime. $CCaN_2$; mol wt 80.10. C 14.99%, Ca 50.03%, N 34.97%. $N\equiv CN=Ca$. Prepn: Kastens, McBurney, *Ind. Eng. Chem.* **43**, 1020 (1951); Franck, Heimann, *Angew. Chem.* **44**, 372 (1931); Owen, *Trans. Faraday Soc.* **57**, 670 (1961); Dedman, Owen, *ibid.* 678.

Commercial grades may occur as grayish-black lumps of powder. While pure calcium cyanamide is nonvolatile and noncombustible, commercial grades may contain small amounts of calcium carbide which will produce acetylene in containers and processing vessels. Other contaminants are carbon, $Ca(OH)_2$, CaO, and $CaCO_3$. Pure calcium cyanamide occurs as glistening, hexagonal crystals belonging to the rhombohedral system. mp ~1340°; d_4^{20} 2.29. Sublimes at 1150-1200°. Heat of formn from $CaC_2 + N_2$: −69.0 kcal/mole (25°). Heat of fusion 1.29 cal/g. Essentially insol in water, but undergoes partial hydrolysis to the sol calcium hydrogen cyanamide, a source of cyanamide ions. No known solvent will bring about soln without decompn.

USE: As fertilizer, defoliant, herbicide, pesticide; manuf and refining of iron; manuf of calcium cyanide, melamine, dicyandiamide.

THERAP CAT (VET): Has been used as an anthelmintic.

1703. Calcium Cyanamide Citrated. Citrated calcium carbimide; CCC; carbimide; Colme; Dipsan; Abstem; Temposil. Contains citric acid in two parts by weight to one part of calcium cyanamide suitably purified for drug use. Used in treatment of alcoholism. Formulations: de Grunigen, Ferguson, **U.S.** pat. **2,998,350** (1961 to Cyanamid and Alcoholism Res. Found., Toronto). Acts by inhibiting aldehyde dehydrogenase: J. A. Smith *et al., J. Am. Med. Assoc.* **165**, 2181 (1957). Effect on cardiovascular system: J. E. Peachey *et al., Clin. Pharmacol. Ther.* **29**, 40 (1981); on liver cells: J. J. Vázquez, S. Cervera, *Lancet* **1**, 361 (1980). Comparison with disulfiram, *q.v.:* M. S. Levy *et al., Am. J. Psychiat.* **123**, 1018 (1967). Review of drug therapy for alcoholism: E. M. Sellers *et al., N. Engl. J. Med.* **305**, 1255 (1981).

THERAP CAT: Alcohol deterrent.

1704. Calcium Cyanide. Cyanogas. C_2CaN_2; mol wt 92.11. C 26.08%, Ca 43.51%, N 30.41%. $Ca(CN)_2$. Prepn: *Gmelin's, Calcium* (8th ed.) **28B**, 173-178, 958-960 (1958). Commercial prepns contain 40-50% $Ca(CN)_2$. Toxicity study: H. F. Smyth *et al., Am. Ind. Hyg. Assoc. J.* **30**, 470 (1969).

Rhombohedric crystals or powder; dec in moist air liberating hydrogen cyanide. *Very poisonous!* Sol in water with gradual liberation of HCN; even very weak acid (CO_2) liberates HCN; sol in alc. *Keep dry.* LD_{50} orally in rats: 39 mg/kg (Smyth).

USE: Fumigant; rodenticide; in stainless-steel manuf; in leaching ores of precious metals; stabilizer for cement.

1705. Calcium Cyclamate. *Cyclohexylsulfamic acid calcium salt; cyclohexanesulfamic acid calcium salt;* calcium cyclohexanesulfamate; calcium cyclohexylsulfamate; cyclamate calcium; Cyclan; Sucaryl Calcium. $C_{12}H_{24}CaN_2O_6S_2$; mol wt 396.54. C 36.35%, H 6.10%, Ca 10.11%, N 7.06%, O 24.21%, S 16.17%. $[C_6H_{11}NHSO_3^-]_2Ca^{2+}$. Prepn: Cummins, Johnson; McQuaid, **U.S.** pats. **2,799,700; 2,804,477**

(both 1957 to du Pont); Freifelder, U.S. pat. **3,082,247** (1963 to Abbott); Birsten, Rosin, U.S. pats. **3,361,798; 3,366,670** (both 1968 to Baldwin-Montrose). Metabolism: Wallace *et al.*, *J. Pharmacol. Exp. Ther.* **175**, 325 (1970); Prosky, O'Dell, *J. Pharm. Sci.* **60**, 1341 (1971); Renwick, Williams, *Biochem. J.* **129**, 869 (1972).

Dihydrate, crystals with pleasant, very sweet taste. Freely sol in water. Practically insol in alc, benzene, chloroform, ether. pH of 10% aq soln 5.5-7.5. Said to be more resistant to cooking temps than saccharin. *See also* Cyclamic Acid.

Note: Consult latest Government regulations on use in foods.

USE: Non-nutritive sweetener.

1706. Calcium Dichromate(VI). Calcium bichromate. $CaCr_2O_7$; mol wt 256.07. Ca 15.65%, Cr 40.61%, O 43.74%. Prepn: Hartford *et al.*, *J. Am. Chem. Soc.* **72**, 3353 (1950).

Trihydrate, bipyramidal orange-red crystals. Nonhygroscopic if pure. Dec on heating above 100° to $CaCrO_4$ and CrO_3. d_4^{30} 2.370. Very sol in water; insol in ether, CCl_4, hydrocarbons; dissolves in alc with immediate reduction of the dichromate and pptn of brown hydrous chromic chromate; dissolves in acetone with subsequent pptn of $CaCrO_4$.

USE: As catalyst; in manuf of $CrCl_3$ and CrO_3; corrosion inhibitor.

1707. Calcium 2-Ethylbutanoate. *2-Ethylbutanoic acid calcium salt; 2-ethylbutyric acid calcium salt;* diethylacetic acid calcium salt; Ethanion. $C_{12}H_{22}CaO_4$; mol wt 270.38. C 53.31%, H 8.20%, Ca 14.82%, O 23.67%. Prepn: *Beilstein* **vol. 2**, 333, 2nd suppl., 291-292, 3rd suppl., 752.

Monohydrate, prismatic twin crystals. Sol in water.

USE: As a stabilizer: Mack, U.S. pat. **2,510,035** (1950 to Advance Solvents & Chem.).

THERAP CAT: Sedative.

1708. Calcium Ferrous Citrate. Ferrous calcium citrate; Ferrocal; Rarical. $C_{12}H_{10}Ca_2FeO_{14}$; mol wt 514.20. C 28.03%, H 1.96%, Ca 15.59%, Fe 10.86%, O 43.56%. Prepn: Opfermann, U.S. pat. **2,691,666** (1954); Oroshnik, U.S. pat. **2,812,344** (1956 to Ortho Pharm.).

Tetrahydrate, tasteless powder, very stable to air oxidation.

THERAP CAT: Hematinic.

1709. Calcium Fluoride. CaF_2; mol wt 78.07. Ca 51.33%, F 48.67%. Occurs in nature as the mineral *fluorite* or *fluorspar*. Prepd from $CaCO_3$ + HF: O. Ruff, *Die Chemie des Fluors* (Berlin, 1920) p 89; Emeleus in *Fluorine Chemistry*, vol. I, J. H. Simons, Ed. (Academic Press, 1950) p 36; Kwasnik in *Handbook of Preparative Inorganic Chemistry*, Vol. 1, G. Brauer, Ed. (Academic Press, New York, 2nd ed., 1963) p 233.

White powder or cubic crystals. When F ions are pptd with Ca^{2+} in the absence of CO_3 ions, a gel is obtained. For the crystals: d 3.18; mp 1403°; bp 2500°; Mohs' hardness: 4. Becomes luminous when heated. Practically insol in water (soly at 18°: 0.0015 g/100 ml); slightly sol in dil mineral acids; is dissolved by concd mineral acids with liberation of HF. LD orally in guinea pigs: > 5 g/kg.

USE: Fluorspar is the main primary source of fluorine and its compds. In ferrous metallurgy it is used as a flux to increase the fluidity of the slag. The steel industry is the largest consumer; the chemical industry, second and glass and ceramics, third. Synthetic fluorspar is used in the optical industry (transmits u.v. rays), and pure calcium fluoride is used as catalyst in dehydration and dehydrogenations. Used to fluoridate drinking water.

1710. Calcium Fluorophosphate. *Phosphorofluoridic acid calcium salt;* calcium monofluorophosphate. $CaFO_3P$; mol wt 138.05. Ca 29.03%, F 13.76%, O 34.77%, P 22.44%. $CaPO_3F$. Prepd from calcium chloride and sodium mono-

fluorophosphate: Rowley, Stuckey, *J. Am. Chem. Soc.* **78**, 4262 (1956).

Dihydrate, monoclinic crystals with a tendency to form twins. Loses fluorine on heating. Soly in water (27°) 0.417 g/100 ml of soln. Practically insoluble in the usual organic solvents.

1711. Calcium Formate. $C_2H_2CaO_4$; mol wt 130.11. C 18.46%, H 1.55%, Ca 30.80%, O 49.19%. $Ca(HCOO)_2$. Prepn from $Ca(OH)_2$ and CO at high temp and pressure: Enderli, U.S. pats. **1,920,851; 1,995,607** (1933, 1935 to Rudolf Koepp); Erasmus, Hamby, U.S. pat. **2,913,318** (1959 to Union Carbide); from $CaCl_2$ and formic acid: Funk, Romer, *Z. Anorg. Allgem. Chem.* **239**, 288 (1938).

Orthorhombic crystals or cryst powder. Slight acetic acid-like odor. d 2.02. Sol in water, practically insol in alc.

USE: Preservative for food, silage; as binder for fine-ore briquets; in drilling fluids and lubricants.

1712. Calcium Gluconate. *D-Gluconic acid calcium salt (2:1);* Calciofon; Calglucon; Ebucin; Glucal; Glucobiogen. $C_{12}H_{22}CaO_{14}$; mol wt 430.38. C 33.49%, H 5.15%, Ca 9.31%, O 52.05%. $Ca[HOCH_2(CHOH)_4COO]_2$.

Odorless, tasteless crystals, granules, or powder. $[\alpha]_D^{20}$ about +6°. Does not lose its water on drying without some decomposition. Slowly sol in 30 parts cold, about 5 parts boiling water; insol in alc or other organic solvents. pH aq soln: 6-7. More concd (20 to 30%) aq solns are easily obtained by the addition of boric acid or similar complex-forming acids. The use of calcium D-saccharate for the prepn of supersatd injectable solns of calcium gluconate is described under Calcium D-Saccharate. Injectable solns of calcium gluconate contg sodium ascorbate are described in **Brit. pat. 495,675** and **Ger. pat. 702,185**. General directions and stability data: Siegrist, *Pharm. Acta Helv.* **24**, 430 (1949).

USE: In sewage purification; in coffee powders to prevent caking.

THERAP CAT: Calcium replenisher.

THERAP CAT (VET): In hypocalcemic states, including bovine milk fever.

1713. Calcium Glycerophosphate. *1,2,3-Propanetriol, mono(dihydrogen phosphate) calcium salt (1:1);* calcium glycerinophosphate; calcium phosphoglycerate; Neurosin. $C_3H_7CaO_6P$; mol wt 210.14. C 17.15%, H 3.36%, Ca 19.07%, O 45.68%, P 14.74%. Three isomers exist: The β-*glycerophosphoric acid calcium salt* (($HOCH_2)_2CHOPO_3Ca$) and D(+)- and L(−)-α-*glycerophosphoric acid calcium salt* ($HOCH_2CH(OH)CH_2OPO_3Ca$). Commercial product is a mixture of calcium β- and DL-α-glycerophosphates: Toal, Phillips, *J. Pharm. Pharmacol.* **1**, 869 (1949). Prepn of the calcium salt of the α- and β-acids: King, Pyman, *J. Chem. Soc.* **105**, 1238 (1914); Toal, Phillips, *loc. cit.* Sepn of the α-acid from β- and polyglycerophosphoric acids via the α-acid calcium salt: Carrara, *Ital. pat.* **460,219** (1950), *C.A.* **46**, 5077a (1952). Protective action against demineralization of dental enamel: T. H. Grenby, J. M. Bull, *Caries Res.* **14**, 210 (1980).

Commercial product, fine, odorless, almost tasteless, slightly hygroscopic powder; dec > 170°. Sol in about 50 parts water; almost insol in alc, boiling water. Soly in water is increased by citric or lactic acid. The aq soln is alkaline. Mixture with calcium lactate, *Calphosan*.

USE: In dentifrices, baking powder, as food stabilizer.

THERAP CAT: Calcium and phosphorus source. Tonic.

THERAP CAT (VET): Has been used as dietary supplement.

1714. Calcium Hexafluorosilicate. Calcium fluosilicate; calcium silicofluoride. CaF_6Si; mol wt 182.15. Ca 22.00%, F 62.58%, Si 15.42%. $CaSiF_6$. Prepn from Ca salt and H_2SiF_6: Moller, Kreth, **Brit. pat. 263,780** (1925).

Dihydrate, powder. d 2.25. Almost insol in cold water; partially dec by hot water; practically insol in acetone. LD orally in guinea pigs: 250 mg/kg, *Handbook of Toxicology*, **vol. 1**, W. S. Spector, Ed. (Saunders, Philadelphia, 1956) pp 58-59.

USE: In wood, rubber, textile industries; flotation agent; insecticide.

1715. Calcium Hydride. CaH_2; mol wt 42.09. Ca 95.21%, H 4.79%. Prepd by direct combination of calcium and hydrogen at 300-400°: **Brit.** pat. **597,055** (1948); by reduction of lime with magnesium in the presence of hydrogen: Gibb, *Trans. Electrochem. Soc.* **93,** 198-211 (1948); from $CaCl_2$ and hydrogen in the presence of sodium: Wade, Alexander, **U.S.** pat. **2,702,740** (1955 to Metal Hydrides). *Reviews:* Halls, *Ind. Chem.* **22,** 680 (1946); Kilb, *USAEC* APEX-485, 57 pp (1959).

Orthorhombic crystals or powder; the commercial product is gray. d 1.7. mp 1.86. Decomposes with water, lower alcohols and carboxylic acids to form hydrogen; moderately powerful condensing agent with ketones and acid esters; more powerful reducing agent toward metal oxides than lithium or sodium hydrides.

USE: To prepare rare metals by reduction of their oxides; as a drying agent for liquids and gases; to generate hydrogen: 1 g of calcium hydride in water liberates 1 liter of hydrogen at STP; in organic syntheses.

1716. Calcium Hydroxide. Calcium hydrate; slaked lime. CaH_2O_2; mol wt 74.09. Ca 54.09%, H 2.72%, O 43.19%. $Ca(OH)_2$. Contains at least 95% $Ca(OH)_2$. Commercial prepn by hydration of lime: W. L. Faith *et al., Industrial Chemicals* (John Wiley, New York, 3rd ed., 1965) pp 483-484. Laboratory prepn by treating an aq soln of a calcium salt with alkali: Ehrlich in *Handbook of Preparative Inorganic Chemistry* vol. 1, G. Brauer (Academic Press, New York, 2nd ed., 1963) p 934. Toxicity study: Smyth *et al., Am. Ind. Hyg. Assoc. J.* **30,** 470 (1969).

Crystals or soft, odorless, granules or powder. Slightly bitter, alkaline taste. Readily absorbs CO_2 from air forming $CaCO_3$. Loses water when ignited; forms CaO. d 2.08-2.34. Slightly sol in water; sol in glycerol, sugar or NH_4Cl solns; sol in acids with evolution of much heat. pH of aq soln satd at 25°: 12.4. *Keep well closed.* LD_{50} orally in rats: 7.34 g/kg (Smyth).

USE: In mortar, plaster, cement and other building and paving materials; in lubricants, drilling fluids, pesticides, fireproofing coatings, water paints; as egg preservative; manuf of paper pulp; in SBR rubber vulcanization; in water treatment; dehairing hides.

THERAP CAT: Astringent.

1717. Calcium Hypochlorite. Losantin. $CaCl_2O_2$; mol wt 142.98. Ca 28.03%, Cl 49.59%, O 22.38%. $Ca(OCl)_2$. Pure product has not been prepd. Commercial product usually contain 50% or more $Ca(OCl)_2$. Preparation of solid product contg 90-94% $Ca(OCl)_2$: Cady, *Inorg. Syn.* **5,** 161 (1957). Impurities include: $Ca(ClO_4)_2$, $CaCl_2$, $CaCO_3$, $Ca(OH)_2$ and water.

USE: Algicide, bactericide, deodorant, disinfectant, fungicide; in sugar refining; oxidizing agent; bleaching agent. Ingredient of *Camporit* which also contains NaCl. Also ingredient of *HTH* and *Perchloron*.

1718. Calcium Hypophosphite. $CaH_4O_4P_2$; mol wt 170.05. Ca 23.57%, H 2.37%, O 37.63%, P 36.43%. $Ca(H_2PO_2)_2$. Prepn: *Gmelin's, Calcium* (8th ed.) **28B,** 1119-1121 (1958).

Monoclinic, prismatic crystals or granular powder. When heated above 300° it evolves spontaneously-inflammable phosphine. Sol in water; slightly sol in glycerol. Practically insol in alcohol. The aq soln is slightly acid. *Incompat.* KI, oxidizers.

USE: As corrosion inhibitor; in nickel plating. Pharmaceutic aid (retards oxidation of ferrous salts).

THERAP CAT: Calcium source.

THERAP CAT (VET): Has been used as a dietary supplement and also as a "nerve tonic".

1719. Calcium Iodate. Lautarite. CaI_2O_6; mol wt 389.88. Ca 10.28%, I 65.10%, O 24.62%. $Ca(IO_3)_2$. Prepd by passing chlorine into a hot soln of lime in which iodine has been dissolved: Bahl, Singh, *J. Indian Chem. Soc.* **17,** 397 (1940).

Nonhygroscopic, monoclinic-prismatic crystals. d_4^{15} 4.519. Stable up to 540°. Sensitive to reducing agents. Soly in water (g/100 ml): 0.10 (0°); 0.95 (100°). More sol in aq solns of iodides and in amino acid solns. Sol in nitric acid. Insol in alcohol.

Monohydrate, cubic crystals. Slightly sol in water.

Hexahydrate, orthorhombic crystals. Slightly sol in water.

USE: Nutritional source of iodine in foods and feedstuffs. More stable in table salts than iodides: *Food Field Reporter,* Aug. 8, 1956; Daum, *C.A.* **51,** 5324 (1957); to improve properties of yeast-leavened bakery products.

THERAP CAT: Antiseptic.

1720. Calcium Iodide. CaI_2; mol wt 293.89. Ca 13.64%, I 86.36%. Prepn: Farr, U.S. pat. **2,415,346** (1947 to Mallinckrodt); Chaigneau, *Bull. Soc. Chim. France* **1957,** 886; *Gmelin's, Calcium* (8th ed.) **28B,** 102, 610-622 (1958). The commercial product usually contains 16-20% water.

Very hygroscopic hexagonal lamella. Becomes yellow and completely insol on exposure to air due to liberation of I_2 and absorption of CO_2. mp 740°; bp 1100°. Very sol in water, methanol, ethanol, acetone; practically insol in ether, dioxane. The aq soln is neutral or slightly alkaline. *Keep tightly closed and protected from light.*

Hexahydrate, hexagonal, thick needles, or plates, or lumps, or powder. Very hygroscopic; becomes yellow in air, mp ~42°. Freely sol in water, alcohol. *Keep tightly closed and protected from light.*

THERAP CAT: Expectorant.

1721. Calcium Iodobehenate. *Iododocosanoic acid calcium salt;* Calioben; Saiodin; Sajodin. $C_{44}H_{84}CaI_2O_4$; mol wt 971.04. C 54.42%, H 8.72%, Ca 4.13%, I 26.14%, O 6.59%. $Ca(C_{21}H_{42}ICO_2)_2$. Prepn: *Beilstein,* vol. 2, 392.

White or yellowish powder; odorless or with slight fat-like odor. Practically insol in water, alcohol, ether; freely sol in chloroform. *Protect from light.*

THERAP CAT: Internally for its iodine action.

THERAP CAT (VET): Has been used internally as a source of iodine.

1722. Calcium Iodostearate. *2-Iodooctadecanoic acid calcium salt;* stearodine. $C_{36}H_{68}CaI_2O_4$; mol wt 858.82. C 50.35%, H 7.98%, Ca 4.67%, I 29.55%, O 7.45%. $Ca(C_{17}H_{34}-ICO_2)_2$. Prepn: *Beilstein,* **vol. 2,** 1st suppl., 177.

Cream-colored, almost odorless powder. Practically insol in water or alcohol; sol in benzene, chloroform, ether.

THERAP CAT: Iodine source.

THERAP CAT (VET): Has been used internally as a source of iodine.

1723. Calcium Lactate. *2-Hydroxypropanoic acid calcium salt.* $C_6H_{10}CaO_6$; mol wt 218.22. C 33.02%, H 4.62%, Ca 18.37%, O 43.99%. $Ca[CH_3CH(OH)COO]_2$. Commercial prepn usually contains about 25% water, and on the anhydr basis it is at least 98% pure. Prepd commercially by neutralization of lactic acid, from fermentation of dextrose, molasses, starch, sugar or whey, with $CaCO_3$: Inskeep *et al., Ind. Eng. Chem.* **44,** 1955 (1952). Clinical use as a supplement: G. M. Day *et al., Pediatric Res.* **9,** 568 (1975). Evaluation as anticaries agent: B. M. Shrestha *et al., Caries Res.* **16,** 12 (1982); M. J. M. Schaeken, J. S. van der Hoeven, *ibid.* **24,** (1990). Clinical pharmacology: M. Goddard *et al., Am. J. Clin. Nutr.* **44,** 653 (1986).

Pentahydrate, odorless, white, slightly efflorescent granules or powder. Becomes anhydr at 120°. pH: 6-8. Sol water. Practically insol in alcohol.

USE: As a preservative in foods and beverages; in dentifrices.

THERAP CAT: Calcium replenisher.

THERAP CAT (VET): Calcium replenisher.

1724. Calcium Levulinate. *4-Oxopentanoic acid calcium salt; levulinic acid calcium salt.* $C_{10}H_{14}CaO_6$; mol wt 270.30. C 44.44%, H 5.22%, Ca 14.83%, O 35.52%. $(CH_3-COCH_2CH_2COO)_2Ca$. Prepn: Cox, Dodds, U.S. pat. **2,-033,909** (1936 to Niacet Chemicals).

Dihydrate, crystals or granular powder. mp 125°. Loses 1 H_2O on drying *in vacuo* at room temp and all H_2O at 50°. Very sol in water; the aq soln is practically neutral.

THERAP CAT: Calcium replenisher.

THERAP CAT (VET): May be used in hypocalcemic states.

1725. Calcium Magnesium Acetate. *Acetic acid, calcium magnesium salt;* CMA; Ice-B-Gon. $C_2H_4O_2.xCa.xMg$. Prepn and use as deicer: S. A. Dunn, R. U. Schenk, *Transport. Res. Rec.* **776,** 12 (1980); *eidem, U.S. Federal Highway*

Administration Report **FHWA-RD-79-109** (1980) pp 17. Prepn by fermentation: C. W. Marynowski *et al., Ind. Eng. Chem. Prod. Res. Dev.* **24,** 457 (1985); via extraction and sorption: H. Reisinger, C. J. King, *Ind. Eng. Chem. Res.* **34,** 845 (1995). Acute toxicity: G. A. Rausina *et al., Acute Toxic. Data* **1,** 87 (1990). Environmental and toxicological evaluation: B. L. McFarland, K. T. O'Reilly in *Chemical Deicers and the Environment,* F. M. D'Itri, Ed. (Lewis Publishers, Chelsea, MI, 1992) pp 193-227. Anti-icing effect: E. E. Adams *et al., ibid.,* pp. 481-493. Degradation in soil: D. W. Ostendorf *et al., J. Environ. Qual.* **22,** 299 (1993). Effects on cement mortar: O. Peterson, *Cem. Concr. Res.* **25,** 617 (1995). Book: *Calcium Magnesium Acetate,* D. L. Wise *et al.,* Eds. (Elsevier Science Publishers, Amsterdam, 1991) pp 511.

Non-flammable. Shows no significant corrosion of steel, aluminum, or zinc. LD_{50} orally in male, female rats: 3240, 3071 mg/kg (Rausina).

USE: Deicing agent for roadways; in fossil fuel combustion for removal of sulfur.

1726. Calcium Mesoxalate. *Mesoxalic acid calcium salt;* calcium ketomalonate; calcium oxomalonate; ketomalonic acid calcium salt; oxomalonic acid calcium salt; Mesoxan. C_3CaO_5; mol wt 156.11. C 23.08%, Ca 25.67%, O 51.24%. Prepn: Scheiber, Hopfer, *Ber.* **53,** 908 (1920); Kobayashi, **Japan.** pat. **4157**('52), *C.A.* **48,** 5212a (1954); Yanagisawa, **Japan.** pat. **7463**('60), *C.A.* **55,** 5880a (1961).

$$\left[O = \begin{array}{c} COO^- \\ \\ COO^- \end{array} \right] Ca^{2+}$$

Crystalline powder, dec 210-220°. Sparingly sol in glacial acetic acid.

Note: May also exist as calcium dihydroxymalonate.

THERAP CAT: Oral hypoglycemic.

1727. Calcium Methionate. *Methanedisulfonic acid calcium salt.* $CH_2CaO_6S_2$; mol wt 214.23. C 5.61%, H 0.94%, Ca 18.71%, O 44.81%, S 29.94%. $CaCH_2(SO_3)_2$. Prepn: *Beilstein* **vol. 1,** 1st suppl., 303; Jenkins, *J. Am. Pharm. Assoc.* **27,** 484 (1938).

Dihydrate, crystals or cryst powder, having a slight odor. Stable in air. Sol in about 2.5 parts water forming a neutral soln (pH 6-7); very slightly sol in alc.

1728. Calcium Molybdate(VI). $CaMoO_4$; mol wt 200.02. Ca 20.04%, Mo 47.97%, O 32.00%. Prepn from sodium molybdate and $CaSO_4$: Carosella, U.S. pat. **2,460,-974** (1949 to U.S. Vanadium); by heating a stoichiometric mixture of CaO or $CaCO_3$ and molybdic acid: Kroger, *Nature* **159,** 674 (1947). Toxicity studies: L. T. Fairhall *et al.,* "The Toxicity of Molybdenum," *U.S. Public Health Service Bulletin No. 293,* Washington D.C. (1945) 36 pp.

Tetragonal crystals. d 4.35. Practically insol in water, alcohol. Sol in concd mineral acids.

USE: In phosphors and luminescent materials.

1729. Calcium Nitrate. CaN_2O_6; mol wt 164.09. Ca 24.42%, N 17.07%, O 58.50%. $Ca(NO_3)_2$. Prepn: *Gmelin's Calcium* (8th ed.) **28B,** 59-69, 341-382 (1956).

Deliquesc granules, mp ~560°. Very sol in water, heat being evolved; freely sol in methanol, ethanol, acetone. Almost insol in concd HNO_3. pH of 5% aq soln 6.0. *Keep well closed.*

Note: $Ca(NO_3)_2$ crystallizes also with $4H_2O$ (30.5%), mp 45°. Technical flake usually contains 28.6% H_2O.

USE: In explosives, fertilizers, matches, pyrotechnics; manuf of incandescent mantles, radio tubes, HNO_3; corrosion inhibitor in diesel fuels.

1730. Calcium Nitrite. CaN_2O_4; mol wt 132.09. Ca 30.34%, N 21.21%, O 48.45%. $Ca(NO_2)_2$. Prepd by reaction of nitric oxide with a mixture of calcium ferrate(III) and calcium nitrate: Ray, Ogg, Jr., *J. Am. Chem. Soc.* **79,** 265 (1957).

White or yellowish, deliquesc, hexagonal crystals. d 2.23. Freely sol in water; slightly sol in alc. *Keep well closed.*

USE: Corrosion inhibitor in lubricants, concrete.

1731. Calcium Oleate. *9-Octadecenoic acid calcium salt; oleic acid calcium salt.* $C_{36}H_{66}CaO_4$; mol wt 603.00. C 71.71%, H 11.03%, Ca 6.65%, O 10.61%. $Ca(C_{18}H_{33}O_2)_2$. Prepn: Harrison, *Biochem. J.* **18,** 1222 (1924); Pink, *J. Chem. Soc.* **1939,** 619.

Pale-yellow transparent solid. Dec above 140°. Slowly absorbs moisture from the air to form the monohydrate. Practically insol in water, alcohol, ether, acetone, petr ether; sol in chloroform, benzene.

USE: Thickening lubricating grease; waterproofing concrete; emulsifier for benzene, kerosene, etc.; in modeling waxes to vary hardness.

1732. Calcium Oxalate. *Ethanedioic acid calcium salt.* C_2CaO_4; mol wt 128.10. C 18.75%, Ca 31.29%, O 49.96%. CaC_2O_4. Prepn from calcium formate: Bredt, U.S. pat. **1,622,991** (1927); from calcium cyanamide: Barsky, Buchanan, *J. Am. Chem. Soc.* **53,** 1270 (1931).

Monohydrate, cubic crystals. Loses all of its water at 200°. When ignited is converted into $CaCO_3$ or CaO without appreciable charring. d 2.2. Practically insol in water or acetic acid; sol in dil HCl or HNO_3.

USE: In ceramic glazes; as carrier for separation of rare earth metals; analysis for calcium: Ingols, Murray, *Anal. Chem.* **21,** 525 (1949).

1733. Calcium Oxide. Lime; burnt lime; calx; quicklime. CaO; mol wt 56.08. Ca 71.47%, O 28.53%. Properly stored lime of commerce contains 90-95% free CaO. Commercial production from limestone: W. L. Faith *et al., Industrial Chemicals* (John Wiley, New York, 3rd ed., 1965) pp 482-487. Lab prepn by ignition of $CaCO_3$: Ehrlich in *Handbook of Preparative Inorganic Chemistry* **vol. 1,** G. Brauer, Ed. (Academic Press, New York, 2nd ed., 1963) p 931. *Review:* R. S. Boynton in Kirk-Othmer *Encyclopedia of Chemical Technology* **vol. 14** (Wiley-Interscience, New York, 3rd ed., 1981) pp 343-382.

Crystals, white or grayish-white lumps, or granular powder; commercial material somctimes has a yellowish or brownish tint, due to iron. mp 2572°; bp 2850°; d 3.32-3.35. Readily absorbs CO_2 and H_2O from air, becoming airslaked. Sol in water forming $Ca(OH)_2$ and generating a large quantity of heat; sol in acids, glycerol, sugar soln; practically insol in alc. *Keep tightly closed and dry.*

Caution: Potential symptoms of overexposure arc irritation of eyes, upper respiratory tract; ulceration or perforation of nasal septum; pneumonia; dermatitis. *See NIOSH Pocket Guide to Chemical Hazards* (DHHS/NIOSH 90-117, 1990) p 56.

USE: In bricks, plaster, mortar, stucco and other building and construction materials; manuf of steel, aluminum, magnesium, and flotation of non-ferrous ores; manuf of glass, paper, Na_2CO_3 (Solvay process), Ca salts and many other industrial chemicals; dehairing hides; clarification of cane and beet sugar juices; in fungicides, insecticides, drilling fluids, lubricants; water and sewage treatment; in laboratory to absorb CO_2 (the combination with NaOH is known as soda-lime, *q.v.*).

1734. Calcium Palmitate. *Hexadecanoic acid calcium salt; palmitic acid calcium salt.* $C_{32}H_{62}CaO_4$; mol wt 550.92. C 69.77%, H 11.34%, Ca 7.27%, O 11.62%. $Ca(C_{16}H_{31}O_2)_2$. Prepn: Harrison, *Biochem. J.* **18,** 1222 (1924).

Powder or rhombic crystals. Dec above 155°. Practically insol in water, alcohol, ether, acetone, petr ether; slightly sol in chloroform, benzene, acetic acid.

USE: Thickening lubricating oils; waterproofing fabrics and lubricating greases; as corrosion inhibitor in halohydrocarbons.

1735. Calcium Permanganate. $CaMn_2O_8$; mol wt 277.95. Ca 14.42%, Mn 39.53%, O 46.05%. $Ca(MnO_4)_2$. Prepn from $KMnO_4$ and $CaCl_2$: Brit. pat. **624,885** (1949 to Boots Pure Drug Co. and T. Hagyard); from $Al(MnO_4)_3$ and $Ca(OH)_2$: Jaskowiak, U.S. pat. **2,504,130** (1950 to Carus Chemical).

Violet or dark-purple, deliquesc crystals. Freely sol in water; dec in alcohol. *Keep tightly closed.*

USE: Antiseptic, disinfectant, deodorizer; with CaF_2 as binder for welding electrode coatings and fluxes.

1736. Calcium Peroxide. Calcium dioxide. CaO_2; mol wt 72.08. Ca 55.60%, O 44.40%. The commercial product usually contains about 60% CaO_2, water, and some Ca-$(OH)_2$ and $CaCO_3$. Prepn: Young, U.S. pat. **2,533,660** (1950 to du Pont); Ehrlich in *Handbook of Preparative Inorganic Chemistry* vol. **1**, G. Brauer, Ed. (Academic Press, New York, 2nd ed., 1963) p 936.

White or yellowish, odorless, almost tasteless powder. Dec in moist air. Slightly sol in water; sol in acids with formation of H_2O_2. *Keep well closed.*

USE: Stabilizer for rubber.

THERAP CAT: Antiseptic.

1737. Calcium Phenolsulfonate. *p-Hydroxybenzenesulfonic acid calcium salt;* calcium sulfocarbolate; calcium sulfophenolate. $C_{12}H_{10}CaO_8S_2$; mol wt 386.42. C 37.30%, H 2.61%, Ca 10.37%, O 33.12%, S 16.60%. $Ca[C_6H_4(OH)SO_3]_2$. Prepn: *Hagers Handb. Pharm. Praxis* vol. **2**, 420 (Berlin, 1930).

Hydrate, odorless cryst powder. Sol in water or alcohol. The aq soln is neutral, and has a bitter, astringent taste.

THERAP CAT (VET): Has been used as an intestinal antiseptic, in dusting powders for ulcers and in ophthalmic solns.

1738. Calcium Phenoxide. Calcium carbolate; calcium phenate; calcium phenolate; calcium phenylate. $C_{12}H_{10}CaO_2$; mol wt 226.29. C 63.69%, H 4.45%, Ca 17.71%, O 14.14%. $Ca(OC_6H_5)_2$. Prepn: Kluge, Drake, U.S. pat. **2,870,134** (1959 to Texas Co.).

Reddish powder. Dec in air. Slightly sol in water or alcohol. *Keep well closed.*

USE: Detergent; additive for motor oils.

1739. Calcium Phosphate, Dibasic. Calcium monohydrogen phosphate; dicalcium orthophosphate; secondary calcium phosphate. $CaHO_4P$; mol wt 136.06. Ca 29.46%, H 0.74%, O 47.04%, P 22.77%. $CaHPO_4$. Occurs in nature as the mineral *monetite.* Prepn from $CaCl_2$ and Na_2HPO_4: Jensen, Rathley, *Inorg. Syn.* **4**, 19, 20 (1953); from $Ca_3(PO_4)_2$ and H_3PO_4: Perloff, Posner, *ibid.* **6**, 16 (1960), where it is an intermediate in the preparation of hydroxyapatite.

Triclinic crystals. At red heat dehydrated to calcium pyrophosphate. Practically insol in water, alcohol. Dihydrate, *brushite.* Monoclinic crystals. Loses water of crystn slowly below 100°. Dehydr at red heat to calcium pyrophosphate. d 2.31. Practically insol in water, alcohol; sol in dil HCl or HNO_3; slightly sol in dil acetic acid.

USE: Chiefly in animal feeds; mineral supplement in cereals and other foods; manuf of glass; in dental products, fertilizers (*see also* Calcium Phosphate, Monobasic).

THERAP CAT: Calcium replenisher.

THERAP CAT (VET): Has been used as a dietary supplement, and as an antacid.

1740. Calcium Phosphate, Monobasic. Acid calcium phosphate; calcium biphosphate; monocalcium orthophosphate; monocalcium phosphate; primary calcium phosphate; "calcium superphosphate". $CaH_4O_8P_2$; mol wt 234.05. Ca 17.12%, H 1.72%, O 54.69%, P 26.47%. $Ca(H_2PO_4)_2$. Commercial prepn for fertilizers by treating pulverized phosphate rock with H_2SO_4 or H_3PO_4: Faith, Keyes & Clark's *Industrial Chemicals*, F. A. Lowenheim, M. K. Moran, Eds. (Wiley-Interscience, New York, 4th ed., 1975) pp 191-200. Laboratory prepn from $CaCO_3$ and H_3PO_4: Jensen, Kathley, *Inorg. Syn.* **4**, 18 (1953).

Monohydrate, large, shining, triclinic plates, cryst powder or granules. Non-hygroscopic when pure, but traces of impurities such as H_3PO_4 cause material to be deliquesc. Strong acid taste. Loses H_2O at 100°, dec at 200°. d_4^{18} 2.220. Moderately sol in water; sol in dil HCl or HNO_3 or acetic acid.

Note: The products obtained from commercial processes are not pure monobasic calcium phosphate. The superphosphate obtained from the H_2SO_4 treatment is about 30% $CaH_4(PO_4)_2.H_2O$, 10% $CaHPO_4$, 45% $CaSO_4$, 10% iron oxide, silica, alumina, etc. and 5% water; it contains 18-21% available P_2O_5. The triple superphosphate obtained from

the H_3PO_4 treatment contains from 43 to 50% available P_2O_5.

USE: Chiefly in fertilizers; as acidulant in baking powders and in wheat flours; mineral supplement for foods and feeds; in enameling.

1741. Calcium Phosphate, Tribasic. Tricalcium orthophosphate; tricalcium phosphate; tertiary calcium phosphate; Calcigenol Simple. $Ca_3O_8P_2$; mol wt 310.18. Ca 38.76%, O 41.27%, P 19.97%. $Ca_3(PO_4)_2$. It is about 96% pure, usually contg an excess of CaO. Occurs in nature as the minerals: *oxydapatit, voelicherite, whitlockite.* The technical product is also known as *"bone ash".* Commercial prepn from phosphate rock: Hignett, Hubbard, *Ind. Eng. Chem.* **38**, 1208 (1946); Elmore, U.S. pat. **2,474,831** (1949 to T.V.A.); Hollingsworth, U.S. pats. **2,556,541** and **2,562,-718** (both 1951 to Coronet Phosphate); Brosheer, Hignett, *Chem. Eng. Rept.* no. 7, 143 pp (1953).

Amorphous, odorless, tasteless powder. mp 1670°. d 3.14. Practically insol in water, alcohol or acetic acid; sol in dil HCl or HNO_3.

USE: Manuf of fertilizers, H_3PO_4 and P compds; manuf milk-glass, polishing and dental powders, porcelains, pottery; enameling; clarifying sugar syrups; in animal feeds; as noncaking agent; in the textile industry.

THERAP CAT: Calcium replenisher.

THERAP CAT (VET): Has been used as a dietary supplement, and as an antacid.

1742. Calcium Phosphide. Photophor. Ca_3P_2; mol wt 182.18. Ca 66.00%, P 34.00%. Prepn: Ehrlich in *Handbook of Preparative Inorganic Chemistry*, vol. **1**, G. Brauer, Ed. (Academic Press, New York, 2nd ed., 1963) p 943.

Red-brown cryst powder or gray lumps. Dec by moist air or water, evolving spontaneously-flammable phosphine. d 2.51; mp ~1600°. *Keep dry and tightly closed.*

USE: For signal fires; in purification of Cu and Cu alloys; as rodenticide.

1743. Calcium Phosphite. $CaHO_3P$; mol wt 120.06. Ca 33.38%, H 0.84%, O 39.98%, P 25.80%. $CaHPO_3$. Prepn: *Gmelin's, Calcium* (8th ed.) **28B**, 1121 (1958).

Monohydrate, crystals. Loses water at 200°; dec above 300°. Slightly sol in water; practically insol in alcohol.

USE: Fertilizers; polymerization catalyst.

1744. Calcium Polycarbophil. WL-140; Carbofil; Mitrolan; Quival; Sorboquel. Calcium salt of a synthetic loosely crosslinked hydrophilic resin of the polycarboxylic type. Review of pharmacology, toxicology, clinical efficacy, and adverse effects: I. E. Danhof, *Pharmacotherapy* **2**, 18-28 (1982).

THERAP CAT: Cathartic.

1745. Calcium Propionate. *Propionic acid calcium salt.* $C_6H_{10}CaO_4$; mol wt 186.22. C 38.70%, H 5.41%, Ca 21.52%, O 34.37%. $Ca(CH_3CH_2COO)_2$. Occurs as mono- or trihydrate. Prepn: *Beilstein* vol. **2**, 238, 2nd suppl., 218, 3rd suppl., 516.

Powder or monoclinic crystals. Sol in water; slightly sol in methanol, ethanol; practically insol in acetone, benzene.

USE: As an inhibitor of molds and other microorganisms in foods, tobacco, pharmaceuticals; in butyl rubber to improve processability and scorching resistance.

THERAP CAT: Antifungal.

1746. Calcium Pyrophosphate. *Calcium diphosphate.* $Ca_2O_7P_2$; mol wt 254.10. Ca 31.55%, O 44.08%, P 24.38%. $Ca_2P_2O_7$. Prepn by ignition of $CaHPO_4$: St. Pierre, *J. Am. Chem. Soc.* **77**, 2197 (1955).

Polymorphous crystals or powder. d 3.09. mp 1353°. Practically insol in water. Sol in dil HCl or HNO_3.

USE: Abrasive; fertilizer; feed supplement; in dentifrices, ceramic ware, china, glass, phosphors.

1747. Calcium D-Saccharate. *D-Glucaric acid calcium salt.* $C_6H_8CaO_8$; mol wt 248.20. C 29.04%, H 3.25%, Ca 16.15%, O 51.57%. $CaC_6H_8O_8$. The normal calcium salt of D-saccharic acid, a dicarboxylic sugar acid derived from the oxidation of D-gluconic acid. Calcium D-saccharate is a true chemical compd and should not be confused with saccharated lime, formerly called "calcium saccharate" and produced by the action of lime upon sugar. Prepn: *Beilstein*, **vol. 3**, 2nd suppl., 378; *Hagers Handb. Pharm. Praxis* **vol. 1**, 755 (Berlin, 1930).

Tetrahydrate, odorless, tasteless crystals or fine white powder. Stable to air. Becomes anhydr upon heating at 100° *in vacuo*. Practically insol in water, alcohol, ether. Sol in dil mineral acids and in calcium gluconate solns.

USE: Pharmaceutic aid (stabilizer for calcium gluconate solns). As plasticizer in cement, concrete, mortar.

1748. Calcium Selenide. CaSe; mol wt 119.04. Ca 33.67%, Se 66.33%. Prepd by reducing CaSeO$_4$ in a stream of H$_2$ at 400-500°: Ehrlich in *Handbook of Preparative Inorganic Chemistry*, **vol. 1**, G. Brauer, Ed. (Academic Press, New York, 2nd ed, 1963) p 939.

White powder. In air may turn red within a few minutes and light brown in a few hours. d 3.82. Decomposed by water. Treatment with HCl produces H$_2$Se gas, and red Se separates.

USE: In electron emitters.

1749. Calcium Silicate. Many different forms of calcium silicate are known. Among the most common forms are CaSiO$_3$, Ca$_2$SiO$_4$ and Ca$_3$SiO$_5$. Usually occur in hydrated form contg various percentages of water of crystallization. Names of calcium silicate minerals are: *afwillite; akermanite; calcium pectolith; centrallasite; crestmoreite; eaklite; foshagite; foshallasite; gjellebaekite; grammite; gyrolite; hillebrandite; larnite; okenite; parawollastonite; pseudo-wollastonite; riversideite; table spate; tobermorite; wollastonite; xonaltite; xonotlite.* Commercial calcium silicate sold for industrial use, such as *Micro-Cell* and *Silene*, is prepared synthetically to control its absorbing power. The usual method of prepn is from lime and diatomaceous earth under carefully controlled conditions: Boss, *Chem. Eng. News* **27**, 677 (1949); Steinour, *Chem. Revs.* **40**, 391 (1947). The commercial product is described here.

White or slightly cream-colored, free-flowing powder. Approximate analysis: CaO 19%, SiO$_2$ 67%, H$_2$O 6 to 8%. d^{25} 2.10. Bulk density: 15-16 lb/cu ft. Absorbs 1 to 2.5 times its weight of liquids and still remains a free-flowing powder. Total absorption power for water about 600%, for mineral oil about 500%. Available surface area: 95 to 175 m^2/g. Ultimate particle size: 0.02 to 0.07 μ. pH of aq slurry 8.0 to 10.0. Practically insol in water. Forms a siliceous gel with mineral acids.

USE: Constituent (produced *in situ*) of lime glass, portland cement; reinforcing filler in elastomers and plastics; absorbent for liquids, gases, vapors; as anti-caking agent, suspension agent, pigment and pigment extender; binder for refractory material; in chromatography; in road construction.

1750. Calcium Stearate. *Octadecanoic acid calcium salt; stearic acid calcium salt.* C$_{36}$H$_{70}$CaO$_4$; mol wt 607.03. C 71.23%, H 11.62%, Ca 6.60%, O 10.54%. Ca(C$_{18}$H$_{35}$O$_4$)$_2$. The commercial prepn also contains palmitate. Prepn: Harrison, *Biochem. J.* **18**, 1222 (1924); Kebrich, Petrot, U.S. pat. **2,650,932** (1953 to National Lead).

Granular, fatty powder. Bulk density ~20 lb/cu ft, mp 147-149° (determined by gradient bar). Practically insol in water, ether, chloroform, acetone, cold alcohol; slightly sol in hot alcohol, in hot vegetable and mineral oils; quite sol in hot pyridine.

USE: For waterproofing fabrics, cement, stucco, explosives; as a releasing agent for plastic molding powders; as a stabilizer for polyvinyl chloride resins; lubricant; in pencils and wax crayons. Food grade calcium stearate, derived from edible tallow, is used as a conditioning agent in certain food and pharmaceutical products.

1751. Calcium Stearyl-2 Lactylate. *Octadecanoic acid 2-(1-carboxyethoxy)-1-methyl-2-oxoethyl ester, calcium salt; stearic acid ester with lactate of lactic acid calcium salt;-* calcium stelate; Verv-Ca. C$_{48}$H$_{86}$CaO$_{12}$; mol wt 895.28. C 64.40%, H 9.68%, Ca 4.48%, O 21.44%. Use in improving the mixing characteristics of flour: Thompson, Buddemeyer, *Cereal Chem.* **31**, 296 (1954); in improving whipping and baking properties of dried egg whites: Gorman, Keith, U.S. pat. **2,919,992** (1960 to Seymour Foods). Metabolism: J. C. Phillips *et al.*, *Food Cosmet. Toxicol.* **19**, 7 (1981).

Free flowing, nonhygroscopic powder. Sparingly sol in water. pH of a 2% aq suspension 4.7.

USE: Dough conditioner in yeast-leavened bakery products; emulsifier in cosmetic and pharmaceutical industry.

1752. Calcium Succinate. *Butanedioic acid calcium salt; succinic acid calcium salt;* Artume. C$_4$H$_4$CaO$_4$; mol wt 156.15. C 30.77%, H 2.58%, Ca 25.67%, O 40.98%. Ca-C$_4$H$_4$O$_4$. Prepn: *Beilstein* **vol. 2**, 607, 2nd suppl., 548, 3rd suppl., 1657.

Trihydrate, needles or granules. Slightly sol in water; practically insol in alcohol; sol in dil acids.

THERAP CAT: Combined with salicylates for rheumatic fever and rheumatoid arthritis.

1753. Calcium Sulfate. CaO$_4$S; mol wt 136.14. Ca 29.44%, O 47.01%, S 23.55%. CaSO$_4$. Review: R. J. Wenk, P. L. Henkels in Kirk-Othmer *Encyclopedia of Chemical Technology*, **vol. 4** (Wiley-Interscience, New York, 3rd ed., 1978) pp 437-448.

The natural form of anhydrous calcium sulfate is known as the mineral *anhydrite*; also as *karstenite, muriacite, anhydrous sulfate of lime, anhydrous gypsum*. Crystals are orthorhombic, color varies, e.g., white with blue, gray or reddish tinge, or brick red. d 2.96. Hardness 3-3.5 (Mohs'). Sol in water (18.75°) 0.2 pts/100 pts. *Insoluble anhydrite* or *dead-burned gypsum* which has the same crystal structure as the mineral is obtained upon complete dehydration of gypsum at above 650°. *Soluble anhydrite* is obtained in granular or powder form by complete dehydration of gypsum at below 300° in an electric oven. Estimated pore space is 38% by volume. Possesses high affinity for water and will absorb 6.6% of its weight of water forming the stable hemihydrate.

Hemihydrate, *dried calcium sulfate; dried gypsum; plaster of Paris; Annalin*. Fine, odorless, tasteless powder. When mixed with water, sets to a hard mass. *Keep well closed*.

Dihydrate, *native calcium sulfate; precipitated calcium sulfate; gypsum; alabaster; selenite; terra alba; satinite; mineral white; satin spar; light spar*. Lumps or powder. d 2.32. It loses only part of its water at 100-150°. Sol in water; very slowly sol in glycerol. Practically insol in most organic solvents.

USE: *Anhydrous:* Insol anhydrite is used in cement formulations and as a paper filler. Soluble anhydride, because of its strong tendency to absorb moisture, is useful as a drying agent for solids, organic liquids and gases; the desiccant used in laboratory and industry is known under the name *Drierite*. This material can be regenerated repeatedly and reused without noticeable decrease in its desiccating efficiency. The *hemihydrate* is used for wall plasters; wallboard; tiles and blocks for the building industry; moldings; statuary; in the paper industry. The *dihydrate* is used in the manuf of portland cement; in soil treatment to neutralize alkali carbonates and to prevent loss of volatile and dissolved nitrogenous compounds by volatilization and leaching; for the manuf of plaster of Paris, artificial marble; as a white pigment, filler or glaze in paints, enamels, pharmaceuticals, paper, insecticide dusts, yeast manuf, water treatment, polishing powders; in the manuf of sulfuric acid, CaC$_2$, (NH$_4$)$_2$SO$_4$, porous polymers. Pharmaceutic aid (in plaster casts).

1754. Calcium Sulfide. CaS; mol wt 72.14. Ca 55.55%, S 44.45%. Pure CaS prepd in the laboratory by heating pure CaCO$_3$ in a stream of H$_2$S + H$_2$ at 1000°: Ehrlich in *Handbook of Preparative Inorganic Chemistry* **vol. 1**, G. Brauer, Ed. (Academic Press, New York, 2nd ed., 1963) p 938. Crude calcium sulfide, erroneously called *sulfurated lime, calcic liver of sulfur, liver of lime, hepar calcis*, made by igniting calcium sulfate with carbonaceous matter. Contains not less than 55% CaS; the balance is calcium sulfate, sulfite and carbonate, and the "ash" from the carbonaceous material. *See Mellor's* **vol III**, p 740 (1928). *Luminous calcium sulfide* or *Canton's phosphorus* made by igniting a mixture of $CaCO_3$ and S with very small quantities of Bi or Mn salts, etc.: Verneuil, *Compt. Rend.* **103**, 600 (1886); *Mellor's, loc. cit.*

White powder if pure; crude and luminous calcium sulfide may be yellowish to pale-gray. Odor of H_2S in moist air; unpleasant alkaline taste. Oxidizes in dry air and dec in moist air. mp > 2000°. d 2.59. Slightly sol in cold, more sol in hot water with partial decompn; freely sol in solns of ammonium salts; practically insol in alcohol; dec even by weak acids, evolving H_2S. *Keep well closed.*

USE: In phosphors; as lubricant additive. Pure CaS used in electron emitters. Luminous CaS used for making luminous paints or varnishes.

THERAP CAT (VET): Has been used in chronic suppurative lesions.

1755. Calcium Sulfite. CaO_3S; mol wt 120.14. Ca 33.36%, O 39.95%, S 26.69%. $CaSO_3$. Prepn: *Gmelin's, Calcium* (8th ed.) **28B**, 107-108, 660-674 (1958).

Dihydrate, crystals or powder. Slowly oxidizes in air to $CaSO_4$. Slightly sol in water, alcohol; sol in SO_2 solns, acids with liberation of SO_2.

USE: Preserving cider and other fruit juices; as disinfectant of brewing vats; antichlor in bleaching textiles; in sugar manuf; in paper pulp cooking; in cement.

1756. Calcium Tartrate. *2,3-Dihydroxybutanedioic acid calcium salt.* $C_4H_4CaO_6$; mol wt 188.15. C 25.53%, H 2.14%, Ca 21.30%, O 51.02%. $CaC_4H_4O_6$. A byproduct of the wine industry. Prepn from wine dregs: Dabul, U.S. pat. **3,114,770** (1963 to Orandi & Massera). *See also* the processes mentioned under L-tartaric acid.

Tetrahydrate, powder. Slightly sol in water (from about 0.04% at 10° to about 0.2% at 85°) or in alcohol; sol in dil HCl or HNO_3.

USE: As preservative for fruits, vegetables, seafoods; in deodorization of fish; as antacid.

1757. Calcium Thiocyanate. Calcium rhodanate; calcium sulfocyanate. $C_2CaN_2S_2$; mol wt 156.25. C 15.37%, Ca 25.65%, N 17.93%, S 41.05%. $Ca(SCN)_2$. Prepn: *Gmelin's, Calcium* (8th ed.) **28B**, 972-976 (1958).

Tetrahydrate, hygroscopic crystals or cryst powder. Dec on heating above 160°. Very sol in water; sol in methanol, ethanol, acetone. *Keep well closed.*

USE: In manuf of acrylonitrile polymers; for parchmentizing; for stiffening of textiles; soln as a solvent for textiles.

1758. Calcium Thioglycollate. *Mercaptoacetic acid calcium derivative;* Depil. $C_2H_2CaO_2S$; mol wt 130.18. C 18.45%, H 1.55%, Ca 30.79%, O 24.58%, S 24.63%. Prepn: Hoshall, *J. Assoc. Offic. Agr. Chem.* **23**, 727 (1940).

Trihydrate, prismatic rod crystals. Odorless or faint mercaptan odor; somewhat astringent and fetid taste. Slowly loses H_2O above 95°, darkens at 220° and partially fuses with decompn at 280-290°. Sol in water; very slightly sol in alcohol, chloroform; practically insol in ether, petr ether, benzene. Solns readily absorb CO_2 from air, pptg $CaCO_3$. *Keep well closed.*

USE: Depilatory; tanning leather; in hair-waving prepns.

1759. Calcium Thiosulfate. Calcium hyposulfite; Tecesal. CaS_2O_3; mol wt 152.21. Ca 26.33%, S 42.13%, O 31.53%. Prepn: Ballezo, Kaufmann, *Monatsh.* **80**, 220 (1949); Levenson, U.S. pat. **2,763,531** (1956 to Kodak).

Hexahydrate, triclinic crystals. When dry, dec on standing forming a yellow crust on surface; more stable if kept damp and stored below 0°. Spontaneously dec at 43-49°. d 1.87. Freely sol in water; practically insol in alcohol. *Keep well closed and in a cool place.*

THERAP CAT: In dermatitis and jaundice due to arsphenamine.

1760. Calcium Tungstate(VI). CaO_4W; mol wt 287.92. Ca 13.92%, O 22.23%, W 63.85%. $CaWO_4$. Occurs in nature as the mineral *scheelite*. The article of commerce is usually made by pptn: Boericke, Boericke, U.S. pat. **2,390,-687** (1945); prepn by heating a stoichiometric mixture of CaO or $CaCO_3$ and tungstic acid: Kroger, *Nature* **159**, 674

(1947); prepn of single crystals: Uitert, Soden, *J. Appl. Phys.* **31**, 328 (1960).

Tetragonal crystals. d 6.06. Practically insol in water. Dec by hot HCl or HNO_3.

USE: Very small crystals have been used for injection into malignant tumors, etc., thus affording by transillumination a means of x-ray treatment; for preparing screens for x-ray observations and photographs; in luminous paints; in scintillation counters.

1761. Caldariomycin. *(1S-trans)-2,2-Dichloro-1,3-cyclopentanediol;* 1,3-dihydroxy-2,2-dichlorocyclopentane. $C_5H_8Cl_2O_2$; mol wt 171.02. C 35.12%, H 4.71%, Cl 41.46%, O 18.71%. Mold metabolite from *Caldariomyces fumago* Woronichin. Isoln and structure: Clutterbuck *et al.*, *Biochem. J.* **34**, 664 (1940). Biosynthesis: Shaw *et al.*, *J. Biol. Chem.* **234**, 2560 (1959); Beckwith, *ibid.* **238**, 3086 (1963); Beckwith, Hager, *ibid.* 3091. Synthesis: *eidem, J. Org. Chem.* **26**, 5206 (1961); Burgstahler *et al., ibid.* **31**, 3516 (1966). Abs configuration: Johnson *et al., J. Am. Chem. Soc.* **90**, 136 (1968).

Fine, colorless needles, mp 121°. Decomposes at temps > 180°; very stable to heat at mod. temps. $[\alpha]_{5461}^{20}$ +59.2° (c = 0.338 in H_2O). Readily sol in cold water, ethanol, ether and most other org solvents. Slightly sol in chloroform.

1762. C-Calebassine. *C*-Toxiferine II; *C*-strychnotoxine. $[C_{40}H_{48}N_4O_2]^{2+}$; mol wt 616.85. From calabash-curare: Wieland *et al., Ann.* **547**, 156 (1941); Karrer, Schmidt, *Helv. Chim. Acta* **29**, 1853 (1946); Zürcher *et al., J. Am. Chem. Soc.* **80**, 1500 (1958). Identity with *C*-toxiferine II and *C*-strychnotoxine: Wieland, Merz, *Ber.* **85**, 731 (1952). Structure: Hesse *et al., Helv. Chim. Acta* **44**, 2211 (1961); Fehlmann *et al., ibid.* **48**, 303 (1965). Synthesis from *C*-dihydrotoxiferine chloride: Bernauer *et al., ibid.* **40**, 1999 (1957); Grdinic *et al., J. Am. Chem. Soc.* **86**, 3357 (1964).

Dichloride, $C_{40}H_{48}Cl_2N_4O_2$, needles from methanol + ether. $[\alpha]_D^{20}$ +72.1° (c = 0.67 in water). uv max (water): 253, 302 nm (log ε 4.37, 3.77).

Dipicrate, $C_{52}H_{52}N_{10}O_{16}$, prisms from acetone + water, mp 215°. Slightly sol in acetone; practically insol in water, methanol, ethanol, dioxane.

THERAP CAT: Muscle relaxant (skeletal).

1763. Calendula. Marigold; Mary-bud; gold-bloom; holligold. Dried, ligulate florets of *Calendula officinalis* L., *Compositae. Habit.* Southern Europe and Levant; cultivated everywhere in gardens. *Constit.* Volatile oil; calendulin; a carotenoid pigments, a saponin which on hydrolysis yields oleanolic acid, bitter principle (caledin). *Refs:* Zimmerman, *Helv. Chim. Acta* **29**, 445 (1946); Gedeon, *Pharmazie* **6**, 547

(1951); **9**, 922 (1954), Kasprzyk, *C.A.* **47**, 6918c (1953); Suchy, Herout, *Coll. Czech. Chem. Commun.* **26**, 890 (1961).
THERAP CAT: Anti-inflammatory (topical).

1764. Californium. Cf; at. no. 98; valence 4, 3, 2. Man-made, radioactive element. No stable nuclides; known isotopes (mass numbers): 239-256. Longest-lived known isotope: 251 ($T_{1/2}$ 898 years, α-emitter). Prepn of isotope ^{245}Cf ($T_{1/2}$ 43.6 min) by bombarding ^{242}Cm with helium ions: S. G. Thompson *et al.*, *Phys. Rev.* **78**, 298 (1950); *eidem, ibid.* **80**, 790. Prepn of ^{249}Cf metal ($T_{1/2}$ 351 years, α-emitter) by reduction of Cf_2O_3 with lanthanum metal: Haire, Baybarz, *J. Inorg. Nucl. Chem.* **36**, 1295 (1974). Medical uses of ^{252}Cf (α-decay $T_{1/2}$ half-life 2.645 years; spontaneous fission $T_{1/2}$ half-life 85.5 years): Seaborg, *Handb. Exp. Pharmakol.* **36**, 929 (1973). Clinical trials in cervical cancer: Y. Maruyama *et al.*, *Int. J. Radiat. Oncol. Biol. Phys.* **11**, 1475 (1985); Y. Maruyama *et al.*, *Cancer* **59**, 1500 (1987). *Reviews:* Cunningham, *J. Chem. Ed.* **36**, 32-37 (1959); M. Haissinsky, J. P. Adloff, *Radiochemical Survey of the Elements* (Elsevier, New York, 1965) pp 28-29; C. Keller, *The Chemistry of the Transuranium Elements* (Verlag Chemie, Weinheim, English Ed., 1971) pp 567-581; Silva, "Trans-Curium Elements" in *MTP Int. Rev. Sci.: Inorg. Chem., Ser. One* vol. 8, A. G. Maddock, Ed. (University Park Press, Baltimore, 1972) pp 71-105; *Comprehensive Inorganic Chemistry* vol.5, J. C. Bailar, Jr. *et al.*, Eds. (Pergamon Press, Oxford, 1973) *passim; Handb. Exp. Pharmakol.* **36**, 689-928 (1973); R. G. Haire in *The Chemistry of the Actinide Elements* vol. 2, J. J. Katz *et al.*, Eds. (Chapman and Hall, New York, 1986) pp 1025-1070. Review of radiobiology and therapeutic applications: Y. Maruyama *et al.*, *Oncology* **35**, 172 (1978).

Metal; two crystalline forms: double hexagonal close-packed α-form, d 15.10, exists below 900°; face-centered cubic β-form, d 8.74 (Katz *et al.*, *loc cit.*, vol. 2, p. 1150). mp 900±30° (Haire, Baybarz). Slowly oxidized in air at room temp; rate increases with increased moisture in air. Oxidized when warmed in air; reacts when heated with nitrogen, hydrogen, or a chalogen. Reacts rapidly with dry hydrogen halides and with aq mineral acids.

Caution: Radiation hazard; handling requires special equipment and shielding facilities (Katz *et al.*, *loc cit.*, vol. 2, p. 1128).

USE: ^{252}Cf as neutron source; startup source for nuclear reactors; in nuclear reactor fuel rod scanners.
THERAP CAT: ^{252}Cf as antineoplastic (radiation source).

1765. Calixarenes. Molecular baskets into which smaller molecules can fit. Name is derived from the Greek "calix" for "chalice" or "cup" and arene for the aryl residues; the name originally referred to the phenolic tetramer linked by methylene groups. The nomenclature uses calix-[n]arene where n is the number of the aryl groups. Modifications to the phenol are indicated by prefixes; whereas total changes of the aryl group are indicated in front of the "arene". Prepn: A. Zinke, E. Ziegler, **Ber.** **77**, 264 (1944); A. Zinke et al., *Monatsh.* **79**, 438 (1948). Modification of basic structures: J. L. Atwood *et al.*, *Angew. Chem. Int. Ed. Engl.* **32**, 1093 (1993). Book: *Calixarenes*, C. D. Gutsche, Ed. (Royal Society of Chemistry, Cambridge, U.K., 1989) 222 pp.

1766. Calmagite. *3-Hydroxy-4-[(2-hydroxy-5-methylphenyl)azo]-1-naphthalenesulfonic acid;* 1-(1-hydroxy-4-methyl-2-phenylazo)-2-naphthol-4-sulfonic acid. $C_{17}H_{14}N_2O_5S$; mol wt 358.37. C 56.98%, H 3.94%, N 7.82%, O 22.32%, S 8.95%. Prepn from 1-amino-2-naphthol-4-sulfonic acid and p-cresol: Lindstrom, Diehl, *Anal. Chem.* **32**, 1123 (1960).

Red crystals from acetone. Sol in water. Absorption max (pH 10.10): 610 nm (ϵ 20300). Functions as acid-base

indicator: Aq solns are bright red at low pH, red at pH 7.1 to 9.1, blue at pH 9.1 to 11.4. The blue color at pH 10 is changed to red by the addition of calcium or magnesium.
USE: As indicator in titration of Ca or Mg with EDTA.

1767. Calmodulin. CaM; calcium-dependent regulator protein; CDR. A calcium-binding multifunctional regulatory protein, ubiquitously distributed in eukaryotic cells. It functions as an intracellular intermediary for calcium ions and activates a number of enzymes involved in fundamental cell processes, such as protein phosphorylation, contractile processes, and metabolism of cyclic nucleotides, of glycogen and of calcium, as well as in other metabolic reactions. Originally discovered as a protein activator of cyclic 3',5'-nucleotide phosphodiesterase: W. Y. Cheung, *Biochem. Biophys. Res. Commun.* **29**, 478 (1967); *idem, Biochim. Biophys. Acta* **191**, 303 (1969); *idem, Biochem. Biophys. Res. Commun.* **38**, 533 (1970); S. Kakiuchi *et al., ibid.* **41**, 1104 (1970). Calcium-binding activity: T. S. Teo, J. H. Wang, *J. Biol. Chem.* **248**, 5950 (1973). Calmodulin is a relatively small, acidic, stable monomer of mol wt 15,000-19,000, lacking cysteine, hydroxyproline, and tryptophan. It has a high content of acidic amino acids and low tyrosine content; almost all calmodulins isolated also contain a single, fully trimethylated lysyl residue. Purification from bovine brain: Y. M. Lin *et al., ibid.* **249**, 4943 (1974); *eidem, Methods Enzymol* **39**, Pt. C, 262 (1974). Amino acid sequence of CaM from bovine brain: T. C. Vanaman in *Calcium Binding Proteins and Calcium Function*, R. H. Wasserman *et al.*, Eds. (Elsevier, New York, 1977) pp 107-116; D. M. Watterson *et al., J. Biol. Chem.* **255**, 962 (1980); from rat testis: J. R. Dedman *et al., ibid.* **253**, 343 (1978); from sea invertebrate, *Renilla reniformis:* T. C. Vanaman, F. Sharief, *Fed. Proc.* **38**, 788 (1979). Radioimmunoassay: R. W. Wallace, W. Y. Cheung, *J. Biol. Chem.* **254**, 6564 (1979); J. C. Chafouleas *et al., ibid.* 10262. CaM has four Ca^{2+}-binding sites; binding to any one of the sites results in a conformational change, which is required for calmodulin to regulate enzyme systems. Conformational transition study: C. B. Klee, *Biochemistry* **16**, 1017 (1977). An important feature of calmodulin's structure is the presence of four homologous internal amino acid sequences, called domains. Each of these domains (one for each bound calcium) are reportedly "E-F Hands", a structural concept described by R. H. Kretsinger in *Calcium Transport in Contraction and Secretions*, E. Carafoli *et al.*, Eds. (Elsevier, Amsterdam, 1975) pp 469-478. Calmodulins obtained from a wide range of phylogenetically different sources are similar in amino acid sequence and in physico-chemical and biological properties; hence, the protein lacks both species and tissue specificity and appears to be structurally and functionally conserved throughout evolution. *Reviews:* W. Y. Cheung, *Science* **207**, 19-27 (1980); A. R. Means, J. R. Dedman, *Nature* **285**, 73-77 (1980); C. B. Klee *et al., Ann. Rev. Biochem.* **49**, 489-515 (1980); A. R. Means, *Recent Progr. Horm. Res.* **37**, 333-367 (1981); Y. M. Lin, *Mol. Cell Biochem.* **45**, 101-112 (1982). Books: *Ann. N.Y. Acad. Sci.* **356**, entitled "Calmodulin and Cell Functions", D. M. Watterson, F. F. Vincenzi, Eds. (1980) 455 pp; *Calcium and Cell Function* vol. 1, W. Y. Cheung, Ed. (Academic Press, New York, 1980) 395 pp.

Isoelectric pt 3.9-4.3. $\epsilon_{276nm}^{1\%}$ 1.8; $\epsilon_{280nm}^{1\%}$ 2.1. Stable when subjected to heat at neutral or acidic pH.

1768. Calomelol. Colloidal calomel; sol calomel. Consists of 80% HgCl and 20% proteins.

Whitish-gray, odorless, tasteless powder. Sol in water to an opalescent suspension; insol in alcohol, ether; pptd by acids, redissolved by alkalies. *Protect from light.*
THERAP CAT: Cathartic.

1769. Calotropin. *[2α(2S,3S,4S,6R),3β,5α]-14-Hydroxy-19-oxo-3,2-[(tetrahydro-3,4-dihydroxy-6-methyl-2H-pyran-2,3-diyl)bis(oxy)]card-20(22)-enolide.* $C_{29}H_{40}O_9$; mol wt 532.63. C 65.40%, H 7.57%, O 27.03%. African arrow poison isolated from milk sap of *Calotropis procera* Dryand., *Asclepiadaceae.* Isolation: Lewin, *Arch. Exp. Path. Pharmakol.* **71**, 142 (1913); Hesse *et al., Ann.* **526**, 252 (1936); **566**, 130 (1950); Rajagopalan *et al., Helv. Chim. Acta* **38**, 1809 (1955). Isoln from *Asclepias curassavica* L., *Asclepiadaceae:* S. M. Kupchan *et al., Science* **146**, 1685 (1964). Structure: G. Hesse, G. Lettenbauer, *Ann.* **623**, 142 (1959); Hesse *et*

al., ibid. **625,** 157, 161 (1959); D. G. H. Crout et al., Tetrahedron Letters **1963,** 63; J. Chem. Soc. **1964,** 2187. Extraction from C. procera R.Br. and toxicity: F. Brüschweiler et
al., Helv. Chim. Acta **52,** 2086 (1969). Revised structure:
eidem, ibid. 2276. Sequestration by larvae of Monarch butterfly Danaus plexippus L.: J. N. Sieber et al., J. Chem. Ecol.
6, 321 (1980). Quantitative analysis of cardenolides in latex
and leaves of C. procera: eidem, Phytochemistry **21,** 2343
(1982). Biosynthesis of labelled compd: M. S. Lee, J. N.
Sieber, ibid. **22,** 923 (1983).

Rectangular platelets from alcohol or ethyl acetate, mp
223° (dec). $[\alpha]_D^{18}$ +66.8° (in methanol). Sol in water, alc.
Practically insol in ether. uv max: 217, 310 nm (log ε 4.21,
1.49). Lethal dose in cats: 0.12 mg/kg (Brüschweiler).

1770. Calumba. Colombo. Root of Jatrorrhiza palmata (DC.) Miers (J. columba Miers), Menispermaceae. Habit.
Eastern Africa. Constit. Columbin, chasmanthin, palmarin
(isomer of chasmanthin), jatrorrhizine, columbic acid, columbamine; contains no tannin: Barton, Elad, J. Chem.
Soc. **1956,** 2085, 2090. Review: Feist, Arzneimittel-Forsch. **1,**
418 (1951).

THERAP CAT (VET): Has been used as a stomachic.

1771. Calusterone. (7β,17β)-17-Hydroxy-7,17-dimethyl-
androst-4-en-3-one; 7β,17α-dimethyltestosterone; U-22550;
Methosarb. $C_{21}H_{32}O_2$; mol wt 316.48. C 79.70%, H
10.19%, O 10.11%. Prepn: Campbell, Babcock, U.S. pats.
3,029,263; 3,341,557 (1962, 1967 both to Upjohn). Epimeric with bolasterone (q.v.). Clinical studies: Gordan et al., J.
Am. Med. Assoc. **219,** 483 (1972).

Crystals from acetone, mp 127-129°. $[\alpha]_D$ +57° (CHCl₃).
uv max (alcohol): 243 nm.

THERAP CAT: Antineoplastic.

1772. Calycanthine. $C_{22}H_{26}N_4$; mol wt 346.48. C
76.27%, H 7.56%, N 16.17%. In Calycanthus floridus L., C.
glaucus Willd.; Chimonanthus praecox (L.) Link, Calycan-
thaceae. First isoln: Eccles, Proc. Am. Pharm. Assoc. **84,**
382 (1888). Extraction procedure: Manske-Marion, Can. J.
Res. **17B,** 293 (1939). Structure: Woodward et al., Proc.
Chem. Soc. **1960,** 76; Hamor et al., ibid. 78; Hamor, Robertson, J. Chem. Soc. **1962,** 194. Configuration: Clayton et al.,
Tetrahedron **18,** 1495 (1962). Synthesis of DL-form: Hendrickson et al., ibid. **20,** 565 (1964); Hall et al., ibid. **23,** 4131
(1967).

Crystals, mp 245° (evac tube). $[\alpha]_D^{18}$ +684° (c = 1.2 in abs
alc). uv max (95% ethanol): 250, 309 nm (log ε 4.28, 3.80).
Alkaline reaction to litmus. Freely sol in alc, chloroform;
sol in ether, acetone, pyridine; slightly sol in water.

Monohydrate, orthorhombic bipyramidal crystals from
water, mp 220°.

Caution: Highly toxic. Can cause violent convulsions,
paralysis, cardiac depression.

1773. Camazepam. Dimethylcarbamic acid 7-chloro-
2,3-dihydro-1-methyl-2-oxo-5-phenyl-1H-1,4-benzodiazepin-3-
yl ester; 7-chloro-1,3-dihydro-3-hydroxy-1-methyl-5-phenyl-
2H-1,4-benzodiazepin-2-one dimethylcarbamate (ester); 7-
chloro-1,3-dihydro-3-(N,N-dimethylcarbamoyl)-1-methyl-5-
phenyl-2H-1,4-benzodiazepin-2-one; SB-5833; Albego. C_{19}-
$H_{18}ClN_3O_3$; mol wt 371.82. C 61.38%, H 4.88%, Cl 9.53%,
N 11.30%, O 12.91%. Prepn: G. Ferrari, C. Casagrande,
Ger. pat. 2,142,181; eidem, **U.S. pat. 3,799,920** (1972, 1974
both to Siphar). Metabolism: F. Marcucci et al., J. Pharm.
Sci. **67,** 1470 (1978). Pharmacology: L. Merlo et al., Arzneimittel-Forsch. **24,** 1759 (1974); R. Ferrini et al., ibid.
2029. Clinical studies: A. Tammaro et al., ibid. **27,** 2177
(1978); S. Carrara et al., Eur. J. Clin. Pharmacol. **13,** 335
(1978).

White crystalline powder from ethyl acetate, mp 173-174°.
Sol in alcohol, moderately sol in water. LD_{50} in mice, rats
(mg/kg): 970, >4000 orally (Ferrini).

Note: This is a controlled substance (depressant) listed in
the U.S. Code of Federal Regulations, Title 21 Part 1308.14
(1995).

THERAP CAT: Anxiolytic.

1774. Cambendazole. [2-(4-Thiazolyl)-1H-benzimid-
azol-5-yl]carbamic acid 1-methylethyl ester; 2-(4-thiazolyl)-5-
benzimidazolecarbamic acid isopropyl ester; 5-isopropoxycarbon-
ylamino-2-(4-thiazolyl)benzimidazole; 5-isopropoxycarbon-
ylaminothiabendazole; isopropyl 2-(4-thiazolyl)-5-benzimid-
azolecarbamate; MK-905; Bonlam; Bovicam; Cambenzole;
Cambet; Equiben; Novazole; Noviben. $C_{14}H_{14}N_4O_2S$; mol
wt 302.36. C 55.61%, H 4.67%, N 18.53%, O 10.58%, S
10.61%. Prepn and activity studies: Hoff, Fisher, **S. Afr.
pat. 68 00,351** (1969 to Merck & Co.), C.A. **72,** 90461q
(1970); Hoff et al., Experientia **26,** 550 (1970). Clinical
studies with sheep, cattle and swine: Egerton et al., Res.
Vet. Sci. **11,** 193, 495, 590 (1970).

Odorless, white crystalline solid, mp 238-240° (dec). Sol
in alcohol, dimethylformamide; sparingly sol in acetone;
slightly sol in benzene; very slightly sol in 0.1M HCl. Prac-

tically insol in isooctane and water (0.02 mg/ml). Stable in acid and base in range of pH 1 to 12. uv max (0.1N HCl): 319, 232 nm ($A_{1cm}^{1\%}$ 740, 670).

THERAP CAT (VET): Anthelmintic.

1775. Camostat. *4-[[4-[(Aminoiminomethyl)amino]-benzeneacetic acid 2-(dimethylamino)-2-oxoethyl ester; N,N-dimethylcarbamoylmethyl-p-(p-guanidinobenzoyl-oxy)phenylacetate.* $C_{20}H_{22}N_4O_5$; mol wt 398.42. C 60.29%, H 5.57%, N 14.06%, O 20.08%. Orally active, non-peptide proteolytic enzyme inhibitor with anti-trypsin and anti-plasmin activities, related structurally to gabexate, q.v. Prepn: S. Fujii et al., Ger. pat. 2,548,886; eidem, U.S. pat. 4,021,-472 (1976, 1977 both to Ono). Enzyme inhibition study: Y. Tamura et al., Biochim. Biophys Acta 484, 417 (1977). Inhibition of exptl tumors in mice: M. Ohkoshi, Gann 72, 959 (1981), C.A. 96, 97304v (1982); M. Ohkoshi, S. Fujii, ibid. 73, 108 (1982), C.A. 96, 155182s (1982). Prevention of acute exptl pancreatitis: S. Takasugi et al., Digestion 24, 36 (1982). Absorption and excretion: S. Hiraku et al., Iyaku-hin Kenkyu 13, 756 (1982), C.A. 97, 120038t (1982).

Monomethanesulfonate, $C_{20}H_{22}N_4O_5 \cdot CH_3SO_3H$, *camostat mesylate, FOY-305, Foipan.* Solid from methanol/ether, mp 150-155°. Sol in water.

THERAP CAT: Protease inhibitor.

1776. Campesterol. *(3β,24R)-Ergost-5-en-3-ol.* $C_{28}H_{48}O$; mol wt 400.69. C 83.93%, H 12.07%, O 3.99%. Small amounts are found in rape-seed oil derived from *Brassica campestris* L., *Cruciferae*, in soybean oil, and in wheat germ oil. Isoln: Fernholz, MacPhillamy, J. Am. Chem. Soc. 63, 1155 (1941). Structure: Fernholz, Ruigh, ibid. 1157. Synthesis: Tarzia et al., Gazz. Chim. Ital. 97, 102 (1967), C.A. 67, 32883q (1967).

Crystals from acetone, mp 157-158°. $[\alpha]_D^{23}$ — 33° (22.5 mg in 5 ml chloroform).
Acetate, $C_{30}H_{50}O_2$, crystals from alc, mp 137-138°. $[\alpha]_D^{23}$ — 35° (28.8 mg in 1 ml chloroform).

1777. Camphene. *2,2-Dimethyl-3-methylenebicyclo-[2.2.1]heptane; 2,2-dimethyl-3-methylenenorbornane; 3,3-di-methyl-2-methylenenorcamphane.* $C_{10}H_{16}$; mol wt 136.24. C 88.16%, H 11.84%. Occurs in many essential oils, such as turpentine (*levo* and *dextro* forms), in cypress oil (*dextro* form), in camphor oil from species of *Lauraceae* (*dextro*), in bergamot oil, in oil of citronella, neroli, ginger, valerian. Reviews on isolation, preparation and properties: J. L. Simonsen, *The Terpenes* vol. II (Cambridge Univ. Press, 1949) pp 280-322; E. Guenther, *The Essential Oils* vol. II (Van Nostrand, 1949) pp 66-70. Synthesis: G. W. Hana, H. Koch, Ber. 111, 2527 (1968).

dl-Form, cubic crystals from alcohol. Large dodecahedra by slow sublimation. Volatilizes on exposure to air. Insipid odor. mp 51-52°. bp_{760} 158.5-159.5°; bp_{100} 92.4°; bp_{16} 55-56°. d_4^{54} 0.8422. n_D^{54} 1.45514. Practically insol in water. Moderately sol in alcohol; sol in ether, cyclohexane, cyclohexene, dioxane, chloroform.
d-Form, mp 52°. $[\alpha]_D^{17}$ +103.5° (c = 9.67 in ether). d_4^{50} 0.8486. n_D^{50} 1.4605.
l-Form, mp 52°. $[\alpha]_D^{21}$ — 119.11° (c = 2.33 in benzene). d_4^{54} 0.8422. n_D^{40} 1.4620.

1778. d-Camphocarboxylic Acid. *4,7,7-Trimethyl-3-oxobicyclo[2.2.1]heptane-2-carboxylic acid; d-2-oxo-3-bornanecarboxylic acid; d-3-camphorcarboxylic acid; d-2-oxo-3-camphanecarboxylic acid; d-3-carboxy-2-bornanone; d-3-carboxy-2-camphanone.* $C_{11}H_{16}O_3$; mol wt 196.25. C 67.32%, H 8.22%, O 24.46%. Prepd by carboxylation of d-camphor: Brühl, Ber. 24, 3373 (1891).

Crystals from benzene, water, ether, or 50% alcohol. mp 127-128°. Sparingly sol in cold water, more sol in warm water; sol in alcohol, ether, chloroform, in about 2 parts boiling benzene. Sparingly sol in cold benzene. Practically insol in cold petr ether; very slightly sol in boiling petr ether. For soln contg 0.38 g in 25 ml solvent: $[\alpha]_D$ +18° (benzene), +60° (alcohol), +73.3° (water).
Ammonium salt, $C_{11}H_{19}NO_3$, *camphydryl, camphor solubilized, Canfoxil.*
Basic bismuth salt, $C_{33}H_{46}Bi_2O_{11}$, *Angimuth.* Prepn: Raiziss, Clemence, U.S. pat. 1,921,638 (1933 to Abbott). Powder. Odor of camphor. Practically insoluble in water. Soluble in methanol, ether, benzene, oils.

THERAP CAT: Basic bismuth salt formerly as antisyphilitic.

1779. Camphor. *1,7,7-Trimethylbicyclo[2.2.1]heptan-2-one; 2-bornanone; 2-camphanone; 2-keto-1,7,7-trimethylnorcamphane; gum camphor; Japan camphor; Formosa camphor; laurel camphor.* $C_{10}H_{16}O$; mol wt 152.24. C 78.90%, H 10.59%, O 10.51%. Occurs in all parts of the camphor tree, *Cinnamomum camphora* T. Nees & Ebermeier, *Lauraceae*. Habit: Java, Sumatra, China (central provinces), Japan, Formosa, Brazil. Obtained by steam distillation from comminuted trees which should be at least 50 years old. Description of various indigenous processes: Gubelmann, Elley, Ind. Eng. Chem. 26, 589 (1934); G. Etzel in Kirk-Othmer Encyclopedia of Chemical Technology, vol. 4 (Wiley, New York, 2nd ed., 1964) pp 54-58. Modern processes start with vinyl chloride and cyclopentadiene to obtain the important intermediate dehydronorbornyl chloride. Review of syntheses: K. Alder in New Methods of Preparative Organic Chemistry (New York, 1948); A. F. Thomas in The Total Synthesis of Natural Products vol. 2, J. ApSimon, Ed. (Wiley-Interscience, New York, 1973) pp 149-154. More than three-fourths of the camphor sold in the U.S. is produced synthetically (usually from pinene), and most is sold in the racemic form, although the U.S.P. specifies the d-form. Configuration: Freudenberg et al., Ann. 594, 76 (1955). Toxicity: A. Smith, G. Margolis, Am. J. Pathol. 30, 857 (1954).

Translucent mass with crystalline fracture. Rhombohedral crystals from alcohol. Cubic crystals by melting and chilling. Familiar fragrant and penetrating odor. Slightly bitter and cooling taste. d_4^{25} 0.992. mp 179.75° (corr., open capillary, 2 mm diam). bp 204°. Sublimes appreciably at room temp and press. *Keep in tight containers away from heat.* At 80° and 12 mm press 14% sublimes within 60 minutes. Very volatile in steam. $[\alpha]_D^{25}$ +41° to +43° (c = 10 in U.S.P. alcohol) according to U.S.P. specif. The water content of the ethanol influences the rotation considerably. $[\alpha]_D^{20}$ +43.8° (c = 7.5 in abs alcohol). uv max (CHCl$_3$): 292 nm. At 25° one gram dissolves in about 800 ml water (giving a colloidal soln), in 1 ml alcohol, 1 ml ether, 0.5 ml chloroform, 0.4 ml benzene, 0.4 ml acetone, 1.5 ml oil of turpentine, 0.5 ml glacial acetic acid. Sol in aniline, nitrobenzene, carbon disulfide, tetralin, decalin, methylhexalin, petr ether, in the higher alcohols, in fixed and volatile oils. Also sol in concd mineral acids in phenol, in liquid NH_3 and in liquid SO_2. Camphor has a peculiar tenacity and cannot be powdered in a mortar unless it is moistened with an organic solvent. Liquefies when triturated with chloral hydrate, menthol, resorcinol, salol, β-naphthol, thymol, phenol, urethan. LD$_{50}$ i.p. in mice: 3000 mg/kg (Smith, Margolis).

Incompat: Incompatible with potassium permanaganate; salts of any kind should not be added to camphor water.

Caution: Ingestion or injection may cause nausea, vomiting, vertigo, mental confusion, delirium, clonic convulsions, coma, respiratory failure, death, *Clinical Toxicology of Commercial Products,* R. E. Gosselin *et al.,* Eds. (Williams & Wilkins, Baltimore, 4th ed., 1976) Section III, pp 77-79.

USE: Excellent plasticizer for cellulose esters and ethers; used in manuf of plastics, esp celluloid; in lacquers and varnishes; in explosives; in pyrotechnics; as moth repellent; in embalming fluids; in manuf cymene; as preservative in pharmaceuticals and cosmetics; in camphorated parachlorophenol and paregoric.

THERAP CAT: Topical anti-infective; topical antipruritic.

THERAP CAT (VET): Has been used internally as a stimulant and carminative; externally as an antipruritic, counterirritant and antiseptic.

1780. Camphoric Acid. *1,2,2-Trimethyl-1,3-cyclopentanedicarboxylic acid;* dextrocamphoric acid. C$_{10}$H$_{16}$O$_4$; mol wt 200.23. C 59.98%, H 8.05%, O 31.96%. By oxidation of camphor: Bredt, *Ber.* **26,** 3047 (1893). Synthesis: Perkin, Thorpe, *J. Chem. Soc.* **89,** 799 (1906); Toivonan, *Acta Chem. Scand.* **2,** 597 (1948).

Leaflets from water, monoclinic prisms from alc, mp 186-188°. d 1.186. $[\alpha]_D^{20}$ +47° to +48° (alc). One gram dissolves in 125 ml water, 10 ml boiling water, 1 ml alc, 20 ml glycerol; sol in chloroform, ether, fats, oils.

1781. d-Camphorsulfonic Acid. *7,7-Dimethyl-2-oxobicyclo[2.2.1]heptane-1-methanesulfonic acid; 2-oxo-10-bornanesulfonic acid;* 10-camphorsulfonic acid; camsylate; camphostyl; β-camphorsulfonic acid; Reychler's acid. C$_{10}$H$_{16}$O$_4$S; mol wt 232.30. C 51.70%, H 6.94%, O 27.55%, S 13.80%. Prepn from powdered camphor, concd H_2SO_4 and acetic anhydride: Reychler, *Bull. Soc. Chim. France* [3] **19,** 120 (1898); Armstrong, Lowry, *J. Chem. Soc.* **81,** 1447 (1902); Lipp, Knapp, *Ber.* **73B,** 915 (1940). Structure: Loudon, *J. Chem. Soc.* **1933,** 823; Komppa, *J. Prakt. Chem.* **162,** 19 (1943).

Prisms from glacial acetic acid or ethyl acetate, dec 193-195°. $[\alpha]_D^{20}$ +43.5° (c = 4.3 in alcohol); $[\alpha]_D^{20}$ +21.5° (c = 4.3 in water). Deliquesc in moist air. Practically insol in ether; slightly sol in glacial acetic acid, ethyl acetate.

Ammonium salt, C$_{10}$H$_{19}$NO$_4$S, needles from water. $[\alpha]_D^{16}$ +20.5° (c = 5 in water). Very sol in water.

Potassium salt, C$_{10}$H$_{15}$KO$_4$S, needles from alcohol. $[\alpha]_D^{16}$ +18.4° (c = 4.4 in water).

USE: Resolution of optically active isomers.

1782. Camphotamide. *3-[(Diethylamino)carbonyl]-1-methylpyridinium salt with 4,7,7-trimethyl-3-oxobicyclo-[2.2.1]heptane-2-sulfonic acid (1:1);* 3-(diethylcarbamoyl)-1-methylpyridinium 2-oxo-3-bornanesulfonate; camphosulfon-yl-*N*-methylpyridine-β-diethylcarboxamide; camphetamide; camphramine; Tonicorine. C$_{21}$H$_{32}$N$_2$O$_5$S; mol wt 424.56. C 59.41%, H 7.60%, N 6.60%, O 18.84%, S 7.55%. Prepd by the condensation of equimolar amounts of camphorsulfonic acid and *N*-methylnikethamide: Lab. Lematte et Boinot in P. Lebeau, M. Janot, *Traité de Pharmacie Chimique* vol. 4, (Paris 1955-56) p 2471; by reacting nicotinic acid with methyl camphorsulfonate and treating with diethylamine: Fr. pat. 812,032 (1937 to Soc. Franc. Recherches Biochim.), *C.A.* **32,** 1052 (1938).

Minute crystals, mp 174-175°. Slight camphor-like odor. Bitter taste with sweet aftertaste. Soluble in water, alcohol, ether. Insol in benzene, other hydrocarbons.

THERAP CAT: CNS stimulant.

1783. Camptothecin. *(S)-4-Ethyl-4-hydroxy-1H-pyrano-[3',4':6,7]indolizino[1,2-b]quinoline-3,14(4H,12H)-dione.* C$_{20}$H$_{16}$N$_2$O$_4$; mol wt 348.36. C 68.96%, H 4.63%, N 8.04%, O 18.37%. Antitumor alkaloid; prototype DNA topoisomerase I inhibitor. Isoln from the stem wood of the Chinese tree, *Camptotheca acuminata* Decsne., *Nyssaceae,* and structure: M. E. Wall *et al., J. Am. Chem. Soc.* **88,** 3888 (1966). Approach to synthesis: Kepler *et al., J. Org. Chem.* **34,** 3853 (1969). Total synthesis: E. J. Corey *et al., ibid.* **40,** 2140 (1975). Total synthesis of racemate: Stork, Schultz, *J. Am. Chem. Soc.* **93,** 4074 (1971); Volkmann *et al., ibid.* 5576; Tang *et al., ibid.* **97,** 159 (1975); J. C. Bradley, G. Buchi, *J. Org. Chem.* **41,** 699 (1976); T. Kametani *et al., J. Chem. Soc. Perkin Trans. I* **1981,** 1563. Pharmacologic and clinical evaluation: Gottlieb *et al., Cancer Chemother. Rep.* **54,** 461 (1970); Gallo *et al., J. Nat. Cancer Inst.* **46,** 789 (1971); S. M. Sieber *et al., Cancer Treat. Rep.* **60,** 1127 (1976). Mechanism of action: Y. H. Hsiang *et al., J. Biol. Chem.* **260,** 14873 (1985). HPLC determn in plasma: J. H. Beijnen *et al., J. Chromatog.* **617,** 111 (1993). *Reviews:* M. Potmesil, *Cancer Res.* **54,** 1431-1439 (1994); M. E. Wall, M. C. Wani, *ibid.* **55,** 753-760 (1995).

Pale yellow needles from methanol + acetonitrile, dec 264-267°. Also reported as mp 275-277° (Volkmann); 287-288° (Stork, Schultz). $[\alpha]_D^{25}$ +31.3° (in chloroform-methanol, 8:2). Exhibits intense blue fluorescence under uv light. uv max: 220, 254, 290, 370 nm (ϵ 37320, 29230, 4980, 19900). Does not form stable salts with acids. Poorly sol in water. Acetate, $C_{22}H_{18}N_2O_5$, crystals, dec 271-274°. uv max: 220, 254, 290, 360-370 nm (ϵ 39010, 28740, 6160, 22000). Chloroacetate, $C_{22}H_{17}ClN_2O_5$, crystals, dec 245-248°.

1784. Camylofine. α-[[2-(Diethylamino)ethyl]amino]benzeneacetic acid 3-methylbutyl ester; N-[2-(diethylamino)ethyl]-2-phenylglycine isopentyl ester; isoamyl α-[N-(β-diethylaminoethyl)amino]phenylacetate; isoamyl N-(β-diethylaminoethyl)-α-aminophenylacetate; α-[(β-diethylaminoethyl)amino]phenylacetic acid isoamyl ester; acamylophenine; Adopon; Avadyl; Belosin; Navadyl; Novospasmin; Sintespasmil; Spasmocan. $C_{19}H_{32}N_2O_2$; mol wt 320.48. C 71.21%, H 10.06%, N 8.74%, O 9.98%. Prepn: Schmeisser et al., **Ger. pat.** 842,206 (1952 to Asta), C.A. **47**, 5445i (1953); **Brit. pat.** 688,331 (1953); Chielmetti, Farm. Sci. Tec. (Pavia) **7**, 625 (1952), C.A. **47**, 11161 (1953); Edwards et al., J. Pharm. Pharmacol. **12**, 179 (1960).

Pale yellow oil, $bp_{1.5}$ 174-178°, bp_4 165-180°. Strongly alkaline reaction.

Dihydrochloride, $C_{19}H_{32}N_2O_2.2HCl$, Avacan. Crystals, mp 174-178°, also reported as mp 172° and 173°. Soluble in water. LD_{50} in mice: 760 mg/kg orally; 1.35 g/kg s.c.; 49.2 mg/kg i.v.

THERAP CAT: Anticholinergic.

1785. Canadine. 5,8,13,13a-Tetrahydro-9,10-dimethoxy-6H-benzo[g]-1,3-benzodioxolo[5,6-a]quinolizine; 9,10-dimethoxy-2,3-(methylenedioxy)berbine; 5,6,13,13a-tetrahydro-9,10-dimethoxy-2,3-(methylenedioxy)-8H-dibenzo[a,g]quinolizine; tetrahydroberberine; xanthopuccine. $C_{20}H_{21}NO_4$; mol wt 339.39. C 70.78%, H 6.24%, N 4.13%, O 18.86%. A protoberberine alkaloid from Corydalis cava (L.), Schweigg & Korte (C. tuberosa DC.), Fumariaceae: Späth, Julian, Ber. **64**, 1131 (1931). Prepn by reduction of berberine: Bersch, Seufert, ibid. **70**, 1121 (1937); Awe, Hertel, Arch. Pharm. **288**, 516 (1955); Russell, J. Am. Chem. Soc. **78**, 3115 (1956). Configuration: Corrodi, Hardegger, Helv. Chim. Acta **39**, 889 (1956). Other syntheses: Kametani et al., J. Chem. Soc. (C) **1969**, 2036; **1971**, 2709; M. Cushman, F. W. Dekon, J. Org. Chem. **44**, 407 (1979); K. Iwasa et al., ibid. **46**, 4744 (1981); N. S. Narasimhan et al., Tetrahedron Letters **22**, 2797 (1981). Pharmacology: F. Sadritdinov, M. B. Sultanov, C.A. **66**, 74714v (1967). Review: R. H. F. Manske, H. L. Holmes, The Alkaloids vol. 4 (Academic Press, New York, 1954) pp 91-92.

LD_{50} in mice (mg/kg): 940 orally; 790 s.c.; 100 i.v. (Sadritdinov, Sultanov).

d-Form, mp 132°. $[\alpha]_D^{15}$ +299° (chloroform). Sol in aq methanol.

l-Form, crystals from methanol, mp 135°. $[\alpha]_D^{22}$ −308°; −317° (c = 0.28; 0.4, both in methanol). Sol in aq methanol.

dl-Form, mp 172° (Späth, Julian), also reported as 163-165° (Cushman, Dekon). uv max (95% ethanol): 209, 284 nm (ϵ 28,300; 5200). Sol in aq methanol.

1786. Canavanine. O-[(Aminoiminomethyl)amino]homoserine; 2-amino-4-(guanidinooxy)butyric acid. $C_5H_{12}N_4O_3$; mol wt 176.18. C 34.09%, H 6.87%, N 31.80%, O 27.24%. $H_2NC(NH)NHOCH_2CH_2CH(NH_2)COOH$. Basic amino acid isolated as L-canavanine from jack beans, Canavalia ensiformis (L.) DC., Leguminosae: Kitagawa, Tomiyama, J. Biochem. (Tokyo) **11**, 265 (1929). Constitutes about 1.5% of the dry weight of alfalfa seeds and sprouts: E. A. Bell Biochem. J. **75**, 618 (1960). Distribution in the plant kingdom: Turner, Harborne, Phytochemistry **6**, 863 (1967). Structure: Gulland, Morris, J. Chem. Soc. **1935**, 763; Kitagawa, Takani, J. Biochem. (Tokyo) **23**, 181 (1936). Configuration: Cadden, Proc. Soc. Exp. Biol. Med. **45**, 224 (1940). Synthesis of racemate: Nyberg, Christensen, J. Am. Chem. Soc. **79**, 1222 (1957); Frankel et al., J. Chem. Soc. **1963**, 3127. Alternate synthesis: Yamada et al., Agr. Biol. Chem. **37**, 2201 (1973). Bears resemblance to arginine, q.v. and is often used as substrate to some enzymes normally acting on arginine. Accordingly it is a potent growth inhibitor of many organisms: Pilcher et al., Proc. Soc. Exp. Biol. Med. **88**, 79 (1955). Toxic to mammals: B. Tschiersch, Pharmazie **17**, 621 (1962). L-Canavanine induces certain hematologic and serologic abnormalities characteristic of systemic lupus erythematosus in monkeys: M. R. Malinow et al., Science **216**, 415 (1982).

Crystals from abs alcohol, mp 184°. $[\alpha]_D^{20}$ +7.9° (c = 3.2). Sulfate, $C_5H_{14}N_4O_7S$, crystals from dil alcohol, dec 172°. $[\alpha]_D^{17}$ +19.4° (c = 2). Freely sol in water. DL-Form, crystals from ethanol, mp 180-182°. Soluble in water. Practically insol in alcohol. Forms a monohydrochloride, mp 190°.

1787. Candelilla Wax. Obtained from the candelilla plants; Euphorbia antisyphilitica Zucc., Euphorbiaceae and E. cerifera Alcocer (which is only doubtfully distinct from E. antisyphilitica) are now the principal sources of candelilla wax. Pedilanthus pavonis Boiss. and P. aphyllus Boiss. are secondary sources, yielding a wax of lower melting range and lower saponification number. Most of the wax is produced in Mexico by immersing the plants in boiling water containing sulfuric acid and skimming off the wax which rises to the surface as described by Dickinson, Am. J. Pharm. **91**, 808 (1919); Hodge, Sineath, Econ. Bot. **10**, 134 (1956). The main constituent is the hydrocarbon hentriacontane. Brief review: C. S. Letcher in Kirk-Othmer Encyclopedia of Chemical Technology vol. 24 (Wiley-Interscience, New York, 3rd ed., 1984) pp 468-469.

Brownish to yellowish-brown, hard, brittle, easily pulverizable lumps. d 0.950-0.990. mp 68-70°. Saponification number 50-65. Acid number 10-20. Iodine number 30-35. Practically insoluble in water. Sparingly sol in alcohol; sol in acetone, benzene, carbon disulfide, decalin, hot petr ether, gasoline, oils, turpentine, hot chloroform, carbon tetrachloride.

USE: Manuf cosmetics, rubber substitutes, furniture and leather polishes, candles, sealing wax, phonograph records; for waterproofing boxes and fabrics; electric insulations; lithographic, printing, stamping and writing inks; molding compositions; sizing paper; hardening other waxes; protective coating for citrus fruits; formerly in chewing gum.

1788. Candesartan. (±)-2-Ethoxy-1-[[2'-(1H-tetrazol-5-yl)[1,1'-biphenyl]-4-yl]methyl]-1H-benzimidazole-7-carboxylic acid; 2-ethoxy-1-[4-[2-(1H-tetrazol-5-yl)phenyl]benzyl]-7-benzimidazolecarboxylic acid; CV-11974. $C_{24}H_{20}N_6O_3$; mol wt 440.46. C 65.45%, H 4.58%, N 19.08%, O 10.90%. Angiotensin II type-1 receptor antagonist. Prepn: T. Naka et al., **Eur. pat.** **Appl.** 459,136; eidem, **U.S. pat.** 5,196,444 (1991, 1993 both to Takeda); K. Kubo et al., J. Med. Chem. **36**, 2182, 2343 (1993). Pharmacology: Y. Shibouta et al., J. Pharmacol. Exp. Ther. **266**, 114 (1993). Clinical pharmacology, pharmacokinetics: T. Ogihara et al., Clin. Ther. **16**, 74 (1994). Pilot study of cilexetil ester in essential hypertension: idem et al., ibid. **15**, 684 (1993).

Colorless crystals from ethyl acetate + methanol, mp 183-185°.

1-[[(Cyclohexyloxy)carbonyl]oxy]ethyl ester, $C_{33}H_{34}N_6O_6$, **candesartan cilexetil, TCV-116**. Ester prodrug; hydrolyzed *in vivo* to the active carboxylic acid. Colorless crystals from ethanol + water, mp 163° (dec).

THERAP CAT: Antihypertensive.

1789. Candicidin. Levorin; Candimon; Vanobid. Heptaene macrolide antifungal antibiotic complex composed of candicidins A, B, C and D (major component). Produced by a strain of *Streptomyces griseus* (Rutgers no. 3570): Lechevalier *et al., Mycologia* **45**, 155 (1953). Methods of isoln: Siminoff, U.S. pat. 2,872,373 (1959 to Penick); Waksman, Lechevalier, U.S. pat. 2,992,162 (1961 to Rutgers Res. & Ed. Found.). Activity: Kligman, Lewis, *Proc. Soc. Exp. Biol. Med.* **82**, 399 (1953); *Lancet* **1954**, 266, 507; Lechevalier, *Presse Med.* **61**, 1327 (1953). Structure of candicidin D: J. Zielinski *et al., Tetrahedron Letters* **1979**, 1791. Biosynthesis studies: Liu *et al., J. Antibiot.* **25**, 116 (1972). Anticholesteremic property and mode of action studies: I. H. Kwon, H. Fisher, *Nutr. Rep. Int.* **9**, 245 (1974); A. K. Singhal *et al., Lipids* **16**, 423 (1981). HPLC comparison of candicidin, hamycin, trichomycin, *q.q.v.:* P. Helboe *et al., J. Chromatog.* **189**, 249 (1980). Toxicity study: L. C. Vining *et al., Antibiot. Ann.* **1954-55**, 980.

Candicidin D

Small yellow needles or rosettes from aq tetrahydrofuran or pyridine/acetic acid/water soln (when pure). Absorption max: 403, 380 ($E^{1\%}_{1cm}$ 1150), 360 nm. Practically insol in water, alcohols, ketones, esters, ethers, hydrocarbons, and other lipophilic solvents. Sol in DMSO, DMF, and lower aliphatic acids. Very sol in 80% aq tetrahydofuran soln. The addn of 5%-25% water to alcohols greatly increases soly. Forms sol salts in alkaline solns. Soly: Marsh, Weiss, *J. Assoc. Offic. Anal. Chem.* **50**, 457 (1967). LD_{50} i.p. in mice: 14 mg/kg (Vining).

Candicidin D, $C_{59}H_{84}N_2O_{18}$, *levorin A_2.* Identity with levorin A_2: Bosshardt, Bickel, *Experientia* **24**, 422 (1968); J. Zielinski *et al., loc. cit.*

THERAP CAT: Antifungal (topical).

1790. Candidin. $C_{47}H_{71}NO_{17}$; mol wt 922.08. C 61.22%, H 7.76%, N 1.52%, O 29.50%. Main component of an antifungal antibiotic complex produced by the soil actinomycete *Streptomyces viridoflavus*: Taber *et al., Antibiot. & Chemother.* **4**, 455 (1954); Vining, Taber, *Can. J. Chem.* **34**, 1163 (1956). The two other active principles are designated as **candidinin** and **candidoin**. Prepn of pure crystalline candidin: Preud'hamme, Vuillemin, Fr. pat. 1,298,345 (1962 to Rhone-Poulenc), *C.A.* **58**, 420c (1963). Structure: Borowski *et al., Tetrahedron Letters* **1971**, 1987. Antifungal activity studies: Solotorovsky *et al., Antibiot. & Chemother.* **8**, 364 (1958).

Golden-yellow needles from methanol + chloroform + water. Does not melt, but darkens slowly above 180°. $[\alpha]_D^{27}$ +363° (c = 0.3 in DMF); $[\alpha]_D^{27}$ +205° (c = 0.3 in glacial acetic acid). uv max (methanol): 406, 383, 362, 347 nm. Practically insol in water and most organic solvents. Moderately sol in glacial acetic acid, pyridine, DMF. LD_{50} in mice (mg/kg): 1.5 i.v.; 7-36 i.p.; 30 s.c.

1791. Canella. White cinnamon; wild cinnamon; false Winter's bark; Bahama white wood; wild canilla. Bark of *Canella alba* Murr. (*C. winterana* Gaertn.), Canellaceae. *Habit.* W. Indies and Florida. *Constit.* Eugenol, cineol, terpenes, caryophyllene, mannitol, resin, cancellin.

USE: As spice and as an addition to smoking tobacco.

1792. Cannabidiol. *2-[3-Methyl-6-(1-methylethenyl)-2-cyclohexen-1-yl]-5-pentyl-1,3-benzenediol; (3R,4R)-2-p-mentha-1,8-dien-3-yl-5-pentylresorcinol.* $C_{21}H_{30}O_2$; mol wt 314.47. C 80.21%, H 9.62%, O 10.18%. Occurs in *Cannabis sativa* L. (*C. sativa* var. *indica Auth.*), Moraceae. Isoln from the marihuana extract of Minnesota wild hemp: Adams *et al., J. Am. Chem. Soc.* **62**, 196, 2194 (1940); U.S. pat. 2,304,669 (1942); Jacob, Todd, *J. Chem. Soc.* **1940**, 649; Russell, Todd, *ibid.* **1942**, 628; Schultz, Haffner, *Arch. Pharm.* **291**, 391 (1958). A physiologically inactive component: Mechoulam *et al., Science* **169**, 611 (1970). Structure: Mechoulam, Shvo, *Tetrahedron* **19**, 2073 (1963). Crystal and molecular structure: T. Ottersen *et al., Acta Chem. Scand. B* **31**, 807 (1977). Synthesis of (−)-form: T. Petrzilka *et al., Helv. Chim. Acta* **52**, 1102 (1969); H. J. Kurth *et al., Z. Naturforsch.* **36B**, 275 (1981); of (±)-form: R. Mechoulam, Y. Gaoni, *J. Am. Chem. Soc.* **87**, 3273 (1965). Abs config: *eidem, ibid.* **93**, 217 (1971).

Pale yellow resin or crystals, mp 66-67°. bp_2 187-190° (bath temp 220°). $bp_{0.001}$ 130°. d_4^{40} 1.040. n_D^{20} 1.5404. $[\alpha]_D^{27}$ −125° (0.066 g in 5 ml 95% ethanol). $[\alpha]_D^{18}$ −129° (c = 0.45 in ethanol). uv max (ethanol): 282, 274 nm (log ε 3.10, 3.12). Practically insol in water or 10% NaOH. Sol in ethanol, methanol, ether, benzene, chloroform, petr ether.

1793. Cannabinol. *6,6,9-Trimethyl-3-pentyl-6H-dibenzo[b,d]pyran-1-ol; 3-amyl-1-hydroxy-6,6,9-trimethyl-6H-dibenzo[b,d]pyran.* $C_{21}H_{26}O_2$; mol wt 310.44. C 81.25%, H 8.44%, O 10.31%. Constituent of the resinous exudate of the female flowers of *Cannabis sativa* L. (*C. sativa* var. *indica* Auth.), Moraceae. Physiologically inactive component of marijuana: Mechoulam *et al., Science* **169**, 611 (1970). Isoln: Wood *et al., J. Chem. Soc.* **69**, 539 (1896); Work *et al., Biochem. J.* **33**, 123 (1939). Structure: Cahn, *J. Chem. Soc.* **1930**, 986; **1931**, 630; **1932**, 1342; **1933**, 1400; Bergel, Vögele, *Ann.* **493**, 250 (1932). Crystal and molecular structure: T. Ottersen *et al., Acta Chem. Scand. B* **31**, 781 (1977). Synthesis: Adams *et al., J. Am. Chem. Soc.* **62**, 2204 (1940); P. C. Meltzer *et al., Synthesis* **1981**, 985; J. Novák, C. A. Salemink, *Tetrahedron Letters* **23**, 253 (1982).

Leaflets from petr ether, mp 76-77°. Sublimes at 4 mm with a bath temp of 180-190°. $bp_{0.05}$ 185°. Insol in water. Sol in methanol, ethanol, aq alkaline solns.

1794. Cannabis. Indian hemp; Indian cannabis; marihuana; marijuana; bhang; ganja; charas; kif; hasach; pot; grass; weed. Dried flowering tops of pistillate plants of *Cannabis sativa* L. (*C. sativa* var. *indica* Auth.), *Moraceae.* *Habit.* Persia, East India, U.S., Central America, cultivated in Europe. *Constit.* Isomeric tetrahydrocannabinols (the most active constituents), cannabinol, cannabidiol. Sepn of constituents by chromatographic methods: Parker *et al., Bull. Narcotics* **20**, 9 (1968). *Reviews:* Adams, *Science* **92**, 115 (1940); Todd, *Experientia* **2**, 55 (1946); Karbe, *Arznei-mittel-Forsch.* **1**, 37 (1951). Status report, including many minor constituents: *Chem. & Eng. News,* July 6, **1970**, pp 30-33. Review of pharmacology: Paton, *Ann. Rev. Pharmacol.* **15**, 191 (1975); of health hazards and therapeutic potential: Council on Scientific Affairs, *J. Am. Med. Assoc.* **246**, 1823-1827 (1981); of human toxicity: G. Nahas, C. Latour, *Med. J. Australia* **156**, 495-497 (1992). Brief summary of Nat. Acad. Sci. report: *Chem. & Eng. News,* March 8, 1982, pp 27-28. Book: *Marijuana: Chemistry, Pharmacology, Metabolism and Clinical Effects,* R. Mechoulam, Ed. (Academic Press, New York, 1973) pp 409. *See also* Hashish and Tetrahydrocannabinols.

Caution: Acute intoxication is frequently due to recreational use by ingestion or by inhalation of smoke. Various psychic responses have been described: impairment of attention and cognitive and psychomotor performance; euphoria, restlessness, confusion, disorientation, delirium, visual and auditory hallucinations, mood changes, drowsiness, dysphoria. In rare cases, elaborate paranoid delusions and severe emotional depression has persisted for days after acute intoxication. Physiological effects observed include elevation of resting heart rate, dry mouth, increased appetite for sweet foods, ataxia, tremor, hyperreflexia; less often headache, nausea, vomiting. Chronic use causes respiratory tract irritation and bronchoconstriction, and may be a cause of lung cancer. *See Clinical Toxicology of Commercial Products,* R. E. Gosselin *et al.,* Eds. (Williams & Wilkins, Baltimore, 5th ed., 1984) Section III, pp 252-258.

Note: This is a controlled substance (hallucinogen) listed in the U.S. Code of Federal Regulations, Title 21 Part 1308.11 (1995).

THERAP CAT (VET): Has been used as a sedative in equine colic.

1795. Canrenone. *(17α)-17-Hydroxy-3-oxopregna-4,6-diene-21-carboxylic acid γ-lactone;* 17α-(2-carboxyethyl)-17β-hydroxyandrosta-4,6-dien-3-one lactone; 17α-(2-carboxyethyl)-17β-hydroxy-3-oxoandrosta-4,6-diene lactone; 6-dehydrotestosterone-17α-propionic acid γ-lactone; 3-oxo-17β-hydroxy-4,6-androstadien-17α-yl)propionic acid γ-lactone; Phanurane. $C_{22}H_{28}O_3$; mol wt 340.46. C 77.61%, H 8.29%, O 14.10%. Aldosterone antagonist. Prepd by de-

hydrogenation of 17-hydroxy-3-oxo-17α-pregn-4-ene-21-carboxylic acid γ-lactone: Cella, Tweit, *J. Org. Chem.* **24**, 1109 (1959); Cella, U.S. pat. **2,900,383** (1959 to Searle).

Crystals from ethyl acetate, mp 149-151°, solidifies and remelts at 165°. $[\alpha]_D$ +24.5° (chloroform). uv max: 283 nm (ε 26700).

Free acid potassium salt, $C_{22}H_{29}KO_4$, *potassium canrenoate, Kanrenol, Soldactone, Venactone.*

THERAP CAT: Diuretic.

1796. Cantharides. Spanish fly; blistering fly; blistering beetle; *Cantharis vesicatoria. Habit.* Southern and Central Europe, mainly upon *Oleaceae* and *Caprifoliaceae. Constit.* 0.6-1% cantharidin, 10-15% fat; resinous substances, acetic and uric acids. Powerful irritant, vesicant, rubefacient. *Review:* Ude, Heeger, *Pharm. Zentralhalle* **82**, 193 (1941).

Caution: Highly toxic by ingestion or absorption from skin and mucous membranes; severe gastroenteritis, nephritis, collapse, death may occur, *Clinical Toxicology of Commercial Products,* R. E. Gosselin *et al.,* Eds. (Williams & Wilkins, Baltimore, 4th ed., 1976) Section II, p 175.

THERAP CAT: Vesicant.

THERAP CAT (VET): Vesicant, counterirritant.

1797. Cantharidin. *Hexahydro-3a,7a-dimethyl-4,7-epoxyisobenzofuran-1,3-dione; 2,3-dimethyl-7-oxabicyclo[2.2.1]-heptane-2,3-dicarboxylic anhydride; exo-1,2-cis-dimethyl-3,6-epoxyhexahydrophthalic anhydride;* cantharides camphor. $C_{10}H_{12}O_4$; mol wt 196.20. C 61.22%, H 6.16%, O 32.62%. Active principle of cantharides *(q.v.)* and other insects, in notorious *"Spanish Fly"* aphrodisiac. Synthesis and stereochemistry: Woodward, Loftfield, *J. Am. Chem. Soc.* **63**, 3167 (1941); Ziegler *et al., Ann.* **551**, 1 (1942). Stereospecific synthesis: Stork, Van Tamelen *et al., J. Am. Chem. Soc.* **75**, 384 (1953). Simple efficient synthesis: W. G. Dauben *et al., ibid.* **102**, 6893 (1980). Crystal structure: M. Zehnder, U. Thewalt, *Helv. Chim. Acta* **60**, 740 (1977).

Orthorhombic plates, scales, mp 218°. Sublimes at about 110° (12 mm Hg, 3-5 mm distance). Insol in cold water, somewhat sol in hot water. One gram dissolves in 40 ml acetone, 65 ml chloroform, 560 ml ether, 150 ml ethyl acetate. Sol in oils.

THERAP CAT: Vesicant.

THERAP CAT (VET): Rubefacient, vesicant, counterirritant.

1798. Canthaxanthin. *β,β-Carotene-4,4'-dione;* 4,4'-dioxo-β-carotene; Food Orange 8; C.I. 40850; Carophyll Red; Orobronze; Roxanthin Red 10; Carotaben plus. $C_{40}H_{52}O_2$; mol wt 564.85. C 85.06%, H 9.28%, O 5.66%. All *trans*-carotenoid pigment widely distributed in nature. Isoln from the edible mushroom *Cantharellus cinnabarinus* Adans. ex Fr., *Agaricaceae:* Haxo, *Bot. Gaz.* **112**, 228 (1950); also isolated from flamingo feathers. Structure and synthesis: Isler *et al., Verh. Naturforsch. Ges. Basel* **67**, 379 (1956); Zeller *et al., Helv. Chim. Acta* **42**, 841 (1959). Alternate syntheses: M. Akhtar, B. C. L. Weedon, *J. Chem. Soc.* **1959**, 4058; R. Rüegg, G. Saucy, U.S. pat. **2,983,752** and J. D. Surmatis, U.S. pat. **3,311,656** (1961, 1967 both to Hoffmann-La Roche); J. D. Surmatis *et al., Helv. Chim. Acta* **53**, 974 (1970); M. Rosenberger *et al., Pure Appl. Chem.* **51**, 871 (1979); *eidem, J. Org. Chem.* **47**, 2130 (1982).

Violet crystals from methylene chloride, dec 217°. Absorption max (cyclohexane): 470 nm ($E_{1cm}^{1\%}$ 2250). Sol in chloroform, oils. Oil solns are more red than those of β-carotene.

USE: Permissible color additive for food and drugs (exempt from certification): *Fed. Reg.* **34**, no. 5 (Jan. 8, 1969). Oral suntanning agent.

1799. Capobenic Acid. *6-[(3,4,5-Trimethoxybenzoyl)-amino]hexanoic acid; 6-(3,4,5-trimethoxybenzamido)hexanoic acid;* 3,4,5-trimethoxybenzoyl-ε-aminocaproic acid; ATBAC; TB-ACA; C-3. $C_{16}H_{23}NO_6$; mol wt 325.36. C 59.07%, H 7.13%, N 4.30%, O 29.50%. Prepn: A. Garzia, Ger. pat. **2,034,192**; *idem,* U.S. pat. **3,697,563** (1971, 1972 both to Ist. Chemioter. Ital.); Bottazzi *et al., Riv. Farmacol. Ter.* **11**, 215 (1971). Pharmacology: Razzaboni *et al., Boll. Soc. Ital. Biol. Sper.* **44**, 1783 (1968). Pharmacology, metabolism and clinical investigation of sodium salt: Greggia *et al., ibid.* **45**, 1447 (1969); Fontanini *et al., Minerva Med.* **60**, 4857 (1969); several authors, *Riv. Farmacol. Ter.* **1** (1970); **2** (1971); **3** (1972).

White, odorless powder, mp 121-123°. uv max (ethanol): 214, 259 nm. Sol in ethanol, acetone, chloroform, alkaline solns. Practically insol in water, ether, carbon tetrachloride. LD_{50} i.p. in rats: 2.5 g/kg (Garzia, 1971).

Sodium salt, $C_{16}H_{22}NNaO_6$, *sodium capobenate, Capben.*

THERAP CAT: Antiarrhythmic (cardiac depressant).

1800. Caprenin. Reduced calorie fat with functional properties similar to cocoa butter. Synthetic triglyceride composed primarily of caprylic and capric acids (derived from coconut and palm-kernel oils) and behenic acid (derived from rapeseed oil). Prepn: A. M. Erhman *et al.,* U.S. pats. **4,888,196** and **5,066,510** (1989, 1991, both to Procter & Gamble). Digestion and absorption in rats: D. R. Webb, R. A. Sanders, *J. Am. Coll. Toxicol.* **10**, 325 (1991). Toxicology and metabolism in animals: D. R. Webb *et al., ibid.* 341. Absorption and caloric value in humans: J. C. Peters *et al., ibid.* 357. Brief description: C. O'Donnell, D. Best, *Prepared Foods* **162**, 41-42 (February, 1993).

Caloric value in humans: 3-5 kcal/g.

USE: Reduced calorie fat substitute in soft candies and confectionary coatings.

1801. Capreomycin. Capastat. Cyclic peptide antibiotic similar to viomycin; produced by *Streptomyces capreolus* NRRL 2773; Herr *et al.,* 140th Am. Chem. Soc. Meet. (Chicago, Sept. 1961), *Abstracts of Papers,* p 49C; *Chem. Eng. News* **39**, 57 (Sept. 18, 1961); Brit. pat. **920,563**; U.S. pat. **3,143,468** (1963, 1964 both to Lilly). Mixture of capreomycins IA, IB, IIA, and IIB in the approx percentages, 25%, 67%, 3%, 6%, resp.: Herr, Redstone, *Ann. N.Y. Acad. Sci.* **135**, 940 (1966). Activity studies: Sutton *et al., ibid.*

Capreomycin IA　R = OH
Capreomycin IB　R = H

947; Lucchesi, *Antibiot. Chemother. (Basel)* **16**, 27 (1970). Proposed structure: Bycroft *et al., Nature* **231**, 301 (1971). Revised structure and total synthesis: T. Shiba *et al., Tetrahedron Letters* **1976**, 3907; S. Nomoto *et al., Tetrahedron* **34**, 921 (1978). Structure of IA and IB: *eidem, J. Antibiot.* **30**, 955 (1977). Toxicity study: Welles *et al., Ann. N.Y. Acad. Sci.* **135**, 960 (1966).

The mixture is a white solid. Sol in water. Practically insol in most organic solvents. pKa in 66% aq DMF: 6.2, 8.2, 10.1, 13.3. Stable in aq soln at pH 4-8; unstable in strongly acidic or strongly basic solns.

Disulfate, *Caprocin, Ogostal.* Solubility data: March, Weiss, *J. Assoc. Offic. Anal. Chem.* **50**, 457 (1967). LD_{50} in mice, rats (mg/kg): 250, 325 i.v.; 514, 1191 s.c. (Welles).

Capreomycin IA, $C_{25}H_{44}N_{14}O_8$, mp 246-248° (dec). $[\alpha]_D^{22}$ −21.9° (c = 0.5 in water). uv max (0.1N HCl): 269 nm (ε 24000); (H_2O): 268 nm (ε 23900); (0.1N NaOH): 287 nm (ε 15900) (all for hydrochloride ethanolate).

Capreomycin IB, $C_{25}H_{44}N_{14}O_7$, mp 253-255° (dec). $[\alpha]_D^{22}$ −44.6° (c = 0.5 in water). uv max (0.1N HCl): 268 nm (ε 22700); (H_2O): 268 nm (ε 22300); (0.1N NaOH): 290 nm (ε 14400) (all for hydrochloride ethanolate).

Note: Capreomycin IIA, capreomycin IIB. Structures corresp to capreomycins IA and IB but lack β-lysine residues.

THERAP CAT: Antibacterial (tuberculostatic).

1802. n-Capric Acid. *Decanoic acid.* $C_{10}H_{20}O_2$; mol wt 172.27. C 69.72%, H 11.70%, O 18.58%. $CH_3(CH_2)_8$-COOH. Prepn from octyl bromide: Shishido *et al., J. Am. Chem. Soc.* **81**, 5817 (1959); Closson, De Pree, U.S. pat. **2,918,494** (1959 to Ethyl Corp.). Recovery from *Cuphea llavea* Llave et Lex, *Lythaceae* seed oil: Miwa *et al.,* U.S. pat. **2,964,546** (1960 to U.S.D.A.). Toxicity study: L. Orö, A. Wretlind, *Acta Pharmacol. Toxicol.* **18**, 141 (1961). Review: *Fatty Acids* Part I, K. S. Markley, Ed. (Interscience, New York, 2nd ed., 1960) pp 34, 39.

Crystalline solid, mp 31.4°. Rancid odor. bp 270°. d_4^{50} 0.8782. n_D^{40} 1.4288. Practically insol in water (0.015 g/100 g at 20°). Sol in ethanol, ether, chloroform, benzene, carbon disulfide. Also sol in dil HNO_3 (d 1.14) from which it precipitates unchanged by addition of water. LD_{50} i.v. in mice: 129 ±5.4 mg/kg (Orö, Wretlind).

USE: Manuf of esters for artificial fruit flavors and perfumes; as an intermediate in other chemical syntheses.

1803. n-Caproic Acid. *Hexanoic acid.* $C_6H_{12}O_2$; mol wt 116.16. C 62.04%, H 10.41%, O 27.55%. $CH_3(CH_2)_4$-COOH. Occurs in milk fats (about 2%), in coconut oil (< 1%), various palm and other oils. Prepn: Vliet *et al., Org. Syn. coll. vol.* II, 417 (1943); Reid, Ruhoff, *ibid.,* 475. Manuf by catalytic reduction of corresponding β-lactone: Caldwell, U.S. pat. **2,484,486** (1949 to Kodak); from oleic acid: Follett, Murray, U.S. pat. **2,580,417** (1952 to Arthur D. Little); from castor oil or a ricinoleate: Steadman, Peterson, U.S. pat. **2,847,432** (1958 to National Res. Corp.); by ozonolysis of tall oil unsaturated fatty acids: Maggiolo, U.S. pat. **2,865,937** (1958 to Welsbach); from 1,3-butadiene and potassium acetate in presence of $NaNH_2$: Schmerling, Toekelt, U.S. pat. **3,075,010** (1963 to Universal Oil Prod.); from cyclohexanol: Bartlett, Lippincott, U.S. pat. **3,121,-728** (1964 to Esso); by catalytic oxidation of n-hexanol: Hay, U.S. pat. **3,173,933** (1965 to General Electric). Toxicity study: H. F. Smyth, C. P. Carpenter, *J. Ind. Hyg. Toxicol.* **26**, 269 (1944). Review: *Fatty Acids,* Part 1, K. S. Markley, Ed. (Interscience, New York, 2nd ed., 1960) pp 34, 37.

Oily liquid, bp 205°. Characteristic goat-like odor. mp −3.4°. d_4^{20} 0.9265. n_D^{20} 1.4163. Slightly soluble in water (1.082 g/100 g); readily soluble in ethanol, ether. LD_{50} orally in rats: 3.0 g/kg (Smyth, Carpenter).

USE: Manuf of esters for artificial flavors, and of hexyl derivatives, especially hexylphenols, hexylresorcinol, etc.

1804. Caproic Aldehyde. *Hexanal;* caproaldehyde; hexaldehyde. $C_6H_{12}O$; mol wt 100.16. C 71.95%, H 12.08%, O 15.97%. $CH_3(CH_2)_4CHO$. Prepn: Bagard, *Bull. Soc. Chim.* **1**, 307 (1907). Toxicity study: Smyth *et al., Arch. Ind. Hyg. Occup. Med.* **10**, 61 (1954).

Liquid. d_4^{20} 0.8335. bp_{760} 131°; bp_{12} 28°. Autooxidizes and polymerizes, especially in the presence of traces of acid. LD_{50} orally in rats: 4.89 g/kg (Smyth).

1805. Caprolactam. *Hexahydro-2H-azepin-2-one;* ε-caprolactam; 2-oxohexamethylenimine; 2-ketohexamethylenimine; aminocaproic lactam. $C_6H_{11}NO$; mol wt 113.16. C 63.69%, H 9.80%, N 12.38%, O 14.14%. Prepn: Wallach, *Ann.* **312,** 187 (1900); **343,** 43 (1905); Ruzicka *et al., Helv. Chim. Acta* **4,** 477 (1921); Eck, Marvel, *J. Biol. Chem.* **106,** 387 (1934); Marvel, Eck, *Org. Syn.* coll. vol. **II,** 371 (1943); Lazier, Rigby, U.S. pat. **2,234,566** (1941 to du Pont); Schlack, U.S. pat. **2,249,177** (1941 to I. G. Farben); **Ger. pats. 739,953** (1943); **745,224** (1943); P. Smith, *J. Am. Chem. Soc.* **70,** 320 (1948); E. Schmitz *et al., J. Prakt. Chem.* **319,** 274 (1977). Purification: Kampschmidt, U.S. pat. **2,786,052** (1957 to Stamicarbon N. V.). Stabilization with alkalies: Indest *et al.,* U.S. pat. **2,884,414** (1959 to Vereinigte Glanzstoff-Fabriken). Toxicity study: H. F. Smyth *et al., Am. Ind. Hyg. Assoc. J.* **30,** 470 (1969). *Reviews: CIOS Repts.* no. 22 and 31, File XXXIII/Synthetic Fiber Developments in Germany, parts I & II; K. Kahr *et al.* in *Ullmann's Encyklopädie der Technischen Chemie* vol. **9,** E. Bartholome *et al.,* Eds. (Verlag Chemie, Weinheim, 4th ed., 1975) pp 96-114.

Hygroscopic leaflets from petr ether, mp 70°. d_4^{75} (liq) 1.02. bp_{50} 180°; bp_3 100°. Viscosity at 78° = 9 centipoises. Flash pt, open cup: 257°F (125°C). Freely sol in water, methanol, ethanol, ether, tetrahydrofurfuryl alcohol, dimethylformamide. Also sol in chlorinated hydrocarbons, cyclohexene, petroleum fractions. A 70% aq soln has d_4^{25} 1.05; n_D^{31} 1.4965; n_D^{40} 1.4935. LD_{50} orally in rats: 2.14 g/kg (Smyth).

USE: Manuf of synthetic fibers of the polyamide type (Perlon); solvent for high mol wt polymers.

1806. Caproyl Chloride. *Hexanoyl chloride.* $C_6H_{11}ClO$; mol wt 134.61. C 53.54%, H 8.24%, Cl 26.34%, O 11.89%. $CH_3(CH_2)_4COCl$. Prepn: Brown, *J. Am. Chem. Soc.* **60,** 1325 (1938). Manuf: Wygant, U.S. pat. **2,806,061** (1957 to Monsanto).

Liquid, bp 151-153°. mp −87.3°. d_4^{15} 0.9805. n_D^{15} 1.4286. Dec by water or alcohol. Sol in ether, chloroform.

1807. Caprylene. *1-Octene;* octylene. C_8H_{16}; mol wt 112.22. C 85.63%, H 14.37%. $CH_3(CH_2)_5CH=CH_2$. Prepn from appropriate alkylmagnesium bromide and allyl bromide or chloride: Geisler, Pilz, *Ber.* **95,** 96 (1962); from formaldehyde or paraformaldehyde and triphenyl(phenylmethylene)phosphorane: Hauser *et al., J. Org. Chem.* **28,** 372 (1963); by catalytic dehydration of 2-octanol: Lundeen, Hoozer, *J. Am. Chem. Soc.* **85,** 2180 (1963).

Liquid, bp 121°, bp_{100} 61.5-61.7°. mp −102°. d_4^{20} 0.7149, d_4^{25} 0.7109. n_D^{20} 1.4087, n_D^{25} 1.4062. Flash pt, open cup: 70°F (21°C). Practically insol in water; misc with alcohol, ether.

1808. Caprylic Acid. *Octanoic acid.* $C_8H_{16}O_2$; mol wt 144.21. C 66.63%, H 11.18%, O 22.19%. $CH_3(CH_2)_6$-COOH. Prepn from 1-heptene: Dupont *et al., Compt. Rend.* **240,** 628 (1955); by oxidation of octanol: Langenbeck, Richter, *Ber.* **89,** 202 (1956). Manuf: Alexander, U.S. pat. **2,821,534** (1958 to GAF); McAlister *et al.,* U.S. pat. **3,053,869** (1962 to Standard Oil Co., Indiana). Antifungal properties: O. Wyss *et al., Arch. Biochem.* **7,** 418 (1945). Toxicity: P. M. Jenner *et al., Food Cosmet. Toxicol.* **2,** 327 (1964). *Review: Fatty Acids,* Part 1, K. S. Markley, Ed. (Interscience, New York, 2nd ed., 1960) pp 34, 38.

Oily liquid, bp 239.7°. Slightly unpleasant rancid taste. mp 16.7°. d_4^{20} 0.910. n_D^{20} 1.4280. Very slightly sol in water (0.068 g/100 g at 20°); freely sol in alcohol, chloroform, ether, carbon disulfide, petr ether, glacial acetic acid. LD_{50} orally in rats: 10,080 mg/kg (Jenner).

USE: An intermediate in manuf of esters used in perfumery; in manuf of dyes, etc.

1809. Caprylic Aldehyde. *Octanal;* caprylaldehyde; octaldehyde. $C_8H_{16}O$; mol wt 128.21. C 74.94%, H 12.58%, O 12.48%. $CH_3(CH_2)_6CHO$. Prepn: Stephen, *J. Chem. Soc.* **127,** 1874 (1925).

Liquid. d_4^{20} 0.821. bp_{760} 163.4°; bp_{20} 72°; bp_9 60°. n_D^{26} 1.41667. Slightly sol in water; misc with alc, ether.

1810. CAPS. *3-(Cyclohexylamino)-1-propanesulfonic acid.* $C_9H_{19}NO_3S$; mol wt 221.32. C 48.84%, H 8.65%, N 6.33%, O 21.69%, S 14.49%. One of the zwitterionic amino acids known as "Good" buffers; active in the pH range of 9.7-11.1. Prepn: H. Dorn, K. Walter, *Z. Chem.* **7,** 151 (1967). Thermodynamic parameters: C. D. McGlothlin, J. Jordan, *Anal. Letters* **9,** 245 (1976). Effects on protein measurement: V. Kaushal, L. D. Barnes, *Anal. Biochem.* **157,** 291 (1986); H. M. Himmel, W. Heller, *J. Clin. Chem. Clin. Biochem.* **25,** 909 (1987). Use as buffer: I. I. Koukli *et al., Analyst* **113,** 603 (1988); R. M. Smith *et al., J. Chromatog.* **514,** 97 (1990).

Colorless needles from methanol, mp 302°-303°. pK (25°): 10.35.

USE: Biological buffer.

1811. Capsaicin. *(E)-N-[(4-Hydroxy-3-methoxyphenyl)-methyl]-8-methyl-6-nonenamide; trans-8-methyl-N-vanillyl-6-nonenamide;* N-(4-hydroxy-3-methoxybenzyl)-8-methylnon-*trans*-6-enamide; Axsain; Mioton; Zostrix. $C_{18}H_{27}NO_3$; mol wt 305.42. C 70.79%, H 4.59%, O 15.72%. Pungent principle in fruit of various species of *Capsicum, Solanaceae.* Isoln from paprika and cayenne: Thresh, *Pharm. J. and Trans.* **7,** 21 (1876); Micko, *Z. Nahr. Genussm.* **1,** 818 (1898). *See Beilstein* **13,** suppl. I, 322. Early structure study: Nelson, *J. Am. Chem. Soc.* **42,** 597 (1920). Synthesis: Späth, Darling, *Ber.* **63,** 737 (1930); L. Crombie *et al., J. Chem. Soc.* **1955,** 1025; O. P. Vig *et al., Indian J. Chem.* **17B,** 558 (1979). Constitution and biosynthesis: D. J. Bennet, E. W. Kirby, *J. Chem. Soc. C* **1968,** 442. Pharmacology: Molnar *et al., Acta Physiol.* **35,** 369 (1969). Capsaicin is a powerful irritant; initial administration causes intense pain. Prolonged treatment causes insensitivity to painful stimuli and induces selective degeneration of certain primary sensory neurons: G. Jancso *et al., Nature* **270,** 741 (1977); R. Gamse, *Arch. Pharmacol.* **320,** 205 (1982); P. Holzer *et al., Neurosci. Letters* **31,** 253 (1982). Neuronal depletion of substance P, *q.v.:* T. M. Jessell *et al., Brain Res.* **152,** 183 (1978); T. L. Yaksh *et al., Science* **206,** 481 (1979). Capsaicin pretreatment also induces long-lasting desensitization of airway mucosa to various mechanical and chemical irritants: J. M. Lundberg, A. Saria, *Nature* **302,** 251 (1983). Preliminary clinical evaluation in chronic postherpetic neuralgia: J. E. Bernstein *et al., J. Am. Acad. Dermatol.* **17,** 93 (1987). *Reviews:* Molnar, *Arzneimittel-Forsch.* **15,** 718 (1965); Walker, Gavern, *Mfg. Chem. Aerosol News* **39** (6), 35 (1968); R. M. Virus, G. F. Gebhart, *Life Sci.* **25,** 1273 (1979); Y. Monsereenusorn *et al., CRC Crit. Rev. Toxicol.* **10,** 321-339 (1982).

Monoclinic, rectangular plates, scales from petr ether, mp 65°. $bp_{0.01}$ 210-220° (air-bath temp). uv max: 227, 281 nm (ε 7000, 2500). Burning taste, one part in 100,000 can be detected by tasting. Practically insol in cold water. Freely sol in alc, ether, benzene, chloroform; slightly sol in CS_2.

USE: As a tool in neurobiological research.

THERAP CAT: Topical analgesic.

1812. Capsanthin. *(3R,3'S,5'R)-3,3'-Dihydroxy-β,κ-caroten-6'-one.* $C_{40}H_{56}O_3$; mol wt 584.88. C 82.14%, H 9.65%, O 8.21%. Carotenoid pigment isolated from paprika *(Capsicum annuum L., Solanaceae):* L. Zechmeister, L. Cholnoky, *Ann.* **454,** 54 (1927); L. Cholnoky *et al., Ann.* **606,** 194 (1957); Warren, Weedon, *J. Chem. Soc.* **1958,** 3972. Structure: Entschel, Karrer, *Helv. Chim. Acta* **43,** 89 (1960); Faigle, Karrer, *ibid.* **44,** 1257, 1904 (1961). Absolute configuration: Bartlett *et al., J. Chem. Soc. (C)* **1969,** 2527. Crystal structure: I. Ueda, W. Nowacki, *Verh. Schweiz. Naturforsch. Ges.* **1971,** 152, *C.A.* **77,** 106449h (1972).

Deep carmine-red needles from petr ether, mp 181-182°. Absorption max 483 nm (ε × 10⁻³ 121). [α]$_{Cd}$ +36° (chloroform). Freely sol in acetone, chloroform. Sol in methanol, ethanol, ether, benzene. Slightly sol in petr ether, CS_2. Diacetate, $C_{44}H_{60}O_5$, red plates from methanol, mp 150°. Dipalmitate, $C_{72}H_{116}O_5$, bordeaux-red plates from benzene + methanol, mp 95°.

1813. Capsicum. Cayenne Pepper. The dried ripe fruit of *Capsicum frutescens* L., *Solanaceae,* known commercially as *African Chillies,* or of *Capsicum annuum* L., var. *conoides* Irish, known as *Tabasco Pepper* or of *Capsicum annuum* var. *longum* Sendt, known as *Louisiana Long Pepper,* or of a hybrid between the Honka variety of Japanese Capsicum and the Old Louisiana Sport Capsicum known as *Louisiana Sport Pepper* (Fam. *Solanaceae). Constit.* Fixed oils; 0.1-1% capsaicin, capsanthin, *q.q.v.* Store in a cool place. Protect from light.

THERAP CAT (VET): Has been used externally as a counter-irritant; internally as a carminative and stomachic.

1814. Captafol. *3a,4,7,7a-Tetrahydro-2-[(1,1,2,2-tetra-chloroethyl)thio]-1H-isoindole-1,3(2H)-dione; N-(1,1,2,2-tet-rachloroethylthio)-4-cyclohexene-1,2-dicarboximide; N-(1,1,-2,2-tetrachloroethylmercapto)-4-cyclohexene-1,2-dicarbox-imide; N-(1,1,2,2-tetrachloroethylthio)-Δ⁴-tetrahydrophthal-imide; N-(1,1,2,2-tetrachloroethylsulfenyl)-cis-4-cyclohex-ene-1,2-dicarboximide; Difolatan.* $C_{10}H_9Cl_4NO_2S$; mol wt 349.06. C 34.41%, H 2.60%, Cl 40.63%, N 4.01%, O 9.17%, S 9.19%. *N*-tetrachloroethyl analog of captan, *q.v.* Prepn: Kohn, **Belg.** pat. **633,205** (1963 to California Res. Corp.), *C.A.* **60,** 15789cd (1964). Toxicology: R. Ben-Dyke *et al., World Rev. of Pest Control* **9,** 119 (1970); G. L. Kennedy, Jr. *et al., Food Cosmet. Toxicol.* **13,** 55 (1975). Carcinogenicity study: S. Tamano *et al., Japan. J. Cancer Res.* **81,** 1222 (1990).

Crystals, mp 160-161°. LD$_{50}$ in rats, rabbits (mg/kg): 2500-6200 orally; 15400 dermally (Ben-Dyke).

Note: The EPA has determined that captafol is oncogenic in mice and rats: *Fed. Reg.* **50,** 1103 (1985).

USE: Agricultural fungicide, especially for potatoes.

1815. Captan. *3a,4,7,7a-Tetrahydro-2-[(trichlorometh-yl)thio]-1H-isoindole-1,3(2H)-dione; N-(trichloromethylthio)-4-cyclohexene-1,2-dicarboximide; N-trichloromethylthio-3a,4,7,7a-tetrahydrophthalimide; N-trichloromethylmercapto-4-cyclohexene-1,2-dicarboximide; N-(trichloromethylmer-capto)-Δ⁴-tetrahydrophthalimide; ENT-26538; SR-406; Merpan; Orthocide-406; Vancide 89.* $C_9H_8Cl_3NO_2S$; mol wt

300.59. C 35.96%, H 2.68%, Cl 35.38%, N 4.66%, O 10.65%, S 10.67%. Prepn: A. R. Kittleson, U.S. pats. **2,553,771; 2,653,155; 2,713,058** (1951, 1953, 1955 all to Standard Oil); *idem, Science* **115,** 84 (1952); *idem, J. Agr. Food Chem.* **1,** 677 (1953). Review of mutagenicity studies: B. A. Bridges, *Mutat. Res.* **32,** 3 (1975); of carcinogenic risk: *IARC Monographs* **30,** 295-318 (1983).

Odorless crystals from CCl_4, mp 178°. d 1.74. Practically insol in water. Soly at 26° in g/100 ml: chloroform 7.78; tetrachloroethane 8.15; cyclohexanone 4.96; dioxane 4.70; benzene 2.13; toluene 0.69; heptane 0.04; ethanol 0.29; ether 0.25. Colorimetric determination: A. R. Kittleson, *Anal. Chem.* **24,** 1173 (1952). LD$_{50}$ orally in rats: 9000 mg/kg (Bridges).

USE: Fungicide; bacteriostat in soap.

1816. Captodiamine. *2-[[[4-(Butylthio)phenyl]phenyl-methyl]thio]-N,N-dimethylethanamine; 2-[p-(butylthio)-α-phenylbenzylthio]-N,N-dimethylethylamine; p-butylmercap-tobenzhydryl β-dimethylaminoethyl sulfide; p-butylthiodi-phenylmethyl 2-dimethylaminoethyl sulfide; captodiam; captodramin.* $C_{21}H_{29}NS_2$; mol wt 359.60. C 70.14%, H 8.13%, N 3.90%, S 17.83%. Prepn: Hübner, Petersen, **U.S.** pat. **2,830,088** (1958). Pharmacological studies: R. Kopf, I. Moller-Nielsen, *Arzneimittel-Forsch.* **8,** 154 (1958).

Hydrochloride, $C_{21}H_{29}NS_2$·HCl, *Covatine, Covatix, Suv-ren.* Crystals, mp 131-132°. LD$_{50}$ (96 hr) in mice, rats (mg/kg): 180, 343 i.p.; 1630, 3800 orally (Kopf, Moller-Nielsen).

THERAP CAT: Anxiolytic.

1817. Captopril. *(S)-1-(3-Mercapto-2-methyl-1-oxoprop-yl)-L-proline; (2S)-1-(3-mercapto-2-methylpropionyl)-L-pro-line; D-2-methyl-3-mercaptopropanoyl-L-proline; SQ-14225; Acediur; Acepril; Aceplus; Alopresin; Acepress; Capoten; Captolane; Captoril; Cesplon; Dilabar; Garranil; Hipertil; Lopirin; Lopril; Tensobon; Tensoprel.* $C_9H_{15}NO_3S$; mol wt 217.29. C 49.75%, H 6.96%, N 6.45%, O 22.09%, S 14.76%. First orally active angiotensin-converting enzyme (ACE) inhibitor. Prepn: M. A. Ondetti, D. W. Cushman, **Ger.** pat. **2,703,828;** *eidem,* U.S. pats. **4,046,889** and **4,105,776** (1977, 1977, 1978 all to Squibb). Design and synthesis: M. A. Ondetti *et al., Science* **196,** 441 (1977); D. W. Cushman *et al., Biochemistry* **16,** 5484 (1977). Improved synthesis: D. H. Nam *et al., J. Pharm. Sci.* **73,** 1843 (1984). Pharma-cology: B. Rubin *et al., Eur. J. Pharmacol.* **51,** 377 (1978); *eidem, Prog. Cardiovasc. Dis.* **21,** 183 (1978). Clinical stud-ies: D. B. Case *et al., ibid.* 195; H. R. Brunner *et al., Ann. Int. Med.* **90,** 19 (1979). Toxicology and metabolism: G. R. Keim in *Captopril and Hypertension,* D. B. Case, Ed. (Plen-um, New York, 1980) p 137. GC/MS determn in biological fluids: T. Ito, Y. Matsuki, *J. Chromatog.* **417,** 79 (1987). Historical review and comprehensive bibliography: Z. P. Horovitz in *Pharmacological and Biochemical Properties of Drug Substances* vol. 3, M. E. Goldberg, Ed. (Am. Pharm. Assoc., Washington, DC, 1981) pp 148-175. Comprehen-sive description: H. Kadin, *Anal. Profiles Drug Subs.* **11,** 79-137 (1982). *Reviews:* Am. Heart J. **104,** pt. 2, 1125-1228 (1982); *Brit. J. Clin. Pharmacol.* **14,** Suppl. 2, 69S-252S (1982). Series of articles on pharmacology and therapeutic efficacy: *Postgrad. Med. J.* **62,** Suppl. 1, 1-191 (1986).

Crystals from ethyl acetate/hexane, mp 103-104° (Ondetti, Cushman). Generally regarded as polymorphic: stable form, mp 106°; unstable form, mp 86° (Florey); also reported as mp 87-88°, resolidifies, second mp 104-105° (Cushman). $[\alpha]_D^{22}$ −131.0° (c = 1.7 in ethanol). pK_1 3.7, pK_2 9.8. Freely sol in water, alc, chloroform, methylene chloride. LD_{50} in mice (mg/kg): 1040 i.v.; 6000 orally (Keim).

Combination with hydrochlorothiazide, *Acezide, Capozide, Captea, Ecazide.*

THERAP CAT: Antihypertensive.

1818. Capuride. *N-(Aminocarbonyl)-2-ethyl-3-methylpentanamide; (2-ethyl-3-methylvaleryl)urea;* (ethyl-*sec*-butylacetyl)urea; (1-ethyl-2-methylpentanoyl)urea; Pacinox. $C_9H_{18}N_2O_2$; mol wt 186.25. C 58.04%, H 9.74%, N 15.04%, O 17.18%. Prepn: Volwiler, Tabern, *J. Am. Chem. Soc.* **58**, 1352 (1936); Adams *et al.,* U.S. pat. **3,282,998** (1966 to Millmaster Onyx).

Crystals from xylene, mp 172°.
THERAP CAT: Hypnotic.

1819. Caramel. Burnt sugar coloring; burnt sugar. Made by heating sugar or glucose, adding small quantities of alkali, alkaline carbonate or a trace of mineral acid during the heating.

Dark-brown, thick liq; pleasant, bitter taste; odor of burnt sugar. d about 1.35. Sol in water, dil alc. Insol in benzene, chloroform, ether, acetone, petr ether, oil turpentine.

USE: Coloring foods, confectionery, galenicals.

1820. Caramiphen. *1-Phenylcyclopentanecarboxylic acid 2-(diethylamino)ethyl ester;* diethylaminoethyl 1-phenylcyclopentane-1-carboxylate. $C_{18}H_{27}NO_2$; mol wt 289.42. C 74.70%, H 9.40%, N 4.84%, O 11.06%. Prepn: **Swiss** pat. **234,452**; H. Martin, F. Häfliger, U.S. pat. **2,404,588** (1945, 1946 both to Geigy). Prepn of the ethanedisulfonate: **Swiss** pat. **272,708** (1951 to Geigy), *C.A.* **46**, 4563i (1952); *Chem. Zentr.* **1952**, 1571. Pharmacology and toxicology of the hydrochloride: C. P. Kraatz *et al., J. Pharmacol. Exp. Ther.* **96**, 42 (1949); of the ethanedisulfonate: J. J. Toner, E. Macko, *ibid.* **106**, 246 (1952). GLC determn in blood: P. Levandoski, T. Flanagan, *J. Pharm. Sci.* **69**, 1353 (1980). Mechanism of antitussive action: E. F. Domino *et al., J. Pharmacol. Exp. Ther.* **233**, 249 (1985).

$bp_{0.07}$ 112-115°.
Ethanedisulfonate, $(C_{18}H_{27}NO_2)_2 \cdot C_2H_6O_6S_2$, *Alcopon, Taoryl, Toryn.* Crystals from acetone, mp 115-116°. More sol in water than the hydrochloride. Sol in alc, pharmaceutical syrups. Mixture with phenylpropanolamine hydrochloride, *Tuss-Ornade.*

Hydrochloride, $C_{18}H_{27}NO_2 \cdot HCl$, *Panparnit, Parpanit.* Crystals, mp 145-146°. Sol in alc; slightly sol in water. LD_{50} i.p. in rats: 209 mg/kg (Kraatz).

THERAP CAT: Anticholinergic. The ethanedisulfonate as antitussive.

1821. Caraway. Dried ripe fruit of *Carum carvi* L., *Umbelliferae. Habit.* Europe, Central and Western Asia; cultivated in England, Russia, U.S. *Constit.* 3.5-7% volatile and fatty oils; resin, sugar, tannin, mucilage.

USE: As a spice in baking. Pharmaceutic aid (flavor).

1822. Carazolol. *1-(9H-Carbazol-4-yloxy)-3-[(1-methylethyl)amino]-2-propanol;* BM-51052; Conducton; Suacron. $C_{18}H_{22}N_2O_2$; mol wt 298.38. C 72.46%, H 7.43%, N 9.39%, O 10.72%. β-Adrenergic blocker. Prepn: H. Leinert *et al.,* Ger. pat. **2,240,599** (1974 to Boehringer, Mann.), *C.A.* **80**, 133455a (1974). Comparative study of cardiac action: W. Bartsch *et al., Arzneimittel-Forsch.* **27**, 1022 (1977). Initial clinico-pharmacological study: E. Chorianopoulos *et al., Herz/Kreislauf* **9**, 965 (1977), *C.A.* **88**, 164460t (1978). Receptor binding studies: G. Kaiser *et al., Arch. Pharmacol.* **305**, 41 (1978); R. B. Innis *et al., Life Sci.* **24**, 2255 (1979). Use in treatment of stress syndrome in pigs: G. Ballarini, F. Guizzardi, *Tierärztl. Umschau* **36**, 171 (1981).

Hydrochloride, $C_{18}H_{23}ClN_2O_2$, crystals, mp 234-235°.
THERAP CAT: Antihypertensive, antianginal, antiarrhythmic.
THERAP CAT (VET): Treatment of stress in pigs.

1823. Carbachol. *2-[(Aminocarbonyl)oxy]-N,N,N-trimethylethanaminium chloride; (2-hydroxyethyl)trimethyl ammonium chloride carbamate;* choline chloride carbamate; carbamylcholine chloride; carbocholine; Carcholin; Miostat; Moryl; Doryl; Coletyl; Lentin. $C_6H_{15}ClN_2O_2$; mol wt 182.65. C 39.46%, H 8.28%, Cl 19.41%, N 15.34%, O 17.52%. $[NH_2COOCH_2CH_2N(CH_3)_3]^+Cl^-$. Prepd by the interaction of β-chloroethylcarbamate and trimethylamine: Kreitmair, *Arch. Exp. Path. Pharmakol.* **164**, 346 (1932); Merck, Ger. pats. **539,329; 553,148; 590,311.** Toxicity: Molitor, *J. Pharmacol. Exp. Ther.* **58**, 337 (1936).

Hard prismatic crystals. Hygroscopic. Odorless, but on standing in an open container develops a faint odor resembling that of an aliphatic amine. mp 200-203° (some decompn); mp 210-212°; mp 204-205° (Kreitmair). One gram dissolves in 1 ml water, 50 ml alcohol, 10 ml methanol. Almost insol in chloroform, ether. The aq soln is stable even when heated and is neutral to litmus. LD_{50} in mice (mg/kg): 15 orally, 0.3 i.v. (Molitor).

Note: The name "Doryl" is also used to designate a high temp insulating resin which has a diphenyl oxide structure linked with methylene groups: Micarta Div., Westinghouse Electric Corp. Ref: *Chem. Week,* Sept. 29, 1962, p 92.

THERAP CAT: Cholinergic, miotic.
THERAP CAT (VET): Parasympathomimetic, used chiefly in large animals, esp for simple colic in the horse.

1824. Carbacrylic Resins. A generic name for cross-linked polyacrylic polycarboxylic ion exchange resins.

Carbacylamine resins, Carbo-Resin. Combination of the following ion exchange resins: carbacrylic resin, potassium carbacrylic resin, and polyaminemethylene resin.

THERAP CAT: Cation-exchange resin.

1825. Carbadox. *(2-Quinoxalinylmethylene)hydrazinecarboxylic acid methyl ester N,N'-dioxide; 3-(2-quinoxalinylmethylene)carbazic acid methyl ester N,N'-dioxide;* methyl 3-(2-quinoxalinylmethylene)carbazate N^1,N^4-dioxide; 2-formylquinoxaline-1,4-dioxide carbomethoxyhydrazone; GS-6244; Fortigro; Mecadox. $C_{11}H_{10}N_4O_4$; mol wt 262.22. C 50.38%, H 3.84%, N 21.37%, O 24.41%. Prepn: Johnston, Belg. pat. **669,353**; idem, U.S. pats. **3,371,090; 3,433,871** (1964, 1968, 1969, all to Pfizer). Animal studies: Thrasher *et al., J. Anim. Sci.* **26**, 911 (1967); Kornegay *et al., ibid.* **27**, 1134 (1968).

Minute yellow crystals, mp 239.5-240°. uv max (water): 236, 251, 303, 366, 373 nm (ε 11000, 10900, 36400, 16100, 16200). Practically insol in water.

THERAP CAT (VET): Antimicrobial.

1826. Carbamazepine. *5H-Dibenz[b,f]azepine-5-car-boxamide;* 5-carbamoyl-5*H*-dibenz[*b,f*]azepine; G-32883; Biston; Calepsin; Carbelan; Epitol; Finlepsin; Sirtal; Staze-pine; Tegretal; Tegretol; Telesmin; Timonil. C₁₅H₁₂N₂O; mol wt 236.27. C 76.25%, H 5.12%, N 11.86% O 6.77%. Prepn: Schindler, U.S. pat. 2,948,718 (1960 to Geigy). Metabolism: P. L. Morselli, A. Frigerio, *Drug Metab. Rev.* **4,** 97 (1975). Review of pharmacokinetics in man: L. Bertilsson, *Clin. Pharmacokinet.* **3,** 128-143 (1978); S. Pynnönen, *Ther. Drug Monit.* **1,** 409-431 (1979). Toxicity: E. G. Stenger, F. C. Roulet, *Med. Exp.* **11,** 191 (1964). Comprehensive description: H. Y. Aboul-Enein, A. A. Al-Badr in *Analytical Profiles of Drug Substances* vol. 9, K. Florey, Ed. (Academic Press, New York, 1980) pp 87-106.

Crystals from abs ethanol + benzene, mp 190-193°. Sol in alcohol, acetone, propylene glycol. Practically insol in water. LD₅₀ orally in mice, rats: 3750, 4025 mg/kg (Stenger, Roulet).

THERAP CAT: Analgesic; anticonvulsant.

1827. Carbamyl Chloride. *Carbamic chloride;* chloroformamide. CH₂ClNO; mol wt 79.49. C 15.11%, H 2.54%, Cl 44.60%, N 17.62%, O 20.13%. H₂NCOCl. Prepd by passing HCl gas over heated cyanuric acid: Gattermann, Rossolymo, *Ber.* **23,** 1190 (1890); Gattermann, *Ber.* **32,** 1117 (1899). From ammonia and phosgene at 400°: Rupe, Labhard, *Ber.* **33,** 236 (1900), *cf.* Gattermann, *Ann.* **244,** 30 (1888).

Liquid. Acrid, offensive odor. Has been obtained cryst, mp about 50°, bp 61-62° (decompn). Reacts violently on contact with water, forming ammonium chloride and carbon dioxide. During storage it gives off HCl and slowly changes to cyanuric acid.

1828. Carbanilic Acid. *Phenylcarbamic acid; N-carb-oxyaniline.* C₇H₇NO₂; mol wt 137.14. C₆H₅NHCOOH. Known only by its derivatives (esters, etc.).

1829. Carbanilide. *N,N′-Diphenylurea;* diphenylcarbamide; 1,3-diphenylurea; *sym*-diphenylurea. C₁₃H₁₂N₂O; mol wt 212.25. C 73.57%, H 5.70%, N 13.20%, O 7.54%. C₆H₅NHCONHC₆H₅. Obtained during the preparation of phenylurea from aniline hydrochloride and urea: Davis, Blanchard, *Org. Syn.* **coll. vol. I,** 453 (2nd ed., 1941). Crystal structure: W. Dannecker *et al., Cryst. Struct. Commun.* **8,** 429 (1979).

Orthorhombic prisms from alc. d 1.239. mp 238°. bp 260° (decompn). Sublimes in current of hydrogen at 220°. Sol in ether, glacial acetic acid. Sparingly sol in water (0.15 g/l), acetone, alcohol, chloroform. Moderately sol in pyridine (69.0 g/l).

1830. Carbarsone. *[4-[(Aminocarbonyl)amino]phenyl]-arsonic acid; N-carbamoylarsanilic acid;* p-ureidobenzene-arsonic acid; *N*-carbamylarsanilic acid; p-carbamidobenzene-arsonic acid; 4-ureido-1-phenylarsonic acid; 4-carbamyl-aminophenylarsonic acid; p-arsonophenylurea; Amabevan;

Ameban; Amibiarson; Arsambide; Carb-O-Sep; Histocarb; Fenarsone; Leucarsone; Aminarsone; Amebarsone. C₇H₉-AsN₂O₄; mol wt 260.08. C 32.33%, H 3.49%, As 28.81%, N 10.77%, O 24.61%. Prepd from the sodium salt of arsanilic acid by treatment with potassium cyanate or cyanogen bromide: **Ger.** pat. 213,155; Stickings, *J. Chem. Soc.* **1928,** 3131; from arsanilic acid by treatment with phosgene: Nakatsu, Kawase, *Ann. Rept. Takamine Lab.* **8,** 44-47 (1956); **Japan.** pat. 4418('58) (to Sankyo).

White powder. mp 174°. Slightly sol in water, alcohol; sol in solns of alkali hydroxides and carbonates. The satd aq soln is acid to litmus. Nearly insol in ether or chloroform. LD₅₀ orally in rats: 510 mg/kg.

THERAP CAT: Antiamebic.

THERAP CAT (VET): Antihistomonad in turkeys.

1831. Carbaryl. *1-Naphthalenol methylcarbamate; methyl carbamic acid 1-naphthyl ester;* 1-naphthyl *N*-methyl-carbamate; ENT-23969; OMS-29; UC-7744; Arylam; Caryl-derm; Clinicide; Derbac; Dicarbam; Ravyon; Seffein; Sevin. C₁₂H₁₁NO₂; mol wt 201.22. C 71.63%, H 5.51%, N 6.96%, O 15.90%. Prepn and description: Haynes *et al., Contrib. Boyce Thompson Inst.* **18,** 507 (1957); Lambrech, U.S. pat. 2,903,478 (1959 to Union Carbide). Metabolism: W. E. Whitehurst *et al., J. Agr. Food Chem.* **11,** 167 (1963); J. B. Houston *et al., Xenobiotica* **5,** 637 (1975). Degradation: D. G. Crosby *et al., J. Agr. Food Chem.* **13,** 204 (1965); D. L. Heywood, *Environ. Qual. Saf.* **4,** 128 (1975). Toxicology: I. Nisbet, D. Miner, *Environment* **13,** 10 (1971). Toxicity: M. Vandekar *et al., Bull. World Health Org.* **44,** 241 (1971). Clinical trial in pediculosis: J. W. Maunder, *Clin. Exp. Dermatol.* **6,** 605 (1981). Review: *Carbamate Insecticides: Chemistry, Biochemistry and Toxicology,* R. J. Kuhr, H. W. Dorough, Eds. (CRC Press, Cleveland, 1976) 301 pp.

Crystals, mp 142°. d₂₀²⁰ 1.232. Moderately sol in DMF, acetone, isophorone, cyclohexanone. Soly in water at 30°: 120 mg/l. Vapor pressure at 25°: <4 × 10⁻⁵ mm Hg. Stable to heat, light, acids; hydrolyzed in alkalies; noncorrosive. LD₅₀ orally in rats: 250 mg/kg (Vandekar).

Caution: Nausea, vomiting, diarrhea, bronchoconstriction, blurring vision, excessive salivation, muscle twitching, cyanosis, convulsions, coma, respiratory failure, *Clinical Toxicology of Commercial Products,* R. E. Gosselin *et al.,* Eds. (Williams & Wilkins, Baltimore, 4th ed., 1976) Section III, pp 79-82.

USE: Contact insecticide.

THERAP CAT: Ectoparasiticide.

THERAP CAT (VET): Ectoparasiticide.

1832. Carbazochrome Salicylate. *2-Hydroxybenzoic acid monosodium salt compd with 2-(1,2,3,6-tetrahydro-3-hydroxy-1-methyl-6-oxo-5H-indol-5-ylidene)hydrazinecarbox-amide (1:1); 3-hydroxy-1-methyl-1,5,6-indolinedione semicar-bazone compd with sodium salicylate;* adrenochrome semi-carbazone compd with sodium salicylate; adrenochrome monosemicarbazone sodium salicylate complex; Adenogen; Adrenosem; Adrestat-F; Statimo. C₁₇H₁₇N₄NaO₆; mol wt 396.34. C 51.52%, H 4.32%, N 14.14%, Na 5.80%, O 24.22%. C₁₀H₁₂N₄O₃·C₇H₅NaO₃. Prepd by refluxing adrenochrome semicarbazone and Na salicylate in 30% methanol: Iwao *et al.,* **Japan.** pat. 57 546 (1957 to Tanabe), *C.A.* **52,** 4693 (1958).

Orange-red powder, mp 196-197.5° (dec). Soly in water (25°): 0.61 mg/ml. Practically insol in ether, chloroform. pH of 10% aq soln 6.7 to 7.3. *See also* Adrenochrome.

THERAP CAT: Hemostatic.

THERAP CAT (VET): Systemic hemostatic for capillary bleeding from increased capillary permeability.

1833. Carbazochrome Sodium Sulfonate. *5-[(Amino-carbonyl)hydrazono]-2,3,5,6-tetrahydro-1-methyl-6-oxo-1H-indole-2-sulfonic acid monosodium salt; 5,6-dihydro-1-methyl-5,6-dioxo-2-indolinesulfonic acid 5-semicarbazone sodium salt;* epinochrome-2-sulfonic acid 5-monosemicarbazone sodium salt; 1-methyl-5-semicarbazono-6-oxo-2,3,5,6-tetrahydroindole-2-sulfonic acid sodium salt; sodium 5,6-dihydro-1-methyl-5,6-dioxo-2-indolinesulfonate 5-semicarbazone; AC-17; Adenaron; Adona; Carbazon; Donaseven; Emex; Odanon; Tazin. $C_{10}H_{11}N_4NaO_5S$; mol wt 322.28. C 37.27%, H 3.44%, N 17.38%, Na 7.13%, O 24.82%, S 9.95%. Prepn: Iwao, *Pharm. Bull. (Tokyo)* **4**, 251 (1956), *C.A.* **51**, 6860f (1957); **Brit. pat. 795,184** (1958 to Gohei Tanabe). Revision of structure from a 3- to a 2-sulfonic acid: Kawazu *et al., J. Heterocycl. Chem.* **10**, 1059 (1973).

Yellow-orange needles from aq methanol, dec 227-228° (free acid dec 195°). Soluble in water.

THERAP CAT: Hemostatic.

1834. Carbazole. *9H-Carbazole;* 9-azafluorene; dibenzopyrrole; diphenylenimine. $C_{12}H_9N$; mol wt 167.21. C 86.20%, H 5.43%, N 8.38%. Prepn: Bunyan, Cadogan, *J. Chem. Soc.* **1963**, 42. Manuf from 2-biphenylamine: Conover, U.S. pat. **2,481,292** (1949 to Monsanto); Voltz *et al.,* U.S. pat. **2,891,965** (1959 to Houdry Process Corp.); Nevitt, Seelig, U.S. pat. **3,085,095** (1963 to Standard Oil, Indiana); from 2-nitrobiphenyl: Larrison, U.S. pat. **2,508,791** (1950 to Allied Chem.); by hydrogenolysis of coal-hydrogenation products: Murray *et al.,* U.S. pat. **2,913,397** (1959 to Union Carbide); from diphenylamine: Grotta, U.S. pat. **2,921,942** (1960 to American-Marietta); Bearse *et al.,* U.S. pat. **3,041,349** (1962 to Martin-Marietta); Nevitt, Seelig, *loc. cit.* Purification: Rottschaefer, U.S. pat. **2,459,135** (1949 to GAF); Insinger, U.S. pat. **2,464,811** (1949 to Koppers Co.). Toxicity study: Eagle, Carlson, *J. Pharmacol. Exp. Ther.* **99**, 450 (1950). Review of prepn and properties: W. C. Sumpter, F. M. Miller, *Heterocyclic Compounds* (Interscience, New York, 1954) pp 70-109.

Crystals from alcohol, benzene, toluene, glacial acetic acid, mp 245°. Sublimes. bp 355°, bp_{147} 200°. d_4^{18} 1.10. Exhibits strong fluorescence and long phosphorescence on exposure to ultraviolet light. Extremely weak base. Insol in water. One gram dissolves in 3 ml quinoline, 6 ml pyridine, 9 ml acetone, 2 ml acetone at 50°, 35 ml ether, 120 ml benzene, 135 ml abs alcohol. Slightly sol in petr ether, chlorinated hydrocarbons, acetic acid. Dissolves in concd H_2SO_4 without dec. KOH fusion yields *N*-potassium salt. LD_{50} orally in rats: > 5 g/kg (Eagle, Carlson).

USE: Important dye intermediate. Used in making photographic plates sensitive to ultraviolet light. Reagent for lignin, carbohydrates, and formaldehyde.

1835. 9-Carbazoleacetic Acid. Carbazyl-*N*-acetic acid; *N,N*-diphenyleneglycine. $C_{14}H_{11}NO_2$; mol wt 225.25. C 74.65%, H 4.92%, N 6.22%, O 14.21%. Prepd by condensing dry potassium carbazole with chloroethyl acetate and saponifying the resulting ester with 35% sodium hydroxide:

Ger. pat. **255,304;** *Chem. Zentr.* **1913,** I, 350; *Frdl.* **11,** 171 (1912-14). *See also:* Seka, *Ber.* **57,** 1527 (1924).

Leaflets from ethyl acetate, mp 215°. Soluble in ether, glacial acetic acid, alcohol, chloroform, xylene. On heating to 240-250° it yields *N*-methylcarbazole.

Ethyl ester, $C_{16}H_{15}NO_2$, crystals from alcohol, mp 97°, sol in organic solvents.

USE: In the detection of nitrates.

1836. Carbendazim. *1H-Benzimidazol-2-ylcarbamic acid methyl ester; 2-benzimidazolecarbamic acid methyl ester;* 2-(methoxycarbonylamino)benzimidazole; methyl 2-benzimidazolecarbamate; carbendazole; BMC; MBC; BCM; BAS-3460; BAS-67054; CTR-6669; HOE-17411; Bavistin; Derosal. $C_9H_9N_3O_2$; mol wt 191.19. C 56.54%, H 4.74%, N 21.98%, O 16.74%. Prepn: H. M. Loux, U.S. pat. **3,010,968** (1961 to du Pont); H. A. Selling *et al., Chem. & Ind. (London)* **1970,** 1625. Activity: G. P. Clemons, H. D. Sisler, *Pestic. Biochem. Physiol.* **1,** 32 (1971). Degradn product of benomyl, *q.v.* Degradn in soil: A. Helweg, *Pestic. Sci.* **8,** 71 (1977).

Light gray powder, mp 302-307° (dec). pKa 4.48. Soly in water at 24°: 8 mg/l at pH 7; 29 mg/l at pH 4. Soly at 24° (mg/l): hexane 0.5; benzene 36; dichloromethane 68; ethanol 300. Slowly decomp in alkaline soln. LD_{50} orally in rats: 6400 mg/kg, *RTECS* **Vol. I,** R. J. Lewis, R. L. Tatken, Eds. (1979) p 229.

USE: Fungicide.

1837. Carbenes. Highly reactive, electron-deficient, divalent carbon intermediates having two unshared electrons, the simplest being *methylene.* Carbenes exist in two states: singlet and triplet. The unshared electrons of singlet carbenes reside in one orbital with paired spins, giving rise to nucleophilic and electrophilic properties. In triplet carbenes, these electrons are found in two orbitals with unpaired spins, and may be considered as diradicals. Carbenes are formed thermally or photochemically from ketenes or diazo compounds. They insert into carbon-hydrogen bonds, add to olefins to form cyclopropanes, dimerize and rearrange. Formation in interstellar space: D. K. Bohme, *Nature* **319,** 473 (1986). Nitrenes, *q.v.,* are nitrogen analogs of carbenes. Books: W. Kirmse, *Carbene Chemistry* (Academic Press, New York, 2nd ed., 1971); *Carbenes* **vol. 1,** M. Jones, R. A. Moss, Eds. (Wiley-Interscience, New York, 1973) 356 pp; **vol. 2** (1975) 373 pp. *Reviews:* R. A. Moss, M. Jones, *Reactive Intermediates* **vol. 1,** M. Jones, R. A. Moss, Eds. (Wiley, New York, 1978) pp 69-116; **vol. 2** (1981) pp 59-133.

1838. Carbenicillin. *[2S-(2α,5α,6β)]-6-[(Carboxyphenylacetyl)amino]-3,3-dimethyl-7-oxo-4-thia-1-azabicyclo-[3.2.0]heptane-2-carboxylic acid; N-(2-carboxy-3,3-dimethyl-7-oxo-4-thia-1-azabicyclo[3.2.0]hept-6-yl)-2-phenylmalonamic acid;* α-carboxybenzylpenicillin; 6-(α-carboxyphenylacetamido)penicillanic acid; α-phenyl(carboxymethylpenicillin). $C_{17}H_{18}N_2O_6S$; mol wt 378.41. C 53.96%, H 4.79%, N 7.40%, O 25.37%, S 8.47%. Semi-synthetic antibiotic related to penicillin. Prepn of monopotassium salt: Hobbs, U.S. pat. **3,142,673** (1964 to Pfizer); of disodium salt: **Belg. pat. 646,991;** Brain, Nayler, U.S. pat. **3,282,926** (1964, 1966 both to Beecham). Activity studies and pharmacology: Naumann, Kempf, *Arzneimittel-Forsch.* **19,** 1222 (1969). Chemistry and mode of action: Butler *et al., J. Infec. Dis.* **122,** *Suppl.,* 81 (1970). Clinical data: Hoffler *et al., ibid.*

1233; Gritz, Naumann, *ibid.* 1237. Toxicity data: E. I. Goldenthal, *Toxicol. Appl. Pharmacol.* **18,** 185 (1971).

Disodium salt, $C_{17}H_{16}N_2Na_2O_6S$, *carbenicillin disodium, BRL-2064, CP-15639-2, Anabactyl, Carbapen, Carbecin, Geopen, Hyoper, Microcillin, Pyocianil, Pyopen.* White powder. LD_{50} i.p. in rats: >2000 mg/kg (Goldenthal).

Phenyl sodium, $C_{23}H_{21}N_2NaO_6S$, *carfecillin sodium, BRL-3475, Gripenin-O, Urocarf, Uticillin.* Prepn and activity studies: Clayton *et al., J. Med. Chem.* **18,** 172 (1975). Crystals from ethanol, $[\alpha]_D^{20}$ +216.2° (H_2O).

THERAP CAT: Antibacterial.

1839. Carbenoxolone. *(3β,20β)-3-(3-Carboxy-1-oxo-propoxy)-11-oxoolean-12-en-29-oic acid; 3β-hydroxy-11-oxo-olean-12-en-30-oic acid hydrogen succinate; 3-O-(β-carboxy-propionyl)-11-oxo-18β-olean-12-en-30-oic acid; 18β-glycyr-rhetic acid hydrogen succinate; carbenoxalone.* $C_{34}H_{50}O_7$; mol wt 570.77. C 71.55%, H 8.83%, O 19.62%. Anti-inflammatory glucocorticoid related to enoxolone, *q.v.* Prepn: Gottfried, Baxendale, **Brit.** pat. **843,133;** U.S. pat. **3,070,623** (1960, 1961 to Biorex). Monograph: *Carbenoxolone Sodium,* J. M. Robson, F. M. Sullivan, Eds. (Butterworths, London, 1969) 263 pp. Symposium on clinical efficacy: *Scand. J. Gastroenterol.* **15,** Suppl. 65, 1-121 (1980). Effect on gastric prostaglandin levels in humans: J. Rask-Madsen *et al., Eur. J. Clin. Invest.* **13,** 351 (1983); P. Minuz *et al., Pharmacol. Res. Commun.* **16,** 875 (1984).

Cream-colored crystals, mp 291-294°. $[\alpha]_D^{20}$ +128° (chloroform).

Disodium salt, $C_{34}H_{48}Na_2O_7$, *Biogastrone, Bioplex, Bioral, Duogastrone, Neogel, Pyrogastrone, Sanodin, Ulcus-Tablinen.* Creamy-white solid. Freely sol in water. LD_{50} in male mice (mg/kg): 198 i.v.; 120 i.p.; in male rats (mg/kg): 3200 orally (Robson, Sullivan).

THERAP CAT: Antiulcerative.

1840. Carbetapentane. *1-Phenylcyclopentanecarboxylic acid 2-(2-diethylaminoethoxy)ethyl ester; 2-(diethylaminoeth-oxy)ethyl 1-phenyl-1-cyclopentanecarboxylate; 2-(diethyl-aminoethoxy)ethyl 1-phenylcyclopentyl-1-carboxylate; 1-phenylcyclopentane-1-carboxylic acid diethylaminoethoxy-ethyl ester; pentoxyverine; pentoxiverin; Atussil.* $C_{20}H_{31}$-NO_3; mol wt 333.47. C 72.04%, H 9.37%, N 4.20%, O 14.39%. Prepn: H. G. Morren, **Brit.** pat. **753,779;** idem, U.S. pat. **2,842,585** (1956, 1958 both to Union Chimique Belge). Antispasmodic activity: D. Wellens, *Arzneimittel-Forsch.* **17,** 495 (1967). Clinical effect on lung function: E. Krieger, *ibid.* **22,** 389 (1972).

$bp_{0.01}$ 165-170°.

Citrate, $C_{20}H_{31}NO_3 \cdot C_6H_8O_7$, *UCB-2543, Antees, Calnathal, Carbetane, Cossym, Fustpentane, Germapect, Pencal, Sedotussin, Toclase, Tosnone, Tuclase.* Crystals, mp 93°. Freely sol in water, chloroform; sol in alcohol, acetone, ethyl acetate. Practically insol in ether, petr ether, benzene.

THERAP CAT: Antitussive.

1841. Carbetocin. *1-Butanoic acid-2-(O-methyl-L-tyro-sine)-1-carbaoxytocin;* 1-butyric acid-2-[3-(p-methoxyphen-yl)-L-alanine]oxytocin; deamino-2-O-methyltyrosine-1-car-baoxytocin; 1-thia-4,7,10,13,16-pentaazacycloeicosane cyclic peptide deriv; 1-desamino-1-monocarba-[2-tyr(OMe)]-OT; (2-O-methyltyrosine)deamino-1-carbaoxytocin; d(COMOT); Decomoton; Depotocin. $C_{45}H_{69}N_{11}O_{12}S$; mol wt 988.18. C 54.70%, H 7.04%, N 15.59%, O 19.43%, S 3.24%. Synthetic carba-analog of oxytocin, *q.v.* Prepn: I. Fric *et al., Coll. Czech. Chem. Commun.* **39,** 1290 (1974); J. H. Cort *et al.,* **Ger.** pat. **2,732,175** (1976 to Czech. Akad. Ved.). Chromatographic properties: M. Lebl, *ibid.* **45,** 2927 (1980). Uterotonic and galactogogic activity: T. Barth *et al., ibid.* 3045; T. Barth *et al., ibid.* **46,** 2441 (1981). Pharmacokinetics in lactating sows: N. Cort *et al., Am. J. Vet. Res.* **42,** 1804 (1981). Use in regulation of bovine labor: Z. Veznik *et al., ibid.* **40,** 425 (1979). Effect on milk let-down in sows: N. Cort *et al., ibid.* **43,** 1283 (1982).

Solid from methanol with ether $[\alpha]_D$ −69.0° (c = 0.25 in 1*M* acetic acid).

THERAP CAT (VET): Oxytocic, stimulates milk let-down.

1842. Carbic Anhydride. *3aα,4,7,7aα-Tetrahydro-4α,-7α-methanoisobenzofuran-1,3-dione; cis-endo-5-norbornene-2,3-dicarboxylic anhydride; endo-cis-bicyclo[2.2.1]hept-5-ene-2,3-dicarboxylic anhydride; 3,6-endomethylene-1,2,3,6-tetrahydro-cis-phthalic anhydride; 3,6-endomethylene-Δ⁴-tetrahydrophthalic anhydride; nadic anhydride;* $C_9H_8O_3$; mol wt 164.16. C 65.85%, H 4.91%, O 29.24%. Prepd by the reaction of maleic anhydride with cyclopentadiene in benzene: Diels, Alder, *Ann.* **460,** 98 (1928). Crystal structure: Destro *et al., Acta Crystallogr.* **25B,** 2465 (1969).

Shiny, orthorhombic crystals from petr ether, mp 164-165°. d 1.417. Converted to equilibrium mixtures with *exo-cis* isomers when heated above mp. Sol in benzene, toluene, acetone, carbon tetrachloride, chloroform, ethanol, ethyl acetate. Slightly sol in petr ether. Reacts with water to form the corresponding acid. Forms the γ-lactone of 5-hydroxy-2,3-norcamphanedicarboxylic acid in 50% H_2SO_4.

1843. Carbidopa. *S-α-Hydrazino-3,4-dihydroxy-α-meth-ylbenzenepropanoic acid monohydrate; (−)-L-α-hydrazino-3,4-dihydroxy-α-methylhydrocinnamic acid monohydrate; α-*

hydrazino-α-methyl-β-(3,4-dihydroxyphenyl)propionic acid monohydrate; L-α-(3,4-dihydroxybenzyl)-α-hydrazinopropionic acid monohydrate; α-methyldopahydrazine; HMD; MK-486; Lodosin; Lodosyn. $C_{10}H_{14}N_2O_4 \cdot H_2O$; mol wt 244.25. C 49.18%, H 6.60%, N 11.47%, O 32.75%. Peripheral decarboxylase inhibitor. Prepn of DL-form: Pfister, **Fr. pat. M1553** (1962 to Merck & Co.), *C.A.* **59**, 12921e (1963); Sletzinger *et al., J. Med. Chem.* **6**, 101 (1963); **Brit. pat. 940,596**; Chemerda *et al.,* U.S. pat. **3,462,536** (1963, 1969 both to Merck & Co.). Synthesis of the L-form: Karady *et al.,* **Ger.** pats. **2,062,285; 2,062,332** (both 1971 to Merck & Co.), *C.A.* **75**, 118122t, 118120r (1971); *eidem, J. Org. Chem.* **36**, 1946, 1949 (1971). Inhibition of dopa decarboxylase: Porter *et al., Biochem. Pharmacol.* **11**, 1067 (1962); Moran, Sourkes, *J. Pharmacol. Exp. Ther.* **148**, 252 (1962); Watanabe *et al., Clin. Pharmacol. Ther.* **11**, 740 (1970). Only the L-form is pharmacologically active: Lotti, Porter, *J. Pharmacol. Exp. Ther.* **172**, 406 (1970).

Crystals from hot water, mp 203-205° (dec). $[\alpha]_D$ —17.3° (methanol). Also reported as mp 208°.

Combination with levodopa, *co-careldopa, Isicom, Nacom, Sinemet.*

DL-Form, tan fluffy crystals, mp 206-208° (dec). uv max (methanol): 282.5 nm (ϵ 2940).

THERAP CAT: In combination with levodopa as antiparkinsonian.

1844. Carbimazole. *2,3-Dihydro-3-methyl-2-thioxo-1H-imidazole-1-carboxylic acid ethyl ester;* 1-ethoxycarbonyl-3-methyl-2-thio-4-imidazoline; ethyl 3-methyl-2-thioimidazoline-1-carboxylate; 1-methyl-3-carbethoxy-2-thioglyoxalone; athyromazole; Neo-mercazole; Neo-Thyreostat. $C_7H_{10}N_2O_2S$; mol wt 186.23. C 45.15%, H 5.41%, N 15.04%, O 17.18%, S 17.22%. Thyroid inhibitor. Prepn: Rimington *et al.,* U.S. pat. **2,671,088**, reissued as **Re 24,505;** U.S. pat. **2,815,349** (1954, 1958, 1957, all to Natl. Res. Dev. Corp.); Baker, *J. Chem. Soc.* **1958**, 2387. Clinical efficacy in Graves' disease: J. Duprey *et al., Presse Méd.* **17**, 1124 (1988).

Crystalline powder with characteristic odor; tasteless at first, followed by bitter taste; mp 122-125°. Sol (at 20°) in 500 parts of water; in 50 parts of ethanol; in 330 parts of ether; in 3 parts of chloroform; in 17 parts of acetone. uv max in 0.1N HCl : water (1:8): 291 nm; in 0.1N H_2SO_4: 227 nm and 291 nm ($E^{1\%}_{1cm}$ 557).

THERAP CAT: Antihyperthyroid.

1845. Carbinoxamine. *2-[(4-Chlorophenyl)-2-pyridinylmethoxy]-N,N-dimethylethanamine; 2-[p-chloro-α-(2-dimethylaminoethoxy)benzyl]pyridine;* paracarbinoxamine. $C_{16}H_{19}ClN_2O$; mol wt 290.79. C 66.09%, H 6.59%, Cl 12.19%, N 9.63%, O 5.50%. Prepn: Tilford, Shelton, U.S. pat. **2,606,-195** (1952 to Wm. S. Merrell); Swain, U.S. pat. **2,800,485** (1957 to McNeil Labs.). Prepn of *l*-form: **Brit. pat. 905,-993** (1962 to McNeil Labs.), *C.A.* **58**, 5644a (1962). Abs config of *l*-form: V. Barouh *et al., J. Med. Chem.* **14**, 834 (1971). Pharmacology and toxicology: R. Cahen, *Ann. Pharm. Franc.* **20**, 463 (1962). GLC determn in serum: D. J. Hoffman *et al., J. Pharm. Sci.* **72**, 1342 (1983).

Liquid, $bp_{0.1}$ 158-162°.

Hydrochloride, $C_{16}H_{19}ClN_2O \cdot HCl$, crystals from isopropanol + ethyl acetate, dec 162-164°. Sol in water.

Maleate, $C_{16}H_{19}ClN_2O \cdot C_4H_4O_4$, *Allergefon, Clistin, Ciberon, Lergefin, Polistin T-Caps.* Bitter crystals from ethyl acetate, mp 117-119°. Freely sol in water, alcohol, chloroform. Very slightly sol in ether. pH of 1% aq soln 4.6-5.1. LD_{50} in mice (mg/kg): 166 i.p. (Cahen).

l-Form, *levocarbinoxamine, rotoxamine, McN-R-73-Z.* $bp_{0.5}$ 143-144°. n_D^{20} 1.5522. $[\alpha]_D^{25}$ —6.8° (c = 2 in methanol).

l-Form *d*-tartrate, $C_{16}H_{19}ClN_2O \cdot C_4H_6O_6$, *Twiston.* Crystals from isopropanol, mp 143-144.5°. $[\alpha]_D^{25}$ +37.2° (c = 20 in methanol).

THERAP CAT: Antihistaminic.

1846. Carbiphene. *α-Ethoxy-N-methyl-N-[2-[methyl(2-phenylethyl)amino]ethyl]-α-phenylbenzeneacetamide; 2-ethoxy-N-methyl-N-[2-(methylphenethylamino)ethyl]-2,2-diphenylacetamide;* etomide (rescinded USAN); etymide. $C_{28}H_{34}N_2O_2$; mol wt 430.59. C 78.10%, H 7.96%, N 6.51%, O 7.43%. Prepn: Krapcho, Turk, *J. Med. Chem.* **6**, 547 (1963).

Hydrochloride, $C_{28}H_{34}N_2O_2 \cdot HCl$, *SQ-10269, Juhalon, Bandol.* Crystals, mp 163-165°.

THERAP CAT: Analgesic.

1847. Carbitol®. *2-(2-Ethoxyethoxy)ethanol;* diethylene glycol monoethyl ether; ethyl digol. $C_6H_{14}O_3$; mol wt 134.18. C 53.71%, H 10.52%, O 35.77%. $CH_3CH_2OCH_2CH_2OCH_2CH_2OH$. Prepn from ethylene oxide and 2-ethoxyethanol in the presence of SO_2: Britton, Sexton, U.S. pat. **2,807,651** (1957 to Dow). Toxicity data: Smyth, Carpenter, *J. Ind. Hyg. Toxicol.* **30**, 63 (1948).

Very hygroscopic liquid, bp 196°. d_4^{25} 0.9855, d_{20}^{20} 1.0273. n_D^{20} 1.4273. Flash pt 96° (Tag open cup). Miscible with acetone, benzene, chloroform, ethanol, ether, pyridine, etc.: Jackson, Drury, *Ind. Eng. Chem.* **51**, 1491 (1959). Also miscible with water. LD_{50} orally in rats: 8.69 g/kg (Smyth, Carpenter).

Acetate, $C_8H_{16}O_4$, liquid, bp 218.5°. mp —25°. d_{20}^{20} 1.0114. n_D^{20} 1.4213. Flash pt 110° (Tag open cup). Miscible with water, alc, ether, most oils. LD_{50} orally in rats: 11 g/kg (Smyth, Carpenter).

USE: As solvent for cellulose esters, in lacquer and thinner formulations, in quick-drying varnishes and enamels, for dyestuffs and wood stains. Acetate is used as a solvent and plasticizer for cellulose esters, gums, resins, etc.

1848. Carbobenzoxy Chloride. *Carbonochloridic acid phenylmethyl ester;* benzyl chloroformate; chloroformic acid benzyl ester; benzylcarbonyl chloride. $C_8H_7ClO_2$; mol wt 170.60. C 56.33%, H 4.14%, Cl 20.78%, O 18.76%. Prepn by action of phosgene absorbed in toluene on benzyl alcohol: Carter *et al., Org. Syn.* **23**, 13 (1943); by reacting carbonyl chloride and benzyl alcohol at —20 to —30°: Farthing, *J. Chem. Soc.* **1950**, 3213.

Oily liquid; acrid odor; lacrimator. bp$_{20}$ 103°; bp$_7$ 85-87°. Dec to CO$_2$ and benzyl chloride upon heating at 100-155°. USE: In peptide synthesis to block the amino group.

1849. Carbocloral. *(2,2,2-Trichloro-1-hydroxyethyl)carbamic acid ethyl ester;* ethyl (2,2,2-trichloro-1-hydroxyethyl)carbamate; chloral-urethane; CI-336; HY-185; Ural; Uraline; Uralium. C$_5$H$_8$Cl$_3$NO$_3$; mol wt 236.48. C 25.40%, H 3.41%, Cl 44.98%, N 5.92%, O 20.30%. Cl$_3$CCH(OH)-NHCOOC$_2$H$_5$. Prepd from chloral and urethane in the presence of HCl: Bischoff, *Ber.* **7,** 628 (1874).

Crystalline powder, mp ~103° with partial decompn. Practically insol in water; freely sol in alcohol, ether.

THERAP CAT: Hypnotic.

1850. Carbocysteine. *S-(Carboxymethyl)-L-cysteine; 3-[(carboxymethyl)thio]alanine;* S-carboxymethylcysteine; AHR-3053; LJ-206; Carbocit; Fluifort; Lisil; Lisomucil; Loviscol; Muciclar; Mucocis; Mucodyne; Mucolase; Mucolex; Mucopront; Mucotab; Mukinyl; Pectox; Pulmoclase; Reomucil; Rhinathiol; Siroxyl; Thiodril; Transbronchin. C$_5$H$_9$NO$_4$S; mol wt 179.20. C 33.51%, H 5.06%, N 7.82%, O 35.71%, S 17.89%. HOOCCH$_2$SCH$_2$CH(NH$_2$)COOH. Prepn: Armstrong, Lewis, *J. Org. Chem.* **16,** 749 (1951); Schöberl, Wagner, *Z. Physiol. Chem.* **304,** 97 (1956); Foye, Verderame, *J. Am. Pharm. Assoc.* **46,** 273 (1957); Goodman *et al., J. Org. Chem.* **23,** 1251 (1958). Pharmacology: Huyen-Vu-Ngoc *et al., C. R. Soc. Biol.* **160,** 1849 (1966); Quevauviller *et al., Therapie* **22,** 485 (1967). Clinical trial in chronic bronchitis: M. Grillage, K. Bernard-Jones, *Brit. J. Clin. Pract.* **39,** 395 (1985). Review of pharmacology and clinical uses: D. T. Brown, *Drug. Intell. Clin. Pharm.* **22,** 603-608 (1988).

L-Form, mp 204-207. [α]$_D^{24}$ 0.5° (1N HCl). DL-Form, spherical aggregates of needles. THERAP CAT: Mucolytic; expectorant.

1851. Carbofuran. *2,3-Dihydro-2,2-dimethyl-7-benzofuranol methylcarbamate; methyl carbamic acid 2,3-dihydro-2,2-dimethyl-7-benzofuranyl ester;* 2,2-dimethyl-2,3-dihydro-7-benzofuranyl-N-methylcarbamate; 2,2-dimethyl-7-coumaranyl N-methylcarbamate; Bay 70143; NIA-10242; Furadan. C$_{12}$H$_{15}$NO$_3$; mol wt 221.26. C 65.14%, H 6.83%, N 6.33%, O 21.69%. Prepn and use as insecticide: **Neth. pat. Appl. 6,407,316** (1964 to Bayer), *C.A.* **63,** 583a (1965); **Neth. pat. Appl. 6,500,340** corresp to W. G. Scharpf, **U.S. pats. 3,474,170-1** (1965, 1969 to FMC); E. F. Orwoll, **U.S. pat. 3,356,690** (1967 to FMC). Metabolism: H. W. Dorough, *J. Agr. Food Chem.* **16,** 319 (1968); J. B. Knaak, *ibid.* **18,** 832 (1970). Toxicity studies: J. S. Tobin, *J. Occup. Med.* **12,** 16 (1970); M. A. Fahmy *et al., J. Agr. Food Chem.* **18,** 793 (1970). Teratogenicity study: K. D. Courtney *et al., J. Environ. Sci. Health* **B20,** 373 (1985).

White crystalline solid, mp 150-153°. Soly in water at 25°: 700 ppm. Unstable in alk. LD$_{50}$ orally in mice: 2 mg/kg (Fahmy, 1970).

Caution: Cholinesterase inhibitor: *Clinical Toxicology of Commercial Products,* R. E. Gosselin *et al.,* Eds. (Williams & Wilkins, Baltimore, 5th ed., 1984) Section II, p 305.

USE: Systemic insecticide, acaricide, nematocide.

1852. Carbohydrazide. *Carbonic dihydrazide;* 1,3-diaminourea. CH$_6$N$_4$O; mol wt 90.08. C 13.33%, H 6.71%, N

62.19%, O 17.76%. NH$_2$NHCONHNH$_2$. Prepd by refluxing diethyl carbonate with hydrazine hydrate: Mohr *et al., Inorg. Syn.* **4,** 32 (1953).

Crystals from water + ethanol, dec 153-154°. Freely sol in water. pH of 1% aq soln ~7.4. Practically insol in alcohol, ether, chloroform, benzene. Forms salts with acids. With nitrous acid it forms the highly explosive carbonyl azide CO(N$_3$)$_2$.

1853. γ-Carboline. *5H-Pyrido[4,3-b]indole; 2H-pyrid-[4,3-b]indole;* 5-carboline. C$_{11}$H$_8$N$_2$; mol wt 168.20. C 78.55%, H 4.79%, N 16.66%. Prepn: Robinson, Thornley, *J. Chem. Soc.* **125,** 2169 (1924). Prepn of derivs: Hörlein, *Ber.* **87,** 463 (1954); C. Ducrocq *et al., J. Heterocycl. Chem.* **12,** 963 (1975). NMR studies: F. Balkau, M. L. Heffernan, *Aust. J. Chem.* **26,** 1501, 1523 (1973).

Monoclinic needles from water, mp 225°. d 1.352. Can be distilled at atmospheric pressure without dec. Strong base. Freely sol in methanol; somewhat less sol in ethanol. Slightly sol in benzene, water.

1854. Carbomycin. Magnamycin. Sixteen-membered-ring macrolide antibiotic complex similar to leucomycin, *q.v.* and erythromycin, *q.v.,* produced by *Streptomyces halstedii.* Isoln and antibacterial activity: F. W. Tanner *et al., Antibiot. & Chemother.* **2,** 441 (1952). Two components have been isolated: Carbomycin A (major) and carbomycin B. Isoln of A: Friedman *et al.,* **U.S. pat. 2,960,438** (1960 to Pfizer); of B: F. A. Hochstein, K. Murai, *J. Am. Chem. Soc.* **76,** 5080 (1954). Structure of A and B: R. B. Woodward, *Angew. Chem.* **69,** 50 (1957); revised structure: M. Kuehne, B. W. Benson, *J. Am. Chem. Soc.* **87,** 4660 (1965); R. B. Woodward *et al., ibid.* 4662. Abs config of A and B: W. D. Celmer, *ibid.* **88,** 5028 (1966). Identity of A with deltamycin A$_4$: Y. Shimauchi *et al., J. Antibiot.* **31,** 270 (1978). Synthesis of B: K. Tatsuta *et al., J. Am. Chem. Soc.* **99,** 5826 (1977). Stereospecific total synthesis of B: *eidem, Tetrahedron Letters* **1980,** 2837. Retrosynthetic studies: K. C. Nicolaou *et al., J. Am. Chem. Soc.* **103,** 1222 (1981). Reviews: D. Vazquez, in *Antibiotics* Vol. 1, D. Gottlieb, P. D. Shaw, Eds. (Springer-Verlag, New York, 1967) pp 366-377; W. Keller-Schierlein in *Fortschr. Chem. Org. Naturst.* **30,** 314-460 (1973).

Carbomycin A

Carbomycin A, C$_{42}$H$_{67}$NO$_{16}$, *(12S,13S)-9-deoxy-12,13-epoxy-12,13-dihydro-9-oxoleucomycin V 3-acetate 4^B-(3-methylbutanoate), magnamycin A, deltamycin A$_4$, M-4209.* Blunt needles from ethanol, mp 214°. [α]$_D^{25}$ −58.6° (chloroform). uv max (abs ethanol): 238, 327 nm (E$_{1cm}^{1\%}$ 185, 0.9). Carbomycin standard is the free base having a potency of 1080 units/mg. For stability of soln data *see* H. L. Martin, *Antibiot. & Chemother.* **3,** 865 (1953). Weak base, pKb 7.2. Solubilities determined by Weiss *et al., ibid.* **7,** 374 (1957) in mg/ml at about 28°: water 0.295; methanol > 20; ethanol > 20. LD$_{50}$ i.v. in mice: 550 mg/kg (Tanner).

Carbomycin B, C$_{42}$H$_{67}$NO$_{15}$, *9-deoxy-9-oxoleucomycin V 3-acetate 4^B-(3-methylbutanoate), magnamycin B.* Colorless anisotropic plates from acetone/water, mp 141-144° (dec), softens at 138°. [α]$_D^{25}$ −35° (c = 1 in chloroform). uv max (abs ethanol): 278 nm ($E_{1cm}^{1\%}$ 276). pKb 7.56. Solubilities in mg/ml at 25°: ethanol 450; water 0.1-0.2.

THERAP CAT: Antibacterial.

THERAP CAT (VET): Antibacterial.

1855. Carbon. C; at. wt 12.01115; at. no. 6; valence 4. Stable isotopes: 12 (98.892%); 13 (1.108%); radioactive isotopes: 9-11; 14-16. Abundance in earth's crust: approx 0.027%. Cosmic abundance: 6 atoms/atom Si. Occurs in 4 allotropic forms: (1) diamond, *q.v.*; (2) graphite, *q.v.* or black lead; (3) amorphous carbon such as coal, lampblack; (4) fullerenes, *see* Buckminsterfullerene, the only molecular form. Comprehensive reviews: P. L. Walker, *Am. Scientist* **50**, 259-293 (June 1962); Holliday *et al.* in *Comprehensive Inorganic Chemistry* vol. 1, J. C. Bailar, Jr. *et al.*, Eds. (Pergamon Press, Oxford, 1973) pp 1173-1294; several authors in Kirk-Othmer *Encyclopedia of Chemical Technology* vol. 4 (Wiley-Interscience, New York, 3rd ed., 1978) pp 556-709.

^{14}C isotope, continuously formed in the earth's atm by the bombardment of nitrogen with cosmic neutrons according to the reaction $^{14}_{7}N + ^{1}_{0}n \rightarrow ^{14}_{6}C + ^{1}_{1}H$. The ^{14}C is rapidly oxidized to CO_2, in this form it penetrates into animals and plants by photosynthesis and metabolism. The ^{14}C content of living matter is estimated at 15.3 disintegrations per minute and per gram of carbon, corresponding to the equilibrium reached between formation of ^{14}C and its exchange with ^{12}C. This equilibrium stops when the plant or animal dies, and the ^{14}C content begins to decrease, because the ^{14}C decays with a half-life of 5760 years. This fact can be used to date organic matter (not more than 40,000 years old) by comparison with the standard 15.3 disintegrations per min per gram: M. Haissinsky, J. P. Adloff, *Radiochemical Survey of the Elements* (Elsevier, New York, 1965) pp 30-32.

1856. Carbon, Amorphous. Carbon black; activated carbon; decolorizing carbon. A quasi-graphitic form of carbon of small particle size. By the term "carbon black" several forms of artificially prepared carbon or charcoal are designated, *e.g.*: (1) *Animal charcoal*, obtained by charring bones, meat, blood, etc.; (2) *Gas black; furnace black; channel black; C.I. 77266*, obtained by incomplete combustion of natural gas; (3) *Lamp black*, obtained by burning various fats, oils, resins, etc., under suitable conditions; (4) *Activated charcoal*, *e.g. Carbomix, Carboraffin, Medicoal, Norit, Opocarbyl, Ultracarbon*, prepd from wood and vegetables. Monograph: H. W. Davidson *et al.*, *Manufactured Carbon* (Pergamon Press, New York, 1968). Reviews: Cohan in *Science of Petroleum* vol. V, Pt 2, B. T. Brooks, A. E. Dunstan, Eds. (Oxford Univ. Press, 1953), pp 79-89; Smisek, Cerny, *Active Carbon* (Elsevier Publishing Co., Amsterdam, 1970).

Note: Carbon black obtained by the impingement or channel process, also known as gas black and channel black, has been banned by the FDA for use as a color additive in foods, drugs and cosmetics.

USE: Pigment for rubber tires; for printing, stenciling and drawing inks; for leather; stove polish, phonograph records, electrical insulating apparatus. Activated charcoal for clarifying, deodorizing, decolorizing and filtering.

THERAP CAT: Activated charcoal as antidote; adsorptive.

THERAP CAT (VET): Internally as an adsorptive in diarrhea; externally in foul wounds.

1857. Carbon Dioxide. Carbonic acid gas; carbonic anhydride. CO_2; mol wt 44.01. C 27.29%, O 72.71%. Occurs in the atms of many planets. In our solar system, e.g., on Venus, the optical layer thickness due to CO_2 is 100,000 cm/atm, but only 220 cm/atm on Earth. Analyses of air in the temperate zones of the Earth show 0.027 to 0.036% (v/v) of CO_2: G. P. Kuiper, *The Atmospheres of the Earth and the Planets* (Univ. of Chicago Press, 1949); Landolt-Bornstein, *Zahlenwerte* vol. III (Springer-Verlag, 6th ed., 1952) pp 59 and 585. Constituent of carbonate type of minerals and products of animal metabolism. Necessary for the respiration cycle of plants and animals. Obtained industrially as a by-product in the manuf of lime during the "burning" of limestone ($CaCO_3$). Also produced by burning coke or other carbonaceous material. In the U.S.A. large amounts are produced by fermentation (Backus process and Reich process). When glucose is fermented by yeast, the chief products are ethyl alcohol and CO_2. Prepd in the

laboratory by dropping acid on a carbonate: E. H. Archibald, *The Preparation of Pure Inorganic Substances* (Wiley, New York, 1932) p 196; Loomis, Walters, *J. Am. Chem. Soc.* **48**, 3103 (1926). Purification: Glemser in *Handbook of Preparative Inorganic Chemistry*, G. Brauer, Ed. (Academic Press, New York, 2nd ed., 1963) p 647. Discovery of a second polymorph of dry ice: L.-G. Liu, *Nature* **303**, 508 (1983). *Reviews:* E. L. Quinn, *J. Chem. Ed.* **7**, 151-162 and 403-419 (1930); J. Kuprianoff, *Die feste Kohlensäure (Trockeneis)* (Enke, Stuttgart, 1939); E. L. Quinn, C. L. Jones, *Carbon Dioxide* (Reinhold, New York, 1947); W. R. Ballou, in Kirk-Othmer *Encyclopedia of Chemical Technology* vol. 4 (Interscience, New York, 3rd ed., 1978) pp 725-742.

Colorless, odorless, noncombustible gas. Faint acid taste. Usually a nonsupporter of combustion, although burning magnesium continues to burn when transferred into a CO_2 atm. Usually marketed in steel cylinders (under sufficient pressure to keep it liquid) or in solid form as *Dry Ice* (compressed carbon dioxide snow, d 1.35). When shipped in steel cylinders, CO_2 is in the form of gas over liquid and at 20° exerts a pressure of 830 psi. Use gloves when handling dry ice, as its temp is at least −78.5°; momentary skin contact with dry ice has caused serious frostbites and blisters. At atmospheric pressures the solid form changes into the gaseous phase without liquefaction. d (gas) 1.527 (air = 1); d (gas) 1.557 ($N_2 = 1$); abs d 0.1146 lb/cu ft at 25°; vol at 25°: 8.76 cu ft/lb. d (gas, 0°) 1.976 g/l at 760 mm; d (liq, 0°) 0.914 at 34.3 atm; d (solid, −56.6°) 1.512. Sublimes at −78.48° (760 mm). $mp_{5.2\ atm}$ −56.6°. Gas is not affected by heat until temp reaches about 2000°. Crit temp 31.3°; crit press 72.9 atm; crit density 0.464. Triple point −56.6° at 5.11 atm. Vapor press at −120°: 10.5 mm; at −100°: 104.2 mm; at −82°: 569.1 mm. Heat of formation 94.05 kcal/mol. Latent heat of vaporization 83.12 g cal/g. Specific heat 0.19 to 0.21 Btu/lb. Soly in water (ml CO_2/100 ml H_2O at 760 mm): 0° = 171; 20° = 88; 60° = 36. More sol at higher pressures. Less sol in alcohol, other neutral organic solvents. Absorbed by alkaline solns with the formation of carbonates.

Caution: Potential symptoms of overexposure are headache, dizziness, restlessness and paresthesia; dyspnea; sweating, malaise; increased heart rate and pulse pressure; elevated blood pressure; coma; asphyxia; convulsions at high concentrations; frostbite (dry ice). *See NIOSH Pocket Guide to Chemical Hazards* (DHHS/NIOSH 90-117, 1990) p 58.

USE: In the carbonation of beverages; manuf of carbonates; in fire prevention and extinction; for inerting flammable materials during manuf, handling and transfer; as propellant in aerosols; as dry ice for refrigeration; to produce harmless smoke or fumes on stage; as rice fumigant; as antiseptic in bacteriology and in the frozen food industry.

THERAP CAT: Respiratory stimulant.

THERAP CAT (VET): Respiratory stimulant (inhalant).

1858. Carbon Diselenide. *Carbon selenide.* CSe_2; mol wt 169.93. C 7.07%, Se 92.93%. Prepd by the action of methylene chloride vapor on heated selenium: Ives *et al., J. Chem. Soc.* **1947**, 1080; or from a mixture of CCl_4 and H_2Se in a stream of N_2 at 500°: Grimm, Metzger, *Ber.* **69**, 1356 (1936); from the elements by electrical discharge on Se vapor in the presence of sugar charcoal: Steudel, *Z. Anorg. Allgem. Chem.* **361**, 195 (1968).

Light-sensitive, golden yellow, strongly refractive, liquid. Odor of rotten radishes. Turns brown to black on storage. d_4^{20} 2.6824; d_4^{25} 2.6626. mp −45.5°. bp 125-126°; $bp_{8.0}$ 10.0°. n_D^{20} 1.845. Heat of formation: 34 kcal/mol. Miscible with carbon tetrachloride, carbon disulfide, toluene, other organic solvents. Practically insol in water. Dec by alc, pyridine.

1859. Carbon Disulfide. Carbon bisulfide; dithiocarbonic anhydride. CS_2; mol wt 76.14. C 15.77%, S 84.23%. Minute amounts occur in coal tar and in crude petroleum. Prepd on an industrial scale by heating charcoal with vaporized sulfur; from sulfur and natural gas: Faith, Keyes & Clark's *Industrial Chemicals*, F. A. Lowenheim, M. K. Moran, Eds. (Wiley-Interscience, New York, 4th ed., 1975) pp 224-229. Laboratory purification: Glemser in *Handbook of Preparative Inorganic Chemistry* vol. 1, G. Brauer, Ed. (Academic Press, New York, 2nd ed., 1963) p 652. Review of production and uses: Bushell, *Chem. & Ind.*

(London) **1961** , 1465; R. W. Timmerman in Kirk-Othmer *Encyclopedia of Chemical Technology* vol. 4 (Wiley-Interscience, New York, 3rd ed., 1978) pp 742-757.

Highly refractive, mobile, very flammable liq. *Poisonous!* The purest distillates ever obtained are reported to have a sweet, pleasing, and ethereal odor, while the usual commercial and reagent grades are foul smelling. Dec on standing for a long time. Burns with a blue flame to CO_2 and SO_2. *Acute fire and explosion hazard,* can be ignited by hot steam pipes. Flash pt, closed cup: $-30°$. Ignition pt: $100°$. Explosive range: 1 to 50% (v/v) in air. d_4^0 1.29272; d_4^{15} 1.27055; d_4^{20} 1.2632; d_4^{30} 1.24817. Vapors sink to the ground. Vapor density 2.67 (air = 1). mp $-111.6°$. $bp_{1.0}$ $-73.8°$; bp_{10} $-44.7°$; bp_{100} $-5.1°$; bp_{400} $+28.0°$; bp_{760} $+46.5°$; $bp_{(2\ atm)}$ $+69.1°$; $bp_{(5\ atm)}$ $+104.8°$. Crit temp $280.0°$; crit press. 72.9 atm. n_D^{15} 1.63189; $n_D^{20.1}$ 1.62803; $n^{23.5}$ 1.62543. Surface tension at 20°: 32.25. Coefficient of viscosity at 20°: 0.363. Heat of vaporization at bp: 84.1 cal/g. Heat of fusion: 1.049 kcal/mole. Heat capacity at 24.3°: 18.17 cal/mole/deg: Brown, Manov, *J. Am. Chem. Soc.* **59**, 500 (1937). Ebullioscopic constant: 2.35°. Dielectric constant at low frequencies: 2.641. Dipole moment: 0.0. Soly in water at 20°: 0.294%. Soly of water in CS_2: < 0.005%. Azeotrope with water bp 42.6°, contains 97.2% CS_2. Misc with anhydr methanol, ethanol, ether, benzene, chloroform, carbon tetrachloride, oils. Can be stored in iron, aluminum, glass, porcelain, Teflon.

Caution: Poisoning usually occurs from inhalation but also may be caused by ingestion and skin absorption. *Acute Toxicity:* euphoria, restlessness, mucous membrane irritation, nausea, vomiting, unconsciousness, terminal convulsions. *Chronic Toxicity:* marked psychic disturbances ranging from extreme irritability to mania with hallucinations, tremors, auditory and visual disturbances, weight loss, blood dyscrasias. Dermal contact with concd solns may cause burning pain, erythema, exfoliation. *See: Clinical Toxicology of Commercial Products,* R. E. Gosselin *et al.,* Eds. (Williams & Wilkins, Baltimore, 4th ed., 1976) Section III, pp 83-86.

USE: In the manuf of rayon, carbon tetrachloride, xanthogenates, soil disinfectants, electronic vacuum tubes. Solvent for phosphorus, sulfur, selenium, bromine, iodine, fats, resins, rubbers.

1860. Carbonic Anhydrase. *Carbonate dihydratase; carbonate hydro-lyase.* Mol wt ~30,000. A small zinc-contg enzyme which catalyzes the hydration of CO_2. Found in higher concns in erythrocytes, renal cortex, and gastric mucosa of mammals; also found in other animal tissues, in plants and in some bacteria. Isoln from bovine erythrocytes: Lindskog, *Biochim. Biophys. Acta* **39**, 218 (1960); from human erythrocytes: Nyman, *ibid.* **52**, 1 (1961); from renal cortex: Höber, *Proc. Soc. Exp. Biol. Med.* **49**, 87 (1942); from gastric mucosa: Davenport, *Physiol. Rev.* **26**, 560 (1946). Human carbonic anhydrase consists of two isoenzymes with distinctly different amino acid sequences and specific activities. The high activity form is called carbonic anhydrase C; the low activity form is called B; modified forms of these two isoenzymes exist: Funakoshi, Deutsch, *J. Biol. Chem.* **243**, 6474 (1968); **244**, 3438 (1969). Amino acid sequence of carbonic anhydrase B: Anderson *et al., Biochem. Biophys. Res. Commun.* **48**, 670 (1972); Lin, Deutsch, *J. Biol. Chem.* **248**, 1885 (1973); sequence of carbonic anhydrase C: Henderson, Henriksson, *Biochem. Biophys. Res. Commun.* **52**, 1388 (1973); Lin, Deutsch, *J. Biol. Chem.* **249**, 2329 (1974). Crystal structure of carbonic anhydrase C: Liljas *et al., Nature New Biol.* **235**, 131 (1972). Catalyzes the reversible reaction of CO_2 and H_2O to HCO_3^- and H^+. Permits CO interchange between blood and tissues. In gastric mucosa, reaction rate is sufficient to neutralize the excess alkalinity produced by the ionization of water and secretion of hydrogen ions: Roughton, Clark in *The Enzymes* vol. 1, part 2, J. B. Sumner, K. Myrbäck, Eds. (Academic Press, New York, 1951) pp 1250-1265. In the kidney, participates in Na^+ transport. Review of physiology: Maren in *Oxygen Affinity of Hemoglobin and Red Cell Acid Base Status, Alfred Benzon Symposium* IV, P. Astrup, M. Roerth, Eds. (Academic Press, New York, 1972) pp 418-433. Review of metal ion function: Prince, Woolley, *Angew. Chem. Int. Ed.* **11**, 408-417 (1972). *Review:* Lind-skog *et al.,* "Carbonic Anhydrase" in *The Enzymes,* vol. **5**, P. D. Boyer, Ed. (Academic Press, New York, 1971) pp 587-665.

1861. Carbon Monoxide. CO; mol wt 28.01. C 42.88%, O 57.12%. Produced on an industrial scale by partial oxidation of hydrocarbon gases from natural gas or by the gasification of coal and coke. Conveniently prepd in the laboratory by heating calcium carbonate with Zn dust: Weinhouse, *J. Am. Chem. Soc.* **70**, 442 (1948); by dehydration of formic acid with H_2SO_4: Gilliland, Blanchard, *Inorg. Syn.* **2**, 81 (1946). Purification of carbon monoxide bought in steel cylinders: A. Klemenc, *Die Behandlung und Reindarstellung von Gasen* (Vienna, 1948) p 160; Glemser in *Handbook of Preparative Inorganic Chemistry* vol. 1, G. Brauer, Ed. (Academic Press, New York, 2nd ed., 1963) p 646. Review of toxic effects in humans: Stewart, *Ann. Rev. Biochem.* **15**, 409-423 (1975). *Review:* C. M. Bartish, G. M. Drissel in Kirk-Othmer *Encyclopedia of Chemical Technology* vol. 4 (Wiley-Interscience, New York, 3rd ed., 1978) pp 772-793.

Highly poisonous, odorless, colorless, tasteless gas. Very flammable, burns in air with a bright blue flame. Ignition pt in air: 700°. mp $-205.0°$. bp $-191.5°$. d_4^{-195} (liq) 0.814. d (gas) 0.968 (air = 1.000). d_4^0 at 760 mm: 1.250 g/liter. The top pressure is 1500 psi. Flammable limits in air: 12 to 75 vol %. Crit press 35 atm, crit temp $-139°$. Heat capacity at 20°: 6.95 cal/mole/°C. Heat value per m³: 3033 kcal. Heat of formation: -26.39 kcal/mol. Dec into carbon and carbon dioxide between 400 and 700°, at lower temp when in contact with catalytic surfaces. Above 800° the equilibrium reaction favors CO formation. Hopcalite, a mixture of the oxides of manganese and copper, catalyzes the decompn at room temp, as does Pd on silica gel. Sparingly sol in water: 3.3 ml/100 ml H_2O at 0°; 2.3 ml/100 ml H_2O at 20°; freely absorbed by a concd soln of cuprous chloride in HCl or in NH_4OH. Appreciably sol in some organic solvents, such as ethyl acetate, $CHCl_3$, acetic acid. The soly in methanol and ethanol is about 7 times as great as the soly in water.

Caution: Combines with the hemoglobin of the blood to form carboxyhemoglobin which is useless as an oxygen carrier. *Toxic symptoms:* Headache, mental dullness, dizziness, weakness, nausea, vomiting, loss of muscular control, increased then decreased pulse and respiratory rates, collapse, unconsciousness, death. *Antidote:* Oxygen. See Patty's *Industrial Hygiene and Toxicology* vol. 2C, G. D. Clayton, F. E. Clayton, Eds. (Wiley-Interscience, New York, 3rd ed., 1982) pp 4114-4124.

USE: As reducing agent in metallurgical operations especially in the Mond process for the recovery of nickel; in organic synthesis especially in the Fischer-Tropsch processes for petroleum-type products and in the oxo reaction; in the manuf of metal carbonyls.

1862. Carbon Nitride. C_3N_4; mol wt 92.06. C 39.14%, N 60.86%. Covalent crystalline solid which exists in two forms α and β. β-form may be harder than diamond. Model of structure and hardness: A. Y. Liu, M. L. Cohen, *Science* **245**, 841 (1989). Synthesis: E. E. Haller *et al.,* **PCT Int. pat. Appl. 91 16,196**; *eidem,* **U.S. pat. 5,110,679** (1991, 1992 both to Regents U. California, Berkeley); C. Niu *et al., Science* **261**, 334 (1993). Hardness evaluation: O. Fukunaga, *Kagaku to Kyoiku* **41**, 325 (1993), *C.A.* **119**, 54572g (1993).

1863. Carbon Suboxide. *1,2-Propadiene-1,3-dione;* tricarbon dioxide. C_3O_2; mol wt 68.03. C 52.96%, O 47.04%. $O=C=C=C=O$. Prepn by thermal decompn of malonic acid: Glemser in *Handbook of Preparative Inorganic Chemistry,* vol. 1, G. Brauer, Ed. (Academic Press, New York, 2nd ed., 1963) p 648. Reactions in organic synthesis: Dashkevich, Beilin, *Russ. Chem. Rev.* **36**, 391 (1967). Comprehensive reviews: Reyerson, Kobe, *Chem. Revs.* **7**, 479 (1930); Vol'kenshtein, *Uspeki Khim.* **4**, 610 (1935); Grauer, *Chimia* **14**, 11 (1960); T. Kappe, E. Ziegler, *Angew. Chem. Int. Ed.* **13**, 491-504 (1974), reprinted in *New Synthetic Methods* vol. 1 (Verlag Chemie, Weinheim, 1975) pp 29-69.

Colorless, highly refractive liquid or colorless gas which burns with a blue, sooty flame. Odor like acrolein and mustard oil. mp $-111.3°$. bp_{760} 6.8°. d_4^0 1.114. n_D^0 1.45384;

n_D^{12} 1.46757. Vapor pressure at 0°: 587-589 mm. Explosive limits, 6 to 30 vol % in air. Dipole moment: 0.7D. Thermodynamic constants: Thompson, *Trans. Faraday Soc.* **37**, 249 (1941). The gas can be stored at pressures of up to 100 mm, but even at these pressures polymerization may occur, giving a red, water-sol product. This invariably occurs at higher pressure or in the liquid state. Polymerization facilitated by presence of P_2O_5. Dec when passed through heated glass tubes, forming a mirror surface. Difficultly sol in carbon disulfide, xylene. With water forms malonic acid quantitatively. Forms malonamide with ammonia.

Caution: In small amounts acts as a lacrimator; in high concns attacks eyes, nose, respiratory organs, producing a feeling of suffocation.

USE: Prepn of malonates; improving dye affinity of fibers.

1864. Carbon Tetrachloride. *Tetrachloromethane;* perchloromethane; Necatorina; Benzinoform. CCl_4; mol wt 153.82. C 7.81%, Cl 92.19%. Obtained from carbon disulfide and chlorine in presence of a catalyst, e.g., $SbCl_5$, Fe filings, or by the chlorination of hydrocarbons: Faith, Keyes & Clark's *Industrial Chemicals*, F. A. Lowenheim, M. K. Moran, Eds. (Wiley-Interscience, New York, 4th ed., 1975) pp 230-234; H. D. DeShon in Kirk-Othmer *Encyclopedia of Chemical Technology* vol. 5 (Wiley-Interscience, New York, 3rd ed., 1979) pp 704-714. Toxicity data: Svirbely, *J. Ind. Hyg. Toxicol.* **29**, 382 (1947); E. Browning, *Toxicity and Metabolism of Industrial Solvents* (Elsevier, New York, 1965) pp 173-188. Use in induction of experimental liver disease: P. Trivedi, A. P. Mowat, *Br. J. Exp. Pathol.* **64**, 25 (1983). Review of carcinogenic risk: *IARC Monographs* **20**, 371-399 (1979); of toxicity studies: R. O. Recknagel *et al.*, *Pharmacol. Ther.* **43**, 139-154 (1989).

Colorless, clear, nonflammable, heavy liquid; characteristic odor. May form phosgene when used to put out electrical fires. *Use only when adequate ventilation is possible.* d_{25}^{25} 1.589. bp 76.7°. mp −23°. n_D^{20} 1.4607. One ml dissolves in 2000 ml water; misc with alcohol, benzene, chloroform, ether, carbon disulfide, petr ether, oils. LC_{50} for mice: 9528 ppm (Svirbely). LD_{50} in rats, mice, dogs (g/kg): 2.92, 12.1-14.4, 2.3 orally; LD_{50} in mice (g/kg): 4.1 i.p., 30.4 s.c. (IARC, 1979).

Caution: Poisoning may occur by inhalation, ingestion or skin absorption. *Acute:* exposure to high concentrations results in depression of the central nervous system manifested as loss of consciousness, dizziness, vertigo, headache, depression, mental confusion, incoordination, and gastrointestinal disturbances such as nausea, vomiting, abdominal pain and diarrhea. Acute exposure may result in renal damage leading to anuria and azotemia, and liver injury. *Chronic:* associated with liver and kidney damage. Skin contact can lead to dermatitis through defatting action. Inhalation may cause pulmonary edema and alveolitis. Concurrent intake with significant amounts of alcohol increases probability of injury. *See Patty's Industrial Hygiene and Toxicology* vol. 2B, G. D. Clayton, F. E. Clayton, Eds. (Wiley-Interscience, New York, 3rd ed., 1981) p 3472-3478; *Clinical Toxicology of Commercial Products*, R. E. Gosselin *et al.*, Eds. (Williams & Wilkins, Baltimore, 5th ed., 1984) Section III, p 101-107. This substance may reasonably be anticipated to be a carcinogen: *Seventh Annual Report on Carcinogens* (PB95-109781, 1994) p 117.

USE: As solvent for oils, fats, lacquers, varnishes, rubber waxes, resins; starting material in manuf of organic compds; grain fumigant. Pharmaceutic aid (solvent). Formerly used as dry cleaning agent and fire extinguisher.

THERAP CAT: Formerly as anthelmintic (Nematodes).

THERAP CAT (VET): Anthelmintic.

1865. Carbon Tetrafluoride. *Tetrafluoromethane;* Freon 14. CF_4; mol wt 88.00. C 13.65%, F 86.35%. Prepd from carbon or carbon monoxide and fluorine: Yost, *Inorg. Syn.* **1**, 34 (1939); Simons, Block, *J. Am. Chem. Soc.* **61**, 2962 (1939); Kwasnik in *Handbook of Preparative Inorganic Chemistry* vol 1, G. Brauer, Ed. (Academic Press, New York, 2nd ed., 1963) p 203. May also be prepd from SiC + F_2: Priest, *Inorg. Syn.* **3**, 178 (1950).

Colorless, odorless gas. Thermally stable. Chemically very inert. d (solid, −195°) 1.98. d (liq, −183°) 1.89. mp −183.6°. bp −127.8°. May be stored in steel cylinders.

Caution: Narcotic in high concns.

USE: Low temp refrigerant; gaseous insulator.

1866. Carbon Tetraiodide. *Tetraiodomethane.* CI_4; mol wt 519.63. C 2.31%, I 97.69%. Prepd by the interaction of carbon tetrachloride and aluminum or calcium iodide: Gustavson, *Ann.* **172**, 173 (1874); boron iodide: Moissan, *Compt. Rend.* **113**, 19 (1891); lithium or calcium iodide: Lantenois, *ibid.* **156**, 1385 (1913); ethyl iodide in the presence of aluminum chloride: Walker, *J. Chem. Soc.* **85**, 1090 (1904); McArthur, Simons, *Inorg. Syn.* **3**, 37 (1950).

Red cubic crystals. Odor of iodine. Dec to iodine and tetraiodoethylene under the influence of light or heat. d_4^{20} 4.32. mp 171°. Sol in benzene, chloroform. Dec by hot alcohol. Practically insol in water, but hydrolyzes slowly in contact with water, forming iodoform and iodine.

1867. *N,N'*-Carbonyldiimidazole. *1,1'-Carbonylbis-1H-imidazole.* $C_7H_6N_4O$; mol wt 162.15. C 51.85%, H 3.73%, N 34.55%, O 9.87%. Prepd from phosgene and imidazole in dry tetrahydrofuran or dry benzene: Staab, *Ann.* **609**, 75 (1957); Anderson, Paul, *J. Am. Chem. Soc.* **80**, 4423 (1958).

Crystals from tetrahydrofuran or benzene, mp 115.5-116°. Should be handled under exclusion of atmospheric moisture. Hydrolyzed by water in a few sec with evolution of CO_2.

USE: In the synthesis of peptides. Reacts readily with carboxylic acids to form acyl imidazoles; subsequent reaction with amines to form amides goes smoothly.

1868. Carbonyl Fluoride. *Carbonic difluoride;* fluophosgene. CF_2O; mol wt 66.01. C 18.20%, F 57.56%, O 24.24%. COF_2. Prepd from CO and F_2 or BrF_3 and CO: Ruff, Miltschitzky, *Z. Anorg. Allgem. Chem.* **221**, 154 (1935); Kwasnik in *Handbook of Preparative Inorganic Chemistry*, vol. 1, G. Brauer, Ed. (Academic Press, New York, 2nd ed., 1963) p 206. Alternate route from CO + AgF_2: Farlow *et al.*, *Inorg. Syn.* **6**, 155 (1960).

Pungent, very hygroscopic gas. d (solid, −190°): 1.388. d (liq, −114°): 1.139. mp −114.0°. bp −83.1°. Heat of formation: 166.6 kcal. Instantly hydrolyzed by water.

Caution: A strong irritant to skin, eyes, mucous membranes, respiratory tract.

1869. Carbophenothion. *Phosphorodithioic acid S-[[(4-chlorophenyl)thio]methyl] O,O-diethyl ester;* S-[[(p-chlorophenyl)thio]methyl] *O,O*-diethyl phosphorodithioate; *O,O*-diethyl *S*-(p-chlorophenylthio)methyl phosphorodithioate; R-1303; Garrathion; Trithion. $C_{11}H_{16}ClO_2PS_3$; mol wt 342.87. C 38.53%, H 4.70%, Cl 10.34%, O 9.33%, P 9.03%, S 28.06%. Prepn: Fancher, U.S. pat. **2,793,224** (1957 to Stauffer). Toxicity study: T. B. Gaines, *Toxicol. Appl. Pharmacol.* **14**, 515 (1969).

Light amber liquid. $bp_{0.01}$ 82°. d_4^{25} 1.271. n_D^{25} 1.5970; n_D^{26} 1.6198 (Fancher). Very low vapor pressure. Practically insol in water. Miscible with vegetable oils and most organic solvents. LD_{50} in female, male rats (mg/kg): 10, 30 orally; 27, 54 dermally (Gaines).

Caution: Cholinesterase inhibitor.

USE: Miticide; insecticide.

1870. Carboplatin. *(SP-4-2)-Diammine[1,1-cyclobutanedicarboxylato(2—)-O,O']platinum; 1,1-cyclobutanedicar-*

boxylic acid platinum complex; cis-diammine(1,1-cyclobu-tanedicarboxylato)platinum(II); CBDCA; JM-8; NSC-241240; Paraplatin. $C_6H_{12}N_2O_4Pt$; mol wt 371.25. C 19.41%, H 3.26%, N 7.55%, O 17.24%, Pt 52.55%. Analog of cisplatin, *q.v.*, with reduced nephrotoxicity. Prepn and antitumor activity: **Neth. pat. Appl. 7,307,863;** M. J. Cleare *et al.,* **U.S. pat. 4,140,707** (1973, 1979 both to Research Corp.). Improved prepn: R. C. Harrison *et al., Inorg. Chim. Acta* **46,** L15 (1980). Crystal structure: S. Neidle *et al., J. Inorg. Biochem.* **13,** 205 (1980). Comparison with other antitumor platinum complexes: M. J. Cleare *et al., Biochimie* **60,** 835 (1978). Early clinical studies: A. H. Calvert *et al., Cancer Chemother. Pharmacol.* **9,** 140 (1982). Clinical pharmacokinetics: S. J. Harland *et al., Cancer Res.* **44,** 1693 (1984); M. J. Egorin *et al., ibid.* 5432. Toxicity, activity in mice, rats, dogs: P. Lelieveld *et al., Eur. J. Cancer Clin. Oncol.* **20,** 1087 (1984). Comparison with cisplatin chemotherapy in advanced seminoma: M. J. Peckham *et al., Brit. J. Cancer* **52,** 7 (1985). In treatment of small cell lung cancer: I. E. Smith *et al., Cancer Treat. Rep.* **69,** 43 (1985).

White crystals, sol in water. LD_{50} in mice (mg/kg): 150 i.p., 140 i.v.; in rats (mg/kg): 85 i.v. (Lelieveld).

THERAP CAT: Antineoplastic.

1871. Carboprost. *(5Z,9α,11α,13E,15S)-9,11,15-Trihy-droxy-15-methylprosta-5,13-dien-1-oic acid; 7-[3,5-dihydroxy-2-(3-hydroxy-3-methyl-1-octenyl)cyclopentyl]-5-heptenoic acid;* (15S)-15-methyl $PGF_{2α}$; U-32921. $C_{21}H_{36}O_5$; mol wt 368.51. C 68.45%, H 9.85%, O 21.71%. Analog of prostaglandin $F_{2α}$, *q.v.* Prepn: G. L. Bundy *et al.,* **Ger. pat. 2,121,980;** G. L. Bundy, **U.S. pat. 3,728,382** (1971, 1973 both to Upjohn); *eidem, Ann. N.Y. Acad. Sci.* **180,** 76 (1971); E. W. Yankee *et al., J. Am. Chem. Soc.* **96,** 5865 (1974). Biological activity: J. R. Weeks *et al., J. Pharmacol. Exp. Ther.* **186,** 67 (1973). Mechanism of action: A. I. Csapo, M. O. Pulkkinen, *Prostaglandins* **18,** 479 (1979). Clinical studies: P. C. Schwallie, K. R. Lamborn, *J. Reprod. Med.* **23,** 289 (1979); M. P. Mapa *et al., Int. J. Gynaecol. Obstet.* **20,** 125 (1982). Teratological study: G. M. Szczech *et al., Advan. Prostaglandin Thromboxane Res.* **4,** 157 (1978).

Tromethamine salt, $C_{25}H_{47}NO_8$, *carboprost trometamol, U-32921E, Hemabate, Prostin/15M.*

Methyl ester, $C_{22}H_{38}O_5$, *carboprost methyl, U-36384.* Crystals from ether/hexane, mp 55-56°. $[α]_D$ +24° (c = 0.81 in ethanol).

THERAP CAT: Oxytocic.

1872. Carboquone. *2-[2-[(Aminocarbonyl)oxy]-1-meth-oxyethyl]-3,6-bis(1-aziridinyl)-5-methyl-2,5-cyclohexadiene-1,4-dione; 2,5-bis(1-aziridinyl)-3-(2-hydroxy-1-methoxyethyl)-6-methyl-p-benzoquinone carbamate (ester);* 2,5-bis(1-aziridinyl)-3-(2-carbamoyloxy-1-methoxyethyl)-6-methyl-1,4-benzoquinone; carbazilquinone; Esquinon. $C_{15}H_{19}N_3O_5$; mol wt 321.33. C 56.07%, H 5.96%, N 13.08%, O 24.90%. Prepn and antitumor activity: Nakao *et al., Chem. Pharm. Bull.* **20,** 1968 (1972); T. Nakamura *et al.,* **Ger. pat. 1,905,-224;** H. Nakao *et al.,* **U.S. pat. 3,631,026** (1969, 1971 both to Sankyo). Effect on tumors in mice: Arakawa *et al., Gann* **61,** 535 (1970). Acute toxicity: H. Masuda, Y. Suzuki, *Oyo Yakuri* **8,** 501 (1974), *C.A.* **81,** 163349g (1974).

Red to reddish-brown crystals, mp 202° (dec). Slightly sol in chloroform, acetone, abs alcohol. Practically insol in water. LD_{50} in male mice, male rats (mg/kg): 6.09, 3.88 i.v.; 3.84, 3.16 i.p.; 30.8, 28.0 orally (Masuda, Suzuki).

THERAP CAT: Antineoplastic.

1873. Carbostyril. *2(1H)-Quinolinone;* 2-hydroxyquinoline; 2-quinolinol; 2(1H)-quinolone; *o*-aminocinnamic acid lactam. C_9H_7NO; mol wt 145.16. C 74.47%, H 4.86%, N 9.65%, O 11.02%. Prepn from quinoline by heating with potassium hydroxide to 225° under anhyd conditions: Tschitschibabin, *Ber.* **56,** 1883 (1923); **Ger. pat. 406,208;** *Chem. Zentr.* **1925,** I, 1536; *Frdl.* **14,** 515. Several other syntheses.

Prisms from methanol, mp 199-200°. Sublimes at atmospheric pressure without dec. A monohydrate has been obtained from a satd aq soln. Kb at 18° = 1.94 × 10^{-9}. Very sparingly sol in water: One gram dissolves in 950 ml H_2O at 22°. Sol in alcohol, ether, dil HCl. Forms easily hydrolyzed Na and K salts.

Compd with 1,3,5-trinitrobenzene, $(C_9H_7NO)_2 \cdot C_6H_3N_3O_6$, yellow needles, mp 178°.

1874. Carboxin. *5,6-Dihydro-2-methyl-N-phenyl-1,4-oxathiin-3-carboxamide;* 2,3-dihydro-5-carboxanilido-6-methyl-1,4-oxathiin; carbathiin; DCMO; Vitavax. $C_{12}H_{13}$-NO_2S; mol wt 235.31. C 61.25%, H 5.57%, N 5.95%, O 13.60%, S 13.63%. Systemic fungicide, effective against loose smut in cereals. Prepn: B. von Schmeling *et al.,* **U.S. pat. 3,249,499** (1966 to U.S. Rubber); M. Kulka *et al.,* **U.S. pat. 3,393,202** (1968 to Uniroyal). Fungicidal activity: B. von Schmeling, M. Kulka, *Science* **152,** 659 (1966). Inhibitor of succinate oxidation: D. E. Mathre, *Phytopathol.* **60,** 671 (1970); G. A. White, *Biochem. Biophys. Res. Commun.* **44,** 1212 (1971); P. C. Mowery *et al., ibid.* **71,** 354 (1976). Residue determn in crops: J. R. Lane, *J. Agr. Food Chem.* **18,** 409 (1970); H. R. Sisken, J. E. Newell, *ibid.* **19,** 738 (1971). Toxicity studies: T. V. Dyadicheva, *Gig. Tr. Prof. Zabol.* **1979**(2), 55, *C.A.* **90,** 146673b (1979).

Crystals, from ethanol or methanol, mp 93-95°. LD_{50} in rats, mice (mg/kg): 430, 3200 orally (Dyadicheva).

USE: Systemic plant fungicide.

1875. β-Carboxyaspartic Acid. *2-Amino-1,1,2-ethane-tricarboxylic acid;* Asa. $C_5H_7NO_6$; mol wt 177.11. C 33.91%, H 3.98%, N 7.91%, O 54.20%. Amino acid found in ribosomal proteins of *E. coli;* homolog of γ-carboxyglutamic acid, *q.v.* Isoln and synthesis of DL-form: J. J. Van Buskirk, W. M. Kirsch, *J. Am. Chem. Soc.* **103,** 3935 (1981). Synthesis of DL-form: E. B. Henson *et al., Tetrahedron* **37,** 2561 (1981). Kinetics of decarboxylation to aspartic acid: P. V. Hauschka *et al., Anal. Chem.* **108,** 57 (1980); M. R. Christy, T. H. Koch, *J. Am. Chem. Soc.* **104,** 1771 (1982); N. E. Dixon, A. M. Sargeson, *ibid.* 6716. Crystal structure and pKa values of DL-zwitterion: B. Richey *et al., Biochemistry* **21,** 4819 (1982).

DL-Form hydrochloride, $C_5H_8ClNO_6$, white crystalline powder. pK_1 0.8 ±0.2, pK_2 2.5 ±0.1, pK_3 4.7 ±0.1, pK_4 10.9 ±0.1.

1876. γ-Carboxyglutamic Acid. *3-Amino-1,1,3-propanetricarboxylic acid;* γ-carboxy-L-glutamic acid; L-γ-carboxyglutamic acid; Gla. $C_6H_9NO_6$; mol wt 191.14. C 37.70%, H 4.75%, N 7.33%, O 50.22%. Amino acid found in blood coagulation proteins (prothrombin, Factor VII, Factor IX, Factor X, *q.q.v.*), plasma proteins and proteins from calcified tissue. Presence of Gla confers metal binding properties to the protein. Discovery in prothrombin: J. Stenflo, *Proc. Nat. Acad. Sci. USA* **71**, 2730 (1974); G. L. Nelsestuen *et al., J. Biol. Chem.* **249**, 6347 (1974). Synthesis: S. Danishefsky *et al., J. Am. Chem. Soc.* **101**, 4385 (1979); of DL-form: H. R. Morris *et al., Biochem. Biophys. Res. Commun.* **62**, 856 (1975); W. Märki, R. Schwyzer, *Helv. Chim. Acta* **58**, 1471 (1975); B. Weinstein *et al., J. Org. Chem.* **41**, 3634 (1976). Resolution of D- and L-forms: W. Märki *et al., Helv. Chim. Acta* **60**, 798 (1977). Biosynthesis by vitamin K-dependent carboxylation of glutamic acid: C. T. Esmon *et al., J. Biol. Chem.* **250**, 4744 (1975). Metal binding properties: G. L. Nelsestuen *et al., Biochem. Biophys. Res. Commun.* **65**, 233 (1975); G. L. Nelsestuen, *J. Biol. Chem.* **251**, 5648 (1976); R. Robertson *et al., ibid.* **253**, 5880 (1978); S. P. Bajaj *et al., ibid.* **257**, 3726 (1982). HPLC determn in proteins, bone, urine: M. Kuwada, K. Katayama, *Anal. Biochem.* **117**, 259 (1981). Reviews: J. Stenflo, J. W. Suttie, *Ann. Rev. Biochem.* **46**, 157-172 (1977); R. E. Olson, J. W. Suttie, *Vitam. Horm.* **35**, 59-108 (1977); J. W. Suttie, *CRC Crit. Rev. Biochem.* **8**, 191-223 (1980); J. P. Burnier *et al., Mol. Cell. Biochem.* **39**, 191-207 (1981).

Crystals, mp 167-167.5° (dec). $[\alpha]_D^{20}$ +35.3° (c = 1 in 6N HCl).
DL-Form, white powder, mp 90-92°.

1877. Carboxymethylcellulose Sodium. *Carboxymethyl ether cellulose sodium salt;* sodium carboxymethylcellulose; sodium cellulose glycolate; Cethylose; CMC; Carmethose; Cel-O-Brandt; Glykocellon; Carbose D; Xylo-Mucine; Tylose MGA; Cellolax; Polycell. R_nOCH_2COONa. Prepd by treating alkali cellulose with sodium chloroacetate: Faith, Keyes & Clark's *Industrial Chemicals*, F. A. Lowenheim, M. K. Moran, Eds. (Wiley-Interscience, New York, 4th ed., 1975) pp 235-238. Review and bibliography: Ott, *Cellulose and Cellulose Derivatives*, New York, 1946 (2nd ed., 1955).

White granules. Soly in water depends on degree of substitution. Water-soluble CMC is available in various viscosities (5-2000 centipoises in 1% soln), and the soly is equally good in hot and cold water (difference from methyl cellulose). Also the presence of metal salts has little effect on the viscosity. Solns are stable between pH 2 and 10. Below pH 2 precipitation of a solid occurs, above pH 10 the viscosity decreases rapidly. pKa 4.30. The free acid is obtained from aq soln at pH 2.5 and may be precipitated with alcohol.
USE: In drilling muds, in detergents as a soil-suspending agent, in resin emulsion paints, adhesives, printing inks, textile sizes, as protective colloid in general. As stabilizer in foods. Pharmaceutic aid (suspending agent; tablet excipient; viscosity-increasing agent).

1878. Carboxypolymethylene. *Carbomer; carbopol;* carboxyvinyl polymer. A vinyl polymer with active carboxyl groups. Description: *Chem. & Eng. News* **36**, 64 (Sept. 29, 1958).

White powder. Highly ionic and slightly acidic. Reacts with fatty amines to form thick and stable emulsions of oils in water.
USE: Thickening, suspending, dispersing, emulsifying agent. In the cosmetic and textile printing fields, in printing inks, in emulsion-based lubricants, in pharmaceuticals, polishes, waxes, paints, waterproof and oil-proof coatings, in industrial specialties.

1879. Carbromal. *N-(Aminocarbonyl)-2-bromo-2-ethylbutanamide;* (α-bromo-α-ethylbutyryl)urea; (α-bromo-α-ethylbutyryl)carbamide; bromodiethylacetylurea; bromodiethylacetylcarbamide; Tildin; Adalin; Planadalin; Diacid; Addisomnol; Bromadal; Uradal; Nyctal. $C_7H_{13}BrN_2O_2$; mol wt 237.10. C 35.46%, H 5.53%, Br 33.70%, N 11.82%, O 13.50%. $(C_2H_5)_2CBrCONHCONH_2$. Prepd by heating urea (to ~50°) with α-bromo-α-ethylbutyryl bromide $(C_2H_5)_2C$-BrCOBr, see Ger. pat. 225,710 (1910); *Frdl.* **10**, 1160; *Chem. Zentr.* **1910**, II, 1008. Patent literature: Slotta, *Grundriss der modernen Arzneistoff-Synthese* (Stuttgart, 1931); H. P. Kaufmann, *Arzneimittel-Synthese* (Berlin, 1953).
Crystals, mp 116-119°. One gram dissolves in about 3000 ml water, 18 ml alcohol, 3 ml chloroform, 14 ml ether. It is very sol in boiling alcohol and dissolves in concd sulfuric, nitric and hydrochloric acids, from which it is precipitated on dilution with water. It is dissolved by solns of alkali hydroxides.
Caution: This substance may be habit forming and is listed in the U.S. Code of Federal Regulations, Title 21 Part 329.1 (1995).
THERAP CAT: Sedative, hypnotic.

1880. Carbubarb. *5-[2-[(Aminocarbonyl)oxy]ethyl]-5-butyl-2,4,6(1H,3H,5H)-pyrimidinetrione; carbamic acid ester with 5-butyl-5-(2-hydroxyethyl)barbituric acid;* 5-butyl-5-carbamoyloxyethylbarbituric acid; 5-butyl-5-(2-hydroxyethyl)-barbituric acid carbamate; carbamic acid ester with 5-(β-hydroxyethyl)-5-butylmalonylurea; 5-(β-hydroxyethyl)-5-butylmalonylurea carbamate; Nogexan. $C_{11}H_{17}N_3O_5$; mol wt 271.27. C 48.70%, H 6.32%, N 15.49%, O 29.49%. Prepn: **Fr.** pat. M1059 (1962 to Consortium Mondial des Grandes Marques), *C.A.* **59**, 7539c (1963); Buzas, **U.S.** pat. 3,150,137 (1964). Purification and toxicity: **Fr.** pat. M2633 (1964 to Interco Fribourg), *C.A.* **62**, 1671e (1965).

Crystals from ethanol, mp 192-194°. LD_{50} s.c. in mice: 1.4 g/kg (**Fr.** pat. M2633).
Note: This is a controlled substance (depressant) listed in the U.S. Code of Federal Regulations, Title 21 Part 1308.13 (1995).
THERAP CAT: Sedative, hypnotic.

1881. Carbutamide. *4-Amino-N-[(butylamino)carbonyl]benzenesulfonamide; 1-butyl-3-sulfanilylurea;* N^1-(butylcarbamoyl)sulfanilamide; N^1-sulfanilyl-N^2-butylurea; N^1-sulfanilyl-N^2-butylcarbamide; N-(4-aminobenzenesulfonyl)-N'-butylurea; aminophenurobutane; BZ-55; U-6987; Bucarban; Glucidoral; Glucofren; Invenol; Nadisan. $C_{11}H_{17}N_3O_3$-S; mol wt 271.34. C 48.69%, H 6.31%, N 15.49%, O 17.69%, S 11.82%. *Ref:* Achelis, Hardebeck, *Deut. Med. Wochenschr.* **80**, 1452 (1955); Achelis *et al., Arch. Exp. Pathol. Pharmakol.* **228**, 163 (1956). Prepn from butylurea and sulfanilamide: Samaniego, **Span.** pat. 229,696 (1956), *C.A.* **51**, 7413 (1957); E. Haack, A. Hagedorn, **U.S.** pat. 2,907,692 (1959 to Boehringer, Mann.).

Crystals, mp 144-145°. Sol in water at pH 5 to 8. LD_{50} s.c. in mice: 3 g/kg (Haack, Hagedorn).

THERAP CAT: Antidiabetic.

1882. Carbuterol. *[5-[2-[(1,1-Dimethylethyl)amino]-1-hydroxyethyl]-2-hydroxyphenyl]urea;* α-(t-butylaminomethyl)-4-hydroxy-3-ureidobenzyl alcohol. $C_{13}H_{21}N_3O_3$; mol wt 267.33. C 58.41%, H 7.92%, N 15.72%, O 17.95%. A β-adrenergic agonist related to isoproterenol, *q.v.*, with selectivity for airway smooth muscle receptors. Prepn: C. Kaiser, S. T. Ross, **Ger. pat.** **2,106,620;** *eidem,* **U.S. pat.** **3,763,-232** (1971, 1973 both to SKF); C. Kaiser *et al., J. Med. Chem.* **17,** 49 (1974). Pharmacology, mechanism of action, toxicity study: J. R. Wardell *et al., J. Pharmacol. Exp. Ther.* **189,** 167 (1974). Analysis in aq soln: L. J. Ravin *et al., J. Pharm. Sci.* **67,** 1523 (1978). Clinical study: T. D. James, H. A. Lyons, *J. Am. Med. Assoc.* **241,** 704 (1979).

Crystals, mp 174-176°. LD_{50} in mice (mg/kg): 32.8 i.v.; 3134.6 orally; in rats: 77.2 i.v. (Wardell).

Hydrochloride, $C_{13}H_{21}N_3O_3 \cdot HCl$, *SKF-40383, Bronsecur, Pirem.* Crystals from ethanol/ether, mp 205-207° (dec).

THERAP CAT: Bronchodilator.

1883. Cardamom Seed. Grains of paradise. Dried ripe seeds of *Elettaria cardamomum* Maton, *Zingiberaceae.* *Habit:* Malabar, cultivated in India, Ceylon, Guatemala. *Constit:* Resin; 2-8% essential oil, 1-2% fixed oil. The essential oil contains eucalyptol (cineol), sabinene, *d,α*-terpineol and acetate, borneol, limonene, terpinene, 1-terpinen-4-ol and its formate and acetate. The fixed oil consists of the glycerides of oleic, stearic, linoleic, palmitic, caprylic and caproic acids. The unsaponifiable matter from the fixed oil contains β-sitosterol. The seeds contain 1.3 γ of vitamin B_1/g. Traces of manganese have also been observed. The ash of the pod including the seeds consists of 24.8% SiO_2, 20.43% Na_2CO_3, 13.3% CaO, 2.54% Cl, 1.5% Al_2O_3 and 4.3% MnO. Other constituents of pods and seeds are pentosans and starch. Description of the cultivation, botany, and the processing of common and oriental cardamom varieties: Viehoever, Sung, *J. Am. Pharm. Assoc.* **26,** 872 (1937); *see also* E. Guenther, *The Essential Oils* **vol. V** (New York, 1952) pp 85-106.

USE: Flavoring baked goods, confectionery, curry powder. In the manufacture of oil of cardamom which is used for flavoring liqueurs. Pharmaceutic aid (flavor).

THERAP CAT: Adjuvant carminative.

THERAP CAT (VET): Has been used as a carminative.

1884. Cardiotoxin. One of the toxic principles from cobra venom. Isoln: Sarkar, *J. Indian Chem. Soc.* **21,** 227 (1947). A single peptide chain composed of 60 amino acid residues cross-linked by four disulfide bridges with leucine and asparagine at the N and C terminals respectively: Narita, Lee, *Biochem. Biophys. Res. Commun.* **41,** 339 (1970). Causes irreversible depolarization of cell membrane and contraction of skeletal and smooth muscle. Shows direct lytic and marked cardiotoxic activities, potentiated by phospholipase A and prevented by polyanions (heparin, RNA, gangliosides). Pharmacology: Lee *et al., Arch. Exp. Pathol. Pharmakol.* **259,** 360 (1968).

1885. 3-Carene. *3,7,7-Trimethylbicyclo[4.1.0]hept-3-ene;* Δ^3-carene; 4,7,7-trimethyl-3-norcarene; isodiprene. $C_{10}H_{16}$; mol wt 136.24. C 88.16%, H 11.84%. Constituent of turpentine. The turpentine from *Pinus sylvestris* L. may

contain as much as 42%, turpentine from *Pinus longifolia* Roxb., *Pinaceae* about 30%. Isoln and structure: Simonsen, *The Terpenes* **vol. II** (Cambridge, 1949) pp 64-72; Guenther, *The Essential Oils* **vol. II** (Van Nostrand, 1949) pp 49-51. Conformation: Acharya, *Tetrahedron Letters* **1966,** 4117. Absorption spectrum: Cole, *J. Chem. Soc.* **1954,** 3807.

d-Form, mobile liquid. Readily oxidized on exposure to air. Sweet and pungent odor, more agreeable than the odor of turpentine. d_{15}^{15} 0.8668; d_{30}^{30} 0.8586. bp_{705} 168-169°. bp_{200} 123-124°. $[\alpha]_D^{20}$ +7.69°. n_D^{30} 1.468. Practically insol in water. Miscible with fat solvents and oils.

d-Form nitrosate, $C_{10}H_{16}N_2O_4$, prepd from *d*-Δ^3-carene with amyl nitrite, acetic and nitric acid, prisms, dec 147.5°.

1886. Carfimate. α-*Ethynylbenzenemethanol carbamate;* α-ethynylbenzyl carbamate; carbamic acid α-ethynylbenzyl ester; α-ethynylbenzyl alcohol carbamate; phenylethynylcarbinol carbamate; Equilium; Nirvotin. $C_{10}H_9NO_2$; mol wt 175.19. C 68.56%, H 5.18%, N 8.00%, O 18.27%. Prepn: **Brit. pat.** **736,340** (1955 to Carlo Erba).

Crystals from alc, mp 86-87°.

THERAP CAT: Sedative; hypnotic.

1887. Cargutocin. *1-Butanoic acid-7-glycine-1,6-dicarbaoxytocin; 1-butyric acid-6-(L-2-aminobutyric acid)-7-glycineoxytocin;* 7-glycine-1,6-aminosuberic acid-oxytocin; [Gly⁷,Asu¹,⁶]oxytocin; Y-5350; Statocin. $C_{42}H_{65}N_{11}O_{12}$; mol wt 916.05. C 55.07%, H 7.15%, N 16.82%, O 20.96%. Analog of oxytocin, *q.v.* Prepn: S. Sakakibara *et al.,* **Ger. pat.** **2,056,298** corresp to **U.S. pat.** **3,749,705** (1971, 1973 both to Yoshitomi). Synthesis and uterotonic activity: T. Yamanaka, S. Sakakibara, *Bull. Chem. Soc. Japan* **47,** 1228 (1974). Pharmacological and biochemical properties: T. Oka *et al., Japan. J. Pharmacol.* **25,** 15 (1975). Comparison with prostaglandin $F_{2\alpha}$ as oxytocic: T. Hashimoto, *Iryo* **30,** 1121 (1976), *C.A.* **87,** 554e (1977). Distribution, metabolism, excretion: Y. Kato *et al., Iyakuhin Kenkyu* **12,** 433 (1981), *C.A.* **95,** 126386k (1981). Teratogenicity studies: Y. Hamada *et al., ibid.* **10,** 26, 41 (1979), *C.A.* **91,** 639u, 83856m (1979).

$[\alpha]_D^{25}$ −44.0° (c = 0.55 in water).

THERAP CAT: Oxytocic.

1888. Carindacillin. *[2S-(2α,5α,6β)]-6-[[3-[(2,3-Dihydro-1H-inden-5-yl)oxy]-1,3-dioxo-2-phenylpropyl]amino]-3,3-dimethyl-7-oxo-4-thia-1-azabicyclo[3.2.0]heptane-2-carboxylic acid; N-(2-carboxy-3,3-dimethyl-7-oxo-4-thia-1-azabicyclo-[3.2.0]hept-6-yl)-2-phenylmalonamic acid 1-(5-indanyl) ester;* 6-[2-(5-indanyloxycarbonyl)phenylacetamido]-3,3-dimethyl-7-oxo-4-thia-1-azabicyclo[3.2.0]heptane-2-carboxylic acid; 1-(5-indanyl) N-(2-carboxy-3,3-dimethyl-7-oxo-4-thia-1-azabicyclo[3.2.0]hept-6-yl)-2-phenylmalonamate; 6-[2-phenyl-2-(5-indanyloxycarbonyl)acetamido]penicillanic acid; 6-(2-

carboxy-2-phenylacetamido)-3,3-dimethyl-7-oxo-4-thia-1-azabicyclo[3.2.0]heptane-2-carboxylic acid 6-(5-indanyl ester); α-(5-indanyloxycarbonyl)benzylpenicillin; carbenicillin indanyl ester; CP-15464. $C_{26}H_{26}N_2O_6S$; mol wt 494.57. C 63.14%, H 5.30%, N 5.66%, O 19.41%, S 6.48%. Semisynthetic antibiotic related to penicillin. Prepn: Butler, **S. Afr.** pat. **69 00,060**; **Ger.** pat. **1,944,376** corresp to U.S. pats. **3,557,090** and **3,574,189** (1969, 1970, 1971, 1971); Nakanishi, **Ger.** pat. **1,959,569** (1970), *C.A.* **73**, 66565b (1970) (all to Pfizer). Activity studies: Butler, *Del. Med. J.* **43**, 366 (1971); English *et al.*, *Antimicrob. Ag. Chemother.* **1**, 185 (1972). Clinical evaluation: Wallace *et al.*, *Antimicrob. Ag. Chemother.* **1970**, 223; Bran *et al.*, *Clin. Pharmacol. Ther.* **12**, 525 (1971); *Indanyl Carbenicillin*, H. Swarz, F. E. Storari, Eds. (Am. Elsevier, New York, 1974) 100 pp.

Sodium salt, $C_{26}H_{25}N_2NaO_6S$, *CP-15464-2*, *Carindapen*, *Geocillin*, *G.U.-Pen*.
THERAP CAT: Antibacterial.

1889. Carisoprodol. *(1-Methylethyl)carbamic acid 2-[[(aminocarbonyl)oxy]methyl]-2-methylpentyl ester; N-isopropyl-2-methyl-2-propyl-1,3-propanediol dicarbamate;* isopropyl meprobamate; isobamate; carisoprodate; Apesan; Arusal; Caprodat; Carisoma; Domarax; Flexal; Flexartal; Miolisodal; Mioril; Rela; Relasom; Sanoma; Soma; Somadril; Somalgit. $C_{12}H_{24}N_2O_4$; mol wt 260.33. C 55.36%, H 9.29%, N 10.76%, O 24.58%. Prepn: Berger, Ludwig, U.S. pat. **2,937,119** (1960 to Carter Prod.). Pharmacology: Berger *et al.*, *J. Pharmacol. Exp. Ther.* **127**, 66 (1959).

Crystals, mp 92-93°. Slightly bitter taste. Very sparingly sol in water: 30 mg/100 ml at 25°; 140 mg/100 ml at 50°. Sol in many common organic solvents. Practically insol in vegetable oils. Stable in dil acids and alkalies. LD_{50} in mice, rats (mg/kg): 2340, 1320 orally; 980, 450 i.p. (Berger).
THERAP CAT: Muscle relaxant (skeletal).

1890. Carlsbad Salt Artificial. A mixture of 1 part potassium sulfate, 9 sodium chloride, 18 sodium bicarbonate and 22 anhydr sodium sulfate. One gram of the mixture in 200 ml water is approx equivalent to an equal vol of Carlsbad water (Sprudel). *Keep dry and well closed.*

Carlsbad salt artificial, effervescent. A mixture of 25 parts of Carlsbad salt artificial, 48 sodium bicarbonate, 17 tartaric acid and 25 citric acid.
THERAP CAT: Cathartic.
THERAP CAT (VET): Purgative.

1891. Carminic Acid. *7-α-D-Glucopyranosyl-9,10-dihydro-3,5,6,8-tetrahydroxy-1-methyl-9,10-dioxo-2-anthracenecarboxylic acid;* C.I. Natural Red 4; C.I. 75470. $C_{22}H_{20}O_{13}$; mol wt 492.39. C 53.66%, H 4.09%, O 42.24%. Glucosidal coloring matter from the scale insect *Coccus cacti* L., *Homoptera* (cochineal). The essential constituent of carmine. Isoln: Schunk, Marchlewski, *Ber.* **27**, 2979 (1894); Dimroth, Scheuer, *Ann.* **399**, 43 (1913). Structure: Dimroth, Kammerer, *Ber.* **53**, 471 (1920); Ali, Haynes, *J. Chem. Soc.* **1959**, 1033; revised structure: Bhatia, Venkataraman, *Indian J. Chem.* **3** (2), 92 (1965). *See also Colour Index* vol. 4 (3rd ed., 1971) p 4632.

Red prisms from alc, no distinct melting point, darkens at 120°. Has a deep red color in water and is yellow to violet in acid solns. uv max (water): 500 nm (ε 6800); (0.02N HCl): 490-500 nm (ε 5800); (0.0001N NaOH): 540 nm (ε 3450). $[\alpha]_{654}^{15}$ +51.6° (water). Sol in water, alc, concd H_2SO_4; slightly sol in ether. Practically insol in petr ether, benzene, chloroform.

Methyl tetra-O-methylcarminate, $C_{27}H_{30}O_{13}$, yellow needles from benzene + petr ether, mp 185-188°.

Aluminum calcium lake, *carmine*. Bright-red, light pieces; easily reduced to powder. Practically insol in cold water or dil acids. Partly sol in hot water; sol in solns of alkali hydroxides or their carbonates giving deep red solns; also sol in borax.

USE: Free acid in color photography; pigment for artists' paints; as bacteriol. stain; rarely now as acid-base indicator or as oxidimetric indicator; as a reagent for aluminum; as a complexing agent for cations. Aluminum calcium lake as dye; in inks; coloring food products and galenicals; in microscopy for making various stains. Approved by the FDA for use in foods and drugs.

1892. Carmofur. *5-Fluoro-N-hexyl-3,4-dihydro-2,4-dioxo-1(2H)-pyrimidinecarboxamide;* 1-(n-hexylcarbamoyl)-5-fluorouracil; HCFU; Mifurol; Yamaful. $C_{11}H_{16}FN_3O_3$; mol wt 257.26. C 51.36%, H 6.27%, F 7.38%, N 16.33%, O 18.66%. Orally active cytostatic deriv of fluorouracil, *q.v.* Prepn: S. Ozaki *et al.*, **Japan. Kokai** 77 78886, *C.A.* **87**, 152265 (1977); *eidem*, **U.S.** pat. **4,071,519** (1977, 1978 both to Mitsui); *eidem*, *Bull. Chem. Soc. Japan* **50**, 2406 (1977). Anti-tumor activity: A. Hoshi *et al.*, *Chem. Pharm. Bull.* **26**, 161 (1978). Determn in body fluids: S. Watanabe *et al.*, *Chemotherapy (Tokyo)* **27**, 778 (1979); O. Nakajima *et al.*, *J. Chromatog.* **225**, 91 (1981). General pharmacology: Z. Henmi *et al.*, *Oyo Yakuri* **19**, 369 (1980), *C.A.* **93**, 125639s (1980). Pharmacokinetics: M. Iigo *et al.*, *Cancer Chemother. Pharmacol.* **4**, 189 (1980). Metabolism: *eidem*, *Xenobiotica* **10**, 847 (1980); T. Kobari *et al.*, *ibid.* **11**, 57 (1981). Phase I clinical study: Y. Koyama, Y. Koyama, *Cancer Treat. Rep.* **64**, 861 (1980). Mutagenicity study: Y. Seino *et al.*, *Cancer Res.* **38**, 2148 (1978). Review: T. Taguchi, *Recent Results Cancer Res.* **70**, 125-132 (1980).

White cryst from ethanol, mp 110-111°. uv max (chloroform): 258 nm (ε 1.16 × 10⁴), S. Ozaki *et al.*, *Bull. Chem. Soc. Japan* **50**, 2406 (1977). Also reported as mp 283° (dec), *eidem*, **U.S.** pat. **4,071,519**.
THERAP CAT: Antineoplastic.

1893. Carmoxirole. *3-[4-(3,6-Dihydro-4-phenyl-1(2H)-pyridinyl)butyl]-1H-indole-5-carboxylic acid.* $C_{24}H_{26}N_2O_2$; mol wt 374.48. C 76.98%, H 7.00%, N 7.48%, O 8.54%. Selective, presynaptic dopamine D_2-receptor agonist. Prepn: H. Böttcher *et al.*, **Ger.** pat. **3,342,632**; *eidem*, **U.S.** pat. **5,256,673** (1985, 1993 both to E. Merck); H. Böttcher, R. Gericke, *Ann.* **1988**, 749. Neurochemical profile: A. F. Haase *et al.*, *Arch. Pharmacol.* **343**, 588 (1991). Pharmacology: G. Haeusler *et al.*, *Eur. Heart J.* **13**, Suppl. D, 129 (1992). Pharmacokinetics and clinical evaluation: W. Meyer *et al.*, *ibid* 121.

Crystals, mp 284-285°.

Hydrochloride, $C_{24}H_{25}N_2O_2.HCl$, *EMD-45609*. Crystals obtained as 0.25 H_2O, mp 298-299°. uv max (methanol): 242, 281 nm.

THERAP CAT: Antihypertensive.

1894. Carmustine. *N,N'-Bis(2-chloroethyl)-N-nitroso-urea;* BCNU; BiCNU; NSC-409962; Becenun; Carmubris; Nitrumon. $C_5H_9Cl_2N_3O_2$; mol wt 214.05. C 28.06%, H 4.24%, Cl 33.13%, N 19.63%, O 14.95%. Chloroethylnitro-sourea derivative with antitumor activity. Similar to chloro-zotocin, lomustine, nimustine, ranimustine, *q.q.v.* Synthesis: Johnston *et al., J. Med. Chem.* **6**, 669 (1963). Properties: Loo *et al., J. Pharm. Sci.* **55**, 492 (1966). Decompn studies as related to antileukemic activity: Montgomery *et al., J. Med. Chem.* **10**, 668 (1967). Antifungal action: Hunt, Pittilo, *Antimicrob. Ag. Chemother.* **1965**, 710. Toxicology studies: Thompson, Larson, *Toxicol. Appl. Pharmacol.* **21**, 405 (1972). Review of pulmonary toxicity: A. C. Smith, *Pharmacol. Ther.* **41**, 443-460 (1989).

Light yellow powder that melts to an oily liquid; mp 30-32°. Both powder and liquid are stable. Dec rapidly in acid and in soln above pH 7. Most stable in petroleum ether or aqueous soln at pH 4. Non-ionized at pH 7 with consequent high lipid solubility. Sol in water up to 4 mg/ml and in 50% ethanol up to 150 mg/ml: DeVita *et al., Cancer Res.* **25**, 1876 (1965). LD_{50} in mice (mg/kg): 19-25 orally, 26 i.p., 24 s.c.; in rats (mg/kg): 30-34 orally (Thompson, Larson).

Note: This substance may reasonably be anticipated to be a carcinogen: *Seventh Annual Report on Carcinogens* (PB95-109781, 1994) p 101.

THERAP CAT: Antineoplastic.

1895. Carnauba Wax. Brazil wax. An exudate from the pores of the leaves of the Brazilian wax palm tree *Copernicia prunifera* (Muell.) H. E. Moore [*Copernicia cerifera* (Arruda da Camara) Mart.], *Palmae.* The botany of the tree and the native wax-collecting procedures are described adequately by A. H. Warth, *The Chemistry and Technology of Waxes* (Reinhold, New York, 1947). The hardness and high-polish capability of this important wax can be ascribed to the presence of esters of hydroxylated unsaturated fatty acids having about 12 carbon atoms in the acid chain. The usual names for the constituents, *i.e.* cerotic acid, melissyl cerotate, carnaubic acid etc., are meaningless. Brief review: C. S. Letcher in Kirk-Othmer *Encyclopedia of Chemical Technology* vol. 24 (Wiley-Interscience, New York, 3rd ed., 1984) pp 469-470.

Hard greenish solid, cryst fracture. Sharp, characteristic, not unpleasant odor upon melting. mp 82-85.5°. d 0.990 to 0.999. Saponification number 78 to 89. Iodine number about 13. n_D^{90} 1.4500. Sparingly soluble in fat solvents at 25°, quite sol at 45°.

USE: Wherever a hard, high-polish wax is desired, *e.g.* in automobile waxes, floor wax emulsions, high quality shoe polishes, in the paper industry (especially for making carbon papers). As a plasticizer in dental impression compounds. To raise the melting point of other waxes; often used together with candelilla wax. The presence of the lower-melting

ouricury wax is considered as an adulteration. Purified and bleached carnauba wax is used for cosmetic materials, such as depilatories and deodorant sticks. In pharmacy as the last stage in tablet coating. Skin sensitization or irritation by carnauba wax seems infrequent.

1896. Carnegine. *(±)-1,2,3,4-Tetrahydro-6,7-dimeth-oxy-1,2-dimethylisoquinoline;* 6,7-dimethoxy-1,2-dimethyl-1,2,3,4-tetrahydroisoquinoline; pectenine. $C_{13}H_{19}NO_2$; mol wt 221.30. C 70.56%, H 8.65%, N 6.33%, O 14.46%. In *Carnegiea gigantea* (Engelm.) Britt. & Rose *(Cereus giganteus* Engelm.), *Cactaceae,* a cactus of Arizona and Mexico. Extraction procedure: Heyl, *Arch. Pharm.* **266**, 668 (1928). Synthesis: Späth, *Ber.* **62**, 1021 (1929); Nakada, Nisgihara, *J. Pharm. Soc. Japan* **64**, 74 (1944). Pharmacology: E. Santi-Soncin, M. Furlanut, *Fitoterapia* **43**, 21 (1972), *C.A.* **80**, 66667f (1974).

Viscous liquid, dec on standing. Distills at 1 mm pressure and 170° air bath temp. Sol in ether, alc, chloroform. LD_{50} i.p. in mice: 15.23 mg/kg (Santi-Soncin, Furlanut).

Hydrochloride monohydrate, $C_{13}H_{19}NO_2.HCl.H_2O$, clusters from dil alc, mp 207° (mp 211° when anhydr). Sol in water, slightly sol in alc.

Hydrobromide monohydrate, $C_{13}H_{19}NO_2.HBr.H_2O$, needles from alc, mp 228°.

Methyliodide, $C_{13}H_{19}NO_2.CH_3I$, needles from methanol, mp 211° (evac tube) after drying at 100° and 10 mm.

1897. Carnidazole. *[2-(2-Methyl-5-nitro-1H-imidazol-1-yl)ethyl]carbamothioic acid O-methyl ester;* O-methyl [2-(2-methyl-5-nitroimidazol-1-yl)ethyl]thiocarbamate; R-25831; Spartrix. $C_8H_{12}N_4O_3S$; mol wt 244.27. C 39.34%, H 4.95%, N 22.94%, O 19.65%, S 13.13%. Prepn and antiprotozoal activity: J. Heeres *et al.,* **Ger.** pat. 2,429,755; *eidem,* U.S. pat. 3,928,374 (both 1975 to Janssen); *eidem, Eur. J. Med. Chem.-Chim. Ther.* **11**, 237 (1976). Crystal structure of monohydrate: N. M. Blaton *et al., Acta Crystallogr.* B35, 753 (1979). Clinical efficacy in vaginal trichomoniasis: A. Notowicz *et al., Brit. J. Vener. Dis.* **53**, 129 (1977); P. Chaudhuri, A. C. Drogendijk, *Eur. J. Obstet. Gynecol. Reprod. Biol.* **10**, 325 (1980).

Crystals from ethanol, mp 142.4° (dec).

THERAP CAT (VET): Antiprotozoal (Trichomonas).

1898. Carnitine. *3-Carboxy-2-hydroxy-N,N,N-trimethyl-1-propanaminium inner salt; (3-carboxy-2-hydroxypropyl)tri-methylammonium hydroxide, inner salt;* γ-amino-β-hydroxy-butyric acid trimethylbetaine; γ-trimethyl-β-hydroxybutyro-betaine; 3-hydroxy-4-(trimethylammonio)butanoate. $C_7H_{15}NO_3$; mol wt 161.20. C 52.16%, H 9.38%, N 8.69%, O 29.78%. $(CH_3)_3N^+CH_2CH(OH)CH_2COO^-$. Essential cofactor of fatty acid metabolism; constituent of striated muscle and liver. Isoln from meat extract: W. Gulewitsch, R. Krimberg, *Z. Physiol. Chem.* **45**, 326 (1905); S. P. Colowick, N. O. Kaplan, *Methods in Enzymology,* vol. **III** (Academic Press, New York, 1957) p 660. Isoln and characterization of Vitamin B_T as carnitine: H. E. Carter *et al., Arch. Biochem. Biophys.* **38**, 405 (1952). Isoln of D-carnitine: S. Friedman *et al., ibid.* **66**, 10 (1957). Synthesis of L-carnitine: M. Tomita, Y. Sendju, *Z. Physiol. Chem.* **169**, 263 (1927); R. Voeffray *et al., Helv. Chim. Acta* **70**, 2058 (1987). Synthesis of DL-carnitine: E. Strack *et al., Ber.* **86**, 525 (1953); H. E. Carter, P. K. Bhattacharyya, *J. Am. Chem. Soc.* **75**,

2503 (1953). Synthesis of DL-, D-, and L-carnitine from epichlorohydrin: E. Strack, I. Lorenz, *Z. Physiol. Chem.* **318**, 129 (1960). Prepn by hydrolysis of γ-dimethylamino-β-hydroxybutyronitrile chloromethylate: Ger. pat. **1,090,676** (1960 to Labaz). Biosynthesis: G. Wolf, C. R. A. Berger, *Arch. Biochem. Biophys.* **92**, 360 (1961). Absolute configuration: Kaneko, Yashida, *Bull. Chem. Soc. Japan* **35**, 1153 (1962). Metabolism: M. E. Mitchell, *Am. J. Clin. Nutr.* **31**, 293 (1978). Use of L-carnitine in hyperlipoproteinemia: C. Carazza, U.S. pat. **4,255,449** (1981); M. T. Ramacci, U.S. pat. **4,315,944** (1982 to Sigma-Tau). Effect on myocardial metabolism in coronary artery disease: R. Ferrari *et al.*, *Clin. Trials J.* **21**, 40 (1984). Clinical evaluation in congestive heart failure: O. Ghidini *et al.*, *Int. J. Clin. Pharmacol. Ther. Toxicol.* **26**, 217 (1988). *Monograph: Recent Research on Carnitine*, G. Wolf, Ed. (MIT Press, Cambridge, Mass., 1965). Book: *Carnitine Biosynthesis, Metabolism, and Functions*, R. A. Frenkel, J. D. McGarry, Eds. (Academic Press, New York, 1980) 356 pp. *Reviews:* G. Fraenkel, S. Friedman, *Vitam. Horm.* **XV**, 73-118 (1957); S. L. De Felice, *Orphan Drugs*, F. E. Karch, Ed. (Marcel Dekker, New York, 1982) pp 33-56. Reviews of nutritional significance and therapeutic applications: C. J. Rebouche, D. J. Paulson, *Ann. Rev. Nutr.* **6**, 41-66 (1986); of clinical pharmacology: J. J. Bahl, R. Bressler, *Ann. Rev. Pharmacol. Toxicol.* **27**, 257-277 (1987); of pharmacokinetics, therapeutic use: K. L. Goa, R . N. Brogden, *Drugs* **34**, 1-24 (1987); of metabolism and enzymology: L. L. Bieber, *Ann. Rev. Biochem.* **57**, 261-283 (1988).

DL-Form, hygroscopic crystalline solid, dec 195-197°.

DL-Form hydrochloride, $C_7H_{15}NO_3.HCl$, *Bicarnesine, Flatistine*. Needles from ethanol, mp 196° (dec). Very sol in water; sol in hot ethanol; slightly sol in cold ethanol. Practically insol in acetone, ether.

D-Form, crystals, dec 210-212°. $[\alpha]_D$ +30.9°. Very sol in water and alcohol. Practically insol in acetone and ether.

D-Form hydrochloride, $C_7H_{15}NO_3.HCl$, crystals, dec 142°.

L-Form, *levocarnitine, vitamin B_T, Cardiogen, Carnitene, Carnicor, Carnitor, Carnum, Carrier, Miocor, Miotonal, Vitacarn.* Crystals from ethanol + acetone, dec 197-198°. Very hygroscopic solid. $[\alpha]_D^{30}$ −23.9° (c = 0.86 in water). Sol in water, hot alcohol. Practically insol in acetone, ether, benzene.

L-Form hydrochloride, $C_7H_{15}NO_3.HCl$, *Lefcar, Metina*. Crystals, dec 142°.

THERAP CAT: L-Form as antihyperlipoproteinemic; DL-hydrochloride as gastric and pancreatic secretion stimulant.

1899. Carnosine. *N-β-Alanyl-ι-histidine*; ignotinc. $C_9H_{14}N_4O_3$; mol wt 226.24. C 47.78%, H 6.24%, N 24.76%, O 21.22%. Naturally occurring dipeptide found in muscle of man and of numerous animals, e.g., pigeons and geese, have *N*-methylcarnosine (anserine, *q.v.*) in their muscles. Isoln: Gulewitsch, Amiradzibi, *Ber.* **33**, 1902 (1900); Wolff, Wilson, *J. Biol. Chem.* **95**, 495 (1932); **109**, 565 (1935). Synthesis from histidine and β-iodo- or β-nitropropionyl chloride: Baumann, Ingvaldsen, *ibid.* **35**, 271 (1918); Barger, Tutin, *Biochem. J.* **12**, 406 (1918). Later syntheses: Sifford, du Vigneaud, *J. Biol. Chem.* **108**, 753 (1935); R. A. Turner, *J. Am. Chem. Soc.* **75**, 2388 (1953); F. J. Vinick, S. Jung, *J. Org. Chem.* **48**, 392 (1983). Crystal structure: H. Itoh *et al.*, *Acta Crystallogr.* **33B**, 2959 (1977). Studies on L-carnosine as a possible neurotransmitter; F. L. Margolis, *Science* **184**, 909 (1974). Possible role of L-carnosine in wound healing: D. E. Fischer *et al.*, *Proc. Soc. Exp. Biol. Med.* **158**, 402 (1978).

L-Form, the natural form, crystals from aqueous ethanol, mp 262° (dec) (Vinick, Jung); also reported as mp 260° (capillary tube) and as mp 308-309° (Dennis bar) (Sifford, du Vigneaud). $[\alpha]_D^{25}$ +21.0° (c = 1.5 in water). pK_1 2.64; pK_2 6.83; pK_3 9.51. Alkaline reaction. One gram dissolves in 3.1 ml water at 25°.

Nitrate, $C_9H_{15}N_5O_6$, crystals, dec 222°. $[\alpha]_D^{20}$ +24.1° (c = 1.5 in water). Very sol in water.

Hydrochloride, $C_9H_{15}ClN_4O_3$, crystals, dec 245°. Very sol in water.

D-Form, crystals, mp 260°. $[\alpha]_D^{28}$ −20.4° (c = 1.5).

1900. Caro's Acid. *Peroxymonosulfuric acid*; sulfomonoperacid; persulfuric acid. H_2O_5S; mol wt 114.08. H 1.77%, O 70.12%, S 28.11%. Dry reagent is prepd by stirring 10 g potassium persulfate into 11 g concd H_2SO_4 for 10 min and adding 30 g finely powdered potassium sulfate; liquid reagent is obtained by triturating potassium persulfate with three times as much (by weight) of H_2SO_4; dil reagent is prepd by stirring 10 g potassium persulfate into 11 g concd H_2SO_4 and adding 50 cc ice: Baeyer, Villiger, *Ber.* **32**, 3625 (1899); another prepn by reacting 90% H_2O_2 with chlorosulfonic acid at −40 to −50°: Ball, Edwards, *J. Am. Chem. Soc.* **78**, 1125 (1956).

The product is a sirupy liquid consisting of about equal amounts of Caro's acid and H_2SO_4. pK_2 of Caro's acid 9.4 ± 0.1. Oxygen is evolved at room temp; should be stored at dry ice temp.

Caution: Can be dangerously unstable, like most peroxides. Description of explosion at Brown University: J. O. Edwards, *Chem. & Eng. News* **33**, 3336 (1955). Explosion at Sun Oil, *ibid.* **38**, 59 (Nov. 21, 1960). May be highly irritating to skin, eyes, mucous membranes.

USE: In prepn of dyes; oxidation of olefins to α-glycols; oxidation of ketones to lactones or esters; treating woolens to prevent felting and shrinking; in bleaching compositions.

1901. α-Carotene. $C_{40}H_{56}$; mol wt 536.88. C 89.49%, H 10.51%. About as widely distributed as its β-isomer, but in smaller amounts. Best sources for both the α- and β-isomers are carrots, palm oil, and green leaves of various species. As a provitamin A it is half as active as β-carotene. Found in the mother liquors after crystallizing β-carotene. Isoln by chromatography: Karrer, Walker, *Helv. Chim. Acta* **16**, 641 (1933). Structure: Kuhn, Lederer, *Ber.* **64**, 1349 (1931); Karrer *et al.*, *Helv. Chim. Acta* **14**, 614 (1931); **16**, 975 (1933). Natural (+)-α-carotene has 6'R configuration: Eugster *et al.*, *ibid.* **52**, 1729 (1969). Synthesis of *dl*-α-carotene: Eugster, Karrer, *ibid.* **38**, 610 (1955); Tscharner *et al.*, *ibid.* **40**, 1676 (1957); Ruegg *et al.*, *ibid.* **44**, 985 (1961).

Deep purple prisms, polyhedra from petr ether or from benzene + methanol, mp 187.50° (evacuated tube). $[\alpha]_{6463}^{18}$ +385° (c = 0.08 in benzene). Absorption max (CHCl₃): 485, 454 nm. More sol than β-carotene. Freely sol in carbon disulfide, chloroform; sol in ether, benzene. Slightly sol in petr ether, alcohol. 100 ml hexane dissolves 294 mg at 0°. Practically insol in water, acids, alkalies. Absorbs oxygen from the air, giving rise to inactive colorless oxidation products. The oxidation in light is autocatalytic. *Store in darkness in sealed ampoules and at low temp (−20°C).*

THERAP CAT: Vitamin A precursor.

THERAP CAT (VET): Vitamin A precursor for all species except cats.

1902. β-Carotene. *β,β-Carotene*; Carotaben; Provatene; Solatene. $C_{40}H_{56}$; mol wt 536.88. C 89.49%, H 10.51%. Most important of the provitamins A. Widely distributed in the plant and animal kingdom. In plants it occurs almost always together with chlorophyll. Isoln from carrots: Willstätter, Escher, *Z. Physiol. Chem.* **64**, 47 (1910); Kuhn, Lederer, *Ber.* **64**, 1349 (1931); Barnett *et al.*, U.S. pat. **2,848,508** (1958). Chromatography: Karrer, Walker, *Helv. Chim. Acta* **16**, 641 (1933). Structure: Willstätter, Mieg, *Ann.* **355**, 1 (1907); Zechmeister *et al.*, *Ber.* **61**, 566 (1928); **66**, 123 (1933); Karrer *et al.*, *Helv. Chim. Acta* **12**, 1142 (1929); **13**, 1084 (1930); **14**, 1033 (1931); Kuhn,

Brockmann, *Ber.* **65,** 894 (1932); **66,** 1319 (1933); **67,** 1408 (1934); *Ann.* **516,** 95 (1935). Crystal structure: Sterling, *Acta Cryst.* **17,** 1224 (1964). Synthesis: Milas *et al., J. Am. Chem. Soc.* **72,** 4844 (1950); Karrer, Eugster, *Compt. Rend.* **250,** 1920 (1950); Inhoffen *et al., Chem. Ztg.* **74,** 285, 309 (1950); Surmatis, Ofner, *J. Org. Chem.* **26,** 1171 (1961); Rüegg *et al., Helv. Chim. Acta* **44,** 985 (1961); Bestmann *et al., Ann.* **1973,** 760; Fischli, Mayer, *Helv. Chim. Acta* **58,** 1584 (1975). Industrial mfg procedure: Isler *et al., Helv. Chim. Acta* **39,** 249 (1956); Isler *et al., U.S.* pat. **2,917,539** (1959 to Hoffmann-La Roche). Microbial production by *Choanephora trispora:* Zajic, U.S. pats. **2,959,521; 2,959,522** and **3,128,236** (1960, 1960 and 1964, all to Grain Processing); Miescher, *U.S.* pat. **3,001,912** (1961 to C.S.C.). *Review:* Fleming, *Selected Organic Syntheses* (John Wiley, London, 1973) pp 70-74.

Deep-purple, hexagonal prisms from benzene + methanol. Red, rhombic, almost square leaflets from petr ether. mp 183° (evacuated tube). Absorption max (chloroform): 497, 466 nm. Less sol than α-carotene. Sol in CS_2, benzene, chloroform. Moderately sol in ether, petr ether, oils. 100 ml hexane dissolve 109 mg at 0°. Very sparingly sol in methanol and ethanol. Practically insol in water, acids, alkalies. Dil solns are yellow. Absorbs oxygen from the air giving rise to inactive, colorless oxidation products. *Keep tightly closed and protected from light. Store at low temp (−20°C).* Commercial crystalline β-carotene has a vitamin A activity of 1.67 million U.S.P. units per gram. The I.U. of 0.6 μg β-carotene is almost exactly equivalent to 0.3 μg vitamin A.

USE: Yellow coloring agent for foods.

THERAP CAT: Vitamin A precursor. Ultraviolet screen.

THERAP CAT (VET): Vitamin A precursor for all species except cats.

1903. γ-Carotene. $C_{40}H_{56}$; mol wt 536.88. C 89.49%, H 10.51%. A rare carotenoid. Has provitamin A activity. Best source is *Penicillium sclerotiorum:* Mase *et al., Arch. Biochem. Biophys.* **68,** 150 (1957). Also present in small proportions in many plant materials, esp fruits containing β-carotene. Chromatographic isoln from crude carotenes: Kuhn, Brockmann, *Ber.* **66,** 407 (1933); Winterstein, *Z. Physiol. Chem.* **219,** 249 (1933). Structure: Kuhn, Brockmann, *loc. cit.; Naturwiss.* **21,** 44 (1933). Synthesis: Garbers *et al., Helv. Chim. Acta* **36,** 1783 (1953); Ruegg *et al., ibid.* **44,** 985 (1961).

Probably crystallizes in polymorphous forms. Synthetic form: red plates, mp 152-153.5°. Absorption max (petr ether): 437, 462, 494 nm ($E_{1cm}^{1\%}$ 2055, 3100, 2720). Natural form: minute, deep-red prisms with bluish luster from benzene + methanol. mp 177.5°. Absorption max (chloroform): 508.5, 475, 446 nm. Somewhat less sol than β-carotene. *Store in darkness in sealed ampoules at low temps (0°C).*

THERAP CAT: Vitamin A precursor.

THERAP CAT (VET): Vitamin A precursor for all species except cats.

1904. δ-Carotene. ϵ,ψ-*Carotene.* $C_{40}H_{56}$; mol wt 536.88. C 89.49%, H 10.51%. Extracted from the fruit of *Gonocaryum pyriforme* Mig., *Icacinaceae:* Winterstein, *Z. Physiol. Chem.* **219,** 249 (1933). Occurs also in carrots and certain varieties of tomatoes. Isoln from tomato mutants: Porter, Murphey, *Arch. Biochem. Biophys.* **32,** 21 (1951). Structure: Kargl, Quackenbush, *ibid.* **88,** 59 (1960). Synthesis: Manchand *et al., J. Chem. Soc.* **1965,** 2019. Ab-

solute configuration: Buchecker, Eugster, *Helv. Chim. Acta* **54,** 327 (1971).

Long orange-red needles from CS_2 + hexane + ethanol, mp 140.5°. $[\alpha]_{Cd}$ +317°; $[\alpha]_D^{25}$ +352° ±16% (hexane). Absorption spectrum: Kargl, Quackenbush, *loc. cit.*

1905. Carotol. *[3R-(3α,3aα,8aα)]-2,3,4,5,8,8a-Hexahydro-6,8a-dimethyl-3-(1-methylethyl)-3a(1H)-azulenol; 2,3,4,-5,8,8a-hexahydro-3-isopropyl-6,8a-dimethyl-3a(1H)-azulenol.* $C_{15}H_{26}O$; mol wt 222.37. C 81.02%, H 11.79%, O 7.19%. From oil of carrot seeds, *Daucus carota* L., *Umbelliferae:* Asahina, Tsukamoto, *J. Pharm. Soc. Japan* **525,** 961 (1925); *C.A.* **20,** 2845 (1926); Sorm *et al., Coll. Czech. Chem. Commun.* **16,** 47 (1951); Pigulevskii, Kovaleva, *Zh. Priklad. Khim.* **32,** 2703 (1959). Structure: Sykora *et al., Coll. Czech. Chem. Commun.* **26,** 788 (1961); Zalkow *et al., J. Org. Chem.* **26,** 981 (1961). Stereochemistry: Levisalles, Rudler, *Bull. Soc. Chim. France* **1964,** 2020; **1967,** 2059. Synthesis of (+)-form: DeBroissia *et al., ibid.* **1972,** 4314; *eidem, Chem. Commun.* **1972,** 855.

(+) - Carotol

Liquid. $bp_{2.5}$ 126°. $[\alpha]_D^{20}$ +30.4°. n_D^{20} 1.4964. d^{20} 0.9624.

1906. Caroverine. *1-[2-(Diethylamino)ethyl]-3-[(4-methoxyphenyl)methyl]-2(1H)-quinoxalinone;* 1-(diethylaminoethyl)-3-(p-methoxybenzyl)dihydro-2-quinoxalone; Spasmium. $C_{22}H_{27}N_3O_2$; mol wt 365.48. C 72.30%, H 7.45%, N 11.50%, O 8.76%. Prepn: Zellner *et al., U.S.* pat. **3,028,384** (1962 to Donau-Pharm.). Polarographic study: P. Pflegel, *Pharmazie* **24,** 667 (1969). Pharmacological study: F. Hahn *et al., Arch. Int. Pharmacodyn. Ther.* **199,** 108 (1972).

Crystals from isopropyl alcohol, mp 69°. $bp_{0.01}$ 202°. Hydrochloride, dec 188°.

THERAP CAT: Antispasmodic.

1907. Caroxazone. *2-Oxo-2H-1,3-benzoxazine-3(4H)-acetamide;* 3-acetamide-4H-1,3-benzoxazin-2-one; 4H-3-methylcarboxamide-3-benzoxazine-2-one; 4H-3-carbamoylmethyl-1,3-benzoxazin-2-one; FI-6654; Timostenil. $C_{10}H_{10}N_2O_3$; mol wt 206.20. C 58.25%, H 4.89%, N 13.59%, O 23.28%. Prepn: **Brit.** pat. **1,115,759;** L. Bernardi *et al., U.S.* pat. **3,427,313** (1965, 1969 both to Soc. Farma. Italia); and pharmacological properties: *eidem, Experientia* **24,** 774 (1968). Pharmacology: G. K. Suchowsky, L. Pegrassi, *Arzneimittel-Forsch.* **19,** 643 (1969); G. K. Suchowsky *et al., Eur. J. Pharmacol.* **6,** 327 (1969). Clinical study: S. Cecchini *et al., J. Int. Med. Res.* **6,** 388 (1978).

Solid, mp 203-205°.
THERAP CAT: Antidepressant.

1908. Carpaine. $C_{28}H_{50}N_2O_4$; mol wt 478.72. C 70.25%, H 10.53%, N 5.85%, O 13.37%. Found in *Carica papaya* L. and in *Vasconcellosia hastata* Caruel, *Caricaceae.* Isoln from papaya leaves: Greshoff, *Mededeel. uit's Lands. Plant., Buitenzorg.* **No. 7,** 5 (1890); Rapoport, Baldridge, *J. Am. Chem. Soc.* **73,** 343 (1951). Reportedly causes brady-cardia, CNS depression: Kakowski, *Arch. Int. Pharm.* **15,** 84 (1905). Structure and stereochemistry: Govindachari *et al., Tetrahedron Letters* **1965,** 1907. Absolute configuration: Coke, Rice, *J. Org. Chem.* **30,** 3420 (1965). Review of struc-tural studies: Govindachari, *J. Indian Chem. Soc.* **45,** 945 (1968).

Monoclinic prisms from acetone, mp 119-120°. Sublimes at 120° under 0.05 mm pressure. $[\alpha]_D^{12}$ +24.7° (c = 1.07 in ethanol). Slightly sol in water. Sol in most organic solvents except petr ether.

1909. Carpetimycins. Carbapenem antibiotics related to thienamycin and olivanic acids, *q.q.v.* Carpetimycins A, B, C and D are known. Prodn of A and B by *Streptomyces* strain No. C-19393, now named *Streptomyces griseus* subsp. *cryophilus,* and antibacterial properties: **Neth.** pat. **Appl. 8,000,628;** A. Imada *et al.,* U.S. pat. **4,518,529** (1980, 1985 to Takeda); A. Imada *et al., J. Antibiot.* **33,** 1417 (1980); by *Streptomyces* strain KC-6643: M. Nakayama *et al., ibid.* 1388. Structure and abs config: S. Harada *et al., J. Anti-biot.* **33,** 1425 (1980); M. Nakayama *et al., ibid.* **34,** 818 (1981). Mode of action study: Y. Nozaki *et al., ibid.* 206. Total synthesis of (±)-carpetimycin A: H. Natsugari *et al., J. Chem. Soc. Perkin Trans. I* **1983,** 403; M. Ihara *et al., Heterocycles* **20,** 2182 (1983); J. D. Buynak, M. N. Rao, *J. Org. Chem.* **51,** 1571 (1986). Synthesis of (−)-carpetimycin A: T. Iimori *et al., J. Am. Chem. Soc.* **105,** 1659 (1983); M. Aratani *et al., Tetrahedron Letters* **26,** 223 (1985); M. Shiba-saki *et al., ibid.* 2217.

Carpetimycin A: R = H
Carpetimycin B: R = SO₃H

Carpetimycin A, $C_{14}H_{18}N_2O_6S$, *[5R-[3[R*(E)],5α,6β]]-3-[[2-(acetylamino)ethenyl]sulfinyl]-6-(1-hydroxy-1-methyleth-yl)-7-oxo-1-azabicyclo[3.2.0]hept-2-ene-2-carboxylic acid, antibiotic Ab 651, antibiotic C-19393-H₂, antibiotic KA-6643-A, C-19393-H₂.* Colorless solid, mp > 145° (dec). $[\alpha]_D^{24}$ −27° (c = 1.7 in water). uv max (water): 240, 288 nm ($E_{1cm}^{1\%}$ 369, 300) (Nakayama, 1980).
Carpetimycin B, $C_{14}H_{18}N_2O_9S_2$, *antibiotic C-19393-S₂, an-tibiotic KA-6643-B, C-19393-S₂.* Colorless solid melting

above 130° (dec). $[\alpha]_D^{24}$ −145° (c = 1 in water). uv max (water): 240, 285 nm ($E_{1cm}^{1\%}$ 357, 305) (Nakayama, 1980).

1910. Carphenazine. *1-[10-[3-[4-(2-Hydroxyethyl)-1-piperazinyl]propyl]-10H-phenothiazin-2-yl]-1-propanone;* 10-[3-[4-(2-hydroxyethyl)-1-piperazinyl]propyl]-3-propionyl-phenothiazine; 2-propionyl-10-[γ-(N′-β-hydroxyethylpiper-azino)propyl]phenothiazine; Proketazine. $C_{24}H_{31}N_3O_2S$; mol wt 425.60. C 67.73%, H 7.34%, N 9.87%, O 7.52%, S 7.53%. Prepn: M. H. Sherlock, N. Sperber, U.S. pat. **2,985,654** (1961 to Schering); R. F. Tislow *et al.,* U.S. pat. **3,023,146** (1962 to Am. Home Products).

Dimaleate, $C_{24}H_{31}N_3O_2S.2C_4H_4O_4$, yellow crystals from isopropanol, mp 175-177° (Tislow). Also reported as crys-tals from methanol, mp 165-167° (Sherlock, Sperber).
THERAP CAT: Antipsychotic.

1911. Carpipramine. *1′-[3-(10,11-Dihydro-5H-dibenz-[b,f]azepin-5-yl)propyl]-[1,4′-bipiperidine]-4′-carboxamide;* 5-[3-(4-piperidino-4-carbamoylpiperidino)propyl]-10,11-dihy-dro-5H-dibenz[b,f]azepine; carbadipimidine; Bay b 4343 b. $C_{28}H_{38}N_4O$; mol wt 446.64. C 75.30%, H 8.58%, N 12.54%, O 3.58%. Prepn: M. Nakanishi, T. Munakata, **Japan.** pat. **66 6752; Brit.** pat. **1,017,986;** Nakanishi *et al.,* U.S. pat. **3,329,683** (1966, 1966, 1967 all to Yoshitomi); Nakanishi *et al., J. Med. Chem.* **13,** 644 (1970). Pharmacology: Naka-nishi *et al., Arzneimittel-Forsch.* **18,** 1435 (1968). Metabo-lism: Nakanishi *et al., J. Pharm. Soc. Japan* **90,** 204 (1970).

Dihydrochloride monohydrate, $C_{28}H_{38}N_4O.2HCl.H_2O$, *Defekton, Prazinil.* Crystals from 90% methanol, mp 260°. Sparingly sol in water. LD_{50} in mice, rats (mg/kg): 28.2, 37.0 i.v.; 136.0, 76.0 i.p.; 2180, 1025 orally (Nakanishi, 1968).
THERAP CAT: Antipsychotic.

1912. Carprofen. *6-Chloro-α-methyl-9H-carbazole-2-acetic acid;* C-5720; Ro-20-5720/000; Imadyl; Rimadyl. $C_{15}H_{12}ClNO_2$; mol wt 273.72. C 65.82%, H 4.42%, Cl 12.95%, N 5.12%, O 11.69%. Prepn: L. Berger, A. J. Corraz, **Ger.** pat. **2,337,340;** *eidem,* U.S. pat. **3,896,145** (1974, 1975 both to Hoffmann-La Roche). Pharmacology: L. O. Randall, H. Baruth, *Arch. Int. Pharmacodyn.* **220,** 94 (1976). Stereospecific assay and disposition: J. M. Kem-merer *et al., J. Pharm. Sci.* **68,** 1274 (1979). Metabolism: F. Rubio *et al., ibid.* **69,** 1245 (1980). Pharmacokinetic and clinical studies in gout: T. F. Yu, J. Perel, *J. Clin. Pharma-col.* **20,** 347 (1980). Use in treatment of rheumatoid arthri-tis: S. Jalava, *Scand. J. Rheumatol.* **1983,** Suppl. 48, 5-12. Review of pharmacology and clinical efficacy: W. M. O'Brien, G. F. Bagby, *Pharmacotherapy* **7,** 16 (1987).

Cryst from chloroform, mp 197-198°. LD$_{50}$ orally in mice: 400 mg/kg (Berger, Corraz).

THERAP CAT: Anti-inflammatory.

1913. Carpronium Chloride. *3-Carboxy-N,N,N-trimethyl-1-propanaminium chloride methyl ester; (3-carboxypropyl)-trimethylammonium chloride methyl ester;* γ-butyrobetaine chloride methyl ester; methyl *N*-trimethyl-γ-aminobutyrate chloride; Actinamin; Furozin. C$_8$H$_{18}$ClNO$_2$; mol wt 195.69. C 49.10%, H 9.27%, Cl 18.12%, N 7.16%, O 16.35%. [CH$_3$-OOCCH$_2$CH$_2$CH$_2$N(CH$_3$)$_3$]$^+$Cl$^-$. Prepn: Strack, Försterling, *Ber.* **76B,** 14 (1943).

Cube-like, hygroscopic crystals from warm acetone, mp 126°. Very sol in water, methanol, ethanol; sol in acetone. Practically insol in ether.

THERAP CAT: Cholinergic.

1914. Carrageenan. Carrageen; carrageenin. Mixture of sulfated polysaccharides extracted from red seaweed *(Rhodophyceae).* Chief sources are *Chondrus crispus* (L.) Stackhouse, and *Gigartina stellata (G. mammillosa)* (Goodenough and Woodward) J. Aghardh, *Gigartinaceae,* found most abundantly in North Atlantic coastal regions from Norway to North Africa. The name carrageenan is derived from the Irish coastal town of Carragheen. The κ and λ structural families are identified based on position of sulfate and the presence/absence of anhydrogalactose. Kappa (κ) family consists of κ,ι,μ,ν carrageenans of which κ and ι are the most prevalent. Characterized by a repeating unit of 4-sulfate-β-D-galactopyranosyl(1→4)-α-D-galactose linked (1→3). The galactose unit varies: 3,6-anhydro-α-D-galactose for κ-form; 3,6-anhydro-α-D-galactose-2-sulfate for ι-form. Due to helical tertiary structure which allows for gelling, κ-family is of major commercial importance. Lambda (λ) family consists of λ and ξ; characterized by a repeating (1→3) linked disaccharide of 2-sulfate-β-D-galactopyranosyl(1→4)-α-D-galactose. The galactose residue is 2,6-sulfated for λ form and 2-sulfated for ξ form. These carrageenans are non-gelling. Conformational study: N. S. Anderson *et al., J. Mol. Biol.* **45,** 85 (1969); and structural analysis: D. A. Rees, E. J. Welsh, *Angew. Chem. Int. Ed.* **16,** 214 (1977); C. Bodeau-Bellion, *Physiol. Veg.* **21,** 785 (1983). Use as phlogistic agent: C. A. Winter *et al., Proc. Soc. Exp. Biol. Med.* **111,** 544 (1962); L. Levy, *Life Sci.* **8,** 601 (1969). Review of effect on inflammation and immunity: H. J. Schwartz in *Inadvert. Modif. Immune Resp.,* **FDA-80-1074,** 109-114 (1980); S. Nicklin, K. Miller, *Food Addit. Contam.* **6,** 425-436 (1989). Review of toxicology: M. L. Weiner, *Agents Actions* **32,** 46-51 (1991). Review of industrial and non-food uses: R. J. Tye, *Carbohydr. Polym.* **10,** 259-280 (1989). *General Reviews*: G. H. Therkelsen in *Industrial Gums,* R. L. Whistler, J. N. BeMiller, Eds. (Academic Press, New York, 3rd ed., 1993) pp 145-180; J. K. Baird, "Gums" in Kirk-Othmer *Encyclopedia of Chemical Technology* vol. 12 (Wiley-Interscience, New York, 4th ed., 1994) pp 842-862.

κ-form. Ability to form thermoreversible gels. Gelation depends on temperature, concentration of counterions and other polysaccharides. Sol in very polar solvents. Sol in hot milk, hot conc. sugar soln; above 60° in water, as Na$^+$ salt in cold water. K$^+$ and Ca^{2+} salts are insol in cold water. Insol in in conc. salt soln; 35% alcohol soln; insol with swelling in cold milk. Optimum stablility occurs in pH 9. Degree of sulphation varys from 25-30%.

ι-form. Ability to form thermoreversible gels. Gelation depends on temperature, concentration of counterions and other polysaccharides. Sol in very polar solvents. Sol in hot milk, hot conc. salt soln; above 60° in water, as Na$^+$ salt in cold water; slightly sol hot conc. sugar soln. K$^+$ and Ca^{2+} salts are insol in cold water. Insol in 35% alcohol soln; insol in cold milk. Optimum stability occurs in pH 9. Degree of sulfation varies from 28-35%. LD$_{50}$ orally in rat: > 5000

mg/kg; dermally in rabbit: > 2000 mg/kg (Weiner). LC50 (4 hr) in rat: > 930.8 ±74.4 mg/m^3 (Weiner).

λ-form. Sol in very polar solvents. Sol in water, milk, hot conc. sugar soln, conc. salt soln., as Na$^+$salt in 35% alcohol soln. Optimum stablility occurs in pH 9. The most highly sulfated from 32-39%.

USE: Gelling, emulsifying, and stabilizing agent and viscosity builder in foods and non-foods, but esp in milk or water systems. Demulcent. To induce experimental edema in laboratory animals.

1915. Carsalam. *2H-1,3-Benzoxazine-2,4(3H)-dione;* carbonylsalicylamide; 2,4-dioxodihydro-5,6-benzo-1,3-oxazine; CSA; Beaprine. C$_8$H$_5$NO$_3$; mol wt 163.13. C 58.90%, H 3.09%, N 8.59%, O 29.42%. Prepn: Shapiro *et al., J. Am. Chem. Soc.* **79,** 2811 (1957); D. N. Dhar, A. K. Bag, *Indian J. Chem.* **21B,** 366 (1982). Crystal and molecular structure: S. Kashino *et al., Acta Crystallogr.* **B34,** 2191 (1978).

Crystals from butanol, mp 229-230°.

THERAP CAT: Analgesic.

1916. Cartap. *Carbamothioic acid, S,S'-[2-(dimethylamino)-1,3-propanediyl] ester;* 1,3-bis(carbamoylthio)-2-*N,N*-(dimethylamino)propane. C$_7$H$_{15}$N$_3$O$_2$S$_2$, mol wt 237.35. C 35.42%, H 6.37%, N 17.70%, O 13.48%, S 27.02%. Insecticide modeled on a toxin isolated from the marine annelid *Lumbrineris heteropoda.* Prepn and insecticidal activity: K. Konishi *et al.,* Fr. pat. **1,452,338;** eidem, U.S. pat. **3,332,943** (1966, 1968 both to Takeda); K. Konishi, *Agr. Biol. Chem.* **34,** 935 (1970). Crystal structure of hydrochloride: C. J. Cheer, F. J. Pickles, *J. Chem. Soc. Perkin Trans. II* **1980,** 1805. Comparative study of insecticidal activity: B. Gumey, N. W. Hussey, *Plant Pathol.* **23,** 127 (1974). Effect on soil enzyme activity: T. Endo *et al., J. Pestic. Sci.* **7,** 101 (1982). Aquatic toxicity study: S. Lakota *et al., Acta Hydrobiol.* **23,** 183 (1981). Chronic toxicity study in mice: Y. Tsubura *et al., Nara Igaku Zasshi* **26,** 368 (1975), *C.A.* **84,** 174837c (1976). *Review:* Y. Kono, *Japan. Pestic. Inf.* **34,** 22 (1978).

Colorless prisms from ethyl acetate, mp 130.5-131° (dec). Monohydrochloride, C$_7$H$_{16}$ClN$_3$O$_2$S$_2$, NTD-2, Padan. Colorless rods from methanol, mp 176° (dec). Sol in water. LD$_{50}$ in mice, rats (mg/kg): 165, 250 orally (Konishi, 1970).

USE: Insecticide.

1917. Carteolol. *5-[3-[(1,1-Dimethylethyl)amino]-2-hydroxypropoxy]-3,4-dihydro-2(1H)-quinolinone;* 5-[3-(*tert*-butylamino)-2-hydroxypropoxy]-3,4-dihydrocarbostyril. C$_{16}$H$_{24}$N$_2$O$_3$; mol wt 292.38. C 65.73%, H 8.27%, N 9.58%, O 16.42%. β-Adrenergic blocker. Prepn: K. Nakagawa *et al.,* Ger. pat. **2,302,027** corresp to U.S. pat. **3,910,924** (1973, 1975 both to Otsuka); eidem, *J. Med. Chem.* **17,** 529 (1974). Pharmacological evaluation in rats: S. Morita *et al., Biochem. Pharmacol.* **25,** 1836 (1976). HPLC determn in plasma and urine: S. Y. Chu, *J. Pharm. Sci.* **67,** 1623 (1978). *In vitro* study: S. Chiba, *Arzneimittel-Forsch.* **29,** 895 (1979). Carcinogenicity study: M. Kurosomi *et al., Farmaco Ed. Prat.* **34,** 202 (1979). Reproduction and toxicity studies: N. Tanaka *et al., J. Toxicol. Sci.* **4,** 47 (1979); M. Tamagawa *et al., ibid.* 59. Acute and subacute toxicity studies: N. Tanaka *et al., Oyo Yakuri* **11,** 165, 173 (1976), *C.A.* **88,** 115336y, 115337z (1978). Series of articles on pharmacology, toxicology and clinical studies: *Arzneimittel-Forsch.* **33,** 277-345 (1983). Acute toxicity data: W. Lang, *ibid.* 290.

Hydrochloride, $C_{16}H_{24}N_2O_3$·HCl, *Abbott 43326, OPC-1085, Arteoptic, Carteol, Cartrol, Endak, Mikelan, Optipress, Tenalet, Tenalin, Teoptic.* Crystals from ethanol, mp 278°. LD_{50} in male mice, rats (mg/kg): 810, 1380 orally; 54.5, 158 i.v.; 380, 400 i.p. (Lang).

THERAP CAT: Antihypertensive, antianginal, antiarrhythmic, antiglaucoma.

1918. Carthamin. *6-β-D-Glucopyranosyl-2-[[3-β-D-glucopyranosyl-2,3,4-trihydroxy-5-[3-(4-hydroxyphenyl)-1-oxo-2-propenyl]-6-oxo-1,4-cyclohexadien-1-yl]methylene]-5,6-dihydroxy-4-[3-(4-hydroxyphenyl)-1-oxo-2-propenyl]-4-cyclohexene-1,3-dione;* carthamic acid; safflor carmine; safflor red; C.I. Natural Red 26; C.I. 75140. $C_{43}H_{42}O_{22}$; mol wt 910.79. C 56.71%, H 4.65%, O 38.65%. Coloring principle from *Carthamus tinctorius* L., *Compositae* (safflower): Kametaka, Perkin, *J. Chem. Soc.* **97**, 1415 (1910). Extraction: Wada, **Japan.** pat. 8943 (1955). Structure: Kuroda, *J. Chem. Soc.* **1930**, 752, 765. Revised structure: H. Obara, J. Onodera, *Chem. Letters* **1979**, 201. Biogenesis: Shimokoriyama, Hattori, *Arch. Biochem. Biophys.* **54**, 93 (1955). *See also: Colour Index* vol. 4 (3rd ed., 1971) p 4625.

Dark red, granular powder with green luster. Slightly sol in water; sol in dil alkali carbonates, alcohol. Practically insol in ether. Solns rapidly dec.

USE: As dye.

1919. Carthamus. Safflower; African saffron; thistle saffron; American saffron; false saffron; bastard saffron; Dyer's saffron. Florets of *Carthamus tinctorius* L. *Compositae. Habit.* Levant; Orient; cultivated extensively in Europe and America. *Constit.* Carthamin, safflor yellow.

USE: In dyeing; surrogate for Spanish saffron; coloring butter, liqueurs, confectionery, cosmetics.

THERAP CAT: Dietary supplement.

1920. Carticaine. *4-Methyl-3-[[1-oxo-2-(propylamino)-propyl]amino]-2-thiophenecarboxylic acid methyl ester;* 3-propylamino-α-propionylamino-2-carbomethoxy-4-methylthiophene; methyl 4-methyl-3-(2-propylaminopropionamido)thiophene-2-carboxylate. $C_{13}H_{20}N_2O_3S$; mol wt 284.38. C 54.91%, H 7.09%, N 9.85%, O 16.88%, S 11.28%. Prepn: H. Ruschig *et al.,* **S. Afr.** pat. **68 04,265**; *eidem,* **U.S.** pat. **3,855,243** (1969, 1974 both to Farbwerke Hoechst). Pharmacological studies: R. Muschaweck, R. Rippel, *Prakt. Anaesth.* **9**, 135 (1974); H. Hofer *et al., ibid.* 157; A. Den Hertog, *Eur. J. Pharmacol.* **26**, 175 (1974). Toxicity studies: C. Baeder *et al., Prakt. Anaesth.* **9**, 147 (1974). Mechanism of action: U. Borchard, H. Drouin, *Eur. J. Pharmacol.* **62**, 73 (1980). *Review:* H. Schneider, *Schweiz. Monatsschr. Zahnheilkd.* **86**, 1188-1194 (1976).

bp$_{0.3}$ 162-167°.

Hydrochloride, $C_{13}H_{20}N_2O_3S$·HCl, *HOE-045, HOE-40045, Ultracain.* White, fine crystals, mp 177-178°. LD_{50} i.v. in mice: 37 mg/kg (Muschaweck, Rippel).

THERAP CAT: Anesthetic (local).

1921. Carubicin. *(8S-cis)-Acetyl-10-[(3-amino-2,3,6-trideoxy-α-L-lyxo-hexopyranosyl)oxy]-7,8,9,10-tetrahydro-1,6,8,-11-tetrahydroxy-5,12-naphthacenedione; (1S,3S)-3-acetyl-1,-2,3,4,6,11-hexahydro-3,5,10,12-tetrahydroxy-6,11-dioxo-1-naphthacenyl 3-amino-2,3,6-trideoxy-α-L-lyxo-hexopyranoside;* 4-O-demethyldaunorubicin; carminomycin; carminomycin I; karminomycin. $C_{26}H_{27}NO_{10}$; mol wt 513.50. C 60.82%, H 5.30%, N 2.73%, O 31.16%. Anthracycline antitumor antibiotic, related to daunorubicin and doxorubicin, q.q.v. Isoln from *Actinomadura carminata:* G. F. Gauze *et al., Antibiotiki* **18**, 675 (1973); M. G. Brazhnikova *et al., ibid.* 678. Antitumor activity: V. A. Shorin *et al., ibid.* 681. Physico-chemical characteristics, structure: M. G. Brazhnikova *et al., J. Antibiot.* **27**, 254 (1974). Pharmacokinetics, pharmacodynamics, toxicity study: L. E. Goldberg *et al., Antibiotiki* **19**, 57 (1974). Production: M. G. Brazhnikova *et al.,* **U.S.S.R.** pat. 508076 (1976 to Inst. Antibiot. Res., USSR), *C.A.* **86**, 15215 (1977). Stereochemistry: *eidem, J. Antibiot.* **29**, 469 (1976). Synthesis from daunomycinone: G. Cassinelli *et al., ibid.* **31**, 178 (1978). Molecular pharmacology: V. H. DuVernay *et al., Cancer Res.* **40**, 387 (1980). Analysis in human serum: S. E. Fandrich, K. A. Pittman, *J. Chromatog.* **223**, 155 (1981). Early clinical studies: L.H. Baker *et al., Cancer Treat. Rep.* **63**, 899 (1979). Embryotoxicity and teratogenicity study: I. Damjanov, A. Celluzzi, *Res. Commun. Chem. Pathol. Pharmacol.* **28**, 497 (1980).

Hydrochloride, $C_{26}H_{27}NO_{10}$·HCl, *NSC-180024.* Crystals from ethanol/benzene. $[\alpha]_D^{20}$ +289°. uv max (ethanol): 236, 255, 462, 478, 492 ($E_{1cm}^{1\%}$ 300), 510, 525 nm. pKa_1 8.00; pKa_2 10.16. Sol in water, methanol. Practically insol in other organic solvents. LD_{50} in mice (mg/kg): 7.3 orally; 1.3 i.v.; 3.7 s.c. (Goldberg).

THERAP CAT: Antineoplastic.

1922. Carumonam. *[2S-[2α,3α(Z)]]-[[[2-[[2-[[(Aminocarbonyl)oxy]methyl]-4-oxo-1-sulfo-3-azetidinyl]amino]-1-(2-amino-4-thiazolyl)-2-oxoethylidene]amino]oxy]acetic acid; (Z)-[[[(2-amino-4-thiazolyl)[[(2S,3S)-2-(hydroxymethyl)-4-oxo-1-sulfo-3-azetidinyl]carbamoyl]methylene]amino]oxy]acetic acid carbamate (ester); (3S,4S)-cis-3-[2-(2-amino-4-thiazolyl)-2-(Z)-carboxymethoxyiminoacetamido]-4-carbamoyloxymethyl-2-azetidinone-1-sulfonic acid.* $C_{12}H_{14}-N_6O_{10}S_2$; mol wt 466.41. C 30.90%, H 3.03%, N 18.02%, O 34.30%, S 13.75%. Synthetic monocyclic β-lactam (monobactam) antibiotic. Prepn: S. Kishimoto *et al.,* **Eur.** pat. **Appl. 93,376**; T. Matsuo *et al.,* **U.S.** pat. **4,572,801** (1983, 1986 both to Takeda); M. Sendai *et al., J. Antibiot.* **38**, 346 (1985). Alternate synthesis: P. S. Manchard *et al., J. Org. Chem.* **53**, 5507 (1988). Comparative *in vitro* antimicrobial activity: R. J. Fass, V. L. Helsel, *Antimicrob. Ag. Chemother.* **28**, 834 (1985); B. R. Smith *et al., ibid.* **29**, 346 (1986); I. M. Hoepelman *et al., Chemotherapy (Basel)* **33**, 103 (1987). β-Lactamase stability: R. L. Then, *ibid.* **30**, 398 (1984). Pharmacokinetics in humans: E. Weidekamm *et al., Antimicrob. Ag. Chemother.* **26**, 898 (1984); C. A. M. McNulty *et al., ibid.* **28**, 425 (1985).

Colorless powder. $[\alpha]_D^{26}$ −45° (c = 1 in DMSO).
Disodium salt, $C_{12}H_{12}N_6Na_2O_{10}S_2$, *AMA-1080, Ro-17-2301, Amasulin, Mobactam.*
THERAP CAT: Antibacterial.

1923. Carvacrol. *2-Methyl-5-(1-methylethyl)phenol;* 2-*p*-cymenol; 2-hydroxy-*p*-cymene; isopropyl-*o*-cresol; isothymol. $C_{10}H_{14}O$; mol wt 150.22. C 79.96%, H 9.39%, O 10.65%. Found in oil of origanum, thyme, marjoram, summer savory: E. Guenther, *The Essential Oils* vol. 2 (Van Nostrand, New York, 1949) p 503; Carpenter, Easter, *J. Org. Chem.* **20**, 401 (1955). Prepn by chlorination of α-pinene with *tert*-butyl hypochlorite: Ritter, Ginsburg, *J. Am. Chem. Soc.* **72**, 2381 (1950); from 2-bromo-*p*-cymol: Strubell, Baumgartel, *Arch. Pharm.* **291**, 66 (1958). Toxicity data: Kochmann, *Arch. Exp. Pathol. Pharmakol.* **161**, 196 (1931).

Liquid; thymol odor. d_4^{20} 0.976; d_4^{25} 0.9751. bp_{760} 237-238°; bp_{18} 118-122°; bp_3 93°. mp ~0°. n_D^{20} 1.52295. uv max (95% ethanol): 277.5 nm (log ε 3.262). Volatile with steam. Practically insol in water. Freely soluble in alc or ether. LD orally in rabbits: 100 mg/kg (Kochmann).
USE: As disinfectant; in organic syntheses.
THERAP CAT: Has been used as anti-infective; anthelmintic (Nematodes).

1924. Carvedilol. *1-(9H-Carbazol-4-yloxy)-3-[[2-(2-methoxyphenoxy)ethyl]amino]-2-propanol;* BM-14190; DQ-2466; Dilatrend; Dimitone; Eucardic; Kredex; Querto. $C_{24}H_{26}N_2O_4$; mol wt 406.48. C 70.92%, H 6.45%, N 6.89%, O 15.74%. Nonselective β-adrenergic blocker with vasodilating activity. Prepn: F. Wiedemann *et al.,* **Ger. pat. 2,815,926;** *eidem,* U.S. pat. **4,503,067** (1979, 1985 both to Boehringer Mannheim). General pharmacological profile: M. Hirohashi *et al., Arzneimittel-Forsch.* **40**, 735 (1990). Clinical pharmacology: E. von Möllendorff *et al., Clin. Pharm. Ther.* **39**, 677 (1986); L. X. Cubeddu *et al., ibid.* **41**, 31 (1987). HPLC determn in biological fluids: K. Reiff, *J. Chromatog.* **413**, 355 (1987). Evaluation in angina: E. A. Rodrigues *et al., Am. J. Cardiol.* **58**, 916 (1986). Series of articles on pharmacology and clinical trials in hypertension: *J. Cardiovasc. Pharmacol.* **19**, Suppl. 1, S1-S146 (1992).

Colorless crystals from ethyl acetate, mp 114-115°.
THERAP CAT: Antihypertensive.

1925. Carvone. *2-Methyl-5-(1-methylethenyl)-2-cyclohexene-1-one; p-mentha-6,8-dien-2-one;* 1-methyl-4-isopropenyl-Δ⁶-cyclohexen-2-one; carvol. $C_{10}H_{14}O$; mol wt 150.22. C 79.96%, H 9.39%, O 10.65%. *d*-Carvone is found in caraway seed and dill seed oils: Schweizer, *J. Prakt.*

Chem. **24**, 257 (1841); Gladstone, *J. Chem. Soc.* **25**, 1 (1872). Isoln of *d*-carvone from mandarin peel oil *(Citrus reticulata* Blanco, *Rutaceae):* Kugler, Kováts, *Helv. Chim. Acta* **46**, 1480 (1963). *l*-Carvone is found in spearmint and kuromoji oils: Kwaenick, *Ber.* **24**, 82 (1891). *dl*-Carvone is found in gingergrass oil: Walbaum, Hüthig, *J. Prakt. Chem.* **71**, 459 (1905). Prepn from α-pinene: Booth, Klein, U.S. pat. **2,796,428** (1957 to Glidden). Structure: Wagner, *Ber.* **27**, 2270 (1894); Tiemann, Semmler, *ibid.* **28**, 2147 (1895). Synthesis of *l*-carvone: Royals, Horne, *J. Am. Chem. Soc.* **73**, 5856 (1951); Shono *et al., Chem. Letters* **1975**, 4330. Synthesis of *dl*-carvone: Suga, *Bull. Chem. Soc. Japan* **31**, 569 (1958); I. Fleming, I. Patterson, *Synthesis* **1979**, 736. Toxicity study: P. M. Jenner *et al., Food Cosmet. Toxicol.* **2**, 327 (1964). *Review:* J. L. Simonsen, *The Terpenes* vol. I (University Press, Cambridge, 2nd ed., 1947) pp 394-408; B. Singaram, J. Verghese, *Perfum. Flavor.* **2**, 47 (1977).

d-Form, liquid, bp_{755} 230°; bp_{5-6} 91°. d_4^{20} 0.965. n_D^{20} 1.4989. $[\alpha]_D^{20}$ +61.2°.
l-Form, liquid, bp_{763} 230-231°. d_{15}^{15} 0.9652. n_D^{20} 1.4988. $[\alpha]_D^{20}$ −62.46°.
dl-Form, liquid, bp 230-231°. d_{15}^{15} 0.9645. n_D^{20} 1.5003. Practically insol in water. Misc with alc. LD_{50} orally in rats: 1640 mg/kg (Jenner).
USE: As oil of caraway; also for flavoring liqueurs; in perfumery and soaps.
THERAP CAT: Carminative.

1926. Caryophyllene. *[1R-(1R*,4E,9S*)]-4,11,11-Trimethyl-8-methylenebicyclo[7.2.0]undec-4-ene;* β-caryophyllene; *trans*-caryophyllene. $C_{15}H_{24}$; mol wt 204.36. C 88.16%, H 11.84%. Sesquiterpenoid occurring in many essential oils and especially in clove oil, the oils from stems and flowers of *Syzygium aromaticum* (L.) Merrill & Perry *(Jambrosa caryophyllus* Niedenzu; *Eugenia caryophyllata* Thunb.), *Myrtaceae.* Occurs in nature as a mixture with isocaryophyllene and α-caryophyllene (humulene, *q.v.).* Isolation of mixture: Schreiner, Kremers, *Pharm. Arch.* **2**, 273, 293 (1899). Existence of isomers: Deussen, *Ann.* **356**, 1 (1907). Structure: Aebi *et al., J. Chem. Soc.* **1953**, 3124; Ramage, Whitehead, *ibid.* **1954**, 4336. Abs config: Barton, Nickon, *ibid.* 4665. Total synthesis of racemic *trans/cis*-forms: Corey *et al., J. Am. Chem. Soc.* **86**, 485 (1964). Rearrangement to isocaryophyllene: Rachlin, **Ger. pat. 2,044,018** (1971 to I.F.F.), *C.A.* **75**, 49364j (1971). *Reviews:* Simonsen, *The Terpenes* vol. III (University Press, Cambridge, 1952) pp 39-71; Barton, de Mayo, *Quart. Rev. (London)* **11**, 197 (1957); Halsall, *ibid.* **16**, 101 (1962).

Caryophyllene   Isocaryophyllene

Liquid. Has a terpene odor about midway between odor of cloves and turpentine. bp_{14} 129-130°; $bp_{9.7}$ 118-119°. $[\alpha]_D^{15}$ −5.2°. n_D^{17} 1.5009; n_D^{15} 1.5030. d_4^{17} 0.9052.
Dihydrochloride, $C_{15}H_{24}$·2HCl, mp 69-70°.
Nitrosochloride, $C_{15}H_{24}$·ClNO. Crystals, dec 159°. $[\alpha]_D^{17}$ −98.07°.
Isocaryophyllene, γ-caryophyllene, cis-caryophyllene. Liquid. bp_{19} 130°; $bp_{14.5}$ 125-125.5°. $[\alpha]_D^{19}$ −26.17°. n_D^{19} 1.4966. d^{19} 0.8995. When treated with hydrogen chloride gives β-caryophyllene dihydrochloride.

Nitrosochloride, $C_{15}H_{24}\cdot ClNO$, exists in two modifications: dec 122°, $[\alpha]_D^{20.5}$ +14.71°, and dec 146°, $[\alpha]_D^{18}$ −33.69°.

USE: In perfumery.

1927. Carzenide. *p-Sulfamoylbenzoic acid; p-sulfamyl-benzoic acid; p-carboxybenzenesulfonamide; p-sulfonamido-benzoic acid; benzoic acid p-sulfamide; Dirnate.* $C_7H_7NO_4S$; mol wt 201.20. C 41.79%, H 3.51%, N 6.96%, O 31.81%, S 15.94%. May be obtained from crude saccharin: Fahlberg, **Ger. pat. 64,624** (1891); from *p*-sulfamoyltoluene: Noyes, *Am. Chem. J.* **7**, 147 (1885); Amazu *et al.*, **Japan. pat. 180,-511** (1949 to Nissan). A metabolite of homosulfanilamide *(q.v.)*: Kakemi *et al.*, *J. Pharm. Soc. Japan* **82**, 1582 (1962). Has some inhibitory effect on carbonic anhydrase: Tanimukai *et al.*, *Biochem. Pharmacol.* **14**, 961 (1965).

Flat, shiny prisms from water, dec 280°. K 3.05×10^{-4} at 18°. Practically insol in cold water, ether, benzene. Freely sol in alcohol.

Sodium salt, $C_7H_6NNaO_4S$, crystals. Neutral. Soly in water: ~20 g/100 ml.

1928. Carzinophilin. Cardinophyllin. Antitumor antibiotic from *Streptomyces sahachiroi.* Isoln: T. Hata *et al.*, *J. Antibiot.* **7A**, 107 (1954); Hata, Sano, **Japan. pat. 7590('56)** (to Kitasato Res. Inst. & Kyowa); **Brit. pat. 777,287** (1957 to Kitasato Res. Inst.). Isoln of active component, carzinophilin A: H. Kamada *et al.*, *J. Antibiot.* **8A**, 187 (1955). Purification: H. Kamada *et al.*, **U.S. pat. 3,044,935** (1962 to Kyowa). Clinical studies: F. Koga, *J. Antibiot.* **7A**, 275 (1954); N. Shimada *et al.*, *ibid.* **8A**, 67 (1955). Mechanism of action: J. W. Lown, K. C. Majumdar, *Can. J. Biochem.* **55**, 630 (1977). Structure studies of carzinophilin A: M. Onda *et al.*, *J. Antibiot.* **22**, 42 (1969); *eidem, Chem Pharm. Bull.* **19**, 2013 (1971). Total structure of carzinophilin A: J. W. Lown, C. C. Hanstock, *J. Am. Chem. Soc.* **104**, 3214 (1982). Synthetic studies on carzinophilin A: M. Shibuya, *Tetrahedron Letters* **24**, 1175 (1983).

Carzinophilin A

Crystals, mp 207° (dec). uv max (water): 217, 290 nm ($E_{1\%}^{1cm}$ 476, 190). Sol in water, alcohols, acetone, butyl acetate, chloroform, benzol. Practically insol in petr ether. Activity is influenced by the change of pH, being most stable at pH 7.0, relatively stable at pH 8-9 and pH 6; loses biological activity quickly when pH is reduced to < 5.0. Activity also influenced by temp change: heating in boiling water for 15 min at pH 7.0 causes activity to drop.

Carzinophilin A, $C_{50}H_{58}N_5O_{18}$, *CZ.* Crystals, mp 217-222° (dec). $[\alpha]_D^{28}$ +57.8° in chloroform. uv max (methanol): 218, 250, 283 nm. Sol in acetone, chloroform, ethyl acetate, butyl acetate, benzene, dioxane, dil alk water; slightly sol in

methanol, ethanol, ether, carbon tetrachloride, petr ether, water. LD_{50} i.v. in mice: 150 γ/kg (Kamada, 1955).

THERAP CAT: Antineoplastic.

1929. Casanthranol. Cantralax; Peristim. Complex mixture of the purified anthranol glycosides derived from *Cascara sagrada (q.v.),* practically devoid of free anthraquinones. Two active fractions have been identified as **casanthranol A** and **casanthranol B.** Upon oxidative hydrolysis, casanthranol A yields emodin, while casanthranol B under similar treatment yields the aglycone, aloe-emodin, and traces of chrysophanic acid. Prepn: Lee, Berger, **U.S. pat. 2,552,896** (1951 to Hoffmann-La Roche).

Component of *Peri-Colace, Casakol.*

THERAP CAT: Cathartic.

1930. Cascara Amara. Honduras bark. Bark of *Picramnia antidesma* Sw., *Simaroubaceae* and related species. *Habit.* West Indies, Mexico.

1931. Cascara Sagrada. Sacred bark; Chittem bark; Chittim bark; Purshiana bark; Persian bark; bearberry bark; bearwood. Dried bark of *Rhamnus purshiana* DC., *Rhamnaceae,* from which a naturally occurring cathartic is extracted. *Habit.* Northern Idaho, west to Northern California. The cathartic properties are primarily due to the presence of *cascarosides,* anthraglycosides which are related to glycosides found in aloe *(q.v.)* and buckthorn. See A. Y. Leung, *Drug Cosmet. Ind.* **121**, 42 (December, 1977). Other constituents are aloins (C-glycosides), O-glycosides, and free anthraquinones: *Analyst* **93**, 749 (1968); *ibid.* **98**, 830 (1973) (Joint Committee Reports). Isoln of anthraquinone aglycones and glycosides: S. C. Yung Su, N. M. Ferguson, *J. Pharm. Sci.* **63**, 899 (1973). Structure of cascarosides A and B: J. W. Fairbairn *et al.*, *J. Pharm. Pharmacol.* **15** (Suppl.), 292T, (1963); F. J. Evans *et al.*, *ibid.* **27** (Suppl.), 91P (1975). Biological evaluation of cascara bark prepns: J. W. Fairbairn, G. E. D. H. Mahran, *ibid.* **5**, 827 (1953).

Commercial prepn, *Cas-Evac.*

THERAP CAT: Cathartic.

THERAP CAT (VET): Laxative.

1932. Cascarilla. Eleuthera bark; sweet-wood bark. Dried bark of *Croton eluteria* (L.) Sw., *Euphorbiaceae. Habit.* Bahama Islands, Cuba, Haiti. *Constit.* About 15% resin; 5.39-6.33% fat; 7.6-8.13% ash; theobromine, volatile oil, cascarillin, betaine, tannin. *Ref:* Farb, *C.A.* **44**, 8060e (1950).

USE: As an addition to smoking tobacco for flavoring.

THERAP CAT: Aromatic bitter.

1933. Cascarillin. *3-(Acetyloxy)-1-[2-(3-furanyl)-2-hydroxyethyl]decahydro-5,6-dihydroxy-2,4a,5-trimethyl-1-naphthalenecarboxaldehyde.* $C_{22}H_{32}O_7$; mol wt 408.49. C 64.69%, H 7.90%, O 27.42%. From bark of *Croton eluteria* (L.) Sw., *Euphorbiaceae.* Isoln: Duval, *J. Pharm.* **8**, 91 (1845); Mylius, *Ber.* **6**, 1051 (1873); Naylor, Littlefield, *Pharm. J.* **57**, 95 (1896). Structure and stereochemistry: McEachan *et al.*, *J. Chem. Soc. (B)* **1966**, 633.

Needles from alc, mp 205°. Very bitter taste. Freely sol in ether, hot alc; slightly sol in water, cold alc, chloroform.

THERAP CAT: Aromatic bitter.

1934. Casein. A mixture of related phosphoproteins occurring in milk and cheese. Present to the extent of 3% in bovine milk. One of the most nutritive milk proteins in that it contains all of the common amino acids and is rich in the essential ones. Produced in mammary tissue from amino acids supplied by the blood. Obtained from milk by removing the cream and acidifying the skimmed milk which causes casein to precipitate. In cheese manufacture, casein is precipitated by the lactic acid formed from the same milk by fermentation. Precipitation by rennet is favored for casein intended for plastics manuf. Curdling by electricity has been described. The prepn of pure casein is described by Hammarsten, *Textbook of Physiological Chemistry* (New York, 7th ed., 1911) p 619, or in Abderhalden's *Handbuch* vol. II (Berlin, 1910) p 384. Alternate prepn: van Slyke, Baker, *J. Biol. Chem.* **35**, 127 (1918); cf. Cohn, Hendry, *Org. Syn.* **coll. vol. II**, 120 (1943). Prepn of casein free from vitamin B$_{12}$ for nutrition experiments: Kissel, U.S. pat. **2,-853,479** (1958 to Natl. Dairy Prods.). The major casein components may be distinguished by electrophoresis and are designated as α-, β-, γ- and K-caseins, in order of decreasing mobility at pH 7. The complete amino acid sequence of bovine β-casein is known and contains 209 residues with an approx. mol wt of 23,600: Ribadeaudumas *et al., Eur. J. Biochem.* **25**, 505 (1972). *Review:* McKenzie, *Advan. Protein Chem.* **22**, pp 75-135 (1967).

White, amorphous powder or granules without odor or taste. Very sparingly sol in water and in nonpolar organic solvents; sol in aqueous solns of alkalies, levorotatory. The isoelectric zone is around pH 4.7; sol in concd HCl with light violet color. Amphoteric; forms salts with both acids and bases. Present in bovine milk as neutral calcium caseinate and in human milk as potassium caseinate. Precipitated from solns satd with metallic salts. Forms a hard, insol plastic with formaldehyde.

Casamino acids, commercial acid-hydrolyzed casein. Hydrolysis is carried out until all the nitrogen in the casein is converted to amino acids or other compounds of relative chemical simplicity. Prepn: Mueller, Miller, *J. Immunol.* **40**, 21 (1941); Mueller, Johnson, *ibid.* 33. Typical analysis: N 10%, NaCl 14%, ash 20%, phosphorus as PO$_4$ 2%, Fe 15 mg/3 g.

Note: Legumin, also known as *avenin* or *vegetable casein*, occurring in beans and nuts, is a globulin resembling casein. Isoln from *Avena sativa* L., *Graminea:* Sanson, *Compt. Rend.* **96**, 75 (1883). Structure: C. J. Bailey, D. Boulter, *Eur. J. Biochem.* **17**, 460 (1970).

USE: In the manuf of molded plastics, adhesives, paints, textile finishes, paper coatings, man-made fibers. Vitamin-free casein is used in diets of animals employed for the biological assay of vitamins. Medicinal grades are used in dietetic prepns and for determining the effectiveness of digestive enzyme prepns contg. pepsin, trypsin, papain. Casamino acids are used in microbial assays and in the prepn of microbiological media.

THERAP CAT: Nutrient.

1935. Casimiroedine. *N*-D-Glucoside of *N*-cinnamoyl-*N*-methylhistamine. C$_{21}$H$_{27}$N$_3$O$_6$; mol wt 417.46. C 60.42%, H 6.52%, N 10.07%, O 23.00%. From fruit, seed and bark of *Casimiroa edulis* Llave et Lex., *Rutaceae:* Power, Callan, *J. Chem. Soc.* **99**, 1993 (1911); Aebi, *Helv. Chim. Acta* **39**, 1495 (1956); Kincl *et al., J. Chem. Soc.* **1956**, 4163; Djerassi *et al., J. Org. Chem.* **21**, 1510 (1956). Structure work: Djerassi *et al., Tetrahedron* **2**, 168a (1958); Raman *et al., Tetrahedron Letters* **1962**, 357.

Needles from methanol, mp 226-228°. $[\alpha]_D^{20}$ −21.3° (c = 1.21 in 5% HCl); $[\alpha]_D$ −33° (aq 1% HCl). uv max (ethanol): 219, 280 nm (log ∈ 4.26, 4.30). Freely sol in dil HCl; dissolves slowly in 2N NaOH or saturated NaHCO$_3$; practically insol in water; sol in hot alc, in amyl alcohol, slightly in benzene, chloroform, ether.

1936. Casimiroin. *6-Methoxy-9-methyl-1,3-dioxolo[4,5-h]quinolin-8(9H)-one;* 1-methyl-4-methoxy-7,8-methylenedioxycarbostyril; 4-methoxy-1-methyl-7,8-methylenedioxy-2-quinolone. C$_{12}$H$_{11}$NO$_4$; mol wt 233.22. C 61.80%, H 4.75%, N 6.01%, O 27.44%. From the seed, trunk, root and bark of *Casimiroa edulis* Llave et Lex., *Rutaceae:* Kincl *et al., J. Chem. Soc.* **1956**, 4163; Iriarte *et al., ibid.* 4170. Structure: Meisels, Sondheimer, *J. Am. Chem. Soc.* **79**, 6328 (1957). Synthesis: Weinstein, Hylton, *Tetrahedron* **20**, 1725 (1964).

Crystals, mp 203°. uv max: 226 nm (log ∈ 4.50). Almost insol in water; sol in alcohol, ether, ethyl acetate.

1937. Cassaidine. *(Dodecahydro-7β,10-dihydroxy-1α,-4bβ,8,8-tetramethyl-2(1H)-phenanthrenylidene)acetic acid 2-(dimethylamino)ethyl ester; (E)-3β,7β-dihydroxy-14α-methyl-podocarpane-Δ^{13,α}-acetic acid 2-(dimethylamino)ethyl ester.* C$_{24}$H$_{41}$NO$_4$; mol wt 407.59. C 70.72%, H 10.14%, N 3.44%, O 15.70%. Cardiotonic principle from the bark of *Erythrophleum guineense* G. Don., *Leguminosae.* Isoln and structure: Ruzicka, Dalma, *Helv. Chim. Acta* **23**, 753 (1940); Engel, *ibid.* **42**, 1127 (1959). Synthesis: Turner *et al., J. Am. Chem. Soc.* **88**, 1766 (1966). Stereochemistry: Clarke *et al., ibid.* **88**, 5865 (1966).

Prisms from acetone + ether, mp 139.5°. $[\alpha]_D^{20}$ −98° (ethanol); $[\alpha]_D^{20}$ −104° (0.1N HCl). Slightly sol in methanol, ethanol, acetone, acetic, chloroform. Practically insol in ether, benzene.

Oxalate, C$_{24}$H$_{41}$NO$_4$.(COOH)$_2$, crystals from methanol+acetone, mp 198-201°. $[\alpha]_D^{17}$ −84° (c = 1.39 in water).

1938. Cassaine. *(Dodecahydro-7β-hydroxy-1α,4bβ,8,8-tetramethyl-10-oxo-2(1H)-phenanthrenylidene)acetic acid 2-(dimethylamino)ethyl ester; (E)-3β-hydroxy-14α-methyl-7-oxopodocarpane-Δ^{13,α}-acetic acid 2-(dimethylamino)ethyl ester.* C$_{24}$H$_{39}$NO$_4$; mol wt 405.58. C 71.07%, H 9.69%, N 3.45%, O 15.78%. Cardiotonic principle from the bark of *Erythrophleum guineense* G. Don., *Leguminosae:* Dalma, *Helv. Chim. Acta* **22**, 1497 (1939). Structure: Turner *et al., Tetrahedron Letters* **1959**, 7; Gensler *et al., J. Am. Chem. Soc.* **81**, 5217 (1959). Absolute configuration: Hauth *et al., Helv. Chim. Acta* **48**, 1087 (1965); Clarke *et al., J. Am. Chem. Soc.* **88**, 5865 (1966). Synthesis: Turner *et al., ibid.* 1766.

Glossy flakes from ether, mp 142.5°. $[\alpha]_D^{23}$ $-110.5°$ (alc); $[\alpha]_D^{23}$ $-101°$ (0.1N HCl). Absorption spectrum: Ruzicka, Dalma, *Helv. Chim. Acta* **22**, 1516 (1939). Sol in methanol, ethanol, acetone, acetic acid, chloroform, ether, benzene.

Hydrochloride hydrate, $C_{24}H_{40}ClNO_4.H_2O$, minute crystals from alcohol + methyl ethyl ketone + ether (1:1:4), mp 220°.

1939. Cassamine. *7-[2-[2-(Dimethylamino)ethoxy]-2-oxoethylidene]tetradecahydro-1,4a,8-trimethyl-9-oxo-1-phenanthrenecarboxylic acid methyl ester; (E)-13-(carboxymethylene)-14α-methyl-7-oxopodocarpan-16-oic acid 13-[2-(dimethylamino)ethyl]methyl ester.* $C_{25}H_{39}NO_5$; mol wt 433.59. C 69.25%, H 9.07%, N 3.23%, O 18.45%. Extracted from the bark of *Erythrophleum guineense* G. Don., *Leguminosae.* Isoln: Engel, Tondeur, *Helv. Chim. Acta* **32**, 2364 (1949). Structure: Arya, Engel, *ibid.* **44**, 1650 (1961).

Crystals from pentane, mp 86-87°. $[\alpha]_D^{20}$ $-56°$ (ethanol). uv max: 225 nm (log ε 4.2). Soluble in petr ether.

Hydrochloride, $C_{25}H_{39}NO_5.HCl$, crystals from acetone, dec 214-217°. $[\alpha]_D^{17}$ $-48°$ (c = 0.65 in water).

Sulfate, $C_{25}H_{39}NO_5.H_2SO_4$, crystals from alcohol + ether, mp 191-194°. $[\alpha]_D^{17}$ $-49.5°$ (c = 0.47 in water).

1940. Cassella's Acid. *7-Hydroxy-2-naphthalenesulfonic acid;* 2-naphthol-7-sulfonic acid; *β-naphthol-δ-monosulfonic acid; β-naphtholsulfonic acid F; F Acid.* $C_{10}H_8O_4S$; mol wt 224.24. C 53.56%, H 3.60%, O 28.54%, S 14.30%. Prepn: Bayer, Duisberger, *Ber.* **20**, 1426 (1887); Green, *J. Chem. Soc.* **55**, 33 (1889).

Needles from HCl, mp 89°. Readily sol in water, alcohol; practically insol in ether, benzene.

USE: Dyestuff intermediate.

1941. Cassella's Acid F. *7-Amino-2-naphthalenesulfonic acid;* 2-naphthylamine-7-sulfonic acid; *β-naphthylamine-δ-sulfonic acid.* $C_{10}H_9NO_3S$; mol wt 223.25. C 53.80%, H 4.06%, N 6.27%, O 21.50%, S 14.36%. Prepn by sulfonation of β-naphthylamine and separation from the 6-amino isomer: Green, *J. Chem. Soc.* **55**, 33 (1889); from 7-hydroxy-2-naphthalenesulfonic acid and ammonia: Green, *loc. cit.;* Wait, U.S. pat. **1,492,497** (1924).

Monohydrate, crystals. Sol in 5040 parts cold water, 350 parts boiling water; sol in glacial acetic acid.

Copper salt, orange-yellow crystals. Sparingly sol in water. *Ref:* Green, Vakil, *J. Chem. Soc.* **113**, 35 (1918).

Note: **Bronner's acid** was first described as *6-amino-2-naphthalenesulfonic acid* or *2-naphthylamine-6-sulfonic acid.* However, this product obtained by sulfonation of β-naphthylamine, was subsequently shown to be a mixture of about equal parts of 6- and 7-amino-2-naphthalenesulfonic acids: Green, *loc. cit.*

USE: Both the title compd and its 6-amino isomer are used in the manuf of azo dyes, *e.g.,* **Brit.** pat. **810,246** (1959 to Bayer).

1942. Cassia Fistula. Cassia pods; drumstick; Indian laburnum; pudding-stick; pudding pipe; purging cassia. Dried fruit of *Cassia fistula* L. (*Cathartocarpus fistula* [L.] Pers.), *Leguminosae. Habit.* Upper Egypt, E. India; cultivated in tropical America and Africa. The pulp of the ripe fruit, **cassia pulp,** is an almost black, viscid mass with a sweetish taste. *Constit.* Hydroxymethylanthraquinones, gum, tannin, albuminoids, about 60% sugars.

THERAP CAT: Cathartic.

1943. Castanea. Chestnut. Leaves of *Castanea dentata* (Marsh.) Borkh., *Fagaceae,* collected in September and October. *Habit.* Southern Europe. There are hardly any chestnut trees left in the U.S. *Constit.* Tannin, gum, albumin, resin.

1944. Castanospermine. *[1S-(1α,6β,7α,8β,8aβ)]-Octahydro-1,6,7,8-indolizinetetrol;* 1,6,7,8-tetrahydroxyoctahydroindolizine; (1S,6S,7R,8R,8aR)-1,6,7,8-tetrahydroxyindolizidine. $C_8H_{15}NO_4$; mol wt 189.21. C 50.78%, H 7.99%, N 7.40%, O 33.82%. Polyhydroxy alkaloid isolated from the seeds of the Australian leguminous tree, *Castanospermum australe.* Inhibits enzymatic glycoside hydrolysis. Isoln of the naturally occurring (+)-form: L. D. Hohenschutz *et al., Phytochemistry* **20**, 811 (1981). Total synthesis and absolute configuration: R. C. Bernotas, B. Ganem, *Tetrahedron Letters* **25**, 165 (1984). Alternate synthesis: H. Hamana *et al., J. Org. Chem.* **52**, 5492 (1987). Inhibition of α- and β-glucosidases: R. Saul *et al., Arch. Biochem. Biophys.* **221**, 593 (1983); *eidem, ibid.* **230**, 668 (1984). Insect antifeedant activity: D. L. Dreyer *et al., J. Chem. Ecol.* **11**, 1045 (1985). Inhibition of HIV infectivity *in vitro:* B. D. Walker *et al., Proc. Natl. Acad. Sci. USA* **84**, 8120 (1987); R. A. Gruters *et al., Nature* **330**, 74 (1987).

Crystals from aq ethanol, mp 212-215° (dec). $[\alpha]_D^{25}$ +79.7° (c = 0.93 in water). pK 6.09.

1945. Castle's Intrinsic Factor. Intrinsic factor; IF. A thermolabile mucoprotein with mol wt about 60,000. Promotes vitamin B_{12} absorption by transporting it through the intestinal wall. Occurs in normal gastric juice, but is deficient in patients with pernicious anemia. Prepd from hog mucosa: Castle *et al., Am. J. Med. Sci.* **178**, 748, 764 (1929); **180**, 305 (1930); Glass *et al., Science* **115**, 101 (1952); Heinle *et al., Trans. Assoc. Am. Phys.* **65**, 214 (1952); Latner *et al., Biochem. J.* **55**, XXIII (1953); Callender *et al., Brit. Med. J.* **I**, 10 (1954); Latner *et al., Lancet* **I**, 497 (1954); Baum, Federman, U.S. pat. **2,912,360** (1959 to Lilly). Purification: Robbins, U.S. pat. **3,008,877** (1961 to Armour); Highley, Ellenbogen, U.S. pats. **3,434,927** and **3,591,678** (1969, 1971, both to Am. Cyanamid). In approx 30% of pernicious anemia patients, antibodies are produced in the serum which combine with IF, thus inhibiting its biological activity. In clinical tests diminished excretion of vitamin B_{12} in the feces is taken as evidence of intrinsic factor activity. Function in the metabolism of vitamin B_{12}: Glass, *Physiol. Rev.* **43**, 529 (1963). *Review:* Gräsbeck, *Progr. Hematol.* **6**, 233 (1969).

Combination with vitamin B_{12}, *Gastrhéma.*

THERAP CAT: Adjuvant in vitamin B_{12} utilization.

1946. Castor Oil. Ricinus oil; oil of Palma Christi; tangantangan oil; Neoloid. Fixed oil obtained by cold-pressing the seeds of *Ricinus communis* L., *Euphorbiaceae.* Triglyceride of fatty acids. Fatty acid composition is approx ricinoleic 87%, oleic 7%, linoleic 3%, palmitic 2%, stearic 1% and dihydroxystearic trace amounts: Binder *et al., J. Am. Oil Chem. Soc.* **39**, 513 (1962). Review and bibliography: Anderson, *J. Philippine Pharm. Assoc.* **42**, 5-16 (1955); Dominguez *et al., J. Chem. Ed.* **20**, 446 (1952); F. C. Naughton *et al.,* in Kirk-Othmer *Encyclopedia of Chemical Technology* **vol. 5** (Wiley-Interscience, New York, 3rd ed., 1979) pp 1-15.

Pale yellow, viscous oil. Slight somewhat characteristic odor. The crude oil tastes slightly acrid with a decidedly nauseating after-taste. Has excellent keeping qualities, does not turn rancid unless subjected to excessive heat. Dextro-rotatory (undil. in sodium light). $d_{15.5}^{15.5}$ 0.961-0.963. Wt of tech grades: 8.1 to 8.9 lbs/gallon. n_D^{25} 1.473-1.477. n_D^{40} 1.466-1.473. Solidif $-10°$ to $-18°$. Viscosity at 25°: 6-8 poises, also expressed as U ± ½ (Gardner-Holdt Scale). Flash pt 445°F (230°C); ignition temp 840°F (449°C). Surface tension (dynes/cm): at 20°, 39.0; at 80°, 35.2. Acid value < 4. Sapon no. 176-187. Iodine no. (Wijs') 81-91. Reichert-Meissl value < 0.5. Polenske value < 0.5. Acetyl value 144-150. Hydroxyl value 161-169. Miscible with abs ethanol, methanol, ether, chloroform, glacial acetic acid. Dissolves in its own vol of petr ether or 95% alcohol. Does not dissolve to any extent in mineral oil, unless mixed with another vegetable oil. When heated to 300° for several hours it polymerizes and becomes miscible with mineral oil.

USE: As an industrial raw material for the prepn of chemical derivs used in coatings, urethane derivs, surfactants and dispersants, cosmetics, lubricants; chief raw material for the production of sebacic acid, a basic ingredient in the production of synthetic resins and fibers; as lubricant in metal drawing, machine lubrication and 2-cycle engine fuels, in hydraulic fluids, rubber preservative and mold lubricants; constituent of embalming fluids; in soap manuf; to impart emollient and lubricant properties to cosmetic prepns; as Turkey-red oil (sulfated castor oil) for dyeing and finishing textiles; as dehydrated castor oil in alkyds, resinous copolymers, varnishes, oil-based paints, enamels, calks and putties; as blown oil (oxidized oil) for plasticizing oilcloth, artificial leather, coated fabrics, and lacquers; to plasticize rosin in the manuf of sticky fly-paper, for nitrocellulose and similar coating systems, hot melts, duplicating and stencil inks, adhesives and laminants; as release and anti-sticking agent in hard candy manuf.

THERAP CAT: Cathartic.

THERAP CAT (VET): Mild purgative, but considered unreliable in adult horses. Emollient.

1947. Castor Oil, Hydrogenated. Opalwax; Castorwax. Mol wt about 932. A hard, white wax, mp 86-88°. Iodine number (Wijs') < 5.0. Extremely insol in water and in the more common organic solvents.

USE: In water-repellent coatings, candles, shoepolish, carbon paper, ointments and cosmetics; for impregnating paper, wood, cloth; for electrical condenser impregnation; as solid lubricant; as a pressure mold release agent in the manuf of formed plastics and rubber goods.

1948. Catalase. Caperase; Equilase; Optidase. Enzymes which promote reactions involving the decompn of hydrogen peroxide to water and oxygen. Although widely distributed among animals, plants, bacteria, and fungi, catalase in mammalian liver and blood has been most intensively studied. Catalase for commercial use obtained from animal liver, bacterial (*Micrococcus lysodeikticus)*, and fungal *(Aspergillus niger)* sources. Isoln from mammalian livers and kidneys: Lolli, Cavanaugh, U.S. pat. **2,703,779** (1955 to Armour); Schroeder *et al., Biochim. Biophys. Acta* **58**, 611 (1962); Dan, U.S. pat. **2,992,167** (1961 to Chr. Hansen's Lab.); from *Aspergillus niger:* Faucett *et al.,* U.S. pat. **3,-102,081** (1963 to Miles Labs.). All catalases so far isolated contain four tetrahedrally arranged subunits of equal size giving an approximate mol wt of 240,000. Each subunit consists of a single polypeptide chain associated with a single prosthetic group, ferric protoporphyrin IX. Amino acid sequence of bovine liver catalase subunit: Schroeder *et al., Arch. Biochem. Biophys.* **131**, 653 (1969). Bovine *hepatocatalase,* a term for catalase obtained from liver, was reported to lower serum cholesterol; the active form was found to contain a catalytic amount of zinc (0.32%): Azarnoff, Curran, *J. Lab. Clin. Med.* **60**, 856 (1962), *cf.* Laporte *et al., Biochem. Pharmacol.* **11**, 670 (1962). *Reviews:* Nicholls, Schonbaum, in *The Enzymes,* **vol. 8,** P. D. Boyer *et al.,* Eds. (Academic Press, New York, 1963) pp 147-225; Deisseroth, Daunce, *Physiol. Rev.* **50**, 319 (1970); Schonbaum, Chance, in *The Enzymes* **vol. 13** (part C), P. D. Boyer, Ed. (Academic Press, New York, 3rd ed., 1976) pp 363-408.

USE: In combination with glucose oxidase, for treatment of food wrappers to prevent oxidative deterioration of food: Sarett, Scott, U.S. pat. **2,765,233** (1956 to Ben L. Sarett). In the removal of traces of peroxide in the process of cold sterilization (preservation of milk and cheese by treatment with hydrogen peroxide). With glucose oxidase, *q.v.,* in food preservation.

1949. Catalposide. *[1aS-(1aα,1bβ,2β,5aβ,6β,6aα)]-1a,- 1b,2,5a,6,6a-Hexahydro-6-[(4-hydroxybenzoyl)oxy]-1a- (hydroxymethyl)oxireno[4,5]cyclopenta[1,2-c]pyran-2-yl-β-D- glucopyranoside.* Catalpin. $C_{22}H_{26}O_{12}$; mol wt 482.44. C 54.77%, H 5.43%, O 39.80%. From unripe fruit of *Catalpa bignonioides* Walt. or *C. ovata* G. Don, *Bignoniaceae:* Claassen, *Am. Chem. J.* **10,** 228 (1888). Historical summary: Bobbitt *et al., J. Org. Chem.* **26,** 3090 (1961). Structure: Lunn *et al., Can. J. Chem.* **40,** 104 (1962). Stereochemistry: Bobbitt *et al., J. Org. Chem.* **31,** 500 (1966); **32,** 1459 (1967).

Crystals from methanol + ethyl acetate, mp 215-216.5°. $[\alpha]_D^{23.2}$ $-184°$ (c = 0.87 in methanol). uv max (ethanol): 260 nm (log ε 4.27); in sodium hydroxide: 303 nm (log ε 4.35). Soluble in water, alcohol; slightly sol in benzene, chloroform, ether.

Hexaacetate, $C_{34}H_{38}O_{18}$, crystals from ethanol, mp 141.5°. $[\alpha]_D^{21.7}$ $-106°$ (c = 0.75 in chloroform).

1950. Catechin. *(2R-trans)-2-(3,4-Dihydroxyphenyl)-3,- 4-dihydro-2H-1-benzopyran-3,5,7-triol; catechol;* 3,3',4',5,7-flavanpentol; catechinic acid; catechuic acid; (+)-cyanidanol-3; dexcyanidanol; cyanidol; Catergen. $C_{15}H_{14}O_6$; mol wt 290.27. C 62.07%, H 4.86%, O 33.07%. Flavonoid found primarily in higher woody plants as (+)-catechin along with *(−)-epicatechin* (cis form). From catechu (gambir and acacia), mahogany wood, etc.: Perkin, Yoshitake, *J. Chem. Soc.* **81**, 1160 (1902); Freudenberg *et al., Ber.* **55**, 1734 (1922). Structure: Freudenberg *et al., Ann.* **444**, 135 (1925). Stereochemistry: Clark-Lewis, *Chem. & Ind. (London)* **1955**, 1218; Hardegger *et al., Helv. Chim. Acta* **40**, 1819 (1957); Clark-Lewis, *J. Chem. Soc.* **1960**, 2433. Pharmacology: Van Cauwenberge *et al., C.R. Soc. Biol.* **165**, 1195 (1971). Metabolism: Das, Sothy, *Biochem. J.* **125**, 417 (1971); and absorption in human: N. P. Das, *Biochem. Pharmacol.* **20**, 3435 (1971). Biosynthesis of epicatechins: Zaprometov, Grisebach, *Z. Naturforsch.* **28c**, 113 (1973). Other catechins such as *afzelechin* and *gallocatechin* and catechin oligomers *(procyanidins)* also exist in nature: Thompson *et al., J. Chem. Soc. Perkin Trans. I* **1972**, 1287. *See also* Bioflavonoids.

Needles from water + acetic acid, occurs as hydrate. mp 93-96°; 175-177° when anhydr. $[\alpha]_D^{18}$ +16° to +18.4°.

dl-Form, needles from water + acetic acid, mp 212-216°. Slightly sol in cold water, ether; sol in hot water, alcohol, glacial acetic acid, acetone. Practically insol in benzene, chloroform, petr ether.

l-form, needles from water + acetic acid, mp 93-96°; 175-177° when anhydr. $[\alpha]_D$ −16.8°.

Note: Catechin is also called catechol (flavan) to distinguish it from catechol (pyrocatechol, *q.v.*).

USE: In dyeing and tanning.

THERAP CAT: Antidiarrheal.

1951. Catechu Black. Cutch; cachou; pegu catechu; cashoo. Extract prepd from heartwood of *Acacia catechu* Willd., *Leguminosae*. *Habit.* India, Hindustan, Ceylon; naturalized in Jamaica. *Constit.* 25-35% catechutannic acid, 2-10% catechin; catechu red, quercetin, gum.

Incompat. Iron compds; gelatin, lime water, zinc sulfate.

USE: Tanning, dyeing fabrics brown and black; staining wood; in toilet preparations.

THERAP CAT: Astringent (diarrheal).

THERAP CAT (VET): Has been used as intestinal astringent.

1952. Catharanthine. *3,4-Didehydroibogamine-18-carboxylic acid methyl ester; 7-ethyl-9,10,12,13-tetrahydro-6,9-methano-5H-pyrido[1′,2′:1,2]azepino[4,5-b]indole-6(6aH)-carboxylic acid methyl ester.* $C_{21}H_{24}N_2O_2$; mol wt 336.43. C 74.97%, H 7.19%, N 8.33%, O 9.51%. Precursor of vinblastine-type alkaloids. Isoln from *Vinca rosea* Linn. (*Catharanthus roseus* G. Don.) *Apocynaceae*: M. Gorman *et al., J. Am. Pharm. Assoc. Sci. Ed.* **48**, 256 (1959); G. H. Svoboda *et al., ibid.* 659. Structure: N. Neuss, M. Gorman, *Tetrahedron Letters* **1961**, 206. Abs config: K. Bláha *et al., ibid.* **1972**, 2763. Synthesis of (±)-form: A. A. Qureshi, A. I. Scott, *Chem. Commun.* **1968**, 947, 948; A. R. Battersby *et al., ibid.* 951; G. Büchi *et al., J. Am. Chem. Soc.* **92**, 999 (1970); J. P. Kutney, F. Bylsma, *Helv. Chim. Acta* **58**, 1672 (1975); B. M. Trost *et al., J. Org. Chem.* **44**, 2052 (1979); T. Imanishi *et al., Tetrahedron Letters* **21**, 3285 (1980).

Crystals from methanol, mp 126-128°. uv max (ethanol): 226, 284, 292 nm (log ε 4.56, 3.92, 3.88). $[\alpha]_D^{27}$ +29.8° (CHCl₃). pKa′ 6.8.

1953. Cathepsins. Intracellular proteinases obtained from animal tissue extracts, the richest sources being liver, kidney and spleen. Located primarily in the lysosomal fraction within the cell. Part of the general enzymic apparatus of animal cells; in most cases they do not specialize in functions characteristic of individual tissues. Review of cathepsins A-C: J. S. Fruton in *The Enzymes* vol. **4**, P. D. Boyer *et al.*, Eds. (Academic Press, New York, 1960) pp 233-241; of cathepsins A-E: M. J. Mycek, *Methods Enzymol.* **19**, 285-315 (1970); of cathepsins B, D and G: several authors, *Res. Monogr. Cell Tissue Physiol.* **2**, 57-89, 181-248 (1977); of cathepsins B, D, G, H, L, N and S: several authors, *Ciba Found. Symp.* **75**, 1-68 (1980).

Cathepsin C, dipeptidyl transferase, dipeptidyl amino peptidase I. Isoln from beef spleen: Tallan *et al., J. Biol. Chem.* **194**, 793 (1952); de la Haba *et al., ibid.* **234**, 316 (1959). Hydrolyzes dipeptidyl amides or esters bearing a free α-amino (or α-imino) group in the N-terminal position, esp. those containing an aromatic amino acid adjacent to the free α-amino group: Planta, Gruber, *Biochim. Biophys. Acta* **53**, 443 (1961); Wurz *et al., Biochemistry* **1**, 19 (1962). Enhances the proteolysis of prothrombin to thrombin and thus plays an important role in blood clotting: Purcell, Barnhart, *Biochim. Biophys. Acta* **78**, 800 (1963).

Cathepsin D. Isoln from bovine spleen: Press *et al., Biochem. J.* **74**, 501 (1960); Webb, *ibid.* **76**, 538 (1960); **84**, 455

(1962). Involved in the catabolism of cartilage and connective tissue: Weston *et al., Nature* **222**, 285 (1969).

Cathepsin G. Isoln from human spleen: G. Starkey *et al., Biochem. J.* **155**, 255 (1976). Chymotrypsin-like enzyme. Catalytic and immunological properties: P. M. Starkey, A. J. Barrett, *ibid.* 273.

1954. Cathinone. *(S)-2-Amino-1-phenyl-1-propanone;* α-aminopropiophenone. $C_9H_{11}NO$; mol wt 149.19. C 72.46%, H 7.43%, N 9.39%, O 10.72%. Pharmacologically active alkaloid extracted from the leaves of khat, *Catha edulis* Forsk., *Celastraceae*. Isoln and identification as active constituent of khat: D. W. Peterson *et al., Life Sci.* **27**, 2143 (1980). Prepn of racemic compd: L. Behr-Bergowski, *Ber.* **30**, 1515 (1897); S. Gabriel, *ibid.* **41**, 1127 (1908). Stereospecific synthesis of (−)-cathinone from norephedrine, characterization as HCl salt: B. D. Berrang *et al., J. Org. Chem.* **47**, 2643 (1982). Similarity of cardiovascular effects with those of (+)-amphetamine: J. D. Kohli, L. J. Goldberg, *J. Pharm. Pharmacol.* **34**, 338 (1982); P. Kalix, *Life Sci.* **32**, 801 (1983). Effect on dopamine metabolism: G. P. Mereu *et al., ibid.* 1383.

Sol in methanol (with racemization), methylene chloride. $[\alpha]_D$ −46.8° (c = 0.24 in methanol).

Hydrochloride, $C_9H_{12}ClNO$, crystals from isopropanol-THF, mp 189-190°.

1955. Catnep. Cataria; catnip; catmint. Herb of *Nepeta cataria* L., *Labiatae*. *Habit.* Europe, Asia; naturalized in U.S. *Constit.* Volatile oil, nepetalactone (*q.v.*), nepetalic acid and related compds, tannin: McElvain, Eisenbraun, *J. Am. Chem. Soc.* **77**, 1599 (1955). Its odor is very attractive to all members of the cat family.

THERAP CAT: Aromatic.

1956. Caulophylline. *(1R)-1,2,3,4,5,6-Hexahydro-3-methyl-1,5-methano-8H-pyrido[1,2-a][1,5]diazocin-8-one; 12-methylcytisine;* N-methylcytisine. $C_{12}H_{16}N_2O$; mol wt 204.27. C 70.56%, H 7.89%, N 13.71%, O 7.83%. Isoln from *Caulophyllum thalictroides* Michx., *Berberidaceae*: Power, Salway, *J. Chem. Soc.* **103**, 191 (1913); from *Thermopsis rhombifolia* Richards., *Leguminosae*: Manske, Marion, *Can. J. Res.* **21B**, 144 (1943); from *Baptisia perfoliata* (L.) R. Br., *Leguminosae*: Marion, Turcotte, *J. Am. Chem. Soc.* **70**, 3253 (1948); from *Cytisus laburnum* L., *Leguminosae*: Pöhm, Galinovsky, *Monatsh.* **84**, 1197 (1953); from *Ormosia stipitata* Schery., *Leguminosae*: Lloyd, Horning, *J. Org. Chem.* **23**, 1074 (1958). Synthesis: Van Tamelen, Baran, *J. Am. Chem. Soc.* **77**, 4944 (1955). Toxicity study: R. B. Barlow, L. J. McLeod, *Brit. J. Pharmacol.* **35**, 161 (1969).

Crystals from ethyl acetate + cyclohexane, mp 140-141°. $[\alpha]_{589}^{25}$ −223° (c = 0.905). Sol in 1.1 parts water, 0.9 part alcohol, 0.5 part methanol, 0.9 part chloroform, 2.8 parts benzene, 2.6 parts acetone. LD₅₀ in mice (mg/kg): 21 i.v.; 51 i.p. (Barlow, McLeod).

1957. Caulophyllum. Blue cohosh; papoose root; squaw root. Dried rhizome and roots of *Caulophyllum thalictroides* Michx., *Berberidaceae*. *Habit.* Canada to North Carolina, Missouri, and Nebraska; Japan. *Constit.* Leontin, caulophylline, saponin, resins.

1958. 2C-B. *4-Bromo-2,5-dimethoxybenzeneethanamine;* 4-bromo-2,5-dimethoxyphenethylamine; BDMPEA;

α-desmethyl DOB; MFT; nexus. $C_{10}H_{14}BrNO_2$; mol wt 260.13. C 46.17%, H 5.42%, Br 30.72%, N 5.38%, O 12.30%. Psychoactive analog of mescaline, q.v. Synthesis and activity: A. T. Shulgin, M. F. Carter, *Psychopharmacol. Commun.* **1**, 93 (1975). Qualitative analysis: F. A. Ragan, Jr. *et al., J. Anal. Toxicol.* **9**, 91 (1985). Pharmacology: R. A. Glennon *et al., Pharmacol. Biochem. Behav.* **30**, 597 (1988); M. Lobos *et al., Gen. Pharmacol.* **23**, 1139 (1992).

Hydrochloride, $C_{10}H_{14}BrNO_2$.HCl. Pale pink crystals from ethanol, mp 237-239° (dec). uv max (dil H_2SO_4): 294 nm.

Note: The DEA has placed this substance into Schedule I of the Controlled Substances Act to avoid an imminent hazard to the public safety: *Fed. Reg.* **59**, 671 (1994).

1959. Ceanothic Acid. *(2α,3β)-3-Hydroxy-A(1),28-dinorlup-20(29)-ene-2,17-dicarboxylic acid;* emmolic acid. $C_{30}H_{46}O_5$; mol wt 486.69. C 74.04%, H 9.53%, O 16.44%. From *Ceanothus americanus* L., *Rhamnaceae* (New Jersey tea). Isoln: Julian *et al., J. Am. Chem. Soc.* **60**, 77 (1938). Identity of ceanothic acid and emmolic acid: Mechoulam, *Chem. & Ind. (London)* **1961**, 1835. Structure: de Mayo, Starratt, *Tetrahedron Letters* **1961**, 259; Mechoulam, *J. Org. Chem.* **27**, 4070 (1962). Stereochemistry: de Mayo, Starratt, *Can. J. Chem.* **40**, 788 (1962). Revised stereochemistry: Eade *et al., Aust. J. Chem.* **24**, 621 (1971).

Crystals from methanol + ether, mp 356-357°. $[\alpha]_D$ +38° (c = 1.20 in ethanol).
Dimethyl ester, $C_{32}H_{50}O_5$, crystals from ether + petr ether, mp 221-223°. $[\alpha]_D$ +41° (chloroform).

1960. Cecropins. A major class of non-specific antibacterial proteins induced in hemolymph of some insects as part of the immune response. First characterized in *Lepidoptera*, particularly *Hyalophora cecropia.* (See also attacins.) Small, basic polypeptides which possess similar amino acid sequences with relatively long hydrophobic regions, mol wt approx 4000 Da. Originally designated *protein P9*, three major peptides have been characterized, cecropins A, B, D; three minor peptides C, E, F are also known. Identification of cell free immune response in *Lepidoptera:* H. Boman *et al., Infect. Immun.* **10**, 136 (1974). Initial identification and isolation of immune response proteins in *H. cecropia:* I. Faye *et al., ibid.* **12**, 1426 (1975). Purification and preliminary characterization of cecropins A and B: D. Hultmark *et al., Eur. J. Biochem.* **106**, 7 (1980); antibacterial spectrum and amino acid sequence: H. Steiner *et al., Nature* **292**, 246 (1981). Partial synthesis of A: R. B. Merrifield *et al., Biochemistry* **21**, 5020 (1982). Total synthesis of A: D. Andreu *et al., Proc. Nat. Acad. Sci. USA* **80**, 6475 (1983); of B: P. van Hofsten *et al., ibid.* **82**, 2240 (1985). Isolation, characterization of cecropins C, D, E, F, and comparison with A and B: D. Hultmark *et al., Eur. J. Biochem.* **127**, 207 (1982); X. Qu *et al., ibid.* 219. Isoln of cecropin-like proteins from flesh fly larvae: M. Okada, S. Natori, *Biochem. J.* **211**, 727 (1983); from tsetse fly: G. P. Kaaya *et al., Insect. Biochem.* **17**, 309 (1987). Cecropin-like proteins designated as *lepidopterans* have been isolated from *Bombyx mori* silkworms: T. Teshima *et al., Tetrahedron Letters* **28**,

4705 (1987). Mechanism of action: H. Steiner *et al., Biochim. Biophys. Acta* **939**, 260 (1988); B. Christensen *et al., Proc. Nat. Acad. Sci. USA* **85**, 5072 (1988). Review of cecropins A and B: H. G. Boman, H. Steiner in *Current Topics in Microbiology and Immunology* **94/95**, W. Henle *et al.*, Eds. (Springer-Verlag, N.Y., 1981) pp 75-91.

1961. Cedrol. *[3R-(3α,3aβ,6α,7β,8aα)]-Octahydro-3,6,-8,8-tetramethyl-1H-3a,7-methanoazulen-6-ol; 8βH-cedran-8-ol;* cedar camphor; cypress camphor. $C_{15}H_{26}O$; mol wt 222.37. C 81.02%, H 11.79%, O 7.19%. Occurs in cedar wood oil from *Juniperus virginiana* L. (*J. sabina* Kook), *Cupressaceae*, in cypress oil from *Cupressus sempervirens: Schimmel's Report* **1904, II**, 20; in the oil from *Juniperus chinensis:* Kondo, *J. Pharm. Soc. Japan* **1907**, 236; and from *Origanum smyrnaeum* L., *Labiatae: Schimmel's Report* **1906**, Oct., 72. Structure: Stork, Breslow, *J. Am. Chem. Soc.* **75**, 3291, 3292 (1953). Stereochemistry and total synthesis: Stork, Clarke, *ibid.* **83**, 3114 (1961). Synthesis of *dl*-form: Corey *et al., ibid.* **91**, 1557 (1969); *eidem, Tetrahedron Letters* **1973**, 3153; E. G. Breitholle, A. G. Fallis, *J. Org. Chem.* **43**, 1964 (1978).

Needles from dil methanol, mp 86-87°. $[\alpha]_D^{28}$ +9.9° (c = 5 in chloroform).
USE: In fragrances.

1962. Cefaclor. *[6R-[6α,7β(R*)]]-7-[(Aminophenylacetyl)amino]-3-chloro-8-oxo-5-thia-1-azabicyclo[4.2.0]oct-2-ene-2-carboxylic acid monohydrate;* 7-(D-2-amino-2-phenylacetamido)-3-chloro-3-cephem-4-carboxylic acid monohydrate; 3-chloro-7-D-(2-phenylglycinamido)-3-cephem-4-carboxylic acid monohydrate; compd 99638; Alfacet; Alfatil; Ceclor; Distaclor; Panacef; Panoral. $C_{15}H_{14}ClN_3O_4S.H_2O$; mol wt 385.83. C 46.70%, H 4.18%, Cl 9.19%, N 10.89%, O 20.73%, S 8.31%. Semi-synthetic cephalosporin antibiotic, related to cephalexin, q.v. Prepn: R. R. Chauvette, **Ger. pat. 2,408,698** (1974 to Lilly), C.A. **82**, 4278n (1975); **U.S. pat. 3,925,372** (1975 to Lilly); R. R. Chauvette, P. A. Pennington, *J. Med. Chem.* **18**, 403 (1975). *In vitro* studies: M. S. Silver *et al., Antimicrob. Ag. Chemother.* **12**, 591 (1977); R. N. Jones, *J. Antibiot.* **30**, 753 (1977); B. R. Meyers, S. Z. Hirschman, *J. Clin. Pharmacol.* **18**, 65 (1978). Metabolism: N. G. Waterman, L. F. Sharfenberger, *Antimicrob. Ag. Chemother.* **14**, 614 (1978). Human pharmacology: G. R. Hodges *et al., ibid.* 454; A. Glynne *et al., J. Antimicrob. Chemother.* **4**, 343 (1978). Clinical studies: J. D. Nelson *et al., Am. J. Dis. Child.* **132**, 992 (1978); B. M. Gray *et al., Antimicrob. Ag. Chemother.* **13**, 988 (1978). Comprehensive description: L. J. Lorenz in *Analytical Profiles of Drug Substances* vol. 9, K. Florey, Ed. (Academic Press, New York, 1980) pp 107-123.

Crystalline solid. uv max (pH 7 buffer): 265 nm (ε 6800). Sol in water. Practically insol in methanol, chloroform, benzene. Solns are stable at pH 2.5-4.5.
THERAP CAT: Antibacterial.

1963. Cefadroxil. *[6R-[6α,7β(R*)]]-7-[[Amino-(4-hydroxyphenyl)acetyl]amino]-3-methyl-8-oxo-5-thia-1-azabicyclo-[4.2.0]oct-2-ene-2-carboxylic acid monohydrate;* 7-[D-(−)-α-amino-α-(4-hydroxyphenyl)acetamido]-3-methyl-3-cephem-4-carboxylic acid monohydrate; p-hydroxycephalexine monohydrate; BL-S578; MJF-11567-3; Baxan; Bidocef; Cefa-Drops; Cefamox; Ceforal; Cephos; Duracef; Duricef; Kefroxil; Oracéfal; Sedral; Ultracef. $C_{16}H_{17}N_3O_5S.H_2O$; mol wt 381.41. C 50.39%, H 5.02%, N 11.02%, O 25.17%, S 8.41%. Semi-synthetic cephalosporin antibiotic. Prepn: **Neth. pat. Appl. 6,812,382;** L. B. Crast, Jr., **U.S. pat. 3,489,-752** (1969, 1970 both to Bristol-Myers); T. Takahashi *et al.*, **Ger. pat. 2,216,113;** *idem*, **U.S. pat. 3,816,253** (1972, 1974, both to Takeda). Prepn of crystalline monohydrate: D. Bouzard *et al.*, **U.S. pat. 4,504,657** (1985 to Bristol-Myers). Antimicrobial activity: R. E. Buck, K. E. Price, *Antimicrob. Ag. Chemother.* **11**, 324 (1977). Pharmacology: M. Pfeffer *et al.*, *ibid.* 331; A. I. Hartstein *et al.*, *ibid.* **12**, 93 (1977). *Review:* *J. Antimicrob. Chemother.* **10**, Suppl. B, 1-162 (1982). Series of articles on clinical trials in respiratory tract infections: *Drugs* **32**, Suppl. 3, 1-56 (1986).

White crystals, mp 197° (dec).
THERAP CAT: Antibacterial.
THERAP CAT (VET): Antibacterial.

1964. Cefamandole. *[6R-[6α,7β(R*)]]-7-[(Hydroxyphenylacetyl)amino]-3-[[(1-methyl-1H-tetrazol-5-yl)thio]methyl]-8-oxo-5-thia-1-azabicyclo[4.2.0]oct-2-ene-2-carboxylic acid; 7-mandelamido-3-[[(1-methyl-1H-tetrazol-5-yl)thio]methyl]-8-oxo-5-thia-1-azabicyclo[4.2.0]oct-2-ene-2-carboxylic acid; 7-D-mandelamido-3-[[(1-methyl-1H-tetrazol-5-yl)thio]methyl]-3-cephem-4-carboxylic acid; 7-D-mandelamido-3-(1-methyl-1,2,3,4-tetrazole-5-thiomethyl)-Δ³-cephem-4-carboxylic acid; CMT; compd 83405.* $C_{18}H_{18}N_6O_5S_2$; mol wt 462.51. C 46.74%, H 3.92%, N 18.17%, O 17.30%, S 13.87%. Broad-spectrum semi-synthetic cephalosporin antibiotic. Prepn: C. W. Ryan, **Ger. pat. 2,018,600;** *idem*, **U.S. pat. 3,641,021** (1970, 1972 to Lilly); J. M. Greene, **Ger. pat. 2,312,997;** *idem*, **U.S. pat. 3,840,531** (1973, 1974 to Lilly). Biological properties: W. E. Wick, D. A. Preston, *Antimicrob. Ag. Chemother.* **1**, 221 (1972). Antibacterial activity: S. Eykyn *et al.*, *ibid.* **3**, 657 (1973); H. C. Neu, *ibid.* **6**, 177 (1974); A. D. Russell, *J. Antimicrob. Chemother.* **1**, 97 (1975). Pharmacologic studies: B. R. Meyers *et al.*, *Antimicrob. Ag. Chemother.* **9**, 140 (1976); R. S. Griffith *et al.*, *ibid.* **10**, 814 (1976). Comprehensive description: R. H. Bishara, E. C. Rickard, *Anal. Profiles Drug Subs.* **9**, 125-154 (1980).

Nafate, $C_{19}H_{17}N_6NaO_6S_2$, *Bergacef, Cedol, Cefam, Cefiran, Cemado, Cemandil, Fado, Kefadol, Kefandol, Lampomandol, Mandokef, Mandol, Mandolsan, Neocefal, Pavecef.* White, odorless needles, mp 190° (dec). uv max (H_2O): 269 nm (ε 10800). pKa 2.6-3.0. Sol in water, methanol. Practically insol in ether, chloroform, benzene, cyclohexane.
THERAP CAT: Antibacterial.

1965. Cefatrizine. *[6R-[6α,7β(R*)]]-7-[[Amino-(4-hydroxyphenyl)acetyl]amino]-8-oxo-3-[(1H-1,2,3-triazol-4-ylthio)methyl]-5-thia-1-azabicyclo[4.2.0]oct-2-ene-2-carboxylic acid; 7-[D-α-amino-α-(4-hydroxyphenyl)acetamido]-3-(1,-2,3-triazol-4(5)-ylthiomethyl)-3-cephem-4-carboxylic acid; BL-S640; SKF-60771; S-640P; Cefaperos.* $C_{18}H_{18}N_6O_5S_2$; mol wt 462.51. C 46.74%, H 3.92%, N 18.17%, O 17.30%, S 13.87%. Orally active semi-synthetic cephalosporin antibiotic. Prepn: G. L. Dunn, J. R. E. Hoover, **Ger. pat.**

2,316,866; *eidem,* **U.S. pat. 3,867,380** (1974, 1975 both to SK&F); D. Willner, L. B. Crast, **Ger. pat 2,364,192** (1974 to Bristol-Myers), *C.A.* **81,** 105538 (1974); G. L. Dunn *et al., J. Antibiot.* **29,** 65 (1976). Prepn of the propylene glycolate: M. A. Kaplan, A. P. Granatek, **Ger. pat. 2,500,385;** *eidem,* **U.S. pat. 3,970,651** (1975, 1976 both to Bristol-Myers). *In vitro* and *in vivo* antibacterial activity and pharmacokinetic behavior: P. Actor *et al., ibid.* **28,** 594 (1975); F. Leitner *et al., Antimicrob. Ag. Chemother.* **7,** 298 (1975). General pharmacological properties: M. Matsuzaki *et al., Japan. J. Antibiot.* **29,** 107 (1976), *C.A.* **85,** 311g (1976). Determn in serum and urine by HPLC: E. Crombez *et al., J. Chromatog.* **173,** 165 (1979). Disposition of ^{14}C-cefatrizine in man: R. C. Gaver, G. Deeb, *Drug Metab. Dispos.* **8,** 157 (1980). Activity and clinical use: A. J. Weinstein, *Drugs* **20,** 137 (1980). Acute toxicity: M. Matsuzaki *et al., Japan. J. Antibiot.* **29,** 612 (1976), *C.A.* **85,** 171724y (1976). Toxicological studies: *ibid.* 639-686, *C.A.* **85,** 171722w-171723x (1976). Series of articles on pharmacokinetics and clinical efficacy: *Drugs Exptl. Clin. Res.* **11,** 441-462 (1985).

Zwitterionic. LD_{50} in male, female mice, male, female rats (mg/kg): 6880, 6410, 4325, 4325 i.p. (Matsuzaki).
Compd with propylene glycol, $C_{18}H_{18}N_6O_5S_2.C_3H_8O_2$, *Brief, Cepticol, Cefotrizin, Faretrizin, Kefoxina, Tamyl, Zitrix.* Rod-like crystals. $[\alpha]_D^{23}$ +55.9° (c = 1% in 1N HCl). uv max: 227, 272 nm.
THERAP CAT: Antibacterial.

1966. Cefazedone. *(6R-trans)-7-[[(3,5-Dichloro-4-oxo-1(4H)-pyridinyl)acetyl]amino]-3-[[(5-methyl-1,3,4-thiadiazol-2-yl)thio]methyl]-8-oxo-5-thia-1-azabicyclo[4.2.0]oct-2-ene-2-carboxylic acid; 3-(5-methyl-1,3,4-thiadiazolyl-2-mercaptomethyl)-7-(3,5-dichloro-4-pyridon-1-ylacetamido)-3-cephem-4-carboxylic acid; EMD-30087.* $C_{18}H_{15}Cl_2N_5O_5S_3$; mol wt 548.45. C 39.42%, H 2.76%, Cl 12.93%, N 12.77%, O 14.59%, S 17.54%. Semi-synthetic cephalosporin antibiotic. Prepn: R. Gerike *et al.,* **Ger. pat. 2,345,402;** *eidem,* **U.S. pat. 4,153,693** (1975, 1979 both to E. Merck, W. Ger.). *See also eidem,* **Ger. pats. 2,621,011, 2,626,026, 2,627,204** (all 1977 to E. Merck, W. Ger.), *C.A.* **88,** 5093, 105383, 121213 (1978). Series of articles on prepn, *in vitro* and *in vivo* activity, kinetics, pharmacology, toxicology, and clinical trials: *Arzneimittel-Forsch.* **29,** 361-462 (1979). Toxicity data: M. von Eberstein *et al., ibid.* 424.

Crystals from acetonitrile.
Sodium salt, $C_{18}H_{14}Cl_2N_5NaO_5S_3$, *Refosporin.* LD_{50} in mice, rats, rabbits, dogs (mg/kg): 6800, 4225, 3200, 3000 i.v. (von Eberstein).
THERAP CAT: Antibacterial.

1967. Cefazolin. *(6R-trans)-3[[(5-Methyl-1,3,4-thiadiazol-2-yl)thio]methyl]-8-oxo-7-[(1H-tetrazol-1-ylacetyl)amino]-5-thia-1-azabicyclo[4.2.0]oct-2-ene-2-carboxylic acid; 7-(1-(1H)-tetrazolylacetamido)-3-[2-(5-methyl-1,3,4-thiadiazolyl)thiomethyl]-Δ³-cephem-4-carboxylic acid; CEZ.* $C_{14}H_{14}N_8O_4S_3$; mol wt 454.51. C 37.00%, H 3.10%, N 24.65%, O 14.08%, S 21.16%. Semi-synthetic antibiotic derived from

7-aminocephalosporanic acid, *q.v.* Prepn: Takano *et al.,* **S. Afr.** pat. **68 04,513** corresp to U.S. pat. **3,516,997** (1969, 1970 to Fujisawa). Synthesis and properties: Kariyone *et al., J. Antibiot.* **23,** 131 (1970). Activity and clinical studies: Nishida *et al., ibid.* **137,** 184; Shibata, Fujii, *Antimicrob. Ag. Chemother.* **1970,** 467. Metabolic studies: Kozatani *et al., Chem. Pharm. Bull.* **20,** 1105 (1972). Toxicology: H. A. Birkhead *et al., J. Infec. Dis.* **128,** Suppl., 379 (1973). Comprehensive description: A. E. Zappala *et al.,* in *Analytical Profiles of Drug Substances* vol. 4, K. Florey, Ed. (Academic Press, New York, 1975) pp 1-20. *Review:* H. Nakano in *Pharmacological and Biochemical Properties of Drug Substances* vol. 1, M. E. Goldberg, Ed. (Am. Pharm. Assoc., Washington, DC, 1977) pp 155-182.

Needles from aq acetone, mp 198-200° (dec). uv max (buffer pH 6.4): 272 nm (ϵ 13150). Easily sol in DMF, pyridine; sol in aq acetone, aq dioxane, aq ethanol; slightly sol in methanol. Practically insol in chloroform, benzene, ether.

Sodium salt, $C_{14}H_{13}N_8NaO_4S_3$, *sodium CEZ, SKF-41558, Acef, Ancef, Atirin, Biazolina, Bor-Cefazol, Cefacidal, Cefamedin, Cefamezin, Cefazil, Cefazina, Elzogram, Firmacef, Gramaxin, Kefzol, Lampocef, Liviclina, Totacef, Zolicef.* White to yellowish-white, odorless crystalline powder with a bitter, salty taste. Crystallizes in α, β, and γ-forms, *see* Kariyone, *loc. cit.* Easily sol in water, slightly sol in methanol, ethanol. Practically insol in benzene, acetone, chloroform. LD_{50} in mice, rats (mg/kg): 3.9, 3.18 i.v.; 6.2, 7.4 i.p. (Birkhead).

THERAP CAT: Antibacterial.

1968. Cefbuperazone. *[6R-[6α,7α,7(2R*,3S*)]]-7-[[2-[[(4-Ethyl-2,3-dioxo-1-piperazinyl)carbonyl]amino]-3-hydroxy-1-oxobutyl]amino]-7-methoxy-3-[[(1-methyl-1H-tetrazol-5-yl)thio]methyl]-8-oxo-5-thia-1-azabicyclo[4.2.0]oct-2-ene-2-carboxylic acid;* (6R,7S)-7-[(2R,3S)-2-(4-ethyl-2,3-dioxo-1-piperazinecarboxamido)-3-hydroxybutyramido]-7-methoxy-3-[[(1-methyl-1H-tetrazol-5-yl)thio]methyl]-8-oxo-5-thia-1-azabicyclo[4.2.0]oct-2-ene-2-carboxylic acid; 7β-[D-α-(4-ethyl-2,3-dioxo-1-piperazinecarboxamido)-β-(S)-hydroxybutanamido]-7α-methoxy-3-[5-(1-methyl-1,2,3,4-tetrazolyl)thiomethyl]-Δ³-cephem-4-carboxylic acid. $C_{22}H_{29}N_9O_9S_2$; mol wt 627.66. C 42.10%, H 4.66%, N 20.08%, O 22.94%, S 10.22%. Broad spectrum injectable cephamycin. Prepn: I. Saikawa *et al.,* **Belg.** pat. **879,217;** *idem,* **U.S.** pat. **4,263,292** (1980, 1981 both to Toyama). Distribution of ^{14}C-cefbuperazone in mice: *idem, Japan. J. Antibiot.* **35,** 2159 (1982), *C.A.* **98,** 11062w (1982). *In vitro* and *in vivo* antibacterial activity: M. Tai *et al., Antimicrob. Ag. Chemother.* **22,** 728 (1982). General pharmacology: A. Takai *et al., Japan. J. Antibiot.* **35,** 2139 (1982), *C.A.* **98,** 65142r (1982). Absorption, distribution, excretion: I. Saikawa *et al., ibid.* 2163, *C.A.* **98,** 100691t (1982). Series of articles on cefbuperazone: *Chemotherapy (Tokyo)* **30,** Suppl. 3 (1982) 986 pp. Stability and degradation studies: S. Takano *et al., Yakugaku Zasshi* **103,** 62 (1983).

Solid, mp 118-120° (dec).

Sodium salt, $C_{22}H_{28}N_9NaO_9S_2$, *T-1982, Keiperazon, Tomiporan.*

THERAP CAT: Antibacterial.

1969. Cefcapene Pivoxil. *[6R-[6α,7β(Z)]]-3-[[(Aminocarbonyl)oxy]methyl]-7-[[2-(2-amino-4-thiazolyl)-1-oxo-2-pentenyl]amino]-8-oxo-5-thia-1-azabicyclo[4.2.0]oct-2-ene-2-carboxylic acid (2,2-dimethyl-1-oxopropoxy)methyl ester;* (6R,7R)-7-[(Z)-2-(2-aminothiazol-4-yl)-2-pentenoylamino]-3-(carbamoyloxymethyl)-3-cephem-4-carboxylic acid pivaloyloxymethyl ester; S-1108; Flumax. $C_{23}H_{29}N_5O_8S_2$; mol wt 567.64. C 48.67%, H 5.15%, N 12.34%, O 22.55%, S 11.30%. Orally absorbed cephalosporin; ester prodrug of the active free acid metabolite, cefcapene. Prepn: Y. Hamashima *et al.,* **Belg.** pat. **904,517;** *idem et al.,* **U.S.** pat. **4,731,-361** (1986, 1988 both to Shionogi); K. Ishikura *et al., J. Antibiot.* **47,** 466 (1994). *In vitro* activity and stability of the free acid: H. C. Neu *et al., Antimicrob. Ag. Chemother.* **36,** 1336 (1992). Metabolism: K. Totsuka *et al., ibid.* 757. Safety and pharmacokinetics: M. Nakashima *et al., ibid.* 762.

Free acid, $C_{17}H_{19}N_5O_6S_2$, *S-1006.*
THERAP CAT: Antibacterial.

1970. Cefclidin. *[6R-[6α,7β(Z)]]-4-(Aminocarbonyl)-1-[[7-[[(5-amino-1,2,4-thiadiazol-3-yl)(methoxyimino)acetyl]amino]-2-carboxy-8-oxo-5-thia-1-azabicyclo[4.2.0]oct-2-en-3-yl]methyl]-1-azoniabicyclo[2.2.2]octane inner salt;* (6R,7R)-3-[(4-carbamoyl-1-quinuclidinio)methyl]-7-[2-(5-amino-1,2,-4-thiadiazol-3-yl)-(Z)-2-methoxyiminoacetamido]-8-oxo-5-thia-1-azabicyclo[4.2.0]oct-1-ene-2-carboxylate; (+)-1-[[(6R,7R)-7-[2-(5-amino-1,2,4-thiadiazol-3-yl)glyoxylamido]-2-carboxy-8-oxo-5-thia-1-azabicyclo[4.2.0]oct-2-en-3-yl]methyl]-4-carbamoylquinuclidinium hydroxide inner salt 7^2-(Z)-(O-methyloxime); 7β-[(Z)-2-(5-amino-1,2,4-thiadiazol-3-yl)-2-methoxyiminoacetamido]-3-[(4-carbamoyl-1-quinuclidinio)methyl-3-cephem-4-carboxylate; E-1040. $C_{21}H_{26}N_8O_6S_2$; mol wt 550.62. C 45.81%, H 4.76%, N 20.35%, O 17.43%, S 11.65%. Broad spectrum fourth generation cephalosporin antibiotic. Prepn: H. Yamauchi *et al.,* **Eur.** pat. **Appl. 188,255;** *eidem,* **U.S.** pat. **4,748,171** (1986, 1988 both to Eisai). Structure activity study: I. Sugiyama *et al., J. Antibiot.* **45,** 103 (1992). Comparative *in vitro* antibacterial spectrum and β-lactamase resistance: N.-A. Watanabe *et al., Antimicrob. Agents Chemother.* **32,** 693 (1988). *In vivo* antipseudomonal activity: T. Toyosawa *et al., Chemotherapy (Tokyo)* **42,** 1242 (1994). Clinical pharmacokinetics: M. Nakashima *et al., J. Clin. Pharmacol.* **29,** 144 (1989). Toxicity study: K. Ogura *et al., Chemotherapy (Tokyo)* **40,** Suppl. 4, 95 (1992).

LD_{50} in male, female mice, rats (mg/kg): > 10000, > 10000, 2236, 2147 i.v.; > 10000 both species orally; > 10000 both species s.c.; > 5000 both species i.m. (Ogura).
THERAP CAT: Antibacterial.

1971. Cefdinir. *[6R-[6α,7β(Z)]]-7-[[(2-Amino-4-thiazolyl)(hydroxyimino)acetyl]amino]-3-ethenyl-8-oxo-5-thia-1-azabicyclo[4.2.0]oct-2-ene-2-carboxylic acid; syn-7-[2-(2-amino-4-thiazolyl)-2-hydroxyiminoacetamido]-3-vinyl-3-*

cephem-4-carboxylic acid; FK-482; BMY-28488. $C_{14}H_{13}N_5$-O_8S_2; mol wt 395.42. C 42.53%, H 3.31%, N 17.71%, O 20.23%, S 16.22%. Cephalosporin antibiotic structurally similar to cefixime, q.v. Prepn: T. Takaya et al., **Belg.** pat. **897,864;** eidem, **U.S.** pat. **4,559,334** (1984, 1985 both to Fujisawa); Y. Inamoto et al., J. Antibiot. **41,** 828 (1988). Antibacterial spectrum in vitro and in vivo: Y. Mine et al., ibid. 1873, 1888. Pharmacokinetics in animals: H. Sakamoto et al., ibid. 1896. Series of articles on antibacterial activity, pharmacology and clinical efficacy: Chemotherapy (Tokyo) **37,** Suppl. 2, 1-1111 (1989).

mp 170° (dec). uv max in pH 7 phosphate buffer: 223, 286 nm (ε 17400, 19700). pKa 9.70.

Sodium salt, $C_{14}H_{12}N_5NaO_5S_2$, dec 220°.

THERAP CAT: Antibacterial.

1972. Cefditoren. [6R-[3(Z),6α,7β(Z)]]-7-[[(2-Amino-4-thiazolyl)(methoxyimino)acetyl]amino]-3-[2-(4-methyl-5-thiazolyl)ethenyl]-8-oxo-5-thia-1-azabicyclo[4.2.0]oct-2-ene-2-carboxylic acid; (+)-(6R,7R)-7-[2-(2-amino-4-thiazolyl)-glyoxylamido]-3-[(Z)-2-(4-methyl-5-thiazolyl)vinyl]-8-oxo-5-thia-1-azabicyclo[4.2.0]oct-2-ene-2-carboxylic acid, 7²-(Z)-(O-methyloxime); 7-[2-methoxyimino-2-(2-aminothiazol-4-yl)acetamido]-3-[2-(4-methylthiazol-5-yl)vinyl]-3-cephem-4-carboxylic acid (syn-isomer, cis-isomer); ME-1206. $C_{19}H_{18}$-$N_6O_5S_3$, mol wt 506.59. C 45.05%, H 3.58%, N 16.59%, O 15.79%, S 18.99%. Broad spectrum, third generation cephalosporin; active metabolite of orally absorbed pivaloyloxymethyl ester prodrug. Prepn: K. Atsumi et al., **Eur.** pat. **Appl.** 175,610; eidem, **U.S.** pat. **4,839,350** (1986, 1989 both to Meiji); K. Sakagami et al., J. Antibiot. **43,** 1047 (1990); idem et al., Chem. Pharm. Bull. **39,** 2433 (1991). In vitro and in vivo activities: A. Tamura et al., Antimicrob. Ag. Chemother. **32,** 1421 (1988). Extensive in vitro activity study: D. Felmingham et al., Drugs Exp. Clin. Res. **20,** 127 (1994). Clinical pharmacokinetics and efficacy in pediatrics: N. Iwai et al., Japan. J. Antibiot. **47,** 181 (1994), C.A. **120,** 260636e (1994).

Sodium salt, sesquihydrate, $C_{19}H_{17}N_6NaO_5S_3 \cdot 1\frac{1}{2} H_2O$, pale yellow crystals from water, mp 195-200° (dec). $[\alpha]_D^{20}$ +121.6° (c = 5 in methanol).

Pivaloyloxymethyl ester, $C_{25}H_{28}N_6O_7S_3$, cefditoren pivoxil, ME-1207, Meiact. Pale yellow powder, mp 127-129°. $[\alpha]_D^{20}$ −48.5° (c = 0.5 in methanol).

THERAP CAT: Antibacterial.

1973. Cefepime. [6R-[6α,7β(Z)]]-1-[[7-[[(2-Amino-4-thiazolyl)(methoxyimino)acetyl]amino]-2-carboxy-8-oxo-5-thia-1-azabicyclo[4.2.0]oct-2-en-3-yl]methyl]-1-methylpyrrolidinium inner salt; 1-[[(6R,7R)-7-[2-(2-amino-4-thiazolyl)-glyoxylamido]-2-carboxy-8-oxo-5-thia-1-azabicyclo[4.2.0]-oct-2-en-3-yl]methyl]-1-methylpyrrolidinium hydroxide inner salt 7²-(Z)-2-(O-methyloxime); 7-[(Z)-2-(2-aminothiazol-4-yl)-2-methoxyiminoacetamido]-3-(1-methylpyrrolidinio)methyl-3-cephem-4-carboxylate; BMY-28142; Maxipime. $C_{19}H_{24}N_6O_5S_2$; mol wt 480.57. C 47.49%, H 5.03%, N 17.49%, O 16.65%, S 13.35%. Semisynthetic, fourth generation cephalosporin antibiotic. Prepn: S. Aburaki et al., **Ger.** pat. **3,307,550;** eidem, **U.S.** pat. **4,406,899** (both 1983 to Bristol-Myers); and antibacterial activity: T. Naito

et al., J. Antibiot. **39,** 1092 (1986). In vitro comparative antimicrobial spectrum: N. J. Khan et al., Antimicrob. Ag. Chemother. **26,** 585 (1984); and β-lactamase stability: H. C. Neu et al., J. Antimicrob. Chemother. **17,** 441 (1986). HPLC determn in plasma and urine: R. H. Barbhaiya et al., Antimicrob. Ag. Chemother. **31,** 55 (1987). Clinical evaluations in infection: N. Clynes et al., Diagn. Microbiol. Infect. Dis. **12,** 257 (1989); S. Oster et al., Antimicrob. Ag. Chemother. **34,** 954 (1990). Review of clinical pharmacokinetics: M. P. Okamoto et al., Clin. Pharmacokinet. **25,** 88-102 (1993).

Colorless powder, mp 150° (dec). uv max (pH 7 phosphate buffer): 235, 257 nm (ε 16700, 16100).

Sulfate, $C_{19}H_{24}N_6O_5S_2 \cdot H_2SO_4$, mp 210° (dec). uv max (pH 7 phosphate buffer): 236, 258 nm (ε 17200, 16900).

Hydrochloride monohydrate, $C_{19}H_{25}ClN_6O_5S_2 \cdot HCl \cdot H_2O$. [6R-[6α,7β(Z)]]-1-[[7-[[(2-Amino-4-thiazolyl)(methoxy-imino)acetyl]amino]-2-carboxy-8-oxo-5-thia-1-azabicyclo-[4.2.0]oct-2-en-3-yl]methyl]-1-methylpyrrolidinium chloride monohydrochloride monohydrate, cefepime hydrochloride, Axepim.

THERAP CAT: Antibacterial.

1974. Cefetamet. [6R-[(6α,7β(Z)]]-7-[[(2-Amino-4-thiazolyl)(methoxyimino)acetyl]amino]-3-methyl-8-oxo-5-thia-1-azabicyclo[4.2.0]oct-2-ene-2-carboxylic acid; (6R,7R)-7-[2-(2-aminothiazol-4-yl)-(Z)-2-(methoxyimino)acetamido]-3-methyl-3-cephem-4-carboxylic acid; (6R,7R)-7-[2-(2-amino-4-thiazolyl)glyoxylamido]-3-methyl-8-oxo-5-thia-1-azabicyclo[4.2.0]oct-2-ene-2-carboxylic acid 7²-(Z)-(O-methyloxime); deacetoxycefotaxime; LY-097964; Ro-15-8074. C_{14}-$H_{15}N_5O_5S_2$; mol wt 397.44. C 42.31%, H 3.80%, N 17.62%, O 20.13%, S 16.14%. Third generation cephalosporin antibiotic. Prepn: R. Heymes, A. Lutz, **Ger.** pat. **2,713,272;** eidem, **U.S.** pat. **4,396,618** (1977, 1983 both to Roussel-UCLAF); R. Bucourt et al., Tetrahedron **34,** 2233 (1978). Prepn of the pivaloyloxymethyl ester: M. Ochiai et al., **Ger.** pat. **2,715,385;** eidem, **U.S.** pat. **4,680,390** (1977, 1987 both to Takeda). Antibacterial spectrum: M. G. Thomas, S. D. R. Lang, Antimicrob. Ag. Chemother. **29,** 945 (1986). HPLC determn in biological fluids: R. Wyss, F. Bucheli, J. Chromatog. **430,** 81 (1988). Pharmacokinetics: K. Stoeckel et al., Curr. Med. Res. Opin. **11,** 432 (1989). Multicenter clinical trial: L. Bernstein-Hahn et al., ibid. 442.

Pivaloyloxymethyl ester, $C_{20}H_{25}N_5O_7S_2$, cefetamet pivoxil, Ro-15-8075, Cefyl, Globocef.

THERAP CAT: Antibacterial.

THERAP CAT (VET): Antibacterial.

1975. Cefixime. [6R-[6α,7β(Z)]]-7-[[(2-Amino-4-thi-azolyl)[(carboxymethoxy)imino]acetyl]amino]-3-ethenyl-8-oxo-5-thia-1-azabicyclo[4.2.0]oct-2-ene-2-carboxylic acid; 7-[2-(2-amino-4-thiazolyl)-2-(carboxymethoxyimino)acet-amido]-3-vinyl-3-cephem-4-carboxylic acid; FK-027; FR-17027; CL-284635; Cefixoral; Cefspan; Cephoral; Oroken; Suprax; Unixime. $C_{16}H_{15}N_5O_7S_2$; mol wt 453.46. C 42.38%, H 3.33%, N 15.44%, O 24.70%, S 14.14%. Orally active, third generation cephalosporin antibiotic. Prepn: T. Takaya et al., **Eur.** pat. **Appl.** 30,630; eidem, **U.S.** pat. **4,409,214** (1981, 1983 both to Fujisawa); H. Yamanaka et al., J. Antibiot. **38,** 1738 (1985). Synthesis and activity of (E)-isomer:

K. Kawabata *et al., Chem. Pharm. Bull.* **34,** 3458 (1986). Mechanism of action: Y. Shigi *et al., J. Antibiot.* **37,** 790 (1984). Comparative antibacterial spectrum *in vitro* and *in vivo:* T. Kamimura *et al., Antimicrob. Ag. Chemother.* **25,** 98 (1984). *In vitro* activity and β-lactamase stability: H. C. Neu *et al., ibid.* **26,** 174 (1984). HPLC determn in human plasma and urine: Y. Tokuma *et al., J. Chromatog.* **311,** 339 (1984). Pharmacokinetics in humans: D. R. P. Guay *et al., Antimicrob. Ag. Chemother.* **30,** 485 (1986). Clinical trial in urinary tract infections: J. Levenstein *et al., S. Afr. Med. J.* **70,** 455 (1986); A. Irvani *et al., Am. J. Med.* **85,** Suppl. 3A, 17 (1988); in respiratory infections: R. Kiani *et al., ibid.* 6.

(*Z*)-Form disodium salt, $C_{16}H_{13}N_5Na_2O_7S_2$, mp > 250°. (*E*)-Form trihydrate, pale yellow solid, mp 218-225° (dec). THERAP CAT: Antibacterial.

1976. Cefmenoxime. *[6R-[6α,7β(Z)]]-7-[[(2-Amino-4-thiazolyl)(methoxyimino)acetyl]amino]-3-[[(1-methyl-1H-tetrazol-5-yl)thio]methyl]-8-oxo-5-thia-1-azabicyclo[4.2.0]oct-2-ene-2-carboxylic acid;* (6R,7R)-7-[2-(2-amino-4-thiazolyl)-glyoxylamido]-3-[[(1-methyl-1H-tetrazol-5-yl)thio]methyl]-8-oxo-5-thia-1-azabicyclo[4.2.0]oct-2-ene-2-carboxylic acid 7²-(*Z*)-(*O*-methyloxime); SCE-1365. $C_{16}H_{17}N_9O_5S_3$; mol wt 511.57. C 37.57%, H 3.35%, N 24.64%, O 15.64%, S 18.80%. Third generation cephalosporin antibiotic; related structurally to cefotaxime and ceftizoxime, *q.q.v.* The name cefmenoxime applies to the isomer having a *syn*-methoxy-imino group. Prepn (unspecified stereochemistry): M. Ochiai *et al.,* **Ger. pat. 2,556,736;** *eidem,* **U.S. pat. 4,098,888** (1976, 1978 both to Takeda); *syn*-isomer: R. Heymes, A. Lutz, **Ger. pat. 2,713,272;** *eidem,* **U.S. pat. 4,476,122** (1983, 1984 both to Roussel-Uclaf). Series of articles on antibacterial activity, absorption, excretion, metabolism, mechanism of action, clinical studies: *Chemotherapy (Tokyo)* **29,** Suppl. 1, 1-998 (1979). *In vitro* and *in vivo* study: K. Tsuchiya *et al., Antimicrob. Ag. Chemother.* **19,** 56 (1981). β-Lactamase stability: K. Okonogi *et al., ibid.* **20,** 171 (1981). Clinical pharmacokinetics study: D. Höffler, P. Koeppe, *Arzneimittel-Forsch.* **33,** 269 (1983). Symposium: *Am. J. Med.* **77**(6A), 1-59 (1984).

Hydrochloride (*syn*-isomer), $(C_{16}H_{17}N_9O_5S_3)_2$.HCl, *Abbott 50192, Bestcall, Cefmax, Cemix, Tacef.* Sodium salt (*syn*-isomer), $C_{16}H_{16}N_9NaO_5S_3$, $[\alpha]_D^{20}$ −13.5° ±1 (c = 1 in water). THERAP CAT: Antibacterial.

1977. Cefmetazole. *(6R-cis)-7-[[[(Cyanomethyl)thio]-acetyl]amino]-7-methoxy-3-[[(1-methyl-1H-tetrazol-5-yl)thio]-methyl]-8-oxo-5-thia-1-azabicyclo[4.2.0]oct-2-ene-2-carboxylic acid;* CS-1170; SKF-83088. $C_{15}H_{17}N_7O_5S_3$; mol wt 471.54. C 38.21%, H 3.63%, N 20.79%, O 16.96%, S 20.40%. Semisynthetic antibiotic derived from cephamycin C, *q.v.* Prepn: H. Nakao *et al.,* **Ger. pat. 2,455,884** (1975 to Sankyo), *C.A.* **83,** 97330g (1975); *eidem, J. Antibiot.* **29,** 554 (1976). See also: J. E. Dolfini, **U.S. pat. 3,920,639** (1975 to Squibb). *In vitro* study: J. V. Uri *et al., ibid.* **31,** 82 (1978). Immunological study: M. Iwata, T. Matuhasi, *Japan. J. Exp. Med.* **48,** 401 (1978). Absorption, distribution, excretion, metabolism in various species: H. Shindo *et al., Chemother. (Tokyo)* **27,** Suppl. 1, 64 (1979). Pharmacology: S. Kobayashi *et al.,*

ibid. **26,** Suppl. 5, 115 (1978). Toxicological studies: H. Masuda *et al., Sankyo Kenkyusho Nempo* **30,** 112 (1978), *C.A.* **90,** 180268h (1979). Comparative clinical study: M. Nishida *et al., J. Antibiot.* **32,** 1319 (1979).

Sodium salt, $C_{15}H_{16}N_7NaO_5S_3$, *Cefmetazon, Metafar, Metazol, Zefazone.* White solid. Very sol in water, methanol, sol in acetone. Practically insol in chloroform. LD_{50} i.v. in rats: > 5000 mg/kg (Masuda). THERAP CAT: Antibacterial.

1978. Cefminox. *[6R-[6α,7α,7(S*)]]-7-[[[(2-Amino-2-carboxyethyl)thio]acetyl]amino]-7-methoxy-3-[[(1-methyl-1H-tetrazol-5-yl)thio]methyl]-8-oxo-5-thia-1-azabicyclo[4.2.0]oct-2-ene-2-carboxylic acid;* 7β-(2-D-amino-2-carboxyethylthio-acetamido)-7α-methoxy-3-[[(1-methyl-1H-tetrazol-5-yl)-thio]methyl]-3-cephem-4-carboxylic acid. $C_{16}H_{21}N_7O_7S_3$; mol wt 519.58. C 36.99%, H 4.07%, N 18.87%, O 21.55%, S 18.51%. Semisynthetic broad spectrum cephamycin antibiotic. Prepn: **Belg. pat. 880,656,** K. Iwamatsu *et al.,* **U.S. pat. 4,357,331** (1980, 1982 both to Meiji Seika). Synthesis, biological activity: *eidem, J. Antibiot.* **36,** 229 (1983). Structure-activity studies: S. Inouye *et al., ibid.* **37,** 1403 (1984). Proposed mechanism of action: T. Tsuruoka *et al., Eur. J. Biochem.* **151,** 209 (1985). *In vitro, in vivo* antibacterial activity compared with that of other cephalosporins: S. Inouye *et al., Antimicrob. Ag. Chemother.* **26,** 722 (1984). Activity against gram-negative infections in mice: T. Watanabe *et al., Drugs Exptl. Clin. Res.* **10,** 293 (1984). Toxicity studies: M. Kurebe *et al., Japan. J. Antibiot.* **37,** 847 (1984). Series of articles on pharmacokinetics, early clinical trials, treatment of urinary tract and respiratory infections: *Chemotherapy (Tokyo)* **32,** Suppl. 5 (1984) 561 pp.

Sodium salt heptahydrate, $C_{16}H_{20}N_7NaO_7S_3.7H_2O$, *MT-141, Meicelin.* Crystals from ethanol-water, mp 90-91°. LD_{50} in male, female mice (mg/kg): 6100, 5200 i.v.; in male, female rats (mg/kg): 6600, 5700 i.v.; 8600, 8550 i.p.; > 15000, > 15000 s.c. or orally (Kurebe). THERAP CAT: Antibacterial.

1979. Cefodizime. *[6R-(6α,7β(Z))]-7-[[(2-Amino-4-thiazolyl)(methoxyimino)acetyl]amino]-3-[[[5-(carboxymeth-yl)-4-methyl-2-thiazolyl]thio]methyl]-8-oxo-5-thia-1-azabicy-clo[4.2.0]oct-2-ene-2-carboxylic acid;* (6R,7R)-7-[2-(2-amino-4-thiazolyl)glyoxylamido]-3-[[[5-(carboxymethyl)-4-methyl-2-thiazolyl]thio]methyl]-8-oxo-5-thia-1-azabicyclo[4.2.0]oct-2-ene-2-carboxylic acid 7²(Z)-(*O*-methyloxime). $C_{20}H_{20}N_6$-O_7S_4; mol wt 584.68. C 41.09%, H 3.45%, N 14.37%, O 19.16%, S 21.94%. Third generation cephalosporin antibiotic with immunomodulating activity; derivative of cefotaxime, *q.v.* Prepn: **Belg. pat. 865,632;** W. Dürckheimer *et al.,* **U.S. pat. 4,278,793** (1978, 1981 both to Hoechst AG); and antibacterial activity: J. Blumbach *et al., J. Antibiot.* **40,** 29 (1987). *In vitro* comparative antibacterial spectrum: R. N. Jones *et al., Antimicrob. Ag. Chemother.* **20,** 760 (1981); and β-lactamase stability: M. Limbert *et al., J. Antibiot.* **37,** 892 (1984). *In vivo* antibacterial activity: N. Klesel *et al., ibid.* 1712. Enhancement of immune response *in vivo:* M. Limbert *et al., ibid.* 1719; and *ex vivo:* A. Fietta *et al., Chemotherapy (Basel)* **34,** 430 (1988). HPLC determn in biological fluids: T. Marunaka *et al., J. Chromatog.* **420,** 329 (1987). Pharmacokinetics: N. Klesel *et al., J. Antibiot.*

37, 901 (1984); in humans: E. E. Dagrosa *et al., Clin. Ther.* **10**, 18 (1987). Clinical study in urogenital gonorrhea: A. H. van der Willigen *et al., Antimicrob. Ag. Chemother.* **32**, 426 (1988). Physical properties, toxicology and structure-activity relationships: A. Bryskier *et al., J. Antimicrob. Chemother.* **26**, Suppl. C, 1 (1990). Series of articles on pharmacology, pharmacokinetics and clinical studies: *Chemotherapy (Tokyo)* **36**, Suppl. 5, 1-980 (1988); and immuno-modulating activity: *J. Antimicrob. Chemother.* **26**, Suppl. C, 1-134 (1990).

$[\alpha]_D^{25}$ −55.9°. pK_1 2.85; pK_2 3.37; pK_3 4.18. uv max (water): 228, 260, 288 nm (log ϵ 4.25, 4.25, 4.20).

Disodium salt, $C_{20}H_{18}N_6Na_2O_7S_4$, *HR-221, THR-221, Kenicef, Timecef*. White to yellow crystalline powder. Soly (g/l): water ~270. LD_{50} in mice, rabbits (mg/kg): 4000-8000 i.v. both species; in rats (mg/kg): 4000-8000 i.v., 15000-17500 s.c., 8000-11000 i.p. (Bryskier).

THERAP CAT: Antibacterial.

1980. Cefonicid. *[6R-[6α,7β(R*)]]-7-[(Hydroxyphenyl-acetyl)amino]-8-oxo-3-[[[1-(sulfomethyl)-1H-tetrazol-5-yl]thio]methyl]-5-thia-1-azabicyclo[4.2.0]oct-2-ene-2-carboxylic acid; (6R,7R)-7-[(R)-mandelamido]-8-oxo-3-[[[1-(sulfometh-yl)-1H-tetrazol-5-yl]thio]methyl]-5-thia-1-azabicyclo[4.2.0]-oct-2-ene-2-carboxylic acid.* $C_{18}H_{18}N_6O_8S_3$; mol wt 542.57 C 39.85%, H 3.34%, N 15.49%, O 23.59%, S 17.73%. Injectable semi-synthetic cephalosporin antibiotic related to cefamandole, *q.v.* Prepn: D. A. Berges, **Ger. pat. 2,611,270;** idem, **U.S. pat. 4,048,311** (1976, 1977 both to Smith Kline); **U.S. pats. 4,093,723, 4,159,373** (1978, 1979 both to Smith Kline). Antibacterial activity, pharmacokinetics: P. Actor *et al., Antimicrob. Ag. Chemother.* **13**, 784 (1978). Series of articles on stability, comparative *in vitro* activity, serum levels: *Curr. Chemother. Infect. Dis., Proc. 11th Int. Congr. Chemother.,* vol. 1 (Am. Soc. Microbiol., Washington, D.C., 1979) pp 246-254. Stability towards β-lactamases: R. Mehta *et al., J. Antibiot.* **34**, 202 (1981). Kinetics and renal handling: D. Pitkin *et al., Clin. Pharmacol. Ther.* **30**, 587 (1981). *In vitro* evaluation vs Group B streptococci: A. S. Bayer *et al., Antimicrob. Ag. Chemother.* **21**, 344 (1982). Review of antibacterial activity, pharmacology, therapeutic use: E. Saltiel, R. N. Brogden, *Drugs* **32**, 222-259 (1986).

Disodium salt, $C_{18}H_{16}N_6Na_2O_8S_3$, *SKF-75073, Cefodie, Monocid, Monocidur, Praticef.*

THERAP CAT: Antibacterial.

1981. Cefoperazone. *[6R-[6α,7β(R*)]]-7-[[[[(4-Ethyl-2,3-dioxo-1-piperazinyl)carbonyl]amino](4-hydroxyphenyl)-acetyl]amino]-3-[[(1-methyl-1H-tetrazol-5-yl)thio]methyl]-8-oxo-5-thia-1-azabicyclo[4.2.0]oct-2-ene-2-carboxylic acid; 7-[D-(−)-α-(4-ethyl-2,3-dioxo-1-piperazinecarboxamido)-α-(4-hydroxyphenyl)acetamido]-3-[[(1-methyl-1H-tetrazol-5-yl)thio]methyl]-3-cephem-4-carboxylic acid.* $C_{25}H_{27}N_9O_8S_2$; mol wt 645.68. C 46.51%, H 4.21%, N 19.52%, O 19.82%, S 9.93%. Broad spectrum third generation cephalosporin antibiotic. Prepn: I. Saikawa *et al.,* **Belg. pat. 837,682;** eidem, **U.S. pat. 4,410,522** (1976, 1983 both to Toyama); eidem, *Yakugaku Zasshi* **99**, 929 (1979). Stability in aq soln: eidem, *ibid.* 1207. *In vitro* activity: M. V. Borobio *et*

al., Antimicrob. Ag. Chemother. **17**, 129 (1980). Kinetics in rats: J. Fabre *et al., Schweiz. Med. Wochenschr.* **110**, 264 (1980); in humans: A. F. Allaz, *ibid.* **109**, 1999 (1979). Review of pharmacology and therapeutic efficacy: R. N. Brogden *et al., Drugs* **22**, 423-460 (1981). Symposium on clinical studies: *ibid.* Suppl. 1, 1-124.

Crystals from acetonitrile/water, mp 169-171° (hydrated). Stable at pH 4.0-7.0; slightly unstable in acid; highly unstable in alkaline soln.

Sodium salt, $C_{25}H_{26}N_9NaO_8S_2$, *CP-52640-2, T-1551, Bio-perazone, Cefazone, Cefobid, Cefobine, Cefobis, Cefogram, Cefoneg, Cefosint, Dardum, Farecef, Kefazon, Novobiocyl, Pathozone, Peracef, Perocef, Tomabef.*

THERAP CAT: Antibacterial.

THERAP CAT (VET): Antibacterial.

1982. Ceforanide. *(6R-trans)-7-[[[2-(Aminomethyl)-phenyl]acetyl]amino]-3-[[[1-(carboxymethyl)-1H-tetrazol-5-yl]thio]methyl]-8-oxo-5-thia-1-azabicyclo[4.2.0]oct-2-ene-2-carboxylic acid; (6R,7R)-7-[2-(α-amino-o-tolyl)acetamido]-3-[[[1-(carboxymethyl)-1H-tetrazol-5-yl]thio]methyl]-8-oxo-5-thia-1-azabicyclo[4.2.0]oct-2-ene-2-carboxylic acid; 7-[o-(aminomethyl)phenylacetamido]-3-[[[1-(carboxymethyl)-1H-tetrazol-5-yl]thio]methyl]-3-cephem-4-carboxylic acid; BL-S786.* $C_{20}H_{21}N_7O_6S_2$; mol wt 519.56. C 46.24%, H 4.07%, N 18.87%, O 18.48%, S 12.34%. Injectable, semi-synthetic cephalosporin antibiotic. Prepn: M. A. Kaplan *et al.,* **Ger. pat. 2,538,804;** W. J. Gottstein *et al.,* **U.S. pats. 4,100,346, 4,172,196** and **4,182,863** (1976, 1978, 1979, 1980 all to Bristol-Myers); W. J. Gottstein *et al., J. Antibiot.* **29**, 1226 (1976). Laboratory evaluation: F. Leitner *et al., Antimicrob. Ag. Chemother.* **10**, 426 (1976). *In vitro* susceptibility comparisons: R. N. Jones *et al., J. Antibiot.* **30**, 576, 583 (1977). Comparative tissue distribution: F. H. Lee *et al., Antimicrob. Ag. Chemother.* **19**, 625 (1981). Pharmacokinetics: E. H. Estey *et al., Clin. Pharmacol. Ther.* **30**, 396 (1981); S. S. Hawkins *et al., ibid.* 468. Clinical study: R. J. Wallace *et al., Antimicrob. Ag. Chemother.* **20**, 648 (1981).

White solid, mp > 150° (dec).

Sodium salt, $C_{20}H_{20}N_7NaO_6S_2$, *Precef.*

THERAP CAT: Antibacterial.

1983. Cefotaxime. *[6R-[6α,7β(Z)]]-3-[(Acetyloxy)meth-yl]-7-[[(2-amino-4-thiazolyl)(methoxyimino)acetyl]amino]-8-oxo-5-thia-1-azabicyclo[4.2.0]oct-2-ene-2-carboxylic acid; (6R,7R)-7-[2-(2-amino-4-thiazolyl)glyoxylamido]-3-(hydr-oxymethyl)-8-oxo-5-thia-1-azabicyclo[4.2.0]oct-2-ene-2-car-boxylate 7²-(Z)-(O-methyloxime) acetate; 7-[2-(2-amino-4-thiazolyl)-2-methoxyiminoacetamido]cephalosporanic acid.* $C_{16}H_{17}N_5O_7S_2$; mol wt 455.47. C 42.19%, H 3.76%, N 15.38%, O 24.59%, S 14.08%. Broad spectrum third generation cephalosporin antibiotic. The name cefotaxime applies to the isomer having a *syn*-methoxyimino group. Prepn (unspecified stereochemistry): M. Ochiai *et al.,* **Ger. pat. 2,556,736;** eidem, **U.S. pat. 4,098,888** (1976, 1978 both to Takeda); *syn*-isomer: R. Heymes, A. Lutz, **Ger. pat. 2,702,-501;** eidem, **U.S. pat. 4,152,432** (1977, 1979 both to Rous-sel-UCLAF). Antibacterial activity: R. Wise, *Antimicrob.*

Ag. Chemother. **14**, 807 (1978); H. W. Van handuyt, M. Pyckavet, *ibid.* **16**, 109 (1979). Human pharmacology: R. Lüthy *et al., ibid.* 127. Symposia: *Drug Therapy* **Suppl.**, 1-108 (Jan. 1981); *Rev. Infect. Dis.* **4**, Suppl., S281-S488 (1982); *Diagn. Microbiol. Infect. Dis.* **2**(3), Suppl., 1S-92S (1984); *Infection* **13**, Suppl. 1, S1-S162 (1985). Comprehensive description: F. J. Muhtadi, M. M. A. Hassan, in *Analytical Profiles of Drug Substances* vol. **11**, K. Florey, Ed. (Academic Press, New York, 1982) pp 139-168. Series of articles on clinical efficacy: *J. Antimicrob. Chemother.* **26**, Suppl. A, 1-83 (1990).

Sodium salt (*syn*-isomer), $C_{16}H_{16}N_5NaO_7S_2$, *HR-756, RU-24756, Cefotax, Chemcef, Claforan, Pretor, Tolycar.* $[\alpha]_D^{20}$ +55°±2 (c = 0.8 in water).

THERAP CAT: Antibacterial.

1984. Cefotetan. *[6R-(6α,7α)]-7-[[[4-(2-Amino-1-carboxy-2-oxoethylidene)-1,3-dithietan-2-yl]carbonyl]amino]-7-methoxy-3-[[(1-methyl-1H-tetrazol-5-yl)thio]methyl]-8-oxo-5-thia-1-azabicyclo[4.2.0]oct-2-ene-2-carboxylic acid;* (6R,7S)-7-[4-(carbamoylcarboxymethylene)-1,3-dithietane-2-carboxamido]-7-methoxy-3-[[(1-methyl-1H-tetrazol-5-yl)thio]methyl]-8-oxo-5-thia-1-azabicyclo[4.2.0]oct-2-ene-2-carboxylic acid. $C_{17}H_{17}N_7O_8S_4$; mol wt 575.63. C 35.47%, H 2.98%, N 17.03%, O 22.24%, S 22.28%. Broad-spectrum injectable semi-synthetic antibiotic derived from cephamycin C, *q.v.* Prepn: M. Iwanami *et al.,* **Ger. pat.** 2,824,559; *eidem,* **U.S. pat.** 4,263,432 (1979, 1981 both to Yamanouchi); *eidem, Chem. Pharm. Bull.* **28**, 2629 (1980). *In vitro* comparison of antibacterial activity: B. Chattopadhyay, J. C. Teli, *J. Antimicrob. Chemother.* **10**, 151 (1982). Toxicity study: K. Imamura *et al., Chemotherapy (Tokyo)* **30**, Suppl. 1, 212 (1982). Series of articles on pharmacology, clinical studies, toxicology: *ibid.* 1-947; *J. Antimicrob. Chemother.* **11**, Suppl. A, 1-236 (1983). Review of antibacterial activity, pharmacokinetics, therapeutic use: A. Ward, D. M. Richards, *Drugs* **30**, 382-426 (1985).

Disodium salt, $C_{17}H_{15}N_7Na_2O_8S_4$, *ICI-156834, YM-09330, Apatef, Cefotan, Ceftenon, Cepan, Darvilen.* LD_{50} in male mice, rats (g/kg): 6.35, 8.48 i.v.; 8.12, 8.37 i.p.; >10 orally and s.c. both species (Imamura).

THERAP CAT: Antibacterial.

1985. Cefotiam. *(6R-trans)-7-[[(2-Amino-4-thiazolyl)-acetyl]amino]-3-[[[1-[2-(dimethylamino)ethyl]-1H-tetrazol-5-yl]-thio]methyl]-8-oxo-5-thia-1-azabicyclo[4.2.0]oct-2-ene-2-carboxylic acid;* 7β-[2-(aminothiazol-4-yl)acetamido]-3-[[[1-(2-dimethylaminoethyl)-1H-tetrazol-5-yl]thio]methyl]ceph-3-em-4-carboxylic acid; SCE-963. $C_{18}H_{23}N_9O_4S_3$; mol wt 525.64. C 41.13%, H 4.41%, N 23.98%, O 12.18%, S 18.30%. Semi-synthetic cephalosporin antibiotic. Prepn: S. Terao *et al.,* **Ger. pat.** 2,458,695; M. Numata *et al.,* **U.S. pat.** 4,080,498 (1975, 1978 both to Takeda); *eidem, J. Antibiot.* **31**, 1262 (1978). Prepn of the crystalline dihydrochloride: K. Naito *et al.,* **Ger. pat.** 2,738,711; *eidem,* **U.S. pat.** 4,146,710 (1978, 1979 both to Takeda). Pharmacodynamics: K. Tsuchiya *et al., ibid.,* 1272. Antibacterial spectrum: *eidem, Antimicrob. Ag. Chemother.* **14**, 557 (1978). Clinical trial in respiratory infections: H. M. Beumer *et al., Int. J. Clin. Pharmacol. Ther. Toxicol.* **23**, 105 (1985).

Dihydrochloride, $C_{18}H_{23}N_9O_4S_3.2HCl$, *Abbott 48999, CGP-14221/E, Halospor, Pansporin, Pansporine, Spizef, Sporidyn.* White to light yellow crystals. Sol in methanol. Slightly sol in ethanol.

1-(Cyclohexyloxycarbonyloxy)ethyl ester dihydrochloride, $C_{27}H_{37}N_9O_7S_3.2HCl$, *cefotiam hexetil hydrochloride, SCE-2174, Pansporin T, Taketiam, Texodil.* Prepn: T. Nishimura *et al.,* **Eur. pat. Appl.** 128,029; *eidem,* **U.S. pat.** 5,120,841 (1984, 1992 both to Takeda); *eidem et al., J. Antibiot.* **40**, 81 (1987). Acute toxicity: S. Chiba *et al., Oyo Yakuri* **35**, 179 (1988), *C.A.* **109**, 31615c (1988). Clinical trial: H. Maier *et al. Arzneimittel-Forsch.* **42**, 980 (1992). Colorless powder. Soly in water at pH 4.5: >1 mg/ml. LD_{50} in male, female mice, male, female rats (mg/kg): 1290, 1150, 1470, 1620 i.p.; >2000 s.c.; >5000 orally (Chiba).

THERAP CAT: Antibacterial.

1986. Cefoxitin. *(6R-cis)-3-[[(Aminocarbonyl)oxy]-methyl]-7-methoxy-8-oxo-7-[(2-thienylacetyl)amino]-5-thia-1-azabicyclo[4.2.0]oct-2-ene-2-carboxylic acid;* 3-(hydroxymethyl)-7-methoxy-8-oxo-7-[2-(2-thienyl)acetamido]-5-thia-1-azabicyclo[4.2.0]oct-2-ene-2-carboxylic acid carbamate (ester); 3-carbamoyloxymethyl-7α-methoxy-7-[2-(2-thienyl)-acetamido]-3-cephem-4-carboxylic acid. $C_{16}H_{17}N_3O_7S_2$; mol wt 447.46. C 44.96%, H 4.01%, N 9.83%, O 26.20%, S 15.00%. Semi-synthetic antibiotic derived from cephamycin C, *q.v.,* possessing high resistance to β-lactamase inactivation. Synthesis: B. G. Christensen *et al.,* **Ger. pats.** 2,129,-675; 2,203,653; *eidem,* **U.S. pat.** 4,297,488 (1971, 1972, 1981 all to Merck & Co.); Karady *et al., J. Am. Chem. Soc.* **94**, 1410 (1972); Ratcliffe, Christensen, *Tetrahedron Letters* **1973**, 4653. Biological evaluation: Wallick, Hendlin, *Antimicrob. Ag. Chemother.* **5**, 25 (1974); Miller *et al., ibid.* 33; Onishi *et al., ibid.* 38; Hamilton, Miller *et al., J. Antibiot.* **27**, 42 (1974). Mode of action: Onishi *et al., Ann. N.Y. Acad. Sci.* **235**, 406 (1974). Toxicity: S. Takayama *et al., Chemotherapy (Tokyo)* **26**, Suppl. 1, 150 (1978). Comprehensive reviews: *J. Antimicrob. Chemother.* **4**, Suppl. B, 1-256 (1978); R. N. Brogden *et al., Drugs* **17**, 1-37 (1979); E. O. Stapley, K. R. Brown, in *Pharmacological and Biochemical Properties of Drug Substances* vol. **3**, M. E. Goldberg, Ed. (Am. Pharm. Assoc., Washington, DC, 1981) pp 262-290. Comprehensive description: G. S. Brenner, *Anal. Profiles Drug Subs.* **11**, 169-195 (1982). Symposium on therapeutic use in anaerobic and aerobic infectons: *Hospital Practice* **25**, Suppl. 4, 1-56 (1990).

Crystals, mp 149-150° (dec). pKa 2.2. Very sol in acetone; sol in aq NaHCO₃; very slightly sol in water. Practically insol in ether and chloroform.

Sodium salt, $C_{16}H_{16}N_3NaO_7S_2$, *MK-306, Betacef, Farmoxin, Mefoxin, Mefoxitin, Merxin, Cenomycin.* White crystals with characteristic odor. $[\alpha]_{589nm}^{25}$ +210° (c = 1 in methanol). Very sol in water; sol in methanol; sparingly sol in ethanol and acetone. Insol in aromatic and aliphatic hydrocarbons. LD_{50} in mice, rats, dogs (g/kg): 5.10, 8.98, >10.0 i.v. (Takayama).

THERAP CAT: Antibacterial.

1987. Cefozopran. *[6R-[6α,7β(Z)]]-1-[[7-[[(5-Amino-1,2,4-thiadiazol-3-yl)(methoxyimino)acetyl]amino]-2-carb-*

oxy-8-oxo-5-thia-1-azabicyclo[4.2.0]oct-2-en-3-yl]methyl]-imidazo[1,2-b]pyridazinium inner salt; 7β-[2-(5-amino-1,2,4-thiadiazol-3-yl)-2(Z)-methoxyiminoacetamido]-3-[(imidazo-[1,2-b]pyridazinium-1-yl)methyl]-3-cephem-4-carboxylate. $C_{19}H_{17}N_9O_5S_2$; mol wt 515.53. C 44.27%, H 3.32%, N 24.45%, O 15.52%, S 12.44%. Prepn: A. Miyake *et al.,* **PCT Int. pat. Appl. 8,605,183;** *eidem,* **U.S. 4,864,022** (1986, 1989 both to Takeda); A. Miyake *et al., J. Antibiot.* **45,** 709 (1992). Antibacterial spectrum: T. Iwahi *et al., Antimicrob. Ag. Chemother.* **36,** 1358 (1992). *In vivo* activity: Y. Iizawa *et al., ibid.* **37,** 100 (1993). Pharmacokinetics: W. Paulfeuerborn *et al., ibid.* 1835. HPLC determn in serum and urine: K. Borner, E. Borner, *J. Chromatog.* **615,** 174 (1993).

Hydrochloride, $C_{19}H_{17}N_9O_5S_2 \cdot HCl$, *SCE-2787.*
THERAP CAT: Antibacterial.

1988. Cefpimizole. *[6R-[6α,7β(R*)]]-1-[[2-Carboxy-7-[[[[(5-carboxy-1H-imidazol-4-yl)carbonyl]amino]phenylacetyl]amino]-8-oxo-5-thia-1-azabicyclo[4.2.0]oct-2-en-3-yl]methyl]-4-(2-sulfoethyl)pyridinium inner salt;* 1-[(6R,7R)-2-carboxy-7-[(R)-2-(5-carboxy-4-imidazolylcarboxamido)-2-phenylacetamido]-8-oxo-5-thia-1-azabicyclo[4.2.0]oct-2-en-3-ylmethyl]pyridino-4-ethylsulfonate; 7-β-[D-(—)-α-(4-carboxyimidazole-5-carboxamido)phenylacetamido]-3-(4-β-sulfoethylpyridinium)methyl-3-cephem-4-carboxylic acid; U-63196; AC-1370. $C_{28}H_{26}N_6O_{10}S_2$; mol wt 670.68. C 50.14%, H 3.91%, N 12.53%, O 23.86%, S 9.56%. Prepn: N. Yasuda *et al.,* **Ger. pat. 2,826,546;** *eidem,* **U.S. pat. 4,217,-450** (1979, 1980 both to Ajinomoto); N. Yasuda *et al., J. Antibiot.* **36,** 242 (1983). *In vitro* antibacterial activity and β-lactamase stability: H. C. Neu, P. Labthavikul, *Antimicrob. Ag. Chemother.* **24,** 375 (1983). Potentiating effect on phagocyte functions: H. Ohnishi *et al., ibid.* **23,** 874 (1983). Toxicity study: S. Hashimoto *et al., Toxicol. Letters* **23,** 135 (1984). Therapeutic efficacy in mice: Y. Obana *et al., J. Antimicrob. Chemother.* **16,** 727 (1985). HPLC determn in human plasma and urine: D. B. Lakings, J. M. Wozniak, *J. Chromatog.* **308,** 261 (1984). Pharmacokinetics in humans: D. B. Lakings *et al., Antimicrob. Ag. Chemother.* **29,** 271 (1986). Efficacy and tolerance in gonorrhea in men: E. T. Sandberg *et al., ibid.* 849.

Monosodium salt, $C_{28}H_{25}N_6NaO_{10}S_2$, *U-63196E, Ajicef, Renilan.* $[α]_D^{20} -28.2°$ (c = 0.5 in water). uv max (water): 257 nm (ε 22400). Sol in water. LD_{50} in male, female mice, male, female rats (g/kg): 2.7, 2.9, 4.2, 3.5 i.v.; 8.2, 6.8, 12.2, 11.5 s.c.; all > 15.0 orally (Hashimoto).
THERAP CAT: Antibacterial.

1989. Cefpiramide. *[6R-[6α,7β(R*)]]-7-[[[[(4-Hydroxy-6-methyl-3-pyridinyl)carbonyl]amino](4-hydroxyphenyl)acetyl]amino]-3-[[(1-methyl-1H-tetrazol-5-yl)thio]methyl]-8-oxo-5-thia-1-azabicyclo[4.2.0]oct-2-ene-2-carboxylic acid;* 7-[(R)-2-(4-hydroxy-6-methylnicotinamido)-2-(p-hydroxyphenyl)acetamido]-3-[[(1-methyl-1H-tetrazol-5-yl)thio]methyl]-8-oxo-5-thia-1-azabicyclo[4.2.0]oct-2-ene-2-carboxylic acid; D-7-[(4-hydroxy-6-methylnicotinamido)-4-hydroxyphenylacetamido]-3-(1-methyltetrazol-5-yl)thiomethylcephem-4-carboxylic acid. $C_{25}H_{24}N_8O_7S_2$; mol wt 612.65. C 49.01%, H

3.95%, N 18.29%, O 18.28%, S 10.47%. Broad spectrum semi-synthetic cephalosporin antibiotic. Prepn: H. Yamada *et al.,* **Belg. pat. 833,063;** *eidem,* **U.S. pats. 4,156,724; 4,-160,087** (1975, 1979, 1979 all to Sumitomo); *eidem,* **Japan. Kokai 79 119,497** (1979 to Sumitomo), *C.A.* **92,** 11042j (1980); I. Isaka *et al.,* **Japan. Kokai 79 30,197** (1979 to Yamanouchi), *C.A.* **91,** 57037a (1979); **Japan. Kokai 84 65,-094** (1984 to Sumitomo), *C.A.* **101,** 110641w (1984). Prepn, NMR data, antibacterial activity: H. Yamada *et al., J. Antibiot.* **36,** 522, 543 (1983). Pharmacokinetics: H. Matsui *et al., Antimicrob. Ag. Chemother.* **22,** 213 (1982); K. Sata *et al., ibid.* **26,** 578 (1984). Metabolism: H. Imasaki *et al., ibid.* **24,** 42 (1983). In vitro activity: H. Wexler *et al., ibid.* **25,** 162 (1984); M. A. Pfaller *et al., ibid.* 368. Comparison with other cephalosporins: N. J. Khan *et al., ibid.* **26,** 585 (1984). Series of articles on *in vitro, in vivo* activity, determn in body fluids, pharmacokinetics, clinical studies: *Chemotherapy (Japan)* **31,** Suppl. 1, 1-842 (1983).

Yellow crystals, mp 213-215° (dec).
Sodium salt, $C_{25}H_{23}N_8NaO_7S_2$, *SM-1652, Wy-44635, Cefpiran, Suncefal, Sepatren.*
THERAP CAT: Antibacterial.

1990. Cefpirome. *[6R-[6α,7β(Z)]]-1-[[7-[[(2-Amino-4-thiazolyl)(methoxyimino)acetyl]amino]-2-carboxy-8-oxo-5-thia-1-azabicyclo[4.2.0]oct-2-en-3-yl]methyl]-6,7-dihydro-5H-cyclopenta[b]pyridinium inner salt;* 3-[(2,3-cyclopenteno-1-pyridinium)methyl]-7-[2-syn-methoxyimino-2-(2-aminothiazol-4-yl)acetamido]ceph-3-em-4-carboxylate; 1-[[(6R,7R)-7-[2-(2-amino-4-thiazolyl)glyoxylamido]-2-carboxy-8-oxo-5-thia-1-azabicyclo[4.2.0]oct-2-en-3-yl]methyl]-6,7-dihydro-5H-1-pyrindinium hydroxide, inner salt, 7^2-(Z)-(O-methyloxime); HR-810. $C_{22}H_{22}N_6O_5S_2$; mol wt 514.59. C 51.35%, H 4.31%, N 16.33%, O 15.55%, S 12.46%. Fourth generation cephalosporin antibiotic. Prepn: R. Lattrell *et al.,* **Brit. pat. Appl. 2,098,216** (1982 to Hoechst). Prepn of crystalline salts: W. Dürckheimer, R. Lattrell, **U.S. pat. 4,609,653** (1986 to Hoechst); and *in vitro* antibacterial activity: G. Seibert *et al., Arzneimittel-Forsch.* **33,** 1084 (1983). Physical properties: R. Lattrell *et al., Proc. 13th Int. Cong. Chemother., Vienna* **4,** part 99, 1 (1983). Comparative *in vitro* antibacterial spectrum and β-lactamase stability: H. C. Neu *et al., Infection* **13,** 146 (1985). HPLC determn in serum and urine: H. Uihlein *et al., ibid.* **16,** 135 (1988). Clinical pharmacokinetics: J. Kavi *et al., J. Antimicrob. Chemother.* **22,** 911 (1988). Series of articles on laboratory and clinical evaluations: *ibid.* **29,** Suppl A 1-104 (1992). Toxicity studies: G. Horstmann, K. Engelbart, *Arzneimittel-Forsch.* **40,** 1250 (1990); H. H. Donaubauer, D. Mayer, *J. Antimicrob. Chemother.* **29,** Suppl. A, 71 (1992).

LD_{50} in mice, rats (g/kg): 1.9-2.4, 1.9-2.15 i.v.; 3.8-4.2, 5.8-6.55 i.p. (Donaubauer, Mayer).
Sulfate, $C_{22}H_{22}N_6O_5S_2 \cdot H_2SO_4$, *Broact, Cefrom, Keitin.* Occurs as the monohydrate, crystals, dec 198-202°. $[α]_D^{25} -4.7°$

(c = 5 in H_2O). uv max: 265 nm (ϵ 21100). Soly (aq buffer, pH 6.5) > 50%.

THERAP CAT: Antibacterial.

1991. Cefpodoxime Proxetil. *[6R-[6α,7β(Z)]]-7-[[(2-Amino-4-thiazolyl)(methoxyimino)acetyl]amino]-3-(methoxy-methyl)-8-oxo-5-thia-1-azabicyclo[4.2.0]oct-2-ene-2-carboxylic acid 1-[[(1-methylethoxy)carbonyl]oxy]ethyl ester;* 1-(isoprop-oxycarbonyloxy)ethyl (6R,7R)-7-[2-(2-amino-4-thiazolyl)-(Z)-2-(methoxyimino)acetamido]-3-methoxymethyl-3-ceph-em-4-carboxylate; CS-807; U-76252; Banan; Cefodox; Ore-lox; Otreon; Vantin. $C_{21}H_{27}N_5O_9S_2$; mol wt 557.61. C 45.23%, H 4.88%, N 12.56%, O 25.82%, S 11.50%. Broad spectrum, orally absorbed third generation cephalosporin, ester prodrug of the active free acid metabolite, cefpodox-ime. Prepn: H. Nakao *et al., Eur. pat. Appl.* **49,118;** *eidem,* **U.S. pat. 4,486,425** (1982, 1984 both to Sankyo). Prepn, pharmacokinetics and NMR analysis: K. Fujimoto *et al., J. Antibiot.* **40,** 370 (1987). *In vitro* and *in vivo* antibacterial activity: Y. Utsui *et al., Antimicrob. Ag. Chemother.* **31,** 1085 (1987). *In vitro* antibacterial spectrum and susceptibil-ity testing of free acid: R. N. Jones, A. L. Barry, *ibid.* **32,** 443 (1988). Series of articles on antibacterial spectrum, pharmacokinetics, toxicology and clinical studies: *Chemotherapy (Tokyo)* **36,** Suppl. 1, 1-1126 (1988). Symposium on pharmacokinetics and clinical efficacy: *J. Antimicrob. Chemother.* **26.** Suppl. E, 1-101 (1990). Review of chemistry, antibacterial activity, toxicity and clinical studies: H. Nakao *et al., Sankyo Kenkyusho Nempo* **39,** 1-44 (1987), *C.A.* **109,** 27503x (1988).

LD_{50} in male, female mice, male, female rats (mg/kg): > 10000, > 10000, > 2000, > 2000 s.c., 3502, 2535, > 4000, > 4000 i.p.; > 8000, > 8000, > 4000, > 4000 orally (Nakao, 1988).

Free acid, $C_{15}H_{17}N_5O_6S_2$, *cefpodoxime, R-3763, U-76253.*

Free acid, sodium salt, $C_{15}H_{16}N_5NaO_6S_2$, *U-76253A, R-3746.*

THERAP CAT: Antibacterial.

1992. Cefprozil. *[6R-[6α,7β(R*)]]-7-[[Amino(4-hydr-oxyphenyl)acetyl]amino]-8-oxo-3-(1-propenyl)-5-thia-1-aza-bicyclo[4.2.0]oct-2-ene-2-carboxylic acid monohydrate;* (6R,-7R)-7-[(R)-2-amino-2-(p-hydroxyphenyl)acetamido]-8-oxo-3-propenyl-5-thia-1-azabicyclo[4.2.0]oct-2-ene-2-carboxylic acid monohydrate; BMY-28100-03-800; Cefzil; Procef. C_{18}-$H_{19}N_3O_5S.H_2O$; mol wt 407.45. C 53.06%, H 5.19%, N 10.31%, O 23.56%, S 7.87%. Semisynthetic oral cephalospo-rin consisting of ~90:10 *Z/E* isomeric mixture. Prepn and antibacterial activity: H. Hoshi *et al.,* **Ger. pat. 3,402,642;** *eidem,* **U.S. pat. 4,520,022** (1984, 1985 both to Bristol-Myers); T. Naito *et al., J. Antibiot.* **40,** 991 (1987). Separa-tion of isomers: M. A. Kaplan *et al.,* **U.S. pat. 4,727,070** (1988 to Bristol-Myers). Comparative *in vitro* antibacterial spectrum and β-lactamase stability: R. N. Jones *et al., Diagn. Microbiol. Infect. Dis.* **9,** 11 (1988). Clinical pharma-cokinetics and tissue penetration: R. H. Barbhaiya *et al., Antimicrob. Ag. Chemother.* **34,** 1204 (1990). Comparative clinical trials: A. Iravani, *ibid.* **35,** 1940 (1991); A. G. Ar-guedas *et al., Pediatr. Infect. Dis. J.* **10,** 375 (1991). HPLC determn in plasma and urine: W. C. Shyu *et al., Pharm. Res.* **8,** 992 (1991). Symposium on clinical use: *Clin. Infect. Dis.* **14,** Suppl. 2, S183-S272 (1992).

Z - isomer

Z-Form, *BMY-28100.* Isolated as the hemihydrate, crys-tals from acetone, mp 218-220° (dec). uv max (pH 7 phos-phate buffer): 228, 279 nm (ϵ 12300, 9800).

E-Form, *BMY-28167, BBS-1067.* Colorless prisms from methanol, mp 230° (dec). uv max (pH 7 phosphate buffer): 228, 292 nm (ϵ 13000, 16900).

THERAP CAT: Antibacterial.

1993. Cefroxadine. *[6R-[6α,7β(R*)]]-7-[(Amino-1,4-cyclohexadien-1-ylacetyl)amino]-3-methoxy-8-oxo-5-thia-1-azabicyclo[4.2.0]oct-2-ene-2-carboxylic acid;* 7-[D-2-amino-2-(1,4-cyclohexadienyl)acetamide]-3-methoxy-3-cephem-4-car-boxylic acid; CGP-9000; Oraspor. $C_{16}H_{19}N_3O_5S$; mol wt 365.41. C 52.59%, H 5.24%, N 11.50%, O 21.89%, S 8.78%. Orally active cephalosporin deriv. Prepn: R. Scartazzini, H. Bickel, **Ger. pat. 2,331,133;** *eidem,* **U.S. pat. 4,073,904** (1974, 1978 both to Ciba-Geigy). Antibiotic activity: O. Zak *et al., J. Antibiot.* **29,** 653 (1976). *In vivo* and *in vitro* microbiological evaluation: K. Yasuda *et al., Antimicrob. Ag. Chemother.* **18,** 105 (1980). Pharmacokinetics: T. Ber-gan, *Chemotherapy* **26,** 225 (1980). Series of articles on activity, toxicity, reproduction studies: *Chemotherapy (Tokyo)* **28,** Suppl 3, 1-335 (1980).

Internal salt, mp 170° (dec). $[\alpha]_D^{20}$ +87° (c = 1.093 in 0.1N HCl). uv max (0.1N HCl): 267 nm (ϵ 6100). LD_{50} in mice (mg/kg): > 6000 orally; 7090 i.p. (Scartazzini, Bickel, 1978).

THERAP CAT: Antibacterial.

1994. Cefsulodin. *[6R-[6α,7β(R*)]]-4-(Aminocarbon-yl)-1-[[2-carboxy-8-oxo-7-[(phenylsulfoacetyl)amino]-5-thia-1-azabicyclo[4.2.0]oct-2-en-3-yl]methyl]pyridinium inner salt;* 7-(α-sulphophenylacetamido)-3-(4'-carbamoylpyridinium)-methyl-3-cephem-4-carboxylic acid. $C_{22}H_{20}N_4O_8S_2$; mol wt 532.55. C 49.62%, H 3.79%, N 10.52%, O 24.03%, S 12.04%. Third generation cephalosporin antibiotic. Prepn: S. Morimoto *et al.,* **Ger. pat. 2,234,280;** *eidem,* **U.S. pat. 4,065,619** (1973, 1977 both to Takeda). Prepn and separa-tion of isomers: H. Nomura *et al., J. Med. Chem.* **17,** 1312 (1974). *In vitro* and *in vivo* antibacterial activity: K. Tsuchiya *et al., Antimicrob. Ag. Chemother.* **13,** 137 (1978). Absorption, distribution, excretion in mice, rats, dogs: *eidem, J. Antibiot.* **31,** 593 (1978). Activity and susceptibil-ity to β-lactamases: A. King *et al., Antimicrob. Ag. Chemo-ther.* **17,** 165 (1980). *In vitro* comparison with other anti-bacterials: H. Grimm, *Arzneimittel-Forsch.* **30,** 1478 (1980). Clinical pharmacology: J. Fuellhaas *et al., Curr. Chemo-ther., Proc. 10th Int. Congr. Chemother.* (Washington, D.C., 1978) **2,** 848-851. Review of activity: H. C. Neu, B. E. Scully, *Rev. Infec. Dis.* **6,** Suppl. 3, S667-S677 (1984). Re-view: A. Bryskier, *Lyon Pharm.* **34,** 343-355 (1983); D. B. Wright, *Drug Intell. Clin. Pharm.* **20,** 845-849 (1986).

Sodium salt, $C_{22}H_{19}N_4NaO_8S_2$, *sulcephalosporin, Abbott 46811, CGP-7174/E, SCE-129, Cefomonil, Monaspor, Pseudomonil, Pseudocef, Pyocefal, Takesulin, Tilmapor, Ulfaret.* Colorless needles from ethanol/water, mp 175° (dec). LD_{50} in mice (mg/kg): > 4000 i.p.; > 15000 orally (Bryskier).

d-Form sodium salt, $[\alpha]_D^{23}$ +16.5° (c = 1.08 in water). uv max (water): 263 nm (ϵ 14600).

l-Form sodium salt, $[\alpha]_D^{23}$ − 16.8° (c = 1.01 in water).

THERAP CAT: Antibacterial.

1995. Ceftazidime. *[6R-[6α,7β(Z)]]-1-[[7-[[(2-Amino-4-thiazolyl)[(1-carboxy-1-methylethoxy)imino]acetyl]amino]-2-carboxy-8-oxo-5-thia-1-azabicyclo[4.2.0]oct-2-en-3-yl]methyl]-pyridinium inner salt;* 1-[[(6R,7R)-7-[2-(2-amino-4-thiazolyl)glyoxylamido]-2-carboxy-8-oxo-5-thia-1-azabicyclo-[4.2.0]oct-2-en-3-yl]methyl]pyridinium hydroxide inner salt 7^2-(Z)-[O-(1-carboxy-1-methylethyl)oxime]; GR-20263. $C_{22}H_{22}N_6O_7S_2$; mol wt 546.58. C 48.34%, H 4.06%, N 15.38%, O 20.49%, S 11.73%. Third generation cephalosporin antibiotic. Prepn: C. H. O'Callaghan *et al., Ger. pat.* **2,921,316**; *eidem,* U.S. pat. **4,258,041** (1979, 1981 both to Glaxo). Prepn of crystalline pentahydrate: A. Brodie, L. A. Wetherill, **Ger.** pat. **3,037,102**; *eidem,* U.S. pat. **4,329,-453** (1981, 1982 both to Glaxo). Chemical and antibacterial properties: *eidem, Antimicrob. Ag. Chemother.* **17**, 876 (1980). Activity vs. *Pseudomonas* and *Enterobacteriaceae:* L. Verbist, J. Verhaegen, *ibid.* 807. *In vitro* comparison with other β-lactams: R. Wise *et al., ibid.* 884; L. Verbist, *ibid.* **19**, 407 (1981). Symposium on clinical studies: *J. Antimicrob. Chemother.* **12**, Suppl. A, 1-414 (1983). Review of antibacterial activity, pharmacokinetics and therapeutic use: D. M. Richards, R. N. Brogden, *Drugs* **29**, 105-161 (1985). Comprehensive description: M. A. Abounassif *et al.,* in *Analytical Profiles of Drug Substances* vol. **19**, K. Florey Ed. (Academic Press, New York, 1990) pp 95-121.

Pentahydrate, $C_{22}H_{22}N_6O_7S_2 \cdot 5H_2O$, *Ceftim, Fortam, Fortaz, Fortum, Glazidim, Kefadim, Kefamin, Kefazim, Modacin, Panzid, Spectrum, Starcef, Tazicef, Tazidime.* Crystalline solid. uv max (pH 6): 257 nm ($E_{1cm}^{1\%}$ 348).

THERAP CAT: Antibacterial.

1996. Cefteram. *[6R-[6α,7β(Z)]]-7-[[(2-Amino-4-thiazolyl)(methoxyimino)acetyl]amino]-3-[(5-methyl-2H-tetrazol-2-yl)methyl]-8-oxo-5-thia-1-azabicyclo[4.2.0]oct-2-ene-2-carboxylic acid;* (+)-(6R,7R)-7-[(Z)-2-(2-amino-4-thiazolyl)-2-(methoxyimino)acetamido]-3-[(5-methyl-2H-tetrazol-2-yl)methyl]-8-oxo-5-thia-1-azabicyclo[4.2.0]oct-2-ene-2-carboxylic acid; 7-[2-(2-aminothiazol-4-yl)-2-syn-methoxyimino-acetamido]-3-[2-(5-methyl-1,2,3,4-tetrazolyl)methyl]-Δ³-cephem-4-carboxylic acid; (+)-(6R,7R)-7-[2-(2-amino-4-thiazolyl)glyoxylamido]-3-[(5-methyl-2H-tetrazol-2-yl)methyl]-8-oxo-5-thia-1-azabicyclo[4.2.0]oct-2-ene-2-carboxylic acid 7^2-(Z)-(O-methyloxime); ceftetrame; Ro-19-5247; T-2525. $C_{16}H_{17}N_9O_5S_2$; mol wt 479.50. C 40.08%, H 3.57%, N 26.29%, O 16.68%, S 13.37%. Third generation, orally active cephalosporin antibiotic. Prepn: **Belg.** pat. **890,499;** H. Sadaki *et al.,* U.S. pat. **4,489,072** (1982, 1984 both to Toyama). *In vitro* antibacterial activity: R. Wise *et al., Anti-*

microb. Ag. Chemother. **29**, 1067 (1986); *in vitro* activity vs *Neisseria gonorrhoeae:* W. R. Bowie *et al., ibid.* **31**, 470 (1987). General pharmacology in animals: S. Hirai *et al., Japan. J. Antibiot.* **39**, 958 (1986), *C.A.* **105**, 183450a (1986). Pharmacokinetics and tissue distribution in animals: I. Saikawa *et al., ibid.* 979, *C.A.* **105**, 183338v (1986). Series of articles on activity, toxicology, pharmacology and clinical efficacy: *Chemotherapy (Tokyo)* **34**, Suppl. 2, 1-984 (1986). Acute toxicity: S. Sato *et al., ibid.* 166.

Crystals from acetone, mp > 200°.

Pivaloyloxymethyl ester, $C_{22}H_{27}N_9O_7S_2$, *cefteram pivoxil, T-2588, Tomiron.* Oral prodrug of cefteram. Crystals, mp 127-128° (dec). LD_{50} in male, female mice, rats (g/kg): > 6.00, 5.86, 5.63, 5.09 i.p.; in both species > 6.00 s.c.; in male mice, rats, dogs (g/kg): > 6.00, > 6.00, > 2.00 orally (Sato).

THERAP CAT: Antibacterial.

1997. Ceftezole. *(6R-trans)-8-Oxo-7-[(1H-tetrazol-1-yl-acetyl)amino]-3-[(1,3,4-thiadiazol-2-ylthio)methyl]-5-thia-1-azabicyclo[4.2.0]oct-2-ene-2-carboxylic acid;* 7-(1H-tetrazol-1-yl)acetamido-3-(1,3,4-thiadiazol-2-ylthio)methylceph-3-em-4-carboxylic acid; CTZ; CG-B3Q. $C_{13}H_{12}N_8O_4S_3$; mol wt 440.49. C 35.45%, H 2.75%, N 25.44%, O 14.53%, S 21.84%. Broad spectrum antibiotic similar to cefazolin, *q.v.* Prepn: T. Takano *et al.,* **S. Afr.** pat. **68 04,513;** *eidem,* U.S. pat. **3,516,997** (1969, 1970 both to Fujisawa). *In vitro* and *in vivo* activity: T. Noto *et al., J. Antibiot.* **29**, 1058 (1976). Pharmacology: Y. Harada *et al., ibid.* 1071; M. Kakimoto *et al., Chemotherapy (Tokyo)* **24**, 722 (1976). Metabolism: K. Koyama *et al., ibid.* 619; Y. Tohira *et al., ibid.* 655. Clinical study: S. Ishiyama *et al., ibid.* 1006. Toxicity and teratogenicity studies: R. Niki *et al., ibid.* 671.

Solid, mp 155° (dec). uv max (phosphate buffer pH 6.4): 273 nm ($E_{1cm}^{1\%}$ 274).

Sodium salt, $C_{13}H_{11}N_8NaO_4S_3$, *FR-10123, Celoslin, Falomesin.* White solid, freely sol in water.

THERAP CAT: Antibacterial.

1998. Ceftibuten. *[6R-[6α,7β(Z)]]-7-[[2-(2-Amino-4-thiazolyl)-4-carboxy-1-oxo-2-butenyl]amino]-8-oxo-5-thia-1-azabicyclo[4.2.0]oct-2-ene-2-carboxylic acid;* 7β-[(Z)-2-(2-aminothiazol-4-yl)-4-carboxy-2-butenoylamino]-3-cephem-4-carboxylic acid; 7432-S; Sch-39720. $C_{15}H_{14}N_4O_6S_2$; mol wt 410.43. C 43.90%, H 3.44%, N 13.65%, O 23.39%, S 15.63%. Orally active third generation cephalosporin antibiotic. Prepn: Y. Hamashima, **Eur.** pat. Appl. **136,721;** *idem,* U.S. pat. **4,634,697** (1985, 1987 both to Shionogi); and *in vitro* antibacterial activity: Y. Hamashima *et al., J. Antibiot.* **40**, 1468 (1987). Industrial synthesis utilizing penicillin G: M. Yoshioka, *Pure Appl. Chem.* **59**, 1041 (1987). Comparative antibacterial spectrum: R. N. Jones, A. L. Barry, *Antimicrob. Ag. Chemother.* **32**, 1576 (1988). Pharmacokinetics in humans: M. Nakashima *et al., J. Clin. Pharmacol.* **28**, 246 (1988).

Dihydrate, $C_{15}H_{14}N_4O_6S_2 \cdot 2H_2O$, *Cedax, Isocef*.
THERAP CAT: Antibacterial.

1999. Ceftiofur. *[6R-[6α,7β(Z)]]-7-[[(2-Amino-4-thia-zolyl)(methoxyimino)acetyl]amino]-3-[[(2-furanylcarbonyl)-thio]methyl]-8-oxo-5-thia-1-azabicyclo[4.2.0]oct-2-ene-2-car-boxylic acid;* (6R,7R)-7-[2-(2-amino-4-thiazolyl)glyoxylami-do]-3-(mercaptomethyl)-8-oxo-5-thia-1-azabicyclo[4.2.0]oct-2-ene-2-carboxylic acid 7²-(Z)-(O-methyloxime) 2-furoate (ester). $C_{19}H_{17}N_5O_7S_3$; mol wt 523.57. C 43.59%, H 3.27%, N 13.38%, O 21.39%, S 18.37%. Broad spectrum, third generation cephalosporin antibiotic. Prepn: B. Labeeuw, A. Salhi, **Eur.** pat. **Appl. 36,812;** *eidem,* **U.S.** pat. **4,464,367** (1981, 1984 both to Sanofi). *In vivo* and *in vitro* antibacterial activity and β-lactamase stability: R. J. Yancey *et al., Am. J. Vet. Res.* **48,** 1050 (1987).

Monosodium salt, $C_{19}H_{16}N_5NaO_7S_3$, *CM-31916, U-64279E, Excenel, Naxcel.*
Monohydrochloride, $C_{19}H_{17}N_5O_7S_3 \cdot HCl$, *U-67279A.*
THERAP CAT (VET): Antibacterial.

2000. Ceftizoxime. *[6R-[6α,7β(Z)]]-7-[[(2-Amino-4-thiazolyl)(methoxyimino)acetyl]amino]-8-oxo-5-thia-1-aza-bicyclo[4.2.0]oct-2-ene-2-carboxylic acid;* (6R,7R)-7-[2-(2-imino-4-thiazolyl)glyoxylamido]-8-oxo-5-thia-1-azabicyclo-[4.2.0]oct-2-ene-2-carboxylic acid 7²-(Z)-(O-methyloxime). $C_{13}H_{13}N_5O_5S_2$; mol wt 383.41. C 40.72%, H 3.42%, N 18.27%, O 20.86%, S 16.73%. Injectable third generation cephalosporin antibiotic related to cefotaxime, *q.v.* Exhibits broad spectrum activity and resistance to β-lactamase hydrolysis. Prepn: T. Takaya *et al.,* **Ger.** pat. **2,810,922;** *eidem,* **U.S.** pat. **4,427,674** (1978, 1984 both to Fujisawa). *In vitro* and *in vivo* antibacterial activity: T. Kamimura *et al., Antimicrob. Ag. Chemother.* **16,** 540 (1979). Comparison with other cephalosporins: M. Nishida *et al., J. Antibiot.* **32,** 1319 (1979). Pharmacokinetics: T. Murakawa *et al., Antimicrob. Ag. Chemother.* **17,** 157 (1980). Series of articles on metabolism, pharmacology, laboratory evaluation: *Arznei-mittel-Forsch.* **30,** 1662-1687 (1980). Toxicity and reproduction studies: K. Fukuhara *et al., ibid.* 1669. Phase I clinical study: N. Nakashima *et al., J. Clin. Pharmacol.* **21,** 388 (1981). *Reviews: J. Antimicrob. Chemother.* **10,** Suppl. C, 1-355 (1982); D. M. Richards, R. C. Heel, *Drugs* **29,** 281-329 (1985).

Crystals, mp 227° (dec).

Sodium salt, $C_{13}H_{12}N_5NaO_5S_2$, *FK-749, FR-13479, SKF-88373, Cefizox, Ceftix, Ceftizon, Epocelin, Eposerin.* LD_{50} in mice, rats (mg/kg): ~6000 i.v. (Fukuhara).
THERAP CAT: Antibacterial.

2001. Ceftriaxone. *[6R-[6α,7β(Z)]]-7-[[(2-Amino-4-thiazolyl)(methoxyimino)acetyl]amino]-8-oxo-3-[[(1,2,5,6-tetrahydro-2-methyl-5,6-dioxo-1,2,4-triazin-3-yl)thio]methyl]-5-thia-1-azabicyclo[4.2.0]oct-2-ene-2-carboxylic acid;* (6R,7R)-7-[2-(2-amino-4-thiazolyl)glyoxylamido]-3-[[(2,5-dihy-dro-6-hydroxy-2-methyl-5-oxo-as-triazin-3-yl)thio]methyl]-8-oxo-5-thia-1-azabicyclo[4.2.0]oct-2-ene-2-carboxylic acid 7²-(Z)-(O-methyloxime); cefatriaxone. $C_{18}H_{18}N_8O_7S_3$; mol wt 554.59. C 38.98%, H 3.27%, N 20.20%, O 20.19%, S 17.35%. Parenteral third generation cephalosporin antibiotic. Prepn: M. Montavon, R. Reiner, **Brit.** pat. **Appl. 2,022,090;** *eidem,* **U.S.** pat. **4,327,210** (1979, 1982 both to Hoffmann-La Roche); R. Reiner *et al., J. Antibiot.* **33,** 783 (1980). *In vitro* and *in vivo* studies: P. Angehrn *et al., Antimicrob. Ag. Chemother.* **18,** 913 (1980). Pharmacokinetics: M. Seddon *et al., ibid.* 240. Determn in plasma, urine, bile: K. H. Trautmann, P. Haefelfinger, *J. High Resolut. Chromatogr. Chromatogr. Commun.* **4,** 54 (1981), *C.A.* **94,** 202437 (1981). Mechanism of action study: R. B. Wright *et al., J. Antibiot.* **34,** 590 (1981). Kinetics in humans: K. Stoekel *et al., Clin. Pharmacol. Ther.* **29,** 650 (1981). Toxicity data: K. Teelman *et al.,* "Experimentelle Toxikologie von Ceftriaxon", in *Ceftriaxon ein neues parenterales Cephalosporin,* Proc. Hahnenklee Symp., 1981, R. Grieshalber, Ed. (Editiones Roche, Basel, 1982) pp 91-111. Review of antibacterial activity, toxicology, pharmacology and clinical studies: D. M. Richards *et al., Drugs* **27,** 469 (1984); T. R. Beam, *Pharmacotherapy* **5,** 237-253 (1985).

Disodium salt hemiheptahydrate, $C_{18}H_{16}N_8Na_2O_7S_3 \cdot 3\frac{1}{2}H_2O$, *Ro-13-9904/001, Rocefin, Rocephin(e).* White cryst powder, mp >155° (dec). $[\alpha]_D^{25}$ -165° (c = 1 in water) (calc for water-free substance). uv max (water): 242, 272 nm (ε 32300, 29530). pKa: ~3 (COOH), 3.2 (NH_3^+), 4.1 (enolic OH). Soly in water at 25°: ~40 g/100 ml. LD_{50} in male, female, mice, rats (mg/kg): 3000, 2800, 2175, 2175 i.v.; >10000 all species orally; >5000 all species s.c. (Teelman).
THERAP CAT: Antibacterial.

2002. Cefuroxime. *[6R-[6α,7β(Z)]]-3-[[(Aminocarbon-yl)oxy]methyl]-7-[[2-furanyl(methoxyimino)acetyl]amino]-8-oxo-5-thia-1-azabicyclo[4.2.0]oct-2-ene-2-carboxylic acid;* (6R,7R)-3-carbamoyloxymethyl-7-[2-(2-furyl)-2-(methoxy-imino)acetamido]-8-oxo-5-thia-1-azabicyclo[4.2.0]oct-2-ene-2-carboxylic acid; (6R,7R)-3-carbamoyloxymethyl-7-[2-(2-furyl)-2-(methoxyimino)acetamido]ceph-3-em-4-carboxylic acid. $C_{16}H_{16}N_4O_8S$; mol wt 424.39. C 45.28%, H 3.80%, N 13.20%, O 30.16%, S 7.56%. Prepn: M. C. Cook *et al.,* **Ger.** pat. **2,439,880;** *eidem,* **U.S.** pat. **3,974,153** (1973, 1976 both to Glaxo). Prepn of the 1-acetoxyethyl ester: M. Gregson, B. Sykes, **Ger.** pat. **2,706,413;** *eidem,* **U.S.** pat. **4,267,320** (1977, 1981 both to Glaxo). *In vitro* studies: C. H. O'Callaghan *et al., Antimicrob. Ag. Chemother.* **9,** 511 (1976); R. N. Jones *et al., ibid.* **12,** 47 (1977). *In vitro* antibacterial activity, human pharmacokinetics: C. H. O'Callaghan *et al., J. Antibiot.* **29,** 29 (1976). Pharmacology: H. Freiesle-ben *et al., Proc. 10th Int. Congr. Chemother.,* Zürich, 1977 (Am. Soc. for Microbiol., Washington, 1978) **II,** pp 873-874. Pharmacokinetics: P. E. Gower, *ibid.* 877-878; J. Kosmidis *et al., ibid.* 875-876. Clinical studies: P. F. Wood *et al., ibid.* 1042-1044; R. Norrby *et al., J. Antimicrob. Chemother.* **3,** 355 (1977). Review of antibacterial activity, pharmacology and therapeutic efficacy: R. N. Brogden *et* al., *Drugs* **17**, 233-266 (1979). Comprehensive description: T. J. Wozniak, J. R. Hicks, *Anal. Profiles Drug Subs.* **20**, 209-236 (1991).

White crystalline solid. $[\alpha]_D^{20}$ +63.7° (c = 1.0 in 0.2*M* pH 7 phosphate buffer). uv max (pH 6 phosphate buffer): 274 nm (ϵ 17600).

Sodium salt, $C_{16}H_{15}N_4NaO_8S$, *Anaptivan, Biociclin, Biofurex, Bioxima, Cefamar, Cefoprim, Cefumax, Cefurex, Cefurin, Curocef, Curoxim, Duxima, Gibicef, Ipacef, Kefurox, Kesint, Lampsporin, Medoxim, Novocef, Spectrazole, Ultroxim, Zinacef.* White solid. $[\alpha]_D^{20}$ +60° (c = 0.91 in water). uv max (water): 274 nm (ϵ 17400). Freely sol in water and buffered solutions; sol in methanol; very slightly sol in ethyl acetate, diethyl ether, octanol, benzene and chloroform. Soly in water: 500 mg/2.5 ml. pKa (water): 2.5; (DMF): 5.1. Solns are stable at room temp for 13 hrs; <10% decompn in 48 hrs at 25° (O'Callaghan, *J. Antibiot.*). 1-Acetoxyethyl ester, $C_{20}H_{22}N_4O_{10}S$, *cefuroxime axetil, CCI-15641, Ceftin, Cefurax, Cepazine, Elobact, Oraxim, Zinat, Zinnat.*

THERAP CAT: Antibacterial.

THERAP CAT (VET): Antibacterial.

2003. Cefuzonam. *[6R-[6α,7β(Z)]]-7-[[(2-Amino-4-thiazolyl)(methoxyimino)acetyl]amino]-8-oxo-3-[(1,2,3-thiadiazol-5-ylthio)methyl]-5-thia-1-azabicyclo[4.2.0]oct-2-ene-2-carboxylic acid;* (−)-(6R,7R)-7-[2-(2-amino-4-thiazolyl)glyoxylamido]-8-oxo-3-[(1,2,3-thiadiazol-5-ylthio)methyl]-5-thia-1-azabicyclo[4.2.0]oct-2-ene-2-carboxylic acid 7^2-(Z)-(O-methyloxime); 7β-[2-(2-aminothiazol-4-yl)-(Z)-2-methoxyiminoacetamido]-3-[(1,2,3-thiadiazol-5-yl)thiomethyl]-ceph-3-em-4-carboxylic acid; CL-118523. $C_{16}H_{15}N_7O_5S_4$; mol wt 513.60. C 37.42%, H 2.94%, N 19.09%, O 15.58%, S 24.97%. Semi-synthetic third generation cephalosporin. Prepn and bactericidal activity: W. V. Curran, A. S. Ross, Fr. pat. **2,488,258**; *eidem,* U.S. pat. **4,399,132** (1982, 1983 both to Am. Cyanamid); *eidem, J. Antibiot.* **36**, 179 (1983). Improved prepn of sodium salt: K. Naito *et al.,* Eur. pat. **145,395** (1985 to Takeda). *In vitro* antibacterial spectrum: M. Hikada *et al., J. Antimicrob. Chemother.* **18**, 585 (1986). Toxicity study: S. Takita, *Chemotherapy (Tokyo)* **34**, Suppl. 3, 96 (1986). Conference proceedings: *Recent Adv. Chemother., Antimicrobial Sect. 2., Proc. 14th Int. Congr. Chemother., Kyoto 1985*, J. Ishigami, Ed. (Univ. Tokyo Press, Tokyo, 1985) pp 887-908. Series of articles: *Chemotherapy (Tokyo)* **34**, Suppl. 3, 1-732 (1986).

White amorphous solid. Absorption max (KBr): 5.62. LD_{50} in female, male mice, female, male rats (mg/kg): 4117, >4800, 4281, 4222 i.v.; 6424, 6783, >8000, >8000 i.p.; all > 10000 orally (Takita).

Sodium salt dihydrate, $C_{16}H_{14}N_7NaO_5S_4\cdot2H_2O$, *L-105 (Lederle), CL-251931, LJC-10305, Cosmosin.* Yellowish crystalline powder. Sol in water.

THERAP CAT: Antibacterial.

2004. Celery Seed. Dried, ripe fruit of *Apium graveolens* L., *Umbelliferae. Habit.* Southern Europe; cultivated everywhere. *Constit.* Volatile and fixed oils; bitter extractive; resin.

2005. Celesticetin. *(S)-2-[(2-Hydroxybenzoyl)oxy]ethyl 6,8-dideoxy-7-O-methyl-6-[[(1-methyl-2-pyrrolidinyl)carbonyl]amino]-1-thio-D-erythro-α-D-galacto-octopyranoside;* α-2-hydroxyethyl 6,8-dideoxy-7-O-methyl-6-(1-methyl-L-2-pyrrolidinecarboxamido)-1-thio-D-erythro-D-galacto-octopyranoside monosalicylate; S-demethyl-3'-depropyl-S-[2-[(2-hydroxybenzoyl)oxy]ethyl]-7-O-methyllincomycin. $C_{24}H_{36}N_2O_9S$; mol wt 528.62. C 54.53%, H 6.86%, N 5.30%, O 27.24%, S 6.07%. Antibiotic substance produced by *Streptomyces caelestis* from a soil obtained from Emigration Canyon near Salt Lake City: DeBoer *et al., Antibiotics Annual* **1954-1955**, 831; Hoeksema *et al., ibid.* 837. Structure: Hoeksema, *J. Am. Chem. Soc.* **86**, 4224 (1964); **90**, 755 (1968).

Amphoteric, amorphous base. $[\alpha]_D^{24}$ +126.6° (c = 0.5 in chloroform). uv max (0.01*N* alcoholic H_2SO_4): 240, 310 nm ($E_{1cm}^{1\%}$ 183.7, 80.6). Soluble in acidic or strongly basic aq solns, practically insol in the pH range of 7 to 10. Soluble in the more polar organic solvents, but not in ether or light hydrocarbons. Active against gram-positive organisms and plant pathogens such as *B. stewartii, P. fascians, X. pruni* at 0.5 γ/ml. The antibacterial activity is rapidly destroyed in aq solns above pH 9 at 24°, but remains at pH 5 to 7 for at least 60 days. Celesticetin is more stable in acid than in basic solns, but is destroyed rapidly in 1*N* HCl at 100°. Cross resistance with erythromycin.

Salicylate, monoclinic tablets, mp 139°. $[\alpha]_D^{24}$ +90.2° (c = 0.5 in H_2O). Sol in water.

Oxalate, crystals, mp 149-154°. $[\alpha]_D^{24}$ +106.6° (c = 0.5 in H_2O). Sol in water.

2006. Celestin Blue. *1-(Aminocarbonyl)-7-(diethylamino)-3,4-dihydroxyphenoxazin-5-ium chloride;* 1-carbamoyl-7-(diethylamino)-3,4-dihydroxyphenazoxonium chloride; C.I. Mordant Blue 14; C.I. 51050; michrome no. 66; Corein 2R. $C_{17}H_{18}ClN_3O_4$; mol wt 363.80. C 56.13%, H 4.99%, Cl 9.75%, N 11.55%, O 17.59%. Prepd by the action of gallamide on *N,N*-diethyl-*p*-nitrosoaniline HCl: Gnehm, Bauer, *J. Prakt. Chem.* **72**, 249 (1905); Grandmougin, Bodmer, *ibid.* **77**, 502 (1908), *see also:* U.S. pats. **534,809** (1895) and **1,227,407** (1917); *Colour Index* vol. 4 (3rd ed., 1971). *Review:* H. J. Conn's *Biological Stains*, R. D. Lillie, Ed. (Williams & Wilkins, Baltimore, 9th ed., 1977) pp 407-408.

Greenish-black powder. With water it yields a reddish-purple soln, which appears blue on extreme diln. The alcoholic soln is blue in all concns. Approximate soly in water 2.0%, abs ethanol 1.5%, Cellosolve 2.25%, ethylene glycol 6.5%, xylene 0.005%.

USE: To dye fabrics a navy-blue color with faint reddish tinge. As nuclear and connective tissue stain. One of the ingredients of Picro-Mallory stain: Lendrum, McFarlane, *J. Pathol. Bacteriol.* **50**, 381 (1940).

2007. Celiprolol. *N'-[3-Acetyl-4-[3-[(1,1-dimethylethyl)-amino]-2-hydroxypropoxy]phenyl]-N,N-diethylurea; N-[3-acetyl-4-(3'-tert-butylamino-2'-hydroxy)propoxy]phenyl-N'-diethylurea; ST-1396.* $C_{20}H_{33}N_3O_4$; mol wt 379.50. C 63.30%, H 8.76%, N 11.07%, O 16.86%. Cardioselective β_1-adrenergic blocker. Prepn: **Belg. pat. 823,411**; G. Zölss *et al.*, **U.S. pat. 4,034,009** (1975, 1977 both to Chemie Linz). Hemodynamic effects: J. Bonelli *et al., Wien Klin. Wochenschr.* **90**, 350 (1978); H. Pittner, *Arch. Pharmacol.* **311**, Suppl., 180 (1980). Series of articles on determn in biological material, pharmacology, toxicology, clinical studies: *Arzneimittel-Forsch.* **33**, 1-79 (1983). Toxicity: W. Wendtlandt, H. Pittner, *ibid.* **41**. Symposium on pharmacology, clinical efficacy and comparison with other β-blockers: *J. Cardiovasc. Pharmacol.* **8**, Suppl. 4, S1-S152 (1986). Comprehensive description: D. J. Mazzo *et al.* in *Analytical Profiles of Drug Substances* vol. 20, K. Florey Ed. (Academic Press, New York, 1991) pp 237-301.

Crystals, mp 110-112°.
Hydrochloride, $C_{20}H_{33}N_3O_4$.HCl, *Celectol, Corliprol, Selecor, Selectol.* White, odorless crystals, mp 197-200° (dec). Soly at ~25°C (g/100 ml): water 15.1, methanol 18.2, ethanol 1.61, chloroform 0.42. uv max (water): 231, 324 nm ($E_{1\%}$ 652, 57); (0.01N HCl): 231, 324 nm ($E_{1\%}$ 660, 60); (0.01N NaOH): 231, 324 nm ($E_{1\%}$ 640, 60); (methanol): 232, 329 nm ($E_{1\%}$ 775, 58). LD_{50} in male mice, rats (mg/kg): 56.2, 68.3 i.v.; 1834, 3826 orally (Wendtlandt, Pittner).

THERAP CAT: Antihypertensive, antianginal.

2008. Cellobiose. *4-O-β-D-Glucopyranosyl-D-glucose; β-cellobiose; cellose; 4-(β-D-glucosido)-D-glucose.* $C_{12}H_{22}O_{11}$; mol wt 342.30. C 42.11%, H 6.48%, O 51.41%. Unit of cellulose and lichenin. Does not occur free in nature, or as glucoside. Prepn from cotton: Braun, *Org. Syn.* coll. vol. **II**, 122, 124 (1943). Prepn from cell-free enzymatic hydrolyzate of cellulose: Whistler, Smart, *J. Am. Chem. Soc.* **75**, 1916 (1953). Structure: Haworth, Hirst, *J. Chem. Soc.* **119**, 193 (1923); Charlton *et al., ibid.* **1926**, 89; Zemplén, *Ber.* **59**, 1254 (1926); Haworth *et al., J. Chem. Soc.* **1927**, 2809; Peterson, Spencer, *J. Am. Chem. Soc.* **49**, 2822 (1927); Helferich *et al., Ber.* **63**, 992 (1930); Hess, Dziengel, *ibid.* **68**, 1594 (1935); Hassid, Ballou in *The Carbohydrates,* W. Pigman, Ed. (Academic Press, New York, 1957) p 490. Synthesis: Haskins *et al., J. Am. Chem. Soc.* **64**, 1289 (1942). Review: Pazur in *The Carbohydrates* vol. **2A**, W. Pigman *et al.,* Eds. (Academic Press, New York, 2nd ed., 1970) pp 109-110; R. G. Edwards, *Dev. Food Carbohyd.* **2**, 229-273 (1980).

Minute crystals from dil alcohol which retain 0.25 to 0.50 mol water after drying in vacuo. Indifferent taste. Dec 225°. Shows mutarotation. $[\alpha]_D^{20}$ +14.2° → +34.6° (15 hrs, c = 8). One gram dissolves in 8 ml water, in 1.5 ml boiling water. Almost insol in abs alc and ether. Reduces Fehling's soln. Hydrolysis with acid or emulsin yields 2 mols β-D-glucose. Not fermented by brewers' yeast, maltase, or invertase.
Octaacetyl-aldehydro-cellobiose, $C_{28}H_{38}O_{19}$, mp 105-110°. $[\alpha]_D^{20}$ +17.7° (c = 3 in chloroform).
Octaacetyl-α-cellobiose, $C_{28}H_{38}O_{19}$, mp 229°. $[\alpha]_D^{20}$ +41° (c = 6 in chloroform).

Octaacetyl-β-cellobiose, mp 202°. $[\alpha]_D^{20}$ −14.7° (c = 5 in chloroform).

2009. Cellocidin. *2-Butynediamide;* acetylenedicarboxamide; acetylenedicarboxylic acid diamide; aquamycin; lenamycin. $C_4H_4N_2O_2$; mol wt 112.09. C 42.86%, H 3.60%, N 24.99%, O 28.55%. $H_2NCOC \equiv CCONH_2$. Antibiotic substance with antibacterial activity. Produced by *Streptomyces chibaensis* from soil collected at Chiba City, Japan: Suzuki *et al., J. Antibiot.* **11**, 81 (1958). Synthesis from dimethyl acetylenedicarboxylate and concd ammonium hydroxide at −10°: Saggiomo, *J. Org. Chem.* **22**, 1171 (1957); Suzuki, Okuma, *J. Antibiot.* **11**, 84 (1958). Identity with lenamycin: Y. Sekizawa, *Meiji Seika Kenkyu Nempo* **1960**, 42, *C.A.* **56**, 14609a (1962). Biosynthesis: E. R. H. Jones, *J. Chem. Soc. Perkin Trans. I* **1973**, 148.
Crystals from dil methanol, dec. 216-218°. uv max (0.1N NaOH): 299 nm ($E_{1cm}^{1\%}$ 290). Sparingly sol in water, methanol, ethanol, acetone, chloroform, glacial acetic acid. Relatively stable in neutral or acid solns, showing no loss of activity at pH 2 to 7 when heated for 10 min at 100°. Unstable in alkaline soln evolving ammonia. LD_{50} i.v. in mice: 11 mg/kg (Suzuki).

2010. Cellophane. Francephane. Transparent, flexible cellulose sheeting made from viscose. The word "cellophane" is not a trademark in the U.S.

2011. Celluloid®. Pyralin; Zylonite. Prepd from nitrocellulose and camphor.
Colorless, amorphous mass. *Flammable.* Prone to spontaneous decompn which is retarded or prevented by the addition of urea, ZnO, $MgCO_3$, diphenylamine, etc. It is rendered less flammable by addition of ammonium phosphate. Softens in boiling water; sol in acetone.
USE: Plastic material for manuf of toilet articles, toys, photographic films; substitute for amber, ivory, ebonite, tortoise shell; also in surgery for bandages and in dentistry as substitute for rubber.

2012. Cellulose. $(C_6H_{10}O_5)_n$. Polysaccharide with the glucose units linked as in cellobiose. Chief constituent of the fiber of plants; cotton is the purest natural form, contg about 90%. Rayon is regenerated cellulose. Books: J. Dorée, *The Methods of Cellulose Chemistry* (Chapman & Hall, London, 1947); T. Lieser, *Kurzes Lehrbuch der Cellulosechemie* (Gebrüder Borntraeger, Berlin, 1953); S. D. Antonovskii, *Chemistry of Wood and Cellulose* (Vsesoyuz. Zaochnyi Lesotekh Instit., Leningrad, 1954); E. Ott *et al., Cellulose and Cellulose Derivatives,* vols. **1-3** (Interscience, New York, 1954, 1955). *Reviews:* Several authors in *Encyclopedia of Polymer Science and Technology* vol. **3**, N. M. Bikales, Ed. (Interscience, New York, 1965) pp 131-539; Shafizadeh, *Pure Appl. Chem.* **35**, 195-208 (1973); A. F. Turbak *et al.* in Kirk-Othmer *Encyclopedia of Chemical Technology* vol. **5** (Wiley-Interscience, New York, 3rd ed., 1979) pp 70-89. Comprehensive review on constitution, conformation, size of molecule, fine structure and superstructure: H. Krässig, *Papier (Darmstadt)* **33**, 9-20 (1979). Review of toxicity studies: R. L. Anderson *et al., Cancer Letters* **63**, 83-92 (1992).
White substance. Practically insol in water or other usual solvents, but is dissolved by concd soln of zinc chloride, by ammoniacal copper hydroxide soln; also by caustic alkali with carbon disulfide.
Microcrystalline form, *Avicel.* Prepn and manuf of crystallite cellulosic aggregates: Battista, *Ind. Eng. Chem.* **42**, 502 (1950); Battista, Smith, U.S. pats. **2,978,446** and **3,141,-875** (1961 to Am. Viscose and 1964 to FMC). Non-fibrous powder. Particle shape: rigid rods. Refractive index: 1.55. Bulk density: 18-19 lb/cubic foot. Practically insol, but dispersible in water; practically insol in and resistant to dil acid; practically insol and inert in organic acids. Partially sol with swelling in dil alkali.
USE: Fibrous form is the basic material for the textile and paper industries. Nitrated it yields nitrocellulose used for manuf of explosives, collodion, lacquers. Basic material also for cellulose acetate, cellulose xanthate. Also used in chromatography and as ion exchange material especially in the form of derivatives such as *DEAE-cellulose* (diethylaminoethyl cellulose) and ECTEOLA-cellulose, *q.v.* Microcrys-

talline forms of cellulose are used as combination binder-disintegrants in tableting, as separatory medium in thin-layer and column chromatography. Colloidal cellulose particles aid in stabilization and emulsification of liquid and foam systems. May be used as pure cellulose raw-material. Incorporation of cellulose crystallite aggregates in foods to reduce caloric content: Battista, **U.S.** pat. **3,023,104** (1962 to American Viscose); also used in food industry as stabilizer, thickener, texturizer.

2013. Cellulose Acetates. Partially acetylated cellulose, *q.v.* Several acetates of cellulose are known, which differ from one another only in the degree of acetylation. In triacetates, no less than 92% of the hydroxyl groups are acetylated. In characterizing the degree of acetylation, percent acetyl value and percent combined acetic acid are used. All cellulose acetates are obtained by treating cellulose with acetic anhydride at various temps for different lengths of time to produce amorphous white solid material in granular, flake, or powder form from which fibers may be formed by extrusion. In the plastics industry, it is usual to acetylate fully and then to lower the acetyl value to 52-56% by partial hydrolysis. Such material when compounded with suitable plasticizers gives a tough thermoplastic product. Manuf: Faith, Keyes & Clark's *Industrial Chemicals*, F. A. Lowenheim, M. K. Moran, Eds. (Wiley-Interscience, New York, 4th ed., 1975) pp 239-243. Toxicity studies: W. C. Thomas *et al., Food Chem. Toxicol.* **29**, 453 (1991). *Review:* G. A. Serad, J. R. Sanders, "Cellulose Acetate and Triacete Fibers" in Kirk-Othmer *Encyclopedia of Chemical Technology* **Vol. 5** (Wiley-Interscience, New York, 3rd ed., 1979) pp 89-117.

R = –OOCCH₃

Cellulose Triacetate

Commercial products do not have sharp melting points. Solubility is affected by the acetyl value; the triacetate is insol in water, alcohol, ether, but sol in glacial acetic acid; the tetraacetate is insol in water, alcohol, ether, glacial acetic acid, methanol; the pentaacetate is insol in water, but sol in alcohol.

USE: Manuf rubber and celluloid substitutes, nonflammable photographic and cinema films, airplane dopes, varnishes and lacquers, filaments, phonograph records; waterproofing fabrics and rendering balloons gas-tight; sizing and finishing fabrics; coating skins; insulating electric wires; tow for cigarette smoke filters.

2014. Cellulose Ethyl Hydroxyethyl Ether. *Ethyl 2-hydroxyethyl ether cellulose;* ethyl hydroxyethyl cellulose;

R is H, –CH₂CH₃, or –CH₂CH₂OH

Etulos. Prepn: Jullander, *C.A.* **48**, 6114g (1954); *idem*, **Ger.** pat. **1,000,367** (1957 to Mo och Domsjö Ab.), *C.A.* **54**, 5088f (1960). Use as laxative: Alm, *C.A.* **50**, 2122i (1956); *idem, Am. J. Dig. Dis.* [NS] **2**, 493 (1957).
THERAP CAT: Cathartic.

2015. Centaurein. *7-(β-D-Glucopyranosyloxy)-5-hydroxy-2-(3-hydroxy-4-methoxyphenyl)-3,6-dimethoxy-4H-1-benzopyran-4-one;* 3',5,7-trihydroxy-3,4',6-trimethoxyflavone 7-β-D-glucoside. $C_{24}H_{26}O_{13}$; mol wt 522.46. C 55.17%, H 5.02%, O 39.81%. Isoln from root of *Centaurea jacea* L., *Compositae:* Bridel, Charaux, *Compt. Rend.* **175**, 833, 1168 (1922). Structure: Farkas *et al., Ber.* **97**, 1666 (1964).

Monohydrate, yellow crystals, mp 208-209°. $[\alpha]_D^{20}$ −76.6° (c = 1.4 in methanol). uv max (methanol): 349, 258 nm (log ε 4.31, 4.30). Sol in hot water, hot alcohol, and hot acetone; practically insol in cold water, chloroform, ether.

2016. Centaury. Minor centaury; lesser centaury; bitter herb. Dried flowering plant of *Centaurium erythracea* Rafn. (syn. *C. umbellatum* Gilib., *Erythraea centaurium* Pers.), *Gentianaceae. Habit.* Europe, North America, Middle East, North Africa. *Constit.* Gentiopicrin, amarogentin, gentisin: Korte, *Ber.* **87**, 1357 (1954); **88**, 704 (1955). *Review:* A. Y. Leung, *Encyclopedia of Common Natural Ingredients* (John Wiley & Sons, New York, 1980) pp 109-110.
Chilean centaury, chanchalagua, canchalagua. Dried flowering plant of *C. cachanlahuen* (Mol) Robins (syn *Gentiana peruviana* Lam.). *Habit.* Chile, Peru.
THERAP CAT: Bitter tonic.

2017. Centaury, American. Rose-pink; bitter-bloom; square-stem rose-gentian. Dried flowering plant of *Sabatia angularis* (L.) Pursh, *Gentianaceae. Habit.* Florida to Louisiana and Oklahoma, northern to southeastern New York; southern Ontario, Michigan, Wisconsin, Missouri. *Constit.* Gentiopicrin, amarogentin, gentisin: Korte, *Ber.* **87**, 1357 (1954); **88**, 704 (1955). Prepn of sedative extracts: Huneker, *Am. J. Pharm.* **43**, 207 (1871).
THERAP CAT: Bitter tonic.

2018. Centchroman. *(±)-(trans)-1-[2-[4-(3,4-Dihydro-7-methoxy-2,2,-dimethyl-3-phenyl-2H-1-benzopyran-4-yl)-phenoxy]ethyl]pyrrolidine; (trans)-1-[2-[p-(7-methoxy-2,2,-dimethyl-3-phenyl-4-chromanyl)phenoxy]ethyl]pyrrolidine;* ormeloxifene; 3,4-*trans*-2,2-dimethyl-3-phenyl-4-[p-(β-pyrrolidinoethoxy)phenyl]7-methoxychroman; *trans*-centchroman. $C_{30}H_{35}NO_3$; mol wt 457.61. C 78.74%, H 7.71%, N 3.06%, O 10.49%. Estrogen antagonist; synthetic nonsteroidal postcoital antifertility agent. Prepd, not claimed: J. W. Bolger, **Ger.** pat. **2,329,201** (1974 to Riker); *idem,* **U.S.** pat. **3,822,287** (1974 to Rexall). Mechanism of action: M. S. Sankaran, M. R. N. Prasad, *Contraception* **9**, 279 (1974). Pharmacology: V. P. Kamboj *et al., Ind. J. Exp. Biol.* **15**, 1144 (1977). Clinical pharmacology: R. Vaidya *et al., ibid.* 1173. Physicochemical properties: R. K. Seth *et al., Ind. J. Pharm. Sci.* **45**, 14 (1983). HPLC determn in serum and clinical pharmacokinetics: J. K. Paliwal *et al., Pharm. Res.* **6**, 1048 (1989). Clinical evaluation in breast cancer: N. C. Misra *et al., Int. J. Cancer* **43**, 781 (1989). Toxicity: I. M. Chak *et al., Ind. J. Exp. Biol.* **15**, 1159 (1977).

Crystals from ether/petr ether, mp 99-101°, sinters ~50°. Hydrochloride, $C_{30}H_{35}NO_3$·HCl, *67/20 CDRI*. White crystals, mp 165-166°. uv max (methanol): 232, 278 nm (ε 3701 at 278 nm). pKa 2.1. Sol in 10 parts chloroform, 20 parts acetone, 60 parts 95% ethanol, 20 parts methanol. Practically insol in water, isobutanol, 0.1*N* HCl, 0.1*N* NaOH. LD$_{50}$ i.p. in mice: 400 mg/kg (Chak).

THERAP CAT: Antiestrogen.

2019. Cephacetrile Sodium. *(6R-trans)-3-[(Acetyloxy)-methyl]-7-[(cyanoacetyl)amino]-8-oxo-5-thia-1-azabicyclo-[4.2.0]oct-2-ene-2-carboxylic acid monosodium salt;* sodium 7-(2-cyanoacetamido)-3-(hydroxymethyl)-8-oxo-5-thia-1-azabicyclo[4.2.0]oct-2-ene-2-carboxylate acetate (ester); sodium 7-(2-cyanoacetamido)cephalosporanic acid; Ciba 36278-Ba; Celospor; Vetimast. $C_{13}H_{12}N_3NaO_6S$; mol wt 361.31. C 43.22%, H 3.35%, N 11.63%, Na 6.36%, O 26.57%, S 8.87%. Semi-synthetic cephalosporin antibiotic. Prepn: **Neth. pat. Appl. 6,600,586**; Bickel *et al.*, **U.S. pat. 3,483,197** (1966, 1969 both to Ciba). Activity studies: Knüsel *et al., Antimicrob. Ag. Chemother.* **1970,** 140; Neu *et al., J. Antibiot.* **25,** 400 (1972). Pharmacokinetics and clinical results: Kradolfer *et al., Schweiz. Med. Wochenschr.* **103,** 711 (1973); Riess *et al., ibid.* 718. Toxicology: Kradolfer *et al., Antimicrob. Ag. Chemother.* **1970,** 150. *Review: Arznei-mittel-Forsch.* **24,** 1446-1533 (1974).

White crystalline powder. LD$_{50}$ in mice (mg/kg): 4500 ± 540 i.v.; 9100 ± 1500 s.c. (Kradolfer).

Free acid, needles from acetone + ether, mp 168-170° (dec). uv max (0.1*N* NaHCO$_3$): 260 nm (ε 9300).

THERAP CAT: Antibacterial.

THERAP CAT (VET): Antibacterial.

2020. Cephaeline. *7′,10,11-Trimethoxyemetan-6′-ol;* desmethylemetine; dihydropsychotrine. $C_{28}H_{38}N_2O_4$; mol wt 466.62. C 72.07%, H 8.21%, N 6.00%, O 13.72%. Next to emetine the most important alkaloid of ipecac, the ground roots of *Uragoga ipecacuanha* (Brot.) Baill. [*Cephaelis ipecacuanha* (Brot.) A. Rich.], *Rubiaceae:* Carr, Pyman, *J. Chem. Soc.* **105,** 1591 (1914); Hesse, *Ann.* **405,** 1 (1914). Structure: Pailer, Porschinski, *Monatsh.* **80,** 101 (1949). Stereochemistry: Van Tamelen *et al., J. Am. Chem. Soc.* **79,** 4817 (1957). Partial synthesis: A. K. Garg, J. R. Gear, *Tetrahedron Letters* **50,** 4377 (1969). Biosynthetic studies: *eidem, ibid.* **1968,** 141. HPLC determn in biological fluids: D. J. Crouch *et al., J. Anal. Toxicol.* **8,** 63 (1984). *Review:* M. Janot in Manske, Holmes, *The Alkaloids* vol. 3 (Academic Press, New York, 1953) pp 363-394.

Needles from ether. Faintly bitter taste. mp 115-116°. $[\alpha]_D^{20}$ —43.4° (c = 2 in chloroform). Practically insol in water. Freely sol in dil hydrochloric acid, dil sulfuric acid, acetic acid, methanol, ethanol, acetone, chloroform. Less sol in ether, petr ether.

Dihydrochloride heptahydrate, $C_{28}H_{40}Cl_2N_2O_4$·$7H_2O$, prisms, sinters at 245°, slowly melts up to 270°. $[\alpha]_D^{20}$ +25.0° (c = 2). Sol in water. Solns turn yellow. Less sol in alc, acetone, chloroform. Practically insol in benzene, ligroin.

Dihydrobromide heptahydrate, $C_{28}H_{40}Br_2N_2O_4$·$7H_2O$, prisms from dil hydrobromic acid, sinters at 266°, slowly melts up to 293°. Sol in water; moderately sol in alcohol, acetone. Practically insol in benzene.

Methyl ether, *see* Emetine.

THERAP CAT: Emetic; antiamebic.

THERAP CAT (VET): Has been used as an emetic and expectorant.

2021. Cephalexin. *[6R-[6α,7β(R*)]]-7-[(Aminophenyl-acetyl)amino]-3-methyl-8-oxo-5-thia-1-azabicyclo[4.2.0]oct-2-ene-2-carboxylic acid;* 7-(D-α-aminophenylacetamido)desace-toxycephalosporanic acid; 7-(D-2-amino-2-phenylacetami-do)-3-methyl-Δ^3-cephem-4-carboxylic acid; Cefadros; Cefaloto; Cefanex; Cefaseptin; Ceporexine; Cex; Derantel; Efalexin; Farexin; Fergon 500; Garasin; Ibilex; Iwalexin; Larixin; Lexibiotico; Llonexina; Madlexin; Mamalexin; Mecilex; Ohlexin; Oracocin; Rinesal; Sencephalin; Sintolexyn; Syncl; Taicelexin; Tokiolexin; Xahl. $C_{16}H_{17}N_3O_4S$; mol wt 347.39. C 55.32%, H 4.93%, N 12.10%, O 18.42%, S 9.23%. Semi-synthetic cephalosporin antibiotic. Prepn: R. B. Morin, B. G. Jackson, **U.S. pats. 3,275,626; 3,507,861** (1966, 1970 both to Lilly); Ryan *et al., J. Med. Chem.* **12,** 310 (1969). Pharmacology and toxicology: Muggleton *et al., Antimicrob. Ag. Chemother.* **1968,** 353; Kind *et al.,* Welles *et al., ibid.* 361, 489. Comprehensive description: L. P. Marrelli, *Anal. Profiles Drug Subs.* **4,** 21-46 (1975). Clinical pharmacology, efficacy, adverse reactions: *Postgrad. Med. J.* **59,** Suppl. 5, 1-56 (1983).

Crystals. uv max: 260 nm (ε 7750). pKa 5.2, 7.3. Monohydrate, *Cefa-Iskia, Cefibacter, Ceporex, Ceporexin, Keforal, Keflet, Keflex, Oracef, Ortisporina, Sartosona, Servispor.* LD$_{50}$ in mice, rats (g/kg): 1.6-4.5, > 5.0 orally; 0.4-1.3, > 3.7 i.p. (Welles).

Monohydrochloride monohydrate, $C_{16}H_{17}N_3O_4S$·HCl.-H_2O, *LY-061188, Keftab.*

Sodium salt, $C_{16}H_{16}N_3NaO_4S$, *Alfaspoven.*

Pivalate ester, *see* Pivcefalexin.

THERAP CAT: Antibacterial.

THERAP CAT (VET): Antibacterial.

2022. Cephalins. Kephalins; phosphatidylethanolamine. Group of phospholipids found in all living organisms. Significant constituent of nervous tissue and brain substance. Cephalins consist of glycerophosphoric acid in which the two free hydroxyls are esterified with long-chain fatty acid residues, and ethanolamine forms an ester linkage with the phosphate group. α-Isomers are derivatives of α-glycerophosphoric acid, *q.v.; β*-isomers are derivatives of *β*-glycerophosphoric acid, *q.v.* Natural products occur in the α-form, while the *β*-form is now recognized to be an artifact. Prepn from commercial "soybean lecithin": Scholfield, Dutton, *Biochem. Prepn.* **5**, 5 (1957); U.S. pat. **2,801,255** (1957 to USDA). Diagnostic use: C. R. Ratliff *et al., Am. J. Gastroenterol.* **55**, 589 (1971). Reviews on natural and synthetic cephalins: E. Baer in *Progress in the Chemistry of Fats and Other Lipids* vol. 6, (MacMillan, New York, 1963) pp 39-44; Van Deenen, de Haas in *Advances in Lipid Research* vol. 2 (Academic Press, New York, 1964) pp 183-189; Verkade, *Bull. Soc. Chim. France* **1963**, 1993.

R and R' can be, but are not necessarily, identical fatty acids.

α - Cephalin

Yellowish amorphous substances; characteristic odor and taste. Practically insol in water, acetone. Freely sol in chloroform, ether; slightly sol in ethanol.

USE: Clinical reagent (liver function test).

THERAP CAT: Hemostatic (local).

2023. Cephaloglycin. *[6R-[6α,7β(R*)]]-3-[(Acetyloxy)methyl]-7-[(aminophenylacetyl)amino]-8-oxo-5-thia-1-aza-bicyclo[4.2.0]oct-2-ene-2-carboxylic acid; 7-(2-amino-2-phenylacetamido)-3-(hydroxymethyl)-8-oxo-5-thia-1-azabicyclo-[4.2.0]oct-2-ene-2-carboxylic acid acetate (ester);* 7-(D-2-amino-2-phenylacetamido)-3-acetoxymethyl-Δ³-cephem-4-carboxylic acid; 7-(D-α-aminophenylacetamido) cephalosporanic acid; Kefglycin. $C_{18}H_{19}N_3O_6S$; mol wt 405.43. C 53.33%, H 4.72%, N 10.36%, O 23.68%, S 7.91%. Semisynthetic cephalosporin antibiotic. Prepn: **Belg.** pat. **635,137;** F. W. Wall *et al.,* U.S. pat. **3,422,103** (1964, 1969 both to Glaxo); **Brit.** pat. **1,017,624** (1966 to Merck & Co.); Spencer *et al., J. Med. Chem.* **9**, 746 (1966); of D- and DL-forms: **Brit.** pat. **985,747** (1965 to Lilly); Kurita *et al., J. Antibiot.* **19A**, 243 (1966). Metabolism and pharmacokinetics in man: J. Haginaka *et al., J. Antibiot.* **33**, 236 (1980).

Dihydrate, *Kafocin.* Crystalline powder, mp 223-250° (dec). uv max (2% DMF): 258 nm ($E_{1cm}^{1\%}$ 166). Isoelectric point and pH of maximum stability: 4.5.

THERAP CAT: Antibacterial.

2024. Cephalonic Acid. *8-Hydroxy-5-oxoophiobola-3,-6,19-trien-25-oic acid;* ophiobolin D. $C_{25}H_{36}O_4$; mol wt 400.56. C 74.96%, H 9.06%, O 15.98%. Minor antibiotic metabolite produced by *Cephalosporium caerulens.* Exhibits weak activity against *Staphylococcus aureus.* Isolation, structure elucidation and physical data: Itai *et al., Tetrahedron Letters* **1967**, 4111; Nozoe *et al., ibid.* 4113. Crystal structure: Itai *et al., Acta Cryst.* **25B**, 872 (1969). Synthetic studies on related compounds: P. C. Dutta, T. K. Das, *Syn. Commun.* **6**, 253 (1976); W. G. Dauben, D. S. Hart, *J. Org. Chem.* **42**, 922 (1977); R. K. Boeckman *et al., ibid.* 3630.

Solid, mp 139°. $[\alpha]_D$ +76.2° (c = 0.54 in chloroform). uv max: 259 nm (ε 11700).

2025. Cephaloridine. *(6R-trans)-1-[[2-Carboxy-8-oxo-7-[(2-thienylacetyl)amino]-5-thia-1-azabicyclo[4.2.0]oct-2-en-3-yl]methyl]pyridinium inner salt;* 1-[(7'-β-[2-(2-thienyl)acet-amido]-8'-oxo-1'-aza-5'-thiabicyclo[4.2.0]oct-2'-en-3'-yl)-methyl]pyridinium-2'-carboxylate; *N*-[7-[(2-thienyl)acet-amido]ceph-3-em-3-ylmethyl]pyridinium-4-carboxylate; *N*-[7-(2'-thienylacetamidoceph-3-ylmethyl)]pyridinium-2-carb-oxylate; cefaloridin; Ceflorin; Cepaloridin; Cer; Faredina; Ceporan; Ceporin; Cilifor; Intrasporin; Keflodin; Keflordin (obsolete); Lloncefal; Sefacin; Cepalorin; Deflorin; Kefspor; Loridine; Ampligram; Floridin. $C_{19}H_{17}N_3O_4S_2$; mol wt 415.49. C 54.92%, H 4.12%, N 10.11%, O 15.40%, S 15.44%. Semi-synthetic cephalosporin antibiotic. Prepn: **Fr.** pat. **1,384,197** (1965 to Glaxo), *C.A.* **63**, 11591c (1965); E. H. Flynn, U.S. pat. **3,449,338** (1969 to Lilly). Toxicity studies: Atkinson *et al., Toxicol. Appl. Pharmacol.* **8**, 398 (1966).

Crystals. Soluble in water. Aq solns are slightly acid (pH 4.5-5); $[\alpha]_D$ +47.7° (c = 1.25 in water). uv max: 239, 252 nm (ε 15160, 13950). pKa 3.2. Aq solns (20% w/v) are stable for four weeks at 4° in the dark. Exposure to sunlight causes discoloration. LD_{50} mice, rats (g/kg): > 15, 2.5-4 orally; in monkeys (g/kg): > 0.2 i.m. (Atkinson).

THERAP CAT: Antibacterial.

THERAP CAT (VET): Antibacterial.

2026. Cephalosporin C. *[6R-[6α,7β(R*)]]-3-[(Acetyloxy)methyl]-7-[(5-amino-5-carboxy-1-oxopentyl)amino]-8-oxo-5-thia-1-azabicyclo[4.2.0]oct-2-ene-2-carboxylic acid; 7-(D-5-amino-5-carboxyvaleramido)-3-(hydroxymethyl)-8-oxo-5-thia-1-azabicyclo[4.2.0]oct-2-ene-2-carboxylic acid acetate.* $C_{16}H_{21}N_3O_8S$; mol wt 415.42. C 46.26%, H 5.10%, N 10.12%, O 30.81%, S 7.72%. Antibiotic substance produced along with penicillin N by a *Cephalosporium* sp. cultivated from sea water near a sewage outlet on the coast of Sardinia: G. Brotzu, *Lav. Ist. Igiene Cagliari* (1948); Newton, Abraham, *Nature* **175**, 548 (1955); *eidem, Biochem. J.* **62**, 651 (1956); *eidem,* **Ger.** pat. **1,014,711** (1957 to Natl. Res. Dev. Corp.); **Brit.** pat. **810,196** (1959); Kelly *et al.,* U.S. pat. **3,082,155** (1963 to Natl. Res. Dev. Corp.); Demain, U.S. pats. **3,116,216/7** (1963 to Merck & Co.). Biosynthesis: Demain, Newkirk, *Appl. Microbiol.* **10**, 321 (1962). Has the same side chain as penicillin N, but the nucleus is different and is named 7-aminocephalosporanic acid: Jago, Heatley, *Brit. J. Pharmacol.* **16**, 170 (1961). Structure: Abraham, Newton, *Biochem. J.* **79**, 377 (1961); Hodgkin, Maslen, *ibid.* 393. In *cephalosporin C_4* the acetoxy group is replaced by a pyridinium group: Hale *et al., ibid.* 403. Transformation to *cephalosporin C_C* for which greater activity is claimed: Abraham *et al.,* U.S. pat. **3,049,541** (1962 to Natl. Res. Dev. Corp.). Total synthesis: Woodward *et al., J. Am. Chem. Soc.* **88**, 852 (1966). Review of cephalosporin C and related compounds: Abraham, *Quart. Rev. Chem. Soc.* **21**, 231 (1967); Abraham, Loder, in *Cephalosporins and Penicillins,* E. H. Flynn, Ed. (Academic Press, New York, 1972) pp 1-26.

Sodium salt, dihydrate, $C_{16}H_{20}N_3NaO_8S.2H_2O$, monoclinic crystals from dil alc. $[\alpha]_D^{16}$ +103° (H_2O). uv max: 260 nm ($E_{1cm}^{1\%}$ 200). pKa values: <2.6, 3.1, 9.8. Sol in water. Practically insol in ethanol, ether. Aq solns are stable at pH 2.5 to 8. Low toxicity. Shows *in vitro* activity against penicillin-resistant staphylococci.

THERAP CAT: Antibacterial.

2027. Cephalosporin P$_1$. *(3α,4α,6α,7β,8α,9β,13α,14β,-16β,17Z)-6,16-Bis(acetyloxy)-3,7-dihydroxy-29-nordammara-17(20),24-dien-21-oic acid.* $C_{33}H_{50}O_8$; mol wt 574.76. C 68.96%, H 8.77%, O 22.27%. Antibiotic substance produced by a *Cephalosporium* sp. cultivated from the sea near Sardinia. Crude cephalosporin P contains at least 5 components: P$_1$, P$_2$, P$_3$, P$_4$, P$_5$, the major active substance being cephalosporin P$_1$. Isoln: Burton, Abraham, *Biochem. J.* **50**, 168 (1951). Structure: Baird *et al., Proc. Chem. Soc.* **1961**, 257; Halsall *et al., ibid.* **1963**, 16; Melera, *Experientia* **19**, 565 (1963); Halsall *et al., Chem. Commun.* **1966**, 685.

Orthorhombic crystals from 50% aq ethanol, $C_{33}H_{50}O_8$·H_2O.½C_2H_5OH. mp 147°. $[\alpha]_D^{20}$ +28° (c = 2.7 in chloroform). uv max: 211 nm (ϵ 9140). Readily sol in chloroform and ethanol; sparingly sol in hexane and water.

2028. Cephalothin. *(6R-trans)-3-[(Acetyloxy)methyl]-8-oxo-7-[(2-thienylacetyl)amino]-5-thia-1-azabicyclo[4.2.0]oct-2-ene-2-carboxylic acid; 3-(hydroxymethyl)-8-oxo-7-[2-(2-thienyl)acetamido]-5-thia-1-azabicyclo[4.2.0]oct-2-ene-2-carboxylic acid acetate; 7-(2-thienylacetamido)cephalosporanic acid; 7-(thiophene-2-acetamido)cephalosporanic acid.* $C_{16}H_{16}N_2O_6S_2$; mol wt 396.44. C 48.47%, H 4.07%, N 7.07%, O 24.21%, S 16.18%. Semi-synthetic cephalosporin antibiotic. Prepn: Chauvette *et al., J. Am. Chem. Soc.* **84**, 3401 (1962); E. H. Flynn, U.S. pat. 3,218,318 (1965 to Lilly); Fr. pat. 1,384,197 (1965 to Glaxo), *C.A.* **63**, 11591c (1965). Bacteriology and pharmacology: Lee, Anderson, *Antimicrob. Ag. Chemother.* **1962**, 695; Walters *et al., ibid.* 706; Naumann, *Arzneimittel-Forsch.* **16**, 1099 (1966). Acute toxicity: M. Kuramoto *et al., Japan. J. Antibiot.* **27**, 746 (1974), *C.A.* **83**, 71972t(1975). Comprehensive description: R. J. Simmons in *Analytical Profiles of Drug Substances* vol. 1, K. Florey, Ed. (Academic Press, New York, 1972) pp 319-341.

mp 160-160.5°. $[\alpha]_D^{20}$ +50° (c = 1.03 in acetonitrile), *see* R. B. Woodward *et al., J. Am. Chem. Soc.* **88**, 852 (1966).

Sodium salt, $C_{16}H_{15}N_2NaO_6S_2$, *Averon-1, Cefalotin, Cemastin, Cephation, Ceporacin, Cepovenin, Chephalotin, Coaxin, Keflin, Lospoven, Microtin, Synclotin, Toricelocin.* mp 204-205°. $[\alpha]_D$ +135° (c = 1.0 in water). uv max: 235, 260 nm (ϵ 12950, 9350). LD$_{50}$ in mice, rats (mg/kg): >20000, >10000 orally; 5670, 7716 i.p. (Kuramoto).

THERAP CAT: Antibacterial.

THERAP CAT (VET): Antibacterial.

2029. Cephamycins. A family of β-lactam antibiotics produced by various *Streptomyces* species. Detection and production: Stapley *et al., Antimicrob. Ag. Chemother.* **2**, 122 (1972). Chemical characterization: Miller *et al., ibid.* 132. Cephamycins A, B, and C have been isolated, the latter being identical to a *Streptomyces clavuligerus* metabolite: Nagarajan *et al., J. Am. Chem. Soc.* **93**, 2308 (1971). Structures: Albers-Schoenberg *et al., Tetrahedron Letters* **1972**, 2911. Antibacterial activity studies: Miller *et al., Antimicrob. Ag. Chemother.* **2**, 281, 287 (1972); Daoust *et al., ibid.* **3**, 254 (1973). Review of syntheses: T. Hiraoka *et al., Heterocycles* **8**, 719 (1977).

Cephamycin C

2030. Cephapirin Sodium. *(6R-trans)-3-[(Acetyloxy)-methyl]-8-oxo-7-[[(4-pyridinylthio)acetyl]amino]-5-thia-1-azabicyclo[4.2.0]oct-2-ene-2-carboxylic acid monosodium salt; 3-(hydroxymethyl)-8-oxo-7-[2-(4-pyridylthio)acetamido]-5-thia-1-azabicyclo[4.2.0]oct-2-ene-2-carboxylic acid acetate monosodium salt; 7-[α-(4-pyridylthio)acetamido]cephalosporanic acid sodium salt; sodium 7-(pyrid-4-ylthioacetamido)cephalosporanate; BL-P-1322; Ambrocef; Brisfirina; Bristocef; Cefadyl; Cefa-Lak; Cefatrexyl; Piricef; ToDay.* $C_{17}H_{16}N_3NaO_6S_2$; mol wt 445.45. C 45.84%, H 3.62%, N 9.43%, Na 5.16%, O 21.55%, S 14.40%. A cephalosporin C antibiotic effective against gram-pos. and gram-neg. bacteria including *Staphylococcus aureus, Escherichia coli, Klebsiella pneumoniae* and *Proteus mirabilis.* Prepn: L. B. Crast, S. Afr. pat. 67 07783; *eidem*, U.S. pat. 3,422,100 (1968, 1970 both to Bristol-Myers). Alternate processes: H. H. Silvestri, D. A. Johnson, U.S. pat. 3,503,967; R. E. Havranek, L. B. Crast, U.S. pat. 3,578,661 (1970, 1971 both to Bristol-Myers); L. B. Crast *et al., J. Med. Chem.* **16**, 1413 (1973). Activity studies: Chisholm *et al., Antimicrob. Ag. Chemother.* **1969**, 244; Axelrod *et al., Appl. Microbiol.* **22**, 904 (1971); Bran *et al., Antimicrob. Ag. Chemother.* **1**, 35 (1972); Wiesner *et al., ibid.* 303.

White crystalline powder, sol in water.

THERAP CAT: Antibacterial.

THERAP CAT (VET): Antibacterial.

2031. Cepharanthine. *6′,12′-Dimethoxy-2,2′-dimethyl-6,7-[methylenebis(oxy)]oxyacanthan.* $C_{37}H_{38}N_2O_6$; mol wt 606.72. C 73.25%, H 6.31%, N 4.62%, O 15.82%. From tubers of *Stephania cephalantha* Hayata, and *Stephania sasahii* Hayata, *Menispermaceae:* Kondo *et al.,* U.S. pat. 2,206,-407 (1940); Kondo, Hasegawa, U.S. pat. 2,248,241 (1941). Structure: Kondo, Keimatsu, *Ber.* **71**, 2553 (1938); Tomita, Sasaki, *Pharm. Bull. (Japan)* **2**, 89, 375 (1954). Total synthesis (*dl*-form): Tomita *et al., Tetrahedron Letters* **1967**, 1201.

Yellow powder, mp 145-155°. Obtained by drying solvated needles from acetone + benzene. $[\alpha]_D^{20}$ +277° (c = 2 in chloroform). Soluble in the usual organic solvents except petr ether.

2032. Cephradine. *[6R-[6α,7β(R*)]]-7-[(Amino-1,4-cyclohexadien-1-ylacetyl)amino]-3-methyl-8-oxo-5-thia-1-azabicyclo[4.2.0]oct-2-ene-2-carboxylic acid;* 7-[D-2-amino-2-(1,4-cyclohexadienyl)acetamido]desacetoxycephalosporanic acid; cefradin; SQ-11436; Anspor; Cefradex; Cefrag; Cefro; Celex; Cesporan; Dimacef; Ecosporina; Eskacef; Lenzacef; Lisacef; Megacef; Samedrin; Sefril; Velocef; Velosef. $C_{16}H_{19}$-N_3O_4S; mol wt 349.41. C 55.00%, H 5.48%, N 12.03%, O 18.32%, S 9.18%. A semisynthetic cephalosporin antibiotic. Prepn and activity data: Weisenborn *et al.*, U.S. pat. 3,485,-819 (1969 to Squibb); Dolfini *et al.*, *J. Med. Chem.* **14,** 117 (1971). Clinical trials: Limson *et al.*, *Curr. Ther. Res.* **14,** 101 (1972); Landa, *ibid.* 496. Comprehensive description: K. Florey, *Anal. Profiles Drug Subs.* **5,** 21-59 (1976). Review: W. E. Brown, J. R. Knill, in *Pharmacological and Biochemical Properties of Drug Substances* vol. 2, M. E. Goldberg, Ed. (Am. Pharm. Assoc., Washington, DC, 1979) pp 279-304.

Monohydrate, *Forticef.* Small, colorless crystals. mp 140-142° (dec). pK_1 2.63; pK_2 7.27. Sol in propylene glycol; slightly sol in acetone, ethanol. Insol in ether, chloroform, benzene, hexane.
Dihydrate, *SQ-22022.* mp 183-185°.
THERAP CAT: Antibacterial.

2033. Ceresin. Purified ozokerite; earth wax; mineral wax; cerosin; cerin. A mixture of hydrocarbons of complex composition purified by treatment with concd H_2SO_4 and filtration through bone-black. Found in Ukraine, Lake Baikal, also in Utah, Texas.
White or yellow, tasteless, waxy cakes. The white is odorless, the yellow has a slight odor. Fracture is very much like that of white wax. d 0.91-0.92. mp 61-78°. It is very stable toward oxidizing agents. Insol in water; sol in 30 parts abs alcohol; sol in benzene, chloroform, petr ether, hot oils.
USE: Substitute for beeswax; for making candles, wax figures; for waxed paper and cloth; in polishes, electrical insulators; waterproofing fabrics; for bottles for hydrofluoric acid; in dentistry for impression and inlay waxes and modeling compounds.

2034. Ceric Fluoride. CeF_4; mol wt 216.11. Ce 64.84%, F 35.16%. Prepd from CeF_3 and F_2: Klemm, Henkel, *Z. Anorg. Allgem. Chem.* **220,** 181 (1934); von Wartenberg, *ibid.* **244,** 343 (1940); Kwasnik in *Handbook of Preparative Inorganic Chemistry,* vol. 1, G. Brauer, Ed. (Academic Press, New York, 2nd ed., 1963) pp 247-248;

from CeO_2 and F_2: Asker, Wylie, *Aust. J. Chem.* **18,** 959 (1965).
Minute crystals, d 4.77. mp > 650°. Practically insol in water; very slowly hydrolyzed by cold water. Thermally stable below 550°. May be reduced to CeF_3 by H_2 at 300°.
Monohydrate, white powder. Sol in acids.
USE: Fluorinating agent.

2035. Ceric Oxide. Ceria. CeO_2; mol wt 172.11. Ce 81.41%, O 18.59%. Usually prepd by ignition of Ce salts Duval, *Anal. Chim. Acta* **1,** 341 (1947); other prepns: Warf, U.S. pat. 2,564,241 (1951 to USAEC); Wilke, *Z. Anorg. Allgem. Chem.* **330,** 164 (1964). Purification: Wetzel in *Handbook of Preparative Inorganic Chemistry,* vol. 2, G. Brauer, Ed. (Academic Press, New York, 2nd ed., 1965) pp 1132-1135. *Review:* Davis, Wayman, *Can. Chem. Process Ind.* **29,** 230 (1945).
Heavy powder or cubic crystals; white if pure, but usually pale yellow; traces of lanthanum, praseodymium, etc., give a pink to reddish-brown color. Practically insol in water, acids. LD_{50} orally in rats: ~1 g/kg, *Toxic Substances List,* H. E. Christensen, Ed. (1972) p 129.
USE: Polishing and decolorizing glass; as opacifier for vitreous enamels; in heat-resistant alloy coatings; in coatings for infrared filters to prevent reflection; in analysis of Ce and oxidimetry; as catalyst for organic reactions.

2036. Ceric Sulfate. CeO_8S_2; mol wt 332.24. Ce 42.17%, O 38.52%, S 19.30%. $Ce(SO_4)_2$. Prepd by heating CeO_2 with concd H_2SO_4: Vanino, *Handbuch der Präparativen Chemie,* vol. 1 (Stuttgart, 3rd ed., 1925) p 755. *Review: Smith, Ceric Sulfate* (G. Frederick Smith Chem. Co., Columbus, Ohio, 2nd ed., 1935) 51 pp. The commercial salt usually contains small amounts of associated rare earths.
Tetrahydrate, yellow to orange powder or orthorhombic crystals. On heating loses water, becoming anhyd at 180-200°; dec above 350°, forming $CeOSO_4$. Sol in a small quantity of water, but dec with much water with separation of a basic salt. Slowly sol in cold, more readily sol in hot mineral acids; sol in dil H_2SO_4.
USE: Oxidizing agent; analytical reagent; in radiation dosimeters.

2037. Cerium. Ce; at. wt 140.115; at. no. 58; valences 3, 4. A rare earth metal, most abundant member of the lanthanide series. Naturally occurring isotopes (mass numbers): 140 (88.48%); 142 (11.08%), radioactive, $T_{1/2}$ > 5 × 10^{16} years; 138 (0.25%); 136 (0.19%). Known artificial radioactive isotopes: 124-135; 137; 139; 141; 143-152. Estimated abundance in the earth's crust: 46-66 ppm. Commercially important sources are the rare earth minerals monazite and bastnaesite; also found in cerite. Discovered by Klaproth, Hisinger and Berzelius in 1803. Can be separated from other rare earths by selective precipitation of ceric (4+) salts from buffered solns (pH 3-4); also by ion exchange techniques. Prepn of metal: Spedding *et al., Ind. Eng. Chem.* **44,** 553 (1952). Review of prepn, properties and compds: *The Rare Earths,* F. H. Spedding, A. H. Daane, Eds. (Krieger, Huntington, N.Y., 1971, reprint of 1961 ed.) 641 pp; Hulet, Bode, "Separation Chemistry of the Lanthanides and Transplutonium Actinides" in *MTP Int. Rev. Sci.: Inorg. Chem., Ser. One,* vol. 7, K. W. Bagnall, Ed. (University Park Press, Baltimore, 1972) pp 1-45; Moeller, "The Lanthanides" in *Comprehensive Inorganic Chemistry,* vol. 4, J. C. Bailar, Jr. *et al.,* Eds. (Pergamon Press, Oxford, 1973) pp 1-101; B. T. Kilbourn in Kirk-Othmer *Encyclopedia of Chemical Technology* vol. 5 (John Wiley & Sons, New York, 4th ed., 1993) pp 728-749; *Chemistry of the Elements,* N. N. Greenwood, A. Earnshaw, Eds. (Pergamon Press, New York, 1984) pp 1423-1449. Brief review of properties: G. T. Seaborg, *Radiochim. Acta* **61,** 115-122 (1993).
Iron-gray, ductile, malleable metal. Only material known to have a solid-solid critical point. Crystalline forms: face-centered cubic α-form, d 6.770, transforms to β-form at −150°; hexagonal β-form transforms to γ-form at −10°; face-centered cubic γ-form transforms to δ-form at 730°; body-centered cubic δ-form exists at > 730°. mp 798°. bp 3433°. Heat of fusion: 5.179 kJ/mol. Heat of sublimation (25°): 422.6 kJ/mol. E^0(aq) Ce^{3+}/Ce −2.48 V (calc). Stable in dry air, but superficially oxidized in moist air; when

finely divided may ignite spontaneously. Slowly dec by cold, rapidly by hot water; sol in dil mineral acids. Ceric salts usually are yellow to orange-red in color and liberate iodine from KI. Cerous salts are usually white and give a white ppt with alkali hydroxides or sulfides, insol in excess of reagent; they also are pptd by ammonium oxalate from cold dil acid sols.

Mischmetall, composed of ~50% Ce, 25% La, 25% other light lanthanides; formed by electrolytic reduction of fused earth chlorides.

USE: In metallurgy as stabilizers in alloys and as an alternative to thorium oxide in welding electrodes. In glass as polishing agent, decolorizer to stabilize impurities, to render glass opaque to near uv radiation, to resist discoloration from strong light or high energy electron bombardment (as in television screens). In ceramics as an opacifying and strengthening agent. Catalysts to impart high cracking activity for crude oil processing, in automotive exhaust control devices, as combustion additive, polymerization initiator, paint drier, polymer stabilizer. As phosphor in fluorescent lamps, cathode ray tubes and thorium dioxide gas mantles.

2038. Ceronapril. *(S)-1-[6-Amino-2-[[hydroxy(4-phenylbutyl)phosphinyl]oxy]-1-oxohexyl]-L-proline;* 1-[(2S)-6-amino-2-hydroxyhexanoyl]-L-proline, hydrogen (4-phenylbutyl)phosphonate (ester); ceranapril; SQ-29852. $C_{21}H_{33}N_2O_6P$; mol wt 440.48. C 57.26%, H 7.55%, N 6.36%, O 21.79%, P 7.03%. Hydroxylphosphonate angiotensin converting enzyme (ACE) inhibitor. Prepn: D. S. Karanewsky, E. W. Petrillo, Jr., **Eur. pat. Appl. 97,534;** *eidem,* **U.S. pat. 4,452,790** (both 1984 to Squibb); *eidem et al., J. Med. Chem.* **31,** 204 (1988). HPLC determn in body fluids: H. Kadin *et al., J. Chromatog.* **487,** 135 (1989). Pharmacology: J. M. DeForrest *et al., J. Cardiovasc. Pharmacol.* **16,** 121 (1990). Radioimmunoassay and pharmacokinetics: J. Tu *et al., Ther. Drug Monit.* **14,** 209 (1992). Clinical evaluation in hypertension: J. Sasaki *et al., Int. J. Clin. Pharmacol. Ther. Toxicol.* **31,** 83 (1993).

Prepd as the monohydrate. Crystals from water, mp 190-195° (dec). $[\alpha]_D$ −47.5° (c = 1 in methanol).
THERAP CAT: Antihypertensive.

2039. Cerous Bromide. Br_3Ce; mol wt 379.83. Br 63.11%, Ce 36.89%. $CeBr_3$. Prepn: Jantsch, Wein, *Monatsh.* **69,** 161 (1936).

Heptahydrate, colorless, deliquesc needles. mp 732°. Sol in water, alcohol. *Keep well closed.*

2040. Cerous Carbonate. $C_3Ce_2O_9$; mol wt 460.26. C 7.83%, Ce 60.89%, O 31.29%. $Ce_2(CO_3)_3$. Prepn: Head, Holley, Jr., *USAEC* **LADC-5579,** 8 pp (1962).

Pentahydrate, powder, or microcryst prisms. Almost insol in water; sol in dil mineral acids and in soln of ammonium salts.

2041. Cerous Chloride. $CeCl_3$; mol wt 246.47. Ce 56.85%, Cl 43.15%. Prepn: Kleinheksel, Kremers, *J. Am. Chem. Soc.* **50,** 959 (1928); Didtschenko, U.S. pat. **2,932,-553** (1960 to Union Carbide); Harmon, Wickers, U.S. pat. **2,982,603** (1961 to USAEC); Taylor, Carter, *J. Inorg. Nucl. Chem.* **24,** 387 (1962). Toxicity study: Graca *et al., Arch. Environ. Health* **5,** 437 (1962).

Very fine powder, mp 822°. d^{25} 3.97. Sol in water, alcohol (exothermic). LD_{50} in mice, guinea pigs (mg/kg): 352, 110 i.p. (Graca).

Heptahydrate, colorless to yellow deliquesc orthorhombic crystals. Begins to lose water above 90°, becomes anhydr by 230°. Very sol in H_2O, alcohol. *Keep well closed.*

USE: In manuf of Ce and Ce salts; with Al or Mg salts as catalyst in the polymerization of olefins.

2042. Cerous Fluoride. CeF_3; mol wt 197.11. Ce 71.08%, F 28.92%. Prepd by the action of HF on CeO_2: von Wartenberg, *Z. Anorg. Allgem. Chem.* **244,** 343 (1940); by the action of F_2 or concd HF on $CeSi_2$: Sterba, *Ann. Chim. Phys.* [8] **2,** 193 (1904); from CCl_2F_2 and CeO_2: Pausewang, Rüdorff, *Z. Anorg. Allgem. Chem.* **369,** 89 (1969).

Hexagonal crystals or powder. d 6.157. mp 1460°. Practically insol in, but slowly hydrolyzed by water.

2043. Cerous Iodide. CeI_3; mol wt 520.83. Ce 26.90%, I 73.10%. Prepn from Ce and HgI_2: Asprey *et al., Inorg. Chem.* **3,** 1137 (1964); from the oxide and HI in the presence of NH_4I: Taylor, Curtis, *J. Inorg. Nucl. Chem.* **24,** 387 (1962).

Bright yellow, orthorhombic crystals. mp 752°. Dec in moist air. Sol in water.

Nonahydrate, white to reddish-white crystals. Very sol in water, the soln readily decomposing with separation of iodine; sol in alc. *Keep well closed and protected from light.*

2044. Cerous Nitrate. CeN_3O_9; mol wt 326.13. Ce 42.96%, N 12.88%, O 44.15%. $Ce(NO_3)_3$. Prepn from Ce sulfate and $Ca(NO_3)_2$: Blumenfeld, U.S. pat. **2,166,702** (1939 to Soc. de Produits Chimiques des Terres Rares); in separation of cerium from other rare earths: Pierce, Butler, *Inorg. Syn.* **2,** 51 (1946); Scargill *et al., J. Inorg. Nucl. Chem.* **4,** 304 (1957). For physical and chemical properties *see USAEC IDO-14504,* 126 pp (1961). Toxicity study: Bruce *et al., Toxicol. Appl. Pharmacol.* **5,** 750 (1963).

Hexahydrate. LD_{50} in rats: 290 mg/kg i.p.; 4.2 g/kg orally (Bruce).

USE: Separation of Ce from other rare earths; catalyst in hydrolysis of phosphoric acid esters; mixture with $Th(NO_3)_4$ (1:99) formerly used in manuf incandescent mantles.

2045. Cerous Oxalate. Sedemesis. $C_6Ce_2O_{12}$; mol wt 544.29. C 13.24%, Ce 51.49%, O 35.27%. $Ce_2(C_2O_4)_3$. Prepn from $Ce(NO_3)_3.6H_2O$ and oxalic acid: Wylie, *J. Chem. Soc.* **1947,** 1687. The article of commerce contains variable amounts of the oxalates of cerium, lanthanum, and other associated elements.

Nonahydrate, white or slightly pink, tasteless powder. Practically insol in water; sol in hot, moderately dil H_2SO_4 or HCl.

THERAP CAT: Antiemetic.

2046. Cerous Sulfate. $Ce_2O_{12}S_3$; mol wt 568.42. Ce 49.30%, O 33.78%, S 16.92%. $Ce_2(SO_4)_3$. Prepn: Blandin, Rerat, *Compt. Rend.* **239,** 1055 (1954); Wendlandt, *J. Inorg. Nucl. Chem.* **7,** 51 (1958).

Octahydrate, orthorhombic octahedral crystals. d 2.87. Loses water on heating, becoming anhydr by 250°; loses SO_3 to form a basic salt above 650°. Soluble in water.

USE: Developing aniline black; said to be superior to vanadium for this purpose.

2047. Cerulenin. *[2R-[2α,3α(4E,7E)]]-3-(1-Oxo-4,7-nonadienyl)oxiranecarboxamide; 2,3-epoxy-4-oxo-7,10-dodecadienamide;* 2,3-epoxy-4-oxo-7,10-dodecadienoylamide; helicocerin. $C_{12}H_{17}NO_3$; mol wt 223.27. C 64.55%, H 7.67%, N 6.27%, O 21.50%. Antifungal antibiotic isolated from *Cephalosporium caerulens; Acryocylindrium oryzae; Helicoceras oryzae:* T. Hata *et al., Japan. J. Bacteriol.* **15,** 1075 (1960); Matsumae *et al., J. Antibiot.* **16A,** 236, 239 (1963); Sano *et al., ibid.* **20A,** 344 (1967); Furuya, Shirasaka, **Japan.** pat. 21,638('70) (to Sankyo), *C.A.* **73,** 108271k (1970). Biological characteristics: Matsumae *et al., J. Antibiot.* **17A,** 1 (1964). Structure: Omura *et al., ibid.* **20A,** 349 (1967). Abs config: *eidem, Chem. Pharm. Bull.* **17,** 2361 (1969); Arison, Omura, *J. Antibiot.* **27,** 28 (1974). Stereoselective synthesis of (+)- and (−)-*tetrahydrocerulenin* and corrected abs config of (+)-cerulenin: H. Ohrui, S. Emoto, *Tetrahedron Letters* **1978,** 2095; J.-R. Pougny, P. Sinay, *ibid.* 3301. Stereoselective synthesis of the (+)-form from D-glucose: N. Sueda *et al., ibid.* **1979,** 2039; M. Pietraszkiewicz, P. Sinay, *ibid.* 4741. Interrupts yeast-type fungi growth by inhibiting the biosynthesis of sterols and fatty acids. Mechanism of action studies: Nomura *et al., J. Biochem.* **71,** 783 (1972); *eidem, J. Antibiot.* **25,** 365 (1972); *see also* D. Vance, *Biochem. Biophys. Res. Commun.* **48,** 649

(1972). Total synthesis of (±)-cerulenin: R. K. Boeckman, Jr., E.W. Thomas, *J. Am. Chem. Soc.* **99**, 2805 (1977); A. A. Jakubowski *et al.*, *Tetrahedron Letters* **1977**, 2399; E. J. Corey, D. R. Williams, *ibid.* 3847; K. Mikami *et al.*, *Chem. Letters* **1981**, 1721; A. A. Jakubowski *et al.*, *J. Org. Chem.* **47**, 1221 (1982).

White needles from benzene, mp 93-94°. bp 120° (10^{-8} mm). $[\alpha]_D^{16}$ +63° (c = 2 in methanol). Stable in neutral and acidic solns. Sol in ethanol, acetone, benzene and most common solvents. Slightly sol in water. Practically insol in petr ether. LD$_{50}$ in mice (mg/kg): 154 i.v.; 211 i.p.; 547 orally (Matsumae 1964).
DL-Form, mp 40-43°.
USE: Biochemical tool.

2048. Ceruletide. *Caerulein;* cerulein; FI-6934. C$_{58}$H$_{73}$N$_{13}$O$_{21}$S$_2$; mol wt 1352.42. C 51.51%, H 5.44%, N 13.46%, O 24.84%, S 4.74%. Cholecystokinin analog; decapeptide discovered in the skins of Australian amphibians: V. Erspamer *et al.*, *Nature* **212**, 204 (1966). Isoln from *Hyla caerulea* and structure: A. Anastasi *et al.*, *Experientia* **23**, 699 (1967); *eidem, Arch. Biochem. Biophys.* **125**, 57 (1968). Synthesis: L. Bernardi *et al.*, *Experientia* **23**, 700 (1967); *eidem,* S. Afr. pat. **67** 04,716; *eidem,* U.S. pat. **3,472,832** (1968, 1969 to Soc. Farm. Italia). Shows hypotensive activity, stimulates smooth muscle and increases gastric, pancreatic, and biliary secretions: V. Erspamer *et al.*, *Experientia* **23**, 702 (1967). Series of articles on metabolism: *Eur. J. Biochem.* **91**, 21-48 (1978). Pharmacodynamics: N. Iwatsuki, O. H. Petersen, *J. Cell. Biol.* **79**, 533 (1978); T. Fujita *et al.*, *Adv. Exp. Med. Biol.* **106**, 147 (1978). Diagnostic use: R. Fujita *et al.*, *Acta Gastroenterol. Belg.* **40**, 167 (1977); C. Monti *et al.*, *Radiology* **129**, 611 (1978). Toxicology: T. Chieli *et al.*, *Toxicol. Appl. Pharmacol.* **23**, 480 (1972). Reviews of pharmacology and clinical applications: G. Bertaccini, *Pharmacol. Rev.* **28**, 127 (1976); and toxicology: M. E. Vincent *et al.*, *Pharmacotherapy* **2**, 223 (1982).

5-oxoPro–Gln–Asp–Tyr–Thr–Gly–Trp–Met–Asp–PheNH$_2$
             |
            SO$_3$H

mp 224-226° (dec). $[\alpha]_D^{20}$ −26° (c = 1 in DMF). Diethylamine salt, *Ceosunin, Takus, Tymtran*. Off-white hygroscopic powder. Sol in DMF and DMSO. Insol in acetone, diethyl ether. LD$_{50}$ i.v. in mice: 1012 mg/kg (Chieli).
THERAP CAT: Stimulant (gastric secretory). Diagnostic aid (pancreatic function); cholecystokinetic agent in cholecystography).

2049. Ceruloplasmin. Caeruloplasmin; ferroxidase. Intensely blue colored copper-containing glycoprotein of the α_2-globulin fraction of mammalian blood; it is the principal copper transport protein and is believed to play an important role in iron mobilization via its ferroxidase activity. First reported by C. G. Holmberg, *Acta Physiol. Scand.* **8**, 227 (1944). Isoln, purification and description of properties of porcine and human ceruloplasmin: C. G. Holmberg, C. B. Laurell, *Acta Chem. Scand.* **2**, 550 (1948). Ceruloplasmin accounts for 90-95% of the circulating copper in normal mammals. Its concentration increases by a factor of 2 to 3 during pregnancy and varies significantly in several diseases and hormonal conditions. Prepd by Cohn cold ethanol fractionation: Cohn *et al.*, *J. Am. Chem. Soc.* **68**, 459 (1946); Steinbuch, Quentin, *Nature* **183**, 323 (1959); and further purified from fraction IV: Sanders *et al.*, *Arch. Biochem. Biophys.* **84**, 60 (1959); U.S. pat. **3,003,918** (1961 to Merck & Co.). Different mol wts have been reported for human ceruloplasmin, ranging from 126,000 to 160,000; the most generally accepted is 134,000 ±3000, cf. L. Ryder, I. Björk, *Biochemistry* **15**, 3411 (1976). Chemical and structural studies of porcine ceruloplasmin: Osaki *et al.*, *J. Biochem. (Tokyo)* **48**, 190 (1960); **50**, 24, 29 (1961); Mukasa *et*

al., *Biochim. Biophys. Acta* **168**, 132 (1968); Matsunaga, Nosoh, *ibid.* **215**, 280 (1970); of human: Kasper, Deutsch, *J. Biol. Chem.* **238**, 2325 (1963); Jamieson, *ibid.* **240**, 2019 (1965); Poillon, Bearn, *Biochim. Biophys. Acta* **127**, 407 (1966); Simons, Bearn, *ibid.* **175**, 260 (1969); Ryden, *Eur. J. Biochem.* **26**, 380 (1972); T. G. Samsonidze *et al.*, *Int. J. Peptide Protein Res.* **14**, 161 (1979); V. N. Zaitsev *et al.*, *Kristallografiya* **25**, 174 (1980), *C.A.* **92**, 210415q (1980). Human metabolism: Kekki *et al.*, *Nature* **209**, 1252 (1966). Reviews of biological role: E. Frieden, H. S. Hsieh, *Adv. Exp. Med. Biol.* **74**, 505-529 (1976); J. M. C. Gutteridge, *Ann. Clin. Chem.* **15**, 293-296 (1978). Comprehensive review: S. H. Laurie, E. S. Mohammed, *Coord. Chem. Rev.* **33**, 279-312 (1980).
Absorption max: 280, 610 nm (E$_{1cm}^{1\%}$ 14.9, 0.68). Electrophoretic mobility (cm/volt/sec): −5.05 × 10^{-5} at pH 7.0 (0.1M phosphate buffer); −5.32 × 10^{-5} at pH 8.6 (0.1M barbital sodium buffer).

2050. Cervicarcin. *1,2,3,4-Tetrahydro-1,3,4,5,10-pentahydroxy-2-methyl-3-[(3-methyloxiranyl)carbonyl]-4a,9a-epoxyanthracen-9(10H)-one; 4a,9a-epoxy-3-(2,3-epoxybutyryl)-1,2,3,4,4a,9a-hexahydro-1,3,4,5,10-pentahydroxy-2-methylanthrone.* C$_{19}$H$_{20}$O$_9$; mol wt 392.36. C 58.16%, H 5.14%, O 36.70%. Antineoplastic antibiotic produced by *Streptomyces ogaensis:* Okuma *et al.*, *J. Antibiot.* **15A**, 152, 247 (1962). Prepn: Sumiki *et al.*, **Japan.** pat. **7400('64)** (to Inst. Phys. & Chem. Res.). Structure: Marumo *et al.*, *J. Am. Chem. Soc.* **86**, 4507 (1964); *eidem, Agr. Biol. Chem.* **32**, 209 (1968). Stereochemistry: *eidem, ibid.* **35**, 1931 (1971). Activity studies: Itakura *et al.*, *J. Antibiot.* **16A**, 231 (1963).

Needles, mp 203-205° (dec). $[\alpha]_D^{20}$ −144° (in ethanol). uv max: 227, 264, 323 nm (ϵ 14200, 7800, 2600). pKa 9.0. Sol in acetone, lower alcohols, acetic acid, pyridine. Moderately sol in ethyl acetate, ethyl ether, chloroform, carbon tetrachloride. Slightly sol in water, benzene. Insol in petr ether, ligroin. Stable in neutral or acidic soln; unstable in alkaline soln. LD$_{50}$ i.p. in mice: 48.5 mg/kg.

2051. Cesium. Caesium. Cs; at. wt 132.9054; at. no. 55; valence 1. Alkali metal. Occurrence in earth's crust: one ppm. One natural isotope: ^{133}Cs; artificial isotopes (mass nos.): 123, 125-132, 134-144. Occurs in nature in the aluminosilicates, *pollucite* and *lepidolite*, and in the borate, *rhodizite*. Discovered by Bunsen and Kirchhoff in 1860. Prepn from pollucite: *Inorg. Syn.* **4**, 5 (1953). ^{137}Cs (T$_{1/2}$ 30 years; β^- 0.514, 1.176 MeV) is an important constituent of radioactive fallout; decays to and reaches radioactive equilibrium with ^{137m}Ba (T$_{1/2}$ 2.554 min; γ 0.662 MeV). Review of cesium and its compounds: Whaley, "Sodium, Potassium, Rubidium, Cesium, and Francium" in *Comprehensive Inorganic Chemistry*, vol. 1, J. C. Bailar, Jr. *et al.*, Eds. (Pergamon Press, Oxford, 1973) pp 369-529; C. T. Williams in Kirk-Othmer *Encyclopedia of Chemical Technology* vol. 5 (Wiley-Interscience, New York, 3rd ed., 1979) pp 327-338.
Silver-white, ductile metal; body-centered cubic structure. Oxidizes rapidly in air; in moist air may ignite spontaneously; colors nonluminous flame reddish-violet. d 1873. mp 28.5°. bp 705°. Sp heat (25°) 0.057 cal/g/°C . Electrical resistivity 36.6 micro-ohm cm at 30°. Reacts with water to form hydroxide with evolution of hydrogen which ignites spontaneously; sol in liquid ammonia. *Keep immersed in mineral oil.*
Caution: Low toxicity in animal expts. Can act as an analog of potassium and thus be potentially harmful. For radioactive ^{137}Cs: max permissible body burden: 30 μcurie; max permissible concn in air: 2 × 10^{-8} μcurie; in water: 2 × 10^{-4} μcurie: *U.S. Bureau of Standards, Handbook 69*, 58 (1959).
USE: In photoelectric cells, as a "getter" in vacuum tubes. The construction and operation of one type of atomic clock

is based on the vibrational frequency (9,192.76 megacy-cles/sec) of the cesium atom. Although ^{60}Co is by far the most commonly used radioisotope for encapsulated energy sources, ^{137}Cs offers competition because as an abundant fission product it is available in large quantities from the reprocessing of fuel elements used in nuclear reactors.

2052. Cesium Bromide. BrCs; mol wt 212.81. Br 37.55%, Cs 62.45%. CsBr. Prepn from pollucite: Thomas, Steton, *Compt. Rend.* **241**, 56 (1955). Toxicity study: Cochran *et al.*, *Arch. Ind. Hyg.* **1**, 637 (1950).
Crystals. d 4.44. mp 636°. bp ~1300°. Very sol in water; sol in alc. Practically insol in acetone. LD$_{50}$ i.p. in rats: 1.4 g/kg (Cochran).
USE: In x-ray fluorescent screens, spectrometer prisms, absorption-cell windows.

2053. Cesium Carbonate. CCs$_2$O$_3$; mol wt 325.82. C 3.69%, Cs 81.58%, O 14.73%. Cs$_2$CO$_3$. Prepn: Suhrmann, Clusius, *Z. Anorg. Allgem. Chem.* **152**, 52 (1926); Thomas, *Ann. Chim.* **6**, 367 (1957); Bernard, *Compt. Rend.* **243**, 1528 (1956).
Very deliquesc crystals; melts at red heat. Extremely sol in water, alcohol; sol in ether.
USE: Catalyst in ethylene oxide polymerization; in coating for spatter-free welding of steel in CO$_2$; in oxide cathode.

2054. Cesium Chloride. ClCs; mol wt 168.36. Cl 21.06%, Cs 78.94%. CsCl. Prepn from carnallite and from pollucite: Donges in *Handbook of Preparative Inorganic Chemistry* vol. **1**, G. Brauer, Ed. (Academic Press, New York, 2nd ed., 1963) pp 951, 955. Toxicity study: Cochran *et al.*, *Arch. Ind. Hyg.* **1**, 637 (1950).
Deliquesc cubic crystals. d 3.99. mp 646°. bp 1303°. Very sol in water; sol in alc. *Keep well closed.* LD$_{50}$ i.p. in rats: 1.5 g/kg (Cochran).
^{131}CsCl, *Cescan-131.*
USE: In the final evacuation of radio and television vacuum tubes; in x-ray fluorescent screens; as radiographic contrast medium; in manuf of cesium.

2055. Cesium Hydroxide. Cesium hydrate. CsHO; mol wt 149.91. Cs 88.66%, H 0.67%, O 10.67%. CsOH. Prepd by electrolysis of Cs salts: Winslow *et al.*, *J. Phys. Colloid Chem.* **51**, 967 (1947); Jolibois, Berger, *Compt. Rend.* **224**, 78 (1947). Alternate prepn: H. Jacobs, B. Harbrecht, *Z. Naturforsch.* **B36**, 270 (1981). Toxicity study: Cochran *et al.*, *Arch. Ind. Hyg.* **1**, 637 (1950).
White or yellowish, fused, very deliquescent, crystalline mass; strongly alkaline reaction. mp 272°. d 3.68. *Corrosive!* Sol in about 0.25 part water, much heat being evolved; sol in alcohol. *Keep tightly closed.* LD$_{50}$ i.p. in rats: 100 mg/kg (Cochran).
USE: In storage-battery electrolytes; as catalyst in the polymerization of cyclic siloxanes.

2056. Cesium Iodide. CsI; mol wt 259.81. Cs 51.15%, I 48.85%. Prepn: *Gmelin's, Cesium* (8th ed.) **25**, pp 188-204 (1938). Toxicity study: Cochran *et al.*, *Arch. Ind. Hyg.* **1**, 637 (1950).
Deliquesc crystals or crystalline powder. d 4.5. mp 621°. bp about 1280°. Very sol in water; sol in ethanol; slightly sol in methanol. Practically insol in acetone. *Keep well closed.* LD$_{50}$ i.p. in rats: 1.4 g/kg (Cochran).
USE: Prisms for infrared spectroscopy; in x-ray fluorescent screens, scintillation counters.

2057. Cesium Nitrate. CsNO$_3$; mol wt 194.91. Cs 68.19%, N 7.19%, O 24.63%. Prepn from pollucite: Watt, *Inorg. Syn.* **4**, 6 (1953). Toxicity study: Cochran *et al.*, *Arch. Ind. Hyg.* **1**, 637 (1950).
White, lustrous hexagonal or cubic prisms; saltpeter taste. d$_4^{20}$ 3.64-3.68. mp 414°; dec at higher temp. Sol in 5 parts cold, 0.5 part boiling water; sol in acetone; very slightly sol in alc. LD$_{50}$ i.p. in rats: 1.2 g/kg (Cochran).
USE: Prepn of other Cs salts.

2058. Cesium Sulfate. Cs$_2$O$_4$S; mol wt 361.87. Cs 73.45%, O 17.69%, S 8.86%. Cs$_2$SO$_4$. Prepn: *Gmelin's, Cesium* (8th ed.) **25**, pp 218-225 (1938).

Orthorhombic or hexagonal prisms. d 4.24. mp 1019°. Very sol in water. Practically insol in alc, acetone, pyridine.
USE: With V or V$_2$O$_5$ as catalyst for SO$_2$ oxidation.

2059. Cetalkonium Chloride. *N-Hexadecyl-N,N-dimethylbenzenemethanaminium chloride; benzylhexadecyldimethylammonium chloride;* cetyldimethylbenzylammonium chloride; hexadecyldimethylbenzylammonium chloride; Banicol; Acetoquat CDAC; Acquat CDAC; Ammonyx G; Zettyn; Ammonyx T; Cetol. C$_{25}$H$_{46}$ClN; mol wt 396.10. C 75.81%, H 11.71%, Cl 8.95%, N 3.54%. Prepn: *Fr.* pat. **771,746** (1934 to I. G. Farben); Piggott, *U.S.* pat. **2,075,958** (1937 to I.C.I.); Westphal, Jerchel, *Ber.* **73**, 1011 (1940). *Review* of this type of quaternary ammonium compounds: J. P. Sisley, P. J. Wood, *Encyclopedia of Surface-Active Agents* (Chem. Publishing Co., New York, 1952) p 96 sqq.

Leaflets from ethyl acetate + petr ether, mp 59°. Sol in water, alc, acetone, ethyl acetate, propylene glycol, sorbitol solns, glycerol, ether, CCl$_4$. pH of aq solns 7.2.
USE: Cationic quaternary ammonium surfactant germicide and fungicide. Used in leather processing, textile dyeing. A mildew preventive in silicone-based water repellents. Compatible with many non-ionic detergents. Active in moderately alkaline solns.
THERAP CAT: Anti-infective (topical).

2060. Cetamolol. *2-[2-[3-[(1,1-Dimethylethyl)amino]-2-hydroxypropoxy]phenoxy]-N-methylacetamide;* 2-[o-[3-(t-butylamino)-2-hydroxypropoxy]phenoxy]-N-methylacetamide; (±)-1-t-butylamino-3-(o-N-methylcarbamoylmethoxyphenoxy)propan-2-ol. C$_{16}$H$_{26}$N$_2$O$_4$; mol wt 310.39. C 61.91%, H 8.44%, N 9.03%, O 20.62%. Cardioselective β_1-adrenergic blocker. Prepn: *Belg.* pat. **767,781**; D. J. Lecount, C. J. Squire, *U.S.* pat. **4,059,622** (1971, 1977 both to ICI). Cardiovascular effects in dogs: G. Beaulieu *et al.*, *Can. J. Physiol. Pharmacol.* **63**, 610 (1984). Determn in human serum: M. D. Stern, *Clin. Biochem.* **17**, 162 (1984). Pharmacology and pharmacodynamics in humans: M. A. Klausner *et al.*, *Curr. Ther. Res.* **36**, 379 (1984); J. Coelho *et al.*, *Brit. J. Clin. Pharmacol.* **19**, 411 (1985). Evaluation in hypertensive patients: W. B. White *et al.*, *Clin. Pharmacol. Ther.* **39**, 664 (1986).

Crystals from benzene-petr ether, mp 96-97°.
Hydrochloride, C$_{16}$H$_{26}$N$_2$O$_4$·HCl, *AI-27303, ICI-72222, Betacor.*
THERAP CAT: Antihypertensive.

2061. Cethexonium Bromide. *N-Hexadecyl-2-hydroxy-N,N-dimethylcyclohexanaminium bromide; hexadecyl(2-hydroxycyclohexyl)dimethylammonium bromide;* cetyldimethyl-(2-hydroxycyclohexyl)ammonium bromide; dimethylcetyl-2-cyclohexanolammonium bromide; N-cetyl-N,N-dimethyl-2-cyclohexanolammonium bromide; Biocidan. C$_{24}$H$_{50}$BrNO; mol wt 448.57. C 64.26%, H 11.24%, Br 17.81%, N 3.12%, O 3.57%. Prepd by the action of cetyl bromide upon 2-dimethylaminocyclohexanol: Mousseron *et al.*, *Bull. Soc. Chim. Biol.* **33**, 369 (1951).

White powder, mp 75°. Sol in water, alcohol, chloroform. Practically insol in petr ether.

THERAP CAT: Antiseptic.

2062. Cetiedil. α-*Cyclohexyl-3-thiopheneacetic acid 2-(hexahydro-1H-azepin-1-yl)ethyl ester;* α-cyclohexyl-α-(3-thienyl)acetic acid 2-hexamethyleneiminoethyl ester. $C_{20}H_{31}NO_2S$; mol wt 349.54. C 68.73%, H 8.94%, N 4.01%, O 9.15%, S 9.17%. Prepn: Pons, Robba, **Fr. pat. 1,460,571** and Pons *et al.*, **Fr. pat. M5504** (1966, 1967, both to Innothera), *C.A.* **68**, 59429d (1968); **71**, 91286c (1969). Prepn and activity: Robba, LeGuen, *Chim. Ther.* **2**, 120 (1967). Antisickling effect: T. Asakura *et al.*, *Proc. Nat. Acad. Sci. USA* **77**, 2955 (1980); L. R. Berkowitz, E. P. Orringer, *J. Clin. Invest.* **68**, 1215 (1981).

Citrate, $C_{20}H_{31}NO_2S.C_6H_8O_7$, *Stratene, Vasocet.* Crystals from ethanol-ether, mp 115°.
Hydrochloride, $C_{20}H_{31}NO_2S.HCl$, crystals from acetonitrile, mp 152° (Robba, LeGuen); also mp 143° (Pons, Robba).

THERAP CAT: Vasodilator (peripheral).

2063. Cetirizine. [2-[4-[(4-Chlorophenyl)phenylmethyl]-1-piperazinyl]ethoxy]acetic acid; [2-[4-(p-chloro-α-phenylbenzyl)-1-piperazinyl]ethoxy]acetic acid. $C_{21}H_{25}ClN_2O_3$; mol wt 388.89. C 64.86%, H 6.48%, Cl 9.12%, N 7.20%, O 12.34%. Nonsedating type histamine H_1-receptor antagonist; major metabolite of hydroxyzine, *q.v.* Prepn: E. Baltes *et al.*, **Eur. pat. Appl. 58,146**; *eidem*, **U.S. pat. 4,525,358** (1982, 1985 both to UCB). Pharmacology: C. De Vos *et al.*, *Ann. Allergy* **59**, 278 (1987); L. Juhlin *et al.*, *J. Allergy Clin. Immunol.* **80**, 599 (1987). Clinical evaluation in asthma: A. Brik *et al.*, *ibid.* 51. Mode of action by eosinophil inhibition: R. Fadel *et al.*, *Clin. Allergy* **17**, 373 (1987). Clinical evaluation of antihistaminic and psychomotor effects: F. M. Gengo *et al.*, *Clin. Pharmacol. Ther.* **42**, 265 (1987).

Crystals from ethanol, mp 110-115°.
Dihydrochloride, $C_{21}H_{25}ClN_2O_3.2HCl$, *P-071, Alerlisin, Formistin, Reactine, Virlix, Zirtek, Zyrtec, Zyrlex.* Crystals from isopropanol, mp 225°.

THERAP CAT: Antihistaminic.

2064. Cetotiamine. *Thiocarbonic acid O-ethyl ester, S-ester with N-[(4-amino-2-methyl-5-pyrimidinyl)methyl]-N-(4-hydroxy-2-mercapto-1-methyl-1-butenyl)formamide ethyl carbonate (ester); O,S-bis(ethoxycarbonyl)thiamine; O,S-dicarbethoxythiamine; DCET.* $C_{18}H_{26}N_4O_6S$; mol wt 426.49. C 50.69%, H 6.14%, N 13.14%, O 22.51%, S 7.52%. Prepn: Takamizawa, Hirai, *Chem. Pharm. Bull.* **10**, 1102 (1962); Takamizawa *et al.*, *ibid.* 1107; Yamamoto *et al.*, *Bitamin* **25**,

472 (1962), *C.A.* **60**, 9773e (1964); **Brit. pat. 944,641** (1963 to Shionogi).

cis-form

Prisms from ethyl acetate + petr ether, mp 113.5-114.5°.
Hydrochloride monohydrate, $C_{18}H_{26}N_4O_6S.HCl.H_2O$, *dicethiamin, Dicetamin.* Crystals from ethyl acetate, dec 122-124°. Sol in water, methanol. Practically insol in ether, benzene.

THERAP CAT: Vitamin B_1 source.

2065. Cetoxime. *N-Hydroxy-2-[phenyl(phenylmethyl)-amino]ethanimidamide; 2-(N-benzylanilino)acetamidoxime;* α-(N-benzyl-N-phenylamino)acetamidoxime. $C_{15}H_{17}N_3O$; mol wt 255.32. C 70.56%, H 6.71%, N 16.46%, O 6.27%. Prepd from (N-benzylanilino)acetonitrile via its thioamide: Benn *et al.*, **Brit. pat. 895,495** (1962 to Boots Pure Drug).

Crystals, mp 107-108°.
Hydrochloride, $C_{15}H_{17}N_3O.HCl$, *Febramine.* Crystals from abs alcohol + ether, mp 164-165°.

THERAP CAT: Antihistaminic.

2066. Cetraric Acid. *9-(Ethoxymethyl)-4-formyl-3,8-dihydroxy-1,6-dimethyl-11-oxo-11H-dibenzo[b,e][1,4]dioxepin-7-carboxylic acid;* cetrarin. $C_{20}H_{18}O_9$; mol wt 402.36. C 59.70%, H 4.51%, O 35.79%. From Iceland moss, *Cetraria islandica (L.)* Ach., *Parmeliaceae.* Isoln: Schnedermann, Knopp, *Ann.* **55**, 144 (1845). Structure: Asahina, Asano, *Ber.* **66**, 893 (1933).

Very bitter prisms from alcohol or acetic acid. Bitterness threshold 1:50,000. Practically insol in hot water, petr ether, benzene, ether, or in cold methanol, alc, acetone and acetic acid. Sol in aq solns of alkalies or their carbonates forming a yellow soln that turns brown on standing.

2067. Cetraxate. *trans-4-[[[4-(Aminomethyl)cyclohexyl]carbonyl]oxy]benzenepropanoic acid; p-hydroxyhydrocinnamic acid trans-(4-aminomethyl)cyclohexanecarboxylate; tranexamic acid p-(2-carboxyethyl)phenyl ester.* $C_{17}H_{23}NO_4$; mol wt 305.37. C 66.86%, H 7.59%, N 4.59%, O 20.96%. Deriv of tranexamic acid, *q.v.* Prepn: Y. Yamamura *et al.*, **Ger. pat. 1,951,061**; *eidem*, **U.S. pat. 3,699,149** (1970, 1972 both to Daiichi); O. Atsuji *et al.*, *J. Med. Chem.* **15**, 247 (1972); S. Kitahara, **Japan. Kokai 73 75547** (1973 to Daiichi), *C.A.* **80**, 59727x (1974). Mechanism of action: Y. Suzuki *et al.*, *Japan. J. Pharmacol.* **29**, 829 (1979), *C.A.* **92**, 88029 (1980). Anti-ulcer effects in rats: T. Hashizume *et al.*, *Arch. Int. Pharmacodyn. Ther.* **240**, 314 (1979). Clinical study: A. Ishimori *et al.*, *Arzneimittel-Forsch.* **29**, 1625 (1979); S. Yamagata, K. Miura, *ibid.* **33**, 1191 (1983).

Crystals from methanol, melts over a range of 200-280°. Hydrochloride, $C_{17}H_{23}NO_4 \cdot HCl$, *DV-1006, Neuer.* Crystals from methanol/ether, mp 238-240°.

THERAP CAT: Antiulcerative.

2068. Cetrimonium Bromide. *N,N,N-Trimethyl-1-hexadecanaminium bromide; hexadecyltrimethylammonium bromide;* cetyltrimethylammonium bromide; Bromat; Cetab; Cetavlon; Cetylamine; C.T.A.B.; Lissolamine V; Micol; Quamonium. $C_{19}H_{42}BrN$; mol wt 364.45. C 62.62%, H 11.62%, Br 21.92%, N 3.84%. $[CH_3(CH_2)_{15}N(CH_3)_3]Br$. Prepd from cetyl bromide and trimethylamine: Shelton *et al., J. Am. Chem. Soc.* **68,** 753 (1946). Toxicity and pharmacology: B. Isomaa, K. Bjondahl, *Acta Pharmacol. Toxicol.* **47,** 17 (1980).

Crystals, mp 237-243°. Soluble in about 10 parts water. Freely sol in alc; sparingly sol in acetone. Practically insol in ether, benzene. Stable in acid soln. LD_{50} in mice, rats (mg/kg): 32.0, 44.0 i.v. (Isomaa, Bjondahl).

p-Toluenesulfonate analog, $C_{26}H_{49}NO_3S$, *cetrimonium tosylate, Cetats.*

Note: Cetrimide is a mixture consisting chiefly of tetradecyltrimethylammonium bromide together with smaller amounts of dodecyltrimethylammonium bromide and cetrimonium bromide.

USE: As cationic detergent and antiseptic; as laboratory reagent.

THERAP CAT: Antiseptic (topical).

THERAP CAT (VET): Antiseptic, cleansing agent.

2069. Cetrimonium Stearate. *N,N,N-Trimethyl-1-hexadecanaminium octadecanoate; hexadecyltrimethylammonium stearate;* cetyltrimethylammonium stearate; trimethylhexadecylammonium stearate; Arquad 16 stearate; Dynafac. $C_{37}H_{77}NO_2$; mol wt 568.02. C 78.24%, H 13.66%, N 2.47%, O 5.63%. $[CH_3(CH_2)_{16}COO][CH_3(CH_2)_{15}N(CH_3)_3]$. Prepn: Gautier *et al., Bull. Soc. Chim. France* **1955,** 634.

Solid, mp 142-143°. Practically insol in water, alcohol.

Note: The commercial product, a waxy solid, also contains other alkyltrimethylammonium stearates, since the hexadecyl chain is derived from soybean fatty acids.

2070. Cetyl Alcohol. *1-Hexadecanol;* ethal; ethol; palmityl alcohol. $C_{16}H_{34}O$; mol wt 242.45. C 79.27%, H 14.14%, O 6.60%. $CH_3(CH_2)_{14}CH_2OH$. Discovered by Chevreul in 1813. Obtained from spermaceti by saponification: Spada, Gavioli, *Farm. Sci. e Tec.* (Pavia) **7,** 435 (1952), *C.A.* **47,** 891c (1953). Prepn from palmitoyl chloride + $NaBH_4$: Caikin, Brown, *J. Am. Chem. Soc.* **71,** 122 (1949); from methylthiopalmitate + Raney Ni: Ruzicka, Prelog, U.S. pat. **2,509,171** (1950 to Ciba); from hexadecyl bromide: Levine, Clippinger, U.S. pat. **3,018,308** (1962 to California Res. Corp.).

White crystals. d 0.811. mp 49°. bp 344°; bp_{15} 190°. n_D^{79} 1.4283. Practically insol in water. Sol in alcohol, chloroform, ether.

Note: The *hexadecyl alcohol* developed by Esso Res. & Eng. Co. for cosmetics is a liquid, primary, branched chain, C_{16} alcohol, made up of an array of isomeric compds maintained in constant proportion by a complex manufacturing process (not from spermaceti): Edman, Lowden, *Drug Cosmet. Ind.* **93,** 631 (Nov. 1963). Liquid, d_{20}^{20} 0.842. bp_{50} 195-205°. Freezes at $< -60°$. Miscible with most alcohols, glycols, esters, ketones, cosmetic oils and aromatics. Immiscible with water.

USE: In cosmetics as emollient, emulsion modifier, coupling agent. Pharmaceutic aid (emulsifying and stiffening agent).

2071. Cetyldimethylethylammonium Bromide. *N-Ethyl-N,N-dimethyl-1-hexadecanaminium bromide; ethylhexadecyldimethylammonium bromide;* ethyl cetab; CDA; Ammonyx DME; Bretol. $C_{20}H_{44}BrN$; mol wt 378.48. C

63.47%, H 11.72%, Br 21.11%, N 3.70%. Cationic germicidal detergent. Prepn and antibacterial activity: R. S. Shelton *et al., J. Am. Chem. Soc.* **68,** 753 (1946).

White powder, mp 178-186°. Soluble in water, alcohol; slightly sol in chloroform, benzene, ether. LD_{50} orally in rats: 500 mg/kg, *RTECS* **Vol. 1,** R. J. Lewis, R. L.Tatken, Eds. (1979) p 107.

USE: Disinfectant; laboratory reagent.

THERAP CAT: Antiseptic (topical).

THERAP CAT (VET): Antiseptic (topical).

2072. Cetyl Lactate. *2-Hydroxypropanoic acid hexadecyl ester; 1-hexadecanol lactate;* lactic acid cetyl ester; lactic acid hexadecyl ester; Ceraphyl 28. $C_{19}H_{38}O_3$; mol wt 314.51. C 72.56%, H 12.18%, O 15.26%. $CH_3CHOH \cdot COOC_{16}H_{33}$. Preparation: Rehberg, Marion, *J. Am. Chem. Soc.* **72,** 1918 (1950).

Waxy solid. mp 41°. $bp_{0.1}$ 132°; bp_1 170°; bp_{10} 219°. n_D^{40} 1.4410; n_D^{50} 1.4370.

USE: Non-ionic emollient. To improve feel and texture of cosmetic and pharmaceutical prepns.

2073. Cetyl Palmitate. *Hexadecanoic acid hexadecyl ester; palmitic acid hexadecyl ester;* hexadecyl palmitate. $C_{32}H_{64}O_2$; mol wt 480.86. C 79.93%, H 13.42%, O 6.65%. $CH_3(CH_2)_{14}COOCH_2(CH_2)_{14}CH_3$. Prepn from palmitoyl chloride and cetyl alcohol in the presence of Mg: Paquot, Bouquet, *Bull. Soc. Chim. France* **1947,** 321; by CrO_3-H_2SO_4 oxidation of cetyl alcohol: Cymerman-Craig, Horning, *J. Org. Chem.* **25,** 2098 (1960). Biosynthesis using inoculum of *Nocardia salmonicolor:* Davis, U.S. pat. **3,169,099** (1965 to Socony Mobil Oil).

Monoclinic leaflets, mp 54°. d^{20} 0.989. n_D^{70} 1.4398. Practically insol in water. Sol in abs alc, ether.

2074. Cetylpyridinium Chloride. *1-Hexadecylpyridinium chloride;* Ceepryn; Cepacol; Dobendan; Medilave; Merocet; Pristacin; Pyrisept. $C_{21}H_{38}ClN$; mol wt 339.99. C 74.19%, H 11.27%, Cl 10.43%, N 4.12%. Pharmacology and toxicology: *J. Pharmacol. Exp. Ther.* **74,** 401 (1942). Review of early literature: C. L. Huyck, *Am. J. Pharm.* **116,** 50 (1944). Toxicity data: J. W. Nelson, S. C. Lyster, *J. Am. Pharm. Assoc.* **35,** 89 (1946).

Monohydrate, *Halset.* White powder, mp 77-83°. Freely sol in water, alcohol, chloroform; very slightly sol in benzene, ether. pH (1% aq soln): 6.0 to 7.0. Surface tension (25°): 43 dyn/cm (0.1% aq soln); 41 dyn/cm (1.0%); 38 dyn/cm (10%). LD_{50} in rats (mg/kg): 250 s.c.; 6 i.p.; 30 i.v.; 200 orally (Nelson, Lyster).

USE: Pharmaceutic aid (preservative).

THERAP CAT: Antiseptic; disinfectant.

THERAP CAT (VET): Topical antiseptic; disinfectant.

2075. Cevadine. *[3β(Z),4α,16β]-4,9-Epoxycevane-3,4,-12,14,16,17,20-heptol 3-(2-methyl-2-butenoate);* veratrine. $C_{32}H_{49}NO_9$; mol wt 591.74. C 64.95%, H 8.35%, N 2.37%, O 24.33%. From seeds of *Schoenocaulon officinale* (Schlecht & Cham.) A. Gray *(Sabadilla officinarum* Brandt), Liliaceae: Poetsch *et al., J. Am. Pharm. Assoc.* **38,** 525 (1949); Ringel, *ibid.* **45,** 433 (1956). Evaluation as insecticide: Ikawa, Link *et al., J. Biol. Chem.* **159,** 517 (1945). Structure: Kupchan, Alfonso, *ibid.* **49,** 242 (1960). Toxicity study: Swiss, Bauer, *Proc. Soc. Exp. Biol. Med.* **76,** 847 (1951). *Review:* Wintersteiner in Graff, *Essays in Biochemistry* (Wiley, New York, 1956) pp 308-321.

Flat needles from ether, decomp 213-214.5°. $[\alpha]_D^{20}$ +12.8° (c = 3.2 in alc). One gram dissolves in about 15 ml alc or ether; slightly sol in water. LD_{50} i.p. in mice: 3.5 mg/kg (Swiss, Bauer).

Aurichloride, fine yellow needles from alc, dec 190°.

Mercurichloride, $C_{32}H_{49}NO_9 \cdot HCl \cdot HgCl_2$, silvery scales, dec 172°.

Caution: Extremely irritating locally, particularly to mucous membranes. Serious poisoning has resulted from local application. Caution must be used in handling. *See also* Veratrine (Mixture).

2076. Cevine. *(3α,4α,16β)-4,9-Epoxycevane-3,4,12,14,-16,17,20-heptol.* $C_{27}H_{43}NO_8$; mol wt 509.64. C 63.63%, H 8.50%, N 2.75%, O 25.11%. By hydrolysis of cevadine. Structure and stereochemistry: Barton *et al., Experientia* **10,** 81 (1954); Kupchan *et al., J. Am. Chem. Soc.* **80,** 1769 (1958); Kupchan *et al., Tetrahedron* **7,** 47 (1959); Eeles, *Tetrahedron Letters* no. 7, 24 (1960). Comparative toxicity: O. Krayer *et al., J. Pharmacol. Exp. Ther.* **82,** 167 (1944); K. Tanaka, *ibid.* **113,** 89 (1955).

Occurs as the hemiheptahydrate, triclinic prisms from dil alc. After drying at 110° it sinters at 165° and becomes a transparent resin at 165-170°: M. Ikawa *et al., J. Biol. Chem.* **159,** 517 (1945). $[\alpha]_D^{17}$ −17.5° (aq alc). Sol in water, alc; slightly sol in ether. LD_{50} i.p. in rats: 87.0 mg/kg (Krayer). LD_{50} s.c. in male mice: 160 mg/kg (Tanaka). Hydrochloride, mp 247°.

2077. Chalcomycin. $C_{35}H_{56}O_{14}$; mol wt 700.82. C 59.98%, H 8.05%, O 31.96%. Antibiotic substance produced from *Streptomyces bikiniensis* NRRL 2737: Frohardt *et al.,* **Ger.** pat. **1,109,835** (1960 to Parke, Davis). Structure: Woo *et al., J. Am. Chem. Soc.* **86,** 2726 (1964). *Review:* Jordan, "Chalcomycin" in *Antibiotics* **I,** D. Gottlieb, P. Shaw, Eds. (Springer-Verlag, New York, 1967) pp 446-450.

Crystals from butanol, mp 121-123°. $[\alpha]_D^{27}$ −43.5° (ethanol). uv max: 218 nm (ϵ 22770). Sol in ethyl acetate, methanol, ethanol, benzene, toluene, chloroform; sparingly sol in carbon tetrachloride, ether, water. Practically insol in petr ether.

2078. Chalcone. *1,3-Diphenyl-2-propen-1-one;* chalkone; benzylideneacetophenone; benzalacetophenone; phenyl styryl ketone. $C_{15}H_{12}O$; mol wt 208.26. C 86.51%, H 5.81%, O 7.68%. $C_6H_5CH=CHCOC_6H_5$. Prepd by the action of NaOH on an alcoholic soln of benzaldehyde and acetophenone: E. P. Kohler, H. M. Chadwell, *Org. Syn. coll. vol.* **I,** 78 (1941).

Pale-yellow orthorhombic prisms from petr ether. mp 57-58°. d_4^{62} 1.0712. bp_{760} 345-348° (slight decompn). bp_{25} 208°. n_D^{62} 1.6458. Freely sol in ether, chloroform, carbon bisulfide, benzene; slightly in alcohol; very sparingly in cold petr ether. *Caution:* Skin irritant.

Note: Exists in other forms (polymorphic) with different mps. Irradiation produces four forms of dimers.

2079. Chalcopyrite. Cupric ferrous sulfide; copper pyrites; yellow copper; Chalkopyrite; Kupferkies. $CuFeS_2$; mol wt 183.52. Cu 34.63%, Fe 30.43%, S 34.94%. Occurs as a mineral. It is an important copper ore found in Canada, Montana, Utah, Arizona, Tennessee, Chile, Peru, Bolivia, Europe, Japan. Prepd synthetically by treating $KFeS_2$ with an ammoniacal soln of Cu^+ ions: Schneider, *J. Prakt. Chem.* [2] **38,** 569 (1888); **56,** 415 (1897); Boon, *Rec. Trav. Chim.* **63,** 69 (1954).

Yellow brass-colored or bronze-colored crystals. Metallic, greenish sheen. Tetragonal crystal system. d 4.1-4.3. mp 950°. Specific heat 0.1291. Hardness 3.5-4.0. Soluble in nitric acid, but dissolves faster in aqua regia. Practically insol in hydrochloric acid.

Note: Copper iron sulfide of the formula $FeS \cdot 2Cu_2S \cdot CuS$ occurs as the mineral *bornite*.

USE: The ore as source of copper. The synthetic material in semiconductor research.

2080. D-Chalcose. *4,6-Dideoxy-3-O-methyl-D-xylo-hexose;* 4,6-dideoxy-3-O-methyl-D-glucose; lankavose. $C_7H_{14}O_4$; mol wt 162.19. C 51.84%, H 8.70%, O 39.46%. Component of chalcomycin and lankomycin, *q.q.v.* Prepn from chalcomycin and structure: Woo *et al., J. Am. Chem. Soc.* **83,** 3352 (1961). Stereochemistry: Woo *et al., ibid.* **84,** 1066 (1962); Foster *et al., J. Chem. Soc.* **1965,** 2318. Synthesis: Kochetkov, Usov, *Tetrahedron Letters* **1963,** 519; McNally, Overend, *Chem. & Ind. (London)* **1964,** 2021; Lawton *et al., Can. J. Chem.* **47,** 2299 (1969). Synthesis of DL-form: R. M. Srivastava, R. K. Brown, *Can. J. Chem.* **48,** 830 (1970); K.Torssell, M. P. Tyagi, *Acta Chem. Scand. B* **31,** 1 (1977); S. Danishefsky, J. F. Kerwin, *J. Org. Chem.* **47,** 1597 (1982).

Crystals, mp 96-99°. $[\alpha]_D^{24}$ +120° (2 min) → +97° (10 min) → +76° (3 hr and 26 hr) (c = 1.5 in water).

Methyl chalcoside, $C_8H_{16}O_4$, crystals, mp 101.5-102°. $[\alpha]_D^{27}$ −21° (c = 2.04 in chloroform).

2081. Chamazulene. *7-Ethyl-1,4-dimethylazulene;* 1,4-dimethyl-7-ethylazulene; dimethulene. $C_{14}H_{16}$; mol wt 184.28. C 91.25%, H 8.75%. Anti-inflammatory principle obtained from chamazulenogenic compds found in chamomile (*Matricaria chamomilla* L.), wormwood (*Artemisia absinthium* L.), and yarrow *(Achillea millefolium* L., *Compositae).* Isoln: Ruzicka, Rudolph, *Helv. Chim. Acta* **9**, 118 (1926); Ruzicka, Haagen-Smit, *ibid.* **14**, 1104 (1931). Structure: Sorm *et al., Coll. Czech. Chem. Commun.* **18**, 527 (1953); Meisels, Weizmann, *J. Am. Chem. Soc.* **75**, 3865 (1953); Suchy *et al., Coll. Czech. Chem. Commun.* **21**, 477 (1956). Synthesis: Mangoni, Bandiera, *Gazz. Chim. Ital.* **90**, 947 (1960); White, Winter, *Tetrahedron Letters* **1963**, 137; D. Mukherjee *et al., J. Am. Chem. Soc.* **101**, 251 (1979). Pharmacology: I. G. Boldina, *Farmakol. Toksikol.* **29**, 672 (1966); I. G. Boldina, M. V. Nazarenko, *C.A.* **67**, 42441h (1967).

Blue oil. bp_{12} 161°; bp_{11} 145°. d_4^{20} 0.9883. uv max: 370 nm (log ϵ 3.7). LD_{50} i.m. in white mice: 3 g/kg (Boldina, Nazarenko).

Trinitrobenzene derivative, $C_{20}H_{19}N_3O_6$, dark violet needles from abs ethanol, mp 131.5-132.5°.

2082. Chamomile. Camomile. The term chamomile refers to the dried flowerheads of either of two distinct plants of the family Asteraceae (Compositae): German chamomile, *Chamomilla recutita* (L.) Rauschert, also known as *Matricaria recutita* L. or *M. chamomilla*, and Roman chamomile, *Chamaemelum nobile* (L.) All. (formerly *Anthemis nobilis* L.). Both chamomiles have been used in traditional medicines as antispasmodic and anti-inflammatory agents. Evaluation of hydroalcoholic extracts for antispasmodic activity: H. B. Forster *et al., Planta Med.* **40**, 309 (1980); for anti-inflammatory activity: A. Tubaro *et al., ibid.* **50**, 359 (1984). Stability of extracts: R. Carle *et al., ibid.* **55**, 540 (1989). HPLC determn of flavones and coumarins in extracts: P. Pietta *et al., J. Chromatog.* **404**, 279 (1987). Reviews: A. Y. Leung, *Encyclopedia of Common Natural Ingredients* (Wiley-Interscience, New York, 1980) pp 110-112; C. Mann, E. J. Staba in *Herbs, Spices, and Medicinal Plants: Recent Advances in Botany, Horticulture, and Pharmacology* vol. 1 (Oryx Press, 1986) pp 235-280.

German chamomile, Hungarian chamomile, wild chamomile, Matricaria. Habit. Europe, northern and western Asia; naturalized in North America. Cultivated in Hungary, Romania, Germany. *Constit.* 0.24-1.9% volatile oil; flavonoids including apigenin, luteolin, quercetin; coumarins such as herniarin, umbelliferone; proazulenes (matricin, matricarin, etc.); phenolic carboxylic acids (anisic, vanillic, syringic, caffeic); polysaccharide mucilage.

German chamomile oil, Matricaria oil. Extracted from flower heads by steam distillation. *Constit.* (−)-α-bisabolol (up to 50%), bisabolol oxides, chamazulene, farnesene. Review of constituents: B. M. Lawrence, *Perfum. Flavor.* **12**, 35-52 (1987). Deep ink-blue oil. d 0.91-0.95. Ester value: <40. Acid value: 5 to 50. Sol in propylene glycol. Insol in glycerin, mineral oil.

Roman chamomile, English chamomile, garden chamomile. Habit. Southern and western Europe; naturalized in North America. Cultivated in Great Britain, Belgium, France, U.S. *Constit.* up to 1.75% volatile oil; up to 0.6% sesquiterpene lactones of the germacranolide type (mainly nobilin, 3-epinobilin); flavonoids (apigenin, luteolin, apiin, etc.); phenolic carboxylic acids (caffeic, ferulic); coumarins; and thiophene derivatives.

Roman chamomile oil, oil of anthemis. Extracted from flowerheads by steam distillation. *Constit.* esters of angelic and tiglic acids (~85%), chamazulene, α-pinene, farnesol. Blue oil. n^{20} 1.44-1.45. $[\alpha]^{20}$ −1 to +4°. Ester value: 250 to 310. Acid value < 15. Sol in mineral oil and most fixed

oils; sol with haziness in propylene glycol. Practically insol in glycerin.

USE: Flowers in herbal teas. Extracts and oils in perfumery; as flavor in foods and alcoholic beverages; hair dye.

THERAP CAT: Carminative; vulnerary. Extracts and oils as topical anti-inflammatory.

2083. Chanoclavine. *[4R-[4α,5β(E)]]-2-Methyl-3-[1,3,-4,5-tetrahydro-4-(methylamino)benz[cd]indol-5-yl]-2-propen-1-ol; (E)-8,9-didehydro-6-methyl-6,7-secoergoline-8-methanol;* chanoclavin-I; secaclavine. $C_{16}H_{20}N_2O$; mol wt 256.35. C 74.97%, H 7.86%, N 10.93%, O 6.24%. Precursor of the tetracyclic ergolines, agroclavine, elymoclavine, and lysergic acid amide, *q.q.v.:* Gröger *et al., Z. Naturforsch.* **21b**, 827 (1966); Floss *et al., Chem. Commun.* **1967**, 105. Occurs in sclerotia of *Claviceps purpurea* (Fries) Tul., *Hypocreaceae*, in *Ipomea tricolor* Cav., *Convolvulaceae* and is one of the active principles of the ancient Aztec drug "Ololiuqui," *Rivea coryrubosa* (L.) Hall. f., *Convolvulaceae:* Hofmann, Tscherter, *Experientia* **16**, 414 (1960). Isoln and structure: Hofmann *et al., Helv. Chim. Acta* **40**, 1358 (1957). Isoln of stereoisomers: Stauffacher, Tscherter, *ibid.* **47**, 2186 (1964). Stereochemistry: Acklin *et al., Chem. Commun.* **1966**, 799. Biosynthetic studies: Floss *et al., J. Am. Chem. Soc.* **90**, 6500 (1968). Total synthesis of (±)-chanoclavine: H. Plieninger *et al., Tetrahedron Letters* **1975**, 1827; H. Plieninger, D. Schmalz, *Ber.* **109**, 2140 (1976); M. Natsume, H. Muratake, *Heterocycles* **16**, 375 (1981).

Thick prisms and polyhedra from acetone or methanol, mp 220-222°. $[\alpha]_D^{20}$ −240° (pyridine), $[\alpha]_D^{20}$ −205° (c = 0.75 in alcohol). uv max: 225, 284, 293 nm (ϵ 4.44, 3.82, 3.72). One gram dissolves in 25 ml boiling methanol, 140 ml boiling acetone, 170 ml boiling ethyl acetate or 350 ml boiling chloroform. Practically insol in water.

Epimer at position 10, *chanoclavine II.*

2084. Chaperonins. Cpns. Subset of molecular chaperone proteins, including the nucleoplasmins and heat shock proteins, that guide the in-vivo transport, folding and assembly of other protein structures but are not themselves components of these structures as part of their biological function. The term "chaperonin" was coined by S. M. Hemmingsen *et al., Nature* **333**, 330 (1988) to describe the function of these proteins. Characterized by large oligomeric cage-like structures with multiple rings of subunits possessing 7,8 or 9-fold symmetry. Ubiquitous conserved proteins which have been identified in most microorganisms and cellular compartments, requires the interaction of two chaperonins to ensure correct folding: J. Martin *et al., Nature* **366**, 228, 279 (1993). Molecular study of mechanism: S. G. Burston *et al., J. Mol. Biol.* **249**, 138 (1995). Review: R. J. Ellis, S. M. van der Vies, *Ann. Rev. Biochem.* **60**, 321-347 (1991). Review of structures and conformations: H. Saibil, S. Wood, *Curr. Opin. Struct. Biol.* **3**, 207-213 (1993). Review of *E. coli* chaperonins, cpn60-cpn10: F. Baneyx, *Ann. N.Y. Acad. Sci.* **745**, 383-394 (1994).

2085. CHAPS. *N,N-Dimethyl-N-(3-sulfopropyl)-3-[[(3α,5β,7α,12α)-3,7,12-trihydroxy-24-oxocholan-24-yl]amino]-1-propanaminium inner salt;* 3-[(3-cholamidopropyl)dimethylammonio]-1-propanesulfonate. $C_{32}H_{58}N_2O_7S$; mol wt 614.89. C 62.51%, H 9.51%, N 4.56%, O 18.21%, S 5.21%. Sulfobetaine zwitterionic derivative of cholic acid, *q.v.* Prepn: L. M. Hjelmeland, *Proc. Nat. Acad. Sci. USA* **77**, 6368 (1980); *idem,* U.S. pat. **4,372,888** (1983 to U.S. Secy. Health and Human Services); and physical properties: L. M. Hjelmeland *et al., Anal. Biochem.* **130**, 72 (1983). Physical characterization: R. E. Stark *et al., J. Phys. Chem.*

88, 6063 (1984). Micellar properties: M. A. Partearroyo *et al., Biochem. Int.* **16**, 259 (1988). Solubilization of membrane proteins: A. J. Bitonti *et al., Biochemistry* **21**, 3650 (1982); W. Wouters *et al., Eur. J. Pharmacol.* **115**, 1 (1985); L. Bancells *et al., Biochem. Pharmacol.* **36**, 2539 (1987); of opioid receptors: J. Simon *et al., J. Neurochem.* **46**, 695 (1986).

Crystals from absolute methanol at 0°.

Hydroxy analog, $C_{32}H_{58}N_2O_8S$, *CHAPSO*. White granular detergent.

USE: Nondenaturing biological detergent.

2086. Chartreusin. *10-[[6-Deoxy-2-O-(6-deoxy-3-O-methyl-α-D-galactopyranosyl)-β-D-galactopyranosyl]oxy]-6-hydroxy-1-methylbenzo[h][1]benzopyrano[5,4,3-cde][1]benzopyran-5,12-dione;* antibiotic X-465A; lambdamycin; NSC-5159. $C_{32}H_{32}O_{14}$; mol wt 640.60. C 60.00%, H 5.03%, O 34.97%. Antibiotic substance produced by *Streptomyces chartreusis* from African soil, also by another *Streptomyces* sp. from North American soil: B. E. Leach *et al., J. Am. Chem. Soc.* **75**, 4011 (1953). Identity with Antibiotic X-465A: J. Berger *et al., ibid.* **80**, 1636 (1958). Structure: L. H. Sternbach *et al., ibid.* 1639; E. Simonitsch *et al., Helv. Chim. Acta* **47**, 1459 (1964); W. Eisenhuth *et al., ibid.* 1475. Anticancer activity: J. P. McGovern *et al., Cancer Res.* **37**, 1666 (1977). Synthesis of *chartarin*, the chartreusin aglycone: T. R. Kelly *et al., J. Am. Chem. Soc.* **102**, 798 (1980); F. M. Hauser, D.W. Combs, *J. Org. Chem.* **45**, 4071 (1980).

Yellow plates from acetone or methylene chloride + ethanol, mp 184-186°. Weak acid. $[\alpha]_D^{25}$ +132.5° (c = 0.2 in pyridine); $[\alpha]_D^{25}$ −33° (c = 0.3 in glacial acetic acid). uv max: 237, 262, 332, 382, 405, 422 nm. Practically insol in water. Sol in acetone.

Sodium salt, gold-colored needles or plates from water. Soly in water: at least 20 mg/ml (pH 9.5). LD_{50} i.v. in mice: 250 mg/kg (Leach).

2087. Charybdotoxin. ChTX; CTX. $C_{176}H_{277}N_{57}O_{55}S_7$; mol wt 4295.95. C 49.21%, H 6.50%, N 18.58%, O 20.48%, S 5.22%. Peptide inhibitor of K^+ channels isolated from the venom of the scorpion *Leiurus quinquestriatus* var. *hebraeus*. Single chain of 4.3 kDa comprised of 37 amino acids with 3 disulfide linkages. Identification: C. Miller *et al., Nature* **313**, 316 (1985). Purification and amino acid sequence: G. Gimenez-Gallego *et al., Proc. Nat. Acad. Sci.* **85**, 3329 (1988). Synthesis: E. E. Sugg *et al., J. Biol. Chem.* **265**, 18745 (1990); P. Lambert *et al., Biochem. Biophys. Res. Commun.* **170**, 684 (1990). NMR structural study: F. Bontems *et al., Eur. J. Biochem.* **196**, 19 (1991). Mechanism of channel block: R. MacKinnon, C. Miller, *J. Gen. Physiol.* **91**, 335 (1988). Characterization of binding sites: J. Vázquez *et al., J. Biol. Chem.* **264**, 20902 (1989); J. Vázquez *et al., ibid.* **265**, 15564 (1990). Use as probe for K^+ channels: M. Garcia, G. Gimenez-Gallego, U.S. pat. **4,960,867** (1990 to Merck & Co.).

Extinction coefficient: 1.52 $(mg/ml)^{-1}cm^{-1}$ at 280 nm; 15.05 $(mg/ml)^{-1}cm^{-1}$ at 215 nm.

USE: Biochemical probe.

2088. Chaulmoogra Oil. Hydnocarpus oil; gynocardia oil. Fixed oil expressed from chaulmoogra, the seeds of *Taraktogenos kurzii* King, *Bixaceae; Hydnocarpus wightiana* Blume or *H. anthelmintica* Pierre, *Flacourtiaceae*. The seeds contain about 50% chaulmoogra oil. *Constit.* Glycerides of chaulmoogric and hydnocarpic acids with small amounts of glycerides of palmitic acid, etc.

Yellow or brownish-yellow oil. Below 25° it is a soft solid; characteristic odor. About d_{25}^{25} 0.95; d_{45}^{45} 0.940. $[\alpha]_D^{20}$ +48° to +60°. Sol in benzene, chloroform, ether, petr ether, slightly in cold alcohol; almost entirely sol in hot alcohol, carbon disulfide.

THERAP CAT: Antibacterial (leprostatic).

2089. Chaulmoogric Acid. *(S)-2-Cyclopentene-1-tridecanoic acid;* D-13-(2-cyclopenten-1-yl)tridecanoic acid; hydnocarpylacetic acid. $C_{18}H_{32}O_2$; mol wt 280.45. C 77.09%, H 11.50%, O 11.41%. Isoln from chaulmoogra oil and characterization: Power, Gornall, *J. Chem. Soc.* **85**, 838, 851 (1904); Barrowcliff, Power, *ibid.* **91**, 557 (1907); Hashimoto, *J. Am. Chem. Soc.* **47**, 2325 (1925); Cole, Cardoso, *ibid.* **61**, 2349, 2351, 3442 (1939). Structure: Shriner, Adams, *ibid.* **47**, 2727 (1925). Synthesis: Noller, Adams, *ibid.* **48**, 1080 (1926); Perkins, Cruz, *ibid.* **49**, 1070 (1927). Synthesis and stereochemistry: Mislow, Steinberg, *ibid.* **77**, 3807 (1955). Activity against *Mycobacterium leprae:* L. Levy, *Am. Rev. Respir. Dis.* **111**, 703 (1975). Biosynthesis: T. Kaneda, *Biochem. Biophys. Res. Commun.* **99**, 1226 (1981).

Shiny leaflets from petr ether or alcohol. mp 68.5°. bp_{20} 247-248°. $[\alpha]_D^{20}$ +60.3° (c = 4 in chloroform). Iodine value 90.5. Freely sol in ether, chloroform, ethyl acetate; sol in other organic fat solvents.

Methyl ester, $C_{19}H_{34}O_2$, prepd by passing HCl gas into a methanol soln of the acid. Needles, mp 22°. bp_{20} 227°. d_{25}^{25} 0.9119. $[\alpha]_D^{15}$ +50° (c = 5 in chloroform).

Ethyl ester, $C_{20}H_{36}O_2$, *chaulmestrol, Moogrol.* Essentially a mixt of the ethyl esters of the unsaturated fatty acids (chaulmoogric and hydnocarpic) of chaulmoogra oil. Pale yellow, clear liquid; slight fruity odor; not unpleasant taste. Iodine no. 90-100. d_{25}^{25} ~0.904. $[\alpha]_D^{25}$ +44.5° (chloroform). Insol in water; misc with alc, chloroform.

THERAP CAT: Ethyl ester in treatment of leprosy, sarcoidosis.

2090. Chavicine. *(Z,Z)-1-[5-(1,3-Benzodioxol-5-yl)-1-oxo-2,4-pentadienyl]piperidine; (Z,Z)-1-piperoylpiperidine.* $C_{17}H_{19}NO_3$; mol wt 285.34. C 71.56%, H 6.71%, N 4.91%, O 16.82%. One of the most active constituents of black pepper: Buchheim, *Arch. Exp. Pathol. Pharmakol.* **5**, 455 (1876); Ott, Eichler, *Ber.* **55**, 2653 (1922); Ott, Lüdemann, *Ber.* **57**, 214 (1924). Stereoisomer of piperine: Lohaus, Gall, *Ann.* **517**, 282 (1935). Loss of pungency of ground pepper during storage attributed to gradual isomerization of chavicine into piperine *(q.v.)*: Newman, *Chem. Prod.* **16**, 379 (1953). Synthesis of chavicine and questioning of its existence as constituent of black pepper: R. Grewe *et al., Ber.* **103**, 3752 (1970). Spectroscopic structural elucidation and prepn: R. DeCleyn, R. Verzele, *Bull. Soc. Chim. Belg.* **84**, 435 (1975).

Yellowish, oily mass; sharp, peppery taste. Sol in alc, ether, petr ether. uv max (methanol): 318 nm (ϵ 16200).

2091. Chavicol. *4-(2-Propenyl)phenol; p-allylphenol; γ-(p-hydroxyphenyl)-α-propylene.* $C_9H_{10}O$; mol wt 134.18. C 80.56%, H 7.51%, O 11.92%. Found together with terpenes in volatile betel oils (from leaves of *Piper betle* L., *Piperaceae*): Eijkman, *Ber.* **22**, 2739 (1889). Prepd from a mixture of estragole, ethyl bromide and magnesium in benzene: Grignard, *Compt. Rend.* **151**, 323 (1910); from 4-allylphenylmagnesium bromide in ether by the action of oxygen: Quelet, *Bull. Soc. Chim. France* [4] **45**, 265 (1929).

Liquid. mp 15.8°. d_4^{15} 1.0203; d_4^{20} 1.0175. bp_{760} 238°; bp_{16} 123°. n_D^{18} 1.5441. Miscible with alcohol, ether, chloroform, petr ether.

Methyl ether, *see* Estragole.

2092. Cheirolin. *1-Isothiocyanato-3-(methylsulfonyl)-propane; isothiocyanic acid 3-(methylsulfonyl)propyl ester; γ-isothiocyanopropyl methyl sulfone; methylpropylsulfone mustard oil.* $C_5H_9NO_2S_2$; mol wt 179.26. C 33.50%, H 5.06%, N 7.81%, O 17.85%, S 35.78%. $CH_3SO_2(CH_2)_3NCS$. From seed of *Cheiranthus cheiri* L., *Cruciferae*: Wagner, *Chem. Ztg.* **32**, 76 (1908); Schneider, *Ber.* **41**, 4466 (1909); **42**, 3416 (1910). Structure: *idem.*, *Ann.* **375**, 207 (1910). Synthesis: Kjaer et al., *Acta Chem. Scand.* **7**, 1370 (1953). Prepn by enzymatic hydrolysis of glucocheirolin: Kjaer, Jensen, *ibid.* **12**, 1746 (1958).

Crystals from ether, mp 47-48°. bp_3 200°. Slightly sol in water, ether; freely sol in alcohol, chloroform, ethyl acetate.

2093. Chelerythrine. *1,2-Dimethoxy-12-methyl[1,3]-benzodioxolo[5,6-c]phenanthridinium; toddaline.* $[C_{21}H_{18}NO_4]^+$. From root of *Chelidonium majus* L., *Papaveraceae*: Kratzmann, *Pharm. Monatsh.* **5**, 161 (1924), *C.A.* **18**, 3406^2 (1924); Platonova et al., *J. Gen. Chem. USSR* **26**, 181 (1956). Structure: Späth, Kuffner, *Ber.* **64**, 1123 (1931); N. Decaudain et al., *Ann. Pharm. Franc.* **35**, 521 (1977). Identity with toddaline: Govindachari, Thyagarajan, *J. Chem. Soc.* **1956**, 769. Synthesis of chelerythrine chloride: Bailey, Worthing, *ibid.* 4535; S. V. Kessar et al., *J. Org. Chem.* **53**, 1708 (1988). Pharmacology: Chelombit'o, Murav'eva, *Aktual. Prob. Farmakol Farm., Vses. Nauch. Konf.* **1971**, 183, *C.A.* **76**, 112g (1972).

Crystals from chloroform + methanol, mp 207° [Manske, *Can. J. Res.* **21B**, 140 (1943)]. The free base is colorless, but its quaternary salts are yellow. Aq solns of base and salts show violet fluorescence. uv spectrum of chelerythrine chloride: Hruban et al., *Coll. Czech. Chem. Commun.* **35**, 3420 (1970).

2094. Chelidonic Acid. *4-Oxo-4H-pyran-2,6-dicarboxylic acid;* 4-oxo-1,4-pyran-2,6-dicarboxylic acid; Jerva acid; jervasic acid. $C_7H_4O_6$; mol wt 184.11. C 45.67%, H 2.19%, O 52.14%. Presence in various plants: Stransky, *Arch. Pharm.* **258**, 56 (1920); Ramstad, *Pharm. Acta Helv.* **28**, 45 (1953). Structure: Verkade, *Rec. Trav. Chim.* **43**, 879 (1924). Synthesis: Riegel, Reinhard, *J. Am. Chem. Soc.* **48**, 1334 (1926); Riegel, Zwilgmeyer, *Org. Syn.* **coll. vol. II**, 126 (1943); Toomey, Riegel, *J. Org. Chem.* **17**, 1492 (1952).

Crystals, dec 257°. uv max (water): 270 nm (ϵ 11500). One gram dissolves in 65 ml water, 26 ml boiling water. Sparingly sol in alc.

Monohydrate, monoclinic crystals from hot alcohol + water. The water of crystn is given up when heated at 102° and then at 160° to constant weight.

Sodium salt, sparingly sol in water.

2095. Chelidonine. *Stylophorin.* $C_{20}H_{19}NO_5$; mol wt 353.37. C 67.98%, H 5.42%, N 3.96%, O 22.64%. Hexahydrobenzophenanthridine alkaloid. Occurs in nature in (+)-form, (−)-form and racemic form. Isoln of (+)-form from root of *Chelidonium majus* L., *Stylophorum diphyllum* (Michx.) Nutt., and *Dicranostigma franchetianum* (Prain) Fedde, *Papaveraceae*: J. M. Probst, *Ann.* **29**, 113 (1839); E. Schmidt, F. Selle, *Arch. Pharm.* **228**, 441 (1890); F. Selle, *ibid.* 96; Manske, *Can. J. Res.* **20B**, 53 (1942); J. Slavik, *Coll. Czech. Chem. Commun.* **20**, 198 (1955); from *Symphoricarpos albus* L., Blake, *Caprifoliaceae*: M. Szaufer et al., *Phytochemistry* **17**, 1446 (1978). Isoln of (−)-form from *Glaucium corniculatum* Curt., *Papaveraceae*: J. Slavik, L. Slavikova, *Coll. Czech. Chem. Commun.* **22**, 279 (1957). Structure: F. von Bruchhausen, H. W. Bersch, *Ber.* **63**, 2520 (1930); E. Späth, F. Kuffner, *ibid.* **64**, 370 (1931); H. W. Bersch, *Arch. Pharm.* **291**, 491 (1958). Identity of (±)-form with diphylline: J. Slavik, *Coll. Czech. Chem. Commun.* **26**, 2933 (1961). Absolute configuration of (+)-*p*-bromobenzoate: N. Takao et al., *Tetrahedron Letters* **1979**, 495. Conformation: M. Cushman, T.-C. Choong, *Heterocycles* **14**, 1935 (1980); M. Sugiura et al., *J. Chem. Soc. Perkin Trans.* II, 175 (1986). Chiroptic properties and absolute configuration of (+)-chelidonine: N. Takao et al., *Arch. Pharm. (Weinheim)* **317**, 223 (1984). Biosynthesis: E. Leete, *J. Am. Chem. Soc.* **85**, 473 (1963). Total synthesis of (±)-form: W. Oppolzer, K. Keller, *ibid.* **93**, 3836 (1971); M. Cushman et al., *J. Org. Chem.* **45**, 5067 (1980); W. Oppolzer, C. Rabbiani, *Helv. Chim. Acta* **66**, 1119 (1983); M. Hanaoka et al., *Chem. Lett.* **1986**, 736. Effect on smooth muscle: P. J. Hanzlik: *J. Pharm. Exp. Ther.* **7**, 99 (1915). Inhibition of reverse transcriptase activity: M. L. Sethi, *Can. J. Pharm. Sci.* **16**, 29 (1981). Cytotoxic effects: M. Cushman et al., loc. cit. Review of pharmacological effects: V. Preininger, "The Biology of Papaveraceae Alkaloids" in *The Alkaloids* **vol. 15**, R. H. F. Manske, Ed. (Academic Press, Orlando, 1975) pp 241-242. General review: V. Simanek, "Benzophenanthridine Alkaloids" *ibid.* **vol. 26**, A. Brossi, Ed. (1985) pp 185-240. Toxicity data: R. C. Anderson, K. K. Chen, *Fed. Proc.* **5**, 163 (1946).

(+)-form

(+)-Form, *[5bR-(5bα,6β,12bα)]-5b,6,7,12b,13,14-hexahydro-13-methyl[1,3]benzodioxolo[5,6-c]-1,3-dioxolo[4,5-i]phenanthridin-6-ol.* Monoclinic prisms from methanol, ethanol or ethanol + chloroform, mp 135-136°. $bp_{0.002}$ 220° (air-bath temp). $[\alpha]_D^{22}$ +115° ±3° (ethanol); $[\alpha]_D^{20}$ +117° (c = 3 in $CHCl_3$). uv max (methanol): 289, 239, 208 nm (log ϵ 3.89, 3.88, 4.69). Sol in alc, chloroform, ether, amyl alc. Practically insol in water. LD_{50} in mice (mg/kg): 34.6 ±2.44 i.v. (Anderson, Chen).

(+)-*O*-Acetylchelidonine, $C_{22}H_{21}NO_6$, crystals from chloroform, mp 184-186°. $[\alpha]_D$ +110°.

(+)-Benzoylchelidonine, $C_{27}H_{23}NO_6$, crystals from chloroform, mp 210-211°.

(−)-Form, crystals from aqueous ethanol, mp 136°. $[\alpha]_D^{22}$ −112° ±3° (c = 0.47 in ethanol).

(±)-Form, *Diphylline.* Crystals from ethanol, mp 215-216°.

2096. Chenodiol. *(3α,5β,7α)-3,7-Dihydroxycholan-24-oic acid; 3α,7α-dihydroxy-5β-cholanic acid;* anthropodesoxycholic acid; gallodesoxycholic acid; 17β-(1-methyl-3-carboxypropyl)etiocholane-3α,7α-diol; chenic acid; chenodeoxycholic acid; CDC; Chendol; Chenix; Chenocedon; Chenocol; Chenodex; Chenofalk; Chenossil; Chenosäure; Cholanorm; Fluibil; Hekbilin; Kebilis; Ulmenide. $C_{24}H_{40}O_4$; mol wt 392.58. C 73.43%, H 10.27%, O 16.30%. A major bile acid in many vertebrates, occurring as the *N*-glycine and/or *N*-taurine conjugate. With other bile acids, forms mixed micelles with lecithin in bile which solubilize cholesterol and thus facilitates its excretion. Facilitates fat absorption in the small intestine by micellar solubilization of fatty acids and monoglycerides. Has cathartic properties since it induces fluid secretion from large intestine. Main constituent of the bile of hens, geese and other fowl; occurs in appreciable amounts in the bile of hamster, hog, guinea pig, bear and man. Epimeric with ursodiol, *q.v.* Isoln: Windhaus *et al., Z. Physiol. Chem.* **140**, 177 (1924); Wieland, Reveney, *ibid.* 186. Configuration: Lettré, *Ber.* **68**, 766 (1935). Prepn from cholic acid: Fieser, Rajagopalan, *J. Am. Chem. Soc.* **72**, 5530 (1950); Hauser *et al., Helv. Chim. Acta* **43**, 1595 (1960); Hofmann, *Acta Chem. Scand.* **17**, 173 (1963). Alternate prepns: Sato, Ikekawa, *J. Org. Chem.* **24**, 1367 (1959); T. Iida, F. C. Chang, *ibid.* **46**, 2786 (1981). Stereoselective total synthesis: T. Kametani *et al., J. Am. Chem. Soc.* **103**, 2890 (1981). Asymmetric total synthesis of (+)-form: *eidem, J. Org. Chem.* **47**, 2331 (1982). Dissolution of cholesterol gallstones: Danzinger *et al., N. Engl. J. Med.* **286**, 1 (1972); Bell *et al., Lancet* **II**, 1213 (1972). Use in long-term treatment of cerebrotendinous xanthomatosis: V. M. Berginer *et al., N. Engl. J. Med.* **311**, 1649 (1984). Monograph on bile acids: *The Bile Acids,* 2 vols., P. P. Nair, D. Kritchevsky, Eds. (Plenum Press, New York, 1971, 1973). Review of pharmacology and therapeutic use of chenodeoxycholic acid: J. H. Iser, A. Sali, *Drugs* **21**, 90-119 (1981). Effect on cholesterol and bile acid metabolism: G. S. Tint *et al., Gastroenterology* **91**, 1007 (1986).

Needles from ethyl acetate + heptane, mp 119°. $[\alpha]_D^{20}$ +11.5° (dioxane). Freely sol in methanol, alc, acetone, acetic acid; more sol in ether and ethyl acetate than deoxycholic acid. Practically insol in water, petr ether, benzene. High solvent power for alkali soaps, but does not form "choleic" acid addition compds as does deoxycholic acid. Forms beautiful cryst salts of Na, K and Ba. While the acid is tasteless, the Na salt tastes slightly sweet at first, then bitter.

Diformate, $C_{28}H_{44}O_6$, clusters of needles from alc; mp with slight effervescence at 137°, upon further heating solidifies again, and finally melts around 172°.

Methyl ester, $C_{25}H_{42}O_4$, fine needles from benzene + heptane, mp 90-91°. $[\alpha]_D^{25}$ +20°.

THERAP CAT: Anticholelithogenic.

2097. CHES. *2-(Cyclohexylamino)ethanesulfonic acid; N-cyclohexyltaurine.* $C_8H_{17}NO_3S$; mol wt 207.29. C 46.35%, H 8.27%, N 6.76%, O 23.15%, S 15.47%. Zwitterionic *N*-substituted aminosulfonic acid in the style of the "Good" buffers; active in the pH range 8.6-10.0. Synthesis: A. Champseix *et al., Bull. Soc. Chim. France* **1985**, 463. Effects of freezing on buffering: D. L. Williams-Smith *et al.,*

Biochem. J. **167**, 593 (1977). Effects on Lowry protein assay: C. Cookson, *Anal. Biochem.* **88**, 340 (1978). Dissociation constants and buffering capacity: M. J. Taylor, Y. Pignat, *Cryobiology* **19**, 99 (1982). Use as buffer: M. G. N. Hartmanis, T. C. Stadtman, *Proc. Nat. Acad. Sci. USA* **84**, 76 (1987); C. Engstrand *et al., Biochim. Biophys. Acta* **1122**, 321 (1992).

Crystals, mp 320°. pKa in water (25°) 9.27 ±0.01; in 20% (w/w) DMSO (25°) 9.10, (0°) 9.76, (−5.5°) 10.01; in 30% (w/w) DMSO (25°) 9.11, (0°) 9.89, (−12°) 10.27.

USE: Biological buffer.

2098. Chicle. The partially evaporated, milky juice from *Manilkara zapotilla* (Jacq.) Gilly *(Achras sapota* L.), Sapotaceae. *Habit.* West Indies, Mexico and Central America.

USE: In the chewing gum industry.

2099. Chimaphila. Pipsissewa; Prince's pine; bitter wintergreen; rheumatism weed; ground holly; pyrola; pine tulip. Dried leaves of *Chimaphila umbellata* (L.) Nutt., Ericaceae. *Habit.* Europe, Asia, North America. *Constit.* Chimaphilin, arbutin, ericolin, urson, tannin, resin.

THERAP CAT: Urinary antiseptic.

2100. Chimaphilin. *2,7-Dimethyl-1,4-naphthalenedione;* 2,7-dimethyl-1,4-naphthoquinone. $C_{12}H_{10}O_2$; mol wt 186.21. C 77.40%, H 5.41%, O 17.18%. From *Chimaphila carymbosa* Pursh. and *Pyrola incarnata* (Fisch.) Fernald, Ericaceae: DiModica *et al., Gazz. Chim. Ital.* **83**, 393 (1953); Inouye, *J. Pharm. Soc. Japan* **76**, 976 (1956). Structure: DiModica, Tira, *Gazz. Chim. Ital.* **86**, 234 (1956).

Yellow needles, mp 113.5-114.5°. uv max (alc): 233, 256, 339 nm. Polymerizes in sunlight to needles, mp 232-233°. Diacetate, crystals from ethanol + water, mp 90.5-91.5°.

2101. Chimonanthine. *1-Demethylcalycanthidine.* $C_{22}H_{26}N_4$; mol wt 346.48. C 76.27%, H 7.56%, N 16.17%. From leaves of *Chimonanthus fragrans* Lindle, Calicanthaceae: Hodson *et al., Proc. Chem. Soc.* **1961**, 465. Structure: Clayton *et al., Tetrahedron* **18**, 1495 (1962). Synthesis: Hendrickson *et al., Proc Chem. Soc.* **1962**, 383; Scott *et al., J. Am. Chem. Soc.* **86**, 302 (1964); Hall *et al., Tetrahedron* **23**, 4131 (1967).

Crystals, mp 188-189°. Weak diacidic base of equivalent weight 173. uv max: 246, 304 nm. Dimethiodide, crystals, dec 235-236°.

2102. Chimyl Alcohol. *(+)-3-(Hexadecyloxy)-1,2-propanediol;* α-hexadecylglyceryl ether; testriol. $C_{19}H_{40}O_3$; mol wt 316.52. C 72.10%, H 12.74%, O 15.16%. $CH_3(CH_2)_{15}O$-$CH_2CHOHCH_2OH$. Isoln from bull and boar testes: Hirano, *J. Pharm. Soc. Japan* **56**, 122 (1936); Prelog *et al.*, *Helv. Chim. Acta* **27**, 674 (1944). Synthesis: Baer, Fischer, *J. Biol. Chem.* **140**, 397 (1941).

Leaflets from hexane, mp 64°. bp$_{0.005}$ 120°. $[\alpha]_D^{20}$ +3.0° (c = 1.16 in chloroform). Sol in acetone, hexane, chloroform.

Bis(phenylurethan), $C_{33}H_{50}N_2O_5$, leaflets from petr ether, mp 97.5-98.5°.

2103. Chinese Wax. The excretion of an insect, *Coccus ceriferus* Fabr., or *C. pela* Westwood, deposited on the twigs and branches of a species of ash tree in Western China. Chief constituent is ceryl cerotate.

White to yellowish-white, practically odorless, tasteless solid. d ~0.93. mp ~92°. Solidif 80-81°. Sapon no. 80-92. Iodine no. 1.4. Acid no. 63. Insol in water; freely sol in benzene, slightly in alcohol or ether.

USE: Manuf candles, leather and furniture polish; treating silk and cotton fabrics; sizing and glazing papers.

2104. Chirata. Chiretta; chirayita; bitter stick; East Indian balmony. Dried plant, *Swertia (Ophelia) chirata* (Roxb.) Buch.-Ham., *Gentianaceae*. Habit. India (Himalaya). *Constit.* Chiratin, ophelic acid.

THERAP CAT: Bitter tonic, antimalarial.

2105. Chitin. $(C_8H_{13}NO_5)_n$. C 47.29%, H 6.45%, N 6.89%, O 39.37%. Cellulose-like biopolymer consisting predominantly of unbranched chains of β-(1→4)-2-acetamido-2-deoxy-D-glucose (also named N-acetyl-D-glucosamine) residues. Found in fungi, yeasts, marine invertebrates and arthropods, where it is a principal component in the exoskeletons. May be regarded as a derivative of cellulose, in which the C-2 hydroxyl groups have been replaced by acetamido residues. Occurrence in lower animals: A. G. Richards, *The Integument of Arthropods* (University of Minnesota Press, Minneapolis, 1951). Occurrence in the plant kingdom: F. von Wettstein, *Handbuch der systematischen Botanik* (F. Deuticke, Leipzig and Vienna, 4th ed., 1933). Typical isoln: Hackman, *Aust. J. Biol. Sci.* **7**, 168 (1954); Horowitz *et al.*, *J. Am. Chem. Soc.* **79**, 5046 (1957). Occurrence in the anthozoan *Stylobates aeneus* Dall, the first sea anemone proved capable of synthesizing chitin: D. F. Dunn, M. H. Liberman, *Science* **221**, 157 (1983). Structure: Dweltz, *Biochim. Biophys. Acta* **44**, 416 (1960); **51**, 283 (1961); Carlstrom, *ibid.* **59**, 361 (1962); Ramachadran, Ramakrishman, *ibid.* **63**, 307 (1962). *Review:* Foster, Webber, *Advan. Carbohyd. Chem.* **15**, 371-393 (1960); C. Jeuniaux, "Chitinous Structures" in *Comprehensive Biochemistry* **vol. 26c**, M. Florkin, E. H. Stotz, Eds. (Elsevier, New York, 1971) pp 595-632. Book: R. A. A. Muzzarelli, *Chitin* (Pergamon Press, New York, 1977). Review of properties and possible novel applications: P. R. Austin *et al.*, *Science* **212**, 749-753 (1981).

Amorphous solid. Practically insol in water, dil acids, dil and concd alkalies, alcohol and other organic solvents; sol in concd HCl, H_2SO_4, 78-97% H_3PO_4, anhydr HCOOH. There are substantial variations in solubility, molecular weight, acetyl values, specific rotation among chitins of different origins and prepared by different methods.

Acetate, $(C_{14}H_{19}NO_{11})_n$. Prepn: Shorigin, Hait, *Ber.* **68**, 971 (1935). Sol in HCOOH, 50% resorcinol, concd HCl, H_2SO_4, HNO_3; practically insol in organic solvents.

Sulfate. Process for sulfating chitin: Cushing, Kratovil, U.S. pat. **2,755,275** (1956 to Abbott). Sulfated chitin has been found to possess anticoagulant properties in laboratory animals: Roth *et al.*, *Am. J. Physiol.* **171**, 761 (1952); *Proc. Soc. Exp. Biol. Med.* **86**, 315 (1954).

USE: Deacylated chitin, *chitosan*, used in water treatment; in photographic emulsions; in improving the dyeability of synthetic fibers and fabrics.

THERAP CAT: Vulnerary.

2106. Chlophedianol. *2-Chloro-α-[2-(dimethylamino)-ethyl]-α-phenylbenzenemethanol; 2-chloro-α-(2-dimethylaminoethyl)benzhydrol;* 1-o-chlorophenyl-1-phenyl-3-dimethyl-amino-1-propanol; 1-phenyl-1-(o-chlorophenyl)-3-dimethyl-aminopropanol; α-(dimethylaminoethyl)-o-chlorobenzhydrol; clofedanol; Tussistop. $C_{17}H_{20}ClNO$; mol wt 289.80. C 70.46%, H 6.96%, Cl 12.23%, N 4.83%, O 5.52%. Prepn: Henecka, Lorenz, Ger. pat. **1,083,277** (1960); Lorenz *et al.*, U.S. pat. **3,031,377** (1962, both to Bayer).

mp 120°. LD$_{50}$ i.v. in mice: 70 mg/kg (Lorenz).

Hydrochloride, $C_{17}H_{20}ClNO \cdot HCl$, *Coldrin, Pectolitan, Refugal, Ulone, ULO.* Crystals, mp 190-191°. Freely sol in water, methanol, ethanol. Sparingly sol in ether, benzene, ethyl acetate. LD$_{50}$ orally in rats: 350 mg/kg; s.c. in mice: 95 mg/kg (Lorenz).

THERAP CAT: Antitussive.

2107. Chloracetyl Chloride. $C_2H_2Cl_2O$; mol wt 112.94. C 21.27%, H 1.78%, Cl 62.78%, O 14.17%. $ClCH_2$-$COCl$. Prepn from chloroacetic acid and benzoyl chloride: Brown, *J. Am. Chem. Soc.* **60**, 1325 (1938); by chlorination of ketene: Erickson, Prill, *J. Org. Chem.* **23**, 141 (1958); Prill, U.S. pat. **2,889,365** (1959 to Monsanto); from chloroacetic acid and pyrocatechylphosphorus trichloride: Gross, Gloede, *Ber.* **96**, 1387 (1963). Manuf from chloroacetic acid and POCl$_2$: MacKenzie, Morris, U.S. pat. **2,848,491** (1958 to Dow); from ketene: Prill, U.S. pat. **2,889,365** (1959 to Monsanto). Physical data: McDonald *et al.*, *J. Chem. Eng. Data* **4**, 311 (1959).

Liquid, very pungent odor. bp 106°. mp −21.77°. d$_4^{20}$ 1.4202. n$_D^{20}$ 1.4541. Dec by water. *Protect from moisture.*

Caution: Irritating to eyes, mucous membranes.

USE: In the synthesis of organic compounds.

2108. Chloracizine. *2-Chloro-10-[3-(diethylamino)-1-oxopropyl]-10H-phenothiazine; 2-chloro-10-(N,N-diethyl-β-alanyl)phenothiazine;* 10-(β-diethylaminopropionyl)-2-chlorophenothiazine; 2-chloro-10-(β-diethylaminopropionyl)-phenothiazine; chloracysin; chlorocyzine; khloratsizin; G-020. $C_{19}H_{21}ClN_2OS$; mol wt 360.91. C 63.23%, H 5.86%, Cl 9.82%, N 7.76%, O 4.43%, S 8.88%. Prepn: Horclois, Metivier, Fr. pat. **1,060,715** (1954 to Rhône Poulenc); **Brit.** pat. **740,932** (1955); Zhuravlev, Gritsenko, *Zhur. Obshch. Khim.* **26**, 3385 (1956), *C.A.* **51**, 9623h (1957). For other prepns in Russian journals see *C.A.* **55**, 6484b, 9425a (1961).

Hydrochloride, $C_{19}H_{21}ClN_2OS \cdot HCl$, crystals from isopropanol, mp 168-169°. Sol in water.

THERAP CAT: Vasodilator (coronary).

2109. Chloral Alcoholate. *2,2,2-Trichloro-1-ethoxy-ethanol;* chloral ethylalcoholate; trichloroacetaldehyde monoethylacetal. $C_4H_7Cl_3O_2$; mol wt 193.46. C 24.83%, H 3.65%, Cl 54.98%, O 16.54%. Prepd by refluxing trichloroacetaldehyde or chloral hydrate with alcohol: Personne,

Compt. Rend. **69**, 1363 (1869); Magnani, McElvain, J. Am. Chem. Soc. **60**, 2212 (1938); Post, J. Org. Chem. **6**, 830 (1941).

Crystals, d 1.143. mp 47.5°. bp$_{760}$ 116°. Less sol in water than chloral hydrate. Sol in organic solvents. With concd H_2SO_4 it forms trichloroacetaldehyde and ethyl sulfate.

2110. Chloral Ammonia. *1-Amino-2,2,2-trichloroethanol.* $C_2H_4Cl_3NO$; mol wt 164.42. C 14.61%, H 2.45%, Cl 64.69%, N 8.52%, O 9.73%. $Cl_3CCH(OH)NH_2$. Prepd from chloral and NH_3: Schiff, Ber. **10**, 165 (1877); Aschan, ibid. **48**, 874 (1915). Believed to be a bimolecular compd, $Cl_3C-CH(OH)NHCH(NH_2)CCl_3.H_2O$ or $C_4H_8Cl_6N_2O_2$: Aschan, loc. cit.

Crystals, mp 62-64° (Schiff, loc. cit.), mp 72-74° (Aschan, loc. cit.). Dissolves slowly in cold water with evolution of NH_3. Sol in ether, benzene.

2111. Chloral Betaine. *1-Carboxy-N,N,N-trimethylmethanaminium inner salt compd with 2,2,2-trichloro-1,1-ethanediol (1:1);* Beta-Chlor; Somilan. $C_7H_{14}Cl_3NO_4$; mol wt 282.55. C 29.76%, H 4.99%, Cl 37.64%, N 4.96%, O 22.65%. Prepd by warming betaine hydrate and chloral hydrate: Petrow et al., **Brit.** pat. **874,246** (1959 to British Drug Houses).

Prisms from water, mp 122.5-124.5°.

Note: This is a controlled substance (depressant) listed in the U.S. Code of Federal Regulations, Title 21 Part 1308.14 (1995).

THERAP CAT: Sedative, hypnotic.

2112. Chloral Formamide. *N-(2,2,2-Trichloro-1-hydroxyethyl)formamide;* chloralamide; chloramide. $C_3H_4Cl_3NO_2$; mol wt 192.43. C 18.73%, H 2.10%, Cl 55.27%, N 7.28%, O 16.63%. $HCONHCH(OH)CCl_3$. Prepn or manuf from chloral and formamide: Bennett, Campbell, Quart. J. Pharm. Pharmacol. **8**, 398 (1955); and herbicidal activity: Reuter, Sehring, **Ger.** pat. **1,168,161** (1964 to Boehringer, Ing.), C.A. **62**, 1024g (1965).

Crystals from water of 30% alc, mp 124-126°, also reported as 115-116° (Bennett, Campbell). Soluble in water; readily sol in alcohol, ether, glycerol, acetone, ethyl acetate. Dec by hot solvents.

Caution: This substance may be habit forming and is listed in the U.S. Code of Federal Regulations, Title 21 Part 329.1 (1995).

THERAP CAT: Sedative, hypnotic.

2113. Chloral Hydrate. *2,2,2-Trichloro-1,1-ethanediol;* trichloroacetaldehyde monohydrate; Escre; Noctec; Nycton; Somnos; Chloraldurat. $C_2H_3Cl_3O_2$; mol wt 165.40. C 14.52%, H 1.83%, Cl 64.30%, O 19.35%. $Cl_3CCH(OH)_2$. Made by adding the required amount of water to trichloroacetaldehyde, q.v. First synthesized by Liebig in 1832. Introduced as hypnotic by Liebreich in 1869. Comprehensive description: J. E. Fairbrother, Anal. Profiles Drug Subs. **2**, 85-143 (1973). Toxicity data: E. Goldenthal, Toxicol. Appl. Pharmacol. **18**, 185 (1971). Mutagenicity: J. Leuschner, F. Leuschner, Arzneimittel-Forsch. **41**, 1101 (1991).

Monohydrate, large, monoclinic plates. Perfect 001 cleavage. Skin irritant. Aromatic, penetrating and sightly acrid odor, slightly bitter, caustic taste. Slowly volatilizes on exposure to air. d 1.91. mp 57° when heated in an open vessel. bp 98° with dissociation into chloral and water. Freely sol in water. One ml of water dissolves the following amounts of chloral hydrate: 2.4 g at 0°; 8.3 g at 25°, 14.3 g at 40°. One gram of chlorate hydrate dissolves in 1.3 ml alcohol, in 2 ml chloroform, in 1.5 ml ether, in 1.4 ml olive oil, in 0.5 glycerol, in 68 g carbon disulfide. Freely sol in acetone, methyl ethyl ketone. Moderately or sparingly sol in turpentine, petr ether, carbon tetrachloride, benzene, toluene. Dissociation in water: Piguet et al., Helv. Chim. Acta **46**, 406 (1963). An alcoholic soln of chloral hydrate (1 in 20) does not at once redden moistened blue litmus paper. Dec by NaOH into chloroform. Reduces ammoniacal $AgNO_3$. Liquefies when triturated with an equal quantity

of menthol, camphor or thymol. LD_{50} orally in rats: 479 mg/kg (Goldenthal).

Caution: May be habit forming. This is a controlled substance (depressant) listed in the U.S. Code of Federal Regulations, Title 21 Parts 329.1 and 1308.14 (1995).

USE: Manuf DDT.

THERAP CAT: Sedative, hypnotic.

THERAP CAT (VET): Anesthetic, sedative, hypnotic.

2114. α-Chloralose. *(R)-1,2-O-(2,2,2-Trichloroethylidene)-α-D-glucofuranose;* α-D-glucochloralose; glucochloral; anhydroglucochloral; chloralosane; Alphakil; Dorcalm; Somio. $C_8H_{11}Cl_3O_6$; mol wt 309.53. C 31.04%, H 3.58%, Cl 34.36%, O 31.01%. Prepd from anhydr glucose and trichloroacetaldehyde (anhydr chloral); β-chloralose is also formed. Prepn: M. Hanroit, C. Richet, C.R. Acad. Sci. **116**, 63 (1893); and structure: Pictet, Reichel, Helv. Chim. Acta **6**, 621 (1923); Freudenberg, Vajda, J. Am. Chem. Soc. **59**, 1955 (1937); cf. Coles et al., ibid. **51**, 519 (1929); White, Hixon, ibid. **55**, 2438 (1933). Crystal structure: T. Taga et al., Acta Crystallogr. **B38**, 1874 (1982). Immobilizing agent in control of depredating birds: Redpath et al., Ann. Appl. Biol. **49**, 77 (1961). Anesthetic for dogs in experimental work: B. G. Bass, N. M. Buckley, Am. J. Physiol. **210**, 854 (1966); R. H. Cox, ibid. **223**, 660 (1972). CNS depressant activity: J. L. Barker, Nature **252**, 52 (1974). GLC determn in rodenticides: J. Theobald, J. Chromatog. **129**, 444 (1976); in tissue: E. Odam et al., Analyst **109**, 1335 (1984). Acute toxicity: E. W. Schafer, Toxicol. Appl. Pharmacol. **21**, 315 (1972). Review of pharmacology: G. U. Balis, R. R. Monroe, Psychopharmacologia **6**, 1-30 (1964); and toxicology: P. Lees, Vet. Rec. **91**, 330-333 (1972). Review as canine anesthetic: H. Holzgrefe et al., Lab. Animal Sci. **37**, 587-595 (1987).

Needles from alcohol or ether, mp 187°. $[\alpha]_D^{22}$ +19° (c = 5 in 98% alc). One gram dissolves in 225 ml water at 15°, in 120 ml at 37°, in 15 ml alc at 25°. Sol in ether, glacial acetic acid; slightly in chloroform. Almost insol in petr ether. LD_{50} orally in mice: 400 mg/kg (Schafer).

β-Chloralose, mp 227-230°, is much less sol in water, alcohol, and ether.

Caution: This substance may be habit forming and is listed in the U.S. Code of Federal Regulations, Title 21 Part 329.1 (1995).

USE: Rodenticide. Control of avian pests.

THERAP CAT: Sedative, hypnotic.

THERAP CAT (VET): Surgical anesthetic for laboratory animals.

2115. Chloramben. *3-Amino-2,5-dichlorobenzoic acid.* $C_7H_5Cl_2NO_2$; mol wt 206.03. C 40.81%, H 2.45%, Cl 34.42%, N 6.80%, O 15.53%. Selective pre-emergence herbicide. Prepn: S. R. McLane et al., **U.S.** pat. **3,014,063** (1961 to Amchem). Activity: A. J. Tafuro, Proc. Northeast Weed Control Conf. **1959**, 423. Degradn in soil: R. E. Wildung et al., Weed Res. **8**, 213 (1968). Carcinogenicity study: J. F. Robens, Vet. Human Toxicol. **22**, 328 (1980).

Cryst, mp 200-201°. Vapor press at 100°: 7 × 10^{-3} mm Hg. Soly in water at 25°: 700 mg/l. Soly in (g/100 ml): acetone 23.3; methanol 22.3; ethanol 17.3; chloroform 0.09; benzene 0.02.

Ammonium salt, $C_7H_8Cl_2N_2O_2$, *ACPM-629, Amiben.* mp 194-197° (dec). Soly in (g/100 gm) at 25°: water 29.6; ethanol 6.9; isopropanol 1.7. Soly in water at 20°: 120 mg/l.

Methyl ester, $C_8H_7Cl_2NO_2$, *Amchem 65-81B, Vegiben.* mp 63-64°. Soly in water at 20°: 120 mg/l.

USE: Herbicide.

2116. Chlorambucil. *4-[Bis(2-chloroethyl)amino]benzenebutanoic acid; 4-[p-[bis(2-chloroethyl)amino]phenyl]butyric acid; γ-[p-di(2-chloroethyl)aminophenyl]butyric acid; N,-N-di-2-chloroethyl-γ-p-aminophenylbutyric acid; chloraminophene; chloroambucil; CB-1348; Amboclorin; Leukeran.* $C_{14}H_{19}Cl_2NO_2$; mol wt 304.22. C 55.27%, H 6.30%, Cl 23.31%, N 4.60%, O 10.52%. Nitrogen mustard deriv. Prepn: W. C. J. Ross *et al.,* **Brit.** pat **727,336;** *eidem,* **U.S.** pat **2,944,079** (1955, 1960 both to Nat. Res. Dev. Corp.); Everett *et al., J. Chem. Soc.* **1953,** 2390. Improved process: Phillips, Mentha, U.S. pat. **3,046,301** (1962 to Burroughs Wellcome). Chemotherapeutic activity: Van Putten *et al., Eur. J. Cancer* **7,** 11 (1971). Toxicity: W. C. J. Ross, *Biochem. Pharmacol.* **13,** 969 (1964). Review of carcinogenicity studies: *IARC Monographs* **9,** 125-134 (1975). Comprehensive description: M. Tariq, A. A. Al-Badr, *Anal. Profiles Drug Subs.* **16,** 85-118 (1987).

Flattened needles from petr ether, mp 64-66°. Sol in ether. Sol at 20° in 1.5 parts alcohol, in 2.5 parts chloroform, in 2 parts acetone. Practically insol in water. LD_{50} i.p. in rats: 58.2 μmole/kg (Ross).

Note: This substance has been listed as a known carcinogen: *Seventh Annual Report on Carcinogens* (PB95-109781, 1994) p 43.

THERAP CAT: Antineoplastic.

THERAP CAT (VET): Antineoplastic agent, esp for leukemia.

2117. Chloramine-B. *N-Chlorobenzenesulfonamide sodium salt; (N-chlorobenzenesulfonamido)sodium;* benzenesulfochloramine; Neomagnol. $C_6H_5ClNNaO_2S$; mol wt 213.62. C 33.74%, H 2.36%, Cl 16.60%, N 6.56%, Na 10.76%, O 14.98%, S 15.01%. $C_6H_5SO_2NNaCl.$ Prepd via benzenesulfonamide: Chattaway, *J. Chem. Soc.* **87,** 145 (1905); Cuiban *et al., Pharmazie* **13,** 407 (1958).

Trihydrate, prisms. Soluble in 20 parts of water, more sol in hot water. Sol in 25 parts of ethanol, yielding a turbid soln. Very sparingly sol in ether, chloroform. An aq soln first turns red litmus paper blue, then gradually bleaches it. Gives a red color wth phenolphthalein.

THERAP CAT: Antibacterial.

THERAP CAT (VET): Antiseptic (topical).

2118. Chloramine-T. *N-Chloro-4-methylbenzenesulfonamide sodium salt; (N-chloro-p-toluenesulfonamido)sodium;* sodium *p*-toluenesulfonchloramide; chloramine; Aktiven; Chloraseptine; Chlorazene; Chlorazone; Euclorina; Gansil; Halamid; Mianine; Tochlorine; Tolamine. $C_7H_7ClNNaO_2S$; mol wt 227.65. C 36.93%, H 3.10%, Cl 15.57%, N 6.15%, Na 10.10%, O 14.06%, S 14.09%. p-$CH_3C_6H_4SO_2NNaCl.$ Prepd via *p*-toluenesulfonamide: Chattaway, *J. Chem. Soc.* **87,** 145 (1905); Inglis, *J. Soc. Chem. Ind. (London)* **37,** 288 (1918); F. J. Welcher, *Organic Analytical Reagents* vol. 4 (Van Nostrand, New York, 1948) pp 316-320.

Trihydrate, prisms. Loses water on drying. Dec slowly on exposure to air. *Keep well closed.* Fairly sol in water. Practically insol in benzene, chloroform, ether. Dec by alc.

USE: Detection of bromate and halogens, F. J. Welcher, *loc. cit.*

THERAP CAT: Antibacterial.

THERAP CAT (VET): Antiseptic (topical).

2119. Chloraminophenamide. *4-Amino-6-chloro-1,3-benzenedisulfonamide; 4-amino-6-chloro-m-benzenedisulfonamide;* 5-chloro-2,4-bis(sulfonamido)aniline; 6-amino-4-chlorobenzene-1,3-disulfonamide; 5-chloro-2,4-disulfamyl-

aniline; Idorese. $C_6H_8ClN_3O_4S_2$; mol wt 285.73. C 25.22%, H 2.82%, Cl 12.41%, N 14.71%, O 22.40%, S 22.44%. Prepn: Novello, Sprague, *J. Am. Chem. Soc.* **79,** 2028 (1957); Novello, U.S. pats. **2,809,194; 2,965,655; 2,965,656** (1957, 1960, 1960 all to Merck & Co.); Novello *et al., J. Org. Chem.* **25,** 965 (1960).

Crystals from aq ethanol, mp 251-252°. uv max (ethanol): 223.5-224.5, 265-266, 312-314 nm (ε 41776, 18633, 3874). Slightly sol in water; more freely sol in alkalies.

THERAP CAT: Diuretic.

2120. Chloramphenicol. *[R-(R*,R*)]-2,2-Dichloro-N-[2-hydroxy-1-(hydroxymethyl)-2-(4-nitrophenyl)ethyl]acetamide;* D-*threo*-N-dichloroacetyl-1-*p*-nitrophenyl-2-amino-1,3-propanediol; D(−)-*threo*-2-dichloroacetamido-1-*p*-nitrophenyl-1,3-propanediol; D-*threo*-N-(1,1'-dihydroxy-1-*p*-nitrophenylisopropyl)dichloroacetamide; Ak-Chlor; Amphicol; Anacetin; Aquamycetin; Chemicetina; Chloramex; Chlorasol; Chloricol; Chlorocid; Chloromycetin; Chloroptic; Cloramfen; Clorocyn; Enicol; Farmicetina; Fenicol; Globenicol; Intramycetin; Kemicetine; Leukomycin; Micoclorina; Mychel; Mycinol; Novomycetin; Ophthochlor; Pantovernil; Paraxin; Quemicetina; Ronphenil; Sintomicetina; Sno Phenicol; Synthomycetin; Tevcocin; Tifomycine; Veticol; Viceton. $C_{11}H_{12}Cl_2N_2O_5$; mol wt 323.13. C 40.89%, H 3.74%, Cl 21.94%, N 8.67%, O 24.76%. Broad spectrum antibiotic obtained from cultures of the soil bacterium *Streptomyces venezuelae:* Bartz, *J. Biol. Chem.* **172,** 445 (1948); Gottlieb *et al., J. Bact.* **55,** 409 (1948); Ehrlich *et al., ibid.* **56,** 467 (1948). Isoln from the moon snail, *Lunatia heros:* C. A. Price *et al., J. Antibiot.* **34,** 118 (1981). Structure: Rebstock *et al., J. Am. Chem. Soc.* **71,** 2458 (1949). Synthesis: Controulis *et al., ibid.* 2463; Long, Troutman, *ibid.* 2469, 2473. *See also* U.S. pats. **2,483,871; 2,483,884; 2,483,892.** Alternate synthesis: Ehrhart *et al., Ber.* **90,** 2088 (1957); **Brit.** pats. **795,131; 796,901,** *C.A.* **53,** 2161 (1959) (both to Chinoin); U.S. pat. **2,839,577** (1958 to Chinoin). Review and evaluation of toxicity studies: *IARC Monographs* **10,** 85-98 (1976). Review of pharmacology and clinical efficacy: I. Shalit, M. I. Marks, *Drugs* **28,** 281-291 (1984). Comprehensive description: D. Szulczewski, F. Eng, *Anal. Profiles Drug Subs.* **4,** 47-90 (1975); A. A. Al-Badr, H. A. El-Obeid, *ibid.* **15,** 701-760 (1986). *Reviews:* Hahn in *Antibiotics,* **vol. 1,** D. Gottlieb, P. D. Shaw, Eds. (Springer-Verlag, New York, 1967) pp 308-330; Pestka, *ibid.* **vol. 3,** J. W. Corcoran, F. E. Hahn, Eds. (1975) pp 370-395; O. Pongs, *ibid.* **vol. 5,** pt. 1, F. E. Hahn, Ed. (1979) pp 26-42.

Needles or elongated plates from water or ethylene dichloride. mp 150.5-151.5°. Sublimes in high vacuum. $[\alpha]_D^{27}$ +18.6° (c = 4.86 in ethanol). $[\alpha]_D^{25}$ −25.5° (ethyl acetate). uv max: 278 nm ($E_{1cm}^{1\%}$ 298). Soly (25°) in water: 2.5 mg/ml; in propylene glycol: 150.8 mg/ml. Very sol in methanol, ethanol, butanol, ethyl acetate, acetone. Fairly sol in ether. Insol in benzene, petr ether, vegetable oils. Soly in 50% acetamide soln about 5%. Additional soly data: Weiss *et al., Antibiot. Chemother.* **7,** 374 (1957). Aq solns are neutral. Neutral and acid solns are stable on heating.

Monosuccinate sodium salt, $C_{15}H_{15}Cl_2N_2NaO_8$, *Protophenicol.* Freely sol in water (about 50% w/w).

Palmitate, *Chlorambon, Chloropal, Clorolifarina.* Prepn: Edgerton, U.S. pat. **2,662,906** (1953 to Parke, Davis). Structure: Edgerton *et al., J. Am. Chem. Soc.* **77,** 27 (1955). Description: Glazko *et al., Antibiot. Chemother.* **2,** 234

(1952). Soly data: Weiss *et al., ibid.* **7,** 374 (1957). Crystals from benzene, mp 90°. Practically tasteless. $[\alpha]_D^{26}$ +24.6° (c = 5 in ethanol). uv max (ethanol): 271 nm ($E_{1cm}^{1\%}$ 179). Very slightly sol in water (1.05 mg/ml at 28°); petr ether (0.225 mg/ml). Freely sol in methanol, ethanol, chloroform, ether, benzene.

Monosuccinate arginine salt, $C_{21}H_{30}Cl_2N_6O_{10}$, *chloramphenicol arginine succinate, Paraxin Succinate A.* mp 135-145° (dec.). *See Japan. Med. Gaz.* 7(10), 15 (1970).

Pantothenate calcium complex (4:1), $C_{62}H_{80}CaCl_8N_{10}O_{30}$, *chloramphenicol pantothenate, Pantofenicol.* Prepn: I. Villax, **Brit.** pats. **866,787, 866,788** and **866,789** (all 1961), I. Villax, **U.S.** pat. **3,078,300** (1963).

THERAP CAT: Antibacterial; antirickettsial.

THERAP CAT (VET): Antibacterial.

2121. Chloranil. *2,3,5,6-Tetrachloro-2,5-cyclohexadiene-1,4-dione; 2,3,5,6-tetrachloro-p-benzoquinone;* tetrachloroquinone; Spergon; Vulklor. $C_6Cl_4O_2$; mol wt 245.88. C 29.31%, Cl 57.68%, O 13.01%. Prepd from *p*-phenylenediamine or phenol by treating with KClO$_3$ and HCl. Because of its great resistance to further oxidation, chloranil is formed as the final product of the chlorate-HCl oxidation of many aromatic compds. For comprehensive list *see* Huntress, *Organic Chlorine Compounds* (New York, 1948). Laboratory procedure starting with phenol or *p*-chlorophenol: Fierz-David, Blangey, *Grundlegende Operationen der Farbenchemie* (Vienna, 5th ed., 1943) p 140.

Golden-yellow platelets from acetic acid or acetone. Monoclinic prisms from benzene or toluene or by sublimation *in vacuo,* mp 290°. Absorption spectrum in chloroform: Lifschitz *et al., Rec. Trav. Chim.* **43,** 276, 658 (1924). Insol in water; almost insol in cold petr ether, cold alcohol. Sol in ether; sparingly sol in chloroform, carbon tetrachloride, carbon disulfide. Solubility data: Dimroth, Bamberger, *Ann.* **438,** 103, 106 (1924). LD$_{50}$ orally in rats: 4.0 g/kg, *Toxic Substances List,* H. E. Christensen, Ed (1973) p 909.

Caution: May be irritating to skin, mucous membranes.

USE: Fungicide. In manuf of chloranil electrodes for pH measurements. As reagent for pamaquine (Plasmochin) in urine (blue color): Schulemann *et al., Chem. Zentr.* **1928, I,** 2193. Two of the chlorine atoms in *para* position are easily substituted, and compds such as 2,5-dianilino-3,6-dichloroquinone are much used in the dye industry.

2122. Chloranilic Acid. *2,5-Dichloro-3,6-dihydroxy-2,5-cyclohexadiene-1,4-dione; 2,5-dichloro-3,6-dihydroxy-p-benzoquinone.* $C_6H_2Cl_2O_4$; mol wt 208.98. C 34.48%, H 0.96%, Cl 33.93%, O 30.62%. Prepd from chloranil by alkaline hydrolysis: Conant, Fieser, *J. Am. Chem. Soc.* **46,** 1866 (1924).

Red crystals, mp 283-284°. Tendency to sublime. Relatively strong dibasic acid. Absorption max (pH 4 in water): 290-340 nm, 520-555 nm.

USE: Reacts with metal cations to form stable salts. Used in spectrophotometry: Hart, *Org. Chem. Bull.* **33,** no. 3 (1961).

2123. Chlorazanil. *N-(4-Chlorophenyl)-1,3,5-triazine-2,4-diamine; 2-amino-4-(p-chloroanilino)-s-triazine; N-(p-chlorophenyl)-2,4-diamino-s-triazine; 2-(p-chloroanilino)-4-amino-s-triazine;* chlorazinil; ASA-226; Diurazine; Triazu-

rol; Orpizin; Daquin; Neo-Urofort; Neurofort. $C_9H_8ClN_5$; mol wt 221.65. C 48.77%, H 3.64%, Cl 15.99%, N 31.60%. Prepn: Clauder, Bulcsu, *Magyar Kém. Folyoirat* **57,** 68 (1951), *C.A.* **46,** 4023g (1952); **Austrian** pat. **168,063** (1951) and **Brit.** pat. **676,024** (1952), *C.A.* **47,** 8097g, 3887i (1953). Synthesis and diuretic activity: Shah *et al., J. Med. Chem.* **11,** 1167 (1968). *See also* Takagi, Ueda, *Chem. Pharm. Bull.* **17,** 1061 (1969).

Crystals, mp 233-234°. Also reported as crystals from ethanol, mp 256-258° (Shah *et al., loc. cit.*).

Hydrochloride, $C_9H_8ClN_5$·HCl, *Doclizid-T, Orpidan.* mp 277-278°.

THERAP CAT: Diuretic.

2124. Chlorbenside. *1-Chloro-4-[[(4-chlorophenyl)-methyl]thio]benzene; p-chlorobenzyl p-chlorophenyl sulfide;* chlorocide; Chlorparacide; Chlorsulphacide; Mitox. $C_{13}H_{10}Cl_2S$; mol wt 269.19. C 58.00%, H 3.74%, Cl 26.34%, S 11.91%. Prepn of analogous compd, *p*-chlorobenzyl *p*-nitrobenzyl sulfide, from *p*-chlorophenylthiol and *p*-nitrobenzyl chloride: Stevenson *et al.,* **Brit.** pat. **738,170** (1955 to Boots Pure Drug). Chemical and biological properties: Cranham *et al., Chem. & Ind. (London)* **1953,** 1206.

Crystals, mp 75-76°. Technical grades may have almond-like odor. d_4^{25} 1.4210. Vapor tension at 30°: 1.21 × 10^{-5} mm. uv max (95% ethanol): 262 nm: Watson, *Chem. & Ind. (London)* **1956,** 349. Soly in water < 1:5000, in ethanol about 2.9%, in kerosene 5-7.5%. Sol in acetone, benzene, toluene, xylene, petr ether. Resistant to acid and alkaline hydrolysis. Strong oxidizing agents convert it to the corresponding sulfoxide and sulfone.

Caution: May be irritating to skin. Kidney and liver injuries have been produced in exptl animals.

USE: Acaricide, esp for the control of eggs and larvae of red spider mites.

2125. Chlorbenzoxamine. *1-[2-[(2-Chlorophenyl)phenylmethoxy]ethyl]-4-[(2-methylphenyl)methyl]piperazine; 1-[2-(o-chloro-α-phenylbenzyloxy)ethyl]-4-o-methylbenzylpiperazine; 1-[2'-(o-chlorobenzhydryloxy)ethyl]-4-(o-methylbenzyl)diethylenediamine; 1-(o-chlorobenzhydryloxyethyl)-4-(o-methylbenzyl)piperazine;* chlorbenzoxyethamine; chlorbenzoxamine. $C_{27}H_{31}ClN_2O$; mol wt 435.01. C 74.55%, H 7.18%, Cl 8.15%, N 6.44%, O 3.68%. Prepn: Morren *et al., Ind. Chim. Belge* **22,** 409 (1957); Morren, **Brit.** pat. **837,986** (1960). Toxicity study: Levis *et al., Arch. Int. Pharmacodyn. Ther.* **118,** 167 (1959).

bp$_{0.01}$ 234-236°; bp$_{0.005}$ 235°.

Dihydrochloride, $C_{27}H_{31}ClN_2O$·2HCl, *UCB-1474, Libratar, Antiulcera Master, Gastomax.* Bitter crystals, mp 197-200°. Freely sol in methanol; sol in ethanol, chloroform, acetic acid; slightly sol in water, acetone. Practically insol in acetonitrile, ether, benzene. LD_{50} in rats (mg/kg): 4000 s.c., 66 i.v.; in mice (mg/kg): 1400 orally (Levis).

THERAP CAT: Anticholinergic.

2126. Chlorbetamide. *2,2-Dichloro-N-[(2,4-dichloro-phenyl)methyl]-N-(2-hydroxyethyl)acetamide; 2,2-dichloro-N-(2,4-dichlorobenzyl)-N-(2-hydroxyethyl)acetamide; N-(2,4-di-chlorobenzyl)-N'-(2-hydroxyethyl)dichloroacetamide; Win-5047; Mantomide; Pontalin.* $C_{11}H_{11}Cl_4NO_2$; mol wt 331.02. C 39.91%, H 3.35%, Cl 42.84%, N 4.23%, O 9.67%. Prepn: Surrey, *J. Am. Chem. Soc.* **76**, 2214 (1954); U.S. pat. **2,732,-402** (1956 to Sterling Drug).

Crystals from benzene + Skellysolve A, mp 112.4-113.4°. Slightly bitter taste. Very slightly sol in water; sol in 95% ethanol to at least 5%.

THERAP CAT: Antiamebic.

2127. Chlorbicyclen. *5,6-Bis(chloromethyl)-1,2,3,4,7,7-hexachlorobicyclo[2.2.1]hept-2-ene; 1,2,3,4,7,7-hexachloro-5,-6-bis(chloromethyl)-2-norbornene; 5,6-bis(chloromethyl)-1,-2,3,4,7,7-hexachloro-2-norbornene; 1,4,5,6,7,7-hexachloro-2,3-bis(chloromethyl)bicyclo[2.2.1]hept-5-ene; Alodan (obsolete).* $C_9H_6Cl_8$; mol wt 397.77. C 27.18%, H 1.52%, Cl 71.30%. Prepn: Fensch, Finkenbrink, Ger. pat. **1,002,341** (1957 to Hoechst); Hoch, Clegg, *J. Am. Chem. Soc.* **81**, 5413 (1959); U.S. pat. **2,951,098** (1960 to Hooker). Synthesis of the *trans*-isomer: I. Schuphan, G. Tölg, *Tetrahedron Letters* **1969**, 3389.

trans-isomer

Solid, mp 104-106°. bp_2 172-176°.
trans-isomer, mp 74-76°.

USE: Formerly as an insecticide and acaricide.

2128. Chlorcyclizine. *1-[(4-Chlorophenyl)phenylmeth-yl]-4-methylpiperazine; 1-(p-chloro-α-phenylbenzyl)-4-methyl-piperazine; 1-(4-chlorobenzhydryl)-4-methylpiperazine; N-methyl-N'-(4-chlorobenzhydryl)piperazine; chlorocyclizine; compd 47-282; Alergicide; Perazyl; Trihistan.* $C_{18}H_{21}ClN_2$; mol wt 300.83. C 71.87%, H 7.04%, Cl 11.78%, N 9.31%. Prepn: Baltzly *et al.*, *J. Org. Chem.* **14**, 775 (1949); Murfitt, Dewing, Brit. pat. **656,043** (1951 to Wellcome Found.); U.S. pat. **2,630,435** (1953 to Burroughs Wellcome). Toxicity study: Castillo *et al.*, *J. Pharmacol. Exp. Ther.* **96**, 388 (1949).

Oil, $bp_{0.1-0.15}$ 137-145°.
Hydrochloride, $C_{18}H_{21}ClN_2 \cdot HCl$, *AH-289, Di-Paralene, Perazil, Histantin.* Crystalline powder, mp 226-227°. One

gram dissolves in about 2 ml water, in 11 ml alc, in about 4 ml chloroform. Practically insol in ether, benzene.
Dihydrochloride, $C_{18}H_{21}ClN_2 \cdot 2HCl$, prisms from alc + ether, mp 216-216.5°. Freely sol in water; sol in alc. Aq soln is acid to litmus. LD_{50} i.p. in mice: 137 mg/kg (Castillo).
Di[(*tert*-butyl)naphthalene sodium sulfonate], $C_{18}H_{21}Cl-N_2 \cdot 2C_{14}H_{15}NaOS$, *Bexedan.*

THERAP CAT: Antihistaminic.

THERAP CAT (VET): The hydrochloride as antihistaminic.

2129. Chlordan(e). *1,2,4,5,6,7,8,8-Octachloro-2,3,3a,4,-7,7a-hexahydro-4,7-methano-1H-indene; 1,2,4,5,6,7,8,8-octa-chloro-3a,4,7,7a-tetrahydro-4,7-methanoindan; 1,2,4,5,6,7,8,-8-octachloro-4,7-methane-3a,4,7,7a-tetrahydroindane; CD-68; Velsicol 1068; Toxichlor; Niran; Octachlor; Ortho-Klor; Synklor; Corodane; Belt.* $C_{10}H_6Cl_8$; mol wt 409.78. C 29.31%, H 1.48%, Cl 69.21%. The commercial product is a mixture containing 60 to 75% of the pure compound and 25 to 40% of related compds. Chlorine content: 64-67%. Prepn: Hyman, Brit. pat. **618,432** (1949); Kleiman, U.S. pat. **2,598,561** (1952 to Velsicol). Toxicity study: R. B. Harbison, *Toxicol. Appl. Pharmacol.* **32**, 443 (1975).

Viscous, amber-colored liquid. Viscosity 69 poises at 25° (about that of 95% glycerol). Viscosity reduced considerably by heating to 120-140°F when it may be sprayed directly. d^{25} 1.59-1.63. n_D^{25} 1.56-1.57. Insol in water. Miscible with aliphatic and aromatic hydrocarbon solvents, including deodorized kerosene. Loses its chlorine in presence of alkaline reagents, and should not be formulated with any solvent, carrier, diluent or emulsifier, which has an alkaline reaction. LD_{50} i.p. in male rats: 343 mg/kg (Harbison).

Note: The EPA has cancelled registrations of pesticides containing this compound with the exception of its use through subsurface ground insertion for termite control and the dipping of roots or tops of non-food plants: *Fed. Reg.* vol. **40**, p 28850 (July 9, 1975).

Caution: Potential symptoms of overexposure are blurred vision; confusion; ataxia, delirium; coughing; abdominal pain, nausea, vomiting and diarrhea; irritability, tremor, convulsions; anuria. *See NIOSH Pocket Guide to Chemical Hazards* (DHHS/NIOSH 90-117, 1990) p 60. Direct contact may cause skin, eye and mucous membrane irritation. *See Patty's Industrial Hygiene and Toxicology* vol. **2B**, G. D. Clayton, F. E. Clayton, Eds. (Wiley-Interscience, New York, 3rd ed., 1981) pp 3718-3725. *See also Clinical Toxicology of Commercial Products*, R. E. Gosselin *et al.*, Eds. (Williams & Wilkins, Baltimore, 5th ed., 1984) Section III, p 108-109.

USE: Insecticide: Stomach poison, contact poison, fumigant. *See* Bussart, Schor, *Soap Sanit. Chemicals* **24**, 126 (1948).

THERAP CAT (VET): Insecticide, acaricide.

2130. Chlordantoin. *5-(1-Ethylpentyl)-3-[(trichloro-methyl)thio]-2,4-imidazolidinedione; 5-(1-ethylpentyl)-3-(tri-chloromethylthio)hydantoin; 5-(1-ethylamyl)-3-(trichloro-methylthio)hydantoin; 3-(trichloromethylthio)-5-(1-ethyl-pentyl)hydantoin; 3-(trichloromethylthio)-5-(1-ethylamyl)-hydantoin.* $C_{11}H_{17}Cl_3N_2O_2S$; mol wt 347.69. C 38.00%, H 4.93%, Cl 30.59%, N 8.06%, O 9.20%, S 9.22%. Prepn of analogous compounds: Kittleson, *Science* **115**, 84 (1952); *cf.* Kupferberg, Doscher, *Antibiot. & Chemother.* **11**, 73 (1961).

Component of *Sporostacin*.
THERAP CAT: Antifungal.

2131. Chlordecone. *1,1a,3,3a,4,5,5,5a,5b,6-Decachloro-octahydro-1,3,4-metheno-2H-cyclobuta[cd]pentalen-2-one;* GC-1189; Kepone. $C_{10}Cl_{10}O$; mol wt 490.64. C 24.48%, Cl 72.26%, O 3.26%. Prepn: Gilbert, Giolito, U.S. pats. **2,616,825** and **2,616,928** (1952 to Allied Chem.); reissued as U.S. pat **Re. 24,435** (1958). Structure: McBee *et al., J. Am. Chem. Soc.* **78**, 1511 (1956). *Review:* Ungnade, McBee, *Chem. Rev.* **58**, 249-320 (1958). Carcinogenicity study: *Am. Ind. Hyg. Assoc. J.* **37**, 680 (1976). Toxicity data: Gaines, *Toxicol. Appl. Pharmacol.* **14**, 515 (1969). Occupational exposure: *Chem. & Eng. News* **54**, 19 (Feb. 2, 1976). Comparative toxicity in man and animals: P. S. Guzelian, *Ann. Rev. Pharmacol. Toxicol.* **22**, 89-113 (1982).

Crystals, dec 350°. Slightly sol in water and hydrocarbon solvents. Sol in alcohols, ketones, acetic acid. LD_{50} orally in rats: 125 mg/kg (Gaines).
Caution: Potential symptoms of overexposure are headache, nervousness, tremor; liver and kidney damage; visual disturbances; ataxia, chest pain, skin erythema; testicular atrophy, low sperm count. *See NIOSH Pocket Guide to Chemical Hazards* (DHHS/NIOSH 94-116, 1994) p 184. *See also Patty's Industrial Hygiene and Toxicology* vol. **2B**, G. D. Clayton, F. E. Clayton, Eds. (John Wiley & Sons, New York, 4th ed., 1994) 1504-1506, 1545-1550. This substance may reasonably be anticipated to be a carcinogen: *Seventh Annual Report on Carcinogens* (PB95, 109781, 1994) p 237.
USE: Formerly as insecticide, fungicide.

2132. Chlordiazepoxide. *7-Chloro-N-methyl-5-phenyl-3H-1,4-benzodiazepin-2-amine 4-oxide; 7-chloro-2-methyl-amino-5-phenyl-3H-1,4-benzodiazepine 4-oxide;* methaminodiazepoxide; clopoxide; Helogaphen; Libritabs; Multum; Risolid; Silibrin; Tropium. $C_{16}H_{14}ClN_3O$; mol wt 299.76. C 64.11%, H 4.71%, Cl 11.83%, N 14.02%, O 5.34%. Prototype of the benzodiazepine anxiolytics. Prepn: Sternbach, U.S. pat. **2,893,992** (1959 to Hoffmann-La Roche); Sternbach, Reeder, *J. Org. Chem.* **26**, 1111 (1961). Comprehensive description: A. MacDonald *et al., Anal. Profiles Drug Subs.* **1**, 15-51 (1972). Review of pharmacokinetics: D. J. Greenblatt *et al., Clin. Pharmacokinet.* **3**, 381-394 (1978); of pharmacology and effect on cognitive function: J. B. Murray, *Genet. Psychol. Monogr.* **109** (2), 167-197 (1984). Clinical trial in alcohol withdrawal: A. K. Burroughs *et al., Alcohol Alcohol.* **20**, 263 (1985).

Light yellow plates from ethanol, mp 236-236.5°. pK 4.8.
Hydrochloride, $C_{16}H_{14}ClN_3O.HCl$, *Ansiacal, A-Poxide, Balance, Benzodiapin, Cebrum, Corax, Disarim, Elenium, Equibral, Labican, Lentotran, Librium, O.C.M., Psichial, Psicoterina, Reliberan, Seren Vita, SK-Lygen, Viansin.* Crystals from methanol, mp 213°. Sol in water.
Note: This is a controlled substance (depressant) listed in the U.S. Code of Federal Regulations, Title 21 Part 1308.14 (1995).
THERAP CAT: Anxiolytic.
THERAP CAT (VET): Tranquilizer.

2133. Chlordimeform. *N'-(4-Chloro-2-methylphenyl)-N,N-dimethylmethanimidamide; N'-(4-chloro-o-tolyl)-N,N-di-methylformamidine;* chlorophenamidine; chlorphenamidine; spanon; CDM; Ciba 8514; Schering 36268; Fundal; Galecron. $C_{10}H_{13}ClN_2$; mol wt 196.68. C 61.07%, H 6.66%, Cl 18.03%, N 14.24%. Member of a class of insecticides possessing the formamidine moiety, effective vs many organophosphate and carbamate resistant pests. Prepn: Arndt, Steinhausen, Ger. pat. **1,172,081** corresp to U.S. pats. **3,378,437** and **3,502,720** (1964, 1968, 1970, all to Schering AG). Acts by interfering with amine regulatory mechanisms: Beeman, Matsumura, *Nature* **242**, 273 (1973); Aziz, Knowles, *ibid.* 417. Review of metabolism: Knowles, *J. Agr. Food Chem.* **18**, 1038 (1970). Degradation products: Witkonton, Ercegovich, *ibid.* **20**, 569 (1972). Toxicity data: C. P. Robinson, P. W. Smith, *J. Toxicol. Environ. Health* **3**, 565 (1977). Toxicological studies: D. S. Folland *et al., J. Am. Med. Assoc.* **13**, 1052 (1978); K. T. Maddy *et al., Toxicol. Letters* **33**, 37 (1986). Review of carcinogenicity studies: *IARC Monographs* **30**, 61-72 (1983).

Colorless crystals, mp 35°. $bp_{0.4}$ 156-157°. n_D^{25} 1.5885. d_4^{25} 1.105. Vapor pressure at 20°: 3.5×10^{-4} mm. Slightly sol in water; easily sol in organic solvents. LD_{50} i.p. in rats: 238 mg/kg (Robinson, Smith).
USE: Acaricide, insecticide.

2134. Chlorendic Anhydride. *4,5,6,7,8,8-Hexachloro-3a,4,7,7a-tetrahydro-4,7-methanoisobenzofuran-1,3-dione; 1,4,5,6,7,7-hexachloro-endo-5-norbornene-2,3-dicarboxylic anhydride;* hexachloroendomethylenetetrahydrophthalic anhydride; 1,4,5,6,7,7-hexachloro-*endo*-bicyclo[2.2.1]hept-5-ene-2,3-dicarboxylic anhydride. $C_9H_2Cl_6O_3$; mol wt 370.83. C 29.15%, H 0.54%, Cl 57.36%, O 12.94%. Prepd from hexachlorocyclopentadiene and maleic anhydride: Herzfeld *et al.,* U.S. pat. **2,606,910** (1952 to Velsicol); Baranauckas *et al.,* U.S. pat. **2,903,463** (1959 to Hooker Chem.). Purification: Zimmer *et al.,* Ger. pat. **1,113,451** (1961 to Hooker Chem.), *C.A.* **57**, 697a (1962).

Crystals, mp 231-235°.
USE: In prepn of polyester resins.

2135. Chlorfenac. *2,3,6-Trichlorobenzeneacetic acid; 2,3,6-trichlorophenylacetic acid;* Fenac; Tri-Fene. $C_8H_5Cl_3O_2$; mol wt 239.48. C 40.12%, H 2.10%, Cl 44.41%, O 13.36%. Prepd by conversion of 2,3,6-trichlorotoluene to the benzyl chloride, the benzyl cyanide, then hydrolysis: **Brit.** pat. **860,310** (1961 to Hooker). Toxicity study: G. W. Bailey, J. L. White, *Residue Rev.* **10**, 97 (1965).

Crystals from benzene, mp 161°. Sol in acetone, alcohol, ether. Practically insol in water. LD_{50} orally in rats: 3000 mg/kg (Bailey, White).

USE: Herbicide.

2136. Chlorfenethol. *4-Chloro-α-(4-chlorophenyl)-α-methylbenzenemethanol; 4,4'-dichloro-α-methylbenzhydrol; p,p'*-dichlorodiphenylmethyl carbinol; 1,1-bis(p-chlorophenyl)ethanol; 1,1-bis(p-chlorophenyl)methyl carbinol; DMC; DCPC; Dimite. $C_{14}H_{12}Cl_2O$; mol wt 267.15. C 62.94%, H 4.53%, Cl 26.54%, O 5.99%. Prepn: Ruthruff *et al.*, U.S. pat **2,430,586** (1947 to Sherwin-Williams); **Brit.** pat. **831,-421** (1960 to Metal & Thermit). Activity: O. Grummitt, *Science* **111**, 361 (1950).

Crystals from Skellysolve C, mp 69-69.5°. Sol in organic solvents. Practically insol in water. LD_{50} orally in rats: 500 mg/kg, *RTECS* **Vol. I**, R. J. Lewis, R. L. Tatken, Eds. (1979) p 225.

Caution: Effects similar to DDT, *q.v.*

USE: Acaricide.

2137. Chlorfenvinphos. *Phosphoric acid 2-chloro-1-(2,-4-dichlorophenyl)ethenyl diethyl ester; O,O-diethyl O-[2-chloro-1-(2,4-dichlorophenyl)vinyl] phosphate; 2,4-dichloro-α-(chloromethylene)benzyl alcohol diethyl phosphate;* CVP; SD-7859; Compd 4072; Birlane; Dermaton; Sapecron; Steladone; Supona. $C_{12}H_{14}Cl_3O_4P$; mol wt 359.57. C 40.08%, H 3.92%, Cl 29.58%, O 17.80%, P 8.61%. Commercial prepn contains at least 90% of the active Z-isomer. Prepn: E. E. Gilbert *et al.*, U.S. pat. **3,003,916** (1961 to Allied Chemical); Phillips *et al.*, U.S. pat. **3,102,842** (1963 to Shell); Whetstone, Phillips *et al.*, *J. Agr. Food Chem.* **14**, 352 (1966). Metabolism: W. F. Chamberlain, D. E. Hopkins, *J. Econ. Entomol.* **55**, 86 (1962); D. H. Hotson *et al.*, *Biochem. J.* **102**, 133 (1967). Toxicity studies: A. M. Ambrose *et al.*, *Toxicol. Appl. Pharmacol.* **17**, 323 (1970); Bunyan *et al.*, *Pestic. Sci.* **2**, 148 (1971).

Z - isomer

Amber liquid, mild odor, $bp_{0.001}$ 120°, $bp_{0.5}$ 167-170°. Vapor press at 25°C: 7.5×10^{-6} mm Hg. n_D^{25} 1.5272. Soly in water at 23°: 145 ppm. Miscible with acetone, ethanol, propylene glycol. Slowly hydrolyzed by water. Corrosive to metal. Toxic to fish. LD_{50} in rats (mg/kg): 6.6 i.v.; 9.66 orally (Ambrose).

USE: Insecticide; acaricide.

2138. Chlorguanide. *N-(4-Chlorophenyl)-N'-(1-methylethyl)imidodicarbonimidic diamide; 1-(p-chlorophenyl)-5-isopropylbiguanide; N^1-p-chlorophenyl-N^5-isopropyldiguanide;* chloroguanide; proguanil. $C_{11}H_{16}ClN_5$; mol wt 253.73. C 52.07%, H 6.36%, Cl 13.97%, N 27.60%. Prepn: Curd, Rose, *J. Chem. Soc.* **1946**, 729; Curd *et al.*, *ibid.* **1948**, 1630; Ainley *et al.*, *ibid.* **1949**, 98. Manuf of acetate: Gailliot, **Fr.** pat. **1,001,548** (1952 to Rhône-Poulenc), *C.A.* **51**, 7411e

(1957). Toxicity study: Schmidt *et al.*, *J. Pharmacol. Exp. Ther.* **90**, 233 (1947).

Rectangular plates from toluene, mp 129°.
Acetate, $C_{11}H_{16}ClN_5 \cdot C_2H_4O_2$, crystals from acetone, mp 189-190°.
Hydrochloride, $C_{11}H_{16}ClN_5 \cdot HCl$, *M-4888, RP-3359, SN-12837, Diguanyl, Drinupal, Guanatol, Palusil, Paludrine, Tirian.* Crystals from water or aq ethanol, mp 243-244°. uv max (alc): 259 nm. Sol in alc; slightly sol in water. Practically insol in chloroform and ether. pH of satd aq soln 5.8-6.3. LD_{50} orally in rats: 200 mg/kg (Schmidt).

THERAP CAT: Antimalarial.

2139. Chlorhexadol. *2-Methyl-4-(2,2,2-trichloro-1-hydroxyethoxy)-2-pentanol;* 2-methyl-2-hydroxy-4-(2,2,2-trichloro-1-hydroxyethoxy)pentane; 2-hydroxy-2-methyl-4-(2,2,2-trichloro-1-hydroxyethoxy)pentane; Chloralodol; Lora (formerly); Mecoral; Medodorm. $C_8H_{15}Cl_3O_3$; mol wt 265.56. C 36.18%, H 5.69%, Cl 40.05%, O 18.07%. Prepd from chloral and 2-methyl-2,4-pentanediol: Christensen, U.S. pat. **2,931,838** (1960 to Danske Medicinal).

Crystals, slightly bitter taste, mp 102-104°. Readily sol in alcohol, chloroform; moderately sol in ether; slightly sol in CCl_4. Hydrolyzes in aq soln.

Note: This is a controlled substance (depressant) listed in the U.S. Code of Federal Regulations, Title 21 Part 1308.13 (1995).

THERAP CAT: Hypnotic.

2140. Chlorhexidine. *N,N''-Bis(4-chlorophenyl)-3,12-diimino-2,4,11,13-tetraazatetradecanediimidamide; 1,1'-hexamethylenebis[5-(p-chlorophenyl)biguanide];* 1,6-bis[N'-(p-chlorophenyl)-N^5-biguanido]hexane; 1,6-bis(N^5-p-chlorophenyl-N'-diguanido)hexane; 1,6-di(4'-chlorophenyldiguanido)hexane; 10040. $C_{22}H_{30}Cl_2N_{10}$; mol wt 505.45. C 52.28%, H 5.98%, Cl 14.03%, N 27.71%. Bisbiguanide with bacteriostatic activity. Prepn: Rose, Swain, *J. Chem. Soc.* **1956**, 4422; *eidem,* U.S. pat. **2,684,924** (1954 to I.C.I.). Antibacterial activity and acute toxicity: G. E. Davies *et al.*, *Brit. J. Pharmacol.* **9**, 192 (1954). Review of toxicology and clinical uses: D. M. Foulkes, *J. Periodont. Res.* **8**, Suppl. 12, 55-60 (1973). Series of articles on clinical efficacy in gingivitis and plaque control: *ibid.* **21**, Suppl. 16, 1-89 (1986).

Crystals from methanol, mp 134°. Strong alkaline reaction. Soly in water at 20°: 0.08% (w/v).

Dihydrochloride, $C_{22}H_{30}Cl_2N_{10} \cdot 2HCl$, *Lisium.* Crystals, dec 260-262°. Soly in water at 20°: 0.06 g/100 ml.

Diacetate, $C_{22}H_{30}Cl_2N_{10} \cdot 2C_2H_4O_2$, *Chlorasept 2000, Nolvasan.* Crystals, mp 154-155°. Neutral reaction. Soly in water at 20°: 1.9 g/100 ml. Aq solns dec when heated above 70°. Soluble in alcohol, glycerol, propylene glycol, polyethylene glycols. LD_{50} orally in mice: 2 g/kg (Davies).

D-Digluconate, *Chlorhexamed, Bacticlens, Corsodyl, Gingisan, Hibiclens, Hibident, Hibidil, Hibiscrub, Hibital, Hibitane, Peridex, pHiso-Med, Plac Out, Rotersept, Secalan, Sterilon, Unisept.* Soly in water at 20°: > 50% (w/v). LD_{50} in mice (mg/kg): 22 i.v.; 1800 orally (Foulkes).

THERAP CAT: Antiseptic; disinfectant.

THERAP CAT (VET): Antiseptic; disinfectant.

2141. Chloric Acid. $ClHO_3$; mol wt 84.46. Cl 41.98%, H 1.19%, O 56.83%. $HClO_3$. Prepd from barium chlorate and sulfuric acid: Lamb *et al.*, *J. Am. Chem. Soc.* **42**, 1643 (1920); from sodium chlorate using ion-exchange resins: Klement, *Z. Anorg. Allgem. Chem.* **260**, 271 (1949).

Known in aq soln only. Aq solns are stable if pure and protected from light. 1% aq soln d_4^{18} 1.0044; 6% soln d_4^{18} 1.0344; 10% soln d_4^{18} 1.0594; 16% soln d_4^{18} 1.0991; 20% soln d_4^{18} 1.1273; 24% soln d_4^{18} 1.1568. A 40% soln corresponds to $HClO_3 \cdot 7H_2O$, d_4^{20} 1.282. If higher concns are attempted by evaporation the soln begins to dec with evolution of chlorine and oxygen and formation of perchloric acid.

Caution: Strongly irritating to skin, mucous membranes.

USE: Oxidizing agent; with H_2SO_3 as catalyst in acrylonitrile polymerization.

2142. Chlorimuron Ethyl. *2-[[[[(4-Chloro-6-methoxy-2-pyrimidinyl)amino]carbonyl]amino]sulfonyl]benzoic acid ethyl ester;* ethyl 2-[[[[(4-chloro-6-methoxypyrimidin-2-yl)amino]carbonyl]amino]sulfonyl]benzoate; DPX-F6025; Classic. $C_{15}H_{15}ClN_4O_6S$; mol wt 414.83. C 43.43%, H 3.64%, Cl 8.55%, N 13.51%, O 23.14%, S 7.73%. Selective sulfonylurea herbicide. Prepn: A. D. Wolf, **Austrian** pat. **8,316,181**; *idem,* U.S. pat. **4,547,215** (1984, 1985 both to Du Pont). Effect on plant growth and pigment synthesis: R. M. Devlin, Z. K. Koszanski, *Proc. Ann. Meet. Northeast. Weed Sci. Soc.* **40**, 115 (1986). Metabolism by plants: H. M. Brown, S. M. Neighbors, *Pestic. Biochem. Physiol.* **29**, 112 (1987). Field trial in soybeans: G. N. Rhodes *et al., Tenn. Farm Home Sci.* **142**, 21 (1987). HPLC determn in crops: J. L. Prince, R. A. Guinivan, *J. Agr. Food Chem.* **36**, 63 (1988). Brief review: J. S. Claus, *Weed Technol.* **1**, 114-115 (1987).

Crystals from butyl chloride, mp 198-201°. Soly (ppm): acetone 71000, acetonitrile 31000, benzene 8000, methylene chloride 153000, water (pH 7) 1200, (pH 6.5) 450, (pH 5) 11. LD_{50} in male, female rats (mg/kg): 4102, 4236 orally (Claus).

USE: Herbicide.

2143. Chlorinated Lime. Bleaching powder. Improperly called *"chloride of lime"* or *"calcium oxychloride"*. A relatively unstable chlorine carrier in solid form; a complex chemical compd of indefinite composition, presumably consisting of varying proportions of $Ca(OCl)_2$, $CaCl_2$, $Ca(OH)_2$ and H_2O in its molecular structure. Maximum available chlorine content approaches 39%. Commercial products usually range between 24% and 37% of available chlorine.

White or grayish-white powder; strong odor of chlorine. On exposure to air it becomes moist and rapidly decomposes. Most of it dissolves in water or alcohol. *Keep dry and tightly closed.*

Caution: Strong solns irritate skin. Inhalation of fumes may cause laryngeal and pulmonary irritation, pulmonary edema, death. Ingestion may produce severe oral, esophageal, gastric irritation.

USE: Bleaching of wood pulp, linen, cotton, straw, oils, soaps, and in laundering; oxidizer in calico printing to obtain white designs on a colored ground; destroying caterpillars; disinfecting drinking water, sewage, etc.; as a decontaminant for mustard gas and similar substances.

THERAP CAT: Disinfectant.

THERAP CAT (VET): Disinfectant for premises. Has been used as a topical antiseptic for superficial wounds.

2144. Chlorindanol. *7-Chloro-2,3-dihydro-1H-inden-4-ol; 7-chloro-4-indanol.* C_9H_9ClO; mol wt 168.62. C 64.11%, H 5.38%, Cl 21.02%, O 9.49%. Prepn: Buck *et al., J. Am. Chem. Soc.* **79**, 3559 (1957); Buck, U.S. pat. **2,990,324** (1961 to Esta Med. Labs.).

Needles from petr ether, mp 91-93°. Absorption spectra: Buck *et al., loc. cit.*

USE: Spermaticide.

2145. Chlorine. Cl; at. wt 35.453; at. no. 17; valences 1 to 7; elemental state: Cl_2. A halogen. Abundance in igneous rock (95% of earth's crust): 0.031% by wt; in seawater: 1.9% by wt (primarily as NaCl). Natural isotopes: 35 (75.53%); 37 (24.47%); seven radioactive isotopes and two isomers are known; radioactive tracer elements: ^{36}Cl ($T_{1/2}$ 3.08 $\times$ 10^5 yrs; β^-, EC); ^{38}Cl ($T_{1/2}$ 37.29 min; β^-); formed in atm by bombardment with cosmic rays. Discovered in 1774 by Scheele; recognized as an element in 1810 by Davy. Produced on a large scale by electrolysis from fused chlorides. The industrial product is about 99.3% pure. Contaminants are traces of bromide, hexachloroethane, hexachlorobenzene, and water. Purification: Fye, Beaver, *J. Am. Chem. Soc.* **63**, 1268 (1941); A. Klemenc, *Die Behandlung und Reindarstellung von Gasen* (Vienna, 2nd ed., 1948) p 153. Lab prepn from MnO_2 and HCl: Schmeisser in *Handbook of Preparative Inorganic Chemistry* vol. **1**, G. Brauer, Ed. (Academic Press, New York, 2nd ed., 1963) p 272. Manuf: Faith, Keyes & Clark's *Industrial Chemicals*, F. A. Lowenheim, M. K. Moran, Eds. (Wiley-Interscience, New York, 4th ed., 1975) pp 244-253. *Reviews: Ciba Review* vol. **12**, no. 139 (Aug. 1960); *Chlorine*, J. S. Sconce, Ed., A.C.S. Monograph Series, no. **154** (Reinhold, New York, 1962) 901 pp; *MTP Int. Rev. Sci.: Inorg. Chem., Ser. One, vol. **3**, V. Gutmann, Ed. (Butterworths, London, 1972); Downs, Adams, "Chlorine, Bromine, Iodine and Astatine" in *Comprehensive Inorganic Chemistry*, vol. **2**, J. C. Bailar, Jr. *et al.*, Eds. (Pergamon Press, Oxford, 1973) pp 1107-1594.

Greenish-yellow, diatomic gas; suffocating odor. mp −101.00°. bp −34.05°, d^{20} at 6.864 atm 1.4085 (liq.); d^{-35} at 0.9949 atm 1.5649 (liq.). Critical temp 144°; critical pressure 76.1 atm. Heat capacity at constant pressure (gas, 25°) 8.11 cal/mole/°C. Marketed in the form of gas over liquid compressed into steel cylinders. *Caution: Never heat cylinders.* Vapor pressure data: Giauque, Powell, *J. Am. Chem. Soc.* **61**, 1970 (1939). Sol in water (25°) with formation of aqueous Cl_2 (0.062 moles/l), HOCl (0.030 moles/l) and Cl^- (0.030 moles/l); total soly: 0.092 moles/l. More sol in alkalies. *See* Chlorinated Lime and Sodium Hypochlorite Soln. Acts as an electron-acceptor in forming complexes with many donor species: Bent, *Chem. Rev.* **68**, 587 (1968). Very reactive; E^0 (aq) $\frac{1}{2}Cl_2/Cl^-$ 1.356 V; dissociation energy (25°): 57.978 kcal; combines readily with all elements except the rare gases (xenon excluded) and nitrogen. Forms explosive mixtures with hydrogen; many finely divided metals will burn in an atm of chlorine. Oxides are strong oxidizing agents and explosive. Monatomic chlorine is unstable under ordinary conditions, however, it can be formed as a result of thermal or optical dissociation, by an electrical discharge or as an intermediate during chemical reactions. LC_{50} (1 hr) in rats, mice: 293 ppm, 137 ppm by inhalation, K. C. Back *et al., Reclassification of Materials Listed as Transportation Health Hazards* (TSA-20-72-3; PB 214-270, 1972).

Caution: Potential symptoms of overexposure are burning of eyes, nose and mouth; lacrimation, rhinorrhea; coughing, choking and substernal pain; nausea, vomiting; headache, dizziness; syncope; pulmonary edema; pneumonia; hypoxemia; dermatitis; eye and skin burns. *See NIOSH Pocket Guide to Chemical Hazards* (DHHS/NIOSH 90-117,

1990) p 62. *See also Patty's Industrial Hygiene and Toxicology* vol. **2B**, G. D. Clayton, F. E. Clayton, Eds. (Wiley-Interscience, New York, 3rd ed., 1981) pp 2954-2965.

USE: Largely for manuf chlorinated lime which is used in bleaching all kinds of fabrics; for purifying water; disinfecting; detinning and dezincing iron; manuf synthetic rubber and plastics, chlorinated hydrocarbons, and a large number of other chemicals. It is an indispensable reagent in synthetic chemistry. Has been used as a military poison gas under the name *bertholite*. ^{36}Cl is considered an extinct nuclide and provides a method of determining the geological age of meteors.

2146. Chlorine Dioxide. Chlorine peroxide. ClO_2; mol wt 67.45. Cl 52.56%, O 47.44%. Prepn from chlorine and sodium chlorite: Derby, Hutchinson, *Inorg. Syn.* **4**, 152 (1953); from potassium chlorate and sulfuric acid: Bodenstein *et al.*, *Z. Anorg. Allgem. Chem.* **147**, 233 (1925); by passing NO_2 through a column of sodium chlorate: Hutchinson, Derby, *Cereal Chem.* **24**, 372 (1947); alternate methods of prepn: Schmeisser in *Handbook of Preparative Inorganic Chemistry* vol. **1**, G. Brauer, Ed. (Academic Press, New York, 2nd ed., 1963) p 301. *Review:* Bedumeau, *Rev. Prod. Chim.* **57**, 173-177, 257-261 (1954). Book: W. J. Masschelein, *Chlorine Dioxide*, R. G. Rice, Ed (Ann Arbor Science Publishers Inc., Ann Arbor, MI, 1979) 189 pp.

Strongly oxidizing, yellow to reddish-yellow gas at room temp. Unpleasant odor similar to that of chlorine and reminiscent of that of nitric acid. Unstable in light; stable in dark if pure, but chlorides catalyze its decompn even in the dark. Reacts violently with organic materials. In concns in excess of 10% at atm pressure easily detonated by sunlight, heat, contact with mercury or carbon monoxide. mp −59°; bp 11°; $d^0(liq)$ 1.642: Cheesman, *J. Chem. Soc.* **1930**, 35. Sol in water (3.01 g/l at 25° and 34.5 mm Hg), with slight hydrolysis to chlorous and chloric acid; sol in alkaline and sulfuric acid solns. Solid is yellowish-red crystalline mass; liquid is reddish-brown.

Caution: Potential symptoms of overexposure are irritation of eyes, nose and throat; coughing, wheezing, bronchitis, pulmonary edema; chronic bronchitis. *See NIOSH Pocket Guide to Chemical Hazards* (DHHS/NIOSH 90-117, 1990) p 62.

USE: Bleaching cellulose, paper-pulp, flour, leather, fats and oils, textiles, beeswax; purification of water; taste and odor control of water; cleaning and detanning leather; manuf of chlorite salts; oxidizing agent; bactericide, antiseptic and deodorizer.

2147. Chlorine Heptoxide. Dichlorine heptoxide; perchloric anhydride. Cl_2O_7; mol wt 182.90. Cl 38.77%, O 61.23%. Prepd by dehydration of perchloric acid with P_2O_5: Michael, Conn, *Am. Chem. J.* **23**, 445 (1900); **25**, 92 (1901); Meyer, Kessler, *Ber.* **54**, 566 (1921); Goodeve, Powney, *J. Chem. Soc.* **1932**, 2078.

Colorless, very volatile oily liquid. d_4^0 1.86. mp −91.5°; bp 82°; $bp_{23.7}$ 0°. Trouton constant 23.4. The most stable oxide of chlorine. *Explodes violently upon concussion or on contact with a flame or iodine.* Dipole moment in CCl_4 at 20° is 0.72. Does not attack wood or paper. Slowly hydrolyzed by water, forming perchloric acid.

Caution: May be irritating to skin, mucous membranes.

USE: Catalyst in cellulose esterification.

2148. Chlorine Monofluoride. Chlorine fluoride. ClF; mol wt 54.45. Cl 65.11%, F 34.89%. Prepd from Cl_2 and F_2 at 400°: Ruff *et al.*, *Z. Anorg. Allgem. Chem.* **176**, 256 (1928); Kwasnik in *Handbook of Preparative Inorganic Chemistry* vol. **1**, G. Brauer, Ed. (Academic Press, New York, 2nd ed., 1963) p 153.

Colorless gas. Slightly yellow when liquid. mp −155.6°. bp −100.1°. Crit temp −14°. d (liq; −108°) 1.67. *Extremely reactive.* Destroys glass instantly, attacks quartz readily in presence of moisture; organic matter bursts into flame on contact; violent reaction with water.

Caution: Extremely corrosive and irritating to skin, eyes, mucous membranes, respiratory tract.

2149. Chlorine Monoxide. Dichlorine monoxide; dichloromonoxide; dichloroxide; hypochlorous anhydride. Cl_2O; mol wt 86.90. Cl 81.59%, O 18.41%. First prepara-

tion by J. L. Gay-Lussac in 1842 from yellow mercuric oxide and chlorine. See also: Cady, *Inorg. Syn.* **5**, 156 (1957); Schmeisser in *Handbook of Preparative Inorganic Chemistry* vol. **1**, G. Brauer, Ed. (Academic Press, New York, 2nd ed., 1963) p 299. Use as a powerful and selective chlorinating agent: F. D. Marsh *et al.*, *J. Am. Chem. Soc.* **104**, 4680 (1982).

Yellowish-brown gas. Disagreeable, penetrating odor. *Explodes on contact with organic matter.* Can also be caused to explode by a spark or by heating. Dec at moderate rate at room temp. mp −120.6°. bp +2.2°. Trouton constant: 22.5. Suffers photochemical and thermal decompn: at 100-140° there is an induction period, followed by a second order reaction, *see* N. V. Sidgwick, *The Chemical Elements and Their Compounds* vol. II (Oxford, 1950) p 1201. One vol of water at 0° dissolves more than 100 vols Cl_2O with formation of HClO; sol in CCl_4. Stored as a liquid or solid at temps below −80°.

Caution: Intensely irritating to eyes, skin, mucous membranes, respiratory tract.

USE: Chlorinating agent.

2150. Chlorine Trifluoride. ClF_3; mol wt 92.45. Cl 38.35%, F 61.65%. Prepd from F_2 and Cl_2 or ClF. May be purified by distillation in suitable steel apparatus: Ruff, Krug, *Z. Anorg. Allgem. Chem.* **190**, 270 (1930); Grisard *et al.*, *J. Am. Chem. Soc.* **73**, 5724 (1951); Kwasnik in *Handbook of Preparative Inorganic Chemistry* vol. **1**, G. Brauer, Ed. (Academic Press, New York, 2nd ed., 1963) p 156. *Reviews:* Kemmitt, Sharp, *Advan. Fluorine Chem.* **4**, 240-241 (1965); Stein, "Physical and Chemical Properties of Halogen Fluorides" in *Halogen Chemistry* vol. **1**, V. Gutmann, Ed. (Academic Press, New York, 1967) pp 133-224; Meinert, *Z. Chem.* **7**, 41-57 (1967).

Corrosive, colorless gas. Somewhat sweet, suffocating odor. mp −76.34°. bp 11.75°. *Extremely reactive.* Glass wool and organic matter burst into flames on contact even with dil vapors. Violently hydrolyzed by water. Attacks quartz if traces of moisture are present. Liquid is yellow-green in color and solid is white.

Caution: Potential symptoms of overexposure are eye and skin burns. *See NIOSH Pocket Guide to Chemical Hazards* (DHHS/NIOSH 90-117, 1990) p 64.

USE: Fluorinating agent; in nuclear reactor fuel processing; incendiary; igniter and propellant for rockets; pyrolysis inhibitor for fluorocarbon polymers.

2151. Chlorisondamine Chloride. *4,5,6,7-Tetrachloro-2,3-dihydro-2-methyl-2-[2-(trimethylammonio)ethyl]-2H-isoindolium dichloride; 4,5,6,7-tetrachloro-2-(2-dimethylaminoethyl)-2-methylisoindolinium chloride methochloride; 4,5,6,7-tetrachloro-2-(2-dimethylaminoethyl)isoindoline dimethylchloride; N-[(2-dimethylammonium)ethyl]-4,5,6,7-tetrachloroisoindolinium dimethochloride; chlorisondamine dimethochloride;* Su-3088; Ecolid; Ecolid chloride. $C_{14}H_{20}Cl_6N_2$; mol wt 429.04. C 39.19%, H 4.70%, Cl 49.58%, N 6.53%. Prepn: Allen, Ocampo, *J. Electrochem. Soc.* **103**, 452, 682 (1956); Huebner, U.S. pat. **3,025,294** (1962 to Ciba).

Non-hygroscopic crystals contg ethanol of crystallization, mp 258-265° (dec). Sol in water, methanol, ethanol. pH of aq solns 4.7 to 6.2. Forms yellow, chloroform-soluble complex with bromcresol green, absorption max 420 nm.

THERAP CAT: Antihypertensive.

2152. Chlormadinone Acetate. *17-(Acetyloxy)-6-chloropregna-4,6-diene-3,20-dione; 6-chloro-17-hydroxypregna-4,6-diene-3,20-dione acetate; 6-chloro-6-dehydro-17α-hydroxyprogesterone acetate; 6-chloro-6-dehydro-17α-acetoxyprogesterone; 17α-acetoxy-6-chloro-6,7-dehydroprogesterone;* Chronosyn; Cyclonorm; Fertiletten; Gestafortin; Lormin;

Luteran; Lutoral; Matrol; Normenon; Menstridyl; Prostal; Traslan; Verton. $C_{23}H_{29}ClO_4$; mol wt 404.93. C 68.22%, H 7.22%, Cl 8.76%, O 15.80%. Orally active progestogen with antiandrogenic activity; has been used in combinations as an oral contraceptive. Prepn: Brückner, **Ger.** pat. **1,075,-114** (1960 to E. Merck, AG); Brückner *et al., Ber.* **94,** 1225 (1961); Sciaky, *Gazz. Chim. Ital.* **91,** 545 (1961); **Brit.** pat. **932,153;** H. J. Ringold, A. Bowers, U.S. pat. **3,485,852** (1963, 1969 both to Syntex). Endocrinological activities: D. M. Brennan, R. J. Kraay, *Acta Endocrinol.* **44,** 367 (1963). Metabolism: S. Honma *et al., Chem. Pharm. Bull.* **25,** 2019 (1977). Clinical evaluation in prostatic carcinoma: R. Nishimura, K. Shida, *Prostate,* Suppl. 1, 27 (1981); in benign prostatic hypertrophy: T. Usui *et al., Acta Urol. Japan* **27,** 327 (1981), *B.A.* **73,** 27225 (1982). Review of carcinogenicity studies: *IARC Monographs* **21,** 365-375 (1979).

Crystals from methanol or ether, mp 212-214°. $[\alpha]_D$ +6° (c = 1 in $CHCl_3$). uv max: 283.5, 286 nm (ε 23400, 22100).

Mixture with ethinyl estradiol, *Amenyl, Lutestral, Menova.*

Mixture with mestranol, *C-Quens, Gestamestrol, Sequens.*

THERAP CAT: Progestogen; antineoplastic (hormonal).

THERAP CAT (VET): Progestogen; estrus regulator.

2153. Chlormequat Chloride. *2-Chloro-N,N,N-trimethylethanaminium chloride; (2-chloroethyl)trimethylammonium chloride;* chlorocholine chloride; choline dichloride; AC-38555; CCC; Cycocel; Cycogan. $C_5H_{13}Cl_2N$; mol wt 158.07. C 37.99%, H 8.29%, Cl 44.86%, N 8.86%. [$ClCH_2$-$CH_2N^+(CH_3)_3$]Cl^-. Prepn: Kauffmann, Vorlander, *Ber.* **43,** 2740 (1910); Freiss, Carville, *J. Am. Chem. Soc.* **76,** 2260 (1954); H. Linser *et al., Austrian* pat. **222,145** (1962 to OSSW), *C.A.* **57,** 7660f (1962). Activity: N. E. Tolbert, *J. Biol. Chem.* **235,** 475 (1960); J. Namokar, *Pesticides* **11,** 53 (1977). Metabolism: R. C. Blinn, *J. Agr. Food Chem.* **15,** 984 (1967). Toxicity study: G. Hennighausen, B. Wiegershausen, *Acta Biol. Med. Ger.* **27,** 663 (1971).

White crystalline solid, mp 245° (dec). Fish-like odor; very hygroscopic. Sol in water, lower alcs. Insol in ether, hydrocarbons. Aq solns corrosive to metal. LD_{50} in mice (mg/kg): 7 i.v., 54 orally (Hennighausen, Wiegershausen).

USE: Plant growth regulator.

2154. Chlormerodrin. *[3-[(Aminocarbonyl)amino]-2-methoxypropyl]chloromercury; 1-[3-(chloromercuri)-2-methoxypropyl]urea;* chlormeroprin; Katonil; Mercoral; Diurone; Mercloran; Percapyl; Merilid; Neohydrin; Oricur. $C_5H_{11}Cl$-HgN_2O_2; mol wt 367.20. C 16.35%, H 3.02%, Cl 9.65%, Hg 54.63%, N 7.63%, O 8.71%. Prepn: R. L. Rowland, U.S. pat. **2,635,982** (1953 to Lakeside); R. L. Rowland *et al., J. Am. Chem. Soc.* **72,** 3595 (1950); E. L. Foreman, U.S. pat. **2,635,983** (1953 to Lakeside). Toxicity data: E. I. Goldenthal, *Toxicol. Appl. Pharmacol.* **18,** 185 (1971).

Crystals from ethanol, mp 152-153°. Bitter, metallic taste. Stable to light and air. Soly in water, methanol: both 1.1 g/100 ml; in ethanol: 0.56 g/100 ml. Sparingly sol in chloroform. Freely sol in alkaline solns. pH of a 0.5% aq soln 4.3-5.0. LD_{50} orally in rats: ~82 mg/kg (Goldenthal).

THERAP CAT: Diuretic.

THERAP CAT (VET): Diuretic.

2155. Chlormezanone. *2-(4-Chlorophenyl)tetrahydro-3-methyl-4H-1,3-thiazin-4-one 1,1-dioxide;* 2-(p-chlorophenyl)-perhydro-3-methyl-1,3-thiazin-4-one 1,1-dioxide; 2-(4-chlorophenyl)-3-methyl-4-metathiazanone 1,1-dioxide; dichloromethazanone; chlormethazanone; Alinam; Banabin-Sintyal; Fenarol; Lobak; Mio-Sed; Rexan; Rilansyl; Rilaquil; Rilassol; Supotran; Suprotan; Tanafol; Trancopal; Trancote; Transanate. $C_{11}H_{12}ClNO_3S$; mol wt 273.74. C 48.27%, H 4.42%, Cl 12.95%, N 5.12%, O 17.53%, S 11.71%. Prepn starting with 4-chlorobenzylidenemethylamine: Surrey *et al., J. Am. Chem. Soc.* **80,** 3469 (1958); **Brit.** pat. **815,203** (1959 to Sterling Drug).

Crystals, mp 116.2-118.2°. Soly in water at 25° less than 0.25% (w/v); in 95% ethanol at 25° less than 1.0% (w/v).

THERAP CAT: Anxiolytic; muscle relaxant (skeletal).

2156. Chlormidazole. *1-[(4-Chlorophenyl)methyl]-2-methyl-1H-benzimidazole; 1-(p-chlorobenzyl)-2-methylbenzimidazole;* 2-methyl-1-(p-chlorobenzyl)benzimidazole; clomidazole. $C_{15}H_{13}ClN_2$; mol wt 256.73. C 70.18%, H 5.10%, Cl 13.81%, N 10.91%. Prepn: Herrling *et al.,* U.S. pat. **2,876,233** (1959 to Grünenthal).

bp$_{12}$ 240-242°.

Monohydrate, mp 67-68°.

Hydrochloride, $C_{15}H_{13}ClN_2$·HCl, *H-115, Diamyceline, Futrican.* mp 227-228°.

THERAP CAT: Antifungal.

2157. Chlornaphazine. *N,N-Bis(2-chloroethyl)-2-naphthylamine;* dichloroethyl-β-naphthylamine; β-naphthyldi-(2-chloroethyl)amine; β-naphthylbis(β-chloroethyl)amine; di(2-chloroethyl)-β-naphthylamine; CB-1048; R-48; Cloronaftina; Erysan. $C_{14}H_{15}Cl_2N$; mol wt 268.19. C 62.70%, H 5.64%, Cl 26.44%, N 5.22%. Prepd from 2-$C_{10}H_7N$-(C_2H_4OH)$_2$ by treatment with $POCl_3$: Ross, *J. Chem. Soc.* **1949,** 183; Ghielmetti, *Farmaco Ed. Sci.* **5,** 275 (1950); **11,** 603 (1956). Mechanism of action studies: Jeney *et al., Kiserl. Orvostud.* **20,** 369 (1968), *C.A.* **70,** 18766j (1969). Review of carcinogenicity studies: *IARC Monographs* **4,** 119-124 (1974).

Platelets from petr ether, mp 54-56°. bp$_5$ 210°. Very sparingly sol in water, glycerol. More sol (in ascending degree) in petr ether, ethanol, olive oil, ether, acetone, benzene.

Note: This substance has been listed as a known carcinogen: *Fourth Annual Report on Carcinogens* (NTP 85-002, 1985) p 44; delisted because no U.S. residents exposed:

Fifth Annual Report on Carcinogens (NTP 89-239, 1989) p 340.

THERAP CAT: Antineoplastic.

2158. Chlornitrofen. *1,3,5-Trichloro-2-(4-nitrophenoxy)benzene;* 4-nitrophenyl-2,4,6-trichlorophenyl ether; 2,4,-6-trichlorophenyl-4'-nitrophenyl ether; CNP; MO-338; MO; Showrone. $C_{12}H_6Cl_3NO_3$; mol wt 318.54. C 45.25%, H 1.90%, Cl 33.39%, N 4.40%, O 15.07%. Prepd, not claimed: **Neth.** pat. **Appl. 6,405,727;** I. Takayuki *et al.,* U.S. pat. **3,316,080** (1964, 1967 both to Mitsui Chemical); and field studies: G. W. Mason, E. G. Brooker, *Proc. 21st N. Z. Weed Pest Control Conf.* 163 (1968). Properties: J. Kanazawa, *Pestic. Sci.* **12,** 417 (1981). Absorption and metabolism in crops: H. Shimotori, S. Kuwatsuka, *J. Pestic. Sci.* **3,** 267 (1978). Degradation studies: M. Oyamada, S. Kuwatsuka, *ibid.* **4,** 157 (1979). GC determn in fish: T. Yamagishi *et al., Bull. Environ. Contam. Toxicol.* **23,** 57 (1979); in river water: H. Kobayashi *et al., J. Chromatog.* **643,** 197 (1993).

Pale yellow needles from ethanol, mp 105-106°. Soly: 0.25 mg/l water. Soly at 20-25°: 764 μg/l water. Vapor pressure (109°): 4.67 × 10⁴ mPa. Partition coefficient (octanol/water): 4709. LC_{50} in carp (48 hr): > 10,000 μg/l; LC_{50} in flea (3 hr): > 40,000 μg/l (Kanazawa).

USE: Herbicide.

2159. Chloroacetaldehyde. 2-Chloro-1-ethanal; monochloroacetaldehyde. C_2H_3ClO; mol wt 78.50. C 30.60%, H 3.85%, Cl 45.16%, O 20.38%. $ClCH_2CHO$. Prepd industrially by carefully controlled chlorination of acetaldehyde: Söll, **Ger.** pat. **844,595** (1943 to Knapsack-Griesheim); Shawinigan Chem. Ltd., **Brit.** pat. **644,914** (1947); Guinot, Tabuteau, *Compt. Rend.* **231,** 234 (1950); **Fr.** pat. **1,012,991** (1950 to Comp. des Prod. Chim. et Electrometallurg.).

Liquid. Acrid, penetrating odor. bp_{760} 85-86°. The anhydr substance polymerizes on standing but reverts to the monomer on distillation.

Hemihydrate, probably $ClCH_2CH(OH)OCH(OH)CH_2Cl$, platelets from water, mp 43-50°, bp 85.5° with decompn into water and chloraldehyde. Sol in water, alcohol, ether.

Caution: Potential symptoms of overexposure are irritation of eyes, skin and mucous membranes; skin burns; eye damage; pulmonary edema; skin, respiratory system sensitization; narcosis, coma. *See NIOSH Pocket Guide to Chemical Hazards* (DHHS/NIOSH 90-117, 1990) p 64.

USE: In the manufacture of 2-aminothiazole; to facilitate bark removal from tree trunks.

2160. Chloroacetamide. *2-Chloroacetamide.* C_2H_4ClNO; mol wt 93.51. C 25.69%, H 4.31%, Cl 37.91%, N 14.98%, O 17.11%. $ClCH_2CONH_2$. Prepn from ethyl chloroacetate and ammonia: Jacobs, Heidelberger, *Org. Syn.* **coll. vol. I,** 153 (2nd ed., 1941); from chloroacetyl chloride and ammonium acetate: Finan, Fothergill, *J. Chem. Soc.* **1962,** 2824.

Crystals from water, mp 119-120°. bp ~225° (dec). Sol in 10 parts water, 10 parts abs alc; very slightly sol in ether. Two isomorphous crystalline modifications: α-form, "stable form", obtained by sublimation and by crystallization from nonpolar solvents; β-form, "unstable form", obtained by quenching the melt or by crystallization from polar solvents, *see* Katayama, *Acta Crystallogr.* **9,** 986 (1956); **10,** 468 (1957); B. Kalyanaraman *et al., J. Cryst. Mol. Struct.* **8,** 175 (1978).

2161. Chloroacetanilide. C_8H_8ClNO; mol wt 169.61. C 56.65%, H 4.75%, Cl 20.90%, N 8.26%, O 9.43%. Prepn of *m-, o-* and *p-*isomers: Roberts *et al., J. Org. Chem.* **24,** 654 (1959). Sepn of *o-* and *p-*isomers: Orton, Bradford, *J. Chem. Soc.* **1927,** 986.

m-Chloroacetanilide, N-(3-chlorophenyl)acetamide, 3'-chloroacetanilide. Needles from 50% glacial acetic acid, mp 77-78°. Readily sol in alcohol, benzene, carbon disulfide; very slightly sol in ligroin. uv max (95% ethanol): 245 nm (log ε 4.19).

o-Chloroacetanilide. Needles from dil glacial acetic acid, mp 87-88°. Sublimes at about 50-60°. Practically insol in water, alkalies; sol in alc; more sol in benzene than corresponding *p-*isomer. uv max (95% ethanol): 240 nm (log ε 4.02).

p-Chloroacetanilide. Orthorhombic crystals from aq glacial acetic acid, alc, or acetone. mp 178-179°. d_4^{22} 1.385. Practically insol in water; readily sol in alc, ether, carbon disulfide; slightly sol in CCl_4, benzene. uv max (95% ethanol): 249 nm (log ε 4.25).

2162. Chloroacetic Acid. Chloroethanoic acid; monochloroacetic acid; MCA. $C_2H_3ClO_2$; mol wt 94.50. C 25.42%, H 3.20%, Cl 37.52%, O 33.86%. $ClCH_2COOH$. Made by chlorination of glacial acetic acid in presence of small amount of sulfur or iodine; also by hydrolysis of trichlorethylene with 90% H_2SO_4. Lab prepn: Gattermann-Wieland, *Praxis des organischen Chemikers* (de Gruyter, Berlin, 40th ed., 1961) pp 109, 110. Manuf: Faith, Keyes & Clark's *Industrial Chemicals,* F. A. Lowenheim, M. K. Moran, Eds. (Wiley-Interscience, New York, 4th ed., 1975) pp 254-257. Toxicity: Woodward *et al., J. Ind. Hyg. Toxicol.* **23,** 78 (1941).

Colorless or white, deliquesc crystals. d 1.580. Exists in α, β and γ forms having mp 63°, 55-56° and 50° respectively. mp for acid of commerce is 61-63°. bp for all forms is 189°. Very sol in water; sol in alcohol, benzene, chloroform, ether. *Keep well closed and in a cool place.*

Sodium salt, $C_2H_2ClNaO_2$, *Monoxone.* White crystalline solid. Soly in water at 20°: 85 g/100 ml. LD_{50} in rats, mice, guinea pigs (mg/kg): 76, 255, 80 orally (Woodward).

Caution: Irritating to skin, mucous membranes.

USE: Herbicide. Manuf various dyes and other organic chemicals.

2163. Chloroacetic Anhydride. *Chloroacetic acid anhydride;* monochloroacetic acid anhydride; *sym-*dichloroacetic anhydride. $C_4H_4Cl_2O_3$; mol wt 170.98. C 28.10%, H 2.36%, Cl 41.47%, O 28.07%. $(ClCH_2CO)_2O$. Prepd by treating monochloroacetic acid in tetrahydrofuran with methoxyacetylene at 15° and distilling the mixture: Eglinton *et al., J. Chem. Soc.* **1954,** 1860; by mixing monochloroacetic acid and HCN with 1,4-dioxane satd with HCl and fractionating the mixture: Krieble, Smellie, U.S. pat. **2,-390,106** (1945); by heating chloroacetyl chloride with monochloroacetic acid in the presence of $AlCl_3$: Strosacker, Schwegler, U.S. pat. **1,713,104** (1929 to Dow).

Prisms from benzene; d_4^{20} 1.5494; mp 46°. bp_{760} 203°; bp_{116} 163°; bp_{62} 149°; bp_{24} 126°; bp_{10} 109-110°; $bp_{0.05}$ 118-120°. Freely sol in ether, chloroform; slightly sol in benzene, practically insol in cold ligroin.

USE: In *N-*acetylation of amino acids in alkaline soln; prepn of cellulose chloroacetates.

2164. Chloroacetone. *1-Chloro-2-propanone;* chloroacetone; monochloroacetone; monochloracetone; acetonyl chloride; chloropropanone; 1-chloro-2-ketopropane; 1-chloro-2-oxopropane. C_3H_5ClO; mol wt 92.52. C 38.94%, H 5.45%, Cl 38.32%, O 17.29%. $ClCH_2COCH_3$. Prepn by the action of chlorine upon diketene: A. B. Boese, Jr., U.S. pat. **2,209,683** (1940 to Carbide and Carbon Chem.); by chlorination of acetone: E. J. Rahrs, U.S. pat. **2,235,562** (1941 to Eastman Kodak); G. H. Morey, U.S. pat. **2,243,484** (1941 to Commercial Solvents Corp.). Stabilization: *idem,* U.S. pat. **2,229,651** (1941 to Commercial Solvents Corp.); E. J. Rahrs, U.S. pat. **2,263,010** (1941 to Eastman Kodak). Forms binary azeotropes with many organic liquids: L. H. Horsley, *Azeotropic Data,* Advances in Chemistry Series No. 6 (Washington, 1952) p 73. Reacts with aryl Grignard reagents to form stilbenes: Huang, *J. Chem. Soc.* **1954,** 2539. Toxicology: E. V. Sargent *et al., Am. Ind. Hyg. Assoc. J.* **47,** 375 (1986).

Liquid. Pungent odor. Lacrimator. Turns dark and resinifies on prolonged exposure to light. May be stabilized by addition of 0.1% H_2O or 1.0% $CaCO_3$. d_4^{25} 1.123; d_4^{15} 1.135. mp −44.5°. bp 119.7°, bp_{50} 61°, bp_{12} 20°. Volatile with steam. Surface tension at 20° = 35.27 dyn/cm. Dipole moment in hexane, 2.36D. Sol in 10 parts water (w/w). Miscible with alcohol, ether, chloroform. LD_{50} (14 day) in mice, rats (mg/kg): 127, 100 orally; LC_{50} (1 hr) in rats (ppm): 262 by inhalation (Sargent).

2,4-Dinitrophenylhydrazone, yellow needles from alc, mp 124°.

Caution: Intensely irritating to eyes, skin, mucous membranes.

USE: Has been proposed as tear gas component for police and military use; in the manuf of couplers for color photography; as enzyme inactivator; intermediate in the manuf of perfumes, antioxidants, drugs; in insecticide formulations; in photopolymerization of vinyl compds. Proposed as catalyst in tetraethyllead production, as selective solvent for separating diolefins.

2165. p-Chloroacetophenone. *1-(4-Chlorophenyl)ethanone; 4'-chloroacetophenone.* C_8H_7ClO; mol wt 154.60. C 62.15%, H 4.56%, Cl 22.93%, O 10.35%. Prepd from chlorobenzene and acetic anhydride in the presence of $AlCl_3$: Adams, Noller, *Org. Syn. coll. vol. I,* 109 (1941). Physical properties: *eidem, ibid.;* Dreisbach, Martin, *Ind. Eng. Chem.* **41,** 2875 (1949).

Liquid, bp_{24} 124-126°, bp 237°. mp 20-21° (Adams, Noller); also reported as 18.4° (Dreisbach, Martin). d_4^{20} 1.192, d_4^{25} 1.188. n_D^{20} 1.555, n_D^{25} 1.553. Practically insol in water. Miscible with alcohol, ether.

Caution: Highly irritating to eyes, mucous membranes.

2166. ω-Chloroacetophenone. *2-Chloro-1-phenylethanone; 2-chloroacetophenone;* α-chloroacetophenone; phenacyl chloride; Chemical Mace; CN. C_8H_7ClO; mol wt 154.60. C 62.15%, H 4.56%, Cl 22.93%, O 10.35%. C_6H_5-$COCH_2Cl$. Chemical warfare agent with lacrimatory properties. Prepn: R. Scholl, H. Korten, *Ber.* **34,** 1902 (1901); H. Rheinboldt, M. Perrier, *J. Am. Chem. Soc.* **69,** 3148 (1947); Schaefer, Sonnenberg, *J. Org. Chem.* **28,** 1128 (1963). Purification: Lofton, U.S. pat. **2,414,418** (1947 to Pennsylvania Coal Prod.). Melting point determination and review of prepn: Macy, *J. Chem. Ed.* **24,** 222 (1947). Comparative toxicity with *o*-chlorobenzylidenemalononitrile, *q. v.:* B. Ballantyne, D. W. Swanston, *Arch. Toxicol.* **40,** 75 (1978).

Crystals from dil alcohol, carbon tetrachloride, or light petr. mp 58-59°; also reported as 54° (Macy) and 56.5° (Rheinboldt, Perrier). bp 244-245°. d^{15} 1.324. Vapor pressure at 20°: 5.4 × 10⁻³ mm Hg. Practically insol in water. Freely sol in alcohol, ether, benzene. LD_{50} in rats (mg/kg): 41 i.v., 36 i.p., 127 orally; LC_{50} in rats: 8750 mg/min/m³ (Ballantyne, Swanston).

Caution: Potential symptoms of overexposure are irritation of eyes, skin and respiratory system; pulmonary edema. *See NIOSH Pocket Guide to Chemical Hazards* (DHHS/NIOSH 90-117, 1990) p 64.

USE: Riot control agent.

2167. Chloroacetyl Isocyanate. Chloroacetic acid anhydride with isocyanic acid. $C_3H_2ClNO_2$; mol wt 119.51. C 30.15%, H 1.69%, Cl 29.67%, N 11.72%, O 26.78%. Cl-CH_2CONCO. Obtained by the reaction of chloroacetamide

with oxalyl chloride: Speziale, Smith, *J. Org. Chem.* **28,** 1808 (1963).

Liquid. bp_{20} 50-55°. $n_D^{21.5}$ 1.4580.

2168. N-(3-Chloroallyl)hexaminium Chloride. *1-(3-Chloro-2-propenyl)-3,5,7-triaza-1-azoniatricyclo[3.3.1.1³,⁷]-decane chloride; 1-(3-chloroallyl)-3,5,7-triaza-1-azoniaadamantane chloride;* Dowicide Q. $C_9H_{16}Cl_2N_4$; mol wt 251.16. C 43.04%, H 6.42%, Cl 28.23%, N 22.31%. Prepn: Scott, Wolf, *Appl. Microbiol.* **10,** 211 (1962).

Cream-colored powder. Freely sol in water (up to 25% w/w). Concd solns are yellow.

USE: Bactericide.

2169. Chloroaniline. C_6H_6ClN; mol wt 127.57. C 56.49%, H 4.74%, Cl 27.79%, N 10.98%. Prepn of *m-, o-* and *p*-isomers: Sidgwick, Rubie, *J. Chem. Soc.* **119,** 1013 (1921). Manufacture of *m-* and *p*-isomers by catalytic hydrogenation of chloronitrobenzene: Pray, Trager, U.S. pat. **2,791,613** (1957 to Columbia-Southern Chem.); by reduction of chloronitrobenzene with NaHS: Latourette *et al.,* U.S. pat. **2,894,035** (1959 to Food Machinery & Chem.). Properties of *o*-isomer: Dreisbach, Martin, *Ind. Eng. Chem.* **41,** 2875 (1949). Toxicity data: H. F. Smyth *et al., Am. Ind. Hyg. Assoc. J.* **23,** 95 (1962).

m-Chloroaniline, 3-chlorobenzenamine. Liquid, bp 230.5°. mp −10.4°. d_4^{22} 1.2150. n_D^{20} 1.5931. Practically insol in water. Sol in most common organic solvents.

o-Chloroaniline, liquid. At 99.61 mol-% purity, bp 208.84°; mp −1.94°. d_4^{22} 1.2114. n_D^{20} 1.5895. Practically insol in water. Sol in acid and in most organic solvents.

p-Chloroaniline, orthorhombic crystals from alcohol or petr ether, mp 72.5°. bp 232°. d_4^{77} 1.169. Sol in hot water; freely sol in alcohol, ether, acetone, carbon disulfide. LD_{50} orally in rats: 0.31 g/kg (Smyth).

2170. Chloroarsenol. *(2-Chloro-1-heptenyl)arsonic acid; 2-chloro-1-heptene-1-arsonic acid.* $C_7H_{14}AsClO_3$; mol wt 256.56. C 32.77%, H 5.50%, As 29.20%, Cl 13.82%, O 18.71%. Prepn: Ger. pat. **296,915** (1917 to Bayer).

Needles from water, mp 115°.

Ammonium salt, $C_7H_{17}AsClNO_3$, *ammonium chloroheptenearsonate.* Freely sol in water. Supplied as a 1% soln made isotonic with 0.7% NaCl under the names: *Solarson; Arsion.*

2171. Chloroazodin. *N,N''-Dichlorodiazenedicarboximidamide;* α,α'-azobis[chloroformamidine]; chlorazodin; dichloroazodicarbonamidine; Azochloramide. $C_2H_4Cl_2N_6$; mol wt 183.00. C 13.13%, H 2.20%, Cl 38.75%, N 45.92%. Prepd by treating an AcOH-NaOAc soln of guanidine nitrate with sodium hypochlorite in the cold: Braz *et al., Appl. Chem. USSR* **17,** 565 (1944), *C.A.* **40,** 2267 (1946). Structure and uv spectra: Kumler, *J. Am. Chem. Soc.* **75,** 3092 (1953).

Bright yellow needles, plates, flakes. Faint chlorine odor, burning taste. Dec explosively at ~155°. Decompn is accelerated by contact with metals. Very slightly sol in water; sparingly sol in alcohol; slightly sol in glycerol, glyceryl triacetate (1:100), vegetable oils, ether, chloroform. Practically insol in carbon tetrachloride and liquid petrolatum. Oil solns are generally made by diluting a triacetin (glyceryl triacetate) soln of chloroazodin with oil. Solns of chloro-azodin in glycerol and in alcohol dec rapidly on warming; all solns dec on exposure to light.

THERAP CAT: Antibacterial.

THERAP CAT (VET): Anesthetic (local).

2172. Chlorobenzene. Monochlorobenzene; benzene chloride. C_6H_5Cl; mol wt 112.56. C 64.03%, H 4.48%, Cl 31.50%. Produced by the chlorination of benzene in the presence of a catalyst. Review of manuf, properties, and use: Faith, Keyes & Clark's *Industrial Chemicals*, F. A. Lowenheim, M. K. Moran, Eds. (Wiley-Interscience, New York, 4th ed., 1975) pp 258-265.

Colorless, very refractive liquid; faint, not unpleasant odor. d^{20} 1.107. bp 131-132°. Solidif $-55°$. mp $-45°$. Flash pt 28°. n_D^{20} 1.5248. Insol in water; freely sol in alcohol, benzene, chloroform, ether.

Caution: Potential symptoms of overexposure are irritation of skin, eyes and nose; drowsiness, incoordination. *See NIOSH Pocket Guide to Chemical Hazards* (DHHS/NIOSH 90-117, 1990) p 64.

USE: Manuf phenol, aniline, DDT; solvent for paints; heat transfer medium.

2173. *p*-Chlorobenzenesulfonic Acid. *4-Chlorobenzene-sulfonic acid;* closylate. $C_6H_5ClO_3S$; mol wt 192.62. C 37.41%, H 2.62%, Cl 18.41%, O 24.92%, S 16.65%. Prepd by heating chlorobenzene with concd sulfuric acid or oleum under dehydrating conditions: Meyer, *Ann.* **433**, 327, 333 (1923).

Monohydrate, crystals from water, mp 67°. The anhydr material is usually obtained as a syrup, bp_{25} 148°. Sol in water, alc. Practically insol in ether, benzene.

Sodium salt monohydrate, $C_6H_4ClNaO_3S.H_2O$, cubic crystals from water. Sol in water.

2174. Chlorobenzilate. *4-Chloro-α-(4-chlorophenyl)-α-hydroxybenzeneacetic acid ethyl ester; 4,4'-dichlorobenzilic acid ethyl ester;* ethyl 4,4'-dichlorobenzilate; ethyl 2-hydroxy-2,2-bis(4-chlorophenyl)acetate; chlorbenzilat; G-23922; compd 338; Acaraben; Akar; Folbex. $C_{16}H_{14}Cl_2O_3$; mol wt 325.19. C 59.10%, H 4.34%, Cl 21.80%, O 14.76%. Prepn: Häfliger, U.S. pat. **2,745,780** (1956 to Geigy); **Brit.** pat. **831,421** (1960 to Metal & Thermit). Toxicity study: T. B. Gaines, *Toxicol. Appl. Pharmacol.* **14**, 515 (1969). Review of carcinogenic risk: *IARC Monographs* **30**, 73-85 (1983).

Viscous liquid. The commercial product is yellow. $bp_{0.04}$ 146-148°. n_D^{20} 1.5727. Vapor press. at 20°: 2.2×10^{-6} mm. Slightly sol in water. Sol in most organic solvents. Incompatible with lime. LD_{50} in male, female rats (mg/kg): 1040, 1220 orally (Gaines).

Caution: Symptoms similar to DDT, *q.v.* *See Clinical Toxicology of Commercial Products,* R. E. Gosselin *et al.,* Eds. (Williams & Wilkins, Baltimore, 5th ed., 1984) Section II, p 283.

USE: Acaricide in spider-mite control; synergist for DDT.

2175. *m*-Chlorobenzoic Acid. *3-Chlorobenzoic acid.* $C_7H_5ClO_2$; mol wt 156.57. C 53.70%, H 3.22%, Cl 22.64%, O 20.44%. Prepn by catalytic oxidation of 1-chloro-3-ethylbenzene: Emerson *et al., J. Am. Chem. Soc.* **71**, 1742 (1949); by chlorination of benzoic acid: Gorvin, *Chem. & Ind. (London)* **1951**, 910; by the von Richter reaction from 1-chloro-4-nitrobenzene and alcoholic KCN: Samuel, *J. Chem. Soc.* **1960**, 1318.

Crystals, mp 158°. d_4^{25} 1.496. Sol in 2850 parts cold water; more sol in hot water; freely sol in alcohol, ether.

Methyl ester, $C_8H_7ClO_2$, *methyl m-chlorobenzoate.* mp 21°, bp 231°.

2176. *o*-Chlorobenzoic Acid. *2-Chlorobenzoic acid.* $C_7H_5ClO_2$; mol wt 156.57. C 53.70%, H 3.22%, Cl 22.64%, O 20.44%. Prepd by $KMnO_4$ oxidation of *o*-chlorotoluene: Clarke, Taylor, *Org. Syn.* **coll. vol. II**, 135 (1943). Crystal structure: Ferguson, Sim, *Acta Cryst.* **12**, 941 (1959), *C.A.* **55**, 24188i (1961).

Monoclinic crystals, mp 142°. d_4^{20} 1.544. Sol in 900 parts cold water; more sol in hot water, freely in alc, ether.

USE: Preservative for glues, paints. Intermediate in the manufacture of fungicides and dyes.

2177. *p*-Chlorobenzoic Acid. Chlorodracylic acid. $C_7H_5ClO_2$; mol wt 156.57. C 53.70%, H 3.22%, Cl 22.64%, O 20.44%. Prepn by catalytic oxidation of 1-chloro-4-ethylbenzene: Emerson *et al., J. Am. Chem. Soc.* **71**, 1742 (1949). Manuf by oxidation of *p*-chlorobenzaldehyde: Shipman, U.S. pat. **3,124,611** (1964 to ICI). Crystal structure: Toussaint, *Acta Cryst.* **4**, 71 (1951), *C.A.* **45**, 6604e (1951); Pollock, Woodward, *ibid.* **7**, 605 (1954), *C.A.* **48**, 13331e (1954).

Triclinic crystals, mp 243°. Sol in 5290 parts water; freely sol in alcohol, ether.

Methyl ester, $C_8H_7ClO_2$, *methyl p-chlorobenzoate,* mp 44°. Sodium salt, $C_7H_4ClNaO_2$, *sodium p-chlorobenzoate.* White, odorless, crystalline powder, freely sol in water.

USE: Sodium salt as a preservative.

2178. *o*-Chlorobenzylidenemalononitrile. *[(2-Chlorophenyl)methylene]propanedinitrile;* o-chlorobenzalmalononitrile; *β,β-dicyano-o-chlorostyrene*; CS. $C_{10}H_5ClN_2$; mol wt 188.62. C 63.68%, H 2.67%, Cl 18.80%, N 14.85%. Lacrimatory chemical warfare agent. Prepn: B.B. Corson, R. W. Stoughton, *J. Am. Chem. Soc.* **50**, 2825 (1928). Pharmacology: R. W. Brimblecombe *et al., Brit. J. Pharmacol.* **44**, 561 (1972). Toxicology: C. L. Punte *et al., Toxicol Appl. Pharmacol.* **4**, 656 (1962); J. R. Gaskins *et al., Arch. Environ. Health* **24**, 449 (1972); B. Ballantyne, S. Callaway, *Med. Sci. Law* **12**, 43 (1972). Comparative toxicity of CS and ω-chloroacetophenone, *q.v.*: B. Ballantyne, D. W. Swanston, *Arch. Toxicol.* **40**, 75 (1978).

White crystalline solid, mp 95-96°, bp 310-315°. Vapor pressure at 20°: 3.4×10^{-5} mm Hg. Sparingly sol in water; sol in acetone, dioxane, methylene chloride, ethyl acetate, benzene. LD_{50} in rats (mg/kg): 28 i.v., 48 i.p.; LC_{50} in rats: 88,480 mg/min/m³ (Ballantyne, Swanston).

Caution: Potential symptoms of overexposure are pain, burning of eyes, lacrimation and conjunctivitis; erythemic eyelids, blepharospasm; irritation of throat, coughing and chest constriction; headache; erythema and vesiculation of skin. *See NIOSH Pocket Guide to Chemical Hazards* (DHHS/NIOSH 90-117, 1990) p 66.

USE: Riot control agent.

2179. p-Chlorobenzylpseudothiuronium Chloride. *Carbamimidothioic acid (4-chlorophenyl)methyl ester monohydrochloride.* $C_8H_{10}Cl_2N_2S$; mol wt 237.15. C 40.52%, H 4.25%, Cl 29.90%, N 11.81%, S 13.52%. Prepd by refluxing a mixture of *p*-chlorobenzyl chloride and thiourea in ethanol: Dewey, Sperry, *J. Am. Chem. Soc.* **61**, 3251 (1939).

Crystals from concd HCl + water, mp 197°.
USE: Identification of organic acids.

2180. Chlorobutanol. *1,1,1-Trichloro-2-methyl-2-propanol; β,β,β-trichloro-tert-butyl alcohol; acetone chloroform; chlorbutol; Chloretone; Coliquifilm; Methaform; Sedaform.* $C_4H_7Cl_3O$; mol wt 177.46. C 27.07%, H 3.98%, Cl 59.93%, O 9.02%. Chlorobutanol may be anhydrous or it may contain up to about one-half molecule of water. Prepn: Fishburn, Watson, *J. Am. Pharm. Assoc.* **28**, 491 (1939); E. Bergman, U.S. pat. **2,446,453** (1948 to Polymerisable Products Ltd.); P. Riegger, H. Richtzenhain, Ger. pat. **1,271,697** (1968 to Dynamit Nobel A.-G.), *C.A.* **69**, 95967g (1968); M. Saljoughian, *Monatsh.* **114**, 813 (1983). Toxicity studies: E. Impens, *Arch. Int. Pharmacodyn. Ther.* **8**, 77 (1901).

Crystals. Camphor odor and taste. A crystal of chlorobutanol placed on a water surface "dances" like a camphor crystal. Sublimes easily. The anhydrous form mp 97° (the hemihydrate mp 78°). bp$_{760}$ 167°; bp$_{246}$ 135°. Easily sol in hot water; one gram dissolves in 1 ml alcohol, 10 ml glycerol; sol in chloroform, ether, acetone, petr ether, glacial acetic acid, oils. MLD orally in dogs, rabbits (mg/kg): 238, 213 (Impens).
Compd with chloral hydrate, mp 65°. Ref: Ger. pat. **151,188** (1904).
Caution: This substance may be habit forming and is listed in the U.S. Code of Federal Regulations, Title 21 Part 329.1 (1995).
USE: Plasticizer for cellulose esters and ethers. Preservative for biological fluids, hypodermic solns, and solns of alkaloids. Pharmaceutic aid (antimicrobial).
THERAP CAT: Analgesic (dental).
THERAP CAT (VET): Has been used as a mild sedative, antipruritic, and antiseptic.

2181. 1-Chloro-2-butene. *α-Chloro-β-butylene; crotyl chloride; γ-methallyl chloride; γ-methylallyl chloride.* C_4H_7Cl; mol wt 90.55. C 53.06%, H 7.79%, Cl 39.15%. $CH_3CH=CHCH_2Cl$. Prepn of *cis-* and *trans-*isomers: Hatch, Nesbitt, *J. Am. Chem. Soc.* **72**, 727 (1950). Manuf: Carlson, U.S. pat. **2,494,034** (1950 to Shell); Montagna, Hess, U.S. pat. **3,055,954** (1962 to Union Carbide). Equilibrium constant for isomerization of 3-chloro-1-butene to 1-chloro-2-butene: Dittmer, Marcantonio, *J. Org. Chem.* **29**, 3473 (1964).
*trans-*Isomer, liquid, bp$_{752}$ 84.8°. n_D^{20} 1.4350, n_D^{25} 1.4327. d$_4^{20}$ 0.9295.
*cis-*Isomer, liquid, bp$_{758}$ 84.1°. n_D^{20} 1.4390. d$_4^{20}$ 0.9426.
Caution: Irritates eyes, respiratory passages.

2182. 3-Chloro-1-butene. *γ-Chloro-α-butylene; α-methallyl chloride; α-methylallyl chloride.* $CH_2=CHCH-ClCH_3$. Empirical formula, etc.: *see* 1-Chloro-2-butene. Prepn of DL-form: Böhme, *Ber.* **71**, 2372 (1938), Curtin,

Gerber, *J. Am. Chem. Soc.* **74**, 4052 (1952); of D(−)-form: Böhme, *loc. cit.;* of L(+)-form: Young, Caserio, *J. Org. Chem.* **26**, 245 (1961). Absolute configuration: Young, Caserio, *loc. cit.* Equilibrium constant for isomerization to 1-chloro-2-butene: Dittmer, Marcantonio, *J. Org. Chem.* **29**, 3473 (1964).
DL-Form, liquid, bp 63.9-64.2°, bp$_{26}$ −5°. d$_4^{20}$ 0.9001. n_D^{20} 1.4150.
D(−)-Form, liquid, bp$_{26}$ −5°. $α_D^{20}$ −2.52° (neat).
L(+)-Form, liquid, $α_D^{25}$ +5.87 (neat).
Caution: Irritates eyes, respiratory passages.

2183. 3-Chloro-d-camphor. *3-Chloro-1,7,7-trimethylbicyclo[2.2.1]heptan-2-one; 3-chloro-d-2-bornanone.* $C_{10}H_{15}ClO$; mol wt 186.68. C 64.34%, H 8.10%, Cl 18.99%, O 8.57%. Of the two isomers, the *endo*-form is more stable than the *exo*-form: Cookson, *J. Chem. Soc.* **1954**, 282. Prepn of *endo*-form: Kipping, Pope, *ibid.* **63**, 548 (1893). Prepn of *exo*-form by isomerization of *endo*-form: Lowry, Steele, *ibid.* **107**, 1382 (1915). uv spectra: Lowry, Owen, *J. Chem. Soc.* **129**, 606 (1926). Crystal structure of *endo*-form: Wiebenga, Krom, *Rec. Trav. Chim.* **65**, 663 (1946). Configuration: Cookson, *loc. cit.;* Kumler *et al., J. Am. Chem. Soc.* **83**, 2711 (1961). *Review: The Terpenes* vol. II, J. L. Simonsen, Ed. (University Press, Cambridge, 2nd ed., 1949) p 397.

endo-form   exo-form

endo-Form, *α-chloro-d-camphor, 3α-chloro-d-camphor, camphor monochlorated.* Monoclinic prisms from alc, mp 94°. Volatile with steam. $[α]_D$ +96.2° (c = 5 in alc). Practically insol in water. Sol in alcohol, chloroform, ether. uv max (cyclohexane): 305 nm (log ε 1.72).
exo-Form, *α'-chloro-d-camphor, 3β-chloro-d-camphor.* Crystals from alc, mp 117°. $[α]_D$ +35° (c = 5 in alc). Readily sol in all ordinary solvents, except water and formamide. Much more sol in 96% alc than *endo*-form. On standing, loses HCl. uv max (cyclohexane): 306 nm (log ε 1.75).

2184. 4-Chloro-m-cresol. *4-Chloro-3-methylphenol; 3-methyl-4-chlorophenol; chlorocresol; parachlorometacresol; 6-chloro-m-cresol; 6-chloro-3-hydroxytoluene; 2-chloro-5-hydroxytoluene.* C_7H_7ClO; mol wt 142.58. C 58.97%, H 4.95%, Cl 24.86%, O 11.22%. Prepd by chlorination of *m*-cresol: Ger. pat. **90,847** (1897 to Kalle), *Frdl.* **4**, 94; Laschinger, U.S. pat. **1,847,566** (1932), *C.A.* **26**, 2471 (1932); Sah, Anderson, *J. Am. Chem. Soc.* **63**, 3165 (1941).

Dimorphous crystals, mp 55.5° and mp 66° (ligroin). Said to be odorless when very pure, but usually a phenolic odor persists. bp 235°. Volatile with steam. One gram dissolves in 260 ml water at 20°, more sol in hot water. Freely sol in alcohol, benzene, chloroform, ether, acetone, petr ether, fixed oils, terpenes, aq alkaline solns. Aq solns turn yellow on exposure to light and air.
USE: Antiseptic, disinfectant, more active in acid than in alkaline solns.
THERAP CAT: Antiseptic (topical).
THERAP CAT (VET): Has been used as a topical antiseptic.

2185. Chlorocyanohydrin. *3,3,3-Trichloro-2-hydroxypropanenitrile; 3,3,3-trichlorolactonitrile; 3,3,3-trichloro-2-hydroxypropionitrile; chloral hydrocyanide.* $C_3H_2Cl_3NO$; mol wt 174.41. C 20.66%, H 1.16%, Cl 60.98%, N 8.03%, O 9.17%. $Cl_3CCH(OH)CN$. Prepd from chloral hydrate and HCN: Chwala, Wassmuth, *Monatsh.* **81**, 843 (1950).

Crystals, mp 61°. Odor of HCN and chloral. Sublimes easily. bp 220°. Freely sol in water, alcohol, ether, and most other common organic solvents; moderately sol in carbon tetrachloride. Practically insol in petr ether.

2186. Chlorodibromomethane. *Dibromochloromethane;* chlorobromoform; monochlorodibromomethane; DBCM. NCI-C55254. $CHBr_2Cl$; mol wt 208.28. C 5.77%, H 0.48%, Br 76.73%, Cl 17.02%. $CHBr_2Cl$. Trihalomethane contaminant found in chlorinated water. Prepn: O. Jacobsen, R. Neumeister, *Ber.* **15**, 599 (1882); M. Fedorynski *et al., Syn. Commun.* **7**, 287 (1977). Toxicity studies: F. J. Bowman *et al., Toxicol. Appl. Pharmacol.* **44**, 213 (1978); Y. Aida *et al., J. Toxicol. Sci.* **17**, 119 (1992). Review of toxicology and human exposure: *Toxicological Profile for Bromoform and Chlorodibromomethane* (PB91-180323, 1990) 121 pp. Review of properties, occurrence, toxicity, and carcinogenicity: *IARC Monographs* **52**, 243-268 (1991).
Liquid. bp 121.3-121.8° (Fedorynski); also reported as bp 123-125° (Jacobsen). d^{15} 2.4450. n_{20}^D 1.5471. LD_{50} in male, female rats (mg/kg): 370, 760 orally (Aida); in male, female mice (mg/kg): 800, 1200 by gavage (Bowman).
USE: Chemical reagent/intermediate in organic synthesis.

2187. 1-Chloro-2,4-dinitrobenzene. 2,4-Dinitro-1-chlorobenzene; 4-chloro-1,3-dinitrobenzene; 6-chloro-1,3-dinitrobenzene. $C_6H_3ClN_2O_4$; mol wt 202.55. C 35.58%, H 1.49%, Cl 17.50%, N 13.83%, O 31.60%. Obtained by nitrating *o*-nitrochlorobenzene: Borsche, Rantscheff, *Ann.* **379**, 152 (1911); Hodgson, Dodgson, *J. Chem. Soc.* **1948**, 1006. The mixture of 2,4- and 2,6-dinitrochlorobenzenes (of which the 2,4-compd is present in largest amount) can be separated with an alkaline ethanol-water soln: Molard, Vaganay, *Mém. Poudres* **39**, 111 (1957). Toxicity study: H. F. Smyth *et al., Am. Ind. Hyg. Assoc. J.* **23**, 95 (1962).

Yellow crystals. *Skin irritant.* d ~1.7. mp 52-54°. bp 315°. Practically insol in water. Sparingly sol in cold, freely sol in hot alc; sol in ether, benzene, CS_2. LD_{50} orally in rats: 1.07 g/kg (Smyth).
Caution: May cause dermatitis of both primary and allergic types.
USE: As a reagent for the detection and determination of nicotinic acid, nicotinamide, and other pyridine compds.

2188. 2-Chloro-1,3-dinitrobenzene. 1-Chloro-2,6-dinitrobenzene; 2,6-dinitro-1-chlorobenzene. $C_6H_3ClN_2O_4$; mol wt 202.55. C 35.58%, H 1.49%, Cl 17.50%, N 13.83%, O 31.60%. Prepd from 2,6-dinitroaniline by the Sandmeyer reaction: Gunstone, Tucker, *Org. Syn.,* coll. vol. IV, 160 (1963).

Yellow crystals from benzene + petr ether or from acetic acid. *Skin irritant.* mp 86-87°. $d_4^{16.5}$ 1.6867. bp 315°. Moderately sol in ether, benzene. Sol in alc. Practically insol in water.
Caution: See 1-Chloro-2,4-dinitrobenzene.

2189. β-Chloroethyl Acetate. *2-Chloroethanol acetate.* $C_4H_7ClO_2$; mol wt 122.55. C 39.20%, H 5.76%, Cl 28.93%, O 26.11%. $CH_3COOCH_2CH_2Cl$. Prepn from ethylene oxide and acetyl chloride: Gustus, Stevens, *J. Am. Chem. Soc.* **55**, 378 (1933).

Liquid, bp 145°. d^{20} 1.1460. n_D^{20} 1.4234. Flash pt 53.89° (closed cup): *Ind. Eng. Chem.* **32**, 880 (1940). Practically insol in water. Sol in alcohol.
USE: Gasoline additive, Henderson, U.S. pat. **3,179,506** (1965 to Shell).

2190. N-(2-Chloroethyl)dibenzylamine Hydrochloride. *N-(2-Chloroethyl)-N-(phenylmethyl)benzenemethanamine hydrochloride;* N,N-dibenzyl-β-chloroethylamine hydrochloride; N,N-dibenzylaminoethyl chloride hydrochloride; Dibenamine hydrochloride. $C_{16}H_{19}Cl_2N$; mol wt 296.24. C 64.87%, H 6.46%, Cl 23.94%, N 4.73%. $(C_6H_5CH_2)_2N$-$CH_2CH_2Cl.HCl$. α-Adrenergic blocker. Prepn: Rabald, Dimroth, **Ger.** pat. **824,208** (1951 to Boehringer, Mann.), *C.A.* **47**, 2206 (1953). Pharmacology and toxicity: M. Nickerson, G. M. Nomaguchi, *J. Pharmacol. Exp. Ther.* **101**, 379 (1951). Use in adrenergic receptor differentiation: R. F. Furchgott, *ibid.* **111**, 265 (1954); in specific receptor labelling: R. D. Green *et al., ibid.* **187**, 524 (1973). *In vivo* studies of protection against hepatotoxic agents: H. M. Maling *et al., Toxicol. Appl. Pharmacol.* **27**, 380 (1974); E. K. Weisburger *et al., ibid.* **28**, 477 (1974); H. M. Maling *et al., Biochem. Pharmacol.* **23**, 1479 (1974).
Crystals, mp 192°, also given as 180-181° (the free base is an oily liquid). Practically insol in water near neutrality. Sol in dil acids (2% at pH 2.1, 1% at pH 2.4 and 0.5% at pH 2.7), in 95% alcohol and in propylene glycol. Stable in acid soln, but rapidly loses activity in neutral or alkaline solns. LD_{50} s.c. in mice: 800 mg/kg (Nickerson, Nomaguchi).

2191. 2-Chloroethyl Vinyl Ether. *(2-Chloroethoxy)ethene.* C_4H_7ClO; mol wt 106.55. C 45.09%, H 6.62%, Cl 33.27%, O 15.02%. $ClCH_2CH_2OCH=CH_2$. Prepd by the action of solid NaOH + triethanolamine upon β,β'-dichlorodiethyl ether: Chitwood, Perkins, U.S. pat. **2,104,717** (1938 to Carbide and Carbon Chem.); *cf.* Cretcher *et al., J. Am. Chem. Soc.* **47**, 1175 (1926). Toxicity study: Smyth *et al., J. Ind. Hyg. Toxicol.* **31**, 60 (1949).
Liquid. d_{15}^{15} 1.0525. bp_{740} 109°. Quite stable to NaOH solns. Even dil acids produce hydrolysis to acetaldehyde and ethylene chlorohydrin [2-chloroethanol]. LD_{50} orally in rats: 250 mg/kg (Smyth).
USE: Mfg anesthetics, sedatives, cellulose ethers.

2192. Chlorofluorocarbons. CFCs; FCCs. Chemically stable series of chlorinated and fluorinated compounds usually with methane or ethane skeleton, marketed under general names such as *Arcton, Freon, Frigen, Genetron.* Known collectively as CFCs, individually identified by a CFC code based on the "rule of 90". (To derive the chemical formula for CFC-12, e.g., add "90" to 12; the resulting number "102" indicates 1 carbon, 0 hydrogen, 2 fluorine yielding the formula CCl_2F_2.) Initial report on suitability as refrigerants: T. Midgley, Jr., A. L. Henne, *Ind. Eng. Chem.* **22**, 542 (1930). Review including physical and chemical properties of various CFCs: E. Heiskel, *Aerosol Rep.* **22**, 403-415 (1983). CFCs do not decompose in the lower atmosphere. Photodecomposition occurs in the stratosphere via absorption of uv radiation and subsequent release of atomic chlorine which can catalyze ozone breakdown. CFC-ozone depletion hypothesis: M. J. Molina, F. S. Rowland, *Nature* **249**, 810 (1974). Reviews focusing on atmospheric chemistry of CFCs, uses, potential hydrogen-substituted replacements, environmental and regulatory issues: J. P. Cohn, *BioScience* **37**, 647-650 (1987); R. Pool, *Science* **242**, 666-668 (1988); F. S. Rowland, *Environ. Conserv.* **15**, 101-115 (1988); L. B. Weisfeld, *Plast. Comp.* **1988**, 15-22, 40-43; F. S. Rowland, *Am. Scientist* **77**, 36-45 (1989). For prepn information *see* dichlorodifluoromethane (CFC-12), trichlorofluoromethane (CFC-11), cryofluorane (CFC-114).
Colorless, essentially odorless, nonflammable, noncorrosive. Miscible with aliphatic, alicyclic and aromatic hydrocarbons, halogenated hydrocarbons, monovalent low molecular alcohols.
Note: Consult latest Government regulations on use as aerosol propellant.
USE: In aerosol propellants (CFC-11, 12, 113); air conditioning; refrigeration (CFC-12); blowing agents for making foam (CFC-11, 12); cleaning fluids (CFC-113); solvents for the electronics industry, bedding and packaging.

2193. Chloroform. *Trichloromethane.* $CHCl_3$; mol wt 119.38. C 10.06%, H 0.84%, Cl 89.09%. Improperly called "formyl trichloride". Made from acetone and bleaching powder by addition of sulfuric acid: $2CH_3COCH_3 + 6CaO-Cl_2 \cdot H_2O \rightarrow 2CHCl_3 + (CH_3COO)_2Ca + 2Ca(OH)_2 + 3Ca-Cl_2 + 6H_2O$. May also be prepd by carefully controlled chlorination of methane: Faith, Keyes & Clark's *Industrial Chemicals,* F. A. Lowenheim, M. K. Moran, Eds. (Wiley-Interscience, New York, 4th ed., 1975) pp 266-269. Has been used as an anesthetic and in pharmaceutical preparations. Toxicity data: H. F. Smyth *et al., Am. Ind. Hyg. Assoc. J.* **23,** 95 (1962); E. T. Kimura *et al., Toxicol. Appl. Pharmacol.* **19,** 699 (1971). Review of toxicology: L. R. Pohl, *Rev. Biochem. Toxicol.* **1,** 79-108 (1979). Review of carcinogenicity studies: *IARC Monographs* **20,** 401-427 (1979). *Review:* M. T. Holbrook in Kirk-Othmer *Encyclopedia of Chemical Technology* vol. 5 (John Wiley & Sons, New York, 4th ed., 1993) pp 1051-1062.

Highly refractive, nonflammable, heavy, very volatile, sweet-tasting liquid; characteristic odor. d_{20}^{20} 1.484. bp 61-62°. mp −63.5°. n_D^{20} 1.4476. Forms a constant boiling mixture with 7% alc, boiling at 59°. d 1.474-1.478 for U.S.P. chloroform contg 0.5-1% ethanol as stabilizer. One ml dissolves in about 200 ml water at 25°. Misc with alcohol, benzene, ether, petr ether, carbon tetrachloride, carbon disulfide, oils. Pure chloroform is light sensitive and reagent grade chloroform usually contains 0.75% ethanol as stabilizer. *Protect from light and keep cool.* LD_{50} (14 day) orally in rats: 2.18 ml/kg (Smyth); 0.9 ml/kg (Kimura).

Caution: Potential symptoms of overexposure are dizziness, mental dullness, nausea and disorientation; headache, fatigue; anesthesia; hepatomegaly; direct contact may cause irritation to eyes and skin. *See NIOSH Pocket Guide to Chemical Hazards* (DHHS/NIOSH 90-117, 1990) p 68. Banned from use in foods, drugs and cosmetics by the FDA. This substance may reasonably be anticipated to be a carcinogen: *Seventh Annual Report on Carcinogens* (PB95-109781, 1994) p 127.

USE: As a solvent for fats, oils, rubber, alkaloids, waxes, gutta-percha, resins; as cleansing agent; in fire extinguishers to lower the freezing temp of carbon tetrachloride; in the rubber industry.

2194. Chlorogenic Acid. *[1S-(1α,3β,4α,5α)]-3-[[3-(3,4-Dihydroxyphenyl)-1-oxo-2-propenyl]oxy]-1,4,5-trihydroxycyclohexanecarboxylic acid; 1,3,4,5-tetrahydroxycyclohexanecarboxylic acid 3-(3,4-dihydroxycinnamate);* 3-caffeoylquinic acid; 3-(3,4-dihydroxycinnamoyl)quinic acid. $C_{16}H_{18}O_9$; mol wt 354.31. C 54.24%, H 5.12%, O 40.64%. Important factor in plant metabolism. Isoln from green coffee beans: Freudenberg, *Ber.* **53,** 237 (1920). Chlorogenic acid and its isomers isochlorogenic acid and neochlorogenic acid occur also in fruit, leaves and other tissues of dicotyledenous plants: Sondheimer, *Arch. Pharm.* **293,** 721 (1960). Forms caffeic acid on hydrolysis: Fiedler, *Arzneimittel-Forsch.* **4,** 41 (1954). Structure: Fischer, Dangschat, *Ber.* **65,** 1037 (1932); Barnes *et al., J. Am. Chem. Soc.* **72,** 4178 (1950); Corse *et al., Tetrahedron* **18,** 1207 (1962). Synthesis: Panizzi *et al., Gazz. Chim. Ital.* **86,** 913 (1956).

Hemihydrate, needles from water. Becomes anhydr at 110°. mp 208°. $[\alpha]_D^{26}$ −35.2° (c = 2.8). pKa (27°) 2.66. R_f values: Fiedler, *loc. cit.* Soly in water at 25° about 4%, much more sol in hot water. Alkaline solns acquire an orange color. Freely sol in alcohol, acetone. Very slightly sol in ethyl acetate. Heating with dil HCl yields caffeic acid. Forms a black compd with iron, said to be responsible for the blackening of cut and cooked potatoes: *Chem. & Ind. (London)* **1958,** 627.

3′-Methyl ether, $C_{17}H_{20}O_9$, *3-feruloylquinic acid.* Crystals from ethyl acetate + petr ether, mp 196-197°. $[\alpha]_D^{25}$ −42.8° (ethanol). uv max (ethanol): 325 nm (ε 19,200).

2195. Chlorogenin. *(3β,5α,6α,25R)-Spirostan-3,6-diol.* $C_{27}H_{44}O_4$; mol wt 432.64. C 74.96%, H 10.25%, O 14.79%. Isoln from bulbs of the California soap plant, amole: *Chlorogalum pomeridianum* (DC.) Kunth, *Liliaceae:* Liang, Noller, *J. Am. Chem. Soc.* **57,** 525 (1935). Chlorogenin occurs in amole as a saponin which kills or stuns fish without rendering them inedible. Structure: Marker, Rohrmann, *ibid.* **61,** 947, 3479 (1939); Marker *et al., ibid.* **62,** 2537, 3006 (1940). On hydrogenation the 3β,6β-isomer (β-chlorogenin) is produced.

Needles from methanol, mp 273-276°. $[\alpha]_{546}^{24}$ −52° (chloroform or isopropanol). Less sol in methanol, more sol in isopropanol than tigogenin.

Diacetate, crystals from dil methanol, mp 154-155°.

Dibenzoate, crystals from methanol + chloroform, mp 200.5-204.5°, $[\alpha]_{546}^{26}$ +9.5° (chloroform).

2196. 1-Chlorohexane. *n*-Hexyl chloride. $C_6H_{13}Cl$; mol wt 120.62. C 59.75%, H 10.86%, Cl 29.39%. $CH_3-(CH_2)_4CH_2Cl$. Prepd from 1-hexanol by treatment with fuming HCl: Henry, *Chem. Zentr.* **1905,** II, 214; with excess $SOCl_2$ or with $PCl_5 + ZnCl_2$: Clark, Streight, *Trans. Roy. Soc. Can.* [3] **23,** III, 77 (1929).

Mobile liquid. d_4^{20} 0.8780. bp_{760} 134°. n_D^{20} 1.4236 (Clark, Streight, *loc. cit.*); n_D^{20} 1.4195 (Mumford, Phillips, *J. Chem. Soc.* **1950,** 75). Insol in water. Refluxing with 10% aq NaOH decomposes 1-chlorohexane to 1-hexanol.

2197. α-Chlorohydrin. *3-Chloro-1,2-propanediol;* 3-chloro-1,2-dihydroxypropane; α-monochlorohydrin; β,β′-dihydroxyisopropyl chloride; glycerol α-monochlorohydrin; 3-chloropropylene glycol; Epibloc. $C_3H_7ClO_2$; mol wt 110.54. C 32.60%, H 6.38%, Cl 32.07%, O 28.95%. $CH_2Cl-CHOHCH_2OH$. Prepd from glycerol and HCl gas: Conant, Quayle, *Org. Syn.* coll. vol. I, 294 (1941). Toxicity study: C. H. Hine *et al., Arch. Ind. Health* **14,** 250 (1956).

Liquid. Sweetish taste. Tendency to turn straw color. d_4^{20} 1.3218. n_D^{20} 1.4831. bp_{760} 213° (dec); bp_{14} 114-120°; bp_{11} 115-117°. Sol in water, alcohol, ether. LD_{50} in mice, rats (g/kg): 0.16, 0.15 orally (Hine).

USE: To lower the freezing point of dynamite; in the manuf of dye intermediates. As rodent chemosterilant.

2198. Chloromethyl Methyl Ether. *Chloromethoxymethane;* methyl chloromethyl ether; monochloromethyl ether; chlorodimethyl ether; CMME. C_2H_5ClO; mol wt 80.51. C 29.84%, H 6.26%, Cl 44.03%, O 19.87%. CH_3O-CH_2Cl. Prepd by passing HCl through a mixture of formalin and methanol: C. S. Marvel, P. K. Porter, *Org. Syn.* coll. vol. I, 377 (1941). *See also Beilstein* **1,** 580 (1918) and supplements. Commercial product usually contaminated by *sym*-dichloromethyl ether, *q.v.* Review of carcinogenic risk: *IARC Monographs* **4,** 239-245 (1974).

Colorless liquid, bp 59°. d_4^{20} 1.0605. n_D^{20} 1.39737.

Caution: Potential symptoms of overexposure are irritation of eyes, skin and mucous membranes; pulmonary edema, pulmonary congestion and pneumonia; burns, necrosis; coughing, wheezing; blood stained sputum; weight loss; bronchial secretions. *See NIOSH Pocket Guide to Chemical Hazards* (DHHS/NIOSH 90-117, 1990) p 68. The technical grade has been listed as a known carcinogen: *Seventh Annual Report on Carcinogens* (PB95-109781, 1994) p 40.

USE: In synthesis of chloromethylated compounds.

2199. 1-Chloro-2-methyl-1-propene. α-Chloroisobutylene; β,β-dimethylvinyl chloride; isocrotyl chloride; 2-methyl-1-propenyl chloride. C_4H_7Cl; mol wt 90.55. C 53.06%, H 7.79%, Cl 39.15%. $(CH_3)_2C=CHCl$. Prepn from iso-butyraldehyde: Kirrmann, *Bull. Soc. Chim. France* **1948**, 163; by isomerization of 3-chloro-2-methylpropene with 80% H_2SO_4: Backhurst *et al., J. Chem. Soc.* **1959**, 2742.

Liquid, bp 68.1°. d_4^{20} 0.9186. n_D^{20} 1.4221.

Caution: This substance may reasonably be anticipated to be a carcinogen: *Seventh Annual Report on Carcinogens* (PB95-109781, 1994) p 185.

USE: In organic syntheses.

2200. 3-Chloro-2-methyl-1-propene. γ-Chloroisobutyl-ene; methallylchloride; β-methallyl chloride; β-methylallyl-chloride. C_4H_7Cl; mol wt 90.55. C 53.06%, H 7.79%, Cl 39.15%. $ClCH_2(CH_3)C=CH_2$. Prepn by chlorination of isobutylene: Burgin *et al., Ind. Eng. Chem.* **31**, 1413 (1939). Manuf: Carlson, U.S. pat. **2,494,034**; Cherniavsky, Brown, U.S. pat. **2,612,530**; Cheney *et al.,* U.S. pat. **2,642,464** (1950, 1952, 1953 all to Shell).

Liquid, bp 71-72°. d_4^{15} 0.9210, d_4^{20} 0.9165 (commercial grade d_{20}^{20} 0.926-0.930). n_D^{15} 1.4318, n_D^{20} 1.4274.

Caution: This substance may reasonably be anticipated to be a carcinogen: *Seventh Annual Report on Carcinogens* (PB95-109781, 1994) p 130.

USE: Insecticide, fumigant. Chemical intermediate in organic syntheses.

2201. 1-Chloronaphthalene. α-Chloronaphthalene. $C_{10}H_7Cl$; mol wt 162.62. C 73.86%, H 4.34%, Cl 21.80%. Prepd by passing chlorine into boiling naphthalene with or without solvent such as chlorobenzene and with or without catalyst such as I_2: DeWitt, Ekeley, *Univ. Colo. Stud.* **18**, 119 (1931), *C.A.* **26**, 2974 (1932). From α-naphthylamine by diazotization and Sandmeyer CuCl reaction: von Auwers, Frühling, *Ann.* **422**, 194, 200, 202 (1921); Hampson, Weissberger, *J. Chem. Soc.* **1936**, 394.

Oily liquid. Volatile with steam. d_4^{20} 1.19382. mp −2.5°. bp_{760} 259.3°; bp_{400} 230.8°; bp_{200} 204.2; bp_{100} 180.4°; bp_{60} 165.6°; bp_{40} 153.2°; bp_{20} 134.4°; bp_{10} 118.6°; bp_5 104.8°; $bp_{1.0}$ 80.6°. n_D^{19} 1.63321. Sol in benzene, petr ether, alcohol. Compound with 2,4,6-trinitro-*m*-cresol, $C_{17}H_{12}ClN_3O_7$, mp 78°.

Styphnate, $C_{16}H_{10}ClN_3O_8$, yellow needles from alcohol, mp 112°.

USE: As immersion liquid in the (microscopic) determn of the refractive index of crystals. Solvent for oils, fat, DDT.

2202. 2-Chloronaphthalene. β-Chloronaphthalene. $C_{10}H_7Cl$; mol wt 162.62. C 73.86%, H 4.34%, Cl 21.80%. Prepd from β-naphthylamine by diazotization and Sandmeyer CuCl reaction: Scheid, *Ber.* **34**, 1813 (1901); van der Kam, *Rec. Trav. Chim.* **45**, 568 (1926); Hampson, Weissberger, *J. Chem. Soc.* **1936**, 394. By chlorination of gaseous naphthalene at 530° in the presence of iodine, ~50% of the product being 1-chloronaphthalene which is sepd by fractional freezing: Britton, Reed, U.S. pat. **1,917,822** (1933).

Monoclinic plates, leaflets from dil alcohol. mp 59.5°. bp_{760} 256°; bp_{60} 161°; bp_{20} 132.6°; bp_{11} 119.7°. Volatile with steam. $n_D^{70.7}$ 1.60787. Sol in alcohol, ether, benzene, chloroform, carbon disulfide.

2203. Chloronitrobenzene. Nitrochlorobenzene. C_6-H_4ClNO_2; mol wt 157.56. C 45.74%, H 2.56%, Cl 22.50%, N 8.89%, O 20.31%. Prepn of *m*-isomer: W. W. Hartman, M. R. Brethen, *Org. Syn.* coll. vol. I (2nd ed., 1964) p 162.

m-Chloronitrobenzene, 1-chloro-3-nitrobenzene, m-nitro-chlorobenzene. Pale-yellow orthorhombic prisms from alcohol. d_4^{20} 1.534. mp 46°. bp_{760} 236°; bp_{12} 117°. Insol in water. Sparingly sol in cold alcohol; freely sol in hot alcohol, chloroform, ether, carbon disulfide, glacial acetic acid.

o-Chloronitrobenzene. Yellow crystals. d 1.305. mp 32-33°. bp 245-246°. Insol in water. Sol in alcohol, benzene, ether.

p-Chloronitrobenzene. Yellow crystals. d 1.520. mp 82-84°. bp 242°. Flash pt 127°. Insol in water. Sparingly sol in cold alcohol, freely in boiling alcohol, ether, carbon disulfide.

Caution: Extremely toxic! Absorbed by inhalation, skin penetration and ingestion. Minute quantities may produce severe methemoglobinemia resulting in cyanosis. Symptoms include lips, tongue and mucous membranes navy blue to black, skin slate gray, lacking signs of cardiac or pulmonary insufficiency; severe headache, nausea, vomiting, dryness of throat. Central nervous symptoms: confusion, ataxia, vertigo, tinnitus, weakness, disorientation, lethargy, drowsiness, coma. Cardiac effects: heart blocks, arrhythmias, shock. Urinary signs: painful micturition, hamaturia, hemoglobinemia, methomoglobinemia, oliguria, mild renal insufficiency. Severe allergic dermatitis frequently occurs after skin contact, *cf. Clinical Toxicology of Commercial Products*, R. E. Gosselin *et al.,* Eds. (Williams & Wilkins, Baltimore, 5th ed., 1984) Section II, p 214, Section III, p 31-36.

USE: In dye chemistry.

2204. Chlorophacinone. *2-[(4-Chlorophenyl)phenyl-acetyl]-1H-indene-1,3(2H)-dione; 2-[(p-chlorophenyl)phenyl-acetyl]-1,3-indandione;* LM-91; Caid; Drat; Liphadione; Quick; Raviac; Rozol. $C_{23}H_{15}ClO_3$; mol wt 374.82. C 73.70%, H 4.03%, Cl 9.46%, O 12.81%. Prepn: Molho *et al.,* U.S. pat. **3,153,612** (1964 to Lipha).

Crystals, mp 140°. Absorption band at 325 nm in acetone soln. Sol in organic solvents; very sparingly sol in water. LD_{50} orally in rats: 20.5 mg/kg.

USE: Anticoagulant rodenticide.

2205. p-Chlorophenacyl Bromide. *2-Bromo-1-(4-chlo-rophenyl)ethanone; 2-bromo-4'-chloroacetophenone;* 4-chloro-ω-bromoacetophenone; α-bromo-p-chloroacetophenone. C_8-H_6BrClO; mol wt 233.49. C 41.15%, H 2.59%, Br 34.22%, Cl 15.18%, O 6.85%. Prepn from bromoacetyl chloride, chlorobenzene, and $AlCl_3$: Collet, *Compt. Rend.* **125**, 718 (1897); by bromination of p-chloroacetophenone: *idem, Bull. Soc. Chim. France* [3] **21**, 69 (1899). Toxicity study: Dat-Xuong *et al., Med. Exp.* **11**(3), 137 (1964).

Needles, mp 96-96.5°. LD_{50} orally in mice: > 2000 mg/kg (Dat-Xuong).

USE: In the prepn of quaternary salts of methenamine and of chlorophenol glyoximes.

2206. Chlorophenol. C_6H_5ClO; mol wt 128.56. C 56.06%, H 3.92%, Cl 27.58%, O 12.45%. Prepn of o-form: Holleman, Rinkes, *Rec. Trav. Chim.* **30**, 48 (1911); Huston, Neely, *J. Am. Chem. Soc.* **57**, 2176 (1935). *cis-trans* equilibria in o-halophenols: Baker, *ibid.* **80**, 3598 (1958). Prepn of p-form: Crawford, Willson, U.S. pat. **1,910,679** (1927 to Monsanto); Neu, *Ber.* **72**, 1505 (1939); Hodgson, Norris, *J. Chem. Soc.* **1949**, Suppl. 1, S181. Manuf: Britton, Keil, U.S. pat. **2,725,402** (1955 to Dow). Prepn of m-form: Acheson, Taylor, *J. Chem. Soc.* **1956**, 4727. Manuf: Stoesser, Gentry; Barnard, Meyer, U.S. pats. **2,835,707** and **2,852,567** (both 1958 to Dow). Toxicity data: Deichmann, *Fed. Proc.* **2**, 76 (1943).

m-Chlorophenol, needles, mp 33.5°. bp 214°. d_4^{45} 1.245, d_4^{78} 1.214. n_D^{40} 1.5565. Dipole moment in benzene, 2.10D. Slightly sol in cold water; sol in alcohol, ether, caustic alkali solns. LD_{50} orally in rats: 0.57 g/kg (Deichmann).

o-Chlorophenol, liquid, bp 175°. mp 9.3°. d_4^{23} 1.2573. n_D^{25} 1.5565; *n* 1.5473. Dipole moment in benzene, 1.33D. Freely sol in alcohol, ether, caustic alkali solns; slightly sol in water. LD_{50} orally in rats: 0.67 g/kg (Deichmann).

p-Chlorophenol, parachlorophenol. Crystals with characteristic phenolic odor, mp 43.2-43.7°. bp 220°. d_4^{78} 1.2238. n_D^{55} 1.5419; n_D^{40} 1.5579. Dipole moment in benzene, 2.22D. Sparingly sol in water, liquid petrolatum; very sol in alcohol, glycerin, ether, chloroform, fixed and volatile oils: *U.S.P.* **XXI**, 1467. LD_{50} orally in rats: 0.67 g/kg (Deichmann).

Caution: May cause tremors, convulsions, dyspnea, coma. Irritating to skin. *See:* Patty's *Industrial Hygiene and Toxicology* Vol. **2A**, G. D. Clayton, F. E. Clayton, Eds. (Wiley-Interscience, New York, 3rd ed., 1981) pp 2612-2615.

THERAP CAT: Parachlorophenol as antiseptic.

2207. Chlorophyll. The green pigment of plants. Higher plants and green algae contain chlorophyll a and chlorophyll b in the approx ratio of 3:1. Chlorophyll c is found together with chlorophyll a in many types of marine algae: Jeffrey, *Biochem. J.* **86**, 313 (1963). Red algae *(Rhodophyta)* contain principally chlorophyll a and also chlorophyll d: Manning, Strain, *J. Biol. Chem.* **151**, 1 (1943). Isoln by chromatography: Tswett, *Ber. Deutsch. Bot. Ges.* **24**, 316, 385 (1906); Willstätter, Stoll, *Investigations on Chlorophyll* (transl by Schertz and Merz: Lancaster, 1928); Schertz, *Ind. Eng. Chem.* **30**, 1073 (1938); Fischer-Orth-Stern, *Die Chemie des Pyrrols* Vol. **II**, part 2 (Leipzig, 1940); Zechmeister, Cholnoky, *Principles and Practice of Chromatography* (New York, 1943). Industrial large-scale isoln processes: Judah *et al., Ind. Eng. Chem.* **46**, 2262 (1954). Structure: Fischer-Orth-Stern, *loc. cit.;* Ficken *et al., J. Chem. Soc.* **1956**, 2273. Total synthesis of chlorophyll a: Woodward *et al., J. Am. Chem. Soc.* **82**, 3800 (1960); *Angew. Chem.* **72**, 651 (1960); Strell *et al., ibid.* 169; Woodward, *Pure Appl. Chem.* **2**, 383 (1963). Abs config of chlorophylls a and b: Brockmann, *Ann.* **754**, 139 (1971); Brockmann, Bode, *Ann.* **1974** (7), 1017. ^{13}C-NMR study of chlorophyll a: S. Lötjönen, P. H. Hynninen, *Org. Mag. Reson.* **16**, 304 (1981); of chlorophyll b: N. Risch, H. Brockmann, *Tetrahedron Letters* **24**, 173 (1983). Review of syntheses: Johnson, *Sci. Progr.* **49**, 77 (1961). Comprehensive reviews with bibliography: Stoll, Wiedemann, *Fortschr. Chem. Org. Naturst.* **1**, 159-254 (1938); *Fortschr. Chem. Forsch.* **2**, 538 (1952); *The Chlorophylls*, L. P. Vernon, G. R. Seely, Eds. (Academic Press, New York, 1966) 679 pp; Inhoffen *et al., Fortschr. Chem. Org. Naturst.* **26**, 284-298 (1968); Inhoffen, *Pure Appl. Chem.* **17**, 443-460 (1968).

	R_1	R_2	R_3
Chlorophyll a	CH_3	CH_2CH_3	X
Chlorophyll b	CHO	CH_2CH_3	X

Chlorophyll a, $C_{55}H_{72}MgN_4O_5$. R = CH_3. Sepn and purification: Anderson, Calvin, *Nature* **194**, 285 (1962). Waxy blue-black microcrystals, usually aggregates of thin, lancet-like leaflets, mp 117-120°. $[\alpha]_D^{20}$ −262° (acetone). Absorption max (ether): 660, 613, 577, 531, 498, 429, 409 nm. Freely sol in ether, ethanol, acetone, chloroform, carbon disulfide, benzene. Sparingly sol in cold methanol. Practically insol in petr ether. The alcoholic soln is blue-green with a deep-red fluorescence.

Chlorophyll b, $C_{55}H_{70}MgN_4O_6$. R = CHO. Waxy blue-black microcrystals. Sinters between 86° and 92°, becomes a viscous liquid at 120-130° and then begins to puff up in large bubbles. $[\alpha]_D^{20}$ −267° (acetone-methanol). Absorption max (ether): 642, 593, 565, 545, 453, 427 nm Sparingly sol in petr ether, ligroin, cold methanol. Freely sol in abs alcohol, ether. The ether soln has a brilliant green color. Solns with other organic solvents are usually green to yellowish-green with red fluorescence.

Chlorophyll c. Structure: Dougherty *et al., J. Am. Chem. Soc.* **88**, 5037 (1966). Reddish-black hexagonal bipyramids or four-sided plates from dil ethanol. Absorption max (acetone): 628, 580, 442 nm ($E_{1cm}^{0.1\%}$ 15.8, 10.7, 115.9). Sol in methanol, ethanol, ethyl acetate; practically insol in ether, acetone.

Chlorophyll d, $C_{54}H_{70}MgN_4O_6$. Chlorophyll a with the 9-vinyl group replaced by a formyl group: Holt, Morley, *Can. J. Chem.* **37**, 507 (1959); Holt, *Can. J. Bot.* **39**, 327 (1961). Absorption max (ether): 686, 445 nm.

The chlorophyll of commerce is an intensely dark-green, aq, alc, or oil soln. Careful alkaline hydrolysis of chlorophyll opens the cyclopentanone ring and replaces the methyl and phytyl ester groups with Na or K; the resulting salts are called *chlorophyllins* and are water sol; e.g. sodium magnesium chlorophyllin, $C_{34}H_{31}N_4Na_3MgO_6$. Acid treatment of chlorophyll removes the Mg replacing it with H_2 which can be replaced with other metals, e.g. iron pheophytin, $C_{55}H_{72}FeN_4O_5$, sol in oil. Comprehensive review with 89 refs: "Chlorophyll Derivatives, Their Chemistry, Commercial Preparation and Uses" by J. C. Kephart in *Econ. Bot.* **9**, 3-38 (1955). Exhaustive analysis and description of commercial chlorophyllin prepns: Strell *et al., Arzneimittel-Forsch.* **5**, 640 (1955); **6**, 8 (1956). Trademarks: *Chloresium, Chlorofolin, Darotol, Ennds, Exodor-Grun, Nullo Chlorophyll, Stozzon-Chlorophyll.*

USE: To color soaps, oils, fats, waxes, confectionery, preserves, liquors, cosmetics, perfumes. Source of phytol. For dyeing leather. As sensitizer for color film. Has been used as antiknock agent in gasoline; as accelerator in the vulcanizing of rubber; in deodorizers.

THERAP CAT: Deodorant.

THERAP CAT (VET): Has been used orally to reduce odors, and topically to promote healing of skin lesions.

2208. Chloropicrin. *Trichloronitromethane;* acquinite; nitrochloroform; Larvacide 100; Picfume. CCl_3NO_2; mol wt 164.37. C 7.31%, Cl 64.70%, N 8.52%, O 19.47%. First prepd in 1848 by Stenhouse from picric acid and bleach powder. Review of prepn, properties, physiological action, and uses: Jackson, *Chem. Rev.* **14**, 251 (1934). Manuf from nitromethane and alkaline hypochlorite: Wilhelm, U.S. pat. **3,106,588** (1963).

Slightly oily liquid, intense odor; bp_{757} 112°. mp −64°; mp −69.2° (corr). d_4^{20} 1.6558, d_4^{25} 1.6483. n_D^{20} 1.4611, n_D^{25} 1.4596. Dipole moment in heptane or benzene, 1.80D. Practically insol in water (0.2272 g and 0.1621 g/100 ml at 0° and 25°); miscible with benzene, abs alc, carbon disulfide; sol in ether.

Caution: Produces severe sensory irritation in upper respiratory passages. Has strong lacrimatory properties, produces increased sensitivity after frequent exposures. Orally causes severe nausea, vomiting, colic, diarrhea. Potent skin irritant: Fairhall, *Industrial Toxicology* (Hafner, New York, 1969) p 196.

USE: Disinfecting cereals and grains; in synthesis, esp in manuf of methyl violet; fumigant; soil insecticide; war gas.

2209. Chloroprednisone. *(6α)-Chloro-17,21-dihydroxy-pregna-1,4-diene-3,11,20-trione;* 6α-chloro-1,4-pregnadiene-17α,21-diol-3,11,20-trione. $C_{21}H_{25}ClO_5$; mol wt 392.88. C 64.20%, H 6.41%, Cl 9.02%, O 20.36%. Prepn: Batres *et al.,* **Ger. pat. 1,079,042** (1960 to Syntex). Prepn of the 21-acetate by selenium dioxide dehydrogenation of 6α-chloro-cortisone 21-acetate: Ringold *et al., J. Am. Chem. Soc.* **80**, 6464 (1958); Ringold, Rosenkranz, U.S. pat. **2,957,895** (1960 to Syntex).

21-Acetate, $C_{23}H_{27}ClO_6$, *Topilan.* Crystals from acetone + hexane, mp 217-219°. uv max (ethanol): 237 nm (log ε 4.19). $[\alpha]_D$ +144° (chloroform).

THERAP CAT: Glucocorticoid (topical).

2210. Chloroprocaine Hydrochloride. *4-Amino-2-chlorobenzoic acid 2-(diethylamino)ethyl ester monohydrochloride;* 2-chloro-4-aminobenzoic acid diethylaminoethyl ester hydrochloride; *Nesacaine.* $C_{13}H_{20}Cl_2N_2O_2$; mol wt 307.22. C 50.82%, H 6.56%, Cl 23.08%, N 9.12%, O 10.42%. Description: Hädicke, *Pharm. Zentralh.* **94**, 384 (1955).

Crystals, mp 176-178° (microstage). Bitter taste. Slowly sol in water. Soly in water at 20° about one gram in 22 ml. Aq solns are just acid to litmus and turn yellow on standing. Soly in 95% ethanol about one gram in 100 ml.

THERAP CAT: Anesthetic (local).

2211. β-Chloropropionic Acid. *3-Chloropropanoic acid.* $C_3H_5ClO_2$; mol wt 108.52. C 33.20%, H 4.64%, Cl 32.67%, O 29.49%. CH_2ClCH_2COOH. Prepd by the hydrolysis of ethylene cyanohydrin with hydrochloric acid; by the oxidation of β-chloropropionaldehyde or of trimethylene chlorohydrin by nitric acid: Moureu, Chaux, *Org. Syn.* **8**, 54 (1928); Powell, *ibid.* 58; Paal, Lobeck, *Ber.* **64**, 2142 (1931).

Leaflets from ligroin. Somewhat hygroscopic. mp 41°. bp_{765} 200°; bp_{35} 127°; bp_{25} 124°; bp_{12} 108°. K at 25° = 1.01 × 10^{−4}. Freely sol in water, alc, chloroform, slightly less in ether.

Methyl ester, $C_4H_7ClO_2$, from β-chloropropionic acid with methanol and HCl or from acrylic acid chloride and methanol. d_4^0 1.198. bp 155-157°.

2212. β-Chloropropionitrile. *3-Chloropropanenitrile;* 3-chloropropanonitrile. C_3H_4ClN; mol wt 89.52. C 40.25%, H 4.50%, Cl 39.60%, N 15.65%. $ClCH_2CH_2CN$. Prepd from acrylonitrile and hydrogen chloride or bromide: Moureu, Clarke, *Bull. Soc. Chim. France* [4] **27**, 905 (1920); Stewart, Clarke, *J. Am. Chem. Soc.* **69**, 714 (1947); Shirley, *Preparation of Organic Intermediates* (Wiley, New York, 1951) p 82.

Liquid. Acrid, characteristic odor. *Poisonous!* mp −51°. d_4^{25} 1.1363. bp_{760} 176° (dec); bp_{200} 132° (dec); bp_{50} 95.2°; bp_5 46.0°. Flash pt 168°F (75.5°C). n_D^{25} 1.4341. Begins to dec when heated above 130° evolving HCl. Absorbs strongly in the infrared. Transparent to uv above 220 nm. Soly in water at 25°: 4.5 g/100 ml. Soly of water in β-chloropropionitrile at 25°: 2.2 ml/100 g. Miscible with ethanol, ether, acetone, benzene, carbon tetrachloride. LD_{50} orally in mice, rats: 9, 100 mg/kg, Fassett in *Industrial Hygiene and Toxicology* vol. 2, F. A. Patty, Ed. (Interscience, New York, 2nd ed., 1962) pp 2025-2026.

Caution: Exposure by any route should be avoided. Somewhat less hazardous than acrylonitrile because of lower vapor pressure. Readily penetrates skin to produce systemic cyanide poisoning, death.

USE: In pharmaceutical and polymer synthesis. Combines the reactivity of a nitrile and an alkyl halide. Because of the cyano group the chlorine atom is more reactive than in ordinary alkyl halides.

2213. 6-Chloropurine. $C_5H_3ClN_4$; mol wt 154.56. C 38.86%, H 1.96%, Cl 22.94%, N 36.25%. Prepd by the action of phosphorus oxychloride on hypoxanthine in N,N-dimethylaniline: Bendich *et al., J. Am. Chem. Soc.* **76**, 6073 (1954). Antineoplastic activity *in vivo:* A. C. Sartorelli, B. A. Booth, *Experientia* **21**, 457 (1965).

Blunt needles from water, dec 175-177° (hot stage preheated to 170°). Also reported as not melted at 290°. uv max (pH 5.2): 265 nm (ε 9120); (pH 13): 274 nm (ε 8790). Soly in water: 0.5% at 20°, also reported as 1 g/182 ml at 24°. Sol in ether, dimethylformamide.

2214. Chloropyramine. *N-[(4-Chlorophenyl)methyl]-N',N'-dimethyl-N-2-pyridinyl-1,2-ethanediamine;* 2-[(p-chlorobenzyl)[2-(dimethylamino)ethyl]amino]pyridine; *N*-dimethylaminoethyl-N-p-chlorobenzyl-α-aminopyridine; *N,N*-dimethyl-N'-(p-chlorobenzyl)-N'-(2-pyridyl)ethylenediamine; halopyramine; Chloropyribenzamine. $C_{16}H_{20}ClN_3$; mol wt 289.81. C 66.31%, H 6.96%, Cl 12.23%, N 14.50%. Prepn: Phillips, Cates, U.S. pat. **2,607,778** (1952 to Merck & Co.); Howard, U.S. pat. **2,569,314** (1951 to Am. Cyanamid); **Swiss** pats. **264,754; 266,234; 266,235** (1950); **Brit.** pat. **651,596** (1951) (all to Geigy).

Light yellow, viscous, oily liquid. Pungent odor. $bp_{0.2}$ 154-155°.

Hydrochloride, $C_{16}H_{20}ClN_3 \cdot HCl$, *Alegan S, Synopen, Synpen.* Crystals from acetone, mp 172-174°.

THERAP CAT: Antihistaminic.

2215. Chloroquine. *N^4-(7-Chloro-4-quinolinyl)-N^1,N^1-diethyl-1,4-pentanediamine; 7-chloro-4-(4-diethylamino-1-methylbutylamino)quinoline;* SN-7618; RP-3377; Aralen; Artrichin; Bemaphate; Capquin; Nivaquine B; Résoquine; Reumachlor; Sanoquin. $C_{18}H_{26}ClN_3$; mol wt 319.88. C 67.59%, H 8.19%, Cl 11.08%, N 13.14%. Prepd by the condensation of 4,7-dichloroquinoline with 1-diethylamino-4-aminopentane: **Ger.** pat. **683,692** (1939); H. Andersag *et al.,* **U.S.** pat. **2,233,970** (1941 to Winthrop); Surrey, Hammer, *J. Am. Chem. Soc.* **68**, 113 (1946). *Review:* Hahn in *Antibiotics* vol. 3, J. W. Corcoran, F. E. Hahn, Eds. (Springer-Verlag, New York, 1975) pp 58-78. Comprehensive description: D. D. Hong, *Anal. Profiles Drug Subs.* **5**, 61-85 (1976). Comparative clinical trial with dapsone in rheumatoid arthritis: P. D. Fowler *et al., Ann. Rheum. Dis.* **43**, 200 (1984); with penicillamine: T. Gibson *et al., Brit. J. Rheumatol.* **26**, 279 (1987).

mp 87°.
Diphosphate, $C_{18}H_{26}ClN_3 \cdot 2H_3PO_4$, *Arechin, Avloclor, Imagon, Malaquin, Resochin, Tresochin.* Bitter, colorless crystals. Dimorphic. One modification, mp 193-195°; the other, mp 215-218°. Freely sol in water; pH of 1% soln about 4.5; less sol at neutral and alkaline pH. Stable to heat in solns of pH 4.0 to 6.5. Practically insol in alcohol, benzene, chloroform, ether.
Sulfate, $C_{18}H_{26}ClN_3 \cdot H_2SO_4$, *Nivaquine.*

THERAP CAT: Antimalarial; antiamebic; antirheumatic. Lupus erythematosus suppressant.

2216. Chloroselenic Acid. $ClHO_3Se$; mol wt 163.42. Cl 21.69%, H 0.62%, O 29.37%, Se 48.32%. $HClSeO_3$. Prepd by the reaction of hydrogen chloride and selenium trioxide: Le Geyt *et al., J. Chem. Soc.* **123**, 2870 (1923).
Liquid. Fumes in air. d 2.26. mp —46°. Dec by heat. Sol in water and selenium oxychloride. Dec in alc. Insol in ether, benzene, chloroform, carbon tetrachloride.

2217. N-Chlorosuccinimide. *1-Chloro-2,5-pyrrolidinedione;* succinchlorimide. $C_4H_4ClNO_2$; mol wt 133.53. C 35.98%, H 3.02%, Cl 26.55%, N 10.49%, O 23.96%. Prepn from succinimide: Hirst, Macbeth, *J. Chem. Soc.* **121**, 2169 (1922); Zimmer, Audrieth, *J. Am. Chem. Soc.* **76**, 3856 (1954). Crystal structure: Brown, *Acta Cryst.* **9**, 193 (1956); **14**, 711 (1961). Toxicity data: Stohlman, Smith, *U.S. Publ. Health Repts.* **59**, 541 (1944).

Orthorhombic crystals, mp 150-151°. Odor of chlorine. Acid to litmus (1:50 aq soln). One gram dissolves in about 70 ml water, 150 ml alcohol, 50 ml benzene. Sparingly sol in ether, chloroform, carbon tetrachloride. Liberates iodine from potassium iodide solns, and bromine from sodium bromide solns. MLD orally in rats: 2.7 g/kg (Stohlman, Smith).

USE: Chlorinating agent.

2218. Chlorosulfonic Acid. Sulfuric chlorohydrin; chlorosulfuric acid. $ClHO_3S$; mol wt 116.52. Cl 30.43%, H 0.87%, O 41.19%, S 27.52%. $SO_2(OH)Cl$. Prepd by the reaction of HCl gas with SO_3: Simon, Kratsch, *Z. Anorg. Allgem. Chem.* **242**, 369 (1939); Briggs, **U.S.** pat. **1,442,335**

(1922 to General Chem.). Purification: Kaplar, Shechter, *Inorg. Syn.* **4**, 52 (1953). *Review:* H. O. Burrus in Kirk-Othmer *Encyclopedia of Chemical Technology* vol. 5 (Wiley-Interscience, New York, 3rd ed., 1979) pp 873-880.
Colorless or slightly yellow, very corrosive liquid, causes severe burns; fumes in air; pungent odor. d^{20} 1.76-1.77; d^0_4 1.784; d^{20}_4 1.753. mp —80°. bp_{755} 151-152°; bp_{19} 74-75°; $bp_{2.4}$ 60-64°. n_D^{14} 1.437. On dropping into water dec with explosive violence. *Keep tightly closed.* When used for the prepn of sulfate esters, the common solvent is pyridine. Other solvents are liquid sulfur dioxide and dichloroethane.
Caution: Highly irritating and corrosive to eyes, skin, mucous membranes.

USE: Manuf sulfone compds, saccharin. As chlorosulfonating and condensing agent in organic syntheses.

2219. Chlorothalonil. *2,4,5,6-Tetrachloro-1,3-benzenedicarbonitrile; tetrachloroisophthalonitrile; m*-tetrachlorophthalodinitrile; *2,4,5,6-tetrachloro-1,3-dicyanobenzene; 1,3-dicyano-2,4,5,6-tetrachlorobenzene;* chlorthalonil; DAC-2787; Daconil 2787; Bravo; Forturf; Exotherm Termil; Termil. $C_8Cl_4N_2$; mol wt 265.91. C 36.14%, Cl 53.33%, N 10.53%. Fungicidal properties first described by N. J. Turner *et al., Contrib. Boyce Thompson Inst.* **22**, 303 (1964). Prepn: R. D. Battershell, H. Bluestone, **U.S.** pat. **3,290,353** (1966 to Diamond Alkali); R. M. Bimber, **U.S.** pat. **3,652,-637** (1972 to Diamond Shamrock). Toxicity studies in mice: H. Yoshikawa, K. Kawai, *Ind. Health* **4**, 11 (1966). Review of carcinogenic risk: *IARC Monographs* **30**, 319-328 (1983).

Crystals. d_4^{25} 1.7. mp 250-251°. bp_{760} 350°. Vapor press <0.01 at 40°. Practically insol in water (soly at room temp reported as 0.6 ppm). Soly in organic solvents at 25° (w/w): xylene 8%, cyclohexane 3%, acetone 2%, kerosine <1.0%. LD_{50} orally in rats: >10.0 g/kg (Turner).

USE: Fungicide, bactericide, nematocide. Agricultural and horticultural fungicide.

2220. Chlorothen. *N-[(5-Chloro-2-thienyl)methyl]-N',-N'-dimethyl-N-2-pyridinyl-1,2-ethanediamine; 2-[(5-chloro-2-thenyl)(2-dimethylaminoethyl)amino]pyridine; N,N-dimethyl-N'-(2-pyridyl)-N'-(5-chloro-2-thenyl)ethylenediamine; N,N-dimethyl-N'-(α-pyridyl)-N'-(2-methyl-5-chlorothienyl)-ethylenediamine; N-5-chloro-2-thienylmethyl-N',N'-dimethyl-N-2-pyridylethylenediamine;* chloropyrilene; chloromethapyrilene; chlorothenylpyramine. $C_{14}H_{18}ClN_3S$; mol wt 295.84. C 56.84%, H 6.13%, Cl 11.98%, N 14.20%, S 10.84%. Prepd by the condensation of 5-chloro-2-thenyl chloride and *N,N*-dimethyl-*N'*-(2-pyridyl)ethylenediamine in the presence of sodium or potassium amide: R. C. Clapp *et al., J. Am. Chem. Soc.* **69**, 1549 (1947); L. P. Kyrides, **U.S.** pat. **2,581,869** (1952 to Monsanto). Toxicity data: J. C. Castillo *et al., J. Pharmacol. Exp. Ther.* **96**, 388 (1949).

Liquid. $bp_{1.0}$ 155-156°. Strong base.
Hydrochloride, $C_{14}H_{18}ClN_3S \cdot HCl$, *Thenclor.* Crystals, mp 106-108°. Freely sol in water.
Citrate, $C_{14}H_{18}ClN_3S \cdot C_6H_8O_7$, *Tagathen.* Crystals, mp 112-116°. On further heating gradually solidifies and remelts 125-140° (dec). uv max: 240 nm ($E_{1cm}^{1\%}$ 390-410). One gram dissolves in 35 ml water, in about 65 ml alc. Prac-

tically insol in ether, chloroform, benzene. pH of 1% aq soln 3.9 to 4.1. LD$_{50}$ i.p. in mice: 105 mg/kg (Castillo).

THERAP CAT: Antihistaminic.

2221. Chlorothiazide. *6-Chloro-2H-1,2,4-benzothiadiazine-7-sulfonamide 1,1-dioxide;* 6-chloro-7-sulfamoyl-2H-1,-2,4-benzothiadiazine 1,1-dioxide; 6-chloro-7-sulfamyl-1,2,4-benzothiadiazine 1,1-dioxide; Alurene; Chlorosal; Chlorurit; Chlotride; Clotride; Diuresal; Diuril; Diurilix; Diurite; Exuril; Urinex; Flumen; Minzil; Neo-Dema; Salunil; Saluretil; Saluric; Warduzide; Yadulan. C$_7$H$_6$ClN$_3$O$_4$S$_2$; mol wt 295.73. C 28.43%, H 2.05%, Cl 11.99%, N 14.21%, O 21.64%, S 21.69%. Synthesis: Novello, Sprague, *J. Am. Chem. Soc.* **79,** 2028 (1957); Novello, Hinkley, U.S. pats. **2,809,194; 2,937,169** (1957, 1960, both to Merck & Co.). HPLC determn in urine: R. O. Fullinfaw *et al., J. Chromatog.* **415,** 347 (1987). Comprehensive description: H. G. Brittain, *Anal. Profiles Drug Subs.* **18,** 33-56 (1989).

Crystals, dec 342.5-343°. pKa$_1$ 6.85, pKa$_2$ 9.45. Freely sol in dimethylsulfoxide, dimethyl formamide; slightly sol in methanol, pyridine; very slightly sol in water. Practically insol in diethyl ether, chloroform, benzene. Soly (g/L): water (pH 4) 0; (pH 7) 0.65. Sol in alkaline aq solns with decompn upon standing or heating.

Sodium salt, C$_7$H$_5$ClN$_3$NaO$_4$S$_2$, *6-chloro-7-sulfamoyl-2H-1,2,4-benzothiadiazin-2-ylsodium 1,1-dioxide,* Diuril Lyovac, *Lyovac Diuril.*

THERAP CAT: Diuretic; antihypertensive.

THERAP CAT (VET): Diuretic.

2222. Chlorothricin. C$_{50}$H$_{63}$ClO$_{16}$; mol wt 955.49. C 62.85%, H 6.65%, Cl 3.71%, O 26.79%. Macrolide antibiotic active against gram-positive bacteria. Isolated together with its dechloro analog from Tü 99, a strain of *Streptomyces antibioticus:* Keller-Schierlein *et al., Helv. Chim. Acta* **52,** 127 (1969). Degradation products studies: Muntwyler *et al., ibid.* **53,** 1544 (1970). Structure: Muntwyler, Keller-Schierlein, *ibid.* **55,** 2071 (1972). Non-competitive inhibitor of pyruvate carboxylase: P. W. Schindler, H. Zaehner, *Arch. Mikrobiol.* **82,** 66, (1972); P. W. Schindler, M. C. Scrutton, *ibid.* **55,** 543 (1975). Biosynthetic studies: O. Mascaretti *et al., J. Nat. Prod.* **42,** 455 (1979); *eidem, Biochemistry* **20,** 919 (1981). Partial synthesis of the aglycone,

chlorothricolide: R. E. Ireland, W. J. Thompson, *J. Org. Chem.* **44,** 3041 (1979); R. E. Ireland *et al., ibid.* **46,** 4863 (1981); W. R. Roush, S. E. Hall, *J. Am. Chem. Soc.* **103,** 5200 (1981). Alternate synthetic approach: R. E. Ireland, M. D. Varney, *J. Org. Chem.* **51,** 635 (1986). *Review:* Keller-Schierlein, *Fortschr. Chem. Org. Naturst.* **30,** 394-396 (1973); H. G. Floss, C.-J. Chang, *Antibiotics* **vol. IV,** J. W. Corcoran, Ed. (Springer-Verlag, New York, 1981) pp 193-214.

Colorless crystals from methylene chloride-methyl acetate, mp 206-207°. uv max (alc): 222, 260 nm (log ε 4.20, 3.81); (0.01N alc KOH): 221, 259 nm (log ε 4.09, 3.95). Dibasic acid; pK: 5.01, 7.91. (The above data applies to a 5:1 mixture with dechlorothricin, Muntwyler, Keller-Schierlein, *loc. cit.).* Slightly sol in water; readily sol in organic solvents of medium polarity.

2223. Chlorothymol. *4-Chloro-5-methyl-2-(1-methylethyl)phenol; 6-chlorothymol;* 4-chlorothymol; 1-methyl-3-hydroxy-4-isopropyl-6-chlorobenzene; 6-chloro-4-isopropyl-1-methyl-3-phenol. C$_{10}$H$_{13}$ClO; mol wt 184.67. C 65.04%, H 7.10%, Cl 19.20%, O 8.66%. Made by the action of sulfuryl chloride on thymol in CCl$_4$: Satriana *et al., J. Am. Pharm. Assoc.* **39,** 135 (1950); H. Pahlicke, **Ger.** pat. **905,738** (1954 to Diwag), *C.A.* **52,** 16294i (1958). Antifungal properties: M. Iannarone, *Drug Standards* **25,** 190 (1957).

Crystals. mp 62-64° (59-61° *N.F.* **XII**). One gram dissolves in about 1000 ml water, 0.5 ml alcohol, 2 ml benzene, 2 ml chloroform, 1.5 ml ether, about 10 ml petr ether; also sol in dil aq NaOH.

2224. Chlorotoluene. C$_7$H$_7$Cl; mol wt 126.59. C 66.42%, H 5.57%, Cl 28.01%. Prepn from diazotized toluidine: Neogi, Mitra, *J. Chem. Soc.* **1928,** 1332; Marvel, McElvain, *Org. Syn. coll.* **vol. I,** 170 (1941). Manuf of *o*- and *p*-isomers by catalytic chlorination of toluene: Di Bella, **U.S.** pat. **3,000,975** (1959 to Heyden Newport Chem.). Absorption spectrum of *m*- and *p*-isomers: Baly, *J. Chem. Soc.* **99,** 1704 (1911): of *m*- and *o*-isomers: Purvis, *ibid.* 1704. Metabolism of *o*-isomer by bacteria: P. A. Vandenbergh *et al., Appl. Environ. Microbiol.* **42,** 737 (1981); by rats: G. B. Quistad *et al., J. Agr. Food Chem.* **31,** 1158 (1983). Brief review: S. Gelfand in Kirk-Othmer *Encyclopedia of Chemical Technology* **vol. 5** (Wiley-Interscience, New York, 3rd ed., 1979) pp 819-827.

m-Chlorotoluene, 1-chloro-3-methylbenzene. Liquid, bp 161.75°. d$_4^{18.7}$ 1.0760. mp —47.8°. n$_D^{20}$ 1.5218.

o-Chlorotoluene, 1-chloro-2-methylbenzene. Liquid, bp 158.97°. Vapor harmful. d$_4^{20}$ 1.0826. mp —35.59°. n$_D^{20}$ 1.5258. Volatile with steam. Slightly sol in water; freely sol in alcohol, benzene, chloroform, ether.

p-Chlorotoluene, 1-chloro-4-methylbenzene. Liquid, bp 162.4°. d$_4^{20}$ 1.0697. mp 7.5°. n$_D^{20}$ 1.5211. Slightly sol in water; sol in alcohol, benzene, chloroform, ether.

USE: Solvent; dyestuff intermediate; in organic syntheses.

2225. Chlorotrianisene. *1,1',1''-(1-Chloro-1-ethenyl-2-ylidene)tris[4-methoxybenzene]; chlorotris(p-methoxyphenyl)ethylene;* tri-p-anisylchloroethylene; tris(p-methoxyphenyl)chloroethylene; Hormonisene; Merbentul; Tace. C$_{23}$H$_{21}$-ClO$_3$; mol wt 380.87. C 72.53%, H 5.56%, Cl 9.31%, O 12.60%. Prepd by reacting tri-p-anisylethylene or tri-p-anisylethanol with Cl in an inert solvent: Basford and I.C.I., **Brit.** pat. **561,508** (1944). Synthesis from p-(p-anisoyl)anisole: Shelton, Van Campen, Jr., **U.S.** pat. **2,430,891** (1947 to Merrell).

Crystals from methanol, mp 114-116°. Softens at 108°. uv max (chloroform): 310 nm, ($E_{1cm}^{1\%}$ 423); min 278 nm. Practically insol in water. Soly in alcohol: 0.28 g/100 ml; in ether: 3.6 g/100 ml. Also sol in glacial acetic acid, acetone, chloroform, carbon tetrachloride, benzene, vegetable oils.

THERAP CAT: Estrogen.

2226. Chlorotris(triphenylphosphine)rhodium. Wilkinson's catalyst. $C_{54}H_{45}ClP_3Rh$; mol wt 925.23. C 70.10%, H 4.90%, Cl 3.83%, P 10.04%, Rh 11.12%. $ClRh[P(C_6H_5)_3]_3$. Prepd from excess triphenylphosphine and $RhCl_3.3H_2O$ in ethanol: Young *et al., Chem. Commun.* **1965**, 131; Bennett, Longstaff, *Chem. & Ind. (London)* **1965**, 846; Osborn *et al., J. Chem. Soc. (A)* **1966**, 1711; Osborn, Wilkinson, *Inorg. Syn.* **10**, 67 (1967). Crystal structure: M. J. Bennett, P. B. Donaldson, *Inorg. Chem.* **16**, 655 (1977).

Dark, burgundy-red crystals from ethanol. Orange crystals from concd ethanol soln refluxed for five minutes; converted to red form on further refluxing. mp 157-158° (Osborn *et al.*); 138° (dec) (Bennett, Longstaff). Soly in chloroform, dichloromethane (25°): ~20 g/l; in benzene, toluene (25°): ~2 g/l; much less sol in acetic acid, acetone and other ketones, methanol, lower aliphatic alcohols. Insol in paraffins, cyclohexane. Reacts with donor solvents such as pyridine, dimethyl sulfoxide, acetonitrile.

USE: Homogeneous catalyst.

2227. Chloroxine. *5,7-Dichloro-8-quinolinol;* 5,7-dichloro-8-hydroxyquinoline; Capitrol. $C_9H_5Cl_2NO$; mol wt 214.05. C 50.50%, H 2.35%, Cl 33.13%, N 6.54%, O 7.47%. Prepd by chlorinating 8-quinolinol: Hebebrand, *Ber.* **21**, 2977 (1888), F. J. Welcher, *Organic Analytical Reagents* vol. I (Van Nostrand, 1947) pp 328-329.

Crystals from alc, mp 179-180°. Soluble in benzene, acetone; slightly sol in cold alcohol, acetic acid; readily sol in sodium and potassium hydroxides and in acids, forming yellow solns.

Mixture with 5-chloro-8-quinolinol and 7-chloro-8-quinolinol, *see* Halquinol.

USE: Analytical reagent.

THERAP CAT: Antiseborrheic.

2228. Chloroxylenol. *4-Chloro-3,5-dimethylphenol;* p-chloro-m-xylenol; 4-chloro-3,5-xylenol; parachlorometaxylenol; 2-chloro-m-xylenol; 2-chloro-5-hydroxy-m-xylene; 2-chloro-5-hydroxy-1,3-dimethylbenzene; Benzytol; Dettol. C_8H_9ClO; mol wt 156.61. C 61.35%, H 5.79%, Cl 22.64%, O 10.22%. Prepd by treating 3,5-dimethylphenol with Cl_2 or SO_2Cl_2: Lesser, Gad, *Ber.* **56**, 974, 976 (1923); von Auwers *et al., Chem. Zentr.* **1924**, II, 2267, *C.A.* **19**, 2339 (1925); Gladden, Cocker, **U.S.** pat. **2,350,677** (1944).

Crystals from benzene, mp 115.5°. Phenolic odor. Volatile with steam. bp 246°. One gram dissolves in 3 liters of water at 20°. More stable in hot water. Soluble in 1 part of 95% alcohol, ether, benzene, terpenes, fixed oils, in solns of alkali hydroxides.

USE: Antiseptic and germicide; for mildew prevention. Claimed to be about 60 times as potent as phenol.

THERAP CAT: Antibacterial; antiseptic (topical and urinary).

THERAP CAT (VET): Antiseptic (topical).

2229. Chlorozotocin. *2-[[[(2-Chloroethyl)nitrosoamino]carbonyl]amino]-2-deoxy-D-glucose;* 2-[3-(2-chloroethyl)-3-nitrosoureido]-2-deoxy-D-glucopyranose; 1-(2-chloroethyl)-1-nitroso-3-(D-glucos-2-yl)urea; DCNU; NSC-178248. $C_9H_{16}ClN_3O_7$; mol wt 313.69. C 34.46%, H 5.14%, Cl 11.30%, N 13.40%, O 35.70%. Chloroethylnitrosourea derivative with antitumor activity. Similar to carmustine, lomustine, nimustine, ranimustine, *q.q.v;* 2-chloroethyl analog of streptozotocin, *q.v.* Synthesis: H. D. Burns *et al., Org. Prep. Proced. Int.* **6**, 259 (1974); T. P. Johnston *et al., J. Med. Chem.* **18**, 104 (1975). Pharmacology: T. Anderson *et al., Cancer Res.* **35**, 761 (1975); P.S. Schein *et al., Cancer Treatment Rep.* **60**, 801 (1976). Decomposition in aqueous media: J. A. Montgomery *et al., J. Med. Chem.* **18**, 568 (1975).

Ivory colored crystals, mp 147-148° (dec with the evolution of gas), (Burns, Heindel). Also reported as mp 140-141° (dec), (Johnston). Sol in water.

THERAP CAT: Antineoplastic.

2230. Chlorphenesin. *3-(4-Chlorophenoxy)-1,2-propanediol;* p-chlorophenyl α-glyceryl ether; Adermykon; Mycil. $C_9H_{11}ClO_3$; mol wt 202.64. C 53.35%, H 5.47%, Cl 17.50%, O 23.69%. Prepd by condensing equimol amts of p-chlorophenol and glycidol in the presence of a tertiary amine or a quaternary ammonium salt as catalyst: Bradley, Forrest, **Brit.** pat. **628,497** (1949 to British Drug Houses).

Crystals, mp 77-79°. Soly in water is less than 1%, may be increased by the addition of solubilizers such as ethylurea or propylene glycol: Berger *et al.,* **U.S.** pat. **2,468,423** (1949 to British Drug Houses).

THERAP CAT: Antifungal (topical).

2231. Chlorphenesin Carbamate. *3-(4-Chlorophenoxy)-1,2-propanediol-1-carbamate; carbamic acid 3-(p-chlorophenoxy)-2-hydroxypropyl ester;* 3-(p-chlorophenoxy)-2-hydroxypropyl carbamate; 1,2-propanediol-3-(p-chlorophenoxy)-1-carbamate; Maolate; Rinlaxer. $C_{10}H_{12}ClNO_4$; mol wt 245.66. C 48.89%, H 4.92%, Cl 14.43%, N 5.70%, O 26.05%. Prepn: Collins, Matthews, **U.S.** pat. **3,161,567;** Parker, **U.S.** pat. **3,214,336** (1964, 1965 both to Upjohn).

Consult the Name Index before using this section.

Clinical evaluation of analgesic activity: L. J. Cass, W. S. Frederick, *J. New Drugs* **2**, 366 (1962).

Crystals from benzene + toluene, mp 89-91°. Readily sol in 95% ethanol, acetone, ethyl acetate; fairly readily sol in dioxane. Almost insol in cold water, benzene, cyclohexane. LD_{50} orally in rats: 748 mg/kg; i.v. in mice: 239 mg/kg (Cass, Frederick).

THERAP CAT: Muscle relaxant (skeletal).

THERAP CAT (VET): Muscle relaxant (skeletal).

2232. Chlorpheniramine. *γ-(4-Chlorophenyl)-N,N-dimethyl-2-pyridinepropanamine; 2-[p-chloro-α-(2-dimethylaminoethyl)benzyl]pyridine;* 1-(p-chlorophenyl)-1-(2-pyridyl)-3-dimethylaminopropane; 1-(p-chlorophenyl)-1-(2-pyridyl)-3-N,N-dimethylpropylamine; 3-(p-chlorophenyl)-3-(2-pyridyl)-N,N-dimethylpropylamine; γ-(4-chlorophenyl)-γ-(2-pyridyl)propyldimethylamine; chlorprophenpyridamine; chlorphenamine; Haynon. $C_{16}H_{19}ClN_2$; mol wt 274.79. C 69.93%, H 6.97%, Cl 12.90%, N 10.19%. Synthesis: Sperber *et al.*, U.S. pats. 2,567,245, 2,676,964 (1951, 1954, both to Schering). Prepn of d-form: L. A. Walter, U.S. pat. 3,061,-517 (1962 to Schering). Solutions: Foley, Ilavsky, U.S. pat. 2,766,174 (1956 to Schering). Pharmacology: F. E. Roth, W. M. Govier, *J. Pharmacol. Exp. Ther.* **124**, 347 (1958). Toxicity data: R. B. Smith *et al.*, *Toxicol. Appl. Pharmacol.* **28**, 240 (1974). Comprehensive description: C. G. Eckhart, T. McCorkle, *Anal. Profiles Drug Subs.* **7**, 43-80 (1978).

Oily liquid, $bp_{1.0}$ 142°.

Maleate, $C_{16}H_{19}ClN_2 \cdot C_4H_4O_4$, *Allergisan, Antagonate, Chlor-Trimeton, Chlor-Tripolon, Cloropiril, C-Meton, Histadur, Histaspan, Lorphen, Piriton, Pyridamal-100, Teldrin.* Crystals, mp 130-135°. uv max (water): 261 nm (ε 5760). Soly in mg/ml at 25°: ethanol 330; chloroform 240; water 160; methanol 130. Slightly sol in benzene, ether. pH of a 2% aq soln about 5. LD_{50} orally in mice: 162 mg/kg (Smith).

d-Form, *dexchlorpheniramine, d-chlorpheniramine.* Oily liquid. $[\alpha]_D^{25}$ +49.8° (c = 1 in DMF).

d-Form maleate, *Fortamine, Isomerine, Phenamin, Phendextro, Polamin, Polaramine, Polaronil, Sensidyn.* Crystals from ethyl acetate, mp 113-115°. $[\alpha]_D^{25}$ +44.3° (c = 1 in dimethylformamide). pH of 1% soln 4-5.

THERAP CAT: Antihistaminic.

THERAP CAT (VET): Antihistaminic.

2233. Chlorphenoxamide. *2,2-Dichloro-N-(2-hydroxyethyl)-N-[[4-(4-nitrophenoxy)phenyl]methyl]acetamide; N-(β-hydroxyethyl)-N-[p-(4-nitrophenoxy)benzyl]dichloroacetamide; N-(β-hydroxyethyl)-N-[p-phenoxy-(4'-nitro)benzyl]dichloroacetamide; dichloro-N-(β-hydroxyethyl)-N-[p-(4'-nitrophenoxy)benzyl]acetamide; clefamide; chlorphenoxamide; Mebinol.* $C_{17}H_{16}Cl_2N_2O_5$; mol wt 399.23. C 51.15%, H 4.04%, Cl 17.76%, N 7.02%, O 20.04%. Prepn: Logemann *et al.*, *Farmaco Ed. Sci.* **13**, 139 (1958); U.S. pat. 2,824,894 (1958 to Carlo Erba). Antiamebic activity and toxicity data: Carneri, *Giorn. Mal. Infett. Parassit.* **10**, 850 (1958), *C.A.* **53**, 5519C (1959).

Crystals from 95% ethanol, mp 136-137°. Practically insol in water (soly ~3 γ/ml). Sol in ethanol, acetone, dioxane. LD_{50} in mice (mg/kg): > 5000 orally, 2000 i.p. (Carneri).

Note: Do not confuse chlorphenoxamide with chlorphenoxamine, *q.v.*

THERAP CAT: Antiamebic.

2234. Chlorphenoxamine. *2-[1-(4-Chlorophenyl)-1-phenylethoxy]-N,N-dimethylethanamine; 2-[p-chloro-α-methyl-α-phenylbenzyl)oxy]-N,N-dimethylethylamine;* β-dimethylaminoethyl (p-chloro-α-methylbenzhydryl) ether; [1-(p-chlorophenyl)-1-phenyl]ethyl (β-dimethylaminoethyl) ether; $C_{18}H_{22}ClNO$; mol wt 303.83. C 71.16%, H 7.30%, Cl 11.67%, N 4.61%, O 5.27%. Prepn: Arnold *et al.*, U.S. pat. 2,785,202 (1957 to Asta-Werke). Synthesis: G. Cahiez *et al.*, *Tetrahedron Letters* **29**, 3659 (1988). Pharmacology: *eidem, Arzneimittel-Forsch.* **4**, 189 (1954); Brock *et al., ibid.* 262. Toxicity studies: Kerley *et al., Toxicol. Appl. Pharmacol.* **4**, 638 (1962). Metabolism: C. Koppel *et al., Arzneimittel-Forsch.* **37**, 1062 (1987). GC-MS determn in urine: H. Maurer, K. Pfleger, *J. Chromatog.* **428**, 43 (1988).

$bp_{0.05}$ 150-155°.

Hydrochloride, $C_{18}H_{22}ClNO.HCl$, *1766, Clorevan, Contristamine, Phenoxene, Phenoxine, Systral.* Needles, mp 128°. Soluble in water; aq solns are stable.

Note: Do not confuse chlorphenoxamine with chlorphenoxamide, *q.v.*

THERAP CAT: Anticholinergic.

2235. Chlorphentermine. *4-Chloro-α,α-dimethylbenzeneethanamine; 4-chloro-α,α-dimethylphenethylamine;* α,α-dimethyl-p-chlorophenethylamine; 1-(p-chlorophenyl)-2-methyl-2-propylamine; 1-(p-chlorophenyl)-2-methyl-2-aminopropane; clorfentermina; Lucofen; Teramine. $C_{10}H_{14}ClN$; mol wt 183.68. C 65.39%, H 7.68%, Cl 19.30%, N 7.63%. Prepn: Bachman *et al., J. Am. Chem. Soc.* **76**, 3972 (1954); Ferrari, *Farmaco Ed. Sci.* **15**, 337 (1960); Fr. pats. M1299 and 1,296,132 (1962 to Simes), *C.A.* **58**, 4467e, 3352g (1963); Brit. pat. 906,331 (1962 to Kefalas), *C.A.* **58**, 6654c (1963). Pharmacology: Jun *et al., Can. J. Pharm. Sci.* **4**, 27 (1969); Ciborska *et al., Acta Pol. Pharm.* **26**, 595 (1969); Moeller-Nielsen, Dubnick, *Proc. Int. Symp. Amphetamines Relat. Compounds 1969,* E. Costa, Ed. (Raven Press, New York, 1970) pp 63-73.

Liquid. bp_2 100-102°.

Hydrochloride, $C_{10}H_{14}ClN.HCl$, *Pre-Sate.* Crystals from alcohol + ether, mp 234°. Soly in water: > 20%. pH of 1% aq soln ~5.5. LD_{50} in mice (mg/kg): 270 orally; 267 s.c. (Ciborska).

Note: This is a controlled substance (stimulant) listed in the U.S. Code of Federal Regulations, Title 21 Part 1308.13 (1995).

THERAP CAT: Anorexic.

2236. Chlorproethazine. *2-Chloro-N,N-diethyl-10H-phenothiazine-10-propanamine; 2-chloro-10-(3-diethylamino-*

propyl)phenothiazine; 3-chloro-10-(3-diethylaminopropyl)-phenothiazine; RP-4909. $C_{19}H_{23}ClN_2S$; mol wt 346.92. C 65.78%, H 6.68%, Cl 10.22%, N 8.07%, S 9.24%. Prepn: Buisson *et al.,* U.S. pat. 2,769,002 (1956 to Rhône-Poulenc).

Hydrochloride, $C_{19}H_{23}ClN_2S.HCl$, *Neuriplege.* Crystals, mp 178°. (Free base bp_1 225-240°). Sensitive to light. Soly in water about 1.0 g/60 ml, ethanol about 1.0 g/300 ml, chloroform 1.0 g/5 ml. Practically insol in acetone, ether, benzene. pH of 1% aq soln 4.8.

THERAP CAT: Muscle relaxant (skeletal); antipsychotic.

2237. Chlorproguanil. *N-(3,4-Dichlorophenyl)-N'-(1-methylethyl)imidodicarbonimidic diamide; 1-(3,4-dichloro-phenyl)-5-isopropylbiguanide; N^1-3,4-dichlorophenyl-N^5-iso-propylbiguanide; N^1-3,4-dichlorophenyl-N^5-isopropyldi-guanide;* M-5943. $C_{11}H_{15}Cl_2N_5$; mol wt 288.18. C 45.85%, H 5.25%, Cl 24.60%, N 24.30%. Method of prepn: Crowther *et al., J. Chem. Soc.* **1951**, 1780; Curd *et al.,* U.S. pat. **2,544,827** (1951); Crowther *et al.,* Brit. pat. **667,116** (1952) (both to ICI).

Hydrochloride, $C_{11}H_{15}Cl_2N_5.HCl$, *Lapudrine.* Crystals, mp 246-247°. Soly in water: 1 g/100 ml. Solns may be boiled without dec.

THERAP CAT: Antimalarial.

2238. Chlorpromazine. *2-Chloro-N,N-dimethyl-10H-phenothiazine-10-propanamine; 2-chloro-10-(3-dimethylami-nopropyl)phenothiazine;* 3-chloro-10-(3-dimethylaminopro-pyl)phenothiazine; *N-(3-dimethylaminopropyl)-3-chloro-phenothiazine;* 2601-A; HL-5746; RP-4560; SKF-2601-A; Chlorderazin; Chlorpromados; Contomin; Esmind; Fenac-til; Novomazina; Promactil; Prozil; Plegomazin; Sanopron; Aminazine; Ampliactil; Amplictil; Promazil; Proma; Elma-rin; Wintermin. $C_{17}H_{19}ClN_2S$; mol wt 318.87. C 64.03%, H 6.01%, Cl 11.12%, N 8.79%, S 10.06%. Prepn: Charpentier *et al., Compt. Rend.* **235**, 59 (1952); Charpentier, U.S. pat. **2,645,640** (1953 to Rhône-Poulenc). Effects of neuroleptics on dopamine receptors: N.-E. Anden *et al., Eur. J. Pharma-col.* **11**, 303 (1970). Toxicity study: E. I. Goldenthal, *Toxi-col. Appl. Pharmacol.* **18**, 185 (1971). Review of analytical methods for determn in pharmaceutical prepns: L. F. S. Chagonda, J. S. Millership, *J. Pharm. Biomed. Anal.* **7**, 271-278 (1989). Brief historical review: G. Curzon, *Trends Pharmacol. Sci.* **11**, 61-63 (1990).

Oily liq. Amine odor. Alkaline reaction. $bp_{0.8}$ 200-205°. Maleate, *Clordelazin.*

Hydrochloride, $C_{17}H_{19}ClN_2S.HCl$, *Hebanil, Hibanil, Hibernal, Klorpromex, Largactil, Largaktyl, Megaphen, Pro-macid, Chloractil, Chlorazin, Sonazine, Marazine, Propa-phenin, Taroctyl, Thorazine, Torazina.* Crystals, dec 179-180° (capillary); 194-196° (microblock). uv curve: Neuhoff, Auterhoff, *Arch. Pharm.* **288**, 400 (1955). pH of 5% aq soln

4.0-5.5. One gram dissolves in 2.5 ml water. Sol in metha-nol, ethanol, chloroform. Practically insol in ether, benz-ene. Slightly acid to litmus. LD_{50} orally in rats: 225 mg/kg (Goldenthal).

THERAP CAT: Antiemetic; antipsychotic.

THERAP CAT (VET): Antiemetic; tranquilizer.

2239. Chlorpropamide. *4-Chloro-N-[(propylamino)-carbonyl]benzenesulfonamide; 1-(p-chlorophenylsulfonyl)-3-propylurea;* 1-(p-chlorobenzenesulfonyl)-3-propylurea; *N-*propyl-*N'*-(p-chlorobenzenesulfonyl)urea; P-607; Adiaben; Asucrol; Catanil; Chloronase; Diabechlor; Diabenal; Dia-betoral; Diabinese; Melitase; Millinese; Oradian; Stabinol. $C_{10}H_{13}ClN_2O_3S$; mol wt 276.74. C 43.40%, H 4.73%, Cl 12.81%, N 10.12%, O 17.34%, S 11.59%. Prepn: Marshall, Sigal, *J. Org. Chem.* **23**, 927 (1958); Brit. pat. 853,555; W. M. McLamore, U.S. pat. **3,349,124** (1960, 1967 both to Pfizer); Bauer *et al., J. Org. Chem.* **31**, 3440 (1960). Phar-macology and metabolism: Khurana *et al., Indian J. Med.* **55**, 1084 (1967); Brotherton *et al., Clin. Pharmacol. Ther.* **10**, 505 (1969); Madsen *et al., Eur. J. Pharmacol.* **13**, 374 (1971). Toxicity study: E. I. Goldenthal, *Toxicol. Appl. Pharmacol.* **18**, 185 (1971).

Crystals from dil ethanol, mp 127-129°. uv max (0.01*N* HCl): 232.5 nm (ϵ 16500). Soly in water at pH 6: 2.2 mg/ml. Practically insol at pH 7.3. Sol in alc; moderately sol in chloroform; sparingly sol in ether, benzene. LD_{50} i.p. in rats: 580 mg/kg (Goldenthal).

THERAP CAT: Antidiabetic.

2240. Chlorpropham. *(3-Chlorophenyl)carbamic acid 1-methylethyl ester; m-chlorocarbanilic acid isopropyl ester;* isopropyl *N-(3-chlorophen-yl)carbamate;* chloro-IPC; chloropropham; CIPC; Chlor-IFC; Furloe; Sprout-Nip. $C_{10}H_{12}ClNO_2$; mol wt 213.66. C 56.21%, H 5.66%, Cl 16.59%, N 6.56%, O 14.98%. Prepn: E. D. Witman, U.S. pat. 2,695,225; Strain, U.S. pat. 2,734,-911 (1954, 1956 both to Columbia-Southern Chem.); Brock-way, U.S. pat. 2,806,051 (1957 to B. F. Goodrich). Toxi-cology: E. M. Boyd, E. Carsky, *Arch. Environ. Health* **19**, 621 (1969).

Solid, mp 40.7-41.1°. bp_2 149°. n_D^{20} 1.5388. Commercial product is a liquid. Slightly sol in water; miscible with most oils and organic solvents. LD_{50} orally in rats: 1.2 g/kg (Boyd, Carsky).

USE: Herbicide; plant growth regulator.

2241. Chlorprothixene. *3-(2-Chloro-9H-thioxanthen-9-ylidene)-N,N-dimethyl-1-propanamine; 2-chloro-N,N-dimeth-ylthioxanthene-$\Delta^{9,\gamma}$-propylamine;* 2-chloro-N,N-dimethyl-3-thioxanthen-9-ylidenepropylamine; 2-chloro-9-(3'-dimethyl-aminopropylidene)thioxanthene; α-2-chloro-10-(3-dimethyl-aminopropylidene)thiaxanthene; N-714; Taractan; Truxal; Truxaletten; Tarasan. $C_{18}H_{18}ClNS$; mol wt 315.87. C 68.45%, H 5.74%, Cl 11.22%, N 4.43%, S 10.15%. Prepn: Brit. pat. 829,763 and Sprague, Engelhardt, U.S. pat. 2,-951,082 (both 1960 to Merck & Co.); Brit. pat. 834,143 (1960 to Am. Cyanamid). Comprehensive description: B. C. Rudy, B. Z. Senkowski, in *Analytical Profiles of Drug Substances* vol. 2, K. Florey, Ed. (Academic Press, New York, 1973) pp 63-84.

Pale yellow crystals, mp 97-98°. Practically insol in water; sol in alcohol, ether, chloroform. Incompatible with acids, alkalies, phenobarbital, thiopental sodium, mepazine. LD_{50} orally in rats: 380 mg/kg, *RTECS* Vol. **II**, E. J. Fairchild, Ed. (1977) p 909.

Hydrochloride, $C_{18}H_{18}ClNS \cdot HCl$, crystals, mp 221°. Freely sol in water at pH 6 to 6.5.

THERAP CAT: Antipsychotic.

2242. Chlorpyrifos. *Phosphorothioic acid O,O-diethyl O-(3,5,6-trichloro-2-pyridinyl) ester; O,O-diethyl O-3,5,6-trichloro-2-pyridyl phosphorothioate;* chlorpyrifos-ethyl; Dowco 179; ENT-27311; Dursban; Lorsban; Pyrinex. C_9-$H_{11}Cl_3NO_3PS$; mol wt 350.59. C 30.83%, H 3.16%, Cl 30.34%, N 4.00%, O 13.69%, P 8.83%, S 9.15%. Prepn: R. H. Rigterink, **Fr.** pat. **1,360,901;** idem, **U.S.** pat. **3,244,586** (1964, 1966 both to Dow); R. H. Rigterink, E. E. Kenaga, *J. Agr. Food Chem.* **14,** 394 (1966). Activity: E. E. Kenaga et al., *J. Econ. Entomol.* **58,** 1043 (1965). Metabolism: G. N. Smith et al., *J. Agr. Food Chem.* **15,** 132 (1967); W. H. Gutenmann et al., ibid. **16,** 45 (1968). Toxicity study: Schafer, *Toxicol. Appl. Pharmacol.* **21,** 315 (1972). Review of environmental fate: K. D. Racke, *Rev. Environ. Contam. Toxicol.* **131,** 1-154 (1993).

White granular crystals, mp 41-42°. Vapor pressure at 25°: 1.87×10^{-5} mm Hg. Soly at 25°: water 2 ppm; isooctane 79% w/w; methanol 43% w/w. Soly (g/100g): acetone 650, benzene 790, carbon disulfide 590, carbon tetrachloride 310, chloroform 630, diethyl ether 510, ethanol 63, ethyl acetate > 200, isooctane 79, kerosene 60, methanol 45, methylene chloride 400, propylene glycol 4, toluene 150, 1,1,1-trichloroethane 400, triethylene glycol 5, xylene 400. Readily sol in other org solvents. uv max: 208, 230, 290 nm. LD_{50} orally in rats: 145 mg/kg (Schafer).

O,O-Dimethyl analog, $C_7H_7Cl_3NO_3PS$, *chlorpyrifos-methyl,* Dowco 214, ENT-27520, Reldan. Crystals, mp 45.5-46.5°. Vapor pressure at 25°: 4.22×10^{-5} mm Hg. Soly in water at 25°: 5 mg/l.

USE: Insecticide; acaricide.

2243. Chlorquinaldol. *5,7-Dichloro-2-methyl-8-quinolinol;* 5,7-dichloro-8-quinaldinol; 5,7-dichloro-8-hydroxyquinaldine; 5,7-dichloro-2-methyl-8-hydroxyquinoline; hydroxydichloroquinaldine; chloroquinaldol; Afungil; Quesil; Siogène; Gyno-Sterosan; Gynotherax; Saprosan; Sterosan; Steroxin; Siosteran. $C_{10}H_7Cl_2NO$; mol wt 228.08. C 52.66%, H 3.09%, Cl 31.09%, N 6.14%, O 7.01%. Prepd by chlorination of 8-hydroxyquinaldine with or without formic acid as solvent: Senn, **U.S.** pat. **2,411,670** (1946 to Geigy).

Yellow needles from alc, mp 114-115° (slight decompn). Medicinal odor. uv max (ethanol): 316 nm ($A_{1cm}^{1\%}$ 170); min 280 nm. Practically insol in water. Soly (25°) in ethanol 1.0 g/100 ml of soln; chloroform 5.0 g; acetone 4.0 g; ether 3.0 g; 0.1N NaOH 1.4 g. Also sol in benzene, glacial acetic acid.

THERAP CAT: Antibacterial.

THERAP CAT (VET): Antibacterial; antifungal.

2244. Chlorsulfuron. *2-Chloro-N-[[(4-methoxy-6-methyl-1,3,5-triazin-2-yl)amino]carbonyl]benzenesulfonamide;* DPX-4189; Glean; Telar. $C_{12}H_{12}ClN_5O_4S$; mol wt 357.78. C 40.29%, H 3.38%, Cl 9.91%, N 19.57%, O 17.89%, S 8.96%. Selective pre- and post-emergence herbicide. Prepn: G. Levitt, **Ger.** pat. **2,715,786;** idem, **U.S.** pat. **4,127,405** (1977, 1978 both to du Pont). Activity: G. Levitt et al., *J. Agr. Food Chem.* **29,** 416 (1981). Mode of action: T. B. Ray, *Pestic. Biochem. Physiol.* **17,** 10 (1982).

Crystals from ether, mp 174-178°. Soly in water: 125 ppm. Moderately sol in methylene chloride; less sol in acetone, acetonitrile. Low soly in hydrocarbon solvents. LD_{50} in male, female rats (mg/kg): 5545, 6293 orally (Levitt).

USE: Herbicide.

2245. Chlortetracycline. *[4S-(4α,4aα,5aα,6β,12aα)]-7-Chloro-4-dimethylamino-1,4,4a,5,5a,6,11,12a-octahydro-3,6,-10,12,12a-pentahydroxy-6-methyl-1,11-dioxo-2-naphthacenecarboxamide;* 7-chlorotetracycline; Acronize; Aureocina; Aureomycin; Biomitsin; Centraureo; Chrysomykine; Orospray. $C_{22}H_{23}ClN_2O_8$; mol wt 478.89. C 55.18%, H 4.84%, Cl 7.40%, N 5.85%, O 26.73%. Antibiotic substance isolated from the substrate of *Streptomyces aureofaciens:* Duggan, *Ann. N.Y. Acad. Sci.* **51,** 177 (1948); **U.S.** pat. **2,482,055** (1949 to Am. Cyanamid); Broschard et al., *Science* **109,** 199 (1949). Structure: Stephens et al., *J. Am. Chem. Soc.* **74,** 4976 (1952); **76,** 3568 (1954). Crystal structure: Donohue et al., ibid. **85,** 851 (1963). Absolute configuration: Dobrynin et al., *Tetrahedron Letters* **1962,** 901. Purification: Winterbottom et al., **U.S.** pat. **2,899,422** (1959 to Am. Cyanamid). Improved process: Miller et al.; Goodman, **U.S.** pats. **2,987,449; 3,050,446** (1961, 1962 to Am. Cyanamid). Toxicity: E. I. Goldenthal, *Toxicol. Appl. Pharmacol.* **18,** 185 (1971). Comprehensive description: G. Schwartzman et al., in *Analytical Profiles of Drug Substances* vol. **8,** K. Florey, Ed. (Academic Press, New York, 1979) pp 101-137.

Golden-yellow crystals, mp 168-169°. $[\alpha]_D^{23}$ −275.0° (methanol). uv max (0.1N HCl): 230, 262.5, 367.5 nm; (0.1N NaOH): 255, 285, 345 nm. Soly in water: 0.5-0.6 mg/ml. Very sol in aq solns above pH 8.5. Freely sol in the Cellosolves, dioxane, and Carbitol. Slightly sol in methanol, ethanol, butanol, acetone, ethyl acetate, benzene. Practically insol in ether, petr ether.

Crystal salt, white powder. Insol in water. Marketed as suspension preserved with methyl and propyl p-hydroxybenzoate; stable and fully active for at least one year at room temps.

Hydrochloride, $C_{22}H_{23}ClN_2O_8 \cdot HCl$, *Aureociclina, Isphamycin.* Bitter, yellow rhomboid crystals. Dec above 210°. $[\alpha]_D^{23}$ −240°. Soly at about 28° (mg/ml): water 8.6; methanol 17.4; ethanol 1.7: Weiss et al., *Antibiot. Chemother.* **7,**

374 (1957). Sol in solns of alkali hydroxides and carbonates. Practically insol in acetone, ether, chloroform, dioxane. pH of satd aq soln 2.8-2.9. LD_{50} orally in rats: 10300 mg/kg (Goldenthal).

THERAP CAT: Antibacterial; antiamebic.

THERAP CAT (VET): Antimicrobial.

2246. Chlorthalidone. *2-Chloro-5-(2,3-dihydro-1-hydroxy-3-oxo-1H-isoindol-1-yl)benzenesulfonamide; 2-chloro-5-(1-hydroxy-3-oxo-1-isoindolinyl)benzenesulfonamide;* 3-hydroxy-3-(4-chloro-3-sulfamylphenyl)phthalimidine; 2-chloro-5-(3-hydroxy-1-oxoisoindolin-3-yl)benzenesulfonamide; 1-oxo-3-(3-sulfamyl-4-chlorophenyl)-3-hydroxyisoindoline; 3-(4'-chloro-3'-sulfamoylphenyl)-3-hydroxyphthalimidine; 1-keto-3-(3'-sulfamyl-4'-chlorophenyl)-3-hydroxyisoindoline; chlorphthalidolone; phthalamudine; phthalamodine; G-33182; Higroton; Hydro-long; Hygroton; Thalitone. $C_{14}H_{11}ClN_2O_4S$; mol wt 338.77. C 49.64%, H 3.27%, Cl 10.47%, N 8.27%, O 18.89%, S 9.47%. Prepn: Graf *et al.*, *Helv. Chim. Acta* **42**, 1085 (1959); U.S. pat. **3,055,904** (1962 to Geigy). Activity and side effects: Holtmeier *et al.*, *Med. Welt* **1967**, 1384; Zsoter *et al.*, *J. Pharmacol. Exp. Ther.* **180**, 723 (1972). Metabolism: Beisenherz *et al.*, *Arch. Int. Pharmacodyn. Ther.* **161**, 76 (1966). Comprehensive description: J. M. Singer *et al.*, in *Analytical Profiles of Drug Substances* vol. 14, K. Florey, Ed. (Academic Press, New York, 1985) pp 1-36.

Crystals from 50% acetic acid, dec 224-226°. mp range may extend from 218° to 264° on slow heating. Can form a monohydrate. uv max (methanol): < 220 nm. Soly in water: 12 mg/100 ml (20°); 27 mg/100 ml (37°); in 0.1*N* Na_2CO_3: 577 mg/100 ml (20°); 990 mg/100 ml (37°). More sol in aq solns of NaOH. Soluble in warm ethanol; slightly sol in ether.

THERAP CAT: Diuretic; antihypertensive.

THERAP CAT (VET): Diuretic.

2247. Chlorthenoxazin(e). *2-(2-Chloroethyl)-2,3-dihydro-4H-1,3-benzoxazin-4-one;* 2-(β-chloroethyl)-2,3-dihydro-4-oxo-(benzo-1,3-oxazine); 4-oxo-2-(β-chloroethyl)-2,3-dihydrobenzo-1,3-oxazine; 2-(β-chloroethyl)-2,3-dihydro-4-keto-(benzo-1,3-oxazine); AP-67; Valtorin; Piroxina; Ossazone; Reumagrip; Ossipirina; Apirazin; Valmorin. $C_{10}H_{10}ClNO_2$; mol wt 211.65. C 56.75%, H 4.76%, Cl 16.75%, N 6.62%, O 15.12%. Prepn: Ohnacker, U.S. pat. **2,943,087** (1960 to Thomae).

Crystals from ethanol, mp 146-147° (dec). Sol in chloroform. uv max (chloroform): 297.5 nm. Ingredient of *Trigatan, Fiobrol.*

THERAP CAT: Antipyretic; analgesic.

2248. Chlorthion®. *Phosphorothioic acid O-(3-chloro-4-nitrophenyl) O,O-dimethyl ester; O,O-dimethyl O-(3-chloro-4-nitrophenyl)* thionophosphate; *p*-nitro-*m*-chlorophenyl dimethyl thionophosphate; compd 22/190; Chlorothion. $C_8H_9ClNO_5PS$; mol wt 297.66. C 32.28%, H 3.05%, Cl 11.91%, N 4.71%, O 26.88%, P 10.41%, S 10.77%. Prepn: Schrader, *Angew. Chem.* **66**, 265 (1954); U.S. pat. **2,701,259** (1955 to Bayer). Toxicity study: T. B. Gaines, *Toxicol. Appl. Pharmacol.* **14**, 515 (1969).

The pure compound as yellow crystals, mp 21°. d_4^{20} 1.437. n_D^{20} 1.5661. Commercial product is a yellow oil, $bp_{0.1}$ 125°. Miscible in benzene, alcohol, ether. Practically insol in water. Hydrolyzed by alkali. LD_{50} in male, female rats (mg/kg): 880, 980 orally (Gaines).

Caution: Cholinesterase inhibitor. Symptoms similar to parathion.

USE: Insecticide.

2249. Chlorzoxazone. *5-Chloro-2(3H)-benzoxazolone; 5-chloro-2-benzoxazolol;* 5-chloro-2-hydroxybenzoxazole; 2-hydroxy-5-chlorobenzoxazole; 5-chlorbenzoxazolin-2-one; 5-chlorobenzoxazolidone; Paraflex; Biomioran; Solaxin. $C_7H_4ClNO_2$; mol wt 169.57. C 49.58%, H 2.38%, Cl 20.91%, N 8.26%, O 18.87%. Prepn from 2-amino-5-chlorobenzoxazole: Marsh, U.S. pat. **2,895,877** (1959 to McNeil Labs.). LC determn in plasma: I. L. Honigsberg *et al.*, *J. Pharm. Sci.* **68**, 253 (1979). Fluorometric determn in tablets and biological fluids: J. T. Stewart, C. W. Chan, *ibid.* 910. Clinical studies: R. Herman, *Curr. Ther. Res.* **9**, 537 (1967); J. J. Scheiner, *ibid.* **19**, 51 (1976). Metabolism study: R. Twele, G. Spiteller, *Arzneimittel-Forsch.* **32**, 759 (1982). Toxicity: G. Hofrichter *et al.*, *Arzneimittel-Forsch.* **17**, 242 (1967). Review of clinical studies: J. K. Elenbaas, *Am. J. Hosp. Pharm.* **37**, 1313 (1980). Comprehensive description: J. T. Stewart, C. A. Janicki in *Analytical Profiles of Drug Substances* vol. 16, K. Florey, Ed. (Academic Press, New York, 1987) pp 119-144.

Crystals from acetone, mp 191-191.5°. Sparingly sol in water; sol in methanol, ethanol, isopropanol. Freely sol in aq solns of alkali hydroxides and ammonia. LD_{50} in mice (mg/kg): 3650 orally, 380 i.p. (suspensions); 440 orally, 183 i.p. (solns of Na salt) (Hofrichter).

THERAP CAT: Muscle relaxant (skeletal).

2250. Cholane. $C_{24}H_{42}$; mol wt 330.60. C 87.19%, H 12.81%. Prepn from bisnorcholyl methyl ketone: Wieland *et al.*, *Z. Physiol. Chem.* **161**, 80, 109 (1926); from potassium cholanate: Kazuno *et al.*, *Proc. Japan Acad.* **28**, 416 (1952).

Stout prisms from alc, mp 90°; $bp_{0.001}$ 190°. Sparingly sol in methanol.

2251. Cholanic Acid. *Cholan-24-oic acid;* ursocholanic acid; 17β-(1-methyl-3-carboxypropyl)etiocholane. $C_{24}H_{40}O_2$; mol wt 360.58. C 79.94%, H 11.18%, O 8.87%. Steroidal acid probably formed by the dehydration and hydrogenation of certain bile acids commonly found in animals. Considered to be a chemical trademark certifying the prehistoric presence of some type of animal, *see* Seifert, *Pure Appl. Chem.* **34**, 633 (1973). This status as a natural product of exclusive animal origins now questioned by its isolation from the embryo of the jequirity bean, *Abrus precatorius, Leguminosae:* Mandava *et al.*, *Steroids* **23**, 357 (1974). Prepn from coprostane, cholic acid, lithocholic acid, desoxy- and chenodesoxycholic acid: Wieland, Weil, *Z. Physiol. Chem.* **80**, 287 (1912); Wieland, Boersch, *ibid.* **106**, 193 (1919); Windaus, Neukirchen, *Ber.* **52**, 1915 (1919); Wieland, Vocke, *Z. Physiol. Chem.* **191**, 69 (1930). From 3-ketocholanic acid diethyl thioacetal by hydrogenolysis: Bernstein, Dorfman, U.S. pat. **2,440,660** (1948 to Am. Cyanamid). 5β-Cholanic acid differs from the thermodynamically more stable 5α-isomer, by being *cis* at the A/B steroid ring function rather than *trans*.

Needles from alc, mp 163-164°. $[\alpha]_D^{20}$ +21.7° (chloroform). Sol in chloroform, alcohol, acetic acid. Forms a molecular compd with allocholanic acid, mp 163.5°.

Methyl ester, $C_{25}H_{42}O_2$, mp 86-87°.

Ethyl ester, $C_{26}H_{44}O_2$, needles from 80% alc, mp 93-94°; bp_{12} 273°. $[\alpha]_D^{19}$ +21° (chloroform).

Note: The name cholanic acid was formerly applied to desoxybilianic acid, $C_{24}H_{36}O_7$.

USE: In chemotaxonomical classification.

2252. Cholanthrene. *1,2-Dihydrobenz[j]aceanthrylene.* $C_{20}H_{14}$; mol wt 254.33. C 94.45%, H 5.55%. Synthesis from 1-β-naphthylhydrindene: Cook *et al.*, *J. Chem. Soc.* **1935**, 667; from 4-indanyl-MgBr and 1-naphthoyl chloride: Fieser, Seligman, *J. Am. Chem. Soc.* **57**, 2174 (1935).

Faintly yellow plates from benzene + ether, mp 173°. Sublimes 210-215° at 0.2 mm Hg. Absorption max: Fieser, Hershberg, *J. Am. Chem. Soc.* **60**, 940 (1938). Sol in benzene, xylene, toluene. Slightly sol in methanol. Insol in water. May be solubilized by aq solns of Na desoxycholate.

Molecular complex with 2,4,7-trinitrofluorenone, olive green crystals, mp 245-246°: Orchin, Woolfolk, *J. Am. Chem. Soc.* **68**, 1727 (1946).

2253. Cholecystokinin. Pancreozymin; cholecystokinin-pancreozymin; CCK-PZ. Polypeptide hormone found in the mammalian gastrointestinal tract and brain. Stimulates pancreatic exocrine secretion and growth. May also play a role in appetite satiation, pain perception, and neuronal transmission. First shown to cause gallbladder contraction: Ivy, Oldberg, *Am. J. Physiol.* **86**, 599 (1928). Discovery of a substance, designated as pancreozymin, which promotes secretion of digestive enzymes by the pancreas: Harper, Raper, *J. Physiol. (London)* **102**, 115 (1943). Identity with pancreozymin: Jorpes *et al.*, *Acta Chem. Scand.* **18**, 2408 (1964). The *C*-terminal pentapeptide has been shown to be identical to that of gastrin and caerulein: V. Mutt, J. E. Jorpes, *Eur. J. Biochem.* **6**, 156 (1968); *idem, Biochem. J.* **125**, 57P (1971). Various biologically active, amino-truncated forms have been identified. Cholecystokinin consisting of 33 amino acids (CCK-33) is the predominant gastrointestinal form; CCK-39 and CCK-58 have also been identified. CCK-8 is the predominant CNS form. Identification of CCK in brain: J. J. Vanderhaeghen *et al.*, *Nature* **257**, 604 (1975); G. J. Dockray, *ibid.* **264**, 568 (1976). Distribution and molecular heterogeneity: J. F. Rehfeld, *J. Biol. Chem.* **253**, 4022 (1978). Synthesis of the *C*-terminal dodecapeptide: M. A. Ondetti *et al.*, *J. Am. Chem. Soc.* **92**, 195 (1970). Synthesis of the *N*-terminal hexapeptide of porcine CCK-33: Bodanszky *et al.*, *J. Org. Chem.* **37**, 2303 (1972). Cloning and nucleotide sequence of the human cholecystokinin gene: Y. Takahashi *et al.*, *Proc. Nat. Acad. Sci. USA* **82**, 1931 (1985). Total synthesis of porcine CCK-

33: Y. Kurano, *Chem. Commun.* **1987**, 323; of human CCK-33: N. Fujii *et al., ibid.* **1988**, 324. Proposed role in suppression of food intake: M. A. Della-Fera, C. A. Baile, *Science* **206**, 471 (1979); C. J. Savory, M. J. Gentle, *Experientia* **36**, 1191 (1980); M. A. Della-Fera *et al., Science* **212**, 687 (1981); in regulation of hypothalamic peptides: S. Itoh *et al., Life Sci.* **25**, 1725 (1979); in modulation of catecholaminergic activity: K. Fuxe *et al., Eur. J. Pharmacol.* **67**, 329 (1980). There is also evidence that CCK acts as a specific antagonist of opiate analgesia: P. L. Faris *et al., Science* **219**, 310 (1983). *Reviews:* E. Straus, R. S. Yalow, *Fed. Proc.* **38**, 2320-2324 (1979); V. Mutt, *Biochem. Soc. Trans.* **8**, 11-14 (1980); *idem, Vitam. Horm.* **39**, 231-426 (1982). Review of physiology: G. J. Dockray, *Brit. Med. Bull.* **38**, 253-258 (1982); of role in appetite satiation and pain perception: G. Stacher, *Psychoneuroendocrinology* **11**, 39-48 (1986). Symposium on neuronal CCK: *Ann. N.Y. Acad. Sci.* **448**, 1-697 (1985).

Human CCK - 33

C-Terminal octapeptide, *see* Sincalide.

2254. (5α)-Cholestane. $C_{27}H_{48}$; mol wt 372.68. C 87.02%, H 12.98%. The *trans*-decalin homolog of coprostane, *q.v.* Prepd from cholesteryl chloride: Diels, Linn, *Ber.* **41**, 548 (1908); Windaus, *Ber.* **50**, 136 (1917); Ruzicka *et al., Helv. Chim. Acta* **16**, 327 (1933). Crystal structure: Haner, Norton, *Acta Cryst.* **20**, 930 (1966).

Scales from ether + alcohol. mp 80°. $[\alpha]_D^{20}$ +24.4° or +30.2° (c = 2 in chloroform). n_D^{88} 1.4887. Freely sol in chloroform, ether, benzene, slightly sol in abs alcohol.

2255. Cholestanol. *(3β,5α)-Cholestan-3-ol.* Dihydrocholesterol; 3β-hydroxycholestane; β-cholestanol. $C_{27}H_{48}O$; mol wt 388.68. C 83.44%, H 12.45%, O 4.12%. Occurs in human feces, in gallstones, in eggs. Prepn by reduction of cholesterol: Willstätter, Mayer, *Ber.* **41**, 2199 (1908); Ellis, Gardner, *Biochem. J.* **12**, 72 (1918). From coprostenone: Diels, Abderhalden, *Ber.* **39**, 884 (1906). *See also* Bruce, Ralls, *Org. Syn.* **coll. vol. II**, 191 (1943).

Monohydrate, scales from alc, mp 141.5-142°. $[\alpha]_D^{22}$ +24.2° (c = 1.3 in chloroform). One gram dissolves in about 100 ml alcohol, in 200 ml dry methanol. Freely sol in hot alc, ether, chloroform. Pptd by digitonin.

Methyl ether, $C_{28}H_{50}O$, needles from acetone, mp 82.5-83°. $[\alpha]_D^{20}$ +20.0°.

Acetate, $C_{29}H_{50}O_2$, prisms from ethyl acetate + methanol, mp 111°. $[\alpha]_D^{20}$ +13.3° (c = 2 in chloroform).

2256. Cholesterol. *(3β)-Cholest-5-en-3-ol;* cholesterin. $C_{27}H_{46}O$; mol wt 386.66. C 83.87%, H 11.99%, O 4.14%. Principal sterol of the higher animals. Found in all body tissues, esp in the brain, spinal cord, and in animal fats or oils. Main constituent of gallstones. Prepd commercially from the spinal cord of cattle by petr ether extraction of the nonsaponifiable matter. Also produced from wool grease. Cholesterol from animal organs always contains cholestanol (dihydrocholesterol) and other satd sterols. Purification by repeated bromination: Schoenheimer, *J. Biol. Chem.* **105**, 355 (1934); Fieser, *Org. Syn.*, **coll. vol. IV**, 195 (1963). Laboratory procedure for isoln from gallstones: L. F. Fieser, *Organic Experiments* (Heath, Boston, 3rd ed., 1964) p 70. Total synthesis: Keana, Johnson, *Steroids* **4**, 457 (1964). Reviews and bibliographies: Fieser, Fieser, *Steroids* (Reinhold, New York, Chapman & Hall, London, 1959); Lettré *et al.*, *Ueber Sterine, Gallensäuren und verwandte Naturstoffe* (Stuttgart, 2nd ed., 1955); R. P. Cook, *Cholesterol (Chemistry, Biochemistry and Pathology)* (Academic Press, New York, 1958) 542 pp; J. T. Gwynne, J. F. Strauss, *Endocrine Rev.* **3**, 299-329 (1982).

Monohydrate, pearly leaflets or plates from dil alcohol. Becomes anhydr at 70-80°. When anhydr mp 148.5°. Has been sublimed as orthorhombic needles bp$_{0.5}$ 233°; bp$_{760}$ 360° (some decompn). d 1.03 (monohydrate); d$_{19}^{19}$ 1.052 (anhydr). $[\alpha]_D^{20}$ −31.5° (c = 2 in ether); $[\alpha]_D^{20}$ −39.5° (c = 2 in chloroform). Absorption spectrum: Heilbron *et al.*, *J. Chem. Soc.* **1928**, 47. Practically insol in water (about 0.2 mg/100 ml H$_2$O). Slightly sol in alc (1.29% w/w at 20°), more sol in hot alc (100 g of satd 96% alcoholic soln contains 28 g at 80°). One gram dissolves in 2.8 ml ether, in 4.5 ml chloroform, in 1.5 ml pyridine. Also sol in benzene, petr ether, oils, fats. Soly in aq solns of bile salts: Rosin, *Z. Physiol. Chem.* **124**, 282 (1923). Solubilization: Gemant, *Life Sci.* **1**, 233 (June 1962). Is pptd by digitonin. Gives intense red color with rosaniline in chloroform soln.
Methyl ether, $C_{28}H_{48}O$, crystals from acetone, mp 84°. $[\alpha]_D^{20}$ −45.8° (c = 1.2 in chloroform).
Acetate, $C_{29}H_{48}O_2$, needles from acetone, mp 115-116°. $[\alpha]_D^{20}$ −47.4° (c = 2 in chloroform).
Benzoate, $C_{34}H_{50}O_2$, mp 145.5° (the melt becomes clear at 180°). $[\alpha]_D^{25}$ −13.7° (c = 0.9 in chloroform).
USE: Pharmaceutic aid (emulsifying agent).

2257. Cholestyramine Resin. Colestyramin; Dowex 1-X2-Cl; Cholybar; Colybar; Cuemid; Quantalan; Questran. A synthetic, strongly basic anion exchange resin contg quaternary ammonium functional groups which are attached to a styrene-divinylbenzene copolymer. Main constituent: Polystyrene trimethylbenzylammonium as Cl⁻ anion, also contains divinylbenzene. Av polymeric mol wt > 10⁶; Cl⁻ content 14-17%. Effect of particle size on bile salt binding capacity: L. M. Hagerman *et al.*, *Proc. Soc. Exp. Biol. Med.* **139**, 248 (1972). Review of pharmacology: H. R. Casdorph in *Lipid Pharmacology* Vol. 2(2), R. Pao-

typified structure of main polymeric groups

letti, C. J. Glueck, Eds. (Academic Press, New York, 1976) pp 222-256. Clinical evaluation in chlordecone detoxification: W. J. Cohn *et al.*, *N. Engl. J. Med.* **298**, 243 (1978). Review of pharmacology and therapeutic efficacy: M. Ast, W. H. Frishman, *J. Clin. Pharmacol.* **30**, 99-106 (1990).
White to buff-colored, hygroscopic, fine powder. Odorless or has not more than a slight amine-like odor. Insol in water, alcohol, benzene, chloroform, ether.
THERAP CAT: Ion-exchange resin (bile salts); antihyperlipoproteinemic.

2258. Cholic Acid. *(3α,5β,7α,12α)-3,7,12-Trihydroxycholan-24-oic acid;* cholalic acid; 17β-(1-methyl-3-carboxypropyl)etiocholane-3α,7α,12α-triol; Colalin. $C_{24}H_{40}O_5$; mol wt 408.58. C 70.55%, H 9.87%, O 19.58%. Occurs in conjugation with glycine or taurine in bile of most vertebrates. Extraction from beef bile: Wieland, Siebert, *Z. Physiol. Chem.* **262**, 1 (1939); laboratory procedure: Gattermann-Wieland, *Praxis des Organischen Chemikers* (de Gruyter, Berlin, 40th ed., 1961) p 360. Structure and synthesis: Fieser, Fieser, *Steroids* (Reinhold, New York, 1959).

Monohydrate, plates from dil acetic acid. Bitter taste with sweetish aftertaste. When anhydr mp 198°. $[\alpha]_D^{20}$ +37° (c = 0.6 in alcohol). pK = 6.4. Not precipitated by digitonin. Soly at 15° in water 0.28 g/l; in alcohol 30.56 g/l; in ether 1.22 g/l; in chloroform 5.08 g/l; in benzene 0.36 g/l; in acetone 28.24 g/l; in glacial acetic acid 152.12 g/l. Sol in solns of alkali hydroxides or carbonates.
Methyl ester, $C_{25}H_{42}O_5$, crystals from 95% alcohol + water, mp 155-156°.
Ethyl ester, $C_{26}H_{44}O_5$, crystals from ethyl acetate + petr ether, mp 162-163°.
Sodium salt, $C_{24}H_{39}NaO_5$, *sodium cholate*. Crystals. Soly in water at 15°: > 568.9 g/l. *Caution:* The names "sodium cholate" and "sodium cholcate" are sometimes used for mixtures of bile salts. The term "sodium choleate" is to be preferred for bile salts, see Ox Bile Extract.
Note: The property of forming molecular compds is common to bile acids. For instance a blue molecular compd $(C_{24}H_{40}O_5I)_4$.KI.H$_2$O, may be prepd by mixing an alcoholic soln of cholic acid and a soln of iodine in aq potassium iodide: Barger, Field, *J. Chem. Soc.* **101**, 1404 (1912).
THERAP CAT: Choleretic.

2259. Choline. *2-Hydroxy-N,N,N-trimethylethanaminium; (β-hydroxyethyl)trimethylammonium;* bilineurine. $[C_5H_{14}NO]^+$. $[HOCH_2CH_2N^+(CH_3)_3]$. Basic constituent of lecithin. Found in many plants and animal organs, *e.g.* bile, brain, yolk of eggs, hops, belladonna, strophanthus. Usually made synthetically from trimethylamine and ethylene chlorohydrin or ethylene oxide: Blackett, Soliday, U.S. pat. **2,774,759** (1956 to Am. Cyanamid). *Review:* T. H. Jukes in Kirk-Othmer *Encyclopedia of Chemical Technology* **vol. 6** (Wiley-Interscience, New York, 3rd ed., 1979) pp 19-28.
Hydroxide, $C_5H_{15}NO_2$, *bursine, fagine, gossypine, luridine, sincaline, vidine.* Viscid, strongly alkaline liq, has been crystallized; absorbs CO$_2$ from the air. Very sol in water, alcohol. Insol in ether. *Keep tightly closed.*

2260. Choline Alfoscerate. *(R)-2-[[(2,3-Dihydroxypropoxy)hydroxyphosphinyl]oxy]-N,N,N-trimethylethanaminium hydroxide inner salt;* D-choline hydroxide 2,3-dihydroxypropyl hydrogen phosphate inner salt; L-α-glycerylphosphorylcholine; sn-glycero-3-phosphorylcholine; L-α-GPC; Brezal; Delecit; Gliatilin. $C_8H_{20}NO_6P$; mol wt 257.22. C 37.36%, H 7.84%, N 5.45%, O 37.32%, P 12.04%. Phospholipid; precursor of choline biosynthesis; intermediate in catabolic pathway of phosphatidylcholine. Isoln from bovine pancreas: G. Schmidt *et al., J. Biol. Chem.* **161**, 523 (1945). Synthesis: E. Baer, M. Kates, *J. Am. Chem Soc.* **70**, 1394 (1948); and properties: N. H. Tattrie *et al., Biochem. Prep.* **6**, 16 (1958). HPLC determn: G. Abbiati *et al., J. Chromatog.* **566**, 445 (1991). Metabolism and biodistribution: *idem et al., Eur. J. Drug Metab. Pharmacokinet.* **18**, 173 (1993). Clinical pharmacology: N. Canal *et al., Int. J. Clin. Pharmacol. Ther. Toxicol.* **29**, 103 (1991). Clinical pharmacokinetics: G. Gatti *et al., ibid* **30**, 331 (1992). Clinical study in multi-infarct dementia: L. Frattola *et al., Curr. Ther. Res.* **49**, 683 (1991); in Alzheimer's type dementia: L. Parnetti *et al., Drugs Aging* **3**, 159 (1993).

White crystals, mp 142.5-143°, sinters at 141°. Extremely hygroscopic. Sol in water. $[\alpha]_{25}^{D}$ −2.7° (c = 2.7 in water, pH 2.5). $[\alpha]_{25}^{D}$ −2.8° (c = 2.6 in water, pH 5.8).

THERAP CAT: Nootropic.

2261. Choline Chloride. *2-Hydroxy-N,N,N-trimethylethanaminium chloride;* Biocolina; Hepacholine; Lipotril. $C_5H_{14}ClNO$; mol wt 139.63. C 43.01%, H 10.11%, Cl 25.39%, N 10.03%, O 11.46%. $[HOCH_2CH_2N^+(CH_3)_3]Cl^-$. Prepn: Klein, Kapp, U.S. pat. **2,623,901** (1952 to Nopco Chem.). Polymorphism: Collin, *J. Am. Chem. Soc.* **79**, 6086 (1957). Enhancement of human serial learning: N. Sitaram *et al., Life Sci.* **22**, 1555 (1978); *eidem, Science* **201**, 274 (1978). Toxicity data: Hartung, Cornish, *Toxicol. Appl. Pharmacol.* **12**, 486 (1968).

Deliquesc crystals. Very sol in water or alcohol. The aq soln is practically neutral. *Keep tightly closed.* LD_{50} in rats (g/kg): 0.400 i.p.; 6.64 orally (Hartung, Cornish).

THERAP CAT: Lipotropic.

THERAP CAT (VET): Nutritional factor. Dietary source of choline (lipotropic factor) in poultry.

2262. Choline Dehydrocholate. Dehydrocholic acid salt of choline; Biscolan. $C_{29}H_{48}NO_6$; mol wt 506.70. C 68.74%, H 9.55%, N 2.76%, O 18.95%. Prepn: Giovambattista, *Rev. Asoc. Bioquim. Arg.* **19**, 180 (1954), *C.A.* **49**, 13275 (1955).

Crystals from methanol+ ether, mp 196-198°.

THERAP CAT: Lipotropic.

2263. Choline Dihydrogen Citrate. (2-Hydroxyethyl)-trimethylammonium citrate; Chothyn; Cirrocolina; Citracholine. $C_{11}H_{21}NO_8$; mol wt 295.29. C 44.74%, H 7.17%, N 4.74%, O 43.35%. $[HOCH_2CH_2N^+(CH_3)_3][C_6H_7O_7^-]$. Prepn: Klein *et al., U.S. pat.* **2,870,198** (1959).

Granules, mp 105-107.5°. Acrid taste. Freely sol in water; very slightly sol in alcohol. Practically insol in benzene, chloroform, ether. The pH of a 25% aq soln is 4.25. May be formulated with calcium citrate for better taste and easier tabletting.

THERAP CAT: Lipotropic.

THERAP CAT (VET): Dietary source of choline (lipotropic factor) in poultry.

2264. Choline Esterase. There are at least two choline esterases. *Acetylcholinesterase* is a "specific" choline esterase, hydrolyzing predominantly choline esters, and characterized by high concns in brain, nerve and red blood cells. The other type, called *butyrylcholinesterase,* is a nonspecific choline esterase ("pseudo" choline esterase), hydrolyzing other esters as well as choline esters, and found in blood serum, pancreas and liver, *cf.* Sumner and Somers, *Chemistry and Methods of Enzymes* (New York, 1947). Prepn from dog pancreas: Mendel, Mundell, *Biochem. J.* **37**, 64 (1943); from horse serum: Strelitz, *ibid.* **38**, 86 (1944). *Reviews:* Davies, Green, *Advan. Enzymol.* **20**, 283 (1958); Froede, Wilson, "Acetylcholinesterase" in *The Enzymes,* P. D. Boyer, Ed. (Academic Press, N. Y., 3rd ed., 1971) **vol. 5**, pp 87-114.

Specific choline esterase develops its max activity at pH 7 and at low levels of acetylcholine (less than 2.5 mg %). Both enzymes are inhibited by very small quantities of phy-

sostigmine (eserine). Acetylcholinesterase is inhibited by phosphorus-containing insecticides and nerve gases.

2265. Choline Salicylate. *2-Hydroxy-N,N,N-trimethylethanaminium salt with 2-hydroxybenzoic acid (1:1); (2-hydroxyethyl)trimethylammonium salicylate;* choline salicylic acid salt; salicylic acid choline salt; Actasal; Arthropan; Artrobione; Audax; Mundisal. $C_{12}H_{19}NO_4$; mol wt 241.29. C 59.73%, H 7.94%, N 5.81%, O 26.52%. Prepn from choline chloride and sodium salicylate: Broh-Kahn, *Intern. Rec. Med.* **173**, 219 (Apr. 1960); *cf.* Johnson, **Brit.** pat. **8031** (1919); Broh-Kahn *et al., U.S. pat.* **3,069,321** (1962 to Labs. for Pharmaceut. Dev.). Variations of process: **Belg.** pat. **583,513** (1960 to Mundipharma, AG).

Extremely hygroscopic solid, mp 49.5-50.0°. Very freely sol in water. Also sol in alcohol, acetone, other hydrophilic sol-vents. Practically insol in ether, petr ether, benzene, oils. Aq solns are stable, they contain the compd in the form of its dissociated choline and salicylate ions. pH of 10% aq soln 6.5. Aq solns are easily discolored by minute traces of iron. The addition of acid to aq solns immediately precipitates free salicylic acid, while choline base, readily recognized by its fishy odor, is liberated upon the addition of alkali.

THERAP CAT: Analgesic; antipyretic.

2266. Choline Theophyllinate. *2-Hydroxy-N,N,N-trimethylethanaminium salt with 3,7-dihydro-1,3-dimethyl-1H-purine-2,6-dione;* theophylline cholinate; theophylline salt of choline; oxtriphylline; oxytrimethylline; Sabidal; Teokolin; Teofilcolina; Filoral; Cholinophylline; Choledyl; Theoxylline; Soliphylline. Prepn: Ladenburg *et al., U.S. pat.* **2,776,287** (1957 to Nepera Chem.).

White granules (contains about 60% of anhydr theophylline). One gram dissolves in 1 ml water. Also sold as Cholinophylline Magnesium Glycinate which contains 43% of anhydr theophylline.

THERAP CAT: Bronchodilator.

2267. Chondrillasterol. *(3β,5α,22E,24R)-Stigmasta-7,22-dien-3-ol.* $C_{29}H_{48}O$; mol wt 412.70. C 84.40%, H 11.72%, O 3.88%. Stereoisomeric with α-spinasterol, *q.v.* Isolated from the green alga *Scenedesmus obliquus* (Turpin) Kuetz., *Scenedesmaceae:* Bergmann, Feeney, *J. Org. Chem.* **15**, 812 (1950). Synthesis: W. Sucrow *et al., Phytochemistry* **15**, 1533 (1976); M. Anastasia *et al., J. Chem. Soc. Perkin Trans. I* **1981**, 2561.

Crystals from chloroform-methanol, mp 168-169°. $[\alpha]_D^{24}$ −2° (chloroform).

Acetate, $C_{31}H_{50}O_2$, crystals from chloroform-methanol, mp 174.5-175.5°. $[\alpha]_D^{24}$ −0.7° (chloroform).

Benzoate, $C_{36}H_{52}O_2$, crystals from dioxane, mp 194-195°. $[\alpha]_D^{24}$ +3.9° (chloroform).

2268. Chondrocurine. *6,6'-Dimethoxy-2,2'-dimethyltubocuraran-7',12'-diol;* d-chondocurine; d-tubocurine. $C_{36}H_{38}N_2O_6$; mol wt 594.71. C 72.71%, H 6.44%, N 4.71%, O 16.14%. Isoln from *Chondodendron tomentosum* Ruiz & Pav., *Menispermaceae:* Wintersteiner, Dutcher, *Science* **97**, 467 (1943); King, *J. Chem. Soc.* **1948**, 1945; Bick, Clezy, *ibid.* **1960**, 2402; Boissier *et al., Lloydia* **28**, 191 (1965), *C.A.* **64**, 948b (1966). Structure: Dutcher, *J. Am. Chem. Soc.* **68**, 419 (1946); Hultin, *Acta Chem. Scand.* **15**, 1130 (1961). Configuration: Bick, Clezy, *J. Chem. Soc.* **1953**, 3893; Hultin, *Acta Chem. Scand.* **17**, 753 (1963). Revised structure:

Everett *et al., Chem. Commun.* **1970**, 1020. ^{13}C-NMR: L. Koike *et al., J. Org. Chem.* **46**, 2385 (1981).

Slender needles from methanol, mp 232-234°; also reported as mp 218-220° (Boissier). $[\alpha]_D^{24}$ +200° (c = 0.5 in 0.1N HCl). $[\alpha]_D^{24}$ +105° (c = 0.9 in pyridine).

Sulfate tetrahydrate, $C_{36}H_{38}N_2O_6 \cdot H_2SO_4 \cdot 4H_2O$, rectangular plates from water. Contains 9.43% H_2O. After drying, mp 263-265° (dec). $[\alpha]_D^{24}$ +184° (c = 0.375 in methanol).

Methiodide, mp 270-275°.

Dimethiodide, $C_{38}H_{44}I_2N_2O_6$, pale amorphous precipitate, mp 275° (dec). $[\alpha]_D^{24}$ +184° (c = 0.375 in methanol). uv max (water): 225, 280 nm (ϵ 62000, 7030).

USE: Methiodides have curarizing power.

2269. Chondrofoline. *(1β)-6,6',7'-Trimethoxy-2,2'-dimethyltubocuraran-12'-ol; (R,R)-7-O-methylbebeerine; 7-O-methylcurine.* $C_{37}H_{40}N_2O_6$; mol wt 608.73. C 73.01%, H 6.62%, N 4.60%, O 15.77%. Alkaloid isolated from the leaves of *Chondrodendron platyphyllum* (A. St. Hil.) Miers, *Menispermaceae:* King, *J. Chem. Soc.* **1940**, 737. Structure, stereochemistry, NMR and mass spectrum: Baldas *et al., Chem. Commun.* **1971**, 132. ^{13}C-NMR: L. Koike *et al., J. Org. Chem.* **46**, 2385 (1981).

Solvated plates from methanol, mp about 135°. $[\alpha]_{5461}^{20}$ −281° (0.1N HCl).

Nitrate hydrate, $C_{37}H_{40}N_2O_6 \cdot 2HNO_3 \cdot H_2O$, needles from water, dec 225°.

2270. Chondroitin Sulfate. Chondroitinsulfuric acid; Chonsurid; Structum. Mol wt estimated at 50,000 depending on source and method of prepn: Schubert, *Fed. Proc.* **17**, 1099 (1958). High viscosity mucopolysaccharides (glycosaminoglycans) with *N*-acetylchondrosine as a repeating unit and with one sulfate group per disaccharide unit. These biological polymers act as the flexible connecting matrix between the tough protein filaments in cartilage to form a polymeric system similar to reinforced rubber. Chondroitin 4-sulfate and chondroitin 6-sulfate are the most abundant mucopolysaccharides in the body and occur both in skeletal and soft connective tissue. Isoln: Bray *et al., Biochem. J.* **38**, 142 (1944); Patat, Elias, *Z. Physiol. Chem.* **316**, 1 (1959); Kasavina *et al.,* USSR pat. **157,466** (1962); Wheat, Davidson, *Biochem. Prepns.* **10**, 52 (1963); Haneno, **Japan.** pat. **7650('64)** (to Yasushi Hano). Structure: Davidson, Meyer, *J. Am. Chem. Soc.* **77**, 4796 (1955). Absorption spectrum of A: Orr, *Biochim. Biophys. Acta* **14**, 173 (1954); of B + C: Mathews, *Nature* **181**, 421 (1958). Clinical trials in atherosclerosis: K. Nakazawa, K. Murata, *J. Int. Med. Res.* **6**, 217 (1978); *eidem, Z. Alternsforsch.* **34**, 153 (1979). *Reviews:* K. Meyer, "Chondroitin Sulfates" in *Polysaccharides in Biology,* Trans. 4th Conf. 1958, G. F. Springer, Ed. (Josiah Macy Jr. Foundn., New York, 1959) p 11; Muir, *Am. J. Med.* **47**, 673-690 (1969); Rodén, *Pure Appl. Chem.* **35**, 181-193 (1973). *See also* Chondrosine.

Chondroitin sulfate A   R = SO₃H   R' = H
Chondroitin sulfate C   R = H   R' = SO₃H

Chondroitin 4-sulfate, chondroitin sulfate A, CSA, Atheroitin. $[\alpha]_D$ −28 to −32°.

Chondroitin 4-sulfate disodium salt, Condrosulf, Lacrypos.

Chondroitin 6-sulfate, chondroitin sulfate C. $[\alpha]_D$ −12° to −18°.

Dermatan sulfate, chondroitin sulfate B, β-heparin. Present in soft connective tissue and abundant in skin, arterial walls and heart valves. Differs from chondroitin sulfate A by containing iduronic acid in place of glucuronic acid, its epimer, at carbon atom 5. Pharmacodynamics: A. M. Traini *et al., J. Int. Med. Res.* **22**, 323 (1994). Clinical evaluation in deep vein thrombosis: B. P. Imbimbo *et al., Thromb. Haemost.* **71**, 553 (1994). $[\alpha]_D$ −60° to −70°.

THERAP CAT: Antihyperlipoproteinemic.

2271. Chondrosine. *2-Amino-2-deoxy-3-O-β-D-glucopyranurosyl-D-galactose; β-D-glucopyranosyluronic acid 2-deoxy-2-amino-D-galactose.* $C_{12}H_{21}NO_{11}$; mol wt 355.30. C 40.57%, H 5.96%, N 3.94%, O 49.53%. Disaccharide unit of chondroitins 4-sulfate and 6-sulfate. Isoln: Hebting, *Biochem. Z.* **63**, 353 (1914); Levene, *J. Biol. Chem.* **140**, 267 (1941); Wolfrom *et al., J. Am. Chem. Soc.* **74**, 1491 (1952); Davidson, Meyer, *ibid.* **76**, 5686 (1954). Structure: *eidem, loc. cit.* **77**, 4796 (1955). Synthesis: Takanashi *et al., ibid.* **84**, 3029 (1962). Chondrosine yields on acid hydrolysis chondrosamine (D-galactosamine, *q.v.*). Prepn and derivatives: Stacey, *J. Chem. Soc.* **1944**, 272; Wolfrom, Onodera, *J. Am. Chem. Soc.* **79**, 4737 (1957).

D-Glucuronic acid   Chondrosamine

Crystals from aq ethanol. $[\alpha]_D^{24}$ +40° (0.05 HCl). $[\alpha]_D^{20}$ +39° (water). Dec on heating.

Methyl ester hydrochloride, $C_{13}H_{24}ClNO_{11}$, crystals from hot ethanol, mp 159-161°. $[\alpha]_D^{23}$ +42° (c = 2 in water).

2272. Chonemorphine. *(3β,5α,20S)-N^{20},N^{20}-Dimethylpregnane-3,20-diamine; 3β-amino-20α-dimethylamino-5α-pregnane.* $C_{23}H_{42}N_2$; mol wt 346.60. C 79.70%, H 12.21%, N 8.08%. Antiamebic principle from bark and leaves of *Chonemorpha macrophylla* G. Don, *Ch. fragrans* Alston, and *Ch. penangensis* Ridl., *Apocynaceae:* Greshoff, *Ber.* **23**, 3545 (1890); Das, Pillay, *J. Sci. Ind. Res. (India)* **13B**, 602, 701 (1954); Chatterjee, Das, *Chem. & Ind. (London)* **1959**,

1445. Structure and synthesis: *eidem, ibid.* **1960**, 290; Janot *et al., Bull. Soc. Chim. France* **1962**, 111. Stereochemistry: Chien *et al., J. Org. Chem.* **29**, 315 (1964).

Bitter crystals, mp 149°. $[\alpha]_D$ +25° (chloroform). Sol in chloroform, ether.

N-Benzilidenechonemorphine, $C_{30}H_{46}N_2$, crystals, mp 182-183°. $[\alpha]_D$ +36°. uv max: 248, 277, 287 nm.

N-Acetylchonemorphine, $C_{25}H_{44}N_2O$, crystals, mp 262-265°. $[\alpha]_D$ +17°.

2273. Chorionic Gonadotropin. Choriogonadotropin; CG. Glycoprotein hormone synthesized by chorionic tissue of the placenta; found in blood and urine during pregnancy. Maintains the function of the corpus luteum and stimulates ovarian steroid secretion in the early stages of gestation. Bioactivity closely resembles that of luteinizing hormone. Structurally similar to LH, FSH, and TSH, *q.q.v.*, all are dimers consisting of noncovalently linked α and β subunits. Within a species, the α-subunits of the four hormones are essentially identical, while the β-subunits, although exhibiting varying degrees of homology, are unique and believed to be responsible for the biological specificity of the individual hormones. Produced by trophoblastic cell neoplasms and used in the diagnosis and management of trophoblastic malignancies. Also produced by cancer cells of various histological types. Discovery in human pregnancy urine: S. Aschheim, B. Zondek, *Klin. Wochenschr.* **6**, 1322 (1927); in serum of pregnant mares: H. H. Cole, G. H. Hart, *Am. J. Physiol.* **93**, 57 (1930). Isoln of human CG: B. Zondek, S. Aschheim, *Klin. Wochenschr.* **7**, 831 (1928); S. Gurin *et al., J. Biol. Chem.* **128**, 525 (1939); P. A. Katzman *et al., ibid.* **148**, 501 (1943). Review of early literature: J. A. Loraine, *J. Reprod. Fertil.* **12**, 23-31 (1966). Purification and physicochemical properties: O. P. Bahl, *J. Biol. Chem.* **244**, 567 (1969). Isoln of α and β subunits: N. Swaminathan, O. P. Bahl, *Biochem. Biophys. Res. Commun.* **40**, 422 (1970); F. J. Morgan, R. E. Canfield, *Endocrinology* **88**, 1045 (1971). Amino acid sequence of α-subunit: R. Bellesario *et al., J. Biol. Chem.* **248**, 6796 (1973); of the β-subunit: R. B. Carlsen *et al., ibid.* 6810; F. J. Morgan *et al., ibid.* **250**, 5247 (1975). Crystal structure: A. J. Lapthorn *et al., Nature* **369**, 438 (1994). Purification using HPLC: M. A. Chlenov *et al., J. Chromatog.* **631**, 261 (1993). Comparison of gonadotropins of various species: J. G. Pierce, T. F. Parsons, *Ann. Rev. Biochem.* **50**, 465-495 (1981). Review of biosynthesis: R. O. Hussa, *Endocrine Rev.* **1**, 268-294 (1980). Diagnostic use as a tumor marker: L. A. Cole *et al., Cancer Res.* **48**, 1356 (1988); D. L. Blithe *et al., Trends Endocrinol. Metab.* **1**, 394 (1990). Secretion in normal, nonpregnant humans: W. D. Odell *et al., ibid.* 418. Synthesis and expression by tumor cells: H. F. Acevedo *et al., Cancer* **76**, 1467 (1995). Review of clinical assay methods: R. J. Norman *et al., Ann. Clin. Biochem.* **27**, 183-194 (1990). Review of structure, secretion and bioactivity of equine CG: B. D. Murphy, S. D. Martinuk, *Endocrine Rev.* **12**, 27-44 (1991).

Human Chorionic Gonadotropin, HCG, pregnancy urine extract, Ambinon, Antuitrin S, A.P.L., Choragon, Choriogonin, Chorex, Choron, Coriantin, Coriovis, Corulon, Endocorion, Follutein, Glukor, Gonadotraphon L.H., Gonic, Libigen, Luteogonin B, Physex, Predalon, Pregnesin, Pregnyl, Primogonyl, Profasi, Profasi HP, Progon. Stable in dry form. Freely sol in water; sol in aq glycerol and glycols. Insol in alcohol, acetone, ether. pI 2.95. mol wt ~39500 Da. α subunit consists of 92 amino acids; β subunit, 145 amino acids. Total carbohydrate content: ~30% by wt.

Equine Chorionic Gonadotropin, pregnant mare serum gonadotropin, PMSG, serum gonadotropin, Anteron, Antex, Gestyl, Primantron, Seragon, Seragonin, Serotropin. Sol in water, dil glycerol and glycols. pI 2.60-2.65. α subunit consists of 96 amino acids; β subunit, 149 amino acids. Total carbohydrate content: ~45% by wt.

THERAP CAT: Gonad-stimulating principle.

THERAP CAT (VET): Gonad-stimulating principle.

2274. Chorismic Acid. *(3R-trans)-3-[(1-Carboxyethenyl)oxy]-4-hydroxy-1,5-cyclohexadiene-1-carboxylic acid;* 3-enolpyruvic ether of *trans-*3,4-dihydroxycyclohexa-1,5-diene carboxylic acid; α-(5-carboxy-1,2-dihydro-2-hydroxyphenoxy)acrylic acid. $C_{10}H_{10}O_6$; mol wt 226.19. C 53.10%, H 4.46%, O 42.44%. The first branch point intermediate in the

biosynthesis of aromatic amino acids via the shikimate pathway in bacteria, fungi, and higher plants; naturally occurring as the (−)-form. Its existence was predicted, then discovered during development of a mutant of *A. aerogenes:* M. I. Gibson *et al., Nature* **195**, 1173 (1962); M. I. Gibson, F. Gibson, *Biochim. Biophys. Acta* **65**, 160 (1962). NMR and preliminary structure study: F. Gibson, M. Jackman, *Nature* **198**, 388 (1963). Isoln and metabolism study: M. Gibson, F. Gibson, *Biochem. J.* **90**, 248 (1964). Prepn and characterization of the barium salt: F. Gibson, *ibid.* 256. Structure, relative and abs config, prepn of the monohydrate: J. M. Edwards, L. M. Jackman, *Aust. J. Chem.* **18**, 1227 (1965). Total synthesis of racemic chorismic acid: D. A. McGowan, G. A. Berchtold, *J. Am. Chem. Soc.* **104**, 1153 (1982); B. Ganem *et al., ibid.* 6787; improved synthesis: G. A. Berchtold *et al., ibid.* **105**, 6265 (1983); short formal synthesis: G. H. Posner *et al., J. Org. Chem.* **52**, 4836 (1987). Total synthesis of (−)-form: J. L. Pawlak, G. A. Berchtold, *ibid.* 1765. Potential use in development of herbicides: S. Stinson, *Chem. & Eng. News* **60**, 31 (Dec. 6, 1982). Reviews of chorismic acid in biosynthesis of aromatic amino acids: F. Gibson, J. Pittard, *Bacteriol. Rev.* **32**, 465-492 (1968); R. J. Ife *et al., J. Chem. Soc. Perkin Trans. I* **1976**, 1776-1783; U. Weiss, J. M. Edwards, *The Biosynthesis of Organic Compounds* (Wiley, New York, 1980) pp 134-143. *See also* shikimic acid.

Crystals may be obtained but show marked tendency toward solvent retention, mp 105-108° (dec). $[\alpha]_D^{21}$ −274° (c = 0.16 in water).

Barium salt trihydrate, $C_{10}H_8BaO_6 \cdot 3H_2O$. Unstable. uv max (aq soln): 272 nm (ε 2700).

Monohydrate, colorless crystals from ethyl acetate+ light petroleum, mp 148-149° (dec); from ether + light petroleum, mp 112° (dec); from ethyl acetate + carbon tetrachloride, mp 115° (dec). $[\alpha]_{5890}^{25}$ −295.5° (c = 0.2 in water). uv max (water): 275 nm (ε 2630).

(±)-Form, crystals from ethyl acetate/hexane, mp 139.4-141° (dec).

2275. Chromic Acetate. $C_6H_9CrO_6$; mol wt 229.13. C 31.45%, H 3.96%, Cr 22.69%, O 41.90%. $Cr(CH_3COO)_3$. The commercial material, usually sold as a concd soln of the basic acetate, $Cr(OH)(C_2H_3O_2)_2$, contains Na acetate or Na_2SO_4 impurities. Industrial prepn: Stover; Drew, U.S. pats. **2,650,239; 2,678,328** (1953, 1954 both to Socony-Vacuum Oil). Prepn of hexahydrate: Hein, Herzog in *Handbook of Preparative Inorganic Chemistry* vol. 2, G. Brauer, Ed. (Academic Press, New York, 2nd ed., 1965) p 1371. Review: *Chromium* vol. 1, M. J. Udy, Ed., A.C.S. Monograph Series no. **132** (Reinhold, New York, 1956) pp 229-233.

Hydrate, approx $Cr(C_2H_3O_2)_3 \cdot H_2O$, gray-green powder or violet plates. Slightly sol in water. Practically insol in alc. MLD i.v. in mice: 2.29 g/kg, *Handbook of Toxicology* vol. **1**, W. S. Spector, Ed. (Saunders, Philadelphia, 1956) pp 70-71.

Hexahydrate, **hexaaquochromium triacetate.** Blue-violet needles. Readily sol in water, with partial hydrolysis, giving a soln which is blue under the incident light and red under transmitted light; solvolyzed by alc.

Basic, $Cr(OH)(C_2H_3O_2)_2$, violet cryst powder. Freely sol in water.

USE: As a mordant in dyeing; in tanning; in hardening photographic emulsions; to improve light stability and dye affinity of textiles and polymers; in catalyst for polymerization of olefins; as an oxidation catalyst.

2276. Chromic Bromide. Br_3Cr; mol wt 291.71. Br 82.18%, Cr 17.82%. $CrBr_3$. Prepd by passing Br_2 vapor over Cr powder at 1000°: Hein, Herzog, in *Handbook of*

Preparative Inorganic Chemistry vol. 2, G. Brauer, Ed. (Academic Press, New York, 2nd ed., 1965) p 1341.

Black, lustrous, hexagonal crystals; green in transmitted and reddish in reflected light. Sol in cold water only upon addition of chromous salts; sol in boiling water.

Hexahydrate, at least two isomeric forms exist. *Dibromotetraaquochromium bromide dihydrate*, $[CrBr_2(H_2O)_4]Br \cdot 2H_2O$; green, deliquesc crystals; sol in water, alc. *Hexaaquochromium tribromide*, $[Cr(H_2O)_6]Br_3$; violet, deliquesc crystals; sol in water. Insol in alc. Both forms practically insol in ether.

USE: In catalysts for polymerization of olefins.

2277. Chromic Carbonate. $Cr_2O_3 \cdot xCO_2 \cdot yH_2O$. A basic carbonate of indefinite composition, contg. 50-55% Cr. Obtained by pptg a chromic salt with a sol carbonate or bicarbonate.

Blue-green amorphous powder. Practically insol in water. Slightly sol in CO_2-contg water; sol in mineral acids.

USE: Prepn of chromic salts.

2278. Chromic Chloride. Cl_3Cr; mol wt 158.35. Cl 67.16%, Cr 32.84%. $CrCl_3$. Prepn: Heisig *et al., Inorg. Syn.* **2**, 193 (1946); Pray, *ibid.* **5**, 153 (1953); Vavoulis *et al., ibid.* **6**, 129 (1960); Hein, Herzog in *Handbook of Preparative Inorganic Chemistry* vol. 2, G. Brauer, Ed. (Academic Press, New York, 2nd ed., 1965) pp 1338-1340; prepn of hexahydrates: *eidem, ibid.*, pp 1348-1350.

Violet, lustrous, hexagonal, cryst scales. Greasy feel. mp 1152°; dissociates above 1300°. d^{25} 2.87. The rate of soln in water, acids, organic solvents is extremely slow. Addition of a trace of $CrCl_2$ or wetting agent aid in rapid soln in water, alcohol. Also exists in hygroscopic, sol, peach-blossom colored form: Pray, *loc. cit. Keep tightly closed.* MLD i.v. in mice: 801 mg/kg: *Handbook of Toxicology* vol. 1, W. S. Spector, Ed. (Saunders, Philadelphia, 1956) pp 70-71.

Hexahydrates, several known isomers. The commercially available, dark green salt is *dichlorotetraaquochromium chloride dihydrate*, $[trans-[CrCl_2(H_2O)_4]Cl \cdot 2H_2O]$; monoclinic crystals; d 1.849: Freeman, Dance, *Inorg. Chem.* **4**, 1555 (1965); Morosin, *Acta Cryst.* **21**, 280 (1966). The violet, rhombohedral hydrate is *hexaaquochromium trichloride*, $[[Cr(H_2O)_6]Cl_3]$. A light green isomer is *chloropentaaquochromium dichloride monohydrate*, $[[CrCl(H_2O)_5]Cl_2 \cdot H_2O]$. All are deliquesc in air. Sol in water; dil aq solns are violet, concd aq solns are green. pH of 0.2 molar aq soln 2.4. Sol in alcohol; slightly sol in acetone. Practically insol in ether. *Keep well closed.*

USE: In chromizing; manuf of Cr metal and compds; as catalyst for polymerization of olefins and other organic reactions; as textile mordant; in tanning; in corrosion inhibitors; as waterproofing agent.

2279. Chromic Fluoride. CrF_3; mol wt 108.99. Cr 47.71%, F 52.29%. Prepd by heating $CrCl_3$ in a current of HF: Kwasnik in *Handbook of Preparative Inorganic Chemistry* vol. 1, G. Brauer, Ed. (Academic Press, New York, 2nd ed., 1963) p 257. Review of chromium halides: Fergusson in *Halogen Chemistry* vol. 3, V. Gutmann, Ed. (Academic Press, New York, 1967) pp 227-302.

Dark green needles. d 3.8. mp 1100°. Sublimes at 1100-1200°. Practically insol in water, alcohol. Sol in HCl with violet color.

Trihydrate, *triaquochromium trifluoride*. Green crystals from solns of Cr or $Cr(OH)_3$ in hydrofluoric acid. Sparingly sol in water.

Other hydrates or Werner complexes are discussed in *Chromium* vol. 1, M. J. Udy, Ed., A.C.S. Monograph Series no. 132 (Reinhold, New York, 1956) p 183.

USE: The hydrates in printing and dyeing woolens; coloring and hardening marble; mothproofing woolen fabrics; treating silk; polishing metals; halogenation catalyst.

2280. Chromic Formate. $C_3H_3CrO_6$; mol wt 187.05. C 19.26%, H 1.62%, Cr 27.80%, O 51.32%. $Cr(HCOO)_3$. Prepd from $Cr(OH)_3$ and formic acid: Akhmedli, Negretov, *J. Gen. Chem. USSR* **20**, 2045 (1950).

Hexahydrate, fine green crystals or gray-green powder. Dec above 300°, evolving CO and CO_2. Sol in water; concd aq soln is blue under incident light, red under transmitted light; dil aq soln is green.

USE: In printing cotton in skeins; in leather tanning and waterproofing.

2281. Chromic Hydroxide. *Chromium hydroxide;* chromic oxide gel; chromic oxide, hydrous. CrH_3O_3; mol wt 103.02. Cr 50.47%, H 2.94%, O 46.59%. $Cr(OH)_3$. Occurs only as hydrates. Prepn: Ruthroff, *Inorg. Syn.* **2**, 190 (1946); Hein, Herzog in *Handbook of Preparative Inorganic Chemistry* vol. 2, G. Brauer, Ed. (Academic Press, New York, 2nd ed., 1965) pp 1345-1346.

Trihydrate, $Cr(OH)_3 \cdot 3H_2O$, or $Cr_2O_3 \cdot 9H_2O$, blue-green powder. Practically insol in water. Sol in dil mineral acids when freshly prepd, giving blue or green soln; becomes insol in acids on aging.

$Cr(OH)_3 \cdot nH_2O$, shining, vitrous, jet-black particles. Useful as catalyst in dehydrogenation of alcohols and paraffins, hydrogenation of olefins.

USE: As pigment; in tanning industry; as mordant; as catalyst for organic reactions.

2282. Chromic Nitrate. CrN_3O_9; mol wt 238.01. Cr 21.85%, N 17.65%, O 60.50%. $Cr(NO_3)_3$. Prepn of anhydr salt from N_2O_5 and $Cr(CO)_6$: Addison, Chapma, *J. Chem. Soc.* **1964**, 539; prepn of nonahydrate from $Cr(OH)_3$ and dil HNO_3 or by reducing CrO_3 in the presence of HNO_3: *Chromium* vol. 1, M. J. Udy, Ed., A.C.S. Monograph Series, no. **132** (Reinhold, New York, 1956) pp 203-204. Toxicity study: Smyth *et al., Am. Ind. Hyg. Assoc. J.* **30**, 470 (1969).

Pale green, extremely deliquesc powder. Non-volatile. Dec above 60°. Sol in water, ethyl acetate, DMSO. Practically insol in benzene, CCl_4, $CHCl_3$. Reacts with ether, sometimes vigorously.

Nonahydrate, deep violet, rhombic, monoclinic crystals. mp about 60°; dec above 100°. Sol in water, alcohol. Aq soln slowly becomes green on heating and rapidly recovers the reddish-violet color on cooling. LD_{50} orally in rats: 3.25 g/kg (Smyth).

USE: Prepn of Cr catalyst; in textile printing; corrosion inhibitor.

2283. Chromic Oxide. Anadonis green; chrome green; chrome ocher; chrome oxide green; chromia; chromium sesquioxide; green cinnabar; green oxide of chromium; green rouge; leaf green; oil green; ultramarine green; C.I. Pigment Green 17; C.I. 77288. Cr_2O_3; mol wt 151.99. Cr 68.42%, O 31.58%. Prepd by reaction of sodium dichromate or chromate with sulfur: Copson in *Chromium* vol. 1, M. J. Udy, Ed., ACS Monograph Series no. **132** (Reinhold, New York, 1956) pp 277-278. *Review:* Wiesburg, *Paint Ind. Mag.* **71**(2), 11 (1956). *See also Colour Index* vol. 4 (3rd ed., 1971) p 4662. Use as catalyst: R. Uma, J. C. Kuriacose, *Indian Chem. Mfr.* **8**, 11 (1970). Cytotoxic effects: V. Bianchi *et al., Toxicology* **17**, 219 (1980).

Light to dark green, fine, hexagonal crystals. mp about 2435°. bp about 3000°. d^{25} 5.22. Turns brown on heating but reverts to green color on cooling. Cryst Cr_2O_3 is extremely hard; will scratch quartz, topaz, zircon. Practically insol in water, alc, acetone. Slightly sol in acids, alkalies.

Caution: Trivalent chromium may cause skin irritation: S. Fregert, H. Rorsman, *Arch. Derm.* **90**, 4 (1964).

USE: In abrasives, refractory materials, electric semiconductors; as pigment, particularly in coloring glass; in alloys; printing fabrics and banknotes; as catalyst for organic and inorganic reactions.

2284. Chromic Phosphate. CrO_4P; mol wt 146.97. Cr 35.38%, O 43.55%, P 21.08%. $CrPO_4$. Prepn: Ness *et al., J. Am. Chem. Soc.* **74**, 4685 (1952); Eickhoff, Kebrich, U.S. pat. **2,749,214** (1956 to National Lead); Wegenknecht, Ger. pats. **1,046,597** (1958); **1,056,104** (1959).

Gray-brown to black crystals of amorphous solid. Does not melt by 1800°. $d^{32.5}$ 2.94. Partially oxidizes to CrO_3 on heating in air. Practically insol in water, acetic acid, HCl, aqua regia.

Hemiheptahydrate, *Arnaudon's green, Plessy's green.* Blue-green powder. d 2.15. Practically insol in water. Sol in acids.

Hexahydrate, violet crystals. Loses water gradually on heating, becoming anhydr after one hour at 800° or 3-4 hrs at 500°. d^{14} 2.121. Practically insol in water. Slightly sol in acetic acid solns; readily sol in mineral acids, alkalis, oxalic acid solns.

Radioactive form, $Cr^{32}PO_4$, *Phosphocol.*

USE: Green pigment; in wash primers; in catalysts for dehydrogenation of hydrocarbons or polymerization of olefins.

THERAP CAT: Radioactive form as antineoplastic (radioactive agent).

2285. Chromic Potassium Oxalate. *Tripotassium tris-(ethanedioato)chromate(3—); tripotassium tris(oxalato)chromate(3—);* potassium trioxalatochromate(III); potassium chromic oxalate. $C_6CrK_3O_{12}$; mol wt 433.35. C 16.63%, Cr 12.00%, K 27.07%, O 44.30%. $K_3[Cr(C_2O_4)_3]$. Prepd by treatment of oxalic acid and $K_2C_2O_4$ with $K_2Cr_2O_7$: Bailar, Jr., Jones, *Inorg. Syn.* **1**, 37 (1939); Hein, Herzog in *Handbook of Preparative Inorganic Chemistry* vol. 2, G. Brauer, Ed. (Academic Press, New York, 2nd ed., 1965) p 1372.

Trihydrate, *potassium trioxalatotriaquochromate(III).* Black-green, monoclinic scales with transparent blue edges. Freely sol in water; practically insol in alcohol.

USE: In tanning industry; dyeing chromate colors on wool.

2286. Chromic Potassium Sulfate. Potassium chromic sulfate; potassium disulfatochromate(III). $CrKO_8S_2$; mol wt 283.22. Cr 18.36%, K 13.80%, O 45.19%, S 22.64%. KCr-$(SO_4)_2$. Dodecahydrate produced by reduction of $K_2Cr_2O_7$ with SO_2: Copson in *Chromium* vol. 1, M. J. Udy, Ed., A.C.S. Monograph Series no. 132 (Reinhold, New York, 1956) p 281; electrolytic manuf: Nishihara *et al.*, **Japan.** pat. 2164'(60), *C.A.* **55**, 5200e (1961).

Dodecahydrate, *chrome alum,* $K[Cr(SO_6H_4)_2(H_2O)_2]$.-$6H_2O$: Duval, *Chim. Anal. (Paris)* **44**, 102 (1962), *C.A.* **57**, 9479d (1962). Large, violet-red to black, octahedral, cubic crystals; ruby-red under transmitted light. d^{25} 1.83. mp 89°; at 400° loses all its H_2O. Sol in 4 parts cold, 2 parts boiling water; practically insol in alcohol. The aq soln is violet when cold, green when hot. The violet color returns after a few weeks at room temp.

USE: Mordant for dyeing fabrics uniformly; tanning leather; printing calico; rendering glue and gum insol; manuf ink, other chromium salts; waterproofing fabrics; hardening photographic emulsions.

2287. Chromic Sulfate. $Cr_2O_{12}S_3$; mol wt 392.18. Cr 26.52%, O 48.95%, S 24.53%. $Cr_2(SO_4)_3$. Prepn of anhydr salt by dehydration of hydrated forms: Rollinson, Bailar, Jr., *Inorg. Syn.* **2**, 197 (1946).

Peach-colored solid. d 3.012. Practically insol in water and acids. MLD i.v. in mice: 247 mg/kg, *Handbook of Toxicology* vol. 1, W. S. Spector, Ed. (Saunders, Philadelphia, 1956) pp 70-71.

Hydrates are known in both green and violet modifications, and have several degrees of hydration up to $18H_2O$: Lukaszewski, Redfern, *Nature* **190**, 805 (1961); Udy in *Chromium* vol. 1, M. J. Udy, Ed., A.C.S. Monograph Series no. 132 (Reinhold, New York, 1956) pp 213-217, 288. The technical product comes in the form of a finely granular, dark-green flake or powder approximating the formula Cr_2-$(SO_4)_3$.$10H_2O$. Readily sol in water; almost insol in alc.

Basic chromic sulfates of the type $Cr(OH)SO_4$.nH_2O are of importance in the tanning industry: Udy, *loc. cit.* and pp 278-280, 305-308. Technical grades are available in two degrees of basicity, one-third and one-half, as finely granular dark-green flakes or powder contg about 25% Cr_2O_3. Readily sol in water.

USE: Insolubilization of gelatin; in catalyst prepn; as mordant in textile industry; in tanning of leather; in chrome plating; in manuf of Cr, CrO_3, and Cr alloys; to improve dispersibility of vinyl polymers in water; in manuf of green varnishes, paints, inks, glazes for porcelain.

2288. Chromium. Cr; at. wt 51.9961; at. no. 24; valences 1-6. Group VIB(6). Four naturally occurring isotopes: 50 (4.31%); 52 (83.76%); 53 (9.55%); 54 (2.38%). Artificial radioactive isotopes: 45-49; 51; 55-57; longest-lived isotope is ^{51}Cr ($T_{1/2}$ 27.704 days) prepd by (n,γ) reaction from ^{50}Cr. Abundance in earth's crust: 122 ppm. Isoln from *crocoite* ($PbCrO_4$): L. N. Vauquelin, *J. Mines* **6**, ser. 1, 737 (1787);

idem, Ann. Chemie **70**, 70 (1809). Commercial sources obtained from chrome ore, *chromite* ($FeO.Cr_2O_3$). Reviews of chromium, its alloys and compds: *Chromium*, M. J. Udy, Ed., A.C.S. Monograph Series, no. **132** (Reinhold, New York, 1956) vol. 1, 433 pp; vol. 2, 402 pp; C. L. Rollinson, "Chromium, Molybdenum and Tungsten" in *Comprehensive Inorganic Chemistry* vol. 3, J. C. Bailar, Jr. *et al.*, Eds. (Pergamon Press, Oxford, 1973) pp 623-700; *Chemistry of the Elements*, N. N. Greenwood, A. Earnshaw, Eds. (Pergamon Press, New York, 1984) pp 1167-1210; J. H. Westbrook in Kirk-Othmer *Encyclopedia of Chemical Technology* vol. **6** (Wiley-Interscience, New York, 4th ed., 1993) pp 228-263; B. J. Page, G. W. Loar, *ibid.* pp 263-311. Review of biological function of the Cr(III) ion as essential trace element: Mertz, *Physiol. Rev.* **49**, 163-239 (1969). Review of carcinogenic risk: *IARC Monographs* **2**, 100-125 (1973); *ibid.* **23**, 205-323 (1980); of metabolism and toxicity: M. D. Cohen *et al., Crit. Rev. Toxicol.* **23**, 255-281 (1993); of toxicology and human exposure: *Toxicological Profile for Chromium* (PB93-182434, 1993) 250 pp. Books: *Chromium: Metabolism and Toxicity*, D. Burrows, Ed. (CRC Press, Boca Raton, 1983) 172 pp; *Chromium in the Natural and Human Environments*, J. O. Nriagu, E. Nieboer, Eds. (John Wiley & Sons, New York, 1988) 571 pp.

Steel-gray, lustrous metal; body-centered cubic structure; hard as corundum and less fusible than platinum. Takes a high polish. mp 1903 ±10°. bp 2642°. d^{20} 7.14. Heat capacity (25°): 5.58 cal/mol/deg C. Heat of fusion: 3.5 kcal/mol; heat of vaporization: 81.7 kcal/mol (at bp); heat of sublimation (25°): 94.8 kcal/mol (Rollinson). Also reported as mp 1875°. bp 2680°. d^{20} 7.19. Specific heat (25° C): 23.9 J/mol/deg K. Heat of fusion: 14.6 kJ/mol; heat of vaporization: 305 kJ/mol (at bp) (Westbrook). Resistant to common corroding agents, highly acid resistant. Reacts with dil HCl, H_2SO_4; not with HNO_3. Resists atmospheric attack at ambient temperatures.

Caution: Potential symptom of overexposure to chromium metal by inhalation is histologic fibrosis of the lungs. See *NIOSH Pocket Guide to Chemical Hazards* (DHHS/NIOSH 90-117, 1990) p 70. Chromic acid and chromate salts are irritating to exposed tissues. Toxic effects due to overexposure may include dermatitis, skin ulcers; nasal inflammation, perforation of nasal septum; cancer of the lungs, nasal cavity and paranasal sinus. See *Patty's Industrial Hygiene and Toxicology* vol. 2C, G. D. Clayton, F. E. Clayton, Eds. (John Wiley & Sons, New York, 4th ed., 1994) p 1973-1985. Chromium and certain chromium compounds have been listed as known carcinogens: *Seventh Annual Report on Carcinogens* (PB95-109781, 1994) p 46.

USE: In manuf of chrome-steel or chrome-nickel-steel alloys (stainless steel), nonferrous alloys, heat resistant bricks for refractory furnaces. To greatly increase strength, hardness and resistance of metals to abrasion, corrosion and oxidation. For chrome plating of other metals; leather tanning; as pigment and mordant; wood preservative. Use of ^{51}Cr as diagnostic aid *see* sodium chromate(VI).

2289. Chromium Carbonyl. Chromium hexacarbonyl. C_6CrO_6; mol wt 220.06. C 32.75%, Cr 23.63%, O 43.62%. $Cr(CO)_6$. Prepn from Cr salts and CO in the presence of a Grignard reagent: Owen *et al., Inorg. Syn.* **3**, 156 (1950); in the presence of Mg and ether: Wender, **U.S.** pat. **3,012,858** (1961 to Diamond Alkali); in the presence of Na and diglyme: Podall *et al., J. Am. Chem. Soc.* **83**, 2057 (1961); in the presence of Na and an aromatic hydrocarbon: Pruett, Wyman, **U.S.** pat. **3,053,629** (1962 to Union Carbide); in the presence of I_2 and a nitrile: Wotiz, **U.S.** pat. **3,100,687** (1963 to Diamond Alkali). Toxicity study: Strohmeier, *Z. Naturforsch.* **19b**, 540 (1964).

Orthorhombic, highly refractive crystals. Sublimes at room temp. Sinters at 90°; dec at 130°; explodes at 210°. Burns with a luminous flame. d^{18} 1.77. Vapor pressure (mm): 0.04 (0°); 1.0 (48°); 66.5 (100°). Almost insol in water, ethanol, methanol. Sol in ether, $CHCl_3$, other organic solvents. Solns or impure solid dec by light. LD_{50} i.v. in mice: 100 mg/kg (Strohmeier).

USE: In catalysts for olefin polymerization and isomerization; gasoline additive to increase octane number; prepn of chromous oxide, CrO.

2290. Chromium Dioxide. CrO_2; mol wt 83.99. Cr 61.90%, O 38.10%. Prepn: Wöhler, *Ann.* **111**, 117 (1859); Thamer *et al., J. Am. Chem. Soc.* **79**, 547 (1957); Swoboda *et al., J. Appl. Phys.* **32**, Suppl. no. 3, 374 (1961); Arthur; Arthur, Ingraham, U.S. pats. **2,959,955; 3,117,093** (1960, 1964 both to du Pont). *Reviews:* Hund, *Farbe + Lack* **78**, 11-16 (1972); Rollinson in *Comprehensive Inorganic Chemistry* vol 3, J. C. Bailar, Jr. *et al.*, Eds. (Pergamon Press, Oxford, 1973) pp 689-690.

Black, ferromagnetic crystals; rutile structure. d 4.89. Metastable in air; various temperatures (250-500°) reported for decompn to Cr_2O_3.

USE: In magnetic recording tapes; as catalyst.

2291. Chromium Picolinate. *Tris(2-pyridinecarboxylato-N^1,O^2)chromium;* chromium(III) trispicolinate. $C_{18}H_{12}$-CrN_3O_6; mol wt 418.31. C 51.68%, H 2.89%, Cr 12.43%, N 10.05%, O 22.95%. Biologically active form of chromium. General prepn, chemical and biological properties: G. W. Evans, D. J. Pouchnik, *J. Inorg. Biochem.* **49**, 177 (1993). Clinical effect on insulin metabolism: G. W. Evans, *Int. J. Biosocial Med. Res.* **11**, 163 (1989). Clinical evaluation of lipid-lowering activity: R. I. Press *et al., West. J. Med.* **152**, 41 (1990). Clinical pharmacokinetics: M. L. Gargas *et al., Drug Metab. Disposit.* **22**, 522 (1994). Use as veterinary diet supplement: M. D. Lindemann *et al., J. Animal Sci.* **73**, 457 (1995). Review of clinical pharmacology: M. F. McCarty, *J. Appl. Nutr.* **43**, 58-66 (1991); *idem, J. Optimal Nutr.* **2**, 36-53 (1993).

Lipophilic. Soly: water (pH 7.0) 0.6 mM, chloroform 2.0 mM. uv max (water): 264 nm (a_M 15546 L mol^{-1}cm^{-1}).

USE: Nutritional supplement.

2292. Chromium Tetrafluoride. CrF_4; mol wt 127.99. Cr 40.63%, F 59.37%. Prepd by reaction of F_2 with Cr or $CrCl_3$: von Wartenberg, *Z. Anorg. Allgem. Chem.* **247**, 136 (1941); Clark, Sadana, *Can. J. Chem.* **43**, 50 (1964).

Very dark greenish-black, amorphous solid; on exposure to moist air becomes brown on surface due to hydrolysis. mp (estimated) 277°: Fergusson in *Halogen Chemistry* Vol. 3, V. Gutmann, Ed. (Academic Press, New York, 1967) p 242. bp ~400° evolving steel-blue vapor; sublimes *in vacuo* above 100°; d 2.89. Less reactive than CrF_3, does not readily form Werner complexes; does not react with NH_3, SO_2, SO_3, pyridine. Sol in water with rapid hydrolysis; practically insol in most organic solvents. Can be stored indefinitely *in vacuo* in well dried Pyrex or silica vessels.

2293. Chromium Trioxide. Chromic acid; chromic anhydride. CrO_3; mol wt 99.99. Cr 52.00%, O 48.00%. Produced commercially by the action of concd H_2SO_4 on a soln of chromate or dichromate: Faith, Keyes & Clark, *Industrial Chemicals*, F. A. Lowenheim, M. K. Moran, Eds. (Wiley-Interscience, New York, 4th ed., 1975) pp 270-274.

Dark red, deliquesc bipyramidal prismatic crystals, flakes or granular power. d 2.70. mp 197°. Dec at 250° to Cr_2O_3 and O_2. Very sol in water; sol in H_2SO_4. Powerful oxidizer; oxidizes alcohol and most other organic substances, sometimes with dangerous violence. *Contact with combustible material may cause fire.*

Caution: Potential symptoms of overexposure are respiratory system irritation, nasal septum perforation; liver and kidney damage; eye injury, conjunctivitis; skin ulcers, sensitization dermatitis. *See NIOSH Pocket Guide to Chemical Hazards* (DHHS/NIOSH 90-117, 1990) p 70. Chronic

overexposure may lead to severe liver damage, CNS involvement and lung cancer. *See Clinical Toxicology of Commercial Products*, R. E. Gosselin *et al.*, Eds. (Williams & Wilkins, 5th ed., 1984) Section II, p 108. *See also Patty's Industrial Hygiene and Toxicology* vol. 2C, G. D. Clayton, F. E. Clayton, Eds. (Wiley-Interscience, New York, 4th ed., 1994) p 1973-1985. This substance has been listed as a known carcinogen: *Seventh Annual Report on Carcinogens* (PB95-109781, 1994) p 46.

USE: Chromium plating; copper stripping; aluminum anodizing; corrosion inhibitor; photography; purifying oil and acetylene; hardening microscopical prepns; oxidizing agent in organic chemistry.

THERAP CAT (VET): Has been used in solution as a topical antiseptic and astringent.

2294. Chromocarb. *4-Oxo-4H-1-benzopyran-2-carboxylic acid;* 2-chromonecarboxylic acid; 4-oxo-4H-chromene-2-carboxylic acid; benzo-γ-pyronecarboxylic acid. $C_{10}H_6O_4$; mol wt 190.16. C 63.16%, H 3.18%, O 33.66%. Prepn: S. Ruhemann, H. E. Stapleton, *J. Chem. Soc.* **77**, 1179 (1900); J. Schmutz *et al., Helv. Chim. Acta* **34**, 767 (1951); G. Pifferi *et al., J. Heterocycl. Chem.* **14**, 1257 (1977). Prepn of diethylamine salt: P. A. Tronche, S. Afr. pat. **68 7352;** *eidem,* U.S. pat. **3,816,470** (1969, 1974 both to Ferlux). Pharmacology of diethylamine salt in animals: J. Couquelet *et al., C.R. Soc. Biol.* **164**, 329 (1970); P. Conquet *et al., ibid.* 800. Clinical comparison with dipyridamole of effect on platelet function: A. Vittoria *et al., Curr. Ther. Res.* **35**, 1033 (1984). Clinical trial in diabetes with vascular disease: N. Ciavarella *et al., ibid.* **36**, 293 (1984). Bioavailability in humans: J.-M. Aiache *et al., Biopharm. Drug Dispos.* **7**, 301 (1986).

Colorless needles from alcohol, mp 250-251° (dec) (Ruhemann, Stapleton); also reported as mp 255-256° (Pifferi). uv max: 230, 305 nm (ε 20220, 8075). Sol in alcohol, ammonia. Sparingly sol in water.

Diethylamine, $C_{14}H_{17}NO_4$, *Angiophtal, Campel, Fludarene.* Microcrystalline powder from alcohol + acetone, mp 138°. Sol in water. LD_{50} in mice: ~800 mg/kg i.v.; > 5 g/kg orally (Tronche, 1974).

THERAP CAT: Diethylamine salt as capillary protectant.

2295. Chromomycins. An antibiotic complex produced by *Streptomyces griseus;* composed of chromomycins group A, group B, and C, with chromomycin A_3 as the major component. Isoln of complex and antibacterial activity: M. Shibata *et al., J. Antibiot.* **13B**, 1 (1960), *C.A.* **54**, 22835g (1960). Separation of components: Mizuno, and Tatsuoka *et al., ibid.* **13B**, 329, 332 (1960), *C.A.* **56**, 1745gi (1962); Miyake *et al.*, **Japan.** pat. **7842('60)** and **Ger.** pat. **1,072,775** (both 1960 to Takeda), *C.A.* **55**, 7758b, 26361f (1961); Mizuno, *J. Antibiot.* **16A**, 22 (1963). Structure of A_3: Miyamoto *et al., Tetrahedron Letters* **1964**, 2367, 2371; Tatsuoka *et al., Proc. Japan Acad.* **40**(3), 236 (1964), *C.A.* **61**, 10767h (1964); Miyamoto *et al., Tetrahedron* **23**, 421 (1967). Abs config of A_3: Harada *et al., J. Am. Chem. Soc.* **91**, 5896 (1969). Revised structure of A_3: J. Thiem, B. Meyer, *J. Chem. Soc. Perkin Trans. II* **1979**, 1331. Series of articles on partial syntheses: *Ber.* **113**, 3039-3074 (1980); J. H. Dodd *et al., J. Org. Chem.* **47**, 4045 (1982). Pharmacology and toxicity: M. Slavik, S. K. Carter, *Adv. Pharmacol. Chemother.* **12**, 1 (1975). *Review:* J. D. Skarbek, M. K. Speedie in *Antitumor Compounds of Natural Origin* vol. 1, A. Aszalos, Ed. (CRC Press, Boca Raton, 1981) pp 191-235.

Chromomycin A₃

Chromomycin A₃, $C_{57}H_{82}O_{26}$, *3ᴮ-O-(4-O-acetyl-2,6-dideoxy-3-C-methyl-α-L-arabino-hexopyranosyl)-7-methylolivomycin D, aburamycin B, Toyomycin.* Yellow powder, dec. 185°. $[\alpha]_D^{23}$ —55° (ethanol). uv max (ethanol): 230, 281, 304, 318, 330, 412 nm (log ε 4.39, 4.72, 3.85, 3.92, 3.84, 4.07). LD_{50} in mice (mg/kg): 1.85 i.v.(Slavik, Carter); 1.7 i.p. (Shibata). THERAP CAT: Antineoplastic.

2296.   Chromonar. *[[3-[2-(Diethylamino)ethyl]-4-methyl-2-oxo-2H-1-benzopyran-7-yl]oxy]acetic acid ethyl ester;* 3-(β-diethylaminoethyl)-4-methyl-7-(carbethoxymethoxy)-coumarin; 3-(β-diethylaminoethyl)-4-methyl-7-(carbethoxymethoxy)-2H-1-benzopyran-2-one; ethyl [[3-[2-(diethylamino)ethyl]-4-methyl-2-oxo-2H-1-benzopyran-7-yl]oxy]acetate; carbochromen; carbocromen. $C_{20}H_{27}NO_5$; mol wt 361.44. C 66.46%, H 7.53%, N 3.88%, O 22.13%. Prepn: **Belg. pat. 621,327;** Ritter *et al.,* **U.S. pat. 3,282,938** (1963, 1966 both to Casella Farbwerke Mainkur). Toxicity study: R. E. Nitz, E. Potzch, *Arzneimittel-Forsch.* **13,** 243 (1963). Metabolic studies: Schraven *et al., ibid.* **20,** 1905 (1970). Series of articles on pharmacology: *ibid.* **13,** 243-268 (1963); on distribution: *ibid.* **22,** 479-511 (1972).

Practically insol in water.
Hydrochloride, $C_{20}H_{27}NO_5$·HCl, *Cassella 4489, Antiangor, Intenkordin, Intensain.* Crystalline powder, mp 159-160°. Freely sol in water, alc, methylene chloride, chloroform. Sparingly sol in acetone, methyl ethyl ketone, benzene, ether. Aq solns show blue fluorescence. LD_{50} in mice (g/kg): 6.3 orally; 0.528 i.p.; 0.0355 i.v. (Nitz, Potzch). THERAP CAT: Vasodilator (coronary).

2297.   Chromotrope 2B. *4,5-Dihydroxy-3-[(4-nitrophenyl)azo]-2,7-naphthalenedisulfonic acid disodium salt; C.I. Acid Red 176;* p-nitrobenzeneazochromotropic acid sodium salt; C.I. 16575. $C_{16}H_9N_3Na_2O_{10}S_2$; mol wt 513.37. C 37.43%, H 1.77%, N 8.19%, Na 8.96%, O 31.17%, S 12.49%. Prepn from diazotized p-nitroaniline and chromotropic acid, *see: Colour Index* **vol. 4** (3rd ed., 1971) p 4097.

Reddish-brown powder. Sol in water with a yellowish-red color. Insol in alcohol.
USE: As a dye; as a reagent for boric acid or borates.

2298.   Chromotropic Acid. *4,5-Dihydroxy-2,7-naphthalenedisulfonic acid;* 1,8-dihydroxynaphthalene-3,6-disulfonic acid. $C_{10}H_8O_8S_2$; mol wt 320.30. C 37.50%, H 2.52%, O 39.96%, S 20.02%. Prepn from 4-chloro-5-hydroxy-2,7-naphthalenedisulfonic acid: **Ger. pat. 147,852** (1904 to BASF); alternate methods: *Frdl.* **3,** 460-466.

Dihydrate, white needles, soluble in water.
Disodium salt dihydrate, $C_{10}H_6Na_2O_8S_2$·$2H_2O$, white needles or leaflets. Very sol in water.
USE: Analytical reagent; intermediate for azo dyes.

2299.   Chromous Acetate. $C_4H_6CrO_4$; mol wt 170.09. C 28.25%, H 3.56%, Cr 30.57%, O 37.63%. $Cr(C_2H_3O_2)_2$. Prepn of monohydrate: Balthis, Jr., Bailar, Jr., *Inorg. Syn.* **1,** 122 (1939); Hatfield, *ibid.* **3,** 148 (1950); Kranz, Witkowska, *ibid.* **6,** 144 (1960); Ocone, Block, *ibid.* **8,** 125 (1966). Toxicity study: H. F. Smyth *et al., Am. Ind. Hyg. Assoc. J.* **30,** 470 (1969).
Monohydrate, deep red powder, or monoclinic crystals; composed of dimeric units. Easily oxidized, especially when moist, to chromic acetate. Loses H_2O when dried over P_2O_5 at 100°, changing color to brown. d 1.79. Slightly sol in cold water; readily sol in hot water; sol in and reacts with most acids; slightly sol in alcohol. Practically insoluble in ether. *Keep tightly closed.* LD_{50} orally in rats: 11.26 g/kg (Smyth).
USE: Prepn of other chromous salts; as O_2 absorber in gas analysis.

2300.   Chromous Bromide. Br_2Cr; mol wt 211.80. Br 75.45%, Cr 24.55%. $CrBr_2$. Prepn: Hein, Herzog in *Handbook of Preparative Inorganic Chemistry,* **Vol. 2,** G. Brauer, Ed. (Academic Press, New York, 2nd ed. 1965) p 1340. Review of chromium halides: Fergusson in *Halogen Chemistry,* **Vol. 3,** V. Gutmann, Ed. (Academic Press, New York, 1967) pp 227-302.
White monoclinic crystals, becoming yellow when heated. mp 842°; d_4^{25} 4.236. Stable in dry air, oxidizes in moist air. Sol in water (exothermic) giving blue soln; sol in alcohol. *Keep well closed.*
USE: In chromizing.

2301.   Chromous Chloride. Cl_2Cr; mol wt 122.90. Cl 57.69%, Cr 42.31%. $CrCl_2$. Prepn: Burg, *Inorg. Syn.* **3,** 150 (1950); Hein, Herzog, in *Handbook of Preparative Inorganic Chemistry,* **Vol. 2,** G. Brauer, Ed. (Academic Press, New York, 2nd ed. 1965) pp 1336-1338. Prepn of tetrahydrate: Balthis, Jr., Bailar, Jr., *Inorg. Syn.* **1,** 125 (1939); Holah, Fackler, *ibid.* **10,** 26 (1967). Toxicity study: H. F. Smyth *et al., Am. Ind. Hyg. Assoc. J.* **30,** 470 (1969).
Lustrous needles or fused, fibrous mass.   mp 824°; d_4^{14} 2.751. Very hygroscopic; stable in dry air but oxidizes rapidly if moist. Powerful reducing agent. Soluble in water giving a blue soln. *Keep well closed.* LD_{50} orally in rats: 1.87 g/kg (Smyth).
Tetrahydrate, *tetraaquochromium dichloride.* Bright blue, hygroscopic crystals. Above 38° changes to isomeric green modification. Loses $1H_2O$, forming trihydrate, at 51°. Absorbs O_2 even when dry, forming a greenish-black oxychloride. Sol in water. Almost insol in concd HCl. On

standing in soln it is oxidized by the water with liberation of H_2. *Keep well closed.*

USE: In chromizing; in prepn of Cr metal; in catalysts for organic reactions; as O_2 absorbent; in analysis.

2302. Chromous Fluoride. CrF_2; mol wt 89.99. Cr 57.78%, F 42.22%. Prepd from $CrCl_2$ + HF: Poulenc, *Compt. Rend.* **116**, 254 (1893); Kwasnik in *Handbook of Preparative Inorganic Chemistry*, Vol. 1, G. Brauer, Ed. (Academic Press, New York, 2nd ed., 1963) pp 256-257. Review of other preparative methods: Sturm, *Inorg. Chem.* **1**, 665-672 (1962).

Blue-green, monoclinic crystals with iridescent sheen. d 3.79. mp 894°. Sparingly sol in water; practically insol in alcohol; sol in boiling HCl; not attacked by hot dil H_2SO_4 or HNO_3. Changes to Cr_2O_3 when heated in air.

USE: In chromizing; in catalytic cracking of hydrocarbons; alkylation catalyst; in nuclear reactor fuels.

2303. Chromous Formate. $C_2H_2CrO_4$; mol wt 142.03. C 16.91%, H 1.42%, Cr 36.61%, O 45.06%. $Cr(HCOO)_2$. Prepd from $CrCl_2$ and Na formate: Lux, Illman, *Ber.* **91**, 2143 (1958); Earnshaw *et al., Proc. Chem. Soc.* **1963**, 281. Prepn of blue and red anhydrous forms from hydrated formates: Herzog, Kalies, *Z. Chem.* **5**, 273 (1965).

Hemihydrate, violet to blue crystals. Sol in water.

Monohydrate, red needles. Sol in water to give blue soln.

Hemipentahydrate, large, dark-red cubic crystals. Sol in water, alcohol, ether, acetone.

USE: In baths for Cr electroplating; in catalysts for organic reactions.

2304. Chromous Oxalate. C_2CrO_4; mol wt 140.02. C 17.16%, Cr 37.14%, O 45.71%. CrC_2O_4. Prepn from chromous acetate and oxalic acid: *Chromium*, Vol. 1, M. J. Udy, Ed., A.C.S. Monograph Ser., no. **132** (Reinhold, New York, 1956) p 227; from chromous sulfate and sodium oxalate: Lux, Illmann, *Ber.* **91**, 2143 (1958). Reported as monohydrate (Udy, *loc. cit.*) or dihydrate (Lux, Illman, *loc. cit.*).

Hydrate, yellow to yellowish-green crystalline powder. d 2.468. Most stable of chromous salts. Not appreciably oxidized by moist air. Practically insol in cold water, alcohol; sol in hot water, dil acids.

2305. Chromous Sulfate. CrO_4S; mol wt 148.06. Cr 35.12%, O 43.22%, S 21.66%. $CrSO_4$. Prepn of pentahydrate: Lux, Illman, *Ber.* **91**, 2143 (1958); Holah, Fackler, *Inorg. Syn.* **10**, 26 (1967). The common hydrated form, long considered to be the heptahydrate, has been reported to be the pentahydrate: Lux, Illman, *loc. cit.*; Lux *et al., Ber.* **97**, 503 (1964). Prepn of standard soln for use as an analytical reagent: Lingane, Pecsok, *Anal. Chem.* **20**, 425 (1948).

Pentahydrate, blue crystals. Stable in air if dry. Sol in water; slightly sol in alcohol; practically insol in acetone. Soluble in dil, dec by concd H_2SO_4. Solns are rapidly oxidized by atmospheric oxygen.

USE: Analytical reagent; absorption of O_2 from gas mixtures; dehydrohalogenating and reducing agent.

2306. Chromyl Chloride. *Dichlorodioxochromium;* chromium dioxychloride. Cl_2CrO_2; mol wt 154.90. Cl 45.77%, Cr 33.57%, O 20.66%. CrO_2Cl_2. Prepn from CrO_3 + HCl: Sisler, *Inorg. Syn.* **2**, 205 (1946); Hein, Herzog in *Handbook of Preparative Inorganic Chemistry*, Vol. 2, G. Brauer, Ed. (Academic Press, New York, 2nd ed., 1965) p 1384; and $AlCl_3$: Flesch, Svec, *J. Am. Chem. Soc.* **80**, 3189 (1958); from Cr_2O_3 and $TiCl_4$: Braos, Cohen, U.S. pat. **3,111,380** (1963 to Natl. Distillers & Chem.). Review of chromium halides: Fergusson in *Halogen Chemistry*, Vol. 3, V. Gutmann, Ed. (Academic Press, New York, 1967) pp 227-302.

Deep red liq; appears black under reflected light. Fumes in moist air. mp −96.5°; bp 117°; d_4^{25} 1.91. Slightly less viscous than water. Nonconductor of electricity. Strong oxidizing agent; *can react explosively with combustible organic and inorganic substances.* Hydrolyzes vigorously on contact with water. Reacts vigorously with liq or gaseous ammonia. Sol in CCl_4, CS_2, benzene, nitrobenzene, $CHCl_3$, $POCl_3$. Its soln in CCl_4 is fairly stable. Liquid CrO_2Cl_2 is stable indefinitely in glass, aluminum, stainless steel containers if protected from light and moisture. CrO_2Cl_2 dis-

solves CrO_3, yielding a powerful oxidant. LD_{50} s.c. in mice: ~5.45 mg/kg: *Toxic Substances List,* H. E. Christensen, Ed. (1972) p 145.

Caution: Burns and blisters the skin. *Handle only in well-ventilated hood.*

USE: Catalyst for polymerization of olefins; oxidation of hydrocarbons; in the Etard reaction for production of aldehydes and ketones; in the prepn of various coordination complexes of Cr.

2307. Chromyl Fluoride. *Difluorodioxochromium;* chromium oxyfluoride. CrF_2O_2; mol wt 121.99. Cr 42.62%, F 31.15%, O 26.23%. CrO_2F_2. Prepn of impure product from CrO_2Cl_2 and F_2: von Wartenberg, *Z. Anorg. Allgem. Chem.* **247**, 140 (1941). Prepn from CrO_3 and SeF_4: Bartlett, Robinson, *J. Chem. Soc.* **1961**, 3549; from $K_2Cr_2O_7$ and HF or from Cr, KNO_3 and HF: Wiechert, *Z. Anorg. Allgem. Chem.* **261**, 315 (1950). Review of earlier prepns and prepn from CrO_3 and ClF, COF_2, MoF_6 or WF_6: P. J. Green, G. L. Gard, *Inorg. Chem.* **16**, 1243 (1977). Review of chromium halides: Fergusson in *Halogen Chemistry*, **vol. 3**, V. Gutmann, Ed. (Academic Press, New York, 1967) pp 227-302.

Red-violet to black rhombic or monoclinic crystals; described as a gas or a liq in earlier work with impure samples. Sublimation temp 29.6°; mp 31.6° (885 mm). *Extremely reactive;* etches glass, quartz. Ignites hydrocarbons such as methane or butane at elevated temps; mixtures burn with a bright flame producing fumes of Cr_2O_3 and CrF_3. Can be stored in a sealed aluminum phosphate glass tube or in a Kel-F tube. Stable indefinitely at room temp in the dark. Polymerizes slowly if exposed to visible, ultraviolet or infrared light to a gray-white solid, mp 200°.

USE: Fluorination catalyst; to increase olefin-polymer receptivity for dyes.

2308. Chrysamminic Acid. *1,8-Dihydroxy-2,4,5,7-tetranitro-9,10-anthracenedione;* 1,8-dihydroxy-2,4,5,7-tetranitroanthraquinone; 2,4,5,7-tetranitrochrysazin; chrysammic acid. $C_{14}H_4N_4O_{12}$; mol wt 420.21. C 40.02%, H 0.96%, N 13.33%, O 45.69%. Prepn by heating chrysazin with fuming nitric acid: Liebermann, *Ann.* **183**, 193 (1876). Structure: Robinson, Simonsen, *J. Chem. Soc.* **95**, 1088 (1909).

Golden-yellow, lustrous, bitter leaflets. Almost insol in water; sol in alcohol with deep-red and in ether with yellow color.

Caution: The acid and its salts explode when ignited or heated rapidly in air.

2309. Chrysanthemaxanthin. *(3S,3′R,5R,6′R,8S)-5,8-Epoxy-5,8-dihydro-β,ε-carotene-3,3′-diol;* $C_{40}H_{56}O_3$; mol wt 584.88. C 82.14%, H 9.65%, O 8.21%. Carotenoid pigment. Occurs together with flavoxanthin q.v.. Has same structural formula, mp, rotation, and absorption as flavoxanthin, but differs sterically. Can be separated chromatographically: in a zinc carbonate column the chrysanthemaxanthin zone locates itself below flavoxanthin and above lutein epoxide. Isoln from asters: Karrer, Jucker, *Helv. Chim. Acta* **26**, 626 (1943). Partial synthesis from lutein: *eidem, ibid.* **28**, 300 (1945). For absolute configuration see: H. Cadosch *et al., ibid.* **61**, 783 (1978).

Golden yellow leaflets, mp 184-185°, also reported as mp 178.5-180 (Codosch). $[\alpha]_D^{23}$ +104° ±5% (c = 1.833 mg/ml in chloroform). Absorption max (CHCl$_3$): 459, 430 nm. uv max (ethanol): 204, 251, 298, 311, 399, 422, 448 nm (ϵ 27300, 28800, 10000, 13100, 97700, 150400, 148500). Freely sol in chloroform, benzene, acetone. Less sol in methanol, ethanol. Almost insol in petr ether.

Note: Chrysanthemaxanthin has no vitamin A activity.

2310. Chrysanthemic Acid. *2,2-Dimethyl-3-(2-methyl-1-propenyl)cyclopropanecarboxylic acid;* chrysanthemum-monocarboxylic acid; chrysanthemumic acid. C$_{10}$H$_{16}$O$_2$; mol wt 168.24. C 71.39%, H 9.59%, O 19.02%. Occurs as esters in pyrethrum flowers, *see* Pyrethrin I and Cinerin I, *also* Allethrin (a synthetic product). Isoln and structure: Staudinger, Ruzicka, *Helv. Chim. Acta* **7**, 177, 201 (1924). Synthesis: Staudinger *et al., ibid.* 390; Campbell, Harper, *J. Chem. Soc.* **1945**, 283; M. J. Devos, A. Krief, *Tetrahedron Letters* **1978**, 1845; **1979**, 1891; O. A. Nesmeyanova *et al., Synthesis* **1982**, 296. Asymmetric synthesis: T. Aratani *et al., ibid.* **1977**, 2599. Stereoselective synthesis of *dl-trans*-chrysanthemic acid: Mills *et al., Chem. Commun.* **1971**, 555; *eidem, J. Chem. Soc. Perkin Trans. I* **1973**, 133; S. C. Welch, T. A. Valdes, *J. Org. Chem.* **42**, 2108 (1977); J. P. Genet *et al., Tetrahedron Letters* **21**, 3183 (1980); of *dl-cis*-form: J. Mann, A. Thomas, *ibid.* **27**, 3533 (1986). Synthesis of (−)-*cis*- and (+)-*trans*-forms: Gopichand *et al., Indian J. Chem.* **13**, 433 (1975).

d-trans form

dl-cis-Form, cubic prisms from ethyl acetate, mp 115-116°.

dl-trans-Form, long prisms, mp 54°. Very sol in ethyl acetate. Exhibits a marked negative heat of soln in ethyl acetate or in methanol.

l-trans-Form, elongated prisms, mp 17-21°. $[\alpha]_D^{25}$ −14.01° (c = 1.535 in abs alc).

d-trans-Form, elongated prisms, mp 17-21°. $[\alpha]_D^{25}$ +14.16° (c = 1.554 in abs alc).

2311. Chrysanthenone. *2,7,7-Trimethylbicyclo[3.1.1]-heptan-6-one; 2-pinen-7-one.* C$_{10}$H$_{14}$O; mol wt 150.22. C 79.96%, H 9.39%, O 10.65%. A constituent of the essential oil of *Chrysanthemum sinense* Sabin, *Compositae*. Prepared by photochemical rearrangement of *l*- or *d*-verbenone (*q.v.*): Hurst, Whitham, *J. Chem. Soc.* **1960**, 2864; Erman, *J. Am. Chem. Soc.* **89**, 3828 (1967); Schuster, Widman, *Tetrahedron Letters* **1971**, 3571. Structure: Kotake, Nonaka, *Ann.* **607**, 153 (1957); Blanchard, *Chem. & Ind. (London)* **1958**, 293. Chemistry: Erman, *J. Am. Chem. Soc.* **91**, 779 (1969).

d-Form, herbaceous odor. bp$_{12}$ 88-89°. n_D^{22} 1.4720. $[\alpha]_D$ +37° (c = 2.1 in chloroform). uv max: 290 nm (ϵ 120). Insoluble in water; sol in alcohol.

2312. Chrysarobin. Purified Goa powder; purified araroba. Improperly called *"medicinal chrysophanic acid"*. The name chrysarobin today has two meanings: (1) A commercial product consisting of 70-85% of different anthraquinone derivs extracted from Goa powder (Araroba); (2) A pure substance, (*1,8-dihydroxy-3-methyl-9-anthrone; 3-methyl-1,8,9-anthracenetriol*), reduction product of chrysophanic acid, *q.v.* This pure substance constitutes about 30% of commercial chrysarobin. Isoln of the commercial product from the wood of *Andira araroba* Aguiar [*Vouacapoua araroba* (Aguiar) Lyons], *Leguminosae:* Liebermann, Siedler, *Ann.* **212**, 29 (1882); Hesse; *ibid.* **309**, 53 (1899). Prepn of the pure substance by reduction of chrysophanic acid: Naylor, Gardner, *J. Am. Chem. Soc.* **53**, 4114 (1931).

Commercial product, brownish to orange-yellow, micro-crystalline, odorless, tasteless powder. Irritating to mucous membranes; causes dangerous inflammation if it enters the eye. Very slightly sol in water; 1 g dissolves in 385 ml alcohol, 30 ml benzene, 15 ml chloroform, 160 ml ether, 180 ml carbon disulfide; also sol in fats. Dissolves in alkali hydroxides or in H$_2$SO$_4$ with a red color and is pptd from the H$_2$SO$_4$ soln by diluting with H$_2$O.

Pure substance, C$_{15}$H$_{12}$O$_3$, yellow needles from glacial acetic acid, mp 203.4-204°.

Caution: Ingestion may cause renal damage, severe gastroenteritis. Application to large areas of skin may cause renal irritation by percutaneous absorption. Inadvertent conjunctival contact frequently produces conjunctivitis. Allergic reaction of skin has been reported.

THERAP CAT: Has been used as antipsoriatic.

THERAP CAT (VET): Has been used in ringworm, and noninfectious diseases of the skin.

2313. 6-Chrysenamine. 6-Chrysenylamine; 6-amino-chrysene; 6-chrysylamine; CP-1001; Chrysenex. C$_{18}$H$_{13}$N; mol wt 243.31. C 88.86%, H 5.39%, N 5.76%. Prepd by reduction of 6-nitrochrysene: Newman, Cathcart, *J. Org. Chem.* **5**, 620 (1940).

Leaflets from alc, mp 210-211°. Slightly sol in alcohol, benzene, ethyl acetate.

USE: In biochemical research to produce leukopenia.

2314. Chrysene. 1,2-Benzphenanthrene. C$_{18}$H$_{12}$; mol wt 228.29. C 94.70%, H 5.30%. Occurs in coal tar. Is formed during distillation of coal, in very small amount during distillation or pyrolysis of many fats and oils. Isoln from coal tar: Liebermann, *Ann.* **158**, 299, 307 (1871). Purification by chromatography: Winterstein, Schön, *Z. Physiol. Chem.* **230**, 146 (1934); Winterstein *et al., ibid.* 158. Synthesis by heating H$_2$ and acetylene to 800°: Meyer, *Ber.* **45**, 1633 (1912). From cholesterol on heating with palladium charcoal or activated charcoal: Schmid, Zentner, *Monatsh.* **49**, 96 (1928).

Orthorhombic bipyramidal plates from benzene. d$_4^{20}$ 1.274. mp 254°. Sublimes easily *in vacuo.* bp 448°. Strong fluorescence under ultraviolet light. Absorption spectrum: Marchlewski, Moroz, *Bull. Soc. Chim. France* [4] **33**, 1406 (1923). Slightly sol in alc, ether, carbon bisulfide, glacial acetic acid. At 25° one gram dissolves in 1300 ml abs alc, 480 ml toluene. About 5% is sol in toluene at 100°. Moderately sol in boiling benzene. Insol in water. Chrysene is generally only slightly sol in cold organic solvents, but fairly sol in these solvents when hot, including glacial acetic acid.

2315. Chrysergonic Acid. C$_{32}$H$_{30}$O$_{14}$; mol wt 638.58. C 60.19%, H 4.74%, O 35.08%. Yellow pigment from ergot. Diastereoisomer of secalonic acids, *q.v.* For isoln and structure *see* secalonic acids.

Pale yellow needles from chloroform, mp 268-270°; pale yellow plates from glacial acetic acid, mp 250-257°. $[\alpha]_D^{20}$ −3° to +34° (c = 1.11 in pyridine). uv max (glacial acetic acid): 335 nm (log ε 4.57). One part sol in 1000 parts boiling chloroform; in 800-1000 parts boiling glacial acetic acid.

2316. Chrysin. *5,7-Dihydroxy-2-phenyl-4H-1-benzopyran-4-one; 5,7-dihydroxyflavone;* chrysidenon 1438. $C_{15}H_{10}O_4$; mol wt 254.24. C 70.86%, H 3.96%, O 25.17%. From heartwood of *Pinus monticola* Dougl., *P. excelsa* Wall., and *P. aristata* Engelm., *Pinaceae:* Linstedt, *Acta Chem. Scand.* **3**, 1147, 1375 (1949); **4**, 55 (1950); from bark of *Dolichandrone falcata* Seem., *Bisnomiaceae:* Kincl, *Naturwiss.* **42**, 646 (1955). Synthesis: Seka, Prosche, *Monatsh.* **69**, 284 (1936); Hutchins, Wheeler, *J. Chem. Soc.* **1939**, 91; Teoule *et al., Bull. Soc. Chim. France* **1961**, 546.

Light yellow prisms from methanol, mp 285°. uv max: 270, 329 nm (log ε 4.40, 3.90). Practically insol in water; sol in alkali hydroxide solns; slightly sol in alcohol, chloroform, ether.

Diacetoxychrysin, $C_{19}H_{14}O_6$, crystals from ethanol, mp 194-195°.

Methylchrysin, $C_{16}H_{12}O_4$, *tectochrysin.* It is present as such or in the form of a glucoside in buds of *Populus* spp., *Salicaceae.* Use of tectochrysin as diuretic: Perrault, U.S. pat. 3,155,579 (1964 to Laroche Navarron). Yellow needles, mp 163°. Sol in alcohol, benzene, chloroform.

2317. Chrysoidine. *4-(Phenylazo)-1,3-benzenediamine monohydrochloride;* 4-phenylazo-*m*-phenylenediamine hydrochloride; 2,4-diaminoazobenzene hydrochloride; chrysoidine orange; chrysoidine Y; C.I. 11270; C.I. Basic Orange 2. $C_{12}H_{13}ClN_4$; mol wt 248.71. C 57.95%, H 5.27%, Cl 14.25%, N 22.53%. Prepn: Hofmann, *Ber.* **10**, 213 (1877); Maximoff, U.S. pat. 2,053,095 (1935 to Azodal Co.). Prepn of salt: Ger. pat. 562,392 (1933 to Chem.-Pharm. Fabrik Hubold & Bartsch). *See also Colour Index* vol. **4** (3rd ed., 1971) p 4019; *H. J. Conn's Biological Stains,* R. D. Lillie, Ed. (Williams & Wilkins, Baltimore, 9th ed., 1977) p 87.

Reddish-brown crystalline powder, mp 118-118.5°. Solubility at 15°: water 5.5%; abs ethanol 4.75%; cellosolve 6.0%; anhydr ethylene glycol 9.5%; xylene 0.005%; slightly sol in acetone. Practically insol in benzene. Gives yellow soln in concd H_2SO_4, orange soln in dil H_2SO_4 and in HNO_3.

Citrate, $C_{12}H_{13}ClN_4 \cdot C_6H_8O_7$, *4-phenylazo-m-phenylenediamine hydrochloride citrate, 2,4-diaminoazobenzene hydrochloride citrate, Azoangin, Azohel.* Sol in water (up to 4%), alcohol.

Free base, $C_{12}H_{12}N_4$, *C.I. Solvent Orange 3.*
USE: Dyeing silk and cotton. Biological stain.
THERAP CAT: Citrate as antiseptic.

2318. Chrysophanic Acid. *1,8-Dihydroxy-3-methyl-9,10-anthracenedione; 1,8-dihydroxy-3-methylanthraquinone;* 3-methylchrysazin; chrysophanol. $C_{15}H_{10}O_4$; mol wt 254.24. C 70.86%, H 3.96%, O 25.17%. Occurs in the free state and as glucoside in cascara sagrada, senna and various species of *Rumex* and *Rheum* (rhubarb). Isoln from rhubarb root: Tutin, Clewer, *J. Chem. Soc.* **99**, 946 (1911); Siesto, Bartoli, *Farmaco Ed. Prat.* **12**, 517 (1957); Carelli, Giuliano, *ibid.* 184; from *Penicillium islandicum* Sopp.: Howard, Raistrick, *Biochem. J.* **46**, 49 (1950); from *Chaetonium affine* Corda: Arkley *et al., Croat. Chem. Acta* **29**, 141 (1957), *C.A.* **53**,

1287h (1959). Synthesis: Eder, Widmer, *Helv. Chim. Acta* **5**, 3 (1922); **6**, 419 (1923); Ayyangar *et al., J. Sci. Ind. Res. (India)* **20B**, 493 (1961). Total synthesis: M. E. Jung, J. A. Lowe, *Chem. Commun.* **1978**, 95.

Hexagonal or monoclinic crystals from alcohol or benzene, mp 196°. Sublimes. Absorption max: 226, 256, 278, 288, 436 nm (ε × 10⁻³ 41, 28, 14, 14, 11.8). Practically insol in water. Slightly sol in cold, freely in boiling alc; sol in benzene, chloroform, ether, glacial acetic acid, acetone, solns of alkali hydrides, and in hot solns of alkali carbonates; very slightly sol in petr ether.

Glucoside, $C_{21}H_{20}O_9$, *chrysophanein, chrysophanii.* Fine yellow needles from alc, mp 248-249°. Slightly sol in hot water; sol in pyridine. Practically insol in cold water, chloroform, ether.

2319. Chymopapain. BAX-1526; Chymodiactin; Discase. Proteolytic enzyme which is the major component of the crude latex of *Carica papaya, Caricaceae.* A sulfhydryl enzyme similar to papain, *q.v.,* with respect to substrate specificities, but differing in electrophoretic mobility, stability and solubility. Original crystallization and partial characterization: Jansen, Balls, *J. Biol. Chem.* **137**, 459 (1941); *eidem,* U.S. pat. 2,313,875 (1943). Consists of four components, two of which have molecular wts of about 35,000 and have been isolated and studied. Chymopapain A: Erbata, Yasunobu, *J. Biol. Chem.* **237**, 1086 (1962); chymopapain B: Kunimitsu, Yasunobu, *Biochim. Biophys. Acta* **139**, 405 (1966); Isunoda, Yasunobu, *J. Biol. Chem.* **241**, 4610 (1966). Purification: Stern, U.S. pat. 3,558,433 (1971 to Baxter Labs.). Specificity studies: Ebata, Takahashi, *Biochim. Biophys. Acta* **118**, 201 (1966). *Reviews:* Kunimitsu, Yasunobu, *Methods Enzymol.* vol. XIX, G. E. Perlmann, L. Lorand, Eds. (Academic Press, New York, 1970) pp 244-252; Glazer, Smith, *The Enzymes* vol. III, P. D. Boyer, Ed. (Academic Press, New York, 3rd ed., 1971) pp 537-538. Review of use in herniated disk treatment: M. J. David, *J. Am. Med. Assoc.* **243**, 2043 (1980).

Powder, more sol in aq soln than papain. Extremely stable at pHs as low as 1.8 and sol in satd NaCl above pH 3. Most active at pHs 2.5 to 4.0. uv max (pH 7): 280 nm ($A_{1cm}^{1\%}$ 18.7 (A); 18.4 (B)). The ratio of milk-clotting to proteolytic activity is twice that of papain although the clotting capacities are the same.

USE: In meat tenderizer.
THERAP CAT: Proteolytic enzyme (used in chemonucleolysis).

2320. Chymotrypsins. Avazyme; Catarase; Chymar; Chymetin; Chymolase; Kymo-trypure; Quimar; Quimoral; Quimotrase; Zolyse. A group of major proteolytic enzymes in the pancreatic juice. Produced in the form of inactive chymotrypsinogen by the acinous cells of pancreas, and carried as such by the pancreatic juice into the duodenum where they are activated by trypsin. They split secondary amide or peptide bonds, carboxylic or phenolic ester bonds and even carbon-carbon bonds. Their main function is to hydrolyze peptide bonds during the intestinal digestion of proteins. Cattle pancreatic juice contains nearly equal quantities of two chymotrypsinogens—A which is cationic at pH 8 and B which is anionic. Chymotrypsinogen A is converted by trypsin into π-chymotrypsin. In the presence of large amounts of trypsin, π-chymotrypsin is degraded to δ-chymotrypsin. In the presence of small amounts of trypsin, π-chymotrypsin is autolyzed to α-chymotrypsin. After long standing α-chymotrypsin gives rise to β- and γ-chymotrypsin which differ from it in their crystal habits. Prepn of π- and δ-chymotrypsins: Bettelheim, Neurath, *J. Biol. Chem.* **212**, 241 (1955). Prepn of α-, β-, γ- and δ-chymotrypsins (and chymotrypsin B): Laskowski, *Methods Enzymol.* **2**, 8 (1955). Amino acid sequence of chymotrypsinogen A: Hartley, *Nature* **201,** 1284 (1964); Melorin *et al.,* *Biochim. Biophys. Acta* **130,** 543 (1966); of chymotrypsinogen B: Hartley *et al., Nature* **207,** 1157 (1965); Smillie *et al., ibid.* **218,** 343 (1968). *Reviews:* Bender *et al., J. Polymer Sci.* **49,** 75 (1961); Desnuelle, *The Enzymes* vol. **4,** P. D. Boyer *et al.,* Eds. (Academic Press, New York, 2nd ed., 1960) pp 93-118; Kraut; Blow; Hess, *ibid.* **vol. 3** (3rd ed., 1971) 165, 185, 213.

α-Chymotrypsin, **Alpha-Chymocutan, Chymotase (tabl.), Chymozym, Impral, Kimopsin, Kimoral.** Arrangement of molecules in monoclinic crystal form: Blow *et al., J. Mol. Biol.* **8,** 65 (1964). Structure: Matthews *et al., Nature* **214,** 652 (1967); Blow *et al., ibid.* **221,** 337 (1969). *Review:* Niemann, *Science* **143,** 1287 (1964).

THERAP CAT: Enzyme (proteolytic).

THERAP CAT (VET): Topically for enzymatic debridement.

2321. Ciafos. *Phosphorothioic acid O-(4-cyanophenyl) O,O-dimethyl ester; phosphorothioic acid O,O-dimethyl ester, O-ester with p-hydroxybenzonitrile; O,O*-dimethyl *O*-(4-cyanophenyl) phosphorothioate; *O,O*-dimethyl *O*-(4-cyanophenyl) thionophosphate; cyanophos; Bay 34727; S-4084; Sumitomo S-4084; Cyanox. $C_9H_{10}NO_3PS$; mol wt 243.22. C 44.44%, H 4.14%, N 5.76%, O 19.73%, P 12.73%, S 13.18%. Prepn: Nishizawa, Mizutani, **Japan.** pat. **62 18,-184;** Kuramoto *et al.,* **U.S.** pat. **3,150,040** (1962, 1964 both to Sumitomo); Sakamoto, Nishizawa, *Agr. Biol. Chem.* **26,** 252 (1962); Bernhart, **Ger.** pat. **2,162,344** corresp to U.S. pat. **3,792,132** (1972, 1974 to Stauffer). Pharmacology and metabolism: H. Yamamoto *et al., Oyo Yakuri* **5,** 75 (1971), *C.A.* **76,** 68907a (1972); Wakimura, Miyamoto, *Agr. Biol. Chem.* **35,** 410 (1971). Evaluation as an insecticide: Nishiyawa *et al., ibid.* **25,** 597 (1961); Tamura *et al., ibid.* 773. Evaluation of neurotoxicity: H. Ohkawa *et al., Biochem. Pharmacol.* **29,** 2721 (1980).

Yellow to reddish-yellow transparent liquid. $bp_{0.09}$ 119-120° (dec). mp 14-15°. $n_D^{32.5}$ 1.5404. Very sol in methanol, ethanol, acetone, chloroform. Sparingly sol in *n*-hexane, kerosene; slightly sol in water. Rapidly dec under alkaline conditions and upon exposure to light, *Japan. Med. Gaz.* **11**(7), 7 (1974). LD_{50} in mice (mg/kg): 1000 orally; 880 i.p. (Yamamoto).

Caution: Cholinesterase inhibitor.

USE: Insecticide.

2322. Cichoriin. *7-(β-D-Glucopyranosyloxy)-6-hydroxy-2H-1-benzopyran-2-one;* 6,7-dihydroxycoumarin-7-glucoside. $C_{15}H_{16}O_9$; mol wt 340.29. C 52.95%, H 4.74%, O 42.32%. Isomeric with esculin. In flowers of the chicory plant *(Cichorium intybus* L., *Compositae).* Extraction procedure: Merz, *Arch. Pharm.* **270,** 476 (1932). Synthesis: Head, Robertson, *J. Chem. Soc.* **1939,** 1266; also Merz, *loc. cit. Review:* Sethna, Shah, *Chem. Rev.* **36,** 1 (1945).

Needles with $2H_2O$, mp 213-215° (after drying in desiccator). $[\alpha]_D^{18}$ — 105° (c = 3 in 50% dioxane). Sol in hot water, alc, glacial acetic acid. Insol in ether, petr ether. Sol in dil alkalies with yellow color, but no fluorescence (difference from esculin). Absorption spectrum of soln: Merz, *loc. cit.*

2323. Cicletanine. *(±)-3-(4-Chlorophenyl)-1,3-dihydro-6-methylfuro[3,4-c]pyridin-7-ol;* 1,3-dihydro-3-(4-chlorophenyl)-7-hydroxy-6-methylfuro[3,4-c]pyridine; cicletanide; cycletanide. $C_{14}H_{12}ClNO_2$; mol wt 261.71. C 64.25%, H 4.62%, Cl 13.55%, N 5.35%, O 12.23%. Furopyridine derivative with antihypertensive and diuretic properties. Prepn: A. Esanu, **Belg.** pat. **891,797;** *idem,* **U.S.** pat. **4,383,998** (1982, 1983 both to Soc. Conseils Recher. Appl. Sci.). Studies on mechanism of action: P. Braquet *et al., Lancet* **1,**

1218 (1983); G. R. Elliott *et al., Thromb. Res.* **33,** 549 (1984). Cardiovascular pharmacology in dogs: R. Jouve *et al., J. Cardiovasc. Pharmacol.* **8,** 208 (1986). *In vitro* comparison of antihistaminic activity of isomers: P. Schoeffter *et al., Eur. J. Pharmacol.* **136,** 235 (1987). HPLC determn in biological fluids: G. Cuisinaud *et al., J. Chromatog.* **341,** 97 (1985). Effect on human prostaglandin metabolism: P. Guinot, J. C. Frölich, *Arzneimittel-Forsch.* **35,** 1714 (1985). Effect on potassium ion transport in hypertensive patients: R. Garay *et al., J. Hypertension* **4,** Suppl. 5, S208 (1986). Pharmacokinetics in humans: J. M. Lize *et al., Therapie* **42,** 399 (1987).

Hydrochloride, $C_{14}H_{12}ClNO_2 \cdot HCl$, *BN-1270, Coverine, Justar, Secletan, Tenstaten.* White crystals, mp 219-228°. Insol in water.

THERAP CAT: Antihypertensive.

2324. Ciclonicate. *3-Pyridinecarboxylic acid 3,3,5-trimethylcyclohexyl ester; trans-3,3,5-trimethylcyclohexyl nicotinate;* cyclonicate; P-350; Bled. $C_{15}H_{21}NO_2$; mol wt 247.34. C 72.84%, H 8.56%, N 5.66%, O 12.94%. Deriv of nicotinic acid, *q.v.* Prepn: K. Matsuda *et al.,* **Ger.** pat. **1,910,481** (1969 to Takeda), *C.A.* **72,** 12769g (1970); G. Massaroli, **Ger.** pat. **2,406,849** (1974 to Poli), *C.A.* **82,** 4129 (1975). Chromatographic study: G. Sekules *et al., Boll. Chim. Farm.* **119,** 521 (1980). Antilipolytic activity in rats: D. Faini *et al., Farmaco Ed. Prat.* **36,** 478 (1981). *In vitro* effect on lipolysis: F. Tessari *et al., Farmaco Ed. Sci.* **36,** 1029 (1981). Clinical study: C. Fossati, *Gazz. Med. Ital.* **139,** 43 (1980).

Liq, $bp_{0.6}$ 127-128°.

THERAP CAT: Vasodilator.

2325. Ciclopirox. *6-Cyclohexyl-1-hydroxy-4-methyl-2(1H)-pyridinone.* $C_{12}H_{17}NO_2$; mol wt 207.27. C 69.54%, H 8.27%, N 6.76%, O 15.44%. Broad spectrum antimycotic agent with some antibacterial activity. Prepn: G. Lohaus, W. Dittmar, **S. Afr.** pat. **69 06039;** *eidem,* **U.S.** pat. **3,883,-545** (1970, 1975 both to Hoechst). *In vitro* study: *eidem, Arzneimittel-Forsch.* **23,** 670 (1973). Series of articles on pharmacokinetics, pharmacology, teratology, toxicity studies: *Oyo Yakuri* **9,** 57-115 (1975), *C.A.* **83,** 53159d, 53538b, 53539c, 71844c, 90833q (1975). Series of articles on chemistry, mechanism of action, toxicology, clinical trials: *Arzneimittel-Forsch.* **31,** 1309-1386 (1981). Toxicity data: H. G. Alpermann, E. Schutz, *ibid.* 1328. *Review:* S. G. Jue *et al., Drugs* **29,** 330-341 (1985).

Solid, mp 144°.

Ethanolamine salt (1:1), $C_{14}H_{24}N_2O_3$, *ciclopirox olamine, HOE-296, Batrafen, Brumixol, Ciclochem, Dafnegin, Lo-*

prox, Micoxolamina, Mycoster, Terit. LD$_{50}$ in mice, rats (mg/kg): 2898, 3290 orally (Alpermann, Schutz).

THERAP CAT: Antifungal.

2326. Ciclosidomine. *N-(Cyclohexylcarbonyl)-3-(4-morpholinyl)sydnone imine.* C$_{13}$H$_{20}$N$_4$O$_3$; mol wt 280.33. C 55.70%, H 7.19%, N 19.99%, O 17.12%. Peripheral vasodilator similar to molsidomine, *q.v.* Prepn: **Fr. pat.** **1,551,-013; Brit. pat. 1,198,283** (1968, 1970 both to Boehringer, Ing.). Clinical evaluation: J. Vos, E. J. D. Mees, *Brit. J. Clin. Pharmacol.* **8,** 155 (1979). Effect on blood pressure and heart rate: R. G. Shanks *et al., ibid.* **18,** 232 (1984). Tissue distribution and excretion: M. R. Matteo *et al., Arzneimittel-Forsch.* **35,** 329 (1985).

Hydrochloride, C$_{13}$H$_{20}$N$_4$O$_3$.HCl, *PR-G 138-Cl, Neopres.* Crystalline powder, mp 187°.

THERAP CAT: Antihypertensive.

2327. Cicrotoic Acid. *β-Methylcyclohexaneacrylic acid;* β-cyclohexyl-β-methacrylic acid; 3-cyclohexyl-2-butenoic acid; β-methylcyclohexanepropenoic acid; AD-106; Accroibilc. C$_{10}$H$_{16}$O$_2$; mol wt 168.24. C 71.39%, H 9.59%, O 19.02%. Prepn: Young *et al., J. Org. Chem.* **28,** 928 (1963); Redel, Raymond, **Fr. pat. M4665** (1967 to Soc. Chim. Org. Biol.), *C.A.* **72,** 42949h (1970).

Colorless prisms from methanol-water, mp 85-86°. uv max: 219 nm (ε 12000). LD$_{50}$ orally in mice, rats: 1.925, 2.900 g/kg.

THERAP CAT: Choleretic.

2328. Cicutoxin. *(E,E,E)-(−)-8,10,12-Heptadecatriene-4,6-diyne-1,14-diol.* C$_{17}$H$_{22}$O$_2$; mol wt 258.36. C 79.03%, H 8.58%, O 12.39%. Poisonous principle of the water hemlock *(Cicuta virosa* L., *Umbelliferae);* naturally occurring as the (−)-form. Isoln: Jacobson, *J. Am. Chem. Soc.* **37,** 916 (1915). Structure: Anet *et al., J. Chem. Soc.* **1953,** 309. Synthesis of (±)-form: Hill *et al., ibid.* **1955,** 1770. Antileukemic activity: T. Konoshima, K.-H. Lee, *J. Nat. Prod.* **49,** 1117 (1986). Simple analytical method: R. A. Smith, D. Lewis, *Vet. Hum. Toxicol.* **29,** 240 (1987).

Prisms from ether + petr ether, transformed in air and light to yellow oily resin. mp 54°. [α]$_D^{15}$ −14.5° (c = 1.7 in ethanol). Sol in alc, chloroform, ether, hot water, alkali hydroxides. Practically insol in petr ether. Very toxic.

(±)-Form, crystals from ether + petr ether, mp 67°. uv max (alc): 242, 252, 318.5, 335.5 nm (ε × 10^{-3} 14.6, 21.6, 50.6, 60.3).

2329. Cidofovir. *(S)-[[2-(4-Amino-2-oxo-1(2H)-pyrimidinyl)-1-(hydroxymethyl)ethoxy]methyl]phosphonic acid;* (S)-1-[3-hydroxy-2-(phosphonylmethoxy)propyl]cytosine; (S)-HPMPC; GS-504; Vistide. C$_8$H$_{14}$N$_3$O$_6$P; mol wt 279.19. C 34.42%, H 5.05%, N 15.05%, O 34.38%, P 11.09%. DNA synthesis inhibitor. Prepn: A. Holy *et al., Eur. pat. Appl.* **253,412;** *eidem,* **U.S. pat. 5,142,051** (1988, 1992 both to Ceskoslov. Akad. Ved; Rega Inst.); and activity vs cytomegalovirus: R. Snoeck *et al., Antimicrob. Ag. Chemother.* **32,** 1839 (1988). Syntheses: J. J. Bronson *et al., Nucleosides*

Nucleotides **9,** 745 (1990); P. R. Brodfuehrer *et al., Tetrahedron Letters* **35,** 3243 (1994). Activity vs herpes simplex virus: G. Andrei *et al., Eur. J. Clin. Microbiol. Infect. Dis.* **11,** 143 (1992). Case studies in immunocompromised patients: J. P. Lalezari *et al., J. Infect. Dis.* **170,** 570 (1994); R. Snoeck *et al., Clin. Infect. Dis.* **18,** 570 (1994).

Fluffy white powder, mp 260° (dec). [α]$_D^{20}$ −97.3° (c = 0.80 in water). Monohydrate, uv max (pH 2): 279 nm (ε 13000).

THERAP CAT: Antiviral.

2330. Cifenline. *2-(2,2-Diphenylcyclopropyl)-4,5-dihydro-1H-imidazole;* (±)-2-(2,2-diphenylcyclopropyl)-2-imidazoline; 1-(2-Δ²-imidazolinyl)-2,2-diphenylcyclopropane; cibenzoline; Ro-22-7796; UP-33-901. C$_{18}$H$_{18}$N$_2$; mol wt 262.35. C 82.41%, H 6.92%, N 10.68%. Prepn: **Belg. pat.** **807,630;** J. C. Cognaco, **U.S. pat. 3,903,104** (1974, 1975 both to Hexachemie). Activity and tolerance: D. Herpin *et al., Acta Cardiol.* **36,** 131 (1981). Electrophysiological effects in man: J. F. Thizy *et al., Lyon Med.* **245,** 119 (1981). Clinical trial: D. Herpin *et al., Curr. Ther. Res.* **30,** 742 (1981).

Crystals from petr ether, mp 103-104°. LD$_{100}$ in rats (mg/kg): 64 i.v. (Cognaco).

Succinate, C$_{18}$H$_{18}$N$_2$.C$_4$H$_6$O$_4$, *Cibenol, Cipralan, Exacor.* Crystals from ethanol + ether, mp 165°.

THERAP CAT: Antiarrhythmic.

2331. Cilastatin. *[R-[R*,S*(Z)]]-7-[(2-Amino-2-carboxyethyl)thio]-2-[[(2,2-dimethylcyclopropyl)carbonyl]amino]-2-heptenoic acid;* MK-791. C$_{16}$H$_{26}$N$_2$O$_5$S; mol wt 358.46. C 53.61%, H 7.31%, N 7.81%, O 22.32%, S 8.95%. Prevents renal metabolism of penem and carbapenem antibiotics by specific and reversible dehydropeptidase I inhibition. Synthesis and combination with thienamycins: D. W. Graham *et al., Eur. pat. Appl.* **48,301;** H. Kropp *et al., Eur. pat. Appl.* **48,025** (both 1982 to Merck & Co.), *C.A.* **97,** 145271b, 145270a (1982). Combination with penems: F. M. Kahan, H. Kropp, **Eur. pat. Appl. 72,014** (1983 to Merck & Co.), *C.A.* **99,** 70272h (1983). The articles cited below discuss the activity of cilastatin alone and in combination with imipenem, *q.v.* Dipeptidase inhibition, pharmacokinetics: S. R. Norrby *et al., Antimicrob. Ag. Chemother.* **23,** 300 (1983). Stimulation of granulocyte function: H. Gnarpe *et al., ibid.* **25,** 179 (1984). HPLC determn in serum: C. M. Myers, J. L. Blumer, *ibid.* **26,** 78 (1984). Enhances intrathecal and ocular penetration of imipenem: A. W. Chow *et al., ibid.* **23,** 634 (1983) and K. R. Finlay *et al., Invest. Ophthalmol. Visual Sci.* **24,** 1147 (1983), respectively. In experimental meningitis: D. E. Washburn *et al., J. Antimicrob. Chemother.* **12,** 39 (1983). Series of articles on pharmacokinetics, safety and tolerance and efficacy of cilastatin/imipenem: *ibid.* **12,** Suppl. D, 1-155 (1983); *Infection* **14,** Suppl. 2, S111-S180 (1986).

Sodium salt, $C_{16}H_{25}N_2NaO_5S$, *cilastatin sodium.* Off-white to yellowish-white hygroscopic, amorphous solid. pKa_1 2.0; pKa_2 4.4; pKa_3 9.2. Very sol in water, methanol.

THERAP CAT: Dipeptidase inhibitor (in combination with imipenem).

2332. Cilazapril. *[1S-[1α,9α(R*)]]-9-[[1-(Ethoxycar-bonyl)-3-phenylpropyl]amino]octahydro-10-oxo-6H-pyridazi-no[1,2-a][1,2]diazepine-1-carboxylic acid monohydrate;* (1S,-9S)-9-[[(S)-1-carboxy-3-phenylpropyl]amino]octahydro-10-oxo-6H-pyridazino[1,2-a][1,2]diazepine-1-carboxylic acid 9-ethyl ester monohydrate; Ro-31-2848; Dynorm; Inhibace; Initiss; Justor; Vascase. $C_{27}H_{31}N_3O_5 \cdot H_2O$; mol wt 435.52. C 60.67%, H 7.64%, N 9.65%, O 22.04%. Angiotensin-converting enzyme (ACE) inhibitor. Prepn: M. R. Attwood *et al.,* Ger. pat. **3,317,290**; *eidem,* U.S. pat. **4,512,924** (1983, 1985 both to Hoffmann-La Roche); *eidem, J. Chem. Soc. Perkin Trans. I* **1986,** 1011. Pharmacology of cilazapril and its active diacid metabolite, cilazaprilat: I. L. Natoff *et al., J. Cardiovasc. Pharmacol.* **7,** 569 (1985). Cardiovascular effects in dogs: M. Holck *et al., ibid.* **8,** 99 (1986). Series of articles on clinical pharmacology and pharmacokinetics: *ibid.* **9,** 26-44 (1987). Enzyme immunoassay in human plasma: H. Tanaka *et al., J. Pharm. Sci.* **76,** 224 (1987). Preliminary evaluation in essential hypertension: A. A. Ajayi *et al., Brit. J. Clin. Pharmacol.* **22,** 167 (1986).

Crystals from aq ethanol, mp 95-97°. $[α]_D^{20}$ −62.51° (c = 1% in ethanol).

Diacid, $C_{20}H_{27}N_3O_5$, *cilazaprilat, Ro-31-3113.* Crystals from water, mp 242°. $[α]_D^{20}$ −74.7° (c = 0.5 in 1M NaOH).

THERAP CAT: Antihypertensive.

2333. Ciliary Neurotrophic Factor. CNTF. Neurotrophic factor unrelated to the neurotrophins, *q.v.* Cytosolic peptide produced by Schwann cells in the peripheral nervous system and by astrocytes in the CNS; not expressed by target tissues. Promotes survival of a broad range of neurons and enhances differentiation of neurons and astrocytes. Initial identification as *in vitro* survival factor for chick ciliary neurons: R. Adler *et al., Science* **204,** 1434 (1979). Purification from chick eye: G. Barbin *et al., J. Neurochem.* **43,** 1468 (1984); from rat sciatic nerve: M. Manthorpe *et al., Brain Res.* **367,** 282 (1986). Molecular cloning of rat CNTF: K. A. Stöckli *et al., Nature* **342,** 920 (1989); of rabbit CNTF: L.-F. H. Lin *et al., Science* **246,** 1023 (1989); of human CNTF: A. Lam *et al., Gene* **102,** 271 (1991); A. Negro *et al., Eur. J. Biochem.* **201,** 289 (1991). Effect on motoneuron survival and proposed role as a "lesion factor" released in pathophysiological conditions: M. Sendtner *et al., J. Cell Sci.* **Suppl. 15,** 103 (1991). Review of bioactivity and structural relationship with hematopoietic cytokines: N. Y. Ip, G. D. Yancopoulos, *Prog. Growth Factor Res.* **4,** 139-155 (1992). Review of CNTF receptor: S. Davis, G. D. Yancopoulos, *Curr. Opin. Neurobiol.* **5,** 281-285 (1993). *Review:* M. Manthorpe *et al.* in *Neurotrophic Factors,* S. E. Loughlin, J. H. Fallon, Eds. (Academic Press, San Diego, 1993) pp 443-473.

Human CNTF, 200 amino acid peptide. mol wt ~23,000 Da. pI ~6.0. Exhibits ~83% amino acid homology with rat CNTF; ~87% with rabbit CNTF.

2334. Cilnidipine. *(E)-(±)-1,4-Dihydro-2,6-dimethyl-4-(3-nitrophenyl)-3,5-pyridinedicarboxylic acid 2-methoxyethyl*

3-phenyl-2-propenyl ester; (±)-(E)-cinnamyl 2-methoxyethyl 1,4-dihydro-2,6-dimethyl-4-(m-nitrophenyl)-3,5-pyridine-dicarboxylate; (±)-FRC-8653. $C_{27}H_{28}N_2O_7$; mol wt 492.53. C 65.84%, H 5.73%, N 5.69%, O 22.74%. Dihydropyridine calcium channel blocker. Prepn: T. Kutsuma *et al.,* **Eur. pat. Appl. 161,877;** *eidem,* U.S. pat. **4,672,068** (1985, 1987 both to Fujirebio). Pharmacology: K. Ikeda *et al., Oyo Yakuri* **44,** 433 (1992). Mechanism of action study: M. Hosono *et al., J. Pharmacobio-Dyn.* **15,** 547 (1992). LC-MS determn in plasma: K. Hatada *et al., J. Chromatog* **583,** 116 (1992). Clinical study: M. Ishii, *Japan. Pharmacol. Ther.* **21,** 59 (1993). Acute toxicity study: S. Wada *et al., Yakuri to Chiryo* **20,** Suppl. 7, S1683 (1992), *C.A.* **118,** 32711 (1992).

Crystals from methanol, mp 115.5-116.6°. LD_{50} in male, female mice, rats (mg/kg): > 5000, > 5000, > 5000, 4412 orally; > 5000 all species s.c.; 1845, 2353, 441, 426 i.p. (Wada).

THERAP CAT: Antihypertensive.

2335. Cilostazol. *6-[4-(1-Cyclohexyl-1H-tetrazol-5-yl)-butoxy]-3,4-dihydro-2(1H)-quinolinone;* 6-[4-(1-cyclohexyl-1H-tetrazol-5-yl)butoxy]-3,4-dihydrocarbostyril; 6-[4-(1-cyclohexyl-5-tetrazolyl)butoxy]-1,2,3,4-tetrahydro-2-oxo-quinoline; OPC-13013; Pletaal. $C_{20}H_{27}N_5O_2$; mol wt 369.47. C 65.02%, H 7.37%, N 18.96%, O 8.66%. Prepn: **Belg. pat. 878,548;** T. Nishi, K. Nakagawa, U.S. pat. **4,277,479** (1979, 1981 both to Otsuka); T. Nishi *et al., Chem. Pharm. Bull.* **31,** 1151 (1983). Physical properties: T. Shimizu *et al., Arzneimittel-Forsch.* **35,** 1117 (1985). Series of articles on pharmacology, metabolism, mechanism of action and clinical evaluations: *ibid.* 1125-1208. HPLC determn in plasma: H. Akiyama *et al., J. Chromatog.* **338,** 456 (1985). Series of articles on toxicology: *Iyakuhin Kenkyu* **16,** 1200-1324 (1985), *C.A.* **104,** 81784z-81787c, 102207f-102209h (1986).

Colorless needle-like crystals from methanol, mp 159.4-160.3°. uv max (methanol): 257 nm (ε 15200). Freely sol in acetic acid, chloroform, n-methyl-2-pyrrolidone, DMSO. Practically insol in ether, water, 0.1N HCl, 0.1N NaOH.

THERAP CAT: Antithrombotic.

2336. Cimaterol. *2-Amino-5-[1-hydroxy-2-[(1-methyl-ethyl)amino]ethyl]benzonitrile;* (±)-5-[1-hydroxy-2-(isopropylamino)ethyl]anthranilonitrile; 1-(4-amino-3-cyanophenyl)-2-isopropylaminoethanol; AB-A-663; CL-263780; AC-263780. $C_{12}H_{17}N_3O$; mol wt 219.29. C 65.73%, H 7.81%, N 19.16%, O 7.30%. β-Adrenergic agonist, related to clenbuterol and mabuterol, *q.q.v.* Prepn: G. Engelhardt *et al.,* Ger. pat. **2,261,914** (1974 to Thomae). Pharmacology: G. Engelhardt, *Arzneimittel-Forsch.* **34,** 1625 (1984). Effect on lipid metabolism in sheep: C. Y. Hu *et al., J. Anim. Sci.* **66,** 1393 (1988). Efficacy as a repartitioning agent in livestock: R. W. Jones *et al., ibid.* **61,** 905 (1985); D. H. Beermann *et al., ibid.* **62,** 370 (1986). *Review:* M. Lafontan *et al., Reprod. Nutr. Develop.* **28,** 61-84 (1988).

Crystals, mp 159-161°.

THERAP CAT (VET): Repartitioning agent.

2337. Cimetidine. *N-Cyano-N'-methyl-N''-[2-[[(5-methyl-1H-imidazol-4-yl)methyl]thio]ethyl]guanidine;* SKF-92334; Acibilin; Acinil; Cimal; Cimetag; Cimetum; Edalene; Dyspamet; Eureceptor; Gastromet; Peptol; Tagamet; Tametin; Tratul; Ulcedin; Ulcedine; Ulcerfen; Ulcimet; Ulcofalk; Ulcomedina; Ulcomet; Ulhys. $C_{10}H_{16}N_6S$; mol wt 252.34. C 47.60%, H 6.39%, N 33.30%, S 12.71%. Competitive histamine H_2-receptor antagonist which inhibits gastric acid secretion and reduces pepsin output. Prepn: G. J. Durant *et al.*, **Belg.** pat. 804,144; *eidem*, **U.S.** pat. 3,950,333 (1974, 1976 both to SK&F); P. Kairisalo, E. Honkanen, *Arch. Pharm. (Weinheim)* **316**, 688 (1983). Chemistry, pharmacology and toxicology: R. W. Brimblecombe *et al., J. Int. Med. Res.* **3**, 86 (1975); *eidem, Brit. J. Pharmacol.* **53**, 435p (1975). Use in combination with other drugs for cancer treatment: R. D. Thornes *et al., Lancet* **2**, 328 (1982); S. Borgström *et al., N. Engl. J. Med.* **307**, 1080 (1982); **308**, 591 (1983). Controlled clinical study in treatment of acute upper gastrointestinal tract bleeding: D. Barer *et al., ibid.* 1571; for prevention of recurrent duodenal ulcer: S. Sontag *et al., ibid.* **311**, 689 (1984). Immunoregulatory effect: J. L. Jorizzo *et al., Ann. Int. Med.* **92**, 192 (1980); W. B. White, M. Ballow, *N. Engl. J. Med.* **312**, 198 (1985). Review of effect on endocrine secretion: C. Scarpignato, G. Bertaccini, *Drugs Exp. Clin. Res.* **5**(4-5), 129-140 (1979). Symposium on clinical efficacy, cytoprotection and antifibrinolytic effects: *Scand. J. Gastroenterol.* **21**, Suppl. 121, 1-62 (1986).

Crystals, mp 141-143°. Soly in water at 37°: 1.14%. Soly increased by dil HCl. LD_{50} in mice, rats (mg/kg): 2600, 5000 orally; 150, 106 i.v.; 470, 650 i.p. (Brimblecombe).

Hydrochloride, $C_{10}H_{16}N_6S.HCl$, *Aciloc, Biomag, Brumetidina, Notul.*

THERAP CAT: Antiulcerative.

2338. Cimetropium Bromide. *[7(S)-(1α,2β,4β,5α,7β)]-9-(Cyclopropylmethyl)-7-(3-hydroxy-1-oxo-2-phenylpropoxy)-9-methyl-3-oxa-9-azoniatricyclo[3.3.1.0²·⁴]nonane bromide;* 8-(cyclopropylmethyl)-6β,7β-epoxy-3α-hydroxy-1αH,5αH-tropanium bromide (−)-(S)-tropate; N-cyclopropylmethylscopolamine bromide; DA-3177; Alginor. $C_{21}H_{28}BrNO_4$; mol wt 438.36. C 57.54%, H 6.44%, Br 18.23%, N 3.20%, O 14.60%. Spasmolytic agent with affinity for intestinal muscarinic receptors. Prepn: S. Casadio, A. Donetti, **Ger.** pat. 2,316,728; *eidem*, **U.S.** pat. 3,853,886 (1973, 1974 both to De Angeli). Crystal and molecular structure: G. Giuseppetti *et al., Farmaco, Ed. Sci.* **35**, 231 (1980). Molecular conformation in solution: A. Gallazzi *et al., ibid.* 913. Pharmacology: A. Schiavone *et al., Arzneimittel-Forsch.* **35**,

796 (1985); C. Scarpignato, G. Bianchi Porro, *Int. J. Clin. Pharm. Res.* **5**, 467 (1985). Clinical trials in treatment of irritable bowel syndrome: G. Piai *et al., Clin. Trials J.* **23**, 6 (1986); A. Ferrari *et al., Clin. Ther.* **8**, 320 (1986).

Crystals from acetonitrile, mp 174°. $[\alpha]_D^{20}$ −18.3° (c = 3).

THERAP CAT: Antispasmodic.

2339. Cimicifuga. Black cohosh; black snake root; Actaea; bugbane; bugwort. Dried rhizome and roots of *Cimicifuga racemosa* (L.) Nutt., *Ranunculaceae.* Habit. U.S., Canada. *Constit.* Isoferulic acid, 15-20% cimicifugin, tannin, volatile oil, resin, a sugar.

THERAP CAT: Astringent bitter.

2340. Cimigenol. *(3β,15α,16α,23R,24S)-16,23:16,24-Diepoxy-9,19-cyclolanostane-3,15,25-triol;* cimicifugol. $C_{30}H_{48}O_5$; mol wt 488.71. C 73.73%, H 9.90%, O 16.37%. From the resin of *Cimicifuga racemosa* Nutt. *(Actaea racemosa* L.), *Ranunculaceae:* Corsano, Spano, *Atti Accad. Nazl. Lincei, Rend., Classe Sci. Fiz. Mat. Nat.* **32**, 674 (1962), *C.A.* **58**, 11408e (1963). Structure: Corsano *et al., Chem. Commun.* **1965**, 185; Corsano, Piancatelli, *Ric. Sci.* **37**, 366 (1967); *eidem, Gazz. Chim. Ital.* **99**, 1140 (1969). Identity with cimicifugol: *J. Pharm. Soc. Japan* **87**, 1569 (1967).

Needles from methanol, mp 227.5-228.5°. $[\alpha]_D$ +38° (c = 0.86 in chloroform).

Diacetate, $C_{34}H_{52}O_7$, needles from hexane, mp 202-204°. $[\alpha]_D$ +46.3° (c = 1.07 in chloroform).

Triacetate, $C_{36}H_{54}O_8$, needles from hexane, mp 149.5-150.5°. $[\alpha]_D$ +25.4° (chloroform).

2341. Cinametic Acid. *3-[4-(2-Hydroxyethoxy)-3-methoxyphenyl]-2-propenoic acid;* 4-(2-hydroxyethoxy)-3-methoxycinnamic acid; Transoddi. $C_{12}H_{14}O_5$; mol wt 238.24. C 60.50%, H 5.92%, O 33.58%. Prepn: **Fr.** pat. M6380 (1968 to Anphar), *C.A.* **76**, 140232w (1972).

THERAP CAT: Hepatobiliary regulator.

2342. Cinchomeronic Acid. *3,4-Pyridinedicarboxylic acid.* $C_7H_5NO_4$; mol wt 167.12. C 50.31%, H 3.02%, N 8.38%, O 38.29%. Prepd in 42% yield by boiling quinine with nitric and fuming nitric acid: Ternájgo, *Monatsh.* **21**, 448 (1900); *cf.* Kirpal, *ibid.* **23**, 248 (1902); Kaas, *ibid.* 252; by selenium dioxide oxidation of isoquinoline: Mueller, **U.S.** pat. 2,436,660 (1948 to Allied Chem.).

Crystals from acidulated water, mp 256°. On heating, a small portion sublimes without decompn. Strong acid. Sparingly sol in hot alc, ether, benzene. Insol in chloroform.

Disodium salt dihydrate, $C_7H_3NNa_2O_4.2H_2O$, plates, moderately sol in water, alcohol.

Dimethyl ester, $C_9H_9NO_4$, crystals, mp 141°.

2343. Cinchona. Calisaya bark; Peruvian bark; Cinchona bark; Jesuit's bark. Dried bark of stem or root of various species of *Cinchona*, fam. *Rubiaceae*, largely of *C. officinalis* L. (*C. ledgeriana* Moens) cultivated mostly in Java, of *C. officinalis* L. (*C. calisaya* Wedd.) from Bolivia, of *C. officinalis* L. and *C. micrantha* R. & P. from Peru, of *C. pubescens* Vahl. (*C. succirubra* Pav.) from Ecuador and *C. pitayensis* Wedd. from Colombia. *Habit.* South America, cultivated mostly in Java, also in India. *Constit.* About 35 alkaloids; cinchotannic, quinic and quinovic acids; cinchona red; volatile oils. The alkaloid content varies according to the source of the bark. The cultivated bark contains 7-10% total alkaloids of which about 70% is quinine. *Cinchona succirubra* Pav. and some other varieties contain more cinchonidine and cinchonine and sometimes quinidine than the cultivated. The standardized method for the assay of cinchona, named "Brussels 1949" is published in the journal *De Belgische Chemische Industrie (Ind. Chim. Belge)* 15, 328-338 (1950).

On heating a small portion of the bark in a test tube a characteristic purple vapor is evolved. Dil H_2SO_4 extracts have a blue fluorescence.

THERAP CAT: Antimalarial.

2344. Cinchonamine. *[1S-(1α,2α,4α,5β)]-2-(5-Ethenyl-1-azabicyclo[2.2.2]oct-2-yl)-1H-indole-3-ethanol;* 3-(β-hydroxyethyl)-2-(5-vinyl-2-quinuclidyl)indole. $C_{19}H_{24}N_2O$; mol wt 296.41. C 76.99%, H 8.16%, N 9.45%, O 5.40%. From the bark of *Remijia purdieana* Wedd., *Rubiaceae:* Arnaud, *Compt. Rend.* **93**, 593 (1881); Hesse, *Ann.* **225**, 211 (1884). Prepn by reduction of quinamine with lithium aluminum hydride and structure: Goutarel *et al., Helv. Chim. Acta.* **33**, 150 (1950). Synthesis: Chen *et al., C.A.* **53**, 7219e (1959). Total synthesis: G. Grethe *et al., Helv. Chim. Acta* **59**, 2271 (1976). Stereochemistry: Wenkert, Bringi, *J. Am. Chem. Soc.* **80**, 3484 (1958); Augustine, *Chem. & Ind. (London)* **1959**, 1071; Sawa, Matsumura, *Tetrahedron* **26**, 2923 (1970). Biosynthetic studies: Battersby, Parry, *Chem. Commun.* **1971**, 31. Conversion of chinchona alkaloids of the quinoline series to those of the indole series: Ochiai *et al., C.A.* **59**, 14040h (1963).

Triboluminescent, orthorhombic prisms from methanol, mp 186°, also reported as mp 194°. $[\alpha]_D^{20}$ +123° (c = 0.66 in ethanol). pK in 80% methyl Cellosolve: 8.28. uv max (methanol): 223, 292 nm (log ε 4.60, 3.88). One gram dissolves in about 35 ml alcohol, 105 ml ether. Sol in benzene, chloroform, petr ether, CS_2; practically insol in water. Hydrochloride monohydrate, $C_{19}H_{24}N_2O.HCl.H_2O$, cubic crystals. One gram dissolves in 200 ml water. Sol in alc.

2345. Cinchonidine. *(8α,9R)-Cinchonan-9-ol;* cinchovatine; α-quinidine. $C_{19}H_{22}N_2O$; mol wt 294.40. C 77.52%, H 7.53%, N 9.52%, O 5.43%. Occurs in most varieties of cinchona bark, especially in bark of *Cinchona pubescens* Vahl. (*C. succirubra* Pav.) and *C. pitayensis* Wedd., *Rubiaceae.* Isoln: Leers, *Ann.* **82**, 147 (1852). Structure: Rabe, *ibid.* **365**, 359 (1909). Stereoisomeric with cinchonine:

Koenigs, *ibid.* **347**, 182 (1906). Configuration: Prelog, Zalán, *Helv. Chim. Acta* **27**, 535 (1944); Prelog, Häfliger, *ibid.* **33**, 2021 (1950); Roth, *Pharmazie* **16**, 257 (1961); Lyle, Keefer, *Tetrahedron* **23**, 3253 (1967). Biosynthetic studies: Battersby, Parry, *Chem. Commun.* **1971**, 31.

Orthorhombic plates, prisms from alcohol, mp 210°. $[\alpha]_D^{20}$ −109.2° (alc). uv absorption data: Kamath *et al., Indian J. Chem.* **6**, 510 (1968). Sol in alcohol and chloroform; moderately sol in ether. Practically insol in water. pK_1 5.80, pK_2 10.03. *Protect from light.* LD_{50} in rats: 206 mg/kg.

Dihydrochloride, $C_{19}H_{22}N_2O.2HCl$, white or slightly yellow crystals or powder. Freely sol in water or alcohol. *Protect from light.*

Hydrochloride dihydrate, $C_{19}H_{22}N_2O.HCl.2H_2O$, cryst powder; loses all of its H_2O at 120°. $[\alpha]_D^{20}$ −117.5° (water). Sol in 25 parts cold water, more sol in boiling water; sol in alcohol, chloroform, slightly in ether. The aq soln is practically neutral. *Protect from light.*

Sulfate trihydrate, $(C_{19}H_{22}N_2O)_2.H_2SO_4.3H_2O$, silky, acicular crystals; efflorescent on exposure to air and darkens in light. mp when anhydr ∼240° with decomp. One gram dissolves in 70 ml water, 20 ml hot water, 90 ml alcohol, 40 ml hot alc, 620 ml chloroform. Practically insol in ether. The aq soln is practically neutral. *Protect from light.*

Epicinchonidine, *(8α,9S)-cinchonan-9-ol,* mp 104°, $[\alpha]_D^{20}$ +63° (c = 0.804 in alc): Rabe *et al., Ann.* **492**, 253 (1932).

THERAP CAT: Antimalarial.

2346. Cinchonine. *(9S)-Cinchonan-9-ol.* $C_{19}H_{22}N_2O$; mol wt 294.40. C 77.52%, H 7.53%, N 9.52%, O 5.43%. Occurs in most varieties of cinchona bark, especially in bark of *Cinchona micrantha* R. & P., *Rubiaceae.* Isoln and structure: Rabe, *Ber.* **41**, 63 (1908). Stereoisomeric with cinchonidine: Koenigs, *Ann.* **347**, 182 (1906). Configuration: Prelog, Zalán, *Helv. Chim. Acta* **27**, 535 (1944); Prelog, Häfliger, *ibid.* **33**, 2021 (1950); Roth, *Pharmazie* **16**, 257 (1961); Lyle, Keefer, *Tetrahedron* **23**, 3253 (1967). Crystal and molecular structure: B. Oleksyn *et al., Acta Crystallogr.* **B35**, 440 (1979). Biosynthetic studies: Battersby, Parry, *Chem. Commun.* **1971**, 31. Toxicity study: C. C. Johnson, C. F. Poe, *Acta Pharmacol. Toxicol.* **4**, 265 (1949).

Prisms, needles from alcohol or ether. mp about 265°; begins to sublime at 220°. $[\alpha]_D$ +229° (alcohol). uv absorption data: Kamath *et al., Indian J. Chem.* **6**, 510 (1968). One gram dissolves in 60 ml alcohol, 25 ml boiling alcohol, 110 ml chloroform, 500 ml ether. Practically insol in water. pK_1 5.85, pK_2 9.92. *Protect from light.* LD_{50} i.p. in rats: 152 mg/kg (Johnson, Poe).

Dihydrochloride, $C_{19}H_{22}N_2O.2HCl$, white or faintly yellow crystals or crystalline powder. Freely sol in water or alcohol. *Protect from light.*

Hydrochloride dihydrate, $C_{19}H_{22}N_2O.HCl.2H_2O$, fine crystals. mp when anhydr about 215° with dec. One gram dissolves in 20 ml water, 3.5 ml boiling water, 1.5 ml alcohol, 20 ml chloroform; slightly sol in ether. The aq soln is practically neutral. *Protect from light.*

Sulfate dihydrate, $(C_{19}H_{22}N_2O)_2.H_2SO_4.2H_2O$, lustrous, very bitter crystals. mp when anhydr about 198°. One gram dissolves in 65 ml water, 30 ml hot water, 12.5 ml alc, 7 ml hot alc, 47 ml chloroform; slightly sol in ether. The aq soln is practically neutral.

Epicinchonine, *(9R)-cinchonan-9-ol,* mp 83°, $[\alpha]_D^{22}$ +120.3° (c = 0.806 in alc), Rabe *et al., Ann.* **492**, 253 (1932).

THERAP CAT: Antimalarial.

2347. Cinchophen. *2-Phenyl-4-quinolinecarboxylic acid; 2-phenylcinchoninic acid;* Atophan; Quinophan; Phenoquin; Agotan; Artam; Alutyl; Cinconal; Vantyl; Viophan; Atocin; Mylofanol; Rhematan; Rheumin; Tophol; Polyphlogin; Quinofen; Tophosan. $C_{16}H_{11}NO_2$; mol wt 249.27. C 77.10%, H 4.45%, N 5.62%, O 12.84%. Quinoline derivative formerly used in treatment of chronic gout: A. B. Gutman, *Arthritis Rheum.* **16**, 431 (1973). Prepd by heating pyruvic acid with aniline and benzaldehyde or with benzylidene aniline in abs alcohol: Doebner, Gieseke, *Ann.* **242**, 290 (1887). Also from acetophenone and isatinic acid in alcoholic KOH: Pfitzinger, *J. Prakt. Chem.* **38**, 582 (1882); **56**, 293 (1897). Cinchophen as ulcerogenic agent: T. P. Churchill, F. H. van Wagoner, *Arch. Path.* **13**, 850 (1932); N. Umeda *et al.*, *Toxicol. Appl. Pharmacol.* **18**, 102 (1971); T. H. Stewart *et al.*, *J. Pathol.* **131**, 363 (1980). Mitochondrial toxicity of cinchophen and derivatives: H. Vainio *et al.*, *Biochem. Pharmacol.* **20**, 1589 (1971).

Needles, mp 213-216°. Stable to air, but turns yellow under the influence of light. Slightly bitter taste. Practically insol in water. One gram dissolves in about 400 ml chloroform, in about 100 ml ether and in about 120 ml alcohol.

Sodium salt, water soluble. In combination with sodium salicylate as *Atophanyl, Arthrosan, Cincosal*.

Hydrochloride, $C_{16}H_{11}NO_2$.HCl, *Chloroxyl*. Yellow cryst powder, slightly bitter, astringent taste; mp about 223°. Practically insol in water. Slightly sol in alcohol; sol in chloroform, ether.

Allyl ester, $C_{19}H_{15}NO_2$, *allyl 2-phenylcinchoninate, Atoquinol*. Prepn: Gams, U.S. pat. **1,336,952** (1920). Long needles from alc, mp 30°. bp_{15} 260°. Practically insol in water. Readily sol in alcohol, ether, acetone, oils. Tasteless.

USE: Experimentally to induce ulcers.

THERAP CAT: Formerly as analgesic.

2348. Cinchotoxine. *(3R-cis)-3-(3-Ethenyl-4-piperidinyl)-1-(4-quinolinyl)-1-propanone; 9-deoxy-9-oxo-1,8-secocinchonine;* cinchonicine. $C_{19}H_{22}N_2O$; mol wt 294.40. C 77.52%, H 7.53%, N 9.52%, O 5.43%. A rearrangement product of cinchonine or cinchonidine obtained by boiling with acetic acid: Pasteur, *Compt. Rend.* **37**, 110 (1853). Miller, Rohde, *Ber.* **33**, 3214 (1900). Conversion to cinchonidinone and cinchonine: Rabe, *Ber.* **44**, 2088 (1911); **41**, 62 (1908). Prepn from cinchonine: Prostenik, Prelog, *Helv. Chim. Acta* **26**, 1965 (1943).

Bitter needles. mp 58-60°. $[\alpha]_D^{15}$ +47° in alcohol. Sparingly sol in water; sol in alcohol, ether, chloroform.

2349. Cinepazet. *4-[1-Oxo-3-(3,4,5-trimethoxyphenyl)-2-propenyl]-1-piperazineacetic acid ethyl ester; 4-(3,4,5-trimethoxycinnamoyl)-1-piperazineacetic acid ethyl ester;* ethyl 4-(3,4,5-trimethoxycinnamoyl)-1-piperazineacetate; 4-(3',4',5'-trimethoxycinnamoyl)-1-(ethoxycarbonylmethyl)piperazine; cinepazic acid ethyl ester; ethyl cinepazate. $C_{20}H_{28}N_2O_6$; mol wt 392.45. C 61.21%, H 7.19%, N 7.14%, O 24.46%. Prepn: C. Fauran *et al.*, Brit. pat. **1,168,108**;

eidem, U.S. pat. **3,590,034** (1969, 1971 both to Delalande). Synthesis and structure-activity studies of the hydrochloride: Fauran, Turin, *Chim. Ther.* **4**, 290 (1969); Huguet *et al.*, *ibid.* 293. Pharmacology and toxicology: Pourrias *et al.*, *Therapie* **26**, 845, 859 (1971). Metabolic data: Chasseaud *et al.*, *Arzneimittel-Forsch.* **22**, 2003 (1972).

Solid from isopropyl ether, mp 96°.

Hydrochloride, mp 200° (dec). LD_{50} in mice (mg/kg): 300 i.v., 1300 orally (Fauran, 1969).

Maleate, $C_{20}H_{28}N_2O_6.C_4H_4O_4$, *MD-6753, Vascoril*. Crystals, mp 130°.

THERAP CAT: Antianginal.

2350. Cinepazide. *1-[2-Oxo-2-(1-pyrrolidinyl)ethyl]-4-[1-oxo-3-(3,4,5-trimethoxyphenyl)-2-propenyl]piperazine; 1-[(1-pyrrolidinylcarbonyl)methyl]-4-(3,4,5-trimethoxycinnamoyl)piperazine;* 1-[4-[2-oxo-2-(1-pyrrolidinyl)ethyl]-1-piperazinyl]-3-(3,4,5-trimethoxyphenyl)-2-propen-1-one; 1-[4-[(3',4',5'-trimethoxycinnamoyl)-1-piperazinyl]acetyl]pyrrolidine. $C_{22}H_{31}N_3O_5$; mol wt 417.51. C 63.29%, H 7.48%, N 10.06%, O 19.16%. Prepn: Belg. pat. **730,345**; C. P. Fauran *et al.*, U.S. pat. **3,634,411** (1969, 1972 both to Delalande); C. P. Fauran, M. J. Turin, *Chim. Ther.* **4**, 290 (1969). Pharmacology: G. Huguet *et al.*, *ibid.* 293; E. Marmo *et al.*, *Farmaco Ed. Prat.* **28**, 132 (1973); B. M. Pourrias *et al.*, *Therapie* **29**, 29, 43 (1974).

Maleate, $C_{22}H_{31}N_3O_5.C_4H_4O_4$, *MD-67350, Brendil, Vasodistal*. Crystals from abs ethanol, mp 135°. LD_{50} in mice, rats (mg/kg): 1000, 1310 orally; 617, 414 i.v. (Pourrias).

THERAP CAT: Vasodilator (peripheral).

2351. Cinerins. Active insecticidal constituents of pyrethrum flowers. Isoln: Ward, *Chem. & Ind. (London)* **1953**, 586; Stephenson, *Pyrethrum Post* **5**, no. 4, 22 (1960); *C.A.* **55**, 13753g (1961). Structure: Schechter *et al.*, *J. Am. Chem. Soc.* **71**, 3165 (1949); Godin *et al.*, *J. Chem. Soc. (C)* **1966**, 332. Stereochemistry: Begley *et al.*, *Chem. Commun.* **1972**, 1276. *Review:* Crombie, Elliott, *Fortschr. Chem. Org. Naturst.* **19**, 120-164 (1961).

Cinerin I   R = CH_3
Cinerin II   R = $COOCH_3$

Cinerin I, $C_{20}H_{28}O_3$, *[1R-[1α[S*(Z)],3β]]-2,2-dimethyl-3-(2-methyl-1-propenyl)cyclopropanecarboxylic acid 3-(2-butenyl)-2-methyl-4-oxo-2-cyclopenten-1-yl-ester*. Viscous liquid. Oxidizes rapidly and becomes inactive in presence of air. $bp_{0.008}$ 136-138°. n_D^{20} 1.5064. $[\alpha]_D^{20}$ −22° (hexane). uv max: 222 nm (ε 21400). Practically insol in water. Sol in alcohol, petr ether, kerosene, carbon tetrachloride, ethylene dichloride, nitromethane.

Cinerin II, $C_{21}H_{28}O_5$, *[1R-[1α[S*(Z)],3β(E)]]-3-(3-meth-oxy-2-methyl-3-oxo-1-propenyl)-2,2-dimethylcyclopropanecar-boxylic acid 3-(2-butenyl)-2-methyl-4-oxo-2-cyclopenten-1-yl ester.* Viscous liquid. Oxidizes rapidly and becomes inactive in presence of air. $bp_{0.001}$ 182-184°. n_D^{20} 1.5183. $[α]_D^{16}$ +16° (isooctane). uv max: 229 nm (ε 28700). Practically insol in water. Sol in alcohol, petr ether (less sol than Cinerin I), kerosene, carbon tetrachloride, ethylene dichloride, nitro-methane.

Caution: Direct skin contact and inhalation may cause allergic attacks in sensitive people. Potential symptoms of overexposure are severe dermatitis, asthma, vasomotor rhi-nitis, anaphylactic reactions; numbness of lips and tongue, sneezing, vomiting, diarrhea; headache, restlessness, tinnitis, incoordination, clonic convulsions, stupor, prostration; death due to respiratory paralysis. *See Clinical Toxicology of Commercial Products,* R. E. Gosselin *et al.,* Eds. (Wil-liams & Wilkins, Baltimore, 5th ed., 1984) Section III, pp 352-355.

USE: Insecticide.

2352. Cinitapride. *4-Amino-N-[1-(3-cyclohexen-1-yl-methyl)-4-piperidinyl]-2-ethoxy-5-nitrobenzamide.* $C_{21}H_{30}N_4$-O_4, mol wt 402.49. C 62.67%, H 7.51%, N 13.92%, O 15.90%. Prepn: A. Vega-Noverola *et al.,* **Ger.** pat. **2,751,-139** (1978 to Anphar); *eidem,* **U.S.** pat. **5,026,858** (1991 to Walton). Pharmacology: R. Massingham *et al., J. Auton. Pharmacol.* **5,** 41 (1985). Spectral determn in pharmaceuti-cal prepns: M. Blanco *et al., J. Pharm. Sci.* **82,** 834 (1993). Clinical evaluation in functional dyspepsia, in comparison with metoclopramide, *q.v.:* F. Mora *et al., Ann. Med. In-tern.* **10,** 323 (1993).

Tartrate, *LAS-17177, Cidine.*
THERAP CAT: Gastroprokinetic.

2353. Cinmetacin. *5-Methoxy-2-methyl-1-(1-oxo-3-phenyl-2-propenyl)-1H-indole-3-acetic acid; 1-cinnamoyl-5-methoxy-2-methylindole-3-acetic acid; 1-cinnamoyl-2-methyl-5-methoxy-3-indolylacetic acid; Cindomet; Indolacin.* C_{21}-$H_{19}NO_4$; mol wt 349.39. C 72.19%, H 5.48%, N 4.01%, O 18.32%. Prepn: H. Yamamoto, M. Nakao, **S. Afr.** pat. **67 02,683;** *eidem,* **U.S.** pat. **3,576,800** (1968, 1971 both to Sumitomo); *eidem, J. Med. Chem.* **12,** 176 (1969). Pharma-cology: T. Komatsu *et al., Arzneimittel-Forsch.* **23,** 1690 (1973).

Yellow crystals from acetone-water, mp 170-172°; also re-ported as mp 164-165°. LD_{50} in mice, rats (mg/kg): 360, 590 i.p.; 750, 1020 orally (Komatsu).
THERAP CAT: Anti-inflammatory.

2354. Cinmethylin. *exo-(±)-1-Methyl-4-(1-methyleth-yl)-2-[(2-methylphenyl)methoxy]-7-oxabicyclo[2.2.1]heptane;* (±)-2-exo-(2-methylbenzyloxy)-1-methyl-4-isopropyl-7-oxa-bicyclo[2.2.1]heptane; SD-95481; Cinch. $C_{18}H_{26}O_2$; mol wt 274.40. C 78.79%, H 9.55%, O 11.66%. Pre-emergence grass herbicide. Member of the cineole (eucalyptol, *q.v.*) family. Prepn: G. B. Payne *et al.,* **Eur.** pat. **Appl. 81,893** (1983 to Shell); *eidem,* **U.S.** pat. **4,670,041** (1987 to Du Pont). Physical and chemical properties: B. T. Grayson *et al., Pestic. Sci.* **21,** 143 (1987). Mechanism of action study: M. H. El-Deek, F. D. Hess, *Weed Sci.* **34,** 684 (1986).

Metabolism in rats: P. W. Lee *et al., J. Agric. Food Chem.* **34,** 162 (1986). Mobility in soil: D. R. Wendt *et al., Proc. South. Weed Sci. Soc.* **1987,** 391; and herbicidal activity: P. C. Lolas, A. Galopoulos, *Zizaniology* **1,** 221 (1985). Field studies in tobacco: P. C. Lolas, *Brit. Crop Prot. Conf. - Weeds* **1985,** 841; in soybeans: P. C. Bhowmik, *Weed Sci.* **36,** 678 (1988).

Colorless liquid, bp 313° ±2. d^{20} 1015 kg/m³. Viscosity (20°): 70-90 mPa. Vapor pressure (20°): 7.6 × 10⁻⁵ mm Hg. Moderately volatile. Partition coefficient (*n*-octanol/-water): 6850. Miscible with most organic solvents. Slight-ly sol in water. Soly in water (20°): 63 ± 2 mg/l. LD_{50} orally in rats: 4.5 g/kg; LD_{50} dermally in rabbits: > 2 g/kg (Lee).

USE: Herbicide.

2355. Cinnabarine. *2-Amino-9-hydroxymethyl-3-oxo-3H-phenoxazine-1-carboxylic acid;* 1-carboxy-2-amino-9-hy-droxymethylphenoxazin-3-one; polystictine. $C_{14}H_{10}N_2O_5$; mol wt 286.24. C 58.75%, H 3.52%, N 9.79%, O 27.95%. Antibiotic substance produced by the fungus *Trametes cin-nabarina* Jacq.: Gripenberg, *Acta Chem. Scand.* **5,** 590 (1951); from *Coriolus sanguineus* (Fr.) (*Polystictus cinnabar-inus* Jacq.): Cavill *et al., J. Chem. Soc.* **1953,** 525. Identity with polystictine: Gripenberg *et al., Acta Chem. Scand.* **11,** 1485 (1957). Structure: Gripenberg, *ibid.* **12,** 603 (1958); **13,** 1305 (1959); Cavill *et al., Tetrahedron* **5,** 275 (1959). Syntheses: Weinstein, Brattesani, *J. Heterocyclic Chem.* **4,** 151 (1967); Schäfer, Schlude, *Tetrahedron Letters* **1968,** 2161.

Red needles, dec 320°. Absorption max: 234, 430, 455 nm (log ε 4.2, 4.0, 4.0). Soly in acetone: 0.04% (20°); in dioxane: 0.1% (100°). Practically insol in ethanol, chloro-form and dimethyl sulfoxide. Sol in cold concd HCl.
O-Methyl ether, $C_{15}H_{12}N_2O_5$, orange red prisms from ethyl acetate + petr ether, mp 200-202°.

2356. Cinnamaldehyde. *3-Phenyl-2-propenal;* cinnamic aldehyde; phenylacrolein; cinnamal. C_9H_8O; mol wt 132.16. C 81.79%, H 6.10%, O 12.11%. $C_6H_5CH{=}CHCHO$. Found in Ceylon and Chinese cinnamon oils. Prepn by condensation of benzaldehyde and acetaldehyde: Peine, *Ber.* **17,** 2117 (1884); **Japan.** pat. **163,097** (1944 to Ogawa Chem. Ind.); from 2-chloroallylbenzene: Bert, Dorier, *Compt. Rend.* **191,** 332 (1930); Bert, Annequin, *ibid.* **192,** 1315 (1931); by condensation of styrene with formylmethyl-aniline: **Brit.** pat. **504,125** (1939 to I. G. Farben); by oxida-tion of cinnamyl alc: Holum, *J. Org. Chem.* **26,** 4814 (1961); Traynelis, Hergenrother, *J. Am. Chem. Soc.* **86,** 298 (1964). Isoln from woodrotting fungus, *Stereum subpilea-tum* Berk. & Curt.: Birkinshaw *et al., Biochem. J.* **66,** 188 (1957). Toxicity data: P. M. Jenner *et al., Food Cosmet. Toxicol.* **2,** 327 (1964).

Yellowish oily liquid, strong odor of cinnamon. d_{15}^{25} 1.048-1.052. Volatile with steam. mp −7.5°. $bp_{1.0}$ 76.1°; bp_5 105.8°; bp_{10} 120.0°; bp_{20} 135.7°; bp_{40} 152.2°; bp_{60} 163.7°; bp_{100} 177.7°; bp_{200} 199.3°; bp_{400} 222.4°; bp_{760} 246.0° (some dec). n_D^{20} 1.618-1.623. Dissolves in about 700 parts water, in about 7 vols of 60% alc. Misc with alcohol, ether, chloroform, oils. LD_{50} in rats (mg/kg): 2220 orally (Jenner).

USE: In the flavor and perfume industry.

2357. Cinnamedrine. *α-[1-[Methyl(3-phenyl-2-propenyl)amino]ethyl]benzenemethanol; N-cinnamylephedrine;* α-[1-(*N*-cinnamyl-*N*-methylamino)ethyl]benzyl alcohol. $C_{19}H_{23}NO$; mol wt 281.40. C 81.10%, H 8.24%, N 4.98%, O 5.69%. Smooth muscle relaxant. Prepn: Stolz, Flaecher, U.S. pat. **1,959,392** (1934 to Winthrop Chem.); Welsh, Keenan, *J. Am. Pharm. Assoc., Sci. Ed.* **30**, 123 (1941).

dl-Hydrochloride, $C_{19}H_{23}NO.HCl$, bitter crystals from ethyl acetate + chloroform, mp 180-185° (base, mp 72-78°). Sol in about 80 parts water and 10 parts alc at 20°, more easily sol at higher temps; sol in 50 parts chloroform; sparingly sol in ether, benzene.

THERAP CAT: Antispasmodic.

2358. Cinnamic Acid. *3-Phenyl-2-propenoic acid;* β-phenylacrylic acid. $C_9H_8O_2$; mol wt 148.16. C 72.96%, H 5.44%, O 21.60%. $C_6H_5CH=CHCOOH$. The ordinary synthetic cinnamic acid which is described here is the *trans*-isomer. Occurs free and partly esterified in storax, balsam Peru or Tolu, oil of cinnamon, coca leaves. Isoln: Beilstein, Kuhlberg, *Ann.* **163**, 123 (1872); von Miller, *Ann.* **188**, 196 (1877). Synthesis (Perkin reaction) from benzaldehyde, acetic anhydride, and potassium acetate: *Org. Reactions* **I**, 248 (1942); from oxalyl bromide + styrene: Treibs *et al., Naturwiss.* **45**, 85 (1958); from acetylene + benzaldehyde: Herbetz, *Ber.* **92**, 541 (1959). Prepn of *cis*- and *trans*-isomers: Comte *et al., Compt. Rend.* **245**, 1144 (1957). Isoln from wood-rotting fungus, *Stereum subpileatum* Berk. & Curt.: Birkinshaw *et al., Biochem. J.* **66**, 188 (1957).

Monoclinic crystals, d_4^4 1.2475. mp 133°; bp 300°. Distilling at 146° causes decarboxylation to styrene. K at 25° = 3.5 × 10⁻⁵. uv max (alc): 273 nm. One gram dissolves in about 2000 ml water at 25° (more sol in hot water), in 6 ml alc, 5 ml methanol, 11 ml chloroform. Freely sol in benzene, ether, acetone, glacial acetic acid, carbon disulfide, oils. The alkali salts are sol in water.

Methyl ester, $C_{10}H_{10}O_2$, crystals, odor fruity and balsamic, reminiscent of strawberries. mp 36°. d_0^{36} 1.042. bp 261.9°; bp_{15} 132.5-134°. n_D^{21} 1.5766. Freely sol in alcohol, ether. Practically insol in water. Clearly sol in 80% alc.

Ethyl ester, $C_{11}H_{12}O_2$, *ethyl cinnamate, ethyl phenylacrylate.* Almost colorless, oily liquid, fruity and balsamic odor, reminiscent of cinnamon with an amber note. d_{25}^{25} 1.045-1.048. d_4^{20} 1.049. bp 271°. mp 6-10°. n_D^{20} 1.559-1.561. Miscible with alcohol, ether. Insol in water. Soluble in 3 vols of 70% alc.

n-Butyl ester, $C_{13}H_{16}O_2$, *butyl cinnamate, Eliminoxy.* Liquid, agreeable ethereal odor when pure. bp_{13} 145°. d_4^{18} 1.012. Very sparingly sol in water (< 0.5%). Sol in 95% alc, ether, acetone, chloroform, benzene. Incompatible with alkalies; stable to light, air, and storage temps.

Anhydride, $C_{18}H_{14}O_3$, *cinnamic anhydride.* Crystals, mp 136°. Practically insol in water. Freely sol in warm benzene; slightly sol in alc.

USE: The main use of cinnamic acid is in the manuf of the methyl, ethyl, and benzyl esters for the perfume industry. Ethyl ester used in glass prisms, lenses. Some esters, such as chaulmoogryl and other derivatives, are used in medicine.

2359. Cinnamon. Dried bark of cultivated trees of *Cinnamomum* sp, Lauraceae. *Constit.* Cinnamaldehyde, tannins, 1-4% volatile oil, sugar. *Review:* A. Y. Leung, *Encyclopedia of Common Natural Ingredients* (John Wiley & Sons, New York, 1980) pp 123-126.

Ceylon cinnamon, true cinnamon. Isolated from *C. zeylanicum* Nees. *Habit.* Ceylon, Sumatra, Borneo; cultivated in tropical Africa, America and Asia.

Saigon cinnamon. Isolated from *C. loureirii* Nees. *Habit.* China; cultivated in Java, Sumatra, S. America.

USE: Pharmaceutic aid (flavor).

THERAP CAT: Carminative.

THERAP CAT (VET): Has been used as a carminative and antidiarrheal.

2360. Cinnamoyl Chloride. *3-Phenyl-2-propenoyl chloride.* C_9H_7ClO; mol wt 166.61. C 64.88%, H 4.23%, Cl 21.28%, O 9.60%. $C_6H_5CH=CHCOCl$. Prepn by the action of oxalyl chloride on sodium cinnamate: Adams, Ulich, *J. Am. Chem. Soc.* **42**, 605 (1920).

Yellowish crystals. mp 35-36°; bp_{58} 170-171°; bp_{25} 154°; bp_{16} 147°; bp_2 101°; $d_4^{45.3}$ 1.1617; $n_D^{42.5}$ 1.614.

USE: Titrimetric determination of small amts of water.

2361. Cinnamoylcocaine. *[1R-(exo,exo)]-8-Methyl-3-[(1-oxo-3-phenyl-2-propenyl)oxy]-8-azabicyclo[3.2.1]octane-2-carboxylic acid methyl ester; ecgonine cinnamate methyl ester;* cinnamoylecgonine methyl ester; cinnamylcocaine; cinnamoylmethylecgonine. $C_{19}H_{23}NO_4$; mol wt 329.40. C 69.28%, H 7.04%, N 4.25%, O 19.43%. From leaves of *Erythroxylon coca* Lam., *Erythroxylaceae,* particularly from the Javanese leaves: Liebermann, *Ber.* **22**, 2661 (1889); de Jong, *Rec. Trav. Chim.* **67**, 484 (1948); Hegnauer, Fikenscher, *Pharm. Acta Helv.* **35**, 43 (1960); from *E. monogynum* Roxb.: Chopra, Gosh, *Arch. Pharm.* **276**, 340 (1938).

(E)-form

Needles, mp 121°. $[α]_D$ −4.7° (chloroform). Almost insol in water; sol in alcohol, chloroform, ether.

2362. Cinnamyl Alcohol. *3-Phenyl-2-propen-1-ol;* cinnamic alcohol; styryl carbinol; γ-phenylallyl alcohol. $C_9H_{10}O$; mol wt 134.18. C 80.56%, H 7.51%, O 11.92%. $C_6H_5CH=CHCH_2OH$. Occurs (in the esterified form) in storax and in balsam Peru, cinnamon leaves, hyacinth oil. Obtained by the alkaline hydrolysis of storax. Prepd synthetically by reducing cinnamal diacetate with iron filings and acetic acid; from cinnamaldehyde by Meerwein-Ponndorf reduction with aluminum isopropoxide: Meerwein, Schmidt, *Ann.* **444**, 221 (1925).

Needles or cryst mass. Odor of hyacinth. mp 33°. d_{35}^{35} 1.0397. $bp_{1.0}$ 72.6°; bp_5 102.5°; bp_{10} 117.8°; bp_{20} 133.7°; bp_{40} 151.0°; bp_{60} 162.0°; bp_{100} 177.8°; bp_{200} 199.8°; bp_{400} 224.6°; bp_{760} 250.0°. n_D^{20} 1.58190; n_D^{33} 1.57580. When small amounts of impurities are present as in the natural article (cinnamyl alcohol from storax), it remains fluid at lower temps than the melting point. Minimum congealing points specified by the Essential Oil Assn are: 33.0° for cinnamic alcohol pure; 28.0° for cinnamic alcohol prime; 20.0° for cinnamic alcohol from storax. Is oxidized slowly on exposure to heat, light and air. Sol in water, glycerol. Clearly sol in 3 vols 50% alc. Freely sol in alc, ether, other common organic solvents.

USE: In perfumery; as deodorant in 12.5% soln in glycerol.

2363. Cinnamyl Anthranilate. *2-Aminobenzoic acid 3-phenyl-2-propenyl ester; anthranilic acid cinnamyl ester;* 3-phenyl-2-propenyl anthranilate; cinnamyl *o*-aminobenzoate. $C_{16}H_{15}NO_2$; mol wt 253.30. C 75.87%, H 5.97%, N 5.53%, O 12.63%. Synthetic imitation grape or cherry flavoring agent. Prepn: A. Seldner, *Am. Perfumer Essent. Oil Rev.* **54**, 295 (1949); R. P. Staiger, E. B. Miller, *J. Org. Chem.* **24**, 1214 (1959). Carcinogenicity studies: G. D. Stoner *et al., Cancer Res.* **33**, 3069 (1973); *Fed. Reg.* **45**, 85832 (1980). *Review:* D. L. Opdyke, *Food Cosmet. Toxicol.* **13**, Suppl., 751-752 (1975).

(E)-form

Crystals, mp 61-61.5°.

USE: Flavoring agent in food; fragrance in soaps and perfumes.

2364. Cinnamyl Cinnamate. *3-Phenyl-2-propenoic acid 3-phenyl-2-propenyl ester; cinnamic acid cinnamyl ester; cinnyl cinnamate; styracin.* $C_{18}H_{16}O_2$; mol wt 264.32. C 81.79%, H 6.10%, O 12.11%. $C_6H_5CH=CHCOOCH_2-CH=CHC_6H_5$. From buds of *Populus balsamifer* L., *Salicaceae:* Goris, Canal, *Bull. Soc. Chim. France* **3**, 1982 (1936); from *Lavanga scandens* Buch.-Ham., *Lavangalata:* Baslas, Deshapande, *J. Indian Chem. Soc.* **27**, 379 (1950). Synthesis of *trans*-cinnamyl *trans*-cinnamate: Klemm *et al.*, *Tetrahedron* **20**, 871 (1964).

trans-trans-Form, needles from abs ethanol, mp 44°. uv max (95% ethanol): 216, 223 nm (log ϵ 3.45, 3.25). Practically insol in water. 1 g dissolves in 22 ml cold, 3 ml boiling alcohol, 3 ml ether.

2365. Cinnarizine. *1-(Diphenylmethyl)-4-(3-phenyl-2-propenyl)piperazine; 1-cinnamyl-4-diphenylmethylpiperazine; N-benzhydryl-N'-trans-cinnamylpiperazine; 1-trans-cinnamyl-4-diphenylmethylpiperazine; 1-cinnamyl-4-benzhydrylpiperazine; 1-diphenylmethyl-4-trans-cinnamylpiperazine;* cinnipirine; 516-MD; Aplactan; Aplexal; Apotomin; Artate; Carecin; Cerebolan; Cerepar; Cinaperazine; Cinazyn; Cinnacet; Cinnageron; Corathiem; Denapol; Dimitron; Eglen; Folcodal; Giganten; Glanil; Hilactan; Ixertol; Katoseran; Labyrin; Midronal; Mitronal; Olamin; Processine; Sedatromin; Sepan; Siptazin; Spaderizine; Stugeron; Stutgeron; Stutgin; Toliman. $C_{26}H_{28}N_2$; mol wt 368.52. C 84.74%, H 7.66%, N 7.60%. Calcium channel blocker with anti-allergic and anti-vasoconstricting activity. Prepn: Janssen, U.S. pat. **2,882,271** (1959, Janssen). Pharmacology: Van Nueten, Janssen, *Arch. Int. Pharmacodyn. Ther.* **204**, 37 (1973). Metabolism: Soudijn, van Wijngaarden, *Life Sci.* **7**, 231 (1968). Clinical study in treatment of intermittent claudication: J. H. Barber *et al.*, *Pharmatherapeutica* **2**, 401 (1980). Assessment of calcium antagonist effects: M. Spedding, *Arch. Pharmacol.* **318**, 234 (1982). HPLC determn in blood and plasma: M. Puttemans *et al.*, *J. Liq. Chromatog.* **7**, 2237 (1984). Comparison of clinical pharmacokinetics with other antihistamines: D. M. Paton, D. R. Webster, *Clin. Pharmacol.* **10**, 477-497 (1985).

Hydrochloride, $C_{26}H_{28}N_2 \cdot HCl$, mp 192° (dec). Soly in water: 2 mg/100 ml.

Mixture with vitamin B_6, *Emesazine*.

THERAP CAT: Antihistaminic. Vasodilator (peripheral, cerebral).

2366. Cinnoline. *1,2-Benzodiazine; benzo[c]pyridazine; 1,2-diazanaphthalene; α-phenodiazine.* $C_8H_6N_2$; mol wt 130.15. C 73.83%, H 4.65%, N 21.52%. First prepd by reduction of 4-chlorocinnoline: Busch, Rast, *Ber.* **30**, 521, 524 (1897). Synthesis from methyl anthranilate: Jacobs *et al.*, *J. Am. Chem. Soc.* **68**, 1310 (1946); from cinnoline-4-carboxylic acid: Morley, *J. Chem. Soc.* **1951**, 1971. Crystal and molecular structure: C. Huiszoon *et al.*, *Acta Crystallogr.* **33B**, 1867 (1977).

Pale yellow clusters of crystals from ligroin, mp 38°; also reported as mp 40-41° (Morley). Solvated needles from ether (contg 1 mol ether), mp 24-25°. $bp_{0.35}$ 114°. uv spectrum: Hearn *et al.*, *J. Chem. Soc.* **1951**, 3318. Geranium-like odor. *Poisonous!* Bitter taste resembling the taste of quinine. Must be kept under nitrogen. On exposure to air it has a tendency to liquefy and turn green. Strong base: *ibid.* **1949**, 1356. Freely sol in the usual solvents.

Hydrochloride, $C_8H_6N_2 \cdot HCl$, brownish needles from alcohol + ether, mp 156-160°. Volatilizes around 100° on slow heating. Freely sol in water, alcohol.

2367. Cinobufotalin. *(3β,5β,15β,16β)-16-(Acetyloxy)-14,15-epoxy-3,5-dihydroxybufa-20,22-dienolide; 14,15β-epoxy-3β,5,16β-trihydroxy-5β-bufa-20,22-dienolide 16-acetate.* $C_{26}H_{34}O_7$; mol wt 458.55. C 68.10%, H 7.47%, O 24.42%. Isolated from the Chinese drug Ch'an Su which is prepd from Chinese toads (*Bufo asiaticus* = *bufo gargarizans* Cantor). Isoln: Kotake, Kuwada, *C.A.* **31**, 7065² (1937); Ruckstubl, Meyer, *Helv. Chim. Acta* **40**, 1270 (1957). Structure: Bernoulli *et al.*, *ibid.* **45**, 240 (1962).

Octahedrons from acetone, mp 259-262°. $[\alpha]_D^{20}$ +11°. uv max: 295 nm (log ϵ 3.72).

2368. Cinolazepam. *7-Chloro-5-(2-fluorophenyl)-2,3-dihydro-3-hydroxy-2-oxo-1H-1,4-benzodiazepine-1-propanenitrile;* 1-(2-cyanoethyl)-7-chloro-3-hydroxy-5-(2'-fluorophenyl)-1,3-dihydro-2H-1,4-benzodiazepin-2-one; OX-373; Gerodorm. $C_{18}H_{13}ClFN_3O_2$; mol wt 357.77. C 60.43%, H 3.66%, Cl 9.91%, F 5.31%, N 11.74%, O 8.94%. Prepn: L. H. Schlager, Ger. pat. **2,950,235**; *idem*, U.S. pat. **4,388,313** (1980, 1983 both to Gerot). Benzodiazepine receptor affinity: W. Sieghart, A. Schuster, *Biochem. Pharmacol.* **33**, 4033 (1984). Polarographic and HPLC analyses: H. Oelschläger *et al.*, *Arch. Pharm.* **325**, 65 (1992). Clinical pharmacokinetics: J. Grünberger *et al.*, *Adv. Ther.* **4**, 84 (1987). Clinical evaluation in insomnia: B. Saletu *et al.*, *Int. J. Clin. Pharm. Res.* **7**, 407 (1987).

Crystals from ethanol, mp 190°-193°.

THERAP CAT: Sedative, hypnotic.

2369. Cinoxacin. *1-Ethyl-1,4-dihydro-4-oxo[1,3]dioxolo[4,5-g]cinnoline-3-carboxylic acid;* 1-ethyl-6,7-methylenedioxy-4(1H)-oxocinnoline-3-carboxylic acid; compd 64716; Cinobac; Noxigram; Uronorm. $C_{12}H_{10}N_2O_5$; mol wt 262.22. C 54.97%, H 3.84%, N 10.68%, O 30.51%. Quinolone antibacterial; analog of oxolinic acid, *q.v.* Prepn: W. A. White, **Ger.** pat. **2,005,104;** *idem,* **U.S.** pat. **3,669,965** (1970, 1972 both to Lilly). *In vitro* and *in vivo* study: W. E. Wick *et al., Antimicrob. Ag. Chemother.* **4,** 415 (1973). Pharmacology: J. J. Szwed *et al., J. Antimicrob. Chemother.* **4,** 451 (1978); K. S. Israel *et al., J. Clin. Pharmacol.* **18,** 491 (1978). Metabolism: R. L. Wolen *et al.* in *Stable Isotopes,* T. A. Baillie, Ed. (Univ. Park Press, Baltimore, 1978) pp 113-125. Clinical studies: S. Colleen *et al., J. Antimicrob. Chemother.* **3,** 579 (1977); S. N. Rous, *J. Urol.* **120,** 196 (1978). Toxicity: I. Narama *et al., Chemotherapy (Tokyo)* **28,** Suppl. 4, 406 (1980), *C.A.* **94,** 25007 (1981). Pharmacokinetics: M. Ohkawa *et al., J. Antimicrob. Chemother.* **8,** 447 (1981); R. Barbhaiya *et al., Antimicrob. Ag. Chemother.* **21,** 472 (1982). *Review:* J. M. Scavone *et al., Pharmacotherapy* **2,** 266-271 (1982). Review of pharmacology and therapeutic efficacy: T. S. Sisca *et al., Drugs* **25,** 544-569 (1983).

Light tan crystals, mp 261-262° (dec). Sol in most polar organic solvents. LD_{50} in rats (mg/kg): 4160 orally, 900 i.v. (Narama).

Sodium salt, $C_{12}H_9N_2NaO_5$, white crystalline solid. Sol in aqueous solvents.

THERAP CAT: Antibacterial.

2370. Cinoxate. *3-(4-Methoxyphenyl)-2-propenoic acid 2-ethoxyethyl ester; p-methoxycinnamic acid 2-ethoxyethyl ester;* 2-ethoxyethyl *p*-methoxycinnamate; Give-Tan; Sun-Dare. $C_{14}H_{18}O_4$; mol wt 250.29. C 67.18%, H 7.25%, O 25.57%. Prepn: **Brit.** pat. **856,411** (1960 to Givaudan).

(E)-form

Viscous liq; bp_2 184-187°. May have a slightly yellow tinge. Practically odorless. Solidifies below −25°. d_{25}^{25} 1.102. n_D^{20} 1.5670. Sapon no. 225.5. Practically insol in water (soly ∼0.05%). Soly in glycerol 0.5%, in propylene glycol 5.0%. Miscible with alcohols, esters, vegetable oils.

USE: Sunscreen agent.

2371. Cinromide. *(E)-3-(3-Bromophenyl)-N-ethyl-2-propenamide; trans-3-bromo-N-ethylcinnamamide;* BW-122U; Vumide. $C_{11}H_{12}BrNO$; mol wt 254.13. C 51.99%, H 4.76%, Br 31.44%, N 5.51%, O 6.30%. Cinnamic acid derivative with anti-epileptic activity. Prepn: E. M. Grivsky, **Ger.** pat. **2,535,599** corresp to **U.S.** pat. **4,041,071** (1976, 1977 to Burroughs Wellcome). Pharmacology: F. E. Soroko *et al., J. Pharm. Pharmacol.* **33,** 741 (1981); G. H. Fromm *et al., Epilepsia* **24,** 394 (1983). Pharmacokinetics and metabolism: R. R. Maddox, B. B. Wannamaker, *Curr. Ther. Res.* **32,** 165 (1982). Mass spec determn in plasma and urine: R. J. Perchalski *et al., Anal. Chem.* **54,** 1466 (1982). Controlled clinical trials: J. W. Allen *et al., Epilepsia* **24,** 422 (1983); B. Spilker *et al., Curr. Ther. Res.* **34,** 7 (1983).

Crystals from ethanol-water, mp 89-90°. LD_{50} in mice (mg/kg): 660 ±28 i.p.; 2277 ±250 orally (Soroko).

THERAP CAT: Anticonvulsant.

2372. Cioteronel. *Hexahydro-4-(5-methoxyheptyl)-2(1H)-pentalenone;* 6-(5-methoxy-1-heptyl)bicyclo[3.3.0]octan-3-one; 2-(5-methoxyhept-1-yl)bicyclo[3.3.0]octan-7-one; X-Andron; CPC-10997; Cyoctol; Exandron. $C_{16}H_{28}O_2$; mol wt 252.40. C 76.14%, H 11.18%, O 12.68%. Topically active nonsteroidal androgen-receptor blocker. Prepn: W. J. Kasha, **PCT Int.** pat **Appl. 83 04,019** (1983 to CBD Corp.), *C.A.* **101,** 23037k (1984). Prepn (not claimed) and use: *idem,* **U.S.** pat. **4,689,345** (1987 to CBD Corp.). Antiandrogenic and antineoplastic activity *in vitro:* L. C. Ford *et al., Chemotherapy (Basel)* **31,** 362 (1985). Series of articles on inhibition of human and microbial dihydrotestosterone-receptor binding *in vitro: Rec. Adv. Chemother., Proc. 14th Int. Congr. Chemother. Antimicrob. Sect. 1* (Univ. Tokyo Press, Tokyo, 1985) pp 261-270; 273-274. Cutaneous metabolism in humans: R. A. de Zeeuw *et al., Pharm. Res.* **7,** 638 (1990). Clinical pharmacokinetics: J. W. Wiechers *et al., Int. J. Pharm.* **65,** 77 (1990).

Clear, colorless oil.

THERAP CAT: Antiacne; antialopecia agent.

2373. Ciprofibrate. *2-[4-(2,2-Dichlorocyclopropyl)-phenoxy]-2-methylpropanoic acid;* 2-[4-(2,2-dichlorocyclopropyl)phenoxy]isobutyric acid; Win-35833; Ciprol; Lipanor; Modalim. $C_{13}H_{14}Cl_2O_3$; mol wt 289.16. C 54.00%, H 4.88%, Cl 24.52%, O 16.60%. Hypolipemic agent, related structurally to clofibrate, *q.v.* Prepn: D. K. Phillips, **Ger.** pat. **2,343,606** corresp to **U.S.** pat. **3,948,973** (1974, 1976 both to Sterling). Pharmacokinetics in animals and man: C. Davison *et al., Drug Metab. Dispos.* **3,** 520 (1975). Inhibition of cholesterol biosynthesis in rats: A. Arnold *et al., Atherosclerosis* **32,** 155 (1979). Blood levels, distribution, duration of action: J. Edelson *et al., ibid.* **33,** 351 (1979). Metabolic effects: A. Arnold *et al., J. Pharm. Sci.* **68,** 1557 (1979). Teratology study: H. Tuchman-Duplessis *et al., Toxicology* **12,** 1 (1979).

Pale cream solid from hexane, mp 114-116°.

THERAP CAT: Antihyperlipoproteinemic.

2374. Ciprofloxacin. *1-Cyclopropyl-6-fluoro-1,4-dihydro-4-oxo-7-(1-piperazinyl)-3-quinolinecarboxylic acid;* Bay q 3939. $C_{17}H_{18}FN_3O_3$; mol wt 331.35. C 61.62%, H 5.48%, F 5.73%, N 12.68%, O 14.49%. Fluorinated quinolone antibacterial. Prepn: K. Grohe *et al.,* **Ger.** pat. **3,142,854;** *eidem,* **U.S.** pat. **4,670,444** (1983, 1987 both to Bayer AG); K. Grohe, H. Heitzer, *Ann.* **1987,** 29. Antibacterial spectrum *in vitro:* B. Watt, F. V. Brown, *J. Antimicrob. Chemother.* **17,** 605 (1986); C. M. Bassey *et al., ibid.* 623. HPLC determn in biological fluids: W. Gau *et al., J. Liq. Chromatog.* **8,** 485 (1985). Pharmacokinetics: G. Hoffken *et al., Antimicrob. Ag. Chemother.* **27,** 375 (1985). Clinical trials: C. A. Ramirez *et al., ibid.* **28,** 128 (1985); B. E. Scully *et al., Lancet* **1,** 819 (1986). Symposia on antibacterial spectrum and clinical use: *Am. J. Med.* **82,** Suppl. 4A, 1-404 (1987); *J. Antimicrob. Chemother.* **26,** Suppl. F, 3-193 (1990). Review of clinical safety and efficacy in children: R. Kubin, *Infection* **21,** 413-421 (1993).

Dec 255-257°.

Monohydrochloride monohydrate, $C_{17}H_{18}FN_3O_3 \cdot HCl \cdot H_2O$, *Bay o 9867, Baycip, Ciflox, Ciloxan, Ciprinol, Cipro, Ciprobay, Ciproxan, Ciproxin, Flociprin, Septicide, Velmonit.* Light yellow crystalline powder. mp 318-320°.

THERAP CAT: Antibacterial.

2375. Ciramadol. *3-[(Dimethylamino)(2-hydroxycyclohexyl)methyl]phenol;* (−)-*cis*-2-(α-dimethylamino-*m*-hydroxybenzyl)cyclohexanol; Wy-15705. $C_{15}H_{23}NO_2$; mol wt 249.35. C 72.25%, H 9.30%, N 5.62%, O 12.83%. Analgesic with narcotic agonist-antagonist properties. Prepn: J. P. Yardley, P. B. Russel, **Ger. pat. 2,317,183;** *eidem,* **U.S. pats. 3,928,626, 4,017,637** (1973, 1975 and 1977 all to Am. Home). Synthesis, stereochemistry, biological activity: J. P. Yardley *et al., Experientia* **34,** 1124 (1978). Determn in pharmaceutical dosage forms: H. K. Chan, A. G. Fogg, *Anal. Chim. Acta* **105,** 423 (1979). Metabolism: C. O.Tio *et al., Fed. Proc.* **38,** 2717 (1979). Respiratory effects in man: H. P. Wuest, *Arch. Int. Pharmacodyn. Ther.* **240,** 45 (1979). Analgesic effect in chronic pain: M. J. Staquet, *Curr. Med. Res. Opin.* **6,** 475 (1980).

Cryst from acetone/hexane, mp 191-193°. $[\alpha]_D^{25}$ −46.92° (c = 1.061 in methanol).

Hydrochloride, $C_{15}H_{23}NO_2 \cdot HCl$, *Ciradol.* Cryst from acetone/methanol, mp 255-257° (effervescence). $[\alpha]_D^{25}$ −15.31° (c = 1.067 in methanol).

THERAP CAT: Analgesic.

2376. Circulins. Polypeptide antibiotics related to polymyxins. Produced by *Bacillus circulans,* a microorganism found in soil and dust: Murray, Tetrault, *Proc. Soc. Am. Bact.* **1,** 20 (1948); McLeod, *J. Bacteriol.* **56,** 749 (1948); Murray *et al., ibid.* **57,** 305 (1949). Isoln yields circulins A and B, circulin A being the major component: Tetrault, **U.S. pat. 2,676,133** (1954 to Purdue Res. Found.). *Review:* Vogner, Studer, *Experientia* **22,** 345 (1966).

Circulin A, $C_{53}H_{100}N_{16}O_{13}$. Structure: Fujikawa *et al., Experientia* **21,** 307 (1965). Synthesis: Studer *et al., Helv. Chim. Acta* **49,** 974 (1966).

Sulfate, $C_{53}H_{100}N_{16}O_{13} \cdot 2\frac{1}{2}H_2SO_4$, crystals or amorphous solid, dec 226-228°. $[\alpha]_D^{25}$ −61.6° (c = 1.25). Sol in water; less sol in the lower alcs. Practically insol in acetone and water-immiscible solvents.

Circulin B, $C_{52}H_{98}N_{16}O_{13}$. Structure: Hayashi *et al., Experientia* **24,** 656 (1968).

2377. Cisapride. *cis-4-Amino-5-chloro-N-[1-[3-(4-fluorophenoxy)propyl]-3-methoxy-4-piperidinyl]-2-methoxybenzamide; cis-*4-amino-5-chloro-*N*-[1-[3-(*p*-fluorophenoxy)-propyl]-3-methoxy-4-piperidinyl]-*o*-anisamide. $C_{23}H_{29}ClFN_3O_4$; mol wt 465.95. C 59.29%, H 6.27%, Cl 7.61%, F 4.08%, N 9.02%, O 13.73%. Gastrointestinal prokinetic agent. Prepn: G. Van Daele, **Eur. pat. Appl. 76,530** (1983 to Janssen), *C.A.* **99,** 194812d (1983); and pharmacology: G. Van Daele *et al., Drug Dev. Res.* **8,** 225 (1986). Pharmacokinetics: J. A. Barone *et al., Clin. Pharm.* **6,** 640 (1987). Review of pharmacology and clinical evaluations: A. Reyntjens *et al., Drug. Dev. Res.* **8,** 251 (1986). Series of articles on pharmacology, mechanism of action and clinical efficacy: *Curr. Ther. Res.* **36,** 1029-1070 (1984).

Monohydrate, $C_{23}H_{29}ClFN_3O_4 \cdot H_2O$, *R-51619, Acenalin, Alimix, Cipril, Prepulsid, Propulsin, Risamal.* Crystals from 2-propanol, mp 109.8°.

THERAP CAT: Peristaltic stimulant.

2378. Cisplatin. *(SP-4-2)-Diamminedichloroplatinum; cis-diamminedichloroplatinum; cis*-platinum II; *cis*-DDP; CACP; CPDC; DDP; NSC-119875; Briplatin; Cismaplat; Cisplatyl; Citoplatino; Lederplatin; Neoplatin; Platamine; Platinex; Platiblastin; Platinol; Platinoxan; Platistin; Platosin; Randa. $Cl_2H_6N_2Pt$; mol wt 300.05. Cl 23.63%, H 2.02%, N 9.34%, Pt 65.02%. Prepn: M. Peyrone, *Ann.* **51,** 1 (1845); G. B. Kauffman, D. O. Cowan, *Inorg. Syn.* **7,** 239 (1963); S. C. Dhara, *Indian J. Chem.* **8,** 193 (1970). Originally known as *Peyrone's salt* or *Peyrone's chloride,* it was primarily of interest in the development of coordination theory. Early structural studies: R. Werner, *Z. Anorg. Chem.* **3,** 267 (1893); H. D. K. Drew *et al., J. Chem. Soc.* **1932,** 988. Discovery of anti-tumor activity: B. Rosenberg *et al., Nature* **205,** 698 (1965); **222,** 385 (1972). Use as neoplasm inhibitor: M. L. Tobe *et al.,* **Ger. pat. 2,318,020** (1972 to Rustenburg Platinum Mines Ltd.), *C.A.* **80,** 55897e (1974); M. J. Cleare *et al.,* **Ger. pat. 2,329,485** (1972 to Research Corp.), *C.A.* **81,** 21172v (1974). Mechanism of action: J. J. Roberts, J. M. Pascoe, *Nature* **235,** 282 (1972); A. Khan, J. M. Hill, *Transplant* **13,** 55 (1972); G. L. Cohen *et al., Science* **203,** 1014 (1979). X-ray structure of cisplatin-DNA adduct: S. E. Sherman *et al., Science* **230,** 412 (1985). Inhibition of *in vitro* DNA synthesis: A. L. Pinto, S. J. Lippard, *Proc. Natl. Acad. Sci. USA* **82,** 4616 (1985). Pharmacology: A. Sirica *et al., Proc. Am. Assoc. Cancer Res.* **12,** 4 (1971); C. L. Litterst *et al., Cancer Res.* **36,** 2340 (1976); N. P. Johnson *et al., Chem. Biol. Interact.* **23,** 267 (1978). Metabolism: R. C. Lange *et al., J. Nucl. Med.* **14,** 191 (1973). Clinical studies: J. J. Ochs *et al., Cancer Treat. Rep.* **62,** 239 (1978); H. M. Pinedo *et al., Eur. J. Cancer* **14,** 1149 (1978). Cisplatin has been shown to have trypanocidal effects: M. S. Wysor *et al., Science* **217,** 454 (1982). Toxicology: R. L. Dixon, *Proc. 7th Int. Congr. Chemother.* Vol. 2 (University Park Press, Baltimore, 1972) pp 241-243; R. W. Fleishman *et al., Toxicol. Appl. Pharmacol.* **33,** 320 (1975). *Reviews:* D. D. Von Hoff, M. Rozencweig in *Advances in Pharmacology and Chemotherapy* vol. 16, S. Garattini *et al.,* Eds. (Academic Press, New York, 1979) pp 273-294; S. J. Lippard, *Science* **218,** 1075-1082 (1982). Review of carcinogenicity studies: *IARC Monographs* **26,** 154-161 (1981); of neurotoxicity: R. J. Cersosimo, *Cancer Treat. Rev.* **16,** 195-211 (1989). Comprehensive description: C. M. Riley, L. M. Sternson, *Anal. Profiles Drug Subs.* **14,** 77-105 (1985). Book: *Cisplatin, Current Status and New Developments,* A. W. Prestayko *et al.,* Eds. (Academic Press, New York, 1980) 527 pp.

Deep yellow solid, dec 270°. Soly in water 0.253 g/100 g at 25°; slowly changes to *trans*-form in aq soln. Insol in most common solvents. Sol in DMF. LD_{50} in guinea pigs: 9.7 mg/kg i.p. (Fleishman).

Note: This substance may reasonably be anticipated to be a carcinogen: *Seventh Annual Report on Carcinogens* (PB95-109781, 1994) p 134.

THERAP CAT: Antineoplastic.

2379. Citalopram. *1-[3-(Dimethylamino)propyl]-1-(4-fluorophenyl)-1,3-dihydro-5-isobenzofurancarbonitrile;* 1-[3-(dimethylamino)propyl]-1-(4-fluorophenyl)-5-phthalancarbonitrile; nitalapram; LU-10-171. $C_{20}H_{21}FN_2O$; mol wt 324.40. C 74.05%, H 6.52%, F 5.86%, N 8.64%, O 4.93%. Serotonin uptake inhibitor. Prepn: K. P. Boegesøe, A. S. Toft, **Ger.** pat. **2,657,013;** *eidem,* **U.S.** pat. **4,136,193** (1977, 1979 both to Kefalas); A. J. Bigler *et al., Eur. J. Med. Chem. Chim. Ther.* **12,** 289 (1977). Pharmacology: A. V. Christensen *et al., Eur. J. Pharmacol.* **41,** 153 (1977). Pharmacokinetics: K. F. Overø, *Eur. J. Clin. Pharmacol.* **14,** 69 (1978). HPLC determn in plasma and urine: E. Oyehaug *et al., J. Chromatog.* **308,** 199 (1984). Clinical evaluations in depression: H. Dufour *et al., Int. Clin. Psychopharmacol.* **2,** 225 (1987); L. Timmerman *et al., ibid.* 239. Analgesic effect in diabetic neuropathy: S. H. Sindrup *et al., Clin. Pharmacol. Ther.* **52,** 547 (1992).

bp$_{0.03}$ 175-181°.
Hydrobromide, $C_{20}H_{21}FN_2O$·HBr, *Cipramil.* Crystals from isopropanol, mp 182-183°.

THERAP CAT: Antidepressant.

2380. Citicoline. *Cytidine 5′-(trihydrogen diphosphate) P′-[2-(trimethylammonio)ethyl] ester inner salt;* choline cytidine 5′-pyrophosphate (ester); cytidine diphosphate choline ester; CDP-choline; Audes; Cereb; Citifar; Colite; Corenalin; Cyscholin; Difosfocin; Emicholine-F; Ensign; Haocolin; Hornbest; Neucolis; Nicholin; Nicolin; Niticolin; Reagin; Recognan; Rexort; Sintoclar; Somazina; Suncholin; $C_{14}H_{26}N_4O_{11}P_2$; mol wt 488.33. C 34.43%, H 5.37%, N 11.47%, O 36.04%, P 12.69%. Naturally occurring nucleotide; coenzyme in one of the biosynthetic pathways of lecithin synthesis. Isoln from liver and yeast: Kennedy, Weiss, *J. Am. Chem. Soc.* **77,** 250 (1955); Lieberman *et al., Science* **124,** 81 (1956). Molecular structure: M. A. Viswamitra *et al., Nature* **258,** 497 (1975). Synthesis: Kennedy, *J. Biol. Chem.* **222,** 185 (1956); **Japan.** pats. **64 6541** and **67 1384** (1964, 1967 both to Takeda), *C.A.* **61,** 13406c (1964) and *C.A.* **66,** 85992k (1967); Kikugawa *et al., Chem. Pharm. Bull.* **19,** 1011, 2466 (1971); N. H. Phuong *et al., Bull. Soc. Chim. France* **1979**(pt 2), 518. Biosynthesis: Ewing, Fina-

more, *Biochim. Biophys. Acta* **218,** 463 (1970). *Review:* B. Nilsson, *Clin. Pharmacol. Drug Epidemiol.* **2,** 273-277 (1979). Series of articles on pharmacology, toxicity, distribution, pharmacokinetics and clinical studies: *Arzneimittel-Forsch.* **33,** 1009-1080 (1983). Toxicity: T. Grau *et al., ibid.* 1033.

Amorphous, somewhat hygroscopic powder. $[\alpha]_D^{20}$ +17.2° (H$_2$O); $[\alpha]_D^{25}$ +19.0° (c = 0.86 in H$_2$O). uv max (pH 1): 280 nm (ϵ 12800). Dissolves readily in water to form acidic soln. Practically insol in most organic solvents. pKa 4.4. LD_{50} in mice, rats (mg/kg): 4600 ±335, 4150 ±370 i.v.; both species 8 g/kg orally (Grau).

Sodium salt, $C_{14}H_{25}N_4NaO_{11}P_2$, *Acticolin, Brassel, Cebroton, Cidifos, Flussorex, Gerolin, Logan, Neuroton, Sinkron.*

THERAP CAT: Vasodilator (cerebral).

2381. Citiolone. *2-Acetamido-4-mercaptobutyric acid γ-thiolactone;* N-(tetrahydro-2-oxo-3-thienyl)acetamide; N-acetylhomocysteinethiolactone; α-acetamido-γ-thiobutyrolactone; AHCTL; BO-714; Citiolase; Thioxidrene. $C_6H_9NO_2S$; mol wt 159.21. C 45.27%, H 5.70%, N 8.80%, O 20.10%, S 20.14%. Prepn: Wagner *et al.,* **Ger.** pat. **1,134,683** (1962 to Degussa, vorm Roessler), *C.A.* **58,** 4648c (1963); Nakanishi *et al.,* **Japan.** pat. **16,712('62)** (to Yoshitomi), *C.A.* **59,** 11660h (1963); Takayananagi *et al.,* and Ogino, Yasui, **Japan.** pats. **3420('64)** and **1376('65)** (both to Sumitomo), *C.A.* **61,** 3193h (1964) and **62,** 14822c (1965). Chemistry: Benesch, Benesch, *J. Am. Chem. Soc.* **78,** 1597 (1956); Laliberté *et al., J. Chem. Soc.* **1963,** 2756. Pharmacology: Kirnberger *et al., Arzneimittel-Forsch.* **8,** 72 (1958); Varga *et al., ibid.* **13,** 1867 (1963).

Needles from toluene, mp 111.5-112.5°. uv max: 238 nm (ϵ 4400). LD_{50}: 1.20 g/kg i.v. in mice; 1.95 g/kg i.p. in rats. USE: Photographic antifogging agent: Dersch, **U.S.** pat. **3,068,100** (1962); Weber, **Ger.** pat. **1,164,828** (1964 to Adox Fotowerke Schleussner GmbH), *C.A.* **60,** 14050f (1964); protector against radiation: Langendorff, Koch, *Strahlentherapie* **106,** 451 (1958); Braun *et al., ibid.* **108,** 262 (1959), *C.A.* **52,** 18841h (1958); **53,** 17325e (1959).

THERAP CAT: In treatment of hepatic disorders.

2382. Citraconic Acid. *(Z)-2-Methyl-2-butenedioic acid;* methylmaleic acid. $C_5H_6O_4$; mol wt 130.10. C 46.16%, H 4.65%, O 49.19%. Obtained by carefully heating citric acid at about 175°. Production of citraconic anhydride from itaconic acid: Humphrey, **U.S.** pat. **2,966,498** (1960 to Pfizer). Toxicity study: P. M. Jenner *et al., Food Cosmet. Toxicol.* **2,** 327 (1964).

Hygroscopic monoclinic crystals; characteristic odor. d 1.62. mp ~90° (dec). Freely sol in water, alc, ether, slightly in chloroform. Practically insol in benzene, petr ether. LD_{50} in rats, mice (mg/kg): 1320, 2260 orally (Jenner).

2383. Citral. *3,7-Dimethyl-2,6-octadienal.* $C_{10}H_{16}O$; mol wt 152.24. C 78.90%, H 10.59%, O 10.51%. Constituent (75 to 85%) of oil of lemon grass, the volatile oil of *Cymbopogon citratus* (DC.) Stapf, or of *Cymbopogon flexuosus* (Nees) Stapf, *Gramineae.* Also present to a limited extent in oils of verbena, lemon, and orange. Isoln: Tiemann, Semuler, *Ber.* **26,** 2708 (1893); Tiemann, *ibid.* **31,** 3310, 3317 (1898); Hibbert, Cannon, *J. Am. Chem. Soc.* **46,** 119 (1924). Synthesis based on acetylene chemistry: Kimel, Sax, **U.S.** pat. **2,661,368** (1953 to Hoffmann-La Roche). Synthesis from isoprene: Leets *et al., J. Gen. Chem. USSR* **27,** 1584 (1957), *C.A.* **53,** 4336e (1959). Alternate synthesis: T. Mandai, *Tetrahedron Letters* **22,** 763 (1981); K. Takabe *et al., Syn. Commun.* **13,** 297 (1983). Citral from natural sources is a mixture of two geometric isomers **geranial** and *neral.* Separation of isomers: Naves, *Bull. Soc. Chim. France* **1952**, 521. Comprehensive reviews: J. L. Simonsen, *The Terpenes*, vol. I, 83-100 (1947); E. Guenther, *The Essential Oils*, vol. II, 326-336 (1949); Gildemeister-Hoffmann, *Die Aetherischen Oele*, vol. IV, 307-356 (4th ed., 1956).

Geranial (Citral a)   Neral (Citral b)

Geranial, light oily liquid. Strong lemon odor. $bp_{2.6}$ 92-93°. d_4^{20} 0.8888. n_D^{20} 1.48982. Practically insol in water. Miscible with alc, ether, benzyl benzoate, diethyl phthalate, glycerol, propylene glycol, mineral oil, essential oils.

Neral, light oily liquid. Lemon odor not as intense but sweeter than gerianal. $bp_{2.6}$ 91-92°. d_4^{20} 0.8869. n_D^{20} 1.48690. Solubilities same as gerianal.

USE: In the synthesis of vitamin A, ionone and methylionone. As a flavor, for fortifying lemon oil. In perfumery for its citrus effect in lemon and verbena scents, in cologne odors, in perfumes for colored soaps. Not stable to alkalies and strong acids. Will cause discoloration of white soaps and alkaline cosmetics.

2384. Citramalic Acid. *2-Hydroxy-2-methylbutanedioic acid; 2-methylmalic acid;* 2-hydroxy-2-methylsuccinic acid; α-hydroxypyrotartaric acid. $C_5H_8O_5$; mol wt 148.12. C 40.55%, H 5.44%, O 54.01%. Enzymatic synthesis: Barker, Blair, *Biochem. Prepns.* **9**, 21 (1962). Chemical synthesis: Barker, *ibid.* 25; J. B. Wilkes, R. G. Wall, *J. Org. Chem.* **45**, 247 (1980). Stereoselective synthesis: E. G. J. Staring *et al.*, *Rec. Trav. Chim.* **105**, 374 (1986).

dl-Form, deliquescent monoclinic prisms from ethyl acetate + petr ether, mp 117°. Sublimes. Freely sol in water, acetone. Sol in ethyl acetate, ether. Practically insol in petr ether, benzene.

d-Form, crystals, mp 112.2-112.8°. $[\alpha]_D^{22}$ +23.6° (c = 3 in H_2O).

l-Form, crystals, mp 112-113°. $[\alpha]_D^{20}$ −23.4° (c = 3 in H_2O).

2385. β-Citraurin. *3-Hydroxy-8'-apo-β,ψ-carotenal;* citraurin. $C_{30}H_{40}O_2$; mol wt 432.65. C 83.29%, H 9.32%, O 7.40%. Carotenoid pigment found only in orange peel. Isoln by chromatography: Zechmeister, Tuzson, *Ber.* **69**, 1878 (1936); **70**, 1966 (1937). The peels from 100 kilos of oranges yield about 35 mg. Structure: Zechmeister, Tuzson, *loc. cit.*; Karrer, Solmssen, *Helv. Chim. Acta* **20**, 682 (1937); Karrer *et al., ibid.* 1020; Zechmeister, v. Cholnoky, *Ann.* **530**, 291 (1937); Karrer *et al., Helv. Chim. Acta* **21**, 445 (1938). Abs config: Bartlett *et al., J. Chem. Soc. (C)* **1969**, 2527. Synthesis: H. Pfander *et al., Helv. Chim. Acta* **63**, 1377 (1980).

Thin orange or yellow-colored plates from benzene + petr ether, mp 147°. Absorption max (benzene): 497, 467 nm. Freely sol in acetone, ethanol, ether, benzene, and carbon disulfide. Sparingly sol in petr ether.

2386. Citrazinic Acid. *1,2-Dihydro-6-hydroxy-2-oxo-4-pyridinecarboxylic acid; 2,6-dihydroxyisonicotinic acid;* 2,6-dihydroxy-4-pyridinecarboxylic acid. $C_6H_5NO_4$; mol wt 155.11. C 46.46%, H 3.25%, N 9.03%, O 41.26%. Prepn from citric acid with aq NH_3 at 140-160° under pressure: Bavley, Hamilton, U.S. pat. **2,729,647** (1956 to Pfizer). Purification: Bavley *et al.*, U.S. pat. **2,738,352** (1956 to Pfizer).

Yellowish powder with a greenish tinge; carbonizes above 300° without melting. Ultrapure material which is white or colorless, has been prepared. Almost insol in water; slightly sol in hot HCl; sol in alkali hydroxide or carbonate solns. Alkaline solns turn blue on standing.

2387. Citric Acid. *2-Hydroxy-1,2,3-propanetricarboxylic acid;* β-hydroxytricarballylic acid. $C_6H_8O_7$; mol wt 192.13. C 37.51%, H 4.20%, O 58.29%. Widely distributed in plants and in animal tissues and fluids. Produced by mycological fermentation on an industrial scale using crude sugar solns, such as molasses and strains of *Aspergillus niger:* See review by Von Loesecke, *Chem. & Eng. News* **23**, 1952 (1945); Schweiger, U.S. pat. **2,970,084** (1961 to Miles Labs.); Faith, Keyes & Clark's *Industrial Chemicals*, F. A. Lowenheim, M. K. Moran, Eds. (Wiley-Interscience, New York, 4th ed., 1975) pp 275-279. Also extracted from citrus fruits (lemon juice contains 5 to 8%) and from pineapple waste. *Reviews:* Wilson, *Chem. & Met. Eng.* **29**, 787 (1923); Browne, *Ind. Eng. Chem.* **13**, 81 (1921); Warneford, Hardy, *ibid.* **17**, 1283 (1925); E. F. Bouchard, E. G. Merritt in Kirk-Othmer *Encyclopedia of Chemical Technology* vol. 6 (Wiley-Interscience, New York, 3rd ed., 1979) pp 150-179. Toxicity: Gruber, Halbeisen, *J. Pharmacol. Exp. Ther.* **94**, 65 (1948).

Anhydr form, mp 153°. Crystals are monoclinic holohedra and crystallize from hot concd aq soln. d 1.665. At 25°, pK_1 3.128; pK_2 4.761; pK_3 6.396, Bates, Pinching, *J. Am. Chem. Soc.* **71**, 1274 (1949). Soly in water: 54.0% w/w at 10°; 59.2% at 20°; 64.3% at 30°; 68.6% at 40°; 70.9% at 50°; 73.5% at 60°; 76.2% at 70°; 78.8% at 80°; 81.4% at 90°; 84.0% at 100°.

Monohydrate, orthorhombic crystals from cold aq solns. Pleasant, sour taste. d 1.542. Monohydrate crystals lose water of crystn in dry air or when heated at about 40 to 50°, slightly deliquescent in moist air. Softens at 75°. mp ~100°. pH of 0.1N soln = 2.2. Densities of aqueous soln (15°/15°): 10% = 1.0392; 20% = 1.0805; 30% = 1.1244; 40% = 1.1709; 50% = 1.2204; 60% = 1.2738. Soly in g/100 g satd soln: ether 2.17; chloroform 0.007; amyl alcohol 15.43; amyl acetate 5.98; ethyl acetate 5.28. Soly at 19° in g/100 g solvent: methanol 197; propanol 62.8. LD_{50} i.p. in rats: 975 mg/kg (Gruber, Halbeisen).

Pharmaceutical Incompatibilities: Potassium tartrate, alkali and alkaline earth carbonates and bicarbonates, acetates, sulfides. Dilute aq solns may ferment on standing.

Barium salt heptahydrate, $C_{12}H_{10}Ba_3O_{14}.7H_2O$, *barium citrate.* Powder. Loses all H_2O at 150°. Sol in 1750 parts water; freely sol in dil HCl or HNO_3; practically insol in alcohol.

Ethyl ester, $C_{12}H_{20}O_7$, *ethyl citrate, triethyl citrate.* Bitter, oily liq. d^{20} 1.137. bp_{760} 294°; $bp_{1.0}$ 127°. Viscosity at 25°: 35.2 cps. Pour pt ~10°. n_D^{20} 1.4455. Soly: water ~6.9%; peanut oil 0.8%. Misc with alc, ether.

USE: Acidulant in beverages, confectionery, effervescent salts, in pharmaceutical syrups, elixirs, in effervescent powders and tablets, to adjust the pH of foods and as synergistic antioxidant, in processing cheese. Used in beverages, jellies, jams, preserves and candy to provide tartness. In the manuf of alkyd resins; in esterified form as plasticizer, foam inhibitor. In the manuf of citric acid salts. As sequestering agent to remove trace metals. As mordant to brighten colors; in electroplating; in special inks; in analytical chemistry for determining citrate-soluble P_2O_5; as reagent for albumin, mucin, glucose, bile pigments.

THERAP CAT: Component of anticoagulant citrate solns (citrate dextrose soln; citrate phosphate dextrose soln; citric acid syrup).

2388. Citrinin. *(3R-trans)-4,6-Dihydro-8-hydroxy-3,4,5-trimethyl-6-oxo-3H-2-benzopyran-7-carboxylic acid;* Antimycin. $C_{13}H_{14}O_5$; mol wt 250.25. C 62.39%, H 5.64%, O 31.97%. Antibiotic substance produced by a white spore aspergillus which has been placed under the species name *Aspergillus niveus* (Thorn and Raper). Also produced in small quantities by *Penicillium citrinum:* Hetherington, Raistrick, *Trans. Roy. Soc. London* **B220,** 269 (1931); Raistrick, Smith, *Chem. & Ind (London)* **60,** 828 (1941); Timonin, *Science* **96,** 494 (1942); Timonin, Rovatt, *Can. J. Pub. Health* **35,** 80 (1944). Identity with antimycin: Haese, *Arch. Pharm.* **296,** 227 (1963). Structure: Brown *et al., J. Chem. Soc.* **1949,** 867; Warren *et al., J. Am. Chem. Soc.* **79,** 3812 (1957); Kovac *et al., Nature* **190,** 1104 (1961). Synthesis: Cartwright *et al., J. Chem. Soc.* **1949,** 1563; J. A. Barber *et al., J. Chem. Soc. Perkin Trans. I* **1986,** 2101. Stereochemistry: Cram, *J. Am. Chem. Soc.* **72,** 1001 (1950); Mehta, Whalley, *J. Chem. Soc.* **1963,** 3777; Mathieson, Whalley, *ibid.* **1964,** 4640. Physical characteristics and toxicity: Nagai *et al., Chem. Zentr.* **1958,** 8088, *C.A.* **55,** 1914 (1961). Crystal and molecular structure: Rodig, *Chem. Commun.* **1971,** 1553. Biosynthesis: J. Barber *et al., J. Chem. Soc. Perkin Trans. I* **1981,** 2577; L. Colombo *et al., ibid.* 2594. Physicochemical data: A. E. Pohland *et al., Pure Appl. Chem.* **54,** 2219 (1982). Toxicology: A. M. Ambrose, F. De Eds, *J. Pharmacol. Exp. Ther.* **88,** 173 (1946). Review: Saito *et al.,* "Yellowed Rice Toxins" in *Microbial Toxins,* A. Ciegler, S. Kadis, A. Ajl, Eds. (Academic Press, New York, 1971) **vol. VI,** pp 357-367.

Lemon-yellow needles from alcohol, dec 175°. $[\alpha]_D^{18}$ −37.4° (c = 1.15 in alc.). uv max: 250, 331 nm ($E_{1cm}^{1\%}$ 418). Strong acid. Practically insol in water. Sol in alcohol, dioxane, dilute alkali. Solns change color with changes in pH, from lemon-yellow at pH 4.6 to cherry-red at pH 9.9. *Poisonous!* LD$_{50}$ in mice, rats (mg/kg): 35, 67 i.p. (Ambrose, De Eds).

Methyl citrinin, $C_{14}H_{16}O_5$, plates from benzene, dec 139°. $[\alpha]_D^{18}$ +217.1° (c = 0.38 in acetone). uv max: 260, 334 nm ($E_{1cm}^{1\%}$ 520, 151.6). Sol in hot alcohol; moderately sol in chloroform. Practically insol in petr ether.

Dihydrocitrinin, $C_{13}H_{16}O_5$, prisms from benzene, dec 171°. $[\alpha]_D^{18}$ −18.8° (c = 4.148 in chloroform). uv max: 260, 330 nm ($E_{1cm}^{1\%}$ 400, 100). Sol in alcohol, acetone, chloroform; sparingly sol in benzene, petr ether.

2389. Citromycetin. *8,9-Dihydroxy-2-methyl-4-oxo-4H,5H-pyrano[3,2-c][1]benzopyran-10-carboxylic acid;* frequentic acid. $C_{14}H_{10}O_7$; mol wt 290.23. C 57.94%, H 3.47%, O 38.59%. Antibiotic substance produced by *Penicillium frequentans* Westling and *P. vesiculosum* Bainier and by *Citromyces* spp: Hetherington, Raistrick, *Phil. Trans. Roy. Soc. London, Ser. B,* **220,** 209 (1931); Grove, Brian, *Nature* **167,** 995 (1951). Structure: Robertson *et al., J. Chem. Soc.* **1951,** 2013. Biosynthesis: Birch *et al., ibid.* **1958,** 4576; Money, *Nature* **199,** 592 (1963). Total synthesis: M. Yamauchi *et al., J. Chem. Soc. Perkin Trans. I* **1987,** 395.

Dihydrate, yellow crystals, effervescence at 155°, dec 290-300° (considerable antecedent blackening). Freely sol in ethanol; readily sol in aq sodium carbonate soln; sparingly sol in water, chloroform. Insol in benzene, hexane. Stable to acid and alkali at 100°.

2390. Citronellal. *3,7-Dimethyl-6-octenal.* $C_{10}H_{18}O$; mol wt 154.25. C 77.87%, H 11.76%, O 10.37%. Chief constituent of citronella oil; also found in many other volatile oils, such as lemon, lemon grass, melissa: Tiemann, *Ber.* **32,** 834 (1899); Spoon, *Chem. Weekbl.* **54,** 236 (1958). Structure: Naves, *Bull. Soc. Chim. France* **1951,** 505; Eschinazi, *J. Org. Chem.* **26,** 3072 (1961).

Liquid. bp$_1$ 47°. n_D^{20} 1.4460. $[\alpha]_D^{25}$ +11.50°. d 0.848-0.856. Soluble in alcohols; very slightly sol in water.

α-citronellal, 3,7-Dimethyl-7-octenal, rhodinal. Liquid. bp$_{1.4}$ 51°. n_D^{20} 1.4410. $[\alpha]_D^{20}$ +9.75°.

USE: In soap perfumes; insect repellent.

2391. β-Citronellol. *3,7-Dimethyl-6-octen-1-ol;* 2,6-dimethyl-2-octen-8-ol; citronellol; cephrol. $C_{10}H_{20}O$; mol wt 156.27. C 76.86%, H 12.90%, O 10.24%. *l*-Form is a constituent of rose and geranium oils. *d*-Form occurs in Ceylon and Java citronella oils. History: J. L. Simonsen, L. N. Owen, *The Terpenes* vol. I (University Press, Cambridge, 2nd ed, 1947). Prepn of (±)-form: Adams, Garvey, *J. Am. Chem. Soc.* **48,** 477 (1926); Ofner *et al., Helv. Chim. Acta* **42,** 2577 (1959). Prepn of (+)-form: Rienäcker, Ohloff, *Angew. Chem.* **73,** 240 (1961); Naves, Tullen, *Helv. Chim. Acta* **44,** 1867 (1961); Eschinazi, *J. Org. Chem.* **26,** 3072 (1961); Rienäcker, *Chimia* **27,** 97 (1973); C. G. Overberger, J. L. Weise, *J. Am. Chem. Soc.* **90,** 3525 (1968); T. Sato *et al., Tetrahedron Letters* **1980,** 3377. Prepn of (−)-form: Ohloff, *loc. cit.;* Rienäcker, *loc. cit.;* Shono *et al., Tetrahedron Letters* **1974,** 1295; K. Mori, T. Sugai, *Synthesis* **1982,** 752. Synthesis of (+) or (−)-form from isoprene: Hidai *et al., Chem. Commun.* **1975,** 170. Stereospecific prepn via microbiological *(Saccharomyces cerevisiae)* reduction: P. Gramatica *et al., Experientia* **38,** 775 (1982). Manuf: Woroch *et al.;* Bain; Webb, U.S. pats. **2,990,422; 3,005,845; 3,028,431** (1961, 1961, 1962, all to Glidden); Eschinazi, U.S. pat. **3,052,730** (1962 to Givaudan). Abs config of the (+)-form: Freudenberg, Hohmann, *Ann.* **584,** 54 (1953); Freudenberg, Lwowski, *ibid.* **587,** 213 (1954). NMR, HPLC determn of *R/S* enantiomer ratios: D. Valentine *et al., J. Org. Chem.* **41,** 62 (1976).

R-(+)-β-Citronellol

(+)-Form, oily liquid, bp 224.5°, bp_{10} 108.4°, d_4^{20} 0.8550. n_D^{20} 1.4559. $[\alpha]_D^{20}$ +5.22°. Very slightly sol in water, miscible with with alcohol, ether.

(−)-Form, *β-rhodinol, Levocitrol.* bp_{10} 108-109°. d_4^{18} 1.4576. $[\alpha]_D^{20}$ −4.76°.

(±)-Form, *dihydrogeraniol.* $d_4^{23.5}$ 0.851. $n_D^{23.5}$ 1.454.
See also Rhodinol.
USE: In perfumery.

2392. Citrulline. N^5-*(Aminocarbonyl)-L-ornithine;* δ-ureidonorvaline; α-amino-δ-ureidovaleric acid; $N^δ$-carbamyl-ornithine. $C_6H_{13}N_3O_3$; mol wt 175.19. C 41.14%, H 7.48%, N 23.99%, O 27.40%. $H_2NCONH(CH_2)_3CH(NH_2)COOH$. An amino acid, first isolated from the juice of watermelon, *Citrullus vulgaris* Schrad., *Cucurbitaceae:* Wada, *Biochem. Z.* **224,** 420 (1930); isoln from casein: Wada, *ibid.* **257,** 1 (1933). Synthesis from ornithine through copper complexes: Kurtz, *J. Biol. Chem.* **122,** 477 (1938); by alkaline hydrolysis of arginine: Fox, *ibid.* **123,** 687 (1938); from cyclopentanone oxime: Fox *et al., J. Org. Chem.* **6,** 410 (1941). Crystallization: Matsuda *et al.,* **Japan.** pat. **71 00,174** (1971 to Ajinomoto), *C.A.* **74,** 126056u (1971). Crystal and molecular structure: Naganathan, Venkatesan, *Acta Crystallogr.* **27B,** 1079 (1971); Ashida *et al., ibid.* **28B,** 1367 (1972). Use in asthenia and hepatic insufficiency: **Fr.** pat. **2,198,739** (1974 to Hublot & Vallet), *C.A.* **82,** 144952c (1975). Clinical trial in treatment of lysinuric protein intolerance: J. Rajantie *et al., J. Pediatr.* **97,** 927 (1980); T. O. Carpenter *et al., N. Engl. J. Med.* **312,** 290 (1985).

Prisms from methanol + water, mp 222°. $[\alpha]_D^{20}$ +3.7° (c = 2). pK_1 2.43; pK_2 9.41. Sol in water. Insol in methanol, ethanol.

Hydrochloride, $C_6H_{13}N_3O_3$·HCl, crystals, dec 185°. $[\alpha]_D^{22}$ +17.9° (c = 2).

Malate (salt), $C_6H_{13}N_3O_3$·$C_4H_6O_5$, *Stimol.*
THERAP CAT: Treatment of asthenia.

2393. Citrullol. $C_{22}H_{38}O_4$; mol wt 366.54. C 72.09%, H 10.45%, O 17.46%. From fruit pulp of *Citrullus colocynthis* Schrad., *Cucurbitaceae:* Power, Moore, *J. Chem. Soc.* **97,** 99 (1910); Power, Salway, *ibid.* **103,** 399, 1022 (1913); Khadem, Rahman, *Tetrahedron Letters* **1962,** 1137.

Crystals, mp 282-283°. uv max: 242, 272, 282 nm (log ∈ 2.85, 2.68, 2.68). Sol in pyridine; practically insol in usual organic solvents.

Diacetate, $C_{26}H_{42}O_6$, crystals, mp 162°.

2394. Citrus Red 2. *1-[(2,5-Dimethoxyphenyl)azo]-2-naphthalenol; C.I. Solvent Red 80;* C.I. 12156. $C_{18}H_{16}N_2O_3$; mol wt 308.34. C 70.12%, H 5.23%, N 9.09%, O 15.57%. Prepn: H. W. Elley, H. W. Daudt, U.S. pat. **2,224,904** (1940 to Du Pont). Metabolism: J. L. Radomski, *J. Pharmacol. Exp. Ther.* **134,** 100 (1961); **136,** 378 (1962). Toxicology: M. Sharratt *et al., Food Cosmet. Toxicol.* **4,** 493 (1966). Review of carcinogenicity studies: *IARC Monographs* **8,** 101-106. *See also Colour Index* vol. **4** (3rd ed., 1971) p 4033.

Crystals, mp 155-157°. Slightly sol in water; partially sol in ethanol and vegetable oils.
USE: To color orange skins.

2395. Civet. Zibeth. Unctuous secretion from receptacles between the anus and genitalia of both male and female civet cat. *Constit.* Civetone and similar compds.

Semi-solid, yellowish to brown unctuous substance; unpleasant, subacrid, bitter taste; fusible and burns without leaving much residue. Insol in water; partly sol in hot alcohol or in ether.
USE: As a fixative in perfumery.

2396. Civetone. *(Z)-9-Cycloheptadecen-1-one.* $C_{17}H_{30}$-O; mol wt 250.42. C 81.54%, H 12.07%, O 6.39%. 17-

Membered macrocyclic musk, constituent of civet: Ruzicka, *Helv. Chim. Acta* **9,** 230 (1926); Ruzicka *et al., ibid.* **10,** 695 (1927). Occurs in nature as *cis*-form. Synthesis of *cis*-civetone: Stoll *et al., ibid.* **31,** 543 (1948); J. Tsuji, T. Mondai, *Tetrahedron Letters* **1977,** 3285; E. Seoane *et al., Chem. & Ind. (London)* **1978,** 165. Synthesis of *trans*-form: H. Hunsdiecker, *Ber.* **77,** 185 (1944); H. H. Mathur, S. C. Bhattacharyya, *J. Chem. Soc.* **1968,** 114. Crystal and molecular structure of *cis*-civetone: G. Bernardinelli, R. Gerdil, *Helv. Chim. Acta* **65,** 558 (1982).

(cis)-form

Crystals, mp 31-32°. Musky odor becoming pleasant in extreme dilns. d_4^{33} 0.917. bp_{742} 342°; bp_2 59°. $n_D^{33.4}$ 1.4830.
USE: In perfumery.

2397. Cladribine. *2-Chloro-2'-deoxyadenosine;* 2-chloro-6-amino-9-(2-deoxy-β-D-*erythro*-pentofuranosyl)purine; 2-chlorodeoxyadenosine; 2-CdA; CldAdo; NSC-105014-F; Leustatin. $C_{10}H_{12}ClN_5O_3$; mol wt 285.69. C 42.04%, H 4.23%, Cl 12.41%, N 24.51%, O 16.80%. Substituted purine nucleoside with antileukemic activity. Prepn as intermediate in synthesis of 2-deoxynucleosides: H. Venner, *Ber.* **93,** 140 (1960); M. Ikehara, H. Tada, *J. Am. Chem. Soc.* **85,** 2344 (1963); *eidem, ibid.* **87,** 606 (1965). Synthesis and biological activity: L. F. Christensen *et al., J. Med. Chem.* **15,** 735 (1972). Stereospecific synthesis: Z. Kazimierczuk *et al., J. Am. Chem. Soc.* **106,** 6379 (1984); R. K. Robins, G. R. Revankar, Eur. pat. Appl. **173,059;** *eidem,* U.S. pat. **4,760,137** (1986, 1988 both to Brigham Young Univ.). Specific toxicity to lymphocytes: D. A. Carson *et al., Proc. Nat. Acad. Sci. USA* **77,** 6865 (1980); *eidem, Blood* **62,** 737 (1983). Mechanism of action: S. Seto *et al., J. Clin. Invest.* **75,** 377 (1985). Clinical evaluation in chronic lymphocytic leukemia: L. D. Piro *et al., Blood* **72,** 1069 (1988); in hairy cell leukemia: *eidem, N. Engl. J. Med.* **322,** 1117 (1990).

Crystals from water, softens at 210-215°, solidifies and turns brown (Christensen). Also reported as crystals from ethanol, mp 220° (softens), resolidifies, turns brown and does not melt below 300° (Kazimierczuk). $[\alpha]_D^{25}$ −18.8° (c = 1 in DMF). uv max in 0.1N NaOH: 265 nm; in 0.1N HCl: 265 nm.
THERAP CAT: Antineoplastic.

2398. Clanobutin. *4-[(4-Chlorobenzoyl)(4-methoxyphenyl)amino]butanoic acid; 4-[p-chloro-N-(p-methoxyphenyl)benzamido]butyric acid; N-(p-chlorobenzoyl)-γ-(p-anisidino)butyric acid; Bykahepar.* $C_{18}H_{18}ClNO_4$; mol wt 347.80. C 62.16%, H 5.22%, Cl 10.19%, N 4.03%, O 18.40%. Prepn: K. Klemm *et al.,* Ger. pat. **1,917,036** corresp to U.S. pat. **3,780,095** (1971, 1973 both to Byk-Gulden). Series of articles on synthesis, physical and pharmacological properties: *Arzneimittel-Forsch.* **29,** 1-15 (1979). *In vitro* biochemical study: H. Wolf *et al., Biochem. Pharmacol.* **29,** 1649 (1980).

Effect on bile excretion in rats, dogs: P. Berchtold et al., *Arzneimittel-Forsch.* **30**, 1878 (1980).

Cryst from ethyl acetate, mp 115-116°. pKa 5.04. Soly in water at 37°: 4.02×10^{-2} mol/l at pH 7. LD_{50} in rats (mg/kg): > 2000 orally; 570 i.v. (Klemm).

THERAP CAT: Choleretic.

THERAP CAT (VET): Choleretic; in treatment of piroplasmosis and anaplasmosis.

2399. Clarase®. Diastatic and proteolytic enzymes. Contains mainly amylase and in smaller amounts maltase, protease, peptidase, lipase and other enzymes: Dzialoszynski, *Nature* **160**, 464 (1947). In determination of thiamin and riboflavin in cowpea: J. K. Edijala, *Analyst (London)* **104**, 637 (1979).

USE: In analytical chemistry for digesting food samples (protein solubilizing, amino acid liberating); in industry for starch conversions, removal of starches, manuf sizing materials, paste making.

2400. Clarithromycin. *6-O-Methylerythromycin;* A-56268; TE-031; Biaxin; Clathromycin; Klacid; Klaricid; Macladin; Naxy; Veclam; Zeclar. $C_{38}H_{69}NO_{13}$; mol wt 747.96. C 61.02%, H 9.30%, N 1.87%, O 27.81%. Semisynthetic antibiotic; derivative of erythromycin, q.v. Prepn: Y. Watanabe et al., *Eur. pat. Appl.* **41,355**; *eidem*, U.S. pat. **4,331,803** (1981, 1982 both to Taisho); and *in vitro* antibacterial activity: S. Morimoto et al., *J. Antibiot.* **37**, 187 (1984). *In vitro* and *In vivo* antibacterial activity: P. B. Fernandes et al., *Antimicrob. Ag. Chemother.* **30**, 865 (1986). Comparative antibacterial spectrum *in vitro*: C. Benson et al., *Eur. J. Clin. Microbiol.* **6**, 173 (1987); H. M. Wexler, S. M. Finegold, *ibid.* 492. HPLC determn in biological fluids: D. Croteau et al., *J. Chromatog.* **419**, 205 (1987). Acute toxicity study: S. Abe et al., *Chemotherapy (Tokyo)* **36**, Suppl. 3, 274 (1988). Symposium on pharmacology and comparative clinical studies: *J. Antimicrob. Chemother.* **27**, Suppl. A, 1-124 (1991).

Colorless needles from chloroform + diisopropyl ether (1:2), mp 217-220° (dec). Also reported as crystals from ethanol, mp 222-225° (Morimoto). uv max (CHCl₃): 288 nm (ε 27.9). $[\alpha]_D^{24}$ −90.4° (c = 1 in CHCl₃). Stable at acidic pH. LD_{50} in male, female mice, male, female rats (mg/kg): 2740, 2700, 3470, 2700 orally, 1030, 850, 669, 753 i.p., > 5000 all s.c. (Abe).

THERAP CAT: Antibacterial.

2401. Clathrates. Compounds that are capable of trapping other substances within their own crystal lattices. The cavities of the host molecules are classified as cages, tunnels, or layered types, depending on the way they include guest molecules. The geometry of the cavities limits the guest molecules by size and shape, rather than by chemical similarity with the host molecules. Among common clathrates are *molecular sieves, cyclotriphosphazenes,* and *Dianin's compound,* as well as hydroquinone, cyclodextrins, o-thymotide, and deoxycholic acid, q.q.v. *Cavitands* are organic hosts with enforced (rigid) cavities: D. J. Cram, *Science* **219**, 1177 (1983); R. C. Helgeson et al., *Chem. Commun.* **1983**, 101. Comprehensive book: *Clathrate Compounds,* V. M. Bhatnagar, Ed. (Chemical Pub. Co., New York, 1970) 244 pp. *Reviews:* D. D. MacNicol et al., *Chem. Soc. Rev.* **7**, 65-87 (1978); E. C. Makin, "Clathration" in Kirk-Othmer *Encyclopedia of Chemical Technology* Vol. 6 (Wiley-Interscience, New York, 3rd ed., 1979) pp 178-189.

USE: As complexing agent; stabilizing agent. In analytical separations.

2402. Clavulanic Acid. *3-(2-Hydroxyethylidene)-7-oxo-4-oxa-1-azabicyclo[3.2.0]heptane-2-carboxylic acid;* MM 14151. $C_8H_9NO_5$; mol wt 199.16. C 48.25%, H 4.55%, N 7.03%, O 40.17%. β-Lactamase inhibitor. Antibiotic produced by *Streptomyces clavuligerus;* first reported naturally occurring fused β-lactam containing oxygen. Isoln: M. Cole et al., Ger. pat. **2,517,316** (1975 to Beecham), *C.A.* **84**, 72635t (1976); A. G. Brown et al., *J. Antibiot.* **29**, 668 (1976). Structure, x-ray crystallography: T. T. Howarth et al., *Chem. Commun.* **1976**, 266. Total synthesis of (±)-form: P. H. Bentley et al., *ibid.* **1977**, 748, 905; *eidem, Tetrahedron Letters* **1979**, 1889. β-Lactamase inhibition and antibacterial spectrum: C. Reading, M. Cole, *Antimicrob. Ag. Chemother.* **11**, 852 (1977). Mechanism of action: B. G. Spratt et al., *ibid.* **12**, 406 (1977). Antibacterial activity, pharmacology and clinical efficacy of combination with amoxicillin: A. P. Ball et al., *Lancet* **1**, 620 (1980); R. N. Brogden et al., *Drugs* **22**, 337-362 (1981). *In vitro* and *in vivo* synergism with ticarcillin: R. Sutherland et al., *Am. J. Med.* **79**, Suppl. 5B, 13 (1985).

Combination of potassium salt with amoxicillin trihydrate, *co-amoxiclav, Amoksiklav, Augmentin, Ciblor, Klavocin, Neo-Duplamox.*

Combination of potassium salt with ticarcillin disodium, *Betabactyl, Timentin.*

Methyl ester, $C_9H_{11}NO_5$, oil. $[\alpha]_D^{22}$ +38°.

p-Nitrobenzyl ester, $C_{15}H_{14}N_2O_7$, monoclinic crystals, mp 117.5-118°.

THERAP CAT: Combination with β-lactam antibiotics as antibacterial.

2403. Clazuril. *2-Chloro-α-(4-chlorophenyl)-4-(4,5-dihydro-3,5-dioxo-1,2,4-triazin-2(3H)-yl)benzeneacetonitrile;* (±)-[2-chloro-4-(4,5-dihydro-3,5-dioxo-*as*-triazin-2(3H)-yl)-phenyl](*p*-chlorophenyl)acetonitrile; Appertex. $C_{17}H_{10}Cl_2N_4O_2$; mol wt 373.20. C 54.71%, H 2.70%, Cl 19.00%, N 15.01%, O 8.57%. Prepn: G. M. Boeckx et al., *Eur. pat. Appl.* **170,316**; *eidem*, U.S. pat. **4,631,278** (both 1986 to Janssen). Anticoccidial effect in pigeons: W. Coussement et al., *Res. Vet. Sci.* **45**, 117 (1988).

mp 196.8°.

THERAP CAT (VET): Coccidiostat.

2404. Clebopride. *4-Amino-5-chloro-2-methoxy-N-[1-(phenylmethyl)-4-piperidinyl]benzamide;* 4-amino-*N*-(1-benzyl-4-piperidyl)-5-chloro-*o*-anisamide; *N*-(1'-benzyl-4'-piperidyl)-2-methoxy-4-amino-5-chlorobenzamide. $C_{20}H_{24}ClN_3O_2$; mol wt 373.88. C 64.25%, H 6.47%, Cl 9.48%, N

11.24%, O 8.56%. Dopamine receptor antagonist related to metoclopramide, *q.v.* Prepn: R. G. Spickett *et al.,* **Ger. pat. 2,513,136** corresp to A. V. Noverola *et al.,* **U.S. pat. 4,138,-492** (1975, 1979 both to Anphar); J. Prieto *et al., J. Pharm. Pharmacol.* **29,** 147 (1977). Pharmacological study: J. L. Masso, D. J. Roberts, *ibid.* **32,** 727 (1980). *In vitro* metabolism studies: G. Huizing *et al., Xenobiotica* **10,** 211 (1980); G. Huizing, A. H. Beckett, *ibid.* 593. Pharmacokinetics: J. Segura *et al., J. Pharm. Pharmacol.* **33,** 214 (1981). Comparative study of anti-emetic and gastrointestinal effects: P. Berga *et al., Arch. Farmacol. Toxicol.* **7,** 189 (1981). Efficacy in healing exptl ulcers: Y. Matsuo *et al., Oyo Yakuri* **24,** 251 (1982); S. Okabe *et al., ibid.* 261, *C.A.* **97,** 208081n, 208082p (1982). Determn in urine and plasma by radioimmunoassay: M. Yano *et al., Chem. Pharm. Bull.* **32,** 1491 (1984). Clinical effect in postoperative nausea: D. F. Duarte *et al., Clin. Ther.* **7,** 365 (1985).

Cryst from methanol, mp 194-195°.

Hydrochloride monohydrate, $C_{20}H_{24}ClN_3O_2.HCl.H_2O$, cryst, mp 217-219°. Approx oral LD_{50} in male Swiss mice: >1000 mg/kg (Prieto).

Malate, $C_{20}H_{24}ClN_3O_2.C_4H_6O_5$, *Amicos, Clanzol, Clast, Cleboril, Cleprid, Motilex.*

THERAP CAT: Antiemetic; antispasmodic.

2405. Clemastine. *[R-(R*,R*)]-2-[2-[1-(4-Chlorophenyl)-1-phenylethoxy]ethyl]-1-methylpyrrolidine; (+)-2-[2-[(p-chloro-α-methyl-α-phenylbenzyl)oxy]ethyl]-1-methylpyrrolidine;* 1-methyl-2-[2-(α-methyl-p-chlorobenzhydryloxy)ethyl]pyrrolidine; 1-methyl-2-[2-(methyl-p-chlorodiphenylmethyloxy)ethyl]pyrrolidine; meclastine. $C_{21}H_{26}ClNO$; mol wt 343.90. C 73.35%, H 7.62%, Cl 10.31%, N 4.07%, O 4.65%. Prepn: **Brit. pat. 942,152** (1963 to Sandoz). Synthesis and abs config: A. Ebnöther, H. P. Weber, *Helv. Chim. Acta* **59,** 2462 (1976). Pharmacology and toxicology of the hydrogen fumarate: Weidmann *et al., Boll. Chim. Farm.* **106,** 467 (1967).

Free base, $bp_{0.02}$ 154°. n_D^{22} 1.5582. $[\alpha]_D^{20}$ +33.6° (ethanol).

Hydrogen fumarate, $C_{25}H_{30}ClNO_5$, *HS-592, Aloginan, Alphamin, Anhistan, Fuluminol, Inbestan, Kinotomin, Lacretin, Lecasol, Maikohis, Mallermin-F, Marsthine, Masletine, Piloral, Reconin, Tavegyl, Tavegyl, Tavist, Telgin-G, Trabest, Xolamin.* mp 177-178°. $[\alpha]_D^{21}$ +16.9° (methanol). LD_{50} in mice, rats (mg/kg): 730, 3550 orally; 43, 82 i.v. (Weimann).

THERAP CAT: Antihistaminic.

2406. Clemizole. *1-[(4-Chlorophenyl)methyl]-2-(1-pyrrolidinylmethyl)-1H-benzimidazole; 1-(p-chlorobenzyl)-2-(1-pyrrolidinylmethyl)benzimidazole;* 1-(p-chlorobenzyl)-2-pyrrolidylmethylenebenzimidazole. $C_{19}H_{20}ClN_3$; mol wt 325.84. C 70.04%, H 6.19%, Cl 10.88%, N 12.90%. Prepn: D. Jerchel *et al., Ann.* **575,** 162 (1952); **Brit. pat. 703,272;** M. Schenck, W. Heinz, **U.S. pat. 2,689,853** (both 1954 to Schering A.G.).

Crystals from dil alc, mp 167°. Forms a water-soluble hydrochloride and sulfate. Used to form a repository form of penicillin G (as salt of benzylpenicillinic acid), trademarks: *Neopenyl, Lergopenin, Depocural. Ref:* H. Mückter *et al., Arzneimittel-Forsch.* **4,** 487 (1954).

Hydrochloride, $C_{19}H_{20}ClN_3.HCl$, *Allercur, Histacuran, Klemidox, Reactrol.* White microscopic rods from butanol, mp 239-241°.

THERAP CAT: Antihistaminic.

2407. Clenbuterol. *4-Amino-3,5-dichloro-α-[[(1,1-dimethylethyl)amino]methyl]benzenemethanol; 4-amino-α-[(tert-butylamino)methyl]-3,5-dichlorobenzyl alcohol;* NAB-365; Monores. $C_{12}H_{18}Cl_2N_2O$; mol wt 277.19. C 52.00%, H 6.55%, Cl 25.58%, N 10.11%, O 5.77%. Substituted phenylethanolamine with β_2 sympathomimetic activity. Prepn: J. Keck *et al.,* **S. Afr. pat. 67 05692;** *eidem,* **U.S. pat. 3,536,-712** (1968, 1970 both to Thomae); *eidem, Arzneimittel-Forsch.* **22,** 861 (1972). Series of articles on pharmacology, reproductive toxicology, pharmacokinetics and metabolism: *ibid.* **26,** 1403-1419, 1427-1460 (1976). Toxicity: M. Ueberberg *et al., ibid.* 1420. Clinical studies: M. Tschan *et al., Eur. J. Clin. Pharmacol.* **15,** 159 (1979); C. Pasotti *et al., Int. J. Clin. Pharmacol. Biopharm.* **17,** 176 (1979). Effect on weight gain in lambs: M. E. Nichols *et al., 1991 Animal Science Research Report,* (Oklahoma Agric. Exp. Station, Oklahoma State University, 1991) 303.

Hydrochloride, $C_{12}H_{18}Cl_2N_2O.HCl$, *NAB-365Cl, Spiropent, Ventipulmin.* Colorless microcrystalline powder from isopropyl alc, mp 174-175.5°. Very sol in water, methanol, ethanol, slightly sol in chloroform. Insol in benzene. LD_{50} in mice, rats, guinea pigs (mg/kg): 176, 315, 67.1 orally; 27.6, 35.3, 12.6 i.v. (Ueberberg).

THERAP CAT: Bronchodilator.

THERAP CAT (VET): Bronchodilator; tocolytic.

2408. Clentiazem. *(2S-cis)-3-(Acetyloxy)-8-chloro-5-[2-(dimethylamino)ethyl]-2,3-dihydro-2-(4-methoxyphenyl)-1,5-benzothiazepin-4(5H)-one; (+)-(2S,3S)-8-chloro-5-[2-(dimethylamino)ethyl]-2,3-dihydro-3-hydroxy-2-(p-methoxyphenyl)-1,5-benzothiazepin-4(5H)-one acetate (ester).* $C_{22}H_{25}ClN_2O_4S$; mol wt 448.97. C 58.86%, H 5.61%, Cl 7.90%, N 6.24%, O 14.25%, S 7.14%. Calcium channel blocker; 8-chloro derivative of diltiazem. Prepn: M. Takeda *et al.,* **Eur. pat. Appl. 127,882;** *eidem,* **U.S. pat. 4,567,175** (1984, 1986 both to Tanabe); H. Inoue *et al., J. Med. Chem.* **34,** 675 (1991). Cardiovascular pharmacology: H. Narita *et al., Arzneimittel-Forsch.* **38,** 515 (1988). Clinical pharmacokinetics: A. K. Shah *et al., J. Clin. Pharmacol.* **33,** 354 (1993). Clinical evaluation in hypertension: S. Kawakita *et al., Clin. Cardiol.* **14,** 53 (1991).

Maleate, $C_{22}H_{25}ClN_2O_4S.C_4H_4O_4$, *TA-3090, Logna.* Crystals from ethanol, mp 160.5-161.5°. $[\alpha]_D^{20}$ +76.5° (c = 1 in methanol).

THERAP CAT: Antihypertensive.

2409. 1,6-Cleve's Acid. *5-Amino-2-naphthalenesulfonic acid;* 1-naphthylamine-6-sulfonic acid. $C_{10}H_9NO_3S$; mol wt 223.25. C 53.80%, H 4.06%, N 6.27%, O 21.50%, S 14.36%. Prepn: Erdmann, *Ann.* **275**, 262 (1893).

Crystals. Sol in 1000 parts water. Practically insol in alc, ether.

USE: In manuf of azo dyes.

2410. 1,7-Cleve's Acid. *8-Amino-2-naphthalenesulfonic acid;* 1-naphthylamine-7-sulfonic acid. $C_{10}H_9NO_3S$; mol wt 223.25. C 53.80%, H 4.06%, N 6.27%, O 21.50%, S 14.36%. Prepn: Erdmann, *Ann.* **275**, 262 (1893); Roos *et al.,* U.S. pat. **2,875,243** (1959 to Bayer). Industrial synthesis: Nakahara *et al., J. Syn. Org. Chem. Japan.* **29**, 1129 (1971).

Needles or prisms from water contg $1H_2O$. Sol in 220 parts water; very slightly sol in alcohol, ether.

USE: In manuf of azo dyes.

2411. Clidanac. *(±)-6-Chloro-5-cyclohexyl-2,3-dihydro-1H-indene-1-carboxylic acid;* (±)-6-chloro-5-cyclohexylindan-1-carboxylic acid; TAI-284; Britai; Indanal. $C_{16}H_{19}ClO_2$; mol wt 278.78. C 68.94%, H 6.87%, Cl 12.72%, O 11.48%. Prepn: P. F. Juby *et al.,* **Ger.** pat. **2,004,038;** *eidem,* U.S. pat. **3,565,943** (1970, 1971 both to Bristol-Myers); S. Noguchi *et al., Chem. Pharm. Bull.* **19**, 646 (1971). Synthesis and anti-inflammatory activity: *eidem, ibid.* **22**, 529 (1974); G. R. Allen *et al., J. Med. Chem.* **15**, 934 (1972); and toxicity data: P. F. Juby *et al., ibid.* 1297. Pharmacology: S. Kuzuna *et al., Japan. J. Antibiot.* **24**, 695 (1974), *C.A.* **82**, 80297h (1976). X-ray crystallographic study: K. Kamiya *et al., Chem. Pharm. Bull.* **23**, 1589 (1975). Metabolism: S. Tanayama, Y. Kanai, *Xenobiotica* **7**, 145 (1977). Immunogenicity study: S. Kuzuna *et al., Oyo Yakuri* **19**, 951 (1980), *C.A.* **94**, 76682 (1981). Inhibition of prostaglandin biosynthesis: S. Tamura *et al., J. Pharm. Pharmacol.* **33**, 29 (1981). Antipyretic effect in cancer patients: H. Furue, *Curr. Ther. Res.* **30**, 71 (1981).

Colorless crystals from petr ether, mp 150.5-152.5°. LD_{50} orally in rats: 41 mg/kg (Juby, 1972).

(*S*)-(+)-Form, colorless needles from petr ether, mp 135-136°. $[\alpha]_D^{25}$ −28.7; $[\alpha]_{365}^{25}$ +87.7 (c = 2 in ethanol). LD_{50} orally in rats: 35 mg/kg (Juby, 1972).

(*R*)-(-)-Form, colorless cryst from petr ether, mp 134-135°. $[\alpha]_D^{25}$ −28.2; $[\alpha]_{365}^{25}$ −87.5° (c = 2 in ethanol).

THERAP CAT: Anti-inflammatory; antipyretic.

2412. Clidinium Bromide. *3-[(Hydroxydiphenylacetyl)-oxy]-1-methyl-1-azoniabicyclo[2.2.2]octane bromide; 3-hydroxy-1-methylquinuclidinium bromide benzilate;* 1-methyl-3-benziloyloxyquinuclidinium bromide; 3-benziloyloxy-1-aza-bicyclo[2.2.2]octane methobromide; Ro-2-3773; Quarzan. $C_{22}H_{26}BrNO_3$; mol wt 432.36. C 61.12%, H 6.06%, Br 18.48%, N 3.24%, O 11.10%. Prepn: Sternbach, U.S. pat. **2,648,667** (1953 to Hoffmann-La Roche). Comprehensive description: B. C. Rudy, B. Z. Senkowski, in *Analytical Profiles of Drug Substances* vol. **2**, K. Florey, Ed. (Academic Press, New York, 1973) pp 145-161.

Crystals from methanol + acetone + ether, mp 240-241°.

THERAP CAT: Anticholinergic.

2413. Clinafloxacin. *7-(3-Amino-1-pyrrolidinyl)-8-chloro-1-cyclopropyl-6-fluoro-1,4-dihydro-4-oxo-3-quinoline-carboxylic acid.* $C_{17}H_{17}ClFN_3O_3$; mol wt 365.79. C 55.82%, H 4.68%, Cl 9.69%, F 5.19%, N 11.49%, O 13.12%. Fluorinated quinolone antibacterial. Prepn: S. Suzue *et al.,* **Eur.** pat. **Appl. 195,316** (1986 to Kyorin). Antibacterial spectrum: M. S. Barrett *et al., Diagn. Microbiol. Infect. Dis.* **14**, 389 (1991); S. M. H. Qadri *et al., Chemotherapy (Basel)* **38**, 92 (1992).

Milk-white powder from chloroform-methanol-conc. aq. ammonia, mp 253-258° (dec).

Hydrochloride, $C_{17}H_{17}ClFN_3O_3.HCl$, *AM-1091, CI-960, PD-127391.* Light yellow prisms from methanol, mp 263-265° (dec).

THERAP CAT: Antibacterial.

2414. Clindamycin. *(2S-trans)-Methyl 7-chloro-6,7,8-trideoxy-6-[[(1-methyl-4-propyl-2-pyrrolidinyl)carbonyl]ami-no]-1-thio-L-threo-α-D-galacto-octopyranoside;* 7(*S*)-chloro-7-deoxylincomycin; 7-deoxy-7(*S*)-chlorolincomycin; clinimy-cin (rescinded); U-21251; Antirobe; Cleocin; Dalacin C; Klimicin; Sobelin. $C_{18}H_{33}ClN_2O_5S$; mol wt 424.99. C 50.87%, H 7.83%, Cl 8.34%, N 6.59%, O 18.82%, S 7.55%. Semi-synthetic antibiotic prepd from lincomycin, *q.v.* Prepn: Magerlein *et al., Antimicrob. Ag. Chemother.* **1966,**

727; Birkenmeyer, U.S. pat. **3,475,407**; Kagan, Magerlein, U.S. pats. **3,509,127** and **3,544,551**; Birkenmeyer, Kagan, U.S. pat. **3,513,155** (1969, 1970, 1970, 1970 all to Upjohn). Synthesis and structure: Birkenmeyer, Kagan, *J. Med. Chem.* **13**, 616 (1970). Stability studies: Oesterling, *J. Pharm. Sci.* **59**, 63 (1970). Activity studies: McGehee *et al.*, *Am. J. Med. Sci.* **256**, 279 (1968); D. A. Leigh, *J. Antimicrob. Chemother.* **7**(suppl. A), 3 (1981). Toxicology: J. E. Gray *et al.*, *Toxicol. Appl. Pharmacol.* **21**, 516 (1972). Comprehensive description of the hydrochloride: L. W. Brown, W. F. Beyer in *Analytical Profiles of Drug Substances* vol. **10**, K. Florey, Ed. (Academic Press, New York, 1981) pp 75-91.

Yellow, amorphous solid. $[\alpha]_D$ +214° (chloroform). Hydrochloride monohydrate, $C_{18}H_{33}ClN_2O_5S \cdot HCl \cdot H_2O$, *Dalacine*. White crystals from ethanol-ethyl acetate, mp 141-143°. $[\alpha]_D$ +144° (H_2O). pKa 7.6. Sol in water, pyridine, ethanol, DMF. LD_{50} in mice (mg/kg): 245 i.v.; 361 i.p.; 2618 orally (Gray).
2-Dihydrogen phosphate, $C_{18}H_{34}ClN_2O_8PS$, *U-28508*, *Dalacin T*.
  THERAP CAT: Antibacterial.
  THERAP CAT (VET): Antibacterial.

2415. Clinofibrate. *2,2′-[Cyclohexylidenebis(4,1-phenyleneoxy)]bis[2-methylbutanoic acid]*; 2,2′-(4,4′-cyclohexylidenediphenoxy)-2,2′-dimethyldibutyric acid; S-8527; Lipoclin. $C_{28}H_{36}O_6$; mol wt 468.59. C 71.77%, H 7.74%, O 20.49%. Anti-atherosclerosis agent, related structurally to clofibrate, *q.v.* Prepn: Y. Nakamura *et al.*, **Ger. pat. 2,017,-331** corresp to **U.S. pat. 3,716,583** (1970, 1973 both to Sumitomo). Hypolipidemic activity: K. Toki *et al.*, *Atherosclerosis* **18**, 101 (1973); K. Suzuki *et al.*, *Japan. J. Pharmacol.* **24**, 407 (1974), *C.A.* **82**, 25821z (1975). Effect on cholesterol and lipoprotein metabolism in rats: K. Suzuki, *Biochem. Pharmacol.* **25**, 325 (1976). General pharmacology: H. Nakatani *et al.*, *Oyo Yakuri* **16**, 687 (1978), *C.A.* **90**, 145808 (1979).

Off-white powder, mp 143-146° (dec). Sol in methanol, ethanol, acetone, chloroform, glacial acetic acid. Slightly sol in CCl_4. Practically insol in water. LD_{50} in male mice, rats (mg/kg): 1800, >4000 orally; 255, 205 i.p.; 410, 2200 s.c., *Japan. Med. Gaz.* **18**(8), 7 (1981).
  THERAP CAT: Antihyperlipoproteinemic.

2416. Clioxanide. *2-(Acetyloxy)-N-(4-chlorophenyl)-3,5-diiodobenzamide*; 4′-chloro-3,5-diiodosalicylanilide acetate; acetoxy-4′-chloro-3,5-diiodobenzanilide; CI-633; SYD-230; Tremerad. $C_{15}H_{10}ClI_2NO_3$; mol wt 541.51. C 33.27%, H 1.86%, Cl 6.55%, I 46.87%, N 2.59%, O 8.86%. Prepn from 4′-chloro-3,5-diiodosalicylanilide: A. Campbell *et al.*, *Experientia* **23**, 992 (1967); **Neth. pat. Appl. 6,604,303** (1966 to Parke, Davis), *C.A.* **66**, 55247d (1967), corresp to **Brit. pat. 1,070,516**. Anthelmintic activity: Presidente *et al.*, *Am. J. Vet. Res.* **33**, 1593 (1972).

Needles from aq DMF or acetone, mp 215-216°. uv max (methanol): 262 nm. LD_{50} in sheep (mg/kg): ~420 by rumenal route; 1600 by abomasal route (Campbell).
  THERAP CAT (VET): Anthelmintic (Trematodes).

2417. Clobazam. *7-Chloro-1-methyl-5-phenyl-1H-1,5-benzodiazepine-2,4(3H,5H)-dione*; 1-phenyl-5-methyl-8-chloro-1,2,4,5-tetrahydro-2,4-dioxo-3H-1,5-benzodiazepine; H-4723; HR-376; LM-2717; Frisium; Urbadan; Urbanyl. $C_{16}H_{13}ClN_2O_2$; mol wt 300.74. C 63.90%, H 4.36%, Cl 11.79%, N 9.31%, O 10.64%. Benzodiazepine psychotherapeutic agent in which the nitrogens in the heterocyclic ring are in the 1,5- rather than in the more common 1,4-positions. Prepn: **Belg. pat. 707,667**; S. Rossi, **U.S. pat. 3,984,-398** (1968, 1976 both to Roussel-UCLAF); S. Rossi *et al.*, *Chim. Ind. (Milan)* **51**, 479 (1969); K. H. Weber *et al.*, *Ann.* **756**, 128 (1972). Toxicology: E. Schütz, *Brit. J. Clin. Pharmacol.* **7**, 33S (1979). Review of pharmacology and therapeutic use in anxiety: R. N. Brogden *et al.*, *Drugs* **20**, 161-178 (1980). Symposium on pharmacology: *Drug Dev. Res.* **1982**, Suppl. 1, 1-186. Metabolism: A. G. Borel, F. S. Abbott, *Drug Metab. Dispos.* **21**, 415 (1993). GC determn in serum: D. F. LeGatt, D. P. McIntosh, *Clin. Biochem.* **26**, 159 (1993). Clinical experience in refractory epilepsy: Canadian Clobazam Cooperative Group, *Epilepsia* **32**, 407 (1991). Review of pharmacology and clinical studies in epilepsy: S. D. Shorvon, "Clobazam" in *Antiepileptic Drugs*, R. H. Levy *et al.*, Eds. (Raven Press, New York, 3rd ed., 1989) pp 821-840.

Crystals from 50% ethanol, mp 166-168°.
Note: This is a controlled substance (depressant) listed in the U.S. Code of Federal Regulations, Title 21 Part 1308.14 (1995).
  THERAP CAT: Anxiolytic; anticonvulsant.

2418. Clobenfurol. *α-(4-Chlorophenyl)-2-benzofuranmethanol*; 2-benzofuryl-p-chlorophenylcarbinol; Menacor. $C_{15}H_{11}ClO_2$; mol wt 258.70. C 69.64%, H 4.29%, Cl 13.70%, O 12.37%. Prepn: **Belg. pat. 644,176**; Ghelardoni, Russo, **Brit. pat. 1,160,925** (1964, 1969 both to Menarini), *C.A.* **63**, 8319h (1965); **71**, 124212f (1969); Ghelardoni, Russo, *Boll. Chim. Farm.* **109**, 48 (1970). Pharmacology of the hemisuccinate deriv: Pisanti, Volterra, *Farmaco Ed. Sci.* **26**, 312 (1971).

White, odorless powder, mp 48-49°.

Hemisuccinate, $C_{19}H_{15}ClO_5$, mp 95-97° from petr ether. LD_{50} in mice, rats (mg/kg): 321, 300 i.p. (Pisanti, Volterra).

THERAP CAT: Vasodilator (coronary).

2419. Clobenoside. *Ethyl 5,6-bis-O-[(4-chlorophenyl)-methyl]-3-O-propyl-D-glucofuranoside;* ethyl 3-O-n-propyl-5,6-di-O-(4-chlorobenzyl)-D-glucofuranoside; ZY-15028; Arvigol; Floganol. $C_{25}H_{32}Cl_2O_6$; mol wt 499.43. C 60.12%, H 6.46%, Cl 14.20%, O 19.22%. Congener of tribenoside, *q.v.* Prepn: **Neth. pat. Appl. 6,812,906;** A. Rossi, **U.S. pat. 3,542,761** (1969, 1970 both to Ciba). Pharmacology and bioavailability in cats: G. Hennings *et al., Arzneimittel-Forsch.* **35,** 498 (1985). Effect on glycoprotein biosynthesis *in vitro:* W. Reutter, C. Bauer, *Adv. Enzyme Regul.* **24,** 405 (1985). Mechanism of action study: D. Araujo *et al., Arch. Int. Pharmacodyn.* **277,** 192 (1985).

Faintly yellowish oil, $[\alpha]_D^{20}$ −17° ±1 (c = 1 in chloroform). Poorly sol in water.

THERAP CAT: Vasoprotectant.

2420. Clobenzepam. *7-Chloro-10-[2-(dimethylamino)-ethyl]-5,10-dihydro-11H-dibenzo[b,e][1,4]diazepin-11-one;* 7-chloro-10-(2-dimethylaminoethyl)dibenzo[b,e][1,4]diazepin-11(10H)-one; 7-chloro-10-(β-dimethylaminoethyl)-10,-11-dihydro-11-oxo-5H-dibenzo[b,e][1,4]diazepine. $C_{17}H_{18}ClN_3O$; mol wt 315.80. C 64.66%, H 5.75%, Cl 11.23%, N 13.31%, O 5.07%. Prepn: **Belg. pat.** 611,926 (1962 to Parke, Davis), *C.A.* **57,** 13785a (1962); **Brit. pat.** 961,106; Schmutz, Hunziker, **U.S. pat. 3,419,547** (1964, 1968 both to Wander). Pharmacology: Ackermann *et al., Med. Exp.* **6,** 205 (1962); Hunziker *et al., Arzneimittel-Forsch.* **13,** 324 (1963).

Colorless prisms from acetone-ether, mp 165-166°. uv max (alcohol): 230 nm (32,740).

Hydrochloride, $C_{17}H_{18}ClN_3O \cdot HCl$, *Tarpan.* Colorless prisms from alcohol-ether, mp 225-233°. LD_{50} orally in mice: 330 mg/kg.

THERAP CAT: Antihistaminic.

2421. Clobenzorex. *(+)-N-[(2-Chlorophenyl)methyl]-α-methylbenzeneethanamine; (+)-N-(o-chlorobenzyl)-α-methyl-phenethylamine; d-N-(1-phenyl-2-propyl)-2-chlorobenzyl-amine.* $C_{16}H_{18}ClN$; mol wt 259.78. C 73.98%, H 6.98%, Cl 13.65%, N 5.39%. Prepn and pharmacology: J. R. Boissier, R. Ratouis, **Fr. pat. 1,429,306** (1966 to Soc. Ind. Fabric. Antibiot.), *C.A.* **66,** 46195h (1967); J. R. Boissier *et al., Ann. Pharm. Fran.* **24,** 57 (1966). Synthesis of labelled compound: J. Lintermans *et al., J. Label. Compounds* **6,** 289 (1970). Metabolic studies: M. Thomasset *et al., Experientia* **26,** 692 (1970); B. Glasson *et al., Arzneimittel-Forsch.* **21,** 1985 (1971).

$bp_{0.1}$ 132-134°.

Hydrochloride, $C_{16}H_{18}ClN \cdot HCl$, *Ba-7205, SD-271-12, Dinintel, Rexigen.* Crystals, mp 182-183° (isopropanol). $[\alpha]_D^{20}$ +26.3° (water). Sol in water, ethanol. Slightly sol in methanol, chloroform. LD_{50} i.p. in mice: 103 mg/kg (Boissier *et al.*).

THERAP CAT: Anorexic.

2422. Clobenztropine. *3-[(4-Chlorophenyl)phenylmeth-oxy]-8-methyl-8-azabicyclo[3.2.1]octane; 3-[(p-chloro-α-phen-ylbenzyl)oxy]tropane;* tropine 4-chlorobenzhydryl ether; Teprin. $C_{21}H_{24}ClNO$; mol wt 341.88. C 73.78%, H 7.08%, Cl 10.37%, N 4.10%, O 4.68%. Prepn: Fromer, **U.S. pat. 2,799,680** (1957); Nield, Bosch, **U.S. pat. 2,782,200** (1957 to Schenley Labs.).

Hydrochloride, $C_{21}H_{24}ClNO \cdot HCl$, crystals from isopropanol, mp 215-217°. Sol in water, ethanol. Practically insol in acetone, benzene, ether.

Hydrobromide, $C_{21}H_{24}ClNO \cdot HBr$, mp 197-200°. Sol in hot water, hot ethanol. Practically insol in acetone, benzene, ether.

Methobromide, $C_{21}H_{24}ClNO \cdot CH_3Br$, mp 245-248°. Sol in hot water, in ethanol. Practically insol in acetone, benzene, ether.

Methochloride, $C_{21}H_{24}ClNO \cdot CH_3Cl$, mp 261-263°. Sol in water, ethanol. Practically insol in acetone, benzene, ether.

THERAP CAT: Antihistaminic.

2423. Clobetasol. *(11β,16β)-21-Chloro-9-fluoro-11,17-dihydroxy-16-methylpregna-1,4-diene-3,20-dione.* $C_{22}H_{28}ClFO_4$; mol wt 410.91. C 64.31%, H 6.87%, Cl 8.63%, F 4.62%, O 15.57%. Prepn: Elks *et al.,* **Ger. pat. 1,902,340;** *eidem,* **U.S. pat. 3,721,687** (1969, 1973 both to Glaxo).

17-Propionate, $C_{25}H_{32}ClFO_5$, *GR-2/925, Clobesol, Dermoval, Dermovate, Dermoxin, Dermoxinale, Temovate.* Crystals, mp 195.5-197°. $[\alpha]_D$ +103.8° (c = 1.04 in dioxane). uv max (ethanol): 237 nm (ϵ 15000).

THERAP CAT: Glucocorticoid; anti-inflammatory.

2424. Clobetasone. *(16β)-21-Chloro-9-fluoro-17-hydr-oxy-16-methylpregna-1,4-diene-3,11,20-trione;* 21-chloro-11-dehydrobetamethasone. $C_{22}H_{26}ClFO_4$; mol wt 408.90. C 64.62%, H 6.41%, Cl 8.67%, F 4.65%, O 15.65%. Prepn: Elks *et al.,* **Ger. pat. 1,902,340;** *eidem,* **U.S. pat. 3,721,687** (1969, 1973, both to Glaxo). Activity: Munro, Wilson, *Brit. Med. J.* **3,** 626 (1975). Radioimmunoassay: W. A. Webb, R. V. Brooks, *Acta Endocrinol. Suppl.* **212,** 203 (1977). Toxicology: J. Tamura *et al., J. Toxicol. Sci.* **5,** 45, 177 (1980). Comparative clinical studies: A. Lassus, *Curr. Med. Res. Opin.* **6,** 165 (1979); T. Fredriksson, K. Nordin, *ibid.* 322. Clinical evaluation of ophthalmic form: D. Lloyd-Jones *et al., Brit. J. Ophthalmol.* **65,** 641 (1981); L. A. Eilon, S. R. Walker, *ibid.* 644; K. R. Wilhelmus *et al., ibid.* 699.

17-Butyrate, $C_{26}H_{32}ClFO_5$, *GR-2/1214, Emovate, Eumovate, Molivate (obsolete).* Crystals from methanol, mp 90-100°.

THERAP CAT: Glucocorticoid; anti-inflammatory.

2425. Clobutinol. *4-Chloro-α-[2-(dimethylamino)-1-methylethyl]-α-methylbenzeneethanol; p-chloro-α-[2-(dimethylamino)-1-methylethyl]-α-methylphenethyl alcohol;* 2-(p-chlorobenzyl)-3-dimethylaminomethyl-2-butanol; 1-p-chlorophenyl-2,3-dimethyl-4-dimethylamino-2-butanol. $C_{14}H_{22}$-ClNO; mol wt 255.79. C 65.74%, H 8.67%, Cl 13.86%, N 5.48%, O 6.25%. Prepn: **Brit. pat. 898,010;** Berg, U.S. pat. **3,121,087** (1962, 1964 both to Thomae). Pharmacology: Engelhorn, *Arzneimittel-Forsch.* **10,** 785 (1960). Metabolism: Beisenherz *et al., ibid.* **19,** 79 (1969).

bp$_{12}$ 179-180°.
Hydrochloride, $C_{14}H_{22}ClNO$·HCl, *KAT-256, Biotertussin, Silomat.* Minute crystals, mp 169-170°. Slightly bitter, acidic taste. Numbs the tongue. Sol in water. LD$_{50}$ in mice (mg/kg): 600 orally; 130 i.p. (Engelhorn).

THERAP CAT: Antitussive.

2426. Clobuzarit. *2-[[4'-Chloro(1,1'-biphenyl)-4-yl]-methoxy]-2-methylpropanoic acid; 2-[[p-(p-chlorophenyl)-benzyl]oxy]-2-methylpropionic acid;* ICI-55897; Clozic. C_{17}-H$_{17}ClO_3$; mol wt 304.77. C 67.00%, H 5.62%, Cl 11.63%, O 15.75%. Analog of clofibrate, *q.v.* Prepn: T. Leigh, L. A. McArdle, **Brit.** pat. **1,140,748;** *eidem,* **U.S.** pat. **3,549,690** (1969, 1970 both to ICI). Pharmacological studies: M. E. J. Billingham, *Brit. J. Pharmacol.* **72,** 523 (1981); M. E. J. Billingham *et al., ibid.* 551. Effects on platelet function in atherosclerosis: J. R. O'Brien *et al., Thromb. Haemostasis* **40,** 75 (1978). Comparative clinical evaluation with penicillamine: D. Y. Bulgen *et al., Ann. Rheum. Dis.* **38,** 567 (1979); H. A. Bird *et al., Clin. Exp. Rheumatol.* **1,** 93 (1983).

Crystals from benzene, mp 154-155°.
THERAP CAT: Antirheumatic.

2427. Clocapramine. *1'-[3-(3-Chloro-10,11-dihydro-5H-dibenz[b,f]azepin-5-yl)propyl][1,4'-bipiperidine]-4'-carboxamide;* 3-chloro-5-[3-(4-piperidino-4-carbamoylpiperidino)-propyl]-10,11-dihydro-5H-dibenz[b,f]azepine; 3-chlorocarpipramine; clocarpramine; CCP. $C_{28}H_{37}ClN_4O$; mol wt 481.08. C 69.91%, H 7.75%, Cl 7.37%, N 11.65%, O 3.33%. Prepn: Nakanishi, Tashiro, **Ger.** pat. **1,905,765** corresp to **U.S.** pat. **3,668,210** (1969, 1972, both to Yoshitomi); M. Nakanishi *et al., J. Med. Chem.* **13,** 644 (1970). Pharmacology and toxicity: M. Nakanishi *et al., Arzneimittel-Forsch.*

21, 391 (1971). Effect on metabolism of serotonin and catecholamines: M. Nakanishi, M. Setoguchi, *ibid.* **23,** 806 (1973). Metabolism: M. Nakanishi *et al., J. Pharm. Soc. Japan* **91,** 1042 (1971). HPLC determn in plasma: K. Hikida *et al., Anal. Sci.* **6,** 367 (1990). *Review: Japan. Med. Gaz.* **10**(7), 5 (1973).

Dihydrochloride monohydrate, $C_{28}H_{37}ClN_4O$·2HCl·H$_2$O, *Y-4153, Clofekton.* White, bitter tasting, crystalline powder, mp 267°. Soly in water: < 3.5 g/100 ml. Freely sol in glacial acetic acid. Practically insol in ether, acetone. LD$_{50}$ in mice, rats (mg/kg): 160, 125 i.p.; 2550, 6800 orally (Nakanishi).

THERAP CAT: Antipsychotic.

2428. Clocinizine. *1-[(4-Chlorophenyl)phenylmethyl]-4-(3-phenyl-2-propenyl)piperazine; 1-(p-chloro-α-phenylbenzyl)-4-cinnamylpiperazine;* 1-cinnamyl-4-(p-chloro-α-phenylbenzyl)piperazine; 1-cinnamyl-4-(4-chlorobenzhydryl)piperazine; cliocinizine. $C_{26}H_{27}ClN_2$; mol wt 402.97. C 77.50%, H 6.75%, Cl 8.80%, N 6.95%. Prepn: **Brit. pat. 809,760** (1959 to Janssen).

Hydrochloride, crystals, mp 200-201°. Component of *Denoral.*

THERAP CAT: Antihistaminic.

2429. Cloconazole. *1-[1-[2-[(3-Chlorophenyl)methoxy]-phenyl]ethenyl]-1H-imidazole;* 1-[1-[2-[(3-chlorobenzyl)oxy]-phenyl]vinyl]-1H-imidazole; croconazole. $C_{18}H_{15}ClN_2O$; mol wt 310.78. C 69.57%, H 4.86%, Cl 11.41%, N 9.01%, O 5.15%. Antimycotic vinylimidazole. Prepn and antifungal properties: M. Ogata *et al.,* **Belg.** pat. **883,665;** *eidem,* **U.S.** pat. **4,328,348** (1980, 1982 both to Shionogi); *eidem, J. Med. Chem.* **26,** 768 (1983). Pharmacology: K. Yamamoto *et al., Oyo Yakuri* **27,** 533 (1984), *C.A.* **101,** 16881c (1984). Series of articles on metabolism in rats and rabbits: K. Mizojiri *et al., Iyakuhin Kenkyu* **16,** 177-206 (1985), *C.A.* **103,** 47788y; 47789z; 47790t (1985). Mechanism of action study: T. Hiratani, H. Yamaguchi, *Chemotherapy (Tokyo)* **33,** 579 (1985).

mp 72-73°. Sol in ethyl acetate.

Monohydrochloride, $C_{18}H_{15}ClN_2O \cdot HCl$, *710674-S, Pilzcin.* Crystals from ethyl acetate + acetonitrile, mp 148.5-150°. LD_{50} in rats (mg/kg): 7000 s.c.; 2500 orally (Ogata, 1983).

THERAP CAT: Antifungal (topical).

2430. Clocortolone. *(6α,11β,16α)-9-Chloro-6-fluoro-11,21-dihydroxy-16-methylpregna-1,4-diene-3,20-dione;* 9-chloro-6α-fluoro-16α-methyl-1,4-pregnadiene-11β,21-diol-3,20-dione. $C_{22}H_{28}ClFO_4$; mol wt 410.91. C 64.31%, H 6.87%, Cl 8.63%, F 4.62%, O 15.57%. Prepn: **Neth.** pat. **Appl. 6,412,708** (1965); Kaspar, Philippson, **Ger.** pat. **2,011,559** corresp to U.S. pat. **3,729,495** (1971, 1973) (all to Schering AG). Clinical trials: Wortmann, *Wien. Med. Wochenschr.* **122**, 701 (1972); Baumann, *Praxis* **61**, 536 (1972).

Crystals from methanol-methylene chloride, mp 254° (dec).

21-Acetate, $C_{24}H_{30}ClFO_5$, *SH-818.* Crystals from methanol-methylene chloride, mp 252° (dec).

21-Pivalate, $C_{27}H_{36}ClFO_5$, *CL-68, SH-863, Cloderm, Purantix.*

THERAP CAT: Glucocorticoid.

2431. Clodinafop-Propargyl. *(R)-2-[4-[(5-Chloro-3-fluoro-2-pyridinyl)oxy]phenoxy]propanoic acid 2-propynyl ester;* 2-propynyl *(R)*-2-[4-(5-chloro-3-fluoro-2-pyridyloxy)phenoxy]propionate; CGA-184927; Topik. $C_{17}H_{13}ClFNO_4$; mol wt 349.75. C 58.38%, H 3.75%, Cl 10.14%, F 5.43%, N 4.00%, O 18.30%. Pyridyloxyphenoxy herbicide for use in cereal crops. Prepn: R. Schurter, **Eur.** pat. **Appl. 248,968;** R. Schurter, H. Rempfler, **U.S.** pat. **4,713,109** (both 1987 to Ciba-Geigy). Description: J. Amrein *et al., Proc. Brighton Crop Prot. Conf. – Weeds,* **1989,** 71-76. Review of field trials with the crop safener, cloquintocet, *q.v.:* D. W. Cornes, *ibid.* 729-734; R. D. Welsh, *Proc. 47th N. Z. Plant Prot. Conf.* **1994,** 22-26.

Odorless, crystalline solid, mp 59.3°. $[\alpha]_D^{20}$ +45.4 (c = 2 in acetone). Vapor pressure (20°): 5.3 × 10⁻⁶ Pa. Soly in water (20°): 2.5 ppm. Sol in most organic solvents. LD_{50} in rats (mg/kg): 1829 orally; >2000 dermally. LC_{50} in rats (4 hr): >2325 mg/m³ by inhalation (Amrein).

USE: Herbicide.

2432. Clodronic Acid. *(Dichloromethylene)bisphosphonic acid; (dichloromethylene)diphosphonic acid;* dichloromethanediphosphonic acid; Cl_2MDP; DMDP. $CH_4Cl_2O_6P_2$; mol wt 244.89. C 4.90%, H 1.65%, Cl 28.95%, O 39.20%, P 25.30%. Bone resorption inhibitor structurally related to inorganic pyrophosphoric acid and to etidronic acid, *q.q.v.* Prepn and use as detergent additive: **Belg.** pat. **672,205** (1966 to Procter and Gamble), *C.A.* **67,** 4040u (1967). Prepn: O. T. Quimby *et al., J. Org. Chem.* **32,** 4111 (1967). Effect on hydroxyapatite (durapatite, *q.v.*) crystal aggregation: N. M. Hansen *et al., Biochim. Biophys. Acta* **451,** 549 (1976); S. Bisaz *et al., ibid.* 560. Effect on osteoclast morphology in rats: D. J. Rowe, E. Hausmann, *Calcif. Tissue Res.* **20,** 53 (1976); S. C. Miller, W. S. S. Jee, *Anat. Rec.* **193,** 439 (1979). Effect (with piroxicam, *q.v.*) on osteolytic and metastatic spread of rat prostate adenocar-

cinoma cells: M. Pollard, P. H. Luckert, *Prostate* **8,** 81 (1986). Study of macrophage function with liposome-encapsulated DMDP: E. Claassen, N. Van Rooijen, *J. Microencapsulation* **3,** 109 (1986). Clinical study in Paget's disease: D. L. Douglas *et al., Lancet* **1,** 1043 (1980); in metastatic bone disease: I. Elomaa *et al., ibid.* **1,** 146 (1983); A. Jung *et al., N. Engl. J. Med.* **308,** 1499 (1983); in hypercalcemia of cancer: R. C. Percival *et al., Brit. J. Cancer* **51,** 665 (1985). Symposium on therapeutic efficacy in neoplastic bone disease: *Bone* **8,** Suppl. 1, S1-S86 (1987).

mp 249-251°.

Disodium salt, $CH_2Cl_2Na_2O_6P_2$, *clodronate disodium, DClMDP, Bonefos, Clasteon, Difosfonal, Loron, Mebonat, Ossiten, Ostac.*

THERAP CAT: Calcium regulator.

2433. Clofazimine. *N,5-Bis(4-chlorophenyl)-3,5-dihydro-3-[(1-methylethyl)imino]-2-phenazinamine; 3-(p-chloroanilino)-10-(p-chlorophenyl)-2,10-dihydro-2-(isopropylimino)phenazine;* 2-(4-chloroanilino)-3-isopropylimino-5-(4-chlorophenyl)-3,5-dihydrophenazine; 2-p-chloroanilino-5-p-chlorophenyl-3,5-dihydro-3-isopropyliminophenazine; G-30320; B-663; Lampren(e). $C_{27}H_{22}Cl_2N_4$; mol wt 473.40. C 68.50%, H 4.68%, Cl 14.98%, N 11.83%. Prepn: V. C. Barry *et al.,* **U.S.** pat. **2,948,726** (1960 to Geigy); Barry *et al., Nature* **179,** 1013 (1957); Barry *et al., J. Chem. Soc.* **1958,** 859; Belton *et al., Proc. Roy. Irish Acad. Sect. B* **62,** 9 (1961), *C.A.* **58,** 4556d (1963). Activity studies: Vischer, *Arzneimittel-Forsch.* **18,** 1529 (1968). HPLC determn in plasma: J. H. Peters *et al., J. Chromatog.* **229,** 503 (1982). Toxicity data: Stenger *et al., Arzneimittel-Forsch.* **20,** 794 (1970). Clinical studies: Mathies, *ibid.* 1838; Karat *et al., Brit. Med. J.* **1,** 198 (1970). Review of clinical pharmacokinetics: M. R. Holdiness, *Clin. Pharmacokinet.* **16,** 74-85 (1989); of clinical use: J. C. Garrelts, *DICP Ann. Pharmacother.* **25,** 525-531 (1991). Comprehensive description: V. K. Kapoor in *Analytical Profiles of Drug Substances* vol. 18, K. Florey, Ed. (Academic Press, New York, 1989) pp 91-120; C. M. O'Driscoll, O. I. Corrigan in *Analytical Profiles of Drug Substances and Excipients* vol. 21, H. G. Brittain, Ed. (Academic Press, San Diego, 1992) pp 75-108.

Dark-red crystals, mp 210-212°. uv max (0.01M methanolic HCl): 284, 486 nm (abs. ~1.30, ~0.64). pKa 8.37; also reported as 8.51. Sol in dil acetic acid, DMF. Sol in 15 parts of chloroform, 700 parts of ethanol, 1000 parts of ether. Practically insol in water. Partition coefficient (octanol/water): 7.48; (isooctane/pH 5.15 buffer) at 20°: 5.01; (n-octanol/pH 5.15 buffer) at 20°: 4.30; (n-octanol/pH 5.15 buffer) at 37°: 4.40; (n-octanol/pH 5.15 buffer) at 45°: 4.48; (n-octanol/pH 5.15 buffer) at 55°: 4.54. LD_{50} orally in mice, rats, and guinea pigs: >4 g/kg (Stenger).

THERAP CAT: Antibacterial (tuberculostatic, leprostatic).

2434. Clofenamide. *4-Chloro-1,3-benzenedisulfonamide; 4-chloro-m-benzenedisulfonamide;* 1-chlorobenzene-2,4-disulfonamide; chlorphenamide; monochlorphenamide; Salco; Saltron; Soluran; Aquedux; Haflutan. $C_6H_7ClN_2O_4S_2$; mol wt 270.72. C 26.62%, H 2.61%, Cl 13.10%, N

10.35%, O 23.64%, S 23.69%. Prepn: Olivier, *Rec. Trav. Chim.* **37**, 307 (1918); Davies, Wood, *J. Chem. Soc.* **1928**, 1125.

Needles, mp 206-207° (Davies), mp 217-219° (Olivier). Sol in hot alcohol, water; slightly sol in cold solvents.

THERAP CAT: Diuretic.

2435. Clofentezine. *3,6-Bis(2-chlorophenyl)-1,2,4,5-tetrazine;* bisclofentezin; NC-21314; Apollo. $C_{14}H_8Cl_2N_4$; mol wt 303.15. C 55.47%, H 2.66%, Cl 23.39%, N 18.48%. Contact acaricide active against eggs and early motile stages of phytophagous mites. Prepn: J. H. Parsons, *Eur.* **pat. Appl. 5912;** *idem,* U.S. pat. **4,237,127** (1979, 1980 both to Fisons). Physical properties, toxicity and bioactivity: K. M. G. Bryan *et al., Proc. Brit. Crop Prot. Conf.-Pests Dis.* **1981,** 67. Field trial on apple trees: A. J. Read, *Proc. 36th N.Z. Weed Pest Control Conf.,* 261 (1983). *Review:* F. Rauch, *Def. Veg.* **39,** 11-17 (1985).

Magenta solid. Crystals from ethyl acetate, mp 179-182°. Soly: 50 g/l in chloroform; 2.5 g/l in benzene; <1 g/l in hexane; <1 mg/l in water. LD_{50} in rats, mice (mg/kg): >3200 orally; LC_{50} (96 hr) in rainbow trout: 100 mg/l (Bryan).

USE: Pesticide.

2436. Clofibrate. *2-(4-Chlorophenoxy)-2-methylpropanoic acid ethyl ester;* ethyl 2-(*p*-chlorophenoxy)-2-methylpropionate; ethyl *p*-chlorophenoxyisobutyrate; Amotril; Anparton; Apolan; Artevil; Ateculon; Ateriosan; Atheropront; Atromidin; Atromid-S; Bioscleran; Claripex; Clobren-SF; Clofinit; CPIB; Hyclorate; Liprinal; Neo-Atromid; Normet; Normolipol; Recolip; Regelan; Serotinex; Sklerolip; Skleromexe; Sklero-Tablinen; Ticlobran; Xyduril. $C_{12}H_{15}ClO_3$; mol wt 242.70. C 59.39%, H 6.23%, Cl 14.61%, O 19.78%. Prepn: W. G. M. Jones *et al.,* **Brit.** pat. **860,303;** *eidem,* U.S. pat. **3,262,850** (1961, 1966 both to I.C.I.). Clinical trials in the primary prevention of ischaemic heart disease: *Brit. Heart J.* **40,** 1069 (1978); *Lancet* **2,** 379 (1980). Toxicity data: G. Metz *et al., Arzneimittel-Forsch.* **27,** 1173 (1977). Review of pharmacology: M. Chevais, *Therapie* **35,** 5-22 (1980); of metabolism: M. N. Cayen, *Rev. Drug Metab. Drug Interact.* **3,** 77-103 (1980); of toxicity studies: *IARC Monographs* **24,** 39-58 (1980). Comprehensive description: M. M. A. Hassan, A. A. Elazzouny, *Anal. Profiles Drug Subs.* **11,** 197-224 (1982).

Oil, bp_{20} 148-150°. Practically insol in water. Misc with ethanol, acetone, chloroform, ether. LD_{50} in mice, rats (g/kg): 1.28, 1.65 orally (Metz).

Mixture with androsterone, *Atromid.*

THERAP CAT: Antihyperlipoproteinemic.

2437. Clofibric Acid. *2-(4-Chlorophenoxy)-2-methylpropanoic acid; 2-(p-chlorophenoxy)-2-methylpropionic acid;* α-(*p*-chlorophenoxy)isobutyric acid; chlorophibrinic acid; Arteriohom; Regulipid. $C_{10}H_{11}ClO_3$; mol wt 214.65. C 55.96%, H 5.17%, Cl 16.52%, O 22.36%. Prepn: Galimberti, Defranceschi, *Gazz. Chim. Ital.* **77,** 431 (1947); Gilman, Wilder, *J. Am. Chem. Soc.* **77,** 6644 (1955); Jones *et*

al., **Brit.** pat. **860,303** (1961 to I.C.I.). Prepn and resolution of isomers: Witiak *et al., J. Med. Chem.* **11,** 1086 (1968). Activity thought to be due to its displacement of thyroxin and its inhibition of cholesterol biosynthesis. *See* Chang *et al., Biochem. Pharmacol.* **16,** 2053 (1967); Walsh *et al., Arch. Biochem. Biophys.* **130,** 7 (1969); Witiak *et al., J. Med. Chem.* **14,** 754 (1971). Metabolism: Almirante *et al., Boll. Chim. Farm.* **108,** 292 (1969).

Crystals from water or methanol, mp 118-119°.

Basic aluminum salt, $C_{20}H_{21}AlCl_2O_7$, *Alufibrate, Atherolip, Atherolipin.*

Magnesium salt, $C_{20}H_{20}Cl_2MgO_6$, *magnesium clofibrate, UR-112, Clomag.* White powder, mp 326-328°. Soly in g/100 mg: water 4.5; abs ethanol 0.7; chloroform 0.02.

Pyridoxine salt, $C_{10}H_{11}ClO_3 \cdot C_8H_{11}NO_3$, *pyridoxine p-chlorophenoxyisobutyrate, Claresan.* Prepn: Sarbach *et al.,* **Ger.** pat. **1,915,497** (1970 to Inst. Rech. Sci.), *C.A.* **74,** 13011g (1971).

Ethyl ester, *see* Clofibrate.

3-Pyridylmethyl ester, *see* Clofenpyride.

Etofylline ester, *see* Theofibrate.

THERAP CAT: Antihyperlipoproteinemic.

2438. Cloflucarban. *N-(4-Chlorophenyl)-N'-[4-chloro-3-(trifluoromethyl)phenyl]urea; 4,4'-dichloro-3-(trifluoromethyl)carbanilide;* Irgasan CF3; Irgosan CF3. $C_{14}H_9Cl_2F_3N_2O$; mol wt 349.14. C 48.16%, H 2.60%, Cl 20.31%, F 16.32%, N 8.02%, O 4.58%. Prepn: Schetty *et al.,* U.S. pat. **2,745,874** (1956 to Geigy); Sydor, **Belg.** pat. **617,117** (1962 to Am. Cyanamid), compd included but not specifically described in abstract, *C.A.* **59,** 1542a (1963).

White crystalline solid, mp 214-215°. Insol in water; dissolves well in organic solvents.

THERAP CAT: Disinfectant.

2439. Clofoctol. *2-[(2,4-Dichlorophenyl)methyl]-4-(1,1,3,3-tetramethylbutyl)phenol;* α-(2,4-dichlorophenyl)-4-(1,1,3,3-tetramethylbutyl)-*o*-cresol; Gramplus; Octofène. $C_{21}H_{26}Cl_2O$; mol wt 365.34. C 69.04%, H 7.17%, Cl 19.41%, O 4.38%. Prepn: J. Debat, U.S. pat. **3,830,852** (1974 to Inst. Recherches Chim. Biol. Appl.). Activity against gram positive and gram negative bacteria: A. Buogo, *Drugs Exp. Clin. Res.* **10,** 321 (1984). Pharmacokinetics in rats: M. Del Tacca *et al.* in *Recent Advances in Chemotherapy, Antimicrobial Section 3,* J. Ishigami, Ed. (University of Tokyo Press, 1985) pp 1927-28; in man: *eidem, J. Antimicrob. Chemother.* **19,** 679 (1987). Mechanism of action: J. Combe *et al., J. Pharmacol.* **11,** 411 (1980); F. Yablonsky, G. Simonet, *ibid.* **13,** 515 (1982). Clinical trials in respiratory infections: J. Vialatte, *Diagnostics* **207,** 65 (1978); in infectious bronchopulmonary diseases: R. Danesi, M. Del Tacca, *Int. J. Clin. Pharm. Res.* **5,** 175 (1985); in ear, nose and throat infections: P. L. Ghilardi, A. Casani, *Drugs Exp. Clin. Res.* **11,** 815 (1985).

Crystals from petr ether, mp 78°. LD$_{50}$ orally in male rats: >4 g/kg (Debat).

THERAP CAT: Antibacterial.

2440. Cloforex. *[2-(4-Chlorophenyl)-1,1-dimethyleth-yl]carbamic acid ethyl ester; (p-chloro-α,α-dimethylphenethyl)carbamic acid ethyl ester;* ethyl *(p-chloro-α,α-dimethylphenethyl)carbamate;* ethyl *N-[α,α-dimethyl-β-(p-chlorophenyl)ethyl]carbamate;* Avicol SL; Frenapyl; Oberex. $C_{13}H_{18}ClNO_2$; mol wt 255.74. C 61.05%, H 7.09%, Cl 13.86%, N 5.48%, O 12.51%. Prepn: **Belg.** pat. **619,886** (1963 to Troponwerke Dinklage), *C.A.* **59**, 11359c (1963), corresp to **Brit.** pat. **970,565.** Metabolism: Ryrfeldt, *Acta Pharmacol. Toxicol.* **28**, 391 (1970); Buelow *et al., Arzneimittel-Forsch.* **21**, 86 (1971).

Crystals, mp 52.5-53°. bp$_{0.005}$ 88-90°.

THERAP CAT: Anorexic.

2441. Clomacran. *2-Chloro-9,10-dihydro-N,N-dimethyl-9-acridinepropanamine; 2-chloro-9-[3-(dimethylamino)propyl]acridan;* SKF-14336D. $C_{18}H_{21}ClN_2$; mol wt 300.83. C 71.87%, H 7.04%, Cl 11.78%, N 9.31%. Prepn: C. L. Zirkle, **U.S.** pat. **3,131,190** (1964 to SKF). Effect on *in vivo* histamine metabolism: R. Schayer, M. A. Reilly, *Arch. Int. Pharmacodyn. Ther.* **203**, 123 (1973). Clinical potency in relation to dopamine-binding: P. Seeman *et al., Nature* **261**, 717 (1976). Pharmacology: P. Fowler *et al., Arzneimittel-Forsch.* **27**, 866 (1977). Photooxidation study: W. R. Knappe, *J. Pharm. Sci.* **67**, 318 (1978).

Phosphate, $C_{18}H_{21}ClN_2 \cdot H_3PO_4$, *SKF-14336, Devryl, Olaxin.* LD$_{50}$ in rats, mice (mg/kg): 350, 222 orally (Fowler).

THERAP CAT: Antipsychotic.

2442. Clomestrone. *(16α)-16-Chloro-3-methoxyestra-1,-3,5(10)-trien-17-one;* 16α-chloroestrone 3-methyl ether; Arterolo; Atheran. $C_{19}H_{23}ClO_2$; mol wt 318.84. C 71.57%, H 7.27%, Cl 11.12%, O 10.04%. Prepn: Mueller *et al., J. Am. Chem. Soc.* **80**, 1769 (1958); Mueller, **U.S.** pat. **2,855,411** (1958 to Searle); Ravizza, **Swiss** pat. **343,396** (1959). Alternate route: Labs. Sobio, **Fr.** pat. **1,350,100** (1964).

Crystals from chloroform, mp 179-180°. [α]$_D$ +161° (chloroform).

THERAP CAT: Antihyperlipoproteinemic.

2443. Clometacin. *3-(4-Chlorobenzoyl)-6-methoxy-2-methyl-1H-indole-1-acetic acid;* [2-methyl-3-(p-chlorobenzoyl)-6-methoxyindol-1-yl]acetic acid; clometazin; mindolic acid; R-3959; C-1656; Dupéran. $C_{19}H_{16}ClNO_4$; mol wt 357.79. C 63.78%, H 4.51%, Cl 9.91%, N 3.91%, O 17.89%. Prepn: Allais, Nomine, **Ger.** pat. **1,901,167** (1969 to Roussel-UCLAF), *C.A.* **72**, 43440x (1970).

White powder, mp 242°.

THERAP CAT: Analgesic.

2444. Clomethiazole. *5-(2-Chloroethyl)-4-methylthiazole;* 4-methyl-5-(β-chloroethyl)thiazole; chlorethiazol; chlormethiazole; S.C.T.Z. C_6H_8ClNS; mol wt 161.65. C 44.58%, H 4.99%, Cl 21.93%, N 8.66%, S 19.84%. Prepd by condensation of thioformamide with chloroacetopropyl alcohol: Buchman, *J. Am. Chem. Soc.* **58**, 1803 (1936); by chlorination of 4-methyl-5-(β-ethoxyethyl)thiazole: Clarke, Gurin, *ibid.* **57**, 1876 (1935); by treating 5-(2-hydroxyethyl)-4-methylthiazole with thionyl chloride: Sawa, Ishida, *J. Pharm. Soc. Japan* **76**, 337 (1956); by H$_2$O$_2$ oxidation of 2-mercapto-4-methyl-5-(β-chloroethyl)thiazole: **Swiss** pat. **200,248** (1938 to Hoffmann-La Roche). Prepn of methane and ethanedisulfonate: Charonnat *et al.,* **U.S.** pat. **3,031,-457** (1962).

Oily, viscous liquid. d$_4^{25}$ 1.233. Characteristic disagreeable odor of thiazole compds. bp$_7$ 92°.

Hydrochloride, $C_6H_8ClNS \cdot HCl$, crystals from abs ethanol + ether, mp 130°. Sol in water, alcohol.

Methanedisulfonate, $C_{13}H_{20}Cl_2N_2O_6S_4$, crystals from methanol + ether, mp 120°.

Ethanedisulfonate, $C_{14}H_{22}Cl_2N_2O_6S_4$, *Distraneurin, Hemineurin, Heminevrin.* Crystals from methanol + ether, mp 124°.

USE: Intermediate in some processes of vitamin B$_1$ manufacture.

THERAP CAT: Sedative, hypnotic, anticonvulsant.

2445. Clometocillin. *[2S-(2α,5α,6β)]-6-[[(3,4-Dichlorophenyl)methoxyacetyl]amino]-3,3-dimethyl-7-oxo-4-thia-1-azabicyclo[3.2.0]heptane-2-carboxylic acid;* 3,4-dichloro-α-methoxybenzylpenicillin; 6-(α-methoxy-3,4-dichlorophenylacetamido)penicillanic acid; clometacillin; clomethacillin; no. 356; penicillin 356. $C_{17}H_{18}Cl_2N_2O_5S$; mol wt 433.31. C 47.12%, H 4.19%, Cl 16.36%, N 6.46%, O 18.46%, S 7.40%. Semi-synthetic antibiotic related to penicillin. Prepn: Vanderhaeghe *et al., Antimicrob. Ag. Chemother.* **1961**, 581; **U.S.** pat. **3,007,920** (1961 to Recherche et Ind. Therap.). Activity studies: van Dijck *et al., Antibiot. Chemother.* **12**, 192 (1962).

Sodium salt, $C_{17}H_{17}Cl_2N_2NaO_5S$, *Rixapen.* [α]$_D$ 210-220°. Pure epimers were prepared by fractional crystn. First crop: Crystals, dec 232-235°. [α]$_D^{20}$ +182° (water). Second crop: [α]$_D^{20}$ +207°. Third crop: Crystals, dec 180-183°. [α]$_D^{20}$ +261° (water).

The free acid yields two diastereoisomers: [α]$_D$ +177° and [α]$_D$ +257°.

THERAP CAT: Antibacterial.

2446. Clomiphene. *2-[4-(2-Chloro-1,2-diphenylethen-yl)phenoxy]-N,N-diethylethanamine; 2-[p-(2-chloro-1,2-di-*

phenylvinyl)phenoxy]triethylamine; 2-[*p*-(β-chloro-α-phenyl-styryl)phenoxy]triethylamine; 1-[*p*-(β-diethylaminoethoxy)-phenyl]-1,2-diphenylchloroethylene; clomifene; chloramiphene; MRL-41. $C_{26}H_{28}ClNO$; mol wt 405.97. C 76.92%, H 6.95%, Cl 8.73%, N 3.45%, O 3.94%. Synthetic estrogen agonist-antagonist. Prepn: Allen *et al.,* U.S. pat. **2,914,563** (1959 to Merrell). Stereochemistry of the geometric isomers: Ernst *et al., J. Pharm. Sci.* **65,** 148 (1976). Induction of ovulation: R. B. Greenblatt, *Fertil. Steril.* **12,** 402 (1961). Clinical trial in anovulatory women: J. Garcia *et al., ibid.* **28,** 707 (1977). Use in male infertility: P. J. Sorbie, R. Perez-Marrero, *J. Urol.* **131,** 425 (1984). HPLC determn of isomers in human plasma: C. L. Baustian, T. J. Mikkelson, *J. Pharm. Biomed. Anal.* **4,** 237 (1986).

Enclomiphene

Citrate, $C_{26}H_{28}ClNO.C_6H_8O_7$, *Clomid, Clomphid, Clomivid, Clostilbegyt, Dyneric, Ikaclomine, Pergotime, Serophene.* Crystals, mp 116.5-118°. Slightly sol in water, chloroform; freely sol in methanol; sparingly sol in alcohol. Insol in ether.

cis-Form, *zuclomiphene.*

trans-Form, *enclomiphene.*

THERAP CAT: Gonad-stimulating principle.

2447. Clomipramine. *3-Chloro-10,11-dihydro-N,N-dimethyl-5H-dibenz[b,f]azepine-5-propanamine; 3-chloro-5-[3-(dimethylamino)propyl]-10,11-dihydro-5H-dibenz[b,f]azepine;* 3-chloro-10,11-dihydro-5-(3-dimethylaminopropyl)-5*H*-dibenz[*b,f*]azepine; 5-(γ-dimethylaminopropyl)-3-chloro-iminodibenzyl; chlorimipramine; G-34586. $C_{19}H_{23}ClN_2$; mol wt 314.86. C 72.48%, H 7.36%, Cl 11.26%, N 8.90%. Prepn: P. N. Craig *et al., J. Org. Chem.* **26,** 135 (1961); W. Schindler, H. Dietrich, **Swiss.** pat. **371,799;** *eidem,* U.S. pat. **3,467,650** (1963, 1969 both to Geigy). Crystal and molecular structure of the hydrochloride: M. L. Post, A. S. Horn, *Acta Crystallogr.* **33B,** 2590 (1977). Clinical trial in obsessive-compulsive disorder: J. DeVeaugh-Geiss *et al., Psychopharmacol. Bull.* **25,** 36 (1989).

$bp_{0.3}$ 160-170°.

Hydrochloride, $C_{19}H_{23}ClN_2.HCl$, *Anafranil.* Crystals from acetone-ether/methanol-ether, mp 189-190° (Craig); also reported as crystals from acetone, mp 191.5°-192° (Schindler, Dietrich).

THERAP CAT: Antidepressant.

2448. Clomocycline. *7-Chloro-4-(dimethylamino)-1,4,-4a,5,5a,6,11,12a-octahydro-3,6,10,12,12a-pentahydroxy-N-(hydroxymethyl)-6-methyl-1,11-dioxo-2-naphthacenecarboxamide;* chlormethylenecycline; N^2-(hydroxymethyl)chlortetracycline; *N*-methylol-7-chlortetracycline; Megaclor. $C_{23}H_{25}ClN_2O_9$; mol wt 508.91. C 54.28%, H 4.95%, Cl 6.97%, N 5.50%, O 28.29%. Semi-synthetic broad spectrum antibiotic related to tetracycline, *q.v.* Prepn: Banci, Tubaro, *Boll. Chim. Farm.* **102,** 471 (1963); **Belg.** pat. **628,142** (1963 to AB Leo), *C.A.* **60,** 15805f (1964).

Yellow material. Dec 145-170°. Sensitive to light and air. pH of 2% aq soln 6.3-6.5. Sol in citrate-phosphate buffer, pH 6-8.

THERAP CAT: Antibacterial.

2449. Clonazepam. *5-(2-Chlorophenyl)-1,3-dihydro-7-nitro-2H-1,4-benzodiazepin-2-one;* 7-nitro-5-(2-chlorophenyl)-3*H*-1,4-benzodiazepin-2(1*H*)-one; Ro-5-4023; Clonopin; Iktorivil; Klonopin; Landsen; Rivotril. $C_{15}H_{10}ClN_3O_3$; mol wt 315.72. C 57.07%, H 3.19%, Cl 11.23%, N 13.31%, O 15.20%. Antiepileptic agent with anxiolytic and antimanic properties. Prepn: L. H. Sternbach *et al., J. Med. Chem.* **6,** 261 (1963); O. Keller *et al.,* U.S. pat. **3,121,076** (1964 to Hoffmann-La Roche). Prepd but not claimed: J. Kariss, H. L. Newmark, U.S. pat. **3,116,203** (1963 to Hoffmann-La Roche). Alternate process: A. Focella, A. I. Rachlin, U.S. pat. **3,335,181** (1967 to Hoffmann-La Roche). Pharmacology: Guerrero-Figueroa *et al., Curr. Ther. Res. Clin. Exp.* **11,** 40 (1969); Lechat *et al., Therapie* **25,** 893 (1970); D'Armagnac *et al., ibid.* **26,** 439 (1971). Toxicology: Blum *et al., Arzneimittel-Forsch.* **23,** 377 (1973). Determn of clonazepam and its main metabolites: S. Ebel, H. Schütz, *ibid.* **27,** 325 (1977). Clinical efficacy in acute mania: G. Chouinard, *Psychosomatics* **26**(12), Suppl., 7 (1985); in panic disorders and agitation: R. Fontaine, *ibid.* 13. Comprehensive description: W. C. Winslow, *Anal. Profiles Drug Subs.* **6,** 61-81 (1977).

White crystals from ethanol-methylene chloride, mp 236.5-238.5°. uv max (7.5% methanol in isopropanol): 248, 310 nm (ε 14500, 11600). Soly in mg/ml at 25°: acetone 31; chloroform 15; methanol 8.6; ether 0.7; benzene 0.5; water <0.1. pK_1 1.5, pK_2 10.5. LD_{50} orally in mice: >4000 mg/kg (Blum).

Note: This is a controlled substance (depressant) listed in the U.S. Code of Federal Regulations, Title 21 Part 1308.14 (1995).

THERAP CAT: Anticonvulsant.

2450. Clonidine. *N-(2,6-Dichlorophenyl)-4,5-dihydro-1H-imidazol-2-amine; 2-(2,6-dichloroanilino)-2-imidazoline;* 2,6-dichloro-*N*-2-imidazolidinylidenebenzenamine; 2-(2,6-dichloroanilino)-1,3-diazacyclopentene-(2); 2-[(2,6-dichlorophenyl)imino]-2-imidazoline. $C_9H_9Cl_2N_3$; mol wt 230.10. C 46.98%, H 3.94%, Cl 30.82%, N 18.26%. α₂-Adrenergic agonist. Prepn: Zeile *et al.,* U.S. pat. **3,202,660** (1965 to Boehringer, Ing.). Use in shaving soap formulations: *eidem,* U.S. pat. **3,190,802** (1965 to Boehringer, Ing.). Pharmacology: Bolme, Fuxe, *Eur. J. Pharmacol.* **13,** 168 (1971). Revised structure: L. M. Jackman, T. Jen, *J. Am. Chem. Soc.* **97,** 2811 (1975). GC determn in plasma: P. O. Edlund, *J. Chromatog.* **187,** 161 (1980). Preliminary studies on potential antidepressant activity: D. C. Jimerson *et al., Biol. Psychiatry* **15,** 45 (1980); J. B. Malick, *Prog. Clin. Biol. Res.* **71,** 165 (1981). Exptl use in drug rehabilitation: M. S. Gold, A. C. Pottash, *Ann. N.Y. Acad. Sci.* **362,** 191-202 (1981). Activity as α-adrenoceptor agonist: A. G. Roach *et al., J. Pharmacol. Exp. Ther.* **227,** 421 (1983). Symposium on cardiovascular and psychotropic pharmacology and clinical experience: *Central Blood Pressure Regulation: The Role of α_2-Receptor Stimulation,* K. Hayduk, K. B. Bock, Eds. (Steinkopff Verlag, Darmstadt, 1983) 284 pp. Effects in acute smoking withdrawal syndrome: A. H. Glassman *et al., Science* **226,** 864 (1984); in alcoholism withdrawal: Z. Jraidi *et al., Therapie* **42,** 21 (1987). Clinical trial in Tourette's syndrome: J. F. Leckman *et al., Neurology* **35,** 343 (1985); in cigarette smoking cessation: R. Davison *et al., Clin. Pharmacol. Ther.* **44,** 265 (1988). *Reviews:* A. Walland in *Pharmacological and Biochemical Properties of Drug Substances* vol. 1, M. E. Goldberg, Ed. (Am. Pharm. Assoc., Washington, D.C., 1977) pp 67-107; H. Schmitt, *Handb. Exp. Pharmacol.* **39,** 299-396 (1977); M. C. Houston, *Prog. Cardiovasc. Dis.* **23,** 337-350 (1981). Comprehensive description: M. A. Abounassif *et al.* in *Analytical Profiles of Drug Substances and Excipients* vol. 21, H. G. Brittain, Ed. (Academic Press, San Diego, 1992) pp 109-147.

Crystals, mp 130°.

Hydrochloride, $C_9H_9Cl_2N_3$·HCl, *ST-155, Catapres, Catapresan, Clonistada, Dixarit, Ipotensium, Isoglaucon, Tenso-Timelets.* Crystals, mp 305°. Sol in abs. ethanol; slightly sol in chloroform. Practically insol in ether. One gram is sol in 6 mls water (60°C); about 13 mls water (20°C); about 5.8 mls methanol; about 25 mls ethanol; about 5000 mls chloroform. uv max (water): 213, 271, 302 nm (ε 8290.327, 713.074, 339.876). LD_{50} in mice, rats (mg/kg): 328, 270 orally; 18, 29 i.v. (Walland).

USE: In shaving soaps.

THERAP CAT: Antihypertensive.

2451. Clonitazene. *2-[(4-Chlorophenyl)methyl]-N,N-diethyl-5-nitro-1H-benzimidazole-1-ethanamine; 2-p-chlorobenzyl-1-(2-diethylaminoethyl)-5-nitrobenzimidazole;* 1-(β-diethylaminoethyl)-2-p-chlorobenzyl-5-nitrobenzimidazole. $C_{20}H_{23}ClN_4O_2$; mol wt 386.88. C 62.09%, H 5.99%, Cl 9.16%, N 14.48%, O 8.27%. Prepn: Hunger *et al., Experientia* **13,** 400 (1957); Hoffmann *et al.,* U.S. pat. **2,935,514** (1960 to Ciba).

Crystals, mp 75-76°.

Hydrochloride, $C_{20}H_{24}Cl_2N_4O_2$, mp 238-240°.

Methanesulfonate, $C_{21}H_{27}ClN_4O_5S$, mp 163-166°.

Note: This is a controlled substance (opiate) listed in the U.S. Code of Federal Regulations, Title 21 Part 1308.11 (1995).

THERAP CAT: Analgesic (narcotic).

2452. Clonitrate. *3-Chloro-1,2-propanediol dinitrate;* 3-chloropropyleneglycol dinitrate; dinitrochlorohydrin; Dylate. $C_3H_5ClN_2O_6$; mol wt 200.54. C 17.97%, H 2.51%, Cl 17.68%, N 13.97%, O 47.87%. Prepn from 3-chloro-1,2-propanediol: Henry, *Ann.* **155,** 165 (1870).

CH_2ONO_2
|
CHONO_2
|
CH_2Cl

Slightly yellow liq, bp_{760} 190-195° (some decompn); bp_{15} 121-123°; bp_{10} 117.5°. d^9 1.5112; d^{15} 1.5408. Sol in alc, acetone, chloroform; practically insol in water, acids.

THERAP CAT: Vasodilator (coronary).

2453. Clonixin. *2-[(3-Chloro-2-methylphenyl)amino]-3-pyridinecarboxylic acid;* 2-(3-chloro-*o*-toluidino)nicotinic acid; 2-(2-methyl-3-chloroanilino)nicotinic acid; 2-(3-chloro-2-methylanilino)nicotinic acid; clonixic acid; CBA-93626; Sch-10304; $C_{13}H_{11}ClN_2O_2$; mol wt 262.70. C 59.44%, H 4.22%, Cl 13.50%, N 10.66%, O 12.18%. Nonsteroidal anti-inflammatory agent. Prepn: M. H. Sherlock, N. Sperber, **Belg. pat. 679,271;** *eidem,* U.S. pat. **3,337,570** (1965, 1967 both to Schering). Pharmacology: A. S. Watnick *et al., Arch. Int. Pharmacodyn.* **190,** 78 (1971). Metabolism in humans: B. Katchen *et al., J. Pharmacol. Exp. Ther.* **187,** 152 (1973). Clinical evaluation of analgesic activity: J. S. Finch, T. J. DeKornfeld, *J. Clin. Pharmacol.* **15,** 371 (1971). GC determn in plasma: C. Giachetti *et al., J. High Resolut. Chromatog.* **13,** 789 (1990). Toxicology studies: H. Tanaka *et al., Oyo Yakuri* **7,** 655 (1973), *C.A.* **80,** 44091m (1974).

Crystals from isopropyl acetate, mp 233-235°. Extremely bitter taste. LD_{50} in male mice, rats (mg/kg): 415, 335 orally; 198, 148 i.p.; 296, 325 s.c. (Tanaka).

Lysine salt, $C_{13}H_{11}ClN_2O_2 \cdot C_6H_{14}N_2O_2$, *L-104, Clonix, Deltar, Dolalgial, Dorixina.*

THERAP CAT: Analgesic.

2454. Clopamide. *(cis)-3-(Aminosulfonyl)-4-chloro-N-(2,6-dimethyl-1-piperidinyl)benzamide;* 1-(4-chloro-3-sulfamoylbenzamido)-2,6-dimethylpiperidine; *N-(2',6'-dimethyl-1'-piperidyl)-3-sulfamoyl-4-chlorobenzamide; 4-chloro-N-(2',6'-dimethyl-1'-piperidyl)-3-sulfamoylbenzamide;* chlosudimeprimyl; DT-327; Aduran; Aquex; Brinaldix. $C_{14}H_{20}ClN_3O_3S$; mol wt 345.85. C 48.62%, H 5.83%, Cl 10.25%, N 12.15%, O 13.88%, S 9.27%. Prepn (stereochemistry unspecified): **Belg. pat. 610,039;** E. Jucker, A. J. Lindenmann, U.S. pat. **3,459,756** (1962, 1969 both to Sandoz). Prepn of *cis-* and *trans-*forms: *eidem, Helv. Chim. Acta* **45,** 2316 (1962). HPLC determn: R. T. Sane *et al., J. Chromatog.* **356,** 468 (1986). Pharmacokinetics and diuretic effect in humans: J. J. McNeil *et al., Clin. Pharmacol. Ther.* **42,** 299 (1987). Clinical trial of combination with pindolol, *q.v.,* in essential hypertension: D. Crowder, E. G. M. Cameron, *Curr. Med. Res. Opin.* **6,** 342 (1979).

Mixture with pindolol, *Viskaldix.*

Hydrazine deriv., crystals from methanol + diisopropyl ether, mp 244-246°.

THERAP CAT: Antihypertensive; diuretic.

2455. Clopenthixol. *4-[3-(2-Chloro-9H-thioxanthen-9-ylidene)propyl]-1-piperazineethanol;* 2-chloro-9-[3-[4-(2-hydroxyethyl)-1-piperazinyl]propylidene]thioxanthene. $C_{22}H_{25}ClN_2OS$; mol wt 400.97. C 65.90%, H 6.28%, Cl 8.84%, N 6.99%, O 3.99%, S 8.00%. Thioxanthene neuroleptic. Prepn (configuration not specified): **Belg. pat. 585,338;** P. V. Petersen *et al.,* U.S. pat. **3,116,291** (1960, 1963 both to Kefalas A/S). Prepn of the pharmacologically active *cis*-isomer: **Belg. pat. 816,855;** N. Lassen, U.S. pat. **3,996,211** (1974, 1976 both to Kefalas A/S). Pharmacology: Cazzullo, Andreola, *Acta Neurol. (Naples)* **20,** 162 (1965); Weissman, *Mod. Probl. Pharmacopsychiat.* **2,** 15 (1969); Moeller Nielsen, *ibid.* 23. Metabolism: Khan, *Acta Pharmacol. Toxicol.* **27,** 202 (1969). HPLC determn of isomers in serum: T. Aaes-Jorgensen, *J. Chromatog.* **188**, 239 (1980). Series of articles on pharmacology and clinical studies: *Acta Psychiatr. Scand.* **64**, Suppl. 294, 1-77 (1981).

cis-isomer

Colorless syrup. Sparingly sol in ether. Readily sol in methanol.

Dihydrochloride, $C_{22}H_{25}ClN_2OS.2HCl$, *AY-62021, N-746, Ciatyl, Sordenac, Sordinol.* Crystals from ethanol, mp 250-260° (dec). Freely sol in water; sparingly sol in alcohol. Practically insol in other organic solvents. LD_{50} in male mice (mg/kg): 111 i.v. (Lassen).

cis(Z)-Form, *α-clopenthixol, zuclopenthixol.* Crystals, mp 84-85°.

cis(Z)-Form dihydrochloride, *Cisordinol, Clopixol.* Crystals, mp 250-260° (dec). LD_{50} in male mice (mg/kg): 105 i.v. (Lassen).

THERAP CAT: Antipsychotic.

2456. Cloperastine. *1-[2-[(4-Chlorophenyl)phenylmethoxy]ethyl]piperidine; 1-[2-[(p-chloro-α-phenylbenzyl)oxy]ethyl]piperidine;* p-chlorobenzhydryl 2-(1-piperidyl)ethyl ether. $C_{20}H_{24}ClNO$; mol wt 329.87. C 72.82%, H 7.33%, Cl 10.75%, N 4.25%, O 4.85%. Prepn: **Brit. pat.** 670,622 (1952 to Parke, Davis); Fujie, *Yakugaku Zasshi* **81**, 693 (1961), *C.A.* **55**, 25848f (1961). Prepn of salts: **Brit. pat.** 1,179,945 (1970 to Yoshitomi Pharm.), *C.A.* **72**, 132549g (1970). Pharmacology: Takagi *et al., Yakugaku Zasshi* **85**, 550 (1965); **87**, 907 (1967), *C.A.* **63**, 8932f (1965); **67**, 107101u (1967). Metabolism: Kato, Furuta, *Oyo Yakuri* **5**, 735 (1972).

$bp_{0.06}$ 172-174°, $bp_{0.15}$ 178-180°.

Hydrochloride, $C_{20}H_{24}ClNO.HCl$, *Hustazol, Nitossil, Novotusil, Seki.* Crystals, mp 147.9°.

Methiodide, $C_{20}H_{24}ClNO.CH_3I$, crystals, mp 140°.

THERAP CAT: Antitussive.

2457. Clopidogrel. *(S)-α-(2-Chlorophenyl)-6,7-dihydrothieno[3,2-c]pyridine-5(4H)-acetic acid methyl ester;* methyl (+)-(S)-α-(o-chlorophenyl)-6,7-dihydrothieno[3,2-c]pyridine-5(4H)-acetate; (+)-methyl-α-5-[4,5,6,7-tetrahydro[3,2-c]thienopyridyl]-(2-chlorophenyl)acetate; SR-25990. $C_{16}H_{16}ClNO_2S$; mol wt 321.83. C 59.71%, H 5.01%, Cl 11.02%, N 4.35%, O 9.94%, S 9.96%. Analog of ticlopidine, *q.v.*; inhibits ADP-induced platelet aggregation. Prepn (unspec. stereochem.): D. Aubert *et al.,* **Eur. pat. Appl.** 99,802; *eidem,* **U.S. pat.** 4,529,596 (1984, 1985 both to Sanofi). Prepn of (+)-form: A. Badorc, D. Fréhel, **Eur. pat. Appl.** 281,459; *eidem,* **U.S. pat.** 4,847,265 (1988, 1989 both to Sanofi). Antithrombotic activity: C. M. Samama *et al., Thromb. Haemostasis* **68**, 500 (1992). Clinical pharmacology and mechanism of action: D. C. B. Mills *et al., Arterioscler. Thromb.* **12**, 430 (1992).

Colorless oil. $[\alpha]_D^{20}$ +51.52° (c = 1.61 in methanol).

Hydrogen sulfate, $C_{16}H_{16}ClNO_2S.H_2SO_4$, *SR-25990C, Plavix.* White crystals, mp 184°. $[\alpha]_D^{20}$ +55.10° (c = 1.891 in methanol).

THERAP CAT: Antithrombotic.

2458. Clopidol. *3,5-Dichloro-2,6-dimethyl-4-pyridinol;* meticlorpindol; clopindol; Coyden. $C_7H_7Cl_2NO$; mol wt 192.04. C 43.78%, H 3.67%, Cl 36.92%, N 7.29%, O 8.33%. Prepn: Stevenson, **U.S. pat.** 3,206,358 (1965 to Dow). Metabolism studies in chickens: Smith, *Poultry Sci.* **48**, 420 (1969). Toxicity study: Plisek *et al., Vet. Spofa* **12**, 111 (1970), *C.A.* **74**, 138780p (1971).

Solid, mp > 320°. Practically insol in water. LD_{50} orally in rats: 18 g/kg (Plisek).

THERAP CAT (VET): Coccidiostat.

2459. Clopirac. *1-(4-Chlorophenyl)-2,5-dimethyl-1H-pyrrole-3-acetic acid;* BRL-13856; CP-172AP; Clopiran. $C_{14}H_{14}ClNO_2$; mol wt 263.72. C 63.76%, H 5.35%, Cl 13.44%, N 5.31%, O 12.13%. Prepn: Lambelin *et al.,* **Ger. pat.** 2,261,965 (1973 to Continental Pharma), *C.A.* **79**, 78604a (1973).

THERAP CAT: Anti-inflammatory.

2460. Cloprednol. *(11β)-6-Chloro-11,17,21-trihydroxypregna-1,4,6-triene-3,20-dione;* 6-chloro-$\Delta^{1,4,6}$-pregnatriene-11β,17α,21-triol-3,20-dione; RS-4691; Cloradryn; Novacort; Syntestan. $C_{21}H_{25}ClO_5$; mol wt 392.88. C 64.20%, H 6.41%, Cl 9.02%, O 20.36%. Prepn: H. J. Ringold, G. Rosenkranz, **Fr. pat.** 1,271,981, *C.A.* **58**, 1148a (1963); *eidem,* **U.S. pat.** 3,232,965 (1962, 1966 both to Syntex). Metabolic effects: E. Ortega *et al., J. Clin. Pharmacol.* **16**, 122 (1976). Effects on hypothalamic-pituitary-adrenal axis function: *eidem, J. Int. Med. Res.* **4**, 326 (1976). Bioavailability in humans: E. Mroszczak *et al., J. Pharm. Sci.* **67**, 920 (1978). HPLC determn in plasma: C. Lee *et al., ibid.* **70**, 669 (1981).

Crystals from acetone/hexane.
THERAP CAT: Glucocorticoid.

2461. Cloprostenol. *[1α(Z),2β(1E,3R*),3α,5α]-(±)-7-[2-[4-(3-Chlorophenoxy)-3-hydroxy-1-butenyl]-3,5-dihydroxycyclopentyl]-5-heptenoic acid.* $C_{22}H_{29}ClO_6$; mol wt 424.92. C 62.19%, H 6.88%, Cl 8.34%, O 22.59%. Aryloxymethyl analog of prostaglandin $F_{2\alpha}$, q.v. Prepn: J. Bowler, **Ger. pat. 2,223,365** (1972 to ICI), C.A. **78**, 110692v (1973); D. Binder et al., Prostaglandins **6**, 87 (1974). Synthesis and biological activity: eidem, ibid. **15**, 773 (1978). Short synthesis: N. R. A. Beeley et al., Tetrahedron **37**, Suppl. 9, 411 (1981). Disposition in the rat and marmoset: G. R. Bourne et al., Xenobiotica **9**, 623 (1979). Effect on fertility in cows: N. Baishya et al., Brit. Vet. J. **136**, 227 (1980); on superovulation, fertilization, egg transport in ewes: D. Whyman, R. W. Moore, J. Reprod. Fertil. **60**, 267 (1980).

Sodium salt, $C_{22}H_{28}ClNaO_6$, *ICI-80996, Estrumate, Planate.*
THERAP CAT (VET): In treatment of infertility in sows, gilts. In synchronization of estrus.

2462. Clopyralid. *3,6-Dichloro-2-pyridinecarboxylic acid; 3,6-dichloropicolinic acid;* 3,6-DCP; Dowco 290; Lontrel; Shield; Reclaim. $C_6H_3Cl_2NO_2$; mol wt 192.00. C 37.53%, H 1.57%, Cl 36.93%, N 7.30%, O 16.67%. Systemic post-emergence herbicide for use in food crops and mesquite. Prepn: H. Johnston, **Belg. pat. 644,105**; idem, **U.S. pat. 3,317,549** (1964, 1967 both to Dow). Alternate process: S. D. McGregor, **U.S. pat. 4,087,431** (1978 to Dow). Physicochemical properties, toxicity and herbicidal activity: J. G. Brown, S. D. Uprichard, Proc. Brit. Crop Prot. Conf. - Weeds **1976**, 119. GLC determn: A. J. Pik, G. W. Hodgson, J. Assoc. Offic. Anal. Chem. **59**, 264 (1976). Persistence in soil: A. J. Pik et al., J. Agr. Food Chem. **25**, 1054 (1977). Field trial in mesquite control: R. W. Bovey, R. E. Meyer, Weed Sci. **33**, 349 (1985).

White, odorless crystalline solid, mp 151-152°. Vapor pressure at 25°: 1.2×10^{-5} mm Hg. Soly at 25°: ~1000 ppm in water, >25% w/w in methanol, acetone, xylene. LD_{50} in male, female rats (mg/kg): >5000, 4300 orally; LC_{50} (96 hr) to rainbow trout: 103.5 mg/l (Brown, Uprichard).
USE: Herbicide.

2463. Cloquintocet-Mexyl. *[(5-Chloro-8-quinolinyl)oxy]acetic acid 1-methylhexyl ester;* CGA-185072. $C_{18}H_{22}ClNO_3$; mol wt 335.83. C 64.38%, H 6.60%, Cl 10.56%, N 4.17%, O 14.29%. Safening agent to protect crops from herbicidal injury. Prepn: A. Hubele, **Eur. pat. Appl. 94,349**; idem, **U.S. pat. 4,902,340** (1983, 1990 both to Ciba-Geigy). Description: J. Amrein et al., Proc. Brighton Crop Prot. Conf. - Weeds **1989**, 71-76. Effects on herbicide metabolism in cereals: K. Kreuz et al., Z. Naturforsch. **46c**, 901 (1991).

Odorless crystals, mp 69°. Vapor pressure (20°): 2.5×10^{-6} Pa. Soly in water (20°): 0.8 ppm. Sol in most organic solvents. LD_{50} in rats (mg/kg): >2000 orally; >2000 dermally. LC_{50} (4 hr) in rats: >935 mg/m³ by inhalation (Amrein).
Mixture with clodinafop-propargyl, *Célio.*
USE: Herbicide safener.

2464. Cloranolol. *1-(2,5-Dichlorophenoxy)-3-[(1,1-dimethylethyl)amino]-2-propanol;* 1-(tert-butylamino)-3-(2,5-dichlorophenoxy)-2-propanol. $C_{13}H_{19}Cl_2NO_2$; mol wt 292.20. C 53.44%, H 6.55%, Cl 24.27%, N 4.79%, O 10.95%. β-Adrenergic blocker. Prepn: G. Richter, **Ger. pat. 2,213,044** (1972 to Res. Inst. Pharm. Chem. Hungary), C.A. **78**, 15780p (1973). Effect on plasma renin and blood pressure: I. Tenyi, Curr. Ther. Res. **21**, 823 (1977). Pharmacology: M. I. K. Fekete et al., Arch. Int. Pharmacodyn. **248**, 190 (1980). Electrophysiological characteristics: F. Solti, E. Czako, Int. J. Clin. Pharmacol. Ther. Toxicol. **18**, 229 (1980). Clinical-pharmacological evaluation: E. Török et al., ibid. **200**. GC-MS determn in plasma: E. Tomori, E. Elekes, J. Chromatog. **204**, 355 (1981). Multicenter clinical trial: E. Török et al., Int. J. Clin. Pharmacol. Ther. Toxicol. **23**, 650 (1985).

Crystals, mp 82-83°.
Hydrochloride, $C_{13}H_{19}Cl_2NO_2 \cdot HCl$, *GYKI-41099, Tobanum.* Crystals, mp 210-212°.
THERAP CAT: Antiarrhythmic.

2465. Clorazepic Acid. *7-Chloro-2,3-dihydro-2-oxo-5-phenyl-1H-1,4-benzodiazepine-3-carboxylic acid.* $C_{16}H_{11}ClN_2O_3$; mol wt 314.73. C 61.06%, H 3.52%, Cl 11.26%, N 8.90%, O 15.25%. Prepn: **Neth. pat. Appl. 6,507,637**; J. Schmitt, **U.S. pat. 3,516,988**; reissued as **U.S. pat. Re. 28,315** (1965, 1970, 1975 all to Clin-Byla). Synthesis and activity of the dipotassium salt: J. Schmitt et al., Chim. Ther. **4**, 239 (1969). Solution chemistry: R. Raveux, M. Briot, ibid. 303. Metabolism: P. Gros, R. Raveux, ibid. 312. Toxicity data: M. Brunaud et al., Arzneimittel-Forsch. **20**, 123 (1970). Series of articles on pharmacology and clinical use: ibid, 123-137. HPLC determn in plasma: P. Colin, G. Sirois, J. Chromatog. **273**, 367 (1983). Clinical trial in anxiety: W. W. K. Zung, J. Clin Psychiatry **48**, 13 (1987); in comparison with buspirone, q.v.: K. Rickels et al., Arch. Gen. Psychiatry **45**, 444 (1988). Comprehensive description: J. A. Raihle, V. E. Papendick, Anal. Profiles Drug Subs. **4**, 91-112 (1975).

Dipotassium salt, $C_{16}H_{10}ClKN_2O_3 \cdot KOH$, *7-Chloro-2,3-dihydro-2-oxo-5-phenyl-1H-1,4-benzodiazepine-3-carboxylic*

acid monopotassium salt compd with potassium hydroxide,
clorazepate dipotassium, Abbott 35616, CB-4306, Belseren,
Mendon, Tranxilène, Tranxilium, Transene, Tranxene.
White powder, freely sol in water. Very poorly sol in
ethanol. Practically insol in ether, chloroform. Aq solns
are alkaline to phenolphthalein. uv max (anhydrous product in water): 231, 311 nm (ε 33500, 2450). LD_{50} in mice
(mg/kg): 700 orally; 290 i.p. LD_{50} orally in rats: > 1000
mg/kg (Brunaud).

Monopotassium salt, $C_{16}H_{10}ClKN_2O_4$, *CB-4311, Azene.*

Note: This is a controlled substance (depressant) listed in
the U.S. Code of Federal Regulations, Title 21 Part 1308.14
(1995).

THERAP CAT: Anxiolytic.

2466. Clorexolone. *6-Chloro-2-cyclohexyl-2,3-dihydro-*
3-oxo-1H-isoindole-5-sulfonamide; 6-chloro-2-cyclohexyl-3-
oxo-5-isoindolinesulfonamide; 5-chloro-2-cylohexyl-1-oxo-6-
sulfamoylisoindoline; 5-chloro-2-cyclohexyl-6-sulfamoyliso-
indolin-1-one; 5-chloro-2-cyclohexylphthalimidine-6-sulfon-
amide; M & B 8430; RP-12833; Flonatril; Nefrolan. C_{14}-
$H_{17}ClN_2O_3S$; mol wt 328.82. C 51.14%, H 5.21%, Cl
10.78%, N 8.52%, O 14.60%, S 9.75%. Prepn: **Belg.** pat.
620,654 corresp to G. E. Lee, W. R. Wragg, **U.S.** pat.
3,183,243 (1963, 1965 both to May & Baker).

Prisms, mp 266-268°. Soly in water at 25°: 16 mg/l.
THERAP CAT: Diuretic.

2467. Cloricromen. *[[8-Chloro-3-[2-(diethylamino)eth-*
yl]-4-methyl-2-oxo-2H-1-benzopyran-7-yl]oxy]acetic acid ethyl
ester; ethyl [[8-chloro-3-[2-(diethylamino)ethyl]-4-methyl-2-
oxo-2H-1-benzopyran-7-yl]oxy]acetate; 8-monochloro-3-(β-
diethylaminoethyl)-4-methyl-7-ethoxycarbonylmethoxy-
coumarin; 8-chlorocarbochromen; AD6. $C_{20}H_{26}ClNO_5$; mol
wt 395.88. C 60.68%, H 6.62%, Cl 8.96%, N 3.54%, O
20.21%. Derivative of coumarin, *q.v.* Prepn: F. della Valle,
Belg. pat. **871,315;** *idem,* **U.S.** pat. **4,452,811** (1979, 1984
both to Fidia). *In vivo* studies of activity as coronary vasodilator: F. Aporti *et al., Pharmacol. Res. Commun.* **10,** 469
(1978); as inhibitor of platelet aggregation: M. Prosdocimi
et al., Thromb. Res. **39,** 399 (1985). Pharmacology in
human platelets: R. A. Travagli *et al., ibid.* **54,** 327 (1989).
Mechanism of action: S. Porcellati *et al., Agents Actions* **29,**
364 (1990).

Crystals from ethyl acetate, mp 147-148°.
Hydrochloride, $C_{20}H_{26}ClNO_5·HCl$, *Cromocap, Proendotel.*
mp 219-220°.
THERAP CAT: Antithrombotic; vasodilator (coronary).

2468. Clorindione. *2-(4-Chlorophenyl)-1H-indene-1,3-*
(2H)-dione; 2-(p-chlorophenyl)-1,3-indandione; 2-(4-chloro-
phenyl)indan-1,3-dione; chlophenadione; G-25766; MG-
2552; Indaliton. $C_{15}H_9ClO_2$; mol wt 256.69. C 70.19%, H
3.53%, Cl 13.81%, O 12.47%. Prepn: Cavallini *et al., Far-*
maco Ed. Sci. **10,** 710 (1955); **Brit.** pat. **748,251** (1956 to
Geigy). Pharmacokinetics: Danek, Pogonowska-Wala, *Pol.*
J. Pharmacol. Pharm. **25,** 307 (1973), *C.A.* **79,** 111540b
(1973). Toxicity study: Fontaine *et al., Med. Pharmacol.*
Exp. **17,** 497 (1967).

Dark red needles from alcohol, pale yellow needles from
dil acetic acid, mp 145-146°. Practically insol in water. Sol
in organic solvents except petr ether. LD_{50} orally in mice:
1220 mg/kg (Fontaine).
THERAP CAT: Anticoagulant.

2469. Clorophene. *4-Chloro-2-(phenylmethyl)phenol; 4-*
chloro-α-phenyl-o-cresol; 2-benzyl-4-chlorophenol; 5-chloro-
2-hydroxydiphenylmethane; *o*-benzyl-*p*-chlorophenol; Septiphene; Santophen 1. $C_{13}H_{11}ClO$; mol wt 218.68. C 71.40%,
H 5.07%, Cl 16.21%, O 7.32%. Prepn: Klarmann *et al., J.*
Am. Chem. Soc. **54,** 3315 (1932); Klarmann, Gate, **U.S.** pat.
1,967,825 (to Lehn & Fink); from Na or K phenoxide and
benzyl chloride followed by chlorination of the resulting *o*-
benzylphenol: Kaiser, **Ger.** pat. **703,955** (1941 to Deutsche
Hydrierwerke).

Crystals, mp 48.5°. $bp_{3.5}$ 160-162°. $d_{15.5}^{55}$ 1.186-1.190.
USE: In household, hospital and veterinary disinfectant
preparations.

2470. Clorprenaline. *2-Chloro-α-[[(1-methylethyl)ami-*
no]methyl]benzenemethanol; o-chloro-α-(isopropylamino-
methyl)benzyl alcohol; 1-(*o*-chlorophenyl)-2-isopropylamino-
ethanol; *N*-(β-*o*-chlorophenyl-β-hydroxyethyl)isopropyl-
amine; isoprophenamine; isofrofenamine. $C_{11}H_{16}ClNO$;
mol wt 213.71. C 61.82%, H 7.55%, Cl 16.59%, N 6.55%, O
7.49%. β-Adrenergic agonist. Prepn: Mills, **U.S.** pat.
2,816,059 (1957 to Lilly); Nash, **U.S.** pat. **2,887,509** (1959
to Lilly).

Hydrochloride monohydrate, $C_{11}H_{16}ClNO·HCl·H_2O$,
Broncon, Clopinerin, Conselt, Fusca, Kalutein, Pentadoll,
Restanolon. mp 163-164°.
THERAP CAT: Bronchodilator.

2471. Clorsulon. *4-Amino-6-(trichloroethenyl)-1,3-*
benzenedisulfonamide; 4-amino-6-(trichlorovinyl)-*m*-benz-
enedisulfonamide; MK-401; Curatrem. $C_8H_8Cl_3N_3O_4S_2$;
mol wt 380.66. C 25.24%, H 2.12%, Cl 27.94%, N 11.04%,
O 16.81%, S 16.85%. Benzenedisulfonamide deriv with fas-
ciolicidal activity. Prepn: H. H. Mrozik, **Ger.** pat. **2,556,-**
122 (1976 to Merck & Co.), *C.A.* **85,** 94100n (1976); H.
Mrozik *et al., J. Med. Chem.* **20,** 1225 (1977). GLC determn in biological fluids: W. J. A. Vanden Heuvel *et al., J.*
Agr. Food Chem. **25,** 389 (1977). Efficacy against liver
fluke, *Fasciola hepatica:* D. A. Ostlind *et al., Brit. Vet. J.*
133, 211 (1977); M. D. Schulman *et al., J. Parasitol.* **65,** 555
(1979). *In vitro* inhibition of glycolysis: M. D. Schulman,
D. Valentino, *Exp. Parasitol.* **49,** 206 (1980). Mechanism of
action: M. D. Schulman *et al., Mol. Biochem. Parasitol.* **5,**
133 (1982). Pharmacokinetics: *eidem, J. Parasitol.* **68,** 603
(1982).

Crystals from ether, mp 194-203°. Another crystal form from water, mp 203-205°. uv max (CH₃OH): 325, 267, 227 nm (ε 4530, 17395, 36310). LD₅₀ in mice (mg/kg): 761 i.p.; > 10,000 orally (Ostlind).

THERAP CAT (VET): Anthelmintic (Trematodes).

2472. Clortermine. *2-Chloro-α,α-dimethylbenzeneeth-anamine; o-chloro-α,α-dimethylphenethylamine; 1-(o-chloro-phenyl)-2-methyl-2-propylamine; 1-(o-chlorophenyl)-2-methyl-2-aminopropane.* C₁₀H₁₄ClN; mol wt 183.68. C 65.39%, H 7.68%, Cl 19.30%, N 7.63%. Prepn and use as anorexic agent: **Belg.** pat. **665,244; Brit.** pat. **1,066,616;** Finocchio, Huebner, **U.S.** pat. **3,415,937** (1965, 1967, 1968, all to Ciba). Toxicity data: E. I. Goldenthal, *Toxicol. Appl. Pharmacol.* **18**, 185 (1971).

bp₁₆ 116-118°.

Hydrochloride, C₁₀H₁₄ClN.HCl, *Su-10568, Voranil.* Crystals from ethanol, mp 245-246°. LD₅₀ orally in rats: 332 ±23 mg/kg (Goldenthal).

Note: This is a controlled substance (stimulant) listed in the U.S. Code of Federal Regulations, Title 21 Part 1308.13 (1995).

THERAP CAT: Anorexic.

2473. Closantel. *N-[5-Chloro-4-[(4-chlorophenyl)cyano-methyl]-2-methylphenyl]-2-hydroxy-3,5-diiodobenzamide; R-31520; Flukiver; Seponver.* C₂₂H₁₄Cl₂I₂N₂O₂; mol wt 663.08. C 39.85%, H 2.13%, Cl 10.69%, I 38.28%, N 4.22%, O 4.83%. Salicylanilide derivative. Prepn: M. A. C. Janssen, V. K. Sipido, **Belg.** pat. **839,481;** *eidem,* **U.S.** pat. **4,005,218** (1976, 1977 both to Janssen). Effectiveness against *Taenia pisiformis* in rabbits: R. A. F. Chevis *et al., Vet. Parasitol.* **7**, 333 (1980); against *Ancylostoma caninum:* J. Guerrero *et al., J. Parasitol.* **68**, 616 (1983); against *Fasciola hepatica* in sheep: B. E. Stromberg *et al., ibid.* **70**, 446 (1984). Prolonged effect on *Haemonchus contortus* in sheep: C. A. Hall *et al., Res. Vet. Sci.* **31**, 104 (1981). Acts by uncoupling oxidative phosphorylation: H. Van den Bossche *et al., Arch. Int. Physiol. Biochim.* **87**, 851 (1979); H. J. Kane *et al., Mol. Biochem. Parasitol.* **1**, 347 (1980).

Crystals from methanol, mp 217.8°.
THERAP CAT (VET): Anthelmintic.

2474. Clospirazine. *8-[3-(2-Chloro-10H-phenothiazin-10-yl)propyl]-1-thia-4,8-diazaspiro[4.5]decan-3-one; 8-[3-(2-chloro-10-phenothiazinyl)propyl]-3-oxo-1-thia-4,8-diaza-spiro[4.5]decane; spiclomazine.* C₂₂H₂₄ClN₃OS₂; mol wt 446.04. C 59.24%, H 5.42%, Cl 7.95%, N 9.42%, O 3.59%, S 14.38%. Prepn: Nakanishi *et al.,* **Japan.** pat. **23,091('69), U.S.** pat. **3,574,204** (1969, 1971 both to Yoshitomi). Pharmacological and metabolic studies: Nakanishi *et al., J.*

Pharm. Soc. Japan **90**, 800, 808 (1970); Imamura *et al., ibid.* 813. *Review: Japan. Med. Gaz.* **8**(11), 10 (1971).

mp 154-156° (acetone).

Hydrochloride, *APY-606, Diceplon, Disepron.* White to brownish, tasteless, odorless crystals or powder, mp 262-264° (dec). Sparingly sol in methanol; slightly sol in chloroform; very slightly sol in water. Decomposes slowly when exposed to light. LD₅₀ in mice, rats (mg/kg): 3400, 2950 i.p.; > 5490, > 5490 s.c.; 3800, 4000 orally (Nakanishi, 1970).

THERAP CAT: Antipsychotic.

2475. Clostebol. *(17β)-4-Chloro-17-hydroxyandrost-4-en-3-one;* 4-chlorotestosterone. C₁₉H₂₇ClO₂; mol wt 322.87. C 70.68%, H 8.43%, Cl 10.98%, O 9.91%. Prepn: Camerino *et al., Farmaco Ed. Sci.* **11**, 586 (1956); *eidem, J. Am. Chem. Soc.* **78**, 3540 (1956); Ringold *et al., J. Org. Chem.* **21**, 1432 (1956); Camerino, **U.S.** pat. **2,953,582** (1960 to Farmitalia); Julian, Printy, **U.S.** pat. **2,933,510** (1960 to Julian Labs.). Clinical pharmacology of the acetate: Molinatti *et al., Folia Endocrinol.* **14**, 528 (1961); Krueskemper, Morgner, *Int. Z. Klin. Pharmakol. Ther. Toxikol.* **1**, 455 (1968).

Crystals from acetone + hexane, mp 188-190°. [α]²⁰_D +148° (CHCl₃). uv max (95% ethanol): 256 nm (log ε 4.13).

Acetate, C₂₁H₂₉ClO₃, *Macrobin, Steranabol.* Crystals from methanol, mp 228-230°. [α]_D +118° ±4° (CHCl₃). uv max: 255 nm (ε 13,300). Sol in ethanol.

Note: This is a controlled substance (anabolic steroid) in the U.S. Code of Federal Regulations, Title 21 Part 1308.13, as defined in Part 1308.02 (1995).

THERAP CAT: Anabolic.

2476. Clothiapine. *2-Chloro-11-(4-methyl-1-piperazin-yl)dibenzo[b,f][1,4]thiazepine;* Entumine; Etumine. C₁₈H₁₈-ClN₃S; mol wt 343.88. C 62.87%, H 5.28%, Cl 10.31%, N 12.22%, S 9.32%. Prepn: **Fr.** pat **CAM 51** (1964 to Dr. A. Wander), *C.A.* **61**, 8328h (1964).

Crystals from ether + petr ether, mp 118-120°.
THERAP CAT: Antipsychotic.

2477. Clotiazepam. *5-(2-Chlorophenyl)-7-ethyl-1,3-di-hydro-1-methyl-2H-thieno[2,3-e]-1,4-diazepin-2-one;* Y-6047; Clozan; Rise; Rize; Rizen; Tienor; Trecalmo; Veratran. C₁₆-H₁₅ClN₂OS; mol wt 318.83. C 60.28%, H 4.74%, Cl

11.12%, N 8.79%, O 5.02%, S 10.06%. A thienodiazepine tranquilizer with biological activity similar to the benzodiazepines. Prepn: M. Nakanishi *et al.*, **Ger.** pat. **2,107,356**; *eidem*, U.S. pat. **3,849,405** (1971, 1974 both to Yoshitomi); *eidem*, *J. Med. Chem.* **16**, 214 (1973). Pharmacology: *eidem*, *Arzneimittel-Forsch.* **22**, 1905 (1972). Effects on biogenic amine metabolism: *eidem*, *ibid.* 1914. Metabolism: *eidem*, *Yakugaku Zasshi* **93**, 311 (1973), *C.A.* **79**, 13397r (1973). Clinical studies: S. Sieberns, *Fortschr. Med.* **97**, 1705 (1979).

Crystals from hexane, mp 105-106°. LD$_{50}$ in mice (mg/kg): 440 i.p., 636 orally (Nakanishi).

Note: This is a controlled substance (depressant) listed in the U.S. Code of Federal Regulations, Title 21 Part 1308.14 (1995).

THERAP CAT: Anxiolytic.

2478. Clotrimazole. *1-[(2-Chlorophenyl)diphenylmethyl]-1H-imidazole; 1-(o-chloro-α,α-diphenylbenzyl)imidazole; 1-[α-(2-chlorophenyl)benzhydryl]imidazole; 1-[(o-chlorophenyl)diphenylmethyl]imidazole; diphenyl-(2-chlorophenyl)-1-imidazolylmethane; 1-(o-chlorotrityl)imidazole;* FB-5097; Bay b 5097; Canesten; Canifug; Empecid; Gyne-Lotrimin; Lotrimin; Mono-Baycuten; Mycelex-G; Mycofug; Mycosporin; Pedisafe; Rimazole; Tibatin; Trimysten. C$_{22}$H$_{17}$ClN$_2$; mol wt 344.84. C 76.63%, H 4.97%, Cl 10.28%, N 8.12%. Prepn: K. H. Buechel *et al.*, **S. Afr.** pat. **68 05,392**; *eidem*, U.S. pat. **3,705,172** (1969, 1972 both to Bayer). Pharmacology: Plempel *et al.*, *Antimicrob. Ag. Chemother.* **1969**, 271; *eidem*, *Deut. Med. Wochenschr.* **94**, 1356 (1969). Clinical findings: Oberste-Lehn *et al.*, *ibid.* 1365. Series of articles on prepn, toxicology, pharmacokinetics, clinical studies: *Arzneimittel-Forsch.* **22**, 1260-1272, 1276-1299 (1972). Toxicity: D. Tettenborn, *ibid.* 1276. Comprehensive description: J. G. Hoogerheide, B. E. Wyka, *Anal. Profiles Drug Subs.* **11**, 225-255 (1982).

Crystals, mp 147-149°. A weak base, slightly sol in water, benzene, toluene; sol in acetone, chloroform, ethyl acetate, DMF. Hydrolyzes rapidly upon heating in aq acids. LD$_{50}$ in male mice, rats (mg/kg): 923, 708 orally (Tettenborn). Hydrochloride, C$_{22}$H$_{17}$ClN$_2$.HCl, mp 159°.

THERAP CAT: Antifungal.
THERAP CAT (VET): Antifungal.

2479. Clove. *Caryophyllus.* Dried flower-buds of *Eugenia caryophyllata* Thunb. (*Caryophyllus aromaticus* L.), *Myrtaceae. Habit.* Molucca Islands, Zanzibar, Sumatra, S. America, W. Indies. *Constit.* 15-18% eugenol, caryophyllin, tannin, gum, resin.

USE: Manuf oil of clove, eugenol; in baking; confections. Pharmaceutic aid (flavor).

THERAP CAT: Analgesic (dental).

2480. Cloxacillin. *[2S-(2α,5α,6β)]-6-[[[3-(2-Chlorophenyl)-5-methyl-4-isoxazolyl]carbonyl]amino]-3,3-dimethyl-7-oxo-4-thia-1-azabicyclo[3.2.0]heptane-2-carboxylic acid; [3-(o-chlorophenyl)-5-methyl-4-isoxazolyl]penicillin; [5-methyl-*

3-(*o*-chlorophenyl)-4-isoxazolyl]penicillin; 6-[3-(*o*-chlorophenyl)-5-methyl-4-isoxazolecarboxamido]penicillanic acid. C$_{19}$H$_{18}$ClN$_3$O$_5$S; mol wt 435.89. C 52.35%, H 4.16%, Cl 8.13%, N 9.64%, O 18.35%, S 7.36%. Semi-synthetic antibiotic related to penicillin; chlorinated deriv of oxacillin, *q.v.* Prepn: F. P. Doyle, J. H. C. Nayler, U.S. pat. **2,996,501** (1961); F. P. Doyle *et al.*, *J. Chem. Soc.* **1963**, 5838. Manuf: *Ind. Chem.* **39**, 513 (1963), *C.A.* **60**, 1543a (1964). Properties and pharmacology: J. H. C. Nayler *et al.*, *Nature* **195**, 1264 (1962). Toxicity data: E. I. Goldenthal, *Toxicol. Appl. Pharmacol.* **18**, 185 (1971). Comprehensive description: D. L. Mays in *Analytical Profiles of Drug Substances* vol. 4, K. Florey, Ed. (Academic Press, New York, 1975) pp 113-136.

Sodium monohydrate, C$_{19}$H$_{17}$ClN$_3$NaO$_5$S.H$_2$O, *sodium cloxacillin, BRL-1621, Bactopen, Cloxapen, Cloxypen, Ekvacillin, Gelstaph, Orbenin, Methocillin-S, Prostaphlin-A, Staphobristol-250, Staphybiotic, Tegopen, Tepogen.* Microcryst powder, dec 170°. [α]$_D^{20}$ +163° (c = 1 in water). pH of 1% aq soln = 6.0-7.5. Sol in water, methanol, ethanol, pyridine, ethylene glycol. LD$_{50}$ in rats, mice (mg/kg): 1630 ±112, 1280 ±50 i.p. (Goldenthal).

Combination of acid and sodium salt, *Latocillin.*

Benzathine salt, C$_{54}$H$_{56}$Cl$_2$N$_8$O$_{10}$S$_2$, *Boviclox, Dry-Clox, Noroclox DC, Opticlox, Orbenin Dry Cow, Triclox.*

THERAP CAT: Antibacterial.
THERAP CAT (VET): Antibacterial.

2481. Cloxazolam. *10-Chloro-11b-(2-chlorophenyl)-2,-3,7,11b-tetrahydrooxazolo[3,2-d][1,4]benzodiazepin-6(5H)-one; 7-chloro-5-(2-chlorophenyl)tetrahydrooxazolo[5,4-b]-2,3,4,5-tetrahydro-1H-1,4-benzodiazepin-2-one; 10-chloro-11b-(2-chlorophenyl)-2,3,5,6,7,11b-hexahydrobenzo[6,7]-1,-4-diazepino[5,4-b]oxazol-6-one; 10-chloro-11b-(2-chlorophenyl)-6-oxo-2,3,5,6,7,11b-hexahydrooxazolo[3,2-d][1,4]-benzodiazepine;* CS-370; Betavel; Enadel; Lubalix; Olcadil; Sepazon; Tolestan. C$_{17}$H$_{14}$Cl$_2$N$_2$O$_2$; mol wt 349.22. C 58.47%, H 4.04%, Cl 20.30%, N 8.02%, O 9.16%. Prepn: R. Tachikawa *et al.*, **Ger.** pats. **1,812,252** and **1,952,201**; *eidem*, U.S. pats. **3,772,371** and **3,696,094** (1969, 1970, 1973, 1972 all to Sankyo); T. Miyadera *et al.*, *J. Med. Chem.* **14**, 520 (1971). Pharmacology: T. Kamioka *et al.*, *Arzneimittel-Forsch.* **22**, 884 (1972). Metabolism: Murata *et al.*, *Chem. Pharm. Bull.* **21**, 404 (1973). Multicenter trials and complementary studies: K. A. Fischer-Cornelssen, *Arzneimittel-Forsch.* **31**, 1757 (1981).

Crystals, mp 202-204° (dec). Freely sol in glacial acetic acid; sparingly sol in chloroform; slightly sol in acetone, dehydrated ethanol, ethyl acetate, benzene. Practically insol in water. LD$_{50}$ in mice (g/kg): 3.3 orally; >2.0 i.p. (Kamioka).

Note: This is a controlled substance (depressant) listed in the U.S. Code of Federal Regulations, Title 21 Part 1308.14 (1995).

THERAP CAT: Anxiolytic.

2482. Cloxotestosterone. *(17β)-17-(2,2,2-Trichloro-1-hydroxyethoxy)androst-4-en-3-one;* testosterone 17-chloral hemiacetal. C$_{21}$H$_{29}$Cl$_3$O$_3$; mol wt 435.82. C 57.88%, H

6.71%, Cl 24.40%, O 11.01%. Prepn: P. Borrevang, U.S. pat. **2,933,514** (1960 to Lövens Kemiske Fabrik); *idem, Acta Chem. Scand.* **16**, 883 (1962).

Crystals from ethyl acetate, mp 200-201°. uv max (ethanol): 241 nm (ε 16300).

Acetate, $C_{23}H_{31}Cl_3O_4$, *Caprosem.* Crystals from ethyl acetate, mp 192-193°. uv max (ethanol): 241 nm (ε 16400).

THERAP CAT: Androgen.

2483. Cloxyquin. *5-Chloro-8-quinolinol;* 5-chloro-8-hydroxyquinoline; 5-chloro-8-oxychinolin; cloxiquine; Chlorisept. C_9H_6ClNO; mol wt 179.61. C 60.19%, H 3.37%, Cl 19.74%, N 7.80%, O 8.91%. Prepn: Bratz, Niementowski, *Ber.* **52**, 189 (1919); Weizmann, Bograchov, *J. Am. Chem. Soc.* **69**, 1222 (1947); Manske *et al., Can. J. Res.* **27F**, 359 (1949). Crystal structure: T. Banerjee, N. N. Saha, *Acta Crystallog.* **C42**, 1408 (1986).

Crystals from ethanol, mp 130°. Sparingly sol in cold, dil HCl.

Hydrochloride, $C_9H_6ClNO \cdot HCl$, yellow needles. mp 256-258°.

Acetate (ester), $C_{11}H_8ClNO_2$, 5-chloro-8-acetoxyquinoline.

THERAP CAT: Antibacterial; antifungal.

2484. Clozapine. *8-Chloro-11-(4-methyl-1-piperazinyl)-5H-dibenzo[b,e][1,4]diazepine;* HF-1854; Clozaril; Leponex. $C_{18}H_{19}ClN_4$; mol wt 326.83. C 66.15%, H 5.86%, Cl 10.85%, N 17.14%. Prepn: Fr. pat. **1,334,944** (1963 to Wander); Schmutz, Hunziker, U.S. pat. **3,539,573** (1970); **Neth.** pat. Appl. **293,201** (1965 to Wander), *C.A.* **64**, 8221a (1966); Hunziker *et al., Helv. Chim. Acta* **50**, 1588 (1967). Structure-activity studies: Schmutz *et al., Chim. Ther.* **2**, 424 (1967). Pharmacology: Stille *et al., Farmaco, Ed. Prat.* **26**, 603 (1971). Metabolism: Gauch, Michaelis, *ibid.* 667. Toxicology: Lindt *et al., ibid.* 585. Clinical studies: De Maio, *Arzneimittel-Forsch.* **22**, 919 (1972). Clinical efficacy in treatment-resistant schizophrenia: J. M. Hane *et al., Psychopharmacol. Bull.* **24**, 62 (1988). *Review:* A. C. Sayers, H. A. Amsler, in *Pharmacological and Biochemical Properties of Drug Substances* vol. 1, M. E. Goldberg, Ed. (Am. Pharm. Assoc., Washington, DC, 1977) pp 1-31. Review of pharmacology and clinical use: R. J. Baldessarini, F. R. Frankenburg, *N. Engl. J. Med.* **324**, 746-754 (1991). Overview of mechanism of action: H. Y. Meltzer, *J. Clin. Psychiatry* **55**, Suppl. B, 47-52 (1994).

Yellow crystals from acetone-petr ether, mp 183-184°. uv max (ethanol): 215, 230, 261, 297 nm (ε 27400, 25800, 16800, 10500). LD_{50} in mice, rats (mg/kg): 61, 58 i.v.; 199, 260 orally (Lindt).

THERAP CAT: Antipsychotic.

2485. Clupeine. Protamine found in herring *(Clupea pallasii)* sperm. Isoln from herring testes contg ripe sperm: Kossel, *The Protamines and Histones* (London, 1928); Rasmussen, *Z. Physiol. Chem.* **224**, 97 (1934); Felix, Mager, *ibid.* **249**, 111 (1937); Block *et al., Proc. Soc. Exp. Biol. Med.* **70**, 494 (1949). Separated into two main fractions, Y and Z, and fraction Y separated into Y_I and Y_{II}: Ando, Sawada, *J. Biochem. (Tokyo)* **49**, 252 (1961). Chemical structure of fraction Z: Ando *et al., Biochem. Biophys. Acta* **56**, 628 (1962); Felix, Hashimoto, *Z. Physiol. Chem.* **330**, 205 (1963). Complete amino acid sequence of Z component: Iwai *et al., J. Biochem. (Tokyo)* **69**, 493 (1971); of Y_{II} component: Suzuki, Ando, *ibid.* **72**, 1419 (1972); of Y_I component: *eidem, ibid.* 1433. Solid-phase synthesis of clupeine Z: Yonezawa *et al., C.A.* **79**, 19093k (1973).

White powder, strongly alkaline reaction. pKa 7.4-8.0; pKb 2.9-3.3.

Usually isolated as the sulfate $B \cdot 2H_2SO_4$; white powder, $[\alpha]_D^{22}$ −85.49° (satd aq soln). One gram dissolves in 80 ml water at room temp. Freely sol in hot water, separates from the supersatd soln on cooling as a clear, colorless oil contg 50% H_2O, n_D^{20} 1.4435. Clupeine is split by protaminase, active trypsin and by chymotrypsin. Compds of clupeine with nucleic acids are described by Kossel, *loc. cit.*

2486. Cnicin. *3,4-Dihydroxy-2-methylenebutanoic acid 2,3,3a,4,5,8,9,11a-octahydro-10-(hydroxymethyl)-6-methyl-3-methylene-2-oxocyclodeca[b]furan-4-yl ester; 6α,8α,15-trihydroxygermacra-1(10),4,11(13)-trien-12-oic acid 12,6-lactone 8-(3,4-dihydroxy-2-methylenebutyrate);* cynisin; centaurin. $C_{20}H_{26}O_7$; mol wt 378.42. C 63.48%, H 6.93%, O 29.60%. Bitter principle of *Cnicus benedictus* L., Compositae. Isolation and review: Korte, Bechmann, *Naturwiss.* **45**, 390 (1958). Structure: Suchy *et al., Tetrahedron Letters* no. 10, 5 (1969); *Ber.* **93**, 2449 (1960). Revised structure: Samek *et al., Tetrahedron Letters* **1969**, 2931. Stereochemistry: Tori *et al., J. Chem. Soc. (B)* **1971**, 1084.

Crystals, mp 143°. $[\alpha]_D^{20}$ +158° (c = 2.3 in ethanol). uv max: 220 nm (log ε 4.34). Slightly soluble in water, freely in alcohol. Practically insol in ether.

2487. Coal Tar. Clinitar; Psorigel; T/Gel. A by-product in the destructive distillation of coal. *Constit.* Benzene, toluene, naphthalene, anthracene, xylene, and other aromatic hydrocarbons; phenol, cresol, and other phenol bodies; ammonia, pyridine, and some other organic bases; thiophene. Monographs: *Coal Tar Data Book*, 2nd ed., 1965, compiled and published by The Coal Tar Res. Assoc., Gomersal (Leeds), England; H. G. Franck, G. Collin, *Steinkohlenteer* (Springer Verlag, Heidelberg-New York, 1968) 255 pp. Review of carcinogenicity studies: *IARC Monographs* **35**, 83-159 (1985).

Almost black, thick liq or semisolid; characteristic odor. A small portion of coal tar dissolves in water; all or almost all dissolves in benzene or nitrobenzene; partly dissolves in alcohol, ether, chloroform, carbon disulfide, methanol, acetone, petr ether, or sodium hydroxide soln. Practically insol in water. Sol in 20 parts alcohol; miscible with abs alcohol, acetone, petrolatum, oils and fats.

Purified coal tar, *pixalbol.* Light-yellow, thin, oily liq.

Occupational exposure to soots, tars and certain mineral oils is known to be carcinogenic: *Seventh Annual Report on Carcinogens* (PB95-109781, 1994) p 406.

THERAP CAT: Antieczematic (topical).

THERAP CAT (VET): Topically in skin disorders.

2488. Cobalt. Co; at. wt 58.9332; at. no. 27; valences 1, 2, 3; rarely 4, 5. One naturally occurring isotope: ^{59}Co; artificial, radioactive isotopes: 54-58; 60-64. Widely distributed in nature; abundance in earth's crust 0.001-0.002%. Principle ores include *cobaltite* ($CoS_2.CoAs_2$), *linnaeite* (Co_3S_4), *smaltite* ($CoAs_2$) and erythrite ($3CoO.As_2O_5.8H_2O$). Metal first isolated in 1735 by Brandt. Reviews of prepn: Whittemore in *Rare Metals Handbook*, C. A. Hampel, Ed. (Reinhold, New York, 1956) pp 105-146; Houot, *Ann. Mines* **1969** (April), 9-36. Prepn of high purity metal: Ware in *Ultrapurification of Semiconductor Materials*, M. S. Brooks, J. K. Kennedy, Eds. (Macmillan, New York, 1962) pp 192-204. Cobalt appears to be essential to life. Plays an important part in animal nutrition; the absence of cobalt-contg vitamin B_{12} causes pernicious anemia. The reactor-produced ^{60}Co ($T_{1/2}$ 5.263 years; β^- 0.314 Mev; γ 1.173, 1.332 Mev) is a widely used source of radioactivity: Centre d'Information du Cobalt, *Cobalt Monograph* (Brussels, 1960) 515 pp. Comprehensive reviews of cobalt and its compds: *Cobalt, Its Chemistry, Metallurgy and Uses*, R. S. Young, Ed., A.C.S. Monograph Series, no. 149 (Reinhold, New York, 1960) 424 pp; Nicholls in *Comprehensive Inorganic Chemistry* vol. 3, J. C. Bailar, Jr. *et al*, Eds. (Pergamon Press, Oxford, 1973) pp 1053-1107; F. Planinsek, J. B. Newkirk in Kirk-Othmer *Encyclopedia of Chemical Technology* vol. 6 (Wiley-Interscience, New York, 3rd ed., 1979) pp 481-494.

Gray, hard, magnetic, ductile, somewhat malleable metal. Exists in two allotropic forms. At room temp the hexagonal form is more stable than the cubic form; both forms can exist at room temperature. Stable in air or toward water at ordinary temp. d 8.92. mp 1493°. bp about 3100°. Brinell hardness: 125. Latent heat of fusion 62 cal/g, latent heat of vaporization 1500 cal/g. Specific heat (15-100°): 0.1056 cal/g/°C. Readily sol in dil HNO_3; very slowly attacked by HCl or cold H_2SO_4. The hydrated salts of cobalt are red, and the sol salts form red solns which become blue on adding concd HCl.

Caution: Potential symptoms of overexposure to metal, dust or fumes are coughing, dyspnea and decreased pulmonary function; weight loss; dermatitis; diffuse nodular fibrosis; respiratory hypersensitivity. *See NIOSH Pocket Guide to Chemical Hazards* (DHHS/NIOSH 90-117, 1990) p 72.

USE: For alloys; manuf cobalt salts; in nuclear technology. Since ^{60}Co can be encapsulated compactly, it has replaced radium in exptl medicine and cancer research. Cobalt is also used in the cobalt bomb, a hydrogen bomb surrounded by a cobalt metal shell. When the nuclear explosion occurs ^{60}Co is formed from ^{59}Co by neutron capture. Considered a "dirty bomb" because of long half-life and intense β^- and γ radiation. Max permissible concentration of ^{60}Co in air: $10^{-7}\mu$Ci/cc, *National Bureau of Standards Handbook* 69, 31 (1959).

THERAP CAT: Trace mineral; ^{60}Co as radiation source (antineoplastic).

2489. Cobaltic Acetate. $C_6H_9CoO_6$; mol wt 236.07. C 30.53%, H 3.84%, Co 24.96%, O 40.66%. $Co(C_2H_3O_2)_3$. Prepd by electrolytic oxidation of $Co(C_2H_3O_2)_2.4H_2O$ in glacial acetic acid contg 2% (v/v) water: Sharp, White, *J. Chem. Soc.* **1952**, 110; by oxidation of solns of cobaltous salts by alkaline persulfates in the presence of acetic acid: Peschanski, Wormser, *Compt. Rend.* **252**, 1607 (1961). Generally considered to be a hydroxo-bridged binuclear complex.

Dark-green, very hygroscopic powder or green octahedral crystals. Becomes black and dec on heating to 100°. Sol in water, acetic acid, alcohol, *n*-butanol; dec by mineral acids. Aq solns hydrolyze slowly at room temp, rapidly at 60-70°; reduced by light or $FeSO_4$.

USE: Catalyst for cumene hydroperoxide decompn.

2490. Cobaltic-Cobaltous Oxide. Cobalto-cobaltic oxide; cobaltosic oxide; tricobalt tetroxide. Co_3O_4; mol wt 240.80. Co 73.42%, O 26.58%. Prepn: *Gmelin's, Cobalt*

(8th ed.) **58** (Part A), 231-235 (1932) and supplement, 202, 491-496 (1961). *Review:* de Bie, Doyen, *Cobalt* **15**, 3-13; **16**, 3-15 (1962).

Black or grey octahedral crystals of cubic system. d 6.11. Commercial material is black and contains about 71% Co. Above 900° loses O_2 to form CoO; absorbs O_2 at lower temps but cryst structure is unchanged. Absorbs water but no definite hydrate has been identified. Readily reduced to Co metal by C, CO, or H_2. Practically insol in water. Sol in acids, alkalies.

USE: In enamels; in semiconductors; in grinding wheels.

2491. Cobaltic Fluoride. Cobalt trifluoride. CoF_3; mol wt 115.93. Co 50.84%, F 49.16%. Prepn from F_2, and CoF_2, $CoCl_2$, or Co_2O_3: Priest, *Inorg. Syn.* **3**, 175 (1950); Kwasnik in *Handbook of Preparative Inorganic Chemistry* vol. 1, G. Brauer, Ed. (Academic Press, New York, 2nd ed., 1963) p 268; from ClF_3 and $CoCl_2$: Rochow, Kukin, *J. Am. Chem. Soc.* **74**, 1615 (1952).

Minute, light brown, hexagonal crystals, d 3.88. Discolors rapidly on exposure to moist air. Reacts with water giving off O_2. Comparatively stable to heat, at 600° the fluorine pressure over the solid is less than 0.1 atm. Volatilizes in an F_2 stream at 600-700°. May be stored in hermetically sealed glass, quartz, or metal containers.

USE: Important fluorinating agent, particularly for complete fluorination of hydrocarbons by the Fowler process.

2492. Cobaltic Oxide Monohydrate. *Cobalt hydroxide oxide;* cobaltic hydroxide. $CoHO_2$; mol wt 91.94. Co 64.10%, H 1.10%, O 34.80%. CoO(OH). Alternate formula $Co_2O_3.H_2O$. Prepn: Glemser in *Handbook of Preparative Inorganic Chemistry* vol 2, G. Brauer, Ed. (Academic Press, New York, 2nd ed., 1965) pp 1520-1521; Schrader, Petzold, *Z. Anorg. Allgem. Chem.* **353**, 186 (1967). Existence of anhydrous cobaltic oxide (Co_2O_3) has not been established: Pagel *et al., J. Am. Chem. Soc.* **57**, 2552 (1935); Nicholls in *Comprehensive Inorganic Chemistry* vol 3, J. C. Bailar, Jr. *et al.,* Eds. (Pergamon Press, Oxford, 1973) p 1096. *See also* de Bie, Doyen, *Cobalt* **15**, 3 (1962). Structure: Schrader, Petzold, *loc. cit.;* Delaplane *et al., J. Chem. Phys.* **50**, 1920 (1969).

Dark-brown to black powder; hexagonal crystal structure. Converted to Co_3O_4 on heating to 148-150° in a vacuum. Practically insol in water. Sol in HCl, evolving Cl_2; sol in HNO_3, H_2SO_4.

USE: Oxidation catalysts, in separation of cobalt from nickel.

2493. Cobaltic Potassium Nitrite. *Hexakis(nitrito-N)-cobaltate(3—) tripotassium;* potassium hexanitrocobaltate-(III); potassium cobaltinitrite; potassium nitrocobaltate-(III); cobalt yellow; Fischer's yellow; C.I. Pigment Yellow 40; C.I. 77357. $CoK_3N_6O_{12}$; mol wt 452.26. Co 13.03%, K 25.94%, N 18.58%, O 42.45%. $K_3Co(NO_2)_6$. Incorrectly called *"Indian Yellow"*. Prepd by addition of KNO_2 to a solution of a Co salt: Salyer, Sweet, *Anal. Chem.* **32**, 548 (1962).

Sesquihydrate, yellow, octahedral, cubic crystals. Very slightly sol in water, dil acetic acid. Practically insol in alcohol; dec by mineral acids.

USE: As an oil- and water-color pigment; in painting on glass and porcelain; in coloring rubber; in separation of Co from Ni; in Co analysis.

2494. Cobaltous Acetate. $C_4H_6CoO_4$; mol wt 177.02. C 27.14%, H 3.42%, Co 33.29%, O 36.15%. $Co(C_2H_3O_2)_2$. Prepd commercially from cobaltous hydroxide or carbonate and an excess of dil acetic acid: Morral in Kirk-Othmer *Encyclopedia of Chemical Technology* vol. 5 (Interscience, New York, 2nd ed., 1964) p 737; prepn from powdered Co and acetic acid: Hahl, U.S. pat. **3,133,942** (1964 to BASF); prepn by oxidation of Co carbonyls in the presence of acetic acid: Gwynn *et al.,* U.S. pat. **3,246,024** (1966 to Gulf). *Review:* de Bie, Doyen, *Cobalt* **15**, 3-13; **16**, 3-15 (1962).

Light-pink crystals. Readily sol in water.

Tetrahydrate, *bis(acetato)tetraaquocobalt.* Intense red, monoclinic, prismatic crystals. d 1.705. On heating becomes anhydrous by 140°. Sol in water, alcohols, dil acids, pentyl acetate. pH of 0.2 molar aq soln 6.8.

USE: Bleaching agent and drier for lacquers, varnishes; in anodizing; catalyst for oxidation and esterification; foam stabilizer for malt beverages.

2495. Cobaltous Arsenate. C.I. 77350. $As_2Co_3O_8$; mol wt 454.64. As 32.96%, Co 38.89%, O 28.15%. $Co_3(AsO_4)_2$. Octahydrate occurs in nature as *erythrite* or *cobalt bloom*. Prepn: *Gmelin's, Cobalt* (8th ed.) **58**, (part A), 305 (1932) and supplement, 752 (1961); Charles-Messance *et al.*, *Bull. Soc. Chim. France* **1962**, 574. *See Colour Index* vol. **4** (3rd ed., 1971) p 4664.

Octahydrate, pink to blood-red, monoclinic, fine needles. On heating becomes anhydr by 400°. Dec by 1000° to $Co_6As_2O_{11}$. d 2.9-3.1. Practically insol in water. Sol in dil mineral acids, in NH_4OH.

USE: Painting on glass and porcelain.

2496. Cobaltous Bromide. Cobalt dibromide. Br_2Co; mol wt 218.74. Br 73.06%, Co 26.94%. $CoBr_2$. Prepn of hexahydrate from $CoCO_3$ and HBr: Clark, Buchner, *J. Am. Chem. Soc.* **44**, 230 (1922). Prepn of anhydr: *eidem, loc. cit.*; Watt *et al.*, *ibid.* **77**, 2752 (1955); Wydeven, Gregory, *J. Phys. Chem.* **68**, 3249 (1964).

Bright green solid or lustrous green cryst leaflets. mp 678° (under HBr and N_2); d_4^{25} 4.909. Hygroscopic, forms hexahydrate in air. Readily sol in water, methanol, ethanol, acetone, methyl acetate.

Hexahydrate, red to reddish-purple, deliquesc, prismatic crystals. mp 47-48°. d_4^{25} 2.46. Loses $4H_2O$ at 100° giving the purple dihydrate, and all H_2O by 130°. Sol in water giving red or blue soln depending on concn and temp, in methanol giving red soln, in ethanol, acetone, ether, methyl acetate giving blue solns. *Keep well closed.*

USE: Chiefly in hygrometers; also in catalysts for organic reactions.

2497. Cobaltous Carbonate. $CCoO_3$; mol wt 118.94. C 10.10%, Co 49.55%, O 40.35%. $CoCO_3$. Occurs in nature as the mineral *cobalt spar* or *sphaerocobaltite*. Prepd by heating a soln of a cobaltous salt with Na_2CO_3: Schlessinger, *Inorg. Syn.* **6**, 189 (1963) where it is the starting material for the prepn of trinitrotriamminecobalt. *Review:* de Bie, Doyen, *Cobalt* **15**, 3-13; **16**, 3-15 (1962).

Red powder or rhombohedral crystals. d 4.13. Almost insol in water, alcohol, methyl acetate. Does not react with cold concd HNO_3 or HCl; when heated, dissolves with evolution of CO_2. Oxidized by air or weak oxidizing agents to cobaltic carbonate.

Hexahydrate, pink to violet-red cryst needles. Pptd when excess CO_2 is present during prepn. On heating becomes anhydr by 140°. Stable in air.

Cobaltous carbonate basic, $C_2H_6Co_5O_{12}$, *cobalt carbonate hydroxide.* $Co_5(OH)_6(CO_3)_2$. Pale-red powder, usually containing some water. Practically insol in water; sol in dilute acids and ammonia.

USE: In ceramics; manuf of Co pigments; prepn of Co compds.

THERAP CAT (VET): Nutritional factor. Used in cobalt deficiency in ruminants.

2498. Cobaltous Chloride. Cobalt dichloride. Cl_2Co; mol wt 129.84. Cl 54.61%, Co 45.39%. $CoCl_2$. Prepn of anhydr from Co powder and Cl_2: Osthoff, West, *J. Am. Chem. Soc.* **76**, 4732 (1954); from the acetate and acetyl chloride: Watt *et al.*, *ibid.* **77**, 2752 (1955); by dehydration of the hexahydrate with $SOCl_2$: Hecht, *Z. Anorg. Chem.* **254**, 51 (1947); prepn of the hexahydrate by treating an aqueous soln of a cobaltous salt with HCl: *Cobalt—Its Chemistry, Metallurgy, and Uses*, R. S. Young, Ed., A.C.S. Monograph Series no. **149** (Reinhold, New York, 1960) p 76. *Review:* de Bie, Doyen, *Cobalt* **15**, 3-13; **16**, 3-15 (1962).

Pale-blue hygroscopic leaflets; colorless in very thin layers; turns pink on exposure to moist air. mp 735°; bp 1049°; d_4^{25} 3.367. Dec 400° on long heating in air. Sublimes at 500° in HCl gas, forming iridescent, fluffy, colorless cryst. Sol in water, alcohols, acetone, ether, glycerol, pyridine. LD_{50} in mice, guinea pigs: 80, 55 mg/kg orally, Krasovskii, Fridlyand, *Gig. Sanit.* **36**, 95 (1971); *C.A.* **74**, 138803y (1971).

Hexahydrate, monoclinic crystals. Structure is reported to be $[CoCl_2(H_2O)_4] \cdot 2H_2O$: Mizuno *et al.*, *J. Phys. Soc. Japan* **14**, 383 (1959), *C.A.* **53**, 14630i (1959). Pink to red,

slightly deliquesc, monoclinic, prismatic crystals. mp 87°; d^{20} 1.924. On heating loses $4H_2O$ at 52-56° forming the dihydrate, violet or blue crystals, d_{25}^{25} 2.477, stable unless exposed directly to moisture. Loses another H_2O by 100°, giving monohydrate, violet, hygroscopic, amorphous solid or needles. Remaining H_2O lost at 120-140°. Sol in water, alcohols, acetone, ether, glycerol. pH of 0.2 molar aq soln 4.6. The aq soln is pink to red, but turns blue when heated or when HCl or H_2SO_4 is added. *Keep well closed.* MLD s.c. in rabbits: 200 mg/kg, *Handbook of Toxicology* vol. **1**, W. S. Spector, Ed. (Saunders, Philadelphia, 1956) pp 72-73.

Caution: Large amounts of $CoCl_2$ depress erythrocyte production. May lead to death in children. Other effects include cutaneous flushing, chest pains, dermatitides, tinnitus, nausea and vomiting, nerve deafness, thyroid hyperplasia, myxedema, congestive heart failure. *See* E. Beutler *et al.*, *Clinical Disorders of Iron Metabolism* (Grune & Stratton, New York, 1963) pp 175-178.

USE: Invisible ink; humidity and water indicator; in hygrometers; temp indicator in grinding; in electroplating; for painting on glass and porcelain; prepn of catalysts; fertilizer and feed additive; foam stabilizer in beer; as absorbent for military poison gas and ammonia; in manuf of vitamin B_{12}. Radioactive cobalt chloride, $^{57}CoCl_2$ (half-life 270 days, pure gamma emitter) used in Mössbauer effect (nuclear clock).

THERAP CAT: Hematinic.

THERAP CAT (VET): Nutritional factor. Used in cobalt deficiency in ruminants.

2499. Cobaltous Chromate(III). Cobalt chromite. $CoCr_2O_4$; mol wt 226.92. Co 25.97%, Cr 45.83%, O 28.20%. Prepn: *Gmelin's, Cobalt* (8th ed.) **58**, (part A), 479 (1932) and supplement, 874-876 (1961).

Brilliant greenish-blue powder having a cubic spinel structure. Almost insol in concd HCl and HNO_3.

USE: Green pigment for ceramics.

2500. Cobaltous Cyanide. Cobalt cyanide. C_2CoN_2; mol wt 110.97. C 21.65%, Co 53.11%, N 25.24%. Prepn: Ray, Sahu, *J. Indian Chem. Soc.* **23**, 161 (1946); *Gmelin's, Cobalt* (8th ed.) **58**, (part A), 364 (1932) and supplement, 712 (1961). Structure reported as $Co_3[Co(CN)_6]_2$: P. S. Poskozim *et al.*, *J. Inorg. Nucl. Chem.* **35**, 687 (1973). Prepn and structure as $Co(CN)_2$: D. M. S. Mosha, D. Nicholls, *Inorg. Chim. Acta* **38**, 127 (1980).

Deep-blue, very hygroscopic powder. d_4^{25} 1.872.

Di- to trihydrate, pink to reddish-brown powder or needles. Practically insol in water, acids, methyl acetate; sol in alkali cyanide solns.

USE: In cobalt catalysts.

2501. Cobaltous Fluoride. Cobalt difluoride. CoF_2; mol wt 96.93. Co 60.80%, F 39.20%. Prepd by the action of HF on anhydr $CoCl_2$: Kwasnik in *Handbook of Preparative Inorganic Chemistry*, vol. **1**, G. Brauer, Ed. (Academic Press, New York, 2nd ed., 1963) p 267; on $CoCO_3$: Clark, Buchner, *J. Am. Chem. Soc.* **44**, 230 (1922); on Co: Muetterties, Castle, *J. Inorg. Nucl. Chem.* **18**, 148 (1961).

Rosy-red tetragonal crystals. mp 1100-1200°, forming a red liq. Volatilizes at about 1400°. d 4.43. Sparingly sol in water; readily sol in warm mineral acids. Forms di-, tri-, and tetrahydrates, all sol in water; their aq solns are dec by boiling, forming the oxyfluoride $CoF_2 \cdot CoO \cdot H_2O$.

USE: Catalyst for organic reactions.

2502. Cobaltous Formate. $C_2H_2CoO_4$; mol wt 148.97. C 16.13%, H 1.35%, Co 39.56%, O 42.96%. $Co(HCOO)_2$. Prepn: *Gmelin's, Cobalt* (8th ed.) **58** (part A), 350 (1932) and supplement, 702 (1961).

Dihydrate, red, cryst powder. d^{22} 2.13. Sol in water; almost insol in alcohol. Becomes anhydr at 140°.

USE: In prepn of Co catalysts.

2503. Cobaltous Hydroxide. CoH_2O_2; mol wt 92.95. Co 63.40%, H 2.17%, O 34.43%. $Co(OH)_2$. Prepd from a solution of a cobaltous salt and an alkali hydroxide: Glemser in *Handbook of Preparative Inorganic Chemistry*, vol. **2**, G. Brauer, Ed. (Academic Press, New York, 2nd ed., 1965) p 1521; Weiser, Milligan, *J. Phys. Chem.* **36**, 722 (1932). *Review:* de Bie, Doyen, *Cobalt* **15**, 3-13; **16**, 3-15 (1962).

Blue-green or rose-red powder or microscopic rhombo-hedral crystals; red form is the more stable of the two. d_4^{15} 3.597. Easily oxidized by air or weak oxidizing agents to $Co(OH)_3$. Amphoteric. Loses water on heating, forming CoO at 168° *in vacuo*. Very slightly sol in water; readily sol in acids; practically insol in dil alkalies; sol in ammonia.

USE: Manuf of Co compds; drier for paints; in enhancing drying properties of lithographic printing inks; in storage battery electrode impregnating solns.

2504. Cobaltous Iodide. Cobalt diiodide. CoI_2; mol wt 312.74. Co 18.84%, I 81.16%. Prepn: Clark, Buchner, *J. Am. Chem. Soc.* **44**, 230 (1922); Chaigneau, *Bull. Soc. Chim. France* **1957**, 886; Chaigneau, Chastagnier, *ibid.* **1958**, 1192; Glemser in *Handbook of Preparative Inorganic Chemistry*, vol. 2, G. Brauer, Ed. (Academic Press, New York, 2nd ed., 1965) p 1518.

The anhyd salt exists in two isomorphous forms. α-CoI_2: black, graphite-like solid. mp 515-520° (in high vacuum). d_4^{25} 5.584. Very hygroscopic, becomes blackish-green in air. Sol in water to give pink to red soln. β-CoI_2: ochre-yellow powder. Blackens at 400° and converts to α-form. d_4^{25} 5.45. Very hygroscopic; deliquesc in moist air forming green droplets. Sol in water to give colorless soln which becomes pink on heating.

Hexahydrate, dark red hexagonal prisms. Loses H_2O on heating becoming anhydr by 130°. d 2.90. Loses I_2 on exposure to air and light. Sol in water to give soln which is red below 20°, olive green at 20 to 40°, and green at higher temps. Sol in ethanol (blue soln), ether (blue to green soln), chloroform (blue soln), acetone.

USE: Indicator for moisture and humidity; determination of water in organic solvents; catalyst for organic reactions.

2505. Cobaltous Nitrate. CoN_2O_6; mol wt 182.94. Co 32.21%, N 15.31%, O 52.47%. $Co(NO_3)_2$. Prepn: *Gmelin's, Cobalt* (8th ed.) **58** (part A), 252-262 (1932) and supplement, 515-521 (1961); Weigel *et al., Bull. Soc. Chim. France* **1964**, 836; Addison, Sutton, *J. Chem. Soc.* **1964**, 5553. Toxicity: *Handbook of Toxicology* vol. 1, W. S. Spector, Ed. (Saunders, Philadelphia, 1956) pp 72-73. *Review:* de Bie, Doyen, *Cobalt* **15**, 3-13; **16**, 3-15 (1962).

Pale red powder. Dec at 100-105°. d 2.49. Sol in water. LD orally in rabbits: 250 mg/kg (Spector).

Hexahydrate, red, deliquesc, monoclinic crystals. mp ~55°. Red liq becomes green and dec to the oxide above 74°. d 1.88. Very sol in water, alcohol, most organic solvents. *Keep well closed in a cool place.*

USE: Manuf of cobalt pigments and invisible inks; decorating stoneware and porcelain; prepn of catalysts; production of vitamin B_{12} supplements.

2506. Cobaltous Oxalate. C_2CoO_4; mol wt 146.95. C 16.35%, Co 40.10%, O 43.55%. CoC_2O_4. Prepn: Robin, *Bull. Soc. Chim. France* **1953**, 1078; *Gmelin's, Cobalt* (8th ed.) **58** (part A) p 355 (1932) and supplement, pp 704-706 (1961). *Review:* de Bie, Doyen, *Cobalt* **15**, 3-13; **16**, 3-15 (1962).

d_4^{25} 3.021. Readily absorbs moisture from air to form hydrates.

Dihydrate, light pink microcryst powder or needles. Almost insol in water; slightly sol in acids; almost insol in aq oxalic acid; freely sol in aq ammonia. Dec on heating with aq KOH or Na_2CO_3 soln.

Tetrahydrate, yellowish-pink amorphous powder. Effloresces on exposure to air. Loses water on heating to 100° giving the dihydrate. Very slightly sol in water; slightly sol in acids; readily sol in aq ammonia.

USE: Prepn of Co catalysts, Co metal powder for powder-metallurgical applications; stabilizer for HCN; temperature indicator.

2507. Cobaltous Oxide. CoO; mol wt 74.93. Co 78.65%, O 21.35%. Prepn: Amiel *et al., Compt. Rend.* **259**, 3512 (1964); Wilke, *Z. Anorg. Allgem. Chem.* **330**, 164 (1964). Toxicity study: Smyth *et al., Am. Ind. Hyg. Assoc. J.* **30**, 470 (1969). *Review:* de Bie, Doyen, *Cobalt* **15**, 3-13; **16**, 3-15 (1962).

Powder, or cubic or hexagonal crystals. Color varies from olive green to red, depending on the particle size, but the commercial material is usually dark grey and contains

about 76% Co. mp ~1935°. d 5.7 to 6.7, depending on method of prepn. Readily absorbs O_2 even at room temp. Practically insol in water. Sol in acids or alkalies. Easily reduced to Co by C or CO. Reacts at high temperatures with silica, alumina, zinc oxide to form pigments. LD_{50} orally in rats: 1.70 g/kg (Smyth).

Note: The commercial oxides are usually not definite chemical compds but mixtures of the cobalt oxides.

USE: In pigments for ceramics; glass coloring and decolor-ization; oxidation catalyst for drying oils, fast-drying paints and varnishes; prepn of cobalt-metal catalysts, Co powder for binder in sintered tungsten carbide; in semiconductors.

2508. Cobaltous Phosphate. C.I. Pigment Violet 14; C.I. 77360. $Co_3O_8P_2$; mol wt 366.74. Co 48.21%, O 34.90%, P 16.89%. $Co_3(PO_4)_2$. Prepn from $CoCl_2$ and $(NH_4)_2HPO_4$: Klement, Haselbeck, *Z. Anorg. Allgem. Chem.* **334**, 27 (1964); from $Ca(H_2PO_4)_2$: Vickery, U.S. pat. **2,914,380** (1959 to Horizons). *Review:* de Bie, Doyen, *Cobalt* **15**, 3-13; **16**, 3-15 (1962). *See also Colour Index* vol. 4 (3rd ed., 1971) p 4665.

Octahydrate, pink to lavender amorph powder. d 2.77. Practically insol in water; sol in mineral acids.

USE: In ceramic pigments; in artists' colors, plastic resins.

2509. Cobaltous Potassium Sulfate. Potassium cobal-tosulfate; potassium cobaltous sulfate; potassium disulfato-cobaltate(II). $CoK_2O_8S_2$; mol wt 329.26. Co 17.90%, K 23.75%, O 38.87%, S 19.48%. $K_2Co(SO_4)_2$. Prepn: *Gmelin's, Cobalt* (8th ed.) **58** (part A), 414 (1932) and supplement, 782-786 (1961).

Hexahydrate, red, monoclinic, prismatic crystals. Loses water on heating above 75°, forming the reddish-brown di-hydrate which is stable from 120 to 150°; becomes anhydr (violet solid) at about 200°. d_4^{20} 2.22. Freely sol in water.

2510. Cobaltous Sulfate. CoO_4S; mol wt 155.00. Co 38.02%, O 41.29%, S 20.69%. $CoSO_4$. Hexahydrate occurs in nature as the mineral *bieberite*. Prepn: Clark *et al., J. Am. Chem. Soc.* **42**, 2483 (1920); Hammel, *Ann. Chim.* **11**, 247 (1939); *Gmelin's, Cobalt* (8th ed.) **58**, (part A) 324-336 (1932) and supplement, 628-647 (1961). *Review:* de Bie, Doyen, *Cobalt* **15**, 3-13; **16**, 3-15 (1962).

Red to lavender dimorphic, orthorhombic crystals. d_4^{25} 3.71. Stable to 708°. Dissolves slowly in boiling water.

Monohydrate, rose-colored, monoclinic crystals. Structure reported to be $Co(H_2SO_5)$. d_4^{25} 3.08. Dissolves slowly in boiling water.

Heptahydrate, structure reported to be $[Co(H_2O)_6]$-$[H_2SO_5]$. Pink to red monoclinic, prismatic crystals. On heating dehydrates to the hexahydrate (monoclinic, prismatic crystals) at 41.5°, and to the monohydrate at 71°. d_4^{25} 2.03. Sol in water; slightly sol in methanol, ethanol.

USE: Usual source of water-soluble cobalt since it is the most economical and it shows less tendency to deliquesc or dehydrate than the chloride or nitrate. Used in storage batteries; in Co-electroplating baths; as drier for lithographic inks, varnishes; in ceramics, enamels, glazes to prevent discoloring; in Co pigments for decorating porcelain.

2511. Cobaltous Sulfide. CoS; mol wt 91.00. Co 64.76%, S 35.24%. Prepn: Glemser in *Handbook of Preparative Inorganic Chemistry*, vol. 2, G. Brauer, Ed. (Academic Press, New York, 2nd ed., 1965) p 1523.

Exists in two forms. α-CoS: black, amorphous powder. Forms Co(OH)S in air. Sol in HCl. β-CoS: grey powder or reddish-silver octahedral crystals. mp >1100°; d 5.45. Practically insol in water; sol in acids.

USE: Catalyst for hydrogenation or hydrodesulfurization.

2512. Cobaltous Thiocyanate. Cobaltous rhodanide; cobaltous sulfocyanate. $C_2CoN_2S_2$; mol wt 175.10. C 13.72%, Co 33.66%, N 16.00%, S 36.63%. $Co(SCN)_2$. Prepn: *Gmelin's, Cobalt* (8th ed.) **58** (part A), 380 (1932) and supplement, 720-722 (1961); Schlessinger, *Inorganic Laboratory Preparations* (Chemical Publishing, New York, 1962) p 44.

Yellow-brown powder. Soluble in water to give a rose-colored soln; sol in ethanol, methanol, ether, acetone, $CHCl_3$ to give blue solns.

Trihydrate, violet to violet-brown rhombic crystals; red in transmitted light. Sol in water to give blue soln which becomes pink on dilution; sol in ethanol, ether, acetone to give blue solns.

USE: As humidity indicator.

2513. Cobamamide. *Co-(5'-deoxyadenosine-5') deriv of 3'-ester of cobinamide dihydrogen phosphate (ester) with 5,6-dimethyl-1-α-D-ribofuranosyl-1H-benzimidazole, inner salt;* cobamamidum; coenzyme B_{12}; DBC; adenosyl-B_{12}; 5'-deoxy-adenosyl-B_{12}; 5'-deoxyadenosylcobalamine; dibencozide; dibenzcozamide; dimebenzcozamide; 5,6-dimethylbenzimid-azolylcobamide coenzyme; 5,6-dimethylbenzimidazolyl-cobamide 5'-deoxyadenosine; vitamin B_{12} coenzyme; Ac-timide; Ademide; Anabasi; Betarin; Calomide; Cobalion; Cobaltamin S; Cobanzyme; Cobazymase; Coenzile; Dolo-nevran; Enzicoba; Héraclène; Hi-Fresmin; Hycobal; Indusil; Ripresil; Sabalamin; Xobaline. $C_{72}H_{100}CoN_{18}O_{17}P$; mol wt 1579.60. C 54.75%, H 6.38%, Co 3.73%, N 15.96%, O 17.22%, P 1.96%. The coenzyme is the metabolically active form of vitamin B_{12}: Barker *et al.*, *Proc. Nat. Acad. Sci. USA* **44**, 1093 (1958); Weissbach, Taylor, *Vitam. Horm. (New York)* **26**, 395 (1968). Isoln from a culture of *Propionibacterium shermanii:* Barker *et al.*, *J. Biol. Chem.* **235**, 181, 480 (1960); *Biochem. Prepn.* **10**, 27 (1964). Prepn of aquoco-balamin and 2',3'-O-isopropylidene-5'-O-p-tolylsulfonyl-adenosine: Hogenkamp, Pailes, *ibid.* **12**, 124 (1968); from treatment of a cobalamin-thiol complex: Murakami *et al.*, U.S. pat. **3,461,114** (1969 to Yamanouchi). The coenzyme differs from Vitamin B_{12} by the presence of a 5'-deoxy-adenosyl group in the axial ligand occupied by the cyanide in the vitamin. Structure: Lenhert, Hodgkin, *Nature* **192**, 937 (1961). The nucleoside is linked to the cobalt via the 5'-carbon atom of its deoxyribosyl moiety and the biological and chemical reactivity reside in this C to Co bond: Hogen-kamp *et al.*, *J. Biol. Chem.* **240**, 3641 (1965) and *Fed. Proc.* **25**, 1623 (1966). *Review:* Smith, *Vitamin B_{12}* (Methuen & Co., London, 3rd ed., 1965).

Yellow-orange six-faced crystals which become deep red upon exposure to air. Absorption max (H₂O): 260, 375, 522 nm (A × 10⁻⁶ 34.7, 10.9, 8.0 cm²/mole). Soly in water (24°): 16.4 mmol. Solns of pH < 3.5 are yellow, > 3.5, red. Sol in ethanol, phenol. Practically insol in acetone, ether, dichloroethylene, dioxane. pKa 3.5. Stability studies: Collado, Nieto, *Ann. Pharm. Fr.* **27**, 427 (1969). Highly sensitive to light, to cyanide and moderately sensitive to acid.

Solns are most stable at pH 6-7 stored in the dark. Heating of acid or alkaline solns produces slow inactivation.

THERAP CAT: Vitamin (hematopoietic).
THERAP CAT (VET): Vitamin (hematopoietic).

2514. Cobrotoxin. The main toxic protein in cobra venom. Crystallization and properties: Yang, *J. Biol. Chem.* **240**, 1616 (1965). Composed of a single peptide chain having 62 amino acid residues intramolecularly cross-linked by four disulfide bonds. Amino acid sequence: Yang *et al.*, *Biochem. Biophys. Acta* **188**, 65 (1969). Complete structure: *eidem, ibid.* **214**, 355 (1970). Synthetic study: H. Aoyagi *et al.*, *ibid.* **263**, 823 (1972). *Review:* C. C. Yang, "Biochemical Studies on the Toxic Nature of Snake Venom Cobrotoxin from Formosan Cobra Venom" in *Toxins of Animal and Plant Origin* I, A. de Vries, E. Kochva, Eds. (Gordon and Breach Science Publishers, New York, 1971) pp 205-236. *See also* A. T. Tu, *Venoms: Chemistry and Molecular Biology* (John Wiley and Sons, New York, 1977) pp 174-189.

2515. Coca. Erythroxylon; cuca; hayo; ipado. Dried leaves of *Erythroxylon coca* Lam., *Erythroxylaceae.* *Habit.* Bolivia, Brazil, Peru, cultivated in Java. *Constit.* 0.5-1% alkaloids in the South American, 1.5-2.5% in the Javanese leaves. In the former, the major alkaloid is cocaine; in the latter, there is very little cocaine, the alkaloids consisting chiefly of ecgonine derivatives such as benzoyl ecgonine, methyl ecgonine, etc. The Java leaves also contain small quantities of tropococaine which is apparently absent in the South American leaves. Other "cocaine" alkaloids present are truxillococaine, isatropylcocaine or cocamine, cocaicine.

Note: This is a controlled substance listed in the U.S. Code of Federal Regulations, Title 21 Part 1308.12 (1995).

THERAP CAT: CNS stimulant.

2516. Cocaethylene. *[1R-(exo,exo)]-3-(Benzoyloxy)-8-methyl-8-azabicyclo[3.2.1]octane-2-carboxylic acid ethyl ester;* ecgonine ethyl ester benzoate (ester); *O-benzoyl-l-ecgonine ethyl ester;* ethylbenzoylecgonine; Homocaine. $C_{18}H_{23}NO_4$; mol wt 317.38. C 68.12%, H 7.30%, N 4.41%, O 20.16%. Homolog of cocaine. Prepn: Merck, *Ber.* **18**, 2952 (1885); Einhorn, *ibid.* **21**, 47 (1888).

Prisms from alcohol, mp 109°. Almost insol in water. Sol in alcohol, ether.

THERAP CAT: Anesthetic (local).

2517. Cocaine. *[1R-(exo,exo)]-3-(Benzoyloxy)-8-methyl-8-azabicyclo[3.2.1]octane-2-carboxylic acid methyl ester; 3β-hydroxy-1αH,5αH-tropane-2β-carboxylic acid methyl ester benzoate;* 2β-carbomethoxy-3β-benzoxytropane; ecgonine methyl ester benzoate; *l-cocaine;* β-cocaine; benzoylmethyl-ecgonine. $C_{17}H_{21}NO_4$; mol wt 303.36. C 67.31%, H 6.98%, N 4.62%, O 21.10%. From the leaves of *Erythroxylon coca* Lam. and other species of *Erythroxylon, Erythroxylaceae* or by synthesis. Extraction procedure: Squibb, *Pharm. J.* [3] **15**, 775, 796; **16**, 67 (1885); Emde in *Ullmann's Enzyklo-pädie der Technischen Chemie;* Schwyzer, *Die Fabrikation pharmazeutischer und chemisch-technischer Produkte* (Berlin, 1931). Configuration: Findlay, *J. Am. Chem. Soc.* **76**, 2855 (1954); O. Kovacs *et al.*, *Helv. Chim. Acta* **37**, 892 (1954). Synthesis: R. Willstätter *et al.*, *Ann.* **434**, 111 (1923). Stereospecific synthesis of *dl*-form: J. J. Tufariello *et al.*, *Tetrahedron Letters* **1978**, 1733; *eidem, J. Am. Chem. Soc.* **101**, 2435 (1979). Biosynthesis: E. Leete, S. H. Kim, *J. Am. Chem. Soc.* **110**, 2976 (1988). Absorption spectrum: J. J. Dobbie, J. J. Fox, *J. Chem. Soc.* **103**, 1193 (1913); Fischer, *Arch. Exp. Pathol. Pharmakol.* **170**, 610 (1933). Toxicity: C. L. Rose *et al.*, *J. Lab. Clin. Med.* **15**, 731 (1930). Vapor pressure studies: A. H. Lawrence *et al.*, *Can. J. Chem.* **62**, 1886 (1984). Comprehensive description: F. J. Muhtadi, A. A. Al-Badr, *Anal. Profiles Drug Subs.* **15**, 151-231 (1986).

Monoclinic tablets from alcohol, mp 98°. Volatile, esp above 90°, but the sublimate is not crystalline. $bp_{0.1}$ 187-188°. $[\alpha]_D^{18}$ −35° (50% alcohol); $[\alpha]_D^{20}$ −16° (c = 4 in chloroform). Aq solns are alkaline to litmus. pKa (15°) 8.61. pKb (15°) 5.59. One gram dissolves in 600 ml water, 270 ml water at 80°, 6.5 ml alcohol, 0.7 ml chloroform, 3.5 ml ether, 12 ml oil turpentine, 12 ml olive oil, 30-50 ml liquid petrolatum. Also sol in acetone, ethyl acetate, carbon disulfide. LD_{50} i.v. in rats: 17.5 mg/kg (Rose).

Hydrochloride, $C_{17}H_{21}NO_4 \cdot HCl$, *cocaine muriate.* Crystals, granules, or powder; saline, slightly bitter taste; numbs tongue and lips. mp ∼195°. $[\alpha]_D$ −72° (c = 2 in aq soln pH 4.5). One gram dissolves in 0.4 ml water; 3.2 ml cold, 2 ml hot alcohol; 12.5 ml chloroform. Also sol in glycerol, acetone. Insol in ether or oils. Avoid heat in preparing soln as it decomposes. Preserve in well-closed, light-resistant containers.

Nitrate dihydrate, $C_{17}H_{22}N_2O_7 \cdot 2H_2O$, crystals, mp 58-63°. Freely sol in water or alcohol; slightly sol in ether.

Sulfate, $C_{17}H_{21}NO_4 \cdot H_2SO_4$, white, granular, crystalline powder. Sol in water or alcohol.

Caution: May be habit forming. Cocaine and its derivatives are controlled substances listed in the U.S. Code of Federal Regulations, Title 21 Parts 329.1 and 1308.12 (1995).

THERAP CAT: Anesthetic (local).

THERAP CAT (VET): Topical anesthetic (ophthalmic).

2518. Cocculus. Fish-berry; Indian berry; *Cocculus indicus;* oriental berry. Dried fruit of *Anamirta cocculus* (L.) Wight & Arn., *Menispermaceae. Habit.* East Indies, Malay Archipelago. *Constit.* Menispermine, paramenispermine, about 1% picrotoxin, picrotoxic acid, cocculine alkaloid, about 50% fat. *Poisonous!*

THERAP CAT: CNS and respiratory stimulant.

2519. Cochineal. The dried female insect, *Coccus cacti* L., enclosing the young larvae. *Habit.* Mexico, Central America; cultivated in West Indies, Canary Islands, Algiers, and Southern Spain. About 70,000 insects to 1 lb. *Constit.* About 10% carminic acid, about 2% coccerin (a wax), about 10% fat. The coloring matter—alkali carminate—is contained only in the fatty parts of the insect and in the yolk of the eggs, to the extent of 10-14%.

USE: Coloring food products and toilet preparations; the source of carmine and carminic acid for manuf red and pink inks and lakes.

2520. Cocillana. Dried bark of *Guarea rusbyi* (Britt.) Rusby, *Meliaceae. Habit.* Bolivia. *Constit.* Rusbyine, about 2.5% resins, about 2.5% fat, tannin.

THERAP CAT: Expectorant.

THERAP CAT (VET): Has been used as an expectorant.

2521. Coclaurine. *(S)-1,2,3,4-Tetrahydro-1-[(4-hydroxyphenyl)methyl]-6-methoxy-7-isoquinolinol;* 1-(p-hydroxybenzyl)-6-methoxy-7-hydroxy-1,2,3,4-tetrahydroisoquinoline; machiline. $C_{17}H_{19}NO_3$; mol wt 285.34. C 71.56%, H 6.71%, N 4.91%, O 16.82%. Isolated as the racemate from species of *Machilus (Lauraceae)* and *Cocculus (Menispermaceae).* First isoln from *C. laurifolius* D.C. believed to be of the *d*-form: Kondo, Kondo, *J. Pharm. Soc. Japan* no. **524**, 876 (1925), *C.A.* **20**, 604[7] (1926); *see also* Johns *et al., Aust. J. Chem.* **20**, 1729 (1967). Structure: Kondo, Kondo, *J. Pharm. Soc. Japan* **48**, 1156 (1928); Tomita, Kusuda, *ibid.* **72**, 280 (1952). Synthesis: Kratzl, Billek, *Monatsh.* **82**, 568 (1951); Finkelstein, *J. Am. Chem. Soc.* **73**, 550 (1951). Identity with machiline: Tomita *et al., J. Pharm. Soc. Japan* **83**, 218 (1963), *C.A.* **59**, 2874a (1963). Crystal structure and

absolute configuration: Fridrichsons, Mathieson, *Tetrahedron* **24**, 5785 (1968).

Plates, tablets from alc, mp 220-221°. Sol in hot alc, hot acetone; slightly sol in water, alc, chloroform, ether, acetone; practically insol in benzene, petr ether.

Hydrochloride, $C_{17}H_{19}NO_3 \cdot HCl$, crystals, mp 263-264°.

2522. Cocoa. A powder prepd from the roasted and cured kernels of ripe seeds of *Theobroma cacao* L. and other species of *Theobroma, Sterculiaceae.* For bibliography *see* Cacao Shell.

Brownish powder of chocolate odor and taste.

USE: In nutrient beverages; as flavoring.

2523. Coconut Oil. Copra oil. Expressed oil from kernels of *Cocos nucifera* L., *Palmae. Constit.* Trimyristin, trilaurin, tripalmitin, tristearin; also various other glycerides.

White, semisolid, lard-like fat; stable to air. Remains bland and edible for several years under ordinary storage conditions. d_4^0 0.903. mp 21-25°. n_D^{40} 1.4485-1.4495. Sapon. no. 255-258. Iodine no. 8-9.5. Acid no. not over 6. Surface tension (20°): 33.4 dyn/cm; (80°): 28.4 dyn/cm. Practically insol in water, 95% alc, more sol in abs alc; very sol in chloroform, ether, carbon disulfide. Soly data: Rao, Arnold, *J. Am. Oil Chem. Soc.* **33**, 389 (1956).

USE: Manuf soap, edible fats, chocolate, candies; in baking instead of lard; in candles and night lights; in dyeing cotton; as an ointment base; in hair dressing; in massage.

2524. Codamine. *(S)-1-[(3,4-Dimethoxyphenyl)methyl]-1,2,3,4-tetrahydro-6-methoxy-2-methyl-7-isoquinolinol; 1,2,3,-4-tetrahydro-6-methoxy-2-methyl-1-veratryl-7-isoquinolinol.* $C_{20}H_{25}NO_4$; mol wt 343.42. C 69.95%, H 7.34%, N 4.08%, O 18.64%. Minor opium alkaloid. Constitutes about 0.003% of Turkish opium. Isoln: Hesse, *Ann.* **282**, 213 (1894). Structure: Späth, Epstein, *Ber.* **59B**, 2791 (1926). Synthesis: Schöpf, Thierfelder, *Ann.* **537**, 143 (1939); Billek, *Monatsh.* **87**, 106 (1956).

dl-Form, large, six-sided prisms from ether, mp 127°. Very sol in alcohol, chloroform. Somewhat sol in boiling water. In soln codamine reacts strongly basic. The salts are bitter in taste, whereas the base is said to be tasteless.

2525. Codeine. *(5α,6α)-7,8-Didehydro-4,5-epoxy-3-methoxy-17-methylmorphinan-6-ol;* methylmorphine; morphine monomethyl ether; morphine 3-methyl ether; Codicept. $C_{18}H_{21}NO_3$; mol wt 299.37. C 72.22%, H 7.07%, N 4.68%, O 16.03%. Present in opium from 0.7 to 2.5%, depending on the source, but mostly prepd by methylation of morphine, *q.v.* Discussion of structure and bibliography: Small, Lutz, "Chemistry of the Opium Alkaloids," *U.S. Public Health Reports,* Suppl. No. 103, Washington (1932). Prepn of (+)-codeine and racemate: Goto, Yamamoto, *Proc. Japan Acad.* **30**, 769 (1954), *C.A.* **50**, 1052h (1956); of (−)-form: E. J. Bijsterveld, H. J. Sinnige, *Rec. Trav. Chim.* **95**, 24 (1976); H. C. Beyerman *et al., ibid.* **97**, 127 (1978). Manuf from morphine: W. R. Heumann, *Bulletin on Narcotics* X, 15 (1958). Facile synthesis from thebaine, *q.v.:* W.

G. Dauben *et al.*, *J. Org. Chem.* **44**, 1567 (1979). Toxicity of the hydrochloride: Eddy, Sumwalt, *J. Pharmacol. Exp. Ther.* **67**, 127 (1939). Comprehensive description of codeine and codeine phosphate, *q.v.*: F. J. Muhtadi, M. M. A. Hassan, *Anal. Profiles Drug Subs.* **10**, 93-138 (1981).

Monohydrate, orthorhombic sphenoidal rods or tablets (octahedra) from water or dil alcohol, mp 154-156° (after drying at 80°). Sublimes (when anhydr) at 140-145° under 1.5 mm pressure. Melts to oily drops when heated in an amount of water insufficient for complete soln, crystallizes on cooling. d_4^{20} 1.32. $[\alpha]_D^{15} -136°$ (c = 2 in alcohol), $[\alpha]_D^{15} -112°$ (c = 2 in chloroform). pK (15°) 6.05; pH of satd aq soln 9.8. One gram dissolves in 120 ml water, 60 ml water at 80°, 2 ml alcohol, 1.2 ml hot water, 13 ml benzene, 18 ml ether, 0.5 ml chloroform; freely sol in amyl alcohol, methanol, dil acids. Almost insol in petr ether or in solns of alkali hydroxides.

Acetate, $C_{20}H_{25}NO_5$. Dihydrate, crystals; acetic acid odor. Sol in water, alc. Loses acetic acid on keeping, then becomes incompletely sol in water. *Keep tightly closed.*

Hydrobromide, $C_{18}H_{21}NO_3 \cdot HBr$. Dihydrate, crystals. Anhydr, mp 190-192°. $[\alpha]_D^{22} -96.6°$. One gram dissolves in 60 ml water, 110 ml alcohol. pH about 5.

Hydrochloride, $C_{18}H_{21}NO_3 \cdot HCl$. Dihydrate, small needles, mp ~280° with some decompn. $[\alpha]_D^{22} -108°$. One gram dissolves in 20 ml water, 1 ml boiling water, 180 ml alcohol. pH about 5. LD_{50} s.c. in mice: 300 mg/kg (Eddy, Sumwalt).

Salicylate, $C_{25}H_{27}NO_6$, white, crystalline powder. Slightly sol in water; freely sol in alcohol or ether.

Caution: May be habit forming. This is a controlled substance (opiate) listed in the U.S. Code of Federal Regulations, Title 21 Parts 329.1 and 1308.12 (1995).

THERAP CAT: Analgesic (narcotic); antitussive.

THERAP CAT (VET): Analgesic (narcotic); antitussive.

2526. Codeine Methyl Bromide. Eucodin. $C_{19}H_{24}BrNO_3$; mol wt 394.31. C 57.88%, H 6.13%, Br 20.26%, N 3.55%, O 12.17%. $C_{18}H_{21}NO_3 \cdot CH_3Br$.

Crystals, mp ~260°. Sol in 2-3 parts water, in hot methanol, sparingly in alc. Insol in chloroform, ether.

Caution: May be habit forming. This is a controlled substance (opium derivative) listed in the U.S. Code of Federal Regulations, Title 21 Parts 329.1, 1308.11 (1995).

THERAP CAT: Analgesic (narcotic); antitussive.

2527. Codeine N-Oxide. Genocodeine; genkodein; Codeigene. $C_{18}H_{21}NO_4$; mol wt 315.37. C 68.55%, H 6.71%, N 4.44%, O 20.29%. Prepn from codeine and 30% hydrogen peroxide: Freund, Speyer, *Ber.* **43**, 3313 (1910); Kelentey *et al.*, *Arzneimittel-Forsch.* **7**, 594 (1957).

Platelets from water, mp 231-232°.

Monohydrate, crystals from alc, mp 215°. $[\alpha]_D^{18} -97.1°$ (c = 2.1 in water).

Hydrochloride monohydrate, $C_{18}H_{21}NO_4 \cdot HCl \cdot H_2O$, crystals, loses crystal water at 110°, mp 219-220°. $[\alpha]_D^{20} -105.8°$ (c = 2 in water). One gram dissolves in 9.5 ml water.

Note: This is a controlled substance (opium derivative) listed in the U.S. Code of Federal Regulations, Title 21 Part 1308.11 (1995).

THERAP CAT: Antitussive.

2528. Codeine Phosphate. Galcodine. $C_{18}H_{24}NO_7P$; mol wt 397.36. C 54.41%, H 6.09%, N 3.52%, O 28.18%, P 7.79%.

Hemihydrate, (U.S.P.), fine, white, needle-shaped crystals or cryst powder. Odorless; affected by light. Solns acid to litmus. Freely sol in water; very sol in hot water; slightly sol in alcohol; more sol in boiling alcohol.

Sesquihydrate, very efflorescent, small crystals or cryst powder. One gram dissolves in 2.3 ml water, 0.5 ml water at 80°, 325 ml alcohol, 125 ml boiling alcohol, 4500 ml chloroform, 1875 ml ether. pH of a 2% aq soln: 4.6. *Keep well closed.*

Note: This is a controlled substance (opiate) listed in the U.S. Code of Federal Regulations, Title 21 Part 1308.12 (1995).

THERAP CAT: Analgesic (narcotic); antitussive.

THERAP CAT (VET): Analgesic (narcotic); antitussive.

2529. Codeine Sulfate. $C_{36}H_{44}N_2O_{10}S$; mol wt 696.82. C 62.05%, H 6.36%, N 4.02%, O 22.96%, S 4.60%.

Trihydrate, crystals or cryst powder. One gram dissolves in 30 ml water, 6.5 ml water at 80°, 1300 ml alc. Insol in chloroform or ether. pH: 5.0. Store in airtight containers; protect from light.

Note: This is a controlled substance (opiate) listed in the U.S. Code of Federal Regulations, Title 21 Part 1308.12 (1995).

THERAP CAT: Analgesic (narcotic); antitussive.

THERAP CAT (VET): Analgesic (narcotic); antitussive

2530. Cod Liver Oil. Gaduol; Tunol. The partially destearinated fixed oil expressed from fresh livers of *Gadus morrhua* L., and other species of *Gadidae*. *Constit.* Most important are vitamins A and D, each gram containing at least 850 U.S.P. units vitamin A (255 μg) and at least 85 U.S.P. units vitamin D (2.125 μg); glycerides of palmitic, stearic, etc. acids (~19% saturated fatty acids; remainder unsaturated); cholesterol, batyl alcohol esters. Source of omega-3 fatty acid: N. Haagsma *et al.*, *J. Am. Oil Chem. Soc.* **59**, 117 (1982). Brief description: D. Hilditch, P. Williams, *The Chemical Constitution of Natural Fats* (Wiley-Interscience, New York, 4th ed., 1964) p 43; *Bailey's Industrial Oils & Fat Products* Vol. 1, D. Swern (Wiley-Interscience, New York, 4th ed., 1979) pp 451-453.

Pale-yellow, thin liq; bland, slightly fishy taste and odor. Becomes yellow, acquires a somewhat disagreeable odor on exposure to air and light. d 0.918-0.927. n_D^{40} 1.4705-1.4745. Sapon no. 180-190. Iodine no. 145-180. Acid no. not over 1.2. Slightly sol in alcohol; freely sol in chloroform, ether, carbon disulfide, ethyl acetate, petr ether.

THERAP CAT: Vitamins A and D source.

THERAP CAT (VET): Source of vitamins A and D. Locally to promote healing.

2531. Coenzyme A. CoA. $C_{21}H_{36}N_7O_{16}P_3S$; mol wt 767.54. C 32.86%, H 4.73%, N 12.77%, O 33.35%, P 12.11%, S 4.18%. An essential cofactor in enzymatic acetyl transfer reactions. Synthesized in cells from pantothenate, ATP and cysteine. Found ubiquitously in mammalian cells. Isoln from animal sources: Lipmann *et al.*, *J. Biol. Chem.* **167**, 869 (1947); **186**, 235 (1950). Many microorganisms contain large amounts of the coenzyme. Isoln from *Streptomyces fradiae*: Kaplan, Lipmann, *ibid.* **174**, 37 (1948). Purifications: De Vries *et al.*, *J. Am. Chem. Soc.* **72**, 4838 (1950); Gregory *et al.*, *ibid.* **74**, 854 (1952). Structure: Baddiley *et al.*, *Nature* **171**, 76 (1953). Total synthesis: Moffatt, Khorana, *J. Am. Chem. Soc.* **81**, 1265 (1959); **83**, 663 (1961); Shimizu *et al.*, *Chem. Pharm. Bull.* **13**, 1142 (1965). Reviews: Lipmann, *Bacteriol. Rev.* **17**, 1-16 (1953); Baddiley, *Advan. Enzymol.* **16**, 1 (1955); Jaenicke, Lynen in *The Enzymes* vol. 3, P. D. Boyer *et al.*, Eds. (Academic Press, New York, 2nd ed., 1960) pp 3-103. Review of metabolism: J. D. Robishaw, J. R. Neely, *Am. J. Physiol.* **248**, E1-E9 (1985); of clinical evaluations in hyperlipoproteinemia: A.

Perin, G. Fraticelli, *Int. J. Tiss. Reac.* **13**, 111-114 (1991); of biochemical role in cellular toxicity: E. P. Brass, *Chemico-Biological Interactions* **90**, 203-214 (1994).

White powder; characteristic thiol odor. May be dried *in vacuo* over phosphorus pentoxide at 34°. uv max: 259.5 nm (ϵ 16800). Fairly strong acid. pK 9.6 (thiol); pK 6.4 (secondary phosphate); pK 4.0 (adenine NH_3^+). Soluble in water. Practically insol in ethanol, ether, acetone. Pure dry coenzyme is best stored in evacuated ampuls at room temp. Readily oxidized by air to the catalytically inactive disulfide.

2532. Coffee, Green. Coffee beans. Dry, unroasted seeds of *Coffea arabica* L., *Rubiaceae*. *Habit*. Tropical Africa, especially Abyssinia; cultivated in many tropical countries, *e.g.*, Java, West Indies, Brazil, etc. *Constit*. Caffeine 1-2%, coffee oil 10-15% and sugars about 8%, proteins about 11%, ash about 5%, chlorogenic and caffeic acids about 6%. Other constituents include cellulose, hemicelluloses, trigonelline, tannic acid, volatile oils. Monograph: M. Sivetz, H. E. Foote, *Coffee Processing Technology*, 2 vols (Avi Publ., 1963).

2533. Coffee Oil. Coffee bean oil. A fatty oil found in coffee beans to the extent of about 15% (dry basis). By-product in the manufacture of instant coffees. Obtained in 2.7% yield by petr ether extraction of green Santos beans after previous dewaxing with tetrachloroethane: Schuette *et al., J. Am. Chem. Soc.* **56**, 2085 (1934). Coffee oil seems devoid of linolenic acid, however positive tests for linoleic (25.66%) and oleic (12.36%) acids were obtained.

Greenish-brown oil. Odor characteristic of green coffee beans. d_{25}^{25} 0.9653. n_D^{25} 1.4790. Iodine no. (Wijs) 100.72. Saponification no. 195.53; Reichert-Meissl no. 0.36. Saturated acids 33.60%. Unsaturated fatty acids 38.02%. Unsaponifiable matter 12.63%.

2534. Coherin. A peptide factor present in the bovine neurohypophysis which may be involved in the normal physiological regulation of intestinal motility in mammals. Appears to stimulate the coordinated or "coherent" contraction of the intestine needed to keep food moving down the tract. Mol wt ~4000. Amino acid analysis (residue/mole): lysine 0.59; aspartic acid 3.9; threonine 0.11; serine 0.22; glutamic acid 4.0; proline 5.1; glycine 4.7; alanine 0.90; half-cystine 4.9; isoleucine 3.3; leucine 3.1; tyrosine 2.9; phenylalanine 0.88. Isoln from bovine posterior pituitary and preliminary data: Goodman, Hiatt, *Science* **178**, 419 (1972). Localization of formation site: I. Galasinska-Pomykol, M. Chilimoniuk, *Endokrynol. Pol.* **24**, 429 (1973), *C.A.* **81**, 102499j (1974). Effect on electrical rhythm of dog ileum *in vivo*: R. B. Hiatt *et al., Am. J. Dig. Dis.* **22**, 108 (1977). Effects on idiopathic delayed gastric emptying in humans: E. D. Davidson *et al., Am. J. Gastroenterol.* **74**, 419 (1980).

Relatively thermostable, losing only 10% of its activity when heated in a boiling water bath at pH 3 for 1 hour. Isoelectric point: pH 6.0.

2535. Colchiceine. *(S)-N-(5,6,7,9-Tetrahydro-10-hydroxy-1,2,3-trimethoxy-9-oxobenzo[a]heptalen-7-yl)acetamide;* 7-acetamido-10-hydroxy-1,2,3-trimethoxy-6,7-dihydrobenzo[a]heptalen-9(5H)-one. $C_{21}H_{23}NO_6$; mol wt 385.42. C 65.44%, H 6.01%, N 3.63%, O 24.91%. Deriv of colchicine, *q.v.*, isolated from *Colchicum autumnale* L., *Liliaceae*: Santavy, Macák, *Coll. Czech. Chem. Commun.* **19**, 805 (1954). Structure and synthesis: Nakamura, *Chem. Pharm. Bull.* **10**, 299 (1962). Circular dichroism: J. Hrbek *et al., Coll. Czech. Chem. Commun.* **47**, 2258 (1982). Effect on antibody

response induced *in vitro*: J. Sterzl *et al., Folia Microbiol.* **27**, 256 (1982).

Yellow crystals from dioxane + ether, mp 178-179°. $[\alpha]_D^{25}$ −255.1° (c = 1 in chloroform). uv max (95% ethanol): 351, 244 nm (log ϵ 4.28, 4.51). Slightly sol in water; freely sol in alcohol, chloroform; almost insol in benzene, ether.

Ethyl ether, $C_{23}H_{27}NO_6$, needles from ethyl acetate + ether, mp 135-139°. $[\alpha]_D^{25}$ −129.4° (c = 1 in chloroform). uv max (95% ethanol): 351, 243.5 nm (log ϵ 4.21, 4.45). LD_{50} in mice: 84 mg/kg i.p., *RTECS* vol. 1, R. J. Lewis, R. L. Tatken, Eds. (1980) p 534.

2536. Colchicine. *(S)-N-(5,6,7,9-Tetrahydro-1,2,3,10-tetramethoxy-9-oxobenzo[a]heptalen-7-yl)acetamide.* $C_{22}H_{25}NO_6$; mol wt 399.44. C 66.15%, H 6.31%, N 3.51%, O 24.03%. A major alkaloid of *Colchicum autumnale* L., *Liliaceae*. Extraction procedure: Chemnitius, *J. Prakt. Chem.* [II] **118**, 29 (1928); F. E. Hamerslag, *Technology and Chemistry of Alkaloids* (New York, 1950) pp 66-80. Structure: Dewar, *Nature* **155**, 141 (1945); King *et al., Acta Cryst.* **5**, 437 (1952); Horowitz, Ullyot, *J. Am. Chem. Soc.* **74**, 487 (1952). Crystal structure: L. Lessinger, T. N. Margulis, *Acta Crystallogr.* **B34**, 578 (1978). Total synthesis: Schreiber *et al., Helv. Chim. Acta* **44**, 540 (1961); Van Tamelen *et al., Tetrahedron* **14**, 8 (1961); Nakamura, *Chem. Pharm. Bull.* **8**, 843 (1960); Sunagawa *et al., ibid.* **9**, 81 (1961); **10**, 281 (1962); Scott *et al., Tetrahedron* **21**, 3605 (1965); Woodward, *Harvey Lectures*, Ser. **59** (Academic Press, New York, 1965) p 31; Kotani *et al., Chem. Commun.* **1974**, 300; D. A. Evans *et al., J. Am. Chem. Soc.* **103**, 5813 (1981). Biosynthesis: Leete, *Tetrahedron Letters* **1965**, 333; Battersby *et al., J. Chem. Soc.* **1964**, 4257; Hill, Unrau, *Can. J. Chem.* **43**, 709 (1965). Tubulin-binding activity: J. M. Andreu, S. N. Timasheff, *Proc. Nat. Acad. Sci. USA* **79**, 6753 (1982). Toxicity: S. J. Rosenbloom, F. C. Ferguson, *Toxicol. Appl. Pharmacol.* **13**, 50 (1968); R. P. Beliles, *ibid.* **23**, 537 (1972). Clinical evaluations in cirrhosis of the liver: M. M. Kaplan *et al., N. Engl. J. Med.* **315**, 1448 (1986); D. Kershenobich *et al., ibid.* **318**, 1709 (1988). Bibliography of early literature: Eigsti, *Lloydia* **10**, 65 (1947). Monograph: O. J. Eigsti, P. Dustin, Jr., *Colchicine in Agriculture, Medicine, Biology and Chemistry* (Iowa State College Press, Ames, Iowa, 1955). *Reviews*: Fleming, *Selected Organic Syntheses* (John Wiley, London, 1973) pp 183-207; G. Lagrue *et al., Ann. Med. Interne* **132**, 496-500 (1981); F. D. Malkinson, *Arch. Dermatol.* **118**, 453-457 (1982). Comprehensive description: D. K. Wyatt *et al., Anal. Profiles Drug Subs.* **10**, 139-182 (1981).

Pale yellow scales or powder, mp 142-150°. Darkens on exposure to light. Has been crystallized from ethyl acetate, pale yellow needles, mp 157°. $[\alpha]_D^{17}$ −429° (c = 1.72). $[\alpha]_D^{17}$ −121° (c = 0.9 in chloroform). pK at 20°: 12.35; pH of 0.5% soln: 5.9. uv max (95% ethanol): 350.5, 243 nm (log ϵ 4.22; 4.47). One gram dissolves in 22 ml water, 220 ml ether, 100 ml benzene; freely sol in alcohol or chloroform. Practically insol in petr ether. Forms two cryst compds with chloroform, $B.CHCl_3$ or $B.2CHCl_3$, which do not give up their chloroform unless heated between 60 and 70° for considerable time. LD_{50} in rats (mg/kg): 1.6 i.v. (Rosenbloom, Ferguson); in mice (mg/kg): 4.13 i.v. (Beliles).

USE: In research in plant genetics (for doubling chromosomes).

THERAP CAT: Gout suppressant. Treatment of Familial Mediterranean Fever.

THERAP CAT (VET): Has been used as an antineoplastic.

2537. Colchicum Corm. Meadow saffron; autumn crocus; wild saffron; meadow crocus. Dried corm of *Colchicum autumnale* L., *Liliaceae*. *Habit.* Central and Southern Europe, North Africa. *Constit.* 0.3-0.5% colchicine, colchicein, colchicoresin, starch, etc. Detn of colchicine in colchicine corm: Self, Corfield, *Quart. J. Pharm. Pharmacol.* **5**, 347 (1932).

THERAP CAT: Gout suppressant.

2538. Colestipol. *N-(2-Aminoethyl)-N'-[2-[(2-aminoethyl)amino]ethyl]-1,2-ethanediamine polymer with (chloromethyl)oxirane.* A basic anion exchange resin described as a high molecular wt copolymer of diethylenetriamine and 1-chloro-2,3-epoxypropane (hydrochloride), with approx 1 out of 5 amine nitrogens protonated, for which no structural formula has been assigned and for which no specific mol wt information is available, due to the highly cross-linked and insoluble nature of the material. Prepn: Nelson, Van den Berg, Ger. pat. **1,927,336;** *eidem,* U.S. pat. **3,692,895** (1969, 1971 to Upjohn); Lednicer, Peery, Ger. pat. **2,053,585;** *eidem,* U.S. pat. **3,803,237** (1971, 1974 to Upjohn). Activity studies: Parkinson *et al., Atherosclerosis* **11**, 531 (1970); **17**, 167 (1973); Marmo *et al., G. Arterioscler.* **8**, 229 (1970); Goodman *et al., J. Clin. Invest.* **52**, 2646 (1973). Toxicology: Webster, Bollert, *Toxicol. Appl. Pharmacol.* **28**, 57 (1974). Review of pharmacology and therapeutic efficacy: R. C. Heel *et al., Drugs* **19**, 161-180 (1980).

Hydrochloride, *U-26597A, Cholestabyl, Colestid, Lestid.* LD$_{50}$ in rats (mg/kg): >4000 i.p.; >1000 orally (Webster, Bollert).

THERAP CAT: Antihyperlipoproteinemic.

2539. Colforsin. *[3R-(3α,4aβ,5β,6β,6aα,10α,10aβ,-10bα)]-5-(Acetyloxy)-3-ethenyldodecahydro-6,10,10b-trihydroxy-3,4a,7,7,10a-pentamethyl-1H-naphtho[2,1-b]pyran-1-one;* *7β-acetoxy-8,13-epoxy-1α,6β,9α-trihydroxylabd-14-en-11-one;* forskolin; boforsin (obsolete); HL-362. C$_{22}$H$_{34}$O$_7$; mol wt 410.51. C 64.37%, H 8.35%, O 27.28%. Diterpene isolated from *Coleus forskohlii,* Briq. *Labiatae,* possessing vasodilating and cardiostimulatory properties. Isoln and characterization: S. V. Bhat *et al., Tetrahedron Letters* **1977**, 1669; *eidem,* Ger. pat. **2,557,784;** *eidem,* U.S. pat. **4,088,659** (1977, 1978 both to Hoechst). Synthesis by hydroxylation of 9-deoxyforskolin: N. J. Hrib, *Tetrahedron Letters* **28**, 19 (1987); *see also* F. E. Ziegler *et al., J. Am. Chem. Soc.* **109**, 8115 (1987). Stereocontrolled synthesis of the 3-ring system: S. Hashimoto *et al., Chem. Commun.* **1987**, 24. Total synthesis of (±)-forskolin: S. Hashimoto *et al., J. Am Chem. Soc.* **110**, 3670 (1988); E. J. Corey *et al., ibid.* 3672. Positive inotropic and blood-pressure lowering activity: E. Lindner *et al., Arzneimittel-Forsch.* **28**, 284 (1978). Activates adenylate cyclase: H. Metzger, E. Lindner, *ibid.* **31**, 1248 (1981). Binds to catalytic site of adenylate cyclase: T. Pfeuffer, H. Metzger, *FEBS Letters* **146**, 369 (1982). Isoln lowers intraocular pressure: J. Caprioli, M. Sears, *Lancet* **1**, 958 (1983). In treatment of glaucoma: *eidem,* U.S. pat. **4,476,-140** (1984 to Yale University). *Reviews:* K. B. Seaman, J. W. Daly, *J. Cyclic Nucl. Res.* **1981**, 201; K. B. Seaman, "Forskolin and Adenylate Cyclase: New Opportunities in Drug Design" in *Ann. Rep. Med. Chem.* **19**, D. M. Bailey, Ed. (Academic Press, New York, 1984) pp 293-302.

Colorless crystals from ethyl acetate/petr ether, mp 230-232°. uv max 210, 305 nm (ε 1000, 50). [α]$_D^{25}$ −26.19° (c = 1.68 in CHCl$_3$).

USE: In purification of adenylate cyclase.

2540. Colfosceril Palmitate. *(R)-4-Hydroxy-N,N,N-trimethyl-10-oxo-7-[(1-oxohexadecyl)oxy]-3,5,9-trioxa-4-phosphapentacosan-1-aminium inner salt 4-oxide;* L-α-dipalmitoyl lecithin; dipalmitoyl-L-α-glycerylphosphorylcholine; dipalmitoyl phosphatidylcholine; DPPC. C$_{40}$H$_{80}$NO$_8$P; mol wt 734.05. C 65.45%, H 10.98%, N 1.91%, O 17.44%, P 4.22%. Major phospholipid constituent and primary surface tension lowering component of natural lung surfactant. Isoln from lung extracts: S. J. Thannhauser *et al., J. Biol. Chem.* **166**, 669 (1946); E. S. Brown, *Am. J. Physiol.* **207**, 402 (1964). Enantioselective synthesis: E. Baer, M. Kates, *J. Am. Chem. Soc.* **72**, 942 (1950). Clinical study in combination therapy for neonatal respiratory distress syndrome (RDS): C. J. Morley *et al., Lancet* **1**, 64 (1981). Diagnostic use as predictor of fetal lung maturity: A. Lohninger *et al., Clin. Chem.* **29**, 650 (1983). HPTLC determn in amniotic fluid: J. G. Alvarez *et al., J. Chromatog. B* **665**, 79 (1995). Review of early literature: D. F. Tierney, *Am. J. Physiol.* **257**, L1-L12 (1989); of role in pulmonary function and clinical use: L. M. G. van Golde *et al., Physiol. Rev.* **68**, 374-455 (1988); of structure and function as surface active agent: B. A. Hills, *Br. J. Anaesth.* **65**, 13-29 (1990); of biosynthesis: H. P. Haagsman, L. M. G. van Golde, *Ann. Rev. Physiol.* **53**, 441-464 (1991).

Occurs as monohydrate, crystals from hot diisobutyl ketone, softens at 75-79°, mp 234-235°. [α]$_D^{23}$ +6.6° (c = 4.2 in 1:1 chloroform-methanol). Readily sol in chloroform, hot diisobutyl ketone, hot dioxane. Readily emulsified in water. Soly at 22-23° (g/100 ml): ethanol 1.5, ether 0.02, acetone 0.02, pyridine 1.1, acetic acid 4.0, methanol 1.4.

THERAP CAT: Pulmonary surfactant. Diagnostic aid (fetal lung maturity).

2541. Colicins. Colicine. Antibiotic substances, or complexes of antibiotic substances, which are highly specific, are produced by certain strains of intestinal bacteria, and act upon other related strains: Fredericq, *Ann. Rev. Microbiol.* **11**, 7 (1957). Colicins produced by various strains may differ in many characteristics, the most conspicuous of which are activity spectra and the specificity of resistant mutants. They give the general reactions of proteins, are antigenic, and their activity is destroyed by proteolytic enzymes: Fredericq, *J. Theoret. Biol.* **4**, 159 (1963). Production of colicin A: Barry *et al., Nature* **198**, 211 (1963); of colicin V: Hutton, Goebel, *Proc. Nat. Acad. Sci. USA* **47**, 1498 (1961); of colicins E$_4$, N, P,V$_2$, V$_3$, V$_4$ and V$_5$: Hamon, Péron, *Ann. Inst. Pasteur* **107**, 44 (1964). Mechanism of action of colicins: Nomura, *Proc. Nat. Acad. Sci. USA* **52**, 1514 (1964). *Reviews: idem* in *Antibiotics* vol. 1, D. Gottlieb, P. D. Shaw, Eds. (Springer-Verlag, New York, 1967) pp 696-704; Wendt, *ibid.* vol. **3**, J. W. Corcoran, F. E. Hahn, Eds. (1975) pp 588-605.

2542. Colistin. Polymyxin E; Colimycin; Coly-Mycin; Colisticina; Totazina. Cyclopolypeptide antibiotic produced by *Bacillus colistinus (Aerobacillus colistinus)* isolated from Japanese soil: Koyama, **Japan. pat. 52 1546** (1952). Composed of colistins A, B and C: Suzuki *et al., J. Biochem. (Tokyo)* **54**, 25 (1963). Structure of colistin A: *eidem, ibid.* **173**, 412. Separation into polymyxins E$_1$ and identity of colistin A with polymyxin E$_1$: Wilkinson, Lowe, *J. Chem. Soc.* **1964**, 4107. Synthesis of colistin A: Studer *et al., Helv. Chim. Acta* **48**, 1371 (1965). Toxicity study: Barnett *et al., Brit. J. Pharmacol. Chemother.* **23**, 552 (1964). *Reviews:* Vogler, Studer, *Experientia* **22**, 345 (1966); *Medicamenta* **61** (509), 177-234 (1973).

DAB = α,γ-diaminobutyric acid

Colistin A　　　R = (+)-6-methyloctanoyl
Polymyxin E₂　　R = 6-methylheptanoyl

Colistin sodium methanesulfonate, $C_{58}H_{105}N_{16}Na_5O_{28}S_5$, *colistimethate sodium, Alficetin, Methacolimycin.* The injectable form. Sol in water and stable in the dry form. LD_{50} i.v. in mice: >550 mg/kg (Barnett).

Colistin formaldehyde-sodium bisulfite, crystals, dec 290-295°. Sol in water, slightly sol in methanol, ethanol, acetone, ether, Koyama *et al.*, **Japan.** pat. **57 4898** (1957).

Sulfate, *Malimyxin, Multimycine.*

Acetylcolistin, Koyama, Kurosawa, **Japan.** pat. **60 9313** (1960 to Kayaku Antibiotics Res.).

Colistin A, $C_{53}H_{100}N_{16}O_{13}$, *polymyxin E₁.* $[\alpha]_{5461}^{22}$ −93.3° (2% acetic acid).

Polymyxin E₂, $C_{52}H_{98}N_{16}O_{13}$. $[\alpha]_{5461}^{22}$ −94.5° (2% acetic acid).

THERAP CAT: Antibacterial.

THERAP CAT (VET): Antibacterial.

2543. Collagen. Ossein. Mol wt about 130,000. Polypeptide substance comprising one third of the total protein in mammalian organisms; main constituent of skin, connective tissue, and the organic substance of bones and teeth. Prepd from bones by dissolving the mineral part of the bones with phosphoric acid: Sciallano, **Fr.** pat. **688,104** (1929), *C.A.* **25,** 786 (1931). Isoln of sol collagens from corium tissue: Reizo, Sakata, *Kumamoto Med. J.* **13,** 27 (1960). Purification: Bloch, Oneson, **U.S.** pat. **2,973,302** (1961 to Ethicon). Commercial extraction process: Highberger, **U.S.** pat. **2,979,438** (1961 to United Shoe Machinery). Collagen production in the body is preceded by the formation of a much larger molecule, the biosynthetic precursor *procollagen,* which is degraded by specific enzymes to make collagen. Different types of collagens exist. They are all composed of molecules containing three polypeptide chains, α-chains, arranged in a triple helical conformation. The amino acid sequence of the α-chain is mostly a repeating structure with glycine in every third position and proline or 4-hydroxyproline frequently preceding the glycine residues. Slight differences in the primary structure (amino acid sequence) establish differences between types. Review of structural studies: Tanzer, *Science* **180,** 561-566 (1973); P. Bornstein, H. Sage, *Ann. Rev. Biochem.* **49,** 959 (1980). Collagen is differentiated from the accompanying fibrous proteins (elastin, *q.v.,* and reticulin) by (*1*) its content of proline, hydroxyproline and hydroxylysine, by (*2*) the absence of tryptophan and its low tyrosine and sulfur content, but particularly by (*3*) its high content of polar groups originating from the difunctional amino acids. The polar groups are responsible for the swelling properties leading eventually to dispersion of collagen in dil acid. Denaturation of collagen is the conversion of the rigidly coiled helix to a random coil called gelatin, *q.v.* Description of the native and denatured states of sol collagen: Boedtker, Doty, *J. Am. Chem. Soc.* **78,** 4267 (1956). Biosynthesized in fibroblastic cells. Review of biosynthesis: Bornstein, *Ann. Rev. Biochem.* **43,** 567-603 (1974). Reviews: Gross, *Sci. Am.* **204,** 121 (May 1961); Harrington, von Hippel, *Advan. Protein Chem.* **16,** 1-138 (1961); Grassmann *et al., Fortschr. Chem. Org. Naturst.* **23,** 195-314 (1965); P. Bornstein, W. Traub, "The Chemistry and Biology of Collagen" in *The Proteins,* vol. IV, H. Neurath, R. L. Hill, Eds. (Academic Press, New York, 1979) pp 411-632; D. R. Eyre, *Science* **207,** 1315-1322 (1980); *Methods in Enzymology* vol. 82, L. W. Cunningham, D. W. Frederiksen, Eds. (Academic Press, New York, 1982) pp 3-555. Summary of recent studies of the role of collagen in disease: *Chem. & Eng. News* **60,** 32 (Jan. 25, 1982). Books: A. Veis, *The Macromolecular Chemistry of Gelatin* (Academic Press, New York, 1964) 433 pp; *Biochemistry of Collagen,* G. N. Ramachandran *et al.,* Eds. (Plenum, New York, 1976) 536 pp; P. P. Fietzek in *Collagen in Health and Disease,* M. I. V. Jayson, J. B. Weiss, Eds. (Churchill-Livingstone, New York, 1982).

USE: As fibers in sutures, in leather substitutes; as a gel in photographic emulsions, in coatings; in food casings.

2544. Collagenase. Clostridiopeptidase A; Iruxol; Santyl. Rare proteolytic enzyme capable of digesting native undenatured collagen, *q.v.,* found in certain *Clostridia* bacteria culture filtrates. Crude prepn from *C. histolyticum:* Mandl *et al., J. Clin. Invest.* **32,** 1323 (1953). Purified prepn: Keller, Mandl, *Arch. Biochem. Biophys.* **101,** 81 (1963). Also isolated from culture media of human wound tissue, skin, bone, leukocytes, gingiva, cornea, rheumatoid synovium; from involuting rat uterus and tadpole tailfin tissue. Sepn of fractions A and B having the respective mol wts 105,000 and 57,400: Harper *et al., Biochem. Biophys. Res. Commun.* **18,** 627 (1965). Amino acid composition: Mandl *et al., Biochemistry* **3,** 1737 (1964); Yoshida, Noda, *Biochim. Biophys. Acta* **105,** 562 (1965). Potential use in treatment of herniated discs: Sussman, **Fr.** pat. **2,008,611** (1970 to Worthington Biochemical). Review of early studies: Mandl, *Advan. Enzymol.* **23,** 163-264 (1961). *Reviews: idem, Science* **169,** 1234-1238 (1970); Nordwig, *Advan. Enzymol. Relat. Areas Mol. Biol.* **34,** 155-205 (1971); Lazarus, *Brit. J. Dermatol.* **86,** 193-199 (1972). Book: *Collagenase,* I. Mandl, Ed. (Gordon and Breach, N.Y., 1972) 215 pp.

USE: In investigation of the structure and biosynthesis of collagen; in dispersion of cells for tissue culture studies.

THERAP CAT: Debriding agent.

2545. Collinomycin. *6-[2-(4,9-Dihydro-8-hydroxy-5,7-dimethoxy-4,9-dioxonaphtho[2,3-b]furan-2-yl)ethyl]-7,8-dihydroxy-1-oxo-1H-2-benzopyran-3-carboxylic acid methyl ester;* α-rubromycin. $C_{27}H_{20}O_{12}$; mol wt 536.45. C 60.45%, H 3.76%, O 35.79%. Antibiotic substance produced by *Streptomyces collinus:* Brockmann, Renneberg, *Naturwiss.* **40,** 166 (1953); **Ger.** pat. **918,162** (1954 to Bayer). Identity with α-rubromycin: Brockmann *et al., Tetrahedron Letters* **1966,** 3525. Structure: Brockmann, Zeeck, *Ber.* **103,** 1709 (1970). Shows considerable *in vitro*-activity against *Staph. aureus.*

Orange prisms from chloroform-methanol, mp 280-282°. Moderately sol in chloroform, acetone, dioxane; sparingly sol in ether, low-molecular alcohols. Practically insol in petr ether, water, aq sodium bicarbonate soln.

2546. Collinsonia. Stone-root; knob root; horse balm; richweed. Root of *Collinsonia canadensis* L., *Labiatae. Habit.* North America, from Ontario to Florida and west to Kansas. *Constit.* Resin, saponin, tannin, mucilage.

2547. Collodion. A soln of 4 g pyroxylin (chiefly nitrocellulose) in 100 ml of a mixture of 25 ml alcohol and 75 ml ether. Contains about 70% ether and 24% abs alc by vol.

Colorless, or slightly yellow, clear or slightly opalescent, syrupy liquid. Has the odor of ether. d_{25}^{25} 0.765-0.775. Exposed in thin layers, it evaporates leaving a tough, colorless film. On the addition of water the pyroxylin ppts.

Flexible collodion, mixture of simple collodion with 2% camphor and 3% castor oil (by wt). Yellow syrupy liquid; contains about 67% ether and about 22% abs alcohol by volume.

Styptic collodion, mixture of flexible collodion with 18% tannic acid by wt. Contains about 61% ether and about 21% abs alcohol by volume.

Caution: Highly flammable! Keep tightly closed in a cool place and away from flame!

USE: Collodion in photography; manuf lacquers, patent and artificial leathers, artificial pearls; process engraving; in cements.

THERAP CAT: Collodion and flexible collodion as topical protectant; styptic collodion as hemostatic.

THERAP CAT (VET): Skin protectant.

2548. Collodion, Cantharidal. Blistering collodion; vesicating collodion. One hundred parts represent 60 parts cantharides. Contains about 57% ether and 22% abs alc by vol.

Olive-green, syrupy liquid.

THERAP CAT: Rubefacient.

2549. Colloidal Bismuth Subcitrate. CBS; De-Nol; De-Noltab; Duosol; Ulcerone. Cytoprotective, polymeric bismuth citrato complex; also known as *tripotassium dicitrato bismuthate*. General prepn of colloidal bismuth solution: L. Vanino, cited in *Mellor's* vol. IX, 598 (1929). Manufacturing processes: C. J. McLoughlin, R. B. Himstedt, **Ger. pat. 2,501,787;** P. J. H. Bos *et al.*, Eur. pat. **Appl. 75,992;** *eidem*, **U.S.** pat. **4,801,608** (1975, 1985, 1989 all to Gist-Brocades). Analytical study: D. R. Williams, *J. Inorg. Nucl. Chem.* **39**, 711 (1977). Crystal structure: E. Asato *et al.*, *Chem. Letters* **1992**, 1967. Pharmacology: J. Wieriks *et al.*, *Scand. J. Gastroenterol.* **17**, Suppl. 80, 11 (1982). Modes of action: D. W. R. Hall, *ibid.* **24**, Suppl. 157, 3 (1989); S. P. Lee, *ibid.* **26**, Suppl. 185, 1 (1991). Pharmacokinetics and safety: L. Z. Benet *ibid.* 29. Symposia on pharmacology and clinical efficacy: *ibid.* **21**, Suppl. 122, 1-54 (1986); *Digestion* **37**, Suppl. 2, 1-64 (1987). Clinical trials in *Helicobacter pylori* associated gastritis and ulcer: T. Rokkas *et al.*, *Gut* **29**, 1386 (1988); W. A. de Boer *et al.*, *Am. J. Gastroenterol.* **89**, 1993 (1994).

White, amorphous powder. Sol in water, dil alkali, ammonia. Forms precipitate of bismuth oxychloride and bismuth citrate at pH < 5.

THERAP CAT: Antiulcerative.

2550. Colocynth. Bitter apple; bitter cucumber; bitter gourd. Dried pulp of fruit of *Citrullus colocynthis* Schrad., *Cucurbitaceae*. *Habit.* Mediterranean region, Africa, Syria. *Constit.* Colocynthin, colocynthidin, pectin, albuminoids.

THERAP CAT: Cathartic.

THERAP CAT (VET): Has been used as a purgative.

2551. Colocynthin. *25-(Acetyloxy) 2 (β D glucopyranosyloxy)-16,20-dihydroxy-9-methyl-19-norlanosta-1,5,23-triene-3,11,22-trione; 2-O-β-D-glucopyranosylcucurbitacin E.* $C_{38}H_{54}O_{13}$; mol wt 718.84. C 63.49%, H 7.57%, O 28.93%. Glucoside from fruit of *Citrullus colocynthis* Schrad., *Cucurbitaceae*: Walz, *Neues Jahrb. Pharm.* **9**, 16, 225 (1858); **16**, 10 (1861); Henke, *Arch. Pharm.* **221**, 200 (1883); Naylas, Chappel, *Pharm. J.* **25**, 117 (1907); Lavie *et al.*, *Acta Univ. Int. Contra Cancerum* **15**, 177 (1959); Khadem, Rahman, *Tetrahedron Letters* **1962**, 1137. Other isolns: H. Ripperger, K. Seifert, *Tetrahedron* **31**, 1561 (1975); H. Ripperger, *ibid.* **32**, 1567 (1976). Structure: Khadem, Rahman, *J. Chem. Soc.* **1963**, 4991.

Yellow crystals, mp 158-160°. $[\alpha]_D$ +50° (c = 0.4 in ethanol). uv max: 234-236 nm (log ϵ 4.11). Sol in ethanol, acetone, chloroform; slightly sol in ether, water.

THERAP CAT: Cathartic.

2552. Colony Stimulating Factors. CSFs. Family of hematopoietic growth factors that regulate the production of myeloid cells. Glycoprotein hormones that stimulate cell proliferation, survival, differentiation commitment, and end-cell functional activity; important to the inflammatory and immune response. Cell types capable of producing one or more CSFs include monocytes, T-lymphocytes, endothelial cells, fibroblasts and skin epithelial cells. Originally identified by their ability to stimulate the *in vitro* growth of hematopoietic progenitor cells into colonies in semisolid media: D. H. Pluznick, L. Sachs, *J. Cell Comp. Physiol.* **66**, 319 (1965); T. R. Bradley, D. Metcalf, *Aust. J. Exp. Biol. Med. Sci.* **44**, 287 (1966). Distinguished by the type(s) of progeny cells which result from stimulation by the specific factor, CSFs include: granulocyte-CSF (G-CSF), granulocyte-macrophage CSF (GM-CSF), macrophage-CSF (M-CSF) and multi-CSF or interleukin-3 (IL-3), *q.q.v.* G-CSF and M-CSF are thought to act on late progenitor cells already committed to their respective lineages. GM-CSF and IL-3 act on earlier, pluripotential progenitors. Potential clinical applications include the reduction of myelosuppression during cancer chemotherapy; augmentation of host defense in infection; and differentiation induction in hematologic malignancies. Reviews: D. Metcalf, *Science* **229**, 16-22 (1985); S. C. Clark, R. Kamen, *ibid.* **236**, 1229-1237 (1987); J. D. Griffin, *Oncology* **2**, 15-21 (1988). Role in infectious disease: R. M. Rose, *Sem. Oncol.* **19**, 415-421 (1992). Reviews of biological effects and clinical potential: J. L. Gabrilove, *Cancer Chemother. Biol. Resp. Mod.* **10**, 492-506 (1988); J. E. Groopman *et al.*, *N. Engl. J. Med.* **321**, 1449-1459 (1989); J. A. Hamilton, *Immunol. Today* **14**, 18-24 (1993).

2553. Colostrokinin. A substance occurring in the colostrum from which it is liberated by kallikrein. Colostrokinin lowers blood pressure and contracts the uterus and intestine. Isoln from cow colostrum: Werle, Trautschold, *Z. Biol.* **112**, 169-180 (1960), *C.A.* **55**, 1725 (1961); Yamazaki, Moriya, *Biochem. Pharmacol.* **18**, 2303 (1969). Properties: *eidem, ibid.* 2313.

2554. Colpormon. *(16α)-3,16-Bis(acetyloxy)estra-1,3,5-(10)-trien-17-one; 3,16α-dihydroxyestra-1,3,5(10)-trien-17-one diacetate;* 16α-hydroxyestrone diacetate; 3,16α-diacetoxy-$\Delta^{1,3,5}$-estratrien-17-one; Colpogynon. $C_{22}H_{26}O_5$; mol wt 370.45. C 71.33%, H 7.07%, O 21.59%. Prepn: Leeds *et al.*, *J. Am. Chem. Soc.* **76**, 2943 (1954); Fr. pat. **M1390** (1962 to Lab. Albert Rolland), *C.A.* **58**, 3484c (1963).

Crystals, mp 179-180°. $[\alpha]_D^{28}$ +122° (chloroform).

THERAP CAT: Estrogen.

2555. Coltsfoot. Coughwort. Dried leaves of *Tussilago farfara* L., *Compositae*. *Habit.* Northern Europe and Asia; naturalized in Northeastern U.S. *Constit.* Pectin, bitter glucoside, tannin, volatile oil, resin, saponin, caoutchouc.

2556. Colubrines. $C_{22}H_{24}N_2O_3$; mol wt 364.44. C 72.51%, H 6.64%, N 7.69%, O 13.17%. Minor alkaloids found in *Strychnos nux vomica L.*; naturally occurring in two chemically distinct forms, α and β. Isolation from mother liquor of commercial strychnine extraction: K. Warnat, *Helv. Chim. Acta* **14**, 997 (1931). Isolation and characterization from *S. nux vomica L.*: G. B. Marini-Bettolo *et al.*, *J. Assoc. Offic. Anal. Chem.* **51**, 185 (1968). Structure of α- and β-colubrine: S. P. Findlay, *J. Am. Chem. Soc.* **73**, 3008 (1951); of β-colubrine: P. Rosenmund, *Ber.* **95**, 2639 (1962); of α-colubrine: J. M. Culver, M. Sainsbury, *J. Chem. Res. (S)* **1987**, 304. Synthesis from strychnine: P. Rosenmund, H. Franke, *Ber.* **97**, 1677 (1964). Determn by GLC: N. G. Bisset, P. Fouché, *J. Chromatog.* **37**, 172 (1968); by HPLC: G. M. Iskander *et al.*, *J. Liq. Chromatog.* **5**, 1481 (1982). Chiroptical properties of β-colubrine: J. W. Snow, T. M. Hooker, Jr., *Can. J. Chem.* **56**, 1222 (1978).

α-Colubrine: R₁ = OCH₃ R₂ = H
β-Colubrine: R₁ = H   R₂ = OCH₃

α-Colubrine, 3-methoxystrychnidin-10-one, 11-methoxy-strychnine. Pyramids from ethyl acetate, mp 189-193°. $[\alpha]_D^{20}$ −72.4° (c = 0.9 in alc). uv max (ethanol): 255, 297 nm (log ε 4.03, 3.77). Sol in alc, benzene, chloroform.

β-Colubrine, 2-methoxystrychnidin-10-one, 10-methoxy-strychnine. Crystals from dil alcohol, mp 222°. Very bitter. $[\alpha]_D^{19}$ −107.7° (c = 2.5 in 80% alcohol). $[\alpha]_D^{21}$ −156° (c = 0.042 in chloroform). uv max (ethanol): 262, 297 nm (log ε 4.40, 3.80). Sol in alcohol, benzene, chloroform.

2557. Columbamine. *5,6-Dihydro-2-hydroxy-3,9,10-tri-methoxydibenzo[a,g]quinolizinium; 7,8,13,13α-tetradehydro-2-hydroxy-3,9,10-trimethoxyberbinium.* $[C_{20}H_{20}NO_4]^+$. Found in *Jatrorrhiza palmata* (DC.) Miers (*J. columba* Miers), *Menispermaceae:* Breslau, *Arch. Pharm.* **245**, 586 (1908); Späth, Burger, *Ber.* **59**, 1486 (1926); Reed, *Diss. Abstr.* **23**, 3638 (1963); from *Berberis lambertii* R. N. Parker, *Berberidaceae:* Chatterjee, Banerjee, *J. Indian Chem. Soc.* **30**, 705 (1953). Synthesis from berberine: Cava, Reed, *J. Org. Chem.* **32**, 1640 (1967).

The free base has not been described, but its quaternary salts are known.

Chloride hemipentahydrate, $C_{20}H_{20}ClNO_4 \cdot 2\frac{1}{2}H_2O$, orange-yellow needles, mp 194°. Sol in water, alcohol.

Chloride tetrahydrate, $C_{20}H_{20}ClNO_4 \cdot 4H_2O$, orange-yellow prisms, mp 184°.

Iodide, orange-yellow needles, mp 224°. Sol in water, alcohol.

2558. Columbin. *9-(3-Furanyl)decahydro-4-hydroxy-4a,10a-dimethyl-1,4-etheno-3H,7H-benzo[1,2-c:3,4-c']dipyran-3,7-dione; 15,16-epoxy-1β,4,12-trihydroxy-5,9-dimethyl-17,18-dinor-8βH,9βH,10α-labda-2,13(16),14-triene-19,20-dioic acid 19,1:20,12-dilactone.* $C_{20}H_{22}O_6$; mol wt 358.39. C 67.03%, H 6.19%, O 26.79%. Major bitter principle from the root of *Jatrorrhiza palmata* (DC.) Miers (*J. columba* Miers), *Menispermaceae.* Isoln: Wittstock, *Poggendorff's Ann.* **19**, 298 (1830). Structure: Barton, Elad, *J. Chem. Soc.* **1956**, 2090. Stereochemistry: Cava *et al., Tetrahedron Letters* **1959**, 1; Overton *et al., J. Chem. Soc. (C)* **1966**, 1482; Cheung *et al., ibid. (B)* **1966**, 853.

Very bitter needles from methanol, mp 195°. $[\alpha]_D$ +52.5° (pyridine). uv max (ethanol): 209 nm (log ε 3.78). Practically insol in water. One gram dissolves in 30 ml acetone, 50 ml ethyl acetate, 75 ml methanol; sol also in chloroform, methylene chloride. Bitterness threshold 1:60,000. Easily isomerized to isocolumbin.

Isocolumbin, $C_{20}H_{22}O_6$, needles from methanol, dec 190°. $[\alpha]_D^{17}$ +0.17° (alk alc). uv max (alc): 209 nm (log ε 3.80).

2559. Concanavalin A. ConA. The most extensively investigated member of the lectin family of plant proteins. Unlike most lectins, it lacks covalently bound carbohydrate and therefore is not a glycoprotein. Isolated from jack bean, *Canavalia ensiformis, Papilionatae:* J. B. Sumner, S. F. Howell, *J. Bacteriol.* **32**, 227 (1936). Its function in *C. ensiformis* is unknown, but it agglutinates a variety of somatic and germ line cells through specific interaction with saccharide-containing cell surface receptors and restores the growth pattern of virus-transformed fibroblasts in tissue culture to that of normal cells, *cf.* M. M. Burger, K. D. Noonan, *Nature* **228**, 512 (1970); G. M. Edelman, C. F. Millette, *Proc. Nat. Acad. Sci. USA* **68**, 2436 (1971). Differential toxicity on normal and transformed cells *in vitro* and inhibition of tumor development *in vivo* have also been reported: J. Shoham *et al., Nature* **227**, 1244 (1970). The molecule consists of identical polypeptide subunits of mol wt about 27,000, existing as dimers in soln at pH < 6 and as tetramers at physiologic pH. The proposed amino acid sequence contains 238 residues; con A has also been shown to have binding site for transition metal ions and calcium ions in addition to saccharide binding sites: G. M. Edelman *et al., Proc. Nat. Acad. Sci. USA* **69**, 2580 (1972). The transition ion, usually Mn⁺² or Ca⁺², apparently stabilizes the formation of the specific saccharide binding site: M. Shoham *et al., Biochemistry* **12**, 1914 (1973). Circular dichroism-NMR study of metal binding sites: A. R. Palmer *et al., ibid.* **19**, 5063 (1980). Oligosaccharide binding study: A. Vanlands *et al., Eur. J. Biochem.* **103**, 307 (1980). Use of con A to study immunoregulation of human T cells: D. M. Dwyer, C. Johnson, *Clin. Exp. Immunol.* **46**, 237 (1981); E. L. Larson *et al., Immunobiology* **161**, 5 (1982). For general refs, *see* Lectins.

USE: As a reagent in analytical and preparative biochemistry; as a probe in studies of cell surface membrane dynamics and cell division.

2560. Condurangin. Kondurangin. Bitter principle from the bark of the Condurango vine, *Marsdenia condurango* Reichb. f., *Asclepiadaceae,* also known as *eagle vine, mataperro, condor vine.* Habit. Ecuador, Peru. Isoln: Korte, Korte, *Z. Naturforsch.* **10b**, 223 (1955); Zechner, Zölss, *Sci. Pharm.* **24**, 107 (1956). Separation of fractions: Cellarius, Zechner, *ibid.* **34**, 10 (1966). Proposed structure of the aglycone, *condurangogenin A:* Tschesche *et al., Tetrahedron* **21**, 1777, 1797 (1965); **23**, 1461 (1967).

Crystals from dil methanol, mp 186-188°. uv max (methanol): 277 nm. Soluble in chloroform, methanol; slightly sol in water. Practically insol in ether, petr ether. Bitterness threshold 1:20,000.

THERAP CAT: Astringent.

2561. Conessine. *(3β)-N,N-Dimethylcon-5-enin-3-amine; 3β-(dimethylamino)con-5-enine;* neriine; roquessine; wrightine. $C_{24}H_{40}N_2$; mol wt 356.60. C 80.84%, H 11.31%, N 7.86%. Antiamebic, antifungal principle in Kurchi bark, the bark of *Holarrhena anti-dysenterica* Wall., *Apocynaceae,* native to India; also found in African species, such as *H. africana* A.DC.; *H. congolensis* Stapf; *H. wulfsbergii* Stapf, and *H. febrifuga* Klotzsch, *Apocynaceae*. Isoln: Haines, *Trans. Med. Soc. Bombay* **4**, 28 (1858); Warnecke, *Ber.* **19**, 60 (1886); Bertho, *Arch. Pharm.* **277**, 237 (1939); Siddiqui, Pillay, *J. Indian Chem. Soc.* **9**, 553 (1932). Structure: Haworth *et al., J. Chem. Soc.* **1953**, 1102; Haworth, McKenna, *Chem. & Ind. (London)* **1957**, 1510; Bertho, Götz, *Ann.* **619**, 96 (1958). Discussion of structure in Fieser, Fieser, *Steroids* (New York, 1959), p 858. Synthesis from Δ^5-pregnene-3β,-20β-diol: Barton, Morgan, *Proc. Chem. Soc.* **1961**, 206; *J. Chem. Soc.* **1962**, 622. Total synthesis: Marshall, Johnson, *J. Am. Chem. Soc.* **84**, 1485 (1962); Stork *et al., ibid.* 2018; Nagata *et al., Tetrahedron Letters* **1963**, 869.

Leaflets or broad plates from acetone, mp 127-128.5°. $[\alpha]_D^{20}$ +25.3° (c = 0.7 in abs alc). Sparingly sol in water.

Dihydrochloride monohydrate, $C_{24}H_{40}N_2 \cdot 2HCl \cdot H_2O$, crystals, sol in water, mp > 340°. $[\alpha]_D^{20}$ +9.3° (c = 0.9 in H_2O).

Dihydrobromide, $C_{24}H_{40}N_2 \cdot 2HBr$, crystals, dec 340°. Very bitter taste. $[\alpha]_D^{20}$ +7.0° (c = 5 in H_2O). Sol in water. Slightly sol in 95% ethanol. Very slightly sol in ether. Practically insol in petr ether.

2562. Congo Red. *3,3'-[[1,1'Biphenyl]-4,4'-diylbis(azo)-]bis[4-amino-1-naphthalenesulfonic acid] disodium salt; C.I. Direct Red 28;* sodium diphenyldiazo-bis-α-naphthylamine-sulfonate; C.I. 22120. $C_{32}H_{22}N_6Na_2O_6S_2$; mol wt 696.68. C 55.17%, H 3.18%, N 12.06%, Na 6.60%, O 13.78%, S 9.21%. Prepn: Böttiger, **DRP 28753** (1884); *Frdl.* **1**, 470; Whitehead, *Chem. Trade J.* **77**, 386 (1925); Kline, *J. Chem. Ed.* **15**, 129 (1938). Purification: Bedaux *et al., Pharm. Weekbl.* **98**, 189 (1963). *See also Colour Index* vol. 4 (3rd ed., 1971) p 4166; *H. J. Conn's Biol. Stains*, R. D. Lillie, Ed. (Williams & Wilkins., Baltimore, 9th ed., 1977) p 148. Toxicity: Richardson, Dillon, *Am. J. Med. Sci.* **198**, 73 (1939). Clinical evaluation as diagnostic aid: E. Ouchi *et al., Tohoku J. Exp. Med.* **118**, Suppl., 191 (1976). Use in combination with Sirius red as histological stain: I. R. Hinds, *Lab. Med.* **16**, 366 (1985).

Brownish-red powder. Absorption max (pH 7.3): 488 nm (E$_{1cm}^{1\%}$ 595). Sol in water (yellowish-red) and in ethanol (orange); very slightly sol in acetone. Practically insol in ether. LD_{50} i.v. in rats: 190 mg/kg (Richardson, Dillon).

Congo Red Paper, *Riegel's paper, Herzberg's paper.* Paper charged with a 0.1% aq soln of Congo Red.

USE: As indicator, usually in 0.1% aq soln for estimating free mineral acids, particularly in presence of organic acids. pH: 3.0 blue-violet, 5.0 red. Detecting and estimating free HCl in gastric contents; detecting acidity of papers. As addition to culture media; reagent for bitter-almond water; dye; biological stain.

THERAP CAT: Diagnostic aid (amyloidosis).

2563. Congressane. *Decahydro-3,5,1,7-[1,2,3,4]butane-tetraylnaphthalene;* pentacyclo[7.3.1.1^{4,12}.0^{2,7}.0^{6,11}]tetradecane; diamantane. $C_{14}H_{20}$; mol wt 188.31. C 89.30%, H 10.70%. Isoln from petroleum: Hala *et al., Angew. Chem. Int. Ed.* **5**, 1045 (1966). Synthesis: Cupas *et al., J. Am.*

Chem. Soc. **87**, 917 (1965); Gund *et al., J. Org. Chem.* **39**, 2979 (1974). Structure: Karle, Karle, *J. Am. Chem. Soc.* **87**, 918 (1965).

Crystals, mp 236-237°.

Note: Structure is congress emblem of the International Congress of Pure and Applied Chemistry.

2564. Conhydrine. $[R-(R^*,S^*)]$-α-Ethyl-2-piperidine-methanol; 2-(α-hydroxypropyl)piperidine. $C_8H_{17}NO$; mol wt 143.23. C 67.09%, H 11.96%, N 9.78%, O 11.17%. From seeds of *Conium maculatum* L., *Umbelliferae*. Isoln: Späth, Adler, *Monatsh.* **63**, 127 (1933). Stereochemistry: Hill, *J. Am. Chem. Soc.* **80**, 1609 (1958); Sicher, Ticky, *Coll. Czech. Chem. Commun.* **23**, 2081 (1958).

Crystals from ether, mp 121°. bp 226°. $[\alpha]_D$ +10°. Slightly sol in water; sol in alc, chloroform, ether.

2565. β-Coniceine. *2-(1-Propenyl)piperidine;* α-allyl-piperidine. $C_8H_{15}N$; mol wt 125.21. C 76.74%, H 12.07%, N 11.19%. Prepn from conhydrine: Löffler, Friederick, *Ber.* **42**, 107 (1909); Löffler, Tschunke, *ibid.* 929.

dl-Form, liquid, solidifies around 8°. bp 168°. d$_4^{15}$ 0.8716.
dl-Form hydrochloride, $C_8H_{15}N \cdot HCl$, mp 206°.
d-Form, needles, mp 39°, bp 168°. $[\alpha]_D^{45}$ +50°.
d-Form hydrochloride, mp 182°.
l-Form, crystals, mp 40°, bp 168°. $[\alpha]_D^{45}$ −50°.
l-Form hydrochloride, mp 183°.
All 3 hydrochlorides are sol in water. The free bases are sol in alcohol and ether; very slightly in water.

2566. γ-Coniceine. *2,3,4,5-Tetrahydro-6-propylpyridine;* 2-*n*-propyl-3,4,5,6-tetrahydropyridine; 2-*n*-propyl-Δ^1-piperidine. $C_8H_{15}N$; mol wt 125.21. C 76.74%, H 12.07%, N 11.19%. Easily reduced to *dl*-coniine (*q.v.*). From seeds of *Conium maculatum* L., *Umbelliferae*: Cromwell, *Biochem. J.* **64**, 259 (1956); Fairbairn, Challen, *ibid.* **72**, 556 (1959); Fairbairn, Sewal, *Phytochemistry* **1**, 38 (1961). Structure: Beyerman *et al., Rec. Trav. Chim.* **80**, 513 (1961). Synthesis: Lukes *et al., Coll. Czech. Chem. Commun.* **12**, 641 (1947); Cervinka, *Chem. Listy* **52**, 1145 (1958).

Alkaline liquid, mousy odor. bp 171°; bp$_{15}$ 63°. Volatile with steam. d$_4^{15}$ 0.8753. n$_D^{16}$ 1.4661. Slightly sol in water; freely sol in alcohol, chloroform, ether.

Hydrochloride, $C_8H_{15}N \cdot HCl$, hygroscopic crystals from ether, mp 143°.

2567. Coniferin. *4-(3-Hydroxy-1-propenyl)-2-methoxyphenyl β-D-glucopyranoside;* 4-hydroxy-3-methoxy-1-(γ-hydroxypropenyl)benzene-4-D-glucoside; abietin; laricin. $C_{16}H_{22}O_8$; mol wt 342.35. C 56.14%, H 6.48%, O 37.39%. Principal glucoside of the conifers. Also in comfrey root, sugar beet, and asparagus. Extraction from the cambium layer of fir: Solntsev, *C.A.* **38**, 3780 (1944). Synthesis: Pauley, Feuerstein, *Ber.* **60**, 1031 (1927). Hydrolysis by emulsin yields coniferyl alcohol and D-glucose. Yields lignin-like material by enzymatic dehydrogenation and polymerization: Freudenberg *et al.*, *Ber.* **85**, 641 (1952); Freudenberg, Rasenack, *Ber.* **86**, 756 (1953).

Dihydrate, crystals from water. Anhydr after 4 hrs at 100°. Anhydr coniferin, mp 186°. $[\alpha]_D^{20}$ −68° (c = 0.5). Absorption spectrum: Patterson, Hibbert, *J. Am. Chem. Soc.* **65**, 1862 (1943). One gram dissolves in 200 ml water, freely sol in boiling water; sol in pyridine; sparingly sol in alc. Practically insol in ether.

2568. Coniferyl Alcohol. *4-(3-Hydroxy-1-propenyl)-2-methoxyphenol; 3-(4-hydroxy-3-methoxyphenyl)-2-propen-1-ol;* 4-hydroxy-3-methoxycinnamic alcohol; γ-hydroxyisoeugenol. $C_{10}H_{12}O_3$; mol wt 180.20. C 66.65%, H 6.71%, O 26.64%. In benzoin, especially Siam benzoin, as coniferyl benzoate (up to 78%): Reinitzer, *Arch. Pharm.* **264**, 131 (1926). From coniferin by hydrolysis with emulsin. From ferula aldehyde (coniferyl aldehyde) by the action of fermenting yeast. Small amounts of coniferyl alcohol are found in brandy from beet molasses.

Prisms, mp 74°. Freely sol in ether; moderately sol in alc; almost insol in water; sol in alkalies. Polymerized by dil acids; converted to an amorphous gum. Absorption spectrum: Herzog, Hillmer, *Ber.* **64B**, 1288 (1931).

2569. Coniine. *(S)-2-Propylpiperidine;* cicutine; conicine. $C_8H_{17}N$; mol wt 127.23. C 75.52%, H 13.47%, N 11.01%. Toxic principle of poison hemlock, *Conium maculatum* L., *Umbelliferae:* Ladenburg, *Ber.* **19**, 439 (1886); Koller, *Monatsh.* **47**, 393 (1926). Occurs naturally as the *(S)*-(+)-isomer. Review of early literature: Marion in *The Alkaloids*, **vol. I**, R. H. F. Manske, H. L. Holmes, Eds. (Academic Press, New York, 1950) pp 211-217. Isoln from the pitcher plant, *Sarracenia flava* and insect paralyzing properties: N. V. Mody *et al. Experientia* **32**, 829 (1976). Absolute configuration: J. C. Craig, S. K. Roy, *Tetrahedron* **21**, 401 (1965). Resolution of (±)-form: J. C. Craig, A. R. Pinder, *J. Org. Chem.* **36**, 3648 (1971). Synthesis of *(S)*-(+)-form: K. Aketa *et al., Chem. Pharm. Bull.* **24**, 621 (1976); of *(R)*-(−)-form: D. Lathbury, T. Gallagher, *Chem. Commun.* **1986**, 114; of (±)-form: T. Nagasaka *et al., Heterocycles* **27**, 1685 (1988). Enantiospecific synthesis of (+)- and (−)-forms: L. Guerrier *et al., J. Am. Chem. Soc.* **105**, 7754 (1983).

Colorless alkaline liquid, darkens and polymerizes on exposure to light and air. Mousy odor. mp ∼−2°. bp 166°-166.5°; bp$_{20}$ 65-66°. Volatile with steam. d_4^{20} 0.844-0.848. n_D^{23} 1.4505. $[\alpha]_D^{25}$ +8.4° (c = 4.0 in CHCl$_3$); $[\alpha]_D^{20}$ +14.6° (neat). pK 3.1. One ml dissolves in 90 ml water, less sol in hot water. The base dissolves about 25% water at room temp. Sol in alcohol, ether, acetone, benzene, amyl alcohol; slightly sol in chloroform.

Hydrobromide, $C_8H_{17}N.HBr$, prisms, mp 211°. One gram dissolves in 2 ml water, 3 ml alcohol; sol in chloroform, ether.

Hydrochloride, $C_8H_{17}N.HCl$, rhomboids, mp 221°. Freely sol in water, alcohol, chloroform.
(R)-(−)-Form, liquid, bp$_{756}$ 165°. $[\alpha]_D^{25}$ −8.1° (c = 4.0 in CHCl$_3$); $[\alpha]_D^{23}$ −14.2° (neat).
(±)-Form, bp 200-210°.
Caution: Potential symptoms of overexposure include drowsiness, paresthesias, weakness, ataxia, nausea, profuse salivation, and bradycardia followed by tachycardia. *See: Clinical Toxicology of Commercial Products*, R. E. Gosselin *et al.*, Eds. (Williams & Wilkins, Baltimore, 5th ed., 1984) Section II, pp 249-250.

2570. Conium Fruit. Hemlock; poison hemlock; spotted hemlock; poison parsley; spotted cowbane. Full-grown, but unripe, carefully dried fruit of *Conium maculatum* L., *Umbelliferae*. *Habit.* Europe, Asia, naturalized in U.S. *Constit.* 0.5-1.5% coniine, conhydrine, pseudoconhydrine, methylconiine, ethylpiperidine, coniic acid, volatile and fixed oil. Has antispasmodic activity.

2571. Conjugated Estrogenic Hormones. Amnestrogen; Conestron; Genisis; Premarin. An amorphous preparation contg water-soluble, conjugated forms of mixed estrogens obtained from urine of pregnant mares. The principal estrogen present is sodium estrone sulfate. The total estrogenic potency of the preparation is expressed in terms of an equiv quantity of sodium estrone sulfate. Prepn: Bates, Cohen, U.S. pat. **2,565,115** (1951 to Squibb); Stiller, O'Keefe, U.S. pat. **2,720,483** (1955 to Olin Mathieson).
Note: This substance has been listed as a known carcinogen: *Seventh Annual Report on Carcinogens* (PB95-109781, 1994) p 51.
THERAP CAT: Estrogen.

2572. Conquinamine. $C_{19}H_{24}N_2O_2$; mol wt 312.41. C 73.05%, H 7.74%, N 8.97%, O 10.24%. Stereoisomer of quinamine, *q.v.*, from which it differs in configuration at one or more of the asymmetric centers. From bark of *Cinchona pubescens* Vahl. (*C. succirubra* Pav.), and *C. rosulenta* How., *Rubiaceae*. Isoln: Hesse, *Ber.* **10**, 2158 (1877); *Ann.* **209**, 62 (1881); Oudemans, *ibid.* **38**. Identity with epiquinamine: Culvenor *et al., J. Chem. Soc.* **1950**, 1485.
Prisms, mp 123°. $[\alpha]_D^{15}$ +200° (alcohol). Slightly sol in water; freely sol in alcohol, chloroform, ether.

2573. Convallamarogenin. $C_{27}H_{42}O_4$; mol wt 430.63. C 75.31%, H 9.83%, O 14.86%. Main genin of the glycoside, *convallamarin*. From root of *Convallaria majalis* L., *Liliaceae*. Isoln and structure: Tschesche *et al., Ber.* **94**, 1699 (1961). From *Reineckia carnea* Kunth., *Liliaceae*: Takeda *et al., Tetrahedron* **19**, 759 (1963).

Prisms from chloroform + methanol, mp 259-261°. $[\alpha]_D^{27}$ −79.1° (c = 1.02 in CHCl$_3$).
Diacetate, $C_{31}H_{46}O_6$, prisms, mp 214-215°. $[\alpha]_D^{27}$ −88.2° (c = 1.10 in CHCl$_3$).

2574. Convallaria. Lily of the valley; May lily; Park lily; May blossom. Dried flowers of *Convallaria majalis* L., *Liliaceae*. *Habit.* U.S., Europe, Northern Asia; cultivated in gardens. *Constit.* of dried flowers: Convallatoxin, convallarin, volatile oil; of dried rhizome and roots: Glycosides of convallamarin, convallatoxin, convallarin. Extraction and identification of glycosides: Tschesche, Seehofer, *Ber.* **87**, 1108 (1954); Erbring, Patt, *Arzneimittel-Forsch.* **8**, 554 (1958).
Convallaria glycosides, Convacard, Convallan, Convallen, Convalyt, Convasid. An aqueous extract containing the total glycosides of Convallaria.
THERAP CAT: Cardiotonic.

THERAP CAT (VET): Has been used as a cardiotonic and diuretic.

2575. Convallatoxin. *(3β,5β)-3-[(6-Deoxy-α-L-manno-pyranosyl)oxy]-5,14-dihydroxy-19-oxocard-20(22)-enolide;* strophanthidin α-L-rhamnoside; Convallaton; Corglykon; Korglykon. $C_{29}H_{42}O_{10}$; mol wt 550.65. C 63.26%, H 7.69%, O 29.06%. From blossoms of lily of the valley *(Convallaria majalis* L., *Liliaceae):* Karrer, *Helv. Chim. Acta* **12**, 506 (1929); from *Ornithogalum umbellatum* L., *Liliaceae:* Mrozik *et al., ibid.* **42**, 683 (1959); from *Antiaris toxicaria* Lesch, *Moraceae:* Juslen *et al., ibid.* **46**, 117 (1963). Structure: Tschesche, Haupt, *Ber.* **69**, 459 (1936); Fieser, Jacobsen, *J. Am. Chem. Soc.* **59**, 2335 (1937). Cleavage into L-rhamnose and strophanthidin: Reichstein, Katz, *Pharm. Acta Helv.* **18**, 521 (1943); *see also C.A.* **38**, 2046. Synthesis from strophanthidin and acetobromrhamnose: Reyle *et al., Helv. Chim. Acta* **33**, 1541 (1950); Haede *et al.,* **Ger. pat. 1,933,-090** (1971 to Hoechst). Toxicity study: W. Förster *et al., Arch. Int. Pharmacodyn.* **155**, 165 (1965).

Prisms from methanol + ether. mp 235-242°. $[\alpha]_D^{22}$ −1.7° ± 3° (c = 0.65 in methanol); $[\alpha]_D^{25}$ −9.4° + 3° (c = 0.72 in dioxane). Sol in alcohol, acetone; slightly sol in chloroform, ethyl acetate, and water (1:2000). Practically insol in ether, petr ether. Gives Legal's reaction. LD_{50} in mice, rats (mg/kg): 10.0 i.p., 16.0 i.v. (Förster).

Tri-*O*-acetyl-convallatoxin, $C_{35}H_{48}O_{13}$, needles from acetone + ether, mp 215-238°. $[\alpha]_D^{25}$ −5.5° ± 2° (c = 0.962 in chloroform).

THERAP CAT: Cardiotonic.

2576. Convicine. *6-Amino-5-(β-D-glucopyranosyloxy)-2,4(1H,3H)-pyrimidinedione.* $C_{10}H_{15}N_3O_8$; mol wt 305.24. C 39.35%, H 4.95%, N 13.77%, O 41.93%. From seeds of *Vicia sativa* L., *Leguminosae:* Ritthausen, *J. Prakt. Chem.* [2] **24**, 202 (1881); *Ber.* **29**, 894 (1896). Structure: Johnson, *J. Am. Chem. Soc.* **36**, 337 (1914); Fisher, Johnson, *ibid.* **54**, 2038 (1932); Bendick, Clements, *Biochim. Biophys. Acta* **12**, 462 (1953); Bien *et al., J. Chem. Soc.* (C) **1968**, 496. Synthesis: *eidem, J. Chem. Soc. Perkin Trans. I* **1973**, 1089.

Leaflets from boiling water, dec without melting at 287°. uv max (pH 7): 245, 271 nm (log ε 3.43, 4.16). Sol in hot water, dil NaOH; slightly sol in cold water, alcohol. Practically insol in chloroform, glacial acetic acid.

2577. COP. *2-Methyl-2-propenoic acid oxiranylmethyl ester polymer with ethyl 2-propenoate;* poly(glycidyl methacrylate-co-ethyl acrylate); [P(GMA-co-EA)]; glycidyl methacrylate-ethyl acrylate polymer; COP Resist. $(C_7H_{10}-O_3.C_5H_8O_2)_x$. Negative electron resist. Prepn and characterization: L. F. Thompson *et al., Tech. Pap., Reg. Tech*

Conf., Soc. Plast. Eng. **1973**, (Oct.) 127. Molecular parameters and lithographic performance: L. F. Thompson *et al., J. Vac. Sci. Technol.* **12**, 1280 (1975). Radiation chemistry: E. D. Feit *et al., ACS Org. Coat. Plast. Chem.* **35**, 287 (1975). Use as matrix for deep UV resist materials: C. Kutal, S. K. Weit, *J. Coating Technol.* **62**, 63 (1990); S. K. Weit *et al., Chem. Mater.* **4**, 453 (1992).

USE: In photosensitive materials as polymer matrix. Photoresist.

2578. COP 1. *L-Glutamic acid polymer with L-alanine, L-lysine and L-tyrosine;* copolymer 1; COP 1 (polyamide); Copaxone. Random basic synthetic copolymer of the L-alanine, L-lysine, L-glutamic acid and L-tyrosine in a molar ratio of 6:1.9:4.7:1. Mol wt of 14,000 to 23,000 Da. Prepn: D. Teitelbaum *et al., Eur. J. Immunol.* **1**, 242 (1971). Suppression of exptl. allergic encephalomyelitis (EAE): *eidem, ibid.* **3**, 273 (1973). Inhibition of T-cell lines for myelin basic protein: D. Teitelbaum *et al., Proc. Nat. Acad. Sci. USA* **89**, 137 (1992); of response of major histocompatibility complex: M. Fridkis-Hareli *et al., ibid.* **91**, 4872 (1994). Clinical trial in multiple sclerosis: M. B. Bornstein *et al., Neurology* **41**, 533 (1991); multicenter trial: K. P. Johnson *et al., Neurology* **45**, 1268 (1995). Review of pharmacology and early clinical evaluation: L. A. Rolak, *Clin. Neuropharmacol.* **10**, 389 (1987); suppression of EAE: R. Arnon, *Israel. J. Med. Sci.* **25**, 686-689 (1989).

THERAP CAT: In treatment of multiple sclerosis.

2579. Copaene. *[1R-(1α,2α,6α,7α,8α)]-1,3-Dimethyl-8-(1-methylethyl)tricyclo[4.4.0.0²,⁷]dec-3-ene; (1R,2S,6S,7S,8S)-(−)-8-isopropyl-1,3-dimethyltricyclo[4.4.0.0²,⁷]dec-3-ene;* α-copaene. $C_{15}H_{24}$; mol wt 204.36. C 88.16%, H 11.84%. Tricyclic sesquiterpene occurring in African copaiba balsam oil; in the so-called supa oil from *Sindora supa* Merr. (*S. wallichii* F.-Vill.), *Leguminosae:* Henderson *et al., J. Chem. Soc.* **1926**, 3077. Also in the essential oil of *Phyllocladus trichomanoides* D. Don, *Podocarpaceae:* Briggs, Sutherland, *J. Org. Chem.* **13**, 1 (1948); from *Cyperus articulatus* Michx., *Cyperaceae:* Couchman *et al., Tetrahedron* **20**, 2037 (1964). Structural studies: Kapadia *et al., ibid.* **21**, 607 (1965). Absolute stereostructure: deMao *et al., ibid.* 619. Synthesis: Heathcock, *J. Am. Chem. Soc.* **88**, 4110 (1966); Heathcock *et al., ibid.* **89**, 4133 (1967); Corey, Watt, *ibid.* **95**, 2303 (1973). *Review:* J. Simonsen, D. H. R. Barton, *The Terpenes* **vol. III** (University Press, Cambridge, 1952) pp 88-91. *See also* Ylangene.

Oily liquid, bp 246-251°; bp_{10} 119-120°; d_{15}^{15} 0.9077; n_D^{20} 1.4894. $[\alpha]_D^{22}$ −6.3° (c = 1.20 in chloroform).

2580. Copaiba. Balsam copaiba; balsam capivi; Jesuit's balsam. Oleoresin from South American species of *Copaifera (Copaiba), Leguminosae. Habit.* Brazil, Venezuela, Colombia, especially the Amazon valley and banks of Orinoco. *Constit.* Volatile oil, resin; illuric and metacopaivic acid (in Maracaibo balsam); copaivic and oxycopaivic acids (in Para balsam). Brief description of balsam copaiba and its constituents: J. A. Wenninger *et al., J. Assoc. Offic. Anal. Chem.* **50**, 1304 (1967); D. L. J. Opdyke, *Food Cosmet. Toxicol.* **14**, 687 (1976); of copaiba oil: *idem, ibid.* **11**, 1075 (1973).

Transparent, viscid, pale yellow to brownish-yellow liq; peculiar odor; bitter, acrid, nauseating taste. d 0.930- 0.995. Acid no. 28-95. Insol in water. Sol in benzene, chloroform, ether, oils, carbon disulfide, abs alcohol, petr ether, partly in 95% alcohol. *Incompat.* Mineral acids, magnesia, water.

USE: In varnishes; for removing old oil varnish from oil paintings; manuf photographic paper.

2581. Copal. Resin copal; gum copal; anime (soft copal); kaurie; cowrie. A resin found as a fossil in Zaire, or exuding from various species of *Trachylobium, Hymenaea courbaril* L., etc., *Leguminosae.* Occurs as hard or soft copals. *Habit.* Zanzibar, Mozambique, also S. America, Australia, Philippine Islands and West Indies. *Constit.* Zanzibar copal contains 80% trachylolic acid, 4% isotrachylolic acid, 6% resene and volatile oil. Kaurie copal contains dammaric acid, dammaran and a resin.

Yellowish to yellowish-brown pieces of various sizes; conchoidal fracture; odorless and tasteless. Hard copals are almost insol in usual solvents; soft copals are partly sol in alcohol, chloroform or glacial acetic acid; both copals after having been fused are sol in oil turpentine and linseed oil.

USE: In varnishes and cements; as substitute for amber; manuf oil cloths and linoleum. In dentistry, for modeling compds and cavity varnishes.

2582. Coparaffinate. *Isopar.* A mixture of water-insol isoparaffinic acids partially neutralized with hydroxybenzyl-dialiphatic amines. The water-insol isoparaffinic acids are obtained by oxidation of petroleum hydrocarbons by the passage of a current of oxygen under pressure at an elevated temp in the presence of a metal catalyst. The water-insol mono- and dicarboxylic acids with from 6 to 16 carbon atoms are separated and purified by fractional distillation. The hydroxybenzyldialiphatic amines are combined directly with the isoparaffinic acids or in suitable solvent. The latter is then removed by distillation: C. E. Earle, U.S. pat. **2,-262,720** (1941).

Dark-brown, viscous, oily liq. Odor of burnt petroleum. d 0.970-0.980. Immiscible with water. Freely sol in alc, volatile and fixed oils. Marketed as a 17% ointment, which also contains 4% titanium dioxide in an ointment base consisting of beeswax, cetyl alcohol, lanolin, and petrolatum.

THERAP CAT: Antifungal (topical).

2583. Copper. Cu; at. wt 63.546; at. no. 29; valences 1, 2. Occurrence in the earth's crust: 70 ppm; also present in seawater: 0.001-0.02 ppm. Two naturally occurring isotopes: 63 (69.09%), 65 (30.91%); nine artificial isotopes: 58-62, 64, 66-68. One of the earliest known metals. Found in nature in its native state; also in combined form in several minerals including chalcopyrite, chalcocite, bornite, *q.q.v.,* *tetrahedrite* ($Cu_{12}Sb_4S_{13}$), enargite (Cu_3AsS_4), antlerite. Extraction from ores: Clark-Hawley, *Encyclopedia of Chemistry* (Reinhold, New York, 2nd ed., 1966) p 288. Metallurgy of copper and its alloys: *Metals Reference Book,* **vols. 1, 2,** C. J. Smithells, Ed. (Butterworth's, London, 3rd ed., 1962). A trace element essential to many plants and animals. Occurs in biological complexes such as *pheophytin* (analog of chlorophyll), hemocyanin, tyrosinase and ceruloplasmin, *q.q.v.* Reviews of copper and copper compounds: *Copper,* A. Butts, Ed., A.C.S. Monograph Series, no. **122** (Reinhold, New York, 1954) 936 pp; Massey, "Copper" in *Comprehensive Inorganic Chemistry,* **vol. 3,** J. C. Bailar, Jr. *et al.,* Eds. (Pergamon Press, Oxford, 1973) pp 1-78; W. M. Tuddenham, P. A. Dougall, Kirk-Othmer, *Encyclopedia of Chemical Technology* **vol. 6** (Wiley-Interscience, New York, 3rd ed., 1979) pp 819-869. Book: *Inflammatory Diseases and Copper,* J. R. J. Sorenson, Ed. (Humana Press, Clifton, N J, 1982) 622 pp.

Reddish, lustrous, ductile, malleable metal; face-centered cubic structure; commercially available in the form of ingots, sheets, wire or powder. Becomes dull when exposed to air. In moist air gradually becomes coated with green basic carbonate. d 8.94. mp 1083°. bp 2595°. Mohs' hardness 3.0. Resistivity 1.673 microohm-cm. Heat of fusion 48.9 cal/g; heat of vaporization 1150 cal/g. Heat capacity at constant pressure (solid) 0.092 cal/g/°C (20°), (liq) 0.112 cal/g/°C. E^0 (aq) Cu^+/Cu +0.521 V; E^0 (aq) Cu^{2+}/Cu +0.337 V. Very slowly attacked by cold hydrochloric or dil sulfuric acid; readily by dil nitric acid, and by both hot concd H_2SO_4 and HBr. It is also attacked by acetic and other organic acids. Slowly sol in ammonia water. Water-soluble cupric salts yield with sodium hydroxide a bluish-green precipitate of cupric hydroxide which is changed to black cupric oxide on warming. Potassium ferrocyanide produces a brownish-red precipitate of copper ferrocyanide. Hydrogen sulfide produces in acid solns a black precipitate of cupric sulfide which is sol in soln of sodium cyanide. Aluminum, iron or zinc precipitate metallic copper from its solns.

Caution: Copper itself probably has little or no toxicity, although there are conflicting reports in the literature. Soluble salts, notably copper sulfate, are strong irritants to skin, mucous membranes. Copper oxide fumes can cause metal fume fever. A relationship between copper and hemochromatosis has been reported. *See* E. Browning, *Toxicity of Industrial Metals* (Appleton-Century-Crofts, New York, 2nd ed., 1969) pp 145-152.

USE: Manuf bronzes, brass, other copper alloys, electrical conductors, ammunition, copper salts, works of art.

2584. Copper Phthalocyanine. *(SP-4-1)-[29H,31H-Phthalocyaninato(2—)-$N^{29},N^{30},N^{31},N^{32}$]copper;* C.I. Pigment Blue 15; C.I. 74160. $C_{32}H_{16}CuN_8$; mol wt 576.08. C 66.72%, H 2.80%, Cu 11.03%, N 19.45%. Discoverers: Dandridge *et al.,* **Brit.** pat. **322,169** (1928 to Scottish Dyes). Prepn: Baumann *et al., Angew. Chem.* **68,** 133 (1956); Sanielevici *et al., Rec. Chim. (Bucharest)* **12,** 281 (1961); Raab, Hoernle, **Brit.** pat. **930,150** (1963 to Bayer). Purification: Heinle, Mau, **Ger.** pat. **1,167,307** (1964 to G. Siegle); Sanielevici *et al.,* **Belg.** pat. **657,307** (1965 to Romania, Ministry of Petroleum and Chemical Industry); H. Tomoda *et al., Chem. Letters* **1980,** 1277. Properties: Dent, Linstead, *J. Chem. Soc.* **1934,** 1027; Easton, Smith, *J. Oil Colour Chem. Assoc.* **49,** 614 (1966). *See also Colour Index,* vol. 4 (3rd ed., 1971) p 4618.

Bright blue microcrystals with purple lustre; sol in 98% H_2SO_4 from which it can be almost quantitatively pptd by dilution with water. Practically insol in water, alcohol, and hydrocarbons. Stable toward heat (crystalline and analytically pure sublime obtained at about 580° in low pressure atmosphere of nitrogen or carbon dioxide), alkalies, dilute acids. Dec by hot nitric acid, or dilute acid permanganate to yield phthalimide.

Exists in two forms: the thermodynamically less stable, redder α-form is a better pigment than the more stable greener β-form. In the presence of aromatic solvents, heat, high shear, etc., α-form is converted to β-form.

USE: In inks and paints. Approved by FDA for use in polypropylene sutures.

2585. Coproergostane. *5β-Ergostane;* pseudoergostane. $C_{28}H_{50}$; mol wt 386.71. C 86.97%, H 13.03%. Prepn from cholanic acid: Fernholz, *Ber.* **69,** 1792 (1936).

Needles from acetone. mp 64°. $[\alpha]_D^{19}$ +25.3° (c = 2 in chloroform).

2586. Coprogen. $C_{35}H_{53}FeN_6O_{13}$; mol wt 821.68. C 51.16%, H 6.50%, Fe 6.80%, N 10.23%, O 25.31%. Iron-contg, growth-promoting complex. Isoln from *Penicillium* spp: Hesseltine *et al., J. Am. Chem. Soc.* **74,** 1362 (1952). Pidacks *et al., ibid.* **75,** 6064 (1953); Gaeumann, Vischer,

U.S. pat. **3,297,526** (1967 to Ciba). Isoln from *Aspergillaceae:* Zaehner *et al., Arch. Mikrobiol.* **45,** 119 (1963). Structure: Keller-Schierlein, Diekmann, *Helv. Chim. Acta* **53,** 2035 (1970).

Clusters of brick-red needles from ethanol. Absorption max (ethanol): 217, 250, 440 nm (log ϵ 4.46, 4.22, 3.47). Sol in water, methanol, ethanol (amorphous coprogen only), propanol, benzyl alcohol, Cellosolves. Practically insol in ether, ethyl acetate, chloroform, benzene, Cellosolve esters.

Desferricoprogen, iron-free coprogen, may be obtained by treating an aq solution with 8-quinolinol yielding a light brown powder.

USE: Coprogen as growth-promoting agent in various organisms. Desferricoprogen to bring about excretion of iron.

2587. Coprostane. *5β-Cholestane;* pseudocholestane. $C_{27}H_{48}$; mol wt 372.68. C 87.02%, H 12.98%. The *cis*-decalin homolog of cholestane. Prepn: Wieland, Jacobi, *Ber.* **59,** 2064 (1926); Young *et al., J. Am. Chem. Soc.* **81,** 1452 (1959); Dart, Henbest, *J. Chem. Soc.* **1960,** 3563; Nickon, Bagli, *J. Am. Chem. Soc.* **83,** 1498 (1961); Caglioti, Grasselli, *Chem. & Ind. (London)* **1964,** 153.

Needles from alcohol, mp 72°. $[\alpha]_D^{11}$ +25.1° (c = 2 in CHCl$_3$). n_D^{88} 1.4884. Freely soluble in ether, chloroform; slightly in abs alcohol.

2588. Coprosterol. *(3β,5β)-Cholestan-3-ol;* 3β-coprostanol; stercorin. $C_{27}H_{48}O$; mol wt 388.68. C 83.44%, H 12.45%, O 4.12%. Found in feces of man and of carnivorous animals: Bondzynski, Humnicki, *Z. Physiol. Chem.* **22,** 396 (1896); Wells *et al., Arch. Biochem. Biophys.* **57,** 437 (1955); Samuel *et al., J. Chromatog.* **14,** 508 (1964). Prepn from coprostenone: Ruzicka *et al., Helv. Chim. Acta* **17,** 1407 (1934); Shoppee, Summers, *J. Chem. Soc.* **1950,** 687. From coprostenol: Schönheimer, Evans, *J. Biol. Chem.* **114,** 567 (1936); Ruzicka *et al., Helv. Chim. Acta* **21,** 498 (1938); Agashe, Summers, *J. Chem. Soc.* **1957,** 3107; Dart, Henbest, *ibid.* **1960,** 3563. Prepn from cholesterol: Rosenfeld, Gallagher, *Steroids* **4,** 515 (1964).

Needles from methanol, mp 101°. $[\alpha]_D^{18}$ +28° (c = 1.8 in CHCl$_3$). Freely sol in ether, chloroform, benzene; slightly sol in methanol (one gram dissolves in 145 ml MeOH). Insol in water.

2589. Coptine. From root of *Coptis trifolia* Salisb., *Ranunculaceae:* Gross, *Am. J. Pharm.* **45**[4], 193 (1873); Mollett, Christensen, *J. Am. Pharm. Assoc.* **23,** 310 (1934); from *C. teeta* Wall.: Chatterjee *et al., J. Indian Chem. Soc.* **28,** 97 (1951); from *C. chinensis* Wils.: Schramm, Tang, *Pharmazie* **14,** 405 (1959).

Crystals. Practically insol in water, ammonia, alkali hydroxides; sol in hot alcohol, dil acids.

2590. Coptis. Goldthread. Dried plant of *Coptis trifolia* Salisb., *Ranunculaceae.* Habit. U.S., northern and middle Atlantic States; Canada. *Constit.* Berberine, coptine.

THERAP CAT: Bitter tonic.

2591. Coptisine. *6,7-Dihydrobis[1,3]benzodioxolo[5,6-a:4',5'-g]quinolizinium; 7,8,13,13a-tetrahedro-2,3:9,10-bis-methylenedioxyberbinium;* bis[methylenedioxy]protoberberine. $[C_{19}H_{14}NO_4]^+$; C 71.24%, H 4.40%, N 4.37%, O 19.98%. From root of *Coptis japonica* Makino, *Ranunculaceae.* Extraction procedure: Kitasato, *C.A.* **21,** 2700 (1927). Structure: Späth, Posega, *Ber.* **62,** 1029 (1929); *cf.* Huang-Minlon, *Ber.* **69,** 1744 (1936). Prepn from rhoeageninediol: Klásek *et al., Tetrahedron Letters* **1968,** 4549.

Hydroxide, $C_{19}H_{15}NO_5$, yellowish needles from alc, mp 218°. Absorption spectrum: Kitasato, *Acta Phytochim.* **3,** 175 (1927). Very slightly sol in water; sparingly in alcohol; sol in alkalies.

Chloride, orange prisms, not melted at 300°.

Iodide, yellow needles, dec above 280°.

Sulfate, yellow crystals, insol in water and alcohol (enables separation from berberine and worenine).

2592. Cord Factor. Trehalose 6,6-dimycolate; 6,6'-di-O-mycolyl-α,α-trehalose; (6-O-mycolyl-α-D-glucopyranosyl)-6-O-mycolyl-α-D-glucopyranoside. Toxic glycolipids responsible for the cord formation and the leukotoxic effect of virulent bacilli. The term cord factor is widely used for the natural mixture of trehalose dimycolates produced by virulent *Mycobacteria, Nocardia, Corynebacteria* and attenuated BCG, *q.v.* For precise designation, the strain from which the preparation was isolated must be mentioned. First isolated from *Mycobacterium tuberculosis:* H. Bloch, *J. Exp. Med.* **91,** 197 (1950). Structure: Y. Asselineau, E. Lederer, *Biochim. Biophys. Acta* **17,** 161 (1955); H. Noll *et al., ibid.* **20,** 299 (1956). Synthesis of *Mycobacterium tuberculosis* cord factor and of trehalose diesters with synthetic mycolic acids: T. Gendre, E. Lederer, *Bull. Soc. Chim.* **1956,** 1478; J. Polonsky *et al., Carbohydr. Res.* **65,** 295 (1978). Antitumor activity: A. Bekierkunst *et al., Science* **174,** 1240

(1971); M. V. Pimm *et al.*, *Int. J. Cancer* **24**, 780 (1979). *Review:* E. Lederer, *Springer Semin. Immunopathol.* **2**, 133-148 (1979).

Colorless wax. Yields trehalose and mycolic acid upon alkaline hydrolysis.

2593. Cordycepin. *3′-Deoxyadenosine;* 9-cordyceposidoadenine. $C_{10}H_{13}N_5O_3$; mol wt 251.25. C 47.81%, H 5.22%, N 27.87%, O 19.10%. First reported nucleoside antibiotic. Isoln from culture fluids of *Cordyceps militaris* (Linn.) Link: K. G. Cunningham *et al.*, *J. Chem. Soc.* **1951**, 2299; N. M. Kredich, A. J. Guarino, *Biochim. Biophys. Acta* **41**, 363 (1960). Proposed structure: H. R. Bentley *et al.*, *J. Chem. Soc.* **1951**, 2301. Identity with 3′-deoxyadenosine and revised structure: E. A. Kaczka *et al.*, *Biochem. Biophys. Res. Commun.* **14**, 456 (1964). Biosynthesis: R. Suhadolnik *et al.*, *J. Am. Chem. Soc.* **86**, 948 (1964). Synthesis: A. R. Todd, T. L. Ulbricht, *J. Chem. Soc.* **1960**, 3275; W. W. Lee *et al.*, *J. Am. Chem. Soc.* **83**, 1906 (1961); E. Walton *et al., ibid.* **86**, 2952 (1964); Y. Ito *et al., ibid.* **103**, 6739 (1981). Cordycepin and cordycepin triphosphate have been used extensively in the study of messenger RNA transcription, see H. T. Shigeura, G. E. Boxer, *Biochem. Biophys. Res. Commun.* **17**, 758 (1964); S. Penman *et al., Proc. Nat. Acad. Sci. USA* **67**, 1878 (1970). *Reviews:* J. J. Fox. *et al.*, "Nucleoside Antibiotics" in *Progr. Nucleic Acid Res. Mol. Biol.* **5**, 258-262 (1966); A. J. Guarino, "Cordycepin" in *Antibiotics* **I**, D. Gottlieb, P. Shaw, Eds. (Springer-Verlag, New York, 1967) pp 468-480.

Needles from ethanol, *n*-butanol, *n*-propanol or water. mp 225-226°. $[\alpha]_D^{20}$ −47°. $[\alpha]_D^{27}$ −42°. uv max (ethanol): 260 nm (ϵ 14600). pH aq soln: 7.1.

Triphosphate, $C_{10}H_{13}N_5O_{12}P_3$, *cordycepin-5′-triphosphate, 3′-deoxy ATP, 3′-deoxyadenosine-5′-(tetrahydrogen triphosphate).* Inhibits the final step of RNA biosynthesis by termination of the ribonucleotide chain due to the absence of the 3′-hydroxyl group. Formation by conversion of 3′-deoxyadenosine in Ehrlich ascites tumor: H. Klenow, *Biochim. Biophys. Acta* **76**, 347 (1963). Metabolism in KB cells: H. Shigeura, S. Sampson, *ibid.* **138**, 26 (1967). Synthesis: J. J. Novak, F. Sorm, *Coll. Czech. Chem. Commun.* **38**, 113 (1973); M. Blandin, *J. Carbohyd. Nucl., Nucl.* **3**(5/6), 341 (1976).

2594. Coriamyrtin. *[1aS-(1aα,1bβ,2β,5β,6aβ,7β,7aα,-8S*)]-Hexahydro-1b-hydroxy-6a-methyl-8-(1-methylethenyl)-spiro[2,5-methano-7H-oxireno[3,4]cyclopent[1,2-d]oxepin-7,-2′-oxiran]-3(2H)-one.* $C_{15}H_{18}O_5$; mol wt 278.30. C 64.74%, H 6.52%, O 28.74%. Sesquiterpene of the picrotoxane group. Main toxic principle from leaves and fruit of *Coraria myrtifolia* L., *Coriariaceae:* M. J. Riban, *C. R. Acad. Sci.* **57**, 798 (1863) and *C. japonica* A. Gray, *Coriariaceae:* Kariyone, Sato, *J. Pharm. Soc. Japan* **50**, 106 (1930); Okuda, *Pharm. Bull. (Japan)* **2**, 185 (1954). Series of articles on structure, absolute configuration, stereochemistry and derivatives: T. Okuda, T. Yoshida, *Chem. Pharm. Bull.* **9**, 379, 404 (1961); **15**, 1687, 1697 (1967). Structural relationship to tutin, *q.v.: eidem, Tetrahedron Letters* **1965**, 439. Biosynthetic studies: M. Biollaz, D. Arigoni, *Chem. Commun.* **1969**, 633. Synthetic studies: K. Tanaka *et al., Chem. Pharm. Bull.* **31**, 1943, 1958, 1972 (1983). Pharmacology: E. E. Swanson, K. K. Chen, *J. Pharmacol. Exp. Ther.* **57**, 410 (1936). Structure-activity relationship: C. H. Jarboe *et al., J. Med. Chem.* **11**, 729 (1968).

Bitter, monoclinic prisms, mp 229-230°. $[\alpha]_D^{14}$ +79°. Slightly sol in water, cold alc; freely sol in hot alc, in ether. LD_{50} i.p. in mice: 3 mg/kg (Jarboe).

2595. Coriander. Dried ripe fruit of *Coriandrum sativum* L., *Umbelliferae. Habit.* Asia, Europe. *Constit.* About 1% volatile oil; fixed oils, malic acid, tannin, mucilage.

THERAP CAT: Carminative, aromatic.

2596. Cork. The bark of the cork oak tree, *Quercus suber* L., *Fagaceae* and to a lesser degree of *Quercus occidentalis* Gray, *Fagaceae*, indigenous to the African and European shores of the western Mediterranean. *Constit.* 35-60% suberin, 30-33% cellulose, 27-32% lignin; small amounts of cerin (a wax), fats, inorganic manganese compds, decacrylic acid. Suberin contains phellonic acid, suberic acid, phloionic acid, phloionolic acid, suberolcarboxylic acid, eicosadicarboxylic acid, stearic acid, cortic acid. Used to seal wine casks since antiquity: Theophrastus, *Historia Plantarum* **3**, 16 and 17. Plinius (major), *Historia Naturalis* **16**, 8, 13. *Reviews and monographs:* A. Klauber, *Monographie des Korkes* (Weber, Berlin, 1920); W. Herrmann, *Kork* in *Ullmanns Encyklopädie der Technischen Chemie* vol. **10**, (Urban & Schwarzenberg, Munich, 3rd ed., 1958) pp 634-637; G. B. Cooke, *Cork and the Cork Tree* (Pergamon Press, New York, 1961).

Pale tan, elastic mass. Floats on water. Specific gravity about 0.10 to 0.25 g/ml. Excellent insulator against heat and sound. Stable to heat up to about 100°. Unaffected by water, brine, dil acids, alcohols, bland oils. Attacked by concd mineral acids, strong alkali and ammonia, free halogens, hydrogen peroxide, ozone.

USE: Bottle stoppers, sound deadeners, heat insulators, life preservers, gaskets, linoleum manufacture, cork tile, bulletin boards, dart boards, inlays for shoes.

2597. Corn Oil. Maize oil; Maydol; Mazola. Obtained as a byproduct by wet milling the grain of *Zea mays* L., *Gramineae* for the manuf of corn starch, corn syrup, glucose, dextrins, etc.: E. W. Eckey, *Vegetable Fats and Oils* (Reinhold, New York, 1954). *Constit.* Glycerides of the following fatty acids: Myristic 0.1-1.7%, palmitic 8-12%, stearic 2.5-4.5%, hexadecenoic 0.2-1.6%, oleic 19-49%, linoleic 34-62%. Unsaponifiable fraction: 1-3% (γ-tocopherol 0.1%, the rest is mostly isomeric sitosterols and wax such as myricyl and ceryl alcohols). The crude oil may contain up to 2% phospholipids (vegetable lecithin, inositol esters). The following constants are for the refined product.

Yellow oil. Faint characteristic odor and taste. d_{25}^{25} 0.916-0.921. mp −18 to −10°. Titer 14-20°. Flash pt 610°F (321°C). Ignition pt 740°F (393°C). n_D^{25} 1.470-1.474; n_D^{40} 1.464-1.468. Acid value 2-6. Saponification value 187-196. Iodine value 109-133. Thiocyanogen value 71-77. Hydroxyl value 8-12. Reichert-Meissl value <0.5. Polenske value <0.5. Hehner value 92-96. Classed as a semidrying oil. On prolonged exposure to air it thickens and becomes rancid. Miscible with chloroform, ether, benzene, petr ether. Slightly sol in alc.

USE: As salad and cooking oil; in prepn of margarine. As pharmaceutic aid (solvent). Some use in the preparation of non-yellowing enamel paint.

2598. Corn Steep Liquor. Corn steep water. Cleaned corn grain is washed thoroughly and then steeped for 36 to 40 hours in approx twice its volume of water contg 0.2% SO_2 at a temp of 46-50°C. As the steep water is drawn from the corn it contains between 6 and 9 lbs of solids per 100 lbs of water. The corn should be soft to the touch, but must not be slimy or smeary—in the latter case it was steeped either too hot or too long. The steep liquor is then evaporated [under 25 mm Hg pressure] in a "Swenson" evaporator to what is called heavy steep liquor with a gravity of 16 to 20° Bé and contg 50-60% of solids. In this form it is sold to manufacturers of antibiotics who may have their own specifications as to age, gravity and SO_2 content. See Bowden, Peterson, *Arch. Biochem.* **9**, 387-399 (April 1946); Graefe, *Stärke* **4**, 275-282 (1952).

USE: In production of penicillin, *meso*-inositol; source of phytin.

2599. Cornus. Dogwood; flowering dogwood. Dried root bark of *Cornus florida* L., *Cornaceae*. *Habit.* Eastern U.S. and Ontario. *Constit.* Cornin (Verbenalin, *q.v.*).

THERAP CAT: Astringent bitter.

2600. Coroxon. *Phosphoric acid 3-chloro-4-methyl-2-oxo-2H-1-benzopyran-7-yl diethyl ester; 3-chloro-7-hydroxy-4-methylcoumarin diethyl phosphate;* 3-chloro-4-methylumbelliferone diethyl phosphate; *O,O*-diethyl *O*-(3-chloro-4-methyl-7-coumarinyl) phosphate; *O,O*-diethyl *O*-7-(3-chloro-4-methylumbelliferone) phosphate; coumaphos oxygen analog; Coralox. $C_{14}H_{16}ClO_6P$; mol wt 346.70. C 48.50%, H 4.65%, Cl 10.23%, O 27.69%, P 8.93%. Metabolite of coumaphos, *q.v.* Prepn: Fusco *et al.*, U.S. pat. **2,951,851** (1960 to Montecatini); G. C. Amin, C. M. Christian, *Indian Chem. Manuf.* **12**, 22 (1974). Crystal and molecular structure: M. R. Gifkins, R. A. Jacobson, *J. Agr. Food Chem.* **24**, 232 (1976). Determn in eggs and milk: K. D. White *et al.*, *J. Assoc. Offic. Anal. Chem.* **66**, 1358 (1983). Ophthalmic cholinesterase inhibitor: M. Kadin, *Am. J. Ophthalmol.* **50**, 622 (1960).

Practically insol in water. Somewhat sol in acetone, chloroform, corn oil.

Mixture with phenothiazine, *Coopex*.

THERAP CAT (VET): Anthelmintic.

2601. Corticosterone. *(11β)-11,21-Dihydroxypregn-4-ene-3,20-dione;* 4-pregnene-11β,21-diol-3,20-dione; Kendall's compound B; Reichstein's substance H; compd B. $C_{21}H_{30}O_4$; mol wt 346.47. C 72.80%, H 8.73%, O 18.47%. Isoln from adrenal cortex: Mason *et al.*, *J. Biol. Chem.* **114**, 613 (1936); Reichstein, von Euw, *Helv. Chim. Acta* **21**, 1197 (1938); Jeanloz *et al.*, U.S. pat. **2,676,904** (1954 to Searle); from adrenal extract: Reichstein, U.S. pat. **2,166,877** (1939 to Roche-Organon). Prepn from desoxycholic acid: Wallis, Chakravorty, U.S. pat. **2,341,250** (1944 to Research Corp.); from 11-deoxycorticosterone: Zaffaroni, U.S. pat. **2,671,752** (1954 to Syntex); from cortisone: Oliveto *et al.*, U.S. pat. **2,927,108** (1960 to Schering).

Trigonal plates from acetone, mp 180-182°. $[\alpha]_D^{15}$ +223° (c = 1.1 in alc). uv max: 240 nm. Insol in water. Sol in usual organic solvents. Upon moistening with concd H_2SO_4 it gives an orange-yellow soln which has a strong fluorescence.

21-Acetate, $C_{23}H_{32}O_5$, clusters of needles from acetone + ether, mp 145°. $[\alpha]_D^{20}$ +195° (c = 0.62 in acetone).

THERAP CAT: Glucocorticoid.

2602. Cortisone. *17,21-Dihydroxypregn-4-ene-3,11,20-trione;* 17-hydroxy-11-dehydrocorticosterone; 11-dehydro-17-hydroxycorticosterone; Δ⁴-pregnene-17α,21-diol-3,11,20-trione; Kendall's compound E; Wintersteiner's compound F; Reichstein's substance Fa; KE; Incortin; Cortone; Cortadren; Scheroson; Corlin; Cortogen; Adreson. $C_{21}H_{28}O_5$; mol wt 360.45. C 69.98%, H 7.83%, O 22.19%. Isoln from suprarenal glands: Pfiffner *et al.*, *J. Biol. Chem.* **111**, 585 (1935); **116**, 291 (1936); Mason *et al.*, ibid. **114**, 613 (1936); Reichstein, *Habit. Helv. Chim. Acta* **19**, 1107 (1936); Kuizenga, Cartland, *Endocrinology* **24**, 526 (1939). Synthesis of the monoacetate from desoxycholic acid: Sarett, *J. Biol. Chem.* **162**, 601 (1946). Further development of prepn methods: Meystre, Wettstein, *Experientia* **3**, 185 (1947); *Helv. Chim. Acta* **30**, 1037, 1256 (1947); Reichstein *et al.*, ibid. **26**, 562, 705, 721 (1943); **27**, 821 (1944); Reichstein, U.S. pat. **2,403,683** (1946); Kendall *et al.*, *J. Biol. Chem.* **166**, 345 (1946); Gallagher, U.S. pat. **2,447,325** (1948); Peterson, Murray, *J. Am. Chem. Soc.* **74**, 1871 (1952); Perlman, ibid. 2126; Sarett, ibid. **70**, 1454 (1948); **71**, 2443 (1949); Mattox, Kendall, ibid. **70**, 882 (1948). Stereospecific total synthesis: Sarett *et al.*, ibid. **74**, 4974 (1952).

Rhombohedral platelets from 95% alcohol, mp 220-224° (some decompn) when heated in evac capillary. $[\alpha]_D^{25}$ +209° (c = 1.2 in 95% alcohol); $[\alpha]_{546}^{25}$ +269° (c = 0.125 in benzene); $[\alpha]_{546}^{25}$ +248° (c = 0.1 to 0.2 in alcohol). uv max: 237 nm (ε 1.4×10^4) see Mason *et al.*, *J. Biol. Chem.* **116**, 267 (1936); Wintersteiner, Pfiffner, ibid. 291. Fairly sol in cold methanol, ethanol, acetone; much less sol in ether, benzene, chloroform; slightly sol in water (28 mg/100 ml at 25°). The water soln is neutral. Gives orange-red soln with intense green fluorescence in concd H_2SO_4. Reduces Benedict's soln on heating.

21-Acetate, $C_{23}H_{30}O_6$, *cortisone acetate, Cortistab, Cortelan, Cortisyl Artriona.* Flat needles from acetone; clusters of radiating rods from chloroform. Becomes opaque at 70-100°, mp 235-238° with slight sintering at 230°. $[\alpha]_D^{25}$ +164° (c = 0.5 in acetone), $[\alpha]_D^{25}$ +208 to +217° (dioxane). uv max: 238 nm (ε 1.58×10^4), see Sarett, *J. Biol. Chem.* **162**, 630 (1946). Soly at 25° in water: 2.2 mg/100 ml; in propylene glycol 44 mg/100 ml; in chloroform 182 mg/g. Reduces ammoniacal silver nitrate soln at room temp. Sol in sulfuric acid giving a yellow soln without fluorescence (difference from hydrocortisone acetate).

21-Cyclopentanepropionate, $C_{29}H_{40}O_6$. Needles from diisopropyl ether, mp 158-161°. $[\alpha]_D^{20}$ +190° (chloroform). uv max (ethanol): 239 nm (ε 16350). Sol in ether, glycols, vegetable oils, especially sesame, peanut, and corn oils.

THERAP CAT: Glucocorticoid.

THERAP CAT (VET): Glucocorticoid, anti-inflammatory agent.

2603. Cortivazol. *(11β,16α)-21-(Acetyloxy)-11,17-dihydroxy-6,16-dimethyl-2'-phenyl-2'H-pregna-2,4,6-trieno[3,2-c]pyrazol-20-one;* 1,2,3,3a,3b,7,10,10a,10b,11,12,12a-dodecahydro-1,11-dihydroxy-2,5,10a,12a-tetramethyl-7-phenylcyclopenta[7,8]phenanthro[2,3-c]pyrazol-1-yl hydroxymethyl ketone acetate; 6,16α-dimethyl-11β,17α,21-trihydroxy-2'-phenyl[3,2-c]pyrazolo-4,6-pregnadien-20-one 21-acetate; MK-650; H-3625; Altim; Diaster; Dilaster. $C_{32}H_{38}N_2O_5$; mol wt 530.66. C 72.43%, H 7.22%, N 5.28%, O 15.07%. Prepn: Fried *et al.*, *J. Am. Chem. Soc.* **85**, 236 (1963); Tishler *et al.*, U.S. pats. **3,067,194** and **3,300,483** (1962, 1967 to Merck & Co.).

Double mp 160-165° and 229-230°; $[\alpha]_D^{23}$ +14° (CHCl₃). uv max (methanol): 283, 315 nm (ε 15700, 19000).

THERAP CAT: Glucocorticoid.

2604. Cortol. *(3α,5β,11β,20S)-Pregnane-3,11,17,20,21-pentol.* C₂₁H₃₆O₅; mol wt 368.51. C 68.45%, H 9.85%, O 21.71%. Isolated from human urine after administration of hydrocortisone or ACTH: Fukushima *et al., J. Biol. Chem.* **212**, 449 (1955).

α-Cortol; 5β-pregnane-3α,11β,17,20α,21-pentol. Crystals from ethyl acetate, mp 250.5-254°. $[\alpha]_D^{26}$ +23.7° (ethanol).

α-Cortol 3,20,21-triacetate, C₂₇H₄₂O₈, crystals from benzene + cyclohexane, mp 168-170°.

β-Cortol; 5β-pregnane-3α,11β,17,20β,21-pentol. Crystals from methanol, mp 262-264.5°. $[\alpha]_D^{25}$ +33.3° (ethanol).

2605. Cortolone. *(3α,5β,20S)-3,17,20,21-Tetrahydroxy-pregnan-11-one;* 3α,17α,20α,21-pregnanetetrol-11-one. C₂₁H₃₄O₅; mol wt 366.50. C 68.82%, H 9.35%, O 21.83%. Isolated from human urine after administration of cortisone, hydrocortisone or ACTH: D. K. Fukushima *et al., J. Biol. Chem.* **212**, 449 (1955). Synthesis: L. H. Sarett, *J. Am. Chem. Soc.* **71**, 1169 (1949); A. H. Soloway *et al., ibid.* **76**, 2941 (1954).

Crystals from acetone, mp 208-209°. $[\alpha]_D^{25}$ +44° (c = 0.5 in ethanol); also reported as $[\alpha]_D^{28}$ +34.2° (ethanol).

3,20,21-Triacetate, C₂₇H₄₀O₈, crystals from methanol, mp 214-216° (Fukushima); also reported as mp 213-214° (Sarett). $[\alpha]_D^{28}$ +28° (ethanol); also reported as $[\alpha]_D^{25}$ +18° (c = 1 in acetone).

2606. Corybulbine. *5,8,13,13a-Tetrahydro-2,9,10-trimethoxy-13-methyl-6H-dibenzo[a,g]quinolizin-3-ol; 2,9,10-trimethoxy-13-methyl-13a-berbin-3-ol;* 3-hydroxy-13-methyl-2,9,10-trimethoxyberbine; Corydalis-G. C₂₁H₂₅NO₄; mol wt 355.43. C 70.96%, H 7.09%, N 3.94%, O 18.01%. In *Corydalis cava* (L.) Schweigg & Korte *(C. tuberosa* DC.); *Fumariaceae:* Freund, Josephi, *Ann.* **277**, 1 (1893); in *C. platy-*

carpa Makino, *Fumariaceae:* Manske, *Can. J. Res.* **21B**, 13 (1943). Structure: Späth, Dobrowsky, *Ber.* **58**, 1274 (1925).

Needles from methanol, mp 242°. $[\alpha]_D^{20}$ +303° (c = 1.4 in CHCl₃). Unstable in light. Weakly basic. Sol in acetone, chloroform, hot benzene; sparingly sol in water, alc, ether.

Hydrochloride, C₂₁H₂₅NO₄·HCl, prisms, dec 245-250°.

2607. Corycavamine. *(R)-4,6,7,14-Tetrahydro-5,14-dimethylbis[1,3]benzodioxolo[4,5-c:5',6'-g]azecin-13(5H)-one;* 7,13-dimethyl-2,3:9,10-bis(methylenedioxy)-7,13a-secoberbin-13a-one. C₂₁H₂₁NO₅; mol wt 367.40. C 68.65%, H 5.76%, N 3.81%, O 21.77%. In *Corydalis cava* (L.) Schweigg & Korte *(C. tuberosa* DC.), *Fumariaceae:* Gadamer *et al., Arch. Pharm.* **240**, 81 (1902). Structure: von Bruchhausen, *ibid.* **263**, 570 (1925). Corycavamine and corycavine originally depicted as the keto and enol tautomers, resp.

Rhombic columns from ether and alcohol, mp 149°. Is converted to corycavine at the mp. $[\alpha]_D^{20}$ +167° (chloroform). Sol in alcohol and chloroform.

dl-Form, corycavine. Orthorhombic plates from alc, mp 218°, mp 222° in vac. Very slightly sol in water; sol in abs alcohol and chloroform; unstable to light.

2608. Corycavidine. *(R)-5,7,8,15-Tetrahydro-3,4-dimethoxy-6,15-dimethyl-[1,3]benzodioxolo[5,6-e][2]benzazecin-14(6H)-one.* C₂₂H₂₅NO₅; mol wt 383.44. C 68.91%, H 6.57%, N 3.65%, O 20.86%. In *Corydalis cava* (L.) Schweigg & Korte *(C. tuberosa* DC.), *Fumariaceae:* Gadamer, *Arch. Pharm.* **249**, 30 (1911); von Bruchhausen, *ibid.* **263**, 570 (1925). Synthesis of *dl*-form: Govindachari, Rajadurai, *J. Sci. Ind. Res. (New Delhi)* **16B**, 506 (1957).

Crystals from methanol, mp 212°. $[\alpha]_D^{20}$ +211°. Unstable to light; racemizes at the mp.

dl-Form, prisms from ether, mp 194°.

Hydrochloride, C₂₂H₂₅NO₅·HCl, crystals, easily sol in water and alcohol.

2609. Corydaldine. *3,4-Dihydro-6,7-dimethoxy-1(2H)-isoquinolinone; 3,4-dihydro-6,7-dimethoxyisocarbostyril;* 1,2,-3,4-tetrahydro-6,7-dimethoxy-1(2H)-1-isoquinolone. C₁₁H₁₃NO₃; mol wt 207.23. C 63.76%, H 6.32%, N 6.76%, O 23.16%. Synthesis: Späth, Dobrowsky, *Ber.* **58**, 1274 (1925); Mohunta, Ray, *J. Chem. Soc.* **1934**, 1263; Wiesner *et al., J. Am. Chem. Soc.* **77**, 675 (1955); Brossi *et al., Helv.*

Chim. Acta **43**, 1459 (1960); Mahuzier, Hamon, *Bull. Soc. Chim. France* **1969**, 684.

Monoclinic prisms from water or alcohol, mp 175°. Sol in water, alcohol, ether, benzene, chloroform. Practically insol in petr ether.

2610. Corydaline. *(13S-trans)-5,8,13,13a-Tetrahydro-2,3,9,10-tetramethoxy-13-methyl-6H-dibenzo[a,g]quinolizine; 2,3,9,10-tetramethoxy-13α-methyl-13aβ-berbine; d*-corydaline. $C_{22}H_{27}NO_4$; mol wt 369.46. C 71.52%, H 7.37%, N 3.79%, O 17.32%. Alkaloid isolated from many species of *Corydalis*. Discovery in *C. tuberosa*: Wackenroder, *Kustner's Arch.* **8**, 423 (1826), as cited in: C. Wehmer, *Die Pflanzenstoffe* I (G. Fischer, Jena, 1929) p 389. Isoln from *Corydalis aurea* Willd. and *C. solida* (L.) Swartz, *Fumariaceae*: Manske, *Can. J. Res.* **16B**, 81 (1938); *Can. J. Chem.* **34**, 1 (1956). Structure: von Bruchhausen, Stippler, *Arch. Pharm.* **265**, 152 (1927). Synthesis: Späth, Kruta, *Ber.* **62**, 1024 (1929); T. Kametani *et al., J. Chem. Soc. Perkin I* **1977**, 1151; M. Cushman, F. W. Dekon, *Tetrahedron* **1978**, 1435; Z. Kiparissides *et al., Can. J. Chem.* **58**, 2770 (1980). Stereochemistry: Bersch, *Arch. Pharm.* **291**, 595 (1958); Jeffs, *Experientia* **21**, 690 (1965). Pharmacology: Berezhinskaya, *C.A.* **78**, 119087j (1973). Toxicity study: R. C. Anderson, K. K. Chen, *Fed. Proc.* **5**, 163 (1946).

Prisms from alc, mp 135°. $[\alpha]_D^{20}$ +311° (c = 0.8 in alc). uv max: 396 nm. Sol in chloroform; moderately sol in ether; sparingly sol in methanol, ethanol. Practically insol in water. LD_{50} in mice (mg/kg): 135.5 ± 12.8 i.v. (Anderson, Chen).

Hydrochloride hydrate, $C_{22}H_{27}NO_4 \cdot HCl \cdot H_2O$, four-sided columns, mp 230-240°.

dl-Form, mp 135°. Slightly sol in water.

dl-Mesocorydaline, mp 158°. The (13R-cis)-analog of corydaline. Slightly sol in water.

d-Mesocorydaline, rhombic crystals, mp 152°. $[\alpha]_D^{20}$ +82° (c = 1.4); +180° (c = 3 in chloroform).

l-Mesocorydaline, rhombic crystals, mp 152°. $[\alpha]_D^{20}$ −85° (c = 1.4); −181° (c = 3 in chloroform).

2611. Corydalis. Squirrel corn; turkey corn. Dried tuber of *Dicentra (Bicuculla) cucullaria* (L.) Bernh., or of *Dicentra canadensis* (DC.) Walp., *Fumariaceae*. *Habit.* Ontario to Kentucky and Missouri. *Constit.* Corydaline, bulbocapnine, isocorydine, corytuberine, corycavine, corybulbine, corydine, fumaric acid, acrid resin, protopine.

2612. Corydine. *(S)-5,6,6a,7-Tetrahydro-2,10,11-trimethoxy-6-methyl-4H-dibenzo[de,g]quinolin-1-ol; 2,10,11-trimethoxy-6aα-aporphin-1-ol;* 1-hydroxy-2,10,11-trimethoxyaporphine. $C_{20}H_{23}NO_4$; mol wt 341.41. C 70.36%, H 6.79%, N 4.10%, O 18.75%. The methyl ether of corytuberine. Occurs in *Corydalis cava* (L.) Schweigg & Körte (*C. tuberosa* DC.), *Fumariaceae*, in the dextrorotatory form. Isoln: Gadamer, Ziegenbein, *Arch. Pharm.* **240**, 94 (1902). Structure: Späth, Berger, *Ber.* **64**, 2038 (1931). Synthesis: Hey, Palluel, *J. Chem. Soc.* **1957**, 2926; Arumugam *et al., Ber.* **91**, 40 (1958); Jackson, Martin, *Chem. Commun.* **1965**, 142, 420; *eidem, J. Chem. Soc. (C)* **1966**, 2222. uv spectrum: Shamma, Yao, *J. Org. Chem.* **36**, 3253 (1971).

Tetragonal prisms from ether, mp 149°. $[\alpha]_D^{20}$ +204° (c = 1.6 in chloroform). Freely sol in chloroform, alcohol, ethyl acetate; moderately sol in ether.

dl-Hydrochloride monohydrate, $C_{20}H_{23}NO_4 \cdot HCl \cdot H_2O$, crystals from ethanol + ether, sinters at 205°, dec 228°. Sparingly sol in water.

2613. Corynantheine. *(16E)-16,17,18,19-Tetradehydro-17-methoxycorynan-16-carboxylic acid methyl ester; (E)-16,-17,18,19-tetradehydro-17-methoxy-17,18-secoyohimban-16-carboxylic acid methyl ester.* $C_{22}H_{26}N_2O_3$; mol wt 366.46. C 72.11%, H 7.15%, N 7.64%, O 13.10%. Isolated from *Corynanthe johimbe* K. Schum., and *Pseudocinchona africana* A. Cheval., *Rubiaceae*: Karrer *et al., Helv. Chim. Acta* **9**, 1059 (1926); **35**, 851 (1952). Structure: Janot *et al., ibid.* **34**, 1207 (1951). Stereochemistry: Van Tamelen *et al., J. Am. Chem. Soc.* **79**, 6426 (1957). Total synthesis of *dl*-form: Van Tamelen, Wright, *Tetrahedron Letters* **1964**, 295; *eidem, J. Am. Chem. Soc.* **91**, 7349 (1969). Total synthesis of *d*-form: Autrey, Scullard, *ibid.* **90**, 4917 (1968).

Crystals: α-form mp 103-107°; β-form mp 165-166°. $[\alpha]_D$ +28° (c = 1 in methanol). uv max (methanol): 227, 280, 291 nm (log ε 4.64, 3.82, 3.80).

Hydrochloride, $C_{22}H_{26}N_2O_3 \cdot HCl$, needles from alcohol + ether, mp 194-206°. Also reported as mp 176-179°. $[\alpha]_D^{20}$ +12° (water). Sol in alcohol, sparingly sol in water.

Dihydrocorynantheine, $C_{22}H_{28}N_2O_3$. Total synthesis of *dl*-form: Van Tamelen, Hester, *J. Am. Chem. Soc.* **81**, 3805 (1959); *eidem, ibid.* **91**, 7342 (1969). Plates from alcohol + water, mp 177-177.5°. $[\alpha]_D$ +28°.

Hydrochloride, $C_{22}H_{28}N_2O_3 \cdot HCl$, mp 212-213°. Also reported as crystals from 95% ethanol-ethyl acetate, mp 242.2-243.3° (dec) (sealed tube), Van Tamelen, Hester, *loc. cit.* (1969).

2614. Corynanthine. *(16β,17α)-17-Hydroxyyohimban-16-carboxylic acid methyl ester;* rauhimbine. $C_{21}H_{26}N_2O_3$; mol wt 354.45. C 71.16%, H 7.39%, N 7.90%, O 13.54%. From bark of *Pseudocinchona africana* Chev., *Corynanthe johimbe* K. Schum., *Rubiaceae* and *Rauwolfia serpentina* (L.) Benth., *Apocynaceae*: Raymond-Hamet, *Compt. Rend.* **212**, 305 (1941); Jorio, *Ann. Chim. Farm.* **1939**, 50, *C.A.* **33**, 9306[9] (1939); Le Hir *et al., Ann. Pharm. Franc.* **11**, 546

(1953); Hofmann, *Helv. Chim. Acta* **37**, 314 (1954). Identity with rauhimbine: *idem, ibid.* 849. Structure and stereochemistry: Janot *et al., Bull. Soc. Chim. France* **1952**, 1085; **1961**, 637.

Stout prisms from acetone, dec 225-226°. $[\alpha]_D^{19}$ −85° (c = 0.5 in pyridine). uv max (methanol): 226, 283, 290 nm (log ε 4.56, 3.87, 3.79). Practically insol in water or petr ether. Sol in 40 parts of boiling chloroform, in 60 parts of boiling benzene, in 20 parts of boiling ethyl acetate, in 5 parts of boiling alcohol.

O,N-Dibutyrylcorynanthine hydrochloride, crystals from benzene, mp 208-210°. Prepn: Reiser *et al.,* U.S. pat. **2,975,183** (1961 to Chemische Werke Albert).

O,N-Dipropionylcorynanthine hydrochloride, crystals from isopropanol, mp 236-237°. Prepn: Reiser *et al., loc. cit.*

2615. Corypalmine. *5,8,13,13a-Tetrahydro-2,9,10-trimethoxy-6H-dibenzo[a,g]quinolizin-3-ol; 2,9,10-trimethoxyberbin-3-ol; 3-hydroxy-2,9,10-trimethoxyberbine; tetrahydrojatrorrhizine.* $C_{20}H_{23}NO_4$; mol wt 341.41. C 70.36%, H 6.79%, N 4.10%, O 18.75%. In *Corydalis cava* (L.) Schweigg & Körte *(C. tuberosa* D.C.) and other varieties of *Corydalis, Fumariaceae.* Synthesis of *dl*-corypalmine: Govindachari *et al., Ber.* **92**, 1654 (1959).

d-Form, small crystals, mp 236°. $[\alpha]_D^{16}$ +280° (chloroform). Insol in water; sol in alcohol and chloroform: Späth *et al., Ber.* **56**, 878 (1923); **58**, 2133 (1925); **60**, 383 (1927).

l-Form, crystals from chloroform-methanol, mp 246° (in vac): Manske, *Can. J. Res.* **20B**, 57 (1942); **21B**, 111 (1943).

dl-Form, crystals from methanol, mp 207°.

2616. Corytuberine. *(S)-5,6,6a,7-Tetrahydro-2,10-dimethoxy-6-methyl-4H-dibenzo[de,g]quinoline-1,11-diol; 2,10-dimethoxy-6aα-aporphine-1,11-diol; 1,11-dihydroxy-2,10-dimethoxyaporphine.* $C_{19}H_{21}NO_4$; mol wt 327.38. C 69.71%, H 6.47%, N 4.28%, O 19.55%. In *Corydalis cava* (L.) Schweigg & Körte *(C. tuberosa* DC), in *Dicentra formosa* (Andr.) Walp., *Fumariaceae.* Isoln: Späth, Berger, *Ber.* **64**, 2038 (1931); Manske, *Can. J. Res.* **10**, 521 (1934). Synthesis of (±)-form: Tomita, Kikkawa, **Japan.** pat. 6466('58) (to Shionogi), *C.A.* **54**, 1584i (1960); T. Kametani, M. Ihara, *J. Chem. Soc. Perkin Trans. I* **1980**, 629. Biosynthesis studies: Blaschke, *Biochem. Physiol. Alkaloide, Int. Symp., 4th,* K. Mothes, Ed. (Akad.-Verlag, Berlin, 1972) pp 283-286. Biomimetric total synthesis: T. Kametani *et al., Heterocycles* **5**, 175 (1976).

Pentahydrate, leaflets, plates, turning gray on exposure to light. mp 240° dec (dry). uv max (methanol): 227, 272, 311.5 nm. Sol in alc, hot water; slightly sol in ether, chloroform, ethyl acetate. $[\alpha]_D^{20}$ +283° (alc). The dried crystals are hygroscopic and quickly attract 1 mol H_2O from the air.

Hydrochloride, $C_{19}H_{21}NO_4 \cdot HCl$, crystals from alcohol-ether, dec above 250°; $[\alpha]_D^{20}$ +168°; sparingly sol in water.

Methyl ether, *see* Corydine.

2617. Cosyntropin. *α$^{1\text{-}24}$-Corticotropin; β$^{1\text{-}24}$-corticotropin; tetracosactide; tetracosactrin; Actholain; Cortrosinta; Cortrosyn; Synacthen.* $C_{136}H_{210}N_{40}O_{31}S$; mol wt 2933.48. C 55.68%, H 7.22%, N 19.10%, O 16.91%, S 1.09%. Ser-Tyr-Ser-Met-Glu-His-Phe-Arg-Trp-Gly-Lys-Pro-Val-Gly-Lys-Lys-Arg-Arg-Pro-Val-Lys-Val-Tyr-Pro. Prepn: **Belg.** pat. **617,177** (1962); H. Kappeler, R. Schwyzer, U.S. pat. **3,228,-926** (1966 to Ciba). Structure and synthesis: H. Kappeler, R. Schwyzer, *Helv. Chim. Acta* **44**, 1136 (1961); R. Schwyzer, H. Kappeler, *ibid.* **46**, 1550 (1963).

Hexaacetate tetradecahydrate, $C_{148}H_{228}N_{40}O_{43}S \cdot 14H_2O$. $[\alpha]_D^{22}$ −88 ± 2° (c = 0.511 in 1% acetic acid).

THERAP CAT: Adrenocorticotropic hormone.

2618. Cotarnine. *5,6,7,8-Tetrahydro-4-methoxy-6-methyl-1,3-dioxolo[4,5-g]isoquinolin-5-ol.* $C_{12}H_{15}NO_4$; mol wt 237.26. C 60.75%, H 6.37%, N 5.90%, O 26.97%. Prepd by the oxidation of narcotine with dil nitric acid. Synthesis: Salway, *J. Chem. Soc.* **97**, 1208 (1910). Is tautomeric. Structure studies: Small, Lutz, "Chemistry of the Opium Alkaloids," Supplement No. 103, *Public Health Reports* (Washington, 1932) p 49; Schneider, Müller, *Ann.* **615**, 34 (1958). Absorption spectrum: Csokán, *Z. Anal. Chem.* **124**, 344 (1942).

Small needles from benzene, dec 132-133°. Sol in alcohol, chloroform, ether, benzene; slightly sol in water; sol in dil acids, in ammonia or sodium carbonate soln, but only slightly in potassium hydroxide soln. Aq or alcoholic solns are yellow.

Hydrochloride, $C_{12}H_{15}NO_4 \cdot HCl$, *Secalysat.*

Chloride, $C_{12}H_{14}ClNO_3$, *7,8-dihydro-4-methoxy-6-methyl-1,3-dioxolo[4,5-g]isoquinolinium chloride, cotarnine chloride, cotarninium chloride, Stypticin.* Dihydrate, light-yellow powder; deliquesc in moist air. Sol in about 1 part water, 4 parts alc. *Keep well closed.*

Phthalate, $C_{32}H_{32}N_2O_{10}$, *Styptol.*

THERAP CAT: Hemostatic.

2619. Cotinine. *1-Methyl-5-(3-pyridinyl)-2-pyrrolidinone; N-methyl-2-(3-pyridyl)-5-pyrrolidone.* $C_{10}H_{12}N_2O$; mol wt 176.22. C 68.16%, H 6.86%, N 15.90%, O 9.08%. Nicotine metabolite first described by Pinner, *Arch. Pharm.* **231**, 378 (1893). Isolated from autoxidized nicotine, nicotine treated with hydrogen peroxide, from nicotine irradiated with ultraviolet light: Frankenburg, Vaitekunas, *J. Am. Chem. Soc.* **79**, 149 (1957).

Viscous oil, bp$_6$ 210-211°. Absorption spectra: Frankenburg, Vaitekunas, *loc. cit.*

Fumarate, $(C_{10}H_{12}N_2O)_2 \cdot C_4H_4O_4$, *Scotine.*

THERAP CAT: Antidepressant.

2620. Cotoin. *(2,6-Dihydroxy-4-methoxyphenyl)phenylmethanone; 2,6-dihydroxy-4-methoxybenzophenone.* $C_{14}H_{12}O_4$; mol wt 244.25. C 68.85%, H 4.95%, O 26.20%. Crystalline principle from true coto bark, the bark of *Aniba coto* (Rusby) Kost, *Lauraceae.* Extraction procedure: Jobst, Hesse, *Ann.* **199**, 17 (1879). Isoln from South American rosewood trees, *Aniba rosaeodora* Ducke and *A. duckei* Kost. *Lauraceae:* Gottlieb, Mors, *J. Am. Chem. Soc.* **80**, 2263 (1958). Structure: Karrer, Lichtenstein, *Helv. Chim. Acta* **11**, 789 (1928). Toxicity study: Jodbauer, Kurz, *Biochem. Z.* **74**, 340 (1916).

Yellow needles from benzene, mp 131-132°. Sharp taste. Almost insol in water. Sol in alcohol, methanol, amyl alcohol, benzene, chloroform, ether, carbon bisulfide, acetone, solns of alkali hydroxides and carbonates. Solns are yellow. Slowly reduces Fehling's soln or ammoniacal $AgNO_3$ in the cold. LD s.c. in frogs: 8 mg/kg (Jodbauer, Kurz).

Caution: Irritates mucous membranes.

2621. Cotton-root Bark. Dried root-bark of one or more of the cultivated species of *Gossypium herbaceum* L. and of other species of *Gossypium, Malvaceae. Habit.* Asia (India, China, Arabia), Egypt, U.S., West Indies, S. America, Australia, Spain, etc. *Constit.* Yellow chromogen about 8% of a pale yellow resin, fixed oil, sugar.

THERAP CAT: Oxytocic.

2622. Cottonseed Oil. Fixed oil from seeds of cultivated varieties of *Gossypium herbaceum* L. or of other species of *Gossypium.*

Pale yellow, oily, practically odorless liq. d_{25}^{25} 0.915-0.921. Solidif 0° to −5°. n_D^{40} 1.4645-1.4655. Sapon no.: 190-198. Iodine no.: 105-114. Surface tension (20°) = 35.4 dyn/cm; (80°) = 31.3 dyn/cm. Slightly sol in alcohol; miscible with chloroform, ether, carbon disulfide, petr ether.

USE: Manuf soaps, oleomargarine, hydrogenated fats, lard substitute, glycerol, leather dressings, lubricants, cosmetics; emollient; also as salad and cooking oil; packing fish. Pharmaceutic aid (solvent).

THERAP CAT (VET): Pediculicide, acaricide, laxative, emollient.

2623. Coumachlor. *3-[1-(4-Chlorophenyl)-3-oxobutyl]-4-hydroxy-2H-1-benzopyran-2-one; 3-(α-acetonyl p-chlorobenzyl)-4-hydroxycoumarin;* 3-(α-p-chlorophenyl-β-acetylethyl)-4-hydroxycoumarin; Geigy rodenticide exp. 332; Ratilan; Tomorin. $C_{19}H_{15}ClO_4$; mol wt 342.78. C 66.58%, H 4.41%, Cl 10.34%, O 18.67%. Prepn: F. Litvan, W. Stoll, U.S. pat. **2,648,682** (1953 to J. R. Geigy). Anticoagulant activity: M. Reiff, R. Weismann, *Acta Tropica* **8**, 97 (1951), *C.A.* **50**, 10976d (1956). Comparative toxicology of coumachlor and warfarin, *q.v.:* H. Wanntorp, *Acta Pharmacol. Toxicol.* **16**, Suppl. 2, 123 pp (1959).

Crystals, mp 169-171°. Sol in alc, acetone, chloroform. Slightly sol in benzene, ether. Practically insol in water. Highly toxic to dogs and pigs. LD_{50} orally in rats: 900 mg/kg, *RTECS* Vol. I, R. J. Lewis, R. L. Tatken, Eds. (1979) p 449.

Caution: Similar to warfarin, with delayed actions on prothrombin level and blood clotting, resulting in death by hemorrhage. *Antidote:* Vitamin K_1.

USE: Rodenticide.

2624. Coumafuryl. *3-[1-(2-Furanyl)-3-oxobutyl]-4-hydroxy-2H-1-benzopyran-2-one; 3-(α-acetonylfurfuryl)-4-hydroxycoumarin;* 3-[α-(2-furyl)-β-acetylethyl]-4-hydroxycoumarin; Fumarin. $C_{17}H_{14}O_5$; mol wt 298.30. C 68.45%, H 4.73%, O 26.82%. Prepn: **Brit.** pat. **734,142** (1955 to Norddeutsche Affinerie and C. F. Spress & Sohn).

Crystals, mp 124°.

USE: Rodenticide.

2625. Coumalic Acid. *2-Oxo-1,2H-pyran-5-carboxylic acid;* α-pyrone-5-carboxylic acid. $C_6H_4O_4$; mol wt 140.10. C 51.44%, H 2.88%, O 45.68%. Prepd by heating anhydr malic acid with oleum: v. Pechmann, *Ann.* **264**, 262, 272 (1891); Wiley, Smith, *Org. Syn.* **coll. vol. IV,** 201 (1963).

Bright yellow prisms from methanol, mp 205-210° (partial decompn). bp_{120} 218°; sublimes partially. Sparingly sol in cold water. Dec by boiling water. Sol in alcohol, glacial acetic acid. Slightly sol in ether, acetone, ethyl acetate. Insol in chloroform, benzene, ligroin.

Methyl ester, $C_7H_6O_4$, leaflets from ligroin, mp 73-74°. bp_{60} 178-180°. bp_{760} 250-260°.

Ethyl ester, $C_8H_8O_4$, crystals, mp 36°. bp_{760} 262-265°. Distills without decompn.

2626. Coumaphos. *Phosphorothioic acid O-(3-chloro-4-methyl-2-oxo-2H-1-benzopyran-7-yl) O,O-diethyl ester; 3-chloro-7-hydroxy-4-methylcoumarin O-ester with O,O-diethyl phosphorothioate;* 3-chloro-4-methylumbelliferone O-ester with O,O-diethyl phosphorothioate; O,O-diethyl O-(3-chloro-4-methyl-7-coumarinyl) phosphorothioate; O,O-diethyl O-(3-chloro-4-methylumbelliferone) thiophosphate; coumafos; Bayer 21/199; Asuntol; Baymix; Co-ral; Meldane; Muscatox; Resitox. $C_{14}H_{16}ClO_5PS$; mol wt 362.77. C 46.35%, H 4.45%, Cl 9.77%, O 22.05%, P 8.54%, S 8.84%. Prepn: Schrader, **Ger.** pat. **881,194;** **U.S.** pat. **2,748,146** (1951, 1956 both to Bayer); Krueger *et al., J. Agr. Food Chem.* **7**, 183 (1959). Toxicity: T. B. Gaines, *Toxicol. Appl. Pharmacol.* **14**, 515 (1969).

Crystals, mp 91°. The commercial product may be slightly brownish. Practically insol in water. Somewhat sol in acetone, chloroform, corn oil. Stable in water. LD_{50} in female, male rats (mg/kg): 16, 41 orally (Gaines).

USE: Insecticide, nematocide.

THERAP CAT (VET): Anthelmintic.

2627. Coumaran. *2,3-Dihydrobenzofuran;* cumaran; dihydrocoumarone. C_8H_8O; mol wt 120.15. C 79.97%, H 6.71%, O 13.32%. Synthesis: Bennett, Mahmoud Hafez, *J. Chem. Soc.* **1941**, 287; *cf.* Hurd, Hoffmann, *J. Org. Chem.* **5**, 212 (1940); Rindfusz, *J. Am. Chem. Soc.* **41**, 669 (1919); J. M. Bakke, H. M. Roholdt, *Acta Chem. Scand.* **B34**, 73 (1980).

Oily liq, bp 188-189°, bp_{13} 74-75°. d_4^{25} 1.058, n_D^{20} 1.5426. Sol in alcohol, ether, chloroform, carbon disulfide.

2628. *p*-Coumaric Acid. *3-(4-Hydroxyphenyl)-2-propen-oic acid; p-hydroxycinnamic acid; β-[4-hydroxyphenyl]acrylic acid.* $C_9H_8O_3$; mol wt 164.16. C 65.85%, H 4.91%, O 29.24%. Isoln: Hlasiwetz, *Ann.* **136**, 31 (1865); Bamberger, *Monatsh.* **12**, 459 (1891). Prepn: Eigel, *Ber.* **20**, 2528 (1887); Konek, Pacsu, *Ber.* **51**, 856 (1918). uv data: Wheeler, Covarrubias, *J. Org. Chem.* **28**, 2015 (1963).

Needles, mp 210-213° (reported in early lit as 206°). Crystallizes in anhydrous form from conc hot aq soln, but as monohydrate from dil aq soln on slow cooling. Slightly sol in cold water; sol in hot water and alc, ether. Practically insol in benzene and ligroin. uv max (95% ethanol): 223, 286 nm (ε 14,450, 19,000).

2629. Coumarilic Acid. *2-Benzofurancarboxylic acid;* coumarone-2-carboxylic acid. $C_9H_6O_3$; mol wt 162.14. C 66.67%, H 3.73%, O 29.60%. Synthesis from coumarin: Fuson *et al., Org. Syn.* coll. vol. III, 209 (1955). *Review:* Sethna, Shah, *Chem. Rev.* **36**, 1 (1945).

Needles from water, bitter taste. mp 192-193°. bp 310-315° with slight decompn. Sol in boiling water, in alcohol; slightly sol in chloroform, carbon disulfide.

2630. Coumarin. *2H-1-Benzopyran-2-one; 1,2-benzopyrone; cis-o-coumarinic acid lactone; cumarin; coumarinic anhydride;* tonka bean camphor. $C_9H_6O_2$; mol wt 146.15. C 73.97%, H 4.14%, O 21.90%. In tonka beans, lavender oil, woodruff *(Asperula* species), in sweet clover *(Melilotus).* Crystal and molecular structure: *Chem. Commun., Univ. Stockholm* **1976**, 21. Toxicity: Jenner *et al., Food Cosmet. Toxicol.* **2**, 327 (1964). *Review:* Sethna, Shah, *Chem. Rev.* **36**, 1 (1945); W. C. Meuly in Kirk-Othmer *Encyclopedia of Chemical Technology* vol. 7 (Wiley-Interscience, New York, 3rd ed., 1979) pp 196-206.

Orthorhombic, rectangular plates. Pleasant, fragrant odor resembling that of vanilla beans, burning taste. mp 68-70°. bp 297-299°. bp_5 139°. One gram dissolves in 400 ml cold, 50 ml boiling water. Freely sol in alc, chloroform, ether, oils; also sol in alkali hydroxide solns. LD_{50} orally in rats, guinea pigs: 680, 202 mg/kg (Jenner).
USE: Pharmaceutic aid (flavor).

2631. Coumarin-3-carboxylic Acid. *2-Oxo-2H-1-benzopyran-3-carboxylic acid.* $C_{10}H_6O_4$; mol wt 190.16. C 63.16%, H 3.18%, O 33.66%. By heating salicylaldehyde with malonic acid in glacial acetic acid. *Review:* Sethna, Shah, *Chem. Rev.* **36**, 1 (1945).

Needles from water, mp 188° (dec). Slightly sol in water; sol in alc, alkalies. Insol in ether, benzene, petr ether.

2632. Coumestrol. *3,9-Dihydroxy-6H-benzofuro[3,2-c]-[1]benzopyran-6-one; 2-(2,4-dihydroxyphenyl)-6-hydroxy-3-benzofurancarboxylic acid δ-lactone;* 7',6-dihydroxycoumarino(3',4',3,2)coumarone. $C_{15}H_8O_5$; mol wt 268.23. C 67.17%, H 3.01%, O 29.82%. An estrogenic factor occurring naturally in forage crops, esp in ladino clover *(Trifolium repens* L.), strawberry clover *(T. fragiferum* L.) and alfalfa

(Medicago sativa L., *Leguminosae).* Isoln: Bickoff *et al., J. Agr. Food Chem.* **6**, 536 (1958); Bickoff, Booth, U.S. pat. **2,890,116** (1959 to U.S.A.). Structure: Bickoff *et al., J. Am. Chem. Soc.* **80**, 3969 (1958). Synthesis: Emerson, Bickoff, *ibid.* 4381; U.S. pat. **2,884,427** (1959 to U.S.A.); Jurd, *Tetrahedron Letters* **1963**, 1151; Kappe, Brandner, *Z. Naturforsch.* **29B**, 292 (1974). Biosynthesis: Grisebach, Barz, *Chem. & Ind. (London)* **1963**, 690.

Crystals, mp 385°. Sublimes at 325°; sublimes in high vacuum at about 175°. uv max (methanol): 208, 243, 343 nm. Exhibits bright blue fluorescence in neutral or acid soln, greenish-yellow fluorescence in strong alkali. Absorption and fluorescence spectra: O. S. Wolfbeis, K. Schaffner, *Photochem. Photobiol.* **32**, 143 (1980). Practically insol in water at acid and neutral pH, in ether. Sparingly sol in water at alkaline pH (pH 11-12); slightly sol in methanol, chloroform, ether; very slightly sol in carbon tetrachloride, benzene.
Diacetate, $C_{19}H_{12}O_7$, crystals from acetic acid, mp 237°.
Dimethyl ether, $C_{17}H_{12}O_5$, crystals from methanol, mp 198°.

2633. Coumetarol. *3,3'-(2-Methoxyethylidene)bis[4-hydroxy-2H-1-benzopyran-2-one]; 3,3'-(2-methoxyethylidene)bis-[4-hydroxycoumarin];* 1-methoxy-2,2-bis[4-hydroxy-3-coumarinyl]ethane; 4,4'-dihydroxy-3,3'-(2-methoxyethylidene)-dicoumarin; 2,2-bis[4'-hydroxy-3'-coumarinyl]ethyl methyl ether; cumetharol; cumethoxaethane; PH-137; Dicoumoxyl; Dicumoxane. $C_{21}H_{16}O_7$; mol wt 380.35. C 66.31%, H 4.24%, O 29.45%. Prepn: Veldstra *et al., Rec. Trav. Chim.* **72**, 358 (1953); **Brit.** pat. **736,388** and Neth. pat. **80,005** (both 1955 to N. V. Amsterdamsche Chininefabriek), *C.A.* **50**, 10795a, 16874b (1956). Stereodynamics: Laruelle, *Tetrahedron Letters* **1970**, 2235. Has antivitamin K action. Activity studies: de Jongh, Kok, *Arch. Int. Pharmacodyn.* **94**, 470 (1953); Gilgenkrantz *et al., Ann. Med. Nancy* **3**, 115 (1964). Series of articles on clinical studies: Hutinel, *Lyon Med.* **23**, 1569-1604 (1964).

Crystals from ethanol or methanol, mp 156-157°.
THERAP CAT: Anticoagulant.

2634. Coumingine. *[2S-(2α,4aα,4bβ,7E,8β,8aα,10aβ)]-3-Hydroxy-3-methylbutanoic acid 7-[2-[2-(dimethylamino)-ethoxy]-2-oxoethylidene]tetradecahydro-1,1,4a,8-tetramethyl-9-oxo-2-phenanthrenyl ester.* $C_{29}H_{47}NO_6$; mol wt 505.70. C 68.88%, H 9.37%, N 2.77%, O 18.98%. Cardiotonic principle from the bark of *Erythrophleum couminga* Baillon, *Leguminosae.* Isoln: Ruzicka *et al., Helv. Chim. Acta* **24**, 63 (1941). Structure: Ruzicka *et al., ibid.* 1449.

Consult the Name Index before using this section.

Thin shiny needles from ether, mp 142°. $[\alpha]_D^{20}$ —70°. Sol in methanol, ethanol, acetone, acetic acid, benzene, chloroform. Practically insol in water, hexane, ether.

Hydrochloride, $C_{29}H_{47}NO_6 \cdot HCl$, crystals from ethanol + ether, mp 195°. Sol in ethanol, water.

2635. Coumithoate. *Phosphorothioic acid O,O-diethyl O-(7,8,9,10-tetrahydro-6-oxo-6H-dibenzo[b,d]pyran-3-yl) ester; 2-(2,4-dihydroxyphenyl)-1-cyclohexene-1-carboxylic acid δ-lactone O,O-diethylphosphorothioate; O,O-diethyl O-7-hydroxy-3,4-tetramethylenecoumarinyl phosphorothioate; O,O-diethylthiophosphoric ester of 3,4-tetramethyleneumbelliferone; 3,4-tetramethyleneumbelliferone O,O-diethylthiophosphoric ester; 7-hydroxy-3,4-tetramethylenecoumarin O,O-diethylthiophosphate; Dithion; Dition.* $C_{17}H_{21}O_5PS$; mol wt 368.39. C 55.43%, H 5.75%, O 21.72%, P 8.41%, S 8.70%. Prepn: Fusco *et al.*, U.S. pat. **2,860,085** (1958 to Montecatini). Toxicity study: R. Ben-Dyke *et al.*, *World Rev. Pest Contr.* **9**, 119 (1970).

Needles from benzene + petr ether, mp 88-88.5°. Practically insol in water. Has only limited solubility in organic solvents. LD_{50} orally in rats: 67 mg/kg (Ben-Dyke).

Caution: A cholinesterase inhibitor. Symptoms similar to parathion.

USE: Insecticide.

2636. Crataegus. Hawthorn; English hawthorn; haws; haw apple; aubépine. Berries, flowers, and leaves of *Crataegus oxyacantha* L., Rosaceae. *Habit.* All temperate zones, abundant in Europe, a garden escape in North America. *Constit.* Triterpene acids (oleanolic, ursolic, crataegolic); purines; anthocyanin type pigments and flavone deriv (pelargonin, quercitrin); choline; acetylcholine; trimethylamine; chlorogenic acid; caffeic acid; ascorbic acid; a growth hormone for caterpillars; a substance named RN 30/9: Hahn *et al.*, *Arzneimittel-Forsch.* **10**, 825 (1960). Some commercial extracts are: *Curtacrat; Crataegus-Kreussler; Esbericard.* Review and bibliography: F. Berger, *Handbuch der Drogenkunde* vol. 3 (Vienna, 1952) pp 202-211.

THERAP CAT: Cardiotonic, vasodilator (coronary).

2637. Creatine. *N-(Aminoiminomethyl)-N-methylglycine; N-amidinosarcosine; (α-methylguanido)acetic acid; N-methyl-N-guanylglycine; methylglycocyamine.* $C_4H_9N_3O_2$; mol wt 131.13. C 36.64%, H 6.92%, N 32.04%, O 24.40%. Present in muscular tissue of many vertebrates. Commercially isolated from meat extracts. Small amounts occur in the blood, but it is not found in normal urine from adults. The greater part of creatine in muscle is combined with phosphoric acid as phosphocreatine. Produced by liver and kidneys by the transfer of the guanidine moiety of arginine to glycine which is then methylated to give creatine. *In vitro* synthesis by liver and kidney tissues: Borsook, Dubnoff, *J. Biol. Chem.* **134**, 635 (1940). Review and bibliography: Peters, Van Slyke, *Quantitative Clinical Chemistry* Vol. I, *Interpretations* (Baltimore, 2nd ed., 1946). Synthesis by heating cyanamide with sarcosine: Strecker, *Jahresber.*

Chem. **1868**, 686; *cf.* Volhard, *Z. Chem.* **5**, 318 (1869); Paulmann, *Arch. Pharm.* **232**, 638 (1894); Bergmann, Zervas, *Z. Physiol. Chem.* **173**, 80 (1928); King *J. Chem. Soc.* **1930**, 2374.

Monohydrate, monoclinic prisms from water. Becomes anhydr at 100°; dec 303°. Neutral reaction to litmus. pKb (20°) 11.02. Adsorption on various chromatographic agents: Grettie, Williams, *J. Am. Chem. Soc.* **50**, 671 (1928). Absorption spectrum: Abderhalden, Haas, *Z. Physiol. Chem.* **164**, 7 (1927). One gram of the monohydrate dissolves in 75 ml water, in about 9 liters alcohol. Insol in ether. In aq soln creatinine is formed, and aq and alkaline solns contain an equilibrium mixture of creatine and creatinine, while in acid solns the formation of creatinine is complete: Cannan, Shore, *Biochem. J.* **22**, 924 (1928). The ratio of the molar concn of creatine to creatinine in water and in various buffer solns is given by Edgar and Shiver, *J. Am. Chem. Soc.* **47**, 1179 (1925).

2638. Creatinine. *2-Amino-1,5-dihydro-1-methyl-4H-imidazol-4-one; 2-amino-1-methyl-4-imidazolidinone; 1-methylhydantoin-2-imide; 1-methylglycocyamidine.* $C_4H_7N_3O$; mol wt 113.12. C 42.47%, H 6.24%, N 37.15%, O 14.14%. The end product of creatine catabolism. Normal constituent of urine; daily output about 25 mg/kg of body weight. Also found together with creatine in muscle tissues and blood. Occurs in all soils and in grain seeds and other vegetable matter. Has been found in certain fish and in crab meat extract. Isoln from urine: Maly, *Ann.* **159**, 279 (1871); Folin, *J. Biol. Chem.* **17**, 463 (1914); Benedict, *ibid.* **18**, 183 (1914). The isoln from urine is tedious, and creatinine is usually prepd from commercial creatine by treatment with HCl: Edgar, Hinegardner, *ibid.* **56**, 881 (1923); *Org. Syn.* **4**, 15 (1925).

Monoclinic plates. Leaflets from water. Dec about 300°. pKb (40°) 10.45. Sol in 12 parts water; slightly sol in alc. Practically insol in acetone, ether, chloroform.

2639. Creolin. Pearson's creolin. Prepd from refined coal-tar oils. Approx composition: Tar acids and oils 75-77%; emulsifying soaps 15-17%; water 8-10%.

Dark brown liquid, characteristic odor resembling that of phenol. Phenol coefficient (against *B. typhosus*) about 10. d 1.02-1.04. Forms stable milky emulsions when dil with much water. Miscible with a small amount of water; also miscible with alcohol, ether, chloroform.

USE: As a general industrial and household disinfectant and deodorant in 1-3% emulsion in water.

THERAP CAT (VET): Antiseptic, parasiticide. Do not use on cats.

2640. Creosol. *2-Methoxy-4-methylphenol; 2-methoxy-p-cresol; 4-methylguaiacol; 3-methoxy-4-hydroxytoluene; 4-hydroxy-3-methoxy-1-methylbenzene.* $C_8H_{10}O_2$; mol wt 138.17. C 69.55%, H 7.30%, O 23.16%. Occurs in beechwood tar. It is one of the active constituents of creosote. Obtained by the Clemmensen reduction of vanillin using amalgamated zinc and toluene as auxiliary solvent: Fletcher, Tarbell, *J. Am. Chem. Soc.* **65**, 1431 (1943); Schwarz, Hering, *Org. Syn.*, coll. vol. IV, 203 (1963).

OH, OCH$_3$, CH$_3$ (structure)

Colorless to yellowish, strongly refractive, aromatic liq. d_4^{25} 1.092. bp$_{760}$ 220°; bp$_{15}$ 105°; bp$_4$ 79°. mp 5.5°. n_D^{25} 1.5353. Slightly sol in water; miscible with alcohol, benzene, chloroform, ether, glacial acetic acid.

Note: Not to be confused with cresol.

2641. Creosote, Coal Tar. Coal tar creosote. A distillate of coal tar produced by high temp carbonization of bituminous coal. *Constit.* Liq and solid aromatic hydrocarbons, tar acids (up to 3%) and tar bases. History and composition: Roche, *J. Forest Prod. Res. Soc.* **2,** 75 (1952). Characterization by GLC: F. H. M. Nestler, *Anal. Chem.* **46,** 46 (1974). GC-MS analysis in treated railroad ties: W. Rotard, W. Mailahn, *ibid.* **59,** 65 (1987). Review of constituents, uses and carcinogenic risk: *IARC Monographs* **35,** 83-159 (1985); of toxicology and human exposure: *Toxicological Profile for Creosote* (PB91-180349, 1990) 125 p.

Translucent brown to black, oily liq. Flammable. Characteristic sharp, penetrating smoky odor; burning caustic taste. Heavier than water. *Typical specification:* $d_{15.5}^{38.0}$ 1.06. Distillation ranges: Up to 210° not >5%; up to 235° not >25% nor <5%; up to 270° not <20%; up to 355° not >85% nor <60%. Flash pt 165°F (75°C); ignition temp 637°F (335°C). Practically insol in water.

Caution: Readily absorbed through GI tract and skin. Potential symptoms following oral overexposure are intense GI irritation and congestion. Symptoms of systemic poisoning include salivation, vomiting, respiratory difficulties, thready pulse, vertigo, headache, loss of pupillary reflexes, hypothermia, cyanosis and mild convulsions. Direct contact may cause skin, eye or mucous membrane irritation, burning and itching, erythema, papular and vesicular eruptions, keratoconjunctivitis. *See* Patty's *Industrial Hygiene and Toxicology* Vol. **2B,** G. D. Clayton, F. E. Clayton, Eds. (John Wiley & Sons, Inc., New York, 4th ed., 1994) pp 1602-1605. Occupational exposure to soots, tars and certain mineral oils is known to be carcinogenic: *Seventh Annual Report on Carcinogens* (PB95, 109781, 1994) p 406.

USE: Wood preservative, disinfectant, insecticide.

THERAP CAT (VET): Has been used as an anthelmintic.

2642. Creosote, Wood. Wood creosote; beechwood creosote; Creasote. Liquid obtained from wood tars by distillation; composed chiefly of guaiacol and creosol, *q.q.v.* Acute toxicity: T. Miyazato *et al., Oyo Yakuri* **21,** 899 (1981), *C.A.* **96,** 28311h (1982). Chronic toxicity and carcinogenicity studies: *eidem, ibid.* **28,** 909, 925 (1984), *C.A.* **102,** 41238b, 57380c (1985). *See also: Clinical Toxicology of Commercial Products,* R. E. Gosselin *et al.,* Eds. (Williams & Wilkins, Baltimore, 5th ed., 1984) Section II, p 192.

Almost colorless or yellowish, very refractive, oily liq; characteristic smoky odor; caustic, burning taste. d_{25}^{25} below 1.076. Begins to boil at about 203° and at least 90% by vol distills between 203-220°. Does not solidify at −20°. Sol in 150-200 parts water, in glycerol, glacial acetic acid, fixed alkali hydroxide solns; miscible with alcohol, chloroform, ether, oils. *Incompat:* Acacia, albumin; cupric, ferric, gold and silver salts; oxidizers.

Calcium deriv., *calcium creosotate, Calcreose.* A mixture of calcium compds of creosote constituents. Dark-brown powder, creosote odor, sharp phenolic taste. Partly sol in water; soln deposits Ca(OH)$_2$ and CaCO$_3$ on standing.

Benzoate, yellowish liquid. Insol in water. Freely sol in alcohol or ether.

Carbonate, colorless to yellowish, clear, viscid, oily liquid; odorless and tasteless or slight odor and taste of creosote. d_{25}^{25} not below 1.145. Insol in water. Freely sol in alcohol; sol in petr ether, fixed oils. Miscible with benzene, chloroform.

Oleate, *Oleocreosote.* Yellowish, oily liquid. d 0.950. Insol in water. Sol in benzene, chloroform, ether.

Phosphate, *Phosote.* Mixture of the phosphoric acid esters of creosote. Yellowish, almost odorless, viscid mass. d_{15}^{15} 1.19. bp 230-235° with decompn. Insol in water. Sol in alcohol.

Valerate, *Eosote.* Colorless to yellowish liquid. bp about 240°. Insol in water. Sol in alcohol, ether.

THERAP CAT: Antiseptic; expectorant.

THERAP CAT (VET): Antiseptic, parasiticide, deodorant; has been used as expectorant, gastric sedative and gastrointestinal antiseptic. Do not use in cats.

2643. Creosotic Acid. Mixture of the isomeric cresotic acids.

White or reddish-white powder. Slightly sol in water; sol in alcohol, ether, alkali hydroxide solns.

USE: Disinfectant generally and in veterinary medicine in 1:500 aq or soap soln; manuf dyes and artificial tannin.

2644. Creslan®. Fiber X-54; Exlan. A copolymer of acrylonitrile possibly with acrylamide or a substituted acrylamide. *Ref:* R. W. Moncrieff, *Man-made Fibres* (John Wiley, New York, 1963) p 483.

Off-white fiber. Specific gravity 1.17. Sticks at 210°. Shrinkage in boiling water, 1%. Sunlight has negligible effect. Resistance to acids: good except to mineral acids. Resistance to alkalies: fair to dilute, poor to concd. Good resistance to cleaning solvents. Superior to its forerunner, *fiber X-51,* in the ease with which it can be dyed.

USE: Recommended for those uses that acrylics generally fill. It is claimed that it can be durably pleated.

2645. Cresol. Cresylic acid; cresylol; tricresol. C$_7$H$_8$O; mol wt 108.14. C 77.75%, H 7.46%, O 14.80%. Mixture of the three isomeric cresols, in which the *m*-isomer predominates. Obtained from coal tar: Paulsen, U.S. pat. **2,998,457** (1962 to Ashland Oil & Ref.). Usually contains a few per cent phenol. Prepn by sulfonation of toluene: Englund *et al., Ind. Eng. Chem.* **45,** 189 (1953); by oxidation of toluene: Braunwarth, Winsted, U.S. pat. **2,994,722** (1961 to Pure Oil). Toxicity study: W. B. Deichmann, S. Witherup, *J. Pharmacol. Exp. Ther.* **80,** 233 (1944). Review of manuf processes: Faith, Keyes & Clark's *Industrial Chemicals,* F. A. Lowenheim, M. K. Moran, Eds. (Wiley-Interscience, New York, 4th ed., 1975) pp 285-293.

OH, CH$_3$ (structure)

Colorless, yellowish, brownish-yellow or pinkish liq; phenolic odor; becomes darker with age and on exposure to light. *Poisonous!* d_{25}^{25} 1.030-1.038. Not less than 90% by vol distills between 195-205°. Soluble in about 50 parts water; miscible with alcohol, benzene, ether, glycerol, petr ether; also sol in solns of fixed alkali hydroxides. A soln in water is neutral to bromocresol purple. *Protect from light.*

m-Cresol, 3-methylphenol. Obtained from coal tar: Maesawa, Kurakano, **Japan.** pat. **8929**('55) (to Osaka Gas), *C.A.* **52,** 1231d (1958); Macak, Rehak, *Brennstoff Chem.* **43,** 80 (1962). Prepn from toluene: Toland, U.S. pat. **2,760,991** (1956 to California Res. Corp.); by oxidation of *o*- or *p*-toluic acid: Kaeding *et al., Ind. Eng. Chem.* **53,** 805 (1961). Colorless or yellowish liquid; phenolic odor. d_4^{20} 1.034. bp 202°. mp 11-12°. Flash pt, closed cup: 187°F (86°C). n_D^{20} 1.5398. Sol in about 40 parts water, in solns of fixed alkali hydroxides; miscible with alc, chloroform, ether. LD$_{50}$ orally in rats: 2.02 g/kg (Deichmann, Witherup).

o-Cresol, 2-methylphenol, o-cresylic acid, o-hydroxytoluene. Prepn from *m*-toluic acid: Toland, U.S. pat. **2,766,294** (1956 to California Res. Corp.); Barnard, Meyer, U.S. pat. **2,852,567** (1958 to Dow). Crystals or liq, becoming dark with age and exposure to air and light; phenolic odor. d_4^{20} 1.047. bp 191-192°. mp 30°. Flash pt 81-83°. n_D^{20} 1.553. Sol in about 40 parts water, in solns of the fixed alkali hydroxides. Miscible with alcohol, chloroform, ether. *Protect from light.* LD$_{50}$ orally in rats: 1.35 g/kg (Deichmann, Witherup).

p-Cresol, 4-methylphenol. Obtained from coal tar. Laboratory prepn from *p*-toluenesulfonic acid by fusion with potassium hydroxide: W. W. Hartman, *Org. Syn. coll. vol.* **I**, 175 (2nd ed., 1941); from toluene: Braunwarth, U.S. pat. **3,046,305** (1962 to Pure Oil). Crystals. Phenolic odor. d_4^{20} 1.0341. mp 35.5°. bp_{760} 201.8°; bp_{200} 179.4°; bp_{100} 140.0°; bp_{40} 117.7°; bp_{20} 102.3°; bp_{10} 88.6°; bp_5 76.5°; $bp_{1.0}$ 53.0°. Flash pt (closed cup) 86°C (187°F). Volatile in steam. n_D^{20} 1.5395. 100 ml water dissolves about 2.5 g at 50°, about 5 g at 100°. Sol in aq solns of alkali hydroxides; in the usual organic solvents. LD_{50} orally in rats: 1.8 g/kg (Deichmann, Witherup).

Caution: Potential symptoms of overexposure are CNS effects; confusion, depression and respiratory failure; dyspnea, irregular rapid respiration and weak pulse; skin and eye burns; dermatitis; lung, liver and kidney damage. *See NIOSH Pocket Guide to Chemical Hazards* (DHHS/NIOSH 90-117, 1990) p 74. Symptoms of intoxication are similar to those produced by phenol, *q.v.* Acute exposure may cause muscular weakness, gastroenteric disturbances, severe depression, collapse and death. Chronic exposure may cause digestive disturbances, liver and kidney damage, and skin eruptions. Cresol has a marked corrosive action on tissues, producing burns and dermatitis. *See* Patty's *Industrial Hygiene and Toxicology* vol. 2A, G. D. Clayton, F. E. Clayton, Eds. (Wiley-Interscience, New York, 3rd ed., 1981) pp 2597-2601.

USE: For making synthetic resins; in disinfectants and fumigants; as industrial solvent; *m*-cresol in photographic developers, explosives.

THERAP CAT: Disinfectant.

THERAP CAT (VET): Local antiseptic, parasiticide, disinfectant; has been used as an intestinal antiseptic.

2646. o-Cresolphthalein. *3,3-Bis(4-hydroxy-3-methylphenyl)-1(3H)-isobenzofuranone; 3,3′-dimethylphenolphthalein.* $C_{22}H_{18}O_4$; mol wt 346.38. C 76.29%, H 5.24%, O 18.48%. Prepn: F. J. Welcher, *Organic Analytical Reagents* vol. 4 (Van Nostrand, New York, 1948) p 482; Hubacher *et al., J. Am. Pharm. Assoc.* **42**, 23 (1953).

Crystals from alc, mp 223°. Slightly sol in water; sol in alcohol. pK 9.4.

Note: The sodium salt exists in the quinone form. *See* Phenolphthalein Sodium.

USE: As indicator: pH range, 8.2 colorless, 9.8 red. Analytical reagent, Welcher, *loc. cit.*

2647. Cresol Red. *4,4′-(3H-2,1-Benzoxathiol-3-ylidene)bis(2-methylphenol) S,S-dioxide; α-hydroxy-α,α-bis(4-hydroxy-m-tolyl)-o-toluenesulfonic acid γ-sultone; o-cresolsulfonphthalein.* $C_{21}H_{18}O_5S$; mol wt 382.44. C 65.95%, H 4.74%, O 20.92%, S 8.38%. Prepn: Sohon, *Am. Chem. J.* **20**, 265 (1898), *Beilstein* vol. **19**, 91; 1st suppl., 650; Lubs, Clark, *J. Wash. Acad. Sci.* **5**, 609 (1915), **6**, 481 (1916).

Tautomerism: Ramart-Lucas, *Bull. Soc. Chim. France* [5] **10**, 282 (1943).

Reddish-brown cryst powder. Can be recrystallized from glacial acetic acid; sol in alc, water, dil acid (yellow soln), dil alkalies (purple soln): Lubs, Acree, *J. Am. Chem. Soc.* **38**, 2772 (1916). pK 8.3: Sager *et al., ibid.* **70**, 732 (1948).

USE: As indicator. pH range: 7.2 yellow, 8.8 red, 2-3 orange to amber.

2648. m-Cresotic Acid. *2-Hydroxy-4-methylbenzoic acid; 2,4-cresotic acid; m-homosalicylic acid; m-cresotinic acid; 2-hydroxy-p-toluic acid; γ-cresotic acid.* $C_8H_8O_3$; mol wt 152.15. C 63.15%, H 5.30%, O 31.55%. Prepn: Prelog *et al., Helv. Chim. Acta* **30**, 675 (1947); Baine *et al., J. Org. Chem.* **19**, 510 (1954); Hauptschein *et al., J. Am. Chem. Soc.* **77**, 2284 (1955).

Crystals or leaflets, mp 177°; volatile with steam. Soly as of the *o*-acid.

Caution: Toxicity is similar to salicylic acid, *q.v.*

USE: In manuf of dyes.

2649. o-Cresotic Acid. *2-Hydroxy-3-methylbenzoic acid; 2,3-cresotic acid; o-homosalicylic acid; o-cresotinic acid; 2-hydroxy-m-toluic acid; 3-methylsalicylic acid.* $C_8H_8O_3$; mol wt 152.15. C 63.15%, H 5.30%, O 31.55%. Prepn: Baine *et al., J. Org. Chem.* **19**, 510 (1954); Jones, *Chem. & Ind. (London)* **1958**, 228; Blicke, McCarty, *J. Org. Chem.* **24**, 1061 (1959); Wessely *et al., Ber.* **93**, 2840 (1960).

White to slightly reddish, odorless cystals, mp 165-166°; volatile with steam. Slightly sol in cold, more sol in hot water; sol in chloroform, alcohol, ether, alkali hydroxides.

Caution: Toxicity is similar to salicylic acid.

USE: In manuf of dyes.

2650. p-Cresotic Acid. *2-Hydroxy-5-methylbenzoic acid; 2,5-cresotic acid; p-homosalicylic acid; p-cresotinic acid; 6-hydroxy-m-toluic acid.* $C_8H_8O_3$; mol wt 152.15. C 63.15%, H 5.30%, O 31.55%. Prepn: Cameron *et al., J. Org. Chem.* **15**, 233 (1950); Baine *et al., ibid.* **19**, 510 (1954); Thomas *et al., J. Am. Chem. Soc.* **80**, 5864 (1958); Blicke, McCarty, *J. Org. Chem.* **24**, 1061 (1959).

White or slightly reddish, almost odorless crystals, mp 151°. Volatile with steam. Sublimes with partial decompn. Soly as of the *o*-acid.

Caution: Toxicity is similar to salicylic acid.

2651. m-Cresyl Acetate. *Acetic acid 3-methylphenyl ester; m-tolyl acetate; m-cresol acetic acid ester; acetic acid m-cresol ester; acetylmetacresol; metacresol acetate; Cresatin; Cresatin-Sulzberger; Cresatin Metacresylacetate; Metacresylacetate-Sulzberger; Kresatin.* $C_9H_{10}O_2$; mol wt 150.18. C 71.98%, H 6.71%, O 21.31%. Prepd by refluxing *m*-cresol with acetic anhydride: Claus, Hirsch, *J. Prakt. Chem.* [2] **39**, 62 (1889); from *m*-cresol and acetyl chloride: Eijkman, *Chem. Weekbl.* **1**, 453 (1904); *see also* U.S. pat.

1,031,971 (1912 to N. Sulzberger). Purification: Lebeau-Janot, *Traité Pharm. Chim.* **II**, 882 (Paris, 1955).

Oily liq. Volatile with steam. Characteristic phenolic odor with a reminiscence of acetone. d_4^{26} 1.048. bp 212°; bp$_{13}$ 99°. uv max (methanol): 262.5, 269.5 nm. Practically insol in water, glycerol. Miscible with alc, ether, chloroform, petr ether, benzene. Soly in liq petrolatum about 5%. Also sol in cottonseed oil.

THERAP CAT: Antiseptic (topical), antifungal.

THERAP CAT (VET): Antiseptic (topical), fungicide.

2652. *o*-Cresyl Acetate. *Acetic acid 2-methylphenyl ester; acetic acid o-tolyl ester;* o-cresol acetate; acetyl-o-cresol; o-cresylic acetate; o-tolyl acetate. $C_9H_{10}O_2$; mol wt 150.18. C 71.98%, H 6.71%, O 21.31%. $CH_3C_6H_4OCOCH_3$.
Liquid. bp ~208°. Almost insol in cold, sol in hot water; freely sol in usual organic solvents, in oils.

2653. CRF. *Corticotropin-releasing factor;* CRH; corticoliberin(e); corticotropin-releasing hormone. Hypothalamic substance that stimulates secretion of corticotropin (ACTH, *q.v.*) and β-endorphin from the pituitary. First direct evidence for its presence in hypothalami: R. Guillemin, B. Rosenberg, *Endocrinology* **57**, 599 (1955); M. Saffran, A. V. Schally, *Can. J. Biochem. Physiol.* **33**, 408 (1955). Purification and characterization of a 41-residue ovine hypothalamic CRF that is highly potent in stimulating secretion of corticotropin and β-endorphin-like immunoactivities: W. Vale *et al., Science* **213**, 1394 (1981). Primary structure: J. Spiess *et al., Proc. Nat. Acad. Sci. USA* **78**, 6517 (1981). When centrally administered, CRF activates the sympathetic nervous system and may therefore function as a key hormone in stress mobilization of the organism, *cf.* W. Vale *et al., loc. cit.;* M. Brown *et al., Life Sci.* **30**, 207 (1982). CRF also stimulates α-MSH secretion and cyclic AMP accumulation in rat pars intermedia cells: H. Meunier *et al., ibid.* **31**, 2129 (1982). A functional relationship between CRF and dynorphin-related opioid peptides has also been suggested: K. A. Roth *et al., Science* **219**, 189 (1983). *Reviews:* A. V. Schally *et al., Recent Progr. Horm. Res.* **24**, 497-588 (1968); R. Burgus, R. Guillemin, *Ann. Rev. Biochem.* **39**, 499-526 (1970); M. Saffran in *Hypothalamic Peptide Hormones and Pituitary Regulation*, J. C. Porter, Ed. (Plenum Press, New York, 1977) pp 225-235; W.Vale *et al.*, in *The Role of Peptides in Neuronal Function*, J. L. Barker, J. G. Smith, Eds. (Dekker, New York, 1980) pp 432-454; several authors in *Polypeptide Hormones*, R. F. Beers, E. G. Bassett, Eds. (Raven Press, New York, 1980) pp 165-271; B. Lutz-Bucher *et al., J. Physiol.* **77**, 939-950 (1981); N. Yasuda *et al., Endocrine Rev.* **3**, 123-140 (1982); E. Stark, G. B. Makara in *Hormonally Active Brain Peptides: Structure and Function*, K. W. McKerns, V. Pantic, Eds. (Plenum Press, New York, 1982) pp 157-179.
Activity is destroyed by trypsin; unaffected by thioglycolate, pepsin, chymotrypsin.

2654. Crilanomer. *Starch polymer with 2-propenenitrile;* Super-Slurper; Clinagel; Intrasite. Hydrolyzed starchpolyacrylonitrile graft copolymers capable of absorbing 2000 times their initial weight of liquid. Prepn: L. A. Guglielmelli *et al., J. Appl. Polym. Sci.* **13**, 2007 (1969). Continuous production: Z. Reyes *et al., Ind. Eng. Chem. Process Des. Develop.* **12**, 62 (1973); manufacturing process: M. O. Weaver *et al., Stärke* **29**, 413 (1977). Analysis: R. J. Dennenberg, T.P. Abbott, *J. Polym. Sci., Polym. Letters* **14**, 693 (1976). Use in wound dressings: W. R. Spence, U.S. pat. **4,226,232** (1980 to Spenco Med). Clinical evaluation in leg ulceration: M. Hornemann, G. Raptis, *Z. Hautkr.* **62**, 41 (1987); B. Esch, G. Raptis, *ibid.* **64**, 1135 (1989). Brief description: H.-G. Elias, *Polym. News* **4**, 26 (1977).

Dry powder.

USE: In dressings for wounds. Absorption of liquids in products ranging from disposable diapers to seed coatings.

2655. Crimidine. *2-Chloro-N,N,6-trimethyl-4-pyrimidinamine; 2-chloro-4-(dimethylamino)-6-methylpyrimidine;* W-491; Castrix. $C_7H_{10}ClN_3$; mol wt 171.63. C 48.99%, H 5.87%, Cl 20.66%, N 24.48%. Prepd from an appropriate 2,4-dihalopyrimidine and dimethylamine: Westphal, U.S. pat. **2,219,858** (1940 to Winthrop).

Commercial product is a brown waxy solid, mp 87°. Practically insol in water; sol in most organic solvents. *Poisonous!* LD$_{50}$ orally in female mice: 1.2 mg/kg. Pyridoxine was found to be an excellent antidote for mice poisoned with crimidine: Karlog, Knudsen, *Nature* **200**, 790 (1963).
Caution: May cause serious CNS damage leading to fatal convulsions.

USE: Rodenticide.

2656. Croceic Acid. *7-Hydroxy-1-naphthalenesulfonic acid;* 2-naphthol-8-sulfonic acid; Bayer's acid. $C_{10}H_8O_4S$; mol wt 224.24. C 53.56%, H 3.60%, O 28.54%, S 14.30%. Prepn: Nietzki, Zübelen, *Ber.* **22**, 453 (1889); Forster, Keyworth, *J. Soc. Chem. Ind. (London)* **46**, 26T (1927).

On evaporation of an aq soln, the free acid dec into sulfuric acid and β-naphthol. However, very concd solns of the acid can be obtained. The normal sodium salt is sol in water, and readily sol in alcohol, crystallizing with 2 mols of alcohol. *Ref:* Forster, Keyworth, *loc. cit.*

2657. Crocetin. *8,8'-Diapo-ψ,ψ-carotenedioic acid; trans*-crocetin. $C_{20}H_{24}O_4$; mol wt 328.41. C 73.15%, H 7.37%, O 19.49%. Carotenoid-dicarboxylic acid isolated from *Crocus sativus* L.; *C. albiflorus* var. *neapolitanus* Hort.; *C. luteus* Lam., *Iridaceae.* Extraction procedure and structure: Jucker, Karrer, *Carotinoide* (Basel, 1948) p 282.

Brick-red rhombs from acetic anhydride, mp 285°. Absorption max (pyridine): 464, 436, 411 nm. Sol in pyridine, in very dil NaOH solns. Very sparingly sol in water, and in the usual organic solvents except pyridine and similar organic bases. Forms a solid dipyridyl salt.
Dimethyl ester, $C_{22}H_{28}O_4$, brick-red elongated leaflets, mp 222.5°. Total synthesis: Buchta, Andree, *Naturwiss.* **46**, 74 (1959).
Di-gentiobiose ester, $C_{44}H_{64}O_{24}$, *crocin.* One of the yellow-red pigments of saffron. Isoln and structure: Karrer, Salomon, *Helv. Chim. Acta* **10**, 397 (1927); **11**, 513, 711 (1928); Karrer *et al., ibid.* **12**, 985 (1929); **13**, 392 (1930); Kuhn, Winterstein, *Ber.* **67**, 344 (1934); Reichstein, *Angew. Chem.* **74**, 887 (1962). Hydrated brownish-red needles from methanol, mp 186° (effervescence). Absorption max (methanol): 464, 434 nm. Freely sol in hot water giving an orange-colored soln. Sparingly sol in abs alcohol, ether, other organic solvents.

2658. Cromolyn. *5,5'-[(2-Hydroxy-1,3-propanediyl)bis-(oxy)]bis[4-oxo-4H-1-benzopyran-2-carboxylic acid]; 5,5'-[(2-hydroxytrimethylene)dioxy]bis(4-oxo-4H-1-benzopyran-2-car-boxylic acid];* 5,5'-(2-hydroxytrimethylenedioxy)bis(4-oxo-chromene-2-carboxylic acid); 1,3-bis(2-carboxychromon-5-yloxy)-2-hydroxypropane; 1,3-di(2-carboxy-4-oxochromen-5-yloxy)propan-2-ol; cromoglycic acid; Duracroman. C_{23}-$H_{16}O_{11}$; mol wt 468.37. C 58.98%, H 3.44%, O 37.58%. Chromone complex which blocks mast cell degranulation. Prepn: Neth. pat. Appl. **6,603,997;** C. Fitzmaurice, T. B. Lee, U.S. pat. **3,419,578** (1966, 1968 both to Fisons). Metabolism: M. J. Ashton *et al., Toxicol. Appl. Pharmacol.* **26,** 319 (1973). Mechanism of action: T. C. Theoharides *et al., Science* **207,** 80 (1980); R. G. Alvarez *et al., Agents Actions* **11,** 94 (1981). Toxicology in primates: J. E. Beach *et al., Toxicol. Appl. Pharmacol.* **57,** 367 (1981). Clinical study in allergic conjunctivitis: G. A. Friday *et al., Am. J. Ophthalmol.* **95,** 169 (1983). Review of pharmacology and clinical use: G. G. Shapiro, P. Konig, *Pharmacotherapy* **5,** 156-170 (1985). *Review:* J. S. G. Cox in *Pharmacological and Biochemical Properties of Drug Substances* **vol. 1,** M. E. Goldberg, Ed. (Am. Pharm. Assoc., Washington, DC, 1977) pp 277-310.

mp 241-242° (dec).
Monohydrate, colorless crystals from ethanol + ether, mp 216-217° (dec).
Disodium salt, $C_{23}H_{14}Na_2O_{11}$, *cromolyn sodium, disodium cromoglycate, DSCG, FPL-670, Aarane, Alercrom, Alerion, Allergocrom, Colimune, Cromovet, Fivent, Gastrofrenal, Inostral, Intal, Introl, Irtan, Lomudal, Lomupren, Lomusol, Lomuspray, Nalcrom, Nalcron, Nasalcrom, Nasmil, Opticrom, Opticron, Rynacrom, Sofro, Vividrin.* Hygroscopic. Freely sol in water (100 mg/ml at 20°). Practically insol in chloroform and alcohol. LD_{50} in mice, rats (mg/kg): > 8000 orally (Cox).
THERAP CAT: Antiasthmatic; antiallergic.
THERAP CAT (VET): Antiallergic.

2659. Cropropamide. *N-[1-[(Dimethylamino)carbonyl]-propyl]-N-propyl-2-butenamide; N-(1-dimethylcarbamoylprop-yl)-N-propylcrotonamide;* α-(N'-crotonyl-N'-propyl)amino-N,N-dimethylbutyramide. $C_{13}H_{24}N_2O_2$; mol wt 240.35. C 64.97%, H 10.06%, N 11.66%, O 13.31%. Prepn: Martin, Gysin, U.S. pat. **2,447,587** (1948 to Geigy).

Liquid, $bp_{0.25}$ 128-130°. Easily sol in water, ether.
Combination with crotethamide, *prethcamide, Micoren, Respirot.*
THERAP CAT: Analgesic. Combination as respiratory stimulant.

2660. Crotamine. One of the poisonous principles found in the venom of Brazilian rattlesnakes, *Crotalus terrificus crotaminicus* Goncalves, *Crotalidae.* A polypeptide having mol wt 10,000-15,000 and isoelectric point at pH 10.3. Purification and properties: Goncalves in *Venoms,* E. E. Buckley, N. Porges, Eds. (Publ. no. **44** of the Am. Assoc. Adv. Science, Washington, D.C., 1956) pp 261-274. Pharmacology: Moussatché *et al., ibid.* pp 275-279.

2661. Crotamiton. *N-Ethyl-N-(2-methylphenyl)-2-bu-tenamide; N-ethyl-o-crotonotoluidide;* crotonyl-N-ethyl-o-toluidine; Crotamitex; Eurax; Euraxil; Veteusan. $C_{13}H_{17}$-NO; mol wt 203.28. C 76.81%, H 8.43%, N 6.89%, O 7.87%. Prepd by treating crotonic acid o-toluidide with

diethyl sulfate: **Swiss** pats. **253,472-3; Brit.** pat. **615,137** (1949 to Geigy).

Yellowish oil, bp_{13} 153-155°. Sol in methanol, ethanol.
USE: Fungicide, insecticide.
THERAP CAT: Scabicide.
THERAP CAT (VET): Scabicide; antipruritic.

2662. Crotethamide. *N-[1-[(Dimethylamino)carbonyl]-propyl]-N-ethyl-2-butenamide; N-(1-dimethylcarbamoylprop-yl)-N-ethylcrotonamide;* α-(N'-crotonyl-N'-ethyl)amino-N,N-dimethylbutyramide. $C_{12}H_{22}N_2O_2$; mol wt 226.32. C 63.69%, H 9.80%, N 12.38%, O 14.14%. Prepn: Martin, Gysin, U.S. pat. **2,447,587** (1948 to Geigy).

Liquid, $bp_{0.03}$ 132-134°. Easily sol in water, ether.
Combination with copropamide, *see* Cropropamide.
THERAP CAT: Analgesic; combination as respiratory stimulant.

2663. Crotonaldehyde. *2-Butenal;* crotonic aldehyde; β-methylacrolein. C_4H_6O; mol wt 70.09. C 68.55%, H 8.63%, O 22.83%. CH$_3$CH=CHCHO. The crotonaldehyde of commerce has the *trans*-configuration: Young, *J. Am. Chem. Soc.* **54,** 2498 (1932). Obtained by adding aldol to a boiling dil acid soln and removing the crotonaldehyde by distillation: v. Auwers, Eisenlohr, *J. Prakt. Chem.* **82,** 115 (1910); Hibbert, *J. Am. Chem. Soc.* **37,** 1759 (1915). Alternate synthesis: J. Smidt *et al., Angew. Chem.* **71,** 176 (1959). Prepn of *cis*-crotonaldehyde: M. Perrier, F. Rouessac, *Nouv. J. Chim.* **1,** 367 (1977). Toxicity studies: H. F. Smyth, C. P. Carpenter, *J. Ind. Hyg. Toxicol.* **26,** 269 (1944). *Review:* W. F. Baxter in Kirk-Othmer *Encyclopedia of Chemical Technology* **vol. 7** (Wiley-Interscience, New York, 3rd ed., 1979) pp 207-218.
Flammable liq. *Lacrimator!* Vapors extremely irritating. Use with adequate ventilation. Keep away from heat and open flame. Keep container closed. mp −76.5°; bp 104.0°; d_{20}^{20} 0.853; $n_D^{17.3}$ 1.4384. Flash pt when anhydr: 13° (55°F) (open cup). Explosive limits in air 2.95-15.5% v/v. Vapor density 2.41 (air = 1). Heat capacity: 0.7 cal/g/°C; heat of vaporization: 8.62 cal/mole. Commercial crotonaldehyde (93%) is stabilized with water (a solid phase separates out at −5°). Readily resinifies to a dimer when pure, slowly oxidizes to crotonic acid. Soly in water (g/100 g) at 20°: 18.1; at 5°: 19.2. Soly of water in crotonaldehyde (g/100 g) at 20°: 9.5; at 5°: 8.0. LD_{50} orally in rats: 0.3 g/kg (Smyth, Carpenter).
Caution: Potential symptoms of overexposure are irritation of eyes and respiratory system. *See NIOSH Pocket Guide to Chemical Hazards* (DHHS/NIOSH 90-117, 1990) p 76.
USE: Manuf of butyl alcohol, butyraldehyde, quinaldine. As warning agent in fuel gases, in locating breaks and leaks in pipes. Minor amounts are used in the manuf of maleic acid, crotyl alcohol, butyl chloral hydrate, and in rubber accelerators. In organic syntheses; as solvent in purification of mineral oils, manuf of resins, rubber antioxidants, insecticides. In chemical warfare.

2664. Crotonic Acid. *(E)-2-Butenoic acid;* β-methyl-acrylic acid; α-crotonic acid; solid crotonic acid; *trans*-crotonic acid. $C_4H_6O_2$; mol wt 86.09. C 55.81%, H 7.02%, O 37.17%. Has been found in clay soil in Texas; formed during the dry distillation of wood. Obtained on a commercial scale exclusively by oxygen- or air-oxidation of croton-aldehyde: Kennedy, U.S. pat. **2,413,235** (1946 to Shawinigan Chem.); Leupold, PB report 70249 (1942); Matthews, B.I.O.S. report 758 (1946). Laboratory procedure using alkaline silver oxide: Young, *J. Am. Chem. Soc.* **54**, 2498 (1932); from acetaldehyde and malonic acid in pyridine: v. Auwers, *Ann.* **432**, 46 (1923); Backer, Bloemen, *Rec. Trav. Chim.* **45**, 102 (1926); Florence, *Bull. Soc. Chim France* [4] **41**, 440 (1927); Letch, Linstead, *J. Chem. Soc.* **1932**, 454. Toxicity study: Smyth, Carpenter, *J. Ind. Hyg. Toxicol.* **26**, 269 (1944). *Review:* W. F. Baxter in Kirk-Othmer *Encyclopedia of Chemical Technology* **vol. 7** (Wiley-Interscience, New York, 3rd ed., 1979) pp 218-226.

Monoclinic needles, prisms from water or ligroin. d_4^{15} 1.018; d_4^{80} 0.964. mp 71.6°. bp$_{10}$ 80.0°; bp$_{20}$ 93.0°; bp$_{40}$ 107.8°; bp$_{60}$ 116.7°; bp$_{100}$ 128.0°; bp$_{200}$ 146.0°; bp$_{400}$ 165.5°; bp$_{760}$ 185.0°. n_D^{80} 1.4228. K at 25° = 2.0×10^{-5}. Soly in water (g/l) at 0°: 41.5; at 10°: 54.6; at 20°: 76.1; at 25°: 94; at 30°: 122; at 40°: 656. Soly in ethanol at 25°: 52.5% w/w; acetone: 53.0% w/w; toluene: 37.5% w/w. LD$_{50}$ orally in rats: 1.0 g/kg (Smyth, Carpenter). *Caution:* Explosions have occurred.

Methyl ester, $C_5H_8O_2$, liq, bp 121°; d^{20} 0.9444; n_D^{20} 1.4242. Ethyl ester, $C_6H_{10}O_2$, liq, bp 138°; d_4^{20} 0.9175; n_D^{20} 1.4245. Vinyl ester, $C_6H_8O_2$, liq, bp 133°; d_4^{20} 0.9410; n_D^{20} 1.450.

USE: In the manuf of copolymers with vinyl acetate used in lacquers and paper sizing; in the manuf of softening agents for synthetic rubber. In medicinal chemistry, e.g., in the manuf of DL-threonine, vitamin A.

2665. Croton Oil. Fixed oil expressed from seeds of *Croton tiglium* L., *Euphorbiaceae*. *Constit.* Croton resin—a powerful vesicant; glycerides of stearic, palmitic, myristic, lauric, tiglic, etc., acids; crotin; phorbol *(q.v.)* esters. *Review:* Hecker, Schmidt, *Fortschr. Chem. Org. Naturst.* **31**, 377 (1974).

Pale-yellow or brownish-yellow, rather viscid liq; slight disagreeable odor. *Poisonous!* d_{25}^{25} 0.935-0.950. n_D^{40} 1.470-1.473. Sapon no. 205-220. Iodine no. 102-118. Slightly sol in alcohol, increasing with age. Freely sol in chloroform, ether, oils, petr ether, carbon disulfide, glacial acetic acid. *Keep well closed and protected from light.*

USE: Has been used as an irritant and cocarcinogen in cancer research.

THERAP CAT: Cathartic, counterirritant.

THERAP CAT (VET): Formerly used as a counterirritant and a purgative.

2666. Crotoxin. Neurotoxin. One of the toxic principles isolated from rattlesnake venom: Slotta, Fraenkel-Conrat, *Ber.* **71**, 1076 (1938). A polypeptide complex of probably two components, one acidic and one basic, having mol wts of ~9000 and 12,000 resp. The subunits do not retain the toxicity of crotoxin but show synergistic action in combination: Hendon, Fraenkel-Conrat, *Proc. Nat. Acad. Sci. USA* **68**, 1560 (1971). Amino acid analyses and fractionation studies: Fraenkel-Conrat, Singer, *Arch. Biochem. Biophys.* **60**, 64 (1956); Rübsamen *et al.*, *Arch. Pharmacol.* **270**, 274 (1971); Horst *et al.*, *Biochem. Biophys. Res. Commun.* **46**, 1042 (1972). Review of early chemical studies: Fraenkel-Conrat, Singer, *Publ. Am. Assoc. Advan. Sci.* **44**, 259 (1956). *Monograph:* Behringwerk-Mitteilungen Sonderband, *Die Giftschlangen der Erde* (N. G. Elwert, Marburg, 1963) 464 pp. Pharmacology: Brazil, Excell, *J. Physiol. (London)* **212**, 34P (1970); Brazil *et al.*, *ibid.* **234**, 63P (1973); Habermann *et al.*, *Arch. Pharmacol.* **273**, 313 (1972). *See also* Crotamine.

Caution: When injected by snake bite, local pain, redness, hemorrhage and necrosis result. Systemic effects are dizziness, sensory and motor depression, collapse, shock. May be fatal. *Antidote:* Rattlesnake antivenin.

2667. Crotoxyphos. *(E)-3-[(Dimethoxyphosphinyl)-oxy]-2-butenoic acid 1-phenylethyl ester; 3-hydroxycrotonic*

acid α-methylbenzyl ester dimethyl phosphate; O,O-dimethyl *O*-[1-methyl-2-(1-phenylcarbethoxy)vinyl] phosphate; dimethyl 2-(α-methylbenzyloxycarbonyl)-1-methylvinyl phosphate; ENT-24717; SD-4294; Ciodrin. $C_{14}H_{19}O_6P$; mol wt 314.28. C 53.51%, H 6.09%, O 30.55%, P 9.86%. Prepn: **Brit.** pat. **855,238** (1960 to Bataafse Petroleum Maatschappij N.V.); Whetstone, Stiles, U.S. pat. **2,982,686** (1961 to Shell). Activity: C. P. Weidenback, R. L. Younger, *J. Econ. Entomol.* **55**, 793 (1962). Metabolism and degradn: K. I. Beynon *et al.*, *Residue Rev.* **47**, 55 (1973). Toxicity study: T. B. Gaines, *Toxicol. Appl. Pharmacol.* **14**, 515 (1969).

Light straw-colored liq, bp$_{0.03}$ 135°. n_D^{25} 1.4988. d^{25} 1.19. Vapor pressure at 20°: 1.4×10^{-5} mm Hg. Soly in water at room temp: 0.1%. Slightly sol in kerosine, saturated hydrocarbons. Sol in acetone, chloroform, ethanol, highly chlorinated hydrocarbons. LD$_{50}$ in female, male rats (mg/kg): 74, 110 orally (Gaines).

Caution: A cholinesterase inhibitor.

USE: Insecticide for external use on livestock.

2668. Crotyl Alcohol. *2-Buten-1-ol;* crotonyl alcohol. C_4H_8O; mol wt 72.11. C 66.63%, H 11.18%, O 22.19%. $CH_3CH=CHCH_2OH$. Prepn by reduction of crotonaldehyde: Nystrom, Brown, *J. Am. Chem. Soc.* **69**, 1197 (1947); Olivier, Young, *ibid.* **81**, 5811 (1959); Smith, Holm, U.S. pat. **2,761,883**; Finch, Furman, U.S. pat. **2,763,696** (both 1956 to Shell); Foreman, U.S. pat. **3,109,865** (1963 to Standard Oil Co., Ohio). Prepn of *cis-* and *trans*-forms: Hatch, Nesbitt, *J. Am. Chem. Soc.* **72**, 727 (1950); Hiskey *et al.*, *J. Org. Chem.* **21**, 429 (1956); of *cis*-form only: Clarke, Crombie, *Chem. & Ind. (London)* **1957**, 143. Toxicity study: H. F. Smyth *et al.*, *Am. Ind. Hyg. Assoc. J.* **23**, 95 (1962).

Liquid, bp 122°. d_4^{20} 0.8532. n_D^{20} 1.4285. Sol in 6 parts water; misc with alc. LD$_{50}$ orally in rats: 0.93 ml/kg (Smyth).

cis-Form, liquid, bp 123.6°. mp −90.15°. d_4^{20} 0.8662. n_D^{20} 1.4342.

trans-Form, liquid, bp 121.2°. d_4^{25} 0.8454. n_D^{20} 1.4289, n_D^{25} 1.4270.

2669. Crown Ethers. Crown compounds. Macrocyclic polyethers with the repeating unit $(-CH_2-CH_2-O-)_n$, where *n* is greater than 2. Crown compounds with other heteroatoms (N, S, P) are known. Prepd by C. J. Pedersen, *J. Am. Chem. Soc.* **89**, 2495, 7017 (1967). Described as "crown" ethers due to appearance of space-filling models and ability to "crown" cations. Proposed nomenclature lists non-ring substituents, number of atoms in ring, the class (crown), and the number of heteroatoms in the ring; e.g. dibenzo-18-crown-6. Bicyclic crowns are called *cryptates*. Crowns complex with cations and solubilize inorganic reagents in organic solvents. Selectivity results from the definite size of the crown cavity which admits only cations of corresponding ionic radii. The stability of complexes depends on the size of the cation and the size of the polyether ring. Chiral crowns are used in optical resolution of enantiomers. *Reviews:* D. J. Cram, J. M. Cram, *Science* **183**, 803-809 (1974); *eidem, Accounts Chem. Res.* **11**, 8-14 (1978);

Dibenzo-18-crown-6

J. J. Christensen *et al., Chem. Rev.* **74**, 351-384 (1974); G. W. Gokel, H. D. Durst, *Synthesis* **1976**, 168-184; A. C. Knipe, *J. Chem. Ed.* **53**, 618-622 (1976); V. Prelog, *Pure Appl. Chem.* **50**, 893-904 (1978). Historical overview: C. J. Pedersen, *Science* **241**, 536-540 (1988).

18-Crown-6, $C_{12}H_{24}O_6$, *1,4,7,10,13,16-hexaoxacyclooctadecane.* Prepn: G. W. Gokel *et al., Org. Syn.* **57**, 30 (1977). Crystals from acetonitrile, mp 38-39.5°.

Dibenzo-18-crown-6, $C_{20}H_{24}O_6$, *2,3,11,12-dibenzo-1,4,7,10,-13,16-hexaoxacyclooctadeca-2,11-diene.* The first crown ether. Prepn: C. J. Pedersen, U.S. pat. **3,687,978** (1972 to Du Pont). White crystals from acetone, mp 164°, bp$_{679}$ 380-384°.

Dicyclohexano-18-crown-6, $C_{20}H_{36}O_6$, *dicyclohexyl-18-crown-6.* Approximate lethal dose in rats (mg/kg): 300 orally; 130 skin absorption (Pedersen).

Caution: Causes eye, skin irritation in test animals.

USE: In organic synthesis as phase transfer reagents, catalysts and in sepn of enantiomers.

2670. CRP. C-Reactive Protein. Homogeneous acute phase protein, found in man and most other animals. Discovered as a result of its precipitation with C-polysaccharide (CPS) in sera of patients with infections and inflammatory disease: W. S. Tillet, T. Francis, *J. Exp. Med.* **52**, 561 (1930). Characterization as a protein and identification of its calcium ion requirement for interaction with CPS: T. J. Abernathy, O. T. Avery, *ibid.* **73**, 173 (1941). CRP is a trace serum protein consisting of single subunits of mol wt about 21,000. In serum and in the purified state, the subunits aggregate as cyclic pentamers having mol wts of 110,000-144,000. The concentrations of C-reactive protein increase up to 1000-fold in inflammatory conditions and in tissue necrosis. In addition, it can initiate reactions of agglutination, precipitation, and opsonization for phagocytosis and can activate the complement system. Other biological activities with platelets and lymphocytes have also been described, but the full biological functions of CRP have not yet been completely elucidated. Review of CRP and the acute phase response: H. Gewurz *et al., Advan. Int. Med.* **27**, 345-372 (1982). Review of structure and function: M. B. Pepys, *Eur. J. Rheumatol. Inflammation* **5**, 386-397 (1982). Book: *Ann. N.Y. Acad. Sci.* **389**, entitled "C-Reactive Protein and the Plasma Protein Response to Tissue Injury", I. Kushner *et al.,* Eds. (1982) 482 pp.

2671. Crufomate. *Methylphosphoramidic acid 2-chloro-4-(1,1-dimethylethyl)phenyl methyl ester; methylphosphoramidic acid 4-tert-butyl-2-chlorophenyl methyl ester; 4-tert-butyl-2-chlorophenyl N-methyl O-methylphosphoramidate;* Dowco 132; Montrel; Ruelene. $C_{12}H_{19}ClNO_3P$; mol wt 291.71. C 49.41%, H 6.56%, Cl 12.15%, N 4.80%, O 16.45%, P 10.62%. Prepn: Wasco *et al.,* U.S. pat. **2,929,762** (1960 to Dow). Toxicity study: T. B. Gaines, *Toxicol. Appl. Pharmacol.* **14**, 515 (1969).

Crystals from petr ether, mp 60-60.5°. Relatively stable in aq dispersions at pH 7 or below. Commercial product is a yellow oil, bp$_{0.01}$ 117-118°. n_D^{20} 1.5142. Sol in acetone, benzene, carbon tetrachloride. Practically insol in water, light petroleum. LD$_{50}$ in male, female rats (mg/kg): 635, 460 orally (Gaines).

USE: Insecticide.

THERAP CAT (VET): Anthelmintic.

2672. Cryofluorane. *1,2-Dichloro-1,1,2,2-tetrafluoroethane;* Freon 114; Frigen 114; Arcton 114. $C_2Cl_2F_4$; mol wt 170.92. C 14.05%, Cl 41.48%, F 44.46%. $ClCF_2CClF_2$. Prepn: Henne, *Organic Reactions* **2**, 49 (1944). Colorless, practically odorless, noncorrosive, nonirritating, nonflammable gas. Faint, ether-like odor in high concentrations. d_{liq}^0 1.5312. mp −94°. bp$_{760}$ +4.1°. n_D^0 1.3092. Critical temp 145.7°; crit pressure 474 lb/sq in. abs. Practi-

cally insol in water. Sol in alcohol, ether. Absorbs less than 0.0025% water. Has little, if any, anesthetic or toxic effect on humans, but toxic substances may be formed on contact with a flame or hot metal surface.

Caution: Potential symptoms of overexposure are irritation of respiratory system; asphyxia; cardiac arrhythmias, cardiac arrest. *See NIOSH Pocket Guide to Chemical Hazards* (DHHS/NIOSH 90-117, 1990) p 90.

Note: Consult latest Government regulations on use as aerosol propellant.

USE: Refrigerant, aerosol propellant.

2673. Cryolite. Kryolith; ice spar; sodium aluminum fluoride. AlF_6Na_3; mol wt 209.94. Al 12.85%, F 54.30%, Na 32.85%. $3NaF.AlF_3$. A mineral; large deposits exist in Greenland and in the Urals. Now prepd synthetically for the aluminum industry. Aluminum oxide is sol and dissociable in molten cryolite, and electrodeposition of aluminum metal is thus possible. Synthetic cryolite is usually made from $NaAlO_2$, $NaHCO_3$ and NaF. See the review by Vogel, *Chemiker-Ztg.* **75**, 603 (1951).

Snow-white, semi-opaque masses, vitreous fracture. Seldom monoclinic crystals with cubic habit, perfect cleavage (001) (110) (101). The natural form may be colored reddish or brown or even black, but loses this discoloration on heating. d 2.95. Mohs' hardness 2.5-3. Fuses fairly easily, mp 1000°. Sol in concd H_2SO_4. Dec by boiling with aq alkali hydroxides or aq calcium hydroxide. LD$_{50}$ orally in rats: 200 mg/kg, *RTECS Vol. II*, R. J. Lewis, R. L. Tatken, Eds. (1979) p 531.

Caution: Potential symptoms of overexposure are irritation of eyes and respiratory system; nausea, abdominal pain, diarrhea; excess salivation, thirst and sweating; stiff spine; dermatitis; calcification of ligaments of ribs and pelvis. *See NIOSH Pocket Guide to Chemical Hazards* (DHHS/NIOSH 90-117, 1990) p 116.

USE: In the aluminum and fluorine industry. Deville, *Ann. Chim. Phys.* [3] **61**, 329 (1861), found that fused cryolite is dissociated by the passage of electric current, with deposition of aluminum metal. Although most aluminum is prepd from bauxite, the presence of cryolite is necessary, since the process operates on the principle mentioned.

2674. Cryptenamine Tannates. Unitensen Tannate. A prepn of ester alkaloids from *Veratrum viride* Ait., *Liliaceae,* obtained by non-aqueous benzene-triethylamine extraction process: Cavallito, U.S. pat. **2,789,977** (1957 to Irwin, Neisler). Contains hypotensive alkaloids protoveratrines A and B, germitrine, neogermitrine, germerine, germidine; also jervine, rubijervine, isorubijervine, germbudine: Kupchan, Gruenfeld, *J. Am. Pharm. Assoc.* **48**, 727 (1959).

Cryptamine tannate is a tan, amorphous powder. Sol in alcohol and slightly sol in water.

Cryptenamine itself is a white, amorphous powder which induces sneezing. Practically insol in water. Freely sol in alcohol. Readily sol in dil acid soln. $[\alpha]_D^{25}$ −25 to −40° (ethanol).

THERAP CAT: Antihypertensive.

2675. Cryptopine. *4,6,7,13-Tetrahydro-9,10-dimethoxy-5-methylbenzo[e]-1,3-dioxolo[4,5-l][2]benzazecin-12(5H)-one;* cryptocavine. $C_{21}H_{23}NO_5$; mol wt 369.42. C 68.28%, H 6.28%, N 3.79%, O 21.65%. Occurs in opium (0.003-0.03%). Has been found in *Corydalis sempervirens* (L.) Pers., and in *Dicentra* spp., *Fumariaceae:* Manske, *Can. J. Res.* **8**, 407 (1933); **7**, 265 (1932); **15B**, 274 (1937). Structure: Perkin, *J. Chem. Soc.* **115**, 713 (1919). Synthesis: Haworth, Perkin, *ibid.* **1926**, 1769. Crystal structure: S. R. Hall, F. R. Ahmed, *Acta Crystallogr.*, Sect. B., **24**, 346 (1968).

Crystallizes as six-sided prisms or plates from benzene, mp 220-221°. d 1.35. Almost insol in water and ether. Soluble in chloroform, acetic acid. One part dissolves in 455 parts alc or 80 parts boiling alc. Sparingly sol in most other organic solvents. May be crystallized from hot alc, benzene, petr ether, methyl ethyl ketone, isoamyl alcohol, acetophenone, pyridine, or an alcohol-pyridine mixture. The salts tend to oil out, but can be crystallized by warming, redissolving, and slow cooling.

2676. Cryptoxanthin. *(3R)-β,β-Caroten-3-ol;* cryptoxanthol; caricaxanthin; hydroxy-β-carotene; β-caroten-3-ol; all-*trans*-cryptoxanthin. $C_{40}H_{56}O$; mol wt 552.88. C 86.90%, H 10.21%, O 2.89%. Carotenoid pigment with vitamin A activity, best isolated from petals and berries of *Physalis* species: Kuhn, Grundmann, *Ber.* **66**, 1746 (1933). Widely distributed, also was isolated from orange rind, *Carica papaya* L., Caricaceae, *Cucurbita pepo* L., Cucurbitaceae, *Zea mays* L., Gramineae, egg yolk, butter, bovine blood serum. Structure: Kuhn, Grundmann, *Ber.* **66**, 1746 (1933); Karrer, Schlientz, *Helv. Chim. Acta* **17**, 55 (1934); Cholnoky *et al., Ann.* **616**, 207 (1958). Total synthesis: Isler *et al., Helv. Chim. Acta* **40**, 456 (1957); Loeber *et al., J. Chem. Soc. (C)* **1971**, 404.

all-trans-form

Red plates with metallic luster from ether + methanol. mp 158-159° (racemate); 169° (natural). Absorption max: 452, 480 nm ($E_{1cm}^{1\%}$ 2380, 2080). Freely sol in chloroform, benzene, pyridine, carbon disulfide. Less sol in ligroin, petr ether, alcohol, methanol.

Monoacetate, $C_{42}H_{58}O_2$; deep red leaflets, mp 117.5°.

2677. Crystallins. Major water-soluble structural proteins found in fiber cells of vertebrate eye lenses; account for the transparency of the lens. Heterogeneous family composed of four groups α, β, γ, δ which have been separated on the basis of size, charge, immunological properties and source. α, β, δ crystallins occur in avian and reptilian lenses while α, β, γ crystallins occur in all other lenses. Several other minor forms have also been identified. Identification and isolation of α, β, γ: C. T. Mörner, *Z. Physiol. Chem.* **18**, 61 (1894); of δ: M. Rabaey, *Exp. Eye Res.* **1**, 310 (1962). Review of early isolation procedures: A. Spector, *Invest. Ophthalmol.* **3**, 182-193 (1964). Structural studies of α: A. F. Van Dam, G. Ten Cates, *Biochim. Biophys. Acta* **121**, 183 (1966); H. A. Kramps *et al., Eur. J. Biochem.* **50**, 503 (1975); of β: H. P. C. Driessen *et al., ibid.* **121**, 83 (1981); G. Wistow *et al., FEBS Letters* **133**, 9 (1981); of γ: T. Blundell *et al., Nature* **289**, 771 (1981); L. Summers *et al., Peptide Protein Rev.* **3**, 147 (1984); of δ: J. Horwitz, J. Piatigorsky, *Biochim. Biophys. Acta* **624**, 21 (1980); L. A. Williams *et al., ibid.* **708**, 49 (1982). Short-range spatial ordering is essential for lens transparency: M. Delaye, A. Tardieu, *Nature* **302**, 415 (1983). Photoinduced aggregation; implications for cataract formation: U. P. Andley *et al., Photochem. Photobiol.* **40**, 343 (1984); K. Mandal *et al., ibid.* **47**, 583 (1988); M. Kono *et al., ibid.* 593. Role of crystallin aging in cataract formation: R. J. Siezen *et al., Proc. Nat. Acad. Sci. USA* **82**, 1701 (1985); H. J. Hoenders, G. J. H. Bessems, *Lens Res.* **3**, 281 (1986). Evolution and expression of crystallin genes: J. Piatigorsky, *Cell* **38**, 620 (1984); W. W. de Jong, W. Hendriks, *J. Mol. Evol.* **24**, 121 (1986). *Reviews:* A. Spector, *Invest. Ophthalmol.* **4**, 579-591 (1965); W. W. de Jong, *Co-ord. Regul. Gene Expression - Proc. 2nd Int. Workshop*, R. M. Clayton, Ed. (Plenum Press, New York, 1986) pp 281-291; G. J. Wistow, J. Piatigorsky, *Ann. Rev. Biochem.* **57**, 479-504 (1988). Review of δ-crystallin: J. Piatigorsky, *Mol. Cell. Biochem.* **59**, 33-56 (1984). Book: *Molecular and Cellular Biology of the Eye Lens*, H. Bloemendal, Ed. (John Wiley & Sons, New York, 1981).

α-Crystallin. High molecular wt multimeric (35-50 subunits) polypeptide for which two major subunit forms exist $αA_1$, $αA_2$; M_r unit 20 kDa. Most acidic crystallin, pI 4.8-5.0.

β-Crystallin. Dioctamer for which several (6-7) subunit forms exist; M_r unit 23-25 kDa. pI 5.7-7.0.
γ-Crystallin. Monomer, M_r 20 kDa. pI 7.1-8.1.
δ-Crystallin. Tetramer for which two possible subunit forms exist; M_r (monomer) 48-50 kDa pI 4.9-5.3.

2678. Cubane. *Pentacyclo[4.2.0.0²,⁵.0³,⁸.0⁴,⁷]octane.* C_8H_8; mol wt 104.15. C 92.26%, H 7.74%. Prepn from bromocyclopentadienone dimer: Eaton, Cole, *J. Am. Chem. Soc.* **86**, 3157 (1964). Structure: Fleischer, *ibid.* 3889.

Rhombic crystals from methanol followed by sublimation at just above room temperature and atmospheric pressure, mp 130-131° (sealed capillary). Thermally unstable. Dec at 200°.

2679. Cubeb. Tailed pepper; Java pepper. Dried, unripe, nearly full-grown fruit of *Piper cubeba* L.f., Piperaceae. *Habit.* Southern Asia, Java, Borneo, Sumatra; cultivated in W. Indies and Ceylon. Berries are globular, 4-5 mm in diameter; blackish-gray; intern. whitish and hollow; strong, spicy odor; aromatic, pungent taste. *Constit.* 10-18% volatile oil, cubebin, about 1% cubebic acid, resin, fat, wax. Total ether extractive about 20%.

2680. Cubebin. *3,4-Bis(1,3-benzodioxol-5-ylmethyl)tetrahydro-2-furanol; tetrahydro-3,4-dipiperonyl-2-furanol; 5-hydroxy-3,4-bis(3,4-methylenedioxybenzyl)tetrahydrofuran.* $C_{20}H_{20}O_6$; mol wt 356.38. C 67.41%, H 5.66%, O 26.94%. From cubeb. Isoln: Capitaine, Soubeiran, *Ann.* **31**, 190 (1839). Structure: Haworth, Kelly, *Chem. & Ind. (London)* **1936**, 901; Haensel *et al., Arch. Pharm.* **300**, 559 (1967). Synthesis: Batterbee *et al., J. Chem. Soc. (C)* **1969**, 2470.

Slender prisms from methanol, mp 131-132°. $[α]_D^{14}$ —17° (acetone). Practically insol in water. Sol in alcohol, chloroform, ether.

Note: "Cubebine" is the French designation for ethereal extract of cubeb.

2681. Cucurbitacins. A group of tetracyclic triterpenes, commonly referred to as "bitter principles of cucurbits", which have antineoplastic and anti-gibberellin activity. They are isolated from various spp of cucurbitaceous plants known since antiquity for their beneficial and toxic properties. The plants have been used as vermifuges, emetics, narcotics, and antimalarials and have been implicated in sporadic livestock poisoning in S. Africa. Seventeen cucurbitacins have been isolated, most from plants of the *Cucurbitaceae* family, but also from plants of *Begoniaceae*, *Cruciferae*, *Datisceae*, *Euphorbiaceae*, and *Scrophulariaceae*. Cucurbitacins B and E are the most commonly identified. Isoln of cucurbitacins A, B, C, D, F: P. R. Enslin, *J. Sci. Food Agr.* **5**, 410 (1954); of G, H, J, K, L: *idem, ibid.* **8**, 673 (1957); of E and I: D. Lavie, S. J. Szinai, *J. Am. Chem. Soc.* **80**, 707 (1958); of O, P, Q and antitumor activity: S. Kupchan *et al., J. Org. Chem.* **35**, 2891 (1970). Structures of A, B, C, D, E, I: W. T. DeKock *et al., J. Chem. Soc.* **1963**, 3828; of B, D, F: D. Lavie *et al., Chem. & Ind. (London)* **1959**, 951; of G, H: C. W. Holzapfel, P. R. Enslin, *J. S. Afr. Chem. Inst.* **17**, 142 (1964), *C.A.* **62**, 10467c (1965); of J, K, L: P. R. Enslin, K. B. Norton, *J. Chem. Soc.* **1964**, 529. Stereochemistry of B, D, E, F, I: D. Lavie *et al., J. Org. Chem.* **28**, 1790 (1963). ¹³C-NMR study of cucurbitacins: V. V. Velde, D. Lavie, *Tetrahedron* **39**, 317 (1983). Use as plant growth regulators: J. Guha, S. P. Sen, *Nature (New Biol.)* **244**, 273 (1973). Toxicity studies: J. LeMen *et al., Chim. Ther.* **4**, 459 (1969); D. Albert *et al., Chim. Ther.* **5**, 205 (1970). *Reviews:* D. Lavie, E. Glotter, *Fortschr. Chem. Org. Naturst.* **29**, 308-357 (1971); J. Guha, S. P. Sen, *Plant Biochem. J.* **2**, 12-28 (1975); A. Shrotria, *Botanica* **26**, 28-31 (1976).

Cucurbitacin B

Cucurbitacin B, $C_{32}H_{46}O_8$, *(2β,9β,10α,16α,23E)-25-(acetyl-oxy)-2,16,20-trihydroxy-9-methyl-19-norlanosta-5,23-diene-3,-11,22-trione, 1,2-dihydro-α-elaterin.* Crystals from abs ethanol, mp 184-186°. $[\alpha]_D^{25}$ +88° (c = 1.55 in ethanol). LD_{10} orally in mice: 5 mg/kg (LeMen).

Cucurbitacin E, $C_{32}H_{44}O_8$, *(9β,10α,16α,23E)-25-(acetyl-oxy)-2,16,20-trihydroxy-9-methyl-19-norlanosta-1,5,23-triene-3,11,22-trione, α-elaterin.* White hexagonal plates from methanol, mp 232-233° (dec). $[\alpha]_D$ −59° (c = 0.7 in chloroform). uv max (chloroform): 234, 267 nm (ε 11700, 8350). LD_{50} orally in mice: 340 mg/kg (Albert).

2682. Cudbear. Crottle. Common names for the lichen *Ochrolechia tartarea* L., *Lecanoraceae* and for the coloring matter from this and other lichens, especially *Lecanoraceae* and *Roccellaceae.* A source of litmus, *q.v.* Acids in lichens (e.g. lecanoric acid) hydrolyze to orcinol, *q.v.*, which, in the presence of ammonia can be oxidized to the dye orcein, *q.v.* The dyes *French Purple, Persio, Orchil,* and *Orseilles* derive from the salts of orcein: *Colour Index* **Vol. 3** (3rd ed., 1971) p 3241.
Purplish-red powder. Imparts fine purplish-red color to alcohol, acid and neutral liquids. The purple tinge may be covered by adding a few drops of burnt-sugar coloring.
USE: Color for syrups, elixirs etc.

2683. Cumene. *(1-Methylethyl)benzene;* cumol; iso-propylbenzene. C_9H_{12}; mol wt 120.19. C 89.94%, H 10.06%. Manuf: Faith, Keyes & Clark's *Industrial Chemicals,* F. A. Lowenheim, M. K. Moran, Eds. (Wiley-Interscience, New York, 4th ed., 1975) pp 294-297. Toxicity study: Smyth *et al., Arch. Ind. Hyg. Occup. Med.* **4,** 119 (1951). Review: D. J. Ward in Kirk-Othmer *Encyclopedia of Chemical Technology* vol. 7 (Wiley-Interscience, New York, 3rd ed., 1979) pp 286-290.

Colorless liquid. d_4^{20} 0.862. bp 152-153°. Flash pt, closed cup: 102°F (39°C). n_D^{20} 1.4914: Hirschler, Faulconer, *J. Am. Chem. Soc.* **68,** 210 (1946). Insol in water. Sol in alcohol and many other organic solvents. LD_{50} orally in rats: 2.91 g/kg (Smyth).
Caution: Potential symptoms of overexposure are irritation of eyes and mucous membranes; headache; dermatitis; narcosis, coma. *See NIOSH Pocket Guide to Chemical Hazards* (DHHS/NIOSH 90-117, 1990) p 76.
USE: In manuf of phenol, acetone, acetophenone, α-methylstyrene.

2684. Cumic Acid. *4-(1-Methylethyl)benzoic acid; p-*isopropylbenzoic acid; cuminic acid. $C_{10}H_{12}O_2$; mol wt 164.20. C 73.15%, H 7.37%, O 19.49%.

Triclinic crystals. d^4 1.162. mp 115-117°. Sparingly sol in water; sol in alcohol, ether; also sol in concd H_2SO_4 without decompn, and pptd unchanged by adding water.

2685. Cumic Alcohol. *4-(1-Methylethyl)benzenemetha-nol;* cuminol; cuminyl alcohol; *p-*isopropylbenzyl alcohol. $C_{10}H_{14}O$; mol wt 150.22. C 79.96%, H 9.39%, O 10.65%. Found in caraway seed. Prepd by the reduction of cuminal; also by treating a suspension of cuminylmagnesium chloride in ether with dry oxygen.

Liquid; intense, persistent, caraway-like odor and burning, aromatic taste. d^{15} 0.981. bp 248.4°. n_D^{24} 1.522. Insol in water; miscible with alcohol, ether.

2686. Cumidine. *4-(1-Methylethyl)benzenamine;* 4-aminocumene; *p-*isopropylaniline; 4-amino-1-isopropylbenz-ene; β-(4-aminophenyl)propane. $C_9H_{13}N$; mol wt 135.21. C 79.95%, H 9.69%, N 10.36%. Prepn from aniline and isopropyl alcohol: Louis, *Ber.* **16,** 111 (1883); by nitration and subsequent reduction of cumene: Constam, Goldschmidt, *Ber.* **21,** 1157 (1898); Vavon, Callier, *Bull. Soc. Chim. France* [4] **41,** 677 (1927).

Liquid. bp_{745} 226-227°. d 0.9526. Insol in water.
Hydrochloride, $C_9H_{13}N·HCl$, prisms from water or alc.
Sulfate, $C_9H_{13}N·\frac{1}{2}H_2SO_4$, leaflets, mp 205°. Sol in hot water or alcohol, less sol in cold water, ether.
USE: In the determination of tungsten.

2687. Cuminaldehyde. *4-(1-Methylethyl)benzaldehyde; p-isopropylbenzaldehyde;* cuminal; cumaldehyde. $C_{10}H_{12}O$; mol wt 148.20. C 81.04%, H 8.16%, O 10.80%. $(CH_3)_2CH-C_6H_4CHO$. Constituent of eucalyptus, myrrh, cassia, cumin, and other essential oils; prepared synthetically from *p-*isopropylbenzoyl chloride. Alternate synthesis by formylation of isopropylbenzene under pressure: Crounse, *J. Am. Chem. Soc.* **71,** 1263 (1949). Toxicity study: P. M. Jenner *et al., Food. Cosmet. Toxicol.* **2,** 327 (1964).

Colorless to yellowish, oily liq; strong persistent odor; acrid, burning taste. d^{20} 0.978. bp_{760} 235-236°; bp_{35} 131-135°. n_D^{20} 1.5301. Practically insol in water. Sol in alc, ether. LD_{50} orally in rats: 1390 mg/kg (Jenner).
Thiosemicarbazone, $C_{11}H_{15}N_3S$, *SHCH-58.* Prepn: P. P. T. Sah, T. C. Daniels, *Rec. Trav. Chim.* **69,** 1545 (1950); J. Bernstein *et al., J. Am. Chem. Soc.* **73,** 906 (1951).
Stout platelets from methanol, mp 147°. Practically insol in water. Moderately sol in alc.
USE: In perfumery.

2688. Cupferron. *N-Hydroxy-N-nitrosobenzenamine ammonium salt; N*-nitrosophenylhydroxylamine ammonium salt. $C_6H_9N_3O_2$; mol wt 155.16. C 46.45%, H 5.85%, N 27.08%, O 20.62%. Prepd from phenylhydroxylamine by treating with $NaNO_2$ at 0° in presence of HCl, filtering, dissolving in ether and treating with NH_3, or by treating an ether soln of phenylhydroxylamine with butyl or amyl nitrite and NH_3 in the cold: Marvel, Kamm, *Org. Syn.* **4**, 19 (1925).

Crystals, mp 163-164°. Freely sol in water or alc. *Keep in well-closed containers to which a piece of ammonium carbonate has been added.*

Note: This substance may reasonably be anticipated to be a carcinogen: *Seventh Annual Report on Carcinogens* (PB95-109781, 1994) p 137.

USE: As a reagent for separating Cu and Fe from other metals. It ppts iron quantitatively from strongly acid soln; as a quantitative reagent for vanadates with which it gives a dark-red ppt sol in alkali soln, and for Ti with which it forms a yellow ppt; also suitable for the colorimetric estimation of Al.

2689. Cupreine. *(8α,9R)-Cinchonan-6',9-diol;* hydroxycinchonidine; ultraquinine. $C_{19}H_{22}N_2O_2$; mol wt 310.40. C 73.52%, H 7.14%, N 9.03%, O 10.31%. In bark and seeds of *Remijia pedunculata* Flueck., *Rubiaceae.* Isoln: Howard, Hodgkin, *J. Chem. Soc.* **41**, 16 (1882). Separated from homoquinine, *q.v.*, by treatment with NaOH: Hesse, *Ann.* **230**, 55 (1885). Reduction to *hydrocupreine:* Giemsa, Halberkann, *Ber.* **51**, 1325 (1918). Conversion into cinchonidine: King, *J. Chem. Soc.* **1946**, 523.

Monoclinic plates from alcohol, mp 202°. $[\alpha]_D^{17}$ −176°, (methanol). pK_1 6.57. Sol in alcohol, NaOH soln, but not in NH_4OH. Sparingly sol in water, ether, chloroform, benzene, petr ether. The neutral salts give yellow aq solns; the acid salts remain colorless in aq soln.

Dihydrate, monoclinic prisms, pseudo-orthorhombic by twinning.

Hydrochloride monohydrate, $C_{19}H_{22}N_2O_2 \cdot HCl \cdot H_2O$, needles, $[\alpha]_D^{17}$ −157°.

Dihydrochloride dihydrate, $C_{19}H_{22}N_2O_2 \cdot 2HCl \cdot 2H_2O$, crystals, $[\alpha]_D^{17}$ −211°.

Hydrogen sulfate hexahydrate, $(C_{19}H_{22}N_2O_2)_2 \cdot H_2SO_4 \cdot 6H_2O$, plates, tablets, when anhydr mp 257° dec, $[\alpha]_D^{22}$ −198°. Sol in 815 parts water.

Diacetate, $C_{23}H_{26}N_2O_4$, scales, plates, mp 88°.

Monomethyl ether, *see* Quinine.

2690. Cupric Acetate. Crystallized verdigris; neutralized verdigris. $C_4H_6CuO_4$; mol wt 181.64. C 26.45%, H 3.33%, Cu 34.99%, O 35.23%. $Cu(CH_3COO)_2$. Prepd by the action of acetic acid on CuO or $CuCO_3$: Winter *et al.,* in Kirk-Othmer *Encyclopedia of Chemical Technology,* **vol. 6** (Interscience, New York, 2nd ed., 1965) p 267. Toxicity study: H. F. Smyth *et al., Am. Ind. Hyg. Assoc. J.* **30**, 470 (1969).

Monohydrate, dark green, monoclinic crystals; exists in dimeric units. mp 115°; dec at 240°; d 1.882. Sol in water, alcohol; slightly sol in ether, glycerol. LD_{50} orally in rats: 0.71 g/kg (Smyth).

USE: Intermediate in manuf of Paris green; as fungicide; as catalyst for organic reactions, including rubber aging; in textile dyeing; as pigment for ceramics.

2691. Cupric Acetate, Basic. Cupric subacetate. Several complexes are known with different cupric acetate:cu-

pric hydroxide:water ratios. *Common verdigris* or *green verdigris* is a 2:1:5 complex; *blue verdigris* or *French verdigris* a 1:1:5 complex; other complexes exist with ratios 1:2:0 and 1:3:2. Review of prepn and properties: Gauthier, *Compt. Rend.* **242**, 644-647 (1956); *idem, Rev. Gen. Sci. Pures Appl. Bull. Assoc. Franc. Avance. Sci.* **66**, 67-78 (1957); Schweizer, Muehlethaler, *Farbe + Lack* **74**, 1159-1173 (1968).

The complexes exist as blue crystals or blue to green powders. Slightly sol in water, alc; sol in dil acids, ammonia.

USE: In manuf of Paris green and other pigments and insecticides; in dyeing and printing fabrics; as fungicide and mold-preventative; preparing gilder's wax in fire gilding; in artificial flowers; as water- or oil-color pigment.

2692. Cupric Acetoarsenite. *(Acetato)trimetaarsenitodicopper; Paris green;* copper acetate arsenite; emerald green; French green; imperial green; mineral green; Mitis green; parrot green; Schweinfurt green; Vienna green; C.I. Pigment Green 21; C.I. 77410. $C_4H_6As_6Cu_4O_{16}$; mol wt 1013.80. C 4.74%, H 0.60%, As 44.34%, Cu 25.07%, O 25.25%. $Cu(C_2H_3O_2)_2 \cdot 3Cu(AsO_2)_2$. Usually contains some water. Prepn: Serciron, U.S. pat. **2,159,864** (1939); Krefft, U.S. pat. **2,268,123** (1944 to Chemical Marketing); Glemser, Sauer in *Handbook of Preparative Inorganic Chemistry,* **vol. 2**, G. Brauer, Ed. (Academic Press, New York, 2nd ed., 1965) p 1027. *See also Colour Index* vol. 4 (3rd ed., 1971) p 4667. Toxicity study: T. B. Gaines, *Toxicol. Appl. Pharmacol.* **14**, 515 (1969).

Emerald green, crystalline powder. *Poisonous!* Stable to air, light. Practically insol in water. Dec on prolonged heating in water. Unstable in acids, bases, and toward H_2S. LD_{50} orally in female rats: 100 mg/kg (Gaines).

Caution: Toxicity primarily result of arsenic content. Gastric disturbance, tremors or muscular cramps, and peripheral neuritis. Local effects on the skin, mucous membranes and conjunctiva.

USE: Insecticide; wood preservative; as pigment, particularly for ships and submarines.

2693. Cupric Arsenite. *Arsonic acid copper(2+) salt (1:1); arsenious acid copper(2+) salt (1:1);* Scheele's green. Compound of variable composition. Usually considered to be $CuHAsO_3$; mol wt 187.46. Also reported as $Cu(AsO_2)_2$: Bhadraver, *Bull. Chem. Soc. Japan* **35**, 1770 (1962). Prepd by reaction of a cupric salt with an alkaline or ammoniacal arsenite: R. N. Kust in Kirk-Othmer *Encyclopedia of Chemical Technology,* vol. 7 (Interscience, New York, 3rd ed., 1979) p 102. *See also: Colour Index* vol. 4 (3rd ed., 1971) p 4667.

Yellowish-green powder. *Poisonous!* Practically insol in water, alcohol; sol in acids, ammonia.

USE: As pigment, wood preservative, insecticide, fungicide, rodenticide.

2694. Cupric Borate. A material of indefinite composition prepd by the reaction of a borax soln with $CuSO_4$: Barnard, Jr. in *Copper,* A. Butts, Ed., ACS Monograph, **no. 122** (Reinhold, New York, 1954) p 794.

Blue to blue-green solid. Extremely hard material.

USE: As ceramic color, pigment in oil paint; in dehydrogenation catalysis; as insulating and sealing medium for lead wires in lamps and mercury rectifiers, as fireproofing agent and preservative for wood; as fungicide, insecticide, parasiticide.

Caution: Symptoms of acute toxicity include vomiting, diarrhea, tremors, convulsions, shock, dermatitis; renal, hepatic and CNS degeneration. Symptoms of chronic ingestion include skin disorders, kidney lesions.

2695. Cupric Bromide. Br_2Cu; mol wt 223.35. Br 71.55%, Cu 28.45%. $CuBr_2$. Prepn: Watt *et al., J. Am. Chem. Soc.* **77**, 2752 (1955); Glemser, Sauer in *Handbook of Preparative Inorganic Chemistry,* **vol. 2**, G. Brauer, Ed. (Academic Press, New York, 2nd ed., 1965) p 1009.

Almost black, iodine-like, deliquesc monoclinic crystals or cryst powder. mp 498°; bp 900°; d_4^{20} 4.710. Very sol in water; sol in alcohol, acetone, ammonia; practically insol in benzene, ether, concd H_2SO_4. *Keep well closed.*

USE: As intensifier in photography; as brominating agent in organic synthesis; as humidity indicator; as wood preservative; in solid-electrolyte battery; as stabilizer for acetylated polyformaldehyde.

2696. Cupric Butyrate. $C_8H_{14}CuO_4$; mol wt 237.74. C 40.42%, H 5.94%, Cu 26.73%, O 26.92%. $Cu(CH_3CH_2CH_2COO)_2$. Prepn from a cupric salt and butyric acid or Na butyrate: Graddon, *J. Inorg. Nucl. Chem.* **17,** 222 (1955); Martin, Waterman, *J. Chem. Soc.* **1957,** 2545; from powdered Cu and butyric acid: Hahl, U.S. pat. 3,133,942 (1964 to BASF).

Monohydrate, large, dark green, monoclinic, hexagonal plates. Becomes dull and disintegrates after several days exposure to air. Sol in water, dioxane, benzene; slightly sol in chloroform, alcohol.

USE: In catalysts; in lubricants.

2697. Cupric Carbonate, Basic. *Copper carbonate hydroxide;* cupric subcarbonate; Bremen blue; Bremen green. $CH_2Cu_2O_5$; mol wt 221.12. C 5.43%, H 0.91%, Cu 57.48%, O 36.18%. $CuCO_3.Cu(OH)_2$. Occurs in nature as the mineral *malachite* (dark green monoclinic crystals). Prepd by adding $CuSO_4$ soln to Na_2CO_3 soln: Winter *et al.,* in Kirk-Othmer *Encyclopedia of Chemical Technology,* **vol. 6** (Interscience, New York, 2nd ed., 1965) p 270. The commercial product usually contains 50-55% Cu.

Green to blue amorphous powder or dark-green monoclinic crystals. Practically insol in water, alcohol; sol in dil acids, ammonia.

Note: Another form of basic copper carbonate, $2CuCO_3.Cu(OH)_2$, occurs in nature as the mineral *azurite* or *chessylite.*

See also Burgundy Mixture.

USE: As seed treatment fungicide; in pyrotechnics; as paint and varnish pigment; in animal and poultry feeds; in sweetening of petrol sour crude stock; in manuf of other Cu salts.

THERAP CAT (VET): Nutritional factor. Used in copper deficiency in ruminants.

2698. Cupric Chlorate. Cl_2CuO_6; mol wt 230.45. Cl 30.77%, Cu 27.58%, O 41.66%. $Cu(ClO_3)_2$. Prepn: *Gmelin's, Copper* (8th ed.) **60B,** 332-334 (1958).

Hexahydrate, blue to green, deliquesc, octahedral crystals. mp 65°. Dec at 100°. Very sol in water, alcohol. *Keep well closed and out of contact with organic matter.*

USE: Mordant in dyeing and printing of textiles.

2699. Cupric Chloride. Cl_2Cu; mol wt 134.45. Cl 52.74%, Cu 47.26%. $CuCl_2$. Prepn: Pray, *Inorg. Syn.* **5,** 153 (1957); *Gmelin's, Copper* (8th ed.) **60B,** 253-295 (1958); Glemser, Sauer in *Handbook of Preparative Inorganic Chemistry,* **vol. 2,** G. Brauer, Ed. (Academic Press, New York, 2nd ed., 1965) p 1008.

Yellow to brown, deliquesc, microcryst powder. Partially dec above 300° to CuCl + Cl_2. mp (extrapolated) 630°. mp of 498° usually reported for this compd is the melting pt of a mixture of $CuCl_2$ and CuCl. d_4^4 3.39. Forms the dihydrate in moist air. Sol in water, alc, acetone. *Keep well closed.*

Dihydrate, green to blue powder or orthorhombic, bipyramidal crystals. Deliquesc in moist air, efflorescent in dry air. Water loss occurs from 70-200°: Bell, Coultard, *J. Chem. Soc. (A)* **1968,** 1417. mp ~100°; d 2.51. Freely sol in water, methanol, ethanol; moderately sol in acetone, ethyl acetate; slightly sol in ether. The aq soln is acid to litmus; pH of 0.2 molar soln 3.6.

Caution: Irritating to skin, mucous membranes.

USE: As catalyst for organic and inorganic reactions; in petroleum industry as deodorizing, desulfurizing, and purifying agent; as mordant for dyeing and printing textiles; as oxidizing agent for aniline dyestuffs; in indelible, invisible and laundry-marking inks; in metallurgy in wet process for recovering mercury from ores, in refining Cu, Ag, Au; in tinting-baths for Fe, Sn; in electrotyping baths for plating Cu on Al; in photography as a fixer, desensitizer and reagent; in producing color in pyrotechnic compositions; in manuf of acrylonitrile, fast black (melanin); in pigments for glass, ceramics; as feed-additive, wood preservative, disinfectant.

2700. Cupric Chromate(VI). Neutral cupric chromate. $CrCuO_4$; mol wt 179.54. Cr 28.96%, Cu 35.39%, O 35.65%. $CuCrO_4$. Prepn from $CuCO_3$, Na_2CrO_4 and CrO_3: Briggs,

J. Chem. Soc. **1929,** 242; from CuO and CrO_3: Campbell, Lemaire, *Can. J. Res.* **25B,** 243 (1947); from $Cu(OH)_2$ and $K_2Cr_2O_7$: Winter *et al.,* in Kirk-Othmer *Encyclopedia of Chemical Technology* vol. 6 (Interscience, New York, 2nd ed., 1965) p 270.

Reddish-brown crystals. Gradually dec to $CuCr_2O_4$ above 400°. Practically insol in water; sol in acids.

Basic cupric chromates. Several cupric chromate-cupric hydroxide compounds are known. $CuCrO_4.Cu(OH)_2$ forms yellow, copper-red or chocolate-brown to lilac crystals; $CuCrO_4.2Cu(OH)_2$, a light brown powder; $2CuCrO_4.3Cu(OH)_2$, yellow to yellowish-brown crystals. Prepn: Campbell, Lemaire, *loc. cit.;* Winter *et al., loc. cit.* All are practically insol in water.

USE: In fungicides, seed protectants, and wood preservatives; as mordant in dyeing textiles; in protecting textiles against insects and microorganisms.

2701. Cupric Chromite. Cupric chromate(III). Cr_2CuO_4; mol wt 231.54. Cr 44.91%, Cu 27.45%, O 27.64%. $CuCr_2O_4$. Prepd by heating $CuCrO_4$ at 400°: Stroupe, *J. Am. Chem. Soc.* **71,** 569 (1949); or by decompn of $(NH_4)_2Cu(CrO_4)_2.2NH_3$ at 700°: Whipple, Wold, *J. Inorg. Nucl. Chem.* **24,** 23 (1962).

Grayish-black to black tetragonal crystals. Dec to $CuCrO_2$ and CrO_3 above 900°. Practically insol in water, dil acids, concd HCl.

Copper-chromium oxide. A mixture of $CuCr_2O_4$ and CuO formed on decompn at temps below 400°, of orange copper ammonium chromate complex prepd from $Cu(NO_3)_2$, $Na_2Cr_2O_7$ and NH_4OH: Adkins *et al., J. Am. Chem. Soc.* **72,** 2626 (1950); Wagner in *Handbook of Preparative Inorganic Chemistry* vol. 2, G. Brauer, Ed. (Academic Press, New York, 2nd ed., 1965) p 1672. Fine, brownish-black to black powder. Stable in atmospheric O_2 and moisture. Practically insol in water, dil acids.

USE: Selective hydrogenation catalysts.

2702. Cupric Citrate. *2-Hydroxy-1,2,3-propanetricarboxylic acid copper salt (1:2);* Cuprocitrol. $C_6H_4Cu_2O_7$; mol wt 315.19. C 22.86%, H 1.28%, Cu 40.32%, O 35.53%. Prepd by the interaction of hot aq solns of $CuSO_4$ and Na citrate: *U.S.D.* 25th ed., p 1653.

Hemipentahydrate, green or bluish-green, odorless, cryst powder. Loses its water at about 100°. Slightly sol in water; sol in ammonia, dil acids; slightly sol in cold, freely in hot solns of alkali citrates.

THERAP CAT: Astringent, antiseptic.

2703. Cupric Ferrocyanide. *Hexakis(cyano-C)ferrate-(4−) copper(2+) (1:2); hexacyanoferrate(4−) dicopper(2+);* cupric hexacyanoferrate(II); Hatchett's brown; C.I. Pigment Brown 9. $C_6Cu_2FeN_6$; mol wt 339.04. C 21.26%, Cu 37.49%, Fe 16.47%, N 24.79%. $Cu_2Fe(CN)_6$. Prepd by addition of $K_4Fe(CN)_6$ to an aqueous soln of a sol cupric salt: Weiser *et al., J. Phys. Chem.* **42,** 945 (1938); **46,** 99 (1942).

Reddish-brown powder or cubic crystals; ppts as a colloid or gel. Practically insol in water, dil acids, most organic solvents. Sol in NH_4OH, solns of alkali cyanides.

USE: As pigment; in photographic toning baths; to lower electric resistance of soil and electrode-to-soil contacts.

2704. Cupric Fluoride. CuF_2; mol wt 101.54. Cu 62.58%, F 37.42%. Prepd by fluorinating Cu or a Cu salt with F_2: von Wartenberg, *Z. Anorg. Allgem. Chem.* **241,** 381 (1939); Jache, Cady, *J. Phys. Chem.* **56,** 1106 (1952); Haendler *et al., J. Am. Chem. Soc.* **76,** 2178 (1954); Ritter, Smith, *J. Phys. Chem.* **70,** 805 (1966); **71,** 2036 (1971).

Monoclinic crystals; turns blue in moist air due to formation of dihydrate. mp ~785° (N_2 atm), 950° (HF atm); d 4.85. Soly in water (20°) 4.7 g/100 ml; hydrolyzed to CuFOH in hot water. May be stored in sealed glass ampuls.

Dihydrate, blue monoclinic crystals. Dec above 130°. d_4^{25} 2.934. Slightly sol in cold water, hydrolyzed to oxyfluoride in hot water.

USE: Anhydr in cathodes in nonaq galvanic cells; high temperature fluorinating agent; dihydrate as flux in casting gray iron.

2705. Cupric Formate. Tubercuprose. $C_2H_2CuO_4$; mol wt 153.58. C 15.64%, H 1.31%, Cu 41.38%, O 41.67%. $Cu(HCOO)_2$. Prepd from cupric carbonate and formic acid: Martin, Waterman, *J. Chem. Soc.* **1959**, 1359.

Three forms of anhydrous compd exist: powder-blue, turquoise, or royal blue crystals. Sol in water. Practically insol in most organic solvents.

Dihydrate, very pale blue, monoclinic needles. Loses $2H_2O$ on standing in air. Turquoise, anhydrous modification formed by dehydration at 100° *in vacuo* over phosphoric oxide. Sol in water.

Tetrahydrate, large, light-blue, monoclinic, holohedral prisms. Powder-blue modification formed by dehydration over $CaCl_2$ under reduced press. Sol in water; very slightly sol in alc. Practically insol in most organic solvents.

USE: As antibacterial agent in the treatment of cellulose.

2706. Cupric Gluconate. $C_{12}H_{22}CuO_{14}$; mol wt 453.84. C 31.76%, H 4.89%, Cu 14.00%, O 49.35%. $Cu[CH_2OH-(CHOH)_4CO_2]_2$. Prepd from gluconic acid and basic cupric carbonate: Suzuki *et al.*, **Japan.** pat. **2889('63)** (to Dainippon), *C.A.* **59**, 11264c (1963).

Hydrate, light blue to bluish-green, odorless crystals or cryst powder. Astringent taste. Soly in water 30 g/100 ml at 25°; slightly sol in alc. Practically insol in most other organic solvents.

USE: In dietary supplements as a readily assimilable form of copper; as oral deodorant.

2707. Cupric Glycinate. *Bis(glycinato)copper;* cupric aminoacetate; glycine copper complex; glycocoll-copper. $C_4H_8CuN_2O_4$; mol wt 211.66. C 22.70%, H 3.81%, Cu 30.02%, N 13.23%, O 30.24%. $(H_2NCH_2COO)_2Cu$. Prepn from glycine and a cupric salt: Tomita, *Bull. Chem. Soc. Japan* **34**, 280 (1960).

Monohydrate, long, deep-blue, rhombic needles. Loses H_2O at 123°, chars at 213°, and dec with gas evolution at 228°. Sol in water; slightly sol in alcohol.

Dihydrate, light blue, powdery crystals. Loses one H_2O at 103°, remaining H_2O at about 140°. Dec with gas evolution at about 225°. Sol in water.

USE: In photometric analysis for copper.

THERAP CAT (VET): Nutritional factor. Given parenterally for copper deficiency in ruminants.

2708. Cupric Hexafluorosilicate. Cupric fluosilicate; cupric silicofluoride. CuF_6Si; mol wt 205.62. Cu 30.90%, F 55.44%, Si 13.66%. $CuSiF_6$. Prepn: Worthington, Haring, *Ind. Eng. Chem., Anal. Ed.* **3**, 7 (1931).

Tetrahydrate, blue, monoclinic, efflorescent crystals. d 2.56: Clark *et al.*, *Can. J. Chem.* **47**, 3859 (1969). Readily sol in water. *Keep well closed.*

USE: Dyeing and hardening white marble; treating plant diseases.

2709. Cupric Hydroxide. Copper hydrate; hydrated cupric oxide. CuH_2O_2; mol wt 97.56. Cu 65.13%, H 2.07%, O 32.80%. $Cu(OH)_2$. Commercial prepn: Furness, U.S. pat. **1,800,828** (1931 to Cellosilk); *idem*, U.S. reissue pat. **24,324** (1957 to Copper Research); Rowe, U.S. pat. **2,536,-096** (1951 to Mountain Copper); laboratory prepn: Weiser *et al.*, *J. Am. Chem. Soc.* **64**, 503 (1942); Gauthier, *Bull. Soc. Chim. France* **1960**, 353.

Blue to blue-green gel or light blue cryst powder. Stability is dependent on the method of prepn; may dec to black CuO on standing a few days or on heating. d 3.37. Practically insol in water. Sol in concd alkali when freshly pptd; sol in acids, NH_4OH.

USE: In manuf of rayon, battery electrodes, other Cu salts; as mordant in dyeing; as pigment; in fungicides, insecticides; as feed additive; in treating and staining paper; in prepn of Schweitzer's reagent; in catalysts.

2710. Cupric Nitrate. CuN_2O_6; mol wt 187.56. Cu 33.88%, N 14.94%, O 51.18%. $Cu(NO_3)_2$. Prepn: *Gmelin's, Copper* (8th ed.) **60B**, 164-179 (1958); Addison, Hathaway, *J. Chem. Soc.* **1958**, 3099. Toxicity: H. F. Smyth *et al.*, *Am. Ind. Hyg. Assoc. J.* **30**, 470 (1969).

Large, blue-green, deliquesc, orthorhombic crystals. Sublimes at 150-225°. mp 255-256°. Sol in water, ethyl acetate,

dioxane; dissolves in and reacts vigorously with ether. *Keep well closed.*

Trihydrate, *gerhardite.* Blue, deliquesc, rhombic plates. mp 114.5°; d 2.05. Freely sol in water, alcohol. Practically insol in ethyl acetate. pH of 0.2 molar aq soln 4.0. *Keep well closed.* LD_{50} orally in rats: 0.94 g/kg (Smyth).

Hexahydrate, blue, deliquesc prismatic crystals. Decomposes to trihydrate at 26.4°. d 2.07. Freely sol in water; sol in alc. *Keep well closed.*

Caution: Irritating to skin, mucous membranes.

USE: In light-sensitive reproductive papers; as ceramic color; as mordant and oxidizing agent in textile dyeing and printing; as reagent for burnishing iron, for giving a black "antique" finish to copper, for coloring zinc brown; in nickel-plating baths; in aluminum brighteners; in wood-preservatives, fungicides, herbicides; in pyrotechnic compositions; as catalyst component in solid rocket fuel; as nitrating agent for aromatic organosilicon compds; as catalyst for organic reactions.

2711. Cupric Oleate. *9-Octadecenoic acid copper salt.* $C_{36}H_{66}CuO_4$; mol wt 626.46. C 69.02%, H 10.62%, Cu 10.14%, O 10.22%. $(C_{17}H_{33}COO)_2Cu$. Prepd from $CuSO_4$ and K oleate: Nelson, Pink, *J. Chem. Soc.* **1954**, 4412.

Blue to green solid. Practically insol in water. Slightly sol in alcohol; sol in ether.

Note: Ingredient of *Cuprex*, a hydrocarbon mixture used as a topical pediculicide.

USE: In antifouling compositions; as emulsifier and dispersing agent; as antioxidant in lubricating oils; as combustion-improver in fuel oils; as stabilizer for amide polymers; as catalyst.

2712. Cupric Oxalate. *Ethanedioic acid copper salt.* C_2CuO_4; mol wt 151.57. C 15.85%, Cu 41.93%, O 42.22%. CuC_2O_4. Prepd by reaction of $CuSO_4$ with oxalic acid: David, *Bull. Soc. Chim. France* **1960**, 719. Usually contains some water.

Blue-white powder. Loses any hydrated water by 200°; dec in air at 310° to CuO. Practically insol in water, alcohol, ether, acetic acid; sol in NH_4OH.

USE: As catalyst for organic reactions; as stabilizer for acetylated polyformaldehyde; in anticaries compositions; in seed treatments to repel birds and rodents.

2713. Cupric Oxide. Black copper oxide. CuO; mol wt 79.55. Cu 79.89%, O 20.11%. Occurs in nature as the minerals *tenorite* (triclinic crystals) and *paramelaconite* (tetrahedral, cubic crystals). Prepn: Glemser, Sauer in *Handbook of Preparative Inorganic Chemistry*, vol. 2, G. Brauer, Ed. (Academic Press, New York, 2nd ed., 1965) p 1012.

Black to brownish-black amorphous or cryst powder or granules. d_4^{14} 6.315. Practically insol in water, alc. Soly study: Hayward *et al.*, *Nature* **215**, 730 (1967). Sol in dil acids, alkali cyanides, $(NH_4)_2CO_3$ soln; slowly sol in NH_3.

USE: As pigment in glass, ceramics, enamels, porcelain glazes, artificial gems; in manuf of rayon, other Cu compds; in sweetening petr gases; in galvanic electrodes; as flux in Cu metallurgy; in correcting Cu deficiencies in soil; as optical-glass polishing agent; to impart flux-and abrasion-resistance to glass fibers; in antifouling paints, pyrotechnic compositions; welding fluxes for bronze; as exciter in phosphor mixtures; as catalyst for organic reactions.

2714. Cupric Perchlorate. Cl_2CuO_8; mol wt 262.45. Cl 27.02%, Cu 24.21%, O 48.77%. $Cu(ClO_4)_2$. Prepd from $Cu(NO_3)_2$ and perchloric acid or nitrosyl perchlorate: Caven, Bryce, *J. Chem. Soc.* **1934**, 514; Hathaway, Underhill, *ibid.* **1960**, 648; **1961**, 3091; Hathaway, *Proc. Chem. Soc.* **1958**, 344.

Very pale green, hygroscopic crystals. Volatilizes on heating. Thermally stable to 130°; above 130° dec to a basic perchlorate. mp about 230-240°. Soluble in water, ether, dioxane, ethyl acetate. Practically insol in benzene, CCl_4, hexane.

Hexahydrate, deep blue, monoclinic crystals. Freely sol in water, methanol, ethanol, acetic acid, acetic anhydride, acetone; slightly sol in ether, ethyl acetate.

USE: Analytical reagent: Kolb, *Ind. Eng. Chem., Anal. Ed.* **11**, 197 (1939); **16**, 38 (1944); Serjeant, *Nature* **186**, 963

(1960). Also in copper electrodeposition; in catalysts for combustion and propellants.

2715. Cupric p-Phenolsulfonate. *p-Hydroxybenzenesulfonic acid copper salt;* cupric sulfocarbolate; Cupriaseptol. $C_{12}H_{10}CuO_8S_2$; mol wt 409.88. C 35.16%, H 2.46%, Cu 15.50%, O 31.23%, S 15.65%. $[C_6H_4(OH)SO_3]_2Cu$. Prepd by double decomposition of $CuSO_4$ and Ba *p*-phenolsulfonate: Lagenbeck, Mahrwald, *Ann.* **605**, 111 (1957).

Hexahydrate, blue to green, rhombic prisms or needles. Soluble in water, alcohol.

USE: In baths for Cu electroplating; esterification catalyst.

2716. Cupric Phosphate. $Cu_3O_8P_2$; mol wt 380.58. Cu 50.09%, O 33.63%, P 16.28%. $Cu_3(PO_4)_2$. Basic salts occur in nature as the minerals: *cornetite, dihydrite, libethenite, phosphorochalcite, pseudolibethenite, pseudomalachite, tagilite.* Prepn from $CuSO_4$ and $(NH_4)_2HPO_4$: Klement, Haselbeck, *Z. Anorg. Allgem. Chem.* **334**, 27 (1964).

Trihydrate, blue or olive orthorhombic crystals or blue to bluish-green powder. Dec on heating. Practically insol in cold water. Slightly sol in hot water; sol in acids, NH_4OH.

USE: Fungicide; catalyst for organic reactions, fertilizer; emulsifier; corrosion inhibitor for H_3PO_4; protectant for metal surfaces against oxidation.

2717. Cupric Salicylate. *2-Hydroxybenzoic acid copper salt.* $C_{14}H_{10}CuO_6$; mol wt 337.78. C 49.78%, H 2.98%, Cu 18.81%, O 28.42%. $Cu[C_6H_4(OH)COO]_2$. Prepd from reaction of $CuSO_4$ with Na salicylate: Hanic, Michalov, *Acta Cryst.* **13**, 299 (1960); Inoue et al., *Inorg. Chem.* **3**, 239 (1964).

Brown, blue-green or green powder. Absorbs water in moist air to form the tetrahydrate.

Tetrahydrate, pale-blue needles or blue-green plates. Blue-green modification loses water on standing in air. Sol in water, alcohol, NH_4OH.

USE: In the separation of monoolefinic from di- or polyolefinic hydrocarbons.

2718. Cupric Selenate. CuO_4Se; mol wt 206.50. Cu 30.77%, O 30.99%, Se 38.24%. $CuSeO_4$. Prepd by the action of selenic acid on cupric carbonate: Klein, *Ann. Chim.* **14**, 263 (1940).

Pentahydrate, light-blue triclinic crystals. Loses water above 80°, forming the monohydrate at 150-220° and becoming anhyd by 265°; dec to the selenite and basic selenite at ~480° and to CuO at ~700°. d 2.56. Sol in water; very slightly sol in acetone. Practically insol in alcohol.

USE: In coloring Cu or Cu alloys black.

2719. Cupric Selenide. CuSe; mol wt 142.51. Cu 44.59%, Se 55.41%. Occurs in nature as the mineral *klockmannite.* Prepd by reduction of cupric selenite with hydrazine: Benzing et al., *J. Am. Chem. Soc.* **80**, 2657 (1958); Kulifay, *J. Inorg. Nucl. Chem.* **25**, 75 (1965).

Blue-black to greenish-black prismatic needles or hexagonal plates. Dec at a dull red heat. d 6.0-6.6. Sol in HCl with H_2Se evolved, in H_2SO_4 with SO_2 evolved; oxidized to $CuSeO_3$ by HNO_3.

USE: Catalyst in Kjeldahl digestions; in semiconductors.

2720. Cupric Selenite. CuO_3Se; mol wt 190.50. Cu 33.36%, O 25.20%, Se 41.45%. $CuSeO_3$. Occurs in nature as the dihydrate, *chalcomenite.* Prepd by the action of selenious acid or an alkaline selenite on a cupric salt: Rocchiccioli, *Compt. Rend.* **247**, 1108 (1958).

Dihydrate, blue, orthorhombic or monoclinic crystals. Becomes anhyd by 265°. Dec to $CuSeO_3 \cdot CuO$ above 460° and to CuO above 660°. d 3.31. Practically insol in water, H_2SeO_3. Sol in acids, NH_4OH.

2721. Cupric Stearate. *Octadecanoic acid copper salt.* $C_{36}H_{70}CuO_4$; mol wt 630.50. C 68.58%, H 11.19%, Cu 10.08%, O 10.15%. $(C_{17}H_{35}COO)_2Cu$. Prepd by metathesis of alcohol cupric acetate with an alcohol soln of stearic acid: Martin, Waterman, *J. Chem Soc.* **1957**, 2545; Rai, Mehrotra, *J. Inorg. Nucl. Chem.* **21**, 311 (1961).

Pale-blue to blue-green amorphous powder. Considered to be dimeric. mp ~250°: Grant, *Can. J. Chem.* **42**, 951 (1964). Practically insol in water, methanol, ethanol, acetone, ether. Sol in pyridine, dioxane, acetic anhydride; sol

in hot but practically insol in cold benzene, toluene, CCl_4, $CHCl_3$, ethyl acetate.

USE: In antifouling paints; in textile and wood preservatives; in rubber aging; as catalyst.

2722. Cupric Sulfate. CuO_4S; mol wt 159.61. Cu 39.81%, O 40.10%, S 20.09%. $CuSO_4$. Occurs in nature as the mineral *hydrocyanite.* Commercial preparation of pentahydrate: Faith, Keyes & Clark's *Industrial Chemicals,* F. A. Lowenheim, M. K. Moran, Eds. (Wiley-Interscience, New York, 4th ed., 1975) pp 280-284; other prepns: *Gmelin's, Copper* (8th ed.) **60B**, 491-560 (1958). Toxicity: Smyth et al., *Am. Ind. Hyg. Assoc. J.* **30**, 470 (1969).

Grayish-white to greenish-white rhombic crystals or amorphous powder. On heating dec above 560°. d 3.6. Hygroscopic. Sol in water. Practically insol in alcohol. *Keep tightly closed.*

Monohydrate, *dried cupric sulfate.* Hygroscopic, almost white powder. Sol in water. Practically insol in alcohol. *Keep tightly closed.*

Pentahydrate, *bluestone, blue vitriol, Roman vitriol, Salzburg vitriol.* Occurs in nature as the mineral *chalcanthite.* Large, blue or ultramarine, triclinic crystals or blue granules or light-blue powder. Slowly efflorescent in air. Loses $2H_2O$ at 30°, 2 more H_2O at 110°; becomes anhyd by 250°. $d_4^{15.6}$ 2.286. Very sol in water; sol in methanol, glycerol; slightly sol in ethanol. pH of 0.2 molar aq soln 4.0. LD_{50} orally in rats: 960 mg/kg (Smyth).

Caution: A strong irritant. *See also* Copper.

USE: Anhyd salt for detecting and removing trace amounts of water from alcohols and other organic compds; as fungicide. Pentahydrate as agricultural fungicide, algicide, bactericide, herbicide; food and fertilizer additive; in insecticide mixtures; in manuf of other Cu salts; as mordant in textile dyeing; in prepn of azo dyes; in preserving hides; in tanning leather; in preserving wood; in electroplating solns; as battery electrolyte; in laundry and metal-marking inks; in petroleum refining; as flotation agent; pigment in paints, varnishes and other materials; in mordant baths for intensifying photographic negatives; in pyrotechnic compositions; in water-resistant adhesives for wood; in metal coloring and tinting baths; in antirusting compositions for radiator and heating systems; as reagent toner in photography and photoengraving; etc.

THERAP CAT: Antidote to phosphorus; antifungal (topical).

THERAP CAT (VET): Nutritional factor. Used in copper deficiency of ruminants. Has also been used as an anthelmintic, emetic, fungicide.

2723. Cupric Sulfate, Basic. *Copper hydroxide sulfate;* cupric subsulfate; Basi-Cop; Cuproxat. Salts of variable compositions of cupric sulfate and cupric hydroxide or oxide. Occurs in nature as the mineral *dolerophane,* $CuSO_4 \cdot$ CuO. *Reviews:* Frear, *Chemistry of the Pesticides* (Van Nostrand, New York, 3rd ed., 1955) pp 316-323; idem, *Agricultural Chemistry,* **vol. 2** (Van Nostrand, New York, 1951) pp 524-525; *Gmelin's, Copper* (8th ed.) **60B**, 579-589 (1958).

Copper sulfate tribasic, $Cu_4H_6O_{10}S$. Occurs in nature as the mineral *brochantite;* monohydrate as the mineral *langite.* Initial product formed during preparation of *Bordeaux mixture* from $CuSO_4$ and $Ca(OH)_2$: Frear, *loc. cit.* Also prepd from $CuSO_4$ and $Cu(OH)_2$: Denk, Leschhorn, *Z. Anorg. Allgem. Chem.* **336**, 58 (1965). Very finely divided, light blue, gelatinous particles. Practically insol in water. Sol in plant and mineral acids, NH_4OH.

Copper sulfate dibasic, $Cu_3H_4O_8S$. Occurs in nature as the mineral *antlerite.* Blue-green, rhombic, bipyramidal crystals. Practically insol in water.

See also Burgundy Mixture.

USE: Fungicide for plants, seed treatment.

2724. Cupric Sulfide. CuS; mol wt 95.61. Cu 66.46%, S 33.54%. Occurs in nature as the mineral *covellite,* or *indigo copper* (blue, hexagonal or monoclinic crystals). Prepn: Glemser, Sauer in *Handbook of Preparative Inorganic Chemistry,* **vol. 2,** G. Brauer, Ed. (Academic Press, New York, 2nd ed., 1965) p 1017.

Black powder. Stable in air when dry, oxidized to $CuSO_4$ by moist air. Practically insol in water, alcohol, dil acids, alkalies. Sol in KCN soln, NH_4OH, hot HNO_3.

USE: In antifouling paints; in prepn of mixed catalysts; in development of aniline black dye in textile printing.

2725. Cupric Tartrate. *2,3-Dihydroxybutanedioic acid copper salt.* $C_4H_4CuO_6$; mol wt 211.62. C 22.70%, H 1.91%, Cu 30.03%, O 45.36%. Prepd by reaction of K tartrate with $Cu(NO_3)_2$: Kirschner, Kiesling, *J. Am. Chem. Soc.* **82**, 4174 (1960).

Trihydrate, green to blue, odorless powder. Slightly sol in water; sol in acids, alkali solns.

USE: In baths for Cu electroplating.

2726. Cupric Tungstate(VI). Cupric wolframate. CuO_4W; mol wt 311.38. Cu 20.41%, O 20.55%, W 59.04%. $CuWO_4$. Prepn: Kosek *et al.*, *Coll. Czech. Chem. Commun.* **24**, 2034 (1959).

Dihydrate, light green powder. Becomes brown to greyish-yellow on heating, with loss of H_2O. Practically insol in water. Slightly sol in acetic acid; sol in NH_3, H_3PO_4; dec by mineral acids.

USE: In semiconductors, nuclear reactors; as catalyst for polyester formation.

2727. Cuprobam. *(Dimethyldithiocarbamato)copper compd with copper chloride;* dimethyldithiocarbamic acid copper salt complex with copper chloride; cuprous dimethyldithiocarbamate cuprous chloride complex; cuprobamé. Prepd as monocuprous chloride complex [$(CH_3)_2NCSSCu$.-$CuCl$]: Couillard, Racine, U.S. pat. 3,037,041 (1962 to Roussel-UCLAF). Prepd as dicuprous chloride complex [$(CH_3)_2NCSSCu.2CuCl$]: **Brit.** pat. 825,901 (1959 to UCLAF).

Both forms are practically insol in water.

USE: Fungicide, molluscicide.

2728. Cuprous Acetate. *Acetic acid copper(1+) salt.* $C_2H_3CuO_2$; mol wt 122.59. C 19.60%, H 2.47%, Cu 51.84%, O 26.10%. CH_3COOCu. Obtained as a sublimate by heating cupric acetate *in vacuo* to temps above 220°: Angel, Harcourt, *J. Chem. Soc.* **81**, 1385 (1902); prepn from cuprous oxide and acetic acid-acetic anhydride: Shimizu, Weller, *J. Am. Chem. Soc.* **74**, 4469 (1952). Crystal structure: T. Ogura *et al.*, *J. Am. Chem. Soc.* **95**, 949 (1973).

Transparent, leafy crystals. Volatilizes on heating; dec on strong heating. Rapidly hydrolyzed by water with the formation of yellow Cu_2O.

2729. Cuprous Bromide. $BrCu$; mol wt 143.45. Br 55.70%, Cu 44.30%. $CuBr$. Prepn: Briggs, *J. Chem. Soc.* **127**, 496 (1925); Keller, Wycoff, *Inorg. Syn.* **2**, 3 (1946); Wagner, Wagner, *J. Chem. Phys.* **26**, 1597 (1957).

White powder or cubic crystals (zinc blende structure); turns green to dark-blue on exposure to sunlight. mp 504°; bp 1345°; d_4^4 4.72. Slightly sol in cold water, dec in hot water, HNO_3. Sol in HCl, HBr, NH_4OH with formation of complexes. Practically insol in acetone, concd H_2SO_4. *Keep tightly closed in a dark place.*

USE: As catalyst for organic reactions.

2730. Cuprous Chloride. Cu-lyt. $ClCu$; mol wt 99.00. Cl 35.81%, Cu 64.19%. $CuCl$. Occurs in nature as the mineral *nantokite* (colorless to grey cubic crystals). Prepn: Keller, Wycoff, *Inorg. Syn.* **2**, 1 (1946); Glemser, Sauer in *Handbook of Preparative Inorganic Chemistry*, **vol. 2**, G. Brauer, Ed. (Academic Press, New York, 2nd ed., 1965) p 1005.

White cryst powder or cubic crystals (zinc-blende structure); stable to air and light if dry, but in presence of moisture turns green on exposure to air and blue to brown on exposure to light. mp 430°; d_4^{25} 4.14. Sparingly sol in water with partial decompn; practically insol in alcohol, acetone; sol in concd HCl, concd NH_4OH with formation of complexes. Solns oxidize rapidly in air. *Keep tightly closed and protected from light.*

USE: As catalyst for organic reactions; catalyst, decolorizer and desulfuring agent in petr industry; in denitration of cellulose; as condensing agent for soaps, fats and oils; in gas analysis to absorb carbon monoxide.

2731. Cuprous Cyanide. Cupricin. $CCuN$; mol wt 89.56. C 13.41%, Cu 70.95%, N 15.64%. $CuCN$. Commercial prepns: Winter *et al.*, Kirk-Othmer *Encyclopedia of Chemical Technology*, **vol. 6** (Interscience, New York, 2nd ed., 1965) p 271; laboratory prepns: Barber, *J. Chem. Soc.* **1943**, 79; Norberg, Jacobson, *Acta Chem. Scand.* **3**, 174 (1949); Vaughan, McCane, *J. Am. Chem. Soc.* **76**, 2504 (1954).

White to cream-colored powder, colorless or dark green orthorhombic crystals, or dark red monoclinic crystals. *Poisonous!* mp 474°. Practically insol in water, alcohol, cold dil acids; sol in NH_4OH; dec by HNO_3, boiling dil HCl. Sol in alkali cyanide solns because of formation of stable cyanocuprate(I) ions.

USE: In electroplating Cu or Fe; as insecticide, fungicide; as antifouling agent in marine paints; as polymerization catalyst.

2732. Cuprous Iodide. Hydro-Giene. CuI; mol wt 190.45. Cu 33.37%, I 66.63%. Occurs in nature as the mineral *marshite* (red-brown crystals). Prepn: Kaufman, Pinnel, *Inorg. Syn.* **6**, 3 (1960); Glemser, Sauer in *Handbook of Preparative Inorganic Chemistry*, **vol. 2**, G. Brauer, Ed. (Academic Press, New York, 2nd ed., 1965) p 1007.

Dense powder or cubic crystals (zinc-blende structure). Less photosensitive than CuBr or CuCl. mp 588-606°; bp ~1290°; d_4^{25} 5.63. Extremely insol in water; practically insol in dil acids, alcohol; sol in aq solns of NH_3, alkali cyanides, thiosulfates, iodides; dec by concd H_2SO_4 and HNO_3.

USE: As catalyst in organic reactions; as ice-nucleating chemical; as coating in cathode-ray tubes; as source of iodine in animal feeds; with HgI_2 as temp indicator; bactericide.

2733. Cuprous Mercuric Iodide. *Dicopper tetraiodomercurate(2—);* cuprous tetraiodomercurate(II); mercuric cuprous iodide. Cu_2HgI_4; mol wt 835.30. Cu 15.22%, Hg 24.01%, I 60.77%.

Deep-red, thermochromic cryst powder. *Poison!* Practically insol in water, alcohol.

USE: For detecting overheating of machine bearings, etc., the red color changes to brownish-black at 60-70° and again becomes red on cooling.

2734. Cuprous Oxide. Red copper oxide; C.I. 77402; Perenex; Yellow Cuprocide; Copper-Sandoz; Caocobre. Cu_2O; mol wt 143.09. Cu 88.82%, O 11.18%. Occurs in nature as the mineral *cuprite* (red to reddish-brown octahedral or cubic crystals). Prepd commercially by furnace reduction of mixtures of copper oxides with Cu: Drapeau, Johnson, U.S. pats. **2,758,014; 2,891,842** (1956, 1959 both to Glidden); by decompn of copper ammonium carbonate: Rowe, U.S. pat. **2,474,497**; Klein, U.S. pat. **2,474,533** (both 1949 to Lake Chemical); Rowe, U.S. pat. **2,536,096;** Munn, U.S. pat. **2,670,273** (1951, 1954 both to Mountain Copper); by treatment of $Cu(OH)_2$ with SO_2: Rowe, U.S. pat. **2,665,-192** (1954 to Mountain Chemical); or by electrolysis of an aq soln of NaCl between Cu electrodes: Arend, *Paint Technology* **13**, 265 (1948). Laboratory prepns: Glemser, Sauer in *Handbook of Preparative Inorganic Chemistry*, **vol. 2**, G. Brauer, Ed. (Academic Press, New York, 2nd ed., 1965) p 1011. Toxicity study: H. F. Smyth *et al.*, *Am. Ind. Hyg. Assoc. J.* **30**, 470 (1969).

Cubic crystals or microcrystalline powder. Color may be yellow, red, or brown depending on the method of prepn and the particle size. Stable in dry air; gradually oxidizes in moist air to CuO. mp 1232°; d_4^{25} 6.0. Practically insol in water. Sol in NH_4OH; in HCl forming CuCl which dissolves in excess HCl. With dil H_2SO_4 or dil HNO_3, the cupric salt is formed and half the copper is pptd as the metal. LD_{50} orally in rats: 0.47 g/kg (Smyth).

USE: Fungicide; antiseptic for fishnets; in antifouling paints for marine use; in photoelectric cells; as red pigment for glass, ceramic glazes; in brazing pastes; in rectifiers; as catalyst.

2735. Cuprous Potassium Cyanide. *Potassium bis(cyano-C)cuprate(1—); potassium dicyanocuprate(I);* potassiocuprous cyanide; potassium cyanocuprate(I). C_2CuKN_2; mol wt 154.68. C 15.53%, Cu 41.08%, K 25.28%, N 18.11%. $KCu(CN)_2$. Prepd by evaporation of an aq soln of CuCN and KCN: Staritzky, Walker, *Anal. Chem.* **28**, 419 (1956).

Monoclinic, prismatic crystals. d 2.38. Practically insol in water; dec to CuCN by cold dil H_2SO_4 or heating in water; sol in DMSO.

USE: In electroplating copper and brass.

2736. Cuprous Selenide. Berzeline; selenkupfer. Cu_2-Se; mol wt 206.05. Cu 61.68%, Se 38.32%. Occurs in nature as the mineral **berzelianite** which in addition to copper and selenium contains 4.73-8.50% silver, 0.35-0.54% iron, up to 0.38% thallium and traces of lead and magnesium. Prepd by reduction of an ammoniacal soln of a cupric salt and H_2SeO_3 with hydrazine: Benzing *et al.*, *J. Am. Chem. Soc.* **80**, 2657 (1958); Kulifay, *J. Inorg. Nucl. Chem.* **25**, 75 (1963); alternate methods: Glemser, Sauer in *Handbook of Preparative Inorganic Chemistry*, **vol. 2**, G. Brauer, Ed. (Academic Press, New York, 2nd ed., 1965) p 1019.

Blue-black to black, tetragonal or cubic crystals with a metallic luster. mp 1113°. d_4^{21} 6.84. Sol in HCl with evolution of H_2Se, in H_2SO_4 with evolution of SO_2, in KCN soln; oxidized by HNO_3 to $CuSeO_3$.

USE: In semiconductors.

2737. Cuprous Sulfide. Cu_2S; mol wt 159.16. Cu 79.85%, S 20.15%. Occurs in nature as the mineral **chalcocite** also called **copper glance** (grey, black, green, blue, or violet rhombic crystals). Prepn: Glemser, Sauer in *Handbook of Preparative Inorganic Chemistry*, **vol. 2**, G. Brauer, Ed. (Academic Press, New York, 2nd ed., 1965) p 1016.

Blue to grayish-black lustrous powder, granules, or orthorhombic crystals. mp ~1100°. d_4^{20} 5.6. On heating in the absence of air it forms Cu and CuS; in the presence of air it forms CuO, $CuSO_4$ and SO_2. Practically insol in water, acetic acid; very sparingly sol in HCl; dec by HNO_3, concd H_2SO_4. Partially sol in NH_4OH, readily in cyanide solns due to complex ion formation.

USE: In luminous paints; in electrodes for thermoelements; in preparation of $CuSO_4$; in solid-lubricant compositions; as catalyst.

2738. Cuprous Sulfite. Cu_2O_3S; mol wt 207.16. Cu 61.35%, O 23.17%, S 15.48%. Cu_2SO_3. Prepn and review: Dasent, Morrison, *J. Inorg. Nucl. Chem.* **26**, 1122 (1964).

Hemihydrate, *Etard's salt.* White to pale-yellow hexagonal crystals. Slightly sol in water; sol in HCl, NH_4OH, alkali solns. Practically insol in ether, alcohol.

Cupro-cupric sulfate, $Cu_3O_6S_2$, *Chevreul's salt.* Dihydrate, red microcryst powder or prismatic crystals. Practically insol in water, alcohol. Sol in NH_4OH, HCl.

Rogojski's salt. Formerly considered to be $Cu_2SO_3.H_2O$; shown to be an equimolar mixture of Chevreul's salt and metallic copper (Dasent, Morrison, *loc. cit.*). Brick-red solid.

USE: As fungicide for grape vines; in dyeing polyacrylic fibers; polymerization catalyst. Chevreul's salt is a selective molluscicide.

2739. Cuprous Thiocyanate. Cuprous sulfocyanate; cuprous sulfocyanide. CCuNS; mol wt 121.63. C 9.88%, Cu 52.25%, N 11.52%, S 26.36%. CuSCN. Prepn: Demmerle *et al.*, *Ind. Eng. Chem.* **42**, 2 (1950); Newman, *Analyst* **88**, 500 (1963).

White to yellow amorphous powder. d 2.85. Practically insol in water, dil acids, alc, acetone; sol in NH_4OH, ether, solns of alkali thiocyanates; dec by concd mineral acids.

USE: In marine antifouling paints; in primer compositions for explosives industry.

2740. Cuproxoline. *Tetrahydrogen bis[8-hydroxy-5,7-quinolinedisulfonato(3—)-N^1,O^8]cuprate(4—) compd with N-ethylethanamine (1:4); 8-hydroxy-5,7-quinolinedisulfonic acid copper derivative compound with diethylamine;* cupric bis[8-hydroxyquinoline di(diethylammonium sulfonate)]; copper DOS; Dicuprene; Cujec; Cuprimyl. $C_{34}H_{56}CuN_6O_{14}$-S_4; mol wt 964.66. C 42.33%, H 5.85%, Cu 6.59%, N 8.71%, O 23.22%, S 13.30%. Clinical efficacy in treatment of arthritis: W. C. Kuzell *et al.*, *Ann. Rheum. Dis.* **10**, 328 (1951). Veterinary use as parenteral copper supplement: I. J. Cunningham, *N. Z. Vet. J.* **7**, 15 (1959); N. F. Suttle, *Vet. Rec.* **109**, 304 (1981); G. J. Judson *et al.*, *Aust. Vet. J.* **61**, 40 (1984). Review of use of organic copper complexes in treatment of rheumatoid and degenerative diseases: J. R. J. Sorenson, W. Hangarter, *Inflammation* **2**, 217-238 (1977).

Review of veterinary use: W. M. Allen, C. B. Mallinson, *Vet. Rec.* **114**, 451-454 (1984).

Green platelets. Sol in water, giving a dark-green soln. A 10% aq soln is almost neutral and can be sterilized by autoclaving. LD_{50} i.m. in rats: 126 mg/kg (Kuzell).

THERAP CAT: Antirheumatic.

THERAP CAT (VET): Copper supplement.

2741. Curare. Ourari; urari; woorari; woorali; wourara. Rendering of an Indian name given to the unstandardized extracts derived mainly from the bark of various spp. of Strychnos and Chondodendron, prepared for use as arrow poisons by Indians in the Amazon and Orinoco valleys, and in the Guianas; the physiologically active principle of which is (+)-tubocurarine chloride (*q.v.*). Three kinds of curare have appeared in commerce, distinguished by the kind of containers in which they were packed: *Tube cucare* or *bamboo curare, pot curare,* and *gourd curare* or *calabash curare.* Listing of members of the family *Menispermaceae*—botanical components of tube and bamboo curare and of pot curare: Krukoff, Moldenke, *Studies of American Menispermaceae, with special reference to species used in the preparation of arrow poisons* in *Brittonia* **3**, 1-74 (1938), also suppl. no. 1-5. (Note that the curare available for medical use under the name Intocostrin is a physiologically standardized extract from *Chondodendron tomentosum* R. & P., *Menispermaceae*). Listing of *Strychnos* spp. *Loganiaceae*—botanical components of calabash curare (and of pot curare): Krukoff, Monachino, *The American Species of Strychnos* in *Brittonia* **4**, 248-322 (1942), also suppl. no. 1-6. Alkaloids from calabash curare and bark of *Strychnos* spp.: P. Karrer, *J. Pharm. Pharmacol.* **8**, 161-164 (1956); Schmid, Karrer, *Helv. Chim. Acta* **29**, 1853 (1946); **30**, 1162 (1947); Marino-Bettolo, *Festschrift Arthur Stoll* (Birkhäuser-Verlag, Basel 1957) pp 257-280. Review of history, chemistry, and use of arrow-poison curare: McIntyre, *Curare* (Chicago, 1947); Bovet *et al.*, *Curare and Curare-like Agents* (Van Nostrand, Princeton, 1959).

Curare is sol in water and in dil alcohol. Stable aq solns (in ampuls) are standardized to contain 20 units per ml.

THERAP CAT: Muscle relaxant (skeletal).

THERAP CAT (VET): Muscle relaxant.

2742. *C*-Curarine I. $[C_{40}H_{44}N_4O]^{2+}$. Isoln from calabash-curare: Wieland, Pistor, *Ann.* **536**, 68 (1938); Karrer, Schmid, *Helv. Chim. Acta* **29**, 1853 (1946); Zürcher *et al.*, *J. Am. Chem. Soc.* **80**, 1500 (1958). Structure: Nagyvàry *et al.*, *Tetrahedron* **14**, 138 (1961); Grdinic *et al.*, *J. Am. Chem. Soc.* **86**, 3357 (1964); Jones, Nowacki, *Chem. Commun.* **1972**, 805. Prepn of *C*-dihydrotoxiferine chloride: Bernauer *et al.*, *Helv. Chim. Acta* **40**, 1999 (1957).

Dichloride, $C_{40}H_{44}Cl_2N_4O$, needles from methanol + ether, mp > 350°. $[\alpha]_D^{20}$ +73.6° (c = 1 in water). uv max (95% alc): 260, 296 nm (log ε 4.41, 4.07). Sol in water, alc; practically insol in ether, acetone.

2743. C-Curarine III. C-Fluorocurarine. $[C_{20}H_{23}N_2-O]^+$. From calabash-curare: Wieland et al., Ann. **547**, 140 (1941); Schmid et al., Helv. Chim. Acta **35**, 1864 (1952); Zürcher et al., J. Am. Chem. Soc. **80**, 1500 (1958). Identity with C-fluorocurarine: Philipsborn et al., Helv. Chim. Acta **41**, 1257 (1958). Structure: Philipsborn et al., ibid. **42**, 461 (1959). Synthesis from C-curarine I: Boekelheide et al., J. Am. Chem. Soc. **81**, 2256 (1959); from strychnine: Fritz et al., Ann. **663**, 150 (1963).

Chloride, $C_{20}H_{23}ClN_2O$, rods from abs alcohol, dec 270-274°. $[\alpha]_D^{20}$ −937° (water). uv max (methanol): 240, 300, 360 nm (log ε 4.00, 3.60, 4.23).

2744. Curcumin. (E,E)-1,7-Bis(4-hydroxy-3-methoxyphenyl)-1,6-heptadiene-3,5-dione; turmeric yellow; diferuloylmethane; C.I. 75300; C.I. Natural Yellow 3. $C_{21}H_{20}O_6$; mol wt 368.39. C 68.47%, H 5.47%, O 26.06%. Natural dyestuff from root of Curcuma longa L., Zingiberaceae. Isoln: Vogel, Ann. **44**, 297 (1842); Perkin, Phipps, J. Chem. Soc. (Trans.) **85**, I, 64 (1904); Rao, Shintre, J. Soc. Chem. Ind. **47**, 54T (1928). Synthesis: Lampe, Ber. **51**, 1347 (1918). Production: Stieglitz, Horn, Ger. pat. 859,145 (1952 to Hoechst). Biosynthesis studies: Roughley, Whiting, Tetrahedron Letters **1971**, 3741. Chromatography: Srinivasan, J. Pharm. Pharmacol. **5**, 448 (1953). See also: H. J. Conn's Biological Stains, R. D. Lillie, Ed. (Williams & Wilkins, Baltimore, 9th ed., 1977) pp 474-476. Pharmacology and anti-inflammatory activity: Srimal, Dhawan, ibid. **25**, 447 (1973).

Orange-yellow, cryst powder, mp 183°. Insol in water, ether. Sol in alcohol, glacial acetic acid. Gives a brownish-red color with alkali; a light-yellow color with acids.
USE: For preparing curcuma paper, pH range 8-9. In the detection of boron.

2745. Curine. (1β)-6,6'-Dimethoxy-2,2'-dimethyltubocuraran-7',12'-diol; l-bebeerine. $C_{36}H_{38}N_2O_6$; mol wt 594.71. C 72.71%, H 6.44%, N 4.71%, O 16.14%. From tubocurare (Chondodendron tomentosum R. & P., Menispermaceae): Boehm, Arch. Pharm. **235**, 660 (1897); Späth et al., Ber. **61**, 1698 (1928). Structure: Späth, Kuffner, Ber. **67**, 55 (1934); Faltis et al., Ber. **69**, 1269 (1936); King, J. Chem. Soc. **1939**, 1157. Configuration: Bick, Clezy, ibid. **1953**, 3893; Hultin, Acta Chem. Scand. **17**, 753 (1963). See Bebeerine for the structural formula.

Efflorescent crystals from methanol. mp 213°; mp 221° in vacuo; four sided prisms from benzene containing one mol benzene, mp 161°. $[\alpha]_D^{20}$ −328° (pyridine). Sol in benzene, chloroform, pyridine.
Hydrochloride, $C_{36}H_{38}N_2O_6 \cdot HCl$, crystals, dec 273°, sol in water and alcohol.
Dihydrochloride, $C_{36}H_{38}N_2O_6 \cdot 2HCl$, crystals.
Methiodide, $C_{36}H_{38}N_2O_6 \cdot CH_3I$, crystals, mp 253°.

2746. Curium. Cm; at. no. 96; valence 3, 4. Manmade radioactive element. No stable nuclides; known isotopes (mass numbers): 238-251. Longest-lived known isotope: 247 ($T_{1/2}$ 1.56 × 10⁷ years, rel. at. mass 247.0703, α-decay). Prepn of first isotope, ²⁴²Cm ($T_{1/2}$ 162.9 days, α-decay), in 1944 by G. T. Seaborg et al. in The Transuranium Elements, G. T. Seaborg et al., Eds. (McGraw-Hill, New York, 1949) pp 1554-1571. Prepn of metal: J. C. Wolfmann et al., J. Am. Chem. Soc. **73**, 493 (1951). Properties: Cunningham, Wallman, J. Inorg. Nucl. Chem. **26**, 271 (1964); Smith et al., J. Chem. Phys. **50**, 5066 (1969). Reviews: M. Haissinsky, J.-P. Adloff, Radiochemical Survey of the Elements (Elsevier, New York, 1965) pp 41-43; C. Keller, The Chemistry of the Transuranium Elements (Verlag Chemie, Weinheim, English Ed., 1971) pp 529-551; Comprehensive Inorganic Chemistry vol 5, J. C. Bailar, Jr. et al., Eds. (Pergamon Press, Oxford, 1973) passim; Handb. Exp. Pharmakol. **36**, 689-928 (1973); P. G. Eller, R. A. Penneman in The Chemistry of the Actinide Elements vol. 2, J. J. Katz et al., Eds. (Chapman and Hall, New York, 1986) pp 962-988.

Silvery, lustrous, malleable metal. Two allotropic modifications: double hexagonal close-packed α-form, d 13.51, transforms to β-form at ~1277°; face-centered cubic β-form, d 12.9, exists from ~1277° to mp. mp 1345°. bp (calc) 3110°. Metal dissolves rapidly in dilute acid solns; oxidizes rapidly in the presence of traces of oxygen. Pyrophoric when finely divided. Chemistry of trivalent state similar to that of trivalent lanthanides.

Caution: Radiation hazard; handling requires special equipment and sheilding facilities (Katz et al., loc. cit. vol. 2, p 1128). Max allowable concn of sol ²⁴⁴Cm in air: 3 × 10⁻¹² µCi/cc; of insol ²⁴⁴Cm: 3 × 10⁻¹¹ µCi/cc, National Bureau of Standards Handbook 69, 89 (1959).

USE: ²⁴²Cm and ²⁴⁴Cm as power sources in radionuclide batteries for space and medical applications. ²⁴²Cm as radioactive heat source. ²⁴⁸Cm in accelerator studies to form superheavy elements.

2747. Curvularin. 4,5,6,7,8,9-Hexahydro-11,13-dihydroxy-4-methyl-2H-3-benzoxacyclododecin-2,10(1H)-dione. $C_{16}H_{20}O_5$; mol wt 292.33. C 65.74%, H 6.90%, O 27.37%. Mold metabolite from a species of Convalaria: Musgrave, J. Chem. Soc. **1956**, 4301; from Penicillium steckii: Fennell et al., Chem. & Ind. (London) **1959**, 1382. Structure: Birch et al., J. Chem. Soc. **1959**, 3146. Synthesis: H. Gerlach, Helv. Chim. Acta **60**, 3039 (1977); of dl-O,O-dimethyl ether: Baker et al., J. Chem. Soc. (C) **1967**, 1913; T. Takahashi et al., Tetrahedron Letters **21**, 3885 (1980); H. H. Wasserman, R. J. Gambale, ibid. **22**, 4849 (1981).

Plates from hot benzene + methanol, mp 206-206.5°. $[\alpha]_D^{18}$ −36.3° (c = 3.8 in ethanol). uv max (ethanol): 223, 272, 304.5 nm (ε 11300; 6350; 5100). Sol in ethanol, methanol, dioxane, acetone, pyridine, concd H_2SO_4; moderately sol in acetic acid, ether; sparingly sol in benzene, petr ether, chloroform, water. Gives yellow solns with aq ammonia, sodium carbonate, and sodium hydroxide. Alkaline solns darken rapidly in air, eventually becoming purple.

Dibenzoate, $C_{30}H_{28}O_7$, needles from benzene + petr ether, mp 133-134°. $[\alpha]_D^{18}$ −10.8° (c = 1.9 in chloroform). uv max (ethanol): 236 nm (ε 36700).

O,O-Dimethyl ether, $C_{18}H_{24}O_5$, short rods from aq ethanol, mp 72°. $[\alpha]_D^{18}$ −2.9° (c = 2.7 in chloroform). uv max (ethanol): 223, 267.5 nm (ε 10500; 5100).

2748. Cuscohygrine. *1,3-Bis(1-methyl-2-pyrrolidinyl)-2-propanone;* cuskhygrine; belladarine. $C_{13}H_{24}N_2O$; mol wt 224.35. C 69.60%, H 10.78%, N 12.49%, O 7.13%. In coca leaves of various origin. Found in crude hygrine. Readily converted to hygrine by acids and bases. Isoln: Liebermann, *Ber.* **22**, 679 (1898); Liebermann, Cybulski, *ibid.* **28**, 578 (1895). Identity with belladarine: Steinegger, Phokas, *Pharm. Acta Helv.* **30**, 441 (1955). Structure: Hess, Fink, *Ber.* **53**, 794 (1920); Sohl, Shriner, *J. Am. Chem. Soc.* **55**, 3828 (1933); Rapoport, Jorgensen, *J. Org. Chem.* **14**, 664 (1949). Synthesis: Späth, Tuppy, *Monatsh.* **79**, 119 (1948); Galinovsky et al., *ibid.* **82**, 551 (1951). Enzymatic synthesis: Tuppy, Faltaous, *ibid.* **91**, 167 (1960). Stereochemistry: Galinovsky, Zuber, *ibid.* **84**, 798 (1953). Biosynthesis: E. Leete, *Chem. Commun.* **1980**, 1170.

Oily liquid, bp_{23} 169-170°; bp_{14} 152°; bp_2 118-125°. d_4^{20} 0.9733. n_D^{20} 1.4832. Miscible with water. Sol in alcohol, ether, benzene.

Hemiheptahydrate, needles, mp 40°.

Hydrobromide, $C_{13}H_{24}N_2O.2HBr$, prisms from alc, mp 234°.

2749. Cuspareine. *2-[2-(3,4-Dimethoxyphenyl)ethyl]-1,-2,3,4-tetrahydro-1-methylquinoline.* $C_{20}H_{25}NO_2$; mol wt 311.42. C 77.14%, H 8.09%, N 4.50%, O 10.27%. From Angostura bark (*Cusparia trifoliata* (Willd.) Engl., *Rutaceae):* Tröger, Runne, *Arch. Pharm.* **249**, 174 (1911); Späth, Pikl, *Monatsh.* **55**, 352 (1930). Structure: Schläger, Leeb, *ibid.* **81**, 714 (1950). Synthesis: Stanek, *ibid.* **88**, 250 (1957).

Crystals from ether, mp 56°. $[\alpha]_D^{20}$ −20.4° (c = 6.8 in abs ethanol).

Methyliodide, $C_{20}H_{25}NO_2.CH_3I$, crystals, mp 156°.

2750. Cusparine. *2-[2-(1,3-Benzodioxol-5-yl)ethyl]-4-methoxyquinoline.* $C_{19}H_{17}NO_3$; mol wt 307.35. C 74.25%, H 5.58%, N 4.56%, O 15.62%. From Angostura bark: Tröger, Runne, *Arch. Pharm.* **249**, 174 (1911); Späth, Pikl, *Monatsh.* **55**, 352 (1930). Synthesis from 2-methyl-4-methoxyquinoline and piperonal: Späth, Brunner, *Ber.* **57**, 1243 (1924).

Apparently trimorphous. White or yellow needles from petr ether, mp 92°, or amber prisms, mp 110-122°. Freely sol in organic solvents; practically insol in water.

2751. Cyacetacide. *Cyanoacetic acid hydrazide;* malononitrile hydrazide; cyanoacetohydrazide; cyanazide; cyanizide; cyanacethydrazine; cyanacethydrazide; cyanoethydrazide; cyanacetylhydrazide; Dictyzide; Mackreazid; Armazal; Reacid; Reazide; Hidacian; Leandin; Neohydrazid; Dictycide. $C_3H_5N_3O$; mol wt 99.09. C 36.36%, H 5.09%, N 42.41%, O 16.15%. $NCCH_2CONHNH_2$. Prepd by boiling ethyl or methyl cyanoacetate with hydrazine hydrate in alc: Rothenburg, *Ber.* **27**, 687 (1894); Darapsky, Hillers, *J. Prakt. Chem.* **92**[2], 313 (1915); Klosa, *Arch. Pharm.* **288**, 453 (1955). Prepn of dosage forms: Muset, U.S. pat. **2,849,369** (1958 to Labs OM).

Stout prisms from alc, mp 114.5-115° (evac tube). Rapidly dec by heat. Very slightly acid to litmus. Very sol in water. Sol in alcohol. Practically insol in ether. Aq solns become discolored after a few days and the pH becomes alkaline.

Hydrochloride, $C_3H_5N_3O.HCl$, crystals, mp 145°. Freely sol in water. Acid reaction. Aq solns are more stable than those of the free hydrazide.

THERAP CAT: Antibacterial (tuberculostatic).

THERAP CAT (VET): Anthelmintic (lungworms).

2752. Cyamelide. *1,3,5-Trioxane-2,4,6-triimine;* insoluble cyanuric acid. $C_3H_3N_3O_3$; mol wt 129.08. C 27.92%, H 2.34%, N 32.55%, O 37.19%. Formed by polymerization of cyanic acid (gaseous or liquid). Prepd by digesting equal parts of potassium cyanate and oxalic acid: Liebig, Wöhler, *Berzelius Jahresber.* **11**, 85. Crude cyamelide contains cyanuric acid which is removed by washing with water: Senier, Walsh, *J. Chem. Soc.* **81**, 290 (1902). Mechanism of polymerization: Werner, Fearon, *ibid.* **117**, 1358 (1920).

White amorphous powder. Practically insol in water. Insol in organic solvents. Slightly sol in ammonia water, in concd acids, in concd NaOH with salt formation.

2753. Cyamemazine. *10-[3-(Dimethylamino)-2-methylpropyl]-10H-phenothiazine-2-carbonitrile;* 2-cyano-10-(3-dimethylamino-2-methylpropyl)phenothiazine; 10-(3-dimethylamino-2-methylpropyl)-3-cyanophenothiazine; cyamepromazine; RP-7204; TH-2602; Kyamepromazine; Cianatil; Tercian. $C_{19}H_{21}N_3S$; mol wt 323.46. C 70.55%, H 6.54%, N 12.99%, S 9.91%. Prepn: Jacob, Robert, U.S. pat. **2,877,224** (1959 to Rhône-Poulenc). Synthesis: Craig et al., *J. Org. Chem.* **26**, 1138 (1961). Biotransformation studies: Robinson, *J. Pharm. Pharmacol.* **18**, 19 (1966). Chemistry: Kiger, Kiger, *Ann. Pharm. Franc.* **23**, 489 (1965).

Yellow oil, $bp_{0.2-0.5}$ 205-220°; also reported as yellow powder, mp 89-96° (Kiger). Practically insol in water. Sol in ethanol and in organic solvents.

Maleate, $C_{23}H_{25}N_3O_4S$, pale yellow plates from methanol-ethanol, mp 196-197° (Craig).

THERAP CAT: Antipsychotic.

2754. Cyanamide. Carbodiimide; hydrogen cyanamide; carbimide; cyanogenamide; amidocyanogen. CH_2N_2; mol wt 42.04. C 28.57%, H 4.80%, N 66.63%. $H_2NC{\equiv}N$. Prepd commercially by continuous carbonation of calcium cyanamide in water according to the equations: 2CaNCN

+ $2H_2O \rightarrow Ca(HNCN)_2 + Ca(OH)_2$; $Ca(HNCN)_2 + CO_2 + H_2O \rightarrow 2H_2NCN + CaCO_3$. *Ref.* Brochure "Cyanamide," edited by Process Chemicals Dept., Am. Cyanamid (New York, Aug. 1959) p 19.

Deliquescent, orthorhombic, elongated, six-sided tablets from dimethyl phthalate. d_4^{20} 1.282. mp 45-46°. $bp_{0.5}$ 83°. Formn of dicyandiamide begins at 122°. Dipole moment in benzene at 20°: 3.8. Cryoscopic constant (water) 26.8-28.4. Sp ht 0.547 cal/g/°C between 0° and 39°. Heat of formation 14.05 kcal/mole (25°); heat of combustion −176.4 kcal/mole (25°); heat of fusion 2.1 kcal/mole. Heat of vaporization 16.4 kcal/mole. Soly (g/100 g soln) in water at 15°: 77.5, at 43°: 100; in butanol at 20°: 28.8, in methyl ethyl ketone: 50.5, in ethyl acetate: 42.4. Sol in alcohols, phenols, amines, ethers, ketones. Very sparingly sol in benzene, halogenated hydrocarbons. Practically insol in cyclohexane. Solid cyanamide should be stored in a cool, dry place. Polymerizes at 122°. Optimum pH for storage of solns is ∼4. Attacks various metals. Solns can be stored in glass provided they are stabilized with phosphoric, acetic, sulfuric, or boric acid. LD_{50} in rats: 125 mg/kg orally.

Note: Term "cyanamide" is also used to designate calcium cyanamide.

Caution: Very irritating and caustic; produces severe dermatitis on moist skin. Inhalation may cause irritation of mucous membranes. Ingestion or inhalation: transitory intense redness of face, headache, vertigo, increased respirations, tachycardia, hypotension. Does not produce "cyanide" effect.

2755. Cyanazine. *2-[[4-Chloro-6-(ethylamino)-1,3,5-triazin-2-yl]amino]-2-methylpropanenitrile; 2-[[4-chloro-6-(ethylamino)-s-triazin-2-yl]amino]-2-methylpropionitrile;* SD-15418; WL-19805; DW-3418; Bladex; Fortrol. $C_9H_{13}ClN_6$; mol wt 240.70. C 44.91%, H 5.44%, Cl 14.73%, N 34.92%. Selective pre-emergence herbicide. Prepn: **Brit.** pat. **1,132,-306;** W. Schwarze, **U.S.** pat. **3,505,325** (1968, 1970 both to Degussa). Activity: T. Chapman *et al., Proc. Brit. Weed Control Conf., 9th* **2,** 1018 (1968). Metabolism: J. V. Crayford, D. H. Hutson, *Pestic. Biochem. Physiol.* **2,** 295 (1975). Persistence in soil: J. T. Majka, T. L. Lavy, *Weed Sci.* **25,** 401 (1977).

White crystals, mp 167.5-169°. Vapor pressure at 20°: 1.6 × 10⁻⁹ mmHg. Soly in water at 25°: 171 mg/l. Soly at 25° (g/l): benzene 15, chloroform 210, ethanol 45, hexane 15. LD_{50} in rats, mice (mg/kg): 182, 380 orally (Chapman).

USE: Herbicide.

2756. Cyanic Acid. Hydrogen cyanate. CHNO; mol wt 43.03. C 27.92%, H 2.34%, N 32.55%, O 37.19%. N≡COH. Best prepd in the laboratory by dry distillation of cyanuric acid: Linhard, *Z. Anorg. Allgem. Chem.* **236,** 200 (1938). *See also* Isocyanic Acid.

Liquid. *Acrid odor, strong lacrimator and vesicant.* d_4^{-20} 1.140°; d_4^{-20} 1.156. mp −86°. bp_{760} 23.5°. Heat of formation −36.5 kcal/mol. Dipole moment 1.592. K at 20° = 2.2 × 10⁻⁴. Rapid heating of the liq may result in an explosion. Polymerizes on standing forming cyamelide and cyanuric acid. Sol in water with decompn to carbon dioxide and ammonia. Dil solns in ice water may be kept for several hours. Dil solns in ether, benzene, toluene can be kept for several weeks.

Caution: Highly irritating to eyes, skin, and mucous membranes. *See Clinical Toxicology of Commercial Products,* R. E. Gosselin *et al.,* Eds. (Williams & Wilkins, Baltimore, 5th ed., 1984) Section II, p 110.

USE: In formation of some cyanates.

2757. Cyanidin Chloride. *2-(3,4-Dihydroxyphenyl)-3,5,7-trihydroxy-1-benzopyrylium chloride; 3,3',4',5,7-pentahydroxyflavylium chloride; 3,3',4',5,7-pentahydroxy-2-phenylbenzopyrylium chloride.* $C_{15}H_{11}ClO_6$; mol wt 322.70. C

55.83%, H 3.44%, Cl 10.99%, O 29.75%. Prepd by acid hydrolysis of cyanin chloride: Willstätter, Everest, *Ann.* **401,** 189 (1913). Isoln from bananas: Simmonds, *Nature* **173,** 402 (1954). Prepn by reduction of quercetin: Bauer *et al., Chem. & Ind. (London)* **1954,** 433; King, White, *J. Chem. Soc.* **1957,** 3901. Structure: Willstätter, Mallison, *Ann.* **408,** 147 (1915). Synthesis: Willstätter *et al., Ber.* **57,** 1938 (1924); Robertson, Robinson, *J. Chem. Soc.* **1928,** 1528. Biosynthesis: Fritsch *et al., Z. Naturforsch.* **26b,** 581 (1971). *See also* Bioflavonoids.

Metallic brownish-red needles (monohydrate) from dil alcoholic HCl. The anhydr compd does not melt below 300°. Absorption max (methanolic HCl): 535 nm. Freely sol in alcohol and in amyl alcohol giving a violet soln. Sol in sodium carbonate soln with blue color. Sparingly sol in dil HCl or H_2SO_4.

3-Glucoside, $C_{21}H_{21}ClO_{11}$, *chrysanthemin, asterin, kuromanin.* From winter aster *(Chrysanthemum indicum* L., Compositae): Willstätter, Bolton, *Ann.* **412,** 136 (1917); from wild strawberries *(Fragaria vesca* L., Rosaceae): Sondheimer, Karash, *Nature* **178,** 648 (1956); from sweet cherries *(Prunus avium* L., Rosaceae): Li, Wagenknecht *ibid.* **182,** 657 (1958). Identity with asterin: Robinson, Willstätter, *Ber.* **61,** 2503 (1928). Structure and synthesis: Murakami *et al., J. Chem. Soc.* **1931,** 2665. Reddish-brown plates or prisms with a metallic shine from dil alcoholic HCl, dec 205° without melting. Absorption max (methanolic HCl): 525 nm. Sol in alc with strong fluorescence and in sodium carbonate with violet-blue color.

3-Rhamnoglucoside, $C_{27}H_{31}ClO_{15}$, *keracyanin, antirrhinin, sambucin, Meralop, Meralops.* From skin of sweet cherries *(Prunus avium* L,. Rosaceae): Willstätter, Zollinger, *Ann.* **412,** 164 (1917); from sour cherries *(Prunus cerasus* L., Rosaceae): Li, Wagenknecht, *J. Am. Chem. Soc.* **78,** 979 (1956). Structure: Robertson, Robinson, *J. Chem. Soc.* **1927,** 2196. Prepn by reduction of quercetin-3-rhamnoglucoside: Bauer *et al., Chem. & Ind. (London)* **1954,** 433. Reduces time to adjust to darkness: F. Trimarchi *et al., Min. Oftalmol.* **18,** 143 (1977). Mechanism of action: F. Trimarchi *et al., Ann. Ottalmol. Clin. Ocul.* **105,** 111 (1979). Red needles from dil HCl or dark prisms from dil methanolic HCl. Absorption max (ethanolic HCl): 532, 333, 282 nm. Sol in hot water, alcohol.

3-Galactoside, $C_{21}H_{21}ClO_{11}$, *idaein, idein.* From cranberries *(Vaccinium vitis idaea* L., Ericaceae): Willstätter, Mallison, *Ann.* **408,** 15 (1915); from Winesap apple *(Pyrus malus* Linn., Rosaceae): Duncan, Dustman, *J. Am. Chem. Soc.* **58,** 1511 (1936). Structure and synthesis: Grove, Robinson, *J. Chem. Soc.* **1931,** 2722. Red needles with bronze metallic luster from dil ethanolic HCl, dec 210-212°. Sol in water, ethanol, methanol, dil HCl. Practically insol in 7% HCl.

3,5-Diglucoside, $C_{27}H_{31}ClO_{16}$, *cyanin, shisonin A.* From cornflower *(Centaurea cyanus* L., Compositae): Willstätter, Everest, *loc. cit.* Structure: Léon, Robinson, *J. Chem. Soc.* **1932,** 221. Synthesis: Robinson, Todd, *ibid.* 2488. Plates with a metallic luster from dil alcoholic HCl, mp 203-204°. $[\alpha]_D$ −258° (in 0.05% HCl). Absorption max (methanolic HCl): 522 nm.

3-Sophoroside, $C_{27}H_{31}ClO_{16}$, *mecocyanin.* From the flowers of *Papaver rhoeas* L., Papaveraceae): Willstätter, Weil, *Ann.* **412,** 231 (1917); from sour cherries: Li, Wagenknecht, *J. Am. Chem. Soc., loc. cit.* Structure: Harborne, *Experientia* **19,** 7 (1963); *Phytochem.* **2,** 85 (1963). (Alternate structure: Grove *et al., J. Chem. Soc.* **1934,** 1608). Dark-red crystals from dil HCl + HOAc or dark-red needles from 2% alcoholic HCl. Absorption max (methanolic HCl): 523 nm. Sol in water. Pptd by glacial acetic acid or acetone.

THERAP CAT: 3-Rhamnoglucoside in treatment of night blindness.

2758. Cyanoacetamide. Malonamide nitrile. $C_3H_4N_2$-O; mol wt 84.08. C 42.86%, H 4.80%, N 33.32%, O 19.03%. $CNCH_2CONH_2$. Prepd by the action of aq or alcoholic ammonia on cyanoacetic ester: Hesse, *Am. Chem. J.* **18**, 724 (1896); Thole, Thorpe, *J. Chem. Soc.* **99**, 429 (1911); Ott, Löpmann, *Ber.* **55**, 1258 (1922); Corson *et al.*, *Org. Syn.* **coll. vol. I** (2nd ed., 1941) p 179.

Needles from alc, mp 119.5°. One gram dissolves in 6.5 ml of cold water. Soly in 100 ml of 95% alc: 1.3 g at 0°; 3.1 g at 26°; 7.0 g at 44°; 14.0 g at 62°; 21.5 g at 71°.

USE: In organic syntheses (*e.g.*, starting material for vitamin B$_6$ synthesis).

2759. Cyanoacetic Acid. Malonic mononitrile. C_3H_3-NO_2; mol wt 85.06. C 42.36%, H 3.55%, N 16.47%, O 37.62%. $N{\equiv}CCH_2COOH$. Prepd from chloroacetic acid and NaCN: Ruggli, Businger, *Helv. Chim. Acta* **25**, 35 (1942); Lapworth, Baker, *Org. Syn.* **coll. vol. I**, 181 (1941); **Brit.** pat. **824,640** (1959); Eaker, **U.S.** pat. **2,539,238** (1951 to Monsanto).

Hygroscopic crystals, mp 66°; dec at 160° into CO_2 and acetonitrile. bp$_{15}$ 108°. Sol in water, alcohol, ether, slightly in benzene, chloroform. *Keep well closed.*

USE: Synthesis of intermediates; manuf barbital.

2760. Cyanofenphos. *Phenylphosphonothioic acid O-(4-cyanophenyl) O-ethyl ester; phenylphosphonothioic acid O-ethyl ester O-ester with p-hydroxybenzonitrile;* CYP; S-4087; Surecide. $C_{15}H_{14}NO_2PS$; mol wt 303.32. C 59.40%, H 4.65%, N 4.62%, O 10.55%, P 10.21%, S 10.57%. Prepn: S. Kuramoto *et al.*, **Japan.** pat. **2870('63)**, *C.A.* **60**, 455c (1964), **Brit.** pat. **929,738**, *C.A.* **59**, 14025a (1963) (both to Sumitomo); **Belg.** pat. **628,298**, *C.A.* **60**, 12055h (1964), (1963 to Stauffer). Insecticidal activity: S. Tamura *et al.*, *Agr. Biol. Chem.* **25**, 773 (1961).

Crystalline solid, mp 83°. n_D^{25} 1.5839. Soly in water at 30°: 6 ppm. Moderately sol in ketones and aromatic solvents. Vapor press at 25°: 1.32×10^{-5} mm Hg. Toxic to fish and bees. LD$_{50}$ orally in mice: 43.7 mg/kg.

USE: Insecticide.

2761. Cyanogen. *Ethanedinitrile;* dicyan; oxalic acid dinitrile. C_2N_2; mol wt 52.04. C 46.16%, N 53.84%. $N{\equiv}C{-}C{\equiv}N$. Usually prepd by adding an aq soln of sodium or potassium cyanide to an aq soln of copper(II) sulfate or chloride. Procedure: Janz, *Inorg. Syn.* **5**, 43 (1957). Alternate prepn from HCN by the use of CuO: Fierce, Millikan, **U.S.** pat. **2,841,472** (1958 to Pure Oil); from HCN and NO_2: Fierce, Sander, *Ind Eng. Chem.* **53**, 985 (1961). *Review:* Brotherton, Lynn, *Chem. Rev.* **59**, 841-883 (1959).

Highly poisonous gas. Almond-like odor. Acrid and pungent when in lethal concns. Burns with pink flame having a bluish border. mp −27.9° (also reported as −34.4°). bp −21.17°. $d_4^{-21.17}$ 0.9537. Heat of vaporization (liquid) 5.778 kcal/mole. Above 500° polymerizes to insol paracyanogen $(CN)_n$. One vol of water dissolves about 4 vols of cyanogen gas. Also sol in alcohol, ether. Slowly hydrolyzed in aq soln giving oxalic acid and ammonia.

Caution: Toxic effects similar to those of hydrogen cyanide *(q.v.).*

2762. Cyanogen Azide. Carbon pernitride. CN_4; mol wt 68.04. C 17.65%, N 82.35%. $N^-{=}N^+{=}N{-}C{\equiv}N$. Prepd by suspending NaN_3 in dry acetonitrile and distilling cyanogen chloride into the cooled suspension: Marsh, Hermes, *J. Am. Chem. Soc.* **86**, 4506 (1964); see also *Chem. & Eng. News* **42**, 51 (Oct. 26, 1964); Marsh, **U.S.** pat. **3,410,-658** (1968); Marsh, *J. Org. Chem.* **37**, 2966 (1972). Spectrum, structure, dipole moment: Costain, Kroto, *Can. J.*

Phys. **50**, 1453 (1972); Almenningen *et al.*, *Acta Chem. Scand.* **27**, 1531 (1973).

Clear, colorless, oily liquid. *Caution: The pure azide detonates violently upon thermal, electrical or mechanical shock!* Can be handled relatively safely in solvents. Half-life of a 27% soln in acetonitrile is 15 days at room temp, more stable at lower temps.

USE: In organic synthesis. Has a versatility and scope of chemical reactivity that is very broad and useful, *e.g.* reacts with alkanes to give primary alkylcyanamides: Anastassiou *et al.*, *J. Am. Chem. Soc.* **87**, 2296 (1965).

2763. Cyanogen Bromide. Bromine cyanide. CBrN; mol wt 105.92. C 11.34%, Br 75.44%, N 13.22%. $BrC{\equiv}N$. Prepd from potassium or sodium cyanide and bromine: Hartman, Dreger, *Org Syn.* **11**, 30 (1931). Industrial prepn from sodium bromide, sodium chlorate, and sodium cyanide in 30% H_2SO_4: Ewan, *Chem. Zentr.* **1907**, I, 591; Göpner, *ibid.* **1908**, I, 1807; Grignard, Crouzier, *Bull. Soc. Chim. France* [4] **29**, 214 (1921).

Needles, cubes; volatile at ordinary temps. *Vapors are highly irritant and very poisonous!* d_4^{20} 2.015. mp 52°. bp 61-62°. Freely sol in water, alcohol, ether. Aqueous solns of alkalies dec it to alkali cyanide and alkali bromide. Pure cyanogen bromide, completely dried by distillation over sodium, may be stored in a desiccator for several months. Impure material dec rapidly and tends to explode.

Caution: Toxic effects similar to those of hydrogen cyanide *(q.v.).*

2764. Cyanogen Chloride. CClN; mol wt 61.47. C 19.54%, Cl 57.67%, N 22.79%. CNCl. By action of Cl on HCN. Toxicity data: *Handbook of Toxicology* **vol. 1**, W. S. Spector, Ed. (Saunders, Philadelphia, 1956) pp 82-83.

Liquid. *Vapors are highly irritant and very poisonous!* d 1.186. bp 13.8°. mp −6°. Sol in water, alc, ether. LD s.c. in rabbits: 20 mg/kg (Spector).

Caution: Toxic effects similar to those of hydrogen cyanide, *q.v.*

USE: In chemical synthesis. Military poison gas.

2765. Cyanogen Iodide. Iodine cyanide. CIN; mol wt 152.92. C 7.85%, I 82.99%, N 9.16%. CNI. Prepd by the action of iodine on sodium cyanide: Bak, Hillebert, *Org. Syn.* **coll. vol. IV**, 207 (1963).

White needles; very pungent odor; acrid taste. *Very poisonous!* mp 146-147°. Sol in water, alcohol, ether, volatile oils. LD orally in cats: 18 mg/kg, *Handbook of Toxicology* **vol. 1**, W. S. Spector, Ed. (Saunders, Philadelphia, 1956) pp 82-83.

Caution: Causes convulsions, paralysis, death from respiratory failure. Lacks inhalation hazard.

USE: Generally for destroying all lower forms of life. In taxidermy for preserving insects, butterflies, etc.

2766. Cyanopsin. Photoreceptor protein found in the retinal cone cells of fresh water and migratory fish, lampreys, and certain amphibians; corresponds to the rod pigment, porphyropsin, *q.v.* Absorption maximum ~620 nm. Composed of the chromophore 11-*cis* 3-dehydroretinal, *q.v.,* bound to a photopsin, the specific protein component of cone cells (*see* Opsins). Proposed existence as a visual pigment and *in vitro* synthesis from 3-dehydroretinal and chicken photopsin: G. Wald, *Science* **118**, 505 (1953). Demonstration of natural occurrence by microspectrophotometry: P. Liebman, G. Entine, *Nature* **216**, 501 (1967). Methods for prepn and assay: R. Hubbard *et al.*, "Methodology of Vitamin A and Visual Pigments" in *Methods in Enzymology* **Vol. 18**, D. B. McCormick, L. D. Wright, Eds. (Academic Press, New York, 1971) pp 615-653. Photochemistry and biological activity: G. Wald, *Science* **162**, 230 (1968). A trichromatic cone system utilizing 3-dehydroretinal as chromophore has been noted to occur in certain amphibians and fish. Three photochemically distinct pigments, corresponding to those of retinal-based color vision, have been identified: P. A. Liebman in *Handbook of Sensory Physiology* **Vol. VII(1)**, H. J. A. Dartnall, Ed. (Springer-Verlag, New York, 1972) pp 481-528. A visual system is generally based on one type of chromophore combined with various opsins. However, pigments utilizing both types of chromophore have been found to coexist in the retina of some of these species. Demonstration of paired-pigment cone systems: E. R. Loew, H. J. A. Dartnall, *Vision Res.* **11**, 551 (1976); A. T. C. Tsin *et al.*, *ibid.* **21**, 943 (1981). Absorption spectrum: J. E. M. Mooij, T. J. T. P. van den Berg, *ibid.* 23 (1983).

2767. Cyanuric Acid. *1,3,5-Triazine-2,4,6(1H,3H,5H)-trione; sym-triazinetriol;* 2,4,6-trihydroxy-1,3,5-triazine; tricyanic acid; trihydroxycyanidine. $C_3H_3N_3O_3$; mol wt 129.08. C 27.92%, H 2.34%, N 32.55%, O 37.19%. Prepn: Schmidt, *J. Prakt. Chem.* [2] **5**, 41-52 (1872); Venable, Moore, *J. Am. Chem. Soc.* **39**, 1752 (1917). HPLC determn: P. Beilstein *et al.*, *J. Agric. Food Chem.* **29**, 1132 (1981); in urine and pool water: T. V. Briggle *et al.*, *J. Assoc. Off. Anal. Chem.* **64**, 1222 (1981). Review of prepn, properties and uses: *The Chemistry of Heterocyclic Compounds*, A. Weissberger, Ed., **vol. 13**, entitled *s-Triazines and Derivatives* by E. M. Smolin, L. Rapoport (Interscience, New York, 1959) pp 17-48; of chemistry, toxicity and antibacterial activity as applied to swimming pools: E. Canelli, *Am. J. Public Health* **64**, 155-162 (1974); F. W. Linda, R. C. Hollenbach, *J. Environ. Health* **40**, 324-329 (1978); of toxicology: B. G. Hammond *et al.*, *Environ. Health Perspect.* **69**, 287-292 (1986). *Review:* J. A. Wojtowicz in Kirk Othmer *Encyclopedia of Chemical Technology* vol. **7** (John Wiley & Sons, 4th ed., 1993) pp 834-851.

White crystalline solid; does not melt up to 330°, sublimes and dissociates to isocyanic acid at higher temps. pKa_1 6.88; pKa_2 11.40; pKa_3 13.5. uv max (0.1*M* HCl): 214 nm (log ε 3.38). uv max (0.1*M* phosphate buffer pH 7): 214 nm (log ε 4.00). uv max (0.1*M* NaOH) 213 nm (log ε 4.64). Slightly sol in common organic solvents such as acetone, benzene, diethyl ether, ethanol, hexane. Soly (%): DMF 7.2; DMSO 17.4. Soly in water (%): 0.2 at 25°; 2.6 at 90°; 10.0 at 150°. Soly in 96% H_2SO_4 (25°): 14.1%. Sol in hot alcohols, pyridine, concd HCl without decompn; in aq solns of NaOH and KOH. Insol in cold methanol, ether, acetone, benzene, chloroform. Forms dihydrate in aq soln; crystallizes as colorless monoclinic prisms that effloresce in dry air. d^{25} 1.75 (anhydr); d^{25} 1.66 (dihydrate). Predominant tautomeric form in solid state and soln is keto form, *isocyanuric acid*; in basic soln the enol form is more stable. LD_{50} orally in rats: > 5.00 g/kg (Canelli).
USE: Convenient lab source of cyanic acid gas. In prepn of melamine, sponge rubber, herbicides, dyes, resin, antimicrobial agents. As stabilizer and disinfectant in swimming pool water.

2768. Cycasin. *(Methyl-ONN-azoxy)methyl β-D-glucopyranoside;* methylazoxymethanol β-D-glucoside; β-D-glucosyloxyazoxymethane. $C_8H_{16}N_2O_7$; mol wt 252.22. C 38.10%, H 6.39%, N 11.11%, O 44.40%. Toxic substance from seeds of *Cycas revoluta* Thumb. and *C. circinalis* L., Cycadaceae from Guam: Nishida *et al.*, *Bull. Agr. Chem. Soc. Japan* **19**, 77 (1955), *C.A.* **50**, 13756g (1956); Riggs, *Chem. & Ind. (London)* **1956**, 926; Matsumoto, Strong, *Arch. Biochem. Biophys.* **101**, 299 (1963). Biological effects, metabolism and mechanism of action of cycasin and its aglycone, *methylazoxymethanol:* Spatz, *Ann. N.Y. Acad. Sci.* **163**, 848 (1969). Structure: Korsch, Riggs, *Tetrahedron Letters* **1964**, 523. Toxicology: Laqueur, Spatz, *Cancer Res.* **28**, 2262 (1968). Carcinogenicity and neurotoxicity studies:

I. Hirono, *Fed. Proc.* **31**, 1493 (1972). Review of carcinogenicity studies: *IARC Monographs* **10**, 121-138 (1976).
Needles from water + acetone + ether, dec 154°. $[\alpha]_D^{18}$ −44° (c = 0.62). uv max (0.4*M* H_2SO_4): 217 nm. LD_{50} in rats, mice, hamsters, rabbits, guinea pigs (mg/kg): 562, 500, <250, ~30, <20 orally (Hirono).
Tetraacetate, $C_{16}H_{24}N_2O_{11}$, plates from acetone + petr ether, mp 137°. $[\alpha]_D^{18}$ −27° (c = 0.98 in $CHCl_3$).
Note: Cycasin has been listed as reasonably anticipated to be a carcinogen: *Fourth Annual Report on Carcinogens* (NTP 85-002, 1985) p 64; delisted because no U.S. residents exposed: *Fifth Annual Report on Carcinogens* (NTP 89-239, 1989) p 340.

2769. Cyclacillin. *[2S-(2α,5α,6β)]-6-[[(1-Aminocyclohexyl)carbonyl]amino]-3,3-dimethyl-7-oxo-4-thia-1-azabicyclo[3.2.0]heptane-2-carboxylic acid;* 6-(1-aminocyclohexane-carboxamido)penicillanic acid; (1-aminocyclohexyl)penicillin; ciclacillin; Wy-4508; Calthor; Citosarin; Cyclapen; Syngacillin; Ultracillin; Vastcillin; Vatracin; Vipicil; Wyvital. $C_{15}H_{23}N_3O_4S$; mol wt 341.43. C 52.77%, H 6.79%, N 12.31%, O 18.74%, S 9.39%. Semi-synthetic antibiotic related to penicillin. Prepn: Alburn *et al.*, U.S. pat. **3,194,802**; Robinson, Nescio, Ger. pat. **1,800,698** corresp to U.S. pat. **3,478,018** (1965; 1969, all to Am. Home Prods.). Activity studies: Rosenman *et al.*, *Antimicrob. Ag. Chemother.* **1967**, 590; Hopper *et al.*, *ibid.* 597; Yurchenco *et al.*, *ibid.* 602; *Chemotherapy* **15**, 209 (1970). Metabolic studies: Poole, *J. Pharm. Sci.* **59**, 1255 (1970); Tucker *et al.*, *Toxicol. Appl. Pharmacol.* **19**, 361 (1971). Physicochemical data: Alburn *et al.*, *Antimicrob. Ag. Chemother.* **1967**, 586; Hou, Poole, *J. Pharm. Sci.* **58**, 1510 (1969); Poole, Bahal, *ibid.* **59**, 1265 (1970).

Crystals, mp 182-183° (anhydrate) (Hou, Poole); also reported as mp 156-158° (dec) (Alburn *et al.*). $[\alpha]_D^{25}$ 268° (water). pK_1, pK_2 in water: 2.68, 7.50; in 50% dioxane: 4.16, 7.04. Stable in acid. Soly at 38°: about 29 mg/ml.
THERAP CAT: Antibacterial.

2770. Cyclamic Acid. *Cyclohexylsulfamic acid; cyclohexanesulfamic acid;* Hexamic Acid. $C_6H_{13}NO_3S$; mol wt 179.24. C 40.21%, H 7.31%, N 7.81%, O 26.78%, S 17.89%. Prepn: Audrieth, Sveda, U.S. pat. **2,275,125** (1942 to du Pont); *J. Org. Chem.* **9**, 89 (1944); Thompson, U.S. pat. **2,800,501** (1957 to du Pont); Shah, Bernsen, U.S. pat. **3,-361,799** (1968 to Abbott). Prepn of the Na salt: Robinson, U.S. pat. **2,383,617** (1945 to du Pont). Other prepns and metabolism: *See* Calcium Cyclamate. Sweetness of the sodium salt discovered by Michael Sveda at the University of Illinois in 1937. Toxicity: Taylor *et al.*, *Food Cosmet. Toxicol.* **6**, 313 (1968). Review of long-term toxicity and carcinogenicity of sodium cyclamate in mice: Brantom *et al.*, *Food Cosmet. Toxicol.* **11**, 735 (1973).

Sweet-sour crystals. mp 169-170°. Fairly strong acid. Very sparingly soluble in water. Slowly hydrolyzed by hot water.
Sodium salt, $C_6H_{12}NNaO_3S$, *sodium cyclohexylsulfamate, cyclamate sodium, sodium cyclamate, Assugrin, Sucaryl Sodium, Sucrosa.* Crystals. Pleasantly sweet to the taste. Freely sol in water. About 30 times as sweet as refined cane sugar. Sweetness is still easily perceptible at a dilution of 1:10,000 (sugar 1:140; saccharin 1:50,000). pH of 10% aq soln between 5.5 and 7.5. Practically insol in alcohol, ether,

benzene, chloroform. LD_{50} orally in mice, rats: 17.0, 15.25 g/kg (Taylor).

Note: Consult latest Government regulations on use in food.

USE: Non-nutritive sweetener.

2771. Cyclandelate. *α-Hydroxybenzeneacetic acid 3,3,-5-trimethylcyclohexyl ester; mandelic acid 3,3,5-trimethylcyclohexyl ester;* 3,3,5-trimethylcyclohexyl mandelate; 3,5,5-trimethylcyclohexyl amygdalate; 3,3,5-trimethylcyclohexanol α-phenyl-α-hydroxyacetate; BS-572; Cyclergine; Cyclobral; Cyclolyt; Cyclomandol; Cyclospasmol; Natil; Novodil; Perebral; Spasmocyclon. $C_{17}H_{24}O_3$; mol wt 276.38. C 73.88%, H 8.75%, O 17.37%. Prepn: N. Brock *et al., Arzneimittel-Forsch.* **2**, 165 (1952); A. B. H. Funcke *et al., ibid.* **3**, 503 (1953); van Sluis, *Chem. Products* **17**, 375 (1954); **Brit.** pat. **707,227**; W. T. Nauta, **U.S.** pat. **2,707,193** (1954, 1955 both to Brocades-Stheeman). Purification: D. Flitter, **U.S.** pat. **3,663,597** (1972 to Am. Home Prods.). Pharmacokinetics: A. Orr, J. R. Whittier, *Int. J. Nucl. Med. Biol.* **1**, 205 (1974). GLC determn in human plasma: G. Andermann, M. Dietz, *J. Chromatog.* **223**, 365 (1981). Clinical trial in chronic cerebrovascular disease: S. Bassi *et al., Brit. J. Clin. Prac.* **38**, 344 (1984). Review of efficacy as cerebral and peripheral vasodilator: C. B. Blakemore, *ibid.* **34**, Suppl., 3-9 (1984). Symposium on calcium modulation and clinical effects: *Drugs* **33**, Suppl. 2, 1-141 (1987). Comprehensive description: C. M. Shearer, *Anal. Profiles Drug Subs. Excip.* **21**, 149-168 (1992).

Crystals, mp 55.0-56.5°. bp_{14} 192-194°. uv max (ethanol): 269, 258, 251 nm (ε 1575, 2020, 1630). Very sol in chloroform, methanol; freely sol in acetonitrile, ethyl acetate, DMF, toluene. Insol in water.

THERAP CAT: Vasodilator (peripheral, cerebral).

2772. Cyclarbamate. *1,1-Cyclopentanedimethanol bis-(phenylcarbamate); 1,1-cyclopentanedimethanol dicarbanilate;* cyclopentane-1,1-dimethanol bis(phenylcarbamate); 1,1-dihydroxymethylcyclopentane *N,N'*-diphenylcarbamate; cyclopentaphene; BSM-906M; Calmalone; Casmalon. $C_{21}H_{24}N_2O_4$; mol wt 368.43. C 68.46%, H 6.57%, N 7.60%, O 17.37%. Prepn: Bourdais, *Bull. Soc. Chim. France* **1962**, 266; E. Rosenberg, **Brit.** pat. **904,410** corresp to **U.S.** pat. **3,067,240** (both 1962 to Lab. Cassenne).

Needles from methanol, mp 151-152°. Practically insol in water. Slightly sol in alcohol, glycerol, propylene glycol.

THERAP CAT: Anxiolytic; muscle relaxant (skeletal).

2773. Cyclazocine. *3-(Cyclopropylmethyl)-1,2,3,4,5,6-hexahydro-6,11-dimethyl-2,6-methano-3-benzazocin-8-ol;* Win-20740; NSC-107429. $C_{18}H_{25}NO$; mol wt 271.40. C 79.66%, H 9.28%, N 5.16%, O 5.90%. Analgesic with mixed narcotic agonist-antagonist properties. Prepn: Archer, **Belg.** pat. **611,000** (1962 to Sterling Drug), *C.A.* **58**, 2439c (1963). Crystal structure: I. L. Karle *et al., Acta Crystallogr.* **B25**, 1469 (1969). Supraspinal analgesic effects: S. Sasson, C. Kornetsky, *Life Sci.* **38**, 21 (1986). Evaluation in anorexia and motor disruption in rats: J. W. Henck *et al., Pharmacol. Biochem. Behav.* **22**, 671 (1985). Determn by HPLC: I. Jane, A. McKinnon, *J. Chromatog.* **323**, 191 (1985). Review of use in treatment of opiate addiction: M. J. Goldstein, *Int. J. Addict.* **15**, 939 (1980).

Crystals from methanol, mp 201-204°.

THERAP CAT: Narcotic antagonist.

2774. Cyclethrin. *2,2-Dimethyl-3-(2-methyl-1-propenyl)cyclopropanecarboxylic acid 3-(2-cyclopenten-1-yl)-2-methyl-4-oxo-2-cyclopenten-1-yl ester;* 3-(2-cyclopentenyl)-2-methyl-4-oxo-2-cyclopentenyl ester of chrysanthemummonocarboxylic acid; chrysanthemummonocarboxylic acid ester with 3-(2-cyclopenten-1-yl)-2-methyl-4-oxo-2-cyclopenten-1-ol. $C_{21}H_{28}O_3$; mol wt 328.45. C 76.79%, H 8.59%, O 14.61%. Prepn and insecticidal activity: H. L. Haynes *et al., Contrib. Boyce Thompson Inst.* **18**, 1 (1954); H. R. Guest, H. A. Stansbury, **U.S.** pat. **2,891,888** (1959 to Union Carbide). Product is a mixture of isomers including four racemic forms or eight optical and geometric isomers which have not yet been isolated and evaluated. Toxicology: C. P. Carpenter *et al., Arch. Ind. Hyg. Occup. Med.* **10**, 162 (1954).

Liquid. n_D^{30} 1.5120. d_{20}^{20} 1.033. LD_{50} orally in rats: 1.4-2.8 g/kg (Carpenter).

USE: Insecticide for flies, roaches, and grain pests.

2775. Cyclexanone. *2-(1-Cyclopenten-1-yl)-2-[2-(4-morpholinyl)ethyl]cyclopentanone;* 2-(Δ-1'-cyclopentenyl)-2-(β-tetrahydroparoxazinoethyl)cyclopentanone; 1-(β-N-morpholinoethyl)-1-cyclopentenylcyclopentane-2-one; pentethylcyclanone. $C_{16}H_{25}NO_2$; mol wt 263.38. C 72.97%, H 9.57%, N 5.32%, O 12.15%. Prepd from 2-cyclopentylidenecyclopentanone and *N-(β-chloroethyl)morpholine*: Ueberwasser, **Ger.** pat. **1,059,901**; **Brit.** pat. **878,677** (1959, 1961 both to Ciba).

Liquid, $bp_{0.035}$ 112°.

Hydrochloride, $C_{16}H_{25}NO_2$·HCl, *Exopan, Exopon.* Crystals from aq ethanol, mp 209°.

Methiodide, $C_{16}H_{25}NO_2$·CH_3I, crystals, mp 106°.

THERAP CAT: Antitussive.

2776. Cyclexedrine. *N,β-Dimethylcyclohexaneethanamine;* cyclohexylisopropylmethylamine; 2-cyclohexyl-*N*-methylpropylamine; 1-cyclohexyl-1-methyl-2-methylaminoethane; ethylhexedrine; isopropylhexedrine. $C_{10}H_{21}N$; mol wt 155.28. C 77.35%, H 13.63%, N 9.02%. Prepn: **Ger.** pat. **970,480** (1958 to Knoll).

Hydrochloride, $C_{10}H_{21}N$·HCl, *Eventin.* Crystalline powder, bitter taste. mp 138-140°. Soluble in water, ethanol, chloroform; slightly sol in ether. LD_{50} orally in mice: 304 mg/kg.

THERAP CAT: Sympathomimetic; anorexic.

2777. Cyclic AMP. *Adenosine cyclic 3',5'-(hydrogen phosphate);* adenosine 3',5'-cyclic monophosphate; adenosine 3',5'-cyclic phosphate; adenosine 3',5'-monophosphate; adenosine 3',5'-phosphate; cyclic adenosine 3',5'-monophosphate; acrasin; 3',5'-AMP; cAMP. $C_{10}H_{12}N_5O_6P$; mol wt 329.21. C 36.48%, H 3.67%, N 21.27%, O 29.16%, P 9.41%. Key intracellular regulator of a number of cellular processes; found in most animal cells, in bacteria, and in some higher plants. First isoln and identification: Rall *et al., J. Biol. Chem.* **224,** 463 (1957); Sutherland, Rall, *ibid.* **232,** 1077 (1958). Molecular structure and conformation: Cook *et al., J. Am. Chem. Soc.* **79,** 3607 (1957); Lipkin *et al., ibid.* **81,** 6198 (1959); Watenpaugh *et al., Science* **159,** 206 (1968). Syntheses: Smith *et al., J. Am. Chem. Soc.* **83,** 698 (1961); Borden, Smith, *J. Org. Chem.* **31,** 3247 (1966). Physical data: D. Lipkin *et al., J. Am. Chem. Soc.* **81,** 6075 (1959). Functions as a mediator of hormone-action for a variety of hormones such as epinephrine, glucagon and ACTH (*q.v.*). Activates phosphorylation of proteins by protein kinases. Defined as a "second messenger" because of its response to hormones ("first messengers"). Converted from adenosine triphosphate (ATP) by the enzyme *adenylate cyclase.* Deactivated by *cyclic nucleotide phosphodiesterases* which convert it to 5'-adenylic acid. Reviews of biochemical model: Sutherland *et al., J. Biol. Chem.* **237,** 1220-1243 (1962); Robison *et al., Ann. Rev. Biochem.* **37,** 149 (1968); Jost, Rickenberg, *ibid.* **40,** 741 (1971); Sutherland, *J. Am. Med. Assoc.* **214,** 1281 (1970); Pastan, Perlman, *Nature New Biol.* **229,** 5 (1971); G. A. Robison *et al.,* Eds., *Ann. N.Y. Acad. Sci.* **185** (1971); Losert, *Pharmazie* **28,** 351 (1973). *See also* Cyclic GMP. Identity with acrasin from cellular slime molds (*Dictyostelium* species), where it acts as a "first" rather than a "second messenger": Konijn *et al., Proc. Nat. Acad. Sci. USA* **58,** 1152 (1967). Review of quantitative methods: *Methods in Molecular Biology,* **vol.** 3 entitled "Methods in Cyclic Nucleotide Research", Mark Chasin, Ed. (Marcel Dekker, New York, 1972). Books: G. A. Robison *et al., Cyclic AMP* (Academic Press, New York, 1971); *Advan. Cyclic Nucleotide Res.* vols. 1, 2, 3, P. Greengard, G. A. Robison, Eds. (Raven Press, New York, 1972, 1973); *Handb. Exp. Pharmacol.* **58** entitled "Cyclic Nucleotides", Pt I: Biochemistry and Pt II: Physiology & Pharmacology, J. A. Nathanson, J. W. Kobabian, Eds. (Springer-Verlag, New York, 1982) 736 and 1000 pp resp.

Hydrate crystallizes from water as platelets with pearly luster, mp 219-220° (with gas evolution). $[\alpha]_D$ −51.3° (c = 0.67). uv max (pH 2): 256 nm (ε 14500); (pH 7): 258 nm (ε 14650). Heat stable; resistant to inactivation by acid or alkali. Found in concns of 10^{-7} to 10^{-6} moles/kg in living cells.

2778. Cyclic GMP. *Guanosine cyclic 3',5'-(hydrogen phosphate);* cyclic guanosine 3',5'-monophosphate; guanosine 3',5'-cyclic monophosphate; guanosine 3',5'-cyclic phosphate; guanosine 3',5'-monophosphate; 3',5'-GMP; cGMP. $C_{10}H_{12}N_5O_7P$; mol wt 345.21. C 34.79%, H 3.50%, N 20.29%, O 32.44%, P 8.97%. A cellular regulatory agent which has been described as a "second messenger". *See* Cyclic AMP. First synthesized because of the interest in cyclic nucleotides generated by cyclic AMP: Smith *et al., J.*

Am. Chem. Soc. **83,** 698 (1961); Borden, Smith, *J. Org. Chem.* **31,** 3247 (1966). Has subsequently been found in animal and bacterial cells in concentrations of 10^{-8} to 10^{-6} moles/kg. First isoln from rat urine: Ashman *et al., Biochem. Biophys. Res. Commun.* **11,** 330 (1963). Structure and conformation: Chwang, Sundaralingam, *Nature, New Biol.* **244,** 136 (1973). Proposed as an antagonist of cyclic AMP in bidirectional systems such as contraction-relaxation or glycogen synthesis-glycogen breakdown. Cyclic GMP levels increase in response to a variety of hormones including acetylcholine, insulin and oxytocin. Formed by conversion of guanosine triphosphate by the enzyme *guanylate cyclase* and hydrolyzed by cyclic nucleotide phosphodiesterases. Found to activate specific protein kinases. Reviews of biochemical model: Goldberg *et al.,* in *Pharmacology and the Future of Man* **vol.** 5, R. A. Maxwell, G. H. Acheson, Eds. (Karger, New York, 1973) pp 146-169; *eidem* in *Advan. Cyclic Nucleotide Res.* **vol.** 3, P. Greengard, G. A. Robison, Eds. (Raven Press, New York, 1973) pp 155-223; Kolata, *Science* **182,** 149-151 (1973); *Nature* **246,** 186 (1973); N. D. Goldberg, M. K. Haddox, *Ann. Rev. Biochem.* **46,** 823-896 (1977). Book: *Handb. Exp. Pharmacol.* **58,** entitled "Cyclic Nucleotides", Pt. I: Biochemistry and Pt. II: Physiology & Pharmacology, J. A. Nathanson, J. W. Kobabian, Eds. (Springer-Verlag, New York, 1982) 736 and 1000 pp resp.

Calcium salt, $C_{20}H_{22}CaN_{10}O_{14}P_2$, decahydrate. uv max (pH 1): 256.5 nm (ε 11350); (pH 7): 254 nm (ε 12950); (pH 12): 262 nm (ε 12400).

2779. Cyclizine. *1-Diphenylmethyl-4-methylpiperazine; N*-benzhydryl-*N'*-methylpiperazine; *N*-methyl-*N'*-benzhydrylpiperazine; (*N*-benzhydryl)(*N'*-methyl)diethylenediamine; Compd 47-83; Wellcome prepn 47-83; Marzine; Marezine; Nautazine; Neo-Devomit. $C_{18}H_{22}N_2$; mol wt 266.39. C 81.16%, H 8.32%, N 10.52%. Prepd by the action of benzhydryl chloride on *N*-methyl piperazine: Baltzly *et al., J. Org. Chem.* **14,** 775 (1949); U.S. pat. **2,630,-435** (1953 to Burroughs Wellcome). Metabolism: R. Kuntzman *et al., Ann. N.Y. Acad. Sci.* **226,** 131 (1973). Comprehensive description: S. A. Benezra in *Analytical Profiles of Drug Substances* **vol.** 6, K. Florey, Ed. (Academic Press, New York, 1977) pp 83-97. Clinical antiemetic efficacy: G. Rowlands, W. J. Currie, *Brit. J. Clin. Pract.* **30,** 197 (1976); W. N. Chestnutt *et al., Eur. J. Anaesthesiol.* **3,** 27 (1986).

Light-sensitive crystals from petr ether, mp 105.5-107.5°. uv max (0.1N HCl): 269, 263, 258, 225 (ε 540, 742, 694, 11300). Soly in g/ml at 25°: chloroform 1.1; ether 0.17; ethanol 0.17; water <0.001. LD_{50} orally in mice: 147 mg/kg, *RTECS* vol. II, E. J. Fairchild, Ed. (1977) p 716.

Hydrochloride, $C_{18}H_{22}N_2 \cdot HCl$, *Valoid.* Crystals, moderately sol in water. *Protect from light.*

THERAP CAT: Antiemetic.

THERAP CAT (VET): Antiemetic.

2780. Cyclobarbital. *5-(1-Cyclohexen-1-yl)-5-ethyl-2,4,-6(1H,3H,5H)-pyrimidinetrione; 5-(1-cyclohexen-1-yl)-5-ethyl-barbituric acid;* cyclobarbitone; hexemal; tetrahydropheno-barbital; Cavonyl; Cyclodorm; Cyklodorm; Fanodormo; Irifan; Namuron; Palinum; Phanodorm; Phanodorn; Philo-dorm; Prälumin; Pro-Sonil; Sonaform. $C_{12}H_{16}N_2O_3$; mol wt 236.27. C 61.00%, H 6.83%, N 11.86%, O 20.31%. Prepn: **Brit. pat. 231,150** (1924 to Bayer); Schulemann, Meisenburg, **U.S. pat. 1,690,796** (1929 to Winthrop); Eckstein, *Przemysl Chem.* **9**, 390 (1953), *C.A.* **49**, 11668c (1955). Toxicity data: G. Hofrichter *et al., Arzneimittel-Forsch.* **17**, 242 (1967).

Shiny crystals, insipid bitter taste, mp 171-174°. Very slightly sol in cold water, appreciably sol in boiling water. One gram dissolves in 5 ml alcohol, 20 ml ether. LD_{50} in mice, rats (mg/kg): 350, 290 i.p. (Hofrichter).

Calcium salt, $C_{24}H_{30}CaN_4O_6$, *hexemal calcium, Itridal, Kollerdormfix, Pronox.* Minute crystals. Appreciably bitter taste. Slightly basic reaction. Soly in water about 1 part in 70 parts (w/w), in 95% alc about 1 part in 500 parts. Practically insol in ether, chloroform.

Caution: May be habit forming. This is a controlled substance (depressant) listed in the U.S. Code of Federal Regulations, Title 21 Parts 329.1 and 1308.13 (1995).

THERAP CAT: Sedative, hypnotic.

THERAP CAT (VET): Has been used as an anesthetic and sedative.

2781. Cyclobendazole. *[5-(Cyclopropylcarbonyl)-1H-benzimidazol-2-yl]carbamic acid methyl ester;* ciclobenda-zole; CC-2481; R-17147; Haptocil. $C_{13}H_{13}N_3O_3$; mol wt 259.26. C 60.23%, H 5.05%, N 16.21%, O 18.51%. Prepn: J. L. Van Gelder *et al.,* **Ger. pat. 2,029,637** corresp to **U.S. pat. 3,657,267** (1971, 1972 both to Janssen). Synthesis and anthelmintic activity: A. H. M. Raeymaekers *et al., Arzneimittel-Forsch.* **28**, 586 (1978). Embryotoxic and antimitotic properties: P. Delatour, Y. Richard, *Therapie* **31**, 505 (1976). Pharmacokinetics: R. R. Brodie *et al., Arzneimittel-Forsch.* **27**, 593 (1977). Biotransformation: B. C. Mayo *et al., Drug Metab. Dispos.* **6**, 518 (1978). Clinical evaluation: A. Degrémont, E. Stahel, *Schweiz. Med. Wochenschr.* **108**, 1430 (1978).

Crystals from acetic acid, mp 250.5°.

THERAP CAT: Anthelmintic (Nematodes).

2782. Cyclobenzaprine. *3-(5H-Dibenzo[a,d]cyclohep-ten-5-ylidene)-N,N-dimethyl-1-propanamine; N,N-dimethyl-5H-dibenzo[a,d]cyclohepten-Δ⁵,ˠ-propylamine;* 5-(3-dimethyl-aminopropylidene)dibenzo[a,e]cycloheptatriene; 1-(3-di-methylaminopropylidene)-2,3:6,7-dibenzo-4-suberene; pro-heptatriene; MK-130; Ro-4-1577; RP-9715. $C_{20}H_{21}N$; mol wt 275.39. C 87.23%, H 7.69%, N 5.09%. Prepn: **Brit. pat. 858,187** (1961 to Hoffmann-La Roche); Villani *et al., J. Med. Pharm. Chem.* **5**, 373 (1962); Winthrop *et al., J. Org. Chem.* **27**, 230 (1962). Pharmacology: C. D. Barnes, W. L. Adams, *Neuropharmacology* **17**, 445 (1978); N. N. Share, *ibid.* 721; and toxicology: J. Metysova *et al., Arch. Int.*

Pharmacodyn. Ther. **144**, 481 (1963). Metabolism: G. Belvedere *et al., Biomed. Mass Spectrom.* **1**, 329 (1974); H. B. Hucker *et al., Drug Metab. Dispos.* **6**, 184 (1978). Bio-availability: *eidem, J. Clin. Pharmacol.* **17**, 719 (1977). Clinical studies: J. V. Basmajian, *Arch. Phys. Med. Rehabil.* **5**, 58 (1978); B. R. Brown, J. Womble, *J. Am. Med. Assoc.* **240**, 1151 (1978). Comprehensive description: M. L. Cotton, G. R. B. Down in *Analytical Profiles of Drug Substances* **vol. 17**, K. Florey, Ed. (Academic Press, New York, 1988) pp 41-72.

bp_1 175-180°. uv max: 224, 289 nm (log ε 4.57, 4.02), (Villani *et al.*)

Hydrochloride, $C_{20}H_{21}N \cdot HCl$, *Flexeril, Flexiban.* Use as muscle relaxant: N. N. Share, **Fr. pat. 2,100,873** (1972 to Frosst), *C.A.* **78**, 47801n (1973). Crystals from isopropanol, mp 216-218°. Soly in water: > 20 g/100 ml. Freely sol in water, methanol, ethanol; sparingly sol in isopropanol; slightly sol in chloroform, methylene chloride. Practically insol in hydrocarbons. uv max: 226, 295 nm (ε 52300, 12000). LD_{50} in mice (mg/kg): 35 i.v., 250 orally (Mcty-sova).

THERAP CAT: Muscle relaxant (skeletal).

2783. Cyclobutane. Tetramethylene. C_4H_8; mol wt 56.11. C 85.63%, H 14.37%. Prepd by hydrogenaton of cyclobutene in the presence of nickel at 100°: Willstätter, Bruce, *Ber.* **40**, 3988 (1907); *see also* Heisig, *J. Am. Chem. Soc.* **63**, 1698 (1941); by pyrolysis of ether, along with other products: Peytral, *Bull. Soc. Chim. France* [4] **35**, 964 (1924). In 39% yield from cyclobutanecarboxylic acid: Cason, Way, *J. Org. Chem.* **14**, 31 (1949), for the prepn of this acid *see* Heisig, Stodola, *Org. Syn.* **coll. vol. III**, 213 (1955).

Gas. Burns with a luminous flame. mp −80°. bp_{741} +13.08°. d^0 0.7038; d^{-5} 0.7185. n_D^0 1.37520. Insol in water; freely sol in alcohol, acetone.

2784. Cyclobutyrol. *α-Ethyl-1-hydroxycyclohexaneace-tic acid;* 1-cyclohexanol-α-butyric acid; α-(1-hydroxycyclo-hexyl)butyric acid; 1-hydroxy-α-ethylcyclohexylacetic acid. $C_{10}H_{18}O_3$; mol wt 186.25. C 64.49%, H 9.74%, O 25.77%. Prepd by hydrolysis of the ethyl ester obtained by the Reformatski reaction: Kon, Naravanan, *J. Chem. Soc.* **1927**, 1536; Kandiah, Linstead, *ibid.* **1929**, 2150; Maillard *et al., Bull. Soc. Chim. France* **1958**, 244; *eidem,* **Ger. pat. 1,094,-254** corresp to **U.S. pat. 3,065,134** (1959, 1962 to Lab. Jacques Logeais). Activity studies: Roquet, Jousse, *Arch. Int. Pharmacodyn.* **125**, 172 (1960).

Colorless crystals from ether-petr ether, mp 81-82°. Formerly reported as viscous oil. $d_4^{18.8}$ 1.0010. bp_{24} 164°; bp_{16} 167-170°; $n_D^{18.8}$ 1.4680 (Kon, Naravanan, *loc. cit.*). Slightly sol in water, petr ether. Very sol in alcohols, acetone, dioxane, chloroform, ether. Completely sol in aq alkalies at pH 7.5-8.

Sodium salt, $C_{10}H_{17}NaO_3$, *Bilimix, Bis-Bil, Colepan, Dimene, Epa-Bon, Hebucol, Maricolene, Tachicol, Tri-Bil, Tri-*

bilina. White crystalline, slightly hygroscopic powder, mp 299-300°.

THERAP CAT: Choleretic.

2785. Cyclobuxine D. *(3β,5α,16α,20S)-14-Methyl-3,20-bis(methylamino)-4-methylene-9,19-cyclopregnan-16-ol;* cyclobuxine. $C_{25}H_{42}N_2O$; mol wt 386.62. C 77.67%, H 10.95%, N 7.25%, O 4.14%. From *Buxus sempervirens* L., *Buxaceae:* E. Schlittler *et al., Helv. Chim. Acta* **32**, 2209 (1949). Probably identical with Alkaloid A: Heusler, Schlittler, *ibid.* 2226. Structure: Brown, Kupchan, *J. Am. Chem. Soc.* **84**, 4590 (1962). Stereochemistry: *eidem, ibid.* **86**, 4424 (1964).

Crystals, dec 245-247°. $[α]_D^{23}$ +98° (chloroform).

Dihydrobromide, $C_{25}H_{42}N_2O.2HBr$, crystals, dec 288-292°.

N,N'-Dimethylcyclobuxine D, $C_{27}H_{46}N_2O$, crystals, dec 204-205°. $[α]_D^{25}$ +99° (chloroform).

2786. Cyclocumarol. *3,4-Dihydro-2-methoxy-2-methyl-4-phenyl-2H,5H-pyrano[3,2-c][1]benzopyran-5-one;* 2-methyl-2-methoxy-4-phenyl-5-oxodihydropyrano[3,2-c][1]benzopyran; 3,4-(2'-methyl-2'-methoxy-4'-phenyl)dihydropyranocoumarin; 4-hydroxycoumarin anticoagulant no 63; BL-5; Cumopyran; Cumopyrin. $C_{20}H_{18}O_4$; mol wt 322.36. C 74.52%, H 5.63%, O 19.85%. Prepn: M. Ikawa *et al., J. Am. Chem. Soc.* **66**, 902 (1944); M. A. Stahmann, *et al.,* U.S. pat. 2,427,579 (1947 to Wisconsin Alumni Res. Found.).

Crystals from ethanol, mp 166°. One gram dissolves in 100 ml water. Slightly sol in alcohol, vegetable oil.

THERAP CAT: Anticoagulant.

2787. Cyclodextrins. Cycloamyloses; cycloglucans; Schardinger dextrins. Naturally occurring clathrates, *q.v.* Obtained from the action of *Bacillus macerans* amylase on starch to form homogeneous cyclic α-(1→4) linked D-glucopyranose units. α-, β- and γ-cyclodextrins are composed of six, seven and eight units, resp. Isoln of β-cyclodextrin: A. Villiers, *Compt. Rend.* **112**, 536 (1891); of α- and β-cyclodextrin: F. Schardinger, *Z. Untersuch. Nahr. Genussm.* **6**, 865 (1903); of γ-cyclodextrin: K. Freudenberg, R. Jacobi, *Ann.* **518**, 102 (1935). X-ray mol wt determn: D. French, R. E. Rundle, *J. Am. Chem. Soc.* **64**, 1651 (1942); D. French *et al., ibid.* **72**, 5150 (1950). Have hydrophobic cavities; form inclusion compds with organic substances, salts and halogens in the solid state or in aq solns. Model for enzyme action: I. Tabushi, *Acc. Chem. Res.* **15**, 66 (1982). *Reviews:* D. French, *Adv. Carbohydr. Chem.* **12**, 189-260 (1957); J. A. Thoma, L. Stewart, *Starch: Chemistry and Technology* vol. **1**, R. L. Whistler, E. F. Paschall, Eds. (Academic Press, New York, 1965) pp 209-249; F. Cramer, H. Hettler, *Naturwiss.* **154**, 625-632 (1967). Book: M. L. Bender, M. Komiyama, *Cyclodextrin Chemistry* (Springer-Verlag, New York, 1978) 96 pp.

α-Cyclodextrin

α-Cyclodextrin, $C_{36}H_{60}O_{30}$, *cyclohexaamylose.* Hexagonal plates or blade-shaped needles. $[α]_D$ +150.5°.

β-Cyclodextrin, $C_{42}H_{70}O_{35}$, *cycloheptaamylose.* Parallelogram shaped crystals. $[α]_D$ +162.0°.

γ-Cyclodextrin, $C_{48}H_{80}O_{40}$, *cyclooctaamylose.* Square plates or rectangular rods. $[α]_D$ +177.4°.

USE: As complexing agent; in study of enzyme action.

2788. Cyclodrine. *1-Hydroxy-α-phenylcyclopentaneacetic acid 2-diethylaminoethyl ester;* β-diethylaminoethyl (1-hydroxycyclopentyl)phenylacetate; 2-phenyl-2-(1-hydroxycyclopentyl)acetic acid β-(diethylamino)ethyl ester. $C_{19}H_{29}NO_3$; mol wt 319.44. C 71.44%, H 9.15%, N 4.38%, O 15.03%. Prepn: Blicke, U.S. pat. 2,922,795 (1960 to Univ. of Michigan). Activity studies: Lands, Luduena, *J. Pharmacol. Exp. Ther.* **117**, 331 (1956); Karczman, Long, *ibid.* **123**, 230 (1958).

Hydrochloride, $C_{19}H_{29}NO_3.HCl$, *GT-92, Cyclopent.* Crystals from isopropanol-isopropyl ether, mp 133-135°.

THERAP CAT: Anticholinergic.

2789. Cyclofenil. *4-[[4-(Acetyloxy)phenyl]cyclohexylidenemethyl]phenol acetate;* α-cyclohexylidene-α-(p-hydroxyphenyl)-p-cresol diacetate; bis(p-acetoxyphenyl)cyclohexylidenemethane; 4,4'-diacetoxybenzhydrylidenecyclohexane; F-6066; H-3452; ICI-48213; Fertodur; Neoclym; Ondonid; Ondogyne; Rehibin; Sanocrisin; Sexadieno; Sexovid. $C_{23}H_{24}O_4$; mol wt 364.44. C 75.80%, H 6.64%, O 17.56%. Prepn: Miquel *et al., J. Med. Chem.* **6**, 774 (1963); Olsson *et al.,* U.S. pat. 3,287,397 (1966). Pharmacology: Hiramatsu *et al., Oyo Yakuri* **6**, 1045 (1972), *C.A.* **78**, 131963z (1973).

Crystals from ethanol, mp 135-136°. uv max (ethanol): 247 nm (ε 17000).

Free diol, $C_{19}H_{20}O_2$, *F-6060.* mp 235-236°.
THERAP CAT: Gonad-stimulating principle.

2790. Cycloguanil. *1-(4-Chlorophenyl)-1,6-dihydro-6,6-dimethyl-1,3,5-triazine-2,4-diamine; 4,6-diamino-1-(p-chlorophenyl)-1,2-dihydro-2,2-dimethyl-s-triazine;* 2,4-diamino-1-p-chlorophenyl-1,6-dihydro-6,6-dimethyl-1,3,5-triazine; 1-p-chlorophenyl-2,4-diamino-6,6-dimethyl-1,6-dihydro-1,3,5-triazine; chlorazine (Russian); chlorguanide triazine; TCl; M-10580; D-20. $C_{11}H_{14}ClN_5$; mol wt 251.72. C 52.49%, H 5.61%, Cl 14.08%, N 27.82%. Metabolic product formed from the antimalarial drug chlorguanide: Carrington *et al., Nature* **168**, 1080 (1951). Prepn: Modest *et al., J. Am. Chem. Soc.* **74**, 855 (1952); Loo, *ibid.* **76**, 5096 (1954); Carrington *et al., J. Chem. Soc.* **1954**, 1017; Bami, *J. Sci. Ind. Res. (India)* **14C**, 231 (1955); Modest, *J. Org. Chem.* **21**, 1 (1956); U.S. pat. **2,900,385** (1959 to Children's Cancer Res. Found.); Elslager, Worth, U.S. pat. **3,074,947** (1963 to Parke, Davis).

Prisms from chloroform + ether, mp 146°. uv max (water): 241 nm (log ε 4.11).

Hydrochloride, $C_{11}H_{14}ClN_5 \cdot HCl$, prisms from water, mp 210-215°. uv max (water): 241 nm (log ε 4.12).

Dihydrochloride, $C_{11}H_{14}ClN_5 \cdot 2HCl$, crystals from acetone, mp 190-196°.

Pamoate, $C_{45}H_{44}Cl_2N_{10}O_6$, *cycloguanil embonate, CI-501, Camolar.* Yellow crystals, mp 231-234°. Soly in water: 0.003%.

THERAP CAT: Antimalarial.

2791. Cycloheptanone. Suberone; ketoheptamethylene; ketocycloheptane. $C_7H_{12}O$; mol wt 112.17. C 74.95%, H 10.78%, O 14.26%. Prepn from cyclohexanol with 1-(aminomethyl)cyclohexanol as the intermediate: Blicke *et al., J. Am. Chem. Soc.* **74**, 2924 (1952). Other methods: *Org. Syn.* **coll. vol. IV**, 221-228 (1963).

Liquid. d_4^{20} 0.9490. bp_{760} 179-181°. bp_{16} 66-70°. n_D^{20} 1.4608. Dipole moment 2.98. Practically insol in water. Freely sol in alcohol. Sol in ether.

Oxime, $C_7H_{13}NO$, *suberoxime.* mp 23°, bp_{20} 152°.

2792. Cyclohexane. Hexahydrobenzene; hexamethylene; hexanaphthene. C_6H_{12}; mol wt 84.16. C 85.63%, H 14.37%. Occurs in petr (0.5-1.0%). Obtained in the distillation of petr or by hydrogenation of benzene. In the distillation of petr the C_4-400°F boiling range naphthas are fractionated to obtain a C_5-200°F naphtha contg 10-14% cyclohexane which on superfractionation yields an 85% concentrate (which is sold as such); further purification necessitates isomerization of pentanes to cyclohexane, heat cracking for removing open chain hydrocarbons and sulfuric acid treatment to remove aromatic compds. The hydrogenation of benzene is done in the liq phase at 150° using Raney nickel catalyst and at least 10 atm H_2 pressure: Sabatier, *Ind. Eng. Chem.* **18**, 1005 (1926). Review and bibliography: Sachanen, *Chemical Constituents of Petroleum* (New York, 1945). Prepn of high purity cyclohexane: Seyer *et al., Ind. Eng. Chem.* **31**, 759 (1939). Cyclohexane can exist in two interconvertible conformations, the boat and the chair. In the chair form its 12 extracyclic bonds fall into two classes: six lie parallel to the main axis of symmetry and are designated "axial", while six extend radially outward at ±109.5° angles to the axis and are designated as "equatorial", Barton *et al., Nature* **172**, 1096 (1954); *Science* **119**, 49 (1954). Solubility:

F. P. Schwarz, *Anal. Chem.* **52**, 10 (1980). Toxicity: Lazarew, *Arch. Exp. Pathol. Pharmacol.* **143**, 223 (1929). Physical properties and methods of purification: L. Scheflan, M. B. Jacobs, *The Handbook of Organic Solvents* (Van Nostrand, 1953) p 233; *Techniques of Chemistry*, A. Weissberger, Ed., **vol. II**, 3rd ed., entitled *"Organic Solvents"* by J. D. Riddick, W. B. Bunger (Wiley-Interscience, New York, 1970) p 592. *Review:* M. L. Campbell in Kirk-Othmer *Encyclopedia of Chemical Technology*, **vol. 12** (Wiley-Interscience, New York, 3rd ed., 1980) pp 931-937.

boat   chair

Flammable liq. Solvent odor. Pungent when impure. d_4^{20} 0.7781; d_4^{80} 0.7206. mp +6.47°. bp_{760} 80.7°; bp_{400} 60.8°; bp_{200} 42.0°; bp_{100} 25.5°; bp_{60} 14.7°; bp_{40} 6.7°. n_D^{20} 1.4264. Flash pt, closed cup: 1°F (−18°C). Flammability limits in air 1.3-8.4% v/v. Soly in water at 23.5°C (w/w): 0.0052%. 100 ml of methanol dissolves 57 grams at 20°C; miscible with ethanol, ethyl ether, acetone, benzene, carbon tetrachloride. LC in mice: ~60-70 mg/l air (Lazarew).

Caution: Potential symptoms of overexposure are irritation of eyes and respiratory system; drowsiness; dermatitis; narcosis, coma. *See NIOSH Pocket Guide to Chemical Hazards* (DHHS/NIOSH 90-117, 1990) p 76.

USE: Solvent for lacquers and resins. Paint and varnish remover. In the extraction of essential oils. In analytical chemistry for mol wt determinations (cryoscopic constant 20.3). In the manuf of adipic acid, benzene, cyclohexyl chloride, nitrocyclohexane, cyclohexanol and cyclohexanone. In the manuf of solid fuel for camp stoves. In fungicidal formulations (possesses slight fungicidal action). In the industrial recrystn of steroids.

2793. Cyclohexanecarboxylic Acid. Hexahydrobenzoic acid. $C_7H_{12}O_2$; mol wt 128.17. C 65.60%, H 9.44%, O 24.97%. Prepn from anisic acid: Lumsden, *J. Chem. Soc.* **87**, 90 (1905); from 2-chlorocycloheptanone: Gutsche, *J. Am. Chem. Soc.* **71**, 3513 (1949); by carbonation of cyclohexylmagnesium chloride: Wagner, Moore, *ibid.* **72**, 974 (1950); from cyclohexane + KI + active Ni: Reppe *et al., Ann.* **582**, 38 (1953); by oxidation of cycloheptanone: Payne, Smith, *J. Org. Chem.* **22**, 1680 (1957); from cyclohexane + HCOOH or CO_2: McKursick *et al., J. Am. Chem. Soc.* **82**, 723 (1960); McKursick, U.S. pat. **2,940,913** (1960 to du Pont).

Liquid. bp 232.5°; bp_{20} 131°; bp_8 110°; $bp_{<1}$ 63-67°. Crystallizes, on cooling, in monoclinic prisms, mp 29°. Odorless but when liq or in soln has a valerian odor. d_4^{15} 1.0480. n_D^{20} 1.4530. Soly in 100 g water at 15°: 0.201 g. Sol in most organic solvents.

Methyl ester, $C_8H_{14}O_2$, fragrant liq, bp 183°. d_4^{15} 0.9954.

USE: Solubilizer for vulcanized rubber; clarifier for mineral oil; in insecticide formulations.

2794. Cyclohexanol. Hexalin; hexahydrophenol. $C_6H_{12}O$; mol wt 100.16. C 71.95%, H 12.08%, O 15.97%. Obtained by oxidation of cyclohexane or hydrogenation of phenol: Faith, Keyes & Clark's *Industrial Chemicals*, F. A. Lowenheim, M. K. Moran, Eds. (Wiley-Interscience, New York, 4th ed., 1975) pp 304-309. Toxicity study: H. F. Smyth *et al., Am. Ind. Hyg. Assoc. J.* **23**, 95 (1962).

Hygroscopic crystals; camphor odor. d^{20} 0.962. mp 23-25°. bp 161°. Flash pt, closed cup: 154°F (68°C); n_D^{22} 1.465. At 20° soly in water: 3.6% (w/w); soly of water in cyclohexanol: 11% (w/w). Miscible with ethanol, ethyl acetate, linseed oil, petr solvents, aromatic hydrocarbons. LD_{50} orally in rats: 2.06 g/kg (Smyth).

Caution: Potential symptoms of overexposure are irritation of eyes, nose, throat and skin; narcosis. *See NIOSH Pocket Guide to Chemical Hazards* (DHHS/NIOSH 90-117, 1990) p 78.

USE: Solvent for alkyd resins, alcohol-sol phenolic resins, ethyl cellulose. Manuf celluloid; finishing textiles; insecticides.

2795. Cyclohexanone. Ketohexamethylene; pimelic ketone; Hytrol O; Anone; Nadone. $C_6H_{10}O$; mol wt 98.14. C 73.43%, H 10.27%, O 16.30%. Obtained from cyclohexanol by catalytic dehydrogenation or by oxidation (which yields cyclohexanone and adipic acid) or from cyclohexane by oxidation (yielding cyclohexanone and cyclohexanol): **Brit. pat.** 310,055 (1928 to Schering-Kahlbaum); **U.S. pats.** 2,223,493; 2,223,494; 2,285,914 (1940, 1940, 1942 to Du Pont). Review of manuf processes: Faith, Keyes & Clark's *Industrial Chemicals,* F. A. Lowenheim, M. K. Moran, Eds. (Wiley-Interscience, New York, 4th ed., 1975) pp 304-309. Toxicity: Smyth *et al., Am. Ind. Hyg. Assoc. J.* **30,** 470 (1969).

Oily liq. Odor reminiscent of peppermint and acetone. d_4^{20} 0.9478; d_4^{25} 0.9421. mp −32.1°. bp_{760} 155.6°; bp_{400} 132.5°; bp_{200} 110.3°; bp_{100} 90.4°; bp_{60} 77.5°; bp_{40} 67.8°; bp_{20} 52.5°; bp_{10} 38.7°; bp_5 26.4°; $bp_{1.0}$ 1.4°. n_D^{20} 1.4507. Flash pt closed cup: 147°F (63°C). Soly in water: 150 g/l at 10°; 50 g/l at 30°. Soly of water in cyclohexanone: 87 g/l at 20°. Sol in alcohol, ether and in other common organic solvents. LD_{50} orally in rats: 1.62 ml/kg (Smyth).

2,4-Dinitrophenylhydrazone, golden yellow needles, mp 160°. uv max (CHCl₃): 262 nm (ε 22500); (0.01N NaOH): 435 nm (ε 19000).

Caution: Potential symptoms of overexposure are irritation of eyes and mucous membranes; headache; narcosis, coma; dermatitis. *See NIOSH Pocket Guide to Chemical Hazards* (DHHS/NIOSH 90-117, 1990) p 78.

USE: Solvent for cellulose acetate, nitrocellulose, natural resins, vinyl resins, crude rubber, waxes, fats, shellac, DDT. In the production of adipic acid for nylon. In the prepn of cyclohexanone resins. 2,4-Dinitrophenylhydrazone deriv as standard in elemental analysis.

2796. Cyclohexene. 1,2,3,4-Tetrahydrobenzene. C_6H_{10}; mol wt 82.15. C 87.73%, H 12.27%. Occurs in coal tar. Prepd by dehydration of cyclohexanol at high temps over various catalysts. Lab prepn using H_2SO_4 as dehydrating agent: Coleman, Johnstone, *Org. Syn.* **5,** 33 (1925); Wagner, *J. Chem. Ed.* **10,** 113 (1933); by distn of cyclohexanol over silica gel or alumina: Hershberg, Ruhoff, *Org. Syn.* **17,** 25 (1937).

Liquid. d_4^{20} 0.8098; d^{50} 0.7823; d^{100} 0.7355. mp −103.5°. bp_{760} 83°. n_D^{20} 1.4465; n_D^{27} 1.4428. Absorption spectrum: Purvis, *Proc. Cambridge Phil. Soc.* **23,** 588 (1927); *Chem. Zentr.* **1927,** II, 379; *cf.* Hartley, Dobbie, *J. Chem. Soc.* **77,** 846 (1900).

Caution: Potential symptoms of overexposure are irritation of eyes, respiratory system and skin; drowsiness. *See NIOSH Pocket Guide to Chemical Hazards* (DHHS/NIOSH 90-117, 1990) p 78.

USE: Alkylation component. In the manuf of adipic acid, maleic acid, hexahydrobenzoic acid and aldehyde. To pre-

pare butadiene in the laboratory. Has been suggested for the synthesis of maleic acid and as stabilizer for high octane gasoline.

2797. Cycloheximide. *[1S-[1α(S*),3α,5β]]-4-[2-(3,5-Dimethyl-2-oxocyclohexyl)-2-hydroxyethyl]-2,6-piperidinedione; 3-[2-(3,5-dimethyl-2-oxocyclohexyl)-2-hydroxyethyl]glutarimide;* naramycin A; NSC-185; U-4527; Actidione. $C_{15}H_{23}NO_4$; mol wt 281.35. C 64.04%, H 8.24%, N 4.98%, O 22.75%. Antibiotic substance isolated from the beers of streptomycin-producing strains of *Streptomyces griseus:* B. E. Leach *et al., J. Am. Chem. Soc.* **69,** 474 (1947); J. H. Ford, B. E. Leach, *ibid.* **70,** 1223 (1948); A. J. Whiffen *et al.,* U.S. pat. **2,574,519** (1951 to Upjohn). Production, assay and antibiotic activity: A. J. Whiffen, *J. Bacteriol.* **56,** 283 (1948). Improved production method: Kominek, U.S. pats. **3,915,802-3** (both 1975 to Upjohn). Structure: E. C. Kornfeld *et al., J. Am. Chem. Soc.* **71,** 150 (1949). Absolute configuration: E. J. Eisenbraun *et al., ibid.* **80,** 1261 (1958); F. Johnson, N. A. Starkovsky, *Tetrahedron Letters* **1962,** 1173; F. Johnson *et al., J. Am. Chem. Soc.* **87,** 4612 (1965). Synthesis of *dl-* and *l-*forms: F. Johnson *et al., ibid.* **88,** 149 (1966).

Plates from amyl acetate or water or 30% methanol, mp 119.5-121° (Whiffen, 1951); also reported as mp 115-116° (Johnson, 1966). $[\alpha]_D^{29}$ −3.38° (c = 9.47 in methanol); $[\alpha]_D^{25}$ +6.8° (c = 2 in H₂O). Soly at 2°: water 2.1 g/100 ml, amyl acetate 7 g/100 ml. Also sol in chloroform, ether, acetone, methanol, ethanol, other common organic solvents except satd hydrocarbons. Relatively heat-stable, acid-stable, destroyed by boiling in aq soln at pH 7 for 1 hr, but shows no loss of activity after 15 min boiling. At pH 2 it is not destroyed by boiling for 1 hr. Rapidly inactivated at room temp by dil alkali with the formation of a volatile, fragrant ketone, 2,4-dimethylcyclohexanone. Extremely repellent to rats. LD_{50} i.v. in mice: 150 mg/kg (Leach).

Acetate, $C_{17}H_{25}NO_5$, glistening plates from methanol, mp 148-149°. $[\alpha]_D^{25}$ +22° (c = 2.3 in methanol).

USE: Fungicide; plant growth regulator. As protein synthesis inhibitor.

2798. Cyclohexylamine. *Cyclohexanamine;* aminocyclohexane; hexahydroaniline. $C_6H_{13}N$; mol wt 99.18. C 72.66%, H 13.21%, N 14.12%. Prepd by the catalytic hydrogenation of aniline at elevated temps and pressures. Fractionation of the crude reaction product yields cyclohexylamine, unchanged aniline, and a high-boiling residue contg N-phenylcyclohexylamine (cyclohexylaniline) and dicyclohexylamine. Review and bibliography: Carswell, Morrill, *Ind. Eng. Chem.* **29,** 1247 (1937). Toxicity data: H. F. Smyth *et al., Am. Ind. Hyg. Assoc. J.* **30,** 470 (1969).

Liquid. Strong, fishy, amine odor. d_{25}^{25} 0.8647. mp −17.7°. bp_{760} 134.5°; bp_{500} 118.9°; bp_{300} 102.5°; bp_{100} 72.0°; bp_{50} 56.0°; bp_{30} 45.1°; bp_{25} 41.3°; bp_{20} 36.4°; bp_{15} 30.5°. n_D^{25} 1.4565. Strong base. Completely misc with water and with common organic solvents, including alcohols, ethers, ketones, esters, aliphatic hydrocarbons, aromatic hydrocarbons and their chlorinated derivatives. On distillation with water, cyclohexylamine forms an azeotropic mixture contg 44.2% cyclohexylamine by weight, bp_{760} 96.4°. Reacts with excess ammonia and zinc chloride at 350° to produce α-picoline: Nordt, *Conversion of Hexahydroaniline into Pyridine Bases,* Off. Pub. Bd., Report PB 704 (1941). LD_{50} orally in rats: 0.71 ml/kg (Smyth).

Caution: Can cause irritation and sensitization. High concs cause nausea and narcotic effects.

USE: In organic synthesis, manuf insecticides, plasticizers, corrosion inhibitors, rubber chemicals, dyestuffs, emulsifying agents, dry-cleaning soaps, acid gas absorbents.

2799. Cyclohexyl Bromide. *Bromocyclohexane.* $C_6H_{11}Br$; mol wt 163.06. C 44.20%, H 6.80%, Br 49.00%.
Liquid; penetrating odor. d_{15}^{15} 1.329. bp 163-165°. n_D^{15} 1.4956. Insol in water. Sol in alcohol, etc.

2800. Cyclohexylcarbinol. *Cyclohexanemethanol;* cyclohexanecarbinol; hydroxymethylcyclohexane; hexahydrobenzyl alcohol. $C_7H_{14}O$; mol wt 114.19. C 73.63%, H 12.36%, O 14.01%. Prepd from cyclohexylmagnesium bromide and formaldehyde in ether: Gilman, Catlin, *Org. Syn.* **coll. vol. I** (2nd ed., 1941) p 188; Hiers, Adams, *J. Am. Chem. Soc.* **48,** 2385 (1926); Marvel *et al., ibid.* **50,** 2810 (1928). For prepn of cyclohexylmagnesium bromide *see Org. Syn., loc. cit.;* Gilman, Zoellner, *J. Am. Chem. Soc.* **53,** 1945 (1931).

Liquid. Slight odor of camphor. d_4^{25} 0.9215. bp_{784} 184-186°; bp_{23} 91-92°. n_D^{25} 1.4640. Sol in alc, ether.
Acetate, bp_{740} 199-201°.

2801. Cyclohexyl Chloride. *Chlorocyclohexane.* $C_6H_{11}Cl$; mol wt 118.61. C 60.76%, H 9.35%, Cl 29.89%.
Colorless liquid; suffocating odor. d_4^{20} 1.000. mp −44°. bp 142°. n_D^{20} 1.4626. Insol in water. Sol in alcohol, etc.

2802. 2-Cyclohexyl-4,6-dinitrophenol. 2,4-Dinitro-6-cyclohexylphenol; dinitro-o-cyclohexylphenol; DNOCHP; SN-46. $C_{12}H_{14}N_2O_5$; mol wt 266.25. C 54.13%, H 5.30%, N 10.52%, O 30.05%. Prepd from phenol, H_2SO_4, and cyclohexene or cyclohexanol: **Brit.** pat. 620,026 (1949 to Pest Control); from o-cyclohexylphenol by nitration in chloroform: Baroni, Kleinan, *Monatsh.* **68,** 251 (1936).

Crystals, mp 106.5-107.5°. Soly in petr oils 2.5 to 6%. Sol in benzene, DMF; very slightly sol in water.
Sodium salt, $C_{12}H_{13}N_2NaO_5$, *2-cyclohexyl-4,6-dinitrophenate, Anobesina.* Crystals, sol in water, alcohol.
USE: Insecticide, esp in control of citrus red mite.

2803. Cycloleucine. *1-Aminocyclopentanecarboxylic acid;* ACPC; NSC-1026. $C_6H_{11}NO_2$; mol wt 129.16. C 55.80%, H 8.58%, N 10.84%, O 24.77%. Synthetic amino acid thought to act as a valine antagonist. Prepn: Zelinsky, Stadnikoff, *Z. Physiol. Chem.* **75,** 350 (1911); Connors, Ross, *J. Chem. Soc.* **1960,** 2119; Cremlyn, *ibid.* **1962,** 3977; Sudo, Ichihara, *Bull. Chem. Soc. Japan* **36,** 34 (1963). Manuf: Neth. pat. Appl. 6,607,754 (1966 to Rohm & Haas), *C.A.* **67,** 73159b (1967). *Review:* Ross *et al., J. Med. Pharm. Chem.* **3,** 1 (1961). Immunopharmacology: Rosenthale *et al., J. Pharmacol. Exp. Ther.* **180,** 501 (1972); Brambilla *et al., Cancer Chemother. Rep., Part 1,* **56,** 579 (1972).

Crystals from ethanol-water, mp 330° (dec). Soly in water approx 5 g/100 ml. Forms stable metal salts.
Hydrochloride, prismic crystals from water, mp 274° (dec). LD_{50} orally in mice, rats: 309, 260-290 mg/kg.
USE: Biological testing material for immunosuppressive properties in mice; for amino acid transport studies.

2804. Cyclomethycaine. *4-(Cyclohexyloxy)benzoic acid 3-(2-methyl-1-piperidinyl)propyl ester; p-cyclohexyloxybenzoic acid ester of N-(3-hydroxypropyl)pipecoline; 3-(2-methylpiperidino)propyl p-cyclohexyloxybenzoate; Surfacaine; Surfathesin; Topocaine.* $C_{22}H_{33}NO_3$; mol wt 359.51. C 73.50%, H 9.25%, N 3.90%, O 13.35%. Prepn: McElvain, Carney, *J. Am. Chem. Soc.* **68,** 2592 (1946), *eidem,* U.S. pat. **2,439,818** (1948). Toxicity data: Schmidt *et al., Toxicol. Appl. Pharmacol.* **1,** 454 (1959).

Hydrochloride, $C_{22}H_{33}NO_3 \cdot HCl$, crystals, dec 178-180°. Soly in water: slightly > 1 g/100 ml.
Sulfate, $(C_{22}H_{33}NO_3)_2 \cdot H_2SO_4$, crystals, mod sol in water.
THERAP CAT: Topical anesthetic.
THERAP CAT (VET): Surface anesthetic.

2805. Cyclonite. *Hexahydro-1,3,5-trinitro-1,3,5-triazine; sym*-trimethylenetrinitramine; cyclotrimethylenetrinitramine; 1,3,5-trinitrohexahydro-*s*-triazine; T_4; RDX; Hexogen. $C_3H_6N_6O_6$; mol wt 222.12. C 16.22%, H 2.72%, N 37.84%, O 43.22%. Prepd by treating methenamine with fuming nitric acid: Henning, **Ger.** pat. **104,280** (1899); Hale, *J. Am. Chem. Soc.* **47,** 2754 (1925). Two moles of cyclonite can be obtained from one mole of methenamine, if ammonium nitrate and acetic anhydride are added: Bachmann, Sheehan, *J. Am. Chem. Soc.* **71,** 1842 (1949). Manuf by nitrolysis of hexamethylenetetramine: Ruth, U.S. pat. **3,049,543** (1962 to Olin Mathieson). Physical data relating to detonation, explosion, and blasting action: M. A. Cook, *Science of High Explosives,* ACS Monograph no. 139 (Reinhold, New York, 1958). Structure: Wood, *Proc. Int. Meet. Mol. Spectrosc., 4th, Bologna, 1959,* **2,** 955 (1962).

Orthorhombic crystals from acetone, mp 205-206°. d_4^{20} 1.82. One gram dissolves in 25 ml acetone. Slightly sol in methanol, ether, ethyl acetate, glacial acetic acid. Practically insol in water, carbon tetrachloride, carbon disulfide.
Mixture with pentaerythritoltetranitrate, *semtex.* Plastic explosive manufactured in two forms, *Semtex H* and *Semtex A,* which differ in the proportion of the two components RDX and pentaerythritoltetranitrate (PETN), *q.v.* Used in the destruction of PanAm Flight 103 in Dec. 1988 over Lockerbie, Scotland. Identification by MS: J. Yinon, *Can. Soc. Forens. Sci. J.* **21,** 46 (1988); by time of flight-MS: A. Marshall *et al., Rapid Commun. Mass Spec.* **8,** 521 (1994). Description of incidents of illicit use: A. W. Feraday in *Adv. Anal. Detect. Explosives,* J. Yinon, Ed. (Kluwer Academic Publishers, Netherlands, 1993) pp 67-72. Analysis of Semtex H and A: J. R. Hobbs, *ibid* pp. 409-427.

Consult the Name Index before using this section.

Caution: Can cause epileptiform seizures, Kaplan *et al.,* *Arch. Environ. Health* **10,** 877 (1965).

USE: High explosive; rat poison.

2806. Cyclonium Iodide. *1-[(2-Cyclohexyl-2-phenyl-1,3-dioxolan-4-yl)methyl]-1-methylpiperidinium iodide;* N-methyl-N-[(2-cyclohexyl-2-phenyl-1,3-dioxolan-4-yl)-methyl]piperidinium iodide; 1-(2-phenyl-2-cyclohexyl-1,3-dioxolan-4-yl)methyl-1-methylpiperidinium iodide; ciclonium iodide; oxapium iodide; SH-100; Esperan. $C_{22}H_{34}INO_2$; mol wt 471.42. C 56.05%, H 7.27%, I 26.92%, N 2.97%, O 6.79%. Prepn and separation of isomers: Kimura *et al.,* **Japan.** pat. **6585('66),** *C.A.* **65,** 7191 (1966) and Saikawa, **Japan.** pat. **16,475('70),** *C.A.* **73,** 45355t (1970) (both to Toyama). Stability studies: Shiuchi, Miyazaki, *Bunseki Kagaku (Japan Analyst)* **18,** 455 (1969), *C.A.* **71,** 74097z (1969). Incompatibility studies: Kunita *et al.,Yakuzaigaku (Pharmacology)* **28,** 84 (1968), *C.A.* **69,** 99335x (1968). Anticholinergic activity and acute toxicity: J. Faff *et al.,* *Pol. J. Pharmacol. Pharm.* **30,** 493 (1978). *Review: Japan. Med. Gaz.* **8**(9), 11 (1971).

α-Form, white crystals from monochlorobenzene or isopropyl alcohol, mp 195-197°. Insol in trichloroethylene.

β-Form, white crystals from monochlorobenzene, mp 150-152°. Sol in hot trichloroethylene.

Both sol in methanol, ethanol, chloroform and tetrachlorethane. Hardly sol in benzene, toluene, xylene and water. LD_{50} i.p. in mice: 106 mg/kg (Faff).

THERAP CAT: Antispasmodic.

2807. Cyclopentadiene. *1,3-Cyclopentadiene.* C_5H_6; mol wt 66.10. C 90.85%, H 9.15%. Obtained from the distillates produced in carbonization of coal, esp from the foreruns of coke-oven light oil: Ward, **U.S.** pat. **2,211,038** (1940); process involving the liquefaction of coke-oven gas: Horclois, *Chim. Ind. (Paris),* special no. April 1934, pp 357-363. Also obtained during the cracking of petr hydrocarbons: Tropsch *et al., Ind. Eng. Chem.* **30,** 169 (1938); *cf.* Dedussenko, *J. Gen. Chem. USSR* **7,** 1467 (1937); *Chem. Zentr.* **109,** I, 793 (1938). Synthesis by passing vaporized cyclopentane over activated alumina and oxides of molybdenum, chromium, or vanadium: Grosse *et al., Ind. Eng. Chem.* **32,** 309 (1940); **U.S.** pats. **2,157,202; 2,157,203; 2,-157,939.** Lab prepn by depolymerization of *dicyclopentadiene:* Moffett, *Org. Syn.* coll. vol. IV, 238 (1963). Toxicity study: Smyth *et al., Arch. Ind. Hyg. Occup. Med.* **10,** 61 (1954). *Review* and discussion of structure: Wilson, Wells, *Chem. Rev.* **34,** 1 (1944). *Review:* M. Fefer, A. B. Small in Kirk-Othmer *Encyclopedia of Chemical Technology* **vol. 7** (Wiley-Interscience, New York, 3rd ed., 1979) pp 417-429.

Liquid. d_4^0 0.8235; d_4^{10} 0.8131; d_4^{20} 0.8021; d_4^{25} 0.7966; d_4^{30} 0.7914. mp −85°. bp_{760} 41.5-42.0°. n_D^{16} 1.44632. Absorption spectrum: Pickett *et al., J. Am. Chem. Soc.* **63,** 1073 (1941). Insol in water. Miscible with alc, ether, benzene, carbon tetrachloride. Sol in carbon disulfide, aniline, acetic acid, liquid petrolatum. Cyclopentadiene polymerizes to dicyclopentadiene on standing. Polymerization is accelerated by the presence of peroxides or trichloroacetic acid. The dimer is cryst, mp 32.5°, with a camphor-like odor, having the structure of a partially hydrogenated indene contg a bridged methylene group. It is a more convenient form in which to handle cyclopentadiene, and is easily depolymerized by distilling at atmospheric pressure. LD_{50} of dimer orally in rats: 0.82 g/kg (Smyth).

Caution: Potential symptoms of overexposure are irritation of eyes and nose. *See NIOSH Pocket Guide to Chemical Hazards* (DHHS/NIOSH 90-117, 1990) p 78.

USE: Manuf resins; in organic synthesis as the diene in the Diels-Alder reaction producing sesquiterpenes, synthetic alkaloids, camphors.

2808. Cyclopentamine. *N,α-Dimethylcyclopentaneethanamine; N,α-dimethylcyclopentaneethylamine;* 1-cyclopentyl-2-methylaminopropane; β-(methylamino)propylcyclopentane; cyclopentadrine; Sinos; Cyklosal; Cyclonarol; Clopane. $C_9H_{19}N$; mol wt 141.26. C 76.53%, H 13.56%, N 9.92%. Prepn: Rohrmann, **U.S.** pat. **2,520,015** (1950 to Lilly).

Hydrochloride, $C_9H_{19}N.HCl$, crystals, mp 113-115° (base bp_{30} 83-86°, n_D^{25} 1.4500). Bitter taste. Freely sol in water.

THERAP CAT: Adrenergic (vasoconstrictor); nasal decongestant.

2809. Cyclopentane. Pentamethylene. C_5H_{10}; mol wt 70.13. C 85.63%, H 14.37%. Occurs in petroleum. Found in petr ether fractions. Prepd by cracking cyclohexane in the presence of alumina at high temps and pressure: Haensel, Ipatieff, *Ind. Eng. Chem.* **35,** 632 (1943); by reduction of cyclopentadiene: David *et al., Bull. Soc. Chim. France* [5] **11,** 561 (1944).

Mobile, flammable liq. mp −94.4°. bp 49.3°. d_4^{20} 0.7460. n_D^{20} 1.4068. Insol in water. Miscible with other hydrocarbon solvents, alcohol, ether. Lethal concn for mice in air: 38,000 ppm, *Handbook of Toxicology* **vol. 1,** W. S. Spector, Ed. (Saunders, Philadelphia, 1956) pp 330-331.

2810. Cyclopentanol. Cyclopentyl alcohol; hydroxycyclopentane. $C_5H_{10}O$; mol wt 86.13. C 69.72%, H 11.70%, O 18.58%. Prepd by modified Meerwein-Pondorf-Verley reduction of cyclopentanone in the presence of aluminum isopropoxide and sodium hydroxide: Truett, Moulton, *J. Am. Chem. Soc.* **73,** 5913 (1951); by catalytic hydrogenation of pure cyclopentanone with copper chromite at 150° and 150 atm: Kögl, Ultée, *Rec. Trav. Chim.* **69,** 1576 (1950); by catalytic hydrogenation of cyclopentanone with platinum oxide and platinum black at 2-3 atm: Noller, Adams, *J. Am. Chem. Soc.* **48,** 1084 (1926); by hydration of cyclopentene in aq H_2SO_4: Hepp, **U.S.** pat. **2,414,646** (1947 to Phillips Petr.); by reduction of cyclopentanone with $LiAlH_4$ in ether at room temp: Nystrom, Brown, *J. Am. Chem. Soc.* **69,** 1197 (1947).

Liquid, odor of amyl alcohol. d_4^0 0.96253; d_4^{15} 0.95078; d_4^{20} 0.9488; d_4^{30} 0.93908. mp −19°. bp 140.85°. Flash pt 124°F. n_D^{15} 1.45512; n_D^{20} 1.4520. Sparingly sol in water. Sol in ethanol.

2811. Cyclopentanone. Ketocyclopentane; ketopentamethylene; adipic ketone. C_5H_8O; mol wt 84.12. C 71.39%, H 9.59%, O 19.02%. Prepd by heating adipic acid to 285-295° in the presence of barium hydroxide, distilling, extracting with ether and fractionating: Thorpe, Kon, *Org. Syn.* coll. vol. I, 192 (2nd ed., 1941).

Liquid. Agreeable odor, somewhat like peppermint. d_4^{18} 0.9509. mp $-58.2°$. bp$_{760}$ 130.6°; bp$_{10}$ 23-24°. Flash pt 85°F. n_D^{20} 1.4366. Slightly sol in water. Miscible with alc, ether. Polymerizes easily, esp in presence of acids.

2812. 1,2-Cyclopentenophenanthrene. $C_{17}H_{14}$; mol wt 218.30. C 93.54%, H 6.46%. Prepn: Ruzicka *et al.*, *Helv. Chim. Acta* **16**, 838 (1933), *cf.* Kon, *J. Chem. Soc.* **1933**, 1081; Cook, Hewett, *Chem. & Ind. (London)* **52**, 451 (1933); Hawthorne, Robinson, *J. Chem. Soc.* **1936**, 763; Bachmann, Kloetzel, *J. Am. Chem. Soc.* **59**, 2207 (1937).

Needles from alc or petr ether, mp 135-136°. Absorption spectrum: Mayneord, Roe, *Proc. Roy. Soc. London* **A152**, 299 (1935).

2813. Cyclopenthiazide. *6-Chloro-3-(cyclopentylmethyl)-3,4-dihydro-2H-1,2,4-benzothiadiazine-7-sulfonamide 1,1-dioxide; 6-chloro-3-cyclopentylmethyl-3,4-dihydro-7-sulfamoyl-2H-1,2,4-benzothiadiazine 1,1-dioxide; 3-cyclopentylmethyl-6-chloro-7-sulfamyl-3,4-dihydro-1,2,4-benzothiadiazine 1,1-dioxide; cyclomethiazide; tsiklometiazid;* Su-8341; Navidrex; Navidrix; Salimid. $C_{13}H_{18}ClN_3O_4S_2$; mol wt 379.89. C 41.10%, H 4.78%, Cl 9.33%, N 11.06%, O 16.85%, S 16.88%. Prepn: Whitehead *et al.*, *J. Org. Chem.* **26**, 2814 (1961); Belg. pat. 587,225 (1960 to Ciba). Pharmacology: Barrett *et al.*, *Arch. Int. Pharmacodyn.* **131**, 325 (1961).

Crystals from dil alc, mp 230°. LD$_{50}$ i.v. in rats, mice: 142, 232 mg/kg.
THERAP CAT: Antihypertensive.

2814. Cyclopentobarbital. *5-(2-Cyclopenten-1-yl)-5-(2-propenyl)-2,4,6(1H,3H,5H)-pyrimidinetrione; 5-allyl-5-(2-cyclopenten-1-yl)barbituric acid;* Cyclopal. $C_{12}H_{14}N_2O_3$; mol wt 234.25. C 61.53%, H 6.02%, N 11.96%, O 20.49%. Prepd by condensing the ethyl ester of cyclopentenylmalonic acid with urea: Compagnie de Bethune: Ger. pat. 589,947 (1930), Frdl. **19**, 1198; Brit. pat. 349,455; Fr. pat. 38,680.

Crystals, mp 139-140°. Bitter taste. Slightly sol in cold water; moderately sol in hot water. Freely sol in alcohol, organic solvents.
Sodium salt, $C_{12}H_{13}N_2NaO_3$, *Cyclopal Sodium.* White powder, very sol in water.

Caution: May be habit forming. This is a controlled substance (depressant) listed in the U.S. Code of Federal Regulations, Title 21 Parts 329.1 and 1308.13 (1995).
THERAP CAT: Sedative, hypnotic.

2815. Cyclopentolate. *α-(1-Hydroxycyclopentyl)benzeneacetic acid 2-(dimethylamino)ethyl ester; 1-hydroxy-α-phenylcyclopentaneacetic acid 2-(dimethylamino)ethyl ester;* 2-dimethylaminoethyl 1-hydroxy-α-phenylcyclopentaneacetate; β-dimethylaminoethyl (1-hydroxycyclopentyl)phenylacetate; 2-phenyl-2-(1-hydroxycyclopentyl)ethanoic acid β-(dimethylamino)ethyl ester. $C_{17}H_{25}NO_3$; mol wt 291.39. C 70.07%, H 8.65%, N 4.81%, O 16.47%. Ophthalmic anticholinergic. Prepn: Treves, U.S. pat. 2,554,511 (1951 to Schieffelin).

Hydrochloride, $C_{17}H_{25}NO_3 \cdot HCl$, *Ak-Pentolate, Alnide, Mydplegic, Cyclogyl, Mydrilate, Zyklolat.* Crystals from ethyl acetate, mp 137-141°. pH of 1% aq soln: 5.0-5.4. Freely sol in water, alcohol. Practically insol in ether.
THERAP CAT: Mydriatic.

2816. Cyclophosphamide. *N,N-Bis(2-chloroethyl)tetrahydro-2H-1,3,2-oxazaphosphorin-2-amine 2-oxide monohydrate; 2-[bis(2-chloroethyl)amino]tetrahydro-2H-1,3,2-oxazaphosphorine 2-oxide monohydrate;* 1-bis(2-chloroethyl)-amino-1-oxo-2-aza-5-oxaphosphoridin monohydrate; bis(2-chloroethyl)phosphamide cyclic propanolamide ester monohydrate; bis(2-chloroethyl)phosphoramide cyclic propanolamide ester monohydrate; *N,N-bis(β-chloroethyl)-N',O-propylenephosphoric acid ester diamide monohydrate; N,N-bis-(β-chloroethyl)-N',O-trimethylenephosphoric acid ester diamide monohydrate;* cyclophosphane; cytophosphane; B-518; Cycloblastin; Cyclostin; Cytoxan; Endoxan; Procytox; Sendoxan. $C_7H_{15}Cl_2N_2O_2P \cdot H_2O$; mol wt 279.10. C 30.12%, H 6.14%, Cl 25.40%, N 10.04%, O 17.20%, P 11.10%. Prepn: H. Arnold *et al.*, *Naturwiss.* **45**, 64 (1957); *eidem, Nature* **181**, 931 (1958); H. Arnold, F. Bourseaux, *Angew. Chem.* **70**, 539 (1958). Absolute configuration of enantiomers: I. L. Karle *et al.*, *J. Am. Chem. Soc.* **99**, 4803 (1977); D. A. Adamaik *et al.*, *Angew. Chem. Int. Ed.* **16**, 330 (1977). Optical resolution study: T. Kawashima *et al.*, *J. Org. Chem.* **43**, 1111 (1978). Asymmetric synthesis of enantiomers: T. Sato *et al.*, *ibid.* **48**, 98 (1983). Metabolism: Connors *et al.*, *Biochem. Pharmacol.* **21**, 1373 (1972); Cox *et al.*, *ibid.* **25**, 993 (1976). Use in combination with azathioprine and hydroxychloroquine, *q.q.v.*, for the treatment of rheumatoid arthritis: D. J. McCarty, G. F. Carrera, *J. Am. Med. Assoc.* **248**, 1718 (1982). Clinical study with ACTH in multiple sclerosis patients: S. L. Hauser *et al.*, *N. Engl. J. Med.* **308**, 173 (1983). Toxicology: A. G. Wheeler *et al.*, *Toxicol. Appl. Pharmacol.* **4**, 324 (1962). Reviews of carcinogenicity studies: *IARC Monographs* **9**, 135-156 (1975); of clinical toxicology: L. H. Fraiser *et al.*, *Drugs* **42**, 781-795 (1991). Review of field trials as defleecing agent for sheep: M. H. Fahmy, Y. Moride, *Anim. Breed. Abstr.* **52**, 7-19 (1984).

mp 41-45°. Soly in water: 40 g/l. Slightly sol in alcohol, benzene, ethylene glycol, carbon tetrachloride, dioxane; sparingly sol in ether and acetone. LD$_{50}$ orally in mice, rats: 350, 94 mg/kg (Wheeler).

Note: Cyclophosphamide has been listed as a known carcinogen: *Seventh Annual Report on Carcinogens* (PB95-109781, 1994) p 53.

USE: Proposed as defleecing agent for sheep.

THERAP CAT: Antineoplastic.

2817. Cyclopropane. Trimethylene. C_3H_6; mol wt 42.08. C 85.63%, H 14.37%. Prepd by reduction of 1,2-dibromocyclopropane with zinc and alcohol: Lott, Christiansen, *J. Am. Pharm. Assoc.* **19**, 341 (1930); Schlatter, *J. Am. Chem. Soc.* **63**, 1733 (1941); from 1,3-dibromopropane with zinc and alcohol in absence of water: Ashdown *et al., ibid.* **58**, 850 (1936); by the action of sodium vapor on 1,3-dibromopropane: Bawn, Hunter, *Trans. Faraday Soc.* **34**, 608 (1938); from 1,3-dichloropropane heated with an excess of zinc dust, iodine and ethanol in 80% yield: Hass *et al., Ind. Eng. Chem.* **28**, 1178 (1936); from 1,3-dichloropropane with zinc in presence of sodium iodide: U.S. pat. 2,102,556 (1937); *cf.* U.S. pat. 2,098,239; Brit. pat. 498,225; U.S. pats. 2,211,787; 2,235,679; 2,235,762; 2,240,513; 2,240,514; 2,242,235. From ethylene by the reaction with methylene iodide and a zinc-copper couple in 29% yield: Simmons, Smith, *J. Am. Chem. Soc.* **80**, 5323 (1958); *Chem. & Eng. News* **36**, 40 (Dec 8, 1958).

Flammable gas. Characteristic odor resembling that of petr ether. mp $-127°$. bp $-33°$. Liquefies at 4-6 atms. One liter of cyclopropane (at 1 atm, 0°) weighs 1.879 g. One vol of cyclopropane dissolves in about 2.7 vols of water at 15°. Freely sol in alcohol, ether. Sol in fixed oils. Concd H_2SO_4 absorbs the gas readily.

Caution: Mixture of cyclopropane with oxygen or air may explode when brought in contact with a flame or other causes of ignition. Explosive limits (% by vol in air), lower: 2.41; upper: 10.3. The explosibility is greater than that of other anesthetic-oxygen mixtures because of the comparatively larger amounts of oxygen that are compatible with cyclopropane anesthesia. Rich oxygen mixtures are therefore to be avoided.

THERAP CAT: Anesthetic (inhalation).

THERAP CAT (VET): Anesthetic (inhalation).

2818. Cyclopropyl Methyl Ether. *Methoxycyclopropane;* cyprome ether. C_4H_8O; mol wt 72.11. C 66.63%, H 11.18%, O 22.19%. Prepd from 1,3-dibromo-2-methoxypropane by reduction with Zn: Olson *et al., J. Am. Chem. Soc.* **69**, 2454 (1947); *see also* Krantz, Jr. *et al., J. Pharmacol.* **69**, 207 (1940); Krantz, Jr., Drake, U.S. pat. 2,330,979 (1944); Haller, U.S. pat. 2,424,029 (1947).

Mobile, flammable liq. Odor similar to that of cyclopropane. d_4^{25} 0.786. mp $-119°$; bp 44.7°; n_D^{20} 1.3802. 100 ml of water dissolve 5.5 g of cyprome ether at 25°. Oil-water coefficient: 6.7 (ethyl ether = 4.5). Vapor press. at 26° (measured in nitrometer): 414 mm (ethyl ether = 555 mm). Explosive mixture, lower limit for air or oxygen: 2.0% v/v (ethyl ether vapor = 2.5 v/v).

Caution: Narcotic in high concns.

USE: Insecticide.

2819. Cyclorphan. *17-(Cyclopropylmethyl)morphinan-3-ol;* (−)-3-hydroxy-*N*-cyclopropylmethylmorphinan. $C_{20}H_{27}NO$; mol wt 297.44. C 80.76%, H 9.15%, N 4.71%, O 5.38%. Prepn: Gates, Montzka, *J. Med. Chem.* **7**, 127 (1964); Gates, U.S. pat. 3,285,922 (1966 to Res. Corp.). Pharmacology: Harris *et al., Arch. Int. Pharmacodyn. Ther.* **165**, 112 (1967).

Crystals from ethyl acetate, mp 187.5-189°. $[\alpha]_D^{30}$ $-120°$ (c = 2.26). LD_{50} in mice, rats (mg/kg): 24, 23 i.v. (Gates, 1966).

2820. Cycloserine. *D-4-Amino-3-isoxazolidinone;* D-4-amino-3-isoxazolidone; orientomycin; PA-94; 106-7; Closina; Farmiserina; Micoserina; Oxamycin; Seromycin. $C_3H_6N_2O_2$; mol wt 102.09. C 35.29%, H 5.92%, N 27.44%, O 31.34%. Antibiotic substance produced by *Streptomyces garyphalus* sive *orchidaceus:* Kuehl, Jr., *et al., J. Am. Chem. Soc.* **77**, 2344 (1955); Hidy *et al., ibid.* 2345; Shull, Sardinas, *Antibiot. & Chemother.* **5**, 398 (1955); Shull *et al.,* U.S. pat. 2,773,878 (1956 to Pfizer); Harned, U.S. pat. 2,789,983 (1957 to Commercial Solvents); Brit. pat. 768,007 (1957 to Commercial Solvents), *C.A.* **51**, 10847e (1957); U.S. pat. 3,124,590 (1964 to Commercial Solvents); Howe, U.S. pat. 2,845,433 (1958 to Merck & Co.). Synthesis: Stammer *et al., J. Am. Chem. Soc.* **77**, 2346 (1955); Peck, U.S. pat. 2,772,280 (1956 to Merck & Co.); Plattner *et al., Helv. Chim. Acta* **40**, 1531 (1957); Holly, Stammer, U.S. pat. 2,840,565 (1958 to Merck & Co.). Prepn of crystalline calcium and magnesium salts: Harris *et al.,* U.S. pat. 2,832,788 (1958 to Merck & Co.). HPLC determn in plasma and urine: D. G. Musson *et al., J. Chromatog.* **414**, 121 (1987). Comprehensive description: J. W. Lamb in *Analytical Profiles of Drug Substances* vol. 1, K. Florey, Ed. (Academic Press, New York, 1972) pp 53-64; H. A. El-Obeid, A. A. Al-Badr, *ibid* vol. 18, (1989) pp 567-597.

Crystals, decomp 155-156°. $[\alpha]_D^{23}$ $+116°$ (c = 1.17); $[\alpha]_{546}^{25}$ $+137°$ (c = 5 in 2N NaOH). uv max: 226 nm ($E_{1cm}^{1\%}$ 402). Sol in water, slightly sol in methanol, propylene glycol. Aq solns have a pH around 6. Forms salts with acids and bases. Neutral or acid solns are unstable. Aq solns buffered to pH 10 with sodium carbonate can be stored without loss for one week at refrigerator temps.

THERAP CAT: Antibacterial (tuberculostatic).

2821. Cyclosporins. A group of nonpolar cyclic oligopeptides with immunosuppressant activity; produced by *Tolypocladium inflatum Gams* (formerly designated as *Trichoderma polysporum* [Link ex Pers.] Rifai) and other fungi imperfecti. The major component, cyclosporin A, has been identified along with several other minor metabolites, cyclosporins B through I. A number of synthetic analogs have also been prepared. Isoln of A and C, structure of A: A. Rüegger *et al., Helv. Chim. Acta* **59**, 1075 (1976); M. Dreyfuss *et al., Eur. J. Appl. Microbiol.* **3**, 125 (1976). Crystal and molecular structure of the iodo deriv of A: T. J. Petcher *et al., Helv. Chim. Acta* **59**, 1480 (1976). Structure of C: R. Traber *et al., ibid.* **60**, 1247 (1977). Production of A and C: E. Härri *et al.,* U.S. pat. 4,117,118 (1978 to Sandoz). Isoln, characterization and antifungal activity of B, D, E, structures of A through D: R. Traber *et al., Helv. Chim. Acta* **60**, 1568 (1977). Isoln and structures of E, F, G, H, I: *eidem, ibid.* **65**, 1655 (1982). Total synthesis of cyclosporin A and analogs: R. Wenger, *Transplant. Proc.* **15**(4), Suppl. 1, 2230 (1983). Biosynthetic studies: H. Kobel *et al., Experientia* **39**, 873 (1983); R. Zocher *et al., Phytochemistry* **29**, 549 (1984). *In vivo* immunosuppressant activity: J. F. Borel *et al., Agents Actions* **6**, 468 (1976); *eidem, Immunology* **32**, 1017 (1977); R. Y. Calne, *Clin. Exp. Immunol.* **35**, 1 (1979). Selective effect on T-cells: D. J. White *et al., Transplantation* **27**, 55 (1979); M. Y. Gordon, J. W. Singer, *Nature* **279**,

433 (1979). Inhibition of γ-interferon synthesis: G. H. Reem *et al., Science* **221**, 63 (1983). Immunosuppressive profile of cyclosporin G: P. C. Hiestand *et al., Transplant. Proc.* **17**, 1362 (1985); *eidem, Immunology* **55**, 249 (1985). Comparative inhibition of cytokine production by A and G: R. M. McKenna *et al., Transplantation* **47**, 343 (1989). HPLC determn of A and G in human serum: P. E. Wallemacq, M. Lesne, *J. Chromatog.* **413**, 131 (1987). Clinical studies of A in graft-versus-host disease: R. Y. Calne *et al., Lancet* **2**, 1323 (1978); R. L. Powles *et al., ibid.* 1327; R. L. Powles *et al., ibid.* **1**, 327 (1980); P. J. Tutschka *et al., Blood* **61**, 318 (1983); in myasthenia gravis: R. S. A. Tindall *et al., N. Engl. J. Med.* **316**, 719 (1987). Toxicological evaluation of cyclosporin A: B. Ryffel *et al., Arch. Toxicol.* **53**, 107 (1983). Comparative toxicological study of A and G: J. I. Duncan *et al., Transplantation* **42**, 395 (1986). Symposium on therapeutic use in dermatology: *J. Am. Acad. Dermatol.* **23**, part 2, 1241-1334 (1990). Review of cyclosporin A: *Progr. Allergy* vol. **38**, K. Ishizaka *et al.,* Eds., entitled "Ciclosporin" by J. F. Borel, Ed. (Karger, Basel, 1986) 474 pp. Comprehensive description: M. M. Hassan, M. A. Al-Yahya, *Anal. Profiles Drug Subs.* **16**, 145-206 (1987). Review of mechanism of action studies: A. Aszalos, *J. Med.* **19**, 297-316 (1988); of use in rheumatoid arthritis: W. B. Harrison, B. von Graffenried, *Agents Actions* **24**, Suppl. 236-253 (1988). Review of genotoxicity of cyclosporin A: A. F. Olshan *et al., Mutat. Res.* **317**, 163-173 (1994).

Cyclosporin A

Cyclosporin A, C$_{62}$H$_{111}$N$_{11}$O$_{12}$, *cyclosporine, ciclosporin, 27-400, Sandimmun(e), Neoral.* White prismatic needles from acetone at −15°. mp 148-151°. [α]$_D^{20}$ −244° (c = 0.6 in chloroform); [α]$_D^{20}$ −189° (c = 0.5 in methanol). Sol in methanol, ethanol, acetone, ether, chloroform; slightly sol in water and saturated hydrocarbons. LD$_{50}$ in mice, rats, rabbits (mg/kg): 107, 25, > 10 i.v.; 2329, 1480, > 1000 orally (Ryffel).
Cyclosporin B, C$_{61}$H$_{109}$N$_{11}$O$_{12}$, *Ala²-cyclosporine.* Amorphous white powder, mp 149-152°. [α]$_D^{20}$ −238° (c = 0.62 in chloroform). [α]$_D^{20}$ −168° (c = 0.56 in methanol).
Cyclosporin C, C$_{62}$H$_{111}$N$_{11}$O$_{13}$, *Thr²-cyclosporine.* Colorless prismatic needles from acetone at −15°. mp 152-155°. [α]$_D^{20}$ −255° (c = 0.5 in chloroform). [α]$_D^{20}$ −182° (c = 0.5 in methanol). Solubility similar to cyclosporin A.
Cyclosporin D, C$_{63}$H$_{113}$N$_{11}$O$_{12}$, *Val²-cyclosporine.* Colorless prisms from acetone at −15°. mp 148-151°. [α]$_D^{20}$ −245° (c = 0.52 in CHCl$_3$). [α]$_D^{20}$ −211° (c = 0.51 in methanol).
Cyclosporin G, C$_{63}$H$_{113}$N$_{11}$O$_{12}$, *7-L-norvaline cyclosporin A, Nva²-cyclosporine.* Colorless, polyhedric crystals from ether + petroleum ether, mp 196-197°. [α]$_D^{20}$ −245° (c = 1.0 in chloroform). [α]$_D^{20}$ −191° (c = 1.04 in methanol).
THERAP CAT: Cyclosporin A as immunosuppressant.

2822. Cyclothiazide. *3-Bicyclo[2.2.1]hept-5-en-2-yl-6-chloro-3,4-dihydro-2H-1,2,4-benzothiadiazine-7-sulfonamide 1,1-dioxide; 6-chloro-3,4-dihydro-3-(2-norbornen-5-yl)-2H-1,2,4-benzothiadiazine-7-sulfonamide 1,1-dioxide;* 6-chloro-3,4-dihydro-3-(2-norbornen-5-yl)-7-sulfamoyl-1,2,4-benzothiadiazine; Beveno; Divanil; Divanon (obsolete); 6-chloro-3-(2-norbornen-5-yl)-7-sulfamyl-3,4-dihydro-1,2,4-benzothiadiazine 1,1-dioxide; Lilly 35483; Aquirel; Anhydron; Doburil; Fluidil. Lilly 35483; Aquirel; Anhydron; Doburil; Fluidil. C$_{14}$H$_{16}$ClN$_3$O$_4$S$_2$; mol wt 389.88. C 43.13%, H 4.14%, Cl 9.09%, N 10.78%, O 16.41%, S 16.45%. Prepn: C. W. Whitehead *et al., J. Org. Chem.* **26**, 2814 (1961); E. Müller, K. Hasspacher, U.S. pat. **3,275,625** (1966 to Boehringer, Ing.). Compre-

hensive description: C. D. Wentling in *Analytical Profiles of Drug Substances* vol. **1**, K. Florey, Ed. (Academic Press, New York, 1972) pp 65-77.

Crystals from dil alc, mp 234°.
THERAP CAT: Diuretic; antihypertensive.
THERAP CAT (VET): Diuretic.

2823. Cyclovalone. *2,6-Divanillylidenecyclohexanone;* 2,6-bis(4-hydroxy-3-methoxybenzylidene)cyclohexanone; divanillalcyclohexanone; Beveno; Divanil; Divanon (obsolete); DVC. C$_{22}$H$_{22}$O$_5$; mol wt 366.41. C 72.12%, H 6.05%, O 21.83%. Prepn: Vorländer, Koch, *Ber.* **62**, 534 (1929); Rumpel, **Austrian** pat. **180,258** (1954 to Chem-Pharm. Fabrik Waldheim).

Crystals from acetic acid, mp 178-179°. Soluble in alcohol, water.
THERAP CAT: Choleretic, cholagogue.

2824. Cycothiamin(e). *N-[(4-Amino-2-methyl-5-pyrimidinyl)methyl]-N-[1-(2-oxo-1,3-oxathian-4-ylidene)ethyl]formamide;* carbothiamine; cyclocarbothiamine; cycotiamine; CCT; Cometamine; Commetamin. C$_{13}$H$_{16}$N$_4$O$_3$S; mol wt 308.36. C 50.64%, H 5.23%, N 18.17%, O 15.57%, S 10.40%. Prepn: **Neth.** pat. **Appl. 6,511,842** corresp. to Murakami *et al.,* **U.S.** pat. **3,324,124** (1966, 1967 to Yamanouchi Seiyaku).

cis-form

Crystals from water, dec 175.5°. LD$_{50}$ orally in mice: 13390 mg/kg.
Hydrochloride, C$_{13}$H$_{16}$N$_4$O$_3$S.HCl, crystals from methanol+ ethyl acetate, or from ethylene dichloride, dec 179-180°.
THERAP CAT: Oral thiamine therapy.

2825. Cycrimine Hydrochloride. *α-Cyclopentyl-α-phenyl-1-piperidinepropanol hydrochloride;* 1-phenyl-1-cyclopentyl-3-piperidino-1-propanol hydrochloride; compd 08958; Pagitane hydrochloride. C$_{19}$H$_{30}$ClNO; mol wt 323.91. C 70.46%, H 9.34%, Cl 10.95%, N 4.32%, O 4.94%. Prepd by the addition of a Grignard reagent to the corresponding β-aminoketone: Denton *et al., J. Am. Chem. Soc.* **72**, 3795 (1950). *See also* Trihexyphenidyl Hydrochloride.

Crystals, dec 241-244°. Very bitter taste. Soly (g/100 ml) at 25° in water: 0.6; in ethanol: 2.0; in chloroform: 3.0. pH of a 0.5% aq soln: 4.9-5.4.

THERAP CAT: Anticholinergic.

2826. Cyfluthrin. *3-(2,2-Dichloroethenyl)-2,2-dimethyl-cyclopropanecarboxylic acid cyano(4-fluoro-3-phenoxyphenyl)methyl ester; (R,S)-α-cyano-4-fluoro-3-phenoxybenzyl-(1R,S)-cis,trans-3-(2,2-dichlorovinyl)-2,2-dimethylcyclopropanecarboxylate;* cyfoxylate; FCR-1272; BAY FCR 1272; Baythroid. $C_{22}H_{18}Cl_2FNO_3$; mol wt 434.29. C 60.84%, H 4.18%, Cl 16.33%, F 4.37%, N 3.23%, O 11.05%. Synthetic pyrethroid insecticide. Commercial product is mixture of 8 isomers, the *(1R)*-isomers primarily responsible for the bioactivity. Prepn of racemic mixture: R. Fuchs *et al.*, Ger. pat. **2,709,264;** *eidem,* U.S. pat. **4,218,469** (1978, 1980 both to Bayer AG). Prepn of stereoisomers: *eidem,* Eur. pat. Appl. **22,970;** *eidem,* U.S. pat. **4,287,208** (both 1981 to Bayer AG). Chiral phase HPLC separation of enantiomers: R. A. Chapman, *J. Chromatog.* **258,** 175 (1983). Metabolism by cell suspension cultures: U. Preiss *et al.*, *Chemosphere* **13,** 861 (1984). Field evaluation against cotton insect pests: J. A. Durant, *J. Agric. Entomol.* **1,** 201 (1984). Review of chemistry, bioactivity and field studies: I. Hammann, R. Fuchs, *Pflanzenshutz-Nachr.* **34,** 121-151 (1981). Review of formulations and potential uses: W. Behrenz *et al.*, *ibid.* **36,** 127-176 (1983).

(1R,3R,αR)-form

Yellowish-brown oil, n_D^{23} 1.5511. Soly in water (20°): 1-2 × 10^{-6} g/l. LD_{50} in male, female rats, male, female mice (mg/kg): 500-800, 1200, 300, 600 orally in Lutrol (Behrenz). LC_{50} (96 hr) in rainbow trout: 0.0006 mg/l (Hammann).

(1R,3R,αR)-Form, colorless oil. $[\alpha]_D^{20}$ −15.0° (c = 1.0 in $CHCl_3$).

(1R,3R,αS)-Form, crystals, mp 50-52°. $[\alpha]_D^{20}$ +24.5° (c = 1.0 in $CHCl_3$).

(1R,3S,αS)-Form, crystals from *m*-hexane mp 68-69°. $[\alpha]_D^{20}$ −2.1° (c = 1.0 in $CHCl_3$).

USE: Agricultural insecticide.

2827. Cyhalothrin. *3-(2-Chloro-3,3,3-trifluoro-1-propenyl)-2,2-dimethylcyclopropanecarboxylic acid cyano(3-phenoxyphenyl)methyl ester;* Grenade; $C_{23}H_{19}ClF_3NO_3$; mol wt 449.86. C 61.41%, H 4.26%, Cl 7.88%, F 12.67%, N 3.11%, O 10.67%. Synthetic pyrethroid insecticide. Prepn: R. K. Huff, Ger. pat. **2,802,962;** *idem,* U.S. pat. **4,183,948** (1977, 1980 both to ICI). Activity: P. D. Bently *et al.*, *Pestic. Sci.* **11,** 156 (1980); V. K. Stubbs *et al.*, *Austr. Vet. J.* **59,** 152 (1982).

(Z)-(1S)-cis-form

Commercial product exists as a mixture consisting of 95% *cis*-isomers.

USE: Insecticide; acaricide.

2828. Cyheptamide. *10,11-Dihydro-5H-dibenzo[a,d]-cycloheptene-5-carboxamide;* dibenzo[a,d][1,4]cycloheptadiene-5-carboxamide; AY-8682. $C_{16}H_{15}NO$; mol wt 237.30. C 80.98%, H 6.37%, N 5.90%, O 6.74%. Prepn: M. A. Davis *et al.*, Fr. pat. **1,355,829** (1964 to Ayerst, McKenna & Harrison); M. A. Davis, S. O. Winthrop, U.S. pat. **3,242,-212** (1966 to Am. Home Prods.). Prepn and anticonvulsant activity: M. A. Davis *et al.*, *J. Med. Chem.* **7,** 88 (1964).

Physical characteristics: H. J. Doorenbos *et al.*, *Pharm. Weekblad* **104,** 732 (1969). Pharmacology: A. B. H. Funcke *et al.*, *Arch. Int. Pharmacodyn.* **187,** 174 (1970). Toxicological studies: C. J. van Eeken *et al.*, *ibid.* **188,** 79 (1970). Metabolism in animals and man: M. Kraml *et al.*, *Biochem. Pharmacol.* **20,** 2327 (1971). Comparison of potency with other anticonvulsants: G. L. Jones *et al.*, *J. Pharm. Sci.* **70,** 618 (1981). Elucidation of structure by x-ray diffraction, mode of action: P. W. Codding *et al.*, *J. Med. Chem.* **27,** 649 (1984).

Long needles from acetonitrile, mp 193-194°. Sol in chloroform; sparingly sol in methanol, acetone; slightly sol in ethanol, ether. Practically insol in water. LD_{50} in mice (g/kg): 4.2-5.2 orally; 2.4-2.6 i.p. (Funcke).

2829. Cyhexatin. *Tricyclohexylhydroxystannane;* TCTH; tricyclohexylstannol; tricyclohexyltin hydroxide; ENT-27395; Dowco 213; Plictran. $C_{18}H_{34}OSn$; mol wt 385.18. C 56.13%, H 8.90%, O 4.15%, Sn 30.82%. Prepn: E. E. Kenaga, U.S. pat. **3,264,177** (1966 to Dow). Activity studies: W. E. Allison *et al.*, *J. Econ. Entomol.* **61,** 1254 (1968); L. R. Jeppson *et al.*, *ibid.*, 1502. Biochemical mode of action: S. Ahmad, C. O. Knowles, *Comp. Gen. Pharmacol.* **3,** 125 (1972). Biodegradation: E. H. Blair, *Environ. Qual. Safety Suppl.* **3,** 406 (1975). Toxicity data: T. B. Gaines, R. E. Linder, *Fundam. Appl. Toxicol.* **7,** 299 (1986). Acute toxicity studies in livestock: J. H. Johnson *et al.*, *Toxicol. Appl. Pharmacol.* **31,** 66 (1975).

Whitish crystalline powder, mp 195-198°. Insol in water. Slightly sol in most organic solvents. Degraded when exposed in thin layers to uv light. Non-toxic to bees, toxic to fish. LD_{50} in adult male, female rats (mg/kg): 779, 826 orally (Gaines, Linder).

Caution: Irritating to eyes.

USE: Acaricide.

2830. Cymarin. *3β-[(2,6-Dideoxy-3-O-methyl-β-D-ribohexopyranosyl)oxy]-5β,14-dihydroxy-19-oxocard-20(22)-enolide;* K-strophanthin-α; Alvonal MR. $C_{30}H_{44}O_9$; mol wt 548.67. C 65.67%, H 8.08%, O 26.24%. A glycoside of cymarose, the aglucon being strophanthidin. Isoln from *Strophanthus kombé* Oliv., *Apocynaceae:* Jacobs, Hoffmann, *J. Biol. Chem.* **67,** 609 (1926); Stoll *et al.*, *Helv. Chim. Acta* **20,** 1484 (1937); from *Adonis vernalis* L., and *A. chrysocyathus* Hook., *Ranunculaceae:* Reichstein, Rosenmund, *Pharm. Acta Helv.* **15,** 150 (1940); Pitra, Cekan, *Coll. Czech. Chem. Commun.* **26,** 1551 (1961); Abubakirov, Yamatova, *Zh. Obshch. Khim.* **31,** 2424 (1961); from *Pentopetia androsaemifolia* Decne., *Asclepiadaceae:* Wyss *et al.*, *Helv. Chim. Acta* **43,** 664 (1960); from *Castilloa elastica* Cerv., *Moraceae:* Adams, Wilkinson, *J. Pharm. Pharmacol.* **13,** 279 (1961). Structure: Kochetkov *et al.*, *Doklady Akad. Nauk SSSR* **136,** 613 (1961). Toxicity data: V. G. Vogel, E. Kluge, *Arzneimittel-Forsch.* **11,** 848 (1961).

Needles from methanol, mp 148°. $[\alpha]_D^{20}$ +39.2° (methanol); $[\alpha]_D^{22}$ +39.0° (c = 1.7 in chloroform). Sol in methanol, chloroform. Practically insol in water. LD_{50} i.v. in rats: 24.8±1.8 mg/kg (Vogel, Kluge).

Sesquihydrate, hexagonal plates from dil ethanol, mp 184-185°.

Monoacetylcymarin, $C_{32}H_{46}O_{10}$, needles from dil methanol, mp 175-176°. $[\alpha]_D^{22}$ +45.1° (ethanol). Sol in chloroform. Practically insol in water.

THERAP CAT: Cardiotonic.

2831. Cymarose. *2,6-Dideoxy-3-O-methyl-ribo-hexose;* 3-methyldigitoxose. $C_7H_{14}O_4$; mol wt 162.19. C 51.84%, H 8.70%, O 39.46%. By hydrolysis of a number of cardiac glycosides from the *Apocynaceae* family. From *Apocynum cannabinum* L., *A. androsaemifolium* L., *A. venetum* L.: Windaus, Hermanns, *Ber.* **48**, 979 (1915); Trabert, *Arzneimittel-Forsch.* **10**, 197 (1960); from *Strophanthus kombé* Oliv., *Periploca graeca* L.: Jacobs, Hoffmann, *J. Biol. Chem.* **67**, 609 (1926); **79**, 519 (1928); from *S. emini* Aschers & Pax: Lamb, Smith, *J. Chem. Soc.* **1936** 442; also from *Castilla elastica* Cerv., *Moraceae:* Adams, Wilkinson, *J. Pharm. Pharmacol.* **13**, 279 (1961). Structure: S. F. Dyke, *The Carbohydrates* (Interscience, New York, 1960) p 104. Synthesis: Bollinger, Ulrich, *Helv. Chim. Acta* **35**, 93 (1952). Review: R. C. Elderfield, "The Carbohydrate Components of Cardiac Glycosides" in W. W. Pigman, M. L. Wolfrom, *Advances in Carbohydrate Chemistry* vol. I (Academic Press, New York, 1945) pp 147-173.

Prisms from ether + petr ether, mp 93-94°. $[\alpha]_D^{20}$ +54.7° (c = 3.2 in water after 24 hrs). Sol in water, alcohol, acetone; sparingly sol in ether, petr ether, chloroform, benzene. Reduces Fehling soln. Does not form an osazone.

2832. Cymene. *Methyl(1-methylethyl)benzene;* isopropyltoluene; methylisopropylbenzene. $C_{10}H_{14}$; mol wt 134.22. C 89.49%, H 10.51%. Usually prepd by alkylation of toluene; *m-, o-* and *p*-isomers obtained: Allen, Yats, *J. Am. Chem. Soc.* **83**, 2799 (1961). Purification and properties of the three isomers: Streiff *et al., Anal. Chem.* **27**, 411 (1955). Separation of the three isomers by gas chromatography: Rihani, Froment, *J. Chromatog.* **18**, 150 (1965). Toxicity study: P. M. Jenner *et al., Food Cosmet. Toxicol.* **2**, 327 (1964).

m-Cymene, 1-methyl-3-(1-methylethyl)benzene. Liquid, bp 175.14°. mp −63.75°. d_4^{20} 0.8610, d_4^{25} 0.8570. n_D^{20} 1.4930, n_D^{25} 1.4906. Practically insol in water. Miscible with alcohol, ether.

o-Cymene, 1-methyl-2-(1-methylethyl)benzene. Liquid, bp 178.15°. mp −71.54°; also reported as −75.24° and −81.53° for two unstable solid forms (Streiff). d_4^{20} 0.8766, d_4^{25} 0.8726. n_D^{20} 1.5006, n_D^{25} 1.4982. Practically insol in water. Miscible with organic solvents.

p-Cymene, 1-methyl-4-(1-methylethyl)benzene, Dolcymene. Occurs in a number of essential oils. Liquid, bp 177.10°. mp −67.94°. d_4^{20} 0.8573, d_4^{25} 0.8533. n_D^{20} 1.4909, n_D^{25} 1.4885. Flash pt, closed cup: 117°F (47°C). Practically insol in water. Misc with alcohol, ether. LD_{50} orally in rats: 4750 mg/kg (Jenner).

2833. Cymiazole. *2,4-Dimethyl-N-(3-methyl-2(3H)-thiazolylidene)benzenamine;* 2-(2′,4′-dimethylphenylimino)-3-methylthiazoline; N-(3-methyl-4-thiazolin-2-ylidene)-2,4-xylidene; xymiazole; CGA-50439; Tifatol. $C_{12}H_{14}N_2S$; mol wt 218.32. C 66.02%, H 6.46%, N 12.83%, S 14.69%. Prepn: D. Duerr, W. D. Traber, Ger. pat. 2,619,724 (1976 to Ciba-Geigy), C.A. **86**, 72627h (1977). Activity: R. M. Immler *et al., Proc. Brit. Crop Prot. Conf.-Pests Dis.* **1977**, 383.

USE: Acaricide.
THERAP CAT (VET): Ectoparasiticide.

2834. Cynanchogenin. *(3β,12β,14β,17α)-12-[(3,4-Dimethyl-1-oxo-2-pentenyl)oxy]-3,8,14-trihydroxypregn-5-en-20-one.* $C_{28}H_{42}O_6$; mol wt 474.64. C 70.86%, H 8.92%, O 20.23%. From root of *Cynanchum caudatum* Max., *Asclepiadaceae:* Mitsuhashi, Shimizu, *Chem. Pharm. Bull.* **8**, 313, 318 (1960). Proposed structure: *eidem, Steroids* **2**, 373 (1963). Revised structure: *eidem, Tetrahedron* **24**, 4143 (1968). ^{13}C-NMR studies: T. Yamagishi *et al., Tetrahedron Letters* **1973**, 3527, 3531.

Fine needles from ether + petr ether, mp 165-167°. $[\alpha]_D^{22}$ −39.5° (c = 1.24 in 90% ethanol). uv max (ethanol): 218 nm (log ε 4).

Monoacetate, $C_{30}H_{44}O_7$, crystals from petr ether + benzene, mp 150°.

2835. Cynarin(e). *[1R-(1α,3α,4α,5β)]-1,3-Bis[[3-(3,4-dihydroxyphenyl)-1-oxo-2-propenyl]oxy]-4,5-dihydroxycyclohexanecarboxylic acid;* 1,3-dicaffeoylquinic acid; 1,5-dicaffeylquinic acid; 3,4-dihydroxycinnamic acid 1-carboxy-4,5-dihydroxy-1,3-cyclohexylene ester; caffeic acid 1-carboxy-4,5-dihydroxy-1,3-cyclohexylene ester; 1-carboxy-4,5-dihydroxy-1,3-cyclohexylenebis(3,4-dihydroxycinnamate); quinic acid 1,5-dicaffeic ester; Cinarine; Listrocol; Plemocil. $C_{25}H_{24}O_{12}$; mol wt 516.46. C 58.14%, H 4.68%, O 37.17%. Active principle of artichoke, *Cynara scolymus* L., *Compositae:* L. Panizzi, M. L. Scarpati, *Gazz. Chim. Ital.* **84**, 792 (1954). Synthesis of originally proposed structure: L. Panizzi *et al., ibid.* 806; L. Panizzi *et al., U.S. pats.* **2,863,-909** (1958, 1963 both to Farmitalia); Alberti *et al., U.S. pat.* **2,918,477** (1959). Revised structure: L. Panizzi, M. L. Scarpati, *Gazz. Chim. Ital.* **95**, 71 (1965). NMR confirmation of structure: I. Horman *et al., J. Agr. Food Chem.* **32**, 538 (1984). HPLC determn in pharmaceutics: A. Bettero, *Boll. Chim. Farm.* **120**, 49 (1981). Inhibition of fatty acid mobilization: P. Dorigo, G. Fassina, *Pharmacol. Res. Commun.* **2**, 109 (1970). Reduction of cholesterol levels in rats: J. Wojcicki, *Drug Alcohol Depend.* **3**, 143 (1978). Clinical trial in hyperlipemic syndrome: M. Montini *et al., Arzneimittel-Forsch.* **25**, 1311 (1975).

Crystals from dil acetic acid. Sweet taste. mp 225-227°. $[\alpha]_D^{25}$ —59° (c = 2 in methanol). uv max (methanol): 326 nm ($E_{1cm}^{1\%}$ 616). Sparingly sol in cold, more in boiling water; sol in glacial acetic acid, alcohols.

THERAP CAT: Choleretic.

2836. Cypermethrin. *3-(2,2-Dichloroethenyl)-2,2-dimethylcyclopropanecarboxylic acid cyano(3-phenoxyphenyl)-methyl ester;* (±)-α-cyano-3-phenoxybenzyl-(±)-*cis,trans*-3-(2,2-dichlorovinyl)-2,2-dimethylcyclopropane carboxylate; NRDC-149; FMC-30980; PP-383; Agrothrin; Ammo; Arrivo; Barricade; Cymbush; Cynoff; Cypercare; Cyperkill; Cypersect; Demon; Dysect; Ectomin; Ectopor; Fastac; Flectron; Nurelle; Polytrin; Ripcord; Rycopel; Sherpa; Topclip Parasol. $C_{22}H_{19}Cl_2NO_3$; mol wt 416.30. C 63.47%, H 4.60%, Cl 17.03%, N 3.36%, O 11.53%. Potent synthetic pyrethroid insecticide, usually exists as a mixture of cis and trans isomers. Prepn of racemic mixture: M. Elliot *et al.,* **Ger. pat. 2,326,077;** *eidem,* **U.S. pat. 4,024,163** (1974, 1977 to NRDC). Activity: *eidem, Pestic. Sci.* **6,** 537 (1975); M. H. Breese, *ibid.* **8,** 264 (1977). Soil degradation: T. R. Roberts, M. E. Standen, *ibid.* 305; D. D. Kaufman *et al., J. Agr. Food Chem.* **29,** 239 (1981). Metabolism: T. Shono *et al., ibid.* **27,** 316 (1979); M. J. Crawford *et al., ibid.* **29,** 130 (1981). Analysis: A. Sapiets *et al., Anal. Methods Pestic. Plant Growth Regul.* **13,** 33-51 (1984). Repellent activity against flies on cattle: J. A. Shemanchuk, *Pestic. Sci.* **12,** 412 (1981); J. E. Hillerton *et al., Brit. Vet. J.* **141,** 160 (1985). Acute toxicity study: F. Cantalamessa, *Arch. Toxicol.* **67,** 510 (1993).

cis-form

Commercial product is a mixture of eight isomers. Viscous semi-solid, mp 60-80°. Insol in water. Sol in methanol, acetone, xylene, methylene dichloride. LD_{50} in 8 day old rats, adult male rats (mg/kg): 14.9, 250.0 orally (Cantalamessa).

USE: Insecticide.

THERAP CAT (VET): Ectoparasiticide.

2837. Cyphenothrin. *2,2-Dimethyl-3-(2-methyl-1-propenyl)cyclopropanecarboxylic acid cyano(3-phenoxyphenyl)-methyl ester;* (RS)-α-cyano-3-phenoxybenzyl (1R)-*cis,trans*-chrysanthemate; α-cyano-*m*-phenoxybenzyl 2,2-dimethyl-3-(2-methylpropenyl)cyclopropanecarboxylate; 3-phenoxy-α-cyanobenzylchrysanthemate; S-2703; S-2703 Forte; Gokilaht. $C_{24}H_{25}NO_3$; mol wt 375.47. C 76.77%, H 6.71%, N 3.73%, O 12.78%. Synthetic pyrethroid. Prepn: T. Matsuo *et al.,* **Ger. pat. 2,231,312;** *eidem,* **U.S. pat. 3,835,176** (1973, 1974 to Sumitomo). Insecticidal activity: M. Elliot *et al., Pestic. Sci.* **13,** 407 (1982); T. Matsunaga *et al., Pesticide Chemistry: Human Welfare and the Environment* **2,** J. Miyamoto *et al.,* Eds (Pergamon Press, Oxford, 1983) pp 231-238; T. Itoh *et al., J. Am. Mosquito Cont.* **2,** 503 (1986). Mechanism of action: T. Narahashi, *Neurobehav. Toxicol. Teratol.* **4,** 753 (1982). Series of articles on stereoselective syntheses of labelled *cis* and *trans* forms: H. Kanamaru *et al., Radioisotopes* **35,** 103, 109, 242 (1986).

cis-form

Yellowish viscous liquid. d_{25}^{25} 1.083. Vapor pressure at 30°: 3.11 × 10^{-6} mm Hg. Viscosity at 30°: 808.8 cP.

Note: Commercial product is a mixture (35:65) of *cis* and *trans* forms.

USE: Insecticide.

2838. Cyprenorphine. *(5α,7α)-17-(Cyclopropylmethyl)-4,5-epoxy-3-hydroxy-6-methoxy-α,α-dimethyl-6,14-ethenomorphinan-7-methanol;* N-(cyclopropylmethyl)-7,8-dihydro-7α-(1-hydroxy-1-methylethyl)-O^6-methyl-6,14-endo-ethenonormorphine; N-(cyclopropylmethyl)-6,14-*endo*-ethenonor-7α-(2-hydroxy-2-propyl)tetrahydronororipavine; N-(cyclopropylmethyl)-6,14-*endo*-ethenotetrahydronororipavine; $C_{26}H_{33}NO_4$; mol wt 423.55. C 73.73%, H 7.85%, N 3.31%, O 15.11%. Narcotic antagonist closely related to diprenorphine, *q.v.* Prepn (7α- or 7β-linkage unspecified): **Belg. pat. 629,070** (1963 to J. F. Macfarlan), *C.A.* **61,** 13364g (1964); K. W. Bentley, **U.S. pat. 3,474,101** (1969 to Reckitt & Sons); of 7α: K. W. Bentley, D. G. Hardy, *J. Am. Chem. Soc.* **89,** 3281 (1967). Activity in rats: K. W. Bentley *et al., Nature* **206,** 102 (1965). Use in large animals: M. R. Jainudeen, *Vet. Rec.* **89,** 686 (1971); K. R. Presnell *et al., J. Wildl. Dis.* **9,** 336 (1973). Binding to neuroleptic receptors: A. Czlonkowski *et al., Life. Sci.* **22,** 953 (1978). Determn by HPLC: I. Jane, A. McKinnon, *J. Chromatog.* **323,** 191 (1985).

Prisms from methanol, mp 234° (7α- or 7β-linkage unspecified).

Hydrochloride, $C_{26}H_{33}NO_4 \cdot HCl$, *M-285.* Crystals from methanol, mp 248°.

Note: This is a controlled substance (opium derivative) listed in the U.S. Code of Federal Regulations, Title 21 Part 1308.11 (1995).

THERAP CAT (VET): Etorphine antagonist.

2839. Cypripedium. Lady's slipper; American valerian; nerve root; yellow moccasin flower; Noah's ark. Dried rhizome and roots of *Calypso bulbosa* (L.) Oakes (*Cypripedium bulbosum* L.), *Cypripedium pubescens* Willd., or of *Cypripedium parviflorum* Salisb., *Orchidaceae. Habit.* Nova Scotia south to Alabama, west to Nebraska and Missouri. *Constit.* Volatile oil, volatile acid, tannin, resins.

THERAP CAT: Sedative.

2840. Cyproconazole. *α-(4-Chlorophenyl)-α-(1-cyclopropylethyl)-1H-1,2,4-triazole-1-ethanol;* (2RS,3RS)-2-(4-chlorophenyl)-3-cyclopropyl-1-(1H-1,2,4-triazol-1-yl)butan-2-ol; SAN-619F; SN-108266; Alto; Atemi 50 SL. $C_{15}H_{18}ClN_3O$; mol wt 291.78. C 61.75%, H 6.22%, Cl 12.15%, N 14.40%, O 5.48%. Broad spectrum triazole antifungal; sterol demethylase inhibitor. Prepn: F. Schaub, **Ger. pat. 3,406,993;** *idem,* **U.S. pat. 4,664,696** (1984, 1987 both to Sandoz). Mode of action: M. S. Fuller *et al., Pest. Biochem. Physiol.* **36,** 115 (1990). Synergism with herbicides: H. R. Kataria, U. Gisi, *Crop Prot.* **9,** 403 (1990). Field trials in winter wheat: J. P. Devey *et al., Brighton Crop Prot. Conf. - Pests and Dis.* **1990,** 801; in cereals in combination with prochloraz, *q.v.:* R. I. Harris *et al., ibid.* 813. *Review:* U. Gisi *et al., Proc. Brit. Crop Prot. Conf. - Pests and Dis.* **1986,** 33-40.

Colorless crystals from hexane/CH_2Cl_2, mp 103-105°. Soly at 25° (%w/w): water 0.0140 ±0.0004; acetone >23; ethanol >23; xylene 12; DMSO >18. Vapor pressure at 20°: 2.6 × 10^{-7} torr. Partition coefficient: 819 ±195. LD_{50} in male and female rats (mg/kg): 1020, 1330 orally; in rats: 2000 dermally. LD_{50} in carp: 18.9 mg/l; in trout 7.2 mg/l; in bluegill sunfish 6.0 mg/l in water. LD_{50} in bobwhite quail (mg/kg): 150 orally; LD_{50} (8 day dietary) in bobwhite quail, mallard duck: 816, 1197 mg/kg (Gisi, 1986).

Combination with prochloraz, *Sportak-Delta*.

USE: Agricultural fungicide.

2841. Cyprodinil. *4-Cyclopropyl-6-methyl-N-phenyl-2-pyrimidinamine;* N-(4-cyclopropyl-6-methylpyrimidin-2-yl)-aniline; 2-phenylamino-4-methyl-6-cyclopropylpyrimidine; CGA-219417; Unix. $C_{14}H_{15}N_3$; mol wt 225.29. C 74.64%, H 6.71%, N 18.65%. Methionine biosynthesis inhibitor. Prepn: A. Hubele, **Eur.** pat. **Appl. 310,550**; *idem*, **U.S.** pat. **5,153,200** (1989, 1992 both to Ciba-Geigy). Comprehensive description: U. J. Heye *et al.*, *Crop Prot.* **13**, 541-549 (1994). Mechanism of action: P. Masner *et al.*, *Pest. Sci.* **42**, 163 (1994). Review of field trials: N. J. Leadbitter *et al.*, *Brit. Crop Prot. Council Monogr.* **57**, 73-78 (1994).

White crystalline solid, mp 75.9°. vapor pressure (25°): 5.1 × 10^{-4} Pa (crystal modif. A); 4.7 × 10^{-4} Pa (crystal modif. B). Log P (*n*-octanol/water): 3.9 (pH 5.0); 4.0 (pH 7.0); 4.0 (pH 9.0). Soly in water at 25° (mg/l): 20 (pH 5.0); 13 (pH7.0); 15 (pH 9.0). LD_{50} orally or dermally in rats: >2000 mg/kg (Heye).

USE: Agricultural fungicide.

2842. Cyproheptadine. *4-(5H-Dibenzo[a,d]cyclohepten-5-ylidene)-1-methylpiperidine;* 1-methyl-4-(5H-dibenzo[a,d]-cycloheptenylidene)piperidine; 5-(1-methylpiperidylidene-4)-5H-dibenzo[a,d]cycloheptene; 1-methyl-4-(5-dibenzo[a,e]-cycloheptatrienylidene)piperidine; Periactinol. $C_{21}H_{21}N$; mol wt 287.40. C 87.76%, H 7.36%, N 4.87%. Prepn: Engelhardt, **U.S.** pat. **3,014,911** (1961 to Merck & Co.); Engelhardt *et al.*, *J. Med. Chem.* **8**, 829 (1965). Metabolism: C. C. Porter *et al.*, *Drug Metab. Dispos.* **3**, 189 (1975). Toxicity study: J. J. Loux *et al.*, *Arzneimittel-Forsch.* **28**, 1644 (1978). Comprehensive description: H. Y. Aboul-Enein, A. A. Al-Badr, *Anal. Profiles Drug Subs.* **9**, 155-179 (1980).

Crystals from dil ethanol, mp 112.3-113.3°.

Hydrochloride sesquihydrate, $C_{21}H_{22}ClN \cdot 1\frac{1}{2}H_2O$, *Anarexol, Antegan, Ifrasarl, Nuran, Periactin, Vimicon, Cipractin, Peritol*. Crystals from abs ethanol + ether, dec 252.6-253.6°. uv max (0.1N H_2SO_4): 224, 285 nm ($E_{1cm}^{1\%}$ 1656, 355). 1.0 g is sol in 1.5 ml methanol, 16 ml chloroform, 35 ml alc, 275 ml H_2O. Practically insol in ether. LD_{50} orally in mice: 74.2 mg/kg (Loux).

Hydrochloride monohydrate, crystals, mp 214-216°. Soly in water: ~5 mg/ml.

THERAP CAT: Antihistaminic; antipruritic.

2843. Cyproquinate. *6,7-Bis(cyclopropylmethoxy)-4-hydroxy-3-quinolinecarboxylic acid ethyl ester;* ethyl 6,7-bis-(cyclopropylmethoxy)-4-hydroxyquinoline-3-carboxylate; cyproquinidate; cyproxyquine; Su-18137; Coxytrol. $C_{20}H_{23}$-NO_5; mol wt 357.41. C 67.21%, H 6.49%, N 3.92%, O 22.38%. Prepn by ring closure of corresp dialkyl (alkoxyphenylamino)methylenemalonates: Mizzoni *et al.*, *Experientia* **24**, 1188 (1968); Mizzoni, DeStevens, S. Afr. pat. **67 05,655** (1968 to Ciba), *C.A.* **70**, 47316t (1969). Review of structure and biology: Mizzoni *et al.*, *J. Med. Chem.* **13**, 870 (1970).

Crystals from DMF, mp 288-288.5° (dec).

THERAP CAT (VET): Coccidiostat.

2844. Cyproterone. *(1β,2β)-6-Chloro-1,2-dihydro-17-hydroxy-3'H-cyclopropa[1,2]pregna-1,4,6-triene-3,20-dione;* 6-chloro-17-hydroxy-1α,2α-methylenepregna-4,6-diene-3,20-dione; 6-chloro-6-dehydro-17α-hydroxy-1,2α-methyleneprogesterone; 6-chloro-1,2α-methylene-4,6-pregnadien-17α-ol-3,20-dione; SH-881; SH-80881. $C_{22}H_{27}ClO_3$; mol wt 374.91. C 70.48%, H 7.26%, Cl 9.46%, O 12.80%. Prepn of free alcohol: Wiechert, Neumann, **Ger.** pat. **1,189,991**, *C.A.* **63**, 1842e (1965); Wiechert, **U.S.** pat. **3,234,093** (1965, 1966 both to Schering AG). Biodynamics in man: Gerhards *et al.*, *Arzneimittel-Forsch.* **23**, 1550 (1973). The free alcohol is an anti-androgen; the acetate is both an anti-androgen and a progestogen. Effect on hormone secretion and on spermatogenesis in man: L. Moltz *et al.*, *Contraception* **21**, 393 (1980). Review of pharmacology and clinical studies (acetate) on acne and hirsutism in women: J. Hammerstein *et al.*, *J. Steroid Biochem.* **19**, 591 (1983).

Crystals from ethyl acetate, mp 237.5-240°.

Acetate, $C_{24}H_{29}ClO_4$, *CPA, SH-714, Androcur, Cyprostat*. Crystals from diisopropyl ether, mp 200-201°. uv max (methanol): 281 nm (ε 17280).

Mixture of acetate with ethinyl estradiol, *Diane 35, Dianette*.

THERAP CAT: The acetate as antiandrogen; combination with estrogen in treatment of acne.

2845. Cyromazine. *N-Cyclopropyl-1,3,5-triazine-2,4,6-triamine;* 2-cyclopropylamino-4,6-diamino-s-triazine; CGA-72662; Citation; Larvadex; Trigard; Vetrazin. $C_6H_{10}N_6$; mol wt 166.19. C 43.36%, H 6.07%, N 50.57%. Insect growth regulator. Prepn: H. U. Brechbuhler *et al.*, **Ger.** pat. **2,736,876**; *eidem*, **U.S.** pat. **4,225,598** (1978, 1980 to Ciba-Geigy). Activity: R. D. Hall, M. C. Foehse, *J. Econ.*

Entomol. **73,** 564 (1980); R. J. Hart *et al., Aust. Vet. J.* **59,** 104 (1982).

Crystals, mp 219-222°.

USE: Insecticide.

THERAP CAT (VET): Ectoparasiticide.

2846. Cystamine. *2,2'-Dithiobisethanamine; 2,2'-dithiobis[ethylamine]; β,β'*-diaminodiethyl disulfide; bis[β-aminoethyl]disulfide; decarboxycystine. $C_4H_{12}N_2S_2$; mol wt 152.28. C 31.55%, H 7.94%, N 18.40%, S 42.11%. (H_2N-$CH_2CH_2)_2S_2$. Forms when cystine is distilled: Neuberg, Ascher, *Chem. Zentr.* **1907,** II, 1156. Prepd by H_2O_2 oxidation of $H_2NCH_2CH_2SH$: Mills, Jr., Bogert, *J. Am. Chem. Soc.* **62,** 1173 (1940); Barnett, *J. Chem. Soc.* **1944,** 5.

Viscous oil. Cannot be distilled without decompn (even in high vacuum). Freely sol in water. Sol in alc.

Dihydrochloride, $C_4H_{12}N_2S_2.2HCl$, needles from methanol, dec 203-214°. Sol in water.

2847. L-Cystathionine. *(R)-S-(2-Amino-2-carboxyethyl)-L-homocysteine; L-2-amino-4-[(2-amino-2-carboxyethyl)-thio]butyric acid.* $C_7H_{14}N_2O_4S$; mol wt 222.27. C 37.83%, H 6.35%, N 12.60%, O 28.79%, S 14.43%. An amino acid whose homocysteine and serine moieties are both L in configuration. Intermediate in transsulfuration whereby the mammal transfers the sulfur of methionine via homocysteine to cysteine: Hope, *Proc. Intern. Congr. Biochem. 4th, Vienna, 1958,* **13,** 63 (1960). Preparation: du Vigneaud *et al., J. Biol. Chem.* **143,** 59 (1942); Weiss, Stekol, *J. Am. Chem. Soc.* **73,** 2497 (1951); Schöberl, Täuber, *Ann.* **599,** 23 (1956). Isolation from human brain: Tallau *et al., J. Biol. Chem.* **230,** 707 (1958). Abs config: G. Jung *et al., Eur. J. Biochem.* **35,** 436 (1973); C.-S. Chen *et al., Biochim. Biophys. Acta* **538,** 534 (1978). *Review:* J. P. Greenstein, M. Winitz, *Chemistry of the Amino Acids* vol. 3 (John Wiley, N. Y., 1961) p 2682.

Crystals (rectangular parallelepipeds), dec 312° (darkening at 270°). $[\alpha]_D^{20}$ +23.7° (1*N* HCl).

2848. Cysteamine. *2-Aminoethanethiol;* mercaptamine; β-mercaptoethylamine; 2-aminoethyl mercaptan; thioethanolamine; decarboxycysteine; MEA; mercamine; L-1573; Becaptan; Lambratene (formerly). C_2H_7NS; mol wt 77.15. C 31.14%, H 9.15%, N 18.16%, S 41.56%. $HSCH_2CH_2NH_2$. A sulfhydryl compound with a variety of biological effects. Prepn: Gabriel, Leupold, *Ber.* **31,** 2837 (1898); Knorr, Rössler, *ibid.* **36,** 1281 (1903); Mills, Jr., Bogart, *J. Am. Chem. Soc.* **62,** 1173 (1940); Wenker, *ibid.* **57,** 2328 (1935); D. A. Shirley, *Preparation of Organic Intermediates* (Wiley, New York, 1951) p 189. Use in treatment of paracetamol (acetaminophen) poisoning: L. F. Prescott *et al., Lancet* **2,** 109 (1976); A. L. Harris, *Brit. Med. J.* **284,** 825 (1982). Effects in nephropathic cystinosis: M. Yudkoff *et al., N. Engl. J. Med.* **304,** 141 (1981). Radioprotective effects: R. P. Bird, *Radiat. Res.* **72,** 290 (1980); C. J. Koch, R. L. Howell, *ibid.* **87,** 265 (1981). Cysteamine has been shown to be a duodenal ulcerogen in rats: H. Selye, S. Szabo, *Nature*

244, 458 (1973); S. Szabo, *Am. J. Path.* **93,** 273 (1978); P. Kirkegaard *et al., Scand. J. Gastroenterol.* **15,** 621 (1980). *Review:* S. Szabo, *Lab. Invest.* **51,** 121 (1984). It has also been found to deplete somatostatin concentration: S. Szabo, S. Reichlein, *Endocrinology* **109,** 2255 (1981); S. M. Sagar *et al., J. Neuroscience* **2,** 225 (1982). In pituitary tissue, cysteamine is a potent depletor of prolactin concentrations *in vivo* and *in vitro:* W. J. Millard *et al., Science* **217,** 452 (1982). Toxicity studies: E. Beccari *et al., Arzneimittel-Forsch.* **5,** 421 (1955); D. L. Klayman *et al., J. Med. Chem.* **12,** 510 (1969); P. K. Srivastava, L. Field, *ibid.* **18,** 798 (1975).

Crystals by sublimation *in vacuo.* Disagreeable odor. mp 97-98.5°. Oxidizes to cystamine on standing in air. Freely sol in water, alkaline reaction. LD_{50} in mice (mg/kg): 625 orally; 250 i.p. (Klayman); (Srivastava, Field).

Hydrochloride, $C_2H_7NS.HCl$, crystals from alc, mp 70.2-70.7°. Sol in water, alcohol. LD_{50} (cg/kg): 23.19 i.p. in rats; 14.95 i.v. in rabbits (Beccari).

USE: Experimentally as a radioprotective agent and to produce acute and chronic duodenal ulcers in rats.

THERAP CAT: Antidote to acetaminophen.

2849. Cysteic Acid. *3-Sulfoalanine;* α-amino-β-sulfopropionic acid. $C_3H_7NO_5S$; mol wt 169.16. C 21.30%, H 4.17%, N 8.28%, O 47.29%, S 18.96%. $HOOCCH(NH_2)$-CH_2SO_3H. Has been isolated from human hair oxidized with permanganate: Lissizin, *Z. Physiol. Chem.* **173,** 309 (1928). Occurs normally in the outer part of the sheep's fleece, where the wool is exposed to light and weather: Martin, Synge, *Advances in Protein Chemistry* **2,** 3 (1945). Prepd from cystine or cysteine by oxidation with bromine in water: Friedmann, *Beiträge zur Chemischen Physiologie und Pathologie* **3,** 25, 38; Gortner, Hoffman, *J. Biol. Chem.* **72,** 435 (1927).

L-Form, octahedra or needles from dil alc (also forms a monohydrate, prismatic needles). When anhydr, dec 260°. $[\alpha]_D^{20}$ +8.66° (1.85 g in 25 ml). pKa_1 (25°): 1.89; pKa_2 8.7; pKb about 12.7. Soluble in water. Insol in alcohol.

DL-Form, crystals, dec 245°.

2850. Cysteine. *L-Cysteine;* Cys; C; β-mercaptoalanine; *(R)*-2-amino-3-mercaptopropanoic acid; 2-amino-3-mercaptopropionic acid; α-amino-β-thiolpropionic acid; half-cystine; thioserine. $C_3H_7NO_2S$; mol wt 121.16. C 29.74%, H 5.82%, N 11.56%, O 26.41%, S 26.47%. A non-essential amino acid in human development. Readily oxides to form a dimeric amino acid, cystine, *q.v.,* in which the two Cys are linked via a disulfide bridge, a common structural feature in proteins. Early chemistry and biochemistry: *Amino Acids and Proteins,* D. M. Greenberg, Ed. (Charles C. Thomas, Springfield, IL, 1951) 950 pp., *passim;* J. P. Greenstein, M. Winitz, *Chemistry of the Amino Acids* vol 1-3 (John Wiley and Sons, Inc., New York, 1961) pp. 1879-1928, *passim.* Simple synthesis of racemic cysteine: V. J. Martens *et al., Angew. Chem. Int. Ed.* **20,** 668 (1981). Determn in proteins: J. G. Hoogerheide, C. M. Campbell, *Anal. Biochem.* **201,** 146 (1992); D. Atherton *et al., ibid.* **212,** 98 (1993). Review of biosynthesis: N. M. Kredich *et al., CIBA Found. Symp.* (Netherlands) **72,** 87-99 (1980). Review of transport in mammalian cells: S. Bannai, *Biochim. Biophys. Acta* **779,** 289-306 (1984). Review of effects on acrylonitrile toxicity: D. E. Nerland *et al., Drug Metabolism. Rev.* **20,** 233-246 (1989). Review of thermodynamics and kinetics: T. R. Ralph *et al., J. Electroanal. Chem.* **375,** 1-15 (1994); of electrosynthesis: *eidem, ibid.* 17-27. Review of role in chemoand radioprotectant strategies: J. C. Roberts, *Amino Acids* **8,** 113-124 (1995).

Crystals. $[\alpha]_D^{25}$ +6.5° (5*N* HCl); $[\alpha]_D^{25}$ +13.0° (glacial acetic acid). pK_1 1.71; pK_2 8.33; pK_3 10.78. Absorption spectrum: Abderhalden, Rossner, *Z. Physiol. Chem.* **178,** 160 (1928). Freely sol in water, alcohol, acetic acid, ammonia water. Insol in ether, acetone, ethyl acetate, benzene, carbon disulfide, carbon tetrachloride. In neutral or slightly alkaline aq solns it is oxidized to cystine by air. More stable in acidic solns.

Hydrochloride, $C_3H_7NO_2S.HCl$, crystals, dec 175-178°. $[\alpha]_D^{25}$ +5.0° (5N HCl); $[\alpha]_D^{25}$ +10.0° (glacial acetic acid). Sol in water, alcohol, acetone; the aq soln is acid. *Keep tightly closed.* Decomposes and oxidizes slowly; hygroscopic.

USE: As dough conditioner.

THERAP CAT (VET): Has been used as a detoxicant.

2851. Cystine. *L-Cystine; [R-(R*,R*)]-3,3'-dithiobis[2-aminopropanoic acid]; dicysteine; β,β'-dithiodialanine; α-diamino-β-dithiolactic acid; β,β'-diamino-β,β'-dicarboxydiethyl disulfide; bis(β-amino-β-carboxyethyl) disulfide. Gelucystine.* $C_6H_{12}N_2O_4S_2$; mol wt 240.30. C 29.99%, H 5.03%, N 11.66%, O 26.63%, S 26.69%. Non-essential amino acid for human development. Formed by the dimerization of two cysteines, *q.v.* through the sulfur. These disulfide bridges occur both within and between polypeptides; often found in extracelluar proteins. First amino acid described in 1810 by Wollaston. Isoln from horn hydrolysate: K. A. H. Mörner, *Z. Physiol. Chem.* **28**, 595 (1899). Early chemistry and biochemistry: *Amino Acids and Proteins,* D. M. Greenberg, Ed. (Charles C. Thomas, Springfield, IL, 1951) 950 pp. *passim*; J. P. Greenstein, M. Winitz, *Chemistry of the Amino Acids* vols **1-3** (John Wiley and Sons, Inc., New York, 1961) pp. 1879-1928, *passim*. Distribution in protein: R. C. Fahey *et al., J. Mol. Evol.* **10**, 155 (1977). Review of lysosomal transport including pathophysiology: W. Gahl in *Pathophysiol. Lysosomal Transp.,* J. G. Thoene, Ed. (CRC Press, Boca Raton, FL, 1992) pp 45-71. Review of thermodynamics and kinetics: T. R. Ralph *et al., J. Electroanal. Chem.* **375**, 1-15 (1994); of electrosynthesis: *eidem, ibid.* 17-27.

Hexagonal tablets from water, dec 260-261° (sealed tube). $[\alpha]_D^{20}$ −223.4° (1.0N HCl). pK_1 1; pK_2 2.1; pK_3 8.02; pK_4 8.71 at 35°. Soly in water (g/l) at 25°: 0.112; at 50°: 0.239; at 75°: 0.523; at 100°: 1.142. Quite sol in aq solns below pH 2 or above pH 8. Insol in alc. Absorption spectrum: Marchlewski, Nowotonowna, *Bull. Soc. Chim. France* [4] **39**, 163, 166 (1926).

D-Form, crystals. $[\alpha]_D^{20}$ +223° (1.0N HCl). Soly in water at 25°: 0.057 g/l.

DL-Form, crystals. Soly in water at 25°: 0.057 g/l.

meso-Form, crystals. Soly in water: 0.056 g/l.

2852. Cystine S-Dioxides. $C_6H_{12}N_2O_6S_2$; mol wt 272.30. C 26.47%, H 4.44%, N 10.29%, O 35.25%, S 23.55%. So-called *cystine disulfoxide* as prepd by Lavine, Toennies, *J. Biol. Chem.* **113**, 571 (1936) and Emilozzi, Pichat, *Bull. Soc. Chim. France* **1959**, 1887 has been shown to be a mixture of two isomers, *cystine S,S-dioxide* (structure I) and *cystine S,S'-dioxide* (structure II). The isomers can be separated by adding 0.25N HCl, filtering of the undissolved S,S-dioxide, and pptg the dissolved S,S'-dioxide with 8N NH4OH. The pH should never exceed 6, since both isomers start to dec at pH 6-7. *Ref:* Utzinger, *Experientia* **17**, 374 (1961).

S,S-dioxide, *2-amino-2-carboxyethyl 2-amino-2-carboxyethanethiosulfonate; L-cystine thiosulfonate.* Main product of oxidation of cystine hydrochloride with a mixture of hydrogen peroxide and formic acid (Utzinger). Dec on heating, becoming partly viscous at ∼205°. Practically insol in water.

S,S'-dioxide, *sym-L-cystine disulfoxide, 3,3'-disulfinyldialanine.* Main product of oxidation of cystine perchlorate in acetonitrile with perbenzoic acid in chloroform (Utzinger). Dec on heating, remaining dry and darkening slowly up to 240°. Slightly more sol in water than the S,S-dioxide.

2853. Cytarabine. *4-Amino-1-β-D-arabinofuranosyl-2(1H)-pyrimidinone; 1-β-D-arabinofuranosylcytosine; β-cytosine arabinoside; aracytidine;* CHX-3311; U-19920; Alexan; Arabitin; Aracytine; Ara-C; Cytarbel; Cytosar; Erpalfa; Iretin; Udicil. $C_9H_{13}N_3O_5$; mol wt 243.22. C 44.45%, H 5.39%, N 17.28%, O 32.89%. Prepn: Walwick *et al., Proc. Chem. Soc.* **1959**, 84; Hunter, U.S. pat. **3,116,282** (1963 to Upjohn); Shen *et al., J. Org. Chem.* **30**, 835 (1965); Fromageot, Reese, *Tetrahedron Letters* **1966**, 3499; Roberts, Dekker, *J. Org. Chem.* **32**, 816 (1967); prebiosynthetic route: *Chem. & Eng. News* **47**, 30 (Oct. 6, 1969). Conversion of cytidylic acid into aracytidine-3'-phosphate: Nagyvary, Tapiero, *Tetrahedron Letters* **1969**, 3481. Efficient synthesis of cytarabine: Hessler, *J. Org. Chem.* **41**, 1828 (1976). Mechanism of action studies: Eridani, *Haematologica* **57**, 341 (1972); Gray, Renis, *J. Theor. Biol.* **39**, 623 (1973). Clinical pharmacology and toxicology: R. C. Donehower *et al., Cancer Treat. Rep.* **70**, 1059 (1986). Symposium on clinical pharmacology, pharmacokinetics and efficacy in leukemia: *Scand. J. Haematol.* **36**, Suppl. 44, 1-74 (1986).

Prisms from 50% ethanol, mp 212-213°. $[\alpha]_D^{23}$ +158°; $[\alpha]_D^{24}$ +153° (c = 0.5 in water). uv max at pH 2: 281.0, 212.5 nm (ε 13171, 10230); at pH 12: 272.5 nm (ε 9259).

THERAP CAT: Antineoplastic; antiviral.

2854. Cythioate. *Phosphorothioic acid O-[4-(aminosulfonyl)phenyl] O,O-dimethyl ester; phosphorothioic acid O,O-dimethyl ester O-ester with p-hydroxybenzenesulfonamide; O,O-dimethyl O-p-sulfamoylphenylphosphorothioate;* CL-26691; ENT-25640; Proban. $C_8H_{12}NO_5PS_2$; mol wt 297.29. C 32.32%, H 4.07%, N 4.71%, O 26.91%, P 10.42%, S 21.57%. Prepn: G. Berkelhammer, U.S. pat. **3,005,004** and R. I. Hewitt, G. Berkelhammer, U.S. pat. **3,179,560** (1961, 1965 to Am. Cyanamid). Pharmacodynamics: H. G. Smith, R. L. Goulding, *J. Econ. Entomol.* **63**, 1640 (1970). Efficacy as ectoparasiticide: C. P. Doval, I. Gupta, *Indian Vet. J.* **55**, 890 (1978); P. M. Bowen, N. J. Caldwell, *Vet. Med. Small Anim. Clinic.* **77**, 79 (1982). Toxicity data: E. E. Kenaga, W. E. Allison, *Bull. Entomol. Soc. Am.* **15**, 85 (1969).

Crystals, mp 70-71°. n_D^{25} 1.5346. LD_{50} orally in rats: 160 mg/kg (Kenaga, Allison).

USE: Insecticide.

THERAP CAT (VET): Ectoparasiticide.

2855. Cytidine. 4-Amino-1-β-D-ribofuranosyl-2-(1H)-pyrimidinone; cytosine riboside; 1-β-D-ribofuranosylcytosine. $C_9H_{13}N_3O_5$; mol wt 243.22. C 44.45%, H 5.39%, N 17.28%, O 32.89%. Constituent of nucleic acids. Isoln from yeast nucleic acid: Levene, Jacobs, *Ber.* **43**, 3154 (1910); Levene, La Forge, *ibid.* **45**, 608 (1912). Sepn from other nucleosides by ion-exchange chromatography: Cohn in Chargaff-Davidson, *The Nucleic Acids* vol. I (New York, 1955) p 211. Synthesis: Howard *et al., J. Chem. Soc.* **1947**,

1052. Crystal structure: Furberg *et al.*, *Acta Cryst.* **18**, 313 (1965). *Review: Basic Principles in Nucleic Acid Chemistry* vol. **1**, P. O. P. Ts'o, Ed. (Academic Press, New York, 1974) *passim*. *See also* Nucleic Acids.

Long needles from 90% ethanol, dec 220-230°. $[\alpha]_D^{25}$ +31° (c = 0.7 in water). Freely sol in water, less sol in alcohol. pK (amino, cationic) 4.22; pK (sugar, anionic) 12.5. uv max (pH 8.2): 271 nm (ϵ 9100); (pH 2.2): 280 nm (ϵ 13400), Voet *et al.*, *Biopolymers* **1**, 193 (1963).
Sulfate, $(C_9H_{13}N_3O_5)_2.H_2SO_4$, long prismatic needles, mp 224-225° (dec with effervescence). $[\alpha]_{589}^{25}$ +34°; $[\alpha]_{546}^{25}$ +43°.

2856. 2'-Cytidylic Acid. Cytidine-2'-monophosphate; cytidylic acid *a*; 2'-cytidinephosphoric acid; cytidine-2'-phosphate; 2'-cytosylic acid; 2'-CMP. $C_9H_{14}N_3O_8P$; mol wt 323.20. C 33.45%, H 4.37%, N 13.00%, O 39.60%, P 9.58%. Ribonuclease inhibitor. Prepn from yeast ribonucleic acid: Cohn, Carter, *J. Am. Chem. Soc.* **72**, 2606 (1950); Loring *et al.*, *J. Biol. Chem.* **196**, 807 (1952); by phosphorylation of $N^6,O^{3'},O^{5'}$-tribenzoylcytidine: Rammler, Khorana, *J. Am. Chem. Soc.* **84**, 3112 (1962). Crystal structure and conformation of the trihydrate: Kartha *et al.*, *Science* **179**, 495 (1973). *Reviews: See* Cytidine; Nucleic Acids.

Crystals, dec 238-240°. $[\alpha]_D^{20}$ +20.7° (water); $[\alpha]_D^{25}$ — 3° (aq NaOH adjusted to pH 10). uv max at pH 2: 278 nm (ϵ 12.7); at pH 12: 272 nm (ϵ 8.6). pKa$_1$ 0.8; pKa$_2$ 4.36; pKa$_3$ 6.17. Sparingly sol in water. Less sol than 3'-cytidylic acid.

2857. 3'-Cytidylic Acid. Cytidine-3'-monophosphate; cytidylic acid *b*; 3'-cytidinephosphoric acid; cytidine-3'-phosphate; 3'-cytosylic acid; 3'-CMP. $C_9H_{14}N_3O_8P$; mol wt 323.20. C 33.45%, H 4.37%, N 13.00%, O 39.60%, P 9.58%. Ribonuclease inhibitor. Prepn from yeast ribonucleic acid: see refs under 2'-Cytidylic Acid. Prepn by pancreatic-ribonuclease-catalyzed ring opening of dicyclohexylguanidinium cytidine 2',3'-cyclic phosphate: Lohrmann, Khorana, *J. Am. Chem. Soc.* **86**, 4188 (1964). X-ray analysis studies of orthorhombic form: Alver, Furberg, *Acta Chem. Scand.* **13**, 910 (1959); Sundaralingam, Jensen, *J. Mol. Biol.* **13**, 914 (1965); of monoclinic form: Bugg, Marsh, *ibid.* **25**, 67 (1967). *Reviews: See* Cytidine; Nucleic Acids.

Crystals, dec 232-234°. $[\alpha]_D^{20}$ +49.4° (c = 1 in water). $[\alpha]_D^{25}$ +50.0° (c = 1 in aq NaOH adjusted to pH 10). uv max at pH 2: 279 nm (ϵ 13.0); at pH 12: 272 nm (ϵ 8.90). pKa$_1$ 0.8; pKa$_2$ 4.28; pKa$_3$ 6.0. Moderately sol in water; more sol than 2'-cytidylic acid.

2858. Cytisine. *(1R)-1,2,3,4,5,6-Hexahydro-1,5-methano-8H-pyrido-[1,2-a][1,5]diazocin-8-one;* baptitoxine; sophorine; ulexine; Cytiton. $C_{11}H_{14}N_2O$; mol wt 190.25. C 69.45%, H 7.42%, N 14.72%, O 8.41%. Toxic principle in seed of *Laburnum anagyroides* Medik. and other *Leguminosae*. Extraction: Ing, *J. Chem. Soc.* **1931**, 2200; Späth, Galinovsky, *Ber.* **65**, 1526 (1932); **66**, 1338 (1933); Lecoq, *Bull. Soc. Chim. France* **10**, 153 (1943). Structure: Ing, *J. Chem. Soc.* **1932**, 2778. Synthesis: Bohlmann *et al.*, *Angew. Chem.* **67**, 708 (1955); Van Tamelen, Baran, *J. Am. Chem. Soc.* **77**, 4944 (1955). Absolute configuration: Okuda *et al.*, *Chem. & Ind. (London)* **1961**, 1751. Pharmacological properties: R. B. Barlow, L. J. McLeod, *Brit. J. Pharmacol.* **35**, 161 (1969).

Orthorhombic prisms from acetone, mp 152-153°. Sublimes. bp$_2$ 218°. $[\alpha]_D^{17}$ — 120°. pK$_1$ 6.11; pK$_2$ 13.08. Sol in 1.3 parts water, 13 parts acetone, 1.3 parts methanol, 3.5 parts alcohol, 30 parts benzene, 10 parts ethyl acetate, 2.0 parts chloroform. Practically insol in petr ether. LD$_{50}$ in mice (mg/kg): 1.73 i.v.; 9.4 i.p.; 101 orally (Barlow, McLeod).
Hydrochloride, $C_{11}H_{14}N_2O.HCl$, deliquescent crystals, sol in water and alcohol. pH of 0.1 molar aq soln 4.3.
Nitrate monohydrate, $C_{11}H_{14}N_2O.HNO_3.H_2O$, needles or leaflets, $[\alpha]_D$ —81.5°. Sol in water, slightly sol in alcohol. Practically insol in ether.

2859. Cytochalasins. A class of mold metabolites exhibiting a number of unusual and varied effects on animal cells. More than twenty cytochalasins are known, isolated from several different mold spp. All are characterized by a highly substituted hydrogenated isoindole ring of known configuration to which is fused a macrocyclic ring. This ring may vary from 11 to 14 atoms in size and may be either a carbocycle or a lactone. Isoln and structure of cytochalasins A, B, C, D: Aldridge *et al.*, *J. Chem. Soc. (C)* **1967**, 1667; *eidem, Chem. Commun.* **1967**, 26; of E and F: Aldridge *et al.*, *ibid.* **1972**, 148. Revised structures of cytochalasins E and F: Aldridge *et al.*, *ibid.* **1973**, 551; Büchi *et al.*, *J. Am. Chem. Soc.* **95**, 5423 (1973). Isoln of H, also known as *paspaline-P* or *kodocytochalasin-1*: G. S. Pendse, *Experientia* **30**, 107 (1974). Structure of H: S. A Patwardhan *et al.*, *Phytochemistry* **13**, 1985 (1974); M. A. Beno *et al.*, *J. Am. Chem. Soc.* **99**, 4123 (1977); x-ray crystal and molecular structure: J. A. McMillan *et al.*, *Chem. Commun.* **1977**, 105. Isoln of K, L, M from *Chalara microspora* and proposed structures: T. Fex, *Tetrahedron Letters* **22**, 2703 (1981). Isoln of E and K from *Aspergillus clavatus* and proposed alternate structure of K: P. S. Steyn *et al.*, *J.*

Chem. Soc. Perkins Trans. I **1982**, 541. Partial synthesis of A and B: S. Masamune *et al., J. Am. Chem. Soc.* **99**, 6756 (1977). Total synthesis of B: G. Stork *et al., ibid.* **100**, 7775 (1978). Major biological effects are the blockage of cytoplasmic cleavage by blocking formation of contractile microfilament structures, resulting in multinucleate cell formation, the reversible inhibition of cell movement, and the induction of nuclear extrusion: Carter, *Nature* **213**, 261 (1967); Krishan, *J. Cell. Biol.* **54**, 657 (1972); E. D. Korn, *Physiol. Rev.* **62**, 703 (1982). Correlation between effects of cytochalasins on cellular structures and cellular events and those on actin *in vitro:* I. Yahara *et al., J. Cell Biol.* **92**, 69 (1982). Other reported effects include the inhibition of glucose transport, of thyroid secretion, of growth hormone release, of phagocytosis, and of platelet aggregation and clot contraction. *See* D. A. Hume *et al., Nature* **272**, 359 (1978). Nomenclature: M. Binder *et al., J. Chem. Soc. Perkin Trans. I* **1973**, 1146. *Reviews:* M. Binder, C. Tamm, *Angew. Chem. Int. Ed.* **12**, 370 (1973); R. B. Herbert in *The Alkaloids,* vol. 7, J. E. Saxton, Ed. (The Chemical Society, London, 1977) pp 29-30; W. G. Thilly *et al., Front. Biol.* **46**, 53-64 (1978); L. V. Domnina *et al., Proc. Nat. Acad. Sci. USA* **79**, 7754-7757 (1982); W. Siess *et al., ibid.* 7709-7713.

Cytochalasin B

Cytochalasin B. C$_{29}$H$_{37}$NO$_5$. (7S,13E,16R,20R,21E)-7,20-Dihydroxy-16-methyl-10-phenyl-24-oxo[14]cytochalasa-6(12),13,21-triene-1,23-dione; (E,E)-16-benzyl-6,7,8,9,10,12a,-13,14,15,15a,16,17-dodecahydro-5,13-dihydroxy-9,15-dimethyl-14-methylene-2H-oxacyclotetradec[2,3-d]isoindole-2,18-(5H)-dione; phomin. The most important and biologically studied of the cytochalasins. Formerly isolated from cultures of a *Phoma* sp. and called phomin: Rothweiler, Tamm, *Experientia* **22**, 750 (1966). Physicochemical data: A. E. Pohland *et al., Pure Appl. Chem.* **54**, 2219 (1982). Felted needles from acetone, mp 218-221°. Completely stable under normal conditions. Solutions in DMSO show no decrease in potency when stored at 4° for three years (Aldrich data sheet). Solubility (mg/ml at 24°): acetone 10.3; ethanol 35.4; DMSO 371; DMF 492. Insol in water.

USE: As tools in cytological research and in characterization of polymerization properties of actin, *q. v.*

2860. Cytochrome c. Myohematin; hematin-protein; Cromoci; Cytorest; Landrax. Hemoprotein in which the catalytically active prosthetic group is a derivative of iron protoporphyrin IX. Occurs in the cells of all aerobic organisms. Found in animal cells in the mitochondrial protein-lipid complex. Can be readily converted to corresponding hemochromogen by replacement of the protein with an organic base, usually pyridine. Because of their ready fluctuation within the cell between the ferrous and ferric states, they are all efficient biological electron-transporters and play a vital role in cellular oxidations in both plants and animals. Generally regarded as universal catalysts of respiration, forming the essential electron-bridge between the respirable substrates and oxygen: Slater, Nord, *Advan. Enzymol. Relat. Subj. Biochem.* **20**, 147 (1958); James, Leech, *Endeavour* **19**, 108 (1960). Isoln of cytochrome c from bakers' yeast: Keilin, *Proc. Roy. Soc. (London) Ser. B* **106**, 418 (1930); Hagikara *et al., Nature* **178**, 629 (1956); from ox heart: Kirby *et al., Acta Chem. Scand.* **10**, 148 (1956); from beef heart: Paleus, *ibid.* **14**, 1743 (1960); Morrison *et al., Biochim. Biophys. Acta* **41**, 334 (1960); from horse heart: Nozaki *et al., J. Biochem. (Tokyo)* **44**, 453 (1957); from fish heart: Hagikara *et al., Nature* **179**, 249 (1957); from wheat germ: Hagikara *et al., ibid.* **181**, 1950 (1958); from human heart: Matsuba, Vasunobu, *J. Biol. Chem.* **236**, 1701 (1961).

Isoln of cytochrome c$_1$ from submitochondrial particles: Green *et al., Biochim. Biophys. Acta* **31**, 34 (1955); from beef heart: Sekugu, *J. Biochem. (Tokyo)* **48**, 214 (1960). Prepn of cytochrome c$_2$ from *Rhodospirillum rubrum* and cytochrome c$_3$ from *Desulfovibrio desulfuricans:* Hario, Kamen, *Biochim. Biophys. Acta* **48**, 266 (1961). Crystn of cytochrome c: Butt, *Proc. Roy. Soc. (London) Ser. B,* **156**, 429 (1962). Improved prepn from mammalian sources: Margoliash, Walasek, U.S. pat. **3,342,796** (1967 to Abbott). Structure consists of a single polypeptide chain of 104 or more amino acid residues with the heme group attached through cysteine residues at positions 14 and 17. Sequence varies among species although a band of 11 residues in positions 70 through 80 are identical in all plants and animals studied. First elucidation of amino acid sequence was that of horse heart cytochrome c: Margoliash *et al., Nature* **192**, 1125 (1961); *J. Biol. Chem.* **237**, 2161 (1962); of human heart cytochrome c: Matsubara, Smith, *ibid.* **237**, 3575 (1962). Sequences are known for a great number of mammalian species, *see* review by Margoliash, Schejter, *Advan. Protein Chem.* **21**, 113 (1966). Homologies of bacterial and mammalian cytochrome c: Dickerson, *J. Mol. Biol.* **57**, 1 (1971). Synthetic studies: Moroder *et al., Biopolymers* **12**, 493-534, 693-750 (1973). *Reviews:* Lemberg, Legge, *Hematin Compounds and Bile Pigments* (New York, 1949); Morton, *Rev. Pure Appl. Chem.* **8**, 161 (1958); Morrison, *Ann. Rev. Biochem.* **30**, 11 (1960); Margoliash, *ibid.,* 549; Margoliash, Schejter, *loc. cit.;* several authors in *Structure and Function of Cytochromes,* Okunuki *et al.,* Eds. (University Park Press, Baltimore, 1968); Harbury, Marks, in *Inorganic Biochemistry* vol. 2, G. L. Eichhorn, Ed. (Elsevier, New York, 1973) pp 902-954.

Reduced form crystallizes as separate needles; oxidized form as rosettes. Mol wt about 13000. Extinction ratio: $E_{550\ red}/E_{580\ ox} = 1.15$-1.25. E$_1^{1\%}$ 550 nm: 22.6-23.1. $E_0' - E_0'$ (quinone) -0.06 log [Ferricyt/Ferrocyt] = +0.251 v. Absorption max reduced form, *Ferrocytochrome c:* 416, 520, 550 nm; oxidized form, *Ferricytochrome c:* 280, 361, 410, 529 nm. Contains one heme per mole weight of 12,000.

Cytochrome c$_1$. Identity with *cytochrome e:* Kellin, Hartree, *Nature* **176**, 200 (1955). Extinction ratio: $E_{418\ red}/E_{411\ ox} = 1.27$. Extinction coefficient, $E_{553\ nm\ red} = 15.3 \times 10^3$ M^{-1} cm^{-1}. $E_0' = E_0'$ (quinone) -0.06 log [Ferricyt/Ferrocyt] = ±0.223 v. Acidic. Non-autoxidizable. Isoelectric pt: pH 3.6. Absorption max of reduced form: 418, 523, 553 nm; oxidized form: 278, 411, 523 nm. Reduced by sodium dithionate, potassium borohydride, ascorbic acid, sodium thioglycolate, cysteine. Stable at 0°. Denatured and coagulated by heating at 50° for 5 min.

Cytochrome c$_2$. Needles changing to squares. Mol wt about 13,000. Extinction ratio: $E_{550\ red}/E_{275\ ox} = 1.14$. Absorption max of reduced form: 272, 316, 415, 521, 550 nm; oxidized form: 275, 357, 410, 520-530 nm.

Cytochrome c$_3$. Needles. Mol wt 11,300. Subject to rapid autoxidation. Total extinction: $A_{552\ nm}^{red}$ 34.2, $A_{570\ nm}^{red}$ 1.90, $A_{280\ nm}^{ox}$ 11.62. Absorption max of reduced form: 418, 522, 552 nm; oxidized form: 349, 409 nm. Contains 2 hemes per mole weight of 12,000.

2861. Cytochromes P$_{450}$. P$_{450}$. Hemoproteins involved in enzyme hydroxylation, demethylation, and *N*-oxidation; present in a number of mammalian tissues but in highest concentrations in adrenal and liver microsomes. Has also been found in insects, yeast, plants, and bacteria. Existence of mitochondrial P$_{450}$ first reported by Harding *et al., Biochim. Biophys. Acta* **92**, 415 (1964). Extraction and purification: Mitani, Horie, *J. Biochem.* **65**, 269 (1969). Acts as a monooxygenase in catalyzing metabolic functions; influences the rate of drug metabolism. Name derived from the characteristic absorption at 450 nm. *Reviews:* several authors in *Oxidases and Redox Systems* vol. 2, King *et al.,* Eds. (University Park Press, Baltimore, 1971); J. T. Groves, *Advances in Inorganic Biochemistry,* vol. 1, G. L. Eichhorn, L. G. Marzilla, Eds. (Elsevier/North-Holland, New York, 1979) pp 119-145; several authors in *Xenobiotica* **12**, 671-800 (1982). Books: *Cytochromes P-450 & B5—Structure, Function & Interaction,* D. Y. Cooper *et al.,* Eds. (Plenum, New York, 1975) 554 pp; *Cytochrome P-450,* R. Sato, T. Omura, Eds. (Academic Press, New York, 1978) 233 pp; *Biochemistry, Biophysics and Regulation of Cytochrome P-*

450, J. A. Gustafsson *et al.*, Eds. (Elsevier/North-Holland, New York, 1980) 626 pp.

2862. Cytohemin. *[SP-5-13-(E,E)]-Chloro[7-ethenyl-17-formyl-12-(1-hydroxy-5,9,13-trimethyl-4,8,12-tetradecatrienyl)-3,8,13-trimethyl-21H,23H-porphine-2,18-dipropanoato-(4—)-N²¹,N²²,N²³,N²⁴]ferrate(2—) dihydrogen; chloro[dihydrogen-3-formyl-8-(1-hydroxy-5,9,13-trimethyltetradecyl)-7,-12,17-trimethyl-13-vinyl-2,18-porphinedipropionato(2—)]-iron.* $C_{49}H_{56}ClFeN_4O_6$; mol wt 888.30. C 66.25%, H 6.35%, Cl 3.99%, Fe 6.29%, N 6.31%, O 10.81%. Isoln from heart muscle: Warburg, Gewitz, *Z. Physiol. Chem.* **288**, 1 (1951); Lemberg, *Biochem. Z.* **338**, 97 (1963). Structure: M. Grassl *et al.*, *ibid.* **337**, 35 (1963); **338**, 771 (1963). Side chain appears to be a mixture of saturated and unsaturated forms: R. Sayffert *et al.*, *3rd Proc. Colloq. Hemes Hemoproteins, 1966*, 45-59. Unsaturated form drawn.

Amorphous powder.

2863. Cytolipin H. *1-O-(4-O-β-D-Galactopyranosyl-β-D-glucopyranosyl)ceramide;* lactosylceramide; ceramide-β-lactoside; cerebronylsphingosylglucosidogalactoside; Galβ1→4Glcβ1→1cer. A glycosphingolipid containing lactose and a ceramide (*N*-acyl fatty acid deriv of a sphingosine). Fatty acid composition is variable but is primarily palmitic, behenic and lignoceric acids. First isolated from human epidermoid carcinoma cells, later found to be one of the major neutrophil glycolipids serving as a neutrophil-differentiation marker. Isoln: M. M. Rapport *et al.*, *Cancer* **12**, 438 (1959). Simplified prepn from ox spleen: M. M. Rapport *et al.*, *J. Biol. Chem.* **237**, 1056 (1962). Structure elucidation: A. C. Schram *et al.*, *Nature* **197**, 1074 (1963). Synthesis: D. Shapiro, E. S. Rachaman, *ibid.* **201**, 878 (1964); K. C. Nicolaou *et al.*, *J. Am. Chem. Soc.* **110**, 7910 (1988). Serological activity: M. M. Rapport, L. Graf, *Nature* **201**, 879 (1964); and effect of fatty acid chain length: L. Graf, M. M. Rapport, *Chem. Phys. Lipids* **13**, 367 (1974). Characterization and distribution of glycolipids in leukemic

and nonleukemic blood cells: J. Hildebrand *et al.*, *J. Lipid Res.* **12**, 361 (1971); in neutrophils: B. A. Macher, J. C. Klock, *J. Biol. Chem.* **255**, 2092 (1980). Role as neutrophil cell surface marker: F. W. Symington *et al.*, *J. Immunol.* **134**, 2498 (1985). Intracellular localization: *eidem*, *J. Biol. Chem.* **262**, 11356 (1987). TLC determn: K. Ogawa *et al.*, *J. Chromatog.* **426**, 188 (1988). Clinical implications in inflammatory bowel disease: C. R. Stevens *et al.*, *Gut* **29**, 580 (1988). Review of structure, organization and role of glycolipids in the cell surface membrane: S. Hakomori, *Ann. Rev. Biochem.* **50**, 733-764 (1981).

Coral-like clusters from pyridine + acetone, mp 230-240° with sintering at 180-190°. $[\alpha]_D^{24}$ —10.8° (c = 2 in pyridine). Practically insol in water, cold methanol, acetone, cold acetic acid, ether, acetonitrile. Soluble in pyridine, chloroform, hot methanol, hot acetic acid. Dissolves readily in chloroform + methanol mixtures of high chloroform content.

2864. Cytosine. *4-Amino-2(1H)-pyrimidinone;* 4-amino-2-oxo-1,2-dihydropyrimidine; 4-amino-2-pyrimidinol; 4-amino-2-hydroxypyrimidine. $C_4H_5N_3O$; mol wt 111.10. C 43.24%, H 4.54%, N 37.82%, O 14.40%. Widely distributed in nature; constituent of nucleic acids. Isoln by hydrolysis of thymus nucleic acids: Levene, Bass, *Nucleic Acids* (New York, 1931) p 57. Syntheses: Hilbert, Johnson, *J. Am. Chem. Soc.* **52**, 1152 (1930); Hilbert *et al.*, *ibid.* **57**, 552 (1935); *cf.* Hunter, Hlynka, *Biochem. J.* **31**, 486 (1937); Ballweg, *Tetrahedron Letters* **1968**, 2171; David, Lubineau, *Bull. Soc. Chim. France* **1969**, 816. Synthesis from carbon dioxide and ammonia on kaolinite: G. R. Harvey *et al.*, *Naturwiss.* **58**, 624 (1971). Exists in keto form. Structure: Barker, Marsh, *Acta Crystallogr.* **17**, 1581 (1964); of monohydrate: Jeffrey, Kinoshita, *ibid.* **16**, 20 (1963); R. J. McClure, B. M. Craven, *ibid.* **B29**, 1234 (1973). Tautomerism: Y. P. Wong, *J. Am. Chem. Soc.* **95**, 3511 (1973). *Review:* Ts'o, "Bases, Nucleosides and Nucleotides" in *Basic Principles of Nucleic Acid Chemistry* vol. 1, P. O. P. Ts'o, Ed. (Academic Press, New York, 1974) pp 453-584. *See also* Nucleic Acids.

Monohydrate, lustrous monoclinic or triclinic platelets from water. Anhydr at 100°; brown around 300°; dec 320-325°. uv max (pH 8.8): 196.5, 267 nm ($\epsilon \times 10^{-3}$ 22.5, 6.1). pK_1 4.60; pK_2 12.16. One gram dissolves in 130 ml water. Slightly sol in alcohol, insol in ether. Gives red color when dissolved in soln of sodium hypochlorite to which a drop of NH_4OH is added. Forms salts with acids.

D

2865. 2,4-D.
(2,4-Dichlorophenoxy)acetic acid; Trinoxol. $C_8H_6Cl_2O_3$; mol wt 221.04. C 43.47%, H 2.74%, Cl 32.08%, O 21.71%. Prepd from 2,4-dichlorophenol and monochloroacetic acid in aq NaOH: Pokorny, *J. Am. Chem. Soc.* **63**, 1768 (1941); Foster, **Brit.** pat. **573,476** (1945); by chlorination of molten phenoxyacetic acid: Manske, **U.S.** pat. **2,471,575** (1949 to U.S. Rubber); from 2,4-dichlorophenol, sodium, and ethyl chloroacetate followed by hydrolysis of the ester: Haskelberg, *J. Org. Chem.* **12**, 426 (1947). Toxicology: V. K. Rowe, T. A. Hymas, *Am. J. Vet. Res.* **15**, 622 (1954); J. M. Way, *Residue Rev.* **26**, 37 (1969).

Crystals from benzene, mp 138°. $bp_{0.4}$ 160°. Almost insol in water. Sol in organic solvents. LD_{50} in mice, rats (mg/kg): 368, 375 orally (Rowe, Hymas).
Sodium salt, $C_8H_5Cl_2NaO_3$, needles from alc, dec 215°. Soly in water: 3.5%.
Isopropyl ester, $C_{11}H_{12}Cl_2O_3$, *Esteron 44*. Liquid, d_{25}^{25} 1.255-1.270. Solidifies at +5°. Sol in oil.
Butyl ester, $C_{12}H_{14}Cl_2O_3$, *Lironox, Weedone*. Prepn: Nagel, **Ger.** pat. **1,144,288** (1963 to Bayer), *C.A.* **59**, 6315e (1963).
Amine salts. Sol in water. The following amines form useful salts: Dimethylamine, isopropylamine, triethylamine, diethanolamine, triethanolamine. These salts are much more sol than the sodium salt. A small amount of sequestering agent, such as ethylenediaminetetraacetic acid, is generally added to prevent complex formation in hard water.
Note: The name Hedonal was formerly a synonym for carbamic acid 1-methylbutyl ester.
Caution: Potential symptoms of overexposure are weakness, stupor, hyporeflexia and muscle twitch; convulsions; dermatitis. *See NIOSH Pocket Guide to Chemical Hazards* (DHHS/NIOSH 90-117, 1990) p 80.
USE: Herbicide. To increase latex output of old rubber trees.

2866. Dacarbazine.
5-(3,3-Dimethyl-1-triazenyl)-1H-imidazole-4-carboxamide; 5(or 4)-(dimethyltriazeno)imidazole-4(or 5)-carboxamide; DIC; DTIC; NSC-45388; Dacatic; DTIC-Dome; Deticene. $C_6H_{10}N_6O$; mol wt 182.19. C 39.56%, H 5.53%, N 46.13%, O 8.78%. First synthesized in 1959 at the Southern Research Institute. Prepn: Shealy *et al., J. Org. Chem.* **27**, 2150 (1962); Hano *et al., Gann* **59**, 207 (1968), *C.A.* **69**, 42527g (1968). Antitumor activity: Shealy *et al., Biochem. Pharmacol.* **11**, 674 (1962); Hano *et al., Gann* **56**, 417 (1965), *C.A.* **63**, 18856g (1965). Mechanism of action studies: Saunders, Schultz, *Biochem. Pharmacol.* **19**, 911 (1970). Metabolism: Skibba *et al., ibid.* 2043; Mizuno, Humphrey, *Cancer Chemother. Rep., Part 1* **56**, 465 (1972). *Review:* Carter, Friedman, *Europ. J. Cancer* **8**, 85-92 (1972), *see also* the series of articles on history, activity, mechanism of action and clinical studies: *Cancer Treatment Rep.* **60**, 123-214 (1976).

Ivory microcrystalline substance; explosive decomp 250-255°. Also reported as mp 205°, Hano *et al., loc. cit.* (1965). uv max (0.1N HCl): 223 nm (7500); (pH 7): 237 nm (11200); both solns protected from light. Stable in neutral soln in absence of light.

Note: This substance may reasonably be anticipated to be a carcinogen: *Seventh Annual Report on Carcinogens* (PB95-109781, 1994) p 138.
THERAP CAT: Antineoplastic.

2867. Dactinomycin.
Actinomycin D; actinomycin-[thr-val-pro-sar-meval]; meractinomycin; actinomycin A_{IV}; actinomycin IV; actinomycin C_1; actinomycin I_1; actinomycin X_1; NSC-3053; Cosmegen. $C_{62}H_{86}N_{12}O_{16}$; mol wt 1255.44. C 59.32%, H 6.90%, N 13.39%, O 20.39%. Antibiotic substance belonging to the actinomycin complex, produced by several *Streptomyces* spp. Historical background of actinomycins and chemistry, toxicology, pharmacology of actinomycin D: *Ann. N.Y. Acad. Sci.* **89**, 285-485 (1960). Isoln from broth cultures of *S. parvulus:* Manaker *et al., Antibiot. Ann.* **1954/55**, 853. Structure: Bullock, Johnson, *J. Chem. Soc.* **1957**, 3280. Synthesis: Brockmann, Manegold, *Naturwiss.* **51**, 383 (1964); Brockmann, Lackner, *ibid.* 384, 435 (1964); Brockmann *et al.,* **Ger.** pat. **1,172,680** (1964 to Bayer); *eidem, Ber.* **101**, 1312 (1968); Meienhofer, *J. Am. Chem. Soc.* **92**, 3771 (1970); T. Tanaka *et al., Bull. Chem. Soc. Japan* **53**, 1352 (1980); K. Nakajima *et al., Pept. Chem.* **19**, 143-148 (1981). Conformation from NMR: Conti, DeSantis, *Nature* **227**, 1239 (1970); Lackner, *Ber.* **104**, 3653 (1971). Toxicity data: F. S. Philips *et al., Ann. N.Y. Acad. Sci.* **89**, 348 (1960). Review and evaluation of studies of carcinogenic action in laboratory animals: *IARC Monographs* **12**, 29-41 (1976).

Trihydrate, bright red, rhomboid prisms from abs alc, dec 241.5-243°. $[\alpha]_D^{28}$ −315° (c = 0.25 in methanol). Abs max (methanol): 244, 441 nm ($A_{1cm}^{1\%}$ 281, 206). Dil solns are very sensitive to light. Sol in alc, propylene glycol, water + glycol mixtures. LD_{50} orally in mice, rats: 13.0, 7.2 mg/kg (Philips).
THERAP CAT: Antineoplastic.

2868. Daidzein.
7-Hydroxy-3-(4-hydroxyphenyl)-4H-1-benzopyran-4-one; 4',7-dihydroxyisoflavone. $C_{15}H_{10}O_4$; mol wt 254.24. C 70.86%, H 3.96%, O 25.17%. The aglucon of daidzin, prepd by hydrolysis with HCl in methanol: Walz, *Ann.* **489**, 118 (1931). Isoln from red clover: Wong, *J. Sci. Food Agr.* **13**, 304 (1962); from the mold *Micromonospora halophytica:* Ganguly, Sarre, *Chem. & Ind. (London)* **1970**, 201. Synthesis: Baker *et al., J. Chem. Soc.* **1933**, 274; Wessely *et al., Ber.* **66**, 685 (1933); Mahal *et al., J. Chem. Soc.* **1934**, 1769; Baker *et al., ibid.* **1953**, 1852; Farkas, *Ber.* **90**, 2940 (1957). Biosynthetic study: Grisebach, Zilg, *Z. Naturforsch.* **23b**, 494 (1968).

Pale-yellow prisms from dil alc, dec 315-323°. uv max: 250 nm (log ε 4.44). Sol in alcohol, ether.
7-Glucoside, $C_{21}H_{20}O_9$, *daidzin*. Isoln from soybean meal *(Soja hispida):* Walz, *loc. cit.* Synthesis: Farkas, Várady, *Ber.* **92**, 819 (1959). Monohydrate, needles from water, dec 234-236°. $[\alpha]_D^{20}$ −36.° (0.02N NaOH). Sol in water and aq alcohol.

2869. Dalapon.
2,2-Dichloropropanoic acid; α,α-dichloropropionic acid. $C_3H_4Cl_2O_2$; mol wt 142.97. C 25.20%, H 2.82%, Cl 49.59%, O 22.38%. Prepn: **Brit.** pat.

752,761 (1956 to Dow); Kowolik, Fisher, **Brit.** pat. **892,584** (1962 to British Celanese); of free acid and sodium salt: Smeykal, Pallutz, *Chem. Tech. (Leipzig)* **15** (11), 654 (1963), *C.A.* **60**, 6741c (1964); of sodium salt: Barrons, **U.S.** pat. **2,642,354** and Maylott, Meyer, **U.S.** pat. **3,007,964** (1953, 1961, both to Dow). Acute toxicity: T. B. Gaines, R. E. Linder, *Fundam. Appl. Toxicol.* **7**, 299 (1986). Toxicity data for the sodium salt: G. W. Bailey, J. L. White, *Residue Rev.* **10**, 97 (1965).

Liquid, bp_{20} 98-99°. d^{20} 1.4014. n_D^{20} 1.4551. LD_{50} in male, female rats (mg/kg): 7126, 6936 orally (Gaines, Linder).

Sodium salt, $C_3H_3Cl_2NaO_2$, *Dowpon, Radapon, Basfapon B.* Crystals, dec 174-176°. Solubility in water at 25°, 45 g/100 ml; aq solns hydrolyze above 70°. Corrosive to iron.

USE: Sodium salt as herbicide.

2870. Dalteparin. Tedelparin. Low molecular weight fragment of heparin, *q.v.*, prepd by nitrous acid depolymerization of porcine mucosal heparin. Mean mol wt 4000-6000 daltons. Prepn: U. Lindahl *et al.*, **PCT Int.** pat. **Appl. 80 1383;** *eidem*, **U.S.** pat. **4,303,651** (1980, 1981 both to Kabi AB). *See also:* U. Lindahl *et al.*, *Proc. Nat. Acad. Sci. USA* **76**, 3198 (1979). Exhibits antithrombotic activity comparable to standard heparin but with diminished hemorrhagic effects: E. Holmer *et al.*, *Thromb. Res.* **18**, 861 (1980). Pharmacokinetics in humans: G. Bratt *et al.*, *Thromb. Haemostasis* **53**, 208 (1985); G. Bratt *et al.*, *Thromb. Res.* **42**, 613 (1986). Clinical comparison with heparin in deep venous thrombosis: G. Bratt *et al.*, *Thromb. Haemostasis* **54**, 813 (1985); in perioperative thromboprophylaxis: M. Koller *et al.*, *ibid.* **56**, 243 (1986). Symposia on pharmacology and clinical efficacy: *Haemostasis* **16**, Suppl. 2, 1-71 (1986); *Sem. Thromb. Hemostas.* **16**, Suppl., 1-76 (1990).

Sodium salt, *Kabi 2165, FR-860, Boxol, Fragmin.*

THERAP CAT: Antithrombotic.

2871. Daltroban. *4-[2-[[(4-Chlorophenyl)sulfonyl]amino]ethyl]benzeneacetic acid; [p-[2-(p-chlorobenzenesulfonamido)ethyl]phenyl]acetic acid;* BM-13505; SKF-96148. $C_{16}H_{16}ClNO_4S$; mol wt 353.83. C 54.31%, H 4.56%, Cl 10.02%, N 3.96%, O 18.09%, S 9.06%. Thromboxane A_2-receptor antagonist. Prepn: E. C. Witte *et al.*, **Ger.** pat. **3,000,377;** *eidem*, **U.S.** pat. **4,443,477** (1981, 1984 both to Boehringer, Mann.). Receptor binding study: A. Yanagisawa *et al.*, *Eur. J. Pharmacol.* **133**, 89 (1987). Pharmacological evaluations in animals: D. J. Lefer *et al.*, *Arch. Int. Pharmacodyn.* **287**, 89 (1987); E. F. Smith, III, J. McDonald, *Pharmacology* **36**, 340 (1988); P. Löbel *et al.*, *Biomed. Biochem. Acta* **47**, S 86 (1988). GC determn in biological fluids: V. Uebis, *J. Chromatog.* **419**, 345 (1987).

THERAP CAT: Antithrombotic.

2872. Damar. Gum Damar; Dammar; resin Damar. Resinous exudate from a species of *Shorea, Dipterocarpaceae.* Habit. East Indies, Philippines. *Constit.* Volatile oil, resins, bitter substance.

Yellowish-white, roundish, or stalactite shaped, friable masses; semi-transparent, conchoidal fracture; varying degrees of hardness. d 1.04-1.12. mp ∼120°. Insol in water. Sol in alcohol, chloroform, ether, carbon disulfide, oil rosemary; partly sol in oil turpentine.

USE: In plasters, varnishes, lacquers, etc. A soln of the purified material in chloroform or xylene is used for preserving animal and vegetable specimens for microscopy.

2873. Damascenine. *3-Methoxy-2-(methylamino)benzoic acid methyl ester; 2-(methylamino)-m-anisic acid methyl ester;* methyl 3-methoxy-2-(methylamino)benzoate; methyldamascenine; nigelline. $C_{10}H_{13}NO_3$; mol wt 195.22. C 61.53%, H 6.71%, N 7.17%, O 24.59%. Odoriferous principle of oil of Nigella from seeds of *Nigella damascena* L. and *Nigella arvensis* L., *Ranunculaceae.* Early syntheses: Ewins, *J. Chem. Soc.* **101**, 544 (1912); Kaufman, Rothlin, *Ber.* **49**, 578 (1916); Sornet, *Manuf. Chem.* **5**, 87 (1924); Keller, Schulze, *Arch. Pharm.* **263**, 481 (1925). Improved synthesis: Mutschler, *ibid.* **298**, 861 (1965); M. Thoinet *et al.*, *Ann. Pharm. Franc.* **36**, 337 (1978). Toxicity study: Bekemeier *et al.*, *Arch. Int. Pharmacodyn. Ther.* **168**, 199 (1967).

Prisms from abs. alc. Nutmeg-like odor, mp 27-29°. bp 270° (slight dec), bp_{10} 147°. Volatile with steam. Insol in water. Freely sol in alc, ether, chloroform, petr ether, oils.

Hydrochloride, deliquesc prisms, mp 156°. (Hydrate, mp 122°). LD_{50} orally in mice: 1800 mg/kg (Bekemeier).

2874. Daminozide. *Butanedioic acid mono(2,2-dimethylhydrazide); N-(dimethylamino)succinic acid;* succinic acid 2,2-dimethylhydrazide; B-9; B-995; Alar; B-Nine; Kylar. $C_6H_{12}N_2O_3$; mol wt 160.17. C 44.99%, H 7.55%, N 17.49%, O 29.97%. $HOOCCH_2CH_2CONHN(CH_3)_2$. Prepn: **Belg.** pat. **613,799** *C.A.* **57**, 10281d (1962); H. A. Hageman, W. L. Hubbard, **U.S.** pat. **3,257,414** (1962, 1966 both to U.S. Rubber). Effect in controlling growth in apples: F. W. Southwick *et al.*, *Proc. Am. Soc. Hort. Sci.* **92**, 71 (1968); D. W. Greene *et al.*, *Fruit Var. J.* **40**, 41 (1986). *In vitro* and *in vivo* mechanism of action studies: K. Ryugo, R. M. Sachs, *J. Am. Soc. Hort. Sci.* **94**, 529 (1969). Hydrolysis to 1,1-dimethylhydrazine, *q.v.*, and determn in food products: W. H. Newsome, *J. Agr. Food Chem.* **28**, 319 (1980); M. K. Conditt, J. R. Baumgardner, *J. Assoc. Offic. Anal. Chem.* **71**, 735 (1988). Discussion of toxicological evaluation: D. Campt, *EPA Journal* **13**, 32 (1987).

Crystals, mp 154-155°. Soly 10% in water, 2.5% in acetone, 5% in methanol.

USE: Plant growth regulator.

2875. Danazol. *(17α)-Pregna-2,4-dien-20-yno[2,3-d]-isoxazol-17-ol;* 1-ethynyl-2,3,3a,3b,4,5,10,10a,10b,11,12,12a-dodecahydro-10a,12a-dimethyl-1H-cyclopenta[7,8]phenanthro[3,2-d]isoxazol-1-ol; 17α-ethynyl-17β-hydroxy-4-androsteno[2,3-d]isoxazole; Win-17757; Bonzol; Chronogyn; Cyclomen; Danocrine; Danol; Danoval; Ladogal; Winobanin. $C_{22}H_{27}NO_2$; mol wt 337.46. C 78.30%, H 8.06%, N 4.15%, O 9.48%. Anterior pituitary supressant. Anabolic steroid deriv of ethisterone, *q.v.*, with mild androgenic side effects (an impeded androgen). Prepn: **Brit.** pat. **905,844** (1962 to Sterling Drug), *C.A.* **58**, 6895c (1963); Manson *et al.*, *J. Med. Chem.* **6**, 1 (1963); Clinton, Manson, **U.S.** pat. **3,135,743** (1964 to Sterling Drug). Activity studies: Sherins *et al.*, *J. Clin. Endocrinol. Metab.* **32**, 522 (1971); Dmowski *et al.*, *Fert. Steril.* **22**, 9 (1971). Clinical studies in endometriosis and other endocrine disorders: R. B. Greenblatt *et al.*, *ibid.* 102. Series of articles on pharmacology, pharmacokinetics and clinical use: *Drugs* **19**, 321-372 (1980). Use in idiopathic thrombocytopenic purpura: Y. S.

Ahn *et al.*, *N. Engl. J. Med.* **308**, 1396 (1983); in hemophilia: H. R. Gralnick *et al.*, *ibid.* 1393.

Crystals from acetone, mp 224.4-226.8°. $[\alpha]_D^{25}$ +7.5° (ethanol); $[\alpha]_D^{25}$ +21.9° (chloroform). uv max (ethanol): 286 nm (ϵ 11300).

THERAP CAT: Antigonadotropin.

2876. Danofloxacin. *(1S)-1-Cyclopropyl-6-fluoro-1,4-dihydro-7-(5-methyl-2,5-diazabicyclo[2.2.1]hept-2-yl)-4-oxo-3-quinolinecarboxylic acid.* $C_{19}H_{20}FN_3O_3$; mol wt 357.38. C 63.86%, H 5.64%, F 5.32%, N 11.76%, O 13.43%. Fluorinated quinolone antibacterial. Prepn: M. R. Jefson, P. R. McGuirk, **Eur.** pat. **Appl.** 215,650; *eidem*, **U.S.** pat. 4,861,779 (1987, 1989 both to Pfizer). *See also:* F. R. Busch *et al.*, **Eur.** pat. **Appl.** 342,849 (1989 to Pfizer), *C.A.* **113**, 6177m (1990). Evaluation in treatment of pneumonia in cattle: W. T. R. Grimshaw *et al.*, *Deutsch. Tierarztl. Wochenschr.* **97**, 529 (1990); C. J. Giles *et al.*, *Vet. Rec.* **128**, 296 (1991).

mp 268-272°.

Monomethanesulfonate, $C_{19}H_{20}FN_3O_3 \cdot CH_4O_3S$, *CP-76136-27, Advocin.*

THERAP CAT (VET): Antibacterial.

2877. Dansyl Chloride. *5-(Dimethylamino)-1-naphthalenesulfonyl chloride.* $C_{12}H_{12}ClNO_2S$; mol wt 269.75. C 53.43%, H 4.48%, Cl 13.14%, N 5.19%, O 11.86%, S 11.89%. Reagent for fluorescent labelling of amines, amino acids and proteins. Synthesis: G. Weber, *Biochem. J.* **51**, 155 (1952); A. Mendel, *J. Chem. Eng. Data* **15**, 340 (1970). In amino acid analysis: J. Airhart *et al.*, *Anal. Biochem.* **53**, 132 (1973). In HPLC analysis of phenols: R. M. Cassidy, D. S. LeGay, *J. Chromatog. Sci.* **12**, 85 (1974). In protein microassays: T. Kinoshita *et al.*, *Anal. Biochem.* **66**, 104 (1975). In measurement of desquamation: R. Marks *et al.*, *Brit. J. Dermatol.* **111**, 265 (1984).

Yellow-orange crystals from hexane, mp 66.5-68°. Sol in acetone, pyridine, benzene, dioxane. Insol in water.

USE: In fluorescent labelling, column chromatography, HPLC.

2878. Danthron. *1,8-Dihydroxy-9,10-anthracenedione; 1,8-dihydroxyanthraquinone;* chrysazin; dantron; Altan; Antrapurol; Diaquone; Dorbane; Istizin; Modane. $C_{14}H_8O_4$; mol wt 240.22. C 70.00%, H 3.36%, O 26.64%. Prepd from 1,8-anthraquinone potassium disulfonate: Fierz-David, Blangey, *Farbenchemie* (Vienna, 5th ed., 1943) pp 224-225; Kozlov, *Doklady Akad. Nauk SSSR* **61**, 281 (1948). Mutation study: J. P. Brown, R. J. Brown, *Mutat. Res.* **40**, 203 (1976).

Orange needles from alc, mp 193-197°. Sublimes. Absorption max: 430, 250 nm (log ϵ 4.35, 4.60). Almost insol in water (6.5 $\times$ 10^{-6} mols/l at 25°), in alcohol (1:2000); moderately sol in ether (1:500), in chloroform; sol in 10 parts hot glacial acetic acid. Very slightly sol in aq solns of alkali hydroxides: about 0.8 g dissolves in 100 ml $0.5N$ NaOH. (Disodium salt is reported to have a water soly of 0.05%.) LD_{50} in mice: 500 mg/kg i.p., *RTECS* Vol. I, R. J. Lewis, R. L. Tatken, Eds. (1979) p 142.

USE: Important intermediate in the manuf of alizarin and indanthrene dyestuffs. Forms insol Ca, Ba, Pb lakes and has been proposed as analytical reagent.

THERAP CAT: Cathartic.

THERAP CAT (VET): Purgative.

2879. Dantrolene. *1-[[[5-(4-Nitrophenyl)-2-furanyl]methylene]amino]-2,4-imidazolidinedione; 1-[[5-(p-nitrophenyl)furfurylidene]amino]hydantoin.* $C_{14}H_{10}N_4O_5$; mol wt 314.26. C 53.51%, H 3.21%, N 17.83%, O 25.46%. Prepn: **Neth.** pat. **Appl.** 6,612,588 corresp to Davis, Snyder, **U.S.** pat. 3,415,821 (1967, 1968 both to Norwich Pharmacal); Snyder *et al.*, *J. Med. Chem.* **10**, 807 (1967); Frimm *et al.*, *Chem. Zvesti* **23**, 916 (1969). Pharmacology and mechanism of action: Ellis, Carpenter, *Arch. Pharmacol.* **275**, 83 (1972); Ellis *et al.*, *J. Pharm. Sci.* **62**, 948 (1973). Metabolic studies: Cox *et al.*, *ibid.* **58**, 987 (1969). Review of pharmacology and use in malignant hyperthermia, neuroleptic malignant syndrome, muscle spasticity: A. Ward *et al.*, *Drugs* **32**, 130-168 (1986).

Crystals from aqueous DMF, mp 279-280°.

Sodium salt hemiheptahydrate, $C_{14}H_9N_4NaO_5 \cdot 3\frac{1}{2}H_2O$, *F-440, Dantamacrin, Dantrium.* Orange powder. Slightly soluble in water; more sol in alkaline soln.

Sodium salt tetrahydrate, $C_{14}H_9N_4NaO_5 \cdot 4H_2O$, *Dantrix.*

THERAP CAT: Muscle relaxant (skeletal).

2880. Daphnandrine. *6,6',12'-Trimethoxy-2-methyloxyacanthan-7-ol.* $C_{36}H_{38}N_2O_6$; mol wt 594.71. C 72.71%, H 6.44%, N 4.71%, O 16.14%. From bark of *Daphnandra micrantha* Benth., *Monimiaceae.* Isoln: Pyman, *J. Chem. Soc.* **105**, 1679 (1914); Bick *et al.*, *ibid.* **1953**, 695. Structure: Bick *et al.*, *ibid.* **1949**, 2767; **1960**, 4928.

Needles from methanol, dec 270°. $[\alpha]_D^{16}$ +480° (c = 1.2 in chloroform). uv max (methanol): 284 nm (log ϵ 3.92). Practically insol in hot water, ethyl acetate, acetone, ether, petr ether; sparingly sol in hot methanol, ethanol, xylene, cold chloroform; sol in hot chloroform.

Hydrochloride hemihydrate, $C_{36}H_{38}N_2O_6 \cdot 2HCl \cdot \frac{1}{2}H_2O$, prisms from water, dec 282°. $[\alpha]_D$ +400° (c = 1.12). Very sol in hot water; sparingly sol in cold water.

Hydrobromide monohydrate, $C_{36}H_{38}N_2O_6.2HBr.H_2O$, prisms from water, dec 291°. $[\alpha]_D +386°$ (c = 3.8). Very sol in hot water; sparingly sol in cold water.

2881. Daphnetin. *7,8-Dihydroxy-2H-1-benzopyran-2-one; 7,8-dihydroxycoumarin.* $C_9H_6O_4$; mol wt 178.14. C 60.68%, H 3.39%, O 35.92%. The aglucon of daphnin. By boiling daphnin with dil mineral acids; by enzymatic hydrolysis of daphnin; by sublimation from daphnin; by synthesis: v. Pechmann, *Ber.* **17**, 934 (1884); Gatterman, Köbner, *Ber.* **32**, 287 (1899); Baker, Savage, *J. Chem. Soc.* **1938**, 1602; Späth, Galinovsky, *Ber.* **70B**, 235 (1937); Sethna, Shah, *Chem. Rev.* **36**, 1 (1945).

Crystals from dil alc, mp 256° (dec). Sublimes. Sol in boiling water, hot dil alcohol, hot glacial acetic acid; very sparingly sol in ether, in carbon bisulfide, in chloroform, in benzene. Sol in alkali and alkali carbonate solns giving a yellow color. Aq solns give a green color with ferric chloride which turns red on addition of sodium carbonate.

2882. Daphnin. *7-β-D-Glucopyranosyloxy-8-hydroxy-2H-1-benzopyran-2-one; 7,8-dihydroxycoumarin 7-β-D-glucoside; daphnetin 7-β-D-glucoside; 7-glucosido-8-hydroxycoumarin.* $C_{15}H_{16}O_9$; mol wt 340.29. C 52.95%, H 4.74%, O 42.32%. In bark and flower of *Daphne* spp. Extraction from fresh bark of *D. mezereum* L., *Thymelaeaceae*: *Beilstein* **XXXI**, 248. Structure: Wessely, Sturm, *Ber.* **63**, 1299 (1930); Hattori, *Chem. Zentr.* **1930** II, 2138. Synthesis: Gandini, *Gazz. Chim. Ital.* **70**, 611 (1940), *C.A.* **35**, 1394 (1941).

Monohydrate, prisms from water. Anhydr after drying at 80-90° at 2 mm Hg. When anhydrous, dec 215° (Wessely, Sturm); dec 224° (Hattori); dec 229° (older data). Optical rotations of monohydrate: $[\alpha]_D^{22} -124°$ (15 mg in 4 ml MeOH); after 2 recrystns from water: $[\alpha]_D^{22} -115°$ (abs MeOH) (Wessely, Sturm); $[\alpha]_D^{15} -84.2°$ (MeOH) (Hattori). Sol in hot water, alcohol; sol in aq solns of alkalies and alkali carbonates with yellow color. Insol in ether. Hydrolyzed by dilute acid yielding 7,8-dihydroxycoumarin and D-glucose. Also split by emulsin.

2883. Daphnoline. *6,6'-Dimethoxy-2-methyloxyacanthan-7,12'-diol;* trilobamine. $C_{35}H_{36}N_2O_6$; mol wt 580.68. C 72.40%, H 6.25%, N 4.82%, O 16.53%. From bark of *Daphnandra micrantha* Benth., *Monimiaceae.* Isoln: Pyman, *J. Chem. Soc.* **105**, 1679 (1914). Identity with trilobamine: Bick *et al., ibid.* **1949**, 2767. Structure: Bick *et al., ibid.* **1960**, 4928.

Crystals from chloroform, mp 194-196°. $[\alpha]_D +459°$ (c = 0.3 in chloroform). uv max (methanol): 285 nm (log ε 3.9). Practically insol in water, ethyl acetate, acetone, ether, petr ether; sparingly sol in methanol, ethanol, xylene, hot chloroform sol in dil acids, cold aq 5% sodium hydroxide.

Hydrochloride, $C_{35}H_{36}N_2O_6.2HCl$, crystals from aq ethanol, dec 290°. $[\alpha]_D +355°$ (c = 2.1 in water). Sol in cold water, hot alcohol.

Hydrobromide monohydrate, $C_{35}H_{36}N_2O_6.2HBr.H_2O$, microscopic needles from water, dec 286°. Very sol in hot water; sparingly sol in cold water.

2884. Dapiprazole. *5,6,7,8-Tetrahydro-3-[2-[4-(2-methylphenyl)-1-piperazinyl]ethyl]-1,2,4-triazolo[4,3-a]pyridine; 5,6,7,8-tetrahydro-3-[2-(4-o-tolyl-1-piperazinyl)ethyl]-s-triazolo[4,3-a]pyridine.* $C_{19}H_{27}N_5$; mol wt 325.46. C 70.12%, H 8.36%, N 21.52%. α-Adrenergic blocker. Prepn: B. Silvestrini, L. Baiocchi, **Ger. pat. 2,915,318;** *eidem,* **U.S. pat. 4,252,721** (1979, 1981 both to Angelini Francesco). Series of articles on general pharmacology: *Arzneimittel-Forsch.* **32**, 674-681 (1982); and toxicity: B. Silvestrini *et al., ibid.* 668. Ocular pharmacokinetics in rabbits: P. Valeri *et al., Pharmacol. Res. Commun.* **18**, 1093 (1986). Topical effects in human eye: A. Reibaldi *et al., Acta Ther.* **10**, 381 (1984); C. Malpassi *et al., ibid* **12**, 55 (1986). Clinical trial in reversal of mydriasis: R. W. Allinson *et al., Ann. Ophthalmol.* **22**, 131 (1990).

mp 158-160°.

Monohydrochloride, $C_{19}H_{27}N_5.HCl$, *AF-2139, Glamidolo, Reversil, Rev-Eyes.* Crystals from absolute ethanol, mp 206-207°. LD_{50} in mice (mg/kg): 260 i.p. (Silvestrini).

THERAP CAT: Antiglaucoma; miotic.

2885. Dapsone. *4,4'-Sulfonylbisbenzeneamine; 4,4'-sulfonyldianiline;* bis(4-aminophenyl)sulfone; 4,4'-diaminodiphenyl sulfone; DDS; diaphenylsulfone; DADPS; 1358F; Avlosulfon; Croysulfone; Diphenasone; Disulone; Dumitone; Eporal; Novophone; Sulfona-Mae; Sulphadione; Udolac. $C_{12}H_{12}N_2O_2S$; mol wt 248.31. C 58.05%, H 4.87%, N 11.28%, O 12.89%, S 12.91%. Prepn: **Fr. pat. 829,926** (1938 to I. G. Farbenind.), *C.A.* **33**, 1761 (1939); Buckles, *J. Chem. Ed.* **31**, 36 (1954); Ferry *et al., Org. Syn.* **coll. vol. III**, 239 (1955). Antibacterial action and metabolism: J. Francis, A. Spinks, *Brit. J. Pharmacol.* **5**, 565 (1950). Mechanism of toxic action: Wu, DuBois, *Arch. Int. Pharmacodyn. Ther.* **183**, 36 (1970). Comprehensive description: C. E. Orzech *et al., Anal. Profiles Drug Subs.* **5**, 87-114 (1976). Review of pharmacology and therapeutic use: J. Uetrecht, *Clin. Dermatol.* **7**, 111-120 (1989).

Crystals from 95% ethanol. mp 175-176° (also a higher melting form, mp 180.5°). pKb 13.0. Soluble in alcohol, methanol, acetone, dil hydrochloric acid. Practically insol in water.

USE: Hardening agent in the curing of epoxy resins.

THERAP CAT: Antibacterial (leprostatic); in treatment of dermatitis herpetiformis.

THERAP CAT (VET): Antibacterial; coccidostat.

2886. Darvan®. Vinylidene cyanide-vinyl acetate copolymer; Darlan; Travis. *Methylenemalononitrile copolymer with vinyl acetate* contg about equal parts of each component. Polymerization of vinylidene cyanide with vinyl acetate: Gilbert *et al., J. Am. Chem. Soc.* **78**, 1669 (1956); Gilbert, Miller, **U.S. pat. 2,615,866** and Sayre, **U.S. pat. 2,975,-158** (1952, 1961, both to B. F. Goodrich). *Review:* R. W. Moncrieff, *Man-Made Fibres* (John Wiley, New York, 4th ed., 1963) pp 488-494. Approximate structure:

A wool-like fiber, sp gr 1.18. Softens at about 175°. Flammability similar to cotton and viscose. Insol in acetone, methylene chloride. Good resistance to sunlight, clothes-moth and carpet-beetle attack, and mildew; moderate resistance to 10% H_2SO_4 at 100° for hrs, and to 0.5% caustic soda at 43° for 7 days. Shrinks about 1% when immersed in boiling water for 3 minutes. Difficult to dye. Relatively resistant to pilling.

USE: Knitted garments, pile coats, and in blends for tropical suitings and men's shirtings.

2887. Datiscetin. *3,5,7-Trihydroxy-2-(2-hydroxyphenyl)-4H-1-benzopyran-4-one; 2',3,5,7-tetrahydroxyflavone;* 2'-hydroxycrysidenolon 1493. $C_{15}H_{10}O_6$; mol wt 286.24. C 62.94%, H 3.52%, O 33.54%. Cleavage product of datiscin: Stenhouse, *Ann.* **98**, 166 (1856). From roots of *Datisca cannabina* L., *Datiscaceae:* Marchlewski, *Biochem. Z.* **3**, 286 (1907). Structure: Leskiewicz, Marchlewski, *Ber.* **47**, 1599 (1914). Synthesis: Kalf, Robinson, *J. Chem. Soc.* **127**, 1968 (1925); Simpson, Wally, *ibid.* **1955**, 166.

Pale yellow needles from alc, mp 271°. uv max (96% alc): 264.0, 375.0 nm (log ε 4.265, 4.005). Sol in alcohol, ether, other organic solvents; practically insol in water. Sol in alkaline solns with yellow color.

2888. Datura. Dried leaves and dried seeds of *Datura fastuosa* L., var. *alba* Nees (also leaves from *D. metel* L.), *Solanaceae. Habit.* India. *Constit.* About 0.5% hyoscine, with traces of hyoscyamine and atropine; the seeds contain also resin and fixed oil. Comprehensive monograph: A. F. Blakeslee, *The Genus Datura* (Ronald Press, New York, 1959) 289 pp.

THERAP CAT: Anticholinergic.

2889. Daucol. *Octahydro-6,8a-dimethyl-3-(1-methylethyl)-1H-3a,6-epoxyazulen-7-ol; 2,3,4,5,6,7a,8,8a-octahydro-3-isopropyl-6,8a-dimethyl-1H-3a,6-epoxyazulen-7-ol.* $C_{15}H_{26}O_2$; mol wt 238.37. C 75.58%, H 10.99%, O 13.42%. From oil of carrot seeds, *Daucus carota* L., *Umbelliferae:* Richter, *Arch. Pharm.* **247**, 391 (1909). Structure: Sykora *et al.,* *Coll. Czech. Chem. Commun.* **26**, 788 (1961); Zalkow *et al.,* *J. Org. Chem.* **26**, 981 (1961). Stereochemistry: Levisalles, Rudler, *Bull. Soc. Chim. France* **1964**, 2020; **1967**, 2059. Synthesis of (−)-form: DeBroissia *et al., Chem. Commun.* **1972**, 855; *eidem, Bull. Soc. Chim. France* **1972**, 4314.

(−)-Daucol

Crystals (from petr ether at −30°), mp 113-115°. bp$_2$ 124-132°. $[\alpha]_D^{20}$ −16.9° (c = 2.76 in ethanol).

2890. Daunorubicin. *(8S-cis)-8-Acetyl-10-[(3-amino-2,3,6-trideoxy-α-L-lyxo-hexopyranosyl)oxy]-7,8,9,10-tetrahydro-6,8,11-trihydroxy-1-methoxy-5,12-naphthacenedione;* daunomycin; leukaemomycin C; rubidomycin; RP-13057; Ceru-

bidin. $C_{27}H_{29}NO_{10}$; mol wt 527.53. C 61.47%, H 5.54%, N 2.66%, O 30.33%. Anthracycline antibiotic related to the rhodomycins, *q.v.* Isolated from fermentation broths of *Streptomyces peucetius:* G. Cassinelli, P. Orezzi, *Giorn. Microbiol.* **11**, 167 (1963), *C.A.* **62**, 9482b (1965); A. Di Marco *et al., Nature* **201**, 706 (1964); *eidem,* **Belg.** pat. **639,-897;** *eidem,* **U.S.** pat. **4,012,284** (1964, 1977 both to Soc. Farmaceut. Italia); S. Pinnert *et al.,* **U.S.** pat. **3,997,662** (1976 to Rhone-Poulenc). Daunorubicin is a glycoside formed by a tetracyclic aglycone, *daunomycinone,* ($C_{21}H_{18}$-O_8) and an amino sugar, *daunosamine,* ($C_6H_{13}NO_3$), 3-amino-2,3,6-trideoxy-L-*lyxo*-hexose: F. Arcamone *et al., J. Am. Chem. Soc.* **86**, 5334, 5335 (1964); R. H. Iwamoto *et al., Tetrahedron Letters* **1968**, 3891. Absolute stereochemistry: F. Arcamone *et al., Gazz. Chim. Ital.* **100**, 949-989 (1970). Identity with rubidomycin: G. L. Tong *et al., J. Pharm. Sci.* **56**, 1691 (1967). Synthesis of daunosamine: J. P. Marsh *et al., Chem. Commun.* **1967**, 973; T. Yamaguchi, M. Kojimo, *Carbohyd. Res.* **59**, 343 (1977); P. M. Wovkulich, M. R. Uskokovic, *J. Am. Chem. Soc.* **103**, 3956 (1981); of daunomycinone: C. M. Wong *et al., Can. J. Chem.* **51**, 466 (1973); J. S. Swenton, P. W. Reynolds, *J. Am. Chem. Soc.* **100**, 6188 (1978); K. Krohn, K. Tolkiehn, *Ber.* **112**, 3453 (1979); F. M. Hauser, S. Prasanna, *J. Am. Chem. Soc.* **103**, 6378 (1981). Total synthesis of daunorubicin: E. M. Acton *et al., J. Med. Chem.* **17**, 659 (1974). Purification: E. Oppici *et al.,* **Belg.** pat. **898,506;** *eidem,* **Brit.** pat. **Appl. 2,133,005** (both 1984 to Farmitalia). Toxicity data: A. Di Marco *et al., Cancer Chemother. Rep.* (part 1) **53**, 33 (1969). Review of properties, biosynthesis, fermentation: R. J. White, R. M. Stroshane, *Drugs Pharm. Sci.* **22**, 569-594 (1984); of carcinogenic action in laboratory animals: *IARC Monographs* **10**, 145-152 (1976); of toxicology: R. J. Maral *et al., Cancer Treat. Rep.* **65**, Suppl. 4, 9-18 (1981); of use in treatment of solid tumors: R. B. Weiss *et al., ibid.* 25-28; of interactions with nucleic acids: S. Neidle, M. R. Sanderson, in *Molecular Aspects of Anti-cancer Drug Action,* S. Neidle, M. J. Waring, Eds. (Verlag-Chemie, Florida, 1983) pp 35-55; of mechanism of cytotoxicity: H. S. Schwartz, *ibid.* pp 93-125; of metabolism and clinical pharmacokinetics: C. E. Riggs, Jr., *Sem. Oncol.* **11**, Suppl. 3, 2-11 (1984). *Review:* A. DiMarco *et al., Antibiotics* vol. 3, J. W. Corcoran, F. E. Hahn, Eds. (Springer Verlag, New York, 1975) pp 101-128.

Daunomycinone

Daunosamine

mp 208-209°. LD$_{50}$ in mice, rats (mg/kg): 20, 13 i.v.; 5, 8 i.p. (DiMarco, 1977).

Hydrochloride, $C_{27}H_{29}NO_{10}$·HCl, *Cérubidine, Daunoblastina, Ondena.* Thin red needles, dec 188-190°. $[\alpha]_D^{20}$ +248 ± 5° (c = 0.05-0.1 in methanol). Sol in water, methanol, aq alcohols. Practically insol in chloroform, ether, benzene. Color of aq soln changes from pink at acid pH to blue at alkaline pH. Absorption max (methanol): 234, 252, 290, 480, 495, and 532 nm (E$_{1cm}^{1\%}$ 665, 462, 153, 214, 218, and 112). LD$_{50}$ in mice (mg/kg): 26 i.v. (DiMarco, 1969).

THERAP CAT: Antineoplastic.

2891. Dauricine. *[R-(R*,R*)]-4-[(1,2,3,4-Tetrahydro-6,7-dimethoxy-2-methyl-1-isoquinolinyl)methyl]-2-[4-[(1,2,3,4-tetrahydro-6,7-dimethoxy-2-methyl-1-isoquinolinyl)methyl]phenoxy]phenol.* $C_{38}H_{44}N_2O_6$; mol wt 624.78. C 73.05%, H 7.10%, N 4.48%, O 15.36%. From *Menispermum dauricum* DC., and *M. canadense* L., *Menispermaceae:* Kondo, Narita, *J. Pharm. Soc. Japan* no. 542, 279 (1927), *C.A.* **21**, 2700 (1927); Manske, *Can. J. Res.* **21B**, 17 (1943). Structure: Kondo *et al., Ber.* **68**, 519 (1935); Kondo, Tomita, *Arch. Pharm.* **274**, 65 (1936); Inubushi, Niwa, *J. Pharm. Soc.*

Japan **72**, 762 (1952); *C.A.* **47**, 6430e (1953). Total synthesis: Kametani, Fukumoto, *J. Chem. Soc.* **1965**, Suppl. 2, 6141.

Slightly yellow amorphous base, mp 115°. $[\alpha]_D^{11}$ −139° in methanol. Sol in alc, acetone, benzene; slightly sol in ether. Dimethiodide, $C_{38}H_{44}N_2O_6.2CH_3I$, needles, mp 204°. $[\alpha]_D^{20}$ −114°.

2892. Dazomet. *Tetrahydro-3,5-dimethyl-2H-1,3,5-thiadiazine-2-thione;* 2-thio-3,5-dimethyltetrahydro-1,3,5-thiadiazine; 3,5-dimethyl-2-thionotetrahydro-1,3,5-thiadiazine; dimethylformocarbothialdine; DMTT; Crag 974; Mylone. $C_5H_{10}N_2S_2$; mol wt 162.28. C 37.01%, H 6.21%, N 17.26%, S 39.52%. Prepd by the action of carbon disulfide on trimethyltrimethylenetriimine: Delépine, *Bull. Soc. Chim.* **15**, 891 (1897); from formaldehyde and methylammonium methyldithiocarbamate: Bodendorf, *J. Prakt. Chem.* **126**, 233 (1930); Ainley *et al., J. Chem. Soc.* **1944**, 147. Prepd commercially by the reaction of carbon disulfide, methylamine and caustic soda: *Chem. Week* **79**, no. 18, 82-83 (1956).

Needles from benzene, mp 106-107°. Soluble in alc with decompn. Dec by water, dil acids. uv max (cyclohexane): 242, 289 nm (ϵ 7150, 9900). LD_{50} orally in rats: 363 mg/kg, *Toxic Substances List*, H. E. Christensen, Ed. (1974) p 757.
USE: Soil fungicide, nematocide, weed killer.

2893. 2,4-DB. *4-(2,4-Dichlorophenoxy)butanoic acid;* 4-(2,4-dichlorophenoxy)butyric acid; Butyrac; Legumex D. $C_{10}H_{10}Cl_2O_3$; mol wt 249.09. C 48.22%, H 4.05%, Cl 28.47%, O 19.27%. Selective post-emergence, translocated herbicide. First described as a plant growth regulator: M. E. Synerholm and P. W. Zimmerman, *Contrib. Boyce Thompson Inst.* **14**, 369 (1947). Prepn: G. W. Kitchingman, A. C. Tucker, **Brit.** pats. **804,565** and **883,255** (1958, 1961, both to ICI). Similar in activity to 2,4-D, *q.v.,* but more selective. Mechanism of selectivity studies: D. L. King, D. E. Bayer, *Proc. West Soc. Weed Sci.* **25**, 37 (1972).

White crystals, mp 117-119°. Soly in water at 25°: 46 ppm. Sol in acetone, ethanol, diethyl ether. Slightly sol in benzene, toluene, kerosene.
Sodium salt, $C_{10}H_9Cl_2NaO_3$, *Embutox.* Water soluble; hard water will show precipitate as the calcium and magnesium salts.
Dimethylamine salt, $C_{12}H_{17}Cl_2NO_3$, *Butoxone.* Water sol; calcium and magnesium salts precipitate in hard water.
USE: Herbicide.

2894. DBHBT. *Mercaptoacetic acid [3,5-bis(1,1-dimethylethyl)-4-hydroxyphenyl]methyl ester;* 3,5-di-*tert*-butyl-4-hydroxybenzylthioglycollate. $C_{17}H_{26}O_3S$; mol wt 310.46.

C 65.77%, H 8.44%, O 15.46%, S 10.33%. Thermal stabilizer. Prepn: K. S. Cottman, U.S. pat. **4,020,042** (1977 to Goodyear); and stabilization of acrylonitrile-butadiene-styrene (ABS) terpolymer: E. G. Kolawole, J. B. Adeniyi, *Eur. Polym. J.* **18**, 469 (1982). Stabilization of polystyrene and poly(methyl methacylate): E. G. Kolawole, *J. Appl. Polym. Sci.* **27**, 3437 (1982); of Nigerian natural rubber: G. C. Ebi, E. G. Kolawole, *ibid.* **45**, 2239 (1992).

White crystals from benzene, mp 107.5-108.5° (Cottman); also reported as mp 110° (Kolawole, Adeniyi).
USE: Antioxidant.

2895. DBMC. *2,4-Bis(1,1-dimethylethyl)-5-methylphenol; 4,6-di-tert-butyl-m-cresol;* 3-methyl-4,6-di-*tert*-butylphenol. $C_{15}H_{24}O$; mol wt 220.35. C 81.76%, H 10.98%, O 7.26%. Prepn: Weinrich, *Ind. Eng. Chem.* **35**, 264 (1943).

Crystals, mp 62.1°. bp 282°. bp_{100} 211°, bp_{20} 167°. d_4^{80} 0.912. Practically insol in water, ethylene glycol; sol in alcohol, benzene, carbon tetrachloride, ether, acetone.
USE: Intermediate in the production of rubber chemicals modified phenolic resins, synthetic musks of the ambrette type.

2896. DCPA. *2,3,5,6-Tetrachloro-1,4-benzenedicarboxylic acid dimethyl ester; 2,3,5,6-tetrachloroterephthalic acid dimethyl ester;* dimethyl 2,3,5,6-tetrachloroterephthalate; chlorthal-methyl; Dacthal; Rid. $C_{10}H_6Cl_4O_4$; mol wt 331.97. C 36.18%, H 1.82%, Cl 42.72%, O 19.28%. Prepn: Lindemann, U.S. pat. **2,923,634** (1960 to Diamond Alkali). Toxicity data: G. W. Bailey, J. L. White, *Residue Rev.* **10**, 97 (1965).

Crystals from methanol, mp 155-156°. Solubility: <5% in water; >5% in acetone, cyclohexanone, xylene. LD_{50} orally in rats: >3000 mg/kg (Bailey, White).
USE: Pre-emergent herbicide.

2897. DDD (Analytical). *6,6'-Dithiobis-2-naphthalenol; 6,6'-dithiodi-2-naphthol;* bis(6-hydroxy-2-naphthyl) disulfide; 2,2'-dihydroxy-6,6'-dinaphthyl disulfide; 6,6'-dithiobis(2-naphthol). $C_{20}H_{14}O_2S_2$; mol wt 350.46. C 68.54%, H 4.03%, O 9.13%, S 18.30%. Prepn: Zincke, Dereser, *Ber.* **51**, 352 (1918).

Leaflets, mp 221-222°.
Note: Not to be confused with the insecticide, *p,p'*-DDD [1,1-dichloro-2,2-bis(*p*-chlorophenyl)ethane].
USE: For determination of protein-bound sulfhydryl groups.

2898. DDT. *1,1'-(2,2,2-Trichloroethylidene)bis[4-chlorobenzene]; 1,1,1-trichloro-2,2-bis(p-chlorophenyl)ethane;* α,-α-bis(*p*-chlorophenyl)-β,β,β-trichlorethane; dichlorodiphenyltrichloroethane; chlorophenothane; clofenotane; dicophane; pentachlorin; *p,p'*-DDT; Agritan; Gesapon; Gesarex; Gesarol; Guesapon; Neocid. $C_{14}H_9Cl_5$; mol wt 354.49. C 47.44%, H 2.56%, Cl 50.01%. Polychlorinated nondegradable pesticide. Prepn: Zeidler, *Ber.* **7**, 1180 (1874); P. Müller, **U.S. pat. 2,329,074** (1944 to Geigy); Rueggeberg, Torrans, *Ind. Eng. Chem.* **38**, 211 (1946); Cook *et al., ibid.* **39**, 868, 1683 (1947). Convenient lab procedures: Bailes, *J. Chem. Ed.* **22**, 122 (1945); Ginsburg, *Science* **108**, 339 (1948). Large scale production: Mosher *et al., Ind. Eng. Chem.* **38**, 916 (1946). Chemical composition of technical grade: H. L Haller *et al., J. Am. Chem. Soc.* **67**, 1591 (1945). Activity: P. Müller, *Helv Chim. Acta* **29**, 1560 (1946). Comprehensive monograph (in English and German): *DDT,* P. Müller, Ed., 3 vols (Birkhäuser Verlag, Basel and Stuttgart, 1955). Toxicity data: Gaines, *Toxicol. Appl. Pharmacol.* **2**, 88 (1960); **14**, 515 (1969). Review of toxicology and human exposure: *Toxicological Profile for 4,4'-DDT, 4,4'-DDE, 4,4'-DDD* (PB95-100137, 1994) 192 p.

Biaxial elongated tablets, needles from 95% alc. mp 108.5-109°. uv max (95% alc): 236 nm. Vapor pressure at 20° = 1.5×10^{-7} mm Hg. Practically insol in water, dil acids, alkalies. Soly (g/100 ml): acetone 58; benzene 78; benzyl benzoate 42; carbon tetrachloride 45; chlorobenzene 74; cyclohexanone 116; 95% alc 2; ethyl ether 28; gasoline 10; isopropanol 3; kerosene 8-10; morpholine 75; peanut oil 11; pine oil 10-16; tetralin 61; tributyl phosphate 50; freely sol in pyridine, dioxane. The soly in organic solvents increases sharply with a rise in temp. Resistant to destruction by light and oxidation. Its unusual stability has resulted in difficulties in residue removal from water, soil and foodstuffs. Should not be kept in iron containers and should not be mixed with iron and aluminum salts nor with alkaline substances. High storage temps should also be avoided. Setting point of technical grade: 88.6-91.4°. LD_{50} in male, female rats (mg/kg): 113, 118 orally (Gaines, 1960).

Caution: Poisoning may occur by ingestion or by absorption through skin or respiratory tract. Potential symptoms of overexposure are numbness, paresthesias, malaise, headache, sore throat, fatigue, weakness, coarse tremors, convulsions and coma; death due to respiratory failure. *See Clinical Toxicology of Commercial Products,* R. E. Gosselin *et al.,* Eds. (Williams & Wilkins, Baltimore, 5th ed., 1984) Section III, pp 134-138. This substance may reasonably be anticipated to be a carcinogen: *Seventh Annual Report on Carcinogens* (PB95-109781, 1994) p 139.

USE: Contact insecticide.

THERAP CAT: Ectoparasiticide; pediculicide.

2899. Deaminooxytocin. *1-(3-Mercaptopropanoic acid)oxytocin;* demoxytocin; desaminooxytocin; ODA-914; Sandopart. $C_{43}H_{65}N_{11}O_{12}S_2$; mol wt 992.19. C 52.05%, H 6.60%, N 15.53%, O 19.35%, S 6.46%. Highly potent analog of the posterior pituitary hormone oxytocin *q.v.,* differing in structure at the 1 position where the free amino group in the half-cystine residue is replaced by hydrogen. Synthesis and biological activity: du Vigneaud *et al., J. Biol. Chem.* **235**, P.C. 64 (1960); Hope *et al., ibid.* **237**, 1563 (1962); Takashima *et al., J. Am. Chem. Soc.* **90**, 1323 (1968). Isoln and purifn: Ferrier *et al., J. Biol. Chem.* **240**, 4264 (1965). Structure-activity correlation: Jarvis, du Vigneaud, *Science* **143**, 545 (1964). Synthesis of D-isomer: G. Flouret, V. du Vigneaud, *J. Med. Chem.* **14**, 556 (1971). Crystal structure analysis: S. P. Wood *et al., Science* **232**, 633 (1986). Pharmacology and comparison with oxytocin: Chan, du Vigneaud, *Endocrinology* **71**, 977 (1962). *Review: Advan. Exp. Med. Biol.* **vol. 2**, 53-104 (1968).

L-Isomer, white plate-like crystals from water, mp 179° (Ferrier); 182-183° (Takashima). For amorphous powder; $[\alpha]_D^{20}$ −88.3° (Ferrier); $[\alpha]_D^{21}$ −107° (Hope); $[\alpha]_D^{25}$ −95.1° (Takashima); all (c = 0.5 in 1N acetic acid). Anhydrous form is very hygroscopic.

D-Isomer, white, fluffy powder, $[\alpha]_D^{20}$ +104° (c = 0.5 in 1N acetic acid).

THERAP CAT: Oxytocic.

2900. Deanol. *2-(Dimethylamino)ethanol;* β-dimethylaminoethyl alcohol; *N,N*-dimethyl-2-hydroxyethylamine; $C_4H_{11}NO$; mol wt 89.14. C 53.90%, H 12.44%, N 15.71%, O 17.95%. $(CH_3)_2NCH_2CH_2OH$. Prepd from equimolar amounts of ethylene oxide and dimethylamine: Knorr, *Ber.* **37**, 3508 (1904); Hanhart, Ingold, *J. Chem. Soc.* **1927**, 1012. Clinical trial in hyperkinetic children: N. Coleman *et al., Psychosomatics* **17**, 68 (1976); in involuntary movement disorders: D. E. Casey, *Dis. Nerv. Syst.* **38**, 7 (1977).

Liquid. d_4^{20} 0.8866. bp_{758} 135°. n_D^{20} 1.43. Miscible with water, alcohol, ether.

Aceglumate, $C_4H_{11}NO.C_7H_{11}NO_5$, *Clérégil, Risatarun.* Prepn: **Fr. pat. M2487** (1964 to Interco Fribourg), *C.A.* **61**, 12085c (1964). Sol in water.

Acetamidobenzoate, $C_4H_{11}NO.C_9H_9NO_3$, *Deaner, Pabenol.* Prepn: **Brit. pat. 879,259** (1957 to Riker Labs); A. Lasslo *et al., J. Am. Pharm. Assoc. Sci. Ed.* **48**, 345 (1959). Crystals from absolute ethanol + ethyl acetate, mp 159-161.5°. Sol in water.

Bitartrate, $C_4H_{11}NO.C_4H_6O_6$, *Liparon.* Crystals, sol in water.

Hemisuccinate, $C_4H_{11}NO.C_4H_6O_4$, *Tonibral, Rischiaril.* White, hygroscopic crystalline powder. Very soluble in water, ethanol; sparingly sol in chloroform, benzene.

THERAP CAT: CNS stimulant.

2901. Debrisoquin. *3,4-Dihydro-2(1H)-isoquinolinecarboximidamide; 3,4-dihydro-2(1H)-isoquinolinecarboxamidine;* 2-amidino-1,2,3,4-tetrahydroisoquinoline; isocaramidine. $C_{10}H_{13}N_3$; mol wt 175.23. C 68.54%, H 7.48%, N 23.98%. Prepn of derivs: W. Wenner, **Belg. pat. 629,007;** *idem,* **U.S. pat. 3,157,573** (1963, 1964 both to Hoffmann-La Roche); *idem, J. Med. Chem.* **8**, 125 (1965). Pharmacology: Moe *et al., Curr. Ther. Res. Clin. Exp.* **6**, 299 (1964); Abrams *et al., J. New Drugs* **4**, 268 (1964). Possible mechanism of action: Medina *et al., Biochem. Pharmacol.* **18**, 891 (1969); Pocelinko *et al., Int. Z. Klin. Pharmakol. Ther. Toxicol.* **2**, 13 (1969). Toxicity study: E. I. Goldenthal, *Toxicol. Appl. Pharmacol.* **18**, 185 (1971).

Sulfate, $2(C_{10}H_{13}N_3).H_2SO_4$, *Ro-5-3307/1, Declinax, Tendor.* Crystals, mp 278-280°, 284-285° or 266-268° (H_2O), depending on humidity. Freely sol in water. LD_{50} in neonate, adult rats (mg/kg): 88±18, 1580±163 orally (Goldenthal).

THERAP CAT: Antihypertensive.

2902. Decaborane(14). Decaboron tetradecahydride. $B_{10}H_{14}$; mol wt 122.22. B 88.45%, H 11.55%. Prepd by the pyrolysis of diborane(6). Reviews of chemistry: Griffo, *Diss. Abstr.* **22**, 2976 (1962); Stanko *et al., Usp. Khim.* **34**(6), 1011-1039 (1965). Review of structure and properties: Campbell, Jr., in *Progress in Boron Chemistry,* Steinberg, McCloskey, Eds., **vol. 1** (Macmillan, New York, 1964) pp 173-188. Reviews of toxicity: Levinskas, "Toxicology of Boron Compounds" in *Boron, Metallo-Boron Compounds and Boranes,* R. M. Adams, Ed. (Interscience, New York, 1964) pp 693-737; E. Browning, *Toxicity of Industrial Metals* (Appleton-Century-Crofts, New York, 2nd ed.,

1969) pp 92-97. General reviews: Stock, *Hydrides of Boron and Silicon* (Cornell Univ. Press, Ithaca, 1933) *passim;* Siegel, Mack, *J. Chem. Ed.* **34,** 314-317 (1957); Major, *Chem. Eng. Progr.* **54**(3), 49-54 (1958); Lipscomb, *Boron Hydrides* (Benjamin, New York, 1963) *passim;* Adams in *Boron, Metallo-Boron Compounds and Boranes, loc. cit.,* pp 647-653; Hawthorn in *Advan. Inorg. Chem. Radiochem.* **5,** 307-345 (1963); Greenwood in *Comprehensive Inorganic Chemistry,* vol. 1, J. C. Bailar, Jr. *et al.,* Eds. (Pergamon Press, Oxford, 1973) pp 818-837.

Orthorhombic crystals. mp 99.6-99.7°; bp 213°; bp_{19} 100°. d_4^{25} 0.94; d_4^{100} (liq) 0.78. Stable indefinitely at room temp; decomp slowly into B + H_2 at 300°. Heat of fusion: 7.8 kcal/mole; of sublimation: 18.33 kcal/mole; of vaporization: 11.6 kcal/mole. Slightly sol in cold water; hydrolyzes in hot water. Sol in ethyl acetate, 1-bromopropane, ethyl silicate, carbon disulfide, benzene, alcohol, acetic anhydride, acetic acid, ethyl borate, carbon tetrachloride. *Highly reactive.* Reacts with amides, acetone, butyraldehyde, acetonitrile at room temp. Decaborane mixtures with carbon tetrachloride are dangerously shock sensitive.

Caution: Potential symptoms of overexposure are dizziness, headache, nausea, lightheadedness and drowsiness; incoordination, local muscle spasms, tremor and convulsions; fatigue. *See NIOSH Pocket Guide to Chemical Hazards* (DHHS/NIOSH 90-117, 1990) p 80. See also *Patty's Industrial Hygiene and Toxicology* vol. **2B,** G. D. Clayton, F. E. Clayton, Eds. (Wiley-Interscience, New York, 3rd ed., 1981) 2987-2990.

USE: In rocket propellants; as catalyst in olefin polymerization.

2903. Decalin®. *Decahydronaphthalene;* perhydronaphthalene; bicyclo[4.4.0]decane; naphthalane; naphthane; Dec; DeKalin. $C_{10}H_{18}$; mol wt 138.25. C 86.88%, H 13.12%. Occurs in two forms: *cis* and *trans.* Hydrogenation of naphthalene in glacial acetic acid in the presence of platinum catalyst at 25° and 130 atm yields a mixture of 77% *cis*-Decalin and 23% *trans*-Decalin. Hydrogenation of Tetralin under the same conditions yields almost only *cis*-Decalin: Baker, Schuetz, *J.Am. Chem. Soc.* **69,** 1250 (1947). Hydrogenation of $\Delta^{4a(8a)}$-octalin in ethanol in the presence of platinum yields *trans*-Decalin as the main component: Linstead *et al., J. Chem. Soc.* **1937,** 1136. Separation of *trans*- and *cis*-forms: Seyer, Walker, *J. Am. Chem. Soc.* **60,** 2125 (1938). Prepn of *trans*-Decalin from *cis*-Decalin: Zelinsky, *Ber.* **65,** 1299 (1932). Conformation studies: Barton, *J. Chem. Soc.* **1948,** 340; Moniz, Dixon, *J. Am. Chem. Soc.* **83,** 1671 (1961).

trans   cis

Liquid. Slight odor resembling menthol. Pure Decalin does not smell of naphthalene. Volatile with steam. The commercial product may be practically all *trans*-Decalin, or a mixture contg up to 60% *cis*-Decalin. The commercial mixture has a flash pt (closed cup) of about 136°F (58°C). Autoignition temp 504°F. Insol in water. Very sol in alcohol, methanol, ether, chloroform. Miscible with propyl and isopropyl alcohol; miscible with most ketones and esters. LD_{50} orally in rats: 4.2 g/kg. Lethal concn for rats in air: 500 ppm, Smyth *et al., Arch. Ind. Hyg. Occup. Med.* **4,** 119 (1951).

cis-Form, mp −43.26°. bp 195.7°; bp_9 67.0°. d_4^{20} 0.8963. n_D^{20} 1.48113.

trans-Form, mp −30.4°. bp 187.25°; bp_9 62.0°. d_4^{20} 0.8700. n_D^{20} 1.46968.

USE: Solvent for naphthalene, fats, resins, oils, waxes; used instead of turpentine in lacquers, shoe polishes, floor waxes. In motor fuel and lubricants. As patent fuel in stoves.

2904. Decamethonium Bromide. *N,N,N,N′,N′,N′-Hexamethyl-1,10-decanediaminium dibromide; decamethylenebis[trimethylammonium bromide];* C-10; Syncurine. C_{16}-

$H_{38}Br_2N_2$; mol wt 418.30. C 45.94%, H 9.16%, Br 38.20%, N 6.70%. $[(CH_3)_3N^+(CH_2)_{10}N^+(CH_3)_3].2Br^-$. Prepn: Blomquist *et al., J. Am. Chem. Soc.* **81,** 678 (1959).

Crystals from methanol + acetone, dec 268-270°. Freely sol in water, alcohol; very slightly sol in chloroform. Practically insol in ether. Aq solns are stable and may be sterilized by autoclaving.

THERAP CAT: Muscle relaxant (skeletal).

2905. Decamethylcyclopentasiloxane. $C_{10}H_{30}O_5Si_5$; mol wt 370.77. C 32.39%, H 8.16%, O 21.58%, Si 37.87%. Isolated from the hydrolysis product of dimethyldichlorosilane: Patnode, Wilcock, *J. Am. Chem. Soc.* **68,** 358 (1946).

Oily liq. mp −38°. bp 210°; bp_{20} 101°. d 0.9593. n_D^{20} 1.3982.

2906. Decamethylene Glycol. *1,10-Decanediol.* C_{10}-$H_{22}O_2$; mol wt 174.28. C 68.92%, H 12.72%, O 18.36%. $HO(CH_2)_{10}OH$. Prepd by the reduction of dimethyl or diethyl sebacate with sodium and ethyl alcohol: Bouveault, Blanc, *Compt. Rend.* **137,** 329 (1903); *Bull. Soc. Chim.* [3] **31,** 1205 (1904); *Ger. pat.* 164,294; *Frdl.* **8,** 1260 (1905); Chuit, *Helv. Chim. Acta* **9,** 264 (1926); Manske, *Org. Syn.* **14,** 20 (1934). By catalytic hydrogenation of sebacic esters: Folkers, Adkins, *J. Am. Chem. Soc.* **54,** 1146 (1932).

Needles from water or dil alcohol. mp 74°. bp_{20} 192°; bp_{11} 179°; bp_8 170°. Freely sol in alcohol, warm ether; almost insol in petr ether and water.

Diethyl ether, $C_{14}H_{30}O_2$, *1,10-diethoxydecane.* d_0^0 0.850; bp 260°.

2907. Decamethyltetrasiloxane. $C_{10}H_{30}O_3Si_4$; mol wt 310.69. C 38.66%, H 9.73%, O 15.45%, Si 36.16%. Prepn: Patnode, Wilcock, *J. Am. Chem. Soc.* **68,** 358 (1946).

Liquid. bp 194°. d 0.8536. n_D^{20} 1.3895. mp ~−70°. Stable. Inert to most chemical reagents and rubber. Maintains about the same viscosity over a wide temp range. Sol in benzene and the lighter hydrocarbons; slightly sol in alc and heavy hydrocarbons.

USE: As a basis for silicone oils or fluids designed to withstand extremes of temp; as a foam suppressant in petr lubricating oil.

2908. Dechlorane® Plus. *1,2,3,4,7,8,9,10,13,13,14,14-Dodecachloro-1,4,4a,5,6,6a,7,10,10a,11,12,12a-dodecahydro-1,4:7,10-dimethanodibenzo[a,e]cyclooctene;* bis(hexachlorocyclopentadieno)cyclooctane; $C_{18}H_{12}Cl_{12}$; mol wt 653.73. C 33.07%, H 1.85%, Cl 65.08%. Polychlorinated hydrocarbon. Prepn by Diels-Alder reaction between cyclooctadiene and hexachlorocyclopentadiene: K. Ziegler, H. Froitzheim-Kühlhorn, *Ann.* **589,** 157 (1954). Use as fire-retardant for plastics: E. V. Gouinlock, Jr., U.S. pat. 3,382,204 (1965 to Hooker). Accumulation in juvenile Atlantic salmon: V. Zitko, *Chemosphere* **9,** 73 (1980). Brief review of fire retardant props: T. J. Machmer, *Life Prop. Protect., Fire Retard. Chem. Assoc., Semi-Annu. Meet.* **1977,** 156-174.

Colorless crystals, mp > 325°. Soluble in *o*-dichlorobenzene.

USE: Fire retardant for plastics.

2909. Decimemide. *4-(Decyloxy)-3,5-dimethoxybenzamide;* EGYT-1050; Denegyt. $C_{19}H_{31}NO_4$; mol wt 337.46. C 67.63%, H 9.26%, N 4.15%, O 18.96%. Prepn: **Neth. pat. Appl. 6,605,460** corresp to E. Kasztreiner *et al.,* **U.S. pat. 3,432,549** (1966, 1969 both to EGYT). Pharmacology: E. Kasztreiner, *Magy. Kém. Lapja* **27**, 370 (1972), *C.A.* **78**, 66697z (1973); I. Kiraly *et al., Acta. Physiol. Acad. Sci. Hung.* **47**, 200 (1976). Metabolism study: G. Molnar *et al., Acta Pharm. Hung.* **48**, Suppl., 58 (1978). Toxicity data: E. Usdin, D. H. Efron, Eds., *Psychotropic Drugs and Related Compounds,* (NIH, Washington, D. C., 2nd Ed., 1972) p 377.

Cryst from methanol, mp 121-122°. LD_{50} in mice, rats (mg/kg): 2950, 1650 orally (Udsin, Efron).

THERAP CAT: Anticonvulsant.

2910. Decoquinate. *6-Decyloxy-7-ethoxy-4-hydroxy-3-quinolinecarboxylic acid ethyl ester;* ethyl 6-(*n*-decyloxy)-7-ethoxy-4-hydroxyquinoline-3-carboxylate; M & B 15497; Deccox. $C_{24}H_{35}NO_5$; mol wt 417.55. C 69.04%, H 8.45%, N 3.35%, O 19.16%. Description of compd: Ball *et al., Chem. & Ind. (London)* **1968**, 56; **Belg. pat. 698,305** (1966 to May & Baker). Metabolism and residue studies: Craine *et al., J. Agr. Food Chem.* **19**, 1228 (1971); Kouba *et al., ibid.* 1234.

THERAP CAT (VET): Coccidiostat.

2911. *n*-Decyl Alcohol. *1-Decanol;* nonylcarbinol. $C_{10}H_{22}O$; mol wt 158.28. C 75.88%, H 14.01%, O 10.11%. $CH_3(CH_2)_8CH_2OH$. Prepn from caprinaldehyde: Krafft, *Ber.* **16**, 1717 (1883); from capric acid methyl ester: Bouveault, Blanc, *Bull. Soc. Chim.* [3] **31**, 674 (1904); from 1-chlorodecane: Schultz, *Ber.* **42**, 3611 (1909); from nonylmagnesium bromide and formaldehyde: Yohe, Adams, *J. Am. Chem. Soc.* **50**, 1507 (1928). Manuf: Faith, Keyes & Clark's *Industrial Chemicals,* F. A. Lowenheim, M. K. Moran, Eds. (Wiley-Interscience, New York, 4th ed., 1975) pp 310-317.

Moderately viscous, strongly refractive liq. Solidifies forming rectangular plates or leaflets; mp 6.4°. bp_{760} 232.9°; bp_{15} 115-120°; bp_8 109.5°. d_4^{20} 0.8297. n_D^{20} 1.43587. Insol in water. Sol in alcohol, ether.

USE: In the manuf of plasticizers, synthetic lubricants, petroleum additives, herbicides, surface active agents, solvents. Has moderate antifoaming capacity.

2912. Deet. *N,N-Diethyl-3-methylbenzamide; N,N-diethyl-m-toluamide;* M-Det; *m*-DETA; ENT-20218; Autan; *m*-Delphene; Detamide; Dieltamid; Flypel; Metadelphene;

Off; Repel. $C_{12}H_{17}NO$; mol wt 191.27. C 75.35%, H 8.96%, N 7.32%, O 8.36%. Prepn from *m*-toluoyl chloride and diethylamine in benzene or ether: Maxim, *Bull. Soc. Chim. Romania* **11**, 29 (1929); *Chem. Zentr.* **1929**, II, 2324, *C.A.* **24**, 94 (1930); *Beilstein* **9**, II, 325; McCabe *et al., J. Org. Chem.* **19**, 493, 496 (1954). Activity: I. H. Gilbert *et al., J. Econ. Entomol.* **48**, 741 (1955). Toxicity study: Ambrose *et al., Toxicol. Appl. Pharmacol.* **1**, 97 (1959). Metabolism: L. Blomquist, W. Thorsell, *Acta Pharmacol. Toxicol.* **41**, 235 (1977). HPLC determn including metabolites: J. M. Yeung, W. G. Taylor, *Drug Metab. Dispos.* **16**, 600 (1988). Review of toxicity: J. R. Clem *et al., Ann. Phamacother.* **27**, 289-293 (1993).

Liquid, d_4^{20} 0.996. bp_{19} 160°; $bp_{1.0}$ 111°. n_D^{25} 1.5206. Practically insol in water. Freely sol in alcohol, ether, benzene. Sparingly sol in petr ether. LD_{50} orally in rats: ~2 g/kg (Ambrose).

Caution: Potential symptoms of toxicity due to acute or chronic overexposure are hypotension, bradycardia; erythema, bullous eruptions, contact urticaria, anaphylaxis; confusion, slurred speech, muscle cramping, insomnia, tremor, clonic jerking, psychosis, seizures, coma (Clcm). Direct contact may cause eye irritation: *See Clinical Toxicology of Commercial Products,* R. E. Gosselin *et al.,* Eds. (Williams & Wilkins, Baltimore, 5th ed., 1984) Section II, p 346.

USE: Insect repellent.

2913. Defensins. A class of low molecular weight cationic peptides which have *in vitro* microbiocidal activity against various bacteria, fungi, and viruses, originally isolated from mammalian neutrophils. Possess similar arginine-rich, cysteine-rich structures of 29-35 amino acids with a conserved core of 8-11 residues. Characterized by six cysteines comprising 3 disulfide linkages. Abbreviated NP (neutrophil peptide) where NP is used for rabbit, HNP for human, GPNP for guinea pig etc.; defensins represent a major protein component of granulocytes. Initial identification studies: H. I. Zeya, J. K. Spitznagel, *Science* **142**, 1085 (1961); *eidem, J. Bacteriol.* **91**, 750 (1966). Isoln, antimicrobial activities and amino acid sequence of six NPs: M. E. Selsted *et al., Infect. Immun.* **45**, 150 (1984); M. E. Selsted *et al., J. Biol. Chem.* **260**, 4579 (1985); of three HNPs: T. Ganz *et al., J. Clin. Invest.* **76**, 1427 (1985); M. E. Selsted *et al., ibid.* 1436. Crystal structure of HNP-3: C. P. Hill *et al., Science* **251**, 1481 (1991). Direct inactivation of enveloped viruses: K. A. Daher *et al., J. Virol.* **60**, 1068 (1986). *In vitro* cytolysis of tumor cells: A. Lichtenstein *et al., Blood* **68**, 1407 (1986). Mechanism of cytolysis: *eidem, J. Immunol.* **140**, 2686 (1988). Identification and isoln of insect defensins from haemolymph of dipteran: J. A. Hoffmann, D. Hoffmann, *Res. Immun.* **141**, 910 (1990). Structure determn of insect forms: P. Lepage *et al., Eur. J. Biochem.* **196**, 735 (1991). Review of insect forms: J. A. Hoffmann, C. Hetru, *Immunol. Today* **13**, 411-415 (1992). *Review:* B. L. Kagan *et al., Toxicology* **87**, 131-149 (1994); including insect forms: T. Ganz, R. I. Lehrer, *Pharmac. Ther.* **66**, 191-205 (1995).

2914. Deferoxamine. *N'-[5-[[4-[[5-(Acetylhydroxyamino)pentyl]amino]-1,4-dioxobutyl]hydroxyamino]pentyl]-N-(5-aminopentyl)-N-hydroxybutanediamide; N-[5-[3-[[5-aminopentyl]hydroxycarbamoyl]propionamido]pentyl]-3-[[5-(N-hydroxyacetamido)pentyl]carbamoyl]propionohydroxamic acid;* 1-amino-6,17-dihydroxy-7,10,18,21-tetraoxo-27-(*N*-acetylhydroxylamino)-6,11,17,22-tetraazaheptaeicosane; desferrioxamine B. $C_{25}H_{48}N_6O_8$; mol wt 560.69. C 53.55%, H 8.63%, N 14.99%, O 22.83%. Natural product forming iron complexes, isolated from *Streptomyces pilosus* Ettlinger *et al.* Prepn from ferrioxamine B and structure: H. Bickel *et al., Helv. Chim. Acta* **43**, 2129 (1960). Syntheses: V. Prelog, A. Walser, *ibid.* **45**, 631 (1962); **Belg. pat. 609,053;** E. Gaeumann *et al.,* **U.S. pat. 3,471,476** (1962, 1969 both to

Ciba). Derivatives: H. Bickel *et al., Helv. Chim. Acta* **46**, 1385 (1963). Clinical efficacy in thallasemia major (Cooley's anemia): B. Modell *et al., Brit. Med. J.* **284**, 1081 (1982). Treatment of hemodialysis-induced aluminum accumulation in brain: P. Ackril *et al, Lancet* **2**, 692 (1980); in bone: H. H. Malluche *et al., N. Engl. J. Med.* **311**, 140 (1984). Neurotoxicity study: N. F. Olivieri *et al., ibid.* **314**, 869 (1986). Review of clinical experience: R. Propper, D. Nathan, *Ann. Rev. Med.* **33**, 509-519 (1982).

Monohydrate, crystals from dil alcohol, mp 138-140°. Sol in water at 20°: 1.2%.

Hydrochloride, $C_{25}H_{48}N_6O_8 \cdot HCl$, *Ba-29837*. Crystals from slightly acidic methanol, mp 172-175°.

Methanesulfonate, $C_{25}H_{48}N_6O_8 \cdot CH_3SO_3H$, *desferrioxamine mesylate, DFOM, Desferal*. Crystals from dil alcohol, mp 148-149°. Sol in water at 20°: >20%.

N-Acetyl derivative, $C_{27}H_{50}N_6O_9$, *N-acetyldeferoxamine, N-acetyldesferrioxamine B*. Crystals from *n*-propanol, mp 180-182°.

Iron complex, $C_{25}H_{45}FeN_6O_8$, *Ferrioxamine B*. A natural microbial growth factor. Isoln: H. Bickel *et al., Helv. Chim. Acta* **43**, 2118 (1960); E. Gaeumann *et al.*, U.S. pats. **3,118,823** and **3,153,621** (both 1964 to Ciba).

Note: Reviews on iron-containing metabolites with growth stimulating properties, *sideramines*, or with antibiotic properties, *sideromycins, ferrimycins*: H. Bickel *et al., Experientia* **16**, 129-133 (1960); V. Prelog, *Pure Appl. Chem.* **6**, 327-338 (1963); F. Knüsel, J. Nüesch, *Nature* **206**, 674-676 (1965); T. F. Emery, *Advan. Enzymol. Relat. Areas Mol. Biol.* **35**, 135-185 (1971); Knüsel, Zimmermann, in *Antibiotics*, vol. 3, J. W. Corcoran, F. E. Hahn, Eds. (Springer-Verlag, New York, 1975) pp 653-667.

THERAP CAT: Parenteral chelating agent (iron and aluminum). Methanesulfonate as antidote to iron poisoning.

2915. Defibrotide. Fraction P; defibrinotide; Dasovas; Noravid; Prociclide. Sodium salt of polydeoxyribonucleotide extract from mammalian lung which stimulates fibrinolysis. mol wt 45,000-50,000 daltons. Prepn: A. Butti *et al.*, **Ger.** pat. **2,154,278**; *eidem*, U.S. pat. **3,899,481** (1972, 1975 both to Crinos). Antithrombotic activity: R. Niada *et al., Thromb. Res.* **23**, 233 (1981). Pharmacokinetics and profibrinolytic activity in rabbits: R. Pescador *et al., ibid.* **30**, 1 (1983). Clinical trial as antithrombotic: S. Coccheri *et al., Int. J. Clin. Pharm. Res.* **2**, 227 (1982). Clinical evaluation in thrombotic microangiopathy: V. Bonomini *et al., Nephron* **40**, 195 (1985). Symposium on pharmacology and clinical efficacy: O. N. Ulutin, Ed., *Haemostasis* **16**, Suppl. 1, 1-68 (1986).

THERAP CAT: Antithrombotic.

2916. Deflazacort. *(11β,16β)-21-(Acetyloxy)-11-hydroxy-2'-methyl-5'H-pregna-1,4-dieno[17,16-d]oxazole-3,20-dione; 11β,21-dihydroxy-2'-methyl-5'βH-pregna-1,4-dieno[17,16-d]oxazole-3,20-dione 2l-acetate;* pregna-1,4-diene-11β,21-diol-3,20-dione[17α,16α-d]-2'-methyloxazoline 21-acetate; oxazacort; azacort; DL-458-IT; L-5458; Calcort; Deflan; Dezacor; Flantadin; Lantadin. $C_{25}H_{31}NO_6$; mol wt 441.52. C 68.01%, H 7.08%, N 3.17%, O 21.74%. Systemic corticosteroid; oxazoline derivative of prednisolone, *q.v.* Prepn: G. Nathansohn, G. Winters, **Belg.** pat. **679,820**; *eidem*, **Brit.** pat. **1,077,393**; *eidem*, U.S. pat. **3,436,389** (1966, 1967, 1969 all to Lepetit); G. Nathansohn *et al., J. Med. Chem.* **10**, 799 (1967). Pharmacology: P. Schiatti *et al., Arzneimittel-Forsch.* **30**, 1543 (1980). Pharmacokinetics: A. Assandri *et al., Eur. J. Drug Metab. Pharmacokinet.* **5**, 207 (1980); *eidem, Adv. Exp. Med. Biol.* **171**, 9 (1984). Immunosuppressive activity in humans: B. H. Hahn *et al., J. Rheumatol.* **8**, 783 (1981). Effect on human mineral metabolism: T. J. Hahn *et al., Calcif. Tissue Int.* **31**, 109 (1980); on human glucose metabolism: P. Cavallo-Perin *et al., Eur. J. Clin. Pharmacol.* **26**, 357 (1984). Clinical comparison with prednisone, *q.v.*, in rheumatoid arthritis and lupus: B. Imbimbo *et al., Adv. Exp. Med. Biol.* **171**, 241 (1984).

Crystals from acetone-hexane, mp 255-256.5°. $[\alpha]_D$ +62.3° (c = 0.5 in chloroform). uv max (methanol): 241-242 nm ($E_{1cm}^{1\%}$ 352.5). LD_{50} orally in mice: 5200 mg/kg (Schiatti).

THERAP CAT: Anti-inflammatory; glucocorticoid.

2917. Defosfamide. *N,N-Bis(2-chloroethyl)-N'-(3-hydroxypropyl)phosphorodiamidic acid 2-chloroethyl ester;* 2-chloroethyl *N,N*-bis(2-chloroethyl)-*N'*-(3-hydroxypropyl)phosphorodiamidate; *N,N,O*-tris(2-chloroethyl)-*N'*-(3-hydroxypropyl)phosphoric acid diamide; desmofosfamide; B-612; Mitarson. $C_9H_{20}Cl_3N_2O_3P$; mol wt 341.60. C 31.64%, H 5.90%, Cl 31.14%, N 8.20%, O 14.05%, P 9.07%. Prepn: Arnold *et al,* **Ger.** pat. **1,061,780** (1959 to Asta-Werke).

Oily liquid. Sp gr 1.3675. Practically insol in water. Sol in organic solvents.

THERAP CAT: Antineoplastic.

2918. Deguelin. *(7aS-cis)-13,13a-Dihydro-9,10-dimethoxy-3,3-dimethyl-3H-bis[1]benzopyrano[3,4-b:6',5'-e]pyran-7(7aH)-one.* $C_{23}H_{22}O_6$; mol wt 394.42. C 70.04%, H 5.62%, O 24.34%. From roots of *Tephrosia toxicaria* Pers., leaves of *Tephrosia vogelii* Hook f., *Leguminosae*, and from derris and cubé roots: H. L. Haller, F. B. LaForge, *J. Am. Chem. Soc.* **56**, 2415 (1934); E. Brierly, H. J. Smith, *J. Pharm. Pharmacol.* **20**, 840 (1968). Structure: Clark, *J. Am. Chem. Soc.* **54**, 3000 (1932). Stereoselective synthesis: P. B. Anzeveno, *J. Org. Chem.* **44**,2578 (1979). Total synthesis of racemate: Fukami *et al., Agr. Biol. Chem. (Tokyo)* **25**, 252 (1961); Omokawa, Yamashita, *ibid.* **38**, 1731 (1974). Stereochemical studies: Djerassi *et al., J. Chem. Soc.* **1961**, 1448.

Yellow oil, $[\alpha]_D^{27}$ −97.2° (c = 0.2 in benzene). (±)-Form, pale green crystals, mp 171°. Practically insol in water. Sol in alc.

Caution: Irritates skin. Inhalation may cause fatal pulmonary damage.

USE: Insecticide.

2919. Dehydroacetic Acid. *3-Acetyl-6-methyl-2H-pyran-2,4(3H)-dione;* 2-acetyl-5-hydroxy-3-oxo-4-hexenoic acid δ-lactone; methylacetopyronone; DHA. $C_8H_8O_4$; mol wt 168.15. C 57.14%, H 4.80%, O 38.06%. Polymerization product of ketene: Steele *et al., J. Org. Chem.* **14**, 460 (1949). From ethyl acetoacetate: Arndt, *Org. Syn.* **coll. vol. III**, 231 (1955). Toxicity study: H. C. Spencer *et al., J. Pharmacol. Exp. Ther.* **99**, 57 (1950). Series of articles on toxicity studies, pharmacology, mechanism of action, absorption, distribution, renal action: *ibid.* 57-111.

White to cream crystalline powder, mp 109-111° (sublimes). bp 269.9°. Soly (w/w at 25°): in acetone 22%; benzene 18%; methanol 5%; carbon tetrachloride 3%; U.S.P. ethanol 3%; ether 5%; glycerol <0.1%; *n*-heptane 0.7%; olive oil 1.6%; propylene glycol 1.7%; water <0.1%. LD_{50} orally in rats: 1000 mg/kg (Spencer).

Sodium salt hydrate, $C_8H_7NaO_4.H_2O$, *DHA-S*. Tasteless white powder. Soly (w/w at 25°): in water 33%; propylene glycol 48%; olive oil <0.1%; methanol 14%; U.S.P. ethanol 1%; *n*-heptane <0.1%; glycerol 15%; ether <0.1%; carbon tetrachloride <0.1%; benzene <0.1%; acetone 0.2%. LD_{50} orally in rats: 570 mg/kg (Spencer).

USE: In organic syntheses; as plasticizer, compatible with nitrocellulose, polystyrene, methacrylate, vinylite resins; as fungicide and bactericide; in antienzyme toothpastes; to reduce pickle bloating.

2920. Dehydroascorbic Acid. L-*threo*-2,3-Hexodiulosonic acid γ-lactone. $C_6H_6O_6$; mol wt 174.11. C 41.39%, H 3.47%, O 55.14%. The reversibly oxidized form of ascorbic acid. Prepd by the action of benzoquinone on ascorbic acid: Ohle, Erlbach, *Ber.* **67**, 555 (1934); Moll, Wieters, *E. Merck's Jahresber.* **50**, 65 (1936); by the action of iodine: Herbert *et al., J. Chem. Soc.* **1933**, 1270; Kenyon, Munro, *ibid.* **1948**, 158; by oxidn with *peri*-naphthindan-2,3,4-trione hydrate: Moubasher, *J. Biol. Chem.* **176**, 533 (1948); *see also* Müller-Mulot, *Z. Physiol. Chem.* **351**, 52 (1970). Isomerization and formn of derivs: Egge, *Tetrahedron Letters* **1969**, 801. Structure studies: Teichmann, Ziebarth, *Z. Prakt. Chem.* **33**, 124 (1966). Toxicity studies: Gaudiano *et al., Boll. Soc. Ital. Biol. Sper.* **43**, 674 (1967).

Fine needles, dec 225°. Sol in water at 60°. In soln the two carbonyl groups (in position 2 and 3) assume the hydrated form $-C(OH)_2-C(OH)_2-$. Practically neutral reaction. pKa: 3.90. $[\alpha]_D^{20}$ +56°. Aq solns are much less stable than those of ascorbic acid. Detailed stability data: Bogdanski, Bogdanska, *Bull. Acad. Polonaise Sci.* (II° classe) **3**, 41 (1955). *See also* Velisek *et al., Coll. Czech. Chem. Commun.* **37**, 1465 (1972). Undecomposed dehydroascorbic acid in soln is easily converted to ascorbic acid by reduction with sulfurous acid. Has same antiscorbutic activity in humans as ascorbic acid (upon oral ingestion).

2921. (3β)-7-Dehydrocholesterol. *Cholesta-5,7-dien-3-ol*; provitamin D_3. $C_{27}H_{44}O$; mol wt 384.65. C 84.31%, H 11.53%, O 4.16%. Occurs in higher animals and in man. Isoln from the horned snail, *Buccinum undatum:* Windaus *et al., Z. Physiol. Chem.* **241**, 100 (1936); from pigskin: Windaus, Bock, *ibid.* **245**, 168 (1937); Boer *et al., Proc. Koninkl. Akad. Wetenschappen Amsterdam* **39**, 622 (1936). Synthesis: Windaus *et al., Ann.* **520**, 98 (1935); Rosenberg, Tinker, U.S. pat. **2,215,727**; Buisman *et al., Rec. Trav. Chim.* **66**, 83 (1947); P. N. Confalone *et al., J. Org. Chem.* **46**, 1030 (1981). Irradiation with ultraviolet light produces vitamin D_3; also lumisterol$_3$ and tachysterol$_3$: Windaus *et al., Ann.* **533**, 118 (1938).

Hydrated plates from ether-methanol. The water of crystn is held tenaciously: Schenck *et al., Ber.* **69**, 2705 (1936). When anhydr mp 150-151°. $[\alpha]_D^{20}$ −113.6° (c = 1 in chloroform), Koch, Koch, *J. Biol. Chem.* **116**, 757 (1936); $[\alpha]_D^{20}$ −127.1° in benzene (Boer). All provitamins D have the same uv maxima at 260, 270, 281, 293.5 nm. Insol in water, sol in the usual organic solvents.

Acetate, $C_{29}H_{46}O_2$, crystals from methanol, mp 129-130°; $[\alpha]_D^{20}$ −85.3° (c = 1.2 in benzene).

Benzoate, $C_{34}H_{48}O_2$, plates from chloroform-acetone, mp 139-140°; $[\alpha]_D^{20}$ −53.2° in chloroform.

2922. Dehydrocholic Acid. *(5β)-3,7,12-Trioxocholan-24-oic acid;* 3,7,12-triketocholanic acid; Acolen; Bilidren; Bilostat; Cholan-DH; Cholagon; Cholepatin; Chologon; Decholin; Dehychol; Deidrocolico Vita; Didrocolo; Erebile; Felacrinos; Procholon. $C_{24}H_{34}O_5$; mol wt 402.53. C 71.61%, H 8.51%, O 19.87%. Prepd from cholic acid by oxidation with CrO_3 in glacial acetic acid: Hammarsten in *Abderhalden's Handbuch der Biol. Arbeitsmethoden*, Abt. I, Teil 6, p 238 (1925); by using chlorine instead of CrO_3: Hinkley, Singleton, U.S. pat. **2,966,499** (1960 to Merck & Co.).

Bitter crystals from acetone, mp 237°. $[\alpha]_D^{20}$ +26° (c = 1.4 in alc). Soly at 15° (g/l): water 0.18; alcohol 3.3; ether 0.46; chloroform 9.04; benzene 1.04; acetone 7.76; ethyl acetate 7.4; glacial acetic acid 7.42.

Sodium salt, $C_{24}H_{33}NaO_5$, *sodium dehydrocholate, Carachol, Dycholium, Suprachol.* Very sol in water. pH of aq soln about 9.0.

THERAP CAT: Choleretic.

2923. 11-Dehydrocorticosterone. *21-Hydroxypregn-4-ene-3,11,20-trione;* Δ⁴-pregnen-21-ol-3,11,20-trione; 17-(1-keto-2-hydroxyethyl)-Δ⁴-androsten-3,11-dione; Kendall's compound A. $C_{21}H_{28}O_4$; mol wt 344.45. C 73.23%, H 8.19%, O 18.58%. Found in adrenal cortex. 1000 lbs of beef glands yield 333 mg: Kendall, *Cold Spring Harbor Symposia Quant. Biol.* **5**, 299 (1937). Isoln procedure: Mason *et al., J. Biol. Chem.* **114**, 613 (1936); *see also* Kendall *et al., Proc. Staff Meet. Mayo Clinic* **12**, 136 (1937). Prepn from corticosterone 21-acetate: Reichstein, *Helv. Chim. Acta* **20**, 953 (1937); from desoxycholic acid: Lardon, Reichstein, *ibid.* **26**, 747 (1943); Gallagher, *Recent Progress in Hormone Research* (New York, 1946) p 83; from 3α-acetoxy-11-ketobisnorcholanic acid: Sarett, *J. Biol. Chem.* **162**, 601 (1946); *J. Am. Chem. Soc.* **68**, 2478 (1946); from 3α-acetyl-11-ketolithocholic acid methyl ester: Wettstein, Meystre, *Helv. Chim. Acta* **30**, 1262 (1947). Manuf: Wettstein, U.S. pat. **2,778,776** (1957 to Ciba).

Large prisms from aq acetone, mp 178-180°. Can be distilled in high vacuum. $[\alpha]_{546}^{25}$ +299°, also given as +347° (c = 0.23 in benzene); $[\alpha]_D^{25}$ +258° (alc). Relatively good soly in benzene.

Acetate, $C_{23}H_{30}O_5$, needles from alc, mp 179-181°. $[\alpha]_{546}^{18}$ +285° (dioxane); $[\alpha]_D^{18}$ +233.7° (dioxane).

2924. Dehydroemetine. *2,3-Didehydro-6',7',10,11-tetramethoxyemetan;* 2,3-dehydroemetine; 2-dehydroemetine. $C_{29}H_{38}N_2O_4$; mol wt 478.63. C 72.77%, H 8.00%, N 5.85%, O 13.37%. Synthetic analog of emetine, *q.v.;* the (−)-form is therapeutically active. Synthesis of racemic 2,3-dehydroemetine: Brossi *et al., Helv. Chim. Acta* **42**, 772 (1959). Stereospecific synthesis of (±)-form and (±)-*2,3-dehydroisoemetine,* the (1'β)-isomer: Clark *et al., J. Chem. Soc.* **1962**, 2479. Separation of the four possible isomers: Brossi, Burkhardt, *Experientia* **18**, 211 (1962). Absolute configuration and bioactivity of isomers: A. Brossi *et al., Helv. Chim. Acta* **45**, 2219 (1962). Prepn of pure (−)-form: A. Brossi, **Belg. pat. 629,898;** *eidem,* **U.S. pat. 3,311,633** (1963, 1967 both to Hoffmann-La Roche); N. Whittaker *J. Chem. Soc. (C)* **1969**, 94. Antiprotozoal activity *in vitro:* G. H. Al-Khateeb *et al., Chemotherapy (Basel)* **23**, 267 (1977). Review of use in amebiasis: G. Woolfe, *Prog. Drug. Res.* **8**, 12 (1965).

Crystals from isopropyl ether, mp 94-96°. $[\alpha]_D$ −183°. (−)-Form dihydrobromide, $C_{29}H_{38}N_2O_4.2HBr$, crystals, mp 243-245°. $[\alpha]_D$ −97° (methanol). uv max (alc): 282 nm (E 7300).

(+)-Form dihydrobromide, crystals, mp 241-243°. $[\alpha]_D$ +95° (methanol). uv max (alc): 282 nm (E 7350).

(±)-Form dihydrochloride, $C_{29}H_{38}N_2O_4.2HCl$, *Dametin, Mebadin.* Crystals from ethanol + ether, mp 235°.

(±)-2,3-Dehydroisoemetine hydrochloride, crystals from ethanol + ether, mp 220-225°.

(−)-2,3-Dehydroisoemetine dihydrobromide, $C_{29}H_{38}N_2O_4.2HBr$, crystals, mp 257-260°. $[\alpha]_D$ −107° (methanol). uv max (alc): 285 nm (ε 7400).

(+)-2,3-Dehydroisoemetine dihydrobromide, crystals, mp 257-258°. $[\alpha]_D$ +109° (methanol). uv max (alc): 285 nm (ε 7450).

THERAP CAT: Antiamebic.

2925. Dehydroergosterol. *(3β,22E)-Ergosta-5,7,9(11),-22-tetraen-3-ol.* $C_{28}H_{42}O$; mol wt 394.64. C 85.22%, H 10.73%, O 4.05%. Prepd from ergosterol: Windhaus, Linsert, *Ann.* **465**, 148 (1928); Callow, Rosenheim, *J. Chem. Soc.* **1933**, 387. From isopyrocalciferol: Windhaus, Dimroth, *Ber.* **70**, 376 (1937).

Solvated plates from alcohol, needles from ether. When dry, mp 146°. $bp_{0.5}$ 230°. $[\alpha]_D^{15}$ +149.2° (c = 1.9 in chloroform). Absorption spectrum: Morton, de Gouveia, *J. Chem. Soc.* **1934**, 916. One gram dissolves in 800 ml methanol. Freely sol in ether, chloroform, benzene.

Acetate, $C_{30}H_{44}O_2$, mp 146°. $[\alpha]_D^{16}$ +204° (c = 1.1 in chloroform).

Methyl ether, $C_{29}H_{44}O$, mp 106°. $[\alpha]_D^{20}$ +166° (c = 2 in chloroform).

2926. 3-Dehydroretinal. *3,4-Didehydroretinal;* 3,7-dimethyl-9-(2,6,6-trimethyl-1,3-cyclohexadien-1-yl)-2,4,6,8-nonatetraenal; retinal 2; retinene 2; vitamin A_2 aldehyde. $C_{20}H_{26}O$; mol wt 282.43. C 85.06%, H 9.28%, O 5.66%. Carotenoid component of the visual pigments of fresh-water and migratory fish, lampreys and certain amphibians. Of the 16 possible stereoisomers, 7 have been synthesized. The 11-*cis*-isomer is the chromophore of porphyropsin and cyanopsin, *q.q.v.* Isoln from retinas of fresh-water fish: G. Wald, *Nature* **139**, 1017 (1937). Recognition as vitamin A_2 aldehyde: R. A. Morton, *ibid.* **153**, 69 (1944). Prepn by the oxidation of vitamin A_2: G. Wald, *Biochim. Biophys. Acta* **4**, 215 (1950). Synthesis of *trans*-isomer: K. R. Farrar *et al., J. Chem. Soc.* **1952**, 1414; of stereoisomers: U. Schwieter *et al., Helv. Chim. Acta* **45**, 517, 528, 541 (1962); R. S. H. Liu *et al., J. Am. Chem. Soc.* **99**, 8095 (1977). HPLC separation: K. Tsukida *et al., J. Chromatog.* **192**, 395 (1980); T. Suzuki, M. Makino-Tasaka, *Anal. Biochem.* **129**, 111 (1983). Biological activity is thought to be analogous with that of retinal, *q.v.:* G. Wald, *J. Gen. Physiol.* **22**, 775 (1939); *idem, Fed. Proc.* **12**, 606 (1953). *Reviews:* G. Wald, *Science* **162**, 230-239 (1968); R. Hubbard *et al.,* "Methodology of Vitamin A and Visual Pigments" in *Methods of Enzymology* **18**, D. B. McCormick, L. D. Wright, Eds. (Academic Press, New York, 1971) pp 615-653. Book: *The Retinoids,* **Vol. 1-2,** M. B. Sporn *et al.,* Eds. (Academic Press, New York, 1984).

trans-Isomer, orange-red prisms from pentane, mp 77-78. uv max (ethanol): 401 nm ($E_{1cm}^{1\%}$ 1470).

11-*cis*-Isomer, oil from petr ether-ether. uv max (ethanol): 393 nm ($E_{1cm}^{1\%}$ 882).

USE: As tool in biological energy transduction research.

2927. 7-Dehydrositosterol. *(3β)-Stigmasta-5,7-dien-3-ol.* $C_{29}H_{48}O$; mol wt 412.70. C 84.40%, H 11.72%, O 3.88%. From soy bean oil β-sitosterol: Wunderlich, *Z. Physiol. Chem.* **241**, 116 (1936). Starting with cholesterol: **U.S. pat. 2,386,635** (1945).

Platelets from alc, browns on contact with air, mp 144-145°. [α]$_D^{21}$ −116° (c = 2 in chloroform). Sparingly sol in methanol, somewhat more in alcohol, freely sol in the other usual organic solvents. Insol in water.

Acetate, C$_{31}$H$_{50}$O$_2$, mp 151-152°. [α]$_D^{21}$ −71° (c = 2 in chloroform).

2928. Delapril. (S)-N-(2,3-Dihydro-1H-inden-2-yl)-N-[N-[1-(ethoxycarbonyl)-3-phenylpropyl]-L-alanyl]glycine; N-[N-[(S)-1-(ethoxycarbonyl)-3-phenylpropyl]-L-alanyl]-N-(indan-2-yl)glycine; ethyl (S)-2-[[(S)-1-[(carboxymethyl)-2-indanylcarbamoyl]ethyl]amino]-4-phenylbutyrate; alindapril; indalapril. C$_{26}$H$_{32}$N$_2$O$_5$; mol wt 452.55. C 69.01%, H 7.13%, N 6.19%, O 17.68%. Angiotensin-converting enzyme (ACE) inhibitor. Prepn: Y. Oka et al., Eur. pat. Appl. 51,391; eidem, U.S. pat. 4,385,051 (1982, 1983 both to Takeda); J. T. Suh et al., Eur. J. Med. Chem. - Chim. Ther. 20, 563 (1985); A. Miyake et al., Chem. Pharm. Bull. 34, 2852 (1986). Pharmacology in animals: Y. Inada et al., Japan. J. Pharmacol. 42, 1 (1986); eidem, ibid., 99. Pharmacology and metabolism in humans: T. Ogihara et al., Curr. Ther. Res. 41, 809 (1987). Pharmacokinetics and evaluation in hypertension: H. Shionoiri et al., Clin. Pharmacol. Ther. 41, 74 (1987).

Hydrochloride, C$_{26}$H$_{32}$N$_2$O$_5$·HCl, CV-3317, REV-6000A, Adecut, Cupressin. Colorless plates from acetone + hydrochloric acid, mp 166-170° (dec). [α]$_D^{22}$ +18.5° (c = 1 in methanol).

THERAP CAT: Antihypertensive.

2929. Delavirdine. 1-[3-[(1-Methylethyl)amino]-2-pyridinyl]-4-[[5-[(methylsulfonyl)amino]-1H-indol-2-yl]carbonyl]piperazine; 1-(5-methanesulfonamido-1H-indol-2-ylcarbonyl)-4-[3-(1-methylethylamino)pyridinyl]piperazine; 1-[3-(isopropylamino)-2-pyridyl]-4-[(5-methanesulfonamidoindol-2-yl)carbonyl]piperazine; U-90152. C$_{22}$H$_{28}$N$_6$O$_3$S; mol wt 456.57. C 57.88%, H 6.18%, N 18.41%, O 10.51%, S 7.02%. Bisheteroarylpiperazine (BHAP) reverse transcriptase inhibitor. Prepn: D. L. Romero et al., PCT Int. pat. Appl. 91 09,849 (1991 to Upjohn). Structure-activity studies: D. L. Romero et al., J. Med. Chem. 36, 1505 (1993). In vitro activity vs HIV-1: T. J. Dueweke et al., Antimicrob. Ag. Chemother. 37, 1127 (1993); P. J. Pagano, K.-T. Chong,

J. Infect. Dis. 171, 61 (1995). Enzyme kinetic studies: I. W. Althaus et al., Biochem. Pharmacol. 47, 2017 (1994).
Crystals from ethyl acetate + hexane, mp 226-228°.
Monomethanesulfonate, C$_{22}$H$_{28}$N$_6$O$_3$S.CH$_4$O$_3$S, delavirdine mesylate, U-90152S, Rescriptor.

THERAP CAT: Antiviral.

2930. Delmadinone Acetate. 17-(Acetyloxy)-6-chloropregna-1,4,6-triene-3,20-dione; 6-chloro-17-hydroxypregna-1,4,6-triene-3,20-dione acetate; 1,6-bisdehydro-6-chloro-17α-acetoxyprogesterone; Δ^1-chlormadinone acetate; RS-1301; Delminal; Estrex; Tardastrex; Tarden; Zenadrex. C$_{23}$H$_{27}$ClO$_4$; mol wt 402.92. C 68.56%, H 6.75%, Cl 8.80%, O 15.88%. Prepn: Ringold et al., J. Am. Chem. Soc. 81, 3485 (1959); Brit. pat. 890,315 (1962 to Upjohn), C.A. 57, 5996c (1962); 76, 14813y (1972); Campbell, Babcock, Ger. pat. 1,243,682 (1967 to Upjohn), C.A. 67, 100343r (1967). Activity studies: Dorfman, Kincl, Steroids 1, 185 (1963); Weichert et al., Arzneimittel-Forsch. 17, 1103 (1967); Gerber et al., J. Small Animal Pract. 14, 151 (1973).

Crystals, mp 168-170°. [α]$_D$ −83° (chloroform). uv max (ethanol): 229, 258, 297 nm (log ε 4.00, 4.00, 4.03).

THERAP CAT: Progestogen; antiandrogen; antiestrogen.
THERAP CAT (VET): Progestogen; antiandrogen; antiestrogen.

2931. Delphinidin. 3,5,7-Trihydroxy-2-(3,4,5-trihydroxyphenyl)-1-benzopyrylium chloride; 3,3',4',5,5',7-hexahydroxyflavylium chloride; 3,3',4',5,5',7-hexahydroxy-2-phenylbenzopyrylium chloride; delphinidol. C$_{15}$H$_{11}$ClO$_7$; mol wt 338.70. C 53.19%, H 3.27%, Cl 10.47%, O 33.07%. The aglucone of delphinin: Willstätter, Mieg, Ann. 408, 61 (1915); the aglucone of violanin: Willstätter, Weil, ibid. 412, 178 (1917). Occurrence in plants: Acheson, Can. J. Biochem. Physiol. 37, 1 (1959). Synthesis: Pratt, Robinson, J. Chem. Soc. 127, 166 (1925); Bradley et al., ibid. 1930, 793.

Chocolate-brown prisms or needles with metallic luster from 5% HCl. Crystallizes from water with 1, 2 or 4 mols H$_2$O. When anhydr does not melt below 350°. Absorption max (methanolic HCl): 544 nm. Sol in methanol, ethanol, ethyl acetate.

3-Glucoside, C$_{21}$H$_{21}$ClO$_{12}$, myrtillin-a. From the whortleberry (Vaccinium myrtillus L., Ericaceae): Willstätter, Zollinger, Ann. 408, 83 (1915); 412, 195 (1916); from the pansy (Viola tricolor L., Violaceae): Karrer, de Meuron, Helv. Chim. Acta 16, 292 (1933). Structure and synthesis: Reynolds, Robinson, J. Chem. Soc. 1934, 1039. Deep purple crystals from dil HCl. Absorption max (methanolic HCl): 535 nm.

3,5-Diglucoside, C$_{27}$H$_{31}$ClO$_{17}$, 3,5-bis(glucosyloxy)-3',4',5',7-tetrahydroxyflavylium chloride, delphin, delphoside, hyacin. From flowers of Salvia patens Cav., Labiatae, structure and synthesis: Reynolds et al., J. Chem. Soc. 1934, 1235; from flowers of Verbena hybrida Hart., Verbenaceae: Scott-Moncrieff, Sturgess, Biochem. J. 34, 268 (1940); from grape skins: Bockian *et al., J. Agr. Food Chem.* **3**, 695 (1955). Identity with hyacin: Saito *et al., Proc. Japan. Acad.* **36**, 340 (1960). Leaflets with bronze luster from dil HCl. Absorption max (methanolic HCl): 534 nm. Practically insol in cold water, alcohol, dil acids; sol in hot dil HCl.

Compound with glucose + hydroxybenzoic acid, $C_{41}H_{39}ClO_{21}$, *delphinin.* From *Delphinium consolida* L., *Ranunculaceae:* Willstätter, Mieg, *loc. cit.* Dark red-brown plates or prisms from 3% HCl. dec 200-203°. Sol in water with decompn, dil alcohol, acetone, boiling alcohol.

Compound with glucose + rhamnose + *p*-hydroxycinnamic acid, $C_{36}H_{37}ClO_{18}$, *violanin.* From flowers of *Viola tricolor* L., *Violaceae:* Willstätter, Weil, *loc. cit.* Structure: Karrer, de Meuron, *loc. cit.* Bluish-violet plates with green metallic luster from methanolic HCl + ether. Sol in amyl alcohol, in 0.15-0.5% HCl; less sol in 1% HCl; slightly sol in water; practically insol in 5-12% HCl.

2932. Delphinine. *(1α,6α,14α,16β)-1,6,16-Trimethoxy-4-(methoxymethyl)-20-methylaconitane-8,13,14-triol 8-acetate 14-benzoate.* $C_{33}H_{45}NO_9$; mol wt 599.72. C 66.09%, H 7.56%, N 2.34%, O 24.01%. A toxic alkaloid from seeds of *Delphinium staphisagria* L., *Ranunculaceae.* Isoln: Markwood, *J. Am. Pharm. Assoc.* **16**, 928 (1927). Purification: Schneider, *Arch. Pharm.* **283**, 283 (1950). Structure: Jacobs, Pelletier, *J. Am. Chem. Soc.* **78**, 3542 (1956); Wiesner *et al., Can. J. Chem.* **47**, 2734 (1969). Configuration: Wiesner *et al., Tetrahedron Letters* **1959**, no. 3, 12; **1960**, no. 15, 23; Gilman, Marion, *ibid.* **1962**, 923; Birnbaum *et al., ibid.* **1971**, 867. Synthesis: Wiesner *et al., Can. J. Chem.* **50**, 1925 (1972).

Orthorhombic, six-sided plates from alc, mp 197.5-199°. $[\alpha]_D^{20}$ +25° (ethanol). Practically insol in water. Sol in 25 parts alcohol, 20 parts chloroform, 10 parts ether.

Hydrochloride, $C_{33}H_{45}NO_9$.HCl, crystals, dec 214°.

2933. Delsoline. *(1α,6β,14α,16β)-20-Ethyl-6,14,16-trimethoxy-4-(methoxymethyl)aconitane-1,7,8-triol.* $C_{25}H_{41}NO_7$; mol wt 467.60. C 64.22%, H 8.84%, N 3.00%, O 23.95%. From seeds of *Dephinium consolida* L., *Ranunculaceae:* Markwood, *J. Am. Pharm. Assoc.* **13**, 696 (1924); Cionga, Iliescu, *Ber.* **74**, 1031 (1941); Marion, Edwards, *J. Am. Chem. Soc.* **69**, 2010 (1947). Structure: Anet *et al., Can. J. Chem.* **35**, 397 (1957); Skaric, Marion, *J. Am. Chem. Soc.* **80**, 4434 (1958); *eidem, Can. J. Chem.* **38**, 2433 (1960); **39**, 1579 (1961).

Delsoline   R = CH_3
Delcosine   R = H

Prisms from methanol, mp 213-216.5° (when immersed at 205°). $[\alpha]_D^{22}$ +53.4° (c = 2.04 in chloroform). Slightly sol in water, sol in alcohol or chloroform.

Delcosine, $C_{24}H_{39}NO_7$. Alkaloid also found in *D. consolida.* mp 203-204°, $[\alpha]_D^{25}$ +56.8° (c = 2.01 in chloroform). Quite sol in water.

USE: Glycosidal dyestuff.

2934. Deltamethrin. *[1R-[1α(S*),3α]]-3-(2,2-Dibromoethenyl)-2,2-dimethylcyclopropanecarboxylic acid cyano(3-phenoxyphenyl)methyl ester;* *(S)*-α-cyano-3-phenoxybenzyl-*(1R)*-*cis*-3-(2,2-dibromovinyl)-2,2-dimethylcyclopropane carboxylate; *decamethrin;* *esbecythrin;* FMC-45498; NRDC-161; RU-22974; Butox; Decis; K-Othrine. $C_{22}H_{19}Br_2NO_3$; mol wt 505.21. C 52.30%, H 3.79%, Br 31.63%, N 2.77%, O 9.50%. Potent synthetic pyrethroid insecticide. Prepn of racemic mixture: M. Elliot *et al.,* Ger. pat. **2,439,-177** (1975 to NRDC), *C.A.* **83**, 73519z (1975); of decamethrin and isomers: *eidem, Pestic. Sci.* **9**, 105 (1978). Activity: *eidem, Nature* **248**, 710 (1974); *eidem, Pestic. Sci.* **9**, 112 (1978). Absolute configuration: J. D. Owen, *J. Chem. Soc. Perkin Trans. I* **1975**, 1865. Photochemistry: L. O. Ruzo *et al., J. Agr. Food Chem.* **25**, 1385 (1977). Metabolism: *eidem, ibid.* **26**, 918 (1978). Toxicology: R. Kavlock *et al., J. Environ. Pathol. Toxicol.* **2**, 751 (1979). Pharmacological effects on central nervous system: P. H. Chanh *et al., Arzneimittel-Forsch.* **34**, 175 (1984).

Crystals, mp 98-101°. Sol in ethanol, acetone, dioxane. Insol in water. LD_{50} in female rats (mg/kg): 31 orally; 4 i.v. (Kavlock).

USE: Insecticide.

2935. Demanyl Phosphate. *Mono[2-(dimethylamino)-ethyl]phosphoric acid ester; 2-dimethylaminoethanol dihydrogen phosphate;* phosphoryldimethylaminoethanol; phosphoryldimethylcolamine; phosphoryldimethylethanolamine; P-DMEA; Panclar. $C_4H_{12}NO_4P$; mol wt 169.12. C 28.41%, H 7.15%, N 8.28%, O 37.84%, P 18.31%. $(CH_3)_2NCH_2$-$CH_2OPO_3H_2$. Prepn: Aron-Samuel *et al.,* Fr. pat. **992,802** (1951), *C.A.* **51**, 458e (1957); Cherbuliez, Rabinowitz, *Helv. Chim. Acta* **41**, 1168 (1958). Brit. pat. **861,463;** U.S. pat. **3,098,072** (1961, 1963 to Ciba). The phosphate ester of a possible choline precursor. Has been isolated from a choline-requiring mutant of *Neurospora crassa,* strain 47904: Wolf, Nyc, *J. Biol. Chem.* **234**, 1068 (1959). Biosynthetic studies: Ansell, Chojnacki, *Nature* **196**, 545 (1962); *eidem, Biochem. J.* **98**, 303 (1966).

Crystals, mp 175-176°.
Monohydrate, mp 78-81°.

THERAP CAT: Psychotonic.

2936. Demecarium Bromide. *3,3'-[1,10-Decanediylbis-[(methylimino)carbonyloxy]]bis[N,N,N-trimethylbenzenaminium]dibromide; (m-hydroxyphenyl)trimethylammonium bromide, decamethylenebis(methylcarbamate);* N,N'-bis[3-trimethylammoniumphenoxycarbonyl]-N,N'-dimethyldecamethylenediamine dibromide; decamethylenebis[m-dimethylaminophenyl N-methylcarbamate] dimethobromide; decamethylenebis[N-methylcarbamic acid m-dimethylaminophenyl ester] bromomethylate; BC-48; Tosmilen; Humorsol. $C_{32}H_{52}Br_2N_4O_4$; mol wt 716.60. C 53.64%, H 7.31%, Br 22.30%, N 7.82%, O 8.93%. Prepn: Schmid, U.S. pat. **2,789,981** (1957 to Oesterreichische Stickstoffwerke).

White, slightly hygroscopic powder, dec 162-167°. Freely sol in water, alc. Sparingly sol in acetone. Insol in ether. Aq solns are neutral, stable, and may be sterilized by heat.

THERAP CAT: Cholinergic (ophthalmic).

2937. Demeclocycline. *[4S-(4α,4aα,5aα,6β,12aα)]-7-Chloro-4-(dimethylamino)-1,4,4a,5,5a,6,11,12a-octahydro-3,6,10,12,12a-pentahydroxy-1,11-dioxo-2-naphthacenecarboxamide; 7-chloro-6-demethyltetracycline;* demethylchlortetracycline (obsolete); RP-10192; Bioterciclin; Declomycin; Deganol; Ledermycin; Periciclina. $C_{21}H_{21}ClN_2O_8$; mol wt 464.86. C 54.26%, H 4.55%, Cl 7.63%, N 6.03%, O 27.53%. Antibiotic related to tetracycline produced by *Streptomyces aureofaciens.* Prepn: McCormick *et al., J. Am. Chem. Soc.* **79,** 4561 (1957); U.S. pat. **2,878,289** (1959 to Am. Cyanamid). Improved fermentation processes: Szumski; Goodman, Matrishin; Goodman, U.S. pats. **3,012,946** and **3,-019,172; 3,050,446** (1961 and 1962 to Am. Cyanamid); Fr. pat. **1,344,645** (1963 to Merck & Co.); Neidleman, U.S. pat. **3,154,476** (1964 to Olin Mathieson). Abs config: Dobrynin *et al., Tetrahedron Letters* **1962,** 901. Toxicity data (hydrochloride): E. I. Goldenthal, *Toxicol. Appl. Pharmacol.* **18,** 185 (1971).

Sesquihydrate, mp 174-178° (dec). $[\alpha]_D^{25}$ −258° (c = 0.5 in 0.1N H_2SO_4). Solubilities: Marsh, Weiss, *J. Assoc. Offic. Anal. Chem.* **50,** 457 (1967).

Hydrochloride, *Clortetrin, Demetraciclina, Detravis, Meciclin, Mexocine.* LD_{50} orally in rats: 2372 mg/kg (Goldenthal).

THERAP CAT: Antibacterial.

THERAP CAT (VET): Antibacterial.

2938. Demecolcine. *6,7-Dihydro-1,2,3,10-tetramethoxy-7-(methylamino)benzo[a]heptalen-9(5H)-one; N-deacetyl-N-methylcolchicine; N-methyl-N-desacetylcolchicine;* colchamine; Santavy's substance F; Colcemid; Omaine. $C_{21}H_{25}NO_5$; mol wt 371.43. C 67.91%, H 6.78%, N 3.77%, O 21.54%. Isoln from *Colchicum autumnale* L., *Liliaceae:* Santavy, *Pharm. Acta Helv.* **25,** 248 (1950); Santavy, Reichstein, *Helv. Chim. Acta* **33,** 1606 (1950); Schlittler, Uffer, **Ger.** pat. **936,268** (1955 to Ciba), *C.A.* **53,** 1396 (1959). Structure: Santavy *et al., Helv. Chim. Acta* **36,** 1319 (1953). Synthesis: Uffer *et al., ibid.* **37,** 18 (1954); H. G. Capraro, A. Brossi, *ibid.* **62,** 965 (1979).

Pale yellow prisms from ethyl acetate + ether, mp 186°. $[\alpha]_D^{20}$ −129.0° (c = 1 in chloroform). uv max (alc): 245, 355 nm (log ε 4.55, 4.24). Basic reaction. Soluble in acidified water, in alcohol, ether, chloroform, benzene.

THERAP CAT: Antineoplastic.

2939. Demegestone. *17-Methyl-19-norpregna-4,9-diene-3,20-dione;* 17α-methyl-19-nor-Δ^{4,9}-pregnadiene-3,20-dione; 17α-methyl-Δ^9-19-norprogesterone; R-2453; Lutionex. $C_{21}H_{28}O_2$; mol wt 312.45. C 80.73%, H 9.03%, O 10.24%. Prepn: **Neth.** pat. **Appl. 6,517,141** corresp to Vignau *et al.,* and Joly *et al.,* U.S. pats. **3,453,267** and **3,547,959** (1966, 1969, 1970, all to Roussel-UCLAF); Velluz *et al., Tetrahe-*

dron, **Suppl 8,** part II, 495 (1966); Joly *et al., Bull. Soc. Chim. France* **1973,** 2694.

Crystals from ether, mp 106°. $[\alpha]_D$ −275° (c = 0.5 in ethanol). uv max (ethanol): 214, 302 nm (ε 6350, 21000).

THERAP CAT: Progestogen.

2940. Demeton. *Phosphorothioic acid O,O-diethyl O-[2-(ethylthio)ethyl] ester mixture with O,O-diethyl S-[2-(ethylthio)ethyl]phosphorothioate;* mercaptophos; Bayer 8169; E-1059; Systox. $C_8H_{19}O_3PS_2$; mol wt 258.34. C 37.19%, H 7.41%, O 18.58%, P 11.99%, S 24.82%. Cholinesterase inhibitor. Isomeric mixture consisting of demeton-O and demeton-S (*O,O-*diethyl *O*(and *S*)-ethylmercaptoethyl thiophosphates). Another name for demeton-S is *Isosystox.* Prepn: Schrader, U.S. pats. **2,571,989** and **2,597,534** (1951, 1952 both to Bayer); Gardner, Heath, *Anal. Chem.* **25,** 1849 (1953). Toxicity study: T. B. Gaines, *Toxicol. Appl. Pharmacol.* **14,** 515 (1969).

Oily liq. Faint odor. bp_2 134°. Virtually insol in water. Sol in ethanol, propylene glycol, toluene and similar hydrocarbons. Absorbed by plants, rendering the foliage and plant fluids toxic to insects. LD_{50} in female, male rats (mg/kg): 2.5, 6.2 orally; 8.2, 14 dermally (Gaines).

Caution: Potential symptoms of overexposure are miosis, aching eyes, rhinorrhea and headache; chest tightening, wheezing, laryngeal spasms, salivation and cyanosis; anorexia, nausea, vomiting, abdominal cramps and diarrhea; local sweating; muscle fasiculation, weakness and paralysis; giddiness, confusion and ataxia; convulsions, coma; low blood pressure; cardiac irregularities; irritation of eyes and skin. *See NIOSH Pocket Guide to Chemical Hazards* (DHHS/NIOSH 90-117, 1990) p 80.

USE: Insecticide.

2941. Demexiptiline. *5H-Dibenzo[a,d]cyclohepten-5-one O-[2-(methylamino)ethyl]oxime.* $C_{18}H_{18}N_2O$; mol wt 278.35. C 77.67%, H 6.52%, N 10.06%, O 5.75%. Prepn: S. Schütz *et al.,* **Ger.** pat. **1,247,302;** *eidem,* U.S. pat. **3,963,778** (1967, 1976 both to Bayer). Synthesis and pharmacological properties: G. Aichinger *et al., Arzneimittel-Forsch.* **19,** 838 (1969); S. Rossi *et al., Farmaco Ed. Sci.* **24,** 685 (1969).

Colorless oil.

Hydrochloride, $C_{18}H_{18}N_2O \cdot HCl$, *LM-2909, Deparon.* Crystals from isopropanol, mp 232-233°. LD_{50} in rats (mg/kg): 330 orally, 25 i.v. (Rossi).

THERAP CAT: Antidepressant.

2942. Denatonium Benzoate. *N-[2-[(2,6-Dimethylphenyl)amino]-2-oxoethyl]-N,N-diethylbenzenemethanaminium benzoate; benzyldiethyl[(2,6-xylylcarbamoyl)methyl]ammonium benzoate; lignocaine benzyl benzoate; Bitrex.* $C_{28}H_{34}N_2O_3$; mol wt 446.59. C 75.31%, H 7.67%, N 6.27%, O 10.75%. Prepn: J. E. Hay, U.S. pat. **3,080,327** (1963 to T. & H. Smith Ltd.); *idem,* U.S. pat. **3,268,575** (1966 to Edinburgh Pharm. Ind.). HPLC determn in rapeseed oil: C. E. Damon, B. C. Pettitt, Jr., *J. Chromatog.* **195**, 243 (1980). Brief review of substance history and uses: H. A. S. Payne, *Chem. & Ind. (London)* **22**, 721-723 (1988).

Crystals from isopropyl alcohol + ethyl acetate, mp 166-170°. Among the most bitter substances known to man. Soluble in water, alcohol; sparingly sol in acetone. Practically insol in ether.

USE: Pharmaceutic aid (alcohol denaturant; flavor). Added to toxic substances as a deterrent to accidental ingestion. Can replace brucine or quassin as denaturant for ethyl alcohol.

2943. Denopamine. *(R)-α-[[[2-(3,4-Dimethoxyphenyl)-ethyl]amino]methyl]-4-hydroxybenzenemethanol; (−)-(R)-α-[[(3,4-dimethoxyphenethyl)amino]methyl]-p-hydroxybenzyl alcohol; (−)-(R)-1-(4-hydroxyphenyl)-2-(3,4-dimethoxy-phenethylamino)ethanol;* TA-064; Carguto; Kalgut. $C_{18}H_{23}NO_4$; mol wt 317.38. C 68.12%, H 7.30%, N 4.41%, O 20.16%. Selective β_1-adrenoceptor agonist with positive inotropic activity. Prepn: M. Ikezaki *et al.*, **Belg. pat. 833,731;** *eidem,* **U.S. pat. 4,032,575** (1976, 1977 both to Tanabe Seiyaku); N. Umino *et al., Chem. Pharm. Bull.* **27**, 1479 (1979). Effect on carbohydrate and lipid metabolism in rats: M. Inamasu *et al., Biochem. Pharmacol.* **33**, 2171 (1984). Cardiovascular pharmacology in animals: T. Nagao *et al., Japan. J. Pharmacol.* **35**, 415 (1984); in humans: M. Kino *et al., Am. J. Cardiol.* **51**, 802 (1983). Metabolism in humans: T. Suzuki *et al., Drug Metabol. Dispos.* **11**, 377 (1983). GC-MS determn in human urine: *eidem, Chem. Pharm. Bull.* **33**, 2549 (1985). Binding affinity and selectivity: K. Naito *et al., Japan. J. Pharmacol.* **38**, 235 (1985). Clinical efficacy in congestive cardiomyopathy: J. Thormann *et al., Am. Heart J.* **110**, 426 (1985). Series of articles on pharmacokinetics, metabolism and disposition in animals: *Arzneimittel-Forsch.* **36**, 643-667 (1986).

l-Form hydrochloride, $C_{18}H_{23}NO_4$·HCl, crystals from isopropanol, mp 138-139.5°. $[\alpha]_D^{25}$ −38.0° (c = 1 in methanol). *dl*-Form hydrochloride, crystals from isopropanol + isopropyl ether, mp 164-167°.

THERAP CAT: Cardiotonic.

2944. Denopterin. *N-[4-[[1-(2-Amino-1,4-dihydro-4-oxo-6-pteridinyl)ethyl]methylamino]benzoyl]-L-glutamic acid;* 9,10-dimethylpteroylglutamic acid; N^{10},9-dimethylfolic acid; Dimetfol. $C_{21}H_{23}N_7O_6$; mol wt 469.46. C 53.73%, H 4.94%, N 20.89%, O 20.45%. Prepn: Hultquist *et al., J. Am. Chem. Soc.* **71**, 619 (1949); Hultquist, Smith, **Brit. pat. 667,098** (1952 to Am. Cyanamid).

Yellow-orange microcrystals. uv max (0.1N NaOH): 253, 305, 365 nm; (0.1N HCl): 315 nm.

THERAP CAT: Antineoplastic.

2945. 6-Deoxy-L-ascorbic Acid. $C_6H_8O_5$; mol wt 160.13. C 45.01%, H 5.04%, O 49.96%. Prepn starting with 2,3-monoacetone-L-sorbomethylose: Müller, Reichstein,

Helv. Chim. Acta **21**, 273 (1938). Alternative procedure: **Swiss.** pats. **203,549** and **209,587** (1939 and 1940, both to Hoffmann-La Roche).

Stout prisms from ethyl acetate, mp 168°. Sublimes at 10^{-3} mm and 160°. $[\alpha]_D^{22}$ +36.7° (0.1N HCl). Freely sol in water; sol in acetone, alcohol; sparingly sol in ethyl acetate. Practically insol in ether.

2946. Deoxycholic Acid. *(3α,5β,12α)-3,12-Dihydroxy-5-cholan-24-oic acid;* 17β-(1-methyl-3-carboxypropyl)etiocholane-3α,12α-diol; desoxycholic acid. $C_{24}H_{40}O_4$; mol wt 392.58. C 73.43%, H 10.27%, O 16.30%. Lacks the C-7 hydroxy group of cholic acid. Occurs in bile of man, ox, sheep, dog, goat, and rabbit. Isoln: Wieland, Siebert, *Z. Physiol. Chem.* **262**, 1 (1939). From cholic acid: Sifferd, **U.S. pat. 2,765,316** (1956 to Armour). Structure and synthesis: L. F. Fieser, M. Fieser, *Steroids* (Reinhold, New York, 1959). Laboratory prepn from ox bile: Gattermann-Wieland, *Praxis des organischen Chemikers* (de Gruyter, Berlin, 40th ed., 1961) p 361. Forms molecular coordination compds (so-called *choleic acids*) with many substances. Complexes wth fatty acids have been studies extensively: Sobotka, Goldberg, *Biochem. J.* **26**, 555 (1932); Sobotka, *Chem. Rev.* **15**, 362 (1934). Surface activity: *idem, Chemistry of the Steroids,* New York (1938).

Crystals from alc, mp 176-178°. Not precipitated by digitonin. $[\alpha]_D^{20}$ +55° (alc). pK 6.58. Soly at 15° in water 0.24 g/l; in alcohol 220.7 g/l; in ether 1.16 g/l; in chloroform 2.94 g/l; in benzene 0.12 g/l; in acetone 10.46 g/l; in glacial acetic acid 9.06 g/l. Sol in solns of alkali hydroxides or carbonates.

Sodium salt, $C_{24}H_{39}NaO_4$, *sodium deoxycholate.* Soly in water at 15°: > 333 g/l.

THERAP CAT: Choleretic.

2947. Deoxycorticosterone. *21-Hydroxypregn-4-ene-3,-20-dione;* 4-pregnen-21-ol-3,20-dione; 21-hydroxyprogesterone; desoxycorticosterone; 11-deoxycorticosterone; cortexone; desoxycortone; Kendall's desoxy compound B; Reichstein's substance Q. $C_{21}H_{30}O_3$; mol wt 330.47. C 76.33%, H 9.15%, O 14.52%. Occurs in adrenal cortex: Reichstein, von Euw, *Helv. Chim. Acta* **21**, 1197 (1938); Steiger, Reichstein, *ibid.* **20**, 1164 (1937). Numerous prepns from other steroids: Schindler *et al., ibid.* **24**, 371 (1941); Reichstein, **Ger. pat. 875,353** (1953 to Schering); Bockmühl *et al.,* **Ger. pat. 871,153** (1953 to Hoechst); Wettstein *et al.,* **U.S. pat. 2,778,776** (1957 to Ciba); **Dutch pat. 89,575** (1958 to Organon); Kaspar *et al.,* **Ger. pat. 1,028,572** (1958 to Schering). Isoln from the prothoracal glands of the water beetle, *Dytiscus marginalis:* Schildknecht *et al., Angew. Chem.* **78**, 392 (1966).

Plates from ether, mp 141-142°. $[\alpha]_D^{22}$ +178° (alc). uv max: 240 nm. Freely sol in alcohol, acetone.

Acetate, *see* Deoxycortiscosterone Acetate.

Tetraacetyl-β-D-glucoside, $C_{35}H_{48}O_{12}$, clusters of needles from 50% alc, mp 176-176.5°. $[\alpha]_D$ +80° (c = 0.515 in chloroform). Believed to occur in nature in this glycosidic form: Johnson, *J. Am. Chem. Soc.* **63**, 3238 (1941).

THERAP CAT: Mineralocorticoid.

2948. Deoxycorticosterone Acetate. *21-Acetyloxy-pregn-4-ene-3,20-dione;* 21-hydroxypregn-4-ene-3,20-dione 21-acetate; 11-deoxycorticosterone acetate; cortexone acetate; desoxycorticosterone acetate; deoxycortone acetate; desoxycortone acetate; DCA; Cortate; Cortiron; Decosteron; Doca; Dorcostrin; Percorten; Syncortyl. $C_{23}H_{32}O_4$; mol wt 372.50. C 74.16%, H 8.66%, O 17.18%.

Orthorhombic needles from alcohol, stable in air, mp 154-160°. Sublimes in high vacuum. $[\alpha]_D$ +168 to +176° (dioxane). Slightly sol in alcohol, methanol, acetone, ether, dioxane, propylene glycol (10 mg/ml), vegetable oils. Practically insol in water.

THERAP CAT: Mineralocorticoid.

THERAP CAT (VET): Adrenocortical steroid, mineralocorticoid.

2949. Deoxydihydrostreptomycin. *O-2-Deoxy-2-(methylamino)-α-L-glucopyranosyl-(1→2)-O-3,5-dideoxy-3-(hydroxymethyl)-α-L-arabinofuranosyl-(1→4)-N,N'-bis(aminoiminomethyl)-D-streptamine;* dihydrodeoxystreptomycin; Desoxymycin; Mervastrept. $C_{21}H_{41}N_7O_{11}$; mol wt 567.60. C 44.44%, H 7.28%, N 17.27%, O 31.01%. Reduction product of streptomycin. Prepn and structure: Ikeda, *C.A.* **50**, 13765g (1956); Yabuta *et al.*, U.S. pat. **2,837,510** (1958); Gailliot, Baget, Fr. pat. **1,198,042** (1959 to Rhône-Poulenc). TLC separation: H. Heding, *Acta Chem. Scand.* **24**, 3086 (1970). Comparative antibacterial activity of streptomycin analogs: H. Heding, O. Lützen, *J. Antibiot.* **25**, 287 (1972).

R = CH₂OH
R' = NHCH₃

Needles, dec 190-195°. $[\alpha]_D^{20}$ −102.5° (water).
Sulfate, $[\alpha]_D^{20}$ −78° (c = 2).
THERAP CAT: Antibacterial (tuberculostatic).

2950. Deoxyepinephrine. *4-[2-(Methylamino)ethyl]-1,-2-benzenediol; 4-[2-(methylamino)ethyl]pyrocatechol;* 3,4-dihydroxyphenylethylamine; methyl[β-(3,4-dihydroxyphenyl)-ethyl]amine; desoxyepinephrine; Epinine. $C_9H_{13}NO_2$; mol wt 167.21. C 64.65%, H 7.84%, N 8.38%, O 19.14%. Prepd by heating HCl and 1-keto-6,7-dimethoxy-2-methyltetrahydroisoquinoline obtained from laudanosine or papaverine: Pyman, *J. Chem. Soc.* **95**, 1266, 1610 (1909). Synthesis from veratrole: Kindler, Peschke, *Arch. Pharm.* **270**, 340 (1932); Kindler, Hesse, ibid. **271**, 439 (1933). From methylhomoveratrylamine: Buck, *J. Am. Chem. Soc.* **52**, 4119 (1930); Bretschneider, *Monatsh.* **76**, 335 (1947).

Crystals, mp 188-189°.
Hydrochloride, $C_9H_{13}NO_2 \cdot HCl$, crystals. Sol in water, alcohol.
THERAP CAT: The hydrochloride as adrenergic; vasoconstrictor.

2951. 2-Deoxy-D-glucose. *2-Deoxy-D-arabino-hexose;* D-arabino-2-desoxyhexose; 2-deoxyglucose; 2-DG; Ba 2758. $C_6H_{12}O_5$; mol wt 164.16. C 43.90%, H 7.37%, O 48.73%. Antimetabolite of glucose, *q.v.*, with antiviral activity. Synthesis: M. Bergmann *et al.*, *Ber.* **55**, 158 (1922); **56**, 1052 (1923); J. C. Sowden, H. O. L. Fischer, *J. Am. Chem. Soc.* **69**, 1048 (1947); H. R. Bolliger, *Helv. Chim. Acta* **34**, 989 (1954); H. R. Bolliger, M. D. Schmid, ibid. 1597, 1671; H. R. Bolliger, "2-Deoxy-D-*arabino*-hexose (2-Deoxy-D-glucose)" in *Methods in Carbohydrate Chemistry* vol. I, R. L. Whistler, M. L. Wolfrom, Eds. (Academic Press, New York, 1962) pp 186-189. Inhibition of influenza virus multiplication: E. D. Kilbourne, *Nature* **183**, 271 (1959). Effects on herpes simplex virus: R. J. Courtney *et al.*, *Virology* **52**, 447 (1973). Mechanism of action studies: M. R. Steiner *et al.*, *Biochem. Biophys. Res. Commun.* **61**, 745 (1974); F. K. Ray *et al.*, *Virology* **58**, 118 (1978). Use in human genital herpes infections: H. A. Blough, R. L. Giuntoli, *J. Am. Med. Assoc.* **241**, 2798 (1979); L. Corey, K. K. Holmes, ibid. **243**, 29 (1980). Effect vs respiratory syncytial viral infections in calves: S. B. Mohanty *et al.*, *Am. J. Vet. Res.* **42**, 336 (1981).

α-2-Deoxy-D-glucose

Cryst from acetone or butanone, mp 142-144°. $[\alpha]_D^{17.5}$ +38.3° (35 min) → +45.9° (c = 0.52 in water); +22.8° (24 hrs) → +80.8° (c = 0.57 in pyridine).
α-Form, cryst from isopropanol, mp 134-136°. $[\alpha]_D^{26}$ +156° → +103° (c = 0.9 in pyridine).
USE: Exptlly as an antiviral agent.

2952. 1-Deoxynojirimycin. *[2R-(2α,3β,4α,5β)]-2-(Hydroxymethyl)-3,4,5-piperidinetriol;* D-5-amino-1,5-dideoxyglucopyranose; moranoline; 1,5-dideoxy-1,5-imino-D-glucitol; (2R,3R,4R,5S)-2-hydroxymethyl-3,4,5-trihydroxy-piperidine; Bay n 5595; S-GI. $C_6H_{13}NO_4$; mol wt 163.17. C 44.17%, H 8.03%, N 8.58%, O 39.22%. α-Glucosidase inhibitor. Prepd by reduction of nojirimycin: S. Inouye *et al.*, *J. Antibiot.* **19A**, 290 (1966); S. Inouye *et al.*, *Tetrahedron* **23**, 2125 (1968). Isoln from mulberry, *Morus* spp. (*Moraceae*): M. Yagi *et al.*, *Nippon Nogei Kagaku Kaishi* **50**, 571 (1976), *C.A.* **86**, 167851r (1977). Production by *Bacillus subtilis* DSM704: D. C. Stein *et al.*, *Appl. Environ. Microbiol.* **48**, 280 (1984). Total synthesis: H. Setoi *et al.*, *Chem. Pharm. Bull.* **34**, 2642 (1986); H. J. G. Broxterman *et al.*,

Rec. Trav. Chim. **106,** 571 (1987). HPLC determn: M. D. Cole *et al., J. Chromatog.* **445,** 295 (1988). Inhibition of α-glucosidase: Y. Yoshikuni, *Agr. Biol. Chem.* **52,** 121 (1988). Antiviral activity: P. S. Sunkara *et al., Biochem. Biophys. Res. Commun.* **148,** 206 (1987). Inhibition of glycoprotein synthesis and secretion: N. Peyrieras *et al., EMBO J.* **2,** 823 (1983); V. Gross *et al., Biochem. J.* **236,** 853 (1986); M. Bollen *et al., Biochem. Pharmacol.* **37,** 905 (1988).

Prisms from water/ethanol, mp 195°. $[\alpha]_D^{21}$ +47° (c = 1.045 in water); $[\alpha]_D^{20}$ +46.7° (c = 0.2 in water); $[\alpha]_D^{20}$ +36° (c = 0.02 in water). pKa 6.6.

2953. Deoxyribonuclease I. Deoxyribonuclease (pancreatic); pancreatic dornase; pancreatic desoxyribonuclease; DNase I; Deanase; Dinase; Dornavac. Single polypeptide chain with two disulfide bridges; 260 amino acid residues; mol wt ~37 kD. Endonuclease enzyme that hydrolyzes DNA. Prepn from beef pancreas: M. Kunitz, *Science* **108,** 19 (1948); Baumgarten, Johnson, U.S. pat. **2,801,956** (1957 to Merck & Co.). Prepn of a stabilized form: *eidem,* U.S. pat. **3,042,587** (1962 to Merck & Co.). Properties of purified DNase I: Price *et al., J. Biol. Chem.* **244,** 917 (1969). Reviews of purification and characterization: M. Laskowski in *The Enzymes* vol. 4, P. D. Boyer, Ed. (Academic Press, New York, 3rd ed., 1971) pp 289-311; S. Moore, *ibid.* **vol. 14** 281-296 (1981); T.-H. Liao, *Molec. Cell Biochem.* **34,** 15-22 (1981).

Recombinant human form, *deoxyribonuclease (human clone 18-1 protein moiety), rhDNase I, dornase alfa, Pulmozyme.* Cloning and expression of human form and *in vitro* reduction of sputum viscosity: S. Shak *et al., Proc. Nat. Acad. Sci. USA* **87,** 9188 (1990); S. Shak, **PCT Int.** pat. **Appl. 7,572** (1990 to Genentech). HPLC purification: J. Cacia *et al., J. Chromatog.* **634,** 229 (1993). Clinical study in cystic fibrosis: H. J. Fuchs *et al., N. Engl. J. Med.* **331,** 637 (1994). Expert report on pharmacology and toxicology: J. D. Green, *Hum. Exp. Toxicol.* **13,** Suppl. 1, S1-S42 (1994).

THERAP CAT: Debriding agent. Recombinant form in treatment of cystic fibrosis (reduction of sputum viscosity).

THERAP CAT (VET): Debriding agent.

2954. Deoxyribonucleic Acid. Desoxyribonucleic acid; DNA; thymus nucleic acid; Desoxiribon; Eucytol. Polynucleotide; essential component of chromosomes in cell nuclei. In its role as the carrier of genetic information, DNA must have two functions: be exactly reproducible in order to transmit its genetic information to future generations; contain information, in chemical code, to direct the development of the cell according to its inheritance. Reviews of biological function: Hotchkiss in *The Nucleic Acids* vol. 2, E. Chargaff, J. N. Davidson, Eds. (Academic Press, New York, 1955) pp 435-473; Crick, *Nature* **227,** 561 (1970); J. N. Davidson, *The Biochemistry of Nucleic Acids* (Academic Press, New York, 7th ed., 1972) pp 6-28. The purine and pyrimidine bases of the nucleosides are primarily adenine, guanine, cytosine and thymine; the sugar is D-2-deoxyribose. The nucleosides are linked together by phosphates in diester linkage from the 3'-hydroxyl of one sugar to the 5'-hydroxyl of the next. The repeating sugar-phosphate linkage forms the backbone of the single polynucleotide strand which is the primary structure of DNA. Chemical analyses of DNA from different species show that the purine content is equal to the pyrimidine content; adenine content equal to thymine; guanine equal to cytosine: Chargaff, *Experientia* **6,** 201 (1950); *idem, Fed. Proc.* **10,** 654 (1951). In the Watson-Crick model of its secondary structure (based on chemical analysis and x-ray studies), DNA consists of two polynucleotide chains forming right-handed helices coiled around the same axis with the sequence of atoms in the two sugar-phosphate backbones running in opposite direction. Two major families of right-handed helix were proposed.

A-DNA and *B-DNA,* each having its own intrinsic restrictions on chain-folding and structure. B-DNA is believed to be the predominant form in biological systems. The purine and pyrimidine bases are inside the helical structure formed by the sugar phosphate backbones; those on one chain form hydrogen bonds to those on the other. Adenine in one chain is always bonded to thymine in the complementary chain by hydrogen bonds; similarly guanine is bonded to cytosine. The linear sequence of bases in one strand completely determines the sequence in the complementary strand. Thus each strand can serve as a template for the replication of the original DNA molecule: Watson, Crick, *Nature* **171,** 737, 964 (1953). X-ray studies: Wilkins *et al., ibid.* 738; Marvin *et al., J. Mol. Biol.* **3,** 547 (1961); Fuller *et al., ibid.* **12,** 60 (1965). DNA also acts as a template in the formation of ribonucleic acids, *q.v.,* which play a fundamental role in the synthesis of proteins in the cell.

Another form of DNA, termed *Z-DNA,* is also known. Its structure is an antiparallel double helix with Watson-Crick base pairing, but it is a left-handed helix with the ribose-phosphate backbone following a zig-zag course. Molecular structure, atomic resolution x-ray crystallographic analysis: A. H.-J. Wang *et al., Nature* **282,** 680 (1979). First identification of Z-DNA in material of biological origin: A. Nordheim *et al., ibid.* **294,** 417 (1981). Studies of B- and Z-DNA: D. J. Patel *et al., Proc. Nat. Acad. Sci. USA* **79,** 1413 (1982). Comparison of A-, B-, and Z-DNA: R. E. Dickerson *et al., Science* **216,** 475 (1982). Demonstration of Z-DNA immunoreactivity in rat tissues: G. Morgenegg *et al., Nature* **309,** 540 (1983). *See also:* Nucleic Acids.

2955. D-2-Deoxyribose. *2-Deoxy-D-erythro-pentose;* desoxyribose; D-2-deoxyarabinose; D-2-ribodesose; D-*erythro*-2-deoxypentose; thyminose. $C_5H_{10}O_4$; mol wt 134.13. C 44.77%, H 7.51%, O 47.71%. Isoln from deoxyribonucleic acid by acidic hydrolysis of purine deoxyribonucleosides which have been isolated by ion-exchange resin chromatography: Laland, Overend, *Acta Chem. Scand.* **8,** 192 (1954). Synthesis: Felton, Freudenberg, *J. Am. Chem. Soc.* **57,** 1637 (1935); Deriaz *et al., J. Chem. Soc.* **1949,** 1879, 2836; Hough, *Chem. & Ind. (London)* **1951,** 406; Sowden, *Biochem. Prepn.* **5,** 75 (1957); I. Ziderman, E. Dimant, *J. Org. Chem.* **32,** 1267 (1967); J. R. Hauske, H. Rapoport, *ibid.* **44,** 2472 (1979); T. Harada, T. Mukaiyama, *Chem. Letters* **1981,** 1109. *Review:* Overend, Stacey, in Chargaff-Davidson, *Nucleic Acids* vol. 1, E. Chargaff, N. J. Davidson, Eds. (Academic Press, New York, 1955) pp 1-80. *See also* Nucleic Acids.

Crystals from isopropanol, mp 91°. Shows mutarotation. Final $[\alpha]_D^{22}$ $-56.2°$ (H_2O). Sol in water, pyridine. Slightly sol in alc.

1,3,4-Triacetate, $C_{11}H_{16}O_7$, needles from methanol, mp 98°. $[\alpha]_D^{23}$ $-171.8°$ (c = 0.56 in chloroform): Allerton, Overend, *J. Chem. Soc.* **1951**, 1480.

3,4,5-Triacetate, oily liq, $bp_{0.001}$ 105°. $[\alpha]_D^{21}$ $+3.4°$ (c = 4.57 in pyridine): Zinner *et al.*, *Ber.* **90**, 2696 (1957).

1,3,4-Tribenzoate, small white nodules from ethanol, mp 127°. $[\alpha]_D^{23}$ $-65°$ (c = 1.02 in chloroform) (Allerton, Overend). Probably a mixture of the two anomeric 2-deoxy-D-ribopyranose tribenzoates: Pedersen *et al.*, *J. Am. Chem. Soc.* **82**, 3425 (1960).

3,4,5-Tribenzoate, fine needles from ethyl acetate + petr ether, mp 118-119°. $[\alpha]_D^{18}$ $-2.8°$ (c = 1.44 in pyridine) (Zinner).

2956. 2-Deoxystreptamine. 2-deoxy-1,3-*myo*-inosadiamine; 1,3-diamino-4,5,6-trihydroxycyclohexane. $C_6H_{14}N_2O_3$; mol wt 162.19. C 44.43%, H 8.70%, N 17.27%, O 29.59%. Important component of several aminocyclitol antibiotics. Production by acid hydrolysis of neamine, *q.v.* and structure: F. A. Kuehl *et al.*, *J. Am. Chem. Soc.* **73**, 881 (1952). Also obtained from hydrolysis products of kanamycins, paromomycins, gentamicins, *q.q.v.* and other aminoglycoside antibiotics. See K. L. Rinehart, *J. Infec. Dis.* **119**, 345 (1969); S. Hanessian, T. H. Haskell, *The Carbohydrates*, vol. IIA, W. Pigman, D. Horton, Eds. (Academic Press, New York, 2nd ed., 1970) pp 159-172. Configuration: J. Daly *et al.*, *J. Am. Chem. Soc.* **82**, 5928 (1960); H. E. Carter *et al.*, *ibid.* **83**, 3723 (1961); R. U. Lemieux, R. J. Cushley, *Can. J. Chem.* **41**, 858 (1963). Synthesis: M. Nakajima *et al.*, *Tetrahedron Letters* **1964**, 967; T. Suami *et al.*, *ibid.* **1967**, 2671; S. Ogawa *et al.*, *J. Org. Chem.* **39**, 812 (1974); H. Pinzbach *et al.*, *Angew. Chem. Int. Ed.* **14**, 223 (1975); S. Ogawa *et al.*, *J. Org. Chem.* **42**, 3083 (1977).

Colorless crystals from ethanol, mp 225-228°.

Dihydrochloride, $C_6H_{14}N_2O_3 \cdot 2HCl$. Crystals, mp 325° (dec), H. Hitomi *et al.*, *Chem. Pharm. Bull.* **9**, 340 (1961).

2957. Deoxyuridine. *2′-Deoxyuridine;* 1-(2-deoxy-β-D-erythro-pentofuranosyl)uracil; 1-(2-deoxy-β-D-ribofuranosyl)uracil; uracil deoxyriboside. $C_9H_{12}N_2O_5$; mol wt 228.20. C 47.37%, H 5.30%, N 12.28%, O 35.05%. Prepn: Dekker, Todd, *Nature* **166**, 557 (1950); Brown *et al.*, *J. Chem. Soc.* **1958**, 3035; Prystas *et al.*, *Coll. Czech. Chem. Commun.* **28**, 3140 (1963); Smejkal *et al.*, *ibid.* **31**, 291 (1966). Conformation: Rahman, Wilson, *Nature* **232**, 333 (1971).

Needles from abs alc or 95% alc, mp 163° (Dekker, Todd) and 167° (Brown *et al.*). $[\alpha]_D^{22}$ $+50°$ (c = 1.1 in *N* NaOH).

α-Anomer, *1-(2-deoxy-α-D-erythro-pentofuranosyl)uracil; 1-(2-deoxy-α-D-ribofuranosyl)uracil.* Prepn: Prystas *et al.*, *loc. cit.*

2958. Deptropine. endo-3-[(10,11-Dihydro-5H-dibenzo-[a,d]cyclohepten-5-yl)oxy]-8-methyl-8-azabicyclo[3.2.1]octane; 3α-[(10,11-dihydro-5H-dibenzo[a,d]cyclohepten-5-yl)oxy]-1αH,5αH-tropane; dibenzheptropine. $C_{23}H_{27}NO$; mol wt 333.47. C 82.84%, H 8.16%, N 4.20%, O 4.80%. Prepn: Van der Stelt *et al.*, *J. Med. Pharm. Chem.* **4**, 335 (1961).

Pharmacology: Funcke *et al.*, *Arch. Int. Pharmacodyn.* **148**, 135 (1964); Timmerman *et al.*, *ibid.* **187**, 291 (1970). Metabolism: Hespe *et al.*, *ibid.* **164**, 397 (1966).

Citrate, $C_{23}H_{27}NO \cdot C_6H_8O_7$, *BS-6987*, *Brontine*. LD_{50} in mice (mg/kg): 32 i.v.; 300 orally (Timmerman).

Maleate, $C_{23}H_{27}NO \cdot C_4H_4O_4$, crystals from ethanol, acetone, ethyl acetate or chloroform, mp 133-136°.

THERAP CAT: Anticholinergic, antihistaminic.

2959. Dequalinium Chloride. *1,1′-(1,10-Decanediyl)-bis-[4-amino-2-methylquinolinium chloride]; 1,1′-decamethylenebis[4-aminoquinaldinium chloride];* BAQD-10; decamine; dekamin; Decatylen; Dekadin; Dequadin Chloride; Dequafungan; Dequavet; Dequavagyn; Eriosept; Evazol; Grocreme; Labosept; Optipect; Phylletten; Polycidine; Sorot. $C_{30}H_{40}Cl_2N_4$; mol wt 527.58. C 68.30%, H 7.64%, Cl 13.44%, N 10.62%. Prepn: Taylor *et al.*, **Brit.** pat. **745,956** (1956 to Allen & Hanburys), *C.A.* **50**, 16878e (1956); Austin *et al.*, *J. Chem. Soc.* **1958**, 1489.

Crystals from ethanol, mp 326° (dec). Soly in water (25°): about one g/200 ml.

Ingredient of *Efisol, Gargilon, Gramipan, Hexalyse, Micrin.*

USE: Bacteriostat.

THERAP CAT: Antiseptic, disinfectant.

THERAP CAT (VET): Antimicrobial.

2960. Dermostatin. Viridofulvin; Dermastatin. Antifungal polyene antibiotic mixture produced by *Streptomyces viridogriseus* Thirum. Comprised of about 43% *Dermostatin A*, $C_{40}H_{64}O_{11}$, and 57% *Dermostatin B*, $C_{41}H_{66}O_{11}$, the first hexaenes to have their structures determined. Preliminary isoln and physico-chemical studies: Bhate *et al.*, *Hindustan Antibiot. Bull.* **4**, 159 (1962), *C.A.* **57**, 13013d (1962); Thirumalachar, Bhate, **Indian** pat. **76,253** (1963 to Hindustan Antibiotics), *C.A.* **59**, 8095b (1963). Purifn: Narasimhachari *et al.*, *Hindustan Antibiot. Bull.* **8**, 111 (1966), *C.A.* **65**, 5301c (1966). Structural studies: Narasimhachari, Swami, *Chemotherapy* **13**, 181 (1968). Revised structure: Pandey *et al.*, *J. Antibiot.* **26**, 475 (1973). Antifungal activity: Menon,

Dermostatin A R = CH₃
Dermostatin B R = CH₂CH₃

Hindustan Antibiot. Bull. **4**, 106 (1962), *C.A.* **57**, 5121c (1962); Gordee, Butler, *J. Antibiot.* **24**, 561 (1971).

Golden-yellow needles, sinters at 180° and darkens at 200°. $[\alpha]_D$ −82° (c = 0.2 in methanol). uv max: 383, 282, 223 nm ($E_{1cm}^{1\%}$ 1000, 212, 130). Sol in aq methanol.

Acetate, prisms from methanol, mp 146-147°. $[\alpha]_D^{23}$ −59.8° (c = 1.37 in chloroform).

THERAP CAT: Antifungal.

2961. Derris Root. Tuba root; Deguelia root. The roots of plants belonging to the genus *Derris,* of which more than 80 species have been described. *Derris elliptica* (Wall.) Benth. and *D. malaccensis* Prain, *Leguminosae* are cultivated in British Malaya and the Dutch East Indies. Toxicity data: Haag *et al., Proc. Soc. Exp. Biol. Med.* **54**, 140 (1943).

LD_{50} orally in mice: 350 mg/kg (as the dry powdered root (Haag)).

USE: As insecticide; as a source of rotenone and rotenoid compds.

2962. Desaspidin BB. *2-[[2,4-Dihydroxy-6-methoxy-3-(1-oxobutyl)phenyl]methyl]-3,5-dihydroxy-4,4-dimethyl-6-(1-oxobutyl)-2,5-cyclohexadien-1-one; 3'-[(5-butyryl-2,4-dihydroxy-3,3-dimethyl-6-oxo-1,4-cyclohexadien-1-yl)methyl]-2',6'-dihydroxy-4'-methoxybutyrophenone;* desaspidin; rosapin. $C_{24}H_{30}O_8$; mol wt 446.50. C 64.56%, H 6.77%, O 28.67%. Isoln from the rhizome of *Dryopteris austriaca* (Jacq.) Wojnar, *Polypodiaceae:* Aebi *et al., Helv. Chim. Acta* **40**, 266 (1957); from *D. caucasica* (A. Br.) Fraser-Jenkins et Corley: Widén *et al., ibid.* **56**, 831 (1973). Structure: Aebi *et al., ibid.* **40**, 572 (1957). Toxicity study: Airaksinan *et al., Acta Pharmacol. Toxicol.* **25**, 33 (1967).

Crystals from ether + petr ether, mp 150-150.5°. uv max: 230, 274 nm (ε 23000, 16800) in cyclohexane. Freely sol in ether, benzene, acetone. Practically insol in methanol, ethanol, petr ether. LD_{50} orally in mice: 340 mg/kg (Airaksinan).

2963. Desatrine. *[3β,4α,6α,7α,15α(R),16β]-4,9-Epoxycevane-3,4,6,7,14,15,16,20-octol 6,7-diacetate 15-(2-methylbutanoate);* protoverine 6,7-diacetate 15-(2-methylbutyrate); protoverine 6,7-diacetate 15(*l*)-2'-methylbutyrate. $C_{36}H_{55}NO_{12}$; mol wt 693.83. C 62.32%, H 7.99%, N 2.02%, O 27.67%. Prepd from protoveratrine B: Kupchan, Ayres, *J. Am. Chem. Soc.* **82**, 2252 (1960); Kupchan, U.S. pats. **3,-009,917** and **3,066,144** (1962 to Wisconsin Alumni Res. Found.). Structure: Kupchan, Ayres, *loc. cit.*

Needles from acetone + petr ether, dec 232-233°. $[\alpha]_D^{26}$ −46° (c = 0.95 in pyridine).

USE: Esters, *e.g., desatrine-3-(N,N-diethylaminoacetate),* used as insecticides.

2964. Deserpidine. *(3β,16β,17α,18β,20α)-17-Methoxy-18-[(3,4,5-trimethoxybenzoyl)oxy]-3,20-yohimban-16-carboxylic acid methyl ester;* 11-desmethoxyreserpine; canescine; recanescine; Harmonyl; Raunormine. $C_{32}H_{38}N_2O_8$; mol wt 578.66. C 66.42%, H 6.62%, N 4.84%, O 22.12%. Isoln from roots of *Rauwolfia canescens* L., *Apocynaceae:* A. Stoll, A. Hofmann, *J. Am. Chem. Soc.* **77**, 820 (1955); M. W. Klohs *et al., ibid.* 4084; N. Neuss *et al., ibid.* 4087; H. B. MacPhillamy *et al., ibid.* 4335; E. Schlittler *et al., Experientia* **11**, 64 (1955); **Brit.** pat. **791,241** (1958 to Penick); **Brit.** pat. **809,912**; P. R. Ulshafer, U.S. pat. **2,982,769** (1959, 1961 both to Ciba). Stereochemistry: C. F. Huebner *et al., Experientia* **11**, 303 (1955); P. E. Aldrich *et al., J. Am. Chem. Soc.* **81**, 2481 (1959). Synthesis: L. Bláha *et al., Coll. Czech. Chem. Commun.* **25**, 237 (1960); **Can.** pat. **678,216** (1964 to Roussel-UCLAF).

Exists in three cryst forms from methanol: α-form, mp 228-232°; β-form, mp 230-232°; and γ-form, mp 138° and 226-232° with resolidification at 175°. $[\alpha]_D^{20}$ −163° (c = 0.5 in pyridine). pKa 6.68 in 40% methanol. uv max (ethanol): 218, 272, 290 nm (log ε 4.79, 4.26, 4.07).

Hydrochloride, $C_{32}H_{38}N_2O_8 \cdot HCl$, thin rectangular plates from acetone, dec 253-256°.

Nitrate, $C_{32}H_{38}N_2O_8 \cdot HNO_3$ crystals, dec 254-260°.

Oxalate, $C_{32}H_{38}N_2O_8 \cdot C_2H_2O_4$, crystals, dec 239-243°.

10-Methoxy deriv, $C_{33}H_{40}N_2O_9$, *10-methoxydeserpine, methoserpidine, Decaserpyl.*

THERAP CAT: Antihypertensive.

2965. Desflurane. *2-(Difluoromethoxy)-1,1,1,2-tetrafluoroethane;* (±)-2-difluoromethyl 1,2,2,2-tetrafluoroethyl ether; I-653; Suprane. $C_3H_2F_6O$. mol wt 168.04. C 21.44%, H 1.20%, F 67.84%, O 9.52%. $CHF_2OCHFCF_3$. Prepd but not claimed: J. P. Russell *et al.*, U.S. pat. **3,897,502** (1975 to Airco). Prepn and use as anesthetic: R. C. Terrell, U.S. pat. **4,762,856** (1988 to BOC). Series of articles on pharmacology and physical properties: E. I. Eger II *et al., Anesth. Analg.* **66**, 971-985 (1987); and toxicity studies: *eidem, ibid.* 1227, 1230. Series of articles on clinical pharmacology and kinetics: *Anesthesiology* **74**, 412-439; 479-498. *Review:* R. M. Jones, *Brit. J. Anaesthesia* **65**, 527-536 (1990).

Volatile liquid. Slight non-pungent odor. Nonflammable, soda lime stable. bp 23.5°. d 1.44. Vapor pressure (20°): 88.53 kPa. Vapor pressure (22-23°): ~700mm. Partition coefficient (37°): blood/gas 0.424 ±0.024; saline/gas 0.225 ±0.002; oil/gas 18.7 ±1.1.

THERAP CAT: Anesthetic (inhalation).

2966. Desipramine. *10,11-Dihydro-N-methyl-5H-dibenz[b,f]azepine-5-propanamine; 10,11-dihydro-5-[3-(methylamino)propyl]-5H-dibenz[b,f]azepine;* 5-(γ-methylaminopropyl)iminodibenzyl; N-(3-methylaminopropyl)iminobibenzyl; desmethylimipramine; norimipramine. $C_{18}H_{22}N_2$; mol wt 266.39. C 81.16%, H 8.32%, N 10.52%. Prepn: **Brit.** pat. **908,788**; **Belg.** pat. **614,616** (both 1962 to Geigy); J. H. Biel, C. I. Judd, **Brit.** pat. **980,231** (1965 to Lakeside); *eidem,* U.S. pat. **3,454,554** (1969 to Colgate Palmolive). Chemistry, pharmacology: E. Eriksoo, O. Rohte, *Arzneimittel-Forsch.* **20**, 1561 (1970). Clinical response, plasma levels, pharmacokinetics: P. D. Hrdina, Y. D. Lapierre, *Prog. Neuropsychopharmacol.* **4**, 591 (1980). Mechanism of action study: S. A. Checkley *et al., Brit. J. Psychiatry* **138**, 248 (1981). Efficacy in depression: W. Z. Potter *et al., Psychopharmacol. Bull.* **17**, 26 (1981); J. W. Stewart *et al., ibid.* 136. Teratological study: L. Aeppli, *Arzneimittel-Forsch.* **19**, 1617 (1969). Carcinogenicity and mutagenicity study: H. Kubinski *et al., Mutat. Res.* **89**, 95 (1981). Has been found to be one of the most effective cmpds described for *in vitro* and *in vivo* reversal of chloroquine, *q.v.*, resistance in *Plasmodium falciparum:* A. J. Bitonti *et al., Science* **242**, 1301 (1988).

bp$_{0.02}$ 172-174°. uv max: 213, 252 nm (log ε 4.39, 3.93). Hydrochloride, C$_{18}$H$_{22}$N$_2$.HCl, *G-35020, JB-8181, NSC-114901, Norpramin, Nortimil, Pertofran, Pertofrane.* Crystals from methanol ether, mp 215-216°. Sol in water. LD$_{50}$ in mice, rats (mg/kg): 500, 385 orally; 94, 48 i.p.; 420, 183 s.c. (Eriksoo, Rohte).

THERAP CAT: Antidepressant.

2967. Deslanoside. *(3β,5β,12β)-3-[(O-β-D-Glucopyrano-syl-(1→4)-O-2,6-dideoxy-β-D-ribo-hexopyranosyl-(1→4)-O-2,-6-dideoxy-β-D-ribo-hexopyranosyl-(1→4)-2,6-dideoxy-β-D-ribo-hexopyranosyl)oxy]-12,14-dihydroxycard-20(22)-enolide; deacetyllanatoside C;* desacetyldigilanide C; Purpurea glycoside C; Cedilanid-D; Desace; Desaci; Lanimerck (ampuls). C$_{47}$H$_{74}$O$_{19}$; mol wt 943.09. C 59.86%, H 7.91%, O 32.23%. From leaves of *Digitalis lanata* Ehrh., *Scrophulariaceae:* Stoll, Kreis, *Helv. Chim. Acta* **16**, 1049, 1390 (1933). Prepn by alkaline degradation of lanatoside C: Kroszczynski *et al., Acta Polon. Pharm.* **21**, 357 (1964), *C.A.* **62**, 10292b (1965).

β-D-glucose–(β-D-digitoxose)$_3$–O

Crystals from methanol, dec 265-268°. [α]$_D^{20}$ +12° (c = 1.084 in 75% alc). One part dissolves in 5000 parts water, 200 parts methanol and 2500 parts ethanol; very slightly sol in chloroform. Practically insol in ether.

THERAP CAT: Cardiotonic.

2968. Deslorelin. *6-D-Tryptophan-9-(N-ethyl-L-prolin-amide)-10-deglycinamideluteinizing hormone-releasing factor (pig);* [D-Trp^6,des-Gly^{10}]-LH-RH ethylamide; Somagard. C$_{64}$H$_{83}$N$_{17}$O$_{12}$; mol wt 1282.47. C 59.94%, H 6.52%, N 18.57%, O 14.97%. Synthetic nonapeptide agonist analog of LH-RH, *q.v.* Prepn: D. H. Coy *et al., J. Med. Chem.* **19**, 423 (1976); W. Kornreich *et al., Int. J. Pept. Protein Res.* **25**, 414 (1985). Pharmacology: K. Sundaram *et al., Life Sci.* **28**, 83 (1981). Pharmacokinetics and metabolism: B. Candas *et al., J. Clin. Endocrinol. Metab.* **70**, 1046 (1990). Clinical inhibition of spermatogenesis: R. Linde *et al., N. Engl. J. Med.* **305**, 663 (1981). Clinical trial in combination with flutamide, *q.v.*, for prostatic carcinoma: A. Dupont *et al., Eur. J. Cancer Clin. Oncol.* **24**, 659 (1988). Clinical evaluation in precocious puberty: P. K. Manasco *et al., J. Pediatrics* **115**, 105 (1989).

5-oxoPro–His–Trp–Ser–Tyr–D-Trp–Leu–Arg–ProNHCH$_2$CH$_3$

[α]$_D^{24}$ −61° (c = 0.37 in 0.1M acetic acid).
THERAP CAT: Treatment of precocious puberty.

2969. Desmopressin. *1-(3-Mercaptopropanoic acid)-8-D-arginine vasopressin; 8-D-arginine-1-(3-mercaptopropanoic acid) vasopressin;* 1-desamino-8-D-arginine vasopressin; Adiuretin SD; DAV Ritter; DDAVP; Desmospray; Minirin. C$_{46}$H$_{64}$N$_{14}$O$_{12}$S$_2$; mol wt 1069.23. C 51.67%, H 6.03%, N 18.34%, O 17.96%, S 6.00%. An analog of vasopressin possessing high antidiuretic activity. Prepn: M. Zaoral *et al., Fr. pat.* **1,540,536;** *eidem,* U.S. pat. **3,497,491** (1968, 1970 both to Ceskoslovenska Akad. Ved). Syntheses: R. L. Huguenin, R. A. Boissonnas, *Helv. Chim. Acta* **49**, 695 (1966); M. Zaoral *et al., Coll. Czech. Chem. Commun.* **32**, 1250 (1967). Clinical studies: I. Vavra *et al., Lancet* **1**, 948 (1968); K. E. Andersson, B. Arner, *Acta Med. Scand.* **192**, 21 (1972). Antidiuretic action in rats: I. Vavra *et al., J. Pharmacol. Exp. Ther.* **188**, 241 (1974).

S–CH$_2$–CH$_2$–C–Tyr–Phe–Gln–Asn–Cys–Pro–D-Arg–GlyNH$_2$
∥
O

[α]$_D^{25}$ +85.5° ± 2° (calculated for the free peptide). Monoacetate, C$_{46}$H$_{64}$N$_{14}$O$_{12}$S$_2$.C$_2$H$_4$O$_2$, *Octostim.*
THERAP CAT: Antidiuretic.

2970. Desmosterol. *3β-Cholesta-5,24-dien-3-ol;* 24-dehydrocholesterol; desmesterol. C$_{27}$H$_{44}$O; mol wt 384.65. C 84.31%, H 11.53%, O 4.16%. Has been isolated from chick embryos and the skin of rats. Stokes *et al., J. Biol. Chem.* **220**, 415 (1956). From barnacles: Fagerlund, Idler, *J. Am. Chem. Soc.* **79**, 6473 (1957); from red algae: Idler *et al., Steroids* **11**, 465 (1968). Synthesis: Dasgupta *et al., J. Org. Chem.* **39**, 1658 (1974); M. A. Apfel, *ibid.* **43**, 2284 (1978). Stereospecific synthesis: M. Koreeda *et al., ibid.* **45**, 1172 (1980); S. Takano *et al., Chem. Commun.* **1983**, 760.

Platelets from methanol, mp 121.5°, also reported as 119-119.5°. [α]$_D^{27}$ −41.0° (c = 1 in chloroform).

2971. Desogestrel. *(17α)-13-Ethyl-11-methylene-18,19-dinorpregn-4-en-20-yn-17-ol;* 17α-ethynyl-18-methyl-11-methylene-Δ^4-estren-17β-ol; Org-2969. C$_{22}$H$_{30}$O; mol wt 310.48. C 85.11%, H 9.74%, O 5.15%. Progestogen with low androgenic potency. Prepn: A. J. van den Broek, **Ger.** pat. **2,361,120;** *idem,* U.S. pat. **3,927,046** (1974, 1975 both to Akzo); A. J. van den Broek *et al., Rec. Trav. Chim.* **94**, 36 (1975). Biological effects: L. Viinikka *et al., Acta Endocrinol.* **83**, 429 (1976). Endocrinological studies in animals: J. van der Vies, J. de Visser, *Arzneimittel-Forsch.* **33**, 231 (1983). Radioimmunoassay: L. Viinikka, *J. Steroid Biochem.* **9**, 979 (1978). Metabolism, pharmacokinetics in humans: L. Viinikka *et al., Eur. J. Clin. Pharmacol.* **15**, 349 (1979). Receptor binding study: E. W. Bergink *et al., J. Steroid Biochem.* **14**, 175 (1981). Clinical study: M. J. Weijers, *Clin. Ther.* **4**, 359 (1982).

Crystals, mp 109-110°. [α]$_D^{20}$ +55° (chloroform).

Mixture with ethinyl estradiol, *Cyclosa, Desogen, Dicromil, Marvelon 150/30, Mercilon, Ortho-Cept, Oviol, Varnoline.*

THERAP CAT: Progestogen. In combination with estrogen as oral contraceptive.

2972. Desomorphine. *4,5a-Epoxy-17-methylmorphinan-3-ol;* dihydrodesoxymorphine-D; Permonid. $C_{17}H_{21}NO_2$; mol wt 271.36. C 75.25%, H 7.80%, N 5.16%, O 11.79%. Synthesis, physical data: L. F. Small *et al., J. Am. Chem. Soc.* **55**, 3863 (1933).

Rectangular plates from acetone-water. Sublimes in high vacuum between 140° and 170°. (Anhydr base mp 189°). $[\alpha]_D^{28}$ −77° (c = 1.6 in methanol). Sol in acetone, ethyl acetate. Alcoholic solns darken and acquire a reddish color.

Hemihydrate, crystals from ethyl acetate, mp 162-164°. $[\alpha]_D^{28}$ −79° (c = 1.08 in ethyl acetate).

Hydrochloride, $C_{17}H_{21}NO_2$.HCl, crystals from alcohol, $[\alpha]_D^{27}$ −67° (c = 0.90). Very sol in water.

Hydrogen sulfate dihydrate, $(C_{17}H_{21}NO_2)_2.H_2SO_4.2H_2O$, scales, plates from water, brown at 210°, unmelted at 230°. Sol in 40 parts water; freely sol in hot water. Soln is stable to boiling. $[\alpha]_D^{29}$ −60° (c = 1.43).

Note: This is a controlled substance (opium derivative) listed in the U.S. Code of Federal Regulations, Title 21 Part 1308.11 (1995).

THERAP CAT: Analgesic (narcotic).

2973. Desonide. *(11β,16α)-11,21-Dihydroxy-16,17-[(1-methylethylidene)bis(oxy)]pregna-1,4-diene-3,20-dione; 11β,16α,17,21-tetrahydroxypregna-1,4-diene-3,20-dione cyclic 16,-17-acetal with acetone;* 16α-hydroxyprednisolone-16α,17-acetonide; 11β,21-dihydroxy-16α,17-isopropylidenedioxy-1,-4-pregnadiene-3,20-dione; 16α-hydroxy-Δ¹-hydrocortisone-16α,17α-acetonide; 16α,17α-isopropylidenedioxyprednisolone; prednacinolone; D-2083; Locapred; Sterax; Steroderm; Topifug; Tridesilon. $C_{24}H_{32}O_6$; mol wt 416.51. C 69.21%, H 7.74%, O 23.05%. Prepn and activity: Bernstein *et al., J. Am. Chem. Soc.* **81**, 4573 (1959); Bernstein, Allen, U.S. pat. **2,990,401** (1961 to Am. Cyanamid); Lee *et al.* and Diassi, Principe, U.S. pats. **3,536,586** and **3,549,498** (both 1970 to Squibb). Pharmacological and toxicological studies: Mascitelli-Coriandoli, Fraia, *Arzneimittel-Forsch.* **20**, 111 (1970); Phillips *et al., Toxicol. Appl. Pharmacol.* **20**, 522 (1971). Structure-activity studies: Ringler *et al., Proc. Soc. Exp. Biol. Med.* **107**, 451 (1961). Chemical and physical properties: Mantica *et al., Arzneimittel-Forsch.* **20**, 109 (1970).

Small plates or white to off-white odorless powder, mp 274-275° from methanol (Mantica); also reported as mp 263-266° from ethyl acetate-petr ether (Bernstein). $[\alpha]_D^{25}$ +123° (c = 0.5 in DMF). uv max: 242 nm ($E_{1cm}^{1\%}$ 356). LD_{50} in rats (mg/kg): 93 s.c. (Phillips).

THERAP CAT: Anti-inflammatory.

2974. Desosamine. *3,4,6-Trideoxy-3-(dimethylamino)-D-xylo-hexose;* 4-dimethylaminotetrahydro-6-methylpyran-2,3-diol; picrocine. $C_8H_{17}NO_3$; mol wt 175.23. C 54.84%, H 9.78%, N 7.99%, O 27.39%. Sugar component of several

macrolide antibiotics, such as the erythromycins. Isoln: Clark, *Antibiot. & Chemother.* **3**, 663 (1953). Identity with picrocine: Brockmann *et al., Ber.* **87**, 856 (1954). Stereochemistry: Woo *et al., Tetrahedron Letters* **1962**, 735. Configuration: Bolton *et al., Chem. & Ind. (London)* **1962**, 1945; Foster *et al., J. Chem. Soc.* **1965**, 2318. Synthesis of DL-form: Korte *et al., Tetrahedron* **18**, 657 (1962); Newman, *Chem. & Ind. (London)* **1963**, 372; *J. Org. Chem.* **29**, 1461 (1964). Stereospecific synthesis of the natural D-form: Richardson, *J. Chem. Soc.* **1964**, 5364; of the L-form: Baer, Chiu, *Can. J. Chem.* **52**, 122 (1974).

Crystals, mp 86-87°.

Hydrochloride, $C_8H_{17}NO_3$.HCl, needles from ethanol + acetone, dec 191-193°. $[\alpha]_D^{20}$ +49.5° (c = 10.0); $[\alpha]_D^{20}$ +53.4° (c = 2.1 in ethanol).

2975. Desoximetasone. *(11β,16α)-9-Fluoro-11,21-dihydroxy-16-methylpregna-1,4-diene-3,20-dione;* 9α-fluoro-16α-methyl-17-desoxyprednisolone; 9α-fluoro-16α-methyl-Δ¹-corticosterone; desoxymethasone; A-41-304; R-2113; HOE-304; Esperson; Stiedex; Topicorte; Topisolon. $C_{22}H_{29}FO_4$; mol wt 376.47. C 70.19%, H 7.76%, F 5.05%, O 17.00%. Prepn: R. Joly *et al., Fr. pat.* **1,296,544;** *eidem,* U.S. pat. **3,099,654** (1962, 1963 to Roussel-UCLAF); Belg. pat. **614,-196;** Kieslich *et al.,* U.S. pat. **3,232,839** (1962, 1966 to Schering AG); R. Joly *et al., Arzneimittel-Forsch.* **24**, 1 (1974). Activity studies: Branceni *et al., Steroids* **6**, 451 (1965); Schröder *et al., Arzneimittel-Forsch.* **24**, 3 (1974). NMR data: Lukacs *et al., C. R. Acad. Sci. Ser. C* **274**, 1458 (1972). Review of pharmacology and therapeutic efficacy in dermatoses: R. C. Heel *et al., Drugs* **16**, 302-321 (1978).

Crystals from ethyl acetate, mp 217°. $[\alpha]_D$ +109° (chloroform). uv max: 238 nm (ε 15750). Sol in alcohol, acetone, chloroform, hot ethyl acetate; slightly sol in ether, benzene. Insol in water, dil aq acids and alkalies.

THERAP CAT: Anti-inflammatory; glucocorticoid.

2976. 6-Desoxy-D-glucosamine. *2-Amino-2,6-dideoxy-D-glucose;* 6-deoxy-D-glucosamine; 2,6-didesoxy-2-amino-D-glucose. $C_6H_{13}NO_4$; mol wt 163.17. C 44.17%, H 8.03%, N 8.58%, O 39.22%. May be prepd by tosylation of the primary alcohol group of a suitable derivative of D-glucosamine, followed by acetylation and replacement of the tosyl group by iodine, and catalytic cleavage of the halide with Raney nickel in the presence of triethylamine: Morel, *Helv. Chim. Acta* **41**, 1501 (1958).

Hydrochloride, $C_6H_{13}NO_4$.HCl, crystals from abs ethanol + ether, dec 172-173°. Shows mutarotation. Final value: $[\alpha]_D^{20}$ +55° (H_2O).

1,3,4-Tri-O-acetyl-N-acetyl-6-desoxy-β-D-glucosamine, $C_{14}H_{21}NO_8$. Crystals from alc, mp 209-210°. $[\alpha]_D^{20}$ +17.5° (chloroform).

USE: In the investigation of amino acid metabolism.

2977. 11-Desoxy-17-hydroxycorticosterone. *17,21-Dihydroxypregn-4-ene-3,20-dione;* 17-hydroxydesoxycorticosterone; 4-pregnene-17α,21-diol-3,20-dione; 17-(1-keto-2-hydroxyethyl)-4-androsten-17α-ol-3-one; Reichstein's Substance S; 11-desoxycortisone; cortexolone. $C_{21}H_{30}O_4$; mol wt 346.47. C 72.80%, H 8.73%, O 18.47%. Isoln and partial synthesis: Reichstein, von Euw, *Helv. Chim. Acta* **21**, 1197 (1938); Reichstein, *ibid.* 1490; Reichstein, von Euw, *ibid.* **23**, 1258 (1940). Partial synthesis involving the chromic acid oxidation of 4-pregnene-17α,20,21-triol-3-one 21-monoacetate: Sarett, *J. Biol. Chem.* **162**, 627 (1946). Prepn from 3α-formoxy-17α-hydroxypregnan-20-one: Gallagher *et al., J. Am. Chem. Soc.* **71**, 3262 (1949); from 16,17-oxido-5-pregnen-3β-ol-20-one acetate: Julian *et al., ibid.* 3574; from 5-pregnen-3β-ol-20-one: Julian *et al., ibid.* **72**, 5145 (1950).

Fine, glistening plates from ether. mp 212.8-216.8°; gives no depression of melting point when mixed with cortisone. Very sparingly sol in water, ether. Sol in acetone, methanol, alcohol. Gives a carmine-red fluorescence reaction with concd. H_2SO_4. Reduces ammoniacal silver nitrate soln at room temp. uv max: 242 nm ($E_{1cm}^{1\%}$ 500). Oxidation with chromic acid in glacial acetic acid yields 4-androstene-3,17-dione.

Acetate, $C_{23}H_{32}O_5$, mp 237.2-240.2° (sinters at 230°); $[\alpha]_D^{24}$ +116° (acetone); uv max (methanol): 242 nm ($E_{1cm}^{1\%}$ 448). When dissolved in concd sulfuric acid produces a typical scarlet color.

2978. Desthiobiotin. *(4R-cis)-5-Methyl-2-oxo-4-imidazolidinehexanoic acid; 5-methyl-2-oxo-4-imidazolidinecaproic acid;* ε-(4-methyl-5-imidazolidone-2)-caproic acid; 4-methyl-5-(ω-carboxyamyl)imidazolidone-2. $C_{10}H_{18}N_2O_3$; mol wt 214.26. C 56.06%, H 8.47%, N 13.07%, O 22.40%. Prepd from biotin by hydrogenolysis of the sulfide linkage: du Vigneaud *et al., J. Biol. Chem.* **146**, 475 (1942). Improved prepn procedure: Melville *et al., Science* **98**, 497 (1943). Synthesis: Melville, *J. Am. Chem. Soc.* **66**, 1422 (1944); Wood, du Vigneaud, *ibid.* **67**, 210 (1945); Duschinsky, Dolan, *ibid.* 2079; Bourquin *et al., Helv. Chim. Acta* **28**, 528 (1954).

Long needles from water. mp 156-158°. $[\alpha]_D^{21}$ +10.7° (c = 2). Sol in water.

Methyl ester, $C_{11}H_{20}N_2O_3$, crystals from methanol. Can be sublimed at 10^{-5} mm and 100°. mp 69-70°. $[\alpha]_D^{28}$ +2.6° (c = 2 in chloroform). Readily sol in chloroform.

2979. Destomycin A. *O-6-Amino-6-deoxy-L-glycero-D-galacto-heptopyranosylidene-(1→2-3)-O-β-D-talopyranosyl-(1→5)-2-deoxy-N'-methyl-D-streptamine; 5-O-[2,3-O-[6-(1-amino-2-hydroxyethyl)tetrahydro-3,4,5-trihydroxy-2H-pyran-2-ylidene]-β-D-talopyranosyl]-2-deoxy-N³-methyl-D-streptamine;* Destonate 20. $C_{20}H_{37}O_3O_{13}$; mol wt 527.53. C 45.54%, H 7.07%, N 7.97%, O 39.43%. Aminoglycoside antibiotic, member of the orthosomycin family. Destomycins A, B and C are known; A is the major component. Isoln of A and B from culture broth of *Streptomyces rimofaciens:* S.-I. Kondo *et al., J. Antibiot.* **18A**, 38 (1965); of C:

eidem, ibid. **28**, 83 (1975). Prepn: *eidem,* **Japan.** pat. **67 07598** (1967 to Meiji Seika Kaisha). Structure of A: *eidem, J. Antibiot.* **19A**, 139 (1966); of A and B: *eidem, ibid.* **28**, 79 (1975). Stereochemical study: S. Horito *et al., Bull. Chem. Soc. Japan* **54**, 2147 (1981). Synthesis of destomycin C: J. Yoshimura *et al., Chem. Letters* **1985**, 1335; J.-I. Tamura *et al., Carbohydr. Res.* **174**, 181 (1988). Anthelmintic activity in poultry: I. Sawada, *Kiseichugaku Zasshi* **21**, 45 (1972), *C.A.* **77**, 135668j (1972). HPLC determn in pork samples: K. Nakaya *et al., Shokukin Eiseigaku Zasshi* **28**, 487 (1987), *C.A.* **109**, 5359c (1988).

White powder, mp 180-190° (dec). $[\alpha]_D^{22}$ +7° (c = 2). Freely sol in water, lower alcohols. Insol or poorly sol in most organic solvents. LD_{50} in mice (after 2-week observation) (mg/kg): 5-10 i.v.; 50-100 orally (Kondo, 1965).

THERAP CAT (VET): Broad spectrum antimicrobial; anthelmintic.

2980. Detaxtran. *Dextran 2-(diethylamino)ethyl ether;* diethylaminoethyldextran; DEAE-dextran; basic dextran. Mol wt ~500,000. Cationic derivative of dextran, *q.v.,* with antilipemic activity. Prepn: W. McKernan, C. R. Ricketts, *Chem. & Ind. (London)* **1959**, 1490; E. Antonini *et al., Giorn. Biochem.* **14**, 88 (1965); *C.A.* **63**, 4375h (1965). Effect on platelet aggregation: R. Brossmer, Th. Pfliederer, *Naturwiss.* **53**, 464 (1966); A. Larcon *et al., Experientia* **28**, 1096 (1972). Use to lower hypercholesterolemia: T. M. Parkinson, U.S. pat. **3,627,872** (1971 to Upjohn); to reduce blood lipid levels: F. Kuzuya *et al.,* U.S. pat. **3,851,057** (1974 to Meito Sangyo). Clinical trial in hyperlipoproteinemia: F. Pupita, A. Barone, *Int. J. Clin. Pharm. Res.* **3**, 287 (1983).

Powder, soluble in water, saline.

Hydrochloride, *Dexide, Pulsar, Lipalt.*

THERAP CAT: Antihyperlipoproteinemic.

2981. Detomidine. *4-[(2,3-Dimethylphenyl)methyl]-1H-imidazole;* 4-(2',3'-dimethylbenzyl)imidazole. $C_{12}H_{14}N_2$; mol wt 186.26. C 77.38%, H 7.58%, N 15.04%. α₂-Adrenoceptor agonist with sedative and analgesic activity. Prepn: A. J. Karjalayne, K. O. A. Kurkela, Eur. pat. Appl. **24,829**; *eidem,* U.S. pat. **4,443,466** (1981, 1984 both to Farmos). Physical studies: E. Laine *et al., Acta Pharm. Suec.* **20**, 451 (1983). Crystal structure: L. H. J. Lajunen *et al., ibid.* **21**, 163 (1984). Pharmacology: R. Virtanen, L. Nyman, *Eur. J. Pharmacol.* **108**, 163 (1985); R. Virtanen, E. MacDonald, *ibid.* **115**, 277 (1985). Mechanism of action: *eidem, J. Vet. Pharmacol. Ther.* **8**, 30 (1985).

Crystals from acetone, mp 114-116°. LD_{50} i.v. in mice: 35 mg/kg (Karjalayne, Kurkela).

Hydrochloride, $C_{12}H_{14}N_2 \cdot HCl$, *Domosedan.* Crystals, mp 160°. Converts reversibly to monohydrate at room temp, 80% humidity.

THERAP CAT (VET): Sedative.

2982. Detoxin Complex. DX.C. A group of selective antagonists of blasticidin S, *q.v.,* produced by *Streptomyces caespitosus* var. *detoxicus* 7072 GC₁. The complex counteracts the inhibitory action of blasticidin S (BS) against *Bacillus cereus* but not against *Piricularia oryzal,* the causative agent of rice blast disease. DX.C also depresses the phytotoxicity of BS to rice plants, and a combination of the two showed less eye irritation in animals than BS alone. Isoln, production, biological properties: H. Yonehara *et al., J. Antibiot.* **21**, 369 (1968); *eidem, Agr. Biol. Chem.* **37**, 2771 (1973). Separation of the complex yields eight groups of detoxins designated A through H, the major components being C and D: N. Otake *et al., J. Antibiot.* **21**, 371 (1968); *eidem, Agr. Biol. Chem.* **37**, 2777 (1973). Detoxin D_1, the major component of the D group, has been shown to have the highest specific activity. It is a unique depsipeptide containing the amino acid *detoxinine* (*β,3-dihydroxy-2-pyrrolidinepropanoic acid*). Structure of D_1: K. Kakenuma *et al., Tetrahedron Letters* **1972**, 2509. ^{13}C-NMR: T. Ogita *et al., Agr. Biol. Chem.* **42**, 2403 (1978). Structures of the minor components of the D group: N. Otake *et al., J. Antibiot.* **27**, 484 (1974). Structures of detoxins A_1, B_1, B_3, C_1, C_2, C_3, E_1: N. Otake *et al., Experientia* **37**, 926 (1981); T. Ogita *et al., Agr. Biol. Chem.* **45**, 2605 (1981). Effect of D on BS uptake in *B. cereus*: A. Shimazu *et al., Experientia* **37**, 365 (1981).

Detoxin D_1 R = CH(CH$_3$)CH$_2$CH$_3$
Detoxin C_1 R = CH$_3$

Light colored fine powder.

Detoxin D_1, $C_{28}H_{41}N_3O_8$, *[2S-[1(R*),2α[R*,(R*)],3α]]-N-(2-methyl-1-oxobutyl)-L-phenylalanine 1-[3-(acetyloxy)-1-(2-amino-3-methyl-1-oxobutyl)-2-pyrrolidinyl]-2-carboxyethyl ester*. Fine crystalline powder, mp 156-158°. $[α]_D^{25}$ −16° (c = 1 in methanol). uv max (methanol): 253, 258, 265, 268 nm ($E_{1cm}^{1\%}$ 3.1, 3.58, 2.77, 1.85). Amphoteric. pKa 4.0, 8.0.

Detoxin C_1, $C_{25}H_{35}N_3O_8$, *[2S-[1(R*),2α(R*),3α]]-N-acetyl-L-phenylalanine 1-[3-(acetyloxy)-1-(2-amino-3-methyl-1-oxobutyl)-2-pyrrolidinyl]-2-carboxyethyl ester, detoxin C_{a1}*. Microneedles from water, mp 142-144°. Amphoteric. pKa 8.0, 3.9. $[α]_D^{25}$ −23° (c = 1 in methanol). uv max (methanol): 248, 253, 259, 265, 269 nm ($E_{1cm}^{1\%}$ 3.7, 3.18, 2.65, 2.8, 1.85).

USE: In mixtures with blasticidin S, to decrease phytotoxicity to rice plants.

2983. Deuterium. Heavy hydrogen. ^2H or D. Exists in the diatomic state; mol wt 4.028. A stable, non-radioactive isotope of hydrogen used in nuclear work. May be obtained by electrolysis of heavy water (obtained by the H_2S/H_2O exchange process) or by fractional distillation of liq hydrogen: Urey *et al., Phys. Rev.* **39**, 164 (1932); Spevack, U.S. pat. **2,787,526** (1957 to USAEC). The hydrogen bomb contains lithium deuteride (LiD) as explosive and plutonium (Pu) as initiator. After detonation by Pu the following reaction sequence takes place (nuclear fusion): ^{6}Li + D = 2^{4}He; ^{6}Li + n = ^{4}He + T; ^{6}Li + T = 2^{4}He + n (n = neutrons, coming from Pu in the second reaction). Since the explosion does not start by itself, there is no critical mass and no limit to the size of the bomb. If controlled thermonuclear fusions can be achieved, deuterium gas will become increasingly important. *Review:* Mackay, Dove, in *Comprehensive Inorganic Chemistry*, vol. **1**, J. C. Bailar, Jr. *et al.*, Eds. (Pergamon Press, Oxford, 1973) pp 77-116; J. J. Katz in Kirk-Othmer *Encyclopedia of Chemical Technology* vol. **7** (Wiley-Interscience, New York, 3rd ed., 1979) pp 539-553.

Colorless, odorless, flammable gas (*see* Hydrogen). d (liq, 20.4°K) 0.169. bp −249.49° (23.67°K). mp −254.43° (18.73°K) at 128.5 mm (triple point). Crit temp −234.75°. Crit press. 16.432 atm. Flammable limits in air: 5-75%.

Calculation of vapor press. from the triple point to the crit point: Friedman *et al., J. Am. Chem. Soc.* **73**, 1310 (1951).

Caution: Considered non-toxic (except for its asphyxiating and flammable properties) as used in laboratories and nuclear power plants. Stunts growth of mammals when drunk regularly as heavy water: J. F. Thomson, *Biological Effects of Deuterium* (Pergamon Press, 1964).

USE: Used extensively in small amts as tracer in the establishment of rates and kinetics of chemical reactions: Wiberg, *Chem. Revs.* **55**, 713-743 (1955).

2984. Deuterium Oxide. Heavy water; water-d_2. D_2O; mol wt 20.028. D 20.11%, O 79.89%.

More associated than H_2O. mp 3.81°; triple point temp 3.82°. bp 101.42°. Critical temp 371.5°. d^{25} 1.1044. Temp of max density 11.23°; d$^{11.23}$ 1.1059. Sp heat of liq (4-25°) 1.028 cal/g/°C. Heat of fusion 1.501 kcal/mole: Long, Kemp, *J. Am. Chem. Soc.* **58**, 1829 (1936). Heat of evapn 9.917 kcal/mole. Dielectric const (25°): 78.06, Vidulich *et al., J. Phys. Chem.* **71**, 656 (1966). Dipole moment in benzene (25°) 1.78, in dioxane 1.87. The surface tension is very slightly smaller and the viscosity at 25° is 1.23 times as great as that of water. On mixing with water, heat is evolved. Reacts with hydrogen at 100° in the presence of 0.2 to $1N$ NaOH forming DOH + HD + OD⁻. Ionization const 1.95 × 10⁻¹⁵: W. F. K. Wynne-Jones, *Trans. Faraday Soc.* **32**, 1397 (1936). pK at 25° 14.955 (molarity scale); 16.653 (mole fraction scale): A. K. Covington *et al., J. Phys. Chem.* **70**, 3820 (1966). *See also* Water.

USE: To study chemical reaction rates and mechanisms. The cross section of deuterium for the capture of thermal neutrons is very low which makes it useful, in the form of heavy water, as a neutron moderator in nuclear reactors. Produces a considerable decrease in neutron energy per collision.

2985. Devarda's Metal. Devarda's alloy. Consists of 50 parts Cu, 45 Al and 5 Zn.

Gray powder. Partly sol in HCl with residue of Cu.

USE: Reducing agent for determination of nitrogen in nitrates and nitrites.

2986. Dexamethasone. (*11β,16α*)-9-*Fluoro-11,17,21-trihydroxy-16-methylpregna-1,4-diene-3,20-dione;* 9α-fluoro-16α-methylprednisolone; 16α-methyl-9α-fluoro-1,4-pregnadiene-11β,17α,21-triol-3,20-dione; 16α-methyl-9α-fluoroprednisolone; 1-dehydro-16α-methyl-9α-fluorohydrocortisone; 16α-methyl-9α-fluoro-Δ¹-hydrocortisone; hexadecadrol; Aeroseb-Dex; Corson; Cortisumman; Decacort; Decaderm; Decadron; Decalix; Decasone; Dekacort; Deltafluorene; Deronil; Deseronil; Dexacortal; Dexacortin; Dexafarma; Dexa-Mamallet; Dexameth; Dexamonozon; Dexapos; Dexa-sine; Dexasone; Dexinoral; Dinormon; Fluormone; Isopto-Dex; Lokalison F; Loverine; Luxazone; Maxidex; Millicorten; Pet-Derm III. $C_{22}H_{29}FO_5$; mol wt 392.47. C 67.33%, H 7.45%, F 4.84%, O 20.38%. Prepn: Arth *et al., J. Am. Chem. Soc.* **80**, 3161 (1958); Oliveto *et al., ibid.* 4431; Muller *et al.*, U.S. pat. **3,007,923** (1961 to Lab. Franc. Chimiothér.); Arth *et al.*, **Ger.** pat. **1,113,690** (1961 to Merck & Co.); **Brit.** pat. **869,511** (to Upjohn). Pharmacology in guinea pigs: L. Zicha *et al., Arzneimittel-Forsch.* **10**, 831 (1960). Clinical trial of anti-emetic efficacy in chemotherapy-induced nausea: M. Markman *et al., N. Engl. J. Med.* **311**, 549 (1984). Use as diagnostic aid in depression: B. J. Carroll *et al., Arch. Gen. Psychiat.* **38**, 15 (1981); P. P. Hubain *et al., Neuropsychobiology* **16**, 57 (1986); M. Ansseau *et al., ibid.* 68. Comprehensive description: E. M. Cohen, *Anal. Profiles Drug Subs.* **2**, 163-197 (1973). Reviews of diagnostic use in Cushing's Syndrome: L. Crapo, *Metabo-*

lism **28,** 955 (1979); in depression: L. Braddock, *Brit. J. Psychiat.* **148,** 363 (1986).

Crystals from ether, mp 262-264°; mp 268-271° (Arth, 1961). $[\alpha]_D^{25}$ +77.5° (dioxane). Soly in water (25°): 10 mg/100 ml. Sol in acetone, ethanol, chloroform.

21-Acetate, $C_{24}H_{31}FO_6$, *Decadronal, Decadron-LA, Dectancyl, Dexacortisyl.* Crystals, mp 215-221° (Arth, 1958); mp 229-231° (Oliveto); mp 238-240° (Arth, 1961). $[\alpha]_D^{25}$ +73° (chloroform) (Arth, 1958); $[\alpha]_D$ +77.6° (Oliveto). uv max: 239 nm (ϵ 14900).

21-(3,3-Dimethylbutyrate), $C_{28}H_{39}FO_6$, *dexamethasone tert-butylacetate, Decadron TBA.*

21-Phosphate disodium salt, $C_{22}H_{28}FNa_2O_8P$, *dexamethasone 21-(dihydrogen phosphate) disodium salt, dexamethasone sodium phosphate, Ak-Dex, Baldex, Colvasone, Dalalone, Dexabene, Dezone, Fortecortin, Hexadrol, Oradexon, Orgadrone, Solu-Decadron, Soldesam.* Used as an injectable form of dexamethasone. Prepn: Chemerda *et al.,* U.S. pat. **2,939,873** (1960 to Merck & Co.); Irmscher, *Chem. & Ind. (London)* **1961,** 1035. Crystals, mp 233-235°. $[\alpha]_D$ +57° (water). Also reported as $[\alpha]_D^{25}$ +74° ±4° (calcd on water-free and alcohol-free basis, concn of 10 mg/ml): *USP XIX,* p 124. uv max (ethanol): 238-239 nm (ϵ 14000). Sol in water.

21-Diethylaminoacetate, $C_{28}H_{41}FNO_6$, *Solu-Forte-Cortin.*

21-Isonicotinate, $C_{28}H_{32}FNO_6$, *dexamethasone 21-(4-pyridinecarboxylate), Auxiloson, Auxisone, Voren.* Crystals, mp 250-252°. $[\alpha]_D^{27}$ +183.5° (dioxane).

17,21-Dipropionate, $C_{28}H_{37}FO_7$, *THS-101, Methaderm.*

21-Palmitate, $C_{38}H_{59}FO_6$, *Limethasone.*

THERAP CAT: Glucocorticoid; anti-inflammatory. Diagnostic aid (Cushing's Syndrome, depression).

THERAP CAT (VET): Glucocorticoid; anti-inflammatory.

2987. Dexetimide. *(S)-3-Phenyl-1'-(phenylmethyl)-[3,-4'-bipiperidine]-2,6-dione; (S)-(+)-2-(1-benzyl-4-piperidyl)-2-phenylglutarimide; (+)-3-(1-benzyl-4-piperidyl)-3-phenylpiperidine-2,6-dione; (+)-1-benzyl-4-(2,6-dioxo-3-phenyl-3-piperidyl)piperidine; dextrobenzetimide; dexbenzetimide.* $C_{23}H_{26}N_2O_2$; mol wt 362.47. C 76.21%, H 7.23%, N 7.73%, O 8.83%. The dextroenantiomer responsible for the pharmacological activity of racemic benzetimide, *q.v.* Resolution of isomers and comparative pharmacology: Janssen *et al., Arzneimittel-Forsch.* **21,** 1365 (1971). Abs config studies: van Wijngaarden *et al., Life Sci.* **9,** part 1, 1289 (1970); Spek *et al., Nature* **232,** 575 (1971). Clinical trials: De Smedt *et al., J. Clin. Pharmacol.* **10,** 207 (1970).

Crystals, mp 181-183°. $[\alpha]_D^{20}$ +125° (chloroform).

Hydrochloride, $C_{23}H_{26}N_2O_2\cdot HCl$, *R-16470, Tremblex.* Crystals, mp 270-275°. $[\alpha]_D^{20}$ +125° (methanol). LD$_{50}$ i.v. in rats: 45 mg/kg (Janssen).

THERAP CAT: Anticholinergic; antiparkinsonian.

2988. Dexpanthenol. *2,4-Dihydroxy-N-(3-hydroxypropyl)-3,3-dimethylbutanamide; D(+)-α,γ-dihydroxy-N-(3-hydroxypropyl)-β,β-dimethylbutyramide; pantothenylol; N-pantoyl-3-propanolamine; panthenol; pantothenyl alcohol; Alcopan-250; Intrapan; Pantenyl; Panthoderm; Motilyn; Bepanthen; Cozyme; Ilopan; Urupan.* $C_9H_{19}NO_4$; mol wt 205.25. C 52.67%, H 9.33%, N 6.82%, O 31.18%. Prepd by the addition of propanolamine to optically active α,γ-dihydroxy-β,β-dimethylbutyrolactone: Schnider, *Jubilee vol. Emil Barell* **1946,** 85; **Swiss** pat. **227,706** (1943); **Brit.** pat. **582,156** (1946); U.S. pat. **2,413,077** (1946 to Hoffmann-La Roche). Only the D(+)-form has vitamin activity.

Viscous, somewhat hygroscopic liq. Slightly bitter taste. d_{20}^{20} 1.2. bp$_{0.02}$ 118-120°. Easily dec on distn. $[\alpha]_D^{20}$ +29.5° (c = 5). n_D^{20} 1.497. Freely sol in water, alcohol, methanol. Slightly sol in ether. Natural pH about 9.5. Reasonably stable to usual sterilization time and temp in aq solns adjusted to pH 3.0-4.0, but long heating causes racemization. Hydrolyzed by alkali and strong acid. Usually more stable than salts of pantothenic acid if pH can be adjusted between 3 and 5. For add'l stability data *see* Rubin, *J. Am. Pharm. Assoc., Sci. Ed.* **37,** 502 (1948). Aq solns can be stabilized with pantolactone: U.S. pat. **2,898,373** (1959).

dl-Form, *panthenol.*

THERAP CAT: Cholinergic; *dl*-form as vitamin.

THERAP CAT (VET): Nutritional factor. Dietary source of pantothenic acid.

2989. Dextran. Dextraven; Gentran; Hemodex; Intradex; Macrose; Onkotin; Plavolex; Polyglucin; Promit. A term applied to polysaccharides produced by bacteria growing on a sucrose substrate, contg a backbone of D-glucose units linked predominantly α-D(1→6). Several organisms produce dextrans but only *Leuconostoc mesenteroides* and *L. dextranicum (Lactobacteriaceae)* have been used commercially. Chemical and physical properties of the dextrans vary with the methods of production. Native dextrans usually have high mol wt; lower mol wt clinical dextrans usually prepared by depolymerization of native dextrans or by synthesis. All dextrans are composed exclusively of α-D-glucopyranosyl units, differing only in degree of branching and chain length. Prepn: Tarr, Hibbert, *Can. J. Res.* **5,** 414 (1931); Novak, Stoycos, U.S. pat. **2,841,578** (1958 to Commonwealth Eng. of Ohio). Enzymic synthesis: Sugg, Hehre, *J. Immunol.* **43,** 119 (1942); Behrens, Ringpfeil, U.S. pat. **3,044,940** (1962 to Serum Werk Bernburg). The crude dextran may be isolated from the culture by precipitation with methanol. Continuous dialysis process: Shurter, U.S. pat. **2,717,853** (1955 to C.S.C.). Elimination of pyrogens: Levi, Lozinski, U.S. pat. **2,762,727** (1956 to Frosst). Method of producing clinical dextran: Novak, Witt, U.S. pat. **2,972,567** (1961). Structure studies: Fowler *et al., Can. J. Res.* **B15,** 486 (1937); Fairhead *et al., ibid.* **B16,** 151 (1938); Peat *et al., J. Chem. Soc.* **1939,** 581; Goldstein, Whelan, *ibid.* **1962,** 170, 176. ¹³C-NMR structure study: F. R. Seymour *et al., Carbohyd. Res.* **51,** 179 (1976). *Reviews:* Evans, Hibbert, *Advan. Carbohyd. Chem.* **2,** 204 (1946); Neely, *ibid.* **15,** 341 (1960); Ricketts, *Progr. Org. Chem.* **5,** 73 (1961); Murphy, Whistler, in *Industrial Gums,* R. L. Whistler, Ed. (Academic Press, New York, 2nd ed., 1973) pp 513-542.

Dextran 40, LMD, LMWD, LVD, Eudextran, Gentran 40, Rheomacrodex, Rheotran. Produced by action of *L. mesenteroides* on sucrose; average mol wt: 40,000.

Dextran 70, Hyskon, Macrodex. Average mol wt: 70,000.

Dextran 75, Gentran 75. Average mol wt: 75,000.

USE: In soft center confections, as a partial substitute for barley malt. Mixed ethers and esters of dextran can be used in lacquers.

THERAP CAT: Plasma volume expander; Dextran 40 also as blood flow adjuvant.

THERAP CAT (VET): Plasma extender.

2990. Dextranase. α-1,6-Glucan 6-glucanohydrolase; E.C. 3.2.1.11. Enzyme which hydrolyzes the α-1→6 glucosidic linkages of the bacterial polysaccharide dextran. *Endodextranases* (dextranases which preferentially split glucosidic linkages remote from end groups) are secreted by various molds and a few bacteria. *Exodextranases* occur

predominantly in mammalian tissues. Prepn: from *Penicillium lilacinum, P. funiculosum,* and *Verticillium coccorum,* Nordström, Hultin, *Svensk. Kem. Tid.* **60**, 283 (1948), *C.A.* **43**, 3050i (1949); from *Aspergillus,* Carlson, Carlson, *Science* **115**, 43 (1952), *eidem,* U.S. pat. 2,709,150 (1955 to Enzymatic Chemicals), *eidem,* U.S. pat. 2,716,084, and Novak, Stoycos, U.S. pat. 2,841,578 (1955 and 1958 both to Commonwealth Eng. of Ohio); from *P. lilacinum, P. funiculosum, P. verruculosum,* and *Spicaria violacea,* Tsuchiya et al., U.S. pat. 2,742,399 and Corman, Tsuchiya, U.S. pat. 2,-776,925 (1956 and 1957 both to U.S. Secy. Agr.); from *P. lilacinum,* Charles, Farrell, *Can. J. Microbiol.* **3**, 239 (1957); from *Lactobacillus bifidus,* Bailey, Clarke, *Biochem. J.* **72**, 49 (1959). Tested as dental caries-control agent in hamsters: Fitzgerald et al., *J. Am. Dental Assoc.* **76**, 301 (1968). *Review:* E. H. Fischer, E. A. Stein, "Cleavage of *O*- and *S*-Glycosidic Bonds (Survey)" in *The Enzymes,* vol. 4, P. D. Boyer et al., Eds. (Academic Press, New York, 2nd ed., 1960) pp 304-307.
USE: In prepn of dextran for clinical use; in dentifrices.

2991. Dextran Iron Complex. Iron-dextran complex; Fenate; Imferon; Ironorm. Prepn: London, Twigg, U.S. pat. 2,820,740 (1958 to Benger); U.S. reissue 24,642 (1959); Herb, U.S. pat. 2,885,393 (1959 to Laros). Acute toxicity: Beliles, *Toxicol. Appl. Pharmacol.* **23**, 537 (1972). Review of carcinogenicity studies: *IARC Monographs* **2**, 161-178 (1973).
LD_{50} i.v. in mice: 2.24 g Fe/kg (Beliles).
Note: This substance may reasonably be anticipated to be a carcinogen: *Seventh Annual Report on Carcinogens* (PB95-109781, 1994) p 236.
THERAP CAT: Hematinic.
THERAP CAT (VET): Nutritional factor (parenteral). Used in iron deficiency anemia, chiefly in pigs.

2992. Dextranomer. *Dextran 2,3-dihydroxypropyl 2-hydroxy-1,3-propanediyl ethers;* Debrisan; Debrisorb. Three-dimensional hydrophilic network of a dextran polymer, linked by cross-chains of epichlorohydrin, *q.v.;* it absorbs moisture and small molecules from suppurating wounds. Chronic tissue response to implantation: J. Falk, G. Tollerz, *Clin. Ther.* **1**(3), 185 (1977). Potential allergic contact sensitization in guinea pigs: G. Jonsson, *ibid.* **1**(4), 260 (1978). Efficacy in treatment of ulcers and wounds: J. Soul, *Brit. J. Clin. Pract.* **32**, 172 (1978); P. N. Sawyer et al., *Surgery* **85**, 201 (1979); S. Di Mascio, *Am. J. Nurs.* **79**, 684 (1979). *Review:* R. C. Heel et al., *Drugs* **18**, 89-102 (1979).

Insol in all solvents; stable in water, salt solns and in alkaline and weakly acidic soln.
THERAP CAT: Vulnerary.

2993. Dextran Sulfate Sodium. Dextran sulfuric acid ester sodium salt; Asuro; Colyonal; Dexulate; Dextrarine; MDS. Heparin-like polysaccharide containing approx. 17% S with up to three sulfate groups per glucose molecule. Mol wt ranges from 4,000-500,000 Da; variations in mol wt are associated with differences in biological activity. Prepn, properties and anticoagulant activity: A. Grönwall et al., *Upsala Läkareforen. Förh.* **50**, 397 (1945); C. R. Ricketts,

Biochem. J. **51**, 129 (1952). Evaluation of toxicity as a function of mol wt: K. W. Walton, *Brit. J. Pharmacol.* **9**, 1 (1954); of carcinogenicity: I. Hirono et al., *Cancer Letters* **18**, 29 (1983). Use in serum HDL cholesterol determn: P. R. Finley et al., *Clin. Chem.* **24**, 931 (1978); G. R. Warnick et al., *ibid.* **28**, 1379 (1982). Antiscrapie effect: B. Ehlers, H. Diringer, *J. Gen. Virol.* **65**, 1325 (1984); R. H. Kimberlin, C. A. Walker, *Antimicrob. Ag. Chemother.* **30**, 409 (1986). Anti-HIV-1 activity *in vitro:* H. Mitsuya et al., *Science* **240**, 646 (1988); M. Baba et al., *Proc. Nat. Acad. Sci. USA* **85**, 6132 (1988).
White powder from alcohol + ether. Freely sol in water. Activity about 17 international heparin units/mg. Aq solns must be buffered (*e.g.,* with sodium bicarbonate) to prevent dec during autoclaving.
USE: Clinical reagent (HDL cholesterol determn).
THERAP CAT: Anticoagulant; antihyperlipoproteinemic.

2994. Dextri-Maltose®. Maltose and dextrins obtained by enzymic action of barley malt on corn flour.
Light, amorphous powder. Readily sol in water or milk. One leveled tablespoonful (8 grams) supplies 27 calories.
USE: As carbohydrate modifier for use with milk and milk products in infants' formulas.

2995. Dextrin. Pyrodextrin; torrefaction dextrin. $(C_6H_{10}O_5)_n \cdot xH_2O$. Produced by the dry heating of unmodified starches. The term also includes products resulting from enzyme or acid-catalyzed hydrolysis of wet starch. *Review:* R. W. Satterthwaite, D. J. Iwinski, in *Industrial Gums,* R. L. Whistler, Ed. (Academic Press, New York, 2nd ed., 1973) pp 577-599.
British gum, starch gum. Produced at high temp in the absence of acid. Dark brown color, odorous. High viscosity; very sol in cold water. Does not reduce Fehling's soln; gives reddish-brown color with iodine.
Canary dextrin, yellow dextrin. Hydrolyzed at high temp for long period of time in the presence of small amts of acid. Light brown to yellow color, slight odor. Low viscosity; very sol in cold water.
White dextrin. Hydrolyzed at low temp for short period of time in the presence of large amts of acid. White color, odorless. Slightly sol in cold water giving a red color with iodine. Very sol in hot water giving a blue color with iodine.
USE: Excipient for dry extracts and pills; for preparing emulsions and dry bandages; for thickening dye pastes and mordants used in printing fabrics in fast colors; sizing paper and fabrics; printing tapestries; preparing felt; manuf printer's inks, glues and mucilage; polishing cereals; in matches, fireworks, and explosives.

2996. Dextroamphetamine Sulfate. (*S*)-*α-Methylbenzeneethanamine sulfate; d-α-methylphenethylamine sulfate; d*-amphetamine sulfate; *d*-1-phenyl-2-aminopropane sulfate; *d*-β-phenylisopropylamine sulfate; Dexampex; Dexedrine Sulfate; Afatin; Dexamphetamine; *d*-Amfetasul; Domafate; Obesedrin; Dexten; Maxiton; Sympamin; Simpamina-D; Albemap; Dadex; Ardex; Dexalone; Amsustain; Betafedrina; *d*-Betaphedrine; Diocurb. $C_{18}H_{28}N_2O_4S$; mol wt 368.50. C 58.67%, H 7.66%, N 7.60%, O 17.37%, S 8.70%. [C_6-$H_5CH_2CH(NH_2)CH_3]_2 \cdot H_2SO_4$. See Amphetamine for structural formula. Prepn by resolution of amphetamine with *d*-tartaric acid followed by treatment with 10% H_2SO_4: Temmler, **Brit.** pat. 508,757 (1939); Nabenhauer, U.S. pat. 2,276,508 (1942 to SK&F); Magidson, Garkusha, *J. Gen. Chem. (USSR)* **11**, 339 (1941); from D-phenylalanine: D. B. Repke et al., *J. Pharm. Sci.* **67**, 1167 (1978). Toxicity data: E. J. Warawa et al., *J. Med. Chem.* **18**, 71 (1975).
Plates, rods [different from amphetamine sulfate crystals, see Keenan, *J. Am. Pharm. Assoc.* **37**, 519 (1948)]. Slightly bitter taste followed by a sensation of numbness. $[\alpha]_D^{20}$ +21.8° (c = 2). mp >300°. One part dissolves in ~10 parts of water, in 500 parts of 95% alcohol. pH 5% aq soln: 5.0 to 6.0. LD_{50} orally in mice: 10 mg/kg (Warawa).
Compounds formed from dextroamphetamine and carboxymethylcellulose are marketed as *Carboxyphen* (mol wt 568,000) and *Bontril* (~40 mg = 15 mg of dextroamphetamine sulfate).

Note: This is a controlled substance (stimulant) listed in the U.S. Code of Federal Regulations, Title 21 Part 1308.12 (1995).

THERAP CAT: CNS stimulant; anorexic.

THERAP CAT (VET): Sympathomimetic, CNS stimulant.

2997. Dextromoramide. *(S)-1-[3-Methyl-4-(4-morpholinyl)-1-oxo-2,2-diphenylbutyl]pyrrolidine; (+)-1-(3-methyl-4-morpholino-2,2-diphenylbutyryl)pyrrolidine;* 4-[2-methyl-4-oxo-3,3-diphenyl-4-(1-pyrrolidinyl)butyl]morpholine; *d*-2,2-diphenyl-3-methyl-4-morpholinobutyrylpyrrolidine; pyrrolamidol; R-875; SKF-5137; *d*-Moramid(e); Palfium; Palphium; Jetrium; Dimorlin. $C_{25}H_{32}N_2O_2$; mol wt 392.54. C 76.50%, H 8.22%, N 7.14%, O 8.15%. Synthesis: Janssen, *J. Am. Chem. Soc.* **78**, 3862 (1956); **Brit. pat. 822,055** (1959 to Janssen).

Crystals, mp 180-184°. $[\alpha]_D^{20}$ +25.5° (c = 5 in benzene). uv max (0.01*N* isopropanol-HCl): 254, 260, 264 nm. Practically insol in water. Soly in 0.1*N* HCl: 1:25 (w/v). Soly (g/100 ml) in ethanol 50; in methanol 40; in acetone 50; in ethyl acetate 40; in benzene 5; in chloroform 5. Sol in ether. Bitartrate, $C_{25}H_{32}N_2O_2 \cdot C_4H_6O_6$, minute crystals, bitter taste. Dec 189-192°. Soly (w/v) at 25°: Water 20%, chloroform 30%, methanol 40%, ethanol 100%, acetone 100%.

Note: This is a controlled substance (opiate) listed in the U.S. Code of Federal Regulations, Title 21 Part 1308.11 (1995).

THERAP CAT: Analgesic (narcotic).

2998. Dezocine. *[5R-(5α,11α,13S*)]-13-Amino-5,6,7,8,-9,10,11,12-octahydro-5-methyl-5,11-methanobenzocyclodecen-3-ol;* Wy-16225; Dalgan. $C_{16}H_{23}NO$; mol wt 245.36. C 78.32%, H 9.45%, N 5.71%, O 6.52%. Synthetic opiate agonist-antagonist. Prepn: M. E. Freed *et al.,* **Ger. pat. 2,159,324** (1972 to Am. Home Products); *C.A.* **79**, 53094w (1973); *eidem, J. Med. Chem.* **16**, 595 (1973). Pharmacology: J. L. Malis *et al., J. Pharmacol. Exp. Ther.* **194**, 488 (1975). Metabolism: S. F. Sisenwine, C. O. Tio, *Drug Metab. Dispos.* **9**, 37 (1981). Pharmacokinetics: S. F. Sisenwine *et al., ibid.* **10**, 366 (1982). Clinical studies: W. Oosterlinck, A. Verbaeys, *Curr. Med. Res. Opin.* **6**, 472 (1980); J. W. Downing *et al., Brit. J. Anaesth.* **53**, 59 (1981).

Hydrobromide salt, $C_{15}H_{23}NO \cdot HBr$, crystalline powder, mp 269-270°. pH (2% aq soln): 4.6. Soly in water: > 20 mg/ml. LD_{50} in female mice, rats (mg/kg): 313, 232 orally; 129, 74 i.m. (Malis).

THERAP CAT: Analgesic (narcotic).

2999. DFDD. *1,1'-(2,2-Dichloroethylidene)bis[4-fluorobenzene]; 1,1-dichloro-2,2-bis(p-fluorophenyl)ethane;* difluorodiphenyldichloroethane. $C_{14}H_{10}Cl_2F_2$; mol wt 287.14. C 58.56%, H 3.51%, Cl 24.69%, F 13.23%. Prepn: Müller, *Helv. Chim. Acta* **9**, 1560 (1946).

Crystals from petr ether, mp 77°.

USE: Contact insecticide.

3000. DFDT. *1,1'-(2,2,2-Trichloroethylidene)bis[4-fluorobenzene]; 1,1,1-trichloro-2,2-bis(p-fluorophenyl)ethane;* difluorodiphenyltrichloroethane; fluorogesarol; HO-2474; Gix. $C_{14}H_9Cl_3F_2$; mol wt 321.58. C 52.29%, H 2.82%, Cl 33.07%, F 11.82%. Prepn: Sumerford, *J. Am. Pharm. Assoc.* **36**, 127 (1947); Cross, **U.S. pat. 2,581,174** (1952). Toxicity: Piekarski, Holz, *Arch. Exp. Pathol. Pharmakol.* **210**, 71 (1950).

Crystals from aq ethanol, mp 44-45°. Practically insol in water; very sol in oils, most organic solvents.

Caution: Effects similar to DDT [1,1,1-trichloro-2,2-bis-(*p*-chlorophenyl)ethane], *q.v.*

USE: Contact insecticide.

3001. Dhurrin. *(S)-α-(β-D-Glucopyranosyloxy)-4-hydroxybenzeneacetonitrile;* p-hydroxymandelonitrile-β-D-glucoside; β-D-glucopyranosyloxy-L-p-hydroxymandelonitrile. $C_{14}H_{17}NO_7$; mol wt 311.29. C 54.02%, H 5.50%, N 4.50%, O 35.98%. From the fruit of young Egyptian plants of *Sorghum vulgare* Pers., *Gramineae:* Dunstan, Henry, *Chem. News* **85**, 301 (1902). Biosynthesis: Koukol *et al., J. Biol. Chem.* **237**, 3223 (1962). Absolute configuration: Towers *et al., Tetrahedron* **20**, 71 (1964). Allied to amygdalin (*q.v.*) and the mandelonitrile glucosides.

Leaflets from water, prisms from alcohol, dec 200°. $[\alpha]_D^{20}$ −65° (alc). uv max: 228 nm (ε 13000). Sol in water, alcohol. Alkali labile.

Pentaacetate, $C_{24}H_{27}NO_{12}$, needles from ethyl acetate + petr ether, mp 132-132.5°. $[\alpha]_D^{20}$ −50.5° (c = 0.24 in alc).

Caution: Yields HCN on hydrolysis.

3002. Diaboline. *1-Acetyl-19,20-didehydro-17,18-epoxy-curan-17-ol; N-acetyl-Wieland-Gumlich aldehyde.* $C_{21}H_{24}N_2O_3$; mol wt 352.43. C 71.57%, H 6.86%, N 7.95%, O 13.62%. From *Strychnos diaboli* Sandw., *Loganiaceae:* King, *J. Chem. Soc.* **1949**, 955; Bader *et al., Helv. Chim. Acta* **36**, 1256 (1953); Casinovi *et al., Nature* **193**, 1178 (1962). Structure: Battersby, Hodson, *Proc. Chem. Soc.* **1959**, 123. Synthesis: Deyrup *et al., Helv. Chim Acta* **45**, 2266 (1962). ^{13}C-NMR study: E. Wenkert *et al., J. Org. Chem.* **43**, 1099 (1978).

Needles from ether, mp 187°. $[\alpha]_D^{22}$ +37.8° (c = 1.72 in chloroform). uv max (ethanol): 249 nm (log ε 4.06).

Hydrochloride monohydrate, $C_{21}H_{24}N_2O_3 \cdot HCl \cdot H_2O$, needles from alcohol + ethyl acetate, mp 300°. For anhydr, $[\alpha]_{5461}^{20}$ +184° (c = 0.57 in water).

3003. Diacerein. *4,5-Bis(acetyloxy)-9,10-dihydro-9,10-dioxo-2-anthracenecarboxylic acid; 9,10-dihydro-4,5-dihydr-*

oxy-9,10-dioxo-2-anthroic acid diacetate; 1,8-diacetoxy-3-carboxyanthraquinone; diacerhein; diacetylrhein; DAR; SF-277; Artrodar; Fisiodar. $C_{19}H_{12}O_8$; mol wt 368.30. C 61.96%, H 3.28%, O 34.75%. Diacetyl derivative of rhein, *q.v.* Demonstrates anti-arthritic activity without inhibiting prostaglandin synthesis. Prepn: A. Tschirch, K. Heuberger, *Arch. Pharm.* **240**, 596 (1902); V. K. Murty *et al., Tetrahedron* **23**, 515 (1967). Effect on prostaglandin release: P. Pomarelli *et al., Farmaco Ed. Sci.* **35**, 836 (1980). Inhibition of proteases: L. Raimondi *et al., Pharmacol. Res. Commun.* **14**, 103 (1982). Use in arthritis: C. A. Friedmann, **Ger.** pat. 2,711,493; *idem,* **U.S.** pat. 4,244,968 (1977, 1981 both to Proter). Preliminary clinical study in osteoarthritis: A. G. Kay *et al., Curr. Med. Res. Opin.* **6**, 548 (1980). Use in multiple sclerosis, amyotrophic lateral sclerosis: C. A. Friedmann, **U.S.** pat. 4,346,103 (1982); *idem,* **Japan. Kokai** 83 225,015 (1983 to Proter), *C.A.* **100**, 144991e (1984). Clinical trial in osteoarthritis: S. Adami *et al., Clin. Ter.* **112**, 439 (1985).

Yellow plates from acetic acid, mp 217-218°.
THERAP CAT: Antiarthritic.

3004. Diacetazotol. *N-Acetyl-N-[2-methyl-4-[(2-methylphenyl)azo]phenyl]acetamide; 4"-(o-tolylazo)-o-diacetotoluidide;* diacetylaminoazotoluene; *N,N*-diacetyl-*o*-tolyazo-*o*-toluidine; 2,3′-dimethyl-4′-(diacetylamino)azobenzene; Dermagan; Diacetotoluide; Dimazon; Pellidol. $C_{18}H_{19}N_3O_2$; mol wt 309.37. C 69.88%, H 6.19%, N 13.58%, O 10.34%. Prepd by treating aminoazotoluene with acetic anhydride or with acetyl chloride in the presence of sodium acetate: **Ger.** pat. 253,884 (1912 to Kalle); Ettel, Hebky, *Coll. Czech. Chem. Commun.* **13**, 161 (1948). Biological properties: Rerábek *et al., C.A.* **46**, 7234i (1952). Allergenic properties: Jirasek *et al., Cesk. Dermatol.* **41**, 38 (1966).

Crystallizes in 2 modifications: Brick-red needles, mp 65°, or stout, red prisms, mp 75° (resembling crystals of potassium dichromate). The German Pharmacopoeia recognizes only a material with mp 74-76° (yellowish-red powder). Insol in water; sol in alcohol, benzene, chloroform, ether, acetone, fixed oils, fats, petrolatum.
Mixture with albumin and tetraiodopyrrole, *Azodolen.*
THERAP CAT: Topical dermatologic.
THERAP CAT (VET): Has been used in ointment or dusting powder to stimulate wound epitheliazation.

3005. Diacetin. *1,2,3-Propanetriol diacetate;* glycerol diacetate; glyceryl diacetate. $C_7H_{12}O_5$; mol wt 176.17. C 47.73%, H 6.87%, O 45.41%. Commercial material is probably a mixture of glycerol 1,2- and 1,3-diacetates. Prepn of the 1,2- and 1,3-compounds: Wegscheider, Zmerzlikar, *Monatsh.* **34**, 1061 (1913); of the 1,2-compd: Golendeev, *J. Gen. Chem. (USSR)* **6**, 1841 (1936); of the 1,3-compd: Lagenbeck, Bollow, *Naturwiss.* **42**, 389 (1955).

Hygroscopic liquid, bp 259°. d_4^{16} 1.184. n_D^{20} 1.44. Sol in water, alc, ether, benzene; practically insol in CS_2.
dl-1,2-Compd, *1,2-diacetin.* bp_{12} 140-142°, bp_{40} 172-173.5°. d_4^{15} 1.1173.
1,3-Compd, *1,3-diacetin.* bp_{40} 172-174°. d^{15} 1.179. n_D^{20} 1.4395.
USE: Technical grade as a plasticizer and softening agent, and as a solvent.

3006. Diacetonamine. *4-Amino-4-methyl-2-pentanone.* $C_6H_{13}NO$; mol wt 115.18. C 62.57%, H 11.38%, N 12.16%, O 13.89%. $CH_3COCH_2C(CH_3)_2NH_2$. Prepd from mesityl oxide with aq NH_3: Haeseler, *J. Am. Chem. Soc.* **47**, 1195 (1925); Orthner, *Ann.* **456**, 245 (1927); Haeseler, *Org. Syn.* **coll. vol. I** (2nd ed., 1941) p 196. From mesityl oxide with liquid ammonia: Smith, Adkins, *J. Am. Chem. Soc.* **60**, 408 (1938).
Liq, $bp_{0.2}$ 25°. Amine odor. Lighter than water. Strongly alkaline. Sol in water, but aq solns prepd in the cold become turbid on heating; miscible with alc, ether. Dec easily forming mesityl oxide and NH_3. Decompn is prone to take place in aq solns or during distillation.
Acid oxalate monohydrate, $C_6H_{13}NO.C_2H_2O_4.H_2O$, spear-shaped crystals, mp 126-127°.

3007. Diacetone Acrylamide. *N-(1,1-Dimethyl-3-oxobutyl)-2-propenamide; N-(1,1-dimethyl-3-oxobutyl)acrylamide; N-[2-(2-methyl-4-oxopentyl)]acrylamide.* $C_9H_{15}NO_2$; mol wt 169.22. C 63.88%, H 8.93%, N 8.28%, O 18.91%. Highly reactive vinyl monomer. Prepd by the reaction of acrylonitrile with acetone in presence of conc sulfuric acid: Coleman *et al., J. Polym. Sci.* **3A**, 1601 (1965); Coleman, **U.S.** pat. 3,277,056 (1966 to Lubrizol). Toxicity data: K. Hashimoto *et al., Arch. Toxicol.* **47**, 179 (1981). *Review: idem,* "N-Oxoalkylacrylamides" in *Encyclopedia of Polymer Science and Technology,* vol. **15**, N. M. Bikales, Exec. Ed. (Wiley-Interscience, New York, 1971) pp 353-364.

White crystalline solid, mp 57-58°. $bp_{0.1-0.3}$ 93-100°; bp_8 120°. Hygroscopic. Has long shelf life; shows no polymerization on prolonged storage. A 50% neutral water soln is stable at pH 7.5-7.7 for up to six months. Similar to acrylic esters in polymerization behavior. Forms high mol wt polymers in bulk, solution, or emulsion. Esp suitable for electropolymerization. LD_{50} orally in mice: 7.7 mmol/kg (Hashimoto).
USE: In the manufacture of coatings, laminates, sealers, adhesives, lubricating oils.

3008. Diacetone Alcohol. *4-Hydroxy-4-methyl-2-pentanone;* Pyranton. $C_6H_{12}O_2$; mol wt 116.16. C 62.04%, H 10.41%, O 27.55%. $(CH_3)_2C(OH)CH_2COCH_3$. Prepd by the action of barium hydroxide or of calcium hydroxide on acetone: Conant, Tuttle, *Org. Syn.* vol. **1**, p 45 (1921); **coll. vol. I**, 193; *cf.* Jacquemain, *Compt. Rend.* **196**, 1622 (1933). Industrial process using KOH and acetone: Schmitt, Disteldorf, **Ger.** pat. 1,052,970 (1959 to Hibernia). Toxicity study: Smyth, Carpenter, *J. Ind. Hyg. Toxicol.* **30**, 63 (1948).
Flammable liq; faint, pleasant odor. d_4^{25} 0.9306 (0.940 for tech grades). bp_{760} 167.9°. bp_{100} 108.2°. bp_{20} 72.0°. bp_{10} 58.8°. $bp_{1.0}$ 22.0°. mp —44°; n_D^{20} 1.4232: Lantz, *J. Am. Chem. Soc.* **62**, 3260 (1940). Flash pt reagent grade: 66°C (151°F). Flash pt commercial grades: 8°C (48°F) (closed cup); 13°C (55°F) (open cup). Vapor density 4.00 (air = 1.00). Miscible with water, alcohol, ether, and other solvents. Is dec by prolonged exposure to alkalies and by

distillation at atmospheric pressure. LD_{50} orally in rats: 4.0 g/kg (Smyth).

Caution: Potential symptoms of overexposure are irritation of eyes, nose, throat and skin; corneal tissue damage; narcosis. *See NIOSH Pocket Guide to Chemical Hazards* (DHHS/NIOSH 90-117, 1990) p 82.

USE: Solvent for cellulose acetate, nitrocellulose, celluloid, fats, oils, waxes, resins. As a preservative in pharmaceutical prepns. In some antifreeze solns and in hydraulic fluids.

3009. Diacetoneglucose. *1,2:5,6-Bis-O-(1-methylethylidene)-D-glucofuranose; 1,2:5,6-diisopropylidene-D-glucose;* 1,2:5,6-diisopropylidene-D-glucofuranose. $C_{12}H_{20}O_6$; mol wt 260.29. C 55.37%, H 7.74%, O 36.88%. Prepd by shaking glucose wih acetone contg 1% HCl: Fischer, Rund, *Ber.* **49**, 90, 93 (1916). Improved procedure (90% yield) using 85% H_3PO_4 and anhydr $ZnCl_2$ in a mechanically stirred suspension of glucose in acetone: Glen *et al.*, U.S. pat. **2,715,121** (1955 to Am. Home Prod.).

Needles from ether or petr ether. Bitter taste. mp 110-111°. Sublimes. $[\alpha]_D^{20}$ −18.5° (c = 5). Reduces Fehling's soln. One part dissolves in about 7 parts of boiling water and in about 200 parts of boiling petr ether. Freely sol in chloroform, alc, acetone, warm ether. Precipitated from aq soln by NaOH soln. Is not split by yeast or emulsin.

3010. Diacetyl. *2,3-Butanedione;* biacetyl; dimethyl diketone; dimethylglyoxal; 2,3-diketobutane. $C_4H_6O_2$; mol wt 86.09. C 55.81%, H 7.02%, O 37.17%. $CH_3COCOCH_3$. Found in bay and other oils; also in butter. Made from methyl ethyl ketone by converting to the isonitroso compd and then decomposing to diacetyl by hydrolysis with HCl; by special fermentation of glucose *via* methylacetylcarbinol: Suomalainen, Jännes, *Nature* **157**, 336 (1946). Toxicity study: P. M. Jenner *et al.*, *Food Cosmet. Toxicol.* **2**, 327 (1964).

Yellowish-green liq; quinone odor; the vapors have a Cl-like odor. d_4^{15} 0.990. bp 88°. n_D^{18} 1.3933. Sol in ~4 parts water; misc with alcohol, ether. LD_{50} orally in rats: 1580 mg/kg (Jenner).

USE: Carrier of aroma of butter, vinegar, coffee, and other foods.

3011. Diacetyldihydromorphine. *(5α,6α)-4,5-Epoxy-17-methylmorphinan-3,6-diol diacetate (ester);* dihydroheroin; paralaudin. $C_{21}H_{25}NO_5$; mol wt 371.43. C 67.91%, H 6.78%, N 3.77%, O 21.54%. Prepn from dihydromorphine: Eddy, Howes, *J. Pharmacol.* **53**, 430 (1935); Small, Mallonee, *J. Org. Chem.* **12**, 558-566 (1947); K. W. Bentley, *The Chemistry of the Morphine Alkaloids* (Oxford, 1954).

Needles, mp 165-167°. Slightly sol in water; sol in chloroform, alcohol, ether.

Hydrochloride, $C_{21}H_{25}NO_5 \cdot HCl$, flakes, scales. mp 215-218°. $[\alpha]_D^{29}$ −59° (c = 1.2). Very sol in water.

3012. Diacetylmorphine. *(5α,6α)-7,8-Didehydro-4,5-epoxy-17-methylmorphinan-3,6-diol diacetate (ester);* heroin; diamorphine; acetomorphine. $C_{21}H_{23}NO_5$; mol wt 369.42. C 68.28%, H 6.28%, N 3.79%, O 21.65%. Narcotic analgesic prepared from morphine and acetic anhydride: O. Hesse, *Ann.* **220**, 203 (1883); from morphine and acetyl chloride: Small, Lutz, *Chemistry of the Opium Alkaloids*, Supplement No. 103, Public Health Reports, Washington (1932); K. W. Bentley, *The Chemistry of the Morphine Alkaloids* (Oxford, 1954). Comparison with acetylmorphine: N. B. Eddy, H. A. Howes, *J. Pharmacol. Exp. Ther.* **53**, 430 (1935). Pharmacodynamics: J. G. Umans, C. E. Inturrisi, *ibid.* **218**, 409 (1981). Toxicity: *eidem, Eur. J. Pharmacol.* **85**, 317 (1982). Vapor pressure studies: A. H. Lawrence *et al., Can. J. Chem.* **42**, 1886 (1984). Comprehensive description: D. K. Wyatt, L. T. Grady, *Anal. Profiles Drug Subs.* **10**, 357-403 (1981).

Orthorhombic plates, tablets from ethyl acetate. mp 173°. bp_{12} 272-274°. $[\alpha]_D^{25}$ −166° (c − 1.49 in methanol). One gram dissolves in 1.5 ml chloroform, 31 ml alcohol, 100 ml ether, 1700 ml water. Slightly sol in ammonia or sodium carbonate soln, sol in alkalies, dec by boiling with water. Turns pink and emits acetic odor on prolonged exposure to air. LD_{50} i.v. in mice: 59 μmol/kg (Umans, Inturrisi).

Hydrochloride monohydrate, $C_{21}H_{23}NO_5 \cdot HCl \cdot H_2O$, fine crystals, mp 243-244°. $[\alpha]_D^{24}$ −156° (c = 1.044). Sol in 2 parts water, 11 parts alcohol. Insol in ether.

Methyl iodide, $C_{21}H_{23}NO_5 \cdot CH_3I$, needles, mp 252°, $[\alpha]_D^{15}$ −107° (c = 0.896).

Note: "China White" has been used as a term for very pure Southeast Asian heroin. This term has also been erroneously used to refer to *3-methylfentanyl* and α-methylfentanyl, *q.v.*, which are potent derivs of fentanyl, *q.v. See:* S. Stinson, *Chem. & Eng. News,* **59**, 71 (Jan. 19, 1981).

Caution: May be habit forming. This is a controlled substance (opium derivative) listed in the U.S. Code of Federal Regulations, Title 21 Parts 329.1 and 1308.11 (1995).

3013. Dialifor. *Phosphorodithioic acid S-[2-chloro-1-(1,3-dihydro-1,3-dioxo-2H-isoindol-2-yl)ethyl] O,O-diethyl ester; phosphorodithioic acid O,O-diethyl ester S-ester with N-(2-chloro-1-mercaptoethyl)phthalimide; O,O-diethyl S-(2-chloro-1-phthalimidoethyl)phosphorodithioate;* dialifos; Hercules 14503; Torak. $C_{14}H_{17}ClNO_4PS_2$; mol wt 393.85. C 42.69%, H 4.35%, Cl 9.00%, N 3.56%, O 16.25%, P 7.86%, S 16.28%. Prepn: Jamison, U.S. pat. **3,355,353** (1967 to Hercules). Acute toxicity: T. B. Gaines, R. E. Linder, *Fundam. Appl. Toxicol.* **7**, 299 (1986).

White crystalline solid, mp 67-69°. Also reported as oil; when recrystallized from toluene + hexane, mp 62-64°. Insoluble in water. Generally sol in aromatic hydrocarbons, ethers, esters, ketones. LD_{50} in adult male, female rats (mg/kg): 24, 6 orally (Gaines, Linder).

USE: Insecticide; acaricide.

3014. Diallate. *Bis(1-methylethyl)carbamothioic acid S-(2,3-dichloro-2-propenyl) ester; diisopropylthiocarbamic acid S-2,3-dichloroallyl ester; S-2,3-dichloroallyl diisopropylthiocarbamate;* DATC; CP-15336; Avadex. $C_{10}H_{17}Cl_2NOS$; mol wt 270.22. C 44.45%, H 6.34%, Cl 26.24%, N 5.18%, O

5.92%, S 11.87%. Prepn: **Brit.** pat. **882,111;** M. W. Harman, J. J. D'Amico, U.S. pat. **3,330,823** (1961, 1967 both to Monsanto). Toxicity data: Bailey, White, *Residue Rev.* **10,** 97 (1965). Mutagenicity studies: F. DeLorenzo *et al., Cancer Res.* **38,** 13 (1978); I. Schuphan *et al., Science* **205,** 1013 (1979). Herbicidal activity: R. Grover *et al., Weed Res.* **19,** 363 (1979). GLC determn in milk and plant tissue: L. W. Cook *et al., J. Assoc. Offic. Anal. Chem.* **65,** 215 (1982). Review of carcinogenic risk: *IARC Monographs* **30,** 235-244 (1983).

(Z)-form

Brown liquid, bp$_9$ 150°. Very slightly sol in water (40 ppm); sol in acetone, benzene, chloroform, ether, heptane. LD$_{50}$ orally in rats: 395 mg/kg (Bailey, White).

USE: Herbicide.

3015. Diallylamine. *N-2-Propenyl-2-propen-1-amine;* di-2-propenylamine. $C_6H_{11}N$; mol wt 97.16. C 74.17%, H 11.41%, N 14.42%. $(CH_2=CHCH_2)_2NH$. Prepd from allylamine and allyl bromide: Ladenburg, *Ber.* **14,** 1879 (1881); from allylamine and allyl chloride: Liebermann, Hagen, *Ber.* **16,** 1641 (1883); from diallylcyanamide by refluxing with dil H_2SO_4: Vliet, *J. Am. Chem. Soc.* **46,** 1307 (1924); *Org. Syn.* **coll. vol. I** (2nd ed.,1941) p 201. Toxicity study: H. F. Smyth *et al., Am. Ind. Hyg. Assoc. J.* **23,** 95 (1962).

Liquid. bp 112°. LD$_{50}$ orally in rats: 0.65 g/kg (Smyth).

3016. Diallylcyanamide. *Di-2-propenylcyanamide; N*-cyanodiallylamine. $C_7H_{10}N_2$; mol wt 122.17. C 68.82%, H 8.25%, N 22.93%. $(CH_2=CHCH_2)_2NCN$. Prepd by the action of allyl bromide on disodium cyanamide: Vliet, *J. Am. Chem. Soc.* **46,** 1307 (1924); *Org. Syn.* **coll. vol. I,** 203 (1941); North, U.S. pat. **1,659,793** (1929).

Liquid. bp$_{90}$ 140-145°; bp$_{57}$ 128-133°; bp$_{18}$ 105-110°. Insol in water, sol in the usual organic solvents.

3017. Diamfenetide. *N,N'-[Oxybis(2,1-ethanediyloxy-4,-1-phenylene)]bisacetamide; β,β'-oxybis[p-acetophenetidide];* β,β'-bis(4-acetamidophenyloxy)ethyl ether; diamphenethide; Coriban. $C_{20}H_{24}N_2O_5$; mol wt 372.42. C 64.50%, H 6.50%, N 7.52%, O 21.48%. Prepn: Harfenist, **Ger.** pat. **2,143,570** (1972), *C.A.* **77,** 19404w (1972). Activity: Dickerson *et al., Brit. Vet. J.* **127,** xl (1971); Kingsbury, Rowlands, *ibid.* **128,** 235 (1972); Armour, Corba, *Vet. Rec.* **91,** 211 (1972). *Review:* Rowlands, *Pestic. Sci.* **4,** 883 (1973).

THERAP CAT (VET): Anthelmintic (trematodes).

3018. 2,4-Diaminoanisole. *4-Methoxy-1,3-benzenediamine;* 4-methoxy-*m*-phenylenediamine; 2,4-DAA; 4-MMPD; C.I. 76050; C.I. Oxidation Base 12. $C_7H_{10}N_2O$; mol wt 138.17. C 60.85%, H 7.29%, N 20.27%, O 11.58%. Prepd by reduction of 2,4-dinitroanisole with iron and acetic acid: **Ger.** pat. **258,653** (1912 to BASF), *Frdl.* **11,** 392 (1912-14). Alternate prepn: K. Fries, *Ann.* **454,** 147 (1927). Prepn of hydrochloride: F. Kehrmann, *Ber.* **50,** 562 (1917). Mutagenicity studies: B. N. Ames *et al., Proc. Nat. Acad. Sci. USA* **72,** 2423 (1975); D. J. N. Hossack, J. C. Richardson, *Experientia* **33,** 377 (1977); W. G. H. Blijleven, *Mutat. Res.-Genetic Toxicol. Testing* **48,** 181 (1977). Toxicity studies: C. Burnett *et al., Food Cosmet. Toxicol.* **13,** 353 (1975); *eidem, J. Toxicol. Environ. Health* **1,** 1027 (1976); *eidem, ibid.* **2,** 657 (1977); G. K. Lloyd *et al., Food Cosmet. Toxicol.* **15,** 607 (1977).

Needles from ether, mp 67-68°. Darkens on exposure to light. LD$_{50}$ of an aq soln containing 0.05% Na_2SO_3: 460 mg/kg orally in rats (Lloyd).

Sulfate, $C_7H_{10}N_2O.H_2SO_4$, *4-MMPDS.* Off white to violet powder. Sol in water and ethanol. When heated to decomposition, it emits very toxic fumes of nitrogen oxides and sulfur oxides. LD$_{50}$ in rats (mg/kg): 372 i.p.; >4000 orally (Burnett, 1977).

Caution: Potential symptoms of toxicity in exptl animals are skin irritation; thyroid and liver changes; teratogenic effects. *See NIOSH Pocket Guide to Chemical Hazards* (DHHS/NIOSH 94-116, 1994) p 90. The sulfate may reasonably be anticipated to be a carcinogen: *Seventh Annual Report on Carcinogens* (PB95, 109781, 1994) p 142.

USE: In prepn of dyes, esp hair and fur dyes; intermediates in the production of C. I. Basic Brown 2; as corrosion inhibitor for steel.

3019. p-Diaminoazobenzene. *4,4'-Azobisbenzenamine; 4,4'-azodianiline; p*-azoaniline; 4,4'-diaminoazobenzene. $C_{12}H_{12}N_4$; mol wt 212.25. C 67.91%, H 5.70%, N 26.40%. Prepn: Witt, Kopetschini, *Ber.* **45,** 1136 (1912); Brode *et al., J. Am. Chem. Soc.* **77,** 2762 (1955). Lab procedure starting with *p*-aminoacetanilide; Santurri *et al., Org. Syn.* **40,** 18 (1960).

Golden-yellow needles, dec 238-241°. Absorption max (ethanol): 400 nm. Slightly soluble in water, benzene, petr ether; freely sol in alcohol.

3020. 3,5-Diaminobenzoic Acid. $C_7H_8N_2O_2$; mol wt 152.15. C 55.26%, H 5.30%, N 18.41%, O 21.03%. Prepn from 3,5-dinitrobenzoic acid: Griess, *Ann.* **154,** 327 (1870).

Monohydrate, needles. Loses its water of crystn at about 110°, mp 228°. Slightly sol in water; the aqueous soln dec on standing. Sol in alcohol or ether.

Ethyl ester, $C_9H_{12}N_2O_2$, prisms from ether, mp 84°. Moderately sol in hot water. Sol in alcohol, benzene, other organic solvents.

USE: Detection and determination of nitrites.

3021. 2,6-Diamino-2'-butyloxy-3,5'-azopyridine. *3-[(6-Butoxy-3-pyridinyl)azo]-2,6-pyridinediamine;* 2',6'-diamino-2-butyloxy-5,5'-azopyridine; 2',6'-diamino-2-butoxy-5,5'-azopyridine; 2,6-diamino-6'-butoxy-3,3'-azopyridine; 2,6-diamino-2'-butoxy-3,5'-azopyridine; 2,6-diamino-3-(6-butoxy-3-pyridylazo)pyridine; Niazo; Neotropin; Savapyrin. $C_{14}H_{18}N_6O$; mol wt 286.34. C 58.73%, H 6.34%, N 29.35%, O 5.59%. Prepn: Dohrn, Diedrich, U.S. pat. **1,862,361** (1932 to Schering-Kahlbaum, A.G.).

Orange-red crystals. mp 129°. Slightly sol in water, freely in organic solvents.

THERAP CAT: Urinary antiseptic.

3022. 4,4'-Diaminodiphenylamine. *N-(4-Aminophenyl)-1,4-benzenediamine; p-p'-diaminodiphenylamine.* $C_{12}H_{13}N_3$; mol wt 199.26. C 72.34%, H 6.58%, N 21.09%. Prepn from *p*-phenylenediamine and aniline hydrochloride: Nietzki, *Ber.* **16**, 474 (1883); from *p*-aminoazobenzene: Barbier, Sisley, *Bull. Soc. Chim. France* [3] **33**, 1232 (1905); by reducing bis(*p*-nitrophenyl)nitroxide with tin chloride and hydrochloric acid: Wieland, Roth, *Ber.* **53**, 213, 226 (1920).

Leaflets from water, mp 158°. Oxidized to bromanil and bromonitromethane by bromine and concd nitric acid.

USE: Dyeing fur: **Ger.** pats. **367,690; 371,231;** manuf dyes. In the detection of hydrogen cyanide.

3023. p,p'-Diaminodiphenylmethane. *4,4'-Methylenebis[benzenamine]; 4,4'-methylenedianiline.* $C_{13}H_{14}N_2$; mol wt 198.27. C 78.75%, H 7.12%, N 14.13%. Prepn from aniline and formaldehyde: Scanlon, *J. Am. Chem. Soc.* **57**, 887 (1935); by hydrogenolysis of *p,p'*-diaminobenzophenone with LiAlH₄: Conover, Tarbell, *ibid.* **72**, 3586 (1950).

Crystals from water or benzene. mp 91.5-92°. bp₇₆₈ 398-399°; bp₁₈ 257°; bp₁₅ 249-253°; bp₉ 232°. Slightly sol in cold water. Very sol in alcohol, benzene, ether.

Caution: Hepatotoxic. Causative agent in "Epping Jaundice"; symptoms include jaundice, weakness, abdominal pain, nausea and/or vomiting, anorexia, fever, chills: H. Kopelman *et al., Brit. Med. J.* **1**, 514 (1966); D. B. McGill, J. D. Motto, *N. Engl. J. Med.* **291**, 278 (1974). This substance and its dihydrochloride may reasonably be anticipated to be carcinogens: *Seventh Annual Report on Carcinogens* (PB95-109781, 1994) p 250.

USE: As curing agent for epoxy resins and urethane elastomers; intermediate in prepn of polyurethanes, Spandex fibers; in production of polyamides; in the determination of tungsten and sulfates; in prepn of azo dyes; as corrosion inhibitor.

3024. 2,4-Diamino-6-hydroxypyrimidine. *2,6-Diamino-4(1H)-pyrimidinone.* $C_4H_6N_4O$; mol wt 126.12. C 38.09%, H 4.80%, N 44.42%, O 12.69%. Prepn from guanidine hydrochloride and cyanoacetic ester: Traube, *Ber.* **33**, 1371 (1900).

Crystallizes from water as the monohydrate. When anhydrous, dec 286°.

USE: In the detection of nitrites and nitrates.

3025. 2,3-Diaminophenazine. *2,3-Phenazinediamine.* $C_{12}H_{10}N_4$; mol wt 210.24. C 68.56%, H 4.79%, N 26.65%. Prepn from *o*-phenylenediamine: Ullmann, Mauthner, *Ber.* **35**, 4304 (1902); Knoevenagel, *J. Prakt. Chem.* [2] **89**, 25 (1914); Richter *Ber.* **44**, 3469 (1911).

Brown to yellow needles. On careful heating sublimes forming yellow leaflets. mp 264°. Sol in alcohol, benzene. Yields 2-aminophenazine and small amounts of phenazine when heated with zinc dust.

USE: Detection of bismuth, cadmium, lead, copper, and mercury.

3026. 2,4-Diaminophenol. $C_6H_8N_2O$; mol wt 124.14. C 58.05%, H 6.50%, N 22.57%, O 12.89%. Prepn by reduction of 2,4-dinitrophenol: Braude *et al., J. Chem. Soc.* **1954**, 3586; Neilson *et al., ibid.* **1962**, 371; from *m*-nitroaniline: Bean, **U.S. pat. 2,525,515** (1950 to Eastman Kodak).

Crystals, dec 78-80°. Somewhat sol in alcohol, acetone; sparingly sol in ether, chloroform, petr ether; readily sol in acids, alkalies.

Dihydrochloride. $C_6H_8N_2O.2HCl$, *Acrol, Amidol.* Crystals, mp 205°. Soly in water at 15°: 27.5 g/100 ml. Slightly sol in alcohol.

USE: Dihydrochloride as photographic developer, in fur and hair dyeing, in test for formaldehyde and ammonia.

3027. 2,3-Diaminopropionic Acid. *3-Aminoalanine;* 2,3-diaminopropanoic acid; α,β-diaminopropionic acid. $C_3H_8N_2O_2$; mol wt 104.11. C 34.61%, H 7.75%, N 26.91%, O 30.74%. Prepd by treating 2,3-dibromopropionic acid with ammonia: Klebs, *Z. Physiol. Chem.* **19**, 314 (1894); Winterstein, Küng, *ibid.* **59**, 146 (1909); Neuberg, Silbermann, *Ber.* **37**, 341 (1904); Frankland, *J. Chem. Soc.* **97**, 1318 (1910); by treating 2-amino-3-chloropropionic acid with ammonia under pressure: Fisher *et al., Ber.* **40**, 3717 (1907); by the reduction of cyanooximoacetic acid esters: Godefroi, **U.S. pat. 2,738,363** (1956 to Parke, Davis). Configuration: Karrer *et al., Helv. Chim. Acta* **9**, 314 (1926). *See also* Bergmann, Grafe, *Z. Physiol. Chem.* **187**, 187 (1930).

DL-Form, radially arranged cryst masses. Hygroscopic, also absorbs CO_2 from air. Begins to melt at 97°, completely liq at 110-120°. Sol in water, yielding alkaline solns. Practically insol in alcohol, ether.

Hydrochloride, $C_3H_8N_2O_2.HCl$, needles or leaflets, mp 236-237°. One gram dissolves in 12 ml water, practically insol in alcohol.

Sulfate hemihydrate, $(C_3H_8N_2O_2)_2.H_2SO_4.\frac{1}{2}H_2O$, six-sided leaflets from water, dec 233-234°. One gram dissolves in 31 ml water at 20°.

Methyl ester dihydrochloride, $C_4H_{10}N_2O_2.2HCl$, crystals, dec 170°. Freely sol in water.

Ethyl ester dihydrochloride, $C_5H_{12}N_2O_2.2HCl$, crystals, mp 164.5-165° (some decompn). Freely sol in water.

L-Form hydrochloride, $C_3H_8N_2O_2.HCl$, needles from water, $[\alpha]_D^{20}$ +25.0° (c = 5 in 1.0N HCl).

D-Form hydrochloride, $C_3H_8N_2O_2.HCl$, needles from water, $[\alpha]_D^{20}$ −25.3° (c = 5 in 1.0N HCl).

USE: The salts as growth inhibitors for microorganisms.

3028. 2,6-Diaminopurine. *1H-Purine-2,6-diamine;* 2,6-diamino-9H-purine. $C_5H_6N_6$; mol wt 150.14. C 40.00%, H 4.03%, N 55.97%. Antagonist of naturally occurring purines: Hitchings, Elion in *Metabolic Inhibitors,* vol. I, R. M. Hochster, J. H. Quastel, Eds. (Academic Press, New York, 1963) pp 216-232. Molecular and electronic structure: Veillard, Pullman, *J. Theoret. Biol.* **4**, 37 (1963). First synthesized by Traube, *Ber.* **37**, 4547 (1904) by cyclization of 2,4,5,6-tetraaminopyrimidine. Prepn from guanidine carbonate in 70% yield and refs to earlier prepns: Taylor *et al., J. Am. Chem. Soc.* **81**, 2442 (1959).

Crystals from ethanol + water, mp 302°. uv max (pH 1.9): 241, 282 nm (log ε 3.98, 4.00).

3029. 3,4-Diaminopyridine. *3,4-Pyridinediamine;* 3,4-DAP. $C_5H_7N_3$; mol wt 109.13. C 55.03%, H 6.47%, N 38.50%. Potassium channel blocker; antagonizes nondepolarizing neuromuscular blockade. Prepn: O. Bremer, *Ann.* **518**, 274 (1935); J. W. Clark-Lewis, R. P. Singh, *J. Chem. Soc.* **1962**, 2379; J. B. Campbell *et al., J. Heterocycl. Chem.* **23**, 669 (1986). HPLC determn in serum: J. Leslie, C. T. Bever, *J. Chromatog.* **496**, 214 (1989). Acute toxicity: P. Lechat *et al., Ann. Pharm. Franc.* **26**, 345 (1968). Effect on neuromuscular transmission: J. Molgo *et al., Eur. J. Pharmacol.* **61**, 25 (1980); R. H. Thomsen, D. F. Wilson, *J. Pharmacol. Exp. Ther.* **227**, 260 (1983). Evaluation in human botulism: A. P. Ball *et al., Quart. J. Med.* **48**, 473 (1979). Clinical evaluations in myasthenic syndromes: K. M. McEvoy *et al., N. Engl. J. Med.* **321**, 1567 (1989); J. Palace *et al., J. Neurol. Neurosurg. Psychiatr.* **54**, 1069 (1991); in multiple sclerosis: C. T. Bever, Jr. *et al., Ann. Neurol.* **27**, 421 (1990); *idem, ibid.* **36**, S118 (1994).

Needles from water, mp 220° (Clark-Lewis, Singh); also reported as white to beige crystals from water, mp 218-219° (Campbell). Readily sol in water, alcohol; slightly sol in ether. LD_{50} i.v. in mice: 13 mg/kg (Lechat).

USE: Intermediate in synthesis of heterocyclic compds.

3030. Diamond. A crystalline form of carbon. Mined as a mineral, principally in South Africa. (Non-commercial) synthesis from other carbon compds (e.g., lignin) by means of elevated temperatures (about 2700°) and pressures (about 800,000 lbs/sq inch): Desch, *Nature* **152**, 148 (1943); Neuhaus, *Angew. Chem.* **66**, 525 (1954); Hall, *Chem. Eng. News* **33**, 718 (1955); Bridgman, *Sci. Amer.* **1955**, 46; Hall, *J. Chem. Ed.* **38**, 484 (1961); Bundy, *Ann. N.Y. Acad. Sci.* **vol. 105**, art 17, pp 951-982 (1964). Books: S. Tolansky, *History and Use of Diamond* (London, 1962) 166 pp; R. Berman, *Physical Properties of Diamond* (Oxford, 1965) 442 pp.

Face-centered cubic crystal lattice. Burns when heated with a hot enough flame (over 800°, oxygen torch). d_4^{25} 3.513. n_D^{20} 2.4173. Hardness = 10 (Mohs' scale). Sp heat at 100°K: 0.606 cal/g-atom/°K. Entropy at 298.16°K: 0.5684 cal/g-atom/°K. Band gap energy: 6.7 ev. Dielectric constant 5.7. Electron mobility: ~1800 cm²/v-sec. Hole mobility: 1200 cm²/v-sec. Can be pulverized in a steel mortar. Attacked by laboratory-type cleaning soln (potassium dichromate + concd H_2SO_4). In the jewelry trade the unit of weight for diamonds is one carat = 200 mg. *Ref: Wall Street J.* **164**, no. 36, p 10 (Aug 19, 1964).

USE: Jewelry. Polishing, grinding, cutting glass, bearings for delicate instruments; manuf dies for tungsten wire and similar hard wires; making styli for recorder heads, longlasting phonograph needles. In semiconductor research.

3031. Diamond Ink. Etching ink. A mixture of HF, $BaSO_4$ and fluorides.

Milky-white liq with a heavy sediment. Shake well before using and warm gently in a lead dish. *Keep in plastic, hardrubber or intern. paraffin-coated bottles.*

USE: For etching glass.

3032. Diampromide. *N-[2-[Methyl(2-phenylethyl)amino]propyl]-N-phenylpropanamide; N-[2-(methylphenethylamino)propyl]propionanilide.* $C_{21}H_{28}N_2O$; mol wt 324.47. C 77.74%, H 8.70%, N 8.63%, O 4.93%. Synthesis: Wright *et*

al., J. Am. Chem. Soc. **81**, 1518 (1959); U.S. pat. **2,944,081** (1960 to American Cyanamid).

Liquid, $bp_{0.5}$ 174-178°. n_D^{26} 1.546.

Sulfate, $C_{21}H_{30}N_2O_5S$, crystals from ethanol + ether, mp 110-111°.

Note: This is a controlled substance (opiate) listed in the U.S. Code of Federal Regulations, Title 21 Part 1308.11 (1995).

THERAP CAT: Analgesic (narcotic).

3033. Diamthazole Dihydrochloride. *6-[2-(Diethylamino)ethoxy]-N,N-dimethyl-2-benzothiazolamine dihydrochloride; 6-[2-(diethylamino)ethoxy]-2-dimethylaminobenzothiazole dihydrochloride; 2-dimethylamino-6-(β-diethylaminoethoxy)benzothiazole dihydrochloride; dimazole dihydrochloride; Ro-2-2453; Asterol Dihydrochloride; Atelor.* $C_{15}H_{25}Cl_2N_3OS$; mol wt 366.35. C 49.18%, H 6.88%, Cl 19.35%, N 11.47%, O 4.37%, S 8.75%. Prepn: Steiger, Keller, U.S. pat. **2,578,757** (1951 to Hoffmann-La Roche).

Crystals, dec 269°, mp 240-243°. Freely sol in water, methanol, ethanol. A 5% aq soln has a pH of ~2.

THERAP CAT: Antifungal.

3034. Diamyl Sodium Sulfosuccinate. *Sulfobutanedioic acid 1,4-dipentyl ester sodium salt; sulfosuccinic acid dipentyl ester sodium salt;* Aerosol AY; Alphasol AY. $C_{14}H_{25}NaO_7S$; mol wt 360.40. C 46.66%, H 6.99%, Na 6.38%, O 31.08%, S 8.90%. The amyl or 1-methylbutyl diester of the monosodium salt of sulfosuccinic acid or a mixture of both. Wetting agent prepd by the action of the appropriate alcohols on maleic anhydride followed by addtion of sodium bisulfite: Jaeger, U.S. pats. **2,028,091** and **2,176,423** (1936, 1939 to Am. Cyanamid).

Available as a mixture of white, hard pellets and powder. Soly in water at 25° = 392 g/liter; at 70° = 502 g/liter. Maximum concn of electrolyte soln in which 1% of the wetting agent is sol: 3% NaCl; 2-4% NH_4Cl (turbid); 3-6% $(NH_4)_2HPO_4$ (turbid); 4% $NaNO_3$ (slightly turbid); 15% Na_2SO_4 (very slightly turbid). Also sol in pine oil, oleic acid, acetone, hot kerosene, carbon tetrachloride, glycerol, hot olive oil; insol in liq petrolatum. Surface tension in water: 0.001% = 69.4 dyn/cm; 0.02% = 68.3 dyn/cm; 0.1% = 50.2 dyn/cm; 0.25% = 41.6 dyn/cm; 1% = 29.2 dyn/cm. Interfacial tension 1% in water *vs.* liquid petrolatum: 5 seconds = 7.55 dyn/cm; 30 seconds = 7.37 dyn/cm; 15 minutes = 7.03 dyn/cm. Interfacial tension 0.1% in water *vs.* liquid petrolatum: 5 seconds = 29.5 dyn/cm; 30 seconds = 28.6 dyn/cm; 15 minutes = 27.5 dyn/cm. Stable in acid and neutral solns, hydrolyzes in alkaline solns.

USE: As emulsifier in emulsion polymerization and as a wetting agent.

3035. 1,2-Dianilinoethane. *N,N'-Diphenyl-1,2-ethanediamine; N,N'-diphenylethylenediamine; N,N'-diphenyl-α,ω-*

diaminoethane; *sym*-diphenylethylenediamine. $C_{14}H_{16}N_2$; mol wt 212.29. C 79.21%, H 7.60%, N 13.20%. Prepd by heating aniline with dichloroethane: Büchi *et al.*, *Helv. Chim. Acta* **39**, 950 (1956); by heating aniline with dibromoethane: Wanzlick, Löchel, *Ber.* **86**, 1463 (1953).

Crystals from dil ethanol, mp 67.5°; (also reported as the monohydrate). bp$_{12}$ 228-230°. Very sol in alc, ether.

USE: Identification of aldehydes; stabilizer for resins and rubber. Intermediate in the manufacture of antihistamines.

3036. Dianisidine. *3,3'-Dimethoxy-[1,1'-biphenyl]-4,4'-diamine; 3,3'-dimethoxybenzidine;* 3,3'-dimethoxy-4,4'-diaminobiphenyl. $C_{14}H_{16}N_2O_2$; mol wt 244.29. C 68.83%, H 6.60%, N 11.47%, O 13.10%. Prepn: Meier, Böhler, *Ber.* **89**, 2301 (1956); Sogn, U.S. pat. **2,794,047** (1957 to Allied Chem.). Manuf of stabilized product: Cashion, U.S. pat. **2,966,519** (1960 to Allied Chem.). Review of carcinogenicity studies: *IARC Monographs* **4**, 41-47 (1974).

Crystals, becoming violet, mp 137-138°. Practically insol in water. Sol in alcohol, benzene, ether.

Note: This substance and its dihydrochloride may reasonably be anticipated to be carcinogens: *Seventh Annual Report on Carcinogens* (PB95-109781, 1994) p 175.

USE: Manuf of azo dyes.

3037. Diathymosulfone. *4,4'-[Sulfonylbis(4,1-phenyleneazo)]bis[5-methyl-2-(1-methylethyl)]phenol; 6,6'-[sulfonylbis(p-phenyleneazo)]dithymol;* bis[p-(4-hydroxy-2-methyl-5-isopropylphenylazo)phenyl] sulfone; 4,4'-bis(p-isopropyl-m-cresylazo)diphenyl sulfone; thymol sulfone; Diatox. $C_{32}H_{34}N_4O_4S$; mol wt 570.71. C 67.35%, H 6.00%, N 9.82%, O 11.21%, S 5.62%. Prepn: **Brit. pat. 758,744** (1956 to Lab. Laborec); Chiarlo *et al.*, *Arch. Ital. Sci. Farmacol.* **7**, 167 (1957), *C.A.* **51**, 16949c (1957).

Red-brown microcrystalline powder from ethanol + water, mp 222-224°. Absorption max (alcohol): 400 nm. Practically insol in water. Sol in alkali solns with red color, in dioxane, acetone; less sol in alcohol, ether.

Di-silver salt, $C_{32}H_{32}Ag_2N_4O_4S$, *thymolated silver sulfone.*

THERAP CAT: Antibacterial (leprostatic).

3038. Diatretyne I. *(E)-8-Amino-8-oxo-2-octene-4,6-diynoic acid; trans-7-carbamoyl-2-heptene-4,6-diynoic acid;* 7-carbamidohepta-2-ene-4,6-diynoic acid; diatretyne amide. $C_8H_5NO_3$; mol wt 163.13. C 58.90%, H 3.09%, N 8.59%, O 29.42%. HOOC—CH=CHC≡CC≡CCONH$_2$. Antibiotic substance produced by the fungus *Clitocybe diatreta*: Anchel, *J. Am. Chem. Soc.* **75**, 4621 (1953); *Arch. Biochem. Biophys.* **78**, 100 (1958). Structure: Bu'Lock *et al.*, *Chem. & Ind. (London)* **1954**, 990. Synthesis: Ashworth *et al.*, *J. Chem. Soc.* **1958**, 950; Prevost *et al.*, *Bull. Soc. Chim. France* **1961**, 2171.

Crystals from methanol, explodes 198°. uv max (95% ethanol): 224, 274, 291, 309.5 nm (ε × 10⁻³ 37.7, 16.5, 23, 19).

3039. Diatretyne II. *7-Cyano-2-heptene-4,6-diynoic acid;* diatretyne nitrile; nudic acid B. $C_8H_3NO_2$; mol wt 145.12. C 66.21%, H 2.08%, N 9.65%, O 22.05%. HOOC-CH=CHC≡CC≡CCN. Antibiotic substance produced by the basidiomycete *Tricholoma nudum* and by the fungus *Clitocybe diatreta*. Isoln: Anchel, *J. Am. Chem. Soc.* **74**, 1588 (1952). Structure: Anchel, *Science* **121**, 607 (1955). Identity of nudic acid B and diatretyne II: Heatly, Stephenson, *Nature* **179**, 1078 (1957). Synthesis: Ashworth *et al.*, *J. Chem. Soc.* **1958**, 950.

Short needles from ether + light petr, dec 179-180°. Darkens and turns black within 2 hrs on exposure to sunlight. Sol in the usual organic solvents except hexane; dissolves slowly in sodium carbonate soln; practically insol in water and dil acids.

3040. Diatrizoate Sodium. *3,5-Bis(acetylamino)-2,4,6-triiodobenzoic acid sodium salt; 3,5-diacetamido-2,4,6-triiodobenzoic acid sodium salt;* sodium 3,5-diacetamido-2,4,6-triiodobenzoate; urografic acid sodium salt; sodium amidotrizoate; sodium diatrizoate; Hypaque Sodium; Triognost. $C_{11}H_8I_3N_2NaO_4$; mol wt 635.90. C 20.78%, H 1.27%, I 59.87%, N 4.41%, Na 3.62%, O 10.06%. Prepn of the acid: Larsen *et al.*, *J. Am. Chem. Soc.* **78**, 3210 (1956). Description: Langecker *et al.*, *Arch. Exp. Pathol. Pharmakol.* **222**, 584 (1954). Use in separation of leukocytes from whole blood: R. J. Perper *et al.*, *J. Lab. Clin. Med.* **72**, 842 (1968); L. P. Bignold, A. Ferrante, *J. Immunol. Meth.* **96**, 29 (1987); J. R. Kalmar *et al.*, ibid. **110**, 275 (1988). Comprehensive description: H. H. Lerner in *Analytical Profiles of Drug Substances* **vol. 4**, K. Florey, Ed. (Academic Press, New York, 1975) pp 137-167.

Rhombic needles, slightly salty taste, mp 261-262° (dec). Soly in water at 20°: 60 g/100 ml. pH of a 50% aq soln 7.0 to 7.5. On chilling solns may become cloudy or form a precipitate which disappears at room temp. Solns may be sterilized by autoclaving. Should be protected from light. LD$_{50}$ i.v. in rats: 14.7 g/kg (Langecker).

Free acid, crystals from dil DMF, mp > 300°.

Note: See also Meglumine Diatrizoate.

USE: Density gradient reagent for blood cell separation.

THERAP CAT: Diagnostic aid (radiopaque medium).

3041. Diaveridine. *5-[(3,4-Dimethoxyphenyl)methyl]-2,4-pyrimidinediamine; 2,4-diamino-5-veratrylpyrimidine;* 2,4-diamino-5-(3',4'-dimethoxybenzyl)pyrimidine. $C_{13}H_{16}N_4O_2$; mol wt 260.30. C 59.99%, H 6.20%, N 21.52%, O 12.29%. Prepn: Falco *et al.*, *J. Am. Chem. Soc.* **73**, 3758 (1951). Hitchings, Falco, U.S. pat. **2,624,732** (1953 to Burroughs Wellcome); Stenbuck *et al.*, *J. Org. Chem.* **28**, 1983 (1963); Hoffer, U.S. pat. **3,341,541** (1967 to Hoffmann-La Roche); M. Hoffer *et al.*, *J. Med. Chem.* **14**, 462 (1971). Crystal structure: T. F. Koetzle, G. Williams, *Acta Crystallogr.* **B34**, 323 (1978).

Crystals, mp 233°.

Ingredient of *Darvisul* which also contains sulfaquinoxaline.

THERAP CAT (VET): Antiprotozoan (in combination with sulfaquinoxaline).

3042. Diazepam. *7-Chloro-1,3-dihydro-1-methyl-5-phenyl-2H-1,4-benzodiazepin-2-one;* 7-chloro-1-methyl-5-phenyl-3H-1,4-benzodiazepin-2(1H)-one; methyl diazepinone; diacepin; LA-III; Ro-5-2807; Wy-3467; NSC-77518; Apaurin; Atensine; Atilen; Bialzepam; Calmpose; Ceregulart; Dialar; Diazemuls; Dipam; Eridan; Eurosan; Evacalm; Faustan; Gewacalm; Horizon; Lamra; Lembrol; Levium; Mandrozep; Neurolytril; Noan; Novazam; Paceum; Pacitran; Paxate; Paxel (obsolete); Pro-Pam; Q-Pam; Relanium; Sedapam; Seduxen; Servizepam; Setonil; Solis; Stesolid; Tranquase; Tranquo-Puren; Tranquo-Tablinen; Unisedil; Valaxona; Valiquid; Valium; Valrelease; Vival; Vivol. $C_{16}H_{13}ClN_2O$; mol wt 284.74. C 67.49%, H 4.60%, Cl 12.45%, N 9.84%, O 5.62%. Prepn: Sternbach, Reeder, *J. Org. Chem.* **26**, 4936 (1961); Reeder, Sternbach, U.S. pat. **3,371,085** (1968 to Hoffmann-La Roche); prepd but not claimed: *eidem,* U.S. pats. **3,109,843** and **3,136,815** (1963, 1964 to Hoffmann-La Roche); M. Gates, *J. Org. Chem.* **45**, 1675 (1980); M. Ishikura *et al., ibid.* **47**, 2456 (1982). Purification: Chase, U.S. pat. **3,102,116** (1963 to Hoffmann-La Roche). Pharmacology: Hudson, Wolpert, *Arch. Int. Pharmacodyn. Ther.* **186**, 388 (1970). Metabolism: Randall *et al., Curr. Ther. Res. Clin. Exp.* **7**, 590 (1965); Van der Kleijn, *Ann. N.Y. Acad. Sci.* **179**, 115 (1971). Toxicity: E. I. Goldenthal, *Toxicol. Appl. Pharmacol.* **18**, 185 (1971). Binding study in rat brain: R. F. Squires, C. Braestrup, *Nature* **266**, 732 (1977). Comprehensive description: A. MacDonald *et al., Anal. Profiles Drug Subs.* **1**, 79-99 (1972). Clinical pharmacokinetics: M. Mandelli *et al., Clin. Pharmacokinet.* **3**, 72 (1978).

Plates from acetone + petr ether, mp 125-126°. pK_a 3.4. Sol in chloroform, DMF, benzene, acetone, alc; slightly sol in water. LD_{50} orally in rats: 710 mg/kg (Goldenthal).

Note: This is a controlled substance (depressant) listed in the U.S. Code of Federal Regulations, Title 21 Part 1308.14 (1995).

THERAP CAT: Anxiolytic; muscle relaxant (skeletal).

3043. Diazinon. *Phosphorothioic acid O,O-diethyl O-[6-methyl-2-(1-methylethyl)-4-pyrimidinyl] ester;* thiophosphoric acid 2-isopropyl-4-methyl-6-pyrimidyl diethyl ester; O,O-diethyl O-2-isopropyl-4-methyl-6-pyrimidyl thiophosphate; diethyl 2-isopropyl-4-methyl-6-pyrimidyl thionophosphate; dimpylate; G-24480; Antigal; Basudin; Diazol; D-z-n; Garden Tox; HelfaCat; HelfaDog; Neocidol; Parasitex; Sarolex; Spectracide; Taberdog; Tabergat. $C_{12}H_{21}N_2O_3PS$; mol wt 304.35. C 47.36%, H 6.95%, N 9.20%, O 15.77%, P 10.18%, S 10.54%. Prepn: Gysin, Margot, U.S. pat. **2,754,243** (1956 to Geigy). Insecticidal properties: Gasser, *Z. Naturforsch.* **8b**, 225 (1953). Toxicity data: T. B. Gaines, *Toxicol. Appl. Pharmacol.* **14**, 515 (1969).

Liquid. Faint ester-like odor. d_4^{20} 1.116-1.118. $bp_{0.002}$ 83-84°. n_D^{20} 1.4978-1.4981. Vapor pressure at 20°: 1.4×10^{-4}; at 40°: 1.1×10^{-3} (~5 times vapor pressure of parathion). Volatility at 20°: 2.4 mg/m³; at 40°: 17.6 mg/m³. Decomposes above 120°. Soly in water at 20°: 0.004%. Miscible with alcohol, ether, petr ether, cyclohexane, benzene, and similar hydrocarbons. More stable in alkaline formulations, then when at neutral or acid pH. LD_{50} in male, female rats (mg/kg): 250, 285 orally (Gaines).

Caution: Cholinesterase inhibitor: *Clinical Toxicology of Commercial Products,* R. E. Gosselin *et al.,* Eds. (Williams & Wilkins, Baltimore, 5th ed., 1984) Section II, p 294.

USE: Insecticide.

THERAP CAT (VET): Ectoparasiticide.

3044. Diaziquone. *[2,5-Bis(1-aziridinyl)-3,6-dioxo-1,4-cyclohexadiene-1,4-diyl]biscarbamic acid diethyl ester; 2,5-bis(1-aziridinyl)-3,6-dioxo-1,4-cyclohexadiene-1,4-dicarbamic acid diethyl ester;* 2,5-diaziridinyl-3,6-bis(ethoxycarbonyl-amino)-1,4-benzoquinone; aziridinylbenzoquinone; AZQ; CI-904; NSC-182986. $C_{16}H_{20}N_4O_6$; mol wt 364.36. C 52.74%, H 5.53%, N 15.38%, O 26.35%. Quinone-containing lipophilic alkylating agent. Prepn as intermediate in dyestuffs: S. Petersen *et al.,* U.S. pat. **2,913,453** (1959 to Schenley Ind.). Synthesis and antitumor activity: A. H. Khan, J. S. Driscoll, *J. Med. Chem.* **19**, 313 (1976); J. S. Driscoll *et al.,* U.S. pat. **4,146,622** (1979 to U.S. Gov't). *In vivo* antitumor studies: J. S. Driscoll *et al., J. Pharm. Sci.* **68**, 185 (1979). Intracerebral penetration and tissue distribution in humans: N. Savaraj *et al., J. Neuro-Oncol.* **1**, 15 (1983). LC determn: B. A. Allen *et al., J. Chromatog.* **222**, 146 (1981). Pharmacokinetics: S. Zimm *et al., Cancer Res.* **44**, 1698 (1984). Clinical studies in treatment of CNS neoplasms: R. T. Eagan *et al., J. Neuro-Oncol.* **5**, 309 (1987); E. Tapazoglou *et al., Am. J. Clin. Oncol.* **11**, 474 (1988). Review: J. F. Bender *et al., Invest. New Drugs* **1**, 71-84 (1983).

Orange needles from ethanol, mp 230° (dec); also reported as yellowish-brown crystals from ethanol decomposing at temperatures above 250° (Petersen). uv max (methanol): 340 nm (log ε 4.17). Soly in water: 0.5 mg/ml. LD_{50} in mice (mg/m²): 30.9 i.v. (Bender).

THERAP CAT: Antineoplastic.

3045. Diazoacetic Ester. Ethyl diazoacetate; ethyl diazoethanoate. $C_4H_6N_2O_2$; mol wt 114.10. C 42.11%, H 5.30%, N 24.55%, O 28.04%. $^-N=N^+=CHCOOC_2H_5$. Prepd by the action of sodium nitrite on glycine ethyl ester hydrochloride: Curtius *J. Prakt. Chem.* [2] **38**, 396 (1888); Silberrad, *J. Chem. Soc.* **81**, 600 (1902); Womack, Nelson, *Org. Syn.* **24**, 56 (1944).

Yellow oil. Pungent odor. Very volatile. Explosive. *Distillation, even under reduced pressure, is dangerous.* mp −22°. $d_4^{17.6}$ 1.0852. bp_5 42°; bp_{12} 45°; bp_{88} 85-86°; bp_{720} 140-141°. $n_D^{17.6}$ 1.4588. Slightly sol in water. Neutral reaction. Volatile with steam, ether, and benzene vapors. Miscible with alcohol, benzene, ligroin, ether. Explodes on contact with concd H_2SO_4.

3046. Diazoaminobenzene. *1,3-Diphenyl-1-triazene;* anilinoazobenzene; benzeneazoaniline. $C_{12}H_{11}N_3$; mol wt 197.24. C 73.07%, H 5.62%, N 21.30%. Made by diazotizing aniline dissolved in HCl with $NaNO_2$ and then adding a concd soln of sodium acetate: Hartman, Dickey, *Org. Syn.* coll. vol. II, 163 (1943).

Golden-yellow, small crystals. mp 98°; explodes when heated to 150°. Insol in water; freely sol in benzene, ether, hot alc.

3047. p-Diazobenzenesulfonic Acid. *4-Sulfobenzenediazonium inner salt;* sulfanilic acid diazide. $C_6H_4N_2O_3S$; mol wt 184.18. C 39.13%, H 2.19%, N 15.21%, O 26.06%, S 17.41%. Prepn by diazotization of sulfanilic acid and structure: Whetsel *et al., J. Am. Chem. Soc.* **78**, 3360 (1956); Bugg *et al., Acta Cryst.* **17**, 767 (1964).

White or slightly red crystals; when dry, easily explodes by heat, a blow, or on rubbing; hence marketed as a paste. Slightly sol in cold water, alcohol; freely sol in hot water, dil alkalies, HCl.

USE: As reagent for phenol, fecal matter in water, glucose, aldehydes, and albumin; also in manuf of azo and other dyes.

3048. Diazomethane. Azimethylene. CH_2N_2; mol wt 42.04. C 28.57%, H 4.80%, N 66.63%. $CH_2=N^+=N^-$. Prepd from chloroform and hydrazine by reaction with potassium hydroxide: Staudinger, Kupfer, *Ber.* **45**, 501 (1912); McManus *et al., J. Org. Chem.* **33**, 4272 (1968); from KOH and nitrosomethylurea: Dessaux, Durand, *Bull. Soc. Chim. France* **1963**, 41. These methods yield gaseous diazomethane. The following procedures yield ether solns of diazomethane. From *N*-nitroso-β-methylaminoisobutyl methyl ketone in ether and isopropanol by reaction with sodium isopropoxide or from the same ketone in ether by reaction with sodium cyclohexoxide: Redemann *et al., Org. Syn.* **25**, 28 (1945). By KOH saponification of nitrosomethylurea in ether: Arndt, *Org. Syn.* **coll. vol. II**, 165 (1943), or of nitrosomethylurethan in ether: von Pechmann, *Ber.* **27**, 1888 (1894); **28**, 855 (1895); Meerwein, Burneleit, *Ber.* **61**, 1845 (1928). In the laboratory diazomethane may be prepd most simply by the action of alkali on the commercially available *N*-methyl-*N*-nitroso-*N'*-nitroguanidine, *cf.* McKay, *J. Am. Chem. Soc.* **70**, 1974 (1948); McKay *et al., Can. J. Res.* **28**, 683 (1950); Fieser and Fieser, *Organic Chemistry* (Reinhold, New York, 3rd ed., 1956) p 176. Alternate intermediate: *p*-tolylsulfonylmethylnitrosamide ("Diazald" Aldrich Chemical Co.), see De Boer, Backer, *Org. Syn.* **coll. vol. IV**, 250 (1963).

Very toxic yellow gas. *Explosive!* Use safety screen. Insidious poison, a well-ventilated hood is absolutely necessary, avoid vapor. mp −145°; bp −23°. Undil liq and concd solns may explode violently, especially if impurities are present. Gaseous diazomethane may explode on heating to 100° or on rough glass surfaces. Ground glass apparatus and glass stirrers with glass sleeve bearings where grinding may occur, should not be used. Alkali metals also produce explosions with diazomethane. Sol in ether, dioxane. Such solns dec only slowly at low temps. Decompn is more rapid if alcohols or water are present. Copper powder causes active decompn with the evolution of nitrogen and the formation of insol white flakes of polymethylene $(CH_2)_x$. Solid calcium chloride or boiling stones have the same effect. This phenomenon appears to occur always during the action of diazomethane on solid substances.

Caution: Potential symptoms of overexposure are coughing, shortness of breath; headache; flushing of skin, fever; chest pain, pulmonary edema, pneumonitis; eye irritation; asthma; fatigue. *See NIOSH Pocket Guide to Chemical Hazards* (DHHS/NIOSH 90-117, 1990) p 82. May act as a respiratory sensitizer. *See Patty's Industrial Hygiene and Toxicology* vol. 2A, G. D. Clayton, F. E. Clayton, Eds. (Wiley-Interscience, New York, 3rd ed., 1981) pp 2784-2786.

USE: Powerful methylating agent for acidic compds such as carboxylic acids, phenols, enols. For syntheses with diazomethane see the reviews by Smith, *Chem. Revs.* **23**, 193 (1938); Eistert, *Z. Angew. Chem.* **54**, 99, 124 (1941) translated by Spangler in *Newer Methods of Preparative Organic Chemistry* (New York, 1948) p 513; J. S. Pizey, *Synthetic Reagents* vol. 2 (John Wiley, New York, 1974) pp 65-142.

3049. 6-Diazo-5-oxo-L-norleucine. DON. $C_6H_9N_3O_3$; mol wt 171.16. C 42.11%, H 5.30%, N 24.55%, O 28.04%. $N_2CHCOCH_2CH_2CH(NH_2)COOH$. Antitumor antibiotic produced by an unidentified species of *Streptomyces* from Peruvian soil: Dion *et al., J. Am. Chem. Soc.* **78**, 3075 (1956); Ehrlich *et al., Antibiot. & Chemother.* **6**, 487 (1956). Several methods of synthesis: De Wald, Moore, *ibid.* **80**, 3941 (1958); *eidem*, U.S. pat. **2,965,634** (1960 to Parke, Davis). *Review:* Pittillo, Hunt, *Antibiotics* vol 1, D. Gottlieb, P. D. Shaw, Eds. (Springer-Verlag, New York, 1967) pp 481-493.

Pale yellow crystals from dil ethanol, dec 145-155°. $[\alpha]_D^{26}$ +21° (c = 5.4). uv max (phosphate buffer): 244, 275 nm ($E_{1cm}^{1\%}$ 376, 683). Freely sol in water. The pH of aq solns should be maintained between 4.5 and 6.5. Also sol in aq solns of methanol, ethanol, acetone. Sparingly sol in absolute alcs.

THERAP CAT: Antineoplastic.

3050. 5-Diazouracil. *5-Diazo-2,4(1H,3H)-pyrimidinedione*; DU; 2,4-dioxo-5-diazopyrimidine; 5-diazo-2,4-dioxopyrimidine. $C_4H_2N_4O_2$; mol wt 138.09. C 34.79%, H 1.46%, N 40.57%, O 23.17%. Prepd by diazotization of 5-aminouracil: Johnson *et al., Ber.* **64**, 2629 (1931). Revised structure: Thurber, Townsend, *J. Heterocycl. Chem.* **9**, 629 (1972). Has significant activity against gram-positive and gram-negative bacteria *in vivo*: Hunt, Pittillo, *Appl. Microbiol.* **16**, 1792 (1968).

Crystals, conflagrates at 198°. The stout prisms obtained from ice water should be white, not yellow or red. Sensitive to light, temp and air. Best stored in evacuated, refrigerated ampuls. Acid reactions. IR spectrum: Band at 4.57 μ. Forms alcoholates, also a red monohydrate, $C_4H_4N_4O_3$, from which a potassium salt, $KC_4H_3N_4O_3$, is obtained which is slightly sol in water with neutral reaction.

USE: In cancer research.

3051. Diazoxide. *7-Chloro-3-methyl-2H-1,2,4-benzothiadiazine 1,1-dioxide*; 3-methyl-7-chloro-1,2,4-benzothiadiazine 1,1-dioxide; SRG-95213; Eudemine Injection; Proglicem; Hyperstat; Hypertonalum; Mutabase; Proglycem. $C_8H_7ClN_2O_2S$; mol wt 230.67. C 41.66%, H 3.06%, Cl 15.37%, N 12.14%, O 13.87%, S 13.90%. Prepn: A. A. Rubin *et al., Science* **133**, 2067 (1961); J. G. Topliss *et al.*, U.S. pat. **2,986,573**; *eidem*, U.S. pat. **3,345,365** (1961, 1967 both to Schering); Raffa, Monzani, *Farmaco Ed. Sci.* **17**, 244 (1962). Crystal and molecular structure: G. Bandoli, M. Nicolini, *J. Cryst. Mol. Struct.* **7**, 229 (1978).

Crystals from dil alc, mp 330-331°. uv max (methanol): 268 nm (ε 11300). Sol in alcohol and alkaline solns. Insol in water.

THERAP CAT: Antihypertensive.

3052. Dibekacin. *O-3-Amino-3-deoxy-α-D-glucopyranosyl-(1→6)-O-[2,6-diamino-2,3,4,6-tetradeoxy-α-D-erythrohexopyranosyl-(1→4)]-2-deoxy-D-streptamine*; DKB; 3',4'-dideoxykanamycin B; debecacin. $C_{18}H_{37}N_5O_8$; mol wt 451.52. C 47.88%, H 8.26%, N 15.51%, O 28.35%. Semisynthetic analog of kanamycin, *q.v.*, effective against kanamycin-resistant bacteria. Prepn: Umezawa *et al., J. Antibiot.* **24**, 485 (1971); *eidem, Bull. Chem. Soc. Japan* **45**, 3624 (1972); Umezawa *et al.*, Ger. pat. **2,135,191** (1972 to Microbiochem. Res. Found.). Improved synthesis: T. Yoneta *et al., Bull. Chem. Soc. Japan* **52**, 1131 (1979). Antibacterial activity: A. G. Paradelis *et al., Antimicrob. Ag Chemother.*

14, 514 (1978). Metabolic studies: Shimizu, *Japan. J. Antibiot.* **26,** 522 (1973). Toxicology: Koeda *et al., ibid.* 221. Pharmacokinetics and acute toxicity: I. Komiya *et al., J. Pharmacobio.-Dyn.* **4,** 356 (1981). HPLC determn in serum: H. Kubo *et al., Antimicrob. Ag. Chemother.* **28,** 521 (1985). *Review:* P. Noone, *Drugs* **27,** 548-578 (1984).

$[\alpha]_D^{20}$ +132° (c = 0.65). LD_{50} in mice (mg/kg): 61.0-68.0 i.v., 373.0-380.0 i.m. (Komiya).

Sulfate, *Débékacyl, Icacine, Kappabi, Orbicin, Panamicin, Panimycin, Tokocin.* White or yellowish-white powder. Slightly bitter taste. Sol in water. Practically insol in ethanol, acetone, and other organic solvents.

THERAP CAT: Antibacterial.

3053. Dibenzalacetone. *1,5-Diphenyl-1,4-pentadien-3-one;* dibenzylidene acetone; distyryl ketone. $C_{17}H_{14}O$; mol wt 234.30. C 87.15%, H 6.02%, O 6.83%. C_6H_5CH=CH-COCH=CHC_6H_5. Prepn from benzaldehyde + acetone: Conrad, Dolliver, *Org. Syn.* **12,** 22 (1939); Haslam, U.S. pat. **2,719,863** (1955 to du Pont); Tokár *et al., Acta Chim. Acad. Sci. Hung.* **19,** 83 (1959). Prepn of geometrical isomers: Dinwiddie *et al., J. Org. Chem.* **27,** 327 (1962).

*trans-trans-*Form, crystals from hot ethyl acetate, mp 110-111°. uv max: 330 nm (ε 34,300). Practically insol in water. Slightly sol in alc, ether; sol in acetone, chloroform.
*cis-trans-*Form, light yellow needles from ethanol, mp 60°. uv max: 295 nm (ε 20,000).
*cis-cis-*Form, yellow oil, $bp_{0.02}$ 130°. uv max: 287 nm (ε 11,000).

USE: In sun protection preparations.

3054. 1,2:5,6-Dibenzanthracene. *Dibenz[a,h]anthracene.* $C_{22}H_{14}$; mol wt 278.35. C 94.93%, H 5.07%. From methyl dinaphthyl ketone: Clar, *Ber.* **62,** 350, 1378 (1929); Fieser, Dietz, *ibid.* 1827; Bachmann, *J. Org. Chem.* **1,** 347 (1937). Separation by chromatography: Winterstein, Schön, *Z. Physiol. Chem.* **230,** 146 (1934). Absorption spectrum: Mayneord, Roe, *Proc. Roy. Soc. London* **A152,** 299 (1935). *Review:* E. Clar, *Polycyclic Hydrocarbons,* 2 vols. (Academic Press, New York, 1964). Review of carcinogenicity studies: *IARC Monographs* **3,** 178-196 (1973).

Plates, leaflets from acetic acid. Crystals may be monoclinic or orthorhombic. Sublimes. mp 266°. Sol in petr ether, benzene, toluene, xylene, oils, other organic solvents; slightly sol in alcohol and ether. Insol in water.

Note: This substance may reasonably be anticipated to be a carcinogen: *Seventh Annual Report on Carcinogens* (PB95-109781, 1994) p 328.

3055. Dibenzepin. *10-[2-(Dimethylamino)ethyl]-5,10-dihydro-5-methyl-11H-dibenzo[b,e][1,4]diazepin-11-one;* 5-methyl-10β-dimethylaminoethyl-10,11-dihydro-11-oxo-dibenzo[b,e][1,4]diazepine; HF-1927. $C_{18}H_{21}N_3O$; mol wt 295.38. C 73.19%, H 7.17%, N 14.23%, O 5.42%. Prepn: F. Hunziker *et al., Arzneimittel-Forsch.* **13,** 324 (1963); **Brit. pat. 961,106;** J. Schmutz, F. Hunziker, U.S. pats. **3,312,689** and **3,419,547** (1964, 1967, 1968 all to Wander). Metabolism: H. Lehner *et al., Arzneimittel-Forsch.* **17,** 185 (1967).

Comprehensive description: A. Egli, W. R. Michaelis in *Analytical Profiles of Drug Substances* **vol. 9,** K. Florey, Ed. (Academic Press, New York, 1980) pp 181-206.

mp 116-117°. $bp_{0.01}$ 185°.
Hydrochloride, $C_{18}H_{21}N_3O.HCl$, *Neodalit, Noveril.* Crystals, mp 238°. uv max (0.1N HCl): 204, 220 nm (log ε 4.530, 4.458). pKa 8.25. Sol in water, alcohol, chloroform. LD_{50} orally in mice: 215 mg/kg (Hunziker).

THERAP CAT: Antidepressant.

3056. Dibenzoylmethane. *1,3-Diphenyl-1,3-propanedione;* phenyl-α-hydroxystyryl ketone; ω-benzoylacetophenone; phenyl phenacyl ketone; γ-hydroxychalcone. $C_{15}H_{12}O_2$; mol wt 224.26. C 80.34%, H 5.39%, O 14.27%. $C_6H_5COCH_2COC_6H_5$. Prepd by the action of sodium methylate on benzylideneacetophenone dibromide: Allen *et al., Org. Syn. coll. vol. I* (2nd ed., 1941) p 205.

Orthorhombic plates, tablets from petr ether, mp 80°. bp_{18} 219-221°. 100 parts alc dissolve 4.43 parts at 19°. Sol in ether, chloroform, aq NaOH. Equilibrium mixtures in methanol, alcohol, and glacial acetic acid contain 96-98% of the enol form.

Monoxime, $C_{15}H_{13}NO_2$, prepd by adding 4 mols hydroxylamine to 1 mol dibenzoylmethane in alcohol. Prisms from ether, mp 165°. Sparingly sol in cold alcohol.

3057. 2,3:6,7-Dibenzphenanthrene. *Pentaphene;* dibenzo[b,h]phenanthrene; β,β'-dibenzphenanthrene; naphtho-2',3'-1,2-anthracene. $C_{22}H_{14}$; mol wt 278.35. C 94.93%, H 5.07%. Prepn by pyrolysis of 1,2- or 1,4-di-o-toluylbenzene: Clar *et al., Ber.* **62,** 947 (1924); Clar, John, *Ber.* **64,** 986 (1931); Winterstein, Schön, *Z. Physiol. Chem.* **230,** 146 (1934); from phthalic anhydride + o-tolylmagnesium bromide: Clar, Stewart, *J. Chem. Soc.* **1951,** 3215; from 2-benzoylbenzoic acid: Marsili, Isola, *Tetrahedron Letters* **1965,** 3023; Franck, Zander, *Ber.* **99,** 396 (1966).

Yellowish-green needles or leaflets from xylene, mp 257°. Absorption max (alc): 423.5, 412, 399, 379, 356, 345, 329, 314.5, 302, 289.5, 257.5, 245 nm. Sol in benzene, xylene; sparingly sol in alcohol, ether. Solns exhibit blue fluorescence in daylight.

Dipicrate, $C_{22}H_{14}.2C_6H_3N_3O_7$, orange needles, mp 184°.

3058. Dibenzylamine. *N-(Phenylmethyl)benzenemethanamine.* $C_{14}H_{15}N$; mol wt 197.28. C 85.24%, H 7.66%, N 7.10%. $C_6H_5CH_2NHCH_2C_6H_5$. Prepd by the action of alcoholic ammonia on benzyl chloride: Mason, *J. Chem. Soc.* **63,** 1312 (1893).

Oil; ammoniacal odor. bp 300° with partial decompn. Practically insol in water; sol in alcohol or ether.

USE: Detection of cobalt, cyanate and iron.

3059. Dibenzyl Chlorophosphonate. *Phosphorochloridic acid bisphenylmethyl ester; benzyl phosphorochloridate;* dibenzylchlorophosphate; DBPCl; dibenzylphosphoryl chloride. $C_{14}H_{14}ClO_3P$; mol wt 296.69. C 56.68%, H 4.76%, Cl 11.95%, O 16.18%, P 10.44%. Prepd by adding sulfuryl chloride to dibenzyl phosphite at 17-19° in CCl_4 diluent: Atherton *et al., J. Chem. Soc.* **1948,** 1106; by adding chlorine to dibenzyl phosphite in CCl_4 at below 0°: Atherton *et al.*, U.S. pat. 2,490,573 (1949 to Hoffmann-La Roche).

Thick oil, dec on standing or distillation. Identified by its reaction with ammonia or amines to form corresponding dibenzyl aminophosphonates.

USE: In solns with inert solvents such as chloroform, carbon tetrachloride, benzene and ether for the phosphorylation of nucleosides and amino acids.

3060. Dibenzyl Disulfide. *Bis(phenylmethyl) disulfide;* benzyl disulfide; α-(benzyldithio)toluene; di(phenylmethyl)-disulfide. $C_{14}H_{14}S_2$; mol wt 246.40. C 68.25%, H 5.73%, S 26.03%. $C_6H_5CH_2SSCH_2C_6H_5$. Prepd by the reaction of benzyl chloride with sodium disulfide or polysulfide: Blanksma, *Rec. Trav. Chim.* 20, 137 (1901); Moran, Crandall, U.S. pat. 2,113,092 (1938); Wojcik, U.S. pat. 2,185,007 (1939).

Leaflets from alcohol, mp 71-72°. Another modification mp 69-70°. Dec > 270°. Practically insol in water. Sol in ether, benzene, hot methanol, hot ethanol.

USE: Antioxidant in rubber compounding, stabilizer for petr fractions, additive to silicone oils. The soly in oils is increased by the presence of benzyl alcohol.

3061. Dibenzyl Phosphite. *Phosphonic acid bis(phenylmethyl) ester;* dibenzyl hydrogen phosphite; benzyl phosphite. $C_{14}H_{15}O_3P$; mol wt 262.25. C 64.12%, H 5.77%, O 18.30%, P 11.81%. $(HO)P(OCH_2C_6H_5)_2$. Prepd by adding dropwise a mixture of dimethylaniline and benzyl alc to a soln of phosphorus trichloride in benzene: Atherton *et al.*, *J. Chem. Soc.* 1945, 382; U.S. pat. 2,490,573 (1949 to Hoffmann-La Roche).

Liquid. mp 0 to +5°. $bp_{0.1}$ 160-164°; (decompn); $bp_{0.01}$ 110-120°; also reported as $bp_{0.001}$ 120-130°. n_D^{18} 1.5521.

USE: In preparation of *N*-phosphorylated amines.

3062. Diborane(6). Boroethane; diboron hexahydride. B_2H_6; mol wt 27.67. B 78.14%, H 21.86%. Review of methods of prepn: Adams in *Borax to Boranes, Advances Chem. Ser.* No. 32 (American Chemical Society, 1961) pp 60-68. Review of structure and properties: Campbell, Jr. in *Progress in Boron Chemistry,* Steinberg, McCloskey, Eds., vol. 1 (Macmillan, New York, 1964) pp 167-184. Reviews of reaction chemistry: Schenker, *Angew. Chem.* 73, 81-107 (1961); Long in *Advan. Inorg. Chem. Radiochem.* 16, 201-296 (1974). Review of toxicity: *see* Decaborane(14). General reviews: Stock, *Hydrides of Boron and Silicon* (Cornell Univ. Press, Ithaca, 1933), *passim;* Siegel, Mack, *J. Chem. Ed.* 34, 314-317 (1957); Major, *Chem. Eng. Progr.* 54(3), 49-54 (1958); Mikhailov, *Usp. Khim.* 31, 417-451 (1962); *Russian Chem. Rev.* (Eng. Ed.) 31, 224-235 (1962); Adams, *Boron, Metalloboron Compounds and Boranes* (Interscience, New York, 1964) pp 555-605; Lipscomb, *Boron Hydrides* (Benjamin, New York, 1963) *passim;* Long, *Progr. Inorg. Chem.* 15, 1-99 (1972); Greenwood in *Comprehensive Inorganic Chemistry* vol. 1, J. C. Bailar, Jr., *et al.*, Eds. (Pergamon Press, Oxford, 1973) pp 763-783.

Colorless, flammable gas; repulsive, sickly-sweet odor. mp −165°; bp −92.5°. d^{-112} 0.447; $d^{-29.6}$ 0.33; $d^{15.0}$ 0.210. Critical temp 16.7°; critical press. 39.5 atm; Cp at 25°: 13.60 cal/mole/°C. Dec at red heat to B + H_2, at lower temps to H_2 and other boron hydrides. Spontaneous ignition temp in air about 40-50°; presence of contaminants may lower the temp limit so that ignition or detonation of diborane(6)-air mixtures may occur at or below room temp. Hydrolyzes in water to H_2 + H_3BO_3. Sol in CS_2. Reacts with NH_3 to form diborane diammoniate; reacts slowly with Br_2 and explosively with Cl_2 to form boron halides; reacts with hydrocarbons or organoboron compds to give alkyl- or arylboron compds; reacts with metal alkyls to form metal borohydrides; reacts with strong electron pair donors to form borane addn compds.

Caution: Potential symptoms of overexposure are tightening of chest, precordial pain, shortness of breath, nonproductive cough and nausea; headache, lightheadedness, vertigo, chills, fever, fatigue, weakness, tremor and muscle fasiculation. *See NIOSH Pocket Guide to Chemical Hazards* (DHHS/NIOSH 90-117, 1990) p 82. *See also Patty's Industrial Hygiene and Toxicology* vol. 2B, G. D. Clayton, F. E. Clayton, Eds. (Wiley-Interscience, New York, 3rd ed., 1981) pp 2987, 2990-2992.

USE: As catalyst for olefin polymerization; as rubber vulcanizer; as reducing agent; as flame-speed accelerator; in rocket propellants; in intermediate in prepn of the boron hydrides; in conversion of olefins to trialkylboranes and primary alcohols; as a doping gas.

3063. Diboron Tetrachloride. *Tetrachlorodiborane(4); boron chloride.* B_2Cl_4; mol wt 163.43. B 13.23%, Cl 86.77%. Prepd by passing gaseous boron trichloride through flow discharge between mercury or copper electrodes: Urry *et al.*, *J. Am. Chem. Soc.* 76, 5293 (1954); Wartik *et al.*, *Inorg. Syn.* 10, 118 (1967); by passing boron trichloride vapor over boron monoxide at 200°: McCloskey *et al.*, *J. Am. Chem. Soc.* 83, 4750 (1961). Infrared and Raman spectra: Linevsky *et al.*, *J. Am. Chem. Soc.* 75, 3287 (1953). Review of boron halides: Massey, *Advan. Inorg. Chem. Radiochem.* 10, 1-152 (1967).

Liquid, ignites in air. mp −92.6°. bp_1 −63.5°; bp_{45} 0°; bp (calc) 65.5°.

3064. Diboron Tetrahydroxide. *Tetrahydroxydiborane-(4); hypoboric acid;* sub-boric acid. $B_2H_4O_4$; mol wt 89.65. B 24.12%, H 4.50%, O 71.38%. $B_2(OH)_4$. Prepd by action of steam on diboron tetrachloride: Wartik, Apple, *J. Am. Chem. Soc.* 77, 6400 (1955), in quantitative yield by aq hydrolysis of tetraethoxydiboron at 0°: McCloskey *et al.*, *ibid.* 83, 4750 (1961).

Minute crystals, starts to lose water at about 90° *in vacuo.*

3065. 9,10-Dibromoanthracene. $C_{14}H_8Br_2$; mol wt 336.03. C 50.04%, H 2.40%, Br 47.56%. Made by brominating a suspension of anthracene in CCl_4: Heilbron, Heaton, *Org. Syn.* 3, 41 (1923).

Yellow needles from xylene, mp 226°. Sublimes. Absorption max: Conrad-Billroth, *Z. Physik. Chem. (Leipzig)* 33B, 133 (1936). Insol in water; slightly sol in alcohol, ether, cold benzene; sol in hot benzene, hot toluene. Oxidation gives anthraquinone.

Addn compd with *sym*-trinitrobenzene, $C_{20}H_{11}Br_2N_3O_6$, mp 179°.

3066. *p*-Dibromobenzene. $C_6H_4Br_2$; mol wt 235.91. C 30.55%, H 1.71%, Br 67.74%. Prepn from diazotized *p*-bromoaniline: Fry, Grote, *J. Am. Chem. Soc.* 48, 710 (1926); by catalytic bromination of bromobenzene: Ferguson *et al.*, *ibid.* 76, 1250 (1954).

Crystals, mp 87.31°. bp 220.40°. $d^{99.6}$ 0.9641. $n_D^{99.3}$ 1.5743. Practically insol in water; sol in about 70 parts alcohol; sol in benzene, chloroform; very sol in ether.

3067. α,α'-Dibromo-*d*-camphor. *d-3,3-Dibromo-1,7,7-trimethylbicyclo[2.2.1]heptan-2-one; d-3,3-dibromo-2-bornanone;* dibromated camphor. $C_{10}H_{14}Br_2O$; mol wt 310.03. C 38.74%, H 4.55%, Br 51.55%, O 5.16%. Prepd by brominating α-bromocamphor: Lowry, *J. Chem. Soc.* 73, 569 (1898). Conformation: Brutcher *et al.*, *J. Am. Chem. Soc.* 81, 4915 (1959). *Review:* J. L. Simonsen, L. N. Owen, *The Terpenes,*

Consult the Name Index before using this section.

vol. 2 (Cambridge University Press, London, 2nd ed., 1949) pp 401, 405.

Crystals from alcohol, benzene, or petr ether, mp 61°. Not very stable and readily assumes yellow color. Volatile with steam. $d_4^{21.6}$ 1.854. $[\alpha]_D^{20}$ +39.6° (c = 17 in benzene). Practically insol in water; sol in about 5 parts abs alcohol; readily sol in ether, benzene, ethyl acetate, petr ether. uv max (cyclohexane): 323 nm (log ε 1.88): Lowry, Owen, *J. Chem. Soc.* **1926**, 606.

3068. Dibromochloropropane. *1,2-Dibromo-3-chloropropane;* 3-chloro-1,2-dibromopropane; DBCP; OS-1897; Fumazone; Nemafume; Nemagon. $C_3H_5Br_2Cl$; mol wt 236.33. C 15.25%, H 2.13%, Br 67.62%, Cl 15.00%. $ClCH_2$-$CHBrCH_2Br$. Prepn: Darmstädter, *Ann.* **152**, 320 (1869). Activity: C. W. McBeth, G. B. Bergeson, *Plant Dis. Rep.* **39**, 223 (1955). Toxicity study: T. R. Torkelson *et al.,* *Toxicol. Appl. Pharmacol.* **3**, 545 (1961). Carcinogenicity studies: W. A. Olsen *et al., J. Nat. Cancer Inst.* **51**, 1993 (1973); E. K. Weisburger, *Environ. Health Perspect.* **21**, 7 (1977). Mutagenicity study: H. S. Rosenkranz, *Bull. Environ. Contam. Toxicol.* **14**, 8 (1975).
Brown liq; pungent odor. bp 196°; bp_{16} 78°; $bp_{0.8}$ 21°. d^{14} 2.093. n_D^{14} 1.553. Vapor press at 21°: 0.8 mm Hg. Slightly sol in water; misc with oils, dichloropropane, isopropyl alcohol. LD_{50} in rats, mice (g/kg): 0.17, 0.26 orally (Torkelson).
Caution: Potential symptoms of overexposure are drowsiness; nausea, vomiting; irritation of eyes, nose, throat and skin; pulmonary edema. *See NIOSH Pocket Guide to Chemical Hazards* (DHHS/NIOSH 90-117, 1990) p 82. *See also Clinical Toxicology of Commercial Products,* R. E. Gosselin *et al.,* Eds. (Williams & Wilkins, Baltimore, 5th ed., 1984) Section II, pp 167-168. This substance may reasonably be anticipated to be a carcinogen: *Seventh Annual Report on Carcinogens* (PB95-109781, 1994) p 146.
USE: Soil fumigant; nematocide.

3069. 1,2-Dibromo-2,4-dicyanobutane. *2-Bromo-2-(bromomethyl)pentanedinitrile;* 2-bromo-2-(bromomethyl)-glutaronitrile; Tektamer 38. $C_6H_6Br_2N_2$; mol wt 265.94. C 27.10%, H 2.27%, Br 60.09%, N 10.53%. Prepn: N. Grier, S. J. Lederer, **Ger.** pat. **2,164,723** corresp to U.S. pat. **3,-833,731** (1972, 1974 to Merck & Co.).

Crystals from ethanol, mp 51.2-52.5°. Mildly pungent odor. Very sol in DMF, acetone, chloroform, ethyl acetate, benzene. Sol in methanol, ethanol, ether. Insol in water.
USE: Preservative for latex paints, adhesives, latex emulsions, dispersed pigments, joint cements, metal working fluids.

3070. 4′,5′-Dibromofluorescein. *4′,5′-Dibromo-3′,6′-dihydroxyspiro[isobenzofuran-1(3H),9′-[9H]xanthen]-3-one;* 4,5-dibromo-3,6-fluorandiol; D & C Orange No. 5; C.I. Solvent Red 72; C.I. 45370:1. $C_{20}H_{10}Br_2O_5$; mol wt 490.10. C 49.01%, H 2.06%, Br 32.61%, O 16.32%. Prepn from fluorescein, *q.v.:* A. Baeyer, *Ann.* **183**, 1 (1876); M. A. Phillips, *J. Chem. Soc.* **1932**, 724. Metabolism: J. M. Webb *et al., J. Pharmacol. Exp. Ther.* **137**, 141 (1962). *See also: Colour Index* vol. 4 (3rd ed., 1971) p 4425.

Red plates, mp 285°. Slightly sol in water (orange with faint yellow fluorescence); sol in ethanol (orange with greenish-yellow fluorescence) and in acetone (pink with yellow fluorescence). Red-yellow soln in conc H_2SO_4, turning yellow-brown with orange ppt on dilution. Absorption spectra: R. C. Gibbs, C. V. Shapiro, *J. Am. Chem. Soc.* **51**, 1769 (1929).
Disodium salt, $C_{20}H_8Br_2Na_2O_5$, *C.I. Acid Orange 11.*
USE: D & C Orange No. 5 permitted for use in lipsticks, mouthwashes and dentifrices: *Fed. Regist.* **47**, 49632 (1982).

3071. Dibromogallic Acid. *2,6-Dibromo-3,4,5-trihydroxybenzoic acid; 2,6-dibromogallic acid;* gallobromol. C_7H_4-Br_2O_5; mol wt 327.91. C 25.64%, H 1.23%, Br 48.73%, O 24.40%. Prepd by bromination of gallic acid: Etti, *Ber.* **11**, 1879 (1878).

Light-brown powder or crystals; crystallizes from water with one H_2O which it loses at 120°. mp 150°, also stated as ~140° with decompn. Soluble in 8 parts cold water, 0.5 part boiling water; freely sol in alcohol, ether; practically insol in chloroform.

3072. 3,5-Dibromo-4-hydroxybenzenesulfonic Acid. 2,6-Dibromophenol-4-sulfonic acid; *p*-hydroxy-3,5-dibromobenzenesulfonic acid. $C_6H_4Br_2O_4S$; mol wt 331.97. C 21.71%, H 1.21%, Br 48.14%, O 19.28%, S 9.66%. Prepn: Senhofer, *Ann.* **156**, 103 (1870); Armstrong, Brown, *J. Chem. Soc.* **25**, 858 (1872). Prepn of the sodium salt: Cannell, *J. Am. Chem. Soc.* **79**, 2927 (1957).

Rectangular platelets. Freely sol in water, alcohol; less sol in ether.
Sodium salt dihydrate, $C_6H_3Br_2NaO_4S.2H_2O$, *Dibromol.* Needles. Freely sol in hot water, less sol in cold water.
THERAP CAT: Disinfectant.

3073. Dibromopropamidine. *4,4′-[1,3-Propanediylbis-(oxy)]bis(3-bromobenzenecarboximidamide); 4,4′-(trimethylenedioxy)bis(3-bromobenzamidine);* 2′,2′′-dibromo-4′,4′′-di-amidino-1,3-diphenoxypropane. $C_{17}H_{18}Br_2N_4O_2$; mol wt 470.16. C 43.43%, H 3.86%, Br 33.99%, N 11.92%, O 6.81%. Prepn: S. S. Berg, G. Newbery, **Brit.** pat. **598,911** (1948 to May & Baker); *eidem, J. Chem. Soc.* **1949**, 642. Antibacterial activity: R. Wien *et al., Lancet* **254**, 711 (1948); A. D. Russell, J. R. Furr, *Int. J. Pharm.* **34**, 115 (1986). Mode of action: W. Woodside, *Microbios* **8**, 23 (1973). Clinical use in *Acanthamoeba* keratitis: J. J. Wiens, W. B. Jackson, *Can. J. Ophthalmol.* **23**, 107 (1988). HPLC determn in cosmetics: B. Wyhowski de Bukanski, M. O. Masse, *Int. J. Cosmet. Sci.* **6**, 283 (1984).

Isethionate, $C_{21}H_{30}Br_2N_4O_{10}S_2$, **Brolene Ointment, Bruli-dine.** Prismatic needles from ethanol, mp 226°. Freely sol in water. One gram dissolves in 2 ml of water at 20°, in 60 g of 95% alcohol at 20°. Sol in glycerol. Practically insol in ether, chloroform, fixed oils, liquid petrolatum. Aq solns are very slightly acidic. Solns may be sterilized by heating at 100° for 30 min, but they should not be stored for more than a few days, and then only in neutral glass containers since some hydrolysis occurs with the formation of sparingly sol urea derivs. *Incompat.* with chlorides, sulfates, and many organic anions, all of which form sparingly sol salts.

USE: Preservative in cosmetics.

THERAP CAT: Antiseptic; antiamebic.

THERAP CAT (VET): Antiseptic, antimicrobial.

3074. 2,3-Dibromopropene. 2,3-Dibromopropylene; 2-bromoallyl bromide; α-bromoallyl bromide; α-epidibromo-hydrin. $C_3H_4Br_2$; mol wt 199.87. C 18.03%, H 2.02%, Br 79.95%. $BrCH_2CBr=CH_2$. Prepd by the action of sodium hydroxide on 1,2,3-tribromopropane: Lespieau, Bourguel, *Org. Syn.* **5**, 49 (1925); *cf.* Tapley, Giesy, *J. Am. Pharm. Assoc.* **15**, 173 (1926).

Liquid. d_4^{20} 1.9336. bp_{760} 140-143°; bp_{75} 75-76°; bp_{18} 42-43°. n_D^{20} 1.5157. May be distilled under atmospheric pressure with very slight decompn, but becomes highly colored on standing in glass-stoppered bottle.

3075. 2,6-Dibromoquinone-4-chlorimide. *2,6-Dibromo-4-(chloroimino)-2,5-cyclohexadien-1-one; 2,6-dibromo-N-chloro-p-benzoquinoneimine;* 2,6-dibromo-N-chloroquinon-imine; BQC reagent. $C_6H_2Br_2ClNO$; mol wt 299.35. C 24.07%, H 0.67%, Br 53.39%, Cl 11.84%, N 4.68%, O 5.34%. Prepn: Hartman *et al., Org. Syn.* **coll. vol. II,** 175 (1943).

Yellow prisms from alc or glacial acetic acid, mp 83°. Sol in about 17,000 parts water; moderately sol in hot alc, sol glacial acetic acid. Gives a blue color with alk phenol soln.

USE: As a reagent for phenol and phosphatases: Ljung-gren, *Proc. 13th Int. Dairy Congr.* **3**, 1319 (1953).

3076. 3,5-Dibromosalicylaldehyde. *3,5-Dibromo-2-hydroxybenzaldehyde;* Dalyde. $C_7H_4Br_2O_2$; mol wt 279.92. C 30.04%, H 1.44%, Br 57.09%, O 11.43%. Prepd by bromi-nation of salicylaldehyde in glacial acetic acid with cooling: Wentworth, Brady, *J. Chem. Soc.* **117**, 1043 (1920); Brew-ster, *J. Am. Chem. Soc.* **46**, 2464 (1924); by bromination in the presence of sodium acetate in glacial acetic acid at 50°: Lindemann, Forth, *Ann.* **435**, 223 (1924).

Pale yellow prisms, mp 86°. Sublimes forming leaflets and needles. Volatile with steam. Sparingly sol in water.

Aq solns are yellow. Readily sol in ether, benzene, chloro-form, hot petr ether, alcohol, glacial acetic acid.

THERAP CAT: Antibacterial.

3077. 3,5-Dibromosalicylic Acid. *3,5-Dibromo-2-hydr-oxybenzoic acid.* $C_7H_4Br_2O_3$; mol wt 295.91. C 28.41%, H 1.36%, Br 54.00%, O 16.22%. Prepd from salicylic acid and Br_2: Earle, Jackson, *J. Am. Chem. Soc.* **28**, 111 (1906); Robertson, *J. Chem. Soc.* **82**, 1481 (1902); and HOBr: Leu-lier, Pinet, *Bull. Soc. Chim. France* [4] **41**, 1362 (1927).

Needles, mp 223°. Sparingly sol in H_2O; sol in alc, ether.

3078. 2,3-Dibromosuccinic Acid. *2,3-Dibromobutane-dioic acid;* α,α'-dibromosuccinic acid; *sym*-dibromosuccinic acid. $C_4H_4Br_2O_4$; mol wt 275.88. C 17.41%, H 1.46%, Br 57.93%, O 23.20%. Prepn of *dl*-form by bromination of maleic acid: McKenzie, *J. Chem. Soc.* **101**, 1196 (1912); Young *et al., J. Am. Chem. Soc.* **61**, 1640 (1939); of *meso*-form by bromination of fumaric acid: *eidem, ibid.;* Rhine-smith, *Org. Syn.* **coll. vol. II,** 177 (1943). Resolution of *dl*-form: McKenzie, *loc. cit.*

dl-Form, *threo-2,3-dibromosuccinic acid.* Crystals, mp 167°. Very sol in cold water.

d-Form, crystalline powder from ethyl acetate + CCl_4, mp 157-158°. $[\alpha]_D^{18}$ +64.4° (c = 5 in water), +135.8° (c = 5 in alc), +147.8° (c = 5 in ethyl acetate).

l-Form, needles from benzene, decomp 157-158°. $[\alpha]_D^{13}$ −148.0° (c = 5.8 in ethyl acetate). Very sol in cold water; sol in ethyl acetate, acetone, methanol, ethanol; sparingly sol in chloroform, petr ether, CCl_4.

meso-Form, *erythro-2,3-dibromosuccinic acid.* Cryst from water, dec 255-256°, also reported as mp 270-273°. Sol in 50 parts cold water, more sol in hot water; sol in alcohol, ether; sparingly sol in chloroform.

3079. 3,5-Dibromo-L-tyrosine. β-(3,5-dibromo-4-hydr-oxyphenyl)alanine; Biotiren; Bromotiren. $C_9H_9Br_2NO_3$; mol wt 338.98. C 31.89%, H 2.68%, Br 47.14%, N 4.13%, O 14.16%. Obtained upon saponification of gorgonin, a sub-stance isolated from coral *(Primnoa lepadifera)* stems: Mör-ner, *Z. Physiol. Chem.* **88**, 139, 152 (1913). Prepd by the action of bromine vapor on tyrosine: Gorup-Besanez, *Ann.* **125**, 281 (1863). By treating tyrosine in aq HBr with Br: Aloy, Rabaut, *Bull. Soc. Chim. France* [4], **3**, 392 (1908); Zeynek, *Z. Physiol. Chem.* **114**, 275 (1921). Clinical evalua-tion in thyroid function: A. Isidori, *Int. J. Clin. Pharmacol. Biopharm.* **16**, 180 (1978).

L-Form dihydrate, efflorescent needles, plates (ortho-rhombic) from water. When anhydr, dec 245°. $[\alpha]_D^{20}$ +1.3° (c = 5 in 4% HCl). pK_1 2.17; pK_2 6.45; pK_3 7.60. One gram dissolves in 250 ml water at 25°, in 30 ml at boiling temp. Slightly sol in alcohol; insol in ether. Freely sol in alkalies and in dil mineral acids. Forms HBr and HCl salts; stable in boiling water.

DL-Form, efflorescent prisms, platelets from water proba-bly contg 1 mol water of crystn. Dec at about 245°. Aq solns are acid to litmus. Behaves as a monobasic acid when titrated with NaOH and phenolphthalein. One gram dissolves in about 590 ml water at 20°.

THERAP CAT: Thyroid inhibitor.

3080. Dibromsalicil. *Bis(5-bromo-2-hydroxyphenyl)ethanedione; 5,5'-dibromosalicil;* DBS; 5,5'-dibromo-2,2'-dihydroxybenzil; 5,5'-dibromo-2,2'-dihydroxybibenzoyl; Dibrosal. $C_{14}H_8Br_2O_4$; mol wt 400.02. C 42.04%, H 2.02%, Br 39.95%, O 16.00%. Prepd by heating dibromodimethoxybenzil in nitrobenzene in the presence of aluminum chloride: Kuhn *et al., Ber.* **76**, 900 (1943).

Flexible, lemon-colored needles or yellow powder from glacial acetic acid. Odorless, tasteless. mp 212-213°. Practically insol in water. Freely sol in alcohol, benzene. Sol in acetone, ether. Sol in dil NaOH yielding solns of an intense deep yellow color.

USE: Antiseptic, germicide.

3081. Dibucaine Hydrochloride. *2-Butoxy-N-[2-(diethylamino)ethyl]-4-quinolinecarboxamide monohydrochloride; 2-butoxy-N-(2-diethylaminoethyl)cinchoninamide hydrochloride;* benzolin; Nupercaine Hydrochloride; Percaine; Cincaine; Sovcaine; Cinchocaine. $C_{20}H_{30}ClN_3O_2$; mol wt 379.93. C 63.23%, H 7.96%, Cl 9.33%, N 11.06%, O 8.42%. Prepn: Miescher, U.S. pat. **1,825,623** (1931 to Ciba). Comprehensive description: G. R. Padmanabhan in *Analytical Profiles of Drug Substances* vol. 12, K. Florey, Ed. (Academic Press, New York, 1983) pp 105-134.

Hygroscopic crystals. Dec 90-98°. (The base mp 64°.) uv max (1N HCl): 247, 320 nm (ε 24700, 8810). Sol in 0.5 part water; freely sol in alcohol, acetone, chloroform; slightly sol in benzene, ethyl acetate and toluene (on warming); insol in ether and in oils. The aq soln, about 1 in 20, is faintly alkaline to litmus.

THERAP CAT: Anesthetic (local).

3082. Dibutoline Sulfate. *2-[[(Dibutylamino)carbonyl]oxy]-N-ethyl-N,N-dimethylethanaminium sulfate (2:1); bis-[ethyl(2-hydroxyethyl)dimethylammonium] sulfate bis(dibutylcarbamate);* (2-dibutyl-carbamyloxyethyl)dimethylethylammonium sulfate; dimethylethyl-β-hydroxyethylammonium sulfate dibutylurethan; Dibuline Sulfate. $C_{30}H_{66}N_4$-O_8S; mol wt 642.94. C 56.04%, H 10.35%, N 8.71%, O 19.91%, S 4.99%. Prepd by treating dibutoline iodide with Ag_2SO_4: Swan, White, U.S. pats. **2,408,893** (1946); **2,432,049** (1947).

Extremely hygroscopic powder. Dec 166°. Usually furnished as a colorless aqueous soln. Also sol in benzene. A surface-active agent; a 5% aq soln (pH 6.75-7.50) has an air-water interfacial tension of 42 dynes/cm at 25°; also reported as 47 dynes/cm: Swan, White, *Am. J. Ophthalmol.* **27**, 933 (1944). Ampuled aq solns may be stored at room temp for at least one year without precipitation, darkening,

change of pH, or loss of potency. Heating of aq solns at 100° causes decompn. Sterilization by filtration is recommended.

THERAP CAT: Anticholinergic.

3083. n-Dibutylamine. *N-Butyl-1-butanamine.* C_8-$H_{19}N$; mol wt 129.25. C 74.35%, H 14.82%, N 10.84%. $(C_4H_9)_2NH$. Prepn from butyl bromide and ammonia with separation of the mono-, di-, and tributylamines formed: Werner, *J. Chem. Soc.* **115**, 1010 (1919). Manuf: Engel, Hoog, U.S. pat. **2,574,693** (1951 to Shell); Davies *et al.,* U.S. pat. **2,609,394** (1952 to I.C.I.); Lemon, Myerly, U.S. pat. **3,022,349** (1962 to Union Carbide); Brois, Rutkowski, U.S. pat. **3,147,310** (1964 to Esso). Toxicity study: H. F. Smyth *et al., Arch. Ind. Hyg. Occup. Med.* **10**, 61 (1954).

Liquid, bp 159-160°. mp −60° to −59°. d_4^{20} 0.7601. n_D^{20} 1.4177. Flash pt, open cup: 135°F (57°C). Sol in water, alc. LD_{50} orally in rats: 550 mg/kg (Smyth).

3084. Di-*tert*-butyl Ether. *2,2'-Oxybis[2-methylpropane].* $C_8H_{18}O$; mol wt 130.23. C 73.78%, H 13.93%, O 12.29%. $[(CH_3)_3C]_2O$. Prepn: R. J. Moore, G. J. O'Donnell, **Brit.** pat. **652,809** (1951 to Bataafsche Petrol.); *C.A.* **46**, 1023d (1952). Synthesis from *t*-butyl chloride and silver carbonate: J. Erickson, W. M. Ashton, *J. Am. Chem. Soc.* **63**, 1769 (1941); from *t*-butyl perbenzoates and Grignard reagents: S. O. Lawesson, N. C. Yang, *ibid.* **81**, 4230 (1959); by alkylation of *t*-butyl alcohol with trimethylcarbenium fluoroantimonate: G. A. Olah *et al., Synthesis* **1975**, 315; from lithium *tert*-butoxide with aromatic sulfonyl chlorides: H. Masada *et al., Tetrahedron Letters* **1979**, 1315.

Clear liquid with camphorous odor. bp 106.5-107.0°. n_D^{20} 1.3949; d_4^{20} 0.7658. Decomposes in the presence of acids.

USE: Gasoline additive.

3085. Di-*tert*-butyl Malonate. *Propanedioic acid bis-(1,1-dimethylethyl) ester;* malonic acid di-*tert*-butyl ester. $C_{11}H_{20}O_4$; mol wt 216.28. C 61.09%, H 9.32%, O 29.59%. $(CH_3)_3COOCCH_2COOC(CH_3)_3$. Prepd from malonic acid, liq isobutylene and sulfuric acid: McCloskey *et al., Org. Syn.* **34**, 26 (1954).

Liquid. mp −6.1 to −5.9°. bp_{31} 112-115°; bp_{10} 93°; bp_1 65-67°. n_D^{20} 1.4184; $n_D^{24.2}$ 1.4161; n_D^{25} 1.4158-1.4161.

USE: In the prepn of ketones.

3086. 2,6-Di-*tert*-butylpyridine. *2,6-Bis(1,1-dimethylethyl)pyridine.* $C_{13}H_{21}N$; mol wt 191.32. C 81.62%, H 11.06%, N 7.32%. Prepd by reacting *tert*-butyllithium with 2-*tert*-butylpyridine: Brown, Kanner, *J. Am. Chem. Soc.* **75**, 3865 (1953); U.S. pat. **2,780,626** (1957 to Research Corp.).

Liquid, mp 2.2°. bp_{23} 100-101°. n_D^{20} 1.5733. pKa 3.58. The base neutralizes HCl but does not react with alkyl halides or boron trifluoride. Undergoes nuclear sulfonation with sulfur trioxide forming a sulfonic acid, $C_{13}H_{21}NSO_3$, mp 310° (decompn).

Chloroaurate, $C_{13}H_{22}NAuCl_4$, mp 184.2-184.5°.

USE: Has been proposed as an additive for lubricating oil, gasoline and for stabilizing Cl-containing polymers.

3087. Dibutyl Succinate. *Butanedioic acid dibutyl ester; succinic acid dibutyl ester;* di-*n*-butyl succinate; Tabutrex; Tabatrex. $C_{12}H_{22}O_4$; mol wt 230.30. C 62.58%, H 9.63%, O 27.79%. $CH_3(CH_2)_3OOCCH_2CH_2COO(CH_2)_3$-$CH_3$. Prepn from butyl alcohol and succinic acid: Vogel, *J. Chem. Soc.* **1948**, 624.

Liquid, bp 274.5°, bp_3 121°. mp −29.0°. d_4^{20} 0.9768. n_D^{20} 1.4299. LD_{50} orally in rats: 8 g/kg.

USE: Insect repellent esp against biting flies of cattle and household ants and roaches.

3088. Di-*tert*-butyl Succinate. *Butanedioic acid bis(1,-1-dimethylethyl) ester;* succinic acid di-*tert*-butyl ester. C_{12}-$H_{22}O_4$; mol wt 230.30. C 62.58%, H 9.63%, O 27.79%. $(CH_3)_3COOCCH_2CH_2COOC(CH_3)_3$. Prepd from succinic acid, liq isobutylene, and sulfuric acid in dioxane: McCloskey *et al., Org. Syn.* **coll. vol. IV**, 263 (1963).

Liquid, mp 36-37°. bp$_9$ 109-110°; bp$_7$ 105-107°.
USE: In Stobbe condensation of ketones.

3089. Dibutyltin Dilaurate. *Dibutylbis[(1-oxododecyl)-oxy]stannane; dibutylbis(lauroyloxy)tin;* Butynorate; Davainex; Tinostat. C$_{32}$H$_{64}$O$_4$Sn; mol wt 631.57. C 60.86%, H 10.21%, O 10.13%, Sn 18.80%. (C$_4$H$_9$)$_2$Sn[OOC(CH$_2$)$_{10}$-CH$_3$]$_2$. Prepn: Sheverdina *et al.*, *Khim. Prom.* **1962** (10), 707, *C.A.* **59**, 8776f (1963); **Fr.** pat. **1,320,473** (1963 to Noury & van der Lande), *C.A.* **59**, 8789b (1963), corresp to **Brit.** pat. **975,369**. Compd claimed but not described: Eberly, **U.S.** pat. **2,560,034** (1951 to Firestone).
Soft crystals or yellow liq, mp 22-24° (Sheverdina). n_D^{20} 1.4683, (**Fr.** pat. **1,320,473**). Practically insol in water, methanol. Soluble in petr ether, benzene, acetone, carbon tetrachloride, ether, organic esters.
USE: Stabilizer for polyvinyl chloride resins. Catalyst for curing certain silicones.
THERAP CAT (VET): Anthelmintic (vs. chicken tapeworms).

3090. Dicamba. *3,6-Dichloro-2-methoxybenzoic acid; 3,6-dichloro-o-anisic acid;* dianat; Velsicol 58-CS-11; Banvel D; Mediben. C$_8$H$_6$-Cl$_2$O$_3$; mol wt 221.04. C 43.47%, H 2.74%, Cl 32.08%, O 21.71%. Sometimes misnamed *methoxydichlorobenzoate.* Prepn: S. B. Richter, **U.S.** pat. **3,013,054** (1961 to Velsicol). Metabolism: D. D. Oehler, G. W. Ivie, *J. Agr. Food Chem.* **28**, 685 (1980). Toxicity: Bailey, White, *Residue Rev.* **10**, 97 (1965).

Crystals from pentane, mp 114-116°. Vapor pressure at 100°: 3.75 × 10^{-3} mm Hg. Sol in ethanol and acetone. Very slightly sol in water. LD$_{50}$ orally in rats: 1040 mg/kg (Bailey, White). One of the ingredients of *Banlene* and of *Cambilene.*
USE: Herbicide.

3091. Dicapthon. *Phosphorothioic acid O-(2-chloro-4-nitrophenyl) O,O-dimethyl ester; O,O-*dimethyl *O-*(2-chloro-4-nitrophenyl)phosphorothioate; *O,O-*dimethyl-*O-*(2-chloro-4-nitrophenyl)thionophosphate; *p*-nitro-*o*-chlorophenyl dimethyl thionophosphate; isochlorothion; American Cyanamid 4124; insecticide ACC 4124; Dicaptan. C$_8$H$_9$ClNO$_5$PS; mol wt 297.66. C 32.28%, H 3.05%, Cl 11.91%, N 4.71%, O 26.88%, P 10.41%, S 10.77%. Prepn: J. H. Fletcher, **U.S.** pat. **2,664,437**; Geoghegan, McPherson, **U.S.** pat. **2,784,207** (1953, 1957 both to Am. Cyanamid); Schrader, *Angew. Chem.* **66**, 265 (1954); *idem, Die Entwicklung neuer insektizider Phosphorsäure-Ester* (Verlag Chemie, 3rd ed., 1963) pp 289-291. Toxicity study: T. B. Gaines, *Toxicol. Appl. Pharmacol.* **14**, 515 (1969).

Crystals from methanol, mp 53°. Sol in acetone, cyclohexanone, ethyl acetate, toluene, xylene, ethylene glycol, propylene glycol, some oils. Practically insol in water. LD$_{50}$ in male, female rats (mg/kg): 400, 330 orally; 790, 1250 dermally (Gaines).
Caution: Cholinesterase inhibitor.
USE: Insecticide.

3092. Dicentrine. *6,7,7a,8-Tetrahydro-10,11-dimethoxy-7-methyl-5H-benzo[g]-1,3-benzodioxolo[6,5,4-de]quinoline; 9,10-dimethoxy-1,2-(methylenedioxy)aporphine;* 1,2-methylenedioxy-9,10-dimethoxyaporphine. C$_{20}$H$_{21}$NO$_4$; mol wt 339.39. C 70.78%, H 6.24%, N 4.13%, O 18.86%. *d*-Form prevalent in nature. Found in *Dicentra pusilla* Sieb. & Zucc., *Fumariaceae* and several other *Dicentra* species. Re-

lated to actinodaphnine, *q.v.*, and laurotetanine, *q.v.* Isoln: Y. Asahina, *Arch. Pharm.* **247**, 201 (1909). Isoln of *l*-form from *Duguetia* A. St. Hil., *Annonaceae:* Casagrande, Ferrari, *Farmaco Ed. Sci.* **25**, 442 (1970). Synthesis of *dl*-form: Haworth *et al.*, *J. Chem. Soc.* **127**, 2018 (1925); Cava *et al.*, *Tetrahedron* **29**, 2245 (1973). Resolution of isomers: Haworth *et al.*, *J. Chem. Soc.* **129**, 29 (1926).

d-Dicentrine

dl-Form, prisms from methanol, mp 178-179°. Freely sol in chloroform, ethyl acetate, acetone, benzene, hot alcohol. Moderately sol in ether, cold alcohol. Absorption max: Girardet, *J. Chem. Soc.* **1931**, 2630.
d-Form, long prisms from ether. mp 169°. [α]$_D^{17}$ +64.1° (c = 1.433 in chloroform). Freely sol in alcohol, ethyl acetate, benzene.
l-Form, long prisms from ether. mp 169°. [α]$_D^{17}$ −63.5° (c = 1.70 in chloroform).
dl-Hydrochloride, C$_{20}$H$_{21}$NO$_4$.HCl, small needles from water, dec 263-265°.
dl-Methiodide monohydrate, C$_{20}$H$_{21}$NO$_4$.CH$_3$I.H$_2$O, plates, mp 228-229°.

3093. Dichlobenil. *2,6-Dichlorobenzonitrile;* H-133; Niagara 5006; Casoron. C$_7$H$_3$Cl$_2$N; mol wt 172.01. C 48.88%, H 1.76%, Cl 41.22%, N 8.14%. Prepn: Reich, *Bull. Soc. Chim. France* [4] **21**, 217 (1917); Norris, Klemka, *J. Am. Chem. Soc.* **62**, 1432 (1940); Chang *et al.*, *C.A.* **53**, 6134d (1959); Koopman, *Rec. Trav. Chim.* **80**, 1075 (1961); Hackmann, ten Haken **Brit.** pat. **861,899**; Higson, **Brit.** pat. **862,937** (both 1961 to Shell). Use as herbicide: Koopman, Daams, **U.S.** pat. **3,027,248** (1962 to N. A. Phillips). Fate in crops, soil and animals: Beynon, Wright, *Residue Rev.* **43**, 23 (1972); Veroop, *ibid.* 55. Toxicity data: G. W. Bailey, J. L. White, *ibid.* **10**, 97 (1965).

Crystals from petr ether, mp 144-145°. Soly in water at 25°: 25 ppm; at 20°: 18 ppm. Vapor pressure at 20°: 3 × 10^{-6} mm Hg; at 25°: 5 × 10^{-4} mm. Absorption spectrum see Koopman, *loc. cit.* LD$_{50}$ in rats, mice (mg/kg): 2710, 6800 orally (Bailey, White).
USE: Herbicide.

3094. Dichlofenthion. *Phosphorothioic acid O-2,4-dichlorophenyl O,O-diethyl ester; O-*2,4-dichlorophenyl *O,O-*diethyl phosphorothioate; *O,O-*diethyl *O-*2,4-dichlorophenyl phosphorothioate; VC-13 Nemacide. C$_{10}$H$_{13}$Cl$_2$O$_3$PS; mol wt 315.16. C 38.11%, H 4.16%, Cl 22.50%, O 15.23%, P 9.83%, S 10.17%. Cholinesterase inhibitor. Prepd from *O,O-*diethyl phosphorochloridothioate, 2,4-dichlorophenol and NaOH: Smithey, Jr., **U.S.** pat. **3,004,054** (1960 to Virginia-Carolina Chemical). Toxicity data: T. B. Gaines, R. E. Linder, *Fund. Appl. Toxicol.* **7**, 299 (1986).

Liquid, $bp_{0.1}$ 164-169°. n_D^{25} 1.5291. d_{20} 1.300. Slightly sol in water; miscible with most organic solvents. LD_{50} in male, female rats (mg/kg): 185, 172 orally; 576, 355 dermally (Gaines, Linder).

USE: Nematocide, insecticide.

3095. Dichlofluanid. *1,1-Dichloro-N-[(dimethylamino)-sulfonyl]-1-fluoro-N-phenylmethanesulfenamide; N-[(dichlorofluoromethyl)thio]-N',N'-dimethyl-N-phenylsulfamide;* Bayer 47531; KUE 13032c; Elvaron; Euparen(e). $C_9H_{11}Cl_2FN_2O_2S_2$; mol wt 333.23. C 32.44%, H 3.33%, Cl 21.28%, F 5.70%, N 8.41%, O 9.60%, S 19.25%. Prepn: E. Klauke *et al.,* **Belg. pat. 609,868** corresp to U.S. pats. **3,285,929, 3,341,403** (1962, 1966, 1967, all to Bayer). Activity: E. Kuhle *et al., Angew. Chem.* **76,** 807 (1964); F. Grewe, *Phytiat.-Phytopharm.* **17,** 47 (1968). UV degradation: T. Clark, D. Watkins, *Pestic. Sci.* **9,** 225 (1978).

White powder, mp 105.0-105.6°. Vapor press at 20°: 1×10^{-6} mm Hg. Insol in water. Sol in acetone, methanol, xylene. Light sensitive; decomp by strong alkaline media. Toxic to fish, harmless to bees. LD_{50} orally in mice: 1250 mg/kg, I. Petersone, D. Berzina, *C.A.* **85,** 117680t (1976).

USE: Fungicide.

3096. Dichlone. *2,3-Dichloro-1,4-naphthalenedione; 2,3-dichloro-1,4-naphthoquinone;* USR-604; Phygon; Phygon Paste; Phygon XL. $C_{10}H_4Cl_2O_2$; mol wt 227.05. C 52.90%, H 1.78%, Cl 31.23%, O 14.09%. Prepd by chlorination of 1,4-naphthoquinone: Bertheim, *Ber.* **34,** 1554 (1901); Brass, Köhler, *Ber.* **55,** 2554 (1922); Sjöstrand, U.S. pat. **2,975,196** (1961 to Svenska Oljeslageri Aktiebolaget); by oxidation of 2,3-dichloro-5,8-dihydro-1,4-naphthohydroquinone: Gaertner, U.S. pat. **2,750,427** (1956 to Monsanto); by oxidation of 2,3-dichloro-*p*-benzoquinone: Grinev *et al., Zh. Obshch. Khim.* **29,** 90 (1959). Toxicity data: G. W. Bailey, J. L. White, *Residue Rev.* **10,** 97 (1965). *Review:* Ter Horst, *Ind. Eng. Chem.* **35,** 1255 (1943).

Golden yellow needles or leaflets from alc, mp 193°. Sublimes. Practically insol in water (soly about 1 part in 10 million parts H_2O). Soly in xylene and *o*-dichlorobenzene about 4%. Moderately sol in acetone, ether, benzene, dioxane. LD_{50} in rats: 1300 mg/kg (Bailey, White). *Caution:* Irritating to skin, mucous membranes; CNS depressant: *Clinical Toxicology of Commercial Products,* R. E. Gosselin *et al.,* Eds. (Williams & Wilkins, Baltimore, 5th ed., 1984) Section II, p 318.

USE: Fungicide for agriculture and textiles; herbicide.

3097. Dichloralphenazone. *1,2-Dihydro-1,5-dimethyl-2-phenyl-3H-pyrazol-3-one compd with 2,2,2-trichloro-1,1-ethanediol(1:2); Dichloralantipyrine.* Bihypnal; Bonadorm; Dormwell; Sedor; Sominat; Welldorm. $C_{15}H_{18}Cl_6N_2O_5$; mol wt 519.03. C 34.71%, H 3.50%, Cl 40.98%, N 5.40%, O 15.41%. $[CCl_3CH(OH)]_2 \cdot C_{11}H_{12}N_2O$. Prepn: Pfeiffer, Seidel, *Z. Physiol. Chem.* **178,** 97 (1928).

Small prismatic needles from water, mp 68°.

THERAP CAT: Sedative, hypnotic.

3098. Dichloramine T. *N,N-Dichloro-4-methylbenzenesulfonamide; N,N-dichloro-p-toluenesulfonamide; p-*toluenesulfonic acid dichloramide. $C_7H_7Cl_2NO_2S$; mol wt 240.11.

C 35.02%, H 2.94%, Cl 29.53%, N 5.83%, O 13.33%, S 13.35%. Prepn from *p*-toluenesulfonic acid: Soper, *J. Chem. Soc.* **125,** 1899 (1924); van Andel, U.S. pat. **2,495,489** (1950 to Shell).

Prisms from chloroform + petr ether, mp 83°. Strong chlorine odor; dec on exposure to air with loss of chlorine. mp about 80°. Almost insol in water, dec by alcohol when warmed; one gram dissolves in about 1 ml benzene, 1 ml chloroform, 2.5 ml carbon tetrachloride; sol in eucalyptol, chlorinated paraffin hydrocarbons, glacial acetic acid; slightly sol in petr ether. *Keep well closed and protected from light.*

USE: Germicide.

THERAP CAT: Antibacterial.

3099. Dichlorisone. *(11β)-9,11-Dichloro-17,21-dihydroxypregna-1,4-diene-3,20-dione; 9α,11β-dichloro-1,4-pregnadiene-17α,21-diol-3,20-dione; 9α,11β-dichloro analog of prednisolone;* Diloderm; Disoderm. $C_{21}H_{26}Cl_2O_4$; mol wt 413.34. C 61.02%, H 6.34%, Cl 17.15%, O 15.48%. Prepn: Robinson *et al., J. Am. Chem. Soc.* **81,** 2191 (1959); Gould *et al.,* U.S. pat. **2,894,963** (1959 to Schering).

Crystals from acetone, dec 238-241°. $[\alpha]_D^{20}$ +134° (pyridine). uv max (methanol): 237 nm (ε 15400).

21-Acetate, $C_{23}H_{28}Cl_2O_5$, *Astroderm.* Crystals from acetone, dec 246-253°. $[\alpha]_D^{125}$ +162° (dioxane). uv max (methanol): 237 nm (ε 15000).

THERAP CAT: Antipruritic (topical).

3100. Dichloroacetic Acid. Bichloracetic acid; dichlorethanoic acid; DCA. $C_2H_2Cl_2O_2$; mol wt 128.94. C 18.63%, H 1.56%, Cl 54.99%, O 24.82%. $CHCl_2COOH$. Transformation of chloral to dichloroacetic acid: Wallach, *Ann.* **173,** 288 (1874); Frantzen, Fikentscher, *ibid.* **623,** 68 (1959); Rosenblum *et al., Chem. & Ind. (London)* **1960,** 718. Toxicity: Smyth *et al., Arch. Ind. Hyg. Occup. Med.* **4,** 119 (1951); P. W. Stacpoole *et al., N. Engl. J. Med.* **300,** 372 (1979); J. L. Cicmanec *et al., Fundam. Appl. Toxicol.* **17,** 376 (1991). Mutagenicity studies: V. Herbert *et al., Am. J. Clin. Nutr.* **33,** 1179 (1980). Use in treatment of lactic acidosis: P. W. Stacpoole *et al., N. Engl. J. Med.* **309,** 390 (1983).

Liquid; pungent odor. d_2^{20} 1.563. bp 193-194°. Apparently occurs in two cryst forms, mp 9.7° and −4°. n_D^{22} 1.4659. Miscible with water, alc, ether. LD_{50} orally in rats: 2.82 g/kg (Smyth).

Ethyl ester, $C_4H_6Cl_2O_2$, liquid, bp 158.3-158.7°. d_4^{20} 1.282. n_D^{20} 1.4386. Slightly sol in water; miscible with alcohol, ether.

THERAP CAT: Caustic; keratolytic; topical astringent.

3101. 1,1-Dichloroacetone. *1,1-Dichloro-2-propanone; α,α-dichloroacetone; uns-dichloroacetone; dichloromethyl methyl ketone.* $C_3H_4Cl_2O$; mol wt 126.97. C 28.38%, H 3.18%, Cl 55.84%, O 12.60%. $CH_3COCHCl_2$. Prepn: Borsche, Fittig, *Ann.* **133,** 113 (1865).

Oily liq. d_{15}^{18} 1.305. bp 120°. Slightly sol in water; sol in alcohol; miscible with ether. Boiling with Na_2CO_3 gives acrylic acid.

3102. 1,3-Dichloroacetone. *1,3-Dichloro-2-propanone; α,γ-dichloroacetone; sym-dichloroacetone;* bis(chloromethyl) ketone. $C_3H_4Cl_2O$; mol wt 126.97. C 28.38%, H 3.18%,

Cl 55.84%, O 12.60%. CH$_2$ClCOCH$_2$Cl. Prepd by the oxidation of dichlorohydrin with sodium dichromate: Conant, Quayle, *Org. Syn.* **2**, 13 (1922).

Plates, needles on distillation. mp 45°. bp 173°. d$_4^{46}$ 1.3826. n$_D^{46}$ 1.47144. Sol in water, very sol in alc, ether.

Caution: Lacrimator, vesicant.

3103. 2,2-Dichloroacetyl Chloride. *Dichloroethanoyl chloride.* C$_2$HCl$_3$O; mol wt 147.39. C 16.30%, H 0.68%, Cl 72.16%, O 10.86%. Cl$_2$CHCOCl. Prepd from pentachloroethane: Ott, *Ger. pat.* **362,748** (1922 to Weiler ter Meer); from chloroform and carbon monoxide in presence of aluminum chloride at high pressure (21% yield): Frank *et al., Ind. Eng. Chem.* **41**, 2061 (1949). Toxicity study: Smyth *et al., Arch. Ind. Hyg. Occup. Med.* **4**, 119 (1951).

Liquid. Fumes in air; acrid, penetrating odor. d$_4^{16}$ 1.5315. bp$_{760}$ 107-108°; bp$_{739}$ 106-107°. n$_D^{16}$ 1.4638. Dec upon contact with water, alcohol. Miscible with ether. LD$_{50}$ orally in rats: 2.46 g/kg (Smyth).

Caution: Irritating to eyes, mucous membranes.

3104. 3,4-Dichloroaniline. *3,4-Dichlorobenzeneamine.* C$_6$H$_5$Cl$_2$N; mol wt 162.02. C 44.48%, H 3.11%, Cl 43.76%, N 8.65%. Prepd by chlorination of *p*-chloroaniline with AlCl$_3$—HCl: Suthers *et al., J. Org. Chem.* **27**, 447 (1962). Manuf by catalytic reduction of 3,4-dichloro-1-nitrobenzene: Dietzler, Keil, **U.S. pat. 3,067,253** (1962 to Dow).

Crystals, mp 71-72°. bp 272°. Practically insol in water. Very sol in alcohol, ether; slightly sol in benzene.

3105. m-Dichlorobenzene. *1,3-Dichlorobenzene.* C$_6$H$_4$Cl$_2$; mol wt 147.00. C 49.02%, H 2.74%, Cl 48.23%. Prepn by Sandmeyer procedure from the appropriate chloroaniline, and, along with *o*- and *p*-dichlorobenzenes, by chlorination of chlorobenzene: Engelsma *et al., Rec. Trav. Chim.* **76**, 325 (1957). Separation of mixture contg *m*-, *o*-, and *p*-dichlorobenzenes by distillation and crystn: Mueller, Wolz, **Fr. pat. 1,374,863** (1964 to Bayer), *C.A.* **62**, 4936e (1965), corresp to **Brit. pat. 999,845.**

Liquid, bp 173°. mp −24.76°. d$_4^{20}$ 1.2884, d$_4^{25}$ 1.2828. n$_D^{20}$ 1.5459. Practically insol in water. Sol in alcohol, ether.

3106. o-Dichlorobenzene. *1,2-Dichlorobenzene;* orthodichlorobenzene. C$_6$H$_4$Cl$_2$; mol wt 147.00. C 49.02%, H 2.74%, Cl 48.23%. Prepn and separation from *m*- and *p*-dichlorobenzenes, *see m*-isomer above. Manuf: Faith, Keyes & Clark's *Industrial Chemicals*, F. A. Lowenheim, M. K. Moran, Eds. (Wiley-Interscience, New York, 4th ed., 1975) pp 258-265.

Liquid, bp 180.5°. mp −17.03°. d$_4^{20}$ 1.3059, d$_4^{25}$ 1.3003. n$_D^{20}$ 1.5515, n$_D^{25}$ 1.5491. Flash pt, closed cup: 151°F (66°C). Practically insol in water. Miscible with alc, ether, benzene.

Caution: Potential symptoms of overexposure are irritation of nose and eyes; liver and kidney damage; skin blisters. *See NIOSH Pocket Guide to Chemical Hazards* (DHHS/NIOSH 90-117, 1990) p 84. Exposure to high concentrations may result in CNS depression. *See Patty's Industrial Hygiene and Toxicology* vol. **2B**, G. D. Clayton, F. E. Clayton, Eds. (Wiley-Interscience, New York, 3rd ed., 1981) pp 3611-3617.

USE: Solvent for waxes, gums, resins, tars, rubbers, oils, asphalts; insecticide for termites and locust borers; fumigant; removing sulfur from illuminating gas; as degreasing agent for metals, leather, wool; as ingredient of metal pol-

ishes; as heat transfer medium; as intermediate in the manuf of dyes.

3107. p-Dichlorobenzene. Paracide; PDB; Paradichlorobenzene; Para-zene; Di-chloricide; Paramoth. C$_6$H$_4$Cl$_2$; mol wt 147.00. C 49.02%, H 2.74%, Cl 48.23%. Prepn and separation from *m*- and *o*-dichlorobenzene, *see m*-isomer above. Crystal structure of triclinic form (β-modification): Housty, Clastre, *Acta Cryst.* **10**, 695 (1957); of monoclinic form (α-modification) and transformation to triclinic form: Panatoni *et al., Gazz. Chim. Ital.* **93**, 813 (1963). Manuf: Faith, Keyes & Clark's *Industrial Chemicals*, F. A. Lowenheim, M. K. Moran, Eds. (Wiley-Interscience, New Yok, 4th ed., 1975) pp 258-265.

Volatile crystals with a characteristic penetrating odor. Sublimes at ordinary temps. mp 53.5° (α-modification), 54° (β-modification). bp 174.12°. n$_D^{60}$ 1.5285. Flash pt (closed cup) 150°F. Practically insol in water. Sol in alcohol, ether, benzene, chloroform, carbon disulfide. Non-corrosive; nonstaining. LD$_{50}$ orally in rats: 500 mg/kg, *Toxic Substances List,* H. E. Christensen, Ed. (1973) p 321.

Caution: Potential symptoms of overexposure are headache; eye irritation, periorbital swelling; profuse rhinitis; anorexia, nausea and vomiting; weight loss, jaundice and cirrhosis. *See NIOSH Pocket Guide to Chemical Hazards* (DHHS/NIOSH 90-117, 1990) p 84. Direct contact with fumes or solid may cause skin irritation. *See Patty's Industrial Hygiene and Toxicology* vol. **2B**, G. D. Clayton, F. E. Clayton, Eds. (Wiley-Interscience, New York, 3rd ed., 1981) pp 3617-3626. This substance may reasonably be anticipated to be a carcinogen: *Seventh Annual Report on Carcinogens* (PB95-109781, 1994) p 152.

USE: Insecticidal fumigant. Popular for domestic use against clothes moths; as room deodorant. Intermediate in production of plastics for electronic components.

3108. 2,2'-Dichlorobenzidine. *2,2'-Dichloro[1,1'-biphenyl]-4,4'-diamine.* C$_{12}$H$_{10}$Cl$_2$N$_2$; mol wt 253.13. C 56.94%, H 3.98%, Cl 28.01%, N 11.07%. Prepn from *m*-chloronitrobenzene: Laubenheimer, *Ber.* **8**, 1625 (1875); Cain, May, *J. Chem. Soc.* **97**, 723 (1910).

Needles from water or prisms from alc, mp 165°. Almost insol in water; moderately sol in alc; readily sol in ether.

Hydrochloride, C$_{12}$H$_{10}$Cl$_2$N$_2$.2HCl, leaflets from water, moderately sol in water.

USE: Manuf azo dyes.

3109. 3,3'-Dichlorobenzidine. *3,3-Dichloro-(1,1'-biphenyl)-4,4'-diamine;* 3,3'-dichloro-4,4'-biphenyldiamine; DCB. C$_{12}$H$_{10}$Cl$_2$N$_2$; mol wt 253.13. C 56.94%, H 3.98%, Cl 28.01%, N 11.07%. Prepn from *o*-chloronitrobenzene: Cohn, *Ber.* **33**, 3552 (1900). Review of carcinogenicity studies: *IARC Monographs* **4**, 49-55 (1974); of bioactivation and toxicology: M. M. Iba, *Drug Metab. Rev.* **21**, 377-400 (1990).

Needles from alc or benzene, mp 132-133°. Almost insol in water. Readily sol in alc, benzene, glacial acetic acid.

Dihydrochloride, C$_{12}$H$_{10}$Cl$_2$N$_2$.2HCl, needles. Slightly sol in water, readily in alc.

Caution: Potential symptoms of overexposure are allergic skin reaction, sensitization dermatitis; headache, dizziness; caustic burns; frequent urination, dysuria; hematuria; gastrointestinal upsets; upper respiratory infection. *See NIOSH Pocket Guide to Chemical Hazards* (DHHS/NIOSH 90-117, 1990) p 86. 3,3'-Dichlorobenzidine and its dihydrochloride may reasonably be anticipated to be a carcinogen: *Seventh Annual Report on Carcinogens* (PB95-109781, 1994) p 155.

USE: Manuf azo dyes; as an intermediate for the Benzidine Yellow pigments.

3110. Dichlorobenzyl Alcohol. *2,4-Dichlorobenzene-methanol;* Dybenal. $C_7H_6Cl_2O$; mol wt 177.03. C 47.49%, H 3.42%, Cl 40.05%, O 9.04%. Prepn: Van de Lande, *Rec. Trav. Chim.* **51,** 98 (1932); Metayer, Dat-Xuong, *Bull. Soc. Chim. France* **1954,** 615.

Crystals, mp 59.5°.

THERAP CAT: Antiseptic.

3111. 1,1-Dichloro-2,2-bis(p-chlorophenyl)ethane. *1,-1'-(2,2-Dichloroethylidene)bis[4-chlorobenzene];* tetrachloro-diphenylethane; TDE; dichlorodiphenyldichloroethane; DDD; *p,p'*-DDD; *p,p'*-TDE; Rhothane. $C_{14}H_{10}Cl_4$; mol wt 320.04. C 52.54%, H 3.15%, Cl 44.31%. Non-degradable pesticide; a component of technical grade DDT. Prepd by condensing dichloroacetaldehyde with chlorobenzene: Haller *et al., J. Am. Chem. Soc.* **67,** 1596, 1600 (1945). Toxicity study: T. B. Gaines, *Toxicol. Appl. Pharmacol.* **14,** 515 (1969).

Crystals, mp 109-110°. The chemical properties and solys are similar to those of DDT. LD_{50} orally in rats: >4000 mg/kg (Gaines). *Caution:* Slightly irritating to skin. *Acute toxicity symptoms:* lethargy but no convulsions. Estimated fatal oral dose 5 g/kg body wt. *Chronic toxicity symptoms:* atrophy of adrenal cortex, liver damage. Symptoms similar to DDT, *q.v.*

USE: Insecticide.

3112. Dichloro(2-chlorovinyl)arsine. *(2-Chloroethenyl)arsonous dichloride;* 2-chlorovinyldichloroarsine; chlorovinylarsine dichloride; Lewisite. $C_2H_2AsCl_3$; mol wt 207.32. C 11.59%, H 0.97%, As 36.14%, Cl 51.30%. ClCH=CH-AsCl$_2$. Obtained together with bis(2-chlorovinyl)chloroarsine and tris(2-chlorovinyl)arsine upon passing acetylene into a mixture of arsenic trichloride and aluminum chloride and fractionating the product in a current of HCl gas: Wieland, *Ann.* **431,** 38 (1923); Lewis, Perkins, *Ind. Eng. Chem.* **15,** 290 (1923); Lewis, Stiegler, *J. Am. Chem. Soc.* **47,** 2546 (1925); Gibson, Johnson, *J. Chem. Soc.* **1931,** 754; Jarman in *Advances in Chemistry Series* **23,** 328-337 (1959).

Liquid. Faint odor of geranium. *Vesicant. War gas. Respiratory and systemic poison.* Solidifies at −13°. mp 0.1°. bp$_{12.5}$ 76-77°; bp$_{26}$ 93°; bp$_{760}$ 190° (dec). d$_4^{20}$ 1.888. Vapor pressure at 0°: 0.087 mm Hg; at 20°: 0.395 mm Hg. Sol in the ordinary organic solvents. Insol in water, dil mineral acids. Hydrolyzed by alkalies. Neutralized and inactivated by bleaching powder, sodium hypochlorite. *Caution:* Extremely toxic! Produces severe vesication, even through rubber. If left on skin, as little as 0.5 ml may give rise to sufficient absorption to produce severe systemic effects; 2 ml may cause death. *Antidote:* Dimercaprol (British Anti-Lewisite), *q.v.*

3113. 2,3-Dichloro-5,6-dicyanobenzoquinone. *4,5-Dichloro-3,6-dioxo-1,4-cyclohexadiene-1,2-dicarbonitrile;* DDQ. $C_8Cl_2N_2O_2$; mol wt 227.01. C 42.33%, Cl 31.24%, N 12.34%, O 14.10%. Prepn: Thiele, Günther, *Ann.* **349,** 45 (1906); Walker, Waugh, *J. Org. Chem.* **30,** 3240 (1965).

Yellow to orange crystals, mp 213.5-215°. Dec in water. Sol in benzene, dioxane, acetic acid. Slightly sol in chloroform, methylene chloride.

USE: Oxidizing agent, especially in steroid synthesis.

3114. Dichlorodifluoromethane. Difluorodichloro-methane; Arcton 12; Freon 12; Frigen 12; Genetron 12; Halon; Isotron 2. CCl_2F_2; mol wt 120.91. C 9.93%, Cl 58.64%, F 31.42%. Prepn: Henne, *Organic Reactions* **2,** 49 (1944); Swarts, *Chem. Zentr.* **1907, II,** 581. Manuf: Faith, Keyes & Clark's *Industrial Chemicals,* F. A. Lowenheim, M. K. Moran, Eds. (Wiley-Interscience, New York, 4th ed., 1975) pp 325-330.

Colorless, practically odorless, noncorrosive, nonirritating, nonflammable gas. Faint, ether-like odor in high concentrations. Stable up to 550°. d$_{liq}^{-29.8}$ 1.486. mp −158°. bp$_{760}$ −29.8°; bp$_{2atm}$ −12.2°; bp$_{5atm}$ +16.1°; bp$_{10atm}$ +42.4°; bp$_{20atm}$ +74.0°; bp$_{30atm}$ +95.6°. Critical temp 111.5°; critical pressure 39.6 atm (582 lb/sq inch). Dipole moment 0.51. Insol in water. Sol in alcohol, ether. Absorbs less than 0.0025% water. Has little, if any, anesthetic or toxic action, but toxic substances may be formed on contact with a flame or hot metal surface. *Caution:* Potential symptoms of overexposure are dizziness, tremors, unconsciousness, cardiac arrhythmias and cardiac arrest. *See NIOSH Pocket Guide to Chemical Hazards* (DHHS/NIOSH 90-117, 1990) p 86. *Note:* Consult latest Government regulations on use as aerosol propellant.

USE: Refrigerant, aerosol propellant.

3115. 1,3-Dichloro-5,5-dimethylhydantoin. *1,3-Dichloro-5,5-dimethyl-2,4-imidazolidinedione;* Dactin; Halane. $C_5H_6Cl_2N_2O_2$; mol wt 197.02. C 30.48%, H 3.07%, Cl 35.99%, N 14.22%, O 16.24%. Prepd by passing chlorine through an aq soln of 5,5-dimethylhydantoin: Biltz, Slotta, *J. Prakt. Chem.* **113,** 248 (1926); Orazi, Orio, *Annales Asoc. Quim. Argentina* **41,** 153 (1953), *C.A.* **48,** 13634 (1954). Structure: Bogaert-Verhoogen, Martin, *Bull. Soc. Chim. Belges* **58,** 567 (1949). Purification by dissolving in concd H_2SO_4 and diluting with ice-water: Lorenz, U.S. pat. **2,828,308** (1958 to Purex).

Four-sided, pointed prisms from chloroform, d$_{20}^{20}$ 1.5. mp 132°. Sublimes at 100°. Turns brown and conflagrates at 212° (after melting at 132°). The dry crystals [combined available chlorine 77.6% (theory)] may be stored without much loss of available chlorine. After 14 weeks at 60° the loss was 1.5% Cl compared with a loss of 37.5% suffered by 70% calcium hypochlorite. On contact with water and esp hot water hypochlorous acid is liberated. At pH 9 nitrogen chloride is formed. Soly in water at 25° 0.21%; at 60° 0.60%. pH of aq soln about 4.4. Freely sol in chlorinated and highly polar solvents at 25°: Chloroform 14%, methylene chloride 30.0%, carbon tetrachloride 12.5%, ethylene dichloride 32.0%, *sym*-tetrachlorethane 17.0%, benzene 9.2%. *Caution:* Potential symptoms of overexposure are irritation of eyes, mucous membranes and respiratory system. *See NIOSH Pocket Guide to Chemical Hazards* (DHHS/NIOSH 90-117, 1990) p 86.

USE: Chlorinating agent, disinfectant, industrial deodorant. In water treatment. Active ingredient of powder laundry bleaches such as Sage's Dry Bleach, Colgate's Pruf. Intermediate for amino acids, drugs, insecticides. Stabilizer for vinyl chloride polymers. Polymerization catalyst.

3116. sym-Dichloroethyl Ether. *1,1'-Oxybis[2-chloroethane];* bis(2-chloroethyl) ether; β,β'-dichloroethyl ether; DCEE; Chlorex. $C_4H_8Cl_2O$; mol wt 143.01. C 33.59%, H 5.64%, Cl 49.58%, O 11.19%. [ClCH$_2$CH$_2$]$_2$O. Prepn: O. Kamm, J. H. Waldo, *J. Am. Chem. Soc.* **43**, 2223 (1921). Toxicity: H. F. Smyth, Jr., C. P. Carpenter, *J. Ind. Hyg. Toxicol.* **30**, 66 (1948). Evaluation of carcinogenic risk: *IARC Monographs* **9**, 117 (1975).

Colorless, clear liq; pungent odor. Vapor harmful. d$_{20}^{20}$ 1.22. bp 178°. mp −50°. n$_D^{20}$ 1.457. Flash pt, closed cup: 145°F (63°C), Buckman Laboratories, Inc. literature; also reported as 131°F (55°C), *Ind. Eng. Chem.* **32**, 880 (1940). Insol in water. Sol in most organic solvents. It dissolves oils, fats, greases, etc. LD$_{50}$ orally in rats: 75 mg/kg (Smyth, Carpenter).

Caution: Potential symptoms of overexposure are lacrimation; irritation of nose and throat, coughing; nausea, vomiting. Direct contact may cause skin and eye irritation. Potential occupational carcinogen. *See NIOSH Pocket Guide to Chemical Hazards* (DHHS/NIOSH 90-117, 1990) p 88.

USE: Reagent for organic synthesis; solvent. Has been used as a scouring agent for textiles; as soil fumigant.

3117. 4',5'-Dichlorofluorescein. *4',5'-Dichloro-3',6'-dihydroxyspiro[isobenzofuran-1(3H),9'-[9H]xanthen]-3-one;* D & C Orange no. 8; 4,5-dichloro-3,6-fluorandiol. $C_{20}H_{10}$-Cl$_2$O$_5$; mol wt 401.20. C 59.88%, H 2.51%, Cl 17.67%, O 19.94%. Prepd by condensation of 2-chlororesorcinol with phthalic anhydride: *Colour Index* vol. 4 (3rd ed., 1971) p 4424.

Orange powder, insol in water and in dil acids. Sol in alcohol and in dil alkali yielding solns which are orange with a greenish-yellow fluorescence. Slightly sol in glycerol and the glycols. Practically insol in oils, fats and waxes. If a soln of the dye in strong alkali is heated, a violet color is produced. *Caution:* Delisted by FDA for use in drugs and cosmetics.

Disodium salt, $C_{20}H_8Cl_2Na_2O_5$, *C.I. Solvent Orange 32*, *C.I.* 45365.

Note: Structure depicted as the open tautomer.

3118. 2,6-Dichloroindophenol Sodium. *2,6-Dichloro-4-[(4-hydroxyphenyl)imino]-2,5-cyclohexadien-1-one sodium salt;* sodium 2,6-dichloroindophenol; sodium 2,6-dichloro-N-(p-hydroxyphenyl)-p-benzoquinone imine; 2,6-dichloro-1,4-benzoquinone-4-(4-hydroxyanil) sodium; 2,6-dichlorophenol-indophenol sodium; Tillman's reagent. $C_{12}H_6Cl_2$-NNaO$_2$; mol wt 290.08. C 49.69%, H 2.08%, Cl 24.44%, N 4.83%, Na 7.93%, O 11.03%. Prepd from 2,6-dichloro-p-quinone-4-chlorimide and phenol in alkaline soln: Gibbs et al., *Publ. Health Repts.* **40**, 650 (1925); Tillmans et al., *Z. Unters. Lebensm.* **56**, 273 (1932).

Dark green powder. May contain up to 2 mols H$_2$O. Freely sol in water, alc. The aq soln is deep blue, changed to red by acids. It liberates iodine from KI in acid solns.

USE: Analytical reagent in the determination of ascorbic acid (vitamin C) which reduces the dye to a colorless hydroxy compd.

3119. sym-Dichloromethyl Ether. *Oxybis[chloromethane];* bis(chloromethyl) ether; BCME. $C_2H_4Cl_2O$; mol wt 114.96. C 20.90%, H 3.51%, Cl 61.68%, O 13.92%. (CH$_2$-Cl)$_2$O. Review of carcinogenicity studies: *IARC Monographs* **4**, 231-238 (1974).

Colorless liquid; suffocating odor. d$_4^{20}$ 1.315. bp 106°. n$_D^{20}$ 1.4346. Unstable in moist air; dec by water into HCl and formaldehyde.

Caution: Potential symptoms of overexposure are irritation of eyes, skin and mucous membranes of respiratory system; pulmonary congestion, edema; corneal damage, necrosis; reduced pulmonary function, coughing, dyspnea and wheezing; blood stained sputum, bronchial secretions. *See NIOSH Pocket Guide to Chemical Hazards* (DHHS/NIOSH 90-117, 1990) p 68. This substance has been listed as a known carcinogen: *Seventh Annual Report on Carcinogens* (PB95-109781, 1994) p 40.

USE: Manuf of plastics and ion exchange resins.

3120. Dichlorophen. *2,2'-Methylenebis[4-chlorophenol];* 2,2'-dihydroxy-5,5'-dichlorodiphenylmethane; 5,5'-dichloro-2,2'-dihydroxydiphenylmethane; bis[5-chloro-2-hydroxyphenyl]methane; di[5-chloro-2-hydroxyphenyl]methane; dichlorophene; G-4; Anthiphen; Dicestal; Didroxane; Di-phenthane-70; Parabis; Plath-Lyse; Preventol G-D; Teniathane; Teniatol; Wespuril. $C_{13}H_{10}Cl_2O_2$; mol wt 269.13. C 58.02%, H 3.75%, Cl 26.35%, O 11.89%. Prepn by adding a CH$_2$O yielding reagent, e.g. aq HCHO, to p chlorophenol: Gump, Luthy, U.S. pat. **2,334,408** (1944 to Burton T. Bush). Acute toxicity: T. B. Gaines, R. E. Linder, *Fundam. Appl. Toxicol.* **7**, 299 (1986).

Crystals from toluene, mp 177-178°. Practically insol in water. Sparingly sol in toluene. One gram dissolves in 1 g of 95% ethanol, in less than 1 g of ether. Also sol in methanol, isopropyl ether, petr ether. Sol (with decompn) in alkaline aq solns. LD$_{50}$ in adult male, female rats (mg/kg): 1506, 1683 orally (Gaines, Linder).

USE: Agricultural fungicide; antimicrobial; germicide in soaps, shampoos, etc.

THERAP CAT: Anthelmintic (Cestodes).

THERAP CAT (VET): Anthelmintic, antiprotozoan, fungicide.

3121. Dichlorophenarsine Hydrochloride. *(3-Amino-4-hydroxyphenyl)arsonous dichloride monohydrochloride; (3-amino-4-hydroxyphenyl)dichloroarsine hydrochloride;* RP-2591; Arseclor; Chlorarsen; Chlorarsol; Clorarsen; Dichloro-Mapharsen; Filarsen; Fontarsol; Halarsol. $C_6H_7AsCl_3$-NO; mol wt 290.41. C 24.82%, H 2.43%, As 25.80%, Cl 36.62%, N 4.82%, O 5.51%. Prepd by the action of arsenic trichloride on o-aminophenol, or from the corresp arsine oxide and an excess of HCl: Ger. pat. **281,101**; from arsphenamine by treatment with mercuric chloride and HCl gas: Binz, Bauer, *Z. Angew. Chem.* **34**, 261 (1921). Toxicity: Beck, *Proc. Soc. Exp. Biol. Med.* **78**, 392 (1951).

White hygroscopic powder. mp 200°. Readily sol in water yielding a soln of oxophenarsine. Usually marketed mixed with 3.33 times its weight of sodium citrate-carbonate as buffer to make a soln of pH 6.2 isotonic with human blood. LD_{50} i.p. in mice: 41 mg/kg (Beck).

THERAP CAT: Formerly as antisyphilitic.

THERAP CAT (VET): Has been used as a filaricide.

3122. 2,4-Dichlorophenol. $C_6H_4Cl_2O$; mol wt 163.00. C 44.21%, H 2.47%, Cl 43.50%, O 9.82%. Key intermediate in synthesis of the herbicide 2,4-D, *q.v.* Prepn by chlorination of phenol: M. Kohn, S. Sussmann, *Monatsh.* **46**, 575 (1926); B. O. Pray, D. N. Sukow, U.S. pat. **2,759,981** (1956 to Columbia-Southern Chem. Corp.); by chlorination of *o*- or *p*-chlorophenol: L. G. Groves *et al., J. Chem. Soc.* **1929**, 512; from 2,4-dinitrophenol: A. Ghosh, *J. Indian Chem. Soc.* **28**, 155 (1951). ^1H NMR study: J. B. Rowbotham, T. Schaefer, *Can. J. Chem.* **52**, 3037 (1974).

Needle-like crystals, mp 45°. bp 209-211°. Volatile with steam. Sol in CCl_4.

3123. 2,6-Dichlorophenol. $C_6H_4Cl_2O$; mol wt 163.00. C 44.21%, H 2.47%, Cl 43.50%, O 9.82%. Synthesis from ethyl 3,5-dichloro-4-hydroxybenzoate: D. S. Tarbell *et al., Org. Syn.* **coll. vol. III**, 267. Crystal structure: C. Bavoux, P. Michel, *Acta Crystallogr.* **30B**, 2043 (1974). Isoln and identification as sex pheromone of lone star tick: R. S. Berger, *Science* **177**, 704 (1972).

White crystals from petr ether, mp 64.5-65.5°.

3124. 1,3-Dichloro-2-propanol. *α-Dichlorohydrin; sym-glycerol dichlorohydrin; glycerol α,γ-dichlorohydrin; sym-dichloroisopropyl alcohol.* $C_3H_6Cl_2O$; mol wt 128.99. C 27.94%, H 4.69%, Cl 54.97%, O 12.40%. $CH_2ClCHOH-CH_2Cl$. Prepd from glycerol, acetic acid and HCl gas: Conant, Quayle, *Org. Syn.* **2**, 29 (1922); **coll. vol. I** (2nd ed.) p 292. Toxicity study: Smyth *et al., Am. Ind. Hyg. Assoc. J.* **23**, 95 (1962).

Liquid. Ethereal odor. d_4^{17} 1.3506. mp $-4°$. bp_{760} 174.3°; bp_{100} 114.8°; bp_{40} 93°; bp_{20} 78°; bp_5 52°; $bp_{1.0}$ 28°. n_D^{17} 1.480245. Sol in 10 parts water; miscible with alcohol, ether. LD_{50} orally in rats: 110 mg/kg, Smyth *et al., Am. Ind. Hyg. Assoc. J.* **23**, 95 (1962).

USE: Solvent for hard resins and nitrocellulose; manuf photographic and Zapon lacquer; cement for celluloid; binder for water colors.

3125. 1,3-Dichloropropene. *α,γ-Dichloropropylene; 1,3-dichloropropylene; γ-chloroallyl chloride; 1,3-D; Telone II.* $C_3H_4Cl_2$; mol wt 110.97. C 32.47%, H 3.63%, Cl 63.90%. $ClCH_2CH=CHCl$. Prepd from 1,2,3-trichloropropane with NaOH: Friedel, Silva, *Compt. Rend.* **75**, 81 (1872); Reboul, *Ann. Chim.* [3] **60**, 37 (1860); from 3-chloro-2-propen-1-ol with PCl_3: Kirrmann *et al., Bull. Soc. Chim. France* [5] **1**, 860 (1934); from acrolein with PCl_5: van Romburgh, *ibid.* [2] **36**, 549 (1881); from 3,3-dichloropropene by isomerization with conc HCl: *idem, ibid.* The usual industrial prepn is from 1,3-dichloro-2-propanol by dehydration with $POCl_3$ or with P_2O_5 in benzene: Hurd, Webb, *J. Am. Chem. Soc.* **58**, 2191 (1936). Prepn of stereoisomers: L. F. Hatch, A. C. Moore, *ibid.* **66**, 285 (1944). Configuration: H. A. Smith, W. H. King, *ibid.* **70**, 3528 (1948); L. F. Hatch, R. H. Perry, *ibid.* **71**, 3262 (1949). Persistence in soil: I. J. Thomason, M. V. McKenry, *Hilgardia* **42**, 393 (1974). GC determn of isomers in rat blood: P. E. Kastl, E. A. Hermann, *J. Chromatog.* **265**, 277 (1983). Toxicity study: T. R. Torkelson, F. Oyen, *Am. Ind. Hyg. Assoc. J.* **38**, 217 (1977).

Liquid with chloroform odor. d^{25} 1.220. bp 108°. n_D^{22} 1.4735. Technical product is mixture of *cis*- and *trans*-isomers. LD_{50} (92% technical product) in male, female rats (mg/kg): 713, 470 orally (10% soln in corn oil); in rabbits (mg/kg): 504 dermally (Torkelson, Oyen).

Mixture with 1,2-dichloropropane, *D-D, Vidden D, Telone.*

cis-Form, bp 104.3°. d_4^{20} 1.224. n_4^{20} 1.4682.

trans-Form, bp 112.0°. d_4^{20} 1.217. n_4^{20} 1.4730.

Caution: Irritating to skin, eyes, and mucous membranes; liver and kidney injury have been produced in exptl animals, *see Clinical Toxicology of Commercial Products,* R. E. Gosselin *et al.,* Eds. (Williams & Wilkins, Baltimore, 5th Ed., 1984) Section III, p 141. This substance may reasonably be anticipated to be a carcinogen: *Seventh Annual Report on Carcinogens* (PB95-109781, 1994) p 164.

USE: Soil fumigant; nematocide. In manuf of pesticides.

3126. Dichloroxylenol. *2,4-Dichloro-3,5-dimethylphenol; 2,4-dichloro-3,5-xylenol;* DCMX. $C_8H_8Cl_2O$; mol wt 191.06. C 50.29%, H 4.22%, Cl 37.11%, O 8.37%. Prepn: Lesser, Gad, *Ber.* **56**, 963 (1923); Jones, *J. Chem. Soc.* **1941**, 267; Gemmell, *Mfg. Chemist* **23**, 63 (1952).

Crystals from petr ether, mp 95-96°. Sublimes. Volatile with steam. Soly in water: one part in 5000. Soly at 15° in 100 parts of solvent: benzene, 14; toluene, 15; acetone, 73; diethyl ketone, 59; petr ether, 4; chloroform, 25; CCl_4, 10.

USE: Bacteriostat in soaps; mold inhibitor, preservative.

3127. Dichlorphenamide. *4,5-Dichloro-1,3-benzenedisulfonamide;* 1,3-disulfamyl-4,5-dichlorobenzene; 4,5-dichloro-1,3-disulfamoylbenzene; Antidrasi; Daranide; Oratrol. $C_6H_6Cl_2N_2O_4S_2$; mol wt 305.16. C 23.62%, H 1.98%, Cl 23.24%, N 9.18%, O 20.97%, S 21.02%. Prepn: Schultz, U.S. pat. **2,835,702** (1958 to Merck & Co.).

Needles from DMSO + water, mp 239-241° (patent, mp 228.5-229°). Practically insol in water. Sol in alkaline solns.

THERAP CAT: Carbonic anhydrase inhibitor.

3128. Dichlorprop. *2-(2,4-Dichlorophenoxy)propanoic acid; 2-(2,4-dichlorophenoxy)propionic acid;* 2,4-DP; dichloroprop; Cornox RK; Hedonal DP. $C_9H_8Cl_2O_3$; mol wt 235.07. C 45.99%, H 3.43%, Cl 30.16%, O 20.42%. Selective pre- and post-emergence herbicide. Exists in two optically active forms; only the (+)-isomer shows biological activity. First prepd and described as a plant growth regulator: M. E. Synerholm, P. W. Zimmerman, *Contrib. Boyce Thompson Inst.* **14**, 91 (1945). Commercial prepn: H. A. Stevenson, R. F. Brookes, **Brit.** pat. **822,199** (1959 to Boots). Commercial products contain approx equal amounts of the two isomers.

Colorless crystals, mp 117-118°. Soly in water at 20°: 350 ppm. Readily sol in organic solvents. Corrosive to metals in the presence of water. Commercial prepns may contain the isooctyl or butyl esters or aq solns of the potassium salt.

LD$_{50}$ orally in rats and mice: 800, 400 mg/kg, *RTECS* Vol. II, E. J. Fairchild, Ed. (1977) p 770.

USE: Herbicide.

3129. Dichlorvos. *Phosphoric acid 2,2-dichloroethenyl dimethyl ester; phosphoric acid 2,2-dichlorovinyl dimethyl ester; O,O-dimethyl O-(2,2-dichlorovinyl) phosphate;* dichlorophos; dichlorovos; DDVP; SD-1750; Astrobot; Atgard; Canogard; Dedevap; Dichlorman; Divipan; Equigard; Equigel; Estrosol; Herkol; Nogos; Nuvan; Task; Vapona; Verdisol. C$_4$H$_7$Cl$_2$O$_4$P; mol wt 220.98. C 21.74%, H 3.19%, Cl 32.09%, O 28.96%, P 14.02%. Cholinesterase inhibitor. Prepn: Whetstone, Harman, U.S. pat. **2,956,073** (1960 to Shell). Toxicity: T. B. Gaines, *Toxicol. Appl. Pharmacol.* **14**, 515 (1969). Review of carcinogenicity studies: M. D. Reuber, *Clin. Toxicol.* **18**, 47-84 (1981).

Liquid. Practically non-flammable. d$_4^{25}$ 1.415. bp$_{20}$ 140°; bp$_{1.0}$ 84°; bp$_{0.5}$ 72°; bp$_{0.01}$ 30°. n$_D^{25}$ 1.451. Vapor pressure at 20°: 1.2 × 10^{-2} mm Hg. Misc with alcohol and most nonpolar solvents. Soly in water: about 1 g/100 ml; in glycerol: about 0.5 g/100 ml. LD$_{50}$ orally in male, female rats: 80, 56 mg/kg (Gaines).

Caution: Potential symptoms of overexposure are miosis, aching eyes; rhinorrhea; headache; chest tightening, wheezing, laryngeal spasms and salivation; cyanosis; anorexia, nausea, vomiting and diarrhea; sweating; muscle fasiculation, paralysis, giddiness and ataxia; convulsions; low blood pressure, cardiac irregularities; irritation of eyes and skin. *See NIOSH Pocket Guide to Chemical Hazards* (DHHS/NIOSH 90-117, 1990) p 90.

USE: Insecticide.

THERAP CAT (VET): Anthelmintic, insecticide.

3130. Diclazuril. *2,6-Dichloro-α-(4-chlorophenyl)-4-(4,5-dihydro-3,5-dioxo-1,2,4-triazin-2(3H)-yl)benzeneacetonitrile;* (p-chlorophenyl)[2,6-dichloro-4-(4,5-dihydro-3,5-dioxo-as-triazin-2(3H)-yl)phenyl]acetonitrile; R-64433; Clinacox. C$_{17}$H$_9$Cl$_3$N$_4$O$_2$; mol wt 407.64. C 50.09%, H 2.23%, Cl 26.09%, N 13.74%, O 7.85%. Nucleotide analog with broad-spectrum anticoccidial activity. Prepn: G. M. Boeckx *et al., Eur.* pat. **Appl. 170,316**; *eidem,* U.S. pat. **4,631,278** (both 1986 to Janssen). Mechanism of action studies: L. Maes *et al., J. Parasitol.* **74**, 931 (1988); A. Verheyen *et al., ibid.* 939. Efficacy against *Eimeria* species in turkeys: O. Vanparijs *et al., Avian Dis.* **33**, 479 (1989); in chickens: O. Vanparijs *et al., Poultry Sci.* **69**, 60 (1990). Preliminary evaluation in coccidial enteritis in AIDS patients: K. Kayembe *et al., Lancet* **1**, 1397 (1989).

mp 290.5°.

THERAP CAT (VET): Coccidiostat.

3131. Diclobutrazol. *(R*,R*)-(±)-β-[(2,4-Dichlorophenyl)methyl]-α-(1,1-dimethylethyl)-1H-1,2,4-triazole-1-ethanol;* 1-(2,4-dichlorophenyl)-4,4-dimethyl-2-(1,2,4-triazol-1-yl)pentan-3-ol; 1-t-butyl-2-(1,2,4-triazol-1-yl)-2-(2′,4′-dichlorobenzyl)ethanol; dichlobutrazol; PP-296; Vigil. C$_{15}$H$_{19}$Cl$_2$N$_3$O; mol wt 328.24. C 54.89%, H 5.83%, Cl 21.60%, N 12.80%, O 4.87%. Prepn: **Belg.** pat. **857,836**; S. Balasubramanyan, M. C. Shephard, U.S. pat. **4,243,405** (1978, 1981 both to ICI). Activity, physical properties, and toxicity: K. J. Bent, A. M. Skidmore, *Proc. Brit. Crop Prot. Conf. - Pests Dis.* **1979**, 477. Fungicidal activity: B. Ram *et al., Test Agrochem. Cultiv.* **3**, 38 (1982). Regulation of plant

growth: J. B. Shanks, *Proc. Plant Growth Reg. Soc. Am.* **9**, 68 (1982). Mechanism of action: P. Gadher *et al., Pest. Biochem. Physiol.* **19**, 1 (1983); T. E. Wiggins, B. C. Baldwin, *Pestic. Sci.* **15**, 206 (1984). Resolution of enantiomers by GC: R. S. Burden *et al., J. Chromatog.* **391**, 273 (1987).

(S,S)-form

Odorless, colorless, crystalline solid, mp 147-149°. Vapor pressure at 20°: 1 × 10^{-8} to 2 × 10^{-8} mm Hg. Soly in water at 20°: 9 mg/l. Sol up to 50 g/l in methanol, ethanol, acetone, and chloroform. LD$_{50}$ in rats (mg/kg): about 4000 orally; >1000 dermally; in mallard ducks: >9000 orally (Bent, Skidmore).

USE: Agricultural fungicide.

3132. Diclofenac Sodium. *2-[(2,6-Dichlorophenyl)amino]benzeneacetic acid monosodium salt; [o-(2,6-dichloroanilino)phenyl]acetic acid sodium salt;* sodium [o-[(2,6-dichlorophenyl)amino]phenyl]acetate; GP-45840; Allvoran; Assaren; Benfofen; Cataflam; Delphimix; Dichronic; Diclobenin; Diclo-Phlogont; Diclo-Puren; Diclord; Dicloreum; Dolobasan; Duravolten; Ecofenac; Effekton; Kriplex; Neriodin; Novapirina; Primofenac; Prophenatin; Rhumalgan; Tsudohmin; Valetan; Voldal; Voltaren; Voltarol; Xenid. C$_{14}$H$_{10}$Cl$_2$NNaO$_2$; mol wt 318.13. C 52.86%, H 3.17%, Cl 22.29%, N 4.40%, Na 7.23%, O 10.06%. Prepn: **Neth.** pat. **Appl. 6,604,752**; A. Sallmann, R. Pfister, U.S. pat. **3,558,690** (1966, 1971 both to Geigy). Pharmacology: Renaud, Lecompte, *Thromb. Diath. Haemorrh.* **24**, 577 (1970), *C.A.* **74**, 86215m (1971); Krupp *et al., Experientia* **29**, 450 (1973). HPLC determn in plasma and urine: J. Godbillon *et al., J. Chromatog.* **338**, 151 (1985). Symposium on pharmacology and clinical experience: *Semin. Arthritis Rheum.* **15**, Suppl. 1, 57-110 (1985); on pharmacology, efficacy and safety: *Am. J. Med.* **80**, Suppl. 4B, 1-87 (1986). Comprehensive description: C. M. Adeyeye, P-K. Li, *Anal. Profiles Drug Subs.* **19**, 123-144 (1990).

Crystals from water, mp 283-285°. uv max (methanol) 283 nm (ε 1.05 × 10^5); (phosphate buffer, pH 7.2) 276 nm (ε 1.01 × 10^5). Soly at 25°C (mg/ml): deionized water (pH 5.2) >9; methanol >24; acetone 6; acetonitrile <1; cyclohexane <1; HCl (pH 1.1) <1; phosphate buffer (pH 7.2) 6. pKa 4. Partition coefficient (n-octanol/aq. buffer): 13.4. LD$_{50}$ in mice, rats (mg/kg): ~390, 150 orally (Krupp).

Free acid, C$_{14}$H$_{11}$ClNO$_2$, crystals from ether-petr ether, mp 156-158°.

THERAP CAT: Anti-inflammatory.

3133. Diclofop-Methyl. *2-[4-(2,4-Dichlorophenoxy)phenoxy]propanoic acid methyl ester;* methyl 2-[4-(2,4-dichlorophenoxy)phenoxy]propionate; HOE-23408; Hoelon; Hoegrass; Il(l)oxan. C$_{16}$H$_{14}$Cl$_2$O$_4$; mol wt 341.19. C 56.33%, H 4.14%, Cl 20.78%, O 18.76%. Selective postemergence herbicide. Prepn: W. Becker *et al.,* **Ger.** pat. **2,223,894** corresp to U.S. pat. **3,954,442** (1973, 1976 to Hoechst). Soil degradn: A. E. Smith, *J. Agr. Food Chem.* **25**, 893 (1977); **27**, 1145 (1979).

mp 39-41°. bp$_{0.1}$ 175-177°. Soly in water: 0.3 mg/100 ml. Soly (g/100 ml): acetone 249; ethanol 11; xylene 253.
USE: Herbicide.

3134. Dicloxacillin. *[2S-(2α,5α,6β)]-6-[[[3-(2,6-Di-chlorophenyl)-5-methyl-4-isoxazolyl]carbonyl]amino]-3,3-dimethyl-7-oxo-4-thia-1-azabicyclo[3.2.0]heptane-2-carboxylic acid;* 6-[3-(2,6-dichlorophenyl)-5-methyl-4-isoxazolecarbox-amido]penicillanic acid; 3-(2,6-dichlorophenyl)-5-methyl-4-isoxazolylpenicillin; BRL-1702; Maclicine. $C_{19}H_{17}Cl_2N_3O_5S$; mol wt 470.33. C 48.52%, H 3.64%, Cl 15.08%, N 8.93%, O 17.01%, S 6.82%. Semi-synthetic antibiotic related to penicillin. Prepn: Naylor, **Brit.** pat. **978,299** corresp to **U.S.** pat. **3,239,507** (1965, 1969 to Beecham). Toxicity data: C. Gloxhuber et al., *Arzneimittel-Forsch.* **15**, 322 (1965). Series of articles on chemistry, pharmacology and toxicology: *ibid.* 322-348; Matsuzaki, *Japan J. Antibiot.* **21**, 274, 284 (1968), *C.A.* **70**, 95267z, 95265x (1969).

Sodium salt monohydrate, $C_{19}H_{16}Cl_2N_3NaO_5S.H_2O$, *sodium dicloxacillin monohydrate, P-1011, Brispen, Constaphyl, Dichlor-Stapenor, Diclocil, Dycill, Dynapen, Noxaben, Pathocil, Pen-Sint, Stampen, Syntarpen, Veracillin.* Crystals, dec 222-225°. [α]$_D^{20}$ 127.2° (water). Sol in water, methanol; less sol in butanol; slightly sol in acetone and the usual organic solvents. LD$_{50}$ in mice (g/kg): 0.9 i.v.; in rats (g/kg): 0.63 i.p.; > 5 orally (Gloxhuber).
THERAP CAT: Antibacterial.
THERAP CAT (VET): Antibacterial.

3135. Dicobalt Octacarbonyl. *Di-μ-carbonylhexacarbonyldicobalt;* octacarbonyldicobalt; cobalt tetracarbonyl; cobalt octacarbonyl. $C_8Co_2O_8$; mol wt 341.95. C 28.10%, Co 34.47%, O 37.43%. Co$_2$(CO)$_8$. Prepn: Mond et al., *J. Chem. Soc.* **97**, 798 (1910); Gilmont, Blanchard, *Inorg. Syn.* **2**, 238 (1946); Wender et al., *ibid.* **5**, 190 (1957); Chini et al., *Chim. Ind. (Milan)* **55**, 120 (1973). Crystal structure: Sumner et al., *Acta Cryst.* **17**, 732 (1964). Two isomeric forms exist in soln: Noack, *Spectrochim. Acta* **19**, 1925 (1963); Bor, *ibid.* 2065; Noack, *Helv. Chim. Acta* **47**, 1064, 1555 (1964). Toxicity studies: Kincaid et al., *Arch. Ind. Hyg. Occup. Med.* **10**, 210 (1954); Brief et al., *Arch. Environ. Health* **23**, 373 (1971); Kalekin, *C.A.* **78**, 67751z (1973); Spiridonova, Shabalina, *Gig. Sanit.* **1973**, 73, *C.A.* **78**, 119835b (1973). Reviews of prepn, properties and chemistry of cobalt carbonyls: I. Wender et al., *U.S. Bur. Mines, Bull.* **600**, 83 pp (1960); *Organic Syntheses via Metal Carbonyls* **vol. 1**, I. Wender, P. Pino, Eds. (Interscience, New York, 1968) *passim;* Chalk, Harrod, "Catalysis by Cobalt Carbonyls" in *Advances in Organometallic Chemistry* **6**, 119-170 (1968); Nicholls in *Comprehensive Inorganic Chemistry* **vol. 3**, J. C. Bailar, Jr. et al., Eds. (Pergamon Press, Oxford, 1973) pp 1059-1064.
Orange platelets obtained by vacuum sublimation. d 1.87. mp 51°; dec above 52°. Stable in an atm of H$_2$ and CO; dec on exposure to air. Insol in water. Sol in organic solvents such as ether, alcohol, CS$_2$, naphtha. Slowly attacked by HCl, H$_2$SO$_4$, more rapidly by HNO$_3$, Br$_2$. LD$_{50}$ in mice, rats (mg/kg): 377.7, 753.8 by gavage (Spiridonova, Shabalina).

Cobalt carbonyl hydride, C_4HCoO_4, *cobalt hydrocarbonyl, tetracarbonylhydridocobalt, tetracarbonylhydrocobalt.* Unstable catalytic intermediate; active species in some reactions catalyzed by Co$_2$(CO)$_8$. Prepn: Gilmont, Blanchard, *loc. cit.,* Wender et al., *loc. cit.* See also reviews above. Yellow, toxic, foul-smelling liquid. mp −26°. Dec at room temp to Co$_2$(CO)$_8$. Sparingly sol in water (3 × 10^{-3} moles/l); behaves as a strong acid. Readily oxidized in air.
USE: Catalyst for hydroformylation, hydrogenation, hydrosilation, isomerization, carboxylation, carbonylation and polymerization reactions.

3136. Dicofol. *4-Chloro-α-(4-chlorophenyl)-α-(trichloromethyl)benzenemethanol; 4,4'-dichloro-α-(trichloromethyl)-benzhydrol;* 1,1-bis(p-chlorophenyl)-2,2,2-trichloroethanol; di(p-chlorophenyl)trichloromethylcarbinol; DTMC; ENT-23648; FW-293; Acarin; Kelthane; Mitigan. $C_{14}H_9Cl_5O$; mol wt 370.49. C 45.39%, H 2.45%, Cl 47.85%, O 4.32%. Prepn: Wilson, Craig, **U.S.** pat. **2,812,280;** Wilson, Wolffe, **U.S.** pat. **2,812,362** (both 1957 to Rohm & Haas); Bergmann, Kaluszyner, *J. Org. Chem.* **23**, 1306 (1958); **Brit.** pat. **831,421** (1960 to Metal & Thermit Corp.). Miticidal activity: J. S. Baker, F. B. Maughan, *J. Econ. Entomol.* **49**, 458 (1956). Toxicology: R. B. Smith et al., *Toxicol. Appl. Pharmacol.* **1**, 119 (1959). Metabolism: J. R. Brown et al., *ibid.* **15**, 30 (1969); K. Tabata et al., *Appl. Entomol. Zool.* **14**, 490 (1979). Review of carcinogenic risk: *IARC Monographs* **30**, 87-101 (1983).

Crystals from petr ether, mp 77-78°. uv max (ethanol): 226, 258, 266, 276 nm (4.43, 2.82, 2.85, 2.60). Practically insol in water. Sol in most aliphatic and aromatic solvents. LD$_{50}$ in rats (mg/kg): 1495 orally; 1150 i.p. (Brown).
USE: Acaricide.

3137. Dicrotophos. *Phosphoric acid 3-(dimethylamino)-1-methyl-3-oxo-1-propenyl dimethyl ester; phosphoric acid dimethyl ester, ester with cis-3-hydroxy-N,N-dimethylcrotonamide;* 3-(dimethoxyphosphinyloxy)-N,N-dimethyl-cis-crotonamide; dimethyl 2-dimethylcarbamoyl-1-methylvinyl phosphate; dimethyl 1-dimethylcarbamoyl-1-propen-2-yl phosphate; C-709; ENT-24482; SD-3562; Bidrin; Carbicron; Ektafos. $C_8H_{16}NO_5P$; mol wt 237.19. C 40.51%, H 6.80%, N 5.91%, O 33.73%, P 13.06%. Compound included in general patent claim, prepn of corresp diethyl ester described: Whetstone, Stiles, **U.S.** pat. **2,802,855** (1957 to Shell). Activity: R. A. Corey, *J. Econ. Entomol.* **58**, 112 (1965). Metabolism and degradn: R. I. Beynon et al., *Residue Rev.* **47**, 55 (1973).

Liquid, commercial grade is brown. bp$_{760}$ 400°. d$_{15}^{15}$ 1.216. n$_D^{23}$ 1.468. Misc with water, ethanol, xylene; somewhat sol in kerosene. Dec at 90° after 7 days; at 75° after 31 days. LD$_{50}$ in female, male rats: 16, 21 mg/kg orally; 42, 43 mg/kg dermally, T. B. Gaines, *Toxicol. Appl. Pharmacol.* **14**, 515 (1969).
Caution: Highly toxic. A cholinesterase inhibitor.
USE: Insecticide.

3138. Dicryl. *N-(3,4-Dichlorophenyl)-2-methyl-2-propenamide; 3',4'-dichloro-2-methylacrylanilide;* N-(3,4-dichlorophenyl)methacrylamide; chloranocryl; Niagara 4556. $C_{10}H_9Cl_2NO$; mol wt 230.09. C 52.20%, H 3.94%, Cl 30.82%, N 6.09%, O 6.95%. Post-emergence herbicide. Prepn from 3,4-dichloroaniline with methacryloyl chloride in the presence of triethylamine. Prepd but not claimed: Thompson,

U.S. pat. **3,169,850** (1965). Toxicity study: G. W. Bailey, J. L. White, *Residue Rev.* **10**, 97 (1965).

Crystals from ethanol + petr ether, mp 128°. Practically insol in water. Soluble in acetone, alcohol, isophorone, dimethyl sulfoxide. LD_{50} orally in rats: 3160 mg/kg (Bailey, White).

USE: Herbicide.

3139. Dictamnine. *4-Methoxyfuro[2,3-b]quinoline;* dictamine. $C_{12}H_9NO_2$; mol wt 199.21. C 72.35%, H 4.55%, N 7.03%, O 16.06%. Alkaloid from root of *Dictamnus albus* Linn., *Skimmia repens* Nakai, *Aegle marmelos* Correa, *Zanthoxylum alatum* Roxb., *Rutaceae:* Thoms, *Ber.* **33**, 68 (1923); Asahina *et al., ibid.* **63**, 2045 (1930); Chatterjee, Roy, *J. Indian Chem. Soc.* **36**, 267 (1959); Deb *et al., ibid.* **39**, 493 (1962). Structure: M. F. Brundon, N. J. McCorkindale, *Chem. & Ind. (London)* **1956**, 1091. Synthesis: Tuppy, Bohm, *Angew. Chem.* **68**, 388 (1956); M. Sato *et al., Chem. Pharm. Bull. (Tokyo)* **35**, 1319 (1987).

Prisms from alcohol, ethyl acetate, benzene + ethyl acetate, mp 133°. Sol in hot alcohol, chloroform; slightly sol in ether. Practically insol in water.

Hydrochloride, $C_{12}H_9NO_2 \cdot HCl$, needles from alcohol, dec 170°.

3140. Dicumarol. *3,3'-Methylenebis[4-hydroxy-2H-1-benzopyran-2-one]; 3,3'-methylenebis[4-hydroxycoumarin];* bishydroxycoumarin (rescinded); Dicoumarol; Dicoumarin; Dicumol; Dufalone; Melitoxin. $C_{19}H_{12}O_6$; mol wt 336.30. C 67.86%, H 3.60%, O 28.54%. Originally isolated from spoiled sweet clover (improperly cured Melilotus hay): Link *et al., J. Biol. Chem.* **138**, 21, 513, 529 (1941); **142**, 941 (1942); now prepd synthetically by the action of formaldehyde on 4-hydroxycoumarin (obtained as the Na derivative by the action of sodium metal on methyl acetylsalicylate): Link, *Fed. Proc.* **4**, 176 (1945), and patents (The Wisconsin Alumni Res. Found.). Toxicity data: Rose *et al., Proc. Soc. Exp. Biol. Med.* **50**, 228 (1942).

Minute crystals. Slight pleasant odor. Has a slightly bitter taste. mp 287-293°. Sol in aq alkaline solns, in pyridine and similar organic bases. Slightly sol in benzene and chloroform. Practically insol in water, alcohol, ether. LD_{50} orally in rats: 541.6 mg/kg (Rose).

THERAP CAT: Anticoagulant.

3141. Dicyanine. *1-Ethyl-2-[3-(1-ethyl-2-methyl-4(1H)-quinolinylidene)-1-propenyl]-4-methylquinolinium iodide;* 1-ethyl-2-[3-(1-ethyl-4(1H)-quinaldylidene)propenyl]lepidinium iodide; 1-ethyl-2-[3-(1-ethyl-2-methyl-4(1H)-quinolylidene)propenyl]-4-methylquinolinium iodide; 2',4-dimethyl-1,1'-diethyl-2,4'-carbocyanine iodide. $C_{27}H_{29}IN_2$; mol wt 508.45. C 63.78%, H 5.75%, I 24.96%, N 5.51%. Prepn by J. A. Aeschlimann as reported by Mills, Odams, *J. Chem. Soc.* **125**, 1913 (1924). See also *Beilstein*, 2nd suppl. **vol. 23**, p 284. Has photoconductive properties that increase sharply with a rise in temp and is classed as an O_2-photoconductor: Meier, *Z. Phys. Chem. (Leipzig)* **208**, 325 (1958). Gives

a color reaction with magnesium ions: Babenko, *Zhur. Anal. Khim.* **13**, 496 (1958).

Olive-green crystals with metallic sheen from methanol. Crystals are usually solvated. Drying at 110° expels methanol of crystn. Decomp 244-252°. Absorption max (methanol): 603.5, 655.5 nm ($A_{1cm}^{1\%}$ 63, 218). Sol in methanol: about 2%. Methanol solns appear deep blue when viewed under incident daylight. Soly in water about 0.2%. Aq solns are dichroic and may appear brownish green. Also sol in abs ethanol (0.5%), in ethylene glycol (1.5%), and in Cellosolve (2.0%). Practically insol in benzene, xylene.

USE: In color photography (sensitizer for extreme reds). Diagnostic stain for cytodiagnosis of ruptured fetal membranes.

3142. Dicyanodiamide. *Cyanoguanidine.* $C_2H_4N_4$; mol wt 84.08. C 28.57%, H 4.80%, N 66.63%. Prepared by controlled polymerization of cyanamide in water in the presence of ammonia, alkaline earth hydroxides, or other suitable bases: Grube, Nitsche, *Z. Angew. Chem.* **27**, 368 (1914); Pinck, Hetherington, *Ind. Eng. Chem.* **27**, 834 (1935); Pinck, *Inorg. Syn.* **3**, 43 (1950).

Monoclinic prismatic crystals from water or alcohol, mp 209.5°. d_4^{25} 1.400. Eutectic with cyanamide at 35.6° (15% dicyanodiamide). Specific heat 0.456 at 0-204°. Neutral reaction. Soly in water at 13°: 2.26%, more sol in hot water. Solns above 80° dec slowly, yielding ammonia. Soly in abs ethanol at 13°: 1.26%; in ether: 0.01%. Insol in benzene, chloroform. Sol in liq ammonia.

USE: In the plastics industry (manuf of melamine). In the pharmaceutical industry (barbiturates, guanidine derivs).

3143. Dicyanodiamidine Sulfate. *(Aminoiminomethyl)urea sulfate (2:1);* biuretamidine sulfate; carbamylguanidine sulfate; guanylurea sulfate. $C_4H_{14}N_8O_6S$; mol wt 302.27. C 15.89%, H 4.67%, N 37.07%, O 31.76%, S 10.61%. Prepd by heating dicyanodiamide with H_2SO_4 and H_2O.

Dihydrate, white needles. At 110° loses its water. Sol in about 20 parts cold, 3 parts boiling water; slightly sol in alc.

USE: For detecting and determining Ni and its separation from Co and other metals.

3144. 9-Dicyanomethylene-2,4,7-trinitrofluorene. *(2,-4,7-trinitro-9H-fluoren-9-ylidene)propanedinitrile;* DTF. $C_{16}H_5N_5O_6$; mol wt 363.25. C 52.91%, H 1.39%, N 19.28%, O 26.43%. Prepn from 2,4,7-trinitrofluorenone: Mukherjee, Levasseur, *J. Org. Chem.* **30**, 644 (1965).

Yellow crystals from acetonitrile, mp 266-268°. uv max (dichloromethane): 365 nm (log ε 4.38).

USE: Forms charge-transfer complexes with aromatic hydrocarbons and amines.

3145. Dicyclohexylamine. *N-Cyclohexylcyclohexanamine;* dodecahydrodiphenylamine. $C_{12}H_{23}N$; mol wt 181.32. C 79.49%, H 12.79%, N 7.72%. Prepn by hydrogenation of equimolar amounts of cyclohexanone and cyclohexylamine: C. F. Winans, H. Adkins, *J. Org. Chem.* **54**, 306 (1932); R. M. Robinson, W. C. Braaten, U.S. pat. **3,154,580** (1964 to Abbott). Forms crystalline derivatives with *N*-protected amino acids: E. Klieger *et al., Ann.* **640**, 157 (1961). *Review:* T. S. Carswell, H. L. Morrill, *Ind. Eng. Chem.* **29**, 1265 (1937).

Liquid. Faint fishy odor. d_{25}^{25} 0.9104. mp −0.1°. pKa 10.4. bp_{760} 255.8°; bp_{300} 214.5°; bp_{200} 199.0°; bp_{100} 174.4°; bp_{50} 154.3°; bp_{25} 135.4°; bp_{11} 121°; bp_4 99.3°; $bp_{1.0}$ 83°. n_D^{25} 1.4823. Flash pt = 110°. Strong base. Sparingly sol in water; sol in the usual organic solvents. Miscible with cyclohexylamine. Readily forms adducts with solvents.

Caution: A skin irritant and possible sensitizer.

USE: Industrial solvent; corrosion inhibitor.

3146. Dicyclohexylcarbodiimide. *N,N′-Methanetetraylbiscyclohexanamine;* carbodicyclohexylimide; DCC; DCCI. $C_{13}H_{22}N_2$; mol wt 206.33. C 75.68%, H 10.75%, N 13.58%. Coupling agent in peptide synthesis: Sheehan, Hess, *J. Am. Chem. Soc.* **77**, 1067 (1955); Khorana, *Chem. & Ind. (London)* **1955**, 1087. Prepn: Schmidt *et al., Ber.* **71**, 1933 (1938); Schmidt, Schnegg, U.S. pat. **2,656,383** (1953 to Bayer); Stevens *et al., J. Org. Chem.* **32**, 2895 (1967); Bestmann *et al., Ann.* **718**, 24 (1968).

Cryst mass, mp 35-36°. bp_{11} 154-156°; $bp_{0.5}$ 98-100°.

Caution: Contact allergen.

USE: In the synthesis of peptides.

3147. Dicyclomine Hydrochloride. *[1,1′-Bicyclohexyl]-1-carboxylic acid 2-(diethylamino)ethyl ester hydrochloride;* diethylaminocarbethoxybicyclohexyl hydrochloride; β-diethylaminoethyl-1-cyclohexylhexahydrobenzoate hydrochloride; β-diethylaminoethyl 1-cyclohexylcyclohexanecarboxylate hydrochloride; bis(cyclohexyl)carboxylic acid diethylaminoethyl ester hydrochloride; dicycloverin hydrochloride; Atumin; Benacol; Bentomine; Bentyl Hydrochloride; Bentylol Hydrochloride; Diocyl Hydrochloride; Di-Syntramine; Dyspas; Mamiesan; Merbentyl; Procyclomin; Wyovin Hydrochloride. $C_{19}H_{36}ClNO_2$; mol wt 345.95. C 65.97%, H 10.49%, Cl 10.25%, N 4.05%, O 9.25%. Prepn: C. H. Tilford *et al., J. Am. Chem. Soc.* **69**, 2903 (1947); M. G. Van Campen, C. H. Tilford, U.S. pat. **2,474,796** (1949 to Merrell).

Crystals from butanone, mp 164-166°. Solubility in water about 25%.

THERAP CAT: Anticholinergic.

3148. Didanosine. *2′,3′-Dideoxyinosine;* dideoxyinosine; ddI; ddIno; NSC-612049; Videx. $C_{10}H_{12}N_4O_3$; mol wt 236.23. C 50.84%, H 5.12%, N 23.72%, O 20.32%. Hypoxanthine nucleoside with antiviral activity; metabolic product of dideoxyadenosine, *q.v.* Enzymatic prepn from dideoxyadenosine: W. Plunkett, S. S. Cohen, *Cancer Res.* **35**, 1547 (1975). Synthesis: G. W. Koszalka, T. A. Krenitsky, Eur. pat. *Appl.* **206,497** (1986 to Wellcome Found.); R. R. Webb *et al., Nucleosides Nucleotides* **7**, 147 (1988). *In vitro* inhibition of HIV-I: H. Mitsuya, S. Broder, *Proc. Nat. Acad. Sci. USA* **83**, 1911 (1986); of HIV-II: D. D. Richman, *Antimicrob. Ag. Chemother.* **31**, 1879 (1987). Antiretroviral spectrum *in vitro:* J. E. Dahlberg *et al., Proc. Nat. Acad. Sci. USA* **84**, 2469 (1987). Cellular pharmacology in cloned human T-cells: G. Ahluwalia *et al., Biochem. Pharmacol.* **36**, 3797 (1987). HPLC determn in human plasma: G. Ray, E. Murrill, *Anal. Letters* **20**, 1815 (1987). Clinical evaluations in AIDS: J. S. Lambert *et al., N. Engl. J. Med.* **322**, 1333 (1990); T. P. Cooley *et al., ibid.* 1340.

White solid, mp 160-163°. uv max 248 nm (pH 2); 254 nm (pH 12).

THERAP CAT: Antiviral.

3149. Didecyldimethylammonium Chloride. *N-Decyl-N,N-dimethyl-1-decanaminium chloride;* dimethyldidecylammonium chloride; Bardac 2250/2280; BTC-1010; Dodigen 1881; Querton 210CL. $C_{22}H_{48}ClN$; mol wt 362.08. C 72.98%, H 13.36%, Cl 9.79%, N 3.87%. Quaternary alkyl ammonium compound with fungicidal and bactericidal activity. Prepn: A. W. Ralston *et al., J. Org. Chem.* **13**, 186 (1948). Conductivity of aqueous soln: *eidem, J. Am. Chem. Soc.* **70**, 977 (1948). Mothproofing agent: A. M. Schwartz, C. A. Rader, U.S. pat. **3,746,767** (1971 to USDA). In carpet dyeing: F. F. Bartsch, R. Feigin, U.S. pat. **3,758,269** (1971 to Sybron). Germicidal activity: R. J. Wright, Ger. pat. **2,810,998;** *eidem,* U.S. pat. **4,272,395** (1978, 1981 both to Unilever). Ion-pair extracting agent in carbapenem isoln: J. D. Hood *et al., J. Antibiot.* **32**, 295 (1979). In frothflotation recovery of metals: V. Petrovich, U.S. pat. **4,225,428** (1980). Eliminates ice-nucleating effect of bacteria in plants: E. A. Youngman, R. C. Schnell, Eur. pat. *Appl.* **37,593;** *eidem,* U.S. pat. **4,311,517** (1981, 1982 both to Shell). Breaks oil-in-water emulsions created during oil-spill recovery procedures: J. Newcombe, U.S. pat. **4,374,734** (1983 to Cities Service). Fungidical wood preservative: A. F. Preston: *J. Am. Oil Chem. Soc.* **60**, 567 (1983).

Extremely hygroscopic. Sol in acetone, extremely sol in benzene. Insol in hexane.

Caution: Causes eye damage and skin irritation. Harmful if swallowed.

USE: General purpose disinfectant, sanitizer; mildew preventative in commercial laundries; water treatment in cooling towers and oil field flood waters; wood preservative.

3150. Didemnins. Biologically active depsipeptides extracted from a Caribbean tunicate (sea squirt) family *Didemnidae*, genus *Trididemnum*. First marine natural product to enter clinical trials as antineoplastic. The three components, didemnins A, B and C are weakly basic compds with antiviral, antitumor activity. Didemnin A is the most abundant, whereas didemnin B is generally the most active. Isoln and bioactivity: K. L. Rinehart, Jr. *et al., Science* **212**, 933 (1981). Extraction and purification: K. L. Rinehart, Jr., **Eur. pat. Appl. 48,149;** *idem,* **U.S. pat. 4,493,796** (1982, 1985 both to Univ. Illinois). Structure determn: K. L. Rinehart, Jr. *et al., J. Am. Chem. Soc.* **103**, 1857 (1981). Revised structure and total synthesis: K. L. Rinehart, Jr. *et al., ibid.* **109**, 6846 (1987). *See also:* U. Schmidt *et al., Tetrahedron Letters* **29**, 3057 (1988); *eidem, ibid.* 4407. Crystal structure of B: M. B. Hossain *et al., Proc. Nat. Acad. Sci. USA* **85**, 4118 (1988). Efficient total synthesis of A and B: Y. Hamada *et al., J. Am. Chem. Soc.* **111**, 669 (1989). Mechanism of action study: L. H. Li *et al., Cancer Letters* **23**, 279 (1984). *In vitro* and *in vivo* immunosuppressive activity of B: D. W. Montgomery, C. F. Zukowski, *Transplantation* **40**, 49 (1985). HPLC determn in biological fluids: J. N. Hartshorn *et al., J. Liq. Chromatog.* **9**, 1489 (1986). Clinical pharmacology, pharmacokinetics: F. A. Dorr *et al., Eur. J. Cancer Clin. Oncol.* **24**, 1699 (1988). Review of bioactivity of A and B: K. L. Rinehart, Jr. *et al., Fed. Proc.* **42**, 87-90 (1983); of activity and toxicology of B: H. G. Chun *et al., Invest. New Drugs* **4**, 279-284 (1986); of biological activity, structure and synthesis: W.-R. Li, M. M. Joullié, *Stud. Nat. Prod. Chem.* **10**, 241-302 (1992).

Didemnin A: R = H

Didemnin B: R = CH₃CHOHCO-N⟨pyrrolidine⟩-CO-

Didemnin C: R = CH₃CHOHCO-

Didemnin A, $C_{49}H_{78}N_6O_{12}$. Greenish-white solid. Sol in methanol, ethanol, dioxane, chloroform. Insol in water.

Didemnin B, $C_{57}H_{89}N_7O_{15}$, *NSC-325319.* Yellow-white amorphous solid. Sol in methanol, ethanol, dioxane, ethyl acetate, chloroform. Insol in water.

Didemnin C, $C_{52}H_{82}N_6O_{14}$, oil. Sol in methanol, ethanol, isopropanol, dioxane, ethyl acetate and chloroform. Sparingly sol in toluene. Insol in water.

3151. Dideoxyadenosine. *2′,3′-Dideoxyadenosine;* 6-amino-9-(2′,3′-dideoxy-β-D-*glycero*-pentofuranosyl)purine; DDA; ddAdo. $C_{10}H_{13}N_5O_2$; mol wt 235.25. C 51.06%, H 5.57%, N 29.77%, O 13.60%. Purine nucleoside analog with antiviral (retrovirus) activity. Prepn: M. J. Robins, R. K. Robins, *J. Am. Chem. Soc.* **86**, 3585 (1964); G. L. Tong *et al., J. Org. Chem.* **30**, 2854 (1965); J. R. McCarthy *et al., J.*

Am. Chem. Soc. **88**, 1549 (1966). *In vitro* inhibition of bacterial DNA synthesis: L. Toji, S. S. Cohen, *Proc. Nat. Acad. Sci. USA* **63**, 871 (1969). Antibacterial activity in mice: G. Beskid *et al., Antimicrob. Ag. Chemother.* **19**, 424 (1981). Inhibition of *in vitro* infectivity of HIV-1 virus: H. Mitsuya, S. Broder, *Proc. Nat. Acad. Sci. USA* **83**, 1911 (1986). HPLC determn in biological fluids: P. A. Blau *et al., J. Chromatog.* **420**, 1 (1987). Metabolism by cloned human T-cells: D. A. Cooney *et al., Biochem. Pharmacol.* **36**, 1765 (1987).

Crystals from ethanol, mp 184-186°. $[\alpha]_D^{25}$ −25.2° (c = 1.01 in water). uv max in methanol: 259.5 nm (ε 14800).

THERAP CAT: Antiviral.

3152. Dieldrin. *(1aα,2β,2aα,3β,6β,6aα,7β,7aα)-3,4,5,-6,9,9-Hexachloro-1a,2,2a,3,6,6a,7,7a-octahydro-2,7:3,6-dimethanonaphth[2,3-b]oxirene; 1,2,3,4,10,10-hexachloro-6,7-epoxy-1,4,4a,5,6,7,8,8a-octahydro-endo,exo-1,4:5,8-dimethanonaphthalene;* insecticide no. 497; HEOD; ENT-16225; compd 497; Octalox. $C_{12}H_8Cl_6O$; mol wt 380.91. C 37.84%, H 2.12%, Cl 55.84%, O 4.20%. Activity: C. W. Kearns *et al., J. Econ. Entomol.* **42**, 127 (1949). Prepn from aldrin, *q.v.:* Soloway, **U.S. pat. 2,676,131** (1954 to Shell); Payne, Smith, **U.S pat. 2,776,301** (1957 to Shell). Synthesis: Korte, Rechmeier, *Ann.* **656**, 131 (1962). Metabolism: M. K. Baldwin *et al., Chem. & Ind. (London)* **1970**, 595; C. T. Bedford, D. H. Hutson, *ibid.* **1976**, 440. Toxicity data: Gaines, *Toxicol. Appl. Pharmacol.* **14**, 515 (1969). A stereoisomer of endrin, *q.v.*

Crystals, mp 176-177°. Vapor press at 20°: 3.1 × 10⁻⁶ mm Hg. Practically insol in water. Moderately sol in common organic solvents except aliphatic petr solvents and methyl alcohol. Stable in org and inorg alkalies and acids commonly used in agriculture. Affected by strong mineral acids. LD_{50} orally in rats: 46 mg/kg (Gaines).

Caution: Readily absorbed through skin. Causes malaise, headache, nausea, vomiting, dizziness, tremors, clonic and tonic convulsions, coma, respiratory failure, *cf. Clinical Toxicology of Commercial Products,* R. E. Gosselin *et al.,* Eds. (Williams & Wilkins, Baltimore, 4th ed., 1976) Section III, pp 121-124.

USE: Formerly as insecticide; manuf and use has been discontinued in the U.S.

3153. Dienestrol. *4,4′-(1,2-Diethylidene-1,2-ethanediyl)bisphenol; 4,4′-(diethylideneethylene)diphenol;* 3,4-bis(p-hydroxyphenyl)-2,4-hexadiene; 4,4′-dihydroxy-γ,δ-diphenyl-β,δ-hexadiene; di(p-oxyphenyl)-2,4-hexadiene; dienoestrol; estrodienol; Cycladiene; Dienol; Dinovex; DV; Estroral; Gynefollin; Hormofemin; Oestrasid; Oestrodiene; Oestroral; Restrol; Retalon; Synestrol. $C_{18}H_{18}O_2$; mol wt 266.34. C 81.17%, H 6.81%, O 12.01%. Synthesis: Dodds *et al., Proc. Roy. Soc.* **127B**, 162 (1939); Hobday, Short, *J. Chem. Soc.* **1943**, 609; Short, Hobday, **U.S. pat. 2,464,203** (1949 to Boots); Adler, **U.S. pat. 2,465,505** (1949 to Hoffmann-La Roche). Configuration: Koch, *Nature* **161**, 309 (1948); Lane, Spialter, *J. Am. Chem. Soc.* **73**, 4408 (1951).

H3C
OH
HO
CH3

(trans-trans)-form

Minute needles from dil alcohol, mp 227-228°. Sublimes at 130° and 1 mm Hg. The sublimate mp 231-234°. Freely sol in alcohol, methanol, ether, acetone, propylene glycol; sol in chloroform, aq solns of alkali hydroxides; sol in vegetable oils after warming, but crystallizes out on standing. Practically insol in water, dil acids.

Diacetate, $C_{22}H_{22}O_4$, *Lipamone, Retalon-Oral.* Prisms from alc, mp 119-120°.

Note: Isodienestrol, mp 179°, has no estrogenic activity.

THERAP CAT: Estrogen.

THERAP CAT (VET): Estrogenic hormone therapy.

3154. Dienochlor. *1,1′,2,2′,3,3′,4,4′,5,5′-Decachlorobi-2,4-cyclopentadien-1-yl;* bis(pentachloro-2,4-cylopentadien-1-yl); decachlor; HRS-16; Pentac. $C_{10}Cl_{10}$; mol wt 474.64. C 25.31%, Cl 74.69%. Acaricide to control mites on ornamental plants. Prepn from hexachlorocyclopentadiene, characterization: E. T. McBee *et al., J. Am. Chem. Soc.* **77,** 4375 (1955). Prepn and rubber vulcanizing props: E. C. Ladd, U.S. pat. **2,732,409** (1956 to U.S. Rubber). Prepn and insecticidal props: J. T. Rucker, U.S. pat. **2,934,470** (1960 to Hooker). Effectiveness as miticide: W. W. Allen *et al., J. Econ. Entomol.* **57,** 187 (1964). Crystal structure: G. Smith *et al., J. Chem. Soc. Perkin Trans. II* **1976,** 796. Photodegradation: G. B. Quistad, K. M. Mulholland, *J. Agr. Food Chem.* **31,** 621 (1983). Effect on oogenesis of Mexican bean beetle: P. R. Hughes, M. A. Penton, *J. Econ. Entomol.* **76,** 1156 (1983).

Cl   Cl
Cl
Cl   Cl
Cl   Cl
Cl   Cl   Cl
Cl

Yellow prisms from petr ether, mp 121.5-122°. uv max: 330 nm (ε 2950). Stable to alkali.

USE: Miticide.

3155. Diethadione. *5,5-Diethyldihydro-2H-1,3-oxazine-2,4(3H)-dione;* 5,5-diethyl-2,3,5,6-tetrahydro-4*H*-1,3-oxazine-2,4-dione; Dioxone; Ledosten; Persisten; Tocèn. $C_8H_{13}NO_3$; mol wt 171.20. C 56.13%, H 7.65%, N 8.18%, O 28.04%. Prepn: Testa *et al., J. Org. Chem.* **24,** 1928 (1959); Brit. pat. **855,244** (1960 to Lepetit, SpA). Toxicity study: Maffii *et al., J. Pharm. Pharmacol.* **13,** 244 (1961).

O   O
O
H3C   NH
CH3   O

Crystals from ether, mp 97-98°. LD_{50} in mice, rats (mg/kg): 52, 31.6 i.p.; 130, 70.5 orally (Maffii).

THERAP CAT: Analeptic.

3156. Diethanolamine. *2,2′-Iminobisethanol; 2,2′-iminodiethanol;* diethylolamine; bis(hydroxyethyl)amine; 2,2′-dihydroxydiethylamine. $C_4H_{11}NO_2$; mol wt 105.14. C 45.70%, H 10.55%, N 13.32%, O 30.44%. $(HOCH_2CH_2)_2$-NH. Produced along with mono- and triethanolamine by ammonolysis of ethylene oxide. *See refs under* Ethanolamine. Toxicity study: H. F. Smyth *et al., J. Ind. Hyg. Toxicol.* **23,** 259 (1941).

Deliquescent prisms, mp 28°. Usually offered as a viscous liquid. Mild ammoniacal odor. d_4^{30} 1.0881; d_4^{60} 1.0693. One

U.S. gallon weighs 9.09 lbs at 30°. Viscosity at 30°: 351.9 cp; at 60°: 53.85 cp. bp_{760} 268.8°; bp_{150} 217°; $bp_{0.01}$ 20°. Strong base: pH of 0.1*N* aq soln: 11.0. n_D^{30} 1.4753. Dipole moment 2.81. Flash pt 300°F. Miscible with water, methanol, acetone. Soly at 25° in benzene: 4.2%, in ether: 0.8%, in carbon tetrachloride: <0.1%, in *n*-heptane: <0.1%. LD_{50} orally in rats: 12.76 g/kg (Smyth).

USE: To scrub gases as indicated under ethanolamine. Diethanolamine can be used with cracking gases and coal or oil gases which contain carbonyl sulfide that would react with monoethanolamine. As rubber chemicals intermediate. In the manuf of surface active agents used in textile specialties, herbicides, petr demulsifiers. As emulsifier and dispersing agent in various agricultural chemicals, cosmetics, and pharmaceuticals. In the production of lubricants for the textile industry. As humectant and softening agent. In organic syntheses.

3157. Diethazine. *N,N-Diethyl-10H-phenothiazine-10-ethanamine; 10-(2-diethylaminoethyl)phenothiazine; N-*(di-ethylaminoethyl)thiodiphenylamine; *N-*(2′-diethylaminoethyl)dibenzoparathiazine; RP-2987; Deparkin; Dinezin; Dolisina; Eazaminum; Ethylemin; Parkazin. $C_{18}H_{22}N_2S$; mol wt 298.45. C 72.44%, H 7.43%, N 9.39%, S 10.74%. Prepd by reacting 10-phenothiazineethyl chloride with diethylamine in presence of copper powder or by reacting diethylaminoethyl chloride with phenothiazine: Charpentier, *Compt. Rend.* **225,** 306 (1947); Huttrer, *Enzymologia* **12,** 293 (1948); Charpentier, U.S. pat. **2,530,451** (1950 to Rhône-Poulenc); Berg, Ashley, U.S. pat. **2,607,773** (1952 to Rhône-Poulenc). Toxicity study: Bovet *et al., Therapie* **2,** 115 (1947).

H3C   N   CH3
N
S

Oily liquid. bp_{4-5} 195-208°; $bp_{0.4-0.5}$ 167-175°.

Hydrochloride, $C_{18}H_{22}N_2S.HCl$, *Antipar, Aparkazin, Casantin, Diparcol, Latibon, Thiantan, Thiontan.* Crystals, mp 184-186°. Burning taste, producing a temporary numbness of the tongue. One part dissolves in about 5 parts water, 6 parts ethanol, 5 parts chloroform. Practically insol in ether. pH of 10% aq soln 5.0-5.3. LD_{50} orally in mice: 450 mg/kg (Bovet).

THERAP CAT: Anticholinergic, antiparkinsonian.

3158. Diethofencarb. *(3,4-Diethoxyphenyl)carbamic acid 1-methylethyl ester;* isopropyl 3,4-diethoxycarbanilate; isopropyl *N-*(3,4-diethoxyphenyl)carbamate; S-165; S-1605; S-32165; Powmil. $C_{14}H_{21}NO_4$; mol wt 267.33. C 62.90%, H 7.92%, N 5.24%, O 23.94%. Prepn: H. Noguchi *et al., Eur. pat. Appl.* **78,663;** *eidem,* U.S. pat. **4,608,385** (1983, 1986 both to Sumitomo). *In vitro* and greenhouse antifungal activities: M. L. Gullino *et al., Tag.-Ber. Akad. Landwirtsch. Wiss.* **253,** 165 (1987); M. L. Gullino, A. Garibaldi, *Med. Fac. Landbouww. Rijksuniv. Gent,* **52,** 895 (1987). Structure-activity study: J. Takahashi *et al., J. Chromatog.* **436,** 316 (1988). Mechanism of action study: M. Fujimura *et al., ACS Symposium Series* **421,** 224-236 (1990). Metabolism in rats: K. Shiba *et al., J. Pestic. Sci.* **15,** 395 (1990). *Review:* M. Fujimura *et al., Japan. Pestic. Info.* **57,** 7-11 (1990).

H3C   O
H
N   O   CH3
H3C   O   O   CH3

Crystalline solid, mp 101.3°. Vapor pressure at 20°: 6.3 × 10⁻⁵, at 25°: 1.1 × 10⁻⁴ mm Hg. Soluble in most organic solvents. Soly at 25 ±2° in water: 26.6 ±0.3 mg/l. LD_{50} in male and female rats (mg/kg): >5000, >5000 orally; >5000, >5000 dermally. LC_{50} (48 hrs) in carp: 13.5

mg/l; (3 hrs) in *Daphnia pulex*: > 10 mg/l (Fujimura, review).

USE: Agricultural fungicide.

3159. Diethylacetic Acid. *2-Ethylbutanoic acid.* $C_6H_{12}O_2$; mol wt 116.16. C 62.04%, H 10.41%, O 27.55%. $(CH_3CH_2)_2CHCOOH$.

Colorless liq; odor somewhat resembling that of caproic acid. d_4^{18} 0.920. bp 194-195°, also stated as 190°. mp ~ −15°. n_D^{10} 1.4179. Slightly sol in water; freely sol in alcohol or ether.

3160. Diethylamine. *N-Ethylethanamine.* $C_4H_{11}N$; mol wt 73.14. C 65.69%, H 15.16%, N 19.15%. $(C_2H_5)_2$-NH. Prepn from ethyl iodide and NH_3 with separation of mono-, di-, and triethylamines formed: Watt, Otto, *J. Am. Chem. Soc.* **69**, 836 (1947). Manuf from ethanol and NH_3, obtained along with mono- and triethylamines: Davies *et al.*, U.S. pat. **2,609,394** (1952 to ICI); Lemon, Myerly, U.S. pat. **3,022,349** (1962 to Union Carbide). Toxicity data: Smyth *et al.*, *Arch. Ind. Hyg. Occup. Med.* **4**, 119 (1951).

Flammable, strongly alkaline liq, bp 55.5°. mp −50°. d_4^{20} 0.7074. n_D^{20} 1.3864. Flash pt < 20°F. Forms a hydrate, $B_2.H_2O$, mp −19°. Miscible with water, alc. It is usually supplied as a soln. *Keep well closed.* LD_{50} orally in rats: 540 mg/kg (Smyth).

Hydrochloride, $C_4H_{11}N.HCl$, crystals from alcohol + ether, mp 226°. Hygroscopic. bp 320-330°. d_4^1 1.048. Sol in water, alcohol, chloroform. Practically insol in ether.

Caution: Potential symptoms of overexposure are eye, skin and respiratory system irritation. *See NIOSH Pocket Guide to Chemical Hazards* (DHHS/NIOSH 90-117, 1990) p 90. *See also* Patty's *Industrial Hygiene and Toxicology* vol. **2B**, G. D. Clayton, F. E. Clayton, Eds. (Wiley-Interscience, New York, 3rd ed., 1981) p 3149.

USE: In the rubber and petroleum industry; in flotation agents; in resins; dyes, pharmaceuticals.

3161. 2-Diethylaminoethanol. β-Diethylaminoethyl alcohol; 2-hydroxytriethylamine. $C_6H_{15}NO$; mol wt 117.19. C 61.49%, H 12.90%, N 11.95%, O 13.65%. $(C_2H_5)_2NCH_2$-CH_2OH. Prepd by the action of ethylene chlorohydrin on diethylamine: Ladenburg, *Ber.* **14**, 1878 (1881); Soderman, Johnson, *J. Am. Chem. Soc.* **47**, 1394 (1925); Hartman, *Org. Syn.* **coll. vol. II**, 183 (1943); by the action of ethylene oxide on diethylamine: Horne, Shriner, *J. Am. Chem. Soc.* **54**, 2928 (1932); Headlee *et al.*, *ibid.* **55**, 1066 (1933).

Liquid. d^{25} 0.8800. bp_{760} 163°; bp_{80} 100°; bp_{10} 55°. n_D^{25} 1.4389. Sol in water, alcohol, ether, benzene.

p-Nitrophenylurethan, crystals, mp 59-60°.

Caution: Potential symptoms of overexposure are nausea, vomiting; irritation of respiratory system, skin and eyes. *See NIOSH Pocket Guide to Chemical Hazards* (DHHS/NIOSH 90-117, 1990) p 92.

3162. Diethylaniline. *N,N-Diethylbenzenamine.* $C_{10}H_{15}N$; mol wt 149.24. C 80.48%, H 10.13%, N 9.39%. Prepd by ethylation of aniline: Whitman, U.S. pat. **2,501,-556** (1950 to du Pont); Voltz, *J. Org. Chem.* **22**, 48 (1957); Closson *et al.*, *ibid.* **22**, 646 (1957); Horyna, Cerny, *Coll. Czech. Chem. Commun.* **21**, 906 (1956); from bromobenzene, sodium amide + diethylamine: Bunnett, Brotherton, *J. Org. Chem.* **22**, 832 (1957).

Colorless to yellow liq. d_4^{25} 0.9302. bp 215-216°; bp_3 62-66°. mp −38°. n_D^{24} 1.5394. Volatile with steam. uv max (isooctane): 303, 259 nm ($\epsilon \times 10^{-3}$ 2.37, 16.7). Slightly sol in alcohol, chloroform, ether. One gram dissolves in 70 ml water at 12°.

USE: As dyestuff intermediate, in organic syntheses.

3163. *N,N*-Diethylbenzhydrylamine. *N,N-Diethyl-α-phenylbenzenemethanamine;* diethylaminodiphenylmethane. $C_{17}H_{21}N$; mol wt 239.36. C 85.31%, H 8.84%, N 5.85%.

Prepd by addn of an excess of phenylmagnesium bromide to *N,N*-diethylformamide and treating the reaction mixture with ammonium chloride: Maxim, Mavrodineanu, *Bull. Soc. Chim. France* [5] **3**, 1084 (1936). Proposed as an antihistaminic: Capraro, *C.A.* **41**, 6989 (1947).

Viscous liquid, bp_{17} 170°. Solidifies on deep cooling, then mp 56° (Maxim, Mavrodineanu); mp 58-59° [Sommelet, *Compt. Rend.* **175**, 1149 (1922)]; mp 61° [Titov, *C.A.* **43**, 4217 (1949)].

USE: In the detection of nitrates.

3164. Diethylbromoacetamide. *2-Bromo-2-ethylbutanamide; 2-bromo-2,2-diethylacetamide;* carbromide. C_6H_{12}-$BrNO$; mol wt 194.07. C 37.13%, H 6.23%, Br 41.17%, N 7.22%, O 8.24%. $(C_2H_5)_2CBrCONH_2$. Prepn: **Ger.** pat. **158,220** (1905 to Kalle).

Crystalline powder; camphor odor; bitter, cooling taste. mp 67°. Sol in 115 parts cold water; sol in alc, benzene, ether, oils. Freely sol in warm water.

Note: Formerly marketed as *Neuronal.*

Caution: This substance may be habit forming and is listed in the U.S. Code of Federal Regulations, Title 21 Part 329.1 (1995).

THERAP CAT: Sedative, hypnotic.

3165. Diethylcarbamazine. *N,N-Diethyl-4-methyl-1-piperazinecarboxamide;* 1-diethylcarbamoyl-4-methylpiperazine; carbamazine; 1-diethylcarbamyl-4-methylpiperazine; 84L; RP-3799; Carbilazine; Caricide; Cypip; Ethodryl; Notézine; Spatonin. $C_{10}H_{21}N_3O$; mol wt 199.30. C 60.27%, H 10.62%, N 21.08%, O 8.03%. Prepn: Kushner *et al.*, *J. Org. Chem.* **13**, 151 (1948); Kushner, Brancone, U.S. pats. **2,467,893; 2,467,895** (1949 to Am. Cyanamid). Pharmacology and toxicology: Harned *et al.*, *Ann. N.Y. Acad. Sci.* **50**, 141 (1948). GC determn in blood: S. Nene *et al.*, *J. Chromatog.* **308**, 334 (1984). Mode of action: J.-Y. Cesbron *et al.*, *Nature* **325**, 533 (1987). Review of pharmacology, mechanisms of action and clinical uses: C. D. MacKenzie, M. A. Kron, *Trop. Dis. Bull.* **82**(10), R1-R37 (1985).

Crystals, mp 47-49°. bp_3 108.5-111°.

Hydrochloride, $C_{10}H_{21}N_3O.HCl$, crystals from acetone, mp 156.5-157°. Very sol in water; sol in chloroform, dioxane.

Citrate (hydrogen citrate), $C_{10}H_{21}N_3O.C_6H_8O_7$, *Banocide, Dec, Dirocide, Filaribits, Filazine, Franocide, Hetrazan, Loxuran, Longicid.* Crystals, mp 141-143°. Freely sol in water (> 75% at 20°). Sparingly sol in cold alc; freely sol in hot alc. Practically insol in benzene, acetone, ether, chloroform. LD_{50} orally in rats: 1.38 g/kg (Harned).

Phosphate, *Ditrazin.* Crystals, freely sol in water.

THERAP CAT: Anthelmintic (Nematodes).

THERAP CAT (VET): Citrate as an anthelmintic.

3166. *N,N*-Diethylcarbanilide. *N,N'-Diethyl-N,N'-di-phenylurea; sym*-diethyldiphenylurea; centralite I; ethyl centralite. $C_{17}H_{20}N_2O$; mol wt 268.36. C 76.09%, H 7.51%, N 10.44%, O 5.96%. $C_2H_5(C_6H_5)NCON(C_6H_5)C_2H_5$. Prepn from ethylaniline and phosgene: Michler, *Ber.* **9**, 712 (1876). Alternate synthesis: N. R. Ayyangar *et al.*, *Chem. & Ind. (London)* **1988**, 599.

Crystals from alc, mp 79°. Insol in water.
USE: As stabilizer of smokeless explosives.

3167. Diethyl Carbitol®. *1,1'-Oxybis(2-ethoxy)ethane;* diethylene glycol diethyl ether. $C_8H_{18}O_3$; mol wt 162.23. C 59.23%, H 11.18%, O 29.59%. $(C_2H_5OCH_2CH_2)_2O$. Toxicity data: H. F. Smyth *et al., J. Ind. Hyg. Toxicol.* **23**, 259 (1941).
Liquid. d_4^{20} 0.907. bp 188°. Very sol in water, alcohol, and other organic solvents. LD_{50} orally in rats: 4.97 g/kg (Smyth).
USE: Solvent; high boiling reaction medium.

3168. Diethylene Glycol. *2,2'-Oxybisethanol;* 2,2'-oxydiethanol. $C_4H_{10}O_3$; mol wt 106.12. C 45.27%, H 9.50%, O 45.23%. $HOCH_2CH_2OCH_2CH_2OH$. Manuf from ethylene oxide and glycol: Faith, Keyes & Clark's *Industrial Chemicals,* F. A. Lowenheim, M. K. Moran, Eds. (Wiley-Interscience, New York, 4th ed., 1975) pp 397-402. Toxicity study: H. F. Smyth *et al., J. Ind. Hyg. Toxicol.* **23**, 259 (1941).
Colorless, hygroscopic, practically odorless liq; sharply sweetish taste. d_{20}^{20} 1.118. Solidifies at −10.45° (when pure). mp −6.5°. bp 244-245°. n_D^{20} 1.4475. Flash pt, open cup: 290°F (143°C). Misc with water, alc, ether, acetone, ethylene glycol. Practically insol in benzene, carbon tetrachloride. LD_{50} in rats, guinea pigs (g/kg): 20.76, 13.21 orally (Smyth).
Caution: Symptoms on ingestion similar to ethylene glycol, *q.v.* Fatal poisoning resulted from its use as a solvent in an elixir. *See* E. Browning, *Toxicity and Metabolism of Industrial Solvents* (Elsevier, New York, 1965) pp 624-628, 686-690; Patty's *Industrial Hygiene and Toxicology* vol. **2C**, G. D. Clayton, F. E. Clayton, Eds. (Wiley-Interscience, New York, 3rd ed., 1982) pp 3832-3838.
USE: In antifreeze soln for sprinkler systems, water seals for gas tanks, etc. (water with 40% diethylene glycol freezes at −18°; with 50% at −28°); as lubricating and finishing agent for wool, worsted, cotton, rayon, and silk; as solvent for vat dyes; in composition corks, glues, gelatin, casein, and pastes to prevent drying out.

3169. Diethylene Glycol Monolaurate. *Dodecanoic acid 2-(2-hydroxyethoxy)ethyl ester;* diethylene glycol laurate; diglycol laurate; Glaurin. $C_{16}H_{32}O_4$; mol wt 288.43. C 66.63%, H 11.18%, O 22.19%. Prepared by controlled esterification of diethylene glycol with lauric acid.

Oily liq. d_4^{25} 0.9572; d_{20}^{20} 0.963-0.968. mp 17-18°. bp ~270° (some dec). May be heated to 250° without decompn. Practically insol in water. Sol in methanol, ethanol, benzene, toluene, chlorinated hydrocarbons, acetone, ethyl acetate, cottonseed oil. Slightly sol in petr naphtha.
USE: Dispersing agent, emulsifier, plasticizer.

3170. Diethyl Ketone. *3-Pentanone;* dimethylacetone; propione; methacetone. $C_5H_{10}O$; mol wt 86.13. C 69.72%, H 11.70%, O 18.58%. $CH_3CH_2COCH_2CH_3$.
Liquid, acetone odor. d_4^{19} 0.816. bp 101.5°. mp −42°. n_D^{25} 1.3905. Sol in about 25 parts water; miscible with alcohol, ether. LD_{50} orally in rats: 2.1 g/kg, Smyth *et al., Arch. Ind. Hyg. Occup. Med.* **10**, 61 (1954).

3171. Diethylmagnesium. Magnesium diethyl. C_4H_{10}-Mg; mol wt 82.43. C 58.29%, H 12.23%, Mg 29.49%. $Mg(C_2H_5)_2$. Prepd by the action of magnesium metal on mercury diethyl in ether: Schlenk, Jr., *Ber.* **64**, 734, 736 (1931).
Solvated crystals from ether, $MgEt_2 \cdot Et_2O$, plates, rods. mp 0°. Liquid at room temp. Spontaneously flammable in air. Violent explosion on contact with water. Loses its ether of crystn on heating *in vacuo.* Not volatile in high vac up to 250° when decompn sets in. Sol in ether. Dec by alc and ammonia. Will glow and catch fire even in CO_2. Must be handled in high vac, or under dry nitrogen or hydrogen.

3172. Diethyl Maleate. *(Z)-2-Butenedioic acid, diethyl ester; maleic acid, diethyl ester;* ethyl maleate. $C_8H_{12}O_4$; mol wt 172.18. C 55.81%, H 7.02%, O 37.17%. Prepn from

maleic acid and ethanol: V. M. Mitchovitch, *Bull. Soc. Chim. France* **4**, 1667 (1937). Physical properties: G. H. Jeffrey, A. I. Vogel, *J. Chem. Soc.* **1948**, 658. Copolymerization with styrene: F. M. Lewis *et al., J. Am. Chem. Soc.* **70**, 1519, 1529 (1948). In Diels-Alder reactions: N. L. Bauld *et al., Tetrahedron Letters* **1972**, 2443. In Michael additions: G. Arsenault *et al., Chem. Commun.* **1983**, 437. Conjugation with glutathione, *q.v.:* E. Boyland, L. F. Chasseaud, *Biochem. J.* **104**, 95 (1967); *eidem, ibid.* **109**, 651 (1968). Toxicity: H. F. Smyth *et al., J. Ind. Hyg. Toxicol.* **31**, 60 (1949).

Oil, bp_{758} 219.5°. d_4^{20} 1.0674. n_D^{20} 1.4402. Insol in water. Soluble in ethanol, ether. LD_{50} orally in rats: 0.30 g/kg (Smyth).
USE: In organic synthesis.

3173. Diethylmalonic Acid. *Diethylpropanedioic acid;* 3,3-pentanedicarboxylic acid. $C_7H_{12}O_4$; mol wt 160.17. C 52.49%, H 7.55%, O 39.96%. Prepn: Daimler, *Ann.* **249**, 173 (1888); Speck, *J. Am. Chem. Soc.* **74**, 2876 (1952). Spectral studies: J. L. Delarbre *et al., J. Raman. Spectrosc.* **16**, 11 (1985). Crystal structure: A. Dubourg *et al., Acta Crystallogr.* **C44**, 1987 (1988).

Crystals, mp 127°. Dec at 170-180° with liberation of CO_2 and formation of diethylacetic acid. Very sol in water; freely sol in alcohol, ether, slightly in chloroform.

3174. Diethyl Oxalate. *Ethanedioic acid diethyl ester;* ethyl oxalate; diethyl ethanedioate; oxalic acid diethyl ester. $C_6H_{10}O_4$; mol wt 146.14. C 49.31%, H 6.90%, O 43.79%. $C_2H_5OOCCOOC_2H_5$. Prepn: Clarke, Davis, *Org. Syn.;* coll. vol. **I**, 256 (2nd ed., 1941); Kenyon, *ibid.* 261.
Liquid. d_4^{20} 1.0785. mp −40.6°; bp_{760} 185.7°; bp_{100} 130.8°; bp_{20} 96.8°; $bp_{1.0}$ 47°. n_D^{20} 1.41011. Flash pt, open cup: 168°F (75°C). Sparingly sol in water which dec it gradually; miscible with alcohol, ether, and other usual org solvents.
USE: Manuf phenobarbital, ethylbenzyl malonate, triethylamine, and similar chemicals, plastics, dyestuff intermediates. Solvent for cellulose esters, perfumes.

3175. Diethylpropion. *2-(Diethylamino)-1-phenyl-1-propanone; 2-diethylaminopropiophenone;* α-benzoyltriethylamine; amfepramone. $C_{13}H_{19}NO$; mol wt 205.30. C 76.06%, H 9.33%, N 6.82%, O 7.79%. Prepn: Hyde *et al., J. Am. Chem. Soc.* **50**, 2287 (1928); Schutte, U.S. pat. **3,001,-910** (1961 to Temmler-Werke). Pharmacology: Cahen *et al., Therapie* **17**, 373-412 (1962).

Hydrochloride, $C_{13}H_{19}NO \cdot HCl$, *Anfamon, Anorex, Danylen, Dobesin, Frekentine, Keramik, Keramin, Magrene, Modulor, Moderatan, Parabolin, Prefamone, Regenon, Tenuate, Tenuate Dospan, Tepanil, Tylinal.* Crystals, dec 168°.
Note: Regenon and Danylen also contain trace elements. Component of *Derfon.*

Note: This is a controlled substance (stimulant) listed in the U.S. Code of Federal Regulations, Title 21 Part 1308.14 (1995).

THERAP CAT: Anorexic.

3176. Diethylsilane. $C_4H_{12}Si$; mol wt 88.22. C 54.46%, H 13.71%, Si 31.83%. $(C_2H_5)_2SiH_2$. Prepd from ethylene and silane: Hurd, U.S. pat. **2,537,763** (1951); Fritz, *J. Naturforsch.* **7b,** 207 (1952); White, Rochow, *J. Am. Chem. Soc.* **76,** 3897 (1954); Clasen, *Angew. Chem.* **70,** 180 (1958).

Liquid. Stable to air when pure. bp_{741} 56°; d_4^{20} 0.6843; n_D^{20} 1.3921. Quickly oxidized by Ag_2O and HgO. Oxidation can be slowed by soln in petr ether or by cooling to $-80°$.

3177. Diethylstilbestrol. *(E)-4,4'-(1,2-Diethyl-1,2-ethenediyl)bisphenol;* α,α'*-diethylstilbenediol;* stilbestrol; stilboestrol; 3,4-bis(*p*-hydroxyphenyl)-3-hexene; 4,4'-dihydroxy-α,β-diethylstilbene; DES; Antigestil; Bufon; Cyren A; Domestrol; Estrobene; Estrosyn; Fonatol; Grafestrol; Makarol; Micrest; Milestrol; Neo-Oestranol I; Oestrogenine; Oestromenin; Oestromensyl; Oestromon; Palestrol; Serral; Sexocretin; Sibol; Stilbetin; Stilboefral; Stilboestroform; Stilkap; Synestrin (tablets); Synthoestrin; Vagestrol. $C_{18}H_{20}O_2$; mol wt 268.36. C 80.56%, H 7.51%, O 11.92%. Prepn: E. C. Dodds *et al., Nature* **141,** 247 (1938); *eidem, Proc. Roy. Soc. London, Ser. B* **127,** 140 (1939). HPLC determn in plasma and urine: A. T. Rhys Williams *et al., J. Chromatog.* **235,** 461 (1982). Comprehensive review of the early literature on synthetic estrogens: U. V. Solmssen, *Chem. Rev.* **37,** 481-598 (1945). Review of metabolism: P. W. Aschbacher, *J. Toxicol. Environ. Health,* Suppl. 1, 45-59 (1976); M. Metzler, *Crit. Rev. Biochem.* **10,** 171-212 (1981). Review of toxicology attributed to its use in the prevention of miscarriage in humans, and as a growth promotant in livestock: K. E. McMartin *et al., J. Environ. Pathol. Toxicol.* **1,** 279-313 (1978). Book: *Developmental Effects of Diethylstilbestrol (DES) in Pregnancy,* A. L. Herbst, H. A. Bern, Eds. (Thieme-Verlag, New York, 1981) 203 pp. Comprehensive description: A. A. Al-Badr, A. G. Mekkawi, *Anal. Profiles Drug Subs.* **19,** 145-192 (1990).

Small plates from benzene, mp 169-172°. uv max (0.1*N* NaOH): 259 nm ($E_{1cm}^{1\%}$ 764). Almost insol in water. Sol in alcohol, ether, chloroform, fatty oils, dil hydroxides.

Dipalmitate, $C_{50}H_{80}O_4$, *stilpalmitate.* Plates from hot alcohol, mp 77-78°.

Diphosphate, *see* Fosfestrol.

Dipropionate, $C_{24}H_{28}O_4$, *Clinestrol, Cyren B, Estilben, Estrobene DP, Euvestin, Neo-Oestranol II, Orestol, Pabestrol D, sin-estrol, Stilbestronate, Stilboestrol DP, Stilbofax, Stilronate, Synestrin (ampuls), Syntestrine, Willestrol.* Plates from methanol, mp 104°. Sol in organic solvents and in vegetable oils.

Dimethyl ether, *see* Dimestrol.

Disulfate, $C_{18}H_{20}O_8S_2$, *Idroestril.* Voluminous white cryst powder from dil HCl. Sparingly sol in water.

Monobenzyl ether, $C_{25}H_{26}O_2$, *Monozol.*

Monomethyl ether, *see* Mestilbol.

Caution: Banned by the FDA in 1979 as a growth promotant in livestock. Diethylstilbestrol has been listed as a known carcinogen: *Seventh Annual Report on Carcinogens* (PB95-109781, 1994) p 55.

THERAP CAT: Estrogen.

THERAP CAT (VET): Formerly in estrogenic hormone therapy.

3178. Diethyl Sulfate. *Sulfuric acid diethyl ester.* $C_4H_{10}O_4S$; mol wt 154.19. C 31.16%, H 6.54%, O 41.51%, S 20.80%. $(C_2H_5)_2SO_4$. Prepd from ethanol + sulfuric acid; by absorption of ethylene in sulfuric acid; from diethyl ether and fuming sulfuric acid. Review of prepn and uses: R. Page, J. A. John in *Ethylene and its Industrial Derivatives,* S.

A. Miller, Ed. (Benn, London, 1969) pp 774-787. Toxicity data: H. F. Smyth *et al., J. Ind. Hyg. Toxicol.* **31,** 60 (1949). Review of carcinogenicity studies: *IARC Monographs* **4,** 277-281 (1974). *Reviews:* C. M. Suter, *The Organic Chemistry of Sulfur* (John Wiley, 1944) pp 62-65; E. E. Gilbert, *Sulfonation and Related Reactions* (Interscience, New York, 1965).

Colorless, oily liquid; peppermint odor. *Poisonous!* Darkens with age. bp 209.5° with decompn; bp_{15} 96°; bp_5 75°. mp $-25°$. d_D^{23} 1.1774. n_D^{20} 1.40037, A. I. Vogel, D. M. Cowan, *J. Chem. Soc.* **1943,** 16. Flash pt: 220°F. Vapor density (air = 1) 5.31. Vapor pressure at 47° = 1 mm. Practically insol in water, gradually dec by it. Rapidly dec by hot water into monoethyl sulfate and alcohol. Miscible with alcohol, ether. LD_{50} orally in rats: 0.88 g/kg (Smyth).

Note: Diethyl sulfate may reasonably be anticipated to be a carcinogen: *Seventh Annual Report on Carcinogens* (PB95-109781, 1994) p 172.

USE: Chiefly as an ethylating agent; as an accelerator in the sulfation of ethylene; in some sulfonations.

3179. Diethylzinc. Zinc diethyl. $C_4H_{10}Zn$; mol wt 123.51. C 38.90%, H 8.16%, Zn 52.94%. $Zn(C_2H_5)_2$. Prepd by the interaction of zinc and ethyl iodide: Simonowitch, *Chem. Zentr.* **1899, I,** 1066; Lachmann, *Am. Chem. J.* **24,** 32 (1900); Dennis, Hance, *J. Am. Chem. Soc.* **47,** 370 (1925); from zinc and ethyl bromide, ethyl iodide: C. R. Noller, *Org. Syn. coll. vol.* II, 184. Used in synthesis of cyclopropanes: J. Furukawa *et al., Tetrahedron* **24,** 53 (1968); *eidem, Tetrahedron Letters* **1968,** 3495; in synthesis of ketocarbenes: L. T. Scott, W. D. Cotton, *J. Am. Chem. Soc.* **95,** 2708 (1973); in ring expansion of arenes: S. Miyano, H. Hashimoto, *Chem. Commun.* **1973,** 216; in preservation of papers: J. C. Williams, G. B. Kelly Jr., U.S. pat. **3,969,549** (1976 to U.S.A.); *eidem, Advan. Chem. Ser.* **193,** 109 (1981).

Mobile liq. Stable in sealed tube and carbon dioxide. d_4^{20} 1.2065; d_4^8 1.245; mp $-28°$; bp_{760} 118°; bp_{30} 27°; $n_{H\alpha}^{20}$ 1.4936. *Ignites in air,* burns with a blue flame, giving off a peculiar, garlic-like odor. Miscible with ether, petr ether, benzene, other hydrocarbons.

USE: In organic synthesis; in preservation on archival papers.

3180. Difemerine. α*-Hydroxy-α-phenylbenzeneacetic acid 2-(dimethylamino)-2-methylpropyl ester;* 2-(dimethylamino)-1,1-dimethylethyl benzilate. $C_{20}H_{25}NO_3$; mol wt 327.42. C 73.37%, H 7.70%, N 4.28%, O 14.66%. Prepn: S. G. Kuznetsov, A. G. El'tsov, *Zh. Obshch. Khim.* **32,** 511 (1962), *C.A.* **58,** 470d (1963); C. Hoffmann, Fr. pat. **M3406** (1965 to Labs. U.P.S.A.), *C.A.* **63,** 13163ab (1965). Toxicity: S. N. Golikov, S. G. Kuznetsov, *C.A.* **61,** 8790b (1964). Comparison of side effects with those of atropine: L. Moser, P. V. Lundt, *Med. Welt* **31,** 1795 (1980). Crystal structure: A. Carpy *et al., Acta Crystallogr.* **B35,** 882 (1979).

Solid, mp 78-78.3°.

Hydrochloride, $C_{20}H_{25}NO_3$·HCl, *Luostyl.* Crystals from isopropanol, mp 182°.

THERAP CAT: Anticholinergic; antispasmodic.

3181. Difenamizole. *2-(Dimethylamino)-N-(1,3-diphenyl-1H-pyrazol-5-yl)propanamide;* 1,3-diphenyl-5-(2-dimethylaminopropionamido)pyrazole; AP-14; Pasalin. $C_{20}H_{22}N_4O$; mol wt 334.42. C 71.83%, H 6.63%, N 16.75%, O 4.78%. Prepn: Kameyama, Niwa, **Japan.** pat. **68 06,621** (1968 to Takeda), *C.A.* **69,** 106704x (1968). Physicochemical properties and stability: Uda, Yashiki, *Takeda Kenkyusho Ho* **29,** 42 (1970), *C.A.* **73,** 28847n (1970). Pharmacology: Fukuda *et al., ibid.* **30,** 389 (1971); Saji *et al., ibid.* 406; Nagawa *et al., ibid.* 427, *C.A.* **76,** 108025v, 108021r, 108023t (1972), resp. Toxicity studies: Tanaka *et al., Oyo Yakuri*

(Pharmacometrics) **6**, 279 (1972). Mechanism of action studies: Yasuhara *et al., ibid.* 313. Acute toxicity: M. Hayashi *et al., Gifu Daigaku Igakubu Kiyo* **20**, 360 (1972), *C.A.* **78**, 92505e (1973). *Review:* Japan. *Med. Gaz.* **10**(11), 12 (1973).

White to pale yellow, odorless powder, mp 123-128°. Also reported as mp 120-122°. Freely sol in acetone, chloroform, benzene. Practically insol in water. Stable to heat, humidity and sunlight for 3 months. LD_{50} in mice (mg/kg): 103 i.v.; 186 i.p.; 525 s.c.; 560 orally (Hayashi).

THERAP CAT: Analgesic, anti-inflammatory.

3182. Difenoconazole. *1-[[2-[2-Chloro-4-(4-chlorophenoxy)phenyl]-4-methyl-1,3-dioxolan-2-yl]methyl]-1H-1,2,-4-triazole; cis,trans*-3-chloro-4-[4-methyl-2-(1*H*-1,2,4-triazol-1-ylmethyl)-1,3-dioxolan-2-yl]phenyl 4-chlorophenyl ether; CGA-169374; Dividend; Geyser; Plandom; Score. $C_{19}H_{17}Cl_2N_3O_3$; mol wt 406.27. C 56.17%, H 4.22%, Cl 17.45%, N 10.34%, O 11.81%. Ergosterol biosynthesis inhibitor. Prepn: A. Hubele, P. Riebli, **Brit. pat. Appl. 2,098,607** (1982 to Ciba-Geigy). Properties and antifungal activity: W. Ruess *et al., Brighton Crop Prot. Conf. - Pests Dis.* **1988**, 543. Field studies: A. J. Leadbeater *et al., ibid.* 917; A. B. Bassi, Jr. *et al., Brit. Crop Prot. Council Monogr.* **57**, 91 (1994). Uptake and translocation in crops: H. Dahmen, T. Staub, *Plant Dis.* **76**, 523 (1992).

White crystalline solid, mp 76°. Vapor pressure (20°): 1.2 × 10⁻⁷ Pa. Soly (20°): 5 mg/l in water. Very soluble in most organic solvents. Partition coefficient: log P 4.3. LD_{50} in rats (mg/kg): 1453 orally; in rabbits: >2010 dermally (Ruess).

USE: Agricultural fungicide.

3183. Difenoxin. *1-(3-Cyano-3,3-diphenylpropyl)-4-phenyl-4-piperidinecarboxylic acid; 1-(3-cyano-3,3-diphenyl-propyl)-4-phenylisonipecotic acid;* difenoxilic acid; difenoxylic acid; McN-JR-15403-11; Lyspafen. $C_{28}H_{28}N_2O_2$; mol wt 424.54. C 79.22%, H 6.65%, N 6.60%, O 7.54%. Active metabolite of diphenoxylate, *q.v.,* from which it may be prepared by hydrolysis of the ethyl ester: Soudijn, van Wijngaarden, **Ger. pat. 1,953,342** corresp to **U.S. pat. 3,646,207** (1970, 1972 to Janssen). Series of articles on pharmacology, toxicology, metabolism: *Arzneimittel-Forsch.* **22**, 513-531 (1972). Toxicity data: Niemegeers *et al., ibid.* 516.

Hydrochloride, $C_{28}H_{28}N_2O_2$·HCl, *R-15403*. White amorphous power, mp 290°. Very slightly sol in water (0.023%); sparingly sol in chloroform, tetrahydrofuran, dimethylacetamide, DMSO. Stable, can be stored for several years under normal conditions; not hygroscopic; not affected by light. LD_{50} orally in rats: 149 mg/kg (Niemegeers).

Note: This is a controlled substance (opiate) listed in the U.S. Code of Federal Regulations, Title 21 Part 1308.11 (1995).

THERAP CAT: Antiperistaltic; antidiarrheal.

3184. Difenpiramide. *N-2-Pyridinyl-[1,1′-biphenyl]-4-acetamide;* diphenpyramide; Z-876; Difenax. $C_{19}H_{16}N_2O$; mol wt 288.35. C 79.14%, H 5.59%, N 9.72%, O 5.55%. Prepn: F. Tenconi, R. M. Tagliabue, **Ger. pat. 2,325,309** (1973 to Zambeletti); L. Molteni *et al.,* **U.S. pat. 3,868,380** (1975). Pharmacological properties: S. Caliari *et al., Arzneimittel-Forsch.* **27**, 2086 (1977). Physicochemical profile: A. Trebbi, G. Filippi, *Boll. Chim. Farm.* **118**, 729 (1979). Therapeutic use: C. Ortolani *et al., Clin. Ter.* **89**, 391 (1979).

Crystals, mp 122-124°. LD_{50} in male mice, rats (mg/kg): 2590, 2075 orally; 1421, 1396 i.p. (Caliari).

THERAP CAT: Anti-inflammatory.

3185. Difenzoquat. *1,2-Dimethyl-3,5-diphenyl-1H-pyrazolium.* $[C_{17}H_{17}N_2]^+$. Post-emergence herbicide for food crops; also effective vs mildew. Prepn of salts: B. Walworth, E. Klingsberg, **Ger. pat. 2,260,485;** *eidem,* **U.S. pat. 3,882,142** (1973, 1975 both to Am. Cyanamid); B. Cross *et al., J. Heterocycl. Chem.* **17**, 905 (1980). Physical properties and herbicidal activity: N. E. Shafer, *Proc. 12th Brit. Weed Control Conf.* **1974**, 831. Effect vs mildew: R. J. Winfield *et al., Proc. Brit. Crop. Prot. Conf. - Pests Dis.* **1977**, 57. HPLC determn in commercial formulations: C. Barry, R. K. Pike, *J. Assoc. Off. Anal. Chem.* **63**, 64 (1980); in water: M. C. Carneiro *et al., J. Chromatog. A* **669**, 217 (1994).

Methyl sulfate, $C_{17}H_{17}N_2$·CH_3O_4S, *AC-84777, CL-84777, Avenge, Finaven.* Crystals from acetone, mp 146-148° (Cross). Soly in water (25°): ~70%. Relatively insol in non-polar solvents. LD_{50} (technical grade) orally in rats: 470 mg/kg; dermally in rabbits: 3540 mg/kg (Shafer).

USE: Herbicide; agricultural fungicide.

3186. Diflorasone. *(6α,11β,16β)-6,9-Difluoro-11,17,21-trihydroxy-16-methylpregna-1,4-diene-3,20-dione; 6α,9α-di-fluoro-16β-methyl-Δ¹,⁴-pregnadiene-11β,17α,21-triol-3,20-dione; 6α,9α-difluoro-16β-methylprednisolone.* $C_{22}H_{28}F_2O_5$;

mol wt 410.46. C 64.38%, H 6.88%, F 9.26%, O 19.49%. The 16β-analog of flumethasone, *q.v.* Prepn of free alcohol and 21-acetate: **Brit. pat. 881,334** (1961 to Pfizer), *C.A.* **56**, 15586c (1962); **Brit. pat. 898,293**; F. H. Lincoln *et al., U.S.* pat. **3,557,158** (1962, 1971 both to Upjohn); **Brit. pat. 912,-015** (1962 to Merck & Co.). Prepn of the 17,21-diacetate: D. E. Ayer *et al.*, **Ger. pat. 2,308,731**; *eidem*, **U.S. pat. 3,980,778** (1973, 1976 both to Upjohn). Proposed mechanism of action: S. Hammarstrom *et al., Science* **197**, 994 (1977). Pharmacology: S. Wickrema *et al., J. Invest. Dermatol.* **71**, 372 (1978). Clinical study: S. M. Bluefarb *et al., J. Int. Med. Res.* **4**, 454 (1976).

Diacetate, $C_{26}H_{32}F_2O_7$, *U-34865, Dermaflor, Diacort, Difulal, Florone, Maxiflor, Psorcon, Soriflor.* Crystals from ethyl acetate-Skellysolve C and acetone-methanol, mp 221-223° (dec). uv max (alc): 238 nm (ε 17250). $[\alpha]_D$ +61° (chloroform).

THERAP CAT: Anti-inflammatory (topical); glucocorticoid.

3187. Difloxacin. *6-Fluoro-1-(4-fluorophenyl)-1,4-dihydro-7-(4-methyl-1-piperazinyl)-4-oxo-3-quinolinecarboxylic acid;* $C_{21}H_{19}F_2N_3O_3$; mol wt 399.40. C 63.15%, H 4.79%, F 9.51%, N 10.52%, O 12.02%. Fluoroquinolone antibacterial structurally related to norfloxacin, *q.v.* Prepn: D. T. W. Chu, **Eur. pat. Appl. 131,839**; *idem*, **U.S. pat. 4,730,000** (1985, 1988 both to Abbott); D. T. W. Chu *et al., J. Med. Chem.* **28**, 1558 (1985); H. Narita *et al.*, **Japan. Kokai 85 237,069** (1985 to Toyama). Antibacterial spectrum *in vitro:* J. M. Stamm *et al., Antimicrob. Ag. Chemother.* **29**, 193 (1986); *in vivo:* P. B. Fernandes *et al., ibid.* 201. Activity vs anaerobic bacteria: *eidem, J. Antimicrob. Chemother.* **18**, 693 (1986). HPLC determn in biological fluids: G. R. Granneman, L. T. Sennello, *J. Chromatog.* **413**, 199 (1987). Pharmacokinetics in humans: G. R. Granneman *et al., Antimicrob. Ag. Chemother.* **30**, 689 (1986).

Monohydrochloride, $C_{21}H_{19}F_2N_3O_3 \cdot HCl$, *Abbott 56619, A-56619.* Crystals, mp > 275°.

THERAP CAT: Antibacterial.

3188. Diflubenzuron. *N-[[(4-Chlorophenyl)amino]carbonyl]-2,6-difluorobenzamide;* 1-(4-chlorophenyl)-3-(2,6-difluorobenzoyl)urea; difluron; DU-112307; PH-60-40; TH-6040; ENT-29054; OMS-1804; Dimilin; Duphacid; Micromite. $C_{14}H_9ClF_2N_2O_2$; mol wt 310.69. C 54.12%, H 2.92%, Cl 11.41%, F 12.23%, N 9.02%, O 10.30%. Substituted benzoylphenylurea, inhibits chitin deposition in insect cuticle. Prepn: K. Wellinga, R. Mulder, **Ger. pat. 2,123,236**; *eidem*, **U.S. pat. 3,748,356** (1971, 1973 both to Philips). Insecticidal activity: K. Wellinga *et al., J. Agr. Food Chem.* **21**, 993 (1973); R. Mulder, M. J. Gijswijt, *Pestic. Sci.* **4**, 737 (1973). Mode of action: E. Hunter, J. F. V. Vincent, *Experientia* **30**, 1432 (1974); S. M. Meola, R. T. Mayer, *Science* **207**, 985 (1980). Field trials against locusts and grass-

hoppers: A. Bouaichi *et al., Crop Protect.* **13**, 53, 60 (1994). Brief review: J. L. Marx, *Science* **197**, 1170-1172 (1977). Review of activity, environmental fate, mode of action: A. Verloop, C. D. Ferrell, *ACS Symposium Series* **37**, 237-270 (1977); of chemistry and analyses: B. Rabenort *et al., Anal. Methods Pestic. Plant Growth Regul.* **10**, 57-72 (1978).

White to yellowish brown crystalline solid, mp 210°-230° (technical grade); mp 239° (pure compd). Soly in water ≃0.2 ppm at 20°. Moderate to good soly in polar to very polar solvents; e.g. acetone 0.65 gm/100 ml at 20°. Partition coefficient in dichloromethane (pH 5.6): 50; in 1-octanol-water: ~5000. LD_{50} in mice, rats (formulation with 50% kaolin) (g/kg): 4.64, > 10 orally (Mulder, Gijswijt).

USE: Insecticide.

3189. Diflucortolone. *(6α, 11β, 16α)-6,9-Difluoro-11,21-dihydroxy-16-methylpregna-1,4-diene-3,20-dione;* 6α,-9α-difluoro-16α-methyl-1,4-pregnadiene-11β,21-diol-3,20-dione; 6α,9α-difluoro-16α-methyl-1-dehydrocorticosterone. $C_{22}H_{28}F_2O_4$; mol wt 394.46. C 66.99%, H 7.15%, F 9.63%, O 16.22%. The 9α-fluoro deriv of fluocortolone, *q.v.* Prepn: **Belg. pat. 639,708**; K. Kieslich *et al., U.S. pat. 3,426,128** (1964, 1969, both to Schering AG); *eidem, Arzneimittel-Forsch.* **26**, 1462 (1976). Toxicity: P. Gunzel *et al., ibid.* 1476. Series of articles on pharmacology, toxicity, metabolism and clinical studies: *ibid.* **26**, 1463-1513 (1976).

Crystals from ethyl acetate-ether, mp 240-244°. uv max: 237 nm (ε 16600) (Kieslich, 1969). Also reported as mp 248-249°. $[\alpha]_D^{22}$ +111° (methanol) (Kieslich, 1969).

21-Valerate, $C_{27}H_{36}F_2O_5$, *Neribas, Neriforte, Nerisona, Nerisone, Temetex, Texmeten.* Crystals, mp 195-195.5°. $[\alpha]_D^{22}$ +100.8° (dioxane). Approx LD_{50} in mice, rats: > 4, 3.1 g/kg orally; 180, 13 mg/kg s.c.; 450, 98 mg/kg i.p. (Gunzel).

THERAP CAT: Anti-inflammatory.

3190. Diflunisal. *2',4'-Difluoro-4-hydroxy-[1,1'-biphenyl]-3-carboxylic acid; 2',4'-difluoro-4-hydroxy-3-biphenylcarboxylic acid;* 2',4'-difluoro-4-hydroxy-[1',1-diphenyl]-3-carboxylic acid; 2-(hydroxy)-5-(2,4-difluorophenyl)benzoic acid; 5-(2,4-difluorophenyl)salicylic acid; MK-647; Adomal; Difludol; Dolisal; Dolobid; Dolobis; Flovacil; Fluniget; Fluodonil; Flustar. $C_{13}H_8F_2O_3$; mol wt 250.20. C 62.41%, H 3.22%, F 15.19%, O 19.18%. Prepn: Ruyle *et al.*, **S. Afr. pat. 67 01,021** (1968 to Merck & Co.), *C.A.* **70**, 106209k (1969); *eidem*, **Fr. pat. 1,522,570** corresp to **U.S. pat. 3,714,-226** (1968, 1973, both to Merck & Co.); J. Hannah *et al., J. Med. Chem.* **21**, 1093 (1978). Metabolism: P. J. DeSchepper *et al., Brit. J. Clin. Pharmacol.* **4**, 645P (1977). Clinical studies: J. A. Wojtulewski *et al., Curr. Med. Res. Opin.* **5**, 562 (1978); J. A. Hicklin, *ibid.* 572; in sickle cell anemia: E. A. Oyewo, *Clin. Trials J.* **24**, 249 (1987). Review of chemistry, pharmacology, clinical pharmacology: S. L. Steelman *et al., ibid.* 506-514. Comprehensive description: M. Cotton, R. A. Hux, *Anal. Profiles Drug Subs.* **14**, 491-526 (1985). *Review:* **Brit. J. Clin. Pharmacol. 4**, Suppl. 1, 1S-52S (1977); C. A. Winter *et al.*, in *Pharmacological and Biochemical Properties of Drug Substances* vol. 3, M. E. Goldberg, Ed. (Am. Pharm. Assoc., Washington, DC, 1981) pp 291-

323. Comprehensive review of pharmacological properties and therapeutic use in pain: R. N. Brogden *et al.*, *Drugs* **19**, 84-106 (1980); *Pharmacotherapy* **3**, no. 2, pt. 2, 1S-82S (1983).

Crystals from toluene, mp 210-211°. Sparingly sol in water. LD₅₀ orally in female mice: 439 mg/kg (Stone).
THERAP CAT: Analgesic; anti-inflammatory.

3191. 2,4-Difluoroaniline. *2,4-Difluorobenzenamine.* $C_6H_5F_2N$; mol wt 129.11. C 55.82%, H 3.90%, F 29.43%, N 10.85%. Prepn: Swarts, *Rec. Trav. Chim.* **35**, 164 (1916); Kutepov, Rozanova, *Zh. Obshch. Khim.* **27**, 2848 (1957), *C.A.* **52**, 8067 (1958); Nad *et al.*, *C.A.* **53**, 14976i (1959).

mp −7.5°. bp₇₅₃ 169.5°. n_D^{25} 1.5043. Specific gravity at 25°: 10.72 lb/gal. Flash pt 158°F (70°C).
USE: In organic syntheses.

3192. p-Difluorobenzene. $C_6H_4F_2$; mol wt 114.09. C 63.16%, H 3.53%, F 33.30%. Prepd by diazotizing p-fluoroaniline in hydrofluoric acid and heating the diazonium salt to 120°: Swarts, *Chem. Zentr.* **1913**, II, 760.

Liquid, pungent, aromatic odor. mp −23.7°. bp 88.82°. d²⁰ 1.17006. $n_D^{18.9}$ 1.44219.

3193. 4,4′-Difluorodiphenyl. *4,4′-Difluoro-1,1′-biphenyl.* $C_{12}H_8F_2$; mol wt 190.19. C 75.78%, H 4.24%, F 19.98%. NMR spectrum: T. K. Halstead *et al.*, *J. Chem. Soc. Faraday Trans. 2*, **1981**, 1817.

Crystalline powder; aromatic odor. d 1.04. mp 92-95°. bp 254-255°. Insol in water. Freely sol in alc, chloroform, ether, oils.

3194. Difluprednate. *(6α,11β)-21-(Acetyloxy)-6,9-difluoro-11-hydroxy-17-(1-oxobutoxy)pregna-1,4-diene-3,20-dione; 6α,9-difluoro-11β,17,21-trihydroxypregna-1,4-diene-3,20-dione 21-acetate 17-butyrate; 6α,9α-difluoroprednisolone-21-acetate-17-butyrate;* CM-9155; W-6309; Epitopic; Myser. $C_{27}H_{34}F_2O_7$; mol wt 508.56. C 63.77%, H 6.74%, F 7.47%, O 22.02%. Prepn: A. Ercoli, R. Gardi, S. Afr. pat. 68 0386; A. Ercoli *et al.*, U.S. pat. 3,780,177 (1968, 1973 both to Warner-Lambert). Anti-inflammatory properties: G. Di Pasquale *et al.*, *Steroids* **16**, 663, 679 (1970); R. Gardi *et al.*, *J. Med. Chem.* **15**, 556 (1972). Pharmacological study: J. Navarro *et al.*, *Arzneimittel-Forsch.* **28**, 2302 (1978). Vasoconstrictive activity in man: A. Barbier *et al.*, *Therapie* **33**, 607 (1978). Bioavailability study: C. Lafille *et al.*, *Dermatologica* **159**, 277 (1979).

Crystals from methylene chloride/ether/petr ether, mp 191-194°. $[\alpha]_D^{22}$ +31.7° (c = 0.5 in dioxane). uv max (ethanol): 237-238 nm (E$_{1cm}^{1\%}$ 320).
THERAP CAT: Anti-inflammatory.

3195. Digallic Acid. *3,4-Dihydroxy-5-[(3,4,5-trihydroxybenzoyl)oxy]benzoic acid;* gallic acid 5,6-dihydroxy-3-carboxyphenyl ester; 4,5-dihydroxybenzoic acid monogallate; gallic acid 3-monogallate; *m*-digallic acid; *m*-galloylgallic acid. $C_{14}H_{10}O_9$; mol wt 322.23. C 52.18%, H 3.13%, O 44.69%. Isoln from Aleppo gallotannin and Chinese gallotannin: Nierensten, *Ber.* **43**, 628 (1910). Synthesis: Fischer, Freudenberg, *Ber.* **46**, 1128 (1913). Structure: Nierenstein *et al.*, *J. Am. Chem. Soc.* **47**, 846 (1925). Determn by HPLC: P. Delahaye, M. Verzele, *J. Chromatog.* **265**, 363 (1983); in tannic acids: M. Verzele *et al.*, *Bull. Soc. Chim. Belg.* **92**, 181 (1983). Role in corrosion inhibition: A. M. Beccaria, E. D. Mor, *Brit. Corrosion J.* **11**, 156 (1976).

Hydrated needles from alc + water, anhydr at 110°; dec 280°. Sol in 1900 parts water at 25°; in 50-60 parts boiling water; sol in methanol, ethanol, acetone; sparingly sol in ether, glacial acetic acid. Upon addn of KCN to an aq soln a transitory pink color appears, which returns on shaking.
Note: In pharmaceutical literature the name digallic acid is frequently confused with tannic acid.

3196. Digalogenin. *(3β,5α,15β,25R)-Spirostan-3,15-diol.* $C_{27}H_{44}O_4$; mol wt 432.64. C 74.96%, H 10.25%, O 14.79%. Aglycon of digalonin. Isoln from saponin mixture and partial synthesis from digitogenin: Tschesche, Wulff, *Ber.* **94**, 2019 (1961).

Crystals from methanol, mp 218.5-220.5°. $[\alpha]_D^{21}$ −75° (CHCl₃). IR spectrum: Tschesche, Wulff *loc. cit.* Sol in chloroform.

3197. Diginatigenin. *(3β,5β,12β,16β)-3,12,14,16-Tetrahydroxycard-20(22)-enolide;* 12-hydroxygitoxigenin; 16-hydroxydigoxigenin. $C_{23}H_{34}O_6$; mol wt 406.52. C 67.96%, H 8.43%, O 23.61%. Aglycone of diginatin from *Digitalis lanata* Ehrk., *Scrophulariaceae:* Murphy, *J. Am. Pharm. Assoc. Sci. Ed.* **44**, 719 (1955). Aglycone of lanatoside D:

Angliker *et al., Ann.* **607**, 131 (1957). Obtained by micro-biological conversion of gitoxigenin: Tamm, Gubler, *Helv. Chim. Acta* **41**, 1762 (1958); Okado *et al., Chem. Pharm. Bull.* **8**, 530 (1960). Structure: Linde *et al., Helv. Chim. Acta* **42**, 2040 (1959); Okado *et al., Chem. Pharm. Bull.* **8**, 535 (1960).

Needles from water, mp 157°. $[\alpha]_D^{20}$ +34° (methanol). Absorption max (ethanol): 318 nm (log ϵ 4.18); (98% w/w H_2SO_4): 230, 310, 390, 490 nm ($E_{1cm}^{1\%}$ 160, 130, 210, 85).

3198. Diginatin. *(3β,5β,12β,16β)-3-[(O-2,6-Dideoxy-β-D-ribo-hexopyranosyl-(1→4)-O-2,6-dideoxy-β-D-ribo-hexopyranosyl-(1→4)-2,6-dideoxy-β-D-ribo-hexopyranosyl)oxy]-12,14,16-trihydroxy-card-20(22)-enolide.* $C_{41}H_{64}O_{15}$; mol wt 796.95. C 61.79%, H 8.09%, O 30.11%. Cardiac glycoside from *Digitalis lanata* Ehrk., *Scrophulariaceae*: Murphy, *J. Am. Pharm. Assoc. Sci. Ed.* **44**, 719 (1955); Angliker *et al., Ann.* **607**, 131 (1957); von Wartburg *et al., Experientia* **14**, 439 (1958). Acid hydrolysis yields 1 mol diginatigenin + 3 mols D-digitoxose. The sugar residue is attached to the hydroxyl group at C-3 of the aglycon.

Prisms from acetone, mp 251-253°. Somewhat hygroscopic. $[\alpha]_D^{20}$ +20.5° ±2°; $[\alpha]_{5463}^{20}$ +26.3° ±2° (c = 0.6 in methanol). Absorption max (98% w/w H_2SO_4): 230, 310, 390, 480, 560 nm ($E_{1cm}^{1\%}$ 185, 175, 315, 230, 120). Sol in alcohol, dil alcohol, dioxane; slightly sol in acetone. One part sol in 1000 parts water and in 2000 parts chloroform at 25°.

Acetyldiginatin-α, $C_{43}H_{66}O_{16}$, flat plates from methanol + chloroform + ether, mp 199-202°. $[\alpha]_D^{20}$ +30.1° (c = 0.6 in pyridine). uv max: 218 nm (log ϵ 4.1).

3199. Diginin. *(3β,12α,14β,17α,20S)-3-[(2,6-Dideoxy-3-O-methyl-D-lyxohexopyranosyl)oxy]-12,20-epoxypregn-5-ene-11,15-dione*3β-(diginosyloxy)-12α,20α-epoxy-14β,17α-pregn-5-ene-11,15-dione. $C_{28}H_{40}O_7$; mol wt 488.62. C 68.83%, H 8.25%, O 22.92%. Isoln from leaves of *Digitalis purpurea* L., *Scrophulariaceae*: Karrer in *E. Barell Festschrift* (1936) p 238; *Chem. Zentr.* **1936**, II, 2727. Isoln: Shoppee, Reichstein, *Helv. Chim. Acta* **23**, 975 (1940); Shoppee, *ibid.* **27**, 426 (1944). Structure: Shoppee *et al., J. Chem. Soc.* **1962**, 3610; Tschesche, Brügmann, *Tetrahedron* **20**, 1469 (1964). Mild hydrolysis yields *diginigenin* and *diginose.* Attempted partial synthesis of diginigenin: Tschesche, Schwinum, *Ber.* **100**, 464 (1967); Tschesche, Müller-Albrecht, *Ber.* **103**, 350 (1970).

Stout prisms from dil alc. Indistinct melting range: 155-183°. $[\alpha]_D^{14}$ −223° (c = 2.3 in chloroform). uv max (ethanol): 309 nm (log ϵ 1.94). Freely sol in chloroform; slightly sol in ether, acetone, ethyl acetate, carbon tetrachloride. Practically insol in water. Positive Legal's test. The Keller reaction produces a brilliant yellow band with diginigenin.

3200. Digitalin. *(3β,5β,16β)-3-[(6-Deoxy-4-O-β-D-glucopyranosyl-3-O-methyl-β-D-galactopyranosyl)oxy]-14,16-dihydroxycard-20(22)-enolide;* digitalinum verum; digitalinum true; Schmiedeberg's digitalin; Diginorgin. $C_{36}H_{56}O_{14}$; mol wt 712.83. C 60.66%, H 7.92%, O 31.42%. Obtained from seeds of *Digitalis purpurea* L., *Scrophulariaceae,* and from roots of *Adenium honghel* A. DC., *Apocynaceae:* Schmiedeberg, *Arch. Exp. Path. Pharmakol.* **3**, 16 (1874); Windaus, Haack, *Ber.* **62**, 475 (1929); Hunger, Reichstein, *Helv. Chim. Acta* **33**, 76 (1950); Sasakawa, *J. Pharm. Soc. Japan* **74**, 474 (1954); Miyatake *et al., Pharm. Bull. (Tokyo)* **5**, 157 (1957).

Crystals from methanol + ether and methanol + water, mp 240-243°. $[\alpha]_D^{20}$ −1.1° (c = 0.894 in methanol). Slightly sol in water, chloroform or ether; sol in alcohol.

16-Acetyldigitalinum verum, $C_{38}H_{58}O_{15}$. Prepn: Miyatake *et al.,* U.S. pat. 3,023,147 (1962). Amorphous powder. Freely sol in water, alcohol, acetone. Practically insol in benzene, ether. uv max: 217 nm (log ϵ 4.16). $[\alpha]_D^{26}$ −21.1° (methanol).

THERAP CAT: Cardiotonic.

3201. Digitalis. Foxglove; fairy gloves; purple foxglove; Digifortis; Digitora; Neodigitalis, Pil-Digis. Dried leaves of *Digitalis purpurea* L., *Scrophulariaceae.* Habit. Southern and central Europe, cultivated in the U.S. *Constit.* Digitoxin (0.2-0.4%), digitonin, digitalin, antirhinic acid, digitalosmin, digitoflavone, inositol, pectin. One U.S.P. digitalis unit represents the potency of 0.1 g of the U.S.P. Digitalis Reference Standard. Toxicology: A. P. Somlyo, *Am. J. Cardiol.* **5**, 523 (1960). Review of mechanism of action and clinical use: T. W. Smith, *N. Engl. J. Med.* **318**, 358-365 (1988). Book: C. Fisch, B. Surawicz, *Digitalis* (Grune & Stratton, New York, 1969) 244 pp.

Caution: The therapeutic dose is close to the toxic dose which causes anorexia, nausea, salivation, vomiting, diarrhea, headache, drowsiness, disorientation, delirium, hallucinations; death may result, *Clinical Toxicology of Commercial Products,* R. E. Gosselin *et al.,* Eds. (Williams & Wilkins, Baltimore, 4th ed., 1976) Section III, pp 124-133.

THERAP CAT: Cardiotonic.

THERAP CAT (VET): Cardiotonic.

3202. Digitalose. *6-Deoxy-3-O-methylgalactose;* 3-methyl-D-fucose. $C_7H_{14}O_5$; mol wt 178.19. C 47.19%, H 7.92%, O 44.90%. First obtained by Kiliani, *Ber.* **25**, 2116 (1892), by hydrolysis of *Digitalinum verum,* a glycoside of gitoxigenin with glucose and digitalose. By hydrolysis of a glycoside from the seeds of *Strophanthus eminii* Aschers & Pax., *Apocynaceae:* Lamb, Smith, *J. Chem. Soc.* **1936**, 442. Configuration: Schmidt *et al., Naturwiss.* **31**, 247 (1943); *Ann.* **555**, 26 (1944). Elderfield in *Advances in Carbohydrate Chemistry* vol. I (New York, 1945) p 150. Prepn by isoln and hydrolysis of emicymarin from *Strophanthus eminii* seeds: Elderfield, *ibid.* 156. Synthesis: Reber, Reichstein, *Helv. Chim. Acta* **29**, 343 (1946). Schmidt, Wernicke, *Ann.* **558**, 70 (1947).

Rosettes of needles from ethyl acetate. mp 106° when freshly prepd, mp 115° after 3 days, mp 119° after 4 mos. $[\alpha]_{5461}^{27}$ +109° (15 min c = 1.7) → $[\alpha]_{5461}^{22}$ +126° or $[\alpha]_{D}^{22}$ +106° (17 hrs). By the Bertrand micro method, 1 mg digitalose has the same reducing power as 9.32 mg anhydr glucose. Soluble in water.

3203. Digitogenin. *(2α,3β,5α,15β,25R)-Spirostan-2,3,15-triol.* $C_{27}H_{44}O_5$; mol wt 448.64. C 72.28%, H 9.89%, O 17.83%. The aglycon of digitonin: Tschesche, *Ber.* **68,** 1090 (1935); Tschesche, Hagedorn, *Ber.* **69,** 797 (1936). Structure and stereochemistry: Marker *et al., J. Am. Chem. Soc.* **64,** 1843 (1942); Meystre *et al., Helv. Chim. Acta* **38,** 381 (1955). Klass *et al., J. Am. Chem. Soc.* **77,** 3829 (1955); and IR spectra: Djerassi *et al., ibid.* **78,** 3166 (1956).

Needles from alc, dec 296°. $[\alpha]_D^{19}$ −81° (c = 1.4 in chloroform). Practically insol in water. Sol in 30 parts chloroform, 35 parts boiling alc, 100 parts alc at 20°.

2,3-Diacetate, crystals from methanol-chloroform, mp 241.5-242°. $[\alpha]_D$ −104°.

Triacetate, minute needles from ether, mp 190°.

3204. Digitonin. Digitin. $C_{56}H_{92}O_{29}$; mol wt 1229.33. C 54.71%, H 7.54%, O 37.74%. Obtained from the seeds of *Digitalis purpurea* L., Scrophulariaceae. Extraction procedure: Gisvold, *J. Am. Pharm. Assoc.* **23,** 664 (1934). Purification of commercial digitonin and its separation into two fractions: G. Ruhenstroth-Bauer, P. M. Breitenfeld, *Z. Physiol. Chem.* **302,** 111 (1955). Structure: Tschesche, Wulff, *Tetrahedron* **19,** 621 (1963). Use as clinical reagent: H. H. Leffler, *Am. J. Clin. Pathol.* **31,** 310 (1959).

Crystals from alc, sinters 225°. Indistinct mp 235-240°. $[\alpha]_D^{20}$ −54° (0.45 g in 15.8 ml methanol). One gram dissolves in 57 ml abs alc, in 220 ml 95% alc. Practically insol in water, forming a soapy suspension. Also practically insol in chloroform, ether.

USE: Clinical reagent (cholesterol determination).

3205. Digitoxigenin. *(3β,5β)-3,14-Dihydroxycard-20-(22)-enolide;* Δ²⁰:²²-3,14,21-trihydroxynorcholenic acid lactone; cerberigenin; echujetin; evonogenin; Thevetigenin. $C_{23}H_{34}O_4$; mol wt 374.52. C 73.76%, H 9.15%, O 17.09%.

The aglycon of digitoxin, thevetin, cerberin, echujin, evomonosid. Prepn by refluxing digitoxin in a mixture of water + alcohol + HCl: Cloetta, *Arch. Exp. Pathol. Pharmakol.* **88,** 113 (1920); Stoll, Kreiss, *Helv. Chim. Acta* **17,** 592 (1934). Structure: Jacobs, Hoffmann, *J. Biol. Chem.* **67,** 333 (1926); Jacobs, Gustus, *ibid.* **78,** 573; **79,** 533 (1928); **82,** 402 (1929); Jacobs, Elderfield, *ibid.* **108,** 497 (1935); Elderfield, *Chem. Rev.* **17,** 187 (1935). Stereochemistry: Meyer, *Helv. Chim. Acta* **30,** 1976 (1947). Synthesis: Danieli *et al., Tetrahedron* **22,** 3189 (1966); W. Fritsch *et al., Ann.* **1974,** 621; S. F. Donovan *et al., Tetrahedron Letters* **1979,** 3287; R. Marinibettolo *et al., Can. J. Chem.* **59,** 1403 (1981); T. Milkova *et al., Tetrahedron Letters* **23,** 413 (1982). Biosynthesis from neriifolin, *q.v.:* A. Cruz *et al., J. Org. Chem.* **42,** 3580 (1977).

Stout prisms from 40% alc, mp 253°. $[\alpha]_D^{17}$ +19.1° (c = 1.36 in methanol). Sol in alc, chloroform, acetone; slightly sol in ethyl acetate; very sparingly sol in ether, water.

3-Acetyldigitoxigenin, $C_{25}H_{36}O_5$, hexagonal plates from acetone + ether, mp 222-227°. $[\alpha]_D^{17}$ +21.4° (c = 1.02 in chloroform).

3206. Digitoxin. *(3β,5β)-3-[(O-2,6-Dideoxy-β-D-ribo-hexopyranosyl-(1→4)-O-2,6-dideoxy-β-D-ribo-hexopyranosyl-(1→4)-2,6-dideoxy-β-D-ribo-hexopyranosyl)oxy]-14-hydroxy-card-20(22)-enolide;* digitalin, crystalline; digitophyllin; Cardigin; Carditoxin; Coramedan; Cristapurat; Crystodigin; Digicor; Digilong; Digimerck; Digimed; Digipural; Ditaven; Digisidin; Digitaline Nativele; Lanatoxin; Myodigin; Purodigin; Purpurid; Tardigal; Unidigin. $C_{41}H_{64}O_{13}$; mol wt 764.95. C 64.38%, H 8.43%, O 27.19%. Secondary glycoside from *Digitalis purpurea* L., Scrophulariaceae. Extracted from the dried leaves with 50% alc. Extraction procedure: Cloetta, *Arch. Exp. Pathol. Pharmakol.* **112,** 261 (1926). Purification: Windaus, Freese, *Ber.* **58,** 2503 (1925). One kilo leaves yield about 6 grams pure digitoxin. Identity with digitophyllin: Cloetta, *Arch. Exp. Pathol. Pharmakol.* **88,** 113 (1920). Structure: *See* reviews by Elderfield in *Chem. Rev.* **17,** 187 (1935), and Stoll, *The Cardiac Glycosides, London* (1937); also Shoppee in *Ann. Rev. Biochem.* **11,** 103 (1942). Acid hydrolysis yields 1 mol digitoxigenin + 3 mol digitoxose. The sugar residue is attached to the hydroxyl group at C-3 of the aglycon. Structure of the sugar moiety: Lichti *et al., Helv. Chim. Acta* **45,** 868 (1962). Digitoxin U.S.P. is either pure digitoxin or a mixture of cardioactive glycosides obtained from *Digitalis purpurea* and consisting chiefly of digitoxin. The potency of digitoxin U.S.P. corresp to the potency of an equal wt of U.S.P. Digitoxin Reference Standard. A deviation of 20% is permitted. The physical characteristics given below are those of the pure compd. Toxicity data: Foerster *et al., Arch. Int. Pharmacodyn. Ther.* **159,** 1 (1966). Comprehensive description: I. M. Jakovljevic, *Anal. Profiles Drug Subs.* **3,** 149-172 (1974).

Very small elongated, rectangular plates from dil alc. May contain ½ or 1 mol H_2O or EtOH which is given up at 118° *in vacuo.* When anhydrous, mp 256-257°. $[\alpha]_D^{20}$ +4.8° (c = 1.2 in dioxane). One gram dissolves in about 40 ml chloroform, about 60 ml alcohol, about 400 ml ethyl acetate. Also sol in acetone, amyl alcohol, pyridine, sparingly sol in ether, petr ether, water (1 g/100 liter at 20°). LD_{50} in guinea pigs, cats (mg/kg): 60.0, 0.18 orally (Foerster).

Acetyl derivatives, *see* Acetyldigitoxins.

THERAP CAT: Cardiotonic.
THERAP CAT (VET): Cardiotonic.

3207. Digitoxose. *2,6-Dideoxy-D-ribo-hexose;* 2-desoxy-D-altromethylose; 2,6-didesoxy-D-allose. $C_6H_{12}O_4$; mol wt 148.16. C 48.64%, H 8.16%, O 43.20%. Obtained by mild acid hydrolysis of the glycosides digitoxin, gitoxin and digoxin: Cloetta, *Arch. Exp. Pathol. Pharmakol.* **88**, 113 (1920); **112**, 261 (1926); Windaus, Stein, *Ber.* **61**, 2436 (1928); Kraft, *Arch. Pharm.* **250**, 118 (1912); Mannich *et al.,* ibid. **268**, 453 (1930); Smith, *J. Chem. Soc.* **1930**, 508; **1931**, 23. Configuration: Micheel, *Ber.* **63**, 347 (1930). Structure: S. F. Dyke, *The Carbohydrates* (Interscience, New York, 1960) p 104. Synthesis: Gut, Prins, *Helv. Chim. Acta* **30**, 1223 (1947); Bolliger, Ulrich, ibid. **35**, 93 (1952). Stereochemical study: S. Tsukamoto *et al., J. Chem. Soc. Perkin Trans. I* **1988**, 2621. Review: R. C. Elderfield in W. W. Pigman, M. L. Wolfrom, *Advances in Carbohydrate Chemistry* vol. I (Academic Press, New York, 1945) pp 159-164.

Crystals from methanol + ether, from ethyl acetate or from acetone + ether, mp 112°. $[\alpha]_D^{17}$ +46.3° (in water); $[\alpha]_D^{20}$ +39.1° (in methanol); $[\alpha]_D^{18}$ +27.9° → +43.3° (after 24 hrs in pyridine). Freely sol in water; sol in acetone, ethanol. Practically insol in ether.

3208. Diglyme. *1,1'-Oxybis[2-methoxyethane]; bis(2-methoxyethyl) ether;* diethylene glycol dimethyl ether. $C_6H_{14}O_3$; mol wt 134.18. C 53.71%, H 10.52%, O 35.77%. $(CH_3OCH_2CH_2)_2O$. Prepn: Cretcher, Pittenger, *J. Am. Chem. Soc.* **47**, 163 (1925); Gallaugher, Hibbert, ibid. **58**, 813 (1936).

Liquid, bp_{760} 162°; bp_{200} 116°; bp_{35} 75°; bp_3 20°. d_4^{20} 0.9451. mp −68°. Flash pt (open cup) 158°F (70°C). n_D^{20} 1.4097. Miscible with water, alcohol, ether, hydrocarbon solvents.

USE: Solvent; reaction medium for Grignard and similar syntheses.

3209. Digoxigenin. *(3β,5β,12β)-3,12,14-Trihydroxy-card-20(22)-enolide;* $\Delta^{20:22}$-3β,12β,14,21-tetrahydroxynorcholenic acid lactone; lanadigenin. $C_{23}H_{34}O_5$; mol wt 390.52. C 70.74%, H 8.78%, O 20.48%. The aglycone of digoxin. By hydrolysis of digoxin: Smith, *J. Chem. Soc.* **1930**, 508. From *Digitalis orientalis* L. and *D. lanata* Ehrh., Scrophulariaceae: Mannick, Schneider, *Arch. Pharm.* **279**, 223 (1941); Pataki *et al., Helv. Chim. Acta* **36**, 1295 (1953). Structure: Meyer, Reichstein, *Experientia* **9**, 253 (1953); Cardwell, Smith, ibid. 367; eidem, *J. Chem. Soc.* **1954**, 2012. Synthesis: P. Welzel, H. Stein, *Tetrahedron Letters* **22**, 3385 (1981).

Dihydrate, prismatic rods from dil alc. Anhydr as stout prisms from ethyl acetate, mp 222°. $[\alpha]_{546}^{20}$ +27.0° (c = 1.77 in methanol). Although it is a 3β-alcohol, it is not precipitated by digitonin (Pataki).

3,12-Diacetyldigoxigenin, prisms from dil methanol, mp 222-223°. $[\alpha]_{546}^{20}$ +61.3° (c = 2 in methanol).

3210. Digoxin. *(3β,5β,12β)-3-[(O-2,6-Dideoxy-β-D-ribo-hexopyranosyl-(1→4)-O-2,6-dideoxy-β-D-ribo-hexopyranosyl-*

(1→4)-2,6-dideoxy-β-D-ribo-hexopyranosyl)oxy]-12,14-dihydroxycard-20(22)-enolide; Cordioxil; Davoxin; Digacin; Dilanacin; Dixina; Dokim; Dynamos; Eudigox; Lanacordin; Lanicor; Lanoxin; LenoxiCaps; Lenoxin; Longdigox; Neo-Dioxanin; Rougoxin; Stillacor; Vanoxin. $C_{41}H_{64}O_{14}$; mol wt 780.95. C 63.06%, H 8.26%, O 28.68%. Secondary glycoside from *Digitalis lanata* Ehrh., or *D. orientalis* Lam., Scrophulariaceae: S. Smith, *J. Chem. Soc.* **1930**, 508; Stoll, *The Cardiac Glycosides* (London, 1937); M. M. Dhar *et al.,* Indian pat. **62,497** (1958 to Council Sci. Indust. Res.), *C.A.* **53**, 653b (1959). *See also* ref under Digoxigenin. Acid hydrolysis of digoxin yields 1 mol digoxigenin + 3 mols digitoxose. The sugar residue is attached to the hydroxyl group at C-3 of the aglycon. Clinical pharmacokinetics: J. K. Aronson, *Clin. Pharmacokinet.* **5**, 137 (1980). Comprehensive description: P. R. B. Foss, S. A. Benezra, *Anal. Profiles Drug Subs.* **9**, 207-243 (1980).

Radially arranged, four- and five-sided triclinic plates from dil alcohol or dil pyridine, decomp 230-265°. $[\alpha]_{D}^{25}$ +13.4° to 13.8° (c = 10 in pyridine). uv max (ethanol): 220 nm (ε 12800). Sol in dil alcohol, pyridine, or mixt of chloroform and alcohol. Almost insol in ether, acetone, ethyl acetate, chloroform, water. More sol in hot 80% alcohol than gitoxin.

β-Methyldigoxin, $C_{42}H_{66}O_{14}$*, medigoxin, metildigoxin, 4'''-O-methyldigoxin, 3β,12β,14β-trihydroxy-5β-card-20(22)-enolide-3-(4'''-O-methyltridigitoxoside), Cardiolan, Lanirapid, Lanitop.* Obtained by the O-methylation of digoxin: F. Kaiser *et al.,* S. Afr. pat. **68 06,079;** eidem, U.S. pat. **3,538,078** (1969, 1970 both to Boehringer, Mann.). Pharmacology and toxicity studies: W. Schaumann, R. Wegerle, *Arzneimittel-Forsch.* **21**, 225 (1971); H. Czerwek *et al.,* ibid. 231. Crystals, mp 227-231°. LD_{50} in rats, mice (mg/kg): 4.8, 4.9 i.v.; 6.2, 4.8 i.p.; 8.3, 7.8 orally (Czerwek).

α-Acetyldigoxin, $C_{43}H_{66}O_{15}$*, Lanatilin, Decurdil, Dioxanin, Sandolanid.* Obtained by enzymatic hydrolysis of digilanide. Prisms from methanol + chloroform, dec 225°. $[\alpha]_D^{20}$ +18.9° (pyridine). Very sparingly sol in ethyl acetate.

β-Acetyldigoxin, Cor-Puren, Novodigal, Agolanid, Kardiamed. Needles from alcohol + chloroform, dec 240°. $[\alpha]_D^{20}$ +30.4° (c = 1.2 in alc). More sol in ethyl acetate than the α-form.

THERAP CAT: Cardiotonic.
THERAP CAT (VET): Cardiotonic.

3211. Dihexyverine. *[1,1'-Bicyclohexyl]-1-carboxylic acid 2-(1-piperidinyl)ethyl ester;* 1-cyclohexylcyclohexanecarboxylic acid β-piperidinoethyl ester; dicyclohexylacetic acid 2-piperidinoethyl ester; dihexiverine. $C_{20}H_{35}NO_2$; mol wt 321.50. C 74.72%, H 10.97%, N 4.36%, O 9.95%. Prepn: Kopp, Tchoubar, *Bull. Soc. Chim. France* **1952**, 84. Pharmacology: Buchel *et al., Arch. Sci. Physiol.* **16**, 57 (1962); **21**, 537 (1967); Guyonneau, *Med. Exp.* **6**, 245 (1962).

Hydrochloride, $C_{20}H_{35}NO_2$.HCl, *JL-1078, Metaspas, Diverine, Olimplex, Seclin, Dispas, Neospasmina, Spasmodex, Spasmalex.* Crystals, mp 175° (copper block). mp 200° (capillary). LD_{50} i.p. in mice: 212 mg/kg (Buchel).
THERAP CAT: Anticholinergic.

3212. Dihydralazine. *2,3-Dihydro-1,4-phthalazinedione dihydrazone; 1,4-dihydrazinophthalazine.* $C_8H_{10}N_6$; mol wt 190.21. C 50.52%, H 5.30%, N 44.18%. Vasodilator. Prepn: J. Druey, U.S. pat. **2,484,785** (1949 to Ciba); J. Druey, B. H. Ringier, *Helv. Chim. Acta* **34**, 195 (1951); Zerweck, Kunze, U.S. pat. **2,786,839** (1957 to Cassella Farbw.). Pharmacology: P. A. Van Zwieten, *Arzneimittel-Forsch.* **18**, 79 (1968). Mechanism of action: I. W. Reimann *et al., Clin. Sci.* **61**, 319s (1981); eidem, *Clin. Exp. Pharmacol. Physiol.* **12**, 79 (1985). Cerebrovascular effects in rats: L. M. Auer *et al., Acta Med. Scand.* **Suppl. 678**, 73 (1983); D. I. Barry *et al., Stroke* **15**, 102 (1984). Hemodynamic responses in man: G. G. Belz, *Clin. Pharmacol. Ther.* **37**, 48 (1985). Clinical studies: A. Salvadeo *et al.*, *Int. J. Clin. Pharmacol. Ther. Toxicol.* **19**, 372 (1981); P. I. Salmela *et al.*, *Ann. Clin. Res.* **13**, 433 (1981). Toxicity data: G. Steiner *et al.*, *J. Med. Chem.* **24**, 59 (1981).

Orange needles from water, dec about 180°. LD_{50} i.p. in rats: 1084 μmol/kg (Steiner).

Sulfate, $C_8H_{10}N_6 \cdot H_2SO_4$, *Pressunic*. Needles, dec 233°.

Sulfate hemipentahydrate, *Depressan, Dihyzin, Nepresol, Népressol*.

Methanesulfonate, $C_8H_{10}N_6 \cdot CH_3SO_3H$, *dihydralazine mesylate, Nepresol Inject.*

THERAP CAT: Antihypertensive.

3213. Dihydrazine Sulfate. $H_{10}N_4O_4S$; mol wt 162.17. H 6.22%, N 34.55%, O 39.46%, S 19.77%. $(N_2H_4)_2 \cdot H_2SO_4$.

White crystalline flakes. Apparent density 55 lbs/cu ft. mp about 104°, decomposes about 180°. Soluble in water (25°): 202 g/100 ml H_2O. Relatively insol in most organic solvents. *See also* Hydrazine Sulfate.

Caution: Irritating to eyes, skin, mucous membranes.

3214. Dihydrocodeine. (*5α,6α*)-*4,5-Epoxy-3-methoxy-17-methylmorphinan-6-ol*; 6-hydroxy-3-methoxy-*N*-methyl-4,5-epoxymorphinan; dihydroneopine; drocode; Codhydrine; Dehacodin; DF 118; DH-codeine; Didrate; Dihydrin; Hydrocodin; Nadeine; Novicodin; Paracodin; Parzone; Rapacodin. $C_{18}H_{23}NO_3$; mol wt 301.39. C 71.73%, H 7.69%, N 4.65%, O 15.93%. Prepd by reduction of codeine or neopine: Skita, Franck, *Ber.* **44**, 2862 (1911); Wieland, Koralek, *Ann.* **433**, 269 (1923); Stein, *Pharmazie* **10**, 180 (1955).

Crystals from methanol + water, mp 112-113°; bp$_{15}$ 248°.

Bitartrate (acid tartrate), $C_{18}H_{23}NO_3 \cdot C_4H_6O_6$, *Fortuss, DHC, Dico*. Crystals from methanol, contains 66.8% dihydrocodeine when completely anhyd. mp 192-193° (Stein). The commercial medicinal grade usually melts at 186-190°. $[\alpha]_D^{25}$ −72° to −75° (c = 1 in H_2O). One gram dissolves in 4.5 ml water. Sparingly sol in alcohol. Insol in ether.

Note: This is a controlled substance (opiate) listed in the U.S. Code of Federal Regulations, Title 21 Part 1308.12 (1995).

THERAP CAT: Analgesic (narcotic). Antitussive.

3215. Dihydrocodeinone Enol Acetate. (*5α*)-*6,7-Didehydro-4,5-epoxy-3-methoxy-17-methylmorphinan-6-ol acetate (ester); demethyldihydrothebaine acetate (ester);* acetyldemethyldihydrothebaine; acetyldihydrocodeinone; thebacon. $C_{20}H_{23}NO_4$; mol wt 341.41. C 70.36%, H 6.79%, N 4.10%, O 18.75%. Prepn by refluxing dihydrocodeinone with acetic anhydride and anhyd sodium acetate: Small *et al., J. Org. Chem.* **3**, 204 (1938).

Needles from methanol, mp 154°. Practically insol in water. Soluble in most organic solvents.

Hydrochloride, $C_{20}H_{23}NO_4 \cdot HCl$, *Acedicon*. Crystals, mp 132-135° (dec), very sol in water, stable in boiling water.

Note: This is a controlled substance (opium derivative) listed in the U.S. Code of Federal Regulations, Title 21 Part 1308.11 (1995).

THERAP CAT: Analgesic (narcotic). Antitussive.

3216. Dihydroequilin. (*17β*)-*Estra-1,3,5(10),7-tetraene-3,17-diol*. $C_{18}H_{22}O_2$; mol wt 270.37. C 79.96%, H 8.20%, O 11.84%. Occurs in the 17α- and 17β-forms. Prepn of β-form by reduction of equilin with sodium in alc: David, *Acta Brev. Nederland* **4**, 63 (1934); Serini *et al., U.S. pat.* 2,221,340 (1940 to Schering). Isoln of β-form from pregnant mares' urine: Gaudry, Glen, *Ind. Chim. Belge, Suppl.* **2**, 435 (1959). Prepn of α-form by reduction of equilin with aluminum isopropoxide: Carol *et al., J. Biol. Chem.* **185**, 267 (1950). Isoln of α-form from pregnant mares' urine: Glen *et al., Nature* **177**, 753 (1956). α-Form was formerly called β-dihydroequilin and vice versa: Banes *et al., J. Biol. Chem.* **187**, 557 (1950).

β-Form

Crystals from actone, mp 174.5-174.6°. $[\alpha]_D^{20}$ +220° (dioxane).

17α-Form, plates from 30% ethanol, mp 205.5-205.6°. $[\alpha]_D^{20}$ +213° (ethanol).

3217. Dihydroergotamine. *9,10-Dihydro-12'-hydroxy-2'-methyl-5'-(phenylmethyl)ergotaman-3',6',18-trione.* $C_{33}H_{37}N_5O_5$; mol wt 583.69. C 67.91%, H 6.39%, N 12.00%, O 13.71%. α-Adrenergic blocker with selective venoconstrictor properties. Also binds to serotonin 5HT$_1$-receptors. Prepn from ergotamine: Stoll, Hofmann, *Helv. Chim. Acta* **26**, 2070 (1943). Clinical pharmacology: H. de Marées *et al., Eur. J. Clin. Pharmacol.* **30**, 685 (1986). Neurotransmitter receptor binding study: B. G. McCarthy, S. J. Peroutka, *Headache* **29**, 420 (1989). Clinical trials in migraine: P. Winner *et al., ibid.* **33**, 471 (1993); D. Ziegler *et al., Neurology* **44**, 447 (1994).

Strongly refractive prisms from dil acetone, contg 2 mols acetone and 2 mols water of crystn. mp 239°. $[\alpha]_D^{20}$ −64°; $[\alpha]_{546}^{20}$ −79° (c = 0.5 in pyridine). Insol in water. Sparingly sol in methanol, ethanol, chloroform, benzene.

Tartrate, $(C_{33}H_{37}N_5O_5)_2 \cdot C_4H_6O_6$, *Divegal*. Six-sided plates from methanol, dec 210-215°.

Methanesulfonate, $C_{33}H_{37}N_5O_5 \cdot CH_3SO_3H$, *dihydroergotamine mesylate, Agit, Angionorm, Dergotamine, DET MS, D.H.E. 45, Diergo, Dihydergot, Dirgotarl, Endophleban, Ergomimet, Ergont, Ergotonin, Ikaran, Migranal, Morena, Orstanorm, Séglor, Tonopres, Verladyn*. Large prisms from 95% alc. mp 230-235°. Moderately sol in water.

THERAP CAT: Antimigraine.

3218. Dihydro-β-erythroidine. (*3β*)-*1,6-Didehydro-14,17-dihydro-3-methoxy-16(15H)-oxaerythrinan-15-one; 12,13-didehydro-2,7,13,14-tetrahydro-α-erythroidine.* $C_{16}H_{21}NO_3$; mol wt 275.35. C 69.79%, H 7.69%, N 5.09%, O 17.43%. The more active of the two isomers that constitute the principal alkaloidal fraction of seeds from several *Erythrina* spp. In contrast to curare, *q.v.*, it retains its ability to block neuromuscular transmission even when administered orally: K. Unna, J. G. Greslin, *J. Pharmacol.* **80**, 53 (1944); K. Unna *et al.*, *ibid.* 39. Prepd by catalytic hydrogenation of β-erythroidine or a salt of β-erythroidine: Folkers, Koniuszy, U.S. pat. **2,370,651** (1945 to Merck & Co.). Structure: Boekelheide *et al.*, *J. Am. Chem. Soc.* **75**, 2550 (1953). Configuration: Hanson, *Proc. Chem. Soc.* **1963**, 52.

Crystals from anhydr ethyl ether, dec 85-86°. $[\alpha]_D^{25}$ +102.5°. Soluble in ethanol. Alkaline hydrolysis yields sodium dihydro-β-erythroidinate.

Hydrochloride, crystals from abs ethanol, mp 238°. $[\alpha]_D^{25}$ +124.7°.

Hydrobromide, crystals from abs ethanol or abs ethanol + abs ether, dec 242°. Bitter taste. $[\alpha]_D^{25}$ +106-107.5°. Soly in water 0.8 g/ml; in ethanol 0.5 g/100 ml.

3219. Dihydroisocodeine. *(5α,6β)-4,5-Epoxy-3-methoxy-17-methylmorphinan-6-ol;* DHIC. $C_{18}H_{23}NO_3$; mol wt 301.39. C 71.73%, H 7.69%, N 4.65%, O 15.93%. Differs from dihydrocodeine only by the spatial arrangement of the —CHOH— group. Prepn by catalytic reduction of isocodeine: Speyer, Krauss, *Ann.* **432**, 233 (1923); Lutz, Small, *J. Am. Chem. Soc.* **54**, 4724 (1932); Rapoport, Payne, *J. Org. Chem.* **15**, 1097 (1950); by epimerization of dihydrocodeine: Baizer, U.S. pat. **2,774,762** (1956 to N.Y. Quinine and Chem. Works).

Prisms from ethanol, mp 199-200°.

Acid tartrate, $C_{18}H_{23}NO_3 \cdot C_4H_6O_6$, crystals, mp 192°, $[\alpha]_D^{29}$ −65.3°.

Acid tartrate trihydrate, $C_{18}H_{23}NO_3 \cdot C_4H_6O_6 \cdot 3H_2O$, crystals from water, mp about 180°. $[\alpha]_D^{26}$ −62.4° (c = 1.94). Soly at 24° = 4.5 g/100 ml H_2O.

Methiodide, $C_{18}H_{23}NO_3 \cdot CH_3I$, crystals from ethanol, dec 269-272°.

3220. Dihydromorphine. *(5α,6α)-4,5-Epoxy-17-methylmorphinan-3,6-diol.* $C_{17}H_{21}NO_3$; mol wt 287.36. C 71.06%, H 7.37%, N 4.87%, O 16.70%. Prepd by hydrogenation of morphine or opium; by demethylation of tetrahydrothebaine; from dihydrocodeine. *Ref:* Small, Lutz, *Chemistry of the Opium Alkaloids*, Suppl. No. 103, Public Health Reports, Washington (1932); Eddy, Reid *J. Pharmacol.* **52**, 468 (1934); K. W. Bentley, *The Chemistry of the Morphine Alkaloids* (Oxford, 1954).

Monohydrate, prisms from chloroform, mp (dry) 157° (dec). Insoluble in water. Sol in acetone, alcohol, chloroform, dil acids.

Hydrochloride, $C_{17}H_{21}NO_3 \cdot HCl$, *paramorphan, paramorfan.* Prismatic crystals, unmelted at 280° *in vacuo*. $[\alpha]_D^{25}$ −112° (c = 1.6). Very sol in water; sparingly sol in abs alcohol.

Hydriodide, $C_{17}H_{21}NO_3 \cdot HI$, crystals, soften at 270°. mp 275°.

Caution: May be habit forming. This is a controlled substance (opium derivative) listed in the U.S. Code of

Federal Regulations, Title 21 Parts 329.1 and 1308.11 (1995).

THERAP CAT: Analgesic (narcotic).

3221. Dihydroresorcinol. *1,3-Cyclohexanedione;* hydroresorcinol. $C_6H_8O_2$; mol wt 112.13. C 64.27%, H 7.19%, O 28.54%.

Crystals, mp about 105° with decompn. Sol in water, alcohol, chloroform, acetone, boiling benzene; slightly sol in ether, carbon disulfide, petr ether.

3222. Dihydrostreptomycin. *O-2-Deoxy-2-(methylamino)-α-L-glucopyranosyl-(1→2)-O-5-deoxy-3-C-(hydroxymethyl)-α-L-lyxofuranosyl-(1→4)-N,N′-bis(aminoiminomethyl)-D-streptamine;* DHSM; DST; Abiocine; Vibriomycin. $C_{21}H_{41}N_7O_{12}$; mol wt 583.60. C 43.22%, H 7.08%, N 16.80%, O 32.90%. Semi-synthetic antibiotic prepd by reduction of streptomycin: Bartz *et al.*, *J. Am. Chem. Soc.* **68**, 2163 (1946); Fried, Wintersteiner, *ibid.* **69**, 79 (1947); Peck, U.S. pat. **2,498,574** (1950 to Merck & Co.); Carboni, Regna, U.S. pat. **2,522,858** (1950 to Pfizer); Levy, U.S. pat. **2,663,-685** (1953 to Schenley); Dolliver, Semenoff, U.S. pat. **2,717,-236** (1955 to Olin Mathieson); Kaplan, U.S. pat. **2,790,792** (1957 to Bristol); Sokol, Popino, U.S. pat. **2,784,181** (1957 to Cyanamid); Jurist, U.S. pat. **2,945,850** (1960 to Olin Mathieson). Isoln from fermentation broth of *Streptomyces humidus:* Nakazawa *et al.*, and Tatsuoka *et al.*, U.S. pats. **2,931,756** and **2,950,277** (both 1960 to Takeda). Crystn of free base (obtained from the sulfate): Rhodehamel *et al.*, *Science* **111**, 233 (1950). Crystn of the hydrochloride: Wolf *et al.*, *ibid.* **109**, 515 (1949); Wolf, U.S. pats. **2,590,139; 2,594,245** (both 1952 to Merck & Co.). Crystn of the sulfate: Solomons, Regna, *Science* **109**, 515 (1949); Wolf *et al.*, *ibid.* 515; Wolf, U.S. pats. **2,590,140; 2,590,141** (both 1952 to Merck & Co.); R. B. Peet, U.S. pat. **2,640,054** (1953 to Heyden Chem.); Katz, U.S. pat. **2,744,892** (1956 to Schenley). Total synthesis: Umezawa *et al.*, *J. Am. Chem. Soc.* **96**, 920 (1974); *eidem, Bull. Chem. Soc. Japan* **48**, 563 (1975); T. Yamasaki *et al.*, *J. Antibiot.* **31**, 1233 (1978).

R = CH3
R′ = CH2OH

Hydrate, crystals from aq acetone. Chars at 240°, turning black up to 300° without melting.

Trihydrochloride, $C_{21}H_{41}N_7O_{12} \cdot 3HCl$, amorphous solid, dec 190-195°, or crystals from methanol. Soly in methanol at 25°: 45 mg/ml for crystalline form; >1 g/ml for amorphous form. Crystals contain methanol of crystn which is lost on heating at 100°C. $[\alpha]_D^{25}$ −95° (1% soln).

Sesquisulfate, $(C_{21}H_{41}N_7O_{12})_2 \cdot 3H_2SO_4$, *Didromycin, Double-mycin, Sol-Mycin, Streptomagma.* Amorphous solid; crystalline form from water + (methanol, methyl ethyl ketone, or other low boiling solvent). Crystals, dec 255-

265°, also reported as dec 250°. $[\alpha]_D^{25}$ —88.5° (1% soln). Crystals show very little hygroscopicity in contrast to amorphous form. Both forms are very sol in water. Solubility in 50% methanol + water: 0.8 mg/ml for crystals; 100 mg/ml for amorphous form. Soly at 28° (mg/ml): >20 in water; 0.35 in methanol; 0.10 in ethanol: Weiss *et al.*, *Antibiot. Chemother.* **7**, 374 (1957). Each milligram contains 800 micrograms of dihydrostreptomycin base on a potency basis.

Pantothenate, $C_{21}H_{41}N_7O_{12}\cdot C_9H_{17}NO_5$, *Didrothenat, Pantostrep.*

THERAP CAT: Antibacterial (tuberculostatic).

THERAP CAT (VET): Antibacterial.

3223. Dihydrotachysterol. *(3β,5E,7E,10α,22E)-9,10-Secoergosta-5,7,22-trien-3-ol;* dichystrolum; anti-tetany substance 10; AT 10; Antitanil; Calcamine; Dygratyl; Dihydral; Hytakerol; Parterol; Tachyrol. $C_{28}H_{46}O$; mol wt 398.67. C 84.36%, H 11.63%, O 4.01%. Prepn by reduction of tachysterol: Windaus *et al.*, *Ann.* **499**, 198 (1932); v. Werder, U.S. pat. **2,228,491** (1941 to Winthrop), *cf. Z. Physiol. Chem.* **260**, 119 (1939). Activity similar to parathyroid extract.

Needles from 90% methanol, mp 125-127°. $[\alpha]_D^{22}$ +97.5° (chloroform). uv max: 242, 251, 261 nm ($E_{1cm}^{1\%}$ 870, 1010, 650). Insol in water. Easily sol in organic solvents.

THERAP CAT: Calcium regulator.

3224. Dihydrothebaine. *(5α)-6,7-Didehydro-4,5-epoxy-3,6-dimethoxy-17-methylmorphinan.* $C_{19}H_{23}NO_3$; mol wt 313.40. C 72.82%, H 7.40%, N 4.47%, O 15.32%. Prepd by reduction of thebaine: Small *et al.*, *J. Am. Chem. Soc.* **58**, 1457 (1936); K. W. Bentley, *The Chemistry of the Morphine Alkaloids* (Oxford, 1954) p 204.

Prisms from ethyl acetate, mp 162-163°. $[\alpha]_D^{20}$ —267° (c = 1.02 in benzene). Sol in alcohol, benzene, ethyl acetate. Insol in water, alkalies.

Methyl iodide, $C_{20}H_{23}NO_4\cdot CH_3I$, rods from alc, mp 257°.

3225. Dihydroxyacetone. *1,3-Dihydroxy-2-propanone;* 1,3-dihydroxydimethyl ketone; Protosol; Ketochromin. $C_3H_6O_3$; mol wt 90.08. C 40.00%, H 6.71%, O 53.28%. $HOCH_2COCH_2OH$. Produced from glycerol by *Acetobacter* sp. under aerobic conditions: Bernhauer, Schoen, *Z. Physiol. Chem.* **177**, 107 (1928); Rutten, *Rec. Trav. Chim.* **70**, 449 (1951); Bousfield *et al.*, *J. Inst. Brewing* **53**, 258 (1947); U.S. pat. **2,948,658** (1960 to Baxter). *Review:* Chem. & Eng. News **38**, 62 (Feb. 22, 1960), 54 (Aug. 15, 1960).

Cryst powder; fairly hygroscopic; characteristic odor; sweet, cooling taste. mp about 75-80°. The normal form is a dimer, slowly sol in 1 part water, 15 parts alcohol. When

freshly prepd reverts rapidly to monomer in soln. The monomer is very sol in water, alcohol, ether, acetone.

USE: Artificial tanning agent; active ingredient in *Man-Tan; Oxatone; Tanorama; Q.T. (Quick Tan); Tan Tone; Magic Tan.* Formulations: Barmak, *Am. Perfum.*, March 1960, p 57; Andreadis, Miklean, U.S. pat. **2,949,403** (1960).

3226. Dihydroxyaluminum Acetylsalicylate. *Dihydroxy(acetylsalicylato)aluminum;* dihydroxyaluminum aspirin. $C_9H_9AlO_6$; mol wt 240.15. C 45.01%, H 3.78%, Al 11.24%, O 39.97%. Prepn: Beekman, U.S. pat. **2,698,332** (1954 to Reheis); Schenck *et al.*, U.S. pat. **2,918,485** (1959 to Keystone Chemurgic).

May contain one or more molecules of $Al(OH)_3$. Stable in aq suspension at neutral pH, dec below pH 4.

THERAP CAT: Analgesic, antipyretic.

3227. Dihydroxyaluminum Aminoacetate. *(Glycinato-N,O)dihydroxyaluminum;* dihydroxy(glycinato)aluminum; aluminum aminoacetate (basic); aluminum dihydroxyaminoacetate; aluminum glycinate; glycine, aluminum salt; Ada; Alamine; Alminate; Alubasine; Alzinox; Aspogen; Dimothyn; Doraxamin; Elcosal; Robalate. $C_2H_6AlNO_4$; mol wt 135.06. C 17.79%, H 4.48%, Al 19.98%, N 10.37%, O 47.39%. $NH_2CH_2COOAl(OH)_2$. Prepd by adding a soln of aluminum isopropoxide to an aq soln of glycine: Krantz *et al.*, *J. Pharmacol.* **82**, 247 (1944).

Very fine powder. Bland taste. Insol in water. Mixes easily with water forming suspensions which do not separate readily. At 25° the pH of the aq suspension (1 in 25) is 7.4. One gram consumes 200 ml of 0.1N HCl under conditions of the pharmacopoeial test for aluminum hydroxide gel.

THERAP CAT: Antacid.

3228. Dihydroxyaluminum Sodium Carbonate. *[Carbonato(1—)-O]dihydroxyaluminum monosodium salt;* aluminum sodium carbonate hydroxide; Kompensan; Minicid. CH_2AlNaO_5; mol wt 144.00. C 8.34%, H 1.40%, Al 18.74%, Na 15.97%, O 55.56%. $(HO)_2AlOCO_2Na$. Prepd by the reaction between an aluminum alkoxide and $NaHCO_3$ in water: Grote, U.S. pat. **2,783,179** (1957 to Chattanooga Medicine).

Amorphous powder or poorly formed crystals. May contain 10 to 11% H_2O. d 2.144. pH of water suspension: 9.7. Acid consuming power: 230 ml 0.1N HCl/gram.

THERAP CAT: Antacid.

3229. Dihydroxymaleic Acid. *(Z)-2,3-Dihydroxy-2-butenedioic acid;* 1,2-dihydroxyethylenedicarboxylic acid. $C_4H_4O_6$; mol wt 148.07. C 32.45%, H 2.72%, O 64.83%. Prepn from tartaric acid: Fenton, *J. Chem. Soc.* **87**, 811 (1905); Nef, *Ann.* **357**, 291 (1907).

Plates from water, usually crystallizes as the dihydrate. The anhydr acid dec 155° without melting. Slightly sol in cold water, ether, acetic acid; more sol in alcohol. Aq solns are unstable.

USE: In the detection of fluorides and of titanium. Has been proposed as antioxidant for frozen foods.

3230. 9,10-Dihydroxystearic Acid. *9,10-Dihydroxyoctadecanoic acid.* $C_{18}H_{36}O_4$; mol wt 316.48. C 68.31%, H 11.47%, O 20.22%. $CH_3(CH_2)_7CH(OH)CH(OH)(CH_2)_7COOH$.

White, odorless, tasteless, lustrous crystals; fatty feel. mp 132-136°. Insol in water. Sol in hot alc or acetone, slightly in ether.

USE: Manuf cosmetic and toilet preparations.

3231. Dihydroxytartaric Acid. *Tetrahydroxybutanedioic acid;* diketosuccinic acid. $C_4H_6O_8$; mol wt 182.09. C

26.39%, H 3.32%, O 70.29%. HOOCC(OH)$_2$C(OH)$_2$-COOH.

White, cryst powder. mp ~114-115° with decompn. Very sol in water; the aq soln dec on heating.

USE: As a reagent for the determination of sodium.

3232. 2,4-Diiodoaniline. *2,4-Diiodobenzenamine.* C$_6$-H$_5$I$_2$N; mol wt 344.92. C 20.89%, H 1.46%, I 73.58%, N 4.06%.

Brown crystals. d 2.75. mp 95-96°. Slightly sol in cold water, moderately in hot water or in alcohol; freely sol in chloroform, ether, acetone, carbon disulfide, boiling alc.

3233. 4′,5′-Diiodofluorescein. *3′,6′-Dihydroxy-4′,5′-diiodospiro[isobenzofuran-1(3H),9′-[9H]xanthen]-3-one;* hydroxydiiodo-o-carboxyphenylfluorone; D & C Orange No. 10; C.I. Solvent Red 73; C.I. 45425:1. C$_{20}$H$_{10}$I$_2$O$_5$; mol wt 584.11. C 41.13%, H 1.73%, I 43.45%, O 13.70%. Prepd by reaction of fluorescein with I$_2$ and HIO$_4$ or ICl in alkali: *Colour Index* vol. 4 (3rd ed., 1971) p 4427.

Orange-red powder. Slightly sol in water; sol in alcohol, in alkali hydroxide solns.

Sodium salt, C$_{20}$H$_8$I$_2$Na$_2$O$_5$, *Erythrosine Extra Yellowish, D & C Orange No. 11, C.I. Acid Red 95, C.I. 45425.*

Note: Structure depicted as the open tautomer.

USE: As an adsorption indicator in the determn of iodides in the presence of chlorides and bromides. The salts in dyeing and printing cotton, in printing half-silk, in dyeing jute, straw, etc.

3234. 3,5-Diiodosalicylic Acid. *2-Hydroxy-3,5-diiodobenzoic acid.* C$_7$H$_4$I$_2$O$_3$; mol wt 389.92. C 21.56%, H 1.03%, I 65.09%, O 12.31%. Prepn from salicylic acid and ICl: Woollett, Johnson, *Org. Syn.* coll. vol. II, 343 (1943); and iodine + H$_2$O$_2$: Jurd, *Aust. J. Sci. Res.* 3A, 587 (1950).

Colorless, odorless needles or slightly yellow cryst powder. Sweetish, bitter taste. mp 235-236° dec. Sol in 5200 parts water at 25°; freely sol in alcohol, ether and most other organic solvents. Practically insol in CHCl$_3$, benzene.

Bismuth salt, *Bijosal:* Pollano, *Minerva Med.* **1930 I**, 786; *C.A.* **24**, 4854^7 (1930).

Ethyl ester, C$_9$H$_8$I$_2$O$_3$, *ethyl 3,5-diiodosalicylate.* Crystals, mp 133°. Practically insol in H$_2$O. Sol in alc, ether, benzene.

USE: As I$_2$ source in foods; in animal feeds as growth promotant.

3235. 3,5-Diiodothyronine. *O-(4-Hydroxyphenyl)-3,5-diiodotyrosine; 3-[4-(p-hydroxyphenoxy)-3,5-diiodophenyl]-alanine.* C$_{15}$H$_{13}$I$_2$NO$_4$; mol wt 525.08. C 34.31%, H 2.50%, I 48.34%, N 2.67%, O 12.19%. Prepn of DL-form: Harington, McCartney, *Biochem. J.* **21**, 852 (1927); Borrows *et al.,*

J. Chem. Soc. **1949,** Suppl. Issue No. 1, S199, S204; Siedel *et al.,* U.S. pat. **2,894,977** (1959 to Hoechst). Prepn of L(+)-form: Harington, *Biochem. J.* **22,** 1429 (1928); Chambers *et al., J. Chem. Soc.* **1949,** 3424; Hillmann, U.S. pat. **2,886,592** (1959); as the hydrochloride: Anthony, U.S. pat. **2,950,315** (1960 to Baxter); Meltzer, U.S. pat. **3,102,136** (1963 to Warner-Lambert). Prepn of D(−)-form: Harington, *loc. cit.;* Elks, Waller, *J. Chem. Soc.* **1952,** 2366.

DL-Form, plates, dec 256-257°.

L(+)-Form, crystals, dec 256°. [α]$_D^{22}$ +26° (c = 1.06 in a 1:2 mixture of 1N HCl + alcohol).

L(+)-Form hydrochloride, crystals, mp 235-240°.

D(−)-Form, crystals, dec 256° (Harington), 265° (Elks, Waller). [α]$_D^{20}$ −27.1° (c = 1 in 1N alc HCl:alc, 1:2 by vol).

USE: An intermediate in manuf of thyroxine.

3236. 3,5-Diiodotyrosine. *3,5-Diiodo-4-hydroxy-β-phenylalanine; iodogorgoic acid; Agontan.* C$_9$H$_9$I$_2$NO$_3$; mol wt 432.98. C 24.97%, H 2.10%, I 58.62%, N 3.23%, O 11.09%. Found in the skeletal proteins of corals, sponges, and other marine organisms. Isoln from *Gorgonia cavollini* as the DL acid: Drechsel, *Z. Biologie* **33**, 99 ; *Jahresber. Tierchemie* **1896,** 574; Henze, *Z. Physiol. Chem.* **38,** 71 (1903); isoln from the common sponge: Wheeler, Mendel, *J. Biol. Chem.* **7,** 1 (1909-10); Low, *J. Marine Res.* **10,** 239 (1951); from marine algae: Coulson, *Chem. & Ind. (London)* **1953,** 997; prepn from L- or DL-tyrosine: Henze, *J. Physiol. Chem.* **51,** 67 (1907); Borrows *et al., J. Chem. Soc.* **1949,** Suppl. Issue no. 1, S185; Jurd, *J. Am. Chem. Soc.* **77,** 5747 (1955); Boyle, Zlatkis, U.S. pat. **2,835,700** (1958 to Basic, Inc.).

L-Form, bunches of needles from water or 70% alcohol, dec 213°. [α]$_D^{20}$ +2.89° (0.246 g in 5 g 4% HCl); [α]$_D^{20}$ +2.27° (0.227 g in 5 g 25% NH$_3$). pK$_1$ 2.12; pK$_2$ 6.48; pK$_3$ 7.82. Soly in water (g/l): at 0° = 0.204; at 25° = 0.617; at 50° = 1.862; at 75° = 5.62; at 100° = 17.00. On boiling with dil alc the crystals swell, and after prolonged boiling a gelatinous precipitate is formed.

DL-Form, double wedge crystals from 70% alc, rectangular plates from water. Dec about 200°. Soly in water (g/l): at 0° = 0.149; at 25° = 0.340; at 50° = 0.773 . On cooling a boiled hydroalcoholic soln, no gelatinous precip is formed.

THERAP CAT: Thyroid inhibitor.

3237. Diisoamylamine. *3-Methyl-N-(3-methylbutyl)-1-butanamine;* isodiamylamine. C$_{10}$H$_{23}$N; mol wt 157.30. C 76.36%, H 14.74%, N 8.90%. (CH$_3$)$_2$CHCH$_2$CH$_2$NHCH$_2$-CH$_2$CH(CH$_3$)$_2$.

Liquid. d$_4^{20}$ 0.767. bp 188°. mp −44°. n$_D^{21}$ 1.4229. Slightly sol in water; sol in alcohol, chloroform, ether.

Hydrochloride, C$_{10}$H$_{23}$N.HCl. Vitreous mass, mp 276°. Freely sol in water, alcohol, ether.

Caution: Irritating to skin, mucous membranes. Has pressor effect.

3238. Diisobutyl Sodium Sulfosuccinate. *Sulfosuccinic acid diisobutyl ester S-sodium salt;* Aerosol IB; Alphasol IB. C$_{12}$H$_{21}$NaO$_7$S; mol wt 332.35. C 43.37%, H 6.37%, Na 6.92%, O 33.70%, S 9.65%. The isobutyl or butyl or 1-methylpropyl diester of the monosodium salt of sulfosuccinic acid or a mixture of all three esters. Wetting agent prepd by the action of the appropriate alcohols on maleic anhydride followed by addition of sodium bisulfite: Jaeger, U.S. pats. **2,028,091; 2,176,423; Brit.** pat. **446,568; Fr.** pat. **776,-495.**

The mixture of the three esters is available as a white, powder-like, easily grindable material. Soly in water at 25°: 760 g/l; at 60°: 804 g/l. Maximum concn of electrolyte soln in which 1% of the wetting agent is sol: 20% NaCl; 20% NH_4Cl; 30% $NaNO_3$; 20% Na_2SO_4. Also sol in glycerol, pine oil, oleic acid. Insol in acetone, kerosene, liq petrolatum, carbon tetrachloride, 2B ethanol, benzene, olive oil. Surface tension (dynes/cm) in water: 0.001% = 72.2; 0.1% = 67.5; 0.5% = 53.4; 1% = 49.1. Interfacial tension 1% in water vs. liq petrolatum: 10 seconds = 32.5 dynes/cm; 30 seconds = 32.3 dynes/cm; 15 minutes = 31.2 dynes/cm. Stable in acid and neutral solns; hydrolyzes in alkaline solns.

Caution: Irritating to eyes, mucous membranes.

USE: Wetting agent.

3239. Diisopromine. *N,N-Bis(1-methylethyl)-γ-phenylbenzenepropanamine; N,N*-diisopropyl-3,3-diphenylpropylamine; 1,1-diphenyl-3-diisopropylaminopropane; disopromine. $C_{21}H_{29}N$; mol wt 295.47. C 85.37%, H 9.89%, N 4.74%. Prepn: **Brit. pat. 808,158** (1959 to N. V. Nederland. Comb. voor Chem. Ind. and Janssen, N. V.).

Hydrochloride, $C_{21}H_{29}N.HCl$, *Agofell, Bilagol.* Crystals, mp 175-176°. uv max: 254, 259.5, 268.8 nm. Soluble in water, methanol, ethanol, chloroform. Practically insol in ether, aq solns of alkalies and ammonia.

THERAP CAT: Antispasmodic.

3240. Diisopropylamine. *N-(1-Methylethyl)-2-propanamine;* DIPA. $C_6H_{15}N$; mol wt 101.19. C 71.22%, H 14.94%, N 13.84%. $[(CH_3)_2CH]_2NH$. Prepn: A. Siersch, *Ann.* **148**, 263 (1868); M. Van der Zande, *Rec. Trav. Chim.* **8**, 202 (1889). Toxicity study: H. F. Smyth *et al., Arch. Ind. Hyg. Occup. Med.* **10**, 61 (1954).

Liquid; characteristic odor; strongly alkaline. d^{22} 0.722. bp 84°. Flash pt, open cup: 21°F (−6°C). Sol in water, alc. LD_{50} orally in rats: 0.77 g/kg (Smyth).

Lithium salt, *LDA, lithiodiisopropylamine, lithium diisopropylamide.* Review of chemistry: *Reagents for Organic Synthesis* vol. **8**, M. Fieser, Ed. (Wiley-Interscience, New York, 1980) pp 292-299. Powder, melts with decompn. Air and moisture sensitive.

Caution: Potential symptoms of overexposure are nausea, vomiting; headache; eye irritation, visual disturbance; pulmonary irritation. Direct contact may cause skin irritation. *See NIOSH Pocket Guide to Chemical Hazards* (DHHS/NIOSH 90-117, 1990) p 94.

USE: In organic synthesis.

3241. Diisopropylamine Dichloroacetate. *Dichloroacetic acid compd with N-(1-methylethyl)-2-propanamine (1:1);* dichloroacetic acid diisopropylammonium salt; diisopropylammonium dichloroacetate; diisopropylamine dichloroethanoate; DADA; DIPA-DCA; IS-401; Dapocel; Dedyl; DIEDI; Disotat; Kalodil; Oxypangam; Tensicor. $C_8H_{17}Cl_2NO_2$; mol wt 230.13. C 41.75%, H 7.45%, Cl 30.81%, N 6.09%, O 13.90%. Prepn: **Brit. pat. 862,248** (1961 to Italseber). Pharmacology: V. A. E. Kraushaar *et al., Arzneimittel-Forsch.* **13**, 109 (1963); P. W. Stacpoole, *J. Clin. Pharmacol. J. New Drugs* **9**, 282 (1969). Mutagenicity study: M. D. Gelernt, V. Herbert, *Nutr. Cancer* **3**, 129 (1982). Constituent of some products sold in the U.S. as "pangamic acid".

Crystals, mp 119-121°. Soly in water: > 50%. LD_{50} orally in mice: 1700 mg/kg (Kraushaar).

THERAP CAT: Vasodilator, hypotensive.

3242. Diisopropyl Paraoxon. *Bis(1-methylethyl)phosphoric acid 4-nitrophenyl ester;* diisopropyl p-nitrophenyl phosphate; Mioticol; Propicol. $C_{12}H_{18}NO_6P$; mol wt 303.25. C 47.53%, H 5.98%, N 4.62%, O 31.66%, P 10.21%. Prepn: **Brit. pat. 687,596** (1953 to Albright & Wilson).

THERAP CAT: Cholinergic (ophthalmic).

3243. Dikegulac. *2,3:4,6-Bis-O-(methylethylidene)-α-L-xylo-2-hexulofuranosonic acid; α-2,3:4,6-di-O-isopropylidene-L-xylo-hexulofuranosonic acid;* di-O-isopropylidene-2-keto-L-gulonic acid; diacetone-2-ketogulonic acid; diacetone-2-oxo-L-gulonic acid; oxogulonic acid diacetonide. $C_{12}H_{18}O_7$; mol wt 274.27. C 52.55%, H 6.61%, O 40.83%. Intermediate in the manuf of ascorbic acid, q.v. Manufacturing process: G. M. Jaffe, E. J. Pleven, **Ger. pat. 2,123,621**; *eidem,* **U.S. pat 3,832,355** (1970, 1974 both to Hoffmann-La Roche). Use as a plant growth regulator: W. Szkrybalo, **Ger. pat. 2,339,239**; *eidem,* **U.S. pat. 4,337,080** (1974, 1982 both to Hoffmann-La Roche); P. Bocian *et al., Nature* **258**, 142 (1975). Physicochemical properties, toxicity and growth retardant effect: W. H. de Silva *et al., Proc. Brit. Crop. Prot. Conf.-Weeds* **1976**, 349. Activity: S. S. Purohit, *Comp. Physiol. Ecol.* **4**, 264 (1979); **6**, 261 (1981).

Sodium salt, $C_{12}H_{17}NaO_7$, *dikegulac-sodium, Ro-7-6145, Atrinal.* Powder, mp > 300°. Vapor pressure at 25°: < 10^{-10} mm Hg. Soly at 20° (g/l): water 590; methanol 390; ethanol 230; chloroform 60; acetone < 10; hexane < 10; cyclohexanone < 10. LD_{50} in mice, male, female rats (mg/kg): 19500, 31000, 18000 orally; LC_{50} (96 hr) in bluegill sunfish, rainbow trout (ppm): > 10000, > 5000 (de Silva).

USE: Plant growth regulator; herbicide.

3244. Dilazep. *3,4,5-Trimethoxybenzoic acid (tetrahydro-1H-1,4-diazepine-1,4(5H)-diyl)di-3,1-propanediyl ester; 3,4,5-trimethoxybenzoic acid diester with tetrahydro-1H-1,4-diazepine-1,4(5H)-dipropanol;* 1,4-bis[3-(3,4,5-trimethoxybenzoyloxy)propyl]perhydro-1,4-diazepine; *N,N'*-bis[3-(3,4,5-trimethoxybenzoyloxy)propyl]homopiperazine; *N,N'*-(bis-ω-hydroxypropyl)homopiperazine 3,4,5-trimethoxybenzoate (diester). $C_{31}H_{44}N_2O_{10}$; mol wt 604.70. C 61.57%, H 7.33%, N 4.63%, O 26.46%. Prepn: **Brit. pat. 1,107,470** corresp to Arnold *et al.,* **U.S. pat. 3,532,685** (1968, 1970 both to Asta-Werke). Series of articles on pharmacology and metabolism: *Arzneimittel-Forsch.* **22**, 639-666 (1972). Toxicology: H. H. Abel *et al., ibid.* 667. Clinical results: Messerich, *Med. Welt.* **1972**, 563.

Dihydrochloride, $C_{31}H_{44}N_2O_{10}\cdot 2HCl$, *Asta C 4898, Comelian, Cormelian, Labitan.* Crystals from ethanol, mp 194-198° (monohydrate). LD_{50} in male mice, male rats (mg/kg): 26.6, 19.1 i.v.; 161, 90.1 i.p.; 3740, >2150 orally (Abel).

THERAP CAT: Vasodilator (coronary).

3245. Dilevalol. *[R-(R*,R*)]-2-Hydroxy-5-[1-hydroxy-2-[(1-methyl-3-phenylpropyl)amino]ethyl]benzamide; (R,R)-*labetalol; Sch-19927; Dilevalon; Levadil; Unicard. $C_{19}H_{24}$-N_2O_3; mol wt 328.41. C 69.49%, H 7.37%, N 8.53%, O 14.62%. Non-selective β-adrenergic blocker with vasodilating activity; active isomer of labetalol, *q.v.* For prepn of racemate *see* labetalol. Synthesis and preliminary pharmacology: J. E. Clifton *et al., J. Med. Chem.* **25**, 670 (1982); E. H. Gold *et al.,* ibid. 1363. Absolute configuration: P. Murray-Rust *et al., Acta Crystallogr.* **C40**, 825 (1984). Adrenoceptor blocking properties in comparison with labetalol: E. J. Sybertz *et al., J. Pharmacol. Exp. Ther.* **218**, 435 (1981). Antihypertensive and hemodynamic effects in rats and dogs: T. Baum *et al.,* ibid. 444; and cardiac effects: J. J. Lynch *et al.,* ibid. **239**, 719 (1986). HPLC determn in biological fluids: K. B. Alton *et al., J. Chromatog.* **425**, 363 (1988). Clinical evaluation in hypertension: J. Soberman *et al., J. Clin. Hypertens.* **3**, 271 (1987); J. D. Wallin *et al., Arch. Intern. Med.* **148**, 534 (1988). Symposium on pharmacology and clinical efficacy: *J. Cardiovasc. Pharmacol.* **11**, Suppl. 2, S1-S45 (1988).

Gum. $[\alpha]_D$ −21.7°.
Hydrochloride, $C_{19}H_{24}N_2O_3\cdot HCl$, polymorphic crystals from ethanol, mp 133-134° (dec), mp 192-193.5° (dec). $[\alpha]_D^{26}$ −30.6° (c = 1.0 in ethanol).

THERAP CAT: Antihypertensive.

3246. Diloxanide. *2,2-Dichloro-N-(4-hydroxyphenyl)-N-methylacetamide; 2,2-dichloro-4'-hydroxy-N-methylacetanilide;* N-dichloroacet-4-hydroxy-N-methylanilide; dichloroacet-4-hydroxy-N-methylanilide; 4-hydroxy-N-methyldichloroacetanilide; Entamide; Ame-Boots. $C_9H_9Cl_2NO_2$; mol wt 234.08. C 46.18%, H 3.88%, Cl 30.29%, N 5.98%, O 13.67%. Prepn: Oxley *et al.,* U.S. pat. **2,912,438** (1959 to Boots Pure Drug).

Crystals from ethyl acetate, mp 175°.
Furoate, *diloxanide 2-furoic acid ester, Furamide, Histomibal, Miforon.* Do not confuse with the amide of furoic acid.

THERAP CAT: Antiamebic.

3247. Diltiazem. *(2S-cis)-3-(Acetyloxy)-5-[2-(dimethylamino)ethyl]-2,3-dihydro-2-(4-methoxyphenyl)-1,5-benzothiazepin-4(5H)-one; (+)-cis-5-[2-(dimethylamino)ethyl]-2,3-dihydro-3-hydroxy-2-(p-methoxyphenyl)-1,5-benzothiazepin-4(5H)-one acetate (ester).* $C_{22}H_{26}N_2O_4S$; mol wt 414.53. C 63.75%, H 6.32%, N 6.76%, O 15.44%, S 7.74%. Calcium channel blocker with coronary vasodilating activity. Prepn: H. Kugita *et al.,* Ger. pat. **1,805,714;** *eidem,* U.S. pat. **3,562,257** (1969, 1971 both to Tanabe Seiyaku); *eidem,*

Chem. Pharm. Bull. **19**, 595 (1971). Resolution of optical isomers: H. Inoue *et al., Yakugaku Zasshi* **93**, 729 (1973), *C.A.* **79**, 66331w (1974); Japan. Kokai 59 20,273 (1984 to Tanabe Seiyaku), *C.A.* **101**, 38486e (1984). Stereospecific synthesis: K. Igarashi, T. Honma, Ger. pat. **3,415,035;** *eidem,* U.S. pat. **4,552,695** (1984, 1985 both to Shionogi). Vasodilating action is stereospecific for the *d-cis*-isomer. Structure-activity studies: Sato *et al., Arzneimittel-Forsch.* **21**, 1338 (1971); T. Nagao *et al., Chem. Pharm. Bull.* **21**, 92 (1973). Pharmacology and toxicity: *eidem, Japan. J. Pharmacol.* **22**, 467 (1972). Metabolism: Meshi *et al., Chem. Pharm. Bull.* **19**, 1546 (1971). Comparative study in variant angina: D. D. Waters *et al., Am. J. Cardiol.* **47**, 179 (1981). Clinical effect in reinfarction: R. S. Gibson *et al., N. Engl. J. Med.* **315**, 423 (1986). Review of pharmacology and efficacy: M. Chaffman, R. N. Brogden, *Drugs* **29**, 387-454 (1985). Symposium: *Acta Pharmacol. Toxicol.* **57**, Suppl. II, 1-73 (1985).

d-cis Form hydrochloride, $C_{22}H_{26}N_2O_4S\cdot HCl$, *CRD-401, Adizem, Altiazem, Anginyl, Angizem, Britiazim, Bruzem, Calcicard, Cardizem, Citizem, Cormax, Deltazen, Diladel, Dilpral, Dilrene, Dilzem, Dilzene, Herbesser, Masdil, Tildiem.* Fine needles from ethanol-isopropanol, mp 207.5-212°. $[\alpha]_D^{24}$ +98.3 ± 1.4° (c = 1.002 in methanol). Freely sol in water, methanol, chloroform; slightly sol in abs ethanol. Practically insol in benzene. LD_{50} in male, female mice, male, female rats (mg/kg): 61, 58, 38, 39 i.v.; 260, 280, 520, 550 s.c.; 740, 640, 560, 610 orally (Nagao, 1972).

THERAP CAT: Antianginal; antihypertensive; antiarrhythmic (class IV).

3248. Dimecrotic Acid. *3-(2,4-Dimethoxyphenyl)-2-butenoic acid; 2,4-dimethoxy-β-methylcinnamic acid; 3-(2,4-dimethoxyphenyl)crotonic acid.* $C_{12}H_{14}O_4$; mol wt 222.24. C 64.85%, H 6.35%, O 28.80%. Prepn: Ger. pat. **1,915,023** (1969 to Unicler), *C.A.* **72**, 66630y (1970). TLC determn in pharmaceutics: M.-F. Etcheverry *et al., Sci. Tech. Pharm.* **7**, 51 (1978).

mp 149°.
Magnesium salt, $C_{24}H_{26}MgO_8$, *Fisiobil, Hepadial.* mp 135°. LD_{50} in mice, rats (mg/kg): 1.3, 1.0 i.p. (Ger. pat.).

THERAP CAT: Choleretic.

3249. Dimefline. *8-[(Dimethylamino)methyl]-7-methoxy-3-methyl-2-phenyl-4H-1-benzopyran-4-one; 8-[(dimethylamino)methyl]-7-methoxy-3-methylflavone;* 8-dimethylaminomethyl-7-methoxy-3-methyl-2-phenylchromone. $C_{20}H_{21}$-NO_3; mol wt 323.39. C 74.28%, H 6.55%, N 4.33%, O 14.84%. Prepn: Da Re *et al., Arzneimittel-Forsch.* **10**, 800 (1960); Brit. pat. **882,537** corresp to Da Re, U.S. pat. **3,-147,258** (1961, 1964 both to Recordati). Pharmacology: Setnikar *et al., J. Pharmacol. Exp. Ther.* **128**, 176 (1960); *eidem, J. Med. Pharm. Chem.* **3**, 471 (1961).

Hydrochloride, $C_{20}H_{21}NO_3 \cdot HCl$, *Rec-7-0267, Remeflin.* Crystals from alcohol + ether, dec 213-214°.

THERAP CAT: Respiratory stimulant.

3250. Dimefox. *Tetramethylphosphorodiamidic fluoride;* tetramethyldiamidophosphoric fluoride; bis(dimethyl-amido)phosphoryl fluoride; fluophosphoric acid di(dimethylamide); bisdimethylaminofluorophosphine oxide; bis(dimethylamido)fluorophosphate; Pestox XIV (obsolete); Terrasytam. $C_4H_{12}FN_2OP$; mol wt 154.12. C 31.17%, H 7.85%, F 12.33%, N 18.18%, O 10.38%, P 20.10%. Prepd by fluorination of bis(dimethylamido)phosphoryl chloride: Schrader, *Angew. Chem.,* suppl. **62,** 18 (1951); *B.I.O.S. Final Rept.* no. **714** (1947); Holmsted, *Acta Physiol. Scand.,* suppl. **90,** 33 (1951); Heap, Saunders, *J. Chem. Soc.* **1948,** 1313; **Brit.** pat. **692,446** (1953); McCombie *et al.,* **U.S.** pat. **2,489,-917** (1949). Toxicity data: A. J. Okinaka *et al., J. Pharmacol. Exp. Ther.* **112,** 231 (1954). *Review:* Kilby, *Chem. & Ind. (London)* **1953,** 856.

Liquid. Fishy odor. d_4^{20} 1.1151. $bp_{4.0}$ 67°; bp_{15} 86°. n_D^{20} 1.4267. Freely sol in water, ether, benzene. Aq solns are stable. LD_{50} in rats (mg/kg): 5.0 i.p.; 7.5 orally (Okinaka).

Caution: A highly toxic cholinesterase inhibitor. Symptoms similar to parathion, *q.v.: Clinical Toxicology of Commercial Products,* R. E. Gosselin *et al.,* Eds. (Williams & Wilkins, Baltimore, 5th ed., 1984) Section II, p 299.

USE: Insecticide; acaricide.

3251. Dimemorfan. *(9α,13α,14α)-3,17-Dimethylmorphinan; d-3-methyl-N-methylmorphinan; AT-17.* $C_{18}H_{25}N$; mol wt 255.40. C 84.65%, H 9.87%, N 5.48%. Prepn: Murakami *et al.,* **Ger.** pat. **2,128,607;** *eidem,* **U.S.** pat. **3,786,054** (1971, 1974 both to Yamanouchi); *eidem, Chem. Pharm. Bull.* **20,** 1706 (1972). Pharmacology: Ida, Fujii, *Oyo Yakuri* **6,** 1207 (1972); Y. Kasé *et al., Arzneimittel-Forsch.* **26,** 353 (1976).

Pale yellow oil, $bp_{0.3}$ 130-136°; or white crystals from acetone, mp 90-93°.

Phosphate, $C_{18}H_{25}N \cdot H_3PO_4$, *Astomin.* White to yellowish-white, odorless, bitter crystals, mp 267-269°. $[\alpha]_D^{23}$ +25.7° (c = 0.5 in methanol). Freely sol in glacial acetic acid; sparingly sol in water, methanol. Practically insol in ethanol, acetone, chloroform, benzene, ether. LD_{50} in mice (mg/kg): 223 s.c.; 475 orally (Kasé).

THERAP CAT: Antitussive.

3252. Dimenhydrinate. *8-Chloro-3,7-dihydro-1,3-dimethyl-1H-purine-2,6-dione compd with 2-(diphenylmethoxy)-N,N-dimethylethanamine (1:1); 2-(benzhydryloxy)-N,N-dimethylethylamine 8-chlorotheophyllinate;* β-dimethyl-aminoethyl benzhydryl ether 1,3-dimethyl-8-chloroxanthine; diphenhydramine 8-chlorotheophyllinate; *N,N*-dimethyl-2-diphenylmethoxyethylamine 8-chlorotheophyllinate; *O*-benzhydryldimethylaminoethanol 8-chlorotheophyllinate; chloranautine; Amosyt; Anautine; Andramine; Antemin; Diamarin; Dimate; Dramamine; Dramarin; Dramocen; Dramyl; Emedyl; Emes; Epha; Gravol; Menhydrinate; Reidamine; Removine; Travel-Gum; Travelin; Travelmin; Vomex A; Xamamina; Faston. $C_{24}H_{28}ClN_5O_3$; mol wt 469.97. C 61.34%, H 6.01%, Cl 7.54%, N 14.90%, O 10.21%. Prepn: Cusic, *Science* **109,** 574 (1949); Cusic, **U.S.** pats. **2,499,058** and **2,534,813** (1950 to Searle).

Crystals, mp 102-107°. Freely sol in alcohol, chloroform. Soluble in benzene. Soly in water: about 3 mg/ml. Almost insol in ether. A satd aq soln has a pH between 6.8 and 7.3.

THERAP CAT: Antiemetic.

THERAP CAT (VET): Has been used as an antiemetic.

3253. Dimenoxadol. *α-Ethoxy-α-phenylbenzeneacetic acid 2-(dimethylamino)ethyl ester; ethoxydiphenylacetic acid 2-dimethylaminoethyl ester; α,α-diphenyl-α-ethoxyacetic acid β-dimethylaminoethyl ester; 2-dimethylaminoethyl ethoxydiphenylacetate; Lokarin.* $C_{20}H_{25}NO_3$; mol wt 327.42. C 73.37%, H 7.70%, N 4.28%, O 14.66%. Prepd by the reaction of diphenylethoxyacetic acid with dimethylchloroethylamine in the presence of isopropanol: Klosa, *Arch. Pharm.* **288,** 42 (1955). Improved prepn: C. N. Yakhontov *et al., Pharm. Chem. J.* **8,** 189 (1974). HPLC determn: I. Jane, A. McKinnon, *J. Chromatog.* **323,** 191 (1985).

Hydrochloride, $C_{20}H_{25}NO_3 \cdot HCl$, long needles, mp 170-172°. Sol in water and alcohols; only slightly sol in chloroform and acetone. Insol in ether, ethyl acetate, benzene. pH (2% aq soln) 4.0-5.5. LD_{50} i.v. in rats: 66 mg/kg (Yakhontov).

Note: This is a controlled substance (opiate) listed in the U.S. Code of Federal Regulations, Title 21 Part 1308.11 (1995).

THERAP CAT: Analgesic (narcotic).

3254. Dimepheptanol. *β-[2-(Dimethylamino)propyl]-α-ethyl-β-phenylbenzeneethanol; 6-(dimethylamino)-4,4-diphenyl-3-heptanol; 2-dimethylamino-4,4-diphenyl-5-heptanol; 3-hydroxy-4,4-diphenyl-6-dimethylaminoheptane; bimethadol; Methadol; Pangerin.* $C_{21}H_{29}NO$; mol wt 311.47. C 80.98%, H 9.38%, N 4.50%, O 5.14%. Prepn of α-isomers by catalytic or lithium aluminum hydride reduction of methadone: Speeter *et al., J. Am. Chem. Soc.* **71,** 57 (1949); Pohland *et al., ibid.* 460. Prepn of β-isomers by sodium propanol reduction of methadone: Eddy *et al., J. Org. Chem.* **17,** 321 (1952); Clark, **U.S.** pats. **2,565,592** and **2,668,814** (1951 and 1954 to Merck & Co.); Speeter, **U.S.** pat. **2,649,445** (1953 to Bristol). Stereochemistry: Portoghese, Williams, *152nd Am. Chem. Soc. Meet.* (New York, Sept. 1966) Abstracts of Papers, p 5.

β-*dl*-Form, mp 127-128°.

β-*dl*-form Hydrochloride, rods from acetone, mp 210-212°.

β-*d*-Form, prisms from alc + water, mp 106-107°; $[\alpha]_D^{20}$ +178° (c = 0.63 in alc).

β-*d*-form Hydrochloride, plates from acetone + ether, mp 206-208°. $[\alpha]_D^{20}$ +73.9° (c = 0.69).

β-*l*-Form, prisms, mp 105-107°. $[\alpha]_D^{20}$ −178° (c = 1.04 in alc).

β-*l*-form Hydrochloride, mp 206-208°. $[\alpha]_D^{20}$ −74° (c = 0.94).

α-*dl*-form Hydrochloride, mp 192-193°.

α-*d*-form Hydrochloride, mp 169-171°. $[\alpha]_D^{25}$ +34° (c = 0.26).

α-*l*-form Hydrochloride, mp 169-171°. $[\alpha]_D^{25}$ −35° (c = 0.30).

Note: This is a controlled substance (opiate) listed in the U.S. Code of Federal Regulations, Title 21 Part 1308.11 (1995).

THERAP CAT: Analgesic (narcotic).

3255. Dimercaprol. *2,3-Dimercapto-1-propanol;* 1,2-dithioglycerol; British Anti-Lewisite; BAL; Dicaptol; Sulfactin. $C_3H_8OS_2$; mol wt 124.23. C 29.01%, H 6.49%, O 12.88%, S 51.62%. $CH_2SHCHSHCH_2OH$. Developed as an anti-gas warfare agent. Effective in stopping the toxic action of Lewisite upon the pyruvate oxidase system in the brain: Peters *et al.*, *Nature* **156**, 616 (1945); Waters, Stock, *Science* **102**, 601 (1945). Prepd by the bromination of allyl alcohol to glycerol dibromohydrin followed by reaction with sodium hydrosulfide under pressure. Also prepd by hydrogenation of hydroxypropylene trisulfide: U.S. pat. **2,-402,665** (1946 to DuPont). Stereospecific synthesis: Anisuzzaman, Owen, *J. Chem. Soc. (C)* **1967**, 1021. Toxicity study: P. Zvirblis, R. I. Ellin, *Toxicol. Appl. Pharmacol.* **36**, 297 (1976).

Viscous oily liq. Pungent offensive odor of mercaptans. d_4^{25} 1.2385; $bp_{0.2}$ 60°; $bp_{5.6}$ 100°; bp_{15} 120°; bp_{25} 130°; bp_{40} 140°; n_D^{25} 1.5720. 8.7 grams dissolve in 100 ml water (dec). Sol in vegetable oils. Marketed as a 10% soln in peanut oil with 20% benzyl benzoate (stabilizer). LD_{50} i.m. in rats: 86.7 mg/kg (Zvirblis, Ellin).

THERAP CAT: Antidote (to arsenic, gold and mercury poisoning).

THERAP CAT (VET): Chelating agent. Detoxicant for heavy metal poisoning.

3256. 2,3-Dimercapto-1-propanesulfonic Acid. 2,3-Dithiolpropanesulfonic acid. $C_3H_8O_3S_3$; mol wt 188.29. C 19.14%, H 4.28%, O 25.49%, S 51.09%. $HSCH_2CH(SH)$-CH_2SO_3H. Chelating agent related to BAL (dimercaprol, *q.v.*). Prepn: N. S. Johary, L. N. Owen, *J. Chem. Soc.* **1955**, 1307; V. E. Petrun'kin, *Ukrain. Khim. Zh.* **22**, 603 (1956); *C.A.* **51**, 5692h (1957). Distribution and excretion in rats: B. Gabard, *Arch. Toxicol.* **39**, 289 (1978). Metabolism study: B. Gabard, R. Walser, *J. Toxicol. Environ. Health* **5**, 759 (1979). Use in removal of internally deposited gold: B. Gabard, *Brit. J. Pharmacol.* **68**, 607 (1980). Protection of mice vs lethal effects of sodium arsenite: H. V. Aposhian *et al.*, *Toxicol. Appl. Pharmacol.* **61**, 385 (1981). Pharmacokinetics in dogs: P. Wiedemann *et al.*, *Biopharm. Drug Dispos.* **3**, 267 (1982). Anti-lewisite activity and stability: H. V. Aposhian *et al.*, *Life Sci.* **31**, 2149 (1982). Toxicological studies: F. Planas-Bohne *et al.*, *Arzneimittel-Forsch.* **30**, 1291 (1980).

Sodium salt, $C_3H_7NaO_3S_3$, *DMPS, unitiol, Dimaval.* Leaflets, mp 235°. LD_{50} i.p. in mice: 5.22 mmol/kg (Aposhian, 1981).

THERAP CAT: Antidote (heavy metal poisoning).

3257. Dimestrol. *(E)-1,1'-(1,2-Diethyl-1,2-ethenediyl)-bis[4-methoxybenzene];* α,α'-diethyl-4,4'-dimethoxystilbene; 4,4'-dimethoxy-α,β-diethylstilbene; stilbestrol dimethyl ether; 3,4-bis(*p*-methoxyphenyl)-3-hexene; 3,4-dianisyl-3-hexene; Depot-Oestromenine; Depot-Oestromon; Synthila. $C_{20}H_{24}O_2$; mol wt 296.41. C 81.04%, H 8.16%, O 10.80%. Dimethyl ether of diethylstilbestrol, *q.v.* Prepd from 3,3-dianisyl-4-hexanone: Sisido, Nozaki, *J. Am. Chem. Soc.* **70**, 777 (1948). Alternate prepns: Dodds *et al.*, *Nature* **142**, 211, 247 (1938); Reid, Wilson, *J. Am. Chem. Soc.* **64**, 1625 (1942). Improved stereospecific prepn: T. Hiyama, H. Nozaki, *Bull. Chem. Soc. Japan* **46**, 2248 (1973). Crystal structure: G. Ruban, P. Luger, *Acta Crystallogr.* **B31**, 2658 (1975). Biological activities: Y. Inamori *et al.*, *Chem. Pharm. Bull.* **33**, 4478 (1985). Review and bibliography: Solmssen, *Chem. Rev.* **37**, 481 (1945). *See also* Mestilbol.

Crystals from petr ether, mp 124°. Less sol than the monomethyl ether, mestilbol. Practically insol in water. Sol in alcohol; freely sol in acetone, ether; also sol in dil aq or alcoholic solns of alkali hydroxides and in vegetable oils.

THERAP CAT: Estrogen.

3258. Dimetacrine. *N,N,9,9-Tetramethyl-10(9H)-acridinepropanamine;* 10-[3-(dimethylamino)propyl]-9,9-dimethylacridan; 5,5-dimethyl-10-dimethylaminopropylacridan; dimethacrine. $C_{20}H_{26}N_2$; mol wt 294.44. C 81.59%, H 8.90%, N 9.51%. Prepn: Holm, **Brit. pat. 933,875** (1963 to Kefalas), *C.A.* **60**, 510a (1964); Molnar, Wagner-Jauregg, *Helv. Chim. Acta* **48**, 1782 (1965); **Fr. pat. 1,438,357** (1966 to Siegfried), *C.A.* **66**, 10856k (1967). Pharmacology and metabolism: Schatz *et al.*, *Arzneimittel-Forsch.* **18**, 862 (1968); Ishitani *et al.*, *Japan. J. Pharmacol.* **20**, 432 (1970).

Free base, bp_1 200°.

Hydrochloride, $C_{20}H_{26}N_2 \cdot HCl$, crystals from abs alc + acetone, mp 151-154°.

Tartrate, *SD-709, Istonil, Linostil.*

THERAP CAT: Antidepressant.

3259. Dimetan®. *Dimethylcarbamic acid 5,5-dimethyl-3-oxo-1-cyclohexen-1-yl ester;* 5,5-dimethyldihydroresorcinol dimethylcarbamate; G-19258. $C_{11}H_{17}NO_3$; mol wt 211.26. C 62.54%, H 8.11%, N 6.63%, O 22.72%. Prepd by the action of dimethylcarbamic acid chloride on 5,5-dimethyldihydroresorcinol: **Swiss. pat. 272,772** (1951); Gysin, **U.S. pat. 2,592,890** (1952 to Geigy). As insecticide: Wiesmann *et al.*, *Experientia* **7**, 117 (1951); Ferguson, Alexander, *J. Agr. Food Chem.* **1**, 888 (1953); Müller, Spindler, *Experientia* **10**, 91 (1954).

Crystals from cyclohexane, mp 45-46°. bp_{11} 170-180°; $bp_{0.3}$ 122-124°. Soly in water at 20° = 3.15 g/100 ml. Freely sol in alc, acetone, ether. Moderately sol in petr ether, gasoline, cyclohexane.

Caution: A cholinesterase inhibitor. Similar in action to physostigmine.

USE: Insecticide (aphicide).

3260. Dimethadione. *5,5-Dimethyl-2,4-oxazolidinedione;* DMO; AC-1198; BAX-1400Z; NSC-30152; Euprac-tone. $C_5H_7NO_3$; mol wt 129.12. C 46.51%, H 5.46%, N 10.85%, O 37.17%. Active metabolite of trimethadione, *q.v.* Prepn: F. Urech, *Ber.* **13**, 485 (1880); R. W. Stoughton, *J. Am. Chem. Soc.* **63**, 2376 (1941). Anticonvulsant activity: C. D. Withrow *et al., J. Pharmacol. Exp. Ther.* **161**, 335 (1968); in comparison with trimethadione: H. Ferngren, *Acta Pharmacol. Toxicol.* **26**, 177 (1968). Use in measurement of intracellular pH and cellular pH gradients: S. Addanki *et al., J. Biol. Chem.* **243**, 2337 (1968); V. Ehrhardt, *Biochim. Biophys. Acta* **775**, 182 (1984); J. B. Arnold *et al., J. Cereb. Blood Flow Metab.* **5**, 369 (1985). GLC determn: W. Gazdzik, W. Kmiotek, *J. Chromatog.* **378**, 482 (1986); LC determn in serum: E. Tanaka, S. Misawa, *J. Chromatog.* **413**, 376 (1987).

Crystals, mp 76-77°. Weak organic acid, pKa (37°) 6.13. LD_{50} i.v. in mice: 450 mg/kg (Stoughton).

USE: *In vivo* measurement of intracellular pH.

THERAP CAT: Anticonvulsant.

3261. Dimethazan. *7-[2-(Dimethylamino)ethyl]-3,7-di-hydro-1,3-dimethyl-1H-purine-2,6-dione; 7-(2-dimethyl-aminoethyl)theophylline;* 1,3-dimethyl-7-(2-dimethylamino-ethyl)xanthine; Elidin. $C_{11}H_{17}N_5O_2$; mol wt 251.29. C 52.58%, H 6.82%, N 27.87%, O 12.73%. Prepn from theophylline and dimethyl-2-chloroethylamine: Moussalli *et al.,* **Brit. pat.** 669,070 (1952), *C.A.* **47**, 5435i (1953); Klosa, *Arch. Pharm.* **288**, 301 (1955). Pharmacology: Balatre, Merlen, *Therapie* **11**, 1146 (1956); Yago, *Japan. Circ. J.* **26**, 407 (1962).

Solid, mp 95°.
Hydrochloride, $C_{11}H_{17}N_5O_2 \cdot HCl$, mp 260°.
THERAP CAT: Antidepressant.

3262. Dimethazone. *2-[(2-Chlorophenyl)methyl]-4,4-di-methyl-3-isoxazolidinone;* FMC-57020; Command. $C_{12}H_{14}$-$ClNO_2$; mol wt 239.70. C 60.13%, H 5.89%, Cl 14.79%, N 5.84%, O 13.35%. Pre-emergence broad-spectrum herbicide. Prepn and phytotoxic activity: J. H. Chang, **Belg.** pat. **889,040**; *idem,* **U.S.** pat. **4,405,357** (1982, 1983 both to FMC). Use on soybeans: M. P. Mascianica, *Proc. Ann. Meet. Northeast Weed Sci. Soc.* **39**, 25 (1985); on white potatoes: C. C. Kupatt *et al., ibid.* 166. Comparison with alachlor, metolachlor, *q.q.v.:* G. R. Marion *et al., ibid.* 147.

Oil.
USE: Pre-emergent herbicide.

3263. Dimethenamid. *2-Chloro-N-(2,4-dimethyl-3-thienyl)-N-(2-methoxy-1-methylethyl)acetamide;* SAN-582H; Frontier. $C_{12}H_{18}ClNO_2S$; mol wt 275.80. C 52.26%, H 6.58%, Cl 12.85%, N 5.08%, O 11.60%, S 11.63%. Preemergence herbicide for use in food crops. Prepn: K. Seckinger *et al.,* **Brit.** pat. **Appl. 2,114,566;** eidem, **U.S.** pat. **4,666,502** (1983, 1987 both to Sandoz). Comprehensive description: J. Harr *et al., Proc. Brit. Crop Prot. Conf. - Weeds* **1991**, 87-92. Review of field trials and persistance in soil: A. Rahman, T. K. James, *Proc. 45th N. Z. Plant Prot. Conf.* **1992**, 84-88.

Yellowish-brown, viscous liquid. Odorless to weak tar-like odor. Mixture of 4 stereoisomers. $bp_{26.7\ Pa}$ 127°. vapor pressure (25°): 36.7 mPa. d^{25} 1.187. Soly (25°): water 1174 mg/l; heptane 28.2 g/100 g; isooctane 22.0 g/100 g; ether, kerosene, ethanol > 50%. LD_{50} in rats (mg/kg): 1570 orally; > 2000 dermally; LC_{50} in bluegill sunfish, rainbow trout (mg/l): 6.4, 2.6 (Harr).

USE: Herbicide.

3264. Dimethicone. Dimethyl polysiloxane; dimeticon; polydimethylsiloxane; α-(trimethylsilyl)-ω-methylpoly[oxy-(dimethylsilylene)]. Silicone oil consisting of a mixture of fully methylated linear siloxane polymers end-blocked with trimethylsiloxy units. Prepn of homologous liquid methyl siloxane polymers: J. F. Hyde, **U.S.** pat. **2,441,098** (1948 to Corning Glass). Dimethicone mixed with silicon dioxide is known as simethicone. Clinical evaluation of simethicone in functional upper GI disease: J. E. Bernstein, A. M. Kasich, *J. Clin. Pharmacol.* **14**, 617 (1974). Stability of simethicone + antacid mixtures in pharmaceutical preparations: J. A. Rider, *Curr. Ther. Res.* **52**, 681 (1992). Review of use of dimethicones in skin and hair products: A. Disapio, P. Fridd, *Int. J. Cosmet. Sci.* **10**, 75-89 (1988). Review of medical use in soft-tissue augmentation: V. J. Selmanowitz, N. Orentreich, *J. Dermatol. Surg. Oncol.* **3**, 597-611 (1977); D. M. Duffy, *Adv. Dermatol.* **5**, 93-109 (1990). *See also:* Silicones.

Clear colorless liquids. Viscosity increases with degree of polymerization. Immiscible with water, alcohol. Miscible with chloroform, ether.

Mixture with silicon dioxide, *simethicone, activated dimethicone, Antifoam A, Baros, Colicon, Endo-Paractol, Gas-X, Infacol, Lefax, Mylicon, Phasil, Phazyme, sab simplex, Silain.* Average number of dimethylsiloxane units is 200 to 350. mol wt 14000-21000. Gray, translucent, viscous liquid. Insol in water, alcohol.

Dimethicone 350, polydimethylsiloxane having a viscosity of 350 centistokes at 25°. d 0.965-0.973. n_D^{25} 1.4013-1.4053.

USE: Oleaginous ointment base, skin protectant, antifoaming agent; prosthetic aid (soft tissue).

THERAP CAT: Antiflatulent.
THERAP CAT (VET): Antibloating agent.

3265. Dimethindene. *N,N-Dimethyl-3-[1-(2-pyridinyl)-ethyl]-1H-indene-2-ethanamine; 2-[1-[2-[2-(dimethylamino)-ethyl]inden-3-yl]ethyl]pyridine; 3-[α-(2'-pyridyl)ethyl]-2-(β-dimethylaminoethyl)indene; dimethpyrindene.* $C_{20}H_{24}N_2$; mol wt 292.42. C 82.15%, H 8.27%, N 9.58%. Synthesis: Huebner *et al., J. Am. Chem. Soc.* **82**, 2077 (1960); Huebner, U.S. pat. **2,970,149** (1961 to Ciba). Pharmacology: W. E. Barrett *et al., Toxicol. Appl. Pharmacol.* **3**, 534 (1961).

Maleate, $C_{20}H_{24}N_2 \cdot C_4H_4O_4$, *Fenistil, Fenostil, Forhistal.* Crystals, mp 159-161°. LD_{50} in rats (mg/kg): 26.8 i.v.; 618.2 orally (Barrett).
THERAP CAT: Antihistaminic.

3266. Dimethirimol. *5-Butyl-2-(dimethylamino)-6-methyl-4(1H)-pyrimidinone; 5-butyl-2-(dimethylamino)-6-methyl-4-pyrimidinol; 2-dimethylamino-4-methyl-5-n-butyl-6-hydroxypyrimidine; PP-675; Milcurb.* $C_{11}H_{19}N_3O$; mol wt 209.29. C 63.13%, H 9.15%, N 20.08%, O 7.64%. Prepn: B. K. Snell *et al.,* S. Afr. pat. **67 01,373**; *eidem,* U.S. pat. **3,980,781** (1968, 1976 both to ICI). Systemic fungicidal activity: R. S. Elias, *Nature* **219**, 1160 (1968); K. J. Bent, *Ann. Appl. Biol.* **66**, 103 (1970). Metabolism and mode of action: A. Calderbank, *Acta Phytopathol.* **6**, 355 (1971). Metabolism: H. Bratt *et al., Food Cosmet. Toxicol.* **10**, 489 (1972).

Needles from ethanol, mp 102°. Vapor pressure at 30°: 1.1×10^{-5} mm Hg. uv max (methanol): 229, 304 nm (ε 15500, 7700). Soly in (g/l) at 25°: water 1.2; acetone 45; chloroform 1200; ethanol 65; xylene 360. LD_{50} in female rats (mg/kg): 200-400 i.p., >4000 orally (Elias).
USE: Fungicide.

3267. Dimethisoquin. *2-[(3-Butyl-1-isoquinolinyl)oxy]-N,N-dimethylethanamine; 3-butyl-1-[2-(dimethylamino)eth-oxy]isoquinoline; 1-(β-dimethylaminoethoxy)-3-n-butyliso-quinoline; quinisocaine.* $C_{17}H_{24}N_2O$; mol wt 272.39. C 74.96%, H 8.88%, N 10.28%, O 5.87%. Prepn: Wilson *et al., J. Am. Chem. Soc.* **71**, 937 (1949); Anderson *et al., J. Am. Pharm. Assoc. Sci. Ed.* **41**, 643 (1952); Ullyot, U.S. pat. **2,612,503** (1952 to SK & F). Clinical studies: Robinson: *Southern Med. J.* **50**, 367 (1957). Toxicity study: Fellows, Macko, *J. Pharmacol. Exp. Ther.* **103**, 306 (1951).

Liquid, bp$_3$ 155-157°. n_D^{20} 1.5486.
Hydrochloride, $C_{17}H_{24}N_2O \cdot HCl$, *Isochinol, Pruralgan, Pruralgin, Quotane.* Crystals, sol in water, alcohol, chloroform. Practically insol in ether. pH of 1% aq soln: ~4.2. LD_{50} i.p. in rats: 45-50 mg/kg (Fellows, Macko).

THERAP CAT: Anesthetic (topical).
THERAP CAT (VET): Anesthetic (topical).

3268. Dimethisterone. *(6α,17β)-17-Hydroxy-6-methyl-17-(1-propynyl)androst-4-en-3-one; 6α-methyl-17-(1-propyn-yl)testosterone; 6α,21-dimethyl-17β-hydroxy-17α-pregn-4-en-20-yn-3-one; 6α,21-dimethylethisterone; 17α-ethynyl-6α,21-dimethyltestosterone; 17α-ethynyl-17-hydroxy-6α,21-di-methylandrost-4-en-3-one; Secrosteron.* $C_{23}H_{32}O_2$; mol wt 340.51. C 81.13%, H 9.47%, O 9.40%. Orally active progestogen; formerly used in combinations as oral contraceptive. Prepn: S. P. Barton *et al., J. Chem. Soc.* **1959**, 1957; *eidem,* U.S. pat. **2,939,819** (1960 to Brit. Drug Houses). Androgenic effects: H. F. L. Scholer, A. M. de Wachter, *Acta Endocrinol.* **38**, 128 (1961). Progestational activity: G. K. Suchowsky, G. Baldratti, *J. Endocrinol.* **30**, 159 (1964). Evaluation of risk in endometrial cancer: N. S. Weiss *et al., N. Engl. J. Med.* **302**, 551 (1980). Review of carcinogenicity studies: *IARC Monographs* **21**, 377-385 (1979).

Crystals, mp 102°. $[\alpha]_D^{20}$ +10° (c = 1 in chloroform). uv max (isopropanol): 240 nm ($E_{1cm}^{1\%}$ 450). Practically insol in water. Sol in ethanol; slightly sol in acetone, chloroform.
Mixture with ethinyl estradiol, *Oracon, Ovin, Tova.*
THERAP CAT: Progestogen.

3269. Dimethoate. *Phosphorodithioic acid O,O-dimeth-yl S-[2-(methylamino)-2-oxoethyl] ester; phosphorodithioic acid O,O-dimethyl ester, ester with 2-mercapto-N-methylacet-amide; O,O-dimethyl S-methylcarbamoylmethyl phosphoro-dithioate; American Cyanamid 12880; Cygon; Fostion MM; Perfekthion; Rogor; Roxion.* $C_5H_{12}NO_3PS_2$; mol wt 229.26. C 26.20%, H 5.28%, N 6.11%, O 20.94%, P 13.51%, S 27.97%. Prepn: Cassaday *et al.,* Young, U.S. pats. **2,494,-283** and **2,996,531** (1950, 1961, both to Am. Cyanamid); Brit. pat. **791,824** (1958 to Montecatini), *C.A.* **52**, 18222 (1958). Toxicity data: E. W. Schafer, *Toxicol. Appl. Pharmacol.* **21**, 315 (1972). Teratogenicity study: K. D. Court-ney *et al., J. Environ. Sci. Health* **B20**, 373 (1985).

Crystals, mp 52-52.5°. d^{65} 1.277. n_D^{65} 1.5334. Burns readily on contact with flame. Very slightly sol in water. Freely sol in most organic solvents, except saturated hydro-carbons. Stable in aq soln; hydrolyzed by aq alkali. LD_{50} orally in rats: 250 mg/kg (Schafer).
Caution: Cholinesterase inhibitor: *Clinical Toxicology of Commercial Products,* R. E. Gosselin *et al.,* Eds. (Williams & Wilkins, Baltimore, 5th ed., 1984) Section II, p 297.
USE: Systemic and contact insecticide.

3270. Dimethocaine. *3-(Diethylamino)-2,2-dimethyl-1-propanol p-aminobenzoate; 3-diethylamino-2,2-dimethyl-propyl p-aminobenzoate; 1-aminobenzoyl-2,2-dimethyl-3-di-ethylaminopropanol; p-aminobenzoate of diethylaminoneo-pentyl alcohol; Larocaine.* $C_{16}H_{26}N_2O_2$; mol wt 278.39. C 69.03%, H 9.41%, N 10.06%, O 11.49%. Prepn: Mannich, Wilder, *Ber.* **65**, 378 (1932); Mannich, U.S. pat. **1,889,678** (1933).

Hydrochloride, $C_{16}H_{26}N_2O_2 \cdot HCl$, minute crystals or powder, sometimes fine leaflets. Bitter taste. mp 196-197°. One part dissolves in 3 parts water at 20°, more sol in hot water. One part dissolves also in 10 parts of cold alcohol, in 5 parts of boiling alcohol. Practically insol in ether, oils, fats. Aq solns are slightly acid to litmus, and may be sterilized by heating at 105° for 10 minutes. LD s.c. in mice: 300 mg/kg.

THERAP CAT: Anesthetic (local).

3271. Dimethomorph. *4-[3-(4-Chlorophenyl)-3-(3,4-dimethoxyphenyl)-1-oxo-2-propenyl]morpholine;* 3-(4-chlorophenyl)-3-(3,4-dimethoxyphenyl) acrylic acid morpholide; CME-151; Acrobat; Forum. $C_{21}H_{22}ClNO_4$; mol wt 387.86. C 65.03%, H 5.72%, Cl 9.14%, N 3.61%, O 16.50%. Antifungal cinnamic acid derivative; comprised of a mixture of *E*- and *Z*- isomers. Process for prepn: J. Curtze, Eur. pat. Appl. 294,907; *idem*, U.S. pat. 4,933,449 (1988, 1990 both to Shell Int.). Comprehensive description: G. Albert *et al.*, *Proc. Brighton Crop Prot. Conf. - Pests Dis.* **1988**, 17-24. Mechanism of action study: P. J. Kuhn *et al.*, *Mycol. Res.* **95**, 333 (1991). Efficacy vs downy mildew of grapevines: T. Wicks, B. Hall, *Plant Dis.* **74**, 114 (1990). Toxicological evaluation: G. E. Veenstra, D. E. Owen, *Arch. Toxicol.*, **1992**, Suppl 15, 113.

Colorless, odorless, crystalline solid. mp 127-148°. Soly (g/l) at 25°: water < 0.05, acetone 50, toluene 20, methanol 20, dichloromethane 500. LD_{50} in rats (mg/kg): 321 i.p.; 3900 orally; > 5000 dermally; LC_{50} (4-hr inhalation): > 4.2 mg/ml (Veenstra, Owen).

USE: Agricultural fungicide.

3272. Dimethoxanate. *10H-Phenothiazine-10-carboxylic acid 2-[2-(dimethylamino)ethoxy]ethyl ester;* β-dimethylaminoethoxyethyl phenothiazine-10-carboxylate; Tussidin; Cotrane. $C_{19}H_{22}N_2O_3S$; mol wt 358.46. C 63.66%, H 6.19%, N 7.81%, O 13.39%, S 8.95%. Prepn: von Seeman, U.S. pat. 2,778,824 (1957 to American Home Prod.).

Hydrochloride, $C_{19}H_{22}N_2O_3S \cdot HCl$, crystals from anhydr acetone or methanol + ether, mp 161-163° (dec).

Ingredient of *Cothera Syrup.*

THERAP CAT: Antitussive.

3273. Dimethoxane. *2,6-Dimethyl-1,3-dioxan-4-ol acetate;* 6-acetoxy-2,4-dimethyl-*m*-dioxane; 2,6-dimethyl-*m*-dioxan-4-yl acetate; acetomethoxane; Dioxin (obsolete); Giv-Gard DXN; $C_8H_{14}O_4$; mol wt 174.20. C 55.16%, H 8.10%, O 36.74%. Prepn: C. S. Marvel *et al.*, *J. Org. Chem.*

4, 252 (1939); Späth *et al.*, *Ber.* **76**, 57 (1943). Description: *Am. Perfum. Cosmet.* **77**, no. 12, 32-34 (1962), *C.A.* **58**, 8848c (1963).

Liquid, bp_6 74-75°. n_D^{20} 1.4310; d_4^{20} 1.0655. Mustard-like odor. Miscible with water, many organic solvents.

USE: Preservative for cutting oils, resin emulsions, water-based paints, cosmetics, inks. Effective range of concn: 0.03-0.1%. Gasoline additive: Chafetz *et al.*, U.S. pat. 3,036,904 (1962 to Texaco).

3274. 1,2-Dimethoxyethane. Ethylene glycol dimethyl ether; monoglyme; α,β-dimethoxyethane; glyme; Dimethyl Cellosolve. $C_4H_{10}O_2$; mol wt 90.12. C 53.31%, H 11.18%, O 35.51%. $CH_3OCH_2CH_2OCH_3$. Prepd from ethylene glycol monomethyl ether, methyl sulfate and metallic sodium: Kranzfelder, Vogt, *J. Am. Chem. Soc.* **60**, 1714 (1938); from ethylene glycol monomethyl ether, methyl chloride and sodium: Capinjola, *ibid.* **67**, 1615 (1945); from chloromethyl methyl ether in the presence of sodium by Wurtz reaction: Geist, Mason, *ibid.* **76**, 3728 (1954).

Liquid. Sharp ethereal odor. d^{15} 0.86877; d_4^{20} 0.86285; d_4^{33} 0.8602; d_{20}^{20} 0.8692. mp −58° (also reported as mp −71°). bp_{760} 82-83°; $bp_{61.2}$ 20°; bp_{50} 16°; bp_{10} −14°. n_D^{20} 1.3739; n_D^{20} 1.3813. Flash pt 4.5°C (40°F). Miscible with water, alcohol. Sol in hydrocarbon solvents.

USE: Solvent, to facilitate the formation of alkali metal-hydrocarbon adducts; in the Reformatsky reaction with methyl γ-bromocrotonate.

3275. 2,6-Dimethoxyquinone. *2,6-Dimethoxy-2,5-cyclohexadiene-1,4-dione;* 2,6-dimethoxybenzoquinone. $C_8H_8O_4$; mol wt 168.15. C 57.14%, H 4.80%, O 38.06%. Obtained from pyrogallol-1,3-dimethyl ether-2-acetate upon oxidation with potassium dichromate: Hofmann, *Ber.* **11**, 337 (1878); or by heating with nitric acid in alc soln: Graebe, Hess, *Ann.* **340**, 237 (1905); by treating pyrogallol trimethyl ether with nitric acid: Will, *Ber.* **21**, 608 (1888); from syringa acid: Graebe, Martz, *ibid.* 221; from sinapic acid: Gadamer, *Ber.* **30**, 2333 (1897); from antiarole by oxidation with ferric chloride: Baker, *J. Chem. Soc.* **1928**, 1029.

Monoclinic golden-yellow prisms from acetic acid, mp 256°. Sublimes easily. Volatile with steam. Slightly sol in hot water, in ether, in alcohol; freely sol in hot glacial acetic acid and in alkaline solns.

3276. Dimethylacetal. *1,1-Dimethoxyethane; acetaldehyde dimethyl acetal;* ethylidene dimethyl ether. $C_4H_{10}O_2$; mol wt 90.12. C 53.31%, H 11.18%, O 35.51%. $CH_3CH(OCH_3)_2$. Prepn from acetaldehyde and methanol: Meadows, Darwent, *Can. J. Chem.* **30**, 501 (1952); Frevel, Hedelund, U.S. pat. 2,691,684 (1954 to Dow). Toxicity study: H. F. Smyth *et al.*, *J. Ind. Hyg. Toxicol.* **31**, 60 (1949).

Liquid, bp 64.5°. d_4^{20} 0.8516. n_D^{20} 1.3665. Miscible with water, alcohol, chloroform, ether. LD_{50} orally in rats: 6.5 g/kg; lethal concn for rats in air: 16000 ppm (Smyth).

USE: As Mering's mixture which is 2 vol dimethylacetal and 1 vol chloroform.

3277. N,N-Dimethylacetamide. Acetic acid dimethylamide; DMAC. C_4H_9NO; mol wt 87.12. C 55.15%, H 10.41%, N 16.08%, O 18.36%. $CH_3CON(CH_3)_2$. Prepn from tris(dimethylamido)phosphate and acetic anhydride: Dye, U.S. pat. 2,667,510 (1954 to Chemstrand); from acetic anhydride and dimethylformamide: Coppinger, *J. Am.*

Chem. Soc. **76**, 1372 (1954). Toxicity study: W. Bartsch *et al., Arzneimittel-Forsch.* **26**, 1581 (1976).

Liquid. d_4^{25} 0.9366; d_4^{20} 0.9429; d_4^0 0.9599. bp_{760} 163-165°; bp_{80} 96°; bp_{33} 85-87°; bp_{26} 74-74.5°; bp_{15} 66-67°; bp_{12} 62-63°. n_D^{20} 1.4373; also reported as n_D^{20} 1.4230; n_D^{25} 1.4358. Flash pt 66°C (151°F). Miscible with water and most organic solvents. LD_{50} orally in rats: 5.4 ml/kg (Bartsch).

Caution: Potential symptoms of overexposure are jaundice, liver damage; depression, lethargy, hallucinations and delusions; skin irritation. *See NIOSH Pocket Guide to Chemical Hazards* (DHHS/NIOSH 90-117, 1990) p 94.

USE: Solvent for many organic reactions and in industrial applications.

3278. Dimethylamine. *N-Methylmethanamine;* C_2H_7N; mol wt 45.08. C 53.28%, H 15.65%, N 31.07%. $(CH_3)_2NH$. Prepn from methanol + ammonia: Smith, U.S. pat. **2,456,-599** (1948 to Commercial Solvents); Serban, *Rev. Chim. (Bucharest)* **14**, 451 (1963), *C.A.* **60**, 5097b (1964); by catalytic hydrogenation of nitrosodimethylamine: Livering, Maury, **Brit.** pat. **797,483** (1958 to Hercules Powder).

A gas at ordinary temp; characteristic odor. d_4^0 (liq) 0.680. bp 7°, mp −96°. Very sol in water forming a very strong alkaline soln; sol in alcohol or ether. It is usually marketed in compressed liq form in tubes or as a 33% aq soln. LD i.v. in rabbits: 4.0 g/kg.

Hydrochloride, $C_2H_7N.HCl$, deliquesc leaflets, mp 171°. Very sol in water; sol in alcohol, chloroform; practically insol in ether. *Keep well closed.*

Caution: Potential symptoms of overexposure are irritation of nose and throat; sneezing, coughing and dyspnea; pulmonary edema; conjunctivitis; burns to skin and mucous membranes; dermatitis. *See NIOSH Pocket Guide to Chemical Hazards* (DHHS/NIOSH 90-117, 1990) p 94.

USE: As accelerator in vulcanizing rubber, tanning, manuf detergent soaps, or attracting boll weevils to exterminate them. As reagent for Mg.

3279. *p*-Dimethylaminoazobenzene. *N,N-Dimethyl-4-(phenylazo)benzenamine;* butter yellow; methyl yellow; C.I. Solvent Yellow 2; C.I. 11020. $C_{14}H_{15}N_3$; mol wt 225.29. C 74.64%, H 6.71%, N 18.65%. Prepn: *Colour Index* vol. **4** (3rd ed., 1971) p 4014. Review of carcinogenic risk: *IARC Monographs* **8**, 125-146 (1975).

Yellow cryst leaflets, mp 114-117°. Insol in water. Sol in alc, benzene, $CHCl_3$, ether, petr ether, mineral acids, oils.

Caution: Potential symptoms of overexposure are enlarged liver; hepatic and renal dysfunction; contact dermatitis; coughing, wheezing, difficulty breathing; bloody sputum; bronchial secretions; frequent urination, hematuria and dysuria. *See NIOSH Pocket Guide to Chemical Hazards* (DHHS/NIOSH 90-117, 1990) p 94. This substance may reasonably be anticipated to be a carcinogen: *Seventh Annual Report on Carcinogens* (PB95-109781, 1994) p 177.

USE: For determination of free HCl in gastric juice; spot test identification of peroxidized fats; pH indicator (red 2.9, yellow 4.0).

3280. *p*-Dimethylaminobenzaldehyde. 4-Dimethylaminobenzenecarbonal; Ehrlich's reagent. $C_9H_{11}NO$; mol wt 149.19. C 72.46%, H 7.43%, N 9.39%, O 10.72%. Prepd by formylation of dimethylaniline with dimethylformamide: Campaigne, Archer, *Org. Syn.*, **coll. vol. IV**, 331 (1963).

Small granular crystals or leaflets from alcohol + water. Unless extremely pure, the crystals are lemon-colored and may turn pink on exposure to light. mp 74°. bp_{17} 176-177°. Slightly sol in water; sol in alcohol, ether, chloroform, acetic acid, many other organic solvents.

Hydrochloride, crystals, mp 109°.

USE: Manuf dyes; reagent for arsphenamine, anthranilic acid, antipyrine, indole, skatole, indican, tryptophan, albumin, ergot alkaloids, colon bacteria, typhoid coli; for differentiating between serum eruptions and true scarlet fever.

3281. *p*-Dimethylaminobenzalrhodanine. *5-[[4-(Dimethylamino)phenyl]methylene]-2-thioxo-4-thiazolidinone; 5-[p-(dimethylamino)benzylidene]rhodanine.* $C_{12}H_{12}N_2OS_2$; mol wt 264.37. C 54.52%, H 4.58%, N 10.60%, O 6.05%, S 24.26%. Prepn: Mackie, Misra, *J. Chem. Soc.* **1954**, 3919.

Deep red needles from xylene, dec 270°, also reported as dec 246°. Sintering ~260° and melting ~280°. Practically insol in water. Very slightly sol in chloroform, ether, benzene; slightly sol in boiling alc; moderately sol in acetone; sol in strong acids with a yellow color.

USE: As 0.03% soln in acetone for detection of silver, mercury, copper, gold, platinum and palladium ions.

3282. 4-(Dimethylamino)benzoic Acid. $C_9H_{11}NO_2$; mol wt 165.19. C 65.44%, H 6.71%, N 8.48%, O 19.37%. Prepn: Willstätter, Kahn, *Ber.* **37**, 411 (1904); Morton, Stevens, *J. Am. Chem. Soc.* **53**, 4028, 4031 (1931); Gilman, Banner, *ibid.* **62**, 344 (1940); Bowman, Stroud, *J. Chem. Soc.* **1950**, 1342. Use of esters as sunscreening agents: Kreps, Ohlsson, U.S. pat. **3,403,207** (1968); Worthington, **Brit.** pat. **1,162,337** and Fr. pat. **1,566,396** (both 1969 to Van Dyk).

Crystals from water, mp 242.5-243.5°. Soluble in alcohol, HCl and KOH solns, sparingly sol in ether. Practically insol in acetic acid. pKa 6.027; pKb 11.488.

3-Methylbutyl ester, $C_{14}H_{21}NO_2$, *padimate A, Escalol 506*.
2-Ethylhexyl ester, $C_{17}H_{27}NO_2$, *padimate O, Escalol 507, Sundown*.

Glyceryl ester, *see* Glyceryl *p*-Aminobenzoate.

THERAP CAT: Esters as ultraviolet screen.

3283. *p*-Dimethylaminobenzophenone. *[4-(Dimethylamino)phenyl]phenylmethanone.* $C_{15}H_{15}NO$; mol wt 225.29. C 79.97%, H 6.71%, N 6.22%, O 7.10%. *p*-$(CH_3)_2NC_6H_4$-COC_6H_5. Prepd from benzanilide, dimethylaniline and phosphorus oxychloride: **Ger.** pat. **41,751**, *Frdl.* **1**, 44 (1887); Meisenheimer *et al., Ann.* **423**, 84 (1921); Hurd, Webb, *Org. Syn.* **7**, 24 (1927).

Leaflets from alc, mp 92-93°. Insol in water. Slightly sol in cold alcohol; freely sol in hot alcohol and in ether. Very weak base, sol in concd mineral acids from which it is precipitated by water.

syn-Oxime, $C_{15}H_{16}N_2O$, prisms from H_2O + alc, mp 163°.
anti-Oxime, mp 176°.

3284. *N,N*-Dimethylaniline. *N,N-Dimethylbenzenamine;* dimethylphenylamine. $C_8H_{11}N$; mol wt 121.18. C 79.29%, H 9.15%, N 11.56%. Made by heating aniline, methyl alcohol, and H_2SO_4 under pressure, the sulfate formed being converted by NaOH to the free base. Toxicity data: H. F. Smyth *et al., Am. Ind. Hyg. Assoc. J.* **23**, 95 (1962).

Oily liq. *Poisonous!* d_4^{20} 0.956. bp 192-194°. mp 2°. Flash pt 61°. n_D^{20} 1.5582. Insol in water. Freely sol in alcohol, chloroform, ether. LD_{50} orally in rats: 1.41 ml/kg (Smyth).

Caution: Potential symptoms of overexposure are anoxia, cyanosis, weakness, dizziness and ataxia. *See NIOSH Pocket Guide to Chemical Hazards* (DHHS/NIOSH 90-117, 1990) p 96.

USE: Solvent; manuf vanillin, Michler's ketone, methyl violet and other dyes. As reagent for methanol, methyl furfural, H_2O_2, nitrate, alcohol, formaldehyde.

3285.   9,10-Dimethyl-1,2-benzanthracene. *7,12-Dimethylbenz[a]anthracene;* 1,4-dimethyl-2,3-benzphenanthrene. $C_{20}H_{16}$; mol wt 256.35. C 93.71%, H 6.29%. Synthesis from *o*-(α-methyl-α-1-naphthyl)toluic acid: Newman, *J. Am. Chem. Soc.* **60**, 1141 (1938). From 9,10-dimethyl-9,10-dihydroxy-9,10-dihydro-1,2-benzanthracene: Bachmann, Chemerda, *J. Am. Chem. Soc.* **60**, 1023 (1938). From 9,10-dimethyl-9,10-dimethoxy-9,10-dihydro-1,2-benzanthracene: *eidem, ibid.* **61**, 2358 (1939). From 9-methyl-1,2-benzanthracen-10-one by a Reformatsky reaction: Mikhailov, Chernova *J. Gen. Chem. (USSR)* **9**, 2171 (1939), *C.A.* **34**, 4068 (1940).

Plates, leaflets from acetone-alc, faint greenish-yellow tinge. mp 122-123°. Maximum fluorescence at 440 nm: *C.A.* **38**, 1276 (1944). Freely sol in benzene; moderately sol in acetone; slightly sol in alcohol. Insol in water. May be solubilized in water by purines such as caffeine, tetramethyluric acid. The nucleosides, adenosine, and guanosine also show a solvent action.

3286.   5,6-Dimethylbenzimidazole. $C_9H_{10}N_2$; mol wt 146.19. C 73.94%, H 6.89%, N 19.16%. Obtained by acid hydrolysis of vitamin B_{12}: Brink, Folkers, *J. Am. Chem. Soc.* **71**, 2951 (1949); **72**, 4442 (1950). Synthesis by refluxing 4,5-diamino-1,2-dimethylbenzene and formic acid in 4*N* HCl: Hobrecker, *Ber.* **5**, 921 (1872); Phillips, *J. Chem. Soc.* **1928**, 2393.

Crystals from ether, mp 205-206°. Sublimes at 140° and 3 mm. uv max (95% ethanol acidified with HCl to be 0.01*N*): 274.5, 284 nm. Alkaline to litmus. Soluble in water, chloroform, ether; freely sol in dil acids.

3287.   p,α-Dimethylbenzyl Alcohol. *α,4-Dimethylbenzenemethanol;* *p*-tolylmethylcarbinol; methyl-*p*-tolylcarbinol; 4-(α-hydroxyethyl)toluene; 4-methyl-α-phenethyl alcohol; 1-*p*-tolyl-1-ethanol. $C_9H_{12}O$; mol wt 136.19. C 79.37%, H 8.88%, O 11.75%. Constituent of the essential oil from *Curcuma longa* L., *Zingiberaceae* and related plants: Dieterle, Kaiser, *Arch. Pharm.* **271**, 337 (1933). Prepd from *p*-tolylmagnesium bromide and acetaldehyde in ether: v. Auwers, Kolligs, *Ber.* **55**, 42 (1922); Eisenlohr, Schulz, *Ber.* **57**, 1816 (1924); from 4-methylacetophenone: Gastaldi, Cherchi, *Gazz. Chim. Ital.* **45**, II, 274 (1915).

Viscous liq. Odor somewhat like that of menthol. The natural product is probably levorotatory. Constants for the synthetic *dl*-form: $d_4^{15.5}$ 0.9668. bp$_{756}$ 219°; bp$_{14}$ 134°; bp$_{11}$

115-116°. Very sparingly sol in water. Miscible with abs alcohol, ether. Also sol in isopropanol, liq petrolatum.

Phenylurethan, $C_{16}H_{17}NO_2$, crystals, mp 97.5°.

3288.   2,3-Dimethyl-1,3-butadiene. Diisopropenyl; β,γ-dimethyl-Δ-α,γ-butadiene. C_6H_{10}; mol wt 82.15. C 87.73%, H 12.27%. Occurs in the 66-70° fraction of Puertollano shale oil. Prepd by slow distillation of anhydr pinacol in the presence of hydrobromic acid: Allen, Bell, *Org. Syn.* **coll. vol. III,** 312; by rapid distillation of anhydr pinacol over alumina at 420-470°: Newton, Coburn, *ibid.* 313.

Liquid. d_{20}^{20} 0.7273; d_4^{20} 0.7267; d_4^{25} 0.7222.   mp −76°. bp$_{769}$ 69.2°; bp$_{760}$ 68.8°. n_D^{25} 1.4362.

USE: In the manuf of synthetic rubber and polymers.

3289.   Dimethylcadmium. Cadmium dimethyl. C_2H_6-Cd; mol wt 142.48. C 16.86%, H 4.24%, Cd 78.90%. Cd-$(CH_3)_2$. Prepn from a Grignard reagent and a cadmium halide: Krause, *Ber.* **50**, 1813 (1917); Gilman, Nelson, *Rec. Trav. Chim.* **55**, 518 (1936); Anderson, Taylor, *J. Phys. Chem.* **56**, 161 (1952).

Liquid. Disagreeable odor. May be kept in a sealed tube. mp −4.5°; bp$_{758}$ 105.5°; $d_4^{17.9}$ 1.9846; n_D 1.5488. Dec with explosive violence when heated above 150°. Catches fire when dropped on filter paper and produces dense clouds of first white, then brown cadmium oxide smoke. When thrown into water, it sinks to the bottom in large drops, which dec in a series of sudden explosive jerks, with crackling sounds. Sol in hydrocarbons.

USE: In organic synthesis; as polymerization catalyst.

3290.   Dimethyl Carbate. *(endo,endo)-Bicyclo[2.2.1]-hept-5-ene-2,3-dicarboxylic acid dimethyl ester; cis-5-norbornene-2,3-dicarboxylic acid dimethyl ester; cis-3,6-endomethylene-Δ⁴-tetrahydrophthalic acid dimethyl ester; dimalone.* $C_{11}H_{14}O_4$; mol wt 210.23. C 62.85%, H 6.71%, O 30.44%. Prepd by a Diels-Alder condensation of cyclopentadiene and maleic anhydride, followed by esterification of resulting anhydride: Bode, *Ber.* **70**, 1167 (1937); Morgan *et al., J. Am. Chem. Soc.* **66**, 404 (1944).

Crystals, mp 38° when very pure. The usual product is a viscous, syrupy liquid. bp$_{12.5}$ 137°; bp$_9$ 130°; bp$_{1.5}$ 115°. d_4^{21} 1.164. n_D^{20} 1.4852. Practically insol in water. Soluble in the usual organic solvents. Solubility in kerosene and mineral oils ∼6%. LD_{50} orally in mice, rats: 1.4, 1.0 g/kg, *Toxic Substances List,* H. E. Christensen, Ed. (1974) p 539.

Caution: Can cause CNS excitation followed by depression.

USE: Insect repellent.

3291.   5,5-Dimethyl-1,3-cyclohexanedione. 1,1-Dimethyl-3,5-diketocyclohexane; 1,1-dimethyl-3,5-cyclohexanedione; dimethyldihydroresorcinol; dimedone; methone. $C_8H_{12}O_2$; mol wt 140.18. C 68.55%, H 8.63%, O 22.83%. Prepd from mesityl oxide and diethyl malonate: Shriner, Todd, *Org. Syn.* **coll. vol. II,** 200 (1943).

Needles from water, prisms from alcohol + ether. Dec 148-150°. Slightly volatile with steam (50 ml of a distillate contained 0.016 g). The dry crystals may be stored in a brown bottle for several years at room temp. Aq solns oxidize on exposure to air and light or upon prolonged storage in the dark. 100 ml of a satd aq soln contains 0.401 g at 19°; 0.416 g at 25°; 1.185 g at 50°; 3.020 g at 80°; 3.837 g at 90°. Monobasic acid in water. pK (25°): 5.15. Dipole moment 3.46. Also sol in methanol, ethanol, chloroform, benzene, acetic acid, and in 50% ethanol-water mixture.

USE: For the separation and identification of aldehydes. Gives insol condensation products with aldehydes, but not with ketones. The use of piperidine as a catalyst is recommended.

3292. N,N-Dimethylformamide. DMF; DMFA. C_3-H_7NO; mol wt 73.09. C 49.30%, H 9.65%, N 19.16%, O 21.89%. $HCON(CH_3)_2$. Prepd from dimethylamine and formic acid: Mitchell, Reid, *J. Am. Chem. Soc.* **53**, 1879 (1931); Brown, *J. Appl. Chem. (London)* **1**, Suppl. Issue no. 2, S159 (1951); Campbell, U.S. pat. **3,015,674** (1962 to Commercial Solvents); Surman, U.S. pat **3,072,725** (1963 to du Pont); from dimethylamine + HCN: Benneville *et al., J. Org. Chem.* **21**, 772 (1956); from HCN + methanol: Fukuoka, Kominami, *Chem. Tech.* **1972** (Nov.), 640. Toxicity study: W. Bartsch *et al., Arzneimittel-Forsch.* **26**, 1581 (1976). Reviews of chemical uses: R. S. Kittila, *Dimethylformamide Chemical Uses* (du Pont, Wilmington, 1967) 264 pp and Suppl. (1973) 148 pp; J. S. Pizey, *Synthetic Reagents* **Vol. 1** (John Wiley, New York, 1974) pp 4-99; C. L. Eberling in Kirk-Othmer *Encyclopedia of Chemical Technology* **vol. 11** (Wiley-Interscience, New York, 3rd ed., 1980) pp 263-268. Review of carcinogenic risk: *IARC Monographs* **47**, 171-197 (1989).

Colorless to very slightly yellow liquid. Faint amine odor. mp −61°. bp_{760} 153°; bp_{39} 76°; $bp_{3.7}$ 25°. d^{25} 0.9445. n_D^{25} 1.42803. Flash pt, open cup: 153°F (67°C). Misc with water and most common organic solvents. pH of 0.5 molar soln in H_2O − 6.7. LD_{50} in mice, rats (ml/kg): 6.8, 7.6 orally; 6.2, 4.7 i.p. (Bartsch).

Caution: Potential symptoms of overexposure are nausea, vomiting and colic; liver damage, hepatomegaly; high blood pressure; facial flush; dermatitis. *See NIOSH Pocket Guide to Chemical Hazards* (DHHS/NIOSH 90-117, 1990) p 96.

USE: Solvent for liqs and gases. In the synthesis of organic compounds. Solvent for Orlon and similar polyacrylic fibers. Wherever a solvent with a slow rate of evaporation is required. Has been termed the universal organic solvent.

3293. N,N-Dimethylglycine. (Dimethylamino)acetic acid; *N*-methylsarcosine; DMG. $C_4H_9NO_2$; mol wt 103.12. C 46.59%, H 8.80%, N 13.58%, O 31.03%. $(CH_3)_2NCH_2$-COOH. Prepn: H. T. Clarke *et al., J. Am. Chem. Soc.* **55**, 4571 (1933); D. E. Pearson, J. D. Bruton, *ibid.* **73**, 864 (1951). Mutagenicity study: N. Colman *et al., Proc. Soc. Exp. Biol. Med.* **164**, 9 (1980). Constituent of some products sold in the U.S. as "pangamic acid".

Hygroscopic crystals, mp 157-160°.

Hydrochloride, $C_4H_9NO_2.HCl$, mp 189-190°. Sol in alc, water. Insol in chloroform, acetone.

3294. N,N-Dimethylglycine Hydrazide Hydrochloride. Girard's reagent D. $C_4H_{12}ClN_3O$; mol wt 153.61. C 31.28%, H 7.87%, Cl 23.08%, N 27.35%, O 10.42%. $(CH_3)_2$-$NCH_2CONHNH_2.HCl$. Prepd from *N,N*-dimethylglycine ethyl ester and hydrazine hydrate in abs alc: Viscontini, Meier, *Helv. Chim. Acta* **33**, 1773 (1950).

Crystals from alc, mp 181°.

Dihydrochloride, $C_4H_{11}N_3O.2HCl$, mp 214.5°.

USE: Reagent for aldehydes and ketones.

3295. Dimethylglyoxime. 2,3-*Butanedionedioxime;* diacetyldioxime; Chugaev's reagent. $C_4H_8N_2O_2$; mol wt 116.12. C 41.37%, H 6.94%, N 24.12%, O 27.56%. Prepn: Semon, Damerell, *Org. Syn. coll. vol. II,* 204 (1943). Manuf: Kamlet, U.S. pat. **2,732,404** (1956 to National Distillers Product Corp.). Crystal structure: *Anal. Chem.* **21**, 1428 (1949); Merritt, Lanterman, *Acta Cryst.* **5**, 811 (1952).

Triclinic crystals from alc + water, mp 238-240°, also reported as melting 242° and 246°. Practically insol in water. Sol in alc, ether, pyridine, acetone.

USE: Detection and determination of Ni and its separation from Co and many other metals. Forms a scarlet red ppt with Ni even in dil solns. Separation of Pd from Sn, Au, Rh, and Ir. Also to detect Bi with which it forms a bright yellow color and ppt.

3296. 1,1-Dimethylhydrazine. *unsym*-Dimethylhydrazine; *asym*-dimethylhydrazine; *N,N*-dimethylhydrazine; UDMH; Dimazine. $C_2H_8N_2$; mol wt 60.10. C 39.97%, H 13.42%, N 46.61%. $(CH_3)_2NNH_2$. Prepd industrially by the reaction of dimethylamine and chloramine; by reduction of nitrosodimethylamine (obtained by treating a dimethylamine salt with sodium nitrite): Hatt, *Org. Syn. coll. vol. II*, 211 (1943). Toxicity studies: Witkin, *Arch. Ind. Health* **13**, 34 (1956); Cornish, Hartung, *Toxicol. Appl. Pharmacol.* **15**, 62 (1969). Review of carcinogenicity studies: *IARC Monographs* **4**, 137-143 (1974).

Flammable, hygroscopic, mobile liq. Fumes in air and gradually turns yellow. Characteristic ammonia-like fishy odor of aliphatic hydrazines. d^{22}_2 0.791; d^{25}_{25} 0.782. mp −58°. bp_{760} 63.9°. $n_D^{22.3}$ 1.40753. Miscible with water with evolution of heat. Also miscible with alcohol, ether, dimethylformamide, hydrocarbons. LD_{50} in mice, rats (mg/kg): 265, 122 orally; 250, 119 i.v. (Witkin).

Hydrochloride, $C_2H_8N_2.HCl$, hygroscopic crystals from abs ethanol, mp 83°. Sol in water, ethanol. Practically insol in ether.

Caution: Potential symptoms of overexposure are irritation of eyes and skin; choking, chest pain and dyspnea; lethargy; nausea; anoxia; convulsions; liver injury. *See NIOSH Pocket Guide to Chemical Hazards* (DHHS/NIOSH 90-117, 1990) p 96. *See also* Patty's *Industrial Hygiene and Toxicology* vol. 2A, G. D. Clayton, F. E. Clayton, Eds. (Wiley-Interscience, New York, 3rd ed., 1981) pp 2801-2803. 1,1-Dimethylhydrazine may reasonably be anticipated to be a carcinogen: *Seventh Annual Report on Carcinogens* (PB95-109781, 1994) p 182.

USE: The base in rocket fuel formulations.

3297. 1,2-Dimethylhydrazine. *N,N'*-Dimethylhydrazine; *sym*-dimethylhydrazine; SDMH. $C_2H_8N_2$; mol wt 60.10. C 39.97%, H 13.42%, N 46.61%. $CH_3NHNHCH_3$. Prepn from dibenzoylhydrazine: Folpmers, *Rec. Trav. Chim.* **34**, 34 (1915); Hatt, *Org. Syn. coll. vol. II*, 208 (1943). Electrosynthesis from nitromethane: Iversen, *Ber.* **104**, 2195 (1971). Toxicity study: Witkin, *Arch. Ind. Health* **13**, 34, (1956). Review and evaluation of studies of carcinogenicity in laboratory animals: *IARC Monographs* **4**, 145-150 (1974).

Flammable, hygroscopic, mobile liquid. Fumes in air and gradually turns yellow. *Corrosive to skin!* Characteristic ammoniacal odor of aliphatic hydrazines. d_4^{20} 0.8274. bp_{753} 81°. n_D^{20} 1.4209. Miscible with water with much evolution of heat. Also miscible with alcohol, ether, dimethylformamide, hydrocarbons. LD_{50} in mice, rats (mg/kg): 36, 160 orally; 29, 175 i.v. (Witkin).

Dihydrochloride, $C_2H_8N_2.2HCl$, extremely hygroscopic prisms, dec 168°. Freely sol in water, ethanol.

Caution: For toxic symptoms *see* 1,1-Dimethylhydrazine.

3298. 1,3-Dimethyl-2-imidazolidinone. *N,N'*-Dimethylethyleneurea; DMI. $C_5H_{10}N_2O$; mol wt 114.15. C 52.61%, H 8.83%, N 24.54%, O 14.02%. Aprotic basic solvent. Prepn: W. R. Boon, *J. Chem. Soc.* **1947**, 307; M. W. Farlow, U.S. pat. **2,422,400** (1947 to du Pont); E. V. Dehmlow, Y. R. Rao, *Syn. Commun.* **18**, 487 (1988). Prepn and pharmacological activity: E. J. Lien, W. D. Kumler, *J. Med. Chem.* **11**, 214 (1968). Prepn and spectral properties: H. Kohn *et al., J. Org. Chem.* **42**, 941 (1977). Evaluation as replacement solvent for HMPA, *q.v.:* C.-C. Lo, P.-M. Chao, *J. Chem. Ecol.* **16**, 3245 (1990).

mp 8.2° (Dehmlow, Rao); also reported as mp below −4° (Lo, Chao). bp_{754} 220°. bp_{17} 106-108°. bp_2 67-68°. n_D^{20} 1.4720. d^{25} 1.0519 g/ml. Dipole moment (dioxane, benzene): 4.09 ± 0.01 D, 4.05 ± 01 D. Dielectric constant (25°): 37.6. Viscosity (25°): 1.944 cP. Soly in toluene: > 5%. LD_{50} i.p. in mice: 2840 mg/kg (Lien).

USE: As substitute solvent for HMPA in organic synthesis.

3299. Dimethylmercury. *Mercury, dimethyl;* methyl mercury. C_2H_6Hg; mol wt 230.66. C 10.41%, H 2.62%, Hg 86.96%. $(CH_3)_2Hg$. Environmental contaminant found together with *monomethyl mercury* compounds in fish and birds. Prepns: Buckton, *Ann.* **108**, 103 (1858); **109**, 219 (1859); Jones, Werner, *J. Am. Chem. Soc.* **40**, 1257 (1918); Gilman, Brown, *ibid.* **52**, 3314 (1930); Wade, U.S. pat. 3,-636,020 (1972 to Ventron); H. H. Sisler *et al., J. Org. Chem.* **45**, 1329 (1980). Synthesized in bottom sediments: Jensen, Jernelöv, *Nature* **223**, 753 (1969). Physical properties: Wilde, *J. Chem. Soc.* **1949**, 72. Formation and degradation studies: Bertilsson, Neujahr, *Biochemistry* **10**, 2805 (1971); DeSimone *et al., Biochim. Biophys. Acta* **304**, 851 (1973); Spangler *et al., Science* **180**, 192 (1973); *eidem, Appl. Microbiol.* **25**, 488 (1973). Neurotoxicity studies: Dales, *Am. J. Med.* **53**, 219 (1972); Skerfving, *Food Cosmet. Toxicol.* **10**, 545 (1972); Kojima, Fujita, *Toxicology* **1**, 43 (1973).

Colorless, volatile, toxic liquid; easily inflammable. bp_{740} 92°. d^{20} 3.1874. n_D^{20} 1.5452. Easily sol in ether, alc. Insol in H_2O.

USE: As inorganic reagent.

3300. Dimethyl(methylene)ammonium Iodide. *N-Methyl-N-methylenemethanaminium iodide;* Eschenmoser's salt. C_3H_8IN; mol wt 185.01. C 19.48%, H 4.36%, I 68.59%, N 7.57%. $H_2C=N(CH_3)_2I$. Mannich type intermediate originally developed to introduce methyl groups into corrin chromophore. Prepn from trimethylamine and diiodomethane: J. Schreiber *et al., Angw. Chem. Int. Ed.* **10**, 330 (1971); from *N,N,N',N'*-tetramethylmethylenediamine: T. A. Bryson *et al., J. Org. Chem.* **45**, 524 (1980). Used in prepn of Mannich bases: J. Hooz, J. N. Bridson, *J. Am. Chem. Soc.* **95**, 602 (1973). In functionalization of indoles: A. P. Kozikowski, H. Isida, *Heterocycles* **14**, 55 (1980). In prepn of α-methylene carbonyls: J. L. Roberts *et al., Tetrahedron Letters* **1977**, 1621; of terminal olefins: *eidem, ibid.* 1299.

Colorless crystals from tetrahydrothiophene dioxide, dec ~240°. Sublimes at 120° at 0.05 torr.

USE: In organic synthesis.

3301. N,N-Dimethyl-1-naphthylamine. *N,N-Dimethyl-1-naphthalenamine;* 1-dimethylaminonaphthalene. $C_{12}H_{13}$-N; mol wt 171.24. C 84.17%, H 7.65%, N 8.18%. Prepn by heating in a closed tube 1-naphthylamine and methyl iodide in methyl alcohol: Landshoff, *Ber.* **11**, 643 (1878); by heating 1-naphthylamine with dimethyl sulfate in sodium hydroxide in presence of pyridine: Germuth, *J. Am. Chem. Soc.* **51**, 1556 (1929). Use: F. G. Germuth, *Ind. Eng. Chem. Anal. Ed.* **1**, 28 (1929).

Oil; aromatic odor. bp_{711} 274.5°, bp_{90} 193°, bp_{69} 184.5°, bp_{13} 139-140°; d_4^4 1.0522, d_{15}^{15} 1.0446, d_{25}^{25} 1.0391; n_D^{20} 1.622. Readily sol in alcohol or ether.

USE: Detection and determination of nitrites.

3302. Dimethylolpropionic Acid. *3-Hydroxy-2-(hydroxymethyl)-2-methylpropanoic acid; 2,2-bis(hydroxymethyl)-*

propionic acid; dihydroxypivalic acid; DMPA. $C_5H_{10}O_4$; mol wt 134.13. C 44.77%, H 7.51%, O 47.71%. Prepn: H. Koch, T. Zerner, *Monatsh.* **22**, 447, 450 (1891); K. E. Wilzbach *et al., J. Am. Chem. Soc.* **70**, 4069 (1948); R. Riemschneider *et al., Monatsh.* **88**, 1099 (1957); H. Vieregge, J. F. Arens, *Rec. Trav. Chim.* **78**, 921 (1959).

Free-flowing granular powder, mp 181-185°. Sol in water, methanol. Slightly sol in acetone. Insol in benzene.

Note: Do not confuse with the herbicide DMPA, *q.v.*

USE: In prepn of water-sol alkyl resins.

3303. Dimethyl-*p*-phenylenediamine. *N,N-Dimethyl-1,4-benzenediamine; p*-aminodimethylaniline. $C_8H_{12}N_2$; mol wt 136.20. C 70.55%, H 8.88%, N 20.57%. Prepn from *p*-nitrosodimethylaniline which is obtained from dimethylaniline: Gattermann-Wieland, *Praxis des Organischen Chemikers* (de Gruyter, Berlin, 40th ed., 1961) p 273. By reduction of methyl orange: A. I. Vogel, *Practical Organic Chemistry* (Longmans, London, 3rd ed., 1959) p 624.

Reddish-violet crystals. mp 53°, also stated as 41°. bp 262°. Sol in water, alcohol, chloroform, ether. *Keep tightly closed and protected from light.*

Dihydrochloride, $C_8H_{12}N_2.2HCl$, white to grayish-white, hygroscopic crystalline powder. Freely sol in water; sol in alcohol.

USE: Dihydrochloride in microscopy and in tests for acetone, uric acid, thallic salts, oxydases, lignin, ozone, H_2O_2, H_2S, Br.

3304. Dimethyl Phthalate. *1,2-Benzenedicarboxylic acid dimethyl ester; phthalic acid dimethyl ester;* methyl phthalate; dimethyl 1,2-benzenedicarboxylate; DMP; Palatinol M; Fermine; Avolin; Mipax. $C_{10}H_{10}O_4$; mol wt 194.19. C 61.85%, H 5.19%, O 32.96%. Prepd industrially from phthalic anhydride and methanol: Faith, Keyes & Clark's *Industrial Chemicals*, F. A. Lowenheim, M. K. Moran, Eds. (Wiley-Interscience, New York, 4th ed., 1975) pp 318-324. Metabolite of *Gibberella fujikuroi:* Cross *et al., J. Chem. Soc.* **1963**, 2937. Toxicity data: J. H. Draize *et al., J. Pharmacol. Exp. Ther.* **93**, 26 (1948).

Oily liq. Slight aromatic odor. $d_{15.6}^{15.6}$ 1.196; d_{20}^{20} 1.1940; d_{25}^{25} 1.189. One gallon weighs 9.93 lbs. Flash pt 295°F (146°C). mp 5.5° (the commercial product freezes around 0°). bp_{760} 283.7°; bp_{400} 257.8°; bp_{200} 232.7°; bp_{100} 210.0°; bp_{60} 194.0°; bp_{40} 182.8°; bp_{20} 164.0°; bp_{10} 147.6°; bp_5 131.8°; $bp_{1.0}$ 100.3°. Vapor pressure at 20° <0.01 mm. The vapor is heavy, d = 6.69 (air = 1). n_D^{20} 1.5168. Viscosity at 25° = 17.2 cps. Heat of vaporization: 93.1 g-cal/g; heat of combustion: 119.7 kg-cal/mole. uv max (ethanol): 277 nm ($E_{1cm}^{1\%}$ 57.7). Miscible with alcohol, ether, chloroform. Practically insol in water (0.43 g/100 ml), petr ether, and other paraffin hydrocarbons. Soly in mineral oil at 20°: 0.34 g/100 g. LD_{50} in mice, rats, guinea pigs (ml/kg): 7.2, 6.9, 2.4 orally (Draize).

Caution: Irritating to mucous membranes, eyes; can cause CNS depression when ingested. Not irritating to or absorbed through skin: *Clinical Toxicology of Commercial Products*, R. E. Gosselin *et al.*, Eds. (Williams & Wilkins, Baltimore, 5th ed., 1984) Section II, p 205.

USE: Solvent and plasticizer for cellulose acetate and cellulose acetate-butyrate compositions. Insect repellent for personal protection against biting insects.

3305. Dimethyl Sulfate. *Sulfuric acid dimethyl ester;* DMS. $C_2H_6O_4S$; mol wt 126.13. C 19.04%, H 4.79%, O 50.74%, S 25.42%. $(CH_3)_2SO_4$. Prepn by distillation of a mixture of oleum and methanol: Guyot, Simon, *Compt. Rend.* **169**, 796 (1919). Technical production from dimethyl ether and SO_3: *BIOS Final Report* No. 986, p 176. *Review:* C. M. Suter, *The Organic Chemistry of Sulfur* (John Wiley, New York, 1944) p 49-61. Use as methylating agent: L. Fieser, M. Fieser, *Reagents for Organic Synthesis* (John Wiley, New York, 1967) pp 293-295. Toxicity data: Smyth *et al., Arch. Ind. Hyg. Occup. Med.* **4**, 119 (1951). Review of toxicology: E. Browning, *Toxicity and Metabolism of Industrial Solvents* (Elsevier, New York, 1965) pp 713-721; of carcinogenicity studies: *IARC Monographs* **4**, 271-276 (1974); of mutagenicity studies: G. R. Hoffmann, *Mutat. Res.* **75**, 63-129 (1980).

Colorless oily liq. *Poisonous. Corrosive.* bp ~188° (with dec); bp_{15} 76°. mp −27°. d_4^{20} 1.3322. n_D^{20} 1.3874. Flash pt 182°F (83°C). Soly in water 2.8 g/100 ml at 18°. Hydrolysis is rapid at or above this temp. Vapor density 4.35. Sol in ether, dioxane, acetone, aromatic hydrocarbons. Sparingly sol in carbon disulfide, aliphatic hydrocarbons. LD_{50} orally in rats: 440 mg/kg (Smyth).

Caution: Vapors and liquid can be absorbed through the skin or respiratory tract, causing local and systemic toxicity. Potential symptoms of overexposure are irritation of eyes, nose, throat; headache; giddiness; photophobia; periorbital edema; dysphonia, aphonia, dysphagia; cough, chest pain; cyanosis; vomiting; diarrhea, dysuria; icterus, albuminuria, hematuria. Liver and kidneys may be affected. Direct contact may cause skin irritation or tissue destruction. *See* NIOSH *Pocket Guide to Chemical Hazards* (DHHS/NIOSH 90-117, 1990) p 98; P. Grandjean, "Skin Penetration: Hazardous Chemicals at Work" (Taylor & Francis, New York, 1990) pp 173-174. This substance may reasonably be anticipated to be a carcinogen: *Seventh Annual Report on Carcinogens* (PB95-109781, 1994) p 183.

USE: Methylating agent in the manuf of many organic chemicals. War gas.

3306. 2,4-Dimethylsulfolane. *Tetrahydro-2,4-dimethylthiophene 1,1-dioxide;* 2,4-dimethylthiacyclopentane 1,1-dioxide; 2,4-dimethyltetramethylene sulfone; 2,4-dimethylcyclotetramethylene sulfone. $C_6H_{12}O_2S$; mol wt 148.23. C 48.62%, H 8.16%, O 21.59%, S 21.63%. Prepd by catalytic hydrogenation of 2,4-dimethyl-3-sulfolene: Morris, Melchior, U.S. pat. **2,451,298** (1948 to Shell).

Colorless to yellow liquid. d_4^{20} 1.1362. n_D^{20} 1.4733. bp_5 123.3; bp 280-281° (some decompn). Miscible with lower aromatic hydrocarbons; partially miscible with naphthenes, olefins, paraffins. Limited miscibility with water.

USE: Solvent for liquid-liquid and vapor-liquid extraction processes.

3307. Dimethyl Sulfone. *Sulfonylbismethane;* $DMSO_2$; methyl sulfone; methylsulfonylmethane. $C_2H_6O_2S$; mol wt 94.13. C 25.52%, H 6.42%, O 33.99%, S 34.06%. CH_3SO_2-CH_3. Has been found in primitive plants such as *Equisetum arvense* L., *Equisetaceae* and in the adrenal cortex of cattle: Pfiffner, North, *J. Biol. Chem.* **134**, 781 (1940). Easily prepd by oxidation of dimethyl sulfide: Douglas, *J. Am. Chem. Soc.* **68**, 1072 (1946); McAllan *et al., ibid.* **73**, 3627 (1951).

Crystals, mp 109°. bp_{760} 238°. Sublimes at 13 mm and 90° to 100°. Infrared absorption (solid) 7600-8700 nm. Dipole moment 4.44 (vapor). Freely sol in water, methanol, ethanol, acetone. Sparingly sol in ether.

USE: High temp solvent for many inorganic and organic substances.

3308. Dimethyl Sulfoxide. *Sulfinylbismethane; methyl sulfoxide;* DMSO; SQ-9453; DMS-70; DMS-90; Deltan; Demasorb; Demavet; Demeso; Dolicur; Domoso; Dromisol; Gamasol 90; Hyadur; Kemsol; Rimso-50; Sclerosol; Somipront; Syntexan. C_2H_6OS; mol wt 78.14. C 30.74%, H 7.74%, O 20.48%, S 41.04%. $(CH_3)_2SO$. Prepd by air-oxidation of dimethyl sulfide in the presence of nitrogen oxides: Smedslund, U.S. pat. **2,581,050** (1952 to Aktiebolaget Centrallaboratorium); Coma, Gerttula, U.S. pat. **3,045,051** (1962 to Crown Zellerbach). Usually obtained as a byproduct of wood pulp manuf for the paper and allied industries: Robbins, *Chem. Eng.* **68**, No. 13, 100 (1961). Purification: Traynelis *et al., J. Org. Chem.* **27**, 2377 (1962). Forms stable coordination complexes with metals: Meek *et al., J. Am. Chem. Soc.* **82**, 6013 (1960). Toxicity data: Brown *et al., J. Pharm. Pharmacol.* **15**, 688 (1963); Willson *et al., Toxicol. Appl. Pharmacol.* **7**, 104 (1965); W. Bartsch *et al., Arzneimittel-Forsch.* **26**, 1581 (1976). Reviews of pharmacology and toxicology: Jacob, Wood, *Arzneimittel-Forsch.* **17**, 1553-1560 (1967); David, *Ann. Rev. Pharmacol.* **12**, 353-374 (1972); of use in organic chemistry: Agami, *Bull. Soc. Chim. France* **1965**, 1021-1039. *Reviews:* Ranky, Nelson, "Dimethyl Sulfoxide" in *Organic Sulfur Compounds,* vol. 1, N. Kharasch, Ed. (Pergamon Press, New York, 1961) pp 170-182; *Ann. N.Y. Acad. Sci.* **411**, entitled "Biological Actions and Medical Applications of Dimethyl Sulfoxide", J. C. de la Torre, Ed. (1983) pp 1-402.

Very hygroscopic liquid. Practically no odor or color. Slightly bitter taste with sweet after-taste. d_4^{20} 1.100. mp 18.45° (supercools easily). bp 189°; bp_{17} 83°; $bp_{5.11}$ 56.6°; $bp_{2.03}$ 47.4°; $bp_{0.70}$ 30°; $bp_{0.37}$ 20°. Flash pt, open cup: 203°F (95°C). n_D^{20} 1.4795; n_D^{21} 1.4787. Viscosity (27°) 1.1 cp. Specific heat 0.7 cal/g (liq). Dielectric constant: 45. Sol in water, ethanol, acetone, ether, benzene, chloroform. LD_{50} orally in rats: 17.9 ml/kg (Bartsch).

Caution: Skin contact results in primary irritation with redness, itching and sometimes scaling. Urticarial wheals are not uncommon. Corneal opacities have been produced in exptl animals.

USE: Solvent for acetylene, sulfur dioxide and other gases. In organic reactions. As antifreeze or hydraulic fluid when mixed with water. Solvent for Orlon. As paint and varnish remover. Dissolves some hydrocarbons more than others, *see* data sheets issued by Crown Zellerbach Corp., Camas, Wash.

THERAP CAT: Anti-inflammatory (topical).

THERAP CAT (VET): Proposed as analgesic, anti-inflammatory agent, and penetrant carrier to enhance absorption.

3309. Dimethylthiambutene. *N,N-Dimethyl-4,4-di-2-thienyl-3-buten-2-amine; N,N,1-trimethyl-3,3-di-2-thienyl-allylamine;* 3-dimethylamino-1,1-bis(2-thienyl)-1-butene; 3-dimethylamino-1,1-di(2'-thienyl)but-1-ene; 338C48; NIH-4542; Ohton; Aminobutene; Dimethibutin; Kobaton; Takaton. $C_{14}H_{17}NS_2$; mol wt 263.43. C 63.83%, H 6.50%, N 5.32%, S 24.35%. Prepn: Grignard reaction of ethyl β-dimethylaminobutyrate with 2-thienyllithium and dehydration of the resulting 3-dimethylamino-1,1-di-2-thienylbutanol with HCl: Adamson, *J. Chem. Soc.* **1950**, 885; U.S. pat. **2,561,899** (1951 to Burroughs Wellcome); Brit. pat. **657,301** (1951 to Wellcome Foundation). Toxicity data: Y. Kase *et al., Chem. Pharm. Bull.* **7**, 372 (1959).

Viscous, dark brown or yellow oil, $bp_{0.05}$ 123-125°; bp_3 157-158°. Sol in chloroform, ether. LD_{50} in mice (mg/10 kg): 1.21 s.c.; in dogs (mg/kg): 20.3 i.v. (Kase).

Hydrochloride, $C_{14}H_{18}ClNS_2$, may be recrystallized from mixt of ethyl acetate and ethanol, mp 168-169°. Sol in water, chloroform. LD_{50} in mice (mg/10 kg): 1.37 s.c.; in dogs (mg/kg): 20.3 i.v. (Kase).

Note: This is a controlled substance (opiate) listed in the U.S. Code of Federal Regulations, Title 21 Part 1308.11 (1995).

THERAP CAT: Analgesic (narcotic).

3310. *N,N'*-Dimethylthiourea. Dimethylthiocarbamide. $C_3H_8N_2S$; mol wt 104.18. C 34.59%, H 7.74%, N 26.89%, S 30.78%. $CH_3NHCSNHCH_3$.

Colorless, exceedingly deliquesc crystals. mp 60-62°. Very sol in water, alcohol, acetone; sparingly sol in benzene, ether, carbon disulfide, very slightly in petr ether. *Keep tightly closed.*

3311. *N,N*-Dimethyltryptamine. *N,N-Dimethyl-1H-indole-3-ethanamine; 3-[2-(dimethylamino)ethyl]indole;* DMT. $C_{12}H_{16}N_2$; mol wt 188.27. C 76.55%, H 8.57%, N 14.88%. Occurs naturally in plants with hallucinogenic properties. Isoln from the leaves of *Prestonia amazonica* (Benth.) Macbride (*Haemadictyon amazonicum* Spruce & Benth.), *Apocynaceae:* Hockstein, Paradies, *J. Am. Chem. Soc.* **79**, 5735 (1957). Metabolism: St. Szára, *Experientia* **12**, 441 (1956); B. R. Sitaram *et al., Biochem. Pharmacol.* **36**, 1509 (1987). Relationship between hallucinogenic activity and electronic configuration: Snyder, Merril, *Proc. Nat. Acad. Sci. USA* **54**, 258 (1965). Synthesis: I. Fleming, M. Woolias, *J. Chem. Soc. Perkins Trans. I* **1979**, 829. Differential interaction with serotonin receptor subtypes: A. V. Deliganis *et al., Biochem. Pharmacol.* **41**, 1739 (1991). Fluorometric detection: O. Jules *et al., Anal. Chim. Acta* **169**, 355 (1985). Review of biosynthesis, metabolism and bioactivity: S. A. Barker *et al., Int. Rev. Neurobiol.* **22**, 83-110 (1981).

Crystals, mp 44.6-46.8°; also reported as plates from ethanol and light petroleum, mp 46° (Fleming, Woolias). bp 60-80°. pKa 8.68 (ethanol-water). Freely sol in dil acetic and dil mineral acids.

Methiodide, mp 216-217°.

Note: This is a controlled substance (hallucinogen) listed in the U.S. Code of Federal Regulations, Title 21 Part 1308.11 (1995).

3312. Dimethylzinc. Zinc dimethyl. C_2H_6Zn; mol wt 95.46. C 25.16%, H 6.34%, Zn 68.50%. $Zn(CH_3)_2$. Prepd by the interaction of zinc and methyl iodide: Simonowitch, *Chem. Zentr.* **1899, I**, 1066; Bamford *et al., J. Chem. Soc.* **1946**, 468.

Mobile liquid. Stable in sealed tube and under CO_2. $d_4^{10.5}$ 1.386; mp −40°; bp 46°. *Ignites in air,* burns with a blue flame, giving off a peculiar, garlic-like odor. Very slow oxidation with traces of air produces methylzinc methylate, CH_3ZnOCH_3. Sol in ether, miscible with hydrocarbons.

3313. Dimetilan. *Dimethylcarbamic acid 1-[(dimethylamino)carbonyl]-5-methyl-1H-pyrazol-3-yl ester; dimethylcarbamic acid ester with 3-hydroxy-N,N,5-trimethylpyrazole-1-carboxamide;* 2-dimethylcarbamoyl-3-methyl-5-pyrazolyl dimethylcarbamate; G-22870; GS-13332; Snip. $C_{10}H_{16}N_4O_3$; mol wt 240.26. C 49.99%, H 6.71%, N 23.32%, O 19.98%. Prepn: T. Grauer, H. Urwyler, U.S. pat. **3,452,043** (1969 to Geigy). Toxicity data: E. F. Edson, *Pharm. J.* **185**, 361 (1960). Insecticidal activity: B. H. Wilson, *J. Econ. Entomol.* **61**, 1764 (1968). Persistence in water: O. M. Aly *et al., Water Res.* **5**, 1191 (1971). HPLC determn: G. Blaicher *et al., Chromatographia* **13**, 438 (1980).

Colorless solid, mp 68-71° (tech: yellow to reddish-brown solid, mp 55-65°). Sol in water, chloroform, DMF, ethanol, acetone, xylene, other org solvents. Boiling range 200-210° (at 13 mm Hg). Hydrolyzed by acid and alkali. LD_{50} in rats (mg/kg): 25 orally; 600-700 dermally (Edson).

USE: Insecticide.

3314. Dimetofrine. *4-Hydroxy-3,5-dimethoxy-α-[(methylamino)methyl]benzenemethanol;* 1-(3,5-dimethoxy-4-hydroxyphenyl)-2-(methylamino)ethanol; dimethophrine; dimetrophine. $C_{11}H_{17}NO_4$; mol wt 227.26. C 58.14%, H 7.54%, N 6.16%, O 28.16%. Selective α_1-adrenergic receptor agonist. Prepn: **Brit.** pat. **1,145,637;** M. Giani *et al.,* U.S. pat. **3,646,144** (1969, 1972 both to Zambeletti). Pharmacology: Scrollini *et al., Atti Accad. Med. Lomb.* **25**, 193, 203 (1970), *C.A.* **76**, 54270a, 54425e (1972). Pharmacokinetics and metabolism: P. Boselli *et al., Arzneimittel-Forsch.* **26**, 2038 (1976); M. S. Benedetti *et al., ibid.* **27**, 158 (1977). TLC determn: G. Musumarra *et al., J. Chromatog.* **350**, 151 (1985). Double blind clinical evaluations: V. Baldrighi *et al., Curr. Med. Res. Opin.* **9**, 78 (1984); U. Marini *et al., ibid.* 265.

mp 178° (dec). pK 9.71.

Hydrochloride, $C_{11}H_{17}NO_4 \cdot HCl$, *Pressamina.* mp 171-173°.

THERAP CAT: Antihypotensive.

3315. Dimetridazole. *1,2-Dimethyl-5-nitro-1H-imidazole;* RP-8595; Emtryl; Unizole. $C_5H_7N_3O_2$; mol wt 141.13. C 42.55%, H 5.00%, N 29.77%, O 22.67%. Prepn: Bhagwat, Pyman, *J. Chem. Soc.* **127**, 1832 (1925).

Needles from water, mp 138-139°. Freely sol in alcohol; sparingly sol in cold water, ether.

Hydrochloride, $C_5H_7N_3O_2 \cdot HCl$, prisms from dil HCl, mp 195°. Freely sol in water, alcohol; sparingly sol in acetone. Also used as the dihydrogen phosphate, freely sol in water.

THERAP CAT (VET): Antiprotozoal (Histomonas).

3316. Diminazene Aceturate. *N-Acetylglycine compd with 4,4'-(1-triazene-1,3-diyl)bis(benzenecarboximidamide) (2:1); N-acetylglycine compd with 4,4'-(diazoamino)dibenzamidine;* diminazene diaceturate; 4,4'-(diazoamino)dibenzamidine diaceturate; 1,3-bis(p-amidinophenyl)triazene bis-(N-acetylglycinate); 1,3-bis[4-guanylphenyl]triazene diaceturate; 4,4'-diamidinodiazoaminobenzene diaceturate; p,p'-diguanyldiazoaminobenzene diaceturate; Azidin; Ganasag; Berenil. $C_{22}H_{29}N_9O_6$; mol wt 515.53. C 51.26%, H 5.67%, N 24.45%, O 18.62%. Prepn: Brodersen *et al.,* U.S. pat. **2,838,485** (1958 to Hoechst); Stavrovskaya, Drusvyatskaya, *C.A.* **59**, 15199a (1963). Mode of action: Festy *et al., C.R. Acad. Sci. Ser. D* **271**, 730 (1970). Chemotherapeutic characteristics: Schmulevich *et al., Veterinariya (Moscow)* **38**, 23 (1961), *C.A.* **56**, 13512b (1962). *Review:* Newton in *Antibiotics* **vol. 3**, J. W. Corcoran, F. E. Hahn, Eds. (Springer-Verlag, New York, 1975) pp 34-47.

Yellow solid, dec 217°. Sol in 14 parts water (20°); slightly sol in alcohol; very slightly sol in ether, chloroform.

Dilactate, $C_{20}H_{27}N_7O_6$, *Babesin.*

THERAP CAT (VET): Antiprotozoal (Trypanosoma, Babesia).

3317. Dimorpholamine. *N,N'-1,2-Ethanediylbis[N-butyl-4-morpholinecarboxamide]; N,N'-dibutyl-N,N'-dicarboxymorpholideethylenediamine; N,N'-dibutyl-N,N'-dicarboxyethylene diaminemorpholide;* TH-1064; Amipan T; Théraleptique; Theraptique. $C_{20}H_{38}N_4O_4$; mol wt 398.55. C 60.27%, H 9.61%, N 14.06%, O 16.06%. Prepn: Boon, *J. Chem. Soc.* **1947,** 307; *idem,* U.S. pats. **2,398,283** and **2,-409,829** (both 1946 to I.C.I.). Toxicity data: Y. Kasé *et al., Arch. Int. Pharmacodyn. Ther.* **163,** 133 (1966).

Crystals from petr ether, mp 41-42°. bp$_{0.4}$ 229°. Solubility in water: 50%. LD$_{50}$ in mice (mg/10 g): 0.54 i.v.; 0.80 i.p.; 1.04 s.c.; 3.80 orally (Kasé).

THERAP CAT: Analeptic.

3318. Dimoxyline. *1-(4-Ethoxy-3-methoxybenzyl)-6,7-dimethoxy-3-methylisoquinoline;* 6,7-dimethoxy-3-methyl-1-(3'-methoxy-4'-ethoxybenzyl)isoquinoline; 6,7-dimethoxy-1-(4'-ethoxy-3'-methoxybenzyl)-3-methylisoquinoline; dioxyline; Paveril. $C_{22}H_{25}NO_4$; mol wt 367.44. C 71.91%, H 6.86%, N 3.81%, O 17.42%. Prepn: Shepard, U.S. pat. **2,728,769** (1955 to Lilly).

Crystals from dil ethanol, mp 124-125°.

Hydrochloride, $C_{22}H_{25}NO_4 \cdot HCl$, crystals, dec 196-208°.

Phosphate, $C_{22}H_{25}NO_4 \cdot H_3PO_4$, crystals, dec 197-199°. More sol than the hydrochloride. LD$_{50}$ i.v. in mice: 112.7 mg/kg.

THERAP CAT: Vasodilator.

3319. β,β'-Dinaphthylamine. *N-2-Naphthalenyl-2-naphthalenamine.* $C_{20}H_{15}N$; mol wt 269.35. C 89.19%, H 5.61%, N 5.20%. Prepn from β-naphthol: Merz, Weith, *Ber.* **13,** 1300 (1880), Benz, *Ber.* **16,** 15 (1883); from β-naphthylamine: Klopsch, *Ber.* **18,** 1586 (1885); Liebermann, Jacobson, *Ann.* **211,** 43 (1882).

Glistening silvery leaves from benzene, mp 170.5°. Slightly sol in boiling alc; readily sol in boiling glacial acetic acid.

USE: Detection of nitrites, nitrates and chlorates: Sa, *Anales Farm. Bioquim. (Buenos Aires)* **5,** 111 (1934), *C.A.* **30,** 6672 (1936).

3320. Diniconazole. *(E)-(±)-β-[(2,4-Dichlorophenyl)-methylene]-α-(1,1-dimethylethyl)-1H-1,2,4-triazole-1-ethanol; (E)-1-(2,4-dichlorophenyl)-4,4-dimethyl-2-(1,2,4-triazol-1-yl)-1-penten-3-ol;* S-3308-10; S-3308L; XE-779L; Ortho Spotless; Spotless; Sumi-8. $C_{15}H_{17}Cl_2N_3O$; mol wt 326.23. C 55.23%, H 5.25%, Cl 21.74%, N 12.88%, O 4.90%. Prepn: Y. Funaki *et al.,* **Ger. pat. 3,010,560;** *eidem,* U.S. pat. **4,-554,007;** resolution of enantiomers: Y. Funaki *et al.,* **Eur. pat. Appl. 54,431;** *eidem,* U.S. pat. **4,435,203** (1980, 1985, 1982, 1984 all to Sumitomo). Metabolism of racemate in rats: N. Isobe *et al., Nippon Noyaku Gakkaishi* **10,** 475 (1985), *C.A.* **104,** 103703b (1986); of enantiomers: *eidem, ibid.* **12,** 421 (1987), *C.A.* **108,** 1715x (1988). Mechanism of sterol inhibition: Y. Yoshida *et al., Biochem. Biophys. Res. Commun.* **137,** 513 (1986); T. Katagi, *J. Agr. Food Chem.* **36,** 344 (1988). Fungicidal activity of the (−)-form and plant growth regulating activity of the (+)-form: C. S. Kvien *et al., J. Plant Growth Regul.* **6,** 233 (1987); D. J. Daigle *et al., Proc. 14th Ann. Mtg. Plant Growth Regul. Soc. Am.* **1987,** p 108. Field trials as foliar fungicide: A. R. Biggs, J. Warner, *Can. J. Plant Pathol.* **9,** 41 (1987); A. K. Hagan *et al., J. Environ. Hort.* **6,** 67 (1988); against soilborne diseases: A. S. Csinos *et al., Appl. Agr. Res.* **2,** 113 (1987). Comprehensive description: H. Takano, *Japan. Pest. Info.* **49,** 18 (1986).

Crystals from isopropanol, mp 148-149°. Vapor pressure at 20°: 2.2 × 10^{-5} mm Hg. Soly at 25° (% v/w): water 4.01; at 23°: acetone 9.5; methanol 9.5; xylene 1.4. Stable to heat, sunlight and moisture. LD$_{50}$ in male, female rats (mg/kg): 639, 474 orally; > 5000, > 5000 dermally (Takano).

(−)-Form, crystals from carbon tetrachloride/n-hexane, mp 160-161°. $[\alpha]_D^{24}$ −31.7° (c = 1 in CHCl$_3$).

(+)-Form, crystals from carbon tetrachloride/n-hexane, mp 160-161°. $[\alpha]_D^{24}$ +26° (c = 1 in CHCl$_3$).

USE: Agricultural fungicide.

3321. Dinitolmide. *2-Methyl-3,5-dinitrobenzamide; 3,5-dinitro-o-toluamide;* 3,5-dinitro-2-methylbenzamide; Zoalene; Zoamix. $C_8H_7N_3O_5$; mol wt 225.16. C 42.68%, H 3.13%, N 18.66%, O 35.53%. Prepn: McGookin *et al., J. Soc. Chem. Ind.* **59,** 92 (1940); Harris *et al.,* U.S. pat. **2,937,-204** (1960 to Dow). Activity studies: Peterson, *Poultry Sci.* **39,** 739 (1960); Joyner, *Res. Vet. Sci.* **1,** 363 (1960). Metabolism: Smith *et al., J. Agr. Food Chem.* **11,** 253 (1963); Smith, *Anal. Biochem.* **7,** 461 (1964).

Crystals from dil alc, mp 181°.

THERAP CAT (VET): Coccidiostat.

3322. 2,4-Dinitroaniline. *2,4-Dinitrobenzenamine.* $C_6H_5N_3O_4$; mol wt 183.12. C 39.35%, H 2.75%, N 22.95%, O 34.95%. Prepn: Wells, Allen, *Org. Syn. coll. vol.* **II,** 221 (1943). Liquid chromatographic determn: A. L. Scher, *J. Assoc. Off. Anal. Chem.* **68,** 474 (1985). Metabolism and disposition in rats: H. B. Matthews *et al., Xenobiotica* **16,** 1 (1986). Toxicity study: E. G. Feldmann, W. O. Foye, *J. Am. Pharm. Assoc. Sci. Ed.* **48,** 419 (1959).

NH₂ / NO₂ / NO₂ (structure)

Yellow needles from dil acetone, greenish-yellow plates from alcohol. mp 187.5-188°. pKa 18.46. Practically insol in cold water, Very sparingly sol in boiling water. 5.8 parts dissolve in 1000 parts of 88% alc at 18°; one part dissolves in 132.6 parts of 95% alc at 21°.

Caution: May be irritating to skin, mucous membranes.
USE: In prepn of azo dyes.

3323. 2,6-Dinitroaniline. *2,6-Dinitrobenzenamine.* $C_6H_5N_3O_4$; mol wt 183.12. C 39.35%, H 2.75%, N 22.95%, O 34.95%. Synthesis starting with chlorobenzene, fuming sulfuric acid, and potassium nitrate: Schultz, *Org. Syn.* coll. vol. IV, 364 (1963).

O₂N / NH₂ / NO₂ (structure)

Light orange needles from alc, mp 139-140°. Soly in 95% ethanol about 0.4 g/100 ml. Also sol in ether, hot benzene. Practically insol in water, petr ether. Dipole moment: 1.9.

3324. 2,4-Dinitrobenzaldehyde. $C_7H_4N_2O_5$; mol wt 196.12. C 42.87%, H 2.06%, N 14.28%, O 40.79%. Prepn: *p*-Nitrosodimethylaniline hydrochloride, obtained by reacting dimethylaniline and sodium nitrite, is condensed with 2,4-dinitrotoluene in the presence of sodium carbonate. The product, dinitrobenzylidene-*p*-aminodimethylaniline is split under steam agitation with HCl: Bennett, Bell, *Org. Syn.* coll. vol. II, 223 (1943).

HC=O / NO₂ / NO₂ (structure)

Yellow to light brown crystals, mp 72°. bp_{10-20} 190-210°. Freely sol in alc, ether, benzene, slightly sol in petr ether, water.
USE: In prepn of Schiff bases.

3325. Dinitrobenzene. $C_6H_4N_2O_4$; mol wt 168.11. C 42.87%, H 2.40%, N 16.66%, O 38.07%.

NO₂ / NO₂ (structure)

m-Dinitrobenzene. Yellowish crystals. mp 89-90°. bp 300-303°. Volatile with steam. One gram dissolves in 2000 ml cold water, 320 ml boiling water, 37 ml alc, 20 ml boiling alcohol; freely sol in benzene, chloroform, ethyl acetate.
o-Dinitrobenzene. White crystals. d 1.57. mp 118°. bp 319°. Volatile with steam. One gram dissolves in 6600 ml cold water, 2700 ml boiling water, about 60 ml alc, 3 ml boiling alc, 20 ml benzene; freely sol in chloroform, ethyl acetate.
p-Dinitrobenzene. White crystals; sublimable. d 1.63. mp 173-174°. bp 299°. Volatile with steam. One gram dissolves in 12,500 ml cold water, 555 ml boiling water, 300 ml alcohol; sparingly soluble in benzene, chloroform, ethyl

acetate. LD orally in cats: 29.4 mg/kg, *Handbook of Toxicology* vol. 1, W. S. Spector, Ed. (Saunders, Philadelphia, 1956) pp 116-117.

3326. 2,4-Dinitrobenzenesulfenyl Chloride. $C_6H_3Cl-N_2O_4S$; mol wt 234.62. C 30.72%, H 1.29%, Cl 15.11%, N 11.94%, O 27.28%, S 13.67%. Prepd from bis(2,4-dinitrophenyl)disulfide and chlorine in ethylene bromide: Kharasch *et al., J. Am. Chem. Soc.* **69**, 1612 (1947); **72**, 1796 (1950); *J. Chem. Ed.* **33**, 585 (1956); from 2,4-dinitrophenol and chlorine in benzene: Perold, Snyman, *J. Am. Chem. Soc.* **73**, 2379 (1951); from 2,4-dinitrophenylbenzyl sulfide and sulfuryl chloride in ethylene chloride: Kharasch, Langford, *Org. Syn.* **44**, 47 (1964). *Review:* Kharasch, *Organic Sulfur Compounds* vol. 1 (Pergamon Press, New York, 1961) pp 375-396.

S-Cl / NO₂ / NO₂ (structure)

Yellow crystals from carbon tetrachloride, mp 96°. Sol in glacial acetic acid, methylene chloride, ethylene chloride, trichlorethylene, benzene, xylene. Somewhat less sol in carbon tetrachloride. Insol in ether. Reacts with alc (even in the cold) to produce ethyl 2,4-dinitrobenzenesulfenate.
USE: Analytical reagent for the characterization of organic compds: Kharasch, *J. Chem. Ed.* **33**. 585 (1956); Langford, Lawson, *ibid.* **34**, 510 (1957).

3327. 3,4-Dinitrobenzoic Acid. $C_7H_4N_2O_6$; mol wt 212.12. C 39.64%, H 1.90%, N 13.21%, O 45.26%. Prepd by treating 3-nitro-4-aminotoluene with Caro's acid and oxidizing the 3-nitro-4-nitrosotoluene with potassium dichromate: Langley, *Org. Syn.* **22**, 47 (1942); by oxidation of 3,4-dinitrotoluene: Goldstein, Voegeli, *Helv. Chim. Acta.* **26**, 475 (1943).

COOH / NO₂ / NO₂ (structure)

Crystals from water + alc. Bitter taste. mp 166°; (mp 165.5-166.5°, Goldstein). 0.673 g dissolves in 100 parts water at 25°. Freely sol in alc, ether and hot water. Sublimes.
USE: In quantitative sugar analysis.

3328. 3,5-Dinitrobenzoic Acid. $C_7H_4N_2O_6$; mol wt 212.12. C 39.64%, H 1.90%, N 13.21%, O 45.26%. Prepd by nitration of benzoic acid with fuming nitric acid: Brewster *et al., Org. Syn.* **22**, 48 (1942); Saunders *et al., Biochem. J.* **36**, 368 (1942). Purification: Jensen *et al., Chem. Weekbl.* **43**, 731 (1947), *C.A.* **42**, 2307d (1948).

COOH / O₂N / NO₂ (structure)

Monoclinic prismatic crystals from alc. mp 205-207°, (mp 206.5-207.2° purified). One gram dissolves in 53 parts of boiling water; much less sol in cold water. Very sol in alc and glacial acetic acid; sparingly sol in ether, carbon disulfide and benzene. Sublimes.
USE: Identification of alcohols, alkyl halides; chromatographic determination of the essential oil constituents.

3329. 3,5-Dinitrobenzoyl Chloride. $C_7H_3ClN_2O_5$; mol wt 230.56. C 36.47%, H 1.31%, Cl 15.38%, N 12.15%, O

34.70%. Prepd from 3,5-dinitrobenzoic acid and PCl₅ at 120-130°: Saunders *et al., Biochem. J.* **36**, 368 (1942).

Needles from light petr (bp 40-60°), mp 69.5°. bp₁₀₋₁₂ 196°.

Caution: A strong irritant.

USE: Identification of amino acids; characterization of the alcohol part of acetals and ketals without prior hydrolysis and separation of the alcohol.

3330. 4,4'-Dinitrocarbanilide. *N,N'-Bis(4-nitrophenyl)-urea;* DNC; 4,4'-dinitrodiphenylurea. C₁₃H₁₀N₄O₅; mol wt 302.25. C 51.66%, H 3.33%, N 18.54%, O 26.47%. Prepd by heating 4-nitroaniline and diphenyl carbonate for 3 hrs at 200°: Vittenet, *Bull. Soc. Chim. France* [3] **21**, 149 (1899); by nitration of diphenylurea with 68.28% HNO₃ in concd H₂SO₄ at 0°: Kogan, Kutepov, *J. Gen. Chem. USSR* **21**, 1297 (1951), *C.A.* **46**, 2003d (1952); by the action of dil HNO₃ on diphenylurea at up to 100°: *eidem,* U.S.S.R. pat. **78,379** (1949), *C.A.* **48**, 7056i (1954).

Yellow needles from alc, mp 312° (dec). Sublimes above 310°. Moderately sol in boiling nitrobenzene; sparingly sol in boiling alc, somewhat more sol in a mixture of glacial acetic acid and nitric acid. Practically insol in acetone, chloroform, benzene, dioxane, acetic acid, ether and linseed oil. Solubilities in g/100 ml solvent at 25°: water 2 × 10⁻⁶; ethanol 0.007; ethyl acetate 0.015; petr ether 0.0; xylene <0.01; methyl Cellosolve 0.1; dimethyl acetamide 0.14; dimethyl sulfoxide 0.47.

3331. Dinitrocresol. *2-Methyl-4,6-dinitrophenol; 4,6-dinitro-o-cresol;* 3,5-dinitro-2-hydroxytoluene; 3,5-dinitro-o-cresol; DN; DNC; DNOC; Antinonnin; Detal; Dinitrol; Elgetol; K III; K IV; Ditrosol; Prokarbol; Effusan; Lipan; Selinon; Sinox; Dekrysil. C₇H₆N₂O₅; mol wt 198.14. C 42.43%, H 3.05%, N 14.14%, O 40.37%. Prepd by sulfonation of o-cresol followed by controlled nitration: Noelting, de Salis, *Ber.* **14**, 987 (1881); Bovini, *Chem. Zentr.* **1928**, II, 112; Bures, *C.A.* **22**, 63 (1928); Datta, Varma, *J. Indian Chem. Soc.* **4**, 321 (1927); Monti, Cianetti, *Gazz. Chim. Ital.* **67**, 628 (1937).

Yellow prisms from alc, mp 87.5°. Moderately volatile with steam. Sparingly sol in water; readily sol in alkaline aq solns, in ether, acetone, alcohol (about 10%); sparingly sol in petr ether. LD₅₀ orally in rats: 30 mg/kg, *Toxic Substances List,* H. E. Christensen, Ed. (1973) p 718.

Sodium salt, C₇H₅N₂O₅Na, red powder. Hydrate, yellow needles very freely sol in water.

The NH₄, K, Ca salts also are sol in water.

Caution: Potential symptoms of overexposure to dinitrocresol are a sense of well being; headache, fever, lassitude, profuse sweating, excessive thirst, tachycardia, hyperpnea, coughing, shortness of breath and coma. *See NIOSH Pocket Guide to Chemical Hazards* (DHHS/NIOSH 90-117, 1990) p 98. *See also Clinical Toxicology of Commercial Products,*

R. E. Gosselin *et al.,* Eds. (Williams & Wilkins, Baltimore, 5th ed., 1984) Section II, p 196.

USE: Selective herbicide, insecticide (ovicidal spray for dormant fruit trees).

3332. 3,7-Dinitro-5-oxophenothiazine. *3,7-Dinitro-10H-phenothiazine 5-oxide;* dinitrodiphenylamine sulfoxide. C₁₂H₇N₃O₅S; mol wt 305.27. C 47.21%, H 2.31%, N 13.76%, O 26.21%, S 10.50%. Prepd from thiodiphenylamine and nitric acid: Buchanan, Schryver, *Brit. Food J.* **11**, 101 (1908), *C.A.* **3**, 2329 (1909); Kehrmann, Nossenko, *Ber.* **46**, 2818 (1913).

Yellowish leaflets from glacial acetic acid, dec 260°. Insol in water and in the usual organic solvents. Sol in ammonia water and in aq solns of alkalies, forming bluish-red solns.

USE: In the determination of tin.

3333. 2,4-Dinitrophenol. α-Dinitrophenol; Aldifen. C₆H₄N₂O₅; mol wt 184.11. C 39.14%, H 2.19%, N 15.22%, O 43.45%. Prepd by the action of NaOH on 1-chloro-2,4-dinitrobenzene. Toxicity data: E. W. Schafer, *Toxicol. Appl. Pharmacol.* **21**, 315 (1972). Crystal and molecular structure: F. Iwasaki, Y. Kawano, *Acta Crystallogr.* **33B**, 2455 (1977).

Yellowish to yellow orthorhombic crystals. d 1.683. mp 112-114°. Sublimes when carefully heated. Volatile with steam. Very sparingly sol in cold water; soly in water (g/100 g of satd soln): at 54.5° = 0.137; at 75.8° = 0.301; at 87.4° = 0.587; at 96.2° = 1.22. Soly at 15° (g/100 g soln): Ethyl acetate 15.55; acetone 35.90; chloroform 5.39; pyridine 20.08; carbon tetrachloride 0.423; toluene 6.36. Also sol in alcohol and benzene. Sol in aq alkaline solns. Forms crystalline sodium salt which is sol in water. LD₅₀ orally in rats: 30 mg/kg (Schafer).

Caution: Highly toxic material. Readily absorbed through intact skin. Vapors absorbed through respiratory tract. Produces marked increase in metabolism and temp; profuse sweating, nausea, vomiting, collapse, death. May cause dermatitis, cataracts, wt loss, granulocytopenia, polyneuropathy, exfoliative dermatitis, *Clinical Toxicology of Commercial Products,* R. E. Gosselin *et al.,* Eds. (Williams & Wilkins, Baltimore, 4th ed., 1976) Section III, pp 134-137.

USE: Manuf dyes, diaminophenol, etc; wood preservative; insecticide; also as indicator. pH range: 2.6 colorless, 4.4 yellow; as a reagent for the detection of potassium and ammonium ions.

3334. 2,5-Dinitrophenol. Yellow crystals. mp 108°; also stated as 104°. Slightly sol in water or cold alcohol; sol in hot alcohol, ether, fixed alkali hydroxides.

USE: Manuf dyes and organic chemicals, as indicator. pH range: 4.0 colorless, 5.4 yellow.

3335. 2,6-Dinitrophenol. Light yellow crystals. mp 63-64°. Slightly sol in cold water or cold alc; freely sol in chloroform, ether, boiling alc, fixed alkali hydroxide solns.

USE: As of the preceding. pH range: 2.0 colorless, 4.0 yellow.

3336. 2,4-Dinitrophenylhydrazine. C₆H₆N₄O₄; mol wt 198.14. C 36.37%, H 3.05%, N 28.28%, O 32.30%. C₆H₃-(NO₂)₂NHNH₂.

Red, cryst powder. mp ~200°. Slightly sol in water or alcohol; sol in moderately dil inorganic acids; readily sol in diglyme.

USE: For the determination of aldehydes and ketones.

3337.   2,4-Dinitroresorcinol. *2,4-Dinitro-1,3-benzene-diol.* $C_6H_4N_2O_6$; mol wt 200.11. C 36.01%, H 2.01%, N 14.00%, O 47.97%.

Yellow crystals, mp 146-148°. Explodes when strongly heated. Sublimes partially undec. Very slightly sol in water or cold alcohol; sol in solns of fixed alkali hydroxides.

USE: For dyeing fabrics mordanted with iron a green color. As a reagent for Co (brown-red ppt) and for Fe (olive-green color).

3338.   5,5'-Dinitrosalicil. *Bis(2-hydroxy-5-nitrophenyl)-ethanedione;* 2,2'-dihydroxy-5,5'-dinitrobenzil; DNS-55. $C_{14}H_8N_2O_8$; mol wt 332.23. C 50.61%, H 2.43%, N 8.43%, O 38.53%. Prepd by oxidation of 2,2'-dimethoxybenzoin, followed by nitration and demethylation of the resulting 2,2'-dimethoxybenzil: Moureu *et al., Bull. Soc. Chim. France* 1951 M, 741; 1955 M, 1155.

Pale yellow needles from acetone, dec 304°. Soly in water about 0.1 mg/ml H_2O. Sparingly sol in the usual organic solvents. Sol in aq 10% NaOH solns with decompn to sodium 5-nitrosalicylate.

3339.   Dinobuton. *Carbonic acid 1-methylethyl 2-(1-methylpropyl)-4,6-dinitrophenyl ester; carbonic acid 2-sec-butyl-4,6-dinitrophenyl isopropyl ester;* 2-*sec*-butyl-4,6-dinitrophenyl isopropyl carbonate; dinitro-*sec*-butylphenyl isopropyl carbonate; isopropyl 2,4-dinitro-6-*sec*-butylphenyl carbonate; Acrex; Dessin; Sytasol. $C_{14}H_{18}N_2O_7$; mol wt 326.31. C 51.53%, H 5.56%, N 8.59%, O 34.32%. Prepn: Pianka, Polton, U.S. pats. **3,234,082; 3,234,260** (both 1966 to Murphy Chem.). Metabolism studies: Bandal, Casida, *J. Agr. Food Chem.* **20,** 1235 (1972). Toxicity study: T. B. Gaines, *Toxicol. Appl. Pharmacol.* **14,** 515 (1969).

Crystals from methanol or petr ether, mp 56-57°. LD$_{50}$ in male, female rats (mg/kg): 59, 71 orally (Gaines).

USE: Miticide.

3340.   Dinocap. DNOCP; CR-1639; ENT-24727; Karathane; Arathane; Mildex; Isocothane; Crotothane. $C_{18}H_{24}N_2O_6$; mol wt 364.40. C 59.33%, H 6.64%, N 7.69%, O 26.34%. Originally thought to be 2-(1-methylheptyl)-4,6-dinitrophenyl butenoate, it is actually a mixture of 2(or 4)-octyl-4,6(or 2,6)-dinitrophenyl butenoate in which the octyl is a mixture of 1-methylheptyl, 1-ethylhexyl, and 1-propylpentyl. Prepn: W. F. Hester *et al.,* U.S. pat. **2,526,660**

(1950 to Rohm & Haas). Fungicidal activity and chemical constitution: A. H. M. Kirby *et al., Ann. Appl. Biol.* **57,** 211 (1966); R. J. W. Byrde *et al., ibid.* 223. Mitocidal activity: N. D. Rishi, A. Q. Rather, *J. Entomol. Res.* **7,** 39 (1983). Degradation in soil: W. Mittelstaedt, F. Fuhr, *J. Agr. Food Chem.* **32,** 1151 (1984). Persistence on crops: B. D. Ripley *et al., Can. J. Plant Sci.* **65,** 229 (1985). HPLC residue analysis in crops: D. Liang *et al., J. Chromatog.* **387,** 385 (1987). Acute and chronic toxicity: P. S. Larson *et al., Arch. Int. Pharmacodyn. Ther.* **119,** 31 (1959). Evaluation of developmental toxicity in rodents of technical grade and selected isomers: L. E. Gray, Jr. *et al., Teratogen. Carcinogen. Mutagen.* **6,** 33 (1986); J. M. Rogers *et al., ibid.* **7,** 341 (1987).

$$R = -CH(CH_2)_5CH_3 \quad or \quad -CH(CH_2)_4CH_3 \quad or \quad -CH(CH_2)_3CH_3$$
$$\qquad\quad CH_3 \qquad\qquad\qquad CH_2CH_3 \qquad\qquad\quad (CH_2)_2CH_3$$

Dark brown liquid. bp$_{0.05}$ 138-140°. Incompatible with oil, oil based sprays, and lime-sulfur mixtures. LD$_{50}$ in male rats (mg/kg): 23 i.v.; 980 orally (Larson).

Caution: Skin irritant: *Clinical Toxicology of Commercial Products,* R. E. Gosselin *et al.,* Eds. (Williams & Wilkins, Baltimore, 5th ed., 1984) Section II, p 197. The EPA has determined that dinocap is teratogenic in rabbits: *Fed. Reg.* **50,** 1119 (1985).

USE: Acaricide; fungicide.

3341.   Dinoseb. *2-(1-Methylpropyl)-4,6-dinitrophenol;* 2-sec-butyl-4,6-dinitrophenol; DNBP; ENT-1122; WSX-8365; Chemox PE; Dow General; Premerge; Subitex; Caldon; Basanite. $C_{10}H_{12}N_2O_5$; mol wt 240.22. C 50.00%, H 5.04%, N 11.66%, O 33.30%. Prepn: L. E. Mills, B. L. Fayerweather, U.S. pat. **2,192,197** (1940 to Dow). Activity: A. S. Crafts, *Science* **101,** 417 (1945). Metabolism: S. K. Bandal, J. E. Casida, *J. Agr. Food Chem.* **20,** 1235 (1972); K. Ingebrigtsen, A. Froeslie, *Acta Pharmacol. Toxicol.* **46,** 326 (1980). Toxicology: R. G. Bough *et al., Toxicol. Appl. Pharmacol.* **7,** 353 (1965); T. B. Gaines, R. E. Linder, *Fundam. Appl. Toxicol.* **7,** 299 (1986).

Orange-brown viscous liquid, mp 38-42°. pKa 4.62. Sol in most organic solvents. LD$_{50}$ in adult male, female rats (mg/kg): 27, 28 orally (Gaines, Linder).

Acetate, $C_{12}H_{14}N_2O_6$, *HOE-2904, Aretit, Ivosit.* Brown oil, vinegar-like odor, mp 26-27°. Vapor press at 20°: 6 × 10^{-4} mm Hg. Sol in aromatics.

Ammonium salt, $C_{10}H_{15}N_3O_5$, *Chemox Selective, Dow Selective, Sinox W.*

Triethanolamine salt, $C_{16}H_{27}N_3O_8$, *DN-289, Chemox DN, Gebutox.*

Methacrylate, see Binapacryl.

USE: Herbicide; insecticide; miticide.

3342.   Dinosterol. *(3β,4α,5α,22E)-4,23-Dimethylergost-22-en-3-ol;* 4α-methyl-5α(*H*)-Δ^{22}-23,24-dimethylcholesten-3β-ol; Black Sea sterol. $C_{30}H_{52}O$; mol wt 428.74. C 84.04%, H 12.22%, O 3.73%. Biogenetically important marine sterol isolated from the toxic dinoflagellate, *Gonyaulax tamarensis.* Isoln and structure determn: Y. Shimizu *et al., J. Am. Chem. Soc.* **98,** 1059 (1976). Stereochemistry: J. Finer *et al., J. Org. Chem.* **43,** 1990 (1978). Identity with the Black Sea sterol: J. J. Boon *et al., Nature* **277,** 125 (1979). Stereospecific synthesis: A. Y. L. Shu, C. Djerassi, *Tetrahedron Letters* **22**, 4627 (1981).

Needles from methanol-chloroform, mp 220-222°. $[\alpha]_D$ ±5° (c = 0.6 in $CHCl_3$). Also reported as mp 211-214° (Shu, Djerassi). $[\alpha]_D^{20}$ −2.2° (in $CHCl_3$).

3343. Dinsed. *N,N′-1,2-Ethanediylbis[3-nitrobenzene-sulfonamide]; N,N′-ethylenebis[3-nitrobenzenesulfonamide]; N,N′*-ethylenebis[*m*-nitrobenzenesulfonamide]; *N,N′*-bis[3-nitrobenzenesulfonyl]ethylenediamine; di-*m*-nitrobenzene-sulfonylethylenediamine. $C_{14}H_{14}N_4O_8S_2$; mol wt 430.42. C 39.07%, H 3.28%, N 13.02%, O 29.74%, S 14.90%. Prepn: Chattaway, *J. Chem. Soc.* **87**, 381 (1905).

Faintly yellow platelets from glacial acetic acid, mp 189-191° (some decompn).

THERAP CAT (VET): Coccidiostat.

3344. Diopterin. *N-[N-[4-[[(2-Amino-1,4-dihydro-4-oxo-6-pteridinyl)methyl]amino]benzoyl]-L-α-glutamyl]-L-glutamic acid;* pteroyl-α-glutamylglutamic acid; pteroyl-diglutamic acid; PDGA. $C_{24}H_{26}N_8O_9$; mol wt 570.52. C 50.53%, H 4.59%, N 19.64%, O 25.24%. Folic acid analog. Synthesis: Mowat *et al., J. Am. Chem. Soc.* **70**, 1096 (1948); Hutchings, and Geraci, U.S. pats. **2,500,825** and **2,501,168** (both 1950 to Am. Cyanamid); Geraci, U.S. pat. **2,766,240** (1956 to Aries Labs.).

Precipitate from hot water. Sol in dil NaOH solns.

3345. Dioscin. *(25R)-Spirost-5-en-3β-yl O-6-deoxy-α-L-mannopyranosyl-(1→2)-O-[6-deoxy-α-L-mannopyranosyl-(1→4)]-β-D-glucopyranoside;* diosgenin bis-α-L-rhamnopy-ranosyl-(1→2 and 1→4)-β-D-glucopyranoside. $C_{45}H_{72}O_{16}$; mol wt 869.06. C 62.19%, H 8.35%, O 29.46%. From *Dioscorea tokoro* Mal., *Dioscoreaceae.* Isoln: Tsukamoto *et al., Pharm. Bull. (Tokyo)* **4**, 35 (1956); Heitz, *Compt. Rend.* **248**, 283 (1958). Structure: Kawasaki, Yamauchi, *Chem. Pharm. Bull.* **10**, 703 (1962).

Crystals, decomp 275-277°. $[\alpha]_D^{13}$ −115° (c = 0.373 in ethanol).

3346. Dioscorea. Wild yam; colic root; rheumatism root. Dried rhizome of *Dioscorea villosa* L., *Dioscoreaceae. Habit.* North America. *Constit.* Saponin, acrid resin. Account of the nature, origins, cultivation and utilization of the useful members of the *Dioscoreaceae:* D. G. Coursey, *Yams* (London, Longmans, 1967) 230 pp. *See also* Yam, Mexican.

3347. Dioscorine. *[1R-(1α,4α,5α)]-2,4′-Dimethylspiro-[2-azabicyclo[2.2.2]octane-5,2′-[2H]pyran]-6′(3′H)-one.* $C_{13}H_{19}NO_2$; mol wt 221.30. C 70.56%, H 8.65%, N 6.33%, O 14.46%. Found in the tubers of *Dioscorea hirsuta* Blume and *D. hispida* Dennst., *Dioscoreaceae.* H. W. Schutte, *Chem. Zentr.* **68**, II, 130 (1897); M. K. Gorter, *Rec. Trav. Chim.* **30**, 161 (1911); A. R. Pinder, *Nature* **168**, 1090 (1951); idem, *J. Chem. Soc.* **1952**, 2236. Structure: W. A. M. Davies *et al., Chem. & Ind. (London)* **1961**, 1410; eidem, *Tetrahedron* **18**, 405 (1962); Morris, A. R. Pinder, *J. Chem. Soc.* **1963**, 1841. Synthesis: Page, A. R. Pinder, *ibid.* **1964**, 4811. Absolute configuration: A. F. Beecham *et al., Tetrahedron Letters* **1969**, 3745. Biosynthetic studies: E. Leete, A. R. Pinder, *Chem. Commun.* **1971**, 1499; E. Leete, *J. Am. Chem. Soc.* **99**, 648 (1977). Pharmacology: A. R. Pinder, *J. Chem. Soc.* **1953**, 1826. Evaluation of toxic components: J. Webster *et al., J. Agr. Food Chem.* **32**, 1087 (1984).

Greenish-yellow prisms from ether, mp 54-55°. $[\alpha]_D^{18}$ −35.0° (c = 3.4 in chloroform). uv max (methanol): 215 nm (ε 10160). Distills unchanged *in vacuo.* Sol in water, alcohol, acetone, chloroform; slightly sol in ether, benzene, petr ether.

Hydrochloride, $C_{13}H_{19}NO_2$.HCl, needles from alcohol + ether, dec 210-211°.

3348. Diosgenin. *(3β,25R)-Spirost-5-en-3-ol;* nitogenin. $C_{27}H_{42}O_3$; mol wt 414.63. C 78.21%, H 10.21%, O 11.58%. Aglycone of saponin dioscin. From *Dioscorea tokoro* Makino, *Dioscoreaceae:* Tsukamoto, *J. Pharm. Soc. Japan* **56**, 135 (1936); from rhizomes of *Trillium erectum* L., *Liliaceae* and *D. villosa* L., *Dioscoreaceae:* Marker *et al., J. Am. Chem. Soc.* **62**, 2542 (1940). Plant sources for diosgenin: Marker *et al., ibid.* **65**, 1199 (1943); Wall *et al., J. Am. Pharm. Assoc. Sci. Ed.* **46**, 653 (1957). Identity with nitogenin: Marker *et al., J. Am. Chem. Soc.* **65**, 1248 (1943). Structure: Marker *et al., ibid.* **62**, 2525 (1940). Configuration at C_{25}: James, *J. Chem. Soc.* **1955**, 637. Synthesis: Mazur *et al., J. Am. Chem. Soc.* **82**, 5889 (1960); Kessar *et al., Tetrahedron Letters* **1966**, 4319. Obtained commercially from *Dioscorea composita* Hemsl. and *D. terpinapensis* Uline (see under Barbasco). Isoln from barbasco varieties: Julian, U.S. pat. **3,019,220** (1962 to Julian Labs.).

Crystals from acetone, mp 204-207°. $[\alpha]_D^{25}$ −129° (c = 1.4 in CHCl$_3$). Sol in the usual organic solvents, in acetic acid.

Acetate, $C_{29}H_{44}O_4$, crystals from acetic acid, mp 198°. $[\alpha]_D^{20}$ −119° (pyridine).

USE: Can be converted to pregnenolone and progesterone: Marker *et al.*, *J. Am. Chem. Soc.* **69**, 2167 (1947).

3349. Diosmetin. *5,7-Dihydroxy-2-(3-hydroxy-4-methoxyphenyl)-4H-1-benzopyran-4-one; 3′,5,7-trihydroxy-4′-methoxyflavone;* cyanidenon-4′-methyl ether 1479; luteolin-4′-methyl ether. $C_{16}H_{12}O_6$; mol wt 300.27. C 64.00%, H 4.03%, O 31.97%. Aglycone of diosmin, *q.v.* Prepn from diosmin isolated from various plant sources: O. A. Oesterle, G. Wander, *Helv. Chim. Acta* **8**, 519 (1925); isoln from lemons *(Citrus limon* Linn., *Rutaceae):* R. M. Horowitz, *J. Org. Chem.* **21**, 1184 (1956). Synthesis and structural elucidation: G. Zemplén, R. Bognár, *Ber.* **76**, 452 (1943). Synthesis: A. Lovecy *et al.*, *J. Chem. Soc.* **1930**, 817; N. B. Lorette *et al.*, *J. Org. Chem.* **16**, 930 (1951); J. H. Looker, M. J. Holm, *ibid.* **24**, 1019 (1959). HPLC determn in biological fluids: D. Baylocq *et al.*, *Ann. Pharmaceut. Franc.* **41**, 115 (1983).

Hemimethanolate, $C_{16}H_{12}O_6 \cdot \frac{1}{2}CH_3OH$, yellow needles from alcohol/ethyl acetate, sinters at 248°. mp 253-254°. Also reported as small yellow needles from methanol, mp 258-259° (Horowitz). uv max: 345, 268, 253 nm (log ε 4.32, 4.25, 4.28).

Triacetate, $C_{22}H_{18}O_9$, colorless needles from methanol, mp 195-196°.

3350. Diosmin. *7-[[6-O-(6-Deoxy-α-L-mannopyranosyl)-β-D-glucopyranosyl]oxy]-5-hydroxy-2-(3-hydroxy-4-methoxyphenyl)-4H-1-benzopyran-4-one; 3′,5,7-trihydroxy-4′-methoxyflavone-7-rutinoside;* 5-hydroxy-2-(3-hydroxy-4-methoxyphenyl)-7-(O^6-α-L-rhamnopyranosyl-β-D-glucopyranosyloxy)chromen-4-one; 5-hydroxy-2-(3-hydroxy-4-methoxyphenyl)-7-β-rutinosyloxy-4H-chromen-4-one; diosmetin 7-β-rutinoside; barosmin; buchu resin; Diosmil; Diosven; Diovenor; Flebosmil; Flebosten; Hemerven; Insuven; Litosmil; Tovene; Varinon; Ven-Detrex; Venosmine. $C_{28}H_{32}O_{15}$; mol wt 608.55. C 55.26%, H 5.30%, O 39.44%. Naturally occurring flavonic glycoside; rhamnoglycoside of diosmetin, *q.v.* Isolation from various plant sources: O. A. Oesterle, G. Wander, *Helv. Chim. Acta* **8**, 519 (1925). Elucidation of structure: G. Zemplén, R. Bognár, *Ber.* **76**, 452 (1943). Prepn from hesperidin, *q.v.: eidem, ibid.;* N. B. Lorette *et al.*, *J. Org. Chem.* **16**, 930 (1951). Isoln from lemon peel *(Citrus limon* Linn. *Rutaceae):* R. M. Horowitz, *J. Org. Chem.* **21**, 1184 (1956); from *Zanthoxylum avicennae, Rutaceae:* H. R. Arthur *et al.*, *J. Chem. Soc.* **1956**, 632; H. R. Arthur *et al.*, *ibid.* **1959**, 4007; from flowers of *Sophora microphylla* Ait. *Leguminosae:* L. H. Briggs *et al.*, *ibid.* **1960**, 1955. Toxicology studies: H. Heusser, W. Osswald,

Arch. Farmacol. Toxicol. **3**, 33 (1977). NMR spectrum: J. L. Nieto, A. M. Gutierrez, *Spectroscop. Letters* **19**, 427 (1986). Mechanism of action: C. Boudet, L. Peyrin, *Arch. Int. Pharmacodyn.* **283**, 312 (1986). Pharmacology: J. R. Caseley-Smith, J. R. Caseley-Smith, *Agents Actions* **17**, 1 (1985); M. Damon *et al.*, *Arzneimittel-Forsch.* **37**, 1149 (1987). HPLC determn in biological fluids: D. Baylocq *et al., Ann. Pharmaceut. Franc.* **41**, 115 (1983). Clinical study in post-phlebitic ulcers: M. C. Nguyen, K. Morere, *Gaz. Med.* **92**, 71 (1985); in acute hemorrhoids: A. Tajana *et al., Minerva Med.* **79**, 387 (1988). Clinical trial in chronic venous insufficiency: R. Laurent *et al., Int. Angiol.* **7**, Suppl. 2, 39 (1988).

Monohydrate, $C_{28}H_{32}O_{15} \cdot H_2O$, mp 275-277° (dec) (Zemplén). Also reported as fine needles from aq pyridine or aq DMF, mp 283° (dec) (Briggs). uv max (ethanol): 255, 268, 345 nm (log ε 4.28, 4.25, 4.30). Practically insol in water, alcohol.

Flavonoid extract, *Daflon, Flebopex, Flebotropin.*

THERAP CAT: Capillary protectant.

3351. Diosphenol. *2-Hydroxy-3-methyl-6-(1-methylethyl)-2-cyclohexen-1-one; 1-methyl-4-isopropyl-1-cyclohexen-2-ol-3-one;* 2-hydroxypiperitone; 1-*p*-menthen-2-ol-3-one; Buchu camphor; Barosma camphor. $C_{10}H_{16}O_2$; mol wt 168.24. C 71.39%, H 9.59%, O 19.02%. The crystalline portion of an oil obtained from Buchu leaves which come from various species of *Barosma,* such as *B. betulina* Bartl. & Wendl., *B. serratifolia* (Curt.) Willd. and *B. crenulata* (L.) Hook., *Rutaceae:* Flückiger, *Pharm. J.* **11**, 174, 219 (1880); Spica, *Gazz. Chim. Ital.* **15**, 195 (1885); Shimoyama, *Arch. Pharm.* **226**, 403 (1888); Kondakov *et al., J. Prakt. Chem.* [II] **54**, 433 (1896); [II] **63**, 49 (1901). Structure: Semmler, McKenzie, *Ber.* **39**, 1160 (1906).

Crystals, mp 83°. Sublimes. bp$_{760}$ 233° (dec); bp$_{10}$ 109°. $d_4^{99.2}$ 0.9542. $n_D^{99.8}$ 1.4607. Absorption spectrum: Lowry, Lishmund, *J. Chem. Soc.* **1935**, 1313; Gillan *et al., ibid.* **1941**, 62. Sparingly sol in water; moderately sol in alcohol; sol in ether, chloroform, carbon disulfide.

3352. Dioxadrol. *2-(2,2-Diphenyl-1,3-dioxolan-4-yl)piperidine;* 2,2-diphenyl-4-(2-piperidyl)-1,3-dioxolane. $C_{20}H_{23}NO_2$; mol wt 309.41. C 77.64%, H 7.49%, N 4.53%, O 10.34%. Prepn of *dl*-forms and resolution of α-racemates: Hardie, Halverstadt, **Belg.** pat. **613,262**; U.S. pat. **3,262,938** (1962, 1966 both to Cutter Labs.); W. R. Hardie *et al., J. Med. Chem.* **9**, 127 (1966).

Oily liquid.

Hydrochloride, $C_{20}H_{23}NO_2 \cdot HCl$, *Rydar*. Crystals from methanol, mp 256-260°. LD_{50} orally in mice: 240 mg/kg (Hardie).

d-Form hydrochloride, *dexoxadrol hydrochloride, Relane*. Crystals, dec 254°. $[\alpha]_D^{20}$ +34° (c = 2 in methanol). LD_{50} orally in mice: 340 mg/kg (Hardie).

l-Form hydrochloride, *levoxadrol hydrochloride, Levoxan*. Crystals, mp 248-254°. $[\alpha]_D^{20}$ −34.5° (c = 2 in methanol). LD_{50} orally in mice: 230 mg/kg (Hardie).

THERAP CAT: Base as antidepressant; *d*-form hydrochloride as stimulant (central), analgesic; *l*-form hydrochloride as anesthetic (local), relaxant (smooth muscle).

3353. Dioxane. 1,4-Diethylene dioxide. $C_4H_8O_2$; mol wt 88.11. C 54.53%, H 9.15%, O 36.32%. Prepd by distilling ethylene glycol with dil H_2SO_4. Monograph: W. Stumpf, *Chemie und Anwendungen des 1,4-Dioxans* (Verlag Chemie, 1956). Toxicity data: E. P. Laug *et al., J. Ind. Hyg. Toxicol.* **21**, 173 (1939). Carcinogenicity studies: M. F. Argus *et al., J. Nat. Cancer Inst.* **35**, 949 (1965); R. J. Kociba *et al., Toxicol. Appl. Pharmacol.* **30**, 275 (1974).

Flammable liq; faint pleasant odor. Vapor harmful. d_4^{20} 1.0329. mp 11.80°. bp_{760} 101.1°; bp_{400} 81.8°; bp_{200} 62.3°; bp_{100} 45.1°; bp_{60} 33.8°; bp_{40} 25.2°; bp_{20} 12°. Cryoscopic constant 4.83. Trouton constant 21.90. Heat of combustion: 581 kcal/mol. Heat of fusion: 2.98 kcal/mol. Specific heat at 20° = 0.0370 kcal/mol/°C. Viscosity at 25° = 0.0120 poise. Crit temp 312°. Crit press 50.7 atm. Flash pt 5-18°C. n_D^{20} 1.4175. Dipole moment: zero. Azeotropic mixture with water: 81.6% dioxane, bp 87.8°. Azeotropic mixture with ethanol: 9.3% dioxane, bp 78.1°. Sol in water and the usual organic solvents. Tends to form explosive peroxides, especially when anhydr. LD_{50} in mice, rats (ml/kg): 5.7, 5.2 orally (Laug).

Caution: Potential symptoms of overexposure are drowsiness, headache; nausea, vomiting; irritation of eyes, nose and throat; liver damage; kidney failure; skin irritation. *See NIOSH Pocket Guide to Chemical Hazards* (DHHS/NIOSH 90-117, 1990) p 100. This substance may reasonably be anticipated to be a carcinogen: *Seventh Annual Report on Carcinogens* (PB95-109781, 1994) p 186.

USE: Stabilizer in chlorinated solvents. Solvent for cellulose acetate, ethyl cellulose, benzyl cellulose, resins, oils, waxes, oil and spirit-sol dyes, and many other organic as well as some inorganic compds.

3354. Dioxaphetyl Butyrate. α,α-*Diphenyl-4-morpholinebutanoic acid ethyl ester;* ethyl 4-morpholino-2,2-diphenylbutyrate; 4-morpholino-2,2-diphenylbutyric acid ethyl ester; Amidalgon; Spasmoxal(e). $C_{22}H_{27}NO_3$; mol wt 353.46. C 74.76%, H 7.70%, N 3.96%, O 13.58%. Prepn: Specter *et al., J. Am. Chem. Soc.* **71**, 57 (1949); Dupré *et al., J. Chem. Soc.* **1949**, 500; Bockmühl, Ehrhart, U.S. pat. **2,230,774** (1940 to Winthrop).

Hydrochloride, $C_{22}H_{27}NO_3 \cdot HCl$, mp 168-169°.

Note: This is a controlled substance (opiate) listed in the U.S. Code of Federal Regulations, Title 21 Part 1308.11 (1995).

THERAP CAT: Analgesic (narcotic). Antispasmodic.

3355. Dioxathion. *Phosphorodithioic acid S,S'-1,4-dioxane-2,3-diyl O,O,O',O'-tetraethyl ester;* 2,3-p-dioxanedithiol S,S-bis(O,O-diethyl phosphorodithioate); AC-528;

ENT-22879; Hercules 528; Delnav; Navadel. $C_{12}H_{26}O_6P_2S_4$; mol wt 456.55. C 31.57%, H 5.74%, O 21.03%, P 13.57%, S 28.09%. Prepn: Diveley, Lohr, U.S. pat. **2,725,328**; Speck, U.S. pat. **2,815,350** (1955, 1957 both to Hercules); Diveley *et al., J. Am. Chem. Soc.* **81**, 139 (1959). Toxicity study: T. B. Gaines, *Toxicol. Appl. Pharmacol.* **14**, 515 (1969).

Tan liquid, d_4^{26} 1.257. mp −20°. n_D^{20} 1.5420. Practically insol in water. Partly sol in hexane. Hydrolyzed by alkali and by heating. LD_{50} in female, male rats (mg/kg): 23, 43 orally; 63, 235 dermally (Gaines).

USE: Insecticide; acaricide.

3356. Dioxethedrine. *4-[2-(Ethylamino)-1-hydroxypropyl]-1,2-benzenediol; N-ethyl-3,4-dihydroxynorephedrine;* α-(1-ethylaminoethyl)protocatechuyl alcohol; 2-ethylamino-1-(3',4'-dihydroxyphenyl)-1-propanol; 1-(3',4'-dihydroxyphenyl)-2-ethylamino-1-propanol; C-247. $C_{11}H_{17}NO_3$; mol wt 211.26. C 62.54%, H 8.11%, N 6.63%, O 22.72%. β-Adrenergic agonist. Prepd by catalytic hydrogenation of the corresponding aminoketone: Lespagnol, Cuingnet, *Ann. Pharm. Franc.* **18**, 445 (1960). Only one of the two possible isomers, believed to be the "erythro" form, was isolated.

Hydrochloride, $C_{11}H_{17}NO_3 \cdot HCl$, crystals from methanol + ether, mp 212-214°.

One of the ingredients of *Bexol*.

THERAP CAT: Bronchodilator.

3357. Dioxybenzone. *(2-Hydroxy-4-methoxyphenyl)(2-hydroxyphenyl)methanone; 2,2'-dihydroxy-4-methoxybenzophenone;* 4-methoxy-2,2'-dihydroxybenzophenone; benzophenone-8; Cyasorb UV 24 (obsolete); Spectra-Sorb UV 24. $C_{14}H_{12}O_4$; mol wt 244.25. C 68.85%, H 4.95%, O 26.20%. Prepn: Hardy *et al.*, U.S. pat. **2,853,521** (1958 to Am. Cyanamid).

Yellow powder, mp 68°. Soly in g/100 ml at 25°: ethanol 21.8; isopropanol 17; propylene glycol 6.2; ethylene glycol 3.0; n-hexane 1.5.

THERAP CAT: Ultraviolet screen.

3358. Dioxypyramidon. *(Dimethylamino)oxoacetic acid 2-acetyl-2-methyl-1-phenylhydrazide; 1-acetyl-1,5,5-trimethyl-2-phenylsemioxamazide;* α-(dimethylamidooxalyl)-β,-β-methylacetylphenylhydrazine; dioxoaminopyrine; dioxoaminopyrine. $C_{13}H_{17}N_3O_3$; mol wt 263.30. C 59.30%, H 6.51%, N 15.96%, O 18.23%. Prepd from aminopyrine by oxidation with 30% hydrogen peroxide: Charonnat, Delaby, *Compt. Rend.* **189**, 850 (1929); Valyashko, Depeshko, *Zh. Obshch. Khim.* **20**, 1667 (1950).

Orthorhombic, translucent prisms from water. Somewhat bitter taste. mp 105.5° (softens at 96°). bp$_2$ 194-201°. uv max (hexane): 248.5 nm (ε 16000). Soly in water at 20° = 7.69 g/100 ml; at 37° = 48.2 g/100 ml. Also sol in alcohol. Hydrolyzed by alkali.

3359. 2,5-Di-*tert*-pentylhydroquinone. *2,5-Bis(1,1-dimethylpropyl)-1,4-benzenediol;* 2,5-di-*tert*-amylhydroquinone; 2,5-bis(1,1-dimethylpropyl)hydroquinone; Santovar A. C$_{16}$H$_{26}$O$_2$; mol wt 250.38. C 76.75%, H 10.47%, O 12.78%. Prepn: Erickson, **Brit.** pat. **596,461** (1948 to Mathieson Alkali Works).

Crystals, mp 179.4-180.4°.

USE: As a staining protector in rubber.

3360. Diperodon Hydrochloride. *3-(1-Piperidinyl)-1,2-propanediol bis(phenylcarbamate) (ester) monohydrochloride;* 3-piperidino-1,2-propanediol dicarbanilate hydrochloride; 3-(1-piperidyl)-1,2-propanediol dicarbanilate hydrochloride; di(phenylurethan) of 1-piperidinepropane-2,3-diol hydrochloride; Diothane Hydrochloride; Proctodon. C$_{22}$H$_{28}$ClN$_3$O$_4$; mol wt 433.93. C 60.89%, H 6.50%, Cl 8.17%, N 9.68%, O 14.75%. Prepn: T. H. Rider, *J. Am. Chem. Soc.* **52**, 2115 (1930); *idem,* **U.S.** pat. **2,004,132** (1935). Stability of solns: E. S. Cook, T. H. Rider, *J. Am. Pharm. Assoc.* **26**, 222 (1937). Pharmacology: *idem, J. Pharmacol.* **64**, 1 (1938). X-ray crystallographic study: R. C. Sullivan, K. P. O'Brien, *Bull. Narcotics* **20**, 31 (1968). IR spectra: *idem, ibid.* **22**, 35 (1970).

Crystals. Bitter taste, followed by a sense of numbness. Stable in air at ordinary temps. Dec 195-200°. Sol in alcohol. Slightly sol in water (1:100), acetone and ethyl acetate. The aq soln is faintly acid to litmus. Extemporaneously prepd aq solns should be used promptly to avoid precipitation of the free base by traces of alkali. The soly in water is increased by the addn of sodium chloride. Insol in benzene or ether.

THERAP CAT: Anesthetic (local).

3361. Diphemanil Methylsulfate. *4-(Diphenylmethylene)-1,1-dimethylpiperidinium methyl sulfate; p-(α-phenylbenzylidene)-1,1-dimethylpiperidinium methylsulfate;* N,N-dimethyl-4-piperidylidene-1,1-diphenylmethane methylsulfate; Demotil; Diphenatil; Nivelona; Prantal; Variton. C$_{21}$H$_{27}$NO$_4$S; mol wt 389.52. C 64.76%, H 6.99%, N 3.60%, O 16.43%, S 8.23%. Prepn: N. Sperber *et al., J. Am. Chem. Soc.* **73**, 5010 (1951); N. Sperber *et al., U.S.* pat. **2,739,968** (1956 to Schering). Pharmacological properties and toxicity data: S. Margolin *et al., Proc. Soc. Exp. Biol. Med.* **78**, 576 (1951).

Crystals, mp 194-195°. Soluble in water. LD$_{50}$ in rats, mice, guinea pigs (mg/kg): 1107, 64, 404 orally (Margolin).

THERAP CAT: Anticholinergic.

THERAP CAT (VET): Anticholinergic.

3362. Diphemethoxidine. *2-(Diphenylmethyl)-1-piperidineethanol;* 2-benzhydryl-1-piperidineethanol; 1-(β-hydroxyethyl)-2-(diphenylmethyl)piperidine; Ba-18189; Cleofil. C$_{20}$H$_{25}$NO; mol wt 295.42. C 81.31%, H 8.53%, N 4.74%, O 5.42%. Prepn: **Brit.** pat. **861,815** (1961 to Ciba), *C.A.* **55**, 15517h (1961).

mp 106-107°. bp$_{0.1}$ 180-181°. Hydrochloride, mp 166-167°.

THERAP CAT: Anorexic.

3363. Diphenadione. *2-(Diphenylacetyl)-1H-indene-1,3-(2H)-dione;* 2-diphenylacetyl-1,3-diketohydrindene; diphacinone; U-1363; Dipaxin; Oragulant; Solvan; Didandin; Diphacin. C$_{23}$H$_{16}$O$_3$; mol wt 340.38. C 81.16%, H 4.74%, O 14.10%. Prepd by the condensation of dimethyl phthalate with diphenylacetone in an inert solvent in the presence of sodium methoxide: Thomas, **U.S.** pat. **2,672,483** (1954 to Upjohn).

Pale yellow crystals from ethanol, mp 146-147°. Acid reaction. Practically insol in water. Slightly sol in benzene, hot ethanol. Sol in acetone, acetic acid. Forms a sodium salt which is sparingly sol in water.

USE: Rodenticide.

THERAP CAT: Anticoagulant.

3364. Diphenamid. N,N-Dimethyl-α-phenylbenzeneacetamide; *N,N-dimethyl-2,2-diphenylacetamide;* N,N-dimethyl-α,α-diphenyl acetamide; 2,2-diphenyl-N,N-dimethylacetamide; L-34314; Dymid; Enide. C$_{16}$H$_{17}$NO; mol wt 239.32. C 80.30%, H 7.16%, N 5.85%, O 6.69%. Selective pre-emergence herbicide. Prepn: Cheney *et al., J. Org. Chem.* **17**, 770 (1952). Toxicity study: G. W. Bailey, J. L. White, *Residue Rev.* **10**, 97 (1965).

Crystals from ethyl acetate, mp 134.5-135.5°. Soluble in water, acetone, dimethylformamide, xylene, phenyl Cellosolve. LD_{50} orally in rats: 700 mg/kg (Bailey, White).

USE: Herbicide.

3365. Diphenane. *4-(Phenylmethyl)phenol carbamate; carbamic acid α-phenyltolyl ester; p-benzylphenyl carbamate; p-benzylphenylurethan; carbamic acid p-benzylphenyl ester; p-hydroxydiphenylmethane carbamate; α-phenyl-p-cresyl carbamate; diphenan; Butolan; Butolen; Carphenol; Oxylan; Palafuge; Parabencil.* $C_{14}H_{13}NO_2$; mol wt 227.26. C 73.99%, H 5.77%, N 6.16%, O 14.08%. Prepd from α-phenyl-p-cresol, $COCl_2$, and NH_4OH: Kropp, U.S. pat. **1,252,-452** (1818).

Crystals from alcohol, mp 147-150°. Practically insol in water. Sparingly sol in 90% ethanol; sol in methanol, abs alcohol, chloroform, ether, benzene.

THERAP CAT: Anthelmintic (Nematodes).

3366. Diphencyprone. *2,3-Diphenyl-2-cyclopropen-1-one;* 1,2-diphenylcyclopropenone; DPC. $C_{15}H_{10}O$; mol wt 206.24. C 87.36%, H 4.89%, O 7.76%. Contact allergen, possibly the smallest neutral aromatic specie. Prepn: R. Breslow *et al., J. Am. Chem. Soc.* **81,** 247 (1959). Crystal sructure: H. L. Ammon, *ibid.* **95,** 7093 (1973). Strain energy: A. Greenberg *et al., ibid.* **105,** 6855 (1983). Mechanism of photodissociation: Y. Hirata, N. Mataga, *Chem. Phys. Letters* **193,** 287 (1992). Report as contact sensitizer: B. M. Hausen, J. Stute, *Chem. Ind.* **1980,** 699. Clinical evaluation in alopecia totalis: B. Monk, *Clin. Exp. Dermatol.* **14,** 154 (1989); of children with alopecia areata: S. MacDonald Hull *et al., Brit. J. Dermatol.* **125,** 164 (1991). Review of clinical efficacy: E. Hoting, A. Boehm, *ibid.* **127,** 625-629 (1992).

Crystallizes as monohydrate from cyclohexane, mp 87-90°. d 1.202 g/cm³. uv max (CH_3CN): 297, 282, 226, 220 nm (log ε 4.3, 4.25, 4.13, 4.16). mp (anhydrous): 118-120°.

THERAP CAT: Antialopecia agent.

3367. Diphenhydramine. *2-Diphenylmethoxy-N,N-dimethylethanamine;* 2-(benzhydryloxy)-*N,N*-dimethylethylamine; β-dimethylaminoethyl benzhydryl ether; *O*-benzhydryldimethylaminoethanol; β-dimethylaminoethanol diphenylmethyl ether; α-(2-dimethylaminoethoxy)diphenylmethane; benzhydramine. $C_{17}H_{21}NO$; mol wt 255.36. C 79.96%, H 8.29%, N 5.49%, O 6.27%. Prepn: Rieveschl, Jr., U.S. pats. **2,421,714; 2,427,878** (both 1947 to Parke, Davis), *cf.* U.S. pat. **2,397,799** (1946). Toxicity: Gruhzit, Fisken, *J. Pharmacol. Exp. Ther.* **89,** 227 (1947). Comprehensive description of the hydrochloride: I. J. Holcomb, S. A. Fusari, *Anal. Profiles Drug Subs.* **3,** 173-232 (1974).

bp$_{2.0}$ 150-165°.

Hydrochloride, $C_{17}H_{21}NO.HCl$, *Alledryl, Allergina, Amidryl, Bagodryl, Bax, Benadryl, Benocten, Benodine, Benzantin, Dibondrin, Dihydral, Diphantine, Dolestan, Fenylhist, Halbmond, Histacyl, Noctomin, S 8, Sedopretten, Sekundal-D, Syntedril, Wehydryl.* Crystals from abs alcohol + ether, mp 166-170°. Bitter taste. Slowly darkens on exposure to light. Stable under ordinary conditions. One gram dissolves in 1 ml water, 2 ml alcohol, 2 ml chloroform, 50 ml acetone. Very slightly sol in benzene, ether. pH of 1% aq soln about 5.5. The aq soln forms a pink precipitate with satd Reinecke's salt soln. LD_{50} orally in rats: 500 mg/kg (Gruhzit, Fisken).

Compd with *p*-sulfonamide benzoate, *Neobenadol.* See **Japan.** pats. **203,405** and **243/64** (to Taisho).

THERAP CAT: Antihistaminic.

THERAP CAT (VET): Antihistaminic. Also in motion sickness.

3368. Diphenic Acid. *[1,1'-Biphenyl]-2,2'-dicarboxylic acid; o,o'-bibenzoic acid.* $C_{14}H_{10}O_4$; mol wt 242.23. C 69.42%, H 4.16%, O 26.42%. Prepd from diazotized anthranilic acid by treatment with a cuproammonia-sulfite reducing agent: Vorländer, Meyer, *Ann.* **320,** 122 (1902); Atkinson, Lawler, *Org. Syn. coll. vol.* **I** (2nd ed, 1941) p 222. By chromic acid oxidation of phenanthrenequinone: Roberts, Johnson, *J. Am. Chem. Soc.* **47,** 1399 (1925). By heating potassium *o*-bromobenzoate with copper powder: Hurtley, *J. Chem. Soc.* **1929,** 1870.

Monoclinic prismatic rods upon slow cooling from water, leaflets from hot water, needles by careful sublimation. mp 228-229°. The satd aq soln is 0.0052N at 25°; soluble in the usual organic solvents.

Dimethyl ester, $C_{16}H_{14}O_4$, monoclinic prismatic plates, tablets from methanol, mp 73.5°; bp$_{14}$ 204-206°.

Diethyl ester, $C_{18}H_{18}O_4$, cubes from alcoholic HCl, mp 42°.

3369. Diphenidol. *α,α-Diphenyl-1-piperidinebutanol;* 1,1-diphenyl-4-piperidino-1-butanol; diphenyl-[3-(1-piperidyl)propyl]carbinol; defenidol; SKF-478. $C_{21}H_{27}NO$; mol wt 309.45. C 81.51%, H 8.79%, N 4.53%, O 5.17%. Prepn: Miescher, Marxer, U.S. pat. **2,411,664** (1946 to Ciba); Barrett, Wilkinson, **Brit.** pat. **683,950** (1952 to Wellcome Foundation), *C.A.* **48,** 2112e (1954). Structure-activity studies: Gautier *et al., Med. Pharmacol. Exp.* **13,** 325 (1965). Clinical studies: Cutt *et al., Aerosp. Med.* **39,** 682 (1968); Benson, *ibid.* **40,** 589 (1969). Acute toxicity: E. I. Goldenthal, *Toxicol. Appl. Pharmacol.* **18,** 185 (1971); *Japan Med. Gaz.* **10**(7), 6 (1973).

Needles from petr ether, mp 104-105°. LD_{50} s.c. in rats: 50 mg/kg (Goldenthal).

Hydrochloride, $C_{21}H_{27}NO.HCl$, *SKF-478-A*, *Ansmin, Cefadol, Celmidol, Difenidolin, Maniol, Mecalmin, Pineroro, Satanolon, Tenesdol, Vontrol, Wansar.* Crystals from chloroform + ethyl acetate, mp 212-214°. Freely sol in methanol; sol in water, chloroform. Practically insol in ether, benzene, petr ether. LD_{50} in mice, rats (mg/kg): 430, 515 orally; 105, 82 i.p.; 37, 29 i.v. *(Japan Med. Gaz.).*

THERAP CAT: Antiemetic.

3370. Diphenolic Acid. *4-Hydroxy-γ-(4-hydroxyphenyl)-γ-methylbenzenebutanoic acid;* 4,4-bis[4'-hydroxyphenyl]pentanoic acid; γ,γ-bis-(p-hydroxyphenyl)valeric acid; DPA. $C_{17}H_{18}O_4$; mol wt 286.33. C 71.31%, H 6.34%, O 22.35%. Prepd by condensing 2.25-4.0 moles of phenol with one mole of levulinic acid in the presence of HCl: **Brit.** pat. **768,206** (1957 to S. C. Johnson & Son); from one mole of phenol, 0.5 mole levulinic acid and HCl, H_2SO_4 or H_3PO_4: Bader, Kantowicz, *J. Am. Chem. Soc.* **76**, 4465 (1954); Bader, **U.S.** pat. **2,933,520** (1960 to S. C. Johnson & Son).

Crystals from hot water. Higher melting modification, mp 171-172°. Appreciably sol in hot water; sol in acetone, acetic acid, ethanol, isopropanol, methyl ethyl ketone.

USE: Intermediate for surface coatings, lubricating oil additives, cosmetics, surfactants, plasticizers, textile chemicals.

3371. Diphenoxylate. *1-(3-Cyano-3,3-diphenylpropyl)-4-phenyl-4-piperidinecarboxylic acid ethyl ester; 1-(3-cyano-3,3-diphenylpropyl)-4-phenylisonipecotic acid ethyl ester;* ethyl 1-(3-cyano-3,3-diphenylpropyl)-4-phenylisonipecotate; ethyl 1-(3-cyano-3,3-diphenylpropyl)-4-phenyl-4-piperidine-carboxylate; 2,2-diphenyl-4-(4-carbethoxy-4-phenylpiperidino)butyronitrile; R-1132. $C_{30}H_{32}N_2O_2$; mol wt 452.60. C 79.61%, H 7.13%, N 6.19%, O 7.07%. Prepn: Janssen, **U.S.** pat. **2,898,340** (1959). Pharmacokinetics and metabolism: Karim *et al., J. Pharmacol. Exp. Ther.* **177**, 546 (1971); *Clin. Pharmacol. Ther.* **13**, 407 (1972). *See also* Difenoxin, the active metabolite of diphenoxylate. Comprehensive description: D. D. Hung, *Anal. Profiles Drug Subs.* **7**, 149-169 (1978).

Hydrochloride, $C_{30}H_{32}N_2O_2.HCl$, crystals, mp 220.5-222°. uv max (methanol): 252, 258, 264 nm. Soly in mg/ml at 25°: acetic acid 500; DMF 500; chloroform 360; methanol > 50; ethanol 3; water 0.8; hexane 0.5.

Mixture of hydrochloride with atropine sulfate, *Lomotil, Diarsed, Reasec.*

Note: This is a controlled substance (opiate) listed in the U.S. Code of Federal Regulations, Title 21 Part 1308.12 (1995).

THERAP CAT: Antiperistaltic; antidiarrheal.

3372. Diphenyl. *1,1'-Biphenyl;* bibenzene; phenylbenzene. $C_{12}H_{10}$; mol wt 154.21. C 93.46%, H 6.54%. Toxicity data: Deichmann *et al., J. Ind. Hyg. Toxicol.* **29**, 1 (1947). *Review:* W. C. Weaver *et al.,* in Kirk-Othmer *Encyclopedia of Chemical Technology* vol. **7** (Wiley-Interscience, New York, 3rd ed., 1979) pp 782-793.

Colorless leaflets; pleasant, peculiar odor. d 1.041. mp 69-71°. bp 254-255°. n_D^{77} 1.588. Insol in water. Sol in alc, ether. LD_{50} orally in rats: 3280 mg/kg (Deichmann).

Caution: Potential symptoms of overexposure are irritation of throat and eyes; headache, nausea, fatigue and numb limbs; liver damage. *See NIOSH Pocket Guide to Chemical Hazards* (DHHS/ NIOSH 90-117, 1990) p 100.

USE: As heat transfer agent; fungistat for oranges (applied to inside of shipping container or wrappers); in organic syntheses.

3373. Diphenylacetamide. Acetyldiphenylamine. $C_{14}H_{13}NO$; mol wt 211.26. C 79.59%, H 6.20%, N 6.63%, O 7.57%. $CH_3CON(C_6H_5)_2$.

White, cryst powder. mp 103°. Sublimes without decomposition. Slightly sol in water; sol in alcohol, ether.

3374. Diphenylacetic Acid. *α-Phenylbenzeneacetic acid;* diphenylmethane-α-carboxylic acid. $C_{14}H_{12}O_2$; mol wt 212.25. C 79.23%, H 5.70%, O 15.08%. $(C_6H_5)_2CHCOOH$. Prepd by the reduction of benzilic acid with hydriodic acid and red phosphorus: C. S. Marvel *et al., Org. Syn.* **coll. vol. I**, 224 (2nd ed., 1941). Ecologically and economically improved synthesis from benzilic acid: P. Strazzolini *et al., Syn. Commun.* **17**, 1919 (1987). Toxicity study: G. W. Bailey, J. L. White, *Residue Rev.,* **10**, 97 (1965).

Small plates from alcohol, needles from water, mp 148°. Sublimes. Sol in hot water, alcohol, ether, chloroform.

Methyl ester, $C_{15}H_{14}O_2$, crystals from dil alcohol, mp 60°. Ethyl ester, $C_{16}H_{16}O_2$, crystals from alcohol, mp 58°. bp_{15} 178°.

Amide, $C_{14}H_{13}NO$, mp 167-168°.

Nitrile, $C_{14}H_{11}N$, *diphenylacetonitrile,* α-cyanodiphenyl-methane, diphenatrile, Dipan. mp 76°. bp_{12} 181°. Sol in alcohol, ether, propylene glycol. LD_{50} orally in rats: 3500 mg/kg (Bailey, White).

Chloride, $C_{14}H_{11}ClO$, crystals from petr ether, mp 57°; bp_{15} 178°. Very sol in benzene. Sol in hot ligroin, in hot isopropyl ether.

Anhydride, $C_{28}H_{22}O_3$, mp 98°; bp_{15} 220°.

USE: Nitrile as herbicide.

3375. Diphenylamine. *N-Phenylbenzeneamine.* $C_{12}H_{11}N$; mol wt 169.23. C 85.17%, H 6.55%, N 8.28%. $C_6H_5NH-C_6H_5$. Made by heating aniline with aniline hydrochloride.

Crystals; floral odor. d 1.16. mp 53-54°. bp 302°. Flash pt 153°. Discolors in light. One gram dissolves in 2.2 ml alcohol, 4.5 ml propyl alcohol; freely sol in benzene, ether, glacial acetic acid, carbon disulfide. Insol in water. Forms salts with strong acids. *Keep protected from light.*

Hydrochloride, $C_{12}H_{11}N.HCl$, crystals, turn blue in air. Freely sol in water, alcohol.

Sulfate, $C_{12}H_{11}N.H_2SO_4$, white to yellowish powder, mp 123-125°. Practically insol in water. Sol in alc, in H_2SO_4.

Caution: May be irritating to mucous membranes. Methemoglobinemia has been produced experimentally. Symptoms similar to aniline, but diphenylamine is less toxic.

USE: Manuf dyes; stabilizing nitrocellulose explosives and celluloid. In anal. chem for the detection of NO_3, ClO_3 and other oxidizing substances with which, in the presence of H_2SO_4, it gives a deep-blue color.

THERAP CAT (VET): Topically in antiscrewworm mixtures. In tests for nitrate or nitrite poisoning.

3376. Diphenylamine-2,2'-dicarboxylic Acid. *2,2'-Iminobis[benzoic acid].* $C_{14}H_{11}NO_4$; mol wt 257.25. C 65.37%, H 4.31%, N 5.44%, O 24.88%. Prepn from the alkali salts of anthranilic acid and of 2-chlorobenzoic acid in presence of copper: Ullmann, *Ann.* **355**, 352 (1907).

Yellow crystals from alcohol, mp 296-297° (dec). Insoluble in water. Very slightly sol in alcohol, ether, chloroform, glacial acetic acid.

USE: Instead of diphenylamine in reactions with oxidizing agents. May be used as an oxidation-reduction indicator in strongly acid solns.

3377. N,N'-Diphenylbenzidine. *N,N'-Diphenyl-[1,1'-biphenyl]-4,4'-diamine.* $C_{24}H_{20}N_2$; mol wt 336.44. C 85.68%, H 5.99%, N 8.33%.

Leaflets or plates, mp 242°. Insol in water. Freely sol in boiling toluene or ethyl acetate, slightly in alc, acetone. Is more sensitive to oxidizing substances than diphenylamine. *Keep protected from light.*

USE: *See* diphenylamine.

3378. 1,4-Diphenyl-1,3-butadiene. *1,1'-(1,3-Butadiene-1,4-diyl)bisbenzene;* distyryl; bistyryl; 1,4-diphenylerythrene. $C_{16}H_{14}$; mol wt 206.29. C 93.16%, H 6.84%. C_6H_5-CH=CHCH=CHC$_6$H$_5$. Prepd by the condensation of phenylacetic acid and cinnamic aldehyde: Kuhn, Winterstein, *Helv. Chim. Acta* **11**, 103 (1928); Corson, *Org. Syn.* **16**, 28 (1936). This method yields the *trans,trans*-form.

cis,cis-Form, leaflets, needles, mp 70.5°. Changes to the *trans,trans*-form under the influence of light. $d_4^{100.6}$ 0.9697. $n_{H\alpha}^{100.6}$ 1.61831; $n_{H\beta}^{100.6}$ 1.66748; $n_{He}^{100.6}$ 1.63473. Soluble in ether, chloroform, benzene, petr ether, hot glacial acetic acid, slightly sol in alcohol.

cis,trans-Form, obtained only in the absence of light. Oily liquid or crystals, mp 88°. $bp_{0.1}$ 133-135°. d_4^{22} 0.9974. $n_{H\alpha}^{22.2}$ 1.59679; $n_{H\beta}^{22.2}$ 1.62830; $n_{He}^{22.2}$ 1.60532.

trans,trans-Form, crystals, mp 149.7°. bp_{720} 350°. Soluble in alcohol; sparingly sol in ether.

3379. sym-Diphenylcarbazide. *2,2'-Diphenylcarbonic dihydrazide.* $C_{13}H_{14}N_4O$; mol wt 242.28. C 64.45%, H 5.82%, N 23.12%, O 6.60%. C_6H_5NHNHCONHNHC$_6$H$_5$.

White, crystal powder, gradually becomes pink. mp 168-171°. Very slightly sol in water; sol in hot alcohol, acetone, glacial acetic acid. *Keep protected from light.*

USE: As indicator in titrating Fe; for the colorimetric determination of Cr, detection of Cd, Hg, Mg, aldehydes, emetine.

3380. Diphenylcarbazone. *Phenyldiazenecarboxylic acid 2-phenylhydrazide.* $C_{13}H_{12}N_4O$; mol wt 240.26. C 64.99%, H 5.03%, N 23.32%, O 6.66%. C_6H_5N=NCONH-NHC$_6$H$_5$.

Orange-red needles. mp about 157° with decompn. Insol in water. Sol in alcohol, chloroform, benzene.

USE: As a sensitive reagent for Hg, with which it gives a blue color.

3381. 1,1-Diphenylethene. *1,1'-Ethenylidenebis[benzene]; unsym-diphenylethylene;* α,α-diphenylethylene; α-methylene-diphenylethylene. $C_{14}H_{12}$; mol wt 180.25. C 93.29%, H 6.71%. $(C_6H_5)_2$C=CH$_2$. Prepd by the action of phenylmagnesium bromide on ethyl acetate in ether, followed by treatment of the reaction product with ammonium chloride soln: Allen, Converse, *Org. Syn.* coll. vol. I (2nd ed., 1941) p 226. Absorption spectrum: Lardy, *J. Chim. Phys.* **21**, 361 (1924).

Liquid, mp 8.2°. d^{20} 1.0232. bp_{760} 277.0°; bp_{400} 249.8°; bp_{200} 222.8°; bp_{100} 198.6°; bp_{60} 183.4°; bp_{40} 170.8°; bp_{20} 151.8°; bp_{10} 135.0°; bp_5 119.6°; $bp_{1.0}$ 87.4°. n_D^{20} 1.60849.

3382. N-(1,2-Diphenylethyl)nicotinamide. *N-(1,2-Diphenylethyl)-3-pyridinecarboxamide;* 1-nicotinylamino-1,2-diphenylethane; nicofetamide; nicofetamide; NC-1065; Lyspamin. $C_{20}H_{18}$-N$_2$O; mol wt 302.38. C 79.44%, H 6.00%, N 9.26%, O 5.29%. Prepn: Suter, U.S. pat. **2,483,250** (1949 to Cilag).

Needles from ethanol, mp 159°. Readily sol in 5N HCl; moderately sol in methanol, acetone, ether, benzene; slightly sol in ethanol. Practically insol in water and alkalies.

THERAP CAT: Anticholinergic.

3383. 1,3-Diphenylguanidine. *N,N'-Diphenylguanidine; sym*-diphenylguanidine; Melaniline; Vulkazit. $C_{13}H_{13}$-N$_3$; mol wt 211.27. C 73.91%, H 6.20%, N 19.89%. Prepn: Naunton, *J. Soc. Chem. Ind. (London)* **45**, 376T (1926); Macholdt-Erdniss, *Ber.* **91**, 1992 (1958); Ferris, Schutz, *J. Org. Chem.* **28**, 71 (1963). Pharmacology: P. Valade *et al., C.R. Soc. Biol.* **143**, 815 (1949).

Crystals from ether, mp 150°. Dec at about 170°. d 1.13. Sparingly sol in water; sol in alcohol, chloroform, hot benzene, hot toluene; readily sol in dil mineral acids. Aq soln is strongly alkaline. MLD s.c. in guinea pigs: 200 mg/kg; i.v. in dogs: 25 mg/kg (Valade).

Phthalate, $C_{34}H_{32}N_6O_4$, *Guantal.* Obtained as the hemihydrate, blue-white to light gray powder. mp 178°. d_4^{25} 1.20. Sol in alcohol. Practically insol in benzene, gasoline.

USE: Recommended as a primary material for standardizing acids. Free base and phthalate as accelerators for vulcanization of rubber.

3384. 1,1-Diphenylhydrazine. $C_{12}H_{12}N_2$; mol wt 184.24. C 78.23%, H 6.56%, N 15.20%. $(C_6H_5)_2$NNH$_2$.

Yellow crystals. d 1.19. mp 34.5°; also reported as mp 44°. bp_{40} 220°. Insol in water. Freely sol in alc, ether.

Hydrochloride, $C_{12}H_{12}N_2$.HCl, white to grayish white, crystalline powder. Slightly sol in water; freely sol in alc.

USE: Hydrochloride as reagent for arabinose and lactose.

3385. Diphenylketene. *Diphenylethenone.* $C_{14}H_{10}O$; mol wt 194.23. C 86.57%, H 5.19%, O 8.24%. $(C_6H_5)_2$-C=C=O. Prepd from benzil monohydrazone: Smith, Hoehn, *Org. Syn.* coll. vol. III, 356 (1955).

Reddish-yellow liquid. Best stored under nitrogen. The addition of hydroquinone helps to retard polymerization. Solidifies in refrigerator. $d_4^{13.7}$ 1.1107. bp_{760} 265-270° (dec); bp_{12} 146°; $bp_{3.5}$ 119-121°. Should be distilled under reduced pressure (3 to 5 mm) in an atm of nitrogen. $n_D^{14.1}$ 1.615. Dipole moment in benzene at 25° = −1.9.

3386. Diphenylmagnesium. Magnesium diphenyl. $C_{12}H_{10}Mg$; mol wt 178.52. C 80.74%, H 5.65%, Mg 13.61%. $Mg(C_6H_5)_2$. Prepd by the action of magnesium on mercury diphenyl: Schlenk, Jr., *Ber.* **64**, 736 (1931); Gilman, Brown, *Rec. Trav. Chim.* **49**, 202 (1930).

Solvated, feathery crystals from ether. Loses its ether *in vacuo* at around 37°. Dec 280°, dissociating into magnesium and diphenyl. Extremely reactive; catches fire in moist, although not in dry, air. Violently dec by water. The etherate is sol in benzene, but the dry compd is not.

3387. Diphenylmethane. *1,1'-Methylenebis[benzene];* benzylbenzene; ditan. $C_{13}H_{12}$; mol wt 168.24. C 92.81%, H 7.19%. $(C_6H_5)_2$CH$_2$. Prepd from methylene chloride and benzene with aluminum chloride as catalyst: Friedel, Crafts, *Bull. Soc. Chim.* [2] **41**, 324 (1884); from benzyl chloride and benzene: Hartman, Phillips, *Org. Syn.* **14**, 34 (1934); L. F. Fieser, *Experiments in Organic Chemistry* (Boston, 3rd ed., 1955) p 157.

Orthorhombic needles. Odor of oranges. d_4^{10} 1.3421 (solid). mp 25.9°. d_4^{26} 1.0008 (liq). bp_{760} 264.5°; bp_{400}

237.5°; bp$_{200}$ 210.7°; bp$_{100}$ 186.3°; bp$_{60}$ 170.2°; bp$_{40}$ 157.8°; bp$_{20}$ 139.8°; bp$_{10}$ 122.8°; bp$_{5}$ 107.4°; bp$_{1.0}$ 76.0°. n_D^{20} 1.57683. Absorption spectrum in hexane: Castille, *Bull. Soc. Chim. Belg.* **36**, 296; *Chem. Zentr.* **1927**, I, 1126; in alcohol: Ondorff, *J. Am. Chem. Soc.* **49**, 1541 (1927). Freely sol in alcohol, ether, chloroform, hexane, benzene. Insol in liq ammonia.

3388. *N,N'*-Diphenyl-*p*-phenylenediamine. *N,N-Diphenyl-1,4-benzenediamine;* 1,4-dianilinobenzene; DPPD. C$_{18}$H$_{16}$N$_2$; mol wt 260.34. C 83.04%, H 6.19%, N 10.76%. Prepd by condensing hydroquinone or *p*-aminophenol with aniline: Calm, *Ber.* **16**, 2805 (1883); Clemens, Magoffin, **U.S. pat. 2,503,712** (1950 to Kodak). Purification: Pecherer, **U.S. pat. 2,833,824** (1958 to Hoffmann-La Roche).

Colorless leaflets from alcohol. Commercial grades are greenish-brown. d 1.20. mp 150-151° (uncorr., Pecherer). bp$_{0.5}$ 220-225°. Sol in monochlorobenzene, benzene, DMF, ether, chloroform, acetone, ethyl acetate, isopropyl acetate, glacial acetic acid. Slightly sol in alcohol. Almost insol in petr ether, water.

USE: Polymerization inhibitor. Antioxidant for rubber, petr oils, feedstuffs.

THERAP CAT (VET): Antioxidant for feedstuffs.

3389. 1,1-Diphenyl-2-picrylhydrazyl (Free Radical). *2,2-Diphenyl-1-(2,4,6-trinitrophenyl)hydrazyl;* DPPH. Prepn: Goldschmidt, Renn, *Ber.* **55B**, 628 (1922); Lyons, Watson, *J. Polymer Sci.* **18**, 141 (1955); Arbuzov, Valitova, *Zh. Obshch. Khim.* **27**, 2354 (1957). Structure: Poirier *et al., J. Org. Chem.* **17**, 1437 (1952).

Large, dark violet prisms from benzene + petr ether, mp 127-129° (dec). mp also reported as 132-133°.

USE: Analytical reagent for reducing substances: Schenck *et al., Tetrahedron Letters* **1967**, 193.

3390. Diphenylpyraline. *4-(Diphenylmethoxy)-1-methylpiperidine;* diphenylpyrilene; 4-benzhydryloxy-*N*-methylpiperidine; 1-methyl-4-piperidyl benzhydryl ether; 1-methyl-4-hydroxypiperidine benzhydryl ether; P-253. C$_{19}$H$_{23}$NO; mol wt 281.40. C 81.10%, H 8.24%, N 4.98%, O 5.69%. Prepn: L. H. Knox, R. Kapp, **U.S. pat. 2,479,843** (1949 to Nopco). Pharmacokinetics: G. Graham, A. G. Bolt, *J. Pharmacokinet. Biopharm.* **2**, 191 (1974). Clinical trial in allergic rhinitis: H. Puhakka *et al., J. Int. Med. Res.* **5**, 37 (1977). GC determn in blood and urine: H. Hattori *et al., J. Chromatog.* **581**, 213 (1992).

Hydrochloride, C$_{19}$H$_{23}$NO.HCl, *Anginosan, Belfene, Dayfen, Diafen, Hispril, Histryl, Histyn, Kolton* (jelly), *Lergoban, Lergobine.* Crystals from isopropanol + ether, mp

206°. Sol in water, ethanol, isopropanol. Practically insol in ether, benzene.

Hydrobromide, C$_{19}$H$_{23}$NO.HBr. Prepn: R. F. Phillips, **U.S. pat. 2,595,405** (1952 to Merck & Co.). Crystals, mp 201-202°. Soluble in water, ethanol.

8-Chlorotheophyllinate, C$_{19}$H$_{23}$NO.C$_7$H$_7$ClN$_4$O$_2$, *piprinhydrinate, diphenylpyraline theoclate, Kolton (tabl.), Mepedyl.* Prepn: W. A. Schuler, Ger. pat. **934,890** (1955 to Promonta), *C.A.* **52**, 20065i (1958). White to off-white crystalline powder, mp 174-176°. Sparingly sol in water. Freely sol in alcohol.

THERAP CAT: Antihistaminic.

3391. *sym*-Diphenylpyrophosphorodiamidic Acid. *N,N'-Diphenyl-P,P'-diamidodiphosphoric acid.* C$_{12}$H$_{14}$N$_2$O$_5$P$_2$; mol wt 328.20. C 43.92%, H 4.30%, N 8.54%, O 24.37%, P 18.87%. Prepd from *O*-benzylphenylphosphoramidic acid: Boger, Friedman, *J. Am. Chem. Soc.* **80**, 2583 (1958).

Dicyclohexylammonium salt, C$_{24}$H$_{40}$N$_4$O$_5$P$_2$, mp 223-224°. Sol in water. No spontaneous hydrolysis at pH 4.0 to pH 11.5 after two hours at 37°.

Dipotassium salt, C$_{12}$H$_{12}$K$_2$N$_2$O$_5$P$_2$, crystals.

USE: Colorimetric estimation of enzymes.

3392. Diphenyl Sulfone. *1,1'-Sulfonylbisbenzene; phenyl sulfone;* sulfobenzide. C$_{12}$H$_{10}$O$_2$S; mol wt 218.28. C 66.03%, H 4.62%, O 14.66%, S 14.69%. (C$_6$H$_5$)$_2$SO$_2$. Prepd by sulfonation of benzene with sulfuric acid; formed as a by-product on production of benzenesulfonic acid and benzenesulfonyl chloride.

White monoclinic prisms or leaflets, mp 128-129°. bp 378-379°. Insoluble in cold water. Slightly sol in boiling water; sol in hot alcohol, in benzene. LD$_{50}$ orally in rats: > 2 g/kg, *Residue Revs.* **36**, 240 (1971).

USE: Ovicide; acaricide.

3393. *sym*-Diphenylthiourea. *N,N'-Diphenylthiourea; thiocarbanilide;* sulfocarbanilide. C$_{13}$H$_{12}$N$_2$S; mol wt 228.32. C 68.39%, H 5.30%, N 12.27%, S 14.04%. C$_6$H$_5$-NHCSNHC$_6$H$_5$. Prepn from aniline and carbon disulfide: Fry, *J. Am. Chem. Soc.* **35**, 1539 (1913); Stasse, **U.S. pat. 2,435,295** (1948 to Allied Chem. & Dye); from potassium ethylxanthate and aniline: Aravindakshan *et al., Indian J. Chem.* **1**(9), 395 (1963). Physical properties and toxicity data: P. J. Hanzlik, A. Irvine, *J. Pharmacol. Exp. Ther.* **17**, 349 (1921).

White crystalline solid, mp 153-154°. d 1.32. Distinct bitter taste. Practically insol in water. Sol in alcohol, ether, chloroform. MLD orally in rabbits: 1.5 g/kg (Hanzlik, Irvine).

USE: Vulcanizing accelerator; sulfur dyes.

3394. Diphetarsone. *[1,2-Ethanediylbis(imino-4,1-phenylene)]bis[arsonic acid] disodium salt; N,N'-ethylenediarsanilic acid disodium salt;* bis[*p*-arsonophenylamino]-1,2-ethane disodium; *N,N'*-bis(*p*-arsonophenyl)ethylenediamine disodium; RP-4763; Amebarsin; Bémarsal; Rodameb. C$_{14}$H$_{16}$As$_2$N$_2$Na$_2$O$_6$; mol wt 504.11. C 33.36%, H 3.20%, As 29.72%, N 5.56%, Na 9.12%, O 19.04%. Prepn: Hamilton, *J. Am. Chem. Soc.* **45**, 2751 (1923).

Decahydrate, powder, no definite melting range. Freely sol in water; slightly sol in boiling 95% ethanol. Practically insol in abs ethanol, acetone, chloroform. pH of a 2% aq soln: 6 to 8. Precipitated from aq solns by acids.

THERAP CAT: Antiamebic.

3395. Diphosgene. *Carbonochloridic acid trichloromethyl ester; chloroformic acid trichloromethyl ester;* trichloromethylchloroformate; trichloromethylcarbonochloridate. $C_2Cl_4O_2$; mol wt 197.83. C 12.14%, Cl 71.68%, O 16.17%. $ClCOOCCl_3$. Used in synthesis as substitute for phosgene, *q.v.* Prepn by photochlorination of methyl formate: W. Hentschel, *J. Prakt. Chem.* **36**, 209 (1887); F. Grignard *et al., C.R. Acad. Sci.* **169**, 1074 (1919); *eidem, ibid.* 1143; by photochlorination of methyl chloroformate: A. Kling *et al., ibid.* 1046; K. Kurita, Y. Iwakura, *Org. Syn.* **59**, 195 (1980). Effect on air-blood barrier in rabbits, toxicity: E. Klika, A. Mysliveckova, *Folia Morphol.* **19**, 5 (1971), *C.A.* **75**, 3433c (1971). Use in synthesis of isocyanides: G. Skorna, J. Ugi, *Angew. Chem. Int. Ed.* **16**, 259 (1977).

Suffocating liquid. bp 128°. bp_{50} 49°. d_{15} 1.664. n_D^{22} 1.45664. Stable at room temp, decomposes to phosgene at ~300°C. LC_{100} in rabbits (10-20 minutes exposure to vapor): 0.9 mg/l air (Klika, Mysliveckova).

USE: In organic synthesis; as war gas.

3396. Dipicrylamine. *2,4,6-Trinitro-N-(2,4,6-trinitrophenyl)benzenamine;* 2,4,6,2',4',6'-hexanitrodiphenylamine. $C_{12}H_5N_7O_{12}$; mol wt 439.21. C 32.82%, H 1.15%, N 22.32%, O 43.71%. *Ref:* Winkel, Maas, *Angew. Chem.* **49**, 827 (1936).

Yellow prisms. Has a tendency to explode. mp about 238° with dec. Insol in water, acetone, alcohol or ether. Sol in alkalies, glacial acetic acid.

USE: Reagent for gravimetric determination of potassium.

3397. Dipin. *1,4-Bis[bis(1-aziridinyl)phosphinyl]piperazine; 1,4-piperazinediylbis[bis(1-aziridinyl)phosphine oxide];* piperazine-1,4-bis(N,N'-diethylenephosphonediamide); tetraethyleneimidopiperazine-N,N'-diphosphoric acid. $C_{12}H_{24}N_6$-O_2P_2; mol wt 346.31. C 41.62%, H 6.99%, N 24.27%, O 9.24%, P 17.89%. Alkylating agent; exptl antineoplastic. Prepn: Kropacheva *et al., Zh. Obshch. Khim.* **30**, 3584 (1960), *C.A.* **55**, 18695c (1961). Spectrophotometric determn: L. Kh. Kartashova, E. M. Salomatin, *Farmatsiya (Moscow)* **33**, 72 (1983), *C.A.* **102**, 119736j (1985). Metabolism in rats: V. V. Chistyakov *et al., Khim.-Farm. Zh.* **21**, 398 (1987), *C.A.* **107**, 146734s (1987).

Crystals, mp 187°.

3398. Dipipanone. *4,4-Diphenyl-6-(1-piperidinyl)-3-heptanone; dl-4,4-diphenyl-6-piperidinoheptan-3-one;* 6-piperidino-4,4-diphenylheptan-3-one; 2-(1-piperidino)-4,4-diphenyl-5-heptanone; phenylpiperone; 378C48; Hoechst 10805; Piperidyl Amidone; Pamedon(e); Fenpidon; Pipadone. $C_{24}H_{31}NO$; mol wt 349.52. C 82.48%, H 8.94%, N 4.01%, O 4.58%. Prepn: Ofner, Walton, *J. Chem. Soc.* **1950**, 2158; **Brit.** pat. 654,975 (1951 to Wellcome Foundation). Variations of the synthesis using phenyllithium as condensing agent: Kazuhiko, Kubota, *Kumamoto Med. J.* **12**, 304-307 (1960).

Hydrochloride, $C_{24}H_{31}NO.HCl$, minute prisms from wet alcohol-ether, mp 123-126°. Also reported as 126-127°.

Hydrobromide, $C_{24}H_{31}NO.HBr$, crystals from water, mp 103-106°.

Note: This is a controlled substance (opiate) listed in the U.S. Code of Federal Regulations, Title 21 Part 1308.11 (1995).

THERAP CAT: Analgesic (narcotic).

3399. Dipiproverine. *α-Phenyl-1-piperidineacetic acid 2-(1-piperidinyl)ethyl ester;* piperidinoethyl α-piperidinophenylacetate. $C_{20}H_{30}N_2O_2$; mol wt 330.47. C 72.69%, H 9.15%, N 8.48%, O 9.68%. Prepn: Klosa, *Pharmazie* **8**, 723 (1953); *idem, ibid.* **11**, 201 (1956); Najer *et al., Bull. Soc. Chim. France* **1958**, 355. Pharmacology: Vaccari *et al., Clin. Ter.* **32**, 483 (1965).

bp_1 188-190°.

Dihydrochloride, $C_{20}H_{30}N_2O_2.2HCl$, *LD-935, Lévospasme, Spasmonal.* Crystals from isopropanol, mp 212-214° (Wunderlich), 238° (Najer). Freely sol in water. Soly in water at 20°, 100°: 184, 323 g/100 ml. Soly in isopropanol at 20°, 50°, 80°: 5, 12.5, 27 g/100 g. Initial pH of a 2.5% soln: 3.35.

THERAP CAT: Anticholinergic.

3400. Dipivefrin. *2,2-Dimethylpropanoic acid 4-[1-hydroxy-2-(methylamino)ethyl]-1,2-phenylene ester;* (±)-3,4-dihydroxy-α-[(methylamino)methyl]benzyl alcohol 3,4-dipivalate; 1-(3',4'-dipivaloyloxyphenyl)-2-methylamino-1-ethanol; dipivalyl epinephrine; DPE. $C_{19}H_{29}NO_5$; mol wt 351.44. C 64.93%, H 8.32%, N 3.99%, O 22.76%. Dipivalyl ester of epinephrine, *q.v.* Prepn: D. Henschler *et al.,* **Ger.** pat. 2,152,058; *eidem,* **U.S.** pat. 4,085,270 (1973, 1978 both to Klinge); A. Hussain, J. E. Truelove, **Ger.** pat. 2,343,657; *eidem,* **U.S.** pats. 3,809,714 and 3,839,584 (all 1974 to Interx). *In vitro* study: A. H. Neufeld, E. D. Page, *Invest. Ophthalmol. Vis. Sci.* **16**, 1118 (1977). Pharmacology: B. C. Wang *et al., J. Pharmacol. Exp. Ther.* **203**, 442 (1977). Effects on intraocular pressure in dogs: R. M. Gwin *et al., Am. J. Vet. Res.* **39**, 83 (1978). Metabolism: I. Abramovsky, J. S. Mindel, *Arch. Ophthalmol.* **97**, 1937 (1979). Clinical study: M. A. Kass *et al., ibid.* 1865. General pharmacology, toxicology and clinical experience in glaucoma: D. A. McClure, *ACS Symposium Series* **14**, 224-235 (1975).

Crystals from ether, mp 146-147°.

Hydrochloride, $C_{19}H_{29}NO_5 \cdot HCl$, *Diopine, d Epifrin, Diphemin, Pivalephrine, Propine.* Crystals from ethyl acetate, mp 158-159°. Sol in water and ethanol.

THERAP CAT: Adrenergic (ophthalmic); antiglaucoma.

3401. Diploicin. *2,4,7,9-Tetrachloro-3-hydroxy-8-methoxy-1,6-dimethyl-11H-dibenzo[b,e][1,4]dioxepin-11-one;* 3,5-dichloro-6-[(3,5-dichloro-6-hydroxy-4-methoxy-o-tolyl)oxy]-4,2-cresotic acid ε-lactone; 5,6'-dimethyl-2',3-dihydroxy-4'-methoxy-2,3',4,5'-tetrachloro-4-carboxydiphenyl ether 2',6-lactone. $C_{16}H_{10}Cl_4O_5$; mol wt 424.06. C 45.32%, H 2.38%, Cl 33.44%, O 18.86%. Antibiotic isolated from the lichen *Buellia canescens* (Dicks.) De Not. [*Diploicia canescens* (Dicks.) Massal.], *Lecideaceae:* Zopf, *Ann.* **336**, 58 (1904); Nolan, *Sci. Proc. Roy. Dublin Soc.* **21**, 67 (1934); Nolan et al., *Chem. & Ind. (London)* **1935**, 577; Barry, *Nature* **158**, 131 (1946). Structure: Nolan et al., *Sci. Proc. Roy. Dublin Soc.* **24**, 319 (1948). Synthesis: Ollis, *ibid.* **27**, 161 (1956); Brown et al., *Proc. Chem. Soc.* **1960**, 393; Hendrickson, Ramsay, *Chem. Commun.* **1968**, 1101; Hendrickson et al., *J. Am. Chem. Soc.* **94**, 6834 (1972); P. D. Djura et al., *J. Chem. Soc., Perkin Trans I,* **1976**, 147.

White needles from methanol, mp 233-234°. uv max: 270 nm (log ε 3.79).

Acetate, $C_{18}H_{12}Cl_4O_6$, mp 234-235°.

3402. Diponium Bromide. *2-[(Dicyclopentylacetyl)oxy]-N,N,N-triethylethanaminium bromide; triethyl(2-hydroxyethyl)ammonium bromide dicyclopentylacetate;* diethylaminoethyl α,α-dicyclopentylacetate ethobromide; dicyclopentylacetic acid β-diethylaminoethyl ester ethobromide; dipenine bromide; SA-267; Unospaston. $C_{20}H_{38}BrNO_2$; mol wt 404.43. C 59.40%, H 9.47%, Br 19.76%, N 3.46%, O 7.91%. Prepn: Barrelet et al., *Ind. Chim. Belg. Suppl.* **2**, 428 (1959); eidem, Swiss pat. **344,054** (1960 to Siegfried), *C.A.* **55**, 418b (1961). Pharmacology: G. Stille, *Arzneimittel-Forsch.* **10**, 911 (1960); G. Häusler, U. Jahn, *ibid.* **15**, 878 (1965).

Crystals, mp 185-186°. Freely sol in water. LD_{50} in mice (mg/kg): 570 orally; 88 i.p.; 6.2 i.v. (Stille); in rats (mg/kg): 6.6 i.v.; 780 orally (Häusler, Jahn).

THERAP CAT: Antispasmodic.

3403. Diprenorphine. *(5α,7α)-17-(Cyclopropylmethyl)-4,5-epoxy-18,19-dihydro-3-hydroxy-6-methoxy-α,α-dimethyl-6,14-ethenomorphinan-7-methanol;* 21-cyclopropyl-6,7,8,14-tetrahydro-7α-(1-hydroxy-1-methylethyl)-6,14-endo-ethanooripavine; N-(cyclopropylmethyl)-19-methylnororvinol; M-5050; RX-5050M. $C_{26}H_{35}NO_4$; mol wt 425.57. C 73.38%, H 8.29%, N 3.29%, O 15.04%. Closely related to cyprenorphine, *q.v.* Prepn: K. W. Bentley, D. G. Hardy, *J. Am. Chem. Soc.* **89**, 3281 (1967). Activity in rats: G. F. Blane, *J. Pharm. Pharmacol.* **19**, 367 (1967); G. F. Blane, D. Dugdall, *ibid.* **20**, 547 (1968). Use as etorphine antagonist in dogs: M. Grange et al., *Rev. Med. Vet.* **124**, 899 (1973); in large animals: B. T. Alford et al., *J. Am. Vet. Med. Assoc.* **164**, 702 (1974). Binding to opiate receptors: C. B. Pert et al., *Life Sci.* **16**, 1849 (1975); J. Pearson et al., *ibid.* **26**, 1047 (1980); to μ opiate receptors: J. J. Frost et al., *ibid.* **38**, 1597 (1986). Treatment of experimental stroke in cats: D. S.

Baskin et al., *Neuropeptides* **5**, 307 (1985); eidem, *J. Neurosurg.* **64**, 99 (1986). Determn by HPLC: I. Jane, A. McKinnon, *J. Chromatog.* **323**, 191 (1985). Toxicity data: N. S. Duggett et al., *Toxicol. Appl. Pharmacol.* **31**, 141 (1977).

Crystals from methanol, mp 185°.

Hydrochloride, $C_{26}H_{35}NO_4 \cdot HCl$, *Revivon.* LD_{50} s.c. in mice: 316.0 ± 20 mg/kg (Duggett).

THERAP CAT (VET): Narcotic antagonist.

3404. Dipropalin. *4-Methyl-2,6-dinitro-N,N-dipropylbenzenamine; 2,6-dinitro-N,N-dipropyl-p-toluidine; N,N-dipropyl-2,6-dinitro-4-methylaniline; 2,6-dinitro-N,N-dipropyl-4-methylaniline; 3,5-dinitro-4-dipropylaminotoluene;* L-35355. $C_{13}H_{19}N_3O_4$; mol wt 281.31. C 55.51%, H 6.81%, N 14.94%, O 22.75%. Prepd from 2,6-dinitro-p-cresol, p-toluenesulfonate and dipropylamine: Hantzsch, *Ber.* **43**, 1662 (1910).

Yellow crystals, mp 80°.

USE: Herbicide.

3405. Dipropetryn. *6-(Ethylthio)-N,N'-bis(1-methylethyl)-1,3,5-triazine-2,4-diamine; 2-(ethylthio)-4,6-bis(isopropylamino)-s-triazine;* GS-16068; Cotofor; Sancap. $C_{11}H_{21}N_5S$; mol wt 255.39. C 51.73%, H 8.29%, N 27.42%, S 12.56%. Pre-emergence herbicide. Prepn: Neth. pat. *Appl.* **6,414,460** corresp to H. Yamamoto, T. Namekawa, U.S. pat. **3,326,912** (1965, 1967 to Nippon Kayaku). Absorption and translocation in plants: E. Basler et al., *Weed Sci.* **26**, 358 (1978).

Powder, mp 104-106°. Vapor pressure at 20°: 7.3×10^{-7} mm Hg. Soly in water at 20°: 16 mg/l. Soluble in organic solvents.

USE: Herbicide.

3406. n-Dipropylamine. *N-Propyl-1-propanamine.* $C_6H_{15}N$; mol wt 101.19. C 71.22%, H 14.94%, N 13.84%. $(CH_3CH_2CH_2)_2NH$.

Colorless liq; odor of ammonia. d_4^{20} 0.738. bp 110°. mp −63°. n_D^{20} 1.40455. Freely sol in water or alcohol. Forms a hydrate with H_2O. LD_{50} orally in rats: 0.93 g/kg, H. F. Smyth et al., *Am. Ind. Hyg. Assoc. J.* **23**, 95 (1962).

3407. Dipropylene Glycol Monomethyl Ether. *1 (or 2)-(2-Methoxymethylethoxy)propanol;* DPGME; DPM; Arcosolv DPM; Dowanol DPM; Poly-Solv DPM. $C_7H_{16}O_3$; mol wt 148.20. C 56.73%, H 10.88%, O 32.39%. Commercial product, prepd from propylene oxide and methanol, may consist of as many as four structural isomers. Prepn of major isomer, *1-(2-methoxy-1-methylethoxy)-2-propanol*, depicted below: A. R. Sexton, E. C. Britton, *J. Am. Chem. Soc.* **75**, 4357 (1953). GC determn in urine: B. Hubner *et al.*, *Fresenius J. Anal. Chem.* **342**, 746 (1992). Toxicity studies: H. F. Smyth, Jr. *et al.*, *Am. Ind. Hyg. Assoc. J.* **23**, 95 (1962); T. D. Landry, B. L. Yano, *Fund. Appl. Toxicol.* **4**, 612 (1984). Review of toxicology: *Patty's Industrial Hygiene and Toxicology* vol. **2D**, G. D. Clayton, F. E. Clayton, Eds. (Wiley-Interscience, New York, 4th ed., 1994) pp 2882-2886; G. Johansson, *NEG and NIOSH Basis for an Occupational Health Standard: Propylene Glycol Ethers and Their Acetates* (PB91-220749, 1991) 47pp.

Colorless liquid. bp_{760} 189.6°. fp −80°. d_4^{25} 0.948. n_D^{25} 1.419. Flash pt, open cup: 185°F. Vapor pressure at 25°: 0.41 mm Hg. Misc with water, benzene. LD_{50} orally in rats, dermally in rabbits: 5.66, 10.0 (ml/kg) (Smyth).

USE: As solvent for automotive fluids, cleaners, dyes, coatings, inks, waxes, adhesives, agricultural products, insect repellents, and cosmetics; chemical intermediate.

3408. Dipropyl Ketone. *4-Heptanone;* butyrone. $C_7H_{14}O$; mol wt 114.19. C 73.63%, H 12.36%, O 14.01%. $CH_3CH_2CH_2COCH_2CH_2CH_3$. Made by passing butyric acid over precipitated $CaCO_3$ at 450°.
Colorless, very refractive liquid; penetrating odor; burning taste. d_4^{15} 0.821. bp 144°. mp −32.6°. n_D^{22} 1.4073. Insol in water; miscible with alcohol, ether.

3409. 5,5-Dipropyl-2,4-oxazolidinedione. Propazone. $C_9H_{15}NO_3$; mol wt 185.22. C 58.36%, H 8.16%, N 7.56%, O 25.91%. Prepd by refluxing dipropylglycolamide and ethylchlorocarbonate with toluene: Stoughton, *J. Am. Chem. Soc.* **63**, 2376 (1941); *see also* Altwegg, Ebin, U.S. pat. **1,375,949** (1921).

Crystals, mp 42-43°. bp_3 141-143°.
Sodium salt, white powder, very sol in water.

3410. Dipyridamole. *2,2',2'',2'''-[(4,8-Di-1-piperidinylpyrimido[5,4-d]pyrimidine-2,6-diyl)dinitrilo]tetrakisethanol;* 2,6-bis(diethanolamino)-4,8-dipiperidinopyrimido-[5,4-*d*]pyrimidine; NSC-515776; RA-8; Anginal; Cardoxin; Cleridium; Coridil; Coronarine; Curantyl; Dipyridan; Gulliostin; Natyl; Peridamol; Persantine; Piroan; Prandiol; Protangix. $C_{24}H_{40}N_8O_4$; mol wt 504.63. C 57.12%, H 7.99%, N 22.21%, O 12.68%. Prepn: **Brit.** pat. **807,826**; F. G. Fischer, *et al.*, **U.S.** pat. **3,031,450** (1959, 1962 both to Thomae). Activity studies: Saraf, Seth, *Indian J. Physiol. Pharmacol.* **15**, 135 (1971). Toxicological study: F. Takenaka *et al.*, *Arzneimittel-Forsch.* **22**, 892 (1972). Review of diagnostic use in echocardiography: M. B. Buchalter *et al.*, *Postgrad. Med. J.* **66**, 531-535 (1990); in ^{201}Tl cardiac imaging: S. G. Beer *et al.*, *Am. J. Cardiol.* **67**, Suppl., 18D-26D (1991).

Deep yellow needles from ethyl acetate, mp 163°. Bitter taste. Slightly sol in H_2O; sol in dil acid having a pH of 3.3 or below; very sol in methanol, ethanol, chloroform; not too sol in acetone, benzene, ethyl acetate. Solns are yellow and show strong blue-green fluorescence. LD_{50} in rats: 8.4 g/kg orally; 208 mg/kg i.v. (Takenaka).

THERAP CAT: Vasodilator (coronary).

3411. α,α'-Dipyridyl. *2,2'-Bipyridine;* 2,2'-dipyridyl; CI-588. $C_{10}H_8N_2$; mol wt 156.19. C 76.90%, H 5.16%, N 17.94%. Prepn from pyridine: Smith, *J. Am. Chem. Soc.* **46**, 414 (1924). Manuf: Freeman, Ghosh, U.S. pat. **2,962,502** (1960 to I.C.I.). Absorption spectrum: P. Krumholz, *J. Am. Chem. Soc.* **73**, 3487 (1951). Pharmacologic study: Bass *et al.*, *J. Pharmacol. Exp. Ther.* **152**, 104 (1966). Toxicity data: R. W. Grady *et al.*, *ibid.* **196**, 478 (1976).

Crystals from dil alcohol, mp 69.7°. bp 272-273°. Sol in about 200 parts water; very sol in alcohol, ether, benzene, chloroform, petr ether. LD_{50} i.p. in mice: 200 mg/kg (Grady).

USE: As a reagent for the determination of iron.

3412. γ,γ'-Dipyridyl. *4,4'-Bipyridine;* 4,4'-dipyridyl. $C_{10}H_8N_2$; mol wt 156.19. C 76.90%, H 5.16%, N 17.94%. Prepn: Dimroth, Frister, *Ber.* **55**, 3695 (1922); Smith, *J. Am. Chem. Soc.* **46**, 414 (1924).

Dihydrate, bitter needles from water, mp 73°. Anhydr form, mp 111-112°. bp_{760} 304.8°. Freely sol in alc, benzene, chloroform, ether; slightly sol in water. Absorption spectrum: Krumholz, *J. Am. Chem. Soc.* **73**, 3487 (1951).
Dihydrochloride, $C_{10}H_8N_2.2HCl$, prisms from water. Soluble in water; practically insol in ether.

3413. Dipyrocetyl. *2,3-Bis(acetyloxy)benzoic acid;* o-pyrocatechuic acid diacetate; diacetylpyrocatechol-3-carboxylic acid; 2,3-diacetoxybenzoic acid; Movirene; Artromialgina. $C_{11}H_{10}O_6$; mol wt 238.20. C 55.47%, H 4.23%, O 40.30%. Prepn: Rietz, U.S. pat. **1,140,716** (1914); Simokoriyama, *Bull. Chem. Soc. Japan* **16**, 284 (1941).

mp 148-150° (Rietz); mp 146-170° (Simokoriyama). Sol in organic solvents. Practically insol in water.

THERAP CAT: Analgesic, antipyretic.

3414. Dipyrone. *[(2,3-Dihydro-1,5-dimethyl-3-oxo-2-phenyl-1H-pyrazol-4-yl)methylamino]methanesulfonic acid sodium salt monohydrate; (antipyrinylmethylamino)methane-sulfonic acid sodium salt;* 1-phenyl-2,3-dimethyl-5-pyrazolone-4-methylaminomethanesulfonate sodium; noraminopyrine methanesulfonate sodium; 4-methylamino-1,5-dimethyl-2-phenyl-3-pyrazolone sodium methanesulfonate; sodium methylaminoantipyrine methanesulfonate; methylmelubrin; methampyrone; metamizol; analgin; sulpyrin; Alginodia; Algocalmin; Bonpyrin; Conmel; Divarine; Dolazon; D-Pron; Dya-Tron; Espyre; Farmolisina; Feverall; Fevonil; Keypyrone; Metilon; Minalgin; Narone; Nartate; Nevralgina; Nolotil; Novacid; Novaldin; Novalgin; Novemina; Novil; Paralgin; Pyralgin; Pyril; Pyrilgin; Pyrojec; Tega-Pyrone; Unagen. $C_{13}H_{16}N_3NaO_4S.H_2O$; mol wt 351.36. C 44.44%, H 5.16%, N 11.96%, Na 6.54%, O 22.77%, S 9.13%. Prepd by methylating the amino group of sulfamipyrine, then treating with formaldehyde sodium bisulfite soln: **Ger. pats. 254,711; 259,503; 259,577** (all 1911 to Hoechst).

Minute crystals from alc. Sol in water (1 g/1.5 ml), methanol. Less so in ethanol. Practically insol in ether, acetone, benzene, chloroform. Aq solns are neutral; may acquire a yellow discoloration without apparent potency loss.

Magnesium salt, $C_{26}H_{32}MgN_6O_8S_2$, *Magnopyrol.*

THERAP CAT: Analgesic; antipyretic.

THERAP CAT (VET): Analgesic; antispasmodic.

3415. Diquat Dibromide. *6,7-Dihydrodipyrido[1,2-a:2',1'-c]pyrazinediium dibromide* 1,1'-ethylene-2,2'-dipyridylium dibromide; FB/2; Aquacide; Reglone. $C_{12}H_{12}Br_2N_2$; mol wt 344.05. C 41.89%, H 3.52%, Br 46.45%, N 8.14%. Prepn: Fielden *et al., U.S. pat.* **2,823,987** (1958 to ICI). Herbicidal properties: Brian *et al., Nature* **181**, 446 (1958). Fate in the environment and toxicity: Simsiman *et al., Residue Rev.* **62**, 131-174 (1976). *Review:* Akhavein, Linscott, *ibid.* **23**, 97-145 (1968).

Monohydrate, pale yellow crystals from water, mp < 320° (dec). Also reported as mp 335-340°. uv max: 308.31 nm (ϵ 18000). Soly in water at 20°: 70%. Insoluble in organic solvents; slightly sol in alcohol. Stable in acid or neutral soln. LD_{50} orally in rats, mice, rabbits, guinea pigs: 231, 125, 101, 100 mg/kg, *Toxic Substances List,* H. E. Christensen, Ed. (1973) p 381.

Caution: Minimal absorption from gastrointestinal tract although distension and irritation occur after large doses in exptl animals: Dalgaard-Mikkelsen, Poulsen, *Pharmacol. Rev.* **14**, 225 (1962).

USE: Contact herbicide used also to produce desiccation and defoliation.

3416. 1,3-Di-6-quinolylurea. *N,N'-Di-6-quinolinylurea; sym-*di-(6-quinolyl)urea; bis(6-quinolyl)urea; 6,6'-diquinolylurea; Babesan. $C_{19}H_{14}N_4O$; mol wt 314.35. C 72.60%, H 4.49%, N 17.82%, O 5.09%. Synthesis starting with 6-aminoquinoline: Schönhöfer, Henecka, **Ger. pat. 583,207** (1933 to I. G. Farben); Haskelberg, *J. Org. Chem.* **12**, 434 (1947); Reuter, *Aust. Chem. Inst. J. & Proc.* **16**, 164 (1949).

Crystals from pyridine, mp 262°. Sol in dil acid.

Bismethosulfate, $C_{23}H_{26}N_4O_9S_2$, *6,6'-ureylenebis[1,1'-dimethylquinolinium] sulfate, quinuronium sulfate, 6,6'-ureylenebis(1-methylquinolinium)bis(methosulfate), dimethyl quinolyl methylsulfate urea, SN-5870, Acaprin, Zothelone, Baburan, Pirevan, Pyroplasmin, Atral.* Yellow crystals from methanol, dec 237°. Freely sol in water.

THERAP CAT (VET): Antiprotozoal *(Babesia).*

3417. Diresorcinol. *[1,1'-Biphenyl]-3,3',5,5'-tetrol;* 3,5,-3',5'-tetrahydroxydiphenyl. $C_{12}H_{10}O_4$; mol wt 218.21. C 66.05%, H 4.62%, O 29.33%.

White to slightly yellowish cryst powder. mp 310°; becomes anhydr at 100°. Sparingly sol in cold water; less sol in hot water or alcohol.

3418. Dirithromycin. *[9S(R)]-9-Deoxo-11-deoxy-9,11-[imino[2-(2-methoxyethoxy)ethylidene]oxy]erythromycin;* LY-237216; AS-E 136; Dynabac; Noriclan; Nortron; Valodin. $C_{42}H_{78}N_2O_{14}$; mol wt 835.09. C 60.41%, H 9.41%, N 3.35%, O 26.82%. Semi-synthetic derivative of erythromycin, *q.v.* Prepn: **Belg. pat. 840,431** (1976 to Thomae); R. Maier *et al., U.S. pat.* **4,048,306** (1977 to Boehringer, Ing.). Synthesis, ^1H- and ^{13}C-NMR, and antimicrobial evaluation: F. T. Counter *et al., Antimicrob. Ag. Chemother.* **35**, 1116 (1991). X-ray structure determn: P. Luger, R. Maier, *J. Cryst. Mol. Struct.* **9**, 329 (1979). HPLC determn in plasma: G. W. Whitaker, T. D. Lindstrom, *J. Liq. Chromatog.* **11**, 3011 (1988). Symposium on antibacterial activity, pharmacology, and clinical experience: *J. Antimicrob. Chemother.* **31**, Suppl. C, 1-185 (1993).

Crystals from ethanol/water, mp 186-189° (dec) (Counter). pKa 9.0 in 66% aq dimethyl fluoride. LD_{50} in mice (g/kg): > 1 s.c.; > 1 orally (Maier).

THERAP CAT: Antibacterial.

3419. Disilane. Disilicoethane; disilicon hexahydride; silicoethane; disilicane. H_6Si_2; mol wt 62.22. H 9.72%, Si 90.28%. Si_2H_6. Obtained by separation of mixed silanes prepd from magnesium silicide and hydrochloric acid: Moissan, Smiles, *Compt. Rend.* **134**, 569, 1549 (1902); Stock, Somiesky, *Ber.* **49**, 111 (1916); **54B**, 524 (1921); **56B**, 247 (1923); Culberston, *U.S. pat.* **2,551,571** (1951 to Union Carbide); prepd by conversion of silane to higher silanes in an ozonizer type electric discharge: Spanier, MacDiarmid, *Inorg. Chem.* **1**, 432 (1962).

Gas; repulsive odor. d_4^{-25} 0.686. mp −132.5°. bp −14.5°. Dec 300°. Ignites spontaneously in air. Slowly dec in water. Explodes on contact with sulfur hexafluoride; reacts vigorously with carbon tetrachloride and chloroform. Sol in carbon disulfide, ethyl alcohol, benzene, and ethyl silicate. Potassium hydroxide liberates hydrogen.

Caution: A powerful irritant.

3420. Disodium Dihydrogen Hypophosphate. Disodium dihydrogen subphosphate; sodium dihydrogen hypophosphate; sodium acid hypophosphate. $H_2Na_2O_6P_2$; mol wt 205.94. H 0.98%, Na 22.33%, O 46.61%, P 30.08%. $Na_2H_2P_2O_6$. Prepd by the oxidation of red phosphorus with sodium chlorite in water: Leininger, Chulski, *Inorg. Syn.* **4**, 68 (1953).

Hexahydrate, monoclinic plates from water, d 1.849. Becomes anhydr at 110°. mp 250°. Soly of the hexahydrate in water at 25°: 2.0 g/100 ml; at 100°: 25.0 g/100 ml. Insol in alc. Aq solns have an acid reaction.

3421. Disodium Phenyl Phosphate. $C_6H_5Na_2O_4P$; mol wt 218.06. C 33.05%, H 2.31%, Na 21.09%, O 29.35%, P 14.20%. $C_6H_5PO_4Na_2$.

White, cryst powder. Freely sol in water, sparingly in alcohol; insol in acetone or ether.

USE: Reagent in the testing of milk for proper pasteurization and for the presence of unpasteurized milk.

3422. Disofenin. *N-[2-[[2,6-Bis(1-methylethyl)phenyl]amino]2-oxoethyl]-N-(carboxymethyl)glycine;* [[[(2,6-diisopropylphenyl)carbamoyl]methyl]imino]diacetic acid; (2,6-diisopropylacetanilido)iminodiacetic acid; *N*-(carboxymethyl)-*N*-[2-(2,6-diisopropylphenyl)amino]-2-oxoethylglycine; DISIDA. $C_{18}H_{26}N_2O_5$; mol wt 350.41. C 61.70%, H 7.48%, N 7.99%, O 22.83%. Iminodiacetic acid analog. Prepn of acid and ^{99m}Tc-complex: M. de Schrijver, **Ger.** pat. **2,723,-605**; *idem*, **U.S.** pat. **4,316,883** (1977, 1982 both to Solco Basel). High yield synthesis of acid: M. P. Best *et al.*, *Aust. J. Chem.* **35**, 2371 (1982). Biodistribution: G. Subramanian *et al.*, *Nuklearmedizin, Suppl.* **16**, 136 (1978). Chromatographic determn of radiochemical purity: A. M. Zimmer *et al.*, *Eur. J. Nucl. Med.* **7**, 88 (1982). Chemical and biological properties: B. Zmbova *et al.*, *Appl. Radiat. Isot.* **38**, 35 (1987). Pharmacokinetics: I. A. Fraser *et al.*, *Eur. J. Nucl. Med.* **14**, 431 (1988); J. E. Love *et al.*, *ibid.* 436. Clinical evaluation as hepatobiliary imaging agent: R. C. Stadalnik *et al.*, *Radiology* **140**, 797 (1981).

White crystals, mp 196-198° (Zmbova). Also reported as crystals from water, mp 191-192.5° (Best); and as mp 166-175° (de Schrijver). Insol in water.

^{99m}Tc complex, ^{99m}Tc diisopropyl IDA, ^{99m}Tc-DISIDA, *Hepatolite.* Distribution coefficient (ethylene dichloride/water): 0.0086.

THERAP CAT: ^{99m}Tc-complex as diagnostic aid (radioactive imaging agent).

3423. Disophenol. *2,6-Diiodo-4-nitrophenol;* DNP; Ancylol. $C_6H_3I_2NO_3$; mol wt 390.90. C 18.44%, H 0.77%, I 64.93%, N 3.58%, O 12.28%. Prepd by iodination of *p*-nitrophenol with iodine−potassium iodide soln in aq ammonia: Datta, Prosad, *J. Am. Chem. Soc.* **39**, 446 (1917). As an anthelmintic: Thorson *et al.*, U.S. pat. **3,081,224** (1963 to Am. Cyanamid). Toxicity studies: J. A. Kaiser, *Toxicol. Appl. Pharmacol.* **6**, 232 (1964); Fowler, *Brit. Vet. J.* **127**, 304 (1971). Comparative field trial in sheep: C. A. Hall *et al.*, *Res. Vet. Sci.* **31**, 104 (1981).

Light yellow, feathery crystals from glacial acetic acid, mp 157°. Freely sol in alcohol, very sparingly sol in water. LD_{50} in rats, mice (mg/kg): 170, 212 orally; 105, 88 i.v.; 105, 107 i.p.; 122, 110 s.c. (Kaiser).

THERAP CAT (VET): Anthelmintic (hookworm).

3424. Disopyramide. *α-[2-[Bis(1-methylethyl)amino]ethyl]-α-phenyl-2-pyridineacetamide;* *α-[2-(diisopropylamino)ethyl]-α-phenyl-2-pyridineacetamide;* 4-(diisopropylamino)-2-phenyl-2-(2-pyridyl)butyramide; H-3292; SC-7031; Dicorantil; Isorythm; Lispine; Ritmodan; Rythmodan. $C_{21}H_{29}N_3O$; mol wt 339.48. C 74.30%, H 8.61%, N 12.38%, O 4.71%. Prepn: Cusic, Sause, **Belg.** pat. **617,730**, *C.A.* **58**, 12522e (1963) and **U.S.** pat. **3,225,054** (1962 and 1965, both to Searle). Synthesis and biological activity: Adelstein, *J. Med. Chem.* **16**, 309 (1973). Pharmacology: *J. Am. Med. Assoc.* **213**, 697 (1970); Mathur, *Am. Heart J.* **84**, 764 (1972); B. Befeler, R. Lazzara, *Heart Lung* **9**, 475 (1980). Conformation-activity study: J. L. Czeisler, R. M. El-Rashidy, *J. Pharm. Sci.* **74**, 750 (1985). Toxicity: P. C. Ruenitz, C. M. Mokler, *J. Med. Chem.* **22**, 1142 (1979). *Reviews:* R. R. Dean *et al.*, in *Pharmacological and Biochemical Properties of Drug Substances* vol. 2, M. E. Goldberg, Ed. (Am. Pharm. Assoc., Washington, DC, 1979) pp 165-185; E. H. Taylor, A. A. Pappas, *Ann. Clin. Lab. Sci.* **16**, 289-295 (1986).

Crystals from hexane, mp 94.5-95.0°. pKa 10.2; also reported as 10.45. LD_{50} i.p. in mice: 517 μmol/kg (Ruenitz, Mokler).

Phosphate, $C_{21}H_{29}N_3O.H_3PO_4$, SC-13957, *Dirythmin SA, Diso-Duriles, Norpace, Rythmodul.*

THERAP CAT: Antiarrhythmic (class IA).

3425. Disparlure. *(2S-cis)-2-Decyl-3-(5-methylhexyl)oxirane; cis-7,8-epoxy-2-methyloctadecane;* ENT-34886; Disrupt. $C_{19}H_{38}O$; mol wt 282.51. C 80.78%, H 13.56%, O 5.66%. The potent sex pheromone of the gypsy moth *Lymantria dispar* (formerly *Porthetria dispar).* Isoln and characterization: B. A. Bierl *et al.*, *Science* **170**, 87 (1970); eidem, *J. Econ. Entomol.* **65**, 659 (1972). Synthesis: S. Iwaki *et al.*, *J. Am. Chem. Soc.* **96**, 7842 (1974); D. G. Farnum *et al.*, *Tetrahedron Letters* **1977**, 4009; K. Mori *et al.*, *Tetrahedron* **35**, 833 (1979); B. E. Rossiter *et al.*, *J. Am. Chem. Soc.* **103**, 464 (1981); K. Mori, T. Ebata, *Tetrahedron Letters* **22**, 4281 (1981). Synthesis of (±)-form: K. Eiter, *Angew. Chem. Int. Ed.* **11**, 60 (1972); idem, **Ger.** pat. **2,145,454**; idem, **U.S.** pat. **3,975,409** (1973, 1976 both to Bayer); H. J. Bestmann *et al.*, *Ber.* **109**, 3375 (1976). Activity of (+)- and (±)-forms: R. T. Cardé, R. P. Webster, *J. Chem. Ecol.* **5**, 935 (1979). Toxicity study: M. Beroza *et al.*, *Toxicol. Appl. Pharmacol.* **31**, 421 (1975).

Viscous colorless oil, bp$_{0.25}$ 146-148°. [α]$_D^{23}$ +0.8° ±0.2° (c = 10 in CCl$_4$). n$_D^{23}$ 1.4450. LD$_{50}$ orally in rats: > 34600 mg/kg (Beroza).

USE: Insect sex attractant.

3426. Distigmine Bromide. *3,3'-[1,6-Hexanediylbis-[(methylimino)carbonyl]oxy]bis[1-methylpyridinium]dibromide; 3-hydroxy-1-methylpyridinium bromide hexamethylenebis[methylcarbamate];* hexamethylenebis[methylcarbamic acid] ester of 3-hydroxy-1-methylpyridinium bromide; hexamethylenebis[N-methylcarbamic acid ester bromomethylate]; hexamethylenebis[N-methylcarbaminoyl-1-methyl-3-hydroxypyridinium bromide]; hexamarium; BC-51; Ubretid. C$_{22}$H$_{32}$Br$_2$N$_4$O$_4$; mol wt 576.33. C 45.85%, H 5.60%, Br 27.73%, N 9.72%, O 11.10%. Prepn: Schmid, U.S. pat. 2,789,981 (1957 to Oesterr. Stickstoffwerke). Pharmacology: Hertting *et al., Arzneimittel-Forsch.* **18,** 479 (1968); Hohenegger, Lindner, *Wien. Klin. Wochenschr.* **81,** 823 (1969). Stability studies: Suzuki, Tanimura, *Yakugaku Zasshi* **90,** 762 (1970).

Crystals, dec 149°.

THERAP CAT: Cholinesterase inhibitor.

3427. Disulfamide. *4-Chloro-6-methyl-1,3-benzenedisulfonamide; 5-chlorotoluene-2,4-disulfonamide;* 2-chloro-4-methylbenzene-1,5-disulfonamide; 5-chloro-2,4-disulfamyltoluene; disulphamide; Disamide; Natirene 25. C$_7$H$_9$ClN$_2$O$_4$S$_2$; mol wt 284.74. C 29.53%, H 3.19%, Cl 12.45%, N 9.84%, O 22.48%, S 22.52%. Prepn: **Brit.** pat. 851,287 (1960 to Brit. Drug Houses). Pharmacology: David, Fellowes, *J. Pharm. Pharmacol.* **12,** 65 (1960).

Non-hygroscopic cryst powder, mp 260°. uv max (ethanol): 285 nm (ε 805). Practically insol in water. Slightly sol in ethanol. Soly in abs ethanol at 25° = 1.89% w/w; in 90% ethanol = 2.23% w/w; in isopropanol = 0.35%; in chloroform = 0.001%. Sol in cold NaOH solns.

THERAP CAT: Diuretic.

3428. Disulfiram. *Tetraethylthioperoxydicarbonic diamide; bis(diethylthiocarbamoyl) disulfide;* tetraethylthiuram disulfide; bis(diethylthiocarbamyl) disulfide; teturamin; TTD; Cronetal; Abstensil; Stopetyl; Contralin; Antadix; Antietanol; Exhoran; Ethyl Thiurad; Antabuse; Etabus; Abstinyl; Thiuranide; Esperal; Tetradine; Noxal; Tetraetil. C$_{10}$H$_{20}$N$_2$S$_4$; mol wt 296.55. C 40.50%, H 6.80%, N 9.45%, S 43.25%. Prepn: Bailey, **U.S.** pat. 1,796,977 (1931 to Roessler and Hasslacher); Adams, Newser, **U.S.** pat. 1,782,-111 (1931 to Naugatuck); cf. Cummings, Simmons, *Ind. Eng. Chem.* **20,** 1173 (1928). Toxicity study: Child, Cramp, *Acta Pharmacol. Toxicol.* **8,** 305 (1952). Comprehensive description: N. G. Nash, R. D. Daley, *Anal. Profiles Drug Subs.* **4,** 168-191 (1975).

Crystals, mp 70°. d 1.30. Practically insol in water (0.02 g/100 ml). Sol in alcohol (3.82 g/100 ml), in ether (7.14 g/100 ml), also sol in acetone, benzene, chloroform, carbon disulfide. LD$_{50}$ orally in rats: 8.6 g/kg (Child, Cramp).

Caution: Therapy should not be instituted for at least 12 hours after ingestion of alcohol. Ingestion of alc after disulfiram administration causes intense vasodilatation of face and neck, tachycardia and tachypnea followed by nausea, vomiting, pallor, hypotension. Occasionally convulsions, cardiac arrhythmias, myocardial infarction may occur. High doses of alc may cause dizziness, headache, dyspnea, unconsciousness, death, cf. *Clinical Toxicology of Commercial Products,* R. E. Gosselin *et al.,* Eds. (Williams & Wilkins, Baltimore, 4th ed., 1976) Section III, pp 137-140.

USE: Rubber accelerator; vulcanizer; seed disinfectant; fungicide.

THERAP CAT: Alcohol deterrent.

3429. Disulfoton. *Phosphorodithioic acid O,O-diethyl S-[2-(ethylthio)ethyl] ester; O,O*-diethyl-S-ethylmercaptoethyl dithiophosphate; thiodemeton; dithiodemeton; Bay 19639; ENT-23347; Dithiosystox; Disyston; Frumin AL; Solvirex. C$_8$H$_{19}$O$_2$PS$_3$; mol wt 274.41. C 35.02%, H 6.98%, O 11.66%, P 11.29%, S 35.06%. Homolog of phorate, *q.v.* Prepn: Schrader, Lorenz, U.S. pat. 2,759,010; **Brit.** pat. 797,307 (1956, 1958 both to Bayer). Toxicity study: T. B. Gaines, *Toxicol. Appl. Pharmacol.* **14,** 515 (1969).

Colorless oil, bp$_{0.01}$ 108°, bp$_{1.5}$ 132-133°. d$_4^{20}$ 1.144. n$_D^{20}$ 1.5348. Vapor pressure at 20°: 1.8 × 10^{-4} mm Hg. Insol in water. LD$_{50}$ in female, male rats (mg/kg): 2.3, 6.8 orally; 6, 15 dermally (Gaines).

USE: Insecticide.

3430. Disul-sodium. *2-(2,4-Dichlorophenoxy)ethanol hydrogen sulfate sodium salt;* sodium 2,4-dichlorophenoxyethyl sulfate; 2,4-dichlorophenoxyethyl hydrogen sulfate sodium salt; 2,4-DES-Na; SES; sesone; Crag Herbicide-1; Crag Sesone. C$_8$H$_7$Cl$_2$NaO$_5$S; mol wt 309.10. C 31.09%, H 2.28%, Cl 22.94%, Na 7.44%, O 25.88%, S 10.37%. Prepn: J. A. Lambrech, U.S. pat. 2,573,769 (1951 to Union Carbide). Toxicity study: G. W. Bailey, J. L. White, *Residue Rev.* **10,** 97 (1965).

Crystals, mp 245° (dec). Soly in water (25°): 25.5%. Insol in most organic solvents except methanol. LD$_{50}$ orally in rats: 730 mg/kg (Bailey, White).

USE: Herbicide.

3431. Dita Bark. Alstonia cortex; Australian fever bark. The dried bark of *Alstonia scholaris* (L.) R. Br. and of *A. constricta* F. Muell., *Apocynaceae. Habit.* India, Ceylon, Philippine Islands, Australia. *Constit.* The alkaloids echitamine, echitenine, alsonine, porphyrine; echicerin—a nonnitrogenous substance; echitin, alstonidine and echitein.

3432. Ditazol. *2,2'-[(4,5-Diphenyl-2-oxazolyl)imino]-bisethanol;* N-(4,5-diphenyloxazol-2-yl)diethanolamine; 2-[bis(β-hydroxyethyl)amino]-4,5-diphenyloxazole; 2,2'-dihydroxy-N-(4,5-diphenyloxazol-2-yl)diethylamine; diethamphenazol; S-222; Ageroplas. C$_{19}$H$_{20}$N$_2$O$_3$; mol wt 324.38. C 70.35%, H 6.21%, N 8.64%, O 14.80%. Prepn: Marchetti *et al., J. Med. Chem.* **11,** 1092 (1968); **Fr.** pat. 1,538,009 corresp to Marchetti, U.S. pat. 3,557,135 (1968, 1971 to Ist. Farmaceut. Serono). Series of articles on pharmacology and toxicology: L. Caprino *et al., Arzneimittel-Forsch.* **23,** 1272-1291 (1973). Metabolism: Marchetti *et al., ibid.* 1291.

Monohydrate, crystals from ethyl ether-petr ether, mp 96-98°. LD_{50} in mice, rats (mg/kg): 9621, 11380 orally; 3390, 7770 i.p. (Caprino).

THERAP CAT: Anti-inflammatory.

3433. Dithianone. *5,10-Dihydro-5,10-dioxonaphtho[2,-3-b]-1,4-dithiin-2,3-dicarbonitrile;* 1,4-dithiaanthraquinone-2,3-dicarbonitrile; 2,3-dicyano-1,4-dithiaanthraquinone; IT-931; MV-119A; Delan. $C_{14}H_4N_2O_2S_2$; mol wt 296.33. C 56.75%, H 1.36%, N 9.45%, O 10.80%, S 21.64%. Prepn: van Schoor *et al.,* U.S. pat. **2,976,296** (1961 to E. Merck). *Review:* E. Amadori, W. Heupt, *Anal. Methods Pestic. Plant Growth Regul.* **10,** 181-187 (1978).

Brown odorless crystals, mp 225°. Practically insol in water. Sol in dioxane, chlorobenzene, chloroform. LD_{50} orally in rats, guinea pigs: 638, 110 mg/kg (Amadori, Heupt).

USE: Fungicide.

3434. Dithiazanine Iodide. *3-Ethyl-2-[5-(3-ethyl-2(3H)-benzothiazolylidene)-1,3-pentadienyl]benzothiazolium iodide;* 3,3'-diethylthiadicarbocyanine iodide; Abminthic; Anelmid; Anguifugan; Delvex; Dejo; Déselmine; Dilombrin; Dizan; Nectocyd; Partel; Telmicid; Telmid. $C_{23}H_{23}IN_2S_2$; mol wt 518.49. C 53.28%, H 4.47%, I 24.48%, N 5.40%, S 12.37%. Prepn from 1-methylbenzothiazole ethiodide and β-(ethylmercapto)acrolein diethyl acetal: Kendall, Edwards, U.S. pat. **2,412,815** (1946 to Ilford).

Green needles from methanol, dec 248°. Practically insol in water. Can be solubilized with polyvinylpyrrolidone: **Can.** pat. **676,636** (1963 to GAF).

USE: Sensitizer for photographic emulsions.

THERAP CAT: Anthelmintic (Nematodes).

THERAP CAT (VET): Anthelmintic.

3435. 2,2'-Dithiobis[benzothiazole]. MBTS; 2,2'-dibenzothiazyl disulfide; benzothiazyl disulfide; dibenzthiazyl disulfide; mercaptobenzthiazyl ether; Thiofide. $C_{14}H_8N_2S_4$; mol wt 332.50. C 50.57%, H 2.43%, N 8.43%, S 38.58%. Formed by oxidation of 2-mercaptobenzothiazole.

Pale yellow needles from benzene. d 1.50. mp 180°; also reported as mp 186°; the commercial product, mp 168° min. Insoluble in water. Solubility at 25° (g/100 ml): in alcohol < 0.2 ; in acetone < 0.5; in benzene < 0.5; in carbon tetra-

chloride < 0.2 (somewhat more in chloroform); in ether < 0.2; in naphtha < 0.5.

USE: As accelerator in the rubber industry.

3436. 2,4-Dithiobiuret. *Thioimidodicarbonic diamide.* $C_2H_5N_3S_2$; mol wt 135.21. C 17.77%, H 3.73%, N 31.08%, S 47.43%. $NH_2CSNHCSNH_2$. Prepd by the interaction of dicyandiamide and hydrogen sulfide in inert solvents under pressure: Sperry, U.S. pat. **2,371,112** (1945 to Am. Cyanamid). *Review:* Kurzer, *Chem. Rev.* **56,** 138-144, 179-181 (1956).

Monoclinic or triclinic crystals. *Poisonous when ingested!* Apparent bulk density 1.522 g/ml at 30°. Dec 181°. uv max: 225, 280 nm (log ε 2.0, 2.2). Solubility at 27° = 0.27 g/100 ml water; 2.2 g/100 g ethanol; 16 g/100 g acetone; about 34 g/100 g Cellosolve. Solubility in boiling water about 8.0%. pH of satd aq soln at 30° = 5.8. Soluble in alkalies with formation of water-soluble salts. Solubility in 1% NaOH = 3.6 g/100 g soln; in 5% NaOH = 16 g/100 g; in 10% NaOH = 29 g/100 g.

Caution: Limited animal expts suggest high toxicity due to respiratory paralysis.

USE: Plasticizer, rubber accelerator, intermediate in resin manuf, in making insecticides and rodenticides. Can be used to delay the wilting of flowers.

3437. 4,4'-Dithiodimorpholine. *4,4'-Dithiobis[morpholine];* morpholine *N,N'-*disulfide; dimorpholine *N,N'-*disulfide. $C_8H_{16}N_2O_2S_2$; mol wt 236.36. C 40.65%, H 6.82%, N 11.85%, O 13.54%, S 27.13%. Prepn: Blake, *J. Am. Chem. Soc.* **65,** 1267 (1943); *idem,* U.S. pat. **2,343,524** (1944 to Monsanto); Harman, U.S. pat. **2,766,236** and **Brit.** pat. **774,570** (1956, 1957, both to Monsanto). Lambrech *et al.,* U.S. pat. **2,902,402** (1959 to Pfizer).

Crystals, mp 124-125°.

USE: As staining protector in rubber; in vulcanization of rubber. As fungicide: Ladd, U.S. pat. **2,429,097** (1949 to U.S. Rubber).

3438. 3,3-Dithiodipyridine Dihydrochloride. *3,3'-Dithiobis[pyridine] dihydrochloride;* 3,3-dipyridyl disulfide dihydrochloride. $C_{10}H_{10}Cl_2N_2S_2$; mol wt 293.24. C 40.96%, H 3.44%, Cl 24.18%, N 9.55%, S 21.87%. Prepd from 3-pyridinesulfonyl chloride and Na_2SO_3: Gibbs, Penny, **Brit.** pat. **582,638** (1946).

Crystals from abs alc, mp 183°. Sol in water.

3439. Dithiopyr. *2-(Difluoromethyl)-4-(2-methylpropyl)-6-(trifluoromethyl)-3,5-pyridinedicarbothioic acid S,S-dimethyl ester;* S,S-dimethyl 2-(difluoromethyl)-4-isobutyl-6-trifluoromethyl-3,5-pyridinedicarbothioate; MON-15100; MON-7200; Dimension. $C_{15}H_{16}F_5NO_2S_2$; mol wt 401.42. C 44.88%, H 4.02%, F 23.66%, N 3.49%, O 7.97%, S 15.98%. Selective herbicide for weed control in turfgrass and selected crops. Prepn: L. F. Lee, Eur. pat. **Appl. 133,612**; *idem,* U.S. pat. **4,692,184** (1985, 1987 both to Monsanto). Herbicidal properties: M. Fujiyama, S. Yamane, *Shokubutsu no Kagaku Chosetsu* **24,** 49 (1989), *C.A.* **111,** 227154n (1989).

Colorless, crystalline solid with faint odor, mp 65°. vapor pressure at 25°: 4×10^{-6} mm Hg. Soly in water (20°):

1.38 ppm. LD_{50} orally in rats: > 5000 mg/kg (Fujiyama, Yamane).

USE: Herbicide.

3440. Dithiosalicylic Acid. *2-Hydroxybenzenecarbodithioic acid.* $C_7H_6OS_2$; mol wt 170.26. C 49.38%, H 3.55%, O 9.40%, S 37.67%. Prepn from salicylaldehyde and ammonium polysulfide: Bruni, Levi, *Atti Accad. Naz. Lincei* **32**, i, 5 (1923), *C.A.* **18**, 2694 (1924); *Beilstein*, 2nd suppl, **vol. 10**, 78.

Orange-yellow needles, mp 48-50°. Moderately sol in water; sol in alcohol, ether, benzene, methanol.

3441. 1,4-Dithiothreitol. *(R*,R*)-1,4-Dimercapto-2,3-butanediol;* Cleland's reagent; *threo*-2,3-dihydroxy-1,4-dithiolbutane. $C_4H_{10}O_2S_2$; mol wt 154.25. C 31.15%, H 6.53%, O 20.74%, S 41.58%. $HSCH_2(CHOH)_2CH_2SH$. Prepn: Evans *et al.*, *J. Chem. Soc.* **1949**, 253.
Slightly hygroscopic needles from ether, mp 42-43°. Can be sublimed at 37° and 0.005 mm press; bp_2 125-130°. Redox potential: −0.33 volts at pH 7. Freely sol in water, ethanol, acetone, ethyl acetate, chloroform, ether.

USE: Protective agent for SH groups: Cleland, *Biochemistry* **3**, 480 (1964), *see also* "Cleland's Reagent" a detailed brochure and bibliography published by Calbiochem, Los Angeles.

3442. Dithizone. *Phenyldiazenecarbothioic acid 2-phenylhydrazide; (phenylazo)thioformic acid 2-phenylhydrazide;* diphenylthiocarbazone. $C_{13}H_{12}N_4S$; mol wt 256.33. C 60.91%, H 4.72%, N 21.86%, S 12.51%. $C_6H_5N=NCSNH-NHC_6H_5$. Prepn: E. Fischer, E. Besthorn, *Ann.* **212**, 316 (1882). Use as an analytical reagent: A. Stock, E. Pohland, *Angew. Chem.* **39**, 791 (1926); H. Fischer, *ibid.* **46**, 442 (1933); **50**, 919 (1937); *Organic Reagents for Metals* (Hopkins & Williams, 1934). Crystal structure: M. Laing, *J. Chem. Soc. Perkin Trans. 2* **1977**, 1248. Conformation in soln: A. T. Hutton, H. M. Irving, *Chem. Commun.* **1981**, 735. Monograph: *Analytical Sciences Monograph No. 5* entitled "Dithizone", H. M. Irving, Ed. (Chemical Society, London, 1977) 112 pp. *Review: idem, CRC Crit. Rev. Anal. Chem.* **8**, 321-405 (1980).
Bluish-black cryst powder, mp 168° (dec). Insol in water. Sparingly sol in alc, freely in carbon tetrachloride, chloroform. Solns are not stable, but may be preserved by a layer of aq SO_2.
Caution: Has caused glycosuria and ocular injury in exptl animals.

USE: Sensitive reagent for several heavy metals, Co, Cu, Pb, and Hg; esp for the estimation of minute amounts of Pb.

3443. Ditiocarb Sodium. *Diethylcarbamodithioic acid sodium salt; diethyldithiocarbamic acid sodium salt;* diethyldithiocarbamate sodium; dithiocarb; DTC; DDC; DEDC; DDTC; DeDTC; Imuthiol. $C_5H_{10}NNaS_2$; mol wt 171.26. C 35.07%, H 5.89%, N 8.18%, Na 13.42%, S 37.45%. $(C_2H_5)_2$-NCS$_2$Na. Chelating agent with strong affinity for Hg, Cu, Ni and Zn. Also active as T-cell specific immunostimulant. Prepn: A. M. Clifford, J. G. Lichty, *J. Am. Chem. Soc.* **54**, 1163 (1932); A. L. Klebanskii, L. P. Fomina, *Zh. Obshch. Khim.* **30**, 794 (1960). Absorption spectrum: H. P. Koch, *J. Chem. Soc.* **1949**, 401. Inhibition of superoxide dismutase in mice: R. E. Heikkila *et al.*, *J. Biol. Chem.* **251**, 2182 (1976); of cisplatin nephrotoxicity in rats: R. F. Borch *et al.*, *Proc. Nat. Acad. Sci. USA* **77**, 5441 (1980). Clinical use in acute nickel carbonyl poisoning: F. W. Sunderman, F. W. Sunderman Jr., *Am. J. Med. Sci.* **236**, 26 (1958); F. W. Sunderman, *Ann. Clin. Lab. Sci.* **9**, 1 (1979); in acute cadmium poisoning: G. R. Gale *et al.*, *ibid.* **11**, 476 (1981) and **13**, 207 (1983). Specific effects on T-cell regulation: A. Pompidou *et al.*, *Int. J. Immunopharmacol.* **7**, 561 (1985). Clinical evaluation in T-cell deficient diseases: E. Lemarie *et al.*, *Meth. Find. Exp. Clin. Pharmacol.* **8**, 51 (1986); in

AIDS-related complex: A. Pompidou *et al.*, *Comp. Immun. Microbiol. Infect. Dis.* **9**, 263 (1986). Review of pharmacology, toxicity and clinical uses: G. Renoux, *J. Pharmacol.* **13**, Suppl. 1, 95-134 (1982); of immunopharmacology and use in cancer immunotherapy: G. Renoux *et al.*, *Advan. Exp. Med. Biol.* **166**, 223-239 (1983).
Trihydrate, $C_5H_{10}NNaS_2\cdot3H_2O$, thin, irregular plate-like crystals from acetone, mp 94-102°. Also reported as 90-92° (Sunderman, Sunderman). Freely sol in water; sol in ethanol, methanol, acetone. Insol in ether, benzene. The aq soln is alkaline to litmus and phenolphthalein and slowly dec. (pH of 10% aq soln is 11.6 at room temp). The addition of an acid to the aq soln produces a white turbidity due to the liberation of carbon disulfide. uv max (ethanol): 257, 290 nm (ϵ 1200, 13000). LD_{50} orally in rats, mice: 2830, 1870 mg/kg; i.v. in mice: > 1000 mg/kg (Renoux, 1982).

USE: For colorimetric determination of small quantities of copper and for its separation from other metals.

THERAP CAT: Immunomodulator. Chelating agent (Wilson's disease). Antidote (nickel, cadmium poisoning).

3444. 1,1-Di-p-tolylethane. *1,1′-Ethylidenebis[4-methylbenzene];* α,α-di-p-tolylethane; *asym*-di-p-tolylethane; 4,4′-α-trimethyldiphenylmethane. $C_{16}H_{18}$; mol wt 210.32. C 91.37%, H 8.63%. Prepd by passing acetylene into a mixture of toluene and concd H_2SO_4 in presence of mercuric sulfate: Reichert, Nieuwland, *Org. Syn.* **IV**, 23 (1925).

Oily, highly refractive liquid. Aromatic odor. d_4^{20} 0.974. Not solid at −20°. bp 295-300°; bp_{12} 155-157°. Soluble in acetic acid.

3445. 1,2-Di-p-tolylethane. *1,1′-(1,2-Ethanediyl)bis[4-methylbenzene];* α,β-di-p-tolylethane; *sym*-di-p-tolylethane; 4,4′-dimethyldibenzyl. $C_{16}H_{18}$; mol wt 210.32. C 91.37%, H 8.63%. Prepd by passing p-xylene vapor over red hot platinum wires: Meyer, Hofmann, *Monatsh.* **37**, 690 (1916).

Leaflets from methanol; plates from ligroin. mp 82°; bp_{730} 296-298°; bp_{18} 178°. Sol in benzene, moderately sol in alcohol, petr ether.

3446. p-Ditolylmercury. *Bis(4-methylphenyl)mercury.* $C_{14}H_{14}Hg$; mol wt 382.86. C 43.92%, H 3.69%, Hg 52.39%. $(CH_3C_6H_4)_2Hg$. Prepd by the action of sodium iodide on p-tolylmercuric chloride in ethanol: Whitmore *et al.*, *Org. Syn.* **3**, 65 (1923); by boiling p-bromotoluene with sodium amalgam in xylene in the presence of ethyl acetate as a catalyst: Dreher, Otto, *Ann.* **154**, 171 (1870); LaCoste, Michaelis, *ibid.* **201**, 246 (1880); *Compt. Rend.* **68**, 1298 (1869).
Needles from xylene. mp 238°. Insol in water. Slightly sol in cold alcohol; sol in hot benzene, xylene, chloroform, carbon disulfide.

3447. Diuron. *N′-(3,4-Dichlorophenyl)-N,N-dimethylurea;* 1,1-dimethyl-3-(3,4-dichlorophenyl)urea; Diurex; Karmex; Urox D. $C_9H_{10}Cl_2N_2O$; mol wt 233.10. C 46.38%, H 4.32%, Cl 30.42%, N 12.02%, O 6.86%. Prepn: Jones, U.S. pat. **2,768,971** (1956 to I.C.I.). Toxicity data: E. M. Boyd, V. Krupa, *J. Agr. Food Chem.* **18**, 1104 (1970).

Crystals, mp 158-159°. Soly in water at 25°: 42 ppm. Very low soly in hydrocarbon solvents. Vapor pressure at 50°: 3.1 × 10⁻⁶ mm Hg. LD₅₀ orally in rats: 437 mg/kg (Boyd, Krupa).

Caution: Repeated doses produce anemia in rats and methemoglobinemia if the compound is hydrolyzed *in vivo* to dichloroaniline: du Pont, *Condensed Technical Information*, April 1961.

USE: Pre-emergent herbicide.

3448. Divicine. *2,6-Diamino-1,6-dihydro-4,5-pyrimidinedione; 2,6-diamino-4,5-pyrimidinediol;* 2,6-diamine-5-hydroxy-4(3H)-pyrimidinone; 2,4-diamino-5,6-dihydroxypyrimidine. $C_4H_6N_4O_2$; mol wt 142.12. C 33.81%, H 4.26%, N 39.42%, O 22.52%. From vicine by heating with dil H_2SO_4: Ritthausen, *Ber.* **29**, 2108 (1896); Levene, Senior, *J. Biol. Chem.* **25**, 607 (1916). Structure: Bendich, Clements, *Biochim. Biophys. Acta* **12**, 462 (1953). Synthesis: Davall, Laney, *J. Chem. Soc.* **1956**, 2124; Chesterfield *et al.,* *ibid.* **1964**, 1001.

Brownish needles, dec above 280°. One gram dissolves in 100 ml boiling water, in about 350 ml cold water. Soluble in 10% KOH.

Diacetate, $C_8H_{10}N_4O_4$, crystals from water, dec 309-312°.

3449. Dixanthogen. *Thioperoxydicarbonic acid diethyl ester; dithiobis[thioformic acid] O,O-diethyl ester; O,O-diethyl dithiobis[thioformate];* bisethylxanthogen; diethyl xanthogenate; ethylxanthic disulfide; preparation K; EXD; Auligen; Aulinogen; Bexide; Herbisan; Lenisarin; Sulfasan. $C_6H_{10}O_2S_4$; mol wt 242.41. C 29.73%, H 4.16%, O 13.20%, S 52.91%. Prepn: Losse, Wottgen, *J. Prakt. Chem.* **13**, 260 (1961). Metabolic studies: Gutenmann, Lisk, *J. Agr. Food Chem.* **19**, 200 (1971). Toxicity: G. W. Bailey, J. L. White, *Residue Rev.* **10**, 97 (1965).

Yellow needles, mp 28-32°. Onion-like odor. Soly in alcohol: 2 g/100 ml. Freely sol in benzene, ether, petr ether, oils. Almost insol in water. LD₅₀ orally in rats: 480 mg/kg (Bailey, White).

USE: Insecticide formulations; herbicide.
THERAP CAT: Ectoparasiticide.

3450. Dixyrazine. *2-[2-[4-[2-Methyl-3-(10H-phenothiazin-10-yl)propyl]-1-piperazinyl]ethoxy]ethanol;* 10-[2-methyl-3-(4-hydroxyethoxyethyl-1-piperazinyl)propyl]phenothiazine; 1-[2-(2-hydroxyethoxy)ethyl]-4-[2-methyl-3-(10-phenothiazinyl)propyl]piperazine; UCB-3412; Esocalm; Esucos. $C_{24}H_{33}N_3O_2S$; mol wt 427.61. C 67.41%, H 7.78%, N 9.83%, O 7.48%, S 7.50%. Prepn: Morren, **Brit.** pat. **861,-420** (1961).

Dihydrochloride, $C_{24}H_{33}N_3O_2S.2HCl$, crystals from isopropanol, mp 192°.

THERAP CAT: Antipsychotic.

3451. Dizocilpine. *10,11-Dihydro-5-methyl-5H-dibenzo[a,d]cyclohepten-5,10-imine;* (+)-5-methyl-10,11-dihydro-5H-dibenzo[a,d]cyclohepten-5,10-imine. $C_{16}H_{15}N$; mol wt 221.30. C 86.84%, H 6.83%, N 6.33%. Non-competitive NMDA, *q.v.,* receptor antagonist. Prepn of racemate: M. E. Christy *et al., J. Org. Chem.* **44**, 3117 (1979); T. R. Lamanec *et al., ibid.* **53**, 1768 (1988); and of enantiomers: **Belg.** pat. **882,361;** P. Anderson *et al.,* U.S. pat. **4,399,141** (1980, 1983 both to Merck & Co.). Series of papers on pharmacological activities: B. V. Clineschmidt *et al., Drug. Devel. Res.* **2**, 123-163 (1982). NMDA receptor binding studies: E. H. F. Wong *et al., Proc. Nat. Acad. Sci. USA* **83**, 7104 (1986); A. C. Foster, E. H. F. Wong, *Brit. J. Pharmacol.* **91**, 403 (1987); J. E. Huettner, B. P. Bean, *Proc. Nat. Acad. Sci. USA* **85**, 1307 (1988). Metabolism in animals: H. B. Hucker *et al., Drug. Metab. Dispos.* **11**, 54 (1983). Pharmacokinetics: A. S. Troupin *et al., Curr. Probl. Epilepsy* **4**, 191 (1986). Protective effects against excitotoxic amino acid-induced neuronal degeneration: G. N. Woodruff *et al., Neuropharmacology* **26**, 903 (1987); A. C. Foster *et al., Neuroscience Letters* **76**, 307 (1987); A. C. Foster *et al., J. Neurosci.* **8**, 4745 (1988); specifically in ischemia: R. Gill *et al., Neuroscience* **25**, 847 (1988); E. Ozyurt *et al., J. Cereb. Blood Flow Metab.* **8**, 138 (1988). Anticonvulsant activity in models of epilepsy: K. Sato *et al., Brain Res.* **463**, 12 (1988); M. E. Gilbert, *ibid.* **90**. Clinical evaluation in attention deficit disorder: F. W. Reimherr *et al., Psychopharmacol. Bull.* **22**, 237 (1986).

White solid from cyclohexane, mp 68.5-69°. [α]²⁰₅₈₉ +161.4° (c = 0.038 g/2 ml ethanol).
Maleate, $C_{16}H_{15}N.C_4H_4O_4$, *MK-801, Neurogard.* Crystalline solid, mp 208.5-210°. [α]²⁰_D +114° (c = 0.0128 g/2 ml ethanol).
(−)-Form, white solid from cyclohexane, mp 68.5-69.5°. [α]₅₈₉ −160.8° (c = 0.032 g/2 ml ethanol).

USE: Biochemical probe for NMDA receptors.

3452. Djenkolic Acid. *S,S'-Methylenebis-L-cysteine; 3,3'-(methylenedithio)dialanine;* 3,3'-methylenedithiobis(2-aminopropanoic acid); L-cysteine thioacetal of formaldehyde. $C_7H_{14}N_2O_4S_2$; mol wt 254.33. C 33.06%, H 11.01%, O 25.16%, S 25.22%. $CH_2[SCH_2CH(NH_2)COOH]_2$. An amino acid isolated from the djenkol bean *(Pithecolobium lobatum* Benth., *Leguminosae):* van Veen, Hyman, *Rec. Trav. Chim.* **54**, 493 (1935). Synthesis from 1 mol formaldehyde and 2 mols L-cysteine in strong HCl solution: Armstrong, du Vigneaud, *J. Biol. Chem.* **168**, 373 (1947); *cf.* du Vigneaud, Patterson, *ibid.* **114**, 633 (1936); Middlebrook, Phillips, *Biochem. J.* **41**, 218 (1947). Synthesis of homologs: Frankel, Gertner, *J. Chem. Soc.* **1960**, 898.

Rosettes of needles of various lengths. Gradually dec between 300 and 350°. [α]²⁰·⁵_D −65.0° (1.0N HCl); [α]²⁵_D −47.5° (c = 2 in 1.0N HCl). Very sparingly sol in cold water. Soly in boiling water about 1 in 200. Readily sol in aq solns of alkalies or acids.

Monohydrochloride, $C_7H_{14}N_2O_4S_2 \cdot HCl$, slender prisms, dec 250-300°. Sol in water.

Dibenzoyldjenkolic acid, $C_{21}H_{22}N_2O_6S_2$, crystals, mp 87.5-89°.

3453. DMPA. *(1-Methylethyl)phosphoramidothioic acid O-(2,4-dichlorophenyl) O-methyl ester; isopropylphosphoramidothioic acid O-2,4-(dichlorophenyl) O-methyl ester;* O-(2,4-dichlorophenyl) O-methyl isopropylphosphoramidothioate; K-22023; Dow 1329; ENT-25647; OMS-115; Dowco 118; Zytron; $C_{10}H_{14}Cl_2NO_2PS$; mol wt 314.17. C 38.23%, H 4.49%, Cl 22.57%, N 4.46%, O 10.19%, P 9.86%, S 10.21%. Prepn: Blair *et al., J. Agr. Food Chem.* **11**, 237 (1963). Synthesis of the optical isomers: Seiber, Tolkmith, *Tetrahedron* **25**, 381 (1969). Use as a herbicide: Leasure, U.S. pat. **3,074,790** (1963 to Dow); as a plant growth regulator: Holmsen, *Weed Sci.* **17**, 187 (1969). Neurotoxicity in chickens: B. M. Francis *et al., J. Environ. Sci. Health* **B15**, 313 (1980). Toxicity study: E. W. Schafer, *Toxicol. Appl. Pharmacol.* **21**, 315 (1972).

Solid, mp 51.4°. Vapor pressure at 150°: 2 mm. Slightly sol in water (5 ppm); freely sol in acetone, benzene, carbon tetrachloride. LD_{50} orally in rats: 270-360 mg/kg (Schafer).

Note: DMPA® is dimethylolpropionic acid, *q.v.*

USE: Herbicide; plant growth regulator.

3454. DMPO. *3,4-Dihydro-2,2-dimethyl-2H-pyrrole 1-oxide; 5,5-dimethyl-1-pyrroline 1-oxide.* $C_6H_{11}NO$; mol wt 113.16. C 63.69%, H 9.80%, N 12.38%, O 14.14%. Nitrone spin trap. Prepn: R. Bonnett *et al., J. Chem. Soc.* **1959**, 2094; D. L. Haire *et al., J. Org. Chem.* **51**, 4298 (1986); and purification: E. G. Janzen *et al., Chem.-Biol. Interactions* **70**, 167 (1989). uv absorption: L. S. Kaminsky, M. Lamchen, *J. Chem. Soc.* **1968**, 1085. Use as spin trap in biological systems: A. I. Cederbaum, G. Cohen, *Arch. Biochem. Biophys.* **204**, 397 (1980); A. S. W. Li *et al., Biochem. Biophys. Res. Commun.* **146**, 1191 (1987); for neutrophil activation studies: B. E. Britigan, D. R. Hamill, *Arch. Biochem. Biophys.* **275**, 72 (1989); *eidem, Free Rad. Biol. Med.* **8**, 459 (1990); in reperfusion injury: C. M. Arroyo *et al., FEBS Letters* **221**, 101 (1987); A. Tosaki, P. Braquet, *Am. Heart J.* **120**, 819 (1990). Use in HPLC quantitation of oxygen free radicals *in vivo*: P. S. Rao *et al., Chromatographia* **30**, 19 (1990).

Nitrone as white crystalline solid. Very hygroscopic. Turns yellow with time even at −20°C sealed under vacuum and protected from light. uv max in cyclohexane: 246 nm (ε 9000); ethanol: 234 nm (ε 7700); water: 226 nm (ε 8600); N-hydrochloric acid: 226 nm (ε 7000); ether: 246 nm.

USE: Spin trap agent for biological systems.

3455. Dobesilate Calcium. *2,5-Dihydroxybenzenesulfonic acid calcium salt;* calcium dobesilate; hydroquinone calcium sulfonate; Dexium; Doxium. $C_{12}H_{10}CaO_{10}S_2$; mol wt 418.42. C 34.45%, H 2.41%, Ca 9.58%, O 38.24%, S 15.33%. Prepn: **Span.** pat. **335,945** (1967 to Labs. Esteve), *C.A.* **69**, 106253z (1968); **Fr.** pat. **M6163**; A. Esteve-Subirana, U.S. pat. **3,509,207** (1968, 1970 both to Labs. OM). Clinical trials: Berson, *Praxis* **59**, 1305 (1970); **61**, 52 (1972). Metabolism: A. Benakis *et al., Therapie* **29**, 211 (1974).

White, powdery crystals from water, mp > 300° (dec). Color deepens to pink upon exposure to air. Very soluble in water and alcohol. Practically insol in ether, benzene, chloroform. LD_{50} in mice: 700 mg/kg (Esteve-Subirana).

THERAP CAT: Vasotropic.

3456. Dobutamine. *(±)-4-[2-[[3-(4-Hydroxyphenyl)-1-methylpropyl]amino]ethyl]-1,2-benzenediol; (±)-4-[2-[[3-(p-hydroxyphenyl)-1-methylpropyl]amino]ethyl]pyrocatechol; 3,-4-dihydroxy-N-[3-(4-hydroxyphenyl)-1-methylpropyl]-β-phenylethylamine;* Compound 81929. $C_{18}H_{23}NO_3$; mol wt 301.39. C 71.73%, H 7.69%, N 4.65%, O 15.93%. $β_1$-Adrenergic agonist; derivative of dopamine, *q.v.* Prepn: Tuttle, Mills, **Ger.** pat. **2,317,710** corresp to U.S. pat. **3,987,200** (1973, 1976 to Lilly). Pharmacology: R. Weber, R. R. Tuttle in *Pharmacological and Biochemical Properties of Drug Substances* vol. 1, M. E. Goldberg, Ed. (Am. Pharm. Assoc., Washington, DC, 1977) pp 109-124; E. H. Sonnenblick *et al., N. Engl. J. Med.* **300**, 17 (1979). Comprehensive description: R. H. Bishara, H. B. Long in *Analytical Profiles of Drug Substances* vol. 8, K. Florey, Ed. (Academic Press, New York, 1979) pp 139-158.

Hydrochloride, $C_{18}H_{23}NO_3 \cdot HCl$, *Inotrex (obsolete)*, *Dobutrex*. Crystals, mp 184-186°. uv max (methanol): 281, 223 nm (ε 4768, 14400). pKa 9.45. Rapidly oxidized at pH 11-13. LD_{50} i.v. in mice: ∼73 mg/kg (Weber, Tuttle).

THERAP CAT: Cardiotonic.

3457. Docarpamine. *(S)-4-[2-[[2-(Acetylamino)-4-(methylthio)-1-oxobutyl]amino]ethyl]-1,2-phenylene diethyl ester;* (−)-(S)-2-acetamido-N-(3,4-dihydroxyphenethyl)-4-(methylthio)butyramide bis(ethylcarbonate) (ester); N-(N-acetyl-L-methionyl)-O,O-bis(ethoxycarbonyl)dopamine; N-(N-acetyl-L-methionyl)-3,4-diethoxycarboxyphenethylamine; TA-870; Tanadopa. $C_{21}H_{30}N_2O_8S$; mol wt 470.54. C 53.60%, H 6.43%, N 5.95%, O 27.20%, S 6.81%. Dopamine prodrug. Prepn: T. Kiguchi *et al., Eur.* pat. **Appl. 7,441**; *eidem,* U.S. pat. **4,228,183** (both 1980 to Tanabe Seiyaku). Pharmacology: I. Yamaguchi *et al., J. Cardiovasc. Pharmacol.* **13**, 879 (1989). Mechanism of action: S. Nishiyama *et al., ibid.* **14**, 175 (1989). Pharmacokinetics: M. Yoshikawa *et al., Drug Metab. Dispos.* **16**, 754 (1988). Metabolism in humans: *eidem, ibid.* **18**, 212 (1990). HPLC determn in biological fluids: *eidem, Biomed. Chromatog.* **4**, 181 (1990). *Review:* S. Nishiyama *et al., Cardiovasc. Drug Rev.* **10**, 101-116 (1992).

Crystals form ethyl acetate/n-hexane, mp 85-90°. $[\alpha]_D^{20}$ −15.6° (c = 2 in methanol). Also reported as crystalline powder, mp 105-108° (Nishiyama, 1992). Slightly sol in water, readily sol in ethanol. LD_{50} in male, female rats (mg/kg): 1000-1400, ~1000 s.c.; in rats, dogs (mg/kg): > 2000 orally (Nishiyama, 1992).

THERAP CAT: Cardiotonic.

3458. Docetaxel. *[2aR-[2aα,4β,4aβ,6β,9α,(αR*,βS*),-11α,12α,12aα,12bα]]-β-[[(1,1-Dimethylethoxy)carbonyl]-amino]-α-hydroxybenzenepropanoic acid 12b-(acetyloxy)-12-(benzoyloxy)-2a,3,4,4a,5,6,9,10,11,12,12a,12b-dodecahydro-4,-6,11-trihydroxy-4a,8,13,13-tetramethyl-5-oxo-7,11-methano-1H-cyclodeca[3,4]benz[1,2-b]oxet-9-yl ester;* N-debenzoyl-N-(*tert*-butoxycarbonyl)-10-deacetyltaxol; NSC-628503; RP-56976; Taxotere. $C_{43}H_{53}NO_{14}$; mol wt 807.89. C 63.93%, H 6.61%, N 1.73%, O 27.73%. Semisynthetic derivative of paclitaxel, *q.v.*, prepd using a natural precursor, *10-deacetyl-baccatin III*, extracted from the needles of the European yew tree, *Taxus baccata* L., Taxaceae. Antimitotic agent that promotes the assembly of microtubules and inhibits their depolymerization to free tubulin. Prepn: M. Colin *et al.*, Eur. pat. Appl. 253,738; *eidem*, U.S. pat. 4,814,470 (1988, 1989 both to Rhône-Poulenc Sante); L. Mangatal *et al.*, Tetrahedron 45, 4177 (1989). Synthesis of the side chain: J.-N. Denis *et al.*, J. Org. Chem. 56, 6939 (1991). Structure-activity study: F. Guéritte-Voegelein *et al.*, J. Med. Chem. 34, 992 (1991). Cytotoxic activity and mechanism of action: I. Ringel, S. B. Horwitz, J. Natl. Cancer Inst. 83, 288 (1991). In vivo antitumor activity: M.-C. Bissery *et al.*, Cancer Res. 51, 4845 (1991). HPLC determn: J. C. Vergniol *et al.*, J. Chromatog. 582, 273 (1992). Clinical pharmacokinetics and toxicology: J.-M. Extra *et al.*, Cancer Res. 53, 1037 (1993). Clinical evaluation: H. Burris *et al.*, J. Clin. Oncol. 11, 950 (1993).

mp 232°. $[\alpha]_D$ −36° (c = 0.74 in ethanol). uv max: 230, 275, 283 nm (ε 14800, 1730, 1670).

THERAP CAT: Antineoplastic.

3459. Docusate Calcium. *Sulfobutanedioic acid 1,4-bis(2-ethylhexyl)ester calcium salt; bis[2-ethylhexyl]calcium sulfosuccinate;* calcium dioctyl sulfosuccinate; dioctyl calcium sulfosuccinate; Surfak. $C_{40}H_{74}CaO_{14}S_2$; mol wt 883.23. C 54.40%, H 8.44%, Ca 4.54%, O 25.36%, S 7.26%. Prepd from dioctyl sodium sulfosuccinate dissolved in isopropanol and from calcium chloride dissolved in methanol: Klotz, U.S. pat. 3,035,973 (1962 to Lloyd Brothers).

White precipitate. Sol in mineral and vegetable oils, liq polyethylene glycol. Practically insol in glycerol. Claimed to have greater surface-active wetting properties than the sodium salt.

Ingredient of *Doxidan* which also contains danthron.

THERAP CAT: Stool softener.

3460. Docusate Sodium. *Sulfobutanedioic acid 1,4-bis-(2-ethylhexyl) ester sodium salt; sulfosuccinic acid 1,4-bis(2-ethylhexyl) ester S-sodium salt;* bis(2-ethylhexyl)sodium sulfosuccinate; dioctyl sodium sulfosuccinate; sodium dioctyl sulfosuccinate; DSS; Aerosol OT; Colace; Comfolax; Co-

prola; Dioctylal; Dioctyl; Diotilan; Disonate; Disposaject; Doxinate; Doxol; Dulcivac; Jamylène; Molatoc; Molcer; Molofac; Nevax; Regutol; Soliwax; Velmol; Waxsol; Yal. $C_{20}H_{37}NaO_7S$; mol wt 444.57. C 54.03%, H 8.39%, Na 5.17%, O 25.19%, S 7.21%. Prepn: Jaeger, U.S. pats. 2,-028,091; 2,176,423 (1936, 1939, both to Am. Cyanamid). Structure and wetting power: Caryl, Ind. Eng. Chem. 33, 731 (1941). Comprehensive description: S. Ahuja, J. Cohen, Anal. Profiles Drug Subs. 2, 199-219 (1973); 12, 713-720 (1983). For structure see Docusate calcium.

Available as wax-like solid, usually in rolls of tissue-thin material; also as 50-75% solns in various solvents. Soly in water (g/l): 15 (25°), 23 (40°), 30 (50°), 55 (70°). Sol in CCl_4, petr ether, naphtha, xylene, dibutyl phthalate, liq petrolatum, acetone, alcohol, vegetable oils. Very sol in water + alcohol, water + water-miscible organic solvents. Stable in acid and neutral solns; hydrolyzes in alkaline solns.

Note: Ingredient of the laxative *Peri-Colace* which also contains casanthranol.

Docusate potassium, $C_{20}H_{37}KO_7S$, *Rectalad.*

USE: Sodium salt as pharmaceutic aid (surfactant); as wetting agent in industrial, pharmaceutical, cosmetic and food applications; dispersing and solubilizing agent in foods; adjuvant in tablet formation.

THERAP CAT: Stool softener.

THERAP CAT (VET): Stool softener.

3461. Dodecahedrane. *Hexadecahydro-5,2,1,6,3,4-[2,3]-butanediyl[1,4]diylidenedipentaleno[2,1,6-cde:2',1',6'-gha]-pentalene;* pentagonal dodecahedrane. $C_{20}H_{20}$; mol wt 260.38. C 92.26%, H 7.74%. Classical uniform convex polyhedrane. Theoretical studies: O. Ermer, Angew. Chem. Int. Ed. 16, 411 (1977); R. L. Disch, J. M. Schulman, J. Am. Chem. Soc. 103, 3297 (1981). Review of synthetic studies: P. E. Eaton, Tetrahedron 35, 2189-2223 (1979). Total synthesis: R. J. Ternansky *et al.*, J. Am. Chem. Soc. 104, 4503 (1982).

Crystals from benzene, mp > 450°.

3462. Dodecamethylcyclohexasiloxane. $C_{12}H_{36}O_6Si_6$; mol wt 444.93. C 32.39%, H 8.16%, O 21.58%, Si 37.87%. Isolated from the hydrolysis product of dimethyldichlorosilane: Patnode, Wilcock, J. Am. Chem. Soc. 68, 358 (1946).

Oily liquid. mp −3°. bp 245°. bp_{20} 128°. d 0.9762. n_D^{20} 1.4015.

3463. Dodecamethylpentasiloxane. $C_{12}H_{36}O_4Si_5$; mol wt 384.84. C 37.45%, H 9.43%, O 16.63%, Si 36.49%. Prepn: Patnode, Wilcock, J. Am. Chem. Soc. 68, 358 (1946).

Liquid. bp_{710} 229°. d 0.8755. n_D^{20} 1.3925. mp ~ −80°. Stable. Inert to most chemical reagents and rubber. Maintains about the same viscosity over a wide temperature range. Sol in benzene and the lighter hydrocarbons; slightly sol in alcohol and the heavy hydrocarbons.

USE: As a basis for silicone oils or fluids designed to withstand extremes of temperature; as a foam suppressant in petr lubricating oil.

3464. 1-Dodecanol. Dodecyl alcohol; lauryl alcohol. $C_{12}H_{26}O$; mol wt 186.34. C 77.35%, H 14.06%, O 8.59%. $CH_3(CH_2)_{11}OH$. Prepd by the reduction of esters of lauric acid with sodium and abs alcohol: Levene, Allen, *J. Biol. Chem.* **27**, 443 (1916); Ford, Marvel, *Org. Syn.* **10**, 62 (1930), or by high pressure hydrogenation of the esters using copper chromite catalyst: Adkins, Folkers, *J. Am. Chem. Soc.* **53**, 1095 (1931).

Leaflets from dil alc, mp 24°. d_4^{24} 0.8309 (liq); d_4^{40} 0.8201; d_4^{99} 0.7781. bp_{760} 259°; bp_{400} 235.7°; bp_{200} 213°; bp_{100} 192°; bp_{60} 177.8°; bp_{40} 167.2°; bp_{20} 150°; bp_{10} 134.7°; bp_5 120.2°; $bp_{1.0}$ 91.0°. Insol in water; sol in alcohol and ether.

Phenylurethan, $C_{19}H_{31}NO_2$, needles from dilute methanol, mp 84°.

USE: Manuf of sulfuric acid esters which are used as wetting agents.

3465. [S-(all-E)]-3-(1,3,5,7,9-Dodecapentaenyloxy)-1,2-propanediol. Fecapentaene 12. $C_{15}H_{22}O_3$; mol wt 250.34. C 71.97%, H 8.86%, O 19.17%. A potent mutagen detected in human feces: W. R. Bruce *et al., Cold Spring Harbor Conf. Cell Proliferation* **3**, 1641 (1977). Isoln and characterization: T. D. Wilkins *et al., Am. J. Clin. Nutr.* **33**, 2513 (1980). Structure: N. Hirai *et al., J. Am. Chem. Soc.* **104**, 6149 (1982). Implicated in human colon cancer: *Chem. & Eng. News* **60**, 22 (Sept. 27, 1982); T. H. Maugh, *Science* **218**, 363 (1982).

Unstable, decomp rapidly in presence of air or acid. uv max: 325, 345, 365 nm. Sol in chloroform, benzene, ether. Insol in water.

3466. Dodecarbonium Chloride. *N-[2-(Dodecylamino)-2-oxoethyl]-N,N-dimethylbenzenemethanaminium chloride; benzyl[(dodecylcarbamoylmethyl)dimethyl]ammonium chloride; N,N-dimethyl-N-benzyl-N-chloro-N'-dodecylglycinamide; (dodecylcarbamoylmethyl)benzyldimethylammonium chloride;* Straminol; Urolocide. $C_{23}H_{41}ClN_2O$; mol wt 397.04. C 69.58%, H 10.41%, Cl 8.93%, N 7.06%, O 4.03%. Preparation: Hentrich *et al.,* U.S. pat. **2,295,655** (1942 to Patchem, AG).

Bitter crystals, mp 147-148°. Sol in water, alcohol. Insol in ether, acetone, benzene. LD_{50} orally in rats: 100 mg/kg.
THERAP CAT: Antiseptic, disinfectant.

3467. Dodemorph. *4-Cyclododecyl-2,6-dimethylmorpholine.* $C_{18}H_{35}NO$; mol wt 281.48. C 76.81%, H 12.53%, N 4.98%, O 5.68%. Prepn: W. Sanne *et al.,* Belg. pat. **614,214;** *eidem,* U.S. pat. **3,686,399** (1962, 1972 both to BASF); K.-H. König *et al., Angew. Chem. Int. Ed.* **4**, 336 (1965). Activity: J. Kradel, E. H. Pommer, *Proc. Brit. Insectic. Fungic. Conf., 4th* **1967**, 170. Toxicity data: D. Marchand, G. Serra, *Def. Veg.* **27**, 144 (1973), *C.A.* **79**, 133569v (1973).

Oil, $bp_{1.5}$ 161-162°. n_D^{25} 1.4907.

Acetate, $C_{18}H_{35}NO.C_2H_4O_2$, *cyclomorph, BAS-238F, Meltatox, Milban.* Yellow liquid, d 0.93. Misc with water. LD_{50} i.p. in mice: 100 mg/kg (Marchand, Serra).
USE: Fungicide.

3468. Dodine. *Dodecylguanidine monoacetate;* AC-5223; Carpene; Cyprex; Melprex. $C_{15}H_{33}N_3O_2$; mol wt 287.45. C 62.68%, H 11.57%, N 14.62%, O 11.13%. Prepn: G. Lamb, U.S. pats. **2,867,562; 2,921,881** (1959, 1960, both to Am. Cyanamid).

Slightly waxy solid, mp 136°. Sol in alcohol, hot water; slightly sol in other solvents. LD_{50} orally in rats: 566 mg/kg, *RTECS* **Vol. I**, R. J. Lewis, R. L. Tatken, Eds. (1979) p 726.
Caution: Strong solns irritating to skin, mucous membranes. Ingestion causes vomiting, diarrhea.
USE: Fungicide.

3469. Dofetilide. *N-[4-[2-[Methyl[2-[4-[(methylsulfonyl)amino]phenoxy]ethyl]amino]ethyl]phenyl]methanesulfonamide;* 1-(4-methanesulfonamidophenoxy)-2-[N-(4-methanesulfonamidophenethyl)-N-methylamino]ethane; UK-68798. $C_{19}H_{27}N_3O_5S_2$; mol wt 441.57. C 51.68%, H 6.16%, N 9.52%, O 18.12%, S 14.52%. Potassium channel blocker. Prepn: J. E. Arrowsmith *et al., Eur. pat.* **Appl. 245,997;** P. E. Cross *et al.,* U.S. pat. **4,959,366** (1987, 1990 both to Pfizer); *idem et al., J. Med. Chem.* **33**, 1151 (1990). HPLC determn in urine: D. K. Walker *et al., J. Chromatog.* **568**, 475 (1991). Mechanism of action study: D. Carmeliet, *J. Pharmacol. Exp. Ther.* **262**, 809 (1992). Clinical evaluation: M. L. Sedgwick *et al., Am. J. Cardiol.* **69**, 513 (1992). Pharmacodynamics and pharmacokinetics: T. C. K. Tham *et al., J. Cardiovasc. Pharmacol.* **21**, 507 (1993). *Review:* H. S. Rasmussen *et al., ibid.* **20**, Suppl. 2, S96-S105 (1992).

Crystals from ethyl acetate/methanol (10:1), mp 147-149° (Cross); from hexane/ethyl acetate, mp 151-152° (Arrowsmith). Also reported as white crystalline solid, mp 161° (Rasmussen). pKa 7.0, 9.0, 9.6. Distribution coefficient (pH 7.4): 0.96.
THERAP CAT: Antiarrhythmic (class III).

3470. Doisynolic Acid. *1-Ethyl-1,2,3,4,4a,9,10,10a-octahydro-7-hydroxy-2-methyl-2-phenanthrenecarboxylic acid;* 3-hydroxy-16,17-secoestra-1,3,5(10)-trien-17-oic acid. $C_{18}H_{24}O_3$; mol wt 288.39. C 74.97%, H 8.39%, O 16.64%. Prepd by alkali fusion of estrone: MacCorquodale *et al., J. Biol. Chem.* **99**, 327 (1933); **101**, 753 (1933); Heer, Miescher, *Helv. Chim. Acta* **28**, 156 (1945). Sixteen isomers are possible. If the hydrogen atom at position 4a is α- and the hydrogen atom at position 10a is β-, the acid belongs to the A series. If the hydrogen at 4a is β- and the hydrogen at 10a is α-, the acid belongs to the B series. If both hydrogens are α-, the acid belongs to the C series. If both hydrogens are β-, the acid belongs to the D series: Miescher, *Chem. Rev.* **43**, 367 (1948). If the hydrogen atom at position 1 and the methyl group at position 2 have either the α,α- or β,β-orientation, the acid is designated *cis;* if they are α,β- the acid is designated *trans:* L. F. Fieser, M. Fieser, *Steroids* (Reinhold, New York, 1959) pp 487-495. Synthesis of *dl-cis* and *dl-trans* of the A series and *dl-cis* of the B series: Anner, Miescher, *Helv. Chim. Acta* **29**, 1889 (1946); **30**, 1422 (1947). Synthesis of *dl-cis* of the C series: Jilck, Protiva, *Coll. Czech. Chem. Commun.* **23**, 692 (1958). Stereochemistry of *d-cis*-form: Iriarte, Crabbe, *Chem. Commun.* **1972**, 1110. *Reviews:* Shoppee, *Annual Reports on the Progress of Chemistry* **44**, 190 (1948); Miescher in *Recent*

Progress in Hormone Research **vol. 3,** G. Pincus Ed. (Academic Press, New York, 1948) pp 47-69.

B-type acid

dl-cis Acid of the A series: Crystals from methanol, mp 181-182°. Appears to have the highest estrogenic potency.

dl-trans Acid of the A series: Plates from methanol, mp 175-177°.

dl-cis Acid of the B series: Crystals from methanol, mp 212-214°.

dl-cis Acid of the C series: Crystals from methanol, mp 113-117°.

3471. Dolasetron. *1H-Indole-3-carboxylic acid (2α,6α,-8α,9aβ)-octahydro-3-oxo-2,6-methano-2H-quinolizin-8-yl ester; endo*-hexahydro-8-(3-indolylcarbonyloxy)-2,6-methano-2*H*-quinolizin-3(4*H*)-one; MDL-73147. $C_{19}H_{20}N_2O_3$; mol wt 324.38. C 70.35%, H 6.21%, N 8.64%, O 14.80%. Bridged pseudopelletierine derivative; 5-HT$_3$ specific receptor antagonist. Prepn: M. W. Gittos, **Eur. pat. Appl. 266,-730** (1988 to Merrell Dow). Binding study: P. H. Boeijinga *et al., Eur. J. Pharmacol.* **219,** 9 (1992). Pharmacology: R. C. Miller *et al., Drug Develop. Res.* **28,** 87 (1993). Clinical pharmacokinetics: H. Boxenbaum *et al., Biopharm. Drug. Dispos.* **13,** 693 (1992); *eidem, ibid.* **14,** 131 (1993). GC-MS and LC determn in plasma: T. A. Gillespie *et al., J. Pharmaceut. Biomed. Anal.* **11,** 955 (1993). Clinical evaluation as antiemetic: V. Kirchner *et al., Ann. Oncol.* **4,** 481 (1993).

Methanesulfonate, $C_{19}H_{20}N_2O_3 \cdot CH_3SO_3H$, *dolasetron mesylate, MDL-73147EF, Anzemet.* Crystalline solid, mp 278°.

THERAP CAT: Antiemetic; antimigraine.

3472. Domesticine. *(S)-5,6,6a,7-Tetrahydro-2-methoxy-6-methyl-4H-benzo[de][1,3]benzodioxolo[5,6-g]quinolin-1-ol; 2-methoxy-9,10-(methylenedioxy)-6aα-aporphin-1-ol; 1-hydr-oxy-2-methoxy-9,10-methylenedioxyaporphine.* $C_{19}H_{19}NO_4$; mol wt 325.36. C 70.14%, H 5.89%, N 4.30%, O 19.67%. In *Nandina domestica* Thunb., *Berberidaceae.* Isoln: Kitasato, Shishido, *Ann.* **527,** 176 (1937). Syntheses: Govindachari *et al., Indian J. Chem.* **7,** 841 (1969); Kessar *et al., ibid.* **8,** 468 (1970); Kametani *et al., J. Chem. Soc. (C)* **1971,** 2446, 2712; *eidem, Chem. Pharm. Bull.* **21,** 766 (1973); Horii *et al., ibid.* **22,** 583 (1974); Hoshino *et al., ibid.* **23,** 2048 (1975).

From methanol + water, mp 115-116°; from abs methanol or benzene, mp 84-85°; dried at 60° over P_2O_5, mp 152-153°. Easily oxidized in air. uv max (ethanol): 221, 283, 310 nm (log ε 4.56, 4.01, 4.17). Very sol in chloroform, sol in hot alc, ethyl acetate, acetic acid, alkalies; slightly sol in ether. Practically insol in water.

Methyl ether, *nantenine, domestine, epidicentrine.* mp 139°. $[\alpha]_D^{18}$ +102° (c = 0.528 in chloroform).

dl-Form, needles from methanol, mp 185-186° (dec).

3473. Domiodol. *2-(Iodomethyl)-1,3-dioxolane-4-methanol; 4-hydroxymethyl-2-iodomethyl-1,3-dioxolane; MG-13608; Mucolitico.* $C_5H_9IO_3$; mol wt 244.03. C 24.61%, H 3.72%, I 52.00%, O 19.67%. Organic iodide mixture of *cis* and *trans* isomers. Prepn: M. Carissimi *et al.*, **Ger. pat. 2,610,704;** *eidem,* **U.S. pat. 4,085,223** (1976, 1978 both to Maggioni). Prepn, expectorant and mucolytic activities: G. Cantarelli *et al., Farmaco Ed. Prat.* **34,** 393 (1979). Pharmacology, toxicity: K. Kogi *et al., Arznemittel-Forsch.* **33,** 1281 (1983). Separation of isomers and comparison of their pharmacological activity: M. Riva *et al., ibid.* 1091. Metabolism and tissue distribution in rats: T. Ohtsuki *et al., Farmaco Ed. Prat.* **39,** 291 (1984). Comparative study with *S*-carboxymethylcysteine, *q.v.,* in chronic obstructive lung disease: L. Casali *et al., Int. J. Clin. Pharmacol. Ther. Toxicol.* **20,** 554 (1982).

LD$_{50}$ in male, female mice (mg/kg): 79, 89 i.p.; 140, 145 orally (Riva).

cis-Form: bp$_{0.2}$ 106-108°. LD$_{50}$ orally in male mice: 135 mg/kg (Riva).

trans-Form: bp$_{0.2}$ 114-116°. LD$_{50}$ orally in male mice: 150 mg/kg (Riva).

THERAP CAT: Mucolytic.

3474. Domiphen Bromide. *N,N-Dimethyl-N-(2-phen-oxyethyl)-1-dodecanaminium bromide; dodecyldimethyl(2-phenoxyethyl)ammonium bromide;* (β-phenoxyethyl)dimeth-yldodecylammonium bromide; PDDB; phenododecinium bromide; NSC-39415; Bradosol Bromide; Oradol; Modicare; Neo-Bradoral. $C_{22}H_{40}BrNO$; mol wt 414.47. C 63.75%, H 9.73%, Br 19.28%, N 3.38%, O 3.86%. Prepd by heating phenoxyethyldimethylamine with dodecyl bromide: Hartmann, Bosshard, **U.S. pat. 2,581,336** (1952 to Ciba).

Crystals, mild characteristic odor, bitter taste, mp 112-113°. Freely sol in water (100 g/100 ml), much less sol at low temps. Sol in ethanol, acetone, ethyl acetate, chloroform; very slightly sol in benzene. Aq solns are clear, colorless, and foam profusely on shaking. pH of 10% soln: 6.42; of 1% commercial product at 25°: 5.5; of 0.1%: 6.8. Surface tension values at 25° (by the capillary rise method) range from 26.75 dynes/cm (10% soln) to 22.08 dynes/cm (0.1% soln). Incompatible with soap.

THERAP CAT: Anti-infective (topical).

3475. Domoic Acid. *[2S-[2α,3β,4β(1Z,3E,5S*)]]-2-Carboxy-4-(5-carboxy-1-methyl-1,3-hexadienyl)-3-pyrrolidineacetic acid.* $C_{15}H_{21}NO_6$; mol wt 311.33. C 57.87%, H 6.80%, N 4.50%, O 30.83%. Excitatory amino acid isolated from the red alga *Chondria armata* Okamura, *Rhodomelaceae,* known in Japanese as "domoi". Structural analog of kainic acid, *q.v.* Shown to be responsible for amnesic shellfish poisoning associated with ingestion of certain cultured blue mussels. Isoln: T. Takemoto, K. Daigo, *Chem. Pharm. Bull.* **6,** 578 (1958); *eidem, Arch. Pharm.* **293,** 627 (1960). Series of articles on isoln, structure, and anthelminthic activity: K. Daigo, *Yakugaku Zasshi* **79,** 350-364 (1959), *C.A.* **53,** 14218 (1959). Structural studies: T. Takemoto *et al., ibid.* **86,** 874 (1966), *C.A.* **66,** 28604m (1967). Total synthesis and revised structure: Y. Ohfune, M. Tomita, *J. Am. Chem. Soc.* **104,** 3511 (1982). Neuroexcitatory activity: T. J. Biscoe *et al., Nature* **255,** 166 (1975); R. Zaczek, J. T. Coyle, *Neuropharmacology* **21,** 15 (1982). Effect on kainate receptor activation: G. Debonnel *et al., Can. J. Physiol. Pharmacol.* **67,** 29, 904 (1989). Identification in toxic mussel extracts: J. L. C. Wright *et al., Can. J. Chem.* **67,** 481 (1989). LC determn in shellfish products: J. F. Lawrence *et al., J. Chromatog.* **462,** 349, 419 (1989). *Review:* M. A. Quilliam, J. L. C. Wright, *Anal. Chem.* **61,** 1053A-1059A (1989).

Dihydrate, mp 217° (dec). $[\alpha]_D^{12}$ −109.6° (c = 1.314 in water). uv max: 242 nm (log ε 4.24). pKa in water: 2.10, 3.72, 4.93, 9.82. Sol in water, acetic acid. Insol in methanol, ethanol, chloroform, acetone, benzene.

3476. Domperidone. *5-Chloro-1-[1-[3-(2,3-dihydro-2-oxo-1H-benzimidazol-1-yl)propyl]-4-piperidinyl]-1,3-dihydro-2H-benzimidazol-2-one; 5-chloro-1-[1-[3-(2-oxo-1-benzimidazolinyl)propyl]-4-piperidyl]-2-benzimidazolinone; R-33812; Euciton; Evoxin; Gastronorm; Mod; Motilium; Nauzelin; Peridon; Peridys.* $C_{22}H_{24}ClN_5O_2$; mol wt 425.92. C 62.04%, H 5.68%, Cl 8.32%, N 16.44%, O 7.51%. A novel *in vitro* dopamine antagonist with antinauseant properties. Prepn: J. Vandenberk *et al.,* **Ger. pat. 2,632,870;** *eidem,* **U.S. pat. 4,066,772** (1977, 1978 both to Janssen). Pharmacology: C. Ennis *et al., J. Pharm. Pharmacol.* **31,** Suppl., 14P (1979). Gastrokinetic properties: J. M. Van Neuten *et al., Life Sci.* **23,** 453 (1978). ^{3}H-domperidone studies: M. P. Martres *et al., ibid.* 1781; M. Baudry *et al., Arch. Pharmacol.* **308,** 231 (1979). Clinical studies: A. J. Reyntjens *et al., Arzneimittel-Forsch.* **28,** 1194 (1978); D. B. Wilson, J. W. Dundee, *Anaesthesia* **34,** 765 (1979). Review of pharmacology, pharmacokinetics and therapeutic efficacy: R. N. Brogden *et al., Drugs* **24,** 360-400 (1982).

Crystals from DMF/water, mp 242.5°.
THERAP CAT: Antiemetic.

3477. Donovan's Solution. An arsenic and mercuric iodides solution prepd from 1 g each of AsI_3 and HgI_2 plus 0.9 g sodium bicarbonate in water to make 100 ml.

Clear, colorless or pale yellow, becoming darker with age. Miscible with water, alcohol.

Incompat. Alkalies, alkaloids or their salts.

THERAP CAT (VET): Has been used in chronic diseases of the skin.

3478. Dopa. *3-Hydroxytyrosine; 3-(3,4-dihydroxyphenyl)alanine; β-(3,4-dihydroxyphenyl)-α-alanine; 2-amino-3-(3,4-dihydroxyphenyl)propanoic acid.* $C_9H_{11}NO_4$; mol wt 197.19. C 54.82%, H 5.62%, N 7.10%, O 32.45%. An amino acid found in seedlings, pods and beans of *Vicia faba* L. (broad beans) *Stizolobium deeringianum* L. (velvet beans) *Leguminosae:* Torquati, *Arch. farm. sper.* **15,** 213, 308 (1913), *C.A.* **7,** 2774 (1913); Sealock, *Biochemical Preparations* **vol. 1,** 25 (1949). Absolute configuration: Guggenheim, *Z. Physiol. Chem.* **88,** 276 (1913). Prepn of DL-, D-, and L-dopa: Yamada *et al., Chem. Pharm. Bull.* **10,** 693 (1962).

DL-Form, prisms from water or from aq $NaHSO_3$ soln, dec 270-272°. Soly in water: 144 mg/40 ml. Readily sol in dil acids and alkalies; slightly sol in benzene, carbon disulfide. Practically insol in abs alcohol, ether, glacial acetic acid, petr ether, chloroform. Oxidizes readily.

D-Form, needles from water, dec 276-278°. $[\alpha]_D^{11}$ +13.0° (c = 5.27 in 1N HCl). Soly in water: 66 mg/40 ml.

L-Form, *see* Levodopa.

3479. Dopamine. *4-(2-Aminoethyl)-1,2-benzenediol; 4-(2-aminoethyl)pyrocatechol; 3-hydroxytyramine; 3,4-dihydroxyphenethylamine; α-(3,4-dihydroxyphenyl)-β-aminoethane.* $C_8H_{11}NO_2$; mol wt 153.18. C 62.73%, H 7.24%, N 9.14%, O 20.89%. Endogenous catecholamine with α and β-adrenergic activity. Isoln from *Hermidium alipes* (S. Watson) *Nyctaginaceae:* Buelow, Gisvold, *J. Am. Pharm. Assoc.* **33,** 270 (1944). Prepn from aminotyramine: Waser, Sommer, *Helv. Chim. Acta* **6,** 61 (1923). From homoveratrylamine: Schöpf, Bayeler, *Ann.* **513,** 196 (1934); Hahn, Stiehl, *Ber.* **69,** 2640 (1936). Comprehensive description of the hydrochloride: J. E. Carter *et al., Anal. Profiles Drug Subs.* **11,** 257-272 (1982). Review of pharmacology and clinical efficacy in oliguria: J. F. Dasta, M. G. Kirby, *Pharmacotherapy* **6,** 304 (1986).

Free base, stout prisms, highly sensitive to oxygen; discolors quickly.

Hydrochloride, $C_8H_{11}NO_2$·HCl, *ASL-279, Cardiosteril, Dopastat, Dynatra, Inovan, Inotropin.* Rosettes of needles from water, dec 241°; may be recrystallized from methanol + ether. Freely sol in water; sol in methanol, in hot 95% ethanol; in aq solns of alkali hydroxides. Practically insol in ether, petr ether, chloroform, benzene, toluene.

Hydrobromide, $C_8H_{11}NO_2$·HBr, crystals, dec 210-214°.

THERAP CAT: Cardiotonic; antihypotensive.

3480. Dopan. *5-[Bis(2-chloroethyl)amino]-6-methyl-2,4(1H,3H)-pyrimidinedione; 5-[bis(2-chloroethyl)amino]-6-methyluracil; 6-methyl-5-[bis(2-chloroethyl)amino]uracil; 4-methyl-5-[bis(β-chloroethyl)amino]uracil; 2,6-dihydroxy-4-methyl-5-bis[2-chloroethyl]aminopyrimidine; Elderfield pyrimidine mustard; NSC-23436.* $C_9H_{13}Cl_2N_3O_2$; mol wt 266.13. C 40.62%, H 4.92%, Cl 26.64%, N 15.79%, O 12.02%. Description: Larionova, Platonova, *Voprosy Onkologii* **1,** no. 5, 36 (1955); *C.A.* **51,** 6862 (1957). Prepn: Nemets *et al.,* **U.S.S.R. pat. 116,912** (1959), *C.A.* **53,** 17438i (1959). Outline of synthesis of the demethyl compd: Petering, Lyttle, *Chem. & Eng. News* **36,** 47 (Sept. 22, 1958). Monograph: L. F. Larionov, *Cancer Chemotherapy* (Pergamon Press, New York, 1965).

Consult the Name Index before using this section.

Snow-white crystals, dec 178-179°. Practically insol in cold water, acetone and benzene. Slightly sol in alc.

USE: In anticancer research: Louis, *J. Chronic Diseases* **15**, 273-281 (March 1962).

3481. Dopastin. *[S-(E)]-N-[2-(Hydroxynitrosoamino)-3-methylbutyl]-2-butenamide; N-(2-nitrosohydroxylamino-3-methylbutyl)crotonamide.* $C_9H_{17}N_3O_3$; mol wt 215.25. C 50.22%, H 7.96%, N 19.52%, O 22.30%. Dopamine β-hydroxylase inhibitor isolated from a culture filtrate of a bacterium of the *Pseudomonas* genus. Prodn, isoln and characterization: Iinuma *et al., Agr. Biol. Chem.* **38**, 2093 (1974). Structure and synthesis: Iinuma *et al., ibid.* 2099. Biochemical and biological studies: Iinuma *et al., ibid.* 2107.

Crystals from acetone + *n*-hexane or from ethyl ether, mp 116-119°. Monobasic acid, pKa 5.1. $[\alpha]_D^{21}$ —246° (c = 0.5 in alcohol). uv max (0.01N HCl): 215 nm (ε 20900).

Copper salt, $C_9H_{16}N_3O_3 \cdot \frac{1}{2}Cu$, dark blue needles from ethanol, mp 142-144°.

3482. Dopexamine. *4-[2-[[6-[(2-Phenylethyl)amino]-hexyl]amino]ethyl]-1,2-benzenediol; 4-[2-[[6-(phenethylamino)hexyl]amino]ethyl]pyrocatechol; FPL-60278; Dopacard.* $C_{22}H_{32}N_2O_2$; mol wt 356.51. C 74.12%, H 9.05%, N 7.86%, O 8.98%. Dopamine receptor and β₂-adrenoreceptor agonist. Prepn: J. B. Farmer *et al.,* **Eur.** pat. Appl. 72,061 (1983 to Fisons). Unlike dopamine, dopexamine has little or no activity at the α- and β₁-adrenoceptors: R. A. Brown *et al., Brit. J. Pharmacol.* **85**, 599 (1985). Cardiovascular activity in dogs: *eidem, ibid.* 609. Hemodynamic effects in chronic congestive heart failure: J. R. Dawson *et al., Brit. Heart J.* **54**, 313 (1985); G. Svensson *et al., Eur. Heart J.* **7**, 697 (1986). Human renovascular effects: F. Magrini *et al., Eur. J. Clin. Pharmacol.* **32**, 1 (1987). Symposium on pharmacology and clinical efficacy: *Am. J. Cardiol.* **62**, Suppl., 1C-88C (1988).

Dihydrochloride, $C_{22}H_{32}N_2O_2 \cdot 2HCl$, *dopexamine hydrochloride, FPL-60278AR.*

Dihydrobromide, $C_{22}H_{32}N_2O_2 \cdot 2HBr$, crystals from ethanol, mp 227-228°.

THERAP CAT: Cardiotonic.

3483. Doramectin. *25-Cyclohexyl-5-O-demethyl-25-de-(1-methylpropyl)avermectin A₁ₐ; 25-cyclohexylavermectin B₁; UK-67994; Dectomax.* $C_{50}H_{74}O_{14}$; mol wt 899.13. C 66.79%, H 8.30%, O 24.91%. Mutational biosynthetic antiparasitic antibiotic structurally related to the avermectins, *q.v.*. Isoln: S. P. Gibson *et al., Eur. pat. Appl.* 214,731; *eidem, U.S. pat.* 5,089,480 (1987, 1992 both to Pfizer). Mutational biosynthesis: C. J. Dutton *et al., J. Antibiot.* **44**, 357 (1991). Pharmacokinetics and efficacy in cattle: *Vet. Parasitol.* **49**, 1-119 (1993).

mp 116°-119°.

THERAP CAT (VET): Antiparasitic.

3484. Dorzolamide. *(4S-trans)-4-(Ethylamino)-5,6-dihydro-6-methyl-4H-thieno[2,3-b]thiopyran-2-sulfonamide 7,7-dioxide.* $C_{10}H_{16}N_2O_4S_3$; mol wt 324.45. C 37.02%, H 4.97%, N 8.63%, O 19.73%, S 29.65%. Carbonic anhydrase inhibitor. Prepn: J. J. Baldwin *et al., Eur. pat. Appl.* 296,-879; *eidem, U.S. pat.* 4,797,413 (1988, 1989 both to Merck & Co.). Mechanism of action study: R.-F. Wang *et al., Arch. Ophthalmol.* **109**, 1297 (1991). HPLC determn in plasma and urine: B. K. Matuszewski, M. L. Constanzer, *Chirality* **4**, 515 (1992). Clinical evaluations in glaucoma and ocular hypertension: E. A. Lippa *et al., Ophthalmology* **98**, 308 (1991); E. A. Lippa *et al., Arch. Ophthalmol.* **110**, 495 (1992).

Hydrochloride, $C_{10}H_{16}N_2O_4S_3 \cdot HCl$, *MK-507, Trusopt.* mp 283-285°. $[\alpha]_D^{24}$ —8.34° (c = 1 in methanol). Sol in water.

THERAP CAT: Antiglaucoma agent.

3485. Dothiepin. *3-Dibenzo[b,e]thiepin-11(6H)-ylidene-N,N-dimethyl-1-propanamine; N,N-dimethyldibenzo[b,e]thiepin-Δ¹¹(6H),γ-propylamine;* 11-(3-dimethylaminopropylidene)-6,11-dihydrodibenzo[b,e]thiepin; dosulepin. $C_{19}H_{21}NS$; mol wt 295.45. C 77.24%, H 7.16%, N 4.74%, S 10.85%. Prepn: M. Protiva *et al., Experientia* **18**, 326 (1962); F. Gadient *et al., Helv. Chim. Acta* **45**, 1860 (1962); K. Stach, H. Spingler, *Monatsh.* **93**, 896 (1962); **Belg.** pat. 618,591; M. Protiva *et al., U.S. pat.* 3,527,766 (1962, 1970 both to SPOFA); M. Rajsner, M. Protiva, *Cesk. Farm.* **11**, 404 (1962), *C.A.* **59**, 2772g (1963). Synthesis of isomers: M. Rajsner *et al., ibid.* **34**, 1963 (1969). Pharmacology: J. Metysova *et al., Arzneimittel-Forsch.* **13**, 1039 (1963); O. Benesova *et al., ibid.* **14**, 100 (1964). Metabolism: O. Horesovsky *et al., Biochem. Pharmacol.* **16**, 2421 (1967).

bp₀.₀₅ 171-172°. mp 55-57°.

Hydrochloride, $C_{19}H_{21}NS.HCl$, *Altapin, Depresym, Prothiaden.* Crystals from ethanol-ether, mp 218-221°. uv max (methanol): 232, 260, 309 nm (log ϵ 4.41, 3.97, 3.53).

THERAP CAT: Antidepressant.

3486. Dowicide 9®. *4-Chloro-2-cyclopentylphenol.* $C_{11}H_{13}ClO$; mol wt 196.68. C 67.18%, H 6.66%, Cl 18.03%, O 8.13%. Prepn: **Ger. pat.** 615,448 (1935 to Hoffmann-La Roche), *C.A.* **29**, 6248[1] (1935); Pajeau, Begue, *Bull. Soc. Chim. France* **1962**, 1923. Bactericidal properties: **Neth. pat. Appl.** 6,513,777 corresp to Lorah, **U.S. pat.** 3,323,988 (1966, 1967 both to Dow).

bp_{18} 181-185°; bp_{11} 160-162°.
USE: Germicide.

3487. Doxacurium Chloride. *[1α,2β(1'S*,2'R*)]-2,2'-[(1,4-Dioxo-1,4-butanediyl)bis(oxy-3,1-propanediyl)]bis[1,2,-3,4-tetrahydro-6,7,8-trimethoxy-2-methyl-1-[(3,4,5-trimethoxyphenyl)methyl]isoquinolinium] dichloride;* (1R,2S;1S,-2R)-1,2,3,4-tetrahydro-2-(3-hydroxypropyl)-6,7,8-trimethoxy-2-methyl-1-(3,4,5-trimethoxybenzyl)isoquinolinium chloride succinate (2:1); *trans,trans*-2,2'-[dimethylenebis(carbonyloxytrimethylene)]bis[1,2,3,4-tetrahydro-6,7,8-trimethoxy-2-methyl-1-(3,4,5-trimethoxybenzyl)isoquinolinium] dichloride; BW-A938U; Nuromax. $C_{56}H_{78}Cl_2N_2O_{16}$; mol wt 1106.14. C 60.81%, H 7.11%, Cl 6.41%, N 2.53%, O 23.14%. Neuromuscular blocking agent. Prepn: H. A. El-Sayad *et al.,* **Eur. pat. Appl.** 54,309 (1982 to Wellcome Found.); *eidem,* **U.S. pat.** 4,701,460 (1987 to Burroughs Wellcome). Clinical pharmacology: S. J. Basta *et al., Anesthesiology* **69**, 478 (1988). Clinical evaluation during anesthesia: R. P. F. Scott, J. Norman, *Brit. J. Anaesth.* **62**, 373 (1989).

Amorphous solid. Sol in water.
THERAP CAT: Muscle relaxant (skeletal).

3488. Doxapram. *1-Ethyl-4-[2-(4-morpholinyl)ethyl]-3,3-diphenyl-2-pyrrolidinone.* $C_{24}H_{30}N_2O_2$; mol wt 378.51. C 76.16%, H 7.99%, N 7.40%, O 8.45%. Prepn: C. D. Lunsford, A. D. Cale, **Belg. pat.** 613,734; *eidem,* **U.S. pat.** 3,192,206 (1962, 1965 both to A. H. Robins); Lunsford *et al., J. Med. Chem.* **7**, 302 (1964). Clinical comparison with mepixanox in chronic bronchopulmonary disease: M. Parziale *et al., G. Ital. Mal Torace* **38**, 323 (1984). Toxicity data: E. I. Goldenthal, *Toxicol. Appl. Pharmacol.* **18**, 185 (1971).

Hydrochloride monohydrate, $C_{24}H_{30}N_2O_2.HCl.H_2O$, *AHR-619, Dopram, Doxapril, Stimulexin.* Crystals from isopropyl ether, mp 217-219°. Bitter taste. Sol in water; sparingly sol in alc, slightly sol in chloroform. LD_{50} orally in rats: 261 mg/kg (Goldenthal).

Benzoate, $C_{31}H_{36}N_2O_4$, crystals, mp 123-124°.

THERAP CAT: Respiratory stimulant.
THERAP CAT (VET): Respiratory stimulant.

3489. Doxazosin. *1-(4-Amino-6,7-dimethoxy-2-quinazolinyl)-4-[(2,3-dihydro-1,4-benzodioxin-2-yl)carbonyl]-piperazine;* 4-amino-2-[4-(1,4-benzodioxan-2-carbonyl)-piperazin-1-yl]-6,7-dimethoxyquinazoline; UK-33274. $C_{23}H_{25}N_5O_5$; mol wt 451.48. C 61.19%, H 5.58%, N 15.51%, O 17.72%. Selective α_1-adrenergic blocker related to prazosin, *q.v.* Prepn: S. F. Campbell, **Ger. pat.** 2,847,623; *idem,* **U.S. pat.** 4,188,390 (1979, 1980 both to Pfizer). Cardiovascular pharmacology in animals: P. B. Timmermans *et al., Arch. Int. Pharmacodyn. Ther.* **245**, 218 (1980). HPLC determn in plasma: M. G. Cowlishaw, J. R. Sharman, *J. Chromatog.* **344**, 403 (1985). Pharmacokinetics in humans: H. L. Elliott *et al., Brit. J. Clin. Pharmacol.* **13**, 699 (1982). Comparison with prazosin in hypertension: P. W. de Leeuw *et al., Eur. J. Clin. Pharmacol.* **23**, 397 (1982). *In vitro* effect on low density lipoprotein-receptor activity: T. P. Leren, *Acta Pharmacol. Toxicol.* **56**, 269 (1985). Symposium on pharmacology and clinical efficacy: *Brit. J. Clin. Pharmacol.* **21**, Suppl. 1, 1S-92S (1986); *Am. J. Cardiol.* **59**, 1G-104G (1987).

Monohydrochloride, $C_{23}H_{25}N_5O_5.HCl$, crystals, mp 289-290°.

Monomethanesulfonate, $C_{23}H_{25}N_5O_5.CH_3SO_3H$, *doxazosin mesylate, UK-33274-27, Alfadil, Cardenalin, Cardular, Cardura, Carduran, Diblocin, Normothen, Supressin.*

THERAP CAT: Antihypertensive.

3490. Doxefazepam. *7-Chloro-5-(2-fluorophenyl)-1,3-dihydro-3-hydroxy-1-(2-hydroxyethyl)-2H-1,4-benzodiazepin-2-one;* N-1-hydroxyethyl-3-hydroxyflurazepam; SAS-643; Doxans. $C_{17}H_{14}ClFN_2O_3$; mol wt 348.76. C 58.55%, H 4.05%, Cl 10.17%, F 5.45%, N 8.03%, O 13.76%. Prepn: G. F. Tamagnone *et al., Arzneimittel-Forsch.* **25**, 720 (1975); *eidem, J. Pharm. Pharmacol.* **26**, 566 (1974); F. De Marchi, G. F. Tamagnone, **Ger. pat.** 2,338,058 (1974 to Schiapparelli), *C.A.* **80**, 121016v (1974). Comparison with flurazepam, *q.v.,* and pharmacology, toxicity studies: M. Babbini *et al., Arzneimittel-Forsch.* **25**, 1294 (1975). Structure-activity studies: *eidem, Pharmacol. Res. Commun.* **7**, 337 (1975). GC determn in blood and tissue: F. Marcucci *et al., J. Chromatog.* **198**, 180 (1980); in plasma, urine: S. Mardente *et al., Therap. Drug Monit.* **3**, 351 (1981). Quantitative EEG, behavioral effects: W. G. Sanita *et al., ibid.* 341. Effect on sleep: G. Rodriguez *et al., Neuropsychobiology* **11**, 133 (1984).

Crystals from methylene chloride/petr ether, mp 138-140°. LD_{50} in rats, mice (mg/kg): 2550, >1500 orally; 586, 774 i.p. (Babbini).

THERAP CAT: Sedative, hypnotic.

3491. Doxenitoin. *5,5-Diphenyl-4-imidazolidinone;* 5,-5-diphenyltetrahydroglyoxalin-4-one; tetrahydro-4-oxo-5,5-

diphenylglyoxaline; SKF-2599; Glior. $C_{15}H_{14}N_2O$; mol wt 238.29. C 75.61%, H 5.92%, N 11.76%, O 6.71%. Prepd by the reduction of diphenylthiohydantoin with sodium: Biltz, Seydel, *Ann.* **391**, 218 (1912); with Raney nickel: Whalley *et al., J. Am. Chem. Soc.* **77**, 745 (1955); Goodman, U.S. pat. **2,744,852** (1956).

Stout plates from methanol, mp 183° (Goodman); crystals from alc, mp 185.5-186.5° (Biltz). Moderately sol in glacial acetic acid; less sol in alc, ethyl acetate, benzene, chloroform. Practically insol in water, ligroin.

Hydrochloride, $C_{15}H_{14}N_2O.HCl$, dec 205-206°.

THERAP CAT: Anticonvulsant.

3492. Doxepin. *3-Dibenz[b,e]oxepin-11(6H)-ylidene-N,-N-dimethyl-1-propanamine; N,N-dimethyldibenz[b,e]oxepin-$\Delta^{11(6H)},\gamma$-propylamine;* 11-(3-dimethylaminopropylidene)-6,11-dihydrodibenz[b,e]oxepin; P-3693A. $C_{19}H_{21}NO$; mol wt 279.38. C 81.68%, H 7.58%, N 5.01%, O 5.73%. Prepn of mixture of *cis-* and *trans-*isomers: K. Stach, F. Bickelhaupt, *Monatsh.* **93**, 896 (1962); F. Bickelhaupt *et al., ibid.* **95**, 485 (1964); Neth. pat. **Appl. 6,407,758;** K. Stach, U.S. pat. **3,438,981** (1965, 1969 both to Boehringer Mann.); and separation and activity of isomers: B. M. Bloom, J. R. Tretter, Belg. pat. **641,498;** *eidem,* U.S. pat. **3,420,851** (1964, 1969 both to Pfizer). Pharmacology: A. Ribbentrop, W. Schaumann, *Arzneimittel-Forsch.* **15**, 863 (1965). Metabolism in animals: D. C. Hobbs, *Biochem. Pharmacol.* **18**, 1941 (1969). Determn in plasma by GC/MS: T. P. Davis *et al., J. Chromatog.* **273**, 436 (1983); by HPLC: T. Emm, L. J. Lesko, *ibid.* **419**, 445 (1987). Clinical study in depression: K. Rickels *et al., Arch. Gen. Psychiat.* **42**, 134 (1985). Comparative clinical trial with cimetidine, *q.v.,* in treatment of ulcer: R. K. Shrivastava *et al., Clin. Ther.* **7**, 181 (1985). Review of pharmacology and therapeutic efficacy: R. M. Pinder *et al., Drugs* **13**, 161 (1977).

Oily liq consisting of a mixture of *cis-* and *trans-*isomers. $bp_{0.03}$ 154-157°, $bp_{0.2}$ 260-270°. LD_{50} in mice, rats (mg/kg): 26, 16 i.v.; 79, 182 i.p.; 135, 147 orally (Ribbentrop, Schaumann).

Hydrochloride, $C_{19}H_{21}NO.HCl$, *Adapin, Aponal, Curatin, Quitaxon, Sinequan.* Crystals, mp 184-186°, 188-189°. (A *cis-trans* mixture of approx 1:5).

Maleate, crystals, mp 161-164°, 168-169°.

*trans-*Form hydrochloride, mp 192-193°.

*cis-*Form hydrochloride, *cidoxepin hydrochloride, P-4599.* Crystals, mp 209-210.5°.

THERAP CAT: Antidepressant.

THERAP CAT (VET): Antipruritic.

3493. Doxifluridine. *5'-Deoxy-5-fluorouridine;* 1-(β-D-5'-deoxyribofuranosyl)-5-fluorouracil; 5'-DFUR; 5'-dFUrd; Ro-21-9738; Flutron; Furtulon. $C_9H_{11}FN_2O_5$; mol wt 246.20. C 43.91%, H 4.50%, F 7.72%, N 11.38%, O 32.49%. Fluorinated pyrimidine nucleoside with cytostatic activity. Prepn: A. F. Cook, U.S. pat. **4,071,680** (1978 to Hoffmann-La Roche); H. Hrebabecky, J. Beranek, *Nucleic Acids Res.* **5**, 1029 (1978); A. F. Cook *et al., J. Med. Chem.* **22**, 1330 (1979). Stereospecific synthesis: J. Kiss *et al., Helv. Chim. Acta* **65**, 1522 (1982). Mechanism of action studies: H.-R. Hartmann, A. Matter, *Cancer Res.* **42**, 2412 (1982); R. D.

Armstrong *et al., Cancer Chemother. Pharmacol.* **11**, 102 (1983). Kinetics and metabolism in humans: J.-P. Sommadossi *et al., Cancer Res.* **43**, 930 (1983). Clinical trials in colorectal carcinoma: R. Abele *et al., J. Clin. Oncol.* **1**, 750 (1983); S. D. Fossa *et al., Cancer Chemother. Pharmacol.* **15**, 161 (1985). Series of articles on animal toxicology: *Yakuri to Chiryo* **13**, Suppl. 2, 221-430 (1985); acute toxicity: M. Shimizu *et al., ibid.* **209**, C.A. **104**, 14673z-14678e (1986). Evaluation of neurotoxicity in humans: M. S. Heier, S. D. Fossa, *Acta Neurol. Scand.* **73**, 449 (1986).

Crystals from ethyl acetate, mp 189-190° (Cook). Also reported as crystals from 2-propanol, mp 186-188° (Hrebabecky, Beranek); needles from methanol + ethyl acetate, mp 192-193° (Kiss). pKa 7.4. $[\alpha]_D^{25}$ +18.4° (c = 0.419 in water). uv max (in methanol): 268-269 nm (ε 8550). LD_{50} (14 day) in mice or rats (mg/kg): > 1000 i.v.: > 2000 s.c.; (in male, female mice, male, female rats (mg/kg): > 5000, > 5000, 3471, 3390 orally (Shimizu).

THERAP CAT: Antineoplastic.

3494. Doxofylline. *7-(1,3-Dioxolan-2-ylmethyl)-3,7-dihydro 1,3 dimethyl-1H-purine-2,6-dione; 7-(1,3-dioxolan-2-ylmethyl)theophylline;* 2-(7'-theophyllinemethyl)-1,3-dioxolane; doxophylline; dioxyfilline; ABC-12/3; Ansimar; Maxivent; Ventax. $C_{11}H_{14}N_4O_4$; mol wt 266.26. C 49.62%, H 5.30%, N 21.04%, O 24.04%. Prepn: U. Avico *et al., Farmaco Ed. Sci.* **17**, 73 (1962). Use as bronchodilator: Belg. pat. **868,556;** J. S. Franzone, T. Tamietto, U.S. pat. **4,187,-308** (1978, 1980 to Istituto Biologico Chemioterapico ABC). Pharmacology: J. S. Franzone *et al., Farmaco Ed. Sci.* **36**, 201 (1981). Pharmacodynamics and toxicity in rats: J. S. Franzone *et al., ibid.* **220**. HPLC detcrmn in pharmaccutical compositions: C. Badini *et al., Farmaco Ed. Prat.* **37**, 320 (1982). Clinical trial in obstructive pneumopathy: C. Bucca *et al., Int. J. Clin. Pharm. Res.* **11**, Suppl 1, 101 (1982).

Crystals, mp 144-145.5°. Sol in water, acetone, ethyl acetate, benzene, chloroform, dioxane, hot methanol or hot ethanol. Practically insol in ethyl ether or petr ether. LD_{50} in mice (mg/kg): 841 orally; 215.6 i.v.; in rats: 1022.4 orally, 445 i.p. (Franzone).

THERAP CAT: Bronchodilator.

3495. Doxorubicin. *(8S-cis)-10-[(3-Amino-2,3,6-trideoxy-α-L-lyxo-hexopyranosyl)oxy]-7,8,9,10-tetrahydro-6,8,11-trihydroxy-8-(hydroxyacetyl)-1-methoxy-5,12-naphthacene-dione;* 14-hydroxydaunomycin; NSC-123127; FI-106. $C_{27}H_{29}NO_{11}$; mol wt 543.53. C 59.67%, H 5.38%, N 2.58%, O 32.38%. Anthracycline antibiotic isolated from *Streptomyces peucetius* var *caesius:* F. Arcamone *et al.,* S. Afr. pat. **68 02378** and U.S. pat. **3,590,028** (1968 and 1971 to Farmitalia); *eidem, Biotechnol. Bioeng.* **11**, 1101 (1969). Synthesis of derivs: F. Arcamone *et al.,* Ger. pat. **1,917,874** (1969 to Farmitalia), C.A. **73**, 45799r (1970). Structural studies: F. Arcamone *et al., Tetrahedron Letters* **1969**, 1007. Synthesis from daunomycin, *q.v.: eidem, Chim. Ind. (Milan)* **51**, 834

(1969); *see also:* E. M. Acton *et al., J. Med. Chem.* **17,** 659 (1974); from 7-deoxydaunomycinone: T. H. Smith *et al.,* U.S. pat. **4,012,448** (1977 to Stanford Res. Inst.). Biochemical comparison with daunomycin: Wang *et al., Proc. Am. Assoc. Cancer Res.* **12,** No. 62, 77 (1971). In acid environment doxorubicin breaks up into a water-insol aglycone, *adriamycinone* ($C_{21}H_{18}O_9$), and a water-sol basic, reducing aminosugar, *daunosamine* ($C_6H_{13}NO_3$), 3-amino-2,3,6-trideoxy-L-lyxohexose: A. Di Marco *et al., Cancer Chemother. Rep.* (part 1) **53,** 33 (1969). Total synthesis of adriamycinone: F. Suzuki *et al., J. Am. Chem. Soc.* **100,** 2272 (1978); regiospecific synthesis: J. S. Swenton, P. W. Reynolds, *ibid.* 6188; of daunosamine: P. M. Wovkulich, M. R. Uskonovic, *Tetrahedron* **41,** 3455 (1985). Pharmacokinetic and chemotherapeutic studies: E. Arena *et al., Arzneimittel-Forsch.* **21,** 1258 (1971). Purification: E. Oppici *et al.,* **Belg.** pat. **898,-506;** *eidem,* **Brit.** pat. **Appl. 2,133,005** (both 1984 to Farmitalia). As modulator of immune response in mice: E. Mihich, M. J. Ehrke, *Transplant. Proc.* **16,** 499 (1984). Doxorubicin's cytotoxicity appears to be due to its ability to intercalate with DNA, interact with plasma membranes and take part in oxidation-reduction reactions: T. R. Tritton, G. Yee, *Science* **217,** 248 (1982); H. Simpkins *et al., Cancer Res.* **44,** 613 (1984); R. S. Youngman, E. F. Elstner, *Arch. Biochem. Biophys.* **231,** 424 (1984). In treatment of cancer of the bladder: M. Pavone-Macaluso *et al., Urology* **23,** 40 (1984); breast: D. C. Tormey *et al., Am. J. Clin. Oncol.* **7,** 231 (1984); prostate: H. Scher *et al., J. Urol.* **131,** 1099 (1984). Toxicology: C. Bertazzoli *et al., Experientia* **26,** 389 (1970); *eidem, Toxicol. Appl. Pharmacol.* **21,** 287 (1972); R. D. Olson *et al., Life Sci.* **29,** 1393 (1981). Review of properties, biosynthesis, fermentation: R. J. White, R. M. Stroshane, *Drugs Pharm. Sci.* **22,** 569-594 (1984); of efficacy: H. L. Davis, T. E. Davis, *Cancer Treat. Rep.* **63,** 809-815 (1979). *Review:* R. H. Blum. S. K. Carter, *Ann. Int. Med.* **80,** 249-259 (1974); G. Aubel-Sadron, D. Londos-Gagliardi, *Biochimie* **66,** 333-352 (1984). Comprehensive description: A. Vigevani, M. J. Williamson, *Anal. Profiles Drug Subs.* **9,** 245-274 (1980). Book: *Doxorubicin,* F. Arcamone, Ed. (Academic Press, New York, 1981) 354 pp.

Adriamycinone

Daunosamine

mp 229-231°.
Hydrochloride, $C_{27}H_{29}NO_{11} \cdot HCl$, *Adriacin, Adriblastina, Adriamycin.* Orange-red colored thin needles, mp 204-205° (dec). $[\alpha]_D^{20}$ +248° (c = 0.1 in methanol). Absorption/uv max (methanol): 233, 252, 288, 479, 496, 529 nm. Sol in water, methanol, aq alcohols. Practically insol in acetone, benzene, chloroform, ethyl ether and petroleum ether. Aq solns are yellow-orange at acid pHs, orange-red at neutral pHs and violet-blue at pH >9. Aq soln unchanged after one month at 5°, but unstable at higher temperatures or at either acid or alkaline pHs. LD_{50} i.v. in mice: 21.1 mg/kg (Bertazzoli, 1970).
Note: Doxorubicin may reasonably be anticipated to be a carcinogen: *Seventh Annual Report on Carcinogens* (PB95-109781, 1994) p 86.
THERAP CAT: Antineoplastic.

3496. Doxycycline. *[4S-(4α,4aα,5α,5aα,6α,12aα)]-4-(Dimethylamino)-1,4,4a,5,5a,6,11,12a-octahydro-3,5,10,12,-12a-pentahydroxy-6-methyl-1,11-dioxo-2-naphthacenecarboxamide monohydrate;* α-6-deoxy-5-hydroxytetracycline monohydrate; α-6-deoxyoxytetracycline monohydrate; 5-hydroxy-α-6-deoxytetracycline monohydrate; GS-3065; Jenacyclin; Supracyclin; Vibramycin. $C_{22}H_{24}N_2O_8 \cdot H_2O$; mol wt 462.46. C 57.14%, H 5.67%, N 6.06%, O 31.14%. Prepn of

family of 6-deoxytetracyclines: C. R. Stephens *et al., J. Am. Chem. Soc.* **80,** 5324 (1958). *See also:* McCormick, Jensen, U.S. pat. **3,019,260** (1962 to Am. Cyanamid). Prepn, separation and configuration of 6α- and 6β-epimers: M. S. von Wittenau *et al., J. Am. Chem. Soc.* **84,** 2645 (1962); C. R. Stephens *et al., ibid.* **85,** 2643 (1963). Prepn of 6α-deoxyoxytetracycline: R. K. Blackwood *et al.,* U.S. pat. **3,200,149** (1965 to Pfizer). ^{1}H-NMR study: M. S. von Wittenau, R. K. Blackwood, *J. Org. Chem.* **31,** 613 (1966). Biological properties: English, *Proc. Soc. Exp. Biol. Med.* **122,** 1107 (1966). Pharmacology: Fabre, *Chemotherapia* **11,** 73 (1966); Gibaldi, *ibid.* **12,** 265 (1967). Toxicity of hyclate: Goldenthal, *Toxicol. Appl. Pharmacol.* **18,** 185 (1971). Clinical trial in prophylaxis of leptospirosis: E. T. Takafuji *et al., N. Engl. J. Med.* **310,** 497 (1984). *Review:* C. Edwards in *Pharmacological and Biochemical Properties of Drug Substances* vol. 2, M. E. Goldberg, Ed. (Am. Pharm. Assoc., Washington, DC, 1979) pp 305-332.

Hydrochloride hemiethanolate hemihydrate, $C_{22}H_{25}Cl\text{-}N_2O_8 \cdot \frac{1}{2}C_2H_6O \cdot \frac{1}{2}H_2O$, *doxycycline hyclate, Azudoxat, Bassado, Clinofug, Diocimex, Doryx, Doxatet, Doxicrisol, Doxychel hyclate, Doxylar, Doxytem, Duradoxal, Granudoxy, Hydramycin, Mespafin, Nordox, Paldomycin, Retens, Ronaxan, Sigadoxin, Spanor, Tetradox, Unacil, Vibramycin Hyclate, Vibra-Tabs, Vibraveineuse, Vibravenös, Zadorin.* Light yellow, crystalline powder from ethanol + HCl. Chars without melting at about 201°. $[\alpha]_D^{25}$ −110° (c = 1 in 0.01N methanolic HCl). uv max (0.01N methanolic HCl): 267, 351 nm (log ε 4.24, 4.12). Sol in water. The alcohol and water of crystallization are lost by drying at 100° under reduced pressure. More active biologically than the corresponding 6β-epimer hydrochloride (Wittenau, 1962). LD_{50} i.p. in rats: 262 mg/kg (Goldenthal).
Sodium metaphosphate (3:1), $3(C_{22}H_{24}N_2O_8) \cdot NaPO_3 \cdot (HPO_3)_3$, *doxycycline fosfatex, Sigacyclat.*
THERAP CAT: Antibacterial.
THERAP CAT (VET): Antibacterial.

3497. Doxylamine. *N,N-Dimethyl-2-[1-phenyl-1-(2-pyridinyl)ethoxy]ethanamine; 2-[α-(2-dimethylaminoethoxy)-α-methylbenzyl]pyridine;* phenyl-2-pyridylmethyl-β-N,N-dimethylaminoethyl ether; 2-dimethylaminoethoxyphenylmethyl-2-picoline. $C_{17}H_{22}N_2O$; mol wt 270.37. C 75.52%, H 8.20%, N 10.36%, O 5.92%. Prepd from phenyl-2-pyridylmethylcarbinol and β-N,N-dimethylaminoethyl chloride in the presence of sodamide in xylene: Sperber *et al., J. Am. Chem. Soc.* **71,** 887 (1949). GC determn: H. C. Thompson *et al., J. Chromatog. Sci.* **20,** 373 (1982). Pharmacology, antihistaminic activity: B. B. Brown, H. Werner, *J. Lab. Clin. Med.* **33,** 325 (1948). Hypnotic efficacy: F. Sjöqvist, L. Lasagna, *Clin. Pharmacol. Ther.* **8,** 48 (1967). Chronic toxicity study of the succinate: C. D. Jackson, B. Blackwell, *J. Am. Coll. Toxicol.* **12,** 1 (1993). *Review:* T. J. Haley, *Dangerous Prop. Ind. Mater. Rep.* **2,** 17 (1982).

Liquid, $bp_{0.5}$ 137-141°. Sol in acids. Slightly volatile, darkens on exposure to light.
Succinate, $C_{17}H_{22}N_2O \cdot C_4H_6O_4$, *Mereprine, Alsadorm, Decapryn succinate, Gittalun, Hoggar N, Sedaplus, Unisom.* Crystals, mp 100-104°, sol in water. One gram dissolves in

1 ml water, 2 ml alcohol, 2 ml chloroform. Slightly sol in benzene and ether. pH (1% aq soln): 4.9 to 5.1. LD_{50} in mice, rabbits (mg/kg): 470, 250 orally; 62, 49 i.v.; in mice, male rats, female rats (mg/kg): 460, 440, 445 s.c. (Brown, Werner).

Note: A combination with pyridoxine hydrochloride, *q.v.*, has been marketed as *Bendectin* for nausea of pregnancy. Prior to 1976, Bendectin also contained dicyclomine, *q.v.* Discussion of Bendectin and the issue of teratogenicity: J. F. Cordero *et al.*, *J. Am. Med. Assoc.* **245**, 2307 (1981); corr. *ibid.* **247**, 2234 (1982); L. B. Holmes, *Teratology* **27**, 277 (1983); L. J. Sheffield, R. Batagol, *Med. J. Aust.* **143**, 143 (1985).

THERAP CAT: Antihistaminic; hypnotic.

THERAP CAT (VET): Antihistaminic.

3498. Dragon's Blood. A resinous secretion found on the fruits of *Daemonorops propinquus* Becc., *D. draco* Blume, and probably other species of *Daemonorops, Palmae* (Rattan palms). *Habit.* Sumatra, Borneo, India. *Constit.* About 55% of a red resin contg about 12-15% of bright-yellow, amorphous dracoresene; 2-3% white amorphous dracoalban. Isoln of the main coloring matter, dracorubin: Brockmann, Haase, *Ber.* **69**, 1950 (1936). Chemical studies of resin pigments: Olaniyi *et al.*, *J. Chem. Soc. Perkin Trans. I* **1973**, 179.

Red sticks, pieces, or cakes; vitreous fracture; makes a bright-crimson powder; odorless and almost tasteless. mp ~120° with sublimation of some benzoic acid. Insol in water; sol in alcohol.

USE: For coloring lacquers and varnishes; occasionally for coloring plasters; in photoengraving on zinc to protect metal parts against etching.

3499. Drazoxolon. *3-Methyl-4-[(2-chlorophenyl)hydrazone]-4,5-isoxazoledione;* 4-(2-chlorophenylhydrazono)-3-methyl-5(4H)-isoxazolone; 3-methyl-4-(o-chlorophenylhydrazono)-5-isoxazolone; PP-781; Ganocide; Mil-Col; Saisan. $C_{10}H_8ClN_3O_2$; mol wt 237.65. C 50.54%, H 3.39%, Cl 14.92%, N 17.68%, O 13.46%. Prepn: Green, Roberts, **Brit. pat. 1,049,103** (1966 to ICI). Structure: Summers, Shields, *Chem. & Ind. (London)* **1964**, 1264. Soil degradation studies: Anderson, Horsgood, *Soil Biol. Biochem.* **3**, 271 (1971). Hydrolysis: Lehtonen *et al.*, *Pestic. Sci.* **3**, 357 (1972). Toxicology: D. G. Clark, T. F. McElligott, *Food Cosmet. Toxicol.* **7**, 481 (1969). *Review:* Yuen, *Anal. Methods Pestic. Plant Growth Regul.* **7**, 665-673 (1973).

Yellow crystals, possessing faint odor, from methanol-benzene, mp 168°. Practically insol in water, acids, and aliphatic hydrocarbons. Sol in alkali, aromatic hydrocarbons (4%), chloroform (about 10%), ethanol (1%), ketones (5%). Stable to dilute acids and forms stable salts with alkalies. LD_{50} orally in female rats, mice: 126, 129 mg/kg (Clark, McElligott).

USE: Fungicide.

3500. Drimenin. *[5aS-(5aα,9aβ,9bα)]-5,5a,6,7,8,9,9a,-9b-Octahydro-6,6,9a-trimethylnaphtho[1,2-c]furan-1(3H)-one.* $C_{15}H_{22}O_2$; mol wt 234.34. C 76.88%, H 9.46%, O 13.65%. From bark of *Drimys winteri* Forst., *Magnoliaceae:* Appel, Dohr, *Scientia* **25**, 137 (1958); *C.A.* **54**, 4663f (1960). Structure and stereochemistry: Appel *et al.*, *J. Chem. Soc.* **1960**, 4685. Synthesis: Wenkert, Strike, *J. Am. Chem. Soc.* **86**, 2044 (1964); Yamagawa, *et al.*, *Synthesis* **1970**, 257; M. Jallali-Naini *et al.*, *Tetrahedron Letters* **22**, 2995 (1981).

Crystals from methanol and sublimed at 110°/0.1 mm, mp 133°. $[\alpha]_D$ −42° (c = 0.76 in benzene); $[\alpha]_D^{25}$ −35.8° (chloroform). Soluble in organic solvents; practically insol in water, acids, bases.

3501. Drofenine. *α-Cyclohexylbenzeneacetic acid 2-(diethylamino)ethyl ester; α-phenylcyclohexaneacetic acid 2-(diethylamino)ethyl ester;* 2-diethylaminoethyl α-phenylcyclohexaneacetate; hexahydroadiphenine. $C_{20}H_{31}NO_2$; mol wt 317.47. C 75.67%, H 9.84%, N 4.41%, O 10.08%. Prepn: Miescher, Hoffman, **U.S. pats. 2,265,184-5** (both 1941 to Ciba). Pharmacology and toxicity: Fleisch *et al.*, *Arzneimittel-Forsch.* **11**, 1119 (1961).

$bp_{0.15}$ 158°.

Hydrochloride, $C_{20}H_{31}NO_2$·HCl, *Trasentine-A, Trasentine 6-H.* Crystals from alc + petr ether, mp 145-147°. Freely sol in water; very sparingly sol in alc, ether. A 5% aq soln is neutral to litmus. LD_{50} i.v. in mice: 65.6 mg/kg (Fleisch).

THERAP CAT: Antispasmodic.

3502. Droloxifene. *(E)-3-[1-[4-[2-(Dimethylamino)ethoxy]phenyl]-2-phenyl-1-butenyl]phenol; (E)-α-[p-[2-(dimethylamino)ethoxy]phenyl]-α'-ethyl-3-stilbenol; (E)-1-[4'-(2-dimethylaminoethoxy)phenyl]-1-(3-hydroxyphenyl)-2-phenyl-but-1-ene;* 3-hydroxytamoxifen. $C_{26}H_{29}NO_2$; mol wt 387.52. C 80.59%, H 7.54%, N 3.61%, O 8.26%. Estrogen-receptor antagonist. Prepn: H. Schickaneder *et al.*, *Eur. pat.* **Appl. 54,168**; eidem, **U.S. pat. 5,047,431** (1982, 1991 both to Klinge); P. C. Ruenitz *et al.*, *J. Med. Chem.* **25**, 1056 (1982). Symposium on pharmacology and clinical studies: *Am. J. Clin. Oncol.* **14**, Suppl. 2, S1-S63 (1991). Clinical trial in breast cancer: P. F. Bruning, *Eur. J. Cancer* **28A**, 1404 (1992).

Colorless crystals from ether, mp 162-163° (Schickaneder); also reported as white crystals from benzene-hexane, mp 160-162° (Ruenitz).

Citrate, $C_{26}H_{29}NO_2$·$C_6H_8O_7$, *K-21060E, FK-435.* Slightly sol in water, methanol, ethanol. Insol in chloroform.

THERAP CAT: Antineoplastic (hormonal).

3503. Drometrizole. *2-(2H-Benzotriazol-2-yl)-4-methylphenol; 2-(2H-benzotriazol-2-yl)-p-cresol;* 2-(2'-hydroxy-5'-methylphenyl)benzotriazole; Tinuvin P. $C_{13}H_{11}N_3O$; mol wt 225.25. C 69.32%, H 4.92%, N 18.65%, O 7.10%. Prepn: Heller *et al.*, **U.S. pats. 3,004,896; 3,189,615** (1961, 1965 to Geigy). Stabilization mechanism in acrylonitrile-butadiene-styrene (ABS) latex: J. B. Adeniyi, *J. Polym. Mater.* **3**, 25 (1986). HPLC determn in plastic: B. Niklasson, B. Björkner, *Contact Dermat.* **21**, 330 (1989). Review of uses: Dunn, Fogg, *J. Appl. Polymer Sci.* **2**, 367 (1959); of toxicology and cosmetic use: CTFA, *J. Am. Coll. Toxicol.* **5**, 455-470 (1986).

Minute crystals, mp 131-133°. bp$_{10}$ 225°. Sol in ethyl acetate, acetone, caprolactam solns, dioctylphthalate, oleyl alcohol, hot petrolatum. Stable to conditions and chemicals used in polymerization or compounding of plastics.

USE: An ultraviolet light absorber for stabilizing plastics and other organic materials against discoloration and deterioration.

THERAP CAT: Ultraviolet screen.

3504. Dromostanolone. *(2α,5α,17β)-17-Hydroxy-2-methyl-androstan-3-one; 17β-hydroxy-2α-methyl-5α-androstan-3-one;* 2α-methylandrostan-17β-ol-3-one; 2α-methyldihydrotestosterone; drostanolone. $C_{20}H_{32}O_2$; mol wt 304.47. C 78.90%, H 10.59%, O 10.51%. Synthetic estrogen antagonist. Prepn: H. J. Ringold *et al., J. Am. Chem. Soc.* **81**, 427 (1959); H. J. Ringold, G. Rosenkranz, U.S. pat. **3,118,915** (1964 to Syntex). GC-MS determn of urinary metabolites: D. DeBoer *et al., J. Steroid Biochem. Molec. Biol.* **42**, 411 (1992).

Crystals from acetone/hexane, mp 149-153°. [α]$_D$ +32° (ethanol).

Propionate, $C_{23}H_{36}O_3$, *NSC-12198, Drolban, Emdisterone, Masterid, Masteril, Masterone, Permastril.* Crystals from hexane, mp 126-130°. [α]$_D$ +24°.

Note: This is a controlled substance (anabolic steroid) in the U.S. Code of Federal Regulations, Title 21 Part 1308.13, as defined in Part 1308.02 (1995).

THERAP CAT: Antineoplastic.

3505. Droperidol. *1-[1-[4-(4-Fluorophenyl)-4-oxobutyl]-1,2,3,6-tetrahydro-4-pyridinyl]-1,3-dihydro-2H-benzimidazol-2-one;* 1-[1-[3-(p-fluorobenzoyl)propyl]-1,2,3,6-tetrahydro-4-pyridyl]-2-benzimidazolinone; 1-[1-[4-(p-fluorophenyl)-4-oxobutyl]-1,2,3,6-tetrahydro-4-pyridyl]-2-benzimidazolinone; dehydrobenzperidol; R-4749; Dridol; Droleptan; Inapsine. $C_{22}H_{22}FN_3O_2$; mol wt 379.43. C 69.64%, H 5.84%, F 5.01%, N 11.07%, O 8.43%. Prepn: P. A. J. Janssen, J. F. Gardocki, U.S. pat. **3,161,645** (1964 to Janssen). Pharmacology and toxicity: J. Yelnosky *et al., Toxicol. Appl. Pharmacol.* **6**, 37 (1964). Comprehensive description: C. A. Janicki, R. K. Gilpin in *Analytical Profiles of Drug Substances* vol. 7, K. Florey, Ed. (Academic Press, New York, 1978) pp 171-192.

Hydrate, crystals, mp 145-146.5°. uv max (9:1, 0.1M HCl:methanol): 245, 280 nm (ε 15600, 7500). Soly at 25° (g/100 ml): chloroform 40; DMF 17; benzene 0.55; methanol 0.41; ethanol 0.34; ether 0.24; 0.1M HCl 0.15; water <0.001. pKa 7.64. Heat and light sensitive. LD$_{50}$ in mice (mg/kg): 125 s.c.; 43 i.v. (Yelnosky).

Note: Ingredient of *Thalamonal, Innovar.*

THERAP CAT: Antipsychotic.

THERAP CAT (VET): Tranquilizer.

3506. Droprenilamine. *N-(2-Cyclohexyl-1-methylethyl)-γ-phenylbenzenepropanamine;* (±)-*N*-(3,3-diphenylpropyl)-α-methylcyclohexaneethylamine; droprenilamine. $C_{24}H_{33}N$; mol wt 335.53. C 85.91%, H 9.91%, N 4.17%. Vasodilator with antiarrhythmic and hypotensive activity. Prepn: M. Carissimi *et al.,* **Ger.** pat. **2,521,113** (1976 to Maggioni), *C.A.* **84**, 164388t (1976). Synthesis and pharmacological study: G. Carenini *et al., Arzneimittel-Forsch.* **26**, 2127 (1976). Hemodynamic and metabolic effects: R. J. Marshall, J. R. Parratt, *Brit. J. Pharmacol.* **59**, 311 (1977). Clinical evaluation: F. Rengo *et al., G. Ital. Cardiol.* **10**, 473 (1980).

Hydrochloride, $C_{24}H_{33}N.HCl$, *MG-8926, Valcor.* Crystals from isopropanol, mp 175-176°. LD$_{50}$ in mice (mg/kg): 2850 orally; 68 i.p. (Carenini).

THERAP CAT: Vasodilator (coronary).

3507. Dropropizine. *3-(4-Phenyl-1-piperazinyl)-1,2-propanediol;* 1-phenyl-4-(2,3-dihydroxypropyl)piperazine; 1-(2,3-dihydroxypropyl)-4-phenylpiperazine; 1-phenyl-4-(2,3-dihydroxypropyl)diethylenediamine; UCB-1967; Ribex. $C_{13}H_{20}N_2O_2$; mol wt 236.31. C 66.07%, H 8.53%, N 11.85%, O 13.54%. Cough suppressive phenylpiperazine derivative. Prepn: H. G. Morren, **Belg.** pat. **601,394;** *eidem,* U.S. pat. **3,163,649** (1961, 1964 to UCB); H. Howell *et al., J. Org. Chem.* **27**, 1709 (1962); J. Bourdais, *Bull. Soc. Chim. France* **1968**, 3246. Prepn of enantiomers: M. Borsa *et al.,* **Eur.** pat. **Appl. 147,847;** *eidem,* U.S. pat. **4,699,911** (1985, 1987 both to Dompé). Pharmacology: K. Cartwright, J. L. Paterson, *J. Pharm. Pharmacol.* **23**, Suppl., 247S (1971). Controlled clinical trials: G. C. Moreo *et al., Gazz. Med. Ital.* **140**, 409 (1981); A. Ravetta, M. Ravetta, *ibid.* **141**, 531 (1982). Chronic oral toxicity: P. R. B. Noel, *Arzneimittel-Forsch.* **19**, 1246 (1969).

Crystals from benzene, mp 105° (Morren); also reported as mp 108° (Bourdais). LD$_{50}$ in rats (mg/kg): 200 i.v., 750 orally (Morren).

S-Form, *levodropropizine, Danka, Levotuss, Rapitux.* White solid from acetone, mp 98-100°. [α]$_D^{25}$ −10° (c = 1.0 in ethanol). LD$_{50}$ in mice, rats (mg/kg): 1287.2, 886.6 orally; 408.0, 401.3 i.p. (Borsa).

R-Form, crystals from acetone, mp 104-105°. [α]$_D^{25}$ +9.7° (c = 1.0 in ethanol). LD$_{50}$ in mice, rats (mg/kg): 871.7, 721.3 orally; 319.2, 363.4 i.p. (Borsa).

THERAP CAT: Antitussive.

3508. Drosera. Common sundew; round-leafed sundew; youthwort. Air-dried, flowering plant, *Drosera rotundifolia* L., frequently mixed with closely allied species *D. anglica* Hudson and *D. longifolia* L., *Droseraceae.* *Habit.* Europe, Asia, North America, south to Florida. *Constit.* Malic and citric acids, resin, tannin.

3509. Drosophilin A. *2,3,5,6-Tetrachloro-4-methoxy-phenol; p-methoxytetrachlorophenol.* $C_7H_4Cl_4O_2$; mol wt 261.92. C 32.10%, H 1.54%, Cl 54.14%, O 12.22%. Antibiotic substance isolated from the basidiomycete *Drosophila subatrata* (Batsch ex Fr.) Quel., grown in a corn-steep medium: Kavanagh, Hervey, *Proc. Nat. Acad. Sci. USA* **38**, 555 (1952). Prepn from tetrachloro-*p*-anisidine: Bures, Hutter, *Casopis Ceskoslov. Lekarnictva* **11**, 29, 57 (1931); *C.A.* **25**, 5153; *Chem. Zentr.* **1931**, II, 225; from 1,4-dimethoxytetrachlorobenzene: Anchel, *J. Am. Chem. Soc.* **74**, 2943 (1952).

Crystals from hexane or dil alcohol, mp 114°. uv max: 301 nm. Sparingly sol in water; sol in hexane, benzene, acetone, alcohol, chloroform, ether.

3510. Drospirenone. *[6R-(6α,7α,8β,9α,10β,13β,14α,-15α,16α,17β)]-1,3′,4′,6,7,8,9,10,11,12,13,14,15,16,20,21-Hexadecahydro-10,13-dimethylspiro[17H-dicyclopropa[6,7;15,16]cyclopenta[a]phenanthrene-17,2′(5′H)-furan]-3,5′(2H)-dione; 6β,7β,15β,16β-dimethylene-3-oxo-4-androstene-[17(β-1′)-spiro-5′]perhydrofuran-2′-one; 6β,7β,15β,16β-dimethylen-3-oxo-17α-pregn-4-ene-21,17-carbolactone; dihydrospirorenone; ZK-30595.* $C_{24}H_{30}O_3$; mol wt 366.50. C 78.65%, H 8.25%, O 13.10%. Synthetic progestogen exhibiting antimineralocorticoid and antiandrogenic activity. Prepn: R. Wiechert et al., **Ger. pat. 2,652,761**; eidem, **U.S. pat. 4,129,-564** (both 1978 to Schering AG); D. Bittler et al., *Angew. Chem.* **94**, 718 (1982). Pharmacological profile: P. Muhn et al., *Contraception* **51**, 99 (1995). HPLC determn in human plasma: W. Krause, U. Jakobs, *J. Chromatog.* **230**, 37 (1982). Clinical evaluation: W. Oelkers et al., *J. Clin. Endocrinol. Metab.* **73**, 837 (1991). Review of synthesis: H. Laurent et al., *J. Steroid. Biochem.* **19**, 771-776 (1983).

mp 201.3°. $[α]_D^{22}$ −182° (c = 0.5 in chloroform). uv (methanol): 265 nm (ε 19000).

THERAP CAT: Progestogen.

3511. Drotebanol. *(6β)-3,4-Dimethoxy-17-methylmorphinan-6,14-diol; 6β,14-dihydroxy-3,4-dimethoxy-N-methylmorphinan; 14-hydroxydihydro-6β-thebainol 4-methyl ether; dihydro-14-hydroxy-4-O-methyl-6β-thebainol; dihydro-14-hydroxy-6β-thebainol 4-methyl ether; oxymethebanol; Metebanyl.* $C_{19}H_{27}NO_4$; mol wt 333.43. C 68.44%, H 8.16%, N 4.20%, O 19.19%. Prepn from thebaine: **Japan. pat. 26,726('64)** (to Sankyo), *C.A.* **62**, 11869b (1965). Metabolism: H. Shindo et al., *Yakugaku Zasshi* **90**, 36 (1970), *C.A.* **72**, 88520q (1970). Pharmacology: S. Kobayashi et al., *Arzneimittel-Forsch.* **20**, 43 (1970). Dependence potential in rhesus monkeys: T. Yanagita et al., *Bull. Narc.* **29**, 33 (1977).

Minute crystals, very slightly hygroscopic, slightly bitter taste. mp 165.5-166.5°. Slightly sol in water. Freely sol in ethanol, chloroform. Sol in acetone, benzene, toluene. Very slightly sol in ether. LD_{50} in mice (mg/kg): 1300 orally, 1150 s.c., 91 i.v. (Kobayashi).

Note: This is a controlled substance (opium derivative) listed in the U.S. Code of Federal Regulations, Title 21 Part 1308.11 (1995).

THERAP CAT: Antitussive.

3512. Droxicam. *5-Methyl-3-(2-pyridinyl)-2H,5H-1,3-oxazino[5,6-c][1,2]benzothiazine-2,4(3H)-dione 6,6-dioxide; E-3128; Dobenam; Droxar; Ombolan.* $C_{16}H_{11}N_3O_5S$; mol wt 357.35. C 53.78%, H 3.10%, N 11.76%, O 22.39%, S 8.97%. Prodrug of piroxicam, *q.v.* Prepn: J. E. Soler, **Fr. pat. 2,528,433**; idem, **U.S. pat. 4,563,452** (1983, 1986 both to Provesan SA). Spectroscopic structural analysis: J. Frigola, *J. Chem. Soc. Perkins Trans. II*, **1988**, 241. Pharmacology: A. J. Farré et al., *Methods Find. Exp. Clin. Pharmacol.* **8**, 407 (1986). Metabolism and pharmacokinetics in animals: A. Esteve et al., ibid. 423. Inhibition of platelet aggregation: J. Esteve et al., ibid. **9**, 209 (1987). Evaluation of ulcerogenic effects in comparison with piroxicam: G. Palacios et al., ibid. 353. Clinical pharmacology: L. Martinez et al., ibid. **10**, 729 (1988). Brief review of pharmacology: J. Esteve et al., *Gen. Pharmacol.* **19**, 49-54 (1988).

Crystals from acetone. mp 259-261°. LD_{50} orally in male, female mice, and male, female rats: 6192, 8841, 1434, 1994 mg/kg (Soler, 1986).

THERAP CAT: Anti-inflammatory.

3513. Droxidopa. *threo-β,3-Dihydroxy-L-tyrosine; L-threo-3-(3,4-dihydroxyphenyl)serine; (−)-(2S,3R)-2-amino-3-hydroxy-3-(3,4-dihydroxyphenyl)propionic acid; threo-dopaserine; L-threo-DOPS; L-DOPS; SM-5688; Dops.* $C_9H_{11}NO_5$; mol wt 213.19. C 50.71%, H 5.20%, N 6.57%, O 37.52%. Synthetic amino acid precursor of norepinephrine, *q.v.* Prepn of racemate: K. W. Rosenmund, H. Dornsaft, *Ber.* **52B**, 1734 (1919). Separation and resolution of diastereomers: B. Hegedüs et al., *Helv. Chim. Acta* **58**, 147 (1975); B. Hegedüs, A. Krasso, **U.S. pat. 3,920,728** (1975 to Hoffmann-La Roche). Improved process for prepn: N. Ohashi et al., **U.S. pat. 4,319,040** (1982 to Sumitomo). Pharmacology of stereoisomers: G. Bartholini et al., *J. Pharmacol. Exp. Ther.* **193**, 523 (1975). Clinical pharmacology of L-threo-form and clinical evaluation in familial amyloid polyneuropathy (FAP): T. Suzuki et al., *Eur. J. Clin. Pharmacol.* **17**, 429 (1980). Reversed-phase chromatography determn in plasma and urine: F. Boomsma et al., *J. Chromatog.* **427**, 219 (1988). Pharmacokinetics in FAP: T. Suzuki et al., *Eur. J. Clin. Pharmacol.* **23**, 463 (1982); in parkinsonism: T. Suzuki et al., *Neurology* **34**, 1464 (1984). Metabolism to norepinephrine: T. Suzuki et al., *Life Sci.* **36**, 435 (1985). Clinical studies in Parkinson's disease: N. Ogawa et al., *J. Med.* **16**, 525 (1985); H. Narabayashi et al., *Advan. Neurol.* **45**, 593 (1986).

Crystals from ethanol and ether, mp 232-235° (dec). $[\alpha]_D^{20}$ −39° (c = 1 in 1N aq HCl). Also cited as crystals from water and L-ascorbic acid, mp 229-232° (dec) (Ohashi). $[\alpha]_D^{20}$ −42.0° (c = 1 in 1N aq HCl).

THERAP CAT: Antiparkinsonian.

3514. DSIP. *Delta sleep-inducing peptide (rabbit);* delta sleep peptide; delta sleep factor. $C_{35}H_{48}N_{10}O_{15}$; mol wt 848.82. C 49.53%, H 5.70%, N 16.50%, O 28.27%. A nonapeptide that shows enhancement and induction of delta (slow-wave) and spindle EEG patterns. Its occurrence was suspected during dialysis of cerebral venous blood of rabbits during sleep induced by electrical stimulation of the thalamus: M. Monnier, L. Hösli, *Science* **146**, 796 (1964). Initial isoln: *eidem, Pflügers Arch.* **282**, 60 (1965). Isoln, characterization: G. A. Schoenenberger *et al., Experientia* **28**, 919 (1972). Amino acid sequence, synthesis of DSIP and analogs: G. A. Schoenenberger, M. Monnier, *Proc. Nat. Acad. Sci. USA* **74**, 1282 (1977). Solid phase synthesis: Y. P. Shvachkin *et al., Zh. Obshch. Khim.* **51**, 719 (1981), *C.A.* **95**, 43644s (1981). Rapid liquid phase synthesis: S. Nozaki, I. Muramatsu, *Bull. Chem. Soc. Japan* **55**, 2165 (1982). HPLC separation: M. Dizaroglu *et al., J. Chromatog.* **237**, 417 (1982). Effect on human sleep: D. Schneider-Helmert *et al., Lancet* **1**, 1256 (1981); *eidem, Int. J. Clin. Pharmacol. Ther. Toxicol.* **19**, 341 (1981); D. Schneider-Helmert, G. A. Schoenenberger, *Experientia* **37**, 913 (1981).

Trp–Ala–Gly–Gly–Asp–Ala–Ser–Gly–Glu

3515. DTBP. *Bis(1,1-Dimethylethyl) peroxide;* di-*tert*-butyl peroxide. $C_8H_{18}O_2$; mol wt 146.23. C 65.71%, H 12.41%, O 21.88%. $(CH_3)_3COOC(CH_3)_3$.

Flammable liq; d_4^{20} 0.7940; mp −40°; bp_{284} 80°; n_D^{20} 1.3890. Flash pt (Tag open cup) 65°F (18°C). Soluble in organic solvents, in most resin monomers and in partial polymers. Soly in water about 0.01%.

USE: As polymerization catalyst.

3516. Dulcamara. Bittersweet; woody nightshade; scarlet berry. Dried stems of *Solanum dulcamara* L., *Solanaceae. Habit.* Europe, Western Asia, Northern Africa, natural in U.S. *Constit.* Solaniceine (about 1%), dulcamarin, dulcamaric and dulcamaretic acids.

3517. Dulcin. *(4-Ethoxyphenyl)urea;* p-phenetolcarbamide; p-phenetylurea; Sucrol; Valzin. $C_9H_{12}N_2O_2$; mol wt 180.21. C 59.99%, H 6.71%, N 15.55%, O 17.76%. Made by treating p-phenetidine with phosgene and then with ammonia: Berlinerblau, *J. Prakt. Chem.* **30**, 103 (1883); from p-phenetidine and urea: Kurzer, *Org. Syn.* **coll. vol. IV,** 52 (1963).

Lustrous needles; very sweet taste—about 250 times as sweet as cane sugar. mp 173-174°. Sol in 800 parts cold water, 50 parts boiling water, 25 parts alcohol.

USE: Non-nutritive sweetener.

3518. Duloxetine. *(S)-N-Methyl-γ-(1-naphthalenyloxy)-2-thiophenepropanamine;* (+)-(S)-N-methyl-γ-(1-naphthyloxy)-2-thiophenepropylamine; (+)-N-methyl-3-(1-naphthalenyloxy)-3-(2-thienyl)propanamine; LY-248686. $C_{18}H_{19}$-NOS; mol wt 297.42. C 72.69%, H 6.44%, N 4.71%, O 5.38%, S 10.78%. Dual serotonin and norepinephrine uptake inhibitor. Prepn: D. W. Robertson *et al., Eur.* pat. **Appl. 273,658;** *eidem,* U.S. pat. **5,023,269** (1988, 1991 both to Lilly); and abs config: J. Deeter *et al., Tetrahedron Let-*

ters **31,** 7101 (1990). Improved process: R. A. Berglund, U.S. pat. **5,362,886** (1994 to Lilly). Pharmacology: D. T. Wong *et al., Neuropsychopharmacology* **8,** 23 (1993). Neurochemical effects *in vivo:* R. W. Fuller *et al., J. Pharmacol. Exp. Ther.* **269,** 132 (1994). Determn of chiral purity: E. C. Rickard, R. J. Bopp, *J. Chromatog. A* **680,** 609 (1994).

Hydrochloride, $C_{18}H_{19}NOS.HCl$. White solid. pKa in DMF-water (66:34): 9.6.

THERAP CAT: Antidepressant.

3519. Durapatite. *Hydroxylapatite;* calcium phosphate hydroxide; calcium orthophosphate basic; hydroxyapatite; Alveograf; Ossopan; Periograf. $3Ca_3(PO_4)_2.Ca(OH)_2$ or $Ca_{10}(PO_4)_6(OH)_2$. Also considered as pentacalcium monohydroxyorthophosphate $Ca_5(OH)(PO_4)_3$. Calcd as $Ca_{10}H_2$-$O_{26}P_6$; Ca 39.89%, H 0.20%, O 41.41%, P 18.50%. Occurs as a mineral in phosphate rock. Constitutes the mineral portion of bone. Prepn from $Ca(NO_3)_2$ and KH_2PO_4: Warington, *J. Chem. Soc.* **26,** 983 (1873); Rathje, *Ber.* **74,** 342 (1941); Hayek in *Handbook of Preparative Inorganic Chemistry,* G. Brauer, Ed. (Academic Press, 2nd ed., 1963) p 545; from calcium phosphate, dibasic: Perloff, Posner, *Inorg. Syn.* **6,** 16 (1960); from $Ca(NO_3)_2.4H_2O$ and $(NH_4)_3PO_4$ plus NH_4OH: Hayek, Newesely, *ibid.* **7,** 63 (1963). Formation and structure of synthetic bone hydroxyapatites: A. S. Posner *et al., Prog. Cryst. Growth Charact.* **3,** 3 (1980).

Hexagonal needles arranged in rosettes. Dec above 1100°. Practically insol in water, even when freshly prepd. Crystallographic data: a_0 9.425; C_0 6.935; C_0/a_0 0.736.

USE: Prosthetic aid (artificial bone and teeth).

THERAP CAT: Calcium supplement; phosphorus supplement.

3520. Durene. *1,2,4,5-Tetramethylbenzene;* Durol. C_{10}-H_{14}; mol wt 134.22. C 89.49%, H 10.51%. Occurs in coal tar. Usually prepd from xylene and methyl chloride in the presence of $AlCl_3$: Smith, *Org. Syn.* **vol. 10,** 32 (1930); *cf.* Smith, Dobrovolny, *J. Am. Chem. Soc.* **48,** 1413 (1926).

Scales with camphor-like odor from alcohol. d_4^{81} 0.84. mp 80°. bp 191-193°. Sublimes and is volatile with steam. Insol in water; freely sol in alcohol, ether, benzene.

3521. Durohydroquinone. *2,3,5,6-Tetramethyl-1,4-benzenediol;* tetramethyl-p-hydroquinone; dihydroxydurene. $C_{10}H_{14}O_2$; mol wt 166.22. C 72.26%, H 8.49%, O 19.25%. For prepn *see* refs under Duroquinone.

Needles from alcohol. mp 233°. Begins to sinter at 220°. Sparingly sol in ether. Treatment with ferric chloride yields duroquinone.

Diacetyldurohydroquinone, needles from alc, mp 207°.

3522. Duroquinone. *2,3,5,6-Tetramethyl-2,5-cyclohexadiene-1,4-dione;* tetramethyl-p-benzoquinone. $C_{10}H_{12}O_2$; mol wt 164.20. C 73.15%, H 7.37%, O 19.49%. Prepn by reduction of dinitrodurene: Smith, *Org. Syn.* **vol. 10,** 40 (1930); Smith, Dobrovolny, *J. Am. Chem. Soc.* **48,** 1420 (1926); by condensation of 2,3-diketopentane with itself in presence of alkalies: von Pechmann, *Ber.* **21,** 1420 (1888); by the action of alkalies on 3,3-dichloropentan-2-one: Faworsky, *J. Prakt. Chem.* [2] **51,** 538 (1895).

Yellow needles from alc, mp 111-112°. Sublimes. Volatile with steam. Insol in water; sol in alcohol, benzene, ether, hot petr ether.

3523. Dyclonine. *1-(4-Butoxyphenyl)-3-(1-piperidinyl)-1-propanone; 3-piperidino-4'-butoxypropiophenone;* β-piperidinoethyl-4-butoxyphenyl ketone; 4-butoxy-β-piperidinopropiophenone; 4-*n*-butoxy-β-(1-piperidyl)propiophenone; 4-butoxyphenyl piperidineethyl ketone; 2-(1-piperidyl)ethyl *p*-butoxyphenyl ketone. $C_{18}H_{27}NO_2$; mol wt 289.42. C 74.70%, H 9.40%, N 4.84%, O 11.06%. Prepd from *p*-butoxyacetophenone by condensation with formaldehyde and piperidine hydrochloride: Pofft, *Chem. Tech. (Berlin)* **4,** 241 (1952), *C.A.* **47,** 10531 (1953). *Compare* Falicain.

Hydrochloride, $C_{18}H_{27}NO_2 \cdot HCl$, **Dyclone, Tanaclone.** Crystals, mp 175-176°. Sol in water, alc, acetone. Phenol coefficient 3.6.

THERAP CAT: Anesthetic (local).

3524. Dydrogesterone. *(9β,10α)-Pregna-4,6-diene-3,20-dione; 10α-pregna-4,6-diene-3,20-dione;* 6-dehydro-*retro*-progesterone; 10α-isopregnenone; Dufaston; Duphaston; Gestatron; Gynorest; Prodel; Retrone. $C_{21}H_{28}O_2$; mol wt 312.45. C 80.73%, H 9.03%, O 10.24%. Prepn: Westerhof, Reerink, *Rec. Trav. Chim.* **79,** 771 (1960); Rappoldt, Westerhof, *ibid.* **80,** 43 (1961).

Crystals from acetone + hexane, mp 169-170°. $[\alpha]_D^{25}$ −484.5° (chloroform). uv max: 286.5 nm (ε 26400).

THERAP CAT: Progestogen.

3525. Dymanthine. *N,N-Dimethyl-1-octadecanamine; N,N-dimethyloctadecylamine;* dimantine. $C_{20}H_{43}N$; mol wt 297.57. C 80.73%, H 14.57%, N 4.71%. $CH_3(CH_2)_{17}N-(CH_3)_2$. Prepd from octadecylamine and formaldehyde: Reck *et al., J. Org. Chem.* **12,** 517 (1947).

mp 22.89°.

Hydrochloride, $C_{20}H_{43}N \cdot HCl$, **GS-1339, NSC-5547, Thelmesan.**

THERAP CAT: Anthelmintic (Nematodes).

THERAP CAT (VET): Anthelmintic (Nematodes).

3526. Dymixal®. Aldrich Dye Mixture. A mixture of three dyes contg 46.1% or 1.5 parts of crystal violet (hexamethyl *p*-rosaniline HCl), 30.8% or 1.0 part of brilliant green (diaminotetraethyltriphenyl carbinol anhydride sulfate), 23.1% or 0.75 part of acriflavine neutral. It may be

prepd by mechanical mixing of the three dyes in their solid state.

Sol in water, forming a neutral greenish-blue soln which shows a yellow fluorescence under ultraviolet light.

THERAP CAT: Topical for burns.

THERAP CAT (VET): Topically for burns (in soln or jelly).

3527. Dynel. Staple fiber copolymer of 60% vinyl chloride and 40% acrylonitrile wet spun from acetone, stretched hot as much as 1300% and then annealed by heat treatment: E. E. Stout, *Introduction to Textiles* (John Wiley, New York, 1960) pp 198-201; R. W. Moncrieff, *Man-made Fibres* (John Wiley, New York, 1963) pp 411-420; Kennedy, "Modacrylic Fibers" in *Encyclopedia of Polymer Science and Technology* **vol. 8,** N. M. Bikales, Ed. (Interscience, New York, 1968) pp 812-839.

Light cream fiber which can be bleached nearly white. Specific gravity 1.31. Tenacity ranges from 2.0 to 3.5 g/denier. Elongation is 30-40%; hygroscopicity is 0.4% under standard conditions. Has extremely good chemical resistance. Acetone is the best solvent; cyclohexanone and dimethylformamide also have some solvent action. Acetic anhydride, acetaldehyde, aniline, ethylene dichloride, and methyl ethyl ketone all plasticize or swell dynel. Resistant to clothes moths' larvae, to carpet beetles, and to mildew and fungus. Will burn in a flame, but if the flame is removed it is self-extinguishing. Resistant to water and non-felting and non-shrinking below the boil. Hot water delusters dynel. Must be ironed with the lowest iron setting and a dry cotton cover over the fabric. May be heat-set in permanent pleats. Can be dyed readily and may be solution-dyed. Resistant to perspiration and to salt-water deterioration.

USE: In apparel and household furnishings; simulated fur coats; chemically resistant clothing. In making wigs and doll hair; the hair can be washed, combed, set, and in some instances, redyed.

3528. Dynorphin. Extremely potent, widely distributed neuropeptide that has 17 amino acid residues and contains leu⁵-enkephalin as its NH_2-terminal sequence. Its name is derived from "dynamis", the Greek word for power, and endorphin, the name applied to the group of opioid peptides to which it belongs. Initially isolated from porcine pituitaries and termed *slow-reversing endorphin:* L. I. Lowney *et al., Life Sci.* **24,** 2377 (1979). Purification, description of properties and amino acid sequence of the first 13 residues: A. Goldstein *et al., Proc. Nat. Acad. Sci. USA* **76,** 6666 (1979). Complete amino acid sequence of the heptadecapeptide from porcine pituitary: *eidem, ibid.* **78,** 7219 (1981). Isoln from porcine duodenum and identity with pituitary dynorphin: S. Tachibana *et al., Nature* **295,** 339 (1982). Synthesis of porcine dynorphin₁-₁₃: M. Wakimasu *et al., Chem. Pharm. Bull.* **29,** 2592 (1981). Soln conformation: R. Maroun, W. L. Mattice, *Biochem. Biophys. Res. Commun.* **103,** 442 (1981). Radioimmunoassay: V. E. Ghazarossian *et al., Life Sci.* **27,** 75 (1980). Comparison of distribution of dynorphin and enkephalin systems in brain: S. J. Watson *et al., Science* **218,** 1134 (1982). Behavioral effects of dynorphin₁-₁₃ in mice and rats: J. M. Walker *et al., Peptides* **1,** 341 (1980); H. Zwiers *et al., Life Sci.* **28,** 2545 (1981). Opiate activity and receptor selectivity: M. Wuester *et al., Neurosci. Letters* **20,** 79 (1980); C. Chavkin, A. Goldstein *Nature* **291,** 591 (1981). In the guinea pig ileum bioassay it has been shown to be 700 times more potent than leu⁵-enkephalin and its agonist effects are 1/13th as sensitive to naloxone antagonism: *eidem, Proc. Nat. Acad. Sci. USA* **78,** 6543 (1981). Dynorphin₁-₁₃ has been proposed as the specific endogenous ligand of the *kappa* opioid receptor (*cf.* endorphins): C. Chavkin *et al., Science* **215,** 413 (1982). It has also been suggested that dynorphin₁-₈ or dynorphin₁-₉ may be transmitters or modulators at the kappa binding site and dynorphins 1-13 and 1-17 may act at a distance from the release site: A. D. Corbett *et al., Nature* **299,** 79 (1982). A possible regulatory role of dynorphin on morphine and β-endorphin-induced analgesia has been proposed: F. Tulunay *et al., J. Pharmacol. Exp. Ther.* **219,** 296 (1981); E. C. Petrie *et al., Peptides* **3,** 41 (1982). Several non-opiate effects have also been described: J. M. Walker *et* al., Science **218**, 1136 (1982); R. Przewlocki et al., ibid. **219**, 71 (1983).

Tyr–Gly–Gly–Phe–Leu–Arg–Arg–Ile–Arg–Pro–Lys–Leu–Lys–Trp–Asp–Asn–Gln

Porcine Dynorphin

Dynorphin$_{1-13}$, C$_{75}$H$_{126}$N$_{24}$O$_{15}$, 1-13-dynorphin (pig). [α]$_D^{23}$ −62.9° (c = 0.5 in 1% acetic acid).

3529. Dyphylline. 7-(2,3-Dihydroxypropyl)-3,7-dihydro-1,3-dimethyl-1H-purine-2,6-dione; 7-(2,3-dihydroxypropyl)theophylline; (1,2-dihydroxy-3-propyl)theophylline; glyphylline; glyfyllin; diprophylline; AFI-phyllin; Astmamasit; Asthmolysin; Astrophyllin; Circair; Coronarin; Cor-Theophyllin; Dilor; Hiphyllin; Hyphylline; Lufyllin; Neostenovasan; Neothylline; Neotilina; Neo-Vasophylline; Neutrafil; Neutraphylline; Prophyllen; Silbephylline; Solufilin; Solufyllin; Theal ampules; Thefylan. C$_{10}$H$_{14}$N$_4$O$_4$; mol wt 254.25. C 47.24%, H 5.55%, N 22.04%, O 25.17%. Prepn: Jones, Maney, U.S. pat. **2,575,344** (1951 to State Univ. of Iowa); Roth, Arch. Pharm. **292**, 234 (1959). Toxicity data: R. A. Al'tshuler, Med. Prom. SSSR **16**, 57 (1962), C.A. **61**, 6230g (1964).

Extremely bitter crystals from alcohol, mp 158°. uv max (0.001% in H$_2$O): 273 nm (A$_{1cm}^1$ 361). Freely sol in water; one gram dissolves in 3 ml H$_2$O at 25°. Soly (g/100 ml): alc 2; chloroform 1. pH (1% aq soln) 6.6 to 7.3. LD$_{50}$ in mice (mg/kg): 3400 orally; 1430 s.c. (Al'tshuler).

THERAP CAT: Bronchodilator.

3530. Dypnone. 1,3-Diphenyl-2-buten-1-one; β-methylchalcone. C$_{16}$H$_{14}$O; mol wt 222.29. C 86.45%, H 6.35%, O 7.20%. Prepd by self-condensation of acetophenone in the presence of aluminum tert-butoxide: Adkins, Cox, J. Am. Chem. Soc. **60**, 1151 (1938); W. Wayne, H. Adkins, Org. Syn. coll. vol. III, 367 (1955). Use as sunscreen: Ind. Eng. Chem. **46**, 15A (July 1954). Toxicity study: H. F. Smyth et al., J. Ind. Hyg. Toxicol. **31**, 60 (1949).

Liquid. d$_4^{15}$ 1.1080. n$_D^{20}$ 1.6343. bp$_{760}$ 340-345° (partial decompn); bp$_{22}$ 225°; bp$_{1.0}$ 150-155°. Insol in water. Sol in alcohol, ether. LD$_{50}$ orally in rats: 3.6 g/kg (Smyth).

USE: Sunscreen.

3531. Dysprosium. Dy; at. wt 162.50; at. no. 66; valences 2, 3, 4. A lanthanide; belongs to the yttrium group of rare earth metals. Naturally occurring isotopes (mass numbers): 156 (0.06%); 158 (0.10%); 160 (2.34%); 161 (18.9%); 162 (25.5%); 163 (24.9%); 164 (28.2%); known artificial radioactive isotopes: 145-155; 157; 159; 165-167. Abundance in earth's crust: ∼4.5 ppm. Occurs in gadoli-

nite, xenotime, samarskite and other rare earth minerals. Discovered in 1886 by Lecoq de Boisbaudran. Sepn: Urbain, Compt. Rend. **139**, 736 (1904); **141**, 521 (1905); **142**, 785 (1906); Spedding et al., J. Am. Chem. Soc. **69**, 2812 (1947); **76**, 2557 (1954). Prepn of salt: Engle, Balke, ibid. **39**, 53 (1917); prepn of a Dy-Al alloy: Schumacher, Harris, ibid. **48**, 3108 (1926). Toxicity studies: Bruce et al., Toxicol. Appl. Pharmacol. **5**, 750 (1962); Haley et al., ibid. **8**, 37 (1966). Reviews of prepn, properties and compds: The Rare Earths, F. H. Spedding, A. H. Daane, Eds. (Krieger, Huntington, N.Y., 1971, reprint of 1961 ed.) 641 pp; Hulet, Bode, "Separation Chemistry of the Lanthanides and Transplutonium Actinides" in MTP Int. Rev. Sci.: Inorg. Chem., Ser. One, vol. 7, K. W. Bagnall, Ed. (Univ. Park Press, Baltimore, 1972) pp 1-45; Moeller, "The Lanthanides" in Comprehensive Inorganic Chemistry, vol. 4, J. C. Bailar, Jr. et al., Eds. (Pergamon Press, Oxford, 1973) pp 1-101; F. H. Spedding in Kirk Othmer Encyclopedia of Chemical Technology vol. 19 (John Wiley & Sons, New York, 3rd ed., 1982) pp 833-854; Chemistry of the Elements, N. N. Greenwood, A. Earnshaw, Eds. (Pergamon Press, New York, 1984) pp 1423-1449. Brief review of properties: G. T. Seaborg, Radiochim. Acta **61**, 115-122 (1993).

Silver metal; tarnishes in moist air. Hexagonal close-packed crystals. d 8.5500. mp 1412°. bp 2567°. Heat of fusion: 10.782 kJ/mol. Heat of sublimation (25°): 290.4 kJ/mol. Forms greenish-yellow salts.

Oxide, Dy$_2$O$_3$, dysprosia. White substance. Prepd by heating the oxalate or sulfate. d^{27} 7.81.

Hydroxide, Dy(OH)$_3$, a gelatinous precipitate. Prepd by adding ammonia to an aq soln of a dysprosium salt; forms a blue colloidal soln.

Chloride, DyCl$_3$, obtained by passing S$_2$Cl$_2$ over heated dysprosia. A hexahydrate is formed from the aq soln. The anhyd chloride, yellow shining crystals, d 3.67; mp 680°. LD$_{50}$ in mice: 585 mg/kg i.p.; 7.65 g/kg orally (Haley).

Sulfate, Dy$_2$(SO$_4$)$_3$, octahydrate, yellow crystals. Prepd by dissolving the oxide in sulfuric acid and precipitating with alcohol; stable in air at 110°, dehydrated at 360°.

Nitrate, Dy(NO$_3$)$_3$, pentahydrate, melts at 88.6° in its water of crystn. Sol in water. LD$_{50}$ (hexahydrate) in rats: 295 mg/kg i.p.; 3.1 g/kg orally (Bruce).

USE: Oxide used in control rods of some nuclear power reactors.

3532. Dystrophin. Protein product of the human Duchenne muscular dystrophy (DMD) gene. Mol wt approx 400,000 daltons. Present in very small amounts in normal muscle (approx 0.002% of total muscle protein) but either absent or abnormal in muscular dystrophy patients. Thought to be an intracellular structural component of the plasma membrane system in normal muscle fibers. Complete cloning of DMD cDNA: M. Koenig et al., Cell **50**, 509 (1987). Isoln of dystrophin from mouse and human muscle tissue: E. P. Hoffman et al., ibid. **51**, 919 (1987). Subcellular localization studies: E. P. Hoffman et al., Nature **330**, 754 (1987); C. M. Knudson et al., J. Biol. Chem. **263**, 8480 (1988); E. E. Zubrzycka-Gaarn et al., Nature **333**, 466 (1988); S. C. Watkins et al., ibid. 863. Amino acid sequence and structural similarity to α-actinin and to the cytoskeletal protein, spectrin: M. Koenig et al., Cell **53**, 219 (1988). Characterization of DMD gene expression in normal and diseased human muscle: M. Oronzi Scott et al., Science **239**, 1418 (1988); in animal muscle and brain: U. Nudel et al., Nature **331**, 635 (1988). Differentiation of muscle and brain DMD mRNA: U. Nudel et al., ibid. **337**, 76 (1989). Correlation of clinical phenotype with dystrophin abnormalities: E. P. Hoffman et al., N. Engl. J. Med. **318**, 1363 (1988). Use of dystrophin cDNA for prenatal diagnosis and carrier detection in muscular dystrophy: B. T. Darras et al., Am. J. Med. Genetics **29**, 713 (1988).

E

3533. E-2020. *2,3-Dihydro-5,6-dimethoxy-2-[[1-(phenylmethyl)-4-piperidinyl]methyl]-1H-inden-1-one hydrochloride;* 5,6-dimethoxy-2-[[1-(phenylmethyl)-4-piperidinyl]methyl]-2,3-dihydro-1H-inden-1-one hydrochloride; 1-benzyl-4-[(5,6-dimethoxy-1-indanon-2-yl)methyl]piperidine hydrochloride. $C_{24}H_{29}NO_3 \cdot HCl$; mol wt 415.96. C 69.30%, H 7.27%, N 3.37%, O 11.54%, Cl 8.52%. Acetylcholinesterase inhibitor. Prepn: H. Sugimoto *et al.*, **Eur.** pat. **Appl.** 296,-560, *eidem*, **U.S.** pat. **4,895,841** (1988, 1990 both to Eisai). Synthesis: Y. Iimura *et al.*, *J. Label. Compd. Radiopharm.* **27**, 835 (1989). Resolution of enantiomers: J. Haginaka, C. Seyama, *J. Chromatog. B.* **577**, 95 (1992). Effect on learning and memory tasks in animal models: G. R. Dawson, S. D. Iversen, *Behaviour. Brain Res.* **57**, 143 (1993). Clinical pharmacokinetics: A. Ohnishi *et al.*, *J. Clin. Pharmacol.* **33**, 1086 (1993). Effect on acetylcholinesterase activity: A. Galli *et al.*, *Eur. J. Pharmacol.* **270**, 189 (1994).

Crystals from methanol/IPE, mp 211-212° (dec).
THERAP CAT: Cognition enhancer.

3534. Ebastine. *1-[4-(1,1-Dimethylethyl)phenyl]-4-[4-(diphenylmethoxy)-1-piperidinyl]-1-butanone;* 4'-*tert*-butyl-4-[4-(diphenylmethoxy)piperidino]butyrophenone; 4-diphenylmethoxy-1-[3-(4-*tert*-butylbenzoyl)propyl]piperidine; LAS-W-090; Bastel; Ebastel. $C_{32}H_{39}NO_2$; mol wt 469.67. C 81.83%, H 8.37%, N 2.98%, O 6.81%. Nonsedating type histamine H_1-receptor antagonist. Prepn: J. M. P. Soto *et al.*, **Eur.** pat. **Appl. 134,124**; *eidem*, **U.S.** pat. **4,550,116** (both 1985 to Fordonal). Metabolized *in vivo* to carebastine, its active carboxylic acid metabolite. Clinical pharmacology, pharmacokinetics: J. Vincent *et al.*, *Brit. J. Clin. Pharmacol.* **26**, 497 (1988). Effect on psychomotor performance: J. Vincent *et al.*, *ibid.* 503.

Fumarate, $C_{32}H_{39}NO_2 \cdot C_4H_4O_4$, crystals from ethanol, mp 197-198°.
THERAP CAT: Antihistaminic.

3535. Ebimar®. Polysaccharide C 16. Degraded carrageenan, *q.v.* Sulfated polysaccharide, extracted from *Chondrus crispus* (L.) Stackhouse, *Gigartinaceae*. Prepn: Anderson, *J. Pharm. Pharmacol.* **13**, 139 (1961); Anderson, Hargreaves, **Brit.** pat. **840,623** (1960 to Evans Medical).
Off-white powder containing 28-30% unchanged bound sulfate. Typical data: S 9.4%, Ca 0.2%, Na 7%, K 1.3%. $[\alpha]_D$ +34°. Freely sol in water and acid. Aq soln less viscous than the aq soln of the parent carrageenan.
THERAP CAT: Antacid.

3536. Ebrotidine. *N-[[[2-[[[2-[(Aminoiminomethyl)amino]-4-thiazolyl]methyl]thio]ethyl]amino]methylene]-4-bromobenzenesulfonamide;* N-p-bromobenzenesulfonyl-N'-[2-[[[2-[(aminoiminomethyl)amino]-4-thiazolyl]methyl]thio]ethyl]formamidine; p-bromo-N-[[[2-[[[2-[(diaminomethylene)amino]-4-thiazolyl]methyl]thio]ethyl]amino]methylene]benzenesulfonamide; FI-3542; Ebrodin; Ulsanic. $C_{14}H_{17}BrN_6O_2S_3$; mol wt 477.43. C 35.22%, H 3.59%, Br 16.74%, N 17.60%, O 6.70%, S 20.15%. Histamine H_2-

receptor antagonist. Prepn: R. Foguet *et al.*, **Eur.** pat. **Appl. 159,012**; *eidem*, **U.S.** pat. **4,728,655** (1985, 1988 both to Ferrer); L. Anglada *et al.*, *Eur. J. Med. Chem.* **23**, 97 (1988). Clinical pharmacology: S. J. Konturek *et al.*, *Scand. J. Gastroenterol.* **27**, 438 (1992); M. Maczka *et al.*, *J. Physiol. Pharmacol.* **43**, 139 (1992). Clinical evaluation in gastroesophageal reflux disease: E. Sito *et al.*, *ibid.* **44**, 259 (1993). HPLC determn in urine and metabolism study: E. Rozman *et al.*, *J. Pharm. Sci.* **83**, 252 (1994).

Crystals from ethyl acetate. mp 142.5-146°.
THERAP CAT: Antiulcerative.

3537. Eburnamonine. *Eburnamenin-14(15H)-one;* (+)-cis-eburnamonine. $C_{19}H_{22}N_2O$; mol wt 294.40. C 77.52%, H 7.53%, N 9.52%, O 5.43%. One of the *Vinca* alkaloids, naturally occurring as the (+)- and (±)-forms. Isoln of the (+)-form from *Hunteria eburnea* Pinchon, *Apocynaceae*: M. F. Bartlett *et al.*, *Compt. Rend.* **249**, 1259 (1959); of the (±)-form from *Vinca minor* L., *Apocynaceae*: J. Mokry *et al.*, *Experientia* **17**, 354 (1961). The (−)-form is obtained by acid hydrolysis of vincamine, *q.v.*: J. Trajanek *et al.*, *Tetrahedron Letters* **1961**, 702; O. Clauder *et al.*, *ibid.* **1962**, 1147; *eidem*, **Hung.** pat. **151,295** (1964 to Gedeon Richter), *C.A.* **60**, 14558e (1964). Structure and synthesis of the (±)-form: M. F. Bartlett, W. I. Taylor, *Tetrahedron Letters* **20**, 20 (1959); *eidem*, *J. Am. Chem. Soc.* **82**, 5941 (1960); E. Wenkert, D. Wickberg, *ibid.* **87**, 1580 (1965). Short synthesis of the (±)-form: E. Wenkert *et al.*, *ibid.* **100**, 4893 (1978); high yield total synthesis: J. L. Hermann *et al.*, *ibid.* **101**, 1540 (1979); T. Imanishi *et al.*, *Chem. Pharm. Bull.* **30**, 1521 (1982). Synthesis of the (−)-form: D. Cartier *et al.*, *Bull. Soc. Chim. France* **1976**, 1961. Structural and biogenetic relationship to vincamine: J. Mokry *et al.*, *Tetrahedron Letters* **1962**, 433. Pharmacology of the (−)-form: P. Lacroix *et al.*, *Arzneimittel-Forsch.* **29**, 1094 (1979). Series of articles on the pharmacodynamics, metabolism, and therapeutic use of the (−)-form: *Eur. Neurol.* **17**, Suppl. 1, 1-172 (1978). *Reviews:* W. I. Taylor in *The Alkaloids* vol. **8**, R. H. F. Manske, Ed. (Academic Press, New York, 1965) pp 253-259; *idem, ibid.* vol. **11**, pp 108-110 (1968).

Cryst, mp 174°. $[\alpha]_D^{25}$ +89° (chloroform). uv max: 241, 268, 296, 302 nm (ε 19800, 10200, 4800, 4800).
(±)-Form, (±)-*eburnamenin-14(15H)-one*, *vincanorine*. Crystals from methanol, mp 201-202.5°. uv max (ethanol): 227, 287, 294 nm (log ε 4.49, 3.89, 3.87).
(−)-Form, (3α,16α)-*eburnamenin-14(15H)-one*, *16-oxo-eburnane*, *vincamone*, CH-846, *Cervoxan*. Solid, mp 168-170°; also reported as mp 177-178° (Cartier). $[\alpha]_D^{25}$ −102° (chloroform) (Clauder); $[\alpha]_D$ −100° (c = 0.783 in chloroform) (Cartier). uv max: 205, 240, 265, 290, 300 nm (log ε 4.28, 4.16, 3.90, 3.59, 3.57).
(−)-Form phosphate, *Eburnal*.
THERAP CAT: The (−)-form as vasodilator.

3538. Ecabet. *[1R-(1α,4aβ,10aα)]-1,2,3,4,4a,9,10,10a-Octahydro-1,4a-dimethyl-7-(1-methylethyl)-6-sulpho-1-phenanthrenecarboxylic acid;* dehydro-6-sulfoabietic acid; 12-sulfodehydroabietic acid. $C_{20}H_{28}O_5S$; mol wt 380.51. C 63.13%, H 7.42%, O 21.02%, S 8.43%. Prepn: L. F. Fieser, W. P. Campbell, *J. Am. Chem. Soc.* **60**, 2631 (1938); H. Wada *et al.*, *Chem. Pharm. Bull.* **33**, 1472 (1985). Metabolism in animals: Y. Ito *et al.*, *J. Pharmacobio-Dyn.* **14**,

533 (1991). Efficacy in healing expt ulcers in rats: Y. Onoda *et al.*, *Arzneimittel-Forsch.* **41**, 546 (1991). Gastroprotective effects in rats in comparison with sucralfate, *q.v.*: M. Kinoshita *et al.*, *Digest. Dis. Sci.* **40**, 661 (1995). Bactericidal activity against *Helicobacter pylori:* K. Shibata *et al.*, *Antimicrob. Ag. Chemother.* **39**, 1295 (1995).

As the hemihydrate, $[\alpha]_D^{25}$ +72.4 (c = 2.5 in alcohol).

Sodium salt, $C_{20}H_{27}NaO_5S$, *(+)-(1R,4aS,10aR)-1,2,3,4,5a,-9,10,10a-octahydro-1,4a-dimethyl-7-(1-methylethyl)-6-sulfo-1-phenanthrenecarboxylic acid 6-sodium salt, TA-2711, Gastron.* Occurs as pentahydrate, mp > 300°. $[\alpha]_D^{20}$ +59.4° (c = 0.5).

THERAP CAT: Antiulcerative.

3539. Ecdysteroids. Polyhydroxylated steroids formerly known as ecdysones. Originally identified as insect molting hormones controlling the pupation of insects. Later shown to be involved in many complex developmental processes in metamorphosis, differentiation and reproduction. Ecdysteroids have been detected in invertebrate species of several phyla belonging to the Protostomia and in some plant species. The terms *zooecdysteroids* and *phytoecdysteroids* are used to distinguish ecdysteroids isolated from animal species from those of plant origin. Nomenclature: T. W. Goodwin *et al.*, *Nature* **272**, 122 (1978). The two major ecdysteroids isolated are ecdysone and 20-hydroxyecdysone. Configuration: M. Koreeda *et al.*, *J. Am. Chem. Soc.* **93**, 4084 (1971). Chromosomal action: M. Ashburner, *Nature* **285**, 435 (1980). GC and HPLC determn: R. P. Evershed *et al.*, *J. Chromatog.* **390**, 357 (1987). Reviews: Kilby, *Discovery* **18**, 13 (1957); P. Karlson, *Angew. Chem. Int. Ed.* **2**, 175-182 (1963); K. Nakanishi, *Pure Appl. Chem.* **25**, 167-195 (1971); M. Koreeda, B. A. Teicher, *Anal. Biochem. Insects* **1977**, 207-240; P. Karlson, *Dev. Endocrinol.* **7**, 1-11 (1980). *Book:* *Ecdysone: From Chemistry to Mode of Action*, J. Koolman, Ed. (Thieme, New York, 1989) 482 pp.

Ecdysone. $C_{27}H_{44}O_6$. *(2β,3β,5β,22R)-2,3,14,22,25-Pentahydroxycholest-7-en-6-one, α-ecdysone.* Secreted by ecdysial tissues, then transformed to more active compound, 20-hydroxyecdysone. First isoln from silkworm moths, *Bombyx mori:* A. Butenandt, P. Karlson, *Z. Naturforsch.* **9b**, 389 (1954). Isoln from rhizomes of *Polypodium vulgare* L.: G. Heinrich, H. Hoffmeister, *Experientia* **23**, 995 (1967); from bracken fern, *Pteridinium aquilinum:* J. N. Kaplanis *et al.*, *Science* **157**, 1436 (1967). Structure: P. Karlson *et al.*, *Ber.* **98**, 2394 (1965). Configuration: R. Huber, W. Hoppe, *ibid.* 2403. Synthesis: U. Kerb *et al.*, *Helv. Chim. Acta* **49**, 1601 (1966); J. B. Siddall *et al.*, *J. Am. Chem. Soc.* **88**, 379, 862 (1966); H. Mori *et al.*, *Chem. Pharm. Bull.* **16**, 563 (1968). mp 238-239°. $[\alpha]_D^{29}$ +62°. uv max: 243 nm (ε 11600).

20-Hydroxyecdysone. $C_{27}H_{44}O_7$. *(2β,3β,5β,22R,)-2,3,14,-20,22,25-Hexahydroxycholest-7-en-6-one, β-ecdysone, ecdysterone, crustecdysone, isoinokosterone, polypodine A.* Most

widely occurring ecdysteroid in both plant and animal species. Isoln from *B. mori:* P. Hocks, R. Wiechert, *Tetrahedron Letters* **1966**, 2989; from seawater crayfish, *Jasus lalandei:* F. Hampshire, D. H. S. Horn, *Chem. Commun.* **1966**, 37. Isolns from plant sources, *Achyranthes fauriei:* T. Takemoto *et al.*, *Yakugaku Zasshi* **87**, 325 (1967); *P. elatus:* M. N. Galbraith, D. H. S. Horn, *Chem. Commun.* **1966**, 905; *P. vulgare:* J. Jizba *et al.*, *Tetrahedron Letters* **1967**, 1689. Isoln from parasitic helminths: H. H. Rees, J. G. Mercer, *Adv. Invertebr. Reprod.* **4**, 173 (1986). Configuration: Dammeier, Hoppe, *Ber.* **104**, 1660 (1971). Synthesis: G. Hüppi, J. B. Siddall, *J. Am. Chem. Soc.* **89**, 6790 (1967); U. Kerb *et al.*, *Tetrahedron Letters* **1968**, 4277; H. Mori, K. Shibata, *Chem. Pharm. Bull.* **17**, 1970 (1969). Total synthesis: T. Kametani *et al.*, *Tetrahedron Letters* **21**, 4855 (1980). From methanol-ethyl acetate, mp 240-242°. uv max: 243 nm (ε 10300). Unstable in alkaline soln.

3540. Ecgonidine. *(1R)-8-Methyl-8-azabicyclo[3.2.1]-oct-2-ene-2-carboxylic acid; 2-tropidinecarboxylic acid;* anhydroecgonin. $C_9H_{13}NO_2$; mol wt 167.21. C 64.65%, H 7.84%, N 8.38%, O 19.14%. Prepn from ecgonine and structure: Findlay, *J. Am. Chem. Soc.* **75**, 1033 (1953). Prepn and formation of *l*-form: de Jong, *Rec. Trav. Chim.* **42**, 980 (1923), **66**, 99 (1947). Synthesis of *dl*-form: Grundmann, Ottmann, *Ann.* **605**, 24 (1957); U.S. pat. 2,783,235 (1957 to Olin Mathieson).

dl-Form, crystals from alcohol + ether, dec 235-236°. Soluble in water; sparingly sol in alcohol.

l-Form, crystals from abs alcohol, mp 235°. $[\alpha]_D^{14}$ −84.6° (c = 1.7).

THERAP CAT: Anesthetic (topical).

3541. Ecgonine. *[1R-(exo,exo)]-3-Hydroxy-8-methyl-8-azabicyclo[3.2.1]octane-2-carboxylic acid; 3β-hydroxy-1αH,-5αH-tropane-2β-carboxylic acid.* $C_9H_{15}NO_3$; mol wt 185.22. C 58.36%, H 8.16%, N 7.56%, O 25.91%. The principal part of the cocaine molecule; naturally occurring as the *l*-form. Obtained by hydrolysis of cocaine: Willstätter *et al.*, *Ann.* **434**, 111 (1923); Bell, Archer, *J. Am. Chem. Soc.* **82**, 4642 (1960). From coca leaves: de Jong, *Rec. Trav. Chim.* **59**, 687 (1940). Structure: Gadamer, John, *Arch. Pharm.* **259**, 227 (1921). Stereochemistry: Fodor, *Nature* **170**, 278 (1952); Fodor, Kovács, *J. Chem. Soc.* **1953**, 724. Synthesis: Willstätter, Bommer, *Ann.* **422**, 15 (1920). *Review:* Stoll, Jucker, *Angew. Chem.* **66**, 376 (1954).

l-Form monohydrate, triboluminescent, monoclinic prisms from alc, mp 198° (anhydr, dec 205°). $[\alpha]_D^{15}$ −45° (c = 5). Neutral to litmus. pKa 11.11; pKb 11.22. One gram dissolves in 5 ml water, 67 ml alc, 20 ml methanol, 75 ml ethyl acetate. Sparingly sol in acetone, ether, benzene, chloroform, petr ether.

Hydrochloride, $C_9H_{15}NO_3 \cdot HCl$, triclinic plates from water, mp 246°. $[\alpha]_D^{15}$ −59° (c = 10). Sol in water; slightly in alc.

dl-Form trihydrate, plates from 90% alcohol, mp 93-118° (anhydr dec 212°).

Note: This is a controlled substance listed in the U.S. Code of Federal Regulations, Title 21 Part 1308.12 (1995).

THERAP CAT: Anesthetic (topical).

3542. Echinacea. Cone flower. Dried rhizome and roots of *Echinacea pallida* (Nutt.) Britt. or *E. angustifolia* DC. *(Brauneria pallida* Nutt.) Britt., *Compositae. Habit.*

Saskatchewan to Texas and Alabama. *Constit.* Inulin, sucrose, betaine, two isomeric 2-methyltetradecadienes, echinacein (neoherculin, α-sanshool), echinacoside (a caffeic acid glycoside), resins, fatty acids. *Refs:* Stoll, *Helv. Chim. Acta* **33**, 1877 (1950); Jacobson, *Science* **120**, 1028 (1954); Crombie, Taylor, *J. Chem. Soc.* **1957**, 2760.

Note: A commercial extract of *Echinaceae* spp, *echinacin,* has been used experimentally as a hyaluronidase antagonist: Koch, Haase, *Arzneimittel-Forsch.* **2**, 464 (1952); Büsing, *ibid.* **5**, 320 (1955).

3543. Echinenone. β,β-*Caroten-4-one;* 4-oxo-β-carotene; 4-keto-β-carotene; aphanin; myoxanthin. $C_{40}H_{54}O$; mol wt 550.87. C 87.22%, H 9.88%, O 2.90%. Carotenoid pigment occurring in algae. Isoln from *Aphanizomenon flosaqua:* Tischer, *Z. Physiol. Chem.* **251**, 109 (1938); **260**, 257 (1939); from *Paracentrotus lividus:* Lederer, *Compt. Rend.* **201**, 300 (1935); from *Oscillatoria rubrescens:* Heilbron, Lythgoe, *J. Chem. Soc.* **1936**, 1376. Structure: Goodwin, Taha, *Biochem. J.* **47**, 244 (1950); Goodwin, *ibid.* **63**, 481 (1956); Ganguly *et al., Arch. Biochem. Biophys.* **60**, 345 (1956). Synthesis: Akhtar, Weedon, *J. Chem. Soc.* **1958**, 3986; **1959**, 4058.

Orange-red crystals from benzene + methanol, mp 178-180°. Absorption max (chloroform): 472-478 nm. Freely sol in carbon disulfide, chloroform, benzene. Slightly sol in pyridine, ether. Practically insol in methanol. Provitamin A activity 54% of that of all-*trans*-β-carotene.

Oxime, $C_{40}H_{55}NO$, red crystals from benzene, mp 208°. Absorption max (chloroform): 464-468 nm.

3544. Echinochrome A. 2-*Ethyl-3,5,6,7,8-pentahydroxy-1,4-naphthalenedione;* 2-ethyl-3,5,6,7,8-pentahydroxy-1,4-naphthoquinone. $C_{12}H_{10}O_7$; mol wt 266.21. C 54.14%, H 3.79%, O 42.07%. The red pigment of the eggs of the sea urchin (*Arbacia pustulosa*). Isoln and structure: Kuhn, Wallenfels, *Ber.* **72**, 1407 (1939). Synthesis from 2-ethyl-1,3,4-trimethoxybenzene and dibenzoyloxymaleic anhydride: Wallenfels, Gauhe, *Ber.* **76**, 325 (1943).

Deep-red needles from dioxane-water, mp 220° (some decompn). Sublimes at 10^{-4} mm at 120°. Absorption max (chloroform): 533, 497, 462 nm. Very sparingly sol in cold water, yet more than sufficiently sol for chemotaxic effects. Readily sol in carbon disulfide, ether, chloroform, benzene, concd H_2SO_4.

3545. Echinomycin. Quinomycin A. $C_{51}H_{64}N_{12}O_{12}S_2$; mol wt 1101.27. C 55.62%, H 5.86%, N 15.26%, O 17.43%, S 5.82%. A *"quinoxaline antibiotic"* similar to the triostins, *q.v.:* Kuroya, Ishida, *J. Antibiot.* **14A**, 324 (1961). Powerful, selective inhibitor of nucleic acid synthesis *in vitro*. Produced by *Streptomyces echinatus* from soil of Cuanza (Angola): Corbaz *et al., Helv. Chim. Acta* **40**, 199 (1957). Structure: Keller-Schierlein *et al., ibid.* **42**, 305 (1959). Revised structure: Dell *et al., J. Am. Chem. Soc.* **97**, 2497 (1975); D. G. Martin *et al., J. Antibiot.* **28**, 332 (1975). Identity with quinomycin A: Katagiri, Sugiura, *Antimicrob. Ag. Chemother.* **1961**, 162. Biosynthesis: Arif *et al., Indian J. Biochem.* **7**, 193 (1970). Isoln and properties of the other quinomycins: Yoshida, Katagiri, *J. Antibiot.* **14A**, 330 (1961); Otsuka, Shoji, *ibid.* **19A**, 128 (1966); *eidem, Tetrahedron* **23**, 1535 (1967). Conformation in soln: H. T. Cheung *et al., J. Am. Chem. Soc.* **100**, 46 (1978). Review of chemis-

try and biochemistry of echinomycin: M. J. Waring in *Antibiotics,* **vol.** 5(pt. 2), F. E. Hahn, Ed. (Springer-Verlag, New York, 1979) pp 173-194.

Slightly hygroscopic crystals, mp 217-218°. $[\alpha]_D^{20}$ −310° (c = 0.86 in chloroform). uv max (methanol): 243, 320 nm ($E_{1cm}^{1\%}$ 622, 100). Has lipophile solubilities. Easily sol in chloroform, dioxane. Insol in water, petr ether, hexane.

3546. Echinopsine. 1-*Methyl-4(1H)-quinolinone;* 1-*methyl-4(1H)-quinolone;* 1,4-dihydro-1-methyl-4-oxoquinoline; *N*-methyl-4-quinolone. $C_{10}H_9NO$; mol wt 159.19. C 75.45%, H 5.70%, N 8.80%, O 10.05%. From *Echinops ritro* L. and other spp of *Echinops, Compositae:* Greshoff, *Rec. Trav. Chim.* **19**, 360 (1900); Ban'kovskii *et al., Dokl. Akad. Nauk SSSR* **148**, 1073 (1963). Structure: Späth, Kolbe, *Monatsh.* **43**, 469 (1923). Synthesis: Kondo, Ikawa, *J. Pharm. Soc. Japan* **51**, 702 (1931); Allison *et al., J. Chem. Soc.* **1954**, 403; King, Abramo, *J. Org. Chem.* **23**, 1609 (1958); Kamiya, *Chem. Pharm. Bull.* **10**, 669 (1962). Simple synthesis: J. R. Merchant, V. Shankaranarayan, *Chem. & Ind. (London)* **1979**, 320.

Needles from benzene, mp 152°. One gram dissolves in about 60 ml water, 6 ml boiling water. Soluble in alcohol, chloroform, hot benzene; slightly sol in ether. LD_{100} in mice: 600 mg/kg.

Hydrochloride, $C_{10}H_9NO.HCl$, crystals, mp 185-186°.

3547. Echinuline. (3S-cis)-3-[[2-(1,1-*Dimethyl-2-propenyl)-5,7-bis(3-methyl-2-butenyl)-1H-indol-3-yl]methyl]-6-methyl-2,5-piperazinedione.* $C_{29}H_{39}N_3O_2$; mol wt 461.65. C 75.45%, H 8.52%, N 9.10%, O 6.93%. From the dry mycelia of *Aspergillus echinulatus:* Quilico, Panizzi, *Ber.* **76**, 348 (1943). Structure: Casnati *et al., Gazz. Chim. Ital.* **92**, 105 (1962); Romanet *et al., Bull. Soc. Chim. France* **1963**, 1048. Absolute configuration: Nakashima, Slater, *Tetrahedron Letters* **1967**, 4433; **1971**, 2649. Synthetic studies: Houghton, Saxton, *J. Chem. Soc. (C)* **1969**, 595, 1003; Takamatsu *et al., Tetrahedron Letters* **1971**, 4661. Total synthesis of optically active *cis* isomer: *eidem, ibid.* **4665**; S. Inoue *et al., J. Pharm. Soc. Japan* **97**, 558 (1977). Epimerization of echinuline with triethylamine in ethanol gives the *trans* isomer, *epi-echinuline,* Westley *et al., Anal. Chem.* **40**, 1888 (1968).

Needles from butanol, mp 242-243°. $[\alpha]_D^{20}$ −26.0° (chloroform). uv max (ethanol): 230, 279, 286 nm (log ϵ 4.60, 3.98, 3.96). Sol in glacial acetic acid, chloroform, pyridine, dioxane; less sol in warm alcohol, warm butanol; slightly sol in benzene, ether, petr ether, carbon tetrachloride, acetone, cold alcohol.

3548. Echitamine. *(3β,16R)-3,17-Dihydroxy-16-(methoxycarbonyl)-4-methyl-2,4(1H)-cyclo-3,4-secoakuammilanium;* ditaine. $[C_{22}H_{29}N_2O_4]$+; mol wt 385.48. From the bark of *Alstonia scholaris* (L.) R.Br., *A. congensis* Engl., *Apocynaceae; A. neriifolia*: Hesse, *Ann.* **203**, 150 (1880); Goodson, Henry, *J. Chem. Soc.* **127**, 1640 (1925); Goodson, *ibid.* **1932**, 2626; Chakravarti *et al., Bull. Calcutta Sch. Trop. Med.* **16**, 81 (1968); *C.A.* **71**, 3529f (1969); Roy, Chatterjee, *J. Ind. Chem. Soc.* **45**, 21 (1968). Structure: Hamilton *et al., Proc. Chem. Soc.* **1961**, 63; *J. Chem. Soc.* **1962**, 5061. Abs config: Manohar, Ramaseshan, *Tetrahedron Letters* **1961**, 814. Synthetic studies: Fritz, Fischer, *Tetrahedron* **20**, 1737 (1964); Dolby, Esfandiari, *J. Org. Chem.* **37**, 43 (1972). *Review:* Govindachari, *J. Indian Chem. Soc.* **45**, 945-957 (1968).

Hydroxide, $C_{22}H_{30}N_2O_5$, white crystals, mp 206°. $[\alpha]_D^{20}$ −29° (alcohol). Sol in water, alcohol, chloroform, ether.

Chloride, $C_{22}H_{29}O_4N_2^+Cl^-$, long needles from water, mp 295°. $[\alpha]_D^{15}$ −58°. uv max (ethanol): 235, 295 nm (log ϵ 3.93, 3.55).

3549. Echothiophate Iodide. *2-[(Diethoxyphosphinyl)-thio]-N,N,N-trimethylethanaminium iodide; (2-mercaptoethyl)trimethylammonium iodide O,O-diethyl phosphorothioate;* diethoxyphosphinylthiocholine iodide; 2-diethoxyphosphinylthioethyltrimethylammonium iodide; O,O-diethyl S-(2-trimethylammoniumethyl)phosphorothioate iodide; S-β-dimethylaminoethyl-O,O-diethylthionophosphate methiodide; ecothiopate iodide; 217-MI; Phospholine Iodide. $C_9H_{23}I-NO_3PS$; mol wt 383.23. C 28.21%, H 6.05%, I 33.11%, N 3.65%, O 12.52%, P 8.08%, S 8.37%. Prepn: Tammelin, *Acta Chem. Scand.* **11**, 1340 (1957); Fitch, U.S. pat. **2,911,430** (1959 to Campbell Pharmaceuticals). Comprehensive description: R. D. Daley in *Analytical Profiles of Drug Substances* vol. 3, K. Florey, Ed. (Academic Press, New York, 1974) pp 233-251.

Crystals, mp 138° (Tammelin); mp 124-124.5° (Fitch). Sol in water, alcohol, chloroform.
THERAP CAT: Cholinergic (ophthalmic).

3550. Econazole. *1-[2-[(4-Chlorophenyl)methoxy]-2-(2,4-dichlorophenyl)ethyl]-1H-imidazole; 1-[2,4-dichloro-β-[(p-chlorobenzyl)oxy]phenethyl]imidazole;* SQ-13050. $C_{18}-H_{15}Cl_3N_2O$; mol wt 381.69. C 56.64%, H 3.96%, Cl 27.87%, N 7.34%, O 4.19%. Prepn: Godefroi *et al., J. Med. Chem.* **12**, 784 (1969); Godefroi, Heeres, **Ger.** pat. **1,940,388** corresp to U.S. pat. **3,717,655** (1970, 1973 both to Janssen). Biological and toxicological properties: Thienpont *et al., Arzneimittel-Forsch.* **25**, 224 (1975). Review of antifungal activity and therapeutic efficacy: R. C. Heel *et al., Drugs* **16**, 177-201 (1978).

mp 86.8°.
Nitrate, $C_{18}H_{15}Cl_3N_2O \cdot HNO_3$, *R-14827, Epi-Pevaryl, Gyno-Pevaryl, Ifenec, Micofugal, Micogin, Palavale, Pargin, Pevaryl, Spectazole.* Crystals from a mixture of 2-propanol, methanol and diisopropyl ether, mp 162°. Very slightly sol in water; very slightly to slightly sol in most common organic solvents. LD_{50} in mice, rats (mg/kg): 462.7, 667.7 orally (Thienpont).
THERAP CAT: Antifungal.

3551. ECTEOLA-Cellulose. Cellulose-ECTEOLA. Anion-exchange material prepd from cellulose, epichlorohydrin, and triethanolamine: Peterson, Sober, *J. Am. Chem. Soc.* **78**, 751 (1956); Veder, *J. Chromatog.* **10**, 507 (1963).

3552. Ectylurea. *(Z)-N-(Aminocarbonyl)-2-ethyl-2-butenamide; (2-ethylcrotonoyl)urea; (α-ethylcrotonoyl)urea; (α-ethylcrotonyl)carbamide;* Astyn; Ektyl; Levanil; Nostal; Nostyn; Cronil; Neuroprocin; Pacetyn. $C_7H_{12}N_2O_2$; mol wt 156.18. C 53.83%, H 7.74%, N 17.94%, O 20.49%. Prepd by treating (2-bromo-2-ethylbutyryl)urea with dil boiling NaOH: Newbery, *J. Chem. Soc.* **127**, 295 (1925). Prepn from carbromal and isopropanol by refluxing with silver oxide: Fancher, U.S. pats. **2,854,379, 2,931,832** (1958, 1960 to Miles Labs.). Pharmacology and toxicity: M. H. Pindell *et al., Fed. Proc.* **12**, 357 (1953).

Needles from benzene, mp 198°, also reported as mp 191-193° (lower melting form, mp 158°). Very slightly sol in ether, hot alcohol, other organic solvents. Sol in concd aq solns of alkalies and acids. LD_{50} in rats (mg/kg): 2500 orally; 900 i.p. (Pindell).
THERAP CAT: Sedative; hypnotic.

3553. Edatrexate. *N-[4-[1-[(2,4-Diamino-6-pteridinyl)methyl]propyl]benzoyl]-L-glutamic acid;* 10-ethyl-10-deazaaminopterin; 10-EdAM; 10-EDAAM; CGP-30694. $C_{22}H_{25}-N_7O_5$; mol wt 467.48. C 56.52%, H 5.39%, N 20.97%, O 17.11%. Dihydrofolate reductase inhibitor. Prepn: F. M. Sirotnak, J. I. DeGraw, **Fr.** pat. **2,464,956** (1981 to SRI International); J. I. DeGraw, Jr., F. M. Sirotnak, U.S. pat. **4,369,319** (1983). Synthesis: J. I. DeGraw *et al., J. Med. Chem.* **25**, 1227 (1982); and resolution of diastereomers: *eidem, ibid.* **29**, 1056 (1986). HPLC determn: J. J. Kinahan *et al., Anal. Biochem.* **150**, 203 (1985); O. Van Tellingen *et al., J. Chromatog.* **529**, 135 (1990). Pharmacology and toxicology: M. P. Fanucchi *et al., Cancer Res.* **47**, 2334 (1987). Clinical evaluation in non-small cell lung cancer: J. S. Lee *et al, Invest. New Drugs* **8**, 299 (1990). Review of pharmacology and antitumor activity: F. M. Sirotnak *et al., NCI Monograph* **5**, 127-131 (1987).

Prepd as 1.75 H$_2$O, uv max (pH 13): 255, 370 nm (ϵ 30731, 7582).

THERAP CAT: Antineoplastic.

3554. Edestin. Globular protein originally obtained from hemp seed *(Cannabis sativa):* Osborne, *Am. Chem. J.* **14**, 662 (1892); Stockwell *et al., Proc. Seed Protein Conf., New Orleans* **1963**, 56, *C.A.* **60**, 3090d (1964). Approx mol wt 300,000. Since the passage of the Marihuana Tax Act in the U.S.A. commercial hemp seed must be heat-treated to destroy its narcotic properties, a process which also destroys the edestin. Very closely related proteins may be obtained from seeds of *Cucurbita pepo* L. (pumpkin), *C. moschata* Duchesne (squash), *C. maxima* Duchesne (squash), *Citrullus vulgaris* Schrader (watermelon), *Cucumis melo* L. (cantaloupe), and *Cucumis sativus* L., *Cucurbitaceae* (cucumber). Isoln from *Cucurbita pepo:* Vickery *et al., Biochem. Prepn.* **2**, 5 (1952). Extraction from hemp seed, prepn of crystalline form and x-ray data: J. Drenth, E. W. Wiebenga, *Rec. Trav. Chim.* **74**, 813 (1955). The amino acid compositions of these globulins are slightly different from hemp seed globulin and from each other: Smith *et al., J. Biol. Chem.* **164**, 159 (1946); Smith, Greene, *ibid.* **167**, 833 (1947); **172**, 111 (1948). Structure studies: Hall, *ibid.* **185**, 45 (1959); Cleemann, Kratky, *Z. Naturforsch.* **15b**, 526 (1960); Dlouhá *et al., Coll. Czech. Chem. Commun.* **28**, 2779 (1963); **29**, 1835 (1964). Electron microscopy and optical diffraction: A. M. H. Schepman *et al., Biochim. Biophys. Acta* **271**, 279 (1972). Electron spin resonance studies: L. J. Dimmey, W. Gordy, *Proc. Nat. Acad. Sci. USA* **77**, 343 (1980).

Octahedral crystals. Completely sol to a clear soln in 10% salt soln. Sol in dil mineral acids. Forms a water-soluble hydrochloride.

3555. Edetate Calcium Disodium. *[[N,N'-1,2-Ethanediylbis[N-(carboxymethyl)glycinato]](4—)-N,N',O,O',O^N,O$^{N'}$]-calciate(2—)disodium; [(ethylenedinitrilo)tetraacetato]calciate(2—) disodium;* ethylenediaminetetraacetic acid calcium disodium chelate; calcium disodium (ethylenedinitrilo)-tetraacetate; calcium disodium ethylenediaminetetraacetate; EDTA calcium; edathamil calcium disodium; calcium disodium edetate; edetic acid calcium disodium salt; sodium calciumedetate; Calcitetracemate Disodium; Calcium Disodium Versenate; Ledclair; Mosatil; Antallin; Sormetal; Versene CA. C$_{10}$H$_{12}$CaN$_2$Na$_2$O$_8$; mol wt 374.27. C 32.09%, H 3.23%, Ca 10.71%, N 7.48%, Na 12.29%, O 34.20%. Prepn: Astakhov, Kiseleva, *Zh. Obshch. Khim.* **20**, 1780 (1950), *C.A.* **45**, 2409 (1951).

Tetrahydrate, powder. Sol in water: at 30° a 0.1*M* soln can be prepd (pH ~7). Practically insol in organic solvents. Exchanges its calcium for lead or other heavy metal ions forming water-sol complexes of the heavy metals.

USE: Color retention agent in foods, flavoring agent.

THERAP CAT: Chelating agent (metal).

THERAP CAT (VET): Chelating agent in lead poisoning.

3556. Edetate Disodium. *N,N'-1,2-Ethanediylbis[N-(carboxymethyl)glycine] disodium salt; (ethylenedinitrilo)-tetraacetic acid disodium salt;* ethylenediaminetetraacetic

acid disodium salt; ethylenebis(iminodiacetic acid) disodium salt; edetic acid disodium salt; edathamil disodium; disodium edathamil; EDTA disodium; tetracemate disodium; disodium ethylenediaminetctraacetate; disodium edetate; Chelaplex III; Endrate disodium; Sequestrene NA 2; Sodium Versenate; Titriplex III; Versene disodium salt. C$_{10}$H$_{14}$N$_2$Na$_2$O$_8$; mol wt 336.21. C 35.72%, H 4.20%, N 8.33%, Na 13.68%, O 38.07%. Prepn: Bersworth, U.S. pat. **2,407,645** (1946 to Martin Dennis); Bersworth, U.S. pat. **2,461,519** (1949). Toxicity: J. E. Wynn *et al., Toxicol. Appl. Pharmacol.* **16**, 807 (1970). *Review:* Biermans, Henrard, *Ind. Chim.* **39**, 6 (1952). For structure see Edetate calcium disodium.

Dihydrate, crystals, mp 252° (dec). pH about 5.3. Sol in water. Has characteristics of weak acid, displacing CO$_2$ from carbonates and reacting with metals to form hydrogen. LD$_{50}$ orally in rats: 2 g/kg (Wynn).

USE: Sequestering agent. Pharmaceutic aid (chelating agent).

THERAP CAT: Chelating agent (metal).

THERAP CAT (VET): Anticoagulant.

3557. Edetate Sodium. *N,N'-1,2-Ethanediylbis[N-(carboxymethyl)glycine] tetrasodium salt; (ethylenedinitrilo)tetraacetic acid tetrasodium salt;* ethylenediaminetetraacetic acid tetrasodium salt; sodium edetate; tetrasodium ethylenediaminetetraacetate; ethylenebis(iminodiacetic acid) tetrasodium salt; tetrasodium ethylenebis(iminodiacetate); EDTA tetrasodium; edetic acid tetrasodium salt; tetracemate tetrasodium; tetrasodium cdetate; tetracemin; Endrate Tetrasodium; Questex; Vcrscnc; Sequestrene; Tetrine; Kalex; Trilon B; Komplexon; Nullapon; Aquamollin; Complexone; Distol 8; Irgalon; Calsol; Syntes 12a; Tyclarosol; Nervanaid B. C$_{10}$H$_{12}$N$_2$Na$_4$O$_8$; mol wt 380.17. C 31.59%, H 3.18%, N 7.37%, Na 24.19%, O 33.67%. (NaOOCCH$_2$)$_2$-NCH$_2$CH$_2$N(CH$_2$COONa)$_2$. Prepn: Bersworth, U.S. pat. **2,407,645** (1946 to Martin Dennis Co.); *idem.,* U.S. pat. **2,461,519** (1949). For bibliography and applications *see "Keys to Chelation"* issued by Dow, Midland, Mich.

Powder, mp >300°. Apparent density: 6.9 lb/gallon. Very sol in water (about 103 g/100 ml). pH of 1% soln 11.3. Reacts with most divalent and trivalent metallic ions forming sol metal chelates. Less sol in alcohol than the potassium salt.

USE: Sequestering agent: one gram complexes 215 mg CaCO$_3$. Usually added to pharmaceuticals in the form of the calcium disodium salt to prevent calcium-depleting action in the body.

THERAP CAT: Chelating agent.

3558. Edetate Trisodium. *N,N'-1,2-Ethanediylbis[N-(carboxymethyl)glycine] trisodium salt; (ethylenedinitrilo)tetraacetic acid trisodium salt;* EDTA trisodium; ethylenediaminetetraacetic acid trisodium salt; trisodium ethylenediaminetetraacetate; trisodium edetate; edetic acid trisodium salt; Limclair; Versene-9; Sequestrene NA 3. C$_{10}$H$_{13}$N$_2$-Na$_3$O$_8$; mol wt 358.19. C 33.53%, H 3.66%, N 7.82%, Na 19.25%, O 35.73%. Prepn: Bersworth, U.S. pat. **2,461,519** (1949). *Review:* Biermans, Henrard, *Ind. Chim.* **39**, 6 (1952). For structure see Edetate calcium disodium.

Monohydrate, crystals from water, mp >300°. More sol in water than the corresp disodium salt or free acid. pH of 1% aq soln 9.3. One gram complexes at least 242 mg CaCO$_3$.

THERAP CAT: Chelating agent.

3559. Edetic Acid. *N,N'-1,2-Ethanediylbis[N-(carboxymethyl)glycine]; (ethylenedinitrilo)tetraacetic acid;* ethylenediaminetetraacetic acid; edathamil; EDTA; Havidote; Versene Acid. C$_{10}$H$_{16}$N$_2$O$_8$; mol wt 292.25. C 41.10%, H 5.52%, N 9.59%, O 43.80%. (HOOCCH$_2$)$_2$NCH$_2$CH$_2$N-(CH$_2$COOH)$_2$. Prepn: Münz, U.S. pat. **2,130,505** (1938 to General Aniline); Bersworth, U.S. pat. **2,407,645** (1946 to Martin Dennis Co.); Bersworth, U.S. pat. **2,461,519** (1949); Smith *et al., J. Org. Chem.* **14**, 355 (1949). Prepn of α- and β-form of crystals: Le Blanc, Spell, *J. Phys. Chem.* **64**, 949 (1960). *Review:* Biermans, Henrard, *Ind. Chim.* **39**, 6 (1952).

Crystals from water, dec 220°. Soly in water at 25°: 0.50 g/l, Palei, Udal'tsova, *C.A.* **56**, 13620c (1962). The free acid is less stable than its salts, and tends to decarboxylate when heated to temps of 150°. Stable on storage and on boiling in aq soln.

Cobalt salt, *Kelocyanor.*

Dipotassium salt dihydrate, $C_{10}H_{14}K_2N_2O_8 \cdot 2H_2O$, *edetate dipotassium, EDTA dipotassium.*

USE: As antioxidant in foods. Pharmaceutic aid (chelating agent).

THERAP CAT (VET): In treatment of lead and heavy metal poisoning of farm animals.

3560. Edifenphos. *Phosphorodithioic acid O-ethyl S,S-diphenyl ester; O-ethyl S,S-diphenyl phosphorodithioate;* EDDP; ediphenphos; Bayer 78418; Hinosan. $C_{14}H_{15}O_2PS_2$; mol wt 310.38. C 54.18%, H 4.87%, O 10.31%, P 9.98%, S 20.66%. Prepn: **Neth. pat. Appl. 6,611,860;** G. Schrader *et al., U.S. pat. 3,499,951* (1967, 1970 both to Bayer). Degradn in soil: C. Tomizawa *et al., J. Environ. Sci. Health* **B11,** 231 (1976). Metabolism: I. Ueyama *et al., Agr. Biol. Chem.* **42,** 885 (1978). Mechanism of action: O. Kodama *et al., ibid.* **44,** 1015 (1980). Toxicology: T. S. Chen *et al., Toxicol. Appl. Pharmacol.* **23,** 519 (1972).

Clear yellow to light-brown liquid, $bp_{0.01}$ 154°. d_4^{20} 1.23; n_D^{22} 1.61. Practically insol in water. Sol in acetone, xylene. LD_{50} in female, male rats (mg/kg): 25.5, 66.5 i.p. (Chen).

USE: Fungicide.

3561. Edoxudine. *2'-Deoxy-5-ethyluridine;* 5-ethyl-2'-deoxyuridine; 5-ethyl-3-(2'-deoxyribosyl)uracil; EDU; EUDR; ORF-15817; Aedurid; Edurid. $C_{11}H_{16}N_2O_5$; mol wt 256.26. C 51.56%, H 6.29%, N 10.93%, O 31.22%. Substituted uracil with anti-herpes activity. Prepn and properties: K. K. Gauri, **Brit. pat. 1,170,565;** *eidem,* **U.S. pat. 3,553,-192** (1968, 1971 both to Robugen). Synthesis via organopalladium intermediates: D. E. Bergstrom, J. L. Ruth, *J. Am. Chem. Soc.* **98,** 1587 (1976); D. E. Bergstrom, M. K. Ogawa, *ibid.* **100,** 8106 (1978). Physical properties: M. Swierkowski, D. Shugar, *J. Med. Chem.* **12,** 533 (1969). Antiviral activity: E. De Clercq, D. Shugar, *Biochem. Pharmacol.* **24,** 1073 (1975). Preferential inhibition of herpes simplex virus type 2: C.-Z. Teh, S. L. Sacks, *Antimicrob. Ag. Chemother.* **23,** 637 (1983). Synergistic effect with interferon-β, *q.v.,* on herpes simplex virus: C. Janz, R. Wigand, *Arch. Virol.* **73,** 135 (1982). Inhibitory effect on growth of murine leukemia cells: J. Balzarini *et al., Invest. New Drugs* **2,** 35 (1984).

Long clear needles from acetone, mp 152-153°. uv max: 267 nm (ε 9610) at pH 2, 267 nm (ε 7280) at pH 1. pKa: 9.98.

THERAP CAT: Antiviral (herpes simplex).

3562. Edrophonium Chloride. *N-Ethyl-3-hydroxy-N,N-dimethylbenzenaminium chloride; ethyl(m-hydroxyphenyl)-dimethylammonium chloride;* (3-hydroxyphenyl)dimethylethylammonium chloride; 3-hydroxy-N,N-dimethyl-N-ethylanilinium chloride; Antirex; Enlon; Reversol; Tensilon. $C_{10}H_{16}ClNO$; mol wt 201.70. C 59.55%, H 8.00%, Cl 17.58%, N 6.94%, O 7.93%. Cholinesterase inhibitor. Prepn: J. A. Aeschlimann, A. Stempel, **U.S. pat. 2,647,924** (1953 to Hoffmann-La Roche). LC determn in biological fluids: M. G. M. DeRuyter *et al., J. Chromatog.* **183,** 193 (1980). Clin-

ical evaluation in reversal of neuromuscular blockade: D. M. Fisher *et al., Anesthesiology* **61,** 428 (1984). Clinical study in supraventricular tachycardia: J. Frieden *et al., Am. J. Cardiol.* **27,** 294 (1971); J. D. Cantwell *et al., Arch. Int. Med.* **130,** 221 (1972). Diagnostic use in myasthenia gravis: G. A. Nicholson *et al., Clin. Exp. Neurol.* **19,** 45 (1983); I. M. Williams *et al., ibid.* **22,** 1 (1986); in esophageal chest pain: J. E. Richter *et al., Ann. Int. Med.* **103,** 14 (1985); C. A. Lee *et al., Dig. Dis. Sci.* **32,** 682 (1987).

Crystals from isopropanol, dec 162-163°. pH of 1% aq soln 4-5. Very sol in water; freely sol in alcohol. Insol in chloroform, ether.

Bromide, $C_{10}H_{16}BrNO$, *edrophone bromide, Ro-2-3198.* Crystals from ethanol + ether, dec 151-152°. Bitter taste. Sol in water (more than 10%). Moderately sol in alcohol. Practically insol in ether. Solns are stable.

THERAP CAT: Cholinergic; antidote to curare principles. Diagnostic aid (myasthenia gravis; esophageal chest pain).

3563. EEDQ. *2-Ethoxy-1(2H)-quinolinecarboxylic acid ethyl ester; N-carbethoxy-2-ethoxy-1,2-dihydroquinoline; N-ethoxycarbonyl-2-ethoxy-1,2-dihydroquinoline; BC-681.* $C_{14}H_{17}NO_3$; mol wt 247.29. C 68.00%, H 6.93%, N 5.66%, O 19.41%. Coupling agent used in the synthesis of peptides: Belleau, Malek, *J. Am. Chem. Soc.* **90,** 1651 (1968); Yajima, Kawatani, *Chem. Pharm. Bull.* **19,** 1905 (1971); Sipos, Gaston, *Synthesis* **1971,** 321. Preparation: Weinberg, **U.S. pats. 3,389,142** and **3,452,140** (1968, 1969, to Bristol-Myers). Has CNS depressant activity: Belleau, *J. Am. Chem. Soc.* **90,** 823 (1968). Pharmacological studies: Martel *et al., Can. J. Physiol. Pharmacol.* **47,** 909 (1969); Chang *et al., Pharmacol. Res. Commun.* **2,** 63 (1970); Weissman, Muren, *J. Med. Chem.* **14,** 49 (1971).

mp 56-57°. $bp_{0.1}$ 125-128°.
USE: In the synthesis of peptides.

3564. Eflornithine. *2-(Difluoromethyl)-DL-ornithine;* α-difluoromethylornithine; DFMO; RMI-71782. $C_6H_{12}F_2N_2O_2$; mol wt 182.17. C 39.56%, H 6.64%, F 20.86%, N 15.38%, O 17.57%. Irreversible inhibitor of ornithine decarboxylase. Prepn: B. W. Metcalf *et al., J. Am. Chem. Soc.* **100,** 2551 (1978); P. Bey *et al., J. Org. Chem.* **44,** 2732 (1979). Antiproliferative effects on cultured tumor cells: P. S. Mamont *et al., Biochem. Biophys. Res. Commun.* **81,** 58 (1978); on Ehrlich ascites cells in rats: L. Alhonen-Hongisto *et al., Acta Chem. Scand.* **B33,** 559 (1979). Inhibition of polyamine biosynthesis: E. Hölttä *et al., Biochem. J.* **178,** 109 (1979). Antitrypanosomal activity in mice: C. J. Bacchi *et al., Science* **210,** 332 (1980). Pharmacokinetics in humans: K. D. Haegele *et al., Clin. Pharmacol. Ther.* **30,** 210 (1981). Clinical evaluations in *Pneumocystis carinii* pneumonia: J. A. Golden *et al., West. J. Med.* **141,** 613 (1984); in trypanosomiasis: S. Van Nieuwenhove *et al., Trans. Royal Soc. Trop. Med. Hyg.* **79,** 692 (1985); in cancer chemotherapy: M. D. Abeloff *et al., Cancer Treat. Rep.* **70,** 843 (1986); V. A. Levin *et al., ibid.* **71,** 459 (1987).

Hydrochloride monohydrate, $C_6H_{12}F_2N_2O_2 \cdot HCl \cdot H_2O$, *Ornidyl.* Crystals from ethanol/water, mp 183°.

THERAP CAT: Antineoplastic; antipneumocystis; antiprotozoal (Trypanosoma).

3565. Efloxate. *[(4-Oxo-2-phenyl-4H-1-benzopyran-7-yl)oxy]acetic acid ethyl ester;* 7-flavone ethyl hydroxyacetate; ethyl flavon-7-yloxyacetate; ethyl 7-flavonoxyacetate; 7-flavonoxyacetic acid ethyl ester; oxyflavil; Re-1-0185; Recordil. $C_{19}H_{16}O_5$; mol wt 324.33. C 70.36%, H 4.97%, O 24.67%. Prepn: Colleoni, Setnikar, *Farmaco Ed. Sci.* **13**, 561 (1958); **Brit.** pats. **803,372, 824,547** (1958, 1959 to Recordati); Da Re, Colleoni, *Ann. Chim. (Rome)* **49**, 1632 (1959).

Crystals from 50% ethanol, mp 123-124°. Soluble in the usual organic solvents; slightly sol in water. LD_{50} i.p. in rats: 3200 mg/kg.

THERAP CAT: Vasodilator (coronary).

3566. Efonidipine. *5-(5,5-Dimethyl-1,3,2-dioxaphosphorinan-2-yl)-1,4-dihydro-2,6-dimethyl-4-(3-nitrophenyl)-3-pyridinecarboxylic acid 2-[phenyl(phenylmethyl)amino]ethyl ester, P-oxide;* 2-(N-benzylanilino)ethyl(±)-1,4-dihydro-2,6-dimethyl-4-(m-nitrophenyl)-5-phosphononicotinate, cyclic 2,2-dimethyltrimethylene ester. $C_{34}H_{38}N_3O_7P$; mol wt 631.67. C 64.65%, H 6.06%, N 6.65%, O 17.73%, P 4.90%. Dihydropyridine calcium channel blocker. Prepn: K. Seto *et al.*, **PCT Int.** pat. **Appl. 8,704,439;** *idem et al.*, **U.S.** pat. **4,885,284** (1987, 1989 both to Nissan); and crystal structure: R. Sakoda *et al.*, *Chem. Pharm. Bull.* **40**, 2362 (1992). Stereoselective synthesis of enantiomers and crystal structure of (S)-form: *idem et al., ibid.* 2377. Pharmacology: C. Shudo *et al.*, *J. Pharm. Pharmacol.* **45**, 525 (1993). Mechanism of action study: T. Yamashita *et al.*, *Japan. J. Pharmacol.* **57**, 337 (1991). Clinical study: T. Saito *et al.*, *Curr. Ther. Res.* **52**, 113 (1992).

Crystals from ethyl acetate, mp 169-170° (Sakoda); also reported as mp 155-156° (Seto).

Hydrochloride, $C_{34}H_{38}N_3O_7P \cdot HCl$. LD_{50} in mice (mg/kg): > 600 orally (Seto).

Hydrochloride ethanol, $C_{34}H_{38}N_3O_7P \cdot C_2H_5OH \cdot HCl$, *NZ-105, Landel.* Yellow crystals from aq ethanol, mp 151° (dec).

(S)- or (R)-Form, pale yellow crystals from ethanol, mp 190-192°. $[\alpha]_D^{25}$ + or −7.0° resp (c = 0.50 in chloroform).

THERAP CAT: Antihypertensive.

3567. Efrotomycin. *31-O-[6-Deoxy-4-O-(6-deoxy-2,4-di-O-methyl-α-L-mannopyranosyl)-3-O-methyl-β-D-allopyranosyl]-1-methylmocimycin; 31-O-[6-deoxy-4-O-(6-deoxy-2,4-di-O-methylhexopyranosyl)-3-O-methylhexopyranosyl]-1-methylmocimycin;* FR-02A; MK-621; Producil. $C_{59}H_{88}N_2O_{20}$; mol wt 1145.35. C 61.87%, H 7.74%, N 2.45%, O 27.94%. Antibiotic produced by *Streptomyces lactamdurans* NRRL 3802: R. G. Wax, W. M. Maiese, **Ger.** pat. **2,450,813** (1975 to Merck & Co.), *C.A.* **83**, 145755y (1975); R. G. Wax *et al., J. Antibiot.* **29**, 670 (1976). *In vitro* and *in vivo* activity: B. M. Frost *et al., ibid.* 1083; **32**, 626 (1979). Production and

growth promoting activity: W. M. Maiese, R. G. Wax, **U.S.** pat. **4,024,251** (1977 to Merck & Co.). Synergism with bottromycin, *q.v.:* B. M. Frost *et al., J. Antibiot.* **32**, 1046 (1979). Structure: R. S. Dewey *et al., ibid.* **38**, 1691 (1985). Stereospecific total synthesis: R. E. Dolle, K. C. Nicolaou, *J. Am. Chem. Soc.* **107**, 1691, 1695 (1985). HPLC determn in feeds: J. D. Strong, *Analyst* **111**, 853 (1986). Effect on gain and feed efficiency in swine: A. G. Foster *et al., J. Anim. Sci.* **65**, 877 (1987).

Pale yellow solid. uv max (pH 7): 232, 327 nm ($E_{1cm}^{1\%}$ 464, 216). LD_{50} in mice (g/kg): > 4 orally; > 2 s.c. (Frost).

THERAP CAT (VET): Growth stimulant.

3568. EGCG. *3,4,5-Trihydroxybenzoic acid, (2R-cis)-3,4-dihydro-5,7-dihydroxy-2-(3,4,5-trihydroxyphenyl)-2H-1-benzopyran-3-yl ester;* (−)-epigallocatechin 3-O-gallate; (−)-epigallocatechol gallate. $C_{22}H_{18}O_{11}$; mol wt 458.38. C 57.65%, H 3.96%, O 38.39%. Polyphenolic constituent of tea; inhibits tumor promotion. Initial identification and isoln from green tea: M. Tsujimura, *Bull. Agr. Chem. Soc. Japan.* **6**, 70 (1930); *C.A.* **25**, 3637 (1931); and crystallization: L. Vuataz *et al., J. Chromatog.* **2**, 173 (1959). Oxidation during tea fermentation: P. Coggon *et al., J. Agr. Food Chem.* **21**, 727 (1973). HPLC/MS extraction from black tea: R. G. Bailey *et al., J. Sci. Food Agric.* **66**, 203 (1994). HPLC determn in plasma and urine: M.-J. Lee *et al., Cancer Epidemiol. Biomark. Prev.* **4**, 393 (1995). Antitumor promoting activity: S. Yoshizawa *et al., Phytother. Res.* **1**, 44 (1987); T. Yamane *et al., Cancer Res.* **55**, 2081 (1995). Inhibition of metastasis in mice: S. Taniguchi *et al., Cancer Letters* **65**, 51 (1992). Brief review of early work: E. A. H. Roberts, *J. Sci Food Agric.* **3**, 193-198 (1952).

White crystals from water, mp 218°. $[\alpha]_D$ −185° ±2° (ethanol). uv max (ethanol): 275 nm (ϵ 11500).

3569. EGF-Urogastrone. EGF-URO. Related polypeptides that are both potent stimulators of cellular proliferation and inhibitors of gastric acid secretion. *Urogastrone* was originally detected as an antisecretory agent during experiments on human urine: J. S. Gray *et al., Science* **89**, 489 (1939); M. H. F. Friedman *et al., Proc. Soc. Exp. Biol. Med.* **41**, 509 (1935). Isoln: J. S. Gray *et al., Endocrinol.* **30**, 129 (1942); R. A. Gregory, *J. Physiol.* **129**, 528 (1955). Improved procedures led to the isoln and amino acid sequence determn of two polypeptides, β-*urogastrone* and γ-*urogas-*

trone: H. Gregory, *Nature* **257**, 325 (1975). These two peptides contain 3 disulfide bonds and consist of 53 and 52 amino acid residues, respectively. The only difference between them is the absence of a C-terminal arginine residue in the γ-peptide. *Epidermal growth factor,* or *EGF,* was isolated from submaxillary glands of male mice after first being detected during purification of a nerve growth promoting protein: S. Cohen, *Proc. Nat. Acad. Sci. USA* **46**, 302 (1960); idem, *J. Biol. Chem.* **237**, 1555 (1962). The primary structure of mouse EGF (mEGF) was found to be a 53 amino acid polypeptide containing 3 disulfide bonds: C. R. Savage *et al., ibid.* **247**, 7612 (1972); **248**, 7669 (1973). Sequence of an mEGF cDNA clone that predicts the synthesis of EGF as a large protein precursor of 1,168 amino acids: A. Gray *et al., Nature* **303**, 722 (1983). mEGF causes premature eye opening and stimulation of epithelial cell tissue growth when injected daily into newborn mice: S. Cohen, *loc. cit.* (1962); S. Cohen, J. M. Taylor, *Recent Progr. Horm. Res.* **30**, 533 (1974). The structural similarities between EGF and urogastrone led to experiments that established the gastric acid-inhibitory activity of mEGF and the proliferative activity of urogastrone, thus showing that both polypeptides have the same intrinsic biological activities, *cf.* H. Gregory, *loc. cit.* It has also been shown that both mEGF and urogastrone can share the same receptor sites in human tissue with almost equal affinities: M. D. Hollenberg, H. Gregory, *Life Sci.* **20**, 267 (1976). Human EGF has been isolated from urine: S. Cohen, G. Carpenter, *Proc. Nat. Acad. Sci. USA* **72**, 1317 (1975). The similarity of its amino acid composition and physicochemical properties to urogastrone has suggested the identity of these polypeptides. Although the mouse and human polypeptides are chemically distinct (16 differences in amino acid sequence), their identical intrinsic biological activities and their ability to share receptor sites with similar affinities has led to the use of the combined term EGF-urogastrone. *Reviews:* S. Cohen, R. Savage, *Recent Progr. Horm. Res.* **30**, 551-574 (1974); M. D. Hollenberg, *Vitam. Horm.* **37**, 69-110 (1979); D. Gospodarowicz, *Ann. Rev. Physiol.* **43**, 251-263 (1981); P. Walker, *J. Endocrinol. Invest.* **5**, 183-196 (1982); M. Das, *Int. Rev. Cytol.* **78**, 233-256 (1982).

Mouse EGF-URO. Mol wt 6041. uv max: 280 nm ($E_{1cm}^{1\%}$ 30.9). Isoelectric pt 4.60. Heat stable and non-dialyzable. Biological activity stable in boiling water but destroyed by heating in dil acid or alkali. Incubation with chymotrypsin or a bacterial protease also destroys biological activity.

Human EGF-URO, anthelone, anthelone U, uroanthelone, uroenterone. Mol wt 6201. Isoelectric pt 4.5. Very sol in water. Sol in methanol, ethylene glycol.

3570. Egg Oil. Oil of egg yolk. Obtained from fresh egg yolks by extraction with ethylene dichloride. Contains fatty glycerides, cholesterol, lecithin. The glyceride fraction is a mixture of the glycerides of satd and unsatd fatty acids. Palmitic, stearic, oleic, linoleic and clupanodonic acids have been isolated from both the glyceride and lecithin fractions. Preparation: Levin, Lerman, *J. Am. Oil Chem. Soc.* **28**, 441 (1951); U.S. pat. 2,503,312 (to Vio-Bin Corp.).

Dark oil. d 0.95. n_D^{20} 1.4790. Sol in the common organic solvents. Miscible with other oils. Insol in water, but disperses readily when shaken with it to form emulsions.

USE: In the formulation of hydrophilic ointment bases for medicinal ointments and cosmetic creams: Bandelin, Tuschhoff, *J. Am. Pharm. Assoc. [Pract. Pharm. Ed.]* **14**, 106, 120 (1953); *Schimmel-Briefs,* no. 235 (Oct. 1954).

3571. Eicosamethylnonasiloxane. $C_{20}H_{60}O_8Si_9$; mol wt 681.46. C 35.25%, H 8.87%, O 18.78%, Si 37.09%. Prepd by the reaction of hexamethyldisiloxane with octamethylcyclotetrasiloxane and sulfuric acid; by varying the proportions of these reactants, other methylpolysiloxanes of a desired average molecular weight may be prepd: Patnode, Wilcock, *J. Am. Chem. Soc.* **68**, 362 (1946).

$$(H_3C)_3Si-O-\left[\begin{array}{c} CH_3 \\ | \\ Si \\ | \\ CH_3 \end{array}-O\right]_7 Si(CH_3)_3$$

Liquid. $bp_{4.9}$ 173°. d 0.918. n_D^{20} 1.3980. Stable. Inert to most chemical reagents and rubber. Maintains about the same viscosity over a wide temp range. Sol in benzene and the lighter hydrocarbons; slightly sol in alc and the heavy hydrocarbons.

USE: As a basis for silicone oils or fluids designed to withstand extremes of temp; as a foam suppressant in petr lubricating oil.

3572. (all-Z)-5,8,11,14,17-Eicosapentaenoic Acid. all-cis-5,8,11,14,17-Eicosapentaenoic acid; all-cis-fatty acid 20:5 omega-3; EPA. $C_{20}H_{30}O_2$; mol wt 302.46. C 79.42%, H 10.00%, O 10.58%. Important polyunsaturated fatty acid of the marine food chain that serves as a precursor for the prostaglandin-3 and thromboxane-3 families. It differs from arachidonic acid, *q.v.* (the eicosatetraenoic acid that is a precursor for the prostaglandin and thromboxane-2 families) by the extra double bond between the third and fourth carbons from the "methyl end" of the molecule. Isoln from cod liver oil: E. Klenk, D. Eberhagen, *Z. Physiol. Chem.* **307**, 42 (1957). Enzymatic conversion to prostaglandin E_3: S. Bergström *et al., J. Biol. Chem.* **239**, PC 4006 (1964). Effects on role of platelets in thrombosis: K. C. Srivastava *et al., Biochem. Exp. Biol.* **16**, 317 (1980). Effects on prostacyclin-like material in human umbilical vasculature: J. Dyerberg, K. A. Jorgensen, *Artery* **8**, 12 (1980). A possible relationship between diets rich in EPA in marine oils and low rates of ischemic heart disease has been proposed: H. O. Bang *et al., Am. J. Clin. Nutr.* **33**, 2657 (1980); J. Dyerberg, *Philos. Trans. Roy. Soc. London [Biol.]* **294**, 373 (1981). Clinical evaluation of lipid lowering effect: Y. Nagakawa *et al., Atherosclerosis* **47**, 71 (1983); of use in rheumatoid arthritis: J. M. Kremer *et al., Ann. Int. Med.* **106**, 497 (1987); of use in Raynaud's phenomenon: R. A. DiGiacomo *et al., Am. J. Med.* **86**, 158 (1989).

Colorless oil. n_D^{20} 1.49865.
Ethyl ester, $C_{22}H_{34}O_2$, *Eskima.*
THERAP CAT: Antihyperlipoproteinemic.

3573. Einsteinium. Es; at. no. 99; valence 3, 2. Manmade radioactive element. No stable nuclides; known isotopes (mass numbers): 243-256. Longest-lived known isotope: 252 ($T_{1/2}$ 472 days, α-emitter, rel. at. mass 252.083). ^{253}Es ($T_{1/2}$ 20.47 days, α-emitter) originally discovered in debris from a thermonuclear test explosion in Nov. 1952. Reviews of discovery: A. Ghiorso *et al., Phys. Rev.* **99**, 1048-1049 (1955); A. Ghiorso, G. T. Seaborg, *Sci. Am.* **195** (6), 67-80 (1956). Prepn of isotopes: A. Chetham-Strode, Jr., L. W. Holm, *Phys. Rev.* **104**, 1314 (1956); B. G. Harvey *et al., ibid.* 1315; G. T. Seaborg, *J. Chem. Ed.* **36**, 38 (1959). *Reviews:* C. Keller, *The Chemistry of the Transuranium Elements* (Verlag Chemie, Weinheim, English Ed., 1971) pp 583-589, Silva, "Trans-Curium Elements" in *MTP Int. Rev. Sci.: Inorg. Chem., Ser. One* vol. **8**, A. G. Maddock, Ed. (University Park Press, Baltimore, 1972) pp 71-105; *Comprehensive Inorganic Chemistry* vol. **5**, J. C. Bailar, Jr. *et al.,* Eds. (Pergamon Press, Oxford, 1973) *passim; Handb. Exp. Pharmakol.* **36**, 7689-928 (1973); E. K. Hulet in *The Chemistry of the Actinide Elements* vol. **2**, J. J. Katz *et al.,* Eds. (Chapman and Hall, New York, 1986) pp 1071-1085; *The Elements Beyond Uranium,* G. T. Seaborg, W. D. Loveland, Eds. (John Wiley & Sons, Inc., New York, 1990) pp 28-38.

Metal; two crystalline forms: double hexagonal close-packed α-form exists below 860°; face-centered cubic β-form, d 8.84. mp 860 ±50°.

Caution: Radiation hazard; handling requires special equipment and shielding facilities (Katz *et al., loc cit.,* vol. 2, p. 1128).

3574. Elaidic Acid. *(E)-9-Octadecenoic acid.* $C_{18}H_{34}O_2$; mol wt 282.47. C 76.54%, H 12.13%, O 11.33%. Stereoisomer of oleic acid, *q.v.* Prepd by the action of HNO_2 on oleic acid. ^{13}C-NMR: J. G. Batchelor *et al., J. Org. Chem.* **39**, 1698 (1974).

H₃C(CH₂)₇ ⌇ (CH₂)₇COOH

White leaflets. d⁷⁹ 0.851. mp 44-45°; also stated as 51°. bp₁₀₀ 288°; bp₁₅ 234°. n_D^{100} 1.4308.

3575. Elaiomycin. *4-Methoxy-3-(1-octenyl-ONN-azoxy)-2-butanol;* (2*S*,3*S*)-4-methoxy-3-(1'-*cis*-octenyl-*cis*-azoxy)-2-butanol; D-*threo*-4-methoxy-3-(1-octenylazoxy)-2-butanol. $C_{13}H_{26}N_2O_3$; mol wt 258.36. C 60.44%, H 10.14%, N 10.84%, O 18.58%. Antibiotic substance produced by *Streptomyces hepaticus:* Haskell *et al., Antibiot. & Chemother.* **4,** 141 (1954); Ehrlich *et al., ibid.* 338; Anderson *et al., ibid.* **6,** 100 (1956). Inhibits the growth of *Mycobacterium tuberculosis* var. *hominis in vitro:* Karlson, *ibid.* **12,** 446 (1962). Reported to have carcinogenic activity: Schoental, *Nature* **221,** 765 (1969). Structure: Stevens *et al., J. Am. Chem. Soc.* **78,** 3229 (1956); **80,** 6088 (1958). Configuration: Stevens *et al., ibid.* **81,** 1435 (1959); McGahren, Kunstmann, *ibid.* **92,** 1587 (1970). Total synthesis: R. A. Moss, M. Matsuo, *ibid* **99,** 1643 (1977). Biosynthesis: R. J. Parry *et al., ibid.* **104,** 339 (1982).

[chemical structure diagram]

Pale yellow oil. Stable to air. $[\alpha]_D^{26}$ +38.4° (c = 2.8 in abs ethanol). n_D^{25} 1.4798. uv max: 237.5 nm ($F_{1cm}^{1\%}$ 428). Spar ingly sol in water; sol in practically all of the common organic solvents. Stable in neutral or slightly acidic aq solns. Dec into a yellow product when dissolved in 0.1*N* NaOH.

Acetate, $C_{15}H_{28}N_2O_4$, oily liquid, bp₀.₅ 84-90°. $[\alpha]_D^{27}$ +25.3° (c = 3 in alc). uv max (alc): 237.5 nm (ε 11000).

3576. Elastase. Pancreatopeptidase E; Elaszym. A protease of broad specificity, obtained from dried pancreas. Mol wt about 25,000. Capable of hydrolyzing proteins at *N*-terminal peptide bonds of aliphatic residues. Breaks down elastin, the specific protein of elastic fibers. Digests other proteins, such as fibrin, hemoglobin, albumin, soybean proteins and casein. Isoln by an adsorption method: Bagdy, Banga, *Acta Physiol. Acad. Sci. Hung.* **11,** 371 (1957); *cf.* Lewis *et al., J. Biol. Chem.* **222,** 705 (1956); Miller, Blum, *ibid.* **218,** 131 (1956). Porcine elastase is a single polypeptide chain of 240 amino acid residues: Shotton, Hartley, *Biochem. J.* **131,** 643 (1973).

White, lyophilized powder of slightly yellowish shade. Isoelectric point (pI): 9.5. Optimum pH between 8.1 and 8.8 according to the buffer employed. At refrigerator temps (4 to 6°) the thawed enzyme remains stable for several days at pH values between 5 and 10. Rapidly inactivated in highly acid media.

3577. Elastin. Elastic, load-bearing protein fibers of animal connective tissue, particularly the ligaments of the vertebrae and the walls of the large arteries. Elastin is an insoluble, highly cross-linked hydrophobic protein, rich in nonpolar amino acid residues, such as valine, leucine, isoleucine, and phenylalanine. On the average, about every third residue is glycine and about every ninth residue is a prolyl residue. Unlike collagen, *q.v.*, which is rich in hydroxyproline, elastin contains only modest amounts. At least two types of elastin are distinguishable: type I elastin isolated from *liga-mentum nuchae*, aorta or skin and type II elastin isolated from cartilage. The most ready source of high purity elastin is the *ligamentum nuchae* of the larger ruminants: Partridge *et al., Biochem. J.* **61,** 11 (1955). Amino acid composition studies: Petruska, Sandberg, *Biochem. Biophys. Res. Commun.* **33,** 222 (1968). Its elastic properties are brought about by a cross-linked structure. Identification of new cross-linking amino acids in elastin: Partridge *et al., Biochem. J.* **93,** 30C (1964); Franzblau *et al., Biochem. Biophys. Res. Commun.* **21,** 575 (1965); Starcher *et al., Biochemistry* **6,** 2425 (1967). Molecular model: Gray *et al., Nature* **246,** 461 (1973). *Review:* Partridge, *Advan. Protein Chem.* **17,** 227-302 (1962); C. Franzblau, "Elastin" in *Com-*

prehensive Biochemistry vol. **26c,** M. Florkin, E. H. Stotz, Eds. (Elsevier, New York, 1971) pp 659-712; Ross, Bornstein, *Sci. Amer.* **224,** 44 (June, 1971). Book (in English): I. Banga, *Structure and Function of Elastin and Collagen* (Akademiai Kiado, Budapest, 1967). Series of articles on structure, biosynthesis, and immunology: *Methods in Enzymology* vol. **82,** L. W. Cunningham, D. W. Frederiksen, Eds. (Academic Press, New York, 1982) pp 559-765.

Purified elastin has a pale yellow color and a bluish fluorescence in uv light. Resists acid and alkaline hydrolysis. Practically insol in a wide range of hydrogen-bond-breaking solvents at temps up to 100° and swells, but does not dissolve, in phenolic solvents. It appears practically impossible to bring elastin into soln except by hydrolytic reagents capable of rupturing peptide bonds. Elastin is one of only a few polymeric substances which, in the presence of water, exist in a form with rubber-like extensibility and low modulus of elasticity. Enzymes which dissolve fibers of the insol protein elastin undamaged and free from contamination are called elastases.

3578. Elcatonin. *1-Butanoic acid-26-L-aspartic acid-27-L-valine-29-L-alanine-1,7-dicarbacalcitonin (salmon);* [1,4,7,-10,13,16-hexaazacyclotricosane cyclic peptide deriv; [aminosuberic acid 1,7]-eel calcitonin; [ASU¹,⁷]-E-CT; carbocalcitonin; HC-58; Calcinil; Carbicalcin; Elcitonin; Turbocalcin. $C_{148}H_{244}N_{42}O_{47}$; mol wt 3363.82. C 52.85%, H 7.31%, N 17.49%, O 22.35%. Synthetic analog of eel calcitonin; amino acid sequence also similar to that of salmon calcitonin. Polypeptide hormone more effective in the reduction of plasma calcium. More stable than natural calcitonins because of the absence of disulfide bridges. Prepn: S. Sakakibara *et al.,* Ger. pat. **2,616,399;** *eidem,* U.S. pat. **4,086,221** (1976, 1978 both to Toyo Jozo); T. Morikawa *et al., Experientia* **32,** 1104 (1976). *In vitro* growth inhibition of human breast cancer cells: Y. Iwasaki *et al., Biochem. Biophys. Res. Commun.* **110,** 235 (1983). Clinical pharmacology: T. Ishioka, *Pharmatherapeutica* **1,** 625 (1977). Clinical study in Paget's disease: A. Caniggia *et al., Minerva Med.* **74,** 993 (1983); in osteoporosis: Italian Osteoporosis Network, *Curr. Ther. Res.* **45,** 502 (1989).

[chemical structure / peptide sequence diagram]

THERAP CAT: Calcium regulator. Treatment of Paget's disease.

3579. Eledoisin. $C_{54}H_{85}N_{13}O_{15}S$; mol wt 1188.41. C 54.58%, H 7.21%, N 15.32%, O 20.19%, S 2.70%. A hendecapeptide from the posterior salivary glands of eledone spp (small octopus spp): Anastasi, Erspamer, *Brit. J. Pharmacol.* **19,** 326 (1962); *eidem, Experientia* **18,** 58 (1962). Synthesis: Sandrin, Boissonnas, *ibid.* 59; *eidem, Helv. Chim. Acta* **47,** 1294 (1964); Lubke *et al., Ann.* **679,** 195 (1964); **Brit.** pat. **984,810** (1965 to Farmitalia); solid-phase synthesis: P. G. Pietta *et al., J. Org. Chem.* **39,** 44 (1974). Its physiologic action resembles that of the other tachykinins, substance P, *q.v.,* and physalaemin, *q.v.* Stimulates extravascular smooth muscle, acts as a potent vasodilator and hypotensive agent; in certain species causes salivation, and increases capillary permeability. Pharmacology: Erspamer, Erspamer, *Brit. J. Pharmacol.* **19,** 337 (1962); Sicuteri *et al., Experientia* **19,** 44 (1963); G. Bertaccini, *Pharmacol. Rev.* **28,** 127 (1976). *Review:* Erdös, *Advan. Pharmacol.* **4,** 64-72 (1966).

5-oxoPro–Pro–Ser–Lys–Asp–Ala–Phe–Ile–Gly–Leu–MetNH₂

Sesquihydrate, powder, dec 230°. $[\alpha]_D^{22}$ −44° (c = 1 in 95% acetic acid). Slowly loses activity when incubated in blood.

Trifluoroacetate, $C_{54}H_{85}N_{13}O_{15}S.C_2HF_3O_2$, Eloisin.

THERAP CAT: Hypotensive; stimulator of lacrimal secretion.

3580. Element 104. *Unnilquadium;* Dubnium (IUPAC proposed name); kurchatovium; Rutherfordium (ACS proposed name). Proposed symbols: Unq, Db (IUPAC), Ku, Rf (ACS). Initial production of element 104 claimed by two groups. Prepn of isotope $^{260}104$ ($T_{1/2}$ ~300 msec; spontaneous fission) by ^{242}Pu (^{22}Ne,4n): G. N. Flerov *et al., At. Energ. (USSR)* **17**, 310 (1964), *C.A.* **62**, 189d (1965); *eidem, Phys. Letters* **13**, 73 (1964); I. Zvara *et al., Radiokhimiya* **11**, 161, 163 (1969). Prepn of two α-emitting isotopes, $^{257}104$ ($T_{1/2}$ 4.5 seconds) by ^{249}Cf(^{12}C,4n) and $^{259}104$ ($T_{1/2}$ 3 seconds) by ^{249}Cf (^{13}C,3n): A. Ghiorso *et al., Phys. Rev. Letters* **22**, 1317 (1969). Ten isotopes reported; mass numbers 253-262. Longest-lived known isotope, $^{261}104$, α-emitter, $T_{1/2}$ 65 ±10 seconds; produced by ^{248}Cm(^{18}O,5n): A. Ghiorso *et al., Phys. Letters B* **32**, 95 (1970). Chemical properties of isotopes and revised $T_{1/2}$ 21 ±1 msec for $^{260}104$: L. P. Somerville *et al., Phys. Rev.* **C31**, 1801-1815 (1985). Solution chemistry: K. R. Czerwinski *et al., Radiochim. Acta* **64**, 23 (1994); *eidem, ibid.* 29. Reviews of history, prepn and properties: C. Keller, *The Chemistry of the Transuranium Elements* (Verlag Chemie, Weinheim, English Ed., 1971) pp 613-618; Silva, "Trans-Curium Elements" in *MTP Int. Rev. Sci.: Inorg. Chem., Ser. One* vol. **8**, A. G. Maddock, Ed. (University Park Press, Baltimore, 1972) pp 71-105; R. J. Silva in *The Chemistry of the Actinide Elements* vol. **2**, J. J. Katz *et al.*, Eds. (Chapman and Hall, New York, 1986) pp 1103-1106; E. K. Hyde *et al., Radiochim. Acta* **42**, 57-102 (1987); Transfermium Working Group, *Pure Appl. Chem.* **65**, 1757-1814 (1993); M. Schädel, *Inst. Phys. Conf. Ser.* **132**, 413-422 (1993). Review of chemistry: D. C. Hoffman, *Proc. of Robert A. Welch Found. Conf. on Chem. Res., XXXIV, Fifty Years with Transuranium Elements* (Houston, Texas, 1990) pp 255-276.

3581. Element 105. *Unnilpentium;* Joliotium (IUPAC proposed name); Hahnium (ACS proposed name); nielsbohrium; eka-tantalum; bohrium. Proposed symbols: Unp, Jl (IUPAC); Ha (ACS); Ns. Prepn of α-emitting $^{260}105$ isotope ($T_{1/2}$ 1.6 ±0.3 sec) by ^{249}Cf (^{15}N,4n): A. Ghiorso *et al., Phys. Rev. Letters* **24**, 1498 (1970). Prepn of $^{261}105$ isotope (decay by spontaneous fission, $T_{1/2}$ 1.8 ±0.6 sec) by ^{243}Am (^{22}Ne,4n): G. N. Flerov *et al., At. Energy (USSR)* **29**, 243 (1970); *eidem, Nucl. Phys. A* **160**, 181 (1971); *eidem, Proc. Int. Conf. Heavy Ion Phys.* **1971** pp 125-143, *C.A.* **80**, 139758y (1974). Prepn of α-emitting isotopes $^{261}105$ ($T_{1/2}$ 1.7 sec) by ^{249}Bk (^{16}O,4n) or by ^{250}Cf (^{15}N,4n); $^{262}105$ ($T_{1/2}$ 40.0 ±10 sec, longest-lived known isotope) by ^{249}Bk(^{18}O,5n): A. Ghiorso *et al., Phys. Rev.* **C4**, 1850 (1971); and revised $T_{1/2}$ 34.1 ±4.6 sec for $^{262}105$: C. E. Bemis, Jr. *et al., Phys. Rev. Letters* **39**, 1246 (1977). Discussion of conflicting claims of discovery: Holcomb, *Science* **168**, 810 (1970); G. N. Flerov, *ibid.* **170**, 15 (1970); A. Ghiorso, *ibid.* **171**, 127 (1971). Solution chemistry: K. E. Gregorich *et al., Radiochim. Acta* **43**, 223 (1988); M. K. Gober *et al., ibid.* **57**, 77 (1992); and identification of $^{263}105$ isotope ($T_{1/2}$ 27 sec, α-emitter) by ^{249}Bk(^{18}O,4n): M. Schädel *et al., ibid.* 85. Reviews of history, prepn and properties: C. Keller, *The Chemistry of the Transuranium Elements* (Verlag Chemie, Weinheim, English Ed., 1971) pp 619-622; Silva, "Trans-Curium Elements" in *MTP Int. Rev. Sci.: Inorg. Chem., Ser. One* vol. **8**, A. G. Maddock, Ed. (University Park Press, Baltimore, 1972) pp 71-105; R. J. Silva in *The Chemistry of the Actinide Elements* vol. **2**, J. J. Katz *et al.*, Eds. (Chapman and Hall, New York, 1986) pp 1106-1109; E. K. Hyde *et al., Radiochim. Acta* **42**, 57-102 (1987); Transfermium Working Group, *Pure Appl. Chem.* **65**, 1757-1814 (1993). Review of chemistry: D. C. Hoffman, *Proc. Robert A. Welch Found. Conf. on Chem. Res., XXXIV, Fifty Years with Transuranium Elements* (Houston, Texas, 1990) pp 255-276.

3582. Element 106. *Unnilhexium.* Rutherfordium (IUPAC proposed name); Seaborgium (IUPAC proposed name). Proposed symbols: Unh; Rf (IUPAC); Sg (ACS). Prepn of isotope $^{263}106$ (α-emitter; $T_{1/2}$ 0.9 ±0.2 seconds;

longest-lived known isotope) by ^{249}Cf (^{18}O,4n): A. Ghiorso *et al., Phys. Rev. Lett.* **33**, 1490 (1974). Four known isotopes: 259-261, 263. Confirmation of discovery: K. E. Gregorich *et al., ibid.* **72**, 1423 (1994). Review of history, prepn and properties: R. J. Silva in *The Chemistry of the Actinide Elements* vol. **2**, J. J. Katz *et al.*, Eds. (Chapman and Hall, New York, 1986) pp 1108-1110; P. Armbruster, *Proc. Int. Sch. Phys. "Enrico Fermi"*, 222-240 (1986); G. N. Flerov, G. M. Ter-Akopian, *Prog. Particle Nucl. Phys.* **19**, 197-239 (1987); G. T. Seaborg, W. D. Loveland, *The Elements Beyond Uranium* (John Wiley & Sons, Inc., New York, 1990) p 54-56; Transfermium Working Group, *Pure Appl. Chem.* **65**, 1757-1814 (1993).

3583. Element 107. *Unnilseptium.* Bohrium (IUPAC proposed name); Neilsbohrium (ACS proposed name). Proposed symbols: Uns; Bh (IUPAC); Ns (ACS). Three known isotopes: 261, 262, 262m. Prepn of α-emitting isotope $^{261}107$ by ^{209}Bi(^{54}Cr,2n); decay by spontaneous fission, $T_{1/2}$ 1-2 msec: Y. T. Organessian *et al., Nucl. Phys. A* **273**, 505 (1976); G. N. Flerov, *C.A.* **87**, 173830v (1977); A. S. Iljinov *et al., Report* **JINR-E-7-9686** (1976), *C.A.* **90**, 62354k (1979). Prepn of isotopes including $^{262}107$ ($T_{1/2}$ 102 ±26 msec, longest-lived known isotope, α-emitter) by ^{209}Bi (^{54}Cr,1n): G. Münzenberg, *et al., Z. Phys.* **A300**, 107 (1981); and revised $T_{1/2}$ 11.8 msec for $^{261}107$: *eidem, ibid.* **A333**, 163 (1989). Review of history, prepn and properties: R. J. Silva in *The Chemistry of the Actinide Elements* vol. **2**, J. J. Katz *et al.*, Eds. (Chapman and Hall, New York, 1986) pp 1110-1112; N. Flerov, G. M. Ter-Akopian *Prog. Particle Nucl. Phys.* **19**, 197-239 (1987); G. T. Seaborg, W. D. Loveland, *The Elements Beyond Uranium* (John Wiley & Sons, Inc., New York, 1990) p 56-57.

3584. Element 108. *Unniloctium.* Hahnium (IUPAC proposed name); hassium (ACS proposed name). Proposed symbols: Uno, Hn (IUPAC), Hs (ACS). Prepn of isotope $^{265}108$ ($T_{1/2}$ ~1.8 msec, α-decay) by ^{208}Pb(^{58}Fe,n): G. Münzenberg *et al., Z. Physik.* **A317**, 235 (1984); of isotopes $^{263}108$ ($T_{1/2}$ ~1.9 sec) by ^{209}Bi (^{55}Mn,n) and $^{264}108$ by ^{207}Pb (^{58}Fe,n) or ^{208}Pb(^{58}Fe,2n): Y. T. Oganessian *et al., ibid.* **A319**, 215 (1984); of isotope $^{264}108$ ($T_{1/2}$ ~76 μsec, α-decay) by ^{207}Pb(^{58}Fe,n): G. Münzenberg *et al., ibid.* **A324**, 489 (1986). Review of synthesis: *eidem, ibid.* **A328**, 49-59 (1987); G. N. Flerov, G. M. Ter-Akopian, *Prog. Particle Nucl. Phys.* **19**, 197-239 (1987); G. T. Seaborg in *Transuranium Elements: A Half Century*, L. R. Morss, J. Fuger, Eds. (Am. Chem. Soc., Washington DC, 1992) p 40-41.

3585. Element 109. *Unnilennium.* Meitnerium (IUPAC and ACS proposed name). Proposed symbols: Une, Mt. Prepn of isotope $^{266}109$ ($T_{1/2}$ ~3.5 msec, α-decay) by ^{209}Bi(^{58}Fe,n): G. Münzenberg *et al., Z. Physik.* **A309**, 89 (1982); *eidem, ibid.* **A315**, 145 (1984). Review of synthesis: G. N. Flerov, G. M. Ter-Akopian, *Prog. Particle Nucl. Phys.* **19**, 197-239 (1987); G. T. Seaborg in *Transuranium Elements: A Half Century*, L. R. Morss, J. Fuger, Eds. (Am. Chem. Soc., Washington DC, 1992) p 41-42.

3586. Elenolide. *[S-(E)]-4-(1-Formyl-1-propenyl)-3,4-dihydro-2-oxo-2H-pyran-5-carboxylic acid methyl ester.* $C_{11}H_{12}O_5$; mol wt 224.21. C 58.93%, H 5.39%, O 35.68%. Hypotensive lactone from fruits, bark and leaves of the olive tree, *Olea europaea* L., Oleaceae: Veer *et al., Rec. Trav. Chim.* **76**, 839 (1957); **Brit. pat. 789,427** (1958 to Organon); U.S. pat. **3,033,877** (1962 to Organon). Structure: Panizzi *et al., Gazz. Chim. Ital.* **90**, 1464 (1960); Beyerman *et al., Bull. Soc. Chim. France* **1961**, 1812. Synthesis from secologanin: L. F. Tietze, H. C. Uzar, *Angew. Chem. Int. Ed.* **18**, 539 (1979).

Needles from hot alc, mp 155.5°. [α]$_D^{20}$ +369° (chloroform). uv max (ethanol): 225, 317 nm (log ε 4.29, 1.75).

3587. Elgodipine. (±)-4-(1,3-Benzodioxol-4-yl)-1,4-dihydro-2,6-dimethyl-3,5-pyridinedicarboxylic acid 2-[[(4-fluorophenyl)methyl]methylamino]ethyl 1-methylethyl ester; 2-[(p-fluorobenzyl)methylamino]ethyl isopropyl (±)- 1,4-dihydro-2,6-dimethyl-4-[2,3-(methylenedioxy)phenyl]-3,5-pyridinedicarboxylate. $C_{29}H_{33}FN_2O_6$; mol wt 524.59. C 66.40%, H 6.34%, F 3.62%, N 5.34%, O 18.30%. Dihydropyridine calcium channel blocker. Prepn: C. F. Torija, J. A. Galiano Ramos, *Eur.* pat. *Appl.* 302,980; *eidem*, U.S. pat. **4,952,592** (1989, 1990 both to Inst. Invest. Desarr. Quimicobiol.). *In vitro* and *in vivo* cardiovascular effects: J. Tamargo *et al.*, *Arzneimittel-Forsch.* **41**, 895 (1991). Hemodynamic profile: C. Drieu la Rochelle *et al.*, *Brit. J. Pharmacol.* **111**, 49 (1994). Clinical evaluation in coronary artery disease: H. Suryapranata *et al.*, *Am. J. Cardiol.* **69**, 1171 (1992). Clinical evaluation in chronic stable angina: D. U. Acharya *et al.*, *Eur. Heart J.* **15**, 665 (1994).

Hydrochloride, $C_{29}H_{33}FN_2O_6$·HCl, *IQB-875.* Crystals from acetone, mp 194-195°. LD$_{50}$ i.p. in mice: 30-40 mg/kg (Torija, Galiano Ramos).

THERAP CAT: Antianginal.

3588. Ellagic Acid. 2,3,7,8-Tetrahydroxy[1]benzopyrano[5,4,3-cde][1]benzopyran-5,10-dione; 4,4′,5,5′,6,6′-hexahydrodiphenic acid 2,6,2′,6′-dilactone; benzoaric acid; Lagistase. $C_{14}H_6O_8$; mol wt 302.20. C 55.64%, H 2.00%, O 42.35%. Occurs free or combined in galls. Isoln from the kino of *Eucalyptus maculata* Hook and *E. hemipholia* F. Muell., *Myrtaceae:* Gell *et al.*, *Aust. J. Chem.* **11**, 372 (1958); Hills, Carle, *ibid.* **16**, 147 (1963). Prepd by sodium persulfate oxidation of gallic acid or by acid hydrolysis of crude tannin from walnuts: Perkin, Nierenstein, *J. Chem. Soc.* **87**, 1412 (1905); Jurd, *J. Am. Chem. Soc.* **78**, 3445 (1956); **79**, 6043 (1957). Purification: Fr. pat. **1,478,523** (1967 to Prod. Chim. Celluloses Rey), *C.A.* **68**, 78267r (1968). Physicochemical properties: Press, Hardcastle, *J. Appl. Chem.* **19**, 247 (1969). Pharmacology: Bhargava *et al.*, *J. Pharm. Sci.* **57**, 1728 (1968). Ellagic acid is a potent antagonist of the mutagenicity of bay-region diol epoxides of several aromatic hydrocarbons. Inhibition of the mutagenicity of the ultimate carcinogenic metabolite of benzo[a]pyrene: A. W. Wood *et al.*, *Proc. Nat. Acad. Sci. USA* **79**, 5513 (1982). Study of the reaction between this metabolite and ellagic acid: J. M. Sayer *et al.*, *J. Am. Chem. Soc.* **104**, 5562 (1982). Brief discussion of ellagic acid as a prototype of a new class of cancer-preventing drugs: J. Fox, *Chem. & Eng. News* **60**, 26 (Oct. 25, 1982).

Cream colored needles from pyridine, mp > 360°. uv max (ethanol): 366, 255 nm (log ε 3.93, 4.60). Slightly sol in water or alcohol; sol in alkalies, in pyridine. Practically insol in ether.

Tetraacetate, $C_{22}H_{14}O_{12}$, needles from acetic anhydride, mp 340°.

THERAP CAT: Hemostatic.

3589. Ellipticine. 5,11-Dimethyl-6H-pyrido[4,3-b]carbazole. $C_{17}H_{14}N_2$; mol wt 246.31. C 82.90%, H 5.73%, N 11.37%. Antitumor alkaloid isolated from *Ochrosia elliptica* Labill., *O. sandwicensis* A. DC., *O. viellardii, O. silvatica, Apocynaceae:* Goodwin *et al.*, *J. Am. Chem. Soc.* **81**, 1903 (1959); Kan Fan *et al.*, *Phytochemistry* **9**, 1351 (1970); Cosson, Schmid, *ibid.* 1353. Structure and synthesis: Woodward *et al.*, *J. Am. Chem. Soc.* **81**, 4434 (1959). Recent syntheses: J. Bergman, R. Carlsson, *Tetrahedron Letters* **1977**, 4663; D. A. Taylor *et al.*, *J. Chem. Res. (M)* **1979**, 4801; S. Kano *et al.*, *J. Org. Chem.* **46**, 2979 (1981); R. Besselievre, H. P. Husson, *Tetrahedron* **37**, 241 (1981). Toxicity study: Rakieten *et al.*, *U.S. Govt. Res. Develop. Rep.* **67**, 38 (1967), *C.A.* **68**, 20827g (1968). Review of syntheses: M. Sainsbury, *Synthesis* **1977**, 437-448; R. Barone, M. Chanon, *Heterocycles* **16**, 1357-1365 (1981). Review of antineoplastic activity: K. W. Kohn *et al.*, *Antibiotics* vol. 5 (pt. 2), F. E. Hahn, Ed. (Springer-Verlag, New York, 1979) pp 195-213.

Bright yellow needles from ethyl acetate, mp 311-315° (dec). uv max: 239, 277, 286, 294, 332, 382, 400 nm (log ε 4.23, 4.61, 4.76, 4.74, 3.65, 3.61, 3.53). LD$_{50}$ in mice (mg/kg): 19.5-22.4 i.v.; 178-204 orally (Rakieten).

3590. Elliptinium Acetate. 9-Hydroxy-2,5,11-trimethyl-6H-pyrido[4,3-b]carbazolium acetate (salt); 9-hydroxy-2-methylellipticinium acetate; HME; NSC-264137; Celiptium. $C_{20}H_{20}N_3O$; mol wt 336.39. C 71.41%, H 5.99%, N 8.33%, O 14.27%. Deriv of ellipticine, *q.v.*, with anticancer activity. Prepn: J. LePecq *et al.*, *Ger.* pat. **2,618,223**; J. B. LePecq, C. Paoletti, U.S. pat. **4,310,667** (1976, 1982 both to Agence Nat. Valorisation Recher.). Activity vs L 1210 mouse leukemia: *eidem*, *Compt. Rend. Ser. D.* **281**, 1365 (1975). HPLC determn: G. Muzard, J. B. LePecq, *J. Chromatog.* **169**, 446 (1979). Metabolism, disposition: N. Van-Bac *et al.*, *Cancer Treat. Rep.* **64**, 879 (1980). Antitumor activity, pharmacology, toxicity study: C. Paoletti *et al.*, *Recent Results Cancer Res.* **74**, 107 (1980). Use in treatment of breast cancers: J. Rouesse *et al.*, *Bull. Cancer* **68**, 437 (1981). Cytochemical and autoradiographic study: N. Sales, E. Puvion, *Eur. J. Cancer Clin. Oncol.* **18**, 291 (1982).

THERAP CAT: Antineoplastic.

3591. Elliptone. 12,12a-Dihydro-8,9-dimethoxy[1]benzopyrano[3,4-b]furo[2,3-h][1]benzopyran-6(6aH)-one; derride. $C_{20}H_{16}O_6$; mol wt 352.34. C 68.18%, H 4.58%, O 27.25%. Found in derris resins of low rotenone content. Isoln of naturally occurring (−)-form from *Derris elliptica* (Wall.) Benth., *Leguminosae:* Buckley, *J. Soc. Chem. Ind.* **55**, 285T (1936); Harper, *Chem. & Ind. (London)* **57**, 1059 (1938); **58**, 292 (1939); *J. Chem. Soc.* **1939**, 1099; *see also* Meyer, Koolhaas, *Rec. Trav. Chim.* **58**, 207 (1939). Structure: Harper, *J. Chem. Soc.* **1939**, 1424; **1942**, 593. Comparison with similar substances: Seiferle, Frear, *Ind. Eng. Chem.* **40**, 683 (1948). Total synthesis of (±)-form: Fukami *et al., Agr. Biol. Chem.* **29**, 82 (1965), *C.A.* **62**, 13115e (1963). Synthesis of (−)-form from rotenone *(q.v.):* P. B. Anzeveno, *J. Heterocycl. Chem.* **16**, 1643 (1979).

Needles from ethanol, mp 160° (Harper), also reported as mp 177-178° (Anzeveno). $[\alpha]_D^{20}$ −18° (benzene); $[\alpha]_D^{20}$ +55° (acetone).

3592. Eltoprazine. *1-(2,3-Dihydro-1,4-benzodioxin-5-yl)piperazine;* 1-(1,4-benzodioxan-5-yl)piperazine. $C_{12}H_{16}$-N_2O_2; mol wt 220.27. C 65.43%, H 7.32%, N 12.72%, O 14.53%. Selective serotonin $5HT_1$ agonist; psychotropic agent with antiaggressive activity. Prepn: J. Hartog *et al.,* **Eur. pat. Appls. 138,280; 189,612** (1985, 1986 both to Duphar). Prepd (not claimed): J. Hartog *et al.,* U.S. pat. **4,833,142** (1989 to Duphar). Crystal structure: M. L. Verdonk *et al., Acta Crystallog.* **C48**, 2271 (1992). Series of articles on pharmacology, receptor binding and mechanism of action: *Drug Metabol. Drug Interact.* **8**, 1-186 (1990). Pharmacokinetics: M. H. de Vries *et al., Eur. J. Clin. Pharmacol.* **41**, 485 (1991). Clinical evaluation in self-injury: W. M. A. Verhoeven *et al., Lancet* **340**, 1037 (1992); in violent psychiatric patients: J. Tiihonen *et al., ibid.* **341**, 307 (1993).

Hydrochloride, $C_{12}H_{16}N_2O_2 \cdot HCl$, *DU-28853.* Crystals from ethanol, mp 256-258°. Soly in water: 19 g/100 ml.
THERAP CAT: Serenic.

3593. Elymoclavine. *8,9-Didehydro-6-methylergoline-8-methanol.* $C_{16}H_{18}N_2O$; mol wt 254.33. C 75.56%, H 7.13%, N 11.01%, O 6.29%. An ergot alkaloid obtained from cultures of fungi parasitic on *Elymus mollis* Trin.: Abe *et al.,* U.S. pat. **2,835,675** (1958). Found in fungi parasitic on *Pennisetum typhoideum* Rich.: Stoll *et al., Helv. Chim. Acta* **37**, 1815 (1954). Structure and stereochemistry: Schreier, *Helv. Chim. Acta* **41**, 1984 (1958). Biosyntheses utilizing *Claviceps* cultures: Naidoo *et al., Chem. Commun.* **1970**, 472; Ogunlana *et al., ibid.* 775; Seiler *et al., ibid.* 1394; Cavender, Anderson, *Biochim. Biophys. Acta* **208**, 345 (1970).

Monoclinic prisms from methanol, 248-252° (dec). $[\alpha]_D^{20}$ −59° (c = 0.1 in ethanol) (Abe); $[\alpha]_D^{20}$ −152° (c = 0.9 in pyridine) (Stoll). uv max: 227, 283, 293 nm (log ε 4.31, 3.84, 3.76). Fairly sol in water with alkaline reaction. Sol in pyridine; very slightly sol in organic solvents.

3594. Emamectin. (4″*R*)-5-*O*-demethyl-4″deoxy-4″-(methylamino)avermectin A_{1a} and (4″*R*)-5-*O*-demethyl-25-de(1-methylpropyl)-4″-deoxy-4″-(methylamino)-25-(1-methylethyl)avermectin A_{1a} (9:1); Affirm; Ban-Lep; Proclaim. Mixture of semi-synthetic avermectins *q.v.,* consisting (9:1) of *4″-epi-(methylamino)-4″-deoxyavermectin B_{1a},* ($C_{49}H_{75}NO_{13}$) and *4″-epi-(methylamino)-4″-deoxyavermectin B_{1b},* ($C_{48}H_{73}NO_{13}$). Prepn: H. Mrozik, U.S. pat. **4,874,749** (1989 to Merck & Co.); and insecticidal activity: H. Mrozik *et al., Experientia* **45**, 315 (1989). Synthesis: R. J. Cvetovich *et al., J. Org. Chem.* **59**, 7704 (1994). *Review:* M. H. Fisher, *ACS Symposium Series* **524**, 169-182 (1993); H. Mrozik, *ibid.* **551**, 54-73 (1994).

Component B_{1a} R = CH_2CH_3
Component B_{1b} R = CH_3

LC_{50} (48 hr) in *Daphnia magna:* 2.9 μg/l. LC_{50} in rainbow trout, bluegill sunfish (mg/l) at 24 hrs: > 1.0, 0.75; at 48 hrs: > 1.0, 0.45; at 96 hrs: 0.67, 0.24.

Benzoate, *4″-deoxy-4″-epi-N-methylaminoavermectin B_1 benzoate, MK-244.* Exists as the anhydrous and various hydrated forms having different crystal morphologies. Amt of water is nonstoichiometric. 1.1% form is an off white crystalline powder, mp 141-146°. $[\alpha]_D$ −6.9° (c = 0.5% in methanol). uv max (10% methanol in water by volume): 244 nm ($A_1^{1\%}$ = 327 corr). log P for *n*-octanol/aq phosphate: 3.0 (pH 5.1); 5.0 (pH 7.0). pKa 4.2; 7.6. Soly 0.32 mg/ml at pH 5 (aq); 0.024 mg/ml at pH 7 (aq). Freely sol in chloroform and methanol. Insol in hexane.
USE: Insecticide.

3595. Embelin. *2,5-Dihydroxy-3-undecyl-2,5-cyclohexadiene-1,4-dione; 2,5-dihydroxy-3-undecyl-p-benzoquinone;* "Embelic acid". $C_{17}H_{26}O_4$; mol wt 294.39. C 69.36%, H 8.90%, O 21.74%. From fruit of *Embelia ribes* Burm., *Myrsinaceae.* Isoln: Heffter, Feuerstein, *Arch. Pharm.* **238**, 15 (1900). Structure and synthesis: Fieser, Chamberlin, *J. Am. Chem. Soc.* **70**, 71 (1948).

Glistening orange plates from alcohol, benzene or acetic acid, mp 142-143°. Soluble in the usual hot organic solvents or in alkali hydroxide solns; very slightly sol in petr ether. Practically insol in water.

Diammonium salt, $C_{17}H_{34}N_2O_4$, *ammonium embelate.* Grayish-violet powder. Sol in water, dil alcohol. *Caution:* Handle with care; irritates mucous membranes and may cause prolonged and violent sneezing!

Disemicarbazone, $C_{19}H_{32}N_6O_4$, pale-brown needles from dil alcohol, dec 236°.

Dioxime, $C_{17}H_{28}N_2O_4$, pale-yellow needles from acetic acid, mp 278°.
THERAP CAT: Ammonium embelate as an anthelmintic (Cestodes).

3596. Embramine. *2-[1-(4-Bromophenyl)-1-phenyleth-oxy]-N,N-dimethylethanamine; 2-[(p-bromo-α-methyl-α-phen-*

ylbenzyl)oxy]-N,N-dimethylethylamine; [2-(1-p-bromophenyl-1-phenylethoxy)ethyl]dimethylamine; 2-[1-(4-bromodiphenyl)ethoxy]-N,N-dimethylethylamine; p-bromo-α-methyl-benzhydryl 2-dimethylaminoethyl ether; 1-(p-bromophenyl)-1-phenyl-1-(2-dimethylaminoethoxy)ethane; β-dimethyl-aminoethyl p-bromo-α-methylbenzhydryl ether. $C_{18}H_{22}Br$-NO; mol wt 348.28. C 62.08%, H 6.37%, Br 22.94%, N 4.02%, O 4.59%. Prepn: L. Novak, M. Protiva, *Coll. Czech. Chem. Commun.* **24**, 3966 (1959); M. Protiva, L. Novak, **Ger.** pat. **1,240,877** (1967 to SPOFA), *C.A.* **67**, 73338j (1967).

bp$_{0.3}$ 135-140°. bp$_1$ 163-170°.
Hydrochloride, $C_{18}H_{22}BrNO.HCl$, *mebrophenhydramine, Bromadryl, Mebryl.* Crystals from acetone + ether, mp 150-152°. LD$_{50}$ in mice (mg/kg): 80 i.v.; 330 orally (Protiva, Novak).

THERAP CAT: Antihistaminic.

3597. Emedastine. *1-(2-Ethoxyethyl)-2-(hexahydro-4-methyl-1H-1,4-diazepin-1-yl)-1H-benzimidazole; 1-[2-(ethoxy)ethyl]-2-(4-methyl-1-homopiperazinyl)benzimidazole.* $C_{17}H_{26}N_4O$; mol wt 302.42. C 67.52%, H 8.67%, N 18.53%, O 5.29%. Histamine H$_1$-receptor antagonist. Prepn: R. Iemura *et al.,* **Eur.** pat. **79,545;** *eidem,* **U.S.** pat. **4,430,343** (1983, 1984 both to Kanebo); *eidem, J. Med. Chem.* **29,** 1178 (1986). GC determn in plasma: T. Hamada *et al., Chem. Pharm. Bull.* **34,** 1168 (1986). Series of articles on pharmacology: *Arzneimittel-Forsch.* **34**, 801-818 (1984). Toxicity data: T. Fukuda *et al., ibid.* 801.

Difumarate, $C_{17}H_{26}N_4O.2C_4H_4O_4$, *KB-2413, LY-188695.* Crystals from ethanol, mp 148-151°. LD$_{50}$ orally in guinea pigs: 744 mg/kg (Fukuda).

THERAP CAT: Antihistaminic.

3598. Emepronium Bromide. *N-Ethyl-N,N,α-trimethyl-γ-phenylbenzenepropanaminium bromide; ethyldimethyl(1-methyl-3,3-diphenylpropyl)ammonium bromide; ethyl(3,3-diphenyl-1-methylpropyl)dimethylammonium bromide; (1-methyl-3,3-diphenylpropyl)dimethylethylammonium bromide; Cetiprin; Restenacht; Ripirin; Uro-Ripirin.* $C_{20}H_{28}$-BrN; mol wt 362.35. C 66.29%, H 7.79%, Br 22.05%, N 3.87%. Prepn: Carlsson, **Swedish** pat. **136,606** (1952 to Aktiebolaget Recip.), *C.A.* **48**, 729a (1954).

Crystals, mp 204°.
THERAP CAT: Antispasmodic.

3599. Emetamine. *1',2',3',4'-Tetradehydro-6',7',10,11-tetramethoxyemetan.* $C_{29}H_{36}N_2O_4$; mol wt 476.62. C

73.08%, H 7.61%, N 5.88%, O 13.43%. Occurs in small amounts in ipecac. Isoln: Pyman, *J. Chem. Soc.* **111**, 419 (1917). Structure: Battersby, *Spec. Publ. Chem. Soc. (London)* **3**, 36 (1955); Battersby *et al., J. Chem. Soc.* **1959**, 1744. Synthesis: Evstigneeva *et al., Proc. Acad. Sci. USSR* **117**, 227 (1957); Battersby *et al., J. Chem. Soc.* **1961**, 3899.

Crystals from ether, mp 142-143°. uv max (ethanol): 236, 283 nm (log ε 4.85, 3.86).

3600. Emetine. *6',7',10,11-Tetramethoxyemetan;* cephaeline methyl ether. $C_{29}H_{40}N_2O_4$; mol wt 480.65. C 72.47%, H 8.39%, N 5.83%, O 13.31%. Principal alkaloid of ipecac, the ground roots of *Uragoga ipecacuanha* (Brot.) Baill. *Rubiaceae*; occurs naturally as (−)-form. Extraction procedures: E. Merck, *BIOS* Final Rpt. **766** (1947); F. E. Hamerslag, *Technology and Chemistry of Alkaloids* (New York, 1950) chapter XI; Stoll, Jucker in *Ullmann's Encyklopädie der Technischen Chemie* 3, Aufl., Bd. III (München-Berlin, 1953) p 231; *Beilstein* 2nd suppl., **vol. 23**, 449. Structure: Robinson, *Nature* **162**, 524 (1948); Janot, *Bull. Soc. Chim. France* **1949**, 185. Stereochemistry: Battersby, *Chem. & Ind. (London)* **1958**, 1324; Battersby, Garrett, *J. Chem. Soc.* **1959**, 3512; Van Tamelen *et al., J. Am. Chem. Soc.* **81**, 6214 (1959). Total synthesis: Evstigneeva *et al., Proc. Acad. Sci. USSR* **75**, 539 (1950); Van Tamelen *et al., J. Am. Chem. Soc.* **91**, 7359 (1969); T. Kametani *et al., J. Chem. Soc. Perkin Trans. I* **1979**, 1211. Alternate syntheses: Burgstahler, Bithos, *J. Am. Chem. Soc.* **82**, 5446 (1960); Openshaw, Whittaker, *J. Chem. Soc.* **1963**, 1461; S. Takano *et al., J. Org. Chem.* **43**, 4169 (1978); T. Fujii, S. Yoshifuji, *Tetrahedron* **36**, 1539 (1980). *See also* Clark *et al.,* **U.S.** pat. **3,102,118** (1963 to Glaxo). Review of syntheses: M. Shamma, *The Isoquinoline Alkaloids* (Academic Press, New York, 1972) pp 426-457. Toxicity: Radomski *et al., J. Pharmacol. Exp. Ther.* **104**, 421 (1952); of dihydrochloride: Child *et al., J. Pharm. Pharmacol.* **16**, 65 (1964). *General review:* Grollman, Jarkovsky, in *Antibiotics* **Vol. 3**, J. W. Corcoran, F. E. Hahn, Eds. (Springer-Verlag, New York, 1975) pp 420-435. Comprehensive description: L. V. Feyns, L. T. Grady in *Analytical Profiles of Drug Substances* **vol. 10**, K. Florey, Ed. (Academic Press, New York, 1981) pp 289-335.

White, amorphous powder. mp 74°. Turns yellow on exposure to light and heat. [α]$_D^{20}$ −50° (c = 2 in CHCl$_3$). Strong alkaline reaction. pK$_1$ 5.77; pK$_2$ 6.64. Freely sol in methanol, ethanol, acetone, ethyl acetate, ether, chloroform. Sparingly sol in water, petr ether. Moderately sol in dil ammonium hydroxide, but sparingly in solns of KOH and NaOH. Absorption spectrum: Bayard, *Chem. Zentr.* **1943**, **II**, 305. LD$_{50}$ i.p. in rats: 12.1 mg/kg (Radomski).

Dihydrochloride, $C_{29}H_{40}N_2O_4$.2HCl, *emetine hydrochloride, Hemometina.* Contains water of crystallization varying from 3 to 8 H_2O. Clusters of needles after drying at 105°, mp 235-255° (dec). $[\alpha]_D$ +11° (c = 1) to $[\alpha]_D$ +21° (c = 8), calculated for the anhydrous salt. One gram of the hydrated salt dissolves in about 7 ml water. pH of aq soln (1 g in 50 ml) 5.6. Sol in alcohol. Solid and solutions turn yellow on exposure to light or heat. LD_{50} (calculated as base) in mice (mg/kg): 32 s.c.; 30 orally (Child).

Acetarsone salt, *Arsemétine.*

THERAP CAT: Antiamebic.

THERAP CAT (VET): Hydrochloride has been used as an antiamebic and in lung worm infection.

3601. Emitefur. *3-[[3-(Ethoxymethyl)-5-fluoro-3,6-dihydro-2,6-dioxo-1(2H)-pyrimidinyl]carbonyl]benzoic acid 6-(benzoyloxy)-3-cyano-2-pyridinyl ester;* 3-[3-(6-benzoyloxy-3-cyano-2-pyridyloxycarbonyl)benzoyl]-1-(ethoxymethyl)-5-fluorouracil; BOF-A2. $C_{28}H_{19}FN_4O_8$; mol wt 558.48. C 60.22%, H 3.43%, F 3.40%, N 10.03%, O 22.92%. Bifunctional prodrug composed of a derivative of 5-fluorouracil (5-FU) and *3-cyano-2,6-dihydroxypyridine (CNDP),* an inhibitor of 5-FU degradation by dihydrouracil dehydrogenase. Prepn: S. Fujii, **Ger. pat. 3,709,699**; *idem,* **U.S. pat. 4,983,-609** (1988, 1991 both to Otsuka); M. Hirohashi *et al., Chem. Pharm. Bull.* **41,** 1498 (1993). Antitumor activity: S. Fujii *et al., Japan. J. Cancer Res.* **80,** 173 (1989). Mechanism of action: T. Okayasu *et al., ibid.* **85,** 101 (1994). Clinical evaluation in lung cancer: Y. Nakai *et al., Acta Oncol.* **33,** 523 (1994). Toxicity study: H. Kinoshita *et al., Yakuri to Chiryo* **22,** 81 (1994), *C.A.* **121,** 99197 (1994).

Crystals, mp 162-164°. LD_{50} in mice, male rats, female rats (mg/kg): > 5000, 1850, 1934 orally (Kiroshita).

THERAP CAT: Antineoplastic.

3602. Emodin. *1,3,8-Trihydroxy-6-methyl-9,10-anthracenedione; 1,3,8-trihydroxy-6-methylanthraquinone;* 4,5,7-trihydroxy-2-methylanthraquinone; frangula emodin; rheum emodin; archin; frangulic acid. $C_{15}H_{10}O_5$; mol wt 270.24. C 66.67%, H 3.73%, O 29.60%. Occurs mostly as the rhamnoside (*see* Frangulin) in rhubarb root, in alder buckthorn (*Rhamnus frangula* L.), in *Cascara sagrada (Rhamnus purshiana* DC., *Rhamnaceae),* also in *Rumex* and in other *Polygonaceae.* Isoln from rhubarb root: Tutin, Clewer, *J. Chem. Soc.* **99,** 946 (1911); Carelli, Giuliano, *Farmaco Ed. Prat.* **12,** 184 (1957); from bark of alder buckthorn: Bridel, Charaux, *Bull. Soc. Chim. Biol.* **15,** 648 (1933). Identity with archin: Chaudhry *et al., J. Sci. Ind. Res. (India)* **9B,** No. 6, 142 (1950), *C.A.* **44,** 9396h (1950). Synthesis from 3,5-dinitrophthalic anhydride and *m*-cresol: Elder, Widmer, *Helv. Chim. Acta* **6,** 966 (1923); from 2-methylanthraquinone: Ayyangar *et al., J. Sci. Ind. Res. (India)* **20B,** 493 (1961), *C.A.* **57,** 8514b (1962).

Orange needles from alc or by sublimation at 12 mm. mp 256-257°. Absorption max (ethanol): 222, 252, 265, 289, 437 nm (log ϵ 4.55, 4.26, 4.27, 4.34, 4.10). Practically insol in water; sol in alc, aq alkali hydroxide solns (cherry-red color), Na_2CO_3 and NH_3 solns. Soly at 25° (g/100 ml of satd soln): ether 0.140; chloroform 0.071; carbon tetrachloride 0.010; carbon bisulfide 0.009; benzene 0.041.

3-Methyl ether, $C_{16}H_{12}O_5$, *1,8-dihydroxy-3-methoxy-6-methylanthraquinone, rheochrysidin.* Brick-red, monoclinic needles, mp 207°. Occurs naturally as *physcione* or *parietin.* Trimethyl ether, $C_{18}H_{16}O_5$, pale yellow needles, mp 225°. *Note: See also* Aloe-emodin.

THERAP CAT: Cathartic.

3603. Emorfazone. *4-Ethoxy-2-methyl-5-(4-morpholinyl)-3(2H)-pyridazinone;* M-73101; Nandron; Pentoyl. $C_{11}H_{17}N_3O_3$; mol wt 239.27. C 55.22%, H 7.16%, N 17.56%, O 20.06%. Prepn: K. Satoda *et al.,* **Japan. pat. 72 24,030** (1972 to Morishita), *C.A.* **77,** 164732f (1972); M. Takaya *et al., J. Med. Chem.* **22,** 53 (1979). Metabolism studies: T. Hayashi *et al., Chem. Pharm. Bull.* **26,** 3124 (1978); **27,** 317 (1979). General pharmacological study: M. Sato *et al., Nippon Yakurigaku Zasshi* **75,** 291 (1979), *C.A.* **91,** 117292 (1979). Mechanism of action studies: M. Sato, A. Yamaguchi, *Arzneimittel-Forsch.* **32,** 379 (1982). Antigenicity study: M. Sato *et al., Oyo Yakuri* **18,** 65 (1979), *C.A.* **92,** 104391 (1980). Toxicity studies: *eidem, ibid.* **16,** 1011 (1978), *C.A.* **90,** 197741 (1979); C. Onodera *et al., J. Toxicol. Sci.* **4,** 229 (1979); K. Shimpo *et al., ibid.* 255.

Cryst from methanol/isopropyl ether, mp 89-91°. LD_{50} i.p. in mice: 700 mg/kg (Takaya).

THERAP CAT: Anti-inflammatory; analgesic.

3604. Emylcamate. *3-Methyl-3-pentanol carbamate; 1-ethyl-1-methylpropyl carbamate;* diethyl methyl carbinol urethan; 3-methyl-3-pentyl carbamate; methyl diethyl carbinol urethan; *tert*-hexanol carbamate; Kabi 925; JD-91; Nuncital; Restetal; Statran; Striatran. $C_7H_{15}NO_2$; mol wt 145.20. C 57.90%, H 10.41%, N 9.65%, O 22.04%. Prepn: **Ger. pat. 245,491** (1912 to Chinifabr. Zimmer); **Ger. pat. 254,472** (1912 to E. Merck); Melander, Hanshoff, **U.S. pat. 2,972,564** (1961 to Kabi).

Needles from 30% ethanol, mp 56-58.5°. $bp_{1.0}$ 35°; $bp_{0.7}$ 24°. Slight odor of camphor. Soly in water: 4.0 mg/ml. Freely sol in alcohol, ether, benzene, glycol ethers.

THERAP CAT: Anxiolytic.

3605. Enalapril. *(S)-1-[N-[1-(Ethoxycarbonyl)-3-phenylpropyl]-L-alanyl]-L-proline;* 1-[N-[(S)-1-carboxy-3-phenylpropyl]-L-alanyl]-L-proline 1'-ethyl ester. $C_{20}H_{28}N_2O_5$; mol wt 376.45. C 63.81%, H 7.50%, N 7.44%, O 21.25%. Angiotensin-converting enzyme (ACE) inhibitor; de-esterified *in vivo* to its active diacid metabolite, enalaprilat, *q.v.* Prepn: A. A. Patchett *et al., Nature* **288,** 280 (1980); *eidem,* **Eur. pat. Appl. 12,401**; E. E. Harris *et al.,* **U.S. pat. 4,374,-829** (1980, 1983 both to Merck & Co.). Pharmacology: D. M. Gross *et al., J. Pharmacol. Exp. Ther.* **216,** 552 (1981); C. S. Sweet *et al., ibid.* 558. Bioavailability and metabolism: E. H. Ulm, *Drug Metab. Rev.* **14,** 99 (1983). Comprehensive description: D. P. Ip, G. S. Brenner, *Anal. Profiles Drug Subs.* **16,** 207-243 (1987). Clinical trial in congestive heart failure: Consensus Trial Study Group, *N. Engl. J. Med.* **316,** 1429 (1987). Review of clinical experience in hypertension: H. J. Gomez *et al., J. Cardiovasc. Pharmacol.* **15,** Suppl. 3, S26-S29 (1990); of clinical pharmacokinetics: R. J. MacFadyen *et al., Clin. Pharmacokinet.* **25,** 274-282 (1993); of combination with hydrochlorothiazide: P. L. Malini, *Adv. Ther.* **10,** 253-262 (1993).

Maleate, $C_{20}H_{28}N_2O_5 \cdot C_4H_4O_4$, *MK-421, Amprace, Bitensil, Cardiovet, Enacard, Enaloc, Enapren, Glioten, Hipoartel, Innovace, Lotrial, Olivin, Pres, Renitec, Reniten, Renivace, Vasotec, Xanef.* White to off-white crystalline powder, mp 143-144.5°. Soly (g/ml): water 0.025; alcohol 0.08; methanol 0.20. $[\alpha]_D^{25}$ −42.2° (c = 1 in methanol). pH (1% water) 2.6. pKa_1 3.0; pKa_2 (25°) 5.4.

Mixture of maleate with hydrochlorothiazide, *Acesistem, Co-Renitec, Innozide, Renacor, Vaseretic, Xynertec.*

THERAP CAT: Antihypertensive.

THERAP CAT (VET): In treatment of heart failure in dogs.

3606. Enalaprilat. *(S)-1-[N-(1-Carboxy-3-phenylpropyl)-L-alanyl]-L-proline dihydrate;* enalaprilic acid; MK-422; Vasotec IV. $C_{18}H_{24}N_2O_5 \cdot 2H_2O$; mol wt 384.43. C 56.24%, H 7.34%, N 7.29%, O 29.13%. Nonsulfhydryl dipeptide angiotensin converting enzyme (ACE) inhibitor. Prepn: A. A. Patchett et al., Nature **288**, 280 (1980); eidem, **Eur. pat. Appl. 12,401**; E. E. Harris et al., U.S. pat. **4,374,829** (1980, 1983 both to Merck & Co.). Active metabolite of enalapril, q.v.: D. J. Tocco et al., Drug Metab. Dispos. **10**, 15 (1982). Pharmacology: D. M. Gross et al., J. Pharmacol. Exp. Ther. **216**, 552 (1981); M. A. Nelson et al., Clin. Exp. Pharmacol. Physiol. **Suppl. 7**, 87 (1982); T. A. Unger et al., Biochem. Pharmacol. **31**, 3063 (1982). Bioavailability: M. L. Cohen et al., J. Pharmacol. Exp. Ther. **226**, 192 (1983). Kinetics of enzyme inhibition: C. H. Reynolds, Biochem. Pharmacol. **33**, 1273 (1984). Pharmacokinetics: K. S. Pang et al., Drug Metab. Dispos. **12**, 309 (1984). Acute hemodynamic effects in essential hypertension: A. C. Simon et al., Clin. Pharm. Ther. **43**, 49 (1988). Clinical evaluation in severe and malignant hypertension: D. J. DiPette et al., ibid. **38**, 199 (1985).

Needles from H_2O, mp 148-151°. $[\alpha]_D$ −67.0° (0.1M HCl).

THERAP CAT: Antihypertensive.

3607. Enallylpropymal. *1-Methyl-5-(1-methylethyl)-5-(2-propenyl)-2,4,6(1H,3H,5H)-pyrimidinetrione; 5-allyl-5-isopropyl-1-methylbarbituric acid;* N-methyl-5-allyl-5-isopropylbarbituric acid. $C_{11}H_{16}N_2O_3$; mol wt 224.26. C 58.91%, H 7.19%, N 12.49%, O 21.40%. Prepn: Schnider, U.S. pat. **2,072,829** and Brit. pat. **454,779** (1937, 1936, both to Hoffmann-La Roche).

Crystals, mp 56-57°. bp_{12} 176-178°. Soluble in organic solvents.

Sodium salt, $C_{11}H_{15}N_2NaO_3$, *Narconumal.* Sol in water.

Caution: May be habit forming. This is a controlled substance (depressant) listed in the U.S. Code of Federal Regulations, Title 21 Parts 329.1 and 1308.13 (1995).

THERAP CAT: Sedative, hypnotic.

3608. Enanthotoxin. *(E,E,E)-2,8,10-Heptadecatriene-4,6-diyne-1,14-diol;* oenanthotoxin. $C_{17}H_{22}O_2$; mol wt 258.36. C 79.03%, H 8.58%, O 12.39%. The poisonous principle of *Oenanthe crocata* L., *Umbelliferae,* the hemlock water dropwort, a toxic plant known since 1746 and believed to be the most poisonous plant in England. Isoln: Clarke et al., J. Pharm. Pharmacol. **1**, 377 (1949). Structure studies: Anet et al., Chem. & Ind. (London) **1952**, 757; J. Chem. Soc. **1953**, 309. Synthesis of DL-form: Bohlmann, Viehe, Ber. **88**, 1245 (1955). Fluorimetric determn: B. Del Castillo et al., Ital. J. Biochem. **29**, 233 (1980). Blocking of sodium current and intramembrane charge movement: J. M. Dubois, M. F. Schneider, Nature **289**, 685 (1981); eidem, Toxicon **20**, 49 (1982). Toxicity study: M. P. Martinez-Honduvilla et al., Arch. Farmacol. Toxicol. **7**, 197 (1981).

Large prisms (natural) or star-shaped crystals (synthetic). Unstable, dec by light and air to brown insol resin. mp 87° (natural), 68° (synthetic). $[\alpha]_D^{15}$ +30.5° (c = 2.0 in methanol). uv max: 213, 252, 267, 281, 296, 315.5, 337.5 ($\epsilon \times 10^{-3}$ 17.5, 33, 29, 17.5, 30.5, 40, 29). Practically insol in water, petr ether, alkalies, dil mineral acids. Readily sol in chloroform, ethanol, methanol, ether, benzene. Average LD i.p. in mice: 0.83 mg/kg (Clarke). LD_{50} i.p. in rats: 2.94 mg/kg (Martinez-Honduvilla).

Caution: Very toxic. May cause convulsions, death.

3609. Encainide. *(±)-4-Methoxy-N-[2-[2-(1-methyl-2-piperidinyl)ethyl]phenyl]benzamide;* (±)-2'-[2-(1-methyl-2-piperidyl)ethyl]-p-anisanilide; 4-methoxy-2'-[2-(1-methyl-2-piperidyl)ethyl]benzanilide. $C_{22}H_{28}N_2O_2$; mol wt 352.48. C 74.97%, H 8.01%, N 7.95%, O 9.08%. Anti-arrhythmic benzanilide derivative. Prepn: S. J. Dykstra, J. L. Minielli, **Ger. pat. 2,210,154** (1972 to Bristol-Myers), C.A. **78**, 4138j (1973); eidem, U.S. pat. **3,931,195** (1976 to Mead Johnson); S. J. Dykstra et al., J. Med. Chem. **16**, 1015 (1973). Antiarrhythmic pharmacology in animals: J. E. Byrne et al., J. Pharmacol. Exp. Ther. **200**, 147 (1977). Clinical pharmacology and efficacy in chronic ventricular arrhythmia: D. M. Roden et al., N. Engl. J. Med. **302**, 877 (1980). Electrophysiology and effects on cardiac conduction: M. Sami et al., Am. J. Cardiol. **44**, 526 (1979). Hemodynamic effects: M. Sami et al., ibid. **52**, 507 (1983). Adverse effects: R. A. Winkle et al., Am. Heart J. **102**, 857 (1981). Comparison with other Class I anti-arrhythmic agents: A. Pottage, Am. J. Cardiol. **52**, 24C (1983). Series of articles on pharmacology, pharmacokinetics, metabolism, clinical safety and efficacy: ibid. **58**(5), 1C-116C (1986). Review: L. B. Mitchell, R. A. Winkle in New Drugs Annual: Cardiovascular Drugs, vol. **1**, A. Scriabine, Ed. (Raven Press, New York, 1983) pp 93-107.

Hydrochloride, $C_{22}H_{28}N_2O_2 \cdot HCl$, *MJ-9067, Enkade, Enkaid.* Crystals, mp 131.5-132.5°. Freely sol in water; slightly sol in ethanol. Insol in heptane. LD_{50} in mice, dogs (mg/kg): 86, 43 orally; 16, 17 i.v. (Mitchell, Winkle).

THERAP CAT: Antiarrhythmic (class IC).

3610. Enciprazine. *(±)-4-(2-Methoxyphenyl)-α-[(3,4,5-trimethoxyphenoxy)methyl]-1-piperazineethanol;* (±)-1-[3-(3,4,5-trimethoxyphenoxy)-2-hydroxypropyl]-4-(2-methoxyphenyl)piperazine. $C_{23}H_{32}N_2O_6$; mol wt 432.52. C 63.87%,

H 7.46%, N 6.48%, O 22.19%. Prepn: A. Kleemann *et al.,* **Ger.** pat. **2,814,168;** *eidem,* **U.S.** pat. **4,335,126** (1978, 1982 both to Degussa). Synthesis and pharmacology: J. Engel *et al., J. Med. Chem.* **33,** 2976 (1990). Metabolism in rats: E. Benfenati *et al., J. Pharm. Pharmacol.* **39,** 312 (1987). Clinical evaluation as anxiolytic: G. Scheibe *et al., Arzneimittel-Forsch.* **40,** 644 (1990); E. Schweizer *et al., Psychopharmacology Bull.* **26,** 215 (1990).

Dihydrochloride, $C_{23}H_{32}N_2O_6 \cdot 2HCl$, *D-13112, Wy-48624.* Colorless crystals, mp 196-197°.

THERAP CAT: Anxiolytic.

3611. Endiandric Acids. A group of novel polycyclic compounds isolated from leaves of the 'Dorrigo Plum', *Endiandra introrsa* C. T. White, *Lauraceae,* a large tree occurring in rain forests of Australia. Although they have a number of asymmetric centers, endiandric acids occur in nature in racemic rather than enantiomeric forms. This fact led to a proposed hypothesis for the "biogenesis" of these compounds in nature from achiral precursors by a series of non-enzymatic electrocyclizations called the "endiandric acid cascade". Isoln from *E. introrsa* leaves (obtained in 1958) and structure of endiandric acid A: W. M. Bandaranayake *et al., Chem. Commun.* **1980,** 162. Postulated electrocyclic reactions, prediction of endiandric acid D as a natural product: *eidem, ibid.* 902. X-ray crystallographic structural elucidation of A and detailed report of isoln: *eidem, Aust. J. Chem.* **34,** 1655 (1981). Isoln and structure of endiandric acid B: *eidem, ibid.* **35,** 557 (1982); of C: *eidem, ibid.* 567. Stereocontrolled total synthesis of A and B, description of "endiandric acid cascade": K. C. Nicolaou *et al., J. Am. Chem. Soc.* **104,** 5555 (1982). Stereocontrolled total synthesis of endiandric acids C-G: *eidem, ibid.* 5557. Synthesis of precursors and thermal studies: *eidem, ibid.* 5558, 5560.

Endiandric acid A
R = CH₂COOH

Endiandric acid C

Endiandric acid B
R = CH₂CH=CHCOOH (E)

Endiandric Acid A, $C_{21}H_{22}O_2$, *(1α,1aα,2aβ,5α,5aα,7aα,-7bα,7cα)-(±)-1a,2,2a,5,5a,7a,7b,7c-octahydro-5-phenyl-1H-cyclobut[bc]acenaphthylene-1-acetic acid, 2-(6'-phenyltetra-cyclo[5,4,2,0³,¹³,0¹⁰,¹²]trideca-4',8'-dien-11'-yl)acetic acid.* Rods from aq ethanol, mp 147-149°. uv max (95% ethanol): 242, 255, 261, 268, 286 nm (log ε 2.19, 2.36, 2.45, 2.32, 1.45). pKa 5.1, 5.0. Gives a yellow color with tetranitromethane.

Endiandric Acid B, $C_{23}H_{24}O_2$, *[1α(E),1aα,2aβ,5α,5aα,-7aα,7bα,7cα]-(±)-4-(1a,2,2a,5,5a,7a,7b,7c-octahydro-5-phen-yl-1H-cyclobut[bc]acenaphthylen-1-yl)-2-butenoic acid, (E)-4-(6'-phenyltetracyclo[5,4,2,0³,¹³,0¹⁰,¹²]trideca-4',8'-dien-11'-yl)-but-2-enoic acid.* Rosettes from aq ethanol and chloroform/petr ether, mp 163-165°. uv max (95% ethanol): 252, 258, 262, 265, 269 nm (log ε 3.11, 3.08, 3.07, 3.01, 2.97).

Endiandric Acid C, $C_{23}H_{24}O_2$, *[1α,(2E,4E),1aα,3β,3aα,6β,-6aα,6bα,7S*]-(±)-1,1a,2,3,3a,6,6a,6b-octahydro-1-(5-phenyl-2,4-pentadienyl)-3,6-methanocyclobut[cd]indene-7-carboxylic acid, 4-[(E,E)-5'-phenylpenta-2',4'-dien-1'-yl]tetracyclo[5,4,0,-0²,⁵,0³,⁹]undec-10-ene-8-carboxylic acid.* Cryst from ethanol and methanol, mp 125-132° (variable and with dec). uv max (95% ethanol): 222, 228, 236, 280, 288 nm (log ε 4.10, 4.05, 3.89, 4.51, 4.53).

3612. Endobenzyline Bromide. *2-[(Bicyclo[2.2.1]hept-5-en-2-ylhydroxyphenylacetyl)oxy]-N,N,N-trimethylethana-minium bromide; choline bromide α-phenyl-5-norbornene-2-glycolate; N,N-dimethylaminoethyl α-(bicyclo-[2.2.1]-5-heptenyl)mandelate methyl bromide; α-(2,5-endo-methyl-ene-Δ³-cyclohexenyl)mandelic acid β-dimethylaminoethyl ester methyl bromide; α-phenyl-5-norbornene-2-glycolic acid 2-dimethylaminoethyl ester methyl bromide; endoben-ziline bromide; PC-1238; Ulcyn.* $C_{20}H_{28}BrNO_3$; mol wt 410.35. C 58.54%, H 6.88%, Br 19.47%, N 3.41%, O 11.70%. Prepn: Kottler, Ohnacker, U.S. pat. **2,884,426** (1959 to Thomae).

Crystals from isopropanol, mp 170°.

THERAP CAT: Anticholinergic.

3613. Endorphins. Generic name derived from the term "endogenous morphine" and applied to a group of neuropeptides that are endogenous ligands of the opiate receptors. They are found in brain, pituitary gland and peripheral tissues of all vertebrates; the effects of endorphins on cells resemble those of opiates such as morphine. Their existence was postulated as a result of the discovery of stereospecific binding sites for narcotic analgesics in animal brain, *cf.* A. Goldstein *et al., Proc. Nat. Acad. Sci. USA* **68,** 1742 (1971); C. B. Pert, S. H. Snyder, *Science* **179,** 1011 (1973); E. J. Simon *et al., Proc. Nat. Acad. Sci. USA* **70,** 1947 (1973). Three types of stereochemically related opiate receptors have been postulated, *mu, delta, kappa:* W. R. Martin *et al., J. Pharm. Exp. Ther.* **197,** 517 (1976); P. E. Gilbert, W. R. Martin, *ibid.* **198,** 66 (1976); J. A. H. Lord *et al., Nature* **267,** 495 (1977). Of the endorphin group of peptides, several important members are known: α-, β- and γ-endorphins (β-endorphin being the most potent), α-neo-endorphin, β-neo-endorphin, dynorphin, *q.v.,* and met-enkephalin and leu-enkephalin, two naturally occurring pentapeptides belonging to the endorphin class. The amino acid sequences of α-, β- and γ-endorphins are contained within the sequence 61-91 of β-lipotropin, *q.v.,* and, in each, the amino terminal pentapeptide sequence is that of met-enkephalin. Dynorphin and α- and β-neo-endorphins have common terminal leu-enkephalin sequences. Initial isoln of active oligopeptide: L. Terenius, A. Wahlstrom, *Acta Pharmacol. Toxicol.* **35,** Suppl. 1, 55 (1974); J. Hughes, *Brain Res.* **88,** 295 (1975). Isoln, characterization and synthesis of two active pentapeptides (*enkephalins*): J. Hughes *et al., Life Sci.* **16,** 1753 (1975); G. W. Pasternak *et al., ibid.* 1765; J. Hughes *et al., Nature* **258,** 577 (1975); J. D. Bower *et al., J. Chem. Soc. Perkin Trans. I* **1976,** 2488. Isoln and properties of β-endorphin from bovine pituitary: H. Tesche-macher *et al., Life Sci.* **16,** 1771 (1975); B. M. Cox *et al., ibid.* 1777. Identity of β-endorphin with carboxy terminal amino acid sequence (61-91) of β-lipotropin: C. H. Li, D. Chang, *Proc. Nat. Acad. Sci. USA* **73,** 1145 (1976); B. M. Cox *et al., ibid.* 1821. Production of biologically active β-endorphin via expression of cloned gene sequences by *E. coli:* J. Shine *et al., Nature* **285,** 456 (1980). For a review of isoln, amino acid sequences, and synthesis of β-endorphins from various species, *see Hormonal Proteins and Peptides* **vol. X,** entitled "β-Endorphin", C. H. Li, Ed. (Academic Press, New York, 1981) pp 2-30.

Isoln and structure of *α-endorphin* or *LPH (61-76)*: R. Guillemin *et al.*, *Compt. Rend. Ser. D* **282**, 783 (1976); isoln, structure, synthesis of α- and γ-endorphin or *LPH (61-77)*: N. Ling *et al.*, *Proc. Nat. Acad. Sci. USA* **73**, 3942 (1976). High yield synthesis of met⁵-enkephalin: B. J. Dhotre *et al.*, *J. Ind. Chem. Soc.* **55**, 1128 (1978). Rapid synthesis of leu⁵-enkephalin: E. Vilkas *et al.*, *Int. J. Peptide Protein Res.* **15**, 29 (1980). Isoln of *α-neo-endorphin*: K. Kagawa *et al.*, *Biochem. Biophys. Res. Commun.* **86**, 153 (1979). Amino acid sequence: *eidem*, *ibid.* **99**, 871 (1981). Isoln, purification, amino acid sequence of *β-neo-endorphin*: N. Minamino *et al.*, *ibid.* 864. Metabolism and physiological effects of endorphins: H. W. Kosterlitz, J. Hughes, *Life Sci.* **17**, 91 (1975); L.-F. Tseng *et al.*, *Proc. Nat. Acad. Sci. USA* **73**, 4187 (1976). Although the physiological role of the endorphins has not been completely elucidated, they are known to block inhibitory pathways in the vertebrate CNS and their possible effects on pain perception, addictive states, and psychiatric disorders have been investigated: D. T. Krieger, A. S. Liotta, *Science* **205**, 366 (1979); R. A. Nicoll *et al.*, *Nature* **287**, 22 (1980). Studies have also suggested that endorphins may be involved in gluco-regulation: M. Feldman *et al.*, *N. Engl. J. Med.* **308**, 350 (1983). *Reviews:* A. Goldstein, *Science* **193**, 1081-1086 (1976); *idem*, *Harvey Lectures* **79**, 291-314 (1979); R. A. North, *Life Sci.* **24**, 1527-1546 (1979); C. R. Beddell *et al.*, *Progr. Med. Chem.* **17**, 1-39 (1980); several authors in *Advan. Biochem. Pharmacol.* **22**, 145-642 (1980); M. S. Gold *et al.*, *Med. Res. Rev.* **2**, 211-246 (1982); S. Zakarian, D. G. Smyth, *Biochem. J.* **202**, 561-571 (1982). *Books: Opiates and Endogenous Opioid Peptides*, H. W. Kosterlitz, Ed. (Elsevier, New York, 1976) 466 pp; *Endorphins: Proceedings*, L. Graf *et al.*, Eds. (Elsevier, New York, 1979) 471 pp; *Endorphins and Opiate Antagonists in Psychiatric Research*, N. Shah, A. G. Donald, Eds. (Plenum, New York, 1982) 423 pp.

```
 61
−Tyr−Gly−Gly−Phe−Met−Thr−Ser−Glu−Lys−Ser−Gln−Thr−Pro−Leu−Val
                                                          |
                                                         Thr
                                                          |
                                                         Leu
                                                          |
 91
HO−Glu−Gly−Lys−Lys−Tyr−Ala−Asn−Lys−Ile−Ile−Ala−Asn−Lys−Phe
```

Human β-Endorphin

β-Endorphin, *β-lipotropin C-fragment*, *β-lipotropin (61-91)*, *LPH (61-91)*.
Met⁵-Enkephalin, $C_{27}H_{35}N_5O_7S$, *methionine⁵-enkephalin*, *met⁵-E*, L-*tyrosylglycylglycyl-L-phenylalanyl-L-methionine*, *LPH (61-55)*. Needles from hot methanol, mp 196-198°. $[α]_{589}^{22}$ −21.9° (c = 1 in DMF); +14° (c = 1 in N HCl).
Leu⁵-Enkephalin, $C_{28}H_{37}N_5O_7$, *leucine⁵-enkephalin*, *leu⁵-E*, L-*tyrosylglycylglycyl-L-phenylalanyl-L-leucine*. White crystalline solid, mp 206° (dec). $[α]_{589}^{22}$ −23.4° (c = 1 in DMF); +18.3° (c = 1 in N HCl).

3614. Endosulfan. *6,7,8,9,10,10-Hexachloro-1,5,5a,6,9,-9a-hexahydro-6,9-methano-2,4,3-benzodioxathiepin 3-oxide; 1,4,5,6,7,7-hexachloro-5-norbornene-2,3-dimethanol cyclic sulfite; 1,2,3,4,7,7-hexachlorobicyclo[2.2.1]-2-heptene-5,6-bisoxymethylene sulfite;* chlorthiepin; Malix; Thiodan; Thionex. $C_9H_6Cl_6O_3S$; mol wt 406.93. C 26.56%, H 1.49%, Cl 52.27%, O 11.80%, S 7.88%. Prepd by reaction of hexachlorocyclopentadiene with *cis*-butene-1,4-diol to form the bicyclic dialcohol, followed by esterification and cyclization with $SOCl_2$: Geering, Nelson, U.S. pat. 2,983,732 (1961 to Hooker). Configuration studies: Reimschneider, Wuscherpfenning, *Z. Naturforsch.* **17b**, 585 (1962); Forman *et al.*, *J.*

Org. Chem. **30**, 169 (1965). Toxicity: T. B. Gaines, *Toxicol. Appl. Pharmacol.* **14**, 515 (1969).
Commercial product brown crystals. mp 70-100° (pure mp 106°). Practically insol in water. Sol in most organic solvents. Stable toward dil mineral acids; hydrolyzed rapidly by alkalies. Commercial product is a mixture of α-isomer, mp 108-110°, and β-isomer, mp 208-210°. LD_{50} orally in male, female rats: 43, 18 mg/kg (Gaines).
USE: Insecticide.

3615. Endothall. *7-Oxabicyclo[2.2.1]heptane-2,3-dicarboxylic acid;* 3,6-endoxohexahydrophthalic acid; endothal. $C_8H_{10}O_5$; mol wt 186.16. C 51.61%, H 5.41%, O 42.97%. Prepn: Olin, U.S. pat. 2,550,494 (1951 to Sharples Chemicals). Of the three possible isomers, the *exo-cis* form has the greatest biological activity. Fate in the environment and toxicity: Simsiman *et al.*, *Residue Rev.* **62**, 131-174 (1976); T. B. Gaines, R. E. Linder, *Fundam. Appl. Toxicol.* **7**, 299 (1986).

Solubility of the acid in g/100 g at 20°: water 10; acetone 7; benzene 0.01; methanol 28. Converted to the anhydride at 90°. LD_{50} in adult male, female rats (mg/kg): 57, 46 orally (Gaines, Linder).
Dipotassium salt, $C_8H_8K_2O_5$, *Aquathol, Aquathol K*.
Caution: May be very irritating to skin, eyes, mucous membranes. Ingestion may cause severe G.I. inflammation with erosion: *Clinical Toxicology of Commercial Products*, R. E. Gosselin *et al.*, Eds. (Williams & Wilkins, Baltimore, 5th ed., 1984) Section II, p 342.
USE: Herbicide; defoliant.

3616. Endothelin. Endothelin-1; ET-1. Potent vasoconstrictor peptide produced by mammalian vascular endothelial cells; present in low concentrations in normal serum. Composed of 21 amino acid residues; mol wt ~2500 Da. Isoln, characterization and cloning of porcine endothelin: M. Yanagisawa *et al.*, *Nature* **332**, 411 (1988). Structural identity of porcine and human endothelin: Y. Itoh *et al.*, *FEBS Letters* **231**, 440 (1988). Structural similarity with the snake venom toxins, *sarafotoxins S6*: C. Y. Lee, V. A. Chiappinelli, *Nature* **335**, 303 (1988); C. Takasaki *et al.*, *ibid.* 303. Identification in serum by radioimmunoassay: K. Ando *et al.*, *FEBS Letters* **245**, 164 (1989). Two additional isoforms, known as *endothelin-2* and *endothelin-3*, have been identified. Structures and comparative pharmacology of human endothelin isopeptides: A. Inoue *et al.*, *Proc. Nat. Acad. Sci. USA* **86**, 2863 (1989). Mechanism of action study: Y. Hirata *et al.*, *Biochem. Biophys. Res. Commun.* **154**, 868 (1988). Possible role in acute renal failure: J. D. Firth *et al.*, *Lancet* **2**, 1179 (1988). *In vivo* effects on pulmonary and systemic hemodynamics: H. L. Lippton *et al.*, *J. Appl. Physiol.* **66**, 1008 (1989). *Review:* T. Masaki, *J. Cardiovasc. Pharmacol.* **13**, Suppl. 5, S1-S4 (1989).

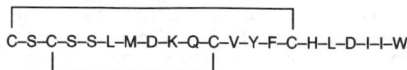

3617. Endralazine. *6-Benzoyl-5,6,7,8-tetrahydropyrido-[4,3-c]pyridazin-3(2H)-one hydrazone;* 6-benzoyl-3-hydrazino-5,6,7,8-tetrahydropyrido[4,3-c]pyridazine. $C_{14}H_{15}N_5O$; mol wt 269.31. C 62.44%, H 5.61%, N 26.01%, O 5.94%. Prepn: E. Schenker, Ger. pat. 2,221,808 corresp to U.S. pat. 3,838,125 (1972, 1974 both to Sandoz); E. Schenker, R. Salzmann, *Arzneimittel-Forsch.* **29**, 1835 (1979). Pharmacology: R. Salzmann *et al.*, *ibid.* 1843. Hemodynamic study: H. U. Lehmann *et al.*, *Z. Cardiol.* **66**, 203 (1977). HPLC determn in human plasma: P. A. Reece *et al.*, *J. Chromatog.* **225**, 151 (1981). Clinical pharmacology: F. C. Reubi, *Eur. J. Clin. Pharmacol.* **13**, 185 (1978). Clinical study in hypertension: H. U. Lehmann *et al.*, *Med. Welt* **29**, 1007 (1978).

Crystals from acetonitrile/water, mp 220-223° (dec). Methanesulfonate, $C_{14}H_{15}N_5O \cdot CH_3SO_4$, *endralazine mesylate, BQ 22-708, Miretilan*. Cryst, mp 185-188° (dec).

THERAP CAT: Antihypertensive.

3618. Endrin. *(1α,2β,2aβ,3α,6α,6aβ,7β,7aα)-3,4,5,6,-9,9-Hexachloro-1a,2,2a,3,6,6a,7,7a-octahydro-2,7:3,6-dimethanonaphth[2,3-b]oxirene; 1,2,3,4,10,10-hexachloro-6,7-epoxy-1,4,4a,5,6,7,8,8a-octahydro-endo,endo-1,4:5,8-dimethanonaphthalene;* mendrin; nendrin; hexadrin; experimental insecticide no. 269; compd 269; ENT-17251. $C_{12}H_8Cl_6O$; mol wt 380.91. C 37.84%, H 2.12%, Cl 55.84%, O 4.20%. A stereoisomer of dieldrin, *q.v.* Prepn from isodrin: H. Bluestone, U.S. pat. **2,676,132** (1954 to Shell); Marks, U.S. pat. **2,899,446** (1959 to Velsicol). Metabolism: M. K. Baldwin *et al., Pestic. Sci.* **7**, 575 (1976). Carcinogenicity studies: M. D. Rueber, *Sci. Total Environ.* **12**, 101 (1979). Toxicity study: T. B. Gaines, *Toxicol. Appl. Pharmacol.* **14**, 515 (1969).

Crystals, dec 245°. Vapor pressure at 25°: 2×10^{-7} mm Hg. Soly in g/100 ml solvent at 25°: acetone 17, benzene 13.8, carbon tetrachloride 3.3, hexane 7.1, xylene 18.3. LD_{50} in female, male rats (mg/kg): 7.5, 18 orally (Gaines). Toxic to fish.

Caution: Potential symptoms of overexposure are epileptiform convulsions; stupor, headache and dizziness; abdominal discomfort, nausea and vomiting; insomnia; aggressiveness, confusion; lethargy, weakness; anorexia. *See NIOSH Pocket Guide to Chemical Hazards* (DHHS/NIOSH 90-117, 1990) p 102.

USE: Formerly as insecticide; manuf and use has been discontinued in the U.S.

3619. Enduracidin. Enramycin. Cyclodepsipeptide antibiotic produced by *Streptomyces fungicidicus,* strain no. B5477, from a soil sample collected in Nishinomiya, Hyogo Prefecture, Japan: Higashide *et al., J. Antibiot.* **21,** 126 (1968). Originally thought to be a single compound. Isoln and characterization: Asai *et al., ibid.* 138; **Fr.** pat. **1,514,-139** corresp to **Brit.** pat. **1,163,270** (1968, 1969 to Takeda). Manufacturing process: **Japan. Kokai 82 79896** (1982 to Takeda), *C.A.* **97,** 108525s (1982). Separation into two components, enduracidins A and B: Hori *et al., Chem. Pharm. Bull.* **21,** 1171 (1973). The two components are each composed of seventeen amino acids, sixteen of which form a macrocyclic peptide lactone, and only differ by one methylene group in their fatty acid moieties. Structural studies: Hori *et al., ibid.* 1175. Final structure: Iwasaki *et al., ibid.* 1184. Antibacterial activity: Goto *et al., J. Antibiot.* **21,** 119 (1968); Tsuchiya *et al., ibid.* 147; M. Kawakami *et al., ibid.* **24,** 583 (1971). Toxicology: H. Yokotani *et al., Takeda Kenkyusho Nempo* **28,** 76 (1969), *C.A.* **72,** 77300s (1970).

Monohydrochloride, *Enradin.* Yellowish powder, dec 234-238°. Slightly sol in water, methanol. Practically insol in acetone, ethanol, chloroform, benzene. LD_{50} in mice and rats (mg/kg): >10000 orally; >5000 i.m.; >5000 s.c. (Yokotani).

Enduracidin A hydrochloride, $C_{107}H_{138}Cl_2N_{26}O_{31} \cdot HCl$, mp 240-245°. $[\alpha]_D^{23}$ +92° (c = 0.5 in DMF). uv max (0.1N HCl): 231, 272 nm.

Enduracidin B hydrochloride, $C_{108}H_{140}Cl_2N_{26}O_{31} \cdot HCl$, mp 238-241°. $[\alpha]_D^{23}$ +92° (c = 0.5 in DMF). uv max (0.1N HCl): 231, 272 nm.

THERAP CAT: Antibacterial.

3620. Enfenamic Acid. *2-[(2-Phenylethyl)amino]benzoic acid; N*-phenethyl anthranilic acid; RH-8; Tromaril. $C_{15}H_{15}NO_2$; mol wt 241.29. C 74.67%, H 6.27%, N 5.80%, O 13.26%. Nonsteroidal anti-inflammatory, antipyretic, analgesic. Prepn: S. R. Hashim *et al.,* **Indian** pats. **103,066** and **114,805** (both 1974 to Council Sci. Ind. Res.). Chemistry and pharmacology: P. Sisoda *et al., Proc. Symp. on CNS Drugs* (Council of Scientific and Industrial Research, New Delhi, 1966) pp 238-248. Antithrombotic activity in rheumatoid arthritis and myocardial infarction: S. R. Manikeri *et al., Ind. J. Med. Res.* **71,** 438 (1980). Natriuretic activity: N. Kshirsagar *et al., Brit. J. Clin. Pharmacol.* **9,** 530 (1980). Clinical trial in nephrotic syndrome: *eidem, Ind. J. Med. Res.* **70,** 801 (1979); as post-partum analgesic: S. Tandon *et al., Ind. Med. Gaz.* **116,** 182 (1982).

Crystals, mp 116-117°. Insol in water. LD_{50} in mice (mg/kg): 575 i.p., >2000 orally (Sisoda).

THERAP CAT: Anti-inflammatory; analgesic.

3621. Enflurane. *2-Chloro-1-(difluoromethoxy)-1,1,2-trifluoroethane; 2-chloro-1,1,2-trifluoroethyl difluoromethyl ether;* methylflurether; compd 347; NSC-115944; Alyrane; Efrane; Ethrane. $C_3H_2ClF_5O$; mol wt 184.49. C 19.53%, H 1.09%, Cl 19.22%, F 51.49%, O 8.67%. CHF_2OCF_2CHClF. Prepn by fluorination of the corresp dichloromethyl ether: R. C. Terrell, **Brit.** pat. **1,138,406;** *idem,* **U.S.** pats. **3,469,-011; 3,527,813** (1969, 1969, 1970 all to Air Reduction). Synthesis and anesthetic properties: R. C. Terrell *et al., J. Med. Chem.* **14,** 517 (1971); **15,** 604 (1972). Enantiomeric resolution: J. Meinwald *et al., Science* **251,** 560 (1991).

Stable, volatile, non-flammable liq. bp 56.5°. n_D^{20} 1.3025. d_{25}^{25} 1.5167. Does not degrade in the presence of alkali or light. Miscible with other organic liquids incl. fats and oils.

THERAP CAT: Anesthetic (inhalation).

3622. Enilconazole. *1-[2-(2,4-Dichlorophenyl)-2-(2-propenyloxy)ethyl]-1H-imidazole;* (±)-*1-[β-(allyloxy)-2,4-dichlorophenethyl]imidazole;* imazalil; R-23979; Clinafarm; Imaverol. $C_{14}H_{14}Cl_2N_2O$; mol wt 297.18. C 56.58%, H 4.75%, Cl 23.86%, N 9.43%, O 5.38%. Broad spectrum antimycotic with agricultural and veterinary applications. Prepn: E. F. Godefroi *et al.,* **Ger.** pat. **2,063,857** corresp to U.S. pat. **3,658,813** (1971, 1972 both to Janssen). Efficacy vs citrus decay: P. R. Harding, *Plant Dis. Rep.* **60,** 643 (1976); S. Ben-Yehoshua *et al., Pestic. Sci.* **12,** 485 (1981). Biological and toxicological properties: D. Thienpont *et al., Arzneimittel-Forsch.* **31,** 309 (1981). Residue analysis in treated grapefruit: E. R. Stein *et al., J. Environ. Sci. Health [B]* **16,** 427 (1981).

Solidified oil. Slightly sol in organic solvents; poorly sol in water.

USE: As a disinfectant for stable and kennel equipment; exptly as an agricultural fungicide.

THERAP CAT: Antifungal.

THERAP CAT (VET): Antifungal.

3623. Enniatins. Ionophore antibiotics produced by the fungus *Fusarium orthoceras* var. *enniatum* and other *Fusaria:* Gaumann *et al., Experientia* **3,** 202 (1947); Plattner, Nager, *ibid.* 325; Plattner *et al., Helv. Chim. Acta* **31**, 594 (1948). They are cyclic depsihexapeptides related structurally to valinomycin, *q.v.* Enniatins A, B, and C are known; A and B are obtained from natural sources and C is a synthetic homolog. Structure and synthesis of enniatin A: Quitt *et al., ibid.* **46**, 1715 (1963); **47**, 166 (1964); of enniatin B: Plattner *et al., ibid.* **46**, 927 (1963); Shemyakin *et al., Tetrahedron Letters* **1963**, 885. Conformation of enniatin B: Ovchinnikov *et al., Biochem. Biophys. Res. Commun.* **37**, 668 (1969). Synthesis of C: *eidem, Izvest. Akad. Nauk SSSR, Ser. Khim.* **1964**, 1912, *C.A.* **62**, 28246b (1965). Enniatins are able to form "sandwich" complexes with cations, a factor used to explain their ability to induce permeability in biological membranes to the complexed ion, *cf.* V. T. Ivanov, *Ann. N.Y. Acad. Sci.* **264**, 221 (1975). Review: Y. A. Ovchinnikov, V. T. Ivanov, "The Cyclic Peptides: Structure, Conformation, and Function" in *The Proteins* **vol. V**, H. Neurath, R. L. Hill, Eds. (Academic Press, New York, 3rd ed., 1982) pp 365-373, 516-529.

Enniatin A	R = CH(CH₃)CH₂CH₃
Enniatin B	R = CH(CH₃)₂
Enniatin C	R = CH₂CH(CH₃)₂

Enniatin A, $C_{36}H_{63}N_3O_9$. Long needles from ethanol + water, mp 122-122.5°. Very slowly sublimes at 10^{-4} mm and 127-128° (oil bath temp). $[\alpha]_D^{18}$ −91.9° (c = 0.926 in chloroform). Sol in ether, benzene, ethyl acetate; sparingly sol in water. Solns are stable to heat, inactivated by alkali.

Enniatin B, $C_{33}H_{57}N_3O_9$. Crystals from petr ether, mp 175-175.5°. $[\alpha]_D^{20}$ −107.9° (c = 0.63 in chloroform). Very sol in organic solvents; slightly sol in petr ether. Practically insol in water.

Enniatin C, $C_{34}H_{59}N_3O_9$, crystals, mp 160-161°. $[\alpha]_D^{23}$ −24°.

3624. Enocitabine. *N-(1-β-D-Arabinofuranosyl-1,2-dihydro-2-oxo-4-pyrimidinyl)docosanamide;* N(4)-behenoyl-1-β-D-arabinofuranosylcytosine; behenoylcytosine arabinoside; BH-AC; NSC-239336; Sunrabin. $C_{31}H_{55}N_3O_6$; mol wt 565.79. C 65.81%, H 9.80%, N 7.43%, O 16.97%. Deriv of cytarabine, *q.v.* Prepn: T. Ishida *et al., Ger. pat.* **2,426,304**; *eidem,* U.S. pat. **3,991,045** (1975, 1976 both to Asahi); M. Akiyama *et al., Chem. Pharm. Bull.* **26**, 981 (1978). Antitumor activity: M. Aoshima *et al., Cancer Res.* **36**, 2726 (1976); *eidem, ibid.* **37**, 2481 (1977). Distribution, excretion:

M. Fukama *et al., Gan to Kagaku Ryoho* **7**, 2109 (1980), *C.A.* **95**, 35186 (1981). Effect and mode of action: T. Kataoka, Y. Sakurai, *Recent Results Cancer Res.* **70**, 147 (1980). Pharmacological and clinical studies: K. Yamada *et al., ibid.* 219.

Crystals from DMSO, mp 141-142°. $[\alpha]_D$ +70° (c = 1 in THF, 22°). uv max (isopropyl alcohol): 216, 248, 303 nm (ε 16400, 15200, 8200).

THERAP CAT: Antineoplastic.

3625. Enoxacin. *1-Ethyl-6-fluoro-1,4-dihydro-4-oxo-7-(1-piperazinyl)-1,8-naphthyridine-3-carboxylic acid;* AT-2266; CI-919; PD-107779; Abenox; Bactidan; Comprecin; Flumark; Penetrex. $C_{15}H_{17}FN_4O_3$; mol wt 320.32. C 56.24%, H 5.35%, F 5.93%, N 17.49%, O 14.98%. Fluorinated quinolone antibacterial; nalidixic acid analog. Prepn and antibacterial activity: J. Matsumoto *et al.,* **Eur.** pat. **Appl. 9425**; *eidem,* U.S. pat. **4,352,803**; *eidem,* U.S. pat. **4,359,578** (1980, 1982, 1982 all to Dainippon and Roger Bellon); J. Matsumoto *et al., J. Med. Chem.* **27**, 292 (1984). HPLC determn in plasma, urine: T. B. Vree *et al., J. Chromatog.* **343**, 449 (1985). *In vitro* comparison with other quinolones and azaquinolones: S. Selwyn, M. Bakhtiar, *Drugs Exptl. Clin. Res.* **10**, 653 (1984). Pharmacokinetics, therapeutic efficacy: K. G. Naber *et al., Infection* **13**, 219 (1985). Efficacy in treatment of urinary tract infections: R. R. Bailey, B. A. Peddie, *N.Z. Med. J.* **98**, 286 (1985). Toxicology studies: H. Senda *et al., Chemotherapy (Tokyo)* **32**, Suppl. 3, 192 (1984). Series of articles on activity, pharmacology, metabolism, early clinical studies: *ibid.* 1-1093; *J. Antimicrob. Chemother.* **14**, Suppl. C, 1-96 (1984); on activity, pharmacokinetics, clinical safety and efficacy: *Infection* **14**, Suppl. 3, S183-S220 (1986).

Crystals from ethanol/methylene chloride, mp 220-224°. LD_{50} in male, female mice, male, female rats (mg/kg): 327, 391, 236, 294 i.v.; 1237, 1320, >2000, >2000 s.c.; all >5000 orally (Senda).

Sesquihydrate, $C_{15}H_{17}FN_4O_3 \cdot 1½H_2O$, Enoxen, Enoxor, *Gyramid.*

THERAP CAT: Antibacterial.

3626. Enoxaparin. PK-10169; RP-54563; Klexane; Lovenox; Ultraparin. Low molecular weight fragment of heparin, *q.v.,* having a 4-eno pyranosuronate sodium structure at the non-reducing end of the chain. Prepd by depolymerization of the benzylic ester of porcine mucosal heparin. Mean mol wt 4000-6000 daltons. Prepn: J. Mardiguian, **Eur.** pat. **Appl. 40,144** (1981 to Pharmindustrie), *C.A.* **96**, 218191s (1982). Determn in human plasma: L. Bara *et al., Haemostasis* **17**, 127 (1987). Symposium on pharmacology and clinical efficacy: *ibid.* **16**, 69-188 (1986). Review of pharmacodynamics, pharmacokinetics and clinical efficacy: M. M. Buckley, E. M. Sorkin, *Drugs* **44**, 465-497 (1992).

THERAP CAT: Antithrombotic.

3627. Enoximone. *1,3-Dihydro-4-methyl-5-[4-(methylthio)benzoyl]-2H-imidazol-2-one;* 4-methyl-5-[p-(methylthio)benzoyl]-4-imidazolin-2-one; fenoximone; MDL-17043; RMI-17043; Perfane; Perfan. $C_{12}H_{12}N_2O_2S$; mol wt 248.31. C 58.05%, H 4.87%, N 11.28%, O 12.89%, S 12.91%. Selective phosphodiesterase inhibitor with vasodilating and positive inotropic activity. Prepn: **Belg.** pat. **883,856** (1980 to Richardson-Merrell); R. A. Schnettler *et al.,* U.S. pat. **4,405,635** (1983 to Merrell-Dow); *eidem, J. Med. Chem.* **25**, 1477 (1982). Cardiovascular pharmacology in animals: R. C. Dage *et al., J. Cardiovasc. Pharmacol.* **4**, 500 (1982); L. E. Roebel *et al., ibid.* 721. *In vitro* mechanism of action studies: T. Kariya *et al., ibid.* 509; H. S. Ahn *et al., Biochem. Pharmacol.* **35**, 1113 (1986). Acute hemodynamic effects in congestive heart failure: B. F. Uretsky *et al., Circulation* **67**, 823 (1983). HPLC determn in plasma: K. Y. Chan *et al., J. Chromatog.* **272**, 396 (1983). Pharmacokinetics in humans: R. G. Alken *et al., Clin. Pharmacol. Ther.* **36**, 209 (1984). Symposium on pharmacology and clinical efficacy in congestive heart failure: *Am. J. Cardiol.* **60**, 1C-90C (1987).

Crystals from isopropanol + water, mp 255-258° (dec).
THERAP CAT: Cardiotonic.

3628. Enoxolone. *(3β,20β)-3-Hydroxy-11-oxoolean-12-en-29-oic acid; 3β-hydroxy-11-oxoolean-12-en-30-oic acid;* glycyrrhetic acid; 18β-glycyrrhetinic acid; uralenic acid; Arthrodont; Biosone; P.O. 12. $C_{30}H_{46}O_4$; mol wt 470.69. C 76.55%, H 9.85%, O 13.60%. From glycyrrhizic acid, *q.v.* Structure: Ruzicka *et al., Helv. Chim. Acta* **26**, 2143, 2278 (1943). Stereochemistry: Beaton, Spring, *J. Chem. Soc.* **1955**, 3126. Prepn from shredded licorice root: Mer, *Am. Perfum. Aromat.* **74**(6), 39 (1959). Manuf: **Brit.** pat. **833,-184** (1960 to Carlo Erba). Identity with uralenic acid: Belous *et al., Zh. Obshch. Khim.* **35**, 401 (1965). Metabolism: Parke *et al., J. Pharm. Pharmacol.* **15**, 500 (1963). Mechanism of action: Helbing, Berntsen, *Pharm. Weekbl.* **100**, 1438 (1965).

Needles from alcohol + petr ether, mp 296°. $[\alpha]_D^{21}$ +86° (alc); $[\alpha]_D^{20}$ +145.5° (dioxane); $[\alpha]_D^{20}$ +163° (chloroform). Freely sol in chloroform, dioxane. Soluble in alcohol, pyridine, acetic acid. Practically insol in petr ether.
18α-Hydrogen form, platelets from dil alcohol, mp 335°. $[\alpha]_D^{20}$ +140° (alcohol); $[\alpha]_D$ +98° (c = 0.1 in chloroform). Sol in alcohol, dioxane, chloroform.
THERAP CAT: Anti-inflammatory (topical).

3629. Enprostil. *[1α,2β(1E,3R*),3α]-7-[3-Hydroxy-2-(3-hydroxy-4-phenoxy-1-butenyl)-5-oxocyclopentyl]-4,5-heptadienoic acid methyl ester; (dl)-9-keto-11α,15α-dihydroxy-16-phenoxy-17,18,19,20-tetranorprosta-4,5,13-trans-trienoic acid methyl ester;* RS-84135; Camleed; Fundyl; Gardrin(e); Syngard. $C_{23}H_{28}O_6$; mol wt 400.47. C 68.98%, H 7.05%, O 23.97%. Prostaglandin E_2 derivative with anti-ulcer activity. Prepn: A. R. Van Horn *et al.,* U.S. pat. **4,178,457** (1979 to Syntex). Pharmacokinetics: D. R. Stanski *et al., Clin. Pharmacol. Ther.* **31**, 273 (1982). Cytoprotective effect against aspirin-induced gastromucosal injury in humans: M. M. Cohen *et al., Gastroenterology* **88**, 382 (1985). Clinical comparison with cimetidine of antisecretory and antigastrin activity in duodenal ulcer: V. Mahachai *et al., ibid.* **89**, 555 (1985); of efficacy in ulcer healing: L. Carling *et al., Scand. J. Gastroenterol.* **22**, 325 (1987). Series of articles on pharmacology, clinical efficacy and safety: *Am. J. Med.* **81**, Suppl. 2A, 1-88 (1986).

White to off-white waxy solid. Softens at 30°, liquid at 46°. Very slightly sol in water. Sol in alcohol, propylene glycol, propylene carbonate. uv max (methanol): 220, 265, 271, 277 nm (log ε 4.01, 3.14, 3.24, 3.16).
THERAP CAT: Antiulcerative. Antisecretory agent.

3630. Enrofloxacin. *1-Cyclopropyl-7-(4-ethyl-1-piperazinyl)-6-fluoro-1,4-dihydro-4-oxo-3-quinolinecarboxylic acid;* CFPQ; Bay Vp 2674; Baytril. $C_{19}H_{22}FN_3O_3$; mol wt 359.40. C 63.50%, H 6.17%, F 5.29%, N 11.69%, O 13.36%. Fluorinated quinolone antibacterial. Prepn: K. Grohe *et al.,* **Ger.** pat. **3,142,854**; *eidem,* U.S. pat. **4,670,444** (1983, 1987 both to Bayer AG); K. Grohe, H. Heitzer, *Ann.* **1987**, 29. Use as plant fungicide: K. Grohe *et al.,* U.S. pat. **4,563,459** (1986 to Bayer AG). Pharmacokinetics in calves: J. N. Davidson *et al., Proc. West. Pharmacol. Soc.* **29**, 129 (1986); in chickens: G. M. Conzelman *et al., ibid.* **30**, 393 (1987). Spectrofluorometric determn of residues in poultry tissues: T. B. Waggoner, M. C. Bowman, *J. Assoc. Offic. Anal. Chem.* **70**, 813 (1987). Toxicology and physical properties: P. Altreuther, *Vet. Med. Rev.* **2**, 87 (1987). Series of articles on pharmacology, *in vitro* antibacterial activity and field trials: *ibid.* 90-140.

Pale yellow crystals, mp 219-221°. Slightly sol in water at pH 7. LD_{50} in male, female mice (mg/kg): > 5000, 4336 orally; ~200, ~200 i.v.; in male rats, male rabbits (mg/kg): > 5000, 500-800 orally (Altreuther).
THERAP CAT (VET): Antibacterial.

3631. Enterobactin. *N,N',N''-(2,6,10-Trioxo-1,5,9-trioxacyclododecane-3,7,11-triyl)tris[2,3-dihydroxy]benzamide; N,N',N''-(2,6,10-trioxo-1,5,9-trioxacyclododecane-3,7,11-triyl)tris-o-pyrocatechuamide;* enterochelin. $C_{30}H_{27}N_3O_{15}$; mol wt 669.56. C 53.82%, H 4.06%, N 6.28%, O 35.84%. Physiologically active macrocyclic iron sequestering agent of the phenolate type, involved in microbial transport and metabolism of iron; it is overproduced by *E. coli* and related enteric bacteria under low-iron stress. Isoln from culture medium of *Salmonella typhimurium* LT2 and structure: J. R. Pollack, J. B. Neilands, *Biochem. Biophys. Res. Commun.* **38**, 989 (1970); also produced by *Escherichia coli* and *Aerobacter aerogenes:* I. G. O'Brien, F. Gibson, *Biochim. Biophys. Acta* **215**, 393 (1970). Conformation: M. Leinás *et al., Biochemistry* **12**, 3836 (1973). Total synthesis: E. J. Corey, S. Bhattacharyya, *Tetrahedron Letters* **1977**, 3919. Synthesis of enterobactin and enantioenterobactin: W. H. Rastetter *et al., J. Org. Chem.* **45**, 5011 (1980); **46**, 3579 (1981). Biosynthetic study: J. R. Pollack *et al., J. Bacteriol.* **104**, 635 (1970). Stability constants and electrochemical behavior of ferric enterobactin: K. N. Raymond, *J. Am. Chem. Soc.* **101**, 6097 (1979). Kinetics and mechanism of iron removal from transferrin: C. J. Carrano *et al., ibid.* 5401. *Review:* J. B. Neilands in *Inorganic Biochemistry* **1**, G. Eichorn, Ed. (Elsevier, New York, 1973) pp 167-202; *idem, Bioinorganic Chemistry* II, K. N. Raymond, Ed. (A.C.S., Washington, 1977) pp 3-33.

Crystals from ethanol/water, mp 202-203°. $[\alpha]_D^{25}$ +7.40° (ethanol). uv max (ethyl acetate): 316 nm (ϵ 9390). Sol in acetone, dioxane, dimethyl sulfoxide, methanol. Practically insol in water, soly increases at pH 7 to 8. Neutral on paper electrophoresis at pH 5. Displays bright blue fluorescence under uv light.

Ferric enterobactin, $[C_{30}H_{21}FeN_3O_{15}]^{3-}$. Anionic complex of enterobactin. uv max: 495 nm (ϵ 5600).

USE: As a growth-promoting agent in various organisms.

3632. Enterogastrone. Anthelone E; enteroanthelone; Duosan; Ileogastrone. An inhibitor of gastric secretion, apparently different from urogastrone. Obtained from extracts of the intestinal mucosa of mammals: Gray, Ivy, *Cold Spring Harbor Sympos. Quant. Biol.* **5**, 405 (1937); Ivy, Greengard, U.S. pat. **2,477,541** (1949 to Research Corp.). Comparison with urogastrone: Friedman, *Vitam. Horm.* (*New York*) **9**, 313 (1951); Gregory, *J. Physiol.* **129**, 528 (1955).

Amorphous buff-colored powder. Freely sol in water.

THERAP CAT: Gastric secretion inhibitor.

3633. Enterolactone. *trans-Dihydro-3,4-bis[(3-hydroxyphenyl)methyl]-2(3H)-furanone; trans-(±)-2,3-bis(3'-hydroxybenzyl)-γ-butyrolactone;* HPMF; HBBL; compd 180/442. $C_{18}H_{18}O_4$; mol wt 298.34. C 72.47%, H 6.08%, O 21.45%. Phenolic compound having 2,3-dibenzylbutane skeleton. First lignan, *q.v.,* found in animal species. Determn in female urine from humans, vervet monkeys and characterization: S. R. Stitch *et al., Nature* **287**, 738 (1980); K. D. R. Setchell *et al., ibid.* 740. Structure determn by GC-MS: *eidem, Biochem. J.* **197**, 447 (1981); *eidem, Biomed. Mass Spectrom.* **10**, 227 (1983). ¹H-NMR, X-ray crystal structure: G. Cooley *et al., J. Chem. Soc. Perkin Trans. I* **1984**, 489. Synthesis: M. B. Groen, J. Leemhuis, *Tetrahedron Letters* **21**, 5043 (1980); G. Cooley *et al., ibid.* **22**, 349 (1981); A. Pelter *et al., ibid.* 1549; P. A. Ganeshpure, R. Stevenson, *Chem. & Ind.* (*London*) **1981**, 778; A. Pelter *et al., J. Chem. Soc. Perkin Trans. I* **1983**, 643. Lignan excretion peaks during luteal phase of menstrual cycle: K. D. R. Setchell *et al., J. Steroid Biochem.* **12**, 375 (1980). Excretion as glucuronide conjugate: M. Axelson, K. D. R. Setchell, *FEBS Letters* **122**, 49 (1980). Biosynthesis by intestinal bacteria: *eidem, ibid.* **123**, 337 (1981). Effect of diet on lignan excretion: H. Aldercreutz *et al., Med. Biol.* **59**, 337 (1981).

Gum, mp 141-143°. uv max (ethanol): 227, 261 nm (log ϵ 4.66, 4.64).

3634. Enteromycin. *3-[[(Methyl-aci-nitro)acetyl]amino]-2-propenoic acid; 3-[2-(methyl-aci-nitro)acetamido]acrylic acid; N-(O-methyl-aci-nitroacetyl)-3-aminoacrylic acid;* seligocidin. $C_6H_8N_2O_5$; mol wt 188.14. C 38.30%, H 4.29%, N 14.89%, O 42.52%. Antibiotic substance isolated from *Streptomyces albireticuli:* Nakazawa, *C.A.* **51**, 10647i (1957); from *S. achromogenes:* Herr *et al., Antimicrob. Ag. Chemother.* **1964**, 100. Structure: Mizuno, *Bull. Chem. Soc. Japan* **34**, 1419, 1425, 1631, 1633 (1961). Process for production: Shibata *et al.,* **Japan.** pat. **5747('58)** (to Takeda), *C.A.* **53**, 16481f (1959). Identification of carboxamide: S. E. DeVoe *et al., Antimicrob. Ag. Chemother.* **1964**, 105; L. A. Mitscher *et al., Tetrahedron* **21**, 267 (1965).

Two interchangeable crystal forms from methanol: needles or fine prisms. Indefinite mp. pKa 4.3. Unstable to heat, acid and alkali. Sparingly sol in water and other solvents. uv max (methanol): 230, 275, 298 nm (ϵ 9000, 13000, 16000).

Carboxamide, $C_6H_9N_3O_4$, rosettes from glacial acetic acid, dec 155°. uv max (methanol): 230, 300 nm ($E_{1cm}^{1\%}$ 650, 820). Sol in glacial acetic acid, dimethyl sulfoxide, dimethylformamide; slightly sol in methanol. Practically insol in the usual organic solvents.

Methyl ester, $C_7H_{10}N_2O_5$, crystals from ethanol, mp 141°.

3635. Entprol. *1,1',1'',1'''-(1,2-Ethanediyldinitrilo)-tetrakis[2-propanol]; 1,1',1'',1'''-(ethylenedinitrilo)tetra-2-propanol; N,N,N',N'-tetrakis(2-hydroxypropyl)ethylenediamine;* Quadrol. $C_{14}H_{32}N_2O_4$; mol wt 292.42. C 57.50%, H 11.03%, N 9.58%, O 21.89%. $[CH_3CH(OH)CH_2]_2NCH_2-CH_2N[CH_2CH(OH)CH_3]_2$. Prepd by the action of propylene oxide on ethylenediamine: Lundsted, Schulz, U.S. pat. **2,697,118** (1954 to Wyandotte).

Viscous liq. bp$_{1.0}$ 190°. Miscible with water. Sol in methanol, ethanol, toluene, ethylene glycol, perchloroethylene.

USE: Cross-linking agent and catalyst in the manuf of urethan foams; in epoxy resin curing; as complexing agent, humectant, emulsifier, plasticizer.

3636. Enviomycin. *(R)-1-(threo-4-Hydroxy-L-3,6-diaminohexanoic acid)-6-[L-2-(2-amino-1,4,5,6-tetrahydro-4-pyrimidinyl)glycine]viomycin;* tuberactinomycin N. $C_{25}H_{43}N_{13}$-O_{10}; mol wt 685.70. C 43.79%, H 6.32%, N 26.56%, O 23.33%. Polypeptide antibiotic produced by a mutant of *Streptomyces griseoverticillatus* var. *tuberacticus.* Isoln and characterization: T. Ando *et al., J. Antibiot.* **24**, 680 (1971); A. Nagata *et al.,* in *Advan. Antimicrob. Antineopl. Chemother.* **vol. 1/2,** M. Hejzlar *et al.,* Eds. (University Park Press, Baltimore, 1972) pp 1039-1041. Prepn: J. Abe *et al.,* **Ger.** pat. **2,133,181;** *eidem,* U.S. pat. **3,892,732** (1972, 1975 both to Toyo Brewing Co., Ltd.). Structural studies: H. Yoshioka *et al., Tetrahedron Letters* **1971**, 2043; T. Wakamiya, T. Shiba, *J. Antibiot.* **28**, 292 (1975). Pharmacology: H. Hamakawa *et al., Oyo Yakuri* **8**, 817 (1974), *C.A.* **82**, 51530y (1975). Metabolism: T. Shimizu *et al., Japan. J. Antibiot.* **27**, 279 (1974). Total synthesis: T. Shiba *et al., J. Antibiot.* **32**, 1078 (1979). For structure *see* Tuberactinomycin.

Hydrochloride, $C_{25}H_{43}N_{13}O_{10}\cdot3HCl$, white crystalline powder, mp >245° (dec). $[\alpha]_D^{21}$ −19.1°. uv max (water, 0.1N HCl): 268 nm ($E_{1cm}^{1\%}$ 342); uv max (0.1N NaOH): 288 nm ($E_{1cm}^{1\%}$ 215). Very sol in water, slightly sol in methanol, ethanol. Insol in common organic solvents.

Sulfate, $C_{50}H_{92}N_{26}O_{32}S_3$, *Tuberactin.* White crystalline powder. Sol in water, slightly sol in usual organic solvents. LD$_{50}$ in mice, rats (mg/kg): 485, 680 i.v. (Hamakawa).

THERAP CAT: Antibacterial (tuberculostatic).

3637. Enviroxime. *6-[(Hydroxyimino)phenylmethyl]-1-[(1-methylethyl)sulfonyl]-1H-benzimidazol-2-amine;* (*E*)-2-amino-6-benzoyl-1-(isopropylsulfonyl)benzimidazole oxime; *anti*-2-amino-1-(isopropylsulfonyl)-6-(α-hydroxyiminobenzyl)benzimidazole; LY-122772. $C_{17}H_{18}N_4O_3S$; mol wt

358.42. C 56.97%, H 5.06%, N 15.63%, O 13.39%, S 8.95%. Benzimidazole deriv that inhibits rhinovirus multiplication. Prepn: C. J. Paget *et al.*, **Ger. pat. 2,638,551** correspds to U.S. pat. **4,118,742** (1977, 1978 both to Lilly). Synthesis and sepn of *syn* and *anti* isomers: J. H. Wikel *et al.*, *J. Med. Chem.* **23**, 368 (1980). Inhibition of rhinovirus replication in organ culture: D. C. De Long, S. E. Reed, *J. Infect. Dis.* **141**, 87 (1980). Metabolic studies: J. F. Quay *et al.*, *Fed. Proc.* **39**, 3079 (1980); C. J. Parli *et al.*, *ibid.* 214. Activity vs rhinovirus infection in man: R. J. Phillpotts *et al.*, *Lancet* **1**, 1342 (1981). Prophylactic activity: F. G. Haden, J. M. Gwaltney, *Antimicrob. Ag. Chemother.* **21**, 892 (1982).

Cryst from acetonitrile, mp 198-199°. uv max (methanol): 218, 290 nm (ε 45600, 27100).

(*Z*)-isomer, *zinviroxime*. Cryst from methanol, mp 182-183°. uv max (methanol): 254, 285 nm (ε 20800, 13200).

3638. Eosine I Bluish. *4',5'-Dibromo-3',6'-dihydroxy-2',7'-dinitrospiro[isobenzofuran-1(3H),9'-[9H]xanthen]-3-one disodium salt; 4',5'-dibromo-2',7'-dinitrofluorescein disodium salt;* hydroxydibromodinitro-*o*-carboxyphenylfluorone sodium; C.I. Acid Red 91; C.I. 45400. $C_{20}H_6Br_2N_2Na_2O_9$; mol wt 624.06. C 38.49%, H 0.97%, Br 25.61%, N 4.49%, Na 7.37%, O 23.07%. Prepd by nitration of 4',5'-dibromo-fluorescein: *Colour Index* vol. **4** (3rd ed., 1971) p 4427.

Red powder. Freely sol in water with green fluorescence; sol in alc.

USE: Dyeing wool, cotton, and paper. In histology as a stain for epithelia, muscular fibers, nuclei, etc.

3639. Eosine Yellowish—(YS). *2',4',5',7'-Tetrabromo-3',6'-dihydroxyspiro[isobenzofuran-1(3H),9'-[9H]xanthen]-3-one disodium salt; 2',4',5',7'-tetrabromofluorescein;* bromo-eosine; tetrabromofluorescein sol; bromofluoresceic acid; eosin; eosine; eosin Y; D & C Red No. 22; C.I. Acid Red 87; C.I. 45380. $C_{20}H_6Br_4Na_2O_5$; mol wt 691.86. C 34.72%, H 0.87%, Br 46.20%, Na 6.65%, O 11.56%. Xanthene dye capable of photodynamic generation of singlet oxygen. Prepd by bromination of fluorescein: *Colour Index* vol. **4** (3rd ed., 1971) p 4426. Photodynamic injury to leaf tissue and inhibition of photosynthesis: J. P. Knox, A. D. Dodge, *Planta* **164**, 22; 30 (1985). *Review:* H. J. Conn's *Biological Stains*, R. D. Lillie, Ed. (Williams & Wilkins, Baltimore, 9th ed., 1977), pp 342-344.

Red crystals with bluish tinge, or brownish-red powder. Freely sol in water, less in alcohol. Insol in ether. The concd aq soln is deep brownish-red, the dilute (1:500) soln is yellowish-red with greenish fluorescence; the alcoholic soln exhibits a strong green fluorescence.

Free acid, $C_{20}H_8Br_4O_5$, *eosine acid, D & C Red No. 21, C.I. Solvent Red 43, C.I. 45380:2.* Bright yellowish red. Sol in acetone, ethanol. Insol in mineral oil, toluene.

USE: In microbiological differential media. Lipstick and nail-polish coloring; dyeing wool, silk and paper; in red inks; as biological stain. Approved by FDA for use in drugs and cosmetics except for use in eye area: *Fed. Reg.* **47**, 53843 (1982).

3640. Epalrestat. *(E,E)-5-(2-Methyl-3-phenyl-2-propen-ylidene)-4-oxo-2-thioxo-3-thiazolidineacetic acid;* 5-[(E,E)-β-methylcinnamylidene]-4-oxo-2-thioxo-3-thiazolidineacetic acid; 3-carboxymethyl-5-(2-methylcinnamylidene)rhoda-nine; ONO-2235; Kinedak; Sorbistat. $C_{15}H_{13}NO_3S_2$; mol wt 319.41. C 56.41%, H 4.10%, N 4.39%, O 15.03%, S 20.08%. Aldose reductase inhibitor. Prepn: T. Tadao *et al.*, **Eur. pat. Appl. 47,109;** *eidem*, U.S. pat. **4,464,382** (1982, 1984 both to Ono Pharmaceutical). Inhibitory effect on aldose reductase *in vitro:* H. Terashima *et al.*, *J. Pharmacol. Exp. Ther.* **229**, 226 (1983). Effect on motor nerve conduction and sorbitol levels in diabetic rats: R. Kikkawa *et al.*, *Diabetologia* **24**, 290 (1983). Effect in rats on peripheral nerve dysfunction: *eidem, Metabolism* **33**, 212 (1984); on retinal microangiopathy: K. Kojima *et al.*, *Japan. J. Ophthalmol.* **29**, 99 (1985).

Crystals from ethanol-water, mp 210-217°.

N-Methyl-D-glucamine salt, $C_{22}H_{30}N_2O_8S_2$, crystals from methanol, mp 163-165°.

THERAP CAT: Treatment of diabetic neuropathy.

3641. Epanolol. *N-[2-[[3-(2-Cyanophenoxy)-2-hydroxy-propyl]amino]ethyl]-4-hydroxybenzeneacetamide;* 1-(2-cyano-phenoxy)-3-β-(4-hydroxyphenylacetamido)ethylamino-2-propanol; ICI-141292; Visacor. $C_{20}H_{23}N_3O_4$; mol wt 369.42. C 65.03%, H 6.28%, N 11.37%, O 17.32%. Cardio-selective β₁-adrenergic blocker with intrinsic sympatho-mimetic activity. Prepn: L. H. Smith, **Ger. pat. 2,362,568;** *idem,* U.S. pat. **4,167,581** (1974, 1979 both to ICI). Phar-macology in dogs: H. J. Smith *et al., J. Pharm. Exp. Ther.* **226**, 211 (1983); in humans: T. H. Pringle *et al., Brit. J. Clin. Pharmacol.* **21**, 249 (1986). Pharmacokinetics and bioavailability in dogs: J. McAinsh *et al., Eur. J. Drug Metabol. Pharmacokinet.* **9**, 129 (1984). Preliminary clinical trial in hypertension: B. Dahlof *et al., Brit. J. Clin. Pharma-col.* **18**, 831 (1984). Comparison with atenolol, *q.v.*, of bron-choconstrictor effect in asthmatics: S. Groth *et al., Eur. J. Clin. Pharmacol.* **30**, 653 (1986).

Crystals from acetonitrile, mp 118-120°.

THERAP CAT: Antihypertensive; antianginal.

3642. Eperisone. *1-(4-Ethylphenyl)-2-methyl-3-(1-piperidinyl)-1-propanone; 4'-ethyl-2-methyl-3-piperidino-propiophenone.* $C_{17}H_{25}NO$; mol wt 259.39. C 78.72%, H 9.71%, N 5.40%, O 6.17%. Spasmolytic agent related struc-turally to tolperisone, *q.v.* Prepn: E. Morita *et al.*, **Ger. pat. 2,458,638;** E. Morita, T. Kanai, U.S. pat. **4,181,803** (1975, 1980 both to Eisai). Pharmacological study: K. Tanaka *et al., Nippon Yakurigaku Zasshi* **77**, 511 (1981), *C.A.* **95**, 35471t (1981). Absorption, distribution, excretion in rats and guinea pigs: T. Fujita *et al., Oyo Yakuri* **21**, 835 (1981),

C.A. **96,** 28193w (1982). Toxicity study: H. Miyagawa *et al., ibid.* 939, C.A. **96,** 79722a (1982).

Hydrochloride, $C_{17}H_{25}NO.HCl$, *EMPP, E-646, Mional, Myonal.* Needles from isopropanol, mp 170-172°. LD_{50} in male S.D. rats, Wistar rats, mice (mg/kg): 1300, 1850, 1024 orally (Miyagawa).

THERAP CAT: Muscle relaxant (skeletal).

3643. Ephedra Equisetina. Ma Huang. Stems and leaves of *Ephedra equisetina* Bunge, *E. sinica* Stapf. and other species of *Ephedra, Gnetaceae.* Indigenous to China and India. It contains 0.75 to over 1% ephedrine; it is the source of natural ephedrine. Contains also variable quantities of pseudoephedrine.

3644. Ephedra Nevadensis. Cay note; canutillo; whorehouse tea; tapopote; teamsters' tea. Leaves and branches of *Ephedra nevadensis* S. Wats. (*E. antisyphilitica* C. A. Mey.), *Gnetaceae.* Habit. U.S. (Calif., Nevada). Contains little or no ephedrine.

3645. Ephedrine. α-[1-(Methylamino)ethyl]benzene-methanol; α-[1-(methylamino)ethyl]benzyl alcohol; 2-methylamino-1-phenyl-1-propanol; 1-phenyl-1-hydroxy-2-methylaminopropane; 1-phenyl-2-methylaminopropanol; α-hydroxy-β-methylaminopropylbenzene. $C_{10}H_{15}NO$; mol wt 165.24. C 72.69%, H 9.15%, N 8.48%, O 9.68%. α- and β-Adrenergic agonist. Occurs in Ma Huang (*Ephedra vulgaris, E. sinica* Stapf., *E. equisetina* Bunge, *Gnetaceae*) and in several other Ephedra spp. Extraction procedure: Rymill, McDonald, *Quart. J. Pharm. Pharmacol.* **10,** 463 (1937). Isomeric forms include *d-* and *l-*ephedrine as well as *d-* and *l-*pseudoephedrine with *l-*ephedrine and *d-*pseudoephedrine as the naturally occurring isomers. Syntheses: Späth, Göhring, *Monatsh.* **41,** 319 (1920); Manske, Johnson, *J. Am. Chem. Soc.* **51,** 580, 1906 (1929). Stereochemistry: Freudenberg *et al., ibid.* **54,** 234 (1932); Freudenberg, Nikolai, *Ann.* **510,** 223 (1934); Witkop, Foltz, *J. Am. Chem. Soc.* **79,** 197 (1957); Pfanz, Kirchner, *Ann.* **614,** 149 (1958). Sepn of isomers: Paris *et al., Ann. Pharm. Franc.* **25,** 177 (1967). Mechanism of action of *l-*ephedrine: Drudi-Baracco *et al., Compt. Rend. Soc. Biol.* **158,** 259 (1964). Toxicity: M. D. Fairchild, G. A. Alles, *J. Pharmacol. Exp. Ther.* **158,** 135 (1967). Review of new methods and developments in stereochemistry: Fodor, *Recent Develop. Chem. Nat. Carbon Compounds* **1,** 15-160 (1965). Comprehensive description: S. A. Benezra, J. W. McRae in *Analytical Profiles of Drug Substances* **vol. 8,** K. Florey, Ed. (Academic Press, New York, 1979) pp 489-507.

*dl-*Form, *racemic ephedrine, racephedrine.* Crystals, mp 79°. Sol in water, alcohol, ether, chloroform, oils.

*dl-*Form hydrochloride, $C_{10}H_{15}NO.HCl$, *racephedrine hydrochloride, Ephetonin.* Crystals, mp 187-188°. One gram dissolves in 4 ml water, in about 40 ml of 95% alc at 20°. Practically insol in ether. pH about 6.

*dl-*Form sulfate, $(C_{10}H_{15}NO)_2.H_2SO_4$, *racephedrine sulfate.* Crystals, mp 247°. Sol in water and alcohol. pH about 6.

*l-*Form, L-*erythro-2-(methylamino)-1-phenylpropan-1-ol.* Waxy solid, crystals or granules. Soapy feel. Gradually dec on exposure to light. May contain up to ½ mole H_2O (5.2%). Anhydr material is hygroscopic. mp 34°. Absorption of water raises the mp to 40°. bp 255°. pH of aq solution (1 in 200) 10.8. One gram dissolves in about 20 ml water, 0.2 ml alcohol; sol in chloroform, ether, oils. *Keep*

well closed in a cool place. Symposium on sympathomimetic agents: *Ind. Eng. Chem.* **37,** 116-148 (1945).

*l-*Form hydrochloride, *Ephedral, Sanedrine.* Orthorhombic needles, affected by light, mp 216-220°. $[\alpha]_D^{25}$ −33 to −35.5° (c = 5). pH of aq soln (1 in 200) 5.9. One gram dissolves in 3 ml water, 14 ml alcohol. Practically insol in ether, chloroform.

*l-*Form sulfate, orthorhombic needles, affected by light, mp 245° (dec). $[\alpha]_D^{25}$ −29.5 to −32.0° (c = 5). One gram dissolves in 1.2 ml water, 95 ml alcohol. Freely sol in hot alcohol. pH about 6.

*dl-*Pseudoephedrine, DL-*threo-2-(methylamino)-1-phenyl-propan-1-ol.* Crystals, mp 118°.

d-Pseudoephedrine, d-ψ-ephedrine, d-isoephedrine. Rhombic tablets from ether, mp 119°. $[\alpha]_D^{20}$ +51° (c = 0.6 in alc). pH of aq soln (1 in 200) 10.8. Sparingly sol in water (differs from *l-*ephedrine). Freely sol in alc or ether.

*d-*Pseudoephedrine hydrochloride, *Galpseud, Novafed, Rhinalair, Otrinol, Sinufed, Sudafed, Symptom 2.* Needles, mp 181-182°. $[\alpha]_D^{20}$ +62° (c = 0.8). uv max (ethanol): 208, 251, 257, 264 nm (ε 8300, 161, 201, 161). pKa 9.22. pH of aq soln (1 in 200) 5.9. Sol in water, alc, chloroform. LD_{50} i.p. in mice: 1.0 mmole/kg (Fairchild, Alles).

*d-*Pseudoephedrine sulfate, *Afrinol.*

THERAP CAT: *l-*Form as bronchodilator. *d-*Pseudoephedrine as decongestant.

THERAP CAT (VET): Sympathomimetic, has been used to counteract hypotension associated with anesthesia; as a mydriatic; in allergic reactions and as a CNS stimulant.

3646. Epiandrosterone. (3β,5α)-3-Hydroxyandrostan-17-one; 3β-hydroxy-17-androstanone; isoandrosterone; 3β-androstanol-17-one; 3β-hydroxyetioallocholan-17-one. C_{19}-$H_{30}O_2$; mol wt 290.45. C 78.57%, H 10.41%, O 11.02%. Present in normal human urine as a minor constituent, it is a less active androgen than androsterone. Synthesis: Cardwell *et al., J. Chem. Soc.* **1953,** 361; Johnson *et al., J. Am. Chem. Soc.* **75,** 2275 (1953); Johnson *et al., ibid.* **78,** 6331 (1956). *Review:* R. I. Dorfman, R. A. Shipley, *Androgens* (Wiley, New York, 1956).

*dl-*Form, crystals, mp 161-162°. Gives off a musk-like odor when hot.

*d-*Form, crystals from ethyl acetate + petr ether, mp 174.5°. $[\alpha]_D^{20}$ +88° (in methanol). Precipitated by digitonin. Practically insol in water. Soluble in organic solvents.

Acetate, $C_{21}H_{32}O_3$, stout prisms, mp 103-104°. $[\alpha]_D^{18}$ +68.5° (chloroform).

Benzoate, $C_{26}H_{34}O_3$, crystals, mp 210-212°.

3647. Epibatidine. *exo-2-(6-Chloro-3-pyridinyl)-7-aza-bicyclo[2.2.1]heptane.* $C_{11}H_{13}ClN_2$; mol wt 208.69. C 63.31%, H 6.28%, Cl 16.99%, N 13.42%. First naturally occurring non-opioid, organochlorine, alkaloid analgesic; isolated from skin of the Ecuadoran poison frog, *Epipedobates tricolor.* Isoln and structural determn: T. F. Spande *et al., J. Am. Chem. Soc.* **114,** 3475 (1992). First natural product identified which contains the 7-azabicyclo[2.2.1]heptane ring system. Synthesis of (±)-form: C. A. Broka, *Tetrahedron Letters* **1993,** 3251.

Basic, relatively polar, colorless oil. uv max (methanol): 217 nm, shoulder 250-280 nm.

3648. Epichlorohydrin. *Chloromethyloxirane; dl-α*-epichlorohydrin; 1-chloro-2,3-epoxypropane; γ-chloropropylene oxide. C_3H_5ClO; mol wt 92.52. C 38.94%, H 5.45%, Cl 38.32%, O 17.29%. Prepn: H. T. Clarke, W. W. Hartman, *Org. Syn.* **coll. vol. I,** 233 (2nd ed., 1941); G. Braun, *ibid.* **coll. vol. II,** 256 (1943). Manuf: Faith, Keyes & Clark's *Industrial Chemicals,* F. A. Lowenheim, M. K. Moran, Eds. (Wiley-Interscience, New York, 4th ed., 1975) pp 335-338. Toxicity data: H. F. Smyth, C. P. Carpenter, *J. Ind. Hyg. Toxicol.* **30,** 63 (1948).

Liquid. d_4^{20} 1.1812; d_4^{25} 1.1750; d_4^{50} 1.1436; d_4^{75} 1.1101. bp_{760} 117.9°; bp_{400} 98.0°; bp_{200} 79.3°; bp_{100} 62.0°; bp_{40} 42.0°; bp_{10} 16.6°; $bp_{1.0}$ −16.5°. mp −25.6°. $n_D^{11.6}$ 1.44195; n_D^{16} 1.43969; n_D^{25} 1.43585. Flash pt, open cup: 105°F (40°C). Insol in water. Misc with alcohol, ether, chloroform, trichloroethylene, carbon tetrachloride; immiscible with petr hydrocarbons. LD_{50} orally in rats: 0.09 g/kg (Smyth, Carpenter).

Caution: Potential symptoms of overexposure are nausea, vomiting; abdominal pain; respiratory distress, coughing; cyanosis; irritation of eyes and skin with deep pain. *See NIOSH Pocket Guide to Chemical Hazards* (DHHS/NIOSH 90-117, 1990) p 102. *See also* Patty's *Industrial Hygiene and Toxicology* **vol. 2A,** G. D. Clayton, F. E. Clayton, Eds. (Wiley-Interscience, New York, 3rd ed., 1981) pp 2242-2247. This substance may reasonably be anticipated to be a carcinogen: *Seventh Annual Report on Carcinogens* (PB95-109781, 1994) p 193.

USE: Solvent for natural and synthetic resins, gums, cellulose esters and ethers, paints, varnishes, nail enamels and lacquers, cement for Celluloid.

3649. Epicholestanol. *(3α,5α)-Cholestan-3-ol;* 3α-hydroxycholestane; ε-cholestanol. $C_{27}H_{48}O$; mol wt 388.68. C 83.44%, H 12.45%, O 4.12%. The 3α-hydroxy epimer of cholestanol. Prepn from cholestanone: Ruzicka, *Helv. Chim. Acta* **17,** 1407 (1934); *cf.* Marker, *et al., J. Am. Chem. Soc.* **57,** 2359 (1935); Barnett *et al., J. Chem. Soc.* **1940,** 1390.

Needles from alcohol, mp 185-186°. $[\alpha]_D^{20}$ +34° (c = 1.7 in chloroform). Less sol than cholestanol. Not precipitated by digitonin.

Acetate, $C_{29}H_{50}O_2$, mp 95.5-96°. Crystals from methanol. Benzoate, $C_{34}H_{52}O_2$, mp 102-103°. $[\alpha]_{546}$ +27.2° in chloroform.

3650. Epicholesterol. *(3α)-Cholest-5-en-3-ol.* $C_{27}H_{46}O$; mol wt 386.66. C 83.87%, H 11.99%, O 4.14%. The 3α-hydroxy epimer of cholesterol. Prepd from 8-oxocholesteryl chloride or from cholest-5-en-3-one or by passing O_2 into a soln of cholesteryl Mg chloride: Ruzicka, Goldberg, *Helv. Chim. Acta* **19,** 1407 (1936); Marker, *et al., J. Am. Chem. Soc.* **58,** 481 (1936); Marker, U.S. pat. **2,117,355** (1938); Barnett *et al., J. Chem. Soc.* **1940,** 1390.

Crystals from alcohol, mp 141.5°. $[\alpha]_D^{30}$ −35° (c = 1 in alcohol). Not precipitated by digitonin.

Acetate, $C_{29}H_{48}O_2$, crystals from methanol, mp 85°.

3651. Epicillin. *[2S-[2α,5α,6β(S*)]]-6-[(Amino-1,4-cyclohexadien-1-ylacetyl)amino]-3,3-dimethyl-7-oxo-4-thia-1-azabicyclo[3.2.0]heptane-2-carboxylic acid;* 6-[D-α-amino-2-(1,4-cyclohexadien-1-yl)acetamido]penicillanic acid; D-α-amino-(1,4-cyclohexadien-1-yl)methylpenicillin; SQ-11302; Dexacilina; Dexacillin; Omnisan; Spectacillin. $C_{16}H_{21}N_3O_4$-S; mol wt 351.43. C 54.68%, H 6.02%, N 11.96%, O 18.21%, S 9.12%. Semi-synthetic antibiotic related to penicillin. Prepn: Weisenborn *et al.,* U.S. pat. **3,485,819** (1969 to Squibb); Dolfini *et al., J. Med. Chem.* **14,** 117 (1971). Activity studies: Basch *et al., Infec. Immunity* **4,** 44 (1971); Gadebusch *et al., ibid.* 50. Clinical trials: Woodruff *et al., N.Y. State J. Med.* **71,** 1087 (1971); Beck *et al., Curr. Ther. Res.* **13,** 530 (1971); Reyes, *ibid.* 602; Landa, *ibid.* 654.

Crystals, dec 202° (hemihydrate).
THERAP CAT: Antibacterial.

3652. 16-Epiestriol. *(16β,17β)-Estra-1,3,5(10)-triene-3,16,17-triol;* $Δ^{1,3,5}$-estratriene-3,16β,17β-triol; 16-epioestriol; Actriol. $C_{18}H_{24}O_3$; mol wt 288.39. C 74.97%, H 8.39%, O 16.64%. Isoln from urine of pregnant women: Marrian, Bauld, *Biochem. J.* **59,** 136 (1955); Watson, Marrian, *ibid.* **63,** 64 (1956). From human placenta: Diczfalusy, Halla, *Acta Endocrinol.* **27,** 303 (1958). Prepn by reduction of 3,16β-diacetoxy-1,3,5(10)-estratrien-17-one: Biggerstaff, Gallagher, *J. Org. Chem.* **22,** 1220 (1957).

Crystals from methanol + benzene, mp 289-291°. $[\alpha]_D^{15}$ +76° (c = 0.297 in ethanol).

Epiestriol-3-allyl ether, crystals, mp 156-156.5°. Prepn: Huffman, U.S. pat. **3,002,009** (1961 to Lasdon Found.).

3653. Epimedii Herba. Yinyanghuo. Chinese herbal medicine made from the dried aerial parts of *Epimedium* sp. (*Berberidaceae*). Used as a tonic and in the treatment of rheumatic disease and hypertension. Principal constituents are prenylflavone glycosides such as icariin, epimedosides, sagittatosides, and baohuosides. Isoln of icariin from *E. macranthum*: S. Akai, *J. Pharm. Soc. Japan* **55,** 537 (1935); *C.A.* **29,** 5850 (1935). Revised structure: Y. Tokuoka *et al., Yakugaku Zasshi* **95,** 825 (1975), *C.A.* **83,** 111172r (1975). HPLC determn of constituents: M. Mizuno *et al., Chem. Pharm. Bull.* **36,** 3487 (1988); Y. Ito *et al., J. Chromatog.* **456,** 392 (1988). Immunoregulatory effects of icariin *in vitro*: W. He *et al., Arzneimittel-Forsch.* **45,** 910 (1995). Review: W. Tang, G. Eisenbrand in *Chinese Drugs of Plant Origin* (Springer-Verlag, Berlin, 1992) pp 491-498.

Icariin

Icariin, $C_{33}H_{40}O_{15}$, *3-[(6-deoxy-α-L-mannopyranosyl)oxy]-7-(β-D-glucopyranosyloxy)-5-hydroxy-2-(4-methoxyphenyl)-8-(3-methyl-2-butenyl)-4H-1-benzopyran-4-one, 4'-O-methyl-8-γ,γ-dimethylallylkaempferol-3-rhamnoside-7-glucoside.* Sesquihydrate, mp 231.5°. $[\alpha]_D^{15}$ −87.09° (in pyridine). Sol in pyridine. Insol in water, alcohol, chloroform, acetone, methanol, ethyl acetate.

3654. Epimestrol. *(16α,17α)-3-Methoxyestra-1,3,5(10)-triene-16,17-diol;* 3-methoxy-17-epiestriol; Org-817; NSC-55975; Stimovul. $C_{19}H_{26}O_3$; mol wt 302.41. C 75.46%, H 8.67%, O 15.87%. Estriol deriv, which stimulates ovulation. Prepn: J. de Visser, **Dutch** pat. **95,275** (1960 to Organon), *C.A.* **55,** 22378d (1961); L. Caglioti, M. Magi, *Tetrahedron Letters* **1962,** 1261; *eidem, Tetrahedron* **19,** 1127 (1963). ^{13}C-NMR study: T. A. Wittstruck, K. I. H. Williams, *J. Org. Chem.* **38,** 1542 (1973). Proton NMR and configurational study: B. Schoenecker *et al., J. Prakt. Chem.* **319,** 419 (1977). Clinical chemistry: G. Falkay *et al., Clin. Chim. Acta* **44,** 209 (1973). Comparative study in rats: D. Bentue-Ferrer *et al., Ann. Endocrinol.* **41,** 203 (1980). Effects in normal men: M. Luisi *et al., Reproduccion* **4,** 77 (1980), *C.A.* **93,** 89260 (1980).

Cryst from acetone/heptane, mp 158-160°. $[\alpha]_D^{20}$ +48° (chloroform).

THERAP CAT: Anterior pituitary activator.

3655. Epinastine. *9,13b-Dihydro-1H-dibenz[c,f]imidazo[1,5-a]azepin-3-amine;* 3-amino-9,13b-dihydro-1*H*-dibenz[c,f]imidazo[1,5-a]azepine; WAL-801. $C_{16}H_{15}N_3$; mol wt 249.32. C 77.08%, H 6.06%, N 16.85%. Tetracyclic, non-sedating histamine H_1 receptor antagonist. Prepn: G. Walther *et al.,* **Brit.** pat. **Appl. 2,071,095;** *eidem,* U.S. pat. **4,313,931** (1981, 1982 both to Boehringer, Ing.); and receptor binding studies: *idem et al., Arzneimittel.-Forsch.* **40,** 440 (1990). Pharmacology: A. Fügner *et al., ibid.* **38,** 1446 (1988). Clinical evaluation: W. S. Adamus *et al., ibid.* **37,** 569 (1987). Mechanism of action: K. Tasaka *et al., Oyo Yakuri* **39,** 365 (1990). Toxicology: J. Nishikawa *et al., ibid.* **42,** 151 (1991).

Crystals from acetonitrile, mp 205-208°. log P_a (octanol/aqueous buffer, pH 7.4, 20°) −0.70. pKa 11.2.

Hydrobromide, crystals from methanol/ethyl actate, mp 284-286°.

Hydrochloride, $C_{16}H_{15}N_3$·HCl, *WAL-801CL, Alesion.* Crystals from methanol/ether, mp 273-275°. Freely sol in water. LD_{50} in male, female rats (mg/kg): 314, 192 orally; 17, 22 i.v. (Nishikawa).

THERAP CAT: Antihistaminic.

3656. Epinephrine. *(R)-4-[1-Hydroxy-2-(methylamino)-ethyl]-1,2-benzenediol;* *(−)-3,4-dihydroxy-α-[(methylamino)-methyl]benzyl alcohol;* *l*-1-(3,4-dihydroxyphenyl)-2-(methyl-amino)ethanol; *l*-3,4-dihydroxy-1-[1-hydroxy-2-(methylami-no)ethyl]benzene; *l*-methylaminoethanolcatechol; adrenaline; levorenin; Bronkaid Mist; Epiglaufrin; Eppy; Glaupo-sine; Primatene Mist; Simplene; Sus-phrine; Suprarenaline. $C_9H_{13}NO_3$; mol wt 183.21. C 59.00%, H 7.15%, N 7.65%, O 26.20%. Endogenous catecholamine with combined α- and β-agonist activity. Principal sympathomimetic hormone produced by the adrenal medulla. Isoln from animal adrenal glands: Takamine, *J. Soc. Chem. Ind.* **20,** 746 (1901); Aldrich, *Am. J. Physiol.* **5,** 457 (1901). Synthesis of *dl*-form: Stolz, *Ber.* **37,** 4149 (1904); Payne, *Ind. Chem.* **37,** 523 (1961). Historic review of syntheses: Loewe, *Arzneimittel-Forsch.* **4,** 583 (1954). Resolution of *dl*-form: Flächer, *Z. Physiol. Chem.* **58,** 189 (1908). Configuration: Pratesi *et al., J. Chem. Soc.* **1958,** 2069. Acute toxicity: A. M. Lands *et al., J. Pharmacol. Exp. Ther.* **90,** 110 (1947). HPLC determn in plasma and urine: C. R. Benedict, *J. Chromatog.* **385,** 369 (1987). Comprehensive description: D. H. Szulczewski, W.-H. Hong, *Anal. Profiles Drug Subs.* **7,** 193-229 (1978). Physiologic review: Malmejac, *Physiol. Rev.* **44,** 186 (1964). Review of biosynthesis: L. A. Pohorecky, R. J. Wurtman, *Pharmacol. Rev.* **23,** 1-35 (1971); of pharmacology and clinical use in cardiopulmonary resuscitation: N. A. Paradis, E. M. Koscove, *Ann. Emerg. Med.* **19,** 1288-1301 (1990); P. Hebert *et al., J. Emerg. Med.* **9,** 487-495 (1991).

Minute crystals, gradually browning on exposure to light and air. mp 211-212°. mp ~215° (dec) when rapidly heated. $[\alpha]_D^{25}$ −50.0° to −53.5° (in 0.6*N* HCl). Very slightly sol in water, alc. Readily sol in aq solns of mineral acids, NaOH, and KOH. Insol in aq solns of ammonia and of the alkali carbonates. Insol in chloroform, ether, acetone, oils. LD_{50} i.p. in mice: 4 mg/kg (Lands).

Hydrochloride, $C_9H_{13}NO_3$·HCl, *Adrenalin, Epifrin, Glaucon, Suprarenin.*

d-Bitartrate, $C_9H_{13}NO_3$·$C_4H_6O_6$, *Asmatane Mist, Asthmahaler, Epitrate, Medihaler-Epi.* Crystals, mp 147-154° (some dec). Darkens slowly on exposure to air and light. One gram dissolves in about 3 ml water. Slightly sol in alc.

3,4-Cyclic borate, $C_9H_{12}BNO_4$, *epinephryl borate, Epinal, Eppy/N.*

dl-Form, *racepinephrine.* Sparingly sol in water, alcohol.

dl-Form hydrochloride, *Asthmanefrin, Vaponefrin.* Crystals from alcohol, mp 157°. Readily sol in water; sparingly sol in abs alc.

THERAP CAT: Bronchodilator; cardiostimulant; mydriatic; antiglaucoma.

THERAP CAT (VET): Vasoconstrictor; cardiostimulant.

3657. Epiquinidine. *(9R)-6'-Methoxycinchonan-9-ol.* $C_{20}H_{24}N_2O_2$; mol wt 324.42. C 74.05%, H 7.46%, N 8.63%, O 9.86%. Occurs in cinchona bark. Isoln from quinoidine: Dirscherl, Thron, *Ann.* **521,** 48 (1938). By epimerization of quinine or quinidine: Rabe, Höter, *J. Prakt. Chem.* **154,** 66 (1940). Stereochemistry: Prelog, Zalán, *Helv. Chim. Acta* **27,** 535, 545 (1944); Prelog, Häfliger, *ibid.* **33,** 2021 (1950); Roth, *Pharmazie* **16,** 257 (1961). Synthesis: Grethe *et al., Helv. Chim. Acta* **55,** 1044 (1972); Gutzwiller, Uskokovic, *ibid.* **56,** 1494 (1973).

Consult the Name Index before using this section.

Lustrous leaflets from ether, mp 111-113°. $[\alpha]_D^{25}$ +107.8° (c = 1.02 in ethanol). Freely sol in alcohol; moderately sol in ether. Shows more blue fluorescence in H_2SO_4 than quinidine or quinine. Forms a double sulfate with epiquinine.

Dihydrochloride, $C_{20}H_{24}N_2O_2$·2HCl, crystals from alc, dec 195-196°. $[\alpha]_D^{20}$ +46° (c = 0.8 in 99% alc).

Neutral dibenzoyl-*d*-tartrate, $(C_{20}H_{24}N_2O_2)_2$·$C_{18}H_{14}O_8$, crystals from alcohol or acetone, dec 166-167°. $[\alpha]_D^{21}$ +3.7° (4:1 alcohol + chloroform).

3658. Epiquinine. *(8α,9S)-6'-Methoxycinchonan-9-ol;* $C_{20}H_{24}N_2O_2$; mol wt 324.42. C 74.05%, H 7.46%, N 8.63%, O 9.86%. For refs *see* Epiquinidine.

Viscous oil, $[\alpha]_D^{22}$ +43° (c = 0.95 in 99% alc). Freely sol in organic solvents. Shows more blue fluorescence in H_2SO_4 than quinine. Forms a double sulfate with epiquinidine.

Dihydrochloride, $C_{20}H_{24}N_2O_2$·2HCl, crystals from acetone, dec 196°. $[\alpha]_D^{21}$ +33° (c = 0.8 in 99% alc).

Neutral dibenzoyl-*d*-tartrate, $(C_{20}H_{24}N_2O_2)_2$·$C_{18}H_{14}O_8$, crystals from acetone, dec 159°, $[\alpha]_D^{25}$ −24.3 (c = 0.93 in ethanol).

3659. Epirizole. *4-Methoxy-2-(5-methoxy-3-methyl-1H-pyrazol-1-yl)-6-methylpyrimidine;* 2-(3-methyl-5-methoxy-1-pyrazolyl)-4-methoxy-6-methylpyrimidine; 2-(3-methoxy-5-methylpyrazol-2-yl)-4-methoxy-6-methylpyrimidine; 1-(4-methoxy-6-methyl-2-pyrimidinyl)-3-methyl-5-methoxypyrazole; mepirizole; DA-398; Mebron. $C_{11}H_{14}N_4O_2$; mol wt 234.26. C 56.40%, H 6.02%, N 23.92%, O 13.66%. Prepn: Naito *et al.,* **S. Afr.** pat. **67 04,936** (1968 to Daiichi Seiyaku), *C.A.* **70,** 57876q (1969); Naito *et al., Chem. Pharm. Bull.* **17,** 1467 (1969). Pharmacology and metabolism: Oshima *et al., ibid.* 1492; Takabatake *et al., ibid.* **18,** 1900 (1970). Toxicity data: H. Ogura *et al., J. Med. Chem.* **15,** 923 (1972).

Minute, white or cream-colored crystals from isopropyl ether, mp 90-92°. Characteristic odor, bitter taste. Sparingly sol in water. Sol in dil acids; freely sol in ethanol, benz-

ene, dichloroethane. Also sol in ether, acetone. LD_{50} orally in mice: 820 mg/kg (Ogura).

THERAP CAT: Analgesic, antipyretic, anti-inflammatory.

3660. Epirubicin. *(8S-cis)-10-[(3-Amino-2,3,6-trideoxy-α-L-arabino-hexopyranosyl)oxy]-7,8,9,10-tetrahydro-6,8,11-trihydroxy-8-(hydroxyacetyl)-1-methoxy-5,12-naphthacenedione;* 3-glycoloyl-1,2,3,4,6,11-hexahydro-3,5,12-trihydroxy-10-methoxy-6,11-dioxo-1-naphthacenyl-3-amino-2,3,6-tridioxy-α-L-arabino-hexopyranoside; 4'-epidoxorubicin; 4'-epiadriamycin; pidorubicin; 4'-epi-DX; IMI-28. $C_{27}H_{29}NO_{11}$; mol wt 543.53. C 59.67%, H 5.38%, N 2.58%, O 32.38%. Analog of the anthracycline antibiotic doxorubicin, *q.v.,* differing only in the position of the C-4 hydroxy group of the sugar moiety. Prepn: F. Arcamone *et al.,* **Ger.** pat. **2,510,-866;** *eidem,* **U.S.** pat. **4,058,519** (1975, 1977 both to Soc. Farmaceut. Italia); *eidem, J. Med. Chem.* **18,** 703 (1975); S. Penco, *Process Biochem.,* **15**(5), 12 (1980). Purification: E. Oppici *et al.,* **Belg.** pat. **898,506;** *eidem,* **Brit.** pat. **Appl. 2,133,005** (both 1984 to Farmitalia Carlo Erba). Comparison with doxorubicin of *in vitro* activity: A. Di Marco *et al., Cancer Res.* **36,** 1962 (1976). Tissue distribution: C. Italia *et al., Brit. J. Cancer* **47,** 545 (1983); F. Arcamone *et al., Cancer Chemother. Pharmacol.* **12,** 157 (1984). HPLC determn in biological fluids: P. E. Deesen, B. Leyland-Jones, *Drug. Metab. Dispos.* **12,** 9 (1984); G. Cassinelli *et al., ibid.* 506. Clinical trials in solid tumors: K. Kolaric *et al., J. Cancer Res. Clin. Oncol.* **106,** 148 (1983). Review of activity in experimental tumors: A. Goldin *et al., Invest. New Drugs* **3,** 3-21 (1985). Review of pharmacology and clinical efficacy: F. Ganzina, *Cancer Treat. Rev.* **10,** 1-22 (1983); R. J. Cersosimo, W. K. Hong, *J. Clin. Oncol.* **4,** 425-439 (1986); P. Hurteloup, F. Ganzina, *Drugs Exp. Clin. Res.* **13,** 233-246 (1986). Review of epirubicin and other anthracycline antineoplastics: R. B. Weiss *et al., Cancer Chemother. Pharmacol.* **18,** 185-97 (1986).

Hydrochloride, $C_{27}H_{29}NO_{11}$·HCl, *Farmorubicin, Pharmorubicin.* Red-orange crystals, mp 185° (dec). $[\alpha]_D^{20}$ +274° (c = 0.01 in methanol). Solution should be protected from sunlight.

THERAP CAT: Antineoplastic.

3661. Epithiazide. *6-Chloro-3,4-dihydro-3-[[(2,2,2-trifluoroethyl)thio]methyl]-2H-1,2,4-benzothiadiazine-7-sulfonamide 1,1-dioxide;* 3-[(2,2,2-trifluoroethylthio)methyl]-6-chloro-3,4-dihydro-2H-1,2,4-benzothiadiazine-7-sulfonamide 1,1-dioxide; 6-chloro-3,4-dihydro-7-sulfamoyl-3-(2,2,-2-trifluoroethylthiomethyl)benzo-1,2,4-thiadiazine 1,1-dioxide; P-2105; Thiaver. $C_{10}H_{11}ClF_3N_3O_4S_3$; mol wt 425.86. C 28.20%, H 2.60%, Cl 8.32%, F 13.38%, N 9.87%, O 15.03%, S 22.59%. Prepn: McManus, **U.S.** pat. **3,009,911** (1961 to Pfizer).

Crystals from acetone + benzene, mp 206-207°.

THERAP CAT: Antihypertensive, diuretic.

3662. Epitiostanol. *(2α,3α,5α,17β)-2,3-Epithioandrostan-17-ol;* 10275-S; Thiodrol. $C_{19}H_{30}OS$; mol wt 306.51. C 74.45%, H 9.87%, O 5.22%, S 10.46%. Episulfide deriv of androstane, *q.v.* Prepn: **Brit.** pat. **977,599;** T. Komeno,

U.S. pat. **3,230,215** (1964, 1966 both to Shionogi); K. Takeda *et al.*, *Tetrahedron* **1965**, 329; P. D. Klimstra *et al.*, *J. Med. Chem.* **9**, 693 (1966). Antitumor effect in mice: A. Matsuzawa, T. Yamamoto, *Cancer Res.* **37**, 4408 (1977). Teratogenicity study: T. Minesita *et al.*, *Oyo Yakuri* **7**, 723 (1973), *C.A.* **80**, 116474p (1974). Toxicity study: *eidem, ibid.* 805, *C.A.* **80**, 66865u (1974). Use in treatment of breast cancer: M. Fujimoro *et al.*, *Cancer* **31**, 789 (1973).

Crystals from acetone, mp 127-128°. $[\alpha]_D^{27.5}$ +24.4° (c = 1.054 in chloroform). uv max (alcohol): 262 nm. LD_{50} in mice, rats (mg/kg): 1, 5 i.p. (Minesita, p 805).

THERAP CAT: Antineoplastic.

3663. EPN. *Phenylphosphonothioic acid O-ethyl O-(4-nitrophenyl) ester;* ethyl *p*-nitrophenyl benzenethiophosphonate; *O*-ethyl *O*-*p*-nitrophenyl phenylphosphonothioate. $C_{14}H_{14}NO_4PS$; mol wt 323.31. C 52.01%, H 4.36%, N 4.33%, O 19.79%, P 9.58%, S 9.92%. Cholinesterase inhibitor. Prepn: A G. Jelinek, **U.S.** pat. **2,503,390** (1950 to du Pont). Manufacturing process: N. Shindo *et al.*, **U.S.** pat. **3,327,026** (1967 to Nissan). Insecticidal activity: R. L. McGarr, A. J. Chapman, *J. Econ. Entomol.* **59**, 1529 (1966); D. A. Wolfenbarger *et al.*, *ibid.* **63**, 1568 (1970). Determn by HPLC: J. M. Lasker *et al.*, *Anal. Biochem.* **109**, 369 (1980); X.-D. Ding, I. S. Krull, *J. Agr. Food Chem.* **32**, 622 (1984). Inhibition of acetylcholinesterase: O. Awad *et al.*, *Arzneimittel-Forsch.* **22**, 1035 (1972); O. M. E. Awad, *Enzyme* **32**, 193 (1984). Metabolism in animals: R. L. Chrzanowski, A. G. Jelinek, *J. Agr. Food Chem.* **29**, 580 (1981). Toxicological study: H. C. Hodge *et al.*, *J. Pharmacol. Exp. Ther.* **112**, 29 (1954). Teratogenic evaluation in mice: K. D. Courtney *et al.*, *J. Environ. Sci. Health* **20B**, 373 (1985). Study of delayed neurotoxicity: B. M. Francis, *Bull. Environ. Contam. Toxicol.* **38**, 283 (1987). Toxicity data: T. B. Gaines, *Toxicol. Appl. Pharmacol.* **2**, 88 (1960).

Light yellow oil, aromatic odor. d^{25} 1.268. n_D^{25} 1.6021. Practically insol in water. Miscible with benzene, toluene, xylene, acetone, isopropanol, methanol. LD_{50} in female, male rats (mg/kg): 7.7, 36 orally; 25, 230 dermal (Gaines).

Caution: Potential symptoms of overexposure are miosis, lacrimation and rhinorrhea; headache; tight chest, wheezing and laryngeal spasm; salivation; cyanosis; anorexia, nausea, abdominal cramp and diarrhea; paralysis, convulsions; low blood pressure; cardiac irregularities; irritation of skin and eyes. *See NIOSH Pocket Guide to Chemical Hazards* (DHHS/NIOSH 90-117, 1990) p 102. *See also Clinical Toxicology of Commercial Products,* R. E. Gosselin *et al.*, Eds. (Williams & Wilkins, Baltimore, 5th ed., 1984) Section II, p 294.

USE: Insecticide; acaricide.

3664. Epostane. *(4α,5α,17β)-4,5-Epoxy-3,17-dihydroxy-4,17-dimethylandrost-2-ene-2-carbonitrile; (2α,4α,5α,17β)-4,5-epoxy-17-hydroxy-4,17-dimethyl-3-oxoandrostane-2-carbonitrile;* Win-32729. $C_{22}H_{31}NO_3$; mol wt 357.49. C 73.92%, H 8.74%, N 3.92%, O 13.43%. 3β-Hydroxysteroid dehydrogenase inhibitor; derivative of trilostane, *q.v.* Prepn as ketone tautomer: R. G. Christiansen, **Ger.** pat. **2,855,-091;** *idem,* **U.S.** pat. **4,160,027** (both 1979 to Sterling). Pharmacology and interceptive activity: J. E. Creange *et al.*,

Contraception **24**, 289 (1981); R. G. Christiansen *et al.*, *J. Med. Chem.* **27**, 928 (1984). Clinical pharmacology: Z. M. Van der Spuy *et al.*, *Clin. Endocrinol.* **19**, 521 (1983); N. S. Pattison *et al.*, *Fertil. Steril.* **42**, 875 (1984). Clinical evaluation as abortifacient: L. Birgerson *et al.*, *Contraception* **35**, 111 (1987); M. J. Crooij *et al.*, *N. Engl. J. Med.* **319**, 813 (1988). Comparative clinical study with mifepristone, *q.v.*, as abortifacient: L. Birgerson, V. Odlind, *Fertil. Steril.* **48**, 565 (1987). Induction of labor in swine: P. A. Martin *et al.*, *J. Anim. Sci.* **64**, 497 (1987); in sheep: M. Silver, *Vet. Rec.* **120**, 299 (1987). Regulation of ovulation in sheep: R. Webb, *J. Reprod. Fertil.* **79**, 231 (1987). Abortifacient efficacy in dogs: D. M. Keister *et al.*, *Theriogenology* **30**, 497 (1988).

Crystals, mp 191-194° from DMF/H_2O. $[\alpha]_D^{25}$ +67.4° (c = 1 in pyridine).

THERAP CAT: Interceptive.

THERAP CAT (VET): Interceptive, labor inducer.

3665. Epoxiconazole. *cis-(±)-1-[[3-(2-Chlorophenyl)-2-(4-fluorophenyl)oxiranyl]methyl]-1H-1,2,4-triazole; (2RS,-3RS)-1-[3-(2-chlorophenyl)-2,3-epoxy-2-(4-fluorophenyl)-propyl]-1H-1,2,4-triazole;* BAS-480-F; Opus. $C_{17}H_{13}ClFN_3$-O; mol wt 329.76. C 61.92%, H 3.97%, Cl 10.75%, F 5.76%, N 12.74%, O 4.85%. Ergosterol biosynthesis inhibitor. Prepn: S. Karbach *et al.*, **Eur.** pat. **Appl. 196,038;** *eidem,* **U.S.** pat. **4,906,652** (1986, 1990 both to BASF). Properties and field trials: E. Ammermann *et al.*, *Brighton Crop Prot. Conf. - Pests Dis.* **1990**, 407; R. Saur *et al.*, *ibid.* 831. Mode of action: A. Akers *et al.*, *ibid.* 837. Use in combination with fenpropimorph, *q.v.:* R. Saur *et al.*, *Gesunde Pflanz.* **46**, 61 (1994). Brief review: A. Floquet, N. Martin, *Phytoma* **449**, 54-57 (1993).

Crystals from diisopropyl ether, mp 136.2°. Soly at 20° (g/100 ml): water 6.63 × 10⁻⁴, acetone 18, dichloromethane 14, *n*-heptane <0.1. Partition coefficient (*n*-octanol/water at pH 7) 3.44. LD_{50} in rats (mg/kg): >5000 orally; >2000 dermally; LC_{50} in rats (4 hr): >5.3 mg/l (dust aerosol) (Ammermann).

USE: Agricultural fungicide.

3666. Eprazinone. *3-[4-(2-Ethoxy-2-phenylethyl)-1-piperazinyl]-2-methyl-1-phenyl-1-propanone; 3-[4-(β-ethoxyphenethyl)-1-piperazinyl]-2-methylpropiophenone.* $C_{24}H_{32}N_2$-O_2; mol wt 380.53. C 75.75%, H 8.48%, N 7.36%, O 8.41%. Prepn of the dihydrochloride: R. Y. Mauvernay, **Neth.** pat. **Appl. 6,602,581;** *idem,* **U.S.** pat. **3,448,192** (1966, 1969 both to Riom). Pharmacodynamics: J. Vacher *et al.*, *Arch. Int. Pharmacodyn. Ther.* **165**, 1 (1967). Pharmacological study: Y. Kase *et al.*, *Nippon Yakurigaku Zasshi* **73**, 605 (1977), *C.A.* **88**, 15657 (1978). Efficacy: W. Spitzer, *Fortschr. Med.* **98**, 871 (1980). Metabolism: P. Toffel-Nadolny, W. Gielsdorf, *Arzneimittel-Forsch.* **31**, 719 (1981).

Dihydrochloride, $C_{24}H_{32}N_2O_2 \cdot 2HCl$, *746-CE, Eftapan, Mucitux*. White, bitter-tasting crystalline powder, mp 201° (Toffel-Nadolny, Gielsdorf); also reported as crystals from methanol, mp 160° (Mauvernay). LD_{50} in mice (mg/kg): 729 orally, 38 i.v. (Vacher).

THERAP CAT: Antitussive.

3667. Eprinomectin. *(4″R)-4″-epi-(Acetylamino)-4″-deoxyavermectin B_1; MK-397.* Mixture of semisynthetic avermectins, *q.v.*; containing 90% or more of component B_{1a} and 10% or less of component B_{1b}. Prepn: H. H. Mrozik, U.S. pat. 4,427,663 (1984 to Merck & Co., Inc.). Synthesis: R. J. Cvetovich *et al., J. Org. Chem.* **59,** 7704 (1994) and biological activity: T. K. Jones *et al., J. Agric. Food Chem.* **42,** 1786 (1994). Mechanism of action study: J. P. Arena et al., *J. Parasitol.* **81,** 286 (1995).

component B_{1a}, R = C_2H_5 component B_{1b}, R = CH₃

Crystalline solid, mp 163.3°-165.7°. LD_{50} orally in mice: 24 mg/kg (Jones).

Component B_{1a}, $C_{50}H_{75}NO_{14}$, *(4″R)-4″-(acetylamino)-5-O-demethyl-4″-deoxyavermectin A_{1a}.*

Component B_{1b}, $C_{49}H_{73}NO_{14}$, *(4″R)-4″-(acetylamino)-5-O-demethyl-25-de(1-methylpropyl)-4″-deoxy-25-(1-methylethyl)-avermectin A_{1a}.*

THERAP CAT (VET): Antiparasitic.

3668. Episteride. *(17β)-17-[[(1,1-Dimethylethyl)amino]carbonyl]androsta-3,5-diene-3-carboxylic acid;* 17β-(N-tert-butylcarboxamido)androsta-3,5-diene-3-carboxylic acid; SKF-105657. $C_{25}H_{37}NO_3$; mol wt 399.57. C 75.15%, H 9.33%, N 3.51%, O 12.01%. 5α-Reductase inhibitor. Prepn and *in vitro* activity: D. A. Holt *et al.,* **Eur. pat. Appl.** 289,327; *eidem,* **U.S. pats.** 4,910,226 and 5,017,568 (1988, 1990, 1991 all to SmithKline Beecham); D. A. Holt *et al., J. Med. Chem.* **33,** 943 (1990). Pharmacology: J. C. Lamb *et al., Endocrinology* **130,** 685 (1992). Antitumor effects in rats: J. C. Lamb *et al., Prostate* **21,** 15 (1992). Mechanism of action study: M. A. Levy *et al., Biochemistry* **29,** 2815 (1990). HPLC determn in human plasma: V. K. Boppana *et al., J. Chromatog.* **631,** 251 (1993).

White crystals from ethyl acetate, mp 242-249°. pK 4.8.

THERAP CAT: Treatment of benign prostatic hypertrophy.

3669. Eprosartan. *(E)-α-[[2-Butyl-1-[(4-carboxyphenyl)methyl]-1H-imidazol-5-yl]methylene]-2-thiophenepropanoic acid;* (E)-3-[2-butyl-1-[(4-carboxyphenyl)methyl]-imidazol-5-yl]-2-(2-thienylmethyl)-2-propenoic acid; SKF-108566. $C_{23}H_{24}N_2O_4S$; mol wt 424.52. C 65.07%, H 5.70%, N 6.60%, O 15.08%, S 7.55%. Prototype of the imidazole-acrylic acid angiotensin II receptor antagonists. Prepn: J. A. Finkelstein *et al.,* **Eur. pat. Appl.** 403,159 (1990 to SmithKline Beckman); *eidem,* **U.S. pat.** 5,185,351 (1993 to SmithKline Beecham); J. Weinstock *et al., J. Med. Chem.* **34,** 1514 (1991); R. M. Keenan *et al., ibid.* **36,** 1880 (1993). Pharmacology: R. M. Edwards *et al., J. Pharmacol. Exp. Ther.* **260,** 175 (1992). Antihypertensive activity: D. P. Brooks *et al., Arch. Pharmacol.* **345,** 673 (1992). Receptor binding study: H. T. Schambye *et al., Mol. Pharmacol.* **47,** 425 (1995).

Crystals from methanol, mp 260-261°. Monomethanesulfonate, $C_{23}H_{24}N_2O_4S \cdot CH_3SO_3H$, *eprosartan mesylate, SKF-108566J.*

THERAP CAT: Antihypertensive.

3670. Eprozinol. *4-(2-Methoxy-2-phenylethyl)-α-phenyl-1-piperazinepropanol;* 1-(2-methoxy-2-phenylethyl)-4-(3-hydroxy-3-phenylpropyl)piperazine. $C_{22}H_{30}N_2O_2$; mol wt 354.49. C 74.54%, H 8.53%, N 7.90%, O 9.03%. Prepn: Saunders, **Brit. pat.** 1,188,505 (1970); to R. Y. Mauvernay, N. Busch, **U.S. pat.** 3,705,244 (1972). Pharmacology: Duchene-Marullaz *et al., Therapie* **26,** 155 (1971). Clinical studies: Sors, Dutarte, *ibid.* 163.

Dihydrochloride, $C_{22}H_{30}N_2O_2 \cdot 2HCl$, *Alecor, Brovel, Eupnéron.* White crystalline powder, mp 164°. Sol in water and alcohol. LD_{50} orally in mice: 500 mg/kg (Mauvernay, Busch).

THERAP CAT: Bronchodilator.

3671. Epsiprantel. *2-(Cyclohexylcarbonyl)-2,3,6,7,8,-12b-hexahydropyrazino[2,1-a][2]benzazepin-4(1H)-one;* 2-cyclohexylcarbonyl-4-oxo-1,2,3,4,6,7,12b-octahydropyrazino[2,1-a][2]benzazepine; BRL-38705; Cestex. $C_{20}H_{26}N_2O_2$; mol wt 326.44. C 73.59%, H 8.03%, N 8.58%, O 9.80%. Anthelmintic structurally related to praziquantel, *q.v.* Prepn: R. J. Dorgan, R. L. Elliott, **Eur. pat. Appl.** 134,984; *eidem,* **U.S. pat.** 4,661,489 (1985, 1987 both to Beecham); M. D. Brewer *et al., J. Med. Chem.* **32,** 2058 (1989). Veterinary trials in dogs: R. M. Corwin *et al., Am. J. Vet. Res.* **50,** 1076 (1989); in cats and dogs: B. R. Manger, M. D. Brewer, *Brit. Vet. J.* **145,** 384 (1989).

White crystals from chloroform/petr ether, mp 189-190°.
THERAP CAT (VET): Anthelmintic (Cestodes).

3672. Eptastigmine. *(3aS-cis)-1,2,3,3a,8,8a-Hexahydro-1,3a,8-trimethylpyrrolo[2,3-b]indol-5-yl heptylcarbamic acid ester;* N-demethyl-N-heptylphysostigmine; heptylphysostigmine; heptylstigmine. $C_{21}H_{33}N_3O_2$. mol wt 359.51. C 70.16%, H 9.25%, N 11.69%, O 8.90%. Cholinesterase inhibitor; analog of physostigmine, q.v. Prepn: M. Pomponi et al., **Eur. pat. Appl. 154,864**; M. Brufani et al., **U.S. pat. 4,831,155** (1985, 1989 both to Consiglio Naz. Ricerche). Prepn of tartrate: P. G. Pagella et al., **Eur. pat. Appl. 298,202** (1989 to Mediolanum Farm. and Consiglio Naz. Ricerche). Prepn and pharmacology: M. Marta et al., *Life Sci.* **43**, 1921 (1988). Anticholinesterase activity: P. De Sarno et al., *Neurobiochem. Res.* **14**, 971 (1989). Clinical cholinesterase inhibition: L. K. Unni et al., *Eur. J. Clin. Pharmacol.* **41**, 83 (1991).

Crystals from heptane, mp 60-64°; also reported as 58-62° (Marta). uv max (methanol): 303, 253 nm (ε 3300, 14200). Tartrate, $C_{21}H_{33}N_3O_2 \cdot C_4H_6O_6$, *MF-201*. Microcrystalline powder, mp 122-123°. Soly in water 70%. LD_{50} i.p. in mice: 35 mg/kg (De Sarno).
THERAP CAT: Cholinergic.

3673. Eptazocine. *(1S)-2,3,4,5,6,7-Hexahydro-1,4-dimethyl-1,6-methano-1H-4-benzazonin-10-ol;* (1S)-1,4-dimethyl-10-hydroxy-2,3,4,5,6,7-hexahydro-1,6-methano-1H-4-benzazonine; ST-2121; Sedapain. $C_{15}H_{21}NO$; mol wt 231.34. C 77.88%, H 9.15%, N 6.05%, O 6.92%. Opioid agonist-antagonist analgesic, related to pentazocine, q.v. Prepn: M. Ikeda et al., **Ger. pat. 2,422,309**; eidem, **U.S. pat. 4,008,219** (1974, 1977 both to Nihon Iyakuhin Kogyo). Chromatographic study: I. Hayashi et al., *Iyakuhin Kenkyu* **12**, 442 (1981), *C.A.* **95**, 220196v (1981). Series of articles on pharmacological activities: *Nippon Yakurigaku Zasshi* **78**, 599-645 (1981), *C.A.* **96**, 62879, 79732, 79733 (1982). Opioid receptor binding study: T. Nabeshima et al., *Res. Commun. Chem. Pathol. Pharmacol.* **48**, 173 (1985).

Hydrobromide, $C_{15}H_{22}BrNO$, crystals, mp 207-210°.
THERAP CAT: Analgesic (narcotic).

3674. EPTC. *Dipropylcarbamothioic acid S-ethyl ester;* dipropylthiocarbamic acid S-ethyl ester; S-ethyl dipropylthiocarbamate; R-1608; FDA-1541; Eptam. $C_9H_{19}NOS$; mol wt 189.32. C 57.10%, H 10.12%, N 7.40%, O 8.45%, S 16.94%. Selective pre-planting herbicide. Prepn: **Brit. pat. 808,753**; H. Tilles, J. Antognini, **U.S. pat. 2,913,327** (both 1959 to Stauffer). Activity: J. Antognini et al., *Proc. Northeast.*

States Weed Control Conf. 1957, 3. Metabolism: V. Y. Ong, S. C. Fang, *Toxicol. Appl. Pharmacol.* **17**, 418 (1970); J. P. Hubbell, J. E. Casida, *J. Agr. Food Chem.* **25**, 404 (1977).

Liquid, bp_{20} 127°. d^{30} 0.9546; n_D^{30} 1.4750. Vapor pressure at 35°: 3.4×10^{-2} mm Hg. Soly in water at 20°: 365 mg/l. Miscible with benzene, alc, toluene, xylene. LD_{50} orally in rats: 1631 mg/kg (Antognini).
USE: Herbicide.

3675. Equilenin. *3-Hydroxyestra-1,3,5,7,9-pentaen-17-one;* 11,12,13,14,15,16-hexahydro-3-hydroxy-13-methyl-17H-cyclopenta[a]phenanthren-17-one; 1,3,5:10,6,8-estrapentaen-3-ol-17-one. $C_{18}H_{18}O_2$; mol wt 266.34. C 81.17%, H 6.81%, O 12.01%. Estrogenic steroidal hormone isolated from urine of pregnant mares: Girard et al., *Compt. Rend.* **195**, 981 (1932). Occurs naturally in the d-form. Not found in human urine. Component of Conjugated Estrogenic Hormones, q.v. Isoln by chromatography: Duschinsky, Lederer, *Bull. Soc. Chim. Biol.* **17**, 1534 (1935). Total synthesis: Bachmann et al., *J. Am. Chem. Soc.* **61**, 974 (1939); **62**, 824 (1940); Johnson et al., ibid. **67**, 2274 (1945); **69**, 2942 (1947); **72**, 505 (1950); Hughes, Smith, *Chem. & Ind. (London)* **1960**, 1022; Bailey et al., *Chem. Commun.* **1967**, 1253; Stein et al., *Tetrahedron* **26**, 1917 (1970). Synthesis from estrone: Corbellini et al., *Farmaco Ed. Sci.* **19**, 913 (1964); O. N. Minailova et al., *Zh. Obschch. Khim.* **49**, 2633 (1979); A. R. Daniewski, T. Kowalczyk-Przewloka, *Tetrahedron Letters* **1982**, 2411. Review of synthetic studies: Taub, "Naturally Occurring Aromatic Steroids" in *The Total Synthesis of Natural Products* vol. 2, J. ApSimon, Ed. (John Wiley & Sons, New York, 1973) pp 642-663.

Needles from dil alc, mp 258-259°. Sublimes at 170-180° at 0.01 mm Hg. [α]_D^{16} +87° (12.8 mg made up to 1.8 ml in dioxane). uv max (ethanol): 231, 270, 282, 292, 325, 340 nm. Soly/100 ml alcohol 0.63 (18°); 2.5 g (boiling).
Acetate, $C_{20}H_{20}O_3$, mp 156-157°.
Benzoate, $C_{25}H_{22}O_3$, mp 222-223° (vac).
Methyl ether, $C_{19}H_{20}O_2$, needles from alc, mp 197-198°; mp 193-194° (vac).
THERAP CAT: Estrogen.

3676. Equilin. *3-Hydroxyestra-1,3,5(10),7-tetraen-17-one;* 1,3,5,7-estratetraen-3-ol-17-one. $C_{18}H_{20}O_2$; mol wt 268.36. C 80.56%, H 7.51%, O 11.92%. Steroidal hormone isolated from urine of pregnant mares: Girard et al., *Compt. Rend.* **195**, 981 (1932); Cartland, Meyer, *J. Biol. Chem.* **112**, 9 (1935/6). Structure: Serini, Logemann, *Ber.* **71**, 186 (1938); Pearlman, Wintersteiner, *J. Biol. Chem.* **130**, 35 (1939). Separation from estrone: Serini, Logemann, **U.S. pat. 2,221,340** (1941 to Schering). Synthesis: Zderic et al., *J. Am. Chem. Soc.* **80**, 2596 (1958); Bagli et al., *Tetrahedron Letters* **1964**, 387; Stein et al., ibid. **1966**, 5015; eidem, *Tetrahedron* **26**, 1917 (1970). Prepn from testosterone by bacterial fermentation: Bowers et al., **U.S. pat. 3,067,212** (1962 to Syntex). Prepn by conversion of equilenin: Bailey et al., *Chem. Commun.* **1967**, 1253; Marshall, Deghenghi, *Can. J. Chem.* **47**, 3127 (1969). Component of Conjugated Estrogenic Hormones, q.v. Review of synthetic studies: Taub, "Naturally Occurring Aromatic Steroids" in *The Total Synthesis of Natural Products* vol. 2, J. ApSimon, Ed. (John Wiley & Sons, New York, 1973) pp 664-670.

Consult the Name Index before using this section.

Orthorhombic sphenoidal plates from ethyl acetate, mp 238-240°. $[\alpha]_D^{25}$ +308° (c = 2 in dioxane); $[\alpha]_D^{25}$ +325° (c = 2 in alc). uv max: 283-285 nm. Soluble in alcohol, dioxane, acetone, ethyl acetate, in other organic solvents; sparingly sol in water.

Benzoate, $C_{25}H_{24}O_3$, mp 196-197°.
Methyl ether, $C_{19}H_{22}O_2$, needles from alc, mp 161-162°.
THERAP CAT: Estrogen.

3677. Equol. *3,4-Dihydro-3-(4-hydroxyphenyl)-2H-1-benzopyran-7-ol; 4',7-isoflavandiol; 4',7-dihydroxyisoflavan; 7-hydroxy-3-(4'-hydroxyphenyl)chroman.* $C_{15}H_{14}O_3$; mol wt 242.27. C 74.36%, H 5.82%, O 19.81%. Isoln from mare's urine: G. F. Marrian, G. A. D. Haslewood, *Biochem. J.* **26**, 1226 (1932); from human urine: M. Axelson *et al., ibid.* **201**, 353 (1982). Physical description: *eidem, ibid.* **29**, 1586 (1935); F. Wessely *et al., Monatsh.* **71**, 215 (1938). Structure: F. Wessely, F. Prillinger, *Ber.* **72B**, 629 (1939); E. L. Anderson, G. F. Marrian, *J. Biol. Chem.* **127**, 649 (1939). Abs config: H. Suginome, *Bull. Chem. Soc. Japan* **39**, 409 (1966). Prepn of (±)-form: J. A. Lamberton *et al., Aust. J. Chem.* **31**, 455 (1978).

Crystals from aq ethanol, mp 189-190.5°. $[\alpha]_D$ −21.5°.

3678. Erabutoxins. Neurotoxic principles isolated from venom of the sea-snake, *Laticauda semifasciata* (in Japanese, *erabu-umihebi*). Their action on post-synaptic membrane blocks neuromuscular transmission. Separation and crystn of erabutoxins A and B: Tamiya, Arai, *Biochem. J.* **99**, 624 (1966). Structures are single polypeptide chains containing 62 amino acid residues with four S-S bridges, *erabutoxin A* differing from *erabutoxin B* only in the amino acid residue at position 26: Sato, Tamiya, *ibid.* **122**, 453 (1971); Endo *et al., ibid.* 463. Isoln and structure of the minor component, *erabutoxin C*: Tamiya, Abe, *ibid.* **130**, 547 (1972). Corrected amino acid sequences: N. Maeda, N. Tamiya, *Biochem. J.* **167**, 289 (1977). Three-dimensional structure of B, description of active site: B. W. Low, *Advan. Cytopharmacol.* **3**, 141 (1979). Molecular conformation: M. R. Kimball *et al., Biochem. Biophys. Res. Commun.* **88**, 950 (1979); F. Inagaki *et al., Eur. J. Biochem.* **109**, 129 (1980). Raman spectra: T. Takamatsu *et al., Biochim. Biophys. Acta* **622**, 189 (1980). Immunological studies: A. Tatsuya *et al., Toxicon* **17**, 571 (1979), *C.A.* **92**, 141459t (1980). Effect on transmission in autonomic ganglia: V. A. Chiappinelli *et al., Brain Res.* **211**, 107 (1981).

LD_{50} in mice, rats (g/g): 0.15, 0.07 i.m. (Tamiya, Arai).

3679. Erbium. Er; at. wt 167.26; at. no. 68; valence 3. Rare earth metal of the yttrium group; member of the lanthanide series. Naturally occurring isotopes (mass numbers): 162 (0.14%); 164 (1.61%); 166 (33.6%); 167 (22.95%); 168 (26.8%); 170 (14.9%). Known artificial radioactive isotopes: 147; 148; 150-161; 163; 165; 169; 171-173. Abundance in earth's crust: 2.47-3.5 ppm. Occurs in small quantities in all the rare earth minerals; main sources: xenotime, fergusonite, gadolinite, euxenite, polycrase, blomstrandine. Discovery: Mosander, *Skand. Naturför. Förh.* **3**, 387 (1842); *Phil. Mag.* [3] **23**, 241 (1843). Separation of oxide: James, *J. Am. Chem. Soc.* **32**, 517 (1910); **34**, 757 (1912); Hofmann, Burger, *Ber.* **41**, 308 (1908); Prandtl, *Z. Anorg. Chem.* **198**, 157 (1931). Sepn by ion-exchange: Spedding *et al., J. Am.*

Chem. Soc. **69**, 2812 (1947); **76**, 2557 (1954). Prepn of the metal: Klemm, Bommer, *Z. Anorg. Chem.* **213**, 138 (1937). Prepn of chloride, bromide: Jantsch *et al., Z. Anorg. Chem.* **207**, 353 (1932). Radioactivity induced by neutron bombardment: Sugden, *Nature* **135**, 469 (1935); McLennan, Rann, *ibid.* **136**, 831 (1935). Spectrum: Eder, *Ber. Wien. Akad.* [2a] **124**, 790 (1915); deGramont, *Compt. Rend.* **171**, 1106 (1920); Mott, McDonald, *Trans. Roy. Soc. Canada* [3] **21**, 230 (1927). Natural isotopic composition: Hayden *et al., Phys. Rev.* **77**, 299 (1950). Toxicity study: Haley, *J. Pharm. Sci.* **54**, 663 (1965). Reviews of prepn, properties and compds: *The Rare Earths*, F. H. Spedding, A. H. Daane, Eds. (Krieger, Huntington, N.Y., 1971, reprint of 1961 ed.) 641 pp; Hulet, Bode, "Separation Chemistry of the Lanthanides and Transplutonium Actinides" in *MTP Int. Rev. Sci.: Inorg. Chem., Ser. One* vol. 7, K. W. Bagnall, Ed. (University Park Press, Baltimore, 1972) pp 1-45; Moeller, "The Lanthanides" in *Comprehensive Inorganic Chemistry* vol. 4, J. C. Bailar Jr. *et al.,* Eds. (Pergamon Press, Oxford, 1973) pp 1-101; F. H. Spedding in Kirk Othmer *Encyclopedia of Chemical Technology* vol. 19 (John Wiley & Sons, New York, 3rd ed., 1982) pp 833-854; *Chemistry of the Elements,* N. N. Greenwood, A. Earnshaw, Eds. (Pergamon Press, New York, 1984) pp 1423-1449. Brief review of properties: G. T. Seaborg, *Radiochim. Acta* **61**, 115-122 (1993).

Dark-gray metallic powder; hexagonal close-packed crystal lattice. d 9.066. mp 1529°. bp 2868°. Heat of fusion: 19.90 kJ/mol. Heat of sublimation (25°): 317.10 kJ/mol. E°(aq) Er^{3+}/Er −230 V (calc). Similar to the other rare earth metals, possesses two reduction potentials 1.770 and 1.875 volts (ref to the normal calomel electrode), Noddack, Brukl, *Angew. Chem.* **50**, 362 (1937).

Oxide, Er_2O_3, erbia. Pinkish powder changing into cubic crystals on heating at 1300°; d 8.64; sp heat 0.065; prepd by igniting the oxalate or basic nitrate. Readily sol in acids. Soly in water: 1.28×10^{-5} g-mol/l at 29°.

Hydroxide, $Er(OH)_3$, pale pink gelatinous ppt. Prepd by action of alkali hydroxide on a soln of erbium nitrate.

Chloride, $ErCl_3$, hexahydrate, deliquesc crystals. Sol in water; slightly sol in alcohol. Dehydrated by heating in a stream of hydrogen chloride. Anhydr form is pinkish powder, d 4.1. LD_{50} in mice (mg/kg): 535 i.p.; 6.2 orally (Haley).

Bromide, $ErBr_3$, nonahydrate, deliquesc rose crystals.
Nitrate, $Er(NO_3)_3$, pentahydrate, reddish, deliquesc crystalline solid. Loses 4 mols of water on heating to 130°. LD_{50} of hexahydrate in female rats (mg/kg): 230 i.p.; 35.8 i.v. (Haley).

3680. Erbium Sulfate. $Er_2O_{12}S_3$; mol wt 622.71. Er 53.72%, O 30.83%, S 15.45%. $Er_2(SO_4)_3$. Prepared by slow evaporation of a soln of erbium oxide in sulfuric acid at 20-25°.

Powder. d 3.678. Hygroscopic; dissociates on heating; dissolves in water with evolution of heat; forms a basic sulfate on heating to 850°.

Octahydrate, pink monoclinic crystals; isomorphic with the corresponding yttrium, praseodymium and neodymium salts. d 3.205. Soly in water (parts/100 parts): 16 (20°); 6.53 (40°). On heating loses its water of hydration.

3681. Erbon. *2,2-Dichloropropanoic acid 2-(2,4,5-trichlorophenoxy)ethyl ester; 2,2-dichloropropionic acid 2-(2,4,-5-trichlorophenoxy)ethyl ester; 2-(2,4,5-trichlorophenoxy)-ethyl 2,2-dichloropropionate;* Baron; Novon. $C_{11}H_9Cl_5O_3$; mol wt 366.45. C 36.05%, H 2.48%, Cl 48.37%, O 13.10%. Prepn: Brust, Seinkbeil, U.S. pat. **2,754,324** (1956 to Dow). Toxicity study: G. W. Bailey, J. L. White, *Residue Rev.* **10**, 97 (1965).

Crystals, mp 49-50°. $bp_{0.5}$ 161-164°. d_4^{50} 1.55. Sol in acetone, alcohol, kerosene, xylene. Insol in water. LD_{50} orally in rats: 1120 mg/kg (Bailey, White).

Caution: Irritation on contact with eyes, skin. Repeated contact with concentrated material may cause dermatitis.

USE: Herbicide.

3682. Erdin. (±)-*5,7-Dichloro-4-hydroxy-6'-methoxy-6-methyl-3,4'-dioxospiro[benzofuran-2(3H),1'-[2,5]cyclohexadiene]-2'-carboxylic acid.* $C_{16}H_{10}Cl_2O_7$; mol wt 385.16. C 49.90%, H 2.62%, Cl 18.41%, O 29.08%. Chlorine containing fungal metabolite isolated from *Aspergillus terreus* Thom, as the racemate, together with geodin: H. Raistrick, G. Smith, *Biochem. J.* **30**, 1315 (1936). Characterization and structure of erdin and geodin: P. W. Clutterbuck *et al.*, *ibid.* **31**, 1089 (1937); C. T. Calam *et al.*, *ibid.* **33**, 579 (1939); *eidem, ibid.* **41**, 458 (1947); D. H. Barton, A. I. Scott, *J. Chem. Soc.* **1958**, 1767. Biosynthesis of geodin: Rhodes *et al., Chem. & Ind. (London)* **1962**, 611.

Needles, mp 210-212°. uv max: 284 nm (ε 21000). (+)-Form, prisms from chloroform + light petr, mp 210-212°. [α]$_D$ +149° (c = 0.41 in dioxane). uv max: 284 nm (ε 21000). Reacts as dibasic acid, dec slowly in presence of alkali. Solubilities of erdin are closely similar to those of geodin except that it is only very slightly sol in chloroform.

(+)-Form methyl ester, $C_{17}H_{12}Cl_2O_7$, *geodin.* Prisms from chloroform + ether, mp 228-231°. [α]$_D$ +140° (c = 0.8). uv max: 284 nm (ε 19000). Practically insol in water, petr ether. Very sparingly sol in benzene, ether, somewhat more sol in alc, ethyl acetate; readily sol in chloroform, acetone, dioxane.

3683. Erdmann's Salt. *Ammonium diamminetetrakis-(nitrito-N)cobaltate(1—).* $CoH_{10}N_7O_8$; mol wt 295.05. Co 19.97%, H 3.42%, N 33.23%, O 43.38%. $NH_4[Co(NH_3)_2(NO_2)_4]$. Prepared by evaporating a neutral aq soln of NH_4NO_2 and $CoCl_2$: Erdmann, *J. Prakt. Chem.* **97**, 410 (1866); Gibbs, *Proc. Am. Acad.* **10**, 13 (1875); improved technique: Jörgensen, *Z. Anorg. Chem.* **17**, 477 (1898). Prepn from $CoCO_3$ and NH_4NO_2: Harkins *et al., J. Am. Chem. Soc.* **38**, 2647 (1916).

Dark, reddish-brown, orthorhombic bipyramidal crystals. d_4^{20} 1.972. Sol in water. A 1% aq soln is precipitated by $AgNO_3$, Tl_2SO_4, CsCl, RbCl, cinchonine sulfate.

3684. Erdosteine. [[*2-Oxo-2-[(tetrahydro-2-oxo-3-thienyl)amino]ethyl]thio]acetic acid;* (±)-[[[(tetrahydro-2-oxo-3-thienyl)carbamoyl]methyl]thio]acetic acid; DL-S-[2-[N-3-(2-oxotetrahydrothienyl)acetamido]]thioglycolic acid; *N*-(carboxymethylthioacetyl)homocysteine thiolactone; dithiosteine; RV-144; Secresolv. $C_8H_{11}NO_4S_2$; mol wt 249.31. C 38.54%, H 4.45%, N 5.62%, O 25.67%, S 25.72%. Homocysteine thiolactone derivative with mucomodulating activity. Prepn: J. Gonella, *Eur. pat. Appl.* **61,386**; *idem, U.S. pat.* **4,411,909** (1982, 1983 both to Refarmed); M. Gobetti *et al., Farmaco Ed. Sci.* **41**, 69 (1986). Clinical trial in bronchitis: G. Ricevuti *et al., Thorax* **43**, 585 (1988). Clinical effect on $α_1$-antitrypsin levels in cigarette smokers: M. Vagliasindi, G. B. Fregnan, *Int. J. Clin. Pharmacol. Ther. Toxicol.* **27**, 238 (1989).

Crystals from ethanol, mp 156-158°. pKa 3.71. LD_{50} in mice and rats (g/kg): > 10 orally; > 3.5 i.v. (Gonella).

THERAP CAT: Mucolytic.

3685. Ergocornine. (5'α)-*12'-Hydroxy-2',5'-bis(1-methylethyl)ergotaman-3',6',18-trione.* $C_{31}H_{39}N_5O_5$; mol wt 561.68. C 66.29%, H 7.00%, N 12.47%, O 14.24%. Natural

ergot alkaloid derived from lysergic acid; a member of the ergotoxine group. Isoln from ergot: Stoll, Hofmann, *Helv. Chim. Acta* **26**, 1570 (1943). Structure: Stoll *et al., ibid.* **34**, 1544 (1951). Separation and purification: Stoll, Hofmann, U.S. pat. **2,447,214** (1948 to Sandoz). Synthesis: Stadler *et al., Helv. Chim. Acta* **52**, 1549 (1969).

Solvated, polyhedra from methanol, dec 181° (contains 1 mole methanol). [α]$_D^{20}$ −110° (pyridine); −175° (chloroform). uv max (methanol): 311 nm (log ε 3.91). Soluble in acetone, chloroform, ethyl acetate; slightly sol in ethyl and methyl alcohol. Nearly insol in water.

Ergocornine phosphate, crystals, dec 190-195°.
Ergocornine ethanesulfonate, crystals, dec 209°.

THERAP CAT: See Ergot.

3686. Ergocorninine. *12'-Hydroxy-2',5'α-bis(1-methylethyl)-8α-ergotaman-3',6',18-trione.* $C_{31}H_{39}N_5O_5$; mol wt 561.68. C 66.29%, H 7.00%, N 12.47%, O 14.24%. An alkaloid isomeric with ergocornine, *q.v.* Isoln, structure, separation, purification and synthesis *see* ergocornine.

Crystallizes solvent-free, unlike ergocornine which tends to retain the solvent of crystn. Prisms from alc, dec 228°, [α]$_D^{20}$ +409° (chloroform). uv max (methanol): 240.5, 312.5 (log ε 4.31, 3.92). Soluble in 15 parts boiling ethanol, 25 parts boiling methanol, 30 parts boiling benzene, 30 parts boiling ethyl acetate; freely sol in acetone, chloroform. Practically insol in water. Does not seem to form salts.

THERAP CAT: See Ergot.

3687. Ergocristine. *12'-Hydroxy-2'-(1-methylethyl)-5'-(phenylmethyl)ergotaman-3',6',18-trione.* $C_{35}H_{39}N_5O_5$; mol wt 609.73. C 68.95%, H 6.45%, N 11.49%, O 13.12%. Natural ergot alkaloid derived from lysergic acid; a member of the ergotoxine group. Isoln from ergot: Stoll, Burckhardt, *Z. Physiol. Chem.* **250**, 1 (1937); Stoll, Hofmann, *Helv. Chim. Acta* **26**, 1570 (1943). Structure: Stoll *et al., ibid.* **34**, 1544 (1951). Separation and purification: Stoll, Hofmann, U.S. pat. **2,447,214** (1948 to Sandoz). Synthesis: Stadler *et al., Helv. Chim. Acta* **52**, 1549 (1969).

Orthorhombic crystals with $2C_6H_6$ from benzene. mp 155-157° (dec) (solvent-free base). [α]$_D^{20}$ −183° (chloroform). Very sol in ethyl and methyl alcohol, acetone, chloroform, ethyl acetate. Slightly sol in ether. Practically insol in water, petr ether.

Ergocristine phosphate, crystals, dec 195°.
Ergocristine ethanesulfonate, crystals, dec 207°.

3688. Ergocristinine. *12'-Hydroxy-2'-(1-methylethyl)-5'α-(phenylmethyl)-8α-ergotaman-3',6',18-trione.* $C_{35}H_{39}N_5$-O_5; mol wt 609.73. C 68.95%, H 6.45%, N 11.49%, O 13.12%. An alkaloid isomeric with ergocristine, *q.v.* Isoln, structure, separation and purification: *see* ergocristine.

Crystallizes solvent-free, unlike ergocristine which tends to retain the solvent of crystn. Long prisms from abs alc, mp 226° (dec). $[α]_D^{20}$ +366° (c = 0.68 in chloroform); +471° (c = 0.35 in pyridine). uv max (methanol/methylene chloride): 313 nm (log ε 3.96). Much less sol than ergocristine. Does not seem to form salts.

3689. Ergocryptine. Ergokryptine. $C_{32}H_{41}N_5O_5$; mol wt 575.71. C 66.76%, H 7.18%, N 12.16%, O 13.90%. Two closely related isomers of the ergotoxine group which differ in the peptide portion of the molecule; α-ergocryptine yielding L-leucine upon hydrolysis, β-ergocryptine yielding L-isoleucine. The ergocryptine discussed in the literature prior to 1967 is now referred to as α-ergocryptine. Isoln from ergot: Stoll, Hofmann, *Helv. Chim. Acta* **26**, 1570 (1943). Structure: Stoll *et al., ibid.* **34**, 1544 (1951). Separation and purification: Stoll, Hofmann, U.S. pat. 2,447,214 (1948 to Sandoz). Separation of β-ergocryptine from α-ergocryptine: Schlientz *et al., Experientia* **23**, 991 (1967); *see also eidem, Pharm. Acta Helv.* **43**, 497 (1968). Synthesis of α- and β-ergocryptine: Stadler *et al., Helv. Chim. Acta* **52**, 1549 (1969).

α-Ergocryptine   R = CH₂CH(CH₃)₂
β-Ergocryptine   R = CH(CH₃)CH₂CH₃

α-Ergocryptine, (5'α)-12'-hydroxy-2'-(1-methylethyl)-5'-(2-methylpropyl)ergotaman-3',6',18-trione. Solvated prisms from acetone, benzene, methanol. With MeOH of crystn, mp 212° (dec). $[α]_D^{20}$ −120° (pyridine); −198° (chloroform). uv max (methanol): 241, 312.5 nm (log ε 4.31, 3.95). Freely sol in alcohol, chlorofom. Almost insol in water.

β-Ergocryptine, [5'α(S)]-12'-hydroxy-2'-(1-methylethyl)-5'-(1-methylpropyl)ergotaman-3',6',18-trione. Rectangular plates from benzene, mp 173° (dec). $[α]_D^{20}$ −98° (c = 0.5 in pyridine); −179° (c = 0.5 in chloroform). uv max (methanol): 312 (log ε 3.93).

THERAP CAT: *See* Ergot.

3690. Ergocryptinine. Ergokryptinine. $C_{32}H_{41}N_5O_5$; mol wt 575.71. C 66.76%, H 7.18%, N 12.16%, O 13.90%. Alkaloid pair isomeric with α- and β-ergocryptine, resp., but differing by an α-configuration at C-8. The literature prior to 1967 refers to α-ergocryptinine as ergocryptinine. Isolation, structure, separation and purification: *see* ergocryptine. Production: Abe *et al.,* U.S. pat. 2,835,675 (1958). Prepn of β-ergocryptinine from β-ergocryptine: Schlientz *et al., Experientia* **23**, 991 (1967). Synthesis of α- and β-ergocryptinin: Stadler *et al., Helv. Chim. Acta* **52**, 1549 (1969).

α-Ergocryptinine, *12'-hydroxy-2'-(1-methylethyl)-5'α-(2-methylpropyl)-8α-ergotaman-3',6',18-trione.* Crystallizes solvent-free, unlike ergocryptine which tends to retain the solvent of crystn. Fine needles from methanol, dec 240-242°. $[α]_D^{20}$ +408° (chloroform); +485° (c = 0.5 in pyridine). uv max (methanol): 241.5, 312.5 nm (log ε 4.30, 3.94). Sol in 20 parts boiling ethanol, 50 parts boiling methanol; freely

sol in acetone, chloroform; almost insol in water. Does not seem to form salts.

β-Ergocryptinine, colorless needles from methylene chloride-methanol, mp 217-218° (dec). $[α]_D^{20}$ +421° (chloroform); +497° (pyridine). uv max (methanol): 240.5, 312 nm (log ε 4.31, 3.94). Differs from the α-isomer by the 1-methylpropyl group at the 5'-position.

3691. Ergoflavin. *[1S-[1α,3β,4β,4aα,7(1'R*,3'R*,4'R*,-4'aR*,9'aS*),9aβ]]-1,1',3,3',4,4',9a,9'a-Octahydro-4,4',8,8',-9a,9'a-hexahydroxy-3,3'-dimethyl-[7,7'-bi-1,4a-(epoxymethano)-4aH-xanthene]-9,9',11,11'(2H,2'H)-tetrone; ergochrome CC(2,2').* $C_{30}H_{26}O_{14}$; mol wt 610.53. C 59.02%, H 4.29%, O 36.69%. Principal pigment from ergot: Freeborn, *Pharm. J.* **88**, 568 (1912); Eglinton *et al., J. Chem. Soc.* **1958**, 1833. Structure: McPhail *et al., ibid.* **1966**, Sect. B, 18. Review of ergoflavin and other ergochromes: Franck, Flasch in *Fortschr. Chem. Org. Naturst.* **30**, 151-206 (1973).

Yellow needles from methanol, decomp 350°. $[α]_D^{21}$ +37.5° (c = 1.236 in acetone). uv max: 240, 260, 381 nm ($E_{1cm}^{1\%}$ 350, 346, 130). Sol in acetone, pyridine; moderately sol in methanol, alcohol, ethyl acetate, dioxane; sparingly sol in ether, benzene. Practically insol in 2N aq NaHCO₃.

Hexaacetate, $C_{42}H_{38}O_{20}$, prisms from chloroform + petr ether, dec 248-249°. $[α]_D^{20}$ +61.2° (c = 0.62 in dioxane). uv max: 340, 338 nm ($E_{1cm}^{1\%}$ 343, 61).

3692. Ergoloid Mesylates. *Dihydroergotoxine monomethanesulfonate (salt);* co-dergocrine mesylate; CCK-179; Circanol; Coristin; Dacoren; DCCK; Deapril-ST; Decril; Dulcion; Ergodesit; Ergohydrin; Ergoplus; Hydergine; Lysergin; Novofluen; Orphol; Pérénan; Progeril; Redergin; Sponsin; Trigot. Hydrogenated ergot alkaloids, specifically an equiproportional mixture of dihydroergocornine methanesulfonate, dihydroergocristine methanesulfonate and α- and β-dihydroergocryptine methanesulfonate in the ratio of 1.5-2.5:1. Neuropharmacological investigations: D. M. Loew *et al., Postgrad. Med. J.* **52**, Suppl. 1, 40 (1976).

Dihydroergocornine   R = CH(CH₃)₂
Dihydroergocristine   R = CH₂C₆H₅
Dihydro-α-ergocryptine   R = CH₂CH(CH₃)₂
Dihydro-β-ergocryptine   R = CH(CH₃)CH₂CH₃

Dihydroergocristine methanesulfonate, $C_{36}H_{45}N_5O_8S$, *Decme, Defluina, Enirant, Insibrin, Nehydrin, Simactil, Unergol.*

THERAP CAT: α-Adrenergic blocker (treatment of impaired mental function in the elderly).

3693. Ergometrinine. *[8α(S)]-9,10-Didehydro-N-(2-hydroxy-1-methylethyl)-6-methylergolinecarboxamide;* D-lysergic acid D-propanolamide; ergonovinine. $C_{19}H_{23}N_3O_2$; mol wt 325.41. C 70.13%, H 7.12%, N 12.91%, O 9.83%.

An alkaloid isomeric with ergonovine, *q.v.* Ergometrinine and ergonovine can be interconverted by simple chemical procedures: Smith, Timmis, *J. Chem. Soc.* **1936,** 1166. Chromatographic determn in cereal grains: G. H. Ware *et al., J. Assoc. Off. Anal. Chem.* **69,** 697 (1986).

Stout prisms from acetone, dec 196°. $[\alpha]_D^{20}$ +416° (c = 0.45 in chloroform). pK 7.3. Freely sol in pyridine, moderately in chloroform; slightly sol in alcohol, acetone. Nearly insol in water.

Hydrobromide monohydrate, $C_{19}H_{24}BrN_3O_2.H_2O$, needles from aq acetone + ether, dec 190°.

Hydrochloride monohydrate, $C_{19}H_{24}ClN_3O_2.H_2O$, needles, dec 175-180°.

Mononitrate, $C_{19}H_{24}N_4O_5$, stout prisms from aq MeOH + ether, dec 235°. $[\alpha]_D^{20}$ +282° (c = 1).

Perchlorate, $C_{19}H_{24}ClN_3O_6$, needles, dec 225° (brown at 210°).

3694. Ergonovine. *[8β(S)]-9,10-Didehydro-N-(2-hydroxy-1-methylethyl)-6-methylergoline-8-carboxamide; N-[α-(hydroxymethyl)ethyl]-D-lysergamide;* D-lysergic acid L-2-propanolamide; ergometrine; Ergobasine; Ergotocine; Ergostetrine; Ergotrate; Ergoklinine; Syntometrine. $C_{19}H_{23}N_3O_2$; mol wt 325.41. C 70.13%, H 7.12%, N 12.91%, O 9.83%. From some ergots: A. Stoll *et al.,* U.S. pat. **2,809,920** (1957 to Saul & Co.). Prepn from D-lysergic acid and L(+)-2-amino-1-propanol: A. Stoll, A. Hofmann, *Helv. Chim. Acta* **26,** 944 (1943); Pioch, U.S. pat. **2,736,728** (1956 to Lilly); Patelli, Bernardi, U.S. pat. **3,141,887** (1964 to Farmitalia). Structure: A. Stoll *et al., Helv. Chim. Acta* **34,** 1544 (1951). Total synthesis: E. C. Kornfeld *et al., J. Am. Chem. Soc.* **78,** 3087 (1956). Metabolism: Slaytor, Wright, *J. Med. Pharm. Chem.* **5,** 483 (1962). Comprehensive description of the maleate: V. D. Reif in *Analytical Profiles of Drug Substances* **vol. 11,** K. Florey, Ed. (Academic Press, New York, 1982) pp 273-312. Toxicity data: R. P. Beliles, *Toxicol. Appl. Pharmacol.* **23,** 537 (1972). Clinical use in diagnosis of angina: L. A. DiCarlo, Jr. *et al., Am. J. Cardiol.* **54,** 744 (1984); R. Nordlander, R. Orinius, *Acta Med. Scand.* **221,** 47 (1987).

Tetrahedra from ethyl acetate, fine needles from benzene. Tends to form solvated crystals, mp 162°. $[\alpha]_D^{20}$ +90° (in water). pK 6.8. Freely sol in lower alcohols, ethyl acetate, acetone; more sol in water than the other principal alkaloids of ergot; slightly sol in chloroform.

Hydrochloride, $C_{19}H_{23}N_3O_2.HCl$, needles from ethyl alcohol, dec 246°. $[\alpha]_D^{25}$ +63° (c = 0.9). More sol in water than the hydrobromide.

Maleate, $C_{19}H_{23}N_3O_2.C_4H_4O_4$, *Cornocentin, Ergotrate Maleate, Ermetrine.* Crystals, dec 167°. $[\alpha]_D^{25}$ +48° to +57°. One gram dissolves in 36 ml water, 120 ml alcohol. Nearly insol in ether and chloroform. LD_{50} i.v. in mice: 8.26 mg/kg (Beliles).

Tartrate hydrate, $(C_{19}H_{23}N_3O_2)_2.C_4H_6O_6.H_2O$, *Basergin, Neofemergen.* Crystals, slightly sol in water.

Hydracrylate, $C_{41}H_{52}N_6O_7$, *Ergotrate-H.*

THERAP CAT: Oxytocic.

THERAP CAT (VET): Oxytocic.

3695. Ergosine. *12′-Hydroxy-2′-methyl-5′α-(2-methylpropyl)ergotaman-3′,6′,18-trione.* $C_{30}H_{37}N_5O_5$; mol wt 547.66. C 65.80%, H 6.81%, N 12.79%, O 14.61%. Isoln of ergosine and ergosinine from ergot: Smith, Timmis, *J. Chem. Soc.* **1937,** 396. Structure: Stoll *et al., Helv. Chim. Acta* **34,** 1544 (1951). Stereochemistry: Stoll *et al., ibid.* **37,** 2039 (1954). Total synthesis of ergosine and ergosinine: Stadler *et al., ibid.* **47,** 1911 (1964).

Prisms from ethyl acetate, dec 228°. $[\alpha]_D^{20}$ − 161° (chloroform). Sol in chloroform; fairly sol in methanol, acetone; sparingly sol in ethyl acetate, benzene.

8α-Isomer, $C_{30}H_{37}N_5O_5$, *ergosinine, ergoclavinine.* Prisms from 90% alcohol, aq acetone, benzene or ethyl acetate, dec 228°. Also reported as colorless needles, mp 190-191° (dec), Stadler *et al., loc. cit.* $[\alpha]_D^{20}$ +420° (chloroform); $[\alpha]_D^{20}$ +380° (acetone). Very readily sol in chloroform; readily sol in acetone; less sol in ethyl acetate; sparingly sol in benzene; very sparingly sol in methyl alcohol; almost insol in water.

Note: Ergoclavine is an equimolar mixture of ergosine and ergosinine.

3696. Ergostane. *5α-Ergostane.* $C_{28}H_{50}$; mol wt 386.71. C 86.97%, H 13.03%. Prepn from allocholanic acid: Fernholz, *Ber.* **69,** 1792 (1938).

Scales, plates from ether + methanol, mp 85°. $[\alpha]_D^{20}$ +17° (c = 2 in chloroform).

3697. Ergostanol. *(3β,5α)-Ergostan-3-ol.* $C_{28}H_{50}O$; mol wt 402.70. C 83.51%, H 12.51%, O 3.97%. Prepn by hydrogenation of ergosta-14,22-dien-7-one: Chen, *Ber.* **70,** 1432 (1937).

Crystals, mp 144-145°. $[\alpha]_D^{20}$ +15.9° (c = 1.8 in chloroform). Is precipitated by digitonin.

Acetate, $C_{30}H_{52}O_2$, mp 145°. $[\alpha]_D^{20}$ +6.0° (c = 1.8 in chloroform).

Benzoate, $C_{35}H_{54}O_2$, mp 163-165°.

3698. α-Ergostenol. *(3β,5α)-Ergost-8(14)-en-3-ol.* $C_{28}H_{48}O$; mol wt 400.69. C 83.93%, H 12.07%, O 3.99%. Prepd by hydrogenation of ergosterol acetate in acetic acid soln: Hart *et al., J. Am. Chem. Soc.* **52,** 2016 (1930); from 22-dihydroergosterol: Windaus, Langer, *Ann.* **508,** 105 (1934).

Plates from methanol, mp 130-131°. $[\alpha]_D^{16}$ +11° (c = 0.9 in methanol). $[\alpha]_{546}$ −10.2° (c = 1 in methanol). Precipitated by digitonin.

Acetate, $C_{30}H_{50}O_2$, mp 110°.

Benzoate, $C_{35}H_{52}O_2$, mp 118-120°.

Methyl ether, $C_{29}H_{50}O$, mp 56°. $[\alpha]_D^{20}$ +6.3° (in chloroform).

3699. β-Ergostenol. *(3β,5α)-Ergost-14-en-3-ol.* $C_{28}H_{48}$-O; mol wt 400.69. C 83.93%, H 12.07%, O 3.99%. From α-ergostenol by treatment with HCl-gas in chloroform: Reindel *et al.*, *Ann.* **452**, 34 (1927).

Plates, tablets from alcohol, mp 141°. $[\alpha]_D^{20}$ +21.2° (c = 0.9 in chloroform).

Acetate, $C_{30}H_{50}O_2$, mp 114°. $[\alpha]_D^{20}$ +13° (c = 1.2 in chloroform).

Benzoate, $C_{35}H_{52}O_2$, mp 158-160°. $[\alpha]_D^{20}$ +18.3°, $[\alpha]_{546}^{20}$ +22.5° (c = 1 in ethyl acetate).

3700. γ-Ergostenol. *(3β,5α)-Ergost-7-en-3-ol.* $C_{28}H_{48}$-O; mol wt 400.69. C 83.93%, H 12.07%, O 3.99%. From 22,23-dihydroergosterol: Laucht, *Z. Physiol. Chem.* **237**, 236 (1935); from ergosterol: Wieland, Benend, *Ann.* **554**, 1 (1943); Bladon *et al.*, *J. Chem. Soc.* **1951**, 2402.

Crystals from isopropanol, mp 146°. Seems optically inactive.

Acetate, $C_{30}H_{50}O_2$, mp 157°. $[\alpha]_D^{14}$ −5.3° (c = 0.7 in chloroform).

Benzoate, $C_{35}H_{52}O_2$, mp 179°.

3701. Ergosterol. *(3β,22E)-Ergosta-5,7,22-trien-3-ol;* ergosta-5:6,7:8,22:23-trien-3-ol; ergosterin. $C_{28}H_{44}O$; mol wt 396.66. C 84.79%, H 11.18%, O 4.03%. Most important of the provitamins D. Usually obtained from yeast which synthesizes it from simple sugars such as glucose. Damp yeast yields about 2.5% ergosterol, the variety of the yeast being very important. Isoln procedure: Green *et al.*, U.S. pat. **3,006,932** (1961 to Vitamins Ltd.). When irradiated with uv light, ergosterol develops powerful vitamin D_2, *q.v.*,

activity. Askew *et al.*, in England and Windaus and collaborators in Germany isolated the antirachitic vitamin D_2. The main irradiation products of ergosterol are lumisterol→tachysterol→vitamin D_2. Structure: Chuang, *Ann.* **500**, 270 (1933). Oxidation products: Fuerst, *Arch. Pharm.* **300**, 144 (1967).

Small hydrated plates from alcohol, in hydrated needles from ether. The best crystallized form contains 1½ mol H_2O, mp 168°: Bills, Honeywell, *J. Biol. Chem.* **80**, 15 (1928). Complete removal of water is almost impossible and results in an amorphous mass, melting range 166-183°. $bp_{0.01}$ 250°. $[\alpha]_D^{20}$ −135° (c = 1.2 in $CHCl_3$ calcd as anhydr). $[\alpha]_{546}^{20}$ −171° ($CHCl_3$). uv max (ethanol): 262, 271, 282, 293 nm, Hogness *et al.*, *ibid.* **120**, 239 (1937). Practically insol in water. One gram dissolves in 660 ml alcohol, in 45 ml boiling alcohol, in 70 ml ether, in 39 ml boiling ether, in 31 ml chloroform. Precipitated by digitonin. Affected by light and air, turns yellow. Oxygen forms peroxides and hydrogen may form polyhydro compds.

Acetate, $C_{30}H_{46}O_2$, mp 179°, clear at 181°.

Benzoate, $C_{35}H_{48}O_2$, plates or needles, mp 169-171°. $[\alpha]_D^{23}$ −71° (c = 1.1 in $CHCl_3$), $[\alpha]_{546}$ −88° ($CHCl_3$).

22,23-Dihydro analog, $C_{28}H_{46}O$, *22,23-dihydroergosterol.* Prepn: A. Windhaus, R. Langer, *Ann.* **508**, 105 (1934); D. H. R. Barton *et al.*, *J. Chem. Soc., Perkin Trans. I* **1976**, 821; D. J. Curry *et al.*, *ibid.* **1977**, 822. uv irradiation gives vitamin D_4, *q.v.* Solvated needles from ethyl acetate + methanol, mp 152-153° (dried). $[\alpha]_D^{19}$ −109° ($CHCl_3$) (Windhaus); also reported as needles from chloroform-methanol, mp 128-130°. $[\alpha]_D^{19}$ −121° (c = 0.1). uv max: 262, 272, 282, 294 nm (ε 8000, 11200, 11800, 6800) (Barton).

22,23-Dihydro analog acetate, $C_{30}H_{48}O_2$, mp 157-158°. $[\alpha]_D^{17}$ −75° (c = 2.1 in $CHCl_3$).

THERAP CAT: Vitamin (antirachitic).

3702. Ergot. Secale cornutum; spurred rye. Dried sclerotia of the fungus *Claviceps purpurea* (Fries) Tul., *Hypocreaceae*, parasitic on rye plants. *Habit.* Europe; cultivated in Spain, Germany, and France. Four main classes of ergot alkaloids can be distinguished: clavine alkaloids, lysergic acids, lysergic acid amides and ergot peptide alkaloids. There are ten ergot peptide alkaloids which are ergotamine, ergosine, ergocristine, ergocryptine, ergocornine, ergotaminine, ergosinine, ergocristinine, ergocryptinine, and ergocorninine, the last five alkaloids being isomers of the first five. These allkaloids are typified by a structure consisting of lysergic acid, dimethylpyruvic acid, proline, and phenylalanine joined in amide linkages. In 1943 A. Stoll and A. Hofmann, *Helv. Chim. Acta* **26**, 1570 (1943) made it clear that the ergotoxine reported by G. Barger and F. H. Carr, *J. Chem. Soc.* **91**, 377 (1907) and by F. Kraft, *Arch. Pharm.* **244**, 336 (1906) was but a mixture of ergocristine, ergocryptine, and ergocornine, and that the ergotinine, first reported by C. Tanret, *Compt. Rend.* **81**, 891 (1875), was also a mixture consisting of ergocristinine, ergocryptinine, and ergocorninine. Other constituents of ergot are ergonovine, ergometrinine, ergoclavine, elymoclavine, trimethylamine, putrescine, cadaverine, agmatine, histamine, tyramine, histidine, tyrosine, valine, leucine, betaine, choline, acetylcholine, ergothioneine, 15-30% fatty oil, ergosterol, mannitol, lactic acid, and succinic acid. Production of ergocryptinine and elymoclavine by cultures of fungi parasitic on *Elymus mollis* Trin: Abe *et al.*, U.S. pat. **2,835,675**. Production of ergot alkaloids by saprophytic cultures: Adams, U.S. pat. **3,117,917** (1964 to Miles Labs.). Ergot alkaloid fermentation: W. J. Kelleher, *Adv. Appl. Microbiol.* vol. **11**, 211-244 (1969). Biosynthesis of ergot alkaloids: H. G. Floss, *Tetra-*

Consult the Name Index before using this section.

hedron **32**, 873 (1976). *Reviews:* Gröger in *Microbial Toxins,* **vol. VIII**, S. Kadis *et al.*, Eds. (Academic Press, New York, 1972) pp 321-373; Stadler, Stütz, "The Ergot Alkaloids" in *The Alkaloids* **vol. 15**, R. H. F. Manske, Ed. (Academic Press, New York, 1975) pp 1-40. Books: F. J. Bove, *The Story of Ergot* (Karger, Basel, 1970) 297 pp; *Handb. Exp. Pharmacol.* **49** entitled "Ergot Alkaloids and Related Compounds", B. Berde, H. O. Schild, Eds. (Springer-Verlag, New York, 1978) 1003 pp.

Caution: Ergot Poisoning: Acute: vomiting, diarrhea, thirst, tachycardia, confusion, coma. *Chronic:* CNS symptoms, G.I. disturbances, peripheral circulatory changes, gangrene, *cf. Clinical Toxicology of Commercial Products,* R. E. Gosselin *et al.*, Eds. (Williams & Wilkins, Baltimore, 4th ed., 1976) Section II, p 146.

THERAP CAT: Vasoconstrictor (specific in migraine).

THERAP CAT (VET): Has been used as an oxytocic.

3703. Ergotamine. *12'-Hydroxy-2'-methyl-5'α-(phenylmethyl)ergotaman-3',6',18-trione.* $C_{33}H_{35}N_5O_5$; mol wt 581.67. C 68.14%, H 6.06%, N 12.04%, O 13.75%. Vasoconstrictor found in ergot of Central Europe. Extraction procedure: Stoll, *Helv. Chim. Acta* **28**, 1283 (1945). Structure: Stoll *et al., ibid.* **34**, 1544 (1951). Stereochemistry: Hofmann *et al., ibid.* **46**, 2306 (1963). Total synthesis: Hofmann *et al., Experientia* **17**, 206 (1961). Comprehensive description of the tartrate: B. Kreilgard in *Analytical Profiles of Drug Substances* **vol. 6**, K. Florey, Ed. (Academic Press, New York, 1977) pp 113-159.

Elongated prisms from benzene. Becomes solvent-free only after prolonged heating in a high vacuum. Very hygroscopic. Darkens and dec on exposure to air, heat and light. Dec 212-214°. $[\alpha]_D^{20}$ −160° (chloroform). Sol in about 70 parts methanol, 150 parts acetone, 300 parts alcohol; freely sol in chloroform, pyridine, glacial acetic acid; moderately sol in ethyl acetate; slightly in benzene. Almost insol in water, petr ether.

Hydrochloride, $C_{33}H_{35}N_5O_5$·HCl, rectangular plates from 90% alc, mp 212° (dec). Sol in water-alcohol mixtures; sparingly in water or alcohol.

Succinate, $(C_{33}H_{35}N_5O_5)_2$·$C_4H_6O_4$, *Ergoton-A.*

Tartrate, $(C_{33}H_{35}N_5O_5)_2$·$C_4H_6O_6$, *Ergate, Ergomar, Ergostat, Ergotartrat, Exmigra, Femergin, Gynergen, Lingraine, Lingran.* Solvated crystals, e.g. the dimethanolate, heavy rhombic plates from methanol, mp 203° (dec). $[\alpha]_D^{25}$ −125° to −155° (c = 0.4 in chloroform). One gram dissolves in 500 ml water or 500 ml alc. *Protect from light and heat.* LD$_{50}$ in mice, rats: 62, 80 mg/kg i.v., *RTECS* **Vol. I**, R. J. Lewis, R. L. Tatken, Eds. (1979) p 618.

THERAP CAT: Antimigraine.

THERAP CAT (VET): Tartrate has been used as an oxytocic.

3704. Ergotaminine. *12'-Hydroxy-2'-methyl-5'α-(phenylmethyl)-8α-ergotaman-3',6',18-trione.* $C_{33}H_{35}N_5O_5$; mol wt 581.67. C 68.14%, H 6.06%, N 12.04%, O 13.75%. An alkaloid isomeric with ergotamine, *q.v.* Isoln from ergot: Stoll, *Helv. Chim. Acta* **28**, 1283 (1945); Stoll *et al.,* U.S. pat. **2,809,920** (1957 to Saul & Co.); Tabor, Vining, *Can. J. Microbiol.* **3**, 55 (1957).

Crystallizes solvent-free, unlike ergotamine which tends to retain the solvent of crystn. Thin rhombic plates from methanol, dec 241-243°. $[\alpha]_D^{20}$ +369° (c = 0.5 in chloroform). Much less sol than ergotamine; sol in about 1000 parts boiling alcohol, 1500 parts boiling methanol; fairly sol

in chloroform, pyridine, glacial acetic acid. Does not seem to form salts.

3705. Ergothioneine. *(S)-α-Carboxy-2,3-dihydro-N,N,-N-trimethyl-2-thioxo-1H-imidazole-4-ethanaminium inner salt; [1-carboxy-2-[2-mercaptoimidazol-4-yl]ethyl]trimethyl-ammonium hydroxide inner salt;* L(+)-*ergothioneine;* thioneine; thiolhistidine-betaine; thiasine; sympectothion. $C_9H_{15}N_3O_2S$; mol wt 229.30. C 47.14%, H 6.59%, N 18.33%, O 13.95%, S 13.98%. First discovered in the sclerotia of the ergot fungus, *Claviceps purpurea*, by Tanret, *J. Pharm. Chim.* **30**, 145 (1909). Has since been found to occur in blood, semen and various mammalian tissues, principally liver and kidneys. Biosynthesis from histidine by *Claviceps purpurea* cultures: Heath, Wildy, *Nature* **179**, 196 (1957); also produced by *Neurospora crassa:* Melville *et al., Fed. Proc.* **15**, 314 (1956). Chemical synthesis: Heath *et al., J. Chem. Soc.* **1951**, 2215. Occurrence in the *Linulus polyphemus* L. (king crab): Ackermann, List, *Z. Physiol. Chem.* **313**, 30 (1958). *Review:* Melville, *Vitamins & Hormones* **17**, 155 (1959).

Dihydrate, needles or leaflets from dil ethanol, dec 256-257°. $[\alpha]_D^{20}$ +116.5°; $[\alpha]_D^{27}$ +115° (H$_2$O). uv max (water): 258 nm (ϵ 16000). One gram dissolves in 5 ml at 25°, much more sol in hot water. Slightly sol in hot methanol, hot ethanol, acetone. Practically insol in ether, chloroform, benzene.

3706. Ergotinine. A 1:1:1 mixture of ergocornine, ergocristinine, and ergocryptinine. Isomeric with the ergotoxine mixture, *q.v.* First reported by Tanret, *Compt. Rend.* **81**, 891 (1875). Resolution into three alkaloids: Stoll, Hofmann, *Helv. Chim. Acta* **26**, 1570 (1943).

Crystallizes solvent-free, unlike ergotoxine which tends to retain the solvent of crystn. Long prisms from acetone. mp 229° with decompn. $[\alpha]_D^{20}$ +365° (c = 0.35 in chloroform). Very sol in chloroform. Sol in 25 parts acetone, 420 parts alcohol, 1000 parts abs ether.

3707. Ergotoxine. Ecboline. A mixture of ergocornine, ergocristine, and ergocryptine, *q.q.v.*, with oxytocic activity. *See also* the isomeric ergotinine mixture. Prepn: Barger, Carr, *J. Chem. Soc.* **91**, 337 (1907); Smith, Timmis, *ibid.* **1930**, 1390; Kofler, *Arch. Pharm.* **275**, 455 (1937); Stoll, Hofmann, *Helv. Chim. Acta* **26**, 1570 (1943); Pitra, Sapara, *Cesk. Farm.* **5**, 585 (1956), *C.A.* **51**, 8369f (1957). Toxicity data: H. Kreitmair, *Arch. Exp. Pathol. Pharmakol.* **176**, 171 (1934).

Orthorhombic crystals with 2C$_6$H$_6$ from benzene. The solvent of crystn is given up only after long drying and heating in high vacuum. mp ~190° (dec) (solvent-free base). $[\alpha]_D^{20}$ −197° (chloroform). Very sol in ethyl and methyl alcohol, acetone, chloroform, ethyl acetate; slightly sol in ether. Almost insol in water, petr ether. LD in mice (mg/kg): 0.107 s.c.; 0.032 i.v. (Kreitmair).

Ergotoxine phosphate, clusters of needles, dec 187°. Sol in 320 parts water, 15 parts boiling alcohol.

Ergotoxine ethanesufonate, acicular crystals, dec 209°. Sol in methanol; slightly in alcohol. Almost insol in water. Solid and solns are sensitive to light and air.

3708. Erigeron. *Fleabane; horseweed.* Leaves and tops of *Conyza canadensis* (L.) Cron. *(Erigeron canadensis L.),* Compositae. *Habit.* Northern and central U.S. *Constit.* Volatile oil, tannin, gallic acid.

3709. Eriochrome® Black T. *3-Hydroxy-4-[(1-hydroxy-2-naphthalenyl)azo]-7-nitro-1-naphthalenesulfonic acid monosodium salt; C.I. Mordant Black 11;* C.I. 14645. $C_{20}H_{12}N_3NaO_7S$; mol wt 461.39. C 52.06%, H 2.62%, N 9.11%, Na 4.98%, O 24.27%, S 6.95%. Prepn: Hagenbach, U.S. pat. **790,363** (1904 to Geigy); *Colour Index* **vol. 4** (3rd ed., 1971) p 4067.

Brownish-black powder with a faint metalic sheen. Sol in hot water giving a reddish-brown soln when cold. Violet-brown precipitate with excess HCl. Deep blue, then red in aq soln of NaOH. Sol in concd H_2SO_4 giving a blackish-blue soln which yields a brown precipitate on dilution.

USE: To dye wool from an acid bath reddish-black, which can be converted to blue-black by afterchroming. As indicator in the determination of the total calcium and magnesium content of water.

3710. Eriodictyol. *2-(3,4-Dihydroxyphenyl)-2,3-dihydro-5,7-dihydroxy-4H-1-benzopyran-4-one; 3',4',5,7-tetrahydroxyflavanone.* $C_{15}H_{12}O_6$; mol wt 288.26. C 62.50%, H 4.20%, O 33.30%. Isoln from *Eriodictyon californicum* (H. & A.) Greene, *Hydrophyllaceae:* Geissman, *J. Am. Chem. Soc.* **62**, 3258 (1940); from lemon: Mager, *Z. Physiol. Chem.* **274**, 109 (1942); Horowitz, *J. Am. Chem. Soc.* **79**, 6561 (1957); U.S. pat. **2,857,318** (1958 to U.S. Dept. Agr.). Structure and synthesis: Reichel *et al., Ann.* **550**, 146 (1942); Zemplén *et al., Ber.* **76B**, 1112 (1943); Pew, *J. Org. Chem.* **27**, 2935 (1962). Synthesis: G. Wurm, U. Geres, *Arch. Pharm.* **315**, 183 (1982). *See also* Bioflavanoids.

Needles with 1½ H_2O from dil alc, dec 257° (rapid heating). After drying in vacuo at 100° for 6 hours, dec 267°. uv max (alc): 290, 326 nm (log ε 2.54, 2.16). Sparingly sol in boiling water, hot alcohol, ether, glacial acetic acid; sol in dil alkalies.

7-L-Rhamnoside, $C_{21}H_{22}O_{10}$, *eriodictin.* Isoln from citrin: Bruckner, Szent-Györgyi, *Nature* **138**, 1057 (1936). Structure: Mager, *loc. cit.* Crystals from ethyl acetate, dec 184-186°. $[\alpha]_D^{20}$ −51.5° (in pyridine). Sol in water, acetone, alcohol; practically insol in ether.

3711. Eriodictyon. Yerba santa; consumptive's weed; bear's weed; mountain balm; gum plant. Dried leaves of *Eriodictyon californicum* (H. & A.) Greene, *Hydrophyllaceae.* Habit. U.S. (California). *Constit.* Volatile oil, eriodictyol, homoeriodictyol, chrysoeriodictyol, xanthoeriodictyol, eriodonol, eriodictyonic acid, ericolin, resin.

USE: Pharmaceutic aid (flavor).

3712. Eritadenine. *[R-(R*,R*)]-6-Amino-α,β-dihydroxy-9H-purine-9-butanoic acid; 2(R),3(R)-dihydroxy-4-(9-adenyl)butyric acid* (D-*erythro*-form); lentinacin; lentysine. $C_9H_{11}N_5O_4$; mol wt 253.22. C 42.69%, H 4.38%, N 27.66%, O 25.27%. Isoln from the mushroom *Lentinus edodes* Sing. (Shiitake) and synthesis: Chibata *et al., Experientia* **25**, 1237 (1969). Syntheses: Kamiya *et al., Tetrahedron Letters* **1969**, 4729; *eidem., Chem. Ind.* **1970**, 652; Okumura *et al., Chem. Commun.* **1970**, 1045; Kawazu *et al., ibid.* **1970**, 1047. Synthesis of D-*threo*- and L-*threo*-forms: Hashimoto *et al., Tetrahedron Letters* **1970**, 1359.

Needles from hot water, mp 279° (dec), $[\alpha]_D$ +50° (0.1N NaOH) and +16° (N HCl). uv max (water): 261 nm (ε 14300).

Sodium salt, $C_9H_{10}N_5NaO_4$, leaflets from 70% ethanol, mp 266-268° (dec), $[\alpha]_D^{20}$ +45.5° (H_2O). IR, NMR: Chibata, *loc. cit.*

THERAP CAT: Anticholesteremic.

3713. Erucic Acid. *(Z)-13-Docosenoic acid; Δ^{13}-cis-docosenoic acid.* $C_{22}H_{42}O_2$; mol wt 338.57. C 78.05%, H 12.50%, O 9.45%. A monoethenoid acid found in the seed fats of *Cruciferae* and *Tropaeolaceae.* It constitutes 40 to 50% of the total fatty acids of rapeseed, mustard and wallflower seed, and it represents up to 80% of fatty acids of nasturtium seeds, *cf.* K. S. Markley, *Fatty Acids* Part I (Interscience, New York, 2nd ed., 1960) p 138-139. Prepn of a crude product by alkaline hydrolysis of rapeseed oil: Noller, Talbot, *Org. Syn. coll. vol. II*, 258 (1943). A purer product is obtained by the method of Dorée and Pepper, *J. Chem. Soc.* **1942**, 477, which involves a fractional precipitation and crystallization. Prepn of a pure product by acid soap crystallization: Chobanov *et al., Chem. & Ind. (London)* **1965**, 606. Synthesis: Bowman, *J. Chem. Soc.* **1950**, 177, 325; Bounds *et al., ibid.* **1953**, 2393. Treatment with nitric acid yields the *trans* isomer, brassidic acid, *q.v.*: Dorée, Pepper, *loc. cit.* Brief review: E. Lower, *Manuf. Chem.* **56**(6), 61-63 (1985).

Needles from alcohol, mp 33.8°. Iodine value 74.98; neutralization value 165.72. d_4^{55} 0.860. bp_{760} 381.5° (decompn); bp_{400} 358.8°; bp_{100} 314.4°; bp_{60} 300.2°; bp_{20} 270.6°; bp_5 239.7°; $bp_{1.0}$ 206.7°. n_D^{45} 1.4534; n_D^{65} 1.44794. Insol in water. About 175 g dissolve in 100 ml ethanol and about 160 g dissolve in 100 ml methanol. Very sol in ether.

3714. Erythritol. *(R*,S*)-1,2,3,4-Butanetetrol;* meso-erythritol; tetrahydroxybutane; erythrol; erythrite; anti-erythrite; erythroglucin; phycite. $C_4H_{10}O_4$; mol wt 122.12. C 39.34%, H 8.25%, O 52.41%. Isoln from algae, lichens, grasses: Bamberger, Landsiedl, *Monatsh.* **21**, 571 (1900); Hesse, *J. Prakt. Chem.* **92**, 425 (1915); Hofmann, *Ber.* **7**, 508 (1874). Prepn by *Aspergillus niger*: Yuill, *Nature* **162**, 652 (1948); by *Penicillium herquei*: Galarraga *et al., Biochem. J.* **61**, 456 (1955); from 2-butene-1,4-diol: Reppe, Schnabel, Ger. pat. **734,025** (1943 to I. G. Farbenind.); from periodate-oxidized starch: Jeanes, Hudson, *J. Org. Chem.* **20**, 1565 (1955). Structure: Shimada, *Acta Cryst.* **11**, 748 (1958).

Tetragonal prisms, mp 121.5°. About twice as sweet as sucrose. bp 329-331°. Very sol in water (satd soln contains about 61% w/w); sol in pyridine (satd soln contains 2.5% w/w); slightly sol in alcohol. Practically insol in ether. pKa (18°): 13.903. LD i.v. in dogs: 5.0 g/kg.

3715. Erythritol Anhydride. *(R*,S*)-2,2'-Bioxirane;* meso-1,2:3,4-diepoxybutane; butadiene diepoxide; butadiene dioxide; butanedione; Bioxiran. $C_4H_6O_2$; mol wt 86.09. C 55.81%, H 7.02%, O 37.17%. Prepn from erythrityl chlorohydrin: Przybytek, *Ber.* **17**, 1091 (1884); by refluxing 1,4-dichloro-2,3-butanediol in tetrahydrofuran with NaOH: Reppe, *Ann.* **596**, 141 (1955); *meso*-form from 1,4-dihydroxy-2-butene or from 3,4-epoxy-1-butene and *dl*-form from 1,4-dibromo-2-butene: Beech, *J. Chem. Soc.* **1951**, 2483. Use to prevent microbial spoilage: H. D. Michener, J. C. Lewis, U.S. pat. **2,934,439** (1960 to U.S. Secy of Agr.). Toxicity data: H. F. Smyth *et al., Arch. Ind. Hyg. Occup. Med.* **10**, 61 (1954).

Liquid. d_4^{18} 1.113. bp 138°; bp_{23} 50-51°. *meso*-form, bp_{761} 140-142°. mp −19°. Compd with piperidine, mp 106° (Beech). Misc with water which hydrolyzes it to erythritol. LD_{50} orally in rats: 0.078 g/kg (Smyth).

Note: This substance may reasonably be anticipated to be a carcinogen: *Seventh Annual Report on Carcinogens* (PB95-109781, 1994) p 166.

USE: Curing polymers; crosslinking textile fibers.

3716. Erythrityl Tetranitrate. (*R**,*S**)-*1,2,3,4-Butanetetroltetranitrate;* erythritol tetranitrate; erythrol tetranitrate; tetranitrol; tetranitrin; nitrocrythrite; Cardilate; Cardiloid. $C_4H_6N_4O_{12}$; mol wt 302.11. C 15.90%, H 2.00%, N 18.55%, O 63.55%. Made by nitration of erythritol.

Leaflets from alcohol, mp 61°. Sol in alcohol, ether, glycerol. Insol in water. Reduces Fehling's soln. Explodes on percussion. Its use in pharmaceutical preparations is only in admixture with carbohydrate substances such as lactose. Sold in tablet form only. Such tablets are nonexplosive.

THERAP CAT: Vasodilator (coronary).

THERAP CAT (VET): Vasodilator.

3717. Erythrocentaurin. *3,4-Dihydro-1-oxo-1H-2-benzopyran-5-carboxaldehyde; 5-formyl-3,4-dihydroisocoumarin;* 5-formyl-3,4-dihydro-1*H*-2-benzopyran-1-one. $C_{10}H_8O_3$; mol wt 176.17. C 68.18%, H 4.58%, O 27.25%. From *Centaurium umbellatum* Gilib., (*Erythraea centaurium* Pers.), Gentianaceae or *Swertia japonica* (Maxim.) Makino, Gentianaceae. By hydrolysis of swertiamarin and erytaurin with emulsin. Isoln: Kariyone, Matsushima, *J. Pharm. Soc. Japan* **47**, 25 (1927). Structure: Kubota, Tomita, *Chem. & Ind. (London)* **1958**, 230; Kubota *et al., Tetrahedron Letters* **1961**, 223. Synthesis: Wenkert *et al., J. Org. Chem.* **29**, 2534 (1964).

Long needles, mp 140-141°. Turns red on exposure to sunlight. uv max: 223, 290 nm (log ϵ 4.30, 3.13).

THERAP CAT: Bitter tonic.

3718. α-Erythroidine. (*3β,12β*)-*1,2,6,7-Tetradehydro-12,17-dihydro-3-methoxy-16(15H)-oxaerythrinan-15-one.* $C_{16}H_{19}NO_3$; mol wt 273.33. C 70.31%, H 7.01%, N 5.12%, O 17.56%. Isoln from *Erythrina* spp, Leguminosae: Folkers, Major, U.S. pat. **2,373,952** (1945 to Merck & Co.); Boekelheide, Grundon, *J. Am. Chem. Soc.* **75**, 2563 (1953). Structure: Godfrey *et al., ibid.* **77**, 3342 (1955). Absolute configuration: Hill, Shearer, *J. Org. Chem.* **27**, 921 (1962); Wenzinger, Boekelheide, *Proc. Chem. Soc.* **1963**, 53. Conversion of α- to β-erythroidine: Boekelheide, Morrison, *J. Am. Chem. Soc.* **80**, 3905 (1958). Biosynthesis studies: Leete, Ahmad, *ibid.* **88**, 4722 (1966).

Needles from pentane, mp 58-60°. $[\alpha]_D^{27}$ +136° (c = 0.5 in water). Unstable on exposure to air.

Hydrochloride, $C_{16}H_{20}ClNO_3$, prisms from ethanol, dec 226-228°. $[\alpha]_D^{32}$ +118° (c = 0.5 in water). uv max (ethanol): 224 nm (log ϵ 4.55).

Methiodide, $C_{17}H_{22}INO_3$, yellow prisms from ethanol, mp 219-220°.

3719. β-Erythroidine. (*3β*)-*1,2,6,7-Tetradehydro-14,17-dihydro-3-methoxy-16(15H)-oxaerythrinan-15-one; 12,13-didehydro-13,14-dihydro-α-erythroidine.* $C_{16}H_{19}NO_3$; mol wt 273.33. C 70.31%, H 7.01%, N 5.12%, O 17.56%. Skeletal muscle relaxant. Isolated from the seeds and other plant parts of *Erythrina* spp., Leguminosae: Folkers, Major, *J. Am. Chem. Soc.* **59**, 1580 (1937); *eidem,* U.S. pats. **2,373,-952; 2,385,266; 2,407,713** (1945 and 1946 to Merck & Co.). Structural studies: Koniuszy, Folkers, *J. Am. Chem. Soc.* **72**, 5519 (1950); Boekelheide *et al., ibid.* **75**, 2550 (1953). Absolute configuration: Wenzinger, Boekelheide, *Proc. Chem. Soc.* **1963**, 53; *eidem,* J. Org. Chem. **29**, 1307 (1964). Toxicity data: F. M. Berger, R. P. Schwartz, *J. Pharmacol. Exp. Ther.* **93**, 362 (1948). *Review:* Boekelheide, *Record Chem. Progr.* **16**, 227-239 (1955).

Crystals from abs ethanol, mp 99.5-100°. $[\alpha]_D^{25}$ +88.8°. Sol in water, benzene, chloroform, methanol, ethanol; moderately sol in diethyl ether. Reacts with NaOH to form sodium erythroidinate. LD_{50} i.p. in mice: 24.0 mg/kg (Berger, Schwartz).

Hydrochloride, $C_{16}H_{19}NO_3 \cdot HCl$, small needles from abs ethanol, mp 232° (dec). $[\alpha]_D^{25}$ +10°. Unstable in air and light. Bitter taste. Sol in water, benzene, chloroform, methanol, ethanol; moderately sol in diethyl ether. *Incompat.* with oxidizing agents, alkaline solns, and akaloidal reagents.

Hydrochloride hemihydrate, $C_{16}H_{19}NO_3 \cdot HCl \cdot \frac{1}{2} H_2O$, small needles from abs ethanol, mp 229.5-230° (dec). $[\alpha]_D$ +95°. Has curare-like action which is antagonized by prostigmine.

Hydrobromide, $C_{16}H_{19}NO_3 \cdot HBr$, crystals from abs ethanol, mp 222.5°. $[\alpha]_D^{25}$ +111.2°.

Hydriodide, $C_{16}H_{19}NO_3 \cdot HI$, crystals from abs ethanol, mp 206°. $[\alpha]_D^{25}$ +108.1°.

Methiodide, $C_{16}H_{19}NO_3 \cdot CH_3I$, white prisms from alcohol, mp 211°.

3720. Erythromycin. Erythromycin A; Abomacetin; Ak-Mycin; Aknin; E-Base; EMU; E-Mycin; Eritrocina; Ery Derm; Erymax; Ery-Tab; Erythromast 36; Erythromid; ERYC; Erycen; Erycin; Erycinum; Ermysin; Ilotycin; Inderm; Retcin; Staticin; Stiemycin; Torlamicina. $C_{37}H_{67}NO_{13}$; mol wt 733.94. C 60.55%, H 9.20%, N 1.91%, O 28.34%. Antibiotic substance produced by a strain of *Streptomyces erythreus* (Waksman) Waksman & Henrici, found in a soil sample from the Philippine Archipelago. Isoln: McGuire *et al., Antibiot. & Chemother.* **2**, 281 (1952); Bunch, McGuire, U.S. pat. **2,653,899** (1953 to Lilly); Clark, Jr., U.S. pat. **2,823,203** (1958 to Abbott). Properties: Flynn *et al., J. Am. Chem. Soc.* **76**, 3121 (1954). Solubility data: Weiss *et al., Antibiot. & Chemother.* **7**, 374 (1957). Structure: Wiley *et al., J. Am. Chem. Soc.* **79**, 6062 (1957). Configuration: Hofheinz, Grisebach, *Ber.* **96**, 2867 (1963); Harris *et al., Tetrahedron Letters* **1965**, 679. There are three erythromycins produced during fermentation, designated A,

B, and C; A is the major and most important component. Erythromycins A and B contain the same sugar moieties, desosamine, q.v., and cladinose (3-O-methylmycarose). They differ in position 12 of the aglycone, erythronolide, A having an hydroxyl substituent. Component C contains desosamine and the same aglycone present in A but differs by the presence of mycarose, q.v., instead of cladinose. Structure of B: P. F. Wiley et al., J. Am. Chem. Soc. 79, 6070 (1957); of C: eidem, ibid. 6074. Synthesis of the aglycone, erythronolide B: E. J. Corey et al., ibid. 100, 4618, 4620 (1978); of erythronolide A: eidem, ibid. 101, 7131 (1979). Asymmetric total synthesis of erythromycin A: R. B. Woodward et al., ibid. 103, 3215 (1981). NMR spectrum of A: D. J. Ager, C. K. Sood, Magn. Reson. Chem. 25, 948 (1987). Biosynthesis: Martin, Goldstein, Progr. Antimicrob. Anticancer Chemother., Proc. 6th Int. Congr. Chemother. II, 1112 (1970); Martin et al., Tetrahedron, 31, 1985 (1975). Cloning and expression of clustered biosynthetic genes: R. Stanzak et al., Biotechnology 4, 229 (1986). Reviews: T. J. Perun in Drug Action and Drug Resistance in Bacteria 1, S. Mitsuhashi, Ed. (University Park Press, Baltimore, 1977) pp 123-152; Oleinick in Antibiotics, vol. 3, J. W. Corcoran, F. E. Hahn, Eds. (Springer-Verlag, New York, 1975) pp 396-419; Infection 10, Suppl. 2, S61-S118 (1982). Comprehensive description: W. L. Koch, Anal. Profiles Drug Subs. 8, 159-177 (1979).

Erythromycin A

Hydrated crystals from water, mp 135-140°, resolidifies with second mp 190-193°. Melting point taken after drying at 56° and 8 mm. $[\alpha]_D^{25}$ −78° (c = 1.99 in ethanol). uv max (pH 6.3): 280 nm (ϵ 50). pKa$_1$ 8.8. Basic reaction. Readily forms salts with acids. Soly in water: ~2 mg/ml. Freely sol in alcohols, acetone, chloroform, acetonitrile, ethyl acetate. Moderately sol in ether, ethylene dichloride, amyl acetate.

Ethylsuccinate, $C_{43}H_{75}NO_{16}$, Anamycin, Arpimycin, Durapaediat, E.E.S., E-Mycin E, Eryliquid, Eryped, Erythro ES, Erythro-Holz, Erythroped, Esinol, Monomycin, Paediathrocin, Pediamycin, Refkas, Sigapedil, Wyamycin E. Prepn: Brit. pat. 830,846; R. K. Clark, U.S. pat. 2,967,129 (1960, 1961 both to Abbott). Hydrated crystals from acetone + water, mp 109-110°. $[\alpha]_D$ −42.5°.
THERAP CAT: Antibacterial.
THERAP CAT (VET): Antibacterial.

3721. Erythromycin Acistrate. Erythromycin 2'-acetate octadecanoate (salt); octadecanoic acid compd with erythromycin 2'-acetate; 2'-acetylerythromycin stearate; Erasis. $C_{57}H_{105}NO_{16}$; mol wt 1060.46. C 64.56%, H 9.98%, N 1.32%, O 24.14%. $C_{39}H_{69}NO_{14}\cdot CH_3(CH_2)_{16}COOH$. Ester prodrug of erythromycin, q.v. Prepn: M. L. A. Marvola et al., Belg. pat. 901,411; E. J. Honkanen et al., U.S. pat. 4,599,326 (1985, 1986 both to Orion-yhtymä Oy). Pharmacology, toxicity and clinical studies: J. Antimicrob. Chemother. 21, Suppl D, 1-122 (1988). For structure see Erythromycin.
THERAP CAT: Antibacterial.

3722. Erythromycin Estolate. Erythromycin 2'-propanoate dodecyl sulfate (salt); erythromycin propionate lauryl sulfate; lauryl sulfate of the propionic acid ester of erythromycin; propionylerythromycin lauryl sulfate; Eritroger; Eromycin; Ilosone; Lauromicina; Neo-Erycinum;

PELS; Roxomicina; Stellamicina; Eriscel; Estomicina; Eupragin; Marcoeritrex; Propiocine Enfant; Togiren. $C_{52}H_{97}NO_{18}S$; mol wt 1056.40. C 59.12%, H 9.25%, N 1.33%, O 27.26%, S 3.04%. $C_{40}H_{71}NO_{14}\cdot CH_3(CH_2)_{11}OSO_3H$. Prepn from propionyl erythromycin: Stephens et al., J. Am. Pharm. Assoc. 48, 620 (1959); Bray, Stephens, U.S. pat. 3,000,874 (1961 to Lilly). Toxicity: E. I. Goldenthal, Toxicol. Appl. Pharmacol. 18, 185 (1971). Comprehensive description: J. M. Mann, Anal. Profiles Drug Subs. 1, 101-117 (1972). For structure see Erythromycin.
Long needles, mp 135-140° (dec). pKa 6.9. Soly in water: 0.024 mg/ml; sol in alcohol, acetone, chloroform. LD$_{50}$ orally in rats: > 5000 mg/kg (Goldenthal).
THERAP CAT: Antibacterial.

3723. Erythromycin Glucoheptonate. D-Glycero-D-gulo-heptonic acid compd with erythromycin (1:1); erythromycin glucoheptonic acid salt; erythromycin gluceptate; Ilotycin Gluceptate. $C_{44}H_{81}NO_{21}$; mol wt 960.12. C 55.04%, H 8.50%, N 1.46%, O 34.99%. $C_{37}H_{67}NO_{13}\cdot C_7H_{14}O_8$. Semi-synthetic macrolide antibiotic. Prepn: Shepler, U.S. pat. 2,852,429 (1958 to Lilly).
Crystals, mp 95-140°. Freely sol in water, alcohol, dioxane, acetone, propylene glycol. Practically insol in ether, carbon tetrachloride, toluene, benzene. 2% aq solns are levorotatory and have a pH of 6.0 to 7.5.
THERAP CAT: Antibacterial.

3724. Erythromycin Lactobionate. 4-O-β-D-Galactopyranosyl-D-gluconic acid compd with erythromycin (1:1); Erythrocin Lactobionate. Semi-synthetic macrolide antibiotic prepd from erythromycin base and lactobiono-δ-lactone in water-acetone: Hoffhine, U.S. pat. 2,761,859 (1956 to Abbott); alternate prepn and antibacterial activity: S. K. Dutta, K. S. Basu, U.S. pat. 4,137,397 (1979 to Jadavpur Univ.). Pharmacokinetics: R. L. Parsons et al., J. Int. Med. Res. 8, suppl. 2, 15 (1980).
White, amorphous powder, mp 145-150°. Soly in water about 200 mg/ml. Freely sol in alcohol; slightly sol in ether. pH of a 2% aq soln 6.0-7.5. A 5% stock soln can be prepd with water or glucose soln but not with NaCl isotonic soln, because NaCl at this concn precipitates erythromycin base. Lower concns of inorganic salts are compatible.
THERAP CAT: Antibacterial.
THERAP CAT (VET): Antibacterial.

3725. Erythromycin Propionate. Erythromycin 2'-propanoate; propionylerythromycin; monopropionylerythromycin; Propiocine. $C_{40}H_{71}NO_{14}$; mol wt 790.00. C 60.82%, H 9.06%, N 1.77%, O 28.35%. Semi-synthetic macrolide antibiotic. Prepn: V. C. Stephens, U.S. pat. 2,993,833 (1961 to Lilly); V. C. Stephens, J. W. Conine, Antibiot. Ann. 1958-59, 346. Pharmacology and toxicology: C. Lee et al., ibid. 354. Clinical studies: R. S. Griffith, ibid. 364; D. M. Perry et al., ibid. 375. Pharmacokinetics in humans: M. Ducci et al., Int. J. Clin. Pharmacol. Ther. Toxicol. 19, 494 (1981).
Monohydrate, crystals from acetone + water, mp 122-126°. $[\alpha]_D^{25}$ −81.6° (acetone). pKa: 6.9. Very slightly sol in water. Readily sol in methanol, ethanol, acetone, ethyl acetate, DMF. LD$_{50}$ in mice, rats (g/kg): 2.87, > 5.0 orally; > 5.0, > 5.0 s.c. (Lee).
THERAP CAT: Antibacterial.

3726. Erythromycin Stearate. Erythromycin octadecanoate (salt); Abboticine; Bristamycin; Dowmycin E; Eratrex; Erypar; Eryprim; Erythro S; Erythrocin; Ethril; Ethryn; Gallimycin; Meberyt; Pantomicina; Pfizer-E; SK-Erythromycin; Wemid. $C_{55}H_{103}NO_{15}$; mol wt 1018.42. C 64.87%, H 10.19%, N 1.38%, O 23.57%. Semi-synthetic macrolide antibiotic. Prepn: O. F. Walasek, U.S. pat. 2,881,163 (1959 to Abbott). Bioavailability: A.-S. Malmborg, J. Antimicrob. Chemother. 5, 591 (1979). Pharmacokinetics and tolerance: D. C. Shanson et al., ibid. 14, 157 (1984). Clinical efficacy in respiratory tract infections: J. P. Butler et al., Chemotherapy (Basel) 25, 367 (1979); in acute otitis media: C. Rosen et al., Acta Otolaryngol. (Stockholm) Suppl. 407, 23 (1984).
Note: The prepn of erythromycin stearate (ester) is described in Booth, Murray, U.S. pat. 2,862,921 (1958 to Upjohn).

Crystals. Slightly bitter taste. Slightly sol in alc, ether, chloroform. Practically insol in water.

THERAP CAT: Antibacterial.

THERAP CAT (VET): Antibacterial.

3727. Erythrophlamine. *[1R-(1α,2α,4aα,4bβ,7E,8β,-8aα,10aβ)]-7-[2-[2-(Dimethylamino)ethoxy]-2-oxoethylidene]-tetradecahydro-2-hydroxy-1,4a,8-trimethyl-9-oxo-1-phenan-threnecarboxylic acid methyl ester.* $C_{25}H_{39}NO_6$; mol wt 449.59. C 66.79%, H 8.74%, N 3.12%, O 21.35%. Extracted from the bark of *Erythrophleum guineense* G. Don., and *E. couminga* Baillon, *Leguminosae*. Isoln: Ruzicka *et al.*, *Experientia* **1**, 160 (1945); Engel, Tondeur, *Helv. Chim. Acta* **32**, 2364 (1949). Structure: Mathieson *et al.*, *Experientia* **16**, 404 (1960); Arya, Engel, *Helv. Chim. Acta* **44**, 1650 (1961).

Crystals from ether + petr ether, mp 149-151°. $[\alpha]_D^{20}$ −62.5° (c = 0.911 in alc). uv max: 222 nm (log ε 4.2).

3728. Erythrophleine. *[1S-(1α,4aα,4bβ,7E,8β,8aα,9α,-10aβ)]-Tetradecahydro-9-hydroxy-1,4a,8-trimethyl-7-[2-[2-(methylamino)ethoxy]-2-oxoethylidene]-1-phenanthrenecarb-oxylic acid methyl ester; norcassamidine.* $C_{24}H_{39}NO_5$; mol wt 421.58. C 68.38%, H 9.32%, N 3.32%, O 18.98%. From the bark of *Erythrophleum guineense* G. Don., *Leguminosae*. Isoln: Harnack, *Arch. Pharm.* **234**, 561 (1896). Structural studies: Blount *et al.*, *J. Chem. Soc.* **1940**, 286. Revised structure and identity with norcassamidine: Friedrich-Fiechtl, Spiteller, *Ber.* **104**, 3538 (1971). Activity: Flacke, *J. Pharmacol. Exp. Ther.* **125**, 49 (1959).

Solid from ethanol-ether, mp 115°. $[\alpha]_D^{20}$ −22.5° (c = 0.65 in ethanol). Sol in water and alcohol.

THERAP CAT: Cardiotonic.

3729. Erythropoietin. Erythropoiesis stimulating factor; hemopoietine; ESF; Ep; Epo. Glycoprotein hormone which stimulates red blood cell formation in higher organisms: P. Carnot, C. Deflandre, *Compt. Rend.* **143**, 384 (1906). Produced primarily in the kidneys and to a lesser extent in the liver. Occurs normally in very low concentrations in plasma. Found in urine and plasma in thousand-fold concns under conditions of anemic or hypoxic stress. Isoln from sheep plasma: W. F. White *et al.*, *Recent Progr. Horm. Res.* **16**, 219 (1960); from human urine: J. Espada, A. Gutnisky, *Biochem. Med.* **3**, 475 (1970). Single chain polypeptide which is heavily glycosylated; Epo contains hexoses, hexosamines, and sialic acid in varying amounts. Two forms of human Epo, termed α and β, differing only in carbohydrate content, have been identified. Purification from human urine: T. Miyake *et al.*, *J. Biol. Chem.* **252**, 5558 (1977). Characterization: F. F. Wang *et al.*, *Endocrinology* **116**, 2286 (1985); M. S. Dordal *et al.*, *ibid.* 2293.

Amino acid sequence: P.-H. Lai *et al.*, *J. Biol. Chem.* **261**, 3116 (1986). Produced in response to hypoxia, Epo stimulates maturation of erythroid progenitor cells in the bone marrow. Biogenesis and control of production: C. Peschle, M. Condorelli, *Science* **190**, 910 (1975); A. J. Erslev *et al.*, *Exp. Hematol.* **8**, Suppl. 8, 1 (1980); J. W. Fisher, *Proc. Soc. Exp. Biol. Med.* **173**, 289 (1983). Production by recombinant DNA technology in *E. coli*: S. Lee-Huang, *Proc. Nat. Acad. Sci. USA* **81**, 2708 (1984); in mammalian cells: K. Jacobs *et al.*, *Nature* **313**, 806 (1985); F.-K. Lin *et al.*, *Proc. Nat. Acad. Sci. USA*, **82**, 7580 (1985). Clinical use of recombinant human Epo (*rHuEpo*) in end-stage renal disease: J. W. Fisher *et al.*, *Am. J. Kidney Dis.* **14**, 163 (1989). Review of chemistry, pharmacology and clinical efficacy of rHuEpo: K. K. Flaharty *et al.*, *Clin. Pharm.* **8**, 769-782 (1989). Brief review of native and recombinant forms: J. L. Spivak, *Nephron* **52**, 289-294 (1989). Symposium on biology and clinical experience: *Am. J. Kidney Dis.* **18**, Suppl. 1, 1-116 (1991).

Epoetin alfa, Epoade, Epogen, Eprex, Erypo, Espo, Procrit. Human Epo produced by recombinant technology. Contains 165 amino acids; mol wt 30,400 daltons.

Epoetin beta, Epogin, Marogen, Recormon. Human Epo produced by recombinant technology. Contains 165 amino acids; mol wt ~30,000 daltons.

THERAP CAT: Hematinic.

3730. Erythropterin. *3-(2-Amino-4,5,6,8-tetrahydro-4,-6-dioxo-7(3H)-pteridinylidene)-2-oxopropanoic acid; 2-amino-3,4,5,6-tetrahydro-4,6-dioxo-7-pteridinepyruvic acid.* $C_9H_7N_5O_5$; mol wt 265.19. C 40.76%, H 2.66%, N 26.41%, O 30.17%. Pigment responsible for the red and orange color spots on the wings of butterflies. Has been detected also in the integument of other insects and in *Mycobacterium lacticola.* Isoln: Schöpf, Becker, *Ann.* **524**, 49 (1936). Structure and synthesis: Tschesche, Ende, *Ber.* **91**, 2074 (1958). Revised structure: Pfleiderer, *ibid.* **95**, 2195 (1962); von Philipsborn *et al.*, *Helv. Chim. Acta* **46**, 2592 (1963). Synthesis: Viscontini, Stierlin, *ibid.* 51.

Monohydrate, red microcrystals from 0.5N HCl. R values: *n*-propanol/4% sodium citrate (1:2) 0.54, *n*-butanol/-ethanol/water (2:2:1) 0.00, 4% ammonium acetate 0.17, 4% disodium phosphate 0.00, 4% sodium bicarbonate 0.27, 4% disodium phosphate satd with *n*-butanol 0.26. Abs max (pH 1.0): 450 nm (log ε 4.02).

3731. D-Erythrose. *(R)-2,3,4-Trihydroxybutanal.* $C_4H_8O_4$; mol wt 120.11. C 40.00%, H 6.71%, O 53.28%. From calcium D-arabonate by oxidation with H_2O_2: Ruff, *Ber.* **32**, 3674 (1899); **33**, 1799 (1900). From D-glucose: Perlin, *Methods in Carbohydrate Chemistry* **1** (Academic Press, New York, 1962) p 64. Synthesis of DL-erythrose: Sonogashira, Nakagawa, *Bull. Chem. Soc. Japan* **45**, 2616 (1972).

Syrup. Shows mutarotation. $[\alpha]_D^{20}$ +1° → −14.5° (3 days, c = 11). Sol in water. Slowly reduces cold Fehling's soln. Sodium amalgam reduces it to natural, inactive erythritol.

Consult the Name Index before using this section.

No aldehyde reaction with benzenesulfhydroxamic acid. Not fermented by yeast.

Phenylosazone, $C_{16}H_{18}N_4O_2$, mp 164°.

3732. L-Erythrose. *[S-(R*,R*)]-2,3,4-Trihydroxybutanal.* $C_4H_8O_4$; mol wt 120.11. C 40.00%, H 6.71%, O 53.28%. From calcium L-arabonate by oxidation with H_2O_2: Ruff, Meusser, *Ber.* **34**, 1365 (1901). From L-arabinose oxime: Wohl, *Ber.* **32**, 3667 (1899). From L-arabonamide by treatment with alkaline NaOCl: Weerman, *Rec. Trav. Chim.* **37**, 35 (1918).

Syrup. Sweet taste. Shows mutarotation. $[\alpha]_D^{24}$ +11.5° (8 min) → +15.2° (120 min) → +30.5° (final, c = 3): Felton, Freudenberg, *J. Am. Chem. Soc.* **57**, 1640 (1935). Soluble in water. Heating with HCl yields lactic acid. Oxidation with Br converts it to L-erythronic acid. Reduces Fehling's soln slowly in the cold. Not fermented by yeast.

Phenylosazone, $C_{16}H_{18}N_4O_2$, mp 164°.

3733. D-Erythrose 4-Phosphate. *[R-(R*,R*)]-2,3-Dihydroxy-4-(phosphonooxy)butanal; 4-D-erythrosephosphoric acid.* $C_4H_9O_7P$; mol wt 200.09. C 24.01%, H 4.53%, O 55.97%, P 15.48%. Occurs in minute amounts in muscle flesh of all animals. Important natural intermediate in the Embden-Meyerhof scheme of alcoholic fermentation and glycolysis. Prepn by chemical synthesis: Ballou *et al., J. Am. Chem. Soc.* **77**, 2658, 5967 (1955). Outline: *Chem. & Eng. News* **34**, 2506 (1956).

Obtained in aq soln only. Shows no observable optical rotation. Is condensed with phosphoenolpyruvate by a cell-free extract from an *E. coli* mutant to give a 90% yield. Best stored and transported as the cyclohexylammonium salt of 4-phosphoryl-D-erythrose dimethyl acetal from which it can be prepared readily when needed.

3734. Erythrosine. *3′,6′-Dihydroxy-2′,4′,5′,7′-tetraiodospiro[isobenzofuran-1(3H),9′-[9H]xanthen]-3-one disodium salt;* 2′,4′,5′,7′-tetraiodofluorescein disodium salt; erythrosine BS; erythrosine B; FD & C Red No. 3; C.I. Food Red 14; C.I. Acid Red 51; C.I. 45430. $C_{20}H_6I_4Na_2O_5$; mol wt 879.86. C 27.30%, H 0.69%, I 57.69%, Na 5.23%, O 9.09%. Prepn: Gilliard *et al.,* **Ger.** pat. 108,838 (1899); *Frdl.* **5**, 215; Gomberg, Tabern, *J. Ind. Eng. Chem.* **14**, 1115 (1922); Dolinsky, Jones, *J. Assoc. Offic. Agr. Chem.* **34**, 114 (1951); *Colour Index* vol. **4** (3rd ed., 1971) p 4428. Structure: Holmes, Scanlan, *J. Am. Chem. Soc.* **49**, 1594 (1927). HPLC determn: F. E. Lancaster, J. F. Lawrence, *J. Chromatog.* **388**, 248 (1987). Acute toxicity: K. R. Butterworth *et al., Food Cosmet. Toxicol.* **14**, 525 (1976); S. L. Yankell *et al., J. Periodontol.* **48**, 228 (1977). Carcinogenicity studies in rodents: J. F. Borzelleca *et al., Food Chem. Toxicol.* **25**, 723, 735 (1987).

Brown powder. Absorption max (water): 524 nm; in 95% alcohol: 531 nm. Soluble in water to cherry-red soln; sol in alcohol. HCl added to an aq soln produces a yellowish-brown ppt. Sodium hydroxide produces a red ppt sol in an excess of the reagent. LD_{50} in male, female mice, male, female rats (g/kg): 0.40, 0.32, 0.34, 0.37 i.p.; 6.7, 6.9, 7.4, 6.8 orally (Butterworth). Also reported as LD_{50} in mice, rats (mg/kg): 2558, 2891 orally (Yankell).

Note: Consult latest governmental regulations on use in food, drugs and cosmetics.

USE: Biological stain; color additive.

3735. L-Erythrulose. *(S)-1,3,4-Trihydroxy-2-butanone; L-glycero-tetrulose.* $C_4H_8O_4$; mol wt 120.11. C 40.00%, H 6.71%, O 53.28%. Prepn by bacterial oxidation of *meso-*erythritol: Müller *et al., Helv. Chim. Acta* **20**, 1468 (1937); Whistler, Underköfler, *J. Am. Chem. Soc.* **60**, 2507 (1937). Synthesis from D-fructose: Gorin *et al., J. Chem. Soc.* **1955**, 2699.

Syrup. $[\alpha]_D^{18}$ +11.4° (c = 2.4 in water). Very sensitive to alkali. Soluble in water, abs alc.

Phenylosazone, $C_{16}H_{18}N_4O_2$, crystals from aq ethanol, mp 162°. $[\alpha]_D$ +32° (10 min) → 0° (24 hr, constant) (c = 0.75 in pyridine + ethanol, 3:2 v/v).

3736. Esaprazole. *N-Cyclohexyl-1-piperazineacetamide; 1-piperazinyl-4-methylenecarbonylcyclohexylamine; N-(1-piperazinylacetyl)cyclohexylamine; N-[(N-cyclohexylcarbamoyl)methyl]piperazine;* hexaprazol(e); exaprazole; C-63; Prazol. $C_{12}H_{23}N_3O$; mol wt 225.33. C 63.96%, H 10.29%, N 18.65%, O 7.10%. Prepn: C. Corvi-Mora, **Ger.** pat. 2,702,537; *idem,* **U.S.** pat. 4,123,530 (1977, 1978 both to Camillo Corvi). Metabolism in animals and man: P. Ventura *et al., Farmaco Ed. Prat.* **39**, 190 (1984). Cytoprotective activity: B. Lumachi, R. Scuri, *Drugs Exp. Clin. Res.* **11**, 693 (1985); and pharmacology: R. Scuri *et al., Boll. Chim. Farm.* **123**, 425 (1983). Pharmacology: G. Coruzzi *et al., Farmaco Ed. Prat.* **40**, 313 (1985); G. Coruzzi *et al., ibid.* 324. Antisecretory activity in humans: L. Erembourg *et al., Int. J. Clin. Pharm. Res.* **7**, 95 (1987). HPLC determn in biological fluids: M. A. Girometta *et al., J. Chromatog.* **383**, 85 (1986). Clinical trial as antiulcerative in comparison with cimetidine, *q.v.:* L. Capurso *et al., Clin. Trials J.* **23**, 293 (1986).

Residue, mp 111-112°. $bp_{0.5}$ 190°. LD_{50} in mice (mg/kg): 1974 orally; 271 i.v.; in rats (mg/kg): 3900 orally (Corvi-Mora).

Hydrochloride, $C_{12}H_{23}N_3O \cdot HCl$, *CO-1063.*

THERAP CAT: Antiulcerative.

3737. Escin. Aescin; Aescusan; Reparil. A mixture of saponins occurring in the seed of the horse chestnut tree, *Aesculus hippocastanum* L., Hippocastanaceae: Winterstein, *Z. Physiol. Chem.* **199**, 25 (1931); Steiner, Holtzem in

Paech-Tracey, *Moderne Methoden der Pflanzenanalyse III* (Springer-Verlag, 1955) p 117. Isoln by chromatography and purification: Fiedler, *Arzneimittel-Forsch.* **4**, 213 (1953); using ion-exchange resins: Erbring *et al.*, U.S. pat. 3,238,-190 (1966 to Madaus). Previously thought to be built up from the aglycon escigenin, glucuronic acid, glucose and xylose: Jermstadt, Waaler, *Pharm. Acta Helv.* **28**, 265 (1953); Patt, Winkler, *Arzneimittel-Forsch.* **10**, 273 (1960); Tschesche *et al.*, *Ann.* **669**, 171 (1963). Recent structural studies indicate that the two major glycosides in the mixture are built up from the aglycon, *protoescigenin*, which is acylated at C-22 by acetic acid, and from the sugar moiety, glucuronic acid and two D-glucose molecules. The two aglycons differ only at the C-21 position which is acylated by either angelic acid, *q.v.* or tiglic acid, *q.v.* Structure and stereochemistry: Wulff, Tschesche, *Tetrahedron* **25**, 415 (1969); Wagner *et al.*, *Arzneimittel-Forsch.* **20**, 205 (1970); eidem, *Z. Physiol. Chem.* **351**, 1133 (1970). Early work identified two forms, *α-escin* and *β-escin:* Wagner, Basse, ibid. **320**, 27 (1960). Identity of *prosaponin B* with *β-escin:* Voigtlander, Rosenberg, *Arzneimittel-Forsch.* **13**, 385 (1963). *β-Escin* is the natural form and can be converted to *α-escin:* Wagner, Schlemmer, U.S. pat. 3,450,691 (1969 to Klinge); Wagner *et al.*, *Arzneimittel-Forsch., loc. cit.* Pharmacology: H. Hampel *et al.*, *ibid.* **20**, 209 (1970); Lang, Mennicke, *ibid.* **22**, 1928 (1972). *Review:* Tschesche, Wulff in *Fortschr. Chem. Org. Naturst.* **30**, 461-606 (1973).

R = Tiglic acid or Angelic acid

Major glycosides of Escin

α-Escin, amorphous powder, mp 225-227°. $[\alpha]_D^{25}$ −13.5° (c = 5 in methanol). Very sol in water. Hemolytic index: 1:20,000. LD_{50} in mice, rats, guinea pigs (mg/kg): 3.2, 5.4, 15.2 i.v.; 320, 720, 475 orally (Hampel).

Sodium salt, mp 251-252°.

β-Escin, Flogencyl. Leaflets from dil ethanol, mp 222-223°. $[\alpha]_D^{27}$ −23.7° (c = 5 in methanol). Practically insol in water. Hemolytic index: 1:40,000. LD_{50} in mice, rats, guinea pigs (mg/kg): 1.4, 2.0, 7.2 i.v.; 134, 400, 188 orally (Hampel).

THERAP CAT: In treatment of peripheral vascular disorders.

3738. Esculetin. *6,7-Dihydroxy-2H-1-benzopyran-2-one; 6,7-dihydroxycoumarin;* cichorigenin. $C_9H_6O_4$; mol wt 178.14. C 60.68%, H 3.39%, O 35.92%. The aglucon of esculin and of cichoriin. By hydrolysis of esculin or of cichoriin: Merz, *Arch. Pharm.* **270**, 486 (1932). By synthesis: Gattermann, Köbner, *Ber.* **32**, 288 (1899); Bert, *Compt. Rend.* **214**, 230 (1942). *Review:* Sethna, Shah, *Chem. Rev.* **36**, 1 (1945).

Prisms from glacial acetic acid; leaflets by vacuum sublimation. mp 268-270°. Soluble in dil alkalies with blue fluorescence; moderately sol in hot alcohol and in glacial acetic acid; almost insol in ether, in boiling water.

USE: In filters for absorption of ultraviolet light.

3739. Esculin. *6-(β-D-Glucopyranosyloxy)-7-hydroxy-2H-1-benzopyran-2-one;* 6,7-dihydroxycoumarin 6-glucoside; esculoside; bicolorin; enallachrome; polychrome; Escosyl. $C_{15}H_{16}O_9$; mol wt 340.29. C 52.95%, H 4.74%, O 42.32%. In leaves and bark of horse chestnut tree *Aesculus hippocastanum* L., Hippocastanaceae. Extraction procedure: Tumann, *Chem. Zentr.* **1916**, I, 1277. Synthesis: Merz, Hagemann, *Naturwiss.* **29**, 650 (1941); eidem, *Arch. Pharm.* **282**, 79 (1944); Amiard, Nominé, *Bull. Soc. Chim. France* **1948**, 476.

Sesquihydrate, needles from hot water, mp 204-206°. One gram dissolves in 580 ml water, 13 ml boiling water. Sol in hot alcohol, methanol, pyridine, ethyl acetate, acetic acid. Aq solns show blue fluorescence above pH 5.8. $[\alpha]_D^{19}$ −78.4° (c = 2.5 in 50% dioxane). Absorption spectrum: Merz, *Arch. Pharm.* **270**, 482 (1932); Goodwin, Pollock, *Arch. Biochim. Biophys.* **49**, 1 (1954). Has vitamin P activity.

Pentaacetate dihydrate, needles from methanol, mp 163-164°.

4-Methylesculin, $C_{16}H_{18}O_9$. Prepn: Velluz, Amiard, *Bull. Soc. Chim. France* **1948**, 1109.

THERAP CAT: Skin protectant.

3740. Eseridine. *(4aS-cis)-2,3,4,4a,9,9a-Hexahydro-2,-4a,9-trimethyl-1,2-oxazino[6,5-b]indol-6-ol methylcarbamate ester;* physostigmine aminoxide; eserine aminoxide; eserine oxide. $C_{15}H_{21}N_3O_3$; mol wt 291.35. C 61.84%, H 7.27%, N 14.42%, O 16.47%. From Calabar bean *(Physostigma venenosum* Balf., Leguminosae): Polonovski, Nitzberg, *Bull. Soc. Chim. France* [IV] **17**, 244 (1915); **21**, 191 (1917). Structure: Stedman, Barger, *J. Chem. Soc.* **127**, 247 (1925). Revised structure: Hootele, *Tetrahedron Letters* **1969**, 2713. Stereochemistry: Robinson, Moorcroft, *J. Chem. Soc. (C)* **1970**, 2077. Total synthesis of racemate: K. Shishido *et al.*, *Chem. Commun.* **1986**, 904.

Rectangular prisms from ether, mp 129°. $[\alpha]_D^{15}$ −175° (alc). Weak base, just alkaline to litmus; does not form crystn salts with mineral acids. Almost insol in water. Sol in alc, chloroform, benzene, ether, petr ether, acetone, dil acids.

Salicylate, $C_{22}H_{27}N_3O_6$, *Génésérine 3.*

THERAP CAT: Cholinergic.

3741. Esmolol. *4-[2-Hydroxy-3-[(1-methylethyl)amino]-propoxy]benzenepropanoic acid methyl ester;* (±)-methyl 3-[4-[2-hydroxy-3-(isopropylamino)propoxy]phenyl]propionate; methyl *p*-[2-hydroxy-3-(isopropylamino)propoxy]hydrocinnamate. $C_{16}H_{25}NO_4$; mol wt 295.38. C 65.06%, H 8.53%, N 4.74%, O 21.67%. Cardioselective β-adrenergic blocker. Prepn: E. I. Carlsson *et al.*, Eur. pat. Appl. 41,491 (1982 to A.B. Hässle), *C.A.* **96**, 122391f (1982); P. W. Erhardt *et al.*, Eur. pat. Appl. 53,435; eidem, U.S. pat. 4,593,119 (1982, 1986 both to Am. Hosp. Supply); P. W. Erhardt *et al.*, *J. Med. Chem.* **25**, 1408 (1982). Pharmacology: R. J. Gorczynski *et al., J. Cardiovasc. Pharmacol.* **5**, 668 (1983); *eidem, ibid.* **6**, 1548 (1984). Pharmacokinetics: A. Yacobi *et al., J. Pharm. Sci.* **72**, 711 (1983). GC-MS determn in blood: C. Y. Sum, A. Yacobi, *ibid.* **73**, 1177 (1984). Clinical studies of effect in supraventricular tachycardia: G. Klein *et al., Int. J. Clin. Pharmacol. Ther. Toxicol.* **22**, 112 (1984); R. J. Gray *et al., J. Am. Coll. Cardiol.* **5**, 1451 (1985). Symposium on pharmacology and clinical efficacy: *Am. J. Cardiol.* **56**(11), 1F-62F (1985). Review of pharmacology and clinical efficacy: P. Benfield, E. M. Sorkin, *Drugs* **33**, 392-412 (1987).

Oil, gradually forming crystalline rosettes at room temp, mp 48-50°.

Hydrochloride, $C_{16}H_{25}NO_4 \cdot HCl$, *ASL-8052, Brevibloc.* Crystals from methanol-ether, mp 85-86°.

THERAP CAT: Antiarrhythmic.

3742. Esparto Wax. Spanish grass wax; halfa wax. A wax derived from a tall, tough grass of the Mediterranean region (S. Europe and N. Africa and Libya). The grass is shipped to Scotland, where it is dewaxed and made into fine paper. The wax is a byproduct. Two species of grass are cultivated for their excellent cellulose content: *Stipa tenacissima* L. (*Macrochloa tenacissima* (L.) Kunth.), *Gramineae* and *Lygeum spartum* L., *Gramineae.   Constit.* Esparto wax consists of 15-17% free wax acids, 20-22% alcohols and hydrocarbons, 63-65% esters. The principal hydrocarbon is hentriacontane ($C_{31}H_{64}$, mp 68°). The acids include cerotic, montanic, myricinic ($C_{30}H_{60}O_2$, mp 68°), lacceric ($C_{32}H_{64}O_2$, mp 70.5°) and hydroxy acids: A. H. Warth, *Chemistry and Technology of Waxes* (Reinhold, New York, 1947) p 139. Extraction of wax from grass: **E. Ger.** pat. **74,703** (1970). TLC identification: H. Schmidt, *Amer. Cosmet. Perfum.* **87**, 35 (1972).

Hard, tough wax. d_4^{25} 0.9887. mp 78.1°. Solidifies at 68.8°. Acid value 23.9. Saponification value 69.8. Soly in ethanol (25°): 0.244 g/100 ml; in ethylene chloride (37°): 1.48 g/100 ml.

USE: Substitute or extender for carnauba wax, *q.v.* Blends well with other waxes. Emulsifies easily and imparts smoothness to polishes. Preferred in the manufacture of carbon papers.

3743. Essential Fatty Acids. EFA. Essential dietary constituents belonging to the linoleic acid family. G. O. Burr and M. M. Burr described expts with albino rats which led them to conclude that in animals given a ration extremely poor in fat, a deficiency disease results which can be cured by linoleic or linolenic acids, *q.q.v.,* and arachidonic acid, *q.v.,* or fats contg these acids (such as lecithin from soybeans): *J. Biol. Chem.* **82**, 345 (1929); **86**, 587 (1930). Support for the conception of the essential nature of these two fatty acids was contained in two reports by Evans and Lepkovsky, *J. Biol. Chem.* **96**, 143, 157 (1932). It was stated that growth would not take place unless unsaturated fatty acids with two or more double bonds were present in the ration. In 1934 a series of three papers by Evans, Lepkovsky, and Murphy appeared, in which the name *vitamin F* was given to the above unsaturated fatty acids. There was a short period of commercial exploitation of vitamin F. Following investigations made by the American Medical Association in 1937, the term "vitamin F" became discredited. At the request of the Council on Foods and Nutrition of the American Medical Association, Hansen and Burr discussed the essentiality of fatty acids in human nutrition in their article "Essential Fatty Acids and Human Nutrition," *J. Am. Med. Assoc.* **132**, 855 (Dec. 7, 1946), in which they state: "In regard to clinical observations made thus far with human subjects, there is no evidence to indicate that a lack of the essential fatty acids produces a disturbance in growth, hematuria, kidney lesions, impaired reproduction or

lactation, and sterility, which abnormalities, are attributed to the dietary lack of either linoleic or arachidonic acid in small experimental animals." *See also* the extensive reviews by Deuel and Greenberg in *Fortschr. Chem. Org. Naturst.* **6**, 1-86 (1950); Stangl, *Int. Rev. Vitamin Res.* **23**, 164-207 (1951); Rahm, Holman, *The Vitamins,* W. H. Sebrell, R. S. Harris, Eds., **vol. 3** (Academic Press, New York, 1971) pp 303-339; W. H. Kunan, *Angew. Chem. Int. Ed.* **15**, 61 (1976); J. F. Mead, "Nutrients with Special Functions: Essential Fatty Acids" in *Human Nutrition, A Comprehensive Treatise,* **vol. 3A,** R. Alfin-Slater, D. Kritchevsky, Eds. (Plenum Press, New York, 1980) pp 213-238. Book: *Essential Fatty Acids and Prostaglandins,* R. J. Holman, Ed. (Pergamon Press, New York, 1982) 968 pp.

3744. Estazolam. *8-Chloro-6-phenyl-4H-[1,2,4]triazolo-[4,3-a][1,4]benzodiazepine; 8-chloro-6-phenyl-4H-s-triazolo-[4,3-a][1,4]benzodiazepine;* D-40TA; Cannoc; Esilgan; Eurodin; Julodin; Nemurel; Nuctalon; ProSom; Somnatrol. $C_{16}H_{11}ClN_4$; mol wt 294.74. C 65.20%, H 3.76%, Cl 12.03%, N 19.01%. Prepn: J. B. Hester, **Ger.** pat. **2,012,190;** *idem,* **U.S.** pat **3,701,782** (1970, 1972 to Upjohn); Meguro, Kuwada, *Tetrahedron Letters* **1970**, 4039; *eidem,* **Ger.** pat. **1,955,349** (1971 to Toyama), *C.A.* **74**, 88078t (1971); Tawada *et al.,* **Ger.** pat. **2,114,441** (1971 to Takeda), *C.A.* **76**, 34320p (1972); Hester *et al., J. Med. Chem.* **14**, 1078 (1971). Structure-activity studies: R. Nakajima *et al., Japan. J. Pharmacol.* **21**, 489 (1971). Pharmacology: *eidem, ibid.* 497; *eidem, Takeda Kenkyusho Ho* **31**, 349 (1972), *C.A.* **78**, 24138m (1972). Toxicity: Yokotani *et al., ibid.* **32**, 152 (1973), *C.A.* **79**, 133006j (1973). Metabolism: Tanayama *et al., Xenobiotica* **4**, 33-64 (1974); *see also ibid.* 229, 441. Molecular structure studies: Kamiya *et al., Chem. Pharm. Bull.* **21**, 1520 (1973).

Crystals from ethyl acetate-Skellysolve B hexanes, mp 228-229°. LD_{50} orally in male mice, rats and rabbits: 740, 3200, 300 mg/kg (Yokotani).

Note: This is a controlled substance (depressant) listed in the U.S. Code of Federal Regulations, Title 21 Part 1308.14 (1995).

THERAP CAT: Sedative, hypnotic.

3745. Ester Gums. These are the glyceryl, methyl, and ethyl esters of rosin acids. They are made by heating rosin and the alcohol under pressure until condensation takes place. The gums are insol in water, but sol in amyl acetate, some oils, turpentine, carbon tetrachloride, etc.

USE: Instead of copal, damar, or kauri in making enamels, paints, cellulose lacquers, and particularly with tung oil for waterproof varnishes.

3746. Estradiol. *(17β)-Estra-1,3,5(10)-triene-3,17-diol; β-estradiol;* α-estradiol (obsolete); *cis-*estradiol; 3,17-epidihydroxyestratriene; dihydrofollicular hormone; dihydrofolliculin; dihydroxyestrin; dihydrotheelin; Dimenformon; Diogyn; Estrace; Estraderm; Estroclim; Evorel; Gynoestryl; Macrodiol; Menorest; Oestrogel; Ovocyclin; Ovocylin; Profoliol B; Progynon; Systen; Vagifem; Zumenon. $C_{18}H_{24}O_2$; mol wt 272.39. C 79.37%, H 8.88%, O 11.75%. Potent mammalian estrogenic hormone produced by the ovary. Triggers the production of gonadotropins leading to ovulation. Isoln from follicular fluid of sow ovaries: D. W. MacCorquodale *et al., J. Biol. Chem.* **115**, 435 (1936). Has also been isolated from urine of pregnant women and mares. Prepn from estrone: A. Butenandt, C. Georgens, *Z. Physiol. Chem.* **248**, 129 (1937); F. Hildebrandt *et al.,* **U.S.** pat. **2,096,744** (1938 to Schering); from cholesterol: H. H. Inhoffen, G. Zühlsdorff, *Ber.* **74**, 1914 (1941). Total syntheses: U. Eder *et al.*, *Ber* **109**, 2948 (1976); W. Oppolzer, D. A. Roberts, *Helv. Chim. Acta* **63**, 1703 (1980). Pharmacology: R. W. Lievertz, *Am. J. Obstet. Gynecol.* **156**, 1289 (1987). HPLC determn in plasma and urine: W. Slikker *et al.*, *J. Chromatog.* **224**, 205 (1981). Pharmacokinetics: W. Kuhnz *et al.*, *Arzneimittel-Forsch.* **43**, 966 (1993). Clinical evaluation in osteoporosis: J. W. W. Studd *et al.*, *Brit. J. Obstet. Gynecol.* **101**, 787 (1994). Comprehensive description: E. G. Salole, *Anal. Profiles Drug Subs.* **15**, 283-318 (1986). Physiologic review: H. G. Burger, *Int. J. Gynecol. Obstet.* **1989**, Suppl. 1, 5-9. Review of clinical safety and efficacy in estrogen replacement therapy: A. Cheang *et al.*, *Drug Safety* **9**, 365-379 (1993); H. L. Judd, *J. Reprod. Med.* **39**, 343-352 (1994).

Prisms from 80% alc, stable in air, mp 173-179°. $[\alpha]_D^{25}$ +76 to +83° (dioxane). uv max: 225, 280 nm. Precipitated by digitonin. Almost insol in water. Freely sol in alcohol; sol in acetone, dioxane, other organic solvents; solns of fixed alkali hydroxides; sparingly sol in vegetable oils. 1 mg = 10,000 international units.

3-Benzoate, $C_{25}H_{28}O_3$, *estradiol benzoate, Agofollin, Benzo-Gynoestryl, Benztrone, Pelanin benzoate, Progynon B, Ovahormon Benzoate.* Crystals from alc, mp 191-196°. Stable in air. $[\alpha]_D^{25}$ +58° to +63° (c = 2 in dioxane). Sol in alc, acetone, dioxane; slightly sol in ether, vegetable oils.

17β-Cyclopentanepropanoate, $C_{26}H_{36}O_3$, *estradiol cypionate, ECP, Depgynogen, Depogen, Estrofem.* Prepn: Ott, U.S. pat. **2,611,773** (1952 to Upjohn). Crystals from benzene + petr ether, mp 151-152°. $[\alpha]_D^{25}$ +45° (chloroform). Sol in ether, methanol, benzene, chloroform, peanut oil, cottonseed oil, corn oil, sesame oil.

Dipropionate, $C_{24}H_{32}O_4$, *Ovahormon Depot.* Prepn: K. Miescher, C. Scholz, U.S. pat. **2,205,627** (1940 to Ciba). mp 104-105°.

Hemisuccinate, $C_{22}H_{28}O_5$, *Eutocol.*

17-Heptanoate, $C_{25}H_{36}O_3$, *estradiol enanthate, SQ-16150.* Prepn: Gauthier *et al.*, *Ann. Pharm. Franc.* **16**, 757 (1958). Crystals from diisopropyl ether, mp 94-96°.

17-Undecanoate, $C_{29}H_{44}O_3$, *estradiol undecylate, SQ-9993, Delestrec.*

17-Valerate, $C_{23}H_{32}O_3$, *Climaval, Cyclacur, Delestrogen, Gynogen LA, Pelanin Depot, Progynon Depot, Progynova, Primofol, Valergen.* Prepn: K. Miescher, C. Scholz, U.S. pat. **2,233,025** (1941 to Ciba). Comprehensive description: K. Florey, *Anal. Profiles Drug Subs.* **4**, 192-208 (1975). Crystals, mp 144-145°.

Note: Estradiol may reasonably be anticipated to be a carcinogen: *Seventh Annual Report on Carcinogens* (PB95-109781, 1994) p 195.

THERAP CAT: Estrogen.

THERAP CAT (VET): Estrogen.

3747. α-Estradiol. *(17α)-Estra-1,3,5(10)-triene-3,17-diol;* 1,3,5-estratriene-3,17α-diol; 3,17-dihydroxyestratriene. $C_{18}H_{24}O_2$; mol wt 272.39. C 79.37%, H 8.88%, O 11.75%. Has been isolated from pregnancy urine of mares. Prepn from β-estradiol by inversion of the hydroxyl group at C-17

after tosylation: Allais, Hoffmann, U.S. pat. **2,835,681** (1958).

Needles with ½H_2O from 80% alcohol, mp 220-223°. $[\alpha]_D^{20}$ +53° to +56° (c = 0.9 in dioxane). Not precipitated by digitonin (in 80% alcoholic soln). Soluble in alcohol, acetone, aq alkalies. One gram dissolves in more than 100 ml of boiling benzene. Slightly sol in ether, chloroform. Insol in water, aq dil acids.

Diacetate, $C_{22}H_{28}O_4$, mp 140-142°.

3-Benzoate, $C_{25}H_{28}O_3$, mp 156-157°; also reported as polymorphous: I, mp 63°; II, mp 153°; III, mp 158°.

3748. Estragole. *1-Methoxy-4-(2-propenyl)benzene; p-allylanisole;* chavicol methyl ether; esdragol. $C_{10}H_{12}O$; mol wt 148.20. C 81.04%, H 8.16%, O 10.80%. Main constituent of *tarragon oil (estragon oil)*, the oil from *Artemisia dracunculus* L., *Compositae* (esdragon) where it occurs to an extent of 60-75%: Grimaux, *Bull. Soc. Chim.* [3] **11**, 34 (1894); Daufresne, *ibid.* [4] **3**, 333 (1908). Occurs also in pine oil and in American turpentine oil. Prepn: Tiffeneau, *Compt. Rend.* **139**, 482 (1904); Verley, Ger. pat. **154,654**; D. Wigfield, K. Taymaz, *Tetrahedron Letters* **1973**, 4841; P. Gramatica *et al.*, *Gazz. Chim. Ital.* **104**, 629 (1974). Toxicity study: P. M. Jenner *et al.*, *Food Cosmet. Toxicol.* **2**, 327 (1964).

Liquid. d_4^{21} 0.9645. bp_{764} 216°; bp_{25} 108-114°; bp_{12} 95-96°. $n_D^{17.5}$ 1.5230. Sol in alcohol, chloroform. Forms azeotropic mixtures with water. LD_{50} in rats, mice (mg/kg): 1820, 1250 orally (Jenner).

USE: In perfumes and as flavor in foods and liqueurs.

3749. Estramustine. *(17β)-Estra-1,3,5(10)-triene-3,17-diol 3-[bis(2-chloroethyl)carbamate]; estradiol 3-bis(2-chloroethyl)carbamate;* estra-1,3,5(10)-triene-3,17β-diol 3-[N,N-bis-(2-chloroethyl)carbamate]; Ro-21-8837. $C_{23}H_{31}Cl_2NO_3$; mol wt 440.41. C 62.73%, H 7.09%, Cl 16.10%, N 3.18%, O 10.90%. Estradiol to which nitrogen mustard is bound. Prepn: Belg. pat. **646,319**; Fex *et al.*, U.S. pat. **3,299,104** (1963, 1967 to AB Leo); Niculescu-Duvaz *et al.*, *J. Med. Chem.* **10**, 172 (1967). Clinical results: Anderes, *Praxis* **60**, 1375 (1971); Muntzing, Nilsson, *Z. Krebsforsch. Klin. Onkol.* **77**, 166 (1972).

Crystals from benzene-petr ether, mp 104-105°. $[\alpha]_D^{20}$ +50° (in dioxane). uv max (alcohol): 270.7, 276.5 nm.

17-Phosphate, $C_{23}H_{32}Cl_2NO_6P$, *Estracyt.* mp 155° (dec). $[\alpha]_D^{20}$ +30° (dioxane). Sol in aqueous and alkali solns.

17-(Dihydrogenphosphate) disodium salt, $C_{23}H_{30}Cl_2N$-Na_2O_6P, *Emcyt.*

THERAP CAT: Antineoplastic.

3750. Estriol. *(16α,17β)-Estra-1,3,5(10)-triene-3,16,17-triol;* 1,3,5-estratriene-3β,16α,17β-triol; 3,16α,17β-trihydroxy-$\Delta^{1,3,5}$-estratriene; 16α-hydroxyestradiol; follicular hormone hydrate; oestriol; trihydroxyestrin; Aacifemine; Colpogyn; Destriol; Gynäsan; Hormomed; Klimax E; Klimoral; Oekolp; Ortho-Gynest; Ovesterin; Ovestin; Ovo-Vinces; Theelol; Tridestrin; Triovex. $C_{18}H_{24}O_3$; mol wt 288.39. C 74.97%, H 8.39%, O 16.64%. A metabolite of, and considerably less potent than the hormone estradiol, *q.v.* It is usually the predominant estrogenic metabolite found in urine. During pregnancy the placenta produces relatively large amounts of estriol. Isoln from human pregnancy urine:

Marrian, *Biochem. J.* **23**, 1090, 1233 (1929); probably occurs as a glycuronide: Cohen, Marrian, *ibid.* **29**, 1577 (1935). Isoln from human placenta: Collip, *Brit. Med. J.* **1930**, II, 1080; Collip *et al.*, *Endocrinology* **18**, 71 (1934). Also obtained from plant sources. Isoln from pussywillows: Skarzynski, *Nature* **131**, 766 (1933). Structure: Huffman, Lott, *J. Am. Chem. Soc.* **69**, 1835 (1947). Crystal and molecular structure: Cooper *et al.*, *Acta Crystallogr.* **25B**, 814 (1969). Soly studies: Ruchelman, Howe, *J. Chromatogr. Sci.* **7**, 340 (1969). Partial synthesis: Huffman, *J. Biol. Chem.* **169**, 167 (1947). Syntheses: Huffman, Lott, *J. Am. Chem. Soc.* **71**, 719 (1949); Leeds *et al.*, *ibid.* **76**, 2943 (1954). Prepn of 16,17-bis(sodium hemisuccinate): **Brit. pat. 879,014** (1960 to Organon).

Very small, monoclinic crystals from dil alc. d 1.27. mp 282°. During heating on the microscope heating stage, rearrangement of the crystal structure takes place at 270° and 275° (rate of heating, 4°/min, Kofler microscope heating stage). $[\alpha]_D^{25}$ +58° ±5° (0.04 g in 1 ml dioxane). uv max: 280 nm. Precipitated by digitonin. Practically insol in water. Sol in alcohol, dioxane, chloroform, ether, vegetable oils; freely sol in pyridine, in solns of fixed alkali hydroxides.

Triacetate, $C_{24}H_{30}O_6$, crystals from 90% alc. mp 126°.

16,17-Bis(sodium hemisuccinate), $C_{26}H_{30}Na_2O_9$, *estriol succinate, Orgastyptin, Stiptanon, Synapause.*

THERAP CAT: Estrogen.

THERAP CAT (VET): Estrogenic hormone therapy.

3751. Estrone. *3-Hydroxyestra-1,3,5(10)-trien-17-one;* 1,3,5-estratrien-3-ol-17-one; oestrone; folliculin; follicular hormone; tokokin; thelykinin; ketohydroxyestrin; Hiestrone; Menformon; Glandubolin; Cristallovar; Destrone; Endofolliculina; Estrol; Femidyn; Folikrin; Kolpon; Crinovaryl; Folisan; Disynformon; Hormovarine; Oestroperos; Wynestron; Thelestrin; Kestrone; Estrusol; Estrugenone; Femestrone Inj; Folipex; Follestrine; Follidrin (tablets); Follicunodis; Hormofollin; Oestrin; Ovifollin; Perlatan; Ketodestrin; Theelin. $C_{18}H_{22}O_2$; mol wt 270.37. C 79.96%, H 8.20%, O 11.84%. A metabolite of 17β-estradiol, *q.v.*, possessing considerably less biological activity: isolated as d-form. Occurs in pregnancy urine of women and mares, in follicular liquor of many animals, in human placenta, in urine of bulls and stallions, in palm-kernel oil. Isoln: Butenandt, *Naturwiss.* **17**, 879 (1929); Doisy *et al.*, *Am. J. Physiol.* **90**, 329 (1929). Also isolated from moghat roots and date palm pollen grains: Amin *et al.*, *Phytochemistry* **8**, 295 (1969). Manuf: E. A. Doisy *et al.*, **U.S. pats. 1,967,350, 1,967,351** (1934 both to St. Louis University); Joly *et al.*, **Fr. pat. 1,305,992** (1962 to Roussel-Uclaf). Synthesis of a stereoisomer of estrone: W. E. Bachmann *et al.*, *J. Am. Chem. Soc.* **64**, 974 (1942). Total synthesis of natural estrone: G. Anner, K. Miescher, *Helv. Chim. Acta* **31**, 2173 (1948); **32**, 1957 (1949); **33**, 1379 (1950); W. S. Johnson *et al.*, *J. Am. Chem. Soc.* **72**, 1426 (1950); **74**, 2832 (1952); I. V. Torgov *et al.*, *Doklady Akad. Nauk SSSR* **135**, 73 (1960), *C.A.* **55**, 11462f (1961). Stereochemistry of estrone isomers: W. S. Johnson *et al.*, *J. Am. Chem. Soc.* **80**, 661 (1958). Review of estrone syntheses: L. F. Fieser, M. Fieser, *Steroids* (Reinhold, New York, 1959) pp 495-502; T. B. Windholz, M. Windholz, *Angew. Chem. Int. Ed.* **3**, 353 (1964); Taub, "Naturally Occurring Aromatic Steroids" in *The Total Synthesis of Natural Products* vol. 2, J. ApSimon, Ed. (John Wiley & Sons, New York, 1973) pp 670-725. Recent syntheses: P. A. Bartlett, W. S. Johnson, *J. Am. Chem. Soc.* **95**, 7501 (1973); S. Danishefsky, P. Cain, *ibid.* **98**, 4975 (1976); T. Kametani *et al.*, *ibid.* **99**, 3461 (1977); P. A. Grieco *et al.*, *J. Org. Chem.* **45**, 2247 (1980). Comprehensive description: D. Both, *Anal. Profiles Drug Subs.* **12**, 135-189 (1983).

Crystals from acetone, mp 254.5-256°. $[\alpha]_D^{22}$ +152° (c = 0.995 in $CHCl_3$). uv max (*p*-dioxane): 282, 296 nm (ϵ 2300, 2130); (conc H_2SO_4): 300, 450 nm; (0.1*M* NaOH): 239, 293 nm. Exists in three cryst phases, one monoclinic, the other two orthorhombic. Pptd by digitonin. Soly in water (25°): 0.003 g/100 ml. One gram of estrone dissolves in 250 ml of 96% alcohol at 15°, in 50 ml of boiling alcohol; in 50 ml acetone at 15°, in 110 ml chloroform at 15°, in 145 ml boiling benzene. Sol in dioxane, pyridine, fixed alkali hydroxide solns; slightly sol in ether, vegetable oils.

Acetate, $C_{20}H_{24}O_3$, *Hogival.* Crystals from alcohol, mp 125-127°.

Propionate, $C_{21}H_{26}O_3$, mp 134-135°, **U.S. pat. 2,156,599.**

Sulfate, *Conjugol.* Price, **U.S. pat. 2,917,522** (1959 to Parke, Davis).

Sulfate piperazine salt, $C_{22}H_{32}N_2O_5S$, *estropipate, piperazine estrone sulfate, Harmogen, Ogen, Sulestrex Piperazine.* Comprehensive description: Z. L. Chang, *Anal. Profiles Drug Subs.* **5**, 375-402 (1976). Cryst powder, melts at 190° to a liquid which resolidifies and then decomp at 245°. $[\alpha]_D^{25}$ +87.8° (c = 1 in 0.4% NaOH). uv max (0.04% NaOH): 275, 268 nm (ϵ 838, 851).

dl-Form, crystals from acetone, mp 251-254°.

Methyl ether, $C_{19}H_{24}O_2$. Synthesis: T. A. Bryson, C. J. Reichel, *Tetrahedron Letters* **21**, 2381 (1980). *dl*-Form, crystals from acetone + methanol, mp 143.2-144.2°. *d*-Form, crystals from methanol, mp 164-165°.

Note: Estrone may reasonably be anticipated to be a carcinogen: *Seventh Annual Report on Carcinogens* (PB95-109781, 1994) p 197.

USE: In the prepn of commercial 19-norsteroids.

THERAP CAT: Estrogen.

THERAP CAT (VET): Estrogen.

3752. Etafedrine. *α-[1-(Ethylmethylamino)ethyl]benzenemethanol; l-N-ethylephedrine; l-α-[1-(ethylmethylamino)ethyl]benzyl alcohol; 2-methylethylamino-1-phenyl-1-propanol; Menetryl; Novedrin.* $C_{12}H_{19}NO$; mol wt 193.29. C 74.57%, H 9.91%, N 7.25%, O 8.28%. β-Adrenergic agonist. Prepn: Skita, Keil, *Ber.* **63**, 34 (1930); Ueda *et al.*, *Pharm. Bull. (Tokyo)* **3**, 465 (1955).

Hydrochloride, $C_{12}H_{19}NO\cdot HCl$, *Nethamine.* Crystals from acetone + alcohol, mp 183-184°. One gram dissolves in 1.5 ml water, 8.0 ml alcohol. Aq solns are very stable.

THERAP CAT: Bronchodilator.

3753. Etafenone. *1-[2-[2-(Diethylamino)ethoxy]phenyl]-3-phenyl-1-propanone; 2'-[2-(diethylamino)ethoxy]-3-phenylpropiophenone; o-diethylaminoethoxy-β-phenylpropiophenone; 2'-(β-diethylaminoethoxy)-3-phenylpropiophenone; β-phenyl-o-(diethylaminoethoxy)propiophenone; LG-11457.* $C_{21}H_{27}NO_2$; mol wt 325.45. C 77.50%, H 8.36%, N 4.30%, O 9.83%. Prepn: G. DiPaco, S. C. Tauro, *Ann. Chim. (Rome)* **48**, 1215 (1958); *eidem,* **Ger. pat. 1,265,-758** (1968 to Guidotti), *C.A.* **69**, 58950a (1968). Toxicity data: H.-J. Hapke, W. Sterner, *Arzneimittel-Forsch.* **19**, 1664 (1969). Series of articles on pharmacology, toxicology, metabolism: *ibid.* 1664-1681.

Liquid, bp_{30} 264-268°.

Hydrochloride, $C_{21}H_{27}NO_2 \cdot HCl$, *hetaphenone, Asamedol, Baxacor, Corodilan, Dialicor, Pagano-Cor, Relicor.* Crystals, mp 129-130°. LD_{50} in rats (mg/kg): 716 orally; 20.8 i.v. (Hapke, Sterne).

THERAP CAT: Vasodilator (coronary).

3754. Etamiphyllin. *7-[2-(Diethylamino)ethyl]-3,7-dihydro-1,3-dimethyl-1H-purine-2,6-dione; 7-(2-diethylaminoethyl)theophylline;* 7-(2-diethylaminoethyl)-1,3-dimethylxanthine; 1,3-dimethyl-7-(2-diethylaminoethyl)xanthine; *dietamiphylline; millophyline.* $C_{13}H_{21}N_5O_2$; mol wt 279.34. C 55.90%, H 7.58%, N 25.07%, O 11.46%. Prepn: Quevauviller *et al., Bull. Soc. Chim. Biol.* **31**, 532 (1946); Moussalli *et al.,* **Brit.** pat. **669,070** (1952); Klosa, *Arch. Pharm.* **288**, 301 (1955). Toxicity data on combinations: Baettig, *Arzneimittel-Forsch.* **21**, 354 (1971).

Waxy solid, mp 75°. Very sol in water, acetone; slightly sol in ethanol, ether.

Hydrochloride, $C_{13}H_{21}N_5O_2 \cdot HCl$, crystals, mp 239-241°. Very sol in water, forming stable neutral soln.

Camphorsulfonate, $C_{23}H_{37}N_5O_6S$, *Camphophyline.* Crystals, mp 174°.

Heparinate, *Milhéparine.*

Iodomethylate, *Iodaphyline, Iodafilina, Jodo-Metil-Fillina.*

Methesculetol salt, *see* Metescufylline.

THERAP CAT: Bronchodilator.

3755. Etamycin. Etamycin A; neoviridogrisein IV; viridogrisein. $C_{44}H_{62}N_8O_{11}$; mol wt 879.02. C 60.12%, H 7.11%, N 12.75%, O 20.02%. Monocyclic peptide lactone antibiotic produced by *Streptomyces griseus* and allied *Streptomyces* spp., which shows activity vs gram positive bacteria, *Mycobacterium tuberculosis* and some fungi. Obtained together with griseoviridin, *q.v.:* B. Heinemann *et al., Antibiot. Ann.* **1954-55**, 728; Q. R. Bartz *et al., ibid.* 777. Chemical studies: T. H. Haskell *et al., ibid.* 784. Biological studies: J. Ehrlich *et al., ibid.* 790. Structure: Sheehan *et al., J. Am. Chem. Soc.* **80**, 3349 (1958). Confirmation of structure: Arnold *et al., J. Chem. Soc.* **1958**, 4466. Prodn with griseoviridin: Q. R. Bartz *et al.,* U.S. pat. **3,023,204** (1962

to Parke, Davis). Total synthesis: Sheehan, Ledis, *J. Am. Chem. Soc.* **95**, 875 (1973). Mode of action: Garcia-Mendoza, *Biochim. Biophys. Acta* **97**, 94 (1965). Mass spectra data: Rozynov *et al., Zh. Obshch. Khim.* **39**, 891 (1969). Prodn of congeners by *S. griseoviridus:* C. Chopra *et al., J. Antibiot.* **32**, 392 (1979). Biosynthetic studies of *neoviridogriseins,* homologs of etamycin: Y. Akumura *et al., ibid.* 575, 1002, 1130.

Weakly basic polypeptide, dec 168-170°. $[\alpha]_D^{26}$ +62°, +31° (c = 5 in chloroform, alc). uv max (alc): 304.5 nm (log ε 3.91). pKa about 1 to 2. Soly in water about 1 mg/ml. Also sol in 1.0*N* HCl and 1.0*N* NaOH, in the lower alcohols and ketones, benzene, chloroform, carbon tetrachloride, carbon disulfide, ethyl acetate, peanut oil, ether. Practically insol in petr ether. LD_{50} in mice: > 2 g/kg s.c.; > 3 g/kg orally.

Hydrochloride, crystals, mp 163-170° (some dec). uv max: 304 nm ($E_{1cm}^{1\%}$ 86), after addition of alkali: 335 nm ($E_{1cm}^{1\%}$ 82). Soly in water about 4 mg/ml. Sol in methanol, ethanol, formamide. Slightly sol in acetone, ethyl acetate, or less polar solvents.

3756. Etanidazole. *N-(2-Hydroxyethyl)-2-nitro-1H-imidazole-1-acetamide;* DUP-453; NSC-301467; SR-2508; Radinyl. $C_7H_{10}N_4O_4$; mol wt 214.18. C 39.26%, H 4.71%, N 26.16%, O 29.88%. Oxygen-mimetic which renders hypoxic cells sensitive to chemo- and radiotherapy. Prepn: A. G. Beaman *et al., Antimicrob. Ag. Chemother.* **1967**, 520. *See also:* W. W. Lee *et al.,* U.S. pat. **4,371,540** (1983 to U.S. Sec'y Health Human Serv.). Pharmacology: J. M. Brown *et al., Int. J. Radiat. Oncol. Biol. Phys.* **7**, 695 (1981). HPLC determn in biological materials: R. Ward, P. Workman, *J. Chromatog.* **420**, 223 (1987). Clinical pharmacokinetics and efficacy in radiation therapy: T. H. Wasserman *et al., Radiother. Oncol.* **20**, Suppl. 1, 129 (1991). Evaluation in combination with alkylating agents: L. N. Shulman *et al., Int. J. Radiat. Oncol. Biol. Phys.* **29**, 545 (1994).

Crystals from ethanol, mp 162-163° (Beaman); also reported as white powder from ethyl acetate + methanol, mp 164.5-165° (Lee). Partition coefficient (octanol/water): 0.046. uv max (isopropanol): 313 nm (ε 7800). Soly in isotonic saline: 200 mg/ml.

THERAP CAT: Antineoplastic adjunct (radiosensitizer).

3757. Etaqualone. *3-(2-Ethylphenyl)-2-methyl-4(3H)-quinazolinone;* 2-methyl-3-(o-ethylphenyl)-4-quinazolone; ethinazone; Aolan. $C_{17}H_{16}N_2O$; mol wt 264.33. C 77.25%, H 6.10%, N 10.60%, O 6.05%. Prepd by condensation of *N*-acetylanthranilic acid and 2-ethylaniline with $POCl_3$: **Belg.** pat. **615,282** (1962 to P. Beiersdorf AG), *C.A.* **58**, 13971ef (1963). Structure-activity studies: Parmar *et al., J. Med. Chem.* **12**, 138 (1969).

mp 81°.

Hydrochloride, $C_{17}H_{16}N_2O \cdot HCl$, mp 247°.

THERAP CAT: Sedative, hypnotic.

3758. Eterobarb. *5-Ethyl-1,3-bis(methoxymethyl)-5-phenyl-2,4,6(1H,3H,5H)-pyrimidinetrione; 5-ethyl-1,3-bis-(methoxymethyl)-5-phenylbarbituric acid; N,N'*-dimethoxymethylphenobarbital; eterobarbital; Ex-12-095; Antilon. $C_{16}H_{20}N_2O_5$; mol wt 320.35. C 59.99%, H 6.29%, N 8.74%, O 24.97%. Alkoxymethyl deriv of phenobarbital, *q.v.,* reported to have little or no hypnotic activity. Prepn: C. M.

Samour, J. A. Vita, **Ger.** pat. **1,939,987** corresp to **U.S.** pat. **3,595,862** (1970, 1971 both to Kendall); C. M. Samour *et al., J. Med. Chem.* **14,** 187 (1971); J. Gal, *J. Pharm. Sci.* **68,** 1562 (1979). Spectrophotographic and polarographic analysis: M. Romer *et al., Anal. Chim. Acta* **88,** 261 (1977). Pharmacologic study: J. A. Vida, M. L. Hasker, *J. Med. Chem.* **16,** 602 (1973). Metabolism: M. A. Goldberg *et al., Ann. Neurol.* **5,** 121 (1979). Efficacy in febrile convulsions: R. M. Julien, G. G. Fowler, *Neuropharmacology* **16,** 719 (1977).

Crystals from ethanol, mp 116-118°. LD$_{50}$ in mice: 470 mg/kg orally (Vida, Hooker).

Note: This is a controlled substance (depressant) listed in the U.S. Code of Federal Regulations, Title 21 Part 1308.13 (1995).

THERAP CAT: Anticonvulsant.

3759. Etersalate. *2-(Acetyloxy)benzoic acid 2-[4-(acetylamino)phenoxy]ethyl ester;* 2-(4-acetamidophenoxy)ethyl 2-acetoxybenzoate; eterilate; etherylate; eterylate; Daital. C$_{19}$H$_{19}$NO$_6$; mol wt 357.36. C 63.86%, H 5.36%, N 3.92%, O 26.86%. Deriv of aspirin, *q.v.* Prepn: C. S. Letelier, F. C. Grafulia, **Ger.** pat. **2,538,206;** *eidem,* **U.S.** pat. **4,014,921** (both 1977 to Alter); C. S. Letelier *et al., Arch. Farmacol. Toxicol.* **4,** 153 (1978). Effect on *in vitro* platelet function in human plasma: M. P. Ortega *et al., ibid.* **6,** 31 (1980). *In vivo* effects on primary immune response: I. Barasoain *et al., Immunopharmacology* **2,** 293 (1980). Comparative toxicity study and gastric tolerance: S. Alonso *et al., Arch. Farmacol. Toxicol.* **4,** 349 (1978).

Cryst from methanol, mp 139-141°. LD$_{10}$ orally in rats: 7000 mg/kg (Leterier, Grafulia, 1977).

THERAP CAT: Analgesic; anti-inflammatory; antipyretic.

3760. Ethacridine. *7-Ethoxy-3,9-acridinediamine;* 6,9-diamino-2-ethoxyacridine; 2,5-diamino-7-ethoxyacridine; 2-ethoxy-6,9-diaminoacridine; etakridin. C$_{15}$H$_{15}$N$_3$O; mol wt 253.30. C 71.13%, H 5.97%, N 16.59%, O 6.32%. Prepn: **Ger.** pats. **364,033; 364,037; 393,411; 395,683** (1922); *Frdl.* **14,** 804, 807-812; A. Albert, W. Gledhill, *J. Soc. Chem. Ind.* **61,** 159 (1942). Toxicity: S. D. Rubbo, *Brit. J. Exp. Pathol.* **28,** 1 (1947); N. J. Joshi *et al., Ind. J. Exptl. Biol.* **16,** 1038 (1978). Pharmacokinetics: T. J. Rising *et al., Arzneimittel-Forsch.* **27,** 872 (1977). Use in protein analysis: R. H. Yue *et al., Thromb. Diath. Haemorrh.* **31,** 439 (1974); F. Franek, *Methods in Enzymology* **121,** 631-638 (1986); in serological detection of *Brucella sp.:* J. D. Huber, P. Nicoletti, *Am. J. Vet. Res.* **47,** 1529 (1986). Clinical evaluation in diarrhea: N. Madanagopalan *et al., Curr. Ther. Res.* **18,** 546 (1975); R. Raedsch *et al., Klin. Wochenschr.* **69,** 863 (1991); as an abortifacient: S. Gardo, M. Nagy, *Arch. Gynecol. Obstet.* **247,** 39 (1990).

Orange-yellow crystals from 50% alc. mp 226°.

Lactate monohydrate, C$_{15}$H$_{15}$N$_3$O.C$_3$H$_6$O$_3$.H$_2$O, *acrinol, Rimaon, Rivanol, Vucine, Acrolactine, Ethodin, Metifex.* Light yellow crystals from 90% alcohol + ether. Darkens at 200°, mp 235°. Slowly sol in 15 parts water; sol in 9 parts boiling water; in 110 parts alcohol at 22°, in 100 parts boiling alcohol. Solns are yellow, fluorescent and stable to boiling. LD$_{50}$ s.c. in mice: 0.12 g/kg (Rubbo).

USE: Reagent in serological testing and protein chemistry.

THERAP CAT: Antiseptic. Abortifacient.

3761. Ethacrynic Acid. *[2,3-Dichloro-4-(2-methylene-1-oxobutyl)phenoxy]acetic acid; [2,3-dichloro-4-(2-methylenebutyryl)phenoxy]acetic acid;* [4-(methylenebutyryl)-2,3-dichlorophenoxy]acetic acid; MK-595; Crinuryl; Edecril; Edecrin; Endecril; Hydromedin; Reomax; Taladren; Uregit. C$_{13}$H$_{12}$Cl$_2$O$_4$; mol wt 303.14. C 51.51%, H 3.99%, Cl 23.39%, O 21.11%. Prepn: Schultz *et al., J. Med. Pharm. Chem.* **5,** 660 (1962); E. M. Schultz, J. M. Sprague, **Belg.** pat. **612,755,** *C.A.* **59,** 12712b (1963); *eidem,* **U.S.** pat. **3,255,241** (1962, 1966 both to Merck & Co.). Pharmacology: Beyer *et al., J. Pharmacol. Exp. Ther.* **147,** 1 (1965). Crystal structure: J. Lamotte *et al., Acta Crystallogr.* **B34,** 2636 (1978). Toxicity study: Peck *et al., Fed. Proc.* **23,** 438 (1964). *Reviews:* Kim *et al., Am. J. Cardiol.* **27,** 407-415 (1971); H. E. Williamson, *J. Clin. Pharmacol.* **17,** 663-672 (1977).

Crystals, mp 121-122° (corr). pKa' 3.50. Sparingly sol in water, aq acids; sol in chloroform. LD$_{50}$ in mice (mg/kg): 176 i.v.; 627 orally (Peck).

Sodium salt, C$_{13}$H$_{11}$Cl$_2$NaO$_4$, *ethacrynate sodium, Lyovac Sodium Edecrin.* uv max (water): 225 nm (ϵ 15287). Soly in water at 25°: up to 9%. Solns at pH 7 at 25° stable for short periods of time. Stability decreased with an increase in pH, temperature and time.

THERAP CAT: Diuretic.

THERAP CAT (VET): Diuretic.

3762. Ethadione. *3-Ethyl-5,5-dimethyl-2,4-oxazolidinedione;* 3-ethyl-5,5-dimethyl-2,4-diketooxazolidine; Dimedione; Petidiol; Didione; Petidion; Epinyl; Etydion; Petisan; Neo-Absentol. C$_7$H$_{11}$NO$_3$; mol wt 157.17. C 53.49%, H 7.05%, N 8.91%, O 30.54%. Prepn: Davis, Hook, **Brit.** pat. **626,971** (1949 to Brit. Schering Res. Labs.).

Flat, vitreous prisms from ether, mp 76-77°.

THERAP CAT: Anticonvulsant.

3763. Ethalfluralin. *N-Ethyl-N-(2-methyl-2-propenyl)-2,6-dinitro-4-(trifluoromethyl)benzenamine;* N-ethyl-N-α,α,-α-trifluoro-N-(2-methylallyl)-2,6-dinitro-p-toluidine; N-ethyl-N-methallyl-4-trifluoromethyl-2,6-dinitroaniline; EL-161; Sonalan; Sonalen. C$_{13}$H$_{14}$F$_3$N$_3$O$_4$; mol wt 333.27. C 46.85%, H 4.23%, F 17.10%, N 12.61%, O 19.20%. Selective, pre-emergent herbicide. Prepn: H. D. Porter, **Ger.** pat. **2,511,897;** *idem,* **Brit.** pat. **1,505,249** (1975, 1978 both to Eli Lilly). Activity: G. Skylakakis *et al., Proc. 12th Brit. Weed Contr. Conf.* **2,** 795 (1974). Review of chemistry and analytical methods: E. W. Day, "Ethalfluralin" in *Analyti-*

cal Methods for Pesticides and Plant Growth Regulators Vol. **10**, G. Zweig, J. Sherma, Eds. (Academic Press, New York, 1978) pp 341-352. Persistence in soil: B. J. Hayden, A. E. Smith, *Bull. Environ. Contam. Toxicol.* **25**, 508 (1983).

Yellow crystals from petr ether, mp 54-57°. Readily sol in acetone, acetonitrile, benzene, chloroform, hexane, methanol, xylene. Soly in water (25°): 0.3 ppm. uv max (methanol): 374, 267 nm. Vapor pressure (25°): 8.2×10^{-5} torr. LD_{50} in mice, rats (g/kg): > 10 orally; LC_{50} in bluegill sunfish, rainbow trout, goldfish (ppb): 22, 37, 260 (Day).

USE: Pre-emergent herbicide.

3764. Ethambutol. *2,2′-(1,2-Ethanediyldiimino)bis-1-butanol; (+)-2,2′-(ethylenediimino)di-1-butanol; d-N,N′-bis-(1-hydroxymethylpropyl)ethylenediamine;* EMB. $C_{10}H_{24}N_2O_2$; mol wt 204.31. C 58.79%, H 11.84%, N 13.71%, O 15.66%. Prepn: Wilkinson *et al., J. Am. Chem. Soc.* **83**, 2212 (1961); *eidem, J. Med. Chem.* **5**, 835 (1962). Pharmacology: Kulig *et al., Diss. Pharm. Pharmacol.* **23**, 463 (1971), *C.A.* **76**, 81239d (1972). Comprehensive description: C. S. Lee, L. Z. Benet, *Anal. Profiles Drug Subs.* **7**, 231-249 (1978). Mechanism of action: W. H. Beggs in *Antibiotics* vol. **5**(pt. 1), F. E. Hahn, Ed. (Springer-Verlag, New York, 1979) pp 43-66.

mp 87.5-88.8°. $[\alpha]_D^{25}$ +13.7° (c = 2 in water). Sol in chloroform, methylene chloride; less sol in benzene; sparingly sol in water.

Dihydrochloride, $C_{10}H_{24}N_2O_2 \cdot 2HCl$, *Dexambutol, Ebutol, Etibi, Etapiam, Myambutol, Mycobutol, Sural, Tibutol.* mp 198.5-200.3°; also reported as mp 201.8-202.6°. $[\alpha]_D^{25}$ +7.6° (c = 2 in water). Sol in water, DMSO; sparingly sol in ethanol; difficultly sol in acetone, chloroform.

Mixture with isoniazid methane sulfonate, *(+)-2,2-(ethylenediimino)dibutanol diisoniazide methanesulfonate, isobutol, Isoetam.* Prepn: C. Ferrer-Salat *et al.,* **Belg.** pat. **753,-862;** U.S. pat. **3,718,655** (1971, 1973 both to Lab. Ferrer). White powder, crystallizes as needles, mp 121-122° (dec). Sol in water, methanol, *N,N*-dimethylformamide. Insol in ethanol, ethyl ether, chloroform. uv max: 265 nm. $[\alpha]_D^{20}$ +3° (c = 5% in water). LD_{50} in mice (g/kg): 2.800 orally; 2.210 i.p. (Ferrer-Salat).

THERAP CAT: Antibacterial (tuberculostatic).

3765. Ethamivan. *N,N-Diethyl-4-hydroxy-3-methoxy-benzamide; N,N-diethylvanillamide;* vanillic acid diethylamide; vanillic diethylamide; Vandid; Cardiovanil; Emivan. $C_{12}H_{17}NO_3$; mol wt 223.27. C 64.55%, H 7.67%, N 6.27%, O 21.50%. Prepd by dissolving vanillic acid *O*-acyl amide in NaOH and neutralizing with CO_2 or acid: Kratzl, Kvasnicka, *Monatsh.* **83**, 18 (1952); by refluxing a mixture of vanillic acid, P_2O_5 and powdered glass in xylene and treating with K_2CO_3: *eidem,* U.S. pat. **2,641,612** (1953 to Oesterr. Stickstoffwerke). Toxicity study: Caujolle *et al., Compt. Rend.* **243**, 609 (1956).

Needles from ligroin, mp 95-95.5°. LD_{50} i.p. in rats: 28 mg/kg (Caujolle).

THERAP CAT: CNS stimulant; respiratory stimulant.

3766. Ethamsylate. *2,5-Dihydroxybenzenesulfonic acid compd with N-ethylethanamine;* diethylammonium 2,5-dihydroxybenzenesulfonate; 1-hydroxy-4-oxo-2,5-cyclohexadiene-1-sulfonic acid compd with diethylamine; diethylammonium cyclohexadien-4-ol-1-one-4-sulfonate; cyclonamine; etamsylate; MD-141; E-141; Aglumin; Altodor; Biosinon; Dicynene; Dicynone. $C_{10}H_{17}NO_5S$; mol wt 263.31. C 45.61%, H 6.51%, N 5.32%, O 30.38%, S 12.18%. Prepn: **Brit.** pat. **895,709** (1962 to Labs. OM); **Span.** pat. **279,303** (1962 to Labs. Esteve), *C.A.* **60**, 1652g (1964). Pharmacology: Demars, *Med. Exp.* **4**, 173 (1961); Huguet *et al., Therapie* **24**, 429 (1969). Toxicity: Esteve *et al., Therapie* **15**, 110 (1960).

Crystals from ethanol, mp 125°. LD_{50} i.v. in mice, rats: 800, 1350 mg/kg (Esteve).

THERAP CAT: Hemostatic.

3767. Ethane. Bimethyl; dimethyl; methylmethane; ethyl hydride. C_2H_6; mol wt 30.07. C 79.89%, H 20.11%. CH_3CH_3. Constituent of natural gas (about 9%). Can be recovered from the gases produced during the distillation of crude petroleum. *Review:* Smith, Hanson, *Oil and Gas J.* **44**, no. 10, 119 (1945).

Colorless, odorless gas. *Flammable asphyxiant.* Burns with a faintly luminous flame. d_4^0 1.0493 (air = 1) or 1.3562 g/l. d_4^0 liq 0.446. mp −172°. bp −88°. Crit temp 32°; crit press. 48.2 atm. Heat of combustion: 1727 Btu/cu ft at 25° (one pound of CH_3CH_3 yields 20,420 Btu [net] at 15.56°). Flammable limits in air: 3.2 to 12.5% by vol. Ignition temp (in air): 530°. Soly in water at 20°: 4.7 ml/100 ml H_2O; in alc at 4°: 46 ml/100 ml alc.

Caution: Narcotic in high concns. A simple asphyxiant.

USE: In the manuf of chlorinated derivs; as refrigerant in some two-stage refrigeration systems where relatively low temps are produced; as fuel gas (so called "bottled gas" or "suburban propane" contains about 90% propane, 5% ethane, and 5% butane).

3768. Ethanearsonic Acid. *Ethylarsonic acid;* ethylarsinic acid. $C_2H_7AsO_3$; mol wt 154.00. C 15.60%, H 4.58%, As 48.65%, O 31.17%. $C_2H_5AsO(OH)_2$. Prepd from potassium arsenite and ethyl iodide: Dehn, *Am. Chem. J.* **33**, 131 (1905); from sodium arsenite and ethyl bromide: Quick, Adams, *J. Am. Chem. Soc.* **44**, 810 (1922).

Needles from alc, mp 99.5°. 70 g dissolve in 100 g water at 27°; 112 g dissolve in 100 g water at 40°; 39.4 g dissolve in 100 g of 95% alcohol at 25°.

Disodium salt, $C_2H_5AsNa_2O_3$, *Mon-Arsone.* Minute hygroscopic crystals, very sol in water.

THERAP CAT: Formerly as antisyphilitic.

3769. 1,2-Ethanedisulfonic Acid. 1,2-Ethylenedisulfonic acid. $C_2H_6O_6S_2$; mol wt 190.20. C 12.63%, H 3.18%, O 50.47%, S 33.72%. $HO_3SCH_2CH_2SO_3H$. Preparation of disodium salt from 1,2-dibromoethane and sodium sulfite: McElvain *et al., J. Am. Chem. Soc.* **67**, 1578 (1945); Ohba *et al.,* U.S. pat. **3,022,172** (1962 to Fuji Photo Film). Free acid obtained by conversion of disodium salt to barium salt followed by liberation with H_2SO_4: McElvain *et al., loc. cit.* Review of prepns: Suter, *Organic Chemistry of Sulfur* (Wiley, New York, 1944) p 166; Gilbert, *Sulfonation and Related Reactions* (Interscience, New York, 1965).

Dihydrate, crystals from acetic acid + acetic anhydride, mp 111-112°. Heating at 145° and 1 mm for 6 hours gives anhydr compd, mp 172-174°. Somewhat soluble in anhydr ether, very sol in dioxane.

Sodium salt dihydrate, $C_2H_4Na_2O_6S_2.2H_2O$, monoclinic prisms. Not very readily sol in water. Practically insol in alcohol.

USE: Base in hypnotics, sedatives.

3770. 1,2-Ethanedithiol. Dithioethyleneglycol; ethylenedimercaptan. $C_2H_6S_2$; mol wt 94.20. C 25.50%, H 6.42%, S 68.08%. $HSCH_2CH_2SH$. Prepd by reacting ethanol, thiourea and ethylene dibromide and subsequent alkaline hydrolysis of the ethylenediisothiuronium bromide: Speciale, *Org. Syn.* **30**, 35 (1950), modified procedure: Grogan *et al.*, *J. Org. Chem.* **20**, 50 (1955).

Liquid. $d^{23.5}$ 1.123. bp_{760} 146°; bp_{150} 76-81°; bp_{46} 63°; bp_{24} 51-52°. n_D^{20} 1.5589. Freely sol in alcohol and in alkalies.

Caution: Vapors of ethanedithiol may cause severe headache and nausea.

3771. Ethanethiol. Ethyl mercaptan; mercaptoethane; ethyl sulfhydrate; thioethyl alcohol. C_2H_6S; mol wt 62.14. C 38.66%, H 9.73%, S 51.61%. CH_3CH_2SH. Found in urine of rabbits after ingestion of cabbage. Is formed in vinous fermentation. Occurs in illuminating gas, in "sour" natural gas of W. Texas; in petroleum distillates from which it may be separated by chemical or physical methods: Thompson *et al.*, *Anal. Chem.* **27**, 175 (1955). Prepn from sodium ethylsulfate and KSH: Klason, *Ber.* **20**, 3407 (1887); catalytically from ethanol and hydrogen sulfide: Kramer, Reid, *J. Am. Chem. Soc.* **43**, 880 (1921). Review on occurrence, prepn, properties and reactions: E. Emmet Reid, *Organic Chemistry of Bivalent Sulfur* **vol. I**, 15-261 (Chemical Publishing Co., New York, 1958).

Liquid; penetrating leek-like odor: minimum detectable concn = 1 part in 50 billion parts of air. d_4^0 0.8617; d_4^{25} 0.83147. bp 34.7-35.04°; bp_{400} 17.7°; bp_{200} 1.5°; bp_{100} —13.0°; bp_{10} —50.2°; bp_1 —76.1°. mp —147.97 to —144.4°. n_D^{20} 1.431; n_D^{25} 1.420. Crit temp 225.5°. Crit pressure 54.2 atm. Explosive limits 2.8 and 18.2% by vol of vapor. Min ignition temp in air 299°, in oxygen 261°. Entropy at 25°: 70.77 cal/deg/mole. Heat capacity at 25°: 17.37 cal/deg/mole. Surface tension at 2°: 23.63 dynes/cm; at 16.7°: 21.62 dynes/cm. Viscosity at 20°: 0.003155 g/cm sec. Azeotrope with *n*-pentane (51% ethanethiol) bp 30.46°; with ether (40% ethanethiol) bp 31.50°. Soly in water at 20°: 6.76 g/l or 0.112 moles/l. Sol in alcohol, ether. *Keep tightly closed and in a cool place.*

Octadecahydrate, needles. Practically insol in water, ethanethiol.

Sodium salt, C_2H_5NaS, hydrolyzes instantly in water.

Caution: Potential symptoms of overexposure to ethanethiol are headache, nausea and irritation of mucous membranes. *See NIOSH Pocket Guide to Chemical Hazards* (DHHS/NIOSH 90-117, 1990) p 112.

USE: Odorant for natural gas; intermediate and starting material in manuf of plastics, insecticides, antioxidants.

3772. Ethanolamine. *2-Aminoethanol;* monoethanolamine; *β*-aminoethyl alcohol; 2-hydroxyethylamine; *β*-hydroxyethylamine; ethylolamine; colamine. C_2H_7NO; mol wt 61.08. C 39.33%, H 11.55%, N 22.93%, O 26.19%. $HO-CH_2CH_2NH_2$. Prepd on a large scale by ammonolysis of ethylene oxide: Knorr, *Ber.* **30**, 909 (1897); **Fr.** pat. **650,574** (1928 to I. G. Farben); Reid, Lewis, **U.S.** pat. **1,904,013** (1933 to Carbide & Carbon); Schwoegler, Olin, **U.S.** pat. **2,373,199** (1945 to Sharples). Also from nitromethane and formaldehyde: *Ullmann's Encyklopädie der technischen Chemie* **3**, 102 (3rd ed., 1953). Manuf: Faith, Keyes & Clark's *Industrial Chemicals,* F. A. Lowenheim, M. K. Moran, Eds. (Wiley-Interscience, New York, 4th ed., 1975) pp 339-344. Toxicity: H. F. Smyth *et al.*, *J. Ind. Hyg. Toxicol.* **23**, 259 (1941).

Viscous, hygroscopic liq. Ammoniacal odor. Absorbs CO_2. d_4^{25} 1.0117; d_4^{40} 0.9998; d_4^{60} 0.9844. One gallon weighs 8.45 lbs in the U.S.A. Viscosity at 25°: 18.95 cps; at 60°: 5.03 cps. mp 10.3°. bp_{760} 170.8°; bp_{12} 70-72°. Strong base. pKa at 25°: 9.4. pH of 25% aq soln: 12.1; of 0.1*N* aq soln: 12.05. n_D^{20} 1.4539. Dipole moment 2.27. Flash pt 195°F. Misc with water, methanol, acetone. Soly at 25°: benzene 1.4%; ether 2.1%; carbon tetrachloride 0.2%; *n*-heptane <0.1%. LD_{50} orally in rats: 10.20 g/kg (Smyth).

Hydrochloride, $C_2H_7NO.HCl$, deliquesc crystals from alc, mp 75-77°.

Oleate, $C_2H_7NO.C_{18}H_{34}O_2$, *Antivariz, Esclerosina, Ethamolin.* Use as a sclerosing agent: S. E. Hedberg *et al.*, *Am. J. Surgery* **143**, 426 (1982).

Caution: Potential symptoms of overexposure to ethanolamine are respiratory system, skin and eye irritation; lethargy. *See NIOSH Pocket Guide to Chemical Hazards* (DHHS/NIOSH 90-171, 1990) p 104.

USE: To remove CO_2 and H_2S from natural gas and other gases; in the synthesis of surface active agents; in polishes, hair waving solns, in emulsifiers; as softening agent for hides; dispersing agent for agricultural chemicals. Is reacted with other substances to form an accelerator in the manuf of antibiotics. Pharmaceutic aid (surfactant).

THERAP CAT: Oleate as sclerosing agent.

3773. Ethaverine. *1-[(3,4-Diethoxyphenyl)methyl]-6,7-diethoxyisoquinoline;* 6,7-diethoxy-1-(3,4-diethoxybenzyl)-isoquinoline; ethylpapaverine; Dyscural. $C_{24}H_{29}NO_4$; mol wt 395.50. C 72.89%, H 7.39%, N 3.54%, O 16.18%. Tetraethyl homolog of papaverine. Synthesis: Wolf, **U.S.** pat. **1,962,224** (1934).

Crystals from alcohol + ether, mp 99-101°. Insol in water. Readily sol in hot alc; slightly in ether and chloroform.

Hydrochloride, $C_{24}H_{29}NO_4.HCl$, *Barbonin, Circubid, Diquinol, Ethabid, Isovex, Laverin, Perparin, Perperine.* Crystals from abs alc, mp 186-188° (dec). One gram dissolves in about 40 ml water. pH of 1% soln 3.6; of 0.1% soln 4.6. Prepn of stable solns: Hereld, **U.S.** pat. **2,971,888** (1961).

Acid oxalate, $C_{24}H_{29}NO_4.(COOH)_2$, crystals from alc, soluble in hot water.

Salt with 7-iodo-8-hydroxyquinoline-5-sulfonic acid, **Hung.** pat. **106,906** (1931); v. Issekutz *et al.*, *Arch. Exp. Pathol. Pharmakol.* **164**, 158 (1932).

THERAP CAT: Antispasmodic.

3774. Ethchlorvynol. *1-Chloro-3-ethyl-1-penten-4-yn-3-ol;* 5-chloro-3-ethylpent-1-yn-4-en-3-ol; ethyl *β*-chlorovinyl ethynyl carbinol; ethclorvynol; Placidyl; Arvynol; Serenesil; Roeridorm; Normoson. C_7H_9ClO; mol wt 144.60. C 58.14%, H 6.27%, Cl 24.52%, O 11.06%. Prepn: W. M. McLamore *et al.*, *J. Org. Chem.* **20**, 109 (1955); Bavley, McLamore, **U.S.** pat. **2,746,900** (1956 to Pfizer). Pharmacological studies: S. Y. P'an *et al.*, *J. Pharmacol. Exp. Ther.* **114**, 326 (1955). Pharmacodynamics: Y. C. Martin, *Biochem. Pharmacol.* **16**, 2041 (1967). Evaluation as hypnotic: J. W. Middleton *et al.*, *Curr. Ther. Res.* **8**, 391 (1966); J. H. Pattison *et al.*, *J. Am. Geriatric Soc.* **20**, 398 (1972). Study of metabolites: J. P. Horwitz *et al.*, *Drug Metab. Dispos.* **8**, 77 (1980).

Liquid, pungent aromatic odor. Slowly darkens on exposure to light and air. d_4^{25} 1.065-1.070 (also reported as d_4^{20} 1.070). bp_{760} 173-174° (also reported as bp_{760} 181°); $bp_{0.1}$ 28.5-30°. n_D^{25} 1.4675-1.4800. Immiscible with water; miscible with most organic solvents. LD_{50} in mice (mg/kg): 290 orally; 240 s.c. (P'an).

Note: This is a controlled substance (depressant) listed in the U.S. Code of Federal Regulations, Title 21 Part 1308.14 (1995).

THERAP CAT: Sedative; hypnotic.

3775. Ethebenecid. *4-[(Diethylamino)sulfonyl]benzoic acid; p-(diethylsulfamoyl)benzoic acid; p-carboxy-N,N-dieth-*

ylbenzenesulfonamide; *p*-carboxybenzenesulfo-di-*N*-ethyl-amide; Antidipsin; Longacid; Urelim. $C_{11}H_{15}NO_4S$; mol wt 257.31. C 51.35%, H 5.88%, N 5.44%, O 24.87%, S 12.46%. Prepn: Gilman, Arntzen, *J. Am. Chem. Soc.* **69**, 1537 (1947); Goldberg, Wragg, *J. Chem. Soc.* **1960**, 1408.

Crystals, mp 192-194° (with turbidity).

THERAP CAT: Uricosuric agent; inhibits penicillin excretion.

3776. Ethenzamide. *2-Ethoxybenzamide;* ethbenzamide; 2-ethoxybenzenecarbonamide; salicylamide *o*-ethyl ether; Lucamide; Protopyrin; Trancalgyl. $C_9H_{11}NO_2$; mol wt 165.19. C 65.44%, H 6.71%, N 8.48%, O 19.37%. Prepn: **Brit.** pat. **656,746** (1951 to H. Lundbeck); Shapiro *et al.*, *J. Am. Chem. Soc.* **81**, 3728 (1959); Bryk, Osowski, **Pol.** pat. **47,822** (1961 to Polfa), *C.A.* **61**, 5573f (1964). Metabolism: Davison *et al.*, *J. Pharmacol. Exp. Ther.* **136**, 226 (1962). Toxicity data: Starmer *et al.*, *Toxicol. Appl. Pharmacol.* **19**, 20 (1971).

Crystals from ethyl acetate + hexane, mp 132-134°. LD_{50} orally in mice: 1160 mg/kg (Starmer).

THERAP CAT: Analgesic.

3777. Ethephon. *(2-Chloroethyl)phosphonic acid;* 2-chloroethanephosphonic acid; CEPA; camposan; CEPHA; Cerone; Florel; Ethrel. $C_2H_6ClO_3P$; mol wt 144.49. C 16.62%, H 4.19%, Cl 24.54%, O 33.22%, P 21.44%. $ClCH_2$-$CH_2PO(OH)_2$. Ethylene-releasing agent used to control growth and ripening of fruit crops. Prepn: Kabachnik, Rossiiskaya, *Izvest. Akad. Nauk S.S.S.R., Otdel. Khim. Nauk* **1946**, 403-410, *C.A.* **42**, 7242e (1948); *Beilstein* **EIII 4**, 1780. Toxicity data: G. Hennighausen *et al.*, *Pharmazie* **32**, 181 (1977). Review of use in viticulture: E. Szyjewicz *et al.*, *Am. J. Enol. Viticult.* **35**, 117-123 (1984).

Very hygroscopic needles from benzene (must be dried over P_2O_5). mp 74-75°. Freely sol in water, methanol, acetone, ethylene glycol, propylene glycol. Slightly sol in benzene, toluene. Practically insol in petr ether. Aq solns are stable below pH 3.5. Above pH 3.5 hydrolysis begins with the release of free ethylene. LD_{50} orally in mice: 2850 mg/kg (Hennighausen).

Caution: Spray formulations are quite acidic, about pH 1.0. May be irritating to exposed skin and eyes, or if inhaled.

USE: Plant growth regulator.

3778. Ethiazide. *6-Chloro-3-ethyl-3,4-dihydro-2H-1,2,-4-benzothiadiazine-7-sulfonamide 1,1-dioxide;* 6-chloro-3-ethyl-3,4-dihydro-7-sulfamoylbenzo-1,2,4-thiadiazine 1,1-dioxide; 3-ethyl-3,4-dihydro-6-chloro-7-sulfamyl-1,2,4-benzothiadiazine 1,1-dioxide; acthiazidum. $C_9H_{12}ClN_3O_4S_2$; mol wt 325.80. C 33.18%, H 3.71%, Cl 10.88%, N 12.90%, O 19.64%, S 19.68%. Prepn: Close *et al.*, *J. Am. Chem. Soc.* **82**, 1132 (1960); Topliss *et al.*, *J. Org. Chem.* **26**, 3842 (1961); **Brit.** pat. **861,367** (1961 to Ciba).

Crystals from tetrahydrofuran + chloroform, mp 269-270°.

Component of *Hypertane.*

THERAP CAT: Diuretic.

3779. Ethinamate. *1-Ethynylcyclohexanol carbamate; carbamic acid 1-ethynylcyclohexyl ester;* ethynylcyclohexyl carbamate; Valamin; Valmid; Valmidate. $C_9H_{13}NO_2$; mol wt 167.21. C 64.65%, H 7.84%, N 8.38%, O 19.14%. Prepn: Junkmann, Pfeiffer, U.S. pat. **2,816,910** (1957 to Schering, AG).

Needles from cyclohexane, mp 96-98°. bp_3 118-122°. Shows no optical activity. $n_D^{21.5}$ 1.4441 (20% in propylene glycol). Soly at 25° in water 0.25%, ethanol 35.0%, hexane 2.0%, sesame oil 0.7%, 1,2-propylene glycol 22%. Stable in N H_2SO_4 at 30°, but dec at 40° and in 0.02N acid at 60°. In 0.01N NaOH, CO_2 is split off at 30°. Lowers the surface tension in satd aq soln at 22° by 21%. The partition coefficient in oil/water is 1.48 (that of ethyl carbamate is 0.14).

Note: The trademark Valamin was used formerly to indicate a brand of *tert*-amyl isovalerate.

Note: This is a controlled substance (depressant) listed in the U.S. Code of Federal Regulations, Title 21 Part 1308.14 (1995).

THERAP CAT: Sedative; hypnotic.

3780. Ethinyl Estradiol. *(17α)-19-Norpregna-1,3,5(10)-trien-20-yne-3,17-diol;* 17α-ethynyl-1,3,5(10)-estratriene-3,17β-diol; 17-ethinylestradiol; ethynylestradiol; Diogyn E; Estigyn; Estinyl; Ethy 11; Eticylol; Eticyclin; Etivex; Feminone; Gynolett; Kolpolyn; Linoral; Lynoral; Oradiol; Orestralyn; Primogyn C; Progynon C; Progynon M. $C_{20}H_{24}O_2$; mol wt 296.41. C 81.04%, H 8.16%, O 10.80%. Synthetic steroid with high oral estrogenic potency: Inhoffen, Hohlweg, *Naturwiss.* **26**, 96 (1938). Prepn from estrone: Inhoffen *et al.*, *Ber.* **71**, 1024 (1938). *See also* **Ger.** pat. **702,063**; **Brit.** pat. **516,444**; U.S. pats. **2,243,887; 2,251,939; 2,265,-976; 2,267,257.** Properties: Petit, Muller, *Bull. Soc. Chim. France* **1951**, 121; L. Ehmann, A. Wettstein, *Pharm. Acta Helv.* **25**, 297 (1950). NMR: Hampel, Kraemer, *Ber.* **98**, 3255 (1965). Toxicity: E. I. Goldenthal, *Toxicol. Appl. Pharmacol.* **18**, 185 (1971). Randomized double-blind clinical studies: S. Koetsawang *et al.*, *Contraception* **25**, 231 (1982); A. Sheth *et al.*, *ibid.* 243. Clinical evaluation in gonadal dysgenesis: L. Cuttler *et al.*, *J. Clin. Endocrinol. Metab.* **60**, 1087 (1985). General review: K. W. Thompson, *J. Clin. Pharmacol.* **8**, 1088-1098 (1948). Review of metabolism and pharmacokinetics: K. Fotherby, *Methods Find. Exp. Clin. Pharmacol.* **4**, 133-141 (1982); of carcinogenicity studies: *IARC Monographs* **21**, 233-255 (1979); *ibid.* Suppl. 4, 186-188 (1982).

Hemihydrate, fine needles from methanol + water, mp 141-146°, $[\alpha]_D^{25}$ 0° ± 1° (dioxane). Dehydrates after melting and further heating, mp 182-184°. $[\alpha]_D^{24}$ 3.5° ± 0.5° (c = 2 in dioxane); −29.5° ± 1° (c = 2 in pyridine). uv max (ethanol): 281 nm (ε 2040 ± 60). Practically insol in water. Soly: 1 part in 6 of ethanol, 1 in 4 of ether, 1 in 5 of acetone, 1 in 4 of dioxane, and 1 in 20 of chloroform. Sol in vegetable oils, and in solns of fixed alkali hydroxides. LD_{50} in rats, mice (mg/kg): 2952, 1737 orally (Goldenthal).

3-Acetate, $C_{22}H_{26}O_3$, crystals, mp 152-153°. $[\alpha]_D^{20}$ +3° (chloroform).

3-Benzoate, $C_{27}H_{28}O_3$, needles from methanol, mp 200-202°.

Note: Also used in combination with chlormadinone acetate, desogestrel, ethynodiol, gestodene, lynestrenol, norethindrone or norgestrel, *q.q.v.* Has been used in combination with dimethisterone, medroxyprogesterone, or megestrol acetate, *q.q.v.*

Note: Ethinyl estradiol may reasonably be anticipated to be a carcinogen: *Seventh Annual Report on Carcinogens* (PB95-109781, 1994) p 199.

THERAP CAT: Estrogen. In combination with progestogen as oral contraceptive.

THERAP CAT (VET): Estrogen.

3781. Ethiodized Oil. Ethyl ester of iodinated fatty acid of poppyseed oil; Ethiodol. Contains 37% organically bound iodine. Ref: Hom *et al., J. Am. Pharm. Assoc. Sci. Ed.* **46,** 254 (1957). Use in lymphography: R. M. Paxton *et al., Brit. Med. J.* **I,** 120 (1975).

Amber-colored oil, extremely fluid, sp gr 1.28. Soluble in acetone, chloroform, ether, hexane. Insol in water.

THERAP CAT: Diagnostic aid (radiopaque medium). Antineoplastic when part of the iodine is ^{131}I.

3782. Ethion. *Phosphorodithioic acid S,S'-methylene O,O,O',O'-tetraethyl ester; ethyl methylene phosphorodithioate; O,O,O',O'-tetraethyl S,S'-methylenediphosphorodithioate; bis[S-(diethoxyphosphinothioyl)mercapto]methane;* diethion; ENT-24105; FMC-1240; Niagara 1240; Nialate. $C_9H_{22}O_4P_2S_4$; mol wt 384.48. C 28.12%, H 5.77%, O 16.65%, P 16.11%, S 33.36%. Prepn: Willard, Henahan, U.S. pats. **2,873,228; 3,014,058** (1959, 1961 both to FMC). Toxicity study: T. B. Gaines, *Toxicol. Appl. Pharmacol.* **14,** 515 (1969).

Liquid, mp −12 to −13°. d_4^{20} 1.220. n_D^{20} 1.5490. Vapor pressure: 1.5×10^{-6} mm Hg. Sol in xylene, methylated naphthalene, chloroform, acetone and kerosene + 1% methyl ethyl ketone or benzene; slightly sol in water. LD_{50} in female, male rats (mg/kg): 27, 65 orally; 62, 245 dermally (Gaines).

Caution: Cholinesterase inhibitor.

USE: Insecticide; acaricide.

3783. Ethionamide. *2-Ethyl-4-pyridinecarbothioamide; 2-ethylthioisonicotinamide;* 3-ethylisothionicotinamide; 2-ethylisothionicotinamide; 2-ethyl-4-thiocarbamoylpyridine; α-ethylisonicotinoylthioamide; amidazine; ethioniamide; Bayer 5312; 1314-Th; Nisotin; Trescatyl; Aetina; Ethimide; Iridocin; Tio-Mid. $C_8H_{10}N_2S$; mol wt 166.25. C 57.80%, H 6.06%, N 16.85%, S 19.29%. Prepn: Libermann *et al., Compt. Rend.* **242,** 2409 (1956); *Bull. Soc. Chim. France* **1958,** 687; **Brit. pat. 800,250** (1958 to Chimie et Atomistique). Mechanism of action study: K. Johnsson *et al., J. Am. Chem. Soc.* **117,** 5009 (1995).

Minute yellow crystals from ethanol, dec 164-166°. Very sparingly sol in water, ether. Sparingly sol in methanol, ethanol, propylene glycol. Sol in hot acetone, dichloroethane. Freely sol in pyridine.

THERAP CAT: Antibacterial (tuberculostatic).

3784. Ethionine. *S-Ethyl-L-homocysteine; 2-amino-4-(ethylthio)butyric acid;* α-amino-γ-(ethylmercapto)butyric acid; homocysteine *S*-ethyl ether. $C_6H_{13}NO_2S$; mol wt 163.24. C 44.15%, H 8.03%, N 8.58%, O 19.60%, S 19.64%. $CH_3CH_2SCH_2CH_2CH(NH_2)COOH$. Prepn from ethanethiol and acrolein, purification via Zn salt: Norton, U.S. pat. **2,840,587** (1958 to Dow). Prepn of DL-form and L-

form: Armstrong, Lewis, *J. Org. Chem.* **16,** 749 (1951). Resolution of D- and L-isomers: Greenstein *et al., J. Biol. Chem.* **204,** 307 (1953). *Review:* E. Farber, "Ethionine Carcinogenesis" in A. Haddow, S. Weinhouse, *Advan. Cancer Res.* vol. 7 (Academic Press, New York, 1963) pp 383-474.

Crystals, dec 272-274°. $[\alpha]_D^{24}$ +25.1°. Also reported as $[\alpha]_D^{23}$ +20.1° (1*N* HCl) (Armstrong, Lewis).

DL-Form, crystals, dec 257-260°. Also reported as dec 272-284° (Armstrong, Lewis).

3785. Ethiozin. *4-Amino-6-(1,1-dimethylethyl)-3-(ethylthio)-1,2,4-triazin-5(4H)-one;* 3-ethylthio-4-amino-6-*tert*-butyl-1,2,4-triazine-5-one; ethyl metribuzin; SMY-1500; BAY SMY 1500; Tycor. $C_9H_{16}N_4OS$; mol wt 228.32. C 47.35%, H 7.06%, N 24.54%, O 7.01%, S 14.04%. Selective pre- and post-emergence herbicide; ethylthio analog of metribuzin, *q.v.* Prepn: **Belg. pat. 697,083;** K. Westphal *et al.,* U.S. pat. **3,671,523** (1967, 1972 both to Bayer AG). Improved process: T. Schmidt, **Ger. pat. 3,339,858;** *idem,* U.S. pat. **4,546,178** (both 1985 to Bayer AG). Physicochemical properties, toxicity and herbicidal activity: H. Hack *et al., Proc. Brit. Crop Prot. Conf. - Weeds* **1985,** 35. Field trials in winter wheat: B. Bolton *et al., ibid.* 915; R. Ratliff, T. F. Peeper, *Weed Technol.* **1,** 235 (1987).

Crystalline solid, mp 95°. Soly at 20°: 420 mg/l in water; 2500 mg/l in hexane; >50% in dichloromethane; >25% in toluol; >15% in 2-propanol. Vapor pressure at 20°: 1×10^{-3} Pa. LD_{50} in male, female rats (mg/kg): 2470, 1280 orally (Hack).

USE: Herbicide.

3786. Ethirimol. *5-Butyl-2-(ethylamino)-6-methyl-4(1H)-pyrimidinone;* 2-ethylamino-4-methyl-5-*n*-butyl-6-hydroxypyrimidine; PP-149; Milstem; Milgo; Milcurb Super. $C_{11}H_{19}N_3O$; mol wt 209.29. C 63.13%, H 9.15%, N 20.08%, O 7.64%. Prepn: B. K. Snell *et al.,* **S. Afr. pat. 67 01,373;** *eidem,* U.S. pat. **3,980,781** (1968, 1976 both to ICI). Systemic fungicidal activity: R. M. Bebbington, *Chem. & Ind. (London)* **1969,** 1512; K. J. Bent, *Ann. Appl. Biol.* **66,** 103 (1970). Metabolism and mode of action: A. Calderbank, *Acta Phytopathol.* **6,** 355 (1971).

Crystalline solid, mp 159-160°. Vapor pressure at 25°: 2×10^{-6} mm Hg. Soly in water at 25°: 200 mg/l. Sol in chloroform, trichloroethylene, aq solns of strong acids and bases. Slightly sol in ethanol; sparingly sol in acetone. LD_{50} orally in rats: 4000 mg/kg (Bebbington).

USE: Fungicide.

3787. Ethisterone. *17α-Hydroxypregn-4-en-20-yn-3-one;* 17α-ethynyltestosterone; 17α-ethynyl-17β-hydroxy-4-androsten-3-one; 17α-ethinyltestosterone; 17α-ethynyl-4-androsten-17β-ol-3-one; anhydrohydroxyprogesterone; pregneninolone; Lutocyclin; Lutocylol; Ora-Lutin; Progestoral; Pranone; Syngestrotabs; Trosinone. $C_{21}H_{28}O_2$; mol wt 312.45. C 80.73%, H 9.03%, O 10.24%. Prepd by addition of acetylene to dehydroepiandrosterone followed by an Oppenauer oxidation: Inhoffen *et al., Ber.* **71,** 1024 (1938).

Crystals from ethyl acetate, mp 269-275°. Sublimes in high vacuum at 190-195°. $[\alpha]_D^{23}$ +23.8° (dioxane); $[\alpha]_D^{25}$ −32.0° (pyridine). uv max (methanol): 241 nm ($E_{1cm}^{1\%}$ 513). Practically insol in water. Slightly sol in alcohol, acetone, ether, chloroform, vegetable oils.

THERAP CAT: Progestogen.

3788. Ethofumesate. *2-Ethoxy-2,3-dihydro-3,3-dimethyl-5-benzofuranol methanesulfonate;* NC-8438; Nortron; Tramat. $C_{13}H_{18}O_5S$; mol wt 286.35. C 54.53%, H 6.34%, O 27.94%, S 11.20%. Prepn: Gates *et al.*, U.S. pat. 3,689,507 (1972 to Fisons).

White crystalline solid, mp 70-72°. Vapor press at 25°: 6.45 × 10⁻⁷ mm Hg. Soly in water at 25°: 110 mg/l. Soly (g/kg): ethanol 100; acetone 400; hexane 4. Stable to hydrolysis in water at pH 7. LD_{50} orally in rats: 6400 mg/kg, *RTECS* Vol. II, R. J. Lewis, R. L. Tatken, Eds. (1979) p 50.

USE: Herbicide.

3789. Ethoheptazine. *Hexahydro-1-methyl-4-phenyl-1H-azepine-4-carboxylic acid ethyl ester;* 4-carbethoxy-1-methyl-4-phenylhexamethylenimine; 4-carbethoxy-1-methyl-4-phenylazacycloheptane; ethyl heptazine; Wy-401. $C_{16}H_{23}NO_2$; mol wt 261.36. C 73.53%, H 8.87%, N 5.36%, O 12.24%. Prepd in several steps from 2-phenyl-4-dimethylaminobutyronitrile as a starting material, the final step being the hydrolysis and esterification of 1-methyl-4-phenyl-4-cyanoazacycloheptane: Diamond *et al.*, *J. Org. Chem.* **22**, 399 (1957); Diamond, Bruce, U.S. pat. 2,666,050 (1954 to Am. Home Prod.).

Liquid. d_4^{26} 1.038. bp_1 133-134°; $bp_{0.5}$ 127-129°; $bp_{0.3}$ 128-130°. n_D^{26} 1.5210; n_D^{28} 1.5220.

Hydrochloride, $C_{16}H_{23}NO_2 \cdot HCl$, recrystallized from ethyl acetate + ether, mp 151-153°.

Citrate, $C_{16}H_{23}NO_2 \cdot C_6H_8O_7$, *Zactane.*

Methobromide, $C_{16}H_{23}NO_2 \cdot CH_3Br$, crystals from ether, mp 215-217°.

THERAP CAT: Analgesic.

3790. Ethohexadiol. 2-Ethyl-1,3-hexanediol; 2-ethyl-3-propyl-1,3-propanediol; octylene glycol; Rutgers 612; 6-12. $C_8H_{18}O_2$; mol wt 146.23. C 65.71%, H 12.41%, O 21.88%. Prepn: Kulpinsky, Nord, *J. Org. Chem.* **8**, 256 (1943); B. G. Wilkes, U.S. pat. 2,407,205 (1946 to Carbide & Carbon Chemicals). Activity: P. Granett, H. L. Haynes, *J. Econ. Entomol.* **38**, 671 (1945); W. V. King, *ibid.* **44**, 339 (1951); B. V. Travis, C. N. Smith, *ibid.* 428.

Slightly oily liq. d_{20}^{20} 0.9422. d_{22}^{22} 0.9325. bp_{760} 244.2°; bp_{50} 163°; bp_{10} 129°; bp_3 102°; $bp_{0.5}$ 94-96°. n_D^{22} 1.4530. Vapor pressure at 20° <0.01 mm. Abs viscosity: 271 cps. Soly in water 0.6% w/w; soly of water in 2-ethyl-1,3-hexanediol 10.8% w/w; sol in ethanol, isopropanol, propylene glycol, castor oil. LD_{50} orally in rats, mice: 1400, 1900 mg/kg, *RTECS* Vol. I, R. J. Lewis, R. L. Tatken, Eds. (1979) p 742.

Caution: Moderately irritating to eyes, mucous membranes, but not to skin. Ingestion causes CNS depression.

USE: Insect repellent.

3791. Ethopabate. *4-Acetamido-2-ethoxybenzoic acid methyl ester;* methyl 4-acetamido-2-ethoxybenzoate; 2-ethoxy-4-acetamidobenzoic acid methyl ester; ethyl pabate. $C_{12}H_{15}NO_4$; mol wt 237.26. C 60.75%, H 6.37%, N 5.90%, O 26.97%. Prepn: Grimme, Schmitz, *Ber.* **87**, 179 (1954); Rogers, Clark, Belg. pat. 613,166 and U.S. pat. 3,211,610 (1962 and 1965 to Merck & Co.); Fr. pat. 1,407,055 corresp to U.S. pat. 3,357,978, and Thominet, Brit. pat. 1,019,781 (1965, 1967 and 1966 to Soc. d'Etudes Sci. et Ind. de l'Ile de France). Anticoccidial activity studies: Rogers *et al.*, *Proc. Soc. Exp. Biol. Med.* **117**, 488 (1964). Metabolism in chickens: Buhs *et al.*, *J. Pharmacol. Exp. Ther.* **154**, 357 (1966).

White to pinkish-white, practically odorless crystals from methanol and water, mp 148-149°. uv max (methanol): 298, 267 nm ($A_{1cm}^{1\%}$ 805, 365). Sol in methanol, ethanol, acetone, acetonitrile. Sparingly sol in isopropanol, *p*-dioxane, ethyl acetate, methylene chloride; very slightly sol in water and isooctane.

Mixture with amprolium, *Amprol Plus.*

THERAP CAT (VET): Mixture as coccidiostat.

3792. Ethoprop. *Phosphorodithioic acid O-ethyl S,S-dipropyl ester;* O-ethyl S,S-dipropylphosphorodithioate; ethoprophos; VC9-104; Mocap. $C_8H_{19}O_2PS_2$; mol wt 242.34. C 39.65%, H 7.90%, O 13.20%, P 12.78%, S 26.46%. Prepn: L. E. Goyette, U.S. pat. 3,112,244 (1963 to Virginia-Carolina Chem. Corp.); J. H. Wilson, Jr., U.S. pat. 3,268,393 (1966 to Mobil). Activity: C. R. Harris, J. L. Hitchon, *J. Econ. Entomol.* **63**, 2 (1970). Metabolism: Z. M. Iqbal, R. E. Menzer, *Biochem. Pharmacol.* **21**, 1569 (1972). Soil degradation: J. H. Smelt *et al.*, *Pestic. Sci.* **8**, 147 (1977).

Pale yellow oil, $bp_{0.2}$ 86-91°. d_4^{20} 1.094. Vapor press. at 26°: 3.5 × 10⁻⁴ mm Hg. Soly in water: 750 mg/l. Sol in most organic solvents.

USE: Insecticide; nematocide.

3793. Ethopropazine. *N,N-Diethyl-α-methyl-10H-phenothiazine-10-ethanamine;* 10-(2-diethylaminopropyl)-phenothiazine; 10-(2-diethylamino-2-methylethyl)phenothiazine; 2-diethylamino-1-propyl-N-dibenzoparathiazine; phenopropazine; profenamine; RP-3356; W-483; Isothazine; Isothiazine; Parkin. $C_{19}H_{24}N_2S$; mol wt 312.48. C 73.03%, H 7.74%, N 8.96%, S 10.26%. Prepd from Grignard complexes of diethylaminopropyl halide and phenothiazine: Berg, Ashley, **U.S. pat. 2,607,773** (1952 to Rhône-Poulenc).

Crystals, mp 53-55°. Usually obtained as an oil because of contamination with 10-(2-diethylamino-1-methylethyl)-phenothiazine. LD_{50} orally in mice: 650 mg/kg.

Hydrochloride, $C_{19}H_{24}N_2S.HCl$, *Dibutil, Lysivane, Pardisol, Parphezin, Parphezhin, Parsidol, Parsitan, Parsotil, Rodipal.* Crystals from ethylene dichloride, mp 223-225° (some decompn). Lower melting points reported are caused by admixture with 10-(2-diethylamino-1-methylethyl)phenothiazine-HCl which melts at 166-168°. One gram dissolves in 400 ml water at 20°, in 20 ml water at 40°. Sol in ethanol, chloroform. Soly in abs ethanol at 25° = 1.0 g/30 ml. Sparingly sol in acetone. Practically insol in ether, benzene. pH of a 5% aq soln is about 5.8.

THERAP CAT: Antiparkinsonian; anticholinergic.

3794. Ethosuximide. *3-Ethyl-3-methyl-2,5-pyrrolidinedione; 2-ethyl-2-methylsuccinimide;* α-ethyl-α-methylsuccinimide; *3-methyl-3-ethylpyrrolidine-2,5-dione;* Atysmal; Capitus; Emeside; Epileo Petitmal; Ethymal; Mesentol; Pemal; Peptinimid; Petinimid; Petnidan; Pyknolepsinum; Simatin; Succimal; Suxilep; Suximal; Suxinutin; Zarontin. $C_7H_{11}NO_2$; mol wt 141.17. C 59.56%, H 7.85%, N 9.92%, O 22.67%. Prepn: Sircar, *J. Chem. Soc.* **1927,** 1252. Acute toxicity data: H. Najer *et al., Bull. Soc. Chim. France* **3,** 1119 (1966). Comparative clinical trial with valproic acid, *q.v.,* in epilepsy: S. Sato *et al., Neurology* **32,** 157 (1982). Brief review: S. J. Wallace, *Neurol. Clin.* **4,** 601 (1986).

Crystals from acetone + ether, mp 64-65°. Freely sol in water. LD_{50} in mice (g/kg): 1.65 i.p.; 1.75 orally (Najer).

THERAP CAT: Anticonvulsant.

3795. Ethotoin. *3-Ethyl-5-phenyl-2,4-imidazolidinedione; 3-ethyl-5-phenylhydantoin;* 1-ethyl-2,5-dioxo-4-phenyl-imidazolidine; Peganone. $C_{11}H_{12}N_2O_2$; mol wt 204.23. C 64.69%, H 5.92%, N 13.72%, O 15.67%. Prepd by heating the potassium salt of 5-phenylhydantoin with ethyl bromide in alc at 100° in a sealed tube: Pinner, *Ber.* **21,** 2320 (1888). Use as anticonvulsant: W. J. Close, **U.S. pat. 2,793,157** (1957 to Abbott Labs.). Metabolism: Dudley *et al., J. Pharmacol. Exp. Ther.* **175,** 27 (1970).

Stout prisms from water, mp 94°. Sparingly sol in cold water, more sol in hot water. Freely sol in alcohol, ether, benzene, dil aq solns of alkali hydroxides.

THERAP CAT: Anticonvulsant.

3796. Ethoxazene. *4-[(4-Ethoxyphenyl)azo]-1,3-benzenediamine; 4-[(p-ethoxyphenyl)azo]-m-phenylenediamine;* 2,4-diamino-4'-ethoxyazobenzene; *p*-ethoxy-2,4-diaminoazobenzene; *p*-ethoxychrysoidine; 4-*p*-phenethylazo-*m*-phenylenediamine; SN-612; Carmurit; Cystural. $C_{14}H_{16}N_4O$; mol wt 256.31. C 65.61%, H 6.29%, N 21.86%, O 6.24%.

Hydrochloride, $C_{14}H_{16}N_4O.HCl$, *Serenium.* Reddish powder. Insol in water. Excreted in water-soluble form, giving a red color to urine. *Ref:* Lott and Christiansen, *J. Am. Pharm. Assoc.* **23,** 785 (1934).

USE: Acid-base indicator.

THERAP CAT: Analgesic.

3797. 2-Ethoxyethanol. Ethylene glycol monoethyl ether; Cellosolve; Oxitol. $C_4H_{10}O_2$; mol wt 90.12. C 53.31%, H 11.18%, O 35.51%. $HOCH_2CH_2OC_2H_5$. Manuf from ethylene oxide and ethanol: Faith, Keyes & Clark's *Industrial Chemicals,* F. A. Lowenheim, M. K. Moran, Eds. (Wiley-Interscience, New York, 4th ed., 1975) pp 403-407. Toxicity data: H. F. Smyth *et al., J. Ind. Hyg. Toxicol.* **23,** 259 (1941).

Colorless, practically odorless liq. d_{20}^{20} 0.931. bp 135°. mp −70°. Flash pt, closed cup: 112°F (44°C); open cup: 120°F (49°C). n_D^{25} 1.406. Misc with water, alc, ether, acetone, liq esters. It dissolves many oils, resins, waxes, etc. LD_{50} orally in rats: 3 g/kg (Smyth).

USE: Solvent for nitrocellulose, lacquers and dopes; in varnish removers, cleansing solns, dye baths; finishing leather with water pigments and dye solns; increasing stability of emulsions.

3798. 2-Ethoxyethyl Acetate. *2-Ethoxyethanol acetate;* cellosolve acetate; ethylene glycol monoethyl ether acetate. $C_6H_{12}O_3$; mol wt 132.16. C 54.53%, H 9.15%, O 36.32%. $C_2H_5OCH_2CH_2OOCCH_3$. Toxicity data: Smyth *et al., J. Ind. Hyg. Toxicol.* **23,** 259 (1941).

Colorless liq, pleasant odor. d_{20}^{20} 0.975. bp 156°. Flash pt, open cup: 134°F (56°C). Sol in about 6 parts water. LD_{50} orally in rats: 5.1 g/kg (Smyth).

Caution: Potential symptoms of overexposure are irritation of eyes and nose; vomiting; kidney damage; paralysis. See *NIOSH Pocket Guide to Chemical Hazards* (DHHS/NIOSH 90-117, 1990) p 104.

USE: Automobile lacquers to retard evaporation and impart high gloss.

3799. 2-Ethoxynaphthalene. Ethyl β-naphthyl ether; ethyl β-naphtholate; nerolin "new"; β-naphthyl ethyl ether; Bromelia. $C_{12}H_{12}O$; mol wt 172.23. C 83.69%, H 7.02%, O 9.29%. Prepd by heating β-naphthol with potassium ethyl sulfate, or with alcohol and H_2SO_4: Yokoyama *et al., Yakugaku Zasshi* **78,** 123 (1958); *C.A.* **52,** 10986c (1958); from β-naphthol + *p*-toluenesulfonyl chloride + ethyl alcohol: Drakowzal, Klamann, *Monatsh.* **82,** 588 (1951); from β-naphthol + H_2SO_4 + ethyl acetate: Patai, Bentov, *J. Am. Chem. Soc.* **74,** 6118 (1952); from β-naphthol + triethylphosphate + *p*-toluenesulfonic acid: Bell, **U.S. pat. 2,683,-748** (1954 to Eastman Kodak).

Lustrous crystals, mp 37-38°. bp 28.2°. $n_D^{47.3}$ 1.5932. d_{20}^{20} 1.0640. Practically insol in water. Sol in alcohol, chloroform, ether, carbon disulfide, toluene, petr ether.

USE: In perfumery.

3800. Ethoxyquin. *6-Ethoxy-1,2-dihydro-2,2,4-trimethylquinoline;* 1,2-dihydro-6-ethoxy-2,2,4-trimethylquinoline; EMQ; Santoflex; Santoquin. $C_{14}H_{19}NO$; mol wt 217.31. C 77.38%, H 8.81%, N 6.45%, O 7.36%. Prepn: Knoevenagel, *Ber.* **54,** 1722 (1921); Baird *et al.,* **Brit. pat. 505,113** (1939 to ICI); C. C. Tung, *Tetrahedron* **19,** 1685 (1963). Structure: Cliffe, *J. Chem. Soc.* **1933,** 1327. Metabolism: R. H. Wilson *et al., Agr. Food Chem.* **7,** 206 (1959); J. U. Skaare, E. Solheim, *Xenobiotica* **9,** 649 (1979). Toxicity data: V. I. Piul'skaya *et al., C.A.* **90,** 4601c (1979).

Yellow liquid. bp$_2$ 123-125°. n_D^{25} 1.569-1.672. d_{25}^{25} 1.029-1.031. LD$_{50}$ orally in rats, mice: 1920, 1730 mg/kg (Piul'-skaya).

USE: Antioxidant in feed and food; antidegradation agent for rubber.

3801. Ethoxzolamide. *6-Ethoxy-2-benzothiazolesulfon-amide;* ethoxyzolamide; Cardrase; Ethamide; Glaucotensil; Redupresin. C$_9$H$_{10}$N$_2$O$_3$S$_2$; mol wt 258.32. C 41.85%, H 3.90%, N 10.84%, O 18.58%, S 24.83%. Carbonic anhydrase inhibitor. Prepn: **Brit. pat. 795,174;** J. Korman, **U.S. pat. 2,868,800** (1958, 1959 both to Upjohn).

Crystals from ethyl acetate + Skellysolve B, mp 188-190.5°.

THERAP CAT: Diuretic.
THERAP CAT (VET): Diuretic.

3802. Ethybenztropine. *endo-3-(Diphenylmethoxy)-8-ethyl-8-azabicyclo[3.2.1]octane;* *3α-(diphenylmethoxy)-8-ethyl-1αH,5αH-nortropane;* N-ethylnortropine benzhydryl ether; tropethydrylin; N-ethyl-8-aza-3-bicyclo[3.2.1]octyl benzhydryl ether; N-ethylbenztropine; ethylbenzatropine. C$_{22}$H$_{27}$NO; mol wt 321.46. C 82.20%, H 8.47%, N 4.36%, O 4.98%. Prepn: **Brit. pat. 804,837** (1958 to Sandoz); Bochringer *et al.,* **Brit. pat. 824,875** (1959 to Boehringer, Ing.). Used as hydrochloride or hydrobromide salts.

Hydrochloride, C$_{22}$H$_{27}$NO.HCl, *Ponalid.* Crystals from acetone, mp 190-191°.
Hydrobromide, C$_{22}$H$_{27}$NO.HBr, *Panolid.* Crystals from methanol + ether, mp 226-228°.
THERAP CAT: Anticholinergic.

3803. Ethyl Acetate. *Acetic acid ethyl ester;* acetic ether; vinegar naphtha. C$_4$H$_8$O$_2$; mol wt 88.11. C 54.53%, H 9.15%, O 36.32%. CH$_3$COOC$_2$H$_5$. Obtained by the slow distillation of a mixture of acetic acid, ethyl alc, and sulfuric acid: Alheritiere, Mercier, **U.S. pat. 2,787,636** (1957 to Usines de Melle); Faith, Keyes, & Clark's *Industrial Chemicals,* F. A. Lowenheim, M. K. Moran, Eds. (Wiley-Interscience, New York, 4th ed., 1975) pp 350-354. Toxicity: H. F. Smyth *et al., Am. Ind. Hyg. Assoc. J.* **23,** 95 (1962).

Clear, volatile, flammable liq; characteristic fruity odor; pleasant taste when diluted. Slowly dec by moisture, then acquires an acid reaction. Absorbs water (up to 3.3% w/w). d$_4^{20}$ 0.902; d$_{25}^{25}$ 0.898. bp 77°. mp −83°. Flash pt +7.2° (open cup). Ignition temp 800°F. Explosive limits (% vol in air): 2.2 to 11.5. n_D^{20} 1.3719. Vapor density 3.04 (air = 1). One ml dissolves in 10 ml water at 25°; more sol at lower and less sol at higher temps. Misc with alc, acetone, chloroform, ether. Azeotropic mixture with water (6.1% w/w) bp

70.4°. Azeotropic mixture with water (7.8% w/w) and alc (9.0% w/w) bp 70.3°. *Keep tightly closed in a cool place and away from fire.* LD$_{50}$ orally in rats: 11.3 ml/kg (Smyth).

Caution: Potential symptoms of overexposure are irritation of eyes, nose and throat; narcosis; dermatitis. *See NIOSH Pocket Guide to Chemical Hazards* (DHHS/NIOSH 90-117, 1990) p 104.

USE: Pharmaceutic aid (flavor); artificial fruit essences; solvent for nitrocellulose, varnishes, lacquers, and aeroplane dopes; manuf smokeless powder, artificial leather, photographic films and plates, artificial silk, perfumes; cleaning textiles, etc.

3804. Ethyl Acetoacetate. *3-Oxobutanoic acid ethyl ester;* acetoacetic acid ethyl ester; acetoacetic ester; ethyl 3-oxobutanoate. C$_6$H$_{10}$O$_3$; mol wt 130.14. C 55.37%, H 7.74%, O 36.88%. CH$_3$COCH$_2$COOC$_2$H$_5$. Only the equilibrium mixture of the keto and enol forms is described here. Prepd from ethyl acetate by the action of sodium, sodium ethoxide, sodamide, or calcium: Inglis, Roberts, *Org. Syn.* **coll. vol. I,** 235 (2nd ed., 1941); Hansley, Schott, **U.S. pat. 2,843,623** (1958 to Natl. Distillers); Scheibler, *Ann.* **565,** 176 (1949); Gattermann-Wieland, *Praxis des Organischen Chemikers* (de Gruyter, Berlin, 40th ed., 1961) p 218. Discussion of keto-enol tautomerism: Ward, *J. Chem. Ed.* **39,** 95 (1962). Toxicity study: H. F. Smyth *et al., J. Ind. Hyg. Toxicol.* **31,** 60 (1949).

Liq. Agreeable odor. d$_4^{10}$ 1.0357; d$_4^{17}$ 1.0288; d$_4^{25}$ 1.0213; d$_4^{54}$ 0.9924; d$_4^{75}$ 0.9703. mp −45°. bp$_{760}$ 180.8°; bp$_{400}$ 158.2°; bp$_{200}$ 138.0°; bp$_{60}$ 106°; bp$_{20}$ 81.1°; bp$_5$ 54.0°; bp$_{1.0}$ 28.5°. n_D^{20} 1.41937. Absorption spectrum: Morton, Rosney, *J. Chem. Soc.* **1926,** 711. Flash pt, closed cup: 184°F. Sol in about 35 parts water; misc with the usual organic solvents. LD$_{50}$ orally in rats: 3.98 g/kg (Smyth).

Caution: Moderately irritating to skin, mucous membranes.

3805. Ethyl Acrylate. *2-Propenoic acid ethyl ester;* acrylic acid ethyl ester. C$_5$H$_8$O$_2$; mol wt 100.12. C 59.98%, H 8.05%, O 31.96%. CH$_2$=CHCOOCH$_2$CH$_3$. Prepd from ethylene chlorohydrin or acrylonitrile, ethanol, and sulfuric acid; also by an oxo reaction from acetylene, carbon monoxide, and ethanol in the presence of suitable catalysts. *See* the refs under Methyl Acrylate.

Monomer, liquid, acrid, penetrating odor, retained by clothing. *Lacrimator.* d$_4^{20}$ 0.9405. fp below −72°. bp$_{760}$ 99.4°; bp$_{39.2}$ 20° (polymerizes on distn). n_D^{20} 1.404. Specific heat at −60°: 0.442 cal/g/°C. Heat of vaporization 8.27 kcal/mol; heat of combustion 655.49 kcal/mol. Flash pt, open cup: 60°F (15°C). Vapor density 3.45 (air = 1). Soly in water at 20°: 2 g/100 ml. Soly of water in ethyl acrylate at 20°: 1.5 g/100 g. Sol in alcohol, ether. Azeotropes: 45.0% water = bp 81°; 56.8% ethanol = bp 76°. Easily polymerizes on standing; polymerization process speeded up by heat, light, and peroxides. If pure, the monomer can be stored below +10° without incurring polymerization.

Polymer, transparent, elastic substance. Practically no odor. Little adhesive power. Resists the usual solvents.

Caution: Potential symptoms of overexposure to the monomer are irritation of eyes, respiratory system and skin. *See NIOSH Pocket Guide to Chemical Hazards* (DHHS/NIOSH 90-117, 1990) p 106. *See also Patty's Industrial Hygiene and Toxicology* vol. 2A, G. D. Clayton, F. E. Clayton, Eds. (Wiley-Interscience, New York, 3rd ed., 1981) p 2292-2296. This substance may reasonably be anticipated to be a carcinogen: *Seventh Annual Report on Carcinogens* (PB95-109781, 1994) p 203.

USE: The monomer in the manuf of water emulsion paint vehicles; in production of emulsion-based polymers used in textile and paper coatings, leather finish resins and adhesives. Imparts flexibility to hard films.

3806. Ethyl Alcohol. *Ethanol;* absolute alcohol; anhydrous alcohol; dehydrated alcohol; ethyl hydrate; ethyl hydroxide. C$_2$H$_6$O; mol wt 46.07. C 52.14%, H 13.13%, O 34.73%. C$_2$H$_5$OH. Manuf: by fermentation of starch, sugar, and other carbohydrates; from ethylene, acetylene, sulfite waste liquors, and synthesis gas (CO + H); by hydrolysis of ethyl sulfate, and oxidation of methane. Toxicity: G. S. Wiberg *et al., Toxicol. Appl. Pharmacol.* **16,** 718 (1970). Embryotoxicity in mammals: N. A. Brown *et al.,*

Science **206**, 573 (1979). Possible mechanism for actions of ethanol on the brain: G. Aston-Jones et al., Nature **296**, 857 (1982). Ethanol-induced chromosomal abnormalities in mice: M. H. Kaufman, ibid. **302**, 258 (1983). Disruption of reproductive function in female primates following alcohol self-administration: N. K. Mello et al., Science **221**, 677 (1983). Review of metabolism and toxicity: C. S. Lieber in Reviews in Biochemical Toxicology vol. **5**, E. Hodgson et al., Eds. (Elsevier, New York, 1983) pp 267-312; of pharmacology: L. Pohorecky, J. Brick, Pharmacol. Ther. **36**, 335-427 (1988); of hepatotoxicity: C. S. Lieber, L. M. DeCarli, J. Hepatol. **12**, 394-401 (1991). General reviews: P. Baud, "Ethyl Alcohol Industry" in Grignard, Traité de Chimie Organique vol. **5** (Masson, 1937) pp 841-975; Zabel, Chem. Inds. (now Chem. Week) **64**, 212 (1949); Faith, Keyes & Clark's Industrial Chemicals, F. A. Lowenheim, M. K. Moran, Eds. (Wiley-Interscience, New York, 4th ed., 1975) pp 355-364; P. D. Sherman, P. R. Kavasmaneck, "Ethanol" in Kirk-Othmer Encyclopedia of Chemical Technology vol. **9** (Interscience, New York, 3rd ed., 1980) pp 338-380.

Clear, colorless, very mobile, flammable liquid; pleasant odor; burning taste. Absorbs water rapidly from air. d_4^{20} 0.789. bp 78.5°. mp −114.1°. n_D^{20} 1.361. Flash pt, closed cup: 13°C. Miscible with water and with many organic liquids. Keep tightly closed, cool, and away from flame! LD_{50} in young, old rats (g/kg): 10.6, 7.06 orally (Wiberg).

The terms 95% alcohol and alcohol (when used alone) refer to a binary azeotrope having a distillate composition of 95.57% ethyl alcohol (by wt) and bp 78.15°. Alcohol, USP is specified as containing not less than 92.3% and not more than 93.8% by weight, corresponding to not less than 94.9% and not more than 96.0% by vol of C_2H_5OH at 15.56°. d_{25}^{25} 0.810; d 0.816 at 15.56° (60°F). Diluted alcohol, prepd from equal vols 95% alcohol and water, contains about 41.5% by wt or about 48.9% by vol of C_2H_5OH. d_{25}^{25} 0.931; d 0.936 at 15.56° (60°F). See U.S.P. **XXI**, 22, 1530 (1985).

Caution: Nausea, vomiting, flushing, mental excitement or depression, drowsiness, impaired perception, incoordination, stupor, coma, death may occur, cf. Clinical Toxicology of Commercial Products, R. E. Gosselin et al., Eds. (Williams & Wilkins, Baltimore, 5th ed., 1984) Section III, pp 166-171.

USE: Most ethyl alcohol is used in alcoholic beverages in suitable dilutions. Other uses are as solvent in laboratory and industry, in the manufacture of denatured alcohol, pharmaceuticals (rubbing compds, lotions, tonics, colognes), in perfumery, in organic synthesis. Octane booster in gasoline. Pharmaceutic aid (solvent).

THERAP CAT: Antiseptic.

THERAP CAT (VET): Antiseptic. To destroy nerve tissue. Solvent and dehydrating agent.

3807. Ethyl Alcohol, Denatured. Denatured alcohol. Ethyl alcohol to which has been added some substance or substances which, while allowing the use of the alcohol in the most varied industries and arts, renders it entirely unfit for consumption as a beverage. The most commonly used denaturants, either alone or in combination, are the following: Methanol, camphor, Aldehol, amyl alcohol, gasoline, isopropanol, terpineol, benzene, castor oil, acetone, nicotine, aniline dyes, ether, cadmium iodide, pyridine bases, sulfuric acid, kerosene, diethyl phthalate. Formula 1 is 5 gallons approved wood alcohol added to 100 gal of 95% ethanol. Formula 2B is 0.5 gal benzene added to 100 gal of 95% ethanol. Similarly formula 3A contains 5 gal commercial methanol, formula 6B contains 0.5 gal pyridine bases, formula 12A 5 gal benzene, formula 13A 10 gal ethyl ether, formula 19 4 gal methyl isobutyl ketone and 1 gal kerosene, formula 20 5 gal crude chloroform, formula 23A 10 gal acetone, formula 28 10 gal benzene, formula 28A 1 gal gasoline, formula 30 10 gal methanol, formula 32 5 gal ethyl ether, formula 33 30 lbs methyl violet, formula 35A 5 gal ethyl acetate, formula 39C 1 gal diethyl phthalate; formula 44 contains 20 gal n-butanol. Additional permissible formulas are given in Appendix to Regulations No. 3, Formulae for Completely and Specially Denatured Alcohol, published by the U.S. Treasury Dept., Bureau of Industrial Alcohol. Reprinted in N. A. Lange, Handbook of Chemistry.

Caution: Denaturants, particularly methanol, may modify and increase toxic symptoms caused by ingestion and exposure to fumes.

3808. Ethylamine. Ethanamine; monoethylamine; aminoethane. C_2H_7N; mol wt 45.08. C 53.28%, H 15.65%, N 31.07%. $C_2H_5NH_2$. Prepn from ethyliodide + liq ammonia: Watt, Otto, J. Am. Chem. Soc. **69**, 836 (1947); from ethanol + ammonia: Davies et al., U.S. pat. **2,609,394** (1952 to ICI); Lemon, Myerly, U.S. pat. **3,022,349** (1962 to Union Carbide). Toxicity study: H. F. Smyth et al., Arch. Ind. Hyg. Occup. Med. **10**, 61 (1954).

Flammable, liq; ammonia odor; strong alkaline reaction. d_{15}^{15} 0.689. bp 16.6°. Solidif −80°. Miscible with water, alcohol, ether. Keep tightly closed and in cold place. LD_{50} orally in rats: 0.40 g/kg (Smyth).

Hydrochloride, $C_2H_7N \cdot HCl$, crystals from ethanol + water, mp 110°. d 1.22. Soluble in 0.4 part water; freely sol in alcohol; slightly sol in chloroform or acetone. Practically insol in ether. Keep well closed.

Hydriodide, $C_2H_7N \cdot HI$, hygroscopic crystals, mp 188°. d 2.10. Freely sol in water or alcohol. Practically insol in chloroform, ether. Keep well closed and protected from light.

Oleate, $C_2H_7N \cdot C_{18}H_{34}O_2$, ethanamine (Z)-9-octadecenoate, Etalate. Commercial prepn is a 5% soln with 2% benzyl alcohol as anodyne.

Caution: Potential symptoms of overexposure are irritation of eyes; skin burns; respiratory irritation; dermatitis. See NIOSH Pocket Guide to Chemical Hazards (DHHS/NIOSH 90-117, 1990) p 106.

USE: In resin chemistry; stabilizer for rubber latex; intermediate for dyestuffs, medicinals; in oil refining; in organic syntheses.

THERAP CAT: Oleate as a sclerosing agent.

3809. N-Ethylamphetamine. N-Ethyl-α-methylbenzeneethanamine; N-ethyl-α-methylphenethylamine; N-ethyl-ω-phenylisopropylamine; 2-ethylamino-1-phenylpropane; Adiparthrol; Apetinil. $C_{11}H_{17}N$; mol wt 163.26. C 80.93%, H 10.50%, N 8.58%. Prepn: Keil, Dobke, Ger. pat. **767,263** (1952 to Theodor H. Temmler); Leonard et al., J. Am. Chem. Soc. **80**, 4858 (1958). Separation of isomers: Brit. pat. **814,339** (1959 to Sterling Drug).

bp$_{14}$ 104.5-106°. n_D^{25} 1.4986.

d-Form hydrochloride, $C_{11}H_{17}N \cdot HCl$, mp 154-156°. $[\alpha]_D^{15}$ +17.2° (c = 2 in water).

l-Form hydrochloride, $C_{11}H_{17}N \cdot HCl$, mp 155-156°. $[\alpha]_D^{25}$ −17.3° (c = 2 in water).

Note: This is a controlled substance (stimulant) listed in the U.S. Code of Federal Regulations, Title 21 Part 1308.11 (1995).

THERAP CAT: Anorexic.

3810. Ethyl Amyl Ketone. 5-Methyl-3-heptanone; amyl ethyl ketone; EAK. $C_8H_{16}O$; mol wt 128.21. C 74.94%, H 12.58%, O 12.48%. Review: Buller, Ind. & Eng. Chem. **48**, 1323 (1956).

Liquid. Mild fruity odor. d_{20}^{20} 0.820-0.824. One gallon weighs 6.83 lbs at 20°. bp$_{760}$ 157-162°. Flash pt 59° (138°F). Evaporation rate 0.3 (n-butyl acetate = 1.0). n_D^{15} 1.4195. Slightly miscible with water. Compatible with alcohols, ketones, ethers, many other organic solvents.

Caution: Narcotic in high concns.

USE: Solvent for nitrocellulose-alkyd, nitrocellulose-maleic, and vinyl resins.

3811. Ethylaniline. N-Ethylbenzenamine; ethylphenylamine. $C_8H_{11}N$; mol wt 121.18. C 79.29%, H 9.15%, N

11.56%. Produced by heating aniline hydrochloride and alcohol at 180°.

Very refractive liquid; rapidly becomes brown on exposure to light and air. Aniline-like odor. d_{25}^{25} 0.958. Solidifies below 80°. bp 204.5°. mp −63.5°. n_D^{20} 1.5559. Insol in water; miscible with alc, ether and many other organic solvents. *Keep well closed and protected from light.* LD_{50} orally in rats: 1.1 g/kg.

3812. Ethylbenzene. C_8H_{10}; mol wt 106.17. C 90.51%, H 9.49%. $C_6H_5C_2H_5$. Prepn from acetophenone: Clemmensen, *Ber.* **46**, 1838 (1913); Gattermann-Wieland, *Praxis des organischen Chemikers* (de Gruyter, Berlin, 40th ed., 1961) p 332. By Huang-Minlon modification of Wolff-Kishner reduction: A. I. Vogel, *Practical Organic Chemistry* (Longmans, London, 3rd ed., 1959) p 516. Physical properties: L. C. Gibbons *et al., J. Am. Chem. Soc.* **68**, 1130 (1946). Manuf: Faith, Keyes & Clark's *Industrial Chemicals*, F. A. Lowenheim, M. K. Moran, Eds. (Wiley-Interscience, New York, 4th ed., 1975) pp 365-370. Toxicity: H. F. Smyth *et al., Am. Ind. Hyg. Assoc J.* **23**, 95 (1962).
Colorless liquid; flammable. d_{25}^{25} 0.866. bp 136.25°. mp −95.01°. n_D^{25} 1.4932. Flash pt, closed cup: 64°F (18°C). Practically insol in water. Misc with the usual organic solvents. LD_{50} orally in rats: 5.46 g/kg (Smyth).
Caution: Potential symptoms of overexposure are irritation of eyes and mucous membranes; headache; dermatitis; narcosis, coma. *See NIOSH Pocket Guide to Chemical Hazards* (DHHS/NIOSH 90-117, 1990) p 106.
USE: For conversion to styrene monomer; as resin solvent.

3813. Ethylbenzhydramine. *2-(Diphenylmethoxy)-N,N-diethylethanamine;* 2-(benzhydryloxy)triethylamine; β-diethylaminoethyl benzhydryl ether; 2-(benzhydryloxy)-N,N-diethylethylamine; benzhydryl 2-diethylaminoethyl ether; etanautine. $C_{19}H_{25}NO$; mol wt 283.41. C 80.52%, H 8.89%, N 4.94%, O 5.65%. Prepn: Martin *et al.,* U.S. pat. **2,397,-799** (1946 to Geigy); Rieveschl, U.S. pat. **2,421,714** (1947 to Parke, Davis).

bp$_{0.15}$ 140-142°.
Hydrochloride, $C_{19}H_{25}NO.HCl$, *Antiparkin.* Prisms from alcohol + ethyl acetate, mp 140°. Bitter taste. Freely sol in water; sol in alcohol, acetone, chloroform; very slightly sol in benzene, ether. pH of 1% aq soln about 5.5.
Methyl iodide, $C_{19}H_{25}NO.CH_3I$, *Metropin.*
THERAP CAT: Anticholinergic; antiparkinsonian.

3814. Ethyl Benzoate. *Benzoic acid ethyl ester.* $C_9H_{10}O_2$; mol wt 150.18. C 71.98%, H 6.71%, O 21.31%. $C_6H_5COOC_2H_5$.
Colorless, clear, refractive liq; aromatic odor; vapors cause cough. $d_{}^{25}$ 1.050. bp 211-213°. mp −34°. n_D^{20} 1.506. Almost insol in water; miscible with alcohol, chloroform, ether, petr ether. LD_{50} orally in rats: 6.48 g/kg, Smyth *et al., Arch. Ind. Hyg. Occup. Med.* **10**, 61 (1954).
USE: In perfumery under the name *Essence de Niobe*; in manuf of *Peau d'Espagne*; artificial fruit essence.

3815. Ethyl Benzoylacetate. *β-Oxobenzenepropanoic acid ethyl ester.* $C_{11}H_{12}O_3$; mol wt 192.21. C 68.74%, H 6.29%, O 24.97%. $C_6H_5COCH_2COOC_2H_5$.
Liquid; pleasant odor; becomes yellow on exposure to air and light. d^{15} 1.122. bp 265-270° with decompn. Volatile

with steam. n_D^{20} 1.5338. Insol in water; miscible with alcohol, ether. *Keep protected from air and light.*

3816. α-Ethylbenzyl Alcohol. *α-Ethylbenzenemethanol;* ω-ethylbenzyl alcohol; 1-phenyl-1-propanol; 1-phenylpropyl alcohol; ethyl phenyl carbinol; α-hydroxypropylbenzene; SH-261; Ejibil; Livonal; Phenycholon; Phenicol; Phenychol; Felicur; Felitrope. $C_9H_{12}O$; mol wt 136.19. C 79.37%, H 8.88%, O 11.75%. Prepd from benzaldehyde or from ethyl phenyl ketone: Norris, Cortese, *J. Am. Chem. Soc.* **49**, 2640 (1927). Toxicity study: O. Linét *et al., Arzneimittel-Forsch.* **12**, 347 (1962).

Oily liquid. Weak, ester-like odor. Sweetish, slightly irritating taste. bp$_{760}$ 219°; bp$_{15}$ 107°; bp$_3$ 78°. d_4^{25} 0.9915. n_D^{23} 1.5169. uv max (methanol): 250, 260 nm (ε 173, 114). Misc with methanol, ethanol, ether, benzene, toluene, olive oil. LD_{50} orally in rats: 1.6 ml/kg (Linét).
USE: As heat transfer medium; in perfumery.
THERAP CAT: Choleretic.

3817. Ethylbenzylaniline. *N-Ethyl-N-phenylbenzenemethanamine; N-ethyl-N-phenylbenzylamine.* $C_{15}H_{17}N$; mol wt 211.31. C 85.26%, H 8.11%, N 6.63%. Prepn: Martin, MacQueen, U.S. pat. **1,887,772** (1933 to Dow); Burgstahler, *J. Am. Chem. Soc.* **73**, 3021 (1951).

Light yellow, oily liq. d_4^{19} 1.034. bp$_{710}$ 287° with slight decompn; bp$_{14}$ 170-180°. n_D^{23} 1.5938. Practically insol in water. One ml dissolves in 5.5 ml alcohol; sol in the usual organic solvents.
USE: Manuf dyes; in organic syntheses.

3818. Ethyl Biscoumacetate. *4-Hydroxy-α-(4-hydroxy-2-oxo-2H-1-benzopyran-3-yl)-2-oxo-2H-1-benzopyran-3-acetic acid ethyl ester; bis(4-hydroxy-2-oxo-2H-1-benzopyran-3-yl)-acetic acid ethyl ester;* ethyl bis(4-hydroxycoumarinyl)acetate; ethyldicoumarol acetate; bis-3,3'-(4-hydroxycoumarinyl)acetic acid ethyl ester; 3,3'-carboxymethylene bis(4-hydroxycoumarin)ethyl ester; ethyl 4,4'-dihydroxydicoumarinyl-3,3'-acetate; B.O.E.A.; Dicumacyl; Pelentan; Stabilene; Tromexan; Tromexan Ethyl Acetate. $C_{22}H_{16}O_8$; mol wt 408.36. C 64.71%, H 3.95%, O 31.34%. Prepn: Rosicky, *Cas. Lek. Cesk.* **83**, 1200 (1944); U.S. pats. **2,482,510-12** (all 1949 to Spojené farm. zovody); Stahmann *et al., J. Am. Chem. Soc.* **65**, 2285 *sqq* (1943). Toxicity study: C. M. Gruber *et al., Fed. Proc.* **10**, 303 (1951).

Crystals, dimorphous, mp 177-182° and mp 154-157°. Bitter, persistent taste. Practically insol in water. Sol in 20 parts acetone, also sol in benzene. Slightly sol in alcohol, ether. LD_{50} in mice, rats (g/kg): 0.88, 0.26 orally; 0.32, 1.1 i.p. (Gruber).
THERAP CAT: Anticoagulant.

3819. Ethyl Bromide. *Bromoethane;* monobromoethane; bromic ether; hydrobromic ether. C_2H_5Br; mol wt 108.97. C 22.05%, H 4.63%, Br 73.33%. CH_3CH_2Br. Made by distilling from a mixture of HBr, ethyl alcohol and

H_2SO_4: Kamm, Marvel, *Org. Syn. coll. vol. I,* 29 (1941). By phosphorus and bromine method: Goshorn *et al., ibid.* 36. Absorption spectrum: Hantzsch, *Ber.* **58,** 619 (1925). Physical properties: Mumford, Phillips, *J. Chem. Soc.* **1950,** 75. Toxicity data: E. H. Vernot *et al., Toxicol. Appl. Pharmacol.* **42,** 417 (1977).

Colorless, flammable, volatile liq; ethereal odor; burning taste; becomes yellowish on exposure to air and light. Vapor harmful. d_4^{20} 1.4612; d_4^{25} 1.4515. bp 38.2°. mp −119°. n_D^{20} 1.4242. Soly in water (g/100 g) at 0°: 1.067; 10°: 0.965; 20°: 0.914; 30°: 0.896; miscible with alcohol, ether, chloroform and with other organic solvents. Explosive limits (% by vol in air), lower 6.75, upper 11.25. Auto-ignition temp 952°F (511°C). LC_{50} rats, mice (ppm): 27000, 16200 (Vernot).

Caution: Potential symptoms of overexposure are irritation of eyes, respiratory system and skin; central nervous system depression; pulmonary edema; liver and kidney disease; cardiac arrhythmias; cardiac arrest. *See NIOSH Pocket Guide to Chemical Hazards* (DHHS/NIOSH 90-117, 1990) p 106.

USE: Ethylating agent in organic synthesis; as refrigerant. Formerly used as a topical and inhalation anesthetic.

3820. Ethyl α-Bromopropionate. *2-Bromopropanoic acid ethyl ester.* $C_5H_9BrO_2$; mol wt 181.03. C 33.17%, H 5.01%, Br 44.14%, O 17.68%. $CH_3CHBrCOOC_2H_5$.

Liquid; sharp, pungent odor; becomes yellow on exposure to light. d_{20}^{20} 1.447. bp 159-160°; also stated as 160-165°. n_D^{20} 1.4469. Insol in water; miscible with alcohol, ether. *Protect from light.*

3821. Ethyl *tert*-Butyl Ether. *2-Ethoxy-2-methylpropane; tert-butyl ethyl ether;* ethyl *tert*-butyl oxide; 1,1-dimethylethyl ethyl ether; ethyl 1,1-dimethylethyl ether; ETBE. $C_6H_{14}O$; mol wt 102.18. C 70.53%, H 13.81%, O 15.66%. $(CH_3)_3C(OCH_2CH_3)$. Prepn J. U. Nef, *Ann.* **309,** 126 (1899). Synthesis: J. F. Norris, G. W. Rigby, *J. Am. Chem. Soc.* **54,** 2088 (1932). Physical properties: T. W. Evans, K. R. Edlund, *Ind. Eng. Chem.* **28,** 1186 (1936). Thermal decomposition: N. J. Daly, C. Wentrup, *Aust. J. Chem.* **21,** 1535 (1968). Brief review focusing on use as gasoline additive: M. Iborra *et al., Chemtech.* **18,** 120 (1988).

bp 69-71°. fp −94.0°. Also reported as bp 73.1° (Norris, Rigby). d_4^{25} 0.7364. n_D^{25} 1.3728. Also reported as bp 72.8° (Evans, Edlund). d_4^{15} 0.7456; d_4^{20} 0.7404; d_4^{25} 0.7353, d_4^{30} 0.7300. n_D^{20} 1.3760. Vapor pressure at 25°: 130 mm Hg. Heat vaporization: 74.3 cal/g. Specific heat (liquid) at 25°: 0.51 cal/g/°C. Surface tension at 24°: 19.8 dynes/cm. Soly in water (20°): 1.2 g/100 g soln. Soly of water in compound (20°): 0.5 g/100 g soln.

USE: Gasoline additive.

3822. Ethyl Butyrate. *Butanoic acid ethyl ester; butyric acid ethyl ester;* ethyl *n*-butyrate. $C_6H_{12}O_2$; mol wt 116.16. C 62.04%, H 10.41%, O 27.55%. $CH_3CH_2CH_2CO-OC_2H_5$. Toxicity data: P. M. Jenner *et al., Food Cosmet. Toxicol.* **2,** 327 (1964).

Colorless liq; pineapple odor. d_4^{20} 0.879. bp 120-121°. mp −93°. n_D^{20} 1.400. Flash pt, closed cup: 78°F (25°C); open cup: 85°F (29°C). Sol in about 150 parts water; misc with alcohol, ether. LD_{50} orally in rats: 13,050 mg/kg (Jenner).

USE: Manuf artificial rum; perfumery; the alcoholic soln constitutes the so-called *"pineapple oil"*.

3823. Ethyl Caprate. *Decanoic acid ethyl ester;* ethyl decanoate. $C_{12}H_{24}O_2$; mol wt 200.32. C 71.95%, H 12.08%, O 15.97%. $CH_3(CH_2)_8COOC_2H_5$.

Colorless liq. d^{20} 0.862. bp 243-245°. Insoluble in water; miscible with alcohol, chloroform, ether.

USE: Manuf wine bouquets, cognac essence.

3824. Ethyl Caproate. *Hexanoic acid ethyl ester;* ethyl hexanoate. $C_8H_{16}O_2$; mol wt 144.21. C 66.63%, H 11.18%, O 22.19%. $CH_3(CH_2)_4COOC_2H_5$.

Colorless to yellowish liquid; pleasant odor. d^{20} 0.873. bp 166-167°. Insol in water; miscible with alcohol, ether.

USE: Manuf artificial fruit flavors.

3825. Ethyl Caprylate. *Octanoic acid ethyl ester;* ethyl octanoate; ethyl octylate. $C_{10}H_{20}O_2$; mol wt 172.27. C 69.72%, H 11.70%, O 18.58%. $CH_3(CH_2)_6COOC_2H_5$.

Colorless, clear, very mobile liquid; pleasant, pineapple odor. d^{17} 0.878. bp 207-209°. Insol in water; misc with alc, ether. LD_{50} orally in rats: 25,960 mg/kg, P. M. Jenner *et al., Food Cosmet. Toxicol.* **2,** 327 (1964).

USE: Manuf fruit ethers; constit of enanthic, cocoic, and cognac ethers.

3826. Ethyl β-Carboline-3-carboxylate. *9H-Pyrido[3,-4-b]indole-3-carboxylic acid ethyl ester;* ethyl norharmancarboxylate; β-CCE. $C_{14}H_{12}N_2O_2$; mol wt 240.26. C 69.99%, H 5.03%, N 11.66%, O 13.32%. Deriv of *β-carboline* that is a potent displacer of ^3H-diazepam from brain benzodiazepine receptors. Isoln from human urine and brain and binding site study: C. Braestrup *et al., Proc. Nat. Acad. Sci. USA* **77,** 2288 (1980). Initially thought to be an endogenous ligand for benzodiazepine receptors in mammalian CNS, it is now believed to be formed during isoln and extraction procedures, R. F. Squires *et al.,* in *GABA and Benzodiazepine Receptors,* E. Costa *et al.,* Eds. (Raven Press, New York, 1980) pp 129-138; M. Nielson *et al., J. Neurochem.* **36,** 276 (1981). Synthesis and psychotropic activity: *Japan. Kokai* **81 43283** (to Schering AG), *C.A.* **95,** 115508a (1981); U. Eder *et al., Eur. pat. Appl.* **30,254** (1981 to A/S Ferrosan; Schering AG). β-CCE has been shown to lower seizure threshold and to reverse the sedative effect of flurazepam, *q.v.:* P. J. Cowen *et al., Nature* **290,** 54 (1981). Neurochemical and pharmacological actions of β-CCE and other β-carbolines: M. Cain *et al., J. Med. Chem.* **25,** 1081 (1982). Anxiogenic and convulsant properties: L. Prado de Carvalho *et al., Nature* **301,** 64 (1983).

mp 229-233°. uv max (pH 7): 215, 242, 279 nm.

3-Hydroxymethyl-β-carboline, $C_{12}H_{10}N_2O$, *9H-pyrido[3,4-b]indole-3-methanol, 3-HMC.* Prepn: F. Hamaguchi, S. Ohki, *Heterocycles* **8,** 383 (1977); M. Cain *et al., loc. cit.* Antagonism of anticonvulsant and anxiolytic actions of diazepam: P. Skolnick *et al., Eur. J. Pharmacol.* **68,** 381 (1980). Crystals, mp 225-228°.

USE: As tools for studying benzodiazepine receptors.

3827. Ethyl Carbonate. *Carbonic acid diethyl ester;* diethyl carbonate; Eufin. $C_5H_{10}O_3$; mol wt 118.13. C 50.84%, H 8.53%, O 40.63%. $(C_2H_5O)_2C=O$. Prepn: Palomaa *et al., Ber.* **72,** 313 (1939). Manuf: Mador, Blackham, U.S. pat. **3,114,762** (1963 to Natl. Distillers & Chem.).

Liquid, bp 126°. Pleasant ethereal odor, mp −43°. Flash pt, closed cup: 77°F (25°C). d_4^{20} 0.9764. n_D^{20} 1.3843. Practically insol in water; miscible with alcohol, ether.

USE: Solvent for nitrocellulose; manuf radio tubes; fixing rare earths to cathode elements.

3828. Ethyl Cellulose. *Cellulose ethyl ether;* Ethocel. Prepd from wood pulp or chemical cotton by treatment with alkali and ethylation of the alkali cellulose with ethyl chloride. Review and bibliography: E. Ott, *Cellulose and Cellulose Derivatives* (New York, 2nd ed., 1955).

White granules. Soly is dependent upon the degree of substitution. Commercial ethyl cellulose has an ethoxy content of 43-50%. A 47% product softens at 140° and is sol in ethyl acetate, ethylene dichloride, benzene, toluene, xylene, butyl acetate, acetone, methanol, ethanol, butanol, carbon tetrachloride. To avoid brittleness, ethyl cellulose formulations usually include an antioxidant such as hydroquinone monobenzyl ether, 4-hexylpyrocatechol, or diphenylamine.

USE: In the manuf of plastics and lacquers. Pharmaceutic aid (tablet binder).

3829. Ethyl Chloride. *Chloroethane;* monochloroethanc; chlorethyl; aethylis chloridum; ether chloratus; ether hydrochloric; ether muriatic; Kelene; Chelen; Anodynon; Chloryl Anesthetic; Narcotile. C_2H_5Cl; mol wt 64.51. C 37.24%, H 7.81%, Cl 54.95%. Prepd by the action of chlorine on ethylene in the presence of HCl and light: U.S. pat. **2,393,509** (1946); by the action of chlorine on ethylene in the presence of the chlorides of copper, iron, antimony, and calcium: **Ger.** pat. **298,931;** Bähr, Zieber, *Angew. Chem.* **1930,** 233, 286; by heating alcohol, HCl and $ZnCl_2$: U.S. pat. **2,396,639** (1946). Review of mfg processes: Faith, Keyes & Clark's *Industrial Chemicals,* F. A. Lowenheim, M. K. Moran, Eds. (Wiley-Interscience, New York, 4th ed., 1975) pp 371-375.

Flammable gas at ordinary temp and pressure. Characteristic ethereal odor; burning taste. At low temps or under increased pressure, ethyl chloride is a mobile, very volatile liquid. Keep away from heat, sparks, and open flame. Keep container closed, out of sun, and away from heat. Use with adequate ventilation. Carbon dioxide is recommended for extinguishing ethyl chloride fires. d_4^0 0.9214. Vapor density: 2.22 (air = 1.00). mp −138.7°. bp_{760} 12.3°. When ethyl chloride is liberated at ordinary room temp from its sealed container (usually a tube with automatic closure) it vaporizes at once. $bp_{2\ atm}$ 32.5°; $bp_{10\ atm}$ 92.6°; $bp_{30\ atm}$ 149.5°; $bp_{50\ atm}$ 180.5°. Crit temp = 187.2°; crit press = 52.0 atm. Soly in water: 0.574 g/100 ml at 20°; in alcohol: 48.3 grams/100 ml. Miscible with ether. Ethyl chloride burns with a smoky, greenish flame, producing hydrogen chloride. Flash pt: −50° (−58°F) (closed cup): 43° (−45°F) (open cup). Explosive limits (% by vol in air): lower 3.6; upper 14.8.

Caution: Potential symptoms of overexposure are incoordination, inebriation; abdominal cramps; cardiac arrhythmias, cardiac arrest; liver and kidney damage. *See NIOSH Pocket Guide to Chemical Hazards* (DHHS/NIOSH 90-117, 1990) p 108.

USE: Refrigerant, solvent, alkylating agent, starting point in the manuf of tetraethyl lead: U.S. pat. **1,907,701** (1933).

THERAP CAT: Anesthetic (topical).

THERAP CAT (VET): Anesthetic (topical).

3830. Ethyl Chloroacetate. *Chloroacetic acid ethyl ester.* $C_4H_7ClO_2$; mol wt 122.55. C 39.20%, H 5.76%, Cl 28.93%, O 26.11%. $CH_2ClCOOC_2H_5$. Prepn: Vogel, *J. Chem. Soc.* **1948,** 644.

Liquid; pungent odor. *Vapors irritate the eyes.* d_4^{20} 1.1498. mp −26°. bp 144-146°. Flash pt 54°. n_D^{20} 1.4227. Insol in water; miscible with alcohol, ether.

3831. Ethyl Chloroformate. *Carbonochloridic acid ethyl ester; chloroformic acid ethyl ester;* ethyl chlorocarbonate. $C_3H_5ClO_2$; mol wt 108.52. C 33.20%, H 4.64%, Cl 32.67%, O 29.49%. $ClCOOC_2H_5$. Prepn from phosgene and alc: Dumas, *Ann.* **10,** 277 (1834).

Corrosive, flammable liquid. Flash pt slightly below 61°F. bp 95°. d_4^{20} 1.1403. n_D^{20} 1.3947. Practically insol and gradually decomp by water; miscible with alcohol, benzene, chloroform, ether.

Caution: Vapors strongly irritate eyes, mucous membranes, and skin.

3832. Ethyl α-Chloropropionate. *2-Chloropropanoic acid ethyl ester.* $C_5H_9ClO_2$; mol wt 136.58. C 43.97%, H 6.64%, Cl 25.96%, O 23.43%. $CH_3CHClCOOC_2H_5$.

Liq; pleasant odor. d_4^{20} 1.087. bp 147-148°. n_D^{20} 1.4185. Insol in water; miscible with alcohol, ether.

3833. Ethyl Cyanoacetate. *Cyanoacetic acid ethyl ester;* cyanoacetic ester; ethyl cyanoethanoate; malonic acid ethyl ester nitrile. $C_5H_7NO_2$; mol wt 113.12. C 53.09%, H 6.24%, N 12.38%, O 28.29%. $CNCH_2COOC_2H_5$. Made by the interaction of sodium cyanide and chloroacetic acid and subsequent esterification of the cyanoacetic acid formed: Kohler, Allen, *Org. Syn.* **3,** 53 (1923); Inglis, *ibid.* **8,** 74 (1928).

Liquid. Slight, pleasant odor. d_4^{25} 1.0560; d_4^{50} 1.0306; d_4^{70} 1.0110. mp −22°. bp_{760} 206.0°; bp_{100} 152.8°; bp_{40} 133.8°; bp_{20} 119.8°; bp_{10} 106.0°; bp_5 93.5°; $bp_{1.0}$ 67.8°. $n_D^{20.5}$ 1.41793.

Insol in water. Miscible with alcohol, ether; sol in ammonia water, aq solns of alkalies.

3834. Ethyl Dibunate. *3,6-Bis(1,1-dimethylethyl)-1-naphthalenesulfonic acid ethyl ester; 3,6-di-tert-butyl-1-naphthalenesulfonic acid ethyl ester;* ethyl 3,6-di-*tert*-butyl-1-naphthalenesulfonate; 2,7-di-*tert*-butylnaphthalene-4-sulfonic acid ethyl ester; ethyl 2,7-di-*tert*-butylnaphthalene-4-sulfonate; dibunate ethyl; NDR-304; Neodyne. $C_{20}H_{28}O_3S$; mol wt 348.51. C 68.93%, H 8.10%, O 13.77%, S 9.20%. Prepd from 3,6-di-*tert*-butyl-1-naphthalenesulfonyl chloride and ethanol: Menard *et al., Can. J. Chem.* **39,** 729 (1961). Pharmacology: Shemano, *Arch. Int. Pharmacodyn. Ther.* **165,** 410 (1967).

Crystals from ethanol, mp 138-139°.

THERAP CAT: Antitussive.

3835. Ethyl Diethylmalonate. *Diethylpropanedioic acid diethyl ester.* $C_{11}H_{20}O_4$; mol wt 216.28. C 61.09%, H 9.32%, O 29.59%. $C_2H_5COOC(C_2H_5)_2COOC_2H_5$. Produced by the action of sodium ethylate and ethylbromide on diethyl malonate.

Liquid. d 0.985-0.990. bp 228-230°. n_D^{20} 1.424. Insol in water; miscible with alcohol, ether.

USE: Manuf barbiturates.

3836. Ethyldimethyl-9-octadecenylammonium Bromide. *N-Ethyl-N,N-dimethyl-9-octadecen-1-aminium bromide;* Onyxide. $C_{22}H_{46}BrN$; mol wt 404.52. C 65.32%, H 11.46%, Br 19.75%, N 3.46%. Prepn: Du Bois, U.S. pat. **2,519,747** (1950 to Onyx Oil & Chem.).

trans-form

USE: A cationic surface active agent. Used in the control of algae. Sold as a 75% concentrate in isopropanol or propylene glycol.

3837. Ethylene. *Ethene;* elayl; olefiant gas. C_2H_4; mol wt 28.05. C 85.63%, H 14.37%. $CH_2{=}CH_2$. Plant hormone; occurs in ripening fruit, in illuminating gas (up to 4%). Prepd by decompn of petr gases or by dehydration of alcohol. Lab prepn from alc: Gattermann-Wieland, *Praxis des organischen Chemikers* (de Gruyter, Berlin, 40th ed, 1961) p 98; F. A. Wunder, E. I. Leupold, *Angew. Chem. Int. Ed.* **19,** 126 (1980). Biosynthesis from 1-aminocyclopropane-1-carboxylic acid: D. Adams, S. F. Yang, *Proc. Nat. Acad. Sci. USA* **76,** 170 (1979). Proposed mechanism of biosynthesis: M. C. Pirrung, *J. Am. Chem. Soc.* **105,** 7207 (1983). Toxicity data: Flury, *Arch. Exp. Pathol. Pharmacol.* **138,** 65 (1928). Comprehensive monograph: S. A. Miller, *Ethylene and its Industrial Derivatives* (Ernest Benn Ltd., London, 1969) 1321 pp. Review of ethylene in biological systems: Spencer, *Fortschr. Chem. Org. Naturst.* **27,** 31-80 (1969); of manuf: Faith, Keyes & Clark's *Industrial Chemicals,* F. A. Lowenheim, M. K. Moran, Eds. (Wiley-Interscience, New York, 4th ed., 1975) pp 376-384; of carcinogenic risk: *IARC Monographs,* **60,** 45-71 (1994). *Review:* K. M. Sundaram *et al.,* in Kirk-Othmer *Encyclopedia of Chemical Technology* vol. 9 (Wiley-Interscience, New York, 4th ed., 1994) pp 877-915.

Colorless, flammable gas with faint sweet odor. Burns with a luminous flame. Solidif −181° forming monoclinic prisms. mp −169.4°, bp_{700} −102.4°. Vapor density 0.978 (air = 1.00). One liter of ethylene gas at 760 mm and 0° weighs 1.260 g. $bp_{2\ atm}$ −90.8°; $bp_{10\ atm}$ −52.8°; $bp_{30\ atm}$

−14.2°; bp$_{50\ atm}$ +8.9°. d$_4^{6.5}$ (liquid): 0.30342. Critical temp +9.6°; crit press. 50.7 atm. One volume of ethylene gas dissolves in about 4 vols of water at 0°, in about 9 vols of water at 25°, in 0.5 vol alcohol at 25°, in about 0.05 vol ether at 15.5°. Sol in acetone, benzene. Explosive limits (% vol in air), lower: 3.02; upper: 34. Autoignition temp: 1009°F (+543°C). LC for mice in air: 950,000 ppm (Flury).

Caution: Can act as a simple asphyxiant due to oxygen displacement. Exposure to high concns may cause CNS depression, unconsciousness. *See Patty's Industrial Hygiene and Toxicology* vol. 2B, G. D. Clayton, F. E. Clayton, Eds. (John Wiley & Sons, New York, 4th ed., 1994) p 1241-1244.

USE: Oxyethylene welding and cutting metals; manuf alcohol, mustard gas, and many other organics. Manuf ethylene oxide (for plastics), "Polythene", polystyrene and other plastics. Plant growth regulator; used commercially to accelerate the ripening of various fruits.

THERAP CAT: Anesthetic (inhalation).

THERAP CAT (VET): Has been used as inhalation anesthetic.

3838. Ethylene Bromohydrin. *2-Bromoethanol;* β-bromoethyl alcohol; glycol bromohydrin. C$_2$H$_5$BrO; mol wt 124.97. C 19.22%, H 4.03%, Br 63.94%, O 12.80%. BrCH$_2$-CH$_2$OH. Prepd by the action of hydrobromic acid on ethylene oxide: Thayer *et al., Org. Syn.* **6**, 12 (1926).

Hygroscopic liq. d$_4^0$ 1.7902; d$_4^{15}$ 1.7696; d$_4^{20}$ 1.7629; d$_4^{25}$ 1.7560; d$_4^{30}$ 1.7494. bp$_{750}$ 149-150° (decompn); bp$_{20}$ 56-57°; bp$_{13}$ 48.5°. n$_D^{20}$ 1.49361. Miscible with water; sol in the usual organic solvents, except petr ether. Hydrolysis of aq solns is accelerated by heat, acids, alkalies. Aq solns have a sweet, burning taste.

Caution: Ethylene bromohydrin vapors are an irritant to the eyes and the mucosa.

3839. Ethylene Chlorohydrin. *2-Chloroethanol;* 2-chloroethyl alcohol; glycol chlorohydrin. C$_2$H$_5$ClO; mol wt 80.51. C 29.84%, H 6.26%, Cl 44.03%, O 19.87%. ClCH$_2$-CH$_2$OH. Made from ethylene by action of a hypochlorite. Toxicity data: H. F. Smyth *et al., J. Ind. Hyg. Toxicol.* **23**, 259 (1941). *Review:* G. H. Riesser in Kirk-Othmer *Encyclopedia of Chemical Technology* vol. 5 (Wiley-Interscience, New York, 3rd ed., 1979) pp 848-864.

Colorless liq. *Poisonous!* d$_4^{20}$ 1.197. bp 128-130°. mp −67°. n$_D^{20}$ 1.4419. Flash pt, open cup: 105°F (40°C). Miscible with water, alcohol. LD$_{50}$ orally in rats: 0.095 g/kg (Smyth).

Caution: Potential symptoms of overexposure are irritation of skin and mucous membranes; nausea, vomiting; vertigo, incoordination; numbness; visual disturbance; headache; thirst; delirium; low blood pressure; collapse, shock and coma. *See NIOSH Pocket Guide to Chemical Hazards* (DHHS/NIOSH 90-117, 1990) p 108. *See also* Patty's *Industrial Hygiene and Toxicology* vol. 2C, G. D. Clayton, F. E. Clayton, Eds. (Wiley-Interscience, New York, 4th ed., 1982) pp 4675-4684.

USE: Solvent; manuf insecticides; treating sweet potatoes before planting.

3840. Ethylene Cyanohydrin. *3-Hydroxypropanenitrile;* hydracrylonitrile; glycol cyanohydrin; β-hydroxypropionitrile. C$_3$H$_5$NO; mol wt 71.08. C 50.69%, H 7.09%, N 19.71%, O 22.51%. HOCH$_2$CH$_2$CN. Prepd by the interaction of ethylene chlorohydrin and alkali cyanide: E. C. Kendall, B. McKenzie, *Org. Syn.* coll. vol. I, 256 (2nd ed., 1941). Toxicity study: H. F. Smyth, C. P. Carpenter, *J. Ind. Hyg. Toxicol.* **26**, 269 (1944).

Liquid. d$_4^{25}$ 1.0404. mp −46°. bp$_{760}$ 228° (slight dec); bp$_{200}$ 178°; bp$_{100}$ 157.7°; bp$_{60}$ 144.7°; bp$_{40}$ 134.1°; bp$_{20}$ 117.9°; bp$_{10}$ 102°; bp$_5$ 87.8°; bp$_{1.0}$ 58.7°. n$_D^{25}$ 1.4241. Miscible with water, acetone, methyl ethyl ketone, ethyl alcohol. Slightly sol in ether (2.3% w/w at 15°). Insol in benzene, petr ether, carbon disulfide, carbon tetrachloride. LD$_{50}$ orally in rats: 10.0 g/kg (Smyth, Carpenter).

USE: Solvent for some cellulose esters and many inorganic salts.

3841. Ethylenediamine. *1,2-Ethanediamine;* 1,2-diaminoethane. C$_2$H$_8$N$_2$; mol wt 60.10. C 39.97%, H 13.42%, N 46.61%. H$_2$NCH$_2$CH$_2$NH$_2$. Manuf from ethylene dichloride and ammonia: Faith, Keyes & Clark's *Industrial Chemicals,* F. A. Lowenheim, M. K. Moran, Eds. (Wiley-

Interscience, New York, 4th ed., 1975) pp 385-388. Toxicity: H. F. Smyth *et al., J. Ind. Hyg. Toxicol.* **23**, 259 (1941).

Colorless, clear, thick liq; ammonia odor. d$_4^{25}$ 0.898. bp 116-117°. mp +8.5°. n$_D^{26}$ 1.4540. Flash pt, closed cup: 110°F (43°C). Volatile with steam. Freely sol in water forming a hydrate, also in alcohol; slightly sol in ether; sol in benzene unless insufficiently dried: S. G. Boas-Traube *et al., Nature* **162**, 960 (1948). *Note:* Strongly alkaline and may readily absorb CO$_2$ from air to form a nonvolatile carbonate. Protect against undue exposure to the atmosphere. LD$_{50}$ orally in rats: 1.16 g/kg (Smyth).

Monohydrate, mp 10°; bp 118°.

Dihydrochloride, C$_2$H$_8$N$_2$.2HCl, crystals. Sublimes without melting. Freely sol in water. Practically insol in alcohol. Aq soln is practically neutral.

Caution: Potential symptoms of overexposure to ethylenediamine are nasal irritation; primary irritation; sensitization dermatitis, irritation of respiratory system and asthma; liver and kidney damage. *See NIOSH Pocket Guide to Chemical Hazards* (DHHS/NIOSH 90-117, 1990) p 108.

USE: Solvent for casein, albumin, shellac, and sulfur; emulsifier; stabilizing rubber latex; as inhibitor in antifreeze solns; in textile lubricants. Pharmaceutic aid (aminophylline injection stabilizer).

THERAP CAT (VET): Has been used as a urinary acidifier.

3842. Ethylene Dibromide. *1,2-Dibromoethane; sym-*dibromoethane; ethylene bromide; EDB; Dowfume W 85. C$_2$H$_4$Br$_2$; mol wt 187.86. C 12.79%, H 2.15%, Br 85.07%. BrCH$_2$CH$_2$Br. Made from ethylene and bromine; also from acetylene and HBr. Manuf: Faith, Keyes & Clark's *Industrial Chemicals,* F. A. Lowenheim, M. K. Moran, Eds. (Wiley-Interscience, New York, 4th ed., 1975) pp 389-391. Toxicity data: G. W. Fischer *et al., J. Prakt. Chem.* **320**, 133 (1978). Carcinogenicity studies: W. A. Olson *et al., J. Nat. Cancer Inst.* **51**, 1993 (1973); *Clin. Toxicol.* **14**, 473 (1979). History of controversial use as a fumigant: J. Walsh, *Science* **215**, 1592 (1982).

Heavy liquid; chloroform odor. d$_{25}^{25}$ 2.172. bp 131-132°. mp +9°. n$_D^{20}$ 1.5379. Vapor pressure at 25°: 11 mm Hg. Sol in about 250 parts water; misc with alcohol, ether. *Protect from light.* LD$_{50}$ i.p. in mice: 220 mg/kg (Fischer).

Caution: Potential symptoms of overexposure are irritation of respiratory system and eyes; dermatitis with vesiculation. *See NIOSH Pocket Guide to Chemical Hazards* (DHHS/NIOSH 90-117, 1990) p 110. *See also* Patty's *Industrial Hygiene and Toxicology* vol. 2B, G. D. Clayton, F. E. Clayton, Eds. (Wiley-Interscience, New York, 3rd ed., 1981) p 3497-3502. This substance may reasonably be anticipated to be a carcinogen: *Seventh Annual Report on Carcinogens* (PB95-109781, 1994) p 148.

USE: Fumigant; in anti-knock gasolines.

3843. Ethylene Dichloride. *1,2-Dichloroethane; sym-*dichloroethane; ethylene chloride; EDC; Dutch liquid; Brocide. C$_2$H$_4$Cl$_2$; mol wt 98.96. C 24.27%, H 4.07%, Cl 71.65%. ClCH$_2$CH$_2$Cl. Made from ethylene and chlorine; also from acetylene and HCl: *Beilstein* I, 84 (1918). Manuf: Faith, Keyes & Clark's *Industrial Chemicals,* F. A. Lowenheim, M. K. Moran, Eds. (Wiley-Interscience, New York, 4th ed., 1975) pp 392-396. Toxicity data: Smyth *et al., Am. Ind. Hyg. Assoc. J.* **30**, 470 (1969). Review of carcinogenicity studies: *IARC Monographs* **20**, 429-444 (1979).

Heavy liq; burns with smoky flame; pleasant odor; sweet taste. d$_4^{20}$ 1.2569. bp 83-84°. mp ∼ −40°. Flash pt, closed cup: 56°F (13°C); open cup: 65°F (18°C). n$_D^{20}$ 1.4443. Sol in about 120 parts water; misc with alcohol, chloroform, ether. LD$_{50}$ orally in rats: 770 mg/kg (Smyth).

Caution: Potential symptoms of overexposure are CNS depression; nausea, vomiting; dermatitis; irritation of eyes, corneal opacity. *See NIOSH Pocket Guide to Chemical Hazards* (DHHS/NIOSH 90-117, 1990) p 110. *See also Clinical Toxicology of Commercial Products,* R. E. Gosselin *et al.,* Eds. (Williams & Wilkins, Baltimore, 5th ed., 1984) Section II, pp 163-164. This substance may reasonably be anticipated to be a carcinogen: *Seventh Annual Report on Carcinogens* (PB95-109781, 1994) p 158.

USE: Solvent for fats, oils, waxes, gums, resins, and particularly for rubber; manuf acetyl cellulose, tobacco extract, etc. Also used as fumigant.

3844. Ethylene Glycol. *1,2-Ethanediol.* $C_2H_6O_2$; mol wt 62.07. C 38.70%, H 9.74%, O 51.55%. $HOCH_2CH_2OH$. Prepd on a large scale by the hydration of ethylene oxide. Flowsheets: *Ullmann's Encyklopädie der technischen Chemie,* **3** (3rd ed., 1953) p 137; M. Sittig, *Organic Chemical Process Encyclopedia* (Noyes Dev. Corp., Park Ridge, N.J., 1967) p 265. Manuf from ethylene and oxygen: Faith, Keyes & Clark's *Industrial Chemicals,* F. A. Lowenheim, M. K. Moran, Eds. (Wiley-Interscience, New York, 4th ed., 1975) pp 397-402. Toxicity studies: Smyth *et al., J. Ind. Hyg. Toxicol.* **23,** 259 (1941); Bornmann, *Arzneimittel-Forsch.* **4,** 643 (1954). Monograph: G. O. Curme, F. Johnston, *Glycols* (Reinhold, New York, 1952).

Slightly viscous liq. Sweet taste. *Poisonous! Do not swallow!* Considerably hygroscopic: Absorbs twice its weight of water at 100% relative humidity. d_0^0 1.1274. d_4^{10} 1.1204. d_4^{20} 1.1135. d_4^{30} 1.1065. One gallon weighs 9.3 lbs. Flash pt, open cup: 240°F (115°C). mp −13°. bp_{760} 197.6°; bp_{97} 140°; bp_{18} 100°; $bp_{3.0}$ 70°; $bp_{0.06}$ 20°. n_D^{15} 1.43312; n_D^{25} 1.43063. Viscosity in centipoises: 26 at 15°; 21 at 20°; 17.3 at 25°. Dielectric constant at 20° and 150 meters wavelength: 38.66 esu. Dipole moment 2.20. Spec heat (20°): 0.561 cal/g/°C. Heat of formation −108.1 kcal/mol. Heat of fusion 44.7 cal/g. Heat of vaporization 191 cal/g. Heat of soln −6.5 cal/g of soln at 17° when 37 parts are mixed with 63 parts H_2O (w/w). Parachor 148.9 (theory 152.2). Surface tension at 20° = 48.4 dynes/cm. Miscible with water, lower aliphatic alcohols, glycerol, acetic acid, acetone and similar ketones, aldehydes, pyridine and similar coal tar bases. Slightly sol in ether (1:200). Practically insol in benzene and its homologs, chlorinated hydrocarbons, petr ether, oils. Density and freezing point of ethylene glycol-water mixtures: 10.15% ethylene glycol by wt (d_4^{20}, fp): 1.013, −3.5°; 20.44% ethylene glycol: 1.027, −8°; 29.88% ethylene glycol: 1.040, −15°; 40.23% ethylene glycol: 1.054, −24°; 50.18% ethylene glycol: 1.067, −36°; 58.37% ethylene glycol: 1.0770, −48°. LD_{50} in rats, guinea pigs (g/kg): 8.54, 6.61 orally (Smyth); in mice (ml/kg): 13.79 orally (Bornmann).

Caution: Constitutes a hazard when ingested: *e.g.,* drinking of antifreeze fluid. Transient stimulation of CNS followed by depression; vomiting, drowsiness, coma, respiratory failure, convulsions; renal damage, which may proceed to anuria, uremia, death. Lethal dose in humans about 1.4 ml/kg or 100 ml. *See* E. Browning, *Toxicity and Metabolism of Industrial Solvents* (Elsevier, New York, 1965) pp 594-600; 686-690; Patty's *Industrial Hygiene and Toxicology* **vol. 2C,** G. D. Clayton, F. E. Clayton, Eds. (Wiley-Interscience, New York, 3rd ed., 1982) pp 3817-3832.

USE: Antifreeze in cooling and heating systems. In hydraulic brake fluids. Industrial humectant. Ingredient of electrolytic condensers (where it serves as solvent for boric acid and borates). Solvent in the paint and plastics industries. In the formulation of printers' inks, stamp pad inks, and inks for ball-point pens. Softening agent for cellophane. Stabilizer for soybean foam used to extinguish oil and gasoline fires. In the synthesis of safety explosives, glyoxal, unsatd ester type alkyd resins, plasticizers, elastomers, synthetic fibers (Terylene, Dacron), and synthetic waxes.

3845. Ethylene Glycol Diacetate. *1,2-Ethanediol diacetate;* glycol diacetate; ethylene diacetate. $C_6H_{10}O_4$; mol wt 146.14. C 49.31%, H 6.90%, O 43.79%. $CH_3COOCH_2-CH_2OOCCH_3$. Prepn from ethylene bromide, glacial acetic acid and potassium acetate: Henry, *Bull. Soc. Chim.* [3] **17,** 207 (1897); Gattermann-Wieland, *Praxis des organischen Chemikers* (de Gruyter, Berlin, 40th ed., 1961) p 107; from synthesis gas via homogeneous ruthenium catalysis: J. F. Knifton, *Chem. Commun.* **1981,** 188. Toxicity study: Smyth *et al., J. Ind. Hyg. Toxicol.* **23,** 259 (1941).

Liquid. d 1.104. bp 190-191°. mp −31°. n_D^{20} 1.415. Flash pt, open cup: 205°F (96°C). Sol in 7 parts water; miscible with alc, ether. LD_{50} orally in rats: 6.86 g/kg (Smyth).

USE: Solvent for oils, cellulose esters, explosives, etc.

3846. Ethylene Glycol Dinitrate. *1,2-Ethanediol dinitrate;* glycol dinitrate; nitroglycol; EGDN. $C_2H_4N_2O_6$; mol wt 152.06. C 15.80%, H 2.65%, N 18.42%, O 63.13%.

$O_2NOCH_2CH_2ONO_2$. Explosive! Prepn: L. Henry, *Ann. Chim. (Paris)* **4,** 253 (1872); P. Golding *et al., Tetrahedron* **49,** 7037 (1993). Determn in blood by GC: M. H. Litchfield, *Analyst* **93,** 653 (1968); in explosion debris by HPLC: R. J. Prime, J. Krebs, *Can. Soc. Forens. Sci. J.* **13,** 27 (1980); in vapor phase by ion mobility spectrometry: A. H. Lawrence, P. Neudorfl, *Anal. Chem.* **60,** 104 (1988). Measurement of vapor pressure: P. A. Pella, *J. Chem. Thermodynam.* **9,** 301 (1977). Adsorption kinetics: D. O. Henderson *et al., Appl. Spectrosc.* **47,** 528 (1993). Brief description of properties: *Am. Ind. Hyg. Assoc. J.* **27,** 574-577 (1966); V. Lindner in Kirk Othmer *Encyclopedia of Chemical Technology* **vol. 10,** (John Wiley & Sons, New York, 4th ed., 1993) pp 4, 21-26.

Colorless, odorless liquid with sweetish taste, mp −22.8°. d 1.490 g/cm³. n_D^{20} 1.395. Insol in water, very sol in alcohol and ether. *Combustible and Explosive.*

Caution: Potential symptoms of overexposure are throbbing headache; dizziness; nausea; vomiting; abdominal pain; hypotension; flushing; palpitations; methemoglobinemia; delerium; CNS depression; angina. Direct contact may cause skin irritation. *See NIOSH Pocket Guide to Chemical Hazards* (DHHS/NIOSH 90-117, 1990) p110.

USE: Explosive for mining and fuel industries. Additive to dynamite. Detection of hidden bombs by analysis of ambient air for EGDN.

3847. Ethylene Glycol Monoacetate. *1,2-Ethanediol monoacetate;* glycol-monoacetin. $C_4H_8O_3$; mol wt 104.11. C 46.15%, H 7.75%, O 46.11%. $CH_3COOCH_2CH_2OH$.

Liquid. d 1.108. bp 182°. Miscible with water, alcohol. LD_{50} orally in rats: 8.25 g/kg, Smyth *et al., J. Ind. Hyg. Toxicol.* **23,** 259 (1941).

3848. Ethylene Oxide. *Oxirane;* Anprolene. C_2H_4O; mol wt 44.05. C 54.53%, H 9.15%, O 36.32%. Prepd from ethylene chlorohydrin and KOH: A. Wurtz, *Ann.* **110,** 125 (1859). Manuf by catalytic oxidation of ethylene: T. E. Lefort, **U.S. pat. 1,998,878** (1935 to Société Française de Catalyse Généralisée) reissued as **U.S. pat. 20,370** (1937 to Carbide and Carbon). Physical props: W. H. Perkin, *J. Chem. Soc.* **63,** 488 (1893). Leukemia in workers exposed to ethylene oxide: C. Hogstedt *et al., J. Am. Med. Assoc.* **241,** 1132 (1979). Hazards and handling: L. G. Hess, V. V. Tilton, *Ind. Eng. Chem.* **42,** 1251 (1950). Genotoxicity: F. Agurell *et al., Mutat. Res.* **250,** 229 (1991). Review of toxicology and human exposure: *Toxicological Profile for Ethylene Oxide* (PB91-180554, 1990) 115 pp; of carcinogenic risk: *IARC Monographs* **60,** 73-159 (1994). *Review:* H. C. Schultze in *Glycols,* G. O. Curme, Jr., Ed. (Reinhold, New York, 1952) pp 74-113; J. P. Dever *et al.,* in Kirk-Othmer *Encyclopedia of Chemical Technology* **vol. 9** (John Wiley & Sons, New York, 4th ed., 1994) pp 915-959.

Colorless, flammable gas at ordinary room temp and pressure; liquid below 12°. *Explosive!* bp 10.7°. d_4^4 0.891, d_7^7 0.887, d_{10}^{10} 0.882. mp −111°. n_D^7 1.3597. Sol in water, alcohol, ether.

Caution: Potential symptoms of overexposure are irritation of eyes, nose and throat; peculiar taste; headache; nausea, vomiting and diarrhea; dyspnea, cyanosis and pulmonary edema; drowsiness, weakness and incoordination; EKG abnormalities; burns to eyes and skin; frostbite. *See NIOSH Pocket Guide to Chemical Hazards* (DHHS/NIOSH 90-117, 1990) p 112. This substance may reasonably be anticipated to be a carcinogen: *Seventh Annual Report on Carcinogens* (PB95-109781, 1994) p 205.

USE: Fumigant for foodstuffs and textiles; to sterilize surgical instruments; agricultural fungicide; (commercially as mixture with inert gases). In organic syntheses, esp in the production of ethylene glycol. Starting material for the manuf of acrylonitrile and nonionic surfactants.

3849. Ethylene Thiourea. *2-Imidazolidinethione; imidazoline-2-thiol;* 2-mercaptoimidazoline; ETU; Akrochem ETU-22; NA-22; Robac 22; Sanceller 22; Vulkacit NPV/C. $C_3H_6N_2S$; mol wt 102.16. C 35.27%, H 5.92%, N 27.42%, S 31.39%. Degradation product of ethylenebisdithiocarbamate fungicides such as mancozeb, maneb, zineb, *q.q.v.* Prepn: C. F. H. Allen *et al., Org. Syn.* **coll. vol. III,** 394 (1955); G. Matolcsy, *Ber.* **101,** 522 (1968). Identification as decomposition product: R A. Ludwig *et al., Can. J. Bot.* **32,** 48 (1954); G. Czegledi-Janko, *J. Chromatog.* **31,** 89 (1967). Determn in various foods by GLC: W. H. Newsome, *J. Agr. Food Chem.* **20,** 967 (1972); by GLC, HPLC: J. H. Onley *et al., J. Assoc. Offic. Anal. Chem.* **60,** 1105 (1977); J. H. Onley, *ibid.* 1111. Review of determn methods: P. Bottomley *et al., Residue Rev.* **95,** 45-89 (1985). Metabolism in mice: K. Savolainen, H. Pyysalo, *J. Agr. Food Chem.* **27,** 1177 (1979); in cats and rats: F. Iverson, *Toxicol. Appl. Pharmacol.* **52,** 16 (1980). Aquatic toxicity data: C. J. Van Leeuwen, *Aquat. Toxicol.* **7,** 145 (1985). Goitrogenic activity: S. L. Graham, W. H. Hansen, *Bull. Environ. Contam. Toxicol.* **7,** 19 (1972); D. M. Smith, *Brit. J. Ind. Med.* **41,** 362 (1984). Review of toxicology: L. Fishbein, *J. Toxicol. Environ. Health* **1,** 713-756 (1976). Use in rubber industry as crosslinking agent: J. T. Oetzel, *Rubber World* **172,** 55 (1975); C. Hepburn, M. S. Mahdi, *Kaut. Gummi. Kunstst.* **39,** 629 (1986).

Needles, prisms from alc or amyl alc. mp 203-204°. Soly in 100 ml water: 2 g at 30°, 9 g at 60°, 44 g at 90°. Moderately sol in methanol, ethanol, ethylene glycol, and pyridine. Insol in acetone, ether, chloroform, benzene, ligroin. LD$_{50}$ orally in rats: 1832 mg/kg (Graham, Hansen).

Note: This substance may reasonably be anticipated to be a carcinogen: *Seventh Annual Report on Carcinogens* (PB95-109781, 1994) p 210.

USE: Accelerator in synthetic rubber production.

3850. Ethylenimine. *Aziridine;* azacyclopropane; dimethylenimine. C$_2$H$_5$N; mol wt 43.07. C 55.78%, H 11.70%, N 32.52%. Prepd by treating 2-chloroethylamine hydrochloride with NaOH: Wystrach *et al., J. Am. Chem. Soc.* **77,** 5915 (1955); **78,** 1263 (1956); from β-aminoethylsulfuric acid by reaction with NaOH: Allen *et al., Org. Syn.* **coll. vol. IV,** 433 (1963). Acute toxicity: H. F. Smyth *et al., J. Ind. Hyg. Toxicol.* **23,** 259 (1941). Toxicology: Weightman, Hoyle, *J. Am. Med. Assoc.* **189,** 543 (1964). Review of carcinogenicity studies: *IARC Monographs* **9,** 37 (1975).

Liquid. Intense odor of ammonia. *Poisonous! Handle in hood only!* Strongly alkaline. Polymerizes easily. d$_4^{24}$ 0.8321. mp −73.96°. bp$_{760}$ 56-57°. n_D^{25} 1.412. Vapor press. (20°) 160 mm Hg. Miscible with water. Sol in alc. Can be stored for some time over a few pellets of sodium hydroxide. LD$_{50}$ orally in rats: 15 mg/kg (Smyth).

Caution: Potential symptoms of overexposure are nausea, vomiting; headache, dizziness; pulmonary edema; liver and kidney damage; eye burns; skin sensitization; irritation of nose and throat. *See NIOSH Pocket Guide to Chemical Hazards* (DHHS/NIOSH 90-117, 1990) p 110. This substance has been listed as a carcinogen by OSHA: *Fed. Reg.* **39,** 3757 (1974).

USE: In the manuf of triethylenemelamine.

3851. Ethylestrenol. *(17α)-19-Norpregn-4-en-17-ol;* *17α-ethylestr-4-en-17β-ol;* 17α-ethyl-17β-hydroxy-4-estrene; 17β-hydroxy-17α-ethyl-19-nor-4-androstene; Orabolin; Durabolin-O; Orgaboral; Maxibolin; Orgabolin (obsolete). C$_{20}$H$_{32}$O; mol wt 288.47. C 83.27%, H 11.18%, O 5.55%. Prepn: de Winter *et al., Chem. & Ind. (London)* **1959,** 905; Szpilfogel, de Winter, U.S. pat. **2,878,267** (1959 to Organon).

Crystals, mp 76-78°. [α]$_D$ +31° (chloroform).

Note: This is a controlled substance (anabolic steroid) in the U.S. Code of Federal Regulations, Title 21 Part 1308.13, as defined in Part 1308.02 (1995).

THERAP CAT: Anabolic.

3852. Ethyl Ether. *1,1'-Oxybisethane;* ethoxyethane; ether; diethyl ether; ethyl oxide; diethyl oxide; sulfuric ether; anesthetic ether. C$_4$H$_{10}$O; mol wt 74.12. C 64.82%, H 13.60%, O 21.58%. C$_2$H$_5$OC$_2$H$_5$. Produced on a large scale by dehydration of ethanol or by hydration of ethylene, both processes being carried out in the presence of sulfuric acid. Review of mfg processes: Himmler in *Ullmann's Encyklopädie der Technischen Chemie,* **vol. 5** (1954) pp 777-782; D. E. Keeley in Kirk-Othmer *Encyclopedia of Chemical Technology* **vol. 9** (Wiley-Interscience, New York, 3rd ed., 1980) pp 381-393.

Mobile, very volatile, highly flammable liq. *Explosive!* Vapor heavier than air. Characteristic, sweetish, pungent odor, more agreeable than chloroform. Burning taste. Tends to form explosive peroxides under the influence of air and light, esp when evaporation to dryness is attempted. Peroxides may be removed from ether by shaking with 5% aq ferrous sulfate soln. Addition of naphthols, polyphenols, aromatic amines, and aminophenols has been proposed for the stabilization of ethyl ether. d$_4^0$ 0.7364; d$_4^{10}$ 0.7249; d$_4^{20}$ 0.7134; d$_4^{30}$ 0.7019. Vapor density 2.55 (air = 1.0). mp −116.3° (stable crystals); mp −123.3° (metastable crystals). bp$_{760}$ 34.6°; bp$_{400}$ 17.9°; bp$_{200}$ +2.2°; bp$_{100}$ −11.5°; bp$_{10}$ −48.1°; bp$_{1.0}$ −74.3°. Satd vapor press. at 0°: 184.9 mm; at 10°: 290.8 mm; at 20°: 439.8 mm; at 50°: 1276 mm; at 70°: 2304 mm. Critical temp 192.7°; crit press. 35.6 atm. Flash pt, closed cup: −49°F (−45°C). Air-ether mixtures containing more than 1.85 volume-% of ether vapor, are explosive hazards. Autoignition temp 180-190°. n_D^{15} 1.35555. Dielectric constant at 26.9° and 85.8 kilocycles = 4.197; good insulator. When shaken under absolutely dry condiions ether can generate enough static electricity to start a fire. Surface tension at 20°: 17.06 dynes/cm. Viscosity at 20°: 0.2448 cp. Heat of vaporization at 30°: 89.80 cal/g. Produces considerable coldness on quick evaporation. Heat of formation −907 cal/g; heat of combustion −8.807 kcal/g. Ether is slightly sol in water and water is slightly sol in ether. A satd water soln contains 8.43% (w/w) of ether at 15° and 6.05% (w/w) at 25°. Ether satd with water contains 1.2% H$_2$O at 20°. Soly in water increased by HCl. Sol in concd hydrochloric acid. May explode when brought in contact with anhyd nitric acid. Miscible with lower aliphatic alcohols, benzene, chloroform, petr ether, other fat solvents, many oils. Azeotrope with water (1.3%), bp 34.2°.

Caution: Potential symptoms of overexposure are dizziness; drowsiness; headache, excitedness and narcosis; nausea, vomiting; irritation of eyes, upper respiratory system and skin. *See NIOSH Pocket Guide to Chemical Hazards* (DHHS/NIOSH 90-117, 1990) p 112.

USE: Solvent for waxes, fats, oils, perfumes, alkaloids, gums. Excellent solvent for nitrocellulose when mixed with alc. Important reagent in organic syntheses, esp in Grignard and Wurtz type reactions. Easily removable extractant of active principles (hormones, etc.) from plant and animal tissues. In the manuf of gun powder. As primer for gasoline engines.

THERAP CAT: Anesthetic (inhalation).

THERAP CAT (VET): Anesthetic (inhalation). Has been used orally in colic, subcutaneously as a stimulant.

3853. Ethyl Formate. *Formic acid ethyl ester.* C$_3$H$_6$O$_2$; mol wt 74.08. C 48.64%, H 8.16%, O 43.20%. HCOOC$_2$H$_5$. Toxicity data: H. F. Smyth *et al., Arch. Ind. Hyg. Occup. Med.* **10,** 61 (1954).

Mobile, flammable liq. d_4^{20} 0.917. bp 53-54°. mp −80°. Flash pt, closed cup: −4°F (−20°C). n_D^{20} 1.3597. Sol in about 10 parts water with gradual decompn into free acid and alcohol; miscible with alcohol, ether. *Keep tightly closed and preferably in contact with calcium chloride.* LD_{50} orally in rats: 4.29 g/kg (Smyth).

Caution: Potential symptoms of overexposure are irritation of eyes and upper respiratory system; narcosis. *See NIOSH Pocket Guide to Chemical Hazards* (DHHS/NIOSH 90-117, 1990) p 112.

USE: As flavor for lemonades and essences; for manuf artificial rum and arrac; also as a solvent for nitrocellulose; as fungicide and larvicide for tobacco, cereals, dried fruits, etc.; in organic synthesis.

3854. 2-Ethyl-1-hexanol. 2-Ethylhexyl alcohol. $C_8H_{18}O$; mol wt 130.23. C 73.78%, H 13.93%, O 12.29%. Manuf from butyraldehyde: Faith, Keyes & Clark's *Industrial Chemicals,* F. A. Lowenheim, M. K. Moran, Eds. (Wiley-Interscience, New York, 4th ed., 1975) pp 413-417. Toxicity study: H. F. Smyth *et al., Am. Ind. Hyg. Assoc. J.* **30,** 470 (1969).

Colorless liq. d_{20}^{20} 0.8344. bp 184-185°. n_D^{20} 1.4300. Flash pt 81°. Soluble in about 720 parts water, in many organic solvents. It dissolves about 2.5% its weight of water at 25°. LD_{50} orally in rats: 12.46 ml/kg (Smyth).

USE: Mercerizing textiles; as a solvent for dyes, resins, oils; also claimed to possess antifoaming properties.

3855. Ethylhydrocupreine. *(8α,9R)-6′-Ethoxy-10,11-dihydrocinchonan-9-ol; hydrocupreine ethyl ether;* Optoquine; Numoquin. $C_{21}H_{28}N_2O_2$; mol wt 340.47. C 74.08%, H 8.29%, N 8.23%, O 9.40%. Prepn: Giemsa, Halberkann, *Ber.* **51,** 1325 (1918); Thron, Freund, U.S. pat. **1,062,203** (1913 to Ver. Chininfabr. Zimmer); Muira, *C.A.* **41,** 4582d (1947). Pharmacology: M. I. Smith, B. Fantus, *J. Pharmacol. Exp. Ther.* **8,** 53 (1916).

White bitter cryst powder. mp 123-128° when solvent-free. $[\alpha]_D^{25}$ −136.2° (alcohol). Practically insol in water; sol in alcohol, benzene, chloroform, ether, dil acids, oils, fats.

Hydrochloride, $C_{21}H_{28}N_2O_2$·HCl, *Neumolisina.* Rhombic crystals from acetone + ether, mp 252-254°. $[\alpha]_D^{21}$ −123.6° (water). One gram dissolves in 2 ml water, 5 ml alcohol, 2.5 ml chloroform; sparingly sol in dry acetone. Practically insol in ether, petr ether. Aq solns are sterilizable without decompn. *Protect from light.*

THERAP CAT: Antiseptic.

3856. Ethylidene Chloride. *1,1-Dichloroethane.* $C_2H_4Cl_2$; mol wt 98.96. C 24.27%, H 4.07%, Cl 71.65%. CH_3CHCl_2. Prepd by the action of PCl_5 on acetaldehyde.

Oily liquid; odor and taste as of chloroform. d_4^{20} 1.1757; d_4^{25} 1.1680. bp 57.3°: Mumford, Phillips, *J. Chem. Soc.* **1950,** 75. mp ~−98°. n_D^{20} 1.4167. Sol in about 200 parts water; miscible with alcohol.

Caution: Potential symptoms of overexposure are CNS depression; skin irritation; liver and kidney damage. *See NIOSH Pocket Guide to Chemical Hazards* (DHHS/NIOSH 90-117, 1990) p 86.

3857. Ethylidene Diacetate. *1,1-Ethanediol diacetate;* 1,1-diacetoxyethane. $C_6H_{10}O_4$; mol wt 146.14. C 49.31%, H 6.90%, O 43.79%. $(CH_3COO)_2CHCH_3$. Control of fungal and bacterial growth in crops and animal feedstuffs: D. L. Kensler *et al.,* U.S. pats. **3,931,412** and **4,012,526** (1976 and 1977 to Chevron).

Liq; sharp, fruity odor. d_4^{12} 1.061. bp 167-169°. Slightly sol in water; miscible with alcohol. With NaOH it yields acetaldehyde.

USE: Agricultural fungicide.

3858. Ethylidene Dicoumarol. *3,3′-Ethylidenebis[4-hydroxy-2H-1-benzopyran-2-one];* 3,3′-ethylidenebis(4-hydroxycoumarin); 3,3-ethylidenebis(4-oxycoumarin); Pertrombon. $C_{20}H_{14}O_6$; mol wt 350.33. C 68.57%, H 4.03%, O 27.40%. Prepn: Jansen, Jensen, *Z. Phys. Chem.* **277,** 66 (1943); Sullivan *et al., J. Am. Chem. Soc.* **66,** 2288 (1943); Stahmann *et al., ibid.* **67,** 900 (1944); Austrian pat. **196,874** (1958 to Spofa), *C.A.* **52,** 10212d (1958); Fucik, Czech. pat. **85,918** (1957), *C.A.* **52,** 2930d (1958).

Crystals from ethanol + dioxane, mp 178°. Dibenzoate, $C_{34}H_{22}O_8$, mp 209-210°.

THERAP CAT: Anticoagulant.

3859. Ethyl Iodide. *Iodoethane.* C_2H_5I; mol wt 155.97. C 15.40%, H 3.23%, I 81.37%. CH_3CH_2I. Made by the action of iodine on alcohol in the presence of red phosphorus. Ethyl iodide contains about 98% CH_3CH_2I and about 2% alcohol.

Heavy, clear, very refractive liquid; ethereal odor. When freshly prepared it is colorless, but soon becomes red (on exposure to light and air) due to liberation of iodine. Silver leaf prevents or greatly retards decompn. d_{20}^{20} 1.950. bp 72°. mp −108°. n_D^{15} 1.5168. Sol in 250 parts water with gradual decompn; miscible with alcohol and most organic solvents. *Keep dry and cool, tightly closed, protected from light.*

3860. Ethyl Isobutyrate. *2-Methylpropanoic acid ethyl ester.* $C_6H_{12}O_2$; mol wt 116.16. C 62.04%, H 10.41%, O 27.55%. $(CH_3)_2CHCOOC_2H_5$.

Liquid; aromatic, fruity odor. d_{20}^{20} 0.870. bp 110-111°. mp −88°. n_D^{20} 1.3903. Slightly sol in water; miscible with alcohol, ether.

USE: Manuf flavoring compds and essences.

3861. Ethyl Isothiocyanate. *Isothiocyanatoethane;* ethyl mustard oil. C_3H_5NS; mol wt 87.15. C 41.35%, H 5.78%, N 16.07%, S 36.80%. C_2H_5NCS. Obtained by the action of mercuric chloride on the product of the reaction of carbon disulfide with ethylamine.

Colorless to yellowish liq; pungent odor. d_4^{18} 1.003. bp 130-132°. mp −6°. n_D^{18} 1.5142. Insol in water. Miscible with alcohol, ether.

Caution: Vapors irritate the eyes; blister the skin.

USE: Military poison gas.

3862. Ethyl Isovalerate. *3-Methylbutanoic acid ethyl ester.* $C_7H_{14}O_2$; mol wt 130.19. C 64.58%, H 10.84%, O 24.58%. $(CH_3)_2CHCH_2COOC_2H_5$.

Colorless, oily liq; apple odor. d_{20}^{20} 0.868. bp 135°. mp −99°. n_D^{20} 1.4009. Sol in about 350 parts water; miscible with alcohol, benzene, ether.

USE: In alc soln for flavoring confectionery and beverages.

3863. Ethyl Lactate. *2-Hydroxypropanoic acid ethyl ester;* ethyl α-hydroxypropionate. $C_5H_{10}O_3$; mol wt 118.13. C 50.84%, H 8.53%, O 40.63%. $CH_3CHOHCOOC_2H_5$.

Colorless liquid; characteristic odor. d_4^{14} 1.042. bp 154°. $[\alpha]_D^{14}$ −10°. Flash pt, closed cup: 117°F (47°C). Miscible

with water (with partial decompn), alcohol, ether. *Keep well closed.*
USE: As solvent for nitrocellulose and cellulose acetate.

3864. Ethyl Laurate. *Dodecanoic acid ethyl ester.* $C_{14}H_{28}O_2$; mol wt 228.38. C 73.63%, H 12.36%, O 14.01%. $CH_3(CH_2)_{10}COOC_2H_5$.
Oil. d_{19}^{19} 0.867. mp $-10°$. bp 269°; bp_{25} 163°. n_D 1.4321. Insol in water. Very sol in alcohol, ether.

3865. Ethyl Levulinate. *4-Oxopentanoic acid ethyl ester.* $C_7H_{12}O_3$; mol wt 144.17. C 58.32%, H 8.39%, O 33.29%. $CH_3COCH_2CH_2COOC_2H_5$.
Liquid. d_{20}^{20} 1.012. bp 205-206°. n_D^{20} 1.4229. Freely sol in water; miscible with alcohol.

3866. Ethyl Linoleate. *(Z,Z)-9,12-Octadecadienoic acid ethyl ester;* mandenol. $C_{20}H_{36}O_2$; mol wt 308.50. C 77.87%, H 11.76%, O 10.37%. Prepn from sunflower seed oil: Mc-Cutcheon, *Org. Syn.* **coll. vol. III,** 526 (1955); *see also* Parker *et al., Biochem. Prepn.* **4,** 88 (1955).

Colorless oil. More stable to air oxidation than linoleic acid. $d_4^{15.5}$ 0.8846; d^{20} 0.8919. bp_6 193°; $bp_{2.5}$ 175°; $bp_{0.001}$ 133°. n_D^{20} 1.46753; n_D^{48} 1.4489. Iodine value 162.5. Miscible with dimethylformamide, fat solvents, oils.
USE: In the vitamin industry.

3867. Ethyl Loflazepate. *7-Chloro-5-(2-fluorophenyl)-2,3-dihydro-2-oxo-1H-1,4-benzodiazepine-3-carboxylic acid ethyl ester;* ethyl 7-chloro-5-(o-fluorophenyl)-2,3-dihydro-2-oxo-1H-1,4-benzodiazepine-3-carboxylate; CM-6912; Meilax; Victan. $C_{18}H_{14}ClFN_2O_3$; mol wt 360.77. C 59.93%, H 3.91%, Cl 9.83%, F 5.27%, N 7.76%, O 13.30%. Benzodiazepine tranquilizer. Prepn: J. Hellerbach *et al.,* **Ger.** pat. **2,001,276;** *eidem,* U.S. pat. **3,657,223** (1970, 1972 both to Hoffmann-La Roche). Analysis in plasma and urine: W. Cantreels, J. R. Jeanniot, *Biomed. Mass Spectrom.* **7,** 565 (1980). GC analysis of metabolites: J. P. Cano, *J. Chromatog.* **226,** 413 (1981). Toxicologic evaluation: G. Mozue *et al., Int. J. Clin. Pharmacol. Ther. Toxicol.* **19,** 453 (1981).

Crystals from ether, mp 193-194°.
Note: This is a controlled substance (depressant) listed in the U.S. Code of Federal Regulations, Title 21 Part 1308.14 (1995).
THERAP CAT: Anxiolytic.

3868. N-Ethylmaleimide. *1-Ethyl-1H-pyrrole-2,5-dione.* $C_6H_7NO_2$; mol wt 125.13. C 57.59%, H 5.64%, N 11.19%, O 25.57%. Prepd by heating N-ethylmaleamic acid in paraffin: Marrian, *J. Chem. Soc.* **1949,** 1515.

Crystals, mp 45°. Lacrimator when liquid.
Caution: Strong irritant.
USE: In cancer research (possible antimitotic activity.)

3869. Ethyl Malonate. *Propanedioic acid diethyl ester;* diethyl malonate; malonic ester. $C_7H_{12}O_4$; mol wt 160.17.

C 52.49%, H 7.55%, O 39.96%. $C_2H_5OOCCH_2COOC_2H_5$. Prepd from chloroacetic acid and sodium cyanide followed by esterification with ethanol and H_2SO_4: L. Gattermann, T. Wieland, *Praxis des Organischen Chemikers* (de Gruyter, Berlin, 40th ed., 1961) pp 220-221.
Liquid. Slightly aromatic, pleasant odor. d 1.055. bp 198-199°; bp_{20} 95°. mp $-50°$. n_D^{20} 1.4143. One gram dissolves in about 50 ml of water. Miscible with alcohol, ether.
USE: Manuf of barbiturates.

3870. Ethyl Menthane Carboxamide. *N-Ethyl-5-methyl-2-(1-methylethyl)cyclohexanecarboxamide;* N-ethyl-4-

menthane-3-carboxamide; WS3. $C_{13}H_{25}NO$; mol wt 211.35. C 73.88%, H 11.92%, N 6.63%, O 7.57%. Prepn: **Neth.** pat. **Appl. 7,201,523;** H. R. Watson *et al.,* U.S. pat. **4,193,936** (1972, 1980 both to Wilkinson Sword). Structure-activity study: H. R. Watson *et al., J. Soc. Cosmet. Chem.* **29,** 185 (1978). Brief review: M. A. Parrish, *Manufact. Chem.* **58,** 31-32 (February, 1987).
White to nearly white powder, mp 88°. $[\alpha]_D$ -49 to $-54°$. Insol in water. Sol in most organic solvents. Average oral cooling threshold: 0.2 μg. LD_{50} in mice, rats (g/kg): 5.3, 2.9 orally (Parrish).
USE: Physiological coolant in foods, beverages, toiletries, cosmetics and pharmaceuticals.

3871. Ethylmercuric Chloride. *Chloroethylmercury;* Granosan. C_2H_5ClHg; mol wt 265.10. C 9.06%, H 1.90%, Cl 13.37%, Hg 75.66%. CH_3CH_2HgCl. Prepd from ethylmercuric bromide by treating with methanolic KOH, filtering and neutralizing with HCl: Slotta, Jacobi, *J. Prakt. Chem.* [2] **120,** 249 (1929). Prepn from $HgCl_2$ and tetraethyl lead: Whelen, *Advances in Chemistry Series* **23,** entitled *Metal-Organic Compounds* (ACS, Washington, D.C., 1960) pp 82-86. For the prepn of the bromide from ethylmagnesium bromide and mercuric bromide *see* Slotta, Jacobi, *loc. cit.,* Marvel *et al., J. Am. Chem. Soc.* **47,** 3009 (1925). *Review:* Krause, von Grosse, *Die Chemie der Metall-organischen Verbindungen* (Berlin, 1937).
White, silvery leaflets from ethanol, mp 192°. Sublimes easily. Solubility in water at 18° = 1.4 × 10^{-4} g/100 ml, at 100° = 2.5 × 10^{-4} g/100 ml; in ethanol at 18° = 0.75 g/100 g; at 78° = 3.5 g/100 g; in chloroform at 18° = 2.6 g/100 g. Slightly sol in ether.
Note: A close analog, *chloro-(2-methoxyethyl)mercury,* Ceresan Wet, is also used as seed fungicide.
Caution: Highly toxic. Causes skin burns; is absorbed through the skin. Chronic exposure has caused permanent injury to brain.
USE: Applied at 2% strength (soln or mixed with solids) as a fungicide for treating seeds.

3872. N-(Ethylmercuri)-p-toluenesulfonanilide. *Ethyl-(4-methyl-N-phenylbenzenesulfonamidato-N)mercury; ethyl-(N-phenyl-p-toluenesulfonamido)mercury;* Ceresan M. $C_{15}H_{17}HgNO_2S$; mol wt 475.96. C 37.85%, H 3.60%, Hg 42.14%, N 2.94%, O 6.72%, S 6.74%.

Crystals. Pungent odor, somewhat reminiscent of garlic. Practically insol in water. LD_{50} orally in rats: 100 mg/kg, *Toxic Substances List,* H. E. Christensen, Ed. (1973) p 585.
Caution: Almost as toxic as mercuric chloride *(q.v.);* less irritating to the skin than ethyl mercury phosphate. Symp-

toms confined to CNS and consist of deafness, ataxia, dysarthria, progressive visual deterioration; dysphagia; sphincteric incontinence, mental confusion, stupor, death: Gleason *et al.*, *Clinical Toxicology of Commercial Products* (Williams & Wilkins, Baltimore, 3rd ed., 1969) p 68, sect. II.

USE: To control the various smuts which befall grain. Reduces bulb rot in gladiolus.

3873. Ethyl Methanesulfonate. *Methanesulfonic acid ethyl ester;* ethyl methanesulfonic acid; ethyl mesylate; EMS; NSC-26805. $C_3H_8O_3S$; mol wt 124.16. C 29.02%, H 6.49%, O 38.66%, S 25.83%. $CH_3SO_2OCH_2CH_3$. Prepn: O. C. Billeter, *Ber.* **38**, 2015 (1905). Mutagenicity studies: T. Alderson, *Nature* **207**, 164 (1965); J. B. Jenkins, *Mutat. Res.* **4**, 90 (1967); A. P. Schalet, *ibid.* **49**, 313 (1978). Review of carcinogenicity studies: *IARC Monographs* **7**, 245-252 (1974). Review of comparative mutagenicity of EMS and methyl methanesulfonate, *q.v.:* S. Kondo, *Environ. Sci. Res.* **24**, 743-785 (1981).

Liquid, bp_{761} 213-213.5°; bp_{10} 85-86°. d_4^{22} 1.1452.

Note: This substance may reasonably be anticipated to be a carcinogen: *Seventh Annual Report on Carcinogens* (PB95-109781, 1994) p 213.

USE: Exptlly as mutagen, teratogen and brain carcinogen.

3874. Ethyl Methyl Ether. *Methoxyethane;* methyl ethyl ether. C_3H_8O; mol wt 60.10. C 59.96%, H 13.42%, O 26.62%. $C_2H_5OCH_3$.

Liquid. d_0^0 0.725. bp 10.8°. Sol in water; miscible with alcohol, ether.

3875. Ethylmethylthiambutene. *N-Ethyl-N-methyl-4,4-di-2-thienyl-3-buten-2-amine; N-ethyl-N,1-dimethyl-3,3-di-2-thienyl 2 propenamine,* N-ethyl-N,1-dimethyl-3,3-di-2-thienylallylamine; 3-ethylmethylamino-1,1-di(2'-thienyl)but-1-ene; ethylmethiambutene; 1C50; NIH-5145; Emethibutin. $C_{15}H_{19}NS_2$; mol wt 277.45. C 64.93%, H 6.90%, N 5.05%, S 23.11%. Prepn: Adamson, **U.S. pat. 2,561,899** (1951 to Burroughs Wellcome). Toxicology: N. B. Eddy, D. Leimbach, *J. Pharmacol. Exp. Ther.* **107**, 385 (1953).

$bp_{0.01}$ 110-113°; bp_{18} 75-76°.

Hydrochloride, $C_{15}H_{19}NS_2$·HCl, crystals, mp 137-138°. LD_{50} in mice (mg/kg): 192 orally; 88 s.c. (Eddy, Leimbach).

Note: This is a controlled substance (opiate) listed in the U.S. Code of Federal Regulations, Title 21 Part 1308.11 (1995).

THERAP CAT: Analgesic (narcotic).

3876. Ethylmorphine. *7,8-Didehydro-4,5-epoxy-3-ethoxy-17-methylmorphinan-6-ol.* $C_{19}H_{23}NO_3$; mol wt 313.40. C 72.82%, H 7.40%, N 4.47%, O 15.32%. Prepd by ethylation of morphine: Baizer, Ellner, *J. Am. Pharm. Assoc.* **39**, 581 (1950); Gorecki, *Ann. Pharm. (Poznan)* **7**, 21 (1969). Toxicity data: M. Aurousseau, J. Navarro, *Ann. Pharm. Franc.* **15**, 640 (1957).

Crystals from ethanol, mp 199-201°.

Hydrochloride dihydrate, $C_{19}H_{23}NO_3$·HCl·2H$_2$O, *Codethyline, Dionin.* White to faintly yellow crystalline powder, mp about 123° (dec); anhydrous form, mp about 170° (dec).

One gram dissolves in 10 ml water and in 25 ml alcohol; slightly sol in chloroform, ether. LD_{50} in mice: 264.6 mg/kg s.c.; 0.771 g/kg orally (Aurousseau, Navarro).

Methiodide, $C_{19}H_{23}NO_3$·CH$_3$I, *ethyl-N-methylmorphinium iodide, Trachyl.*

Note: This is a controlled substance (opiate) listed in the U.S. Code of Federal Regulations, Title 21 Part 1308.12 (1995).

THERAP CAT: Analgesic (narcotic); antitussive.

THERAP CAT (VET): Analgesic (narcotic); antitussive; mydriatic.

3877. Ethyl Nitrite. *Nitrous acid ethyl ester;* nitrous ether. $C_2H_5NO_2$; mol wt 75.07. C 32.00%, H 6.71%, N 18.66%, O 42.63%. C_2H_5ONO. Made by the action of sodium nitrite on a mixture of alcohol and H_2SO_4 in the cold. The article of commerce contains 90-95% ethyl nitrite, the remainder is chiefly alcohol as preservative. In the laboratory ethyl nitrite gas is conveniently generated by placing 20 liters of 90% ethanol in a suitable vessel, diluting with 200 liters of water, and while stirring, adding to the dil alc 18.3 kg of nitrosyl chloride at the rate of 2.25 kg per hour—*see also* Chase, **U.S. pat. 2,615,896** (1952).

Colorless or yellowish, clear, flammable, exceedingly volatile liq. Characteristic odor; burning, sweetish taste. d_{15}^{15} 0.90. bp 17°. Slightly sol in water and dec by it; miscible with alcohol, ether. On keeping, it gradually dec, becoming acid and oxides of nitrogen form. Decompn is hastened by air, light, and moisture. *Keep tightly closed, in a cool place, protected from light.*

Caution: May cause methemoglobinemia and hypotension and, in high concns, narcosis.

USE: For preparing spirit nitrous ether by mixing with 21 parts alcohol by weight.

3878. Ethyl Nitrobenzoate. $C_9H_9NO_4$; mol wt 195.17. C 55.39%, H 4.65%, N 7.18%, O 32.79%.

Colorless crystals. Insol in water. Freely sol in alc, ether. *o*-Ester, mp 30°. bp_{18} 173°.

m-Ester, mp 47°. bp about 298°.

p-Ester, mp 57°.

3879. N-Ethyl-N-nitrosourea. *N-Nitroso-N-ethylurea;* ENU. $C_3H_7N_3O_2$; mol wt 117.11. C 30.77%, H 6.02%, N 35.88%, O 27.32%. $C_2H_5N(NO)CONH_2$. Potent mouse mutagen. Prepn: E. A. Werner, *J. Chem. Soc.* **115**, 1093 (1919); F. Arndt, H. Scholz, *Angew. Chem.* **46**, 47 (1933). Carcinogenicity studies: M. F. Rajewsky, R. Goth, *Mol. Base Malig., Sel. Pap. Int. Symp.* **1976**, 2. Mutagenicity studies: W. L. Russell *et al.*, *Proc. Nat. Acad. Sci. USA* **16**, 5818 (1979). Toxicity studies: H. Druckrey *et al.*, *Nature* **210**, 1378 (1966).

Pale buff-yellow, hexagonal plates, mp 103-104°. LD_{50} i.v. in rats: 240 mg/kg (Druckrey).

Note: This substance may reasonably be anticipated to be a carcinogen: *Seventh Annual Report on Carcinogens* (PB95-109781, 1994) p 292.

USE: Exptlly as mutagen; ethylating agent.

3880. Ethylnorepinephrine. *4-(2-Amino-1-hydroxybutyl)-1,2-benzenediol; α-(1-aminopropyl)-3,4-dihydroxybenzyl alcohol; α-(1-aminopropyl)protocatechuyl alcohol; 1-(3,4-dihydroxyphenyl)-2-aminobutanol;* 1-(3,4-dihydroxyphenyl)-1-hydroxy-2-aminobutane; ethylnoradrenaline; ethylnorsuprarenin; E.N.E.; E.N.S.; $C_{10}H_{15}NO_3$; mol wt 197.23. C 60.90%, H 7.67%, N 7.10%, O 24.34%. Prepd by catalytic hydrogenolysis of α-benzhydrylamino-3,4-dibenzyloxybutyrophenone-HCl: Suter, Ruddy, *J. Am. Chem. Soc.* **66**, 747 (1944); **U.S. pat. 2,431,285** (1947 to Winthrop-Stearns).

Hydrochloride, $C_{10}H_{15}NO_3$·HCl, *Bronkephrine.* Crystals, dec 199-200°. Soluble in water. Marketed as an 0.2% aq soln.

THERAP CAT: Bronchodilator.

3881. Ethyl Oenanthate. *Heptanoic acid ethyl ester;* ethyl heptanoate; ethyl *n*-heptoate; oenanthic ether; cognac oil, synthetic; oil of grapes; aether oenanthicus; oleum vitis viniferae. $C_9H_{18}O_2$; mol wt 158.24. C 68.31%, H 11.47%, O 20.22%. $CH_3(CH_2)_5COOC_2H_5$. Obtained commercially either by esterification of heptanoic acid, yielding a high-grade product; or by the direct esterification of coconut oil yielding a mixture of the ethyl esters of lauric, myristic, palmitic, and other higher fatty acids plus ethyl and isoamyl alcohol. For some flavoring applications the latter product is more desirable. Prepn: Rogers, *J. Am. Pharm. Assoc., Sci. Ed.* **12**, 503 (1923). Toxicity: P. M. Jenner *et al., Food Cosmet. Toxicol.* **2**, 327 (1964).

Pure ester, liquid. Fruity, wine-like odor and taste with burning aftertaste. d_4^{15} 0.8723; d_4^{25} 0.8630. bp_{760} 189°; bp_{35} 95°; bp_8 68°. Viscosity at 25°: 0.0111 g/cm/sec. Insol in water. Misc with alcohol, ether, chloroform. Forms an azeotrope with water, the azeotropic mixture boils at 98.5° and contains 72% w/w of ethyl oenanthate. LD_{50} orally in rats: > 34,640 mg/kg (Jenner).

Note: Artificial Ethyl Oenanthate, also known as *artificial cognac essence* or *oil of wine,* is a mixture of amyl and ethyl caprates with ethyl and isoamyl butyrates and caprylates.

USE: In the manuf of liqueurs. Plays an important part in the formulation of raspberry, gooseberry, grape, cherry, apricot, currant, bourbon, and other artificial essences.

3882. Ethyl Oxalacetate. *Oxobutanedioic acid diethyl ester;* diethyl oxalacetate; oxaloacetic ester. $C_8H_{12}O_5$; mol wt 188.18. C 51.06%, H 6.43%, O 42.51%. $C_2H_5OOCCH_2$-$COCOOC_2H_5$. Obtained by the action of sodium on a mixture of ethyl oxalate and ethyl acetate.

Oily liquid. d_4^{20} 1.131. bp_{24} 132°. $n_D^{16.6}$ 1.45614. Insol in water. Miscible with alc, benzene, ether. Dec when boiled at ordinary pressure. With water it gradually dec into alc and acetic and oxalic acids.

Semicarbazone, crystals, mp 162°.

3883. Ethylparaben. *4-Hydroxybenzoic acid ethyl ester;* ethyl *p*-hydroxybenzoate; Nipagin A; Ethyl Parasept; Solbrol A. $C_9H_{10}O_3$; mol wt 166.18. C 65.05%, H 6.07%, O 28.88%. Prepd by esterification of *p*-hydroxybenzoic acid: Cavill, Vincent, *J. Soc. Chem. Ind. (London)* **66**, 175 (1947).

Crystals, mp 116°. bp 297-298° (decompn). Freely sol in alcohol, ether. Soly in water at 20°: 0.070% w/w; at 25°: 0.075% w/w.

USE: Preservative for pharmaceuticals.

3884. Ethyl Pelargonate. *Nonanoic acid ethyl ester;* ethyl nonanoate; wine ether. $C_{11}H_{22}O_2$; mol wt 186.29. C 70.92%, H 11.90%, O 17.18%. $CH_3(CH_2)_7COOC_2H_5$. Toxicity data: P. M. Jenner *et al., Food Cosmet. Toxicol.* **2**, 327 (1964).

Liquid. d_4^{18} 0.866. bp ~220°. mp −44°. Insol in water. Misc with alc, ether. LD_{50} orally in rats: > 43,000 mg/kg (Jenner).

Note: The product of commerce is an alcoholic soln of various essences in form of a yellowish liq, d about 0.823. It is used in production of cognac-like beverages; 1 part ethyl pelargonate and 20 parts alcohol constitute one kind of artificial "Cognac Essence".

3885. Ethyl Phenylacetate. *Benzeneacetic acid ethyl ester;* α-toluic acid ethyl ester. $C_{10}H_{12}O_2$; mol wt 164.20. C 73.15%, H 7.37%, O 19.49%. $C_6H_5CH_2COOC_2H_5$. Prepd by heating benzyl cyanide with alcohol and H_2SO_4: Adams, Thal, *Org. Syn.* **2**, 27 (1922).

Liquid. Pleasant odor. d_4^{20} 1.0333. bp_{760} 226°; bp_{32} 135°; bp_{20} 121°. $n_D^{18.5}$ 1.49921. Absorption spectrum: Baly, Collie, *J. Chem. Soc.* **87**, 1344 (1905); Baly, Tryhorn, *ibid.* **107**, 1063 (1915).

USE: In perfumery.

3886. Ethyl Phosphorochloridite. *Phosphorochloridous acid diethyl ester;* diethylchlorophosphinate; diethylchlorophosphonate; diethylphosphorous acid chloride; ethyl chlorophosphite. $C_4H_{10}ClO_2P$; mol wt 156.55. C 30.69%, H 6.44%, Cl 22.65%, O 20.44%, P 19.79%. $(C_2H_5O)_2PCl$. Prepd by passing chlorine into diethylphosphorous acid in ligroin contg Na wire: Arbusow, Arbusow, *Ber.* **65**, 195 (1932); by heating triethyl phosphite and phosphorus trichloride; or by adding phosphorus trichloride to a soln of ethanol and diethylaniline in ether: Cook *et al., J. Chem. Soc.* **1949**, 2021.

Liquid. Characteristic odor of acid chlorides. Fumes in moist air; reacts violently with water. d_4^{20} 1.0816; d_0^0 1.0962; d_0^{20} 1.0747. bp_{760} 153-155° (also reported as bp_{760} 143-148°); bp_{30} 63-65°; bp_{15} 45-53°; bp_{10} 37-38°. n_D^{20} 1.4350.

USE: In peptide synthesis.

3887. 3-Ethyl-4-picoline. *3-Ethyl-4-methylpyridine;* β-collidine; 3-ethyl-γ-picoline; β-ethyl-γ-methylpyridine. $C_8H_{11}N$; mol wt 121.18. C 79.29%, H 9.15%, N 11.56%. Prepd from 2,6-dichloro-3-ethyl-4-methylpyridine by reduction with HI and red phosphorus: Ruzicka, Fornasir, *Helv. Chim. Acta* **2**, 338 (1919); from 3-acetyl-1,4-dimethyl-1,2,5,-6-tetrahydropyridine: Prelog, Komzak, *Ber.* **74**, 1705 (1941); from γ-methylnicotinic acid: Rabe, Jantzen, *Ber.* **54**, 925 (1921); from β-ethylpyridine: Koenigs, Hoffmann, *Ber.* **58**, 194 (1925); from crotonaldehyde and ammonia: Tschitschibabin, Oparina, *Ber.* **60**, 1877 (1927).

Liquid, aromatic odor. d_4^{17} 0.9286. bp_{753} 195-196°; bp_{23} 88-90°; bp_{12} 76°. Volatile with steam. Sparingly sol in water. Soluble in alcohol, ether, chloroform, dil acids.

Hydrochloride, $C_8H_{11}N$·HCl, hygroscopic platelets, sol in water.

3888. 4-Ethyl-2-picoline. *4-Ethyl-2-methylpyridine;* α-methyl-γ-ethylpyridine; 4-ethyl-α-picoline; α-collidine. $C_8H_{11}N$; mol wt 121.18. C 79.29%, H 9.15%, N 11.56%. Prepd from 2-picoline and acetic anhydride in the presence of zinc dust: H. Maier-Bode, J. Altpeter, *Das Pyridin und Seine Derivate* (Halle, 1934) p 54; Ger. pat. 390,333.

Liquid. d_4^0 0.9291; d_4^{16} 0.9268. bp_{751} 179°. Volatile with steam. Sol in alcohol, ether, benzene, dil acids. Sparingly sol in water.

3889. 5-Ethyl-2-picoline. *5-Ethyl-2-methylpyridine;* aldehyde-collidine; aldehydine; 2-methyl-5-ethylpyridine; 5-ethyl-α-picoline; 3-ethyl-6-methylpyridine. $C_8H_{11}N$; mol wt 121.18. C 79.29%, H 9.15%, N 11.56%. Prepd by heating acetaldehyde ammonia in a double vol of abs alc: Ador, Bayer, *Ann.* **155**, 297 (1870); by heating ammonia water and paraldehyde in presence of ammonium acetate: Dürkopf, Schlaugk, *Ber.* **21**, 294 (1888); Ger. pats. **347,820; 349,184; 349,267;** H. Maier-Bode, J. Altpeter, *Das Pyridin und Seine Derivate* (Halle, 1934) p 56; Frank *et al., J. Am. Chem. Soc.* **68**, 1368 (1946); Dunn, U.S. pat. **2,717,897** (1955 to UCC); R. L. Frank *et al., Org. Syn.* coll. vol. IV, 451 (1963). Continuous process: Takeba *et al.,* U.S. pat. **2,935,513** (1960 to Takeda).

Liquid. Aromatic odor. d_4^{23} 0.9184. bp_{747} 177.8°; bp_{20} 74-75°; bp_{17} 65-66°. n_D^{20} 1.4971. Volatile with steam. Practically insol in water. Sol in alcohol, ether, benzene, dil acids, concd H_2SO_4.

Hydrochloride, $C_8H_{11}N \cdot HCl$, hygroscopic needles, freely sol in water.

3890. 1-Ethyl-3-piperidinol. 1-Ethyl-3-hydroxypiperidine. $C_7H_{15}NO$; mol wt 129.20. C 65.07%, H 11.70%, N 10.84%, O 12.38%. Prepn: Biel et al., J. Am. Chem. Soc. **74**, 1485 (1952); Biel, U.S. pat. **2,802,007** (1957 to Lakeside).

Liquid. bp_{15} 93-95°. n_D^{14} 1.4777.

3891. Ethyl Propionate. *Propanoic acid ethyl ester.* $C_5H_{10}O_2$; mol wt 102.13. C 58.80%, H 9.87%, O 31.33%. $CH_3CH_2COOC_2H_5$.

Colorless liquid; fruity odor. d_4^{20} 0.891. bp 99°. mp −73°. n_D^{20} 1.3844. Flash pt, closed cup: 12°. Sol in about 60 parts water; miscible with alcohol, ether.

3892. 3-Ethylpyridine. "β-Lutidine". C_7H_9N; mol wt 107.16. C 78.46%, H 8.47%, N 13.07%.

Colorless to brownish liquid. d^{23} 0.940. bp 162-165°. n_D^{22} 1.5021. Slightly sol in water; freely sol in alcohol, or ether.

3893. 4-Ethylpyridine. γ-Ethylpyridine. C_7H_9N; mol wt 107.16. C 78.46%, H 8.47%, N 13.07%. Found in California petr: Hackmann et al., Rec. Trav. Chim. **62**, 229 (1943). Best prepd from pyridine by treatment with acetic anhydride and zinc: Dohrn, Horsters, Ger. pat. **390,333** (1924 to Schering); Wibaut, Arens, Rec. Trav. Chim. **60**, 119 (1941); Frank, Smith, Org. Syn. **27**, 38 (1947). Using iron powder: Tenenbaum, Fand, U.S. pat. **2,712,019** (1955 to Nepera Chem.).

Liquid. Obnoxious odor. Turns brown if not very pure. d_4^{22} 0.9404. bp_{750} 169.6-170.0°. n_D^{18} 1.5029. Sparingly sol in water; sol in alcohol, ether.

Caution: Symptoms similar to pyridine, q.v.

3894. Ethyl Salicylate. *2-Hydroxybenzoic acid ethyl ester;* Salicylic acid ethyl ester; salicylic ether; sal ethyl. $C_9H_{10}O_3$; mol wt 166.18. C 65.05%, H 6.07%, O 28.88%. $HOC_6H_4COOC_2H_5$.

Clear, very refractive liquid; pleasant odor resembling that of methyl salicylate. Becomes yellowish-brown on long exposure to air and light. d_4^{20} 1.131. bp 231-234°. mp +1°. n_D^{20} 1.5226. Slightly sol in water; miscible with alc, ether.

Protect from light.

USE: Manuf artificial perfumes.

THERAP CAT (VET): Has been used as a counterirritant.

3895. Ethyl Silicate. *Silicic acid tetraethyl ester;* tetraethyl silicate. $C_8H_{20}O_4Si$; mol wt 208.33. C 46.12%, H 9.68%, O 30.72%, Si 13.48%. $Si(OC_2H_5)_4$. Prepn from abs alcohol and silicon tetrachloride: J. J. Ebelmen, Ann. **57**, 319 (1846); A. W. Dearing, E. E. Reid, J. Am. Chem. Soc. **50**, 3058 (1928); G. Sumrell, E. E. Ham, ibid. **78**, 5573 (1956). Toxicological studies: H. F. Smyth, J. Seaton, J. Ind. Hyg. Toxicol. **22**, 288 (1940); V. H. Rowe et al., ibid. **30**, 332 (1948). Review: H. D. Cogan, C. A. Setterstrom, Ind. Eng. Chem. **39**, 1364 (1947).

Colorless, flammable liquid. mp −77°. bp 165-166°. Flash point 125°F (52°C). d_4^{20} 0.933; n_D^{25} 1.3818. Viscosity 0.6 cps. Practically insol in water, and slowly dec by it. Miscible with alcohol.

Caution: Potential symptoms of overexposure are irritation of skin, eyes and nose. See NIOSH Pocket Guide to Chemical Hazards (DHHS/NIOSH 90-117, 1990) p 114.

USE: In weatherproofing and hardening stone, arresting decay and disintegration; manuf weatherproof and acid-proof mortars and cements. In the "lost wax" process for casting of high-melting alloys.

3896. Ethylstibamine. Bayer 693; Neostibosan; Stibosamine; Astaril. A complex of p-aminobenzenestibonic acid (present largely as a tetramer), p-acetylaminobenzenestibonic acid (present largely as a dimer), antimonic acid and diethylamine in the approx molar ratio of 1:2:1:3. Contains 41 to 44% organic pentavalent antimony (the acetyl content is between 6 and 7%). The manuf requires special care since the compd is both a colloid and a molecular addition compd and slight variations affect its toxicity greatly: Schmidt, Angew. Chem. **43**, 963 (1930); U.S. pat. **1,988,632**, C.A. **29**, 1586 (1935).

Light-yellow to yellowish-brown powder. Freely sol in water forming neutral, colloidal solns. The pH of a 5% soln is between 6.5 and 7.6. A 25% soln is approx isotonic. Solns for injection should be prepd immediately before use with sterile distilled water, as they are unstable on heating or standing.

THERAP CAT: Antiprotozoal (Leishmania).

3897. Ethyl Sulfate. *Sulfuric acid monoethyl ester;* ethyl hydrogen sulfate; ethylsulfuric acid; sulfethylic acid; sulfovinic acid. $C_2H_6O_4S$; mol wt 126.13. C 19.04%, H 4.79%, O 50.74%, S 25.42%. $C_2H_5OSO_2OH$. Prepn: Hamid et al., J. Chem. Soc. **129**, 1098 (1926); Atwood, U.S. pat. **2,683,731** (1954 to Natl. Petro-Chemicals). Review: C. M. Suter, The Organic Chemistry of Sulfur (John Wiley, 1944) p 23-27.

Corrosive.

Barium salt, $C_4H_{10}BaO_8S_2$, barium ethyl sulfate. Dihydrate, lustrous leaflets. Freely soluble in water; slightly sol in alcohol. Not pptd by H_2SO_4 or sulfates.

Calcium salt, $C_4H_{10}CaO_8S_2$, calcium ethyl sulfate, calcium sulfovinate. Dihydrate, crystals. Soluble in water, alcohol.

See also Diethyl Sulfate.

USE: Intermediate in the manuf of ethanol from ethylene.

3898. Ethyl Sulfide. *1,1'-Thiobisethane;* diethyl sulfide; thioethyl ether. $C_4H_{10}S$; mol wt 90.19. C 53.27%, H 11.18%, S 35.55%. $(C_2H_5)_2S$. Made by distillation of sodium ethyl sulfate with a soln of K_2S.

Liquid; ethereal odor. d_4^{20} 0.837. bp 92°. n_D^{20} 1.44233. Insol in water; miscible with alcohol, ether.

USE: Solvent for anhydr mineral salts; in plating baths for coating metals with gold or silver.

3899. 2-(Ethylsulfonyl)ethanol. Ethylsulfonylethyl alcohol. $C_4H_{10}O_3S$; mol wt 138.19. C 34.77%, H 7.29%, O 34.73%, S 23.20%. $CH_3CH_2SO_2CH_2CH_2OH$. Prepn from 2-(ethylthio)ethanol + H_2O_2: Lee, Muller, U.S. pat. **2,805,964** (1957 to Pennsalt Chem.); Shostakovskii et al., Zh. Obshch. Khim. **30**, 1123 (1960).

Hygroscopic crystals from acetone, mp 46°. $bp_{2.5}$ 153°. d_4^{20} 1.25. Flash pt 188°C (371°F). Soluble in water and in many organic solvents. LD_{50} orally in mice: 18 g/kg.

Benzoate, $C_{11}H_{14}O_4S$, needles from alcohol, mp 118°.

USE: When liq, as solvent for polar org substances. Intermediate for pharmaceuticals, plasticizers, solvents. Humectant. Antistatic agent for synthetic fibers and fabrics.

3900. Ethyl Tartrate. *(R)-2,3-Dihydroxybutanedioic acid diethyl ester;* diethyl tartrate. $C_8H_{14}O_6$; mol wt 206.20. C 46.60%, H 6.84%, O 46.56%. $C_2H_5OOCCH(OH)CH(OH)COOC_2H_5$.
Colorless, thick, oily liquid. d_4^{20} 1.204. bp 280°; bp_{11} 150°. mp +17°. $[\alpha]_D^{20}$ 7.5°. n_D^{20} 1.4476. Slightly sol in water; miscible with alcohol or ether.

3901. Ethyl Tartrate, Acid. *(R)-2,3-Dihydroxybutanedioic acid monoethyl ester;* monoethyl tartrate. $C_6H_{10}O_6$; mol wt 178.14. C 40.45%, H 5.66%, O 53.89%. HOOCCH(OH)CH(OH)COOC_2H_5.
Colorless, very hygroscopic crystals. Sweet taste. mp 90°. Sol in water, alcohol; insol in ether. *Keep well closed.*
USE: Printing with Indol Blue and Crystal Fast Blue on fustian, etc.

3902. 2-(Ethylthio)ethanol. β-Hydroxydiethyl sulfide. $C_4H_{10}OS$; mol wt 106.19. C 45.24%, H 9.49%, O 15.07%, S 30.20%. $CH_3CH_2SCH_2CH_2OH$. Prepn from 2-chloroethanol and ethyl mercaptan: Demuth, Meyer, *Ann.* **240**, 310 (1887); Steinkopf *et al., Ber.* **53**, 1010 (1920); from 2-thioethanol and diethyl sulfate: Bergmann, *J. Am. Chem. Soc.* **74**, 829 (1952).
Liquid. d_{20}^{20} 1.015-1.025. bp_{760} 184.5°; bp_{28} 99°. Soluble in ether, other organic solvents.
USE: Intermediate for pesticides, lubricating and cutting oil additives, flotation agents, plasticizers.

3903. Ethyl p-Toluenesulfonate. *4-Methylbenzenesulfonic acid ethyl ester.* $C_9H_{12}O_3S$; mol wt 200.26. C 53.98%, H 6.04%, O 23.97%, S 16.01%. $CH_3C_6H_4SO_3C_2H_5$.
Monoclinic crystals. d 1.17. mp 33°. bp_{15} 173°.
USE: For ethylation.

3904. Ethyl Vanillin. *3-Ethoxy-4-hydroxybenzaldehyde;* ethylprotocatechuic aldehyde; bourbonal; ethylprotal; vanillal; Ethavan; Ethovan. $C_9H_{10}O_3$; mol wt 166.18. C 65.05%, H 6.07%, O 28.88%.

Colorless flakes possessing a finer and more intense vanilla odor and taste than vanillin. mp 77-78°. Sparingly sol in water; sol in alc, ether, glycerol, propylene glycol. Soly in 95% alcohol about 1 g/2 ml. LD_{50} orally in rats: >2000 mg/kg, P. M. Jenner *et al., Food Cosmet. Toxicol.* **2**, 327 (1964).
USE: In flavoring and perfumery.

3905. Ethynodiol. *(3β,17α)-19-Norpregn-4-en-20-yne-3,17-diol;* 17α-ethynyl-19-norandrost-4-ene-3β,17β-diol; 17α-ethynyl-4-estrene-3β,17β-diol; ED. $C_{20}H_{28}O_2$; mol wt 300.44. C 79.96%, H 9.39%, O 10.65%. Prepn: F. B. Colton, U.S. pat. 2,843,609 (1958 to Searle). Prepn of the 3-acetate, 17-acetate, and diacetate: P. D. Klimstra, U.S. pat. 3,176,013 (1965 to Searle); *see also* F. Sondheimer, Y. Klibansky, *Tetrahedron* **5**, 15 (1959). Pharmacokinetics and metabolism: C. J. Lewis *et al., Xenobiotica* **10**, 705 (1980). Comparative clinical study in oral contraceptives: M. H. Briggs, *J. Reprod. Med.* **28**, Suppl. 1, 92 (1983). Review of carcinogenicity studies: *IARC Monographs* **21**, 387-398 (1979). Comprehensive description: E. P. K. Lau, J. L. Sutter, in *Analytical Profiles of Drug Substances* vol. 3, K. Florey, Ed. (Academic Press, New York, 1974) pp 253-279.

Diacetate, $C_{24}H_{32}O_4$, *3β,17β-diacetoxy-17α-ethynyl-4-estrene, SC-11800, Femulen, Luteonorm, Luto-Metrodiol, Metrodiol.* Crystals from methanol + water, mp ~126-127°. $[\alpha]_D$ −72.5° (chloroform).
Diacetate mixture with mestranol, *Luteolas, Metrulen, Ovaras, Ovulen.*
Diacetate mixture with ethinyl estradiol, *Conova, Demulen, Miniluteolas.*
THERAP CAT: Progestogen; in combination with estrogen as oral contraceptive.

3906. Ethynylbenzene. Phenylacetylene. C_8H_6; mol wt 102.14. C 94.08%, H 5.92%. $C_6H_5C≡CH$. Prepd by dropping β-bromostyrene on molten potassium hydroxide and distilling: Hessler, *Org. Syn.* **2**, 67 (1922).
Liquid. d_4^{20} 0.9300. mp −44.8°. bp_{760} 142.4°; bp_{90} 75°; bp_{15} 39°. n_D^{20} 1.5489. Insol in water; miscible with alcohol, ether, other organic solvents.

3907. Etidocaine. *(±)-N-(2,6-Dimethylphenyl)-2-(ethylpropylamino)butanamide;* 2-(ethylpropylamino)-2',6'-butyroxylidide. $C_{17}H_{28}N_2O$; mol wt 276.42. C 73.87%, H 10.21%, N 10.13%, O 5.79%. Prepn: H. J. F. Adams *et al.,* Ger. pat. 2,162,744; *eidem,* U.S. pat. 3,812,147 (1972, 1974 both to Astra). Activity and toxicity studies: *eidem, J. Pharm. Sci.* **61**, 1829 (1972).

Hydrochloride, $C_{17}H_{28}N_2O.HCl$, *W-19053, Duranest.* Crystals from abs ethanol-ether and isopropanol-isopropyl-ether, mp 203-203.5°. LD_{50} in female mice (mg/kg): 6.7 i.v.; 99 s.c. (Adams, 1972).
THERAP CAT: Anesthetic (local).

3908. Etidronic Acid. *(1-Hydroxyethylidene)bisphosphonic acid;* (1-hydroxyethylidene)diphosphonic acid; ethane-1-hydroxy-1,1-diphosphonic acid; EHDP. $C_2H_8O_7P_2$; mol wt 206.03. C 11.66%, H 3.91%, O 54.36%, P 30.07%. Bone resorption inhibitor structurally related to inorganic pyrophosphoric acid and to clodronic acid, *q.q.v.* Also a sequestering and chelating agent. Effect on hydroxyapatite (durapatite, *q.v.*) crystal aggregation: N. M. Hansen *et al., Biochim. Biophys. Acta* **451**, 549 (1976); S. Bisaz *et al., ibid.* 560. Prepn by reaction of phosphorous trichloride, acetic acid, and aqueous alkali: H. von Bayer, K. A. Hoffmann, *Ber.* **30**, 1973 (1897); D. R. Peck, U.S. pat. 3,468,935 (1969 to Albright and Wilson Mfg.). Continuous prepn process: L. Rogovin *et al.,* U.S. pat. 3,400,147 (1968 to Procter and Gamble). Prepn from acetic anhydride and dimethylphosphite: R. R. Irani, R. S. Mitchell, U.S. pat. 3,475,486 (1969 to Monsanto). Prepn and characterization of the acid and disodium salt: F. Kasparek, *Monatsh.* **99**, 2016 (1968); B. Blaser *et al., Z. Anorg. Allgem. Chem.* **381**, 247 (1971). Use as builder for detergents: F. L. Diehl, U.S. pat. 3,159,581 (1964 to Proctor & Gamble). Pharmacology: M. D. Francis *et al., Science* **165**, 1264 (1969); J. Jowsey *et al., J. Lab. Clin. Med.* **76**, 126 (1970); B. S. Strates *et al., Biochim. Biophys. Acta* **244**, 121 (1971). Toxicology: G. A. Nolen, E. V. Buehler, *Toxicol. Appl. Pharmacol.* **18**, 548 (1971). Metabolism: W. R. Michael *et al., ibid.* **21**, 503 (1972). Clinical study in postmenopausal osteoporosis: R. P. Heaney, P. D. Saville, *Clin. Pharmacol. Ther.* **20**, 593 (1976); in Paget's disease: R. D. Altman *et al., N. Engl. J. Med.* **289**, 1379

(1973); R. D. Altman, *Am. J. Med.* **79**, 583 (1985); C. J. Preston *et al., Brit. Med.J.* **292**, 79 (1986); in hypercalcemia of cancer: E. Ryzen *et al., Arch. Intern. Med.* **145**, 449 (1985). Symposium on clinical efficacy in malignancy-related hypercalcemia: *Am. J. Med.* **82**, Suppl 2A, 1-78 (1987).

Crystallizes from water as the monohydrate. pK_1 1.35 ±0.08; pK_2 2.87 ±0.01; pK_3 7.03 ±0.01; pK_4 11.3. Very sol in water (69% at 20°C). Insol in acetic acid.

Disodium salt, $C_2H_6Na_2O_7P_2$, *disodium dihydrogen (1-hydroxyethylidene)bis[phosphonate], etidronate disodium, Calcimux, Didronel, Diphos, Etidron.* Crystallizes from water as the di- or tetrahydrate.

THERAP CAT: Calcium regulator.

3909. Etifelmin. *2-Diphenylmethylenebutylamine;* 2-ethyl-3,3-diphenyl-2-propenylamine; EDPA; Na III. $C_{17}H_{19}N$; mol wt 237.34. C 86.03%, H 8.07%, N 5.90%. Prepn: Heinrich, Heiger, **Ger. pat. 1,122,514** (1962 to Giulini GmbH), *C.A.* **57**, 733c (1962).

Hydrochloride, $C_{17}H_{19}N.HCl$, *Tensinase D.* Crystals, mp 232°. Freely sol in water.

Mixture of hydrochloride and nicotinate, *Gilutensin.*

THERAP CAT: CNS stimulant; antihypotensive.

3910. Etifoxine. *6-Chloro-N-ethyl-4-methyl-4-phenyl-4H-3,1-benzoxazin-2-amine;* 6-chloro-2-(ethylamino)-4-methyl-4-phenyl-4H-3,1-benzoxazine; HOE-36801. $C_{17}H_{17}ClN_2O$; mol wt 300.79. C 67.88%, H 5.70%, Cl 11.79%, N 9.31%, O 5.32%. Psychotropic agent with anxiolytic and anticonvulsant activity. Prepn: **Fr. pat. M7358**; H. Kuch *et al.,* **U.S. pat. 3,725,404** (1969, 1973 both to Hoechst); I. Hoffmann *et al., Arzneimittel-Forsch.* **20**, 975 (1970). Exptl psychopharmacological study: J. R. Boissier *et al., Therapie* **27**, 325 (1972). Analysis of EEG effects: D. Bente *et al., Arzneimittel-Forsch.* **25**, 944 (1975). Evaluation of psychotropic effect: R. Corsico *et al., Psychopharmacologia* **45**, 301 (1976).

Colorless crystals from petr ether, mp 90-92°. uv max (ethanol): 273 nm (ε 21200). LD_{50} orally in mice: 12 g/kg (Hoffmann).

Hydrochloride, $C_{17}H_{17}ClN_2O.HCl$, *Stresam.* Crystals, mp 150-151°.

THERAP CAT: Anxiolytic.

3911. Etilefrin. *α-[(Ethylamino)methyl]-3-hydroxybenzenemethanol; α-[(ethylamino)methyl]-m-hydroxybenzyl alcohol; m-hydroxy-α-(ethylaminomethyl)benzyl alcohol; α-(m-hydroxyphenyl)-β-(ethylamino)ethanol; ethylphenylephrine; etiladrianol.* $C_{10}H_{15}NO_2$; mol wt 181.23. C 66.27%, H 8.34%, N 7.73%, O 17.66%. Prepn: Goto, *J. Pharm. Soc. Japan* **74**, 318 (1954), *C.A.* **49**, 3960 (1955); **Span. pat. 273,-595** (1962 to Labs. Fher S.A.), *C.A.* **60**, 1649e (1964).

dl-Form, crystals, mp 147-148°.

dl-Form hydrochloride, $C_{10}H_{15}NO_2.HCl$, *Apocretin, Circupon, Effontil, Effortil, Effortilvet, Efortil, Ethyl Adrianol, Eti-Puren, Kertasin, Pulsamin, Tonus-forte.* Crystals, mp 121°. Bitter taste. Freely sol in water; sol in alcohol. Practically insol in chloroform.

THERAP CAT: Antihypotensive.

THERAP CAT (VET): Antihypotensive.

3912. Etiocholane. *(5β)-Androstane;* 5-epiandrostane. $C_{19}H_{32}$; mol wt 260.46. C 87.62%, H 12.38%. Parent compd of alkyl substituted etiocholanes, such as pregnane, cholane, coprostane. Prepd from etiocholane-17-one semicarbazone: Butenandt, Dannenbaum, *Z. Physiol. Chem.* **229**, 192 (1934).

Needles from acetone, mp 78-80°.

3913. Etiocholanic Acid. *(5β,17β)-Androstane-17-carboxylic acid;* aetiocholanic acid; etianic acid; etiocholane-17β-carboxylic acid. $C_{20}H_{32}O_2$; mol wt 304.47. C 78.90%, H 10.59%, O 10.51%. Prepn: Wieland *et al., Z. Physiol. Chem.* **161**, 80 (1926); Jacobs, Elderfield, *J. Biol. Chem.* **108**, 497 (1935); Tschesche, *Ber.* **68**, 9 (1935); Steiger, Reichstein, *Helv. Chim. Acta* **21**, 841 (1938).

Needles from glacial acetic acid; elongated leaflets from acetic acid. Sublimes at 0.002 mm press. and 160° bath temp, mp 228-229°. Insol in water; sol in pentane.

Methyl ester, $C_{21}H_{34}O_2$, needles from methanol, mp 99-101°.

Note: The stem name "etianic acid" was proposed by the Subcommittee on Steroid Nomenclature of the National Research Council as a replacement for "etiocholanic acid" in order to avoid the use of the same name for parent hydrocarbons of different carbon content: *J. Am. Chem. Soc.* **74**, 2817 (1952).

3914. Etiocobalamin. *Cobinamide dicyanide;* vitamin B_{12p}; Factor B. $C_{50}H_{72}CoN_{13}O_8$; mol wt 1042.14. C 57.63%, H 6.96%, Co 5.66%, N 17.47%, O 12.28%. Vitamin B_{12} factor obtained by removal of the nucleotide from cyanocobalamin by acid hydrolysis. Isoln from calf feces: Ford, Porter, *Biochem. J.* **51**, V (1952). Prepn by acid hydrolysis: Gant *et al., ibid.* **56**, XXXIV (1954); Friedrich, Bernhauer, *Angew. Chem.* **65**, 627 (1953). Prepn from factor V_{1a} and D(−)-1-amino-2-propanol: Bernhauer *et al., Helv. Chim. Acta* **43**, 696 (1960); Bernhauer, **U.S. pat. 3,072,674** (1963 to Hoffmann-La Roche).

Note: The term etiocobalamin is used by some authors to designate any cobalamin lacking the nucleotide group present in vitamin B$_{12}$.

3915. Etioporphyrin. *2,7,12,18-Tetraethyl-3,8,13,17-tetramethyl-21H,23H-porphine;* 1,3,5,8-tetramethyl-2,4,6,7-tetraethylporphine; etioporphyrin III; mesoetioporphyrin. $C_{32}H_{38}N_4$; mol wt 478.68. C 80.29%, H 8.00%, N 11.70%. Occurs in Bavarian oil shale, crude petr, ozokerite, amber, cannel coal and other hard varieties of bituminous coal: Treibs, *Ann.* **510,** 60 (1934); **517,** 184 (1935); *Angew. Chem.* **49,** 682 (1936). Prepd by decarboxylation of mesoporphyrin IX: Fischer, Treibs, *Ann.* **466,** 191, 206 (1928). Synthesis: Fischer, Stangler, *Ann.* **462,** 265 (1928); Johnson *et al., J. Chem. Soc.* **1959,** 3416. Structure: Abraham *et al., ibid.* **1961,** 3468. *Review:* Rimington, Kennedy, in M. Florkin, H. S. Mason, *Comparative Biochemistry* (Academic Press, New York, 1962) pp 557-614.

Long prismatic needles or butterflies from pyridine or from chloroform-petr ether, mp 360-363°. pKa 18. Absorption max: 246, 269, 396, 497, 532, 566, 620, 645 nm (log ε 3.90, 3.89, 5.22, 4.13, 3.99, 3.81, 3.65, 2.62).

Copper salt, $C_{32}H_{36}N_4Cu$, red needles from pyridine-acetic acid.

Magnesium salt, $C_{32}H_{36}N_4Mg$, crystals from methanol.

3916. Etiproston. *[1R-[1α(Z),2β(E),3α,5α]]-7-[3,5-Dihydroxy-2-[2-[2-(phenoxymethyl)-1,3-dioxolan-2-yl]ethenyl]cyclopentyl]-5-heptenoic acid;* (5Z,13E)-(8R,9S,11R,12R)-9,-11-dihydroxy-15,15-ethylenedioxy-16-phenoxy-17,18,19,20-tetranorprostadienoic acid; 15-deoxy-15,15-ethylenedioxy-16-phenoxy-17,18,19,20-tetranorprostaglandin F$_{2α}$; Prostavet; Vetiprost. $C_{24}H_{32}O_7$; mol wt 432.51. C 66.65%, H 7.46%, O 25.89%. Prostaglandin F$_{2α}$ analog with estrus cycle synchronizing activity. Prepn: W. Skuballa *et al.,* **Ger. pat. 2,434,133;** *eidem,* **U.S. pat. 4,088,775** (1976, 1978 both to Schering AG); and biological activity: W. Skuballa *et al., J. Med. Chem.* **21,** 443 (1978).

Colorless oil.
THERAP CAT (VET): Luteolytic.

3917. Etiroxate. *O-(4-Hydroxy-3,5-diiodophenyl)-3,5-diiodo-α-methyl-*DL-*tyrosine ethyl ester;* DL-*α-methylthyroxine ethyl ester;* CG-635. $C_{18}H_{17}I_4NO_4$; mol wt 818.96. C 26.40%, H 2.09%, I 61.98%, N 1.71%, O 7.81%. Deriv of thyroxine, *q.v.* Prepn: **Neth. pat. Appl. 6,614,150;** H. Kummer, R. Beckmann, **U.S. pat. 3,930,017** (1967, 1975 both to Grünenthal). Animal studies: R. Beckmann, *Arzneimittel-Forsch.* **29,** 499 (1979). Effect on iodine metabolism in man: D. Emrich, *ibid.* **27,** 422 (1977). Use in hyperlipoproteinemia: H. Banz, F. P. Gall, *Fortschr. Med.* **97,** 1942 (1979); *eidem, Med. Klin.* **75,** 51 (1980).

Cryst from ethanol, mp 156-157°.
Hydrochloride, $C_{18}H_{17}I_4NO_4 \cdot HCl$, *Skleronorm.*
THERAP CAT: Antihyperlipoproteinemic.

3918. Etisazol. *N-Ethyl-1,2-benzisothiazol-3-amine;* 3-*(ethylamino)-1,2-benzisothiazole;* Netrosylla. $C_9H_{10}N_2S$; mol wt 178.26. C 60.64%, H 5.65%, N 15.72%, S 17.99%. Prepd by the reaction of diphenyldisulfide-2,2'-dicarbonyl dichloride with ethylamine, followed by treatment with PCl$_5$ and ammonia: Boeshagen, *Ber.* **99,** 2566 (1966). Chemistry studies: Geiger *et al., ibid.* **102,** 1961 (1969); Boeshagen *et al., ibid.* **103,** 3166 (1970).

mp 78°.
Hydrochloride, $C_9H_{10}N_2S \cdot HCl$, *Bay Va 5387, Ectimar.* Crystals from ethanol, mp 171°.
THERAP CAT (VET): Antifungal.

3919. Etizolam. *4-(2-Chlorophenyl)-2-ethyl-9-methyl-6H-thieno[3,2-f][1,2,4]triazolo[4,3-a][1,4]diazepine;* 1-methyl-6-*o*-chlorophenyl-8-ethyl-4H-*s*-triazolo[3,4-*c*]thieno[2,3-*e*]-1,4-diazepine; Y-7131; Depas. $C_{17}H_{15}ClN_4S$; mol wt 342.85. C 59.56%, H 4.41%, Cl 10.34%, N 16.34%, S 9.35%. Prepn: M. Nakanishi *et al.,* **Ger. pat. 2,229,845;** *eidem,* **U.S. pat. 3,904,641** (1972, 1973 both to Yoshitomi). Pharmacology and toxicity studies: T. Tsumagari *et al., Arzneimittel-Forsch.* **28,** 1158 (1978). Effect on monoamine metabolism in brain: M. Setoguchi *et al., ibid.* 1165; on rage responses in cats: T. Fukuda, T. Tsumagari, *Japan. J. Pharmacol.* **33,** 885 (1983).

Crystals from toluene, mp 147-148°. LD_{50} in male, female rats, male, female mice (mg/kg): 3619, 3509, 4358, 4258 orally; 865, 825, 830, 783 i.p.; > 5000 s.c. (Tsumagari).

THERAP CAT: Anxiolytic.

3920. Etodolac. *1,8-Diethyl-1,3,4,9-tetrahydropyrano-[3,4-b]indole-1-acetic acid;* etodolic acid; AY-24236; Edolan; Lodine; Ramodar; Tedolan; Ultradol; Zedolac. $C_{17}H_{21}NO_3$; mol wt 287.36. C 71.06%, H 7.37%, N 4.87%, O 16.70%. Prepn: C. A. Demerson *et al.,* **Ger.** pat. **2,301,525** (1973 to Ayerst); *eidem,* U.S. pat. **3,939,178** (1976 to Am. Home Products); *eidem, J. Med. Chem.* **19,** 391 (1976). Anti-inflammatory and analgesic properties: R. R. Martel, J. Klicius, *Can. J. Physiol. Pharmacol.* **54,** 245 (1976). Metabolic disposition in animals and man: M. N. Cayen *et al., Drug. Metab. Rev.* **12,** 339 (1981); E. S. Ferdinandi *et al., Xenobiotica* **16,** 153 (1986). Mechanism of action: K. Inoue *et al., Arzneimittel-Forsch.* **41,** 235 (1991). Clinical comparison with sulindac in rheumatoid arthritis: G. Jacob *et al., Curr. Ther. Res.* **37,** 1124 (1985).

Crystals from hexane/chloroform, mp 145-148°.

THERAP CAT: Anti-inflammatory; analgesic.

3921. Etodroxizine. *2-[2-[2-[4-[(4-Chlorophenyl)phenylmethyl]-1-piperazinyl]ethoxy]ethoxy]ethanol;* 1-(p-chlorobenzhydryl)-4-[2-[2-(2-hydroxyethoxy)ethoxy]ethyl]piperazine; 1-(p-chloro-α-phenylbenzyl)-4-[2-[2-(2-hydroxyethoxy)ethoxy]ethyl]diethylenediamine; hydrochlorbenzethylamine. $C_{23}H_{31}ClN_2O_3$; mol wt 418.96. C 65.94%, H 7.46%, Cl 8.46%, N 6.69%, O 11.46%. Prepn: Morren, **Brit.** pat. **817,231** (1959). GC determn in plasma: R. Pentz, A. Schutt, *Arch. Toxicol.* **39,** 225 (1978). Clinical evaluations in insomnia: R. Loire, A. Perrin, *Lyon Med.* **219,** 1795 (1968); S. Fedeli, *Bruxelle Med. Belg.* **48,** 517 (1968). Toxicology: M. Giurgea, J. Puigdevall, *Proc. Eur. Soc. Study Drug Toxicity* **9,** 134 (1968).

Liquid, $bp_{0.01}$ 250°.

Dimaleate, $C_{23}H_{31}ClN_2O_3 \cdot 2C_4H_4O_4$, *Indunox, Drimyl.* LD_{50} orally in rats: 920 mg/kg (Giurgea, Puigdevall).

THERAP CAT: Hypnotic.

3922. Etofenamate. *2-[[3-(Trifluoromethyl)phenyl]-amino]benzoic acid 2-(2-hydroxyethoxy)ethyl ester; N-(α,α,α-trifluoro-m-tolyl)anthranilic acid 2-(2-hydroxyethoxy)ethyl ester;* B-577; TV-485; Bayrogel; Glasel; Rheumon gel; Traumon Gel. $C_{18}H_{18}F_3NO_4$; mol wt 369.34. C 58.54%, H 4.91%, F 15.43%, N 3.79%, O 17.33%. Percutaneously ac-

tive antiphlogistic agent. Prepn: K. H. Boltze *et al.,* **Ger.** pat. **1,939,112** corresp to U.S. pat. **3,692,818** (1971, 1972 both to Troponwerke). Series of articles on chemistry, analysis, biochemistry, pharmacology, toxicology, and clinical studies: *Arzneimittel-Forsch.* **27,** 1300-1363 (1977). Toxicity: H. Jacobi *et al., ibid.* 1333. Metabolism and GC determn: *ibid.* **31,** 9-21 (1981).

Pale yellow viscous oil, thermolabile at 180°. $bp_{.001}$ 130-135°. uv max (methanol): 286 ($E^{1\%}_{1cm}$ 423). n^{25}_D 1.564. Sol in lower alcohols, ethyl acetate, acetone, chloroform, ether, benzene. Soly in water at 22°: 0.16 mg/100 ml. LD_{50} in male, female rats (mg/kg): 292, 470 orally; 140, 226 i.v.; 373, 397 i.p.; 643, 568 s.c. (Jacobi).

THERAP CAT: Anti-inflammatory.

3923. Etofibrate. *3-Pyridinecarboxylic acid 2-[2-(4-chlorophenoxy)-2-methyl-1-oxopropoxy]ethyl ester; nicotinic acid 2-hydroxyethyl ester 2-(p-chlorophenoxy)-2-methylpropionate (ester);* ethylene glycol 1-[2-(p-chlorophenoxy)-2-methylpropionate]-2-nicotinate; α-[(p-chlorophenoxy)isobutyroyl]-β-(nicotinoyl)glycol; ethofibrate; Lipo-Merz. $C_{18}H_{18}ClNO_5$; mol wt 363.80. C 59.43%, H 4.99%, Cl 9.75%, N 3.85%, O 21.99%. Glycol diester of nicotinic and clofibric acids, q.q.v. Prepn: A. Scherm, D. Peteri, **Ger.** pat. **1,941,217;** *eidem,* U.S. pat. **3,723,446** (1971, 1973 both to Merz). Lipid lowering effects in rats: W. Sterner, A. Schultz, *Arzneimittel-Forsch.* **24,** 1990 (1974). Metabolism: H. Oelschläger *et al., ibid.* **30,** 984 (1980). *In vitro* effect on human platelet function: M. P. Ortega *et al., Thromb. Res.* **19,** 409 (1980). HPLC determn in plasma: E. R. Garrett, M. R. Gardner, *J. Pharm. Sci.* **71,** 14 (1982). Bioavailability and pharmacokinetics in humans: K. I. Johnson *et al., Arzneimittel-Forsch.* **34,** 1785 (1984). Review of pharmacology and comparison with other hypolipidemic agents: R. Paoletti *et al., Am. J. Cardiol.* **52,** Suppl B, 21B-27B (1983).

Crystalline powder, mp 100°.

THERAP CAT: Antihyperlipoproteinemic.

3924. Etofylline. *3,7-Dihydro-7-(2-hydroxyethyl)-1,3-dimethyl-1H-purine-2,6-dione; 7-(2-hydroxyethyl)theophylline;* oxyethyltheophylline; Oxytheonyl; Oxphylline; Phyllocormin N. $C_9H_{12}N_4O_3$; mol wt 224.22. C 48.21%, H 5.39%, N 24.99%, O 21.41%. Obtained by the action of glycol monochlorohydrin upon theophylline in an alkaline medium: Lespagnol *et al., Bull. Soc. Pharm. Lille* **1948,** no. 2, p 18; Roth, *Arch. Pharm.* **292,** 234 (1959); Fabbrini, Cencioni, *Farmaco Ed. Sci.* **17,** 660 (1962).

Crystals from abs ethanol, mp 158°. Freely sol in water. Moderately sol in alcohol. pH of a 5% aq soln 6.5-7.0. Solns may be sterilized by heating. Pharmacological action as of theophylline.

Combination with theophylline, *Cordalin.*
THERAP CAT: Bronchodilator.

3925. Etofylline Nicotinate. *3-Pyridinecarboxylic acid 2-(1,2,3,6-tetrahydro-1,3-dimethyl-2,6-dioxo-7H-purin-7-yl)-ethyl ester; nicotinic acid ester with 7-(2-hydroxyethyl)theophylline;* 7-(2-hydroxyethyl)theophylline nicotinate; Hesotanol; Hesotin. $C_{15}H_{15}N_5O_4$; mol wt 329.32. C 54.71%, H 4.59%, N 21.27%, O 19.43%. Prepn: Pongratz, Zirm, *Monatsh.* **88**, 330 (1957). Pharmacology: Fischbach, Haas, *Arzneimittel-Forsch.* **17**, 313 (1967).

Fine dendritic needles from abs alcohol, mp 151-152°. Hydrated crystals from water. One gram dissolves in 50 ml boiling water; moderately sol in alcohol.
THERAP CAT: Vasodilator.

3926. Etoglucid. *2,2′-(2,5,8,11-Tetraoxadodecane-1,12-diyl)bisoxirane; 1,2:15,16-diepoxy-4,7,10,13-tetraoxahexadecane;* 1,2-bis[2-(2,3-epoxypropoxy)ethoxy]ethane; triethylene glycol diglycidyl ether; TDE; ethoglucid; ICI-32865; Epodyl. $C_{12}H_{22}O_6$; mol wt 262.30. C 54.95%, H 8.45%, O 36.60%. Prepn: Greenshields *et al.,* **Brit.** pat. **901,876** (1962 to ICI); Blyakhman, *Zh. Org. Khim.* **3**, 1423 (1967), *C.A.* **68**, 59384k (1968); X.-P. Gu *et al., Synthesis* **1985**, 649. Pharmacology and metabolism: Duncan, Snow, *Biochem. J.* **82**, 8P (1962); James, Solheim, *Xenobiotica* **1**, 43 (1971). Clinical evaluation in bladder tumors: M. R. G. Robinson *et al., J. Urol.* **118**, 972 (1977); J. Flamm, F. Grof, *Urologe* **B27**, 26 (1987).

Colorless liquid, mp −15° to −11°. bp$_{0.005}$ 140°; bp$_{0.1}$ 133-149°; bp$_2$ 195-197°. d^{20} 1.1312. n_D^{20} 1.4584.
THERAP CAT: Antineoplastic.

3927. Etomidate. *(R)-1-(1-Phenylethyl)-1H-imidazole-5-carboxylic acid ethyl ester; (R)-(+)-1-(α-methylbenzyl)imidazole-5-carboxylic acid ethyl ester;* R-16659; Amidate; Hypnomidate. $C_{14}H_{16}N_2O_2$; mol wt 244.29. C 68.83%, H 6.60%, N 11.47%, O 13.10%. Prepn of the (±)-form: **Belg.** pat. **662,474** corresp to E. F. Godefroi *et al.,* **U.S.** pat. **3,354,173** (1965, 1967 both to Janssen); *eidem, J. Med. Chem.* **8**, 220 (1965). Prepn of the (R)-(+)-form: L. F. C. Roevens *et al.,* **Ger.** pat. **2,609,573** corresp to U.S. pat. **3,991,072** (both 1976 to Janssen). Metabolism: E. Goetz, *Anaesthesist* **23**, 331 (1974). Pharmacokinetics: M. J. Van Hamme *et al., Anesthesiology* **49**, 274 (1978). Pharmacology: P. A. J. Janssen *et al., Arch. Int. Pharmacodyn. Ther.* **214**, 92 (1975). Clinical study: R. J. Fragen, N. Caldwell, *Anesthesiology,* **50**, 242 (1979). Comprehensive description: Z. L. Chang, J. B. Martin, in *Analytical Profiles of Drug Substances* vol. **12**, K. Florey, Ed. (Academic Press, New York, 1983) pp 191-214.

Crystals from diisopropyl ether, mp 67°. $[\alpha]_D^{20}$ +66° (c = 1 in ethanol). uv max (isopropanol): 240 nm (ϵ 12200). Soly in water at 25°: 0.0045 mg/100 ml. Sol in chloroform, methanol, ethanol, propylene glycol, acetone. LD$_{50}$ in mice, rats (mg/kg): 29.5, 14.8-24.3 i.v. (Janssen).
THERAP CAT: Hypnotic.

3928. Etomidoline. *2-Ethyl-2,3-dihydro-3-[[4-[2-(1-piperidinyl)ethoxy]phenyl]amino]-1H-isoindol-1-one;* 2-ethyl-3-(β-piperidino-p-phenetidino)phthalimidine; K-2680; Smedolin. $C_{23}H_{29}N_3O_2$; mol wt 379.50. C 72.79%, H 7.70%, N 11.07%, O 8.43%. Anticholinergic antispasmodic. Prepn: N. Giraldi, V. Mariotti, **S. Afr.** pat. **6,705,704;** *eidem,* **U.S.** pat. **3,624,206** (1968, 1971 both to Erba). Quantitative evaluation in humans: G. Sacchetti *et al., Farmaco Ed. Prat.* **27**, 708 (1972). Spasmolytic properties: M. Takeda, *Oyo Yakuri* **7**, 903 (1973), *C.A.* **80**, 66678 (1974). Pharmacology: M. Takeda *et al., ibid.* 1073, *C.A.* **81**, 33285 (1974); T. Mukai *et al., Japan. J. Pharmacol.* **31**, 147 (1981), *C.A.* **95**, 18068 (1981).

Crystals from ligroin, mp 106-107°. LD$_{50}$ in mice (mg/kg): 176 orally; 44.8 i.v. (Giraldi, Mariotti, 1971).
THERAP CAT: Antispasmodic.

3929. Etonitazene. *2-[(4-Ethoxyphenyl)methyl]-N,N-diethyl-5-nitro-1H-benzimidazole-1-ethanamine; 1-[(2-diethylamino)ethyl]-2-(p-ethoxybenzyl)-5-nitrobenzimidazole;* 2-p-ethoxybenzyl-1-(2-diethylaminoethyl)-5-nitrobenzimidazole. $C_{22}H_{28}N_4O_3$; mol wt 396.49. C 66.65%, H 7.12%, N 14.13%, O 12.11%. Prepn: Hunger *et al., Experientia* **13**, 400 (1957); Hoffmann *et al.,* **U.S.** pat. **2,935,514** (1960 to Ciba). Crystal structure: C. Humblet *et al., Acta Crystallogr.* **B34**, 3828 (1978). Binds selectively to mu-opioid receptors: M. S. Moolten *et al., Life Sci.* **52**, PL199 (1993).

Hydrochloride, $C_{22}H_{28}N_4O_3$·HCl, mp 162-164°.
Note: This is a controlled substance (opiate) listed in the U.S. Code of Federal Regulations, Title 21 Part 1308.11 (1995).
THERAP CAT: Analgesic (narcotic).

3930. Etoperidone. *2-[3-[4-(3-Chlorophenyl)-1-piper-azinyl]propyl]-4,5-diethyl-2,4-dihydro-3H-1,2,4-triazol-3-one.* $C_{19}H_{28}ClN_5O$; mol wt 377.92. C 60.39%, H 7.47%, Cl 9.38%, N 18.53%, O 4.23%. Psychotropic drug related structurally to trazodone, *q.v.* Prepn: G. Palazzo, **Ger.** pat. **2,351,739;** *idem,* **U.S.** pat. **3,857,845** (both 1974 to Angelini). Pharmacology and acute toxicity: R. Lisciani *et al., Arzneimittel-Forsch.* **28**, 417 (1978); M. T. Ramacci *et al., ibid.* **29**, 294 (1979). Clinical evaluation: P. Bertoletti, G. *Clin. Med.* **58**, 393 (1977). Teratology study: S. Barcellona *et al., Toxicology* **8**, 87 (1977). Acute cardiovascular toxicity study: R. Lisciani *et al., ibid.* **10**, 151 (1978).

Liq, bp$_{0.5}$ 230°.

Hydrochloride, C$_{19}$H$_{28}$ClN$_5$O.HCl, *AF-1191, ST-1191, Axiomin, Depracer, Etoran, Staff, Tropene.* Crystals from isopropanol, mp 197-198°. pH (2% aq soln): 4.1 ± 0.2. Sol in water, alc; slightly sol in chloroform. Practically insol in acetone, benzene, ether. LD$_{50}$ in mice, rats (mg/kg): 580, 720 orally; 72, 62 i.v.; 135, 120 i.p. (Lisciani).

THERAP CAT: Antidepressant.

3931. Etoposide. *9-[(4,6-O-Ethylidene-β-D-glucopyranosyl)oxy]-5,8,8a,9-tetrahydro-5-(4-hydroxy-3,5-dimethoxyphenyl)furo[3′,4′:6,7]naphtho[2,3-d]-1,3-dioxol-6(5aH)-one; 4′-demethylepipodophyllotoxin 9-[4,6-O-ethylidene-β-D-glucopyranoside];* EPEG; NSC-141540; VP-16-213; Lastet; Vepesid. C$_{29}$H$_{32}$O$_{13}$; mol wt 588.57. C 59.18%, H 5.48%, O 35.34%. Semi-synthetic deriv of podophyllotoxin, *q.v.,* related structurally to teniposide, *q.v.* Prepn: C. Keller-Juslén *et al.,* U.S. pat. **3,524,844** (1970 to Sandoz); C. Keller-Juslén *et al., J. Med. Chem.* **14,** 936 (1971). Activity in exptl tumors: H. Stähelin, *Eur. J. Cancer* **9,** 215 (1973). Mechanism of action: S. B. Horwitz, J. D. Loike, *Lloydia* **40,** 82 (1977). Pharmacokinetics: F. R. Pelsor *et al., J. Pharm. Sci.* **67,** 1106 (1978). Clinical studies: R. L. Chard *et al., Cancer Treat. Rep.* **63,** 1755 (1979); R. E. Slayton *et al., ibid.* 2089. Teratogenicity and cytogenicity study: S. M. Sieber *et al., Teratology* **18,** 31 (1978). *Review:* B. F. Issell, S. T. Crooke, *Cancer Treat. Rev.* **6,** 107-124 (1979). Review of pharmacology and clinical experience: P. J. O'Dwyer *et al., N. Engl. J. Med.* **312,** 692-700 (1985). Review of clinical pharmacology, pharmacokinetics and assay methods: P. I. Clark, M. L. Slevin, *Clin. Pharmacokinet.* **12,** 223-252 (1987). Comprehensive description: J. J. M. Holthuis *et al.* in *Analytical Profiles of Drug Substances* **vol. 18,** K. Florey, Ed. (Academic Press, New York, 1989) pp 121-151.

Crystals from methanol, mp 236-251°. [α]$_D^{20}$ −110.5° (c = 0.6 in chloroform). uv max (abs. methanol): 283 nm (ε 4245). pKa 9.8.

THERAP CAT: Antineoplastic.

3932. Etorphine. *[5α,7α(R)]-4,5-Epoxy-3-hydroxy-6-methoxy-α,17-dimethyl-α-propyl-6,14-ethenomorphinan-7-methanol; tetrahydro-7α-(1-hydroxy-1-methylbutyl)-6,14-endo-ethenooripavine; 7,8-dihydro-7α-[1(R)-hydroxy-1-methylbutyl]-O⁶-methyl-6,14-endo-ethenomorphine; 7α-[1(R)-hydroxy-1-methylbutyl]-6,14-endo-ethenotetrahydrooripavine; 19-propylorvinol; tetrahydro-7α-(2-hydroxy-2-pentyl)-6,14-endo-ethenooripavine.* C$_{25}$H$_{33}$NO$_4$; mol wt 411.54. C 72.96%, H 8.08%, N 3.40%, O 15.55%. Narcotic analgesic. Prepn: **Belg. pat. 618,392;** K. W. Bentley, **Brit. pat. 937,214** (1962, 1963 both to J. F. MacFarlan); and structure: K. W. Bentley, D. G. Hardy, *Proc. Chem. Soc.* **1963,** 220. Improved prepn: W. R. Hydro, U.S. pat. **3,763,167** (1973 to U.S. Army). Determn by HPLC: I.

Jane, A. McKinnon, *J. Chromatog.* **323,** 19 (1985). Analgesic activity: G. F. Blane, *J. Pharm. Pharmacol.* **19,** 367 (1967). Pharmacology: Lister, *ibid.* **16,** 364 (1964). Use as tranquilizer in dogs: J. L. Crooks *et al., Vet. Rec.* **87,** 498 (1970); M. Grange *et al., Rev. Med. Vet.* **124,** 899 (1973); in large animals: A. M. Harthoorn, *J. Am. Vet. Med. Assoc.* **149,** 875 (1966); B. T. Alford *et al., ibid.* **164,** 702 (1974).

White crystals from aqueous ethoxyethanol, mp 214-217°. (7β-isomer, melts at 280°).

Hydrochloride, C$_{25}$H$_{33}$NO$_4$.HCl, *M-99.* mp 266-267°. (7β-isomer hydrochloride, dec at 290°)°.

Hydrochloride in combination with acepromazine, *q.v., Immobilon.*

3-Acetate, C$_{27}$H$_{35}$NO$_5$, *acetorphine, O³-acetyl-7,8-dihydro-7α-[1(R)-hydroxy-1-methylbutyl]-O⁶-methyl-6,14-endo-ethenomorphine, 3-O-acetyl-7α-[1(R)-hydroxy-1-methylbutyl]-6,14-endo-ethenotetrahydrooripavine, 3-O-acetyl-17-propylorvinol.* Crystals from methanol, mp 195°.

3-Acetate hydrochloride, C$_{27}$H$_{35}$NO$_5$.HCl, *M-183.* mp 206°.

Note: Acetorphine and etorphine are controlled substances (opium derivatives) listed in the U.S. Code of Federal Regulations, Title 21 Parts 1308.11, 1308.12 (1995). Use with extreme care; very small amounts may bring about respiratory paralysis and death: D. W. Upson, *Upson's Handbook of Clinical Veterinary Pharmacology* (VM Publishing, Bonner Springs, 1980) pp 380-381.

THERAP CAT (VET): Used to immobilize large animals.

3933. Etoxadrol. *[2S-[2α,4β(R*)]]-2-(2-Ethyl-2-phenyl-1,3-dioxolan-4-yl)piperidine; (+)-2-(2-Ethyl-2-phenyl-1,3-dioxolan-4-yl)piperidine; 2-ethyl-2-phenyl-4-(2-piperidyl)-1,3-dioxolanc.* C$_{16}$H$_{23}$NO$_2$; mol wt 261.36. C 73.53%, H 8.87%, N 5.36%, O 12.24%. Prepn of *dl*-form: Hardie *et al., J. Med. Chem.* **9,** 127 (1966); Hardie, Halverstadt, U.S. pat. **3,262,938** (1966 to Cutter Labs). Resolution of the α-racemates: Allen *et al.,* **Ger. Offen.** pat. **2,001,616** (1970 to Cutter Labs), *C.A.* **74,** 13129b (1971). Pharmacological studies: Traber *et al., J. Pharmacol. Exp. Ther.* **175,** 395 (1970); Hidalgo *et al., Anesth. Analg. (Cleveland)* **50,** 231 (1971); Kelly *et al., ibid.* 262.

(+)-Hydrochloride, C$_{16}$H$_{23}$NO$_2$.HCl, *CL-1848C, Thoxan.* Crystals from isopropanol, mp 221.5-222°. [α]$_D^{25}$ +16.63°.

THERAP CAT: Anesthetic (intravenous).

3934. Etozolin. *[3-Methyl-4-oxo-5-(1-piperidinyl)-2-thiazolidinylidene]acetic acid ethyl ester; 3-methyl-4-oxo-5-piperidino-Δ²,α-thiazolidineacetic acid ethyl ester; 2-carbethoxymethylene-3-methyl-5-piperidino-4-thiazolidone;* Gö-687; W-2900A; Elkapin. C$_{13}$H$_{20}$N$_2$O$_3$S; mol wt 284.38. C 54.91%, H 7.09%, N 9.85%, O 16.88%, S 11.28%. A member of a new class of heterocyclic compounds, the 2-methylenethiazolidones; a diuretic with choleretic properties. Prepn: Satzinger, *Ann.* **665,** 150 (1963); U.S. pat. **3,072,653** (1963 to Warner-Lambert). Pharmacology: O. Heidenreich *et al., Arzneimittel-Forsch.* **14,** 1242 (1964). Series of articles on activity, metabolism and toxicology: G. Satzinger *et al.,* *ibid.* **27,** 1742-1817 (1977).

Crystals from methanol, mp 140°. uv max (methanol): 283, 243 nm (log ϵ 4.32, 4.0). LD_{50} i.p. in mice, rats: 1.210, 1.575 g/kg.

Hydrochloride, $C_{13}H_{20}N_2O_3S \cdot HCl$, crystals, mp 158-159°.

Free acid, $C_{11}H_{16}N_2O_3S$, *Gödecke 3282, ozolinone.*

THERAP CAT: Diuretic.

3935. Etretinate. *(all-E)-9-(4-Methoxy-2,3,6-trimethylphenyl)-3,7-dimethyl-2,4,6,8-nonatetraenoic acid ethyl ester;* Ro-10-9359; Tegison; Tigason. $C_{23}H_{30}O_3$; mol wt 354.49. C 77.93%, H 8.53%, O 13.54%. Aromatic analog of retinoic acid, *q.v.* Prepn: W. Bollag *et al.,* Ger. pat. **2,414,619;** *eidem,* U.S. pats. **4,105,681** and **4,215,215** (1974, 1978, 1980 all to Hoffmann-La Roche). Effects on skin metabolism: W. P. Raab, B. M. Gmeiner, *Arch. Dermatol. Res.* **256,** 247 (1976). Toxicity study in organ culture: I. Lasnitzki, *Brit. J. Cancer* **35,** 110 (1977). Pharmacokinetics: R. Haenni, *Dermatologica* **157,** Suppl, 5 (1978). Mutation study: H. J. Juhl *et al., Mutat. Res.* **58,** 317 (1978). HPLC determn in plasma: R. Haenni *et al., J. Chromatog.* **162,** 615 (1979). Toxicology, carcinogenicity and teratogenicity study: J. J. Kamm, *J. Am. Acad. Dermatol.* **6,** 652 (1982). Clinical evaluation in psoriatic arthritis: M. L. Ciompi *et al., Int. J. Tissue React.* **10,** 25 (1988). *Review:* H. Mayer *et al., Experientia* **34,** 1105-1119 (1978). Review of pharmacology and therapeutic efficacy: A. Ward *et al., Drugs* **26,** 9-43 (1983); of clinical pharmacology: A. Vahlquist, O. Rollman, *Dermatologica* **175,** Suppl. 1, 20-27 (1987). Book: *The Retinoids,* Vol. **1-2,** M. B. Sporn *et al.,* Eds. (Academic Press, New York, 1984).

Crystals, mp 104-105°. LD_{50} in mice (1 day): >4000 mg/kg i.p. (Bollag). LD_{50} (20 day) in mice, rats (mg/kg): 1176, >2000 i.p.; >2000, >4000 orally (Kamm).

Free acid, *see* Acitretin.

THERAP CAT: Antipsoriatic.

3936. Etrimfos. *O-(6-Ethoxy-2-ethyl-4-pyrimidinyl)-phosphorothioic acid O,O-dimethyl ester; O,O-*dimethyl *O-*(2-ethyl-4-ethoxypyrimidinyl)-6-thionophosphate; SAN-197; Ekamet; Satisfar. $C_{10}H_{17}N_2O_4PS$; mol wt 292.30. C 41.09%, H 5.86%, N 9.58%, O 21.89%, P 10.60%, S 10.97%. Organophosphorous insecticide with stomach and contact activity against pests of fruit and vegetable crops. Prepn: K. Milzner, F. Reisser, Ger. pat. **2,209,554;** *eidem,* U.S. pat. **3,862,188** (1972, 1975 both to Sandoz). Physical properties and activity: H. J. Knutti, F. Reisser, *Proc. 8th Brit. Insect. Fungicide Conf.* **2,** 675 (1975). Residue analysis and toxicity study: M. C. Bowman *et al., J. Agr. Food Chem.* **26,** 35 (1978). Metabolism in plants: M. Akram *et al., ibid.* 925. Review of analytical methods: J. C. Karapally *et al., Anal. Methods Pestic. Plant Growth Regul.* **11,** 125-138 (1980).

Colorless oil, slight characteristic odor. mp −3.35°. d 1.195; n_D^{20} 1.5068. Vapor pressure at 20°: 6.5×10^{-5} mm Hg. Soly in water at 23-24°: 40 ppm. Miscible with ethanol, dimethyl sulfoxide, ethyl acetate, acetone, ether, chloroform, xylene, hexane. LD_{50} in male rats, mice (mg/kg): 1800, 437 orally; LC_{50} in carp (mg/l): 13.6 (48 hr); 13.3 (96 hr) (Knutti).

USE: Agricultural insecticide.

3937. Etryptamine. *α-Ethyl-1H-indole-3-ethanamine; 3-(2-aminobutyl)indole;* α-ethyltryptamine. $C_{12}H_{16}N_2$; mol wt 188.27. C 76.55%, H 8.57%, N 14.88%. Prepn: Heinzelman *et al., J. Org. Chem.* **25,** 1548 (1960); **Brit. pat. 933,786** (1963 to ICI).

Crystals from ethyl acetate + petr ether, mp 97-99°. uv max: 220.5, 281, 289.5 nm (ϵ 20500, 6000, 5200).

Acetate, $C_{12}H_{16}N_2 \cdot C_2H_4O_2$, *Monase.* Crystals from ethyl acetate + methanol, mp 165-166°.

Hydrochloride, $C_{12}H_{16}N_2 \cdot HCl$, mp 215.5-218°.

THERAP CAT: CNS stimulant.

3938. Etymemazine. *2-Ethyl-N,N,β-trimethyl-10H-phenothiazine-10-propanamine; 10-[3-(dimethylamino)-2-methylpropyl]-2-ethylphenothiazine;* 3-ethyl-10-(3-dimethylamino-2-methylpropyl)phenothiazine; 3-(3-ethyl-10-phenothiazinyl)-2-methyl-1-dimethylaminopropane; ethotrimeprazine; ethylisobutrazine; RP-6484; Diquel. $C_{20}H_{26}N_2S$; mol wt 326.51. C 73.57%, H 8.03%, N 8.58%, S 9.82%. Prepn: Jacob, Robert, U.S. pat. **2,837,518** (1958 to Rhône-Poulenc). Activity studies in pigs: Lambrerth, *Aust. Vet. J.* **44,** 333 (1968). Biotransformation studies: Robinson, *J. Pharm. Pharmacol.* **18,** 19 (1966).

Hydrochloride, $C_{20}H_{26}N_2S \cdot HCl$, *Nuital, Sergetyl.* mp 160-163°. Soluble in water, ethanol, methanol, acetone, chloroform. Practically insol in ether.

THERAP CAT: Antihistaminic.

THERAP CAT (VET): Tranquilizer.

3939. β-Eucaine. *2,2,6-Trimethyl-4-piperidinol benzoate (ester);* α-4-benzoyloxy-2,2,6-trimethylpiperidine; α-vinyldiacetonalkamine benzoate; benzamine; betacaine; eucaine B. $C_{15}H_{21}NO_2$; mol wt 247.34. C 72.84%, H 8.56%, N 5.66%, O 12.94%. Prepn from lower melting form (α-form) of 2,2,6-trimethyl-4-piperidinol and benzoyl chloride: Harries, *Ann.* **417,** 107, *see* p 175 (1918). Stereochemistry: King, *J. Chem. Soc.* **125,** 41 (1924); Stenlake, *J. Pharm. Pharmacol.* **6,** 164 (1954). Configuration: Perks, Russell, *ibid.* **19,** 318 (1966).

Crystals from petr ether, mp 70-71° (corr).

Hydrochloride, crystals from water, mp 277-279° (corr). pKa 9.4. One gram dissolves in 30 ml water, 35 ml alcohol, 6 ml chloroform; more sol in boiling water or boiling alcohol. Aq soln is neutral to litmus. MLD i.v. in rats: 15-25 mg/kg. *Incompat:* Salicylates.

d-Form, crystals from petr ether, mp 57-58° (corr).

d-Form hydrochloride, $[\alpha]_D$ +11.5° (water).

l-Form, crystals from petr ether, mp 57-58° (corr).
l-Form hydrochloride, crystals, mp 244-245° (corr). [α]$_D$ —11.3° (water).

THERAP CAT: Anesthetic (local).

THERAP CAT (VET): Has been used as a local anesthetic.

3940. Eucalyptol. *1,3,3-Trimethyl-2-oxabicyclo[2.2.2]-octane; 1,8-epoxy-p-menthane;* cineole; cajeputol. C$_{10}$H$_{18}$O; mol wt 154.25. C 77.87%, H 11.76%, O 10.37%. The chief constituent of oil of eucalyptus: Cloez, *Ann.* **154**, 372 (1870); Berry, *Australasian J. Pharm.* **1929**, 203; from wormwood oil: Wallach, Brass, *Ann.* **225**, 291 (1884). Identity with cajeputol: Wallach, *ibid.* 314. Identity with cineole: Jahns, *Ber.* **17**, 2941 (1884). Structure: Wallach, *Ann.* **291**, 342 (1896). Biosynthesis: Birch *et al., Tetrahedron Letters,* no. 3, 1 (1959). Activity: Jori *et al., Eur. J. Pharmacol.* **9**, 362 (1970). Cockroach repellent activity: T. H. Maugh, *Science* **218**, 278 (1982).

Colorless liquid; camphor-like odor; spicy, cooling taste. d$_{25}^{25}$ 0.921-0.923. bp 176-177°. mp 1.5°. n$_D^{20}$ 1.455-1.460. Flash pt, closed cup: 118°F (48°C). Practically insol in water; miscible with alcohol, chloroform, ether, glacial acetic acid, oils.

USE: Pharmaceutic aid (flavor).

3941. Eucalyptus. Gum wood; Australian fever tree; blue-gum tree. Dried leaves of *Eucalyptus globulus* Labill., *Myrtaceae;* this is a dwarf species, known as *Malee* in Australia, and richest in oil of Eucalyptus. *Habit.* Australia, cultivated in subtropics, Europe, N. Africa and Southern U.S. *Constit.* 1-3% volatile oil, tannin, resins, eucalyptic acid, rutin.

USE: Pharmaceutic aid (flavor).

3942. Eucalyptus Gum. Eucalyptus kino; red gum. Dried gummy exudation from *Eucalyptus longirostris* F. Muell. (*E. rostrata* Schlecht.) and other species of *Eucalyptus, Myrtaceae. Habit.* Australia. *Constit.* About 45% kino-tannic acid, kino red, glucoside, catechol, pyrocatechol.

THERAP CAT: Astringent, antidiarrheal.

3943. Eucatropine. *α-Hydroxybenzeneacetic acid 1,2,2,-6-tetramethyl-4-piperidinyl ester;* mandelic acid β-1,2,2,6-tetramethyl-4-piperidyl ester; β-4-hydroxy-1,2,2,6-tetramethyl-piperidine mandelic acid ester; *N*-methyl-β-vinyldiacetonalkamine* mandelate; β-4-mandeloyloxy-1,2,2,6-tetramethyl-piperidine; β-1,2,2,6-tetramethyl-4-piperidinol mandelate; Euphthalmine. C$_{17}$H$_{25}$NO$_3$; mol wt 291.39. C 70.07%, H 8.65%, N 4.81%, O 16.47%. Prepn from higher melting form (β-form) of 1,2,2,6-tetramethyl-4-piperidinol and mandelic acid: Harries, *Ann.* **296**, 328 (1897); Kipping, *J. Chem. Soc.* **123**, 3115 (1923). Description: Harries, *Ber.* **31**, 665 (1898).

dl-Form, prisms from petr ether sinters at 108°, mp 113°. Practically insol in cold water, more sol in hot water; readily sol in organic solvents other than petr ether.
dl-Form hydrochloride, crystals from alcohol + ether, sinters at 181°, mp 183-184°. Very sol in water; sol in alcohol (1 g dissolves in 2 ml boiling alcohol), in chloroform. Practically insol in ether. Aq solns are neutral to litmus.

THERAP CAT: *dl*-Form hydrochloride as anticholinergic (ophthalmic).

3944. Eugenol. *2-Methoxy-4-(2-propenyl)phenol; 4-allyl-2-methoxyphenol;* allylguaiacol; eugenic acid; caryophyllic acid. C$_{10}$H$_{12}$O$_2$; mol wt 164.20. C 73.15%, H 7.37%, O 19.49%. Obtained from many natural sources: *Beilstein* **vol. 6**, 961. Prepn: Claisen, *Ann.* **418**, 113 (1919); from oil of cloves: Waterman, Priester, *Rec. Trav. Chim.* **48**, 1272 (1929). Toxicity study: E. C. Hagan *et al., Toxicol. Appl. Pharmacol.* **7**, 18 (1965).

Colorless or pale yellow liquid, bp 255°. Darkens and thickens on exposure to air. Odor of cloves; spicy, pungent taste. mp —9.2° to —9.1°. d$_4^{20}$ 1.0664. n$_D^{20}$ 1.5410. Practically insol in water. Misc with alcohol, chloroform, ether, oils. One ml dissolves in 2 ml 70% alcohol; sol in glacial acetic acid, in aq fixed alkali hydroxide solns. Ferric chloride, potassium permanganate. LD$_{50}$ in rats, mice (mg/kg): 2680, 3000 orally (Hagan).

Benzoate, C$_{17}$H$_{16}$O$_3$, *O-benzoyleugenol.* Crystals, mp 69-70°. bp 360°. Freely sol in benzene, chloroform, ether, hot alcohol. Practically insol in water.

USE: In perfumery instead of oil of cloves; manuf vanillin. As insect attractant.

THERAP CAT: Analgesic (dental).

3945. Euonymus. Wahoo; Indian arrow wood; bitter ash; burning bush; strawberry tree; spindle tree. Dried root bark of *Evonymus atropurpureus* Jacq., *Celastraceae. Habit.* U.S., Ontario to Florida east of Mississippi. *Constit.* Euonymol—an alcohol; euonysterol, atropurpurin, asparagine, citrullol, dulcite; malic, citric and tartaric acids.

THERAP CAT: Cathartic, diuretic.

3946. Euparin. *1-[6-Hydroxy-2-(1-methylethenyl)-5-benzofuranyl]ethanone; 6-hydroxy-2-isopropenyl-5-benzofuranyl methyl ketone;* 5-acetyl-6-hydroxy-2-isopropenylbenzofuran. C$_{13}$H$_{12}$O$_3$; mol wt 216.24. C 72.21%, H 5.59%, O 22.20%. From roots of *Eupatorium purpureum* L., *E. cannabinum* L., *Compositae:* Kamthong, Robertson, *J. Chem. Soc.* **1939**, 925; Jermanowska, *C.A.* **48**, 5848h (1954). Structure: Kamthong, Robertson, *J. Chem. Soc.* **1939**, 933. Synthesis: Ramachandran *et al., J. Org. Chem.* **28**, 2744 (1963). *Review:* Gizycki, *Pharmazie* **6**, 686 (1951).

Yellow needles from alcohol, mp 121-122°. uv max (ethanol): 263, 358 nm (ε 34400, 5900). Sol in alcohol, benzene, chloroform, ether. Practically insol in water, alkali.

Acetate, C$_{15}$H$_{14}$O$_4$, prisms from petr ether, mp 80°.

O-Methyl deriv, C$_{14}$H$_{14}$O$_3$, *O-methyleuparin,* 2-isopropenyl-6-methoxy-5-benzofuranyl methyl ketone. Needles from dil alcohol, mp 76-77°. Sol in the usual organic solvents.

3947. Eupatorin. *5-Hydroxy-2-(3-hydroxy-4-methoxyphenyl)-6,7-dimethoxy-4H-1-benzopyran-4-one; 3',5-dihydroxy-4',6,7-trimethoxyflavone.* C$_{18}$H$_{16}$O$_7$; mol wt 344.32. C 62.79%, H 4.68%, O 32.53%. Isoln from *Eupatorium semiserratum* DC., *Compositae* and structure: Kupchan *et al., J. Pharm. Sci.* **54**, 929 (1965).

Crystals from dioxane + water, mp 196-198°. uv max (alcohol): 243, 254, 274, 342 nm (ε 17400, 19300, 19800, 27700).

3948. Eupatorium. Boneset; thoroughwort. Flowering tops and dried leaves of *Eupatorium perfoliatum* L., *Compositae. Habit.* Canada to Florida and west to Texas and Nebraska. *Constit.* Eupatorin, volatile oil, resin, tannin, sugar, inulin, wax.

3949. Euphorbia. Pill-bearing spurge; snake-weed; cat's hair; Queensland asthma weed; flowery-headed spurge. Dried herb of *Euphorbia hirta* L. or *E. pilulifera* L., *Euphorbiaceae. Habit.* Queensland (Australia), India, widely distributed in tropical countries. *Constit.* Several resins, an unstable glucoside.

3950. Euprocin. *(8α,9R)-10,11-Dihydro-6'-(3-methylbutoxy)cinchonan-9-ol;* hydrocupreine isopentyl ether; isoamylhydrocupreine; isopentylhydrocupreine; eucupine; Eucupin. $C_{24}H_{34}N_2O_2$; mol wt 382.55. C 75.35%, H 8.96%, N 7.32%, O 8.36%. Prepd from hydrocupreine and *p*-toluenesulfonic acid isopentyl ester: Slotta, Behnisch, *Ber.* **66**, 360 (1933).

Almost tasteless powder, mp 152°. Sol in alc, ether, chloroform; practically insol in water. LD s.c. in mice: 300 mg/kg; LD i.v. in rabbits: 13 mg/kg, *Handbook of Toxicology* vol. **1**, W. S. Spector, Ed. (Saunders, Philadelphia, 1956) pp 140-141.

Dihydrochloride monohydrate, $C_{24}H_{34}N_2O_2.2HCl.H_2O$, fine bitter needles. Sol in 15 parts water; freely sol in alc.
THERAP CAT: Anesthetic (local); antiseptic.

3951. Europium. Eu; at. wt 151.965; at. no. 63; valences 2, 3. A lanthanide; belongs to cerium group of rare earth metals. Naturally occurring isotopes (mass numbers): 151 (47.8%); 153 (52.2%); known artificial radioactive isotopes: 138-150; 152; 154-160. Abundance in earth's crust: 1.06-2.1 ppm. Commercially important sources are the rare earth minerals monazite and bastnaesite; also found in gadolinite. Has been detected spectroscopically in the sun and in certain stars. Discovered and prepd as the oxide: Demarçay, *Compt. Rend.* **122**, 728 (1896); **130**, 1019, 1469 (1900); **132**, 1484 (1901). Separation: Spedding *et al.*, *J. Am. Chem. Soc.* **76**, 2557 (1954); by paper chromatography: Lederer, *Nature* **176**, 462 (1953). Toxicity study: Haley, *J. Pharm. Sci.* **54**, 663 (1965). Reviews of prepn, properties and compds: *The Rare Earths*, F. H. Spedding, A. H. Daane, Eds. (Krieger, Huntington, N.Y., 1971, reprint of 1961 ed.) 641 pp; S. P. Sinha, *Europium* (Springer, New York, 1967) 164 pp; Hulet, Bode, "Separation Chemistry of the Lanthanides and Transplutonium Actinides" in *MTP Int. Rev. Sci.: Inorg. Chem., Ser. One, vol. 7*, K. W. Bagnall, Ed. (University Park Press, Baltimore, 1972) pp 1-45; Moeller, "The Lanthanides" in *Comprehensive Inorganic Chemistry*, vol. **4**, J. C. Bailar, Jr. *et al.*, Eds. (Pergamon Press, Oxford, 1973) pp 1-101; F. H. Spedding in Kirk Oth-

mer *Encyclopedia of Chemical Technology* vol. **19** (John Wiley & Sons, New York, 3rd ed., 1982) pp 833-854; *Chemistry of the Elements*, N. N. Greenwood, A. Earnshaw, Eds. (Pergamon Press, New York, 1984) pp 1423-1449. Brief review of properties: G. T. Seaborg, *Radiochim. Acta* **61**, 115-122 (1993).

Body-centered cubic crystal lattice; d 5.244; mp 826°. bp 1429°. Heat of fusion: 9.221 kJ/mol. Heat of sublimation (25°): 144.7 kJ/mol. Sol in liq ammonia. Shows two reduction potentials −0.710 and −2.510 v. (referred to a normal calomel electrode): Noddack, Brukl, *Angew. Chem.* **50**, 362 (1937); gives two definite series of salts, in one the metal is divalent, and in the other it is trivalent.

Sesquioxide, Eu_2O_3, *europia.* Pink powder, d 7.42, prepd by heating the hydroxide, nitrate, oxalate or sulfate at 1600°. The oxide of the divalent metal is prepd by reduction of the sesquioxide at elevated temp.

Hydroxide, $Eu(OH)_3$, prepd by adding ammonia or an alkali hydroxide to a soln of an europic salt.

Europic chloride, $EuCl_3$, greenish-yellow needles; mp 623° in nitrogen (in a closed tube), d^{35} 4.471, prepd by passing sulfur chloride over the heated oxide at 200-500°. LD_{50} of trichloride in mice: 550 mg/kg i.p.; 5 g/kg orally (Haley).

Europous chloride, $EuCl_2$. Prepd by reduction of $EuCl_3$ with hydrogen at 600°. White amorphous powder, sol in water.

Europic sulfate, $Eu_2(SO_4)_3$, octahydrate, a pinkish cryst solid, prepd by dissolving the oxide in sulfuric acid. Soly in water: 2.56 parts per 100 parts at 20°, 1.93 parts per 100 parts at 40°. On heating at 375° yields the anhydr sulfate.

Europic nitrate, $Eu(NO_3)_3$, hexahydrate, mp 85° in its water of crystallization (sealed tube). LD_{50} in rats (mg/kg): 210 i.p.; > 5000 orally (Haley).

Europous sulfate, $EuSO_4$, colorless crystals. Insol in water and in dil acids. Prepd by electrolytic reduction of europic salts.
USE: The salts in cathode ray tube coatings for color television receivers. Eu has a very high cross-section for the capture of thermal neutrons which is of value in the construction of electric atomic power stations. Organic derivs as shift reagents in NMR spectroscopy: C. C. Hinckley, *J. Am. Chem. Soc.* **91**, 5160 (1969); R. E. Sievers, *Nuclear Magnetic Resonance Shift Reagents* (Academic Press, New York, 1973).

3952. Evan's Blue. *6,6'-[(3,3'-Dimethyl[1,1'-biphenyl]-4,4'-diyl)bis(azo)]bis[4-amino-5-hydroxy-1,3-naphthalenedisulfonic acid] tetrasodium salt; C.I. Direct Blue 53;* 4,4'-bis[7-(1-amino-8-hydroxy-2,4-disulfo)naphthylazo]-3,3'-bitolyl tetrasodium salt; C.I. 23860; T-1824; Azovan Blue. $C_{34}H_{24}N_6Na_4O_{14}S_4$; mol wt 960.82. C 42.50%, H 2.52%, N 8.75%, Na 9.57%, O 23.31%, S 13.35%. Prepd by coupling 1 mol of diazotized *o*-tolidine with 2 mols of Chicago acid (1-amino-8-naphthol-2,4-disulfonic acid): Ger. pats. **35,341**; **38,802**; *Frdl.* **1**, 469, 488 (1877-1887); **3,949**; **57,327**; **75,469**; *Frdl.* **3**, 685, 687, 690 (1890-1894); Hartwell, Fieser, *Org. Syn.* **coll. vol. II**, 145 (1943). Diagnostic use: M. H. Nielsen, N. C. Nielsen, *Scand. J. Clin. Lab. Invest.* **14**, 605 (1962); O. Linderkamp *et al.*, *Eur. J. Pediat.* **125**, 135 (1977).

Blue crystals with bronze to green luster. Sol in water, alcohol, acids, alkalies. Indicator changing color near pH 10. Destroyed by strong oxidizing and reducing agents and precipitated from soln by strong concns of neutral salts. Rather stable in aq soln, and may be autoclaved at 15 lbs pressure for 30 min. Dye made up in physiological saline should not be autoclaved.
THERAP CAT: Diagnostic aid (blood volume determination).

3953. Evening Primrose Oil. EPO. Seed oil of the evening primrose, *Oenothera biennis* L., *Onagraceae*, which contains approx 72% linoleic acid and approx 9% γ-linolenic acid, *q.q.v.*, as the two main constituents. Unique among vegetable oils because of its high content of γ-linolenic acid. Effect on prostaglandin biosynthesis in rats: B. A. Schölkens *et al.*, *Prostaglandins, Leukotienes Med.* **8**, 273 (1982). Clinical studies in atopic eczema: C. R. Lovell *et al.*, *Lancet* **1**, 278 (1981); S. Wright, J. L. Burton, *ibid.* **2**, 1120 (1982); P. L. Biagi *et al.*, *Drugs Exp. Clin. Res.* **14**, 285 (1988). Ingredient in cosmetics for aging skin: J. P. Marty, **Ger.** **pat. 3,447,618** (1985 to Roussel-UCLAF), *C.A.* **103**, 146984r (1985). Brief review including discussion of uses: A. J. Barber, *Pharm. J.* **240**, 723-725 (1988). Clear, golden yellow oil. d_{15} 0.9283. n_D^{25} 1.4782. Sapon. no. 287.8. Iodine no. 154.8.

Note: Evening primrose oil products include **Efamol**, **Efamast**, **Epogam**.

USE: Dietary supplement.

THERAP CAT: In treatment of atopic eczema and mastaglia.

3954. Evodiamine. *8,13,13b,14-Tetrahydro-14-methyl-indolo[2',3':3,4]pyrido[2,1-b]quinazolin-5(7H)-one.* $C_{19}H_{17}$-N_3O; mol wt 303.36. C 75.23%, H 5.65%, N 13.85%, O 5.27%. From *Evodia rutaecarpa* Hook. & Thoms and bark of *Zanthoxylum rhetsa* DC., *Rutaceae:* Y. Asahina, K. Kashiwaki, *J. Pharm. Soc. Japan* **1915**, 1293, *C.A.* **10**, 607 (1916); Gopinath *et al.*, *Tetrahedron* **8**, 293 (1960). Structure: Y. Asahina *J. Pharm. Soc. Japan* **1924**, 1; Ohta, *J. Pharm. Soc. Japan* **65**, 15 (1945), *C.A.* **45**, 5697 (1951). Synthesis: Asahina, Ohta, *Ber.* **61B**, 319 (1928); T. Kametani *et al.*, *J. Am. Chem. Soc.* **98**, 6186 (1976); *eidem, Heterocycles* **4**, 23 (1976). Biosynthesis: M. Yamazaki *et al.*, *Tetrahedron Letters* **1966**, 3221; **1967**, 3317. Mass spec.: J. Tamas *et al.*, *Acta Chim. Acad. Sci. Hung.* **89**, 85 (1976).

Yellow plates from alc, mp 278°. $[\alpha]_D^{15}$ +352° (acetone); $[\alpha]_D$ +440° (chloroform). uv max (acetonitrile): 272, 280, 291, 335 nm (log ε 4.06, 4.02, 3.90, 3.30). Sol in acetone; slightly sol in alcohol, ether, chloroform. Practically insol in water, petr ether, benzene. Does not seem to form salts.

3955. Exalamide. *2-(Hexyloxy)benzamide;* HBA; Hyperan. $C_{13}H_{19}NO_2$; mol wt 221.30. C 70.56%, H 8.65%, N 6.33%, O 14.46%. Deriv of salicylamide, *q.v.* Prepn: E. M. Bevin *et al.*, *J. Pharm. Pharmacol.* **4**, 872 (1952); F. J. MacRae, D. E. Seymour, **Brit. pat. 726,786** (1955 to Herts Pharm.), *C.A.* **50**, 5751i (1956); L. V. Coates *et al.*, *J. Pharm. Pharmacol.* **9**, 855 (1957). Metabolism: Y. Kanaiwa *et al.*, *Yakugaku Zasshi* **95**, 411 (1975), *C.A.* **83**, 71428v (1975). General pharmacologic effects: R. Inoki *et al.*, *Oyo Yakuri* **10**, 165 (1975), *C.A.* **88**, 115369 (1976). Comparative study of antifungal efficacy vs clinical isolates: T. Kusunoki, S. Harada, *J. Dermatol.* **11**, 227 (1984).

Cryst from ethanol, mp 71°. Sol in methanol, acetone, chloroform, benzene. Slightly sol in ether. Practically insol in water.

THERAP CAT: Antifungal (topical).

3956. Exaltolide®. *Oxacyclohexadecan-2-one;* 15-hydroxypentadecanoic acid ε-lactone. $C_{15}H_{28}O_2$; mol wt 240.39. C 74.95%, H 11.74%, O 13.31%. Musk compd responsible for the pleasant odor of angelica root oil. Synthesis: Ruzicka, Stoll, *Helv. Chim. Acta* **11**, 1159 (1928); Carnduff *et al.*, *Chem. & Ind. (London)* **1960**, 559; Dhenkne *et al.*, *J. Chem. Soc.* **1962**, 2348; Mathur, Bhattacharya, *ibid.* **1963**, 3505; Becker, Ohloff, *Helv. Chim. Acta* **54**, 2889 (1971). Alternate syntheses: T. Mukaiyama *et al.*, *Chem. Letters*

1977, 441; K. Narasaka *et al.*, *ibid.* **763**; W. H. Kruizinga, R. M. Kellogg, *Chem. Commun.* **1979**, 286.

Thick oil. Odor of amber and musk. $bp_{0.25}$ 110°; bp_{15} 176°. d_4^{20} 0.9549. n_D^{20} 1.4708. When sublimed in vacuum, needles, mp 32°. Sol in alc.

USE: As a fixative in perfumery.

3957. Exametazime. *[R*,R*-(E,E)]-(±)-3,3'-[(2,2-Dimethyl-1,3-propanediyl)diimino]bis-2-butanone dioxime;* *(RR,SS)*-4,8-diaza-3,6,6,9-tetramethylundecane-2,10-dione bisoxime; (±)-(3RS,3'RS)-3,3'-[(2,2-dimethyltrimethylene)-diimino]di-2-butanone dioxime; *d,l*-hexamethylpropyleneamine oxime; *d,l*-HM-PAO; hexametazine; Ceretec. $C_{13}H_{28}$-N_4O_2; mol wt 272.39. C 57.32%, H 10.36%, N 20.57%, O 11.75%. Propyleneamine oxime derivative that forms a neutral, lipophilic complex with ^{99m}Tc which diffuses across cellular membranes, including the blood brain barrier. Used in regional cerebral blood flow imaging. Prepn: L. R. Canning *et al.*, **Eur. pat. Appl. 194,843;** *eidem,* **U.S. pat. 4,789,736** (1986, 1988 both to Amersham). Prepn and radiopharmacology of ^{99m}Tc complex: D. P. Nowotnik *et al.*, *Nucl. Med. Commun.* **6**, 499 (1985); R. D. Neirinckx *et al.*, *J. Nucl. Med.* **28**, 191 (1987). Clinical comparison of isomers: P. F. Sharp *et al.*, *ibid.* **27**, 171 (1986). Biodistribution in humans: K. Nakamura *et al.*, *Eur. J. Nucl. Med.* **15**, 100 (1989). Clinical evaluation in single photon emission tomography (SPECT) of meningiomas: S. Nakano *et al.*, *J. Nucl. Med.* **30**, 1101 (1989). ^{99m}Tc radiolabeling of human platelets: H. J. Danpure, S. Osman, *Nucl. Med. Commun.* **9**, 267 (1988); of human leukocytes: H. J. Danpure *et al.*, *ibid.* **465**. Clinical study of labeled leukocytes for imaging inflammatory lesions: J. H. Reynolds, *Clin. Radiol.* **42**, 195 (1990). Series of articles on basic and clinical studies: *J. Cereb. Blood Flow Metab.* **8**, S1-S126 (1988).

Crystals from ethyl acetate, mp 128-130°.

THERAP CAT: ^{99m}Tc complex as diagnostic aid (radioactive imaging agent).

3958. Exifone. *(2,3,4-Trihydroxyphenyl)(3,4,5-trihydroxyphenyl)methanone;* 2,3,3',4,4',5'-hexahydroxybenzophenone; 4-galloylpyrogallol; Adlone. $C_{13}H_{10}O_7$; mol wt 278.22. C 56.12%, H 3.62%, O 40.25%. Cognition enhancing agent. Prepn: H. Bleuler, A. G. Perkin, *J. Chem. Soc.* **109**, 529 (1916); see also **Ger. pat. 49,149** (1889 to BASF). Prepd but not claimed: J. M. Gazave *et al.*, **Ger. pat. 2,-501,443;** *eidem,* **U.S. pat. 4,015,017** (1975, 1977 both to Pharmascience). Pharmacology: R. D. Porsolt *et al.*, *Arzneimittel-Forsch.* **37**, 388 (1987). Animal studies in memory improvement: R. D. Porsolt *et al.*, *Pharmacol. Biochem. Behav.* **27**, 253 (1987); in antagonism of benzodiazepine-induced amnesia: R. D. Porsolt *et al.*, *Psychopharmacol.* **95**, 291 (1988). Clinical evaluation in senile dementia: F. Piette *et al.*, *Rev. Geriatr.* **11**, 375 (1986); in Parkinson's disease: H. Allain *et al.*, *Fundam. Clin. Pharmacol.* **2**, 1 (1988).

Crystals from aqueous solution, mp 270° (Gazave); also reported as faintly yellow needles from water, mp 272-273° (Bleuler, Perkin). LD_{50} in rats (g/kg): 0.355 i.p.; 1.425 orally (Gazave).

THERAP CAT: Nootropic.

3959. Exiproben. *2-[3-(Hexyloxy)-2-hydroxypropoxy]-benzoic acid.* $C_{16}H_{24}O_5$; mol wt 296.36. C 64.84%, H 8.16%, O 26.99%. Prepn: **Brit.** pat. **933,980** (1963 to Thomae), *C.A.* **60**, 2866a (1964); G. Ohnacher, **U.S.** pat. **3,198,827** (1965 to Boehringer, Ing.). Metabolism study:

A. Pinelli, *Pharmacol. Res. Commun.* **5**, 17 (1973). Choleretic activity: R. Bossa *et al., Fegato* **17**, 21 (1971), *C.A.* **76**, 135808j (1972). Clinical study: N. D'Imperio *et al., Clin. Ter.* **78**, 153 (1976).

Sodium salt, $C_{16}H_{23}NaO_5$, *DCH-21, Etopalin.* Solid, mp 147°.

THERAP CAT: Choleretic.

3960. Exosurf®. *(R)-4-Hydroxy-N,N,N-trimethyl-10-oxo-7-[(1-oxohexadecyl)oxy]-3,5,9-trioxa-4-phosphapentaco-san-1-aminium hydroxide, inner salt, 4-oxide mixture with formaldehyde polymer with oxirane and 4-(1,1,3,3-tetrameth-ylbutyl)phenol and 1-hexadecanol;* Surfexo. $C_{40}H_{80}NO_8P.$-$C_{16}H_{34}O.(C_{14}H_{22}O.C_2H_4O.CH_2O)_x.$ Protein-free synthetic lung surfactant composed of a 13.5:1.5:1 mixture of colfosceril palmitate, cetyl alcohol and tyloxapol, *q.q.v.* Prepn: J. A. Clements, U.S. pat. **5,110,806** (1992 to Univ. of Calif.). Pharmacology: D. J. Durand *et al., J. Pediatr.* **107**, 775 (1985); W. H. Tooley *et al., Am. Rev. Respir. Dis.* **136**, 651 (1987). Clinical trial in neonatal respiratory distress syndrome (RDS): W. Long *et al., J. Pediatr.* **118**, 595 (1991). Symposium on pulmonary mechanics and clinical use in RDS: *ibid.* **120**, Suppl., S1-S50 (1992). Review of pharmacodynamics, clinical efficacy and tolerance: K. L. Dechant, D. Faulds, *Drugs* **42**, 877-894 (1991); of metabolism: R. L. DeAngelis, J. W. Findlay, *Clin. Perinatol.* **20**, 697-710 (1993).

THERAP CAT: Pulmonary surfactant.

F

3961. Factor V. *Blood-coagulation factor V;* proaccelerin; labile factor; accelerator globulin; Ac-globulin. Mol wt above 300,000. Participates in the later stages of blood coagulation. Activated by factor X, *q.v.*, and along with Ca^{2+} and phospholipid forms a complex which converts prothrombin, *q.v.*, to thrombin, *q.v.* Accelerates the interaction of prothrombin, thromboplastin and Ca^{2+} ions: Ware, Seegers, *Fed. Proc.* **7**, 131 (1948). Thrombin produces serum Ac-globulin, *accelerin,* a far more potent accelerator, from plasma Ac-globulin, apparently by catalytic action: Ware *et al., Science* **106**, 618 (1947); Rapaport *et al., Blood* **21**, 221 (1963). Purification: Ware, Seegers, *J. Biol. Chem.* **172**, 699 (1948); Blombäck, Blombäck, *Nature* **198**, 886 (1963). Role in the scheme of blood coagulation: Pavlovsky in *Fibrinogen and Fibrin Turnover of Clotting Factors,* R. B. Hunter *et al.,* Eds. (Schattauer-Verlag, Stuttgart, 1963) pp 443-454. Proaccelerin deficiency is characterized by prolonged Quick's prothrombin time, and cephalin time, abnormal thromboplastin generation test, abnormal prothrombin consumption, and prolonged whole blood clotting time: Owren in *Thrombolytic Activity and Related Phenomena,* I. S. Wright *et al.,* Eds. (Schattauer-Verlag, Stuttgart, 1961) pp 387-391. *Review:* R. W. Colman, *Prog. Hemostasis Thromb.* **3**, 109-143 (1976). Purification, stability, amino acid composition, physical properties of bovine and human factor V: R. W. Colman, R. M. Weinberg, *Methods Enzymol.* **45B**, 107-122 (1976).

3962. Factor VII. *Blood-coagulation factor VII;* proconvertin; co-thromboplastin; serum prothrombin conversion accelerator; SPCA. Enzyme which, in conjunction with tissue factor and Ca^{2+} ions initiates the activation of factor X, *q.v.* Isoln from human plasma or serum: Fantl, Osborn, *Thromb. Diath. Haemorrh.* **8**, 286 (1962); from bovine plasma: J. Jesty, Y. Nemerson, *J. Biol. Chem.* **249**, 509 (1974). Separation from factor X by DEAE cellulose: Hougie, Bunting, in *New Blood Clotting Factors,* I. S. Wright *et al.,* Eds. (Schattauer-Verlag, Stuttgart, 1960) pp 40-42. Purification and structure of bovine factor VII: J. Jesty, Y. Nemerson, *J. Biol. Chem.* **249**, 509 (1974); R. Radcliffe, Y. Nemerson, *ibid.* **250**, 388 (1975); *eidem, Methods Enzymol.* **45B**, 49 (1976). Its deficiency results in retarded prothrombin conversion and an elevated one-stage prothrombin time: Alexander in *Thrombolytic Activity and Related Phenomena,* I. S. Wright *et al.,* Eds. (Schattauer-Verlag, Stuttgart, 1961) pp 392-402. Factor VII deficiency is associated with severe liver disease, vitamin K deficiency and broad spectrum antibiotic therapy, or may follow anticoagulant coumarin therapy.

3963. Factor VIII. *Blood-coagulation factor VIII;* antihemophilic globulin; antihemophilic factor A; AHG; AHF; Factorate; Hemofil; Humafac; Koate-HP; Monoclate-P; Nordiocto; Profilate. Plasma protein which is involved in the intrinsic formation of thromboplastin by the activation of Factor X, *q.v.:* Pavlovsky in *Fibrinogen and Fibrin Turnover of Clotting Factors,* R. B. Hunter *et al.,* Eds. (Schattauer-Verlag, Stuttgart, 1963), pp 443-454. Present at extremely low concn in plasma. Isoln and characterization of human and bovine factor VIII: Hershgold *et al., J. Lab. Clin. Med.* **77**, 185 (1971); Schmer *et al., J. Biol. Chem.* **247**, 2512 (1972); M. E. Legaz, E. W. Davie, *Methods Enzymol.* **45B**, 83 (1976). Deficiency of factor VIII results in classical hemophilia or hemophilia A, a sex-linked bleeding diathesis in which blood clots slowly, prothrombin is converted to thrombin at an abnormally slow rate and a rebleeding phenomenon is frequently seen: Wagner, Brinkhous in *Thrombolytic Activity and Related Phenomena,* I. S. Wright *et al.,* Eds. (Schattauer-Verlag, Stuttgart, 1961) pp 403-407. Molecular immunology of factor VIII: Zimmerman, Edgington, *Ann. Rev. Med.* **25**, 303 (1974). *Reviews:* Ingram, *Advan. Clin. Chem.* **8**, 189 (1965); Barrow, Graham, *Physiol. Rev.* **54**, 23 (1974); E. J. Hershgold, *Prog. Hemostasis Thromb.* **2**, 99-139 (1974); D. L. Aronson, *Semin. Thromb. Hemostasis* **6**, 12-27 (1979).

Relatively stable in liquid, frozen, or lyophilized plasma when blood is properly drawn. Heat-labile in plasma. About half the activity is lost from canine citrated plasma in 10 min at 56°. Poorly adsorbed by $BaSO_4$ from citrated or oxalated plasma.

Recombinant human factor VIII, *Kogenate, Recombinate.*
THERAP CAT: Antihemophilic factor (human).

3964. Factor IX. *Blood-coagulation factor IX;* Christmas factor; PTC; plasma thromboplastin component; antihemophilic factor B; autoprothrombin II. A plasma and serum glycoprotein that participates in the middle phases of blood coagulation. Activated by factor XI and Ca^{2+}. Interacts then with factor VIII, Ca^{2+}, and phospholipid to form the complex which converts factor X into factor X_a. Isoln and characterization of bovine factor IX: K. Fujikawa *et al., Biochemistry* **12**, 4938 (1973); K. Fujikawa, E. W. Davie, *Methods Enzymol.* **45B**, 74 (1976). Deficiency results in a congenital bleeding disorder known as Christmas disease or hemophilia B. *Review:* Macfarlane in *Thrombolytic Activity and Related Phenomena,* I. S. Wright *et al.,* Eds. (Schattauer-Verlag, Stuttgart, 1961) pp 408-415; B. N. Bouma, J. A. Van Mourik, *Haemostasis: Biochemistry, Physiology & Pathology,* D. Ogston, B. Bennett, Eds. (Wiley-Interscience, New York, 1977) pp 56-77.

Factor IX complex (human), Konyne, Profilnine, Proplex T. A mixture of human factors II, VII, IX, and X.
Factor IX (human), *AlphaNine, Mononine.*
THERAP CAT: Hemostatic.

3965. Factor X. *Blood-coagulation factor X;* Stuart-Prower factor; Stuart factor; Prower factor. Enzyme which is activated by a complex of factors IX, VIII, calcium ions and phospholipid, or by factor VII and tissue factor in the intrinsic and extrinsic pathways of blood coagulation resp. Non-physiological activators are trypsin and a specific protein from Russel's viper venom: *see* Fujikawa *et al., Biochemistry* **11**, 4892 (1972). Active factor X apparently acts upon factor V, *q.v.* in the presence of calcium ions and lipid to form a complex which activates the conversion of prothrombin to thrombin. Present in both plasma and serum because neither increases nor decreases during clotting. Presumably synthesized in the liver because level is reduced with coumarin drugs and in liver disease. Isoln and characterization of bovine factor X: *eidem., ibid.* 4882. Factor X_a is a serine protease which, in the presence of factor V, Ca^{2+} ions and phospholipids, activates prothrombin. Severe deficiency causes hemorrhagic state resembling hemophilia. Mild deficiency may result in easy bruising. In severe cases deep muscular and joint hemorrhages occur. Improved purification and bibliography of earlier isoln procedures: Bajaj, Mann, *J. Biol. Chem.* **248**, 7729 (1973). Comparison of the molecular forms generated by the different methods of activation: Radcliffe, Barton, *ibid.* 6788. *Review:* Graham, Hougie in *Thombolytic Activity and Related Phenomena,* I. S. Wright *et al.,* Eds. (Schattauer-Verlag, Stuttgart, 1961) pp 416-420; C. M. Jackson, *Thromb. Diath. Haemorrh., Suppl.* **57**, 197-216 (1974). Reviews on prepn, physical properties and function in the coagulation process: K. Fujikawa, E. W. Davie, *Methods Enzymol.* **45B**, 89 (1976); J. Jesty, Y. Nemerson, *ibid.* p 95.

Stable at room temperature for several days. Adsorbed by $BaSO_4$, $Al(OH)_3$, $Ca_3(PO_4)_2$, Seitz filters, bentonite and DEAE cellulose. At 56° is destroyed rapidly, in serum, less rapidly as citrate eluate. At 65° is destroyed rapidly in any form. Stable between pH 6.1-9.0. Soluble in water. Precipitates from water between pH 8-4. Isoelec pt pH 4.3.

3966. Factor XI. *Blood-coagulation factor XI;* PTA; plasma thromboplastin antecedent. Participates in the initial phases of blood coagulation. In normal plasma, factor XI is present in a precursor form and is converted to an active form, factor XI_a, which in turn activates factor IX to factor IX_a in the presence of Ca^{2+} ions: Ratnoff, Davie, *Biochemistry* **1**, 677 (1962); Schiffman *et al., Blood* **22**, 733 (1963); K. Fujikawa *et al., Biochemistry* **13**, 4508 (1974). Deficiency is characterized by the moderate bleeding symptoms of hemophilia C; by impaired clotting and recalcification times, impaired prothrombin consumption, and delayed thromboplastin formation: Duckert, Soulier in *New Blood Clotting Factors,* I. S. Wright *et al.,* Eds. (Schattauer-

Verlag, Stuttgart, 1960) pp 145-147, 123-131. Review on bovine factor XI: T. Koide, *Methods Enzymol.* **45B**, 65-73 (1976).

3967. Factor XII. *Blood-coagulation factor XII;* Hageman factor; HF. Mol wt about 82,000. An enzyme which circulates in zymogen form in blood and when activated initiates the first of a series of events in coagulation of blood plasma. Enzymatic nature of blood coagulation is described by the "cascade" theory in which clotting factors interact with one another in a stepwise manner, one acting as an enzyme, the other as substrate, in a sequence of reactions leading to the formation of thrombin, *q.v.:* Davie, Ratnoff, *Science* **145**, 1310 (1964); MacFarlane, *Nature* **202**, 498 (1964). Factor XII initiates clotting when blood or plasma comes into contact with glass, collagen, or surface-active agents. Once activated, it interacts with factor XI, *q.v.,* to convert it to an enzyme. Factor XII is a sialoglycoprotein with esterase activity. Its activation by glass may be viewed as a dynamic alteration of the tertiary structure of the protein moiety: Schoenmakers *et al., Biochim. Biophys. Acta* **101**, 166 (1965). Isoln of human and rabbit factor XII in zymogen form: C. G. Cochrane, K. D. Wuepper, *J. Exp. Med.* **134**, 986 (1971); H. Saito *et al., Circ. Res.* **34**, 641 (1974). Highly purified Hageman factor undergoes a change in physical properties during activation: Donaldson, Ratnoff, *Science* **150**, 754 (1965). Factor XII deficiency is termed the Hageman trait. Unlike deficiencies in all other clotting factors, persons with Hageman trait do not show significant bleeding tendencies. *Reviews:* Ratnoff in *New Blood Clotting Factors,* I. S. Wright *et al.,* Eds. (Schattauer-Verlag, Stuttgart, 1960) pp 116-122; Ratnoff *et al.,* in *Thrombolytic Activity and Related Phenomena,* I. S. Wright *et al.,* Eds. (Schattauer-Verlag, Stuttgart, 1961) pp 364-378; several authors in *Recent Advances in Blood Coagulation,* L. Poller, Ed. (J. & A. Churchill, London, 1969); J. Spragg, K. F. Austen, *Compr. Immunol.* **3**, 125-143 (1977); J. H. Griffin, C. G. Cochrane, *Methods Enzymol.* **45B**, 56-65 (1976).

Sedimentation coefficient, $s^\circ_{20,\omega} = 7.08$. Diffusion constant, $D^\circ_{20,\omega} = 7.14 \times 10^{-7}$ cm^2/sec. uv max at pH 7: 280 nm ($E^{1\%}_{1cm}$ 12.0). Isoelectric pt pH 8.0.

3968. Factor XIII. *Blood-coagulation factor XIII;* fibrin-stabilizing factor; FSF; fibrinase; Laki-Lorand factor; LLF; Fibrogamin. Plasma enzyme precursor which, when activated by thrombin in the presence of Ca^{2+}, converts soluble fibrin, *q.v.,* gel to a tough insoluble clot: Laki, Lorand, *Science* **108**, 280 (1948); Lorand, *Nature* **166**, 694 (1950). During the terminal stage of blood-clotting, factor XIII catalyzes the formation of cross-links in the fibrin gel by eliminating NH$_3$ in the transamidation reaction in which the ε-amino group of lysine from a fibrin molecule forms a peptide linkage with a glutaminyl residue from a nearby fibrin monomer: Lorand *et al., Biochem. Biophys. Res. Commun.* **25**, 629 (1966); Fuller, Doolittle, *ibid.* 694. Trypsin, papain, and reptilase also activate factor XIII: Lorand *et al., ibid.* **31**, 222 (1968). Structure contains two each of two types of soluble proteinaceous subunits called the α and β chains; mol wt of subunits each about 80,000: Schwartz *et al., J. Biol. Chem.* **248**, 1395 (1973). Amino acid sequence studies: Holbrook *et al., Biochem. J.* **135**, 901 (1973). Clinical studies: Cucuianu *et al., Thromb. Diath. Haemorrh.* **30**, 480 (1973). *Review:* L. Lorand, *Ann. N.Y. Acad. Sci.* **202**, 6-30 (1972); A. G. Loewy, *ibid.* 41-58; C. G. Curtis, L. Lorand, *Haemostasis: Biochemistry, Physiology & Pathology,* D. Ogston, B. Bennett, Eds. (Wiley-Interscience, New York, 1977) pp 186-201; *eidem, Methods Enzymol.* **45B**, 177-191 (1976).

uv max: 280 nm ($E^{1\%}$ 13.8). Retains activity in plasma kept at 0°; inactivated in a few days at room temperature. Activity destroyed if serum heated at 60° for 10 min. Does not dialyse out from plasma. Inactivated by *Echis coloratus* venom. Pretreatment with compounds containing an SH-group, such as cysteine, reduced glutathione, DL-penicillamine, or mercapto-1-propanol prevents inactivation as well as restores activity after venom treatment.

THERAP CAT: Antihemorrhagic.

3969. Fadrozole. *4-(5,6,7,8-Tetrahydroimidazo[1,5-a]pyridin-5-yl)benzonitrile;* 5-*p*-cyanophenyl-5,6,7,8-tetrahydroimidazo[1,5-a]pyridine. C$_{14}$H$_{13}$N$_3$; mol wt 223.28. C

75.31%, H 5.87%, N 18.82%. Aromatase inhibitor. Prepn: L. J. Browne, **Eur. pat. Appl. 165,904;** *idem,* U.S. pat. **4,617,307** (1985, 1986 both to Ciba-Geigy); L. J. Browne *et al., J. Med. Chem.* **34**, 725 (1991). Effects on estrogen biosynthesis: R. E. Steele *et al., Steroids* **50**, 147 (1987). Pharmacology: K. Schieweck *et al., Cancer Res.* **48**, 834 (1988). GC/MS determn in plasma and urine: R. Ackermann, G. Kaiser, *Biomed. Environ. Mass Spect.* **18**, 558 (1989). Pharmacokinetics: G. M. Kochak *et al., J. Clin. Endocrinol. Metabol.* **71**, 1349 (1990). Clinical evaluation in metastatic breast cancer: J. I. Raats *et al., J. Clin. Oncol.* **10**, 111 (1992).

mp 117-118°.

Hydrochloride, C$_{14}$H$_{13}$N$_3$.HCl, *CGS-16949A.* Crystals from 2-propanol, mp 231-233°. Sol in water.

THERAP CAT: Antineoplastic.

3970. Fagarine. *4,8-Dimethoxyfuro[2,3-b]quinoline;* γ-fagarine; 8-methoxydictamnine. C$_{13}$H$_{11}$NO$_3$; mol wt 229.24. C 68.11%, H 4.84%, N 6.11%, O 20.94%. Antiarrhythmic principle in *Fagara coco* (Gill.) Engl., *Rutaceae.* Isoln: Deulofeu *et al., J. Am. Chem. Soc.* **64**, 2326 (1942); Briggs, Cambi, *Tetrahedron* **2**, 256 (1958). Structure: Berinzaghi *et al., J. Org. Chem.* **10**, 181 (1945). Synthesis: Grundon, McCorkindale, *Chem. & Ind. (London)* **1956**, 1091.

Prisms from alcohol, mp 142°. uv max: 238, 332, 370 nm (log ε 4.76, 3.88, 3.89). Sol in chloroform, benzene, ether; slightly sol in water, petr ether.

Hydrochloride, C$_{13}$H$_{12}$ClNO$_3$, needles from chloroform + ether, mp 158-159°.

3971. Famciclovir. *2-[2-(2-Amino-9H-purin-9-yl)ethyl]-1,3-propanediol diacetate (ester);* 9-[4-acetoxy-3-(acetoxymethyl)but-1-yl]-2-aminopurine; FCV; BRL-42810; Famvir. C$_{14}$H$_{19}$N$_5$O$_4$; mol wt 321.34. C 52.33%, H 5.96%, N 21.79%, O 19.92%. Prodrug of penciclovir, *q.v.* Prepn: R. L. Jarvest, M. R. Harnden, **Austral. pat. Appl. 84 31,973** (1984 to Beecham); M. R. Harnden *et al., J. Med. Chem.* **32**, 1738 (1989). Bioavailability and metabolism: R. A. Vere Hodge *et al., Antimicrob. Ag. Chemother.* **33**, 1765 (1989). HPLC determn in plasma and urine: J. R. McMeekin *et al., Anal. Proc.* **29**, 178 (1992). Review of metabolism and mode of action: R. A. Vere Hodge, *Antiviral Chem. Chemother.* **4**, 67-84 (1993); of antiherpes activity in animals: D. Sutton, E. R. Kern, *ibid.,* Suppl. 1, 37-46 (1993); of clinical pharmacokinetics: S. Daniels, J. J. Schentag, *ibid.,* 57-64.

White shiny plates from ethyl acetate-hexane, mp 102°-104°. uv max (methanol): 222, 244, 309 nm (ε 27500, 4890, 7160). Sol in water (25°): > 25% w/v initially; rapidly pptes as sparingly sol monohydrate (2-3% w/v). Freely sol in acetone, methanol; sparingly sol in ethanol, isopropanol.

THERAP CAT: Antiviral.

3972. Famotidine. *3-[[[2-[(Aminoiminomethyl)amino]-4-thiazolyl]methyl]thio]-N-(aminosulfonyl)propanimidamide;* [1-amino-3-[[[2-[(diaminomethylene)amino]-4-thiazolyl]-methyl]thio]propylidene]sulfamide; *N*-sulfamoyl-3-[(2-guanidinothiazol-4-yl)methylthio]propionamide; YM-11170; MK-208; Amfamox; Dispromil; Famodil; Famodine; Famosan; Famoxal; Fanosin; Fibonel; Ganor; Gaster; Gastridin; Gastropen; Ifada; Lecedil; Motiax; Muclox; Nulcerin; Pepcid; Pepcidina; Pepcidine; Pepdine; Pepdul; Peptan; Ulcetrax; Ulfamid; Ulfinol. $C_8H_{15}N_7O_2S_3$; mol wt 337.45. C 28.47%, H 4.48%, N 29.06%, O 9.48%, S 28.51%. Histamine H_2-receptor antagonist. Prepn, NMR and mass spectral data: H. Yasufumi *et al.*, Belg. pat. **882,071**; *eidem*, U.S. pat. **4,283,408**; Japan. Kokai **81 55383**, *C.A.* **95**, 203930n (1980, 1981, 1981 all to Yamanouchi). Inhibition of gastric acid and pepsin secretion in rats: M. Takeda *et al.*, *Arzneimittel-Forsch.* **32**, 734 (1982); in man: M. Miwa *et al.*, *J. Clin. Pharmacol. Ther. Toxicol.* **22**, 214 (1984). Effect on disposition of antipyrine in liver: Ch. Staiger *et al.*, *Arzneimittel-Forsch.* **34**, 1041 (1984). Chromatographic determn in plasma and urine: W. C. Vincek *et al.*, *J. Chromatog.* **338**, 438 (1985). Pharmacokinetics: T. Takabatke *et al.*, *Eur. J. Clin. Pharmacol.* **28**, 327 (1985). Clinical trial in Zollinger-Ellison syndrome: J. M. Howard *et al.*, *Gastroenterology* **88**, 1026 (1985). Symposia on pharmacology and clinical efficacy: *Am. J. Med.* **81**, Suppl. 4B, 1-64 (1986); *Scand. J. Gastroenterol.* **22**, Suppl. 134, 1-62 (1987).

mp 163-164°. Soly at 20° (%, w/v): 80 in DMF; 50 in acetic acid; 0.3 in methanol; 0.1 in water; <0.01 in ethanol, ethyl acetate, chloroform. LD_{50} i.v. in mice: 244.4 mg/kg (Yasufumi).

THERAP CAT: Antiulcerative.

3973. Famphur. *Phosphorothioic acid O-[4-[(dimethylamino)sulfonyl]phenyl] O,O-dimethyl ester; phosphorothioic acid O,O-dimethyl ester, O-ester with p-hydroxy-N,N-dimethylbenzenesulfonamide;* O,O-dimethyl-O.p-(dimethylsulfamoyl)phenyl phosphorothioate; *p*-hydroxy-*N,N*-dimethylbenzenesulfonamide ester with phosphorothioic acid *O,O*-dimethyl ester; famophos; American Cyanamid 38023; ENT-25644; Warbex. $C_{10}H_{16}NO_5PS_2$; mol wt 325.35. C 36.92%, H 4.96%, N 4.31%, O 24.59%, P 9.52%, S 19.71%. Prepn: Berkelhammer, U.S. pat. **3,005,004** (1961 to Am. Cyanamid). Metabolism: P. E. Gatterdam *et al.*, *J. Agr. Food Chem.* **15**, 845 (1967). Toxicology: E. W. Schafer, *Toxicol. Appl. Pharmacol.* **21**, 315 (1972); W. D. Black *et al.*, *ibid.* **50**, 167 (1979).

Crystals, mp 52.5-53.5°. LD_{50} orally in rats: 35 mg/kg (Schafer).

Caution: Cholinesterase inhibitor. *See* Parathion.

USE: Insecticide.

THERAP CAT (VET): Insecticide (grubicide).

3974. Fampridine. *4-Pyridinamine;* 4-aminopyridine; γ-aminopyridine; 4-pyridylamine; 4-AP; EL-970. $C_5H_6N_2$; mol wt 94.12. C 63.81%, H 6.43%, N 29.76%. Potassium channel blocker. Prepn: A. Kirpal, *Monatsh. Chem.* **23**, 244 (1902); R. Camps, *Arch. Pharm.* **240**, 345 (1902); C. R. Hauser, G. A. Reynolds, *J. Org. Chem.* **15**, 1224 (1950). Prolongs action potential in demyelinated nerve fibers: R. M. Sherratt *et al.*, *Nature* **283**, 570 (1980). HPLC determn

in serum: A. van der Horst *et al.*, *J. Chromatog.* **574**, 166 (1992). Neurophysiology and clinical efficacy in multiple sclerosis: H. A. M. van Diemen *et al.*, *J. Neurol. Sci.* **116**, 220 (1993). Clinical evaluation: C. H. Polman *et al.*, *Arch. Neurol.* **51**, 1136 (1994). Review of pharmacology and efficacy of aminopyridines in MS: C. T. Bever, Jr., *Ann. Neurol.* **36**, S118-S121 (1994).

Crystals from CHCl$_3$, mp 158-159°. Sol in water, alcohol; slightly sol in benzene, ether; very slightly sol in ligroin.

THERAP CAT: In treatment of multiple sclerosis.

3975. Fantofarone. *3,4-Dimethoxy-N-methyl-N-[3-[4-[[2-(1-methylethyl)-1-indolizinyl]sulfonyl]phenoxy]propyl]-benzeneethanamine;* 2-isopropyl-1-[4-[3-[*N*-methyl-*N*-(3,4-dimethoxy-β-phenethyl)amino]propyloxy]benzenesulfonyl]-indolizine; SR-33557. $C_{31}H_{38}N_2O_5S$; mol wt 550.72. C 67.61%, H 6.95%, N 5.09%, O 14.53%, S 5.82%. Indolizinsulfone calcium channel blocker. Prepn: J. Gubin *et al.*, Eur. pat. **235,111**; J. Gubin *et al.*, U.S. pat. **4,957,925** (1987, 1990 both to Sanofi); *eidem*, *J. Med. Chem.* **35**, 981 (1992). Binding studies: A. Schmid *et al.*, *Mol. Pharmacol.* **35**, 766 (1989); B. A. Kenny *et al.*, *Brit. J. Pharmacol.* **108**, 93 (1993). X-ray structure and electronic properties: V. J. Gibon *et al.*, *Eur. J. Med. Chem.* **27**, 485 (1992). Pharmacokinetics: E. Dellissant *et al.*, *Eur. J. Clin. Pharmacol.* **41**, 329 (1991). Review of pharmacology and clinical evaluations: P. Chatelain *et al.*, *Cardiovasc. Drug Reviews* **9**, 123-146 (1991).

Monoclinic colorless sheets from diisopropyl ether / CHCl$_2$, mp 82°-83°. d_z 1.21 g/cm³. Sol in all organic solvents, in water at 0.06%. pKa 8.34 ± 0.05.

THERAP CAT: Antihypertensive.

3976. α-Farnesene. *3,7,11-Trimethyl-1,3,6,10-dodecatetraene;* 2,6,10-trimethyl-2,6,9,11-dodecatetraene; farnesene. $C_{15}H_{24}$; mol wt 204.36. C 88.16%, H 11.84%. Isoln of *(E,E)*-form from natural coating of apples: Huelin, Murray, *Nature* **210**, 1260 (1966); Murray, *Aust. J. Chem.* **22**, 197 (1969); from Dufour's gland in ants: Cavill *et al.*, *Tetrahedron Letters* **1967**, 2201. Isoln of *(Z,E)*-form from oil of perilla: T. Sakai, Y. Hirose, *Bull. Chem. Soc. Japan* **42**, 3615 (1969). Oxidation products of farnesene are believed to cause scald, a serious storage disorder of apples. Four possible geometric isomers. Synthetic studies: Ruzicka, *Helv. Chim. Acta* **6**, 490, 501 (1923); Ruzicka, Capato, *ibid.* **8**, 267 (1925). Configuration of *(E,E)*-form confirmed by synthesis from *trans*-β-farnesene: Brieger *et al.*, *J. Org. Chem.* **34**, 3789 (1969). Stereospecific synthesis of *(E,Z)*- and *(Z,Z)*-isomers: Anet, *Aust. J. Chem.* **23**, 2101 (1970). Synthesis of *(E,E)*-form: Tanaka *et al.*, *J. Am. Chem. Soc.* **97**, 3252 (1975). *(E,E)*- and *(Z,E)*-forms are components of aphid alarm pheromones: J. A. Pickett, D. C. Griffiths, *J. Chem. Ecol.* **6**, 349 (1980); of the trail pheromone of red imported fire ants: R. K. Vandermeer *et al.*, *Tetrahedron Letters* **22**, 1651 (1981).

(E,E)- α-Farnesene

Thin oil. bp$_{12}$ about 125°. d$_4^{20}$ 0.8410. n$_D^{20}$ 1.4836. Practically insol in water. Misc with hydrocarbon solvents. uv max of *(E,E)*-form (alc): 233 nm (ε 27,000); of *(Z,E)*-form (alc): 238 nm (ε 11300).

3977. β-Farnesene. *7,11-Dimethyl-3-methylene-1,6,10-dodecatriene.* C$_{15}$H$_{24}$; mol wt 204.36. C 88.16%, H 11.84%. The naturally occurring *trans* form or *(E)*-isomer is a constituent of various essential oils, *see* F. Sorm *et al., Coll. Czech. Chem. Commun.* **14,** 699 (1949); **15,** 626 (1951); also an alarm pheromone of several aphid species: Bowers *et al., Science* **177,** 1121 (1972); Edwards *et al., Nature* **241,** 126 (1973); W. S. Bowers *et al., J. Insect. Physiol.* **23,** 697 (1977). Isoln from leaves of wild potato, *Solanum berthaultii* Hawkes, *Solanaceae:* R. W. Gibson, J. A. Pickett, *Nature* **302,** 608 (1983). Prepd by dehydration of farnesol: Bhati, *Perfum. Essent. Oil Rec.* **54,** 376 (1963); Brieger, *J. Org. Chem.* **32,** 3720 (1967); alternate syntheses: Tanaka *et al., J. Am. Chem. Soc.* **97,** 3252 (1975); S. Akutagawa *et al., Chem. Letters* **1976,** 485; O. P. Vig *et al., Indian J. Chem. Sect. B* **18,** 33 (1979). Synthesis of *(Z)*-isomer: Anet, *Aust. J. Chem.* **23,** 2101 (1970); O. P. Vig *et al., J. Indian Chem. Soc.* **47,** 999 (1970); *eidem, Indian J. Chem.* **13,** 1244 (1975).

(E)- β-Farnesene

(E)-Form. Oil, bp 124°. d$_4^{20}$ 0.8310. n$_D^{20}$ 1.4870. uv max (hexane): 224 nm (ε 14000).
(Z)-Form. Oil, bp$_{3.4}$ 95-107°. n$_D^{32}$ 1.4780. uv max (hexane): 224 nm (ε 17300).

3978. Farnesol. *3,7,11-Trimethyl-2,6,10-dodecatrien-1-ol.* C$_{15}$H$_{26}$O; mol wt 222.37. C 81.02%, H 11.79%, O 7.19%. Found in oils of citronella, neroli, cyclamen, lemon grass, tuberose, rose, musk, balsam Peru, and tolu. Isoln: Elge, *Chem. Ztg.* **34,** 857 (1910); **37,** 1422 (1913); Kerschbaum, *Ber.* **46,** 1732 (1913); Naves, *Helv. Chim. Acta* **32,** 1798, 2181 (1949); LaFace, *ibid.* **33,** 249 (1950). Synthesis: Ruzicka, *ibid.* **6,** 492 (1923); Ruzicka, Firmenich, *ibid.* **22,** 392 (1939); Nazarov *et al., Zh. Obshch. Khim.* **28,** 1444 (1958); Shvarts, Petrov, *ibid.* **30,** 3598 (1960); Popjak *et al., J. Biol. Chem.* **237,** 56 (1962). Four possible stereoisomers. Stereochemistry: Bates *et al., Chem. & Ind. (London)* **1961,** 1907; *J. Org. Chem.* **28,** 1086 (1963). *trans-trans*-Farnesol is the only stereoisomer present in many essential oils but occurs mixed with *cis-trans*-farnesol in petitgrain oil and several other oils: Naves, *Compt. Rend.* **251,** 900 (1960). Stereospecific synthesis of *trans-trans*-farnesol: Corey *et al., J. Am. Chem. Soc.* **92,** 6637 (1970).

trans-trans-Farnesol. Liquid. bp$_{0.35}$ 111°. n$_D^{25}$ 1.4872. uv max: 192-196 nm (ε 28,500).
Commercial farnesol. bp$_{0.2}$ 110-113°. d$_4^{20}$ 0.8871. n$_D^{20}$ 1.4870.
USE: In perfumery, to emphasize the odor of sweet floral perfumes, such as lilac and cyclamen.

3979. Fast Green FCF. *N-Ethyl-N-[4-[[4-[ethyl[(3-sulfophenyl)methyl]amino]phenyl](4-hydroxy-2-sulfophenyl)-methylene]-2,5-cyclohexadien-1-ylidene]-3-sulfobenzene-methanaminium inner salt, disodium salt;* FD & C Green No. 3; C.I. Food Green 3; C.I. 42053. C$_{37}$H$_{34}$N$_2$Na$_2$O$_{10}$S$_3$; mol wt 808.86. C 54.94%, H 4.24%, N 3.46%, Na 5.68%, O 19.78%, S 11.89%. Prepn: H. Johnson, P. Staub, *Ind. Eng. Chem.* **19,** 497 (1927). Toxicity studies: F. C. Lu, A. Lavalle, *Can. Pharm. J.* **97,** 30 (1964); W. H. Hansen *et al., Food Cosmet. Toxicol.* **4,** 389 (1966). Review of carcinogenicity studies: *IARC Monographs* **16,** 187-197 (1978). See also: *Colour Index* vol. 4 (3rd ed., 1971) p 4383.

Dark green powder or granules with a metallic lustre. Absorption max: 628 nm. Very sol in water; sol in ethanol. Dull orange soln in conc H$_2$SO$_4$, changing to dull green on dilution. Orange soln in conc HCl or conc HNO$_3$. Bright blue soln in 10% aq NaOH. LD$_{50}$ orally in rats: > 2 g/kg (Lu, Lavalle).
USE: Biological stain. Approved by FDA for use in food, drugs and cosmetics excluding use in eye area: *Fed. Regist.* **47,** 52140 (1982).

3980. Fasudil. *Hexahydro-1-(5-isoquinolinylsulfonyl)-1H-1,4-diazepine; N-(5-isoquinolinesulfonyl)-1,4-perhydro-diazepine; 1-(5-isoquinolinesulfonyl)homopiperazine.* C$_{14}$-H$_{17}$N$_3$O$_2$S; mol wt 291.37. C 57.71%, H 5.88%, N 14.42%, O 10.98%, S 11.01%. Intracellular calcium antagonist. Prepn: H. Hidaka, T. Sone, *Eur. pat.* **Appl. 187,371;** *eidem,* U.S. pat. **4,678,783** (1986, 1987 both to Asahi); A. Morikawa *et al., Chem. Pharm. Bill.* **40,** 770 (1992). Pharmacology: T. Asano *et al., Br. J. Pharmacol.* **98,** 1091 (1989). Mechanism of action: *idem et al., J. Pharmacol. Exp. Ther.* **241,** 1033 (1987). Clinical trials: M. Shibuya *et al., J. Neurosurg.* **76,** 571 (1992). Acute toxicity: H. Koga *et al., Yakuri to Chiryo* **20,** S1433 (1992), *C.A.* **117,** 226005b (1992). Review of pharmacology and clinical evaluation: M. Shirotani *et al., Cardiovasc. Drug Rev.* **10,** 333-357 (1992).

Hydrochloride, C$_{14}$H$_{17}$N$_3$O$_2$S.HCl, *AT-877, HA-1077, Eril.* Crystals from water, mp 220.5°. Also reported as white crystalline powder, mp 219.3° (Shirotani). Sol in water up to 2 × 10^{-2}M at pH 5.0-7.0. LD$_{50}$ in mice, rats (mg/kg): 67.5, 59.9 i.v.; 124.5, 123.2 s.c.; 273.9, 335.0 orally (Koga).
THERAP CAT: Vasodilator (cerebral).

3981. Fazadinium Bromide. *1,1'-Azobis[3-methyl-2-phenylimidazo[1,2-a]pyridinium] dibromide;* AH-8165; Fazadon. C$_{28}$H$_{24}$Br$_2$N$_6$; mol wt 604.35. C 55.65%, H 4.00%, Br 26.44%, N 13.91%. A short-acting curarimimetic with rapid onset. Prepn: D. Jack, E. E. Glover, *Ger. pat.* **2,127,355** corresp to U.S. pats. **3,773,746** and **3,849,557** (1971, 1973 and 1974, all to Allen & Hanburys); E. E. Glover, M. Yorke, *J. Chem. Soc. (C)* **1971,** 3280. Pharmacology: L. Bolger *et al., Nature* **238,** 354 (1972); R. T. Brittain, M. B. Tyers, *Brit. J. Anaesth.* **45,** 837 (1973); E. L. Post *et al., Anesthesiology* **42,** 240 (1975); P. T. Hiser *et al., ibid.* 245.

Dihydrate, yellow crystals from water, mp 215-219° (softens at 196°). Also reported as yellow solid from methanol/ethyl acetate, mp 218-220°. uv max (H_2O): 285, 297 nm (log ϵ 4.04, 4.34). LD_{50} in mice: 0.31 mg/kg i.v.

THERAP CAT: Muscle relaxant (skeletal).

3982. Febantel. *[[2-[(Methoxyacetyl)amino]-4-(phenylthio)phenyl]carbonimidoyl]biscarbamic acid dimethyl ester;* dimethyl [[2-(2-methoxyacetamido)-4-(phenylthio)phenyl]-imidocarbonyl]dicarbamate; Bay Vh 5757; Bay h 5757; Rintal. $C_{20}H_{22}N_4O_6S$; mol wt 446.48. C 53.80%, H 4.97%, N 12.55%, O 21.50%, S 7.18%. Prepn: H. Koelling et al., Ger. pat. 2,423,679 corresp to U.S. pat. 3,993,682 (1975, 1976 both to Bayer). Anthelmintic efficacy: H. Wollweber et al., Arzneimittel-Forsch. **28**, 2193 (1977); H. Thomas, Res. Vet. Sci. **25**, 290 (1978). Safety evaluation in horses: J. A. Shmidl, Vet. Med. Small Anim. Clin. **73**, 775 (1978).

Crystals, mp 129-130°.

THERAP CAT (VET): Anthelmintic

3983. Febarbamate. *1-[2-[(Aminocarbonyl)oxy]-3-butoxypropyl]-5-ethyl-5-phenyl-2,4,6(1H,3H,5H)-pyrimidinetrione; 1-(3-butoxy-2-hydroxypropyl)-5-ethyl-5-phenylbarbituric acid carbamate;* 1-(3-butoxy-2-carbamoyloxypropyl)-5-phenyl-5-ethylbarbituric acid; Gö-560; G-Tril; Tymium. $C_{20}H_{27}N_3O_6$; mol wt 405.45. C 59.25%, H 6.71%, N 10.36%, O 23.68%. Prepn: P. Gold-Aubert, E. Gysin, Helv. Chim. Acta **44**, 105 (1961); P. Gold-Aubert, U.S. pat. 3,075,983 (1963 to Sapos). Pharmacology: E. Frommel et al., Helv. Physiol. Acta **19**, 241 (1961). Metabolism in rats: J. Vachta, P. Gold-Aubert, Eur. J. Drug Metab. Pharmacokinet. **7**, 147 (1982); in man: J. Vachta et al., ibid. **8**, 297 (1983).

Syrup; could not be crystallized, dec ~196°. Sol in alcohol, dioxane. Insol in water. LD_{50} orally in mice: 1065 mg/kg (Frommel).

THERAP CAT: Antidepressant; thymoanaleptic.

3984. Febrifugine. *(2S-trans)-3-[3-(3-Hydroxy-2-piperidinyl)-2-oxopropyl]-4(3H)-quinazolinone; 3-[3-(3-hydroxy-2-piperidyl)acetonyl]-4(3H)-quinazolinone;* 3-[β-keto-γ-(3-hydroxy-2-piperidyl)propyl]-4-quinazolone; β-dichroine. $C_{16}H_{19}N_3O_3$; mol wt 301.35. C 63.77%, H 6.36%, N 13.94%, O 15.93%. Isolated from the *Dichroa febrifuga* Lour., *Saxifragaceae* and from the common hydrangea: Koepfli et al., J. Am. Chem. Soc. **71**, 1048 (1949); Ablondi et al., J. Org. Chem. **17**, 14 (1952). Identity with β-dichroine: Fu, Yang,

C.A. **43**, 1530a (1949). Structure: Koepfli et al., J. Am. Chem. Soc. **72**, 3323 (1950). Synthesis: Baker et al., J. Org. Chem. **17**, 132 (1952); **18**, 178 (1953); Koepfli et al., U.S. pat. 2,504,847 (1950 to U.S. Gov't); Baker, Querry, U.S. pat. 2,651,632 (1953 to Am. Cyanamid). Config: Hill, Edwards, Chem. & Ind. (London) **1962**, 858. Revised stereochemistry: Barringer et al., J. Org. Chem. **38**, 1937 (1973).

Dimorphic crystals: needles from ethanol, mp 139-140°; from chloroform, mp 154-156°. $[\alpha]_D^{25}$ +6° (c = 0.5 in chloroform); $[\alpha]_D^{25}$ +28° (c = 0.5 in ethanol). Freely sol in methanol + chloroform, water + ethanol; slightly sol in water, ethanol, acetone, chloroform. Practically insol in ether, benzene, petr ether. LD_{50} orally in white mice: 2.5-3.0 mg/kg (Koepfli, 1949).

Dihydrochloride monohydrate, crystals from 90% ethanol, mp 220-222°. $[\alpha]_D^{25}$ +12.8° (c = 0.8).

THERAP CAT (VET): Coccidiostat.

3985. Febuprol. *1-Butoxy-3-phenoxy-2-propanol;* H-33; K-10033; Valbil; Valbilan. $C_{13}H_{20}O_3$; mol wt 224.30. C 69.61%, H 8.99%, O 21.40%. Prepn: W. F. Minor et al., J. Am. Chem. Soc. **76**, 2993 (1954). Prepn and use as choleretic: H. Hoffmann et al., Ger. pat. 2,120,396; eidem, U.S. pat. 3,839,587 (1972, 1974 both to Klinge). Pharmacology: G. Hofrichter et al., Arzneimittel-Forsch. **24**, 111 (1974). Pharmacokinetics, metabolism: K. Rehm et al., ibid. **26**, 813 (1976). Spasmolytic action: R. Eisenburger, ibid. **27**, 1429 (1977). Clinical studies: C. Wolf et al., Münch. Med. Wochenschr. **118**, 1285 (1976); W. E. Scholing, W. Weinert, ibid. **120**, 143 (1978); U. Ritter, H. J. Kyrein, Arzneimittel-Forsch. **33**, 891 (1983).

Colorless oil, bp_{11} 165°; $bp_{1.0}$ 125-132°. n_D^{20} 1.5004. d_4^{20} 1.027. LD_{50} in mice, rats (mg/kg): 3050, 2370 orally; 436, 400 i.p. (Hofrichter).

THERAP CAT: Choleretic.

3986. Feclemine. *2-(Cyclohexylphenylmethyl)-N,N,N',N'-tetraethyl-1,3-propanediamine;* 1,3-bis(diethylamino)-2-(α-phenyl-α-cyclohexylmethyl)propane; 1-phenyl-1-cyclohexyl-2,2-bis(diethylaminomethyl)ethane; phenecyclamine; phenetamine; UCB-1545; Licaran; Spasmexan. $C_{24}H_{42}N_2$; mol wt 358.61. C 80.38%, H 11.80%, N 7.81%. Prepn: Morren et al., Ind. Chim. Belg. **20**, 733 (1955); Morren, U.S. pat. 2,781,398 (1957 to U.C.B.).

Dihydrochloride, $C_{24}H_{42}N_2\cdot2HCl$, Licabile. Hygroscopic crystals, mp below 70°. $bp_{0.05}$ 143°. Also used as pamoate.

THERAP CAT: Antispasmodic.

3987. Fedotozine. *(R)-α-Ethyl-N,N-dimethyl-α-[[(3,4,5-trimethoxyphenyl)methoxy]methyl]benzenemethanamine;* (+)-(R)-α-ethyl-N,N-dimethyl-α-[[(3,4,5-trimethoxybenzyl)-

oxy]methyl]benzylamine; (+)-(1*R*)-1-phenyl-1-[(3,4,5-trimethoxy)benzyloxymethyl]-*N,N*-dimethyl-*n*-propylamine. $C_{22}H_{31}NO_4$; mol wt 373.49. C 70.75%, H 8.37%, N 3.75%, O 17.13%. Peripheral kappa opioid receptor agonist. Prepn of racemate: **Neth. pat. Appl. 80 03,601**; D. R. Torossian *et al., U.S. pat.* **4,301,163** (both 1981 to Soc. Ind. Prod. Synth.); of isomers: G. G. Aubard *et al.,* **Eur. pat. Appl. 384,088**; *eidem,* **U.S. pat. 5,245,080** (1990, 1993 both to Jouveinal). Mechanism of action and pharmacokinetics in dogs: N. Pascaud *et al., J. Pharm. Pharmacol.* **42**, 546 (1990). Clinical trial in nonulcer dyspepsia: B. Fraitag *et al., Dig. Dis. Sci.* **39**, 1072 (1994).

Pale yellow, viscous oil. $[\alpha]_D$ +16.5° (c = 6 in ethanol). D(−)-Tartrate, $C_{22}H_{31}NO_4 \cdot C_4H_6O_6$, *JO-1196.* Crystals from ethanol, mp 147°. $[\alpha]_D^{25}$ +14.5° (c = 5 in HCl).

THERAP CAT: Gastroprokinetic.

3988. Felbamate. *2-Phenyl-1,3-propanediol dicarbamate;* carbamic acid 2-phenyltrimethylene ester; ADD-03055; W-554; Felbamyl; Felbatol; Taloxa. $C_{11}H_{14}N_2O_4$; mol wt 238.24. C 55.46%, H 5.92%, N 11.76%, O 26.86%. Antiepileptic, structurally similar to meprobamate, *q.v.* Prepn: F. M. Berger, B. J. Ludwig, **U.S. pat. 2,884,444** (1959 to Carter); B. J. Ludwig *et al., J. Med. Chem.* **12**, 462 (1969). Use in treatment of epileptic seizures: R. D. Sofia, **U.S. pat. 4,978,680** (1990 to Carter-Walace). HPLC determn in plasma: R. P. Remmel *et al., Ther. Drug Monitor.* **12**, 90 (1990). Pharmacology: E. A. Swinyard *et al., Epilepsia* **27**, 27 (1986). Metabolism in animals: J. T. Yang *et al., Drug Metab. Dispos.* **19**, 1126 (1991). Penetration of blood brain barrier: E. M. Cornford *et al., Epilepsia* **33**, 944 (1992). Clinical pharmacokinetics: A. J. Wilensky *et al., ibid.* **26**, 602 (1985). Clinical trial in refractory partial seizures: E. Faught *et al., Neurology* **43**, 688 (1993). Efficacy in childhood epileptic encephalopathy (Lennox-Gastaut syndrome): F. J. Ritter *et al., New Engl. J. Med.* **328**, 29 (1993). Review of pharmacology, toxicology and clinical evaluation: R. D. Sofia *et al., Epilepsy Res.* Suppl. 3, 103-108 (1991).

White, odorless powder, mp 151-152°. Sparingly sol in water, methanol, ethanol, acetone and chloroform; freely sol in DMSO, 1-methyl-2-pyrrolidinone and DMF. LD_{50} i.p. in mice: 4000 mg/kg (Ludwig *et al.*).

THERAP CAT: Anticonvulsant.

3989. Felbinac. *(1,1′-Biphenyl)-4-acetic acid; 4-biphenylacetic acid;* 4-carboxymethylbiphenyl; BPAA; LY-61017; L-141; LJC-10141; Napageln; Traxam. $C_{14}H_{12}O_2$; mol wt 212.25. C 79.23%, H 5.70%, O 15.08%. One of the metabolites of fenbufen, *q.v.* Prepn: E. Schwenk, D. Papa, *J. Org. Chem.* **11**, 798 (1946). Improved prepn: G. R. Malone, A. I. Meyers, *ibid.* **39**, 618 (1974). Pharmacology: A. E. Sloboda, A. C. Osterberg, *Inflammation* **1**, 415 (1976). Mode of action study in ocular inflammation: E. L. Tolman *et al., Invest. Ophthalmol.* **15**, 1005 (1976). HPLC determn in serum: J. S. Fleitman *et al., J. Chromatog.* **228**, 372 (1982).

White needles from ethyl ether, mp 164-165°. LD_{50} orally in rats: 164 mg/kg (Sloboda, Osterberg). Ethyl ester, $C_{16}H_{16}O_2$, *LM-001, Daitec.*

THERAP CAT: Anti-inflammatory, analgesic.

3990. Felinine. *S-(3-Hydroxy-1,1-dimethylpropyl)-L-cysteine; L-3-[(3-hydroxy-1,1-dimethylpropyl)thio]alanine;* (−)-felinine. $C_8H_{17}NO_3S$; mol wt 207.29. C 46.35%, H 8.27%, N 6.76%, O 23.15%, S 15.47%. HOOCCH(NH₂)CH₂SC-(CH₃)₂CH₂CH₂OH. Sulfur containing amino acid; occurs as L-(−)-form in cat urine. Isoln: R. G. Westall, *Biochem. J.* **55**, 244 (1953). Synthesis of (±)-felinine: S. Trippett, *J. Chem. Soc.* **1957**, 1929; of (−)-felinine: H. Eggerer, *Ann.* **657**, 212 (1962); A. Schöberl *et al., Ber.* **101**, 373 (1968).

Needles from water + acetone, dec 177°. $[\alpha]_D^{25}$ −11.5° (c = 2.81 in water). Sol in water, ethanol. Practically insol in acetone, ethyl acetate, ether.

(±)-Felinine, crystals from aq ethanol, dec 181°.

3991. Felodipine. *4-(2,3-Dichlorophenyl)-1,4-dihydro-2,6-dimethyl-3,5-pyridinedicarboxylic acid ethyl methyl ester;* H-154/82; Agon; Feloday; Flodil; Hydac; Munobal; Plendil; Prevex; Splendil. $C_{18}H_{19}Cl_2NO_4$; mol wt 384.26. C 56.26%, H 4.98%, Cl 18.45%, N 3.65%, O 16.65%. Dihydropyridine calcium channel blocker marketed as the racemate. Prepn: P. B. Berntsson *et al.,* **Eur. pat. Appl. 7293**; *eidem,* **U.S. pat. 4,264,611** (1980, 1981 both to AB Hassle). Conformation: P. B. Berntsson, R. E. Carter, *Acta Pharm. Suec.* **18**, 221 (1981). Interaction with calmodulin, *q.v.:* S. L. Boström *et al., Nature* **292**, 777 (1981). Diuretic-natriuretic properties: G. F. Dibona *et al., Clin. Res.* **30**, 571A (1982). Hemodynamic effects in coronary patients: P. Decoster *et al., Eur. J. Clin. Invest.* **12**, 43 (1982). Enantioselective determn in plasma: P. A. Soons *et al., J. Chromatog.* **528**, 343 (1990). Pharmacokinetics and pharmacodynamics: B. Edgar *et al., Biopharm. Drug Dispos.* **8**, 235 (1987). Pharmacokinetics, vascular selectivity, effects on hypertension, hemodynamics and renal function: *Drugs* **29**, Suppl. 2, 1-212 (1985).

Crystals from isopropyl ether, mp 145°.

THERAP CAT: Antihypertensive; antianginal.

3992. Felypressin. *Vasopressin 2-(L-phenylalanine)-8-L-lysine;* 2-(phenylalanine)-8-lysine vasopressin; Phe²-Lys⁸-vasopressin; Phe²-Phe²-Lys⁸-oxytocin; PLV-2. Octapressin; $C_{46}H_{65}N_{13}O_{11}S_2$; mol wt 1040.24. C 53.11%, H 6.30%, N 17.50%, O 16.92%, S 6.17%. Prepn: Boissonnas, Guttmann, *Helv. Chim. Acta* **43**, 190 (1960); Meienhofer, du Vigneaud, *J. Am. Chem. Soc.* **82**, 6336 (1960); **Brit. pat. 928,607**, corresp to **U.S. pat. 3,232,923** (1963, 1966, both to Sandoz).

Cys–Phe–Phe–Gln–Asn–Cys–Pro–Lys–GlyNH₂

THERAP CAT: Vasoconstrictor.

3993. Femoxetine. *(3R-trans)-3-[(4-Methoxyphenoxy)methyl]-1-methyl-4-phenylpiperidine;* (+)-*trans*-1-methyl-3-[*p*-(methoxy)phenoxymethyl]-4-phenylpiperidine. $C_{20}H_{25}NO_2$; mol wt 311.42. C 77.14%, H 8.09%, N 4.50%, O 10.27%. Serotonin uptake inhibitor. Prepn: J. A. Christensen, R. F. Squires, **Belg. pat. 810,310** corresp to **U.S. pat.**

3,912,743 (1974, 1975 to A/S Ferrosan). Neuropharmacology: J. B. Lassen *et al., Psychopharmacologia* **42**, 21 (1975). Pharmacology and toxicity: *eidem, Eur. J. Pharmacol.* **32**, 108 (1975). GC determn in body fluids: E. Bechgaard, J. Lund, *J. Chromatog.* **133**, 147 (1977). Metabolism: H. Larsson, J. Lund, *Acta Pharmacol. Toxicol.* **48**, 424 (1981). Bioavailability and pharmacokinetics: H. Mengel *et al., Arzneimittel-Forsch.* **33**, 462 (1983). Double-blind clinical trials in treatment of migraine: P. G. Andersson, E. N. Petersen, *Acta Neurol. Scand.* **64**, 280 (1981); in treatment of endogenous depression: L.-E. Dahl *et al., Acta Psychiatr. Scand.* **66**, 9 (1982); P. N. Reebye *et al., Pharmacopsychiatria* **15**, 164 (1982).

Hydrochloride, $C_{20}H_{25}NO_2 \cdot HCl$, *FG-4963, Malexil.* LD_{50} in female, male mice (mg/kg): 48, 45 i.v.; 941, 723 s.c.; 1408, 1687 orally (Lassen).

THERAP CAT: Antidepressant.

3994. Fenadiazole. *2-(1,3,4-Oxadiazol-2-yl)phenol;* 2-(o-hydroxyphenyl)-1,3,4-oxadiazole; Hypnazol. $C_8H_6N_2O_2$; mol wt 162.15. C 59.26%, H 3.73%, N 17.28%, O 19.73%. Prepn: Maillard *et al., Bull. Soc. Chim. France* **1961**, 529; Vincent *et al. ibid.* **1962**, 1580; Drit. pat. **902,388** and Maillard *et al.,* Fr. pat. M379, *C.A.* **57**, 15251g (1962) (both 1962 to Logeais Labs.).

Crystals from methanol, mp 111-112°. $bp_{0.1}$ 180°. LD_{50} i.p. in mice: 940 mg/kg.

THERAP CAT: Hypnotic.

3995. Fenalamide. *α-[[[2-(Diethylamino)ethyl]amino]-carbonyl]-α-ethylbenzeneacetic acid ethyl ester; N-[2-(diethylamino)ethyl]-2-ethyl-2-phenylmalonamic acid ethyl ester;* ethyl N-[2-(diethylamino)ethyl]-2-ethyl-2-phenylmalonamate; phenylethylmalonic acid (diethylamino)ethylamide ethyl ester; Spasmamide. $C_{19}H_{30}N_2O_3$; mol wt 334.46. C 68.23%, H 9.04%, N 8.38%, O 14.35%. Smooth muscle relaxant. Prepn: Galimberti *et al.,* U.S. pat. 3,025,317 (1962 to Soc. Ital. Prod. Schering).

bp_3 182-188°. Sol in acetone, methanol, ethanol, ethyl acetate, benzene, chloroform, ether, mineral acids. Practically insol in water, alkalies.

Hydrobromide, crystals, dec 74-80°.
Hydrochloride, crystals, dec 71-74°.

THERAP CAT: Antispasmodic.

3996. Fenalcomine. *α-Ethyl-4-[2-[(1-methyl-2-phenylethyl)amino]ethoxy]benzenemethanol;* α-ethyl-p-[2-[(α-methylphenethyl)amino]ethoxy]benzyl alcohol; 2-[N-[β-[4-(α-hydroxypropyl)phenoxy]ethyl]amino]-1-phenylpropane. $C_{20}H_{27}NO_2$; mol wt 313.44. C 76.64%, H 8.68%, N 4.47%,

O 10.21%. Prepn: Pinhas, Fr. pat. M7255 (1969 to Laroche-Navarron), *C.A.* **76**, 14107w (1972). Pharmacology: Pham-Huu-Chanh *et al., Pharmacology* **6**, 137 (1971); *eidem, Arch. Int. Pharmacodyn. Ther.* **194**, 270 (1971).

Hydrochloride, $C_{20}H_{27}NO_2 \cdot HCl$, *Cordoxene.*
THERAP CAT: Cardiac stimulant; anesthetic (local).

3997. Fenamiphos. *(1-Methylethyl)phosphoramidic acid ethyl 3-methyl-4-(methylthio)phenyl ester; isopropylphosphoramidic acid ethyl 4-(methylthio)-m-tolyl ester;* ethyl 4-(methylthio)-m-tolyl isopropylphosphoramidate; ethyl 3-methyl-4-(methylthio)phenyl (1-methylethyl)phosphoramidate; phenamiphos; Bay 68138; SRA-3886; B-68138; Bayer 68138; Nemacur. $C_{13}H_{22}NO_3PS$; mol wt 303.36. C 51.47%, H 7.31%, N 4.62%, O 15.82%, P 10.21%, S 10.57%. Systemic broad spectrum nematocide with anticholinesterase activity. Prepn: H. Kayser, G. Schrader, U.S. pat. 2,978,479 (1961 to Bayer). Nematocidal activity: W. M. Zeck, *Pflanzenschutz-Nach.* **24**, 114 (1971). Mode of action: B. Homeyer, K. Wagner, *Nematologica* **27**, 215 (1981). Effect on biological and chemical activity in soil: S. P. Mathur *et al., J. Environ. Sci. Health* **B15**, 61 (1980); D. J. Ross, T. W. Speir, *Soil. Biol. Biochem.* **17**, 123 (1985). GC determn: M. J. Brown, *J. Agr. Food Chem.* **29**, 1129 (1981). Residual toxicity: D. C. Read, *J. Econ. Entomol.* **69**, 429 (1976). Acute toxicity: E. F. Hill, M. B. Camardese, *Ecotoxicol. Environ. Safety* **8**, 551 (1984); E. W. Schafer, Jr., W. A. Bowles, Jr., *Arch. Environ. Contam. Toxicol.* **14**, 111 (1985). Solubility data: B. T. Bowman, W. W. Sans, *J. Environ. Sci. Health* **B18**, 221 (1983). *Review:* B. Homeyer, *Pflanzenschutz-Nach.* **24**, 48-67 (1971). Review of metabolism and biochemical properties, efficacy: T. B. Waggoner, A. M. Khasawinah, *Residue Review* **53**, 79-97 (1974).

Crystals, mp 49°. Soly in water: 329 μg/ml. LD_{50} in rats, birds (mg/kg): 10-25 (Waggoner, Khasawinah), 2.4 (Hill, Camardese) orally.
USE: Nematocide.

3998. Fenapanil. *α-Butyl-α-phenyl-1H-imidazole-1-propanenitrile;* phenapronil; RH-2161; Sisthane. $C_{16}H_{19}N_3$; mol wt 253.35. C 75.85%, H 7.56%, N 16.59%. Prepn: G. A. Miller *et al.,* Ger. pat. 2,604,047 corresp to U.S. pat. 4,073,921; *eidem,* U.S. pat. 4,225,723 (1976, 1978, 1980 all to Rohm & Haas).

Hydrochloride, mp 160-162°.
USE: Fungicide.

3999. Fenarimol. *α-(2-Chlorophenyl)-α-(4-chlorophenyl)-5-pyrimidinemethanol;* 2,4'-dichloro-α-(pyrimidin-5-yl)-benzhydryl alcohol; EL-222; Bloc; Rimidin(e); Rubigan. $C_{17}H_{12}Cl_2N_2O$; mol wt 331.20. C 61.65%, H 3.65%, Cl 21.41%, N 8.46%, O 4.83%. Prophylactic and curative leaf fungicide; inhibits ergosterol biosynthesis. Prepn: **Neth. pat. Appl. 6,806,106;** H. M. Taylor *et al.,* U.S. pat. 3,818,-

009 (1968, 1974 both to Lilly). Improved prepn: *eidem,* U.S. pat. **3,869,456;** use as fungicide: *eidem,* U.S. pat. **3,887,708** (both 1975 to Lilly). Mode of action, efficacy against various fungi: H. Buchenauer, *Proc. Brit. Crop Prot. Conf.-Pests Dis.* **1977,** 699. Absorption by leaf tissue, local systemic activity: I. F. Brown, H. R. Hall, *ibid.* **1981**(2), 573. In control of apple scab: J. M. Olivier, J. Guillaumes, *Monogr. Brit. Crop Prot. Conf. - Fungic. Crop Prot.* **31,** 485 (1985). Post-infection efficacy: A. L. O'Leary, T. B. Sutton, *Phytopathol.* **76,** 119 (1986). Brief review: J.-M. Beraud *et al., Def. Veg.* **34,** 17 (1980).

White odorless crystals, mp 117-119°. Practically insol in water (13.7 ppm at pH 7). Sol in most organic solvents. Vapor pressure at 25° $< 10^{-7}$ millibar. LD_{50} in mice, rats (mg/kg): 4500, 2500 orally (Beraud).

USE: Plant fungicide.

4000. Fenbendazole. *[5-(Phenylthio)-1H-benzimidazol-2-yl]carbamic acid methyl ester;* 5-(phenylthio)-2-benzimidazolecarbamic acid methyl ester; methyl 5-(phenylthio)-2-benzimidazolecarbamate; HOE-881v; Panacur; Safe-Gard. $C_{15}H_{13}N_3O_2S$; mol wt 299.35. C 60.18%, H 4.38%, N 14.04%, O 10.69%, S 10.71%. Prepn: H. Loewe *et al., Ger.* pat. **2,164,690;** *eidem,* U.S. pat. **3,954,791** (1973, 1976 both to Hoechst); C. Baeder *et al., Experientia* **30,** 753 (1974). Efficacy vs gastrointestinal nematodes in swine: Enigk *et al., Deut. Tierärztl. Wochenschr.* **81,** 177 (1974).

Light brownish-gray, odorless, tasteless crystalline powder, mp 233° (dec). Insol in water; insol or only very slightly sol in the usual solvents. Freely sol in DMSO.

THERAP CAT (VET): Anthelmintic (Nematodes).

4001. Fenbenicillin. *[2S-(2α,5α,6β)]-3,3-Dimethyl-7-oxo-6-[(phenoxyphenylacetyl)amino]-4-thia-1-azabicyclo-[3.2.0]heptane-2-carboxylic acid;* 3,3-dimethyl-7-oxo-6-(2-phenoxy-2-phenylacetamido)-4-thia-1-azabicyclo[3.2.0]heptane-2-carboxylate; α-phenoxybenzylpenicillin; 6-(α-phenoxyphenylacetamido)penicillanic acid; 6-(α-phenoxy-α-phenylacetamido)penicillanate; phenbenicillin. $C_{22}H_{22}N_2O_5$-S; mol wt 426.49. C 61.96%, H 5.20%, N 6.57%, O 18.76%, S 7.52%. Semi-synthetic antibiotic related to penicillin. Prepn: **Brit.** pat. **877,120** (1961 to Beecham). Bacteriological and pharmacological properties: Rollo, Burley, *Brit. Med. J.* **I,** 76 (1962).

Potassium salt, $C_{22}H_{21}KN_2O_5S$, *Penspek (formerly).* Powder, mp 88-95°, dec 120-125°. Readily sol in water. LD_{50} in mice (mg/kg): 225 i.v.; 520 i.p.; 3000 orally (Rollo, Burley).

THERAP CAT: Antibacterial.

4002. Fenbuconazole. α-[2-(4-Chlorophenyl)ethyl]-α-phenyl-1H-1,2,4-triazole-1-propanenitrile; (R,S)-4-(4-chlorophenyl)-2-phenyl-2-[(1H-1,2,4-triazol-1-yl)methyl]butyronitrile; fenethanil; RH-7592; Enable; Govern; Indar. C_{19}-

$H_{17}ClN_4$; mol wt 336.82. C 67.75%, H 5.09%, Cl 10.53%, N 16.63%. Systemic triazole antifungal for food and ornamental crops. Prepn: S. H. Shaber *et al.,* **Ger.** pat. **3,721,786;** *idem,* U.S. pat. **5,087,635** (1988, 1992 both to Rohm & Haas). Comprehensive description: D. Driant *et al., Proc. Brighton Crop Prot. Conf. - Pests Dis.* **1988,** 33-40. Field trials vs *Botrytis* infection in roses: Y. Elad *et al., Crop Prot.* **12,** 69 (1993); vs root rot in barley: P. J. Cotterill, *ibid.* 273.

White crystalline solid, mp 124-126°. vapor pressure (20°): 0.37×10^{-7} torr. Soly in water (25°): 0.2 ppm. Sol in common organic solvents. Insol in aliphatic hydrocarbons. LD_{50} in rats (mg/kg): > 2000 orally; > 5000 dermally (Driant).

USE: Agricultural fungicide.

4003. Fenbufen. γ-*Oxo-[1,1'-biphenyl]-4-butanoic acid;* 3-(4-biphenylylcarbonyl)propionic acid; β-p-phenylbenzoylpropionic acid; diphenyl-4-γ-oxo-γ-butyric acid; 4-(4-biphenylyl)-4-oxobutyric acid; CL-82204; Bufemid; Cinopal; Cinopol; Lederfen. $C_{16}H_{14}O_3$; mol wt 254.29. C 75.57%, H 5.55%, O 18.88%. Prepn: **Fr.** pat. **798,941** (1936 to I. G. Farbenind.), *C.A.* **30,** 7729[4] (1936); D. H. Hey, R. Wilkinson, *J. Chem. Soc.* **1940,** 1030; *cf.* M. Weizmann *et al., Chem. & Ind.* **1940,** 402; W. Reppe *et al., Ann.* **596,** 223 (1955); A. S. Tomcufcik *et al., Ger.* pat. **2,147,111** corresp to U.S. pat. **3,784,701** (1972, 1974 to Am. Cyanamid). Series of articles on chemistry, pharmacology, reproductive toxicology and clinical experience: *Arzneimittel-Forsch.* **30,** 695-720, 725-746 (1980). Toxicity: H. F. Bolte *et al., ibid.* 721. Review of pharmacology and therapeutic use: R. N. Brogden *et al., Drugs* **21,** 1-22 (1981).

Crystals from ethanol, mp 185-187°. LD_{50} in various strains of mice, rats (mg/kg): 795-1673, 200-720 orally; 506-811, 265-575 i.p. (Bolte).

THERAP CAT: Anti-inflammatory.

4004. Fenbutatin Oxide. *Hexakis(2-methyl-2-phenylpropyl)distannoxane;* di[tri-(2-methyl-2-phenylpropyl)tin]-oxide; hexakis(β,β-dimethylphenethyl)distannoxane; SD-14114; Torque; Vendex. $C_{60}H_{78}OSn_2$; mol wt 1052.70. C 68.46%, H 7.47%, O 1.52%, Sn 22.55%. Selective organotin miticide. Prepn: **Ger.** pat. **2,225,666;** C. A. Horne, **U.S.** pat. **3,657,451** (1971, 1972 to Shell Oil Co.). Miticidal activity: L. R. Jeppson *et al., J. Econ. Entomol.* **68,** 707 (1975).

White crystalline powder, mp 138-139°. Soly in g/l at 23°: acetone 6; benzene 140; methylene chloride 380. Insol in water. Converts to the hydroxide in the presence of water. Nontoxic to bees, toxic to fish. LD_{50} in rats: 2630 mg/kg orally, *RTECS Vol.* **I,** R. J. Lewis, R. L. Tatken, Eds. (1980) p 709.

Caution: Corrosive, skin and eye irritatant.
USE: Acaricide.

4005. Fenbutrazate. *α-Ethylbenzeneacetic acid 2-(3-methyl-2-phenyl-4-morpholinyl)ethyl ester; 2-phenylbutyric acid 2-(3-methyl-2-phenylmorpholino)ethyl ester; 2-(3-methyl-2-phenylmorpholino)ethyl 2-phenylbutyrate; phenbutrazate.* $C_{23}H_{29}NO_3$; mol wt 367.49. C 75.17%, H 7.95%, N 3.81%, O 13.06%. Prepn: H. Siemer, O. Hengen, U.S. pat. **3,018,222** (1962 to Ravensberg). Physical properties and pharmacology: O. Hengen, H. Siemer, *Arzneimittel-Forsch.* **5**, 526 (1955).

Honey colored, viscous oil, $bp_{0.05}$ 235-240°. Sol in methanol with neutral reaction.
Hydrochloride, $C_{23}H_{29}NO_3$.HCl, *R-381.* Crystals, mp 154°. Slightly sol in water; sol in ethanol, acetone. Practically insol in ether, benzene. LD_{50} in mice (mg/kg): 3200 orally; 2800 s.c. (Hengen, Seimer).
Hydrochloride mixture with phenmetrazine theoclate, *Cafilon, Filon.*
THERAP CAT: Anorexic.

4006. Fencamfamine. *N-Ethyl-3-phenylbicyclo[2.2.1]-heptan-2-amine; N-ethyl-3-phenyl-2-norbornanamine; 2-ethylamino-3-phenylnorcamphane; 2-phenyl-3-ethylaminonorbornane; 2-ethylamino-3-phenylnorbornane; 2-ethylamino-3-phenylbicyclo[2.2.1]heptane; Euvitol.* $C_{15}H_{21}N$; mol wt 215.34. C 83.67%, H 9.83%, N 6.50%. Prepd from 3-amino-2-phenylnorbornane and acetaldehyde followed by hydrogenation in the presence of PtO_2: Thesing *et al.*, **Ger. pat. 1,110,159** (1961 to E. Merck), *C.A.* **56**, 2352g (1962).

$bp_{0.1}$ 128-131°.
Hydrochloride, $C_{15}H_{21}N$.HCl, *H-610.* Crystals from acetone, mp 192°. Freely sol in water, ethanol, methanol, chloroform; slightly sol in benzene. Practically insol in ether. LD_{50} orally in mice: 135 mg/kg.
Note: The hydrochloride is the principal ingredient of *Reactivan* which also contains vitamins B_1, B_6, B_{12} and C.
THERAP CAT: CNS stimulant.

4007. Fencamine. *3,7-Dihydro-1,3,7-trimethyl-8-[[2-[methyl(1-methyl-2-phenylethyl)amino]ethyl]amino]-1H-purine-2,6-dione; 8-[[2-[methyl(α-methylphenethyl)amino]-ethyl]amino]caffeine; 8-[2-(N,α-dimethyl-β-phenyl)ethylamino]-1,3,7-trimethyl-2,6-dioxopurine; N¹-(1,3,7-trimethyl-2,6-dioxopurin-8-yl)-N²-(1-methyl)phenethyl-N²-methylethylenediamine; N-8-caffeyl-N'-methyl-N'-(α-methylphenethyl)ethylenediamine; phencamine; ST-374.* $C_{20}H_{28}N_6O_2$; mol wt 384.48. C 62.48%, H 7.34%, N 21.86%, O 8.32%. Prepn: **Span. pats. 347,509; 352,077** (both 1969 to Labs. Miquel); Pitarch *et al., Quim. Ind. (Madrid)* **17**, 71 (1971). Pharmacology: *eidem, ibid.* 76. Colorimetric and TLC determn in urine: J. Mallol *et al., Arzneimittel-Forsch.* **24**, 1301 (1974). Toxicity data: E. Usdin, D. H. Efron, *Psychotropic Drugs and Related Compounds* (National Institute of Mental Health, Rockville, MD, 2nd ed., 1972) p 229. Clinical trial in depression: P. G. Quiros e Isla, *Rev. Clin. Espan.* **119**, 437 (1970).

Crystals from methanol, mp 150-152°.
Hydrochloride, $C_{20}H_{28}N_6O_2$.HCl, *Altimina, Sicoclor.* Slightly hygroscopic, bitter crystals, decomp 278-279°. uv max (water): 296 nm ($E_{1cm}^{1\%}$ 398). Sol in water; slightly sol in many organic solvents. LD_{50} in rats, mice (mg/kg): 93, 82 i.p.; 508, 418 orally (Usdin, Efron).
THERAP CAT: Analeptic.

4008. *d*-Fenchone. *(1S)-1,3,3-Trimethylbicyclo[2.2.1]-heptan-2-one; d-1,3,3-trimethyl-2-norbornanone; d-1,3,3-trimethyl-2-norcamphanone.* $C_{10}H_{16}O$; mol wt 152.24. C 78.90%, H 10.59%, O 10.51%. Occurs in fennel oil and in the essential oil of *Lavandula stoechas* L., *Labiatae.* Isoln: Wallach, *Ann.* **263**, 129 (1891); **353**, 209 (1907); **369**, 63 (1909); Shavrygin, *J. Appl. Chem. USSR* **12**, 1201 (1939). Total synthesis: Boyle *et al., Chem. Commun.* **1971**, 395; G. Buchbauer, H. C. Rohner, *Ann.* **1981**, 2093. Toxicity study: P. M. Jenner *et al., Food Cosmet. Toxicol.* **2**, 327 (1964). *Review:* Simonsen, *The Terpenes,* vol. II (Cambridge, 2nd ed., 1949) pp 560-580; D. L. J. Opdyke, *Food Cosmet. Toxicol* **14**, Suppl., 769-771 (1976).

Oily liq. Camphor-like odor. d_4^{18} 0.948. mp 6.1°. bp_{760} 193.5°; bp_{100} 122°; bp_{20} 82°; bp_{15} 66°. $[\alpha]_D^{20}$ +66.9°. n_D^{18} 1.4636. Practically insol in water (pH of satd soln 6.82). Very sol in abs alcohol, ether. LD_{50} orally in rats: 6.16 g/kg (Jenner).
USE: As flavor in foods; in perfumes.
THERAP CAT: Counterirritant.

4009. Fencibutirol. *α-Ethyl-1-hydroxy-4-phenylcyclo-hexaneacetic acid; α-(1-hydroxy-4-phenylcyclohexyl)butyric acid; MG-4833; Biligen; Hepasil; Verecolene.* $C_{16}H_{22}O_3$; mol wt 262.35. C 73.25%, H 8.45%, O 18.30%. Prepn: Carissimi, Ravenna, U.S. pat. **3,027,302** (1962 to Maggioni).

Crystals from ligroin, mp 157°.
THERAP CAT: Choleretic.

4010. Fenclozic Acid. *2-(4-Chlorophenyl)-4-thiazole-acetic acid; 2-(p-chlorophenyl)thiazol-4-ylacetic acid; acidum fenclozicum; ICI-54450; Myalex.* $C_{11}H_8ClNO_2S$; mol wt 253.71. C 52.08%, H 3.18%, Cl 13.97%, N 5.52%, O 12.61%, S 12.64%. Prepn: W. Hepworth, G. J. Stacey, **Neth. Appl. 6,614,130;** *eidem,* U.S. pat. **3,538,107** (1967, 1970 both to ICI; Aries, Fr. pat. **1,561,433** (1969), *C.A.* **72**, 43654y (1970). Pharmacology: W. Hepworth *et al., Nature* **221**, 582 (1969); Newbould, *Brit. J. Pharmacol.* **35**, 487 (1969). Metabolic studies: Foulkes, *J. Pharmacol. Exp. Ther.* **172**, 115, 449 (1970). Clinical evaluation: Chalmers *et al., Ann. Rheum. Dis.* **28**, 590, 595 (1969).

Colorless crystalline solid from ethyl acetate, mp 155-156°. Soluble in most organic solvents; sparingly sol in water. LD_{50} in rats, mice (mg/kg): 850, 1000 orally; 250, 300 i.v. (Hepworth).

THERAP CAT: Anti-inflammatory.

4011. Fendiline. γ-*Phenyl-N-(1-phenylethyl)benzene-propanamine; N-(3,3-diphenylpropyl)-α-methylbenzylamine;* N-(1-phenylethyl)-3,3-diphenylpropylamine. $C_{23}H_{25}N$; mol wt 315.46. C 87.57%, H 7.99%, N 4.44%. Calcium blocking agent. Prepn: **Belg. pat. 621,300;** Harsányi *et al.,* U.S. pat. **3,262,977** (1962, 1966 both to Chinoin); *eidem, J. Med. Chem.* **7,** 623 (1964); Klosa, *J. Prakt. Chem.* **34,** 312 (1966). Prepn and studies of the labelled compound: Volford, Harsányi, *J. Label. Compounds* **9,** 219 (1973). Structure-activity studies: Leszkovszky *et al., Acta Physiol. Acad. Sci. Hung.* **29,** 283 (1966). Pharmacologic properties: W. R. Kukovetz *et al., Arzneimittel-Forsch.* **26,** 1321 (1976); A. Fleckenstein *et al., ibid.* **27,** 562 (1977). Pharmacokinetics and tolerance: R. Weyhenmeyer *et al., ibid.* **37,** 58 (1987). Use in ischemic heart disease: Z. Antaloczy, I. Preda, *Ther. Hung.* **27,** 71 (1979). Assessment of Ca^{2+}-antagonist effects: M. Spedding, *Arch. Pharmacol.* **318,** 234 (1982).

$bp_{0.3}$ 206-210°.

Hydrochloride, $C_{23}H_{25}N \cdot HCl$, *HK-137, Cordan, Fendilar, Sensit.* Almost white or slightly pink powder, mp 204-205°. Very slightly sol in water; easily sol in methanol, ethanol, chloroform. LD_{50} in mice (mg/kg): 14.5 i.v.; 950 orally (Harsányi).

THERAP CAT: Vasodilator (coronary).

4012. Fendosal. *5-(4,5-Dihydro-2-phenyl-3H-benz[e]indol-3-yl)-2-hydroxybenzoic acid;* 5-(4,5-dihydro-2-phenyl-3H-benz[e]indol-3-yl)salicylic acid; 3-(3-carboxy-4-hydroxyphenyl)-2-phenyl-4,5-dihydro-3H-benz[e]indole; HP-129; P-71-0129; Alnovin. $C_{25}H_{19}NO_3$; mol wt 381.43. C 78.72%, H 5.02%, N 3.67%, O 12.58%. Prepn: R. C. Allen, B. Anderson, **Ger. pat. 2,407,671** corresp to U.S. pat. **3,878,-225** (1974, 1975, both to Hoechst); V. B. Anderson *et al., J. Med. Chem.* **19,** 318 (1976). Pharmacology: H. B. Lassman *et al., Agents Actions* **8,** 209 (1978). Clinical study: S. S. Bloomfield *et al., Clin. Pharmacol. Ther.* **23,** 390 (1978).

Crystals from acetic acid, mp 223-225° (dec). LD_{50} in mice, rats: 740, 450 mg/kg orally.

THERAP CAT: Anti-inflammatory.

4013. Fenethazine. *N,N-Dimethyl-10H-phenothiazine-10-ethanamine; 10-(2-dimethylaminoethyl)phenothiazine;* 10-(β-dimethylaminoethyl)phenothiazine; RP-3015; SC-1627; Anergan; Anergen; Ethysine; Etisine; Lisergan; Lysergan;

Phenethazinum; Rutergan. $C_{16}H_{18}N_2S$; mol wt 270.40. C 71.07%, H 6.71%, N 10.36%, S 11.86%. Prepd by the condensation of phenothiazine with dimethylaminoethyl chloride: Charpentier, **U.S. pat. 2,519,886** (1950 to Rhône-Poulenc).

bp_1 183-187°.

Hydrochloride, $C_{16}H_{18}N_2S \cdot HCl$, mp 201-201.5°. LD_{50} i.p. in mice: 115-120 mg/kg.

THERAP CAT: Antihistaminic.

4014. Fenethylline. *3,7-Dihydro-1,3-dimethyl-7-[2-[(1-methyl-2-phenylethyl)amino]ethyl]-1H-purine-2,6-dione; 7-[2-[(α-methylphenethyl)amino]ethyl]theophylline;* 7-[β-(α-methyl-β-phenylethylamino)ethyl]theophylline; 7-(phenylisopropylaminoethyl)theophylline; 7-(3-phenyl-2-propylaminoethyl)theophylline; theophyllineethylamphetamine. $C_{18}H_{23}N_5O_2$; mol wt 341.41. C 63.32%, H 6.79%, N 20.51%, O 9.37%. Prepn from 7-(β-chloroethyl)theophylline and amphetamine: Kholstaedt, Klinger, **Ger. pat. 1,123,329** (1962 to Degussa), *C.A.* **57,** 5933c (1962). Metabolism in man: T. Ellison *et al., Eur. J. Pharmacol.* **13,** 123 (1970).

Hydrochloride, $C_{18}H_{23}N_5O_2 \cdot HCl$, *H-814, Captagon.* Crystals, two different modifications, mp 227-229° and 237-239°.

d-Form hydrochloride, crystals, mp 246-247°.

l-Form hydrochloride, crystals, mp 246-247°.

Note: This is a controlled substance (stimulant) listed in the U.S. Code of Federal Regulations, Title 21 Part 1308.11 (1995).

THERAP CAT: CNS stimulant.

4015. Fenfluramine. *N-Ethyl-α-methyl-3-(trifluoromethyl)benzeneethanamine; N-ethyl-α-methyl-m-(trifluoromethyl)phenethylamine;* 2-ethylamino-1-(3-trifluoromethylphenyl)propane; S-768. $C_{12}H_{16}F_3N$; mol wt 231.26. C 62.32%, H 6.97%, F 24.65%, N 6.06%. Prepn: L. G. Beregi *et al.,* **Fr. pat. M1658;** *eidem,* U.S. pat. **3,198,833** (1963, 1965 both to Sci. Union et Cie Soc. Franc. Recherche Méd.). Prepn of optical isomers: *eidem,* U.S. pat. **3,198,-834** (1965 to Sci. Union et Cie Soc. Franc. Recherche Med.). Pharmacology: *Presse Med.* **71,** 181 (1963). Pharmacology and toxicity of isomers and racemate: J. C. Le Douarec *et al., Arch. Int. Pharmacodyn. Ther.* **161,** 206 (1966). Pharmacokinetics: S. Caccia *et al., Eur. J. Clin. Pharmacol.* **29,** 221 (1985). Clinical trial of dextrofenfluramine in refractory obesity: N. Finer *et al., Curr. Ther. Res.* **38,** 847 (1985). Comprehensive review: Pinder *et al., Drugs* **10,** 241-323 (1975).

bp$_{12}$ 108-112°. LD$_{50}$ i.p. in mice: 144 mg/kg (U.S. pat. 3,198,833).

Hydrochloride, C$_{12}$H$_{16}$F$_3$N.HCl, *Acino, Adipomin, Obedrex, Pesos, Ponderal, Ponderax, Ponderex, Pondimin, Rotondin.* Crystals from ethanol + ether, mp 166°.

d-Form, *dexfenfluramine, dextrofenfluramine.* [α]$_D^{25}$ +9.5° (c = 8 in ethanol). LD$_{50}$ orally in rats: 114.6 mg/kg (Le Douarec).

d-Form hydrochloride, *Adifax, Glypolix, Isomeride.* Crystals from ethyl acetate, mp 160-161°.

l-Form, [α]$_D^{25}$ −9.6° (c = 8 in ethanol). LD$_{50}$ orally in rats: 195 mg/kg (Le Douarec).

l-Form hydrochloride, crystals from ethyl acetate, mp 160-161°.

Note: This is a controlled substance listed in the U.S. Code of Federal Regulations, Title 21 Part 1308.14 (1995).

THERAP CAT: Anorexic.

4016. Fenipentol. *α-Butylbenzenemethanol; α-butylbenzyl alcohol;* phenylbutylcarbinol; 1-phenyl-1-hydroxypentane; phenylpentanol; PC 1; Ph BC; Pancoral. C$_{11}$H$_{16}$O; mol wt 164.25. C 80.44%, H 9.82%, O 9.74%. Synthesis: Fourneau, Puyal, *An. Soc. Espan. Fis. Quim.* **18**, 323 (1920); Adams, Vander Werf, *J. Am. Chem. Soc.* **72**, 4368 (1950); Protiva *et al., Chem. Listy* **46**, 37 (1952). Prepn of the drug: Scheffler, Engelhorn, **Brit.** pat. **915,515** corresp to U.S. pat. **3,084,100** (both 1963 to Thomae). Pharmacology: Engelhorn, *Arzneimittel-Forsch.* **10**, 255 (1960), Beck, Bierwisch, *ibid.* **20**, 693 (1970). Metabolism: Koss *et al., ibid.* **14**, 195 (1964).

Colorless or slightly yellow liquid. bp$_{12}$ 123-124°. n_D^{20} 1.5112. Miscible with organic liquids. Practically insol in water. LD$_{50}$ in mice: 1.03 g/kg i.p.; 3.10 g/kg orally.

THERAP CAT: Choleretic. Also used in treatment of mild chronic pancreatitis.

4017. Fenitrothion. *Phosphorothioic acid O,O-dimethyl O-(3-methyl-4-nitrophenyl) ester; O,O-dimethyl O-4-nitro-m-tolyl phosphorothioate; O,O-dimethyl O-(3-methyl-4-nitrophenyl) phosphorothioate; O,O-dimethyl O-4-nitro-m-tolyl thiophosphate;* MEP; metathion; Bayer 41831; Bayer S 5660; ENT-25715; OMS-45; AC-47300; Accothion; Cyten; Cyfen; Folithion; Sumithion. C$_9$H$_{12}$NO$_5$PS; mol wt 277.24. C 38.99%, H 4.36%, N 5.05%, O 28.86%, P 11.17%, S 11.57%. Prepn: **Belg.** pat. **594,669** (1960 to Sumitomo); **Belg.** pat. **596,091** (1960 to Bayer). Activity: Y. Nishizawa *et al., Agr. Biol. Chem.* **25**, 605 (1961). Toxicity study: G. Schrader, *Angew. Chem.* **73**, 331 (1961).

Yellow oil. bp$_{0.05}$ 118°. n_D^{25} 1.5528. d$_4^{25}$ 1.3227. Vapor press at 20°: 6×10^{-6} mm Hg. uv max: 269.5 nm (ε 6756). Practically insol in water. Low soly in aliphatic hydrocarbons; sol in most organic solvents. LD$_{50}$ orally in rats: 250 mg/kg (Schrader).

Caution: Cholinesterase inhibitor.

USE: Insecticide.

4018. Fennel. Large fennel; sweet fennel. Dried, ripe fruit of cultivated varieties of *Foeniculum vulgare* Mill., *Umbelliferae. Habit.* Southern Europe, Western Asia, widely cultivated. Contains 3-4% volatile oil.

4019. Fenofibrate. *2-[4-(4-Chlorobenzoyl)phenoxy]-2-methylpropanoic acid 1-methylethyl ester;* isopropyl [4'-(p-chlorobenzoyl)-2-phenoxy-2-methyl]propionate; procetofen; procetofene; LF-178; Ankebin; Elasterin; Fenobrate; Fenotard; Lipanthyl; Lipantil; Lipidil; Lipoclar; Lipofene; Liposit; Lipsin; Nolipax; Procetoken; Protolipan; Secalip. C$_{20}$H$_{21}$ClO$_4$; mol wt 360.84. C 66.57%, H 5.87%, Cl 9.83%, O 17.74%. Prepn: A. Mieville, **Ger.** pat. **2,250,327** (1973 to Orchimed), *C.A.* **79**, 53029 (1973), addn to **Ger.** pat. **2,003,-430;** *eidem,* **U.S.** pat. **4,058,552** (1970, 1977 both to Orchimed). Series of articles on synthesis, metabolism, pharmacology and clinical trials: *Arzneimittel-Forsch.* **26**, 885-909 (1976). Toxicity data: R. Sornay *et al., ibid.* 885. GC-MS of major metabolite in humans: L. F. Elsom *et al., J. Chromatog.* **123**, 463 (1976). Efficacy in hyperlipidemias: H. B. Stähelin *et al., Praxis* **68**, 24 (1979). Effect on human biliary lipids: E. M. Grandjean *et al., Schweiz. Med. Wochenschr.* **109**, 601 (1979). Mechanism of action: W. Wülfert *et al., Artery* **9**, 120 (1981).

Crystals from isopropanol, mp 80-81°. Practically insol in water. Slightly sol in methanol, ethanol. Sol in acetone, ether, benzene, chloroform. LD$_{50}$ in mice: 1600 mg/kg orally (Sornay).

THERAP CAT: Antihyperlipoproteinemic.

4020. Fenoldopam. *6-Chloro-2,3,4,5-tetrahydro-1-(4-hydroxyphenyl)-1H-3-benzazepine-7,8-diol;* 6-chloro-7,8-dihydroxy-1-p-hydroxyphenyl-2,3,4,5-tetrahydro-1H-3-benzazepine; SKF-82526. C$_{16}$H$_{16}$ClNO$_3$; mol wt 305.76. C 62.85%, H 5.27%, Cl 11.59%, N 4.58%, O 15.70%. Dopamine D$_1$-receptor agonist. Prepn: J. Weinstock, **Ger.** pat. **2,751,258;** *idem,* **U.S.** pat. **4,197,297** (1978, 1980 both to SmithKline); J. Weinstock *et al., J. Med. Chem.* **23**, 973 (1980). HPLC determn in urine and plasma: V. K. Boppana *et al., J. Chromatog.* **317**, 463 (1984). Clinical pharmacology: R. M. Stote *et al., Clin. Pharmacol. Ther.* **34**, 309 (1983); R. M. Carey *et al., J. Clin. Invest.* **74**, 2198 (1984). Hemodynamic effects in hypertension: H. O. Ventura *et al., Circulation* **69**, 1142 (1984); M. P. Caruana *et al., Brit. J. Clin. Pharmacol.* **24**, 721 (1987). Clinical evaluation in congestive heart failure: G. S. Francis *et al., Am. Heart J.* **116**, 473 (1988).

Hydrobromide, C$_{16}$H$_{16}$ClNO$_3$.HBr, mp 277° (dec).

Methanesulfonate, C$_{16}$H$_{16}$ClNO$_3$.CH$_3$SO$_3$H, *fenoldopam mesylate, SKF-82526-J, Corlopam.* mp 274° (dec).

THERAP CAT: Antihypertensive.

4021. Fenoprofen. *(±)-α-Methyl-3-phenoxybenzeneacetic acid; (±)-m-phenoxyhydratropic acid;* α-dl-2-(3-phenoxyphenyl)propionic acid; Lilly 53838. $C_{15}H_{14}O_3$; mol wt 242.27. C 74.36%, H 5.82%, O 19.81%. Prepn: Marshall, **Fr. pat. 2,015,718** corresp to **U.S. pat. 3,600,437** (1970, 1971 to Lilly). Pharmacology: Rubin *et al., J. Pharm. Sci.* **60,** 1797 (1971); **61,** 800 (1972); Herrmann, *Proc. Soc. Exp. Biol. Med.* **139,** 548 (1972). Metabolism: Rubin *et al., J. Pharmacol. Exp. Ther.* **183,** 449 (1972). Toxicology: J. L. Emmerson *et al., Toxicol. Appl. Pharmacol.* **25,** 444 (1973). Comprehensive description: C. K. Ward, R. E. Schirmer in *Analytical Profiles of Drug Substances* vol. 6, K. Florey, Ed. (Academic Press, New York, 1977) pp 161-182. *Review:* R. N. Brogden *et al., Drugs* **13,** 241-265 (1977); R. Nickander *et al.,* in *Pharmacological and Biochemical Properties of Drug Substances* vol. 1, M. E. Goldberg, Ed. (Am. Pharm. Assoc., Washington, DC, 1977) pp 183-213.

Viscous oil, $bp_{0.11}$ 168-171°. n_D^{25} 1.5742. pKa 7.3.

Calcium salt dihydrate, $C_{30}H_{26}CaO_6.2H_2O$, *Lilly 69323, Fenopron, Fepron, Feprona, Nalfon, Nalgesic, Progesic.* White crystalline powder. Soly in mg/ml at 37°: *n*-hexanol 11; methanol 8; water 2.5; chloroform 0.01. pKa 4.5. Aq solns sensitive to intense uv light. LD_{50} orally in mice: 800 mg/kg (Emmerson).

THERAP CAT: Anti-inflammatory; analgesic.

4022. Fenoterol. *5-[1-Hydroxy-2-[[2-(4-hydroxyphenyl)-1-methylethyl]amino]ethyl]-1,3-benzenediol; 3,5-dihydroxy-α-[[(p-hydroxy-α-methylphenethyl)amino]methyl]benzyl alcohol;* 1-(3,5-dihydroxyphenyl)-1-hydroxy-2-[(4-hydroxyphenyl)isopropylamino]ethane; 1-(*p*-hydroxyphenyl)-2-[[β-hydroxy-β-(3′,5′-dihydroxyphenyl)]ethyl]aminopropane; TH-1165. $C_{17}H_{21}NO_4$; mol wt 303.36. C 67.31%, H 6.98%, N 4.62%, O 21.10%. β2-Adrenergic agonist. Prepn and sepn of stereoisomers: **Belg. pat. 640,433;** Zelle *et al.,* **U.S. pat. 3,341,593** (1962, 1967 to Boehringer, Ing.). Pharmacology: Schuster, Baum, *Arzneimittel-Forsch.* **19,** 1905 (1969); O'Donnell, *Eur. J. Pharmacol.* **12,** 35 (1970). Toxicity: E. I. Goldenthal, *Toxicol. Appl. Pharmacol.* **18,** 185 (1971). Metabolism in mice: S. Kojima *et al., Arzneimittel-Forsch.* **30,** 959 (1980). Clinical evaluation: Tweel, *Ann. Allergy* **29,** 142 (1971); Rebuck, Saunders, *Med. J. Aust.* **1,** 225 (1972). Comparison with ritodrine, *q.v.,* in preterm labor: J. Gerris *et al., Eur. J. Clin. Pharmacol.* **18,** 443 (1980). Review of pharmacology and therapeutic efficacy: R. C. Heel *et al., Drugs* **15,** 3-32 (1978); N. Svedmyr, *Pharmacother.* **5,** 109-126 (1985).

Hydrobromide, $C_{17}H_{21}NO_4.HBr$, *TH-1165a, Airum, Berotec, Dosberotec, Partusisten.* Crystals from methanol-ether, mp 222-223°. LD_{50} in mice (mg/kg): 1100 s.c.; 1990 orally (Goldenthal).

Hydrochloride, $C_{17}H_{21}NO_4.HCl$, mp 183° (acetonitrile-ether).

THERAP CAT: Bronchodilator; tocolytic.

4023. Fenoverine. *10-[[4-(1,3-Benzodioxol-5-ylmethyl)-1-piperazinyl]acetyl]-10H-phenothiazine;* 10-[4-(3,4-dioxymethylenebenzyl)-1-piperazinylacetyl]phenothiazine; 10-[(4-piperonyl-1-piperazinyl)acetyl]phenothiazine; Spasmopriv. $C_{26}H_{25}N_3O_3S$; mol wt 459.57. C 67.95%, H 5.48%, N 9.14%, O 10.44%, S 6.98%. Piperonylpiperazine derivative with spasmolytic activity. Prepn: A. Buzas, R. Pierre, **Fr. pat. 2,092,639** (1972), *C.A.* **77,** 140143p (1972). Pharmacology, clinical studies in gastrointestinal disorders, dysmenorrhea: R. Pierre, R. Roustan, *Med. Interne* **15,** 49 (1980). In

treatment of colon dysfunction: Claudon, *Rev. Fr. Gastro-Enterol.* **198,** 31 (1984).

Crystals from isopropyl ether, mp 141-142°. LD_{50} in mice (g/kg): ~1.50 orally, ~2.50 i.p. (Buzas, Pierre).

THERAP CAT: Antispasmodic.

4024. Fenoxaprop-ethyl. *(±)-2-[4-[(6-Chloro-2-benzoxazolyl)oxy]phenoxy]propanoic acid ethyl ester;* HOE-33171; Acclaim; Furore; Puma; Whip. $C_{18}H_{16}ClNO_5$; mol wt 361.78. C 59.76%, H 4.46%, Cl 9.80%, N 3.87%, O 22.11%. Selective postemergent herbicide to control grassy weeds in broadleaved crops and established turfgrass. Prepn: **Belg. pat. 858,618;** R. Handte *et al.,* **U.S. pat. 4,130,413** (both 1978 to Hoechst). Metabolism in soybeans: O. Wink *et al., J. Agr. Food Chem.* **32,** 187 (1984). Persistence in soil: A. E. Smith, *ibid.* **33,** 483 (1985). Field trials in agricultural crops: G. W. Kerse *et al., Proc. 36th N.Z. Weed Pest Control Conf.,* 265 (1983); in turfgrass: P. H. Dernoeden, J. D. Fry, *Proc. Ann. Meet. Northeast Weed Sci. Soc.* **39,** 282 (1985). Brief description: J. Bieringer *et al., Proc. Brit. Crop Prot. Conf.-Weeds* **1982,** 11-17.

mp 84-85°. bp 200° at 100 Pa. Vapor pressure at 20°: 0.19×10^{-5} Pa. Soly at 20° (%): >0.5 in hexane; >1 in cyclohexane, ethanol, 1-octanol; >20 in ethyl acetate; >30 in toluene; >50 in acetone. Soly in water at 25°: 0.9 mg/l. LD_{50} in male, female rats (mg/kg): 2357, 2500 orally; 739, 864 i.p. (Bieringer).

USE: Postemergent herbicide.

4025. Fenoxazoline. *4,5-Dihydro-2-[[2-(1-methylethyl)-phenoxy]methyl]-1H-imidazole; 2-[(o-cumenyloxy)methyl]-2-imidazoline;* 2-(o-isopropylphenoxymethyl)-2-imidazoline; phenoxazoline. $C_{13}H_{18}N_2O$; mol wt 218.30. C 71.53%, H 8.31%, N 12.83%, O 7.33%. Preparation: **Fr. pat. 1,365,971** (1964 to Lab. Dausse and Soc. BMC). Use as sympathomimetic agent: Giudicelli, **U.S. pat. 3,198,703** (1965 to Lab. Dausse).

Hydrochloride, $C_{13}H_{18}N_2O.HCl$, *Aturgyl, Snup.* Crystals, mp 174°. Sol in water, ethanol.

THERAP CAT: Sympathomimetic.

4026. Fenoxedil. *2-(4-Butoxyphenoxy)-N-(2,5-diethoxyphenyl)-N-[2-(diethylamino)ethyl]acetamide; 2-(p-butoxyphenyl)-N-[2-(diethylamino)ethyl]-2′,5′-diethoxyacetanilide.* $C_{28}H_{42}N_2O_5$; mol wt 486.65. C 69.11%, H 8.70%, N 5.76%, O 16.44%. Prepn: **Ger. pat. 1,964,712;** Thuillier, Geffroy, **U.S. pat. 3,818,021** (1970, 1974 both to CERPHA).

Hydrochloride, $C_{28}H_{42}N_2O_5 \cdot HCl$, *Suplexedil.* mp 140°. LD_{50} in mice (mg/kg): 750 orally, 17 i.v. (Thuillier, Geffroy).

THERAP CAT: Vasodilator.

4027. Fenoxycarb. *[2-(4-Phenoxyphenoxy)ethyl]carbamic acid ethyl ester;* ethyl [2-(4-phenoxyphenoxy)ethyl]carbamate; ABG-6215; Ro-13-5223; Award; Insegar; Logic; Pyctyl; Torus; Varikill. $C_{17}H_{19}NO_4$; mol wt 301.34. C 67.76%, H 6.36%, N 4.65%, O 21.24%. Insect growth regulator. Prepn: U. Fischer *et al.,* **Eur.** pat. **Appl. 4,334;** *eidem,* **U.S.** pat. **4,215,139** (1979, 1980 both to Hoffmann-La Roche). Insecticidal properties: S. Dorn *et al., Z. Pflanzenkr. Pflanzenschutz* **88,** 269 (1981). Field trials, persistence in water, and bioaccumulation in fish: C. H. Schaefer *et al., J. Econ. Entomol.* **80,** 126 (1987).

Crystals from petr ether, mp 53-54°. Soly in water (23°): 5.76 ppm. LD_{50} in rats (mg/kg): 9220 i.p.; 16800 orally (Dorn).

USE: Insecticide.

4028. Fenozolone. *2-(Ethylamino)-5-phenyl-4(5H)-oxazolone;* 2-ethylamino-4-oxo-5-phenyl-Δ^2-oxazoline; 5-phenyl-2-(ethylimino)-4-oxazolidone; 5-phenyl-2-ethylamino-4-oxazolinone; LD-3394; Ordinator. $C_{11}H_{12}N_2O_2$; mol wt 204.23. C 64.69%, H 5.92%, N 13.72%, O 15.67%. Prepn: Najer, Giudicelli, *Bull. Soc. Chim. France* **1961,** 1231; **Belg.** pat. **613,985** and **Ger.** pat. **1,297,108** (1962, 1969 both to Dausse), *C.A.* **57,** 13761a (1962); **71,** 70586k (1969). Exists predominantly in the tautomeric ethylaminooxazolinone form: Najer *et al., Compt. Rend.* **254,** 2173 (1962). Activity studies: Giudicelli *et al., ibid.* 2862.

Crystals from benzene, mp 148°. Stable in alkaline, unstable in acid soln. uv max (alcohol): 221 nm (log ε 4.42). LD_{50} in white mice (g/kg): 0.425 orally; 0.175 i.p. (Belg. pat.).

THERAP CAT: CNS stimulant.

4029. Fenpentadiol. *2-(4-Chlorophenyl)-4-methyl-2,4-pentanediol;* 2-methyl-4-(*p*-chlorophenyl)-2,4-pentanediol; RD-292; Tredum. $C_{12}H_{17}ClO_2$; mol wt 228.72. C 63.02%, H 7.49%, Cl 15.50%, O 13.99%. Prepn: Valette, **Fr.** pat. **M1984** (1963 to Labs Albert Rolland), *C.A.* **60,** 2828h (1964). Psychopharmacology: J. Ginet *et al., Arzneimittel-Forsch.* **21,** 1 (1971). Pharmacokinetics: Y. Dormand, J. C. Levron, *Eur. J. Toxicol.* **5,** 43 (1972). Clinical evaluations: P. Mouren *et al., Marseille Med.* **106,** 555 (1969); P. J. Parquet, *Lille Med.* **18,** 836 (1973). Toxicity: H. Kriegel *et al., Arzneimittel-Forsch.* **21,** 9 (1971).

Crystals from petr ether, mp 76.5°. uv max (0.1% in abs ethanol): 228 nm. LD_{50} in male, female mice, rats (mg/kg): 940, 995, 1200, 1250 orally (Kriegel).

THERAP CAT: Antidepressant.

4030. Fenpiclonil. *4-(2,3-Dichlorophenyl)-1H-pyrrole-3-carbonitrile;* 3-(2,3-dichlorophenyl)-4-cyanopyrrole; 4-cyano-3-(2,3-dichlorophenyl)pyrrole; CGA-142705; Beret; Galbas; Gambit. $C_{11}H_6Cl_2N_2$; mol wt 237.09. C 55.73%, H 2.55%, Cl 29.91%, N 11.82%. Phenylpyrrole fungicide for seed treatment; structurally related to pyrrolnitrin, *q.v.* Prepd (not claimed): K. Ohkuma *et al.,* **Brit.** pat. **Appl. 2,024,824;** *eidem,* **U.S.** pat. **4,229,465** (both 1980 to Nippon Soda). Process: P. Martin, **Eur.** pat. **Appl. 174,910;** *idem,* **U.S.** pat. **4,812,580** (1986, 1989 both to Ciba-Geigy). Comprehensive description: D. Nevill *et al., Brighton Crop Prot. Conf. - Pests Dis.* **1988,** 65-72. Field trials in potatoes: A. J. Leadbeater, W. W. Kirk, *ibid.* **1992,** 657. Mechanism of action study: A. B. K. Jespers, M. A. De Waard, *Pest. Biochem. Physiol.* **49,** 53 (1994).

Colorless, odorless crystals, mp 152.9°. Soly in water (20°): 2 ppm. Partition coefficient (log *P*): 4.3 (*n*-octanol/water). LD_{50} in rats, mice, rabbits (mg/kg): > 5000 orally. LD_{50} dermally in rats: > 2000 mg/kg. LC_{50} (4 hr) in rats: > 1502 mg/m³ by inhalation (Nevill).

USE: Agricultural fungicide.

4031. Fenpiprane. *1-(3,3-Diphenylpropyl)piperidine;* 1,1-diphenyl-3-piperidinopropane; 1-(3,3-biphenylpropyl)-piperidine; 1,1-diphenyl-3-pentamethyleneiminopropane. $C_{20}H_{25}N$; mol wt 279.43. C 85.97%, H 9.02%, N 5.01%. Prepn: Bockmühl, Ehrhart, *Ann.* **561,** 52 (1948); **U.S.** pat. **2,446,522** (1948 to Winthrop-Stearns); Ruddy, Buckley, *J. Am. Chem. Soc.* **72,** 718 (1950); Ruddy, Becker, **U.S.** pat. **2,662,886** (1953 to Winthrop-Stearns).

Crystals, mp 41-42.5°. bp$_8$ 210-220°. Hydrochloride, $C_{20}H_{25}N \cdot HCl$, crystals, mp 216-217°. *Note:* Active ingredient of *Aspasan.*

THERAP CAT: Antiallergic, antispasmodic.

4032. Fenpiverinium Bromide. *1-(4-Amino-4-oxo-3,3-diphenylbutyl)-1-methylpiperidinium bromide;* 1-(3-carbamoyl-3,3-diphenylpropyl)-1-methylpiperidinium bromide; α,α-diphenyl-1-piperidinebutyramide methobromide; α,α-diphenyl-γ-piperidylbutyramide methobromide; 2,2-diphenyl-2-(2-piperidinoethyl)acetamide methyl bromide; (diphenyl-piperidinoethyl)acetamide methyl bromide; fenpipramide methobromide; 12494 Hoechst; Resantin. $C_{22}H_{29}BrN_2O$; mol wt 417.39. C 63.31%, H 7.00%, Br 19.14%, N 6.71%, O 3.83%. Prepn: **Brit.** pat. **708,859** (1954 to Hoechst); Moffett, Aspergren, *J. Am. Chem. Soc.* **79,** 4451 (1957).

Crystals, mp 177.5-178.5°. Dimorphic crystals from iso-propanol + ethyl acetate, mp 216-216.5°. Freely sol in water (water soln is neutral).

THERAP CAT: Antispasmodic.

4033. Fenpropathrin. *2,2,3,3-Tetramethylcyclopropane carboxylic acid cyano(3-phenoxyphenyl)methyl ester;* α-cy-ano-3-phenoxybenzyl 2,2,3,3-tetramethylcyclopropanecarb-oxylate; fenpropanate; S-3206; SD-41706; WL-41706; Danitol; Meothrin; Rody. $C_{22}H_{23}NO_3$; mol wt 349.43. C 75.62%, H 6.63%, N 4.01%, O 13.74%. Synthetic pyrethroid insecticide with repellant and contact activity. Prepn: T. Matsuo *et al.,* Ger. pat. **2,231,312**; *eidem,* U.S. pat. **3,835,-176** (1973, 1974 to Sumitomo). Commercial product is mix-ture of stereoisomers, the (S)-isomer primarily responsible for the bioactivity. Separation of enantiomers by chiral phase HPLC: R. A. Chapman, *J. Chromatog.* **258**, 175 (1983). Pesticidal activity: M. H. Breese, *Pestic. Sci.* **8**, 264 (1977). Metabolism: M. J. Crawford, D. H. Hutson, *ibid.* 579. Degradn in soil: T. R. Roberts, M. E. Standen, *ibid.* 600; R. A. Chapman, *Bull. Environ. Contam. Toxicol.* **26**, 513 (1981). Fish toxicity: J. R. Coats, *ibid.* **23**, 250 (1979). Mammalian toxicity study: R. D. Verschoyle, W. N. Ald-ridge, *Arch. Toxicol.* **45**, 325 (1980).

Pale yellow oil, n_D^{26} 1.5283. LC_{50} (24 hr) in rainbow trout: 76.7 ppb (Coats). LD_{50} in rats (mg/kg): 2.5 i.v. (Ver-schoyle); in male, female rats (mg/kg): 24-36, 18-24 orally (Crawford).

USE: Insecticide, acaricide.

4034. Fenpropidin. *1-[3-[4-(1,1-Dimethylethyl)phenyl]-2-methylpropyl]piperidine;* (RS)-1-[3-(4-tert-butylphenyl)-2-methylpropyl]piperidine; Patrol. $C_{19}H_{31}N$; mol wt 273.46. C 83.45%, H 11.43%, N 5.12%. Piperidine fungicide active against powdery mildews of grain. Prepn: **Belg.** pat. **861,-002**; A. Pfiffner, U.S. pat. **4,241,058** (1978, 1980 both to Hoffmann-La Roche). Improved process: N. Goetz, L. Hupfer, Eur. pat. Appl. **17,893**; *eidem,* U.S. pat. **4,283,534** (1980, 1981 both to BASF). Structure-activity relation-ships: W. Himmele, E.-H. Pommer, *Angew. Chem. Int. Ed.* **19**, 184 (1980). Inhibition of ergosterol biosynthesis: R. I. Baloch *et al.,* *Phytochemistry* **23**, 2219 (1984).

Oil, $bp_{0.2}$ 117°; $bp_{0.045}$ 125°; also reported as $bp_{0.032}$ 104°.

USE: Agricultural fungicide.

4035. Fenpropimorph. *4-[3-[4-(1,1-Dimethylethyl)-phenyl]-2-methylpropyl]-2,6-dimethylmorpholine;* BAS-42100F; Ro-14-3169/000; Corbel; Mistral. $C_{20}H_{33}NO$; mol wt 303.49. C 79.15%, H 10.96%, N 4.62%, O 5.27%. Sys-temic fungicide for control of powdery mildew, rust in ce-real crops. Prepn, fungicidal activity: W. Himmele *et al.,* Ger. pat. **2,656,747** (1978 to BASF), *C.A.* **89**, 109522k

(1978). Field trials in control of cereal diseases: J. C. Atkin *et al., Proc. Brit. Crop Prot. Conf. - Pests Dis.* **1981**, 307. Inhibition of ergosterol biosynthesis: R. I. Baloch *et al., Phytochemistry* **23**, 2219 (1984). Structure-activity relation-ships of 3-phenylpropylamines: W. Himmele, E.-H. Pom-mer, *Angew. Chem. Int. Ed.* **19**, 184 (1980); of substituted morpholines: E.-H. Pommer, *Pestic. Sci.* **15**, 285 (1984). Brief review: K. Bohnen *et al., Proc. Brit. Crop Prot. Conf. - Pests Dis.* **1979**, 541-548.

cis-(−)-S-form

Liquid, $bp_{0.05}$ 120°. Soly in water: 1 gm/l; sol in most organic solvents. Vapor pressure at 20°: 2.5×10^{-5} mm Hg. LD_{50} in male, female rats (mg/kg): 3650, 3420 orally; 4200, 4380 dermally; in male, female mice (mg/kg): 1180, 1270 i.p. (Bohnen).

*cis-*form; (±)-*cis-4-[3-(4-tert-butylphenyl)-2-methylpropyl]-2,6-dimethylmorpholine.*

USE: Systemic fungicide.

4036. Fenproporex. (±)-3-[(1-Methyl-2-phenylethyl)-amino]propanenitrile; (±)-3-[(α-methylphenethyl)amino]pro-pionitrile; (±)-N-2-cyanoethylamphetamine. $C_{12}H_{16}N_2$; mol wt 188.27. C 76.55%, H 8.57%, N 14.88%. Deriv of am-phetamine, *q.v.* Prepn: **Fr.** pat. **M4364** corresp to P. Pohr-bach, J. Blum, U.S. pat. **3,485,924** (1966, 1969 both to Bottu). Pharmacological studies: B. M. Beecham *et al., J. Pharm. Pharmacol.* **23**, 140 (1971); A. H. Beckett *et al., ibid.* **24**, 194 (1972). Peripheral effects in human and rat adipose tissue: M. Dubost *et al., Brit. J. Pharmacol.* **58**, 436P (1976). Chromatographic identification of amphetamine in urine of patients treated with fenproporex: R. B. Sznelvar, *Eur. J. Toxicol. Environ. Hyg.* **8**, 5 (1975). Clinical trial: G. Hertel, W. Fallot-Burghardt, *Fortschr. Med.* **96**, 2380 (1978).

Liquid, bp_2 126-127°.

Hydrochloride, $C_{12}H_{16}N_2 \cdot HCl$, Gacilin, Solvolip. White, cryst, odorless powder from abs ethanol, mp 146°. Bitter taste. Sol in water, 95% ethanol.

THERAP CAT: Anorexic.

4037. Fenprostalene. *[1α,2β(1E,3R*),3α,5α]-(±)-7-[3,5-Dihydroxy-2-(3-hydroxy-4-phenoxy-1-butenyl)cyclopen-tyl]-4,5-heptadienoic acid methyl ester;* (±)-9α,11α,15α-trihy-droxy-16-phenoxy-17,18,19,20-tetranorprosta-4,5,13-*trans*-trienoic acid methyl ester; RS-84043; Bovilene; Porcilene; Synchrocept B. $C_{23}H_{30}O_6$; mol wt 402.49. C 68.64%, H 7.51%, O 23.85%. Synthetic analog of prostaglandin $F_{2α}$, related structurally to prostalene, *q.v.* Prepn: J. M. Muchowski, J. H. Fried, U.S. pat. **3,985,791**; A. R. Van Horn *et al.,* U.S. pat. **4,178,457** (1976, 1979 both to Syntex). Effect on pregnancy in beagles: B. Vickery, G. McRae, *Biol. Reprod.* **22**, 438 (1980). Duration of action study: B. H. Vickery *et al., Prostaglandins Med.* **5**, 93 (1980).

uv max (methanol): 220, 265, 271, 278 nm (log ϵ 3.99, 3.11, 3.23, 3.16).

THERAP CAT (VET): Luteolytic.

4038. Fenpyroximate. *(E)-4-[[[[(1,3-Dimethyl-5-phenoxy-1H-pyrazol-4-yl)methylene]aminoJoxy]methyl]benzoic acid 1,1-dimethylethyl ester; tert-butyl (E)-α-(1,3-dimethyl-5-phenoxypyrazol-4-ylmethyleneaminooxy)-p-toluate;* NNI-850; HOE-555-02A; Sequel. $C_{24}H_{27}N_3O_4$; mol wt 421.50. C 68.39%, H 6.46%, N 9.97%, O 15.18%. Phenoxypyrazole based acaricide. Chemical, biological properties and field evaluation: T. Konno *et al., Proc. Brit. Crop Prot. Conf. - Pests Dis.* **1990**, 71. Mode of action: K. Motoba *et al., Pest. Biochem. Physiol.* **43**, 37 (1992). Control of spider mites: M. Van de Veire, D. Degheele, *Med. Fac. Landbouww. Univ. Gent.* **57**, 925 (1992). Metabolism in rats: H. Nishizawa *et al., J. Pesticide Sci.* **18**, 59 (1993). Soil degradation: Y. Izawa *et al., ibid.* 67.

White crystalline powder, mp 101.1-102.4°. Vapor pressure at 25°: 5.6 × 10^{-8} mm Hg. Soly at 20° (g/l)· water 0.015 × 10^{-3}; at 25°: methanol 15; *n*-hexane 4; xylene 175. Log *P* (octanol/water): 5.01 (20°). LD$_{50}$ in male, female rats (mg/kg): 480, 245 orally; > 2000, > 2000 dermally. LC$_{50}$ (48hr) in carp: 6.1 μg/l; LC$_{50}$ (3hr) in *daphnia pulex*: 85 μg/l (Konno).

USE: Acaricide.

4039. Fenquizone. *7-Chloro-1,2,3,4-tetrahydro-4-oxo-2-phenyl-6-quinazolinesulfonamide; 7-chloro-2-phenyl-6-sulfamyl-1,2,3,4-tetrahydro-4-quinazolinone;* MG-13054. $C_{14}H_{12}ClN_3O_3S$; mol wt 337.79. C 49.78%, H 3.58%, Cl 10.50%, N 12.44%, O 14.21%, S 9.49%. Thiazide-like diuretic with saluretic and hypocalciuric activity. Prepn: M. G. Biressi *et al., Farmaco Ed. Sci.* **24**, 199 (1969). Prepn of potassium salt: M. Carissimi, F. Ravenna, **Belg.** pat. **799,-087** (1973 to Maggioni); *eidem*, **U.S.** pat. **3,870,720** (1975). Gas-liquid chromatography: A. Marzo *et al., J. Chromatog.* **272**, 95 (1983). Metabolism in dogs: D. Ceriani *et al., Farmaco Ed. Sci.* **31**, 379 (1976). Human pharmacokinetics: G. C. Maggi *et al., Arzneimittel-Forsch.* **35**, 994 (1985). Site of action: C. Ferrando *et al., J. Pharm. Pharmacol.* **33**, 219 (1981). Clinical trials: O. Biadi *et al., Drugs Exp. Clin. Res.* 7, 763 (1981); N. Glorioso *et al., Curr. Ther. Res.* **35**, 483 (1984); A. Cupisti *et al., ibid.* **38**, 293 (1985).

mp > 310°. Insol in water.
Monopotassium salt, $C_{14}H_{12}ClKN_3O_3S$, *Idrolone.* White crystalline powder. Sol in water. pH ~9.9 (1% aqueous soln).

THERAP CAT: Diuretic.

4040. Fenretinide. *N-(4-Hydroxyphenyl)retinamide;* 4-HPR; all-*trans-N*-4'-hydroxyretinanilide; 4-(all-*trans*-retinoyl)aminophenol; McN-R-1967. $C_{26}H_{33}NO_2$; mol wt 391.55. C 79.76%, H 8.49%, N 3.58%, O 8.17%. Synthetic analog of vitamin A, *q.v.* Prepn: R. J. Gander, J. A. Gurney, **Belg.** pat. **847,942**; *eidem*, **U.S.** pat. **4,190,594** (1977, 1980 both to J & J); and pharmacology: R. C. Moon *et al., Cancer Res.* **39**, 1339 (1979). Large scale prepn and physical properties: Y. F. Shealy *et al., J. Pharm. Sci.* **73**, 745 (1984). Clinical efficacy and safety in bladder cancer: A. Decensi *et al., Eur.*

J. Cancer Prev. **3**, 377 (1994). Review of pharmacology and clinical evaluation in breast cancer prevention: *idem et al., Oncol. Rep.* **1**, 817-824 (1994); M. A. Cobleigh, *Leukemia* **8**, Suppl. 3, s59-s63 (1994); and oral leukoplakia: A. Costa *et al., Cancer Res.* **54**, Suppl., 2032s-2037s (1994).

Crystallizes in several polymorphic or solvated forms. Crystals from ethanol/water, mp 173-175° (Shealy). Crystals from chloroform/hexanes, mp 162-163°; solvated form obtained by slow crystallization from methanol, mp 178-181° (Moon). uv max (chloroform): 370 nm (ε 44500). uv max (methanol): 362 nm (47900).

THERAP CAT: Antineoplastic.

4041. Fenspiride. *8-(2-Phenylethyl)-1-oxa-3,8-diazaspiro[4.5]decan-2-one;* decaspiride; DESP. $C_{15}H_{20}N_2O_2$; mol wt 260.34. C 69.20%, H 7.74%, N 10.76%, O 12.29%. α-Adrenergic blocker. Prepn: **Neth.** pat. **Appl. 6,504,602**; Regnier *et al.,* **U.S.** pat. **3,399,192** (1965 and 1968, both to Sci. Union et Cie-Soc. Franc. Recherche Méd.). Pharmacology: LeDouarec *et al., Arzneimittel-Forsch.* **19**, 1263 (1969); Duhault *et al., ibid.* **22**, 1947 (1972).

Hydrochloride, $C_{15}H_{20}N_2O_2$.HCl, *NAT-333, NDR-5998A, Decaspir, Fluiden, Pneumorel, Respiride, Tegencia, Viarespan.* Crystals decomp 232-233°. Soluble in water. LD$_{50}$ i.v. in mice: 106 mg/kg; orally in rats: 437 mg/kg (LeDouarec).

THERAP CAT: Bronchodilator.

4042. Fensulfothion. *Phosphorothioic acid O,O-diethyl O-[4-(methylsulfinyl)phenyl] ester; O,O-diethyl O-[p-(methylsulfinyl)phenyl] phosphorothioate;* Bay 25141; Dasanit; Terracur P. $C_{11}H_{17}O_4PS_2$; mol wt 308.36. C 42.85%, H 5.56%, O 20.75%, P 10.04%, S 20.80%. Prepn: Homeyer, Schrader, **Belg.** pat. **666,012** (1965 to Bayer), *C.A.* **64**, 20555f (1966).

Liquid. bp$_{0.01}$ 138-141°. LD$_{50}$ orally in rats: 5 mg/kg.
Note: For application 1 part of fensulfothion may be mixed with 0.5 part xylene, 0.6 part emulsifier added and the mixture diluted with water. An effective concn of 10 to 40 ppm is claimed.

USE: Nematocide, insecticide.

4043. Fentanyl. *N-Phenyl-N-[1-(2-phenylethyl)-4-piperidinyl]propanamide; N-(1-phenethyl-4-piperidyl)propionanilide;* phentanyl; R-4263. $C_{22}H_{28}N_2O$; mol wt 336.48. C 78.53%, H 8.39%, N 8.33%, O 4.75%. Prepn: P. A. J. Janssen, **Fr.** pat. **1,344,366**; *idem*, **U.S.** pat. **3,164,600** (1963, 1965 both to Janssen). Pharmacology: J. F. Gardocki, J. Yelnosky, *Toxicol. Appl. Pharmacol.* **6**, 48 (1964); R. Hess *et al., J.*

Pharmacol. Exp. Ther. **179,** 474 (1971). Effects on cerebral circulation and metabolism in rats: C. Carlsson *et al., Anesthesiology* **57,** 375 (1982). Clinical studies: E. A. Welchew, J. A. Thornton, *Anaesthesia* **37,** 309 (1982); M. J. Stephens *et al., Med. J. Aust.* **1,** 419 (1982).

Crystals, mp 83-84°.

Citrate, $C_{22}H_{28}N_2O.C_6H_8O_7$, *Fentanest, Leptanal, Pentanyl, Sublimaze.* Crystalline powder, mp 149-151°. Bitter taste. One gram dissolves in about 40 ml water. Sol in methanol; sparingly sol in chloroform. LD_{50} in mice (mg/kg): 11.2 i.v.; 62 s.c. (Gardocki, Yelnosky).

Note: This is a controlled substance (opiate) listed in the U.S. Code of Federal Regulations, Title 21 Part 1308.12 (1995).

THERAP CAT: Analgesic (narcotic).

THERAP CAT (VET): Analgesic; tranquilizer.

4044. Fenthion. *Phosphorothioic acid O,O-dimethyl O-[3-methyl-4-(methylthio)phenyl] ester; O,O*-dimethyl *O*-(4-methylmercapto-3-methylphenyl) thionophosphate; *O,O*-dimethyl *O*-(3-methyl-4-methylthiophenyl) thiophosphate; *O,O*-dimethyl *O*-(4-methylthio-3-methylphenyl) thiophosphate; *O,O*-dimethyl *O*-[4-(methylthio)-*m*-tolyl] phosphorothioate; 4-methylmercapto-3-methylphenyl dimethyl thiophosphate; Bayer 29493; ENT-25540; S-1752; Baycid; Baytex; Entex (obsolete); Lebaycid; Mercaptophos; Queletox; Talodex; Tiguvon. $C_{10}H_{15}O_3PS_2$; mol wt 278.33. C 43.15%, H 5.43%, O 17.24%, P 11.13%, S 23.04%. Prepn: Schrader, *Die Entwicklung neuer insektizider Phosphorsäure Ester* (Verlag Chemie, 3rd ed., 1963) pp 298-313; Schlegk, Schrader, **Ger.** pat. **1,116,656,** *C.A.* **56,** 14170g (1962); and **U.S.** pat. **3,042,703** (1961 and 1962, both to Bayer); Nishizawa, Mizutani, **Japan.** pat. **64 15,130** (1964 to Sumitomo), *C.A.* **61,** 16016h (1964). Properties: Jung, *Bull. World Health Org.* **21,** 215 (1959); Schrader, *loc. cit.* Toxicity: Gaines, *Toxicol. Appl. Pharmacol.* **2,** 88 (1960).

Liquid, slight odor of garlic. $bp_{0.01}$ 87° Commercial product, $bp_{0.01}$ 105°. d_4^{20} 1.250. n_D^{20} 1.5698. Vapor pressure at 20°: 3×10^{-5} mm Hg. Thermally stable up to 210°; resistant to alkalies up to pH 9. Readily sol in methanol, ethanol, ether, acetone and many other organic solvents, esp chlorinated hydrocarbons. Practically insol in water (55 mg/l). LD_{50} in male, female rats (mg/kg): 215, 245 orally (Gaines).

Caution: A cholinesterase inhibitor. *See* Parathion.

USE: Insecticide; acaricide.

THERAP CAT (VET): Ectoparasiticide.

4045. Fentiazac. *4-(4-Chlorophenyl)-2-phenyl-5-thiazoleacetic acid;* BR-700; CH-800; Donorest; Flogene; Norvedan. $C_{17}H_{12}ClNO_2S$; mol wt 329.81. C 61.91%, H 3.67%, Cl 10.75%, N 4.25%, O 9.70%, S 9.72%. Prepn: K. Brown, **Brit.** pat. **1,145,884;** *idem,* **U.S.** pat. **3,476,766** (both 1969 to Wyeth). Anti-inflammatory activity: K. Brown *et al., Nature* **219,** 164 (1968). Crystallographic study: R. Destro, *Acta Crystallogr. B* **34,** 959 (1978). Absorption: G. Zanolo *et al., Boll. Chim. Farm.* **119,** 209 (1980). Metabolism: S. Fumero *et al., Arzneimittel-Forsch.* **30,** 1253 (1980). Phar-

macokinetics in humans: M. Quattrini *et al., ibid.* **31,** 1046 (1981). Toxicity study: D. A. Shriver *et al., Toxicol. Appl. Pharmacol.* **42,** 75 (1977). Review of exptl and clinical pharmacology: E. Marmo, *Curr. Med. Res. Opin.* **6,** 53-63 (1979).

Colorless needles from benzene, mp 161-162°. LD_{50} in rats, mice (mg/kg): 661, 692 orally (Marmo).

THERAP CAT: Anti-inflammatory.

4046. Fenticlor. *2,2'-Thiobis[4-chlorophenol];* bis[2-hydroxy-5-chlorophenyl]sulfide; 2,2'-dihydroxy-5,5'-dichlorodiphenyl sulfide; S-7; Novex. $C_{12}H_8Cl_2O_2S$; mol wt 287.17. C 50.19%, H 2.81%, Cl 24.69%, O 11.14%, S 11.17%. Prepn by chlorination of bis[2-hydroxyphenyl]sulfide: Muth, **Ger.** pat. **568,944** (1931 to I. G. Farben); Dunning *et al., J. Am. Chem. Soc.* **53,** 3466 (1931).

Fine needles from toluene, mp 175°. Sol in aq solns of NaOH, in alcohol, in hot benzene.

Note: Novex is also a trade name for a thermoplastic resin.

USE: Fungicide; esp used against *Monosporium apiospermum:* Reifferscheid, Seeliger, *Deut. Med. Wochenschr.* **80,** 1841, 1850 (1955).

THERAP CAT: Anti-infective.

4047. Fenticonazole. *1-[2-(2,4-Dichlorophenyl)-2-[[4-(phenylthio)phenyl]methoxy]ethyl]-1H-imidazole;* 1-[2,4-dichloro-β-[[p-(phenylthio)benzyl]oxy]phenethyl]imidazole; 1-(2,4-dichlorophenyl)-2-(N-imidazolyl)ethyl-4-phenylthiobenzyl ether; 2,4-dichloro-4'-phenylthio-(N-imidazolylmethyl)dibenzyl ether. $C_{24}H_{20}Cl_2N_2OS$; mol wt 455.41. C 63.30%, H 4.43%, Cl 15.57%, N 6.15%, O 3.51%, S 7.04%. Broad spectrum antimycotic; also active as antibacterial. Prepn: D. Nardi *et al.,* **Ger.** pat. **2,917,244;** *eidem,* **U.S.** pat. **4,221,803** (1979, 1980 both to Recordati); *eidem, Arzneimittel-Forsch.* **31,** 2123 (1981). Series of articles on physicochemical properties, *in vitro* and *in vivo* antibacterial and antifungal activity, pharmacology and toxicology: *ibid.* 2127-2154. Toxicity data: G. Graziani *et al., ibid.* 2145. Preliminary clinical comparison with miconazole in vaginal candidiasis: A. Gastaldi, *Curr. Ther. Res.* **38,** 489 (1985). Comparative clinical trials in dermatomycoses: A. Finzi *et al., Mykosen* **29,** 41 (1986); E. M. Kokoschka *et al., ibid.* 45.

Mononitrate, $C_{24}H_{20}Cl_2NS.HNO_3$, *Rec-15-1476, Falvin, Fentiderm, Fentigyn, Lomexin.* Odorless, white crystalline powder, mp 136°. uv max (methanol): 252 nm (ϵ 13894). Soly at 20° (mg/ml): water <0.1; ethyl ether <0.1; ethanol 30; methanol 100; chloroform 300; DMF 600. pKa 6.54.

LD$_{50}$ in mice, male rats, female rats (mg/kg): 1191, 440, 309 i.p.; all > 3000 orally (Graziani).

THERAP CAT: Antifungal (topical).

4048. Fentonium Bromide. *[3(S)-endo]-8-(2-[1,1'-Bi-phenyl]-4-yl-2-oxoethyl)-3-(3-hydroxy-1-oxo-2-phenylpro-poxy)-8-methyl-8-azoniabicyclo[3.2.1]octane bromide; 3α-hydroxy-8-(p-phenylphenacyl)-1αH,5αH-tropanium bromide (−)-tropate; N-(4-phenylphenacyl)-1-hyoscyaminium bro-mide; phentonium bromide; FA-402; Z-326; Ketoscilium;* Ulcesium. C$_{31}$H$_{34}$BrNO$_4$; mol wt 564.52. C 65.96%, H 6.07%, Br 14.15%, N 2.48%, O 11.34%. Prepn: U. Teotino, D. Della Bella, **Brit.** pat. **1,026,640**; *idem,* **U.S.** pat. **3,356,-682** (1966, 1967 both to Whitefin Holding); Teotino *et al., Chim. Ther.* **3,** 453 (1968). Human pharmacology: Azzol-lini *et al., Curr. Ther. Res.* **12,** 734 (1970). Clinical studies: Alberto *et al., Minerva Med.* **62,** 852 (1971).

Crystals, mp 203-205° (dec). [α]$_D^{23}$ −5.68° (c = 5 in DMF). Also reported as mp 193-194°. [α]$_D^{25}$ −4.7° (c = 5 in DMF). LD$_{50}$ in mice (mg/kg): 12.1 i.v.; > 400 s.c. and orally (Teotino, Della Bella).

THERAP CAT: Anticholinergic; antispasmodic.

4049. Fenugreek. Greek Hay. *Trigonella foenumgrae-cum* L., *Leguminosae,* an annual herb, cultivated in South-ern Europe, Northern Africa and India for its seeds which are used in making curry. The seeds contain the alkaloid trigonelline *(q.v.)* and the steroidal sapogenin diosgenin *(q.v.):* Fazli, Hardman, *Trop. Sci.* **10,** 66-78 (1968).

USE: Has been used in making imitation maple syrup and for culinary spices other than curry, flavoring agent.

THERAP CAT (VET): Emollient.

4050. Fenuron. *N,N-Dimethyl-N-phenylurea; N-phenyl-N',N'-dimethylurea;* Dybar. C$_9$H$_{12}$N$_2$O; mol wt 164.21. C 65.83%, H 7.37%, N 17.06%, O 9.74%. C$_6$H$_5$NHCON-(CH$_3$)$_2$. Prepn: Crosby, Niemann, *J. Am. Chem. Soc.* **76,** 4458 (1954); Gilbert, Sorma, **U.S.** pat. **2,729,677** (1956 to Allied Chem.); Jones, **U.S.** pat. **2,768,971** (1956 to ICI); Applegath *et al.,* **U.S.** pat. **2,857,430** (1958 to Monsanto). Prepn of trichloroacetate: Gilbert *et al.,* **U.S.** pat. **2,782,112** (1957 to Allied Chem.). Toxicity data: G. W. Bailey, J. L. White, *Residue Rev.* **10,** 97 (1965).

Crystals, mp 131-133°. Sparingly sol in water (0.29% at 24°); in hydrocarbons. LD$_{50}$ orally in rats: 7500 mg/kg (Bailey, White).

Trichloroacetate, C$_{10}$H$_{14}$Cl$_3$N$_2$O$_3$, *fenuron TCA, Urab.* mp 65-68°. Esp effective in woody plant control.

Caution: If hydrolyzed to aniline can cause methemo-globinemia: *Clinical Toxicology of Commercial Products,* R. E. Gosselin *et al.,* Eds. (Williams & Wilkins, Baltimore, 5th ed., 1984) Section II, p 331.

USE: Herbicide.

4051. Fenvalerate. *4-Chloro-α-(1-methylethyl)benzene-acetic acid cyano(3-phenoxyphenyl)methyl ester; α-cyano-3-phenoxybenzyl α-(4-chlorophenyl)isovalerate; cyano(3-phen-oxyphenyl)methyl 4-chloro-α-(1-methylethyl)benzeneace-tate; α-cyano-3-phenoxybenzyl-2-(4-chlorophenyl)-3-meth-ylbutyrate; phenvalerate; S-5602; SD-43775; WL-43775;* Belmark; Pydrin; Pyridin; Sumicidin; Tirade. C$_{25}$H$_{22}$ClNO$_3$; mol wt 419.91. C 71.51%, H 5.28%, Cl 8.44%, N 3.34%, O 11.43%. A synthetic pyrethroid insecticide without the usual cyclopropane ring. Prepn: **Ger.** pat. **2,335,347**; K. Fujimoto *et al.,* **U.S.** pat. **3,996,244** (1974, 1976, both to

Sumitomo); D. A. Wood, **Ger.** pat. **2,651,341**; *idem,* **U.S.** pat. **4,061,664** (1977, both to Shell). Has two chiral centers giving four possible optical isomers; prepn and comparative insecticidal activity of isomers: M. Hirano *et al.,* **Ger.** pat. **2,737,297**; *idem,* **U.S.** pat. **4,503,071** (1978, 1985 both to Sumitomo). Absolute configuration of most active isomer, esfenvalerate: K. Aketa *et al., Agr. Biol. Chem.* **42,** 895 (1978). GC separation of isomers: G. R. Cayley, B. W. Simpson, *J. Chromatog.* **356,** 123 (1986); determn of esfen-valerate in technical prepn: S. Sakaue *et al., Agr. Biol. Chem.* **51,** 1671 (1987). Comprehensive description of activ-ity and physical properties of esfenvalerate: H. Oo'uchi, *Japan. Pest. Info.* **1985,** 21-24. Comparative soil metabolism of isomers: P.W. Lee *et al., J. Agr. Food Chem.* **35,** 384 (1987).

Clear yellow viscous liquid at 23°. d^{23} 1.17. n$_D^{20}$ 1.5533. Vapor press at 25° = 1.1×10^{-8} mm Hg. Soly at 20° (g/l): acetone, > 450; chloroform, > 450; methanol, > 450; hex-ane, 77. Insol in water. More stable in acidic soln than in alk soln. Dec gradually between 150-300°. No significant breakdown after 100 hrs at 75°. LD$_{50}$ orally in rats: 451 mg/kg (DMSO); > 3200 mg/kg (aqueous suspension), Shell Technical Data Bulletin. Highly toxic to fish and bees.

(S,S)-Isomer, *esfenvalerate, Aα, OMS-3023, S-1844, S-5602α, Sumi-alpha.* White crystalline solid, mp 59-60.2°. [α]$_D^{25}$ −15.0° (c = 2.0 in CH$_3$OH). d$_{23}^{23}$ 1.163. Vapor pres-sure at 20°: 2.63×10^{-7} mm Hg; at 25°: 5.00×10^{-7} mm Hg. Soly (1%): acetonitrile > 60, chloroform > 60, DMF > 60, DMSO > 60, ethyl acetate > 60, acetone > 60, ethyl cellosolve 40-50, n-hexane 1-5, kerosene < 1, methanol 7-10, α-methylnaphthalene 50-60, xylene > 60.

Caution: Eye, skin irritant.

USE: Insecticide.

THERAP CAT (VET): Ectoparasiticide.

4052. Fepradinol. *(±)-α-[[(2-Hydroxy-1,1-dimethyleth-yl)amino]methyl]benzenemethanol; 1-phenyl-2-(α,α-dimeth-ylethanolamino)ethanol.* C$_{12}$H$_{19}$NO$_2$; mol wt 209.29. C 68.87%, H 9.15%, N 6.69%, O 15.29%. Prepn: E. Biekert, J. Sonnenbichler, *Ber.* **95,** 1451 (1962); G. Staibano, C. Pro-tto, *Farmaco. Ed. Sci.* **27,** 929 (1972); G. Staibano, **Belg.** pat. **Appl. 774,629** (1972 to Gentili). Pharmacology: J. M. Masso *et al., J. Pharm. Pharmacol.* **45,** 959 (1993). Mechan-ism of action study: *eidem, Arzneimittel-Forsch.* **44,** 68 (1994). Clinical trial: C. Bestit *et al., Curr. Ther. Res.* **45,** 53 (1989).

White cryst from ethanol, mp 142-143°; also reported as mp 140-142° (Staibano). Cryst from benzene, mp 139° (Biekert). uv max (0.1N HCl): 250.5, 256.4, 262.0 nm (log ε 2.17, 2.28, 2.25).

Hydrochloride, C$_{12}$H$_{19}$NO$_2$.HCl, *Dalgen, Flexidol.*
THERAP CAT: Anti-inflammatory.

4053. Feprazone. *4-(3-Methyl-2-butenyl)-1,2-diphenyl-3,5-pyrazolidinedione; 4-prenyl-1,2-diphenyl-3,5-pyrazoli-dinedione; 4-prenyl-1,2-diphenyl-3,5-dioxopyrazolidine; 4-(β-isoamylenyl)-1,2-diphenyl-3,5-pyrazolidinedione; 4-(2-isopentenyl)-1,2-diphenyl-3,5-pyrazolidinedione; phenyl-prenazone; prenazone; DA-2370; Analud;* Methrazone; Zepelin. C$_{20}$H$_{20}$N$_2$O$_2$; mol wt 320.39. C 74.98%, H 6.29%, N 8.74%, O 9.99%. Non-steroidal anti-inflammatory drug. Prepn: Casadio, Pala, **Ger.** pat. **2,031,238** corresp to U.S. pat. **3,703,528** (1971, 1972 both to Ist. De Angeli); Casadio *et al., Arzneimittel-Forsch.* **22,** 171 (1972). Series of articles on physical-chemical data, pharmacology, toxicology and clinical studies: *ibid.* 174-281. Toxicity data: C. Bianchi, G. Bonardi, *ibid.* 196.

Fine, white odorless and tasteless crystalline powder, mp 156.5° (ethanol). Very sol in acetone, chloroform, DMF; sparingly sol in acetonitrile, benzene, 10% NaOH. Slightly sol in ether, methanol, ethanol, cyclohexane. Practically insol in 10% HCl, 10% acetic acid, water. uv max (ethanol): 246 nm (log ε 4.19); (pH 9-12 buffer): 264 nm (log ε 4.322-4.326). pKa 5.09 ± 0.07. LD_{50} in male mice, rats (mg/kg): 408.8, 386.4 i.p.; 1067, > 2000 orally (Bianchi, Bonardi).

THERAP CAT: Anti-inflammatory.

4054. Ferbam. *Tris(dimethylcarbamodithioato-S,S')-iron; tris(dimethyldithiocarbamato)iron;* dimethyldithiocarbamic acid iron salt; ferric dimethyldithiocarbamate; Carbamate; Ferbeck; Fermate; Ferradow; Karbam Black. $C_9H_{18}FeN_3S_6$; mol wt 416.50. C 25.95%, H 4.36%, Fe 13.41%, N 10.09%, S 46.19%. $[(CH_3)_2NCS_2]_3Fe$. Prepd from Na dimethyldithiocarbamate and a soln of a ferric salt: Tisdale, Williams, U.S. pat. 1,972,961 (1934 to du Pont). Toxicity data: H. C. Hodge *et al., J. Am. Pharm. Assoc.* 41, 662 (1952).

Black solid. mp >180° (dec). Solubility in water 120 ppm, pH of soln 5.0; sol in acetone, chloroform, pyridine, acetonitrile. Tends to decompose on prolonged storage or exposure to heat and moisture. LD_{50} in male mice, rats (mg/kg): 3000, 2700 i.p. (Hodge).

Caution: Potential symptoms of overexposure are irritation of eyes and respiratory tract; dermatitis; gastrointestinal disturbance. *See NIOSH Pocket Guide to Chemical Hazards* (DHHS/NIOSH 90-117, 1990) p 114. *See also Clinical Toxicology of Commercial Products,* R. E. Gosselin *et al.,* Eds. (Williams & Wilkins, Baltimore, 5th ed., 1984) Section II, p 311.

USE: Fungicide.

4055. Fermium. Fm; at. no. 100; valence 3, 2. Manmade, radioactive element. No stable nuclides; known isotopes (mass numbers): 242-259. Longest-lived known isotope: 257 ($T_{1/2}$ 101 days, α-emitter, rel. at. mass 257.0951), produced by neutron bombardment of plutonium (or other elements with higher at. no.). ^{255}Fm ($T_{1/2}$ 20.07 hrs, α-emitter) originally discovered in debris from a thermonuclear test explosion in Nov. 1952. Reviews of discovery: A. Ghiorso *et al., Phys. Rev.* 99, 1048-1049 (1955); A. Ghiorso, G. T. Seaborg, *Sci. Am.* 195(6), 67-80 (1956). *Reviews:* G. T. Seaborg, *J. Chem. Ed.* 36, 38-44 (1959); C. Keller, *The Chemistry of the Transuranium Elements* (Verlag Chemie, Weinheim, English Ed., 1971) pp 591-594; Silva, "Trans-Curium Elements" in *MTP Int. Rev. Sci.: Inorg. Chem.,* Ser. One vol. 8, A. G. Maddock, Ed. (University Park Press, Baltimore, 1972) pp 71-105; *Comprehensive Inorganic Chemistry* vol. 5, J. C. Bailar, Jr. *et al.,* Eds. (Pergamon Press, Oxford, 1973) *passim; Handb. Exp. Pharmakol.* 36, 689-738 (1973); R. J. Silva in *The Chemistry of the Actinide Elements* vol. 2, J. J. Katz *et al.,* Eds. (Chapman and Hall, New York, 1986) pp 1086-1092; *The Elements Beyond Uranium,* G. T. Seaborg, W. D. Loveland, Eds. (John Wiley & Sons, Inc., New York, 1990) pp 28-38.

Caution: Radiation hazard; handling requires special equipment and sheilding facilities (Katz *et al., loc cit.,* vol. 2, p. 1128).

4056. Ferredoxins. A group of electron transfer factors found in plants and bacteria, which are non-heme iron-sulfur proteins and which play an important role in photosynthesis, nitrogen and carbon dioxide fixation, and respira-

tion. They are generally classified by the presence of either 2 or 4 iron atom clusters and an equivalent amount of inorganic or "acid-labile" sulfide bonded to the peptide chain through 4 cysteine sulfhydryl groups. The two-iron ferredoxins are found primarly in plants and blue-green algae and are sometimes referred to as chloroplast or "plant type" ferredoxins; the four-iron ferredoxins are predominant in bacteria. Mol wts of chloroplast ferredoxins are about 12,000; those of bacterial ferredoxins range from about 6,000 to 24,000. Isoln from *Clostridium pasteurianum:* L. E. Mortenson *et al., Biochem. Biophys. Res. Commun.* 7, 448 (1962). Prepn of bacterial ferredoxins: L. E. Mortenson, U.S. pat. 3,344,130 (1960 to duPont). The amino acid sequence of several bacterial and chloroplast ferredoxins has been elucidated. Amino acid sequence of ferredoxin from *Clostridium pasteurianum:* Tanaka *et al., Biochem. Biophys. Res. Commun.* 16, 422 (1964); *eidem, Biochemistry* 5, 1666 (1966); from spinach: Matsubara, Sasaki, *J. Biol. Chem.* 243, 1732 (1968). Synthesis of peptide chain of *C. pasteurianum* ferredoxins: Bayer *et al., Tetrahedron* 24, 4853 (1968); Trakatellis, Schwartz, *Proc. Nat. Acad. Sci. USA* 63, 436 (1969). *Reviews:* Buchanan, *Struct. Bonding* 1, 109 (1966); Malkin, Rabinowitz, *Ann. Rev. Biochem.* 36, 113 (1967); Arnon, *Naturwiss.* 56, 295 (1969); Buchanan, Arnon, *Advan. Enzymol.* 33, 119 (1970); W. Lovenberg, "Ferredoxin and Rubredoxin" in *Microbial Iron Metabolism,* J. B. Neilands, Ed. (Academic Press, New York, 1974) pp 161-182; D. C. Yoch, R. P. Carithers, *Microbiol. Rev.* 43, 384-421 (1979).

Absorption max of bacterial ferredoxins: 280, 385-400 nm; of plant ferredoxins: 280, 325, 420, 463 nm. All show negative oxidation-reduction potentials near that of the hydrogen electrode: −0.390 V *(C. pasteurianum);* −0.420 V (spinach): Tagawa, Arnon, *Biochim. Biophys. Acta* 153, 602 (1968). Autoxidizable. Acidification, as well as treatment with iron-chelating agents or mercurials, results in evolution of H_2S and loss of the visible absorption.

4057. Ferric Acetate, Basic. $C_4H_7FeO_5$; mol wt 190.94. C 25.16%, H 3.70%, Fe 29.25%, O 41.90%. Fe(OH)(CH_3COO)_2. Prepd by heating ferric acetate soln: Casey, Doyle in Kirk-Othmer *Encyclopedia of Chemical Technology,* vol. 12 (Interscience, New York, 2nd ed., 1967) p 24.

Brownish-red scales or amorphous powder; faint acetic odor. Practically insol in water; sol in alcohol, acids. *Protect from light.*

USE: In the textile industry as a mordant in dyeing and printing, and for the weighting of silk and felt; as wood preservative; in leather dyes; as medicine.

4058. Ferric Albuminate. Albumized iron. Combinations of egg albumin and iron contg 17-19% Fe. The exact form in which iron exists in these compds is not known and probably varies with the method of manuf. Structure and composition of compds of Fe and albumin: Heymann, Oppenheimer, *Biochem. Z.* 199, 468 (1928).

Reddish-brown, lustrous granules, or brown powder. Odorless. Freely sol in water; almost insol in alcohol.

THERAP CAT: Hematinic.

4059. Ferric and Ammonium Acetate Solution. Basham's mixture. Contains 0.16-0.20% Fe and about 3.5% ammonium acetate w/v. Prepn: *U.S.D.* 25th ed, p 722.

Clear, reddish-brown liq, having an aromatic odor. Acid to litmus.

THERAP CAT: Hematinic.

THERAP CAT (VET): In iron deficiency.

4060. Ferric Bromide. Br_3Fe; mol wt 295.56. Br 81.11%, Fe 18.89%. FeBr_3. Prepn from Fe and Br_2: Gregory, Thackrey, *J. Am. Chem. Soc.* 72, 3176 (1950); Gregory, *ibid.* 73, 472 (1951); from Fe_2(SO_4)_3 and LiBr: Lieser, Elias, *Z. Anorg. Allgem. Chem.* 316, 208 (1962).

Dark-red or black hexagonal, rhombic plates. Very hygroscopic. Loses some Br_2 on exposure to air and light. Sol in water, alcohol, ether, acetic acid; slightly sol in liq NH_3; aq soln dec to FeBr_2 + Br_2 on boiling. *Keep cool, well-closed and protected from light.*

Caution: Irritant.

USE: Catalyst for organic reactions, particularly in bromination of aromatic compds.

4061. Ferric Chloride. Flores martis. Cl_3Fe; mol wt 162.20. Cl 65.57%, Fe 34.43%. $FeCl_3$. Occurs in nature as the mineral *molysite*. Prepn: Tarr, *Inorg. Syn.* **3**, 191 (1950); Pray, *ibid.* **5**, 153 (1957); Epperson *et al.*, *ibid.* **7**, 163 (1963); Lieser, Elias, *Z. Anorg. Allgem. Chem.* **316**, 208 (1962); Attwood, Shelton, *J. Inorg. Nucl. Chem.* **26**, 1758 (1964); Bardawil *et al.*, *Inorg. Chem.* **3**, 149 (1964). Acute toxicity: C. S. Hosking, *Aust. Paediatr. J.* **6**, 92 (1970). Use as clinical reagent: L. Cassidei *et al.*, *Clin. Chim. Acta* **90**, 121 (1978).

Hexagonal, dark leaflets or plates. Red by transmitted light, green by reflected light; sometimes appears brownish-black. Very hygroscopic. Melts and volatilizes about 300°; bp ~316°; d^{25} 2.90. Vapor density measurements show that it is dimeric at about 400° but monomeric above 750°. Dissociates at high temps to $FeCl_2$ and Cl_2. Readily absorbs water in air to form the hexahydrate. Readily sol in water, alcohol, ether, acetone; slightly sol in CS_2. Practically insol in ethyl acetate. *Keep well closed.*

Hexahydrate, brownish-yellow or orange monoclinic crystals. d 1.82. Structure: *trans*-$[FeCl_2(H_2O)_4]Cl.2H_2O$: Lind, *J. Chem. Phys.* **47**, 990 (1967). Usually slight odor of HCl; very hygroscopic. mp ~37°. Readily sol in water, alcohol, acetone, ether; pH of 0.1 molar aq soln 2.0. *Keep well closed.* LD_{50} i.v. in mice: 0.049 mg Fe/g (Hosking).

Caution: Anhydr form is irritant.

USE: In photoengraving; photography; manuf of other Fe salts, pigments, ink; as catalyst in organic reactions; purifying factory effluents and deodorizing sewage; chlorination of Ag and Cu ores; as mordant in dyeing and printing textiles; oxidizing agent in dye manuf. Clinical reagent (amino acids in urine, esp in phenylketonuria).

THERAP CAT: Hexahydrate as astringent, styptic.

THERAP CAT (VET): Styptic, astringent.

4062. Ferric Chromate(VI). *Chromic acid iron(3+) salt (3:2);* C.I. Pigment Yellow 45; C.I. 77505; Siderin yellow. $Cr_3Fe_2O_{12}$; mol wt 459.67. Cr 33.93%, Fe 24.30%, O 41.77%. $Fe_2(CrO_4)_3$. Preparation: Boericke, Bangert, *U.S. Bur. Mines, Rept. Investigations* **3813**, 19 pp (1945), *C.A.* **39**, 5164[7] (1945).

Yellow powder. Practically insol in water. Sol in HCl.

USE: Pigment for ceramics, glass and enamels.

4063. Ferric Citrate. A combination of iron and citric acid of indefinite composition. Prepn: Belloni, *Gazz. Chim. Ital.* **50**, II, 159 (1920).

Garnet-red, transparent scales or pale-brown powder. Odorless; slight ferruginous taste. Slowly but completely sol in cold water, readily sol in hot water, but diminishing in soly with age. Practically insol in alc.

THERAP CAT: Hematinic.

THERAP CAT (VET): In iron deficiency.

4064. Ferric Ferrocyanide. *Hexakis(cyano-C)ferrate-(4-) iron(3+) (3:4);* Ferric hexacyanoferrate(II); C.I. Pigment Blue 27; C.I. 77510; Berlin blue; Chinese blue; Hamburg blue; mineral blue; Paris blue; Prussian blue. $C_{18}Fe_7$-N_{18}; mol wt 859.23. C 25.16%, Fe 45.50%, N 29.34%. Fe_4-$[Fe(CN)_6]_3$. Usually contains some water and some alkali ferrocyanide. Prepn and list of refs: *Colour Index* vol. 4 (3rd ed, 1971) p 4673.

Dark-blue powder or lumps. Loses all of its water only at 250° with partial decompn. d 1.80. Practically insol in water, dil acids, most organic solvents. When freshly prepd is sol in aq oxalic acid.

USE: As pigment in printing inks, paints, alkyd resin enamels, linoleum, leathercloth, carbon papers, typewriter ribbons, rubbers, plastics, artists' colors; in removal of H_2S from gases.

4065. Ferric Fluoride. F_3Fe; mol wt 112.84. F 50.51%, Fe 49.49%. FeF_3. Prepn from $FeCl_3$ and HF: Kwasnik in *Handbook of Preparative Inorganic Chemistry* vol. 1, G. Brauer, Ed. (Academic Press, New York, 2nd ed., 1963) p 266; Shinn *et al.*, *Inorg. Chem.* **5**, 1927 (1966); from Fe_2O_3 or Fe_2S_3 and SF_4: Oppegard *et al.*, *J. Am. Chem. Soc.* **82**, 3835 (1960); from $FeCl_3$ and CH_3CHF_2: Natta *et al.*, **Brit. pat. 995,186** (1965 to Montecatini).

Green hexagonal crystals. Sublimes > 1000°. d 3.87. Very slightly sol in water; freely sol in dil HF; practically

insol in alcohol, ether, benzene. Soly in liq HF: 0.008 g/100 g HF, Jache, Cady, *J. Phys. Chem.* **56**, 1106 (1952).

USE: As catalyst in organic reactions.

4066. Ferric Formate. $C_3H_3FeO_6$; mol wt 190.90. C 18.88%, H 1.58%, Fe 29.25%, O 50.29%. $Fe(HCOO)_3$. Prepn: Weinland, Reihlin, *Ber.* **46**, 3144 (1913); Starke, *J. Inorg. Nucl. Chem.* **25**, 823 (1963). May contain 1 or 2 H_2O.

Red or yellow microcryst powder. Readily hydrolyzed to basic formates. Sol in water; very slightly sol in alcohol. *Protect from light.*

USE: Preservation of silage.

4067. Ferrichromes. Growth-promoting iron chelates. Ferrichrome and ferrichrome A first isolated from the rust fungus, *Ustilago sphaerogena:* Neilands, *J. Am. Chem. Soc.* **74**, 4846 (1952); Garibaldi, Neilands, *ibid.* **77**, 2429 (1955). Ferrichrome and four related compounds, *ferrichrysin, ferricrocin, ferrirubin* and *ferrirhodin* have also been isolated from *Aspergillaceae* and from *Penicillium resticulosum:* Zähner *et al.*, *Arch. Mikrobiol.* **45**, 119 (1963); Keller-Schierlein, Deer, *Helv. Chim. Acta* **46**, 1907 (1963); Keller-Schierlein, *ibid.*, 1920. Ferrichromes are cyclic hexapeptides containing three small, neutral amino acids and three derivatives of δ-N-hydroxy-L-ornithine [$HONH(CH_2)_3CH(NH_2)$-COOH]. Structure: Rogers, Neilands, *Biochemistry* **3**, 1850 (1964). Crystal and molecular structure of ferrichrome A: Zalkin *et al.*, *J. Am. Chem. Soc.* **88**, 1810 (1966). Synthesis of ferrichrome: Keller-Schierlein, Maurer, *Helv. Chim. Acta* **52**, 603 (1969); Isowa *et al.*, *Bull. Chem. Soc. Japan* **47**, 215 (1974). Reviews of ferrichromes and other hydroxamic acids: Neilands, *Struct. Bonding* **1**, 59-108 (1966); Emery, *Advan. Enzymol. Relat. Areas Mol. Biol.* **35**, 135-185 (1971); J. B. Neilands, *Bioinorganic Chemistry* **11**, K. N. Raymond, Ed. (A.C.S., Washington, 1977), pp 3-32.

| Ferrichrome | R = CH$_3$ | R' = H |
| Ferrichrome A | R = trans-HOOCCH$_2$(CH$_3$)C=CH | R' = CH$_2$OH |

Ferrichrome, $C_{27}H_{42}FeN_9O_{12}$, long, yellow needles from anhydr methanol, shrink and blacken at 240-242° without melting. $[\alpha]_D$ +300° (c = 0.04). Absorption max (methanol): 425 nm ($E^{1\%}_{1cm}$ 39.4). Sol in water and hot methanol; sparingly sol in ethanol, acetone, ether, chloroform.

Ferrichrome A, $C_{41}H_{58}FeN_9O_{20}$, crystals from water. Absorption max (0.1M phosphate buffer; pH 7): 440 nm ($E^{1\%}_{1cm}$ 33.8). Very sol in methanol, ethanol, propanol; slightly sol in hot water. Practically insol in acetone, petr ether, ether, chloroform.

4068. Ferric Hydroxide. *Ferric hydroxide oxide;* hydrated ferric oxide. $FeHO_2$; mol wt 88.85. Fe 62.85%, H 1.13%, O 36.01%. $FeO(OH)$. Occurs in nature as the minerals *goethite* [α-$FeO(OH)$], *lepidocrocite* [γ-$FeO(OH)$], and *limonite* [$FeO(OH).nH_2O$]. Other known allomorphic forms: β-$FeO(OH)$; δ-$FeO(OH)$. The hydroxide $Fe(OH)_3$ is not known. Prepn: Lux in *Handbook of Preparative Inor-*

ganic Chemistry, **vol 2,** G. Brauer, Ed. (Academic Press, New York, 2nd ed., 1965) p 1499. Crystal structure of α-FeO(OH): Sampson, *Acta Cryst.* **25B,** 1683 (1969). *Review:* Bernal *et al., Clay Miner. Bull.* **4,** 15-30 (1959).

Red to brown powder or crystals. Loses H_2O to form Fe_2O_3. d 3.4-3.9. Practically insol in water, alcohol. Sol in mineral acids.

USE: In purifying water; as absorbent in chemical processing; as pigment; as catalyst.

4069. Ferric Hypophosphite. $FeH_6O_6P_3$; mol wt 250.81. Fe 22.27%, H 2.41%, O 38.27%, P 37.05%. Fe$(H_2PO_2)_3$. Prepn: *U.S.D.* 25th ed, p 573.

White or grayish-white powder. Odorless, tasteless. Sol in 2300 parts cold water, 1200 parts boiling water; more sol in water in presence of H_3PO_2; sol in warm concd solns of alkali citrates. *Protect from light.* Should not be heated or triturated with chlorates, nitrates, or other oxidizing agents.

USE: Formerly as dietary supplement for phosphorus.

4070. Ferriclate Calcium Sodium. *Pentaaqua[D-gluconato(4—)-O²,O⁴,O⁵]tetra-μ-hydroxydioxotriferrate(3—) calcium sodium (2:1:4);* monocalcium tetrasodium bis[pentaaquatetra-μ-hydroxy[D-gluconato(4—)]dioxotriferrate(3—)]; Kelfer. $C_{12}H_{44}CaFe_6Na_4O_{36}$; mol wt 1231.57. C 11.70%, H 3.60%, Ca 3.25%, Fe 27.21%, Na 7.47%, O 46.77%.

THERAP CAT: Hematinic.

4071. Ferric Nitrate. FeN_3O_9; mol wt 241.86. Fe 23.09%, N 17.37%, O 59.54%. $Fe(NO_3)_3$. Prepn: *Gmelin's, Iron* (8th ed.) **59,** part B, 161-172 (1932). Toxicity study: H. F. Smyth *et al., Am. Ind. Hyg. Assoc. J.* **30,** 470 (1969).

Nonahydrate, pale-violet to grayish-white, somewhat deliquesc crystals. mp 47°. Dec below 100°. d^{21} 1.68. Freely sol in water, alcohol, acetone; slightly sol in cold concd HNO_3. LD_{50} orally in rats: 3.25 g/kg (Smyth).

USE: As mordant in dyeing, weighting silks, tanning; as reagent in analytical chemistry; as corrosion inhibitor.

4072. Ferric Oxide. Ferric sesquioxide; jeweler's rouge. Fe_2O_3; mol wt 159.69. Fe 69.94%, O 30.06%. α-Form occurs in nature as the mineral *hematite.* γ-Form occurs in nature as the mineral *maghemite;* prepd by dehydration of α-FeO(OH): Giovanoli, Brütsch, *Chimia* **28,** 188 (1974). Prepn of a third allomorphic form, ε-Fe_2O_3: Schrader, Büttner, *Z. Anorg. Allgem. Chem.* **320,** 220 (1963); Trautmann, Forestier, *Compt. Rend.* **261,** 4423 (1965). Color and appearance of Fe_2O_3 are dependent upon the size and shape of the particles and the amount of combined water. Preparation and properties: *Gmelin's, Iron* (8th ed.) **59,** part B, 63-94 (1932); Baudisch, Hartung, *Inorg. Syn.* **1,** 185 (1939); *Ullmann's Encyklopädie der Technischen Chemie* **vol. 6,** 421-423 (1955); Bernal *et al., Clay Miner. Bull.* **4,** 15-30 (1959). Toxicology: L. T. Fairhall, *Industrial Toxicology* (Hafner, New York, 2nd ed., 1969) pp 64-66.

Note: The composition of the substance called δ-Fe_2O_3 is actually FeO(OH) (Bernal *et al.*).

Caution: Potential symptoms of overexposure to dust and fumes are benign pneumoconiosis with x-ray shadows indistinguishable from fibrotic pneumoconiosis. *See NIOSH Pocket Guide to Chemical Hazards* (DHHS/NIOSH 90-117, 1990) p 130; Fairhall, *loc. cit.*

USE: As pigment for rubber, paints, paper, linoleum, ceramics, glass; in paint for ironwork, ship hulls; as polishing agent for glass, precious metals, diamonds; in electrical resistors and semiconductors; in magnets, magnetic tapes; as catalyst; colloidal solns as stain for polysaccharides.

4073. Ferric Oxide, Saccharated. Saccharated iron; iron sugar; Colliron I.V.; Feojectin; Ferrivenin; Ferum Hausmann; Iviron; Neo-Ferrum; Proferrin; Sucrofer. Contains 2.8-3.2% Fe. Prepn: *U.S.D.* 26th ed., p 627 (1967). Prepn of soln contg 2% Fe suitable for i.v. injection: Slack, Wilkinson, *Lancet* **256,** 11 (1949).

Brown powder. Sol in water. Practically insol in alcohol. Solns are unstable in the presence of electrolytes: *Do not mix with physiological saline!*

THERAP CAT: Hematinic.

4074. Ferric Phosphate. FeO_4P; mol wt 150.82. Fe 37.03%, O 42.43%, P 20.54%. $FePO_4$. Occurs in nature as the minerals *beraunite, cacoxenite, dufrenite, koninckite,*

phosphosiderite, strengite. Prepn from $Fe(H_2PO_4)_3$: Remy, Boulle, *Compt. Rend.* **253,** 2699 (1961); from $Fe(CO)_5$ and H_3PO_4: Cate *et al., Soil Sci.* **88**(3), 130 (1959); from phosphate rock: Vickery, **U.S.** pat. **2,914,380** (1959 to Horizons Inc.); from mill scale and H_3PO_4: Alexander, Mathes, **U.S.** pat. **3,070,423** (1962 to Chemetron).

Dihydrate, white, grayish-white, or light pink, orthorhombic or monoclinic crystals or amorphous powder. Loses water above 140°. d 2.87. Practically insol in water. Slowly sol in HNO_3; readily sol in HCl.

USE: As food and feed supplement, particularly in bread enrichment; as fertilizer.

4075. Ferric Pyrophosphate. $Fe_4O_{21}P_6$; mol wt 745.21. Fe 29.98%, O 45.09%, P 24.94%. $Fe_4(P_2O_7)_3$. Prepn: *Gmelin's, Iron* (8th ed.) **59,** part B, 777 (1932); Knight, Kelly, **U.S.** pat. **3,014,784** (1962 to American Oil).

Nonahydrate, yellowish-white powder. Practically insol in water or acetic acid. Sol in mineral acids.

USE: As catalyst; in fireproofing of synthetic fibers; in corrosion-preventing pigments.

THERAP CAT: Hematinic.

4076. Ferric Sodium Edetate. *[[N,N'-Ethanediylbis[N-(carboxymethyl)glycinato]](4—)]-N,N',O,O',Oᴺ,Oᴺ'-ferrate-(1—) sodium; sodium [(ethylenedinitrilo)tetraacetato]ferrate-(1—);* (ethylenedinitrilo)tetraacetic acid sodium salt iron complex; ferric monosodium ethylenediaminetetraacetate; edetic acid sodium iron salt; sodium iron edetate; sodium feredetate; Ferrostrane; Ferrostrene; Sybron. $C_{10}H_{12}FeN_2$-NaO_8; mol wt 367.05. C 32.72%, H 3.30%, Fe 15.21%, N 7.63%, Na 6.26%, O 34.87%. Prepd from disodium ethylenediaminetetraacetic acid and ferric nitrate: Sawyer, McKinnie, *J. Am. Chem. Soc.* **82,** 4191 (1960).

Crystals from water + ethanol.

THERAP CAT: Iron source.

4077. Ferric Sodium Pyrophosphate. Sodium ferric pyrophosphate. $Fe_4Na_8O_{35}P_{10}$. mol wt 1277.02. Fe 17.49%, Na 14.40%, O 43.85%, P 24.25%. Hydrate; Na_8-$Fe_4(P_2O_7)_6$.xH_2O. The commercial product contains 15.5-16.5% Fe and 50.5-52.5% P_2O_5.

White powder. Bulk density 1.4-1.6. Sol in hydrochloric acid. Insol in water.

USE: Food enrichment. Less prone to induce rancidity than orthophosphates.

4078. Ferric Subsulfate Solution. Basic ferric sulfate soln; Monsel's soln. Approx: $Fe_4(OH)_2(SO_4)_5$. Prepn from $FeSO_4$ and HNO_3: *U.S.D.* 25th ed., p 574.

Reddish-brown liquid. Almost odorless; sour, strongly astringent taste. Acid to litmus. Affected by light. d^{25} 1.548. Miscible with water, alcohol. May crystallize or solidify at low temps. *Keep well closed, protected from light, in a warm place.*

USE: As mordant in dyeing textiles.

THERAP CAT: Styptic.

THERAP CAT (VET): Styptic, astringent.

4079. Ferric Sulfate. Ferric persulfate; ferric sesquisulfate; ferric tersulfate. $Fe_2O_{12}S_3$; mol wt 399.88. Fe 27.93%, O 48.01%, S 24.06%. $Fe_2(SO_4)_3$. Prepn: *Gmelin's, Iron* (8th ed.) **59,** part B, 439-462 (1932).

Grayish-white powder, or rhombic or rhombohedral crystals. Very hygroscopic. Commercial product usually contains about 20% water and is yellowish in color. d^{18} 3.097. Slowly sol in water, rapidly sol in the presence of a trace of $FeSO_4$; sparingly sol in alcohol; practically insol in acetone, ethyl acetate. Hydrolyzed slowly in aq soln. *Keep well closed and protected from light.*

USE: In preparation of iron alums, other iron salts and pigments; as coagulant in water purification and sewage treatment; in etching aluminum; in pickling stainless steel and copper; as mordant in textile dyeing and calico printing; in soil conditioners; as polymerization catalyst.

4080. Ferric Tannate. Ferric gallotannate. Variable composition. Contains 8-10% Fe, 70-80% tannin.

Bluish-black powder. Insol in water; sol in dil mineral acids.

USE: In inks.

4081. Ferric Thiocyanate. Ferric sulfocyanate; ferric sulfocyanide. $C_3FeN_3S_3$; mol wt 230.10. C 15.66%, Fe 24.27%, N 18.26%, S 41.81%. $Fe(SCN)_3$. Prepn: *Gmelin's Handb. Anorg. Chem., Iron* (8th ed.) part B, 747-761 (1932).

Sesquihydrate, red, deliquesc crystals. Dec on heating. Sol in water, alcohol, ether, acetone, pyridine, ethyl acetate. Practically insol in $CHCl_3$, CCl_4, CS_2, toluene. *Keep well closed.*

USE: Analytical reagent.

4082. Ferrite. Ferrospinel. A crystalline, usually man-made material, having a spinel structure and consisting essentially of ferric oxide and at least one other metallic oxide which is usually, although not always, divalent in nature. When molded into compressed bodies, the material is characterized by high magnetic permeability. Typified composition: Fe_2O_3 67-70%; ZnO 10-10.5%; MnO_2 20-22.5%; CuO 0.1-10%; Co_3O_4 0.1%. Ferrites are prepd by ceramic techniques. The oxides or carbonates are milled in steel ball mills, and the mixture of very fine particles is dried and prefired in order to obtain a homogeneous end product: Hilpert, *Ber.* **42**, 2248 (1909). Examples of modern techniques: Simpkiss; Harvey, U.S. pats **2,723,238-9** (both 1955 to RCA). Prepn of single crystals: Rooymans, *Colloq. Int. Cent. Nat. Rech. Sci.* No. 205, 151 (1972). Books: Snoek, *New Developments in Ferromagnetic Materials* (Elsevier, New York, 1947); Smit, Wijn, *Ferrites* (John Wiley, New York, 1959); Soohov, *Theory and Applications of Ferrites* (Prentice Hall, 1960); Standley, *Oxide Magnetic Materials* (Clarendon Press, Oxford, 1962); *Ferrites, Proc. Int. Conf.*, Y. Hoshino *et al.*, Eds. (University Park Press, Baltimore, 1971) 671 pp; E. E. Riches, *Ferrites, A Review of Materials and Applications* (Mills and Boon, London, 1972) 88 pp. Reviews with bibliographies: Gorter, *Proc. I.R.E.* **43**, 1945-1973 (1955); Fresh, "Methods of Preparation and Crystal Chemistry of Ferrites," *ibid.* **44**, 1303-1311 (1956); Brailsford, *Magnetic Materials* (John Wiley, New York, 3rd ed., 1960) pp 160-181; Hogen, *Sci. Am.* **202**, 92-104 (1960); Economos in Kirk-Othmer's *Encyclopedia of Chemical Technology* vol. 8 (Interscience, New York, 2nd ed., 1965) pp 881-901; Gray, "Oxide Spinels" in *High Temperature Oxides*, Part IV, A. M. Alper, Ed. (Academic Press, New York, 1971) pp 77-107.

Note: The term "ferrites" has been expanded to mean any oxidic magnetic material.

USE: Magnetic cores for inductors and transformers; microwave devices; information storage; electromechanical transducers: E. E. Riches, *loc. cit.;* Brockman, *Ceram. Ind.* **99**, 24 (1972).

4083. Ferritin. Epadora; Ferrofolin; Ferrol; Ferrosprint; Ferrostar; Sanifer; Sideros; Unifer. Major iron storage protein; found in spleen, liver and intestinal mucosa of vertebrates; widely distributed in the plant and animal kingdoms. Isoln and crystallization of horse spleen ferritin: Laufberger, *Bull. Soc. Chim. Biol.* **19**, 1575 (1937); Granick, *J. Biol. Chem.* **14**, 451 (1942); Crichton *et al., Biochem. J.* **131**, 51 (1973). Consists of a protein shell surrounding a crystalline, hydrated iron oxide/phosphate core. The core may contain up to 4500 Fe^{3+} ions; unfractionated, horse-spleen ferritin contains approx 20% iron on a dry weight basis. The protein shell, *apoferritin,* has a mol wt of ~445,000. Several proposed structures of protein moiety: Harrison, *J. Mol. Biol.* **6**, 404 (1963); Crichton, *Biochem. J.* **126**, 761 (1972); Niitsu *et al., Biochem. Biophys. Res. Commun.* **55**, 1134 (1973). Absorption spectra: Granick, *Chem. Rev.* **38**, 379 (1946). Use of serum ferritin and isoferritins in clinical medicine: J. W. Halliday, L. W. Powell, *Prog. Hematol.* **11**, 229 (1979). *Reviews:* Harrison, Hoy, "Ferritin" in *Inorganic Biochemistry* vol. 1, G. L. Eichhorn, Ed. (Elsevier, New York, 1973) pp 253-277; Crichton, *Angew.*

Chem. Int. Ed. **12**, 57-65 (1973); P. Aisen, I. Listowsky, *Ann. Rev. Biochem.* **49**, 357 (1980).

Red-brown, water-soluble protein. Forms cubic and orthorhombic crystals.

THERAP CAT: Hematinic.

4084. Ferrocene. Dicyclopentadienyliron; biscyclopentadienyliron. $C_{10}H_{10}Fe$; mol wt 186.03. C 64.56%, H 5.42%, Fe 30.02%. Prepn: Kealy, Pauson, *Nature* **168**, 1039 (1951); Pauson, U.S. pat. **2,680,756** (1954 to du Pont); Miller *et al., J. Chem. Soc.* **1952**, 632; Anzilotti, Weinmayr, U.S. pat. **2,791,597** (1957 to du Pont). Other prepn: Pruett, Morehouse, *Advances in Chemistry Series* **23**, 368-371 (1959); Wilkinson, *Org. Syn.* **coll. vol. IV**, 473 (1963); Cordes, **Fr.** pat. **1,341,880** (1963 to BASF), *C.A.* **60**, 6873a (1964). Structure studies: Wilkinson *et al., J. Am. Chem. Soc.* **74**, 2125 (1952); Seibold, Sutton, *J. Chem. Phys.* **23**, 1967 (1955). Synthesis of a helical ferrocene: T. J. Katz, J. Pesti, *J. Am. Chem. Soc.* **104**, 346 (1982). *Reviews:* Rausch *et al., J. Chem. Ed.* **34**, 268-272 (1957); M. Rosenblum, *Chemistry of the Iron Group Metallocenes* (John Wiley, New York, 1965) 241 pp; Bruce, *Organometal. Chem. Rev.* (Sect. B) **10**, 75-122 (1972); B. W. Rockett, G. Marr, *J. Organometal. Chem.* **211**, 215-278 (1981).

Orange needles from methanol or ethanol; odor of camphor. mp 173-174°. Sublimes above 100°. Volatile in steam. Practically insol in water, 10% NaOH, and concd boiling HCl. Sol in alcohol, ether, benzene. Dissolves in dil nitric and concd sulfuric acids forming a deep red soln with blue fluorescence. The molecule is diamagnetic and the dipole moment is effectively zero.

USE: Antiknock additive for gasoline; catalyst.

4085. Ferrocholinate. 2-Hydroxy-N,N,N-trimethylethanaminium (OC-6-44)-triaqua[2-hydroxy-1,2,3-propanetricarboxylato(4—)]ferrate(1—); [hydrogen citrato(3—)]triaquoiron, choline salt; iron choline citrate; Chelafer; Chel-Iron; Ferrolip. $C_{11}H_{24}FeNO_{11}$; mol wt 402.16. C 32.85%, H 6.02%, Fe 13.89%, N 3.48%, O 43.76%. Chelate prepd by interaction of equimolar quantities of choline dihydrogen citrate and freshly prepd $Fe(OH)_3$ or $FeCO_3$: Bandelin, U.S. pat. **2,575,611** (1951 to Flint Eaton); by treatment of a freshly prepd solution of ferric citrate with an equimolar amount of choline: Chakrabarti, Sen, *Chem. & Ind. (London)* **1961**, 1407. Pharmaceutical prepn, characteristics and applications: G. N. Bookwalter *et al., J. Food Sci.* **38**, 618 (1973). Bioavailability in pigs: E. R. Miller *et al., J. Animal Sci.* **52**, 783 (1981).

Greenish-brown, reddish-brown or brown amorphous solid with glistening surface upon fracture. Freely sol in water, yielding stable solns; sol in acids, alkalies. One gram of pharmaceutical grade is equivalent to 120 mg of elemental iron and 360 mg of choline base.

Note: Other combinations of iron, choline and citric acid have been prepd for pharmaceutical use. A 1:2:2 chelate, $C_{22}H_{36}FeN_2O_{16}$, was reported by Chakrabarti and Sen, *loc. cit.,* and a 2:3:3 chelate, $C_{33}H_{57}Fe_2N_3O_{24}$, by Rosenfelder, U.S. pat. **2,865,938** (1958 to H. Rosenstein).

THERAP CAT: Hematinic.

4086. Ferroglycine Sulfate. Ferroglycine sulfate complex; ferrous sulfate glycine complex; ferrous aminoacetosulfate; iron sulfate-glycine complex; glycine-ferrous sulfate complex; Plesmet; Kelferon; Ferronord; Ferro sanol; Glyferro; Pleniron. Prepn from glycine and ferrous sulfate: Rummell, U.S. pats. **2,877,253** and **2,957,806** (1959, 1960 both to Schwarz Arzneimittelfabrik).

THERAP CAT: Hematinic.

4087. Ferrosoferric Oxide. Ferric ferrous oxide; triiron tetraoxide; black iron oxide; magnetic iron oxide; Ethiops iron. Fe_3O_4; mol wt 231.53. Fe 72.36%, O 27.64%. Occurs in nature as the mineral *magnetite* (red-black lumps). Prepn: *Gmelin's, Iron* (8th ed) part B, 36-62 (1932); *Ullmanns Encyklopädie der Technischen Chemie* vol. **6**, 420 (1955). Review: Robl, *Angew. Chem.* **70**, 367 (1958).

Black cubes or amorphous powder. mp 1538°; d 5.2. Oxidized to Fe_2O_3 on heating in air. Practically insol in water. Sol in acids.

USE: Pigment in paints, linoleum, ceramic glazes; in coloring glass; as a polishing compd; in the textile industry; in cathodes; as catalyst.

4088. Ferrous Bromide. Br_2Fe; mol wt 215.65. Br 74.10%, Fe 25.90%. $FeBr_2$. Prepn: Baxter, *Z. Anorg. Chem.* **38**, 236 (1904); Baxter *et al., ibid.* **70**, 333 (1911); Schimmel, *Ber.* **62**, 963 (1929); Kühnl, Ernst, *Z. Anorg. Allgem. Chem.* **317**, 84 (1962).

Light yellow to dark brown hygroscopic crystals. mp 684°. d_4^{25} 4.63. Very sol in water and alcohol. *Keep tightly closed.*

Hexahydrate, pale green to bluish-green rhombic prisms. Forms tetrahydrate at 49°; dihydrate at 83°. Rapidly oxidized in moist air. *Keep tightly closed.*

USE: Polymerization catalyst.

4089. Ferrous Carbonate Mass. Blaud's mass; Vallet's mass; Fecarb. Contains 36-41% $FeCO_3$, the remainder consisting of honey and sugar. Prepn: *U.S.D.* 25th ed., p 575.

Dark greenish-gray to brown, moderately soft mass. Practically insol in water; appreciably sol in water satd with CO_2; sol in dil acids.

THERAP CAT: Hematinic.

THERAP CAT (VET): Has been used in iron deficiency.

4090. Ferrous Carbonate Saccharated. Freshly pptd $FeCO_3$ protected from oxidation by admixture with sugar. Contains not less than 15% $FeCO_3$. Prepn: *U.S.D.* 25th ed., p 575.

Olive-gray to greenish-brown powder. Odorless. Partially sol in water, completely sol in dil mineral acids.

THERAP CAT: Hematinic.

THERAP CAT (VET): In iron deficiency.

4091. Ferrous Chloride. Cl_2Fe; mol wt 126.75. Cl 55.94%, Fe 44.06%. $FeCl_2$. Occurs in nature as the mineral *lawrencite.* Preparation: Kovacic, Brace, *Inorg. Syn.* **6**, 172 (1960); Kühnl, Ernst, *Z. Anorg. Allgem. Chem.* **317**, 84 (1962). Prepn of dihydrate: Gayer, Woontner, *Inorg. Syn.* **5**, 179 (1957).

White rhombohedral crystals; may sometimes have a green tint. Very hygroscopic. mp 674°; bp 1023°; d^{25} 3.16. Can be sublimed in a stream of HCl at about 700°. Forms $FeCl_3$ and Fe_2O_3 on heating in air. Freely sol in water, alc, acetone; slightly sol in benzene; practically insol in ether.

Dihydrate, white monoclinic crystals with pale green tint. Loses 1 H_2O at 120°: Gayer, Woontner, *loc. cit.;* also reported to lose 1 H_2O at 150-160°: Schafer, *Z. Anorg. Allgem. Chem.* **258**, 69 (1949). Sol in water.

Tetrahydrate, pale green to blue-green, monoclinic crystals or cryst powder. Loses $2H_2O$ at 105-115°: Schafer, *loc. cit.* d 1.93. Sol in water, alcohol. The technical product may not be completely sol without the addn of acid. Aq solns are readily oxidized.

Caution: Mild irritant.

USE: In metallurgy; as reducing agent; in pharmaceutical prepns; as mordant in dyeing.

4092. Ferrous Citrate. Several forms of this salt are known. Prepn from citric acid and Fe powder: Oroshnik, Haffcke, U.S. pat. **2,904,573** (1959 to Ortho); and ferrous salts: Carlson, U.S. pat. **3,091,626** (1963 to Scherer).

Monohydrate, *monoferrous acid citrate monohydrate.* Powder. Practically insol in water, alcohol, acetone.

Decahydrate, *triferrous dicitrate decahydrate.* Very slightly colored powder or white crystals. Very stable to air oxi-

dation. If H_2O is removed by vacuum desiccation, the dehydrated ferrous salt rapidly oxidizes to a ferric salt. Practically insol in water, acetone.

THERAP CAT: Hematinic.

4093. Ferrous Fluoride. F_2Fe; mol wt 93.84. F 40.49%, Fe 59.51%. FeF_2. Prepn from $FeCl_2$ and HF gas: Kwasnik in *Handbook of Preparative Inorganic Chemistry,* G. Brauer, Ed. (Academic Press, New York, 2nd ed., 1963) p 266; from Fe powder and liq HF: Muetterties, Castle, *J. Inorg. Nucl. Chem.* **18**, 148 (1961).

Tetragonal crystals (rutile type) or powder. Sublimes at about 1100°. d 4.09. Sparingly sol in water; more sol in dil HF. Practically insol in alcohol, ether, benzene.

USE: As catalyst in organic reactions.

4094. Ferrous Fumarate. Cpiron; Erco-Fer; Ferrofume; Ferronat; Ferrone; Ferrotemp; Ferrum; Fersamal; Firon; Fumafer; Fumar F; Fumiron; Galfer; Heferol; Ircon; Meterfer; One-Iron; Toleron; Tolferain; Tolifer. $C_4H_2FeO_4$; mol wt 169.90. C 28.28%, H 1.19%, Fe 32.87%, O 37.67%. Prepd by mixing hot aq solns of ferrous sulfate and sodium fumarate and separating the resulting slurry by filtration: Bertsch, Lemp, U.S. pat. **2,848,366** (1958 to Mallinckrodt); Prabhakaran, Patel, *Indian J. Chem.* **7**, 266 (1969). The hot soln of sodium fumarate is preferably added to the ferrous sulfate soln. The commercial material contains a minimum of 31.3% total Fe and not less than 2.0% ferric iron. Toxicity: Weaver *et al., Am. J. Med. Sci.* **241**, 296 (1961).

Reddish-orange to reddish-brown, granular powder. d^{25} 2.435. Odorless; almost tasteless. Not melted at 280°. Soly at 25° in water: 0.14 g/100 ml; in alcohol <0.01 g/100 ml. Soly in acid is limited by liberation of fumaric acid: up to 0.45 g can be dissolved in 100 ml of 1.0N HCl; up to 0.6 g in 0.1N HCl. LD_{50} i.p. in mice: 480 mg/kg (Weaver).

THERAP CAT: Hematinic.

4095. Ferrous Gluconate. *Bis(D-gluconato-O^1,O^2)-iron;* Fergon; Ferlucon; Ferronicum; Iromon; Irox; Nionate. $C_{12}H_{22}FeO_{14}$; mol wt 446.14. C 32.31%, H 4.97%, Fe 12.52%, O 50.21%. $Fe[HOCH_2(CHOH)_4CO_2]_2$. Prepd from Ba gluconate and $FeSO_4$. Prepn of isotonic solns: Hammarlund, *Pharm. Acta Helv.* **35**, 593 (1960). Toxicity study: Hoppe *et al., Am. J. Med. Sci.* **230**, 491 (1955).

Dihydrate, yellowish-gray or pale greenish-yellow powder. Slight odor of caramel. Acid to litmus. Sol in water. Practically insol in alcohol. Aq solns are stabilized by the addition of glucose. Extensive stability studies on aq solns: Johnson, Thomas, *J. Pharm. Pharmacol.* **6**, 1037 (1954). LD_{50} in mice: 114 mg/kg i.v.; 3.7 g/kg orally (Hoppe).

USE: As coloring and flavoring in foods.

THERAP CAT: Hematinic.

4096. Ferrous Hydroxide. FeH_2O_2; mol wt 89.86. Fe 62.15%, H 2.24%, O 35.61%. $Fe(OH)_2$. Prepn: Rihl, Fricke, *Z. Anorg. Allgem. Chem.* **251**, 406 (1943); Leussing, Kolthoff, *J. Am. Chem. Soc.* **75**, 2476 (1953). Oxidation products: Feitknecht, Keller, *Z. Anorg. Chem.* **262**, 61 (1950); Mayne, *J. Chem. Soc.* **1953**, 129; Shipko, Douglas, *J. Phys. Chem.* **60**, 1519 (1956).

White, amorphous powder or white to pale green hexagonal crystals. Oxidized on exposure to air; may ignite spontaneously on exposure to air if finely divided. Practically insol in water. More sol in solns of NH_4 salts; sol in concd NaOH soln.

4097. Ferrous Iodide. FeI_2; mol wt 309.65. Fe 18.03%, I 81.97%. Prepn: Chaigneau, *Bull. Soc. Chim. France* **1957**, 886; Lieser, Elias, *Z. Anorg. Allgem. Chem.* **316**, 208 (1962); Lux in *Handbook of Preparative Inorganic Chemistry* vol. 2, G. Brauer, Ed. (Academic Press, New York, 2nd ed., 1965) p 1495.

Large, thin, red-violet crystals or black leaflets. Very hygroscopic. Sol in water, alcohol, ether; aq soln is readily oxidized by air. *Keep tightly closed and protected from light.*

USE: As catalyst for organic reactions.

THERAP CAT: Antitubercular.

THERAP CAT (VET): Source of iron and iodine.

4098. Ferrous Lactate. $C_6H_{10}FeO_6$; mol wt 233.99. C 30.80%, H 4.31%, Fe 23.87%, O 41.03%. $Fe(C_3H_5O_3)_2$. Prepn: *Gmelin's, Iron* (8th ed.) **59**, part B, p 532 (1932).

Prepn of isotonic solns: Hammarlund, *Pharm. Acta Helv.* **35**, 593 (1960).

Trihydrate, greenish-white powder or cryst masses. Slight characteristic odor; mild, sweet and ferruginous taste. Sol in water; freely sol in alkali citrates forming a green soln. Almost insol in alc. On exposure to air it becomes darker and incompletely sol. *Keep well closed and protected from light.* LD in rabbits: 578 mg/kg s.c.; 287 mg/kg i.v., *Handbook of Toxicology* vol. 1, W. S. Spector, Ed. (Saunders, Philadelphia, 1956) pp 142-143.

THERAP CAT: Hematinic.

4099. Ferrous Oxalate. Ferrox. C_2FeO_4; mol wt 143.86. C 16.70%, Fe 38.82%, O 44.48%. FeC_2O_4. Prepn of dihydrate: *Gmelin's, Iron* (8th ed.) **59**, part B, 532-533 (1932); prepn of anhydr salt: Waterman, Vivian, *J. Org. Chem.* **14**, 289 (1949).

Dihydrate, pale yellow, odorless, cryst powder. d 2.28. Dec at 150-160° on heating in air. Slightly sol in water; sol in dil mineral acids.

USE: As a photographic developer for silver bromide-gelatin plates; to impart a greenish-brown tint to optical glass (sunglasses, windshields, railroad car windows), for decorative glassware; pigment for plastics, paints, lacquers.

4100. Ferrous Oxide. FeO; mol wt 71.84. Fe 77.73%, O 22.27%. Prepn: Lux in *Handbook of Preparative Inorganic Chemistry* vol. 2, G. Brauer, Ed. (Academic Press, New York, 2nd ed., 1965) p 1497.

Jet-black powder. mp 1360°; d 5.7. Easily oxidized by air. It is a strong base and readily absorbs CO_2. Practically insol in water and alkalies. Readily sol in acids.

USE: In manuf of green, heat-absorbing glass; in steel manuf; in enamels; as catalyst.

4101. Ferrous Phosphate. $Fe_3O_8P_2$, mol wt 357.48. Fe 46.87%, O 35.81%, P 17.33%. $Fe_3(PO_4)_2$. Prepn of octahydrate: *Gmelin's, Iron* (8th ed.) **59**, part B, 771 (1932). Due to unavoidable oxidation, the article of commerce contains basic ferric phosphate.

Octahydrate, grayish-blue powder or monoclinic crystals. d 2.58. Practically insol in water. Sol in mineral acids. *Keep well closed and protected from light.*

USE: In ceramics, as catalyst.

4102. Ferrous Phosphide. Iron hemiphosphide; diiron phosphide. Fe_2P; mol wt 142.66. Fe 78.29%, P 21.71%. Prepn from Fe and red P: Rundqvist, Jellinek, *Acta Chem. Scand.* **13**, 425 (1959); Cadeville, Meyer, *Compt. Rend.* **252**, 1124 (1961).

Gray, hexagonal needles or blue-gray powder. d 6.85. Ferromagnetic. Insol in water, dil acid, dil alkali. Reacts with hot mineral acids.

4103. Ferrous Selenide. FeSe; mol wt 134.81. Fe 41.43%, Se 58.57%. Prepn from the elements: Gronvold, Westrum, *Acta Chem. Scand.* **13**, 241 (1959).

Black mass with metallic luster. Stable in air. Dec when heated in O_2. d 6.78. Practically insol in water. Sol in HCl with the evolution of H_2Se.

USE: In electrical semiconductors.

4104. Ferrous Succinate. Cerevon; Ferromyn. C_4H_4-FeO_4; mol wt 171.92. C 27.95%, H 2.35%, Fe 32.48%, O 37.23%. $Fe^{2+}(OOCCH_2CH_2COO)^{2-}$. Prepn: Franke, *Ann.* **491**, 30 (1931).

Tetrahydrate, sparingly sol in water. Also obtained as a dihydrate.

THERAP CAT: Hematinic.

4105. Ferrous Sulfate. FeO_4S; mol wt 151.91. Fe 36.76%, O 42.13%, S 21.11%. $FeSO_4$. Hydrates occur in nature as the minerals: *melanterite, siderotil, szomolnikite, tauriscite.* Heptahydrate prepd commercially by the action of H_2SO_4 on Fe: Faith, Keyes & Clark's *Industrial Chemicals,* F. A. Lowenheim, M. K. Moran, Eds. (Wiley-Interscience, New York, 4th ed., 1975) pp 418-421. Crystal structure of heptahydrate: Baur, *Acta Cryst.* **17**, 1167 (1964). Acute toxicity: Hoppe *et al., Am. J. Med. Sci.* **230**, 491 (1955).

Monohydrate, *dried ferrous sulfate, exsiccated ferrous sulfate, Feromax, Feroritard, Ferro-Gradumet, Fespan, Tetucur.*

White to yellow cryst powder. Loses H_2O at about 300°. Dec at higher temps. Sol in water.

Heptahydrate, *copperas, green vitriol, iron vitriol, Feosol, Feospan, Fesofor, Fesotyme, Fero-Gradumet, Fer-in-Sol, Haemofort, Ironate, Mol-Iron, Presfersul, Sulferrous.* Blue-green, monoclinic, odorless crystals or granules. Efflorescent in dry air; oxidizes in moist air forming a brown coating of basic ferric sulfate. Forms tetrahydrate at 56.6° and monohydrate at 65°. d 1.897. Sol in water. Practically insol in alcohol. Aq solns are oxidized slowly by air when cold, rapidly when hot; rate of oxidation increased by addn of alkali or exposure to light. *Incompat.* Alkalies, sol carbonates, Au and Ag salts, Pb acetate, lime water, KI, K and Na tartrate, Na borate, tannin, vegetable astringent infusions and decoctions. LD_{50} in mice: 65 mg/kg i.v.; 1.52 g/kg orally (Hoppe).

Caution: G.I. disturbances (e.g., gastric distress, colic, constipation, diarrhea) may occur. In children, ingestion of large quantities may cause vomiting, hematemesis, hepatic damage, tachycardia, peripheral vascular collapse: *Clinical Toxicology of Commercial Products,* R. E. Gosselin *et al.,* Eds. (Williams & Wilkins, Baltimore, 4th ed., 1976) Sect. III, pp 153-158.

USE: In manufacture of Fe, Fe compds, other sulfates; in Fe electroplating baths; in fertilizer; as food and feed supplement; in radiation dosimeters; as reducing agent in chemical processes; as wood preservative; as weed-killer; in prevention of chlorosis in plants; in other pesticides; in writing ink; in process engraving and lithography; as dye for leather; in etching aluminum; in water treatment; in qualitative analysis ("brown ring" test for nitrates); as polymerization catalyst.

THERAP CAT: Hematinic.

THERAP CAT (VET): In iron deficiency. Astringent.

4106. Ferrous Sulfide. Iron sulfuret. FeS; mol wt 87.91. Fe 63.52%, S 36.48%. Occurs in nature as the minerals: *magnetkies, pyrrhotine, troillite.* Usually prepd from the elements: Lux in *Handbook of Preparative Inorganic Chemistry* vol 2, G. Brauer, Ed. (Academic Press, New York, 2nd ed., 1965) p 1502. The commercial product is ~75-80% FeS.

Colorless hexagonal crystals when pure; usually gray to brownish-black lumps, rods or granular powder. mp 1194°; d 4.84. Trimorphic with transition points at 135° and 325°. Oxidized by moist air to S and Fe_3O_4. Practically insol in water. Soluble in acids with the evolution of H_2S.

USE: As a laboratory source of H_2S; in the ceramic industry; as a paint pigment; in anodes; in lubricant coatings.

4107. Ferrous Thiocyanate. Ferrous sulfocyanate; ferrous sulfocyanide. $C_2FeN_2S_2$; mol wt 172.01. C 13.97%, Fe 32.47%, N 16.29%, S 37.28%. $Fe(SCN)_2$. Prepn: *Gmelin's, Iron* (8th ed) **59**, part B, 743-747 (1932); O'Brien, *Chem. & Eng. News* **33**, 2008 (1955).

Trihydrate, pale green monoclinic prisms. Rapidly oxidized on exposure to air; dec by heat. Freely sol in water, alcohol, ether. *Keep tightly closed and protected from light.*

USE: Indicator for peroxides in organic solns.

4108. Fertilysin. *N,N'-1,8-Octanediylbis[2,2-dichloroacetamide]; N,N'-octamethylenebis(2,2-dichloroacetamide); N,N'-bis(dichloroacetyl)-1,8-octamethylenediamine;* Win-18446. $C_{12}H_{20}Cl_4N_2O_2$; mol wt 366.11. C 39.37%, H 5.51%, Cl 38.73%, N 7.65%, O 8.74%. $Cl_2CHCONH(CH_2)_8NH$-$OCCHCl_2$. Inhibits spermatogenesis without affecting production of gonadotropins in male laboratory animals. Prepn and biological activity: Surrey, Mayer, *J. Med. Pharm. Chem.* **3**, 419 (1961); Surrey, U.S. pat. **3,143,566** (1964 to Sterling Drug). Assessment of antifertility activity in rats: S. Nag, J. J. Ghosh, *J. Steroid Biochem.* **11**(1B), 681 (1979). Teratological study: P. Taleporos *et al., Teratology* **18**, 5 (1978).

Crystals, mp 122.4-123.6°. LD_{50} orally in mice: > 16000 mg/kg.

4109. Fertirelin. *9-(N-Ethyl-L-prolinamide)-10-deglycinamide-luteinizing hormone-releasing factor(pig);* Des-Gly^{10}-NH_2-LH-RH-ethylamide; TAP-031. $C_{55}H_{76}N_{16}O_{12}$; mol wt 1153.31. C 57.28%, H 6.64%, N 19.43%, O 16.65%. Synthetic nonapeptide analog of LH-RH, *q.v.* Prepn: M.

Fujino *et al.*, **Ger. pat. 2,321,174;** *eidem,* U.S. pat. **3,853,837** (1973, 1974 both to Takeda); S. Shinagawa, M. Fujino, *Chem. Pharm. Bull.* **23,** 229 (1975); *see also:* M. Fujino *et al., Biochem. Biophys. Res. Commun.* **49,** 863 (1972). Enzyme immunoassay in bovine plasma: J. Okada, S. Kondo, *ibid.* **33,** 4464 (1985). Field trial in bovine cystic ovarian disease: T. Nakao *et al., Japan. J. Vet. Sci.* **45,** 269 (1983); in induction of ovulation: T. Nakao *et al., Theriogenology* **20,** 111 (1983); D. A. Coleman *et al., ibid.* **30,** 149 (1988).

5-oxoPro–His–Trp–Ser–Tyr–Gly–Leu–Arg–ProNHCH$_2$CH$_3$

Monoacetate, C$_{55}$H$_{76}$N$_{16}$O$_{12}$·C$_2$H$_4$O$_2$, *U-69689E, Conceral, Ovalyse.*
Monoacetate pentahydrate, white, fluffy powder, $[\alpha]_D^{25}$ $-53.6°$ (c = 0.5 in 5% acetic acid).
THERAP CAT (VET): Gonad stimulating principle.

4110. Ferulic Acid. *3-(4-Hydroxy-3-methoxyphenyl)-2-propenoic acid; 4-hydroxy-3-methoxycinnamic acid;* 3-methoxy-4-hydroxycinnamic acid; caffeic acid 3-methyl ether. C$_{10}$H$_{10}$O$_4$; mol wt 194.19. C 61.85%, H 5.19%, O 32.96%. Widely distributed in small amounts in plants. Isoln from *Ferula foetida* Reg. *Umbelliferae:* H. Hlasiwetz, L. Barth, *Ann.* **138,** 61 (1866); from *Pinus laricio* Poir. *Abietineae:* M. Bamberger, *Monatsh.* **12,** 441 (1891); *see also Beilstein* **10,** 436 (1927) and supplements. Prepd by the interaction of vanillin, malonic acid and piperidine in pyridine for 3 weeks, then precipitating with HCl: Vorsatz, *J. Prakt. Chem.* **145,** 265 (1936); Pearl, Beyer, *J. Org. Chem.* **16,** 216 (1951). Sepn of *cis* and *trans* isomers: Comte *et al., Compt. Rend.* **245,** 1144 (1957). ^{13}C NMR study: C. J. Kelley *et al., J. Org. Chem.* **41,** 449 (1976). Discovery as a component of cell walls in wheat and barley: M. G. Smart, T. P. O'Brien, *Aust. J. Plant Physiol.* **6,** 485 (1979). Use as food preservative: T. Tsuchiya, M. Takasawa, **Japan. Kokai 75 18621** (1975 to Kyokuto Shibosan), *C.A.* **83,** 7602v (1975).

trans - Ferulic Acid

cis-Form, yellow oil. uv max (alcohol): 316 nm.
trans-Form, orthorhombic needles from water, mp 174°. uv max (alcohol): 236, 322 nm. Sol in hot water, alcohol, ethyl aetate. Moderately sol in ether. Sparingly sol in petr ether, benzene. Forms a sodium salt.
USE: Food preservative.

4111. Fervenulin. *6,8-Dimethylpyrimido[5,4-e]-1,2,4-triazine-5,7-(6H,8H)-dione;* 6,8-dimethyl-5,7-dioxo-5,6,7,8-tetrahydropyrimido[5,4-e]-*as*-triazine; 1,3-dimethylazalumazine; planomycin. C$_7$H$_7$N$_5$O$_2$; mol wt 193.17. C 43.53%, H 3.65%, N 36.26%, O 16.57%. Antibiotic from culture filtrates of *Streptomyces fervens:* Eble *et al., Antibiot. Ann.* **1959-1960,** 227. Structure: Daves *et al., J. Org. Chem.* **26,** 5256 (1961). Synthesis: Pfleiderer, Schündehütte, *Ann.* **615,** 42 (1958); Daves *et al., J. Am. Chem. Soc.* **84,** 1724 (1962); Yoneda, Nagamatsu, *Bull. Chem. Soc. Japan* **48,** 2884 (1975); Taylor, Sowinski, *J. Org. Chem.* **40,** 2321 (1975); S. Senda *et al., J. Am. Chem. Soc.* **99,** 7358 (1977).

Yellow orthorhombic crystals, mp 178-179°. uv max (ethanol): 238, 275, 340 nm (ε 18,500, 1600, 4200). Sol in practically all of the common organic solvents; sol in cold water to about 2 mg/ml; in hot water to about 40 mg/ml.

Practically insol in hydrocarbons. Labile to alkali; stable to acid.
4-Oxide, C$_7$H$_7$N$_5$O$_3$. Synthesis: M. Ichiba *et al., J. Heterocycl. Chem.* **14,** 175 (1977); K. Senga *et al., Heterocycles* **6,** 273 (1977); synthesis and conversion to fervenulin: M. Ichiba *et al., J. Org. Chem.* **43,** 469 (1978). Crystals from alc, mp 179-180°. uv max (alc): 240, 304 nm (log ε 4.10, 3.21).

4112. Feverfew. Featherfew; featherfoil; midsummer daisy. *Tanacetum parthenium* (L.) Sch. Bip., (formerly *Chrysanthemum parthenium* (L.) Bernh.) *Compositae;* a perennial, strongly aromatic herb found in Britain and the Balkan peninsula. Used medicinally since the Middle Ages as a febrifuge. Constituents include sesquiterpene lactones such as parthenolide, *q.v.:* P. J. Hylands, D. M. Hylands in *Development of Drugs and Modern Medicines,* J. W. Gorrod *et al.,* Eds. (Ellis Horwood, Chichester, 1986) pp 100-104. Inhibition of prostaglandin biosynthesis by feverfew extract: H. O. J. Collier *et al., Lancet* **2,** 922 (1980). Effect on human platelet phospholipase: A. N. Makheja, J. M. Bailey, *ibid.* 1054 (1981); *eidem, Prostaglandins, Leukotrienes Med.* **8,** 653 (1982); J. K. Thakkar *et al., Biochim. Biophys. Acta* **750,** 134 (1983). Inhibition of platelet secretory activity: S. Heptinstall *et al., Lancet* **1,** 1071 (1985); S. Heptinstall *et al., J. Pharm. Pharmacol.* **39,** 459 (1987). Structure and anti-secretory activity study: W. A. Groenewegen *et al., ibid.* **38,** 709 (1986). Clinical trials in migraine using freeze dried feverfew leaves: E. S. Johnson *et al., Brit. Med. J.* **291,** 569 (1985). Use of oil extract in migraine: E. S. Johnson *et al.,* U.S. pat. **4,758,433** (1988 to R. P. Scherer). *Review:* M. I. Berry, *Pharm. J.* **232,** 611-614 (1984). Brief review of activity and possible side effects: C. A. Baldwin *et al., ibid.* **239,** 237-238 (1987).

4113. Fexofenadine. *α,α-Dimethyl-4-[1-hydroxy-4-[4-(hydroxydiphenylmethyl)-1-piperidinyl]butyl]benzeneacetic acid;* carboxyterfenadine; terfenadine carboxylate; MDL-16455; Allegra. C$_{32}$H$_{39}$NO$_4$; mol wt 501.67. C 76.62%, H 7.84%, N 2.79%, O 12.76%. Identification as a metabolite of terfenadine, *q.v.,* and antihistaminic activity: D. A. Garteiz *et al., Arzneimittel-Forsch.* **32,** 1185 (1982). HPLC separation from terfenadine: K. Y. Chan *et al., J. Chromatog.* **571,** 291 (1991); determn in biological fluids: A. Terhechte, G. Blaschke, *ibid. (A)* **694,** 219 (1995). Effects on cardiac K$^+$ channels: D. Rampe *et al., Mol. Pharmacol.* **44,** 1240 (1993). Synthesis: S. H. Kawai *et al., J. Org. Chem.* **59,** 2620 (1994). Use as antihistaminic: B. E. McNutt, **PCT Int. pat. Appl. 95 10278** (1995 to Marion Merrell Dow).

White crystals from methanol, mp 142°-143°.
THERAP CAT: Antihistaminic.

4114. Fialuridine. *1-(2-Deoxy-2-fluoro-β-ᴅ-arabinofuranosyl)-5-iodo-2,4(1H,3H)-pyrimidinedione;* 1-(2-deoxy-2-fluoro-β-ᴅ-arabinofuranosyl)-5-iodouracil; 5-iodo-2'-fluoroarauracil; FIAU. C$_9$H$_{10}$FIN$_2$O$_5$; mol wt 372.09. C 29.05%, H 2.71%, F 5.11%, I 34.11%, N 7.53%, O 21.50%. Exptl antiviral agent; nucleoside analog with antihepatitis B activity. Prepn: K. A. Watanabe *et al., J. Med. Chem.* **22,** 21 (1979). Antiviral activity: J. M. Colacino, C. Lopez, *Antimicrob. Ag. Chemother.* **24,** 505 (1983); K. A. Staschke *et al., Antivir. Res.* **23,** 45 (1994). Clinical pharmacokinetics: R. R. Bowsher *et al., Antimicrob. Ag. Chemother.* **38,** 2134 (1994). Report of trial suspension: S. R. Ahmed, *Lancet* **342,** 166 (1993). Evaluation of mechanism of hepatotoxicity: L. Cui *et al., J. Clin. Invest.* **95,** 555 (1995). Clinical toxicological profile: W. Stevenson *et al., Transpl. Proc.* **27**, 1219 (1995).

Crystals from ethanol, mp 216-217°.

4115. Fibrin. Fibrin monomer is fibrinogen from which one or two peptides have been removed by means of thrombin: Laki, Chandrasekhar, *Nature* **197**, 1267 (1963). The term fibrin is usually applied to polymerized fibrin monomer. Terminal clotting takes place in four steps: (1) fibrinogen hydrolyzes under the influence of thrombin into fibrin and fibrinopeptide fragments; (2) fibrin forms soft clots which can be readily dispersed; (3) thrombin activates fibrin-stabilizing factor, *q.v.*, an enzyme precursor, present in blood plasma; (4) fibrin in the networks cross-links under the influence of the activated FSF to give the final hard clots: *Chem. & Eng. News* **43**, no. 32, 38 (1965); O. D. Ratnoff, B. Bennett, *Science* **179**, 1291 (1973). Fibrin occurs in two principal forms, *fibrin-i*, "insoluble" fibrin, differing from *fibrin-s*, "soluble" fibrin, by urea solubility as well as other characteristics. Fibrin-i is formed through the reaction of a fibrinogen-like plasma protein, FSF, which in the presence of Ca^{2+} converts what would otherwise be a "soluble" weakly bonded gel into a covalently bonded, insol clot: Rosenberg, Carman, *Nature* **204**, 994 (1964). Chemical studies of crosslinking segments: Chen, Doolittle, *Proc. Nat. Acad. Sci. USA* **66**, 472 (1970); *eidem, Biochemistry* **10**, 4486 (1971); Doolittle *et al., Biochem. Biophys. Res. Commun.* **44**, 94 (1971). *Reviews:* W. H. Seegers, *Prothrombin* (Harvard University Press, 1962) 728 pp; Laki, Gladner, *Physiol. Rev.* **44**, 127 (1964); Lorand, *Fed. Proc.* **24**, no. 4, part 1, 784 (1965); A. L. Copley, *Thromb. Res.* **14**, 249 (1979). Review of chemistry: several authors, *Thromb. Diath. Haemorrh., Suppl.* **39** (1970).

4116. Fibrinogen. Factor I; Parenogen. A plasma glycoprotein belonging structurally to the keratin-myosin group. Synthesized and secreted by hepatic parenchymal cells. Present to the extent of 0.3-0.4 g/100 ml in human plasma. Essential to the clotting of blood. Its synthesis is greatly increased during acute inflammatory challenge. The fibrinogen molecule consists of three peptide chains, α (A), β (B), and γ (C), crosslinked by several disulfide bonds. The mol wt of about 400,000 represents a dimeric form of the molecule. Thrombin releases fibrinopeptides A and B from the *N*-terminal ends of the α and β chains of fibrinogen in the formation of fibrin during coagulation. Because fibrinogen is less sol than other plasma proteins it is readily separated by precipitation with sodium chloride: Florkin, *J. Biol. Chem.* **87**, 629 (1930); or with ammonium sulfate: Nanninga, *Arch. Neer. Physiol.* **28**, 241 (1946). Prepn from human plasma: Edsall *et al., J. Am. Chem. Soc.* **69**, 2731 (1947). Purification: Cama *et al., Naturwiss.* **48**, 574 (1961). End group determination: Lorand, Middlebrook, *Science* **118**, 515 (1953). Structure studies: Schauenstein, Hochenegger, *Z. Naturforsch.* **8b**, 473 (1953); Edsall, *J. Polymer Sci.* **12**, 253 (1954). *Reviews:* Seegers, *Physiol. Rev.* **34**, 711 (1954); Lorand, *Fed. Proc.* **24**, no. 4, part 1, 784 (1965); several authors, *Thromb. Diath. Haemorrh., Suppl.* **39** (1970); A. L. Copley, *Thromb. Res.* **14**, 249 (1979). Fibrinogen has been shown to be the receptor for the endogenous lectin (agglutinin) secreted by activated platelets: T. K. Gartner *et al., Nature* **289**, 688 (1981). Review of biosynthesis: G. M. Fuller, D. G. Ritchie, *Ann. N.Y. Acad. Sci.* **389**, 308-322 (1982).

Sparingly sol in water. Aq solns are viscous. Isoelectric point 5.5. Readily denatured by heating to 56° or higher, and by chemical agents such as salicylaldehyde, naphtho-quinone sulfonates, ninhydrin, and alloxan. Small amounts of papain will clot fibrinogen, but larger amounts will digest the clot.

THERAP CAT: Coagulant (clotting factor).

4117. Fibroblast Growth Factor. FGF. Growth stimulatory factor originally isolated from bovine brain and pituitary and found to stimulate DNA synthesis in cultured fibroblast cells. Isoln: D. Gospodarowicz, *Nature* **249**, 123 (1974). Mitogenic effect on cultured cell lines and induction of amphibian limb regeneration *in vivo:* D. Gospodarowicz *et al., Advan. Metab. Disord.* **8**, 301 (1975). Two closely related forms have been identified, known as basic (*bFGF*) and acidic (*aFGF*) fibroblast growth factors, having a total amino acid sequence homology of 55%. Both induce the proliferation and differentiation of a wide variety of cell types, including corneal and vascular endothelial cells, myoblasts, chondrocytes, osteoblasts and glial cells. FGF has neurotrophic and angiogenic activity and may play an important role in the wound healing process. Purification of bFGF from pituitary: D. Gospodarowicz, *J. Biol. Chem.* **250**, 2515 (1975); from brain: D. Gospodarowicz *et al., ibid.* **253**, 3736 (1978). Identification of aFGF from bovine brain: K. A. Thomas *et al., ibid.* **255**, 5517 (1980). Comparison of fibroblast growth factors: S. K. Lemmon *et al., J. Cell Biol.* **95**, 162 (1982). Purification and characterization of aFGF: K. A. Thomas *et al., Proc. Nat. Acad. Sci. USA* **81**, 357 (1984). Identity of bFGF from brain and pituitary: D. Gospodarowicz *et al., ibid.* 6963. Amino acid sequence of bFGF: F. Esch *et al., ibid.* **82**, 6507 (1985); of aFGF: G. Gimenez-Gallego *et al., Science* **230**, 1385 (1985); F. Esch *et al., Biochem. Biophys. Res. Commun.* **133**, 554 (1985). Possible identity of aFGF with *endothelial cell growth factor (ECGF)* and *eye-derived growth factor-II (EDGF-II):* A. B. Schreiber *et al., J. Cell Biol.* **101**, 1623 (1985); of bFGF with *macrophage-derived growth factor (MDGF):* A. Baird *et al., Biochem. Biophys. Res. Commun.* **126**, 358 (1985); of FGFs with *retina-derived endothelial cell growth factors:* A. Baird *et al., Biochemistry* **24**, 7855 (1985). Cloning of cDNA for bovine bFGF: J. A. Abraham *et al., Science* **233**, 545 (1986); for human bFGF: T. Kurokawa *et al., FEBS Letters* **213**, 189 (1987); M. Iwane *et al., Biochem. Biophys. Res. Commun.* **146**, 470 (1987). Expression of a chemically synthesized gene for bioactive bovine aFGF: D. L. Linemeyer *et al., Biotechnology* **5**, 960 (1987). Receptor binding study: G. Neufeld, D. Gospodarowicz: *J. Biol. Chem.* **261**, 5631 (1986).

FGF-like factors have been isolated from several human tumor cell lines: R. R. Lobb *et al., Biochem. Biophys. Res. Commun.* **139**, 861 (1986); M. Klagsbrun *et al., Proc. Nat. Acad. Sci. USA* **83**, 2448 (1986); D. Moscatelli *et al., J. Cell. Physiol.* **129**, 273 (1986). FGF has also been shown to be structurally homologous to the protein products of several oncogenes: C. Dickson, G. Peters, *Nature* **326**, 833 (1987); M. Taira *et al., Proc. Nat. Acad. Sci. USA* **84**, 2980 (1987); P. Delli Bovi *et al., Cell* **50**, 729 (1987). Review of tissue distribution and bioactivity of bFGF: A. Baird *et al., Rec. Prog. Horm. Res.* **42**, 143-205 (1986). Review of structural characterization and biological functions: D. Gospodarowicz *et al., Endocrine Rev.* **8**, 95-114 (1987). Potential role in the control of pituitary and gonad development: D. Gospodarowicz, N. Ferrara, *J. Steroid Biochem.* **32**, 183-191 (1989). *Reviews:* K. A. Thomas, G. Gimenez-Gallego, *Trends Biochem. Sci.* **11**, 81-84 (1986); D. Gospodarowicz *et al., Mol. Cell. Endocrinol.* **46**, 187-204 (1986); K. A. Thomas, *FASEB J.* **1**, 434-440 (1987).

Acidic fibroblast growth factor, pI 5-7. Exists in 2 microheterogeneous forms: *aFGF-1*, a 140 amino acid peptide, mol wt 15,900 daltons, and *aFGF-2*, an amino truncated form lacking 6 amino terminal residues, mol wt 15,200 daltons.

Basic fibroblast growth factor, pI 9.6. 146 amino acid peptide, mol wt ~16,000 daltons. Also exists as an amino truncated form, *des 1-15 bFGF*, lacking the first 15 amino acid residues.

4118. Fibroin. Protein filaments produced by members of the phylum *Arthropoda*, particularly by certain species belonging to the classes *Insecta* (insects) and *Arachnida* (spiders, etc.). Fibroin is the main protein of silk and is secreted by the insect in its silk glands together with *sericin*, the second silk protein, in aq soln and converted into silk by a process called "spinning." Organization of amino acid sequences in silk fibroin of *Bombyx mori* and general review of silk proteins: F. Lucas, K. M. Rudall in *Comprehensive Biochemistry*, M. Florkin, E. M. Stotz, Eds., **vol. 26B** (Elsevier, New York, 1968) pp 475-558. *See also* P. M. Lizardi, *Cell* **18**, 581 (1979); L. P. Gage, R. F. Manning, *J. Biol. Chem.* **255**, 9444 (1980). Chemical and crystalline structures: B. Lotz, F. Colonna Cesari, *Biochimie* **61**, 205 (1979).

Pale yellow mass resembling silk. Insol in water, alcohol, ether, dil alkalies. Sol in concd alkalies, concd mineral acids and in ammoniacal nickel oxide soln.

4119. Fibronectins. α_2-*SB glycoproteins;* α_2-*opsonins;* CIG; CSP; CAF; GAP A; LETS; Zeta protein. High mol wt multifunctional glycoproteins, found on cell surfaces, in body fluids (especially plasma), in soft connective tissue matrices, and in most basement membranes. Although fibronectins apparently function as adhesive ligand-like molecules, the full range of their biological activities and relationships are still being elucidated. Their importance in cell adhesion, oncogenic transformation, reticuloendothelial system function, embryonic differentiation, phagocytosis, hemostasis, and chemotaxis is being studied. Discovered as a result of isoln of a partially purified fraction of human plasma and initially termed "cold-insoluble globulin" or CIG: P. R. Morrison *et al.*, *J. Am. Chem. Soc.* **70**, 3103 (1948). Subsequent studies described various proteins or factors, named according to sources or biological activities, that are now designated as fibronectins. At least two types are known to exist, termed plasma and cellular fibronectin, respectively. Both forms contain subunits of mol wt > 200,000, joined by disulfide bonds. They are similar in amino acid compositions, carbohydrate structures and secondary and tertiary structures; they cannot be distinguished in biological activity in assays of cell interactions with substrates or in opsonic activity for macrophages. They differ in their effects on cell morphology, on alignment of transformed cells and on hemagglutination; they also have differences in solubility and in the number of subunits linked by disulfide bonds. Monoclonal antibody studies have indicated that the two forms are distinct: B. T. Atherton, R. O. Hynes, *Cell* **25**, 133 (1981); K. D. Noonan *et al.*, *J. Supramol. Struct. Cell Biochem.* **5**, Suppl, 302 (1981). Regulation of fibronectin biosynthesis: D. R. Senger *et al.*, *Am. J. Physiol.* **245**, 144 (1983). Structure-function relationships: T. Vartio, A. Vaheri, *Trends Biochem. Sci.* **8**, 442 (1983). Role in cellular adhesion, spreading and cytoskeletal organization: I. Virtanen *et al.*, *Nature* **298**, 660 (1982); in phagocytosis: L. Van de Water *et al.*, *Science* **220**, 201 (1983); in wound healing: G. R. Martin *et al.*, "Regulation of Tissue Structure and Repair by Collagen and Fibronectin" in *The Surgical Wound*, P. Dineen, C. Hildrick-Smith, Eds. (Lea & Febiger, Philadelphia, 1981) pp 110-122. Use in treatment of corneal trophic ulcer therapy: T. Nishida *et al.*, *Arch. Ophthalmol.* **101**, 1046 (1983). Review of role in cellular adhesion: S. K. Akiyama *et al.*, *J. Supramol. Struct. Cell Biochem.* **16**, 345-358 (1981); role in inflammation: D. F. Mosher *et al.*, *Adv. Inflamm. Res.* **2**, 187-207 (1981); C. Bianco, *Ann. N.Y. Acad. Sci.* **408**, 602-609 (1983); actvity in various disease states: S. K. Akiyama, K. M. Yamada, "Fibronectin in Disease" in *Monographs in Pathology* No. 24, N. Kaufman, Ed., entitled "Connective Tissue Diseases", B. M. Wagner *et al.*, Eds. (Williams & Wilkins, Baltimore, 1983). *General Reviews:* E. Pearlstein *et al.*, *Mol. Cell. Biochem.* **29**, 103-128 (1980); M. W. Mosesson, D. L. Amrani, *Blood* **56**, 145-158 (1980); E. Ruoslahti, *J. Oral Pathol.* **10**, 3-13 (1981); R. O. Hynes, K. M. Yamada, *J. Cell Biol.* **95**, 369-377 (1982); R. O. Hynes, *Scientific American* **254**, 42-51 (1986).

4120. Fichtelite. *[1S-(1α,4aα,4bβ,7β,8aα,10aβ)]-Tetradecahydro-1,4a-dimethyl-7-(1-methylethyl)phenanthrene;* 18-norabietane. $C_{19}H_{34}$; mol wt 262.48. C 86.94%, H 13.06%. From decayed wood of conifers: Bromeis, *Ann. Pharm.* **37**, 304 (1841). Structure: L. Ruzicka, E. Waldmann, *Helv. Chim. Acta* **18**, 611 (1935); Crowfoot, *J. Chem. Soc.* **1938**, 1241. Stereochemistry: Burgstahler, Marx, *Tetrahedron Letters* **1964**, 3333. Synthesis from abietic acid: Jensen,

Johnson, *J. Org. Chem.* **32**, 2045 (1967); Burgstahler, Marx, *ibid.* **34**, 1562 (1969). Synthesis of *dl*-form: Johnson *et al.*, *J. Am. Chem. Soc.* **88**, 3859 (1966); **90**, 5872 (1968); D. F. Taber, S. A. Saleh, *ibid.* **102**, 5085 (1980).

Crystals from methanol, mp 45-46°. bp_{43} 235-236°. d_4^{22} 0.9380. n_D^{20} 1.5052. $[\alpha]_D$ +19°.

4121. Ficin. Ficus proteinase; Ficus protease; Debricin; Higueroxyl Delabarre. A proteolytic enzyme of est. mol wt 23,800-25,500 which requires a free sulfhydryl group for activity and as such is a member of a group which includes papain, *q.v.*, and bromelain, *q.v.* Occurs in the latex of tropical trees of the genus *Ficus* subgenus *Pharmacosyce, Moraceae* (Oje trees). The commercial product is a concentrate prepd by filtering and drying the latex of *Ficus glabrata* H. B. & K., *Moraceae*. First crystallized from fresh fig latex: Walti, *J. Am. Chem. Soc.* **60**, 493 (1938). Characterization: Cohen, *Nature* **182**, 659 (1958); Englund *et al.*, *Biochemistry* **7**, 163 (1968). Purification: Gibian, Bratfisch, U.S. pat. 2,950,227 (1960 to Schering AG). Amino acid composition: Wong, Liener, *Biochem. Biophys. Res. Commun.* **17**, 470 (1964); Metrione *et al.*, *Arch. Biochem. Biophys.* **122**, 137 (1967); Husain, Lowe, *Biochem. J.* **117**, 333 (1970). Pharmacodynamics: H. Heisto, M. K. Fagerhol, *Transfusion* **19**, 545 (1979); A. Perkash *et al.*, *ibid.* **20**, 301 (1980). *Reviews:* Liener, Friedenson, *Methods Enzymol.* **19**, 260-273 (1970); Glazer, Smith, *The Enzymes* vol. III, P. D. Boyer, Ed. (Academic Press, New York, 3rd ed., 1971) pp 538-542.

Buff to cream-colored hygroscopic powder. Acrid odor, growing stronger with age. Bulky, approx 3 ml/g, not free-flowing. Appears dry, even with 15% H_2O present. Not completely sol in water, 2-10% insol material. The sol portion of 1 g dissolves in 3 ml H_2O. pH (2% soln): 4.1. Insol in usual organic solvents. Loses about 10-20% activity when stored 1-3 yrs at ordinary temp and atm conditions. Aq solns inactivated at 100°, solid partially inactivated within a few hours. Solns relatively stable between pH 4-8.5; incompatible with iron, copper, aluminum. Gelatin, coagulated egg white, casein, meat and most protein-like material hydrolyzed in aq ficin solns. Ficin is 10-20 times as active as papain in regard to milk clotting; 4-10 times as active, in general. LD_{50} orally in rats and mice: ~10 g/kg; in rabbits and guinea pigs: ~5 g/kg.

Note: The name ficin is currently used to describe both the crude dried latex from different species of the genus *Ficus*, as well as the enzyme itself.

Caution: Handle ficin carefully because of its tissue-dissolving properties. Can cause irritation to skin, eyes, mucous membranes. Large doses by mouth cause purging.

USE: Protein digestant. In the brewing industry as a chillproofing agent in beer. In the cheese industry as a substitute for rennet in the coagulation of milk. In the meat industry as a meat tenderizer and as an agent for removing casings from formed sausage. In the leather industry for the bating of leather. In the textile industry for shrinkproofing wool, for removing gelatin from sized thread, and mixed with amylases and maltases as a spot remover. In the prepn of peptones. For solubilizing protein material in spent grains. Also used in the determination of the Rh factor. Speeds 10 times the agglutination of human blood cells by the Rh factor when in contact with the anti-Rh serum.

THERAP CAT (VET): Has been used as a trichuricide.

4122. Filicinic Acid. *3,5-Dihydroxy-4,4-dimethyl-2,5-cyclohexadien-1-one;* 1,1-dimethylcyclohexane-2,4,6-trione; *gem*-dimethylphloroglucinol; filicinsäure (German). $C_8H_{10}O_3$; mol wt 154.17. C 62.33%, H 6.54%, O 31.13%. Isolated from male fern and by hydrolytic decompn of filixic acid or aspidin: Boehm, *Ann.* **302,** 171 (1898); **329,** 321 (1903). Synthesis: Robertson, Sandrock, *J. Chem. Soc.* **1933,** 1617; Angus *et al., Chem. & Ind. (London)* **1954,** 546; Inagaki *et al., J. Pharm. Soc. Japan* **76,** 1258 (1956); Hoefer, Riedl, *Angew. Chem.* **74,** 501 (1962); *eidem, Ann.* **656,** 127 (1962).

Prisms from water or benzene, mp 214-215° (dec). pK 5.8. Practically insol in water, petr ether. Sol in 70 parts boiling water, 10 parts boiling alcohol; sparingly sol in ether, benzene, glacial acetic acid. Reduces Tollen's reagent.

3,5-Diacetylfilicinic acid, $C_{12}H_{14}O_5$, crystals from methanol, mp 65-66°. pK 5.0. Sol in cold methanol, benzene, ether; slightly sol in hot water.

4123. Filipin. Filimarisin; U-5956; NSC-3364. Polyene antibiotic complex containing at least eight pentaene compounds, which has been resolved into three pure components, filipin II, III (major), and IV, which differ in the number of hydroxyl groups present (8, 9, 9, resp.) and filipin I, a mixture of at least 5 components. Described as a single entity in the earlier literature. Initial isolation from *Streptomyces filipenensis* in Philippine soil: Whitfield *et al., J. Am. Chem. Soc.* **77,** 4799 (1955). Early structural work: Dhar *et al., Proc. Chem. Soc.* **1960,** 310; Djerassi *et al., Tetrahedron Letters* **1961,** 383; Golding, Rickards, *ibid.* **1964,** 2615; Dhar *et al., J. Chem. Soc.* **1964,** 842; Ceder, Ryhage, *Acta Chem. Scand.* **18,** 588 (1964). Separation of filipin complex into components: Bergy, Eble, *J. Antibiot.* **23,** 414 (1970); *see also* Rickards *et al., ibid.* 603. Mechanism of action: R. W. Holz in *Antibiotics* **vol. 5,** pt. 2, F. E. Hahn, Ed. (Springer-Verlag, New York, 1979) pp 313-340. Filipin also interacts specifically with 3β-hydroxysterols (e.g. cholesterol): P. M. Elias *et al., J. Histochem. Cytochem.* **27,** 1247 (1979); *see also* N. J. Severs, H. J. Simons, *Nature* **303,** 637 (1983).

Filipin III

Yellow, feathery needles from chloroform, mp 195-205°. Sensitive to air. $[\alpha]_D^{22}$ −148.3° (c = 0.89 in methanol). uv max (methanol): 322, 338, 355 nm ($E_{1cm}^{1\%}$ 910, 1360, 1330). Freely sol in DMF, pyridine. Also sol in 95% ethanol, methanol, butanol, isopropanol, glacial acetic acid, ether. Practically insol in water, chloroform.

Filipin III. $C_{35}H_{58}O_{11}$, *4,6,8,10,12,14,16,27-octahydroxy-3-(1-hydroxyhexyl)-17,28-dimethyloxacyclooctacosa-17,19,21,-23,25-pentaen-2-one; 15-deoxylagosin.* Isomeric with filipin IV. Crystals from propanol, mp 163-180° (Bergy, Eble). $[\alpha]_D^{25}$ −245° (c = 0.8 in DMF). uv max (methanol): 243, 308, 321, 337, 354 nm ($E_{1cm}^{1\%}$ 62, 413, 851, 1368, 1343).

USE: Produces arthritis rapidly in rabbits: Pras, Weissman, *Drug Trade News,* July 4, 1966, p 40; as a sterol probe in freeze-fracture cytochemistry.

THERAP CAT: Antifungal.

4124. Filixic Acids. Filicin; filicic acid; Filixsäure (German). Natural filixic acid from male fern, *Dryopteris filix-mas,* is a mixture of six homologues, the 3 main com-

ponents (BBB, PBB, PBP) of which are obtained by recrystallization from ethyl acetate. When raw filixic acid is purified by treatments with methanol but without recrystns from ethyl acetate, a product is obtained which seems to contain 3 additional homologues with an analogous composition but with acetyl groups in the side chain. Isoln: Luck, *Ann.* **54,** 119 (1845); Boehm, *ibid.* **318,** 253 (1901). Structure studies: Riedl, *Ann.* **585,** 32 (1954); Chan, Hassal, *Experientia* **13,** 349 (1957). Structure of the three main filixic acids: Penttilä, Sundman, *Acta Chem. Scand.* **17,** 191 (1963).

BBB	$R_1 = R_2 = C_3H_7$
PBB	$R_1 = C_2H_5$ $R_2 = C_3H_7$
PBP	$R_1 = R_2 = C_2H_5$

Natural filixic acid, pale-yellow plates from ethyl acetate, mp 184-185°. uv max: 228, 288 nm (ϵ 41000, 29000). Practically insol in water, methanol, acetone. Sol in chloroform, benzene, warm ethyl acetate, acetic acid; slightly sol in ether.

BBB, $C_{36}H_{44}O_{12}$, crystals from ethyl acetate, mp 172-174°.
PBB, $C_{35}H_{42}O_{12}$, crystals from ethyl acetate, mp 184-186°.
PBP, $C_{34}H_{40}O_{12}$, crystals from ethyl acetate, mp 192-194°.

THERAP CAT (VET): Has been used as an anthelmintic.

4125. Finasteride. *(5α,17β)-N-(1,1-Dimethylethyl)-3-oxo-4-azaandrost-1-ene-17-carboxamide;* 17β-*(N-tert-butylcarbamoyl)-4-aza-5α-androst-1-en-3-one;* MK-906; Andozac; Chibro-Proscar; Finastid; Proscar; Prostide. $C_{23}H_{36}N_2$-O_2; mol wt 372.55. C 74.15%, H 9.74%, N 7.52%, O 8.59%. Inhibitor of 5α-reductase, the enzyme which converts testosterone to the more potent androgen, 5α-dihydrotestosterone. Prepn: G. H. Rasmusson, G. F. Reynolds, *Eur. pat.* **Appl. 155,096;** *eidem,* U.S. pat. **4,760,071** (1985, 1988 both to Merck & Co.); G. H. Rasmusson *et al., J. Med. Chem.* **29,** 2298 (1986); A. Bhattacharya *et al., J. Am. Chem. Soc.* **110,** 3318 (1988). Inhibition of 5α-reductase *in vitro:* T. Liang *et al., Endocrinology* **117,** 571 (1985); *in vivo:* J. R. Brooks *et al., Steroids* **47,** 1 (1986). HPLC determn in plasma and urine and preliminary pharmacokinetics: J. R. Carlin *et al., J. Chromatog.* **427,** 79 (1988). Effects on hair growth in stumptail macaques: L. Rhodes *et al., J. Clin. Endocrinol. Metab.* **79,** 991 (1994). Clinical pharmacology: J. Imperato-McGinley *et al., ibid.* **70,** 777 (1990); G. J. Gormley *et al., ibid.* 1136. Clinical effects on hormone levels in male pattern baldness: A. L. Dallob *et al., ibid.* **79,** 703 (1994).

White to off-white crystalline solid, mp ~257° (dried to anhydrous under N_2). Also reported as mp 252-254° (Rasmusson, 1988). $[\alpha]_D$ −59° (c = 1 in methanol). Freely sol in chloroform, DMSO, ethanol, methanol, n-propanol; sparingly sol in propylene glycol, polyethylene glycol 400; very slightly sol in 0.1N HCl, 0.1N NaOH, water.

THERAP CAT: Treatment of benign prostatic hypertrophy.

4126. Fipexide. *1-(1,3-Benzodioxol-5-ylmethyl)-4-[(4-chlorophenoxy)acetyl]piperazine; 1-[(p-chlorophenoxy)acetyl]-*

Consult the Name Index before using this section.

4-piperonylpiperazine; 1-[2-(4-chlorophenoxy)acetyl]-4-(3,4-methylenedioxybenzyl)piperazine. $C_{20}H_{21}ClN_2O_4$; mol wt 388.85. C 61.78%, H 5.44%, Cl 9.12%, N 7.20%, O 16.46%. Prepn of hydrochloride: A. Buzas, R. Pierre, **Fr.** pat. **M7524** (1970 to Lab Bouchard), *C.A.* **75**, 112865r (1971). Manufacturing process: G. P. Gardini *et al.,* **U.S.** pat. **4,225,714** (1980 to Farmaceutici Geymon.). Improved prepn: G. P. Gardini *et al., Syn. Commun.* **12**, 887 (1982). Determn by TLC: G. Musumarra *et al., J. Chromatog.* **350**, 151 (1985). Toxicology and pharmacology: G. David *et al., Acta Therap.* **11**, 387 (1985). Clinical trial in the elderly: B. Bompani, G. Scali, *Curr. Med. Res. Opin.* **10**, 99 (1986).

Hydrochloride, $C_{20}H_{21}ClN_2O_4 \cdot HCl$, *BP-662, Attentil, Vigilor.* Crystals from ethanol, mp 230-232°. LD_{50} in Swiss mice, Sprague-Dawley rats, Wistar rats (mg/kg): 4150, 4482, 7000 orally; 499, 537, 450 i.p. (David).

THERAP CAT: Nootropic.

4127. Fipronil. *5-Amino-1-[2,6-dichloro-4-(trifluoromethyl)phenyl]-4-[(trifluoromethyl)sulfinyl]-1H-pyrazole-3-carbonitrile;* 5-amino-3-cyano-1-(2,6-dichloro-4-trifluoromethylphenyl)-4-trifluoromethylsulfinylpyrazole; (±)-5-amino-1-(2,6-dichloro-α,α,α-trifluoro-*p*-tolyl)-4-trifluoromethylsulfinylpyrazole-3-carbonitrile; MB-46030; Frontline. $C_{12}H_4Cl_2F_6N_4OS$; mol wt 437.15. C 32.97%, H 0.92%, Cl 16.22%, F 26.08%, N 12.82%, O 3.66%, S 7.34%. GABA-gated chloride channel blocker. Prototype of the phenylpyrazole insecticides known as *fiproles.* Prepn: I. G. Buntain *et al.,* **Eur.** pat. **295,117** (1988 to May & Baker); L. R. Hatton *et al.,* **U.S.** pat. **5,232,940** (1993). Mechanism of action study: L. M. Cole *et al., Pest. Biochem. Physiol.* **46**. 47 (1993). Comprehensive description: F. Colliot *et al., Brighton Crop Prot. Conf.—Pests Dis.* **1992**, 29-34.

White solid, mp 200.5-201°. Vapor pressure (20°): 2.8 × 10^{-9} mm Hg. Log P (*n*-octanol/water): 4.0. Soly: water 2 mg/l; acetone > 50%; corn oil > 10,000 mg/l. LD_{50} in rats (mg/kg): 100 orally; > 2000 dermally (Colliot); in mice (mg/kg): 32 i.p. (Cole).

USE: Pesticide.

4128. Firefly Luciferin. *(S)-4,5-Dihydro-2-(6-hydroxy-2-benzothiazolyl)-4-thiazolecarboxylic acid;* 2-(6-hydroxybenzothiazol-2-yl)-2-thiazoline-4-carboxylic acid; D-(−)-luciferin. $C_{11}H_8N_2O_3S_2$; mol wt 280.33. C 47.13%, H 2.88%, N 9.99%, O 17.12%, S 22.88%. Light emission in the American firefly, *Photinus pyralis,* has been shown to involve the interaction of magnesium ion, oxygen, ATP, the enzyme luciferase, and the oxidizable substrate luciferin. Isoln from fireflies (yield from 15,000 active fireflies about 9 mg): Bitler, McElroy, *Arch. Biochem. Biophys.* **72**, 358 (1957); from Japanese firefly *(Luciola cruciata):* Kishi *et al., Tetrahedron Letters* **1968**, 2847. Structure and synthesis: E. H. White *et al., J. Am. Chem. Soc.* **83**, 2402 (1961); **85**, 337 (1963); *J. Org. Chem.* **30**, 2344 (1965); S. Seto *et al., Bull. Chem. Soc. Japan* **36**, 331 (1963). Configuration: G. E. Blank *et al., Biochem. Biophys. Res. Commun.* **42**, 583 (1971). Measurement of ATP: D. M. Karl, O. Holm-Hansen, *Anal. Biochem.* **75**, 100 (1976). *Review:* L. J. Bowie, *Methods Enzymol.* **57**, 15 (1978).

Pale yellow needles from methanol, dec 189.5-190°. Recrystallizes with difficulty and sublimes with decomposition and decarboxylation. $[\alpha]_D^{22}$ −36° (c = 1.2 in DMF). uv max (H_2O): 268, 327 nm (log ε 3.88, 4.27). Slightly sol in water at pH 6.5. Sol in alkaline aq solns, methanol, acetone, ethyl acetate, DMF. Aqueous solns are sensitive to extremes in pH, esp. in presence of light and oxygen. Racemization occurs rapidly in some solvents. Should be stored for extended periods dry, under nitrogen atmosphere in light-tight containers.

USE: In the assay of ATP.

4129. Fisetin. *2-(3,4-Dihydroxyphenyl)-3,7-dihydroxy-4H-1-benzopyran-4-one; 3,3',4',7-tetrahydroxyflavone;* 5-desoxyquercetin; fisidenolon 1521; C.I. 75620; C.I. Natural Brown 1. $C_{15}H_{10}O_6$; mol wt 286.24. C 62.94%, H 3.52%, O 33.54%. Coloring matter isolated from *Rhus cotinus* L. *Anacardiaceae:* Chevreul, *Lecons Chim. Appl. a la Teint.* **2**, 169 (1833) [*cf. J.* Schmid, *Ber.* **19**, 1734 (1886)]; from *R. rhodanthema* Müll: Perkin, *J. Chem. Soc.* **71**, 1194 (1897); Oyamada, *Ann.* **538**, 44 (1939); from heartwood of *Acacia* spp, *Leguminosae:* Roux, Paulus, *Biochem. J.* **77**, 315 (1960); Roux *et al., ibid.* **78**, 834 (1961). Structure: Seshadri, *Ann. Rev. Biochem.* **20**, 492 (1951). Synthesis: Kostanecki *et al., Ber.* **37**, 784 (1904); Allan, Robinson, *J. Chem. Soc.* **129**, 2334 (1926). Inhibition of aflatoxin cytotoxicity: A. G. Schwartz, W. R. Rate, *J. Environ. Pathol. Toxicol.* **2**, 1021 (1979). Mutagenicity studies: J. P. Brown *et al., Biochem. Soc. Trans.* **5**, 1489 (1977); J. P. Brown, P. S. Dietrich, *Mutat. Res.* **66**, 223 (1979).

Yellow needles from dil alc, dec 330°. uv max (ethanol): 252, 320, 360 nm (log $E_{1cm}^{1\%}$ 2.62, 2.51, 2.73). Sol in alcohol, acetone, acetic acid, solns of fixed alkali hydroxides. Practically insol in water, ether, benzene, chloroform and petr ether. Its soln in dil alcohol NaOH shows a dark green fluorescence. LD_{50} in mice: 180 mg/kg i.v., *RTECS* **Vol. I,** R. J. Lewis, R. L. Tatken, Eds. (1980) p 686.

4130. Flavaspidic Acid. *3,5-Dihydroxy-4,4-dimethyl-2-(1-oxobutyl)-6-[[2,4,6-trihydroxy-3-methyl-5-(1-oxobutyl)-phenyl]methyl]-2,5-cyclohexadien-1-one;* 3'-[(5-butyryl-2,4-dihydroxy-3,3-dimethyl-6-oxo-1,4-cyclohexadien-1-yl)methyl]-5'-methylphlorobutyrophenone; polystichocitrin; Toxifren. $C_{24}H_{30}O_8$; mol wt 446.50. C 64.56%, H 6.77%, O 28.67%. Isolated from the rhizomes of male fern: Boehm, *Ann.* **318**, 253 (1901); **329**, 310 (1903). Structure and synthesis: McGookin *et al., J. Chem. Soc.* **1953**, 1828; Riedl, *Ann.* **585**, 32 (1954); Aebi, *Helv. Chim. Acta* **39**, 153 (1956). Crystal structure of α- and β-forms: Erämetsä, Penttilä, *Acta Chem. Scand.* **24**, 3335 (1970). Toxicity study: Airaksinen *et al., Acta Pharmacol. Toxicol.* **25**, 33 (1967).

α-Form, orthorhombic crystals from methanol or ethanol, mp 92°; solidifies again at 110° and melts again at 156°. LD$_{50}$ orally in mice: 690 mg/kg (Airaksinen).

β-Form, monoclinic crystals from benzene, xylene or acetic acid, mp 156°.

4131. Flavine-Adenine Dinucleotide. *Riboflavine 5'-(trihydrogen diphosphate) 5'→5'-ester with adenosine; adenosine 5'-(trihydrogen diphosphate) 5'→5'-ester with riboflavine;* FAD; riboflavine 5'-adenosine diphosphate; isoalloxazine-adenine dinucleotide; Fademin; Flamitajin; Flanin F; Flavitan. C$_{27}$H$_{33}$N$_9$O$_{15}$P$_2$; mol wt 785.56. C 41.28%, H 4.23%, N 16.05%, O 30.55%, P 7.89%. The prosthetic group of certain flavoproteins including D-amino acid oxidase, glucose oxidase, glycine oxidase, fumaric hydrogenase, histaminase, and xanthine oxidase. Isoln from yeast: Warburg *et al.*, *Biochem. Z.* **297**, 417 (1938). Structure and isoln from liver, kidneys, hearts, muscles: Warburg, Christian, *ibid.* **298**, 150 (1938); isoln from the mycelium of *Eremothecium ashbyii:* Yagi, Tada, *Biochem. Prepn.* **7**, 51 (1959); Masuda *et al.*, U.S. pat. **2,973,305** (1961 to Takeda). Synthesis: Christie *et al.*, *J. Chem. Soc.* **1954**, 46; Huennekens, Kilgour, *J. Am. Chem. Soc.* **77**, 6716 (1955); DeLuca, Kaplan, *J. Biol. Chem.* **223**, 569 (1956); Moffatt, Khorana, *J. Am. Chem. Soc.* **80**, 3756 (1958). Review: A. Holmgren, *Experientia* **36** (Suppl), 149-180 (1980). Review of FAD and other flavin coenzymes: Beinert, *The Enzymes* vol. **2**, P. D. Boyer *et al.*, Eds. (Academic Press, New York, 2nd ed., 1960) pp 339-416; *see also* vol. **XIII Part B** (Academic Press, New York, 3rd ed., 1975), several authors.

Isolated as the barium salt, C$_{27}$H$_{31}$BaN$_9$O$_{15}$P$_2$, small yellow spheres clustered like grapes. Absorption max: 366, 445 nm. The absorption curve is practically identical with that of riboflavine. There is some stronger absorption between 450 and 510 nm resulting in aq solns which are more reddish and less green than those of riboflavine. The appearance of a strong greenish fluorescence indicates decomposition and loss of catalytic activity.

4132. Flavone. *2-Phenyl-4H-1-benzopyran-4-one;* 2-phenylchromone; 2-phenyl-γ-benzopyrone; 2-phenyl-1,4-benzopyrone. C$_{15}$H$_{10}$O$_2$; mol wt 222.24. C 81.07%, H 4.54%, O 14.40%. Prepn: Feuerstein, Kostanecki, *Ber.* **31**, 1757 (1898); Kostanecki, Tambor, *Ber.* **33**, 330 (1900); Kostanecki, Szabranski, *Ber.* **37**, 2634 (1904); Ruhemann, *Ber.* **46**, 2192 (1913); Bogert, Marcus, *J. Am. Chem. Soc.* **41**, 89 (1919). Isoln from *Primula malacoides*, Franch., *Primulaceae* and biological properties: Weller *et al.*, *Antibiot. & Chemother.* **3**, 603 (1953). Syntheses: Wheeler, *Org. Syn.* **coll. vol. IV**, 478 (1963); Y. Ashihara *et al.*, *Bull. Chem. Soc. Japan* **50**, 3298 (1977); A. Banerji, N. C. Goomer, *Synthesis* **1980**, 874.

Crystals from petr ether, mp 99-100°. Practically insol in water. Sol in most organic solvents. Absorption max: 350, 405 nm.

4133. Flavopereirine. *3-Ethyl-12H-indolo[2,3-a]quinolizin-5-ium inner salt;* melinonine G. C$_{17}$H$_{14}$N$_2$; mol wt 246.31. C 82.90%, H 5.73%, N 11.37%. From *Strychnos melinoniana* Baillon., *Loganiaceae:* Bächli *et al.*, *Helv. Chim. Acta* **40**, 1167 (1957); from *Geissospermum vellosii* Baillon., *Apocynaceae:* Rapoport *et al.*, *J. Am. Chem. Soc.* **80**, 1601 (1958). Bertho *et al.*, *Ber.* **91**, 2581 (1958). Structure: Bejar *et al.*, *Compt. Rend.* **244**, 2066 (1957). Synthesis: Le Hir *et al.*, *Bull. Soc. Chim. France* **1958**, 551; Prasad, Swan, *J. Chem. Soc.* **1958**, 2024; G. W. Gribble, D. A. Johnson, *Tetrahedron Letters* **28**, 5259 (1987); of perchlorate: J. Ninomiyo *et al.*, *Heterocycles* **9**, 1527 (1978).

Orange crystals from acetone, mp 233-235°. uv max (ethanol): 230, 238, 248, 294, 351, 390 nm (log ε 4.40, 4.43, 4.39, 4.14, 4.25, 4.14).

Perchlorate, C$_{17}$H$_{15}$ClN$_2$O$_4$, crystals from methanol, mp 307-308°. uv max (0.015N HCl-ethanol): 238, 294, 350, 389 nm (log ε 4.57, 4.22, 4.31, 4.21).

4134. Flavoxanthin. *(3S,3'R,5R,6'R,8R)-5,8-Epoxy-5,8-dihydro-β,ε-carotene-3,3'-diol.* C$_{40}$H$_{56}$O$_3$; mol wt 584.88. C 82.14%, H 9.65%, O 8.21%. Carotenoid pigment; stereoisomer of chrysanthemaxanthin, *q.v.* Often found in plants, but in minute amounts only and never as a principal pigment. Isoln from *Ranunculus acris* L., *Ranunculaceae:* Kuhn, Brockmann, *Z. Physiol. Chem.* **213**, 192 (1932). Structure: Karrer, Rutschmann, *Helv. Chim. Acta* **25**, 1144 (1942). Partial synthesis: Karrer, Jucker, *ibid.* **28**, 300 (1945). Absolute configuration: H. Cadosch *et al.*, *ibid.* **61**, 783 (1978).

Golden-yellow aggregates of prisms, mp 184°. Absorption max (chloroform): 430, 459 nm. [α]$_{Cd}^{20}$ +190° (c = 0.04 in benzene). Freely sol in chloroform, benzene, acetone; less sol in methanol, ethanol. Almost insol in petr ether.

4135. Flavoxate. *3-Methyl-4-oxo-2-phenyl-4H-1-benzopyran-8-carboxylic acid 2-(1-piperidinyl)ethyl ester;* 3-methylflavone-8-carboxylic acid β-piperidinoethyl ester; 2-piperidinoethyl 3-methyl-4-oxo-2-phenyl-4H-1-benzopyran-8-carboxylate; 2-piperidinoethyl 3-methylflavone-8-carboxylate. C$_{24}$H$_{25}$NO$_4$; mol wt 391.47. C 73.64%, H 6.44%, N 3.58%, O 16.35%. Smooth muscle relaxant. Prepn of the hydrochloride: P. Da Re *et al.*, *J. Med. Pharm. Chem.* **2**, 263 (1960); P. Da Re, U.S. pats. **2,921,070** and **3,350,411** (1960 to Recordati and 1967 to Seceph). Pharmacological studies: I. Setnikar *et al.*, *J. Pharmacol. Exp. Ther.* **130**, 356 (1960). Mechanism of action: P. Cazzulani *et al.*, *Arch. Int. Pharmacodyn.* **274**, 189 (1985). Pharmacokinetics in rats: I. Setnikar *et al.*, *Arzneimittel-Forsch.* **25**, 1916 (1975); in man: M. Bertoli *et al.*, *Pharmacol. Res. Commun.* **8**, 417 (1976). *In vivo* activity: C. Pietra, P. Cazzulani, *Farmaco Ed. Prat.* **41**, 267 (1986). Determn in tissues: A. Cova, I. Setnikar,

Arzneimittel-Forsch. **25,** 1707 (1975). Clinical evaluations: D. V. Bradley, R. J. Cazort, *J. Clin. Pharmacol.* **10,** 65 (1970); A. Zanollo, F. Catanzaro, *Urol. Int.* **35,** 176 (1980). Clinical comparison with phenazopyridine, *q.v.:* S. Gould, *Urology* **5,** 612 (1975). Short review: R. Ruffmann, A. Sartini, *Drug Exp. Clin. Res.* **13,** 57 (1987).

Crystals, pK 7.3. Soly in water at 37°: 0.001% (w/v). Sol in ethanol and chloroform. LD_{50} in rats (mg/kg): 1110 orally; 20.8 i.v. (Setnikar, 1975).

Hydrochloride, $C_{24}H_{25}NO_4 \cdot HCl$, *DW-61, Rec-7-0040, Bladderon, Genurin, Patricin, Spasuret, Urispas.* Crystals from ethanol + ether, mp 232-234°. LD_{50} i.v. in rats: 27.4 mg/kg (Cazzulani).

Succinate, $C_{24}H_{25}NO_4 \cdot C_4H_6O_4$, soly in water at 37°: 33.7% (w/v).

THERAP CAT: Antispasmodic.

4136. Flecainide. *N-(2-Piperidinylmethyl)-2,5-bis(2,2,2-trifluoroethoxy)benzamide.* $C_{17}H_{20}F_6N_2O_3$; mol wt 414.35. C 49.28%, H 4.87%, F 27.51%, N 6.76%, O 11.58%. Prepn: E. H. Banitt, W. R. Brown, U.S. pat. **3,900,481** (1975 to Riker); of the acetate: *eidem,* U.S. pat. **4,005,209** (1977 to Riker); E. H. Banitt *et al., J. Med. Chem.* **20,** 821 (1977). Preliminary pharmacological study: J. R. Schmid *et al., Fed. Proc.* **34,** 775 (1975). *In vitro* electrophysiological study: A. B. Hodess *et al., J. Cardiovasc. Pharmacol.* **1,** 427 (1979). Antiarrhythmic effects: P. Somani, *Clin. Pharmacol. Ther.* **27,** 464 (1980). Use in acute exptl myocardial infarction: H. Gülker *et al., Z. Cardiol.* **70,** 124 (1981). Clinical study in ventricular arrhythmias: J. L. Anderson *et al., N. Engl. J. Med.* **305,** 473 (1981). Determn of acetate in human plasma by spectrophotofluorometry: S. F. Chang *et al., Arzneimittel-Forsch.* **33,** 251 (1983). Review of pharmacology and clinical efficacy: D. M. Roden, R. L. Woosley, *N. Engl. J. Med.* **315,** 36-41 (1986). Symposium on clinical experience: *Am. J. Cardiol.* **62,** Suppl., 1D-67D (1988). Comprehensive description: S. Alessi-Severini *et al., Anal. Profiles Drug Subs. Excip.* **21,** 169-195 (1992).

uv max (ethanol): 205, 230, 300 nm ($E_{1cm}^{1\%}$ 521, 219, 59). Monoacetate, $C_{17}H_{20}F_6N_2O_3 \cdot C_2H_4O_2$, *R-818, Almarytm, Apocard, Ecrinal, Tambocor.* White granular solid from isopropyl alcohol/isopropyl ether, mp 145-147°. Soly at 37° (mg/ml): water 48.4, alcohol 300.

THERAP CAT: Antiarrhythmic (class IC).

4137. Fleroxacin. *6,8-Difluoro-1-(2-fluoroethyl)-1,4-dihydro-7-(4-methyl-1-piperazinyl)-4-oxo-3-quinolinecarboxylic acid;* AM-833; Ro-23-6240; Megalocin; Quinodis. $C_{17}H_{18}F_3N_3O_3$; mol wt 369.34. C 55.28%, H 4.91%, F 15.43%, N 11.38%, O 13.00%. Fluorinated quinolone antibacterial. Prepn: **Belg.** pat. **887,574;** T. Irikura *et al.,* **U.S.** pat. **4,398,029** (1981, 1983 both to Kyorin). Antibacterial spectrum *in vitro* and *in vivo:* K. Hirai *et al., Antimicrob. Ag. Chemother.* **29,** 1059 (1986). *In vitro* activity vs anaerobic bacteria: J. Wüst, U. Hardegger, *Eur. J. Clin. Microbiol.* **6,** 688 (1987). HPLC determn in biological fluids: H. Kusajima *et al., J. Chromatog.* **381,** 137 (1986). Pharmacokinetics and bioavailability: E. Weidekamm *et al., Antimicrob. Ag. Chemother.* **31,** 1909 (1987). Preliminary clinical evaluation in gonorrhea: J. B. J. Boerema *et al., J. Antimicrob. Chemother.* **21,** 140 (1988).

Hydrochloride, $C_{17}H_{18}F_3N_3O_3 \cdot HCl$, crystals from water, mp 269-271° (dec).

THERAP CAT: Antibacterial.

4138. Flesinoxan. *(R)-N-[2-[4-[2,3-Dihydro-2-(hydroxymethyl)-1,4-benzodioxin-5-yl]-1-piperazinyl]ethyl]-4-fluorobenzamide;* (+)-*(S)-p-fluoro-N-[2-[4-[2-(hydroxymethyl)-1,4-benzodioxan-5-yl]-1-piperazinyl]ethyl]benzamide.* $C_{22}H_{26}FN_3O_4$; mol wt 415.46. C 63.60%, H 6.31%, F 4.57%, N 10.11%, O 15.40%. Serotonin (5-HT$_{1A}$)-receptor agonist; phenylpiperazine derivative. Prepn: J. Hartog *et al.,* **Eur.** pat. **Appl. 138,280;** *eidem,* U.S. pat. **4,833,142** (1985, 1989 both to Duphar). Synthesis and resolution of enantiomers: M. D. Ennis, N. B. Ghazal, *Tetrahedron Letters* **33,** 6287 (1992). GC determn in plasma and urine: M. P. van Berkel *et al., Methodol. Surveys Biochem. Anal.* **18,** 209 (1988). Pharmacology: J. Schipper *et al., Hum. Psychopharmacol.* **6,** S53 (1991). Clinical evaluation in depression: P. Grof *et al., Int. Clin. Psychopharmacol.* **8,** 167 (1993).

$[\alpha]_D$ +26° (c = 1 in methanol). Hydrochloride, $C_{22}H_{26}FN_3O_4 \cdot HCl$, *DU-29373,* mp 184.5-185.5°. Also reported as crystals from ethanol/ether, mp 183.0-184.0° (Ennis). $[\alpha]_D$ +25° (c = 1 in methanol). Freely sol in water; sparingly sol in ethanol.

THERAP CAT: Anxiolytic.

4139. Flindersine. *2,6-Dihydro-2,2-dimethyl-5H-pyrano[3,2-c]quinolin-5-one;* 2,2'-dimethyl-α-pyrano(5',6',3,4)-2(1H)-quinolone. $C_{14}H_{13}NO_2$; mol wt 227.26. C 73.99%, H 5.77%, N 6.16%, O 14.08%. From wood of *Flindersia australis* R. Br., *Rutaceae:* Matthes, Schreiber, *Ber. Deut. Pharm. Ges.* **24,** 385 (1914). Structure: Brown *et al., Aust. J. Chem.* **7,** 348 (1954). Synthesis: *eidem, ibid.* **9,** 277 (1956); Piozzi *et al., Gazz. Chim. Ital.* **99,** 711 (1969); Bowman *et al., Chem. Commun.* **1970,** 666; Huffman, Hsu, *Tetrahedron Letters* **1972,** 141.

Crystals from methanol, dec 185-186°. uv max (methanol): 235, 333, 350, 365 nm (log ε 4.42, 4.00, 4.10, 3.93). Sol in alc, benzene, chloroform, glacial acetic acid, paraffin, fatty oils, alkali hydroxides, slightly sol in petr ether. Practically insol in water.

N-Methylflindersine, $C_{15}H_{15}NO_2$, crystals from petr ether, mp 84°.

Dihydroflindersine, $C_{14}H_{15}NO_2$, hexagons from ethyl acetate, mp 229°. uv max (methanol): 225, 272, 283, 312 nm (log ε 4.44, 3.94, 3.96, 3.93).

4140. Floctafenine. *2-[[8-(Trifluoromethyl)-4-quinolinyl]amino]benzoic acid 2,3-dihydroxypropyl ester; N-[8-(trifluoromethyl)-4-quinolyl]anthranilic acid 2,3-dihydroxy-*

propyl ester; 4-[o-(2′,3′-dihydroxypropyloxycarbonyl)phenyl]amino-8-trifluoromethylquinoline; 8-trifluoromethyl-7-deschloroglafenine; R-4318; RU-15750; Diralgan; Idalon; Idarac; Novodolan. $C_{20}H_{17}F_3N_2O_4$; mol wt 406.36. C 59.11%, H 4.22%, F 14.03%, N 6.89%, O 15.75%. Prepn: Allais *et al.*, **Ger.** pat. **1,815,467** corresp to **U.S.** pat. **3,644,-368** (1969, 1972, both to Roussel-UCLAF); G. Mouzin *et al.*, *Synthesis* **1980**, 54. Analgesic activity: Allais *et al.*, *Chim. Ther.* **8**, 154 (1973); M. Peterfalvi *et al.*, *Arch. Int. Pharmacodyn. Ther.* **216**, 97 (1975). Clinical investigation: Stenport, *Curr. Ther. Res.* **18**, 303 (1975). Toxicology: R. Glomot *et al.*, *Toxicol. Appl. Pharmacol.* **36**, 173 (1976). Metabolism: R. K. Lynn *et al.*, *J. Clin. Pharmacol.* **19**, 20 (1979).

Crystals from methanol, mp 179-180°. Sol in alcohol, acetone; very slightly sol in ether, chloroform, methylene chloride. Insol in water. LD_{50} in male mice, rats (mg/kg): 3400, 960 orally; 180, 160 i.v. (Glomot).

THERAP CAT: Analgesic.

4141. Flomoxef. *(6R cis)-7-[[[(Difluoromethyl)thio]-acetyl]amino]-3-[[[1-(2-hydroxyethyl)-1H-tetrazol-5-yl]thio]-methyl]-7-methoxy-8-oxo-5-oxa 1-azabicyclo[4.2.0]oct-2-ene-2-carboxylic acid;* 7-β-difluoromethylthioacetamido-7α-methoxy-3-[[1-(2-hydroxyethyl)-1H-tetrazol-5-yl]thiomethyl]-1-oxa-3-cephem-4-carboxylic acid. $C_{15}H_{18}F_2N_6O_7S_2$; mol wt 496.47. C 36.29%, H 3.65%, F 7.65%, N 16.93%, O 22.56%, S 12.92%. Semi-synthetic oxacephalosporin antibiotic. Prepn: T. Tsuji *et al.*, **Belg.** pat. **898,541;** *eidem,* **U.S.** pat. **4,532,233** (1984, 1985 both to Shionogi); *eidem, J. Antibiot.* **38**, 466 (1985). *In vitro* activity and β-lactamase stability: H. C. Neu, N.-X. Chin, *Antimicrob. Ag. Chemother.* **30**, 638 (1986).

Crystals from acetone + methylene chloride, mp 82.5°-87.5°.

Sodium salt, $C_{15}H_{17}F_2N_6NaO_7S_2$, *6315-S, Flumarin.*

THERAP CAT: Antibacterial.

4142. Flopropione. *1-(2,4,6-Trihydroxyphenyl)-1-propanone; 2′,4′,6′-trihydroxypropiophenone;* phloropropiophenone; RP-13907; Argobyl; Cospanon; Flopion; Gallepronin; Gasstenon; Labroda; Labrodax; Pasmus; Profenon; Spasmorin; Spasmoril; Supanate; Supazlun. $C_9H_{10}O_4$; mol wt 182.18. C 59.34%, H 5.53%, O 35.13%. Prepn: Shinoda, *J. Pharm. Soc. Japan* **1927**, 111; Canter *et al.*, *J. Chem. Soc.* **1931**, 1245; Howells, Little, *J. Am. Chem. Soc.* **54**, 2451 (1932).

Monohydrate, needles from water. Anhydr compd, mp 175-176°. Sol in ethanol, ether, ethyl acetate, hot water; very slightly sol in cold water.

THERAP CAT: Antispasmodic.

4143. Florantyrone. *γ-Oxo-8-fluoranthenebutanoic acid;* β-(8-fluoranthoyl)propionic acid; Zanchol. $C_{20}H_{14}O_3$; mol wt 302.33. C 79.46%, H 4.67%, O 15.88%. Prepn: R. R. Burtner, **U.S.** pat. **2,773,091** (1956 to Searle). *See also:* O. E. Fancher, **U.S.** pat. **2,560,425** (1951 to Miles Labs).

Lemon yellow crystals from acetic acid, mp ~208°. Soluble in methanol, ethanol, aq solns of sodium carbonate.

THERAP CAT: Choleretic.

4144. Floredil. *4-[2-(3,5-Diethoxyphenoxy)ethyl]morpholine.* $C_{16}H_{25}NO_4$; mol wt 295.38. C 65.06%, H 8.53%, N 4.74%, O 21.67%. Prepn by refluxing 3,5-diethoxyphenol with 4-(2-chloroethyl)morpholine in ethanol: Lefon, **Ger.** pat. **2,020,464** (1970 to Orsymonde), *C.A.* **74**, 31650u (1971).

Hydrochloride, $C_{16}H_{25}NO_4$.HCl, *Carfonal.*

THERAP CAT: Vasodilator (coronary).

4145. Florfenicol. *[R-(R*,S*)]-2,2-Dichloro-N-[1-(fluoromethyl)-2-hydroxy-2-[4-(methylsulfonyl)phenyl]ethyl]acetamide;* D-*(threo)*-1-p-methylsulfonylphenyl-2-amino-3-fluoro-1-propanol; fluorothiamphenicol; Sch-25298. Aquafen. $C_{12}H_{14}Cl_2FNO_4S$; mol wt 358.22. C 40.24%, H 3.94%, Cl 19.79%, F 5.30%, N 3.91%, O 17.87%, S 8.95%. Fluorinated derivative of thiamphenicol, *q.v.* Prepn: T. Nagabhushan, **Eur.** pat. **Appl. 14,437;** **U.S.** pat. **4,235,892** (both 1980 to Schering). *In vitro* antibacterial activity: H. C. Neu, K. P. Fu, *Antimicrob. Ag. Chemother.* **18**, 311 (1980); V. P. Syriopoulou *et al.*, *ibid.* **19**, 294 (1981). Pharmacokinetics in veal calves: K. J. Varma *et al.*, *J. Vet. Pharmacol. Therap.* **9**, 412 (1986). Therapeutic effect on pseudotuberculosis in yellowtail tuna: N. Yasunaga, S. Yasumoto, *Fish Pathol.* **23**, 1 (1988).

mp 153-154°. Sol in water.

THERAP CAT (VET): Antibacterial.

4146. Flosequinan. *7-Fluoro-1-methyl-3-(methylsulfinyl)-4(1H)-quinolinone;* flosequinon; BTS-49465; Manoplax. $C_{11}H_{10}FNO_2S$; mol wt 239.27. C 55.22%, H 4.21%, F 7.94%, N 5.85%, O 13.37%, S 13.40%. Mixed arterial and venous vasodilator. Prepn: R. V. Davies *et al.*, **Ger.** pat. **3,011,994;** *eidem,* **U.S.** pat. **4,302,460** (1980, 1981 both to Boots). Pharmacology: J. G. Smith, G. T. Kinasewitz, *J. Cardiovasc. Pharmacol.* **8**, 878 (1986); M. F. Sim *et al.*, *Brit. J. Pharmacol.* **94**, 371 (1988). HPLC determn in plasma, serum and urine: M. B. Slegowski *et al.*, *J. Chromatog.* **425**, 227 (1988). Pharmacokinetics and hemodynamics in humans: A. J. Cowley *et al.*, *J. Hypertens.* **2**, Suppl. 3, 547

(1984); R. D. Wynne *et al., Eur. J. Clin. Pharmacol.* **28,** 659 (1985). Preliminary evaluation in hypertension: A. J. Cowley *et al., ibid.* **33,** 203 (1987); in congestive heart failure: P. D. Kessler, M. Packer, *Am. Heart J.* **113,** 137 (1987); P. D. Kessler *et al., J. Cardiovasc. Pharmacol.* **12,** 6 (1988). Clinical study in chronic heart failure: A. J. Cowley *et al., Brit. Med. J.* **297,** 169 (1988).

Crystals, mp 226-228°.

THERAP CAT: Antihypertensive.

4147. Floxacillin. *[2S-(2α,5α,6β)]-6-[[[3-(2-Chloro-6-fluorophenyl)-5-methyl-4-isoxazolyl]carbonyl]amino]-3,3-dimethyl-7-oxo-4-thia-1-azabicyclo[3.2.0]heptane-2-carboxylic acid;* 3-(2-chloro-6-fluorophenyl)-5-methyl-4-isoxazolyl-penicillin; 6-[3-(2-chloro-6-fluorophenyl)-5-methyl-4-isoxazolecarboxamido]penicillanic acid; flucloxacillin; BRL-2039; Abboflox; Flupen; Penplus; Staphlipen. $C_{19}H_{17}$-$ClFN_3O_5S$; mol wt 453.88. C 50.28%, H 3.78%, Cl 7.81%, F 4.19%, N 9.26%, O 17.63%, S 7.06%. Semi-synthetic antibiotic active against penicillin-resistant staphylococci. Halogen-substituted derivative of oxacillin, *q.v.* Prepn: **S. Afr.** pat. **63 4,323** (1964 to Beecham); Nayler, **Brit.** pat. **978,299;** *idem,* **U.S.** pat. **3,239,507** (1964, 1966 both to Beecham). Pharmacology and toxicity: R. Sutherland *et al., Brit. Med. J.* **4,** 460 (1970). Clinical studies: Harding *et al., Clin. Trials J.* **7,** 368 (1970); Qureshi *et al., ibid.* 375.

Sodium monohydrate, $C_{19}H_{16}ClFN_3NaO_5S.H_2O$, *Culpen, Floxapen, Ladropen, Stafoxil, Staphcil, Staphylex.* Freely sol in water. LD$_{50}$ in mice (g/kg): 2.2 s.c., 3.8 orally (Sutherland).

THERAP CAT: Antibacterial.

4148. Floxuridine. *2′-Deoxy-5-fluorouridine;* 1-(2-de-oxy-β-D-ribofuranosyl)-5-fluorouracil; 5-fluoro-2′-deoxy-β-uridine; NSC-27640; FUDR. $C_9H_{11}FN_2O_5$; mol wt 246.20. C 43.91%, H 4.50%, F 7.72%, N 11.38%, O 32.49%. Prepn: Hoffer *et al., J. Am. Chem. Soc.* **81,** 4112 (1959); Heidelberger, Duschinsky, **U.S.** pat. **2,885,396** (1959); Duschinsky *et al.,* **U.S.** pat. **2,970,139** (1961); Hoffer, **U.S.** pats. **2,949,-451; 3,041,335** (1960, 1962 both to Hoffmann-La Roche). Structure: Lemieux, Hoffer, *Can. J. Chem.* **39,** 110 (1961). Crystal and molecular structure: Harris, *Diss. Abstr.* **24,** 4425 (1964); Harris, McIntyre, *Biophys. J.* **4,** 203 (1964). Conformation of furanose ring in molecule: Sundaralingam, *J. Am. Chem. Soc.* **87,** 599 (1965).

Crystals from butyl acetate, mp 150-151° (Hoffer, 1959); also reported as mp 145° (**U.S.** pat. **3,041,335**). $[\alpha]_D$ +37°

(water), +48.6° (DMF). uv max (pH 7.2): 268 nm (ε 7570); (pH 14): 270 nm (ε 6480).

α-Anomer, *1-(2-deoxy-α-D-erythro-pentofuranosyl)-5-fluo-rouracil.* Crystals from butyl acetate, mp 150-151°. $[\alpha]_D^{25}$ −21° (c = 2 in water).

THERAP CAT: Antiviral; antineoplastic.

4149. Fluacizine. *10-[3-(Diethylamino)-1-oxopropyl]-2-(trifluoromethyl)-10H-phenothiazine; 10-(N,N-diethyl-β-alan-yl)-2-(trifluoromethyl)phenothiazine;* fluoracisine; fluoracizine; ftorazicine; Phtorazisin. $C_{20}H_{21}F_3N_2OS$; mol wt 394.46. C 60.90%, H 5.37%, F 14.45%, N 7.10%, O 4.06%, S 8.13%. Phenothiazine derivative with psychotropic activity. Prepn: Y. I. Vikhlyaev *et al.,* **USSR** pat. **360,342** (1972), *C.A.* **78,** 97683w (1973); of hydrochloride: S. V. Zhuravlev *et al.,* **Brit.** pat. **1,191,800** (1970), *C.A.* **73,** 25493h (1970); Y. I. Vikhlyaev *et al.,* **Fr.** pat. **2,035,748** (1971), *C.A.* **75,** 140872j (1971); *eidem,* **Ger.** pat. **1,805,659** (1979), *C.A.* **75,** 49101w (1971). Use as antidepressant: *eidem,* **USSR** pat. **356,992** (1972), *C.A.* **78,** 75888q (1973) (all to Inst. Pharmacol. Chemother., Acad. Med. Sci., USSR). Mass spectrum: S. Morosawa *et al., Org. Mass Spec.* **17,** 309 (1982). Effect on drug metabolism: V. Avakumov *et al., Biochem. Pharmacol.* **27,** 2177 (1978). Pharmacokinetics in animals, man: V. P. Zherdev *et al., Farmakol. Toksikol. (USSR)* **45,** 83 (1982).

Hydrochloride, $C_{20}H_{21}F_3N_2OS.HCl,$ *fluoracyzine, tora-cizin.* White cryst powder, mp 163-165°. Sol in water, warm alcohols.

THERAP CAT: Antidepressant.

4150. Fluanisone. *1-(4-Fluorophenyl)-4-[4-(2-methoxy-phenyl)-1-piperazinyl]-1-butanone; 4′-fluoro-4-[4-(o-methoxy-phenyl)-1-piperazinyl]butyrophenone;* haloanisone; R-2028; Sedalande. $C_{21}H_{25}FN_2O_2$; mol wt 356.44. C 70.76%, H 7.07%, F 5.33%, N 7.86%, O 8.98%. Prepn: Janssen, **U.S.** pat. **2,997,472** (1961). Determn in human plasma: M. P. Quaglio, A. M. Bellini, *Farmaco Ed. Prat.* **36,** 204 (1981). Comparison between biochemical and behavioral effects: G. B. Fregnan, R. Porta, *Arzneimittel-Forsch.* **31,** 70 (1981). Toxicity study: C. Cascio *et al., Farmaco Ed. Sci.* **35,** 605 (1980).

Crystals, mp 67.5-68.5°. Sol in chloroform; sparingly sol in methanol; slightly sol in ether. Practically insol in water. LD$_{50}$ i.p. in mice: 200 mg/kg (Cascio).

Hydrochloride, mp 205-205.5°.

THERAP CAT: Antipsychotic.

4151. Fluazacort. *(11β,16β)-21-(Acetyloxy)-9-fluoro-11-hydroxy-2′-methyl-5′H-pregna-1,4-dieno[17,16-d]oxazole-3,20-dione; 9-fluoro-11β,21-dihydroxy-2′-methyl-5′βH-preg-na-1,4-dieno[17,16-d]oxazole-3,20-dione 21-acetate;* L-6400; Azacortid. $C_{25}H_{30}FNO_6$; mol wt 459.51. C 65.35%, H 6.58%, F 4.13%, N 3.05%, O 20.89%. Prepn of the 21-hydroxy compound: G. Nathansohn *et al., J. Med. Chem.* **10,** 799 (1967); of the 21-hydroxy and fluazacort: *eidem,* **Brit.** pat. **1,119,082;** *eidem,* **U.S.** pat. **3,461,119** (1968, 1969 to Lepetit); *eidem, Steroids* **13,** 383 (1969). Pharmacological and physico-chemical properties: *eidem, ibid.* 365. Percutaneous absorption: J. D. Lewis *et al., Arzneimittel-Forsch.* **25**, 1646 (1975).

Crystals from ethanol, mp 252-255°. $[\alpha]_D$ +54.8° (c = 0.5 in CHCl₃). uv max (methanol): 238-240 nm ($E_{1cm}^{1\%}$ 330).
THERAP CAT: Anti-inflammatory.

4152. Fluazifop-butyl. *2-[4-[[5-(trifluoromethyl)-2-pyridinyl]oxy]phenoxy]propanoic acid butyl ester;* butyl 2-[4-(5-trifluoromethyl-2-pyridyloxy)phenoxy]propionate; PP-009; TF-1169; Fusilade. C₁₉H₂₀F₃NO₄; mol wt 383.37. C 59.53%, H 5.26%, F 14.87%, N 3.65%, O 16.69%. Selective post-emergence herbicide. Prepn: R. Nishiyama *et al.,* **Ger. pat. 2,812,571** (1979 to Ishihara Sangyo Kaisha), *C.A.* **90**, 152017g (1979). Activity: R. E. Plowman *et al., Proc. Brit. Crop Prot. Conf. — Weeds* **1980**, 29; J. R. Finney, P. B. Sutton, *ibid.* 429.

$bp_{0.05}$ 167°.
USE: Herbicide.

4153. Fluazinam. *3-Chloro-N-[3-chloro-2,6-dinitro-4-(trifluoromethyl)phenyl]-5-(trifluoromethyl)-2-pyridinamine;* 3-chloro-N-(3-chloro-5-trifluoromethyl-2-pyridinyl)-α,α,α-trifluoro-2,6-dinitro-p-toluidine; N-(3-chloro-5-trifluoromethyl-2-pyridyl)-2,6-dinitro-3-chloro-4-trifluoromethyl-aniline; IKF-1216; ICIA-192; PP-192; Frowncide; Shirlan. C₁₃H₄Cl₂F₆N₄O₄; mol wt 465.10. C 33.57%, H 0.87%, Cl 15.25%, F 24.51%, N 12.05%, O 13.76%. Blight fungicide with uncoupling activity on oxidative phosphorylation. Prepn: R. Nishiyama *et al.,* **Eur. pat. Appl. 31,257;** *eidem,* **U.S. pat. 4,331,670** (1981, 1982 both to Ishihara Sangyo Kaisha). Mechanism of action study: Z. Guo *et al., Biochim. Biophys. Acta* **1056**, 89 (1991). Field trials in bulbs: B. P. Anema *et al., Meded. Fac. Landbouwwet., Rijksuniv. Gent* **53**, 635 (1988); in potatoes: B. P. Anema *et al., Brighton Crop Prot. Conf. - Pests Dis.* **1992**, 663.

mp 100-102°. pKa 7.11.
USE: Agricultural fungicide.

4154. Flubendazole. *[5-(4-Fluorobenzoyl)-1H-benzimidazol-2-yl]carbamic acid methyl ester;* 5-(p-fluorobenzoyl)-2-benzimidazolecarbamic acid methyl ester; R-17889; Flubenol; Flumoxal; Flumoxane; Fluvermal. C₁₆H₁₂FN₃O₃; mol wt 313.29. C 61.34%, H 3.86%, F 6.06%, N 13.41%, O 15.32%. Fluoro analog of mebendazole, *q.v.* Prepn: J. L. VanGelder *et al.,* **Ger. pat. 2,029,637** corresp to **U.S. pat. 3,657,267** (1971, 1972 both to Janssen); A. H. M. Raeymaekers *et al., Arzneimittel-Forsch.* **28**, 586 (1978). Pharmacological, biological properties: D. Thienpont *et al., ibid.* 605. Anthelmintic activity in rats: O. Vanparijs *et al., Vet.*

Parasitol. **5**, 237 (1979); in pigs: E. Telléz-Giron *et al., Am. J. Trop. Med. Hyg.* **30**, 135 (1981).

Crystals, mp 260°. LD₅₀ in mice, rats, guinea pigs (mg/kg): > 2560 orally (Thienpont).
THERAP CAT (VET): Anthelmintic.

4155. Flubenzimine. *N-[3-Phenyl-4,5-bis[(trifluoromethyl)imino]-2-thiazolidinylidene]benzenamine;* N^2,3-diphenyl-N^4,N^5-bis(trifluoromethyl)thiazolidine-2,4,5-triylidene triamine; SLJ-312; Cropotex. C₁₇H₁₀F₆N₄S; mol wt 416.35. C 49.04%, H 2.42%, F 27.38%, N 13.46%, S 7.70%. Prepn: H.-J. Scholl *et al.,* **Ger. pat. 2,062,348** corresp to **U.S. pat. 3,895,020** (1972, 1975 to Bayer). Mode of action: G. Zoebelein *et al., Pflanzenschutz-Nachr.* **33**, 169 (1980); D. J. Bluett, A. Wainwright, *Proc. Brit. Crop Prot. Conf.-Pests Dis.* **1981**, 75.

Yellow crystals from methanol, mp 118-119°. Vapor pressure at 20°: 1 × 10⁻⁵ mm Hg. Soly in water at 20°: 30 mg/l. LD₅₀ orally in rats: 3750 mg/kg (Bluett, Wainwright).
USE: Acaricide.

4156. Fluchloralin. *N-(2-Chloroethyl)-2,6-dinitro-N-propyl-4-(trifluoromethyl)benzenamine;* N-(2-chloroethyl)-α,α,α-trifluoro-2,6-dinitro-N-propyl-p-toluidine; BAS-392H; Basalin. C₁₂H₁₃ClF₃N₃O₄; mol wt 355.70. C 40.52%, H 3.68%, Cl 9.97%, F 16.02%, N 11.81%, O 17.99%. Soil incorporated, pre-planting herbicide. Prepn: K. H. Karl, K. Kiehs, **Ger. pat. 2,161,879** (1973 to BASF), *C.A.* **79**, 65990y (1973). Photochemistry: G. P. Nilles, M. J. Zabik, *J. Agr. Food Chem.* **22**, 684 (1974). Degradn and metabolism in soil: S. Otto, *Environ. Qual. Saf., Suppl.* **3**, 277 (1975).

Orange-yellow cryst solid, mp 42-43°. Soly in water: 10 ppm.
USE: Herbicide.

4157. Flucloronide. *(6α,11β,16α)-9,11-Dichloro-6-fluoro-21-hydroxy-16,17-[(1-methylethylidene)bis(oxy)]-pregna-1,4-diene-3,20-dione;* 9,11β-dichloro-6α-fluoro-16α,17,21-trihydroxypregna-1,4-dien-3,20-dione cyclic 16,17-acetal with acetone; 9α,11β-dichloro-6α-fluoro-21-hydroxy-16α,17α-(isopropylidenedioxy)pregna-1,4-diene-3,20-dione; 6α-fluoro-9α,11β-dichloro-1,4-pregnadiene-16α,17,21-triol-3,20-dione 16,17-acetonide; fluclorolone acetonide; RS-2252; Topilar. C₂₄H₂₉Cl₂FO₅; mol wt 487.40. C 59.14%, H 6.00%, Cl 14.55%, F 3.90%, O 16.41%. Prepn: Bowers, **U.S. pat. 3,201,391** (1965 to Syntex). Activity studies: Dorfman *et al., Acta Endocrinol.* **49**, 262 (1965); several authors in *Brit.*

J. Dermatol. **82**, suppl. 6, 53-98 (1970). Clinical studies: several authors in *Clin. Trials J.* **8**, 35-63 (1971).

THERAP CAT: Glucocorticoid.

4158. Fluconazole. α-*(2,4-Difluorophenyl)-α-(1H-1,2,-4-triazol-1-ylmethyl)-1H-1,2,4-triazole-1-ethanol;* 2,4-difluoro-α,α-bis(1*H*-1,2,4-triazol-1-ylmethyl)benzyl alcohol; 2-(2,-4-difluorophenyl)-1,3-bis(1*H*-1,2,4-triazol-1-yl)propan-2-ol; UK-49858; Biozolene; Diflucan; Elazor; Triflucan. $C_{13}H_{12}$-F_2N_6O; mol wt 306.27. C 50.98%, H 3.95%, F 12.41%, N 27.44%, O 5.22%. Orally active bistriazole antifungal agent. Prepn: K. Richardson, **Brit.** pat. **2,099,818**; *idem,* U.S. pat. **4,404,216** (1982, 1983 both to Pfizer). Antifungal activity *in vivo:* K. Richardson *et al., Antimicrob. Ag. Chemother.* **27**, 832 (1985); and *in vitro:* T. E. Rogers, J. N. Galgiani, *ibid.* **30**, 418 (1986). Pharmacokinetics: M. J. Humphrey *et al., ibid.* **28**, 648 (1985). GC determn in human plasma and urine: P. R. Wood, M. H. Tarbit, *J. Chromatog.* **383**, 179 (1986). Evaluation as maintenance therapy for cryptococcal meningitis in AIDS patients: S. A. Bozzette *et al., N. Engl. J. Med.* **324**, 580 (1991). Symposium on mode of action, toxicology, animal models and clinical studies: *International Telesymposium on Recent Trends in the Discovery, Development and Evaluation of Antifungal Agents,* **Section 2,** R. A. Fromtling, Ed. (J. R. Prous Science Publ., Barcelona, 1987) pp 77-174.

Crystals from ethyl acetate/hexane, mp 138-140°.
THERAP CAT: Antifungal.

4159. Flucycloxuron. *N-[[[4-[[[[(4-Chlorophenyl)cyclopropylmethylene]amino]oxy]methyl]phenyl]amino]carbonyl]-2,6-difluorobenzamide;* 1-[α-(4-chloro-α-cyclopropylbenzylideneaminooxy)-*p*-tolyl]-3-(2,6-difluorobenzoyl)urea; DU-319722; PH-70-23; UBI-A1335; Andalin. $C_{25}H_{20}ClF_2N_3O_3$; mol wt 483.90. C 62.05%, H 4.17%, Cl 7.33%, F 7.85%, N 8.68%, O 9.92%. Insect growth regulator which interferes with chitin biosynthesis. Prepn: M. S. Brouwer, A. C. Grosscurt, **Eur.** pat. **Appl. 117,320;** *eidem,* U.S. pat. **4,550,-202** (1984, 1985 both to Duphar). Comprehensive description: P. Scheltes *et al., Proc. Brighton Crop Prot. Conf. — Pests Dis.* **1988**, 559. Mechanism of action and insecticidal activity: A. C. Grosscurt *et al., Pest. Sci.* **22**, 51 (1988). Determn in fruit and soil samples: J. van Zijtveld *et al., J. Chromatog.* **600**, 211 (1992).

E-isomer

Odorless crystalline powder. Product consists of 50-80% *E*-isomer and 50-20% Z-isomer. Data given for ~70%:30%

E:Z mixture. mp 143.6°. vapor pressure (20°): <4.4 mPa. Soly (20°): water <1μg/l; xylene 0.2 g/l; cyclohexane 3.3 g/l; ethanol 3.9 g/l. LC_{50} (4 hr) in rats: 3.3 mg/l; LC_{50} (96 hr) in rainbow trout, bluegill sunfish: >100 mg/l; LD_{50} in rats (mg/kg): >5000 orally; >2000 dermally (Scheltes).
USE: Insecticide.

4160. Flucythrinate. *4-(Difluoromethoxy)-α-(1-methylethyl)benzeneacetic acid cyano(3-phenoxyphenyl)methyl ester;* (±)-cyano-(3-phenoxyphenyl)methyl (+)-4-(difluoromethoxy)-α-(1-methylethyl)benzeneacetate; AC-222705; Cybolt; Guardian; Pay-Off; Stock Guard. $C_{26}H_{23}F_2NO_4$; mol wt 451.47. C 69.17%, H 5.13%, F 8.42%, N 3.10%, O 14.18%. Synthetic pyrethroid insecticide with contact and stomach poison activity. Prepn: G. Berkelhammer, V. Kameswaran, **Ger.** pat. **2,757,066;** *eidem,* U.S. pat. **4,199,595** (1977, 1980 both to Am. Cyanamid). Activity: A. F. Saad *et al., Proc. Brit. Crop Prot. Conf.-Pests Dis.* **1981**, 381; K. Wettstein, *ibid.* 563. Efficacy in cattle fly control: S. M. Taylor *et al., Vet. Rec.* **116**, 566 (1985).

Viscous liquid. Sol in acetone, xylene, 2-propanol.
USE: Insecticide.
THERAP CAT (VET): Ectoparasiticide.

4161. Flucytosine. *4-Amino-5-fluoro-2(1H)-pyrimidinone;* 5-fluorocytosine; 5-FC; Ro-2-9915; Alcobon; Ancobon; Ancotil. $C_4H_4FN_3O$; mol wt 129.09. C 37.22%, H 3.12%, F 14.72%, N 32.55%, O 12.39%. Prepn: Duschinsky *et al., J. Am. Chem. Soc.* **79**, 4559 (1957); Heidelberger, Duschinsky, U.S. pat. **2,802,005** (1957); Duschinsky, Heidelberger, U.S. pat. **2,945,038** and Duschinsky, U.S. pat. **3,040,026** (1960, 1962 both to Hoffmann-La Roche); Undheim, Gacek, *Acta Chem. Scand.* **23**, 294 (1969). Patents as a fungicide: Berger, Duschinsky, **Belg.** pat. **628,615** and U.S. pat. **3,368,938** (1963, 1968 both to Hoffmann-La Roche). Activity studies: Grunberg *et al., Antimicrob. Ag. Chemother.* **1963**, 566; Shadomy *et al., ibid.* **1968**, 452; Grunberg *et al., in Vth Int. Congr. Chemother., Proc.,* **vol. IV,** K. Spitzy, Ed. (Verlag Wiener Med. Akad., 1967, Austria) p 69. Metabolic studies: Koechlin *et al., Biochem. Pharmacol.* **15**, 435 (1966). Clinical results: Utz *et al., Antimicrob. Ag. Chemother.* **1968**, 344; Warner *et al., ibid.* **1970**, 473. Comprehensive description: E. H. Waysek, J. H. Johnson in *Analytical Profiles of Drug Substances* **vol. 5,** K. Florey, Ed. (Academic Press, New York, 1976) pp 115-138.

Odorless, white crystalline solid, mp 295-297° (dec). uv max (0.1*N* HCl): 285 nm (ε 8900). Soly in water: 1.5 g/100 ml at 25°C. pKa$_1$: 3.26. LD_{50} in mice (mg/kg): >2000 orally and s.c.; 1190 i.p.; 500 i.v. (Grunberg, 1963).
THERAP CAT: Antifungal.

4162. Fludarabine. *9-β-D-Arabinofuranosyl-2-fluoro-9H-purin-6-amine;* 9-β-D-arabinofuranosyl-2-fluoroadenine; 2-fluorovidarabine; 2-fluoro-9-β-D-arabinofuranosyladenine; 2-F-araA; NSC-118218; NSC-118218-H. $C_{10}H_{12}FN_5O_4$; mol wt 285.23. C 42.11%, H 4.24%, F 6.66%, N 24.55%, O 22.44%. Adenosine deaminase-resistant purine nucleoside antimetabolite. Prepn and *in vitro* cytotoxicity: J. A. Montgomery, K. Hewson, *J. Med. Chem.* **12**, 498 (1969). Improved prepn: J. A. Montgomery *et al., J. Heterocycl. Chem.* **16**, 157 (1979); J. A. Montgomery, U.S. pat. **4,210,-745** (1980 to U.S. Dept. Health, Education and Welfare).

Inhibition of DNA synthesis and *in vivo* antileukemic activity: R. W. Brockman *et al., Biochem. Pharmacol.* **26**, 2193 (1977). Metabolized to 5'-monophosphate: R. W. Brockman *et al., Cancer Res.* **40**, 3610 (1980). HPLC determn in human leukemia cells: V. Gandhi *et al., J. Chromatog.* **413**, 293 (1987). Prepn of 5'-monophosphate: J. A. Montgomery, A. T. Shortnacy, U.S. pat. **4,357,324** (1982 to U.S. Dept. of Health and Human Services). Pharmacokinetics in humans: M. R. Hersh *et al., Cancer Chemother. Pharmacol.* **17**, 277 (1986). Evaluation of therapeutic efficacy and CNS toxicity in acute refractory leukemia: R. P. Warrell, Jr., E. Berman, *J. Clin. Oncol.* **4**, 74 (1986); H. G. Chun *et al., Cancer Treat. Rep.* **70**, 1225 (1986). Series of articles on pharmacology and therapeutic use: *Sem. Oncol.* **17**, Suppl. 8, 1-78 (1990).

Crystals from ethanol + water, mp 260°. $[\alpha]_D^{25}$ +17 ±2.5° (c = 0.1 in ethanol). uv max (pH 1, pH 7, pH 13): 262, 261, 262 nm ($\epsilon \times 10^{-3}$ 13.2, 14.8, 15.0). Sparingly sol in water, organic solvents.

5'-Monophosphate, $C_{10}H_{13}FN_5O_7P$, *2-F-ara-AMP, NSC-328002, NSC-312887, Fludara.* Sol in water.

THERAP CAT: Phosphate as antineoplastic.

4163. Fludeoxyglucose F¹⁸. *2-Deoxy-2-(fluoro-¹⁸F)-α-D-glucopyranose;* 2-[¹⁸F]fluoro-2-deoxy-D-glucose; ¹⁸FDG. $C_6H_{11}^{18}FO_5$. Radioactive glucose analog used clinically to evaluate glucose metabolism by PET. Prepn: T. Ido *et al., J. Labelled Comp. Radiopharm.* **14**, 175 (1978). Review of syntheses: T. J. Tewson, *Nucl. Med. Biol.* **16**, 533-551 (1989). Biodistribution: B. M. Gallagher *et al., J. Nucl. Med.* **19**, 1154 (1978). Cerebral glucose measurement in man: M. Reivich *et al., Circ. Res.* **44**, 127 (1979). Clinical evaluation of metabolic imaging in myocardium: R. J. Gropler *et al., J. Nucl. Med.* **31**, 1749 (1990); in tumors: J. Okada *et al., ibid.* **33**, 325 (1992); N. Y. Tse *et al., Ann. Surg.* **216**, 27 (1992). Reviews of imaging in cerebrovascular disease: K. Herholz, W. D. Heiss, *Sem. Neurol.* **9**, 293-300 (1989); Alzheimer's disease: J. M. Hoffman *et al., Eur. Neurol.* **29**, Suppl. 3, 16-24 (1989); brain tumors: R. E. Coleman *et al., J. Nucl. Med.* **32**, 616-622 (1991).

THERAP CAT: Diagnostic aid (radioactive imaging agent).

4164. Fludiazepam. *7-Chloro-5-(2-fluorophenyl)-1,3-dihydro-1-methyl-2H-1,4-benzodiazepin-2-one;* ID-540; Ro-5-3438; Erispan. $C_{16}H_{12}ClFN_2O$; mol wt 302.74. C 63.48%, H 4.00%, Cl 11.71%, F 6.28%, N 9.25%, O 5.28%. Fluorinated analog of diazepam, *q.v.* Prepn: E. Reeder *et al.,* Ger. pat. **1,136,709** corresp to U.S. pat. **3,051,701** (both 1962 to Hoffmann-LaRoche); L. H. Sternbach *et al., J. Org. Chem.* **27**, 3788 (1962). Synthesis and pharmacology: Y. Asami *et al., Arzneimittel-Forsch.* **24**, 1563 (1974). Pharmacodynamics: M. Nakamura, H. Fukushima, *J. Pharm. Pharmacol.* **30**, 56, 254 (1978); T. Tsuchiya, H. Fukushima, *Eur. J. Pharmacol.* **48**, 421 (1978). Anxiolytic vs. sedative properties: M. Babbini *et al., Life Sci.* **25**, 15 (1979).

Colorless prisms from *n*-hexane/isopropanol, mp 88-92°, (Asami); also reported as mp 69-72° (Sternbach). LD_{50} in mice (mg/kg): 910 orally; 360 i.p.; 1150 s.c. (Asami).

Note: This is a controlled substance (depressant) listed in the U.S. Code of Federal Regulations, Title 21 Part 1308.14 (1995).

THERAP CAT: Anxiolytic.

4165. Fludioxonil. *4-(2,2-Difluoro-1,3-benzodioxol-4-yl)-1H-pyrrole-3-carbonitrile;* 3-(2,2-difluorobenzodioxol-4-yl)-4-cyanopyrrole; CGA-173506; Celest; Maxim; Saphire. $C_{12}H_6F_2N_2O_2$; mol wt 248.19. C 58.07%, H 2.44%, F 15.31%, N 11.29%, O 12.89%. Nonsystemic phenylpyrrole fungicide structurally related to pyrrolnitrin, *q.v.* Prepn: R. Nyfeler, J. Ehrenfreund, Eur. pat. **Appl. 206,999;** *eidem,* U.S. pat. **4,705,800** (1986, 1987 both to Ciba-Geigy). Comprehensive description: K. Gehmann *et al., Proc. Brighton Crop Prot. Conf. — Pests Dis.* **1990**, 399-406. Review of field trials as seed treatment: E. Koch, A. J. Leadbeater, *ibid.* **1992**, 1137-1146.

Colorless, odorless crystals, mp 199.4°. Soly in water (20°): 1.53 ppm. log P (*n*-octanol/water): 2.6. LD_{50} in rats (mg/kg): >2000 orally; >2000 dermally; LC_{50} (4 hr) in rats: >2600 mg/m³ (Gehmann).

USE: Agricultural fungicide.

4166. Fludrocortisone. *(11β)-9-Fluoro-11,17,21-trihydroxypregn-4-ene-3,20-dione;* 9α-fluorohydrocortisone; 9α-fluoro-17-hydroxycorticosterone; 9α-fluorocortisol; fluodrocortisone; fluohydrisone; fluohydrocortisone; Astonin H. $C_{21}H_{29}FO_5$; mol wt 380.46. C 66.30%, H 7.68%, F 4.99%, O 21.03%. Prepn: J. Fried, E. F. Sabo, *J. Am. Chem. Soc.* **76**, 1455 (1954); **Brit.** pat. **792,224;** J. Fried, U.S. pat. **2,852,511** (both 1958 to Olin Mathieson). Series of articles on pharmacology and metabolism: *Arzneimittel-Forsch.* **21**, 1103-1158 (1971). Comprehensive description of the acetate: K. Florey, *Anal. Profiles Drug Subs.* **3** 281-306 (1974).

Crystals, dec 260-262°. $[\alpha]_D^{23}$ +139° (c = 0.55 in 95% ethanol). uv max (ethanol): 239 nm (ϵ 17600). Soly in water: 0.14 mg/ml.

21-Acetate, $C_{23}H_{31}FO_6$, *Alflorone, F-Cortef, Florinef.* Crystals, polymorphic, mp 233-234° (occasionally mp 205-208°, resolidifying on further heating, then mp 226-228°). Crystallizing procedure: R. P. Graber, C. S. Snoddy, U.S. pat. **2,957,013** (1960 to Merck & Co.). $[\alpha]_D^{23}$ +123° (c = 0.64 in chloroform). uv max (ethanol): 238 nm (ϵ 16800).

Soly (mg/ml): water 0.04; acetone 56; alc 20; chloroform 20; ether 4.
THERAP CAT: Mineralocorticoid.
THERAP CAT (VET): Mineralocorticoid.

4167. Flufenamic Acid. *2-[[3-(Trifluoromethyl)phenyl]amino]benzoic acid; N-(α,α,α-trifluoro-m-tolyl)anthranilic acid;* 3'-trifluoromethyldiphenylamine-2-carboxylic acid; CI-440; INF-1837; Achless; Ansatin; Arlef; Fullsafe; Meralen; Paraflu; Parlef; Ristogen; Sastridex; Surika; Tecramine. $C_{14}H_{10}F_3NO_2$; mol wt 281.23. C 59.79%, H 3.58%, F 20.27%, N 4.98%, O 11.38%. Prepn from *m*-aminobenzotrifluoride and *o*-iodobenzoic acid: Wilkinson, Finar, *J. Chem. Soc.* **1948**, 32. Pharmacology: Winder *et al., Arthritis Rheum.* **6**, 36 (1963); **12**, 472 (1969). Antiviral activity: Inglot, *J. Gen. Virol.* **4**, 203 (1969). Toxicity data: Zoni *et al., Farmaco Ed. Sci.* **26**, 191 (1971). Comprehensive description: E. Abignente, P. deCaprariis, in *Analytical Profiles of Drug Substances* vol. **11**, K. Florey, Ed. (Academic Press, New York, 1982) pp 313-346.

Pale yellow needles from 50% alc, mp 125°. LD_{50} in mice: 715 mg/kg orally (Zoni).
Aluminum salt, $C_{42}H_{27}AlF_9N_3O_6$, *aluminum flufenamate, Alfenamin, Opyrin.*
Butyl ester, $C_{18}H_{18}F_3NO_2$, *Fenazol, Combec.*
THERAP CAT: Anti-inflammatory; analgesic.

4168. Fluindione. *2-(4-Fluorophenyl)-1H-indene-1,3-(2H)-dione; 2-(p-fluorophenyl)-1,3-indandione;* LM-123; Previscan. $C_{15}H_9FO_2$; mol wt 240.23. C 75.00%, H 3.78%, F 7.91%, O 13.32%. Prepn: Geiger *et al.,* Ger. pat. **1,130,439** (1962 to USV); *C.A.* **57**, 12403a (1962); Molho, Boschetti, Fr. pats. **1,369,396** and **M 6913** (1964, 1969 to LIPHA), *C.A.* **62**, 3988g (1965); **74**, 141378u (1971). Alternate syntheses: Shapiro *et al., J. Org. Chem.* **26**, 3580 (1961); Hrnciar, Kovalcik, *Chem. Zvesti* **16**, 200 (1962). Activity and toxicity data: Fontaine *et al., Med. Pharmacol. Exp.* **17**, 497 (1967).

Crystals from acetic acid, mp 120°. LD_{50} orally in mice: 240 mg/kg (Fontaine).
THERAP CAT: Anticoagulant.

4169. Flumazenil. *8-Fluoro-5,6-dihydro-5-methyl-6-oxo-4H-imidazo[1,5-a][1,4]benzodiazepine-3-carboxylic acid ethyl ester;* ethyl-8-fluoro-5,6-dihydro-5-methyl-6-oxo-4H-imidazo[1,5-a][1,4]benzodiazepine-3-carboxylate; flumazepil; Ro-15-1788; Anexate; Lanexat; Mazicon; Romazicon. $C_{15}H_{14}FN_3O_3$; mol wt 303.29. C 59.40%, H 4.65%, F 6.26%, N 13.85%, O 15.83%. Imidazodiazepine which selectively blocks the central effects of classic benzodiazepines. Prepn: W. Haefely *et al.,* Eur. pat. Appl. **27,214**; M. Gerecke *et al.,* U.S. pat. **4,316,839** (1981, 1982 both to Hoffmann-La Roche). Specific inhibition of benzodiazepine-receptor binding: W. Hunkeler *et al., Nature* **290**, 514 (1981). Electrophysiological study in animals: P. Pole *et al., Arch. Pharmacol.* **316**, 317 (1981). Antagonist, agonist and inverse agonist properties: S. E. File *et al., Psychopharmacol.* **89**, 113 (1986). HPLC determn in human plasma: U. Timm, M. Zell, *Arzneimittel-Forsch.* **33**, 358 (1983). Clinical reversal of benzodiazepine-induced sedation: A. Darragh *et al., Lancet* **2**, 8 (1981); *eidem, ibid.* 1042; B. Ricou *et al., Brit. J. Anaesth.* **58**, 1005 (1986). Clinical evaluation in drug overdose: G. F. O'Sullivan *et al., Clin. Pharmacol. Ther.* **42**, 254 (1987).

Crystals from alcohol, mp 201-203°. LD_{50} in mice, rats (mg/kg): 4000, 1360 i.p.; 4300, 6000 orally (Hunkeler).
THERAP CAT: Benzodiazepine antagonist.

4170. Flumecinol. *α-Ethyl-α-phenyl-3-(trifluoromethyl)benzenemethanol;* α-ethyl-3-(trifluoromethyl)benzhydrol; RGH-3332; Zixoryn; Zyxorin. $C_{16}H_{15}F_3O$; mol wt 280.29. C 68.56%, H 5.39%, F 20.33%, O 5.71%. Hepatic microsomal enzyme inducer. Prepn: E. Toth *et al.,* Ger. pat. **2,438,399**; *eidem,* U.S. pat. **4,039,589** (1975, 1977 both to Gedeon-Richter). GC determn in biological fluids: I. Klebovich, L. Vereczkey, *J. Chromatog.* **221**, 403 (1980). Toxicity study: M. Ledniczky *et al., Arzneimittel-Forsch.* **28**, 669 (1978). Series of articles on metabolism, CNS effects, enzyme induction, pharmacological properties: *ibid.* 663-679.

Oil, $bp_{0.03}$ 106-108°. uv max (ethanol): 259, 265, 271 nm. LD_{50} orally in adult rats: 2235 mg/kg (Ledniczky).
THERAP CAT: Enzyme inducer (hepatic).

4171. Flumedroxone Acetate. *(6α)-17-(Acetyloxy)-6-(trifluoromethyl)pregn-4-ene-3,20-dione; 17-hydroxy-6α-(trifluoromethyl)pregn-4-ene-3,20-dione acetate;* 17-acetoxy-6α-(trifluoromethyl)progesterone; 17-hydroxy-6α-(trifluoromethyl)progesterone acetate; 6α-(trifluoromethyl)-17-hydroxyprogesterone acetate; WG-537; Demigran. $C_{24}H_{31}F_3O_4$; mol wt 440.50. C 65.44%, H 7.09%, F 12.94%, O 14.53%. Prepn: Godtfredsen, Vangedal, *Acta Chem. Scand.* **15**, 1786 (1961); Godtfredsen, **Brit.** pat. **905,694** (1962 to Lövens Komiske Fabrik). Solubilization study: T. Lovgren *et al., Acta Pharm. Suec.* **15**, 233 (1978).

Solid, mp 206-207°. $[\alpha]_D^{20}$ +30°. uv max (ethanol): 234 nm (ε 15600).
THERAP CAT: Antimigraine.

4172. Flumequine. *9-Fluoro-6,7-dihydro-5-methyl-1-oxo-1H,5H-benzo[ij]quinolizine-2-carboxylic acid;* R-802; Apurone; Fantacin. $C_{14}H_{12}FNO_3$; mol wt 261.25. C 64.36%, H 4.63%, F 7.27%, N 5.36%, O 18.37%. Fluorinated quinolone antibacterial. Prepn: J. F. Gerster, **Ger.** pat. **2,264,163**; *idem,* U.S. pat. **3,896,131** (1973, 1975 both to Riker). *In vitro* study: G. Stilwell *et al., Antimicrob. Ag. Chemother.* **7**, 483 (1975). Bioevaluation: S. R. Rohlfing *et al., J. Antimicrob. Chemother.* **3**, 615 (1977). Activity vs *E. coli:* D. Greenwood, *Antimicrob. Ag. Chemother.* **13**, 479 (1978).

White microcrystalline powder, mp 253-255°. Sol in alkaline solns and alcohol. Insol in water.

THERAP CAT: Antibacterial.

THERAP CAT (VET): Antibacterial.

4173. Flumethasone. *(6α,11β,16α)-6,9-Difluoro-11,17,-21-trihydroxy-16-methylpregna-1,4-diene-3,20-dione;* 6α,9α-difluoro-16α-methylprednisolone; 6α-fluorodexamethasone; Aniprime; Cortexilar; Flucort; Methagon. $C_{22}H_{28}F_2O_5$; mol wt 410.46. C 64.38%, H 6.88%, F 9.26%, O 19.49%. Prepn: **Brit. pat. 902,292;** F. H. Lincoln *et al.,* **U.S. pat. 3,499,016** (1962, 1970 both to Upjohn). Prepn of the 21-acetate: J. A. Edwards *et al., J. Am. Chem. Soc.* **81,** 3156 (1959); **82,** 2318 (1960).

21-Acetate, $C_{24}H_{30}F_2O_6$, crystals from acetone + hexane, mp 260-264°. $[\alpha]_D$ +91°. uv max (ethanol): 237 nm (log ϵ 4.16).

21-Pivalate, $C_{27}H_{36}F_2O_6$, *Locacorten (formerly), Locorten, Lorinden, Losalen.*

THERAP CAT: Glucocorticoid; anti-inflammatory.

THERAP CAT (VET): Glucocorticoid.

4174. Flumethiazide. *6-(Trifluoromethyl)-2H-1,2,4-benzothiadiazine-7-sulfonamide 1,1-dioxide;* 6-trifluoromethyl-7-sulfamoyl-4H-1,2,4-benzothiadiazine 1,1-dioxide; 6-trifluoromethyl-7-sulfamyl-1,2,4-benzothiadiazine 1,1-dioxide; trifluoromethylthiazide; Ademol; Fludemil. $C_8H_6F_3N_3O_4S_2$; mol wt 329.28. C 29.18%, H 1.84%, F 17.31%, N 12.76%, O 19.44%, S 19.48%. Synthesis: Holdrege *et al., J. Am. Chem. Soc.* **81,** 4807 (1959); Yale *et al., ibid.* **82,** 2042 (1960); **U.S. pat. 3,040,042** (1962 to Olin Mathieson); **Brit. pat. 861,809** (1961 to SK&F).

Crystals, dec 305.4-307.8°. uv max: 278 nm ($E_{1cm}^{1\%}$ 335) (50% diglyme + 50% 0.1N HCl). Sparingly sol in water (50 mg/ml in boiling water with decompn); sol in methanol, ethanol, DMF. Practically insol in ethyl acetate, methyl ethyl ketone, benzene, toluene. Unstable in alkaline soln (conversion to precursor α,α,α-trifluoro-3-amino-4,6-disulfamoyltoluene).

THERAP CAT: Carbonic anhydrase inhibitor.

4175. Flumethrin. *3-[2-Chloro-2-(4-chlorophenyl)ethenyl]-2,2-dimethylcyclopropanecarboxylic acid cyano(4-fluoro-3-phenoxyphenyl)methyl ester;* 3'-phenoxy-4'-fluoro-α-cyanobenzyl 2,2-dimethyl-3-[2-(4-chlorophenyl)-2-chlorovinyl]cyclopropane carboxylate; Bayticol; Bayvarol. $C_{28}H_{22}Cl_2FNO_3$; mol wt 510.39. C 65.89%, H 4.34%, Cl 13.89%, F 3.72%, N 2.74%, O 9.40%. Synthetic pyrethroid insecticide. Prepn: R. Fuchs *et al.,* **Ger. pat. 2,730,515** corresp to **U.S.**

pat. **4,276,306** (1979, 1981 to Bayer). Series of articles on laboratory evaluation, efficacy and toxicology: *Vet. Med. Rev.* **2,** 115-139, 158-177 (1982).

Oil, n_D^{25} 1.5831.

USE: Insecticide; acaricide.

THERAP CAT (VET): Ectoparasiticide.

4176. Flumetramide. *6-[4-(Trifluoromethyl)phenyl]-3-morpholinone;* 6-(α,α,α-trifluoro-*p*-tolyl)-3-morpholinone; McN-1546; Duraflex. $C_{11}H_{10}F_3NO_2$; mol wt 245.20. C 53.88%, H 4.11%, F 23.24%, N 5.71%, O 13.05%. Prepn: Gannon, Poos, **U.S. pat. 3,308,121** (1967 to McNeil).

mp 115.5-116.5°.

THERAP CAT: Muscle relaxant (skeletal).

4177. Flumetsulam. *N-(2,6-Difluorophenyl)-5-methyl-[1,2,4]triazolo-[1,5-a]pyrimidine-2-sulfonamide;* DE-498; Broadstrike. $C_{12}H_9F_2N_5O_2S$; mol wt 325.30. C 44.31%, H 2.79%, F 11.68%, N 21.53%, O 9.84%, S 9.86%. Selective inhibitor of acetolactate synthase, a key enzyme in branched chain amino acid biosynthesis. Prepn: W. A. Kleschick *et al.,* **Eur. pat. Appl. 142,152** (1985 to Dow). Improved process: L. H. McKendry, **U.S. pat. 4,910,306** (1990 to Dow). Synthesis, toxicity and herbicidal activity: W. A. Kleschick *et al., J. Agric. Food Chem.* **40,** 1083 (1992). Degradation in soil: R. G. Lehmann *et al., Weed Res.* **33,** 187 (1993). Metabolism in wheat, corn and barley: D. S. Frear *et al., Pestic. Biochem. Physiol* **45,** 178 (1993).

White powder, mp 250-251°. pKa 4.6. Partition coefficient (octanol/water): 0.21. Vapor pressure (25°): 3.7 $\times$ 10^{-13} Pa. Soly in water (g/l): 0.049 (pH 2.5); 5.65 (pH 7). LD_{50} in rats (mg/kg): >5000 orally; in rabbits (mg/kg): >2000 dermally (Kleschick, 1992).

USE: Herbicide.

4178. Flumiclorac. *[2-Chloro-4-fluoro-5-(1,3,4,5,6,7-hexahydro-1,3-dioxo-2H-isoindol-2-yl)phenoxy]acetic acid;* 2-chloro-4-fluoro-5-[(3,4,5,6-tetrahydro)phenoxyacetic acid; [2-chloro-5-(cyclohex-1-ene-1,2-dicarboximido)-4-fluorophenoxy]acetic acid. $C_{16}H_{13}ClFNO_5$; mol wt 353.73. C 54.33%, H 3.70%, Cl 10.02%, F 5.37%, N 3.96%, O 22.62%. Postemergent herbicide for use in food crops. Prepn: E. Nagano *et al.,* **Eur. pat. Appl. 83,055;** *eidem,* **U.S. pat. 4,770,695** (1983, 1988 both to Sumitomo). Comprehensive description: K. Kamoshita *et al.,* in *Pest Management in Soybean,* L. G. Copping *et al.,* Eds. (Elsevier, London, 1992) pp 317-324.

Pentyl ester, $C_{21}H_{23}ClFNO_5$, V-23031, S-23031, *Resource.* mp 88.87-90.13°. Vapor pressure (22.4°): $< 1.0 \times 10^{-7}$ mm Hg. d^{20} 1.3316. Soly (25°): methanol 47.8 g/l; hexane 3.28 g/l; acetone 590 g/l; acetonitrile 589 g/l; water 0.189 mg/l. LD_{50} in rats (mg/kg): > 5000 orally; > 2000 dermally (Kamoshita).

USE: Herbicide.

4179. Flunarizine. *(E)-1-[Bis(4-fluorophenyl)methyl]-4-(3-phenyl-2-propenyl)piperazine; (E)-1-[bis(p-fluorophenyl)-methyl]-4-cinnamylpiperazine;* 1-cinnamyl-4-(di-p-fluoro-benzhydryl)piperazine. $C_{26}H_{26}F_2N_2$; mol wt 404.50. C 77.20%, H 6.48%, F 9.39%, N 6.93%. Calcium channel blocker; fluorinated deriv of cinnarizine, *q.v.* Prepn: P. A. J. Janssen, **Ger.** pat. **1,929,330** and **Fr.** pat. **2,014,487;** *idem,* **U.S.** pat. **3,773,939** (1970, 1970, 1973, all to Janssen). Pharmacology: L. K. Desmedt *et al., Arzneimittel-Forsch.* **25,** 1408 (1975); T. Godfraind, *Eur. J. Pharmacol.* **53,** 273 (1979). Clinical studies: J. Schetz *et al., Curr. Ther. Res. Clin. Exp.* **23,** 131 (1978); G. Rudofsky *et al., Angiology* **30,** 479 (1979). Rheological effects in humans: J. D. Cree *et al., ibid.* 505. Mechanism of calcium blocking activity: T. Godfraind, *Fed. Proc.* **40,** 2866 (1981). Clinical study in retinal vasculopathy: P. Nihard, *Angiology* **33,** 37 (1982). Clinical trial in chronic cerebrovascular disorders: A. Agnoli *et al., Int. J. Clin. Pharm. Res.* **8,** 189 (1988). Series of articles on absorption, distribution, excretion and metabolism: *Arzneimittel-Forsch.* **33,** 1135-1151 (1983). Review of pharmacology, toxicology and clinical studies: B. Holmes *et al., Drugs* **27,** 6 (1984).

Dihydrochloride, $C_{26}H_{26}F_2N_2 \cdot 2HCl$, *Dinaplex, Flugeral, Flunagen, Flunarl, Fluxarten, Gradient, Issium, Mondus, Sibelium.* Crystals from a mixture of 2-propanol/ethanol, mp 251.5°.

THERAP CAT: Vasodilator (cerebral and peripheral).

4180. Flunisolide. *(6α,11β,16α)-6-Fluoro-11,21-dihydroxy-16,17-[(1-methylethylidene)bis(oxy)]pregna-1,4-diene-3,-20-dione; 6α-fluoro-11β,16α,17,21-tetrahydroxypregna-1,4-diene-3,20-dione cyclic 16,17-acetal with acetone;* 6α-fluoro-11β,21-dihydroxy-16α,17-isopropylidenedioxy-Δ^{1,4}-pregnadiene-3,20-dione; RS-3999; Aerobid; Bronalide; Lunis; Nasalide; Rhinalar; Synaclyn; Syntaris. $C_{24}H_{31}FO_6$; mol wt 434.50. C 66.34%, H 7.19%, F 4.37%, O 22.09%. Synthetic fluorinated corticosteroid related to prednisolone, *q.v.* Prepn using *S. roseochromogenes:* **Brit.** pat. **933,867** (1963 to Am. Cyanamid), *C.A.* **60,** 3070f (1964). Prepn using *Cunninghamella blakesleeana* 8688b: H. J. Ringold, G. Rosenkranz, **U.S.** pat. **3,124,571;** H. J. Ringold *et al.,* **U.S.** pat. **3,126,375** (both 1964 to Syntex). Metabolism: N. I. Chu *et al., Drug Metab. Dispos.* **7,** 81 (1979). Use in rhinitis: J. N. Sahay *et al., Clin. Allergy* **9,** 17 (1979); J. K. Sarsfield, G. E. Thomson, *Brit. Med. J.* **2,** 95 (1979). Use in bronchial asthma: D. R. Webb *et al., Ann. Allergy* **42,** 80 (1979). Clinical study in chronic asthma: R. G. Slavin *et al., J. Allergy Clin. Immunol.* **66,** 379 (1980). Review of pharma-

cology and therapeutic efficacy: G. E. Pakes *et al., Drugs* **19,** 397-411 (1980).

21-Acetate, $C_{26}H_{33}FO_7$, *RS-1320.*

THERAP CAT: Glucocorticoid; antiasthmatic.

4181. Flunitrazepam. *5-(2-Fluorophenyl)-1,3-dihydro-1-methyl-7-nitro-2H-1,4-benzodiazepin-2-one;* 1-methyl-7-nitro-5-(2-fluorophenyl)-3H-1,4-benzodiazepin-2(1H)-one; Ro-5-4200; Narcozep; Rohypnol; Roipnol. $C_{16}H_{12}FN_3O_3$; mol wt 313.29. C 61.34%, H 3.86%, F 6.06%, N 13.41%, O 15.32%. Prepn: L. H. Sternbach *et al., J. Med. Chem.* **6,** 261 (1963); J. Kariss, H. L. Newmark, U.S. pats. **3,116,203** and **3,123,529;** O. Keller *et al.,* **U.S.** pat. **3,203,990** (1963, 1964, 1965 all to Hoffmann-La Roche). Industrial prepn and purifn: J. M. Autin, **Fr.** pat. **2,529,203** (1983 to Fabre), *C.A.* **100,** 191913r (1984). Pharmacology and clinical studies: E. Eidelberg *et al., Neurology* **5,** 223 (1965); Schuler *et al., Psychopharmacologia* **27,** 123 (1972); J. M. Monti, H. Altier, *ibid.* **32,** 343 (1973); Schallek *et al., Arch. Int. Pharmacodyn. Ther.* **206,** 161 (1973); Kaplan *et al., J. Pharm. Sci.* **63,** 527 (1974). Review of pharmacology and therapeutic use: M. A. K. Mattila, H. M. Larni, *Drugs* **20,** 353-374 (1980).

Pale yellow needles from methylene chloride-hexane, mp 166-167°. Also reported as crystals from acetonitrile and methanol, mp 170-172°.

Note: This is a controlled substance (depressant) listed in the U.S. Code of Federal Regulations, Title 21 Part 1308.14 (1995).

THERAP CAT: Hypnotic.

4182. Flunixin. *2-[[2-Methyl-3-(trifluoromethyl)phenyl]amino]-3-pyridinecarboxylic acid;* 2-(2-methyl-3-trifluoromethylanilino)nicotinic acid; 2-(α³,α³,α³-trifluoro-2,3-xylidino)nicotinic acid; Sch-14714. $C_{14}H_{11}F_3N_2O_2$; mol wt 296.25. C 56.76%, H 3.74%, F 19.24%, N 9.46%, O 10.80%. Cyclo-oxygenase inhibitor. Prepn: M. H. Sherlock, N. Sperber, **Belg.** pat. **679,271;** *eidem,* **U.S.** pat. **3,337,570** (1966, 1967 both to Schering). Prepn of the acid and meglumine salt: M. H. Sherlock, **Belg.** pat. **812,772;** *idem,* **U.S.** pat. **3,839,344** (both 1974 to Schering). Pharmacodynamics in horses: M. M. Hardee, J. N. Moore, *Res. Vet. Sci.* **40,** 152 (1986); and pharmacokinetics: L. R. Soma *et al., J. Vet. Pharmacol. Ther.* **15,** 292 (1992). GC determn in urine: M. Johansson, E. -L. Anlér, *J. Chromatog.* **427,** 55 (1988). Clinical evaluation in horses: J. W. Houdeshell, P. W. Hennessey, *J. Equine Med. Surg.* **1,** 57 (1977). Experimental use in food-producing animals: M. Kopcha, A. S. Ahl, *J. Am. Vet. Med. Assoc.* **194,** 45 (1989). Comparative toxicological study in horses: C. G. MacAllister *et al., ibid.* **202,** 71 (1993).

Crystals from acetone/hexane, mp 226-228°. pKa' 5.82. Meglumine salt, $C_{14}H_{11}F_3N_2O_2 \cdot C_7H_{17}NO_5$, **Banamine, Finadyne**. Colorless cryst from ethanol/ether, mp 135-137°. Crystals from acetonitrile, mp 136-139°. Sol in water.

THERAP CAT (VET): Anti-inflammatory; analgesic; antipyretic.

4183. Flunoxaprofen. *(+)-2-(4-Fluorophenyl)-α-methyl-5-benzoxazoleacetic acid;* RV-12424; Priaxim. $C_{16}H_{12}$-FNO_3; mol wt 285.27. C 67.37%, H 4.24%, F 6.66%, N 4.91%, O 16.83%. Nonsteroidal anti-inflammatory. Prepn of racemate: D. Evans *et al.*, **Ger. pat. 2,324,443**; *eidem,* **U.S. pat. 3,912,748** (1973, 1975 both to Lilly). Prepn of racemate and anti-inflammatory activity: D. W. Dunwell *et al., J. Med. Chem.* **18**, 53-58 (1975). Improved process for prepn of racemate and enantiomers: F. Mauri, R. Signorini, **Belg. pat. 877,887**; *eidem,* **U.S. pat. 4,304,918** (1979, 1981 both to Ravizza); R. Signorini, A. Verga, **Ger. pat. 3,325,-672** (1984 to Ravizza). HPLC determn of enantiomers in body fluids: S. Pedrazzini *et al., J. Chromatog.* **415**, 214 (1987). Prophylactic and therapeutic treatment of asthma and other immediate hypersensitivity diseases: W. Dawson, **U.S. pat. 4,416,892** (1983 to Lilly).

Crystals from acetone-water or acetic acid, mp 162-164°. $[\alpha]_D^{20}$ +50° (c = 2% in DMF). LD_{50} orally in mice: ~1200 mg/kg (Dunwell).

THERAP CAT: Anti-inflammatory.

4184. Fluoboric Acid. *Hydrogen tetrafluoroborate;* borofluoric acid; hydrofluoboric acid. BF_4H; mol wt 87.81. B 12.31%, F 86.54%, H 1.15%. HBF_4. Prepd from H_3BO_3 + HF: Fichter, Thiele, *Z. Anorg. Allgem. Chem.* **67**, 302 (1910); Mathers *et al., J. Am. Chem. Soc.* **37**, 1516 (1915); Funk, Binder, *Z. Anorg. Allgem. Chem.* **155**, 327; **159**, 121 (1926); Sheintsis, *J. Appl. Chem. USSR* **13**, 1101 (1940); Wamser, Christian, *J. Am. Chem. Soc.* **70**, 1209 (1948); Kwasnik in *Handbook of Preparative Inorganic Chemistry,* **vol. 1**, G. Brauer, Ed. (Academic Press, New York, 2nd ed., 1963) pp 221-222. For other methods of prepn *see* H. S. Booth, D. R. Martin, *Boron Trifluoride and Its Derivatives* (New York, 1949). *Review:* Sharp, *Advan. Fluorine Chem.* **1**, 68-128 (1960).

Liquid. *Poisonous!* bp 130° (decompn). Miscible with water, alcohol. Strong acid. n_D^{20} of a 20% aq soln 1.3284. Heat of formn 388.5 kcal. Undergoes limited hydrolysis in water to form hydroxyfluoborate ions; major product is BF_3OH^-. Pure HBF_4 may be stored in glass vessels at room temps. Forms cryst salts wth metals. The heavy metal salts and $LiBF_4$ and $NaBF_4$ are very sol in water. *See* Potassium Tetrafluoborate about color phenomena appearing in concd solns of fluoboric acid.

Caution: Strong caustic action on skin, mucous membranes. Irritating to eyes, respiratory tract. Plating solns contg fluoborates are considered toxic.

USE: As catalyst for preparing acetals, esterifying cellulose; to clean metal surfaces before welding; to brighten aluminum; as a solute in electrolytes for plating metals such as chromium, iron, nickel, copper, silver, zinc, cadmium, indium, tin, and lead (has a high throwing power). Reagent for sodium in the presence of magnesium and potassium ions; for making stabilized diazo salts (diazonium and tetrazonium fluoborates). An 0.1 to 0.5% soln retards fermentation: Homeyer, *Pharm. Ztg.* **34**, 761 (1889).

4185. Fluocinolone Acetonide. *(6α,11β,16α)-6,9-Difluoro-11,21-dihydroxy-16,17-[(1-methylethylidene)bis(oxy)]-pregna-1,4-diene-3,20-dione; 6α,9α-difluoro-11β,16α,17,21-tetrahydroxypregna-1,4-diene-3,20-dione cyclic 16,17-acetal with acetone;* 6α,9α-difluoro-16α-hydroxyprednisolone 16,-17-acetonide; 6α,9α-difluoro-16α,17α-isopropylidenedioxy-1,4-pregnadiene-3,20-dione; Coriphate; Cortiplastol; Dermalar; Fluonid; Fluovitef; Fluvean; Fluzon; Jellin; Localyn; Synalar; Synamol; Synandone; Synemol; Synotic; Synsac. $C_{24}H_{30}F_2O_6$; mol wt 452.50. C 63.71%, H 6.68%, F 8.40%, O 21.21%. Prepn: J. S. Mills *et al., J. Am. Chem. Soc.* **82**, 3399 (1960); J. S. Mills, A. Bowers, **U.S. pat. 3,014,938**; H. J. Ringold *et al.*, **U.S. pat. 3,126,375** (1961, 1964 both to Syntex). IR spectrum: Sammul *et al., J. Assoc. Offic. Agr. Chem.* **47**, 952 (1964).

Crystals from acetone + hexane, mp 265-266°. $[\alpha]_D$ +95° (chloroform). uv max: 238 nm (log ε 4.21).

THERAP CAT: Glucocorticoid; anti-inflammatory.

THERAP CAT (VET): Adrenocortical steroid (topical use).

4186. Fluocinonide. *(6α,11β,16α)-21-(Acetyloxy)-6,9-difluoro-11-hydroxy-16,17-[(1-methylethylidene)bis(oxy)]-pregna-1,4-diene-3,20-dione; 6α,9-difluoro-11β,16α,17,21-tetrahydroxypregna-1,4-diene-3,20-dione, cyclic 16,17-acetal with acetone, 21-acetate;* 21-acetoxy-6α,9α-difluoro-11β-hydroxy-16α,17α-isopropylidenedioxy-1,4-pregnadiene-3,-20-dione; 16α,17α-isopropylidene-6α-fluorotriamcinolone 21-acetate; fluocinolide (obsolete); fluocinolide acetate (obsolete); fluocinolone acetonide acetate; Biscosal; Dermaplus; Lidex; Metosyn; Synalate (rescinded); Straderm; Topsym; Topsymin; Topsyne; Topsyn. $C_{26}H_{32}F_2O_7$; mol wt 494.53. C 63.15%, H 6.52%, F 7.68%, O 22.65%. Prepn: **Brit. pat. 916,996** (1963 to Olin Mathieson), *C.A.* **59**, 1716b (1963); H. J. Ringold, G. Rosenkranz, **U.S. pat. 3,124,571**; H. J. Ringold *et al.*, **U.S. pat. 3,126,375** (both 1964 to Syntex); J. Fried, **U.S. pat. 3,197,469** (1965 to Pharm. Res. Prod.). NMR: A. D. Cross, P. W. Landis, *J. Am. Chem. Soc.* **86**, 4005 (1964).

Crystals from methanol. mp 308-311°. $[\alpha]_D$ +83° (chloroform). uv max: 237 nm (log ε 4.18).

THERAP CAT: Anti-inflammatory; glucocorticoid.

4187. Fluocortin Butyl. *(6α,11β,16α)-6-Fluoro-11-hydroxy-16-methyl-3,20-dioxopregna-1,4-dien-21-oic acid butyl ester;* butyl 6α-fluoro-11β-hydroxy-16α-methyl-3,20-dioxopregna-1,4-dien-21-oate; SH K 203; Varlane; Vaspit. $C_{26}H_{35}FO_5$; mol wt 446.56. C 69.93%, H 7.90%, F 4.25%, O 17.91%. The butyl ester deriv of fluocortolone-21-acid, a metabolite of fluocortolone, *q.v.*, in humans. Prepn: H. Laurent *et al.*, **Ger. pats. 2,150,268, 2,150,270**; *eidem*, **U.S. pat. 3,824,260** (1973, 1973, 1974, all to Schering AG); H. Laurent *et al., Angew. Chem.* **87**, 70 (1975). Description of this new corticoid structure type, the pregnan-21-oic esters, their activity and methods for synthesis: H. Laurent *et al.*, *J. Steroid Biochem.* **6**, 185 (1975). Toxicity data: P. Günzel *et al.*, *Arzneimittel-Forsch.* **27**, 2217 (1977). Series of articles on synthesis, pharmacology and clinical trials: *ibid.* 2185-2246.

Crystals from acetone/hexane, mp 195.1°. $[\alpha]_D^{25}$ +136° (c = 0.5 in chloroform). uv max (methanol): 242 nm (ϵ 16800). Soluble in chloroform, ethanol; poorly sol in ethyl ether. Insol in water. LD_{50} in mice, rats (g/kg): >4 orally and s.c. (Günzel).

THERAP CAT: Anti-inflammatory.

4188. Fluocortolone. *(6α,11β,16α)-6-Fluoro-11,21-dihydroxy-16-methylpregna-1,4-diene-3,20-dione;* 6α-fluoro-16α-methyl-1-dehydrocorticosterone; 6α-fluoro-16α-methyl-$\Delta^{1,4}$-pregnadiene-11β,21-diol-3,20-dione; Ultralan oral. $C_{22}H_{29}FO_4$; mol wt 376.47. C 70.19%, H 7.76%, F 5.05%, O 17.00%. Prepn: **Belg. pat. 614,196**; Kieslich *et al.*, **U.S. pat. 3,232,839** (1962, 1966 to Schering AG); Doménico *et al.*, *Arzneimittel-Forsch.* **15**, 46 (1965); Kieslich *et al.*, *Ann.* **726**, 168 (1969). Pharmacology: Doménico *et al.*, *loc. cit;* Gillich *et al.*, *Int. Z. Klin. Pharmakol. Ther. Toxicol.* **1**, 197 (1968); Von Schoening *et al.*, *Klin. Wochenschr.* **48**, 1448 (1970). Metabolism: Gerhards *et al.*, *Acta Endocrinol. (Copenhagen)* **68**, 98 (1971). Clinical studies: Breuer *et al.*, *Arzneimittel-Forsch.* **15**, 50 (1965).

Crystals, mp 188-190.5°. Soly (mg/l): 295 in water (37°); 120 in ethanol (20°); 440 in toluene (20°). $[\alpha]_D^{20}$ +100° (dioxane). uv max (methanol): 242 nm (ϵ 16300).

21-Acetate, $C_{24}H_{31}FO_5$, crystals, mp 237-239°. uv max (methanol): 240 nm (ϵ 15860).

21-Hexanoate, $C_{28}H_{39}FO_5$, *fluocortolone 21-caproate, SH-770, Ficoid, Ultralanum.* Crystals, mp 242-245°. $[\alpha]_D^{20}$ +98.5° (dioxane). uv max (methanol): 242 nm (ϵ 16200). Soly (mg/l): 7.8 in water (37°); 450 in ethanol (20°); 440 in toluene (20°).

21-Pivalate, $C_{22}H_{37}FO_5$, *fluocortolone trimethylacetate.* Crystals, mp 187°. Almost insol in water. Sol in chloroform, methanol. Slightly sol in ether.

THERAP CAT: Glucocorticoid.

4189. Fluometuron. *N,N-Dimethyl-N'-[3-(trifluoromethyl)phenyl]urea;* 1,1-dimethyl-3-(α,α,α-trifluoro-m-tolyl)-urea; N-(3-trifluoromethylphenyl)-N',N'-dimethylurea; Ciba 2059; Cotoran; Cottonex; Lanex. $C_{10}H_{11}F_3N_2O$; mol wt 232.21. C 51.73%, H 4.77%, F 24.55%, N 12.06%, O 6.89%. Made by the reaction of dimethylamine on 3-trifluoromethylphenyl isocyanate: Abel, *Chem. & Ind. (London)* **1957**, 1106. Herbicidal preparations: Martin, Aebi, **U.S. pat. 3,134,665** (1964 to Ciba). Toxicity data: R. Ben-Dyke *et al.*, *World Rev. Pest. Control* **9**, 119 (1970). Review of carcinogenic risk: *IARC Monographs* **30**, 245-253 (1983).

Crystals, mp 163-164.5°. Soly in water at 25°: 80 ppm; sol in acetone, ethanol, isopropanol, DMF and other organic solvents. LD_{50} in rats (mg/kg): 8900 orally; in rabbits (mg/kg): >10,000 dermally (Ben-Dyke).

USE: Herbicide.

4190. 9H-Fluorene. *o*-Biphenylenemethane; diphenylenemethane; 2,2'-methylenebiphenyl. $C_{13}H_{10}$; mol wt 166.22. C 93.94%, H 6.06%. Occurs in coal tar (about 1.6%). Isoln: Kruber, *Ber.* **70**, 1556 (1937). Also found in coke-oven tar: Weiss, Downs, *Ind. Eng. Chem.* **15**, 1022 (1923). From acetylene and hydrogen in red-hot tube: Meyer, *Ber.* **45**, 1609 (1912); Meyer, Taeger, *Ber.* **53**, 1261 (1921). From charcoal by boiling with fuming HNO_3: Dimroth, Kerkovins, *Ann.* **399**, 120 (1913). From 2,2'-dibromodiphenylmethane on boiling with hydrazine hydrate in presence of Pd: Busch, Weber, *J. Prakt. Chem.* [2] **146**, 47 (1936). Reactions: Rieveschl Jr., Ray, *Chem. Rev.* **23**, 287 (1938).

Dazzling white leaflets or flakes from alc. d 1.202. Sublimes easily in high vacuum. mp 116-117°. bp 295°. Absorption spectrum: Mayneord, Roe, *Proc. Roy. Soc. London* **A158**, 634 (1937). Freely sol in glacial acetic acid; sol in carbon disulfide, ether, benzene, hot alcohol. Soly data: Mortimer, *J. Am. Chem. Soc.* **45**, 633 (1923).

4191. 9H-Fluorene-2,7-diamine. 2,7-Diaminofluorene. $C_{13}H_{12}N_2$; mol wt 196.25. C 79.56%, H 6.16%, N 14.27%. Prepd by nitrating fluorene and reducing the 2,7-dinitrofluorene formed with tin and hydrochloric acid: Schmidt, Hinderer, *Ber.* **64**, 1793 (1931).

Needles from water, mp 165°. Slightly sol in cold water, more sol in hot water. Readily sol in alc.

Hydrochloride, $C_{13}H_{12}N_2 \cdot HCl$, crystals, readily sol in hot water.

USE: Detection of bromide, chloride, nitrate, persulfate, cadmium, copper, cobalt, zinc.

4192. N-2-Fluorenylacetamide. *N-9H-Fluoren-2-ylacetamide;* 2-acetylaminofluorene; AAF; 2-FAA. $C_{15}H_{13}NO$; mol wt 223.27. C 80.69%, H 5.87%, N 6.27%, O 7.17%. Synthesis: Hayashi, Nakayama, *J. Soc. Chem. Ind. Japan* [Suppl] **36**, 127B (1933). Carcinogenicity studies: R. H. Wilson *et al.*, *Cancer Res.* **1**, 595 (1941). Toxicity studies: Haley *et al.*, *Proc. Soc. Exp. Biol. Med.* **143**, 1117 (1973); **146**, 648 (1974).

Crystals from alcohol + water, mp 194°. uv max: 285 nm. Insol in water. Sol in alcohols, glycols, fat solvents.

Caution: Potential symptoms of overexposure are reduced function of liver, kidneys, bladder and pancreas. Direct contact may cause lung and skin irritation. *See NIOSH Pocket Guide to Chemical Hazards* (DHHS/NIOSH 90-117, 1990) p 32. This substance may reasonably be anticipated to be a carcinogen: *Seventh Annual Report on Carcinogens* (PB95-109781, 1994) p 78.

USE: As a positive control to study the carcinogenicity and mutagenicity of aromatic amines.

4193. Fluorescamine. *4-Phenylspiro[furan-2(3H),1'-(3'H)-isobenzofuran]-3,3'-dione;* 4-phenylspiro[furan-2(3H),-1'-phthalan]-3,3'-dione; Ro-20-7234; Fluram. $C_{17}H_{10}O_4$; mol wt 278.26. C 73.38%, H 3.62%, O 23.00%. Nonfluorescent reagent that reacts readily with primary amines to form highly fluorescent compds: S. Udenfriend *et al.*, *Science* **178**, 871 (1972). Prepn: M. Weigele *et al.*, *J. Am. Chem. Soc.* **94**, 5927 (1972); eidem, *J. Org. Chem.* **41**, 388 (1976). Use as fluorometric reagent: W. Leimgruber, M. Weigele, Ger. pat. **2,350,179** corresp to U.S. pat. **3,830,629** (both 1974 to Hoffmann-La Roche). Review of analytical uses: C. Y. Lai, *Methods Enzymol.* **47**, 236-243 (1977); S. Stein, *Peptides in Neurobiology*, H. Gainer, Ed. (Plenum, New York, 1977) pp 9-37; S. Udenfriend, *Pharmacology* **19**, 223-227 (1979).

mp 154-155°. uv max (ether): 235, 276, 284, 306 nm (ε 25900, 3950, 4100, 3800).

USE: Analytical reagent.

4194. Fluorescein. *3',6'-Dihydroxyspiro[isobenzofuran-1(3H),9'-[9H]xanthen]-3-one;* 9-(o-carboxyphenyl)-6-hydroxy-3H-xanthen-3-one; 3',6'-dihydroxyfluoran; 3',6'-fluorandiol; 9-(o-carboxyphenyl)-6-hydroxy-3-isoxanthenone; resorcinolphthalein; D & C Yellow no. 7; C.I. Solvent Yellow 94; C.I. 45350:1. $C_{20}H_{12}O_5$; mol wt 332.31. C 72.29%, H 3.64%, O 24.07%. Prepd by heating phthalic anhydride with resorcinol: Fischer, Bollmann, *J. Prakt. Chem.* **104**, 123 (1922); McKenna, Sowa, *J. Am. Chem. Soc.* **60**, 124 (1938). Structure: Ramart-Lucas, *Compt. Rend.* **205**, 864 (1937); Nagase *et al.*, *J. Pharm. Soc. Japan* **73**, 1033, 1039 (1953). Review of synthesis, properties and histological use: R. F. Steiner, H. Edelhoch, *Chem. Rev.* **62**, 457 (1962). Use as label in immunoassays: E. F. Ullman *et al.*, *J. Biol. Chem.* **251**, 4172 (1976); Y. Suzuki *et al.*, *Japan. J. Exp. Med.* **49**, 179 (1979). Toxicity studies in fish: L. L. Marking, *Progr. Fish Cult.* **31**, 139 (1969). Toxicity data: S. L. Yankell, J. J. Loux, *J. Periodontol.* **48**, 228 (1977). *See also: Colour Index* vol. 4 (3rd ed., 1971) p 4424; *H. J. Conn's Biological Stains*, R. D. Lillie, Ed. (Williams & Wilkins, Baltimore, 9th ed., 1977) p 337.

Yellowish-red to red powder. mp 314-316° in sealed tube, with decompn. Insol in water, benzene, chloroform, ether. Sol in hot alcohol or glacial acetic acid; also sol in alkali hydroxides or carbonates with a bright green fluorescence appearing red by transmitted light. Absorption max: 493.5, 460 nm.

Disodium salt, $C_{20}H_{10}Na_2O_5$, *soluble fluorescein, resorcinol phthalein sodium, uranin(e), uranine yellow, D & C yellow No. 8, C.I. Acid Yellow 73, C.I. 45350, Ak-Fluor, Fluorescite, Fluorets, Fluor-i-strip, Ful-Glo, Funduscein, Irescein.* Hygroscopic orange-red powder. Freely sol in water with yellowish-red color and intense yellowish-green fluorescence

perceptible down to a dil of 0.02 ppm under uv light. The fluorescence disappears when the soln is made acid, and reappears when the soln is again made neutral or alkaline. Absorption max (water): 493.5 nm. Slightly sol in alc. LD_{50} in mice, rats (mg/kg): 4738, 6721 orally (Yankel, Loux).

USE: In examining subterranean waters. Serves to ascertain source of springs, connections between streams and sea, determining approx vol of water delivered by a spring, detecting source of contamination of drinking water, infiltration of soil with waste waters of factories. Approved by FDA for use in externally applied drugs and cosmetics. Analytical reagent (protein label). Clinical reagent (immuno-histological stain, immuno-fluorescent label).

THERAP CAT: Diagnostic aid (corneal trauma indicator), ophthalmic angiography, contact lens fitting.

THERAP CAT (VET): Diagnostic aid (corneal lesions, intraocular inflammation).

4195. Fluorescein Paper. Zellner's paper. Paper charged with a black, substantive, neutral dye, then impregnated with a fluorescein soln and dried. Prepn and application: Zellner, **Ger.** pat. **124,922** (1901); *Chem. Zentralbl.* **1901,** II, 1032; *Pharm. Zentralh.* **1901,** 521; **1902,** 297; *E. Merck's Jahresber.* **1901,** 161-162.

USE: Exceedingly sensitive to alkalies (1:3,000,000) and particularly to ammonia (1:5,000,000) in spring or well waters; usable with dark or strongly colored liqs.

4196. Fluorescin. *2-(3,6-Dihydroxy-9H-xanthen-9-yl)-benzoic acid;* resorcinolphthalin. $C_{20}H_{14}O_5$; mol wt 334.33. C 71.85%, H 4.22%, O 23.93%. Obtained by heating fluorescein with NaOH and zinc dust. Formation by *Pseudomonas aeruginosa:* King *et al.*, *Can. J. Res.* **26C**, 514 (1948); Totter, Moseley, *J. Bacteriol.* **65**, 45 (1953).

Bright yellow powder, mp 125-127°. Readily oxidizes to fluorescein. Practically insol in water. Sol in alkali hydroxides or carbonates, alcohol, ether. *Keep well closed.*

USE: Reagent for oxidases, peroxides.

4197. Fluoresone. *1-(Ethylsulfonyl)-4-fluorobenzene;* ethyl p-fluorophenyl sulfone; p-fluorophenyl ethyl sulfone; Bripadon; Caducid. $C_8H_9FO_2S$; mol wt 188.22. C 51.05%, H 4.82%, F 10.09%, O 17.00%, S 17.04%. Prepn: G. Thuillier *et al.*, *Compt. Rend.* **248**, 2492 (1959); P. Rumpf, G. Thuillier, **Fr.** pat. **M399** corresp to U.S. pat. **3,084,101** (1962, 1963 both to Centre Nat. Recher. Scient.); A. A. Mignot, P. Rumpf, *Bull. Soc. Chim. France* **1968**, 435. Pharmacology: J. Thuillier *et al.*, *Proc. Meeting Coll. Int. Neuropsychopharmacol., 3rd, Munich* **1962**, 317-326 (Publ. 1964). Clinical evaluation: H. Akimoto, S. Taen, *ibid.* 326. Gas chromatography: E. Marozzi *et al.*, *Farmaco Ed. Prat.* **31**, 180 (1976).

Crystals, mp 41°. LD_{50} orally in mice: 2.5 g/kg (G. Thuillier); also reported as 850 mg/kg (J. Thuillier); 542 mg/kg (Akimoto, Taen).

THERAP CAT: Anticonvulsant; analgesic; anxiolytic.

4198. Fluoridamid. *N-[4-Methyl-3-[[(trifluoromethyl)-sulfonyl]amino]phenyl]acetamide;* 5-acetamido-2-methyltrifluoromethanesulfonanilide; 3-trifluorosulfonamido-p-acetotoluidide; Sustar. $C_{10}H_{11}F_3N_2O_3S$; mol wt 296.27. C

40.54%, H 3.74%, F 19.24%, N 9.46%, O 16.20%, S 10.82%. Prepn: Harrington *et al.,* U.S. pat. 3,639,474 (1972 to Minnesota Mining & Mfg).

mp 175-176.5°.
USE: Plant growth retardant.

4199. Fluorine. F; at. wt 18.9984032; at. no. 9; valence 1; elemental state F_2. A halogen. Occurrence in earth's crust 0.065% by wt. Natural abundance of isotopes: ^{19}F 100%; ^{18}F ($T_{1/2}$ 109.7 min) is prepared in nuclear reactors. Does not occur in elemental state in nature. Most important sources are fluorite, cryolite, and fluorapatite: Finger, "Fluorine Resources and Fluorine Utilization" in *Advances in Fluorine Chemistry,* vol. 2, M. Stacey *et al.,* Eds. (Butterworths, London, 1961) pp 35-54. Discovered in 1771 by Scheele. Isolated in 1886 by electrolyzing a soln of potassium fluoride in anhydr hydrogen fluoride at −23°, using platinum-iridium electrodes: Moissan, *Compt. Rend.* 102, 1543 (1886); 103, 202, 256. Subsequent methods of prepn: Ruff, *Ber.* 69, 181 (1936); Henne, *J. Am. Chem. Soc.* 60, 96 (1938). ^{18}F has been used to study fluorine exchange reactions: Dave, Sowerby, "Isotopic Halogen Exchange Reactions" in *Halogen Chemistry,* vol. 1, V. Gutmann, Ed. (Academic Press, New York, 1967) pp 41-132. Toxicity data: Keplinger, Suissa, *Am. Ind. Hyg. Assoc. J.* 29, 10 (1968). *Reviews:* A. J. Rudge, *The Manufacture and Use of Fluorine and its Compounds* (Oxford University Press, 1962); O'Donnell, "Fluorine" in *Comprehensive Inorganic Chemistry,* vol. 2, J. C. Bailar, Jr. *et al.,* Eds. (Pergamon Press, Oxford, 1973) pp 1009-1106; A. J. Woytek in Kirk-Othmer *Encyclopedia of Chemical Technology* vol. 10 (Wiley-Interscience, New York, 3rd ed., 1980) pp 630-654.

Pale yellow, diatomic gas. mp −219.61° (53.54°K); bp −188.13° (85.02°K); d (liq, −188.13°) 1.5127; vapor pressure data: Hu *et al., J. Am. Chem. Soc.* 75, 5642 (1953); White *et al., ibid.* 76, 2584 (1954). Crit temp: −129°; crit pressure: 55 atm. Most reactive nonmetal; higher oxidation potential than ozone; most electronegative element; E° (calc) ½F/F⁻ 2.9 V. F-F bond weaker than Cl-Cl and Br-Br bonds; enthalpy of dissociation: 37.7 kcal. Reacts vigorously with most oxidizable substances at room temp, frequently with ignition. Combines directly or indirectly, to form fluorides with all the elements except helium, neon and argon. Dec water, giving hydrofluoric acid, HF, oxygen fluoride, OF_2, hydrogen peroxide, oxygen and ozone. Reacts with nitric acid, forming the explosive gas, fluorine nitrate, NO_3F; with sulfuric acid, giving fluorosulfuric acid, $HFSO_3$. Yields the metal fluorides, water, oxygen and oxygen fluoride when made to react with metal hydroxides in the cold. Reacts violently with organic compds, usually with disintegration of the molecule. Under controlled conditions, however, hydrocarbon vapors may be fluorinated with elemental fluorine. Solid fluorine explodes when brought in contact with liquid hydrogen. Under ordinary conditions it does not react directly with oxygen, nor does it react with oxides of sodium, potassium or calcium. LC_{50} (1 hr) inhalation by rats, mice, guinea pigs: 185, 150, 170 ppm (by vol) (Keplinger, Suissa).

Caution: Potential symptoms of overexposure are irritation of eyes, nose and respiratory tract; laryngeal spasms, bronchial spasms; pulmonary edema; eye and skin burns. *See NIOSH Pocket Guide to Chemical Hazards* (DHHS/ NIOSH 90-117, 1990) p 116. Chronic ingestion of high concentrations from water supply can cause mottled enamel of teeth and osteosclerosis. *See Clinical Toxicology of Commercial Products,* R. E. Gosselin *et al.,* Eds. (Williams & Wilkins, Baltimore, 5th ed., 1984) Section III, pp 185-193.

4200. Fluorine Dioxide. Dioxygen difluoride. F_2O_2; mol wt 70.00. F 54.28%, O 45.72%. Prepd from O_2 and F_2: Ruff, Menzel, *Z. Anorg. Allgem. Chem.* 211, 204 (1933); 217, 85 (1934); Goetschel *et al., J. Am. Chem. Soc.* 91, 4702 (1969). Chemical behavior: Streng, *ibid.* 85, 130 (1963). *Review:* Kemmitt, Sharp, *Advan. Fluorine Chem.* 4, 214-215 (1965).

Thermally unstable gas at room temp; begins to dec into fluorine and oxygen at −100°. Pale yellow solid or yellow liq at low temp. mp −154°C (119°K). Early prepns, described as brown gas, red liq and orange solid (mp −163.5°), probably contained other fluorine-oxygen compounds as impurities: Goetschel *et al., loc. cit.* Strong oxidizing and fluorinating agent.

4201. Fluorine Monoxide. Oxygen fluoride; fluorine oxide. F_2O; mol wt 54.00. F 70.37%, O 29.63%. Prepd by passing fluorine slowly through an aq NaOH soln: Yost, *Inorg. Syn.* 1, 109 (1939); Schnizlein *et al., J. Phys. Chem.* 56, 233 (1952). Toxicity data: Darmer *et al., Am. Ind. Hyg. Assoc. J.* 33, 661 (1972). *Review:* Kemmitt, Sharp, *Advan. Fluorine Chem.* 4, 213-314 (1965).

Colorless gas. Yellowish-brown when liq. Peculiar smell. Does not attack glass in the cold. Corrodes mercury. Reacts very slowly with water. The gas may be kept over water unchanged for a month. d (liq; −224°) 1.90. mp −223.8°. bp −145.3°. Trouton constant 20.65. Soly in water (0°) 6.8 ml gas/100 ml H_2O. LC_{50} (1 hr) inhalation by rats, mice: 2.6, 1.5 ppm (Darmer).

Caution: Potential symptoms of overexposure are intractable headache; respiratory system irritation, pulmonary edema; eye and skin burns. *See NIOSH Pocket Guide to Chemical Hazards* (DHHS/NIOSH 90-117, 1990) p 170.

4202. Fluorine Nitrate. *Nitroxy fluoride;* nitrogen trioxyfluoride; nitryl hypofluorite. FNO_3; mol wt 81.00. F 23.45%, N 17.29%, O 59.25%. $FONO_2$. Prepd by the action of fluorine on nitric acid: Cady, *J. Am. Chem. Soc.* 56, 2635 (1934); Ruff, Kwasnik, *Angew. Chem.* 48, 238 (1935); Kwasnik in *Handbook of Preparative Inorganic Chemistry,* vol. 1, G. Brauer, Ed. (Academic Press, New York, 2nd ed., 1963) pp 187-189. *Reviews:* Kemmitt, Sharp, *Advan. Fluorine Chem.* 4, 216-218 (1965); Woolf, *ibid.* 5, 1-30 (1965); Schmutzler, *Angew. Chem. Int. Ed.* 7, 440-455 (1968).

Colorless gas. Moldy, acrid odor. mp −175°. bp −45.9°. d (liq at bp) 1.507. d$^{−193.2}$ (solid) 1.951. Trouton const 20.8. The liq explodes on slight percussion. Hydrolyzed by water to OF_2, O_2, HF and HNO_3. Sol in acetone. Conflagrates on contact with alcohol, ether, aniline. May be stored in vacuum-sealed glass ampuls cooled by liq oxygen. Powerful oxidizing agent.

USE: Oxidizing agent in rocket propellants.

4203. Fluorine Perchlorate. Chlorine tetroxyfluoride. $ClFO_4$; mol wt 118.45. Cl 29.93%, F 16.04%, O 54.03%. $FOClO_3$. Prepd by passing fluorine over cold 72% aq perchloric acid in platinum apparatus: Rohrback, Cady, *J. Am. Chem. Soc.* 69, 677 (1947).

Colorless gas. Pungent, acrid odor. mp −167.3°. bp$_{755}$ −15.9°. Explodes on the slightest provocation, *i.e.,* on contact with rough surfaces, dust, grease, rubber, on melting, distilling, etc.

Caution: Attacks lungs even in traces.

4204. Fluoroacetamide. Fluoroacetic acid amide; monofluoroacetamide; 1081; Fluorakil 100; Fussol. C_2H_4-FNO; mol wt 77.06. C 31.17%, H 5.23%, F 24.65%, N 18.18%, O 20.76%. CH_2FCONH_2. Numerous syntheses, e.g. from fluoroacetyl chloride and NH_3: Truce, *J. Am. Chem. Soc.* 70, 2828 (1948). Mode of action study and toxicity data: F. Matsumura, R. D. O'Brien, *Biochem. Pharmacol.* 12, 1201 (1963).

Crystals. Sublimes on heating. Freely sol in water; sol in acetone; sparingly sol in chloroform. LD_{50} i.p. in mice: 85 mg/kg (Matsumura, O'Brien).

USE: Rodenticide. Insecticide proposed mainly for use on fruits to combat scale insects, aphids, and mites.

4205. Fluoroacetic Acid. Fluoroethanoic acid; gifblaar poison. $C_2H_3FO_2$; mol wt 78.04. C 30.78%, H 3.87%, F 24.34%, O 41.00%. CH_2FCOOH. Occurs in "Gifblaar,"

Dichapetalum cymosum (Hook.) Engl. (*Chailletia cymosa* Hook.), *Dichapetalaceae,* one of the most poisonous plants, habitat South Africa. Prepd from methyl iodoacetate by heating with silver fluoride or mercurous fluoride or from methyl chloroacetate by heating with potassium fluoride, followed by saponification of the methyl ester with baryta: Swarts, *Bull. Soc. Chim.* [3] **15,** 1134 (1896); Gryszkiewicz-Trochimowski *et al., Rec. Trav. Chim.* **66,** 419 (1947). Review and biochemical aspects: Peters, *Endeavour* **13,** 147 (1954); Peters *et al., Biochem. J.* **77,** 17 (1960). Reviews of toxicity: *Clinical Toxicology of Commercial Products,* R. E. Gosselin *et al.,* Eds. (Williams & Wilkins, Baltimore, 4th ed., 1976) Section III, pp 163-166; Gribble, *J. Chem. Ed.* **50,** 460-462 (1973).

Crystals. Burns with a green flame.

Sodium salt, $C_2H_2FNaO_2$, *Compd 1080, 1080, Fratol.* A fine white powder, sol in water. Oral lethal dose estimated to be 2-5 mg/kg body wt.

Methyl ester, $CH_2FCOOCH_3$, liq, odor of ethyl acetate. d_4^{15} 1.1613. bp 104.5°. Sol in water; slightly sol in petr ether.

Ethyl ester, $CH_2FCOOC_2H_5$, liq, odor of ethyl acetate. $d^{20.5}$ 1.0926. bp_{758} 121.6°. $n_D^{20.5}$ 1.3767. Sol in water.

Caution: Extremely toxic. Causes convulsions and ventricular fibrillation. Should never be used where inhalation or food contamination may occur. Locally, it is an irritant.

USE: The sodium salt is used as a water-soluble rodenticide.

4206. 3-Fluoro-D-alanine. *S*-2-Amino-3-fluoropropanoic acid; FA. $C_3H_6FNO_2$; mol wt 107.08. C 33.65%, H 5.65%, F 17.74%, N 13.08%, O 29.88%. Synthesis by *C*-photofluorination of D-alanine or by resolution of the DL-form: J. Kollonitsch, F. M. Kahan; J. Kollonitsch, Ger. pats. **2,136,067; 2,229,245,** *C.A.* **76,** 100053p (1972); **78,** 72586d (1973); *idem,* U.S. pat. **3,839,170** (1972, 1972, 1974 all to Merck & Co.). The 2-deuterated analog has also been prepared. Antibacterial activity: J. Kollonitsch *et al., Nature* **243,** 346 (1973).

Crystals from water, dec 168°. $[\alpha]_D^{20}$ −10° (c = 3 in 1*N* HCl). Stable at acid pH; less stable at pH > 8.

4207. p-Fluoroaniline. *4-Fluorobenzenamine.* C_6H_6- FN; mol wt 111.12. C 64.85%, H 5.44%, F 17.10%, N 12.61%. Prepd by reduction of 1-fluoro-4-nitrobenzene by Raney nickel: Benington *et al., J. Org. Chem.* **18,** 1508 (1953); Finger *et al., J. Am. Chem. Soc.* **81,** 98 (1959); by sulfurated sodium borohydride: Lalancette, Brindle, *Can. J. Chem.* **49,** 2990 (1971).

Liquid. d_4^{20} 1.1725; d_4^{25} 1.1690. mp −1.9°. bp 188°, bp_{20} 86°. n_D^{20} 1.51954. Heat of combustion 780.4 kcal/mol. Very slightly sol in water.

USE: Intermediate in the manuf of herbicides and plant growth regulators.

4208. Fluorobenzene. C_6H_5F; mol wt 96.10. C 74.99%, H 5.24%, F 19.77%. Obtained by warming benzenediazonium chloride with concd HF. Solubility value: P. M. Gross *et al., J. Am. Chem. Soc.* **55,** 650 (1933). Pressure, vol, temp constants: Douslin *et al., J. Am. Chem. Soc.* **80,** 2031 (1958).

Liquid; benzene odor. d_4^{20} 1.024. bp_{760} 84.73°; $bp_{13\ atm}$ 200°; $bp_{38\ atm}$ 275°. mp −40°. n_D^{20} 1.4677. Miscible with alcohol, ether. Sol (30°C): 1.54 g/1000 g water.

4209. p-Fluorobenzoic Acid. $C_7H_5FO_2$; mol wt 140.11. C 60.01%, H 3.60%, F 13.56%, O 22.84%. Prepd by treating *p*-carbethoxybenzenediazonium chloride with fluoboric acid, followed by thermal decompn of the resulting *p*-carb-

ethoxybenzenediazonium fluoborate and by hydrolysis of the ensuing ethyl ester of *p*-fluorobenzoic acid: Schiemann, Winkelmüller, *Org. Syn.* **13,** 52 (1933).

Monoclinic prisms from water. Peculiar sweet taste. mp 182.6°. Sparingly sol in cold water; freely sol in hot water. Soly in water at 32°: 1.1 g/l. Water solns evaporate without leaving a residue. Sol in alc, ether. pK (25°): 3.85.

Ethyl ester, $C_9H_9FO_2$, crystals, mp 26°. bp 210°.

4210. 1-Fluoro-2,4-dinitrobenzene. 2,4-Dinitro-1-fluorobenzene; DNFB; FDNB; Sanger's reagent. $C_6H_3FN_2O_4$; mol wt 186.10. C 38.72%, H 1.62%, F 10.21%, N 15.05%, O 34.39%. Prepn: A. F. Holleman, J. W. Beekman, *Rec. Trav. Chim.* **23,** 225 (1904); Cook, Saunders, *Biochem. J.* **41,** 558 (1947). Use in peptide analysis: Sanger, *Biochem. J.* **39,** 507 (1945); **40,** 261 (1946); **45,** 563 (1949); Porter, Sanger, *ibid.* **42,** 287 (1948). Tumor promoting activity: F. G. Bock *et al., Cancer Res.* **29,** 179 (1969). Mutagenicity study: D. R. Jagannath *et al., Mutat. Res.* **78,** 91 (1980). Review of uses: *Reagents for Organic Synthesis,* L. F. Fieser, M. Fieser, Eds. (Wiley, New York, 1967) pp 321-322.

Pale yellow crystals from ether, mp 26°. $bp_{2.0}$ 137°. Sol in benzene, ether, propylene glycol.

Caution: Vesicant. For proper handling see: J. S. Thompson, O. P. Edmunds, *Ann. Occup. Hyg.* **23,** 27 (1980).

USE: Reagent for labeling a terminal amino acid group; in modified Wohl degradations of aldoses. As hapten.

4211. Fluoroform. *Trifluoromethane.* CHF_3; mol wt 70.01. C 17.16%, H 1.44%, F 81.41%. Prepd from $CHCl_3$ and HF: Meslans, *Compt. Rend.* **110,** 717 (1890); Valentiner, U.S. pat. **643,835** (1900); Ruff, *Ber.* **69,** 299 (1936); Whalley, *J. Soc. Chem. Ind. (London)* **66,** 429 (1947); Kwasnik in *Handbook of Preparative Inorganic Chemistry* vol. 1, G. Brauer, Ed. (Academic Press, New York, 2nd ed., 1963) pp 204-205. Early industrial prepn: Pearlson in *Fluorine Chemistry* vol. I, J. H. Simons, Ed. (Academic Press, New York, 1950) p 467.

Colorless, odorless gas. Stable up to 1150°. Chemically very inert. d (solid) 1.935. d (liq; −100°) 1.52. mp −160°. bp −84.4°. Critical temp 33°, critical pressure 47 atm, critical density 0.516. May be stored over water. Practically non-toxic: N. V. Sidgwick, *The Chemical Elements and Their Compounds,* vol. II (Oxford, 1950) p 1130.

Caution: May be slightly irritating to respiratory tract, and, in high concns, narcotic.

USE: Refrigerant for low temps.

4212. Fluoromethane. Methyl fluoride. CH_3F; mol wt 34.03. C 35.29%, H 8.88%, F 55.82%. Prepn in 82% yield by heating fluorosulfonic acid methyl ester with KF: Zappel, Jonas, Ger. **1,131,197** (1962 to Bayer).

Gas. Agreeable ether-like odor. Burns with evolution of HF, the flame being about as colorless as that of alc: Dumas, Peligot, *Ann.* **15,** 59 (1835). d (liq; −78°) 0.8774. d (gas) 1.1951 (air = 1); d (gas) 1.0813 (oxygen = 1). Dipole moment 1.81. Molecular volume: 22.03. Van der Waals constants: 0.00923; 0.002350. Critical temp 44.9°; crit press. 62.0 atm. Dielectric constant (for wavelengths below 10^4 cm) = 1.00948. mp −141.8°. bp_{872} −75.7°; bp_{760} −78.2°. bp_{143} −103.7°. One hundred vols of water dissolve 166 vols of the gas at 15°. Freely sol in alcohol, ether.

Caution: Narcotic in high concns.

4213. Fluorometholone. *(6α,11β)-9-Fluoro-11,17-dihydroxy-6-methylpregna-1,4-diene-3,20-dione;* 21-desoxy-9α-fluoro-6α-methylprednisolone; 21-desoxy-6α-methyl-9α-fluoroprednisolone; fluormetholon; Cortilet; Delmeson; Effumidex; Fluaton; Flumetholon; FML; Loticort; Oxylone;

Ursnon. $C_{22}H_{29}FO_4$; mol wt 376.47. C 70.19%, H 7.76%, F 5.05%, O 17.00%. Prepn: F. H. Lincoln *et al.*, U.S. pat. **2,867,637** (1959 to Upjohn). Prepn of acetate and other esters: B. J. Magerlein *et al.*, U.S. pat. **3,038,914** (1962 to Upjohn). Pharmacology of acetate: A. Kupferman *et al.*, *Arch. Ophthalmol.* **100**, 640 (1982). HPLC determn: H. Tokunaga *et al.*, *Chem. Pharm. Bull.* **32**, 4012 (1984). Clinical evaluation in ocular inflammation: H. M. Leibowitz *et al.*, *Ann. Ophthalmol.* **16**, 1110 (1984).

Crystals from acetone, mp 292-303°.

17-Acetate, $C_{24}H_{31}FO_5$, crystals from ethyl acetate + hexanes, mp 230-232°. $[\alpha]_D$ +28° (chloroform).

THERAP CAT: Glucocorticoid; anti-inflammatory.

4214. 5-Fluoroorotic Acid. *5-Fluoro-1,2,3,6-tetrahydro-2,6-dioxo-4-pyrimidinecarboxylic acid;* 5-fluoroorotate; 5-fluoro-6-carboxyuracil; 5-FOA; ENT-26398; NSC-31712; Ro-2-9945; WR-152520. $C_5H_3FN_2O_4$; mol wt 174.09. C 34.50%, H 1.74%, F 10.91%, N 16.09%, O 36.76%. Pyrimidine precursor; selectively toxic to yeast cells that synthesize the enzyme orotidine-5'-P decarboxylase. Prepn: R. Duschinsky *et al.*, *J. Am. Chem. Soc.* **79**, 4559 (1957); Brit. pat. **806,584** (1958 to Hoffmann-La Roche); R. Duschinsky, C. Heidelberger, U.S. pat. **2,948,725** (1960). Improved synthesis: D. H. R. Barton *et al.*, *J. Chem Soc. Perkin Trans. I* **1974**, 2095; and mass spectrum: S. N. Alam *et al.*, *Acta Pharm. Suec.* **12**, 375 (1975). Antitumor activity: C. Heidelberger *et al.*, *Nature* **179**, 663 (1957). Inhibition of RNA synthesis: C. T. Garrett *et al.*, *Arch. Biochem. Biophys.* **155**, 342 (1973). Clinical evaluation of antimycotic activity: K. Nikolova *et al.*, *Methods Find. Exp. Clin. Pharmacol.* **9**, 85 (1987). Review of *in vitro* use in the positive selection of genetically transformed yeast cells: J. D. Boeke *et al.*, *Methods Enzymol.* **154**, 164-175 (1987).

Monohydrate, crystals, mp 255° (dec) (Duschinsky, 1957); also reported as 258-259° (Barton). Partially sol in water. uv max (0.1*N* HCl): 284-285 nm (ε 7100). LD$_{50}$ i.p. in male mice: 300 mg/kg (Nikolova).

USE: Research tool in molecular genetics.

4215. *p*-Fluorophenylacetic Acid. *4-Fluorobenzeneacetic acid.* $C_8H_7FO_2$; mol wt 154.14. C 62.34%, H 4.58%, F 12.33%, O 20.76%. Prepd by treating *p*-fluorobenzyl cyanide with sulfuric acid: Olah *et al.*, *J. Org. Chem.* **22**, 879 (1957).

Crystals from chloroform, mp 86°. bp$_2$ 164°.

β-Dimethylaminoethyl ester HCl, crystals, mp 103°.

USE: Intermediate in the manuf of fluorinated anesthetics.

4216. Fluorosalan. *3,5-Dibromo-2-hydroxy-N-[3-(trifluoromethyl)phenyl]benzamide; 3,5-dibromo-α,α,α-trifluoro-m-salicylotoluidide;* 3,5-dibromo-3'-trifluoromethylsalicylanilide; Fluorophene. $C_{14}H_8Br_2F_3NO_2$; mol wt 439.03. C 38.30%, H 1.84%, Br 36.40%, F 12.98%, N 3.19%, O 7.29%. Prepn: Stecker, U.S. pat. **3,041,236** (1962).

THERAP CAT: Disinfectant.

4217. Fluorosulfonic Acid. *Fluorosulfuric acid;* fluosulfonic acid. FHO_3S; mol wt 100.07. F 18.99%, H 1.01%, O 47.96%, S 32.04%. HSO_3F. Prepn: Thorpe, Kirman, *J. Chem. Soc.* **61**, 921 (1892); Meyer, Schramm, *Z. Anorg. Allgem. Chem.* **206**, 25 (1932); Kwasnik in *Handbook of Preparative Inorganic Chemistry*, **vol. 1**, G. Brauer, Ed. (Academic Press, New York, 2nd ed., 1963) pp 177-178. Reviews of properties and chemistry: Gillespie, *Accounts Chem. Res.* **1**, 202-209 (1968); Thompson, "Fluorosulfuric Acid" in *Inorganic Sulfur Chemistry*, G. Nickless, Ed. (Elsevier, New York, 1968) pp 587-606; Jache in *Advan. Inorg. Chem. Radiochem.* **16**, 177-200 (1974).

Colorless liquid; fumes in moist air. d_4^{18} 1.740; d_4^{25} 1.726. mp $-89°$. bp$_{760}$ 163°; bp$_{120}$ 110.0°; bp$_{19}$ 77.0°. Stable to 900°. Considerably more acidic than 100% H_2SO_4. Does not attack glass when anhydr and pure. Violent reaction with water although it is incompletely and reversibly hydrolyzed. Reddish-brown color with acetone. Forms stable salts which are little hydrolyzed by water and which may be recrystallized from water.

Methyl ester, *see* Methyl Fluorosulfonate.

Caution: May be highly irritating to skin, mucous membranes.

USE: Fluorinating agent. Catalyst in alkylation, acylation, polymerization and condensation reactions; in hydrofluorination of olefins; in production of substituted pyridines. In production of petroleum products. *See* Thompson, *loc. cit.* *Magic Acid*, a 1:1 HSO_3F-SbF_5 soln, is used in the study of stable solns of alkyl- and arylcarbonium ions: Olah, *Science* **168**, 1298 (1970).

4218. Fluorotoluene. *Fluoromethylbenzene.* C_7H_7F; mol wt 110.13. C 76.34%, H 6.41%, F 17.25%. Prepn of *o*-, *m*-, *p*-isomers: A. F. Holleman, J. W. Beekman, *Rec. Trav. Chim.* **23**, 225 (1904); A. F. Holleman, *ibid.* **25**, 330 (1906); E. D. Bergmann, S. Berkovic, *J. Org. Chem.* **26**, 919 (1961).

Colorless liquids. Insol in water. Miscible with alc, ether.

o-Fluorotoluene, 1-fluoro-2-methylbenzene. d^{13} 1.004. mp $\sim -80°$. bp 114°. n_D^{20} 1.4704.

m-Fluorotoluene. d^{13} 0.997. mp $-111°$. bp 116°. n_D^{20} 1.4691.

p-Fluorotoluene. Odor of oil of bitter almonds. d^{16} 1.001. bp 116°. n_D^{20} 1.470.

4219. Fluorouracil. *5-Fluoro-2,4(1H,3H)-pyrimidinedione;* 2,4-dioxo-5-fluoropyrimidine; 5-FU; Ro-2-9757; NSC-19893; Adrucil; Arumel; Efudex; Efudix; Fluril; Fluracil; Fluoroplex; Fluroblastin; Fluro Uracil; Timazin. $C_4H_3FN_2O_2$; mol wt 130.08. C 36.93%, H 2.32%, F 14.61%, N 21.54%, O 24.60%. Prepn: Duschinsky *et al.*, *J. Am. Chem. Soc.* **79**, 4559 (1957); Heidelberger, Duschinsky, U.S. pats. **2,802,005** (1957) and **2,885,396** (1959); Barton *et al.*, *J. Org. Chem.* **37**, 329 (1972); O. Miyashita *et al.*, *Chem. Pharm. Bull.* **29**, 3181 (1981). Pharmacokinetics: W. Sadee, C. G. Wong, *Clin. Pharmacokinet.* **2**, 437 (1977). Site of action: F. Maley, G. F. Maley, *FEBS Symp.* **57**, 21 (1979). Comprehensive description: B. C. Rudy, B. Z. Senkowski, *Anal. Profiles Drug Subs.* **2**, 221-244 (1973); S. M. Bayomi, A. A.

Al-Badr, *ibid.* **18**, 599-639 (1989). Review of clinical trials: R. M. Hansen, *Cancer Invest.* **9**, 637-642 (1991).

Crystals from water or methanol-ether, dec 282-283°. Sublimes (0.1 mm) 190-200°. uv max (0.1*N* HCl): 265-266 nm (ε 7070).

Sodium salt, *Effluderm.*

THERAP CAT: Antineoplastic.

4220. Fluosilicic Acid. *Hydrogen hexafluorosilicate;* hexafluosilicic acid; hydrosilicofluoric acid; hydrofluosilicic acid; silicofluoric acid; fluorosilicic acid. F_6H_2Si; mol wt 144.09. F 79.11%, H 1.40%, Si 19.49%. H_2SiF_6. Prepd from HF + SiO_2; also prepd by the action of water on SiF_4; by the action of H_2SO_4 on $BaSiF_6$: Hempel, *Ber.* **18**, 1438 (1885); Baur, Glaessner, *Ber.* **36**, 4215 (1903); Söll, *FIAT-Review* **23**, 257 (1946); Kwasnik in *Handbook of Preparative Inorganic Chemistry* **vol. 1**, G. Brauer, Ed. (Academic Press, New York, 2nd ed., 1963) p 214-215; Lange in *Fluorine Chemistry*, **vol. I**, J. H. Simon, Ed. (New York, 1950) p 129. Commercial grades of fluosilicic acid soln are obtained as a by-product in the superphosphate industry. *Review:* Colton, *J. Chem. Ed.* **35**, 562-563 (1958).

Liquid, when anhydr dissociates almost instantly into SiF_4 and HF. Marketed as aq soln only. A 60-70% soln solidifies around 19°, forming a cryst dihydrate. May be distilled without decompn only as a 13.3% aq soln. Fairly strong acid. Sour, pungent odor. $d_{17.5}^{17.5}$ 5% soln: 1.0407; 10%: 1.0834; 15%: 1.1281; 20%: 1.1748; 25%: 1.2235; 30%: 1.2742; 34%: 1.3162. The more concd solns (but not the anhydr liq) can be stored in glass, although some etching will take place around the surface. Usually stored in iron containers.

Caution: Severe corrosive effect on skin, mucous membranes.

USE: A 1-2% soln is used widely for sterilizing equipment in brewing and bottling establishments. Other concns are used in the electrolytic refining of lead, in electroplating, for hardening cement, crumbling lime or brick work, for the removal of lime from hides during the tanning process, to remove molds, as preservative for timber.

4221. Fluosol DA. Biologically inert fluorocarbon emulsion of perfluorodecalin and perfluorotripropylamine that performs the oxygen-carrying function of red blood cells. Has a particle size of 0.05 to 0.25μ and is stable at room temperature for years. Can be administered regardless of blood type; does not carry hepatitis or other infectious diseases. Prepn: K. Yokoyama *et al.*, *Ger. pat.* **2,630,-586;** *eidem, U.S. pat.* **4,252,827** (1977, 1981 to Green Cross). Hemodynamic and oxygen transport effects: K. K. Tremper *et al.*, *Crit. Care Med.* **8**, 738 (1980). Clinical studies: H. Ohyanagi *et al.*, *Clin. Ther.* **2**, 306 (1979); T. Suyama *et al.*, *Prog. Clin. Biol. Res.* **55**, 609 (1981); T. Mitsuno *et al.*, *Ann. Surg.* **195**, 60 (1982). *Review: Int. Congr. Ser.-Excerpta Med.* **486**, 1-471 (1979). Review of use as red cell substitute: K. C. Lowe, *Comp. Biochem. Physiol.* **87A**, 825-838 (1987).

20% Emulsion, *FDA-20.* Viscosity at 37°: 2.3 cp. pH 7.4.

Perfluorodecalin, $C_{10}F_{18}$, *octadecafluorodecahydronaphthalene, perflunafene, FDC.* Prepn: E. T. McBee, U.S. Bechtol, *Ind. Eng. Chem.* **39**, 380 (1947). mp −10° to −7°. bp 140°. d_4^{20} 1.9456. n_D^{20} 1.3118. Vapor pressure at 37°: 12.7 mm Hg.

Perfluorotripropylamine, $C_9F_{21}N$, *1,1,2,2,3,3,3-heptafluoro-N,N-bis(heptafluoropropyl)-1-propanamine, heneicosafluorotripropylamine, perfluamine, FTPA.* Prepn: R. N. Haszeldine, *J. Chem. Soc.* **1951**, 102. bp 129.5-130.5°. d_4^{25} 1.822. n_D^{25} 1.279. Vapor press. at 37°: 20.0 mm Hg.

USE: Blood substitute.

4222. Fluoxetine. (±)*-N-Methyl-γ-[4-(trifluoromethyl)-phenoxy]benzenepropanamine;* (±)*-N-methyl-3-phenyl-3-*

[(α,α,α-trifluoro-*p*-tolyl)oxy]propylamine; *N*-methyl-3-(*p*-trifluoromethylphenoxy)-3-phenylpropylamine. $C_{17}H_{18}F_3$-NO; mol wt 309.33. C 66.01%, H 5.87%, F 18.43%, N 4.53%, O 5.17%. Serotonin uptake inhibitor. Prepn: B. B. Malloy, K. K. Schmiegel, **Ger. pat. 2,500,110;** *eidem, U.S.* pat. **4,314,081** (1975, 1982 both to Lilly). Pharmacology: D. T. Wong *et al.*, *Life Sci.* **15**, 471 (1974); *eidem, J. Pharmacol. Exp. Ther.* **193**, 804 (1975). GC determn in plasma: J. F. Nash *et al.*, *Clin. Chem.* **28**, 2100 (1982). Interaction with ethanol: L. Lemberger *et al.*, *Clin. Pharmacol. Ther.* **6**, 658 (1985). Clinical trial in depression: L. F. Fabre, L. Crismon, *Curr. Ther. Res.* **37**, 115 (1985). Pharmacology and toxicity data: P. Stark *et al.*, *J. Clin. Psychiatry* **46**, Suppl. **3**(2), 7 (1985). Series of articles on clinical pharmacology and therapeutic efficacy: *ibid.* 14-67. Review of pharmacology, therapeutic efficacy: P. Benfield *et al.*, *Drugs* **32**, 481-508 (1986). Comprehensive description: D. S. Risley, R. J. Bopp in *Analytical Profiles of Drug Substances* **vol. 19**, K. Florey Ed. (Academic Press, New York, 1990) pp 193-219.

Hydrochloride, $C_{17}H_{18}F_3NO\cdot HCl$, *LY-110140, Adofen, Fluctin, Fluoxeren, Fontex, Foxetin, Prozac, Reneuron.* White to off-white crystalline solid. mp 158.4-158.9. Freely sol in methanol, ethanol; sol in acetonitrile, chloroform, acetone; slightly sol in ethyl acetate, dichloromethane, water (with sonication at pH 1.2, 4.5, 7.0). Practically insol in toluene, cyclohexane, hexane. Soly (mg/ml): methanol, ethanol > 100; acetone, acetonitrile, chloroform between 33-100; dichloromethane between 5-10; water between 1-2; ethyl acetate between 2-2.5; toluene, cyclohexane, hexane between 0.5-0.67. Maximum soly in water: 14 mg/ml. uv max (methanol) 227, 264, 268, 275 nm ($E_{1cm}^{1\%}$ 372.0, 29.2, 29.3, 21.5). LD_{50} in mice, rats (mg/kg): 248, 452 orally (Stark).

Oxalate, $C_{19}H_{18}F_3NO_5$, crystals from ethyl acetate-methanol, mp 179-182° (dec).

THERAP CAT: Antidepressant.

4223. Fluoxymesterone. *(11β,17β)-9-Fluoro-11,17-di-hydroxy-17-methylandrost-4-en-3-one;* 11β,17β-dihydroxy-9α-fluoro-17α-methyl-4-androsten-3-one; 9α-fluoro-11β-hydroxy-17α-methyltestosterone; Androsterolo; Androfluorene; Androfluorone; Halotestin; Oratestin; Ora-Testryl; Testoral; Ultandren; $C_{20}H_{29}FO_3$; mol wt 336.45. C 71.40%, H 8.69%, F 5.65%, O 14.27%. Prepn: Herr *et al.*, *J. Am. Chem. Soc.* **78**, 501 (1956). *cf.* Herr, U.S. pat. **2,813,881** (1957 to Upjohn). Comprehensive description: J. Kirschbaum, *Anal. Profiles Drug Subs.* **7**, 251-275 (1978).

Crystals, dec 270°. $[\alpha]_D$ +109° (ethanol). uv max (ethanol): 240 nm (ε 16700). Sol in pyridine; slightly sol in acetone, chloroform; sparingly sol in methanol. Practically insol in water, ether, benzene, hexanes.

Note: This is a controlled substance (anabolic steroid) in the U.S. Code of Federal Regulations, Title 21 Part 1308.13, as defined in Part 1308.02 (1995).

THERAP CAT: Androgen.

4224. Flupentixol. *4-[3-[2-(Trifluoromethyl)-9H-thioxanthen-9-ylidene]propyl]-1-piperazineethanol;* 2-trifluoromethyl-9-[3-[4-(β-hydroxyethyl)-1-piperazinyl]propylidene]-thioxanthene; flupenthixol; N-7009; LC-44. $C_{23}H_{25}F_3N_2OS$;

mol wt 434.53. C 63.58%, H 5.80%, F 13.12%, N 6.45%, O 3.68%, S 7.38%. Neuroleptic agent related structurally to thiothixene, *q.v.* Prepn: P. N. Craig, C. L. Zirkle, U.S. pat. **3,282,930** (1966 to SKF). Prepn of *cis-* and *trans-*forms: **Brit.** pat. **925,538** (1963 to SKF); P. V. Petersen, T. Ammitzboll, **U.S.** pat. **3,681,346** (1972 to Kefalas). Metabolism of decanoate: A. Jorgensen *et al., Acta Pharmacol. Toxicol.* **29,** 339 (1971). α-Flupentixol, the *cis-*isomer, shows greater pharmacological activity than β-flupentixol, the *trans-*isomer: I. Moller Nielsen *et al., Acta Pharmacol. Toxicol.* **33,** 353 (1973). X-ray crystallography of isomers: M. L. Post *et al., Nature* **256,** 342 (1975).

cis-form

Dihydrochloride, $C_{23}H_{25}F_3N_2OS.2HCl$, *Emergil, Fluanxol, Siplarol, Metamin.*

Decanoate, *LU-5-110, Depixol, Fluanxol Dépôt, Viscoleo.*

THERAP CAT: Antipsychotic.

4225. Fluperolone Acetate. *[11β,17α,17(S)]-17-[2-(Acetyloxy)-1-oxopropyl]-9-fluoro-11,17-dihydroxyandrosta-1,4-dien-3-one; 9-fluoro-11β,17α-dihydroxy-17(S)-lactoylandrosta-1,4-dien-3-one 17β-acetate; 9α-fluoro-11β,17α,21-trihydroxy-21-methylpregna-1,4-diene-3,20-dione 21-acetate; 17β-[2-acetoxypropionyl]-9α-fluoro-11β,17-dihydroxyandrosta-1,4-dien-3-one; 21-methyl-9α-fluoroprednisolone acetate;* P-1742; ALAcortril; Methral. $C_{24}H_{31}FO_6$; mol wt 434.50. C 66.34%, H 7.19%, F 4.37%, O 22.09%. Prepn: E. J. Agnello *et al., Experientia* **16,** 357 (1960); *eidem, J. Org. Chem.* **28,** 1531 (1963); E. J. Agnello, G. D. Laubach, **U.S.** pat. **3,234,095** (1966 to Pfizer). Activity studies: Child *et al., Arch. Dermatol.* **97,** 407 (1968).

Crystals, mp 251-253°. $[\alpha]_D$ +87°. uv max: 239 nm (ε 15350).

THERAP CAT: Glucocorticoid; anti-inflammatory.

4226. Fluphenazine. *4-[3-[2-(Trifluoromethyl)-10H-phenothiazin-10-yl]propyl]-1-piperazineethanol;* 1-(2-hydroxyethyl)-4-[3-(3-(trifluoromethyl-10-phenothiazinyl)propyl]-piperazine; 10-[3'-[4''-(β-hydroxyethyl)-1''-piperazinyl]propyl]-3-trifluoromethylphenothiazine; 2-(trifluoromethyl)-10-[3-[1-(β-hydroxyethyl)-4-piperazinyl]propyl]phenothiazine; S-94; SQ-4918. $C_{22}H_{26}F_3N_3OS$; mol wt 437.53. C 60.39%, H 5.99%, F 13.03%, N 9.60%, O 3.66%, S 7.33%. Prepn: H. L. Yale, F. Sowinski, *J. Am. Chem. Soc.* **82,** 2039 (1960); **Brit.** pat. **829,246;** G. E. Ullyot, **U.S.** pat. **3,058,979** (1960, 1962 both to SKF); **Brit.** pat. **833,474** (1960 to Scherico), *C.A.* **54,** 21143e (1960); E. L. Anderson *et al., Arzneimittel-Forsch.* **12,** 937 (1962); H. L. Yale, R. C. Merrill, **U.S.** pat. **3,194,733** (1965 to Olin Mathieson). Metabolism: J. Dreyfuss, A. J. Cohen, *J. Pharm. Sci.* **60,** 826 (1971). Comprehensive description of the enanthate ester: K. Florey, *Anal. Profiles Drug Subs.* **2,** 245-262 (1973); of the dihydrochloride: *idem, ibid.* 263-294; of the decanoate ester: G. Clarke, *ibid.* **9,** 275-294 (1980).

Dark brown viscous oil, $bp_{0.5}$ 268-274°; $bp_{0.3}$ 250-252°.

Dihydrochloride, $C_{22}H_{26}F_3N_3OS.2HCl$, *Anatensol, Dapotum, Lyogen, Moditen, Omca, Pacinol, Permitil, Prolixin, S iqualone, Tensofin, Valamina.* Crystals from abs ethanol, mp 235-237°. Also reported as mp 224.5-226°.

Decanoate, $C_{32}H_{44}F_3N_3O_2S$, *SQ-10733, QD-10733, Dapotum D, Lyogen Depot, Modecate, Prolixin Decanoate, Siqualine.* Pale yellow-orange, viscous liquid. Slowly crystallizes at room temp. mp 30-32°. Very sol in chloroform, ether, cyclohexane, methanol, ethanol. Insol in water.

Enanthate, $C_{29}H_{38}F_3N_3O_2S$, *SQ-16144, Moditen Enanthate, Prolixin Enanthate.* Pale yellow to yellow-orange viscous liquid or oily solid.

THERAP CAT: Antipsychotic.

4227. Flupirtine. *[2-Amino-6-[[(4-fluorophenyl)methyl]amino]-3-pyridinyl]carbamic acid ethyl ester; 2-amino-6-[(p-fluorobenzyl)amino]-3-pyridinecarbamic acid ethyl ester;* D-9998. $C_{15}H_{17}FN_4O_2$; mol wt 304.32. C 59.20%, H 5.63%, F 6.24%, N 18.41%, O 10.51%. Substituted pyridine with central analgesic properties. Prepn: K. Thiele, W. von Bebenburg, **S. Afr.** pat. **69 02,364** (1970 to Degussa); W. von Bebenburg *et al., Chem. Ztg.* **103,** 387 (1979); *eidem, ibid.* **105,** 217 (1981). Prepn of maleate: W. von Bebenburg, S. Pauluhn, **Belg.** pat. **890,331;** *eidem,* **U.S.** pat. **4,-481,205** (1980, 1984 both to Degussa). Comparison of pharmacology with other analgesics: V. Jakovlev *et al., Arzneimittel-forsch.* **35,** 30 (1985). Pharmacokinetic studies: K. Obermeier *et al., ibid.* 60. Effect on driving ability: B. Biehl, *ibid.* 77. Clinical trials in treatment of cancer pain: W. Scheef, D. Wolf-Gruber, *ibid.* 75. Efficacy in treatment of pain after hysterectomy: R. A. Moore *et al., Brit. J. Anaesth.* **55,** 429 (1983). Symposium on pharmacology and clinical efficacy: *Postgrad. Med. J.* **63,** Suppl. 3, 1-113 (1987).

Crystals from isopropanol, mp 115-116°. 5% ethanol soln is colorless, turns green on exposure to air for 20 hours. LD_{50} orally in mice, rats: 617, 1660 mg/kg (Jakovlev).

Hydrochloride, $C_{15}H_{17}FN_4O_2.HCl$, crystals from water, mp 214-215°. When prepd industrially contains intensely blue by-product.

Maleate, $C_{15}H_{17}FN_4O_2.C_4H_4O_4$, *Katadolon.* Colorless crystals from isopropanol, mp 175.5-176°. Formed as mixture of two crystalline forms A and B; mixtures containing 60-90% A are preferred.

THERAP CAT: Analgesic.

4228. Fluprednidene Acetate. *(11β)-21-(Acetyloxy)-9-fluoro-11,17-dihydroxy-16-methylenepregna-1,4-diene-3,20-dione; 9α-fluoro-11β,17,21-trihydroxy-16-methylenepregna-1,-4-diene-3,20-dione 21-acetate; 9α-fluoro-16-methylene-Δ1,4-pregnadiene-11β,17,21-triol-3,20-dione 21-acetate; 16-methylene-9α-fluoroprednisolone 21-acetate; 9α-fluoro-16-methyleneprednisolone 21-acetate;* fluprednylidene 21-acetate; StL-1106; Etacortin; Decoderm sine Gentamycin. $C_{24}H_{29}FO_6$; mol wt 432.49. C 66.65%, H 6.76%, F 4.39%, O 22.20%. Prepn: Taub *et al., J. Org. Chem.* **25,** 2258 (1960); **29,** 3486 (1964); v. Werder *et al., Arzneimittel-Forsch.* **18,** 7 (1968); Wendler, Taub; Taub, Wendler; Taub *et al.;* Wendler *et al.*, U.S. pats. **3,065,239; 3,068,224; 3,068,226** and **3,136,760** (1960 and 1964 to Merck & Co.).

Crystals, mp 231-234°. [α]$_D$ +43° (CHCl$_3$); [α]$_D$ +32° (dioxane) (v. Werder). uv max (methanol): 238 nm (ε 15700).
THERAP CAT: Anti-inflammatory (topical).

4229. Fluprednisolone. *(6α,11β)-6-Fluoro-11,17,21-trihydroxypregna-1,4-diene-3,20-dione;* 6α-fluoroprednisolone; 6α-fluoro-1,4-pregnadiene-11β,17α,21-triol-3,20-dione; 6α-fluoro-1-dehydrohydrocortisone; U-7800; NSC-47439; Alphadrol; Etadrol. C$_{21}$H$_{27}$FO$_5$; mol wt 378.44. C 66.65%, H 7.19%, F 5.02%, O 21.14%. Prepn: Hogg, Spero, U.S. pat. **2,841,600** (1958 to Upjohn); Batres *et al.*, **Ger.** pat. **1,079,042** (1960 to Syntex); Lettré, Hotz, **Ger.** pat. **1,088,-953** (1960 to Bayer).

Crystals, mp 208-213°. [α]$_D$ +92°.
21-Acetate, C$_{23}$H$_{29}$FO$_6$, crystals, mp 235-238°.
THERAP CAT: Glucocorticoid; anti-inflammatory.
THERAP CAT (VET): Anti-inflammatory.

4230. Fluproquazone. *4-(4-Fluorophenyl)-7-methyl-1-(1-methylethyl)-2(1H)-quinazolinone;* 1-isopropyl-7-methyl-4-(p-fluorophenyl)-2(1H)-quinazolinone; RF-46-790; Tormosyl. C$_{18}$H$_{17}$FN$_2$O; mol wt 296.34. C 72.96%, H 5.78%, F 6.41%, N 9.45%, O 5.40%. Fluoro analog of proquazone, *q.v.* Prepn: G. E. Hardtmann *et al.*, **Ger.** pat. **2,230,393** (1973 to Sandoz), *C.A.* **78**, 111363a (1973); G. E. Hardtmann, **U.S.** pat. **3,937,705** (1976 to Sandoz). Analgesic effect: U. Herrmann *et al.*, *Clin. Pharmacol. Ther.* **27**, 379 (1980); B. von Graffenried, E. Nuesch, *Brit. J. Clin. Pharmacol.* **10**, Suppl. 2, 225 (1980). Series of articles on pharmacology, metabolism, analysis, clinical studies, toxicology: *Arzneimittel-Forsch.* **31**, 871-940 (1981). Toxicity: G. Rüttimann *et al.*, *Arzneimittel-Forsch.* **31**, 882 (1981).

Crystals, mp 172-174°. LD$_{50}$ in mice, rats, rabbits (mg/kg): > 5000, > 3000, 742 orally (Rüttimann).
THERAP CAT: Analgesic.

4231. Fluprostenol. *[1α(Z),2β(1E,3R*),3α,5α]-(±)-7-[3,5-Dihydroxy-2-[3-hydroxy-4-[3-(trifluoromethyl)phenoxy]-1-butenyl]cyclopentyl]-5-heptenoic acid;* ICI-81008; E-quimate. C$_{23}$H$_{29}$F$_3$O$_6$; mol wt 458.47. C 60.25%, H 6.38%, F 12.43%, O 20.94%. Potent luteolytic agent related to prostaglandin F$_{2α}$, *q.v.*: Dukes, Walpole, *Nature* **250**, 330 (1974). Prepn: Binder *et al.*, *Prostaglandins* **6**, 87 (1974). Induction of estrus and luteolysis in mares: Cooper, Farr, *Vet. Rec.* **94**, 161 (1974); Berwyn-Jones, Irvine, *N.Z. Vet. J.* **22**, 107 (1974).

Sodium salt, C$_{23}$H$_{28}$F$_3$NaO$_6$, *ICI-80008.*
THERAP CAT (VET): Treatment of infertility in mares.

4232. Flurandrenolide. *(6α,11β,16α)-6-Fluoro-11,21-dihydroxy-16,17-[(1-methylethylidene)bis(oxy)]pregn-4-ene-3,-20-dione;* 6α-fluoro-11β,16α,17,21-tetrahydroxypregn-4-ene-3,20-dione cyclic 16,17-acetal with acetone; 6α-fluoro-16α,-17-isopropylidenedioxy-4-pregnene-11β,21-diol-3,20-dione; 6α-fluoro-11β,16α,17,21-tetrahydroxyprogesterone cyclic 16,17-acetal with acetone; fludroxycortide; fluorandrenolone; flurandrenolone; flurandrenolone acetonide (obsolete); Cordran; Drenison; Drocort; Haelan; Sermaka. C$_{24}$H$_{33}$FO$_6$; mol wt 436.52. C 66.04%, H 7.62%, F 4.35%, O 21.99%. Prepn: Ringold *et al.*, **Ger.** pat. **1,131,213** corresp to U.S. pat. **3,126,375** (1962, 1964 both to Syntex).

Crystals from acetone + hexane, mp 247-255°. [α]$_D$ +140-150° (CHCl$_3$). uv max: 236 nm (log ε 4.17).
THERAP CAT: Glucocorticoid; anti-inflammatory.

4233. Flurazepam. *7-Chloro-1-[2-(diethylamino)ethyl]-5-(2-fluorophenyl)-1,3-dihydro-2H-1,4-benzodiazepin-2-one;* Felmane; Noctosom; Stauroderm. C$_{21}$H$_{23}$ClFN$_3$O; mol wt 387.88. C 65.03%, H 5.98%, Cl 9.14%, F 4.90%, N 10.83%, O 4.12%. Prepn: G. A. Archer *et al.*, **Belg.** pat. **629,005,** *C.A.* **60**, 15896f (1964); **Neth.** pat. **Appl. 6,401,335,** *C.A.* **62**, 5289a (1965); G. A. Archer *et al.*, **U.S.** pat. **3,299,053** (1963, 1964, 1967 all to Hoffmann-La Roche). Manuf process: R. Fryer, L. H. Sternbach, U.S. pat. **3,567,710** (1971 to Hoffmann-La Roche). Synthesis: S. Inaba *et al.*, *Chem. Pharm. Bull.* **19**, 263 (1971). Structure-activity studies: L. H. Sternbach *et al.*, *J. Med. Chem.* **8**, 815 (1965); Armagnac *et al.*, *Therapie* **26**, 439 (1971). Metabolism: Schwartz *et al.*, *J. Med. Chem.* **11**, 770 (1968); *eidem, J. Pharm. Sci.* **59**, 1800 (1970). Pharmacology: L. O. Randall *et al.*, *Arch. Int. Pharmacodyn. Ther.* **178**, 216 (1969). Toxicology: H. C. Rosenberg, *Pharmacol. Biochem. Behav.* **13**, 415 (1980). GC determn in plasma: Z. Salama *et al.*, *Arzneimittel-Forsch.* **38**, 400 (1988). Comprehensive description: B. C. Rudy, B. Z. Senkowski, *Anal. Profiles Drug Subs.* **3**, 307-331 (1974).

White rods from ether-petr ether, mp 77-82°.

Dihydrochloride, $C_{21}H_{23}ClFN_3O.2HCl$, *flurazepam hydrochloride, Ro-5-6901, Benozil, Dalmadorm, Dalmane, Dalmate, Dormodor, Felison, Insumin, Lunipax, Somlan.* Pale yellow crystals from methanol + ether, mp 190-220°. LD_{50} in mice (mg/kg): 290 i.p.; 870 orally; 84 i.v. (Randall).

Note: This is a controlled substance (depressant) listed in the U.S. Code of Federal Regulations, Title 21 Part 1308.14 (1995).

THERAP CAT: Sedative, hypnotic.

4234. Flurbiprofen. *2-Fluoro-α-methyl[1,1'-biphenyl]-4-acetic acid; 2-fluoro-α-methyl-4-biphenylacetic acid;* 2-(2-fluoro-4-biphenylyl)propionic acid; 3-fluoro-4-phenylhydratropic acid; BTS-18322; U-27182; Adfeed; Ansaid; Antadys; Cebutid; Froben; Flurofen; Ocufen; Stayban; Zepolas. $C_{15}H_{13}FO_2$; mol wt 244.27. C 73.76%, H 5.36%, F 7.78%, O 13.10%. Prepn: **Fr.** pat. **M5737;** Adams *et al.*, **U.S.** pat. **3,755,427** (1968, 1973 both to Boots Co., Ltd.). Pharmacology: Chalmers *et al., Ann. Rheum. Dis.* **31**, 319 (1972); *ibid.* **32**, 58 (1973); Glenn *et al., Agents Actions* **3**, 210 (1973); Nishizawa *et al., Thromb. Res.* **3**, 577 (1973). Symposium on pharmacokinetics and clinical efficacy in pain management: *Am. J. Med.* **80**, Suppl. 3A, 1-157 (1986).

Crystals from petr ether, mp 110-111°.

THERAP CAT: Anti-inflammatory; analgesic.

4235. Flurogestone Acetate. *(11β)-17-(Acetyloxy)-9-fluoro-11-hydroxypregn-4-ene-3,20-dione; 9-fluoro-11β,17-dihydroxypregn-4-ene-3,20-dione-17-acetate;* 17α-acetoxy-9α-fluoro-11β-hydroxyprogesterone; 9-fluoro-11β,17-dihydroxyprogesterone 17-acetate; flugestone acetate; SC-9880; Chronogest; Cronolone; Synchronate. $C_{23}H_{31}FO_5$; mol wt 406.49. C 67.96%, H 7.69%, F 4.67%, O 19.68%. Prepn: Bergstrom *et al., J. Am. Chem. Soc.* **81**, 4432 (1959); Bergstrom, Dodson; Bergstrom, Nicholson, **U.S.** pats. **2,892,851; 2,963,498** (1959, 1960 to Searle).

Crystals from benzene + petr ether, or ethyl acetate + petr ether, mp 266-269°. $[\alpha]_D$ +77.6° (in chloroform). uv max (methanol): 238 nm (ϵ 17500).

Component of *Syncro-Mate.*

THERAP CAT: Progestogen.

THERAP CAT (VET): Progestogen. Estrus regulation.

4236. Flurothyl. *1,1'-Oxybis[2,2,2-trifluoroethane]; bis-(2,2,2-trifluoroethyl) ether;* hexafluorodiethyl ether; bis(trifluoroethyl) ether; Indoklon. $C_4H_4F_6O$; mol wt 182.07. C

26.39%, H 2.21%, F 62.61%, O 8.79%. $CF_3CH_2OCH_2CF_3$. Prepn: **Brit.** pat. **814,493** (1959 to Pennsalt Chem.); **Brit.** pat. **889,282** (1962 to Air Reduction). Toxicology in animals: A. Cherkin, *Psychopharmacologia* **15**, 404 (1969). Review of pharmacology: L. Arce, *Psychosomatics* **11**, 358-360 (1970); of clinical trials in psychiatric convulsant therapy: J. G. Small, I. F. Small, *Sem. Psychiatry* **4**, 13-26 (1972).

Colorless, mobile liq, mild ethereal, pleasant odor. d_4^{20} 1.41. bp 63.9°. Practically insol in water. Sol in alcohol.

THERAP CAT: CNS stimulant.

4237. Fluroxene. *(2,2,2-Trifluoroethoxy)ethene; 2,2,2-trifluoroethyl vinyl ether;* Fluoromar. $C_4H_5F_3O$; mol wt 126.08. C 38.11%, H 4.00%, F 45.21%, O 12.69%. $CF_3CH_2-OCH=CH_2$. Prepn: Shukys, **U.S.** pat. **2,830,007** (1958) and Townsend, **U.S.** pat. **2,870,218** (1959 to Air Reduction).

Liquid. bp_{751} 42.5°. n_D^{20} 1.3192. d^{20} 1.135. *Potentially explosive.*

THERAP CAT: Anesthetic (inhalation).

4238. Fluroxypyr. *[(4-Amino-3,5-dichloro-6-fluoro-2-pyridinyl)oxy]acetic acid.* $C_7H_5Cl_2FN_2O_3$; mol wt 255.03. C 32.97%, H 1.98%, Cl 27.80%, F 7.45%, N 10.98%, O 18.82%. Auxin-type herbicide for use in cereals. Prepn: S. D. McGregor, **U.S.** pat. **3,761,486** (1973 to Dow). Description: J. A. Paul *et al., Proc. Brit. Crop Prot. Conf. - Weeds* **1985**, 939. Field trials in grassland: A. R. Thompson, *ibid.* **1987**, 735. Mode of action: G. E. Sanders, K. E. Pallet, *Ann. Appl. Biol.* **111**, 385 (1987). HPLC determn: R. G. Lehmann, J. R. Miller, *J. Chromatog.* **485**, 581 (1989). Persistence in soil: L. F. Bergström *et al., Pestic. Sci.* **29**, 405 (1990). Uptake and metabolism in weeds: R. L. MacDonald *et al., Weed Res.* **34**, 333 (1994).

White crystalline solid, mp 232-233°. Vapor pressure (25°): 9.42×10^{-7} mm Hg. Soly at 20° (g/l): water 0.091; acetone 41.6. LD_{50} orally in rats: 2405 mg/kg; i.p. in male, female rats: 458, 519 mg/kg; percutaneous in rabbits > 5000 mg/kg (Paul).

1-Methylheptyl ester, $C_{15}H_{21}Cl_2FN_2O_3$, *fluroxypyr-meptyl, Dowco 433, Starane.*

USE: Herbicide.

4239. Flurprimidol. *α-(1-Methylethyl)-α-[4-(trifluoromethoxy)phenyl]-5-pyrimidinemethanol; α-isopropyl-α-[p-(trifluoromethoxy)phenyl]-5-pyrimidinemethanol;* Compd 72500; EL-500; Cutless. $C_{15}H_{15}F_3N_2O_2$; mol wt 312.29. C 57.69%, H 4.84%, F 18.25%, N 8.97%, O 10.25%. Growth regulator for use on established turfgrass. Prepn: R. L. Benefiel, **Belg.** pat. **815,245** (1974 to Lilly), *C.A.* **83**, 97354t (1975); R. L. Benefiel, E. V. Krumkalns, **U.S.** pats. **3,967,949, 4,002,628** (1976, 1977 both to Lilly). Growth retardant properties, effect on turfgrass: L. G. Thompson, *Proc. West. Soc. Weed Sci.* **35**, 113 (1982). Effect on Kentucky bluegrass, comparison with other growth retardants: T. L. Watschke, *Proc. Ann. Meet. Northeast. Weed Sci. Soc.* **35**, 322 (1981); N. E. Christians, *J. Am. Soc. Hort. Sci.* **110**, 765 (1985). Long term study of effect on bluegrass-fescue turf: P. H. Dernoeden, *Agron. J.* **76**, 807 (1984). Efficacy in dwarfing southern pine seedlings: R. C. Hare, *Can. J. For. Res.* **14**, 123 (1984).

Non-volatile white crystals, mp 94-96°. Sol in acetone, ethanol, methanol, DMSO, diethyl ether. LD_0 orally in rats: > 500 mg/kg; LD_{50} dermally in rabbits: > 2000 mg/kg (Thompson).

USE: Growth retardant for grasses.

4240. Flusilazole. *1-[[Bis(4-fluorophenyl)methylsilyl]-methyl]-1H-1,2,4-triazole;* bis(4-fluorophenyl)methyl(1H-1,2,4-triazol-1-ylmethyl)silane; fluzilazol; DPX-H6573; Nustar; Olymp; Punch. $C_{16}H_{15}F_2N_3Si$; mol wt 315.40. C 60.93%, H 4.79%, F 12.05%, N 13.32%, Si 8.90%. Sterol-inhibiting, broad spectrum foliar fungicide. Prepn: W. K. Moberg, **Eur. pat Appl. 68,813;** idem, **U.S. pat 4,510,136** (1983, 1985 both to Du Pont); W. K. Moberg et al., Pest. Sci. Biotech., Proc. 6th Int. Congr. Pest. Chem., 57 (1987). Field performance, physical properties and toxicity: T. M. Fort, W. K. Moberg, Proc. Brit. Crop Prot. Conf.-Pest Dis. **1984,** 413. Field trials against apple scab: A. R. Biggs, J. Warner, Can. J. Plant Pathol. **9,** 41 (1987); and mildew: A. A. J. Swait et al., Tests Agrochem. Cultiv. **8,** 42 (1987).

Crystalline solid, mp 55°. Soly: > 2 g/ml in many organic solvents. LD_{50} in male, female rats (mg/kg): 1110, 674 orally; in rabbits > 2000 dermally (Fort, Moberg).

USE: Agricultural fungicide.

4241. Fluspirilene. *8-[4,4-Bis(4-fluorophenyl)butyl]-1-phenyl-1,3,8-triazaspiro[4.5]decan-4-one;* R-6128; Imap; Redeptin. $C_{29}H_{31}F_2N_3O$; mol wt 475.58. C 73.24%, H 6.57%, F 7.99%, N 8.84%, O 3.36%. Prepn: **Belg. pat. 633,914;** P. A. J. Janssen, **U.S. pat. 3,238,216** (1963, 1966 both to Janssen). HPLC determn: A. H. Hikal, H. I. Al-Shoura, J. Liq. Chromatog. **5,** 2205 (1982). Metabolism in rat: J. P. P. Heykants, Life Sci. **8,** Part I, 1029 (1969). Pharmacology: P. A. J. Janssen et al., Arzneimittel-Forsch. **20,** 1689 (1970). Calcium channel blocking activity: R. J. Gould et al., Proc. Nat. Acad. Sci. USA **80,** 5122 (1983); J. P. Galizzi et al., ibid. **83,** 7513 (1986). Clinical trials in schizophrenia: H. Immich et al., Arzneimittel-Forsch. **20,** 1699 (1970); G. Chouinard et al., J. Clin. Psychopharmacol. **6,** 21 (1986).

White to yellowish amorphous or crystalline solid, mp 187.5-190°. Soly in water: 0.015-0.020 mg/ml. LD_{50} i.m. in rats: 146 ±14 mg/kg (Janssen).

THERAP CAT: Antipsychotic.

4242. Flutamide. *2-Methyl-N-[4-nitro-3-(trifluoromethyl)phenyl]propanamide;* α,α,α-trifluoro-2-methyl-4'-nitro-m-propionotoluidide; 4'-nitro-3'-trifluoromethylisobutyranilide; niftolid; Sch-13521; Drogenil; Eulexin; Euflex; Flucinom; Fugerel. $C_{11}H_{11}F_3N_2O_3$; mol wt 276.22. C 47.83%, H 4.01%, F 20.63%, N 10.14%, O 17.38%. Prepn: Baker et al., J. Med. Chem. **10,** 93 (1967); Neri, Topliss, **Ger. pat. 2,130,-450** (1972 to Sherico), C.A. **78,** 58091g (1973); Gold, **Ger. pat. 2,261,293;** idem, **U.S. pat. 3,847,988** (1973, 1974 both to Schering). Activity studies: Neri et al., Endocrinology **91,** 427 (1972); Peets et al., ibid. **94,** 532 (1974); B. A. Gladue, L. G. Clemens, Endocrinology **106,** 1917 (1980). Effect in prostatic cancer: P. C. Sogani, W. F. Whitmore, J. Urol. **122,** 640 (1979).

Crystals from benzene, mp 111.5-112.5°.

THERAP CAT: Antiandrogen; antineoplastic (hormonal).

THERAP CAT (VET): Antiandrogen.

4243. Flutazolam. *10-Chloro-11b-(2-fluorophenyl)-2,3,-7,11b-tetrahydro-7-(2-hydroxyethyl)oxazolo[3,2-d][1,4]benzodiazepin-6(5H)-one;* MS-4101; Ro-7-6102; Coreminal. $C_{19}H_{18}ClFN_2O_3$; mol wt 376.81. C 60.56%, H 4.81%, Cl 9.41%, F 5.04%, N 7.43%, O 12.74%. Tricyclic benzodiazepine deriv. Prepn: **Belg. pat. 765,253;** M. E. Derieg et al., **U.S. pat. 3,905,956** (1971, 1975 both to Hoffmann-La Roche). Physicochemical properties: A. Ito et al., Iyakuhin Kenkyu **9,** 787 (1978), C.A. **89,** 169013e (1979). Psychopharmacological profile: T. Mitsushima, S. Ueki, Folia Pharmacol. Japon. **74,** 959 (1978), C.A. **90,** 180083u (1979). Toxicological studies: K. Ishimura et al., Oyo Yakuri **15,** 1081 (1978); eidem, ibid. **16,** 693 (1978), C.A. **90,** 16379z, 145909p (1979).

White prisms from toluene, mp 142-147°. pKa 5.40. Freely sol in chloroform, ethanol. Moderately sol in acetone, benzene, methanol. Practically insol in water.

THERAP CAT: Anxiolytic.

4244. Fluticasone Propionate. *(6α,11β,16α,17α)-6,9-Difluoro-11-hydroxy-16-methyl-3-oxo-17-(1-oxopropoxy)androsta-1,4-diene-17-carbothioic acid S-(fluoromethyl) ester;* S-fluoromethyl 6α,9α-difluoro-11β-hydroxy-16α-methyl-17α-propionyloxy-3-oxoandrosta-1,4-diene-17β-carbothioate; CCI-18781; Cutivate; Flixonase; Flixotide; Flovent; Flunase. $C_{25}H_{31}F_3O_5S$; mol wt 500.58. C 59.99%, H 6.24%, F 11.39%, O 15.98%, S 6.41%. Derivative of flumethasone, q.v. Prepn: **Neth. pat. Appl. 81 00,707;** G. H. Phillipps et al., **U.S. pat. 4,335,121** (1981, 1982 both to Glaxo). Series of articles on structure-activity, pharmacology and clinical studies: Respir. Med. **84,** Suppl. A, 19-35 (1990). Clinical trials in allergic rhinitis: E. O. Meltzer et al., J. Allergy Clin. Immunol. **86,** 221 (1990). Evaluation in untreated celiac disease: H. C. Mitchison et al., Gut **32,** 260 (1991).

Crystals, mp 272-273° (dec). $[\alpha]_D$ +30° (c = 0.35).

THERAP CAT: Antiallergic; anti-inflammatory.

4245. Flutoprazepam. *7-Chloro-1-(cyclopropylmethyl)-5-(2-fluorophenyl)-1,3-dihydro-2H-1,4-benzodiazepin-2-one;* KB-509; ID-1937; Restar; Restas. $C_{19}H_{16}ClFN_2O$; mol wt 342.80. C 66.57%, H 4.70%, Cl 10.34%, F 5.54%, N 8.17%,

O 4.67%. Benzodiazepine derivative. Prepn: H. Yamamoto *et al.*, **Brit.** pat. **1,253,368;** *eidem,* U.S. pat. **3,925,364** (1971, 1975 both to Sumitomo). New process: T. Okamoto *et al.*, **Ger.** pat. **2,151,540;** *eidem,* U.S. pat. **3,832,344** (1972, 1974 both to Sumitomo). Stability and degradation mechanism in aqueous solns: K. Akimoto *et al.*, *Int. J. Pharmacol.* **3,** 109 (1979). Pharmacokinetics and metabolism: A. Kozaki *et al.*, *Yakugaku Zasshi* **102,** 1177 (1982), *C.A.* **98,** 119080q (1983). Pharmacology of flutoprazepam and metabolites: K. Oki *et al.*, *Arch. Int. J. Pharmacodyn.* **269,** 180 (1984). Acute toxicity study: H. Takebe *et al.*, *Oyo Yakuri* **20,** 1055 (1980), *C.A.* **95,** 73630b (1981).

Crystals, mp 118°-122°. LD_{50} in male, female mice, rats (mg/kg): 2640, 2430, 13760, 10060 orally; 2400, 2110, 2460, 2230 i.p.; > 5000 s.c. all species (Takebe).

THERAP CAT: Anxiolytic.

4246. Flutriafol. *α-(2-Fluorophenyl)-α-(4-fluorophenyl)-1H-1,2,4-triazole-1-ethanol; (RS)-2,4'-difluoro-α-(1H-1,2,4-triazol-1-ylmethyl)benzhydryl alcohol;* flutriafen; R-152450; PP-450; Impact. $C_{16}H_{13}F_2N_3O$; mol wt 301.30. C 63.78%, H 4.35%, F 12.61%, N 13.95%, O 5.31%. Prepn: K. P. Parry *et al.*, **Eur.** pat. **Appl. 15,756** (1980 to ICI). Activity, physical properties, and toxicology: A. M. Skidmore *et al.*, *Proc. 10th Int. Congr. Plant Prot.* **1,** 368 (1983). Mode of action: B. C. Baldwin *et al.*, *Med. Fac. Landbouww. Rijksuniv. Gent* **49,** 303 (1984). Field trials: P. J. Northwood *et al.*, *Proc. 10th Int. Congr. Plant Prot.* **3,** 930 (1983); J. S. Brown, *Crop Prot.* **6,** 157 (1987).

White crystalline solid, mp 130°. Vapor pressure at 20°: 3.00 × 10⁻⁹ mm Hg. Soly at 20° (g/l): acetone 190; dichloromethane 150; hexane 0.30; methanol 69; xylene 12; water (pH 7.0) 0.13. LD_{50} male, female rats (mg/kg): 1140, 1480 orally; rats, rabbits: > 1000, > 2000 percutaneously (Skidmore).

USE: Agricultural fungicide.

4247. Flutrimazole. *1-[(2-Fluorophenyl)(4-fluorophenyl)phenylmethyl]-1H-imidazole; 1-[o-fluoro-α-(p-fluorophenyl)-α-phenylbenzyl]imidazole;* UR-4056. $C_{22}H_{16}F_2N_2$; mol wt 346.38. C 76.29%, H 4.66%, F 10.97%, N 8.09%. Prepn: J. Bartroli, M. Anguita, **Eur.** pat. **Appl. 352,352;** *eidem,* U.S. pat. **5,149,707** (1990, 1992 both to Uriach). Series of articles on synthesis, antifungal activity, pharmacokinetics: *Arzneimittel-Forsch.* **42,** 832-863 (1992). Toxicity study: M. L. Vericat *et al.*, *ibid.* 841.

Crystals from acetonitrile, mp 164°-167°. Also reported as white solid, mp 161-163°. LD_{50} in male, female mice, male, female rats (mg/kg): > 1000, > 1000, 808, 1214 orally; > 2000, > 2000, 1079, 1446 i.p. (Vericat).

THERAP CAT: Antifungal (topical).

4248. Flutropium Bromide. *(endo,syn)-8-(2-Fluoroethyl)-3-[(hydroxydiphenylacetyl)oxy]-8-methyl-8-azoniabicyclo-[3.2.1]octane bromide; (8r)-3α-benziloyloxy-8-(2-fluoroethyl)-8-methyl-8-azoniabicyclo[3.2.1]octane bromide; (8r)-8-(2-fluoroethyl)-3α-hydroxy-1αH,5αH-tropanium bromide benzilate; N-β-fluoroethylnortropine benzilate methobromide; benzilic acid N-β-fluoroethylnortropine ester methobromide;* Ba-598Br; Flubron. $C_{24}H_{29}BrFNO_3$; mol wt 478.40. C 60.26%, H 6.11%, Br 16.70%, F 3.97%, N 2.93%, O 10.03%. Anticholinergic agent. Prepn: R. Banholzer *et al.*, **Ger.** pat. **2,540,633;** *eidem,* U.S. pat. **4,042,700** (both 1977 to Boehringer, Ing.). Synthesis: R. Banholzer *et al.*, *Arzneimittel-Forsch.* **36,** 1161 (1986). Molecular and crystal structure: G. Kiel, *ibid.* 1166. Pharmacology in animals: S. Yanaura *et al.*, *Japan. J. Pharmacol.* **33,** 971 (1983); R. Bauer, A. Fügner, *Arzneimittel-Forsch.* **36,** 1348 (1986). Pharmacodynamics: K. Yoshimura *et al.*, *Iyakuhin Kenkyu* **18,** 240 (1987), *C.A.* **107,** 17197v (1987).

White crystals from acetonitrile, mp 192-193° (dec); from ethanol-acetone, mp 198-199° (dec). LD_{50} in male, female mice, male, female rats (mg/kg): 12.5, 11.0, 16.4, 18.4 i.v.; 760, 810, 830, 740 orally (Bauer, Fügner).

THERAP CAT: Bronchodilator.

4249. Fluvalinate. *N-[2-Chloro-4-(trifluoromethyl)-phenyl]-DL-valine cyano(3-phenoxyphenyl)methyl ester;* ZR-3210; Mavrik. $C_{26}H_{22}ClF_3N_2O_3$; mol wt 502.92. C 62.09%, H 4.41%, Cl 7.05%, F 11.33%, N 5.57%, O 9.54%. Synthetic pyrethroid insecticide without the usual cyclopropane ring. Prepn: C. A. Henrick, B. A. Garcia, **Ger.** pat. **2,812,169** (1978 to Zoecon), *C.A.* **90,** 122072d (1970). Activity: C. A. Henrick *et al.*, *Pestic. Sci.* **11,** 224 (1980).

Yellow-amber liquid. Vapor press. at 25°: 1 × 10⁻⁷ mm Hg. Soly in water: 2.0 ppb. Sol in organic solvents.

USE: Insecticide.

4250. Fluvastatin. *[R*,S*-(E)]-(±)-7-[3-(4-Fluorophenyl)-1-(1-methylethyl)-1H-indol-2-yl]-3,5-dihydroxy-6-heptenoic acid; (±)-(3R*,5S*,6E)-7-[3-(p-fluorophenyl)-1-isopropylindol-2-yl]-3,5-dihydroxy-6-heptenoate;* $C_{24}H_{26}$-

FNO_4; mol wt 411.47. C 70.06%, H 6.37%, F 4.62%, N 3.40%, O 15.55%. Synthetic HMG-CoA reductase inhibitor. Prepn: F. G. Kathawala, **PCT Int.** pat. **Appl. 84 02131;** idem, **U.S.** pat. **4,739,073** (1984, 1988 both to Sandoz). Clinical pharmacology: J. Yuan et al., Atherosclerosis **87,** 147 (1991). Clinical pharmacokinetics: F. L. S. Tse et al., J. Clin. Pharmacol. **32,** 630 (1992). Metabolism in humans: J. G. Dain et al., Drug Metab. Disposit. **21,** 567 (1993). HPLC determn in plasma: G. Kalafsky et al., J. Chromatog. Biomed. Appl. **614,** 307 (1993). Clinical study: E. Leitersdorf et al., Eur. J. Clin. Pharmacol. **45,** 513 (1993). Review of pharmacology and therapeutic use: R. I. Levy et al., Circulation **87,** Suppl. III, III-45—III-53 (1993).

Sodium salt, $C_{24}H_{25}FNNaO_4$, fluindostatin, XU 62-320, Lescol. mp 194-197°.

THERAP CAT: Antihyperlipoproteinemic.

4251. Fluvoxamine. (E)-5-Methoxy-1-[4-(trifluoromethyl)phenyl]-1-pentanone O-(2-aminoethyl)oxime; 5-methoxy-4'-(trifluoromethyl)valerophenone (E)-O-(2-aminoethyl)oxime. $C_{15}H_{21}F_3N_2O_2$; mol wt 318.34. C 56.60%, H 6.65%, F 17.90%, N 8.80%, O 10.05%. Serotonin uptake inhibitor. Prepn: **Neth.** pat. **Appl. 7,503,310;** H. B. A. Welle, V. Claassen, **U.S.** pat. **4,085,225** (1975, 1978 both to Philips-Duphar). Inhibition of 5-HT uptake: V. Claassen et al., Brit. J. Pharmacol. **60,** 505 (1977). HPLC determn in plasma: G. J. De Jong, J. Chromatog. **183,** 203 (1980). Quantitative EEG, psychometric and pharmacokinetic studies in man: B. Salctu et al., J. Neurol. Transm. **49,** 63 (1980). Use in endogenous depression: J. E. De Wilde, D. P. Doogan, J. Affective Disord. **4,** 249 (1982). Effects on clonidine-induced depression in mice: J. Maj et al., J. Neurol. Transm. **55,** 19 (1982). Review: Brit. J. Clin. Pharmacol. **15,** Suppl. 3, 347S-450S (1983). Review of pharmacology, clinical efficacy: P. Benfield, A. Ward, Drugs **32,** 313 (1986).

Maleate, $C_{15}H_{21}F_3N_2O_2 \cdot C_4H_4O_4$, DU-23000, MK-264, Dumirox, Faverin, Fevarin, Floxyfral, Luvox, Maveral. Crystals from acetonitrile, mp 120-121.5°.

THERAP CAT: Antidepressant.

4252. Folescutol. 6,7-Dihydroxy-4-(4-morpholinylmethyl)-2H-1-benzopyran-2-one; 6,7-dihydroxy-4-(morpholinomethyl)coumarin; 4-morpholinomethylescutol; pholescutol. $C_{14}H_{15}NO_5$; mol wt 277.28. C 60.64%, H 5.45%, N 5.05%, O 28.85%. Prepn by the reaction of 4-chloromethyl-6,7-dihydroxycoumarin with morpholine: **Fr.** pat. **M2035** (1963 to Labs. Dausse), C.A. **60,** 4113 (1964). Pharmacological study: J. P. Tarayre et al., Ann. Pharm. Franc. **33,** 467 (1975).

Crystals from 50% ethanol, mp 232°.

Hydrochloride, $C_{14}H_{15}NO_5 \cdot HCl$, **LD-2988,** Covalan. Crystals from 75% ethanol, mp 259-261°.

THERAP CAT: Capillary protectant.

4253. Folic Acid. N-[4-[[(2-Amino-1,4-dihydro-4-oxo-6-pteridinyl)methyl]amino]benzoyl]-L-glutamic acid; N-[p-[[(2-amino-4-hydroxy-6-pteridinyl)methyl]amino]benzoyl]glutamic acid; pteroylglutamic acid; N-(p-[(2-amino-4-hydroxy-pyrimido[4,5-b]pyrazin-6-yl)methylamino]benzoyl)glutamic acid; PGA; liver Lactobacillus casei factor; vitamin Bc; vitamin M; folsäure; Cytofol; Folacin; Foldine; Foliamin; Folicet; Folipac; Folettes; Folsan; Folvite; Incafolic; Millafol. $C_{19}H_{19}N_7O_6$; mol wt 441.40. C 51.70%, H 4.34%, N 22.21%, O 21.75%. Hematopoietic vitamin present, free or combined with one or more additional molecules of L(+)-glutamic acid, in liver, kidney, mushrooms, spinach, yeast, green leaves, grasses: Mitchell et al., J. Am. Chem. Soc. **63,** 2284 (1941). Isoln: Pfiffner et al., ibid. **69,** 1476 (1947); Stokstad et al., ibid. **70,** 3 (1948). Structure: Mowat et al., ibid. 14. Crystal structure: D. Mastropaolo et al., Science **210,** 334 (1980). History of the different folic acid factors: Ann. N. Y. Acad. Sci. **48,** 255-350 (1946). See also reviews by Subbarow et al. in Vitam. Horm. (New York) **3,** 237-296 (1945); Pfiffner, Hogan, ibid. **4,** 1-13 (1946). Several syntheses, see reviews by Gates, Chem. Rev. **41,** 63-95 (1947); Hutchings, Mowat, Vitam. Horm. (New York) **6,** 1-25 (1948); Sletzinger et al., J. Am. Chem. Soc. **77,** 6365 (1955); **U.S.** pats. **2,786,-056; 2,816,109; 2,821,527-8** (all 1958 to Merck & Co.); Sadao Kawanishi, **U.S.** pat. **2,956,057** (1960 to Kongo Kagaku Kabushiki Kaisha). Alternate syntheses: Bieri, Viscontini, Helv. Chim. Acta **56,** 2905 (1973); E. Khalifa et al., ibid. **59,** 242 (1976). Comprehensive reviews: Jaenicki, Kutzbach, Fortschr. Chem. Org. Naturst. **21,** 183-274 (1963); Marchetti, Acta Vitaminol. Enzymol. **25,** 41-64 (1971); F. J. Al-Shammary et al., Anal. Profiles Drug Subs. **19,** 221-259 (1990).

Yellowish-orange crystals. Extremely thin platelets (elongated at two ends) from hot water, no mp. Darkens and chars from about 250°. $[\alpha]_D^{25}$ +23° (c = 0.5 in 0.1N NaOH). uv max (pH 13): 256, 283, 368 nm (log ∈ 4.43, 4.40, 3.96). Very slightly sol in cold water (0.0016 mg/ml at 25°), sol to about 1% in boiling water. Slightly sol in methanol, appreciably less sol in ethanol and butanol. Insol in acetone, chloroform, ether, benzene. Relatively sol in acetic acid, phenol, pyridine, solns of alkali hydroxides and carbonates. Sol in hot dil HCl and H_2SO_4. Injectable solns are prepd by dissolving folic acid in normal sodium bicarbonate soln (which should be sterilized by filtration) or by preparing solns of the sodium or methylglucamine salt. A suspension of 1 g folic acid in 10 ml water has a pH of 4.0-4.8. Aq solns prepd with sodium bicarbonate have a pH between 6.5 and 6.8.

THERAP CAT: Vitamin (hematopoietic).

THERAP CAT (VET): Nutritional factor (dietary requirement in poultry).

4254. Folinic Acid. *N-[4-[[(2-Amino-5-formyl-1,4,5,6,-7,8-hexahydro-4-oxo-6-pteridinyl)methyl]amino]benzoyl]-L-glutamic acid; N-[p-[[(2-amino-5-formyl-5,6,7,8-tetrahydro-4-hydroxy-6-pteridinyl)methyl]amino]benzoyl]glutamic acid; 5-formyl-5,6,7,8-tetrahydropteroyl-L-glutamic acid; 5-formyl-5,6,7,8-tetrahydrofolic acid; CF; citrovorum factor; leucovorin.* $C_{20}H_{23}N_7O_7$; mol wt 473.45. C 50.74%, H 4.90%, N 20.71%, O 23.66%. Intermediate product of the metabolism of folic acid; the active form into which that acid is converted in the body, ascorbic acid being a necessary factor in the conversion process. First reported as the *Leuconostoc citrovorum* 8081 growth factor: H. E. Sauberlich, C. A. Baumann, *J. Biol. Chem.* **176,** 165 (1948). Isoln from houseflies: S. Miller, A. S. Perry, *Life Sci.* **4,** 1573 (1965). Prepn: J. A. Brockman, Jr., *et al., J. Am. Chem. Soc.* **72,** 4325 (1950); E. Khalifa *et al., Helv. Chim. Acta* **63,** 2554 (1980). Isomers: D. B. Cosulich *et al., J. Am. Chem. Soc.* **74,** 4215 (1952). Structure: May *et al., ibid.* **73,** 3067 (1951); Pohland *et al., ibid.* 3247. Manuf: Shive, U.S. pat. **2,741,608** (1956 to Res. Corp.). Stereoselective synthesis: J. Owens, *et al., J. Chem. Soc., Perkin Trans. 1* **7,** 871 (1993). Used as an antidote to folic acid antagonists such as methotrexate, *q.v.,* which block the conversion of folic acid into folinic acid. Review of clinical combination therapy with methotrexate: J. R. Bertino *et al., Ann. N.Y. Acad. Sci.* **186,** 486-495 (1971). Pharmacokinetics: P. F. Nixon, *Clin. Exp. Pharmacol. Physiol.* **Suppl. 5,** 35 (1979). Comprehensive description: L. O. Pont *et al., Anal. Profiles Drug Subs.* **8,** 315-350 (1979). Review of clinical synergy with fluorouracil in cancer: R. J. DeLap, *Yale J. Biol. Med.* **61,** 23-34 (1988).

Crystals, dec 240-250°. uv max (0.1*N* NaOH): 282 nm (% T = 27.0 for 10 mg/l). $[\alpha]_D^{20}$ +14.26° (c = 3.42 as anhydr Ca salt). pKa (3 groups): 3.1, 4.8, and 10.4. Sparingly sol in water. pH of satd aq soln 2.8-3.0 at which pH partial decompn takes place. More stable at neutral or mildly alkaline pH.

Calcium salt pentahydrate, $C_{20}H_{21}CaN_7O_7.5H_2O$, *calcium folinate, NSC-3590, Folaren, Foliben, Lederfolat, Lederfolin, Leucovorin, Leucosar, Rescufolin, Rescuvolin, Tonofolin, Wellcovorin.* Off-white to light beige amorphous, odorless powder, freely sol in water. Practically insol in alc. $[\alpha]_D^{21}$ +14.9° (c = 1 in water).

l-Form calcium salt, *calcium (6S)-folinate, calcium levofolinate, levoleucovorin calcium, Elvorine.* $[\alpha]_D^{20}$ −15.1° (c = 1.82).

THERAP CAT: Antidote to folic acid antagonists; antianemic (folate deficiency).

4255. Folpet. *2-[(Trichloromethyl)thio]-1H-isoindole-1,3(2H)-dione; N-(trichloromethylthio)phthalimide; Phaltan.* $C_9H_4Cl_3NO_2$-S; mol wt 296.56. C 36.45%, H 1.36%, Cl 35.86%, N 4.72%, O 10.79%, S 10.81%. *Cf:* Captan. Prepn: Kittleson, U.S. pat. **2,553,770** (1951 to Standard Oil). Toxicity: T. B. Gaines, R. E. Linder, *Fundam. Appl. Toxicol.* **7,** 299 (1986).

Crystals from benzene, mp 177°. LD_{50} orally in adult male, female rats: > 5000 mg/kg (Gaines, Linder).

Caution: May irritate mucosal surfaces: *Clinical Toxicology of Commercial Products,* R. E. Gosselin *et al.,* Eds.

(Williams & Wilkins, Baltimore, 5th ed., 1984) Section II, p 317.

USE: Agricultural fungicide.

4256. Fomecins. Antibacterial substances produced by several strains of the basidiomycete *Fomes juniperinus* Schrenk. Isoln of fomecin A (major) and fomecin B (minor) from cultures grown in corn steep liquor: M. Anchel *et al., Proc. Nat. Acad. Sci. USA* **38,** 655 (1952). Structures: T. C. McMorris, M. Anchel, *Can. J. Chem.* **42,** 1595 (1964). Synthesis of B: S. M. Al-Mousawi *et al., Bull. Soc. Chim. Belg.* **88,** 883 (1979); of A and B: K. Hayashi *et al., Chem. Pharm. Bull.* **28,** 1971 (1980).

Fomecin A R = CH₂OH
Fomecin B R = CHO

Fomecin A, $C_8H_8O_5$, *2,3,4-trihydroxy-6-(hydroxymethyl)-benzaldehyde.* Cream-colored to orange crystals from ethanol-water, ethanol-benzene, acetone-benzene, or ethyl acetate. Dec above 160° without melting. Optically inactive in ethanol. uv max (ethanol): 241, 304 nm (ϵ 10800, 15300). Weakly acidic. Sparingly sol in water (1 mg/ml); slightly more sol in ethanol, acetone, ethyl acetate. Less sol in chloroform, benzene. Aq solns at neutral or acid pH are stable. The activity is lost around pH 8. Intravenous injection in doses up to 1 mg per mouse (*ca.* 50 m/kg) had no apparent ill effect.

Fomecin B, $C_8H_6O_5$, *3,4,5-trihydroxy-1,2-benzenedicarboxaldehyde.* Yellow needles from ethyl acetate. Darkens on heating; mp ∼230°. uv max (ethanol): 263, 336 nm (ϵ 26400, 9200).

4257. Fomepizole. *4-Methyl-1H-pyrazole;* 4-MP. $C_4H_6N_2$; mol wt 82.11. C 58.52%, H 7.37%, N 34.12%. Alcohol dehydrogenase inhibitor. Prepn: H. Pechmann, E. Burkard, *Ber.* **33,** 3590 (1900); D. S. Noyce *et al., J. Org. Chem.* **20,** 1681 (1955); T. Momose *et al., Heterocycles* **30,** 789 (1990). Inhibition of human liver alcohol dehydrogenase: T.-K. Li, H. Theorell, *Acta Chem. Scand.* **23,** 892 (1969). Toxicity study: G. Magnusson *et al., Experientia* **28,** 1198 (1972). GC determn in plasma and urine: R. Achari, M. Mayersohn, *J. Pharm. Sci.* **73,** 690 (1984). Clinical pharmacology: D. Jacobsen *et al., Alcoholism (New York)* **12,** 516 (1988). Pharmacokinetics: *eidem, Eur. J. Clin. Pharmacol.* **37,** 599 (1989). Case study in ethylene glycol poisoning: F. J. Baud *et al., N. Engl. J. Med.* **319,** 97 (1988). *Review:* J. Likforman *et al., J. Toxicol. Clin. Exp.* **7,** 373-382 (1987).

mp 15.5-18.5°. bp₁₈ₘₘ 98.5-99.5°; bp₇₃₀ 204-205°. n_D^{22} 1.4913. uv max in 95% ethanol: 220 nm (log ϵ 3.47); in 6*N* HCl: 226 nm (log ϵ 3.65). Sol in water, alcohol. LD_{50} (7 days) in mice, rats (mmol/kg): 3.8, 3.8 i.v.; 7.8, 6.5 orally (Magnusson).

THERAP CAT: Antidote to methanol and ethylene glycol poisoning.

4258. Fominoben. *N-[3-Chloro-2-[[methyl-[2-(4-morpholinyl)-2-oxoethyl]amino]methyl]phenyl]benzamide; 3'-chloro-α-[methyl[(morpholinocarbonyl)methyl]amino]-o-benzotoluidide; 3'-chloro-β-[N-methyl-N-[(morpholinocarbonyl)methyl]aminomethyl]benzanilide.* $C_{21}H_{24}ClN_3O_3$; mol wt 401.89. C 62.76%, H 6.02%, Cl 8.82%, N 10.46%, O

11.94%. Prepn: **Fr. pats. 1,482,547** and **Addn. 92,488** (1967, 1968 both to Thomae), *C.A.* **68**, 95494e (1968); **72**, 21938p (1970); Kruger *et al., U.S.* pat. **3,661,903** (1972 to Boehringer, Ing.); *eidem, Arzneimittel-Forsch.* **23**, 290 (1973). Series of articles on pharmacology and clinical findings: *ibid.* 296-375. Metabolism: Zimmer, *ibid.* 1798. Toxicity data: S. Püschmann, R. Engelhorn, *ibid.* 296.

Base, mp 122.5-123°.
Hydrochloride, $C_{21}H_{24}ClN_3O_3 \cdot HCl$, *PB-89, Finaten, Noleptan, Oleptan, Terion, Tussirama.* Crystals, mp 206-208° (dec). Soly in water: 0.1 mg/100 ml; in 0.05N aq tartaric acid: 5 g/100 ml. LD_{50} in mice, rats (mg/kg): 630, 1201 i.p.; 2200, 1250 orally (Püschmann, Engelhorn).
THERAP CAT: Antitussive; respiratory stimulant.

4259. Fomocaine. *4-[3-[4-(Phenoxymethyl)phenyl]propyl]morpholine; 4-[3-(α-phenoxy-p-tolyl)propyl]morpholine;* P-652; Erbocain. $C_{20}H_{25}NO_2$; mol wt 311.42. C 77.14%, H 8.09%, N 4.50%, O 10.27%. Prepn: **Brit. pat. 786,128** (1957 to Promonta), *C.A.* **52**, 9225h (1958); H. Oelschläger, *Arzneimittel-Forsch.* **9**, 313 (1959). Improved synthesis: *eidem, ibid.* **27**, 1625 (1977). Pharmacology: O. Nieschulz *et al., ibid.* **8**, 539 (1958). Metabolism: H. Blume, H. Oelschläger, *ibid.* **28**, 956 (1978). Bioavailability and local anesthetic effect: H. Oelschläger, D. Rothley, *ibid.* **29**, 693 (1979).

Colorless crystals from petr ether, mp 52-53°. $bp_{1.1}$ 238-240°. uv max (ethanol): 220, 269 nm (ε 15820, 1373). LD_{50} i.v. in mice: 175 mg/kg (Nieschulz).
THERAP CAT: Anesthetic (local).

4260. Fonazine. *10-[2-(Dimethylamino)propyl]-N,N-dimethyl-10H-phenothiazine-2-sulfonamide; 10-(2-dimethylaminopropyl)-3-dimethylsulfamidophenothiazine; 3-dimethylsulfamido-10-(2-dimethylaminopropyl)phenothiazine; dimethothiazine; dimethiotazine; dimetiotazine;* RP-8599. $C_{19}H_{25}N_3O_2S_2$; mol wt 391.56. C 58.28%, H 6.44%, N 10.73%, O 8.17%, S 16.38%. Serotonin inhibitor. Prepn: **Brit. pat. 814,512** (1959 to Rhône-Poulenc). Pharmacokinetics and distribution in dog and rat: O. R. W. Lewellen, R. Templeton, *Adv. Biochem. Psychopharmacol.* **9**, 213 (1974). Clinical evaluation in spasticity: W. B. Matthews *et al., Acta Neurol. Scand.* **48**, 635 (1972). HPLC determn in plasma: J. Holt *et al., Brit. J. Clin. Pharmacol.* **13**, 282 (1982).

Hydrochloride, $C_{19}H_{25}N_3O_2S_2 \cdot HCl$, crystals, dec 214°.
Methanesulfonate, $C_{19}H_{25}N_3O_2S_2 \cdot CH_3SO_3H$, *dimethothiazine mesylate, fonazine mesylate, Banistyl, Bonpac, Calsekin, Migristène, Neomestine, Promaquid, Yoristen.*
THERAP CAT: Antimigraine.

4261. Fonofos. *Ethylphosphonodithioic acid O-ethyl S-phenyl ester; O-ethyl S-phenyl ethylphosphonothiolothio-* nate; N-2790; Dyfonate. $C_{10}H_{15}OPS_2$; mol wt 246.33. C 48.76%, H 6.14%, O 6.49%, P 12.57%, S 26.03%. Prepn: Szabo *et al., U.S.* pat. **2,988,474** (1961 to Stauffer); Pitt, Simone, **Ger.** pat. **2,002,629**, *C.A.* **73**, 77392u (1970), corresp to **U.S.** pat. **3,642,960** (1970, 1972, both to Stauffer).

Light yellow liquid, $bp_{0.1}$ 130°. d_{25}^{25} 1.16. n_D^{30} 1.5883. Practically insol in water. Miscible with organic solvents. LD_{50} orally in rats: 3 mg/kg, *Toxic Substances List,* H. E. Christensen, Ed. (1974) p 602.
Caution: Cholinesterase inhibitor.
USE: Soil insecticide.

4262. Formaldehyde. Methanal; oxomethane; oxymethylene; methylene oxide; formic aldehyde; methyl aldehyde. CH_2O; mol wt 30.03. C 40.00%, H 6.71%, O 53.28%. HCHO. Formed by incomplete combustion of many organic substances. Present in coal and wood smoke, esp in smoke as produced for smoking ham and fish. Found in the atm, esp over large cities. Prepd commercially by catalytic vapor phase oxidation of methanol using air as the oxidizing agent and heated silver, copper, alumina, or coke as catalysts. Process using molybdenum iron oxide catalyst: Allyn *et al., U.S.* pat. **2,812,309** (1957) and **U.S.** pat. **2,849,-492** (1958 to Reichhold Chem.). Prepn of stable formaldehyde by heating low molecular polyoxymethylenes with P_2O_5: **Ger.** pat. **1,070,611** (1959 to BASF). Mfg processes: Faith, Keyes & Clark's *Industrial Chemicals,* F. A. Lowenheim, M. K. Moran, Eds. (Wiley-Interscience, New York, 4th ed., 1975) pp 422-429. Prepn of semicarbazone: M. Pomerantz *et al., J. Org. Chem.* **47**, 2217 (1982). Toxicity data: H. F. Smyth *et al., J. Ind. Hyg. Toxicol.* **23**, 259 (1941). Carcinogenicity study: J. A. Swenberg *et al., Cancer Res.* **40**, 3398 (1980). Assessment, regulation and evaluation of carcinogenicity: F. Perera, C. Petito, *Science* **216**, 1285 (1982). *Review:* H. R. Gerberich *et al.,* in Kirk-Othmer *Encyclopedia of Chemical Technology* vol. 11 (John Wiley & Sons, New York, 4th ed., 1994) pp 929-951. Review of toxicology: H. d'A. Heck *et al., Crit. Rev. Toxicol.* **20**, 397-426 (1990); of oral toxicity: P. Restani, C. L. Galli, *ibid.* **21**, 315-328 (1991).

Flammable colorless gas at ordinary temp. Pungent suffocating odor. d 1.067 (air = 1.000). d_4^{-20} 0.815. mp −92°. bp_{760} −19.5°; bp_{400} −33.0°; bp_{200} −46.0°; bp_{100} −57.3°; bp_{60} −65.0°; bp_{40} −70.6°; bp_{20} −79.6°; bp_{10} −88.0°. Ignition temp about 300° (572°F). Very sol in water, up to 55%; sol in alcohol, ether. Very reactive, combines readily with many substances, and polymerizes easily. *See also* Paraformaldehyde.
Formaldehyde solution, Formalin, Formol, Morbicid, Veracur. A soln of about 37% by wt of formaldehyde gas in water, usually with 10-15% methanol added to prevent polymerization. This soln is the full strength and also known as Formalin 100% or Formalin 40 which signifies that it contains 40 grams of formaldehyde within 100 ml of the soln. Colorless liq; pungent odor. On standing, esp in the cold, may become cloudy and on exposure to very low temp a ppt of trioxymethylene is formed. When evaporated, some formaldehyde escapes, but most of it is changed to trioxymethylene. It is a powerful reducing agent especially in presence of alkali. In the air it slowly oxidizes to formic acid. d_{25}^{25} 1.081-1.085. One gallon weighs 9.1 lbs. bp_{760} 96°. n_D^{20} 1.3746. Flash pt 60° (140°F). pH 2.8-4.0. Misc with water, alcohol, acetone. *Keep well closed in a moderately warm place.* LD_{50} orally in rats: 0.80 g/kg (Smyth).
Formaldehyde semicarbazone, $C_2H_5N_3O$, crystals from ethanol, mp 120-121°. uv max (CH_3CN): 227 nm (log ε 3.73).
Caution: Potential symptoms of overexposure are irritation of eyes, nose and throat; lacrimation, burns of the nose and coughing; bronchial spasm, pulmonary irritation. *See NIOSH Pocket Guide to Chemical Hazards* (DHHS/NIOSH

90-117, 1990) p 116. May act as a primary irritant on skin causing erythmatous or eczematous dermatitis. Sensitization can result. *See* Patty's *Industrial Hygiene and Toxicology* vol. 2A, G. D. Clayton, F. E. Clayton, Eds. (Wiley-Interscience, New York, 3rd ed., 1981) pp 2637-2646. Ingestion may cause immediate intense pain in mouth, pharynx and stomach; nausea, vomiting, hematemesis, abdominal pain, diarrhea; pale, clammy skin and other signs of shock; difficult urination, hematuria, anuria; vertigo, convulsions, stupor, coma; death due to respiratory failure. *See* Clinical Toxicology of Commercial Products, R. E. Gosselin *et al.*, Eds. (Williams & Wilkins, New York, 5th ed., 1984) Section III, pp 196-197. Formaldehyde gas may reasonably be anticipated to be a carcinogen: *Seventh Annual Report on Carcinogens* (PB95-109781, 1994) p 214.

USE: In the prodn of amino, phenolic, and polyacetal resins, wood products, plastics, fertilizers and foam insulation. As a textile finish, preservative, stabilizer, disinfectant and antibacterial food additive.

THERAP CAT: Solution as disinfectant.

THERAP CAT (VET): Solution as antiseptic; fumigant; has been used in tympany, diarrhea, mastitis, pneumonia, internal bleeding.

4263. Formaldehyde Sodium Bisulfite. *Hydroxymethanesulfonic acid monosodium salt;* sodium formaldehydebisulfite; methylolsulfonic acid sodium salt. CH_3NaO_4S; mol wt 134.09. C 8.96%, H 2.26%, Na 17.15%, O 47.73%, S 23.91%. $HOCH_2SO_3Na$. Prepn: Skrabal, Skrabal, *Monatsh.* **69**, 17 (1936); Chhabria *et al.*, **Ger.** pat. **1,173,059** (1964 to Schill & Seilacher). Structure: Lauer, Langkammerer, *J. Am. Chem. Soc.* **57**, 2360 (1935); Caughlan, Tartar, *ibid.* **63**, 1266 (1941).

Monohydrate, needles from water. pK (25°): 6.92.

USE: Fixing agent for keratin-containing fibers. Flotation of lead-zinc ores. Protecting color pictures exposed to high humidities against fading and staining.

4264. Formamide. Methanamide; carbamaldehyde. CH_3NO; mol wt 45.04. C 26.67%, H 6.71%, N 31.10%, O 35.52%. $HCONH_2$. Prepd on a large scale from carbon monoxide and ammonia at high press. and temp: Meyer, Orthner, *Ber.* **54**, 1705 (1921); **55**, 857 (1922); Meyer, **Ger.** pat. **390,798** (1924); **392,409** (1924); **414,257** (1925 to BASF); Wietzel, Herbst, **U.S.** pat. **1,843,434** (1932 to I. G. Farben). Toxicity study: Pham-Huu-Chank *et al.*, *Toxicol. Appl. Pharmacol.* **26**, 596 (1973). *Review:* C. L. Eberling in Kirk-Othmer *Encyclopedia of Chemical Technology* vol. **11** (Wiley-Interscience, New York, 3rd ed., 1980) pp 258-263.

Slightly viscous, odorless, colorless liq. Industrial grades may have faint odor of ammonia. mp +2.55°. bp$_{760}$ 210.5° (partial decomp into CO and NH_3 at atm pressure beginning at 180°); bp$_{400}$ 193.5°; bp$_{200}$ 175.5°; bp$_{60}$ 147.0°; bp$_{20}$ 122.5°; bp$_{10}$ 109.5°; bp$_{1.0}$ 70.5°. d_4^{15} 1.13756; d_4^{20} 1.13340; d_4^{30} 1.12483. Dielectric constant ϵ = 84. Viscosity $\eta \times 10^5$ at 15° = 4320, at 30° = 2926. Surface tension γ at 20° = 58.35. n_D^{15} 1.44911; n_D^{20} 1.44754; n_D^{110} 1.4170; n_D^{130} 1.4095. Flash pt, open cup: 310°F (154°C). pH of 0.5 molar aq soln 7.1. Misc with water, methanol, ethanol, acetone, acetic acid, dioxane, ethylene glycol, U.S.P. glycerol, phenol. Very slightly sol in ether, benzene. Dissolves casein, glucose, zein, tannins, starch, lignin, polyvinyl alcohol, cellulose acetate, nylon, the chlorides of copper, lead, zinc, tin, cobalt, iron, aluminum, nickel, the acetates of the alkali metals, some inorganic sulfates and nitrates. LD_{50} in mice, rats (g/kg): 4.6, 5.7 i.p. (Pham-Huu-Chanh).

Caution: Moderately irritating to skin, mucous membranes.

USE: As ionizing solvent, manuf formic esters, hydrocyanic acid by catalytic dehydration, as softener for paper, animal glues, water-sol gums.

4265. Formanilide. *N-Phenylformamide;* formylaniline. C_7H_7NO; mol wt 121.14. C 69.41%, H 5.82%, N 11.56%, O 13.21%. C_6H_5NHCHO. Review on methods of prepn: de Jonge *et al.*, *Rec. Trav. Chim.* **75**, 5 (1956).

Crystals from ether + petr ether, mp 46.6-47.5°. bp 271°; bp$_{120}$ 216°; bp$_{14}$ 166°. uv max (96% ethanol): 242-243 nm (ϵ 13700); in water: 239-240 nm (ϵ 11200); in cyclohexane: 240 nm (ϵ 12600). Soly in water at 20°: 25.4 g/l; at 25°: 28.6 g/l.

4266. Formebolone. *11α,17β-Dihydroxy-17-methyl-3-oxoandrosta-1,4-diene-2-carboxaldehyde;* 2-formyl-17α-methylandrosta-1,4-diene-11α,17β-diol-3-one; 2-formyl-11α-hydroxy-Δ^1-methyltestosterone; formyldienolone; Esiclene. $C_{21}H_{28}O_4$; mol wt 344.45. C 73.23%, H 8.19%, O 18.58%. Prepn: Canonica *et al.*, *Gazz. Chim. Ital.* **95**, 138 (1968); Gomarasca, **Brit.** pat. **1,168,931;** **Fr.** pat. **1,584,960** (1969, 1970 both to Lab. Prod. Biol. Braglia); *C.A.* **72**, 90742g (1970); **73**, 131212a (1970). Pharmacology: *idem, Minerva Med.* **62**, 842 (1971); Vaccari, Livini, *ibid.* 846; De Marchi *et al.*, *Arzneimittel-Forsch.* **23**, 1583 (1973). GC/MS determn of metabolites in urine: R. Massé *et al.*, *Anal. Chim. Acta* **247**, 211 (1991). Toxicological studies: Travella, *Gazz. Int. Med. Chir.* **76**, 400 (1971), *C.A.* **75**, 30792a (1971).

Crystals from ethyl acetate, mp 209-212°. [α]$_D^{25}$ −105° (CHCl$_3$). Water soluble. Approx LD$_{50}$ in rats, mice (mg/kg): 104, 187 i.p.; 270, 293 s.c.; orally in rats: > 1000 (Travella).

11-β-Epimer, mp 95-103° from ethyl ether-petr ether.

Note: This is a controlled substance (anabolic steroid) in the U.S. Code of Federal Regulations, Title 21 Part 1308.13, as defined in Part 1308.02 (1995).

THERAP CAT: Anabolic.

4267. Formestane. *4-Hydroxyandrost-4-ene-3,17-dione;* 4-OHA; CGP-32349; Lentaron. $C_{19}H_{26}O_3$; mol wt 302.41. C 75.46%, H 8.67%, O 15.87%. Selective aromatase inhibitor. Prepn: R. D. Burnett, D. N. Kirk, *J. Chem. Soc. Perkin Trans. I* **1973**, 1830; J. Mann, B. Pietrzak, *ibid.* **1983**, 2681; P. G. Ciattini *et al.*, *Synth. Commun.* **22**, 1949 (1992). Aromatase inhibition and effect on estrogen-dependent processes *in vivo:* A. M. H. Brodie *et al.*, *Endocrinology* **100**, 1684 (1977). Radioimmunoassay in plasma: J. Khubieh *et al.*, *J. Steroid Biochem.* **35**, 377 (1990). GC determn in plasma and urine: P. H. Degen, W. Schneider, *J. Chromatog.* **565**, 67 (1991). Pharmacokinetics and endocrine effects in breast cancer patients: M. Dowsett *et al.*, *Eur. J. Cancer* **28**, 415 (1992). Clinical evaluation in breast cancer: R. C. Coombes, *J. Steroid Biochem. Mol. Biol.* **43**, 145 (1992).

Needles from aq methanol, mp 199-202° (Burnett, Kirk); also reported as crystals from ethyl acetate, mp 203.5-206° (Brodie). uv max (99.5% ethanol): 278 nm (ϵ 11030). [α]$_D^{20}$ +181° (c = 7.7 in chloroform).

THERAP CAT: Antineoplastic (hormonal).

4268. Formic Acid. Ameisensäure (German). CH_2O_2; mol wt 46.03. C 26.10%, H 4.38%, O 69.52%. HCOOH. Observed by S. Fisher in 1670 in the products resulting from the distillation of ants. Acute and chronic toxicity: G. Malorny, *Z. Ernährungswiss.* **9**, 332 (1969). Production: P. G. Jessop *et al.*, *Nature* **368**, 231 (1994). Review of metabolism and toxicology: J. Liesivuori, H. Savolainen, *Pharmacol. Toxicol.* **69**, 157-163 (1991). *Reviews:* A. Aguilo, T. Horlenko, *Hydrocarbon Process., Int. Ed.* **59**, 120-130 (1980); D. J. Drury in Kirk-Othmer *Encyclopedia of Chemical Technology* vol. **11** (Wiley-Interscience, New York, 4th ed., 1994) pp 951-958.

Colorless liq; pungent odor. bp_{760} 100.8°. mp 8.4°. d^{20} 1.220. pKa (20°) 3.75. n_D^{20} 1.3714. Flash point, open cup: 59°. Surface tension at 20°: 37.67 dyn/cm. Viscosity at 20°: 1.784 cP. Dielectric constant at 20°: 57.9. Vapor pressure at 20°: 33.55 mm Hg. Heat of fusion: 3031 cal/mole; of vaporization at 100°: 104 cal/g. Azeotropic mixture with water: 77.5 wt% formic acid, bp_{760} 107.3°. Strong reducing agent. Misc with water, ether, acetone, ethyl acetate, methanol, ethanol. Partially sol in benzene, toluene, xylenes. LD_{50} in mice (mg/kg): 1100 orally; 145 i.v. (Malorny).

Caution: Potential symptoms of overexposure are eye irritation, lacrimation; nasal discharge; throat irritation, coughing and dyspnea; nausea; skin burns, dermatitis. *See NIOSH Pocket Guide to Chemical Hazards* (DHHS/NIOSH 90-117, 1990) p 118.

USE: Silage additive, prophylactic in treatment of animal feeds; acidulant in dyeing of natural and synthetic fibers, leather tanning; coagulating latex in rubber production; chemical synthesis.

THERAP CAT: Counterirritant; astringent.

4269. Formicin. *N-(Hydroxymethyl)acetamide;* formaldehyde acetamide; methylal acetamide. $C_3H_7NO_2$; mol wt 89.09. C 40.44%, H 7.92%, N 15.72%, O 35.92%. CH_3CO-$NHCH_2OH$. Obtained by warming acetamide with formaldehyde and a small amount of potash: Fuchs, *Pharm. Ztg.* **1905,** 803; Ger. pat. **164,610** (1905 to Kalle).

Colorless, very hygroscopic mass. Marketed only as a slightly yellowish, syrupy liq. d 1.14-1.18. Very sol in water, alcohol; sol in chloroform, glycerol. Insol in ether.

USE: Antiseptic, disinfectant esp for surgical instruments.

4270. Formocortal. *(11β,16α)-21-(Acetyloxy)-3-(2-chloroethoxy)-9-fluoro-11-hydroxy-16,17-[(1-methylethylidene)bis(oxy)]-20-oxopregna-3,5-diene-6-carboxaldehyde; 3-(2-chloroethoxy)-9-fluoro-11β,16α,17,21-tetrahydroxy-20-oxopregna-3,5-diene-6-carboxaldehyde, cyclic 16,17-acetal with acetone, 21-acetate;* 3-(2-chloroethoxy)-6-formyl-9α-fluoropregna-3,5-diene-11β,16α,17,21-tetrol-20-one 21-acetate 16α,17α-acetonide; 3-(2-chloroethoxy)-9α-fluoro-6-formyl-11β,21-dihydroxy-16α,17α-isopropylidenedioxypregna-3,5-dien-20-one; fluoroformylon; FI-6341; Cortocin-F; Cutisterol; Deflamene; Fluderma. $C_{29}H_{38}ClFO_8$; mol wt 569.07. C 61.21%, H 6.73%, Cl 6.23%, F 3.34%, O 22.49%. Prepn: Camerino et al., Fr. pat. **1,396,602,** C.A. **63,** 5713c (1965); Neth. pat. Appl. **6,508,458;** U.S. pat. **3,314,945** (1965, 1966, 1967 all to Farmitalia); Baldratti et al., *Experientia* **22,** 468 (1966). Pharmacology: Suchowsky, Baldratti, *Proc. Int. Congr. Horm. Steroids, 2nd, Milan* **1966,** L. Martini, F. Fraschini, M. Motta, Eds. (Excerpta Medica Found., Amsterdam, 1967) pp 536-540; Tajima, Otsu, *Japan. J. Pharmacol.* **62,** 115 (1966), C.A. **67,** 72329v (1967); Baldratti et al., *Arzneimittel-Forsch.* **21,** 1533 (1971). Toxicity studies: Bertazzoli et al., *J. Eur. Toxicol.* **4,** 88 (1971).

White crystalline powder from ether-petroleum ether, mp 180-182°. $[\alpha]_D^{20}$ +26° (chloroform). uv max (ethanol): 216, 324 nm (ϵ 12,100, 17,100). LD_{50} in mice: > 2000 mg/kg orally; 490 mg/kg s.c.; 537 mg/kg i.p. (Bertazzoli).

THERAP CAT: Glucocorticoid.

4271. Formononetin. *7-Hydroxy-3-(4-methoxyphenyl)-4H-1-benzopyran-4-one; 7-hydroxy-4'-methoxyisoflavone;* biochanin B; formononetol; neochanin. $C_{16}H_{12}O_4$; mol wt 268.27. C 71.64%, H 4.51%, O 23.86%. Isoln from soybean meal *(Soja hispida):* Walz, *Ann.* **489,** 118 (1931); from

clover species *Trifolium subterraneum* L. and *T. pratense* L., *Leguminosae* in which it is the major estrogenic factor: Bradbury, White, *J. Chem. Soc.* **1951,** 3447; Bate-Smith, Swain, *Chem. & Ind. (London)* **1953,** 1127. Identity with biochanin B: Bose, *J. Sci. Ind. Res.* **15B,** 325 (1956). Structure: Baker et al., *J. Chem. Soc.* **1933,** 274. Synthesis: Wessely et al., *Ber.* **66,** 685 (1933); Kagal et al., *Tetrahedron Letters* **1962,** 593. Biological half-life in *Cicer arietinum* L., *Papilionatae:* N. Amrhein, E. Diederich, *Naturwiss.* **67,** 40 (1980). HPLC analysis: R. E. Carlson, *J. Chromatog.* **198,** 193 (1980). ^{13}C-NMR study: H. C. Jha et al., *Can. J. Chem.* **58,** 1211 (1980). Mutagenicity study: R. M. Bartholomew et al., *Mutat. Res.* **78,** 317 (1980).

Needles from alcohol, mp 258°. uv max (ethanol): 250, 300 nm (ϵ 27440, 11240).

7-Glucoside, $C_{22}H_{22}O_9$, ononin, 4'-methyldaidzin. From *Ononis spinosa* L., *Leguminosae:* Hlasiwetz, *J. Prakt. Chem.* **65,** 415 (1855). Synthesis: Farkas, Varady, *Ber.* **92,** 819 (1959). Needles from water, $C_{22}H_{22}O_9.H_2O$, mp 210-214°. When anhydr, dec 245°. $[\alpha]_D^{25}$ −24.2° (pyridine). Sol in alcohol; slightly sol in water, ether.

4272. Formoterol. *(R*,R*)-(±)-N-[2-Hydroxy-5-[1-hydroxy-2-[[2-(4-methoxyphenyl)-1-methylethyl]amino]ethyl]phenyl]formamide;* 3-formylamino-4-hydroxy-α-[N-[1-methyl-2-(p-methoxyphenyl)ethyl]aminomethyl]benzyl alcohol; (±)-2'-hydroxy-5'-[(RS)-1-hydroxy-2-[[(RS)-p-methoxy-α-methylphenethyl]amino]ethyl]formanilide. $C_{19}H_{24}N_2O_4$; mol wt 344.41. C 66.26%, H 7.02%, N 8.13%, O 18.58%. β-Adrenoreceptor stimulating catecholamine analog with selective bronchodilator activity. Prepn: M. Murakami et al., Ger. pat. **2,305,092;** eidem, U.S. pat. **3,994,974;** K. Murase et al., Japan. Kokai **75 12,040** (1973, 1976, 1975 all to Yamanouchi); C.A. **83,** 113941q (1975); and separation of racemates (there are two pairs of enantiomers): eidem, *Chem. Pharm. Bull.* **25,** 1368 (1977). The following studies have been done on the "A" racemate fumarate dihydrate. GC-MS determn in urine: H. Kamimura et al., *J. Chromatog.* **229,** 337 (1982). Radioimmunoassay in plasma, urine: K. Yokoi et al., *Life Sci.* **33,** 1665 (1983). Pharmacology: H. Ida, *Arzneimittel-Forsch.* **26,** 839, 1337 (1976). Metabolism in rats, dogs: H. Sasaki et al., *Xenobiotica* **12,** 803 (1982). Anti-allergic activity in mice and guinea pigs: K. Tomioka et al., *Arch. Int. Pharmacodyn.* **250,** 279 (1981). Toxicity: T. Yoshida et al., *Pharmacometrics* **26,** 811 (1983). Review of pharmacology and clinical use: D. Faulds et al., *Drugs* **42,** 115-137 (1991).

Fumarate dihydrate, "A"-racemate; $(C_{19}H_{24}N_2O_4)_2.$-$C_4H_4O_4.2H_2O$, BD-40A, Atock, Foradil. Crystals from 95% isopropyl alcohol, mp 138-140°. LD_{50} in male, female, rats, mice (mg/kg): 3130, 5580, 6700, 8310 orally; 98, 100, 72, 71 i.v.; 1000, 1100, 640, 670 s.c.; 170, 210, 240, 210 i.p. (Yoshida).

Fumarate dihydrate, "B"-racemate; mp 154-155°.

THERAP CAT: Antiasthmatic.

4273. Formothion. *Phosphorodithioic acid S-[2-(formylmethylamino)-2-oxoethyl] O,O-dimethyl ester; phosphorodithioic acid O,O-dimethyl ester S-ester with N-formyl-2-mer-*

capto-N-methylacetamide; S-(N-formyl-N-methylcarbamoyl-methyl) O,O-dimethyl phosphorodithioate; J-38; OMS-968; Aflix; Anthio. $C_6H_{12}NO_4PS_2$; mol wt 257.27. C 28.01%, H 4.70%, N 5.44%, O 24.88%, P 12.04%, S 24.93%. Prepn: Lutz, Schuler, **Brit.** pat. **900,557** (1962 to Sandoz). Toxicity: K. Liesche, Deut. Apoth. **108**, 604 (1968), C.A. **69**, 26287y (1968).

Yellowish liq, mp ~25°. Practically insol in water. Miscible with alcohols, ether, chloroform, benzene. LD_{50} in rats, rabbits (mg/kg): 350, 410 orally (Liesche).
Caution: Cholinesterase inhibitor. *See* Parathion.
USE: Acaricide, systemic insecticide.

4274. Formyl Fluoride. CHFO; mol wt 48.02. C 25.01%, H 2.10%, F 39.57%, O 33.32%. HCOF. Prepn by the interaction of anhydr formic acid, potassium fluoride and benzoyl chloride (16% yield): Nyesmejanov, Kahn, *Ber.* **67**, 203 (1946); from benzoyl fluoride and formic acid (36% yield): Masentshev, *J. Gen. Chem. USSR* **16**, 203 (1946); from benzoyl chloride, formic acid, and KHF_2 (35% yield): Olah *et al., Ber.* **89**, 862 (1956); Olah, Kuhn, *J. Am. Chem. Soc.* **82**, 2381 (1960); from acetic formic anhydride, $CH_3COOOCH$, and anhydr HF (67% yield): Olah, Kuhn, *ibid.* 2380.
Gas at ordinary temps and pressure: bp_{760} −29°. mp −142°. Trouton constant 20.8. Heat of vaporization at −24°: 5.175 kcal.
Caution: Strong irritant.
USE: Acylating agent in organic synthesis.

4275. N²-Formylsulfisomidine. *N-(2,6-Dimethyl-4-pyri-midinyl)-4-(formylamino)benzenesulfonamide; 4′-[(2,6-di-methyl-4-pyrimidinyl)sulfamoyl]formanilide; N-[4-[[(2,6-di-methyl-4-pyrimidinyl)amino]sulfonyl]phenyl]formamide;* formylsulfamethine; FAK III; Wometin. $C_{13}H_{14}N_4O_3S$; mol wt 306.35. C 50.97%, H 4.61%, N 18.29%, O 15.67%, S 10.47%. Prepd by formylation of sulfisomidine: Brand, Rieckhoff, **Ger.** pats. **1,122,511; 1,126,857** (both 1962 to VEB Farbenfabrik Wolfen).

Crystals, mp 248.5-250.5°.
THERAP CAT: Antibacterial.

4276. Fortimicins. Aminoglycoside antibiotic complex produced by *Micromonospora olivoasterospora.* The main components, fortimicins A and B, both exhibit broad-spectrum antibacterial activity. Isoln of A: T. Nara *et al.,* **Japan. Kokai 75 29,789** (to Kyowa); *eidem,* **U.S.** pat. **3,-976,768** (1976 to Abbott). Isoln of B: *eidem,* **Ger.** pat. **2,418,349** (1974 to Kyowa); *eidem,* **U.S.** pat. **3,931,400** (1976 to Abbott). Isoln and biological properties of A and B: *eidem, J. Antibiot.* **30**, 533 (1977); physical and chromatographic properties: R. Okachi *et al., ibid.* 541. Structures of A and B: R. S. Egan *et al., ibid.* 552. *In vitro* activity: R. Girolami, J. S. Stamm, *ibid.* 564. Isoln and properties of minor components, fortimicins C, D, KE: M. Sugimoto *et al., ibid.* **32**, 868 (1979). Synthesis of B: T. Suami, Y. Honda, *Chem. Letters* **1980**, 641; Y. Honda, T. Suami, *Bull. Chem. Soc. Japan* **55**, 1156 (1982).

Fortimicin A, $C_{17}H_{35}N_5O_6$, *4-amino-1-[(aminoacetyl)meth-ylamino]-1,4-dideoxy-3-O-(2,6-diamino-2,3,4,6,7-pentadeoxy-β-L-lyxo-heptopyranosyl)-6-O-methyl-L-chiro-inositol, astro-micin, Abbott 44747.* White amorphous powder, mp >200° (dec). $[\alpha]_D^{25}$ +87.5° (c = 0.1 in water). Sol in water and lower alcohols. Insol in organic solvents. LD_{50} (of the sulfate salt) in mice (mg/kg): 380 i.v.; 400 s.c. (Nara, 1977).
Fortimicin B, $C_{15}H_{32}N_4O_5$, *4-amino-1,4-dideoxy-3-O-(2,6-diamino-2,3,4,6,7-pentadeoxy-β-L-lyxo-heptopyranosyl)-6-O-methyl-1-(methylamino)-L-chiro-inositol.* White amorphous powder, mp 101-103°. $[\alpha]_D^{25}$ +22.2° (c = 0.1 in water). Sol in water and lower alcohols. Insol in organic solvents.
THERAP CAT: Antibacterial.

4277. Foscarnet Sodium. *Dihydroxyphosphinecarbox-ylic acid oxide trisodium salt;* trisodium phosphonoformate; trisodium carboxyphosphate; A-29622; Foscavir. CNa_3O_5P; mol wt 191.95. C 6.26%, Na 35.93%, O 41.68%, P 16.14%. Inhibits viral DNA polymerase and reverse transcriptase. Prepn: P. Nylén, *Ber.* **57B**, 1023 (1924). Crystal structure: R. R. Naqvi *et al., J. Chem. Soc. (A)* **1971**, 2751; S. C. Abrahams, *Acta Crystallogr.* **B28**, 2886 (1972). Use as antiviral agent: B. F. H. Eriksson *et al.,* **Ger.** pat. **2,728,685** corresp to **U.S.** pat. **4,215,113** (1978, 1980 to Astra). Activity against herpes simplex virus: E. Helgstrand *et al., Science* **201**, 819 (1978); J. M. Reno *et al., Antimicrob. Ag. Chemother.* **13**, 188 (1978); S. Alenius *et al., Arch. Virol.* **60**, 197 (1979). Mechanism of inhibition of Epstein-Barr virus replication: A. K. Datta, R. E. Hood, *Virology* **114**, 52 (1981). Preclinical evaluation: B. Oeberg *et al., Int. Congr. Ser.-Excerpta Med.* **571**, 175 (1982). Inhibitory effect on HIV-1 (HTLV-III/LAV) virus replication *in vitro:* P. S. Sarin *et al., Biochem. Pharmacol.* **34**, 4075 (1985).

Hexahydrate, mp >250°. LD_{50} i.p. in mice: 2000-4000 µmol/kg (Eriksson).
THERAP CAT: Antiviral.

4278. Fosetyl Al. *Phosphonic acid monoethyl ester aluminum salt;* aluminum tris(ethyl phosphite); aluminum tris-(O-ethylphosphonate); efosite Al; phosethyl Al; LS-74-783; Aliette. $C_6H_{18}AlO_9P_3$; mol wt 354.11. C 20.35%, H 5.12%, Al 7.62%, O 40.66%, P 26.24%. Systemic fungicide translocated both upwards and downwards in plants. Prepn: V. V. Orlovskii *et al., J. Gen. Chem. U.S.S.R.* **42**, 1924 (1972). Alternate prepn, fungicidal properties: J. Ducret *et al.,* **Ger.** pat. **2,456,627**; *eidem,* **U.S.** pat. **4,139,616** (1975, 1979 both to Rhone-Poulenc); J. Abblard *et al.,* **U.S.** pat. **4,143,059** (1979 to Philagro). Biological properties and field trials against vine mildew and other *Phycomycetes* pathogens: A. Bertrand *et al., Phytiatr.-Phytopharm.* **26**, 3 (1977). Evaluation of carcinogenicity: J. A. Quest *et al., Reg. Toxicol. Pharmacol.* **14**, 3 (1991). Review of properties and fungicidal action on tropical, temperate crops: D. J. Williams *et al., Proc. Brit. Crop Prot. Conf. - Pests Dis.* **1977**, 565.

White odorless crystals, mp > 300°. Soly in water at 20°C: 120 g/l. Practically insol in acetonitrile, propylene glycol (< 80 mg/l). LD_{50} in rats, mice, Japanese quail, rabbits (mg/kg): 5800, 3700, 4997, 2680 orally. LD_{50} i.p. in rats: 550 mg/kg (Williams).

USE: Agricultural fungicide.

4279. Fosfestrol. *(E)-4,4'-(1,2-Diethyl-1,2-ethenediyl)-bisphenol bis(dihydrogen phosphate); α,α'-diethyl-4,4'-stilbenediol diphosphoric acid ester;* diethylstilbestrol 4,4'-diphosphoric ester; diethylstilbestrol diphosphate; diethylstilbestryl diphosphate; diethyldihydroxystilbene diphosphate; stilbestrol diphosphate; Stilphostrol. $C_{18}H_{22}O_8P_2$; mol wt 428.32. C 50.48%, H 5.18%, O 29.88%, P 14.46%. Prepd by treating stilbestrol with phosphorus oxychloride in pyridine: Miescher, Heer, U.S. pat. **2,234,311** (1941); Arnold, U.S. pat. **2,802,854** (1957 to Asta-Werke). Prepn of stable solns at pH 10: Fonner, Collins, U.S. pat. **2,828,244** (1958 to Miles Labs.). Prepn of sodium salts: Dawson, U.S. pat. **2,971,975** (1961 to Miles Labs.). Clinical pharmacokinetics: H. Oelschläger *et al., Arzneimittel-Forsch.* **36**, 1284 (1986). Review of clinical trials in prostate cancer: J-P. Droz *et al, Cancer* **71**, 1123-1130 (1993).

Voluminous white cryst powder from dil HCl. Dec 204-206°. Sparingly sol in water.

Disodium salt, $C_{18}H_{18}Na_2O_8P_2$, crystals. Softens at 190°, uniform melt at 230°. Sol in water. pH of aq soln 6.5. pH also reported as 4.85 to 5.00.

Tetrasodium salt, $C_{18}H_{20}Na_4O_8P_2$, *Cytonal, Honvan, Honvol, ST-52, Stilbostatin.*

THERAP CAT: Antineoplastic (hormonal); estrogen.

4280. Fosfomycin. *(2R-cis)-(3-Methyloxiranyl)phosphonic acid; (−)-(1R,2S)-(1,2-epoxypropyl)phosphonic acid;* fosfonomycin; phosphonomycin; MK-955; Fosfocina. $C_3H_7O_4P$; mol wt 138.06. C 26.10%, H 5.11%, O 46.35%, P 22.44%. Antibiotic produced by *Streptomyces* strains: D. Hendlin *et al., Science* **166**, 122 (1969); B. G. Christensen *et al., ibid.* 123; also produced by *Pseudomonas syringae:* J. Shogi *et al., J. Antibiot.* **39**, 1011 (1986). Isoln: D. Hendlin *et al.,* **Belg. pat. 718,507;** *eidem,* U.S. pat. **3,914,231** (1969, 1975 both to Merck & Co.). Synthesis and resolution: B. G. Christensen *et al.,* **Belg. pats. 723,072; 723,073** (both 1969 to Merck & Co.). Alternate synthesis: Girotra, Wendler, *Tetrahedron Letters* **1969**, 4647; Glamkowski *et al., J. Org. Chem.* **35**, 3510 (1970). Series of articles on characterization, activity and clinical testing: *Antimicrob. Ag. Chemother.* **1969**, 284-351; H. B. Woodruff *et al., Chemotherapy* **23**, Suppl. 1, 1-22 (1977). Mechanism of action: Kahan *et al., Ann. N. Y. Acad. Sci.* **235**, 354 (1974). Pharmacokinetics in humans: M. Goto *et al., Antimicrob. Ag. Chemother.* **20**, 393 (1981). Pharmacology of *cofosfolactamines,* fixed dose combinations of fosfomycin with β-lactam antibiotics: P. Periti, *Drugs Exp. Clin. Res.* **6**, 305 (1980). Clinical trial of cofosfolactamines: T. Barreca *et al., ibid.* **10**, 55 (1984).

Crystals, mp ~94°. Soluble in water.

Benzylammonium salt, $C_{10}H_{16}NO_4P$, mp 170-174°. $[\alpha]_{405}$ −9.1° (c = 5).

Calcium salt monohydrate, *Afos, Biocin, Biofos, Endociclina, Faremicin, Fonofos, Fosfobiotic, Fosfocin, Fosfogram, Fosfotricina, Fosmicin, Foximin, Francital, Gram-Micina, Ipamicina, Lancetina, Lofoxin, Neofocin, Palmofen, Priomicina, Selemicina, Ultramicina, Valemicina.*

Tromethanol, $C_7H_{18}NO_7P$, *(2R-cis)-(3-methyloxiranyl)-phosphonic acid compd with 2-amino-2-(hydroxymethyl)-1,3-propanediol (1:1), fosfomycin trometamol, Monuril.*

THERAP CAT: Antibacterial.

4281. Fosfosal. *2-(Phosphonooxy)benzoic acid; salicylic acid dihydrogen phosphate; o-carboxyphenyl phosphate;* salicyl phosphate; UR-1521; Disdolen. $C_7H_7O_6P$; mol wt 218.10. C 38.55%, H 3.23%, O 44.01%, P 14.20%. Salicylic acid deriv. Prepn: A. C. Marin Moga, E. Francia Barra, **Ger. pat. 2,641,526** (1978 to Uriach), *C.A.* **88**, 136317h (1978). Pharmacology and comparison with aspirin: J. Garcia Rafanell *et al., Arzneimittel-Forsch.* **30**, 1091 (1980). Toxicity studies: M. S. Sanchez *et al., ibid.* 1098. Pharmacokinetics: V. Rimbau *et al., Arch. Farmacol. Toxicol.* **7**, 61 (1981). Double-blind trials in musculoskeletal/articular pain: C. Diaz *et al., Clin. Ther.* **4**, 121 (1981); L. Madera Cat, J. Garcia Rafanell, *Med. Clin.* **76**, 18 (1981).

White solid, mp 168-170°. Sol in water, ethanol, acetone. Insol in non-polar organic solvents. Hydrolyzes in aq soln. LD_{50} in male, female mice, male, female rats at pH 1.0, aq soln (mg/kg): 94, 105, 153, 257 i.v.; 352, 253, 338, 360 i.p.; 1455, 1539, 1104, 1213 orally; at pH 3.5 (mg/kg): 117, 118, 207, 215 i.v.; 1592, 1483, 1085, 1128 i.p.; 1702, 2007, 1685, 2225 orally (Sanchez).

THERAP CAT: Analgesic.

4282. Fosinopril. *[1[S*(R*)],2α,4β]-4-Cyclohexyl-1-[[[2-methyl-1-(1-oxopropoxy)propoxy](4-phenylbutyl)phosphinyl]acetyl]-L-proline; (4S)-4-cyclohexyl-1-[[[(RS)-1-hydroxy-2-methylpropoxy](4-phenylbutyl)phosphinyl]acetyl]-L-proline propionate (ester);* fosenopril. $C_{30}H_{46}NO_7P$; mol wt 563.67. C 63.93%, H 8.23%, N 2.48%, O 19.87%, P 5.50%. Phosphinic acid containing angiotensin converting enzyme (ACE) inhibitor. Prepn: E. W. Petrillo, Jr., U.S. pat. **4,337,201** (1982 to Squibb); of active diacid form: J. Krapcho *et al., J. Med. Chem.* **31**, 1148 (1988). Metabolism and pharmacokinetics: S. M. Singhvi *et al., Brit. J. Clin. Pharmacol.* **25**, 9 (1988). GC determn of diacid: M. Jemal *et al., J. Chromatog.* **345**, 299 (1985). Clinical trial in hypertension: P. A. Sullivan *et al., Am. J. Hypertension* **1**, 280S (1988). Brief description: E. W. Petrillo, Jr., *et al., Clin. Exp. Theory Prac.* **A9**, 235 (1987). Series of articles on pharmacology, pharmacokinetics, and clinical experience: *Drug Invest.* **3**, Suppl. 4, 1-53 (1991).

Sodium salt, $C_{30}H_{45}NNaO_7P$, *SQ-28555, Acecor, Monopril, Secorvas, Staril.*

Diacid, $C_{23}H_{34}NO_5P$, *SQ-27519.* mp 149-153°. $[\alpha]_D$ −24° (c = 1 in methanol).

THERAP CAT: Antihypertensive.

4283. Fospirate. *Dimethylphosphoric acid 3,5,6-trichloro-2-pyridinyl ester; O,O-dimethyl-O-3,5,6-trichloro-2-pyridyl phosphate;* Dowco 217; Torelle. $C_7H_7Cl_3NO_4P$; mol wt 306.47. C 27.43%, H 2.30%, Cl 34.70%, N 4.57%, O 20.88%, P 10.11%. Prepn by reaction of 3,5,6-trihalo-2-pyridinol with phosphoryl chloride: Rigterink, Kenaga, *J. Agr. Food Chem.* **14**, 304 (1966).

Crystals from petr ether, mp 86.5-88°.
THERAP CAT (VET): Anthelmintic.

4284. Fosthietan. *1,3-Dithietan-2-ylidenephosphoramidic acid diethyl ester; phosphonodithioimidocarbonic acid cyclic methylene P,P-diethyl ester;* 2-(diethoxyphosphinylimino)-1,3-dithietane; AC-64475; CL-64475; Nem-A-Tak. $C_6H_{12}NO_3PS_2$; mol wt 241.27. C 29.87%, H 5.01%, N 5.81%, O 19.89%, P 12.84%, S 26.58%. Prepn: R. W. Addor, S. Afr. pat. **68 01,064**; *idem,* U.S. pat. **3,470,207** (1968, 1969 both to Am. Cyanamid); *idem, J. Heterocycl. Chem.* **7**, 381 (1970). Activity: W. K. Whitney, J. L. Aston, *Proc. Brit. Insectic. Fungic. Conf.* **2**, 625 (1975).

Pale yellow oil, mercaptan-like odor. n_D^{25} 1.5348; d_{25} 1.3. Vapor press at 25°: 6.5 × 10⁻⁶ mm Hg. Soly in water at 25°: 50 g/kg. Sol in acetone, chloroform, methanol, toluene. LD_{50} in rats, mice (mg/kg): 5, 18 orally (Whitney, Aston).
USE: Nematocide; insecticide.

4285. Fotemustine. *[1-[[[(2-Chloroethyl)nitrosoamino]carbonyl]amino]ethyl]phosphonic acid diethyl ester;* (±)-diethyl [1-[3-(2-chloroethyl)-3-nitrosoureido]ethyl]phosphonate; 1-[N-(2-chloroethyl)-N-nitrosoureido]ethylphosphonic acid diethyl ester; S-10036; Muphoran. $C_9H_{19}ClN_3O_5P$; mol wt 315.69. C 34.24%, H 6.07%, Cl 11.23%, N 13.31%, O 25.34%, P 9.81%. Amino acid-linked nitrosourea alkylating agent. Prepn: G. Lavielle, C. Cudennec, Fr. pat. **2,536,075**; *eidem,* U.S. pat. **4,567,169** (1984, 1986 both to ADIR). Clinical evaluation in advanced cancers: D. Khayat *et al., Cancer Res.* **47**, 6782 (1987); in disseminated malignant melanoma: D. Khayat *et al., J. Nat. Cancer Inst.* **80**, 1407 (1988).

mp 85°.
THERAP CAT: Antineoplastic.

4286. Francium. Eka-cesium. Fr; at. no. 87. ²²³Fr, *Actinium K,* is the most stable isotope ($T_{1/2}$ 21 min; β-emitter) formed by α-decay of actinium (²²⁷Ac). Found in uranium minerals. First obtained in 1939 from an actinium prepn by Perey, *J. Phys. Radium* **10**, 439 (1939); *J. Chim. Phys.* **43**, 155, 262 (1946); Hyde, Ghiorso, *Phys. Rev.* **90**, 267 (1953); Hyde, *J. Am. Chem. Soc.* **74**, 4181 (1952). Isoln by paper chromatography: Perey, Adloff, *Compt. Rend.* **236**, 1163 (1953). Also obtainable by proton bombardment of thorium. Mass numbers of other known isotopes: 204-213; 217-222; 224. Most electropositive element. Chemical behavior similar to that of other alkali metals. *Reviews:* Hyde, *J. Chem. Ed.* **36**, 15-21 (1959); Whaley, "Sodium, Potassium, Rubidium, Cesium and Francium" in *Comprehensive Inorganic Chemistry* vol. **1**, J. C. Bailar Jr., *et al.*, Eds. (Pergamon Press, Oxford, 1973) pp 369-529.

4287. Frangula. Buckthorn bark; alder buckthorn; black dogwood; berry alder; arrow wood; Persian berries. Dried bark of *Rhamnus frangula* L., *Rhamnaceae. Habit.* Europe, Russian Asia, Mediterranean coast of Africa. *Constit.* Frangulin, emodin, chrysophanic acid.

4288. Frangulin. Franguloside; avornin; Cascarin. In berries, bark, and rootbark of *Rhamnus* spp., especially in alder buckthorn *(Rhamnus frangula* L.), *Rhamnus carthartica* L., and *Rhamnus purshiana* DC. *(Cascara sagrada), Rhamnaceae.* Prepn from bark of alder buckthorn: Bridel, Charaux, *Bull. Soc. Chim. Biol.* **15**, 642 (1933); M. Kubiak, *Herba Pol.* **23**, 217 (1977), *C.A.* **88**, 166260b (1978). Consists of the two glucosides, frangulins A and B, which were originally thought to be isomeric. Structure and synthesis of frangulin A: Hörhammer, Wagner, *Z. Naturforsch.* **27B**, 959 (1972). Structure of frangulin B: Wagner, Demuth, *Tetrahedron Letters* **1972**, 5013.

Frangulin A   R =

Frangulin B   R =

Frangulin A, $C_{21}H_{20}O_9$. *1,3,8-Trihydroxy-6-methylanthraquinone-l-rhamnoside, emodin-l-rhamnoside, rhamnoxanthin.* Crystals, mp 228°. Absorption max: 225, 264, 282, 300, 430 nm (log ε 4.52, 4.28, 4.15, 3.97, 4.05).
Frangulin B, $C_{20}H_{18}O_9$. *6-O-(D-Apiofuranosyl)-1,6,8-trihydroxy-3-methylanthraquinone.* mp 196°.
THERAP CAT: Cathartic.

4289. Fraxetin. *7,8-Dihydroxy-6-methoxy-2H-1-benzopyran-2-one; 7,8-dihydroxy-6-methoxycoumarin.* $C_{10}H_8O_5$; mol wt 208.17. C 57.70%, H 3.87%, O 38.43%. The aglucon of fraxin. Obtained by heating fraxin with dil H_2SO_4. Structure: Wessely, Demmer, *Ber.* **61**, 1279 (1928). Synthesis: Aghoramurthy, Seshadri, *J. Chem. Soc.* **1954**, 3065.

Plates from alc, mp 228°. Turns yellow at 150° and brown at mp. Sol in 10 liters cold water, in 300 ml boiling water; somewhat more sol in alcohol; hardly sol in ether.
Dimethyl ether, $C_{12}H_{12}O_5$, *6,7,8-trimethoxycoumarin.* Orthorhombic bipyramidal crystals, mp 104°. $bp_{0.2}$ 90-100°.

4290. Fraxin. *8-(β-D-Glucopyranosyloxy)-7-hydroxy-6-methoxy-2H-1-benzopyran-2-one;* 7,8-dihydroxy-6-methoxycoumarin-8-β-D-glucoside; 8-(glucosyloxy)-6-methoxyumbelliferone; fraxetin-8-glucoside; fraxoside; paviin. $C_{16}H_{18}O_{10}$; mol wt 370.31. C 51.90%, H 4.90%, O 43.21%. From bark of the common European ash *(Fraxinus excelsior* L.): Salm-Horstmar, *Pogg. Ann.* **100**, 607 (1857); from *F. oxyphylla* Mar., *Oleaceae:* Paris, Stambouli, *Compt. Rend.* **253**, 313 (1961). Identity with paviin: Stokes, *J. Chem. Soc.* **12**, 126 (1859). From the horse chestnut tree *(Aesculus hippocastanum* L., *Hippocastanaceae):* Reppel, *Planta Med.* **4**, 199 (1956); from *Diervilla* spp., *Caprifoliaceae:* Charaux, *J.*

Pharm. Chim. [7] **4**, 248 (1911). Structure: Wessely, Demmer, *Ber.* **62**, 120 (1929).

Yellow hydrated needles from water or dil alc. Slightly bitter, astringent taste. The water of crystn (about 3 mols) is removed at 130° and 0.2 mm Hg. The anhydr substance, mp 205° (rapid heating). Sparingly sol in cold water; freely sol in hot water, hot alcohol. Practically insol in ether. Alkaline solns are sulfur-yellow and on dilution show characteristic blue-green fluorescence.

4291. Fredericamycin A. *(E,E)-6',7'-Dihydro-4,9,9'-trihydroxy-6-methoxy-3'-(1,3-pentadienyl)spiro[2H-benz(f)indene-2,8'-(8H)cyclopent(g)isoquinoline]-1,1',3,5,8(2'H)-pentone;* FCRC-A48; NSC-305623. $C_{30}H_{21}NO_9$; mol wt 539.50. C 66.79%, H 3.92%, N 2.60%, O 26.69%. Antitumor antibiotic produced by *Streptomyces griseus* (FCRC-48); represents new structural class of antibiotics containing a spiro-[4,4]nonane ring system. Isoln together with two biologically inactive components, fredericamycins B and C: R. C. Pandey *et al., J. Antibiot.* **34**, 1389 (1981). Antimicrobial and cytotoxic activity *in vitro* and antitumor activity *in vivo:* D. J. Warnick-Pickle *et al., ibid.* 1402. Prepn and biological activity of water soluble salts: R. Misra, *ibid.* **51**, 976 (1988). Structure determn by x-ray crystallography: R. Misra *et al., J. Am. Chem. Soc.* **104**, 4478 (1982). Spectroscopic and mass spectral characterization: *eidem, J. Antibiot.* **40**, 786 (1987). Synthetic studies: A. V. R. Rao *et al., Chem. Commun.* **1984**, 1119; K. A. Parker *et al., Tetrahedron Letters* **26**, 2181 (1985). Synthesis of (±)-form: T. R. Kelly *et al., J. Am. Chem. Soc.* **108**, 7100 (1986). Biosynthetic study: K. M. Byrne *et al., Biochemistry* **24**, 478 (1985). Mechanism of action study: B. D. Hilton *et al., ibid.* **25**, 5533 (1986).

Thin, platelet-like crystals from acetonitrile + water, mp > 350° (dec). pKa (DMF) 6.80; 8.88. Sol in acetic acid, DMSO, DMF, pyridine; partially sol in acidic methanol, acidic chloroform, acidic ethyl acetate. Insol in water, petr ether, ether. Acts as an indicator: Red in acidic soln, green to blue in basic soln.
Potassium salt, solubility (mg/ml): water (1.0); DMSO (8/1.5); 1:1 DMSO-H_2O (3.0); also readily sol in DMF, dimethylacetamide, pyridine. Sparingly sol in ethyl acetate, acetonitrile, methanol, chloroform. Insol in hexanes, benzene, acetone, ether.

4292. Frenolicin. *[1R-(1α,3β,4aβ,10aβ)]-3,4,5,10-Tetrahydro-9-hydroxy-5,10-dioxo-1-propyl-4a,10a-epoxy-1H-naphtho[2,3-c]pyran-3-acetic acid.* $C_{18}H_{18}O_7$; mol wt 346.34. C 62.42%, H 5.24%, O 32.34%. Antibiotic produced by *Streptomyces fradiae:* Van Meter *et al., Antimicrob. Ag. Ann.* **1960**, 77. Structure: Ellestad *et al., J. Am. Chem. Soc.* **88**, 4109 (1966); **90**, 1325 (1968). Total synthesis: A. Ichihara *et al., Tetrahedron Letters* **21**, 4469 (1980).

Pale yellow needles from benzene, mp 160-161°. $[\alpha]_D^{25}$ −37.7° (c = 1.5 in methanol). uv max (methanol): 234, 362 nm (ε 18300, 5200). Sol in methanol, ethanol, ethyl acetate, acetone, glacial acetic acid, carbon tetrachloride; slightly sol in benzene. Practically insol in water, cyclohexane, petr ether.

4293. Frequentin. *6-(1,3-Heptadienyl)-3,4-dihydroxy-2-oxocyclohexanecarboxaldehyde.* $C_{14}H_{20}O_4$; mol wt 252.31. C 66.65%, H 7.99%, O 25.36%. Mold metabolite with antineoplastic and antibiotic activity; isolated from *Penicillium frequentans* Westling and *P. palitans* Westling: Curtis *et al., Nature* **167**, 557 (1951); Birkinshaw, *Biochem. J.* **51**, 271 (1952). Structure: Sigg, *Helv. Chim. Acta* **46**, 1061 (1963). Effect on Ehrlich ascites carcinoma cells: J. Fuska *et al., Int. Congr. Chemother., Proc., 7th, Prague, 1971* vol. 2 (Univ. Park Press, Baltimore, 1972) pp 97-98. Synthetic study: C. G. Unson, *Diss. Abstr. B* **39**, 1296 (1978).

Needles from boiling water, mp 134.5°. When dissolved, exists as the 1-hydroxymethylene tautomer. $[\alpha]_D^{22}$ +65° (c = 0.95 in $CHCl_3$); $[\alpha]_D^{22}$ +61° (c = 0.96 in methanol). uv max (dioxane): 290, 232 nm (log ε 3.74, 4.52). Sol in acetone, dioxane; less sol in chloroform, ethanol, benzene; slightly sol in carbon tetrachloride, water.

4294. Friedelin. *D:A-Friedooleanan-3-one;* friedelan-3-one. $C_{30}H_{50}O$; mol wt 426.73. C 84.44%, H 11.81%, O 3.75%. Major triterpene constituent of cork, obtained by extraction of ground cork with alc. Isoln: Istrati, Ostrogovich, *Compt. Rend.* **128**, 1581 (1899). Isolated also from *Ceratopetalum apetalum* D. Don, *Cunoniaceae:* Jefferies, *J. Chem. Soc.* **1954**, 473. Structure, sterochemistry: Brownlie *et al., Chem. & Ind. (London)* **1955**, 1156; Kane, Stevenson, *Tetrahedron* **15**, 223 (1961); Stevenson, *J. Org. Chem.* **28**, 188 (1963). Total synthesis of (±)-form: R. E. Ireland, D. M. Walba, *Tetrahedron Letters* **1976**, 1071.

Needles from ethyl acetate or alcohol, mp 263-263.5°. $[\alpha]_D$ −27.8° (chloroform). One gram dissolves in 8.6 ml chloroform, 264 ml 99% alcohol.

4295. Fructose. *D-Fructose;* β-D-fructose; levulose; fruit sugar; Fructosteril; Laevoral; Levugen; Laevosan. $C_6H_{12}O_6$; mol wt 180.16. C 40.00%, H 6.71%, O 53.28%. Occurs in a large number of fruits, honey, and as the sole sugar in bull and human semen: Auerbach, Bodlander, *Angew. Chem.* **36**, 602 (1923); Mann, *Nature* **157**, 79 (1946); Pryde, *ibid.* 660. Prepd by adding abs alcohol to the syrup obtained from the acid hydrolysis of inulin: Bates *et al., Natl. Bur.*

Std. (U.S.), Circ. **C440,** 399 (1942). Prepn from dextrose: Cantor, Hobbs, U.S. pat. **2,354,664** (1944 to Corn Prod. Refining). From sucrose by enzymatic conversion: Koepsell *et al.,* U.S. pat. **2,729,587** (1956 to U.S.A.). Crystal and molecular structure: J. A. Kanters *et al., Acta Crystallogr.* **B33,** 665 (1977). *Review:* Barry, Honeyman, *Advan. Carbohyd. Chem.* **7,** 53-98 (1952); M. Chen, R. L. Whistler, *ibid.* **34,** 285-343 (1977). Occurs in both the furanose and pyranose forms. An aq soln at 20° contains about 20% of the furanose form.

Orthorhombic, bisphenoidal prisms from alc, dec 103-105°. Sweetest of the sugars. Shows mutarotation. $[\alpha]_D^{20}$ $-132°$ → $-92°$ (c = 2). Rapid and anomalous mutarotation involves pyranose-furanose interconversion. The final value is obtained instantly in the presence of hydroxyl ions. pKa (18°): 12.06. Freely sol in water. One gram dissolves in 15 ml alc, in 14 ml methanol. Slightly sol in cold, freely in hot acetone; sol in pyridine, ethylamine, methylamine.

Ferric form, *CB-302, Ferritose.* Prepn: Stitt *et al., Proc. Soc. Exp. Biol. Med.* **110,** 70 (1962); Saltman, Charley, U.S. pats. **3,074,927** and **3,275,514** (1966).

USE: To prevent sandiness in ice cream.

THERAP CAT: Fluid and nutrient replenisher, and ferric form as hematinic.

THERAP CAT (VET): For bovine ketosis.

4296. DL-**Fructose.** α-Acrose; methose. $C_6H_{12}O_6$; mol wt 180.16. C 40.00%, H 6.71%, O 53.28%. Component of formose (polymerization product of formaldehyde): Vogel, *Helv. Chim. Acta* **11,** 370 (1928).

Needles from methanol. d_4^{16} 1.665. mp 129-130° (slow heating). Reduces Fehling's soln.

Phenylosazone, $C_{18}H_{22}N_4O_4$, mp 216-217°.

4297. Fructose-1,6-diphosphate. *D-Fructose 1,6-bis(dihydrogen phosphate); 1,6-D-fructosediphosphoric acid;* hexose diphosphate; Harden-Young ester. $C_6H_{14}O_{12}P_2$; mol wt 340.12. C 21.19%, H 4.15%, O 56.45%, P 18.21%. Formed from fructose-6-phosphate in the presence of ATP, Mg^{2+} and the enzyme phosphohexokinase. Prepn from glucose, mannose, fructose, sucrose by the action of yeasts: A. Harden, *Alcoholic Fermentation* (Longmans, Green & Co., New York, 4th ed., 1932); v. Lebedev, *Biochem. Z.* **36,** 254 (1911); *cf* Ger. pats. **292,817; 293,864; 301,590.** Fructose-1,6-diphosphate is reversibly split in the presence of aldolase forming 1-phosphodihydroxyacetone and 3-phosphoglyceraldehyde: Meyerhof *et al., Biochem. Z.* **286,** 301 (1936). Metabolism regulation study: M. E. Kirtley, M. McKay, *Mol. Cell. Biochem.* **18,** 141 (1977).

Sodium salt monohydrate, *Esafosfina.*

THERAP CAT: Roborant; tonic.

4298. Fructose-6-phosphate. *D-Fructose 6-(dihydrogen phosphate);* D-fructose-6-phosphoric acid; fructose monophosphate; hexose phosphate; hexose monophosphate; Neuberg ester. $C_6H_{13}O_9P$; mol wt 260.14. C 27.70%, H 5.04%, O 55.35%, P 11.91%. Present in animal tissues as an equilibrium mixture with glucose-6-phosphate. The glucose-6-phosphate may be reversibly transformed into fructose-6-phosphate by the enzyme phosphohexose isomerase. Prepn by hydrolysis of 1,6-fructose diphosphate with dil acid: Neuberg, *Biochem. Z.* **88,** 432 (1917); Ger. pat. **334,250** (Bayer); *Chem. Zentr.* **1921,** II, 961; *Frdl.* **13,** 948. Role in metabolic regulation and heat generation: E. A. Newsholme, *Biochem. Soc. Trans.* **4,** 978 (1976).

Very sol in water. $[\alpha]_D^{21}$ +2.5° (c = 3), Meyerhof, Lohmann, *Biochem. Z.* **185,** 117 (1927). $[\alpha]_D$ +1.2° (c = 0.9), Lohmann, *ibid.* **262,** 145 (1933).

The magnesium and zinc salts also are sol in water. The equilibrium mixture of 75-80% glucose-6-phosphate and 20-25% fructose-6-phosphate is called *lactacidogen* or *Embden ester.*

4299. FSH. *Follicle-stimulating hormone;* urofollitrophin; Fertinorm; Follitropin; Luteoantine; Metrodin; Thylakentrin. Mol wt about 36,000. Glycoprotein gonadotropic hormone found in pituitary tissue of mammals that directly regulates the metabolic activity of granulosa cells of the ovary and Sertoli cells of the testis. In the female, FSH induces the maturation of the Graafian follicles of the ovary. The follicular cells which surround the growing ovum develop the capacity to produce estrogens which induce proliferative changes in the walls of the uterus and vagina. The rising titer of estrogen will bring on ovulation and the changeover to progestational activity. In the male, promotes the development of the germinal cells of the testes. Isoln procedures: Fevold *et al., Endocrinology* **26,** 999 (1940); Fraenkel-Conrat *et al., Proc. Soc. Exp. Biol. Med.* **45,** 627 (1940); McShan, Meyer, *J. Biol. Chem.* **135,** 473 (1940); Greep *et al., ibid.* **133,** 289 (1940); Li *et al., Science* **109,** 445 (1949). Prepn of human FSH from pituitaries and from postmenopausal urine: Roos, Gemzell in *Ciba Foundation Study Group* no. 22 (Little, Brown and Co., Boston, 1965). Isoln and a study of some chemical, physical, immunological and biological properties of FSH; evidence of subunit structure: Roos, *Acta Endocrinol. (Copenhagen)* **59,** Suppl. **131** (1968); Reichart; Rathnam, Saxena, in *Gonadotropins,* B. B. Saxena *et al.,* Eds. (Wiley-Interscience, New York, 1972) pp 107-131. Amino acid sequence of human FSH α-subunit: P. Rathnam, B. B. Saxena, *J. Biol. Chem.* **250,** 6735 (1975); of human β-subunit: B. B. Saxena, P. Rathnam, *ibid.* **251,** 993 (1976); of equine α: P. Rathnam *et al., ibid.* **253,** 5355 (1978); of equine β: Y. Fujiki *et al., ibid.* 5363. Physical and hydrodynamic properties: R. J. Ryan *et al., Recent Progr. Horm. Res.* **26,** 105 (1970). Effect on gonadal functions: J. H. Dorrington, D. T. Armstrong, *ibid.* **35,** 301 (1979). *Reviews:* A. R. Means *et al., Ann. Rev. Physiol.* **42,** 59-70 (1980); J. G. Pierce, T. F. Parsons, *Ann. Rev. Biochem.* **50,** 465-495 (1981). *See also* HCG, LH, TSH.

Solid. Isoelectric point 4.5. Soluble in water, physiological saline soln, 50% alc. Soluble in a pH 4.4 acetate buffer contg 20.5% sodium sulfate (different from LH which is insol in this solvent). A 1 mg-% soln in saline may be kept at 60° for 30 min. In solns of pH 7-8 the activity is retained at 75° for 30 min. In 50% alcoholic soln the activity is destroyed at 60° within 15 min.

Note: **Menotropins** is a term applied to an extract of postmenopausal urine containing FSH and LH activity. Marketed under the tradenames *Pergonal, Humegon,* and *Pregova.* Used to treat female sterility. For veterinary use, purified FSH is obtained from the pituitary gland of domesticated animals.

THERAP CAT: Gonad-stimulating principle.

THERAP CAT (VET): In certain reproductive disorders.

4300. Fucosamine. *2-Amino-2,6-dideoxygalactose.* $C_6H_{13}NO_4$; mol wt 163.17. C 44.17%, H 8.03%, N 8.58%, O 39.22%. Isoln of D-form from lipopolysaccharide of *Chromobacterium violaceum:* Crumpton, Davies, *Biochem. J.* **70,** 729 (1958); from *Bacillus licheniformis:* Sharon *et al., ibid.* **93,** 210 (1964). Isoln of L-form from type V *Pneumococcus* capsular polysaccharide: Barker *et al., Nature* **189,** 303 (1961). Synthesis of L-form: Kuhn *et al., Ann.* **628,** 186

(1959); J. Lehmann *et al.*, *Ber.* **112**, 1470 (1979); of D-form: Zehavi, Sharon, *J. Org. Chem.* **29**, 3654 (1964).

α-D-Fucosamine

D-Form hydrochloride, $C_6H_{13}NO_4 \cdot HCl$, crystals from aq acetone, dec 170-175°. $[\alpha]_D^{20}$ +91° (water). Absorption max: 400 nm.

N-Acetyl-D-fucosamine, $C_8H_{15}NO_5$, crystals from ethanol, dec 196-197°. $[\alpha]_D^{22}$ +92° (c = 2 in water).

L-Form hydrochloride, $C_6H_{13}NO_4 \cdot HCl$, rods from methanol + isopropanol, mp 192-193°. $[\alpha]_D^{27}$ −92° (c = 0.89 in water).

N-Acetyl-L-fucosamine, $C_8H_{15}NO_5$, crystals from ethanol, mp 197-198°. $[\alpha]_D^{26}$ −82° (c = 1.46 in water).

4301. D-Fucose. *6-Deoxy-D-galactose;* D-galactomethylose; rhodeose. $C_6H_{12}O_5$; mol wt 164.16. C 43.90%, H 7.37%, O 48.73%. Obtained from glucosides found in various species of *Convolvulaceae, e.g.*, convolvulin, jalapin, β-turpethein. Isoln: Votocek, *Z. Zuckerind. Böhmen* **24**, 249; *Chem. Zentr.* **1901**, I, 1042; **1902**, II, 1361; **1905**, II, 1528. Isoln from jalap resin: Votocek, Bulir, *ibid.* **1906**, I, 1818. Prepn by acid hydrolysis of chartreusin: Sternbach *et al.*, *J. Am. Chem. Soc.* **80**, 1639 (1958).

α-Form, needles from alc. Sweet taste. mp 144°. Shows mutarotation, $[\alpha]_D^{19}$ +127.0° (7 min) → +89.4° (31 min) → +77.2° (71 min) → +76.0° (final value 146 min, c = 10). Soluble in water; moderately sol in alcohol.

Pentaacetate, $C_{16}H_{22}O_{10}$, mp 115.5°.
Oxime, $C_6H_{13}NO_5$, mp 188.5°.
Phenylhydrazone, $C_{12}H_{18}N_2O_4$, mp 172°.
Phenylosazone, $C_{18}H_{22}N_4O_3$, mp 176.5°.

4302. L-Fucose. *6-Deoxy-L-galactose;* L-galactomethylose. $C_6H_{12}O_5$; mol wt 164.16. C 43.90%, H 7.37%, O 48.73%. Occurs in seaweed: *Ascophyllum nodosum* (L.) Ledol. (*Fucus nodosus* L.), *Fucus vesiculosus* L., *F. serratus* L., *F. virsoides* (Don) J. Ag., *Fucaceae*, and in gum tragacanth. Isoln from seaweed: Clark, *J. Biol. Chem.* **54**, 65 (1922); Hockett *et al.*, *J. Am. Chem. Soc.* **61**, 1658 (1939). Manuf from fucoidan: Schweiger, U.S. pat. 3,240,775 (1966 to Kelco). Synthesis from D-galactose: Dejterjuszynski, Flowers, *Carbohydr. Res.* **28**, 144 (1973); from D-glucose: T. Chiba, S. Tejima, *Chem. Pharm. Bull.* **27**, 2838 (1979); from D-mannose: J. Defaye *et al.*, *Carbohydr. Res.* **94**, 131 (1981). Review of chemistry and biochemistry: H. M. Flowers, *Adv. Carbohydr. Chem. Biochem.* **39**, 279-345 (1981).

α-Form, minute needles from abs alcohol, mp 140°. Shows mutarotation. $[\alpha]_D^{20}$ −124.1° (10 min) → −108.0° (20 min) → −91.5° (36 min) → −78.6° (70 min) → −75.6° (final value, 24 hrs, c = 9). Soluble in water and alcohol.

4303. Fucosterol. *(3β,24E)-Stigmasta-5,24(28)-dien-3-ol;* 24-ethylidenecholest-5-en-3β-ol. $C_{29}H_{48}O$; mol wt 412.70. C 84.40%, H 11.72%, O 3.88%. Isoln from *Fucus vesiculosus* L., *Fucaceae:* Heilbron *et al.*, *J. Chem. Soc.* **1934**, 1572; from marine brown algae *(Phaeophyceae):* Tsuda, *Chem. Pharm. Bull.* **6**, 724 (1958). Structure: MacPhillamy, *J. Am. Chem. Soc.* **64**, 1732 (1942). Stereochemistry:

C. Brooks *et al.*, *Steroids* **20**, 487 (1972). Synthesis: Hayazu, *Pharm. Bull.* **5**, 452 (1957).

Crystals from methanol, mp 124°. $[\alpha]_D^{20}$ −38.4° (chloroform). Soluble in most organic solvents.

Acetate, $C_{31}H_{50}O_2$, crystals, mp 120-121°. $[\alpha]_D^{20}$ −45° (chloroform).

4304. Fucoxanthin. *(3S,3'S,5R,5'R,6S,6'R)-3'-(Acetyloxy)-6',7'-didehydro-5,6-epoxy-5,5',6,6',7,8-hexahydro-3,5'-dihydroxy-8-oxo-β,β-carotene; 6',7'-didehydro-5,6-epoxy-4',5,-5',6,7,8-hexahydro-3,3',5'-trihydroxy-8-oxo-α-carotene 3'-acetate;* all-*trans*-fucoxanthin. $C_{42}H_{58}O_6$; mol wt 658.92. C 76.56%, H 8.87%, O 14.57%. Carotenoid pigment found in fresh brown algae (Don) J. Ag., *Fucaceae*. Also found in *Zygnema pectinatum* (Vauch.) Ag., *Zygnemaceae* and in *Polysiphonia nigrescens* (Dillw.) Grev., *Rhodomelaceae*. Isoln: Karrer *et al.*, *Helv. Chim. Acta* **14**, 628 (1931); Willstätter, Page, *Ann.* **404**, 237 (1914). Structure: Bonnet *et al.*, *Chem. Commun.* **1966**, 515; *J. Chem. Soc. (C)* **1969**, 429. Abs config: DeVille *et al.*, *Chem. Commun.* **1969**, 1311; K. Bernhard *et al.*, *Tetrahedron Letters* **1976**, 115.

Needles from ether + petr ether, mp 160°. $[\alpha]_D^{18}$ +72.5° ±9° (chloroform). Abs max (chloroform): 492, 457 nm. Abs max (ethanol): 450 nm ($E_{1cm}^{1\%}$ 1140), Antia, *Can. J. Chem.* **43**, 302 (1965). Freely sol in ethanol; less sol in carbon disulfide; sparingly sol in ether. Practically insol in petr ether. 1.66 g dissolves in 100 g boiling methanol.

4305. Fucus. Bladder-wrack; sea-wrack; bladder fucus; kelpware; black-tang; cut-weed; sea-oak. Dried thallus of *Fucus vesiculosus* L., *F. serratus* L., or *F. siliquosus*, L. *Fucaceae. Habit.* Atlantic and Pacific Oceans. *Constit.* Algin, about 0.01% iodine and some bromine mannite.

4306. Fuller's Earth. Floridin. A nonplastic variety of kaolin containing an aluminum magnesium silicate. The name is derived from an ancient process of cleaning or fulling wool, to remove the oil and dirt particles, with a water slurry of earth or clay. At the present time, the term fuller's earth is applied to any clay that has adequate decolorizing and purifying capacity to be used commercially in oil refining without chemical treatment. It is sometimes considered to be synonymous with *montmorillonite, q.v., kaolinite* ($Al_2O_3 \cdot 2SiO_2 \cdot 2H_2O$) and *Halloysite* ($Al_2O_3 \cdot 2SiO_2 \cdot 4H_2O$). A long list of minerals likely to be found in fuller's earth is given by Porter, *U.S. Geol. Survey, Bull.* **315**, 268 (1907), *C.A.* **1**, 1684 (1907), and the opinion is expressed that fuller's earth results from the decompn of hornblendes and augites rather than from feldspars. Fuller's earth has for its base a series of amorphous, hydrous aluminum silicates that have a rather persistent colloidal (used in its widest sense) structure. It is to this colloidal structure, which is not lost at 130° and possibly higher, that the bleaching power is due.

The bleaching efficiency of fuller's earth is usually increased by treatment with dilute acids.

USE: Decolorizer for oils and other liquids; filtering medium; filler for rubber; in agricultural formulations; also instead of absorbent charcoal.

4307. Fulvoplumierin. *(E,E)-7-(2-Butenylidene)-1,7-dihydro-1-oxocyclopenta[c]pyran-4-carboxylic acid methyl ester;* 3-(2-butenylidene)-2-carboxy-α-(hydroxymethylene)-1,4-cyclopentadiene-1-acetic acid δ-lactone methyl ester; methyl 7-crotonylidenecyclopenta[c]pyran-1-(7H)-one-4-carboxylate. $C_{14}H_{12}O_4$; mol wt 244.25. C 68.85%, H 4.95%, O 26.20%. Occurs together with plumieride and plumericin. Isoln from *Plumeria acutifolia* Poir., *Apocynaceae,* also from roots of *P. rubra* var *alba:* Grumbach *et al., Experientia* **8,** 224 (1952). Structure: Schmid, Bencze, *Helv. Chim. Acta* **36,** 206, 1468 (1953). Stereochemistry: Albers-Schönberg *et al., ibid.* **45,** 1406 (1962). Synthesis: Büchi, Carlson, *J. Am. Chem. Soc.* **90,** 5336 (1968); **91,** 6470 (1969).

Orange needles from chloroform + petr ether, ethyl acetate, or alcohol, dec 151-152°. Sublimes in high vacuum. uv max (ethanol): 272, 365 nm (ε 7,000; 33,700). Sol in chloroform, hot ethyl acetate, benzene, alcohol; less sol in pyridine, acetone. Practically insol in water, petr ether.

4308. Fumagillin. *[3R-[3α,4α(2R*,3R*),5β,6β(all-E)]]-2,4,6,8-Decatetraenedioic acid, mono[5-methoxy-4-[2-methyl-3-(3-methyl-2-butenyl)oxiranyl]-1-oxaspiro[2.5]oct-6-yl] ester; 2,4,6,8-decatetraenedioic acid mono[4-(1,2-epoxy-1,5-dimethyl-4-hexenyl)-5-methoxy-1-oxaspiro[2.5]oct-6-yl] ester;* Amebacilin; Fugillin; Fumadil B; Fumidil. $C_{26}H_{34}O_7$; mol wt 458.55. C 68.10%, H 7.47%, O 24.42%. Antibiotic substance produced by *Aspergillus fumigatus:* T. E. Eble, F. R. Hanson, *Antibiot. & Chemother.* **1,** 55 (1951); F. R. Hanson, T. E. Eble, U.S. pat. **2,652,356** (1953 to Upjohn). M. H. Peterson *et al.,* U.S. pat. **2,803,586** (1957 to Abbott). Purification: D. S. Tarbell *et al., J. Am. Chem. Soc.* **77,** 5613 (1955). Structure: *eidem, ibid.* **82,** 1005 (1960); **83,** 3096 (1961). Stereochemistry: N. J. McCorkindale, J. G. Sime, *Proc. Chem. Soc.* **1961,** 331; J. R. Turner, D. S. Tarbell, *Proc. Nat. Acad. Sci. USA* **48,** 733 (1962). Biosynthesis: A. J. Birch, S. F. Hussain, *J. Chem. Soc. (C)* **1969,** 1473. Total synthesis: E. J. Corey, B. B. Snider, *J. Am. Chem. Soc.* **94,** 2549 (1972). Anti-amebic activity: M. C. McCowen *et al., Science* **113,** 202 (1951). Antineoplastic activity: J. A. DiPaolo *et al., Antibiotics Annual* **1958-1959,** 541. *Review:* Girolami, "Fumagillin" in Kavanagh's *Anal. Microbiol.* (Academic Press, New York, 1963) pp 295-301; B. J. Wilson, "Miscellaneous *Aspergillus* Toxins" in *Microbial Toxins,* vol. VI, A. Ciegler *et al.,* Eds. (Academic Press, New York, 1971) pp 277-281.

Yellow needles from methanol, mp 194-195°. $[\alpha]_D^{25}$ −26.6° (c = 1 in 95% ethanol). Absorptivity: 156.0 at 335 nm and 146.5 at 351 nm (soln of 100 mg in 10 ml chloroform diluted with alcohol to 0.0004% fumagillin and 0.04% chloroform). Practically insol in water, dil acids, satd hydrocarbons. Sol in most other organic solvents, in aq solns

of bicarbonates and alkali hydroxides. Best stored in dark, evacuated ampuls at low temps. Stability data: T. E. Eble, E. R. Garrett, *J. Am. Pharm. Assoc.* **43,** 536 (1954); E. R. Garrett, *ibid.* 539. LD_{50} in mice (mg/kg): ~800 s.c. (Di-Paolo).

Methyl ester, $C_{27}H_{36}O_7$, crystals from dil methanol, mp 147-150°.

THERAP CAT: Formerly used as antiamebic.

THERAP CAT (VET): Antiprotozoal. Control of *Nosema apis* in honey bees.

4309. Fumaric Acid. *(E)-2-Butenedioic acid; trans-1,2-ethylenedicarboxylic acid; allomaleic acid; boletic acid.* $C_4H_4O_4$; mol wt 116.07. C 41.39%, H 3.47%, O 55.14%. Occurs in many plants, *e.g.,* in *Fumaria officinalis* L., *Fumariaceae,* in *Boletus scaber* Bull., *Boletaceae,* and in *Fomes igniarius* (Fries) Kickx., *Polyporaceae.* Essential to vegetable and animal tissue respiration. Prepd industrially from glucose by the action of fungi such as *Rhizopus nigricans:* Foster, Waksman, *J. Am. Chem. Soc.* **61,** 127 (1939). Laboratory prepn by the oxidation of furfural with sodium chlorate in the presence of vanadium pentoxide: Milas, *Org. Syn. coll. vol. II,* 302 (1943). Molecular structure: J. L. Derissen, *J. Mol. Struct.* **38,** 177 (1977). *Review:* W. D. Robinson, R. A. Mount in Kirk-Othmer *Encyclopedia of Chemical Technology* vol. **14** (Wiley-Interscience, New York, 3rd ed., 1981) pp 770-793.

Monoclinic, prismatic needles or leaflets from water. d 1.625. Sublimes at 200°. Sublimes at 165° at 1.7 mm pressure. Partial carbonization and formation of maleic anhydride occur at 230° (open vessel). mp 287° (closed capillary, rapid heating). pK_1 (25°): 3.03; pK_2: 4.54. Absorption spectrum: Macbeth, Stewart, *J. Chem. Soc.* **111,** 830 (1917). Soly in 100 g water at 25°: 0.63 g; at 40°: 1.07 g; at 60°: 2.4 g; at 100°: 9.8 g; in 100 g 95% alcohol at 30°: 5.76 g; in 100 g acetone at 30°: 1.72 g; in 100 g ether at 25°: 0.72 g. Almost insol in olive oil, chloroform, carbon tetrachloride, benzene, xylene, molten camphor, liq ammonia.

Monomethyl ester, $C_5H_6O_4$, prisms from alc, mp 144.5°.

Dimethyl ester, $C_6H_8O_4$, crystals, mp 102°. bp 192°.

USE: Substitute for tartaric acid in beverages and baking powders; as a replacement or partial replacement for citric acid in fruit drinks. As an antioxidant. Manuf polyhydric alcohols, synthetic resins. As mordant in dyeing.

4310. Fumigatin. *3-Hydroxy-2-methoxy-5-methyl-2,5-cyclohexadiene-1,4-dione; 3-hydroxy-2-methoxy-5-methyl-p-benzoquinone;* 6-hydroxy-3-methoxy-p-toluquinone; 3-hydroxy-4-methoxy-2,5-toluquinone. $C_8H_8O_4$; mol wt 168.15. C 57.14%, H 4.80%, O 38.06%. Fungal toxin with antibiotic properties isolated from metabolism soln of *Aspergillus fumigatus* Fres: Anslow, Raistrick, *Biochem. J.* **32,** 687 (1938); Waksman, Geiger, *J. Bact.* **47,** 391 (1944). Synthesis: Baker, Raistrick, *J. Chem. Soc.* **1941,** 670; Posternak, Ruelius, *Helv. Chim. Acta* **26,** 2045 (1943); Seshadri, Venkatasubramanian, *J. Chem. Soc.* **1959,** 1660. Biosynthesis: Pettersson, *Acta Chem. Scand.* **17,** 1323 (1963); Simonart, Verachtert, *Bull. Soc. Chim. Biol.* **49,** 543 (1967). Formation and polarographic assay: J. Lafond-Grellety *et al., Ann. Microbiol.* **129B,** 3 (1978). *Review:* Wilson, "Miscellaneous Aspergillus Toxins," in *Microbial Toxins* vol. VI, A. Ciegler *et al.,* Eds. (Academic Press, New York, 1971) p 281.

Maroon-colored needles or hexagonal plates from petr ether, mp 116°. Sublimes *in vacuo.* Volatile with steam.

Sparingly sol in water, petr ether; freely sol in acetone, ether, chloroform, benzene, ethyl acetate, alcohol.

4311. Fumonisin B₁. *1,2,3-propanetricarboxylic acid 1,-1'-[1-(12-amino-4,9,11-trihydroxy-2-methyltridecyl)-2-(1-methylpentyl)-1,2-ethanediyl] ester;* macrofusine; FB₁. C_{34}-$H_{59}NO_{15}$; mol wt 721.84. C 56.57%, H 8.24%, N 1.94%, O 33.25%. Most prevalent of a family of mycotoxins produced by *Fusarium moniliforme*, a common mold associated with corn; also isolated from other *Fusarium* species. Isolation: W. C. A. Gelderblom *et al., Appl. Environ. Microbiol.* **54**, 1806 (1988). Structure elucidation of family: S. C. Bezuidenhout *et al., Chem. Commun.* **1988**, 743. Causative agent of pulmonary edema in pig: L. R. Harrison *et al., J. Vet. Diagn. Invest.* **2**, 217 (1990). Association of B₁, B₂ with human esophageal cancer: J. P. Rheeder *et al., Phytopathology* **82**, 353 (1992). Metabolism: G. S. Shephard *et al., Toxicon.* **30**, 768 (1992). Toxicity and carcinogenicity in rat: W. C. A. Gelderblom *et al., Carcinogenesis* **12**, 1247 (1991). Toxicology in pig: W. H. Haschek *et al., Mycopathologia* **117**, 83 (1992). LC determn in corn of B series fumonisins: M. E. Stack, R. M. Eppley, *J. Assoc. Offic. Anal. Chem.* **75**, 834 (1992); P. A. Murphy *et al., J. Agric. Food Chem.* **41**, 263 (1993). Review of animal toxicoses: P. F. Ross *et al., Mycopathologia* **117**, 109-114 (1992). *Review:* W. P. Norred, *J. Toxicol. Environ. Health* **38**, 309-328 (1993).

4312. Fungichromin. Antibiotic A 246; cogomycin; lagosin; pentamycin; Cantricin. $C_{35}H_{58}O_{12}$; mol wt 670.84. C 62.67%, H 8.71%, O 28.62%. Antifungal polyene macrolide antibiotic, related structurally to filipin, *q.v.* Isoln from *Streptomyces cellulosae:* A. A. Tytell *et al., Antibiot. Ann.* **1954-1955**, 716. Isoln (as pentamycin) from *S. penticus:* S. Umezawa, Y. Tanaka, *J. Antibiot.* **11A**, 26 (1958). Isoln (as lagosin) from *S. roseoluteus:* C. J. Bessel *et al., U.S. pat.* **3,013,947** (1961 to Glaxo). Isoln from *S. griseus* (FCRC-21): R. C. Pandey *et al., Biomed. Mass. Spectrom.* **7**, 93 (1980). Structures: A. C. Cope, H. E. Johnson, *J. Am. Chem. Soc.* **80**, 1504 (1958); A. C. Cope *et al., ibid.* **84**, 2170 (1962); M. L. Dhar *et al., J. Chem. Soc.* **1964**, 842; M. P. Berry, M. C. Whiting, *ibid.* 862; V. Pozgay *et al., J. Antibiot.* **29**, 472 (1976). Identity of fungichromin with lagosin and cogomycin and comparison of reported physico-chemical constants: R. C. Pandey *et al., J. Antibiot.* **35**, 988 (1982). Biosynthesis and NMR assignment: H. Noguchi *et al., J. Am. Chem. Soc.* **110**, 2938 (1988).

Light yellow cryst. mp 157-162° (dec). $[\alpha]_D^{20}$ −227.7° (c = 0.53 in DMF). uv max (methanol): 357, 338, 322 nm ($E_{cm}^{1\%}$ 1231, 1250, 786). LD₅₀ in mice (mg/kg): 1624 orally; 33.3 i.p. (Umezawa, Tanaka).

THERAP CAT: Antifungal (topical).

4313. Fungisterol. *(3β,5α,22E)-Ergosta-6,8,22-trien-3-ol.* $C_{28}H_{44}O$; mol wt 396.66. C 84.79%, H 11.18%, O 4.03%. Formed from ergosterol by fungi. Isoln from *Penicillium chrysogenum:* A. Saito, *J. Ferment. Technol.* **29**, 457 (1951), *C.A.* **47**, 12507f (1953). Structure: *idem, ibid.* **31**, 141 (1953), *C.A.* **48**, 5276d (1954); *idem, ibid.* **32**, 138, 140 (1954), *C.A.* **49**, 9009i (1955).

Crystals from alcohol + ether + ethyl acetate, mp 147.5°. $[\alpha]_D^{15}$ −21.9° (chloroform).

Acetate, $C_{30}H_{46}O_2$, crystals, mp 158.5°. $[\alpha]_D^{15}$ −15.7° (chloroform).

Note: Term fungisterol is also used for γ-ergostenol, *q.v.*

4314. Funtumine. *(3α,5α)-3-Aminopregnan-20-one;* 3α-amino-20-oxo-5α-pregnane. $C_{21}H_{35}NO$; mol wt 317.52. C 79.44%, H 11.11%, N 4.41%, O 5.04%. Steroidal alkaloid isolated from *Funtumia latifolia* Stapf., Apocynaceae: Janot *et al., Compt. Rend.* **246**, 3076 (1958); from leaves of *Holorrhena febrifuga* Stapf., Apocynaceae: H. Dodoun *et al., Phytochemistry* **12**, 923 (1973). Structure: Janot *et al., Bull. Soc. Chim. France* **1960**, 1640, 1669. Prepn: H. Kapnang *et al., Tetrahedron Letters* **1977**, 3469. *Review:* R. Goutarel, *Bull. Soc. Chim. France* **1960**, 769. Effect on liver carcinogenesis in rats: A. Lacassagne *et al., Compt. Rend. Ser. D* **274**, 2830 (1972).

Prisms from ethyl acetate, mp 126°. $[\alpha]_D$ +95° (c = 1.7 in chloroform). pK 9.18. Sol in the usual organic solvents. LD₅₀ i.v. in mice: 30 mg/kg (Lacassagne).

4315. Furaltadone. *5-(4-Morpholinylmethyl)-3-[[(5-nitro-2-furanyl)methylene]amino]-2-oxazolidinone;* 5-morpholinomethyl-3-(5-nitrofurfurylideneamino)-2-oxazolidinone; 3-(5-nitro-2-furfurylideneamino)-5-(4-morpholinomethyl)-2-oxazolidone; furmethonol; nitrofurmethone; NF-260; Altafur; Altabactina; Furazolin; Ibifur; Medifuran; Nitraldone; Otifuril; Sepsinol; Ultrafur; Valsyn. $C_{13}H_{16}N_4O_6$; mol wt 324.29. C 48.15%, H 4.97%, N 17.28%, O 29.60%. Prepn: Gever, **U.S. pat. 2,802,002** (1957 to Norwich).

Yellow crystals from 95% ethanol, dec 206°. Sparingly sol in water: about 75 mg/100 ml at 25°.

THERAP CAT: Antibacterial.

THERAP CAT (VET): Antibacterial.

4316. Furan. Furfuran; oxole; tetrole; divinylene oxide. C_4H_4O; mol wt 68.08. C 70.57%, H 5.92%, O 23.50%. Occurs in oils obtained by the distillation of rosin contg pine wood. Prepd by decarboxylation of 2-furancarboxylic acid: Wilson, *Org. Syn.* **coll. vol. I** (2nd ed., 1941), p 274. Has been prepd directly from furfural over hot soda-lime or by dropping furfural on a fused mixt of sodium and potassium hydroxides: Hurd *et al., J. Am. Chem. Soc.* **54**, 2532 (1932). Toxicity data: Henderson, *J. Pharmacol. Exp. Ther.* **57**, 394 (1936). Thermodynamic properties: G. B. Guthrie, Jr. *et al., J. Am. Chem. Soc.* **74**, 4662 (1952).

Consult the Name Index before using this section.

Liquid. $d_4^{19.4}$ 0.9371. bp_{760} 31.36°; bp_{758} 32°. n_D^{20} 1.4216. Flash pt, closed cup: $-32°F$ ($-35°C$). Absorption spectrum: Purvis, *J. Chem. Soc.* **97**, 1648, 1655 (1910). Insol in water. Freely sol in alcohol and ether. Stable to alkalies; resinifies on evaporation or when in contact with mineral acids. Lethal concn for rats in air: 30400 ppm (Henderson).

Caution: The vapors are anesthetic. Can be absorbed through skin. *See: Toxicity of Industrial Metals,* E. Browning (Appleton-Century-Crofts, New York, 2nd ed., 1969) p 698.

4317. 2-Furanacrylic Acid. *3-(2-Furanyl)-2-propenoic acid;* β-2-furylacrylic acid; furacrylic acid; 2-furalacetic acid; furfurylidene acetic acid. $C_7H_6O_3$; mol wt 138.12. C 60.87%, H 4.38%, O 34.75%. Prepd by the condensation of furfural with malonic acid in the presence of pyridine. Laboratory procedure: S. Rajagopalan, P. V. A. Raman, *Org. Syn.* **25**, 51 (1945). Large scale procedure: Johnson, *ibid.* **20**, 55 (1940). Acid dissociation determn: C. C. Price, E. A. Dudley, *J. Am. Chem. Soc.* **78**, 68 (1956).

(E)-form

Colorless needles from water, mp 141°. Sublimes in high vacuum at 112°. bp_{760} 286° (also given as 226°); bp_8 117°. pKa in ethanol (25°): 6.49. Volatile in steam. One gram dissolves in 500 ml water at 15°. 100 ml of a satd benzene soln contains 1.14 g at 19°. Sol in alcohol, ether, glacial acetic acid.

Labile form, mp 104°. Can be converted to the stable form by exposure to sunlight in benzene soln contg some iodine.

4318. 2-Furanacrylonitrile. *3-(2-Furanyl)-2-propenenitrile;* 3-(2-furyl)acrylonitrile. C_7H_5NO; mol wt 119.12. C 70.58%, H 4.23%, N 11.76%, O 13.43%. Prepd from furfural and cyanoacetic acid in the presence of ammonium acetate and pyridine: Patterson, *Org. Syn.* **40**, 46 (1960).

(E)-form

Liquid. bp_{760} 95-97°. n_D^{25} 1.5824. Miscible with toluene, dimethylformamide.

4319. Furazabol. *(5α,17β)-17-Methylandrostano[2,3-c]-[1,2,5]oxadiazol-17-ol;* 17β-hydroxy-17α-methyl-5α-androstano[2,3-c]furazan; androfurazanol; furazalon; DH-245; Frazalon; Miotolon; Myotolon. $C_{20}H_{30}N_2O_2$; mol wt 330.47. C 72.69%, H 9.15%, N 8.48%, O 9.68%. Prepn: Ohta *et al.*, **Belg.** pat. **645,743** corresp to **U.S.** pat. **3,245,-988** (1964, 1966 both to Daiichi Seiyaku); Shimizu *et al.*, *Chem. Pharm. Bull.* **13**, 895 (1965); Ohta *et al., ibid.* 1445. Pharmacological studies: Kasahara *et al., ibid.* **13**, 1460 (1965); **16**, 1456, 1460 (1968). Toxicity: *eidem, ibid.* **14**, 285 (1966). Metabolic studies: Takegoshi *et al., ibid.* **20**, 1243 (1972).

Needles from methanol, mp 152-153°. $[\alpha]_D$ +39.4° (c = 1.42 in $CHCl_3$). uv max (ethanol): 217 nm (ε 4300). LD_{50} in mice (g/kg): 2.330 orally; >4 s.c.; 0.494 i.p. (Kasahara, 1966).

THERAP CAT: Anticholesteremic.

4320. Furazolidone. *3-[[(5-Nitro-2-furanyl)methylene]-amino]-2-oxazolidinone; 3-(5-nitrofurfurylideneamino)-2-oxazolidinone;* N-(5-nitro-2-furfurylidene)-3-amino-2-oxazolidone; NF-180; Furovag; Furoxane; Furoxone; Giarlam; Giardil; Medaron; Neftin; Nicolen; Nifulidone; Ortazol; Roptazol; Tikofuran; Topazone. $C_8H_7N_3O_5$; mol wt 225.16. C 42.68%, H 3.13%, N 18.66%, O 35.53%. Prepn: **Brit.** pat. **735,136**; G. Gever, **U.S.** pat. **2,742,462**; G. D. Drake *et al.,* **U.S.** pat. **2,759,931**; G. Gever, C. J. O'Keefe, **U.S.** pat. **2,927,110** (1955, 1956, 1956, 1960 all to Norwich); G. Gever *et al., J. Am. Chem. Soc.* **77**, 2277 (1955). Antimicrobial spectrum and toxicity: J. A. Yurchenco *et al., Antibiot. & Chemother.* **3**, 1035 (1953); G. S. Rogers *et al., ibid.* **6**, 231 (1956). Pharmacology and toxicology: B. H. Ali, *Vet. Res. Commun.* **6**, 1 (1983).

Yellow crystals from DMF, mp 256-257°. Darkens under strong light. Soly in water (pH 6): approx 40 mg/l. Dec by alkali.

THERAP CAT: Anti-infective (topical); topical antiprotozoal (Trichmonas).

THERAP CAT (VET): Antimicrobial.

4321. Furazolium Chloride. *6,7-Dihydro-3-(5-nitro-2-furanyl)-5H-imidazo[2,1-b]thiazol-4-ium chloride;* NF-963; Dermafur. $C_9H_8ClN_3O_3S$; mol wt 273.70. C 39.50%, H 2.95%, Cl 12.95%, N 15.35%, O 17.54%, S 11.72%. Prepn: **Neth.** pat. **Appl. 6,400,380** corresp to Snyder, Jr., **U.S.** pat. **3,169,970** (1964, 1965, both to Norwich Pharmacal); Snyder, Jr., Benjamin, *J. Med. Chem.* **9**, 402 (1966).

Crystals from methanol, dec >250°.

Free base, $C_9H_7N_3O_3S$, crystals from benzene + hexane, dec 171-172°.

THERAP CAT: Antibacterial.

4322. Furcellaran. Furcellaria gum; Danish agar; Burtonite 44. A gum obtained from a seaweed of the *Rhodophyceae,* the red alga *Furcellaria fastigiata,* fam. *Furcellariaceae,* order *Gigartinales.* The weed is found primarily in Northern European waters, especially in the Kattegat (between Sweden and Denmark). The gum is the potassium salt of the sulfuric acid ester of a high molecular weight polysaccharide. Consists mainly of D-galactose, 3,6-anhydro-D-galactose, and the half-ester sulfates of these sugars; one sulfate group occurs for each three or four monomeric units, which are arranged in an alternating sequence of $(1\rightarrow3)$ and $(1\rightarrow4)$-linked units. *Review:* Bjerre-Petersen *et al.,* in *Industrial Gums,* R. L. Whistler, Ed. (Academic Press, New York, 2nd ed., 1973) pp 123-136.

The processed gum is a white, odorless powder. Sol in hot or warm water. Easily dispersed in cold water to a homogeneous suspension without lumps; the furcellaran particles hydrate, swell and become almost invisible but do not dissolve unless heated. Forms agar-like gels at low concns. The strength of the gel can be increased by adding salts, esp potassium salts. Highly viscous. Solns in neutral medium are not adversely affected by prolonged exposure to high heat. However, exposure to heat in acidic media results in rapid hydrolysis and loss of gelling power.

USE: Natural colloid, gelling agent, viscosity control agent used primarily in food products but also in pharmaceuticals. Also in products for diabetics, proprietaries for reducing excess body wt, toothpastes. As carrier for food preservatives, bactericides. In bacteriological culture media.

4323. Furethidine. *4-Phenyl-1-[2-[(tetrahydro-2-furanyl)methoxy]ethyl]-4-piperidinecarboxylic acid ethyl ester; 4-phenyl-1-[2-(tetrahydrofurfuryloxy)ethyl]isonipecotic acid ethyl ester;* ethyl 1-(tetrahydrofurfuryloxyethyl)-4-phenyl-piperidine-4-carboxylate; 1-(2'-tetrahydrofurfuryloxyethyl)-norpethidine. $C_{21}H_{31}NO_4$; mol wt 361.48. C 69.78%, H 8.64%, N 3.87%, O 17.70%. Prepn: Frearson, Stern, **Brit. pat. 797,448** (1958 to J. F. Macfarlan & Co.); Frearson *et al., J. Chem. Soc.* **1960,** 2103.

$bp_{0.5}$ 210°; $bp_{0.3}$ 175-183°. mp ~28°. n_D^{20} 1.5219. pKa 7.48.

Methiodide, $C_{22}H_{34}INO_4$, crystals from ethyl acetate, mp 174°.

4324. Furfural. *2-Furancarboxaldehyde; 2-furaldehyde;* pyromucic aldehyde; artificial oil of ants; "furfurol." $C_5H_4O_2$; mol wt 96.09. C 62.50%, H 4.20%, O 33.30%. Occurs in some essential oils. Prepd industrially from pentosans which are contained in cereal straws and brans. Laboratory prepn from corncobs: R. Adams, V. Voorhees, *Org. Syn.* **coll. vol. I,** 280 (2nd ed., 1941). May also be prepd from pyridine. Toxicity study: P. M. Jenner *et al., Food Cosmet. Toxicol.* **2,** 327 (1964).

Colorless oily liq. Peculiar odor, somewhat resembling the odor of benzaldehyde. Turns yellow to brown on exposure to air and light and resinifies (the polymerization is greatly accelerated by hot alkali). d_4^{25} 1.1563. bp_{760} 161.8°; bp_{100} 103°; bp_{20} 67.8°; $bp_{1.0}$ 18.5° mp −36.5°. Volatile in steam. n_D^{20} 1.5261. Absorption spectrum: Purvis, *J. Chem. Soc.* **97,** 1655 (1910). Sol in 11 parts water; very sol in alcohol, ether. Flash pt, closed cup, 140°F (60°C); open cup, 155°F (68°C). Lower explosive limit: 2.1% by vol in air. Autoignition temp 797°F (392°C). *Keep in airtight container and protect from light.* LD_{50} orally in rats: 127 mg/kg (Jenner).

Caution: Potential symptoms of overexposure are irritation of eyes and upper respiratory system; headache; dermatitis. *See NIOSH Pocket Guide to Chemical Hazards* (DHHS/NIOSH 90-117, 1990) p 118.

USE: In the manufacture of furfural-phenol plastics such as Durite; in solvent refining of petroleum oils; in the prepn of pyromucic acid. As a solvent for nitrated cotton, cellulose acetate, and gums; in the manuf of varnishes; for accelerating vulcanization; as insecticide, fungicide, germicide; as reagent in analytical chemistry. In the synthesis of furan derivatives.

4325. Furfuryl Alcohol. *2-Furanmethanol;* 2-furylcarbinol; 2-furancarbinol; α-furylcarbinol; furfuralcohol; 2-hydroxymethylfuran. $C_5H_6O_2$; mol wt 98.10. C 61.22%, H 6.16%, O 32.62%. Usually prepd from furfural which is obtained by the processing of corncobs. The oil obtained by steam distillation of roasted coffee bean meal consists of 50% furfuryl alcohol after all organic acids have been removed. Laboratory prepn from furfural by the Cannizzaro reaction: Wilson, *Org. Syn.* **coll. vol. I** (2nd ed., 1941) p 276; *cf.* **U.S. pat. 2,041,184** (to Quaker Oats), *C.A.* **30,** 4515

(1936). Has been obtained by yeast reduction of furfural. Prepd industrially by the catalytic reduction of furfural using nickel and Cu-CrO catalysts: Peters, **U.S. pat. 1,906,873** (1933 to Quaker Oats); Wojcik, *Ind. Eng. Chem.* **40,** 210 (1948).

Liquid. Faint burning odor. Bitter taste. *Poisonous!* d_4^{23} 1.1282. bp_{760} 170°; bp_{400} 151.8°; bp_{200} 133.1°; bp_{100} 115.9°; bp_{60} 104.0°; bp_{40} 95.7°; bp_{20} 81.0°; bp_{10} 68.0°; bp_5 56.0°; $bp_{1.0}$ 31.8°. n_D^{23} 1.48515. Miscible with water, but unstable in water. Very sol in alcohol and ether. Easily resinified by acids. Spontaneous ignition temp +490° (915°F). Flash pt +75° (167°F). LD_{50} orally in rats: 275 mg/kg (fed as 2% aq soln), *Handbook of Toxicology* vol. I, W. S. Spector, Ed. (Saunders, Philadelphia, 1956) pp 146-147.

Caution: Potential symptoms of overexposure are dizziness; nausea, diarrhea; diuresis; respiration, body temperature depression; vomiting. *See NIOSH Pocket Guide to Chemical Hazards* (DHHS/NIOSH 90-117, 1990) p 118.

USE: Solvent; manuf wetting agents, resins.

4326. 5-Furfuryl-5-isopropylbarbituric Acid. *5-(2-Furanylmethyl)-5-(1-methylethyl)-2,4,6(1H,3H,5H)-pyrimidinetrione;* 5-isopropyl-5-furfurylbarbituric acid; 5-(2-furylmethyl)-5-isopropylbarbituric acid; Dormovit. $C_{12}H_{14}N_2O_4$; mol wt 250.25. C 57.59%, H 5.64%, N 11.19%, O 25.57%. Prepn: Heilner, **U.S. pat. 2,035,317** (1935 to Wiernik AG).

Crystals, mp 168-170°. Practically insol in water. Sol in hot water, alcohol, ether, chloroform. Aq soln of the sodium salt has a pH of 9.47.

Note: This is a controlled substance listed in the U.S. Code of Federal Regulations Title 21, Part 1308.13 (1995).

THERAP CAT: Sedative, hypnotic.

4327. Furfurylmethylamphetamine. *N-Methyl-N-(1-methyl-2-phenylethyl)-2-furanmethanamine; N-methyl-N-(α-methylphenethyl)furfurylamine;* N-methyl-N-(1-phenyl-2-propyl)furfurylamine; Furfenorex. $C_{15}H_{19}NO$; mol wt 229.32. C 78.56%, H 8.35%, N 6.11%, O 6.98%. Prepn of d- and dl-forms: Boissier, Ratouis, **Fr. pats. M3332** and **1,407,075** (both 1965 to S.I.F.A.), *C.A.* **63,** 13214g, 13215c (1965).

dl-Form, $bp_{0.1}$ 86-88°.
dl-Form cyclohexylsulfamate, $C_{15}H_{19}NO.C_6H_{13}NO_3S$, *Frugalan.* Crystals from acetone + ether, mp 113-114°.
d-Form, $bp_{0.1}$ 103-106°. n_D^{21} 1.5295.
d-Form cyclohexylsulfamate. Crystals from acetone, mp 90-92°. $[\alpha]_D^{20}$ +12.7° (c = 1 in water).

THERAP CAT: Anorexic.

4328. α-Furildioxime. *Di-2-furanylethanedione dioxime.* $C_{10}H_8N_2O_4$; mol wt 220.18. C 54.55%, H 3.66%, N 12.72%, O 29.07%.

Monohydrate, needle-like crystals, mp 166-168°. Very sol in alcohol, ether, slightly sol in benzene, petr ether.

USE: As a reagent, in alcohol soln, for nickel with which it gives an orange-red compd.

4329. 2-Furoic Acid. *2-Furancarboxylic acid;* α-furoic acid; pyromucic acid; Brenzschleimsäure (German). $C_5H_4O_3$; mol wt 112.08. C 53.58%, H 3.60%, O 42.82%. Prepd from furfural by a Cannizzaro reaction: Wilson, *Org. Syn.* coll. vol. I (2nd ed., 1941) p 276; *cf.* U.S. pat. 2,041,184 (to Quaker Oats), *C.A.* **30**, 4515 (1936); Wojcik, *Ind. Eng. Chem.* **40**, 210 (1948); Harrisson, Moyle, *Org. Syn.* coll. vol. IV, 493 (1963). Formed in man and several animal species as a metabolite of furfural and related compds.

Elongated monoclinic prisms from water or by sublimation at 130-140° and 50-60 mm pressure. mp 133-134°. bp_{760} 230-232°; bp_{20} 141-144°. K at 25°: 7.6×10^{-4}. Absorption spectrum: Hartley, Dobbie, *J. Chem. Soc.* **73**, 600 (1898). One gram dissolves in 26 ml water at 15°; in 4 ml boiling water. More sol in alcohol; freely sol in ether.

Ethyl ester, $C_7H_8O_3$, *ethyl furoate, ethyl pyromucate.* Colorless crystals. d_4^{20} 1.117. mp 34-36°. bp_{706} 195°. Insol in water; sol in alcohol.

Methyl ester, $C_6H_6O_3$, *methyl furoate, methyl pyromucate.* Liquid. Agreeable odor. bp 181.3°; d_4^{21} 1.1786; n_D^{21} 1.4871. Slightly sol in water; sol in alcohol.

4330. Furonazide. *4-Pyridinecarboxylic acid [1-(2-furanyl)ethylidene]hydrazide; isonicotinic acid α-methylfurfurylidenehydrazide;* 2-furyl methyl ketone isonicotinoylhydrazone; α-methylfurfurylidenehydrazide of isonicotinic acid; INF; Furilazone; Clitizina; Menazone. $C_{12}H_{11}N_3O_2$; mol wt 229.24. C 62.87%, H 4.84%, N 18.33%, O 13.96%. Prepn: Miyatake *et al., J. Pharm. Soc. Japan* **75**, 1066 (1955).

Crystals, mp 199-201.5°.

THERAP CAT: Antibacterial (tuberculostatic).

4331. Furosemide. *5-(Aminosulfonyl)-4-chloro-2-[(2-furanylmethyl)amino]benzoic acid; 4-chloro-N-furfuryl-5-sulfamoylanthranilic acid;* 4-chloro-N-(2-furylmethyl)-5-sulfamoylanthranilic acid; frusemide; fursemide; LB-502; Aisemide; Beronald; Desdemin; Discoid; Diural; Dryptal; Durafurid; Errolon; Eutensin; Frusetic; Frusid; Fulsix; Fuluvamide; Furesis; Furo-Puren; Furosedon; Hydro-rapid; Impugan; Katlex; Lasilix; Lasix; Lowpston; Macasirool; Mirfat; Nicorol; Odemase; Oedemex; Profemin; Rosemide; Rusyde; Trofurit; Urex. $C_{12}H_{11}ClN_2O_5S$; mol wt 330.75. C 43.58%, H 3.35%, Cl 10.72%, N 8.47%, O 24.19%, S 9.69%. Prepn: K. Stürm *et al.,* **Ger.** pat. 1,122,541; *eidem,* U.S. pat. 3,058,882 (both 1962 to Hoechst). Molecular and crystal structure: J. Lamotte *et al., Acta Crystallogr.* **B34**, 1657 (1978). Review of pharmacokinetics: L. Z. Benet, *J. Pharmacokinet. Biopharm.* **7**, 1-27 (1979); R. E. Cutler, A. D. Blair, *Clin. Pharmacokinet.* **4**, 279-296 (1979). Pharmacodynamics and pharmacokinetics of a slow-release formulation in man: A. Ebihara *et al., Arzneimittel-Forsch.* **33**, 163 (1983). Toxicity: E. I. Goldenthal, *Toxicol. Appl. Pharmacol.* **18**, 185 (1971). Comprehensive description: A. M. Al-Obaid *et al., Anal. Profiles Drug Subs.* **18**, 153-193 (1989).

Crystals from aq ethanol, mp 206°. uv max (95% ethanol): 288, 276, 336 nm ($E_{1cm}^{1\%}$ 945, 588, 144); (0.1N NaOH): 226, 273, 336 nm ($E_{1cm}^{1\%}$ 1147, 557, 133). Slightly sol in water, chloroform, ether. Sol in acetone, methanol, DMF, aq solns above pH 8.0. Less sol in ethanol. *Pharmaceut. Incompat:* Calcium gluconate, ascorbic acid, tetracyclines, urea, epinephrine. LD_{50} orally in female, male rats: 2600, 2820 mg/kg (Goldenthal).

THERAP CAT: Diuretic; antihypertensive.

THERAP CAT (VET): Diuretic.

4332. Furoyl Chloride. *Furancarbonyl chloride.* $C_5H_3ClO_2$; mol wt 130.53. C 46.01%, H 2.32%, Cl 27.16%, O 24.51%. C_4H_3OCOCl.

Colorless liquid; mp −2°. bp 170°. Dec by water or alcohol. Sol in ether.

Caution: Vapors irritate the eyes.

4333. Fursultiamine. *N-[(4-Amino-2-methyl-5-pyrimidinyl)methyl]-N-[4-hydroxy-1-methyl-2-[[(tetrahydro-2-furanyl)methyl]dithio]-1-butenyl]formamide;* thiamine tetrahydrofurfuryl disulfide; Alinamin F; Diteftin; Judolor; TTFD. $C_{17}H_{26}N_4O_3S_2$; mol wt 398.55. C 51.23%, H 6.58%, N 14.06%, O 12.04%, S 16.09%. Prepn: S. Yurugi, T. Fushimi, U.S. pat. 3,016,380 (1962 to Takeda). Metabolism: S. Kikuchi *et al., Eur. J. Pharmacol.* **9**, 367 (1970); C. Mitoma, *Drug Metab. Dispos.* **1**, 698 (1973). Series of papers on function of tetrahydrofurfuryl mercaptan moiety: *J. Biochem.* **74**, 717 (1973). Crystal structure: W. Shin, Y. C. Kim, *Bull. Kor. Chem. Soc.* **7**, 331 (1986).

E - form

Colorless prisms from ethyl acetate, mp 132° (dec). d 1.29. Sparingly sol in water; sol in organic solvents, dil mineral acids. LD_{50} in rats (mg/kg): 2200 orally; 540 i.p. (Yurugi, Fushimi).

THERAP CAT: Vitamin (enzyme cofactor).

4334. Furtrethonium. *N,N,N-Trimethyl-2-furanmethanaminium; furfuryltrimethylammonium;* trimethylfurfurylammonium; furtrimethonium; Furmethide. $[C_8H_{14}NO]^+$. Prepn of salts: Nabenhauer, U.S. pat. 2,185,220 (1938 to SK & F); Weilmuenster, Jordan, *J. Am. Chem. Soc.* **67**, 415 (1945); Khromov-Borisov *et al., J. Gen. Chem. USSR* **24**, 2021 (1954).

Iodide, $C_8H_{14}INO$, *Furamon, Furanol.* Crystals from ethanol + ethyl acetate, mp 116-117° (Weilmuenster, Jordan); also reported as mp 118-120° (Khromov-Borisov). Soluble in water, alcohol. Practically insol in benzene. pH of 1% aq soln, 5.3-6.0.

Benzenesulfonate, $C_{14}H_{19}NO_4S$, *Benzamon.* Crystals from ethanol + butyl acetate, mp 134-135° (Kromov-Borisov). Soluble in water, alcohol.

p-Toluenesulfonate, $C_{15}H_{21}NO_4S$, *Fisostina.*

THERAP CAT: Cholinergic. Iodide as parasympathomimetic.

4335. Fusafungine. S-314; Biofusal; Fusaloyos; Fusarine; Locabiotal. Antibiotic used as an aerosol for local application. Isoln from *Fusarium* species: Couchoud, **Fr. pat. 1,164,181** (1958), *C.A.* **54**, 20074b (1960); Servier, **Belg. pat. 612,474** (1962), *C.A.* **57**, 11320b (1962); *idem*, **Brit. pat. 1,018,626** (1966 to Biofarma), *C.A.* **64**, 16585b (1966). Bacteriological activity: D. Haler, *J. Int. Med. Res.* **5**, 61 (1977); *idem, Vie Med.* **61**, 507 (1980).

Solid, mp 125-129°. Stable up to 180°. Practically insol in water. Sol in glycols and fats.

THERAP CAT: Antibacterial.

4336. Fusaric Acid. *5-Butyl-2-pyridinecarboxylic acid; 5-butylpicolinic acid.* $C_{10}H_{13}NO_2$; mol wt 179.22. C 67.02%, H 7.31%, N 7.82%, O 17.85%. Antibiotic (wilting agent) first isolated from the fungus *Fusarium heterosporium*, Nees: Yabuta *et al.*, *J. Agr. Chem. Soc. Japan* **10**, 1059 (1934). Isoln from other *Fusarium* species and from *Gibberella fujikuroi* and synthesis: Plattner *et al.*, *Helv. Chim. Acta* **37**, 1379 (1954). Prepn: Hardegger, Nikles, *ibid.* **39**, 505 (1956); **40**, 2428 (1957); Schreiber, Adam, *Ber.* **93**, 1848 (1960); Umezawa, Nagatsu, **Ger. pat. 2,005,255** (1970 to Microbiochem. Res. Found.); R. Tschesche, W. Führer, *Ber.* **111**, 3502 (1978). Dopamine β-hydroxylase inhibitor and hypotensive activity: Suda *et al.*, *Chem. Pharm. Bull.* **17**, 2377 (1969); Nagatsu *et al.*, *Biochem. Pharmacol.* **19**, 35 (1970). Toxicity study: Ishii *et al.*, *Arzneimittel-Forsch.* **25**, 55 (1975).

Colorless crystals, mp 96-98°. LD_{30} orally in mice: 230 mg/kg (Ishii).

Copper salt, bluish-violet crystals from water, mp 258-259°.

4337. Fusarubin. *3,4-Dihydro-3,6,9-trihydroxy-7-methoxy-3-methyl-1H-naphtho[2,3-c]pyran-5,10-dione; 5,8-dihydroxy-2-(hydroxymethyl)-6-methoxy-3-(2-oxopropyl)-1,4-naphthalenedione; 3-acetonyl-5,8-dihydroxy-2-(hydroxymethyl)-6-methoxy-1,4-naphthoquinone;* oxyjavanicin. $C_{15}H_{14}O_7$; mol wt 306.27. C 58.83%, H 4.61%, O 36.57%. Isoln from *Fusarium solani*: Ruelius, Gauhe, *Ann.* **569**, 38 (1950). Structure: Hardegger *et al.*, *Helv. Chim. Acta* **47**, 2027 (1964). Efficient total synthesis: Y. Tanouc *et al.*, *Bull. Chem. Soc.* **60**, 2927 (1987).

Red prisms from benzene, dec 218°. Absorption max (ether): 535, 499 nm. Sol in glacial acetic acid, tetrahydrofuran, acetone, dioxane, pyridine; slightly sol in cold chloroform, cold alcohol, ether. Practically insol in carbon disulfide, cyclohexane, cold benzene. Dissolves in dil NaOH with a violet color. Practically insol in bicarbonate soln.

4338. Fuscin. *9,10-Dihydro-5-hydroxy-4,8,8-trimethyl-2H,4H-benzo[1,2-b:4,3-c']dipyran-2,6(8H)-dione.* $C_{15}H_{16}O_5$; mol wt 276.29. C 65.21%, H 5.84%, O 28.95%. Antibacterial pigment produced by the fungus *Oidiodendron fuscom* Robak: Michael, *Biochem. J.* **42**, XL; **43**, 528 (1948). Derivatives and degradation products: Birkinshaw *et al.*, *ibid.* **48**, 66 (1951). Structure and synthesis: Birch, *Chem. & Ind. (London)* **1955**, 682; Barton, Hendrickson, *J. Chem. Soc.* **1956**, 1028. Alternate synthesis: Pyuskyulev *et al.*, *Tetrahedron* **29**, 2849 (1973). Biosynthesis: Birch, *Ciba Foundation Symposium, Quinones Electron Transport* **1960**, 233; Birch *et al.*, *J. Chem. Soc.* **1965**, 1231.

Orange-colored, diamond-shaped plates from alc, mp 230°. uv max: 355, 283 nm (ε 27,600; 1000). Practically insol in water, petr ether. Sol in chloroform, acetone, ethyl acetate, acetic acid, ether, benzene. Slightly sol in cold ethanol, more sol in hot ethanol. Dissolves in aq solns of alkali hydroxides and pyridine with an intense purple color. Easily reduced by hydrosulfite or other reducing agents to the colorless dihydrofuscin, $C_{15}H_{18}O_5$, mp 206°, which is also found in the metabolic soln from the fungus grown on Czapek-Dox medium.

4339. Fusel Oil. A by-product of carbohydrate fermentations to produce ethyl alc. The material varies widely in composition, depending on the fermentation raw material used, but contains chiefly isopentyl alcohol and 2-methyl-1-butanol as well as isobutyl alcohol (20%), *n*-propyl alcohol (3-5%), and small amounts of other alcohols, esters and aldehydes. Described as an oily liq with a disagreeable odor; 60% boils at 122-138°. *Amyl alcohol (commercial)* obtained by chemical treatment and refining of fusel oil contains about 85% isopentyl alcohol and 15% 2-methyl-1-butanol. *Ref: Industrial Chemicals*, W. L. Faith *et al.*, Eds. (John Wiley, New York, 2nd ed., 1957) pp 107-114.

Caution: Commercial amyl alcohol is considerably more toxic than ethyl alcohol. *See also* 1-Pentanol and Isopentyl Alcohol.

4340. Fusidic Acid. *(3α,4α,8α,9β,11α,13α,14β,16β,-17Z)-16-(Acetyloxy)-3,11-dihydroxy-29-nordammara-17(20),-24-dien-21-oic acid; 3α,11α,16β-trihydroxy-29-nor-8α,9β,-13α,14β-dammara-17(20),24-dien-21-oic acid 16-acetate; 3α,-11α,16β-trihydroxy-4α,8,14-trimethyl-18-nor-5α,8α,9β,13α,-14β-cholesta-17(20),24-dien-21-oic acid 16-acetate; 3,11,16-trihydroxy-4,8,10,14-tetramethyl-17-(1'-carboxyisohept-4'-enylidene)cyclopentanoperhydrophenanthrene 16-acetate;* ramycin; Fucithalmic. $C_{31}H_{48}O_6$; mol wt 516.72. C 72.06%, H 9.36%, O 18.58%. Antibiotic substance isolated from the fermentation broth of *Fusidium coccineum*. Isoln and structure: W. O. Godtfredsen *et al.*, *Nature* **193**, 987 (1962); *Lancet* I, 928 (1962); W. O. Godtfredsen, S. Vangedal, *Tetrahedron* **18**, 1029 (1962). Identity with ramycin: H. Vanderhaeghe *et al.*, *Nature* **205**, 710 (1965). Structure: D. Arigoni *et al.*, *Experientia* **19**, 521 (1963). Stereochemistry: W. O. Godtfredsen *et al.*, *Tetrahedron* **21**, 3505 (1965). Synthetic studies: W. G. Dauben *et al.*, *J. Am. Chem. Soc.* **94**, 8593 (1972); R. E. Ireland, U. Hengartner, *ibid.* 3652; M. Tanabe *et al.*, *Tetrahedron Letters* **1977**, 1481. Total synthesis: W. G. Dauben *et al.*, *J. Am. Chem. Soc.* **104**, 303 (1982). *Review:* Tanaka in *Antibiotics* vol. 3, J. W. Corcoran, F. E. Hahn, Eds. (Springer-Verlag, New York, 1975) pp 436-447; W. von Daehne *et al.*, *Advan. Appl. Microbiol.* **25**, 95-146 (1979).

Crystals from ether or benzene, mp 192-193°. $[\alpha]_D^{20}$ −9° (chloroform). uv max: 204 nm (ε 9900). pK: 5.35 in water. Sol in alc, acetone, chloroform, pyridine, dioxane; sparingly sol in water, ether, hexane. LD_{50} in mice (g/kg): 1.2 s.c.; 1.5 orally (Godtfredsen).

Sodium salt, $C_{31}H_{47}NaO_6$, *sodium fusidate, ZN-6, Fucidin, Fucidina, Fucidine, Fucidin Intertulle.* Crystals, sol in water. LD_{50} in mice (g/kg): 0.2 i.v. (Godtfredsen, *Nature* 1962).

THERAP CAT: Antibacterial.

4341. Fustin. *(trans)-2-(3,4-Dihydroxyphenyl)-2,3-dihydro-3,7-dihydroxy-4H-1-benzopyran-4-one; 3,3',4',7-tetrahydroxyflavanone;* dihydrofisetin. $C_{15}H_{12}O_6$; mol wt 288.26. C 62.50%, H 4.20%, O 33.30%. From wood of *Rhus cotinus* L. (Venice sumac) and *R. succedanea* L., *Anacardiaceae:* Schmid, *Ber.* **19,** 1734 (1886); from *Gleditsia triacanthos* L., *Leguminosae:* Chadenson *et al., Compt. Rend.* **249,** 1362 (1955). Structure: Oyamada, *Ann.* **538,** 44 (1939). Stereochemistry: Weinges, *Ann.* **627,** 229 (1959); Roux, Paulus, *Biochem. J.* **77,** 315 (1960); Gaffield, *Tetrahedron* **26,** 4093 (1970).

(±)-Form, crystals, mp 226-228°. $[\alpha]_D^{21}$ −2.4°.

(±)-Form tetraacetate, $C_{23}H_{20}O_{10}$, crystals from methanol, mp 147-148°.

(−)-Form, crystals, mp 216-218°. $[\alpha]_D^{25}$ −26° (c = 2 in 1:1 acetone, water).

(−)-Form tetraacetate, crystals, mp 117-118°. $[\alpha]_D^{23}$ −25.2° (c = 0.8 in tetrachloroethane).

(+)-Form, needles from water, mp 228-229°. $[\alpha]_D^{23}$ +28.3° (c = 0.9 in 1:1 acetone, water).

(+)-Form tetraacetate, crystals from ethanol, mp 116-119°. $[\alpha]_D^{25}$ +24.4° (c = 1.3 in tetrachloroethane).

4342. Fyrol FR-2®. *1,3-Dichloro-2-propanol phosphate (3:1);* phosphoric acid tris(1,3-dichloro-2-propyl)ester; tris-(1,3-dichloroisopropyl)phosphate; tris[2-chloro-1-(chloromethyl)ethyl]phosphate; TCPP; PF-38; Emulsion 212. $C_9H_{15}Cl_6O_4P$; mol wt 430.91. C 25.09%, H 3.51%, Cl 49.36%, O 14.85%, P 7.19%. Flame retardant formerly used in children's sleepwear; once considered as a potential replacement for Tris-BP, *q.v.* Prepn: W. J. Jones *et al., J. Chem. Soc.* **1946,** 824. Use in flameproofing: R. J. Polacek, U.S. pat. **3,041,293** (1962 to Celanese). Thermal studies: K. Paciorek *et al., Am. Ind. Hyg. Assoc. J.* **39,** 633 (1978); N. Inagaki *et al., J. Appl. Polym. Sci.* **21,** 217 (1977); **24,** 1 (1979). Mutagenicity studies: M. D. Gold *et al., Science* **200,** 785 (1978); A. Nakamura *et al., Mutat. Res.* **66,** 373 (1979); D. Brusick *et al., J. Environ. Pathol. Toxicol.* **3,** 207 (1979). Toxicity study: J. K. Piotrowski *et al., Bromatol. Chem. Toksykol.* **9,** 141 (1976), *C.A.* **85,** 104825u (1976).

Viscous liquid, bp_5 236-237°. n_D^{20} 1.5022. Soly in water ∼100 ppm. LD_{50} orally in rats: 1.85 g/kg (Piotrowski).

USE: Flame retardant.

G

4343. Gabapentin. *1-(Aminomethyl)cyclohexaneacetic acid;* CI-945; Gö-3450; GOE-3450; Neurontin. $C_9H_{17}NO_2$; mol wt 171.24. C 63.13%, H 10.01%, N 8.18%, O 18.69%. Amino acid structurally related to γ-aminobutyric acid (GABA), *q.v.*, designed to cross the blood brain barrier. Prepn: G. Satzinger *et al.*, **Ger.** pat. **2,460,891** (1976 to Gödecke); *eidem*, **U.S.** pat. **4,024,175** (1977 to Warner-Lambert). Pharmacokinetics and metabolism: K.-O. Vollmer *et al.*, *Arzneimittel-Forsch.* **36**, 830 (1986). Clinical pharmacology: B. Saletu *et al.*, *Int. J. Clin. Pharmacol. Ther. Toxicol.* **24**, 362 (1986). GC determn in biological fluids: W. D. Hooper *et al.*, *J. Chromatog.* **529**, 167 (1990). Clinical trial as additional therapy in partial epilepsy: The US Gabapentin Study Group No. 5, *Neurology* **43**, 2292 (1993). Review of pharmacology and clinical trials: B. Schmidt, "Gabapentin" in *Antiepileptic Drugs*, R. H. Levy *et al.*, Eds. (Raven Press, New York, 3rd ed., 1989) pp 925-935; K. L. Goa, E. M. Sorkin, *Drugs* **46**, 409-427 (1993).

Crystals from ethanol/ether, mp 162-166° (Satzinger); also reported as mp 165-167° (Schmidt). pKa_1 (25°) 3.68; pKa_2 10.70. Isoelectric point 7.14. Solubility in water at pH 7.4 exceeds 10%.

THERAP CAT: Anticonvulsant.

4344. Gabexate. *4-[[6-[(Aminoiminomethyl)amino]-1-oxohexyl]oxy]benzoic acid ethyl ester; p-hydroxybenzoic acid ethyl ester 6-guanidinohexanoate; p-carbethoxyphenyl ε-guanidinocaproate.* $C_{16}H_{23}N_3O_4$; mol wt 321.38. C 59.80%, H 7.21%, N 13.08%, O 19.91%. Non-peptide proteolytic enzyme inhibitor which also inhibits the hydrolytic effects of thrombin, plasmin, and kallikrein, trypsin but not chymotrypsin; *cf.* aprotinin. Prepn as the *p*-toluenesulfonate salt: S. Fujii, T. Watanabe, **Ger.** pat. **2,050,484**; *eidem*, **U.S.** pat. **3,751,447** (1971, 1973 both to Ono). Enzyme inhibition: M. Muramatu, S. Fujii, *Biochim. Biophys. Acta* **268**, 221 (1972); S. Tamura *et al.*, *ibid.* **484**, 417 (1977). Pharmacology: T. Okegada *et al.*, *Nippon Yakurigaku Zasshi* **71**, 71 (1975), *C.A.* **84**, 218m (1976). Metabolism: M. Sugiyama *et al.*, *Oyo Yakuri* **9**, 733 (1977), *C.A.* **83**, 188145s (1975). Metabolism of inhibitory effect on platelet aggregation: G. Kosaki *et al.*, *Thromb. Res.* **20**, 587 (1980). Beneficial action in traumatic shock: A. M. Lefer *et al.*, *IRCS Med. Sci.: Libr. Compend.* **8**, 278 (1980); in exptl acute pancreatitis: J. R. Wisner *et al.*, *Pancreas* **2**, 181 (1987). Comparative clinical study in acute pancreatitis: N. Tanaka *et al.*, *Adv. Exp. Med. Biol.* **120**, 367 (1979). Teratology and toxicity study: T. Fujita *et al.*, *Oyo Yakuri* **9**, 743 (1975), *C.A.* **83**, 188322x (1975).

Monomethanesulfonate, $C_{16}H_{23}N_3O_4 \cdot CH_3SO_3H$, *gabexate mesylate, FOY, Megacert.* White crystals. Sol in water, ethanol, chloroform. Slightly sol in acetone. Practically insol in ether. pH of soln (1:100): 4.0-5.0. LD_{50} in mice (mg/kg): 8000 orally; 4700 s.c.; 25 i.v. (Fujita).

THERAP CAT: Enzyme inhibitor (proteinase).

4345. Gadodiamide. *[5,8-Bis(carboxymethyl)-11-[2-(methylamino)-2-oxoethyl]-3-oxo-2,5,8,11-tetraazatridecan-13-oato(3—)]gadolinium;* gadolinium diethylenetriamine pentaacetic acid bismethylamide; Gd-DTPA-BMA; S-041.

$C_{16}H_{26}GdN_5O_8$; mol wt 573.66. C 33.50%, H 4.57%, Gd 27.41%, N 12.21%, O 22.31%. Nonionic, low-osmolar paramagnetic gadolinium chelate complex. Prepn: S. C. Quay, **PCT Int.** pat. **Appl. 8,602,841** (1986 to New Salutar); *idem*, **U.S.** pat. **4,687,659** (1987 to Salutar). Pharmacokinetics, hemodynamics and toxicology: E. S. Harpur *et al.*, *Invest. Radiol.* **28**, Suppl. 1, S28 (1993). Clinical pharmacokinetics: M. VanWagoner, D. Worah, *ibid*, S44. Clinical study of CNS imaging: G. Sze *et al.*, *ibid*, **28**, Suppl. 1, S49 (1993); J. Valk *et al.*, *Neuroradiol.* **35**, 173 (1993); of myocardial imaging: M.-C. Dulce *et al.*, *Am. J. Roentgenol.* **160**, 963 (1993). Review of physicochemical properties: C. A. Chang, *Invest. Radiol.* **28**, Suppl. 1, S21-S27 (1993).

Osmolality (37°, 0.1M): 111 μmol/kg water. LD_{50} i.v. in mice: 14 mmol/kg (Chang).

Gadodiamide injection, $C_{16}H_{28}GdN_5O_9 \cdot C_{16}H_{26}CaN_5$-$NaO_8$, *Omniscan.* Osmolality (37°, 0.5M): 789 mOsM/kg water. Viscosity (cP): 2.0 (20°), 1.4 (37°). d^{20} 1.13. Partition coefficient (butanol/water): —2.1. LD_{50} i.v. in mice: 34 mmol/kg (Harpur).

THERAP CAT: Diagnostic aid (MRI contrast agent).

4346. Gadolinium. Gd; at. wt 157.25; at. no. 64; valence 3. A lanthanide; belongs to yttrium group of rare earth metals. Naturally occurring isotopes (mass numbers): 152 (0.20%), radioactive, $T_{1/2}$ 1.08 × 10^{14} years, α-emitter); 154 (2.18%); 155 (14.80%); 156 (20.47%); 157 (15.65%); 158 (24.84%); 160 (21.86%); known artificial radioactive isotopes: 137, 139, 142-151, 153, 159, 161, 162. Abundance in earth's crust: 6.1-6.36 ppm. Sources: samarskite, gadolinite (ytterbite), xenotime, and other rare earth minerals. Discovered by J. C. G. de Marignac in 1880. Prepn of metal: Trombe, *Compt. Rend.* **200**, 459 (1935); *idem, Bull. Soc. Chim. France* **2**, 660 (1935); Klemm, Bommer, *Z. Anorg. Chem.* **231**, 138 (1937). Sepn from other rare earths: F. H. Spedding *et al.*, *J. Am. Chem. Soc.* **69**, 2812 (1947); **76**, 2557 (1954). Spectrum: Albertson, *Phys. Rev.* **47**, 370 (1935); Spedding *et al.*, *J. Chem Phys.* **5**, 33 (1937). Toxicity study: Haley, *J. Pharm. Sci.* **54**, 663 (1965). Reviews of prepn, properties and compds: *The Rare Earths*, F. H. Spedding, A. H. Daane (Krieger, Huntington, N.Y., 1971, reprint of 1961 ed.) 641 pp; Hulet, Bode, "Separation Chemistry of the Lanthanides and Transplutonium Actinides" in *MTP Int. Rev. Sci: Inorg. Chem., Ser. One*, vol. 7, K. W. Bagnall, Ed. (University Park Press, Baltimore, 1971) pp 1-45; T. Moeller, "The Lanthanides" in *Comprehensive Inorganic Chemistry* vol. 4, J. C. Bailar, Jr. *et al.*, Eds. (Pergamon Press, Oxford, 1973) pp 1-101; F. H. Spedding in Kirk Othmer *Encyclopedia of Chemical Technology* vol. 19 (John Wiley & Sons, New York, 3rd ed., 1982) pp 833-854; *Chemistry of the Elements*, N. N. Greenwood, A. Earnshaw, Eds. (Pergamon Press, New York, 1984) pp 1423-1449. Brief review of properties: G. T. Seaborg, *Radiochim. Acta* **61**, 115-122 (1993).

Colorless or faintly yellowish metal; tarnishes in moist air. Crystalline forms: hexagonal close-packed α-form, d 7.886, transforms to β-form at 1262°; body-centered cubic β-form exists at > 1262°. mp 1312°. bp 3273°. Heat of fusion: 10.05 kJ/mol. Heat of sublimation (25°): 397.5 kJ/mol. E°(aq) Gd^{3+}/Gd —2.4 V (calc). Experimental reduction potentials (referred to a normal calomel electrode): —1.810, —1.955 V: Noddack, Brukl, *Angew. Chem.* **50**, 362 (1937).

Oxide, Gd_2O_3, *gadolinia.* Colorless, hygroscopic powder; prepd by igniting the hydroxide, nitrate, carbonate or oxalate, d^{15} 7.407, absorbs CO_2 from the air.

Hydroxide, $Gd(OH)_3$, gelatinous precipitate, prepd by the action of alkali or ammonium hydroxide on a soln of a gadolinium salt. Absorbs CO_2 from the air.

Chloride, $GdCl_3$, white monoclinic crystals, prepd by heating the oxide with excess of ammonium chloride above 200°. d^0 4.52. mp ~609°. Sol in water; forms double salts with platinic and auric chlorides. A hexahydrate, $GdCl_3 \cdot 6H_2O$, deliquesc crystals, d 2.424, is obtained from the aq soln. LD_{50} in mice (mg/kg): 550 i.p.; > 2000 orally (Haley).

Sulfate, $Gd_2(SO_4)_3$, octahydrate, colorless monoclinic crystals. Soly in water decreases with rise in temp. On heating at 400° yields the anhyd sulfate, d 4.139; begins to dec at 500°.

Nitrate, $Gd(NO_3)_3$, hexahydrate, deliquesc triclinic crystals. d 2.332. mp 91°. Sol in water, in alcohol. A pentahydrate, prismatic crystals, mp 92°, d 2.406, very insol, has been prepd. LD_{50} (hexahydrate) in rats (mg/kg): 230 i.p.; > 5000 orally (Haley).

USE: Oxide in control rods of some nuclear power reactors.

4347. Gadopentetic Acid. *[N,N-Bis[2-[bis(carboxymethyl)amino]ethyl]glycinato(5−)]gadolinate(2−) dihydrogen; N,N-*bis[2-[bis(carboxymethyl)amino]ethyl]glycine gadolinium complex; gadolinium diethylenetriaminepentaacetic acid; Gd-DTPA. $C_{14}H_{20}GdN_3O_{10}$; mol wt 547.58. C 30.71%, H 3.68%, Gd 28.72%, N 7.67%, O 29.22%. Evaluation as water-soluble paramagnetic relaxation reagent (PARR) for NMR spectrometry: J. J. Dechter, G. C. Levy, *J. Magnet. Reson.* **39,** 207 (1980); and prepn: T. J. Wenzel *et al., Anal. Chem.* **54,** 615 (1982). Use as diagnostic medium for NMR imaging: H. Gries *et al.,* **Ger.** pat. **3,129,906;** *eidem,* **U.S.** pat. **4,647,447** (1983, 1987 both to Schering AG). Physicochemical properties: H. Gries, H. Miklautz, *Physiol. Chem. Phys. Med. NMR* **16,** 105 (1984); E. Roux, L. De Broe, *J. Belg. Radiol.* **71,** 31 (1988). Evaluation as NMR contrast-enhancing agent in animals: R. C. Brasch *et al., Am. J. Roentgenol.* **142,** 625 (1984). Pharmacokinetics and toxicity: H. J. Weinmann *et al., ibid.* 619; in humans: H. J. Weinmann *et al., Physiol. Chem. Phys. Med. NMR* **16,** 167 (1984). HPLC determn: M. M. Vora *et al., J. Chromatog.* **369,** 187 (1986). Exptl use in imaging reperfused myocardium in dogs: S. Schaefer *et al., J. Am. Coll. Cardiol.* **12,** 1064 (1988). Clinical studies of diagnostic use in brain imaging: R. Felix *et al., Radiology* **156,** 681 (1985); J. P. Stack *et al., Neuroradiology* **30,** 145 (1988). Safety and tolerance in humans: H. A. Goldstein *et al., Radiology* **174,** 17 (1990). Reviews of diagnostic use in intracranial lesions: J. R. Hesselink, G. A. Press, *Radiol. Clin. North Am.* **26,** 873-887 (1988); in spinal disease: G. Sze, *ibid.* 1009-1024.

Dimeglumine salt, $C_{28}H_{54}GdN_5O_{20}$, *SHL-451A, ZK-93035, Magnevist, Resovist.* Freely sol in water. LD_{50} i.v. in rats: 10 mmol/kg (Weinmann).

THERAP CAT: Diagnostic aid (NMR contrast medium).

4348. Gadoteridol. *[10-(2-Hydroxypropyl)-1,4,7,10-tetraazacyclododecane-1,4,7-triacetato(3−)-N^1,N^4,N^7,N^{10},O^1,-O^4,O^7,O^{10}]gadolinium;* 10-(2-hydroxypropyl)-1,4,7,10-tetraazacyclododecane-1,4,7-triacetic acid gadolinium complex; gadolinium(III) 1,4,7-*tris*(carboxymethyl)-10-(2'-hydroxypropyl)-1,4,7,10-tetraazacyclododecane; Gd(HP-DO3A); SQ-32692. $C_{17}H_{29}GdN_4O_7$; mol wt 558.69. C 36.55%, H 5.23%, Gd 28.15%, N 10.03%, O 20.05%. Nonionic, lowosmolar paramagnetic gadolinium (III) chelate. Prepn: M. F. Tweedle *et al., Eur.* pat. **Appl. 292,689;** *eidem,* **U.S.** pat. **4,885,363** (1988, 1989 both to Squibb); D. D. Dischino *et al., Inorg. Chem.* **30,** 1265 (1991). Biodistribution: P. Wedeking *et al., Magn. Reson. Imaging* **10,** 641 (1992). RIA determn in biological fluids: M. D. Ogan *et al., J. Pharm. Sci.* **82,** 475 (1993). Toxicological studies: R. A. Soltys, *Invest. Radiol.* **27,** Suppl. 1, S7 (1992). Symposium on physicochemical properties, pharmacokinetics and clinical use in neurological disease: *ibid,* 1-63.

White solid from methanol/acetone, mp > 225°. Partition coefficient (butanol:water) −1.98.

Gadoteridol injection, *ProHance.* Osmolality (37°): 630 mOsM/kg water. Viscosity (cP): 2.0 (20°), 1.3 (37°). d^{25} 1.140. LD_{50} in mice, rats (mmol/kg): 11-14, > 10 i.v. (Soltys).

THERAP CAT: Diagnostic aid (MRI contrast agent).

4349. Galactaric Acid. Mucic acid; galactosaccharic acid; tetrahydroxyadipic acid; saccharolactic acid; Schleimsäure (German). $C_6H_{10}O_8$; mol wt 210.14. C 34.29%, H 4.80%, O 60.91%. Prepd by oxidation of lactose or of galactose: Kent, Tollens, *Ann.* **227,** 221 (1885); Maurer, Drefahl, *Ber.* **75B,** 1489 (1942). Manuf from wood sawdust: Acree, **Brit.** pat. **160,777** (1921). *Review:* B. A. Lewis *et al.,* "Galactaric Acid and Its Derivatives" in Whistler, Wolfrom, *Methods in Carbohydrate Chemistry,* **vol. II** (Academic Press, New York, 1963) pp 38-46.

Cryst powder, dec ~255° when rapidly heated, also reported as 225°. Soluble in 300 parts cold water, 60 parts boiling water, alkalies; practically insol in alcohol, ether.

Ammonium salt, $(NH_4)_2C_6H_8O_8$, acicular crystals. Soluble in water.

USE: Has been proposed to replace potassium bitartrate in baking powder and for manuf of granular effervescing salts.

4350. Galactitol. Dulcitol; dulcite; dulcose; euonymit; melampyrite; melampyrum; melampyrin. $C_6H_{14}O_6$; mol wt 182.17. C 39.56%, H 7.75%, O 52.70%. Found in dulcite or Madagascar manna (*Melampyrum nemorosum* L.) and in other species of *Melampyrum, Scrophulariaceae,* and *Evonymus atropurpureus* Jacq., *Celastraceae.* Isoln: Hünefeld, *Ann.* **24,** 241 (1837); Bouchardat, *Ann. Chim. Phys.* **27,** 68 (1872); Fischer, Hertz, *Ber.* **25,** 1261 (1892); Rogerson, *J. Chem. Soc.* **101,** 1040 (1912). Prepn by catalytic isomerization of D-glucitol: Wright, Hartmann, *J. Org. Chem.* **26,** 1588 (1961). Synthesis: Lespieau, *Bull. Soc. Chim. France* [5] **1,** 1374 (1934); Delepine, Horeau, *ibid.* **4,** 1524 (1937); Wiemann, Gordon, *ibid.* **1958,** 433. Structure: R. L. Lohmar "The Polyols" in W. Pigman, *The Carbohydrates* (Academic Press, New York, 1957) p 247.

Crystals from methanol + water, mp 188-189°. Slightly sweet taste. d^{20} 1.47. bp_1 275-280°. One gram dissolves in 30 ml water, in 2 ml boiling water. Slightly sol in alc. Ka at 18° = 3.5×10^{-14}.

Hexa-*O*-acetylgalactitol, $C_{18}H_{26}O_{12}$, crystals from ethanol, mp 168-169°.

Hexanitrate, *nitrodulcitol,* mp 94-95°. Has explosive properties: Taylor, Rinkenbach, *J. Franklin Inst.* **204,** 374 (1927).

4351. Galactoflavin. *1-Deoxy-1-(3,4-dihydro-7,8-dimethyl-2,4-dioxobenzo[g]pteridin-10(2H)-yl)-D-galactitol; 7,8-dimethyl-10-(D-galacto-2,3,4,5,6-pentahydroxyhexyl)benzo[g]pteridine-2,4(3H,10H)-dione; 7,8-dimethyl-10-(D-galacto-2,3,-4,5,6-pentahydroxyhexyl)isoalloxazine; 7,8-dimethyl-10-(d-1'-dulcityl)isoalloxazine; 6,7-dimethyl-9-(d-1'-dulcityl)isoalloxazine; 6,7-dimethyl-9-(1-deoxy-D-galactitol-1-yl)isoalloxazine.* $C_{18}H_{22}N_4O_7$; mol wt 406.40. C 53.20%, H 5.46%, N 13.79%, O 27.56%. Prepd from 1-deoxy-1-(3,4-dimethyl-6-phenylazo)anilino-D-galactitol and barbituric acid: Berezovskii, Eremenko, *Zh. Obshch. Khim.* **32,** 4056 (1962), *C.A.* **59,** 736b (1963). Structure: Emerson *et al., J. Biol. Chem.* **160,** 165 (1945). Pharmacology: Lane, Brindley, *Proc. Soc. Exp. Biol. Med.* **116,** 57 (1964). Produces congenital malformations in animals: Nelson *et al., J. Nutr.* **58,** 125 (1956); Miller *et al., J. Biol. Chem.* **237,** 968 (1962); Mackler, *Pediatrics* **43,** 915 (1969).

Yellow crystals, dec 260°. Absorption max: 223, 267, 370, 445 nm (ϵ 2730, 28100, 9100, 10800). Compd has yellow-green fluorescence in water.

USE: Riboflavine antagonist.

4352. D-Galactosamine. *2-Amino-2-deoxy-D-galactose;* chondrosamine; GalN. $C_6H_{13}NO_5$; mol wt 179.17. C 40.22%, H 7.31%, N 7.82%, O 44.65%. Amino sugar isolated from chondroitin sulfate, *q.v.:* P. A. Levene, F. B. La Forge, *J. Biol. Chem.* **18,** 123 (1914). Sepn of α- and β-anomers: P. A. Levene, *ibid.* **57,** 337 (1923). Synthesis: S. P. James *et al., Nature* **156,** 308 (1945) eidem, *J. Chem. Soc.* **1946,** 625; R. Kuhn, W. Kirschenlohr, *Ann.* **600,** 126 (1956); P. A. Gent *et al., J. Chem. Soc. Perkin Trans. 1* **1972,** 277. Chemistry: D. Horton in *The Amino Sugars* Vol. 1A, R. W. Jeanloz, Ed. (Acdemic, New York, 1969) pp 133-145. Inducer of exptl hepatitis: D. Keppler *et al., Exp. Mol. Pathol.* **9,** 279 (1968); K. Decker, D. Keppler in *Progress in Liver Diseases* Vol. IV, H. Popper, F. Schaffner, Eds. (Grune & Stratton, New York, 1972) p 183. Powerful inhibitor of hepatic RNA synthesis: D. Keppler *et al., J. Biol. Chem.* **249,** 211 (1974); T. Anukarahanonta *et al., Eur. J. Cancer* **16,** 1171 (1980).

α-form

Hydrochloride, $C_6H_{14}ClNO_5$, crystals, mp 180° (dec). Shows mutarotation. α-Form: $[\alpha]_D^{23}$ +124° → +93° (water). β-Form: $[\alpha]_D^{23}$ +47° → +93° (water).

4353. D-Galactose. Cerebrose; brain sugar. $C_6H_{12}O_6$; mol wt 180.16. C 40.00%, H 6.71%, O 53.28%. Constituent of many oligo- and polysaccharides occurring in pectins, gums, and mucilages. Prepn: Kent, Tollens, *Ann.* **227,** 224

(1885); E. P. Clark, *J. Biol. Chem.* **47,** 2 (1921). Mutarotation and purification of β-form: C. S. Hudson, E. Yanosky, *J. Am. Chem. Soc.* **39,** 1021 (1917). Structural configuration: J. Pryde, *J. Chem. Soc.* **123,** 1809 (1923); W. Charlton *et al., ibid.* **1926,** 94; W. N. Haworth *et al., ibid.* **1927,** 2428; E. L. Jackson, C. S. Hudson, *J. Am. Chem. Soc.* **59,** 994 (1937); R. M. Hann *et al., ibid.* **66,** 1912 (1944). Isoln in the processing of the red alga, *Porphyra umbilicalis:* S. Peat *et al., J. Chem. Soc.* **1961,** 1590. *Review:* W. Pigman, *The Carbohydrates* (Academic Press, New York, 1957) pp 88-90. Review of diagnostic use: W. J. Schirmer *et al., J. Surg. Res.* **41,** 543 (1986).

α-form

α-Form, prisms from water or ethanol, mp 167°. $[\alpha]_D$ +150.7° → +80.2° (water). Soluble in about 0.5 parts water; freely sol in hot water; final soly in water at 25° = 68%; sol in pyridine; slightly sol in alcohol.

β-Form, crystals, mp 167°. $[\alpha]_D$ +52.8° → +80.2° (water). Sol in 1.7 parts water at 17°.

Monohydrate, prisms from water, mp 118-120°.

Microparticulate form, SH U 454, Echovist. Suspension of galactose microparticle granules in a galactose solution. Prepn: J. S. Rasor, E. G. Tickner, Eur. pat. **Appl. 131,540** (1985 to Schering). Series of articles on *in vivo* use in echocardiography: *Arzneimittel Forsch.* **36,** 1030-1040 (1986). Review of formulations and clinical diagnostic use: R. Schürmann, R. Schlief, *Radiol. Med.* **87,** Suppl. 1, 15-23 (1994).

Transpulmonary microparticulate form, SH U 508A, Levovist. Suspension of galactose microparticle granules containing 0.1% physiologic palmitic acid in a sterile water solution.

THERAP CAT: Diagnostic aid (hepatic function). Microparticulate forms as diagnostic aid (ultrasound contrast agent).

4354. D-Galacturonic Acid. $C_6H_{10}O_7$; mol wt 194.14. C 37.12%, H 5.19%, O 57.69%. Obtained by hydrolysis of pectin where it is present as polygalacturonic acid: Ehrlich, *Chem. Ztg.* **41,** 197 (1917); Ehrlich, Guttmann, *Biochem. Z.* **259,** 100 (1933); Ber. **66,** 220 (1933); Niemann, Link, *J. Biol. Chem.* **95,** 203 (1932); **104,** 743 (1934); Morell, Link, *ibid.* **100,** 385 (1933); Anderson, King, *J. Chem. Soc.* **1961,** 5333. Isoln from mustard seeds: Goering, U.S. pat. **2,987,448** (1961 to Oil Seed Prod.).

α-form

α-Form, monohydrate, needles, mp 159°. $[\alpha]_D^{20}$ +98.0° → +50.9° (water). Soluble in water; slightly sol in hot alcohol. Practically insol in ether.

β-Form, mp 166°. $[\alpha]_D$ +27° → +55.6° (water).

Phenylhydrazone, mp 141°.

4355. Galanga. Galangal; colic root; East India root; Chinese ginger. Dried rhizome of *Alpinia officinarum* Hance, *Zingiberaceae*. Habit. China. Constit. Volatile oil, resin, kaempferid, galangin, dioxyflavanol, galangol.

4356. Galangin. *3,5,7-Trihydroxy-2-phenyl-4H-1-benzopyran-4-one; 3,5,7-trihydroxyflavone;* norizalpinin. $C_{15}H_{10}$-O_5; mol wt 270.24. C 66.67%, H 3.73%, O 29.60%. Isoln from galanga root, *Alpininia officinarum,* Hance and characterization: E. Jahns, *Ber.* **14,** 2807 (1881). Prepn: T. Heap, R. Robinson, *J. Chem. Soc.* **129,** 2336 (1926); J. J.

Chavan, R. Robinson, *ibid.* **1933**, 368. Mutagenicity studies: J. T. MacGregor, L. Jurd, *Mutat. Res.* **54**, 297 (1978); J. P. Brown, P. S. Dietrich, *ibid.* **66**, 223 (1979).

Yellowish needles from ethanol, mp 214-215°. Moderately sol in ethanol, ether; insol in water. Very sol in chloroform, benzene.

4357. Galanthamine. *4a,5,9,10,11,12-Hexahydro-3-methoxy-11-methyl-6H-benzofuro[3a,3,2-ef][2]benzazepin-6-ol;* galantamine; lycoremine; Jilkon. $C_{17}H_{21}NO_3$; mol wt 287.36. C 71.06%, H 7.37%, N 4.87%, O 16.70%. From Caucasian snowdrops, *Galanthus woronowii* Vel., *Amaryllidaceae:* N. F. Proskurnina, A. P. Yakovleva, *J. Gen. Chem.* **22**, 1899 (1952); from *Narcissus* spp: Boit *et al., Ber.* **90**, 725, 2197 (1957). Structure work: Kobayashi *et al., Chem. & Ind. (London)* **1956**, 177. Synthesis and stereochemistry: Barton, Kirby, *Proc. Chem. Soc.* **1960**, 392; *J. Chem. Soc.* **1962**, 806; Williams, Rogers, *Proc. Chem. Soc.* **1964**, 357. Alternate total synthesis: Kametani *et al., J. Chem. Soc. (C)* **1971**, 1043. Asymmetric synthesis of (+)- and (−)-forms from L-tyrosine: K. Shimizu *et al., Heterocycles* **8**, 277 (1977). Biosynthesis studies: D. H. R. Barton *et al., J. Chem. Soc.* **1963**, 4545; W. Döbke, *Heterocycles* **6**, 551 (1977). Pharmacokinetics: D. Mihailova, I. Yamboliev, *Pharmacology* **32**, 301 (1986). Toxicology study: S. L. Friess *et al., Toxicol. Appl. Pharmacol.* **3**, 347 (1961).

Crystals from benzene, mp 126-127°. $[\alpha]_D^{20}$ −118.8° (c = 1.378 in ethanol). Monoacidic base. Fairly sol in hot water; freely sol in alcohol, acetone, chloroform. Less sol in benzene, ether.

Hydrochloride, $C_{17}H_{21}NO_3 \cdot HCl$, crystals from water, dec 256-257°. Sparingly sol in cold, more sol in hot water. Very sparingly sol in alcohol, acetone.

Hydrobromide, $C_{17}H_{21}NO_3 \cdot HBr$, *Nivalin.* Crystals from water, dec 246-247°. $[\alpha]_D^{20}$ −93.1° (c = 0.1015 in 15 ml H_2O). LD_{50} i.v. in mice (mg/kg): 5.2 ± 0.2 (Friess).

THERAP CAT: Cholinesterase inhibitor.

4358. Galegine. *(3-Methyl-2-butenyl)guanidine; N-3,3-dimethylallylguanidine; isoamyleneguanidine.* $C_6H_{13}N_3$; mol wt 127.19. C 56.66%, H 10.30%, N 33.04%. Isoprenoid guanidine deriv from seeds of *Galega officinalis* L., *Leguminosae:* Tanret, *Compt. Rend.* **158**, 1182, 1426 (1914); **159**, 108 (1914); Markovic, Dittertová, *Chem. Zvesti* **9**, 576 (1955), *C.A.* **50**, 8137d (1956). Structure: Barger, White, *Biochem. J.* **17**, 827 (1923). Synthesis: Späth, Spitzy, *Ber.* **58**, 2273 (1925); Babor, Jezo, *Chem. Zvesti* **8**, 18 (1954), *C.A.* **49**, 7495f (1955). Metabolic effects: G. Weitzel *et al., Z. Physiol. Chem.* **353**, 535 (1972). Effects on mitochondria: B. Lotina *et al., Arch. Biochem. Biophys.* **159**, 520 (1973). Biosynthetic study: J. Steiniger, G. Reuter, *Biochem. Physiol. Pflanz.* **166**, 275 (1974). *Review:* Braun, *J. Chem. Ed.* **8**, 2175 (1931).

Hygroscopic, bitter crystals. mp 60-65°. Freely sol in water or alcohol, slightly in ether. *Keep well closed.*

4359. Galipine. *2-[2-(3,4-Dimethoxyphenyl)ethyl]-4-methoxyquinoline.* $C_{20}H_{21}NO_3$; mol wt 323.39. C 74.28%, H 6.55%, N 4.33%, O 14.84%. From Angostura bark *(Cusparia trifoliata* Engl., *Rutaceae):* Körner, Böhringer, *Gazz. Chim. Ital.* **13**, 363 (1883); Tröger, Kroseberg, *Arch. Pharm.* **250**, 494 (1912). Synthesis: Späth, Eberstaller, *Ber.* **57**, 1687 (1924); Späth, Pikl, *Ber.* **62**, 2244 (1929); Schläger, Leeb, *Monatsh.* **81**, 714 (1950).

Prismatic needles from alc, mp 116°. Soluble in alcohol, benzene, chloroform, ether; slightly sol in water, petr ether. The salts are more sol than those of cusparine.

Hydrochloride tetrahydrate, $C_{20}H_{21}NO_3 \cdot HCl \cdot 4H_2O$, plates, become anhydr at 100°, mp 165°.

Methiodide, $C_{20}H_{21}NO_3 \cdot CH_3I$, yellow needles, mp 146°.

4360. Gallacetophenone. *1-(2,3,4-Trihydroxyphenyl)-ethanone; 2',3',4'-trihydroxyacetophenone;* Alizarine yellow C; C.I. 57000. $C_8H_8O_4$; mol wt 168.15. C 57.14%, H 4.80%, O 38.06%. Prepn: Hart, Woodruff, *J. Am. Chem. Soc.* **58**, 1957 (1936); Campbell, Coppinger, U.S. pat. 2,-686,123 (1954 to U.S. Secy. Agr.); Knowles, U.S. pat. 2,-763,691 (1956 to Kodak); Price, Israelstam *J. Org. Chem.* **29**, 2800 (1964).

White to brownish-gray, cryst powder, mp 173°. uv max (methanol): 237, 296 nm (ε 8560, 12,500). Sol in 600 parts cold water, more in hot water; sol in alcohol, ether, soln of sodium acetate.

USE: Antiseptic.

4361. Gallamine Triethiodide. *2,2',2''-[1,2,3-Benzene-triyltris(oxy)]tris[N,N,N-triethylethanaminium] triiodide; [v-phenenyltris(oxyethylene)]tris[triethylammonium triiodide];* 1,2,3-tris(2-triethylammonium ethoxy)benzene triiodide; 1,2,3-tris(2-diethylaminoethoxy)benzene tris(ethyl iodide); tri(β-diethylaminoethoxy)-1,2,3-benzene triiodoethylate; pyrogallol 1,2,3-(diethylaminoethyl ether) tris(ethyl iodide); benzcurine iodide; RP-3697; F-2559; Tricuran; Retensin; Relaxan; Flaxedil. $C_{30}H_{60}I_3N_3O_3$; mol wt 891.54. C 40.42%, H 6.78%, I 42.70%, N 4.71%, O 5.38%. Curaring properties: D. Bovet *et al., Compt. Rend.* **225**, 74 (1947); F. Depierre, *ibid.* 956. Prepn: E. Fourneau, U.S. pat. **2,544,076** (1951 to Rhone-Poulenc). Comparatve clinical pharmacokinetics: W. Buzello, S. Agoston, *Anaesthesist* **27**, 313 (1978). Mode of action: D. Colquhoun, R. E. Sheridan, *Brit. J. Pharmacol.* **66**, 78 (1979); *eidem, Proc. Roy. Soc. London, Ser. B* **211**, 181 (1981). Effects in mammalian and amphibian nerve fibers: K. J. Smith, C. L. Schauf, *Science* **212**, 1170 (1981).

White cryst from acetone/water, mp 152-153° (indefinite). Freely sol in water, alc, dil acetone; sparingly sol in anhydr acetone, ether, benzene, chloroform. *Pharmaceutical Incompat:* Meperidine hydrochloride (solns must not be mixed).

THERAP CAT: Muscle relaxant (skeletal).

THERAP CAT (VET): Muscle relaxant (skeletal).

4362. Gallein. *3′,4′,5′,6′-Tetrahydroxyspiro[isobenzofuran-1(3H),9′-[9H]xanthen]-3-one;* 3′,4′,5′,6′-tetrahydroxyfluoran; 3′,4′,5′,6′-tetrahydroxyspiro[phthalan-1,9′-xanthen]-3-one; pyrogallolphthalein; C.I. 45445; mordant violet 25. $C_{20}H_{12}O_7$; mol wt 364.31. C 65.94%, H 3.32%, O 30.74%. Obtained by heating 1 part phthalic anhydride with 2 parts of pyrogallol or gallic acid: Baeyer, *Ber.* **4**, 457 (1871); Buchka, *Ann.* **209**, 261 (1881). Use as a biological stain: R. D. Lillie *et al., Stain Technol.* **49**, 339 (1974); R. Welsh, *ibid.* **52**, 261 (1977). *See also* H. J. Conn's *Biol. Stains,* R. D. Lillie, Ed. (Williams & Wilkins, Baltimore, 9th ed., 1977) p 351.

Brownish-red powder or crystals with 1½ H_2O, or red crystals with greenish-yellow color when anhydr. Loses the water of crystn at about 180° and blackens above this temp. Does not melt even at 300°. pH 3.8 brownish-yellow; pH 6.6 rose-red. Almost insol in water, benzene, chloroform. Slightly sol in ether; sol in alc, acetone, alkalies.

Disodium salt, $C_{20}H_{10}Na_2O_7$, *alizarin violet.* pH 10.6 rose; pH 13.0 violet.

USE: Clinical reagent (phosphates in urine). Monophosphates give a yellow, dibasic a red, tribasic a violet color. Used in soln of 0.5 g in 100 ml 50% alc; 2-3 drops for 100 ml liq. As sensitive indicator for acids, alkali hydroxides, NH_3, but not for carbonates. Biological stain.

4363. Gallic Acid. *3,4,5-Trihydroxybenzoic acid.* $C_7H_6O_5$; mol wt 170.12. C 49.42%, H 3.55%, O 47.02%. Obtained by alkaline or acid hydrolysis of the tannins from nutgalls; also by enzymatic hydrolysis using spent broths from *Penicillium glaucum* or *Aspergillus niger* which contain tannase: A. G. Perkin, O. Gunnell, *J. Chem. Soc.* **69**, 1303 (1896); Hsias, *Huang Hai* No. 7, 51 (1946), *C.A.* **42**, 3901i (1948); Cochrane, *Econ. Bot.* **2**, 145 (1948); Toth, Henster, *Acta Chim. Acad. Sci. Hung.* **2**, 209 (1952). Prepn from tannin containing materials: Krueger *et al.,* U.S. pat. **2,-723,992** (1955 to Mallinckrodt). Synthesis from aliphatic materials: Shipchandler *et al., J. Chem. Soc. Perkin Trans. I* **1975**, 1400. Biosynthesis: Haslam *et al., J. Chem. Soc.* **1961**, 1854. Study of polymorphic forms: E. Lindpainter, *Mikrochemie* **27**, 21 (1939). Toxicity studies: J. W. Dollahite *et al., Am. J. Vet. Res.* **23**, 1264 (1962).

Needles from abs methanol or chloroform, formerly reported as dec 235-240° (Perkin, Gunnell). Sublimes at 210° giving a stable form with mp 258-265° (dec) and an unstable form mp 225-230° (Lindpainter). One gram dissolves in 87 ml water, 3 ml boiling water, 6 ml alcohol, 100 ml ether, 10 ml glycerol, 5 ml acetone. Practically insol in benzene, chloroform, petr ether. *Protect from light.* LD_{50} in rabbits (g/kg): 5.0 orally (Dollahite).

Methyl ester, $C_8H_8O_5$, *methyl gallate, gallicin.* Monoclinic prisms from methanol, often hydrated or solvated. When dry, mp 202°. Sol in hot water, alcohol, methanol, ether.

Propyl ester, *see* Propyl Gallate.

USE: Manuf gallic acid esters, pyrogallol, inks; as photographic developer; in tanning; in dyeing; in testing for free mineral acids, dihydroxyacetone and alkaloids. Esters as antioxidants.

THERAP CAT: Formerly as astringent, styptic.

THERAP CAT (VET): Has been used as intestinal astringent.

4364. Gallium. Ga; at. wt 69.72; at. no. 31; valences 3, 2, 1. Natural isotopes: 69 (60.2%); 71 (39.8%), artificial radioactive isotopes: 63-68; 70; 72-76. Best source is the mineral germanite, a copper sulfide ore; occurs in very small quantities in zinc blendes, in aluminum clays, found in ores of iron, chromium, manganese; constitutes 5×10^{-4}% of the crust of the earth. Discovered by L. Boisbaudran, *Compt. Rend.* **81**, 493, 1100 (1875); **82**, 163, 1036 (1876); isolated pure by L. Boisbaudran and E. Jungfleisch, *Bull. Soc. Chim.* [2] **31**, 50 (1879). Isoln from rhenium-rich copper schist: Feit, *Angew. Chem.* **46**, 216 (1933). From bauxite: *Chem. Eng. News* **34**, 4300 (1956). Purification by zone melting: *Chem. Ztg.* **80**, 787 (1956). Alternate methods of purification: Gebauhr, U.S. pat. **2,928,731**; Merkel, U.S. pat. **2,927,853** (both 1960 to Siemens-Schuckert). Spectra: L. Boisbaudran *et al.,* cited in *Mellor's* **vol. 5**, 378 (1929). *Reviews:* Wagner, Gitzen, *J. Chem. Ed.* **29**, 162 (1952); Greenwood, *Inorg. Chem. Radiochem.* **5**, 91-134 (1963); Wade, Banister, *Comprehensive Inorganic Chemistry* **vol. 1**, J. C. Bailar, Jr. *et al.,* Eds. (Pergamon Press, Oxford, 1973) pp 997-1000, 1069-1117; P. de la Bretèque in Kirk-Othmer *Encyclopedia of Chemical Technology* **vol. 11** (Wiley-Interscience, New York, 3rd ed., 1980) pp 604-620. Review of diagnostic use of ^{67}Ga citrate in tumor detection: C. Bekerman *et al., Semin. Nucl. Med.* **14**, 296-323 (1984); of pharmacology and clinical antitumor activity of nitrate: B. J. Foster *et al., Cancer Treat. Rep.* **70** 1311-1319 (1986).

Grayish metal; possesses a greenish-blue reflection; tin- or silver-like when molten; has a crystalline orthorhombic texture. mp 29.78°. bp approx 2400°: Cochran, Foster, *J. Electrochem. Soc.* **109**, 144 (1962). Shows a tendency to remain in supercooled state. Contracts on melting; $d^{29.65}$ (solid) 5.9037; $d^{29.8}$ (liq) 6.0947: Richards, Boyer, *J. Am. Chem. Soc.* **43**, 274 (1921). Heat capacity: 0.09 cal/g/°C (0-24°, solid). Latent heat of fusion 19.16 cal/g. Stable in dry air; tarnishes in moist air or oxygen. Reacts with alkalies with evolution of hydrogen; attacked by cold concd hydrochloric acid; rendered passive by hot nitric acid; readily attacked by halogens.

Sesquioxide, Ga_2O_3; obtained by thermal decompn of the salts; exists in five crystalline modifications. β-Form is most stable, mp 1740±20°. d 5.94. On heating with magnesium is violently reduced to the metallic state.

Suboxide, Ga_2O; brown powder; obtained by heating the sesquioxide and the metal at 700°; stable in dry air; dec above 800°; converted to the trivalent state by nitric acid or bromine.

Hydroxide, $Ga(OH)_3$, a gelatinous precipitate; obtained by the action of ammonia or alkali hydroxide on a soln of a gallic salt.

Chloride, $GaCl_3$, colorless needles, mp 78°; prepd by the action of chlorine or hydrogen chloride on the metal. Free energy of formation (ΔG_f°) −25.4 kcal/mole.

Dichloride, $Ga^+[GaCl_4]^-$, mp 175-176° to a colorless liquid. Prepd by heating the metal in a sealed tube with stoichiometric quantities of mercury(I) or II chloride. Heat of fusion: 5.1 kcal/mol; heat of formation: −168.2 kcal/mole.

Sulfate octadecahydrate, $Ga_2(SO_4)_3.18H_2O$, octahedral crystals; formed by dissolving the sesquioxide or the hydroxide in sulfuric acid and precipitating with ether or alcohol; sol in water or 60% alcohol. Prepn: Reinmann, Tanner, *Z. Naturforsch.* **20b**, 71 (1965).

Citrate, $C_6H_5^{67}GaO_7$, *Neoscan*.

Nitrate, GaN_3O_9, *NSC-15200, Ganite*. White crystalline powder. Sol in warm and cold aqueous solutions, absolute alcohol, ether. Clinical study in treatment of cancer related hypercalcemia: R. P. Warrell, Jr. *et al., Ann. Int. Med.* **108**, 669 (1988). LD_{50} in mice, rats, rabbits (mg/kg): 55, 46, 43 i.v. (Foster).

Lactate, $C_9H_{15}GaO_9$, white amorphous powder, mp 270° (dec). Slightly hygroscopic. Soly (g/100 ml): water (20°) 8.9, water (100°) 21.5, 95% ethyl alcohol (20°) 1.5, ether, anhydrous (20°) 0.015, acetone, anhydrous (20°) 0.004. pH 2.7. LD_{50} in rats, rabbits (mg/kg): 47, 43 i.v.; 121, 97 s.c. (Dudley, Levine). Synthesis and physical properties: H. C. Dudley, R. F. Garzoli, *J. Am. Chem. Soc.* **70**, 3942 (1948). Toxicity study: H. C. Dudley, M. D. Levine, *J. Pharmacol. Exp. Ther* **95**, 487 (1949).

Nitrate nonahydrate, $Ga(NO_3)_3.9H_2O$, obtained by dissolving the metal or the oxide in concd nitric acid (Reinmann, Tanner).

Caution: Administration to humans has caused metallic taste, skin rashes, bone marrow depression: H. E. Stokinger in Patty's *Industrial Hygiene and Toxicology* **vol. 2A**, G. D. Clayton, F. E. Clayton, Eds. (Wiley-Interscience, New York, 3rd ed., 1981) pp 1630-1637.

THERAP CAT: ^{67}Ga citrate as diagnostic aid (neoplasm). Nitrate as antineoplastic, antihypercalcemic.

4365. Gallium Arsenide. AsGa; mol wt 144.64. As 51.80%, Ga 48.20%. GaAs. Prepd by passing a mixture of hydrogen and arsenic vapor over gallium(III) oxide at 600°: Goldschmidt, *Skr. Akad. Oslo* **1926**, no. 8, pp 34, 100; *Ber.* **60**, 1289 (1927). Simplified procedure: Juza, Schulz, *Z. Anorg. Allgem. Chem.* **275**, 65 (1954); Minden, *Sylvania Technologist* **11**, no. 1, p 19 (Jan. 1958). Vapor phase crystal growth: McAleer *et al., J. Electrochem. Soc.* **108**, 1168 (1961). Use in high-speed microcircuits: A. L. Robinson, *Science* **219**, 275 (1983).

Cubic crystals. mp 1238°. Dark gray with metallic sheen. Hardness 4.5. d_4^{25} 5.31. Thermal expansion coefficient: 5.9×10^{-6}. Thermal conductivity 0.52 watt units. Specific heat 0.086 cal/g/°C. Molten gallium arsenide attacks quartz, therefore graphite boats or quartz boats, which are carbon coated by pyrolytic decompn of methane, should be used in zone refining. GaAs single crystals have been grown by the Czochralski technique and by the floating zone method. Extensive twinning occurs. Intrinsic electron concn 10^7. Energy gap at room temp 1.38 electron volts. Electron mobility 8800 cm^2/volt sec. Hole mobility 450 cm^2/volt sec. Effective mass for electrons 0.06 m_0. Lattice constant 5.654Å. Dielectric constant 11.1. Intrinsic resistivity at 300°K = 3.7×10^8 ohm-cm. Electron lattice mobility at 300°K = 10,000 cm^2/volt-sec. Intrinsic charge density at 300°K = $1.4 \times 10^6/cm^3$. Electron diffusion constant at 300°K = 310 cm^2/sec. Hole diffusion constant = 11.5 cm^2/sec.

USE: In semiconductor applications (transistors, solar cells, lasers).

4366. Gallium Phosphide. GaP; mol wt 100.70. Ga 69.24%, P 30.76%. Prepn and description: Folberth, Oswald, *Z. Naturforsch.* **9a**, 1050 (1954); Wolff *et al., Phys. Rev.* **94**, 753 (1954); Antell, Effer, *J. Electrochem. Soc.* **106**, 509 (1959); **107**, 252 (1960); Frosch, Derick, *ibid.* **108**, 251 (1961); Addamiano, *J. Am. Chem. Soc.* **82**, 1537 (1960); Gershenzon, Mikulyak, *ibid.* **108**, 548 (1961); Pizzarello, *ibid.* **109**, 226 (1962). Seguin, Gans, U.S. pat. **2,862,787** (1958); Chang, U.S. pat. **2,921,905** (1960 to Westinghouse).

Translucent, amber-colored crystals of the zinc blende type. Greenish-yellow, opaque cryst mass, when unreacted gallium is present. mp 1465°. Dielectric constant 8.4. Energy gap 2.25 ev. Hole mobility 70. Electron mobility 1200. Current density at 4 volts = 25 amps/cm^2 at 25° (in rectifier circuitry).

USE: In semiconductor electronics.

4367. Gallium Trifluoride. F_3Ga; mol wt 126.72. F 44.98%, Ga 55.02%. GaF_3. Prepd by thermal decompn of ammonium hexafluogallate $(NH_4)_3(GaF_6)$: Hannebohn, Klemm, *Z. Anorg. Allgem. Chem.* **229**, 342 (1936); Kwasnik in *Handbook of Preparative Inorganic Chemistry* **vol. 1**, G. Brauer, Ed. (Academic Press, New York, 2nd ed., 1963) pp 227-228; from Ga and HF: Brewer *et al., J. Inorg. Nucl. Chem.* **9**, 56 (1959); Muetterties, Castle, *ibid.* **18**, 148 (1961).

White powder. d 4.47 (after heating to 630° in a current of F_2). mp > 1000°. bp ~950°. Can be sublimed in N_2 current at about 800° without decomp. Soly in water (25°) 0.0024 g/100 ml. Soly in hot dil HCl 0.0028 g/100 ml.

Trihydrate, white substance, mp > 140°. More sol in water than anhydr form.

4368. Gallocyanine. *1-Carboxy-7-(dimethylamino)-3,4-dihydroxyphenoxazin-5-ium chloride; C.I. Mordant Blue 10;* C.I. 51030. $C_{15}H_{13}ClN_2O_5$; mol wt 336.73. C 53.50%, H 3.89%, Cl 10.53%, N 8.32%, O 23.76%. Made by introducing nitrosodimethylaniline hydrochloride into a suspension of gallic acid in boiling methanol: Koechlin, Ger. pat. **19,580** (1881), *Frdl.* **1**, 269; Nietzki, Otto, *Ber.* **21**, 1740 (1888); *Colour Index* **vol. 4** (3rd ed., 1971) p 4460.

Green crystals. Practically insol in cold water. Slightly sol in hot water; sol in alcohol, glacial acetic acid, in alkali carbonates with reddish color, in concd HCl with a blue color which becomes red upon diluting with water.

USE: As a dye; in alkali carbonate soln as a reagent for lead with which it forms a deep violet color.

4369. Gallopamil. α-*[3-[[2-(3,4-Dimethoxyphenyl)ethyl]methylamino]propyl]-3,4,5-trimethoxy-α-(1-methylethyl)-benzeneacetonitrile; 5-[(3,4-dimethoxyphenethyl)methylamino]-2-isopropyl-2-(3,4,5-trimethoxyphenyl)valeronitrile;* α-*isopropyl-α-[(N-methyl-N-homoveratryl)-γ-aminopropyl]-3,4,5-trimethoxyphenylacetonitrile; methoxyverapamil; D-600.* $C_{28}H_{40}N_2O_5$; mol wt 484.64. C 69.39%, H 8.32%, N 5.78%, O 16.51%. Calcium channel blocking agent; methoxy analog of verapamil, *q.v.* Prepn: F. Dengel, Belg. pat. **615,861**; *idem,* U.S. pat. **3,261,859** (1962, 1966 to Knoll). Stereoselective synthesis of enantiomers: L. J. Theodore, W. L. Nelson, *J. Org. Chem.* **52**, 1309 (1987). Pharmacology: H. Haas, E. Busch, *Arzneimittel-Forsch.* **17**, 257 (1967); *idem, ibid.* **18**, 401 (1968); A. Fleckenstein, *ibid.* **20**, 1317 (1970). Clinical trial in stable angina pectoris: N. S. Khurmi *et al., Am. J. Cardiol.* **53**, 684 (1984). Symposium on pharmacology and clinical efficacy: *J. Cardiovasc. Pharmacol.* **20**, Suppl. 7, S1-S95 (1992). Review of pharmacokinetics and therapeutic potential in ischemic heart disease: R. N. Brogden, P. Benfield, *Drugs* **47**, 93-115 (1994).

Pale yellow viscous oil. n_D^{25} 1.5402.

Hydrochloride, $C_{28}H_{40}N_2O_5 \cdot HCl$, *Algocor, Procorum.* mp 145-148°.

d-Form hydrochloride, white crystalline solid from propanol/cyclohexane, mp 160.5-161.5°. $[\alpha]_D^{25}$ +11.7° (c = 5.02 in ethanol).

l-Form hydrochloride, white crystalline solid from propanol/cyclohexane, mp 160.5-161.5°. $[\alpha]_D^{23}$ −11.7° (c = 5.04 in ethanol).

THERAP CAT: Antianginal.

4370. Gamabufotalin. *(3β,5β,11α)-3,11,14-Trihydroxybufa-20,22-dienolide;* gamabufogenin; gamabufagin. $C_{24}H_{34}O_5$; mol wt 402.53. C 71.61%, H 8.51%, O 19.87%. One of the genins found in the venom of a Japanese toad (gama; *Bufo vulgaris formosus*). Isoln and structure: Kotake, *Ann.* **465**, 11 (1928); Wieland, Vocke, *Ann.* **481**, 215 (1930); Chen et al., *J. Pharmacol. Exp. Ther.* **49**, 26 (1933); Kondo, Ohno, *J. Pharm. Soc. Japan* **59**, 186 (1939); Jensen, *J. Am. Chem. Soc.* **59**, 767 (1937); Kuno Meyer, *Helv. Chim. Acta* **32**, 1599 (1949). Also found in Ch'an-Su, A Chinese drug prepared from Chinese toads (*Bufo asiaticus = Bufo gargarizans* Cantor): Ruckstuhl, Meyer, *Helv. Chim. Acta* **40**, 1270 (1957).

Bitter, solvated prisms from alcohol + ether, mp 254°; when dry dec 262-263°. $[\alpha]_D^{18}$ +1.26° (c = 0.793 in methanol). Produces numbness of the tongue. uv max about 300 nm. Very sparingly sol in chloroform, acetone, water. Somewhat more sol in methanol.

4371. Gambir. Catechu; pale catechu; gambir catechu; terra japonica. Dried aqueous extract of leaves and twigs of *Uncaria gambier* (Hunter) Roxb. *(Ouroparia gambier* (Hunter) Baill.), *Rubiaceae. Habit.* Southern Asia. *Constit.* 7-30% catechol, 20-50% catechutannic acid; quercetin, catechu-red, gambir-fluorescein, fixed oil, wax. Contains not less than 70% water-sol substance and not less than 60% is sol in alcohol. *Incompat:* Iron compds, gelatin, lime-water, mercuric chloride, zinc sulfate, alkalies.
USE: Tanning, dyeing fabrics brown or black.
THERAP CAT: Astringent.
THERAP CAT (VET): Astringent (intestinal).

4372. Gamboge. Cambogia. Gum-resin from *Garcinia hanburyi* Hook. f., *Guttiferae. Habit.* East Indies. *Constit.* 70-80% resin, cambogic acid, 15-25% gum.
THERAP CAT: Cathartic.
THERAP CAT (VET): Drastic purgative.

4373. Gambogic Acid. *[1R-[1α,1(Z),3aβ,5α,11β,-14aS*]]-2-Methyl-4-[3a,4,5,7-tetrahydro-8-hydroxy-3,3,11-trimethyl-13-(3-methyl-2-butenyl)-11-(4-methyl-3-pentenyl)-7,15-dioxo-1,5-methano-1H,3H,11H-furo[3,4-g]pyrano[3,2-b]xanthen-1-yl]-2-butenoic acid.* $C_{38}H_{44}O_8$; mol wt 628.76. C 72.59%, H 7.05%, O 20.36%. Principal acidic component of gamboge from the latex of *Garcinia hanburyi* Hook. f., *Guttiferae:* Amorosa, Lipparini, *Ann. Chim. (Rome)* **45**, 977 (1955). Structure: Ollis et al., *Tetrahedron* **21**, 1453 (1965); Ahmed et al., *J. Chem. Soc. (C)* **1966**, 772.

Golden, friable mass. $[\alpha]_D^{20}$ −685° (methanol). uv max (ethanol): 217, 280, 291, 362 nm (ε 26000, 16700, 17000, 14900).

Methyl ester monomethyl ether, $C_{40}H_{48}O_8$, yellow prisms from methanol, mp 130-131°. $[\alpha]_D^{22}$ −560° (c = 0.7 in chloroform). uv max (ethanol): 224, 299 nm (ε 36000, 13600).

Pyridine salt, $C_{43}H_{49}NO_8$, orange needles from ether + petr ether, mp 147-149°. $[\alpha]_D$ −550° (chloroform). uv max (ethanol): 291.5, 359.5 nm (ε 22300, 18100).

4374. Ganciclovir. *2-Amino-1,9-[[2-hydroxy-1-(hydroxymethyl)ethoxy]methyl]-6H-purin-6-one;* 9-[(1,3-dihydroxy-2-propoxy)methyl]guanine; 2'-nor-2'-deoxyguanosine; DHPG; 2'NDG; BIOLF-62; BW-759U; BW-759; BW-759U; RS-21592; $C_9H_{13}N_5O_4$; mol wt 255.23. C 42.35%, H 5.13%, N 27.44%, O 25.07%. Nucleoside analog structurally related to acyclovir, *q.v.* Prepn: J. P. Verheyden, J. C. Martin, U.S. pat. **4,355,032** (1982 to Syntex); K. K. Ogilvie et al., *Can. J. Chem.* **60**, 3005 (1982). Prepn and antiherpetic activity: W. T. Ashton et al., *Biochem. Biophys. Res. Commun.* **108**, 1716 (1982); J. C. Martin et al., *J. Med. Chem.* **26**, 759 (1983). Antiviral spectrum *in vitro:* K. O. Smith et al., *Antimicrob. Ag. Chemother.* **22**, 55 (1982); Y.-C. Cheng et al., *Proc. Nat. Acad. Sci. USA* **80**, 2767 (1983). Inhibitory effect on human cytomegalovirus *in vitro:* E.-C. Mar et al., *Antimicrob. Ag. Chemother.* **24**, 518 (1983); M. J. Tocci et al., *ibid.* **25**, 247 (1984). Comparison with acyclovir in herpes-infected mice: P. Collins, N. M. Oliver, *Antiviral Res.* **5**, 145 (1985). Clinical treatment of cytomegalovirus infection in immunodeficient patients: S. H. Koretz et al., *N. Engl. J. Med.* **314**, 801 (1986); O. L. Raskin et al., *J. Infect. Dis.* **155**, 323 (1987). Symposium on pharmacology and clinical efficacy vs cytomegalovirus: *Rev. Infect. Dis.* **10**, Suppl. 3, S457-S572 (1988).

Crystals from methanol, mp 250° (dec) (Verheyden, Martin); also reported as crystalline monohydrate from water, mp 248-249° (dec) (Ashton); crystals from water, mp > 300° (Martin). uv max (methanol): 254 nm (ε 12880). Soln in water (25°): 4.3 mg/ml at pH 7. LD_{50} i.p. in mice: 1-2 g/kg (Martin).

Sodium salt, $C_9H_{12}N_5NaO_4$, *Cymevan, Cymevene, Cytovene, Denosine.*
THERAP CAT: Antiviral.

4375. Ganglefene. *4-(2-Methylpropoxy)benzoic acid 3-(diethylamino)-1,2-dimethylpropyl ester;* p-isobutoxybenzoic acid 3-(diethylamino)-1,2-dimethylpropyl ester; 1,2-dimethyl-3-diethylaminopropyl p-isobutoxybenzoate; p-isobutoxybenzoic acid α,β-dimethyl-γ-diethylaminopropyl ester. $C_{20}H_{33}NO_3$; mol wt 335.49. C 71.60%, H 9.91%, N 4.18%, O 14.31%. Prepn: A. L. Mndzhoyan et al., U.S.S.R. pat. **115,905** (1958), *C.A.* **53**, 17439a (1959). Pharmacologic studies: M. G. Udel'nov et al., *Farmakol. Toksikol.* **36**, 299 (1973), *C.A.* **79**, 61442y (1973); V. A. Gusel, *Pharmacol.,*

Biochem. Behav. **2**, 1 (1974), *C.A.* **81**, 114675b (1974). Stability study: V. G. Belikov *et al.*, *Khromatogr. Metody Farm.* **1977**, 42, *C.A.* **88**, 141588 (1978).

LD$_{50}$ in mice: 530 mg/kg s.c., *RTECS* Vol. I, R. J. Lewis, R.C. Tatken, Eds. (1980) p 257.

Hydrochloride, C$_{20}$H$_{33}$NO$_3$.HCl, *Gangleron.*

THERAP CAT: Vasodilator (coronary).

4376. Gangliosides. Animal glycosphingolipids occurring in highest concentration in the central nervous system but widely distributed in other tissues. Primarily located in the outer surface of mammalian cell plasma membranes, they are also found in the synaptic membrane of the central nervous system. More than 60 gangliosides have been characterized. Each consists of a fatty acid (often stearic acid) and an oligosaccharide moiety, both attached to sphingosine. Sialic acid, *q.v.*, is the identifying sugar. The sialic acid residues of gangliosides and those of glycoprotein are the main cause of cell surface negative charge. Gangliosides are also thought to be involved in many different cell functions, e.g. metabolism, growth, malignant transformation. Ganglioside nomenclature: L. Svennerholm, "Ganglioside Metabolism" in *Comprehensive Biochemistry* **vol. 18,** M. Florkin, E. H. Stotz, Eds. (Elsevier, Amsterdam, 1970) pp 201-204. Isoln from normal brain and from the brain of patients with Tay-Sachs disease: Klenk, *Z. Physiol. Chem.* **268,** 50 (1941); **273,** 76 (1942); Trams, Lauter, *Biochim. Biophys. Acta* **60,** 350 (1962); Svennerholm, *Acta Chem. Scand.* **17,** 239 (1963). Occurrence of different gangliosides: Kuhn, Egge, *Angew. Chem.* **72,** 805 (1960). Chromatographic studies: Kuhn *et al., ibid.* **73,** 580 (1961); Klenk, Gielen, *Z. Physiol. Chem.* **326,** 144 (1961). Structure of ganglioside G$_I$ (or G$_{M1}$): Kuhn, Wiegandt, *Ber.* **96,** 866 (1963). Structure of gangliosides G$_I$ and G$_{II}$ (or G$_{D1a}$): Kuhn, Egge, *ibid.* **96,** 3338 (1963). Structure of gangliosides G$_{II}$, G$_{III}$ (or G$_{D1b}$) and G$_{IV}$ (or G$_{T1b}$): Kuhn, Wiegandt, *Z. Naturforsch.* **18b,** 541 (1963). Molecular diversity, biological implications of brain and thymus gangliosides: Y. Nogai, M. Iwamori, *Mol. Cell. Biochem.* **29,** 81 (1980). Biosynthesis in tissues: S. Basu *et al., Adv. Exp. Med. Biol.* **125,** 213 (1980). Action as biotransducers of membrane-mediated information: R. O. Brady, P. H. Fishman, *Adv. Enzymol.* **50,** 303 (1979). Reviews: Wiegandt, *Angew. Chem. Int. Ed.* **7,** 87-96 (1968). Collection of articles on structure and function: *Adv. Exp. Med. Biol.* **125,** 1-555 (1980); on structure analysis, function and properties, regeneration and recovery in the nervous system, clinical trials in amyotropic lateral sclerosis and diabetic neuropathy: *ibid.* **174,** 1-649 (1984). Book: *Gangliosides in Neurological and Neuromuscular Function, Development, and Repair,* M. M. Rapport, A. Gorio, Eds. (Raven Press, New York, 1981) 267 pp.

Ganglioside G$_{M1}$

Gangliosides are colorless crystallizable substances which melt with decompn. Insol in non-polar solvents. Soly in polar solvents increases with the size of the sugar residue and the sialic acid content. Forms micelles in aq soln, having mol wt of about 200,000-250,000. Forms molecular

solns in DMF or tetrahydrofuran, having mol wt of 1000-3000.

Ganglioside G$_{M1}$,C$_{73}$H$_{131}$N$_3$O$_{31}$, *siagoside, Sygen.*

Combination of gangliosides G$_{M1}$, G$_{D1a}$, G$_{D1b}$, and G$_{T1b}$, *Cronassial, Gangliovet.*

THERAP CAT: In treatment of peripheral neuropathies.

THERAP CAT VET: In treatment of peripheral neuropathies.

4377. Gardenins. A group of flavones isolated from the resinous exudate of leaf buds of *Gardenia lucida* Roxb. *Rubiaceae,* a small spiny tree occurring throughout India. Isoln of the main flavonoid pigment, gardenin A (originally thought to be the only component and referred to in early literature as "gardenin"): J. Stenhouse, C. G. Groves, *J. Chem. Soc.* **1877,** 552; *Ann.* **200,** 311 (1880); K. J. Balakrishna, T. R. Seshadri, *Proc. Indian Acad. Sci.* **A 27,** 91 (1948). Isoln and structure of gardenins B, C, D, E: A. V. R. Rao *et al., Indian J. Chem.* **8,** 398 (1970). Isoln of B from *Citrus jambhiri* lush. *Rutaceae* and synthesis: B. P. Chaliha *et al., Tetrahedron* **21,** 1441 (1965). Structure of A: P. K. Bose, *J. Indian Chem. Soc.* **22,** 233 (1945). Revised structure: A. V. R. Rao, K. Venkataraman, *Indian J. Chem.* **6,** 677 (1968). Synthesis of A: M. Krishnamurti *et al., ibid.* **8,** 575 (1970); M. Kamalam, A. V. R. Rao, *ibid.* 573; of C: A. J. Kalra *et al., ibid.* **11,** 96 (1973); of E: *eidem, ibid.* 1092. Antineoplastic activity and cytotoxicity: J. Edwards *et al., J. Nat. Prod.* **42,** 85 (1979). ^{13}C-NMR study: M. Iinuma *et al., Chem. Pharm. Bull.* **28,** 708 (1980).

Gardenin A	R = R' = OCH$_3$
Gardenin B	R = R' = H
Gardenin C	R = OH, R' = OCH$_3$
Gardenin D	R = OH, R' = H
Gardenin E	R = R' = OH

Gardenin A, C$_{21}$H$_{22}$O$_9$, *5-hydroxy-6,7,8-trimethoxy-2-(3,4,-5-trimethoxyphenyl)-4H-1-benzopyran-4-one.* Golden yellow needles from ethanol, mp 162-163°. Sol in alcohol, chloroform. All the gardenins give a green color with alcoholic ferric chloride.

Gardenin B, C$_{19}$H$_{18}$O$_7$, *5-O-desmethyltangeretin.* Fine yellow needles from benzene/petr ether, mp 176-177°. uv max (ethanol): 292, 330 nm (log e 4.40, 4.35); (1% AlCl$_3$): 310, 350 nm (log ε 4.48, 4.49).

Gardenin C, C$_{20}$H$_{20}$O$_9$. Yellow flakes from ethyl acetate-petr ether, mp 179-180°. uv max (methanol) 304, 323 nm (log ε 4.18, 4.20).

Gardenin D, C$_{19}$H$_{18}$O$_8$. Cryst, mp 190-192°. uv max (ethanol): 256, 280, 343 nm.

Gardenin E, C$_{19}$H$_{18}$O$_9$. Golden yellow needles from ethyl acetate/petr ether, mp 234-235°. uv max (methanol): 280, 325 nm (log ε 4.37, 4.36).

4378. Gardinol Type Detergents. A mixture of the sodium salts of sulfated fatty alcohols made by reducing the mixed fatty acids of coconut oil or of cottonseed oil, and of fish oils. Sometimes natural waxes such as spermaceti, wool fat, and beeswax are sulfated directly. The mixture of the sulfated alcohols which goes by the commercial name "lauryl alcohol" consists of about 15% mixed C$_8$ and C$_{10}$ (octyl and decyl) alcohols, 40% C$_{12}$ (lauryl or dodecyl) alcohol, 30% C$_{14}$ (myristyl or tetradecyl) alcohol, and 15% mixed C$_{16}$ and C$_{18}$ (cetyl, stearyl, and oleyl) alcohols.

Some of the tradenames designating this type of detergent are: *Gardinol; Duponol; Modinal; Aurinol; Maprofix; Tergavon; Sadopan; Cyclopon; Cyclanon; Sapidan; Lissapol; Teepol. See also* Sodium Lauryl Sulfate.

USE: Detergents, wetting, emulsifying, dispersing agents for cosmetics, as dry cleaning aids, in fungicidal sprays, in metal cleaning, leather processing, in textile manuf.

Caution: May cause local sensitivity reactions.

4379. Gardol®. *N-Methyl-N-(1-oxododecyl)glycine sodium salt; N-lauroylsarcosine sodium salt;* sodium *N*-lauroyl sarcosinate; Medialan LL-99. $C_{15}H_{28}NNaO_3$; mol wt 293.38. C 61.41%, H 9.62%, N 4.77%, Na 7.84%, O 16.36%. Prepn: Jungermann *et al., J. Am. Chem. Soc.* **78,** 172 (1956).

Aq soln, *Medialan LL-33.*
USE: Detergent, foaming agent, antienzyme for dentifrices: King, U.S. pat. **2,689,170** (1954 to Colgate-Palmolive).

4380. Garlic. Allium; Carisano. Fresh bulb of *Allium sativum* L., *Liliaceae. Habit.* Central Asia, Southern Europe, U.S.; cultivated worldwide. *Constit.* Alliin, *q.v.,* volatile and fatty oils, mucilage, albumin. Upon bruising of the bulb, the enzyme allinase converts alliin to allicin, *q.v.,* the source of the garlic odor. Series of articles on cardiovascular effects: *Brit. J. Clin. Pract.* **44,** Suppl. 69, 1-39 (1990). Review of chemistry: E. Block, *Sci. Am.* **252,** 114-119 (March, 1985). Brief reviews: A. Y. Leung, *Encyclopedia of Common Natural Ingredients* (John Wiley & Sons, New York, 1980) pp 176-178; P. Mansell, J. P. D. Reckless, *Brit. Med. J.* **303,** 379-380 (1991).

Powdered form, *Kwai.*
USE: As a spice and seasoning in foods.
THERAP CAT: In treatment of hypertension and hyperlipidemia.

4381. Garryine. $C_{22}H_{33}NO_2$; mol wt 343.51. C 76.92%, H 9.68%, N 4.08%, O 9.32%. From bark of *Garrya veutchii* Kellog, *Garryaceae,* where it occurs together with veatchine and other alkaloids: Oneto, *J. Am. Pharm. Assoc.* **35,** 204 (1946); Wiesner *et al., Can. J. Chem.* **30,** 608 (1952). Structure: Wiesner *et al.: J. Am. Chem. Soc.* **76,** 6068 (1954); Djerassi *et al., ibid.* **77,** 4801 (1955). Stereochemistry: Solo, Pelletier, *Chem. & Ind. (London)* **1960,** 1108. Racemic syntheses and resolution: Masamune, *J. Am. Chem. Soc.* **86,** 290 (1964); Nagata *et al., ibid.* 929, **89,** 1499 (1967); Guthrie *et al., Coll. Czech. Chem. Commun.* **31,** 602 (1966).

Monohydrate, crystals from dil acetone. Bitter taste, mp 74-82°. pK 8.70. $[\alpha]_D^{27.5}$ −84.23° (c = 1.44 in ethanol). Sol in water, alcohol.
Hydrochloride, $C_{22}H_{33}NO_2 \cdot HCl$, crystals from abs ethanol + ether, dec 263-268°. Soluble in water.

4382. Gasoline. Petrol (British); Benzin (German). A mixture of C_4 to C_{12} hydrocarbons. Natural gasoline, obtained by fractional distillation of petroleum, contains mostly saturated hydrocarbons; commercial grades of motor gasoline contain paraffins, olefins, naphthenes, and aromatics, all in substantial concns. Motor gasolines are made chiefly by cracking processes, in which heavier petr fractions are converted into more volatile fractions by thermal or catalytic decompn; have also been made commercially by catalytic high-pressure hydrogenation of soft coal and by catalytic synthesis of hydrocarbons from carbon monoxide and hydrogen. Some gasolines sold in the U.S. contain a minor

proportion of tetraethyllead added to motor gasoline to increase the octane numbers and thereby prevent "knock" in engines in which the gasoline is used as fuel. Knock is the audible manifestation of an excessive rate of pressure rise when the gasoline vapor is ignited under compression in an engine. The relative knocking tendencies of gasolines are measured in terms of "Octane Number," which is defined as the percentage of iso-octane, having "100 Octane No.," to be blended with *n*-heptane, having "0 Octane No." by definition, in order to obtain the same degree of knock as is obtained with the gasoline being rated, under standard conditions in a standardized test engine. Additives such as ethanol, methanol, benzene, toluene, MTBE and MMT, *q.q.v.,* are replacing tetraethyllead, and only a small percentage of leaded gasoline is sold in the U.S. *Review:* J. C. Lane in Kirk-Othmer *Encyclopedia of Chemical Technology* vol. 11 (Wiley-Interscience, New York, 3rd ed., 1984) pp 652-695. Review of toxicity: N. K. Weaver, *Ann. N.Y. Acad. Sci.* **534,** 441-451 (1988); of carcinogenicity: M. A. Mehlman, *Toxicol. Ind. Health* **7,** 143-152 (1990). Symposium on toxicology, exposure and health effects: *Environ. Health Perspec.* **101,** Suppl. 6, 1-212 (1993).

Gasoline is a highly flammable, mobile liq with characteristic odor; evaporates quickly. Flash pt 40-70°C. bp 32-210°C. Explosive limits, vol % in air: lower 1.3, upper 6.0. Insol in water. Freely sol in abs alcohol, ether, chloroform, benzene. Dissolves fats, oils, natural resins. LD_{50} orally in rats: 18.85 ml/kg (Weaver).

Caution: Potential symptoms due to mild overexposure by ingestion may include inebriation, vomiting, vertigo, drowsiness, confusion and fever; severe overexposure by ingestion may result in mild excitation, loss of consciousness, convulsions, cyanosis, congestion, capillary hemorrhaging of the lungs and internal organs, and death due to circulatory failure. When aspirated into the respiratory tract, severe inflammatory reaction and chemical pneumonitis may develop. Direct contact may cause skin and eye irritation. Prolonged skin exposure may result in a chemical burn. Repeated exposure may cause defatting of the skin. Potential symptoms due to acute overexposure by inhalation of vapor may include irritation and burning in the respiratory system, dizziness, anesthesia, bronchopneumonia, asphyxiation; inhalation of vapor at high levels may induce CNS depression, coma, convulsions, myocardial irritability. Potential symptoms due to chronic overexposure by inhalation may include vomiting, diarrhea, insomnia, headache, dizziness, anemia, impairment of renal function, muscle and neurological symptoms. *See Patty's Industrial Hygiene and Toxicology* vol. 2B, G. D. Clayton, F. E. Clayton, Eds. (Wiley-Interscience, New York, 4th ed., 1994) p 1395-1406; Weaver *loc. cit.*
USE: As fuel in internal combusion engines of the spark-ignited, reciprocating type; diluent; finishing agent; industrial solvent.

4383. Gastrins. Gastrointestinal hormones isolated from the mucosal lining of the gastric antrum of various mammalian species. Highly potent gastric secretion stimulants, first discovered by J. S. Edkins: *Proc. Roy. Soc. London* **76B,** 376 (1905). Several gastrins have been identified; they are referred to as "little gastrin", "big gastrin" and "minigastrin". "Little gastrin" or G-17 exists in two forms, 18-34-gastrin I or G-17-I and 18-34-gastrin II or G-17-II, which are heptadecapeptides that are identical in amino acid sequence, with the latter having sulfated tyrosine residues. There are relatively small species differences in amino acid sequences of G-17, although there are differences in the ratio of the non-sulfated to the sulfated forms. This ratio is about 3:4 in hog compared to about 2:1 in man. "Big gastrin" or 1-34-gastrin, also found in two forms, consists of the heptadecapeptides of "little gastrin" extended from their *N*-termini by additional heptadecapeptides with amino acid compositions different from G-17. For each species, the G-17-I and G-17-II portions of the G-34 gastrin are identical. Pairs of shorter gastrins, or "minigastrins" (22-34-gastrin) are C-terminal tetradecapeptides. Structure of porcine "little gastrins": Gregory *et al., Nature* **204,** 391 (1964); human: Bentley *et al., ibid.* **209,** 583 (1966). Synthesis of porcine "little gastrins": Anderson *et al., ibid.* **204,** 933 (1964); human: Beecham *et al., ibid.* **209,** 585 (1966);

human and canine: Agarwal, Kenner, *J. Chem. Soc. (C)* **1969**, 2213; ovine: Agarwal *et al., ibid.* 954; feline: *eidem, Experientia* **25**, 346 (1969). General synthetic method: E. Brown *et al., Chem. Commun.* **1980**, 1093. Isoln of "big gastrins" from Zollinger-Ellison tumor tissue: R. A. Gregory, H. J. Tracy, *Lancet* **2**, 797 (1972). Isoln of "mini-gastrins" from Zollinger-Ellison tumor tissue: *eidem, Gut* **15**, 683 (1974). Amino acid sequences of porcine and human "big gastrins": *eidem,* in *Gastrointestinal Hormones,* J. C. Thompson, Ed. (Univ. Texas Press, Austin, 1975) pp 13-14. Revised sequence of porcine: G. J. Dockray *et al., Bioorg. Chem.* **8**, 465 (1979); of human: A. M. Choudhury *et al., Z. Physiol. Chem.* **361**, 1719 (1980). Synthesis of human: G. Wendlberger *et al., Monatsh.* **112**, 1297 (1981). Structure and synthesis of "minigastrins": R. A. Gregory *et al., Z. Physiol. Chem.* **360**, 73 (1979). Review of chemical studies: Kenner, Sheppard, *Proc. Roy. Soc. London* **170B**, 89 (1968). Review of physiological advances: Gregory, *ibid.* 81. General reviews: Sanders, Schimmel, *Am. J. Med.* **49**, 380 (1970); McGuigan, *Vitam. Horm.* **32**, 47-88 (1974); V. Mutt, *ibid.* **39**, 231-426 (1982).

5-oxo-Pro–Gly–Pro–Trp–Leu–Glu–Glu–Glu–Glu–Glu–Ala–Tyr–Gly–Trp–Met–Asp–PheNH₂

18-34-Gastrin I (human)

4384. Gaultherin. *2-[(6-O-β-D-Xylopyranosyl-β-D-gluco-pyranosyl)oxy]benzoic acid methyl ester;* methyl salicylate-2-glucoxyloside; methyl salicylate-2-primeveroside; mono-tropitin; monotropitoside. $C_{19}H_{26}O_{12}$; mol wt 446.41. C 51.12%, H 5.87%, O 43.01%. In the wintergreen plant, *Gaultheria procumbens* L., in *Monotropa hypopitys* L., *Ericaceae* in *Betula lenta* L., *Betulaceae,* in *Spiraea ulmaria* L. and *S. filipendula* L., *Rosaceae:* Bridel, *Compt. Rend.* **177**, 642 (1923); **179**, 991 (1924); **180**, 1421, 1864 (1925); Bridel, Grillon, *ibid.* **187**, 609 (1928). Synthesis: Robertson, Waters, *J. Chem. Soc.* **1931**, 1881. On hydrolysis with 3% H_2SO_4 gaultherin forms 1 mol methyl salicylate, 1 mol D-glucose and 1 mol D-xylose. Enzymatic hydrolysis gives methyl salicylate and primeverose (glucoxylose).

Needles in star formation from 99% acetone. mp 180°. $[\alpha]_D^{20}$ −58° (c = 2). Sol in water and alcohol, slightly sol in ethyl acetate, acetone. Insol in ether.

Gaultherioside, $C_{13}H_{24}O_{10}$, ethyl primeveroside. mp 185°. $[\alpha]_D$ −58°. Gaultherioside forms glucose, xylose, and EtOH on hydrolysis with 3% H_2SO_4.

Note: The older literature regards gaultherin as methyl-salicylate-D-glucoside.

4385. Gefarnate. *(E,E,E)-5,9,13-Trimethyl-4,8,12-tetra-decatrienoic acid 3,7-dimethyl-2,6-octadienyl ester;* geranyl farnesylacetate; DA-688; Alsanate; Arsanyl; Dixnalate; Gefanil; Gefarnil; Gefarnyl; Gefulcer; Osteol; Salanil; Zac-kal. $C_{27}H_{44}O_2$; mol wt 400.65. C 80.94%, H 11.07%, O 7.99%. Prepn: Cardani *et al., J. Med. Chem.* **6**, 457 (1963); **Belg. pat.** 617,994 (1962 to Ist. de Angeli). Metabolism: G. Coppi *et al., Arzneimittel-Forsch.* **19**, 1519 (1969). Clinical evaluation in duodenal ulcer: C. R. Newman, D. A. Montgomery, *Brit. J. Clin. Pract.* **27**, 85 (1973).

Slightly yellowish liq with weak terpenic smell. $bp_{0.05}$ 165-168°. n_D^{20} 1.4900. uv max: 204 nm ($E_{1cm}^{1\%}$ 486). Sol in alc, ether, dimethylformamide, acetone, fatty oils. Practi-

cally insol in water, formamide, ethylene glycol, propylene glycol, glycerine.

THERAP CAT: Antiulcerative.

4386. Geissoschizoline. *Curan-17-ol;* pereirine. C_{19}-$H_{26}N_2O$; mol wt 298.43. C 76.47%, H 8.78%, N 9.39%, O 5.36%. From bark of *Geissospermum vellosii* Allem., *Apocynaceae:* Hesse, *Ann.* **202**, 141 (1880); Bertho, Moog, *ibid.* **509**, 241 (1934). Identity with pereirine: Bertho, Koll, *Naturwiss.* **48**, 49 (1961). Structure: Janot *et al., Compt. Rend.* **250**, 4383 (1960); Bertho, Koll *Ber.* **94**, 2737 (1961). Synthesis: Hymon, Schmid, *Helv. Chim. Acta* **49**, 2067 (1966); of *dl*-form: Dadson, Harley-Mason, *Chem. Commun.* **1969**, 665; Harley-Mason, Taylor, *ibid.* **1970**, 812.

Coarse crystals from abs methanol, mp 142.5-143°. $[\alpha]_D^{21}$ +32° (ethanol). uv max in (ethanol): 245, 300 nm (log ε 3.93, 3.47). Sol in alcohol, chloroform, ether. Practically insol in water.

Methyl chloride, $C_{19}H_{26}N_2O.CH_3Cl$, rods from abs methanol, mp 297°.

Methiodide, $C_{19}H_{26}N_2O.CH_3I$, rods from abs methanol, mp 254°.

4387. Geissospermine. *(16R,19E)-19,20-Didehydro-16-[(10β,13β,21β)-23-deoxy-21,22-dihydro-11-oxa-12,14-seco-strychnidin-10-yl]corynan-17-oic acid methyl ester; 19,20-di-dehydro-16-(15-ethyl-3a,5,5a,7,8,13a-hexahydro-4H-4,6-etha-no-1H,3H-[1,3]oxazino[3,4,5-lm]pyrrolo[2,3-d]carbazol-1-yl)-corynan-17-oic acid methyl ester;* β,17-epoxy-α-(3-ethylidene-1,2,3,4,6,7,12,12b-octahydroindole[2,3-a]quinolizin-2-yl)-curan-1-propanoic acid methyl ester. $C_{40}H_{48}N_4O_3$; mol wt 632.85. C 75.92%, H 7.65%, N 8.85%, O 7.58%. From bark of *Geissospermum vellosii* Allem., *Apocynaceae:* Hesse, *Ann.* **202**, 141 (1880). Structure: Puisieux, LeHir, *Compt. Rend.* **252**, 902 (1961); Janot, *Tetrahedron* **14**, 113 (1961). Crystal structure: A. Chiaroni, C. Riche, *Acta Crystallogr.* **35B**, 1820 (1979). Pharmacology: M. M. Aurousseau, *Ann. Pharm. Franc.* **19**, 175 (1961). On hydrolysis with HCl splits into geissoschizine and geissoschizoline, *q.v.*

Anhydr form, crystals from abs acetone, mp 213-214° (dec). $[\alpha]_D^{20}$ −101° (ethanol). uv max (methanol): 251, 285, 293 nm (log ε 4.10, 3.91, 3.90). Slightly sol in water, ether; sol in alcohol.

Dihydrate, mp 207° (sinters at 145°).

4388. Gelatin. Gelfoam; Puragel. A heterogeneous mixture of water-soluble proteins of high average mol wt. Gelatin is not found in nature but derived from collagen, *q.v.* by hydrolytic action. Obtained by boiling skin, tendons, ligaments, bones, etc., with water. Approx amino acid content: glycine 26.0-27.0%, alanine 8.7-9.6%, isoleucine 1.4-1.7%, leucine 3.2-3.6%, valine 2.5-2.7%, serine 3.2-3.8%, threonine 1.9-2.2%, proline 14.8-17.6%, phenylalanine

2.2-2.6%, tyrosine 0.49-1.1%, tryptophan 0.0-0.003, methionine 0.6-1.0%, cystine 0.1-0.2%, histidine 0.6-1.0%, arginine 8.6-9.3%, lysine 4.1-5.9%, aspartic acid 5.5-6.8%, glutamic acid 10.2-11.7%, hydroxyproline 12.6-14.4%, hydroxylysine 0.76-1.5%. Analysis: R. E. Neuman, *Arch. Biochem.* **24**, 289 (1949); J. E. Eastoe, *Biochem. J.* **61**, 589 (1955). Nutritionally, gelatin is an incomplete protein lacking tryptophan and contg but small amounts of other important amino acids. Review of the chemistry and structure of collagen with emphasis on its transformation to gelatin: A. Veis, *The Macromolecular Chemistry of Gelatin* (Academic Press, New York, 1964) 433 pp. *Incompat:* Tannin, formaldehyde.

Colorless or slightly yellow, transparent, brittle, practically odorless, tasteless sheets, flakes, or coarse powder. Amphoteric. Swells up and absorbs 5-10 times its weight of water to form a gel in solutions below 35-40°. Sol in hot water, glycerol, acetic acid. Insol in organic solvents.

USE: As stabilizer, thickener and texturizer in food; manuf rubber substitutes, adhesives, cements, lithographic and printing inks, plastic compds, artificial silk, photographic plates and films, matches, light filters for mercury lamps; clarifying agent; in hectographic masters; sizing paper and textiles; for inhibiting crystn in bacteriology, for preparing cultures. Pharmaceutic aid (suspending agent; encapsulating agent; tablet binder; tablet and coating agent).

THERAP CAT (VET): Plasma expander; hemostasis (sponge).

4389. Gellan Gum. Native gellan gum; high acyl gellan gum; polysaccharide S-60; PS-60. Extracellular polysaccharide obtained by aerobic fermentation of *Pseudomonas elodea*. Anionic hydrocolloid composed of acetylated linear tetrasaccharide repeat unit β-D-glucose, β-D-glucuronic acid and α-L-rhamnose in a 2:1:1 molar ratio. Prepn of native form: K. S. Kang, G. T. Veeder, U.S. pat. **4,326,053**; of deacetylated form: K. S. Kang *et al.*, U.S. pat. **4,326,052** (both 1982 to Merck & Co.); R. Moorhouse *et al.*, *ACS Symposium Series* **150**, 111 (1981). Structural studies: M. A. O'Neill *et al.*, *Carbohydr. Res.* **124**, 123 (1983); P.-E. Jansson *et al.*, *ibid.* 135. Gelation properties: H. Grasdalen, O. Smidsrod, *Carbohydr. Polymers* **7**, 371 (1987). Texture profile analysis: G. R. Sanderson *et al.* in *Gums and Stabilizers for the Food Industry* **vol. 4**, G. O. Phillips *et al.*, Eds. (IRL Press, Oxford, 1988) pp 219-229. Determn in food gels: J. K. Baird, W. W. Smith, *Food Hydrocolloids* **3**, 407 (1989); in food products: H. D. Graham, *ibid.* 435 (1990). Use as agar substitute in culture media: D. Shungu *et al.*, *Appl. Environ. Microbiol.* **46**, 840 (1983). Physicochemical properties: M. Milas *et al.*, *Biopolymers* **30**, 451 (1990). Rheological study: P. B. Deasy, K. J. Quigley, *Int. J. Pharmaceutics* **73**, 117 (1991). Comparative review of gelling agents: G. R. Sanderson *et al.*, *Cereal Foods World* **34**, 991-998 (1989). *Reviews:* G. R. Sanderson in *Food Gels*, P. Harris, Ed. (Elsevier Appl. Sci., London, 1990) pp 201-232; W. Gibson in *Thickening and Gelling Agents in Food*, A. Imeson, Ed. (Blackie Academic & Professional, London, 1992) pp 227-249.

Low Acyl form

Readily sol in water. Forms thermoreversible gels.

Low acyl purified gellan gum, deacylated gellan gum, deacylated PS-60, Gelrite, Kelcogel. Unsubstituted tetrasaccharide. White to tan powder, mol wt ~0.5 × 10⁶. Partially sol in cold water; aqueous dispersions dissolve > 70°. Forms thermoreversible gels. Incompat with strong oxidizers. Bulk density ~50 lb/cu ft. LD₅₀ orally in rats: > 5000 mg/kg (Sanderson, 1990).

Note: Approved by U.S. FDA as safe for general use as a food additive in nonstandardized foods (U.S. Code of Federal Regulations, Title 21 Part 172.665, 1993).

USE: Gelling, texturizing, film forming and suspending agent.

4390. Gelsemine. $C_{20}H_{22}N_2O_2$; mol wt 322.41. C 74.51%, H 6.88%, N 8.69%, O 9.92%. CNS stimulant from roots and rhizome of *Gelsemium sempervirens* (L.) Ait., *Loganiaceae*. Isoln: Gerrard, *Pharm. J.* **13**, 641 (1883); Moore, *J. Chem. Soc.* **97**, 2223 (1910); **99**, 1231 (1911); Sayre, Watson, *J. Am. Pharm. Assoc.* **8**, 708 (1919); Chou, *Chinese J. Physiol.* **5**, 131 (1931), *C.A.* **25**, 4085⁶ (1931); Schwarz, Marion, *Can. J. Chem.* **31**, 958 (1953). Structure: Conroy, Chakrabarti, *Tetrahedron Letters* **1959**(4), 6; Lovell *et al.*, *ibid.* 1; Roe, Gates, *Tetrahedron* **11**, 148 (1960). NMR spectroscopic study: Y. Schun, G. A. Cordell, *J. Nat. Prod.* **48**, 969 (1985). Partial syntheses: W. E. Earley *et al.*, *Tetrahedron Letters* **29**, 3781, 3785 (1988).

Crystals from acetone, mp 178°. *Poisonous!* [α]$_D^{20}$ +13° (c = 1.2 in chloroform). pKa 7.75 in 80% methylcellosolve. uv max (methanol): 210, 252, 280 nm (log ε 4.50, 3.87, 3.15). Slightly sol in water; sol in alcohol, benzene, chloroform, ether, acetone, dilute acids.

Hydrochloride, $C_{20}H_{23}ClN_2O_2$, prisms from methanol + ether, mp 326°. [α]$_D$ +5° (c = 1.072 in water). Sol in water, slightly sol in alcohol.

4391. Gelsemium. Yellow jasmine; yellow jessamine; wild woodbine. CNS stimulant from dried rhizome and roots of *Gelsemium sempervirens* (L.) Ait., *Loganiaceae*. *Habit.* Southern U.S. *Constit.* Gelsemine, gelsemoidine, scopoletin, gelsemic acid, volatile oil, resin. *Poisonous!*

4392. Gemcitabine. *2'-Deoxy-2',2'-difluorocytidine;* 1-(2-oxo-4-amino-1,2-dihydropyrimidin-1-yl)-2-deoxy-2,2-difluororibose; dFdC; dFdCyd; LY-188011; Gemzar. $C_9H_{11}F_2N_3O_4$; mol wt 263.20. C 41.07%, H 4.21%, F 14.44%, N 15.97%, O 24.32%. Prepn: L. W. Hertel, **Brit.** pat. **Appl. 2,136,425**; *idem*, U.S. pat. **4,808,614** (1984, 1989 both to Lilly); L. W. Hertel *et al.*, *J. Org. Chem.* **53**, 2406 (1988); T. S. Chou *et al.*, *Synthesis* **1992**, 565. Antitumor activity: L. W. Hertel *et al.*, *Cancer Res.* **50**, 4417 (1990). Mode of action study: V. W. T. Ruiz *et al.*, *Biochem. Pharmacol.* **46**, 762 (1993). Clinical pharmacokinetics and toxicity: J. L. Abbruzzese *et al.*, *J. Clin. Oncol.* **9**, 491 (1991). Review of clinical studies: B. Lund *et al.*, *Cancer Treat. Rev.* **19**, 45-55 (1993).

Crystals from water, pH 8.5. [α]$_{365}$ +425.36°; [α]$_D$ +71.51° (c = 0.96 in methanol). uv max (ethanol): 234, 268 (ε 7810, 8560). LD₁₀ i.v. in rats: 200 mg/m² (Abbruzzese).

Hydrochloride, $C_9H_{11}F_2N_3O_4$·HCl, crystals from water-acetone, mp 287-292° (dec). [α]$_D$ +48°; [α]$_{365}$ +257.9° (c = 1.0 in deuterated water). uv max (water): 232, 268 nm (ε 7960, 9360).

THERAP CAT: Antineoplastic.

4393. Gemeprost. *(2E,11α,13E,15R)-11,15-Dihydroxy-16,16-dimethyl-9-oxoprosta-2,13-dien-1-oic acid methyl ester;* 16,16-dimethyl-*trans*-Δ²-PGE₁ methyl ester; ONO-802; Cergem; Cervagem(e); Preglandin. $C_{23}H_{38}O_5$; mol wt 394.55. C

70.02%, H 9.71%, O 20.28%. Analog of prostaglandin E₁, *q.v.* Prepn: M. Hayashi *et al.*, **Ger.** pat. **2,700,021** corresp to U.S. pat. **4,052,512** (both 1977 to Ono); H. Suga *et al.*, *Prostaglandins* **15**, 907 (1978). Effects on uterine contractility and steroid hormone plasma levels: K. Oshima *et al.*, *J. Reprod. Fertil.* **55**, 353 (1979). Effects on reproductive function: K. Matsumoto *et al.*, *Nippon Yakurigaku Zasshi* **79**, 15 (1982), *C.A.* **96**, 98392 (1982). Use in termination of first trimester pregnancy: O. Reiertsen *et al.*, *Prostaglandins Leukotrienes Med.* **8**, 31 (1982).

THERAP CAT: Abortifacient; oxytocic.

4394. Gemfibrozil. *5-(2,5-Dimethylphenoxy)-2,2-dimethylpentanoic acid; 2,2-dimethyl-5-(2,5-xylyloxy)valeric acid;* CI-719; Decrelip; Genlip; Gevilon; Lipozid; Lipur; Lopid. C₁₅H₂₂O₃; mol wt 250.34. C 71.97%, H 8.86%, O 19.17%. Serum lipid regulating agent. Prepn: P. L. Creger, **Ger.** pat. **1,925,423**; *eidem*, U.S. pat. **3,674,836** (1969, 1972, both to Parke, Davis). Production: O. P. Goel, U.S. pat. **4,126,637** (1978 to Warner-Lambert). Pharmacology: A. H. Kissebach *et al.*, *Atherosclerosis* **24**, 199 (1976); M. T. Kahonen *et al.*, *ibid.* **32**, 47 (1979). Series of articles on metabolism, clinical pharmacology, kinetics and toxicology: *Proc. Roy. Soc. Med.* **69**, Suppl 2, 1-120 (1976). Toxicity data: S. M. Kurtz *et al.*, *ibid.* 15. Controlled clinical trial in hyperlipidemia: J. E. Lewis *et al.*, *Pract. Cardiol.* **9**, 99 (1983). Multicenter clinical trial of long-term effect on coronary heart disease: V. Manninen *et al.*, *J. Am. Med. Assoc.* **260**, 641 (1988).

Crystals from hexane, mp 61-63°. bp₀.₀₂ 158-159°. LD₅₀ in mice, rats (mg/kg): 3162, 4786 orally (Kurtz).
THERAP CAT: Antihyperlipoproteinemic.

4395. Genistein. *5,7-Dihydroxy-3-(4-hydroxyphenyl)-4H-1-benzopyran-4-one;* 4′,5,7-trihydroxyisoflavone; prunetol; genisteol. C₁₅H₁₀O₅; mol wt 270.24. C 66.67%, H 3.73%, O 29.60%. Phytoestrogen found in soy products; the aglucon of genistin and of sophoricoside, *q.v.* Specific protein kinase inhibitor. Prepn from the glucoside by hydrolysis with emulsin: Charaux, Rabaté, *J. Pharm. Chim.* [9] **1**, 404 (1941); by hydrolysis with HCl in methanol: Walter, *J. Am. Chem. Soc.* **63**, 3273 (1941). Isoln from *prunus* spp., *Rosaceae:* Hasegawa *ibid.* **79**, 1738 (1957); from *Podocarpus spicata* R.Br., *Podocarpaceae:* Briggs, Cebalo, *Tetrahedron* **6**, 145 (1959). Structure: Baker, Robinson, *J. Chem. Soc.* **1925**, 1981; **1926**, 2713; Walz, *Ann.* **489**, 118 (1931). Synthesis: Baker, Robinson, *J. Chem. Soc.* **1928**, 3115; Narasimhachari *et al.*, *J. Sci. Ind. Res.* **12**, 287 (1953); Yoder *et al.*, *Proc. Iowa Acad. Sci.* **61**, 271 (1954); Zemplén *et al.*, *Acta Chim. Acad. Sci. Hung.* **19**, 277 (1959). HPLC determn in biological fluids: J. G. Supko, L. R. Phillips, *J. Chromatog. B* **666**, 157 (1995); A. A. Franke *et al*, *Proc. Soc. Exp. Biol. Med.* **208**, 18 (1995). Review of synthesis and isotopic labeling: K. Wähälä *et al.*, *ibid.* 27-32. Series of articles on chemopreventive properties and mechanism of action: *ibid.* 103-115, 120-130; *J. Nutr.* **125**, Suppl. 3, 777S-797S (1995).

Rectangular or six-sided rods from 60% alcohol. Dendritic needles from ether. mp 297-298° (slight decompn). Sol in the usual organic solvents (Walter, *loc. cit.*); practically insol in water; sol in dil alkalies with yellow color. uv max: 262.5 nm (ε 138). Alkaline hydrolysis yields phloroglucinol and *p*-hydroxyphenylacetic acid.

Triacetate, C₁₅H₇O₅(CH₃CO)₃, clusters of needles from alcohol, mp 200-202°. Sol in organic solvents; practically insol in alkalies.

7-D-Glucoside, C₂₁H₂₀O₁₀, *genistin.* For isoln and structure *see* Walter, Hasegawa, Walz, *loc. cit.* Synthesis: Zemplén, Farkas, *Ber.* **76B**, 1110 (1943). Pale yellow plates from 80% ethanol, mp 256°. [α]D²¹ −28° (c = 0.6 in 0.02N NaOH); [α]D²⁶ −21.4° (pyridine). uv max (85% ethanol): 262.5 nm (a 90.5). Practically insol in cold water; slightly sol in hot water, hot ethanol, hot methanol; sol in hot 80% ethanol, hot 80% methanol, hot acetone, pyridine.

4′-Methyl ether, C₁₆H₁₂O₅, *5,7-dihydroxy-4′-methoxyisoflavone; Biochanin A; olmelin.* Isoln from red clover: Pope *et al.*, *Chem. & Ind. (London)* **1953**, 1092; Wong, *J. Sci. Food Agr.* **13**, 304 (1962); from *Andira inermis* (Swartz) H.B.K., *Leguminosae:* Crocker *et al.*, *J. Chem. Soc.* **1962**, 4906. Identity with olmelin: Gakhokidze, *J. Appl. Chem. USSR* **23**, 789 (1950), *C.A.* **46**, 9098i (1952). Structure: Bose, Siddiqui, *J. Sci. Ind. Res.* **9B**, no. 1, 25 (1950). Synthesis: Baker *et al.*, *Nature* **169**, 706 (1952). Yellow needles from methanol, mp 212-216°.

USE: Chemical probe to explore signal transduction pathways.

4396. Genisteine-Alkaloid. *[7S-(7α,7aα,14α,14aα)]-Dodecahydro-7,14-methano-2H,6H-dipyrido[1,2-a:1′,2′-e][1,5]diazocine; 11-isosparteine; l-α-isosparteine.* C₁₅H₂₆N₂; mol wt 234.38. C 76.87%, H 11.18%, N 11.95%. A stereoisomer of sparteine, *q.v.*, isolated from *Cytisus scoparius* (L.) Link, (*Spartium scoparium* L.), *Leguminosae:* Valeur, *Compt. Rend.* **167**, 23, 163 (1918); from *Lupinus caudatus* Kellog, *Leguminosae:* Marion *et al.*, *Can. J. Chem.* **29**, 22 (1951). Identity with *l*-α-isosparteine: Marion, Leonard, *ibid.* 297. Structure: Leonard *et al.*, *J. Am. Chem. Soc.* **77**, 1552 (1955). Absolute configuration: Okuda, Tsuda, *Chem. & Ind. (London)* **1961**, 1115.

Monohydrate, needles from boiling acetone, mp 108-110°. [α]D²² −51.6° (c = 0.7 in abs ethanol).

4397. Genite®. *2,4-Dichlorophenol benzenesulfonate;* benzenesulfonic acid 2,4-dichlorophenyl ester; 2,4-dichlorophenyl benzenesulfonate; EM-923; Genitol 923. C₁₂H₈Cl₂O₃S; mol wt 303.17. C 47.54%, H 2.66%, Cl 23.39%, O 15.83%, S 10.58%. Prepn: Gilbert, U.S. pat. **2,618,583** (1952 to Allied Chemical).

Waxy solid. Practically insol in water; sol in most organic solvents.

USE: Acaricide.

4398. Gentamicin. *Gentamycin.* Antibiotic complex produced by fermentation of *Micromonospora purpurea* or *M. echinospora* and variants thereof: M. J. Weinstein *et al., Antimicrob. Ag. Chemother.* **1963**, 1. Isoln, purification and characterization: J. P. Rosselet *et al., ibid.* 14. Industrial pats.: G. M. Luedemann, M. J. Weinstein; Charney, U.S. pats. **3,091,572; 3,136,704** (1963, 1964, both to Schering). Consists of three closely related components, gentamicins C_1, C_2, C_{1a}, and also gentamicin A which differs from the other members of the complex but is similar to kanamycin C, *q.v.* Separation of gentamicin C components: H. Maehr, C. P. Schaffner, *J. Chromatog.* **30**, 572 (1967); Wagman *et al., ibid.* **34**, 210 (1968). Structures contain 2-deoxystreptamine, *q.v.*, linked to two saccharide units, these being *garosamine* and a *purpurosamine* in the C series gentamicins. Structure studies: D. J. Cooper *et al., J. Chem. Soc. (C)* **1971**, 960, 2876, 3126. Structure of gentamicin A: H. Maehr, C. P. Schaffner, *J. Am. Chem. Soc.* **89**, 6787 (1967); **92** 1697 (1969). Sepn and structures of gentamicins A_1 to A_4: Nagabhushan *et al., J. Org. Chem.* **40**, 2830, 2835 (1975). Synthetic studies: W. Meyer zu Reckendorf, Bischof, *Ber.* **105**, 2546 (1972); M. Chmielewski *et al., Carbohyd. Res.* **70**, 275 (1979). Review of activity, toxicity and clinical pharmacology: J. Black *et al., Antimicrob. Ag. Chemother.* **1963**, 138-147. Comprehensive description: B. E. Rosenkrantz *et al., Anal. Profiles Drug Subs.* **9**, 295-340 (1980). Determn in serum by immunoassay: H. A. Holt *et al., J. Antimicrob. Chemother.* **34**, 747 (1994). Review of clinical use: S. B. Shrimpton *et al., ibid.* **31**, 599-606 (1993).

Gentamicin C_1	$R_1 = R_2 = CH_3$
Gentamicin C_2	$R_1 = CH_3, R_2 = H$
Gentamicin C_{1a}	$R_1 = R_2 = H$

White amorphous powder, mp 102-108°. $[\alpha]_D^{25}$ +146°. Freely sol in water; sol in pyridine, DMF, in acidic media with salt formation; moderately sol in methanol, ethanol, acetone. Practically insol in benzene, halogenated hydrocarbons.

Gentamicin C_1, $C_{21}H_{43}N_5O_7$, mp 94-100°. $[\alpha]_D^{25}$ +158°.
Gentamicin C_2, $C_{20}H_{41}N_5O_7$, mp 107-124°. $[\alpha]_D^{25}$ +160°.
Gentamicin C_{1a}, $C_{19}H_{39}N_5O_7$, *O-3-deoxy-4-C-methyl-3-(methylamino)-β-L-arabinopyranosyl-(1 → 6)-O-[2,6-diamino-2,3,4,6-tetradeoxy-α-D-erythro-hexopyranosyl-(1 → 4)]-2-deoxy-D-streptamine, gentamicin D.*

Gentamicin A, $C_{18}H_{36}N_4O_{10}$, *O-2-amino-2-deoxy-α-D-gluco-pyranosyl-(1 → 4)-O-[3-deoxy-3-(methylamino)-α-D-xylopyranosyl-(1 → 6)]-2-deoxy-D-streptamine.*

Hydrochloride, mp 194-209°. $[\alpha]_D^{25}$ +113°. Freely sol in water, methanol; slightly in ether. Practically insol in other organic solvents.

C complex sulfate, *Alcomicin, Bristagen, Cidomycin, Duragentam, Garamycin, Garasol, Genoptic, Gentacin, Gentak, Gentalline, Gentalyn, Gentibioptal, Genticin, Gentocin, Gentogram, Gent-Ophtal, Gentrasul, Lugacin, Nichogencin, Ophtagram, Pangram, Refobacin, Septopal, Sulmycin, U-gencin.* White, hygroscopic powder, mp 218-237°. $[\alpha]_D^{25}$ +102°. Sol in ethylene glycol, formamide. LD_{50} in mice (mg base/kg): 430 i.p.; 485 s.c.; 75 i.v.; > 9050 orally (Black).

THERAP CAT: Antibacterial.
THERAP CAT (VET): Antibacterial.

4399. Gentian. Yellow gentian; pale gentian; bitter root. Dried rhizome and roots of *Gentiana lutea* L., *Gentianaceae.* Habit. Central and Southern Europe. *Constit.* Gentiin, gentiamarin, gentisin, gentisic acid, gentiopicrin, gentianose (a sugar), pectin.

THERAP CAT: Bitter tonic.
THERAP CAT (VET): Bitter tonic.

4400. Gentianine. *5-Ethenyl-3,4-dihydro-1H-pyrano-[3,4-c]pyridin-1-one;* 4-(2-hydroxyethyl)-5-vinylnicotinic acid δ-lactone; erythricine. $C_{10}H_9NO_2$; mol wt 175.19. C 68.56%, H 5.18%, N 8.00%, O 18.27%. From *Gentiana kirilowi, Gentianaceae:* Proskurnina, *J. Gen. Chem. USSR* **14**, 1148 (1944), *C.A.* **40**, 7213 (1946); from *Anthocleista procera* Afz. and *Fagraea fragrans* Roxb., *Loganiaceae:* Lavie, Taylor-Smith, *Chem. & Ind. (London)* **1963**, 781; Wan, Chow, *J. Pharm. Pharmacol.* **16**, 484 (1964). Isoln from *Slevogtia orientalis* Gris., *Gentianaceae*, structure and synthesis: Govindachari *et al., J. Chem. Soc.* **1957**, 551, 2725.

Needles from ether or petr ether, mp 82-83°. uv max: 220 nm (log ε 4.38). Sol in alkali.

Hydrochloride, $C_{10}H_9NO_2$.HCl, needles from alcohol + ether, dec 169-170°.

Dihydrogentianine, $C_{10}H_{11}NO_2$, crystals from ether + petr ether, mp 74-76°. uv max: 270 nm (log ε 3.4).

4401. Gentian Violet. *N-[4-[Bis[4-(dimethylamino)-phenyl]methylene]-2,5-cyclohexadien-1-ylidene]-N-methyl-methanaminium chloride; C.I. Basic Violet 3;* hexamethylpararosaniline chloride; hexamethyl-*p*-rosaniline chloride; aniline violet; crystal violet; methylrosaniline chloride; C.I. 42555; Adergon; Axuris; Badil; Gentiaverm; Meroxylan; Meroxyl; Pyoktanin; Vianin; Viocid. $C_{25}H_{30}ClN_3$; mol wt 407.99. C 73.60%, H 7.41%, Cl 8.69%, N 10.30%. Prepn and properties: *Colour Index vol. 4* (3rd ed., 1971) p 4391. Toxicity studies: H. C. Hodge *et al., Toxicol. Appl. Pharmacol.* **22**, 1 (1972).

Dark green powder or greenish, glistening pieces with metallic luster. Sol in water, chloroform. Practically insol in ether. One gram dissolves in about 10 ml alcohol, in about 15 ml glycerin. LD_{50} orally in mice, rats: 1.2, 1.0 g/kg (Hodge).

Note: Commercial product, which is usually admixed with *pentamethylpararosaniline chloride* and *tetramethylpararosaniline chloride*, contains not less than 96% gentian violet.

USE: As dye for wood, silk, paper; in inks; as biological stain.

THERAP CAT: Anti-infective (topical). Has been used as anthelmintic (Nematodes).

THERAP CAT (VET): Antimicrobial (topical).

4402. Gentiobiose. *6-O-β-D-Glucopyranosyl-D-glucose;* 6-(β-D-glucosido)-D-glucose; amygdalose. $C_{12}H_{22}O_{11}$; mol wt 342.30. C 42.11%, H 6.48%, O 51.41%. From gentianose by partial hydrolysis with 0.2% H_2SO_4 or with invertin. From D-glucose by enzymatic synthesis with emulsin: Helferich, Lette, *Org. Syn.* **22,** 53 (1942). Prepn and structure: Haworth, Wylam, *J. Chem. Soc.* **123,** 3120 (1923); Hudson, *J. Am. Chem. Soc.* **51,** 1708 (1930). Structure: Hassid, Ballou in W. Pigman, *The Carbohydrates* (Academic Press, New York, 1957) p 492. Synthesis: Helferich, Klein, *Ann.* **450,** 219 (1926); Reynolds, Evans, *J. Am. Chem. Soc.* **60,** 2559 (1938). Hydrolysis with almond emulsin gives 2 mols D-glucose.

α-Form, lentil-shaped crystals with $2CH_3OH$ from methanol. Bitter taste. Very hygroscopic. mp 86°. Shows mutarotation. $[\alpha]_D^{22} +16°$ (3 min) → $+8.3°$ (3½ hrs c = 4). Sol in water, hot methanol, hot 90% alcohol.

β-Form, anhydr crystals from alc, mp 190-195°. Shows mutarotation. $[\alpha]_D^{22} -5.9°$ (6 min) → $+9.6°$ (6 hrs, c = 3). Sol in water, hot methanol, hot 90% alcohol.

4403. Gentiopicrin. *(5R-trans)-5-Ethenyl-6-(β-D-glucopyranosyloxy)-5,6-dihydro-1H,3H-pyrano[3,4-c]pyran-1-one;* gentiopicroside. $C_{16}H_{20}O_9$; mol wt 356.33. C 53.93%, H 5.66%, O 40.41%. The principal bitter glucoside of common gentians, which was isolated in 1862 from *Gentiana lutea,* L., *Gentianaceae:* Canonica *et al., Tetrahedron* **16,** 192 (1961). Isoln from roots of *G. lutea:* Korte, *Ber.* **87,** 512 (1954); Korte *et al., ibid.* **91,** 759 (1958). Structure: Canonica *et al., Tetrahedron Letters* **1960**(24), 7; *Tetrahedron* **16,** 192 (1961). Revised structure: Inouye *et al., Tetrahedron Letters* **1968,** 4429. Absolute configuration: Manitto, Pagnoni, *Gazz. Chim. Ital.* **94,** 229 (1964).

Crystals from anhydr ethyl acetate or abs alcohol, mp 191°. $[\alpha]_D^{20} -199°$ (ethanol). uv max (c = 0.0285 g/l methanol): 270 nm (log ε 3.96).

Hemihydrate, crystals from 50% ethanol, mp 121°.

Tetraacetate, $C_{24}H_{28}O_{13}$, crystals from methanol. mp 138.5-139.5°. uv max (c = 0.0214 g/l methanol): 272 nm (log ε 3.84).

THERAP CAT: Antimalarial.

4404. Gentisic Acid. *2,5-Dihydroxybenzoic acid;* 5-hydroxysalicylic acid. $C_7H_6O_4$; mol wt 154.12. C 54.55%, H 3.92%, O 41.52%. Occurs in gentian: Redgrove, *Pharm. J.* **122,** 324 (1929). Found in urine of dogs after ingestion of salicylates: Neuberg, *Berl. Klin. Wochenschr.* **48,** 799 (1911). Prepd from hydroquinone: Senhofer, Sarlay, *Monatsh.* **2,** 448 (1881); Juch, *ibid.* **26,** 839 (1905); Brunner, *Ann.* **351,** 321 (1907); Zeltner, Landau, **Ger. pat.** 258,887; *Frdl.* **11,** 210; Dyson, *Manual of Organic Chemistry* **vol. I** (London, 1950) p 614; by oxidation of salicylic acid with potassium persulfate: **Ger. pat.** 81,297 (to Schering AG); *Frdl.* **4,** 127; from *p*-hydroxyphenol: Meyer, **U.S. pat. 2,-588,336** (1952 to Monsanto); from 5-bromo-2-hydroxybenzoic acid: Lowenthal, Pepper, *J. Am. Chem. Soc.* **72,** 3292

(1950); by Kolbe synthesis: Clemens, **U.S. pat. 2,816,137** (1957 to Eastman Kodak). Metabolic product of *Penicillium patulum:* Tanenbaum, Bassett, *Biochim. Biophys. Acta* **28,** 21 (1958); of *Polyporus tumulosus* Cooke: Crowden, Ralph, *Aust. J. Chem.* **14,** 475 (1961). Biosynthesis: Gatenback, Linnroth, *Acta Chem. Scand.* **16,** 2298 (1962).

Needles, monoclinic prisms from water, mp 199-200°. Crystals are dimorphic and undergo phase inversion upon heating. The stable phase begins to sublime at 200° and melts at 205°. pK (25°): 2.93. Sol in water (about 1 part in 200 parts H_2O at 5°, much more sol in hot water), alcohol, ether. Practically insol in carbon disulfide, chloroform, benzene.

Sodium salt, $C_7H_5NaO_4$, *sodium gentisate, Gentinatre, Gentisod, Legential, Gentisine U.C.B.* Crystals with 5½ H_2O from water. Rapidly loses 3H_2O on exposure to air, but holds ½H_2O tenaciously even at 100°. Sol in water.

THERAP CAT: Analgesic; anti-inflammatory.

4405. Gentisin. *1,7-Dihydroxy-3-methoxy-9H-xanthen-9-one;* 4,7-dihydroxy-2-methoxyxanthone; gentianic acid; gentianin; gentiin. $C_{14}H_{10}O_5$; mol wt 258.23. C 65.12%, H 3.90%, O 30.98%. From root of *Gentiana lutea* L., *Gentianaceae:* Henry, Caventou, *J. Pharm. Chim.* **1821,** 178; Canonica, Pelizzoni, *Gazz. Chim. Ital.* **85,** 1007 (1955). Structure: Korte, *Ber.* **87,** 1357 (1954). Synthesis: Anand, Venkataraman, *Proc. Indian Acad. Sci.* **25A,** 438 (1947); Rao, Seshadri, *ibid.* **37A,** 710 (1953).

Yellow needles from alc, mp 266-267°. Absorption max (methanol): 260, 275, 315, 410 nm (log ε 4.35, 4.30, 4.10, 3.70). Very slightly sol in water or organic solvents.

Diacetate, $C_{18}H_{14}O_7$, crystals from alc, mp 196-197°. uv max (methanol): 240, 270, 300 nm (log ε 4.58, 4.05, 4.10).

4406. Gentisyl Alcohol. *2-(Hydroxymethyl)-1,4-benzenediol; 2,5-dihydroxybenzyl alcohol.* $C_7H_8O_3$; mol wt 140.14. C 60.00%, H 5.75%, O 34.25%. A metabolic product of *Penicillium patulum:* Birkinshaw *et al., Biochem. J.* **37,** 726 (1943); Brack, *Helv. Chim. Acta* **30,** 1 (1947); Tanenbaum, Bassett, *Biochim. Biophys. Acta* **28,** 21 (1958). Prepn from gentisaldehyde: Birkinshaw, *loc. cit.*

Clusters of needles from chloroform. mp 100°. Sublimes in high vacuum at 70-80° (bath temp). Freely sol in water, alcohol, ether; slightly sol in benzene, chloroform. Insol in petr ether. Upon heating with 2N H_2SO_4, a white amorphous precipitate is obtained. With aq NaOH: yellow color deepening to reddish brown.

4407. Gentrogenin. *(3β,25R)-3-Hydroxyspirost-5-en-12-one;* 20α,22a,25D-spirost-5-en-3β-ol-12-one; botogenin. $C_{27}H_{40}O_4$; mol wt 428.61. C 75.66%, H 9.41%, O 14.93%. Isoln from acid hydrolysis mixture of sapogenins of *Dioscorea mexicana:* Marker, Lopez, *J. Am. Chem. Soc.* **69,** 2397 (1947); from crude sapogenin mixture extracted from tubers of *D. spiculiflora, Dioscoreaceae:* Walens *et al., J. Org. Chem.* **22,** 182 (1957); Townley, **U.S. pat. 2,954,374** (1960 to Schering); from *Heloniopsis orientalis* (Thunb.) C. Tanaka, *Liliaceae:* Okanishi *et al., Chem. Pharm. Bull.* **10**, 1195 (1962). Structure: Walens *et al.,* U.S. pat. 2,830,986 (1958 to USDA).

Rectangular plates from methanol, mp 215-216°. $[\alpha]_D^{28}$ −56° (c = 1.02 in CHCl₃).

Acetate, $C_{29}H_{42}O_5$, crystals from ethyl acetate, mp 227°. $[\alpha]_D^{25}$ −58.1° (c = 1.02 in CHCl₃).

USE: Precursor for synthesis of pharmacologically active steroids.

4408. Geosmin. *[4S-(4α,4aα,8aβ)]-Octahydro-4,8a-dimethyl-4a(2H)-naphthalenol; octahydro-4α,8aβ-dimethyl-4aα(2H)-naphthol;* 1,10-*trans*-dimethyl-*trans*-(9)-decalol. $C_{12}H_{22}O$; mol wt 182.31. C 79.06%, H 12.16%, O 8.78%. Major volatile component of beet essence, also determined to be the potent earthy odor contaminant of fish, beans, water: A. C. Thaysen, *Ann. Appl. Biol.* **23**, 99 (1936); J. Silvey, A. W. Roach, *J. Am. Water Works Assoc.* **56**, 60 (1964); A. A. Rosen *et al., Water Treat. Exam.* **19**(Pt. 2), 106 (1970); R. G. Buttery *et al., J. Agr. Food Chem.* **24**, 419, 1246 (1976). Isoln from *Actinomyetes,* algae: N. N. Gerber, H. A. Lechevalier, *Appl. Microbiol.* **13**, 935 (1965); A. A. Rosen *et al., ibid.* **16**, 178 (1968); R. S. Safferman *et al., Environ. Sci. Technol.* **1**, 429 (1967). Isoln from beets: K. E. Murray, P. A. Bannister, *Chem. & Ind. (London)* **1975**, 973; L. D. Tyler *et al., J. Agr. Food Chem.* **26**, 1466 (1978). Structure: N. N. Gerber, *Biotechnol. Bioeng.* **9**, 321 (1967); *eidem, Tetrahedron Letters* **1968**, 2971. Biological degradation: L. V. Narayan, W. J. Nunez, *J. Am. Water Works Assoc.* **66**(Pt. 1), 532 (1974).

Colorless neutral oil, bp 270°. $[\alpha]_D^{25}$ −16.5°. Threshold odor conc: 0.1 ppb. Decomp in acid to odorless compds.

4409. Gepefrine. *(S)-3-(2-Aminopropyl)phenol;* α-methyl-*m*-tyramine. $C_9H_{13}NO$; mol wt 151.21. C 71.49%, H 8.67%, N 9.26%, O 10.58%. Sympathomimetic isomer of hydroxyamphetamine, *q.v.* Prepn: W. S. Saari *et al., J. Med. Chem.* **11**, 1115 (1968); G. Petrik *et al.,* Ger. pat. **2,-712,860** (1978 to Helopharm), *C.A.* **88**, 136296a (1978). Molecular configuration and biological actions: M. L. Torchiana *et al., Arch. Int. Pharmacodyn. Ther.* **174**, 184 (1968). *In vitro* study of tritium-labeled gepefrine: R. L. Dorris, *Eur. J. Pharmacol.* **35**, 225 (1976). Biochemical-pharmacological study: R. R. Ruffolo *et al., Biochem. Pharmacol.* **25**, 399 (1976).

Crystals from ethanol/hexane, mp 155-158°. $[\alpha]_D^{25}$ +31.8° (c = 2 in methanol). Sublimes at 100° (0.3 mm).

Tartrate, $C_9H_{13}NO\cdot C_4H_6O_6$, *Pressionorm, Wintonin.*
THERAP CAT: Antihypotensive.

4410. Gephyrotoxin. *[1R-(1α,3aβ,5aα,6α(Z),9aα)]-Dodecahydro-6-(2-penten-4-ynyl)pyrrolo[1,2-a]-quinoline-1-ethanol;* HTX D; histrionicotoxin D. $C_{19}H_{29}NO$; mol wt 287.45. C 79.39%, H 10.17%, N 4.87%, O 5.57%. Parent member of a class of tricyclic perhydrobenzoindolizine neurotoxin alkaloids isolated from the skin secretions of the Colombian poison-dart frogs *Dendrobates histrionicus* and *Dendrobates occultator,* family *Dendrobatidae.* The name gephyrotoxin was coined from the Greek word "gephyra" meaning "bridge", and refers to the bridge presumably formed by addition of the nitrogen function to one side of a 2,6-disubstituted piperidine precursor to form the bicyclic indolizine. Isolation of naturally occurring *l*-form: T. Tokuyama *et al., Helv. Chim. Acta* **57**, 2597 (1974). Structure and absolute configuration: J. W. Daly *et al., ibid.* **60**, 1128 (1977). Revised configuration: R. Fujimoto, Y. Kishi, *Tetrahedron Letters* **1981**, 4197. Total synthesis of (±)-form: R. Fujimoto *et al., J. Am. Chem. Soc.* **102**, 7154 (1980); D. J. Hart, *J. Org. Chem.* **46**, 3576 (1981); D. J. Hart, K. Kanai, *J. Am. Chem. Soc.* **105**, 1255 (1983); L. E. Overman *et al., ibid.* **5373**. Complete proton and ¹³C NMR assignment: M. W. Edwards, A. Bax, *ibid.* **108**, 918 (1986). Effect on neuromuscular transmission: C. Souccar *et al., Mol. Pharmacol.* **25**, 384, 395 (1984). Receptor binding study: R. S. Aronstam *et al., Neurochem. Res.* **11**, 1227 (1986). Review including discussion of gephyrotoxin congeners: J. W. Daly, *Fortschr. Chem. Org. Naturst.* **41**, 283-300 (*see also* refs pp 326-340) (1982).

mp 231-232° (dec). uv max (ethanol): 225 nm (ε 8400). $[\alpha]_D^{25}$ −51.5° (c = 1 in ethanol).

4411. Geraniol. *(E)-3,7-Dimethyl-2,6-octadien-1-ol; trans-3,7-dimethyl-2,6-octadien-8-ol;* lemonol. $C_{10}H_{18}O$; mol wt 154.25. C 77.87%, H 11.76%, O 10.37%. An olefinic terpene alcohol constituting the chief part of oil of rose and of oil of palmarosa; also found in many other essential oils such as citronella, lemon grass, etc. Isomeric with linalool. Isoln: Jacobsen, *Ann.* **157**, 234 (1871). Structure: Verley, *Bull. Soc. Chim. France* **25**, 68 (1919); J. L. Simonsen, *The Terpenes* vol. I (University Press, Cambridge, 2nd ed., 1947) pp 40-52. Stereochemistry: Burrell *et al., Proc. Chem. Soc.* **1959**, 263; Bates *et al., J. Org. Chem.* **28**, 1086 (1963). Synthesis: Burrell *et al., J. Chem. Soc. (C)* **1966**, 2144; K. Takabe *et al., Chem. Letters* **1977**, 1025; K. K. Mathew *et al., Indian J. Chem.* **B20**, 340 (1981).

Oily liq. Sweet rose odor. bp_{757} 229-230°; bp_{12} 114-115°. d_4^{20} 0.8894. n_D^{20} 1.4766. uv max: 190-195 nm (ε 18000). Practically insol in water. Miscible with alcohol, ether.

Acetate, $C_{12}H_{20}O_2$, sweet, fragrant liq. bp ~242° with decompn. d_{15}^{15} 0.9174. n_D^{15} 1.4628. Almost insol in water. Very sol in alcohol; miscible with ether.

Butyrate, $C_{14}H_{24}O_2$, liq. Characteristic fragrant odor. bp_{18} 152°. d_4^{17} 0.901. Almost insol in water. Sol in alc, ether.

Formate, $C_{11}H_{18}O_2$, liq. Odor of roses and of green rose leaves. bp_{15} 113-114°. d_4^{20} 0.927. Almost insol in alc, ether.

USE: In perfumery. As insect attractant. Butyrate for compounding artificial attar of rose. Formate as constituent of artificial neroli oil and of artificial orange blossom oil.

4412. Geranium. Cranesbill; storksbill; alum root. Dried rhizome of *Geranium maculatum* L., *Geraniaceae*. *Habit.* Canada and Eastern U.S., south to Georgia. *Constit.* 10-28% tannin, gallic, acid, resin, sugar, pectin.

THERAP CAT: Astringent.

THERAP CAT (VET): Astringent (intestinal).

4413. Geranylhydroquinone. *(E)-2-(3,7-Dimethyl-2,6-octadienyl)-1,4-benzenediol; trans-(3,7-dimethyl-2,6-octadienyl)hydroquinone; trans-1,4-dihydroxy-2-(3,7-dimethyl-2,6-octadienyl)benzene; geranyl-1,4-benzenediol; geroquinol;* Béradia. $C_{16}H_{22}O_2$; mol wt 246.35. C 78.01%, H 9.00%, O 12.99%. Prepn and use: Baranger, *Fr.* pat. **M2694** (1964), *C.A.* **61**, 15940e (1964). Radioprotective activity: G. Rudali, *C. R. Soc. Biol.* **160**, 1365 (1966). Effects on cancers in mice: G. Rudali, L. Menetrier, *Therapie* **22**, 895 (1967). Discovery as a potent contact allergen from trichomes of *Phacelia crenulata* var. *funerea* J. Voss, *Hydrophyllaceae*: G. Reynolds, E. Rodriguez, *Phytochemistry* **18**, 1567 (1979).

Colorless needles from *n*-hexane/ethyl acetate, mp 61-62°, H. Inouye *et al.*, *Ber.* **101**, 4057 (1968).

USE: Exptly as a radioprotective agent.

4414. Germane. Germanium hydride. GeH_4; mol wt 76.64. Ge 94.74%, H 5.26%. Prepd by the action of lithium aluminum hydride on a germanium halide in ether soln: Finholt *et al.*, *J. Am. Chem. Soc.* **69**, 2692 (1947); by reduction of GeO_2 by sodium hydoborate: Griffiths, *Inorg. Chem.* **2**, 375 (1963).

Colorless gas. mp $-165°$, bp $-90°$, d_4^{-142} 1.523. Slightly sol in hot hydrochloric acid, dec in nitric acid.

Caution: Hemolytic gas: E. Browning, *Toxicity of Industrial Metals* (Appleton-Century-Crofts, New York, 2nd ed., 1969) pp 162-163.

4415. Germanium. Ge; at. wt 72.59; at. no. 32; valences 4, 2. Five naturally occurring isotopes: 70 (20.55%); 72 (27.37%); 73 (7.67%); 74 (36.74%); 76 (7.67%); artificial, radioactive isotopes: 65-69; 71; 75; 77; 78. Extent of occurrence in the earth's crust about 0.0007%. Predicted and called ekasilicon by Mendeléeff. Discovered in 1886 by Clemens Winkler: *J. Prakt. Chem.* **34**, 177 (1886). Obtained industrially from the flue dusts of smelters processing zinc-bearing ores: Jaffee *et al.*, *Trans. Electrochem. Soc.* **89**, 277 (1946). Purification by zone refining: Pfann, *J. Metals* **4**, 747 (1952). Physical properties: Hassion *et al.*, *J. Phys. Chem.* **59**, 1076 (1955). Inhalation toxicity studies: J. H. E. Arts *et al.*, *Food Chem. Toxicol.* **28**, 571 (1990). Review and description of modern isolation techniques: Pirest in L. P. Hunter, *Handbook of Semiconductor Electronics* (McGraw-Hill, New York, 1956), section 6. Comprehensive monograph: V. I. Davydov, *Germanium* (Gordon & Breach, New York, 1966) 417 pp. *Reviews:* Rochow in *Comprehensive Inorganic Chemistry* vol. 2, J. C. Bailar, Jr. *et al.*, Eds. (Pergamon Press, Oxford, 1973) p 1-41; J. H. Adams in Kirk-Othmer *Encyclopedia of Chemical Technology* vol. 11 (Wiley-Interscience, New York, 3rd ed., 1980) pp 791-802.

Grayish-white, lustrous, brittle metalloid. Diamond-cubic structure when cryst. Poor conductor of electricity. d_4^{25} 5.323. Reported melting points range from 925-975°; best value 937.2° (Hassion). Vol smaller by a few % when molten. bp 2700°. Thermal expansion coefficient (at ~25°): $6.1 \times 10^{-6}/°C$. Thermal conductivity (at 25°): 0.14 cal/sec cm/°C. Specific heat (0-100°): 0.074 cal/g/°C. Lattice constant at 25°: 5.657×10^{-8} cm. Atoms/cc = 4.42×10^{22}. Volume compressibility: 1.3×10^{-12} cm²/dyn. Dielectric constant: 16. Covalent bond ionization energy at 0°K = 1.2 ev. Band gap: 0.67 ev. Impurity atom ionization energy: ~0.01 ev. Intrinsic resistivity at 300°K = 47 ohm-cm. Electron mobility at 300°K = 3900 cm²/v sec. Hole mobility at 300°K = 1900 cm/v sec. Magnetic (mass)

susceptibility ($\chi \times 10^6$) = -0.12. Intrinsic charge density at 300°K = 2.4×10^{13}. Electron diffusion constant at 300°K = 100. Hole diffusion constant at 300°K = 49. Insol in water, hydrochloric acid, dil alkali hydroxides. Attacked by aqua regia, concd nitric or sulfuric acids, fused alkalies, alkali peroxides, nitrates, or carbonates. Relatively stable, unaffected by air, becomes oxidized above 600°; is slowly oxidized by hydrogen peroxide at room temp, fairly rapidly at 90°; is attacked by hydrogen above 1000°. When finely divided, burns in chlorine or bromine.

USE: In electronics: Manuf rectifying devices (germanium diodes), transistors, in red-fluorescing phosphors; in dental alloys; in the production of glass capable of transmitting infrared radiation. Review of uses: Aldington, Cumming, *Endeavour* **14**, 200-204 (1955); *New Uses for Germanium*, F. I. Metz, Ed. (Midwest Research Institute, 1974) 120 pp.

THERAP CAT (VET): Astringent (intestinal).

4416. Germanium Dichloride. Cl_2Ge; mol wt 143.52. Cl 49.41%, Ge 50.59%. $GeCl_2$. Obtained as residue upon low temp distillation of $GeHCl_3$ (prepd by the reaction of hydrogen chloride with germanium monosulfide): Moulton, Miller, *J. Am. Chem. Soc.* **28**, 2702 (1956).

Unstable substance, dec into polymeric subchlorides even at low temps. Sol in ether, benzene.

4417. Germanium Dioxide. Germanic acid. GeO_2; mol wt 104.61. Ge 69.41%, O 30.59%.

White powder. Sol in about 250 parts cold water, 100 parts boiling water; sol in acids or in solns of the fixed alkalies. LD_{50} i.p. in rats: 750 mg/kg, Rosenfeld, Wallace, *Arch. Ind. Hyg. Occup. Med.* **8**, 466 (1953).

4418. Germanium Tetrachloride. Cl_4Ge; mol wt 214.42. Cl 66.14%, Ge 33.86%. $GeCl_4$. Prepd from Ge and Cl_2 or GeO_2 and HCl: Bauer, Burschkies, *Ber.* **66**, 277 (1933); Foster *et al.*, *Inorg. Syn.* **2**, 109 (1946).

Mobile liq. Fumes in air. Peculiar, acidic odor, but can be distinguished from that of concd HCl. Appreciably volatile at room temp. d_{20}^{20} 1.879. bp_{760} 83.1°. mp $-49.5°$. Hydrolyzed by water with a crackling noise. Is stable, but not very sol in 6N HCl. If more concd HCl is used, the soly of $GeCl_4$ is reduced. Sol in benzene, ether, other organic solvents.

Caution: Fumes irritating to eyes, mucous membranes of respiratory tract.

4419. Germanium Tetrafluoride. F_4Ge; mol wt 148.60. F 51.14%, Ge 48.86%. GeF_4. Prepd by heating $BaGeF_6$ to around 700°: Dennis, Laubengayer, *Z. Physik. Chem.* **130**, 520 (1927); Dennis, *Z. Anorg. Allgem. Chem.* **174**, 119 (1928); Biltz, *ibid.* **207**, 65 (1932); Hoffman, Gutowsky, *Inorg. Syn.* **4**, 147 (1953).

Colorless gas. Odor of garlic. Thermally stable up to about 1000°. d^0 (liq) 2.162; d (solid; $-195°$) 3.148. mp $-15°$ (under 3032 mm pressure). Sublimes at $-36.5°$. On contact with water, hydrolyzes to GeO_2 and H_2GeF_6. Does not attack glass when absolutely dry. Corrodes mercury, grease.

Trihydrate, obtained by slow evaporation of a soln of germanium dioxide in 20% HF. This is said to be water-soluble.

Caution: Highly irritating to eyes, skin, lungs, mucous membranes.

4420. Germine. *(3β,4α,7α,15α,16β)-4,9-Epoxycevane-3,4,7,14,15,16,20-heptol.* $C_{27}H_{43}NO_8$; mol wt 509.64. C 63.63%, H 8.50%, N 2.75%, O 25.11%. Alkamine present in many polyester alkaloids which occur in *Veratrum* and *Zygadenus* species. Isoln: Poethke, *Pharm. Monatsh.* **18**, 77 (1937); *idem*, *Arch. Pharm.* **275**, 571 (1937); Seiferle *et al.*, *J. Econ. Entomol.* **35**, 35 (1942); Klohs *et al.*, *J. Am. Chem. Soc.* **75**, 4925 (1953); Kupchan, Deliwala, *ibid.* **76**, 5545 (1954); Myers *et al.*, *ibid.* **77**, 3348 (1955); **78**, 1621 (1956). Structure and configuration: Kupchan, Narayanan, *ibid.* **81**, 1913 (1959). Comparative toxicity: O. Krayer *et al.*, *J. Pharmacol. Exp. Ther.* **82**, 167 (1944).

Crystals from methanol, mp 221.5-223°. $[\alpha]_D^{25}$ +4.5° (95% ethanol); $[\alpha]_D^{16}$ +23.1° (c = 1.13 in 10% acetic acid). Sol in chloroform, methanol, ethanol, acetone, water; slightly sol in ether. LD_{50} i.v. in mice: 139.0 mg/kg (Krayer).

3-Acetate, $C_{29}H_{45}NO_9$, needles from ether, mp 219-221°. $[\alpha]_D^{23}$ +10° (c = 1.05 in pyridine).

16-Acetate, $C_{29}H_{45}NO_9$, crystals from chloroform, mp 225-227°. $[\alpha]_D^{23}$ −19° (c = 0.93 in pyridine).

3,4,7,15,16-Pentaacetate, $C_{37}H_{53}NO_{13}$, prisms from acetone + petr ether, dec 285-287°. $[\alpha]_D^{23}$ −65° (c = 0.65 in pyridine).

4421. Gestodene. *(17α)-13-Ethyl-17-hydroxy-18,19-dinorpregna-4,15-dien-20-yn-3-one;* 17α-ethynyl-17β-hydroxy-18-methyl-4,15-estradien-3-one; 17α-ethynyl-13-ethyl-17β-hydroxy-4,15-gonadien-3-one; SH B 331. $C_{21}H_{26}O_2$; mol wt 310.44. C 81.25%, H 8.44%, O 10.31%. Orally active gestogen with progesterone-like profile of activity. Prepn: **Belg. pat. 847,090;** H. Hofmeister *et al.,* U.S. pat. **4,081,537** (1977, 1978 both to Schering AG). Pharmacokinetics: B. Duesterberg *et al., Contraception* **24,** 673 (1981); B. Duesterberg, *Steroids* **43,** 43 (1984). Comparative study of anti-aldosterone activity: W. Losert *et al., Arzneimittel-Forsch.* **35,** 459 (1985). Clinical trial as oral contraceptive: G. Hoppe, *Adv. Contracept.* **3,** 159 (1987).

Crystals from acetone-hexane, mp 197.9°.

Mixture with ethinyl estradiol, *Femodene, Femovan, Ginoden, Gynera, Milvane, Minulet, Monodie, Phaeva, Tri-Minulet.*

THERAP CAT: Progestogen. In combination with estrogen as oral contraceptive.

4422. Gestonorone Caproate. *17-[(1-Oxohexyl)oxy]-19-norpregn-4-ene-3,20-dione; 17-Hydroxy-19-norpregn-4-ene-3,-20-dione hexanoate;* 17α-hydroxy-19-norprogesterone caproate; 17β-acetyl-17-hydroxyestr-4-ene-3-one hexanoate; gestronol caproate; SH-582; Depostat. $C_{26}H_{38}O_4$; mol wt 414.59. C 75.32%, H 9.24%, O 15.44%. Prepn: Popper *et al., Arzneimittel-Forsch.* **19,** 352 (1969); Popper *et al.,* Ger. pat. **1,074,582** (1960 to Schering). Pharmacological studies:

Junkmann, *Anglo-Ger. Med. Rev.* **1,** 385 (1962). Metabolism: Breuer, Lisboa, *Acta Endocrinol.* **51,** 114 (1966).

Crystals, mp 123-124°. $[\alpha]_D$ +13° (chloroform). uv max: 239 nm (ε 17540).

THERAP CAT: Progestogen; in treatment of prostatic hypertrophy.

4423. Gestrinone. *(17α)-13-Ethyl-17-hydroxy-18,19-dinorpregna-4,9,11-trien-20-yn-3-one;* 13β-ethyl-17α-ethynyl-17β-hydroxy-4,9,11-gonatrien-3-one; 13β-ethyl-17α-ethynyl-Δ^{4,9,11}-gonatriene-17β-ol-3-one; ethylnorgestrienone; A-46745; R-2323; RU-2323; Dimetriose; Dimetrose; Nemestran; Tridomose. $C_{21}H_{24}O_2$; mol wt 308.42. C 81.78%, H 7.84%, O 10.38%. Steroidal antiestrogen, antiprogestogen, analog of norgestrienone, *q.v.* Prepn: G. Nomine *et al.,* U.S. pat. **3,257,278;** Neth. pat. **Appl. 6,607,609,** C.A. 67, 44029d (1967); D. Bertin, A. Pierdet, U.S. pat. **3,478,067** (1966, 1966, 1969 all to Roussel-UCLAF). Radioimmunoassay in human plasma: J. Frick *et al., Urol. Res.* **5,** 55 (1977). Clinical evaluation of contraceptive efficacy: G. Azadian-Boulanger *et al., Am. J. Obstet. Gynecol.* **125,** 1049 (1976); of use in fibrocystic breast disease: E. M. Coutinho, G. Azadian-Boulanger, *Int. J. Gynaecol. Obstet.* **22,** 363 (1984); in endometriosis: E. J. Thomas, I. D. Cooke, *Brit. Med. J.* **294,** 272 (1987); E. M. Coutinho, G. Azadian-Boulanger, *Fertil. Steril.* **49,** 418 (1988).

Crystals from ethyl acetate and benzene-cyclohexane (1:1), mp 154°. $[\alpha]_D^{20}$ +84.6° (c = 0.41 in methanol).

THERAP CAT: Antigonadotropin.

4424. Ghatti Gum. Gum Ghatti; Indian gum. The gummy exudate from stems of *Anogeissus latifolia* Wall., Combretaceae, abundant in India and Ceylon, *cf.* C. L. Mantell, *The Water-Soluble Gums* (New York, 1947). Name derived from the word *ghats,* meaning passes, and given to the gum because of its ancient mountain transportation routes. Structure is a complex water-soluble polysaccharide occurring as a calcium-magnesium salt; composed of L-arabinose, D-galactose, D-mannose, D-xylose, D-glucuronic acid, in a molar ratio of 10:6:2:1:2, and traces of 6-deoxyhexose: Aspinall *et al., J. Chem. Soc.* **1955,** 1160. Early investigation of chemistry and mol wt: Shaw *et al., Proc. S. Dakota Acad. Sci.* **15,** 46 (1935); **16,** 34 (1936); **17,** 27 (1937); **19,** 130 (1939); **21,** 78 (1941). *Review:* Meer *et al.,* in *Industrial Gums,* R. L. Whistler, Ed. (Academic Press, New York, 2nd ed., 1973) pp 265-271.

Ghatti gum sold in the U.S. usually has been autoclaved in order to make all of the gum water-sol. The *U.S. Dispensatory* (24th ed.) states that gum Ghatti suitable as clinical laboratory reagent "is entirely sol in 5 parts of cold water." Forms a very viscous mucilage, more viscous but less adhesive than acacia. Insol in 90% alcohol. $[\alpha]_D^{25}$ +42° (dil H_2SO_4). Gum ghatti solns may be colored slightly due to traces of pigment remaining in the gum. Does not form a true gel.

USE: As substitute for acacia. As emulsifying agent in pharmaceuticals, oils, waxes.

4425. Ghi. Ghee; samli; clarified butter. Prepd from cream or butter by melting and heating to 122° for about 25 mins. The water evaporates, and lactose, salt, and albuminous substances sink to the bottom. This treatment kills existing microorganisms by heat and deprives incoming microorganisms of moisture and necessary nutrients.

Properly clarified butter is of finely grained consistency, yellow color, mp 30° and retains most of the vitamin A content of the original butter. Keeps much longer than ordinary butter and contains less than 0.7% water. Used in India instead of butter.

4426. Gibberellic Acid. *(1α,2β,4aα,4bβ,10β)-2,4a,7-Trihydroxy-1-methyl-8-methylenegibb-3-ene-1,10-dicarboxylic acid 1,4a-lactone;* gibberellin X; gibberellin A_3. $C_{19}H_{22}O_6$; mol wt 346.38. C 65.88%, H 6.40%, O 27.71%. Plant hormone; most outstanding of the plant-growth promoting metabolites of *Gibberella fujikuroi.* Isoln: P. J. Curtis, B. E. Cross, *Chem. & Ind. (London)* **1954**, 1066; B. E. Cross, *J. Chem. Soc.* **1954**, 4670; P. W. Brian *et al.,* U.S. pat. **2,842,-051**; C. T. Calam, P. J. Curtis, U.S. pat. **2,950,288**; A. J. Birch *et al.,* U.S. pat. **2,977,285** (1958, 1960, 1961, all to ICI). Stereochemistry and structure: G. Stork, H. Newman, *J. Am. Chem. Soc.* **81**, 5518 (1959); B. E. Cross *et al.,* *Proc. Chem. Soc.* **1959**, 302; F. McCapra *et al., ibid.* **1962**, 185; D. C. Aldridge *et al., J. Chem. Soc.* **1963**, 143; P. M. Bourn *et al., ibid.* **1963**, 154. Partial synthesis: E. J. Corey *et al., J. Am. Chem. Soc.* **93**, 7316 (1971). Stereospecific total synthesis: *eidem, ibid.* **100**, 8034 (1978). Promotes growth of seedlings: M. J. Bukovac, S. H. Wittwer, *Quart. Bull. Mich. Agr. Expt. Sta.* **39**, 307 (1956); J. M. Merritt, "Gibberellins for Agriculture", *J. Agr. Food Chem.* **6**, 184 (1958). *Reviews:* B. E. Cross *et al.* in *Gibberellins, Advances in Chemistry Series,* **vol. 28**, 13 (A.C.S., Washington, 1961); series of articles in *Plant Growth Subst., Proc. 7th Int. Conf.,* D. J. Carr, Ed. (Springer-Verlag, Berlin, 1972).

Crystals from ethyl acetate, mp 233-235° (effervescence). $[\alpha]_D^{19}$ +86° (c = 2.12). pK 4.0. Slightly sol in water, ether. Freely sol in methanol, ethanol, acetone. Moderately sol in ethyl acetate. Sol in aq solns of sodium bicarbonate and sodium acetate.

Methyl ester, $C_{20}H_{24}O_6$, needles from benzene + methanol, mp 209-210°. $[\alpha]_D^{20}$ +75° (c = 0.5).

Potassium gibberellate, $C_{19}H_{21}KO_6$, *Gibrel.* White, bulky powder. $[\alpha]_D^{25}$ +43° to +60° (c = 5 in water). pH of 5% soln 5.5-6.5. One gram dissolves in 20 ml water.

USE: Plant growth regulator. Promoting growth of plants, esp the growth of seedlings. Food additive in the malting of barley: *Fed. Reg.* **25**, 2162 (1960).

4427. Gibberellins. GAs. A class of plant growth hormones first isolated in 1938 from cultures of *Gibberella fujikuroi* (Sawada) Wollenweber *(Fusarium moniliforme* Sheldon) the fungus causing Bakanae disease in rice: Yabuta, Sumiki, *J. Agr. Chem. Soc. Japan* **14**, 1526 (1938). Isolated also from higher plants; for source references see review by Lang, *Ann. Rev. Plant Physiol.* **21**, 537 (1970). More than 60 gibberellins are known of which gibberellin A_3 (gibberellic acid, *q.v.*) is the most important. GA_3 and mixtures of GA_4 and GA_7 are available commercially. All gibberellins are diterpenoid acids based on the *gibberellane* skeleton containing the *gibbane* nucleus. Major structural differences lie in the substituents at positions 4a, 7, 8 (gibbane numbering) and the presence or absence of a γ-lactone ring. Total synthesis of racemic gibberellins A_2, A_4, A_9, A_{10}: Mori *et al., Tetrahedron* **25**, 1293 (1969); of gibberellin A_4: A. L. Cossey *et al., Tetrahedron Letters* **1980**, 4383; of gibberellin A_{15}: Nagata *et al., J. Am. Chem. Soc.* **93**, 5740 (1971); of gibberellins A_{15} and A_{37}: E. Fujita *et al., J. Chem. Soc., Perkin Trans. I* **1977**, 611. Stereochemistry: Meguro, Fuzimura, *Tetrahedron Letters* **1968**, 6305. Biosynthesis: Cross *et al., J. Chem. Soc. (C)* **1968**, 1054; Shechter, West, *J. Biol. Chem.* **244**, 3200 (1969). Nomenclature: MacMillan, Takahasni, *Nature* **217**, 170 (1968). Early reviews: Brian *et al., Fortschr. Chem. Org. Naturst.* **18**, 350 (1960); Paleg, *Ann. Rev. Plant Physiol.* **16**, 291 (1965). Recent reviews: Cleland in *The Physiology of Plant Growth and Development,* M. B. Wilkins, Ed. (McGraw-Hill, New York, 1969) pp 49-81; L. Rappaport, "Applications of Gibberellins in Agriculture", in *Plant Growth Subst., Proc. Int. Conf. 10th,* F. K. Skoog, Ed. (Springer-Verlag, Berlin, 1980) pp

377-391; I. D. Railton, *Cell. Biol. Int. Rep.* **6**, 319-337 (1982). Comprehensive synthetic review: E. Fujita, M. Node, *Heterocycles* **7**, 709 (1977).

Gibbane   Gibberellane

USE: Plant growth hormone. For specific agricultural uses *see* review by Turner, *Outlook Agr.* **7**, 14 (1972).

4428. Gibbs Reagent. *2,6-Dichloro-4-(chloroimino)-2,5-cyclohexadien-1-one;* 2,6-dichloro-*p*-benzoquinone-4-chloroimine; *N*,2,6-trichloro-*p*-quinoneimine; 2,6-dichloroquinone chloroimide; *N*,2,6-trichloroquinoneimine; *N*,2,6-trichlorobenzoquinone imine. $C_6H_2Cl_3NO$; mol wt 210.45. C 34.24%, H 0.96%, Cl 50.54%, N 6.66%, O 7.60%. Reagent used to determine the presence of phenols. The reaction with phenols unsubstituted in the *para* position is called the Gibbs Reaction: H. D. Gibbs, *Chem. Rev.* **3**, 291 (1927). Prepn of the reagent: *idem, J. Biol. Chem.* **72**, 649 (1927); G. I. Mikhailov, *Trans. Inst. Pure Chem. Reagents* **16**, 83 (1939). Analysis of the Gibbs color reaction: J. C. Dacre, *Anal. Chem.* **43**, 589 (1971).

Yellow needles from alcohol, mp 65-67°.
USE: In determination of phenols.

4429. Gigantine. *1,2,3,4-Tetrahydro-6,7-dimethoxy-1,2-dimethyl-5-isoquinolinol.* $C_{13}H_{19}NO_3$; mol wt 237.30. C 65.80%, H 8.07%, N 5.90%, O 20.23%. Hallucinogen from the cactus *Carnegia gigantea* Engelm. *Cactaceae:* Hodgkins *et al., Tetrahedron Letters* **1967**, 1321. Structure: Kapadia *et al., Chem. Commun.* **1970**, 856. Synthesis: *eidem, Chem. & Ind. (London)* **1970**, 1593; Choudhury *ibid.* **1971**, 578.

mp 151-152°. $[\alpha]_D^{25}$ +27.1° (c = 2 in chloroform).

4430. Ginger. Dried rhizome of *Zingiber officinale* Roscoe, *Zingiberaceae. Habit.* Southeast Asia; cultivated in India, West Indies, Africa, Australia, most tropical and subtropical countries. *Constit.* Volatile oil (1-3%), containing zingiberene, *ar*-curcumene, β-sesquiphellandrene and β-bisabolene; oleoresin (4-7.5%) containing the pungent gingerols and shogaols; starch (40-60%); proteins (10%); and fats (10%). Analysis: D. W. Connell *et al., Flavour Ind.* **1**, 677 (1970). Clinical trial in hyperemesis of pregnancy: W. Fischer-Rasmussen *et al., Eur. J. Obstet. Gynecol. Reprod. Biol.* **38**, 19 (1990); in postoperative nausea: S. Phillips *et al., Anaesthesia* **48**, 715 (1993). Comprehensive review: V. S. Govindarajan, *CRC Crit. Rev. Food Sci. Nutr.* **17**, 1-96; 189-258 (1982). Review of cultivation, processing and uses: B. M. Lawrence, *Perfumer Flavorist* **9**, 1-40 (1984).
USE: Flavor in foods and beverages.

THERAP CAT: Antiemetic; carminative.

THERAP CAT (VET): Carminative.

4431. [6]-Gingerol. *(S)-5-Hydroxy-1-(4-hydroxy-3-methoxyphenyl)-3-decanone.* $C_{17}H_{26}O_4$; mol wt 294.39. C 69.36%, H 8.90%, O 21.74%. The major phenol and most important of the pungent principles of ginger oil, isolated from rhizome of *Zingiber officinale* Roscoe, *Zingiberaceae:* Thresh, *Pharm. J.* [3] **10**, 171 (1879); **12**, 721 (1881); **14**, 798 (1883); **15**, 208 (1884); Garnett, Grier, *ibid.* [4] **25**, 118 (1907); Nelson, *J. Am. Chem. Soc.* **39**, 1466 (1917). Structure studies: Lapworth *et al., J. Chem. Soc.* **111**, 777 (1917). Review of early literature: Redgrove, *Pharm. J.* **125**, 54 (1930); Jacobs, *Am. Perfumer* **48**, no. 7, 60, 62 (1946). Demonstration of the presence of an homologous series of phenolic ketones in the pungent constituents of ginger, [6]-gingerol being the major member: D. W. Connell, M. D. Sutherland, *Austral. J. Chem.* **22**, 1033 (1969). Synthesis of (±)-form: Hirao *et al., Chem. Pharm. Bull.* **20**, 2287 (1972); K. Banno, T. Mukaiyama, *Bull. Chem. Soc. Japan* **49**, 1453 (1976); P. Denniff, D. A. Whiting, *Chem. Commun.* **1976**, 712; P. Denniff *et al., J. Chem. Soc. Perkin Trans. I* **1981**, 82. Stereoselective synthesis of $S(+)$-form: D. Enders *et al., Ber.* **112**, 3703 (1979). Biosynthesis of (±)-form: P. Denniff, D. A. Whiting, *Chem. Commun.* **1976**, 711; I. Macleod, D. A. Whiting, *ibid.* **1979**, 1152; P. Denniff *et al., J. Chem. Soc. Perkin Trans. I* **1980**, 2637. Physical properties and cardiotonic effects of gingerols: N. Shoji *et al., J. Pharm. Sci.* **71**, 1174 (1982). Mutagenicity studies: H. Nakamura, T. Yamamoto, *Mutat. Res.* **103**, 119 (1982); *eidem, ibid.* **122**, 87 (1983).

Normally obtained as pungent, yellow oil. n_D^{25} 1.5224. uv max (ethanol) 282 nm (ε 2560) (Connell, Sutherland). Crystalline form, mp 30-32°. $[\alpha]_D$ +27.8° (c = 1 in CHCl$_3$). uv max (ethanol) 284 nm (ε 2700). Sol in 50% alcohol, ether, chloroform, benzene; moderately sol in hot petr ether.

4432. Ginkgo. Maidenhair tree; kew tree. *Ginkgo biloba* Linn; the only living member of the *Ginkgoaceae.* Valued as a street and park tree, the ripe fruit gives off a foul odor. The seeds and leaves have been used in Chinese traditional medicine for asthma and vascular disease. *Habit.* China, cultivated in Eastern U.S. and Canada. *Constit.* A wide variety of compounds have been identified in the fruit, leaves and bark including: ginkgolides, *q.v.,* bilobalide, biflavones, flavonol glycosides. Fruit pulp contains toxic principles: ginkgolic acid, ginkgol, bilobol. Analysis of constituents of leaves: S. Furukawa, *Sci. Papers Inst. Phys. Chem. Res. Tokyo* **19**, 27 (1932); **21**, 273 (1933), *C.A.* **27**, 303, 5745 (1933). Isoln and characterization of biflavonyl constituents of ginkgo leaves: W. Baker *et al., J. Chem. Soc.* **1963**, 1477; of phenolic constituents: K. Weinges *et al., Arzneimittel-Forsch.* **18**, 539 (1968). Isoln of ginkgolides, *q.v.,* from root bark: M. Maruyama *et al., Tetrahedron Letters* **1967**, 299-326. HPLC determn of biflavones in leaf extracts: F. Briançon-Scheid *et al., Planta Med.* **49**, 204 (1983); of terpenes: P. G. Pietta *et al., Chromatographia* **29**, 251 (1990). Reviews: W. Tang, G. Eisenbrand, *Chinese Drugs of Plant Origin* (Springer-Verlag, Berlin, 1992) pp 555-565; P. Houghton, *Pharm. J.* **253**, 122-123 (1994).

Leaf extract, *Ginkgo biloba extract, GBE, EGb 761, Rökan, Tanakan, Tebonin.* A defined extract obtained from the leaves: P. Kloss, H. Jaggy, *Ger. pat.* **2,117,429** (1972 to Willmar Schwabe), *C.A.* **78**, 47787 (1973). Clinical use: U. Bauer, *Arzneimittel-Forsch.* **34**, 716 (1984). *Review:* F. V. DeFeudis, *Ginkgo biloba Extract (EGb 761): Pharmacological Activities and Clinical Applications* (Elsevier, Amsterdam, 1991) 187 pp.

THERAP CAT: In treatment of cerebral and peripheral vascular insufficiency.

4433. Ginkgolides. Family of bioactive terpenes isolated from the root bark and leaves of *Ginkgo biloba* L., *Ginkgoaceae.* Specific platelet activating factor (PAF) antagonists. Isoln from leaves: S. Furukawa, *Sci. Papers Inst. Phys. Chem. Res. Tokyo* **19**, 27 (1932), *C.A.* **27**, 303 (1933). Isoln from root bark and characterization: M. Maruyama *et al., Tetrahedron Letters* **1967**, 299; K. Nakanishi, *Pure Appl. Chem.* **14**, 89 (1967); from leaves: N. Sakabe *et al., Chem. Commun.* **1967**, 259; K. Okabe *et al., J. Chem. Soc. (C)* **1967**, 2201. Crystal structure of B: L. Dupont *et al., Acta Cryst.* **C42**, 1759 (1986); of A and C: M. Sbit *et al., ibid.* **C43**, 2377 (1987). Total synthesis of (±)-B: E. J. Corey *et al., J. Am. Chem. Soc.* **110**, 649 (1988); of (±)-A: E. J. Corey, A. K. Ghosh, *Tetrahedron Letters* **29**, 3205 (1988). Enantioselective route to B: E. J. Corey, A. V. Gavai, *ibid.*, 3201. Review of syntheses: E. J. Corey, *Chem. Soc. Rev.* **17**, 111-133 (1988). Proton- and ^{13}C-NMR: C. Roumestand *et al., Tetrahedron* **45**, 1975 (1989). Effect on PAF receptor binding: R. Korth *et al., Eur. J. Pharmacol.* **152**, 101 (1988). *Review:* P. Braquet *et al., Med. Res. Rev.* **11**, 295-355 (1991).

Ginkgolide B

Ginkgolide A, $C_{20}H_{24}O_9$, *BN-52020.* Bitter crystals from ethanol, mp ~300 (dec). $[\alpha]_D^{24}$ −53.4° (c = 1 in ethanol). uv max (ethanol): 219 nm (log ε 2.72).

Ginkgolide B, $C_{20}H_{24}O_{10}$, *(1β)-1-hydroxyginkgolide A, BN-52021.* Review of pharmacology and PAF-antagonism: P. Braquet, R. H. Bourgain, *Adv. Exp. Med. Biol.* **215**, 215-235 (1987). Clinical trial in asthma: K.-H. Hsieh, *Chest* **99**, 877 (1991); in severe sepsis: J. F. Dhainaut *et al., Crit. Care Med.* **22**, 1720 (1994). Bitter crystals from ethanol, mp ~300 (dec). $[\alpha]_D^{24}$ −52.6° (c = 1 in ethanol). uv max (ethanol): 219 nm (log ε 2.64).

Ginkgolide C, $C_{20}H_{24}O_{11}$, *(1α,7β)-1,7-dihydroxyginkgolide A, BN-52022.* Bitter crystals from ethanol, mp ~300 (dec). $[\alpha]_D^{24}$ −14.7° (c = 1 in ethanol). uv max (ethanol): 220 nm (log ε 2.30).

USE: Reference compd for PAF receptors in biological systems.

THERAP CAT: Ginkgolide B in treatment of severe sepsis.

4434. Ginseng. Panax. Root of *Panax ginseng* C.A. Mey., *Araliaceae,* a perennial herb indigenous to Eastern Asia, *P. quinquefolium* L., found in Eastern U.S. and Canada, and *P. pseudoginseng* Wall, found in India, China and Japan. The biologically active constituents are considered to be a series of saponin glycosides known as *ginsenosides, panaxosides* or *panaquilins.* Review of constituents: J. P. Hov, *Comp. Med. East West* **5**, 123-145 (1977). Isoln and identification of ginseng saponin glycosides: S. Shibata *et al., Tetrahedron Letters* **1962** 419; G. B. Elyakov *et al., ibid.* **1964**, 3591; R. Kasai *et al., Chem. Pharm. Bull.* **31**, 2120 (1983). Chemico-pharmacological study: T. Kaku, *Arzneimittel-Forsch.* **25**, 539 (1975). Effect on brain biogenic amines: V. Petkov, *ibid.* **28**, 388 (1978). TLC analysis of saponin content of commercial ginseng products: L. E. Liberti, A. D. Marderosian, *J. Pharm. Sci.* **67**, 1487 (1978). Use in oriental medicine as tonic: K. Chimin Wong, Wu Lien-teh, *History of Chinese Medicine* (Shanghai, 2nd ed., 1936) 906 pp. Comprehensive review of morphology, cultivation and uses: Baranov, *Econ. Bot.* **20**, 403-406 (1966). Review of chemical constituents: J. P. Hou, *Comp. Med. East West* **5**, 123-145 (1977). Brief reviews: W. E. Court, *Pharm. J.* **214**, 180-181 (1975); B. J. Spalding, *Chem. Week* **139**, 19-21 (1986).

Sweet, slightly aromatic taste.
THERAP CAT: Tonic.

4435. Giractide. *1-Glycine-18-L-argininamide-α*[(1-18)]-*corticotropin.* $C_{100}H_{156}N_{34}O_{22}S$; mol wt 2218.62. C 54.14%, H 7.09%, N 21.47%, O 15.87%, S 1.45%. Polypeptide corresp to the first 18 amino acid residues in corticotropin, in which the 1-serine is replaced by glycine. Prepn: H. Otsuka *et al., Bull. Chem. Soc. Japan* **43**, 196 (1970); of the hexaacetate: *eidem,* **Japan.** pat. **70 19,061** (1970 to Shionogi), *C.A.* **73**, 110138r (1970). Adrenocorticotropic potency in man: *J. Clin. Endocrinol. Metab.* **33**, 355 (1971). Fluorometric study: T. Muraki *et al., Experientia* **32**, 1605 (1976). Absorption: M. Hirata *et al., Chem. Pharm. Bull.* **26**, 1061 (1978). Chromatographic studies: S. Terabe *et al., J. Chromatog.* **172**, 163 (1979); M. Schoeneshoefer, A. Fenner, *ibid.* **224**, 472 (1981).

Gly–Tyr–Ser–Met–Glu–His–Phe–Arg–Trp–Gly–Lys–Pro–Val–Gly–Lys–Lys–Arg–ArgNH₂

$[\alpha]_D^{23.5}$ −51.4 ±1.9° (c = 0.472 in 0.1*N* acetic acid). uv max (0.1*N* NaOH): 281, 288 nm (ε 6750, 6490).

Hexaacetate salt, $C_{100}H_{156}N_{34}O_{22}S.6C_2H_4O_2$, *S-50022, Acthormon.*

THERAP CAT: Adrenocorticotropic hormone.

4436. Girard Reagents. Quaternary ammonium acetylhydrazine chlorides which form water-soluble hydrazones with carbonyl compounds. The hydrazones formed can subsequently be hydrolyzed in order to regenerate the original carbonyl compounds. Prepn: A. Girard, G. Sandulesco, **Brit.** pat. **Appl. 6640** (1934); *eidem, Helv. Chim. Acta* **19**, 1095 (1936); A. Girard, *Org. Syn.* **coll. vol. II**, 85 (1943). *Review:* Wheeler, *Chem. Rev.* **62**, 205 (1962). Use of Girard reagents in separation of carbonyl compounds: Schubert, Wehrberger, *Endokrinologie* **48**, 70 (1965); R. E. J. Mitchel, H. C. Birnboim, *Anal. Biochem.* **81**, 47 (1977); W. Holstein, D. Severin, *Erdoel Kohle, Erdgas, Petrochem.* **32**, 487 (1979); *C.A.* **92**, 44207b (1980).

Girard reagent T   Girard reagent P

Girard reagent T, $C_5H_{14}ClN_3O$, *2-hydrazino-N,N,N-trimethyl-2-oxoethanaminium chloride, (carboxymethyl)trimethylammonium chloride hydrazide, betaine hydrazide hydrochloride.* Highly hygroscopic needles. May be stored in well-stoppered containers. Material that has developed an odor should be recrystallized from abs ethanol. mp 192° (slight decomp). Very freely sol in water; sol in about 150 parts of absolute ethanol; more sol in methanol. Also very sol in acetic acid, glycerol, ethylene glycol. Practically insol in organic solvents devoid of hydroxyl groups.
Girard reagent P, $C_7H_{10}ClN_3O$, *1-(2-hydrazino-2-oxoethyl)pyridinium chloride.* Non-hygroscopic crystals from methanol, dec 200°. Less soluble in polar solvents than Girard reagent T.
USE: In the isoln of 17-ketosteroids and other carbonyl compounds.

4437. Gitalin. Gitaligin. An extract of *Digitalis purpurea* L., *Scrophulariaceae,* once believed to be a single amorphous compd, is a stable, relatively constant mixture of digitoxin, gitoxin, *q.q.v.*, and *gitaloxin (16-formylgitoxin, Cristaloxine)* with smaller amounts of several other glycosides and genins, together with minute amounts of other substances. Review of prepn and composition: Mitchell, *Am. J. Pharm.* **136**, 71 (1964).
Amorphous powder. Readily sol in alcohol, slightly (but slowly) sol in water. A satd aq soln has pH of about 7.0 and an intensely bitter taste. *Ref:* N.N.D. 1960, 301.
THERAP CAT: Cardiotonic.
THERAP CAT (VET): Cardiotonic.

4438. Gitogenin. *(2α,3β,5α,25R)-Spirostan-2,3-diol;* digin. $C_{27}H_{44}O_4$; mol wt 432.64. C 74.96%, H 10.25%, O

14.79%. Obtained from gitonin by heating with dil HCl. Structure: Tschesche, *Ber.* **68**, 1090 (1935); Jacobs, Simpson, *J. Biol. Chem.* **110**, 429 (1935); Marker, Rohrmann, *J. Am. Chem. Soc.* **61**, 2724 (1939); **62**, 647 (1940); Noller, Lieberman, *ibid.* **63**, 2131 (1941); Klass *et al, ibid.* **77**, 3829 (1955). Pharmacological study: H. K. Iwamoto *et al., J. Pharmacol. Exp. Ther.* **91**, 130 (1947).

Leaflets from benzene, dec 271.5-275°. $[\alpha]_D^{20}$ −75° (c = 1.02 in CHCl₃). Sol in chloroform, hot alc; sparingly sol in cold ethyl acetate, in ether; practically insol in water. Not precipitated by digitonin.
Diacetate, $C_{31}H_{48}O_6$, long needles from ether + methanol, mp 251-254°. $[\alpha]_D^{20}$ −96° (c = 1.92 in CHCl₃).

4439. F-Gitonin. Gitogenin β-lycotetraoside. $C_{50}H_{82}$-O_{23}; mol wt 1051.19. C 57.13%, H 7.86%, O 35.01%. Leaf saponin from *Digitalis purpurea* L., *Scrophulariaceae:* Kawasaki, Nishioka, *Chem. Pharm. Bull.* **12**, 1311 (1964). Structure: Kawasaki *et al., Tetrahedron* **21**, 299 (1965). A gitogenin tetraglycoside of which the sugar composition is 2 moles D-glucose, 1 mole D-galactose, and 1 mole D-xylose; differs from *gitonin* (the gitogenin tetraglycoside in *D. purpurea* seeds), in which the sugar composition is 2 galactose, 1 glucose, and 1 xylose: Tschesche, Wulff, *Ber.* **94**, 2019 (1961).

Dihydrate, needles from butanol + water, dec 252-255°. $[\alpha]_D^{25}$ −58.5° (c = 0.53 in pyridine).

4440. Gitoxigenin. *(3β,5β,16β)-3,14,16-Trihydroxycard-20(22)-enolide;* $\Delta^{20,22}$-3,14,16,21-tetrahydroxynorcholenic acid lactone. $C_{23}H_{34}O_5$; mol wt 390.52. C 70.74%, H 8.78%, O 20.48%. The aglycon of gitoxin. By refluxing gitoxin in a mixture of water + alcohol + HCl: Smith, *J. Chem. Soc.* **1931**, 23. Structure: Jacobs, Elderfield, *J. Biol. Chem.* **100**, 671 (1933); Elderfield, *Chem. Rev.* **17**, 217 (1935); Henderson, Chen, *J. Med. Pharm. Chem.* **5**, 988 (1962). Configuration: Moore, *Helv. Chim. Acta* **37**, 659 (1954); Repke, Klesczewski, *Arch. Exp. Pathol. Pharmakol.* **239**, 131 (1960). *Cf.* ref under Digoxigenin.

Sesquihydrate, plates from dil alc. After drying at 100° *in vacuo* mp 234°. $[\alpha]_{545}^{20}$ +38.5° (c = 0.68 in methanol). Absorption max (96% H_2SO_4): 310, 485, 520 nm. Slightly sol in alcohol, acetone, ethyl acetate. Treatment with alcoholic HCl yields digitaligenin with loss of $2H_2O$.

3,16-Diacetylgitoxigenin, mp 249-250°.

3,16-Dibenzoylgitoxigenin, mp 262°.

4441. Gitoxin. *(3β,5β,16β)-3-[(O-2,6-Dideoxy-β-D-ribohexopyranosyl-(1→4)-O-2,6-dideoxy-β-D-ribo-hexopyranosyl-(1→4)-2,6-dideoxy-β-D-ribo-hexopyranosyl)oxy]-14,16-dihydroxycard-20(22)-enolide;* anhydrogitalin; bigitalin; pseudodigitoxin. $C_{41}H_{64}O_{14}$; mol wt 780.95. C 63.06%, H 8.26%, O 28.68%. Secondary glycoside mainly from *Digitalis purpurea* L., also from *D. lanata* Ehrh., Scrophulariaceae. Byproduct of digitoxin manuf. Isoln: Kraft, *Arch. Pharm.* **250**, 118 (1912); Cloetta, *Arch. Exp. Pathol. Pharmakol.* **112**, 261 (1926); Smith, *J. Chem. Soc.* **1931**, 23. Purification: McChesney *et al., J. Am. Pharm. Assoc.* **37**, 364 (1948). Acid hydrolysis yields 1 mol gitoxigenin + 3 mols digitoxose, *q.q.v.*: Windaus, Schwarte, *Ber.* **58**, 1515 (1925). *See also* Satoh, Aoyama, *Chem. Pharm. Bull.* **18**, 94 (1970). Pharmacokinetics in humans: M. Lesne, *Int. J. Clin. Pharmacol. Biopharm.* **16**, 456 (1978). Prepn of pentaacetate: **Belg. pat. 668,116; Neth. pat. Appl. 6,506,250** (1965, 1966 both to VEB Arzneimittelwerk Dresden), *C.A.* **65**, 23366 (1966); **67**, 91092d (1967). Toxicity study: Foerster *et al., Arch. Int. Pharmacodyn.* **155**, 165 (1965). *Review:* R. Megges *et al., Pharmazie* **32**, 665-667 (1977).

Stout prisms from chloroform + methanol, dec 285° (rapid heating). $[\alpha]_{546}^{20}$ +3.5° (c = 1.02 in pyridine). Absorption max (98% H_2SO_4): 315, 415, 495, 530 nm ($E_{1cm}^{1\%}$ 275, 185, 430, 505). Almost insol in chloroform, ethyl acetate and acetone. Dissolves in a mixture of chloroform and alcohol or in pyridine or in dil alcohol. Less sol in hot 80% alcohol than digoxin.

α- and β-Acetylgitoxin, $C_{43}H_{66}O_{15}$, obtained by enzymatic hydrolysis of digilanide B. β-Form: Long, thin, hair-like prisms from dil methanol, dec 220-225°. $[\alpha]_D^{20}$ 15.7° (c = 1.28 in pyridine). Sol in 80-100 parts methanol; sparingly sol in water and ether.

Pentaacetate, $C_{51}H_{74}O_{19}$, *pengitoxin, penta-O-acetylgitoxin, Carnacid-Cor, Cordoval, Pentagit.* Mass spectral studies: Blessington, Morton, *Org. Mass Spectrom.* **3**, 95 (1970). Rhomboid crystals, mp 151-155°. $[\alpha]_D^{20}$ 14.0 ±1.5° (c = 1.6

in pyridine). LD_{50} in mice (mg/kg): 6.4 i.p.; in rats (mg/kg): 21.0 i.v. (Foerster).

THERAP CAT: Cardiotonic.

4442. Gladiolic Acid. *2,3-Diformyl-6-methoxy-5-methylbenzoic acid;* 4-methoxy-5-methyl-o-phthalaldehyde-3-carboxylic acid. $C_{11}H_{10}O_5$; mol wt 222.20. C 59.46%, H 4.54%, O 36.00%. Fungistatic antibiotic produced by *Penicillium gladioli* Machacek: Brian *et al., Nature* **157**, 697 (1946). Active only in the form shown; the lactol form, 4-formyl-3-hydroxy-7-methoxy-6-methylphthalide, is inactive. Structure: Grove, *Biochem. J.* **50**, 648 (1952); **54**, 664 (1953); *J. Chem. Soc.* **1952**, 3345; Raistrick, Ross, *Biochem. J.* **50**, 635 (1952). Synthesis: Brown, Newbold, *J. Chem. Soc.* **1954**, 1076.

Silky needles from water, mp 160°. pK 4.4. uv max: 214, 271, 304 nm (ε 18500, 6900, 3200). Sol in hot water.

4443. Glafenine. *2-[(7-Chloro-4-quinolinyl)amino]benzoic acid 2,3-dihydroxypropyl ester; N-(7-chloro-4-quinolyl)-anthranilic acid 2,3-dihydroxypropyl ester;* 4-[(2-carboxyphenyl)amino]-7-chloroquinoline α-monoglyceride; 2-[(7-chloro-4-quinolyl)amino]benzoic acid α-glyceride; 2,3-dihydroxypropyl N-(7-chloro-4-quinolyl)anthranilate; glycerylaminophenaquine; glaphenine; R 1707; Glifan; Glifanan; Privadol. $C_{19}H_{17}ClN_2O_4$; mol wt 372.81. C 61.21%, H 4.60%, Cl 9.51%, N 7.51%, O 17.17%. Prepn: Fr. pat. **M2413;** A. Allais, J. Meier, U.S. pat. **3,232,944** (1964, 1966 both to Roussel-UCLAF); G. Mouzin *et al., Synthesis* **1980**, 54. HPLC determn in plasma: M.-C. Tournet *et al., J. Chromatog.* **224**, 348 (1981). Biotransformation in the rat and man: J. Pottier *et al., Eur. J. Drug Metab. Pharmacokinet.* **4**, 109 (1979). Bioavailability, absorption: F. Moolenaar *et al., Int. J. Pharm.* **4**, 195 (1980). Alkalimetric determn: M. S. Tawakkol *et al., Pharmazie* **36**, 163 (1981). Comprehensive description: A. A. Badwan *et al.,* in *Analytical Profiles of Drug Substances and Excipients* vol. **21**, H. G. Brittain, Ed. (Academic Press, San Diego, 1992) pp 197-232.

Pale yellow prisms, mp 169-170° (Mouzin); also reported as mp 165° (Allais, Meier). Practically insol in water. Slightly sol in abs alcohol, acetone, ether, benzene, chloroform; sol in dil aq alkalies and acids. Soly at 30°C (g/100 ml): hexane < 0.001, water 0.001, chloroform 0.260, acetone 0.297, ethanol 0.700, 0.1N HCl 1.295. pKa (20°C) 7.2. uv max (methanol): 356, 225, 255 nm; (0.1N HCl) 342, 223, 252 nm. LD_{50} orally in mice: > 2 g/kg (Allais, Meier).

THERAP CAT: Analgesic.

4444. Glaucarubin. *[1β,2α,11β,12α,15β(S)]-11,20-Epoxy-1,2,11,12-tetrahydroxy-15-(2-hydroxy-2-methyl-1-oxobutoxy)picras-3-en-16-one; 2-hydroxy-2-methylbutyric acid 4-ester with 1,2,3,3aβ,4,6aβ,7,7aα,10,11,11a,11bα-dodecahydro-1α,2α,4β,10α,11β-pentahydroxy-3α,8,11aβ-trimethyl-5H-1,11cβ-(epoxymethano)phenanthro[10,1-bc]pyran-5-one;* α-kirondrin; Glaumeba. $C_{25}H_{36}O_{10}$; mol wt 496.55. C 60.47%, H 7.31%, O 32.22%. From aceituno meal, the press cake from seeds of *Simaruba glauca* DC, Simarubaceae: Brink, *13th Int. Congr. Pure Appl. Chem.* (Stockholm and Uppsala, 1953); Ham *et al., J. Am. Chem. Soc.* **76**, 6066 (1954); Cuckler *et al., Arch. Int. Pharmacodyn.* **114**, 307

(1958); Shafer, U.S. pat. **2,864,745** (1958 to Merck & Co.). Structure: Polonsky *et al., Bull. Soc. Chim. France* **1964**, 1827. Identity of α- and β-kirondrin with glaucarubinone and glaucarubin, resp: Bourguignon, Polonsky, *Bull. Soc. Chim. Biol.* **46**, 1145 (1964). Crystal structure: Kartha *et al., Nature* **202**, 389 (1964). Configuration: Nyburg *et al., Chem. Commun.* **1965**, 203.

Bitter crystals from methanol, dec 250-255°. $[\alpha]_D^{25}$ +45° (c = 1.7 in pyridine); $[\alpha]_D^{25}$ +69° (c = 0.6 in methanol). Sol in 0.1N NaOH soln. Practically insol in aq sodium bicarbonate solns. Only slightly sol in water (less than 1.8 mg/ml). Prompt acidification of an alkaline soln regenerates glaucarubin, but standing in alkaline soln leads to decomp.

Tetramethoxyglaucarubin, $C_{29}H_{44}O_{10}$, crystals from ether, mp 202-205°. $[\alpha]_D$ +17.7° (c = 0.705 in pyridine); $[\alpha]_D$ +65.5° (c = 0.926 in methanol).

Pentaacetate, $C_{35}H_{46}O_{15}$, crystals from benzene + petr ether, mp 158-160°. $[\alpha]_D$ +49° (c = 0.534 in pyridine).

THERAP CAT: Antiamebic.

4445. Glaucine. *(S)-5,6,6a,7-Tetrahydro-1,2,9,10-tetramethoxy-6-methyl-4H-dibenzo[de,g]quinoline; 1,2,9,10-tetramethoxyaporphine;* boldine dimethyl ether; Bromcholitin; Glauvent. $C_{21}H_{25}NO_4$; mol wt 355.43. C 70.96%, H 7.09%, N 3.94%, O 18.01%. *d*-Form prevalent in nature. Found in *Glaucium flavum* Crantz (*G. luteum* Scop.), *Papaveraceae* and in *Dicentra* and *Corydalis* species, *Fumariaceae*. Isoln: Fischer, *Arch. Pharm.* **239**, 426 (1901). Structure and prepn: Gadamer, *ibid.* **249**, 680 (1911); Späth, Tharrer, *Ber.* **66**, 904 (1933). Configuration: Faltis, Adler, *Arch. Pharm.* **284**, 281 (1951). Synthesis of *dl*-form: Chan, Maitland, *J. Chem. Soc. (C)* **1966**, 753; Jackson, Martin, *ibid.* 2061; Cava *et al., J. Org. Chem.* **35**, 175 (1970). Pharmacology and toxicity of *dl*-glaucine phosphate: Y. Kasé *et al., Arzneimittel-Forsch.* **33**, 936 (1983). Sites of antitussive action: *eidem, ibid.* 947.

d-Glaucine

Orthorhombic plates, prisms from ethyl acetate or ether, mp 120°. $[\alpha]_D^{20}$ +115° (c = 3 in alc). Sol in acetone, alcohol, chloroform, ethyl acetate. Moderately sol in ether, petr ether. Practically insol in water and benzene.

Hydrochloride trihydrate, $C_{21}H_{25}NO_4 \cdot HCl \cdot 3H_2O$, needles. mp 232° (anhydrous). Sol in water, alcohol, chloroform.

Hydrobromide, $C_{21}H_{25}NO_4 \cdot HBr$, crystals, mp 235°. Less sol than the hydrochloride.

dl-Form phosphate, $(C_{21}H_{25}NO_4)_2 \cdot 3H_3PO_4$, *DL-832.* Crystalline powder. LD_{50} in mice (mg/kg): 98 i.v.; 401 orally (Kasé).

4446. Gliadin. A simple protein, one of the prolamins, derived from the gluten of wheat, rye, etc. May contain up to 43% glutamine (*C.A.* **50**, 15792a (1956)). Studies on the physical nature of gliadin: Holme, Briggs, *Cereal Chem.* **36**,

321 (1959). Use of deamidized gliadin in food products: McDonald, U.S. pat. **3,030,211** (1962 to USDA).

Practically insol in water, abs alcohol, and other neutral solvents. Sol in 70-80% alcohol, dil acid, dil alkali.

4447. Glibornuride. *[1S-(endo, endo)]-N-[[(3-Hydroxy-4,7,7-trimethylbicyclo[2.2.1]hept-2-yl)amino]carbonyl]-4-methylbenzenesulfonamide; (1R,2R,3S,4S)-1-(2-hydroxy-3-bornyl)-3-(p-tolylsulfonyl)urea;* D-3-*endo*-p-tosylureidoborneol; 1-(*p*-tolylsulfonyl)-3-(2-*endo*-hydroxy-3-*endo*-D-bornyl)urea; Ro-6-4563; Gluborid; Glutril. $C_{18}H_{26}N_2O_4S$; mol wt 366.48. C 58.99%, H 7.15%, N 7.64%, O 17.46%, S 8.75%. Prepn: Bretschneider *et al., Monatsh.* **100**, 2133 (1969); *eidem,* S. Afr. pat. 67 06,161, *C.A.* **71**, 61601j (1969); S. Afr. pat. 69 02,395 corresp to U.S. pat. **3,654,357** (1968, 1969, 1972, all to Hoffmann-La Roche). Comprehensive review of pharmacology: *Arzneimittel-Forsch.* **22**, 2153-2222 (1972).

Crystals, mp 192-195° (ethanol-water); also reported as 195-198°. $[\alpha]_D$ +63.8° (ethanol).

THERAP CAT: Antidiabetic.

4448. Gliclazide. *N-[[(Hexahydrocyclopenta[c]pyrrol-2(1H)-yl)amino]carbonyl]-4-methylbenzenesulfonamide; 1-(hexahydrocyclopenta[c]pyrrol-2(1H)-yl)-3-(p-tolylsulfonyl)-urea;* N-(4-methylbenzenesulfonyl)-N'-(3-azabicyclo[3.3.0]-oct-3-yl)urea; 1-(3-azabicyclo[3.3.0]oct-3-yl)-3-(*p*-tolylsulfonyl)urea; S-1702; Diamicron; Glimicron; Nordialex. $C_{15}H_{21}N_3O_3S$; mol wt 323.42. C 55.71%, H 6.54%, N 12.99%, O 14.84%, S 9.91%. Prepn: Beregi *et al.,* Fr. pat. **1,510,714** and U.S. pat. **3,501,495** (1968 and 1970, both to Sci. Union & Co.-Soc. Franc. Rech. Med.). Series of articles on pharmacology: *Arzneimittel-Forsch.* **22**, 1686-1695 (1972). Toxicity data: J. Duhault *et al., ibid.* 1682. Review of pharmacology, efficacy: B. Holmes *et al., Drugs* **27**, 301 (1984).

Crystals from anhydrous ethanol, mp 180-182°. LD_{50} orally in mice: > 3 g/kg (Duhault).

THERAP CAT: Antidiabetic.

4449. Glimepiride. *trans-3-Ethyl-2,5-dihydro-4-methyl-N-[2-[4-[[[(4-methylcyclohexyl)amino]carbonyl]amino]sul-fonyl]phenyl]ethyl]-2-oxo-1H-pyrrole-1-carboxamide;* N-[4-[2-(3-ethyl-4-methyl-2-oxo-3-pyrroline-1-carboxamido)-eth-yl]-benzenesulfonyl]-N'-4-methylcyclohexylurea; 1-[4-[2-(3-ethyl-4-methyl-2-oxo-3-pyrroline-1-carboxamido)ethyl]phenylsulfonyl]-3-(4-methylcyclohexyl)urea; HOE-490. $C_{24}H_{34}N_4O_5S$; mol wt 490.62. C 58.75%, H 6.98%, N 11.42%, O 16.31%, S 6.54%. Sulfonylurea. Prepn: R. Weyer *et al.,* Ger. pat. **2,951,135**; *eidem,* U.S. pat. **4,379,785** (1981, 1983 both to Hoechst). Synthesis: R. Weyer, V. Hitzel, *Arzneimittel-Forsch.* **38**, 1079 (1988). Pharmacology: K. Geisen, *ibid.,* 1120. Effects on insulin and glucagon secretion: V. Leclercq-Meyer *et al., Biochem. Pharmacol.* **42**, 1634 (1991). HPLC determn in biological fluids: K. H. Lehr, P. Damm, *J. Chromatog.* **526**, 497 (1990). Clinical pharmacokinetics: K. Ratheiser *et al., Arzneimittel-Forsch.* **43**, 856 (1993). Toxicity study: U. Schollmeier *et al., ibid.* 1038.

mp 207°.
THERAP CAT: Antidiabetic.

4450. Gliotoxin. *[3R-(3α,5aβ,6β,10aα)]-2,3,5a,6-Tetra-hydro-6-hydroxy-3-(hydroxymethyl)-2-methyl-10H-3,10a-epi-dithiopyrazino[1,2-a]indole-1,4-dione.* $C_{13}H_{14}N_2O_4S_2$; mol wt 326.40. C 47.84%, H 4.32%, N 8.58%, O 19.61%, S 19.65%. Antibiotic substance produced by various spp of *Trichoderma, Gladiocladium fimbriatum, Aspergillus fumigatus,* and *Penicillium* spp: Weindling, Emerson, *Phytopathology* **26**, 1068 (1936); **27**, 1175 (1937); Johnson et al., *J. Am. Chem. Soc.* **65**, 2005 (1943); Menzel et al., *J. Biol. Chem.* **152**, 419 (1944). Structure: Bell et al., *J. Am. Chem. Soc.* **80**, 1001 (1958); Beecham et al., *Tetrahedron Letters* **1966**, 3131. Crystallographic data: McCrone, *Anal. Chem.* **26**, 1662 (1954). Biosynthesis: Suhadolnik, Chenoweth, *J. Am. Chem. Soc.* **80**, 4391 (1958); Winstead, Suhadolnik, *ibid.* **82**, 1644 (1960); J. D. M. Herscheid et al., *J. Org. Chem.* **45**, 1885 (1980). Synthetic studies: Poisel, Schmidt, *Ber.* **104**, 1714 (1971); *ibid.* **105**, 625 (1972); Oehler et al., *ibid.* **625**. Total synthesis of *dl*-form: T. Fukuyama, Y. Kishi, *J. Am. Chem. Soc.* **98**, 6723 (1976); T. Fukuyama et al., *Tetrahedron* **37**, 2045 (1981).

Monoclinic needles from methanol or benzene, dec 221°. $[\alpha]_D^{25}$ −290° (c = 0.08 in ethanol). uv max: 270 nm (ε 4500). Soly in mg/ml at 7°: acetic acid 12; acetone 9.0; acetonitrile 10.2; benzene 5.5; carbon tetrachloride 0.8; chloroform 20; dioxane 73 (decompn); dimethylformamide 17; ethyl acetate 8.5; ethanol 4.7; methanol 1.4; pyridine 77; water (30°): 0.07. Sensitive to oxidation and heat; inactivated by heating for 10 min at 100°.
Monoacetate, $C_{15}H_{16}N_2O_5S_2$, isolated from cultures of *Penicillium terlikowski* Zaleski: Johnson et al., *J. Am. Chem. Soc.* **75**, 2110(1953). Orthorhombic crystals from benzene, mp 162-163°.

4451. Glipizide. *N-[2-[4-[[[(Cyclohexylamino)carbonyl]amino]sulfonyl]phenyl]ethyl]-5-methylpyrazinecarboxamide; 1-cyclohexyl-3-[[p-[2-(5-methylpyrazinecarboxamido)ethyl]phenyl]sulfonyl]urea;* glydiazinamide; K-4024; Glibenese; Glucotrol; Mindiab; Minidiab. $C_{21}H_{27}N_5O_4S$; mol wt 445.54. C 56.61%, H 6.11%, N 15.72%, O 14.36%, S 7.20%. Second generation sulfonylurea with hypoglycemic activity. Prepn: Ambrogi, Logemann, **Ger.** pat. **2,012,138** corresp to U.S. pat. **3,669,966** (1970, 1972 both to Carlo Erba); Ambrogi et al., *Arzneimittel-Forsch.* **21**, 200 (1971). Pharmacology: *eidem, ibid.* 208; Marigo et al., *ibid.* 215. Metabolism: Goldaniga et al., *ibid.* **23**, 242 (1973); Fuccella et al., *J. Clin. Pharmacol.* **13**, 68 (1973). Toxicity: Ambrogi et al., *Arzneimittel-Forsch.* **21**, 208 (1971). Review of pharmacology and

therapeutic efficacy: R. N. Brogden et al., *Drugs* **18**, 329-353 (1979); H. E. Lebovitz, *Pharmacother.* **5**, 63-77 (1985).
Crystals from ethanol, mp 208-209°. Also reported as mp 200-203°. LD_{50} in mice, rats (g/kg): >3, 1.2 i.p. (Ambrogi).
THERAP CAT: Antidiabetic.

4452. Gliquidone. *N-[(Cyclohexylamino)carbonyl]-4-[2-(3,4-dihydro-7-methoxy-4,4-dimethyl-1,3-dioxo-2(1H)-isoquinolinyl)ethyl]benzenesulfonamide; 1-cyclohexyl-3-[p-[2-(3,4-dihydro-7-methoxy-4,4-dimethyl-1,3-dioxo-2(1H)-isoquinolyl)ethyl]phenyl]sulfonyl]urea;* 1,2,3,4-tetrahydro-2-[p-(N'-cyclohexylureido-N-sulfonyl)phenethyl]-4,4-dimethyl-7-methoxyisoquinoline-1,3-dione; AR-DF 26; Glurenorm. $C_{27}H_{33}N_3O_6S$; mol wt 527.64. C 61.46%, H 6.30%, N 7.96%, O 18.19%, S 6.08%. Prepn: E. Kutter et al., **Ger.** pat. **2,000,339** (1971 to Thomae); *eidem,* U.S. pat. **3,708,486** (1973 to Boehringer, Ing.); *eidem,* **Ger.** pat. **2,011,126** (1971 to Thomae), *C.A.* **76**, 14359e (1972). Pharmacokinetics and metabolism: Kopitar, *Arzneimittel-Forsch.* **25**, 1455 (1975); Kopitar et al., *ibid.* 1933.

Crystals from boiling methanol, mp 180-182°.
Sodium salt, sinters at 160°. LD_{50} in mice: >2 g/kg orally; 234 mg/kg i.v. (Kutter, 1973).
THERAP CAT: Antidiabetic.

4453. Glisoxepid. *N-[2-[4-[[[[(Hexahydro-1H-azepin-1-yl)amino]carbonyl]amino]sulfonyl]phenyl]ethyl]-5-methyl-3-isoxazolecarboxamide; 1-(hexahydro-1H-azepin-1-yl)-3-[[p-[2-(5-methyl-3-isoxazolecarboxamido)ethyl]phenyl]sulfonyl]-urea;* 4-[4-[β-(5-methylisoxazole-3-carboxamido)ethyl]phenylsulfonyl]-1,1-hexamethylenesemicarbazide; BS-4231; RP-22410; Pro-Diaban. $C_{20}H_{27}N_5O_5S$; mol wt 449.53. C 53.44%, H 6.05%, N 15.58%, O 17.80%, S 7.13%. Prepn: H. Plümpe, W. Puls, **S. Afr.** pat. **68 06,886**; *eidem,* U.S. pat. **3,668,215** (1969, 1972 both to Bayer). Pharmacology: Loubatieres et al., *Compt. Rend. Ser. D* **271**, 1446 (1970); *J. Pharmacol.* **3**, 171, 229 (1972). Toxicity study: Tettenborn, *Arzneimittel-Forsch.* **24**, 409 (1974). Series of articles: *ibid.* 363-452.

Colorless crystals from ethanol, mp 189°. LD_{50} in mice, rats, cats, dogs (g/kg): >10.0, >10.0, >4.0, >2.0 orally; in mice, rats (mg/kg): 283, 196 i.v. (Tettenborn).
THERAP CAT: Antidiabetic.

4454. Globin. The colorless, basic protein of hemoglobin. Formed from tissue protein in the body, the globin part of catabolized hemoglobin is re-used (unlike heme, the prosthetic group of hemoglobin). Prepn from ox hemoglobin: Anson, Mirsky, *J. Gen. Physiol.* **13**, 469 (1930). Globin from normal adult human hemoglobin consists of four polypeptide chains: two α-chains and two β-chains. The α-chain contains 141, the β-chain 146 amino acids. Thus globin contains 574 amino acids and has an approx mol wt of 62,000. Abnormal globins may contain γ- and δ-chains. Structure: Braunitzer et al., *Z. Physiol. Chem.* **325**, 283 (1961); **331**, 1 (1963); Konigsberg et al., *J. Biol. Chem.* **237**,

1549, 2547 (1962); **238**, 2016, 2028 (1963). Review of prepn and properties: Rossi Fanelli *et al., Advan. Protein Chem.* **19**, 124 (1964). *Review:* Braunitzer *et al., ibid.* 1.

Denatures rapidly above 17°. At pH values near neutrality, combines with ferroprotoporphyrin to yield hemoglobin, or with ferriprotoporphyrin to yield methemoglobin.

4455. Glucagon. Glukagon; hyperglycemic-glycogenolytic factor; HG-factor; HGF. Mol wt approx 3550. Polypeptide hormone produced in the alpha cells of the islets of Langerhans in the pancreas. Existence of this hyperglycemic pancreatic factor was originally postulated to explain the initial hyperglycemia following insulin administration. *Refs:* Murlin *et al., J. Biol. Chem.* **56**, 252 (1923); Kimball, Murlin, *ibid.* **58**, 337 (1923); Bürger, Kramer, *Arch. Exp. Pathol. Pharmakol.* **156**, 1 (1930); Bürger, Brandt, *Z. Ges. Exp. Med.* **96**, 375 (1935); Staub *et al., Science* **117**, 628 (1953); Ingle *et al., Proc. Soc. Exp. Biol. Med.* **84**, 232 (1953); Geschwind, Staub, *ibid.* 244; Staub *et al., J. Biol. Chem.* **214**, 619 (1955); **Brit.** pat. **762,885** (1956 to Lilly). Amino acid sequence of porcine, bovine, and human glucagon: Bromer *et al., J. Am. Chem. Soc.* **79**, 2807 (1957); Bromer *et al., J. Biol. Chem.* **246**, 2822 (1971); Thomsen *et al., FEBS Letters* **21**, 315 (1972). Primary structure of glucagon from all species studied thus far is the same, with the exception of the turkey which differs by one amino acid residue. Total synthesis: E. Wünsch, *Z. Naturforsch.* **22B**, 1269 (1967). *See also* E. Wünsch, G. Wendlberger, *Ber.* **101**, 3659 (1968) and preceding articles in the series which describe in detail the prepn of the fragments of glucagon and the assembly of the total sequence. Usually present in increased concns in diabetes. *Reviews:* Behrens, Bromer, *Vitamins and Hormones* **16**, 263-301 (1958); A. C. Beynon, M. J. Geelen, *Vet. Res. Commun.* **5**, 223-236 (1982). Books: P. J. Lefèbvre, R. H. Unger, *Glucagon* (Pergamon Press, New York, 1972) 535 pp; *Glucagon* **Pts. I and II**, P. Lefèbvre, Ed. (Springer, New York, 1983) 700 pp.

1
His–Ser–Gln–Gly–Thr–Phe–Thr–Ser–Asp–Tyr–Ser–Lys–Tyr–Leu–Asp–Ser–Arg

29
Thr–Asn–Met–Leu–Trp–Gln–Val–Phe–Asp–Gln–Ala–Arg

Rhombic dodecahedra. Stable. Practically insol in water. Sol in acidic, basic media, *i.e.* below pH 3 and above pH 9.5.

THERAP CAT: Antihypoglycemic.

4456. Glucametacin. *2-[[[1-(4-Chlorobenzoyl)-5-methoxy-2-methyl-1H-indol-3-yl]acetyl]amino]-2-deoxy-D-glucose;* 2-[2-[1-(*p*-chlorobenzoyl)-5-methoxy-2-methylindol-3-yl]-acetamido]-2-deoxy-D-glucose; glucametacine; glucamethacin; indomethacin glucosamide. $C_{25}H_{27}ClN_2O_8$; mol wt 518.95. C 57.86%, H 5.24%, Cl 6.83%, N 5.40%, O 24.66%. Deriv of indomethacin, *q.v.* Prepn: A. Demetrio *et al.,* **Ger.** pat. **2,223,051** (1973 to SIR Lab. Chem. Biol. SpA), *C.A.* **80**, 83529e (1974). Pharmacological study: E. Paroli *et al., Arzneimittel-Forsch.* **28**, 819 (1978). Clinical studies: P. Petera *et al., Int. J. Clin. Pharmacol.* **15**, 581 (1977); L. Capelli *et al., Curr. Med. Res. Opin.* **7**, 227 (1981).

Monohydrate, $C_{25}H_{27}ClN_2O_8 \cdot H_2O$, *Euminex, Teorema, Teoremac.*

THERAP CAT: Anti-inflammatory.

4457. Glucamine. *1-Amino-1-deoxy-D-glucitol;* 1-amino-1-deoxysorbitol; glycamine; D-glucamine. $C_6H_{15}NO_5$; mol wt 181.19. C 39.77%, H 8.34%, N 7.73%, O 44.15%. Prepn from D-glucose: Holly *et al., J. Am. Chem. Soc.* **72**, 5416 (1950); from *N*-benzylglycamine: Kagan *et al., ibid.* **79**, 3541 (1957); by catalytic reduction of glucose in the presence of hydrazine: Lemieux, **U.S.** pat. **2,830,983** (1958 to Natl. Res. Council, Ottawa). Commercial prepn: Flint, Salzberg, **U.S.** pat. **2,016,962** (1935 to du Pont); Groggins, Stirton, *Ind. Eng. Chem.* **29**, 1358 (1937).

Crystals from methanol, mp 127. Sharp, slightly sweet taste. $[\alpha]_D^{15}$ −7.95° (c = 10 in water). Very sol in water; slightly sol in alcohol. Practically insol in ether.

4458. D-Glucaric Acid. Saccharic acid; D-glucosaccharic acid; D-tetrahydroxyadipic acid. $C_6H_{10}O_8$; mol wt 210.14. C 34.29%, H 4.80%, O 60.91%. Best prepd by nitric acid oxidation of starch; yields as high as 65% are obtained in contrast to much lower yields from glucose or sucrose: Kiliani, *Ber.* **58**, 2344 (1925); Schmidt *et al., ibid.* **70**, 2402 (1937). Prepn from D-glucose: Mehltretter *et al.,* **U.S.** pat. **2,472,168** (1949 to U.S. Secy. Agr.); Truchan, **U.S.** pat. **2,809,989** (1957 to Cowles Chem.); Phillips *et al., J. Chem. Soc.* **1958**, 3522; **Belg.** pat. **615,023** (1962 to Ciba).

Needles from 95% ethanol, mp 125-126°. Shows mutarotation. $[\alpha]_D^{19}$ +6.86° → +20.60° (H₂O). Ka_1 = 1.0 × 10⁻⁵ at 25°. Soluble in water, ethanol. Sparingly sol in ether.

1,4-Lactone, *saccharolactone.* Strong inhibitor of β-glucuronidase: Levvy, *Biochem. J.* **52**, 464 (1952); Boyland, Williams, *ibid.* **64**, 578 (1956). Monohydrate, crystals, mp 90°.

4459. D-Glucoascorbic Acid. *D-arabino-Hept-2-enonic acid γ-lactone;* 3-keto-D-glucoheptonofuranolactone. $C_7H_{10}O_7$; mol wt 206.15. C 40.78%, H 4.89%, O 54.33%. A physiologically inactive homolog of ascorbic acid. Prepn: Ault *et al., J. Chem. Soc.* **1933**, 741; Baird *et al., ibid.* **1934**, 62; Reichstein *et al., Helv. Chim. Acta* **17**, 510 (1934); Stacey, Turton, *J. Chem. Soc.* **1946**, 661; Stedehouder, *Rec. Trav. Chim.* **71**, 831 (1952). Review on analogs of ascorbic acid: Smith, *Advances in Carbohydrate Chemistry* **2**, 79 (1946).

Consult the Name Index before using this section.

Monohydrate, clusters of rod-like crystals with pointed ends from acetone + methanol + petr ether, mp 101-105° (Reichstein); mp 138° (Baird). Becomes anhydr at 70° under high vacuum, mp 191° (Ault). $[\alpha]_D^{20}$ −14° (c = 1 as hydrate); $[\alpha]_D^{20}$ −22° (c = 1 as hydrate in methanol); $[\alpha]_D^{14.5}$ −37.8° (c = 2.41 in 0.01N HCl); $[\alpha]_D^{20}$ −80° (c = 0.75 neutralized with NaOH). pK_1 4.26; pK_2 11.58. Soluble in water; moderately sol in alcohol.

4460. Glucofrangulin. Erbalax-N; Irgalax. $C_{27}H_{30}O_{14}$; mol wt 578.53. C 56.06%, H 5.23%, O 38.72%. Anthraquinone glycoside from bark of *Rhamnus frangula* L., *Rhamnaceae* (alder buckthorn): Casparis, Maeder, *Bull. Soc. Chim. Biol.* **9**, 324 (1927); Cucu, Jarpo, *Pharmazie* **14**, 316 (1959); Knap et al., *Czech.* pat. **110,024** (1964), *C.A.* **61**, 4159e (1964). Consists of two isomers, glucofrangulin A and glucofrangulin B, which differ by the linkage to the sugar moiety at the 3 position of the aglycone. Structure: Schindler, *Helv. Chim. Acta* **29**, 411 (1946); Hörhammer et al., *Naturwiss.* **51**, 310 (1964); Longo et al., *Arch. Pharm.* **297**, 248 (1964); Wagner, Hörhammer, *Naturwiss.* **53**, 585 (1966). Proof of structure of glucofrangulin A and partial synthesis: eidem, *Z. Naturforsch. B* **24**, 1408 (1969).

Glucofrangulin A

Glucofrangulin A, 3-[(6-deoxy-α-L-mannopyranosyl)oxy]-1-(β-D-glucopyranosyloxy)-8-hydroxy-6-methyl-9,10-anthracenedione.

Glucofrangulin A octaacetate, $C_{43}H_{46}O_{22}$, needles from methanol, mp 228-230°. $[\alpha]_D^{20}$ −124° (c = 1.16 in acetone). uv max: 212, 264, 360 nm (log ∈ 4.57, 4.56, 4.26).

THERAP CAT: Cathartic.

4461. α-Glucogallin. *α-D-Glucopyranose 1-(3,4,5-trihydroxybenzoate); α-D-glucopyranose-1-gallate;* 1-galloyl-α-D-glucose. $C_{13}H_{16}O_{10}$; mol wt 332.26. C 46.99%, H 4.85%, O 48.15%. Synthesis: Schmidt, Herok, *Ann.* **587**, 63 (1954); Schmidt, Schmadel, *ibid.* **649**, 149 (1961).

Crystals, dec 179-181°. $[\alpha]_D^{20}$ +83° (c = 3 in methanol). Dihydrate, prisms from water, mp 171-173°. $[\alpha]_D^{25}$ +79.1° (water). Freely sol in water, methanol, ethanol, dioxane, acetic acid. Sparingly soluble in ethyl acetate, ether and acetone.

4462. β-Glucogallin. *β-D-Glucopyranose 1-(3,4,5-trihydroxybenzoate); β-D-glucopyranose-1-gallate;* glucogallic acid; 1-galloyl-β-D-glucose. $C_{13}H_{16}O_{10}$; mol wt 332.26. C 46.99%, H 4.85%, O 48.15%. Glucoside or glucotannoid from chinese rhubarb, *Rheum officinale,* Baill., *Polygonaceae:* Gilson, *Compt. Rend.* **136**, 385 (1903). Structure and synthesis: Fischer, Bergmann, *Ber.* **51**, 1760 (1918); Schmidt, Schmadel, *Ann.* **649**, 149 (1961).

Bitter microscopic prisms from water, methanol or 80% ethanol, mp 207°. $[\alpha]_D^{25}$ −24.5° (c = 1.75 in water). Freely sol in hot water. Sparingly sol in cold water, methanol, ethanol, acetone, ethyl acetate. Practically insol in ether, benzene, chloroform, petr ether.

4463. Glucoheptonic Acid. *D-glycero-D-gulo-Heptonic acid;* α-glucoheptonic acid; glucosemonocarboxylic acid; glucomonocarbonic acid. $C_7H_{14}O_8$; mol wt 226.18. C 37.17%, H 6.24%, O 56.59%. Obtained by treating glucose with HCN yielding a cyanohydrin which is saponified to glucoheptonic acid: Kiliani, *Ber.* **19**, 769 (1886); Fischer, *Ann.* **270**, 71 (1892); Armestar, *C.A.* **45**, 2865 (1951). Process starting with calcium cyanide and glucose: Clevenot, **U.S.** pat. **2,735,866** (1956 to Lab. Clevenot). Diagnostic use of ^{99m}Tc complexes in renal scintigraphy: R. E. Boyd et al., *Brit. J. Radiol.* **46**, 604 (1973); in brain scanning: J. Léveillé et al., *J. Nucl. Med.* **18**, 957 (1977); T. W. Ryerson et al., *Radiology* **127**, 429 (1978). Subacute toxicity study: L. Belbeck et al., *Can. J. Comp. Med.* **45**, 299 (1981).

Lactonizes upon evapn. The lactone forms large sweetish crystals, mp 145-148°. $[\alpha]_D^{20}$ −56.0° (shows mutarotation). Sol in water.

Sodium salt, $C_7H_{13}NaO_8$, *gluceptate sodium, sodium glucoheptonate.* Prepn from corn syrup: Behnke, **U.S.** pat. **3,022,343** (1962 to Pfanstiehl Labs). Crystals (α-form), dec 161°. $[\alpha]_D^{20}$ +6.06° (c = 10 in H_2O). Freely sol in water.

Calcium salt, $C_{14}H_{26}CaO_{16}$, *gluceptate calcium, calcium glucoheptonate, calcium glucosemonocarbonate, calcium glucomonocarbonate, Calciforte, Calheptose.* Prepn from Na salt: Holstein, **U.S.** pat. **3,033,900** (1962 to Pfanstiehl Labs.). Hygroscopic crystals, somewhat acrid taste, dec 200°. Sol in water.

Magnesium salt, $C_{14}H_{26}MgO_{16}$, *magnesium glucoheptonate, magnesium glucosemonocarbonate, magnesium glucomonocarbonate, Navolin.* Prepn: Cipelli, **U.S.** pat. **3,063,896** (1962 to Merck & Co.). Water-sol crystals, pleasant taste.

Complex with ^{99m}Tc, *^{99m}Tc gluceptate, ^{99m}Tc gluheptonate, Glucoscan, TechneScan gluceptate.*

USE: Pharmaceutic aid.

THERAP CAT: ^{99m}Tc complex as diagnostic aid (radioactive imaging agent).

4464. Gluconic Acid. *D-Gluconic acid; dextronic acid; maltonic acid; glyconic acid; glycogenic acid; pentahydroxycaproic acid.* $C_6H_{12}O_7$; mol wt 196.16. C 36.74%, H 6.17%, O 57.09%. May be prepd from glucose by electrolytic oxidation in alkaline medium: Helwig, **U.S.** pat. **1,937,273** (1933); by chemical oxidation with hypobromites: Stoll, Kussmaul, **U.S.** pat. **1,648,368** (1927). At present produced in commercial quantities by the fermentative oxidation of the aldehyde group in glucose from corn using *Aspergillus niger, A. fumaricus, Acetobacter aceti, Penicillium chrysogenum,* and other *Penicillia:* Bernhauer, Schulof, **U.S.** pat. **1,849,053** (1932 to Pfizer); Williams, *Mfg. Chemist* **16**, 239 (1945); Moyer et al., *Ind. Eng. Chem.* **32**, 1379 (1940); Prescott et al., *ibid.* **45**, 338 (1953); van Gelder, **U.S.** pat. **2,916,-**

515 (1959 to Noury & van der Lande, Holland); Roehr, *Naturwiss.* **48,** 478 (1961). Prepn by hydrolysis of α-D-glucose with bromine + H_2SO_4: Foster, Vardheim, *J. Chem. Soc.* **1957,** 989; by γ-irradiation of D-glucose: Phillips *et al., ibid.* **1958,** 3522; Grant, Ward, *ibid.* **1959,** 2871. Prepn of solid: Isbell, U.S. pat. **1,985,255** (1934 to U.S. Secy of Commerce).

$$
\begin{array}{c}
COOH \\
HC\!-\!OH \\
HO\!-\!CH \\
HC\!-\!OH \\
HC\!-\!OH \\
CH_2OH
\end{array}
$$

Crystals, mp 131°. Mild acid taste. $[\alpha]_D^{20}$ −6.7° (c = 1). K at 25° = 2.5 × 10^{-4}. Freely sol in water, slightly sol in alcohol. Insol in ether and most other organic solvents. In aq solns the acid is partially transformed into an equilibrium mixt with gamma and delta gluconolactones. Because of the difficulties of preparing a solid cryst product, gluconic acid of commerce is a 50% aq soln, light amber color, faint odor of vinegar, d_4^{25} 1.24. May be stored in stainless steel drums.

Ammonium salt, $NH_4C_6H_{11}O_7$, needles, deliquesc in moist air and become yellow on exposure to light. Dec 154°. Neutral reaction. Soly in water: 31.6 g/100 ml at 25°. Slightly sol in alc. Practically insol in most other organic solvents. The salt dec on steaming with the formation of ammonia and gluconic acid.

Magnesium salt dihydrate, $C_{12}H_{22}MgO_{14}.2H_2O$, *magnesium gluconate, Almora, Ultra-Mg.*

Zinc complex, $C_{12}H_{22}O_{14}Zn$, *Rubozinc.*

USE: *See* Gluconolactone. Ammonium salt used as latent acid catalyst in textile printing.

THERAP CAT: Magnesium salt as antispasmodic; magnesium replenisher.

4465. Gluconolactone. *D-Gluconic acid δ-lactone;* glucono delta lactone; delta gluconolactone; Fujiglucon. $C_6H_{10}O_6$; mol wt 178.14. C 40.45%, H 5.66%, O 53.89%. Prepn by oxidation of glucose with bromine water: Isbell, Pigman, *J. Res. Nat. Bur. Stand.* **10,** 337 (1933); by oxidation of glucose in *Acetobacter suboxydans:* King; Cheldelin, *Biochem. J.* **68,** 31P (1958). Structure: J. Stanék *et al., The Monosaccharides* (Academic Press, New York, 1963) p 271.

Crystals, dec 153°. Sweet taste (different from gluconic acid). $[\alpha]_D^{20}$ +61.7° (c = 1). Soly in water 59 g/100 ml; in alc about 1 g/100 g. Insol in ether. Hydrolyzed to gluconic acid by water. A freshly prepd 1% aq soln has a pH of 3.6 changing to pH 2.5 within 2 hrs.

USE: Component of many cleaning cmpds because of the sequestering ability of the gluconate radical which remains active in alk solns; in the dairy industry to prevent milkstone; in breweries to prevent beerstone; as latent acid catalyst for acid colloid resins, particularly in textile printing; as a coagulant for tofu.

4466. Glucosamine. *2-Amino-2-deoxy-D-glucose;* chitosamine. $C_6H_{13}NO_5$; mol wt 179.17. C 40.22%, H 7.31%, N 7.82%, O 44.65%. Found in chitin, in mucoproteins, and in mucopolysaccharides. Isoln from chitin: Ledderhose, *Z. Physiol. Chem.* **2,** 213 (1878); Hackman, *Aust. J. Biol. Sci.* **7,** 168 (1954). Synthesis: Fischer, Leuchs, *Ber.* **35,** 3787 (1902); **36,** 24 (1903). Separation of α- and β-forms: Westphal, Holzmann, *ibid.* **75B,** 1274 (1942). Structure: Haworth *et al., J. Chem. Soc.* **1939,** 271; Cutler, Peat, *ibid.* 782; Cox, Jeffrey, *Nature* **143,** 894 (1939). Pharmacokinetics in

dog and man: I. Setnikar *et al., Arzneimittel-Forsch.* **36,** 729 (1986). Clinical trials in arthrosis: Y. Vajarudal, *Clin. Ther.* **3,** 336 (1981); M. J. Tapadinhas *et al., Pharmatherapeutica* **3,** 157 (1982). *Review:* Foster, Stacey, "The Chemistry of the 2-Amino Sugars" in C. S. Hudson *et al., Advan. Carbohyd. Chem.* **vol.** 7 (Academic Press, New York, 1952) pp 247-288.

α-Form, crystals, mp 88°. $[\alpha]_D^{20}$ +100° changing to +47.5° after 30 min (water).

β-Form, needles from methanol, dec 110°. $[\alpha]_D^{20}$ +28° changing to +47.5° after 30 min (water). Very sol in water, sol in about 38 parts boiling methanol; sparingly sol in cold methanol or ethanol. Practically insol in ether, chloroform.

N-Acetylglucosamine, $C_8H_{15}NO_6$, needles from methanol + ether, mp 205°. $[\alpha]_D^{18}$ +64° changing to +40.9° (in water).

Sulfate salt, $C_6H_{13}NO_5.xH_2SO_4$, *Dona.*

USE: Pharmaceutic aid.

THERAP CAT: Antiarthritic.

4467. Glucose. D-Glucose; dextrose; blood sugar; grape sugar; corn sugar; Dextropur; Dextrosol; Glucolin. $C_6H_{12}O_6$; mol wt 180.16. C 40.00%, H 6.71%, O 53.28%. A main source of energy for living organisms. Occurs naturally and in the free state in fruits and other parts of plants. Combined in glucosides, in di- and oligosaccharides, in the polysaccharides cellulose and starch, and in glycogen. Normal human blood contains 0.08-0.1%. Manuf on a large scale from starch: Dean, Gottfried, *Advan. Carbohyd. Chem.* **5,** 127 (1950). Below 50°, α-D-glucose hydrate is the stable cryst form, above 50° the anhydr form is obtained and at still higher temps β-D-glucose is formed: W. Pigman, *The Carbohydrates* (Academic Press, New York, 1957) p 92. Structure: Kjaer, Lindberg, *Acta Chem. Scand.* **13,** 1713 (1959). Conformation: E. Percival, *Structural Carbohydrate Chemistry* (J. Garnet Miller, London, 1962) pp 51-57. Comprehensive monograph: H. Bartelheimer *et al.,* D-*Glucose und verwandte Verbindungen in Medizin und Biologie* (Enke, Stuttgart, 1966) 1126 pp.

α-D-Glucose

α-Form monohydrate, crystals from water, mp 83°. $[\alpha]_D$ +102.0° → +47.9° (water). 0.74 times as sweet as sucrose. One gram dissolves in about 1 ml water and in about 60 ml alcohol. LD i.v. in rabbits: 35 g/kg.

α-Form anhyd, crystals from hot ethanol or water, mp 146°. $[\alpha]_D$ +112.2° → +52.7° (c = 10 in water). The final value is obtained instantly in the presence of hydroxyl ions. Formula for varying concns: $[\alpha]_D^{20}$ +52.5° + 0.0188p (p = g/100 ml). pH of 0.5 molar aq soln 5.9. $d_{17.5}^{17.5}$ of water solns w/v: 5% = 1.019; 10% = 1.038; 20% = 1.076; 30% = 1.113; 40% = 1.149. n_D^{20} 10% soln 1.3479. One gram dissolves in 1.1 ml water at 25°; in 0.8 ml at 30°; in 0.41 ml at 50°; in 0.28 ml at 70°; in 0.18 ml at 90°; in 120 ml methanol at 20°. Very sparingly sol in abs alcohol, ether, acetone; sol in hot glacial acetic acid, pyridine, aniline. LD i.v. in rabbits: 35 g/kg.

β-Form, crystals from hot water + ethanol, from dil acetic acid, or from pyridine, mp 148-155°. $[\alpha]_D$ +18.7° → +52.7° (c = 10 in water).

THERAP CAT: Fluid and nutrient replenisher.

THERAP CAT (VET): Nutrition (usually parenterally), hypoglycemia, ketosis, to counteract hepatotoxins.

4468. Glucose Oxidase. *β*-ᴅ-Glucopyranose aerodehydrogenase; P-FAD; corylophyline; microcide; mikrotsid; notatin. An enzyme obtained from mycelia of fungi, such as *Aspergilli* and *Penicillia;* a typical aerobic dehydrogenase which catalyzes the oxidation of glucose to gluconic acid (molecular oxygen is reduced to hydrogen peroxide). It is a flavoprotein, the prosthetic group being flavine-adenine dinucleotide (FAD). Commercial prepns frequently contain appreciable amounts of another enzyme, catalase, which is desirable for certain uses since it removes hydrogen peroxide aerobically generated by glucose oxidase. Names of some commercial prepns are: *DeeO, Fermcozyme, OxyBan, Ovazyme.* Isoln from *Penicillia* cultures: Coulthard *et al., Biochem. J.* **39,** 24 (1945). Commercial production from *Aspergilli* and *Penicillia:* Goldsmith *et al.,* U.S. pat. **2,926,-122** (1960); from *Aspergillus niger:* Faucett *et al.,* U.S. pat. **3,102,081** (1963 to Miles Labs.). Removal of proteolytic enzymes from glucose oxidase (contg catalase) obtained from *Aspergilli* or *Penicillia* cultures: Ohlmeyer, U.S. pat. **2,940,904** (1960 to Ben L. Sarett). Separation from catalase: Pazur *et al., Biochem. Biophys. Acta* **65,** 369 (1962). Properties: Muller, *Enzymologia* **10,** 40 (1941); Keilin, Hartree, *Biochem. J.* **42,** 221 (1948), **50,** 331 (1952). *Reviews:* L. A. Underkofler "Glucose Oxidase: Production, Properties, Present and Potential Applications" in *Soc. Chem. Ind. (London) Monograph* **no. 11,** 72-86 (1961); R. Bentley, "Glucose Oxidase" in *The Enzymes* vol. 7, P. D. Boyer *et al.,* Eds. (Academic Press, New York, 1963) pp 567-586.

Amorphous powder or crystals. Abs max between 270-280, 375-380, and 450-460 nm (aq soln). Freely sol in water giving yellowish-green solns. Most active at pH 5.5-6.0 and 30-35°. Stable between pH 4.5 and 7.0. Stable to pepsin and trypsin. A glucose oxidase unit is defined as that quantity of enzyme which will cause the uptake of 10 mm³ oxygen per min in a Warburg manometer at 30° in the presence of excess air and excess catalase with a substrate contg 3.3% glucose monohydrate and 0.1*M* phosphate buffer, pH 5.9 with 0.4% sodium dehydroacetate: Scott, *J. Agr. Food Chem.* **1,** 727 (1953).

USE: Mainly in the protection of foods: for the removal of glucose from egg albumin and whole eggs prior to drying. To remove oxygen from canned foods, soft drinks, beer, and stored food. In the manuf of test papers for diabetes control and fertility tests. To stabilize ascorbic acid and vitamin B₁₂ prepns. In combination with catalase, for treatment of food wrappers to prevent oxidative deterioration of food: Sarett, Scott, U.S. pat. **2,765,233** (1956 to Ben L. Sarett).

4469. Glucose-6-phosphate. *ᴅ-Glucose 6-(dihydrogen phosphate);* glucose-6-phosphoric acid; Robison ester. $C_6H_{13}O_9P$; mol wt 260.14. C 27.70%, H 5.04%, O 55.35%, P 11.91%. A normal constituent of resting muscle, probably always existing in equilibrium with fructose-6-phosphate. For the enzymatic conversion from the 1-phosphate *see* α-Glucose-1-phosphate. Isoln from a crude mixture of hexose phosphates, obtained by yeast fermentation: Robison, King, *Biochem. J.* **25,** 323 (1931). Prepn by the action of phosphoglucomutase on α-glucose-1-phosphate: Colowick, Sutherland, *J. Biol. Chem.* **144,** 423 (1942); from acetone glucose: Levene, Raymond, *ibid.* **92,** 757 (1931); by phosphorylation of 1,2,3,4-tetraacetylglucose followed by deacetylation: Fischer, Lardy, *ibid.* **164,** 513 (1946); *Biochem. Prepn.* **2,** 39 (1952). Prepn from starch: de Chatelperron *et al.,* Fr. pat. **1,379,068** (1964), *C.A.* **62,** 9394b (1965).

Barium salt, $C_6H_{11}O_9PO_5Ba$, nonhygroscopic, stable powder. [α]²⁴ +17.9°. Easily sol in water.
Dipotassium salt, $C_6H_{11}O_5PO_4K_2$, precipitate from methanol. [α]²⁴_D +21.2° (c = 1.3). Freely sol in water.

4470. α-Glucose-1-phosphate. *α-ᴅ-Glucopyranose 1-dihydrogenphosphate;* α-Glucose-1-phosphoric acid; α-ᴅ-

glucopyranose-1-phosphate; Cori ester. $C_6H_{13}O_9P$; mol wt 260.14. C 27.70%, H 5.04%, O 55.35%, P 11.91%. Found widely in both plants and animals. In plants it is the immediate precursor of starch, and in animals of glycogen, being also the first product in the breakdown and utilization of these substances. Isoln from muscle and synthesis using trisilver phosphate: Cori *et al., J. Biol. Chem.* **121,** 465 (1937); Krahl, Cori, *Biochem. Prepn.* **1,** 33 (1949). Prepn from α-acetobromglucose + silver diphenyl phosphate: Posternak, *J. Am. Chem. Soc.* **72,** 4824 (1950); by phosphorolysis of starch using phosphorylase and orthophosphate: McCready, Hassid, *Biochem. Prepn.* **4,** 63 (1955). Structure: Wolfrom, Pletcher, *J. Am. Chem. Soc.* **63,** 1050 (1941). Configuration: Wolfrom *et al., ibid.* **64,** 23 (1942); Harmon, *Diss. Abstr.* **24,** 4400 (1964); Beevers, Maconochie, *Acta Cryst.* **18,** 232 (1965).

Free acid, [α]²⁵_D +120°. pK₁ = 1.11; pK₂ = 6.13. Stronger acid than H_3PO_4. Extremely sol in water.
Barium salt trihydrate, nonhygroscopic, stable powder. [α]²⁵_D +75° (c = 1.26). Easily sol in water.
Calcium salt, *Actigam.*
Dipotassium salt dihydrate, crystals from ethanol. [α]²⁰_D +78° (c — 4); [α]²⁰_546 +90° (c = 4). Freely sol in water.
THERAP CAT: Calcium salt as roborant.

4471. Glucosulfone Sodium. *1,1′-[Sulfonylbis(4,1-phenyleneimino)]bis[1-deoxy-1-sulfo-ᴅ-glucitol] disodium salt; p,p′-sulfonyldianiline N,N′-diglucoside disodium disulfonate; p,p′*-sulfonyldianiline-*N,N′*-di-ᴅ-glucose sodium bisulfite compd; *p,p′*-diaminodiphenylsulfone-*N,N′*-di(dextrose sodium sulfonate); disodium *p,p′*-diaminodiphenylsulfone-*N,N′*-diglucose sulfonate; 501-P; Protomin; Promin; Promanide; Angeli's sulfone. $C_{24}H_{34}N_2Na_2O_{18}S_3$; mol wt 780.71. C 36.92%, H 4.39%, N 3.59%, Na 5.89%, O 36.89%, S 12.32%. Prepd by refluxing a mixture of 4,4′-diaminodiphenylsulfone, glucose, sodium bisulfite, and 80% ethanol: Jain *et al., C.A.* **40,** 4687⁷ (1946); Swiss pat. **234,108** (1944 to B. Siegfried), *C.A.* **43,** 4297a (1949).

White, amorphous powder. Soluble in water; slightly sol in alc. Insol in ether, benzene, methanol, ethyl acetate, pyridine. Aq solns may be sterilized by autoclaving. Marketed as an injectable aq soln of a mixture of 88.5% glucosulfone sodium + 11.5% glucose. LD₅₀ orally in mice: 3.93 g/kg.
THERAP CAT: Antibacterial (leprostatic).

4472. *N⁴*-*β*-ᴅ-Glucosylsulfanilamide. *4-(β-ᴅ-Glucopyranosylamino)benzenesulfonamide; N⁴-β-d-*glucosidosulfanilamide; sulfanilamide-*d*-glucoside; *N-p-*sulfamylphenyl-ᴅ-glucosylamine; Prontoglucal; Prontoglukal. $C_{12}H_{18}N_2O_7S$; mol wt 334.35. C 43.11%, H 5.43%, N 8.38%, O 33.50%, S 9.59%. Prepn: Kuhn, Birkofer, *Ber.* **71,** 621 (1938); Bognár, Nánási, *J. Chem. Soc.* **1953,** 1703; Bognár *et al., Magy. Kém. Foly.* **62,** 271 (1956), *C.A.* **54,** 4399g (1960). Structure: Braun *et al., J. Org. Chem.* **7,** 19 (1942).

β-D-glucose—N—⟨structure⟩

Fine needles from aq ethanol, mp 204°. $[\alpha]_D^{22}$ −117° (c = 0.9 in pyridine); $[\alpha]_D^{22}$ −128° (c = 0.9 in water).

2,3,4,6-Tetraacetate, $C_{20}H_{26}N_2O_{11}S$, crystals from ethanol, mp 204°. $[\alpha]_D^{22}$ −81° (pyridine).

$N,N',2,3,4,6$-Hexaacetate, $C_{24}H_{30}N_2O_{13}S$, crystals from ethanol, mp 115°.

THERAP CAT: Antibacterial.

4473. Glucovanillin. *4-(β-D-Glucopyranosyloxy)-3-methoxybenzaldehyde;* vanillin-D-glucoside; avenein; vanilloside. $C_{14}H_{18}O_8$; mol wt 314.29. C 53.50%, H 5.77%, O 40.72%. From green fruit of vanilla: Goris, *Compt. Rend.* **179**, 70 (1924). From coniferin by oxidation with CrO_3: Tremann, *Ber.* **18**, 1595 (1885). Structure and synthesis: Fischer, Raske, *Ber.* **42**, 1475 (1909); Thorpe, Williams, *J. Chem. Soc.* **1937**, 494.

Needles from methanol, mp 189-190°. Bitter taste. $[\alpha]_D^{20}$ −89.9° (water). Soluble in hot water, alcohol. Almost insol in ether.

Tetraacetate, $C_{22}H_{26}O_{12}$, crystals from dil alcohol, mp 142-143°. $[\alpha]_D$ −48.3° (chloroform).

USE: Pharmaceutic aid (flavor).

4474. D-Glucuronic Acid. $C_6H_{10}O_7$; mol wt 194.14. C 37.12%, H 5.19%, O 57.69%. Widely distributed in the plant and animal kingdoms. Usually occurs in "paired" form, *i.e.* as a glycosidic combination with phenols, alcohols, etc. Such glucuronides form in the liver to detoxify poisonous hydroxyl-containing substances. The glucuronides present in normal urine are those of phenol, cresol, and indoxyl. After the ingestion of poisons such as morphine, chloral hydrate, camphor, or turpentine, glucuronides formed with the poison or its hydroxylated derivatives appear in the urine. Review and bibliography: Stacey, *Advan. Carbohyd. Chem.* **2**, 161 (1946); Jones, Smith, *ibid.* **4**, 243 (1949). Structure: Pryde, Williams, *Nature* **128**, 187 (1931); Levene, Meyer, *J. Biol. Chem.* **92**, 257 (1931); Levene, Kreider, *ibid.* **120**, 597 (1937). Review of syntheses: Mehltretter, *Advan. Carbohyd. Chem.* **8**, 231 (1953). Prepn by irradiation of D-glucose in dil aq soln: Phillips *et al., J. Chem. Soc.* **1958**, 3522; by γ-irradiation of aq sucrose soln: Phillips, Moody, *ibid.* **1960**, 762. Electrophoretic sepn of D-glucuronic acid and its C-5 epimer, *L-iduronic acid*: I. Miyamoto, S. Nagase, *Anal. Biochem.* **115**, 308 (1981). Monographs: N. E. Artz, E. M. Osman, *Biochemistry of Glucuronic Acid* (Academic Press, New York, 1950); G. J. Dutton, Ed., *Glucuronic Acid, Free and Combined* (Academic Press, New York, 1966) 629 pp.

β-D-Glucuronic acid

β-Form, needles from alcohol or ethyl acetate. mp 165°. Shows mutarotation: $[\alpha]_D^{24}$ +11.7° → +36.3° (2 hrs, c = 6). Soluble in water, alcohol. Reduces Fehling's soln.

4475. β-Glucuronidase. Glusulase. Glucuronide-splitting enzyme found in liver, spleen, and certain tissues of the endocrine and reproductive systems. Rats seem to have a higher concn than other mammals. Also found in fish liver, snails, mollusks, and some insects. Isoln from rat livers, kidneys, and spleens: Fishman, Talalay, *Science* **105**, 131 (1947). *Review and bibliography:* Fishman, *Advan. Enzymol.* **16**, 361-409 (1955).

White powder. Soluble in water.

USE: In the determination of urinary steroids and of steroid conjugates in blood.

4476. D-Glucuronolactone. *D-Glucuronic acid γ-lactone;* D-glucofuranurono-6,3-lactone; glucurolactone; glucurone; Dicurone; Glucoxy; Guronsan. $C_6H_8O_6$; mol wt 176.13. C 40.92%, H 4.58%, O 54.50%. Found in many plant gums in polymeric combination with other carbohydrates. Important structural constituent of practically all fibrous and connective tissues in the animal organism, *cf.* D-glucuronic acid. Prepd synthetically from many polysaccharides or suitable glucosides where the hydroxyl at carbon 6 may be oxidized while the other sensitive groups are protected. Prepn: Stacey, *J. Chem. Soc.* **1939**, 1529; Hardegger, Spitz, *Helv. Chim. Acta* **33**, 337 (1950); Marsh, *Proc. Biochem. Soc.* [*Biochem. J.*], **50**, XI (1951); Mehltretter *et al., J. Am. Chem. Soc.* **73**, 2424 (1951); Phillips, Moody, *J. Chem. Soc.* **1960**, 762. Structure: J. Stanek *et al., The Monosaccharides* (Academic Press, New York, 1963) p 259. For isoln procedures see the ref under glucuronic acid.

Crystals from ethanol, mp 176-178°. (Commercial grades, mp 172°.) d_4^{30} 1.76. $[\alpha]_D^{25}$ +19.8° (c = 5.19). Soluble in water (26.9 g/100 ml of soln); slightly sol in methanol (2.8 g/100 ml). Very slightly sol in abs ethanol (0.7 g/100 ml), in glacial acetic acid (0.3 g/100 ml). The free acid is more sol than the lactone. At room temp an aq soln of glucuronolactone reaches an equilibrium of about 20% lactone and 80% acid within 2 months. At 100° an equilibrium of 60% lactone and 40% free acid is reached within 2 hrs. Initial pH of 10% aq soln 3.5, after 1 week the pH is about 2.5.

THERAP CAT: Detoxicant.

4477. Glutamic Acid. *L-Glutamic acid;* Glu; E; glutaminic acid; *(S)-2-aminopentanedioic acid;* α-aminoglutaric acid; 1-aminopropane-1,3-dicarboxylic acid; Glutacid; Glutaminol; Glutaton. $C_5H_9NO_4$; mol wt 147.13. C 40.82%, H 6.17%, N 9.52%, O 43.50%. Non-essential amino acid for human development; referred to as an excitatory amino acid (EAA) due to its role in neurotransmission. Isoln from wheat gluten: H. Ritthausen, *J. Prakt. Chem.* **99**, 454 (1866). Early chemistry and biochemistry: *Amino Acids and Proteins,* D. M. Greenberg, Ed. (Charles C. Thomas, Springfield, IL, 1951) 950 pp., *passim;* J. P. Greenstein, M. Winitz, *Chemistry of the Amino Acids* vols **1-3** (John Wiley and Sons, Inc., New York, 1961) pp. 1929-1954, *passim;* C. W. Huffman, W. G. Skelly, *Chem. Rev.* **63**, 625-644 (1963). Interconversion from L-proline: S. Yoshifuji *et al., Tetrahedron Letters* **21**, 2963 (1980). Determn in serum: C. D. Stalikas *et al., Eur. J. Clin. Chem.* **32**, 767 (1994); of naturally occurring levels in food: D. H. Daniels *et al., Food Addit. Contam.* **12**, 21 (1995). Identification as excitatory neurotransmitter: D. R. Curtis *et al., J. Physiol.* **150**, 656 (1960). Review of biosynthesis: A. Hamberger *et al., Adv. Biochem. Psychopharmacol.* **27**, 115-126 (1981); of metabolism and associated disorders: S. B. Prusiner, *Ann. Rev. Med.* **32**, 521-542 (1981). Review of receptor binding: P. A. Briley *et al., Mol. Cell. Biochem.* **39**, 347 (1981). Review as neurotransmitter: B. Engelsen, *Acta Neurol. Scand.* **74**, 337-355 (1986); E. Marmo, *Med. Res. Rev.* **8**, 441-458 (1988). Review of role in learning and memory: W. J. McEntee, T. H. Crook, *Psychopharmacology* **111**, 391-401 (1993); in Parkinson's disease: M. S. Starr, *Synapse* **19**, 264-

293 (1995). *Books:* R. Powell, *Monosodium Glutamate and Glutamic Acid* (Noyes Dev. Corp., Park Ridge, N.J., 1968) 256 pp.; *Glutamic Acid: Advances in Biochemistry & Physiology,* L. J. Filer, Jr. *et al.,* Eds. (Raven, New York, 1979).

Orthorhombic, bisphenoidal crystals from aq alc. d_4^{20} vac 1.538. Melts 160° with conversion to to L-pyrrolidonecarboxylic acid. Sublimes at 200°. $[\alpha]_D^{22.4}$ +31.4° (6N HCl). pK$_1'$ 2.19; pK$_2'$ 4.25; pK$_3'$ 9.67. Soly in water (g/l): 8.64 (25°); 21.86 (50°); 55.32 (75°); 140.00 (100°). Insol in methanol, ethanol, ether, acetone, cold glacial acetic acid and common neutral solvents.

Hydrochloride, $C_5H_9NO_4$·HCl, *Acidulin, Acidoride, Hypochylin, Antalka, Aciglumin, Pepsdol, Glutamidin, Acidogen, Aclor, Gastuloric, Glutan-HCl, Glutasin, Hydrionic, Muriamic.* Orthorhombic bisphenoidal plates, dec 214°. $[\alpha]_D^{22}$ +24.4° (c = 6).

Sodium salt, *see* Monosodium Glutamate.

Magnesium salt hydrobromide monohydrate, $C_{10}H_{17}Br$-MgN_2O_8·H_2O, *magnesium glutamate hydrobromide, magnesium bromoglutamate, Psicosoma, Psycho-Soma, Psychoverlan.*

DL-Form, Orthorhombic crystals from water, 225-227° (dec). d_4^{20} 1.4601. Soly in water (g/l): 20.54 (25°); 49.34 (50°); 118.6 (75°); 284.9 (100°). Sparingly sol in alcohol, ether, petr ether.

D-Form, Shiny leaflets from water. $[\alpha]_D^{20}$ −30.5° (c = 1.00 in 6N HCl).

THERAP CAT: Exptl antiepileptic; hydrochloride as gastric acidifier; magnesium salt hydrobromide as anxiolytic.

4478. L-Glutamic Acid 5-Ethyl Ester. L-Glutamic acid γ-ethyl ester; γ-ethyl L-glutamate; Glutestere. $C_7H_{13}NO_4$; mol wt 175.18. C 47.99%, H 7.48%, N 8.00%, O 36.53%. HOOCCH(NH$_2$)CH$_2$CH$_2$COOC$_2$H$_5$. Prepn from glutamic acid, ethanol, and hydrogen chloride: Hegedus, *Helv. Chim. Acta* **31**, 737 (1948); Pravda, *Coll. Czech. Chem. Commun.* **24**, 2083 (1959); **Czech.** pat. **88,344** (1959), *C.A.* **54**, 8660h (1960).

Hydrochloride, $C_7H_{13}NO_4$·HCl, crystals from ethanol + ether, mp 134-136°.

4479. Glutamine. *L-Glutamine;* Gln; Q; 2-aminoglutaramic acid; (*S*)-2,5-diamino-5-oxopentanoic acid; glutamic acid 5-amide; Cebrogen; Glumin; Levoglutamina; Stimulina. $C_5H_{10}N_2O_3$; mol wt 146.15. C 41.09%, H 6.90%, N 19.17%, O 32.84%. Non-essential amino acid for human development; most abundant free amino acid in plasma and tissue. Isoln from sugarbeet juice: E. Schulze, E. Bosshard, *Ber.* **16**, 312 (1883). Due to ease of conversion to glutamic acid, *q.v.*, it was not isolated from protein until 1932: M. Damodoran *et al., Biochem. J.* **26**, 1704 (1932). Early chemistry and biochemistry: R. M. Archibald, *Chem. Rev.* **37**, 161-208 (1945); *Amino Acids and Proteins,* D. M. Greenberg, Ed. (Charles C. Thomas, Springfield, IL, 1951) 950 pp., *passim;* J. P. Greenstein, M. Winitz, *Chemistry of the Amino Acids* **vol 1-3** (John Wiley and Sons, Inc., New York, 1961) pp. 1929-1954, *passim.* HPLC determn in biotech samples: B. Polanuer *et al., J. Chromatog.* **594**, 173 (1992). Review of nutritive needs: R. J. Smith, D. W. Wilmore, *J. Parent. Enter. Nutr.* **14**, (Suppl.) 94S-99S (1990); and metabolism: W. W. Souba, *J. Nutr. Biochem.* **4**, 2-9 (1993); L. M. Castell *et al., Amino Acids* **7**, 231-243 (1994). Review of metabolism and physiologic implications: R. J. Smith, *J. Parent. Enter. Nutr.* **14**, (Suppl.) 40S-44S (1990); B. Moskovitz *et al., Pharmacol. Res.* **30**, 61-71 (1994).

Fine opaque needles from water or dil ethanol, dec 185-186°. $[\alpha]_D^{20}$ +6.1° (c = 3.6). pK$_1$ 2.17; pK$_2$ 9.13. One gram dissolves in 20.8 ml water at 30°, in 38.5 ml at 18°, in 56.7 ml at 0°. Practically insol in methanol (3.5 mg/100 ml at 25°), ethanol (0.46 mg/100 ml at 23°), ether, benzene, acetone, ethyl acetate, chloroform.

DL-Form. Synthesis: F. E. King, D. A. A. Kidd, *J. Chem. Soc.* **1949**, 3315; G. B. Kline, S. H. Cox, *J. Org. Chem.* **26**, 1854 (1961). Prisms from dil acetone, mp 185-186° (King); mp 173-174.5° (Kline). One gram dissolves in 38.5 ml water at 18°.

4480. Glutaraldehyde. *Pentanedial;* glutaral; glutaric dialdehyde; 1,3-diformylpropane; Cidex; Glutarol; Novaruca; Verucasep. $C_5H_8O_2$; mol wt 100.12. C 59.98%, H 8.05%, O 31.96%. OHCCH$_2$CH$_2$CH$_2$CHO. Prepd by treating 2-ethoxy-3,4-dihydro-2H-pyran with aq HCl: A. C. Cope *et al., Org. Syn.* **coll. vol. IV**, 816 (1963). *See also* Beilstein **1**, 776 (1918) and supplements. Toxicity data: H. F. Smyth *et al., Am. Ind. Hyg. Assoc. J.* **23**, 95 (1962).

Oil, bp$_{760}$ 187-189°; bp$_{50}$ 106-108°; bp$_{10}$ 71-72°. n_D^{25} 1.4338. Sol in water, volatile in steam. Polymerizes in water to a glassy form which regenerates the dialdehyde on vacuum distillation. LD$_{50}$ of 25% soln orally in rats: 2.38 ml/kg; by skin penetration in rabbits: 2.56 ml/kg (Smyth).

Dioxime, $C_5H_{10}N_2O_2$, crystals from water or pyridine, mp 178°. Sublimes. Treatment with hot mineral acids gives pyridine.

USE: In sterilization of endoscopic instruments, thermometers, rubber or plastic equipment which cannot be heat sterilized; as a tanning agent for leather.

Caution: Causes severe eye irritation in rabbits.

THERAP CAT: Antiseptic.

4481. Glutaric Acid. *Pentanedioic acid;* 1,3-propanedicarboxylic acid. $C_5H_8O_4$; mol wt 132.12. C 45.46%, H 6.10%, O 48.44%. COOH(CH$_2$)$_3$COOH. Occurs in green sugar beets; is found in water extracts of crude wool. Manuf from cyclopentanone by oxidative ring fission with hot 50% nitric acid in the presence of vanadium pentoxide. Lab prepn by acid hydrolysis of trimethylene cyanide: Marvel, Tuley, *Org. Syn.* **5**, 69 (1925), or of methylenedimalonic ester: Otterbacher, *ibid.* **10**, 58 (1930). Several new methods: Paris *et al., Org. Syn.* **coll. vol. IV**, 496 (1963); English, Dayan, *ibid.* 499.

Large monoclinic prisms, mp 97.5-98°. d_4^{15} 1.429. bp$_{760}$ 302-304° (very slight decompn); bp$_{20}$ 200°; bp$_{10}$ 195-198°, pK$_1$ (25°): 4.34; pK$_2$: 5.22. n_D^{106} 1.41878. Absorption spectrum: *Compt. Rend.* **189**, 915 (1929). Soly in water (g/l): at 0°: 429; at 20°: 639; at 50°: 957; at 65°: 1118. Freely sol in abs alcohol, ether; sol in benzene, chloroform; slightly sol in petr ether.

Dimethyl ester, $C_7H_{12}O_4$, liquid, faint agreeable odor. d_4^{15} 1.0934. bp$_{752}$ 213.5-214°; bp$_{13}$ 93.5-94.5°. n_D^{20} 1.4246. Very sol in alcohol and ether.

4482. Glutaronitrile. *Pentanedinitrile;* glutaric acid dinitrile; trimethylene dicyanide; trimethylene cyanide. $C_5H_6N_2$; mol wt 94.12. C 63.81%, H 6.43%, N 29.76%. CN-(CH$_2$)$_3$CN. Prepd by the action of potassium cyanide on trimethylene bromide: Marvel, McColm, *Org. Syn.* **5**, 103 (1925).

Viscous liquid. Bitter-sweet taste. d^{23} 0.9888. mp −29°. bp$_{760}$ 286°; bp$_{100}$ 206°; bp$_{60}$ 190°; bp$_{40}$ 176°; bp$_{20}$ 157°; bp$_{10}$ 140°; bp$_5$ 124°; bp$_{1.0}$ 91.3°. n_D^{23} 1.4365. Soluble in water, alcohol, chloroform. Insol in ether, carbon disulfide.

4483. Glutathione. *N-(N-L-γ-Glutamyl-L-cysteinyl)glycine;* L-glutathione; glutathione-SH; GSH; Agifutol S; Copren; Deltathione; Glutamed; Glutasan; Glutathin; Glutathiol; Glutathion; Glutinal; Isethion; Neuthion; Tathiclon; Tathion; Tationil; Triptide. $C_{10}H_{17}N_3O_6S$; mol wt 307.33. C 39.08%, H 5.58%, N 13.67%, O 31.24%, S 10.43%. The major low mol wt thiol compound of the living plant or animal cell. Isoln from yeast: Hopkins, *J. Biol. Chem.* **84**, 269 (1929). Synthesis: du Vigneaud, Miller, *Biochem. Prepn.* **2**, 87 (1952); Goldschmidt *et al., Ber.* **97**, 2434 (1964); Y. Ozawa *et al., Bull. Chem. Soc. Japan* **53**, 2592 (1980). Review of early syntheses: Jeschkeit *et al., Pharmazie* **18**, 658 (1963). Review of metabolism: A. Meister, M. E. Anderson, *Ann. Rev. Biochem.* **52**, 711-760 (1983); of metabolic role in antineoplastic chemotherapy: B. A. Arrick, C. F. Nathan, *Cancer Res.* **44**, 4224-4232 (1984). *Monographs:* S. Colowick *et al., Glutathione* (Academic Press, New York, 1954); *Glutathione,* E. M. Crook, Ed., Biochem. Soc. Symposium No. 17, London, 1958 (Cambridge University Press, 1959); *Glutathione,* L. Flohe *et al.,* Eds. (Academic Press, New York, 1974); *Glutathione: Metabolism & Function,* I. M. Arias, W. B. Jackoby, Eds. (Raven, New York, 1976) .

Crystals from 50% ethanol, mp 195°. $[\alpha]_D^{25}$ −18.9° (c = 4.653). $[\alpha]_D^{27}$ −21° (c = 2.74). pK_1' 2.12; pK_2' 3.53; pK_3' 8.66; pK_4' 9.12. Freely sol in water, dil alcohol, liq ammonia, dimethylformamide.

Disulfide, $C_{20}H_{32}N_6O_{12}S_2$, L-γ-*glutamyl*-L-*cysteinyl*-*glycine disulfide, N,N'-[dithiobis[1-[(carboxymethyl)carbamoyl]ethylene]]diglutamine, GSSG, oxidized glutathione.* Crystals, mp 123°. $[\alpha]_D^{20}$ −108° (c = 2 in water).

4484. Gluten. Protein substance of wheat which is intermixed with the starchy endosperm of the grain. Causes the carbon dioxide produced during dough fermentation to be retained by the dough in a manner which provides the porous and spongy structure of bread. Prepn from wheat: Rist, *Sugar J.* **11**, no. 9, 26 (1949), *C.A.* **43**, 9505c (1949); Christensen, U.S. pat. **2,583,684** (1952 to Gateway Chemurgic). Amino acid composition: Pence *et al., Cereal Chem.* **27**, 335 (1950). *Reviews:* M. J. Blish "Wheat Gluten" in M. L. Anson, J. T. Edsall, *Advan. Protein Chem.* **vol. II** (Academic Press, New York, 1945) pp 337-359; Meredith, *Cereal Sci. Today* **9**, 33, 54 (1964).

Yellowish-gray powder. Practically insol in water. Partly sol in alcohol, dil acids; sol in alkalies.

USE: As adhesive and as substitute for flour.

4485. Glutethimide. *3-Ethyl-3-phenyl-2,6-piperidinedione; 2-ethyl-2-phenylglutarimide; α-ethyl-α-phenylglutarimide; 3-ethyl-3-phenyl-2,6-dioxopiperidine; 3-ethyl-3-phenyl-2,6-diketopiperidine; Elrodorm; Doriden-Sed; Doriden.* $C_{13}H_{15}NO_2$; mol wt 217.27. C 71.87%, H 6.96%, N 6.45%, O 14.73%. Synthesis: Tagmann *et al., Helv. Chim. Acta* **35**, 1541 (1952); Salmon-Legagneur, Neveu, *Compt. Rend.* **234**, 1060 (1952); *Bull. Soc. Chim. France* **1953**, 70; Hoffmann, Tagmann, U.S. pat. **2,673,205** (1954 to Ciba). Resolution: Kukalja *et al., Croat. Chem. Acta* **33**, 41 (1961), *C.A.* **55**, 27193g (1961). Abs config of antipodes: Finch *et al., Experientia* **31**, 1002 (1975). Comprehensive description: H. Y. Aboul-Enein, *Anal. Profiles Drug Subs.* **5**, 139-187 (1976).

dl-Form, crystals from ether or from ethyl acetate + petr ether, mp 84°. uv max (methanol): 251, 257, 263 nm. Freely sol in ethyl acetate, acetone, ether, chloroform; sol in ethanol, methanol. Practically insol in water.

d-Form, crystals, mp 102.5-103°. $[\alpha]_D^{20}$ +176° (methanol).

l-Form, crystals, mp 102-103°. $[\alpha]_D^{20}$ −181° (methanol).

Note: This is a controlled substance (depressant) listed in the U.S. Code of Federal Regulations, Title 21 Part 1308.12 (1995).

THERAP CAT: Sedative; hypnotic.

4486. Glyburide. *5-Chloro-N-[2-[4-[[[(cyclohexylamino)carbonyl]amino]sulfonyl]phenyl]ethyl]-2-methoxybenzamide; 1-[[p-[2-(5-chloro-o-anisamido)ethyl]phenyl]sulfonyl]-*

3-cyclohexylurea; N-[4-(β-(2-methoxy-5-chlorobenzamido)-ethyl)benzosulfonyl]-N'-cyclohexylurea; N¹-[4-[β-(2-methoxy-5-chlorobenzoylamino)ethyl]benzenesulfonyl]-N²-cyclohexylurea; glybenzcyclamide; glibenclamide; HB-419; U-26452; Adiab; Azuglucon; Bastiverit; Dia-basan; Diabeta; Daonil; Duraglucon; Euglucon; Gilemal; Gliben-Puren N; Glidiabet; Glimidstada; Glubate; Glucoremed; Gluco-Tablinen; Glycolande; Lederglib; Libanil; Lisaglucon; Malix; Maninil; Micronase; Praeciglucon. $C_{23}H_{28}ClN_3O_5S$; mol wt 494.01. C 55.92%, H 5.71%, Cl 7.18%, N 8.51%, O 16.19%, S 6.49%. Second generation sulfonylurea with hypoglycemic activity. Prepn: Aumuller *et al., Arzneimittel-Forsch.* **16**, 1640 (1966); Neth. pat. **6,603,398** (1966 to Boehringer, Mann.), *C.A.* **66**, 65289h (1967); Neth. pat. Appl. **6,610,580**; H. Weber *et al.,* U.S. pat. **3,454,635** (1967, 1969 both to Hoechst). Pharmacology: Loubatières, Mariani, *C. R. Acad. Sci. Ser. D* **265**, 643 (1967). Toxicity: Mizukami *et al., Arzneimittel-Forsch.* **19**, 1413 (1969). Series of articles on synthesis, pharmacology, toxicology and clinical studies: *ibid.* 1323-1494. Effect on release of insulin, glucagon and somatostatin: S. Efendic *et al., Proc. Nat. Acad. Sci. USA* **76**, 5901 (1979). Symposium on pharmacology, mechanism of action and clinical trials: *Ann. Clin. Res.* **15**, Suppl. 37, 1-35 (1983). Comprehensive description: P. G. Takla, *Anal. Profiles Drug Subs.* **10**, 337-355 (1981). Review of pharmacology and clinical efficacy: J. M. Feldman, *Pharmacotherapy* **5**, 43-62 (1985).

Crystals from methanol, mp 169-170° (Weber); also reported as mp 172-174° (Aumüller). pKa 5.3. Sparingly sol in water, sol in the usual organic solvents. LD_{50} in rats and mice (g/kg): > 20 orally; > 12.5 i.p.; > 20 s.c. (Mizukami).

THERAP CAT: Antidiabetic.

4487. Glybuthiazol(e). *4-Amino-N-[5-(1,1-dimethylethyl)-1,3,4-thiadiazol-2-yl]benzenesulfonamide; N¹-(5-tert-butyl-1,3,4-thiadiazol-2-yl)sulfanilamide; sulfatertiobutylthiadiazole; 2-(p-aminobenzenesulfamido)-5-tertiobutyl-1,3,4-thiadiazole; glybuthizol; RP-2259; Glipasol.* $C_{12}H_{16}N_4O_2S_2$; mol wt 312.42. C 46.13%, H 5.16%, N 17.93%, O 10.24%, S 20.53%. Prepn: Brit. pat. **828,963** (1960 to Rhône-Poulenc).

Needles from ethanol, mp 221-223°. Sol in ethanol (1.0 g/65 ml); in acetone (1.0 g/15 ml); in DMF (1.0 g/3 ml). Insol in water, ether, benzene. Forms a water-sol Na salt.

THERAP CAT: Antidiabetic.

4488. Glybuzole. *N-[5-(1,1-Dimethylethyl)-1,3,4-thiadiazol-2-yl]benzenesulfonamide; N-(5-tert-butyl-1,3,4-thiadiazol-2-yl)benzenesulfonamide; 2-benzenesulfonamido-5-tert-butyl-1,3,4-thiadiazole; desaglybuzole; TH-1395; RP-7891; AN-1324; Gludiase.* $C_{12}H_{15}N_3O_2S_2$; mol wt 297.40. C 48.46%, H 5.08%, N 14.13%, O 10.76%, S 21.56%. Prepn: Macrae, Drain, Brit. pat. **822,947** (1959 to Smith & Nephew), *C.A.* **54**, 4622h (1960); Fr. pat. **M3389** (1965 to Rhône-Poulenc), *C.A.* **64**, 3553a (1966). Pharmacodynamics: Bargeton *et al., Arch. Int. Pharmacodyn.* **153**, 379 (1965).

Needles, mp 163°. LD$_{50}$ orally in mice: 1315 mg/kg.
THERAP CAT: Antidiabetic.

4489. Glycarsamide. *[4-[(Hydroxyacetyl)amino]phenyl]arsonic acid;* N-glycoloylarsanilic acid; N-glycolylarsanilic acid; p-glycolylaminobenzenearsonic acid; p-glycolylaminophenylarsinic acid; Astryl. C$_8$H$_{10}$AsNO$_5$; mol wt 275.09. C 34.93%, H 3.66%, As 27.24%, N 5.09%, O 29.08%. Prepd by treating arsanilic acid with chloroacetyl chloride and NaOH: Streitwolf *et al.*, **Ger. pat. 513,210; Brit. pat. 344,-532** (1929 to I. G. Farben).

Crystals from dil acid or water, mp 202-203°. Freely sol in hot water, methanol. Sparingly sol in cold water, acetic acid. Practically insol in acetone, benzene.

Sodium salt, C$_8$H$_9$AsNNaO$_5$, *SNGA, Quarcyl, Allegan.* Toxicity and physiological disposition: McChesney *et al.*, *Toxicol. Appl. Pharmacol.* **4**, 14 (1962). Crystals. Noticeable acid odor. Freely sol in water.

Nitroacridine complex, *see* Nitroakridin 3582.
THERAP CAT (VET): Anthelmintic (Strongyloidosis).

4490. Glyceraldehyde. *2,3-Dihydroxypropanal;* glyceric aldehyde; α,β-dihydroxypropionaldehyde. C$_3$H$_6$O$_3$; mol wt 90.08. C 40.00%, H 6.71%, O 53.28%. DL-glyceraldehyde together with its isomer dihydroxyacetone is obtained from glycerol by mild oxidation with hydrogen peroxide and ferrous salts as catalysts: Witzemann, *J. Am. Chem. Soc.* **36**, 2227 (1914). *See also* Org. Syn. **coll. vol. II,** 305 (1943). The equilibrium mixture of glyceraldehyde and dihydroxyacetone is called *glycerose*. The two isomers are convertible into another through a common enediol resulting from the migration of hydrogen atoms (Lobry de Bruyn-van Eckenstein rearrangement). The equilibrium mixture plays an important role in the fermentation of sugars and in the biogenesis of constituents of the animal organism, *cf.* L.F. Fieser, M. Fieser, *Advanced Organic Chemistry* (Reinhold, New York, 1961) pp 78, 284, 405. Glyceraldehyde is the simplest aldose and its D- and L-forms are taken as the configurational reference standard for carbohydrates. The two forms have been obtained through the action of nitrous acid on the corresponding form of 3-amino-2-hydroxypropanal: Wohl, Momber, *Ber.* **47**, 3346 (1914); Pictet, Barbier, *Helv. Chim. Acta* **4**, 924 (1921); *cf.* Baer, Fischer, *Science* **88**, 108 (1938). Prepn of L-glyceraldehyde from L-sorbose and of D-glyceraldehyde from D-fructose: Perlin, *Methods in Carbohydrate Chemistry* **1**, 61 (1962).

L-form   D-form

DL-Form, tasteless crystals from alcohol + ether, d$_{18}^{18}$ 1.455. mp 145°. Distills at 140-150° (bath temp) and 0.8 mm pressure. 100 ml water dissolve 3 g at 18°. Insol in benzene, petr ether, pentane. Osazone, mp 132°.

L-Form, [α]$_D^{25}$ −8.7° (c = 2 in H$_2$O). Its dimethylacetal, bp$_{17-20}$ 126-129°. [α]$_D^{26}$ −20.9° (p = 9.22).

D-Form, [α]$_D^{25}$ +8.7° (c = 2 in H$_2$O). Its dimethylacetal, bp$_{17}$ 127-129°, bp$_{10}$ 123-126°. [α]$_D^{15}$ +21.2° (c = 18).

4491. Glyceraldehyde 3-Phosphate. *2-Hydroxy-3-(phosphonooxy)propanal;* 3-phosphoglyceraldehyde. C$_3$H$_7$O$_6$P; mol wt 170.06. C 21.19%, H 4.15%, O 56.45%, P 18.21%. An intermediate product of carbohydrate metabolism. Prepn of DL-form by enzymatic route: Meyerhof, Junowicz-Kocholaty, *J. Biol. Chem.* **149**, 71 (1943); by reductive cleavage of glyceraldehyde 1-benzyl ether 3-phosphoric acid: Fischer, Baer, *Ber.* **65**, 337, 1040 (1932); by hydrolysis of dimeric gylceraldehyde 1,3-diphosphoric acid: Baer, Fischer, *J. Biol. Chem.* **150**, 213 (1943); by hydrolysis of dimeric glyceraldehyde 1-bromide 3-phosphoric acid: *ibid.* 223. *See also* Baer in *Biochem. Prepn.* **1**, 50 (1949). Prepn of D-form by enzymatic route: Meyerhof, Junowicz-Kocholaty, *loc. cit.*; by synthetic route: Ballou, Fischer, *J. Am. Chem. Soc.* **77**, 3329 (1955).

DL-Form calcium salt dihydrate, C$_3$H$_5$CaO$_6$P.2H$_2$O, crystals, sol in water. Aq solns are not stable, particularly when alkaline. The dioxane addition compd of DL-glyceraldehyde 1-bromide 3-phosphoric acid described by Baer is relatively stable and may be stored in the refrigerator. An aq soln of the Na or K salt of glyceraldehyde-3-phosphoric acid (contg dioxane and bromide ion) is readily obtained by dissolving the dioxane compd in cold water and carefully neutralizing to pH 7.

D-Form calcium salt dihydrate, amorphous powder. [α]$_D^{25}$ +14.5° (c = 1.2 in 0.1N HCl calcd as the free acid). Prepn of calcium-free aq soln: Ballou, Fischer, *loc. cit.*

4492. Glyceric Acid. *2,3-Dihydroxypropanoic acid;* α,β-dihydroxypropionic acid. C$_3$H$_6$O$_4$; mol wt 106.08. C 33.97%, H 5.70%, O 60.33%. CH$_2$OHCH(OH)COOH. Prepn from α,β-dibromopropionic acid by treatment with silver oxide: Karrer, Klarer, *Helv. Chim. Acta* **7**, 931 (1924); from isoserine and nitrous acid: Fischer, Jacobs, *Ber.* **40**, 1069 (1907); from glycerol and nitrous acid: Mulder, *Ber.* **9**, 1902 (1876); Beilstein, *Ann.* **120**, 229 (1861). Dextrorotatory glyceric acid is obtained by the action of *Penicillia* or *Aspergilli* on the DL-form: McKenzie, Harden, *J. Chem. Soc.* **83**, 431 (1903). Levorotatory glyceric acid has been obtained by the oxidation of D(+)-glyceric aldehyde: Wohl, Schellenberg, *Ber.* **55**, 1408 (1922).

DL-Form, syrup, dec on distn. pK (25°): 3.55. Miscible with water, alcohol, acetone. Nearly insol in ether.

L(+)-Form, syrup. Its esters and salts are levorotatory and its salts are much more sol in water than those of the DL-form.

L-Form calcium salt, Ca(C$_3$H$_5$O$_4$)$_2$.2H$_2$O, monoclinic sphenoidal crystals, mp 137°. One gram dissolves in 10 ml water. [α]$_D^{20}$ −14.6° (c = 5).

D(−)-Form, syrup. Its salts are dextrorotatory.

D-Form calcium salt, Ca(C$_3$H$_5$O$_4$)$_2$.2H$_2$O, prisms, mp 138°. [α]$_D^{20}$ +14.5° (c = 5).

4493. Glycerol. *1,2,3-Propanetriol;* glycerin; glycerine; trihydroxypropane; incorporation factor; IFP; Bulbold; Cristal; Glyceol; Ophthalgan. C$_3$H$_8$O$_3$; mol wt 92.09. C 39.13%, H 8.76%, O 52.12%. CH$_2$OHCHOHCH$_2$OH. Obtained from oils and fats as byproduct in the manuf of soaps and fatty acids. During World War I supplementary quantities were produced by the "Protol" fermentation process from sugar, a process based upon the fixation of acetaldehyde by sodium sulfite. Just prior to World War II the synthesis of glycerol from propylene was announced. Production from sugars by fermentation: Onishi, U.S. pat. **3,012,945** (1961 to Noda). Identity with incorporation factor: Kuehl *et al.*, *J. Am. Chem. Soc.* **82**, 2079 (1960). In nucleic acid the incorporation factor may exist as a bound form of glycerol. Acute toxicity: W. Bartsch *et al.*, *Arzneimittel-Forsch.* **26**, 1581 (1976). Reviews and bibliographies: J. W. Lawrie, *Glycerol and the Glycols* (New York, 1928); G. Leffingwell, M. Lesser, *Glycerin* (Brooklyn, 1945); C. S. Miner, N. N. Dalton, *Glycerol* (New York, 1953); C. Lüttgen, *Glyzerin und glyzerinähnliche Stoffe* (Heidelberg, 2nd ed., 1955); J. C. Kern in Kirk-Othmer *Encyclopedia of Chemical Technology* **vol. 11** (Wiley-Interscience, New York, 3rd ed., 1980) pp 921-932.

Syrupy liquid. Sweet warm taste. About 0.6 times as sweet as cane sugar. Absorbs moisture from air; also absorbs H_2S, HCN, SO_2. *Caution: Contact with strong oxidizing agents such as chromium trioxide, potassium chlorate, or potassium permanganate may produce an explosion.* Neutral to litmus. Solidifies after prolonged cooling at 0° forming shiny orthorhombic crystals, mp 17.8°. bp_{760} 290.0° (dec); bp_{400} 263.0°; bp_{200} 240.0°; bp_{100} 220.1°; bp_{60} 208.0°; bp_{20} 182.2°; bp_{10} 167.2°; bp_5 153.8°; $bp_{1.0}$ 125.5°. n_D^{15} 1.4758; n_D^{20} 1.4746; n_D^{25} 1.4730. d_{15}^{15} 1.26557; d_{20}^{20} 1.26362; d_{25}^{25} 1.26201. Flash point, open cup: 350°F (176°C). Specific gravities of 95% aq soln w/w (U.S.P. grade): d_{15}^{15} 1.25270; d_{20}^{20} 1.25075; d_{25}^{25} 1.24910; 90% aq soln w/w: d_{15}^{15} 1.23950; d_{20}^{20} 1.23755; d_{25}^{25} 1.23585; 80% d_{15}^{15} 1.213; 70% d_{15}^{15} 1.185; 60% d_{15}^{15} 1.157; 50% d_{15}^{15} 1.129; 20% d_{15}^{15} 1.049; 5% d_{15}^{15} 1.0122. Viscosity (cps. at 20°): 5% soln 1.143; 10% 1.311; 25% 2.095; 50% 6.050; 60% 10.96; 70% 22.94; 83% 111. Freezing points of aq solns w/w: 10% −1.6°; 30% −9.5°; 50% −23.0°; 66.7% −46.5°; 80% −20.3°; 90% −1.6°. Miscible with water, alcohol. One part dissolves in 11 parts ethyl acetate, in about 500 parts ethyl ether. Insol in benzene, chloroform, carbon tetrachloride, carbon disulfide, petr ether, oils. LD_{50} in rats (ml/kg): > 20 orally; 4.4 i.v. (Bartsch).

USE: As solvent, humectant, plasticizer, emollient, sweetener, in the manuf of nitroglycerol (dynamite), cosmetics, liq soaps, liqueurs, confectioneries, blacking, printing and copying inks, lubricants, elastic glues, lead oxide cements; to keep fabrics pliable; to preserve printing on cotton; for printing rollers, hectographs; to keep frost from windshields; as antifreeze in automobiles, gas meters and hydraulic jacks, in shock absorber fluids. In fermentation nutrients in the production of antibiotics. Pharmaceutic aid (humectant; solvent; vehicle). Leffingwell and Lesser (*op. cit.*) give 1583 different uses.

THERAP CAT: Diagnostic aid (ophthalmic).

THERAP CAT (VET): Emollient, demulcent. As a source of glucose in bovine ketosis.

4494. Glycerol Formal. Methylidinoglycerol; Glicerinformal; Sericosol N. $C_4H_8O_3$; mol wt 104.11. C 46.15%, H 7.75%, O 46.11%. Mixture of isomeric α,α'- and α,β-forms prepd from glycerin and formaldehyde: M. Schultz, B. Tollens, *Ann.* **289**, 20 (1895). Prepn and separation of isomers: H. Hibbert, M. Carter, *J. Am. Chem. Soc.* **50**, 3120 (1928); J. D. van Roon, *Rec. Trav. Chim.* **48**, 173 (1929). Prepn and pharmacology: P. Gimeno, **Span.** pat. **475,962** (1979 to Calipe), *C.A.* **93**, 26445u (1980). Toxicological evaluation for use as pharmaceutical solvent: D. M. Sanderson, *J. Pharm. Pharmacol.* **11**, 150, 446 (1959). Teratogenicity studies: V. Aliverti *et al.*, *Arch. Sci. Biol.* **61**, 89 (1977); E. Giavini, M. Prati, *Acta Anat.* **106**, 203 (1980).

α,α'-form   α,β-form

Liquid, bp_{760} 191-195°, bp_{20} 95-97°. d_4^{20} 1.215; n_D^{20} 1.451. Sol in water, alc, chloroform. pH of 10% soln: 4-6.5. LD_{50} in rats (ml/kg): 8.6 orally, 3.5 i.v. (Gimeno).

α,α'-Form, *1,3-dioxan-5-ol, α,α'-methylene glycerin, α,α'-formaldehyde glycerol, 5-m-dioxanol.* bp 193.8°. d_4^{25} 1.2200. n_D^{25} 1.4527.

α,β-Form, *1,3-dioxolane-4-methanol, α,β-methylene glycerin, α,β-formaldehyde glycerol, 4-(hydroxymethyl)-1,3-dioxolane.* bp 192.5°. d_4^{25} 1.2008. n_D^{25} 1.4469.

USE: Pharmaceutic aid (solvent).

4495. Glycerophosphoric Acid. Phosphoric acid glycerol esters. $C_3H_9O_6P$; mol wt 172.07. C 20.94%, H 5.27%, O 55.79%, P 18.00%. Three isomers exist: *β-glycerophosphoric acid*, $(HOCH_2)_2CHOPO(OH)_2$ and the D(+)- and L(−)-forms of *α-glycerophosphoric acid*, $HOCH_2CH(OH)-CH_2OPO(OH)_2$. The L-α-acid is the naturally occurring

form; the β-acid, present in hydrolyzates of lecithins from natural sources, arises from migration of the phosphoryl group from the α-carbon atom. See review: Dawson, *Ann. Rept. Progr. Chem. (Chem. Soc. London)* **55**, 365 (1958). Prepn by phosphorylation of glycerol results in a mixture of the α- and β-acids: Cherbuliez, Weniger, *Helv. Chim. Acta* **29**, 2006 (1946). Prepn and configuration of L-α-acid: Baer, Fischer, *J. Biol. Chem.* **128**, 491 (1939). Prepn of D-α-acid: *eidem, ibid.* **135**, 321 (1940). Separation of α-acid from β- and polyglycerophosphoric acids: Carrara, **Ital.** pat. **460,-219** (1950), *C.A.* **46**, 5077a (1952).

Absolute acid (commercial mixture of α- and β-acids); clear syrupy liq, mp −25°. d_4^{14} 1.59. Tends to dec during concn; hence, usually marketed as a 25-50% soln. Soluble in water, alcohol.

α-Acid L-form, syrup. Readily sol in water, methanol, ethanol. Practically insol in ether. $[\alpha]_D$ −1.45° (barium salt, c = 10.3 in 2N HCl).

Note: Phosphatidic acids are fatty acid diesters of glycerophosphoric acid.

USE: Absolute acid used to manuf certain glycerophosphates or to impart taste to solns of glycerophosphates which are generally used medicinally. *See also:* Calcium Glycerophosphate.

4496. Glyceryl p-Aminobenzoate. *1,2,3-Propanetriol 1-(4-aminobenzoate); p-aminobenzoic acid monoglyceryl ester;* monoglycerol p-aminobenzoate; Escalol 106. $C_{10}H_{13}NO_4$; mol wt 211.22. C 56.87%, H 6.20%, N 6.63%, O 30.30%. Prepd by controlled esterification of p-aminobenzoic acid with glycerol.

Semisolid, waxy mass or syrup. Faint aromatic odor. Liquefies and congeals very slowly. Soluble in methanol, ethanol, isopropanol, glycerol, propylene glycol. Insol in water, oils, fats.

USE: In cosmetic sunscreen prepns (up to 1%).

4497. Glyceryl Iodide. *3-Iodo-1,2-propanediol;* γ-iodopropyleneglycol; 3-iodo-1,2-dihydroxypropane. $C_3H_7IO_2$; mol wt 201.99. C 17.84%, H 3.49%, I 62.83%, O 15.84%. $ICH_2CH(OH)CH_2OH$. Prepd from 2,3-epoxy-1-propanol: Kratzl *et al.*, *Monatsh.* **93**, 49 (1962).

Crystals, mp 48-49°. Freely sol in water, alcohol.

4498. Glyceryl Monostearate. *Octadecanoic acid monoester with 1,2,3-propanetriol;* Monostearin. The commercial product is a mixture of variable proportions of glyceryl monostearate and glyceryl monopalmitate.

White, wax-like solid or wax-like beads, or flakes, mp 56-58°. Saponification value 164-170. Iodine value not more than 6. Soluble in hot organic solvents such as alcohol, benzene, ether, acetone, mineral or fixed oils. Insol in water, but may be dispersed in hot water with the aid of a small amount of soap or other suitable surface active agent.

USE: In pharmaceutical dispensing, see Green, *J. Am. Pharm. Assoc., Pract. Pharm. Ed.* **7**, 299 (1946).

4499. Glycidol. *Oxiranemethanol; 2,3-epoxy-1-propanol;* 3-hydroxypropylene oxide. $C_3H_6O_2$; mol wt 74.08. C 48.64%, H 8.16%, O 43.20%. Prepd by the action of perbenzoic acid on allyl alc: Prileshajew, *Ber.* **42**, 4813 (1909); from glycerol-1-monochlorohydrin by the action of KOH in alc, or by the action of sodium in ether: Rider, Hill, *J. Am. Chem. Soc.* **52**, 1521 (1930). L-form prepd from L-1-(p-toluenesulfonyl)glycerol: Sowden, Fischer, *ibid.* **64**, 1291 (1942). Acute toxicity data: E. D. Thompson, R. A. Hiles, *Food Cosmet. Toxicol.* **19**, 347 (1981); E. D. Thompson, D. P. Gibson, *Food Chem. Toxicol.* **22**, 665 (1984).

DL-Form, slightly viscous liq. d_4^{25} 1.1143. bp_{760} 167° (decompn); $bp_{2.5}$ 66°; $bp_{0.9}$ 25°. Miscible with water. LD_{50} in female rats (g/kg): 0.42 orally, 0.20 i.p. (Thompson, Hiles). LD_{50} in male, female rats (g/kg): 0.35, 0.21 i.p.; 0.76, 0.64 orally (Thompson, Gibson).

L-Form, $[\alpha]_D^{20}$ +15° (neat).

Caution: Potential symptoms of overexposure are irritation of eyes, nose, throat and skin; narcosis. *See NIOSH Pocket Guide to Chemical Hazards* (DHHS/NIOSH 90-117, 1990) p 118. *See also Patty's Industrial Hygiene and Toxicology* **vol. 2A**, G. D. Clayton, F. E. Clayton, Eds. (John Wiley & Sons, New York, 4th ed., 1994) 422-425. This substance may reasonably be anticipated to be a carcinogen: *Seventh Annual Report on Carcinogens* (PB95-109781, 1994) p 222.

USE: Stabilizer in manuf of vinyl polymers; intermediate in synthesis of glycerol, glycidyl ethers, and amines; additive for oil and synthetic hydraulic fluids; epoxy resin diluent.

4500. Glycine. Gly; G; aminoacetic acid; aminoethanoic acid; glycocoll; Gyn-Hydralin; Glycosthène. $C_2H_5NO_2$; mol wt 75.07. C 32.00%, H 6.71%, N 18.66%, O 42.63%. Non-essential amino acid for human development. Only amino acid with no asymmetric carbon. Major inhibitory neurotransmitter. Isoln from gelatin: H. Braconnot, *Ann. Chim. Phys.* **13**, 113 (1820). Early chemistry and biochemistry: *Amino Acids and Proteins*, D. M. Greenberg, Ed. (Charles C. Thomas, Springfield, IL, 1951) 950 pp., *passim*; J. P. Greenstein, M. Winitz, *Chemistry of the Amino Acids* **vols 1-3**, (John Wiley and Sons, Inc., New York, 1961) pp. 1955-1970, *passim*. Exists in three polymorphic forms, α-, β-, and γ-: Y. Iitaka, *Nature* **183**, 390 (1959). Review of metabolism and radioprotection: S. Capalna, *Rev. Roum. Physiol.* **9**, 17-34 (1972); of metabolism in humans: A. A. Jackson, M. H. Golden, *Clin. Sci.* **58**, 517-522 (1980); in anerobes: J. R. Andreesen, *Anton. Leeuw.* **66**, 223-227 (1994). Review of role in neurotransmission: S. M. Paul in *Psychopharmacology: The Fourth Generation of Progress*, F. E. Bloom, D. J. Kupfer, Eds. (Raven Press, New York, 1995) pp 87-94. *Book: Glycine Neurotransmission*, O. P. Ottersen, J. Storm-Mathisen, Eds. (John Wiley & Sons, Chichester, U.K., 1990) pp 489.

Sweet, monoclinic prisms from alc, starts to dec at 233°, completely sintered at 290°. d 1.1607. pK_1' 2.34; pK_2' 9.60. pH of 0.2 molar soln in H_2O = 4.0. Adsorption on various chromatographic agents: Grettie, Williams, *J. Am. Chem. Soc.* **50**, 671 (1928). Soly in 100 ml water at 25°: 25.0 g; at 50°: 39.1 g; at 75°: 54.4 g; at 100°: 67.2 g. 100 g of abs alc dissolve about 0.06 g. Sol in 164 parts pyridine. Almost insol in ether.

Hydrochloride, $C_2H_5NO_2$.HCl, hygroscopic prisms from HCl, mp 182°.

4501. Glycine Sulfate. Triglycine sulfate. $C_6H_{17}N_3$-$O_{10}S$; mol wt 323.28. C 22.29%, H 5.30%, N 13.00%, O 49.49%, S 9.92%. $(NH_2CH_2COOH)_3.H_2SO_4$. Prepn: Horsford, *Ann.* **60**, 1 (1846); *see also* Matthias, Miller, Remeika, *Phys. Rev.* **104**, 849 (1956). Crystal growing during prepn: Konstantinova *et al.*, *Kristallografiya* **4**, 69-73 and 125-129 (1959).

Orthorhombic crystals. Very freely sol in water. Has ferroelectric properties: Curie point 47°. Spontaneous polarization at room temp: 2.2×10^{-6} coul/cm². Coercive field 220 v/cm.

USE: In electronics research.

4502. Glycinin. Chief protein constituent of soybeans: Osborne, Campbell, *J. Am. Chem. Soc.* **20**, 419 (1898); Smith, Circle, *Ind. Eng. Chem.* **31**, 1282 (1939). *Review:* H. Neurath, K. Bailey, Eds., *The Proteins* **vol. I**, Part A (Academic Press, New York, 1953) pp 208-209, 223; *eidem, ibid.* **vol. II**, part A, (1954) pp 503, 506.

Soluble in water in the pH range 1-4, as well as above 7.

4503. Glycobiarsol. *[[4-[(Hydroxyacetyl)amino]phenyl]arsonato(1−)]oxobismuth; oxo(hydrogen N-glycoloylarsanilato)bismuth*; bismuth glycoloylarsanilate; bismuth *p*-

glycolylaminophenylarsonate; bismuth glycolyl arsanilate; *N*-glycoloylarsanilic acid bismuth deriv; Dysentulin; Milibis; Viasept; Wintodon. $C_8H_9AsBiNO_6$; mol wt 499.06. C 19.25%, H 1.82%, As 15.01%, Bi 41.87%, N 2.81%, O 19.24%. Prepd from bismuth nitrate and sodium *N*-glycoloylarsanilate: *Hagers Handb. Pharm. Praxis* **Band I** (Suppl. 2), 759 (Berlin, 1958).

Yellowish to pink powder. Dec on heating. Very slightly sol in water, alcohol. Practically insol in ether, chloroform, benzene. pH of satd aq soln 2.8-3.5.

THERAP CAT: Antiamebic.

THERAP CAT (VET): For trichomoniasis.

4504. Glycocholic Acid. *N-[(3α,5β,7α,12α)-3,7,12-trihydroxy-24-oxocholan-24-yl]glycine; N-cholylglycine.* C_{26}-$H_{43}NO_6$; mol wt 465.63. C 67.07%, H 9.31%, N 3.01%, O 20.62%. The product of conjugation of cholic acid with glycine; chief ingredient of the bile of herbivorous animals. In the weakly alkaline bile fluid glycocholic acid exists as the sodium salt. Prepn from bile: Hammarsten in *Abderhalden's Handbuch der Biol. Arbeitsmethoden*, Abt. I, Teil 6, p 211 (1925). Prepn from cholic acid: Cortese, *J. Am. Chem. Soc.* **59**, 2532 (1937). Synthesis: Cortese, Bauman, *ibid.* **57**, 1393 (1935); Bergstrom, Norman, *Acta Chem. Scand.* **7**, 1126 (1953). Separation: Antonides, **Brit.** pat. **928,635** (1963 to Armour). Metabolism: Norman, *Scand. J. Gastroenterol.* **5**, 231 (1970).

Sesquihydrate, crystals from 5% alc, mp about 130°. $[\alpha]_D^{23}$ +30.8° (c = 7.5 in 95% ethanol). Anhyd form, mp 165-168°. pK 4.4. Soly in water at 15°: 0.33 g/l; in boiling water: 8.3 g/l. Is hydrolyzed to cholic acid and glycine by acids and alkalies. Forms addition compds with nitrobenzene, aniline, benzyl alcohol, benzaldehyde, triolein.

Sodium salt, $C_{26}H_{42}NNaO_6$, crystals from 95% alcohol + ether, mp 230-240°. $[\alpha]_D^{24}$ +32° (water). Soly at 15° in water > 274 g/l; in alcohol > 340 g/l.

4505. Glycocyamine. *N-(Aminoiminomethyl)glycine; N-amidinoglycine*; guanidineacetic acid; guanidoacetic acid. $C_3H_7N_3O_2$; mol wt 117.11. C 30.77%, H 6.02%, N 35.88%, O 27.32%. Prepd from *S*-ethylthiourea hydrobromide, sodium hydroxide, and glycine: Brand, Brand, *Org. Syn.* **coll. vol. III**, 440 (1955). Crystal and molecular structure: Guha, *Acta Crystallogr.* **B29**, 2163 (1973); J. Berthou *et al.*, *ibid.* **B32**, 1529 (1976).

Crystals, dec 280-284°. Appreciably sol in water.

THERAP CAT: In combination with betaine as cardiotonic.

4506. Glycogen. Animal starch; liver starch. $(C_6H_{10}$-$O_5)_n$; mol wt from about 2.7×10^5 to 3.5×10^6. Reserve carbohydrate of the animal organism. High molecular wt polymer having branched-chain structure composed of D-glucopyranose residues. Distributed through the cell protoplasm. Found esp in the liver and in rested muscle. Occurs also in insects and lower plants including fungi and yeasts. Isoln by alkaline destruction of the other cell constituents: Claude Bernard, *Lecons sur le diabete* (Paris, 1877) p 553; by destruction with trichloroacetic acid: Bell, Young, *Biochem. J.* **28**, 882 (1934); by centrifugation: Meyer, Jeanloz, *Advan. Enzymol.* **3**, 112 (1943); by hydraulic pressure: Stockhausen, Silbereisen, *Biochem. Z.* **287**, 276 (1936). For biological synthesis and lysis from the Cori ester (glucose-1-phosphate) *see* the review and bibliography by Meyer, *Advan. Enzymol.* **3**, 109 (1943); *see also* Nord, *Chem. Rev.* **26**, 423 (1940); Kalckar, *ibid.* **28**, 71 (1941). Isoln from the causal agent of cotton root rot, *Phymatotrichum omnivorum* (Shear) Duggar: Ergle, *J. Am. Chem. Soc.* **69**, 2061 (1947). Studies on linkages: Bahl, Smith, *J. Org. Chem.* **31**, 2915 (1966).

White powder. $[\alpha]_D^{25}$ +196° to +197°. Sol in water with opalescence. Insol in alc. Does not reduce Fehling's soln. With iodine, brown to violet colors are produced.

4507. Glycol Dilaurate. *Dodecanoic acid 1,2-ethanediyl ester;* ethylene dilaurate. $C_{26}H_{50}O_4$; mol wt 426.68. C 73.19%, H 11.81%, O 15.00%. $C_{11}H_{23}COOCH_2CH_2OOC$-$C_{11}H_{23}$.
Colorless, amorphous mass, mp 50-52°. bp_{20} 188°. Insol in alcohol, ether.
USE: In lacquers and varnishes as a plasticizer.

4508. Glycolic Acid. *Hydroxyacetic acid;* hydroxyethanoic acid. $C_2H_4O_3$; mol wt 76.05. C 31.59%, H 5.30%, O 63.11%. $HOCH_2COOH$. Constituent of sugar cane juice. Made by the action of NaOH on monochloroacetic acid; also by electrolytic reduction of oxalic acid. *Review:* Sales brochure on hydroxyacetic acid from E. I. du Pont.
Odorless, somewhat hygroscopic crystals, mp 80°. K at 25°: 1.48×10^{-4}. Soluble in water, methanol, alcohol, acetone, acetic acid, ether. pH of aq solns: 2.5 (0.5%); 2.33 (1.0%); 2.16 (2.0%); 1.91 (5.0%); 1.73 (10.0%). LD_{50} orally in rats: 1.95 g/kg, H. F. Smyth *et al., J. Ind. Hyg. Toxicol.* **23**, 259 (1941).
Caution: Mild irritant to skin, mucous membranes.
USE: In the processing of textiles, leather, and metals; in pH control, and wherever a cheap organic acid is needed, e.g. in the manuf of adhesives, in copper brightening, decontamination cleaning, dyeing, electroplating, in pickling, cleaning and chemical milling of metals.

4509. Glycol Salicylate. *2-Hydroxybenzoic acid 2-hydroxyethyl ester;* monoglycol salicylate; ethylene glycol monosalicylate; 2-hydroxyethyl salicylate; GL-7; Glysal; Norgesic; Phlogont (salve); Spirosal. $C_9H_{10}O_4$; mol wt 182.18. C 59.34%, H 5.53%, O 35.13%. $C_6H_4(OH)$-$COOCH_2CH_2OH$.
Almost colorless, odorless liq. bp_{12} 169-172°. Soluble in about 110 parts water, 8 parts olive oil; very sol in alcohol, benzene, chloroform, ether.
THERAP CAT: Counterirritant, anti-inflammatory (topical).

4510. Glyconiazide. *D-Glucuronic acid γ-lactone 1-[(4-pyridinylcarboxyl)hydrazone];* D-glucuronolactone isonicotinoylhydrazone; isonicotinoylhydrazone of D-glucuronic acid lactone; isonicotinic acid hydrazide hydrazone with glucuronic acid lactone; Galatone; Gatalone; Glucazide; Gluconiazide; Gluronazide; Guidazide; Hydronsan; INH-G; Mycobactyl. $C_{12}H_{13}N_3O_6$; mol wt 295.25. C 48.82%, H 4.44%, N 14.23%, O 32.51%. Prepd by heating isonicotinic acid hydrazide with D-glucuronolactone in methanol: Sah, *J. Am. Chem. Soc.* **75**, 2512 (1953); Sah, U.S. pat. **2,940,899** (1960 to U. of Calif.).

Plates and rods from methanol, needles from abs ethanol. Dec 150-160°. Freely sol in water. Practically insol in cold alc; 1.2 g dissolve in 100 ml methanol at 66°.
THERAP CAT: Antibacterial (tuberculostatic).

4511. Glycopyrrolate. *3-[(Cyclopentylhydroxyphenylacetyl)oxy]-1,1-dimethylpyrrolidinium bromide; 3-hydroxy-1,1-dimethylpyrrolidinium bromide α-cyclopentylmandelate;* α-cyclopentylmandelic acid ester with 3-hydroxy-1,1-dimethylpyrrolidinium bromide; 1-methyl-3-pyrrolidyl α-cyclopentylmandelate methobromide; 1-methyl-3-pyrrolidyl α-phenyl-α-cyclopentylglycolate methobromide; 3-(2-phenyl-2-cyclopentylglycoloyloxy)-1,1-dimethylpyrrolidinium bromide; glycopyrronium bromide; AHR-504; Nodapton; Robanul; Robinul; Tarodyl; Tarodyn. $C_{19}H_{28}BrNO_3$; mol wt 398.34. C 57.29%, H 7.08%, Br 20.06%, N 3.52%, O 12.05%. Prepn: Franko, Lunsford, *J. Med. Pharm. Chem.* **2**, 523 (1960); Lunsford, U.S. pat. **2,956,062** (1960 to A. H. Robins). Pharmacodynamics: E. Kaltiala *et al., J. Pharm. Pharmacol.* **26**, 352 (1974). Toxicology: B. V. Franko *et al., Toxicol. Appl. Pharmacol.* **17**, 361 (1970). Clinical comparison with atropine, *q.v.,* in anaesthetic practice: F. Kongsrud, S. Sponheim, *Acta Anaesth. Scand.* **26**, 620 (1982); A. I. Webb, R. M. McMurphy, *Am. J. Vet. Res.* **48**, 1733 (1987); B. V. G. Malling *et al., Brit. J. Anaesth.* **60**, 426 (1988). Brief review of pharmacology and clinical use: R. K. Mirakhur, J. W. Dundee, *Anaesthesia* **38**, 1195-1204 (1983).

White crystals from butanone, mp 193.2-194.5°. Sol in water. LD_{50} (72 hr.) in female mice, female rats (mg/kg): 107, 196 i.p.; in male rats (mg/kg): 1150 orally (Franko).
THERAP CAT: Anticholinergic.
THERAP CAT (VET): Anticholinergic.

4512. Glycosine. *1-Methyl-2-(phenylmethyl)-4(1H)-quinazolinone;* 2-benzyl-1-methylquinazol-4-one; arborine. $C_{16}H_{14}N_2O$; mol wt 250.30. C 76.78%, H 5.64%, N 11.19%, O 6.39%. Found in the toothbrush plant, *Glycosmis pentaphylla* (Retz.) Corr., and *G. arborea* Corr., *Rutaceae*. Isoln from dried, powdered leaves: Chatterjee, Majumdar, *J. Am. Chem. Soc.* **76**, 2459 (1954). Identity of arborine and glycosine, structure: Chakravarti *et al., Tetrahedron* **16**, 224 (1961). Synthesis: Pakrashi *et al., Indian J. Chem.* **6**, 472 (1968); Ziegler *et al., Monatsh.* **100**, 948 (1969); T. Kametani *et al., Heterocycles* **9**, 1585 (1978).

Rhombohedral prisms from chloroform + ethyl acetate, mp 155-156°. uv max (ethanol): 231, 268, 277, 306 nm. Freely sol in chloroform, ethyl acetate, benzene, ethanol. Sparingly sol in ether.
Hydrochloride, $C_{16}H_{14}N_2O\cdot HCl$, leaflets from 90% ethanol, dec 209-210°.

4513. N-Glycylglycine. $C_4H_8N_2O_3$; mol wt 132.12. C 36.36%, H 6.10%, N 21.20%, O 36.33%. NH_2CH_2CONH-CH_2COOH. The simplest of all peptides. Prepn from 2,5-diketopiperazine: Schott *et al., J. Org. Chem.* **12**, 490 (1947); Greenstein, Winitz, *Chemistry of the Amino Acids* vol. 2, (New York, 1961) p 803. From tritrylglycylglycine: Zervas *et al., J. Am. Chem. Soc.* **78**, 1359 (1956). From phthalylglycylglycine: Sheehan, Frank, *ibid.* **71**, 1856 (1949). From the dicyclohexylamine salt of trifluoroacetylglycylglycine: Weygand, Reiher, *Ber.* **88**, 26 (1955).

Crystals from dil alc. Crystal shape described as small tetrahedral leaves with a lustrous ball in center. Dec 262-264°. pK$_1'$ 3.12; pK$_2'$ 8.17. Heat of combustion: 472.4 kcal/mole. Soluble in hot water; slightly sol in ethanol. Practically insol in ether.

Hydrochloride monohydrate, $C_4H_8N_2O_3.HCl.H_2O$, crystals from water + ethanol.

Ethyl ester hydrochloride, crystals from abs ethanol, dec 182°.

USE: In the synthesis of more complicated peptides.

4514. Glycyrrhiza. Licorice; liquorice; sweet root. Dried rhizome and roots of *Glycyrrhiza glabra* L., var. *typica* Regel & Herder (Spanish licorice), or of *G. glabra* L., var. *glandulifera* (Waldst. & Kit.) Regel & Herder (Russian licorice), or of other varieties of *G. glabra* yielding a yellow and sweet wood, *Leguminosae. Habit.* Southern Europe to Central Asia. *Constit.* 6-14% glycyrrhizin (the glucoside of glycyrrhetic acid), asparagine, sugars, resin. Used chiefly in the form of glycyrrhiza syrup. *Incompat.* Acids, metallic salts.

USE: Extract and syrup as pharmaceutic aids (flavor and flavored vehicles).

4515. Glycyrrhizic Acid. *(3β,20β)-20-Carboxy-11-oxo-30-norolean-12-en-3-yl 2-O-β-D-glucopyranuronosyl-α-D-glucopyranosiduronic acid;* glycyrrhizin; glycyrrhizinic acid; glycyrrhetinic acid glycoside. $C_{42}H_{62}O_{16}$; mol wt 822.94. C 61.30%, H 7.59%, O 31.11%. Extraction from *Glycyrrhiza glabra* L., *Leguminosae:* Karrer, Chao, *Helv. Chim. Acta* **4**, 100 (1921); Ruzicka, Louenberger, *ibid.* **19**, 1402 (1936). From commercial glycyrrhizinum ammoniacale: Tschirch, Cederberg, *Arch. Pharm.* **245**, 97 (1907); Voss *et al., Ber.* **70**, 122 (1937). Revised method of isoln: Conn, Conn, *J. Lab. Clin. Med.* **47**, 20 (1956). Structure: Lythgoe, Trippett, *J. Chem. Soc.* **1950**, 1983. Revised structure: Marsh, Levvy, *Biochem. J.* **63**, 9 (1956). See also: I. Kitagawa *et al., Chem. Pharm. Bull.* **36**, 3710 (1988); T. Hatano *et al., ibid.* **39**, 1238 (1991). *Review:* Nieman, *Chem. Weekbl.* **48**, 213 (1952). Synthesis of derivatives: Brieskorn, Sax, *Arch. Pharm.* **303**, 905 (1970).

Crystals from glacial acetic acid. Intensely sweet taste. $[\alpha]_D^{17}$ +46.2° (c = 1.5 in alc). Freely sol in hot water, alcohol; practically insol in ether.

Ammonium glycyrrhizinate pentahydrate, $C_{42}H_{65}NO_{16}.$5H$_2$O, needles from 75% aqueous ethanol, decomp 212-217°. $[\alpha]_D^{20}$ +46.9° (c = 1.5 in 40% ethanol). uv max: 248 nm (ε 11400). Sol in ammonia water, glacial acetic acid.

Dipotassium salt, $C_{42}H_{60}K_2O_{16}$, *Rizinsan K2 A2.*

4516. Glyhexamide. *N-[(Cyclohexylamino)carbonyl]-2,3-dihydro-1H-indene-5-sulfonamide; 1-cyclohexyl-3-(5-indanylsulfonyl)urea;* SQ-15860; Subose. $C_{16}H_{22}N_2O_3S$; mol wt 322.43. C 59.60%, H 6.88%, N 8.69%, O 14.89%, S 9.95%. Prepd from hydrindene-5-sulfonamide and cyclohexyl isocyanate: Hoehn, Breuer, U.S. pat. **3,097,242** (1963 to Olin Mathieson). Clinical pharmacology: Grinnell *et al., Am. J. Med. Sci.* **253**, 312 (1967).

Crystals from 70% acetone, mp 153-155°.
THERAP CAT: Antidiabetic.

4517. Glymidine. *N-[5-(2-Methoxyethoxy)-2-pyrimidinyl]benzenesulfonamide;* 2-benzenesulfonamido-5-(β-methoxyethoxy)pyrimidine; glycodiazine. $C_{13}H_{15}N_3O_4S$; mol wt 309.35. C 50.48%, H 4.89%, N 13.58%, O 20.69%, S 10.37%. Prepn: **Belg.** pat. **609,270**; H. Priewe *et al.,* **U.S.** pat. **3,275,635** (1962, 1966 both to Schering, AG); Gutsche *et al., Arzneimittel-Forsch.* **14**, 373 (1964). Series of articles on pharmacology: *ibid.* 377-412. Activity: Losert *et al., ibid.* **23**, 1251 (1973). Metabolism: Soyfer *et al., Chim. Ther.* **5**, 441 (1970). Toxicity data: Kramer *et al., Arzneimittel-Forsch.* **14**, 377 (1964).

Crystals, mp 152-154°. Soly in ethanol: 0.91%; in toluene: 0.67%.

Sodium salt, $C_{13}H_{14}N_3NaO_4S$, *SH-717, Glyconormal, Gondafon, Lycanol, Redul.* Crystals, mp 221-226°. Sparingly sol in alc. Soly in water at 37°: 70.5%. LD$_{50}$ in mice, rats (g/kg): 1.48, 2.00 i.v.; 5.30, 2.85 orally (Kramer).

THERAP CAT: Antidiabetic.

4518. Glyodin. *2-Heptadecyl-4,5-dihydro-1H-imidazole monoacetate;* 2-heptadecylglyoxalidine acetate; Crag Fruit Fungicide 341. $C_{22}H_{44}N_2O_2$; mol wt 368.60. C 71.69%, H 12.03%, N 7.60%, O 8.68%. Prepn from stearic acid and ethylenediamine: Kiff, U.S. pat. **2,540,171** (1951 to Union Carbide and Carbon).

Light orange crystals, mp 62-68°. d^{20} 1.035. Insol in water, acetone, toluene. Sol in isopropanol.

Base, soft greasy wax, mp 94°.
USE: Agricultual fungicide.

4519. Glyoxal. *Ethanedial;* biformyl; diformyl; oxalaldehyde. $C_2H_2O_2$; mol wt 58.04. C 41.39%, H 3.47%, O 55.14%. OHCCHO. Prepd by the oxidation of acetaldehyde by nitric or selenious acid: Lubawin, *Ber.* **8**, 768 (1875); Wyss, *Ber.* **10**, 1366 (1877); Kölln, *Ann.* **416**, 230 (1918); Riley *et al., J. Chem. Soc.* **1932**, 1881; Ronzio, Waugh, *Org. Syn. coll. vol. III*, 438 (1955); by hydrolysis of dichlorodioxane: Butler, Cretcher, *J. Am. Chem. Soc.* **54**, 2988 (1932). Review of commercial development: J. F. Bohmfalk *et al., Ind. Eng. Chem.* **43**, 786 (1951). Toxicity study: H. F. Smyth *et al., J. Ind. Hyg. Toxicol.* **23**, 259 (1941). *Review:* A. B. Boese *et al.,* in *Glycols*, G. O. Curme, F. Johnston, Eds. (Reinhold, New York, 1952) pp 125-128.

Yellow prisms or irregular pieces turning white on cooling. d^{20} 1.14. Opaque at 10°, mp 15°. bp$_{776}$ 51°. The vapors are green and burn with a purple flame. *Explosive:* Mixtures with air may explode! $n_D^{20.5}$ 1.3826. Sol in anhyd solvents. pH of a 40% aq soln: 2.1-2.7; d$_4^{20}$ 1.27. Polymerizes quickly on standing, on contact with water (violent reaction), or when dissolved in solvents contg water. The anhyd polymer changes to the monomer on heating. Solns of the monomer are obtained on heating the polymer with anethole, phenetole, safrole, methyl nonyl ketone, or benzaldehyde. LD$_{50}$ in rats, guinea pigs (mg/kg): 2020, 760 orally (Smyth).

Dihydrate, $(OHCCHO)_3.2H_2O$, crystalline powder, nonhygroscopic. More sol in hot water than in cold water. Commercially available in anhyd form as crystalline dihydrate, or as a 40% aq soln which may contain polymerization inhibitors.

Caution: Moderately irritating to skin, mucous membranes.

USE: In textiles, organic synthesis, glues, biocides.

4520. Glyoxal-Sodium Bisulfite. *1,2-Dihydroxy-1,2-ethanedisulfonic acid disodium salt; glyoxal compd with sodium bisulfite.* $C_2H_4Na_2O_8S_2$; mol wt 266.16. C 9.03%, H 1.51%, Na 17.28%, O 48.09%, S 24.10%. Prepn: Ronzio, Waugh, *Org. Syn.* coll. vol. **III**, 438 (1955).

Monohydrate, hard crystals. Faint SO_2 odor. Freely sol in water. Practically insol in alcohol.

4521. Glyoxylic Acid. *Oxoacetic acid; formylformic acid;* glyoxalic acid; oxoethanoic acid. $C_2H_2O_3$; mol wt 74.04. C 32.45%, H 2.72%, O 64.83%. OHCCOOH. Occurs in unripe fruit and in young green leaves; has also been found in very young sugar beets. Prepd by heating dibromoacetic acid with some water: Grimaux, *Bull. Soc. Chim.* [2] **26**, 483; Cramer, *Ber.* **25**, 714 (1892); by electrolytic reduction of oxalic acid: Meyer, *Ber.* **37**, 3592 (1904); by the action of *Aspergillus niger* on calcium acetate, malonic or citric acid: Challenger *et al., J. Chem. Soc.* **1927**, 205, 207. Prepn of glyoxylic acid soln for analytical use: *Beilstein* vol. **III**, 594.

Hemihydrate, crystals from water, mp 70-75°. Also obtained in anhydr form as monoclinic crystals from water, mp 98°. Obnoxious odor. Strong, corrosive acid. K = 4.6 × 10⁻⁴. Deliquesces rapidly and forms a syrup on short exposure to air. Sparingly sol in alc, ether, benzene. Freely sol in water; aq solns tend to acquire a yellowish tint. Attacks most base metals except certain stainless steel alloys.

Monohydrate, crystals, mp ~50°. Highly hygroscopic.
Caution: Irritant, corrosive.

4522. Glyphosate. *N-(Phosphonomethyl)glycine;* MON-0573. $C_3H_8NO_5P$; mol wt 169.07. C 21.31%, H 4.77%, N 8.28%, O 47.31%, P 18.32%. Broad-spectrum post-emergence, translocated herbicide. Prepn: J. E. Franz, **Ger.** pat. **2,152,826;** *idem,* **U.S.** pats. **3,799,758** and **3,853,-530** (1972, 1974, 1974 all to Monsanto). Metabolism and degradation in soil and water: M. L. Rueppel *et al., J. Agr. Food Chem.* **25**, 517 (1977). Toxicity study: E. A. Bababunmi *et al., Toxicol. Appl. Pharmacol.* **45**, 319 (1978).

White solid, mp 230° (dec). Soly in water at 25°: 12 g/l. Insol in most organic solvents. LD_{50} in rats, mice (mg/kg): 4873, 1568 orally (Bababunmi).

Mono(isopropylamine) salt, $C_6H_{17}N_2O_5P$, *MON-2139, Round-up.* Very sol in water.
USE: Herbicide.

4523. Glyphosine. *N,N-Bis(phosphonomethyl)glycine;* N-carboxymethyl-N,N-bis(methylenephosphonic acid)-amine; CP-41845; Polaris. $C_4H_{11}NO_8P_2$; mol wt 263.08. C 18.26%, H 4.21%, N 5.32%, O 48.65%, P 23.55%. Method of prepn: Irani, Moedritzer, **U.S.** pat. **3,288,846** (1966 to Monsanto). As ripening agent for sugar cane: Hamm, **U.S.** pat. **3,556,762** (1971 to Monsanto). Brief review of ripening effect and toxicity: Porter, Ahlrichs, *Hawaii Sugar Technol. Rep.* **30**, 71 (1971) (publ. 1972).

White crystals from aq ethanol. Soly in water at 20°: 248 g/l. Slightly corrosive to metals. LD_{50} orally in mammals:

3925 mg/kg, *RTECS* vol. **1**, R. J. Lewis, Sr., R. L. Tatken, Eds. (1979) p 719.
USE: Chemical ripener.

4524. Glypinamide. *4-Chloro-N-[[(hexahydro-1H-azepin-1-yl)amino]carbonyl]benzenesulfonamide;* 1-(p-chlorophenylsulfonyl)-3-(hexahydro-1H-azepin-1-yl)urea; azepinamide; Parinase. $C_{13}H_{18}ClN_3O_3S$; mol wt 331.82. C 47.06%, H 5.47%, Cl 10.68%, N 12.66%, O 14.46%, S 9.66%. Orally active, sulfonylurea hypoglycemic agent. Prepn: Wright, **Brit.** pat. **887,886** (1962 to Upjohn); Wright, Willette, *J. Med. Pharm. Chem.* **5**, 815 (1962).

Crystals from methyl ethyl ketone, mp 197-198.5°.
THERAP CAT: Antidiabetic.

4525. Goitrin. *(S)-5-Ethenyl-2-oxazolidinethione;* 5-vinyl-2-thiooxazolidone; (−)-5-vinyl-2-oxazolidinethione. C_5H_7NOS; mol wt 129.18. C 46.49%, H 5.46%, N 10.84%, O 12.39%, S 24.82%. An antithyroid compd isolated from seeds of different species of *Brassica, Cruciferae:* Astwood *et al., J. Biol. Chem.* **181**, 121 (1949); Greer, *J. Am. Chem. Soc.* **78**, 1260 (1956). Stereochemistry: Kjaer *et al., Acta Chem. Scand.* **13**, 144 (1959). Activity: Langer *et al., Endokrinologie* **57**, 225 (1971).

Large prisms from ether, mp 50°. $[\alpha]_D^{31}$ −70.5° (c = 2 in methanol). Behaves as a weak acid; pKa 10.5. Stable in alkali, but not in acid.

4526. Gold. Au; at. wt 196.9665; at. no. 79; valences 1, 3. Occurrence in the earth's crust: 0.005 ppm. One natural isotope: 197; artificial isotopes (mass numbers): 177-179, 181, 183, 185-196, 198-203. Probably the first pure metal known to man. Occurs in nature in its native form and in minute quantities in almost all rocks and in seawater. Gold ores include *calavarite,* $(AuTe_2)$, *sylvanite,* [(Ag,Au)-Te_2], *petzite,* [(Ag,Au)$_2$Te]. Methods of mining, extracting and refining: Hull, Stent, in *Modern Chemical Processes,* **Vol. 5** (Reinhold, New York, 1958) pp 60-71. Lab prepn of gold powder from gold pieces: Block, *Inorg. Syn.* **4**, 15 (1953). Chemistry of gold drugs in the treatment of rheumatoid arthritis: D. H. Brown, W. E. Smith, *Chem. Soc. Rev.* **9**, 217 (1980). Use as catalyst in oxidation of organic compds by NO_2: R. E. Sievers, S. A. Nyarady, *J. Am. Chem. Soc.* **107**, 3726 (1985). Least reactive metal at interfaces with gas or liquid: B. Hammer, J. K. Norskov, *Nature* **376**, 238 (1995). *Reviews: Gmelin's Handb. Anorg. Chem., Gold* (8th ed.) **62**, parts 2, 3 (1954); Johnson, Davis, "Gold" in *Comprehensive Inorganic Chemistry,* vol. **3**, J. C. Bailar Jr. *et al.,* Eds. (Pergamon Press, Oxford, 1973) pp 129-186; J. G. Cohn, E. W. Stern in Kirk-Othmer *Encyclopedia of Chemical Technology* vol. **11** (Wiley-Interscience, New York, 3rd ed., 1980) pp 972-995.

Yellow, soft metal; face-centered cubic structure; when prepared by volatilization or precipitation methods, deep violet, purple, or ruby powder, mp 1064.76°; bp 2700°. d 19.3. Hardness (Mohs') 2.5-3.0; (Brinell's) 18.5. Extremely inactive; not attacked by acids, air or oxygen. Superficially attacked by aq halogens at room temp. Reacts with aqua regia, with mixtures contg chlorides, bromides, or iodides if they can generate nascent halogens, with many oxidizing mixtures especially those contg halogens. Also with alkali cyanides, solns of thiocyanates and double cyanides.
USE: In manuf jewelry; in gold plating other metals; as a standard of currency; most frequently alloyed with silver

and copper. For use in medicine, *see* Gold, Radioactive, Colloidal.

4527. Gold, Explosive. "Fulminating gold". An auric compd of nitrogen. Obtained by the action of ammonia on auric chloride or ammonium chloride on auric oxide: Raschig, *Ann.* **235**, 355 (1886); Weitz, *Ann.* **410**, 117 (1915).

Dark brown powder, which explodes on heating or rubbing to give gold, nitrogen, and ammonia. Exact compn of the compd is unknown since it is too explosive to be dried. Therefore, the only elements that can be determined are gold, nitrogen, and chlorine.

4528. Gold Monochloride. Aurous chloride. AuCl; mol wt 232.42. Au 84.75%, Cl 15.25%. Prepd by thermal decompn of gold trichloride: Biltz, Wein, *Z. Anorg. Allgem. Chem.* **148**, 192 (1925); Capella, Schwab, *C.R. Acad. Sci.* **260**, 4337 (1965).

Yellowish powder. d 7.57. Dec at about 289° into gold and Cl_2. Practically insol in water, but slowly dec by it, more rapidly on heating, with formation of gold trichloride and separation of metallic gold; sol in alkali cyanides. With solns of alkali bromides, metallic gold and potassium auribromide are formed.

4529. Gold Monocyanide. Aurous cyanide; gold cyanide. CAuN; mol wt 222.98. C 5.39%, Au 88.33%, N 6.28%. AuCN. May be prepd by decompn of Na[Au-$(CN)_2$] with HCl: Wogrinz, *Metalloberfläche* **8**, B162 (1954).

Yellow, odorless powder; iridescent in sunlight; slowly dec in presence of moisture. When warmed with HCl, HCN is evolved. When ignited, dec into metal Au and CN. d_4^{20} 7.14. Hexagonal when cryst. Practically insol in water, alcohol, ether, dil acids; sol in ammonia, soln of NaCN; dissolved by aqua regia. On warming with concd H_2SO_4 half of the gold separates as metal.

Caution: May liberate HCN.

4530. Gold Monoiodide. Aurous iodide. AuI; mol wt 323.87. Au 60.82%, I 39.18%. Prepn: Weiss, Weiss, *Z. Naturforsch.* **11b**, 604 (1956); Grange, *Bull. Soc. Chim. France* **1964**, 2418.

Yellowish to greenish-yellow powder; dec slowly at ordinary temp, more rapidly at higher. d 8.25. Insol in water; sol in alkali iodide or cyanide solns; dec by warm acids.

4531. Gold Monosulfide. Aurous sulfide. Au_2S; mol wt 426.00. Au 92.47%, S 7.53%. May be prepd from potassium gold cyanide, H_2S and HCl: Hoffmann, Krüss, *Ber.* **20**, 2361 (1887); Glemser, Sauer in *Handbook of Preparative Inorganic Chemistry*, **Vol. 2**, G. Brauer, Ed. (Academic Press, New York, 2nd ed., 1965) pp 1061-1062.

Brownish-black powder. When freshly pptd it forms a colloidal soln in water. Practically insol in water or single acids; sol in aqua regia, alkali cyanide solns.

4532. Gold Monoxide. Aurous oxide. Existence doubtful; probably a mixture of gold and auric oxide: Pollard, *J. Chem. Soc.* **129**, 1347 (1926). "Au_2O" has been prepared by treating $K(AuBr_2)$ with alkali.

Pale gray-violet solid. May be dried at 200°, but gives off oxygen a few degrees above this, and rapidly at 250°. Soly in water (25°): 0.04 mg/l. When freshly precipitated, it dissolves in alkalies, but the soln deposits gold rapidly.

4533. Gold, Radioactive, Colloidal. Radioactive colloidal gold; colloidal gold—^{198}Au; radio-gold (^{198}Au) colloid; gold colloid ^{198}Au; Aurcoloid; Auroscan-198; Aureotope. Colloidal dispersion of radioactive gold (^{198}Au) for parenteral administration.

Particle diameter range, 3-7 mμ. Has half-life of 2.7 days and emits beta and gamma radiation. Stable to heat except autoclaving under pressure. Compatible with saline solns, radiopaque media and other agents; flocculated by polyvalent metal ion.

THERAP CAT: Antineoplastic.

4534. Gold Selenate. Auric selenate. $Au_2O_{12}Se_3$; mol wt 822.81. Au 47.88%, O 23.33%, Se 28.79%. $Au_2(SeO_4)_3$. Prepd from gold and selenic acid: Lenher, *J. Am. Chem. Soc.* **24**, 354 (1902); Caldwell, Eddy, *ibid.* **71**, 2247 (1949).

Small, yellow crystals. Dec in light. Soluble in sulfuric and nitric acids. Insol in water.

4535. Gold Selenide. Auric selenide. Au_2Se_3; mol wt 630.81. Au 62.45%, Se 37.55%. Prepd by adding an alcoholic soln of gold chloride to a soln of hydrogen selenide: Moser, Atynsky, *Monatsh.* **45**, 235 (1925); by passing hydrogen selenide into an aq soln of gold chloride: Uelsmann, *Ann.* **116**, 127 (1860); by combination of elements in solns or melts: Kulifay, *U.S. pat.* **3,026,175** (1962 to Monsanto).

Black amorphous solid. d^{22} 4.65. Dec by heat. Forms mercuric selenide when heated with mercury. Sol in aqua regia, soln of alkali sulfides.

4536. Gold Sodium Thiomalate. *Mercaptobutanedioic acid monogold(1+) sodium salt;* sodium aurothiomalate; Kidon; Myochrysine; Myocrisin; Shiosol; Tauredon. A variable mixture of the mono- ($C_4H_4AuNaO_4S$) and disodium ($C_4H_3AuNa_2O_4S$) salts of gold thiomalic acid. Previously described as the monohydrate of the disodium salt. Prepd by the interaction of sodium thiomalate and a gold halide: *U.S. pat.* **1,994,213** (1935). Mechanism of action: P. E. Lipsky, M. Ziff, *Adv. Inflammation Res.* **3**, 219 (1982). Pharmacokinetics: E. S. Waller *et al., Res. Commun. Chem. Pathol. Pharmacol.* **37**, 33 (1982). Detection in blood by atomic absorption spectroscopy: A. I. A. Rodgers *et al., Anal. Proc.* **19**, 87 (1982).

$$\text{Au} \diagdown_{\text{S}} \diagup \diagdown_{\text{COO}^-}^{\text{COO}^-} \quad \bullet \; x \, Na^+ \quad \bullet \quad (2-x) \, H^+$$

White to yellowish-white powder, metallic taste. Very sol in water. Practically insol in alcohol, ether. Aq solns are colorless to pale yellow. pH of a 5% aq soln: 5.8-6.5.

THERAP CAT: Antirheumatic.

4537. Gold Sodium Thiosulfate. *Bis[monothiosulfato-(2−)-O,S]aurate(3−) trisodium; bis(monothiosulfato)aurate(3−) trisodium;* aurothiosulfate natrium; aurothiosulfate sodium; double thiosulfate of gold and sodium; hyposulfite of gold and sodium; sodium aurothiosulfate; sel de Fordos et Gélis; Auricidine; Aurocidin; Aurolin; Auropex; Auropin; Aurosan; Aurothion; Crisalbine; Crytion; Novacrysin; Sanochrysine; Solfocrisol; Thiochrysine. $AuNa_3O_6S_4$; mol wt 490.20. Au 40.18%, Na 14.07%, O 19.58%, S 26.17%. $Na_3Au(S_2O_3)_2$. Dihydrate prepd by adding small amounts of gold chloride to a concd aq soln of sodium thiosulfate; after each addition of gold chloride a red color appears which must disappear before another addition can be made; the complex is finally precipitated by the addition of alc to the colorless aq soln. All the gold is reduced from the auric to the aurous state: Fordos, Gélis, *Ann. Chim. Phys.* **13**, 394 (1845); Brown, *J. Am. Chem. Soc.* **49**, 958 (1927); *Roger's Inorganic Pharmaceutical Chemistry*, T. O. Soine, C. O. Wilson, Eds. (Lea & Febiger, Philadelphia, 8th ed., 1967) pp 343-345. Crystal structure of dihydrate: Ruben *et al., Inorg. Chem.* **13**, 1836 (1974).

Dihydrate, white glistening crystals, needle-like or prismatic. d 3.09. Darkens slowly on exposure to light. The 2 moles of water of crystn are not given up until heated to 150 or 160°. One gram dissolves in 2 ml water. Insol in alc and most other organic solvents. A 1:20 aq soln is neutral or slightly alkaline to litmus. Aq solns dec on standing and turn yellow. LD i.v. in rabbits: 100 mg/kg.

Caution: Dermatitis, stomatitis, hepatitis, nephritis, G.I. disturbances, hematologic reaction may occur. *Antidote:* Dimercaprol (BAL).

THERAP CAT: Antirheumatic.

4538. Gold Stannate. Aurous stannate; C.I. Pigment Red 109; C.I. 77482; gold-tin precipitate; gold-tin purple; purple of Cassius. Contains gold, tin and oxygen; composition of commercial products varies and in some cases may be a complex mixture of gold and stannic acid. Prepn: *Colour Index*, vol. 4 (3rd ed., 1971) p 4669.

Brown powder. Practically insol in water. Sol in ammonia.

USE: Manuf ruby glass, colored enamels, and painting porcelain.

4539. Gold Tribromide. Auric bromide. $AuBr_3$; mol wt 436.68. Au 45.11%, Br 54.89%. Prepn: Burawoy, Gibson, *J. Chem. Soc.* **1935**, 217.

Brownish-black powder. mp about 160° with decompn. Sol in water with reddish-brown color; also sol in alcohol, glycerol, but gradually reduced by them. *Keep well closed and protected from light.*

4540. Gold Tribromide, Acid. Hydrogen tetrabromoaurate(1—); bromoauric acid. $AuBr_4H$; mol wt 517.59. Au 38.05%, Br 61.75%, H 0.19%. $HAuBr_4$. Prepd from $AuBr_3$ and HBr.

Pentahydrate, dark reddish-brown, odorless, deliquesc crystals or granular masses. mp about 27°. Very sol in water; sol in alcohol. *Keep tightly closed in a cool place and protected from light.*

4541. Gold Trichloride. Auric chloride. $AuCl_3$; mol wt 303.32. Au 64.94%, Cl 35.06%. Conveniently prepd in the laboratory from metallic gold and iodine monochloride: Gutmann, *Z. Anorg. Allgem. Chem.* **264**, 169 (1951).

Dihydrate, dark orange-red crystals, deliquesc in moist air. d_4^{20} 3.9. Sublimes at 180° (760 mm). bp 229°. Decomp 254°. Soluble in water, alcohol, ether. *Keep well closed and protected from light.* LD s.c. in mice: 1.5 g/kg.

4542. Gold Trichloride, Acid. Hydrogen tetrachloroaurate(1—); hydrochloroauric acid; aurochlorohydric acid; chloroauric acid. $AuCl_4H$; mol wt 339.79. Au 57.97%, Cl 41.74%, H 0.30%. $HAuCl_4$. Prepd according to the equation $2Au + 3Cl_2 + 2HCl \rightarrow 2HAuCl_4$: Thomsen, *Ber.* **16**, 1585 (1883); Biltz, Wein, *Z. Anorg. Allgem. Chem.* **148**, 192 (1925).

Tetrahydrate, golden-yellow to reddish-yellow, very hygroscopic and deliquesc monoclinic crystals; readily affected by sunlight. (Also available as brown crystals or cryst masses, contg 50-51% gold, and having the same properties as the yellow crystals.) Has caustic action on the skin (blisters) and then on exposure to light leaves violet-brown spots. Dec on strong heating to Cl_2, HCl, and metallic gold. d about 3.9. Very sol in water, alcohol; sol in ether. *Keep tightly closed and protected from light.*

USE: Photography, gold-plating, gilding glass and porcelain, manuf ruby glass; as a reagent for alkaloids.

4543. Gold Tricyanide. Auric cyanide. C_3AuN_3; mol wt 275.02. C 13.10%, Au 71.62%, N 15.28%. $Au(CN)_3$. Prepn of trihydrate: Chadwick, Sharpe, *Adv. Inorg. Chem. Radiochem.* **8**, 157-158 (1966).

Trihydrate, white, deliquesc crystals; dec at 50°. *Poison!* Very soluble in water, slightly sol in alcohol. *Keep tightly closed and protected from light.*

4544. Gold Trihydroxide. Auric hydroxide. AuH_3O_3; mol wt 247.99. Au 79.43%, H 1.22%, O 19.36%. $Au(OH)_3$. Usually contains about $3H_2O$ and hence about 65% gold. May be prepd in lab according to the equation $2AuHCl_4 + 3Na_2CO_3 + 3H_2O \rightarrow 2Au(OH)_3 + 6NaCl + 2KCl + 3CO_2$: Lydén, *Z. Anorg. Allgem. Chem.* **240**, 157 (1939).

Brown powder; dec by sunlight to metallic gold; also slowly dec with age or at 100°, and completely at 250°. Practically insol in water. Sol in soln of NaCN, in HCl or concd HNO_3. With NH_3 yields gold fulminate which *explodes easily in dry form. Protect from light.*

USE: In gold-plating solns; for decorating porcelains.

4545. Gold Trioxide. Auric oxide; gold sesquioxide; gold oxide. Au_2O_3; mol wt 441.93. Au 89.14%, O 10.86%. Prepd from the hydroxide: Roseveare, Buehner, *J. Am. Chem. Soc.* **49**, 1221 (1927).

Brown powder; begins to evolve oxygen at 110°; at 250° it is entirely dec to metallic gold; also slowly dec by sunlight. Practically insol in water. Sol in HCl, concd HNO_3 and in NaCN soln. *Keep protected from light.*

4546. Gold Trisulfide. Auric sulfide. Au_2S_3; mol wt 490.13. Au 80.37%, S 19.63%. Prepd by treating dry lithium aurichloride $Li(AuCl_4)$ with hydrogen sulfide at —10°. The reaction product consists of HCl, lithium chloride, and auric sulfide. The lithium chloride can be removed by extraction with alcohol. *Ref:* Antony, Lucchesi, *Gazz. Chim. Ital.* **19**, 552 (1889). Prepd from $AuCl_3$ or $HAuCl_4$ and H_2S

in abs ether at low temp: Guthier, Dürrwachter, *Z. Anorg. Allgem. Chem.* **121**, 266 (1922).

Black powder. Heating to 200° dec it into its elements.

USE: In photography.

4547. Goserelin. 6-[O-(1,1-Dimethylethyl)-D-serine]-10-deglycinamideluteinizing hormone-releasing factor (pig) 2-(aminocarbonyl)hydrazide; D-Ser(Bu^t)^{6}Azgly^{10}-gonadorelin; D-Ser(Bu^t)^{6}Azgly^{10}-luliberin; ICI-118630. $C_{59}H_{84}N_{18}O_{14}$; mol wt 1269.43. C 55.82%, H 6.67%, N 19.86%, O 17.65%. Synthetic peptide agonist analog of LH-RH, *q.v.* Prepn: A. S. Dutta *et al.*, *Ger.* pat. 2,720,245; *eidem,* U.S. pat. **4,100,**-**274** (1977, 1978 both to I.C.I.); *eidem, J. Med. Chem.* **21**, 1018 (1978). Effect on ovulation in cattle: P. S. Jackson, B. J. A. Furr, *Res. Vet. Sci.* **34**, 182 (1983). Radioimmunoassay in human serum: R. N. Clayton *et al.*, *Clin. Endocrinol.* **22**, 453 (1985). Pharmacokinetics, endocrinology in humans: T. J. Perren *et al.*, *Cancer Chemother. Pharmacol.* **18**, 39 (1986). Preliminary clinical trials in prostatic carcinoma: J. M. Allen *et al.*, *Brit. Med. J.* **286**, 1607 (1983); K. J. Walker *et al.*, *Lancet* **2**, 413 (1983); S. R. Ahmed *et al.*, *ibid.* 415; in breast cancer: P. N. Plowman *et al.*, *Brit. J. Cancer* **54**, 903 (1986). Suppression of human ovarian activity: C. P. West, D. T. Baird, *Clin. Endocrinol.* **26**, 213 (1987). Review of pharmacokinetics and therapeutic applications: P. Chrisp, K. L. Goa, *Drugs* **41**, 254-288 (1991).

5-oxoPro—His—Trp—Ser—Tyr—D-Ser(t-Bu)—Leu—Arg—Pro—NHNHCONH$_2$

Acetate, $C_{61}H_{88}N_{18}O_{16}$, *Zoladex.*

THERAP CAT: Antineoplastic (hormonal); in treatment of prostatic carcinoma.

4548. Gossyplure. 7,11-Hexadecadien-1-ol acetate. C_{18}-$H_{32}O_2$; mol wt 280.45. C 77.09%, H 11.50%, O 11.41%. Sex pheromone of pink bollworm, *Pectinophora gossypiella* (Saunders): Hummel *et al., Science* **181**, 873 (1973). Isoln and prepn of 1:1 mixture of (Z,Z) and (Z,E) isomers: B. A. Bierl *et al., J. Econ. Entomol.* **67**, 211 (1974). Improved prepn: R. J. Anderson, C. A. Hendrick, U.S. pat. **3,919,329** (1975 to Zoecon); *eidem, J. Am. Chem. Soc.* **97**, 4327 (1975). Stereoselective synthesis of isomers: K. Mori *et al., Agr. Biol. Chem.* **38**, 1551 (1974); *eidem, Tetrahedron* **31**, 1846 (1975); H. Su, P. G. Mahany, *J. Econ. Entomol.* **67**, 319 (1974); H. J. Bestmann *et al., Tetrahedron Letters* **1976**, 353; J. M. Muchowski, M. C. Venuti, *J. Org. Chem.* **46**, 459 (1981). Activity of isomers: H. M. Flint *et al., Environ. Entomol.* **6**, 274 (1977). Degradn: R. D. Henson, *ibid.* 821.

(Z,Z) - form

Yellow liquid. Sol in most org solvents. Extremely flammable.

(Z,Z)-Form, bp 130-132°. n_D^{21} 1.4592.
(Z,E)-Form, bp 132-134°. n_D^{21} 1.4591.
USE: Sex attractant for pink bollworm.

4549. Gossypol. 1,1',6,6',7,7'-Hexahydroxy-3,3'-dimethyl-5,5'-bis(1-methylethyl)[2,2'-binaphthalene]-8,8'-dicarboxaldehyde; 1,1',6,6',7,7'-hexahydroxy-5,5'-diisopropyl-3,3'-dimethyl[2,2'-binaphthalene]-8,8'-dicarboxaldehyde; 2,2'-bis[1,-6,7-trihydroxy-3-methyl-5-isopropyl-8-aldehydonaphthalene]; 2,2'-bis[8-formyl-1,6,7-trihydroxy-5-isopropyl-3-methylnaphthalene]. $C_{30}H_{30}O_8$; mol wt 518.56. C 69.49%, H 5.83%, O 24.68%. Poisonous pigment found in cottonseed. Name derived from the botanical name of the cotton plant, *Gossypium* L., *Malvaceae.* Isoln (0.5% yield): Campbell *et al., J. Am. Chem. Soc.* **59**, 1723, 1726 (1937). Process for removal from cottonseed flour: Vix *et al., J. Am. Oil Chem. Soc.* **48**, 611 (1971). Isoln of (+)-form from *Thespesia populnea, Malvaceae:* King, de Silva, *Tetrahedron Letters* **1968**, 261; Datta *et al., Indian J. Chem.* **10**, 263 (1972). Structure: Adams, *J. Am. Chem. Soc.* **60**, 2193 (1938); Shirley, Dean, *ibid.* **79**, 1205 (1957). Probably exists in three tautomeric forms. Conformation: Wood *et al., Chem. & Ind. (London)* **1969**, 1738. Synthesis: Edwards, *ibid.* **80**, 3798 (1958); *idem, J. Am. Oil Chem. Soc.* **47**, 441 (1970). Toxic to nonruminant animals by reducing the oxygen-carrying capacity of blood. Metabolism in rats: Abou-Donia *et al., Lipids* **5**, 938 (1970). Potential use as a male contraceptive agent: H. Pösö *et al., Lancet* **1**, 885 (1980); T. H. Maugh, *Science* **212**, 314 (1981); in the treatment of genital herpes: V. Wichmann *et al., Am. J. Obstet. Gynecol.* **142**, 593 (1982). Biochemical mechanism of action study: C.-Y. G. Lee *et al., Mol. Cell. Biochem.* **47**, 65 (1982). Toxicity study: El-Nockrashy *et al., J. Am. Oil Chem. Soc.* **40**, 14 (1963). Reviews: Adams *et al., Chem. Rev.* **60**, 555-631 (1960); L. C. Berardi, L. A. Goldblatt, "Gossypol" in *Toxic Constituents of Plant Foodstuffs*, I. E. Liener, Ed. (Academic Press, New York, 2nd ed., 1980) pp 183-237.

At least 3 crystal modifications: mp 184° from ether, mp 199° from chloroform, mp 214° from ligroin. Very slightly sol in petr ether. Sol in methanol, ethanol, ether, chloroform, DMF. Freely sol (with slow decompn) in dil aq solns of ammonia and sodium carbonate. Insol in water. LD₅₀ orally in rats: 2.57 g/kg (El-Nockrashy).

Caution: May be irritating to G.I. tract. In experimental animals large doses cause edema of lungs, shortness of breath, paralysis.

USE: Proposed as rubber antioxidant, stabilizer for vinyl polymers. Potential insecticide.

4550. Gougerotin. *(R)-1-(4-Amino-2-oxo-1(2H)-pyrimidinyl)-1,4-dideoxy-4-[[3-hydroxy-2-[[(methylamino)acetyl]-amino]-1-oxopropyl]amino]-β-D-glucopyranuronamide; 4-amino-2-oxo-1(2H)-pyrimidinyl)-1,4-dideoxy-4-[D-2-[2-(methylamino)acetamido]hydracrylamido]glucopyranuronamide;* 1-[4-deoxy-4-(sarcosyl-D-seryl)amino-β-D-glucopyranuronamide]cytosine; aspiculamycin; asteromycin. C₁₆H₂₅N₇O₈; mol wt 443.42. C 43.34%, H 5.68%, N 22.11%, O 28.87%. Antibiotic substance with antibacterial and antineoplastic activity. Isoln from *Streptomyces gougerotii:* Kanzaki *et al., J. Antibiot.* **15A**, 93 (1962). Identity with asteromycin: Ikeuchi *et al., ibid.* **25**, 548 (1972). Structure: Iwasaki, *Yakugaku Zasshi* **82**, 1358 (1962). Revised structure: Fox *et al., Tetrahedron Letters* **1968**, 6029; Watanabe *et al., Chem. Pharm. Bull.* **17**, 416 (1969). Total synthesis: *eidem, J. Am. Chem. Soc.* **94**, 3272 (1972); Lichtenthaler *et al., Tetrahedron Letters* **1975**, 3527. Identity with aspiculamycin: Lichtenthaler *et al., ibid.* 665. Mechanism of action study: J. C. Lacal *et al., J. Antibiot.* **33**, 441 (1980). Reviews: Clark in *Antibiotics*, vol. 1, D. Gottlieb, P. D. Shaw, Eds. (Springer-Verlag, New York, 1967) pp 278-282; Yukioka, *ibid.* vol. 3, J. W. Corcoran, F. E. Hahn, Eds. (1975) pp 448-458.

Needles, mp 211-217° (dec). $[\alpha]_D^{27}$ +53° (c = 0.8). uv max (water): 267, 235 nm (ε 9400, 9300); in 0.1N HCl: 275 nm (ε 13600); in 0.1N NaOH: 267 nm (ε 9800). LD₅₀ in mice (mg/kg): 57 i.v. (Kanzaki).

4551. G-Proteins. GTP binding proteins. Distinct class of membrane associated *guanine nucleotide binding proteins* characterized by their function as couplers between a wide variety of receptors and their effector molecules in transmembrane signalling pathways. An example is the retinal G-protein, *transducin*, which links the photon receptor, rhodopsin, *q.v.*, to cGMP phosphodiesterase. G-Proteins are heterotrimeric, with apparent mol wt of 100 kDa and composed of α, β, γ subunits. The α subunit contains the guanine nucleotide binding site, possesses GTPase activity, and is specific for each G-protein. β and γ subunits form a noncovalent, membrane attached complex. Reviews: A. M. Spiegel, *Mol. Cell. Endocrinol.* **49**, 1-16 (1987); *idem, Ann. Rep. Med. Chem.* **23**, 235-242 (1988); P. J. Casey, A. G. Gilman, *J. Biol. Chem.* **263**, 2577-2580 (1988); H. R. Bourne, *Nature* **337**, 504-505 (1989); L. Birnbaumer, *Ann. Rev. Pharmacol. Toxicol.* **30**, 675-705 (1990); M. E. Linder, A. G. Gilman, *Sci. Am.* **267**, 56-65 (July, 1992). Review of role in disease: A. C. Dolphin, *Trends Neurosci.* **10**, 53-57 (1987).

4552. Gramicidin S. Gramicidin S (Soviet); gramicidin C (Soviet). C₆₀H₉₂N₁₂O₁₀; mol wt 1141.47. C 63.13%, H 8.12%, N 14.73%, O 14.02%. Cyclic decapeptide antibiotic produced by a strain of *Bacillus brevis.* Isoln: Gause *et al., Compt. Rend. Acad. Sci. USSR* **43**, 217 (1944), *C.A.* **39**, 1195 (1945); Gause, Brazhnikova, *Lancet* **247**, 715 (1944). More closely related to tyrocidines, *q.v.*, in biological and chemical properties than to true gramicidins, *q.v.* Structure: Synge, *Biochem. J* **39**, 363 (1945); Consden *et al., ibid.* **40**, xliii (1946); **41**, 596 (1947); Battersby, Craig, *J. Am. Chem. Soc.* **73**, 1887 (1951); Erlanger, Goode, *Nature*, **174**, 840 (1954). Synthesis and absorption spectrum: Schwyzer, Sieber, *Helv. Chim. Acta* **40**, 624 (1957); Waki, Izuniya, *Bull. Chem. Soc. Japan* **40**, 1687 (1967). Solid phase synthesis: Losse, Neubert, *Tetrahedron Letters* **1970**, 1267; M. Ohno *et al., J. Am. Chem. Soc.* **93**, 5251 (1971). Improved synthesis via a linear pentapeptide: Y. Minematsu *et al., Tetrahedron Letters* **1980**, 2179; via a linear decapeptide: T. Mukaiyama *et al., Chem. Letters* **1981**, 1367. Industrial procedure: Brit. pat. **836,725** (1960 to Ciba). Review: Y. A. Ovchinnikov, V. T. Ivanov, "The Cyclic Peptides: Structure, Conformation, and Function" in *The Proteins* vol. V, H. Neurath, R. L. Hill, Eds. (Academic Press, New York, 3rd ed., 1982) pp 547-555.

Hydrochloride, C₆₀H₉₂N₁₂O₁₀.2HCl, prisms from ethanol + aq HCl, dec 277-278°. $[\alpha]_D^{24}$ −289° (c = 0.43 in 70% ethanol). Freely sol in alcohol; slightly sol in acetone. Practically insol in water, acids, alkalies. LD₅₀ i.p. in rats: 17 mg/kg (Gause, Brazhnikova).

THERAP CAT: Topical antibacterial.

4553. Gramicidin(s). Gramicidin D (Dubos); linear gramicidins; Gramoderm. Polypeptide antibiotic complex first isolated from the mixture tyrothricin *(q.v.)* along with tyrocidine *(q.v.)* from cultures of *Bacillus brevis:* Dubos, Hotchkiss, *J. Exp. Med.* **73**, 629 (1941); *eidem, J. Biol. Chem.* **141**, 155 (1941). Commercial extraction: Baron, U.S. pat. **2,534,541** (1950 to Penick). Commercial preparation is a mixture of the four components, gramicidin A, B, C, and D, comprising about 87.5, 7.1, 5.1, 0.3 percent resp: Gross, Witkop, *Biochemistry* **4**, 2495 (1965). Each of the components A, B, and C consist of 2 chains, one with valine in position 1, comprising 80-95% of the component, and the other with isoleucine in position 1. Structure, characterization, and synthesis of the two isoforms of gramicidin A, *N*-formylvaline and isoleucine: Sarges, Witkop, *J. Am. Chem. Soc.* **86**, 1862 (1964); **87**, 2011, 2020 (1965); Bauer *et al., Biochemistry* **11**, 3266 (1972). Structure of gramicidin B: Sarges, Witkop, *J. Am. Chem. Soc.* **87**, 2027 (1965); of gramicidin C: *eidem, Biochemistry* **4**, 2491 (1965). Synthesis of valine-gramicidin B and C: K. Noda, E. Gross in *Chemistry and Biology of Peptides, Proc. 3rd Am. Peptide Symp.,* J. Meienhofer Ed. (Ann Arbor Science Publishers, Michigan, 1972) pp 241-250. *Review:* Hunter, Schwartz, "Gramicidins" in *Antibiotics* I, S. Gottlieb, P. Shaw, Eds. (Springer-Verlag, New York, 1967) pp 642-648. Comprehensive description: G. A. Brewer in *Analytical Profiles of Drug Substances* vol. 8, K. Florey, Ed. (Academic Press, New York, 1979) pp 179-218.

HC–Val–Gly–Ala–D-Leu–Ala–D-Val–Val–D-Val–Trp–D-Leu–Trp–D-Leu–Trp–D-Leu–Trp–NHCH₂CH₂OH
‖
O

Valine-Gramicidin A

Spear-shaped or lenticular platelets, mp 229-230°. Almost insol in water (0.6 mg/100 ml). Soluble in the lower alcohols, acetic acid, pyridine. Moderately sol in dry acetone and dioxane. Practically insol in ether, hydrocarbons. Tends to form colloidal suspensions in water.

THERAP CAT: Antibacterial.

THERAP CAT (VET): Antimicrobial.

4554. Gramine. *N,N-Dimethyl-1H-indole-3-methanamine; 3-(dimethylaminomethyl)indole;* Donaxine. $C_{11}H_{14}$-N_2; mol wt 174.25. C 75.82%, H 8.10%, N 16.08%. In chlorophyll-deficient mutants of barley: Euler *et al., Z. Physiol. Chem.* **217,** 23 (1933). In the Asiatic reed *Arundo donax* L., *Gramineae:* Orechoff, Norkina, *Ber.* **68,** 436 (1935). From *Acer saccharinum* L. (the Silver Maple) and *A. rubrum* L., *Aceraceae:* Pachter *et al., J. Org. Chem.* **24,** 1285 (1959); Pachter, *J. Am. Pharm. Assoc., Sci. Ed.* **48,** 670 (1959). Synthesis: Kühn, Stein, *Ber.* **70,** 567 (1937). Biosynthesis from tryptophan in barley: Bowden, Marion, *Can. J. Chem.* **29,** 1037 (1951); O'Donovan, Leete, *J. Am. Chem. Soc.* **85,** 461 (1963); Gower, Leete, *ibid.* 3683; *see also* Gross *et al., Tetrahedron Letters* **1971,** 4047.

Shiny, flat needles or plates from acetone, mp 138-139°. Absorption spectrum: Kanakoa *et al., Chem. Pharm. Bull.* **8,** 294 (1960). Sol in alcohol, ether, chloroform; slightly sol in cold acetone. Practically insol in petr ether, water.

Hydrochloride, $C_{11}H_{14}N_2$·HCl, crystals from ethanol + ether, dec 191°. Sol in water.

Methiodide, $C_{11}H_{14}N_2$·CH₃I, crystals from methanol + benzene, mp 168-169°. Sol in water: Geissman, Armen, *J. Am. Chem. Soc.* **74,** 3916 (1952).

4555. Granaticin. *[3aS-(3aα,5α,8α,9α,11β,13bα,15S*)]-3,3a,5,8,11,13b-Hexahydro-7,8,12,15-tetrahydroxy-5,9-dimethyl-8,11-ethanofuro[2,3-e]naphtho[2,3-c:6,7-c']dipyran-2,-6,13(9H)-trione;* antibiotic WR 141; litmomycin. $C_{22}H_{20}O_{10}$; mol wt 444.39. C 59.46%, H 4.54%, O 36.00%. Antibiotic substance produced by *Streptomyces olivaceus* from soil of Portuguese West Africa. Isoln and antibacterial activity: R. Corbaz *et al., Helv. Chim. Acta* **40,** 1262 (1957). Determn by microbiological diffusion assay: A. Ricicova, M. Podojil, *Folia Microbiol.* **10,** 299 (1965). Isoln of granaticin B, the α-L-rhodinoside of granaticin: S. Barcza *et al., Helv. Chim. Acta* **49,** 1736 (1966); Fr. pat. **1,525,993;** W. Keller, H. Zaehner, U.S. pat. **3,836,642** (1968, 1974 both to Ciba-Geigy). Structure of granaticin and granaticin B: W. Keller-Schierlein *et al., Helv. Chim. Acta* **51,** 1257 (1968); M. Brufani, M. Dobler, *ibid.* 1269; *Naturally Occurring Quinones,* R. H. Thomson, Ed. (Academic Press, New York, 2nd ed., 1971) pp 298-302. Identity of granaticin with antibiotic litmomycin: C.-J. Chang *et al., J. Antibiot.* **28,** 156 (1975). Biosynthesis: C. E. Snipes *et al., J. Nat. Prod.* **42,** 627 (1979); *eidem, J. Am. Chem. Soc.* **101,** 701 (1979). Total synthesis of (±)-form: K. Nomura *et al., ibid.* **109,** 3402 (1987); of the natural (−)-form: K. Okazaki *et al., Chem. Commun.* **1989,** 354. Cytotoxic action on carcinoma cells: E. Sturdik, L. Drobnica, *Neoplasma* **30,** 3 (1983). Inhibition of RNA synthesis: A. Ogilvie *et al., Biochem. J.* **152,** 517 (1975); P. Heinstein, *J. Pharm. Sci.* **71,** 197 (1982).

Deep red, garnet-like crystals from acetone, dec 204-206°. Also reported as mp 211-213° (dec). Acts as an indicator: red in acids, blue in alkalies. Absorption max (abs ethanol): 223, 286, 532, 576 nm (log ε 4.58, 3.76, 3.87, 3.75).

Tetraacetylgranaticin, $C_{30}H_{28}O_{14}$, yellow crystals from alc, mp 242-243°. $[\alpha]_D^{20}$ −100° (c = 0.818 in chloroform).

Granaticin B, $C_{28}H_{30}O_{12}$, red crystalline solid from methanolic-HCl, mp 117-119°. $[\alpha]_D^{22}$ 17.2° (c = 0.83 in pyridine). Absorption max (methanol): 223, 285, 527, 566 nm (log ε 4.42, 3.68, 3.76, 3.57).

4556. Grandisol. *(1R-cis)-1-Methyl-2-(1-methylethenyl)cyclobutaneethanol; cis-(+)-2-isopropenyl-1-methylcyclobutaneethanol;* (+)-(1R,2S)-1-(2'-hydroxyethyl)-1-methyl-2-isopropenylcyclobutane. $C_{10}H_{18}O$; mol wt 154.25. C 77.87%, H 11.76%, O 10.37%. Major component of *grandlure,* the sex attractant of the boll weevil *(Anthonomus grandis,* Boheman). Isoln and synthesis: J. H. Tumlinson *et al., Science* **166,** 1010 (1969); *eidem, J. Org. Chem.* **36,** 2616 (1971). Synthesis of optically active grandisol: P. D. Hobbs, P. D. Magnus, *Chem. Commun.* **1974,** 856; *eidem, J. Am. Chem. Soc.* **98,** 4594 (1976); K. Mori, *Tetrahedron* **34,** 915 (1978); of enantiomerically pure grandisol: J. B. Jones *et al., Can. J. Chem.* **60,** 2007 (1982). Synthesis of racemate: B. M. Trost *et al., J. Am. Chem. Soc.* **99,** 3088 (1977). Short stereoselective synthesis of (±)-grandisol: I. Aljancic-Solaja *et al., Helv. Chim. Acta* **70,** 1302 (1987). Review of syntheses: J. A. Katzenellenbogen, *Science* **194,** 139-148 (1976); J. M. Brand *et al., Fortschr. Chem. Org. Naturst.* **37,**18-29 (1979), *see also* refs pp 157-190; K. Mori, "The Synthesis of Insect Pheromones" in *The Total Synthesis of Natural Products* vol. 4, J. ApSimon, Ed. (Wiley-Interscience, New York, 1981) pp 80-85.

Liquid, $bp_{1.0}$ 50-60°. $[\alpha]_D^{21.5}$ +18.5° (c = 1 in hexane). n_D^{20} 1.4748.

4557. Granisetron. *endo-1-Methyl-N-(9-methyl-9-azabicyclo[3.3.1]non-3-yl)-1H-indazole-3-carboxamide.* $C_{18}H_{24}$-N_4O; mol wt 312.41. C 69.20%, H 7.74%, N 17.93%, O 5.12%. Specific serotonin (5HT₃) receptor antagonist. Prepn: F. D. King, *Eur. pat.* **200,444** (1986 to Beecham). 5HT₃ receptor binding study: G. J. Kilpatrick *et al., Nature* **330,** 746 (1987). Clinical evaluations as antiemetic for cancer chemotherapy patients: R. A. Joss *et al., J. Nat. Cancer Inst.* **80,** 1340 (1988); J. Carmichael *et al., Brit. Med. J.* **297,** 110 (1988).

Hydrochloride, $C_{18}H_{24}N_4O.HCl$, *BRL-43694A, Kytril.* mp 290-292°.

THERAP CAT: Antiemetic.

4558. Granulocyte Colony-Stimulating Factor. CSF-β; G-CSF; GM-DF; MGI-2; pluripoietin. Hematopoietic growth factor that stimulates the development of committed progenitor cells to neutrophils and enhances the functional activities of the mature end-cell. Glycoprotein of mol wt 18-22 kDa; produced in response to specific stimulation by a variety of cells including monocytes, fibroblasts and endothelial cells. Murine and human molecules exhibit cross species reactivity. Originally identified as a differentiation factor for murine leukemic cells. Characterization in mouse serum: J. Lotem *et al., Int. J. Cancer* **25**, 763 (1980); A. W. Burgess, D. Metcalf, *ibid.* **26**, 647 (1980). Purification of murine G-CSF: N. A. Nicola *et al., J. Biol. Chem.* **258**, 9017 (1983); of human G-CSF: K. Welte *et al., Proc. Nat. Acad. Sci. USA* **82**, 1526 (1985); N. Nicola *et al., Nature* **314**, 625 (1985); H. Nomura *et al., EMBO J.* **5**, 871 (1986). Production of human G-CSF by recombinant DNA technology: L. M. Souza *et al., Science* **232**, 61 (1986); S. Nagata *et al., Nature* **319**, 415 (1986). Review of clinical potential and comparison with GM-CSF: W. P. Steward, *Lancet* **342**, 153-157 (1993). *Reviews:* G. D. Demetri, J. D. Griffin, *Blood* **78**, 2791-2808 (1991); L. S. Tkatch, D. J. Tweardy, *Lymphokine Cytokine Res.* **12**, 477-488 (1993). Review of role in infectious diseases: D. C. Dale *et al., J. Infect. Dis.* **172**, 1061-1075 (1995).

Filgrastim, N-L-methionylcolony-stimulating factor (human clone 1034), recombinant methionyl human G-CSF, r-metHuG-CSF, KRN-8601, Neupogen. 175 amino acid peptide produced in *E. coli* by recombinant DNA technology; mol wt 18.8 kDa. Differs from endogenous human G-CSF by the addition of an *N*-terminal methionine and the absence of glycosylation. Clinical trial in severe chronic neutropenia: D. C. Dale *et al., Blood* **81**, 2496 (1993). Book: *Filgrastim in Clinical Practice,* G. Morstyn, T. M. Dexter, Eds. (Marcel Dekker, New York, 1994) 351 pp.

Lenograstim, Granocyte, Neutrogin. Glycoprotein with 174 amino acids and 4% sugar chains produced in Chinese hamster ovary cells by recombinant DNA technology; mol wt ~20 kDa. Closely resembles endogenous human G-CSF. Clinical trial in bone-marrow transplantation: C. Gisselbrecht *et al., Lancet* **343**, 696 (1994).

Nartograstim, 1-(N-L-methionyl-L-alanine)-3-L-threonine-4-L-tyrosine-5-L-arginine-17-L-serinecolony-stimulating factor (human clone 1034), marograstim, KW-2228, Noyap. G-CSF mutein of 175 amino acids produced in *E. coli* by recombinant DNA technology; mol wt 18.9 kDa. *In vitro* activity: T. Suzuki *et al., Acta Haematol.* **87**, 181 (1992). ELISA determn in plasma: T. Kuwabara *et al., J. Pharmacobio-Dyn.* **25**, 121 (1992).

THERAP CAT: Hematopoietic stimulant; antineutropenic.

4559. Granulocyte-Macrophage Colony-Stimulating Factor. *Colony-stimulating factor-2;* CSF-2; CSFα; GM-CSF; NIF-T. Hematopoietic growth factor that stimulates the development of neutrophils and macrophages and promotes the proliferation and development of early erythroid, megakaryocytic and eosinophilic progenitor cells. Produced by endothelial cells, monocytes, fibroblasts and T lymphocytes. Inhibits neutrophil migration and enhances the functional activities of the mature end-cells. Purification of murine GM-CSF: A. W. Burgess *et al., J. Biol. Chem.* **252**, 1998 (1977); of human GM-CSF: N. A. Nicola *et al., Blood* **54**, 614 (1979); and identity with neutrophil migration inhibition factor (NIF-T): J. C. Gasson *et al., Science* **226**, 1339 (1984). Partial amino acid sequence of murine: L. G. Sparrow *et al., Proc. Nat. Acad. Sci USA* **82**, 292 (1985). Cloning and expression of murine GM-CSF: N. M. Gough *et al., Nature* **309**, 763 (1984); of human: G. G. Wong *et al., Science* **228**, 810 (1985); F. Lee *et al., Proc. Nat. Acad. Sci. USA* **82**, 4360 (1985); M. A. Cantrell *et al., ibid.* 6250. Pharmacokinetics: W. P. Petros, *Pharmacotherapy* **12**, Suppl 2-2, 32S (1992). Clinical comparison with G-CSF: S. Blackwell, J. Crawford, *ibid.* 20S. *Review:* J. C. Gasson, *Blood* **77**, 1131-1145 (1991). Review of clinical and biological effects: S. G. Louie, B. Jung, *Am. J. Hosp. Pharm.* **50**, Suppl 3, S10-S18 (1993).

Molgramostim, colony-stimulating factor 2 (human clone pHG25 protein moiety reduced), Sch-39300, Leucomax. Nonglycosylated peptide of 127 amino acids produced in *E. coli* by recombinant DNA technology. Mol wt 14.5 kDa. Pharmacokinetics in AIDS and ARC: R. G. Hewitt *et al., Antimicrob. Ag. Chemother.* **37**, 512 (1993).

Regramostim, colony-stimulating factor 2 (human clone pCSF-1 protein moiety reduced) glycoform GMC 89-107. Glycoprotein of 127 amino acids produced in Chinese hamster ovary cells by recombinant DNA technology. Mol wt 21-34 kDa.

Sargramostim, 23-L-leucinecolony-stimulating factor 2 (human clone pHG25 protein moiety), BI 61.012, Leukine, Prokine. Variably glycosylated glycoprotein of 127 amino acids produced in yeast by recombinant DNA technology. Mol wt 15.5-19.5 kDa. Differs from endogenous human GM-CSF by the substitution of a leucine moiety at position 23 of the amino acid sequence. Clinical evaluation in myelodysplasia: W. J. Gradishar *et al., Blood* **80**, 2463 (1992).

THERAP CAT: Hematopoietic stimulant; antineutropenic.

4560. Graphite. Plumbago; black lead; mineral carbon. Obtained by mining, especially in Canada and Ceylon. Monograph: A. R. Ubbelohde, F. A. Lewis, *Graphite and Its Crystal Compounds* (Oxford, 1960). *Review:* Holliday *et al.* in *Comprehensive Inorganic Chemistry* vol. 1, J. C. Bailar, Jr. *et al.,* Eds. (Pergamon Press, Oxford, 1973) pp 1250-1294.

Crystallized carbon with traces of Fe, SiO_2, etc. Usually soft, black scales, crystals rare. d 2.09-2.23. Mohs' hardness — 1.0. Commercial varieties usually withstand temps up to 2820°. Sol in molten iron.

Caution: Potential symptoms of overexposure are coughing, dyspnea, black sputum, pulmonary function impairment and lung fibrosis. *See NIOSH Pocket Guide to Chemical Hazards* (DHHS/NIOSH 90-117, 1990) p 120.

USE: For "lead" pencils, refractory crucibles, stove polish; as pigment, lubricant, graphite cement; for matches and explosives, commutator brushes, anodes, arc-lamp carbons, electroplating; polishing compds, rust and needle-paper; coating for cathode ray tubes; moderator in nuclear piles.

4561. Graphite Fluoride. *Fluorine compd with graphite.* Non-wettable layered solid, CF_x, (0.5 < x < 1.3), obtained by direct fluorination of graphite at high temp. Two covalent forms have been produced by controlling the quality of the graphite and the reaction temp: $(CF)_n$ at (300°-600°) and $(C_2F)_n$ at (350°-400°). Composition and color are a function of the fluorination temp; the F/C ratio increases with increasing temp, and color changes from black to gray to white. Prepn: O. Ruff, O. Bretschneider, *Z. Anorg. Allgem. Chem.* **217**, 1 (1934); and thermal/electrical characteristics: C.-C. Hung *et al., SAMPE Quarter.* **19**, 12 (1988). Evaluation as solid lubricant of $(CF_x)_n$: R. L. Fusaro, H. E. Sliney, *ASLE Trans.* **13**, 56 (1970); of $(C_2F)_n$: H. Miyake *et al., Proc. JSLE Int. Tribiol. Conf.* **2**, 395 (1985). Lubrication mechanism and wear life: R. L. Fusaro, *Wear* **53**, 303 (1979). Brief review of physical properties: Central Glass Co. in *Proc. BMRA Symp.,* Brussels 1983, A. Kozawa, M. Nagayama, Eds. (Battery Mat. Res. Assoc., 1984) 135-141; of prepn, properties and uses: T. Nakajima, N. Watanabe, *Chemtech* **1990**, 426-430. Book: N. Watanabe *et al., Graphite Fluorides,* (Elsevier Press, Amsterdam, 1988), 262 pp.

$(CF)_n$, *poly(carbon monofluoride), Cefbon-CF.* Gray-white (commercial). d 2.58 g/cm³; packed bulk density 0.7 g/cm³. Insol in all solvents. Fluorine content: 61-64 wt%.

$(C_2F)_n$, *poly(dicarbon monofluoride), Cefbon-C_2F.* Dark gray-black (commercial). d 2.79 g/cm³; packed bulk density 0.8 g/cm³. Insol in all solvents. Fluorine content: 49-53 wt%.

USE: Solid lubricant; as cathode material in lithium-fluoride batteries; in controlling wettability of surfaces.

4562. Graphitic Acid. Graphite oxide; graphitic oxide. This material, obtained by oxidation of graphite, was first prepd by Brodie in 1859; Hummers, Offeman, *J. Am. Chem. Soc.* **80**, 1339 (1958). Its composition is not well defined but usually given as $C_4O(OH)$: Aragon de la Cruz, Cowley, *Nature* **196**, 468 (1962), *Acta Cryst.* **16**, 531 (1963). Prepn and manufacture: Hummers, U.S. pat. **2,798,878** (1957 to National Lead); Hummers, Offeman, *loc. cit.;* Ruskin, U.S.

pats. **2,933,381** and **2,944,881** (both 1960 to Union Carbide). Crystal structure: Aragon de la Cruz, Cowley, *loc. cit.*

Very light to dark brown, or yellowish-brown solid.

USE: In rocket propellant mixtures.

4563. Gratiogenin. *(3β,9β,10α,24S)-20,24-Epoxy-3,25-dihydroxy-9-methyl-19-norlanost-5-en-11-one.* $C_{30}H_{48}O_4$; mol wt 472.71. C 76.23%, H 10.23%, O 13.54%. Tetracyclic triterpene closely related to the cucurbitacins, *q.v.* Prepd by hydrolysis of gratioside: Retzlaff, *Arch. Pharm.* **240,** 561 (1902). Structural studies: Tschesche, Heesch, *Ber.* **85,** 1067 (1952). Revised structure: Tschesche *et al., Ann.* **674,** 196 (1964).

Clusters of needles from methanol or ethanol, mp 191-196°. $[\alpha]_D^{20}$ +168° (c = 1.193 in chloroform); $[\alpha]_D^{20}$ +151° (c = 0.863 in methanol). Also reported as mp 202-203°. $[\alpha]_D^{25}$ +175 ± 3° (chloroform), (Tschesche, 1964). Freely sol in methanol, ethanol, chloroform, acetone, pyridine. Practically insol in water, ether, petr ether.

4564. Gratioside. $C_{42}H_{68}O_{14}$; mol wt 796.99. C 63.30%, H 8.60%, O 28.10%. Triterpene glycoside monohydrate from tubers of *Gratiola officinalis* L., *Scrophulariaceae:* Retzlaff, *Arch. Pharm.* **240,** 561 (1902); Tschesche *et al., Ann.* **674,** 196 (1964). Dec on hydrolysis into 1 mol gratiogenin and 2 mol glucose: Tschesche, Heesch, *Ber.* **85,** 1067 (1952).

Monohydrate, clusters of crystals from ethanol, mp 268-274°; $[\alpha]_D^{20}$ +75° (c = 0.825 in alc). Also reported mp 286-289°; $[\alpha]_D^{24}$ 81° (c = 0.83 in ethanol), (Tschesche, 1964). uv max (methanol): 295 nm (log ε 1.84). Slightly sol in water, freely in alcohol. Practically insol in ether.

4565. Gravitol(e). *2-(2-Allyl-6-methoxyphenoxy)triethylamine;* 2-methoxy-6-allylphenol diethylaminoethyl ether; Clavitol; Uterol. $C_{16}H_{25}NO_2$; mol wt 263.38. C 72.97%, H 9.57%, N 5.32%, O 12.15%. Prepn: Slotta, *Grundriss der modernen Arzneistoff-Synthese* (Stuttgart, 1931) p 118; H. Hahl, Ger. pat. **433,182** (1926 to I.G. Farbenind.), *Frdl.* **15,** 1500; *Chem. Zentr.* **1926,** II, 2223; H. Pal'gi, *Zh. Obshch. Khim. Eng. Ed.* **28,** 2275 (1958). Pharmacology: E. Käer, G. Barkan, *Arch. Exp. Pathol. Pharmakol.* **170,** 111 (1933).

Oily liq. bp_{10} 160-161°; bp_{2-3} 141-144°. $n_D^{24.5}$ 1.5075. Slightly sol in water.

Hydrochloride, $C_{16}H_{26}ClNO_2$, crystals, freely sol in water.

4566. Grayanotoxins. Toxic diterpenoids present in leaves of the various species of *Rhododendron, Kalmia,* and *Leucothoe, Ericaceae;* also found in honey from rhododendron flowers. Eighteen grayanotoxins have been isolated, the first three being the most important. Isoln of grayanotoxins I, II, III: Kakisawa *et al., Tetrahedron* **21,** 3091 (1965); of IV and V: Okuno *et al., ibid.* **26,** 4765 (1970); of V, VI, and VII: Hikino *et al., Chem. Pharm. Bull.* **18,** 2357 (1970); of VIII, IX, X, and XI: Hikino *et al., ibid.* **19,** 1289 (1971); of XII and XIII: Hikino *et al., ibid.* **20,** 422 (1972). Approaches to synthesis of the grayanotoxin skeleton: T.

Shiozaki *et al., Tetrahedron Letters* **1972,** 657; T. Kametani *et al., Chem. Pharm. Bull.* **27,** 152 (1979); *eidem, Tetrahedron Letters* **22,** 2379 (1981); *eidem, Tetrahedron* **37,** 3813 (1981). Toxicity study: H. Hikino *et al., Toxicol. Appl. Pharmacol.* **35,** 303 (1976).

	R_1	R_2	R_3
Grayanotoxin I	OH	CH_3	$COCH_3$
Grayanotoxin III	OH	CH_3	H
Grayanotoxin II	$R_1R_2 = =CH_2$		$R_3 = H$

Grayanotoxin I, $C_{22}H_{36}O_7$, *(3β,6β,14R)-grayanotoxane-3,5,6,10,14,16-hexol 14-acetate, G-I, acetylandromedol, andromedotoxin, rhodotoxin, asebotoxin.* R_1 = OH; R_2 = CH_3; R_3 = $COCH_3$. From *Leucothoe grayana* Max., *Ericaceae:* Miyajimi, Takei, *J. Agr. Chem. Soc. Japan* **10,** 1093 (1934); from *Rhododendron maximum* L., *Ericaceae:* Wood *et al., J. Am. Chem. Soc.* **76,** 5689 (1954). Identity with acetylandromedol, andromedotoxin and rhodotoxin: Tallent *et al., ibid.* **79,** 4548 (1957). Stereochemistry: Iwasa, Nakamura, *Tetrahedron Letters* **1969,** 3973; Narayanan *et al., ibid.* **1970,** 3943; Hikino *et al., Chem. Pharm. Bull.* **18,** 1071 (1970). Approach to synthesis: Okuno, Matsumoto, *Tetrahedron Letters* **1969,** 4077. Has hypotensive action: Moran *et al., J. Pharmacol. Exp. Ther.* **110,** 415 (1954). Crystals from ethyl acetate, mp 258-260° to 267-270°, depending on rate of heating. $[\alpha]_D^{25}$ −8.8° (c = 2.3 in ethanol). Sol in hot water, alcohol, acetic acid, hot chloroform; very slightly sol in benzene, ether, petr ether. LD_{50} i.p. in mice: 1.31 mg/kg (Hikino, 1976).

Grayanotoxin II, $C_{20}H_{32}O_5$, *(3β,6β,14R)-grayanotox-10-(20)-ene-3,5,6,14,16-pentol, G-II, deacetylanhydroandromedotoxin.* $R_1R_2 = CH_2=$; R_3 = H. From *L. grayana* Max., *Ericaceae:* Miyajimi, Takei, *loc. cit.* Identity with deacetylanhydroandromedotoxin: Meguri, *Yakagaku Zasshi* **79,** 1060 (1959); *C.A.* **54,** 5599g (1960). Stereochemistry: Iwasa, Nakamura, *loc. cit.;* Kumazawa, Iriye, *Tetrahedron Letters* **1970,** 927; Yasue *et al., Chem. Pharm. Bull.* **18,** 2586 (1970). Synthesis: S. Gasa *et al., Tetrahedron Letters* **1976,** 553. Columns, mp 199-200°. $[\alpha]_D^{28}$ −41.88°. LD_{50} i.p. in mice: 26.1 mg/kg (Hikino, 1976).

Grayanotoxin III, $C_{20}H_{34}O_6$, *(3β,6β,14R)-grayanotoxane-3,5,6,10,14,16-hexol, G-III, deacetylandromedotoxin.* R_1 = OH; R_2 = CH_3; R_3 = H. From *L. grayana* Max., *Ericaceae:* Miyajimi, Takei, *J. Agr. Chem. Soc. Japan* **12,** 947 (1936), *C.A.* **30,** 6747[9] (1936). Stereochemistry: Hikino *et al., Chem. Pharm. Bull.* **18,** 1071 (1970). LD_{50} i.p. in mice: 0.84 mg/kg (Hikino, 1976).

4567. Grepafloxacin. *1-Cyclopropyl-6-fluoro-1,4-dihydro-5-methyl-7-(3-methyl-1-piperazinyl)-4-oxo-3-quinolinecarboxylic acid.* $C_{19}H_{22}FN_3O_3$; mol wt 359.40. C 63.50%, H 6.17%, F 5.29%, N 11.69%, O 13.36%. Fluorinated quinolone antibacterial. Prepn: J. M. Domagala *et al.,* **PCT Int. pat. Appl. 89 06,649;** *eidem,* U.S. pat. **4,920,120** (1989, 1990 both to Warner-Lambert); S. E. Hagen *et al., J. Med. Chem.* **34,** 1155 (1991). Comparative *in vitro* activity: F. Marco *et al., J. Antimicrob. Chemother.* **33,** 647 (1994); R. C. Arduino *et al., ibid.* **34,** 403. HPLC determn in human biological matrices: J. M. Woodcock *et al.*, *FEMS Microbiol. Lett.* **119**, 315 (1994). Clinical pharmacokinetics: J. Child *et al.*, *Antimicrob. Ag. Chemother.* **39**, 513 (1995); P. J. Cook *et al.*, *J. Antimicrob. Chemother.* **35**, 317 (1995).

Dihydrate, mp 190-192°.

Hydrochloride, $C_{19}H_{22}FN_3O_3 \cdot HCl$, *OPC-17116*.

THERAP CAT: Antibacterial.

4568. Grindelia. Gum-plant (of California). Dried leaves and flowering tops of *Grindelia camporum* Greene or of *G. humilis* H. & A. *(G. cuneifolia* Auth.), Compositae. *Habit.* North America (California). *Constit.* Volatile oil, over 20% resin, grindelol, saponin, tannin, robustic acid.

THERAP CAT: Expectorant.

4569. Grindelic Acid. *[1'R-[1'α(S*),4'Aα,8'αβ]]-4,4'a,-5,5',6',7',8',8'a-Octahydro-2',5,5',5',8'a-pentamethylspiro-[furan-2(3H),1'(4'H)-naphthalene]-5-acetic acid; 9,13-epoxy-labd-7-en-15-oic acid.* $C_{20}H_{32}O_3$; mol wt 320.47. C 74.96%, H 10.06%, O 14.98%. From the resin of *Grindelia robusta* Nutt., Compositae. Isoln and structure: Panizzi *et al.*, *Gazz. Chim. Ital.* **92**, 522 (1962). Configuration: Mangoni, Belardini, *ibid.* **92**, 1379 (1962); **93**, 455, 465 (1963). Synthesis: M. Adinolfi, *et al.*, *ibid.* **106**, 625 (1976).

Crystals from acetic acid, mp 100-101°. $[\alpha]_D$ 102.2°. Methyl ester, $C_{21}H_{34}O_3$, crystals from methanol, mp 70-70.5°. $[\alpha]_D$ −134.1° (c = 1.46 in methanol).

4570. Griscein. Antibiotic substance produced by strains of *Streptomyces griseus.* Isoln: Reynolds *et al.*, *Proc. Soc. Exp. Biol. Med.* **64**, 50 (1947); Reynolds, Waksman, *J. Bacteriol.* **55**, 739 (1948). Improved method of isoln: F. A. Kuehl, L. Chaiet, U.S. pat. **2,505,053** (1950 to Merck); F. A. Kuehl *et al.*, *J. Am. Chem. Soc.* **73**, 1770 (1951). Analysis of composition: $C_{40}H_{61}FeN_{10}O_{20}S$. Degradation of griscein by acid hydrolysis yielded 3-methyluracil and at least two amino acids. One of the acids appears to be glutamic acid: F. A. Kuehl *et al., loc. cit.* Probably is a mixture of components; similar or identical to albomycin (*q.v.*): Stapley, Ormond, *Science* **125**, 587 (1957); Turková *et al.*, *Coll. Czech. Chem. Commun.* **31**, 2444 (1966). Toxicity study: V. I. Aksenov, *Veterinariya (Moscow)* **12**, 93 (1974), *C.A.* **83**, 54127d (1975). Biosynthesis: V. V. Kuklin *et al., Antibiotiki* **25**, 403 (1980), *C.A.* **93**, 146173a (1980).

Amorphous red powder. Sol in water; slightly sol in 95% alcohol. Practically insol in abs alcohol, ether, acetone, chloroform, benzene. The activity remains unchanged when an aq soln is heated to 100° for 10 min. LD_{50} in mice (mg/kg): 600 orally; 34 s.c. (Aksenov).

4571. Griseofulvin. *(1'S-trans)-7-Chloro-2',4,6-trimethoxy-6'-methylspiro[benzofuran-2(3H),1'-[2]cyclohexene]-3,4'-dione; 7-chloro-4,6-dimethoxycoumaran-3-one-2-spiro-1'-(2'-methoxy-6'-methylcyclohex-2'-en-4'-one)*; amudane; Curling factor; Fulcin; Fulvicin; Grifulvin; Grisactin; Grisefuline; Grisovin; Gris-PEG; Grysio; Lamoryl; Likuden; Neo-Fulcin; Polygris; Poncyl-FP; Spirofulvin; Sporostatin. $C_{17}H_{17}ClO_6$; mol wt 352.77. C 57.88%, H 4.86%, Cl 10.05%, O 27.21%. Antibiotic substance produced by *Penicillium griseofulvum* Dierckx and by *P. janczewskii* Zal. [same as *P. nigricans* (Banier)Thom]. Isoln: Oxford *et al., Biochem. J.* **33**, 240 (1939); Brian *et al., Trans. Brit. Mycol. Soc.* **29**, 173 (1946); Hockenhull, Dorey *et al.*, U.S. pats. **3,069,328, 3,069,329** (both 1962 to Glaxo). Structure: Grove *et al., Chem. & Ind. (London)* **1951**, 219; *J. Chem. Soc.* **1952**, 3977. Stereochemistry: MacMillan, *ibid.* **1959**, 1823; Brown, Sim, *ibid.* **1963**, 1050. Total synthesis: Brossi *et al., Helv. Chim. Acta* **43**, 1444, 2071 (1960); Taub *et al., Tetrahedron* **19**, 1 (1963); Stork, Tomasz, *J. Am. Chem. Soc.* **86**, 471 (1964); S. Danishefsky, F. J. Walker, *ibid.* **101**, 7018 (1979). Conformation: Levine, Hicks, *Tetrahedron Letters* **1971**, 311.

Crystal structure: G. Malmros *et al., Cryst. Struct. Commun.* **6**, 463 (1977). Review and evaluation of studies of carcinogenic action in laboratory animals: *IARC Monographs* **10**, 153-161 (1976). *Review:* Grove, *Quart. Rev.* **17**, 1 (1963); Huber in *Antibiotics* vol. 3, J. W. Corcoran, F. E. Hahn, Eds. (Springer-Verlag, New York, 1975) pp 606-613. Comprehensive description: E. R. Townley, *Anal. Profiles Drug Subs.* **8**, 219-249 (1979).

Stout octahedra or rhombs from benzene, mp 220°. $[\alpha]_D^{17}$ +370° (satd CHCl₃ soln). uv max: 286, 325 nm. Soly in DMF at 25°: 12 to 14 g/100 ml. Slightly sol in ethanol, methanol, acetone, benzene, CHCl₃, ethyl acetate, acetic acid. Practically insol in water, petr ether.

THERAP CAT: Antifungal.

THERAP CAT (VET): Antifungal.

4572. Griseoviridin. GV. $C_{22}H_{27}N_3O_7S$; mol wt 477.54. C 55.33%, H 5.70%, N 8.80%, O 23.45%, S 6.71%. Antibiotic substance obtained together with Etamycin (viridogrisein) from *Streptomyces griseus:* Bartz *et al., Antibiot. Ann.* **1954-1955**, 777. Characterization: Ames *et al., J. Chem. Soc.* **1955**, 4260; Ames, Bowman, *ibid.* **1955**, 4264. Structure studies: Fallona *et al., J. Am. Chem. Soc.* **84**, 4162 (1962); *eidem, Can. J. Chem.* **42**, 371, 394 (1964). Revised structure: G. I. Birnbaum, S. R. Hall, *J. Am. Chem. Soc.* **98**, 1926 (1976). Absolute configuration: B. W. Bycroft, T. J. King, *J. Chem. Soc. Perkin Trans. I* **1976**, 1996. Production of griseoviridin and viridogrisein: Bartz *et al.*, U.S. pat. **3,023,204** (1962 to Parke, Davis). Synthetic study: A. I. Meyers, R. A. Amos, *J. Am. Chem. Soc.* **102**, 870 (1980).

Polymorphic crystals, dec 158-166°, or 194-200°, or 230-240° depending on the crystal modification. $[\alpha]_D^{27}$ −237° (c = 0.5 in methanol). uv max (methanol): 221 nm ($E^{1\%}_{1cm}$ 870). Sol in pyridine, moderately sol in lower alcohols, sparingly sol in water and nonpolar solvents.

Diacetate, $C_{26}H_{31}N_3O_9S$, needles from methanol + ether + petr ether, dec 137-140°, $[\alpha]_D^{27}$ −230° (c = 0.44 in methanol). uv max (ethanol): 218 nm (ε 41800).

4573. Guaiac. Gum guaiac; resin guaiac; guaiacum. Resin from wood of *Guajacum officinale* L. or *G. sanctum* L., Zygophyllaceae. *Constit.* About 70% α- and β-guaiaconic acids, about 11% guaiacic acid, related compds and guaiaretic acid, 15% vanillin, guaiac yellow, guaiac saponin (guaiacin). Use as clinical reagent for occult blood: R. H. Wilkinson, W. A. F. Penfold, *Lancet* **2**, 847 (1969). Acute toxicity: P. M. Jenner *et al., Food Cosmet. Toxicol.* **2**, 327 (1964).

Brown or greenish-brown, irregular lumps. mp 85-90°. Insol in water. Freely sol in alcohol, chloroform, ether, creosote, soln of chloral hydrate, alkalies; slightly sol in benzene, carbon disulfide. *Incompat:* Of liquid preparations: Mineral acids, acacia, ferric chloride, gold chloride, permanganates, spirit nitrous ether, water. LD_{50} orally in rats: > 5000 mg/kg (Jenner).

USE: Clinical reagent (blood or hemoglobin).

4574.　Guaiac-Copper Sulfate Paper. Schönbein-Pagenstecher's paper. White filter paper impregnated first with an alc soln of guaiac resin, then after drying, with an aq soln of cupric sulfate. Used for detecting HCN, a trace of which colors the paper blue.

4575.　Guaiacol. *2-Methoxyphenol;* methylcatechol; *o*-hydroxyanisole; 1-hydroxy-2-methoxybenzene; Anastil. $C_7H_8O_2$; mol wt 124.14. C 67.73%, H 6.50%, O 25.78%. Isolated from guaiac resin: Sobrero, *Ann.* **48**, 19 (1843); from hardwood tar: McGinness *et al., Tappi* **43**, 1027 (1960). Prepd by mercuric oxide oxidation of lignin: Lewis, Pearl, U.S. pat. 2,433,227 (1947 to Sulphite Prod.); by oxidation of anisole with trifluoroperoxyacetic acid: McClure, Williams, *J. Org. Chem.* **27**, 627 (1962); from acetovanillone + $ZnCl_2$: Read, U.S. pat. 3,057,927 (1962 to Ontario Res. Found.); from the diazonium salt of *o*-anisidine: Herbst, Ger. pat. 1,148,236 (1963 to Hoechst). Toxicity data: Taylor *et al., Toxicol. Appl. Pharmacol.* **6**, 378 (1964).

White or slightly yellow cryst mass or colorless to yellowish, very refractive liquid; characteristic odor. Darkens on exposure to air and light. d (crystals) 1.129; d (liq) ~1.112. Solidif 28°, but may remain liq for a long time even at a much lower temp. bp 204-206°; bp_4 53-55°. One gram dissolves in 60-70 ml water, 1 ml glycerol; miscible with alcohol, chloroform, ether, oils, glacial acetic acid. Slightly sol in petr ether; sol in NaOH soln; with moderately concd KOH it forms a sparingly sol compd. *Protect from light.* LD_{50} orally in rats: 725 mg/kg (Taylor).

Phenylacetate, $C_{15}H_{14}O_3$, *Gujaphenyl, Gunyl.*

THERAP CAT: Expectorant.

THERAP CAT (VET): Expectorant.

4576.　Guaiacol Benzoate. *2-Methoxyphenol benzoate;* benzoylguaiacol; Benzosol. $C_{14}H_{12}O_3$; mol wt 228.25. C 73.67%, H 5.30%, O 21.03%. $CH_3OC_6H_4OOCC_6H_5$. Prepn: Lynch, Moore, *Can. J. Chem.* **40**, 1461 (1962).

Odorless, almost tasteless, cryst powder. mp 57-58°. Slightly sol in water; sol in hot alcohol, ether, chloroform.

THERAP CAT: Expectorant.

4577.　Guaiacol Carbonate. Carbonic acid bis(2-methoxyphenyl) ester; guaiacol carbonic acid neutral ester; carbonic acid guaiacol ether; Duotal. $C_{15}H_{14}O_5$; mol wt 274.27. C 65.69%, H 5.14%, O 29.17%. Prepd by the action of phosgene on a concd soln of guaiacol in aq NaOH: Ger. pat. 58,129 (to Bayer); S. P. Schotz, *Synthetic Organic Compounds* (London, 1925) p 137; J. Schwyzer, *Die Fabrikation Pharmazeutischer und Chemischtechnischer Produkte* (Berlin, 1931) p 212; H. P. Kaufmann, *Arzneimittel-Synthese* (Springer-Verlag, 1953) p 607.

Odorless needles from ethanol, mp 88.1°. Faint taste of guaiacol. Practically insol in water. At room temp 1 g dissolves in 60 ml ethanol, 1 ml chloroform, 18 ml ether. Much more sol in hot alcohol, hot benzene, hot chloroform. Slightly sol in liq fatty acids. At 20°, 100 g of glycerol (d 1.2612) will dissolve 0.043 g of guaiacol carbonate. Difficult to hydrolyze. Upon ingestion much is excreted unchanged.

THERAP CAT: Expectorant.

THERAP CAT (VET): Expectorant.

4578.　Guaiacol Phosphate. $C_{21}H_{21}O_7P$; mol wt 416.37. C 60.58%, H 5.08%, O 26.90%, P 7.44%. $(CH_3OC_6H_4O)_3$-PO. Obtained by the action of $POCl_3$ upon guaiacol sodium: Kucherov, *Zh. Obshch. Khim.* **19**, 126 (1949). Prepn of sodium salt: Auger, Dupuis, *Compt. Rend.* **146**, 1151 (1908).

Crystals, mp 91°. bp_3 275-280°. Practically insol in water, ether. Sol in hot alc, freely in chloroform, toluene, acetone.

Sodium salt, *Novocol.*

THERAP CAT: Expectorant.

4579.　Guaiacol Valerate. $C_{12}H_{16}O_3$; mol wt 208.26. C 69.21%, H 7.74%, O 23.05%. $CH_3OC_6H_4OOCC_4H_9$. Has antibacterial properties: Kellner, Kober, *Arzneimittel-Forsch.* **5**, 224 (1955).

Yellowish liq. d 1.05. bp 265°. Practically insol in water. Freely sol in alcohol, benzene, chloroform, ether.

USE: Detection of copper: Shapiro, *Zh. Analit. Khim.* **17**, 248 (1962).

4580.　Guaiapate. *1-[2-[2-[2-(2-Methoxyphenoxy)ethoxy]ethoxy]ethyl]piperidine;* MG-5454; Klamar. $C_{18}H_{29}NO_4$; mol wt 323.43. C 66.84%, H 9.04%, N 4.33%, O 19.79%. Basic ether deriv of guaiacol, *q.v.* Prepn: M. Carissimi, F. Ravenna, Fr. pat. 1,386,633; *eidem,* U.S. pat. 3,320,254 (1965, 1967 both to Maggioni); M. Carissimi *et al., J. Med. Chem.* **8**, 542 (1965). Pharmacological and clinical studies: G. Barbi, *Gazz. Med. Ital.* **127**, 13 (1968). Analytical study: S. Cavicchi *et al., Boll. Chim. Farm.* **108**, 682 (1969).

Liq, $bp_{0.5}$ 190-193°. LD_{50} i.p. in mice: 254 μmoles/kg (Carissimi).

THERAP CAT: Antitussive.

4581.　Guaiazulene. *1,4-Dimethyl-7-(1-methylethyl)azulene;* 7-isopropyl-1,4-dimethylazulene; *S*-guaiazulene; AZ 8; AZ 8 Beris; Eucazulen; Kessazulen; Vaumigan. $C_{15}H_{18}$; mol wt 198.31. C 90.85%, H 9.15%. Isoln from chamomile oil: Sorm *et al., Coll. Czech. Chem. Commun.* **16**, 626 (1951); from guaiac wood oil: Joos, Swiss pat. 314,487 (1956), *C.A.* **52**, 443b (1958). Total synthesis: Plattner *et al., Helv. Chim. Acta* **32**, 2452 (1949); Sorm *et al., Coll. Czech. Chem. Commun.* **16**, 168 (1951); Jacob *et al., Tetrahedron* **20**, 2821 (1964); J. Mukherjee *et al., J. Am. Chem. Soc.* **101**, 251 (1979). Pharmacokinetics of guaiazulene soluble in animals: H. Mukai *et al., J. Pharmacobio-Dyn.* **8**, 329, 337 (1985). Effect on gastric and duodenal ulcers in rats: S. Okabe *et al., Nippon Yakurigaku Zasshi* **88**, 467 (1986), *C.A.* **106**, 43769 (1987).

Blue oil. bp_{10} 165-170°.

3-Sulfonate sodium salt, $C_{15}H_{17}NaO_3S$, *5-isopropyl-3,8-dimethyl-1-azulenesulfonic acid sodium salt, sodium gualenate, guaiazulene soluble, Azulon.*

Trinitrobenzene deriv, $C_{15}H_{18}\cdot C_6H_3N_3O_6$, violet to black needles from ethanol, mp 151°.

THERAP CAT: Anti-inflammatory; antiulcerative.

4582.　Guaifenesin. *3-(2-Methoxyphenoxy)-1,2-propanediol;* glycerol mono(2-methoxyphenyl) ether; glycerol α-(2-methoxyphenyl) ether; guaiacyl glyceryl ether; glyceryl guaiacyl ether; glycerol guaiacolate; α-glyceryl guaiacol ether; *o*-methoxyphenyl glyceryl ether; 1,2-dihydroxy-3-(2-

methoxyphenoxy)propane; guaiacol glyceryl ether; guaiphenesin; guaiacuran; MY-301; XL-90; Actifed-C; Calmipan; Colrex Expectorant; Equicol; Glycodex; Guaiamar; Guayanesin; Miocaina; Myocaine; Myoscain; Oresol; Oreson; Relaxil G; Reorganin; Respenyl; Resyl; Robitussin; Sirotol; Tenntuss; Tulyn. $C_{10}H_{14}O_4$; mol wt 198.22. C 60.59%, H 7.12%, O 32.29%. Centrally acting muscle relaxant with expectorant properties. Prepn: Marle, *J. Chem. Soc.* **101**, 305 (1912); Yale *et al.*, *J. Am. Chem. Soc.* **72**, 3710 (1950); Roviralta, Astoul, **Span.** pat. **212,920** (1954), *C.A.* **49**, 8332b (1955). Prepn from 2-methoxyphenol and glycidol: W. Merk *et al.*, **Ger.** pat. **3,106,995** corresp to U.S. pat. **4,390,732** (1982, 1983 to Degussa AG). GLC determn in blood: W. R. Maynard, R. B. Bruce, *J. Pharm. Sci.* **59**, 1346 (1970). Clinical use in chronic respiratory disease: D. G. Workman *et al.*, *Curr. Ther. Res.* **7**, 665 (1965). Clinical efficacy as antitussive: J. J. Kuhn *et al.*, *Chest* **82**, 713 (1982). Pharmacokinetics and cardiopulmonary effects in horses: J. A. E. Hubbell *et al.*, *Am. J. Vet. Res.* **41**, 1751 (1980). Use in equine anesthesia: J. L. Grandy, W. N. McDonell, *J. Am. Vet. Med. Assoc.* **176**, 619 (1980); G. J. Brouwer, *Eq. Vet. J.* **17**, 133 (1985).

Minute rhombic prisms from ether, mp 78.5-79°. bp_{19} 215°. Slightly bitter aromatic taste. One gram dissolves in 20 ml water at 25°. Much more sol in hot water; freely sol in ethanol; sol in chloroform, glycerol, propylene glycol, DMF; moderately sol in benzene. Practically insol in petr ether.

Mixture with theophylline (1:1), *guaithylline, Eclabron*.

THERAP CAT: Expectorant; guaithylline also as bronchodilator.

THERAP CAT (VET): Expectorant; muscle relaxant.

4583. Guaiol. *[3S-(3α,5α,8α)]-1,2,3,4,5,6,7,8-Octahydro-α,α,3,8-tetramethyl-5-azulenemethanol;* 3,8-dimethyl-5-(α-hydroxyisopropyl)-Δ⁹-octahydroazulene; champaca camphor; champacol; guaiac alcohol; Guajol. $C_{15}H_{26}O$; mol wt 222.37. C 81.02%, H 11.79%, O 7.19%. A sesquiterpene alc from guaiac wood: *Michelia champaca* L., *Magnoliaceae*: also from oil of wood of *Bulnesia sarmienti* Lorentz, *Zygophyllaceae*. Isoln: Plattner, Lemay, *Helv. Chim. Acta* **23**, 897 (1940). Structure: Plattner, Magyar, *ibid.* **25**, 581 (1942). Stereochemistry: Takeda, Minato, *Tetrahedron Letters* **1960**(22), 33; Minato, *Chem. Pharm. Bull.* **9**, 625 (1961); *idem*, *Tetrahedron* **18**, 365 (1962). Total synthesis of *dl*-form: Buchanan, Young, *Chem. Commun.* **1971**, 643; *eidem*, *J. Chem. Soc. Perkin Trans. I* **1973**, 2404; Marshall *et al.*, *Tetrahedron Letters* **1971**, 885; Marshall, Greene, *J. Org. Chem.* **37**, 982 (1972); Andersen, Uh, *Tetrahedron Letters* **1973**, 2079.

Trigonal pyramidal crystals from alc, mp 91°. d_{20}^{100} 0.9074. bp_{760} 288° (slight decompn); bp_{17} 165°; bp_{10} 148°. $[\alpha]_D^{20}$ −30° (c = 4 in alc). n_D^{100} 1.4716. Sol in alc, ether. Insol in water. Methyl ether, $C_{16}H_{28}O$, liq, d_4^{25} 0.9332. bp_9 142°. $[\alpha]_D^{20}$ −31.8°. $n_D^{18.5}$ 1.4896.

Note: The name "guaiol" is also applied to 1,2-dimethylacrolein, isolated from guaiacum resin.

4584. Guamecycline. *[4S-(4α,4aα,5aα,6β,12aα)]-N-[[4-[[(Aminoiminomethyl)amino]iminomethyl]-1-piperazinyl]methyl]-4-(dimethylamino)-1,4,4a,5,5a,6,11,12a-octahydro-3,-*

6,10,12,12a-pentahydroxy-6-methyl-1,11-dioxo-2-naphthacenecarboxamide; tetrabiguanide; "xanthomycin"; xantomicina. $C_{29}H_{38}N_8O_8$; mol wt 626.67. C 55.58%, H 6.11%, N 17.88%, O 20.42%. Prepn: **Brit.** pat. **1,042,207** (1966 to Societa Prodotti Antibiotici), *C.A.* **65**, 16922g (1966). Clinical studies: O. Restivo, G. Sindoni, *Gazz. Med. Ital.* **127**, 17 (1968).

Dihydrochloride, $C_{29}H_{38}N_8O_8 \cdot 2HCl$, *Terratrex, Xantociclina*.

THERAP CAT: Antibacterial.

4585. Guanabenz. *2-[(2,6-Dichlorophenyl)methylene]-hydrazinecarboximidamide; [(2,6-dichlorobenzylidene)amino]guanidine; N-(2,6-dichlorobenzylidene)-N'-amidinohydrazine;* NSC-68982. $C_8H_8Cl_2N_4$; mol wt 231.08. C 41.58%, H 3.49%, Cl 30.68%, N 24.25%. α₂-Adrenergic agonist. Prepn: J. Yates, E. Haddock, **Brit.** pat. **1,019,120** (1966 to Shell), *C.A.* **64**, 11132h (1966). Use as antihypertensive: W. J. Houlihan *et al.*, **Ger.** pat. **1,804,634** (1969 to Sandoz), *C.A.* **71**, 89976j (1969). Pharmacology: T. Baum *et al.*, *J. Pharmacol. Exp. Ther.* **171**, 276 (1970); E. Lampa *et al.*, *Experientia* **36**, 228 (1980). Disposition of ¹⁴C-guanabenz in humans: R. H. Meacham *et al.*, *Clin. Pharmacol. Ther.* **27**, 44 (1980). Mechanism of action: G. F. DiBona, *J. Cardiovasc. Pharmacol.* **6**, Suppl. 3, S543 (1984). Radioimmunoassay determn in plasma: H. Tatsumi, *Arzneimittel-Forsch.* **34**, 1704 (1984). Clinical studies: A. Reppelli *et al.*, *Boll. Soc. Ital. Cardiol.* **23**, 177 (1978); C. V. Ram *et al.*, *J. Clin. Pharmacol.* **19**, 148 (1979). Clinical studies in opiate withdrawal: F. S. Tennant, R. A. Rawson, *Nat. Inst. Drug Abuse Res. Monogr. Ser.* **49**, 338 (1984); J. T. Murphy, *Drug Intell. Clin. Pharm.* **19**, 32 (1985). Review of pharmacodynamic properties and therapeutic efficacy: B. Holmes *et al.*, *Drugs* **26**, 212-229 (1983). Comprehensive description: C. M. Shearer in *Analytical Profiles of Drug Substances* vol. 15, K. Florey, Ed. (Academic Press, New York, 1986) pp 319-336.

White solid from acetonitrile, mp 227-229° (dec).

Monoacetate, $C_{10}H_{12}Cl_2N_4O_2$, *Wy-8678, Rexitene, Tenelid, Wytensin*. Solid, mp 192.5° (dec). Soly at 25° (mg/ml): water 11; alcohol 50; propylene glycol 100; chloroform 0.6; ethyl acetate 1.

THERAP CAT: Antihypertensive.

4586. Guanacline. *[2-(3,6-Dihydro-4-methyl-1(2H)-pyridinyl)ethyl]guanidine; N-(2-guanidinoethyl)-4-methyl-Δ³-piperidine; N-(2-guanidinoethyl)-4-methyl-1,2,3,6-tetrahydropyridine; 1-(2-guanidinoethyl)-4-methyl-1,2,3,6-tetrahydropicoline;* cyclazenin; FBA-1464. $C_9H_{18}N_4$; mol wt 182.27. C 59.31%, H 9.95%, N 30.74%. Prepn: **Fr.** pat. **M3016** (1965 to Bayer), *C.A.* **62**, 16206b (1965).

Monosulfate, $C_9H_{18}N_4 \cdot H_2SO_4$, *Leron*. Crystals from ethanol + ethyl acetate, dec 185-186°. Unstable in air.

Neutral sulfate, crystals, mp 232.5°. Stable in air.
THERAP CAT: Antihypertensive.

4587. Guanadrel. *(1,4-Dioxaspiro[4.5]dec-2-ylmethyl)-
guanidine.* $C_{10}H_{19}N_3O_2$; mol wt 213.28. C 56.32%, H
8.98%, N 19.70%, O 15.00%. Orally active postganglionic
synpathetic inhibitor. Prepn: W. R. Hardie, J. E. Aaron,
S. Afr. pat. 67 06,328; eidem, U.S. pat. 3,547,951 (1968,
1970 both to Cutter). Antihypertensive activity: L. Hansson et al., Clin. Pharmacol. Ther. 14, 204 (1973). In vitro
adrenergic neuron blocking activity: L. Roller, Aust. J.
Pharm. Sci. 5, 35 (1976). Pharmacologic study: E. M.
Johnson, F. E. Hunter, Biochem. Pharmacol. 28, 1525
(1979). Effect on patients with thyrotoxicosis: S. Ruben-
feld et al., Arch. Intern. Med. 138, 1106 (1978). Review of
pharmacology and efficacy in hypertension: F. A. Finnerty
Jr., R. N. Brogden, Drugs 30, 22-31 (1985).

Sulfate, $(C_{10}H_{19}N_3O_2)_2 \cdot H_2SO_4$, *CL-1388R, U-28288D,
Anarel, Hylorel.* Cryst from methanol/ethanol, mp 213.5-
215°.
THERAP CAT: Antihypertensive.

4588. Guanazodine. *[(Octahydro-2-azocinyl)methyl]-
guanidine;* α-guanidinomethylheptamethylenimine; 1-azacy-
clooct-2-ylmethylguanidine. $C_9H_{20}N_4$; mol wt 184.28. C
58.66%, H 10.94%, N 30.40%. Hypotensive agent, related
structurally to guanethidine, q.v. Prepn of the sulfate salt:
J. Rakoczi et al., Hung. pat. 155,990; eidem, U.S. pat. 3,-
856,778 (1969, 1974 both to EGYT). Pharmacokinetics in
man: T. Past et al., Int. J. Clin. Pharmacol. Ther. Toxicol. 8,
111 (1973). General pharmacology: H. Iwata et al., Oyo
Yakuri 14, 235 (1977), C.A. 88, 16070 (1977). Analytical
study: E. Zollner-Ivan, Acta Pharm. Hung. 48, 76 (1978).
Clinical studies in hypertension: S. Dobi et al., Ther. Hung.
28, 60 (1980); Z. Herpai et al., ibid. 181.

Sulfate monohydrate, $C_9H_{20}N_4 \cdot H_2SO_4 \cdot H_2O$, *EGYT-739,
Calnegyt, Sanegyt.* White crystals from water, mp 239-241°.
LD_{50} in mice (mg/kg): 2450 orally, 700 s.c., 165 i.v. (Rakoczi, 1974).
THERAP CAT: Antihypertensive.

4589. Guanethidine. *[2-(Hexahydro-1(2H)-azocinyl)-
ethyl]guanidine;* [2-(octahydro-1-azocinyl)ethyl]guanidine;
1-(2-guanidinoethyl)octahydroazocine; 2-(1'-azacyclooctyl)-
ethylguanidine; N-(2-perhydroazocin-1-ylethyl)guanidine; 2-
(1-N,N-heptamethylenimino)ethylguanidine; oktadin; oktatenzin; Su-5864; Eutensol; Dopom; Octatensine; Oktatensin;
Sanotensin; Abapresin. $C_{10}H_{22}N_4$; mol wt 198.31. C
60.57%, H 11.18%, N 28.25%. Prepn: Maxwell et al., Experientia 15, 267 (1959); Mull, U.S. pat. 2,928,829 (1960 to
Ciba). Improved process: Mull, U.S. pats. 3,006,913; 3,-
055,882 (1961, 1962 to Ciba).

Sulfate, $(C_{10}H_{22}N_4)_2 \cdot H_2SO_4$, *Guethine, Iporal, Isobarin,
Ismelin.* Crystals from dil ethanol, mp 276-281° (dec).
Monosulfate, $C_{10}H_{22}N_4 \cdot H_2SO_4$, component of *Esimil.*
THERAP CAT: Antihypertensive.

4590. Guanfacine. *N-(Aminoiminomethyl)-2,6-dichlo-
robenzeneacetamide; N-amidino-2-(2,6-dichlorophenyl)acet-*

amide; [(2,6-dichlorophenyl)acetyl]guanidine. $C_9H_9Cl_2N_3O$;
mol wt 246.10. C 43.93%, H 3.69%, Cl 28.81%, N 17.07%,
O 6.50%. Centrally acting α_2-adrenoceptor agonist. Prepn:
J. B. Bream, C. W. Picard, Fr. pat. 1,584,670; eidem, U.S.
pat. 3,632,645 (1969, 1972, both to Wander). Pharmacology: D. M. Coward et al., Arzneimittel-Forsch. 27, 2326
(1977); H. F. Oates et al., Arch. Int. Pharmacodyn. Ther.
231, 148 (1978); eidem, Clin. Exp. Pharmacol. Physiol. 6, 61
(1979). Determn in biological fluids: M. Guerrat et al., J.
Pharm. Sci. 68, 219 (1979). Kinetics: Y. A. Weiss et al.,
Clin. Pharmacol. Ther. 25, 283 (1979). Clinical studies: U.
C. Dubach et al., Arzneimittel-Forsch. 27, 674 (1977); P.
MacCarthy et al., Clin. Exp. Pharmacol. Physiol. 5, 187
(1978). Symposium on pharmacology, pharmacokinetics,
clinical efficacy and comparison with clonidine: Am. J. Cardiol. 57, 1E-61E (1986). Review: Brit. J. Clin. Pharmacol.
10, Suppl. 1, 1S-208S (1980).

White grains from methanol/ether, mp 225-227°.
Hydrochloride, $C_9H_9Cl_2N_3O \cdot HCl$, *BS-100-141, LON-798,
Estulic, Tenex.* White needles, mp 213-216°. LD_{50} in mice:
165 mg/kg orally (Coward).
THERAP CAT: Antihypertensive.

4591. Guanidine. Aminomethanamidine; carbamamidine; carbamidine; aminoformamidine; iminourea. CH_5N_3;
mol wt 59.07. C 20.33%, H 8.53%, N 71.14%. $(NH_2)_2$-
C=NH. Strong organic base existing primarily as guanidinium ion at physiological pH. Found in turnip juice,
mushrooms, corn germ, rice hulls, mussels, earthworms.
Has antiviral, antifungal, antipyretic and muscle stimulatory activity. Prepn of guanidine nitrate from dicyanodiamide and ammonium nitrate: Smith et al., Ind. Eng.
Chem. 23, 1124 (1931); Davis, Org. Syn. coll. vol. I (2nd ed.,
1941) p 302; from SO_2, CO_2, and NH_3: Boivin, U.S. pat.
2,762,843 (1956); from urea: Mackay, U.S. pat. 2,590,257
(1952 to Am. Cyanamid); Craig, Minor, U.S. pat. 3,009,949
(1961 to Deere); Shaver, U.S. pat. 3,108,999 (1963 to Monsanto); from ammonium thiocyanate or thiourea + ammonia: Watt, Makosky, Ind. Eng. Chem. 46, 2599 (1954).
Review of mode of action: D. R. Tershak et al., "Guanidine" in Handbook of Experimental Pharmacology Vol. 61,
G. V. R. Born et al., Eds. entitled "Chemotherapy of Viral
Infections" P. E. Came, L. A. Caliguiri, Eds. (Springer-
Verlag, New York, 1982) pp 343-375. Review of experimental and clinical uses: F. Davidoff, N. Engl. J. Med. 289,
141-146 (1973). General reviews: M. Schenck, Pharmazie
3, 5 (1948); G. Schaefer, "Guanidines and Biguanidines" in
International Encyclopedia of Pharmacology and Therapeutics 107, M. Erecinska, Ed. (Pergamon, Oxford 1981) pp
165-185.

Deliquescent, cryst mass, mp ∼50°. pKa ∼12.5. Absorbs CO_2 from air. Very sol in water, alcohol. On heating
to 160° it is converted to melamine and NH_3. *Keep well
closed.* LD orally in rabbits: 500 mg/kg, Handbook of
Toxicology vol. I, W. S. Spector, Ed. (Saunders, Philadelphia, 1956) p 152.

Hydrochloride, $CH_5N_3 \cdot HCl$, cryst powder. Freely sol in
water, alcohol. Aq soln is neutral.
Nitrate, $CH_5N_3 \cdot HNO_3$, cryst powder, mp 214°. Soluble
in 10 parts water; in alcohol. Aq soln is neutral.

4592. Guanidinium Aluminum Sulfate Hexahydrate.
Guanidine compd with aluminum sulfate; GASH. CH_{18}-
$AlN_3O_{14}S_2$; mol wt 387.28. C 3.10%, H 4.68%, Al 6.97%, N
10.85%, O 57.84%, S 16.56%. Prepd from an aq soln of an
equimolecular mixture of guanidine sulfate and aluminum
sulfate: Ferraboschi, Proc. Cambridge Phil. Soc. 14, 473
(1908); Holden et al., Phys. Rev. 101, 962 (1956); Wieder,
Proc. IRE 45, 1094 (1957). Crystal structure: Schein et al.,
J. Chem. Phys. 47, 5183 (1967). Crystal growth: T. A.
Zarembovskaya, Zh. Fiz. Khim. 45, 2504 (1971). Optical

properties: P. M. Nikolic *et al.*, *Fizika* (Zagreb) **12**, Suppl. 1, 165 (1980).

Large hexagonal plates belonging to the trigonal system with perfect basal cleavage. Has ferroelectric properties.

4593. Guanine. *2-Amino-1,7-dihydro-6H-purin-6-one; 2-aminohypoxanthine.* $C_5H_5N_5O$; mol wt 151.13. C 39.74%, H 3.33%, N 46.34%, O 10.59%. Constituent of nucleic acids; widespread in animal and plant kingdom. First isolated from guano. Syntheses: Fisher, *Ber.* **30**, 2226 (1897); Traube, *ibid.* **33**, 1371 (1900); **Ger. pats.** **134,984** (1903), **158,591** (1903), **162,336** (1904). Prepn of ^{15}N-isotopic guanine following Traube's synthesis: Plentl, Schoenheimer, *J. Biol. Chem.* **153**, 203 (1944). Several desmotropic forms. Crystal structure of hydrochloride monohydrate: Broomhead, *Acta Cryst.* **4**, 92 (1951). *Reviews:* Shapiro, *Progr. Nucleic Acid Res. Mol. Biol.* **8**, 73-112 (1968); Ts'o, "Bases, Nucleosides and Nucleotides" in *Basic Principles in Nucleic Acid Chemistry* vol. 1, P. O. P. Ts'o, Ed. (Academic Press, New York, 1974) pp 453-584. *See also* Nucleic Acids.

Usually amorphous. Small rhombic crystals by slow evaporation of aq soln contg large excess of NH_3. Dec above 360° with partial sublimation. uv max (pH 6.2): 246, 275 nm ($\epsilon \times 10^{-3}$ 10.7, 8.1). Freely sol in ammonia water, aq KOH solns, dil acids; sparingly sol in alcohol, ether. Almost insol in water. pKb 3.22; pKa 9.92, detd at 40°. Many compds with acids, bases and metals have been prepared.

Hydrochloride monohydrate, $C_5H_5N_5O \cdot HCl \cdot H_2O$, cryst powder. Loses H_2O at 100°, HCl at 200°. Practically insol in water, alcohol, ether; sol in acidulated water.

4594. Guano. Bird manure. The dried excrements of sea birds (cormorants) and bats from coastal islands of Peru, Chile, West Indies, and Africa. Usually mixed with feathers and bones. Contains about 9% nitrogen, 6% phosphorus, 2% potassium, and 15-20% moisture. Used as fertilizer.

4595. Guanochlor. *2-[2-(2,6-Dichlorophenoxy)ethyl]hydrazinecarboximidamide; [[2-(2,6-dichlorophenoxy)ethyl]-amino]guanidine.* $C_9H_{12}Cl_2N_4O$; mol wt 263.13. C 41.08%, H 4.60%, Cl 26.95%, N 21.29%, O 6.08%. Prepn of free base and sulfate: **Belg. pat.** **629,613** (1963 to Pfizer), *C.A.* **60**, 14437d (1964).

Sulfate, $(C_9H_{12}Cl_2N_4O)_2 \cdot H_2SO_4$, *Vatensol.* Crystals from water, mp 214°.
THERAP CAT: Antihypertensive.

4596. Guanosine. *2-Amino-9-β-D-ribofuranosyl-9H-purine-6(1H)-one; guanine riboside; vernine.* $C_{10}H_{13}N_5O_5$; mol wt 283.24. C 42.41%, H 4.63%, N 24.73%, O 28.24%. Constituent of nucleic acids. Prepn from yeast nucleic acid: P. A. Levene, L. W. Bass, *Nucleic Acids* (New York, 1931) p 163. Prepn from polynucleotides: P. A. Levene, E. Jorpes, *J. Biol. Chem.* **86**, 389 (1930). Prepn from plants: H. Stendel, E. Peiser in G. Klein, *Handbuch der Pflanzen-*

analyse IV (Vienna, 1933) p 448. Structure: Levene, Tipson, *J. Biol. Chem.* **97**, 491 (1932); Gulland *et al.*, *J. Chem. Soc.* **1934**, 1639; Tsuboi *et al.*, *Biochim. Biophys. Acta* **55**, 1 (1962). Synthesis: Davoll, *J. Chem. Soc.* **1958**, 1593. Tautomerism in aq soln: Miles *et al.*, *Science* **142**, 1458 (1963). Crystal structure and conformation: Bugg *et al.*, *Biochem. Biophys. Res. Commun.* **3**, 436 (1968). *Review: Basic Principles in Nucleic Acid Chemistry* vol. 1, P. O. P. Ts'o, Ed. (Academic Press, New York, 1974) *passim. See also* Nucleic Acids.

Dihydrate, needles from water. Anhydr at 110°. Dec 240° in sealed tube (rapid heating). $[\alpha]_D^{20}$ −61° (in water); $[\alpha]_D^{24}$ −72° (c = 0.96 in 0.1N NaOH). uv max (pH 5.5): 188.3, 252.5 nm ($\epsilon \times 10^{-3}$ 26.8, 13.7), Voet *et al.*, *Biopolymers* **1**, 193 (1963). One gram dissolves in 1320 ml water at 18°, in 33 ml boiling water. Soluble in dil mineral acids, in hot acetic acid, and in dil bases. Insol in alcohol, ether, chloroform, benzene.

4597. Guanoxabenz. *2-[(2,6-Dichlorophenyl)methylene]-N-hydroxyhydrazinecarboximidamide; 1-[(2,6-dichlorobenzylidene)amino]-3-hydroxyguanidine;* Compd 43-663. $C_8H_8Cl_2N_4O$; mol wt 247.08. C 38.89%, H 3.26%, Cl 28.70%, N 22.68%, O 6.48%. Prepn: W. J. Houlihan, R. E. Manning, **Ger. pat.** **1,902,449** corresp to **U.S. pat.** **3,591,636** (1969, 1971 both to Sandoz). Pharmacological study: J. C. Doxey, A. S. Hersom, *Brit. J. Pharmacol.* **70**, 171 (1980).

Hydrochloride, $C_8H_8Cl_2N_4O \cdot HCl$, *Benzerial.* Crystals from ethanol/ether, mp 173-175°.
THERAP CAT: Antihypertensive.

4598. Guanoxan. *[(2,3-Dihydro-1,4-benzodioxin-2-yl)-methyl]guanidine; (1,4-benzodioxan-2-ylmethyl)guanidine; 2-guanidinomethyl-1,4-benzodioxan.* $C_{10}H_{13}N_3O_2$; mol wt 207.23. C 57.96%, H 6.32%, N 20.28%, O 15.44%. Prepn: **Belg. pat.** **632,701**; J. Augstein, S. M. Green, **U.S. pat.** **3,247,221** (1963, 1966, both to Pfizer); Monro, *Chem. & Ind. (London)* **1964**, 1806; Gardner, **Brit.** pat. **996,708** (1965 to SKF), *C.A.* **63**, 14877d (1965); Yu, Shen, *C.A.* **66**, 94969k (1967). Prepn and properties of (+) and (−)-forms: Stenlake *et al.*, *J. Pharm. Pharmacol.* **20** (Suppl), 82 (1968). Pharmacology: Augstein *et al.*, *J. Med. Chem.* **8**, 446 (1965); Cession-Fossion, *Arch. Int. Pharmacodyn. Ther.* **164**, 419 (1966); Vidal-Beretewide *et al.*, *Arzneimittel-Forsch.* **19**, 947 (1969). Metabolic studies: Jack *et al.*, *J. Pharm. Pharmacol.* **23**, 2225 (1971); *Xenobiotica* **2**, 35 (1972).

Crystals, mp 164-165°.
Sulfate, $(C_{10}H_{13}N_3O_2)_2 \cdot H_2SO_4$, *Envacar.*
THERAP CAT: Antihypertensive.

4599. 3'-Guanylic Acid. Guanosine 3'-monophosphate; guanine riboside-3'-phosphoric acid; guanylic acid b.

$C_{10}H_{14}N_5O_8P$; mol wt 363.22. C 33.07%, H 3.88%, N 19.28%, O 35.24%, P 8.53%. From yeast or pancreas. Prepn: P. A. Levene, L. W. Bass, *Nucleic Acids* (New York, 1931) pp 224-227. Structure: Levene, Jorpes, *J. Biol. Chem.* **81**, 579 (1929); Levene, Harris, *ibid.* **95**, 755 (1932); **98**, 9 (1932). Early work probably done on a mixture of 2'- and 3'-guanylic acids; see physical data below. Separation of two isomers: Cohn, *J. Am. Chem. Soc.* **72**, 1471 (1950); Khym, Cohn, *ibid.* **76**, 1818 (1954); *eidem*, *Biol. Prepn.* **5**, 40 (1957). Absorption spectrum: Voet *et al.*, *Biopolymers* **1**, 193 (1963). *Reviews: see* Guanine, Nucleic Acids.

Dihydrate, long prisms from water. The water of crystn is given up at 118° and is taken up again at room temp. When anhydrous, dec 180° (closed tube). $[\alpha]_D^{25}$ $-8°$ (c = 2); $-65°$ (c = 2 in 5% NaOH). Acid to litmus. Soluble in cold water, freely sol in hot water. Boiling with dil mineral acids yields guanine, H_3PO_4, and D-ribose.

Neutral sodium salt, $Na_2C_{10}H_{12}N_5O_8P$, flakes from water, contains 21.1% H_2O. Sol in cold, freely sol in hot water.

Brucine salt heptahydrate, $C_{10}H_{14}N_5O_8P.(C_{23}H_{26}N_2O_4)_2$·7$H_2O$, rectangular leaflets from alc. When anhydr, dec 233-240°. $[\alpha]_D^{20}$ $-26°$ (35% alc). One gram dissolves in 100 ml water.

4600. 5'-Guanylic Acid. Guanosine 5'-monophosphate; GMP; guanosine 5'-phosphate; guanine riboside-5-phosphoric acid. $C_{10}H_{14}N_5O_8P$; mol wt 363.22. C 33.07%, H 3.88%, N 19.28%, O 35.24%, P 8.53%. Nucleotide widely distributed in nature; found in hydrolyzates of RNA. Isolated together with inosinic acid from sardines or yeast extract: Kuninaka *et al.*, *New Food Ind. (Tokyo)* **3**, no. 1, 21 (1961). Also by direct biosynthesis using microorganisms or enzymes: Abrams, Bentley, *Arch. Biochem. Biophys.* **79**, 91 (1959); Magasanik, Karibian, *J. Biol. Chem.* **235**, 2672 (1960); Okumura *et al.*, U.S. pat. **3,249,511** (1966). Chemical synthesis: Michelson, Todd, *J. Chem. Soc.* **1949**, 2483; Chambers *et al.*, *J. Am. Chem. Soc.* **79**, 3747 (1957); Gilham, Tener, *Chem. & Ind. (London)* **1959**, 542; Tener, *J. Am. Chem. Soc.* **83**, 159 (1961); Koransky *et al.*, *Z. Naturforsch.* **17B**, 291 (1962). Prepn of Na salt: Ishibashi, Ito, U.S. pat. **3,190,877** (1965 to Takeda). Monograph on synthesis of nucleotides: G. R. Pettit, *Synthetic Nucleotides* **vol. 1** (Van Nostrand Reinhold, New York, 1972) 252 pp. *Reviews: See* Guanidine; Nucleic Acids.

Microcrystals, dec 190-200°. Sparingly sol in cold water. Barium salt octahydrate, $C_{10}H_{12}N_5O_8PBa.8H_2O$, white powder. uv max (pH 2): 256 nm (ϵ 12400); (pH 12): 260 nm (ϵ 12100).

Disodium salt monohydrate, $C_{10}H_{12}N_5O_8PNa_2.H_2O$, hygroscopic crystals, decomp at about 250°. Characteristic meaty taste. a_M (molar absorbancy): 13.7×10^3 at 252.5 nm (pH 7). Soly in water at 25° about 25 g/100 ml. Practically insol in alcohol, acetone, ether.

USE: The disodium salt as flavor intensifier, like sodium inosinate and sodium glutamate. Said to be more effective than either.

4601. Guaran. Principal polysaccharide from endosperm of guar seeds, *Cyamopsis tetragonaloba* (L.) Taub., *Leguminosae:* Heyne, Whistler, *J. Am. Chem. Soc.* **70**, 2249 (1948). Structure: Whistler, Durso, *ibid.* **74**, 5140 (1952). Configuration: Koleske, Kurath, *J. Polymer Sci.* Pt. A, **2**, 4123 (1964). *Review:* Deuel *et al.*, *Chimia* **8**, 64 (1954).

$[\alpha]_D^{25}$ +53° (1N NaOH). Sol in cold water.

Triacetate, fibrous material, mp 226-227°. Can be formed into strong films which can be elongated 550%. Becomes birefringent and does not develop crystallinity.

USE: In textile and paper industry.

4602. Guar Gum. Guar flour; gum cyamopsis; cyamopsis gum; Burtonite V-7-E; Jaguar; Decorpa; Guarina; Glucotard; Guarem. Mol wt about 220,000. The ground endosperms of *Cyamopsis tetragonolobus* (L.) Taub., *Leguminosae* which is cultivated in India as livestock feed. The water soluble fraction (85%) of guar flour is called guaran which consists of linear chains of $(1\rightarrow 4)$-β-D-mannopyranosyl units with α-D-galactopyranosyl units attached by $(1\rightarrow 6)$ linkages. Ratio of D-galactose to D-mannose is 1:2. Effect on lipid metabolism: D. J. A. Jenkins *et al.*, *Brit. Med. J.* **2**, 1555 (1979); on glucose and lipid levels in diabetic and healthy volunteers: U. Smith, G. Holm, *Atherosclerosis* (Shannon, Ire.) **45**, 1 (1982); on renal tumors in diabetic rats: B. C. Chin *et al.*, *Biomed. Res.* **5**, 273 (1984). As source of fiber in patients with non-insulin dependent diabetes: M. E. McIvor *et al.*, *Am. J. Clin. Nutr.* **41**, 891 (1985). Toxicology studies: S. L. Graham *et al.*, *Food Cosmet. Toxicol.* **19**, 287 (1981). Comprehensive monograph: F. Smith, R. Montgomery, *The Chemistry of Plant Gums and Mucilages* (Reinhold, New York, 1959) 627 pp. *Review:* Goldstein *et al.* in *Industrial Gums*, R. L. Whistler, Ed. (Academic Press, New York, 2nd ed., 1973) p 303-321.

Free flowing powder. Completely sol in cold and hot water; practically insol in oils, greases, hydrocarbons, ketones, esters. Water solns are tasteless, odorless, nontoxic, of a pale, translucent gray color, and neutral. Stable to heat. Has five to eight times the thickening power of starch. Water solns may be converted to a gel by small amounts of borax. Aq dispersions are neutral. *Cf.* "A Comparative Study of Commercially Available Guar Gums" by I. A. Schlakman, A. J. Bartilucci, *Drug Standards* **25**, 149-154 (1957). LD_{50} in male, female rats (g/kg): 7.35, 6.77 orally (Graham).

USE: In paper sizing; as a protective colloid, stabilizer, thickening and film forming agent for cheese, salad dressings, ice cream, soups; as a binding and disintegrating agent in tablet formulations; in pharmaceutical jelly formulations; in suspensions, emulsions, lotions, creams, toothpastes; in the mining industry as a flocculant, as a filtering agent; in water treatment as a coagulant aid.

THERAP CAT: Adjunct to diet, insulin or oral hypoglycemics in control of diabetes.

4603. Guinea Green B. *N-Ethyl-N-[4-[[4-[ethyl[(3-sulfophenyl)methyl]amino]phenyl]phenylmethylene]-2,5-cyclo-*

hexadien-1-ylidene]-3-sulfobenzenemethanaminium inner salt, sodium salt; C.I. Acid Green 3; C.I. Food Green 1; FD & C Green 1; C.I. 42085. $C_{37}H_{35}N_2NaO_6S_2$; mol wt 690.82. C 64.33%, H 5.11%, N 4.06%, Na 3.33%, O 13.90%, S 9.28%. Prepn: Jones *et al., J. Assoc. Offic. Agr. Chem.* **38**, 977 (1955). Toxicity studies: F. C. Lu, A. Lavalle, *Can. Pharm. J.* **97**, 30 (1964); W. H. Hansen *et al., Food Cosmet. Toxicol.* **4**, 389 (1966). *See also: Colour Index* vol. **4** (3rd ed., 1971) p 4385.

A dull, dark green powder, or a bright, crystalline solid. Sol in water to a green soln which becomes brownish-yellow on addn of HCl and blackish-green with NaOH. An excess of NaOH decolorizes the soln. Sparingly sol in alcohol; it dissolves in concd H_2SO_4 to a yellow soln which, when diluted with water, turns first yellowish-red, then green. LD_{50} orally in rats: > 2 g/kg (Lu, Lavalle).

USE: Limited use as a dye for silk and wool fabrics; as biological stain. Delisted by FDA in 1966 for use in foods, drugs, and cosmetics.

4604. D-Gulonic Acid. $C_6H_{12}O_7$; mol wt 196.16. C 36.74%, H 6.17%, O 57.09%. Prepd as the sodium salt by reduction of sodium glucuronate with sodium amalgam in alkaline medium: Fischer, Piloty, *Ber.* **24**, 525 (1891); from D-gulonic acid γ-lactone: Rehorst, Naumann, *ibid.* **77**, 24 (1944).

$[\alpha]_D^{20}$ $-6°$ (10 min) → $-38.6°$ (15 days). The free acid forms the lactone spontaneously. pK (25°): 3.68.

Sodium salt, $C_6H_{11}NaO_7$, crystals. $[\alpha]_D^{20}$ +11.5°. Sol in water.

Calcium salt, $Ca(C_6H_{11}O_7)_2$. $[\alpha]_D^{21}$ $-14.45°$ (c = 1.73). Precipitated from aq soln by alc.

4605. L-Gulonic Acid. Xylosecarboxylic acid. $C_6H_{12}O_7$; mol wt 196.16. C 36.74%, H 6.17%, O 57.09%. Prepd from L-xylose and HCN followed by hydrolysis of the nitrile: Fischer, Stahel, *Ber.* **24**, 529 (1891). Prepn from D-glucuronic acid: **Ger. pat.** 618,907 (1935 to Hoffmann-La Roche); from L-gulonolactone: Ishidate *et al., Chem. Pharm. Bull.* **13**, 173 (1965).

Crystallizes as the lactone on evapn of an aq soln.
Sodium salt, $[\alpha]_D^{20}$ +12.7° (c = 9). Freely sol in water.

4606. D-Gulose. $C_6H_{12}O_6$; mol wt 180.16. C 40.00%, H 6.71%, O 53.28%. Prepd by sodium amalgam reduction of an acid soln of the γ-lactone of D-gulonic acid: Fischer,

Stahel, *Ber.* **24**, 532 (1891); van Ekenstein, Blanksma, *Rec. Trav. Chim.* **27**, 3 (1908). Alternate synthesis: Meyer zu Reckendorf, *Angew. Chem. Int. Ed.* **6**, 177 (1967); idem, *Methods Carbohyd. Chem.* **6**, 129 (1972); R. Köster *et al., Angew. Chem. Int. Ed.* **19**, 547 (1980).

Syrup. Sweet taste. $[\alpha]_D^{20}$ $-20.4°$. Sol in water, slightly sol in alcohol. Not fermentable by yeast.

4607. L-Gulose. $C_6H_{12}O_6$; mol wt 180.16. C 40.00%, H 6.71%, O 53.28%. Prepd by sodium amalgam reduction of an acid soln of the γ-lactone of L-gulonic acid: Fischer, Piloty, *Ber.* **24**, 526 (1891). *See also* van Ekenstein, Blanksma, *Rec. Trav. Chim.* **27**, 3 (1908); Levene, LaForge, *J. Biol. Chem.* **20**, 430 (1915); Talen, *Rec. Trav. Chim.* **44**, 891 (1925); Isbell, *J. Am. Chem. Soc.* **55**, 2167 (1933). Synthesis from D-mannose: Evans, Parrish, *Carbohyd. Res.* **28**, 359 (1973); from D-glucose: D. K. Minster, S. M. Hecht, *J. Org. Chem.* **43**, 3987 (1978).

R = CH₂OH

Syrup. $[\alpha]_D^{20}$ +61.6°. $[\alpha]_D$ +21.3° (c = 4.58) (Evans, Parrish). Freely sol in water; slightly sol in alcohol. Not fermentable by yeast.

4608. Gum Benzoin. Resin benzoin; resin benjamin; gum benjamin. Balsamic resin from *Styrax benzoin* Dryand., known as Sumatra benzoin, or from *S. tonkinensis* (Pierre) Craib, *Styracaceae*, or other species of *Styrax* known as Siam benzoin. *Habit.* Thailand, Cambodia, S. Vietnam, Sumatra, Java, and Sunda Islands. *Constit.* Ethereal oil, free and combined benzoic and cinnamic acids up to 39%, vanillin, coniferyl benzoate, resin (a mixture of benzoresinol and benzoresinotannol) esterified with benzoic acid, styrol, styracin. Not less than 90% of Siam and not less than 75% of Sumatra benzoic is sol in alc (U.S.P.). *Ref:* Reinitzer, *Arch. Pharm.* **264**, 131 (1926); Brans, *Pharm. Weekbl.* **73**, 374 (1936); Freudenberg, Bittner, *Ber.* **83**, 600 (1950).

USE: Preserving ointments; preparing natural benzoic acid; for fumigating pastilles; in perfumery and cosmetics.

THERAP CAT: Topical protectant.

THERAP CAT (VET): Tincture is used topically as an antiseptic and to promote healing; as an inhalant for bronchitis, and orally as an expectorant.

4609. Gum Tragacanth. Tragacanth. Mol wt about 840,000. The dried gummy exudation from *Astragalus gummifer* Labill. (white gavan) or other Asiatic species of *Astragalus, Leguminosae,* found largely in Iran, also in Asia Minor and in Syria. When mixed with water gives a soluble fraction, as a hydrosol, called *tragacanthin* which is a complex mixture of polysaccharides containing D-galacturonic acid, other sugars, and traces of starch and cellulose. The insoluble fraction swells to a gel and consists of 60-70% bassorin, q.v. Structural studies: Norman, *Biochem. J.* **25**, 200 (1931); James, Smith, *J. Chem. Soc.* **1945**, 739, 749; Aspinall, Baillie, *ibid.* **1963**, 1702, 1714. *Reviews:* Beach, in *Natural Plant Hydrocolloids,* Advances in Chemistry Series **11** (A.C.S., Washington, 1954) pp 38-44; Meer *et al.* in *Industrial Gums,* R. L. Whistler, Ed. (Academic Press, New York, 2nd ed., 1973) pp 289-299. *Book:* F. Smith, R. Montgomery, *The Chemistry of Plant Gums and Mucilages* (Reinhold, New York, 1959) 627 pp.

Odorless. Insipid, mucilaginous taste. Acid reaction. One gram requires 0.9 ml 0.1*N* NaOH for neutralization to phenolphthalein: Gabel, *J. Am. Pharm. Assoc.* **23**, 341 (1934). Viscosity of tragacanth mucilages is reduced by adding acid, alkali, and NaCl particularly if the mucilage is heated: Mantell, *The Water-Soluble Gums* (New York,

1947). Maximum initial viscosity of solns at pH 8; maximum stable viscosity near pH 5. Forms a deep yellow stringy precipitate when a soln is boiled with a few drops of 10% aqueous ferric chloride soln. A stringy precipitate formed also on heating a soln with Schweitzer reagent. Tragacanth is entirely insol in alcohol.

USE: In pharmaceutical compounding and dispensing, e.g., to suspend heavy insol powders, as an excipient for tablets and to impart consistence to troches; also in making emulsions and emulsifying agents; as stabilizer, thickener, texturizer in food; in adhesives (mucilages, pastes); in textile sizing, textile printing and general printing inks, and in dyeing with insol color lakes.

4610. Gusperimus. (±)-7-[(Aminoiminomethyl)amino]-N-[2-[[4-[(3-aminopropyl)amino]butyl]amino]-1-hydroxy-2-oxoethyl]heptanamide; (±)-N-[[[4-[(3-aminopropyl)amino]-butyl]carbamoyl]hydroxymethyl]-7-guanidinoheptanamide; 1-amino-19-guanidino-11-hydroxy-4,9,12-triazanonadecane-10,13-dione; deoxyspergualin; (±)-15-deoxyspergualin. $C_{17}H_{37}N_7O_3$; mol wt 387.53. C 52.69%, H 9.62%, N 25.30%, O 12.39%. Synthetic derivative of the antitumor antibiotic spergualin. Prepn: **Belg.** pat. **894,651;** H. Umezawa et al., U.S. pat. **4,518,532** (1983, 1985 both to Microbiochem. Res. Found.); Y. Umeda et al., J. Antibiot. **38,** 886 (1985). Prepn of the active (−)-isomer: H. Iwasawa et al., ibid. **35,** 1665 (1982); H. Umezawa et al., **Eur.** pat. **Appl. 94,632;** eidem, U.S. pat. **4,525,299** (1983, 1985 both to Microbiochem. Res. Found.). Synthesis and bioactivity of isomers: Y. Umeda et al., J. Antibiot. **40,** 1316 (1987). Mechanism of action study: W. E. G. Müller et al., ibid. **1028.** HPLC determn in plasma: R. Nakanuma et al., J. Chromatog. **527,** 208 (1990). Pharmacokinetics: J. F. Muindi et al., Cancer Res. **51,** 3096 (1991). Preliminary evaluation in renal transplant rejection: H. Amemiya et al., Transplant. Proc. **25,** 730 (1993). Series of articles on synthesis and immunomodulating activity: Ann. N.Y. Acad. Sci. **685,** 123-206 (1993).

Trihydrochloride, $C_{17}H_{37}N_7O_3$·3HCl, BMS-181173, NKT-01, NSC-356894, Spanidin. White powder, prepd as the sesquihydrate. Sol in water. pH of 50 mg/ml saline soln: ~4.9. LD_{50} in mice (mg/kg): 25-50 i.p. (Umezawa).

(−)-Form trihydrochloride, prepd as the dihydrate. Colorless syrup, no def mp. $[\alpha]_D^{25}$ −7.3° (c = 1 in H_2O). LD_{50} in mice (mg/kg): 35-40 i.v. or i.p. (Iwasawa).

THERAP CAT: Immunosuppressant.

4611. Gutta-Percha. The purified, coagulated, milky exudate of various trees of the genus Palaquium, Sapotaceae. Habit. Malayan Archipelago. Extensive review: Williams, Econ. Bot. **18,** 5-26 (1964). Defined as a trans isomer of rubber. Rubber has a repeat period of 8.2 Å, whereas α-gutta-percha has 8.7 Å and β-gutta-percha has 4.8 Å. The short period of β-gutta-percha identifies it almost uniquely as an all-trans polyisoprene.

Becomes pliable at 25-30°, plastic at 60°. mp 100° (partial dec). On exposure to air and sunlight, it absorbs oxygen and becomes brittle. Insol in water; partially sol in hot alcohol; 90% or more dissolves in chloroform, carbon disulfide, petr ether, oil of turpentine. Keep under water and protected from light.

USE: Insulator in electrotechnics, as dental cement; in orthopedics for fracture splints; manuf surgical instruments; covering golf balls.

4612. Guvacine. 1,2,5,6-Tetrahydro-3-pyridinecarboxylic acid; 1,2,5,6-tetrahydronicotinic acid. $C_6H_9NO_2$; mol wt 127.14. C 56.68%, H 7.13%, N 11.02%, O 25.17%. From betel nuts, the seeds of Areca catechu L., Palmae. Extraction: Jahns, Ber. **24,** 2615 (1891). Synthesis: Freudenberg, ibid. **51,** 976, 1669 (1918); Hess, Leibbrandt, ibid. **51,** 806; **52,** 206 (1919).

Prisms from water, dec 295°. Neutral to litmus. Sol in water. Almost insol in abs alc, ether, chloroform, benzene.

Hydrochloride, $C_6H_{10}ClNO_2$, needles from water, dec 318°. Sol in water.

Hydrobromide, $C_6H_{10}BrNO_2$, needles from abs alcohol, dec 280°. Almost insol in acetone.

USE: Has been proposed as growth factor for Staphylococcus aureus and Proteus vulgaris instead of nicotinic acid.

4613. Gymnemic Acid. Gymnemin. Antisaccharin principle occurring as the potassium salt (gymnemin) in the leaves of Gymnema sylvestre R.Br. and allied Asclepiadaceae: Hooper, Chem. News **59,** 159 (1889); Mhaskar, Caius, Indian J. Med. Res. No. **16,** 1 (1930); Warren, Pfaffman, J. Appl. Physiol. **14,** 40 (1959); Stöcklin et al., Helv. Chim. Acta **50,** 474 (1967). A complex mixture of at least nine closely related acidic glycosides, the major active component being gymnemic acid A₁: Stöcklin, J. Agr. Food Chem. **17,** 704 (1969); Dateo, Long, ibid. **21,** 899 (1973). Separation of major components: Sinsheimer et al., J. Pharm. Sci. **59,** 622, 629 (1970). Completely obtunds taste for several hours for bitter or sweet, e.g., quinine or sugar, but not for sour, astringent or pungent substances.

The acid is a yellow to brown amorphous, bitter powder. Almost insol in water; sol in alcohol. The potassium salt is a reddish-brown cryst mass, sol in water or alcohol.

4614. Gypsogenin. (3β,4α)-3-Hydroxy-23-oxoolean-12-en-28-oic acid; githagenin; albasapogenin; gypsophilasapogenin. $C_{30}H_{46}O_4$; mol wt 470.69. C 76.55%, H 9.85%, O 13.60%. From Agrostemma githago (L.) Scop., Caryophyllaceae: Wedekind, Krecke, Z. Physiol. Chem. **155,** 122 (1925); from Gypsophila oldhamiana, Caryophyllaceae: Kutani, Karr, J. Pharm. Soc. Japan **64,** 18 (1944), C.A. **45,** 2961d (1951). Structure: Ruzicka, Giacomello, Helv. Chim. Acta **20,** 299 (1937). Identity with githagenin: Kon, Soper, J. Chem. Soc. **1940,** 617.

Needles or leaflets from methanol, mp 274-276°. $[\alpha]_D$ +91° (alc).

Acetate, $C_{32}H_{48}O_5$, square tablets from methanol, mp 176-177°. $[\alpha]_D^{20}$ +78° (CHCl₃).

Methyl ester, $C_{31}H_{48}O_4$, crystals, mp 192°.

H

4615. H3®. Gerovital; Gerovital H3; Gerontex H3; GH 3; Sex-Ex. A factor claimed to exist in commercial procaine hydrochloride prepns according to the Romanian M.D. Anna Aslan [Institute of Geriatrics, Bucharest]. H3 is Dr. Aslan's name for the factor in procaine HCl which she found to be effective in achieving an apparent reversal of phenomena previously considered irreversible, such as those encountered in cerebral arteriosclerosis. *Ref: Publishers' Weekly,* Dec. 7, 1959, p 36; Aslan, *C.A.* **53,** 1465b (1959). Historical review of controversy: P. M. McGrady, *The Youth Doctors* (Coward-McCann, New York, 1968) pp 181-192; A. Hecht, *FDA Consumer* (March 1980) p 17. Clinical study: M. R. Hall *et al., Age Ageing* **12,** 302 (1983); and a response: P. H. Millard, *Br. Med. J.* **289,** 1094 (1984).

Note: Oral version is marketed as *KH3.* Each KH3 capsule contains: Procaine hydrochloride 0.05 g, hematoporphyrin 0.0002 g, magnesium carbonate 0.03 g.

4616. Hachimycin. *Trichomycin;* Cabimicina; Trichonat. Heptaene macrolide antibiotic substance produced by *Streptomyces hachijoensis* from soil of the Pacific Island Hachijo Jima: S. Hosoya *et al., J. Antibiot.* **5,** 564 (1952); T. Yamaguchi, *ibid.* **7A,** 10 (1954). Purification and prepn of insol derivs: S. Hosoya *et al., ibid.* **8A,** 5 (1955). At one time, trichomycin was believed to be identical to candicidin and hamycin, *q.q.v.:* H. Burrows, D. Calam, *J. Chromatog.* **53,** 566 (1970). Subsequent HPLC studies have shown the three polyene antibiotics to be different entities: P. Helboe *et al., ibid.* **189,** 249 (1980). Active against *Trichomonas vaginalis, Treponema pallidum, Trichophyton, Candida,* and yeast, weak activity against *Asporgillus* and *Penicillium:* T. Yamaguchi, *J. Antibiot.* **7A,** 10 (1954). Toxicity: S. Hosoya *et al., J. Antibiot.* **6A,** 98 (1953). *Review:* S. A. Waksman, H. A. Lechevalier, *The Actinomycetes* vol. III (Williams & Wilkins, Baltimore, 1962) pp 397-399.

Yellow crystals. Acid reaction. Forms a water-sol sodium salt. Can be pptd as a water-insol salt by acridine derivs such as acriflavine, by sulfa drugs such as homosulfanilamide, by basic antibiotics such as streptomycin base, by enzymes such as papain and lysozyme, by dyes such as methylene blue, and by metallic salts such as $CaCl_2$. LD_{50} i.p. in mice: 0.05 mg/10-12 g mouse (Hosoya).

THERAP CAT: Antifungal. Antiprotozoal (Trichomonas).

4617. Hadacidin. *N-Formyl-N-hydroxyglycine; N*-formyl-*N*-hydroxyaminoacetic acid; *N*-hydroxyformamidoacetic acid. $C_3H_5NO_4$; mol wt 119.08. C 30.26%, H 4.23%, N 11.76%, O 53.74%. Antitumor antibiotic originally isolated from cultures of *Penicillium frequentans* Westling. Synthesis: Kaczka *et al., Biochemistry* **1,** 340 (1962); Kinnel, Schoenewaldt, **U.S.** pat. **3,154,578** (1964 to Merck & Co.). Biosynthesis: Stevens, Emery, *Biochemistry* **5,** 74 (1966). *Review:* Shigeura, "Hadacidin" in *Antibiotics* I, D. Gottlieb, P. Shaw, Eds. (Springer-Verlag, New York, 1967) pp 451-456.

Unstable crystals, mp 119-120°. Turns brown and liquefies on standing. The decompn products are formic acid and *N*-hydroxyglycine. Dibasic acid, potentiometric titration shows pH peak at 3.5 and 9.1. Soluble in water, methanol, ethanol, acetone, ether.

Monosodium salt, $C_3H_4NNaO_4$, crystals. Easily forms a hydrate, very freely sol in water.

4618. Hafnium. Hf; at. wt 178.49; at. no. 72; valences 4; also 2, 3. Six naturally occurring isotopes: 180 (35.22%); 178 (27.1%); 177 (18.56%); 179 (13.75%); 176 (5.21%); 174 (0.163%); α-emitter, $T_{1/2}$ 2 × 10^{15} years); artificial isotopes: 157; 158; 168-173; 175; 181-183. Abundance in earth's crust: 5 ppm. Found in all zirconium-contg minerals. Discovered in 1923 by Coster and Hevesy: *Nature* **111,** 79 (1923). Extraction from the mineral cyrtolite: Larsen *et al.,*

Inorg. Syn. **3,** 67 (1950). Prepd by thermal decompn of its iodide; by reduction of the tetrachloride or of the hydrofluohafniate with metallic sodium; by reduction of the oxide with a mixture of calcium and sodium: van Arkel, de Boer, *Z. Anorg. Chem.* **148,** 345 (1925); de Boer, Fast, *ibid.* **187,** 193 (1930); **U.S.** pat. **2,741,628.** Spectra: Coster *et al.,* cited by Mellor, *A Comprehensive Treatise on Inorganic and Theoretical Chemistry* 7, 170 (1930). A dibromide and tribromide have been prepd: Schumb, Morehouse, *J. Am. Chem. Soc.* **69,** 2696 (1947). Reviews of hafnium and its compds: Larsen, "Zirconium and Hafnium Chemistry" in *Advan. Inorg. Chem. Radiochem.* **13,** 1-133 (1970); Bradley, Thornton, "Zirconium and Hafnium" in *Comprehensive Inorganic Chemistry,* vol. 3, J. C. Bailar, Jr. *et al.,* Eds. (Pergamon Press, Oxford, 1973) pp 419-490; R. H. Nielson in Kirk-Othmer *Encyclopedia of Chemical Technology* **vol. 12** (Wiley-Interscience, New York, 3rd ed., 1980) pp 67-80.

Highly lustrous, ductile metal of hexagonal cryst structure. mp 2227°; d 13.3. Resembles zirconium and thorium.

Dioxide, HfO_2, obtained by igniting the hydroxide, the oxalate, or the sulfate. d^{20} 9.68. mp 2774°.

Tetrachloride, $HfCl_4$, white cryst mass, obtained by heating the oxide in the presence of chlorine and a reducing agent. Hydrolyzed to hafnyl chloride, $HfOCl_2$, by water. Volatilizes appreciably at 250°.

Sulfate, $Hf(SO_4)_2$, prepd by the action of fuming sulfuric acid on the tetrachloride, dec above 500°.

4619. Halazepam. *7-Chloro-1,3-dihydro-5-phenyl-1-(2,-2,2-trifluoroethyl)-2H-1,4-benzodiazepin-2-one;* Sch-12041; Paxipam. $C_{17}H_{12}ClF_3N_2O$; mol wt 352.74. C 57.89%, H 3.43%, Cl 10.05%, F 16.16%, N 7.94%, O 4.54%. Polyfluoroalkyl analog of diazepam, *q.v.* Prepn: J. G. Topliss, **U.S.** pat. **3,429,874** (1969 to Schering); M. Steinman *et al., J. Med. Chem.* **16,** 1354 (1973). Pharmacology. J. D. Petel *et al., Psychopharmacology* **61,** 25 (1979). Use in schizophrenia: K. Y. Ota *et al., Curr. Ther. Res.* **15,** 327 (1973); in anxiety: K. Rickels *et al., Int. Pharmacopsychiatry* **13,** 118 (1978).

Crystals from acetone-hexane, mp 164-166°. LD_{50} in mice: > 4000 mg/kg orally (Steinman).

Note: This is a controlled substance (depressant) listed in the U.S. Code of Federal Regulations, Title 21 Part 1308.14 (1995).

THERAP CAT: Anxiolytic.

4620. Halazone. *4-[(Dichloroamino)sulfonyl]benzoic acid; p-(dichlorosulfamoyl)benzoic acid; p-sulfondichloramidobenzoic acid; p-carboxybenzenesulfondichloroamide;* Pantocid. $C_7H_5Cl_2NO_4S$; mol wt 270.09. C 31.13%, H 1.87%, Cl 26.25%, N 5.19%, O 23.69%, S 11.87%. Prepd by chlorination of *p*-sulfamylbenzoic acid in alkaline medium. The commercially available product is a mixture of mono- and dichloroamides with the dichloro compd predominating, see Proschko, **Ger.** pat. **492,249;** *Chem. Zentr.* **101,** I, 2630 (1930); **U.S.** pat. **1,697,139,** *C.A.* **23,** 1138 (1929); Claas, **Ger.** pat. **318,899;** *Chem. Zentr.* **91,** IV, 14 (1920). Review of use in water disinfection: O'Connor, Kapoor, *J. Am. Water Works Assoc.* **62,** 80 (1970).

Crystals or white powder, dec about 195°. Odor of chlorine. Slightly sol in water, chloroform (differing from dichloramine-T which is very sol in chloroform). Sol in glacial acetic acid and in solns of alkali hydroxides and of alkali carbonates with the formation of a salt. MLD orally in rats: 3.5 g/kg.

Sodium salt, $C_7H_4Cl_2NNaO_4S$, *Aseptamide, Gynamide.*
USE: Has been proposed for use in the determination of iodine numbers of fats and oils. Used as water disinfectant.

4621. Halcinonide. *(11β,16α)-21-Chloro-9-fluoro-11-hydroxy-16,17-[(1-methylethylidene)bis(oxy)]pregn-4-ene-3,-20-dione; 21-chloro-9-fluoro-11β,16α,17-trihydroxypregn-4-ene-3,20-dione cyclic 16,17-acetal with acetone;* 21-chloro-9α-fluoro-11β-hydroxy-16α,17α-isopropylidenedioxy-4-pregnene-3,20-dione; 9α-fluoro-21-chloro-11β,16α,17α-trihydroxypregn-4-ene-3,20-dione 16,17-acetonide; SQ-18566; Halciderm; Halcimat; Halog. $C_{24}H_{32}ClFO_5$; mol wt 454.97. C 63.36%, H 7.09%, Cl 7.79%, F 4.18%, O 17.58%. Prepn: Bernstein *et al., J. Org. Chem.* **27,** 690 (1962); L. T. Difazio, M. A. Augustine, Ger. pat. 2,355,710; *eidem,* U.S. pat. 3,-892,857 (1972, 1975 both to Squibb). Pharmacological evaluation: Bagatell, Augustine, *Curr. Ther. Res.* **16,** 748 (1974); R. C. Millonig, E. Yiakas, in *Pharmacological and Biochemical Properties of Drug Substances* **vol. 1,** M. E. Goldberg, Ed. (Am. Pharmaceut. Assoc., Washington, DC, 1977) pp 215-231. Comprehensive description: J. Kirshbaum, *Anal. Profiles Drug Subs.* **8,** 251-281 (1979).

Crystals from acetone-petr ether, mp 264-265° (dec). uv max (methanol): 238 nm (ε 16400). $[\alpha]_D^{25}$ +155° (CHCl$_3$). Sol in acetone, chloroform, DMSO. Slightly sol in benzene, ethanol, ethyl ether, methanol. Insol in water, 0.1M HCl, 0.1M NaOH, hexanes. LD$_{50}$ i.p. in mice: 150 mg/kg (Millonig, Yiakas).

THERAP CAT: Anti-inflammatory (topical).

4622. Halethazole. *5-Chloro-2-[p-(diethylaminoethoxy)phenyl]benzothiazole;* 2-[p-(diethylaminoethoxy)phenyl]-5-chlorobenzothiazole; haletazole; Episol. $C_{19}H_{21}ClN_2OS$; mol wt 360.91. C 63.23%, H 5.86%, Cl 9.82%, N 7.76%, O 4.43%, S 8.88%. Prepn: Stephens, U.S. pat. 2,996,512 (1961 to Crookes Labs.).

Crystals from ethanol, mp 93-94°.
Citrate, $C_{19}H_{21}ClN_2OS \cdot C_6H_8O_7$, crystals from water, mp 167°.
THERAP CAT: Antiseptic, antifungal.

4623. Halibut Liver Oil. From the liver of the halibut, *Hippoglossus hippoglossus* L. It usually contains not less than 60,000 U.S.P. vitamin A units and not less than 500 U.S.P. vitamin D units per gram.

Pale yellow liq; slight fishy odor and taste. d 0.922-0.925. n_D^{40} 1.470-1.478. Sapon no. 170-180. Iodine no. 120-136. Unsaponifiable matter 8-13%.

THERAP CAT: Source of vitamins A and D.
THERAP CAT (VET): Source of vitamins A and D.

4624. Halimide®. *Dodecylbenzyltrimethylammonium chloride.* $C_{22}H_{40}ClN$; mol wt 354.02. C 74.64%, H 11.39%, Cl 10.01%, N 3.96%. Prepn: Darragh, Stayner, *Ind. Eng. Chem.* **46,** 254 (1954); Stayner, U.S. pat. 2,692,286 (1954 to Calif. Res. Corp.).

Pale yellow, glassy solid. Phenol coefficient *(S. aureus):* 408; *(S. typhosa):* 367.
USE: Disinfectant; cationic surface active agent.

4625. Halobetasol Propionate. *(6α,11β,16β)-21-Chloro-6,9-difluoro-11-hydroxy-16-methyl-17-(1-oxopropoxy)pregna-1,4-diene-3,20-dione;* 21-chloro-6α,9-difluoro-11β,17-dihydroxy-16β-methylpregna-1,4-diene-3,20-dione 17-propionate; ulobetasol propionate; BMY-30056; CGP-14458; Ultravate. $C_{25}H_{31}ClF_2O_5$; mol wt 484.97. C 61.92%, H 6.44%, Cl 7.31%, F 7.83%, O 16.50%. Trihalogenated corticosteroid structurally related to clobetasol, *q.v.* Prepn: J. Kalvoda, G. Anner, Ger. pat. 2,743,069; *eidem,* U.S. pat. 4,619,921 (1978, 1986 both to Ciba-Geigy). Clinical pharmacology in psoriatic patients: W. A. Watson *et al., Pharmacotherapy* **10,** 107 (1990). Clinical trial in localized vitiligo: L. Xunquan *et al., Int. J. Dermatol.* **29,** 295 (1990).

Crystals from methylene chloride/ether, mp 220-221°.
THERAP CAT: Anti-inflammatory.

4626. Halofantrine. *1,3-Dichloro-α-[2-(dibutylamino)-ethyl]-6-(trifluoromethyl)-9-phenanthrenemethanol;* 1-(1,3-dichloro-6-trifluoromethyl-9-phenanthryl)-3-di(n-butyl)-aminopropanol; γ-(dibutylamino)-1,3-dichloro-6-(trifluoromethyl)-9-phenanthrenepropanol. $C_{26}H_{30}Cl_2F_3NO$; mol wt 500.43. C 62.40%, H 6.04%, Cl 14.17%, F 11.39%, N 2.80%, O 3.20%. Prepn of the hydrochloride: W. T. Colwell *et al., J. Med. Chem.* **15,** 771 (1972); of the β-glycerophosphate salt: J. F. Rossignol, Eur. pat. Appl. 138,374; *idem,* U.S. pat. 4,507,288 (both 1985 to Smithkline Beckman). Antimalarial activity *in vivo:* L. H. Schmidt *et al., Antimicrob. Ag. Chemother.* **14,** 292 (1978). HPLC determn in human blood: J. W. Hines *et al., J. Pharm. Sci.* **74,** 433 (1985). Antimalarial activity in clinically-induced infection in humans: J. Rinehart *et al., Am. J. Trop. Med. Hyg.* **25,** 769 (1976); T. M. Cosgriff *et al., ibid.* **31,** 1075 (1982); in naturally acquired disease: J. P. Coulaud *et al., Trans. Roy. Soc. Trop. Med. Hyg.* **80,** 615 (1986). Review of pharmacology and therapeutic potential: H. M. Bryson, K. L. Goa, *Drugs* **43,** 236-258 (1992).

Hydrochloride, $C_{26}H_{30}Cl_2F_3NO \cdot HCl$, *SKF-102886, WR-171669, Halfan.* Two crystalline forms have been reported: mp 93-96°, mp 203-204°.
β-Glycerophosphate, $C_{29}H_{39}Cl_2F_3NO_7P$, white crystals, mp 60-65°.
THERAP CAT: Antimalarial.

4627. Halofuginone. *trans-(±)-7-Bromo-6-chloro-3-[3-(3-hydroxy-2-piperidinyl)-2-oxopropyl]-4(3H)-quinazolinone;* (±)-trans-7-bromo-6-chloro-3-[3-(3-hydroxy-2-piperidyl)aceto-*

Consult the Name Index before using this section.

nyl]-4(3H)-quinazolinone; 7-bromo-6-chlorofebrifugine; HAL. $C_{16}H_{17}BrClN_3O_3$; mol wt 414.69. C 46.34%, H 4.13%, Br 19.27%, Cl 8.55%, N 10.13%, O 11.57%. Halogenated deriv of febrifugine, *q.v.* Prepn: E. Waletzky *et al.,* U.S. pat. **3,320,124** (1967 to Am. Cyanamid). *In vivo* and *in vitro* activity: J. G. Ryley *et al., Parasitology* **70,** 203 (1975). Toxicity to fresh water organisms: *Bull. Environ. Contam. Toxicol.* **15,** 720 (1976). HPLC determn in chicken feed: A. Anderson *et al., J. Chromatog.* **168,** 471 (1979); in chicken serum: R. C. Beier *et al., J. Liq. Chromatogr.* **17,** 2961 (1994). Evaluation as coccidiostat: P. E. Waibel *et al., Poult. Sci.* **66,** 1629 (1987). Inhibition of collagen synthesis and effect on skin tearing: I. Granot *et al., ibid.* **70,** 1559 (1991).

Hydrobromide, $C_{16}H_{17}BrClN_3O_3 \cdot HBr$, *RU-19110, Stenorol.* Crystals, mp 247° (dec).

THERAP CAT (VET): Antiprotozoal (coccidiostat).

4628. Halometasone. *(6α,11β,16α)-2-Chloro-6,9-difluoro-11,17,21-trihydroxy-16-methylpregna-1,4-diene-3,20-dione;* 2-chloroflumethasone; C-48401-Ba. $C_{22}H_{27}ClF_2O_5$; mol wt 444.90. C 59.39%, H 6.12%, Cl 7.97%, F 8.54%, O 17.98%. Synthetic corticosteroid. Prepn: G. G. Anner *et al.,* Neth. pat. **Appl. 540,244** correp to U.S. pat. **3,652,554;** *eidem,* Swiss pat. **551,399,** *C.A.* **81,** 120864d (1974) (1968, 1972, 1974 all to Ciba-Geigy). Series of clinical trials in eczema, psoriasis, dermatomycoses, bacterial infections: *J. Int. Med. Res.* **11,** Suppl. 1 (1983).

mp 220-222° (dec).
Monohydrate, $C_{22}H_{27}ClF_2O_5 \cdot H_2O$, *Sicorten.*
THERAP CAT: Anti-inflammatory (topical); antipruritic.

4629. Haloperidol. *4-[4-(4-Chlorophenyl)-4-hydroxy-1-piperidinyl]-1-(4-fluorophenyl)-1-butanone; 4-[4-(p-chlorophenyl)-4-hydroxypiperidino]-4'-fluorobutyrophenone;* 4'-fluoro-4-(4-hydroxy-4-p-chlorophenylpiperidino)butyrophenone; 1-(3-p-fluorobenzoylpropyl)-4-p-chlorophenyl-4-hydroxypiperidine; R-1625; Aloperidin; Bioperidolo; Brotopon; Dozic; Einalon S; Eukystol; Haldol; Halosten; Keselan; Linton; Peluces; Serenace; Serenase; Sigaperidol. $C_{21}H_{23}ClFNO_2$; mol wt 375.87. C 67.11%, H 6.17%, Cl 9.43%, F 5.05%, N 3.73%, O 8.51%. Prepn: P. A. J. Janssen *et al., J. Med. Pharm. Chem.* **1,** 281 (1959); P. A. J. Janssen, Belg. pat. **577,977** (1959), *C.A.* **54,** 4630c (1960); *idem,* Brit. pat. **895,309;** *idem,* U.S. pat. **3,438,991** (1962, 1969 both to Janssen). Toxicity: E. I. Goldenthal, *Toxicol. Appl. Pharmacol.* **18,** 185 (1971); A. J. Collins, M. Horlington, *Brit. J. Pharmacol.* **37,** 140 (1969). Metabolism in man: A. Forsman *et al., Curr. Ther. Res., Clin. Exp.* **21,** 606 (1977). Comprehensive description: C. A. Janicki, C. Y. Ko, *Anal. Profiles Drug Subs.* **9,** 341-369 (1980). Review of pharmacology and therapeutic efficacy of decanoate in psychosis: R. Beresford, A. Ward, *Drugs* **33,** 31-49 (1987).

Crystals, mp 148.0-149.4°. uv max (9:1 0.1M HCl:methanol): 247, 221 nm (ε 13300, 15000). pKa 8.3. Soly in water: 1.4 mg/100 ml. Freely sol in chloroform, methanol, acetone, benzene, dil acids. LD_{50} orally in rats: 165 mg/kg (Goldenthal); i.p. in mice: 60 mg/kg (Collins, Horlington).

Hydrochloride, $C_{21}H_{23}ClFNO_2 \cdot HCl$, crystals, mp 226-227.5°. Soly in water: 300 mg/100 ml.

Decanoate, $C_{31}H_{41}ClFNO_3$, *KD-136, Haldol Decanoate, Halomonth, Neoperidole.*

THERAP CAT: Antidyskinetic (in Gilles de la Tourette's disease); antipsychotic.

4630. Halopredone Acetate. *(6β,11β)-17,21-Bis(acetyloxy)-2-bromo-6,9-difluoro-11-hydroxypregna-1,4-diene-3,20-dione;* 2-bromo-6β,9-difluoro-11β,17,21-trihydroxypregna-1,4-diene-3,20-dione 17,21-diacetate; Haloart; Topicon. $C_{25}H_{29}BrF_2O_7$; mol wt 559.40. C 53.68%, H 5.23%, Br 14.28%, F 6.79%, O 20.02%. Prepn: M. Riva, L. Toscano, Ger. pat. **2,508,136;** M. Riva *et al.,* U.S. pat. **4,226,862** (1975, 1980 both to Pierrel); L. Toscano *et al., J. Med. Chem.* **20,** 213 (1977). Physico-chemical and analytical studies: *eidem, Arzneimittel-Forsch.* **27,** 1636 (1977). Pharmacology: A. Bianchetti *et al., ibid.* 2096. Clinical study: B. Palmerio, P. Magnani, *ibid.* 2404. Toxicological study: L. Casilli *et al., ibid.* 2102.

Crystals from benzene, mp 290-292° (dec). $[\alpha]_D^{24}$ −36° (c = 1 in chloroform). uv max (methanol): 246 nm (ε 12500).

THERAP CAT: Anti-inflammatory (topical).

4631. Haloprogin. *1,2,4-Trichloro-5-[(3-iodo-2-propynyl)oxy]benzene; 3-iodo-2-propynyl 2,4,5-trichlorophenyl ether;* 2,4,5-trichlorophenyl γ-iodopropargyl ether; M-1028; Halotex; Mycanden; Mycilan; Polik. $C_9H_4Cl_3IO$; mol wt 361.39. C 29.91%, H 1.12%, Cl 29.43%, I 35.12%, O 4.43%. Prepn: S. Seki *et al., Agr. Biol. Chem. (Tokyo)* **27,** 150 (1963), *C.A.* **58,** 14635g (1963); *eidem,* Japan. pat. **64 19,-791;** *eidem,* U.S. pat. **3,322,813** (1964, 1967 both to Meiji Seika). Pharmacology: S. Seki *et al., Antimicrob. Ag. Chemother.* **1963,** 569; E. F. Harrison *et al., Appl. Microbiol.* **19,** 746 (1970).

White or pale yellow crystals, mp 113-114°; dec 190°. uv max (anhydr ethanol): 288.5, 298.5 nm. Easily soluble in methanol, ethanol; very slightly sol in water. LD_{50} in mice: > 3 g/kg orally and s.c.; 510 mg/kg i.p. (Seki, 1963).

THERAP CAT: Antibacterial.

4632. Halopropane. *3-Bromo-1,1,2,2-tetrafluoropropane;* 1-bromo-2,2,3,3-tetrafluoropropane; 1,1,2,2-tetrafluoro-3-bromopropane; FHD-3; Tebron. $C_3H_3BrF_4$; mol wt 194.95. C 18.48%, H 1.55%, Br 40.99%, F 38.98%. $CHF_2-CF_2CH_2Br$. Prepn: Cohen, U.S. pat. **3,080,430** (1963 to du Pont). Use in the study of cardiac arrhythmias: Katz, *J. Pharmacol. Exp. Ther.* **152**, 88 (1966).

Liquid, bp 74°; d_{20} 1.81; n_D^{25} 1.3558.

THERAP CAT: Anesthetic (inhalation).

4633. Halostachine. *(R)-α-[(Methylamino)methyl]-benzenemethanol;* *(—)-α-(methylaminomethyl)benzyl alcohol;* 1-hydroxy-1-phenyl-2-methylaminoethane; (—)-phenyl-(methylaminomethyl)carbinol; 2-methylamino-1-phenyl-ethanol. $C_9H_{13}NO$; mol wt 151.21. C 71.49%, H 8.67%, N 9.26%, O 10.58%. Isoln from *Halostachis caspica (Halostachis caspia):* Menshikov, Rubinshtein, *J. Gen. Chem. USSR* **13**, 801 (1943), *C.A.* **39**, 1172 (1945). Synthesis: Menshikov, Borodina, *J. Gen. Chem. USSR* **17**, 1569 (1947), *C.A.* **42**, 2245 (1948). Configuration: Lukes *et al.*, *Coll. Czech. Chem. Commun.* **26**, 466 (1961). As a possible cause of ryegrass staggers: Aasen *et al.*, *Aust. J. Agr. Res.* **20**, 71 (1969). TLC determn in rye and fescue grasses: L. P. Bush, J. A. D. Jeffreys, J. Chromatog. **111**, 165 (1975). Vasoactive potential: C. B. Davis *et al.*, *Vet. Hum. Toxicol.* **25**, 6 (1983). One pot synthesis: P. Zandbergen *et al.*, *Tetrahedron* **48**, 3977 (1992).

Crystals, mp 43-45°. $[\alpha]_D^{20}$ —47.03°. Soluble in water, alc, ether.

Hydrochloride, crystals from alcohol, mp 113-114°. $[\alpha]_D^{20}$ —52.46°. Freely sol in water.

4634. Halothane. *2-Bromo-2-chloro-1,1,1-trifluoroethane;* bromochlorotrifluoroethane; 1,1,1-trifluoro-2,2-chlorobromoethane; Fluothane; Rhodialothan. $C_2HBrClF_3$; mol wt 197.38. C 12.17%, H 0.51%, Br 40.48%, Cl 17.96%, F 28.88%. $CF_3CHClBr$. Prepn from a mixt of F_3CCH_2Cl and F_3CCBr_2Cl: Suckling, Raventos, U.S. pat. **2,921,098** (1960 to I.C.I.); by rearrangement of $F_2BrCCHFCl$: Scherer, Kühn, U.S. pat. **2,959,624** (1960 to Hoechst); from BrCl-CHCBrCl₂: Chapman, McGinty, Brit. pat. **805,764** (1958 to I.C.I.); from Br_2ClCCF_3: McGinty, U.S. pat. **3,082,263** (1963 to I.C.I.); Madai, Muller, *J. Prakt. Chem.* **19**, 83 (1963). Enantiomeric resolution: J. Meinwald *et al.*, *Science* **251**, 560 (1991). Comprehensive description: R. D. Daley, *Anal. Profiles Drug Subs.* **1**, 119-147 (1972).

Non-flammable, highly volatile liquid. Characteristic, sweetish, not unpleasant odor. d_4^{20} 1.871. bp 50.2°; bp_{243} 20°. n_D^0 1.3697. Sensitive to light, may be stabilized with 0.01% thymol. Soly in water 0.345%. Miscible with petr ether, other fat solvents.

THERAP CAT: Anesthetic (inhalation).

THERAP CAT (VET): Anesthetic (inhalation).

4635. Haloxazolam. *10-Bromo-11b-(2-fluorophenyl)-2,3,7,11b-tetrahydrooxazolo[3,2-d][1,4]benzodiazepin-6(5H)-one;* CS-430; Somelin. $C_{17}H_{14}BrFN_2O_2$; mol wt 377.21. C 54.13%, H 3.74%, Br 21.18%, F 5.04%, N 7.43%, O 8.48%. Sleep-inducing agent, related structurally to oxazolam and cloxazolam, *q.v.* Prepn of analogous compounds: T. Miyadera *et al.*, *J. Med. Chem.* **14**, 520 (1971). Structure-activity study: M. Yoshimoto *et al.*, *Chem. Pharm. Bull.* **25**, 1378 (1977). Pharmacological studies: T. Kamioka *et al.*, *Arzneimittel-Forsch.* **28**, 838 (1978). Metabolism in animals: R. Hayashi *et al.*, *Oyo Yakuri* **17**, 617 (1979), *C.A.* **91**, 204161s (1979).

Colorless cryst, mp 185°. Sparingly sol in water. LD_{50} in mice: 1850 mg/kg orally (Kamioka).

Note: This is a controlled substance (depressant) listed in the U.S. Code of Federal Regulations, Title 21 Part 1308.14 (1995).

THERAP CAT: Sedative, hypnotic.

4636. Haloxon. *Phosphoric acid bis(2-chloroethyl) 3-chloro-4-methyl-2-oxo-2H-1-benzopyran-7-yl ester; 7-[[bis(2-chloroethoxy)phosphinyl]oxy]-3-chloro-4-methyl-2H-1-benzopyran-2-one; 3-chloro-7-hydroxy-4-methylcoumarin bis(2-chloroethyl) phosphate; O,O-bis(β-chloroethyl) O-(3-chloro-4-methyl-7-coumarinyl) phosphate; 3-chloro-7-hydroxy-4-methyl-2H-1-benzopyran-2-one bis(chloroethyl) phosphate; 3-chloro-4-methylumbelliferone di(2-chloroethyl) phosphate; galoxone; helmirone; 96-H-60;Galloxan; Loxon; Luxon.* $C_{14}H_{16}Cl_3O_6P$; mol wt 415.59. C 40.46%, H 3.40%, Cl 25.59%, O 23.10%, P 7.45%. Prepn: Belg. pat. **610,896** (1962 to Cooper, McDougall & Robertson), *C.A.* **57**, 13729g (1962). Toxicity data: Brown *et al.*, *Nature* **194**, 379 (1962).

Crystals from ethanol, mp 91°. LD_{50} orally in rats: 900 mg/kg (Brown).

THERAP CAT (VET): Anthelmintic.

4637. Halquinol. *5,7-Dichloro-8-quinolinol mixt with 5-chloro-8-quinolinol and 7-chloro-8-quinolinol;* chlorquinol; SQ-16401; CHQ; Halquivet; Quinolor; Quixalin; Quixalud; Tarquinor. Proportions resulting naturally from chlorination of 8-quinolinol. Prepn of products contg 57-74 wt-% 5,7-dichloro-8-quinolinol, 23-40 wt-% 5-chloro-8-quinolinol, and up to 3 wt-% 7-chloro-8-quinolinol: Fr. pat. **1,372,414** (1964 to Olin Mathieson), *C.A.* **62**, 13130c (1965).

THERAP CAT: Anti-infective (topical).

THERAP CAT (VET): Antimicrobial (intestinal infections).

4638. Hamamelis. Witch hazel; winter bloom; snapping hazel; striped alder; spotted alder; tobacco wood. Dried leaves of *Hamamelis virginiana* L., *Hamamelidaceae,* collected in autumn. History of hamamelis extract and distillate: Lloyd, Lloyd, *J. Am. Pharm. Assoc.* **24**, 220 (1935). *Habit.* North America (New England to Minnesota, southward to Louisiana). *Constit.* Hamamelitannin, *q.v.,* gallic acid, volatile oil, bitter principle.

THERAP CAT: Astringent.

THERAP CAT (VET): Has been used as astringent, hemostatic, sedative.

4639. Hamamelitannin. *2-C-[[(3,4,5-Trihydroxybenzoyl)oxy]methyl]-D-ribofuranose 5-(3,4,5-trihydroxybenzoate); 2-C-(hydroxymethyl)-D-ribofuranose 2',5-digallate.* $C_{20}H_{20}O_{14}$; mol wt 484.37. C 49.59%, H 4.16%, O 46.24%. From bark of *Hamamelis virginiana* L., *Hamamelidaceae:* Grüttner, *Arch. Pharm.* **236**, 278 (1898); from bark of *Castanea dentata* (Marsh.) Borkh., *Fagaceae:* Mayer, Kunz, *Naturwiss.* **46**, 206 (1959). Structure: Mayer *et al.*, *Ann.* **688**, 232 (1965). Synthesis: Ezekiel *et al.*, *Carbohyd. Res.* **11**, 233 (1969).

Prisms from water, mp 145-147°. $[\alpha]_D^{19}$ +32.6° (c = 1.5). Hexamethyl ether, $C_{26}H_{32}O_{14}$, needles from dioxane + water, mp 73-76°. $[\alpha]_{578}^{25}$ +10.0° (c = 2.0 in dioxane). Sol in acetone, dioxane, methanol. Practically insol in water.

4640. Hamamelose. *2-C-(Hydroxymethyl)-D-ribose.* C_6-$H_{12}O_6$; mol wt 180.16. C 40.00%, H 6.71%, O 53.28%. From tannin of witch hazel *(Hamamelis virginiana* L., *Hamamelidaceae):* Fischer, Freudenberg, *Ber.* **45**, 2709 (1912); Freudenberg, Peters, *ibid.* **53**, 953 (1920); Anderson, *U.S.-A.E.C.* UCRL-8870, 114 (1959). Structure: Schmidt, *Ann.* **476**, 250 (1929). Configuration: Schmidt, Heintz, *ibid.* **515**, 77 (1934). Synthesis: Novák, Sorm, *Coll. Czech. Chem. Commun.* **30**, 3303 (1965); Paulsen *et al., Ber.* **105**, 1978 (1972). Synthesis of L-hamamelose: Burton *et al., Proc. Chem. Soc.* **1962**, 181. Synthesis and ^1H, ^{13}C NMR study of aqueous equilibrium: W. A. Szarek *et al., Can. J. Chem.* **61**, 461 (1983).

D-Form, crystals from abs ethanol, mp 111°. $[\alpha]_D^{21}$ −7.4° (equilib in water).

L-Form, crystals from ethanol + ethyl acetate, mp 110-111°. $[\alpha]_D^{22}$ +1.3° (3 min) → +7.3° (equilib after 17 min).

4641. Hamycin. Primamycin. Polyene antibiotic complex produced by *Streptomyces pimprina:* M. J. Thirumalachar *et al., Hindustan Antibiot. Bull.* **3**, 136 (1961), *C.A.* **55**, 27515i (1961); M. J. Thirumalachar, U.S. pat. **3,261,751** (1966 to Hindustan Antibiot.). Formerly believed to be identical to trichomycin and candicidin, *q.q.v.:* H. Burrows, D. Calam, *J. Chromatog.* **53**, 566 (1970). Subsequent HPLC studies have shown the three polyene antibiotics to be different entities: P. Helboe *et al., ibid.* **189**, 249 (1980). Structural study showing hamycin to be a mixture of components A, B, C, D: R. C. Pandey, K. L. Rinehart, *16th Interscience Conference on Antimicrob. Ag. Chemother.*, Chicago, 1976, *Abstracts of Papers,* no. 41. Absorption, excretion, tissue concentration: M. G. Phatak *et al., Indian J. Exp. Biol.* **12**, 284 (1974), *C.A.* **82**, 118703j (1975). Toxicity studies: Williams *et al., Antimicrob. Ag. Chemother.* **1964**, 737; antifungal activity: A. C. Parekh, C. V. Dave, *Life Sci.* **19**, 1737 (1976). HPLC study: W. Mechlinski, C. P. Schaffner, *J. Antibiot.* **33**, 591 (1980). Review of clinical use in mycoses: M. J. Thirumalachar, *J. Sci. Ind. Res.* **31**, 542-544 (1972), *C.A.* **79**, 13308 (1973).

Yellow amorphous powder, no definite mp, decomp 160°. $[\alpha]_D^{22}$+216°. uv max (80% methanol): 383 nm ($E_{1cm}^{1\%}$ 916). An amphoteric compd. Almost insol in water, benzene, chloroform, dry lower aliphatic alcohols, ether. Sol in basic solvents such as pyridine, collidine, and in aq lower alcohols. In concd H_2SO_4 gives stable blue color; no coloration with ferric chloride or with HCl. LD_{50} i.v. in mice (using sodium carboxymethyl cellulose as a vehicle): 6.16 mg/kg (Williams). Also reported as LD_{50} i.v. in Swiss mice (using a colloidal suspension): 1.20 mg/kg (Parekh, Dave).

Hamycin A, mp > 300° (dec), $[\alpha]_D^{24}$ +181.1° (c = 0.6 in DMF). uv max: 380 nm ($E_{1cm}^{1\%}$ 989).

THERAP CAT: Antifungal.

4642. Haplophytine. *[15(3aS,7R)]-3,4-Didehydro-19-hydroxy-16,17-dimethoxy-1-methyl-15-(2,3,5,6-tetrahydro-11-hydroxy-4-methyl-1,13-dioxo-1H-3a,7-methanopyrrolo[1,2-a]-[1,3]benzodiazocin-7(4H)-yl)aspidospermidin-21-oic acid γ-lactone.* $C_{37}H_{40}N_4O_7$; mol wt 652.75. C 68.08%, H 6.18%, N 8.58%, O 17.16%. Insecticidal alkaloid from the Mexican plant *Haplophyton cimicidum* A. DC., *Apocynaceae:* Rogers *et al., J. Am. Chem. Soc.* **74**, 1987 (1952); **76**, 2819 (1954). Partial structure: Snyder *et al., ibid.* **80**, 3708 (1958). Structure: Rae *et al., ibid.* **89**, 3061 (1967). Crystal structure and absolute config.: Zacharias, *Acta Crystallogr.* **26B**, 1455 (1970). Synthetic approaches: P. Yates, D. A. Schwartz, *Can. J. Chem.* **61**, 509 (1983). *Review:* Yates *et al., J. Am. Chem. Soc.* **95**, 7842 (1973).

Crystals from ethanol + chloroform, mp 290-293° (rapid heating, starting at 250°). Also reported as mp 300-302° (Yates *et al., loc. cit.*). $[\alpha]_D^{25}$ +109.0° (chloroform). uv max (ethanol): 220, 265, 305 nm (ε 48500, 14300, 4500). Very sol in chloroform, benzene, dioxane, ethyl acetate. Moderately sol in acetone, methanol, somewhat less in ethanol. Practically insol in water, ether, petr ether. Readily sol in dil acids or alkalies.

Dihydrochloride, $C_{27}H_{31}N_3O_5.2HCl$, dec 208-218° (darkens at 200°).

O-Methylhaplophytine, $C_{28}H_{33}N_3O_5$, crystals from ether + ethanol, dec 288-291°. $[\alpha]_D^{24}$ +12° (c = 4.37 in chloroform).

4643. Haptoglobins. α_2-Globulins which combine with hemoglobin to form a weak peroxidase and which constitute about one-quarter of the α_2-globulin fraction of human plasma: Jayle, Conas, *Bull. Soc. Chim. Biol.* **34**, 65 (1952). Prepn: Herman-Boussier *et al., ibid.* **48**, 817, 837 (1960). For each individual there are three genetic types of haptoglobins, each reacting identically with antibodies to each of the others. Structure of this serum glycoprotein is a tetramer composed of α- and β-polypeptide chains, the α-chain varying in the three phenotypes, the β-chains being identical. Amino acid sequence of α-chains: Black, Dixon, *Nature* **218**, 736 (1968); Malchy, Dixon, *Can. J. Biochem.* **51**, 321 (1973). C- and N-terminal sequences of the β-chain: Barnett *et al., Biochemistry* **11**, 1189 (1972); A. Kurosky *et al., Comp. Biochem. Physiol.* **55B**, 453 (1976); *eidem, Biochemistry* **15**, 5326 (1976). Lowered levels of haptoglobins found in individuals with acute hepatitis and pernicious anemia. *Review:* Laurell, Grönvall, *Advan. Clin. Chem.* **5**, 135-172 (1962). Monograph: Kirk, *The Haptoglobin Groups in Man* (S. Karger, New York, 1968) 77 pp. Review on genetics, biochemistry and physiology of human haptoglobins: J. Javid, *Curr. Top. Hematol.* **1**, 151-192 (1978).

4644. Harmaline. *4,9-Dihydro-7-methoxy-1-methyl-3H-pyrido[3,4-b]indole;* 1-methyl-7-methoxy-3,4-dihydro-β-carboline; 3,4-dihydroharmine; harmidine; harmalol methyl ether; *O-methylharmalol.* $C_{13}H_{14}N_2O$; mol wt 214.27. C 72.87%, H 6.59%, N 13.07%, O 7.47%. CNS stimulant from seeds of *Peganum harmala* L., *Zygophyllaceae:* Goebel, *Ann.* **38**, 363 (1841); from *Banisteria caapi* Spruce, *Malpighiaceae:* Hochstein, Paradies, *J. Am. Chem. Soc.* **79**, 5735 (1957). Structure: Manske *et al., J. Chem. Soc.* **1927**, 1. Synthesis: Späth, Lederer, *Ber.* **63**, 120, 2102 (1930); Spenser, *Can. J. Chem.* **37**, 1851 (1959). Identity with harmidine: Robinson, *Chem. & Ind. (London)* **1965**, 605. Pharmacology: Fuentes, Longo, *Neuropharmacology* **10**, 15 (1971). Metabolism: Ho *et al., Biochem. Pharmacol.* **20**, 1313 (1971). Review of structure and synthesis work: Hofmann, *Sv. Farm. Tidskr.* **75**, 933 (1971), *C.A.* **76**, 149609g (1972).

Orthorhombic bipyramidal prisms, tablets from methanol, rhombic octahedra from ethanol, mp 229-231°. Solns fluoresce blue. pK 4.2. uv max (methanol): 218, 260, 376 nm (log ε 4.27, 3.90, 4.02). Slightly sol in water, alcohol, ether; quite sol in hot alcohol, dil acids.

Hydrochloride dihydrate, slender, yellow needles, moderately sol in water, alcohol.

N-Acetylharmaline, needles, mp 204-205°.

4645. Harmalol. *4,9-Dihydro-1-methyl-3H-pyrido[3,4-b]indol-7-ol;* 3,4-dihydro-1-methyl-9*H*-pyrido[3,4-*b*]indol-7-ol. $C_{12}H_{12}N_2O$; mol wt 200.24. C 71.98%, H 6.04%, N 13.99%, O 7.99%. From seeds of *Peganum harmala* L., *Zygophyllaceae:* Göbel, *Ann.* **38**, 363 (1841); Fischer, *Ber.* **18**, 400 (1885); by demethylation of harmaline: Coulthard *et al., Biochem. J.* **27**, 727 (1933).

Trihydrate, red needles from water. Dec 212° (anhydr). Readily sol in hot water, acetone, chloroform, alkali hydroxides, but not carbonates. Aq solns are yellow with green fluorescence.

Lactate monohydrate, bright yellow leaflets, mp 116-120°; when anhydr mp 174-176°. Sol in water and alcohol.

O-Ethylharmalol, brown needles from alc, mp 237-239°.

O-n-Propylharmalol, brown needles from alc, mp 195-197°.

O-n-Butylharmalol, colorless needles from alc, mp 173°.

O-Methylharmalol, *see* Harmaline.

4646. Harman. *1-Methyl-9H-pyrido[3,4-b]indole;* 3-methyl-4-carboline; 2-methyl-β-carboline; aribine; loturine; passiflorin. $C_{12}H_{10}N_2$; mol wt 182.22. C 79.10%, H 5.53%, N 15.37%. From bark of *Sickingia rubra* (Mart.) K. Schum. *(Arariba rubra* Mart.), *Rubiaceae; Symplocus racemosa* Roxb., *Symplocaceae;* and *Passiflora incarnata* L., *Passifloraceae:* Rieth, Wohler, *Ann.* **120**, 247 (1861); Späth, *Monatsh.* **40**, 151 (1920); Neu, *Arzneimittel-Forsch.* **4**, 601 (1954). Structure: Neu, *ibid.* **6**, 94 (1956). Synthesis: Harvey, Robson, *J. Chem. Soc.* **1938**, 97; Snyder *et al., J. Am. Chem. Soc.* **70**, 222 (1948); Clemo, Holt, *J. Chem. Soc.* **1953**, 1313; Kametani *et al., ibid. (C)* **1968**, 1006. Isoln from cigarette smoke: Poindexter, Carpenter, *Chem. & Ind. (London)* **1962**, 176. Toxicity study: E. B. Sigg *et al., Arch. Int. Pharmacodyn.* **149**, 164 (1964).

Bitter orthorhombic crystals from heptane + cyclohexane, mp 237-238°. Exhibits bright blue fluorescence in uv light. pKa's 7.37, 144.6. uv max (methanol): 234, 287, 347 nm (log ε 4.57, 4.21, 3.66). Absorption and fluorescence spectra: O. S. Wolfbeis *et al., Monatsh. Chem.* **113**, 509 (1982). Practically insol in water. Sol in dil acids. LD_{50} i.p. in mice: 50 mg/kg (Sigg).

Hydrochloride, $C_{12}H_{10}N_2 \cdot HCl$, rosettes of needles from ethanol + 20% HCl in water, sublimes at 120-130°.

4647. Harmine. *7-Methoxy-1-methyl-9H-pyrido[3,4-b]-indole;* banisterine; yageine; telepathine; leucoharmine. C_{13}-$H_{12}N_2O$; mol wt 212.25. C 73.57%, H 5.70%, N 13.20%, O 7.54%. CNS stimulant isolated from seeds of *Peganum har-*

mala L., *Zygophyllaceae:* Göbel, *Ann.* **38**, 363 (1841); Reinhard *et al., Phytochemistry* **7**, 503 (1968); from *Banisteria caapi* Spruce, *Malpighiaceae:* Hochstein, Paradies, *J. Am. Chem. Soc.* **79**, 5735 (1957); from *Banisteriopsis inebrians* Morton, *Malpighiaceae:* O'Connell, Lynn, *J. Am. Pharm. Assoc.* **42**, 753 (1953). Structure: Manske *et al., J. Chem. Soc.* **1927**, 1, 240. Synthesis: Späth, Lederer, *Ber.* **63B**, 120 (1930); Akaboro, Saito, *ibid.* 2245; Hahn *et al., Ber.* **67B**, 2031 (1934); *Ann.* **520**, 107, 123 (1935); *Ber.* **71B**, 2163, 2175 (1938); Harvey, Robson, *J. Chem. Soc.* **1938**, 97. Metabolism: Slotkin, DiStefano, *J. Pharmacol. Exp. Ther.* **174**, 456 (1970). Toxicity study: K. K. Chen *et al., J. Pharmacol. Exp. Ther.* **79**, 127 (1943).

Slender, orthorhombic prisms from methanol, mp 261° (dec). Sublimes. pKa 7.70. uv max (methanol): 241, 301, 336 nm (log ε 4.61, 4.21, 3.69). Absorption and fluorescence spectra: O. S. Wolfbeis, E. Fürlinger, *Z. Physikal. Chem.* **129**, 171 (1982). Slightly sol in water, alcohol, chloroform, ether.

Hydrochloride dihydrate, crystals, mp 262° (dec), mp 321° when anhydrous. Sol in 40 parts water, freely sol in hot water. Aq solns have blue fluorescence. LD_{50} i.v. in mice: 38 mg/kg (Chen).

4648. Hashish. Hasach; kif. Purified alcoholic extract of *Cannabis sativa* L. *(Cannabis sativa* var. *indica* Auth.), *Moraceae* deprived of its volatile oil; resin secreted by the flowering tops of the female plant. Principal active constituents are tetrahydrocannabinols, *q.v.* Review and recent advances in the chemistry of hashish: Mechoulam, Gaoni, *Fortschr. Chem. Org. Naturst.* **25**, 175 (1967). *See also* Cannabis.

Caution: May be habit forming. This is a controlled substance (hallucinogen) listed in the U.S. Code of Federal Regulations, Title 21 Parts 329.1 and 1308.11 (1995).

4649. Hasubanonine. *7,8-Didehydro-3,4,7,8-tetramethoxy-17-methylhasubanan-6-one.* $C_{21}H_{27}NO_5$; mol wt 373.45. C 67.54%, H 7.29%, N 3.75%, O 21.42%. Alkaloid having a modified morphinan skeleton with a five membered heterocyclic ring, hitherto unknown in natural sources. (Alkaloids with this unique skeleton are called *hasubanan* alkaloids; *see* S. Shiotani, T. Kometani, *Tetrahedron Letters* **1976**, 767 for synthesis of hasubanan alkaloids). Isolated from *Stephania japonica* Miers, *Menispermaceae:* Kondo *et al., Ann. Rept. ITSUU Lab.* **2**, 35 (1951); Satomi, *ibid.* **3**, 37 (1953); Kondo, Satomi, *ibid.* **8**, 41 (1957); *C.A.* **47**, 5951f (1953); **48**, 2728c (1954); **51**, 17956i (1957). Structure: Tomita *et al., Tetrahedron Letters* **1964**, 2937; *eidem, Chem. Pharm. Bull.* **13**, 538 (1965). Total synthesis: Ibuka *et al., Tetrahedron Letters* **1970**, 4811; *Chem. Pharm. Bull.* **22**, 782 (1974). Biosynthesis: A. R. Battersby *et al., Chem. Commun.* **1974**, 773; *eidem, J. Chem. Soc. Perkin Trans. I* **1981**, 2010, 2016, 2030. *Review:* Y. Inubushi, T. Ibuka in *The Alkaloids,* vol. **XVI**, R. H. F. Manske, Ed. (Academic Press, New York, 1977) pp 393-428.

Prisms from methanol, mp 116°. $[\alpha]_D$ −219° (ethanol).

4650. HCS. *Lactogen (human placental); lactogenic hormone (placental human);* human chorionic somatomammotropin; chorionic growth hormone-prolactin; CGP; human placental lactogen; HPL. Mol wt about 21,000. Growth hormone from human placenta. Exhibits lactogenic as well as growth-promoting activity: Josimovich, MacLearen, *Endocrinology* **71**, 209 (1962); Kaplan, Grumbach, *J. Clin. Endocrinol. Metab.* **24**, 80 (1964). Structure is a single polypeptide chain containing 190 amino acid residues: Li *et al., Science* **173**, 56 (1971); Sherwood *et al., Nature New Biol.* **233**, 59 (1971). Structural revision to 191 amino acid residues and comparison with human growth hormone and human prolactin: Li *et al., Arch. Biochem. Biophys.* **155**, 95 (1973); Niall *et al., Recent Progr. Horm. Res.* **29**, 387 (1973). Purification: Parcells, Dahlgren, U.S. pat. **3,687,833** (1972 to Upjohn). Terminology: Li *et al., Experientia* **24**, 1288 (1968). *Review:* M. Chatterjee, H. N. Munro in *Vitamins and Hormones* vol. **35**, P. L. Munson *et al.*, Eds. (Academic Press, New York, 1977) pp 149-208.

4651. Heart Muscle Extract. Usually prepd from the hearts of calf embryos. *Ref:* Strauss, *Deut. Med. Wochenschr.* **1952**, 1284; Umhau, *Landarzt* **1952**, 100; Kristen, *Deut. Med. J.* **1952**, 250; Willems, *Die Medizinische* **1953**, no. 13; Jasinski, *Cardiologia* **23**, 49 (1953); Brenner, *Klin. Wochenschr.* **1954**, 295; Witzleb *et al., ibid.* 297. Commercial preparations: *Myocardone; Recosen; Rocosenin; Corhormon; Lysomiol; Hormocardiol; Hrezolan; Cordiomon* (inj.).

THERAP CAT: Vasodilator (coronary).

4652. Hecogenin. *(3β,5α,25R)-3-Hydroxyspirostan-12-one.* $C_{27}H_{42}O_4$; mol wt 430.63. C 75.31%, H 9.83%, O 14.86%. A steroidal sapogenin which has been isolated from plants, particularly from numerous *Agave* species. Isoln: Marker *et al., J. Am. Chem. Soc.* **69**, 2167 (1947); Rubin, U.S. pat. **3,303,186** (1967). Synthesis and configuration: Mazur *et al., J. Am. Chem. Soc.* **82**, 5889 (1960). Separation of an optically inactive product from sapogenin mixtures: Cardenas, U.S. pat. **3,013,010** (1961 to Searle). *Review:* L. F. Fieser, M. Fieser, *Steroids* (Reinhold, New York, 1959) pp 667-671 sqq.

Crystals from acetone, mp 264-266°. $[\alpha]_D$ +8° (CHCl$_3$) (Mazur *et al.*); mp 245°, 253°, 268° (Marker *et al.*); mp 240-245°, 245-250°. $[\alpha]_D$ ±0° (CHCl$_3$) (Cardenas).

Pseudohecogenin, $C_{27}H_{42}O_4$, *(3β,5α,25R)-3,26-Dihydroxyfurost-20(22)-en-12-one.* Prepn from hecogenin: Cameron *et al., J. Chem. Soc.* **1955**, 2807; Wall, Serota, U.S. pat. **2,870,143** (1959 to USDA). Plates from aq acetone, mp 190-191°. $[\alpha]_D$ +103° (c = 1.5 in chloroform), +96° (c = 1 in dioxane). uv max (ethanol): 213 nm (ε 6400).

USE: In prepn of steroidal hormones.

4653. Hectorite. Strese & Hofmann's hectorite. Swelling and gelling clay of the montmorillonite group. Approx formula: $Na_{0.67}(Mg,Li)_3Si_8O_{20}(OH,F)_4$. Some analyses show neither Li nor F, whether by oversight or by variation in the mineral is uncertain: M. H. Hey, *Mineral Species and Varieties* (Brit. Museum, London, 2nd ed., 1962) p 205.

Caution: Dust can be irritating to respiratory tract.

USE: In the chill-proofing of beer: Shaler *et al.*, U.S. pat. **3,100,707** (1963 to American Tansul).

4654. Hederagenin. *(3β,4α)-3,23-Dihydroxyolean-12-en-28-oic acid;* caulosapogenin; melanthigenin. $C_{30}H_{48}O_4$; mol wt 472.71. C 76.23%, H 10.23%, O 13.54%. Occurs as glycoside in many saponins, see α-Hederin. Isoln and structure: van der Haar, *Arch. Pharm.* **250**, 424 (1912); Ruzicka *et al., Helv. Chim. Acta* **28**, 380 (1945); Haynes *et al., J.*

Chem. Soc. **1963**, 744. Identity with melanthigenin: Mustafa, Soliman, *ibid.* **1943**, 70. Identity with caulosapogenin: McShefferty, Stenlake, *ibid.* **1956**, 2314. Molecular and crystal structure: R. Roques *et al., Acta Crystallogr.* **B34**, 1634 (1978). *See also* α- and β-Amyrin and Oleanolic Acid.

Crystals from alc, mp 332-334°. $[\alpha]_D^{20}$ +81° (c = 0.7 in pyridine). Freely sol in pyridine; sol in chloroform-alcohol mixtures; slowly sol in alcohol; practically insol in water; sol in dil alcoholic NaOH, but not in water solns of alkalies.
Diacetate, mp 172-174°. $[\alpha]_D^{20}$ +64° (CHCl$_3$).
Methyl ester, mp 240°. $[\alpha]_D^{23}$ +76° (c = 0.8 in CHCl$_3$).

4655. α-Hederin. Helixin (the saponin). $C_{41}H_{66}O_{12}$; mol wt 750.97. C 65.58%, H 8.86%, O 25.57%. Isoln from ivy leaves *(Hedera helix* L., *Araliaceae):* van der Haar, *Arch. Pharm.* **250**, 424 (1912); **251**, 632, 650; idem, *Biochem. Z.* **76**, 335 (1916); idem, *Ber.* **54**, 3142 (1921). Isoln and structure: Tschesche *et al., Z. Naturforsch.* **20b**, 708 (1965).

Precipitated from ethanol by addition of ether, mp 256-259°. $[\alpha]_D^{20}$ +14.5° (c = 0.92 in methanol).

4656. Helenalin. *[3aS-(3aα,4α,4aβ,7aα,8α,9aα)]-3,3a,-4,4a,7a,8,9,9a-Octahydro-4-hydroxy-4a,8-dimethyl-3-methyleneazuleno[6,5-b]furan-2,5-dione; 6α,8β-dihydroxy-4-oxoambrosa-2,11(13)-dien-12-oic acid 12,8-lactone.* $C_{15}H_{18}O_4$; mol wt 262.31. C 68.69%, H 6.92%, O 24.40%. Pseudoguaianolide sesquiterpenoid lactone from *Helenium autumnale* L., *H. amarum* (Raf.) H. Roch, *H. microcephalum* DC., *Compositae.* Isoln: Clark, *J. Am. Chem. Soc.* **58**, 1982 (1936); and characterization: Adams, Herz, *ibid.* **71**, 2546 (1949). Structure: Büchi, Rosenthal, *ibid.* **78**, 3860 (1956); Barton, de Mayo, *Quart. Rev. (London)* **11**, 189 (1957); Herz, *J. Org. Chem.* **27**, 4043 (1962). Abs config studies: Herz, Kagan, *ibid.* **32**, 216 (1967). Synthesis of racemic form: Y. Ohfune *et al., J. Am. Chem. Soc.* **100**, 5946 (1978); M. R. Roberts, R. H. Schlessinger, *ibid.* **101**, 7626 (1979); C. H. Heathcock *et al., ibid.* **104**, 1907 (1982). Toxicity study: D. A. Witzel *et al., Am. J. Vet. Res.* **37**, 859 (1976).

Bitter, sternutative crystals from benzene, mp 167-168°. $[\alpha]_D^{25}$ −102.8° (c = 3.64 in 95% ethanol). uv max: 223 nm (ε 11900). Slightly sol in water; sol in alcohol, chloroform, hot benzene. LD$_{50}$ in mice: 150 mg/kg orally.

Acetylhelenalin, C$_{17}$H$_{20}$O$_5$, crystals from aq methanol, mp 184°. uv max: 221, 316 nm (ε 12600, 61).

4657. Helenynolic Acid. *[S-(E)]-9-Hydroxy-10-octadecen-12-ynoic acid.* C$_{18}$H$_{30}$O$_3$; mol wt 294.43. C 73.43%, H 10.27%, O 16.30%. From seed oil of *Helichrysum bracteatum* Andr., *Compositae:* R. G. Powell *et al., J. Am. Oil. Chem. Soc.* **42**, 165 (1965). Structure: *eidem, J. Org. Chem.* **30**, 610 (1965). Absolute configuration: J. C. Craig *et al., ibid.* 4342. Partial synthesis of racemic helenynolic acid: H. R. S. Conacher, Gunstone, *Lipids* **5**, 137 (1970). Synthesis of (±)-form: T. B. Patrick, G. F. Melm, *J. Org. Chem.* **44**, 645 (1979).

Methyl ester, C$_{19}$H$_{32}$O$_3$. uv max (isooctane): 228, 238 nm (ε 17400, 14300). [α]$_{600}^{26}$ −7° (c = 3.6 in ethanol).

4658. Helicin. *2-(β-D-Glucopyranosyloxy)benzaldehyde;* salicylaldehyde β-D-glucoside. C$_{13}$H$_{16}$O$_7$; mol wt 284.27. C 54.93%, H 5.67%, O 39.40%. Prepn by the oxidation of salicin with dil nitric acid: Schiff, *Ann.* **154**, 19 (1870); from salicylaldehyde + O-tetraacetyl-α-glucosidyl bromide: Robertson, Waters, *J. Chem. Soc.* **1930**, 2729. ORD and stereochemical studies: Tsuzuki *et al., Bull. Chem. Soc. Japan* **44**, 526 (1971).

Needles with 0.75 mol H$_2$O from H$_2$O; mp 175-176° when dried at 100°. [α]$_D^{20}$ −60° (c = 1.4). One gram dissolves in 55 ml water; freely sol in hot water, alcohol. Forms compds with urea, thiourea, certain amino acids.

Tetraacetate, C$_{21}$H$_{24}$O$_{11}$, needles from alc, mp 142°. [α]$_D^{20}$ −37° (acetone).

Note: The same formula is ascribed to *spirein,* found in *Spiraea camtschatica* Pall. and in *S. ulmaria* L., *Rosaceae.* Emulsin hydrolyzes helicin and spirein, yielding D-glucose and salicylaldehyde.

4659. Heliosupine. *2-Methyl-2-butenoic acid 7-[[2,3-dihydroxy-2-(1-hydroxyethyl)-3-methyl-1-oxobutoxy]methyl]-2,-3,5,7a-tetrahydro-1H-pyrrolizin-1-yl ester;* cynoglossophine. C$_{20}$H$_{31}$NO$_7$; mol wt 397.47. C 60.44%, H 7.86%, N 3.52%, O 28.18%. Hepatotoxic pyrrolizidine alkaloid isolated from *Heliotropium supinum* L., *Boraginaceae:* Denisova *et al., Doklady Akad. Nauk SSSR* **93**, 59 (1953); from *Cynoglossum officinale* L., *Boraginaceae:* Man'ko, Borisyuk, *Ukrain. Khim. Zhur.* **23**, 362 (1957), *C.A.* **52**, 2188a (1958); Man'ko, Marchenko, *Khim. Prir. Soedin.* **7**, 537 (1971), *C.A.* **75**, 126598t (1971). Identity with cynoglossophine: Man'ko, *Ukrain. Khim. Zhur.* **25**, 627 (1959), *C.A.* **54**, 12494d (1960). Structure: Crowley, Culvenor, *Aust. J. Chem.* **12**, 694 (1959). Biosynthesis studies: Crout, *J. Chem. Soc. (C)* **1967**, 1233. *Reviews:* see Lasiocarpine.

Colorless gum. [α]$_D^{20}$ −4.3° (c = 5.1 in ethanol). Methyl ether, *see* Lasiocarpine.

4660. Helium. He; at. wt 4.002602; at. no. 2. Group VIIIA(18), also known as Group 0. A noble gas characterized by an electronic structure in which the outer *p* subshell is entirely filled. Natural isotopic mixture (mass numbers): 4 (99.999862%), 3 (1.38 × 10^{-4}%). Known artificial radioactive isotopes: 5-8. Longest-lived isotope: ^{6}He (T$_{1/2}$ 806.7 msec, β-emitter). Abundance in igneous rock of earth's crust: 3 × 10^{-3} ppm by wt; concentration in air: 5.24 ppm by vol. Identified in the spectrum of the sun's chromosphere by Lockyer and Frankland in 1868. Obtained by Hillebrand in 1890 by heating uranium minerals and identified by Ramsay in 1895. Found in natural gas from which it is extracted on a commercial scale. Produced in the decay of radioactive elements: 1 kg of uranium in its conversion into 865 g of lead forms 756 l of helium; also produced in nature by the bombardment of beryllium, lithium, and other light elements with cosmic rays, x-rays and high-speed protons and deuterons. Monograph: G. A. Cook, *Argon, Helium and the Rare Gases* (Interscience, New York, 1961). *Reviews:* Cockett, Smith, "The Monatomic Gases" in *Comprehensive Inorganic Chemistry* vol. 1, J. C. Bailar, Jr. *et al.,* Eds. (Pergamon Press, Oxford, 1973) pp 139-211; E. Cook, *Science* **206**, 1141-1146 (1979); S.-C. Hwang, W. R. Weltmer, Jr. in Kirk Othmer *Encyclopedia of Chemical Technology* vol. 13 (John Wiley & Sons, 4th ed., 1995) pp 1-38; *Chemistry of the Elements,* N. N. Greenwood, A. Earnshaw, Eds. (Pergamon Press, New York, 1984) pp 1042-1059. Review of use in inductively coupled plasma-mass spectrometry: S. F. Durrant, *Fres. J. Anal. Chem.* **347**, 389-392 (1993).

Colorless, odorless, tasteless, nonflammable, monatomic, inert gas; will form compds with highly electronegative elements such as O, F, Cl. Cannot be frozen by lowering the temp at ordinary press.; no triple point. Very slightly sol in water (ml/100 ml): 0.97 at 0°; 1.08 at 50°. Trouton's const 4.64.

^{4}He, critical temp 5.2014°K, critical press 227.5 kPa, critical d 69.64 kg/m^3. Gas: d^0 (101.3 kPa) 0.17850 kg/m^3, d (normal bp) 16.89 kg/m^3. Liquid: normal bp 4.224°K, d (normal bp) 125.0 kg/m^3, heat of vaporization (normal bp) 81.70 J/mol. Two liq forms exist: He I above ∼2.2°K; He II below ∼2.2°K. He II is a superconducting liq; has very low viscosity; superfluid.

^{3}He, critical temp 3.324°K, critical press 116.4 kPa, critical d 41.3 kg/m^3. Gas: d^0 (101.3 kPa) 0.1347 kg/m^3, d (normal bp) 23.64 kg/m^3. Liquid: normal bp 3.191°K, d (normal bp) 58.9 kg/m^3, heat of vaporization (normal bp) 25.48 J/mol.

Caution: Can act as a simple asphyxiant by displacing air. *See: Matheson Gas Data Book* (Matheson Co., Inc., 4th ed., East Rutherford, NJ, 1966) pp 249-253.

USE: Liquid helium (the most volatile liq known) as cryogen for the production of low temps, in MRI machines. Gas in manuf of semiconducting devices; in detection devices for leaks in vacuum systems; in Ne-He lasers; in gas mixtures as the working fluid in plasma devices; in mixtures with Ne and Ar in Geiger counters. *Q-Gas,* a mixture of 98.7% He and 1.3% butane, has been used as a filling for gas-flow Geiger counters. Gas as a shield in gas tungsten-arc welding, in metal processing; substitute for N$_2$ in synthetic breathing gas for deep sea divers and workers in high pressure conditions; coolant in high temp nuclear reactors; carrier gas in gas-liquid and gas-solid chromatography; inert diluent; to create inert atmosphere; to fill balloons and airships, lifting power is 0.93 if hydrogen is taken as 1.00.

4661. Hellebrin. *3-[(6-Deoxy-4-O-β-D-glucopyranosyl-α-L-mannopyranosyl)oxy]-5,14-dihydroxy-19-oxobufa-20,22-dienolide;* hellebrigenin glucorhamnoside. C$_{36}$H$_{52}$O$_{15}$; mol wt 724.80. C 59.66%, H 7.23%, O 33.11%. Cardiac glycoside isolated from rhizome of *Helleborus niger* L., *Ranunculaceae:* Karrer, *Festschrift E. C. Barell* **1936**, 238, *C.A.* **31**, 2348^8 (1937). Structure: Karrer, *Helv. Chim. Acta* **26**, 1353 (1943); Wissner, Kating, *Planta Med.* **20**, 344 (1971).

β-D-glucose—rhamnose-O

Crystals from hot methanol, mp 283-284°. $[\alpha]_D^{20}$ −23.4° (50% methanol). Sol in dil alcohol; less sol in methanol, ethanol; slightly sol in water. Practically insol in ether. MLD in guinea pigs: 0.85 μMol/kg, Klup, *Arch. Exp. Pathol. Pharmakol.* **252**, 314 (1966).

4662. Helminthosporal. *[1R-(endo, anti)]-1,7-Dimethyl-4-(1-methylethyl)bicyclo[3.2.1]oct-6-ene-6,8-dicarboxaldehyde.* $C_{15}H_{22}O_2$; mol wt 234.34. C 76.88%, H 9.46%, O 13.65%. Toxin produced as (−)-form by the fungus *Bipolaris sarokiniana* Shoemaker *(Helminthosporium sativum).* Isoln: deMayo *et al., Can. J. Chem.* **39**, 1608 (1961). Structure: deMayo *et al., J. Am. Chem. Soc.* **84**, 494 (1962). Stereochemistry: deMayo *et al., Can. J. Chem.* **41**, 2996 (1963). Total synthesis and absolute configuration: Corey, Nozoe, *J. Am. Chem. Soc.* **87**, 5728 (1965); (+)-form: E. Piers, H. P. Isenring, *Can. J. Chem.* **55**, 1039 (1977); of (±)-form: M. Yanagiya *et al., Tetrahedron Letters* **1979**, 1761.

Crystals from light petr, mp 56-59°. $[\alpha]_D$ −49° (c = 1.18 in CHCl$_3$). bp$_{0.015}$ 115-120°. uv max (ethanol): 266 nm (ϵ 11000).

Bis-oxime, $C_{15}H_{24}N_2O_2$, crystals from alc, mp 177-182°.

4663. Helminthosporol. *[1R-(endo, anti)]-8-(Hydroxymethyl)-1,7-dimethyl-4-(1-methylethyl)bicyclo[3.2.1]oct-6-ene-6-carboxaldehyde; 8-(hydroxymethyl)-4-isopropyl-1,7-dimethylbicyclo[3.2.1]oct-6-ene-6-carboxaldehyde.* $C_{15}H_{24}O_2$; mol wt 236.35. C 76.23%, H 10.23%, O 13.54%. A plant-growth regulator isolated from *Helminthosporium sativum* as (−)-form. Prepn: Tamura *et al., Agr. Biol. Chem. (Tokyo)* **29**, 216 (1965); Tamura, Sakurai, *ibid.* **28**, 337 (1964); Tamura *et al., ibid.* **27**, 738 (1963). Synthesis of (+)-form: E. Piers, H. P. Isenring, *Can. J. Chem.* **55**, 1039 (1977). *See also* Helminthosporal.

Crystals from *n*-hexane, mp 98°. $[\alpha]_D^{23}$ −28.7° (c = 1.93 in CHCl$_3$). uv max (ethanol): 267 nm.

4664. Helonias. False unicorn; blazing star; starwort. Rhizome and roots of *Chamaelirium luteum* (L.) A. Gray, *Liliaceae.* Habit. Ontario and Eastern U.S.

4665. Helveticoside. *(3β,5β)-[(2,6-Dideoxy-β-D-ribohexopyranosyl)oxy]-5,14-dihydroxy-19-oxocard-20(22)-enolide;* erisimin; erysimin; alleoside A. $C_{29}H_{42}O_9$; mol wt 534.65. C 65.15%, H 7.92%, O 26.93%. A β-glycoside consisting of one mole strophanthidin and one mole D-digitoxose. Isoln from *Erysimum helveticum* (Jacq.) DC. and *Erysimum crepidifolium* Reichenb., *Cruciferae,* and structure: Nagata *et al., Helv. Chim. Acta* **40**, 41 (1957); *eidem, Festschrift Arthur Stoll* **1957**, 715; Gmelin, **Ger.** pat. **1,221,764** (1966). From *Erysimum* spp.: Kowalewski, *Helv. Chim. Acta* **43**, 1314 (1960). Toxicity study: Graebner, Giesel, *Arzneimittel-Forsch.* **22**, 1854 (1972).

Dihydrate, needles from dil methanol, mp 153-157°. $[\alpha]_D^{25}$ +30.7° (c = 1.5 in methanol); $[\alpha]_D^{24}$ +26.0° (c = 1.1 in chloroform). uv max (ethanol): 217, 304 nm (log ϵ 4.27, 1.55). *Violent poison!* LD$_{100}$ i.v. in cats: 0.104 mg/kg (Graebner, Giesel).

3',4'-Diacetate, $C_{33}H_{46}O_{11}$, prisms from methanol + ether, mp 237-244°. $[\alpha]_D^{25}$ +41.3° (chloroform).

4666. Helvolic Acid. *(4α,6β,8α,9β,13α,14β,16β,17Z)-6,16-Bis(acetyloxy)-3,7-dioxo-29-nordammara-1,17(20),24-trien-21-oic acid; (Z)-6β,16β-dihydroxy-3,7-dioxo-29-nor-8α,-9β,13α,14β-dammara-1,17(20),24-trien-21-oic acid diacetate;* fumigacin. $C_{33}H_{44}O_8$; mol wt 568.71. C 69.70%, H 7.80%, O 22.51%. Antibiotic substance of the fusidane class. Produced by *Aspergillus fumigatus:* Waksman *et al., J. Bact.* **45**, 233 (1943); by *A. fumigatus* mut. *helvola:* Chain *et al., Brit. J. Exp. Pathol.* **24**, 108 (1943). Structural studies: Okuda *et al., Chem. Pharm. Bull.* **12**, 121 (1964); Oxley, *Chem. Commun.* **1966**, 729; Okuda *et al., Tetrahedron Letters* **1967**, 2295. Revised structure and stereochemistry: Iwaki *et al., Chem. Commun.* **1970**, 1119. Chemical and biological data: Reshetova, *Antibiotiki* **14**, 554 (1969). Review of literature until 1960. Wilson, "Miscellaneous *Aspergillus* Toxins" in *Microbial Toxins,* vol. VI, A. Ciegler *et al.,* Eds. (Academic Press, New York, 1971) p 265.

Needles from methanol, mp 215°. $[\alpha]_D^{25}$ −121° (chloroform). uv max (ethanol): 231 nm (log ϵ 4.24). Very slightly sol in water; slightly sol in petr ether, methanol, ethanol. More sol in hot methanol, ethanol; sol in chloroform, acetone, ethyl acetate, benzene, pyridine, glacial acetic acid, dioxane, ether.

Methyl ester, $C_{34}H_{46}O_8$, *methyl helvolate.* Crystals from methanol, mp 257°. $[\alpha]_D^{25}$ −140° (chloroform).

4667. Hematein. *6a,7-Dihydro-3,4,6a,10-tetrahydroxybenz[b]indeno[1,2-d]pyran-9(6H)-one;* hydroxybrasilein; hydroxybrazilein. $C_{16}H_{12}O_6$; mol wt 300.27. C 64.00%, H 4.03%, O 31.97%. Not to be confused with hematin. From hematoxylin or logwood extract and NH$_3$ by treatment with air: Engels *et al., J. Chem. Soc.* **93**, 1115 (1908); Rolland,

Teintex **3**, 261, 322, 460 (1938); Justin-Mueller, *Melliand Textilber.* **30**, 26, 63 (1949), *C.A.* **46**, 8375h (1952).

Reddish-brown crystals with yellowish-green metallic luster. mp >200°; also stated as 250° with decompn. Sol in about 1700 parts water; slightly sol in alcohol, ether; insol in benzene, chloroform; freely sol in ammonia with brownish-violet color and in dil NaOH with bright red color. Forms salts with heavy metals.

USE: As an indicator like hematoxylin; for staining animal tissue, particularly cell nuclei.

4668. Hematin. (SP-5-13)-[7,12-Diethenyl-3,8,13,17-tetramethyl-21H,23H-porphine-2,18-dipropanoato(4—)-N^{21},N^{22},N^{23},N^{24}]hydroxyferrate(2—) dihydrogen; [dihydrogen 3,7,-12,17-tetramethyl-8,13-divinyl-2,18-porphinedipropionato-(2—)]hydroxyiron; hydroxy[dihydrogen protoporphyrin IX-ato(2—)]iron; ferriheme hydroxide; ferriporphyrin hydroxide; ferriprotoporphyrin basic; hydroxyhemin; phenodin. $C_{34}H_{33}FeN_4O_5$; mol wt 633.50. C 64.46%, H 5.25%, Fe 8.82%, N 8.84%, O 12.63%. Found in the body in pathological conditions, *e.g.*, after phosgene poisoning, in pernicious anemia. Prepn: Fischer-Orth, *Die Chemie des Pyrrols* **II**, 1, 386 (Leipzig, 1937). Brief review including structure: J. E. Falk, *Porphyrins and Metalloporphyrins* (Elsevier, New York, 1964) pp 17, 23, 46. Toxicology: D. L. Lips *et al.*, *Toxicol. Letters* **2**, 329 (1978). Use in treatment of hepatic porphyria: C. J. Watson *et al.*, *Ann. Intern. Med.* **79**, 80 (1973); G. J. Dhar *et al.*, *ibid.* **83**, 20 (1975); J. M. Lamon *et al.*, *Medicine* **58**, 252 (1979).

Solvated crystals from pyridine, dry at 40° *in vacuo*. Freely sol in dil solns of alkali hydroxides, slightly sol in hot pyridine. Unstable. Absorption max (10% aq NaOH): 580 nm (E_{mM} 10.5). LD_{50} i.v. in rats: 4.32 mg/100 g (Lips).

4669. Hematoporphyrin. 7,12-Bis(1-hydroxyethyl)-3,8,-13,17-tetramethyl-21H,23H-porphine-2,18-dipropanoic acid; 7,12-bis(1-hydroxyethyl)-3,8,13,17-tetramethyl-2,18-porphinedipropionic acid; 1,3,5,8-tetramethyl-2,4-bis(α-hydroxyethyl)porphine-6,7-dipropionic acid; hematoporphyrin IX; Photodyn. $C_{34}H_{38}N_4O_6$; mol wt 598.70. C 68.21%, H 6.40%, N 9.36%, O 16.03%. Prepd from hemin by the action of hydrobromic acid in glacial acetic acid: Fischer-Orth, *Die Chemie des Pyrrols* **II**, 1, 421 (Leipzig, 1937). Review and prepn: J. E. Falk, *Porphyrins and Metalloporphyrins* (Elsevier, New York, 1964) pp 175-177.

Deep red crystals. Absorption max (0.1N KOH): 615.5, 565, 534.4, 499.5 nm. Insol in water. Sol in alc. Sparingly sol in ether, chloroform.

Hydrochloride monohydrate, $C_{34}H_{38}N_4O_6 \cdot HCl \cdot H_2O$, *Sensibion*. Prepn of a stable product: Woods, Steigman, U.S. pat. **2,858,320** (1958 to Baxter Labs.). Dark red crystals.

THERAP CAT: Antidepressant.

4670. Hematoxylin. cis-(+)-7,11b-Dihydrobenz[b]indeno[1,2-d]pyran-3,4,6a,9,10(6H)-pentol; hematoxiline; hydroxybrazilin; hydroxybrasilin. $C_{16}H_{14}O_6$; mol wt 302.28. C 63.57%, H 4.67%, O 31.76%. From the heart-wood of logwood *(Haematoxylon campechianum* Linn., *Leguminosae):* Chevreul, *Ann. Chim. Phys.* **82**, 54, 126 (1810). Structure: Perkin, Robinson, *J. Chem. Soc.* **93**, 489 (1908). Synthesis: Dann, Hofmann, *Angew. Chem.* **75**, 1125 (1963); Morsingh, Robinson, *Tetrahedron* **26**, 281 (1970); Kirkiacharian, Billet, *Bull. Soc. Chim. France* **1972**, 3292. Stereochemistry: Craig *et al.*, *J. Org. Chem.* **30**, 1573 (1965). Spectroscopic and physico-chemical properties: C. Bettinger, H. W. Zimmerman, *Histochemistry* **95**, 279 (1991). Prepn for staining: R. B. Bosma *et al.*, *J. Histotechnol.* **16**, 371 (1993). Use in determn of ruthenium: A.-A. Y. El-Sayed, *Fresenius J. Anal. Chem.* **349**, 830 (1994). *Review:* Robinson, *Bull. Soc. Chim. France* **1958**, 125. Review of theory and application as histological stain: B. B. Hrapchak, *Am. J. Med Tech.* **42**, 371-379 (1976).

Trihydrate, white to yellowish crystals; redden on exposure to light, mp 100-120°; also stated as 140°. Slightly sol in cold water, ether; sol in hot water, hot alc, also in alkali hydroxides, borax, glycerol. Its solns darken on standing.

USE: Chiefly as a stain in microscopy; also in manufacture of ink.

4671. Hematoxylon. Logwood. Heart-wood of *Haematoxylon campechianum* L., *Leguminosae.* *Habit.* Central America; grown in the West Indies. *Constit.* About 10% hematoxylin; tannin; resin.

USE: Largely as dye.

4672. Heme. (SP-4-2)-[7,12-Diethenyl-3,8,13,17-tetramethyl-21H,23H-porphine-2,18-dipropanoato(4—)-N^{21},N^{22},-N^{23},N^{24}]ferrate(2—) dihydrogen; [dihydrogen 3,7,12,17-tetramethyl-8,13-divinyl-2,18-porphinedipropionato(2—)]-iron; 1,3,5,8-tetramethyl-2,4-divinylporphine-6,7-dipropionic acid ferrous complex; ferroheme; heme; protoheme; protoheme IX; reduced hematin; ferroprotoporphyrin. $C_{34}H_{32}FeN_4O_4$; mol wt 616.50. C 66.24%, H 5.23%, Fe 9.06%, N 9.09%, O 10.38%. Heme occurs free in tissues in the presence of certain pathological conditions, and in normal tissues; it occurs as the prosthetic group of a number of hemoproteins. It has been identified as the prosthetic group of hemoglobins, erythrocruorins (the hemoglobin analog of many invertebrates), myoglobins, some peroxidases, catalases, and cytochromes b. It is the color-furnishing portion of hemoglobin. Obtained when a soln of hematin in alkali is reduced in absence of nitrogenous substances: Bertin-Sans, de Moitessier, *Compt. Rend.* **114**, 923 (1892); Dhéré *et al.*, *ibid.* **165**, 515 (1917). Synthesis: Fischer-Orth, *Die Chemie des Pyrrols* **II**, 1, 384 (Leipzig, 1937). Biosynthesis: Shemin, *Naturwiss.* **57**, 185 (1970). *Review:* J. E. Falk, *Porphyrins and Metalloporphyrins* (Elsevier, NewYork, 1964) pp 8, 94, 183. Comprehensive monograph: Chance *et al.*, *Hemes and Hemoproteins* (Academic Press, New York, 1966) 624 pp; *Handbook of Experimental Pharmacology* **vol. 44**, entitled "Heme and Hemoproteins", F. DeMatteis, W. N. Aldridge, Eds. (Springer-Verlag, New York, 1978) 449 pp.

Fine brown needles with a dark violet sheen. Absorption max in phosphate buffer at pH 7: ~550, 575 nm (E_{mM}^{572} 5.5). Sparingly sol in glacial acetic acid; freely sol in the presence of oxygen. Very unstable.

4673. Hementin. Anticoagulant enzyme from the salivary glands of the South American giant leech *Haementeria ghiliarii.* Extraction: R. T. Sawyer *et al.,* U.S. pat. **4,390,-630** (1983 to Univ. California, Berkeley). Anticoagulant, fibrinolytic properties: A. Z. Budzynski *et al., Proc. Soc. Exp. Biol. Med.* **168,** 266 (1981). Purification and identification as single polypeptide chain with mol wt approx 120000: S. M. Malinconico *et al., J. Lab. Clin. Med.* **103,** 44 (1984). Proteolytic degradation of fibrinogen by hementin: *eidem, ibid.* **104,** 842 (1984). Protease inhibition: E. H. Murer *et al., Thromb. Haemostasis* **51,** 24 (1984).

4674. Hemerythrin. Mol wt 108,000. A non-heme, oxygen-carrying protein found in members of four invertebrate phyla: sipunculids, polychaetes, priapulids and brachiopods. Oxygenated form is called *oxyhemerythrin.* Isoln from the sipunculid, *Golfingia goldii* (or *Phascolosoma goldii*): Klotz *et al., Arch. Biochem. Biophys.* **68,** 284 (1957). Consists of eight subunits, each containing two ferrous ions which form a complex with one molecule of oxygen: Klotz, Keresztes-Nagy, *Biochemistry* **2,** 445, 923 (1963). Each subunit is built up from 113 amino acids. Amino acid sequence: Klippenstein *et al., ibid.* **7,** 3868 (1968); Klippenstein, *ibid.* **11,** 372 (1972). Structure of trimeric hemerythrin: J. L. Smith *et al., Nature* **303,** 86 (1983). *Reviews:* Okamura, Klotz, "Hemerythrin" in *Inorganic Biochemistry,* **vol. 1,** G. L. Eichhorn, Ed. (Elsevier, New York, 1973) pp 320-343; Klotz *et al., Science* **192,** 335-344 (1976); J. S. Loehr, T. M. Loehr in *Advances in Inorganic Biochemistry* **vol. 1,** G. L. Eichhorn, L. G. Marzilli, Eds. (Elsevier, New York, 1979) pp 235-252.

Hemerythrin is colorless. Oxyhemerythrin forms violet-pink crystals. Spectral data: Garbett *et al., Arch. Biochem. Biophys.* **135,** 419 (1969).

4675. Hemicelluloses. Large group of polysaccharides found, in association with lignin, in the primary and secondary cell walls of all plants and of some seaweeds. Whether lignin and hemicellulose are chemically bonded or lignin mechanically entraps hemicellulose molecules is still unknown. There are variations in the hemicellulosic composition of plants and even between the organs of the same plant. Hemicellulosic composition of plants changes with growth and maturation and is influenced also by environmental factors during growth. Principal sugar residues present in hemicellulose are: D-xylose, D-glucose, D-galactose, D-mannose, L-arabinose, D-glucuronic acid, D-galacturonic acid, 4-O-methyl-D-glucuronic acid, L-rhamnose and L-fucose. The most ubiquitous and abundant hemicelluloses are the *xylans* which are composed of linear and/or branched chains of β-(1→4) linked D-xylopyranosyl units. Hemicelluloses from woods: T. E. Timell, *Adv. Carbohyd. Chem.* **19,** 247 (1964); *idem,* **20,** 409 (1965). Hemicelluloses from grasses and cereals, K. C. B. Wilke, *ibid.* **36,** 215 (1979).

4676. Hemicholinium. *2,2'-[1,1'-Biphenyl]-4,4'-diylbis-[2-hydroxy-4,4-dimethylmorpholinium]; 2,2'-(4,4'-biphenyl-ylene)bis[2-hydroxy-4,4-dimethylmorpholinium].* $[C_{24}H_{34}-N_2O_4]^{2+}$. Specific, competitive inhibitor of the sodium-dependent high-affinity choline uptake system (HACU) localized on cholinergic presynaptic nerve terminals, resulting in the inhibition of acetylcholine synthesis. Prepn of seco tautomer: J. P. Long, F. W. Schueler, *J. Am. Pharm. Assoc.* **43,** 79 (1954). *In vitro* pharmacology and revised hemiacetal structure: F. W. Schueler, *J. Pharmacol. Exp. Ther.* **115,** 127 (1955). Structure-activity studies: R. P. DiAugustine, V. B. Haarstad, *Biochem. Pharmacol.* **19,** 559 (1970); J. G. Cannon *et al., Pharm. Res.* **5,** 359 (1988). Pharmacology: J. J. Freeman *et al., J. Pharmacol. Exp. Ther.* **210,** 91 (1979); J. J. Hagan *et al., Psychopharmacology* **98,** 347 (1989). Receptor binding studies in human brain: J. Pascual *et al., J. Neurochem.* **54,** 792 (1990); in rat brain: *eidem, Pharmacol Res.* **24,** 345 (1991). Potential uses of tritium labeled compound: T. W. Vickroy *et al.* in *Synth. Appl. Isotop. Labeled Compd. 1985,* R. R. Muccino, Ed. (Elsevier, Amsterdam, 1986) pp 145-150; H. K. Happe, L. C. Murrin, *J. Neurochem.* **60,** 1191 (1993).

Dibromide, $C_{24}H_{34}Br_2N_2O_4$, *hemicholinium-3, HC-3.* mp 180°. Moderately soluble in water, soluble in ethanol. Occurs as the monohydrate, white solid from ethanol/ether, mp 226-228 (dec). LD_{50} i.p. in female mice: 0.048-0.082 mg/kg (DiAugustine, Haarstad).

USE: Research tool as cholinergic probe, to deplete acetylcholine stores.

4677. Hemi-Dewar Biphenyl. *2-Phenylbicyclo[2.2.0]-hexa-2,5-diene.* $C_{12}H_{10}$; mol wt 154.21. C 93.46%, H 6.54%. Prepn by decompn of cyclobutadieneiron tricarbonyl with ceric ammonium nitrate in the presence of phenylacetylene: Burt, Pettit, *Chem. Commun.* **1965,** 517.

Oil. n_D^{26} 1.5834. Upon heating to 90° for several minutes is converted quantitatively into biphenyl.

4678. Hemin. *(SP-5-13)-Chloro[7,12-diethenyl-3,8,13,-17-tetramethyl-21H,23H-porphine-2,18-dipropanoato(4—)-$N^{21},N^{22},N^{23},N^{24}$]ferrate(2—) dihydrogen; chloro[dihydrogen 3,7,12,17-tetramethyl-8,13-divinyl-2,18-porphinedipropionato(2—)]iron;* chlorohemin; 1,3,5,8-tetramethyl-2,4-divinylporphine-6,7-dipropionic acid ferrichloride; Teichmann's crystals; ferriheme chloride; ferriprotoporphyrin chloride; ferriporphyrin chloride. $C_{34}H_{32}ClFeN_4O_4$; mol wt 651.95. C 62.64%, H 4.95%, Cl 5.44%, Fe 8.57%, N 8.59%, O 9.82%. Prepd from hemoglobin soln by heating with acetic acid and sodium chloride. Practical procedure for its prepn from ox blood: Schalfejeff, *J. Russ. Physiol. Chem. Soc.* **1885,** 30; *Ber.* **18,** 232 (1885); Gattermann-Wieland, *Praxis des Organischen Chemikers* 23rd ed., p 407; Fischer-Orth, *Die Chemie des Pyrrols* **II,** 1, 377 (Leipzig, 1937); H. Fischer, *Org. Syn. coll. vol. III,* 442 (1955); Labbe, Nishida, *Biochim. Biophys. Acta* **26,** 437 (1957). Biosynthesis: Karlzeile, *Angew. Chem.* **66,** 729 (1954). *Review:* Stoll, *Experientia* **4,** 6 (1948); H. H. Inhoffen, *Naturwiss.* **55,** 457 (1968); W. S. Caughey in *Inorganic Biochemistry* vol. 2, G. L. Eichhorn, Ed. (Elsevier, New York, 1973) pp 797-831.

Consult the Name Index before using this section.

Long, thin blades from glacial acetic acid or from chloroform-pyridine-acetic acid, appearing brown in transmitted light and steel-blue in reflected light. Sinters at 240° but is not melted at 300°. Absorption spectrum: Fischer-Orth, *loc. cit.* Freely sol in dil ammonia water, also in solns of sodium hydroxide with hematin formation, *i.e.*, the chlorine is displaced by an OH group. Practically insol in carbonate solns, dil acid solns. Sol in strong organic bases such as trimethylamine, *p*-toluidine, dimethylaniline. Sol in concd H_2SO_4 with loss of Fe. Sparingly sol in 70-80% alc. Practically insol but stable in water.

Dimethyl ester, $C_{36}H_{36}ClFeN_4O_4$, needles from benzene, not melted at 300°. Freely sol in acetic acid, benzene, chloroform, acetone.

Diethyl ester, $C_{38}H_{40}ClFeN_4O_4$, crystals, freely sol in chloroform.

4679. Hemipyocyanine. *1-Phenazinol*; α-hydroxyphenazine; 1-hydroxy-5,10-diazoanthracene; pyoxanthose. $C_{12}H_8N_2O$; mol wt 196.21. C 73.46%, H 4.11%, N 14.28%, O 8.15%. Found in old cultures of *Pseudomonas pyocyanea*. Isoln: Fordos, *Compt. Rend.* **56**, 1128 (1863). Prepd from pyocyanine by the action of 2% NaOH: Wrede, Strack, *Z. Physiol. Chem.* **140**, 12 (1924); by the reaction of pyrogallol monomethyl ether and *o*-phenylenediamine followed by demethylation of the formed α-methoxyphenazine: Surrey, *Org. Syn. coll. vol.* III, 753 (1955). Alternate syntheses: Hegedüs, *Helv. Chim. Acta* **33**, 766 (1950); Vivian, *Nature* **178**, 753 (1956).

Yellow needles from benzene, mp 159-160°. Sublimes easily *in vacuo*. Slightly sol in hot water; freely sol in the usual organic solvents except petr ether. Sol in aq alkaline solns with purplish-red color which turns yellow on neutralization. Forms red salts with mineral acids.

4680. Hemisulfur Mustard. *2-(2-Chloroethylthio)ethanol*; 2-chloro-2'-hydroxydiethyl sulfide; 2-chloroethyl 2-hydroxyethyl sulfide; β-chloroethyl β-hydroxyethyl thioether; mustard chlorohydrin; semi-mustard gas; semisulfur mustard. C_4H_9ClOS; mol wt 140.63. C 34.16%, H 6.45%, Cl 25.21%, O 11.38%, S 22.80%. $ClCH_2CH_2SCH_2CH_2OH$. Prepn: Ogston *et al.*, *Trans. Faraday Soc.* **44**, 45 (1948); Grant, Kinsey, *J. Am. Chem. Soc.* **68**, 2075 (1946); Fuson, Ziegler, *J. Org. Chem.* **11**, 510 (1946); Rueggeberg *et al.*, *ibid.* **13**, 110 (1948); Tsou *et al.*, *ibid.* **26**, 4987 (1961).

Oily, somewhat hygroscopic liquid. $bp_{0.6}$ 100° (Fuson); $bp_{0.5-0.75}$ 87°; bp_6 44.5° (Rueggeberg); n_D^{20} 1.5188; $n_D^{24.5}$ 1.5205. Sol in water; miscible with bis(2-chloroethyl) sulfide, thiodiglycol. Unstable, forming sulfonium salts at room temp or in solns with ethanol or chloroform.

4681. Hemocyanins. A non-heme, oxygen carrying copper protein found in arthropods and mollusca of which *keyhole-limpet hemocyanin* (*KLH*) is an example. Dissolved in the hemolymph; not found in blood cells. Mol wt ranges from 4.5×10^5 to 1.3×10^7. One molecule of oxygen is bound by two copper atoms. The oxygenated form, *oxy-*

hemocyanin is blue, while *deoxyhemocyanin* is colorless. The copper can be removed reversibly to form *apohemocyanin*. Structure of snail, *Helix pomatia*, hemocyanin: J. E. Mellema, A. Klug, *Nature* **239**, 146 (1972). The role of copper in hemocyanin: R. Lontie, L. Vanquickenborne, *Met. Ions Biol. Syst.* **3**, 183 (1974). *Reviews:* K. E. van Holde, E. F. J. van Bruggen in *Subunits in Biological Systems* vol. **5**, pt. A, S. N. Timasheff, G. D. Fasman, Eds. (Dekker, New York, 1971) pp 1-53; R. Lontie, R. Witters in *Inorganic Biochemistry*, vol. I, G. L. Eichhorn, Ed. (Elsevier, New York, 1973) pp 344-358; E. F. J. van Bruggen, *Trends Biochem. Sci.* **5**, 185-8 (1980). Review of respiratory function: J. Bonaventura, C. Bonaventura, *Am. Zool.* **20**, 7-17 (1980); C. P. Magnum, *ibid.* 19-38.

USE: KLH as an experimental antigen in animals.

4682. Hemoglobin. Hb; ferrohemoglobin. The major component of red blood cells which transports oxygen from the lungs to body tissues and facilitates the return transport of carbon dioxide. Mammalian hemoglobins have mol wts of about 64,500. Composed of four peptide chains called globins, *q.v.*, each of which is bound to a heme, *q.v.* Normal human hemoglobin is composed of a pair of two identical chains. Iron is coordinated to four pyrrole nitrogens of protoporphyrin IX, and to an imidazole nitrogen of a histidine residue from the globin side of the porphyrin. The sixth coordination position is available for binding with oxygen and other small molecules. Called *oxyhemoglobin*, HbO_2, in the oxygenated form and *carboxyhemoglobin*, HbCO, when oxygen is displaced by carbon monoxide. Binds reversibly with oxygen while the heme iron remains in the ferrous state. Autoxidation is prevented by the cover of hydrophobic groups of the globin. When the iron in hemoglobin is oxidized from the ferrous to the ferric state the compd is called methemoglobin, *q.v.* and is accompanied by a loss of oxygen-binding capacity. Hemoglobin is usually prepd by separating the red blood corpuscles from the lighter plasma by centrifuging; the plasma is siphoned off, and on adding ether to the blood corpuscle paste, the cells burst. After another centrifugation to remove the ruptured cell envelopes, a clear red soln of the protein is obtained [Fieser, Fieser, *Org. Chem.* (New York, 3rd ed., 1956) p 455]. Prepn of cryst HbO_2 from washed horse or dog erythrocytes: Heidelberger, *J. Biol. Chem.* **53**, 31 (1922); *see also* Ferry, Green, *ibid.* **81**, 175 (1929); from human blood: Drabkin, *ibid.* **164**, 703 (1946). Structure studies: Muirhead, Perutz, *Nature* **199**, 633 (1963); Perutz *et al.*, *ibid.* **219**, 131 (1968); Perutz, *Proc. Roy. Soc. London* **173B**, 113 (1969). Respiratory properties of hemoglobin and its function as a carrier of oxygen and carbon dioxide: Peters, Van Slyke, *Quantitative Clinical Chemistry* vol. I (Baltimore, 1932). Mechanism of action: Arnone in *Ann. Rev. Med.* **25**, 123-130 (1974). *Reviews:* Lemberg, Legge, *Hematin Compounds and Bile Pigments* (New York, 1949); F. W. Sunderman, *Hemoglobin—its Precursors and Metabolites* (Lippincott, Philadelphia, 1964); H. Lehmann, R. G. Huntsman, *Man's Hemoglobins* (*ibid.* 1966); Huisman, Schroeder, *New Aspects of the Structure, Function, and Synthesis of Hemoglobins* (Butterworth, London, 1971); M. F. Perutz, *Ann. Rev. Biochem.* **48**, 327-386 (1979); G. Fermi, M. F. Perutz, *Haemoglobin & Myoglobin* (Oxford Univ. Press, New York, 1982) 104 pp; R. E. Dickerson, I. Geis, *Hemoglobin: Structure, Function, Evolution, and Pathology* (Benjamin-Cummings, Menlo Park, Calif., 1983) 176 pp.

Crystal form, solubility, affinity for oxygen, absorption spectra differ quantitatively in hemoglobins of different species, due to the variation in amino acid sequence of the protein moiety since the same heme group is present in all vertebrate and many invertebrate hemoglobins. Human HbO_2 = tetragonal crystals; horse HbO_2 = orthorhombic crystals from citrated blood, monoclinic crystals from oxalated blood. Absorption spectra, *see* Lemberg, Legge, *loc. cit.*, 228. Oxyhemoglobin is an article of commerce where it is called hemoglobin. Brownish-red powder or scales. Soluble in about 7 parts water, slowly sol in glycerol.

4683. Henna. Dried powdered leaves of *Lawsonia alba* Lam., *L. inermis* L., and *L. spinosa* L., Lythraceae. Obtained from North Africa or India. Contains about 1% of lawsone, q.v. Ref: Cox, Analyst **63**, 397 (1938); Talaat, Brit. Med. J. **II**, 944 (1960).

USE: For dyeing hair and nails auburn to red, in the Orient together with "reng," the dried, powdered leaves of the indigo plant, in order to produce darker and even bluish-black shades. For relatively permanent dyeing the pH must be about 5.5; this is achieved by the addition of citric, boric, or adipic acid. Ingredient of many commercial hair rinses.

4684. Heparamine. N-Desulfoheparin. The —NHSO$_3$H groups in heparin are replaced by —NH$_2$ groups. Prepn of sodium salt by methanolysis of sodium heparinate and saponification: Velluz et al., Compt. Rend. **247**, 1521 (1958); from sodium heparinate heated with HCl, neutralized with NaHCO$_3$ and dialyzed against running water: Foster et al., J. Chem. Soc. **1961**, 1204; from sodium heparinate demineralized with Dowex 50 resin: **Brit. pat. 863,-235** (1961 to UCLAF).

Sodium salt, solid, sol in water. $[\alpha]_D^{20}$ +67 ± 2°.

N-Acyl derivatives, heparides. Similar to heparin in clearing alimentary lipemia but are practically devoid of its anticoagulant activity.

4685. Heparin. Heparinic acid; Arteven; Leparan. Glycosaminoglycan with anticoagulant activity. Heterogenous mixture of variably sulfated polysaccharide chains composed of repeating units of D-glucosamine and either L-iduronic or D-glucuronic acids. Mol wt ranges from 6000-30000 Da. Biosynthesized and stored in mast cells of various animal tissues, particularly liver, lung or gut. Commercial heparin is isolated from beef lung or pork intestinal mucosa. Isoln from mammalian tissue: Howell, Am. J. Physiol. **63**, 434 (1922-23); **71**, 553 (1924-25); Korn, J. Biol. Chem. **234**, 1325 (1959); L. B. Jaques, Can. J. Biochem. Physiol. **37**, 1183 (1959); J. A. Bush et al., U.S. pat. **2,884,-358** (1959 to So. Calif. Gland). Purification: G. Nominé et al., U.S. pat. **2,989,438** (1961 to UCLAF); Toccaceli, U.S. pat. **3,016,331** (1962 to Ormonoterapia Richter); L. Roden et al., Methods Enzymol. **26**, 73 (1972). Structural studies: M. L. Wolfrom, J. Am. Chem. Soc. **72**, 5796 (1950); Velluz et al., Compt. Rend. **247**, 1521 (1958); M. L. Wolfrom et al., J. Org. Chem. **29**, 540 (1964). Configuration of glycosidic linkages: M. L. Wolfrom et al., ibid. **31**, 1173 (1966); A. S. Perlin et al., Can. J. Chem. **48**, 2260 (1970); T. Helting, U. Lindahl, J. Biol. Chem. **246**, 5442 (1971). Identification of L-iduronic acid residues: A. S. Perlin et al., Carbohyd. Res. **7**, 369 (1968). Antithrombotic activity results from the binding and activation of antithrombin III, a plasma protein which inhibits several enzymes in the coagulation cascade: R. D. Rosenberg, Fed. Proc. **36**, 10 (1977). Anticoagulant activity is related to the mol wt of the polysaccharide fragments; low molecular weight components exhibit decreased hemorrhagic effects while retaining antithrombin binding ability: L.-O. Andersson et al., Thromb. Res. **15**, 531 (1979); T. W. Barrowcliffe et al., Brit. J. Haematol. **41**, 573 (1979); J. Hirsch et al., Sem. Thromb. Hemostas. **11**, 13 (1985). Characterization of the antithrombin binding site: U. Lindahl et al., Proc. Nat. Acad. Sci. USA **76**, 3198 (1979); J. Choay et al., Thromb. Res. **18**, 573 (1980). Synthesis of the pentasaccharide corresponding to the binding site sequence: eidem, Biochem. Biophys. Res. Commun. **116**, 492 (1983). Symposium on structure, activity and clinical applications: Fed. Proc. **36**, 9-116 (1977). Review of mechanism of action: I. Björk, U. Lindahl, Mol. Cell. Biochem. **48**, 161-182 (1982); of structure-activity relationships and prepn of low mol wt fractions: B. Casu, Advan. Carbohyd. Chem. Biochem. **43**, 51-134 (1985); of biosynthesis: U. Lindahl et al., Trends Biochem. Sci. **11**, 221-225 (1986). Comprehen-

sive description: F. Nachtmann et al., Anal. Profiles Drug Subs. **12**, 215-276 (1983). Overview of clinical results in pulmonary embolism and venous thrombosis: R. Collins et al., N. Engl. J. Med. **318**, 1162-1170 (1988); J. Hirsh, ibid. **324**, 1565-1574 (1991); of clinical studies with low mol wt heparinoids: H. ten Cate et al., Am. J. Hematol. **27**, 146-153 (1988).

Heparin has a rotation of $[\alpha]_D^{20}$ +55°.

Calcium salt, Calciparine, Ecasolv.

Magnesium salt, magnesium heparinate, Cutheparine. Sol in water. Insol in organic solvents.

Potassium salt, Clarin (formerly).

Sodium salt, heparin sodium, Heprinar, Hepsal, Lipo-Hepin, Liquémin, Lipo-Hepinette, Longheparin, Monoparin, Panheprin, Pularin, Liquaemin Sodium, Minihep, Thrombo-Hepin, Thromboliquine, Thrombophob, Unihep. White to grayish-brown amorphous powder. Odorless, hygroscopic. $[\alpha]_D^{25}$ +47° (c = 1.5 in water). One gram dissolves in 20 ml water. Sol in saline soln. Practically insol in alcohol, acetone, benzene, chloroform, ether. pH of 1% aq soln = 6.0 to 7.5. Absorption spectrum: Burson et al., J. Am. Chem. Soc. **78**, 5874 (1956). Ampuled solns may be stored at room temp for at least 12 months. Commercially available ampuled, sterile solns contain 0.5% phenol or chlorobutanol as preservative.

THERAP CAT: Anticoagulant.

THERAP CAT (VET): Anticoagulant.

4686. Hepaxanthin. 5,6-Epoxy-5,6-dihydroretinol; 5,6-monoepoxyvitamin A; vitamin A epoxide; 574-chromogen. C$_{20}$H$_{30}$O$_2$; mol wt 302.46. C 79.42%, H 10.00%, O 10.58%. Isoln from cod liver oil and formation from vitamin A: Karrer, Jucker, Helv. Chim. Acta **28**, 717 (1945); **30**, 559 (1947). Prepn and properties: Jungalwala, Cama, Biochem. J **95**, 17 (1965). See also Lakshmanan et al., ibid. 27; Tsukida et al., J. Vitaminol. **14**, 95 (1968).

Viscous yellow oil. uv max (ethanol): 272 nm. Freely sol in chloroform, ether, benzene. Moderately sol in alcohol. Slightly sol in petr ether.

4687. HEPES. 4-(2-Hydroxyethyl)-1-piperazineethanesulfonic acid; N-(2-hydroxyethyl)piperazine-N'-2-ethanesulfonic acid. C$_8$H$_{18}$N$_2$O$_4$S; mol wt 238.31. C 40.32%, H 7.61%, N 11.76%, O 26.85%, S 13.46%. One of several zwitterionic N-substituted aminosulfonic acids known as "Good" buffers, active in the pH range of 6-8.5. Prepn: N. E. Good et al., Biochemistry **5**, 467 (1966). Buffering characteristics: W. J. Ferguson et al., Anal. Biochem. **104**, 300 (1980). Temperature effects on pKa and pH: R. N. Roy et al., Cryo-Letters **6**, 285 (1985). Interaction with hydroxyl radicals: M. Hicks, J. M. Gebicki, FEBS Letters **199**, 92 (1986). Interference with Lowry protein determination: H. M. Himmel, W. Heller, J. Clin. Chem. Clin. Biochem. **25**, 909 (1987). Use as biological buffer: H. Eagle, Science **174**, 500 (1971); K. V. Rao, Ind. J. Exp. Biol. **15**, 552 (1977); E. D. Lalague et al., J. Microsc. **127**, 307 (1982); G. I. McFadden, M. Melkonian, Phycologia **25**, 551 (1986).

Crystals from alcohol + water, mp 234°. pKa$_1$ $\simeq$3, pKa$_2$ (20°) 7.55. ΔpKa/°C −0.014. Saturated aqueous soln is 2.25M at 0°.

USE: Biological buffer.

4688. Heparonicate. 3-Pyridinecarboxylic acid 2-hexyl-2-[[(3-pyridinylcarbonyl)oxy]methyl]-1,3-propanediyl ester; nicotinic acid triester with 2-hexyl-2-(hydroxymethyl)-1,3-propanediol; 2-hexyl-2-(hydroxymethyl)-1,3-propanediol trinic-

R = H or SO$_3^-$ R' = SO$_3^-$ or COCH$_3$

Antithrombin binding site of Heparin

otinate; 2,2-dihydroxymethyl-*n*-octanol trinicotinate; 1,1,1-trimethylolheptane trinicotinate; 1,1,1-(trihydroxymethyl)-heptane trinicotinate; Megrin. $C_{28}H_{31}N_3O_6$; mol wt 505.57. C 66.52%, H 6.18%, N 8.31%, O 18.99%. Prepn: M. Nakanishi *et al.*, **Neth.** pat. **Appl. 6,514,807;** *eidem,* **U.S.** pat. **3,384,642;** *eidem,* **Japan.** pat. **67 10988** (1966, 1968 and 1967, all to Yoshitomi).

Crystals from ethanol, mp 94-96°.
THERAP CAT: Vasodilator (peripheral).

4689. Heptabarbital. *5-(1-Cyclohepten-1-yl)-5-ethyl-2,-4,6(1H,3H,5H)-pyrimidinetrione; 5-(1-cyclohepten-1-yl)-5-ethylbarbituric acid;* 5-ethyl-5-cycloheptenylbarbituric acid; heptabarb; Heptadorm; Medomin. $C_{13}H_{18}N_2O_3$; mol wt 250.30. C 62.38%, H 7.25%, N 11.19%, O 19.18%. Prepn: **Fr.** pat. **870,714** (1942 to Geigy); Taub, **U.S.** pat. **2,501,551** (1950).

Crystals, mp 174°. Slightly bitter taste. uv max (0.2N NaOH): 218.5, 254 nm. Very sparingly sol in water, more sol in alcohol. At 25° 100 ml of soln contains: 4.0 g in alcohol; 5.7 g in acetone; 1.4 g. in chloroform. Soluble in alkaline solns. Forms water-soluble sodium, magnesium, and calcium salts.
Caution: May be habit forming. This is a controlled substance (depressant) listed in the U.S. Code of Federal Regulations, Title 21 Parts 329.1 and 1308.13 (1995).
THERAP CAT: Sedative, hypnotic.

4690. Heptachlor. *1H-1,4,5,6,7,8,8-Heptachloro-3a,4,-7,7a-tetrahydro-4,7-methanoindene;* E-3314; Velsicol 104; Drinox; Heptamul. $C_{10}H_5Cl_7$; mol wt 373.32. C 32.17%, H 1.35%, Cl 66.48%. Prepn: Bluestone *et al.*, **U.S.** pat. **2,576,-666** (1951 to Julius Hyman); McKenna *et al.*, **U.S.** pat. **2,661,377-8** (1953 to Shell); Kleiman, Tapas, **U.S.** pat. **2,-904,599** (1959 to Velsicol). Toxicity data: T. B. Gaines, *Toxicol. Appl. Pharmacol.* **14,** 515 (1969).

Crystals, mp 95-96°. Vapor pressure at 25° = 3 × 10⁻⁴ mm Hg. Soly in g/100 ml solvent at 27°: acetone 75, benzene 106, carbon tetrachloride 112, cyclohexanone 119, alcohol 4.5, xylene 102. LD_{50} in male, female rats (mg/kg): 100, 162 orally (Gaines).
Caution: Poisoning may occur by ingestion, inhalation, skin contamination. May cause blood dyscrasias and liver necrosis. *See: Clinical Toxicology of Commercial Products,* R. E. Gosselin *et al.,* Eds. (Williams & Wilkins, Baltimore, 5th ed., 1984) Section II, pp 285-286.
Note: The EPA has cancelled registration of pesticides containing this compound with the exception of its use through subsurface ground insertion for termite control and

the dipping of roots or tops of non-food plants, *Fed. Reg.* **vol. 40,** p 28850 (July 9, 1975).
USE: Insecticide for control of cotton boll weevil.

4691. Heptaminol. *6-Amino-2-methyl-2-heptanol;* 2-methyl-6-amino-2-heptanol; 6-methyl-2-amino-6-heptanol. $C_8H_{19}NO$; mol wt 145.25. C 66.16%, H 13.19%, N 9.64%, O 11.02%. Prepn: H. T. F. Givens, R. M. Herbst, **U.S.** pat. **2,457,656** (1948 to Bilhuber); of hydrochloride: J. Doeuvre, J. Poizat, *Compt. Rend.* **224,** 286 (1947). Pharmacokinetics and metabolism: F. Chanoine *et al.*, *Arzneimittel-Forsch.* **31,** 1430 (1981). HPLC determn in biological fluids: R. R. Brodie *et al.*, *J. Chromatog.* **274,** 179 (1983). Clinical trial in hypotension: D. Milon *et al.*, *Fundam. Clin. Pharmacol.* **4,** 695 (1990). Review of mode of action: B. Pourrias *Ann. Pharm. Franc.* **49,** 127-138 (1991).

Colorless liquid, bp₇ 92-93°.
Hydrochloride, $C_8H_{19}NO \cdot HCl$, *RP-2831, Cortensor, Eoden, Hept-a-myl, Heptylon.* Crystals, mp 150°. Freely sol in water. Sol in alcohol. Practically insol in acetone, benzene, ether. pH of 2% aq soln: 4.5-5.5.
5'-Adenylate, *Ampecyclal.*
THERAP CAT: Antihypotensive.

4692. Heptanal. Heptaldehyde; Aldehyde C-7; heptylaldehyde; oenanthal; enanthal; oenanthol; oenanthaldehyde; enanthaldehyde. $C_7H_{14}O$; mol wt 114.19. C 73.63%, H 12.36%, O 14.01%. $CH_3(CH_2)_5CHO$. Obtained by distilling castor oil under reduced pressure: Rogers, *J. Am. Pharm. Assoc., Sci. Ed.* **12,** 503 (1923); Dominguez *et al.*, *J. Chem. Ed.* **29,** 446 (1952). Catalytic dehydration of ricinoleic acid methyl ester yields heptanal as a cleavage product in almost quantitative yield: Panjutin, *Chem. Zentr.* **1928,** II, 747.
Liquid. Penetrating fruity odor. d₄⁰ 0.83423; d₄¹⁵ 0.82162; d₄³⁰ 0.80902. mp −43.3°. bp₇₆₀ 152.8°; bp₃₀ 59.6°; bp₁₀ 42.5°. n_D²⁰ 1.42571. Viscosity at 15°: 0.977 cp; at 30°: 0.791 cp. Surface tension (γ) at 30° = 25.68. Surface tension against water at 30°: 14.41. Heat of combustion (liq) −1062.4 kcal/mol. Slightly sol in water. Misc with alc, ether. Sol in 3 vols of 60% alc.
USE: Manufacture of 1-heptanol; ethyl oenanthate.

4693. Heptanal Sodium Bisulfite. *1-Hydroxy-1-heptanesulfonic acid monosodium salt;* heptaldehyde sodium bisulfite; Hepbisul. $C_7H_{15}NaO_4S$; mol wt 218.25. C 38.52%, H 6.93%, Na 10.53%, O 29.32%, S 14.69%. $CH_3(CH_2)_5$-CHOHSO₃Na. Prepd by shaking a 30% ether soln of heptanal with an excess of 40% aq sodium bisulfite. Soln: Garay, **U.S.** pat. **3,019,161** (1962 to Baker Castor Oil Co., compd and method not claimed). General method of prepn: Lauer, Langkammerer, *J. Am. Chem. Soc.* **57,** 2360 (1935).
Moderately sol in water. Somewhat sol in methanol. Practically insol in ethanol, ether.
USE: In antifungal compositions.

4694. *n*-Heptane. C_7H_{16}; mol wt 100.20. C 83.91%, H 16.09%. $CH_3(CH_2)_5CH_3$. A hydrocarbon from petroleum.
Volatile, flammable liquid. d₄²⁰ 0.684. bp 98.4°. mp −90.7°. Flash pt, open cup: 30°F (−1°C); closed cup: 25°F (−4°C). n_D²⁵ 1.3855. Insol in water; sol in alcohol, chloroform, ether. Lethal concn for mice in air: 15900 ppm, *Handbook of Toxicology* **vol 1,** W. S. Spector, Ed. (Saunders, Philadelphia, 1956) pp 338-339.
Caution: Potential symptoms of overexposure are lightheadedness, giddiness and stupor; no appetite, nausea; dermatitis; chemical pneumonia; unconsciousness. *See NIOSH Pocket Guide to Chemical Hazards* (DHHS/NIOSH 90-117, 1990) p 120.
USE: As standard in testing knock of gasoline engines.

4695. Heptanoic Acid. Enanthic acid; oenanthic acid; oenanthylic acid; *n*-heptoic acid; *n*-heptylic acid. $C_7H_{14}O_2$; mol wt 130.19. C 64.58%, H 10.84%, O 24.58%. CH_3-$(CH_2)_5COOH$. Found in the various fusel oils in appreciable amounts. Has been observed in rancid oils. Prepd by the oxidation of heptaldehyde with potassium permanganate in dil sulfuric acid: Ruhoff, *Org. Syn.* **coll. vol. II**, 315 (1943). Toxicity study: L. Orö, A. Wretlind, *Acta Pharmacol. Toxicol.* **18**, 141 (1961).

Oily liquid. Disagreeable, rancid odor. Faint, tallow-like odor when spectroscopically pure. d_4^0 0.9345; d_4^{15} 0.9222; d_4^{20} 0.9181; d_4^{30} 0.9099. mp $-7.5°$. bp_{760} 223.01°, also reported as 221.9°; bp_{256} 187.5°; bp_{64} 150.8°. Specific heat 0.54 cal/g. Heat of combustion -986.1 cal/g (20°). n_D^{20} 1.42162. Viscosity 3.40 cp at 30°; 0.82 cp at 120°. Interfacial tension against water: 7.0 dynes/cm. pka (25°): 4.4. Soly in water (15°): 0.2419 g/100 ml H_2O. Sol in ethanol, ether, DMF, DMSO. LD_{50} i.v. in mice: 1200±56 mg/kg (Orö, Wretlind).

Methyl ester, $C_8H_{16}O_2$, liquid, d_4^{20} 0.8815; mp $-55.8°$; bp 173.8°. n_D^{20} 1.41152.

Ethyl ester, *see* Ethyl Oenanthate.

4696. 1-Heptanol. *n*-Heptyl alcohol; enanthic alcohol; 1-hydroxyheptane. $C_7H_{16}O$; mol wt 116.20. C 72.35%, H 13.88%, O 13.77%. $CH_3(CH_2)_5CH_2OH$. Prepd from heptaldehyde by reduction with iron filings in dil acetic acid: Clarke, Dreger, *Org. Syn.* **6**, 52 (1926); *cf.* Noller, Bannerot, *ibid.* **14**, 91 (1934); **coll. vol. I** (2nd ed., 1941) p 304. Other methods include the reaction between pentane and ethylene oxide in the presence of anhydr aluminum bromide: I. G. Farbenind., *Fr. pat.* **716,604**, *C.A.* **26**, 2198 (1932); and the action of amyl magnesium bromide on ethylene oxide: Vaughn *et al., J. Am. Chem. Soc.* **55**, 4207 (1933).

Liquid. d_4^{25} 0.8187. mp $-34.6°$. bp_{760} 175.8°; bp_{400} 155.6°; bp_{200} 136.6°; bp_{100} 119.5°; bp_{40} 9.98°; bp_{20} 85.8°; bp_{10} 74.7°; bp_5 64.3°; $bp_{1.0}$ 42.4°. n_D^{25} 1.4224. Water solns are colloidal: Traube, Klein, *Koll. Z.* **29**, 236 (1921), *Chem. Zentr.* **1922**, I, 233. One liter of water dissolves 1.0 g at 18°; 2.85 g at 100°; 5.15 g at 130°. Miscible with alcohol, ether.

4697. 2-Heptanol. Amylmethylcarbinol; 2-hydroxyheptane. $C_7H_{16}O$; mol wt 116.20. C 72.35%, H 13.88%, O 13.77%. $CH_3CH(OH)(CH_2)_4CH_3$. Occurs in oil of cloves: Masson, *Compt. Rend.* **149**, 630 (1907). Prepd by the action of amylmagnesium bromide on acetaldehyde: Henry, de Wael, *Rec. Trav. Chim.* **28**, 446 (1909); by the reduction of methyl amyl ketone with sodium in alcoholic soln: Thoms, Mannich, *Ber.* **36**, 2544 (1903); Pickard, Kenyon, *J. Chem. Soc.* **99**, 58 (1911); **105**, 849 (1914); Whitmore, Otterbacher, *Org. Syn.* **10**, 60 (1930); by the action of *Penicillium palitans* on coconut oil: Stokoe, *Biochem. J.* **22**, 82, 84 (1928). Toxicity data: Smyth *et al., Arch. Ind. Hyg. Occup. Med.* **10**, 61 (1954).

dl-Form, liquid. d^0 0.8344; d^{20} 0.8193. bp_{760} 158-160°, n_D 1.42131. Almost insol in water: 3.5 g/l. Sol in alc, ether, benzene. LD_{50} orally in rats: 2.58 g/kg (Smyth).

d-Form, liquid. d_4^{20} 0.8190; d_4^{35} 0.8050; d_4^{51} 0.7920; d_4^{64} 0.7815; d_4^{110} 0.7417. bp_{20} 73.5°. $[\alpha]_D^{20}$ +11.45° (1.039 g in 20 ml abs ethanol); $[\alpha]_4^{20}$ +13.71° (0.992 in 20 ml benzene).

l-Form, liquid. d_4^{20} 0.8184. bp_{23} 74.5°. $[\alpha]_D^{17}$ $-10.48°$.

4698. 2-Heptanone. Methyl amyl ketone. $C_7H_{14}O$; mol wt 114.19. C 73.63%, H 12.36%, O 14.01%. $CH_3(CH_2)_4COCH_3$. Found in oil of cloves and in cinnamonbark oil. Responsible for the "peppery" odor in cheeses of the Roquefort type: Hammer, Bryant, *Iowa State Coll. J. Sci.* **11**, 281 (1937). Prepd by the ketone decompn of ethyl butylacetoacetate: Drake, Riemenschneider, *J. Am. Chem. Soc.* **52**, 5005 (1930); Dehn, Jackson, *ibid.* **55**, 4285 (1933); Johnson, Hager, *Org. Syn.* **coll. vol. I** (2nd ed., 1941) p 351; by hydration of 1-heptyne and 2-heptyne: Thomas *et al., J. Am. Chem. Soc.* **60**, 719 (1938). Toxicity study: Smyth *et al., Am. Ind. Hyg. Assoc. J.* **23**, 95 (1962).

Liquid. Penetrating fruity odor. d_4^0 0.8324; d_4^{15} 0.8197; d_4^{30} 0.8068. bp_{760} 151.5°; bp_{21} 111°. n_D^{15} 1.41156; n_D^{25} 1.40729. Very slightly sol in water. Sol in alc, ether. LD_{50} orally in rats: 1.67 g/kg (Smyth).

Caution: Potential symptoms of overexposure are irritation of eyes and mucous membranes; headache; narcosis, coma; dermatitis. *See NIOSH Pocket Guide to Chemical Hazards* (DHHS/ NIOSH 90-117, 1990) p 144.

USE: In perfumery as constituent of artificial carnation oils; as industrial solvent.

4699. Heptenophos. *Phosphoric acid 7-chlorobicyclo-[3.2.0]hepta-2,6-dien-6-yl dimethyl ester;* 7-chlorobicyclo-

[3.2.0]hepta-2,6-dien-6-yl dimethylphosphate; HOE-2982; Ragadan; Hostaquick. $C_9H_{12}ClO_4P$; mol wt 250.62. C 43.13%, H 4.83%, Cl 14.15%, O 25.54%, P 12.36%. Systemic and contact insecticide of short persistence. Prepn: B. Böhner, K. Rüfenacht, **S. Afr. pat.** 67 06,947; *eidem,* U.S. pat. 3,600,474 (1968, 1971 both to Geigy). Activity: R. T. Hewson, *Proc. 8th Brit. Insectic. Fungic. Conf.* **2**, 697 (1975); W. Bonin, *ibid.* 705.

Pale amber liquid, $bp_{0.001}$ 94-95°. d_4^{20} 1.294. Vapor pressure at 20°: 7.5×10^{-4} mm Hg. Sol in xylene, acetone, methanol. LD_{50} orally in rats: 96-117 mg/kg (Hewson).

USE: Insecticide.

THERAP CAT (VET): Ectoparasiticide.

4700. Heptoxime. *1,2-Cycloheptanedione dioxime.* $C_7H_{12}N_2O_2$; mol wt 156.18. C 53.83%, H 7.74%, N 17.94%, O 20.49%. Prepn from cycloheptanone: Vander *et al., J. Org. Chem.* **14**, 836 (1949); Belcher *et al., J. Chem. Soc.* **1958**, 2743. *Review:* Banks, Nicholas, *USAEC* ISC-737 (1956).

Crystals from benzene, mp 182°. pK_1: 10.65 ± 0.2; pK_2: 12.21 ±0.2.

Monohydrate, crystals from water, mp 179-180°.

USE: As reagent in quantitative determination of nickel.

4701. D-*manno*-Heptulose. D-*manno*-Ketoheptose. $C_7H_{14}O_7$; mol wt 210.18. C 40.00%, H 6.71%, O 53.28%. Isoln from the wet pulp of the fruit of the avocado tree, *Persea gratissima* Gaertn. f. (*P. americana* Mill.), *Lauraceae*: LaForge, *J. Biol. Chem.* **28**, 511 (1917); Montgomery, Hudson, *J. Am. Chem. Soc.* **61**, 1654 (1939); Richtmyer in *Methods in Carbohydrate Chemistry*, R. L. Whistler *et al.*, Eds. **vol. 1**, 173 (1962).

Large, transparent prisms from methanol, mp 151-152°. $[\alpha]_D^{20}$ +29° (c = 2 in H_2O). Sol in water. Usually accompanied by perseitol. Although perseitol is much more insol in water and methanol than D-*manno*-heptulose, clean separation of the two substances is not always easy and may require patient fractional crystn.

4702. Hercynine. *(S)-α-Carboxy-N,N,N-trimethyl-1H-imidazole-4-ethanaminium inner salt; L-(1-carboxy-2-imidazol-4-ylethyl)trimethyl ammonium hydroxide inner salt;* histidine-betaine; histidine trimethylbetaine. $C_9H_{15}N_3O_2$; mol wt 197.24. C 54.81%, H 7.67%, N 21.30%, O 16.22%. Present in many fungi, especially in *Amanita muscaria* Fr. and *Agaricus compestris* L., *Agaricaceae*. Isoln: Kutscher, *Zentr. Physiol.* **24**, 775 (1910); **26**, 569 (1912); Barger, Ewins, *J. Chem. Soc.* **99**, 2340 (1911); Küng, *Z. Physiol. Chem.* **91**, 249 (1914). Occurrence in the *Limulus polyphemus* L. (king crab), and prepn: Ackermann, List, *ibid.* **313**, 30 (1958). Synthesis: Reinhold *et al., J. Med. Chem.* **11**, 258 (1968). Biosynthesis by fungi and *Actinomycetales*: Genghof, *J. Bacteriol.* **103**, 475 (1969).

White crystals from methanol and ether, mp 237-238° (dec). $[\alpha]_D^{22}$ +44.5° (5N HCl). Sol in water, alc.

4703. Herqueinone. *(7aS-cis)-8,9-Dihydro-4,6,7a-trihydroxy-5-methoxy-1,8,8,9-tetramethyl-3H-phenaleno[1,2-b]furan-3,7(7aH)-dione.* $C_{20}H_{20}O_7$; mol wt 372.37. C 64.51%, H 5.41%, O 30.08%. Red fungal pigment isolated from *Penicillium herquei*: Stodola *et al.*, *Nature* **167**, 773 (1951); Galarraga *et al.*, *Biochem. J.* **61**, 456 (1955); Harman *et al.*, *J. Org. Chem.* **20**, 1260 (1955). Structure: Cason *et al.*, *Tetrahedron* **18**, 839 (1962); Paul, Sim, *Proc. Chem. Soc.* **1962**, 352; Cason *et al.*, *J. Org. Chem.* **35**, 179 (1970); J. S. Brooks, G. A. Morrison, *Tetrahedron Letters* **1970**, 963; *eidem*, *J. Chem. Soc. Perkin Trans. I* **1972**, 421; **1974**, 2114. Crystal structure and abs config: A. Quick *et al.*, *Chem. Commun.* **1980**, 1051. ^{13}C-NMR study: T. Suga *et al.*, *Chem. Letters* **1981** 1063.

Red needles from alc, dec 226°. Sublimes in high vacuum at 175-190°. Absorption max (ethanol): 220, 250, 314, 416 nm (log ε 4.29, 4.09, 4.47, 3.66). Sol in acetone, DMF, 2N NaOH, concd HCl, cold concd H_2SO_4; fairly sol in methanol, ethanol, ether, ethyl acetate, chloroform; slightly sol in benzene, carbon disulfide. Practically insol in petr ether, carbon tetrachloride, water, aq Na_2CO_3.

4704. Hesperetin. *(S)-2,3-Dihydro-5,7-dihydroxy-2-(3-hydroxy-4-methoxyphenyl)-4H-1-benzopyran-4-one; 3',5,7-trihydroxy-4'-methoxyflavanone;* cyanidanon 4'-methyl ether 1626. $C_{16}H_{14}O_6$; mol wt 302.28. C 63.57%, H 4.67%, O 31.76%. The aglucon of hesperidin. Prepd by hydrolysis of hesperidin or by synthesis: Shinoda, Kawagoye, *C.A.* **23**, 2957 (1929); Seka, Prosche, *Monatsh.* **69**, 284 (1936). Sepn of isomers: Arthur *et al.*, *J. Chem. Soc.* **1956**, 632. Structure and configuration: Arakawa, Nakazaki, *Ann.* **636**, 111 (1960). *See also* Bioflavonoids.

(±)-Form, prisms from alc, mp 226-228°. Freely sol in alc, moderately in ether. Slightly sol in water, chloroform, benzene; sol in dil alkalies. Precipitated by carbonates.
Triacetate, $C_{22}H_{20}O_9$, crystals, mp 139-141°.
(−)-Form, plates from alc, mp 216-218°. $[\alpha]_D^{27}$ −37.6° (c = 1.80 in alc).
Triacetate, crystals, mp 130-132°. $[\alpha]_D^{26}$ +21.1° (c = 1.28 in chloroform).

4705. Hesperidin. *(S)-7-[[6-O-(6-Deoxy-α-L-mannopyranosyl)-β-D-glucopyranosyl]oxy]-2,3-dihydro-5-hydroxy-2-(3-hydroxy-4-methoxyphenyl)-4H-1-benzopyran-4-one;* hesperetin 7-rhamnoglucoside; cirantin; hesperetin-7-rutinoside. $C_{28}H_{34}O_{15}$; mol wt 610.57. C 55.08%, H 5.61%, O 39.31%. Predominant flavonoid in lemons and sweet oranges *(Citrus*

sinensis). Extraction procedures: Higby, *J. Am. Pharm. Assoc., Sci. Ed.* **30**, 629 (1941); U.S. pat. **2,421,061** (1947); Baier, U.S. pat. **2,442,110** (1948 to Calif. Fruit Growers Exchange). Structure: King, Robertson, *J. Chem. Soc.* **1931**, 1704; Arthur *et al.*, *ibid.* **1956**, 632; Horowitz, Gentili, *Tetrahedron* **19**, 773 (1963). Synthesis: Zemplen, Bognar, *Ber.* **75**, 1043 (1943); **76**, 773 (1943). Identity with cirantin: Manwaring *et al.*, *Phytochemistry* **7**, 1881 (1968). *See also* Bioflavonoids, Rutinose.

Fine, dendritic needles by precipitation at pH 6-7, mp 258-262° (softens at 250°). $[\alpha]_D^{20}$ −76° (c = 2 in pyridine). One gram dissolves in 50 l water. Sol in formamide, DMF at 60°. Slightly sol in methanol, hot glacial acetic acid. Almost insol in acetone, benzene, chloroform. Freely sol in dil alkalies, pyridine.
Octaacetate, $C_{44}H_{50}O_{23}$, needles from methanol, mp 176-178°. $[\alpha]_D^{20}$ −47.3° (pyridine).

4706. Hetacillin. *[2S-[2α,5α,6β(S*)]]-6-(2,2-Dimethyl-5-oxo-4-phenyl-1-imidazolidinyl)-3,3-dimethyl-7-oxo-4-thia-1-azabicyclo[3.2.0]heptane-2-carboxylic acid;* 6-(2,2-dimethyl-5-oxo-4-phenyl-1-imidazolidinyl)penicillanic acid; phenazacillin; BRL-804; Penplenum; Versapen; Versatrex. $C_{19}H_{23}N_3O_4S$; mol wt 389.48. C 58.59%, H 5.95%, N 10.79%, O 16.43%, S 8.23%. Semi-synthetic antibiotic related to penicillin. Prepn and structure: Hardcastle *et al.*, *J. Org. Chem.* **31**, 897 (1966); Johnson, Panetta, U.S. pat. **3,198,804** (1965 to Bristol-Myers). Pharmacology: Kirby, Kind, *Ann. N.Y. Acad. Sci.* **145**, 291 (1967); Ueda *et al.*, *J. Antibiot.* **20B**, 206 (1967), *C.A.* **69**, 95016w (1968). Stability studies: Saccani, Pansera, *Boll. Chim. Farm.* **107**, 640 (1968). Epimerization at C-6 to *epihetacillin*: Johnson *et al.*, *Tetrahedron Letters* **1968**, 1903.

Rectangular plates from water + methyl isobutyl ketone, dec 182.8-183.9°; also reported as mp 189.2-191.0°. $[\alpha]_D^{25}$ +366° (pyridine). Practically insol in most organic solvents and water; sol in dil aq NaOH soln (pH 7-8), DMF, DMSO, pyridine, methanol (with dec).
Potassium salt, $C_{19}H_{22}KN_3O_4S$, Uropen, Versapen K, Hetacin K, Natacillin.
Methyl ester, $C_{20}H_{25}N_3O_4S$, crystals from carbon tetrachloride, mp 101.5-102°. Sol in most organic solvents.
Epihetacillin, crystals, mp 164-165°. $[\alpha]_D^{23}$ +232° (pyridine).

THERAP CAT: Antibacterial.
THERAP CAT (VET): Antibacterial.

4707. Hetastarch. *Starch 2-hydroxyethyl ether;* hydroxyethyl starch; HES; 6-H.E.S.; Hespan; Hespander; Hestar; Hestat; Hestsol; Plasmasteril; Volex. A starch derivative of undetermined exact composition. Comprised of more than 90% amylopectin and etherified to the extent that an average of 7 to 8 of the OH groups present in every ten D-glucopyranose units of the polymer have been converted to OCH_2CH_2OH groups. Preparation by treating starch with pyridine and ethylene chlorohydrin: Mima, Yokoyama, **Japan.** pat. **06,556**('70) (to Green Cross), *C.A.* **73**, 36822r (1970). Structure evaluation studies: Banks *et al.*,

Brit. J. Pharmacol. **47**, 172 (1973). Physicochemical and biological properties: Tamada *et al., Chem. Pharm. Bull.* **19**, 286 (1971); Banks *et al., Staerke* **24**, 181 (1972). Pharmacology: Irikura *et al., Oyo Yakuri* **6**, 985 sqq (1972). Metabolism: Bogan *et al., Toxicol. Appl. Pharmacol.* **15**, 206 (1969); Ryan *et al., Xenobiotica* **2**, 141 (1972). Toxicity studies: Irikura *et al., Oyo Yakuri* **6**, 1023 (1972). Introduction as a plasma expander: Wiedersheim, *Arch. Int. Pharmacodyn.* **111**, 353 (1957). Review of production and uses: Hjermstad, *Starch: Chem. Technol.* **2**, 423-432 (1967).

R or R' = H or CH₂CH₂OH

Amylopectin derivative. The sequence is frequently interrupted by a unit in which R' is the residue of an additional *O*-hydroxyethylated α-D-glucopyranosyl moiety that constitutes the first unit in a branch or sub-branch of the polymer.

USE: Cryoprotective agent for erythrocytes.

THERAP CAT: Plasma volume expander.

4708. 5-HETE. *[S-(E,Z,Z,Z)]-5-Hydroxy-6,8,11,14-eicosatetraenoic acid;* (S)-5-hydroxy-6,8,11,14-*cis*-eicosatetraenoic acid. $C_{20}H_{32}O_3$; mol wt 320.47. C 74.96%, H 10.06%, O 14.98%. Important intermediate in a series of biosynthetic processes leading from arachidonic acid, *q.v.*, to a number of biologically active compounds. First discovered during the transformation of arachidonic acid by rabbit peritoneal polymorphonuclear leukocytes: P. Borgeat *et al., J. Biol. Chem.* **251**, 7816 (1976). Revised description: *eidem, ibid.* **252**, 8772 (1977). It has also been found during metabolism of arachidonic acid by human peripheral blood polymorphonuclear leukocytes: P. Borgeat, B. Samuelsson, *Proc. Nat. Acad. Sci. USA* **76**, 2148 (1979), and has been shown to be strongly chemotactic for human eosinophils and neutrophils: E. J. Goetzl *et al., Immunology* **39**, 491 (1980); E. J. Goetzl, *N. Engl. J. Med.* **303**, 822 (1980). 5-HETE is formed via *5-HPETE* (*5-hydroperoxy-6,8,11,14-eicosatetraenoic acid*), a precursor of the leukotrienes, *q.v.* Chemical and enzymic syntheses of 5-HETE and 5-HPETE: E. J. Corey *et al., J. Am. Chem. Soc.* **102**, 1435 (1980). Synthesis via phenylselenylation of arachidonic acid: J. E. Baldwin *et al., Tetrahedron* **37**, Suppl. 1, 263 (1981). Large-scale synthesis of 5-HETE: E. J. Corey, S. Hashimoto, *Tetrahedron Letters* **22**, 299 (1981). Biosynthetic study: R. C. Mebane, *Diss. Abstr. B* **42**, 2839 (1982). Selected synthesis of octadeuterated (±)-5-HETE: W. C. Hubbard *et al., Prostaglandins* **23**, 61 (1982). Stereospecific syntheses of 5-(S)- and 5-(R)-HETE and transformation to (±)-5-HPETE: R. Zamboni, J. Rokach, *Tetrahedron Letters* **24**, 999 (1983). Review of syntheses: J. G. Atkinson, J. Rokach, in *Handbook of Prostaglandins and Related Lipids: The Eicosanoids*, A. L. Willis *et al.*, Eds. (CRC Press, Boca Raton, in press).

Methyl ester, $C_{21}H_{34}O_3$, colorless oil. $[\alpha]_D^{23}$ +14.0°; $[\alpha]_{436}^{23}$ +35.7° (c = 2.02 in benzene).

4709. Heteronium Bromide. *3-[(Hydroxyphenyl-2-thienylacetyl)oxy]-1,1-dimethylpyrrolidinium bromide; 3-hydroxy-1,1-dimethylpyrrolidinium bromide α-phenyl-2-thiopheneglycolate;* 3-hydroxy-1,1-dimethylpyrrolidinium bromide α-phenyl-α-(2-thienyl)glycolate; 1-methyl-3-pyrrolidyl phenyl-2-thienylglycolate methobromide; Hetrum bromide. $C_{18}H_{22}BrNO_3S$; mol wt 412.35. C 52.43%, H 5.38%, N 3.40%, O 11.64%, S 7.78%. Prepn: Ryan, Ainsworth, *J. Org. Chem.* **27**, 2901 (1962); of α- and β-diastereoisomers, Ainsworth, Ryan, U.S. pat. **3,138,614** (1964 to

Lilly). Toxicity data: E. I. Goldenthal, *Toxicol. Appl. Pharmacol.* **18**, 185 (1971).

Crystals from abs ethanol; α- and β-diastereoisomers, separated by fractional crystn, have different crystal forms and different infrared absorption spectra. LD_{50} in female, male rats (mg/kg): 3399±624, 3576±954 orally (Goldenthal).

α-Diastereoisomer, crystals from methanol + ethyl acetate, mp 210-211°.

β-Diastereoisomer, crystals from methanol + ethyl acetate, mp 182-184°.

THERAP CAT: Anticholinergic.

4710. Hetolin®. *1-Methyl-4-[3,3,3-tris(4-chlorophenyl)-1-oxopropyl]piperazine; 1-methyl-4-[3,3,3-tris(p-chlorophenyl)propionyl]piperazine;* 1-[β,β,β-tris(p-chlorophenyl)propionyl]-4-methylpiperazine; LZ-544. $C_{26}H_{25}Cl_3N_2O$; mol wt 487.86. C 64.01%, H 5.17%, Cl 21.80%, N 5.74%, O 3.28%. Prepn: **Belg. pat. 634,833** (1964 to Hoechst), *C.A.* **61**, 16076h (1964). Pharmacology: Schorr *et al., Arzneimittel-Forsch.* **14**, 1151 (1964).

Crystals from ethyl acetate, mp 213-215°.

Hydrochloride, $C_{26}H_{25}Cl_3N_2O\cdot HCl$, *Dicroden*. Crystals from ethanol, mp 267-269°. Freely sol in hot water and alcohol. LD_{50} orally in mice: 610 mg/kg.

THERAP CAT (VET): Anthelmintic (flukicide).

4711. Hexaaminecobalt Trichloride. Luteocobaltic chloride; hexammino-cobalt chloride. $Cl_3CoH_{18}N_6$; mol wt 267.47. Cl 39.76%, Co 22.03%, H 6.78%, N 31.42%. [Co-$(NH_3)_6]Cl_3$. Prepd from $CoCl_2$, NH_4Cl, NH_3 and O_2: Bjerrum, McReynolds, *Inorg. Syn.* **2**, 217 (1946).

Wine-red or brownish-orange-red, monoclinic crystals. Sol in water; on long boiling with water, NH_3 is evolved and $Co(OH)_3$ pptd.

USE: As reagent for pyrophosphoric acid, for the estimation of phosphate.

4712. Hexaborane(10). Hexaboron decahydride; borohexane. B_6H_{10}; mol wt 74.95. B 86.55%, H 13.45%. Prepd by the reaction of magnesium boride with hydrochloric or phosphoric acid: Stock, Kuss, *Ber.* **56B**, 789 (1923).

Liquid. mp −62.3°; bp 108°; vapor pressure (0°): 7.5 mm: Burg, Kratzer, *Inorg. Chem.* **1**, 725 (1962). d^0 0.69. Slowly dec at room temp. Hydrolyzes in water after long heating.

4713. Hexacarbacholine Bromide. *N,N,N,N',N',N'-Hexamethyl-4,13-dioxo-3,14-dioxa-5,12-diazahexadecane-1,-16-diaminium dibromide; 2,2'-[1,6-hexanediylbis(iminocarbonyloxy)]bis[N,N,N-trimethylethanaminium] dibromide;* choline bromide hexamethylenedicarbamate; hexamethylenedicarbamic acid choline bromide diester; hexamethylene-1,6-bis(carbamoylcholine bromide); N,N'-hexamethylenebis-[(2-carbamoyloxyethyl)trimethylammonium bromide]; BC-16; Imbretil. $C_{18}H_{40}Br_2N_4O_6$; mol wt 536.35. C 40.31%, H 7.52%, Br 29.80%, N 10.45%, O 11.93%. Prepn: Schmied *et al.*, **Austrian pat. 185,371** (1956); **Ger. pat. 1,021,842** (1958 to Oesterreichische Stickstoffwerke).

Crystals from ethanol, mp 174-176°.

THERAP CAT: Muscle relaxant (skeletal).

4714. Hexachlorobenzene. Perchlorobenzene; Anticarie; Bunt-cure; Bunt-no-more; Julin's carbon chloride. C_6Cl_6; mol wt 284.78. C 25.31%, Cl 74.69%. Prepn: Becke, Sperber, U.S. pat. **2,792,434** (1957 to BASF). Teratogenicity study: K. D. Courtney *et al.*, *Toxicol. Appl. Pharmacol.* **35**, 239 (1976). Carcinogenicity studies: J. R. P. Cabral *et al.*, *Nature* **269**, 510 (1977); D. L. Arnold *et al.*, *Food Chem. Toxicol.* **23**, 779 (1985) (corr. *ibid.* **26**, 169 (1988)). uv spectrum: H. Conrad-Billroth, *Z. Phys. Chem.* **19**, 76 (1932); O. Schnepp, R. Kopelman, *J. Chem. Phys.* **30**, 868 (1959). Toxicology: G. Vettorazzi, *Residue Rev.* **56**, 107 (1975), R. Ockner, R. Schmid, *Nature* **189**, 499 (1961).

Needles. d^{23} 2.044. mp 231°. bp 323-326°. Vapor pressure at 20°: 1.09×10^{-5} mm Hg. Sublimable. Insol in water. Sparingly sol in cold alcohol; sol in benzene, chloroform, ether. Flash point 242°.

Note: Not to be confused with benzene hexachloride, *see* Lindane.

Caution: Potential symptoms of chronic oral overexposure are dermal lesions, hypertrichosis, anorexia, weight loss, focal alopecia, corneal opacity, atrophic hands, hepatomegaly, porphyria, skeletal muscle wasting. *See Clinical Toxicology of Commercial Products*, R. E. Gosselin *et al.*, Eds. (Williams & Wilkins, Baltimore, 5th ed., 1984) Section II, pp 170-171; *Patty's Industrial Hygiene and Toxicology* vol. **2B**, G. D. Clayton, F. E. Clayton, Eds. (John Wiley & Sons, New York, 4th ed., 1994) pp 1477-1492. This substance may reasonably be anticipated to be a carcinogen: *Seventh Annual Report on Carcinogens* (PB95-109781, 1994) p 223.

USE: In organic syntheses. Formerly as agricultural fungicide.

4715. Hexachloroethane. Carbon hexachloride; perchloroethane. C_2Cl_6; mol wt 236.74. C 10.15%, Cl 89.85%. CCl_3CCl_3. Prepn: *Beilstein* **1**, 87 (1918) and suppls. Toxicity data: Barsoum, Saad, *Quart. J. Pharm. Pharmacol.* **7**, 205 (1934).

Crystals; camphorous odor. d 2.09. Readily sublimes without melting. bp 186.8° (triple point). Heat of sublimation 12.2 kcal/mol. Sol in alcohol, benzene, chloroform, ether, oils. Insol in water. MLD i.v. in dogs: 325 mg/kg (Barsoum, Saad).

Caution: Direct contact may cause eye irritation. *See NIOSH Pocket Guide to Chemical Hazards* (DHHS/NIOSH 90-117, 1990) p 122. *See also Patty's Industrial Hygiene and Toxicology* vol. **2E**, G. D. Clayton, F. E. Clayton, Eds. (John Wiley & Sons, New York, 4th ed., 1994) p 4144-4151. This substance may reasonably be anticipated to be a carcinogen: *Seventh Annual Report on Carcinogens* (PB95-109781, 1994) p 226.

USE: In metallurgy for refining aluminum alloys, removing impurities from molten metals, recovering metal from ores or smelting products. Degassing agent for magnesium; to inhibit explosiveness of methane and combustion of ammonium perchlorate. Smoke generator in grenades; in pyrotechnics. Ignition suppressant, in fire extinguishing

fluids, polymer additive, flame-proofing agent, vulcanizing agent. In prodn of synthetic diamonds.

THERAP CAT (VET): Anthelmintic (flukicide).

4716. Hexachlorophene. *2,2'-Methylenebis[3,4,6-trichlorophenol];* 2,2'-dihydroxy-3,3',5,5',6,6'-hexachlorodiphenylmethane; bis(3,5,6-trichloro-2-hydroxyphenyl)methane; AT-7; G-11; Bilevon; Dermadex; Exofene; Hexosan; pHisohex; Surgi-Cen; Surofene. $C_{13}H_6Cl_6O_2$; mol wt 406.91. C 38.37%, H 1.49%, Cl 52.28%, O 7.86%. Prepd by the condensation of 2 mols of 2,4,5-trichlorophenol with 1 mol formaldehyde in the presence of concd sulfuric acid: Gump, U.S. pat. **2,250,480** (1941 to Burton T. Bush). Improved procedures: U.S. pat. **2,435,593** (1948) and **2,812,365** (1957 to Givaudan). Acute toxicity: T. B. Gaines, R. E. Linder, *Fundam. Appl. Toxicol.* **7**, 299 (1986).

Crystals from benzene, mp 164-165°. Practically insol in water. Sol in alcohol, acetone, ether, chloroform, propylene glycol; polyethylene glycols; olive oil; cottonseed oil; dil aq solns of the alkalies. Forms salts with alkalies and alkaline earths. Phenol coefficient ~125 (monopotassium salt). Incompatible with Tweens from bacteriological point of view. LD_{50} in adult male, female rats (mg/kg): 66, 57 orally (Gaines, Linder).

Monophosphate, *Hepadist.*

Caution: Excessive dosage to animals results in symptoms of neurotoxicity. Reversible vacuolar changes mainly affecting the myelin of the brain and spinal cord have been reported. Because of potential neurotoxicity in humans, the FDA has regulated use. *See* Lockhart, *Pediatrics* **50**, 229 (1972).

USE: Chiefly in the manuf of germicidal soaps.

THERAP CAT: Antiseptic; disinfectant.

THERAP CAT (VET): Anthelmintic (flukicide).

4717. Hexacyclonate Sodium. *1-(Hydroxymethyl)cyclohexaneacetic acid sodium salt;* sodium 3,3-pentamethylene-4-hydroxybutyrate; sodium β,β-pentamethylene-γ-hydroxybutyrate; β,β-pentamethylene-γ-hydroxybutyric acid sodium salt; Neuryl. $C_9H_{15}NaO_3$; mol wt 194.21. C 55.66%, H 7.79%, Na 11.84%, O 24.72%. Prepn: Van Wessem, Sakal; Shavel *et al.*, U.S. pats. **2,960,441; 3,007,940** (1960; 1961 to Warner-Lambert).

Monohydrate, platelets from *n*-butanol + benzene, mp 106-108°. The anhyd salt is hygroscopic. Readily sol in water, methanol, ethanol; sparingly sol in ether, acetone.

THERAP CAT: CNS stimulant.

4718. Hexadecyl 3-Hydroxy-2-naphthoate. *3-Hydroxy-2-naphthalenecarboxylic acid hexadecyl ester.* $C_{27}H_{40}O_3$; mol wt 412.61. C 78.60%, H 9.77%, O 11.63%. Prepd by the action of 3-hydroxy-2-naphthoyl chloride on cetyl alc: Oshima, Hayashi, *J. Soc. Chem. Ind. Japan* **44**, 821 (1941).

Greenish-white, flaky crystals, mp 72-73°. Soluble in benzene, glacial acetic acid, petr ether. Sparingly sol in cold alcohol. Insol in water.

USE: As waterproofing agent for rayon.

4719. Hexadimethrine Bromide. *N,N,N',N'-Tetramethyl-1,6-hexanediamine polymer with 1,3-dibromopropane;* polymer of *N,N,N',N'*-tetramethylhexamethylenediamine and trimethylene bromide; poly(*N,N,N',N'*-tetramethyl-*N*-trimethylenehexamethylenediammonium dibromide); Polybrene. $(C_{13}H_{30}Br_2N_2)_x$. Toxicity study: Kimura *et al., Toxicol. Appl. Pharmacol.* **1**, 185 (1959).

White, hygroscopic, amorphous polymer. Soluble in water up to 10%. pH of 1% saline soln 5-9. Stable in soln and when autoclaved. Polymers with mol wt of 5000-10,000 have LD_{50} i.v. in mice: 25-40 mg/kg (Kimura).

THERAP CAT: Heparin antagonist.

4720. Hexaflumuron. *N-[[[3,5-Dichloro-4-(1,1,2,2-tetrafluoroethoxy)phenyl]amino]carbonyl]-2,6-difluorobenzamide;* 1-[3,5-dichloro-4-(1,1,2,2-tetrafluoroethoxy)phenyl]-3-(2,6-difluorobenzoyl)urea; DE-473; XRD-473; Consult; Trueno. $C_{16}H_8Cl_2F_6N_2O_3$; mol wt 461.15. C 41.67%, H 1.75%, Cl 15.38%, F 24.72%, N 6.07%, O 10.41%. Insect growth regulator; inhibits chitin synthesis. Prepn: R. H. Rigterink, R. J. Sbragia, **Eur.** pat. **Appl. 71,279;** *eidem,* U.S. pat. **4,468,405** (1983, 1984 both to Dow). Physical properties and activity: R. J. Sbragia *et al., Proc. 10th Int. Congr. Plant Prot.* **1**, 417 (1983). Chromatographic determn in soil: A. Khoshab, R. Teasdale, *J. Chromatog. A* **660**, 195 (1994). Environmental distribution: D. Yon *et al., Brighton Crop Prot. Conf. - Pests Dis.* **1992**, 907. Field trials in food crops: K. N. Komblas, A. C. Hunter, *Brit. Crop Prot. Conf. - Pests Dis.* **1986**, 907; vs subterranean termites: N.-Y. Su, *J. Econ. Entomol.* **87**, 389 (1994).

White solid, mp 197-199°. Soly in water (23°): 0.7 mg/l. vapor pressure (298° K): 5.87 × 10⁻⁹.

USE: Insecticide.

4721. Hexafluorenium Bromide. *N,N'-Di-9H-fluoren-9-yl-N,N,N',N'-tetramethyl-1,6-hexanediaminium dibromide; hexamethylenebis[9-fluorenyldimethylammonium bromide];* hexamethylenebis(dimethyl-9-fluorenylammonium bromide); Mylaxen. $C_{36}H_{42}Br_2N_2$; mol wt 662.55. C 65.26%, H 6.39%, Br 24.12%, N 4.23%. Prepn: Cavallito *et al., J. Am. Chem. Soc.* **76**, 1862 (1954); Cavallito, Gray, U.S. pat. **2,-783,237** (1957 to Irwin, Neisler).

Crystals from *n*-propanol, mp 188-189°.

THERAP CAT: Muscle relaxant (skeletal); succinylcholine synergist.

4722. Hexafluorobenzene. Perfluorobenzene. C_6F_6; mol wt 186.06. C 38.73%, F 61.27%. Prepn: E. T. McBee *et al., Ind. Eng. Chem.* **39**, 378 (1947); J. A. Godsell *et al., Nature* **178**, 199 (1956). Toxicology: C. F. B. Nhachi, *Toxicology* **39**, 317 (1986). Mechanistic study of metabolite formation: I. M. C. M. Rietjens, J. Vervoort, *Chem. Res. Toxicol.* **5**, 10 (1992).

mp −13 to −11°. bp_{743} 81.0-82.0° (McBee); also reported as bp 80° (Godsell). n_D^{20} 1.3760; n_D^{18} 1.3746. d_4^{25} 1.612.

USE: Solvent; intermediate in chemical synthesis.

4723. Hexalure. *(Z)-7-Hexadecen-1-ol acetate; cis-7-hexadecenyl acetate;* cis-1-acetoxy-7-hexadecene; hexalene. $C_{18}H_{34}O_2$; mol wt 282.47. C 76.54%, H 12.13%, O 11.33%. Synthetic sex attractant for pink bollworm moths, *Pectinophora gossypiella* (Saunders). Discovery and prepn: N. Green *et al., Experientia* **25**, 682 (1969); N. Green, J. C. Keller, **Ger.** pat. **1,960,155;** *eidem,* U.S. pat. **3,586,712** (1970, 1971 both to U.S. Sec. Agric.). Field trials: J. C. Keller *et al., J. Econ. Entomol.* **62**, 1520 (1969). Toxicity study: M. Beroza *et al., Toxicol. Appl. Pharmacol.* **31**, 421 (1975). *See also* Gossyplure, Propylure.

Clear oily liquid, $bp_{0.001}$ 100-104°. n_D^{25} 1.4484. Insol in water. Sol in hexane, ether, acetone, benzene. LD_{50} orally in rats: > 34600 mg/kg (Beroza).

USE: Insect sex attractant.

4724. Hexamethonium. *N,N,N,N',N',N'-Hexamethyl-1,6-hexanediaminium; hexamethylenebis(trimethylammonium);* α,ω-bis(trimethylammonium)hexane; hexathonide; hexamethone. $[C_{12}H_{30}N_2]^{2+}$. $[(CH_3)_3N(CH_2)_6N(CH_3)_3]^{2+}$. Prepn: H. J. Barber, U.S. pat. **2,641,610** (1953 to May & Baker); H. J. Barber, K. Gaimster, *J. Pharm. Pharmacol.* **3**, 663 (1951). Ganglion blocking agent: R. Wien, D. F. J. Mason, *Brit. J. Pharmacol.* **6**, 611 (1951).

Bromide, $C_{12}H_{30}Br_2N_2$, *C-6, Bistrium bromide, Esametina, Gangliostat, Hexameton bromide, Hexanium bromide, Simpatoblock, Vegolysen, Vegolysin.* Hygroscopic crystals, mp 274-276°. Sol in water, alc. Insol in acetone, chloroform, ether. pH of 1% soln: 6.2-7.0.

Chloride, $C_{12}H_{30}Cl_2N_2$, *Bistrium chloride, Chloor-hexaviet, Esomid chloride, Hestrium chloride, Hexameton chloride, Hexone chloride, Hiohex chloride, Methium chloride, Meton.* Hygroscopic crystals, dec 289-292°. Sol in water, alc. Practically insol in chloroform, ether. pH of 10% soln: 5.5-6.5.

Iodide, $C_{12}H_{30}I_2N_2$, *Hexathide.* Hygroscopic cryst powder. Sol in water. Practically insol in alc.

Tartrate, $C_{20}H_{40}N_2O_{12}$, *Vegolysen-T.* Hygroscopic powder, dec 186°. Sol in water. Practically insol in alc. pH of 10% soln: 3.8.

THERAP CAT: Antihypertensive.

4725. Hexamethyldisilazane. *1,1,1-Trimethyl-N-(trimethylsilyl)silanamine;* HMDS. $C_6H_{19}NSi_2$; mol wt 161.39. C 44.65%, H 11.87%, N 8.68%, Si 34.80%. $(CH_3)_3SiNHSi(CH_3)_3$. Silylating agent. Prepn: R. O. Sauer, *J. Am. Chem. Soc.* **66**, 1707 (1944). Surface tension and density as a function of temperature: R. S. Myers, H. L. Clever, *J. Chem. Eng. Data* **14**, 161 (1969). Mass spec. study: J. Tamás, P. Miklos, *Org. Mass Spectrom.* **10**, 859 (1975). Raman, IR and NMR spectra: K. Hamada, H. Morishita, *Spectrosc. Letters* **16**, 717 (1983). Molecular structure and conformation: T. Fjeldberg, *J. Mol. Struct.* **112**, 159 (1984). Use in deactivation of chromatography columns: T. Welsch *et al., J. Chromatog.* **241**, 41 (1982); R. N. Kong *et al., Chromatographia* **18**, 362 (1984). Use as adhesion promoter for photoresists: J. N. Helbert, H. G. Hughes in *Proc. Symp. Adhesion Aspects Polymeric Coatings,* K. L. Mittal, Ed. (Plenum Press, New York, 1983) pp 499-508; D. Freeman, W. Kern, *J. Vac. Sci. Technol.* **7**, 1446 (1989). Physical properties, toxicity and technical data: Petrarch Systems Inc. Product Information Sheets. Brief review of chemistry and applications: R. J. De Pasquale, *Amer. Lab.* **5**(6), 35-39 (1973).

Colorless liquid with an ammonia-like odor. bp 126.0°. d_4^{20} 0.7742. n_D^{20} 1.4080. Flash pt, closed cup: 81°F (27°C). pKa 7.55. Vapor pressure at 50°C: 79 mm. Viscosity: 0.90 ctsk. Ignition pt: 380°. Dielectric constant (1000 Hz): 2.27. Insol in water, reacts slowly. LD_{LO} i.p. in mice: 650 mg/kg (Petrarch Systems Inc. Product Information Sheets).

Caution: Causes severe burns to eyes and irritation of the skin (Petrarch Systems Inc. Product Information Sheets).

USE: Deactivation of chromatographic support materials. In electronic industry as an adhesion promoter for photoresists on silicon.

4726. Hexamethylene Glycol. *1,6-Hexanediol;* 1,6-dihydroxyhexane. $C_6H_{14}O_2$; mol wt 118.18. C 60.98%, H 11.94%, O 27.08%. $HOCH_2(CH_2)_4CH_2OH$. Prepared by reduction of ethyl adipate with copper chromite: Lazier *et al., Org. Syn. coll. vol.* **II,** 325 (1943); from 2,5-tetrahydro-furandimethanol: Utne *et al.,* U.S. pat. **3,070,633** (1962 to Merck & Co.).

Crystals, mp 42.8°. bp_{760} 208°; bp_{10} 134°. n_D^{25} 1.4579. Dipole moment: 2.48 (dioxane). Sol in water, alcohol; sparingly sol in hot ether. LD_{50} orally in rats: 3.73 g/kg, *Toxic Substances List,* H. E. Christensen *et al.,* Ed. (1974) p 399.

Bis(3,5-dinitrobenzoate), $C_{20}H_{18}N_4O_{12}$, crystals from dioxane + ethanol, mp 169-171°.

USE: Intermediate in the production of nylon; to make hexamethylenediamine, polyesters, polyurethans; in gasoline refining; as plasticizer.

4727. Hexamethylolmelamine. *(1,3,5-Triazine-2,4,6-triyltrinitrilo)hexakismethanol; (s-triazine-2,4,6-triyltrinitrilo)hexamethanol;* hexakis(hydroxymethyl)melamine; Resloom M 75. $C_9H_{18}N_6O_6$; mol wt 306.28. C 35.29%, H 5.92%, N 27.44%, O 31.34%. Prepn: Gams *et al., Helv. Chim. Acta* **24,** 302E (1941); Widmer, Fisch, U.S. pat. **2,-387,547** (1945 to Ciba).

Amorphous to cryst, white mass, mp 135-139°. Sol in water. Easily forms insol polymerization products. For the prepn of aq solns double distilled water and hard glass vessels are recommended. A 1.4-2% aq soln is thus possible.

USE: In fireproofing and creaseproofing of cottons, rayons.

4728. Hexamidine. *4,4'-[1,6-Hexanediylbis(oxy)]bis-benzenecarboximidamide;* 4,4'-(hexamethylenedioxy)dibenz-amidine; 4,4'-diamidino-α,ω-diphenoxyhexane. $C_{20}H_{26}N_4O_2$; mol wt 354.45. C 67.77%, H 7.39%, N 15.81%, O 9.03%. Prepn: A. J. Ewins *et al.,* **Brit.** pat. **507,565** (1939 to May & Baker); and trypanocidal activity: J. N. Ashley *et al., J. Chem. Soc.* **1942,** 103. Activity in fibrinolytic systems: J. D. Geratz, *Thromb. Diath. Haemorrh.* **29,** 154 (1973). Antibacterial activity: G. Michel *et al., J. Int. Med. Res.* **14,** 205 (1986). Antifungal activity: M. C. Reynaud, C. Chauve, *Bull. Soc. Fr. Mycol. Med.* **15,** 269 (1986). HPLC determn in pharmaceutics: P. Taylor *et al., J. Pharm. Sci.* **72,** 1477 (1983); in cosmetics: B. Wyhowski de Bukanski, M. O. Masse, *Int. J. Cosmet. Sci.* **6,** 283 (1984). Clinical use as a topical antiseptic: M. J. Fénelon, *Bordeaux Med.* **3,** 867 (1970). Use in treatment of acne: P. Taylor, A. A. Levy, **Eur.** pat. Appl. **93,186** (1983 to Richardson-Vicks).

Isethionate, $C_{24}H_{38}N_4O_{10}S_2$, *RF-2535, Desomedine, Hexomedin, Ophtamedine.* Prisms from HCl, mp 246-247° (dec).

USE: Preservative in cosmetics.

THERAP CAT: Antiseptic (topical).

4729. *n*-Hexane. C_6H_{14}; mol wt 86.18. C 83.63%, H 16.37%. $CH_3(CH_2)_4CH_3$. Chief constituent of petr ether or ligroin. Toxicity data: *Handbook of Toxicology* **vol. 1,** W. S. Spector, Ed. (Saunders, Philadelphia, 1956) pp 338-339; E. T. Kimura *et al., Toxicol. Appl. Pharmacol.* **19,** 699 (1971).

Colorless, very volatile liquid; faint, peculiar odor. d^{20} 0.660. bp 69°. mp −100° to −95°. n_D^{20} 1.375. Insol in water; miscible with alcohol, chloroform, ether. LC for mice in air: ~40000 ppm (Spector). LD_{50} orally in young adult rats: 49.0 ml/kg (Kimura).

Caution: Potential symptoms of overexposure are light-headedness; nausea, headache; numbness of extremities, muscle weakness; irritation of eyes and nose; dermatitis; chemical pneumonia; giddiness. See *NIOSH Pocket Guide to Chemical Hazards* (DHHS/NIOSH 90-117, 1990) p 122.

USE: Determining refractive index of minerals; filling for thermometers instead of mercury, usually with a blue or red dye.

4730. 1,6-Hexanediamine. Hexamethylenediamine; 1,6-diaminohexane. $C_6H_{16}N_2$; mol wt 116.21. C 62.02%, H 13.88%, N 24.11%. $NH_2(CH_2)_6NH_2$. Prepd by reducing adiponitrile with sodium and alcohol: Slotta, Tschesche, *Ber.* **62,** 1404 (1929). Manuf: Faith, Keyes & Clark's *Industrial Chemicals,* F. A. Lowenheim, M. K. Moran, Eds. (Wiley-Interscience, New York, 4th ed., 1975) pp 442-444.

Platelets, leaflets. Sublimes as long needles. Odor of piperidine. Absorbs water and carbon dioxide from air. mp 42°. bp 205°; bp_{20} 100°. Freely sol in water; slightly sol in alcohol, benzene.

Dihydrochloride, $C_6H_{16}N_2.2HCl$, needles from water or alcohol, mp 248°. Freely sol in water.

USE: Intermediate in the manuf of nylon.

4731. Hexanitrohexaazaisowurtzitane. *Octahydro-1,-3,4,7,8,10-hexanitro-5,2,6-(iminomethenimino)-1H-imidazo-[4,5-b]pyrazine;* CL-20; HNIW; 2,4,6,8,10,12-hexanitro-2,4,-6,8,10,12-hexaazatetracyclo[5.5.0.0^{5,9}.0^{3,11}]dodecane. C_6H_6-$N_{12}O_{12}$; mol wt 438.19. C 16.45%, H 1.38%, N 38.36%, O 43.82%. Symmetric polyazacyclic nitramine explosive. Polymorphs α, β, γ, ε, ζ have been identified and confirmed by x-ray diffraction or IR techniques. A sixth polymorph δ has been identified, but not confirmed. First synthesized by A. T. Nielsen in 1987; modified by R. Wardle and J. C. Hinshaw. Gas phase dissociation: J. R. Doyle, Jr., *Org. Mass Spectrom.* **26,** 723 (1991). EPR spectra: M. D. Pace, *J. Phys. Chem.* **95,** 5858 (1991). Identification of δ polymorph by high-pressure phase transition and verification of ζ by ir: T. P. Russell *et al., ibid.* **96,** 5509 (1992). Pressure-temperature phase diagrams: *eidem, ibid.* **97,** 1993 (1993). Thermal decomposition: J. C. Oxley *et al., ibid.* **98,** 7004 (1994). High-pressure/thermal shock matrix isolation deflagration and detonation study: J. K. Rice, T. P. Russell, *Chem. Phys. Letters* **234,** 195 (1995). Review: S. Borman, *Chem. Eng. News* **72,** 18-22 (Jan. 17, 1994).

Explosive! Crystalline.

USE: Rocket propellants. Munitions. Construction or demolition explosive.

4732. 1-Hexanol. *n*-Hexyl alcohol; amylcarbinol; pentylcarbinol; 1-hydroxyhexane. $C_6H_{14}O$; mol wt 102.18. C 70.53%, H 13.81%, O 15.66%. $CH_3(CH_2)_4CH_2OH$. Occurs as the acetate in seeds and fruits of *Heracleum sphondylium* L. and *H. giganteum* Fisch., *Umbelliferae*. Lab prepn by the action of butylmagnesium bromide on ethylene oxide: Dreger, *Org. Syn.* coll. vol. I, 306 (2nd ed., 1941). Reduction of 1,3-hexadienal with iron wire in the presence of nickel acetate: Zeisel, Neuwirth, *Ann.* **433**, 127 (1923); cf. Baumgarten, Glatzel, *Ber.* **59**, 2659 (1926); Kuhn, Hoffer, *Ber.* **63**, 2165 (1930). Industrial prepn by reducing ethyl caproate with sodium in abs alcohol: Bouveault, Blanc, **Ger. pat. 164,294** (1903). Toxicity study: H. F. Smyth *et al., Arch. Ind. Hyg. Occup. Med.* **4**, 119 (1951).

Liquid. d_4^{25} 0.8153; d_4^{35} 0.8082. mp −51.6°. bp_{760} 157°; bp_{400} 138°; bp_{200} 119.6°; bp_{100} 102.8°; bp_{60} 92°; bp_{40} 83.7°; bp_{20} 70.3°; bp_{10} 58.2°; bp_5 47.2°; $bp_{1.0}$ 24.4°. n_D^{25} 1.4162. Absorption spectrum: Massol, Faucon, *Bull. Soc. Chim.* [4] **11**, 932. Flash pt, closed cup: 145°F (63°C). Slightly sol in water; miscible with alcohol, ether. LD_{50} orally in rats: 4.59 g/kg (Smyth).

USE: Manuf antiseptics, hypnotics.

4733. Hexapropymate. *1-(2-Propynyl)cyclohexanol carbamate; carbamic acid 1-(2-propynyl)cyclohexyl ester;* 1-(2-propynyl)cyclohexyl carbamate; 1-carbamoyloxy-1-(2-propynyl)cyclohexane; hexopropynate; L-2103; Modirax; Lunamin; Merinax. $C_{10}H_{15}NO_2$; mol wt 181.23. C 66.27%, H 8.34%, N 7.73%, O 17.66%. Prepn: Lauger *et al., Helv. Chim. Acta* **42**, 2379 (1959); Prost, **U.S. pat. 2,931,828** (1960 to Soc. Belge des Labs. Labaz).

Crystals from petr ether or dioxane, mp 99°. Slight characteristic odor. Sol in ethanol, glycerol, propylene glycol.

THERAP CAT: Sedative, hypnotic.

4734. Hexazinone. *3-Cyclohexyl-6-(dimethylamino)-1-methyl-1,3,5-triazine-2,4(1H,3H)-dione;* DPX-3674; Velpar. $C_{12}H_{20}N_4O_2$; mol wt 252.32. C 57.12%, H 7.99%, N 22.21%, O 12.68%. Broad spectrum, pre- and post-emergence herbicide effective against woody and herbaceous weeds. Prepn: **Neth. pat. Appl. 7,307,218**; J. J. Fuchs, J. B. Wommack, **U.S. pat. 3,850,924**; K. Lin, **U.S. pat. 3,983,116** (1973, 1974, 1976 all to Du Pont). Herbicidal activity: D. A. Allison, T. D. Joyce, *Proc. Brit. Weed Control Conf.,* **1974**, 279; J. D. Riggleman, *Proc. South Weed Sci. Soc.* **31**, 141 (1978). Field trials: S. J. B. Hay, R. G. Jones, *Proc. Symp. Methods Weed Control* **1977**, 139; D. E. Yarborough *et al., Weed Sci.* **34**, 723 (1986). HPLC determn in soil and water: D. C. Bouchard, T. L. Lavy, *J. Chromatog.* **270**, 396 (1983). Residues in ground water: D. G. Neary, *South. J. Appl. Forestry* **7**, 217 (1983). Retention and degradation in soil: K. I. N. Jensen, E. R. Kimball, *Bull. Environ. Contam. Toxicol.* **38**, 232 (1987).

White crystalline solid. mp 97-100.5°. Very soluble in water, 330 g/l at 25°. LD_{50} orally in rats: 1690 mg/kg (Neary).

USE: Herbicide.

4735. Hexazole. *4-Cyclohexyl-3-ethyl-4H-1,2,4-triazole;* 3-ethyl-4-cyclohexyl-4*H*-1,2,4-triazole; Azoman; Triazol Ingelheim; T-156. $C_{10}H_{17}N_3$; mol wt 179.27. C 67.00%, H

9.56%, N 23.44%. Description: Behrens *et al., Klin. Wochenschr.* **16**, 944 (1937).

Stout prisms from ether, mp 89-90°. bp_{10} 227°. Can be distilled in vacuum. Slightly basic reaction. Freely sol in water; also sol in benzene, more sol in chloroform.

4736. Hexedine. *2,6-Bis(2-ethylhexyl)hexahydro-7a-methyl-1H-imidazo[1,5-c]imidazole;* 3,7-bis(2-ethylhexyl)-5-methyl-1,3,7-triazabicyclo[3.3.0]octane; Sterisol (Domestic). $C_{22}H_{45}N_3$; mol wt 351.62. C 75.15%, H 12.90%, N 11.95%. Prepn: McMillan, **Ger. pat. 1,209,249** corresp to **U.S. pat. 3,357,886** (1966, 1967, both to Warner-Lambert). Compound claimed but not described: Senkus, **U.S. pat. 2,393,826** (1946 to CSC).

Liquid, $bp_{0.025}$ 131°. n_D^{22} 1.4660.

THERAP CAT: Antibacterial.

4737. 3-Hexen-1-ol. Leaf alcohol; Blättcralkohol. $C_6H_{12}O$; mol wt 100.16. C 71.95%, H 12.08%, O 15.97%. $CH_3CH_2CH=CHCH_2CH_2OH$. Occurs in leaves of odoriferous plants (including shrubs and trees). Isoln from Japanese oil of peppermint: Walbaum *J. Prakt. Chem.* [2] **96**, 245 (1917); Walbaum, Rosenthal, *Ber. Schimmel* 1929 Jubiläums-Ausg., p 205; from raspberry juice: Bohnsack, *Ber.* **75**, 72 (1942). The natural product has the *cis* configuration: Crombie, Harper, *J. Chem. Soc.* **1950**, 873; Harper, Smith, *ibid.* **1955**, 1512.

Liquid. Strong odor resembling that of isoamyl alc, approaching the odor of green leaves when highly dil. d_{15}^{22} 0.846. bp 156-157°; bp_9 55-56°. n_D^{20} 1.4389.

4738. Hexestrol. *4,4'-(1,2-Diethyl-1,2-ethanediyl)bisphenol; 4,4'-(1,2-diethylethylene)diphenol; meso-*3,4-bis(*p*-hydroxyphenyl)-*n*-hexane; 4,4'-dihydroxy-γ,δ-diphenylhexane; 4,4'-dihydroxy-α,β-diethyldiphenylethane; dihydrodiethylstilbestrol; hexoestrol; Synthovo; Cycloestrol; Hexanoestrol; Hormoestrol; Syntrogène. $C_{18}H_{22}O_2$; mol wt 270.37. C 79.96%, H 8.20%, O 11.84%. From anethole HBr: Campbell *et al., Proc. Roy. Soc.* **B128**, 253 (1940); Bernstein, Wallis, *J. Am. Chem. Soc.* **62**, 2871 (1940); **U.S. pat. 2,357,985** (1944); Buu-Hoi, Hoán, *J. Org. Chem.* **14**, 1023 (1949). By the reduction of Grignard compds with cobaltous chloride: Kharasch *et al., J. Am. Chem. Soc.* **65**, 491 (1943). Proof of meso configuration: Wessely, Welleba, *Ber.* **74**, 777 (1941). Comprehensive description: H. Y. Aboul-Enein *et al.,* in *Analytical Profiles of Drug Substances* vol. 11, K. Florey, Ed. (Academic Press, New York, 1982) pp 347-374.

Needles from benzene, thin plates from dil alc, mp 185-188°. Freely sol in ether; sol in acetone, alcohol, methanol; slightly sol in benzene, chloroform. Sol in vegetable oils upon slight warming, also in dil solns of alkali hydroxides. Practically insol in water and in dil mineral acids.

Diacetate, *Retalon-Lingual.* Crystals, mp 137-139°.

Dipropionate, *Retalon Oleosum.* Crystals from petr ether, mp 127-128°.

Diphosphate, $C_{18}H_{24}O_8P_2$, *hexestrol 4,4'-diphosphoric ester, Cytostatin.* Prepn: **Brit.** pat **593,480** (1947 to Roche); Atherton *et al.,* **U.S.** pat. **2,490,573** (1949 to Hoffmann-La Roche).

THERAP CAT: Estrogen; antineoplastic (hormonal).
THERAP CAT (VET): Estrogen.

4739. Hexestrol Bis(β-diethylaminoethyl ether). *2,2'-[(1,2-Diethyl-1,2-ethanediyl)bis(4,1-phenyleneoxy)]bis[N,N-diethylethanamine]; 2,2'''-[(1,2-diethylethylene)bis(p-phenyleneoxy)]bis(triethylamine);* 4,4'-bis(β-diethylaminoethoxy)-α,β-diethyldiphenylethane; 3,4-bis[p-(β-diethylaminoethoxy)phenyl]hexane; α,α'-diethyl-4,4'-bis(β-diethylaminoethoxy)-bibenzyl. $C_{30}H_{48}N_2O_2$; mol wt 468.72. C 76.87%, H 10.32%, N 5.98%, O 6.83%. Prepn: Lowe *et al., J. Chem. Soc.* **1951,** 3286.

Dihydrochloride, $C_{30}H_{48}N_2O_2 \cdot 2HCl$, *Coralgil, Coralgina.* Needles from alc + ethyl acetate, mp 226-227°. Freely sol in water, methanol, chloroform, hot alc.

THERAP CAT: Vasodilator (coronary).

4740. Hexethal Sodium. *5-Ethyl-5-hexyl-2,4,6(1H,3H,5H)-pyrimidinetrione monosodium salt; sodium 5-ethyl-5-hexylbarbiturate;* Ortal Sodium; Hebaral. $C_{12}H_{19}N_2NaO_3$; mol wt 262.28. C 54.95%, H 7.30%, N 10.68%, Na 8.77%, O 18.30%. Prepn: A. W. Dox, **U.S.** pat. **1,624,546.** Anesthetic properties and toxicity data: C. M. Gruber, J. T. Brundage, *J. Pharmacol. Exp. Ther.* **60,** 439 (1937).

White or slightly yellowish powder, mp 126°. Bitter taste. Very sol in water; sol in alcohol. Insol in ether, benzene. Aq solns are alkaline to litmus. Aq solns are unstable and dec on standing. Boiling evolves ammonia and causes precipitation of a solid. MLD i.p. in rats: 240-250 mg/kg (Gruber, Brundage).

Caution: May be habit forming. This is a controlled substance (depressant) listed in the U.S. Code of Federal Regulations, Title 21 Parts 329.1 and 1308.13 (1995).

THERAP CAT: Sedative; hypnotic.
THERAP CAT (VET): Sedative; hypnotic.

4741. Hexetidine. *1,3-Bis(2-ethylhexyl)hexahydro-5-methyl-5-pyrimidinamine; 5-amino-1,3-bis(2-ethylhexyl)hexahydro-5-methylpyrimidine;* 1,3-bis(β-ethylhexyl)-5-methyl-5-aminohexahydropyrimidine; 5-amino-1,3-di(β-ethylhexyl)hexahydro-5-methylpyrimidine; Glypesin; Hexigel; Hexocil; Hexoral; Hextril; Oraldene; Sterisil; Steri/Sol. $C_{21}H_{45}N_3$; mol wt 339.61. C 74.27%, H 13.36%, N 12.37%. Prepn: Senkus, *J. Am. Chem. Soc.* **68,** 1611 (1946); **U.S.** pat. **2,415,047;** Bell, Necker, **U.S.** pat. **3,054,797** (1947, 1962 both to Comm. Solvents Corp.). Comprehensive description: G. Satzinger *et al., Anal. Profiles Drug Subs.* **7,** 277-295 (1978).

Liquid. d_{20}^{20} 0.8889. $bp_{0.4}$ 160°. n_D^{20} 1.4668. Sol in petr ether, methanol, benzene, acetone, ethanol, n-hexane, chloroform. Practically insol in water. pKa 8.3. Has good thermal stability.

THERAP CAT: Antifungal.
THERAP CAT (VET): Antifungal.

4742. Hexobarbital. *5-(1-Cyclohexen-1-yl)-1,5-dimethyl-2,4,6(1H,3H,5H)-pyrimidinetrione; 5-(1-cyclohexen-1-yl)-1,5-dimethylbarbituric acid;* methylhexabital; methexenyl; enhexymal; hexobarbitone; Citodon; Citopan; Cyclonal; Dorico; Evipal; Evipan; Hexanastab Oral; Noctivane; Sombucaps; Sombulex; Somnalert. $C_{12}H_{16}N_2O_3$; mol wt 236.27. C 61.00%, H 6.83%, N 11.86%, O 20.31%. Prepd by condensation of monomethylurea with methyl cyclohexenylmethylcyanoacetate in abs alcohol in the presence of sodium: **U.S.** pat. **1,947,944;** *cf.* **Ger.** pat. **595,175;** **Fr.** pat. **753,178.** Comparative clinical trial in anesthesia: D. W. Barron *et al., Brit. J. Anaesth.* **38,** 802 (1966). GC determn in human plasma: D. D. Breimer, J. M. Van Rossum, *J. Chromatog.* **88,** 235 (1974). Pharmacokinetics: N. P. E. Vermeulen *et al., Brit. J. Clin. Pharmacol.* **15,** 459 (1983); M. Van der Graaff *et al., Biopharm. Drug Dispos.* **7,** 265 (1986).

Prismatic crystals, practically tasteless, mp 145-147°. Practically insol in water. One gram dissolves in about 3 liters of water. Sol in methanol, hot ethanol, ether, chloroform, acetone, benzene, aq solns of alkali hydroxides, but not carbonates.

Sodium salt, $C_{12}H_{15}N_2NaO_3$, *hexobarbital soluble, Cyclonal Sodium, Dorico Soluble, Evipal Sodium, Evipan Sodium, Hexanastab, Hexenal, Methexenyl Sodium, Noctivane Sodium, Narcosan Soluble, Privenal.* White, bitter, very hygroscopic powder. Very sol in water, freely sol in alcohol, methanol, acetone. Practically insol in chloroform, ether, benzene. A soln in water absorbs carbon dioxide from the air, causing precipitation of the insol free acid. Aq and alcoholic solns are alkaline to litmus and phenolphthalein. The pH of a 10% w/v soln of hexobarbital soluble is about 11.5.

Caution: May be habit forming. This is a controlled substance listed in the U.S. Code of Federal Regulations, Title 21 Parts 329.1 and 1308.13 (1995).

THERAP CAT: Sedative, hypnotic. Anesthetic (intravenous).
THERAP CAT (VET): Sedative, hypnotic.

4743. Hexobendine. *3,4,5-Trimethoxybenzoic acid 1,2-ethanediylbis[(methylimino)-3,1-propanediyl] ester; 3,4,5-trimethoxybenzoic acid diester with 3,3'-[ethylenebis(methylimino)]di-1-propanol; N,N'-dimethyl-N,N'-bis[3-(3',4',5'-trimethoxybenzoxy)propyl]ethylenediamine;* hexabendin. $C_{30}H_{44}N_2O_{10}$; mol wt 592.69. C 60.80%, H 7.48%, N 4.73%, O 26.99%. Prepn: O. Kraupp, K. Schlögl, **Austrian** pat. **231,432** corresp to U.S. pat. **3,267,103** (1964, 1966 to OSSW). Pharmacology: Rudolph *et al., Arzneimittel-Forsch.* **20,** 637 (1970); Kolassa, Pfleger, *Biochem. Pharmacol.* **20,** 490 (1971); H. Rameis *et al., Arzneimittel-Forsch.* **30,** 671 (1980).

mp 75-77°.

Dihydrochloride, $C_{30}H_{44}N_2O_{10} \cdot 2HCl$, *ST-7090, Reoxyl, Ustimon, Andiamine.* Crystals, mp 170-174°. uv max: 267

nm. Freely sol in water, less sol in alc. Practically insol in ether.

THERAP CAT: Vasodilator (coronary).

4744. Hexocyclium Methyl Sulfate. *4-(2-Cyclohexyl-2-hydroxy-2-phenylethyl)-1,1-dimethylpiperazinium methyl sulfate (salt); N-(β-cyclohexyl-β-hydroxy-β-phenylethyl)-N¹-methylpiperazine methosulfate; N-(β-cyclohexyl-β-hydroxy-β-phenylethyl)-N¹-methylpiperazine dimethylsulfate; 4-(β-cyclohexyl-β-hydroxy-β-phenethyl)-1,1-dimethylpiperazinium methyl sulfate; Tral; Tralin.* $C_{21}H_{36}N_2O_5S$; mol wt 428.59. C 58.85%, H 8.47%, N 6.54%, O 18.67%, S 7.48%. Description: Helgren *et al., J. Am. Pharm. Assoc., Sci. Ed.* **46**, 639 (1957). Prepn: Weston, U.S. pat. **2,907,765** (1959 to Abbott).

Crystals, mp 200-210°. uv max (0.1N H_2SO_4): 252, 257, 263 nm. Soly in water about 50% w/v. Slightly sol in chloroform. Insol in ether.

THERAP CAT: Anticholinergic.

4745. Hexoprenaline. *4,4'-[1,6-Hexanediylbis[imino(1-hydroxy-2,1-ethanediyl)]]bis-1,2-benzenediol; α,α'-[hexamethylenebis(iminomethylene)]bis[3,4-dihydroxybenzyl alcohol]; N,N'-bis[2-(3,4-dihydroxyphenyl)-2-hydroxyethyl]hexamethylenediamine;* BYK 1512. $C_{22}H_{32}N_2O_6$; mol wt 420.51. C 62.84%, H 7.67%, N 6.66%, O 22.83%. $β_2$-Adrenergic agonist. Prepn: Schmid *et al.,* **Austrian** pat. **241,436** corresp to U.S. pat. **3,329,709** (1965, 1967, both to OSSW); Schmid *et al.,* **Ger.** pat. **1,215,729** (1966 to Lentia GmbH). Pharmacology: Thiede *et al., Arzneimittel-Forsch.* **21**, 416 (1971). Clinical studies: Schindl, *ibid.* **20**, 1755 (1970); several authors in *Wien. Klin. Wochenschr.* **83**, 75, 80, 101, 114, 117, 130 (1971). Clinical evaluation for interruption of labor: J. Lipshitz *et al., J. Reprod. Med.* **31**, 1023 (1986).

Crystals, mp 162-165° (hemihydrate).
Dihydrochloride, $C_{22}H_{32}N_2O_6$·2HCl, *ST-1512, Ipradol.* Crystals from methanol-ether, mp 197.5-198°.
Sulfate, $C_{22}H_{32}N_2O_6$·H_2SO_4, *Bronalin, Delaprem, Etoscol, Gynipral, Ipradol, Leanol.* Crystals from water-alcohol, mp 222-228°.

THERAP CAT: Bronchodilator; tocolytic.

4746. Hexylcaine Hydrochloride. *1-(Cyclohexylamino)-2-propanol benzoate (ester) hydrochloride;* D-109; Cyclaine. $C_{16}H_{24}ClNO_2$; mol wt 297.82. C 64.53%, H 8.12%, Cl 11.90%, N 4.70%, O 10.74%. Prepn: Cope, Hancock, *J. Am. Chem. Soc.* **66**, 1453 (1944); Cope, U.S. pat. **2,486,374** (1949 to Sharp & Dohme). Pharmacokinetics: R. M. Rodgers *et al., Res. Commun. Chem. Pathol. Pharmacol.* **29**, 99 (1980).

Crystals from abs alc, mp 177-178.5°. Soluble in water to the extent of about 12% w/w. A 1% soln is stable to boiling and autoclaving for sterilization purposes.

THERAP CAT: Anesthetic (local).
THERAP CAT (VET): Anesthetic (local).

4747. 2-Hexyldecanoic Acid. $C_{16}H_{32}O_2$; mol wt 256.43. C 74.94%, H 12.58%, O 12.48%. Prepn: Lederer *et al., Bull. Soc. Chim. France* **1952**, 413.

Viscous oil, $bp_{0.02}$ 140-150°. n_D^{24} 1.4432.
Sodium salt, *Devaricin.* White powder. Sol in water.

THERAP CAT: Sclerosing agent.

4748. Hexylene Glycol. *2-Methyl-2,4-pentanediol; α,α,α'-trimethyltrimethyleneglycol; pinakon.* $C_6H_{14}O_2$; mol wt 118.18. C 60.98%, H 11.94%, O 27.08%. Prepn: Franke, *Monatsh.* **22**, 1067 (1901); Leopold, **Ger.** pat. **486,767** (1925 to I. G. Farben), *Frdl.* **16**, 679; Adkins, Cramer, *J. Am. Chem. Soc.* **52**, 4349 (1930); Arundale, Mikeska, U.S. pat. **2,367,324** (1945 to Standard Oil). Toxicity study: Smyth, Carpenter, *J. Ind. Hyg. Toxicol.* **30**, 63 (1948).

Liquid. Mild sweetish odor. d_{15}^{15} 0.924. bp_{760} 198°; bp_{10} 97°. Flash pt: ca. 93° (200°F). n_D^{20} 1.4276. Dipole moment: 2.8. Viscosity at 20° = 34 cp. Sol in water, alcohol, ether, lower aliphatic hydrocarbons. LD_{50} orally in rats: 4.70 g/kg (Smyth, Carpenter).

Diacetate, $C_{10}H_{18}O_4$. Liquid. bp_{12} 95°.

USE: Cosmetics, hydraulic brake fluids (as coupling agent to castor oil).

4749. Hexyl Methyl Ketone. *2-Octanone;* methyl hexyl ketone. $C_8H_{16}O$; mol wt 128.21. C 74.94%, H 12.58%, O 12.48%. $CH_3(CH_2)_5COCH_3$.

Liquid; apple odor; camphor taste. d_4^{20} 0.820. mp −16°. bp 172-173°. n_D^{20} 1.41512. Insol in water. Miscible with alc, ether.

4750. 4-Hexylresorcinol. *4-Hexyl-1,3-benzenediol;* 4-hexyl-1,3-dihydroxybenzene; ST-37; Ascaryl; Caprokol; Crystoids; Gelovermin; Sucrets; Worm-Agen. $C_{12}H_{18}O_2$; mol wt 194.27. C 74.19%, H 9.34%, O 16.47%. Prepd by reduction of hexanoylresorcinol with zinc amalgam + dil HCl: Dohme *et al., J. Am. Chem. Soc.* **48**, 1688 (1926); Twiss, *ibid.* 2206; **Ger.** pats. **488,419** and **489,117** (1929 to Sharp & Dohme). Toxicity: Lamson *et al., J. Pharmacol. Exp. Ther.* **53**, 198 (1935).

Pale yellow, heavy liq becoming solid on standing at room temp. Needles from benzene or petr ether, mp 67.5-69°. $bp_{6.7}$ 178-180°; bp_{13-14} 198-200°; bp_{760} 333-335°. Pungent odor; sharp astringent taste. Sol in ether, chloroform, acetone, alcohol, vegetable oils; slightly sol in petr ether; sol in about 2000 parts water. LD_{50} orally in rats: 550 mg (Lamson).

Caution: Concd solns can cause irritation of skin, mucous membranes: *Clinical Toxicology of Commercial Products,* R. E. Gosselin *et al.,* Eds. (Williams & Wilkins, Baltimore, 5th ed., 1984) Section II, p 190.

THERAP CAT: Anthelmintic (Nematodes). Antiseptic (topical).
THERAP CAT (VET): Anthelmintic.

4751. Hexythiazox. *trans-5-(4-Chlorophenyl)-N-cyclohexyl-4-methyl-2-oxo-3-thiazolidinecarboxamide; trans-4-*

methyl-5-(4-chlorophenyl)-3-cyclohexylcarbamoyl-2-thia-
zolidone; HTZ; NA-73; DPX-Y5893-9; Nissorun; Acarflor;
C'esar; Savey; Zeldox. $C_{17}H_{21}ClN_2O_2S$; mol wt 352.88. C
57.86%, H 6.00%, Cl 10.05%, N 7.94%, O 9.07%, S 9.09%.
Prepn: I. Iwataki et al., Ger. pat. 3,037,105; eidem, U.S.
pat. 4,442,116 (1981, 1984 to Nippon Soda). Miticidal activity: M. A. Hoy, Y.- L. Ouyang, J. Econ. Entomol. 79,
1377 (1986); C. Welty et al., ibid. 81, 586 (1988). Residue
determn in crops: M. Tokieda et al., J. Pestic. Sci. 12, 711
(1987), C.A. 108, 162770b (1988). Brief description of physical properties, toxicity and efficacy: Nippon Soda Co.,
Ltd., Japan. Pest. Info. 44, 21 (1984).

White odorless crystals, mp 105.5°. Vapor pressure at
20°: 2.54 × 10^{-8} mm Hg. Soly at 20° (g/100 ml): acetone
16, chloroform 137.9, methanol 2.06, n-hexane 0.39, xylene
36.2, acetonitrile 2.86; water 0.5 ppm. LD_{50} in male, female
mice, male, female rats (mg/kg): all >5000 orally; all
>5000 dermally (Nippon Soda Co.).

USE: Acaricide.

4752. Hippuric Acid. N-Benzoylglycine; benzoylaminoacetic acid; benzamidoacetic acid. $C_9H_9NO_3$; mol wt
179.18. C 60.33%, H 5.06%, N 7.82%, O 26.79%. C_6H_5-
CONHCH$_2$COOH. Present in the urine of herbivorous
animals; also in smaller amounts in human urine. Prepd
from benzoyl chloride and glycine in NaOH soln: Ingersoll,
Babcock, Org. Syn. coll. vol. II, 328 (1943).

Crystals. mp 187-188°. One gram dissolves in about 250
ml cold water, 1000 ml chloroform, 400 ml ether, 60 ml
amyl alcohol; slightly sol in cold, freely in hot alcohol or
hot water; also sol in aq soln of sodium phosphate. Practically insol in benzene, carbon disulfide, petr ether.

Ammonium salt, $NH_4C_9H_8NO_3$, crystals. Freely sol in
water; sol in alcohol.

Potassium salt monohydrate, $KC_9H_{10}NO_4.H_2O$, cryst
powder. Very sol in water; sol in alcohol.

4753. Hirsutic Acid C. [1aR-(1aα,2β,3aβ,3bα,5α,6aα,-
7aS*)]-Decahydro-2-hydroxy-3a,5-dimethyl-3-methylenecyclo-
penta[4,5]pentaleno[1,6a-b]oxirene-5-carboxylic acid; 6β,6aβ-
epoxy-2,3,3aα,3b,4,5,6,6a,7,7aα-decahydro-5β-hydroxy-2β,-
3bβ-dimethyl-4-methylene-1H-cyclopenta[a]pentalene-2-car-
boxylic acid. $C_{15}H_{20}O_4$; mol wt 264.32. C 68.16%, H 7.63%,
O 24.21%. Antibiotic metabolite isolated from the fungus
Stereum hirsutum: Heatley et al., Brit. J. Exp. Pathol. 28, 35
(1947). Structure and stereochemistry: Comer et al., Chem.
Commun. 1965, 310; Comer, Trotter, J. Chem. Soc. (B)
1966, 11; Comer et al., Tetrahedron 23, 4761 (1967). Synthesis of (±)-form: Hashimoto et al., Tetrahedron Letters
1974, 3745; M. Yamazaki et al., Chem. Letters 1981, 1245.
Stereosynthesis: B. M. Trost et al., J. Am. Chem. Soc. 101,
1284 (1979); of (+)-form: M. Shibasaki et al., Tetrahedron
Letters 23, 5311 (1982).

Prisms from ethanol, mp 179-182°. $[\alpha]_D^{23}$ +116° (c =
1.05).
Methyl ester, $C_{16}H_{22}O_4$, colorless prisms, mp 161-162°.
$[\alpha]_D^{20}$ +119° (c = 2.25).

4754. Hirudin. Exhirud; Exhirudine; Hirudex. Anticoagulant protein extracted from leeches. The medicinal
leech of Central Europe, Hirudo medicinalis Linn. (Sangui-

suga officinalis Savigny) yields about 3 mg per head. American leeches contain much less hirudin. Several isoforms
have been identified. Isoln: Haycraft, Arch. Exp. Pathol.
Pharmakol. 18, 209 (1884); Bock, ibid. 41, 160 (1898);
Franz, ibid. 49, 342 (1903); Ger. pats. 136,103 (1902); 150,-
805 (1903); Bodong, Arch. Exp. Pathol. Pharmakol. 52, 242
(1905); Abel et al., J. Pharmacol. 5, 302 (1913/14); Marshall, ibid. 7, 157 (1915); Yanagisawa, Yokoi, Proc. Imp.
Acad. Japan 14, 69 (1938); Kirsanov, Bystritskaya, Biokhimiya 5, 596 (1940), C.A. 36, 6308. Initial characterization: F. Markwardt, Naturwiss. 42, 537 (1955). Revised
extraction procedure: Schremmer et al., Monatsh. 87, 87
(1956). Purification: Jutisz et al., Bull. Soc. Chim. Biol. 45,
55 (1963), C.A. 58, 10426c (1963). Hirudin is a single chain
polypeptide containing 65 amino acids; mol wt approximately 7000 daltons. C-Terminal sequence: Llosa et al.,
Biochim. Biophys. Acta 93, 40 (1964); N-terminal sequence:
Graf et al., ibid. 310, 416 (1973). Hirudin is characterized
by a high proportion of dicarboxylic acids, which explains
its acid character, and by the absence of tryptophan, methionine and arginine: Llosa et al., Bull. Soc. Chim. Biol. 45,
69 (1963). Complete amino acid sequence: T. E. Petersen
et al. in Protides of the Biological Fluids, vol. 23, H. Peeters,
Ed. (Pergamon Press, London, 1976) pp 145-149. A specific inhibitor of thrombin, q.v., hirudin inhibits coagulation
in the initial stages of clotting and does not require presence
of other coagulation factors or plasma constituents: F.
Markwardt, Methods Enzymol. 19, 924 (1970). Pharmacokinetics in humans: F. Markwardt et al., Thromb. Haemostasis 52, 160 (1984). Chromogenic assay in plasma: U.
Griessbach et al., Thromb. Res. 37, 347 (1985). Review of
pharmacology: F. Markwardt, Biomed. Biochim. Acta 44,
1007 (1985); and therapeutic potential: P. H. Johnson, Ann.
Rev. Med. 45, 165-177 (1994); of therapeutic use in myocardial infarction: U. Zeymer, K.-L. Neuhaus, Drug Safety 12,
234-239 (1995). Series of articles on development, pharmacology, and therapeutic potential of natural and recombinant hirudins: Semin. Thromb. Hemostasis 15, 261-333
(1989).

Gray or white flakes or powder. pI 3.9. Sol in water,
physiol saline soln, in pyridine. Practically insol in alcohol,
ether, acetone, benzene. Deteriorates on storage in sealed
ampuls, on exposure to heat and when in soln with dil
acids.

Recombinant desulfatohirudin variant 1, 63-desulfohiru-
din (Hirudo medicinalis isoform HV1); desirudin, r-hirudin,
rec-hirudin, ds-hirudin, CGP-39393, Revasc. Identical in
amino acid sequence to natural hirudin variant 1 (HVI)
except that it lacks a sulfate group on Tyr at residue 63.
Production by recombinant DNA technology: M. Liersch
et al., Eur. pat. Appl. 168,342 (1986 to Ciba-Geigy); eidem,
U.S. pat. 5,422,249 (1995 to Ciba-Geigy; UCP Gen-
Pharma); E. Fortkamp et al., DNA 5, 511 (1986); J. Dodt et
al., FEBS Letters 202, 373 (1986). Physicochemical properties: A. Electricwala et al., Thromb. Haemostasis 63, 499
(1990). Clinical pharmacodynamics and pharmacokinetics:
M. Verstraete et al., J. Am. Coll. Cardiol. 22, 1080 (1993).
Comparative clinical evaluation in coronary angioplasty: A.
A. van den Bos et al., Circulation 88, [Pt. 1] 2058 (1993).
Clinical evaluation as adjunctive therapy in acute myocardial infarction: C. P. Cannon et al., J. Am. Coll. Cardiol. 23,
993 (1994).

THERAP CAT: Antithrombotic.

4755. Histaminase. Diamine oxidase; benzylamine oxidase; histamine deaminase; histamine oxidase; E.C. 1.4.3.6.
A copper contg enzyme present in tissues, esp in kidneys
and in the intestinal mucosa. Attacks diamines such as
histamine in the body by oxidative deamination: Best, J.
Physiol. 67, 256 (1929); Zeller, Helv. Chim. Acta 21, 880
(1938); Advan. Enzymol. 2, 93 (1942). Extraction from hog
kidneys: Swedin, Acta Med. Scand. 114, 21 (1943). Identity
with diamine oxidase: Zeller, Fed. Proc. 24, 764 (1965).
Appears to have the general properties of a flavoprotein.
Reviews: E. A. Zeller "Diamine Oxidases" in The Enzymes,
vol. 8, P. D. Boyer et al., Eds. (Academic Press, New York,
2nd ed., 1963) pp 313-335; Buffoni, Pharmacol. Rev. 18,
1163-1199 (1966); Hansson, Scand. J. Clin. Lab. Invest. vol.
31, suppl. 129, 7 (1973).

Has been sold in enteric-coated tablets as *Torantil* (*Torantyl*) and *Metoryl* (*Metoral*).

4756. Histamine. *1H-Imidazole-4-ethanamine; 2-(4-imidazolyl)ethylamine;* 4-imidazoleethylamine; 5-imidazole-ethylamine; β-aminoethylimidazole; β-aminoethylglyoxaline. $C_5H_9N_3$; mol wt 111.15. C 54.03%, H 8.16%, N 37.81%. A potent vasodilator found in normal tissues and blood. Occurs widely in nature as a result of putrefactive processes. Stimulates the secretion of pepsin and acid by the stomach; eating and vagal stimulation cause the release of histamine from gastric mucosa. The flavoprotein diamine oxidase converts histamine to the corresponding aldehyde and ammonia. Some undegraded histamine, in the form of the *N*-acetyl and *N'*-methyl derivs, is excreted in the urine. Produced from histidine by *pneumococci* or *B. coli:* Lévy-Brühl, Ungar, *Ann. Inst. Pasteur* **61**, 828 (1938). Synthesis from imidazolylpropionic acid: Pyman, *J. Chem Soc.* **99**, 668 (1911); Koessler, Hanke, *J. Am. Chem. Soc.* **40**, 1716 (1918); from α-aminobutyrolactone: Garforth, Pyman, *J. Chem. Soc.* **1935**, 489. Diagnostic use as gastric stimulant: O. M. Laudano, *Gastroenterol.* **50**, 653 (1966); in pheochromocytoma: T. Nakai, R. Yamada, *J. Clin. Endocrinol. Met.* **57**, 19 (1983). Role of histamine in induction of drinking by food intake: F. S. Kraly, *Nature* **302**, 65 (1983). *Review:* H. Wetterqvist, *Handb. Exp. Pharmakol.* **18**(pt. 2), 131-150 (1978). Toxicity: K. Nagai *et al., Arzneimittel-Forsch.* **17**, 1575 (1967).

Deliquescent needles from chloroform, mp 83-84°. bp_{18} 209-210°. Freely sol in water, alcohol, hot chloroform. Sparingly sol in ether. LD_{50} i.p. in mice: 2020 mg/kg (Nagai).

Dihydrochloride, $C_5H_9N_3.2HCl$, *Amin-Glaukosan, Imido, Imadyl (obsolete), Ergamine, Peremine.* Prisms from ethanol, mp 244-246°. Freely sol in water, methanol; sol in ethanol.

Phosphate, $C_5H_{15}N_3O_8P_2$, *Histapon.* Prismatic crystals, mp 140°. Readily sol in water.

THERAP CAT: Diagnostic aid (gastric secretion, pheochromocytoma). Antiallergic (hyposensitization therapy).

4757. Histapyrrodine. *N-Phenyl-N-(phenylmethyl)-1-pyrrolidineethanamine; 1-[2-(N-benzylanilino)ethyl]pyrrolidine; N-pyrrolidylethyl-N-phenylbenzylamine; N-benzyl-N-phenylpyrrolidinoethylamine; N-benzyl-N-pyrrolidinoethylaniline.* $C_{19}H_{24}N_2$; mol wt 280.41. C 81.38%, H 8.63%, N 9.99%. Prepn: Hopff *et al.,* U.S. pat. **2,623,880** (1952). Chemical properties: Auterhoff, *Arch. Pharm.* **284**, 123 (1951).

bp_1 198-205°.

Hydrochloride, $C_{19}H_{24}N_2.HCl$, *Calcistin, Domistan, Luvistin.* Crystals from alc, mp 196-197°. Soly in water (18°); 2 g/100 ml.

THERAP CAT: Antihistaminic.

4758. Histidine. *L-Histidine;* His; H; (*S*)-α-amino-1*H*-imidazole-4-propanoic acid; α-amino-4(or 5)-imidazolepropionic acid; glyoxaline-5-alanine. $C_6H_9N_3O_2$; mol wt 155.16. C 46.45%, H 5.85%, N 27.08%, O 20.62%. Essential amino acid for human development; precursor to histamine, *q.v.* and a component of carnosine, *q.v.*. Independently isolated in 1896 by both A. Kossel and S. G. Hedin, *Zeit. Physiol.* **22**, 176, 191 (1896). Early chemistry and biochemistry: *Amino Acids and Proteins*, D. M. Greenberg, Ed.

(Charles C. Thomas, Springfield, IL, 1951) 950 pp., *passim;* J. P. Greenstein, M. Winitz, *Chemistry of the Amino Acids* vols 1-3 (John Wiley and Sons, Inc., New York, 1961) pp. 1971-1995, *passim.* NMR spectra: D. H. Sachs *et al., J. Biol. Chem.* **246**, 6576 (1971). HPLC determn in dipeptides: E. Kasziba *et al., J. Chromatog.* **432**, 315 (1988). Often found at the active site of proteins: W. T. Morgan *et al., J. Biol. Chem.* **268**, 6256 (1993); J.-W. P. Boots *et al., Biochim. Biophys. Acta* **1248**, 27 (1995). Role in proton translocation — a model: J. E. Morgan *et al., J. Bioenerg. Biomembr.* **26**, 599 (1994). Review of metabolism: F. B. Stifel, R. H. Herman, *Am. J. Clin. Nutr.* **24**, 207-217 (1971). Review of role in nutrition: *eidem, ibid.* **25**, 182-185 (1972); L. P. Mercer *et al., Nutrition* **6**, 273-277 (1990).

Sweet needles or plates, dec 287° (softens at 277°). $[\alpha]_D^{20}$ −10.9° (c = 0.77 in 0.5*N* NaOH); $[\alpha]_D^{25}$ −38.95° (c = 0.75 to 3.77); $[\alpha]_D^{25}$ +13.34° (c = 1.00-4.05 in 6.1*N* HCl). pK_1' 1.78; pK_2' 5.97; pK_3' 8.97. Soly in water at 25°: 41.9 g/l. Insol in common neutral solvents except water.

Monohydrochloride, $C_6H_9N_3O_2.HCl$, *Ecristidine, Laristine, Larostidin, Plexamine.* Rhombic crystals, dec 251-252° (also forms a monohydrate, mp 80°, anhydrous, mp at 140°). $[\alpha]_D^{26}$ +8.0° (c = 2 in 3 mols HCl). Fairly sol in water. Insol in alcohol, ether.

DL-Form, quadrilateral plates or tetragonal prisms, dec 285°. Sol in water.

4759. Histones. Small chromosomal proteins (mol wt 12,000-20,000) possessing an open, unfolded structure, attached to DNA of cell nuclei by ionic linkages. First isolated from bird erythrocytes: Kossel, *Z. Physiol. Chem.* **8**, 511 (1884). Prepd also from cell nuclei of calf thymus, spleen, and liver, liver and spleen of leukemic rats, bovine spermatozoa: Hnilica, *Biochem. J.* **82**, 123 (1962); Berry, Mayer, *Exp. Cell Res.* **20**, 116 (1960). Classification is based on relative amounts of lysine and arginine: Histone I is very rich in lysine and has several subtypes; histone II, moderately rich in lysine with two subtypes; histone III, moderately rich in arginine, contains cysteine; histone IV, very rich in arginine and in glycine. Histones of the same type obtained from various plant and animal sources are very similar in amino acid sequence. Amino acid composition of animal histones: Hnilica *et al., Biochim. Biophys. Acta* **124**, 109 (1966); of plant histones and similarity of plant and animal histones: Fambrough, Bonner, *Biochemistry* **5**, 2563 (1966). Complete amino acid sequence of calf thymus histone IV: DeLange *et al., J. Biol. Chem.* **244**, 319 (1969); Ogawa *et al., ibid.* 4387; of histone III: DeLange *et al., Proc. Nat. Acad. Sci. USA* **69**, 882 (1972); *eidem, J. Biol. Chem.* **248**, 3261 (1973); Hooper *et al., ibid.* 3275; of the two subtypes of histone II: Iwai *et al., Nature* **226**, 1056 (1970); Yeoman *et al., J. Biol. Chem.* **247**, 6018 (1972). Toxicity study: Starbuck *et al., Arch. Int. Pharmacodyn. Ther.* **165**, 374 (1967). Review of structural studies: DeLange, *Accounts Chem. Res.* **5**, 368 (1972). Review of histone in the chromosome structure: Bradbury *et al., Ann. N.Y. Acad. Sci.* **222**, 266 (1973). Review of biological functions: Binner, Garrard, *Life Sci.* **14**, 209 (1974). General review: Butler *et al., Progr. Biophys. Mol. Biol.* **18**, 211 (1968); R. J. DeLange, E. L. Smith in *Proteins* vol. **4**, H. Neurath, R. L. Hill, Eds. (Academic, New York, 3rd ed., 1979) pp 134-243.

Histones are susceptible to enzymatic cleavage; sol in an $Hg_2SO_4-H_2SO_4$ medium. Infrared spectra: de Lozé, *Compt. Rend.* **246**, 417 (1958). LD_{50} i.v. in rats of lysine-rich, slightly lysine-rich, and arginine-rich histones: 90, 60-70, 60 mg/kg (Starbuck).

4760. Histrelin. *6-[1-(Phenylmethyl)-D-histidine]-9-(N-ethyl-L-prolinamide)-10-deglycinamide luteinizing hormone-releasing factor (pig);* 5-oxo-L-prolyl-L-histidyl-L-tryptophyl-L-seryl-L-tyrosyl-N^7-benzyl-D-histidyl-L-leucyl-L-arginyl-N-ethyl-L-prolinamide; L-pyroglutamyl-L-histidyl-L-tryptophyl-L-seryl-L-tyrosyl-D-N^{im}-benzyl-histidyl-L-leucyl-L-arginyl-L-proline ethylamide; [(im-Bzl)-D-His[6],Pro[9]-NEt]-gonadotropin-releasing hormone; ORF-17070; RWJ-17070. $C_{66}H_{86}N_{18}O_{12}$; mol wt 1323.52. C 59.90%, H 6.55%, N 19.05%, O 14.51%. Synthetic nonapeptide agonist analog of LH-RH, q.v. Prepn: J. E. F. Rivier, W. W. Vale, Jr., Eur. pat. Appl. 21,620; eidem, U. S. pat. 4,244,946 (both 1981 to Salk Inst.). See also: J. E. Rivier, W. W. Vale, Life Sci. 23, 869 (1978). Pharmacology: Y-Q. Cao et al., Int. J. Androl. 5, 158 (1982); J. W. Gunnet et al., J. Endocrinol. 131, 211 (1991). HPLC purification: J. Rivier et al., J. Chromatog. 288, 303 (1984). Amino acid sequence: C. J. Shaw, M. L. Cotter, Chromatographia 21, 197 (1986). Clinical evaluations in polycystic ovarian disease: K. Steingold et al., J. Clin. Endocrinol. Metab. 65, 773 (1987); in endometriosis: idem et al., Obstet. Gynecol. 69, 403 (1987); in porphyria: K. E. Anderson et al., Arch. Int. Med. 150, 1469 (1990); in PMS: J. F. Mortola et al., J. Clin. Endocrinol. Metab. 72, 252A (1991). Review of pharmacology and therapeutic use in central precocious puberty: L. B. Barradell, D. McTavish, Drugs 45, 570-588 (1993).

5-oxoPro—His—Trp—Ser—Tyr—NH----C---C— Leu—Arg—ProNHCH₂CH₃

$[\alpha]_D^{20}$ −33.9° (c = 1 in acetic acid).
Acetate, $C_{66}H_{86}N_{18}O_{12}\cdot xC_2H_4O_2\cdot yH_2O$, Supprelin.
THERAP CAT: Treatment of central precocious puberty.

4761. HMPA. Hexamethylphosphoric triamide; hexamethylphosphoramide; hexametapol; hempa; HMPT; ENT-50882. $C_6H_{18}N_3OP$; mol wt 179.20. C 40.21%, H 10.12%, N 23.45%, O 8.93%, P 17.28%. $[(CH_3)_2N]_3PO$. Prepn: Saul, Godfrey, U.S. pat. 2,752,392 (1956 to Monsanto); Godfrey, U.S. pat. 2,852,550 (1958 to Monsanto); Miller, Lomonte, U.S. pat. 3,084,190 (1963 to Dow); Vetter, Noeth, Ber. 96, 1308 (1963). Solvation effects: J. E. Dubois, A. Bienvenue, Tetrahedron Letters 1966, 1809. Toxicity studies: T. B. Gaines, Toxicol. Appl. Pharmacol. 14, 515 (1969). Carcinogenicity studies: J. A. Zapp, Science 190, 422 (1975); IARC Monographs 15, 211 (1977). Review: H. Normant, Angew. Chem. Int. Ed. 6, 1047 (1967).
Colorless liquid, completely miscible with water. mp 7.20°. bp₇₆₀ 235°. bp₁₁ 105-107°, bp₆ 97-99°; bp₂.₅ 78°. n_D^{21} 1.4572. d₂₀ 1.03. Dipole moment at 25°: 5.37D. LD₅₀ in male, female rats (mg/kg): 2650, 3360 orally (Gaines).
Note: This substance may reasonably be anticipated to be a carcinogen: Seventh Annual Report on Carcinogens (PB95-109781, 1994) p 229.
USE: Aprotic solvent in organic synthesis. Solvent for polymers, gases; polymerization catalyst; thermal stabilizer in polystyrene; protective additive for polyvinyl and polyolefin resins against uv light degradation. De-icing additive for jet fuels. Chemosterilant for a number of insect pests; chemical mutagen.

4762. Holarrhenine. (3β,12β)-3-(Dimethylamino)con-5-enin-12-ol; 12β-hydroxyconessine. $C_{24}H_{40}N_2O$; mol wt 372.59. C 77.37%, H 10.82%, N 7.52%, O 4.29%. From leaves and bark of Holarrhena congolensis Stapf, Apocynaceae. Isoln: Pyman, J. Chem. Soc. 115, 163 (1919).

Structure: Uffer, Helv. Chim. Acta 39, 1834 (1956); van Hove, Tetrahedron 7, 104 (1959).
Needles from acetic acid or ethyl acetate, mp 197-198°. $[\alpha]_D^{24}$ −7° (chloroform); $[\alpha]_D^{24}$ +9° (alcohol). Practically insol in water. Sol in alcohol, chloroform, slightly in ether, acetone, ethyl acetate.
O-Acetylholarrhenine, $C_{26}H_{42}N_2O_2$, crystals from pentane or acetone, mp 177-178°.

4763. Holmium. Ho; at. wt 164.93032; at. no. 67; valence 3. A rare earth metal of the yttrium group; member of the lanthanide series. Naturally occurring isotope (mass number): 165; known artificial radioactive isotopes: 146-164; 166-170. Estimated abundance in earth's crust 1.15-1.4 ppm; occurs in rare earth minerals. Discovered by Soret in 1878 and independently by Cleve in 1879: Cleve, Compt. Rend. 89, 478, 708 (1879). Purification of salts: Holmberg, Z. Anorg. Chem. 71, 226 (1911). Separation from erbium: Driggs, Hopkins, J. Am. Chem. Soc. 47, 363 (1925); from other rare earths: Spedding et al., ibid. 69, 2812 (1947); 76, 2557 (1954). Absorption spectrum: Severin, Z. Physik 125, 455 (1949). Analysis by means of emission spectra: Smith, Wiggins, Analyst 74, 95 (1949). Toxicity study: Haley et al., Toxicol. Appl. Pharmacol. 8, 37 (1966). Reviews of prepn, properties and compds: The Rare Earths, F. H. Spedding, A. H. Daane, Eds. (Krieger, Huntington, N.Y., 1971, reprint of 1961 ed.) 641 pp; Hulet, Bode, "Separation Chemistry of the Lanthanides and Transplutonium Actinides" in MTP Int. Rev. Sci.: Inorg. Chem., Ser. One, vol. 7, K. W. Bagnall, Ed. (University Park Press, Baltimore, 1972) pp 1-45; Moeller, "The Lanthanides" in Comprehensive Inorganic Chemistry, vol. 4, J. C. Bailar, Jr. et al., Eds. (Pergamon Press, Oxford, 1973) pp 1-101; F. H. Spedding in Kirk Othmer Encyclopedia of Chemical Technology vol. 19, (John Wiley & Sons, New York, 3rd ed., 1982) pp 833-854; Chemistry of the Elements, N. N. Greenwood, A. Earnshaw, Eds. (Pergamon Press, New York, 1984) pp 1423-1449. Brief review of properties: G. T. Seaborg, Radiochim. Acta 61, 115-122 (1993).
Metal; hexagonal close-packed crystals. d 8.7947. mp 1474°. bp 2700°. Heat of fusion: 16.874 kJ/mol. Heat of sublimation (25°): 300.8 kJ/mol. Forms yellow-green salts.
Oxide, Ho_2O_3, holmia. Yellow solid. Obtained by igniting the hydroxide, nitrate, sulfate, oxalate. Dissolves in acids with formation of a yellow salt.
Chloride, $HoCl_3$, bright yellow, crystalline solid, mp 718°. Formed by heating the hydrated salt in a current of hydrogen chloride at 350°. LD₅₀ in mice: 560 mg/kg i.p.; 7.2 g/kg orally (Haley).
Bromide, $HoBr_3$, mp 914°.
Iodide, HoI_3, light-yellow solid. mp 1010° ±10°. Obtained by passing hydrogen iodide over the anhyd chloride at 600°: Jantsch et al., Z. Anorg. Chem. 207, 353 (1932).

4764. Holomycin. N-(4,5-Dihydro-5-oxo-1,2-dithiolo-[4,3-b]pyrrol-6-yl)acetamide; N-(4,5-dihydro-5-oxo-1,2-dithiolo[4,3-b]pyrrol-6-yl)-N-methylformamide; 6-acetamido-1,2-dithiolo[4,3-b]pyrrol-5(4H)-one; N-demethylthiolutin. $C_7H_6N_2O_2S_2$; mol wt 214.27. C 39.24%, H 2.82%, N 13.07%, O 14.93%, S 29.93%. Antibiotic substance produced by a strain of Streptomyces griseus (Krainski) Waksman et Henrici. Isoln and structure: Ettlinger et al., Helv. Chim. Acta 42, 563 (1959). Crystal structure: Jensen, J. Antibiot. 22, 231 (1969). Activity: Von Daehne et al., ibid. 233. Manuf process: Gaumann et al., U.S. pat. 3,014,922 (1961 to Ciba). Total synthesis: Schmidt, Geiger, Ann. 664, 168 (1963); Büchi, Lukas, J. Am. Chem. Soc. 86, 5654 (1964); Hagio, Yoneda, Bull. Chem. Soc. Japan 47, 1484 (1974); J. E. Ellis et al., J. Org. Chem. 42, 2891 (1977).

Consult the Name Index before using this section.

Orange-yellow flakes from methanol + ethyl acetate, dec 268-270. uv max: 245, 302, 290 nm (log ε 3.78, 3.51, 4.05). LD$_{50}$ i.v. in mice: 5-10 mg/kg (Von Daehne).

4765. Homarine. *2-Carboxy-1-methylpyridinium inner salt;* 1-methyl-2-pyridinium carboxylate; picolinic acid *N*-methylbetaine; *N*-methylpicolinic acid. C$_7$H$_7$NO$_2$; mol wt 137.14. C 61.31%, H 5.14%, N 10.21%, O 23.33%. Quaternary ammonium base occurring in tissues of marine animals such as sea urchin, *Arabacia pustulosa,* jellyfish, *Velella spirans,* lugworm, *Arenicola marina.* Isoln: Hoppe-Seyler, *Z. Physiol. Chem.* **222,** 105 (1933). Prepn: Kosower, Patton, *J. Org. Chem.* **26,** 1318 (1961); Quast, Schmitt, *Ann.* **732,** 64 (1970). Review of distribution and function in animals: Brodzicki, *Kosmos (Warsaw)* **16A,** 431-438 (1967), *C.A.* **68,** 18617v (1968).

Crystals from methanol. Does not have a melting point, slowly carbonizes when heated. Absorption spectrum: Gasteiger *et al., Ann. N.Y. Acad. Sci.* **90,** 624 (1962).

Hydrochloride, C$_7$H$_7$NO$_2$.HCl, fine needles, dec 170-175°. Freely sol in water, less sol in methanol and ethanol.

4766. Homatropine. *endo-(±)-α-Hydroxybenzeneacetic acid 8-methyl-8-azabicyclo[3.2.1]oct-3-yl ester; 1αH,5αH-tropan-3α-ol mandelate;* mandelyltropeine; tropine mandelate. C$_{16}$H$_{21}$NO$_3$; mol wt 275.35. C 69.79%, H 7.69%, N 5.09%, O 17.43%. Prepd from mandelic acid and tropine: Ladenburg, *Ann.* **217,** 82 (1883); Chemnitius, *J. Prakt. Chem.* **117,** 142 (1927). Prepn of D(−)- and L(+)-forms: Werner, Miltenberger, *Ann.* **631,** 163 (1960). Comprehensive description: F. J. Muhtadi *et al.* in *Analytical Profiles of Drug Substances* vol. **16,** K. Florey, Ed. (Academic Press, New York, 1987) pp 245-290. Toxicity: R. L. Cahen, K. Tvede, *J. Pharmacol. Exp. Ther.* **105,** 166 (1952).

DL-Form, prisms from ether, mp 99-100°. *Poisonous!* Hygroscopic, but only slightly sol in water; sol in alcohol, benzene, chloroform, ether, acetone, dil acids.

DL-Hydrobromide, C$_{16}$H$_{21}$NO$_3$.HBr, *Homatrisol, Bufopto Homatrocel.* Orthorhombic, bipyramidal prisms from water, mp ~212° (partial decompn). *Poisonous!* One gram dissolves in 6 ml water, 40 ml alcohol (12 ml at 60°), 420 ml chloroform. Insol in ether. pH (1% aq soln): 5.4.

DL-Hydrochloride, C$_{16}$H$_{21}$NO$_3$.HCl, prisms from water, mp 217-220° (dec). *Poisonous!* Freely sol in water, alcohol. pH (1% aq soln): 5.4.

DL-Methylbromide, C$_{16}$H$_{21}$NO$_3$.CH$_3$Br, *Arkitropin, Homapin, Malcotran, Mesopin, Novatrin, Novatropine, Sethyl.* Minute needles from alc + ether, mp 191-192° (slight dec). *Poisonous!* Freely sol in water, dil alc; slightly sol in abs alc. Insol in ether. pH (1% aq soln) 5.9; pH (10% aq soln) 4.5. LD$_{50}$ in mice (mg/kg): 1400 orally, 60 i.p. (Cahen, Tvede).

THERAP CAT: Anticholinergic (ophthalmic).

THERAP CAT (VET): Anticholinergic (ophthalmic).

4767. Homidium. *3,8-Diamino-5-ethyl-6-phenylphenanthridinium;* 2,7-diamino-9-phenyl-10-ethylphenanthridinium; 2,7-diamino-10-ethyl-9-phenylphenanthridinium; ethidium; RD-1572; Novidium; Babidium. [C$_{21}$H$_{20}$N$_3$]$^+$. Prepn: T. I. Watkins, *J. Chem. Soc.* **1952,** 3059; Short *et al., U.S. pat.* **2,662,082** (1953 to Boots Pure Drug). uv spec-

trum: B. Hudson, R. Jacobs, *Biopolymers* **14,** 1309 (1975). Inhibition of DNA synthesis: B. A. Newton, *J. Gen. Microbiol.* **17,** 718 (1957); of DNA polymerase: W. H. Elliott, *Biochem. J.* **86,** 562 (1963). Intercalcation of double-stranded DNA: M. J. Waring, *J. Mol. Biol.* **13,** 269 (1965); J.-B. LePecq, C. Paoletti, *ibid.* **27,** 87 (1967). Mutagenicity studies: J. T. MacGregor, I. J. Johnson, *Mutat. Res.* **48,** 103 (1977); G. S. Probst *et al., Environ. Mutagen.* **3,** 11 (1981). Review of interactions with nucleic acids: J.-B. LePecq in *Methods of Biochemical Analysis* vol. **20,** D. Glick, Ed. (Wiley-Interscience, New York, 1971) pp 41-86; M. Waring in *Antibiotics* vol. **3,** J. W. Corcoran, F. E. Hahn, Eds. (Springer-Verlag, New York, 1975) pp 141-165.

Bromide, C$_{21}$H$_{20}$BrN$_3$, *Dromilac.* Bitter tasting dark red crystals from alc, mp 238-240°. uv max in water: 210, 285, 316, 343 nm (ε 200-500, 5000-10000, 50000, 40000). Sol in 20 parts water and 750 parts chloroform at 20°.

Chloride, C$_{21}$H$_{20}$ClN$_3$, dark red cryst powder. Crystallizes with 1 mol of ethanol. Sol in 5 parts of water at room temp.

USE: In sepn and determn of nucleic acids.

THERAP CAT (VET): Antiprotozoal (Trypanosoma).

4768. Homocamfin. *3-Methyl-5-(1-methylethyl)-2-cyclohexen-1-one; m-menth-6-en-5-one;* methylisopropylcyclohexenone; 5-isopropyl-3-methyl-Δ²-cyclohexenone; Cyclosal; Hexetone. C$_{10}$H$_{16}$O; mol wt 152.24. C 78.90%, H 10.59%, O 10.51%. Prepn: Horning *et al., J. Org. Chem.* **9,** 547 (1944); Whitmore, Roberts, *ibid.* **13,** 31 (1948).

Pale yellow, bitter, oily liq. bp$_{18}$ 127-128°; bp$_{16}$ 124-126°; bp$_{15}$ 120-122°; bp$_{11.5}$ 113-116°; bp$_9$ 108-110°. Slightly sol in water; sol in alc, benzene, ether. Marketed in aq soln each 100 ml containing 10 g of the oil and 25 g sodium salicylate.

THERAP CAT: CNS stimulant.

THERAP CAT (VET): Has been used as a CNS stimulant.

4769. Homochelidonine. *4b,5,6,11b,12,13-Hexahydro-1,2-dimethoxy-12-methyl[1,3]benzodioxolo[5,6-c]phenanthridin-5-ol;* α-homochelidonine. C$_{21}$H$_{23}$NO$_5$; mol wt 369.42. C 68.28%, H 6.28%, N 3.79%, O 21.65%. From root of *Chelidonium majus* L., *Papaveraceae:* Schmidt, Selle, *Arch. Pharm.* **228,** 441 (1890). Structure: Späth, Kuffner, *Ber.* **64,** 1123 (1931); H. W. Bersch, *Arch. Pharm.* **291,** 491 (1958). Total synthesis: I. Ninomiya *et al., Heterocycles* **7,** 137 (1977).

Orthorhombic prisms from ethyl acetate, mp 182°. $[\alpha]_D$ +118° (alc). Very sol in chloroform. Sol in alcohol, acetic acid, ethyl acetate, dil mineral acids; sparingly sol in ether.

4770. Homochlorcyclizine. *1-[(4-Chlorophenyl)phenyl-methyl]hexahydro-4-methyl-1H-1,4-diazepine;* N-(p-chloro-benzhydryl)-N'-methylhomopiperazine; 1-(4-chlorodiphen-ylmethyl)-4-methyl-2,3,4,5,6,7-hexahydro-1,4-diazepine; 1-(p-chloro-α-phenylbenzyl)-4-methylhomopiperazine; SA-97; Curosajin; Homorestar. $C_{19}H_{23}ClN_2$; mol wt 314.86. C 72.48%, H 7.36%, Cl 11.26%, N 8.90%. Prepn: Weston, Sommers, U.S. pat. **2,655,498** (1953 to Abbott); *eidem, J. Am. Chem. Soc.* **76,** 5805 (1954).

Hydrochloride, $C_{19}H_{23}ClN_2$.HCl, *Homoginin.*
Dihydrochloride, $C_{19}H_{23}ClN_2$.2HCl, *Homoclomin.* Crystals from ethanol, mp 227-228°. Base $bp_{0.8}$ 177°; n_D^{25} 1.5804.
THERAP CAT: Serotonin antagonist.

4771. Homocysteine. *2-Amino-4-mercaptobutyric acid.* $C_4H_9NO_2S$; mol wt 135.19. C 35.54%, H 6.71%, N 10.36%, O 23.67%, S 23.72%. HSCH₂CH₂CH(NH₂)COOH. A sulfur containing amino acid, produced by the demethylation of methionine and an intermediate in the biosynthesis of cysteine from methionine. Originally obtained from the liver of mammals. Has also been obtained from cystathionine: Binkley, *Methods Enzymol.* **2,** 314 (1955). D-Homocysteine may be prepd from S-benzyl-D-homocysteine, while L-homocysteine is best obtained from L-homocystine: du Vigneaud, Brown, *Biochem. Prepn.* **5,** 93 (1957), *see also* Patterson, du Vigneaud, *J. Biol. Chem.* **111,** 393 (1935); Riegel, du Vigneaud, *ibid.* **112,** 149 (1935). Origin and biochemical conversions in review by Shapiro, Schlenk, *Advan. Enzymol.* **22,** 264-268 (1960). Use of L-form sodium salt as flavor enhancer similar to monosodium glutamate: Kaneko *et al.,* U.S. pat. **3,259,505** (1966 to Ajinomoto Kabushiki Kaisha).

DL-Form, platelets from dil ethanol, mp 232-233°. pK_1 2.22; pK_2 8.87; pK_3 10.86.
Thiolactone hydrochloride, C_4H_8ClNOS, crystals from abs ethanol. L-form: $[\alpha]_D^{26}$ +21.5° (c = 1); D-form: $[\alpha]_D^{26}$ −21.5° (c = 1).

4772. Homocystine. *4,4'-Dithiobis[2-aminobutanoic acid];* 4,4'-dithiobis[2-aminobutyric acid]. $C_8H_{16}N_2O_4S_2$; mol wt 268.36. C 35.81%, H 6.01%, N 10.44%, O 23.85%, S 23.90%. Synthesis by the "malonate" method: Patterson, du Vigneaud, *J. Biol. Chem.* **111,** 393 (1935); Shiroishi *et al., Rept. Food Res. Inst. (Tokyo)* **5,** 75 (1951); du Vigneaud, Brown, *Biochem. Prepns.* **5,** 93 (1957); from 3,6-bis(β-chloroethyl)-2,5-dioxopiperazine: Snyder, Cannon, *J. Am. Chem Soc.* **66,** 511 (1944).

DL-Form, platelets from water, dec 263-265°. pK_1 1.59; pK_2 2.54; pK_3 8.52; pK_4 9.44.
L-Form, crystals, dec 281-284°. $[\alpha]_D^{21}$ −16° (H₂O); $[\alpha]_D^{26}$ +79° (1.0N HCl).
D-Form, crystals, dec 281-284°. $[\alpha]_D^{21}$ +16° (H₂O); $[\alpha]_D^{26}$ −79° (1.0N HCl).

4773. Homoeriodictyol. *2,3-Dihydro-5,7-dihydroxy-2-(4-hydroxy-3-methoxyphenyl)-4H-1-benzopyran-4-one; 4',5,7-trihydroxy-3'-methoxyflavanone;* eriodictyonone; cyanida-

non-3-methyl ether 1625. $C_{16}H_{14}O_6$; mol wt 302.28. C 63.57%, H 4.67%, O 31.76%. From *Eriodictyon californicum* (H. & A.) Torr., and *E. angustifolium* Nutt., *Hydrophyllaceae:* Geissman, *J. Am. Chem. Soc.* **62,** 3258 (1940); Hadley, Gisvold, *J. Am. Pharm. Assoc.* **33,** 275 (1944). Purification: Seka, Prosche, *Monatsh.* **69,** 284 (1936). Structure: Shinoda, Sato, *J. Pharm. Soc. Japan* **49,** 64 (1929), *C.A.* **23,** 4210 (1929). Synthesis: Farooq *et al., Naturwiss.* **46,** 76 (1959).

Crystals from 70% acetic acid. Needles by sublimation in high vacuum (0.003-0.005 mm Hg) at 190-195°. Plates from dil alc, dec 225° (after drying *in vacuo* at 110°). $[\alpha]_D^{20}$ −28° (alc). uv max (alc): 290, 328 nm (log ε 2.26, 2.33). Moderately sol in alcohol, acetic acid. Nearly insol in water, ethyl acetate; practically insol in benzene, chloroform.
Triacetate, mp 115-116°.

4774. Homofenazine. *Hexahydro-4-[3-[2-(trifluoro-methyl)phenothiazin-10-yl]propyl]-1H-1,4-diazepine-1-ethanol;* 1-(β-hydroxyethyl)-4-[3-(2-trifluoromethylphenothiazin-10-yl)propyl]hexahydro-1,4-diazepine; 1-(2-hydroxyethyl)-4-[3-(3-trifluoromethylphenothiazin-10-yl)propyl]-homopiperazine; 3-trifluoromethyl-10-[3-[4-(2-hydroxyethyl)homopiperazino]propyl]phenothiazine. $C_{23}H_{28}F_3N_3OS$; mol wt 451.56. C 61.18%, H 6.25%, F 12.62%, N 9.31%, O 3.54%, S 7.10%. Prepn: Schuler *et al.,* U.S. pat. **3,040,043** (1962 to Degussa).

bp_1 230-240°.
Dihydrochloride, $C_{23}H_{28}F_3N_3OS$.2HCl, *Pasaden.*
Difumarate, $C_{23}H_{28}F_3N_3OS$.2C₄H₄O₄, crystals, mp 148°.
THERAP CAT: Sedative.

4775. Homogentisic Acid. *2,5-Dihydroxybenzeneacetic acid; 2,5-dihydroxyphenylacetic acid;* 2,5-dihydroxy-α-toluic acid. $C_8H_8O_4$; mol wt 168.15. C 57.14%, H 4.80%, O 38.06%. An important intermediate in metabolism of tyrosine *(q.v.)* and phenylalanine *(q.v.).* Occurs in plants and in the urine of alkaptonurics: Garrod, *Inborn Errors of Metabolism* (Oxford Medical Publications, London, 1923, and later). Isoln from alkaptonuric urine: Mörner, *Z. Physiol. Chem.* **117,** 85 (1921). Synthesis from 2,5-dihydroxymandelic acid or from 2,5-dihydroxyphenylglyoxylic acid by boiling with fuming HI: Neubauer, Flatow, *ibid.* **52,** 395 (1907). Alternate syntheses: L. DeForrest Abbott, J. D. Smith, *J. Biol. Chem.* **179,** 365 (1949); S. B. Bostock, A. H. Renfrew, *Synthesis* **1978,** 66; J. L. Bloomer, K. M. Damodaran, *ibid.* **1980,** 111. Biosynthesis from tyrosine: Davies *et al., J. Chem. Soc.* **1964,** 3126. Metabolic studies: W. E. Knox, M. LeMay-Knox, *Biochem. J.* **49,** 686 (1951); B. N. LaDu, V. G. Zannoni, *J. Biol. Chem.* **217,** 777 (1955).

Monohydrate, prisms from water. Anhydrous leaflets from hot alcohol + chloroform, mp 152°. Freely sol in water, alcohol, ether; insol in chloroform, benzene. Easily dehydrated to the lactone. Aq solns are stable.

Dimethyl ether, $C_{10}H_{12}O_4$, mp 124.5°.

Methyl ester dimethyl ether, $C_{11}H_{14}O_4$, mp 45°.

4776. Homosalate. *2-Hydroxybenzoic acid 3,3,5-trimethylcyclohexyl ester; salicylic acid 3,3,5-trimethylcyclohexyl ester;* 3,3,5-trimethylcyclohexyl salicylate; homomenthyl salicylate; Heliophan. $C_{16}H_{22}O_3$; mol wt 262.35. C 73.25%, H 8.45%, O 18.30%. Prepn: Stockelbach, U.S. pat. **2,369,084** (1945 to Fries Bros.).

bp$_4$ 161-165°. n^{20} 1.516 to 1.518. d$_{25}^{25}$ 1.045.

A component of *Coppertone* products.

USE: Ultraviolet screen.

4777. Homoserine. 2-Amino-4-hydroxybutanoic acid; 2-amino-4-hydroxybutyric acid; α-amino-γ-hydroxy-*n*-butyric acid. $C_4H_9NO_3$; mol wt 119.12. C 40.33%, H 7.62%, N 11.76%, O 40.29%. Principal free amino acid occurring in pea plants: A. I. Virtanen, *Acta Chem. Scand.* **7**, 1423 (1953); J.A. Bakhuis, *Nature* **180**, 713 (1957). Prepn: Fischer, Blumenthal, *Ber.* **40**, 106 (1907); Armstrong *J. Am. Chem. Soc.* **70**, 1756 (1948); Birnbaum, Greenstein, *Arch. Biochem. Biophys.* **42**, 212 (1953); M. Frankel, Y. Knobler, *J. Am. Chem. Soc.* **80**, 3147 (1958). Review of homoserine production by fermentation: T. Nara in *Microbial Prod. Amino Acids,* K. Yamada, Ed. (Wiley, New York, 1972) pp 417-434.

L-form

L-Homoserine, flat prisms from 90% alc. Dec 203°. [M]$_D$ +21.8° (5*N* HCl), [M]$_D$ +14.3° (glacial acetic acid). [α]$_D^{26}$ −8.8° (c = 5 in H_2O); [α]$_D^{26}$ +18.3° (c = 2 in 2*N* HCl). On standing for 8 hrs at 26° the [α]$_D$ of the HCl soln decreases to nearly zero as the corresponding levorotatory-γ-butyrolactone is formed.

L-Homoserine γ-lactone monohydrochloride, prepd by refluxing L-homoserine with 2*N* HCl for 2 hrs, crystals, [α]$_D^{26}$ −27.0° (c = 5).

D-Homoserine, crystals, dec 203°. [α]$_D^{26}$ +8.8° (c = 5).

DL-Homoserine, crystals from dil ethanol, dec 186-187°.

4778. Homovanillic Acid. *4-Hydroxy-3-methoxybenzeneacetic acid; (4-hydroxy-3-methoxyphenyl)acetic acid;* 4-hydroxy-3-methoxy-α-toluic acid. $C_9H_{10}O_4$; mol wt 182.18. C 59.34%, H 5.53%, O 35.13%. Metabolite found in human urine. Prepn: F. Tiemann, N. Nagai, *Ber.* **10**, 201 (1877); F. Mauthner, *Ann.* **370**, 368 (1909); H. E. Fisher, H. Hibbert, *J. Am. Chem. Soc.* **69**, 1208 (1947). End product of dopamine, *q.v.:* K. Shaw *et al., J. Biol. Chem.* **226**, 255 (1957); M. Goodall, H. Alton, *Biochem. Pharmacol.* **25**, 2635 (1968).

White crystals, mp 143°. Sol in water, benzene. Slightly sol in alcohol, ether. Insol in cyclohexane.

4779. HON. *5-hydroxy-4-oxonorvaline; 2-amino-5-hydroxylevulinic acid;* δ-hydroxy-γ-oxo-L-norvaline; 2-amino-

5-hydroxy-4-oxopentanoic acid. $C_5H_9NO_4$; mol wt 147.13. C 40.82%, H 6.17%, N 9.52%, O 43.50%. HOCH$_2$CO-CH$_2$CH(NH$_2$)COOH. Antitubercular antibiotic substance isolated from *Streptomyces akiyoshiensis nova* sp.: Tatsuoka *et al., J. Antibiot.* **14A**, 39 (1961). Synthesis and structure: Miyake, *Chem. Pharm. Bull.* **8**, 1071, 1074, 1079 (1960).

Needles from water + acetone, no definite mp. [α]$_D^{20}$ −6° (water); [α]$_D^{17}$ −8.2° (c = 3.4 in water). pKa 2.91. uv max (water): 271 nm (ε 24). Stable in pure, dry state. Less stable in basic than in acidic solns. Colors red when heated in caustic soln, losing antibiotic activity. The DL-form is half as active as L-HON. LD$_{50}$ in mice (mg/kg): 5200 i.v.; 8000 s.c.; 7600 orally (Tatsuoka).

4780. Honokiol. *3',5-Di-2-propenyl-[1,1'-biphenyl]-2,4'-diol;* 3',5-diallyl-2,4'-biphenyldiol; 5,3'-diallyl-2,4'-dihydroxydiphenyl. $C_{18}H_{18}O_2$; mol wt 266.34. C 81.17%, H 6.81%, O 12.01%. Bioactive principle isolated from the bark of *Magnolia obovata,* Thunb., *Magnoliaceae* and other *Magnolia* species used in Japanese and Chinese traditional medicine. Name derived from "Honoki", the Japanese name of *M. obovata.* Isomeric with magnolol, *q.v.* Identification in methanolic extracts of *M. obovata* bark: M. Fujita *et al., Chem. Pharm. Bull.* **20**, 212 (1972). Isoln from seeds of *M. grandiflora* L.: F. S. El-Feraly, W.-S. Li, *Lloydia* **41**, 442 (1978). Synthesis: T. Takeya *et al., Chem. Pharm. Bull.* **34**, 2066 (1986). CNS depressant effects *in vivo:* K. Watanabe *et al., Planta Med.* **49**, 103 (1983). HPLC determn in *M. officinalis* extracts: T.-H. Tsai, C.-F. Chen, *J. Chromatog.* **598**, 143 (1992).

Colorless needles, mp 86-86.5° (El-Feraly, Li); also reported as mp 87.5° (Fujita). Sol in usual organic solvents and caustic alkali. uv max (ethanol): 294 nm (ε 8200).

4781. Hopantenic Acid. *(R)-4-[(2,4-Dihydroxy-3,3-dimethyl-1-oxobutyl)amino]butanoic acid;* 4-(2,4-dihydroxy-3,3-dimethylbutyramido)butyric acid; D-homopantothenic acid. $C_{10}H_{19}NO_5$; mol wt 233.26. C 51.49%, H 8.21%, N 6.00%, O 34.29%. Homolog of pantothenic acid, *q.v.* Prepn: R. Fuerst, L. L. Li, *Biochem. Biophys. Acta* **86**, 26 (1964); C. M. DeSha, R. Fuerst, *ibid.* 33. Prepn of the calcium salt: Y. Nishizawa *et al.,* **Japan.** pat. **66 732** (1966 to Tanabe), *C.A.* **64**, 12555c (1965). Series of articles on analysis, physicochemical properties, electrophysiology, pharmacology, toxicity studies, teratology: *Bitamin* **33**, 603-632 (1966), *C.A.* **65**, 5949a-g (1966). Colorimetric determn, properties: T. Kodama *et al., J. Vitaminol.* **13**, 298 (1967), *C.A.* **68**, 62755n (1968). Peri- and postnatal studies in rats: Y. Asano *et al., Oyo Yakuri* **19**, 1011 (1980), *C.A.* **94**, 58160 (1981). Mechanism of action: V. M. Arakumov, M. A. Kovler, *Farmakol. Toksikol.* **44**, 30 (1981), *C.A.* **94**, 114577 (1981). Determn in plasma: Y. Umeno *et al., J. Chromatog.* **226**, 333 (1981). Toxicology studies: Y. Nishizawa *et al., J. Vitaminol.* **15**, 26 (1969). Pharmacology and acute toxicity study: Y. Nishizawa, *Med. J. Osaka Univ.* **35**, 41 (1984).

Sol in water; stable at pH 5-6. [α]$_D^{20}$ +23.8°. pKa 4.52 at 25°. LD$_{50}$ in male, female mice, male female rats (mg/kg): 850, 954, 1575, 1458 i.p.; 2063, 2495, 5940, 7348 s.c.; 6297, 7935, 16810, 13350 orally (Nishizawa).

Calcium salt hemihydrate, $C_{20}H_{36}CaNO_5 \cdot \frac{1}{2}H_2O$, *calcium homopantothenate, hopantenate calcium, pantogam, Hopate.* White powder, mp 155-165°. [α]$_D^{20}$ +24.19° (c = 2 in water).

Bitter taste. Sol in water, slightly sol in methanol. Practically insol in organic solvents. Stable at pH 5-6.

THERAP CAT: Cerebral activator.

4782. Hops. Carefully dried strobiles of *Humulus lupulus* L., *Moraceae,* bearing their glandular trichomes. *Habit.* Europe, Asia, North America, cultivated widely. *Constit.* Volatile oil (0.3-1%), lupulone, humulone, xanthohumol, lactaric acid, ceryl alcohol, cerotic acid, soft and hard hop-resins. Review of chemistry and constituents: Stevens, *Chem. Rev.* **67,** 19 (1967).

USE: Beer brewing.

THERAP CAT: Aromatic bitter.

4783. Hordenine. *4-[2-(Dimethylamino)ethyl]phenol; N,N*-dimethyltyramine; *p*-hydroxy-*N,N*-dimethylphenethylamine; anhaline; eremursine; peyocactine. $C_{10}H_{15}NO$; mol wt 165.24. C 72.69%, H 9.15%, N 8.48%, O 9.68%. Isoln from barley germs: Leger, *Compt. Rend.* **142,** 108 (1906); **143,** 234, 916 (1906); **144,** 488 (1907); Erspamer, Falconieri, *Naturwiss.* **39,** 431 (1952). Structure: Leger, *Bull. Soc. Chim. France* [3] **35,** 868 (1906); [4] **1,** 148 (1907); Gaebel, *Arch. Pharm.* **244,** 441 (1906). Synthesis from phenethyl alcohol: Barger, *J. Chem. Soc.* **95,** 2193 (1909); from tyrosine: Raoul, *Compt. Rend.* **204,** 74 (1937); from *p*-(β-hydroxyethyl)anisole: Cheng *et al., J. Am. Chem. Soc.* **73,** 4081 (1951).

Orthorhombic prisms from alcohol or from benzene + petr ether, needles from water, mp 117-118°. bp_{11} 173°. Sublimes 140-150°. Very sol in alcohol, chloroform, ether. 7 grams dissolve in 1000 ml water. Sparingly sol in benzene, toluene, xylene. Practically insol in petr ether.

Hydrochloride, $C_{10}H_{15}NO.HCl$, needles from alcohol, mp 177°. Very sol in water. LD_{50} in mice: 113.5 mg/kg.

4784. Horehound. Hoarhound. *Marrubium vulgare* (Tourn.) L., *Labiatae. Habit.* Europe, Asia, naturalized in the U.S. *Constit.* Bitter principle marrubiin, volatile oil.

THERAP CAT: Expectorant.

4785. Horse-radish. Raphanus rusticanus; Kren; Maliner KrenThe root of *Radicula armoracia* (L.) Robinson [*Cochlearia armoracia* L.; *Armoracia lopathifolia* Gilib.; *Roripa armoracia* Hitch.; *Nasturtium armoracia* Fries], *Cruciferae. Habit.* Europe, naturalized in North America. Contains ascorbic acid and sinigrin which yields allyl isothiocyanate on hydrolysis with peroxidase or myrosinase, an enzyme from black mustard: Stoll, Seebeck, *Helv. Chim. Acta* **31,** 1432 (1948).

Active ingredient of *Rasapen,* a urinary antiseptic.

USE: Condiment.

4786. HPA-23. *Hexaoctacontaoxononaantimonateheneicosatungstate(19—) heptadecaammonium disodium;* ammonium 5-tungsto-2-antimonate; 5-tungsto-2-antimonate; 21-tungsto-9-antimonate. $H_{68}Na_2N_{17}O_{86}Sb_9W_{21}$, mol wt 6685.04. H 1.03%, Na 0.69%, N 3.56%, O 20.58%, Sb 16.39%, W 57.75%. $(NH_4)_{17}Na(NaSb_9W_{21}O_{86})$. One of the condensed mineral heteropolyanions, also called polyoxotungstates. Synthesis: M. Michelon, G. Hervé, *Compt. Rend. Ser. C* **274,** 209 (1972); M. Michelon *et al., J. Inorg. Nucl. Chem.* **42,** 1583 (1980). Crystal structure: J. Fischer *et al., J. Am. Chem. Soc.* **98,** 3050 (1976). Laser Raman studies of intracellular localization: C. Cibert, C. Jasmin, *Biochem. Biophys. Res. Commun.* **108,** 1424 (1982). Antiviral activity in mice: C. Jasmin *et al., J. Nat. Cancer Inst.* **53,** 469 (1974); G. H. Werner *et al., J. Gen. Virol.* **31,** 59 (1976). *In vitro* inhibition of rabies virus: H. Tsiang *et al., J. Gen. Virol.* **40,** 665 (1978). Effects in scrapie model systems: R. H. Kimberlin, C. A. Walker, *Lancet* **2,** 591 (1979); *eidem, Arch. Virol.* **78,** 9 (1983). Structure-activity study: M. Hervé *et al., Biochem. Biophys. Res. Commun.* **116,** 222 (1983). Clinical evaluation in AIDS: W. Rozenbaum *et al.,*

Lancet **1,** 450 (1985); B. L. Moskovitz, *Antimicrob. Ag. Chemother.* **32,** 1300 (1988).

Crystals from water. Solution has pH near 6.7, and is stable around neutrality. LD_{50} i.p. in mice: 750 mg/kg (Werner).

Note: Formerly known as *ammonium antimony tungsten oxide.*

4787. HQNO. *2-Heptyl-4-quinolinol 1-oxide;* 2-heptyl-4-hydroxyquinoline *N*-oxide. $C_{16}H_{21}NO_2$; mol wt 259.35. C 74.10%, H 8.16%, N 5.40%, O 12.34%. Inhibitor of electron transport through the cytochrome bc_1 segment of the respiratory chain. Isoln of the naturally occurring antagonist to dihydrostreptomycin from *Pseudomonas pyocyanea:* Hays *et al., J. Biol. Chem.* **159,** 725 (1948); J. Lightbown, *J. Gen. Microbiol.* **11,** iv (1954). Properties: J. W. Cornforth, A. T. James, *Biochem. J.* **58,** xlviii (1954). Synthesis: *eidem, ibid.* **63,** 124 (1956); D. E. Ames *et al., J. Chem. Soc.* **1956,** 3079. Inhibition of electron transport: J. W. Lightbown, F. L. Jackson, *Biochem. J.* **63,** 130 (1956); M. Avron, *ibid.* **78,** 735 (1961); N. J. Jacobs, M. J. Wolin, *Biochim. Biophys. Acta* **69,** 29 (1963). Effect on proton permeability of the mitochondrial membrane: K. Krab, M. Wikström, *Biochem. J.* **186,** 637 (1980). Sites of inhibition: G. Izzo *et al., FEBS Letters* **93,** 320 (1978); M. Droppa *et al., Z. Naturforsch.* **36C,** 109 (1981).

Crystals from methylethyl ketone, mp 156-157°.

4788. Humic Acids. Allomelanins found in soils, coals, and peat, resulting from the decompn of organic matter, particularly dead plants. Consists of a mixture of complex macromolecules having polymeric phenolic structures with the ability to chelate with metals, esp iron. Review: R. A. Nicolaus, *Melanins* (Hermann, Paris, 1968) pp 147-153; W. Flaig *et al.* in *Soil Components* vol. 1, J. E. Gieseking, Ed. (Springer, New York, 1975) pp 1-211.

Chocolate-brown, dust-like powder. Slightly sol in water, usually with much swelling; sol in alkali hydroxides and carbonates; also sol in hot concd HNO_3 with dark-red color.

USE: In mud baths, drilling muds, pigments for printing inks, fertilizers, growth hormones for plants, transporters of trace minerals in soil: Steelnick, *J. Chem. Ed.* **40,** 379 (1963).

4789. Humulene. *(E,E,E)-2,6,6,9-Tetramethyl-1,4,8-cycloundecatriene;* α-humulene; α-caryophyllene. $C_{15}H_{24}$; mol wt 204.36. C 88.16%, H 11.84%. Sesquiterpenoid isomer of caryophyllene, *q.v.* occurring in many essential oils, especially oil of hops *(Humulus lupulus* L. *Moraceae)* and leaves of *Lindera strychnifolia* (F.) Will *Lauraceae.* Occurs in nature as a mixture with β-humulene. Isolation of mixture: A. C. Chapman, *J. Chem. Soc.* **67,** 54, 780 (1895). Identity with α-caryophyllene: F. Sorm *et al., Coll. Czech. Chem. Commun.* **14,** 693, 699, 716 (1949). Structure: F. Sorm *et al., ibid.* **19,** 570 (1954). Stereochemistry: A. T. McPhail, G. A. Sim, *J. Chem. Soc. (B)* **1966,** 112. Synthesis: E. J. Corey, E. Hanamaka, *J. Am. Chem. Soc.* **89,** 2758 (1967). Stereoselective synthesis: Y. Kitagawa *et al., ibid.* **99,** 3864 (1977); E. J. Corey *et al., Tetrahedron Letters* **34,** 3675 (1993). Chromatographic conversion to β-humulene: V. Benesova *et al., Coll. Czech. Chem. Commun.* **26,** 1832 (1961). *Reviews:* F. Sorm in *Fortschr. Chem. Org. Naturst.* **19,** 1-32 (1961); *Rodd's Chemistry of Carbon Compounds,* S. Coffey, Ed., **vol. IIC** (Elsevier, New York, 2nd ed., 1969) pp 282-283.

Liquid. bp$_5$ 106–107°. n_D^{30} 1.5004, N. P. Damodaran, S. Dev, *Tetrahedron* **24**, 4113 (1968). Also reported as bp$_{10}$ 123°. n_D^{25} 1.5015. d$_4^{25}$ 0.8865, R. P. Hildebrand *et al., Chem. Ind. (London)* **1959**, 489. NMR spectrum: S. Dev *et al., J. Am. Chem. Soc.* **90**, 1246 (1968).

Silver nitrate complex, $C_{15}H_{24}.2AgNO_3$, crystals from aq ethanol, mp 175°.

β-Humulene, (E,E)-1,4,4-trimethyl-8-methylene-1,5-cyclo-undecadiene. Liquid. n_D^{20} 1.5014. d$_4^{20}$ 0.8905.

4790. Humulon. *(R)-3,5,6-Trihydroxy-4,6-bis(3-methyl-2-butenyl)-2-(3-methyl-1-oxobutyl)-2,4-cyclohexadien-1-one;* α-bitter acid; α-lupulic acid; humulone. $C_{21}H_{30}O_5$; mol wt 362.47. C 69.59%, H 8.34%, O 22.07%. Antibiotic constituent of hops (*Humulus lupulus* L., Moraceae). *See also* Lupulon. Isoln from commercial hops: Bungener, *Bull. Soc. Chim.* [2] **45**, 487 (1886); Barth, Lintner, *Ber.* **31**, 2022 (1898); Wollmer, *Ber.* **49**, 780 (1916); Lewis *et al., J. Clin. Invest.* **28**, 916 (1949). Structure: Riedl, *Ber.* **85**, 692 (1952); Carson, *J. Am. Chem. Soc.* **73**, 4652 (1951). Absolute configuration and structure of preferred isomer: De-Keukeleire, Verzele, *Tetrahedron* **26**, 385 (1970).

Crystals from ether, mp 65–66.5°. Bitter taste, esp in alcoholic soln. More stable to air than lupulon. Monobasic acid. $[\alpha]_D^{20}$ −212° (1.0 g in 15.5 g 96% alc). uv max (ethanol): 237, 282 nm (ε 13,760; 8330). Soluble in the usual organic solvents. Slightly sol in boiling water from which it separates as a milky precipitate on cooling. Forms a sodium salt which is readily sol in water. Suffers no loss of bacteriostatic potency against *Staphylococcus aureus* upon autoclaving 40 ppm in phosphate buffer at pH 6.5 or 8.5. The presence of ascorbic acid in low concns extends the duration of bacteriostatic action.

4791. Huperzine A. *[5R-(5α,9β,11E)]-5-Amino-11-ethylidene-5,6,9,10-tetrahydro-7-methyl-5,9-methanocyclo-octa[b]pyridin-2(1H)-one;* selagine; HUP. $C_{15}H_{18}N_2O$; mol wt 242.32. C 74.35%, H 7.49%, N 11.56%, O 6.60%. Reversible alkaloid inhibitor of AChE which crosses the blood-brain barrier. Occurs as the (−)-form in the vegetative part of clubmosses; teas brewed from these mosses have traditionally been used in China to alleviate memory problems. Isolated as selagine from *Lycopodium selago* L., Lycopodiaceae: J. Muszynski, *Quart. J. Pharm. Pharmacol.* **21**, 34 (1948); from *Huperzia serrata* as huperzine: Chin. Co-op Res. Group, *J. Tradit. Chin. Med.* **2**, 45 (1982). Original structure: Z. Valenta *et al., Tetrahedron Letters* **1960**, 26; revised structure as huperzine A: J.-S. Liu *et al., Can. J. Chem.* **64**, 837 1986. Identity with selagine: W. A. Ayer *et al., ibid.* **67**, 1538 (1989). Synthesis of (±)-form: A. P. Kozikowski *et al., J. Org. Chem.* **56**, 4636 (1991); G. Campiani *et al., ibid.* **58**, 7660 (1993). NMR spectra: B. N. Zhou *et al., Phytochemistry* **34**, 1425 (1993). Anticholinesterase activity: Y.-E. Wang *et al., Acta Pharmacol. Sin.* **7**, 110 (1986); binding profile of enantiomers: M. McKinney *et al., Eur. J. Pharmacol.* **203**, 303 (1991); binding specificity: Y. Ashani *et al., Mol. Pharmacol.* **45**, 555 (1994). Clinical evaluation in senile dementia: R.-W. Zhang *et al., Acta Pharmacol. Sin.* **12**, 259 (1991). Brief review of chemistry

and clinical use: D. Bai, *Pure Appl. Chem.* **65**, 1103–1112 (1993).

Monoclinic crystals from acetone, mp 214°–215°. $[\alpha]_D$ −147° (c = 0.36 in CH$_3$OH) (Ayer). Also reported as mp 230°. $[\alpha]_D^{24.5}$ −150.4° (c = 0.498 in MeOH) (Liu). uv max (EtOH): 231, 313 nm (log ε 4.01, 3.89).

THERAP CAT: In treatment of memory disorders.

4792. Hyalobiuronic Acid. *2-Amino-2-deoxy-3-O-β-D-glucopyranuronosyl-D-glucose;* 3-O-(β-D-glucopyranosyluronic acid)-2-amino-2-deoxy-D-glucose. $C_{12}H_{21}NO_{11}$; mol wt 355.30. C 40.57%, H 5.96%, N 3.94%, O 49.53%. Disaccharide unit of hyaluronic acid. Isoln from hyaluronic acid: Rapport *et al., Nature* **168**, 996 (1951). Structure: Weissman, Meyer, *J. Am. Chem. Soc.* **76**, 1753 (1954). Synthesis: Takanashi *et al., ibid.* **84**, 3029 (1962).

Rectangular prisms from water, darken at 190° with no characteristic melting or dec point. pK$_1'$ = 2.6, pK$_2'$ = 7.1. Shows mutarotation: $[\alpha]_D^{20}$ +34° → +30° (c = 1.08 in 0.1N HCl). Sparingly sol in hot water, dilute HCl, dil NaHCO$_3$. Practically insol in water, glacial acetic acid, ethanol, methanol and pyridine.

N-Acetylhyalobiuronic acid, $C_{14}H_{23}NO_{12}$, amorphous. pK' = 3.3. $[\alpha]_D^{24}$ −32° (c = 2.0 in water).

4793. Hyaluronic Acid. Mol wt is within the range of 50,000 to 8 × 10^6 depending on source, methods of prepn, and determination. A natural high viscosity mucopolysaccharide with alternating β (1-3) glucuronidic and β (1-4) glucosaminidic bonds. Found in the umbilical cord, in vitreous humor, in synovial fluid, in pathologic joints, in group A and C hemolytic streptococci and in Wharton's jelly. Isoln and characterization: Meyer, Palmer, *J. Biol. Chem.* **107**, 629 (1934); **114**, 689 (1936); Balazs, *Fed. Proc.* **17**, 1086 (1958); Laurent *et al., Biochim. Biophys. Acta* **42**, 476 (1960). Structure: Weissman, Meyer, *J. Am. Chem. Soc.* **76**, 1753 (1954); Meyer, *Fed. Proc.* **17**, 1075 (1958). Crystal structure of hyaluronate films: Dea *et al., Science* **179**, 560 (1973); Atkins, Sheehan, *ibid.* 562. Possible role in determining blood vessel location in the embryo: R. N. Feinberg, D. C. Beebe, *Science* **220**, 1177 (1983). *Reviews:* Tauber, *Chemistry and Technology of Enzymes* (New York, 1946); Meyer, Rapport in *Advan. Enzymol.* **13**, 199 (1952); Whistler, Olson in *Advan. Carbohyd. Chem.* **12**, 299 (1957). Review of role in various developmental processes: B. P. Toole, *Cell Biology of Extracellular Matrix,* E. D. Hay, Ed. (Plenum Press, New York, 1981) pp 259–288.

Sodium salt, *ARTZ, Connettivina, Equron, Healon, Healonid, Hyacid, Hyalgan, Hyalovet, Hyonate, Ial, Opegan, Provisc, Synacid.* $[\alpha]_D^{25}$ −74° (c = 0.25 in water): Rapport *et al., J. Am. Chem. Soc.* **73**, 2416 (1951). Most viscosity determinations of hyaluronic acid vary from 1-8: Jensen,

Acta Chem. Scand. **7**, 603 (1953). Infrared absorption spectra: Orr, Biochim. Biophys. Acta **14**, 173 (1954).

USE: Surgical aid (ophthalmological).

THERAP CAT (VET): Adjunct in treatment of noninfectious synovitis.

4794. Hyaluronidases. Spreading factor; diffusing factor; invasin; Alidase; Apertase; Diffusin; Enzodase; Harodase; Hyalase; Hyalozima; Hyalidase; Hyasmonta; Hyason; Hyazyme; Infiltrase; Jalovis; Kinaden; Kinetin; Luronase; Permease; Rondase; Ronidase; Thiomucase; Unidasa; Wydase. Enzymes which have in common the cleavage of glycosidic bonds of hyaluronic acid, q.v., and, to a variable degree, of some other acid mucopolysaccharides of connective tissue. The skin is probably the largest store of hyaluronidase in the body; the enzyme although generally present in an inactive form, may be supposed to regulate the velocity of water and metabolite exchange by decreasing the viscosity of the intercellular matrix. Also has a physiological role in fertilization: The sperm is rich in the enzyme and can thus advance better in the cervical canal and reach the ovum. Found in the type II pneumococci, in group A and C hemolytic streptococci, Staphylococcus aureus and Clostridium welchii: Linker et al., J. Biol. Chem. **219**, 13 (1956); in heads of leeches: Linker et al., Nature **180**, 810 (1957); in snake venoms: Favilli, ibid. **145**, 866 (1940); in testes: Hahn, Biochem. Z. **315**, 83 (1943); Högberg, Acta Chem. Scand. **8**, 1098 (1954). Biochemical properties: D. Platt, Arzneimittel-Forsch. **20**, 1836 (1970). Review: Meyer, Rapport in Advan. Enzymol. **13**, 199-236 (1952); Meyer et al., The Enzymes vol. **4**, P. D. Boyer et al., Eds. (Academic Press, New York, 2nd ed., 1960) pp 447-460; Meyer, ibid. vol. **5** (3rd ed., 1971) pp 307-320. Reviews of clinical trials in myocardial infarction: G. S. May et al., Progr. Cardiovasc. Dis. **25**, 335-359 (1983); A. B. Saunders, Emerg. Med. Clin. North Am. **6**, 361-372 (1988). Hyaluronidase manufacturers define their product in terms of turbidity-reducing (TR) units or in viscosity units. Prepd solns for injection usually contain 150 turbidity-reducing units or 500 viscosity units dissolved in 1 ml of isotonic NaCl soln.

USE: Pharmaceutic aid (diffusing agent—s.c. injections).

THERAP CAT: Spreading agent.

THERAP CAT (VET): To promote diffusion, absorption, resorption.

4795. Hycanthone. 1-[[2-(Diethylamino)ethyl]amino]-4-(hydroxymethyl)-9H-thioxanthen-9-one. $C_{20}H_{24}N_2O_2S$; mol wt 356.49. C 67.38%, H 6.79%, N 7.86%, O 8.98%, S 8.99%. Metabolite of lucanthone, q.v.: Rosi et al., Nature (London) **208**, 1005 (1965). Prepn by oxidative fermentation of lucanthone and schistosomicidal activity: Rosi et al., J. Med. Chem. **10**, 867 (1967); Neth. pat. **Appl. 6,410,359**, and Rosi, Peruzzotti, U.S. pats. **3,294,803; 3,312,598** (1965, 1966, 1967 all to Sterling Drug). Alternate synthesis: Laidlaw et al., J. Org. Chem. **38**, 1743 (1973).

Crystals, mp 100.6-102.8°. Absorption max (ethanol): 233, 258, 329, 438 nm (ε 19400, 37000, 9700, 6600). Extremely sensitive to acid.

Hydrochloride, mp 173-176° (dec).

Mesylate, Etrenol.

THERAP CAT: Anthelmintic (Schistosoma).

4796. Hydantoin. 2,4-Imidazolidinedione; 2,4-(3H,-5H)-imidazoledione; glycolylurea. $C_3H_4N_2O_2$; mol wt 100.08. C 36.01%, H 4.03%, N 27.99%, O 31.97%. Prepn: Baeyer, Ann. **130**, 129 (1864). Manuf: Gresham, Schweitzer, U.S. pat. **2,402,134** (1946 to du Pont); White, Wysong,

U.S. pat. **2,663,713** (1953 to Dow). Review: J. H. Bateman in Kirk-Othmer Encyclopedia of Chemical Technology vol. **12** (Wiley-Interscience, New York, 3rd ed., 1980) pp 692-711.

Needles from methanol, mp 220°. Slightly sol in water or ether; sol in alcohol, in solns of fixed alkali hydroxides.

4797. Hydnocarpic Acid. (R)-2-Cyclopentene-1-undecanoic acid; 11-(2-cyclopenten-1-yl)undecanoic acid. $C_{16}H_{28}O_2$; mol wt 252.40. C 76.14%, H 11.18%, O 12.68%. Component of chaulmoogra oil; naturally occurring in d-form. Isoln from seeds of Hydnocarpus wightiana Blume or H. anthelmintica Pierre, Flacourtiaceae, or from the seeds of Taraktogenos kurzii King, Bixaceae: F. B. Power, M. Barrowcliff, J. Chem. Soc. **87**, 884 (1905); ibid. **91**, 557 (1907). Structure: R. L. Shriner, R. Adams, J. Am. Chem. Soc. **47**, 2727 (1925). Synthesis of dl-hydnocarpic acid: D. G. M. Diaper, J. C. Smith, Biochem. J. **42**, 581 (1948). Toxicity: Handbook of Toxicology, vol. **1**, W. S. Spector, Ed. (Saunders, Philadelphia, 1956) pp 274-275. Antimicrobial spectrum: P. L. Jacobsen, L. Levy, Proc. West. Pharmacol. Soc. **15**, 44 (1972). Mechanism of action: eidem, Antimicrob. Ag. Chemother. **3**, 373 (1973). Chromatographic determn in seed oils: W. W. Christie et al., Lipids **24**, 116 (1989).

Colorless, glistening leaflets from petr ether + ethyl acetate, mp 59-60°. $[\alpha]_D$ +68.3° (chloroform). Sparingly sol in usual organic solvents; sol in chloroform.

dl-Form, pearly plates from alcohol, ethyl actate or petr ether + ethyl acetate, mp 59-59.5°.

Sodium salt, sodium hydnocarpate, hydnocarpate sodium, sodium gynocardate. Yellowish powder. Sol in water, alc. The aq soln is alkaline. MLD i.v. in rats: 100-125 mg/kg (Spector).

THERAP CAT: Antibacterial (leprostatic).

4798. Hydracarbazine. 6-Hydrazino-3-pyridazinecarboxamide; 3-hydrazino-6-carbamoylpyridazine; 3-hydrazinopyridazine-6-carboxamide. $C_5H_7N_5O$; mol wt 153.14. C 39.21%, H 4.61%, N 45.73%, O 10.45%. Prepn: Libermann, Rouaix, Bull. Soc. Chim. France **1959**, 1793; Brit. pat. **856,-409** (1960 to Chimie et Atomistique).

Crystals, dec 249-250°.

Note: A component of Normatensyl.

THERAP CAT: Antihypertensive; diuretic.

4799. Hydracrylic Acid. 3-Hydroxypropanoic acid; β-hydroxypropionic acid; ethylene lactic acid. $C_3H_6O_3$; mol wt 90.08. C 40.00%, H 6.71%, O 53.28%. CH_2OHCH_2COOH. Prepd by alkaline hydrolysis of the nitrile: R. R. Read, Org. Syn. coll. vol. **I**, 321 (2nd ed., 1941).

Viscous liq. Strong acid, pK (25°): 4.51. On distn or boiling with 50% H_2SO_4 dec into water and acrylic acid. Very sol in water, sol in alcohol, miscible with ether.

Sodium salt, $NaC_3H_5O_3$, deliquescent crystals, mp 143°.

Calcium salt dihydrate, $Ca(C_3H_5O_3)_2.2H_2O$, prisms, mp 140-145°; freely sol in cold water.

4800. Hydralazine. 1(2H)-Phthalazinone hydrazone; 1-hydrazinophthalazine; Ciba 5968; Präparat 5968; C-5968; Hipoftalin; Hypophthalin; Apresoline. $C_8H_8N_4$; mol wt

160.18. C 59.99%, H 5.03%, N 34.98%. Prepd by the action of hydrazine hydrate on 1-chloro- or 1-phenoxyphthalazine: Hartmann, Druey, U.S. pat. **2,484,029** (1949 to Ciba); Druey, Ringier, *Helv. Chim. Acta* **34**, 204 (1951). Metabolism: Z. H. Israile, P. G. Dayton, *Drug Metab. Rev.* **6**, 283 (1977); K. Schmid *et al.*, *Arzneimittel-Forsch.* **31**, 1143 (1981). Pharmacology: J. L. Cangiano *et al.*, *J. Lab. Clin. Med.* **92**, 516 (1978). Clinical paper: R. F. Albrecht *et al.*, *Int. Anesthesiol. Clin.* **16**, 299 (1978). Acute toxicity: L. Dorigotti *et al.*, *Pharmacol. Res. Commun.* **8**, 295 (1976). Comprehensive description: C. E. Orzech *et al.*, *Anal. Profiles Drug Subs* **8**, 283-314 (1979).

Yellow needles from methanol, mp 172-173° (rapid heating). One gram dissolves in 3 ml 2*N* acetic acid, in 12 ml warm methanol. Forms a red compd (phthalazinylhydrazone) with acetone at 60° in presence of 2*N* acetic acid. LD_{50} in mice, rats (mg/kg): 122, 90 orally; 101, 40 i.p. (Dorigotti).

Hydrochloride, $C_8H_8N_4$·HCl, *Lopress.* Yellow crystals, dec 273°. Soly in water (g/100 ml) at 15°: 3.01; at 25°: 4.42; in 95% ethanol: 0.2 g/100 ml. Very slightly sol in ether, pH of a 2% aq soln 3.5 to 4.5. uv max (0.001% aq soln): 211, 240, 260, 304, 315 nm. Aq solns containing 20 mg/ml may be preserved with 0.5% chlorobutanol.

THERAP CAT: Antihypertensive.

4801. Hydrallostane. *(5α,11β)-11,17,21-Trihydroxypregnane-3,20-dione;* 11β,17α,21-trihydroxyallopregnane-3,-20-dione; allodihydrohydrocortisone; allopregnane-11β,-17α,21-triol-3,20-dione; 4,5α-dihydrocortisol; allodihydro F. $C_{21}H_{32}O_5$; mol wt 364.48. C 69.20%, H 8.85%, O 21.95%. Isoln from beef and hog adrenals: Neher, Wettstein, *Helv. Chim. Acta* **39**, 2062 (1956). Prepn: Pataki *et al.*, *J. Biol. Chem.* **195**, 753 (1952); Fukushima, Daum, *J. Org. Chem.* **26**, 520 (1961); Gould, Oliveto, U.S. pats. **2,783,254; 2,897,216** (1957; 1959 to Schering); Gould, Herzog, U.S. pat. **2,783,226** (1957 to Schering); **Brit.** pat. **742,-888** (1956 to Syntex).

Crystals, mp 234-240°. $[\alpha]_D^{25}$ +83° (acetone). Practically insol in water. Sol in methanol, acetone, chloroform.

21-Acetate, $C_{23}H_{34}O_6$, crystals from hexane + ethyl acetate, mp 211-213°. $[\alpha]_D^{20}$ +69° (chloroform).

21-*tert*-Butylacetate, $C_{27}H_{42}O_6$, crystals from ethanol, mp 256-264° (also crystallizes as a hydrate). Sol in chloroform, oils, fats.

4802. Hydramethylnon. *Tetrahydro-5,5-dimethyl-2(1H)-pyrimidinone[3-[4-(trifluoromethyl)phenyl]-1-[2-[4-(trifluoromethyl)phenyl]ethenyl]-2-propenylidene]hydrazone;* 1,5-bis(α,α,α-trifluoro-*p*-tolyl)-1,4-pentadiene-3-one (1,4,5,6-tetrahydro-5,5-dimethyl-2-pyrimidinyl)hydrazone; AC-217300; Amdro; Combat; Maxforce. $C_{25}H_{24}F_6N_4$; mol wt 494.48. C 60.73%, H 4.89%, F 23.05%, N 11.33%. Slow-activating stomach poison insecticide. Prepn: J. B. Lovell, U.S. pat. **4,087,525** (1975 to Am. Cyanamid). Activity: *idem, Proc. Brit. Crop Prot. Conf.-Pests Dis.* **1979**, 575. Field trials for control of red imported fire ant: D. P. Harlan *et al.*, *Southwest. Entomol.* **6**, 150 (1981); W. A. Banks *et al.*, *ibid.* 158; for control of cockroaches: J. F. Milio *et al.*, *J. Econ. Entomol.* **79**, 1280 (1986).

Crystals from isopropanol, mp 189-191°. Insol in water. Sol in alc, acetone.

USE: Insecticide.

4803. Hydramitrazine. *4,6-Bis(diethylamino)-1,3,5-tri-azin-2(1H)-one hydrazone;* 2,4-bis(diethylamino)-6-hydrazi-no-s-triazine; 2-hydrazino-4,6-bis(diethylamino)-1,3,5-triazine; Meladrazine. $C_{11}H_{23}N_7$; mol wt 253.35. C 52.15%, H 9.15%, N 38.70%. Prepn: Hüni, Staehelin, U.S. pat. **2,824,-103** (1958 to Ciba).

d-Tartrate, $C_{11}H_{23}N_7$·$C_4H_6O_6$, *Lisidonil.*
THERAP CAT: Antispasmodic.

4804. Hydrangea. Seven barks. Dried rhizome and roots of *Hydrangea arborescens* L., *Saxifragaceae.* Habit. Eastern U.S. Constit. The glucoside hydrangin, saponin, resins, fixed and volatile oils, starch.

4805. Hydrargaphen. *[μ-[[3,3'-Methylenebis[2-naph-thalenesulfonato]](2−)]]diphenyldimercury;* phenylmercuric methylenebis(2-naphthyl-3-sulfonic acid); phenylmercury 2,2'-dinaphthylmethane-3,3'-disulphonate; phenylmercuric dinaphthylmethanedisulfonate; methylenedinaphthylene-sulfonic acid bisphenylmercuri salt; phenylmercury methyl-enedinaphthalenesulfonate; bis(phenylmercuri)methylenedi-naphthalenesulfonate; phenyl mercuric Fixtan; Conotrane; P.M.F.; Versotrane; Hydraphen; Penotrane; Fibrotan; Septotan. $C_{33}H_{24}Hg_2O_6S_2$; mol wt 981.86. C 40.37%, H 2.46%, Hg 40.86%, O 9.78%, S 6.53%. Prepn: Bywater, Pritchard, **Brit.** pats. **584,196** and **2,555,114** (1949 and 1951 to Ward, Blenkinsop). Review and properties: Goldberg, *Mfg. Chemist* **22**, 182 (1951); Hopf, *ibid.* **24**, 444 (1953).

Amorphous powder. Practically insol in water but readily goes into colloidal soln in alkali metal dinaphthylmeth-ane disulfonates. The colloidal solns have strong tendency to absorb at interfaces and to form charged hydrated aggregates. LD_{100} orally in mice: 80 mg/kg.

USE: Bactericide and fungicide for the treatment of wool, hides, leather, textiles, timber and wood-pulp, paints, adhesives.

THERAP CAT: Anti-infective (topical).

4806. Hydrastine. *[S-(R*,S*)]-6,7-Dimethoxy-3-(5,6,7,-8-tetrahydro-6-methyl-1,3-dioxolo[4,5-g]isoquinolin-5-yl)-1(3H)-isobenzofuranone;* *l*-β-hydrastine. $C_{21}H_{21}NO_6$; mol wt 383.40. C 65.79%, H 5.52%, N 3.65%, O 25.04%. Isoln of naturally occurring *l-β*-form from *Hydrastis canadensis* L., *Ranunculaceae* together with berberine and canadine. Synthesis of diastereoisomeric mixtures of hydrastines: Hope *et al., J. Chem. Soc.* **1931**, 236; Marshall *et al., ibid.* **1934**, 1315; M. Hanaoka *et al., Chem. Pharm. Bull.* **27**, 1947 (1979); J. R. Falck, S. Manna, *Tetrahedron Letters* **22**, 619 (1981). Resolution to the *l-β*-form: Haworth, Pinder, *J. Chem. Soc.* **1950**, 1776; Haworth *et al., Nature* **165**, 529 (1950). Structure: Knabe, *Arch. Pharm.* **293**, 121 (1960). Abs config: Ohta *et al., Tetrahedron Letters* **1963**, 859; Blaha *et al., Coll. Czech. Chem. Commun.* **29**, 2328 (1964); Snatzke *et al., Tetrahedron* **25**, 5059 (1969). Biosynthesis: Gear, Spenser, *Can. J. Chem.* **41**, 783 (1963).

Orthorhombic prisms from alc, mp 132°. $[\alpha]_D^{20} - 50°$ (c = 0.3 in abs alc). uv max (ethanol): 202, 218, 238, 298, 316 nm (log ϵ 4.79, 4.53, 4.15, 3.86, 3.63). pK 7.8. Freely sol in acetone and benzene; insol in water. The salts hydrolyze easily and do not crystallize well.

Hydrochloride, $C_{21}H_{21}NO_6 \cdot HCl$. Prepn: R. Paech, M. V. Tracy, *Modern Methods of Plant Analysis* vol. **IV** (Springer-Verlag, Berlin, 1955) p 383. Powder, mp 116°. $[\alpha]_D^{17} +127°$ (c = 4 in dil HCl). Very sol in water, alcohol; slightly sol in $CHCl_3$; very slightly in ether. pH (0.5% aq soln): 4.2. Hygroscopic. *Keep well closed.*

THERAP CAT: Hydrochloride formerly as uterine hemostatic, antiseptic.

4807. Hydrastinine. *5,6,7,8-Tetrahydro-6-methyl-1,3-dioxolo[4,5-g]isoquinolin-5-ol;* 1-hydroxy-6,7-methylenedioxy-2-methyl-1,2,3,4-tetrahydroisoquinoline. $C_{11}H_{13}NO_3$; mol wt 207.23. C 63.76%, H 6.32%, N 6.76%, O 23.16%. By oxidation of hydrastine: Freund, Will, *Ber.* **20**, 88 (1887). From berberine: Freund, *Ger. pat.* 241,136 (1910), *C.A.* **6**, 2145 (1912). From cotarnine: Pyman, Remfry, *J. Chem. Soc.* **101**, 1595 (1912); Topchiev, *J. Appl. Chem. USSR* **6**, 529 (1933). From formylhomopiperonylamine: Decker *et al., Ann.* **395**, 299, 321, 328 (1913); Rosenmund, *Ber. Deut. Pharm. Ges.* **29**, 200 (1919). From safrole: Kindler, Peschke, *Arch. Pharm.* **270**, 353 (1932). Structure study: Schneider, Müller, *Ann.* **615**, 34 (1958).

Needles from petr ether, mp 117°. Freely sol in alcohol, chloroform, ether, dil acids; practically insol in cold, moderately sol in hot water. Solns with water or alcohol are yellow and fluorescent, those with nonpolar organic solvents are colorless. pK 2.62. Absorption spectra: Dobbie, Tinkler, *J. Chem. Soc.* **85**, 1005 (1904).

THERAP CAT: The hydrochloride as cardiotonic; uterine hemostatic.

4808. Hydrastis. Golden seal; orange root; yellow root; yellow puccoon; Indian turmeric. Dried rhizome and roots of *Hydrastis canadensis* L., *Ranunculaceae,* contg not less than 2.5% ether-soluble alkaloids. *Habit.* North America. *Constit.* 2-4% hydrastine, 2-3% berberine; canadine, volatile oil, resin.

4809. Hydrazine. Hydrazine anhydrous. H_4N_2; mol wt 32.05. H 12.58%, N 87.42%. H_2NNH_2. Prepn from hydrazine hydrate: Sisler *et al., J. Am. Chem. Soc.* **76**, 3914 (1954); Schenk in *Handbook of Preparative Inorganic Chem-*

istry vol. 1, G. Brauer, Ed. (Academic Press, New York, 1963) pp 469-472. Toxicity data: Witkin, *Arch. Ind. Health* **13**, 34 (1956). Toxicology study: Back, Thomas, *Ann. Rev. Pharmacol.* **10**, 395 (1970). Review of carcinogenicity studies: *IARC Monographs* **4**, 127-136 (1974); of toxicology: R. von Burg, T. Stout, *J. Appl. Toxicol.* **11**, 447-450 (1991). Books: L. F. Audrieth, B. A. Ogg, *The Chemistry of Hydrazine* (Wiley, New York, 1951); C. C. Clark, *Hydrazine* (Mathieson Chem., Baltimore, 1953). *Reviews:* Troyan, *Ind. Eng. Chem.* **45**, 2608-2612 (1953); Zimmer, *Chem. Ztg.* **79**, 599-605 (1955); Hudson *et al.,* "Hydrazine" in *Mellor's* vol. VIII, suppl. II, *Nitrogen* (Part 2), 69-114 (1967); Jones in *Comprehensive Inorganic Chemistry* vol. 2, J. C. Bailar, Jr. *et al.,* Eds. (Pergamon Press, Oxford, 1973) p 250-265; H. W. Schiessl in Kirk-Othmer *Encyclopedia of Chemical Technology* vol. 13 (John Wiley & Sons, New York, 4th ed., 1995) pp 560-606.

Colorless oily liq, fuming in air. Penetrating odor resembling that of ammonia. Burns with violet flame. Explodes during distn if traces of air are present, also affected by uv and metal ion catalysts. Can be stored for years if sealed in glass and kept in a cool, dark place. Flash and fire pt 126°F (52°C). Contracts on freezing. d_4^{-5} 1.146; d_4^0 1.0253; d_4^2 1.024; d_4^{15} 1.011; d_4^{25} 1.0036; d_4^{35} 0.9955. One gallon of commercial product weighs 8.38 lbs. mp 2.0°. bp_{760} 113.5°; bp_{71} 56°; $bp_{5\ atm}$ 170°; $bp_{10\ atm}$ 200°; $bp_{20\ atm}$ 236°. n_D^{25} 1.46979; n_D^{35} 1.46444. Dipole moment 1.83-1.90. Dielectric constant (25°): 51.7. Latent heat of fusion (mp): 3.025 kcal/mole; latent heat of vaporization (bp): 9760 kcal/mole (calc). Crit temp 380°; crit pressure 14 atm. Diacidic base. pK_1 (25°): ~6.05. Forms salts with inorganic acids. Highly polar solvent. Powerful reducing agent. Dissolves many inorganic substances. Misc with water, methyl, ethyl, propyl, isobutyl alcohols. Forms an azeotropic mixture with water, bp_{760} 120.3°, which contains 55 mole-% (68.5 weight-%) N_2H_4. LD_{50} in mice (mg/kg): 57 i.v.; 59 orally (Witkin).

Dihydrochloride, $H_4N_2 \cdot 2HCl$, white crystalline powder, mp 198°. d 1.42. Freely sol in water, slightly in alcohol.

Caution: Potential symptoms of overexposure to hydrazine are irritation of eyes, nose and throat; temporary blindness; dizziness, nausea; dermatitis; burns skin and eyes. *See NIOSH Pocket Guide to Chemical Hazards* (DHHS/NIOSH 90-117, 1990) p 124. *See also* Patty's *Industrial Hygiene and Toxicology,* vol. 2E, G. D. Clayton, F. E. Clayton, Eds. (John Wiley & Sons, Inc., New York, 4th ed., 1994) pp 3435-3441. Hydrazine may reasonably be anticipated to be a carcinogen: *Seventh Annual Report on Carcinogens* (PB95-109781, 1994) p 231.

USE: Chemical intermediate in manuf of agricultural chemicals, spandex fibers and antioxidants. Reducing agent; organic hydrazine derivs; rocket fuel. Dihydrochloride as chlorine scavenger for HCl gas streams.

4810. Hydrazine Hydrate. H_6N_2O; mol wt 50.06. H 12.08%, N 55.96%, O 31.96%. $H_2NNH_2 \cdot H_2O$. Prepd from hydrazine sulfate by the action of NaOH, followed by distn under nitrogen.

Fuming refractive liquid, faint characteristic odor. *Violent poison! Causes delayed eye irritation.* d^{21} 1.03. mp −51.7° or below −65° (two eutectics). bp_{740} 118-119°; bp_{26} 47°. n_D^{20} 1.42842. Strong base, very corrosive, attacks glass, rubber, cork, but not stainless V_2A steel or Allegheny stainless 304 and 347. Molybdenum steels such as Allegheny stainless 316 should not be used. Very powerful reducing agent. Miscible with water and alcohol. Insol in chloroform and ether.

Mixture with methanol, *C-Stuff.*

USE: Reducing agent, solvent for inorganic materials. Manuf "Helman" catalyst, consisting of 80% hydrazine hydrate, 19.5% ethanol, 0.5 to 0.05% copper, used to dec hydrogen peroxide in V-2 type rockets. Mixture with methanol as propellant for rocket engines.

4811. Hydrazine Sulfate. Hydrazinium sulfate; hydrazonium sulfate. $H_6N_2O_4S$; mol wt 130.12. H 4.65%, N 21.53%, O 49.18%, S 24.64%. $H_2NNH_2 \cdot H_2SO_4$. Prepd by Raschig synthesis: $2NH_3 \cdot aq + [Ca(OCl)_2/Na_2CO_3/colloid]$ and treatment with H_2SO_4. Starch, glue, or gelatin are used as colloids, and sodium hypochlorite may be used instead of bleaching powder: Adams, Brown, *Org. Syn.* **2**, 37 (1922);

Audrieth, Nickles, *Inorg. Syn.* **1**, 90 (1939). Industrial prepn by the action of sodium hypochlorite on urea in the presence of NaOH: *B.I.O.S. Final Report* **369**; Moncrieff, *Manuf. Chem.* **18**, 177 (1947). Revised lab procedures: Pfeiffer, Simons, *Ber.* **80**, 127 (1947); Adams, Brown, *Org. Syn.* **coll. vol. I**, 2nd ed. (1941), p 309. Crystal structure: Nitta *et al., Acta Cryst.* **4**, 289 (1951); Jönsson, Hamilton, *ibid.* **26B**, 536 (1970). Review of activity and clinical studies in cancer cachexia: J. Gold, *Nutr. Cancer* **9**, 59-66 (1987).

Orthorhombic crystals. Glass-like plates or prisms. d 1.378: Curtis, Jay, *J. Prakt. Chem.* **39**, 39 (1889); d⁷ 2.016. mp 254°. Sol in about 33 parts water; freely sol in hot water. Insol in alcohol. pH of 0.2 molar aq soln 1.3.

Note: This substance may reasonably be anticipated to be a carcinogen: *Seventh Annual Report on Carcinogens* (PB95-109781, 1994) p 231.

USE: In the gravimetric estimation of nickel, cobalt and cadmium; in the refining of rare metals; as antioxidant in soldering flux for light metals; as reducing agent in the analysis of minerals and slags; in separating polonium from tellurium; in tests for blood; for destroying fungi and molds; in the prepn of hydrazine hydrate.

4812. Hydrazine Tartrate. Hydrazine acid tartrate; hydrazine hydrogen tartrate; hydrazine bitartrate. $C_4H_{10}N_2O_6$; mol wt 182.13. C 26.38%, H 5.53%, N 15.38%, O 52.71%. $H_2NNH_2 \cdot C_4H_6O_6$.

Crystals, mp 182-183°. $[\alpha]_D^{20}$ +22.5°. Soly in water at 0° about 6 g/100 ml. pH of a satd aq soln 3.6.

USE: In chemical deposition of metals (silvering mirrors, etc.): Owen, U.S. pat. **2,801,935** (1957 to Merck & Co.).

4813. 4-Hydrazinobenzenesulfonic Acid. *p*-Sulfophenylhydrazine; phenylhydrazine-*p*-sulfonic acid. $C_6H_8N_2O_3S$; mol wt 188.21. C 38.29%, H 4.28%, N 14.88%, O 25.50%, S 17.04%. Prepn by sulfonation of phenylhydrazine: L. Claisen, P. Roosen, *Ann.* **278**, 296 (1894); by the reduction of *p*-diazobenzenesulfonic acid: Th. Zincke, A. Kuchenbecker, *Ann.* **330**, 1 (1903); L. V. Lazeeva *et al.,* USSR pat. **1,057,-493** (1983 to Tambov Pigment), *C.A.* **100**, 138755q (1984). Used in resoln of 2-pyrazoline cmpds: M. Mukai *et al., Can. J. Chem.* **57**, 360 (1979); in isoln of volatile ketones: W. Treibs, H. Röhnert, *Ber.* **84**, 433 (1951); in analysis of trace amounts of selenium: T. Kawashima *et al., Anal. Chim. Acta* **49**, 443 (1970); *eidem, ibid.* **89**, 65 (1977).

Needles from water, mp 286°. Slightly sol in water, alcohol.

4814. 2-Hydrazinoethanol. 2-Hydroxyethylhydrazine; β-hydroxyethylhydrazine; Omaflora. $C_2H_8N_2O$; mol wt 76.10. C 31.57%, H 10.60%, N 36.81%, O 21.02%. HO-$CH_2CH_2NHNH_2$. Prepn from hydrazine monohydrate and 2-chloroethanol: Gansser, Rumf, *Helv. Chim. Acta* **36**, 1423 (1953); from hydrazine monohydrate and ethylene oxide: Gever, O'Keefe, U.S. pat. **2,660,607** (1953 to Eaton Labs.); from hydrazine and ethylene oxide: Brit. pat. **776,113** (1957 to Olin Mathieson).

Colorless, slightly viscous liquid. d 1.11. One gallon weighs 9.26 lbs. mp −70°. $bp_{17.5}$ 110-130°; bp_{25} 145-153°. Flash pt 224°F (106°C). Misc with water. Sol in the lower alcohols. Slightly sol in ether.

USE: Plant growth regulant.

4815. Hydrazoic Acid. Hydrogen azide; hydronitric acid; triazoic acid; stickstoffwasserstoffsäure (German). HN_3; mol wt 43.03. H 2.34%, N 97.66%. Produced by the action of sulfuric acid on sodium azide: L. F. Audrieth, C. F. Gibbs, *Inorg. Syn.* **1**, 77 (1939); using stearic acid: Günther, Meyer, *Z. Elektrochem.* **41**, 541 (1935). Prepn of water and ether solns of hydrazoic acid: W. S. Frost *et al., J. Am. Chem. Soc.* **55**, 3516 (1933); L. F. Audrieth, C. F. Gibbs, *loc. cit.;* P. W. Schenk in *Handbook of Preparative Inorganic Chemistry* **vol. 1**, G. Brauer, Ed. (Academic Press, New York, 2nd ed., 1963) pp 472-474. GC determn: J. M. Zehner, R.A. Simonaitis, *J. Chrom. Sci.* **14**, 493 (1976). Toxic-

ity study: Graham *et al., J. Ind. Hyg. Toxicol.* **30**, 98 (1948). Review of toxicology: C. F. Reinhardt, M. R. Brittelli, in Patty's *Industrial Hygiene & Toxicology* Vol. 2A, G. D. Clayton, F. E. Clayton, Eds. (Wiley-Interscience, New York, 1981) pp 2779-2784. *Reviews:* Mason in *Mellor's Comprehensive Treatise on Inorganic and Theoretical Chemistry* **vol. VIII**, suppl. II, *Nitrogen* (part II), 1-15 (1967); Jones in *Comprehensive Inorganic Chemistry* **vol. 2**, J. C. Bailar Jr. *et al.,* Eds. (Pergamon Press, Oxford, 1973) pp 276-293.

Mobile liquid. Intolerable pungent odor. *Extremely explosive!* mp −80°. bp 37°. pKa 4.72. LD_{50} in mice (mg/kg): 21.5 i.p. (Graham).

Caution: Acute exposure: eye irritation, cough, headache, fall in blood pressure, weakness, collapse. *Chronic exposure:* hypotension, weakness, palpitation, ataxia (Graham).

USE: Industrially in prepn of heavy metal azides for shell detonators.

4816. Hydrindantin. *2,2'-Dihydroxy-[2,2'-bi-1H-indene]-1,1',3,3'-(2H,2'H)-tetrone; 2,2'-dihydroxy-[2,2'-biindan]-1,1',3,3'-tetrone;* reduced ninhydrin. $C_{18}H_{10}O_6$; mol wt 322.27. C 67.09%, H 3.13%, O 29.79%. Formed by the action of potassium cyanide on ninhydrin: Bruice, Richards, *J. Org. Chem.* **23**, 145 (1958). Convenient prepn by reduction of ninhydrin with ascorbic acid: Moore, Stein, *J. Biol. Chem.* **211**, 907 (1954).

Dihydrate, prisms from acetone, anhydr at 100°, turns reddish-brown at 200°, dec 249-254°. Very sparingly sol in hot water; sol in Methyl Cellosolve. Sol with decompn in aq Na_2CO_3 solns (deep red color), and in NaOH solns (deep blue color). Can be precipitated from carbonate solns by the addn of acid. Deep purple color with ammonia and blue color with amino acids.

USE: Reagent for the photometric determination of amino acids and similar compds.

4817. Hydriodic Acid. A soln of hydrogen iodide in water. Marketed in various concns, *e.g.,* 57% HI, d 1.7; 47%, d 1.5; 10%, d 1.1. Prepd by absorption of hydrogen iodide gas in water or by the action of iodine on hydrogen sulfide according to the eq $H_2S + I_2 \rightarrow 2HI + S$: Heisig, Frykholm, *Inorg. Syn.* **1**, 157 (1939). *See also* Hydrogen Iodide.

Colorless when freshly made, but rapidly turns yellowish or brown on exposure to light and air. This discoloration can be prevented by the addition of about 1.5% hypophosphorous acid (H_3PO_2). Concd solns that have been stored for some time are usually opaque from oxidation, they may be regenerated with hypophosphorous acid: Foster, Nahas, Jr., *Inorg. Syn.* **2**, 210 (1946). *Keep protected from light and air, preferably not above 30°.* Miscible with water or alcohol. Dissolves iodine. The azeotrope (constant-boiling acid) bp_{760} 127°, d 1.70, contains 56.9% HI. Strong, corrosive acid, attacks natural rubber. 0.1 molar soln, pH 1.0.

Caution: Strong irritant.

USE: Reducing agent, manuf of inorganic iodides, pharmaceuticals, disinfectants. The 57% acid is also used for analytical purposes, such as methoxyl determinations.

THERAP CAT: Expectorant.

4818. Hydrobenzoin. *1,2-Diphenyl-1,2-ethanediol;* diphenylethyleneglycol. $C_{14}H_{14}O_2$; mol wt 214.26. C 78.48%, H 6.59%, O 14.93%. Prepn of *d-, l-, dl-* and *meso*-forms: Forst, Zincke, *Ann.* **182**, 262, 275 (1876); Irvine, Weir, *J. Chem. Soc.* **91**, 1390 (1907); Buck, Jenkins, *J. Am. Chem. Soc.* **51**, 2163 (1929); L. F. Fieser, *Organic Experiments* (D. C. Heath, Boston, 1964) pp 210, 214, 216-217, 229-231. Improved method for prepn of *dl*-form from the *meso* isomer: Collet, *Synthesis* **1973**, 664.

dl-Form, crystals from ether + petr ether, mp 120°. Resolved into d- and l-forms by slow, repeated crystallizations from ether.

d-Form, $[\alpha]_D^{20}$ +97.6° (chloroform).

l-Form, $[\alpha]_D^{20}$ — 97.0° (chloroform).

meso-Form, monoclinic leaflets from alcohol or water, mp 139°. Soly in water: 0.25% at 20°, 1.25% at 100°. Freely sol in hot alcohol, chloroform. Infrared absorption (chloroform): 2.82, 2.96μ. Dipole moment 2.67.

4819. Hydrobromic Acid. A soln of hydrogen bromide gas in water. Marketed in various concns, e.g., 50% HBr, d 1.517; 40% HBr, d 1.38; 34%, d 1.31; 10%, d 1.08. Lab prepn according to the eq H_2SO_4 + KBr → $KHSO_4$ + HBr: Heisig, Amdur, *Inorg. Syn.* **1**, 155 (1939). *See also* Hydrogen Bromide.

Colorless or faintly yellow; slowly darkens on exposure to air and light. Miscible with water, alcohol. *Keep protected from light.* When dil hydrobromic acid is distilled, a weaker acid comes over first and when a very concd acid is boiled, HBr gas chiefly distills over first; in both cases a "constant boiling" acid contg about 47.5% HBr remains which distills unchanged at 126°. The bp and compn of the azeotrope varies with pressure: at 100 mm, bp 74.12°, compn 49.80% HBr; at 400 mm, bp 107.00°, compn 48.47%; at 700 mm, 122°, compn 47.74%; at 800 mm, bp 125.79°, compn 47.56%: Bonner *et al., J. Am Chem. Soc.* **55**, 1406 (1933). Aq solns are strongly acid. The satd aq soln contains 68.85% HBr at 0° and 66% at 25°.

Caution: Strong irritant.

USE: The concd acid is used principally in analytical chemistry and organic prepns.

THERAP CAT: Sedative.

THERAP CAT (VET): Has been used as a sedative.

4820. Hydrocarbostyril. *3,4-Dihydro-2(1H)-quinolinone;* 3,4-dihydrocarbostyril; 3,4-dihydro-2-quinolinol; *o*-aminohydrocinnamic acid lactam; 2-oxo-1,2,3,4-tetrahydroquinoline; dihydro-α-quinolone. C_9H_9NO; mol wt 147.18. C 73.45%, H 6.16%, N 9.52%, O 10.87%. Prepd by the catalytic reduction of *o*-nitrocinnamic acid: Blout, Silverman, *J. Am. Chem. Soc.* **66**, 1442 (1944).

Prisms from methanol + water, mp 165-166.5°. Freely sol in alcohol, ether, dimethylformamide. Practically insol in water. Sol in hot aq NaOH solns.

4821. Hydrochloric Acid. Muriatic acid. A soln of hydrogen chloride gas (HCl) in water. Prepn and reviews: *see* Hydrogen Chloride.

Fumes in air. May be colored yellow by traces of iron, chlorine, and organic matter. Reagent grade concd hydrochloric acid contains close to 38.0% HCl. 83 ml of concd HCl poured into sufficient water to make 1 liter yields approx 1.0N HCl. The pH of 1.0N HCl is 0.10; of 0.1N = 1.10; of 0.01N = 2.02; of 0.001N = 3.02; of 0.0001N = 4.01. n_D^{18} (1.0N soln) 1.34168. d_4^{15} 1.05 (10.17% w/w soln); 1.10 (20%); 1.15 (29.57%); 1.20 (39.11%). Freezing pt: — 17.14° (10.81% soln); — 62.25° (20.69%); — 46.2° (31.24%); — 25.4° (39.17%), *Gmelin's, Chlorine* (8th ed.) **6**, 136-137 (1927). Constant boiling azeotrope with water bp_{760} 108.58° contg 20.22% HCl, d_4^{15} 1.096. Boiling weaker or stronger aq solns results in loss of either component until the constant boiling acid is obtained.

Caution: Corrosive burns may result from the inhalation of acid fumes and from skin contact with or the ingestion of strong acid. Symptoms after ingestion or skin contact include immediate pain and ulceration of all membranes and tissues which come in contact with the acid. Ingestion may be associated with nausea, vomiting and intense thirst; corrosion of the stomach may lead within a few hours or a few days to gastric perforation and peritonitis. Late esophageal, gastric and pyloric strictures and stenoses should be anticipated. Contact of conc acid with the eye can cause extensive necrosis of the conjunctiva and corneal epithelium, resulting in perforation or opaque scarring. Chemical pneumonitis can be expected after respiratory exposure to acid vapors or after tracheobronchial aspiration of ingested acid. Death may occur due to complications such as circulatory shock, asphyxia due to glottic or laryngeal edema, perforation of the stomach with peritonitis, gastic hemorrhage, infection or anition due to stricture formation. *See Clinical Toxicology of Commercial Compounds*, R. E. Gosselin *et al.,* Eds. (Williams & Wilkins, Baltimore, 5th ed., 1984) Section III, pp 8-11.

USE: In the production of chlorides; refining ore in the production of tin and tantalum; for the neutralization of basic systems; as laboratory reagent; hydrolyzing of starch and proteins in the prepn of various food products; pickling and cleaning of metal products; as catalyst and solvent in organic syntheses. Also used for oil- and gas-well treament and in removing scale from boilers and heat-exchange equipment. Pharmaceutic aid (acidifier).

THERAP CAT (VET): Has been used as gastric acidifier.

4822. Hydrochlorothiazide. *6-Chloro-3,4-dihydro-2H-1,2,4-benzothiadiazine-7-sulfonamide 1,1-dioxide;* 6-chloro-3,4-dihydro-7-sulfamoyl-2H-1,2,4-benzothiadiazine 1,1-dioxide; 6-chloro-7-sulfamyl-3,4-dihydro-1,2,4-benzothiadiazine 1,1-dioxide; 3,4-dihydrochlorothiazide; chlorsulfonamidodihydrobenzothiadiazine dioxide; chlorosulthiadil; Aquarius; Bremil; Cidrex; Dichlorosal; Dichlotride; Diclotride; Direma; Diu-melusin; Disalunil; Esidrex; Esidrix; Fluvin; Hidroronol; Hydril; Hydro-Aquil; Hydro-Diuril; Hydrosaluric; Hydrothide; Hypothiazide; Ivaugan; Maschitt; Nefrix; Neo-Codema; Neoflumen; Oretic; Panurin; Thiaretic; Thiuretic; Urodiazin; Vetidrex. $C_7H_8ClN_3O_4S_2$; mol wt 297.74. C 28.24%, H 2.71%, Cl 11.91%, N 14.11%, O 21.49%, S 21.54%. Prepn: de Stevens *et al., Experientia* **14**, 463 (1958); Werner *et al., J. Am. Chem. Soc.* **82**, 1161 (1960); Jones, Novello, U.S. pat. **3,025,292** (1962 to Merck & Co.); de Stevens, Werner, U.S. pat. **3,163,645** (1964 to Ciba); Klosa, **Ger.** pat. **1,163,332** (1964); J. S. Irons, T. M. Cook, U.S. pat. **3,164,588** (1965 to Merck & Co.). Purification: Downing, U.S. pat. **3,043,840** (1962 to Merck & Co.). Toxicity data: J. J. Piala *et al., J. Pharmacol. Exp. Ther.* **134**, 273 (1961). Comprehensive description: H. P. Deppeler, *Anal. Profiles Drug Subs.* **10**, 405-441 (1981).

Crystals, mp 273-275°. uv max (methanol + trace HCl): 317, 271, 226 nm ($A_{1cm}^{1\%}$ 130, 654, 1280). pKa 7.9, 9.2. Sol in dil ammonia, or sodium hydroxide; also sol in methanol, ethanol, acetone. Practically insol in water. LD_{50} in mice (mg/kg): 590 i.v.; > 8000 orally (Piala).

THERAP CAT: Diuretic.

THERAP CAT (VET): Diuretic.

4823. Hydrocinchonidine. *(8α,9R)-10,11-Dihydrocinchonan-9-ol;* (—)-dihydrocinchonidine; cinchamidine. $C_{19}H_{24}N_2O$; mol wt 296.41. C 76.99%, H 8.16%, N 9.45%, O 5.40%. Found in cinchona barks: Forst, Böhringer, *Ber.* **14**, 1270 (1881); Hesse, *ibid.* 1683; *idem, Ann.* **214**, 1 (1882); Skita, Nord, *Ber.* **45**, 3312 (1912); Heidelberger, Jacobs, *J. Am. Chem. Soc.* **41**, 817 (1919). Configuration: Ochiai *et al., J. Pharm. Soc. Japan* **67**, 211 (1947). Stereospecific synthesis from secologanin: R. T. Brown, D. Curless, *Tetrahedron Letters* **27**, 6005 (1986). Total synthesis: M. Ihara *et al., J. Chem. Soc. Perkin Trans. I* **1988**, 1277. Stereoisomer of hydrocinchonine, *q.v.*

Needles or leaflets, mp 230°. $[\alpha]_D^{15}$ −98° (alcohol). Practically insol in water. Sol in alcohol, chloroform; slightly sol in ether.

4824. Hydrocinchonine. *(9S)-10,11-Dihydrocinchonan-9-ol;* cinchotine; cinconifine; (+)-dihydrocinchonine; pseudocinchonine; $C_{19}H_{24}N_2O$; mol wt 296.41. C 76.99%, H 8.16%, N 9.45%, O 5.40%. From cinchona barks: Caventou, Willm, *Compt. Rend.* **69,** 284 (1869). Prepn from cinchonine: Hesse, *Ann.* **300,** 46 (1898); Pum, *Monatsh.* **16,** 68 (1895); Arlt, *ibid.* **20,** 426, 439 (1899); Heidelberger, Jacobs, *J. Am. Chem. Soc.* **41,** 817 (1919). Structure: Rabe, *Ber.* **55,** 522 (1922). Conversion of hydroquinidine to hydrocinchonine: King, *J. Chem. Soc.* **1946,** 523. HPLC determn: C.-T. A. Chung, E. J. Staba, *J. Chromatog.* **295,** 276 (1984). Stereospecific synthesis from secologanin: R. T. Brown, D. Curless, *Tetrahedron Letters* **27,** 6005 (1986). Total synthesis: M. Ihara *et al., J. Chem. Soc. Perkin Trans. I* **1988,** 1277.

Prisms, mp 268-269°. $[\alpha]_D^{14}$ +204° (c = 0.6 in alc). Almost insol in water, ether. Sol in alcohol.

Hydrochloride, $C_{19}H_{24}N_2O.HCl$, crystals, mp 220-221°. $[\alpha]_D^{23}$ +155° (c = 0.8 in water).

4825. Hydrocinnamic Acid. *Benzenepropanoic acid;* 3-phenylpropionic acid; β-phenylpropionic acid; benzylacetic acid. $C_9H_{10}O_2$; mol wt 150.18. C 71.98%, H 6.71%, O 21.31%. $C_6H_5CH_2CH_2COOH$. Prepn by reduction of cinnamic acid: Ingersoll, *Org. Syn.* **9,** 42 (1929); from propiophenone: Schwenk, Papa, *J. Org. Chem.* **11,** 798 (1946); by Mauer oxidation of 1-phenyl-3-propanol: Langenbeck, Richter, *Ber.* **89,** 202 (1956); by chromate oxidation of propylbenzene: Reitsema, Allphin, *J. Org. Chem.* **27,** 27 (1962).

White, cryst powder, mp 47-48°. bp 280°; bp_{75} 194-197°; bp_{18} 145-147°; bp_6 125-129°. Sol in 170 parts cold, more sol in hot water, in alcohol, benzene, chloroform, ether, glacial acetic acid, petr ether, carbon disulfide.

4826. Hydrocodone. *4,5-Epoxy-3-methoxy-17-methyl-morphinan-6-one;* dihydrocodeinone; Bekadid; Dicodid. $C_{18}H_{21}NO_3$; mol wt 299.37. C 72.22%, H 7.07%, N 4.68%, O 16.03%. Prepn by hydrogenation of codeinone: Mannich, Löwenheim, *Arch. Pharm.* **258,** 295 (1920); by oxidation of dihydrocodeine, **Ger. pat.** 415,097 (1925 to E. Merck), *Frdl.* **15,** 1518 (1925-1927); by catalytic rearrangement of codeine: **Ger. pat.** 623,821. Industrial prepn from dihydrocodeine: K. Pfister, M. Tishler, **U.S. pat.** 2,715,626 (1955 to Merck & Co.). Pharmacology and toxicity data: N. B. Eddy, J. G. Reid, *J. Pharmacol. Exp. Ther.* **52,** 468 (1934). Brief description: A. Stein, *Pharmazie* **10,** 180 (1955). *Review:* Small, Lutz, "Chemistry of the Opium Alkaloids," Suppl. No. 103, Public Health Reports, Washington (1932).

Prisms from alcohol, mp 198°. Sol in alcohol, dil acids. Insol in water. uv max: 280 nm (ε 1310). LD_{50} s.c. in mice: 85.7 mg/kg (Eddy, Reid).

Hydrochloride monohydrate, $C_{18}H_{21}NO_3.HCl.H_2O$, crystals, mp 185-186° dec. $[\alpha]_D^{27}$ −130° (c = 2.877). Very sol in water.

Bitartrate hemipentahydrate, $C_{18}H_{21}NO_3.C_4H_6O_6.$-2½$H_2O$, *Calmodid, Codinovo, Duodin, Kolikodal, Orthoxycol, Mercodinone, Synkonin, Norgan, Hydrokon.* Needles, mp 118-128°. One gram dissolves in 16 ml water, in 150 g 95% ethanol. Almost insol in ether, chloroform. pH of a 2% aq soln about 3.6.

Hydriodide, $C_{18}H_{21}NO_3.HI$, mp 219-220°.

Methiodide, $C_{18}H_{21}NO_3.CH_3I$, mp 250-255°.

Caution: May be habit forming. This is a controlled substance (opiate) listed in the U.S. Code of Federal Regulations, Title 21 Parts 329.1 and 1308.12 (1995).

THERAP CAT: Analgesic (narcotic); antitussive.

4827. Hydrocortamate. *N,N-Diethylglycine (11β)-11,-17-dihydroxy-3,20-dioxopregn-4-en-21-yl ester;* cortisol 21-ester with *N,N-diethylglycine;* cortisol 21-(N,N-diethyl)glycinate; hydrocortisone 21-diethylaminoacetate; Ulcort. $C_{27}H_{41}NO_6$; mol wt 475.63. C 68.18%, H 8.69%, N 2.94%, O 20.18%. Prepn: Pinson, Laubach, **Ger. pat.** 1,016,708 (1957 to Pfizer), *C.A.* **54,** 22737a (1960); Richter, Schenck, **Ger. pat.** 1,037,451 (1958 to Schering AG), *C.A.* **54,** 22730h (1960).

Crystals from ethyl acetate, mp 162-163°.

Hydrochloride, $C_{27}H_{41}NO_6.HCl$, *ethamicort, Magnacort.* Crystals from ethyl acetate, dec 222°.

THERAP CAT: Glucocorticoid.

4828. Hydrocortisone. *(11β)-11,17,21-Trihydroxy-pregn-4-ene-3,20-dione;* cortisol; 4-pregnene-11β,17α,21-triol-3,20-dione; 17-hydroxycorticosterone; anti-inflammatory hormone; Kendall's compound F; Reichstein's substance M; Aeroseb-HC; Ala-Cort; Anflam; Cetacort; Cort-Dome; Cortef; Cortenema; Cortril; Dermacort; Dermocortal; Dermolate; Dioderm; Efcortelan; Evacort; Ficortril; Hydracort; Hydro-Adreson; Hydrocort; Hydrocortisyl; Hydrocortone; Hytone; Lacticare-HC; Medicort; Mildison; Nutracort; Penecort; Proctocort; Scheroson F; Synacort; Texacort; Timocort; Zenoxone. $C_{21}H_{30}O_5$; mol wt 362.47. C 69.59%, H 8.34%, O 22.07%. Principal glucocorticoid hormone produced by the adrenal cortex. Biosynthesis stimulated by ACTH, *q.v.* Circulates in plasma primarily bound to *corticosteroid-binding globulin*, also known as *transcortin*, and to albumin. Isoln from adrenal glands: Reichstein, *Helv. Chim. Acta* **20,** 953 (1937); Mason *et al., J. Biol. Chem.* **124,** 459 (1938); from urine: Mason, Sprague, *ibid.* **175,** 451 (1948); from blood: Reich *et al., ibid.* **187,** 411 (1950). Configuration: von Euw, Reichstein, *Helv. Chim. Acta* **25,** 988 (1942); **30,** 205 (1947). Synthesis: H. Wendler *et al., J. Am. Chem. Soc.* **72,** 5793 (1950). Biosynthesis by isolated adrenal glands: O. Hechter *et al., Arch. Biochem. Biophys.* **25,** 457 (1950); A. Zaffaroni *et al., J. Am. Chem. Soc.* **73,** 1390 (1951). Prepn by microbial transformation: H. C. Murray, D. H. Peterson, **U.S. pat.** 2,602,769 (1952 to Upjohn); D. R. Colingsworth *et al., J. Biol. Chem.* **203,** 807 (1953). Comprehensive description: K. Florey, *Anal. Profiles Drug Subs.* **12,** 277-324 (1983). Review of clinical use in dermatoses: A. M. Kligman, K. H. Kaidbey, *Cutis* **22,** 232-244 (1978). Review of clinical assays in serum and urine: A. Moore *et al., Ann. Clin. Biochem.* **22,** 435-454 (1985). Physiological role in immunity: W. M. Jefferies,

Med. Hypoth. **34**, 198-208 (1991); in fetal maturation: G. C. Liggins, *Reprod. Fertil. Devel.* **6**, 141-150 (1994).

Bitter-tasting crystalline, striated blocks from abs ethanol or isopropanol, mp 217-220° with some decompn. $[\alpha]_D^{22}$ +167° (abs ethanol). uv max: 242 nm ($E_{1cm}^{1\%}$ 445). Soly (mg/ml) at 25°: water 0.28; ethanol 15.0; methanol 6.2; acetone 9.3; chloroform 1.6; propylene glycol 12.7; ether about 0.35. Sol in concd sulfuric acid with intense green fluorescence.

21-Acetate, $C_{23}H_{32}O_6$, *Colifoam, Colofoam, Cortaid, Cordes, Cortifoam, Efcorlin, Hc45, Hydrocal, Hydrocortistab, Hydrocortone Acetate, Lanacort, Lenirit, Sigmacort, Sintotrat, Velopural.* Monoclinic, sphenoidal, tabular crystals from dil acetone. Tasteless. d_4^{20} 1.289; dec 223°. $[\alpha]_D^{25}$ +166° (c = 0.4 in dioxane); $[\alpha]_D^{25}$ +150.7° (c = 0.5 in acetone). uv max (methanol): 242 nm ($E_{1cm}^{1\%}$ 390). Somewhat hygroscopic. Soly in water: 1 mg/100 ml; in ethanol: 0.45 g/100 ml; in methanol: 3.9 mg/ml; in acetone: 1.1 mg/g; in ether: 0.15 mg/ml. One gram dissolves in about 200 ml chloroform. Very sol in DMF; sol in dioxane.

17-Butyrate, $C_{25}H_{36}O_6$, *Alfason, Laticort, Locoid, Plancol.*

21-Phosphate disodium salt, $C_{21}H_{29}Na_2O_8P$, *hydrocortisone sodium phosphate, Cleiton, Efcortesol, Hydrocortone Phosphate.* White powder. uv max (methanol): 242 nm ($A_{1cm}^{1\%}$ 298-341). $[\alpha]_D^{25}$ +120° (H_2O). Soly in water (25°): > 500 mg/ml. pH of a 1% aq soln: 7.5-8.5.

21-Sodium succinate, $C_{25}H_{33}NaO_8$, *hydrocortisone hemisuccinate sodium salt, A-hydroCort, Buccalsone, Corlan, Efcortelan Soluble, Saxizon, Solu-Cortef, Solu-Glyc.* Amorphous, hygroscopic, white powder, mp 169.0-171.2°. Soly in water: ~500 mg/ml. Similarly sol in methanol, ethanol; sparingly sol in chloroform.

17-Valerate, $C_{26}H_{38}O_6$, *hydrocortisone valerate, Westcort.*

THERAP CAT: Glucocorticoid.

THERAP CAT (VET): Glucocorticoid.

4829. Hydrocotarnine. *5,6,7,8-Tetrahydro-4-methoxy-6-methyl-1,3-dioxolo[4,5-g]isoquinoline;* 8-methoxy-5,6-methylenedioxy-2-methyl-1,2,3,4-tetrahydroisoquinoline. $C_{12}H_{15}NO_3$; mol wt 221.26. C 65.14%, H 6.83%, N 6.33%, O 21.69%. Found in mother liquors from morphine extraction. It is not certain whether it is formed from narcotine during the extraction or whether it exists in the poppy plant. May also be prepd by reduction of cotarnine: Topchiev, *J. Appl. Chem. USSR* **6**, 529 (1933), *C.A.* **28**, 2718 (1934); Schneider, Müller, *Ann.* **615**, 34 (1958); Knabe, *Arch. Pharm.* **292**, 652 (1959). Reduction of hydrocotarnine with sodium in alcohol leads to replacement of the methoxyl group by hydrogen, with formation of hydrohydrastinine. Review and bibliography: Small, Lutz, "Chemistry of the Opium Alkaloids," Suppl. No. 103, *Public Health Reports,* Washington (1932).

Hemihydrate, plates from petr ether, mp 56°. Loses water of crystn at 60°. May be distilled with little decompn at 100°: Hesse, *Ber.* **4**, 693 (1871). Absorption spectrum: Hantzsch, *Ber.* **44**, 1816 (1911); Steiner, *Compt. Rend.* **176**, 244, 1379 (1923); Csokán, *Z. Anal. Chem.* **124**, 344 (1942). Almost insol in water, alkaline solns; sol in alcohol, acetone, chloroform, benzene, ether.

Hydrochloride monohydrate, prisms; sol in water.

Hydrobromide, crystals, mp 237°; sparingly sol in water.

Hydriodide, needles from methanol, mp 196°; sol in hot water.

Methiodide, needles from water, plates from alc, mp 206°.

Methobromide, needles from chloroform, dec 221°.

4830. Hydroflumethiazide. *3,4-Dihydro-6-(trifluoromethyl)-2H-1,2,4-benzothiadiazine-7-sulfonamide 1,1-dioxide;* 6-trifluoromethyl-3,4-dihydro-7-sulfamoyl-2H-1,2,4-benzothiadiazine 1,1-dioxide; 3,4-dihydro-7-sulfamyl-6-trifluoromethyl-1,2,4-benzothiadiazine 1,1-dioxide; trifluoromethylhydrothiazide; dihydroflumethiazide; methforylthiazidine; metflorylthiazidine; Bristab; Bristurin; Di-Ademil; Diucardin; Elodrine; Finuret; Hydol; Hydrenox; Leodrine; NaClex; Olmagran; Rodiuran; Rontyl; Saluron; Sisuril; Vergonil. $C_8H_8F_3N_3O_4S_2$; mol wt 331.30. C 29.00%, H 2.43%, F 17.20%, N 12.68%, O 19.32%, S 19.36%. Synthesis: Holdrege *et al., J. Am. Chem. Soc.* **81**, 4807 (1959); Close *et al., ibid.* **82**, 1132 (1960); Yale *et al., ibid.* 2042 (1960); Novello *et al., J. Org. Chem.* **25**, 970 (1960). Numerous patents, *e.g.,* Lund *et al.,* U.S. pat. **3,254,076** (1966 to Lövens Kemiske Fabrik). Pharmacology: J. J. Piala *et al., J. Pharmacol. Exp. Ther.* **134**, 273 (1961). Comprehensive description: C. E. Orzech *et al., Anal. Profiles Drug Subs.* **7**, 297-317 (1978).

Crystals, mp 272-273°. uv max (methanol): 272.5 nm (log ε 4.286). Soly in mg/ml at 25°: acetone > 100; methanol 58; acetonitrile 43; water 0.3; ether 0.2; benzene < 0.1. pK_1 8.9; pK_2 10.7. Forms water-sol salts with bases. LD_{50} in mice (mg/kg): > 8000 orally, 750 i.v., 6280 i.p. (Piala).

THERAP CAT: Antihypertensive; diuretic.

4831. Hydrofluoric Acid. Fluohydric acid. A soln of hydrogen fluoride gas in water. Obtained by distilling calcium fluoride with H_2SO_4. Usually marketed in concns of about 47% and 53%. d 1.15-1.18. Refs: See Hydrogen Fluoride.

Colorless or almost colorless, fuming liquid. *Poisonous!* Miscible with water. Vapor pressure data: Brosheer *et al., Ind. Eng. Chem.* **39**, 423 (1947). Compn of liq and vapor: Munter *et al., ibid.* 427. A weak acid: pKa 3.19. The 38.2% (w/w HF) soln is a binary azeotrope; bp 112.2°. It attacks glass or stoneware, dissolving the silica. *Keep in plastic, lead, wax, or paraffin paper bottles.*

Caution: Acute effects: External contact—liquid or vapor causes severe irritation of eyes and eyelids which may result in prolonged or permanent visual defects or total destruction of eyes. Skin contact may result in severe burns. Handle with care as it causes painful sores on the skin usually noticed on the next day only. Inhalation—extreme irritation of respiratory tract, pulmonary inflammation, congestion; avoid inhaling the vapors. Ingestion—necrosis of esophagus and stomach with nausea, vomiting, diarrhea, circulatory collapse, death. *Chronic effects:* (from inhalation or ingestion), may cause fluorosis. Symptoms—wt loss, malaise, anemia, leukopenia, discoloration of teeth, osteosclerosis.

USE: Cleaning cast iron, copper, brass; removing efflorescence from brick and stone, or sand particles from metallic castings; working over too heavily weighted silks; frosting, etching glass and enamel; polishing crystal glass; decomposing cellulose; enameling and galvanizing iron; increasing porosity of ceramics. Its salts are used as insecticides and to arrest undesirable fermentation in brewing. Also used in analytical work to determine SiO_2, etc.

4832. Hydrofuramide. *1-(2-Furanyl)-N,N'-bis(2-furanylmethylene)methanediamine; N,N'-difurfurylidene-2-furanmethanediamine;* furfuramide. $C_{15}H_{12}N_2O_3$; mol wt 268.27. C 67.16%, H 4.51%, N 10.44%, O 17.89%. Prepn: Hartley, Dobbie, *J. Chem. Soc.* **73**, 598 (1898); Taniyama, *J. Chem. Soc. Japan,* Ind. Chem. Sect. **51**, 33 (1948); Kapur *et al., J. Sci. Ind. Res. (India)* **19B**, 509 (1960); Kamal *et al., Tetra-*

hedron **19**, 869 (1963). Structure: Soundararajan, Anantakrishnan, *Proc. Indian Acad. Sci.* **38A**, 176 (1953).

Brownish crystals from abs alcohol, mp 117°. bp ~250° with decompn. uv max: 259, 215 nm (log ε 4.18, 4.16). Practically insol in water; freely sol in alc, ether; readily dec by acids.

USE: Vulcanization accelerator.

4833. Hydrogen. Protium. H; at. wt 1.0079; at. no. 1; valence 1; elemental state: H_2. Isotopes: 1 (protium 99.9844%); 2 (deuterium 0.0156%); 3 (tritium, traces only). The most abundant element in the known universe. Occurrence in the earth's atmosphere 0.00005% H_2. First recognized as an element by Cavendish in 1766; named by Lavoisier. Obtained by passing H_2O vapors over heated iron; by electrolysis of water or by action of HCl or H_2SO_4 on Fe or Zn; by hydrolysis of metal hydrides. Produced industrially by electrolysis; from methane or coke and steam. *Reviews: Nouveau Traité de Chimie Minérale* vol. **1**, P. Pascal, Ed. (Masson, Paris, 1956) pp 565-675; Mackay in *Comprehensive Inorganic Chemistry* vol. **1**, J. C. Bailar, Jr. *et al.*, Eds. (Pergamon Press, Oxford, 1973) pp 1-76; B. G. Mandelik, D. S. Newsome in Kirk-Othmer *Encyclopedia of Chemical Technology* vol. 12 (Wiley-Interscience, New York, 3rd ed., 1980) pp 938-982. *See also* Deuterium and Tritium.

Colorless, odorless, tasteless gas; flammable or explosive when mixed with air, oxygen, chlorine, etc. mp −259.2° (13.96°K) at 54 mm (triple point). bp −252.77° (20.39°K). d_{gas} 0.069 (air = 1); d_{liq} 0.0700 (at bp); d_{sol} 0.0763 (13°K). A liter of the gas at 0° weighs 0.08987 g. Crit. temp −239.9°; crit press. 12.8 atm. Sol in about 50 vols of water at 0°.

Caution: No specific toxic action. In high concns can act as a simple asphyxiant.

USE: In oxy-hydrogen blowpipe (welding) and limelight; autogenous welding of steel and other metals; manuf ammonia, synthetic methanol, HCl; hydrogenation of oils, fats, naphthalene, phenol; in balloons and airships; manuf tungsten. In thermonuclear reactions: the hydrogen atom has a single peripheral electron (1s). It ionizes to form protons, deuterons (D) or tritons (T). The ionization potential of the H atom is 13.59 eV. Accelerated protons bring about extremely varied nuclear reactions. Liq hydrogen used in bubble chambers to study subatomic particles; as a coolant.

4834. Hydrogen Bromide. *Anhydrous hydrobromic acid.* BrH; mol wt 80.91. Br 98.75%, H 1.25%. HBr. Prepd commercially by direct combination of the elements at 375° preferably over a catalyst such as platinized silica gel or platinized asbestos: Richards, Hönigschmid, *J. Am. Chem. Soc.* **32**, 1581 (1910); Smyth, Hitchcock, *ibid.* **55**, 1830 (1933); Schneider, Johnson, *Inorg. Syn.* **1**, 152 (1939). Lab procedure from tetrahydronaphthalene and bromine: Müller, *Monatsh.* **49**, 29 (1928); Duncan, *Inorg. Syn.* **1**, 151 (1939); Schmeisser in *Handbook of Preparative Inorganic Chemistry* vol. **1**, G. Brauer, Ed. (Academic Press, New York, 2nd ed., 1963) pp 282-286. Detailed description of laboratory methods of prepn: Houben-Weyl, *Methoden der organischen Chemie*, vol **5/4** (Thieme, Stuttgart, 4th ed., 1960) p 16-20. Review of prepn and properties of HBr and other hydrogen halides: Woolf in *Mellor's Comprehensive Treatise on Inorganic and Theoretical Chemistry* vol. **II**, Suppl I (originally published as Suppl II, part 1) 724-741 (1956); John in *Bromine and its Compounds*, Z. E. Jolles, Ed. (Ernest Benn, London, 1966) pp 81-105; Downs, Adams in *Comprehensive Inorganic Chemistry* vol. **2**, J. C. Bailar Jr., *et al.*, Eds. (Pergamon Press, Oxford, 1973) pp 1280-1329.

Colorless, corrosive, nonflammable gas. Acrid odor. Fumes in moist air forming clouds which have a sour taste.

d 2.71 (air = 1.00). mp −86.9°. bp_{760} −66.8°; $bp_{11.0\ atm}$ −4.8°; $bp_{17.1\ atm}$ +12°; $bp_{30.0\ atm}$ 36°; $bp_{59.2\ atm}$ 70°. Crit temp 89.8°; crit press. 84.5 atm. Sp heat (cal/g/°C): solid (−91°) 0.152; liq 0.176; gas (27°) 0.085. Heat of fusion at mp: 7.44 cal/g. Heat of vaporization at bp 51.3 cal/g. Freely sol in water: One vol H_2O dissolves 600 vols HBr gas at 0°. Also sol in alc. Soly in organic solvents: Fernandes, *J. Chem. Eng. Data* **17**, 377 (1972); Gerrard, *Chem. & Ind. (London)* **1969**, 295; Ahmed *et al.*, *J. Appl. Chem.* **20**, 109 (1970). Aq solns are strongly acid. The satd aq soln contains 68.85% HBr at 0° and 66% at 25°. The boiling point of a constant-boiling mixture is 122.5° at 740 mm and 126° at 760 mm. The composition of the constant-boiling mixture is 47.38% HBr at 752 mm. For complete tables *see* Bonner *et al.*, *J. Am. Chem. Soc.* **55**, 1406 (1943). *See also* Hydrobromic Acid. Anhydr HBr is marketed in steel cylinders in the form of a gas over liquid. LC_{50} in mice, rats: 814, 2858 ppm by inhalation, K. C. Back *et al.*, *Reclassification of Materials Listed as Transportation Health Hazards* (TSA-20-72-3, PB 214-270, 1972).

Caution: Potential symptoms of overexposure are irritation of eyes, nose and throat; skin and eye burns. *See NIOSH Pocket Guide to Chemical Hazards* (DHHS/NIOSH 90-117, 1990) p 124.

USE: Mfg organic and inorganic bromides, hydrobromic acid, as reducing agent and as catalyst in controlled oxidations, in the alkylation of aromatic compds, in the isomerization of conjugated diolefins.

4835. Hydrogen Chloride. *Anhydrous hydrochloric acid.* ClH; mol wt 36.46. Cl 97.24%, H 2.76%. HCl. *See also* Hydrochloric Acid. Produced industrially by the interaction of NaCl and H_2SO_4; from NaCl, SO_2, air and water vapor; by controlled combination of the elements; or as a by-product of the synthesis of chlorinated hydrocarbons: A. C. Cumming, *Hydrochloric Acid and Salt Cake* (Gurney and Jackson, London, 1923); N. A. Laury, *Hydrochloric Acid and Sodium Sulfate* (Chem. Catalog Co., New York, 1927); Maude, *Chem. Eng. Progress* **44**, 179 (1948); Faith, Keyes & Clark's *Industrial Chemicals*, F. A. Lowenheim, M. K. Moran, Eds. (Wiley-Interscience, New York, 4th ed., 1975) pp 454-461. Prepn of pure HCl for research purposes: Hönigschmid *et al.*, *Z. Anorg. Allgem. Chem.* **163**, 315 (1927); Kemp, *J. Chem. Ed.* **37**, 142 (1960); Schmeisser in *Handbook of Preparative Inorganic Chemistry* vol. **1**, G. Brauer, Ed. (Academic Press, New York, 2nd ed., 1963) pp 280-282. Toxicity: K. I. Darmer *et al.*, *Am. Ind. Hyg. Assoc. J.* **35**, 623 (1974). Reviews of prepn and properties: Addison, Lewis in *Mellor's Comprehensive Treatise on Inorganic and Theoretical Chemistry* vol. **II**, suppl. I (originally published as suppl. II, part I) 402-475 (1956); Downs, Adams in *Comprehensive Inorganic Chemistry* vol. **2**, J. C. Bailar, Jr. *et al.*, Eds. (Pergamon Press, Oxford, 1973) pp 1280-1329; D. S. Rosenberg in Kirk-Othmer *Encyclopedia of Chemical Technology* vol. 12 (Wiley-Interscience, New York, 3rd ed., 1980) pp 983-1015.

Colorless, corrosive, nonflammable gas. Characteristic pungent odor. Fumes in air. d 1.268 (air = 1.000). d 1.639 g/l. mp −114.22°. bp_{760} −85.05°; bp_{100} −114.61°; bp_{10} −137.77°; $bp_{1.0}$ −154.37°. Critical temp 51.4°; critical pressure 81.6 atm; critical density 0.42 g/ml. n_D^{20} (liquid under pressure) 1.256. Heat capacity at constant volume (15°): 0.1939 cal/g/°C. Heat capacity at constant pressure (15°): 0.1375 cal/g/°C. Heat of vaporization at −85°: 3860 cal/mole; heat of soln (infinite dilution) −17.88 kcal/mole; heat of formation of gas at 25°: −22.063 kcal/mole. Dielectric constant (gas at 0°) 1.0046; dipole moment 1.07. Soly in water (g/100 g H_2O): 82.3 (0°); 67.3 (30°); 63.3 (40°); 59.6 (50°); 56.1 (60°). Forms a const boiling mixture: 20.22 g/100 g soln; *see* Hydrochloric Acid. Soly in methanol (g/100 g soln): 54.6 (−10°); 51.3 (0°); 47.0 (20°); 43.0 (30°); in ethanol: 45.4 (0°); 42.7 (10°); 41.0 (20°); 38.1 (30°); in ether: 37.52 (−10°); 35.6 (0°); 24.9 (20°); 19.47 (30°). LC_{50} (30 min) in mice, rats: 2142, 5666 ppm (Darmer).

Caution: Potential symptoms of overexposure are inflammation of nose, throat and larynx; coughing, burning of throat and choking; burning of eyes and skin; dermatitis. *See NIOSH Pocket Guide to Chemical Hazards* (DHHS/NIOSH 90-117, 1990) p 126. *See also Patty's Industrial Hygiene and Toxicology* vol. **2B**, G. D. Clayton, F. E. Clayton, Eds. (Wiley-Interscience, New York, 3rd ed., 1981) pp 2959-2961.

USE: In the manuf of pharmaceutical hydrochlorides, vinyl chloride from acetylene, alkyl chlorides from olefins, and arsenious chloride from arsenious oxide. In the chlorination of rubber, as a gaseous flux for babbitting operations. In organic reactions involving isomerization, polymerization, and alkylation. For making chlorine where economical.

4836. Hydrogen Cyanide. *Hydrocyanic acid;* Blausäure (German); prussic acid. CHN; mol wt 27.03. C 44.44%, H 3.73%, N 51.83%. HCN. Prepd on a large scale by the catalytic oxidation of ammonia-methane mixtures (Andrussow Process): *see* Andrussow, *Angew. Chem.* **48**, 593 (1935); Maffezzoni, *Chim. Ind. (Milan)* **34**, 460 (1952); Faith, Keyes & Clark's *Industrial Chemicals*, F. A. Lowenheim, M. K. Moran, Ed. (Wiley-Interscience, New York, 4th ed., 1975) pp 482-486. May also be prepd by the catalytic decompn of formamide. Conveniently prepd in the laboratory by acidifying NaCN or $K_4[Fe(CN)_6]$: Glemser in *Handbook of Preparative Inorganic Chemistry* vol. 1, G. Brauer, Ed. (Academic Press, New York, 2nd ed., 1963) pp 658-660.

Colorless gas or liquid; characteristic odor; very weakly acid (does not redden litmus); burns in air with a blue flame; *intensely poisonous* even when mixed with air. d(gas) 0.941 (air = 1); d(liq) 0.687. mp −13.4°. bp 25.6°. Miscible with water, alc; slightly sol in ether. LC_{50} in rats, mice, dogs: 544 ppm (5 min), 169 ppm (30 min), 300 ppm (3 min) by inhalation, K. C. Back *et al., Reclassification of Materials Listed as Transportation Health Hazards* (TSA-20-72-3; PB214-270, 1972).

Caution: High concn produces tachypnea (causing increased intake of cyanide); then dyspnea, paralysis, unconsciousness, convulsions and respiratory arrest. Headache, vertigo, nausea and vomiting may occur with lesser concns. Chronic exposure over long periods may cause fatigue, weakness. Exposure to 150 ppm for ½ to 1 hr may endanger life. Death may result from a few min exposure to 300 ppm. Average fatal dose: 50 to 60 mg. *Antidote:* Sodium nitrite and sodium thiosulfate. *See Patty's Industrial Hygiene and Toxicology* vol. **2C**, G. D. Clayton, F. E. Clayton, Eds. (Wiley-Interscience, NewYork, 3rd ed., 1982) pp 4850-4853.

USE: The compressed gas is used for exterminating rodents and insects in ships and for killing insects on trees, etc. *Must be handled by specially trained experts.*

4837. Hydrogen Fluoride. *Hydrofluoric acid gas;* fluohydric acid gas; anhydr hydrofluoric acid. FH; mol wt 20.01. F 94.96%, H 5.04%. HF. Obtained by the action of sulfuric acid on fluorspar (calcium fluoride): Faith, Keyes & Clark's *Industrial Chemicals*, F. A. Lowenheim, M. K. Moran, Eds. (Wiley-Interscience, New York, 4th ed., 1975) pp 462-467; prepn of pure HF: Simons, *Inorg. Syn.* **1**, 134 (1939); Shamir, Netzer, *J. Sci. Instrum.* (Ser. 2) **1**, 770 (1968). Exists as hydrogen-bonded polymers: Simons, Hildebrand, *J. Am. Chem. Soc.* **46**, 2183 (1924); Jarry, Davis, *J. Phys. Chem.* **57**, 600 (1953); Atoji, Lipscomb, *Acta Cryst.* **7**, 173 (1954). Reviews of prepn, properties and chemistry: Simons in *Fluorine Chemistry* vol. 1, J. H. Simons, Ed. (Academic Press, New York, 1950); Hyman, Katz, "Liquid Hydrogen Fluoride" in *Non-Aqueous Solvent Systems*, T. C. Waddington, Ed. (Academic Press, New York, 1965) pp 47-81; O'Donnell in *Comprehensive Inorganic Chemistry* vol. 2, J. C. Bailar, Jr. *et al.*, Eds. (Pergamon Press, Oxford, 1973) pp 1038-1054; J. F. Gall in Kirk-Othmer *Encyclopedia of Chemical Technology* vol. **10** (Wiley-Interscience, New York, 3rd ed., 1980) pp 733-753.

Colorless gas. Fumes in air. *Highly irritating, corrosive and poisonous!* d^{34} 1.27 (air = 1); d_0^0 1.002. mp −83.55°: Gillespie, Humphreys, *J. Chem. Soc. (A)* **1970**, 2311. bp 19.51°; bp_{400} +2.5°; bp_{200} −13.2°; bp_{100} −28.2°; bp_{40} −45.0°; bp_{20} −56.0°; bp_5 −74.7°. Very sol in water and alcohol. Slightly sol in ether. Sol in many organic solvents; soly (wt % at 5°): in benzene 2.54; toluene 1.80; *m*-xylene 1.28; tetralin 0.27. Many compds are sol in HF. *See* Simons, *loc. cit.;* Hyman, Katz, *loc. cit.* Anhydr HF is one of the most acidic substances known; Hammett acidity function (H_0) −10.98: Hyman, Katz, *loc. cit.* In aqueous soln, it is a

weak acid: Ka 6.46×10^{-4} mole/l, Vanderborgh, *Talanta* **15**, 1009 (1968). Forms a constant boiling mixture with water, *see* Hydrofluoric Acid. Dissolves silica, silicic acid, glass. Should be stored in steel cylinders. LC_{50} (1 hr) in rats, mice, monkeys: 1278, 500, 1780 ppm by inhalation, K. C. Back *et al., Reclassification of Materials Listed as Transportation Health Hazards* (TSA-20-72-3, PB 214-270, 1972).

Caution: Potential symptoms of overexposure are irritation of eyes, nose and throat; pulmonary edema; skin and eye burns; nasal congestion; bronchitis. *See NIOSH Pocket Guide to Chemical Hazards* (DHHS/NIOSH 90-117, 1990) p 126. *See also Patty's Industrial Hygiene and Toxicology* vol. **2B**, G. D. Clayton, F. E. Clayton, Eds. (Wiley-Interscience, New York, 3rd ed., 1981) pp 2945-2948.

USE: Catalyst, especially in the petroleum industry (paraffin alkylation); in fluorination processes, especially in the aluminum industry; in the manuf of fluorides; for separating uranium isotopes; in making fluorine contg plastics; in dye chemistry.

4838. Hydrogen Iodide. *Anhydrous hydriodic acid.* HI; mol wt 127.91. H 0.79%, I 99.21%. Prepd by catalytic union of the elements: Caley, Burford, *Inorg. Syn.* **1**, 159 (1939); Powell, Campbell, *J. Am. Chem. Soc.* **69**, 1227 (1947). May also be prepd by treating concd hydriodic acid solns with P_2O_5: Schmeisser in *Handbook of Preparative Inorganic Chemistry* vol. 1, G. Brauer, Ed. (Academic Press, New York, 2nd ed., 1963) pp 286-289. Lab prepn: Hoffman, *Inorg. Syn.* **7**, 180 (1963). Purification: A. Klemenc, *Die Behandlung und Reindarstellung von Gasen* (Vienna, 2nd ed., 1948) p 239; Irving, Wilson, *Chem. & Ind. (London)* **1964**, 653. Reviews of prepn and properties of HI and other hydrogen halides: Hills in *Mellor's Comprehensive Treatise on Inorganic and Theoretical Chemistry* vol. **II**, suppl. 1 (originally published as suppl. II, part 1) 857-869 (1956); Downs, Adams, in *Comprehensive Inorganic Chemistry* vol. 2, J. C. Bailar, Jr. *et al.*, Eds. (Pergamon Press, Oxford, 1973) pp 1280-1329.

Colorless, acrid, non-flammable gas. Fumes in moist air. Decomposed by light. mp −50.8°. bp_{760} −35.1°; $bp_{2 atm}$ −18.9°; $bp_{5 atm}$ +7.3°; $bp_{10 atm}$ 32.0°; $bp_{60 atm}$ 127.5°. d^0 5.66; d^{25} 5.23 g/l. Crit temp 151.0°, crit press. 82.0 atm. Sp heat (25°) 0.0545 cal/g/°C. Extremely sol in water. Soly (g/100 g H_2O): 234 (10°); 900 (0°). Soly in organic solvents: Gerrard, *Chem. & Ind. (London)* **1969**, 295; Ahmed *et al., J. Appl. Chem.* **20**, 109 (1970). Forms an azeotrope with water, *see* Hydriodic Acid. Reacts with the lower aliphatic alcohols forming the corresponding iodo compds. Forms a colorless liquid at atm pressure when cooled with dry ice and ether or similar cooling mixture. Attacks natural rubber.

Caution: Strong irritant.

USE: Manuf of hydriodic acid, organic iodo compds, to remove iodine from iodo compds.

4839. Hydrogen Peroxide. Hydrogen dioxide; hydroperoxide; Albone; Hioxyl; Lensan A; Mirasept; Oxysept; Pegasyl. H_2O_2; mol wt 34.01. H 5.93%, O 94.07%. First reported by Thenard in 1818; prepd by treating barium peroxide with acid. Manuf of aqueous solns: Faith, Keyes & Clark's *Industrial Chemicals*, F. A. Lowenheim, M. K. Moran, Eds. (Wiley-Interscience, New York, 4th ed., 1975) pp 487-495; R. Powell, *Hydrogen Peroxide Manufacture* (Noyes Dev. Corp., Park Ridge, N.J., 1968) 221 pp. Production of anhydr hydrogen peroxide by continuous fractional crystn: Crewson, Ryan, U.S. pat. **2,724,640** (1955 to Becco). *Reviews:* W. C. Schumb *et al., Hydrogen Peroxide* A.C.S. Monograph Series no. **128** (Reinhold, New York, 1955) 759 pp; Ebsworth *et al.*, in *Comprehensive Inorganic Chemistry* vol. **2**, J. C. Bailar, Jr. *et al.*, Eds. (Pergamon Press, Oxford, 1973) pp 771-778; J. R. Kirchner in Kirk-Othmer *Encyclopedia of Chemical Technology* vol. **13** (Wiley-Interscience, New York, 3rd ed., 1981) pp 12-38.

Colorless, rather unstable liquid; bitter taste. Distillable in high vacuum. May dec violently if traces of impurities are present. d^0 1.463. mp −0.43°. bp 152°. Misc with water; sol in ether. Insol in petr ether. Decomposed by many organic solvents.

Marketed as a soln in water in concns of 3-90% by wt. Solns of hydrogen peroxide gradually deteriorate and are usually stabilized by the addition of acetanilide or similar organic materials. Agitation or contact with rough surfaces, metals or many other substances accelerates decomposition. Rapidly dec by alkalies, finely divided metals; the presence of mineral acid renders it more stable.

Caution: Potential symptoms of overexposure are irritation of eyes, nose and throat; corneal ulceration; erythema, vesicles on skin; bleaching of hair. *See NIOSH Pocket Guide to Chemical Hazards* (DHHS/NIOSH 90-117, 1990) p 126.

USE: A 90% soln is used in rocket propulsion. As dough conditioner, maturing and bleaching agent in food.

THERAP CAT: Antiseptic; disinfectant.

THERAP CAT (VET): Topical antiseptic and cleansing agent (as a dilute soln).

4840. Hydrogen Peroxide Solution 3%. Hydrogen dioxide soln; oxydol. Contains 2.5-3.5% by wt of H_2O_2 = 8-12 vols oxygen.

Colorless, slightly acid liq. d ~1.00. Foams in the mouth. *Keep protected from light and in a cool place. Incompat:* Alkalies, ammonia and their carbonates, albumin, balsam Peru, phenol, charcoal, chlorides, alkali citrates; ferrous, mercurous or gold salts; hypophosphites, iodides, lime water, permanganates, sulfites, tinctures, and organic matter in general.

USE: In the plastics industry; white discharge printing on indigo-dyed wool; bleaching feathers, hair, silk, straw, ivory, flour, bone, gelatin, and textile fabrics; renovating old paintings, engravings; as oxidizer in manuf dyes; disinfecting water and hides; artificially aging wines, liquors, etc.; refining oils and fats; as antichlor; with paraphenylenediamine as a dye for furs, dead hair, etc.; in photography as hypo eliminant; with NaOH for cleaning metal surfaces, for gilding, silvering, etc. In pharmaceutical prepns, mouthwashes, dentifrices, sanitary lotions.

THERAP CAT: Anti-infective (topical).

THERAP CAT (VET): Topical antiseptic and cleansing agent.

4841. Hydrogen Peroxide Solution 30%. Superoxol. Contains 30% by wt of H_2O_2 = 100 vols of oxygen.

Clear, colorless liquid. *Strong oxidizing agent!* d ~1.11. Miscible with water. Now replacing the 3% soln for industrial uses; diluted to the required strength immediately before use. It also is used for making the 3% soln.

Caution: Avoid contact with skin and eyes—wear rubber gloves and goggles. Avoid contact with combustible materials. Drying of concd product on clothing or other combustible materials may cause fire. In case of contact, immediately flush with plenty of water for at least 15 min; for eyes, get medical attention. Avoid contamination from any source, including metals, dust, etc. Such contamination may cause rapid decompn, generation of large quantities of oxygen gas and high pressures.

Store in original closed container. Be sure that the container vent is working satisfactorily. Do not add any other product to container. When empty, rinse thoroughly with clean water.

4842. Hydrogen Selenide. Selenium hydride. H_2Se; mol wt 80.98. H 2.49%, Se 97.51%. Prepd by heating selenium and hydrogen in a sealed tube at 440°: Hautefeuille, *Bull. Soc. Chim.* [2] **7**, 198 (1867); by passing a mixture of hydrogen and selenium vapor over pumice stone at 440°: Corenwinder, *Ann. Chim. Phys.* [3] **34**, 77 (1852); by warming potassium or ferrous selenide with hydrochloric acid: Berzelius, *Acad. Handl. Stockholm* **39**, 13 (1818); by the action of water on aluminum selenide: Fonzes-Diacon, *Traité de Chimie Minérale, Paris* **1**, 469 (1904); Waitkins, Shutt, *Inorg. Syn.* **2**, 183 (1946).

Gas. Disagreeable odor. d_4^{-42} 2.12. bp −41.3°. Liquefies at 0° under a pressure of 6.6 atm; at 18°, 8.6 atm; at 52°, 21.5 atm; at 100°, 47.1 atm; at the crit temp 137°, 91.0 atm. mp −65.73°. v.p. at −30°, 1.75 atm; v.p. at 0.2°, 4.5 atm; v.p. at 30.8°, 12 atm. K_1 at 25° = 1.30×10^{-4}; K_2 at 25° = 1×10^{-11}. Soly in water (ml/100 ml): 377 (4°); 270 (22.5°). Sol in carbonyl chloride and carbon disulfide. Unites directly with most metals to form metal selenides. Approx LC_{50} (30 min) in guinea pigs: 6 ppm, *Handbook of Toxicology* vol. 1, W. S. Spector, Ed. (Saunders, Philadelphia, 1956) pp 340-341.

Caution: Potential symptoms of overexposure are irritation of eyes, nose and throat; nausea, vomiting and diarrhea; metallic taste, garlic breath; dizziness, lassitude and fatigue. *See NIOSH Pocket Guide to Chemical Hazards* (DHHS/NIOSH 90-117, 1990) p 128.

4843. Hydrogen Sulfide. Sulfureted hydrogen; "hydrosulfuric acid". H_2S; mol wt 34.08. H 5.91%, S 94.09%. Evolved from numerous environmental natural sources such as bacterial decomposition of vegetable and animal proteinaceous material. Occurs naturally as a component of crude petroleum, natural gas, volcanic gas and sulfur springs. Also a pollutant released into the environment as a by-product of a variety of industrial operations. Lab prepn: Bickford, Wilkinson, *Inorg. Syn.* **1**, 111 (1939). Purification: Ward *et al., ibid.* **3**, 14 (1950). Toxicity studies: E. H. Vernot *et al., Toxicol. Appl. Pharmacol.* **42**, 417 (1977); M. F. Tansy *et al., J. Toxicol. Environ. Health* **8**, 71-88 (1981). Review of toxicity and properties: R. O. Beauchamp, Jr., *et al., CRC Crit. Rev. Toxicol.* **13**, 25-97 (1984); of toxicology: R. J. Reiffenstein *et al., Ann. Rev. Pharmacol. Toxicol.* **32**, 109-134 (1992).

Flammable, poisonous gas with characteristic odor of rotten eggs, perceptible in air at concns of 0.02-0.13 ppm, sweetish taste. Burns in air with pale blue flame. Ignition temp 260°. Explosive limits when mixed with air: lower limit 4.3% by vol, upper limit 46% by vol. mp −85.49°; bp −60.33°: Giauque, Blue, *J. Am. Chem. Soc.* **58**, 831 (1936). Heavier than air; 1.5392 g/l (0°; 760 mm). d^{gas} 1.19 (air = 1.00). Vapor pressure 18.75×10^5 Pa. One gram H_2S dissolves in 187 ml water at 10°, in 242 ml water at 20°, in 314 ml water at 30°; in 94.3 ml abs alcohol at 20°; in 48.5 ml ether at 20°. Sol in glycerol, gasoline, kerosene, carbon disulfide, crude oil. Water solns of H_2S are not stable, absorbed oxygen causes the formation of elemental sulfur, and the solns become turbid rapidly. In a 50:50 v/v mixture of glycerol and water the precipitation of sulfur is retarded considerably. pH of freshly prepd satd water soln 4.5. pKa_1 7.04; pKa_2 11.96. LC_{50} in mice, rats (ppm): 634, 712 (1 hr inhalation) (Vernot). LC_{50} in rats (ppm): 444 (4 hr inhalation) (Tansy).

Caution: Highly toxic, can be fatal. Irritant and chemical asphyxiant. Insidious poison, since sense of smell may be fatigued and fail to give warning of high concns. Direct contact with gas may cause irritation of eyes and respiratory tract resulting in keratoconjunctivitis, photophobia, lacrimation, corneal opacity; rhinitis, laryngitis, cough, bronchopneumonia. Direct contact with solution may cause skin irritation, erythema. Potential symptoms of overexposure by inhalation include salivation, GI disturbances; giddiness, headache, vertigo, confusion, unconsciousness; tachypnea, tachycardia, sweating, fatigue. Exposure to very high vapor concentrations may result in systemic intoxication leading to paralysis of respiratory center of brain, apnea and sudden collapse. *See Clinical Toxicology of Commercial Products,* R. E. Gosselin *et al.,* Eds. (Williams & Wilkins, Baltimore, 5th ed., 1984) Section III, pp 198-202; *NIOSH Pocket Guide to Chemical Hazards* (DHHS/NIOSH 94-116, 1994) p 170.

USE: To produce elemental sulfur and sulfuric acid; in manuf of heavy water and other chemicals; in metallurgy; as analytical reagent.

4844. Hydrogen Telluride. H_2Te; mol wt 129.62. H 1.56%, Te 98.44%. Prepd by the action of H_2O or HCl on aluminum telluride; by electrolysis of a 50% soln of sulfuric or phosphoric acid with a Te cathode: Dennis, Anderson, *J. Am. Chem. Soc.* **36**, 882 (1914); Fehér in *Handbook of Preparative Inorganic Chemistry* vol. 1, G. Brauer, Ed. (Academic Press, New York, 2nd ed., 1963) pp 438-441.

Colorless gas. Offensive, garlic-like odor. *Highly poisonous!* mp −49°. bp −2°. d_{liq}^{-12} 2.68. Wt of one liter of the gas: 6.234 g. Liquid H_2Te is dec immediately by light. The dry gas is stable to light, but dec in the presence of dust, traces of moisture, rubber, cork, etc. Sol in water with fairly quick decompn. A satd aq soln is about 0.1N.

Caution: Imparts offensive odor to breath. Symptoms similar to hydrogen selenide, *q.v.*

4845. Hydrogen Tetracarbonylferrate(II). Iron hydrocarbonyl; iron tetracarbonyl dihydride. $C_4H_2FeO_4$; mol wt 169.90. C 28.28%, H 1.19%, Fe 32.87%, O 37.67%. H_2Fe-

(CO)$_4$.   Prepn from Fe(CO)$_5$:   Blanchard, Coleman, *Inorg. Syn.* **2**, 243 (1946); Sternberg *et al., J. Am. Chem. Soc.* **79**, 6116 (1957); Bishop *et al., J. Chem. Soc.* **1959**, 2484.

Colorless crystals, mp −70°; colorless liq, very unstable, dec below −10° on exposure to air.   Slight traces of iron carbonyls impart a red or yellow color.   Gas has extremely nauseating odor.   Sol in alkalies; neutral aq solns of salts unstable at room temp; strongly alkaline solns of salts, e.g. K$_2$Fe(CO)$_4$, are stable at room temp when all traces of air are excluded but dec gradually at 100° to CO and H$_2$.

4846.   Hydrohydrastinine. *5,6,7,8-Tetrahydro-6-methyl-1,3,dioxolo[4,5-g]isoquinoline.* C$_{11}$H$_{13}$NO$_2$; mol wt 191.23. C 69.09%, H 6.85%, N 7.32%, O 16.73%. Prepn from cotarnine, *q.v.:* Topchiev, *J. Appl. Chem. USSR* **6**, 529 (1933), *C.A.* **28**, 2718 (1934); Clayson, *J. Chem. Soc.* **1949**, 2016; Schneider, Müller, *Ann.* **615**, 34 (1958); Knabe, *Arch. Pharm.* **292**, 652 (1959). *See also:* Hydrastinine, Hydrocotarnine.

Crystals from petr ether, mp 66°. bp$_{752}$ 303°. Sol in alc, ether, acetone, benzene, CS$_2$, ethyl acetate.   Absorption spectrum: Dobbie, Tinkler, *J. Chem. Soc.* **85**, 1007 (1904).

Hydrochloride, C$_{11}$H$_{13}$NO$_2$.HCl, crystals from water or alcohol, dec 278°. Sol in water.

Hydrobromide, C$_{11}$H$_{13}$NO$_2$.HBr, needles from water, dec 272°.

Hydriodide, C$_{11}$H$_{13}$NO$_2$.HI, crystals from water, dec 242°.   Sol in water or alc.

Platinichloride, (C$_{11}$H$_{13}$NO$_2$)$_2$.H$_2$PtCl$_6$, yellow tablets, dec 222°. Sparingly sol in alc.

4847.   Hydromorphone. *4,5-Epoxy-3-hydroxy-17-methylmorphinan-6-one;* dihydromorphinone; Dimorphone; Novolaudon. C$_{17}$H$_{19}$NO$_3$; mol wt 285.34. C 71.56%, H 6.71%, N 4.91%, O 16.82%. Prepn by electrolytic reduction of morphine: Takagi, Ueda, *J. Pharm. Soc. Japan* **56**, 44 (1936); Nakamura, *ibid.* **62**, 347 (1942); by oxidation of dihydromorphine: Rapoport *et al., J. Org. Chem.* **15**, 1103 (1950); Homeyer, De la Mater, U.S. pats. **2,628,962** and **2,654,756** (both 1953 to Mallinckrodt); Rapoport, U.S. pat. **2,649,454** (1953 to Univ. of California). Crystal structure: Steinmetz, *Z. Krist.* **67**, 434 (1928). Toxicity data: M. E. Buchwald, G. S. Eadie, *J. Pharm. Exp. Ther.* **71**, 197 (1941). *Review:* King *et al., U.S. Pub. Health Repts.* Suppl. no. 113, 38 pp (1935).

Crystals from ethanol, mp 266–267°. [α]$_D^{25}$ −194° (c = 0.98 in dioxane).

Hydrochloride, C$_{17}$H$_{19}$NO$_3$.HCl, *Dilaudid, Laudicon, Hymorphan.*   Crystals, dec 305–315° (evacuated tube). [α]$_D^{25}$ −133°. Sol in 3 parts water; sparingly sol in alc. LD$_{50}$ in mice (mg/kg): 61-96 i.v. (Buchwald, Eadie).

Caution: May be habit forming. This is a controlled substance (opiate) listed in the U.S. Code of Federal Regulations, Title 21 Parts 329.1 and 1308.12 (1995).

THERAP CAT: Analgesic (narcotic).

4848. Hydrone®. An alloy of 35% Na and 65% Pb. Small, irregular lumps which on contact with water yield pure hydrogen gas (2.6 cu ft per lb); leaves a residue of lead-sponge and NaOH soln.

USE: As a source of hydrogen for laboratory work, portable blowpipes, and burners.

4849.   Hydroorotic Acid. *Hexahydro-2,6-dioxo-4-pyrimidinecarboxylic acid; 4,5-dihydroorotic acid.* C$_5$H$_6$N$_2$O$_4$; mol wt 158.11. C 37.98%, H 3.82%, N 17.72%, O 40.48%. Prepn from carbethoxyasparagine: Miller *et al., J. Am. Chem. Soc.* **75**, 6086 (1953); U.S. pat. **2,773,872** (1956 to Merck & Co.).

L-Form, crystals, dec 266°. [α]$_D^{25.3}$ +33.23° (c = 1.992 in 1% NaHCO$_3$).

D-Form, crystals from water, dec 266°. [α]$_D^{25.3}$ −31.54° (c = 2.01 in 1% NaHCO$_3$).

DL-Form, crystals, dec 259°. Forms a water-sol salt.

USE: The L- and DL-dihydroorotic acids are precursors to the biological pyrimidines, such as thymine, uracil, cytosine and hydroxymethylcytosine.   When employed in large amounts D-dihydroorotic acid is an antimetabolite of L-dihydroorotic acid in pyrimidine utilization.   It has also been found to inhibit the orotic acid utilization of those organisms which have a requirement for it in order to grow.   In general it may be said to inhibit bacterial growth.

4850. Hydroprene. *(E,E)-3,7,11-Trimethyl-2,4-dodecadienoic acid ethyl ester;* ethyl *(2E,4E)*-3,7,11-trimethyldodeca-2,4-dienoate; OMS-1696; SHA-486300; ZR-512; Altozar IGR; Gencor; Gentrol; Mator. C$_{17}$H$_{30}$O$_2$; mol wt 266.42. C 76.64%, H 11.35%, O 12.01%. Juvenile hormone mimic. Prepn: C. A. Henrick, U.S. pat. **4,021,461** (1977 to Zoecon); *idem et al., J. Agric. Food Chem.* **21**, 354 (1973); and insecticidal activity of enantiomers: *idem et al., ibid.* **26**, 542 (1978). Effect on sweet potato weevil: G. M. Ram, A. C. Sekhar, *Ind. J. Comp. Anim. Physiol.* **2**, 87 (1991); cockroach: B. L. Reid, G. W. Bennett, *J. Econ. Entomol.* **87**, 1537 (1994). Effect of (*S*)-enantiomer on cockroach: J. P. Edwards, J. E. Short, *ibid.* **86**, 436 (1993). Review and use in food industries: M. Martinez, *Cereal Food World* **38**, 818-820 (1993).

bp$_{0.03}$ 95°. uv max (hexane): 262 nm (ε 28300). LD$_{50}$ orally in rats: > 34000 mg/kg (Edwards).

(*S*)-form. bp$_{0.05}$ 80°. [α]$_D^{25}$ +2.9° (c = 0.020 in methanol).

USE: Insect growth regulator; insecticide.

4851. Hydroquinidine. *(9S)-10,11-Dihydro-6'-methoxy-cinchonan-9-ol;* dihydroquinidine; hydroconchinine. C$_{20}$H$_{26}$N$_2$O$_2$; mol wt 326.44. C 73.59%, H 8.03%, N 8.58%, O 9.80%. An alkaloid of cinchona, stereoisomeric with hydroquinine.   Usually prepd by hydrogenation of quinidine: Heidelberger, Jacobs, *J. Am. Chem. Soc.* **41**, 826 (1919). Conversion to dihydrocinchonine by removal of the methoxy group: King, *J. Chem. Soc.* **1946**, 523. Manuf pat.: Gutzwiller, Uskokovic, **Ger.** pat. **1,933,599** (1970 to Hoffmann-La Roche), *C.A.* **72**, 90696v (1970). Pharmacology: Cosnier *et al., Therapie* **26**, 97 (1971).

Plates from ether, needles from alcohol, mp 169°. $[\alpha]_D^{20}$ +231° (c = 2.02 in alc); +299° (c = 0.82 in 0.1N H_2SO_4). Readily sol in hot alcohol; slightly sol in water and ether.

Hydrochloride, $C_{20}H_{26}N_2O_2$.HCl, *Serecor*. Rhombic plates, mp 273-274°. $[\alpha]_D^{26}$ +184° (c = 1.3). Freely sol in methanol, chloroform; less readily in water or abs alcohol; difficultly sol in dry acetone.

THERAP CAT: Antiarrhythmic.

4852. Hydroquinine. *(8α,9R)-10,11-Dihydro-6'-meth-oxycinchonan-9-ol;* dihydroquinine. $C_{20}H_{26}N_2O_2$; mol wt 326.44. C 73.59%, H 8.03%, N 8.58%, O 9.80%. An alkaloid of cinchona, found in quinine sulfate mother liquors. Stereoisomeric with hydroquinidine, *q.v.* Usually prepd by careful hydrogenation of quinine: Heidelberger, Jacobs, *J. Am. Chem. Soc.* **41**, 819 (1919). Total synthesis: Rabe *et al., Ber.* **64B**, 2487 (1931). Synthesis of isomers: Rubtsov, *J. Gen. Chem. USSR* **9**, 1493 (1939), *C.A.* **34** 2850 (1940); *ibid.* **13**, 593, 702 (1943), *C.A.* **39**, 705 (1945). LC determn in quinine beverages: L. P. Valenti, *J. Assoc. Offic. Anal. Chem.* **68**, 782 (1985).

Needles from ether or benzene, mp 172°. $[\alpha]_D^{18}$ −142° (alc); $[\alpha]_D^{20}$ −236° (c = 0.82 in 0.1N H_2SO_4). pK_1 = 5.33; K_1 = 4.7×10⁻⁶. Freely sol in acetone, alcohol, chloroform, ether, petr ether; fairly sol in ammonia water. Almost insol in water (290 mg/l).

Hydrochloride hemihydrate, $C_{20}H_{27}ClN_2O_2$.½H_2O, prisms from water, mp 208° (Heidelberger). $[\alpha]_D^{21}$ −124° (c = 1.1). pH of 0.005 molar soln 5.85. Freely sol in water, alcohol, methanol, acetone. Almost insol in ether. Crystallizes also with 2 mols H_2O. Infrared spectrum: Suszko, Dega-Szafran, *Ber.* **62**, 7819b (1965).

THERAP CAT: Depigmentor.

4853. Hydroquinone. *1,4-Benzenediol;* p-dihydroxybenzene; hydroquinol; quinol; Aida; Black and White Bleaching Cream; Eldoquin; Eldopaque; Tecquinol. C_6H_6-O_2; mol wt 110.11. C 65.45%, H 5.49%, O 29.06%. Prepd by the oxidation of aniline: L. Gattermann, T. Wieland, *Die Praxis des Organischen Chemikers* (de Gruyter, Berlin, 40th ed., 1961) p 266; by reduction of quinone: Kitchen, U.S. pat. **1,322,580** (1920); Seyewetz, Miodon, *Bull. Soc. Chim. France* **33**, 449 (1923); by Elbs persulfate oxidation of phenol: Baker, Brown, *J. Chem. Soc.* **1948**, 2303; Forrest, Petrow, *ibid.* **1950**, 2340; from acetylene + CO: Howk, Sauer, U.S. pat. **3,055,949** (1962 to du Pont). Toxicity data: Woodard *et al., Fed. Proc.* **8**, 348 (1949). Toxicology and carcinogenicity study: F. W. Kari *et al., Food. Chem. Toxicol.* **30**, 737 (1992). *Review:* J. Varagnat in Kirk-Othmer *Encyclopedia of Chemical Technology* **vol. 13** (Wiley-Interscience, New York, 3rd ed., 1981) pp 39-69.

Crystals, mp 170-171°. d_{15} 1.332. bp 285-287°. Sol in 14 parts water; freely sol in alcohol, ether, slightly in benzene. Its soln becomes brown in the air due to oxidation; the oxidation is very rapid in presence of alkali. *Keep well closed and protected from light. Handle with caution. Avoid contact with skin and eyes.* LD_{50} orally in rats: 320 mg/kg (Woodard).

Caution: Harmful effects may occur following overexposure to dust and vapors by inhalation or by contact with skin or eyes. Contact with skin may cause dermatitis. Contact with eyes may cause eye irritation, keratitis, discoloration of conjunctiva and corneal changes. Ingestion may cause tinnitus, nausea, dizziness, sense of suffocation, increased respiration rate, vomiting, pallor, muscle twitching, headache, dyspnea, cyanosis, delirium, and collapse. Urine is usually green or brownish green in color and continues to darken on standing. *See* Patty's *Industrial Hygiene and Toxicology* **vol. 2A**, G. D. Clayton, F. E. Clayton, Eds. (Wiley-Interscience, New York, 3rd ed., 1981) pp 2589-2592. *See also NIOSH Pocket Guide to Chemical Hazards* (DHHS/NIOSH 90-117, 1990) p 128.

USE: As photographic reducer and developer; as reagent in the determination of small quantities of phosphate; as antioxidant.

THERAP CAT: Depigmentor.

4854. Hydroxocobalamin. *Cobinamide hydroxide dihydrogen phosphate (ester) inner salt 3'-ester with 5,6-dimethyl-1-α-D-ribofuranosyl-1H-benzimidazole;* α-(5,6-dimethylbenzimidazolyl)hydroxocobamide; vitamin B_{12a}; hydroxocobemine; OHB₁₂; alphaCobione; Alpha-Ruvite; Axlon; Cobalin-H; Codroxomin; Droxomin; Duradoce; Duralta-12; Neo-Betalin 12; Neo-Cytamen; alphaRedisol; Redisol H; Sytobex-H. $C_{62}H_{89}CoN_{13}O_{15}P$; mol wt 1346.37. C 55.31%, H 6.66%, Co 4.38%, N 13.52%, O 17.82%, P 2.30%. Physiological analog of vitamin B_{12} where the CN group is replaced with OH. Exists in aq soln as an equilibrium mixture of the hydroxy isomer and the ionic aqua isomer (aquacobalamin). Precursor of the coenzymes methylcobalamin and cobamamide, *q.q.v.* Prepn: E. A. Kaczka *et al., J. Am. Chem. Soc.* **71**, 1514 (1949); *eidem, ibid.* **73**, 335 (1951); *eidem,* U.S. pat. **2,738,301** (1956 to Merck & Co.). Acids replace the hydroxo group with the anion of the acid: *eidem, Science* **112**, 354 (1950). Prepn from cyanocobalamin: R. Bieganowski, G. Klar, *Chem.-Ztg.* **106**, 235 (1982). Clinical evaluation as antidote to cyanide: J. C. Forsyth *et al., Clin. Toxicol.* **31**, 277 (1993). *Review:* "Vitamin B_{12}" in *Vitamins,* W. Friedrich, Ed. (de Gruyter, Berlin, 1988) pp 837-928.

Dark red, orthorhombic needles or platelets from water + acetone. Darkens at 200°, but not melted at 300°. Absorption max (H_2O): 279, 325, 359, 516, 537 nm (ε 19000, 11400, 20600, 8900, 9500). Moderately sol in water and in the lower aliphatic alcohols. Practically insol in acetone, ether, petr ether, halogenated hydrocarbons, benzene.

Aquacobalamin, $C_{62}H_{90}CoN_{13}O_{15}P$.OH, *cobinamide hydroxide monohydrate dihydrogen phosphate (ester) inner salt 3'-ester with 5,6-dimethyl-1-α-D-ribofuranosyl-1H-benzimidazole,* aquocobalamin, vitamin B_{12b}; vitamin B_{12d}. Absorption max (H_2O): 274, 317, 351, 499, 527 nm (ε 20600, 6100, 26500, 8100, 8500).

Acetatocobalamin, $C_{64}H_{91}CoN_{13}O_{16}P$, *cobinamide acetate (salt) dihydrogen phosphate (ester) inner salt 3'-ester with 5,6-dimethyl-1-α-D-ribofuranosyl-1H-benzimidazole, Depogamma, Docelan, Novidroxin.* Prepn of stable solns: Marcus *et al.,* Fr. pat. **1,336,671** (1963 to Merck & Co.), *C.A.* **60**, 380a (1964).

Nitrocobalamin, $C_{62}H_{88}CoN_{14}O_{16}P$, *cobinamide nitrite (salt) dihydrogen phosphate (ester) inner salt 3'-ester with 5,6-dimethyl-1-α-D-ribofuranosyl-1H-benzimidazole, nitritocobalamin,* vitamin B_{12c}. Prepn: E. A. Kaczka *et al., J. Am. Chem. Soc.* **73**, 3569 (1951). Red crystalline solid. Absorption max in water: 357, 527.5 nm ($E_{1cm}^{1\%}$ 153.2, 59.5); in 0.01N NaOH: 357, 535 nm ($E_{1cm}^{1\%}$ 139, 62.5).

Sulfitocobalamin, $C_{62}H_{89}CoN_{13}O_{17}PS$, *cobinamide sulfite (1:1) (salt) dihydrogen phosphate (ester) inner salt 3'-ester with 5,6-dimethyl-1-α-D-ribofuranosyl-1H-benzimidazole, cobalaminsulfonic acid.* Prepn: Fricke, U.S. pat. **2,721,162** (1955 to Abbott). Absorption max: 275, 365, 418, 516 nm ($E_{1cm}^{1\%}$ 328, 130, 49, 61).

THERAP CAT: Vitamin (hematopoietic).

THERAP CAT (VET): Vitamin (hematopoietic).

4855. Hydroxyamphetamine. *(±)-4-(2-Aminopropyl)-phenol;* dl-p-hydroxy-α-methylphenethylamine; dl-1-p-hydroxyphenyl-2-propylamine; p-hydroxyphenylisopropylamine; α-methyltyramine; Paredrine; Paredrinex; Pulsoton. C_9H_{13}-NO; mol wt 151.21. C 71.49%, H 8.67%, N 9.26%, O 10.58%. Prepn from oxime of p-methoxyphenyl acetone:

Mannich, Jacobsohn, *Ber.* **43**, 189 (1910), **Ger. pat. 243,546.** From *p*-nitrobenzyl chloride and a salt of nitroethane: Hoover, Hass, *J. Org. Chem.* **12**, 501 (1947).

Crystals (rosettes) from benzene, mp 125-126°. Sol in water, alcohol, chloroform, ethyl acetate.

Iodide, $C_9H_{14}INO$, stout prisms, mp 155°. Freely sol in water, alcohol, acetone.

Hydrochloride, $C_9H_{14}ClNO$, crystals from HCl, mp 171-172°. Sol in water, alcohol. Practically insol in ether.

Hydrobromide, $C_9H_{14}BrNO$, crystals. Freely sol in water, alcohol, acetone. Sold as Paredrine Hydrobromide Aqueous—a 1% aq soln made isotonic with sodium chloride and preserved with sodium ethylmercuri thiosalicylate. Also sold as Paredrine Hydrobromide Ophthalmic 1%, with boric acid—a 1% aq soln made tear-isotonic with 2% boric acid and preserved with sodium ethylmercuri thiosalicylate.

THERAP CAT: Adrenergic (ophthalmic); mydriatic.

4856. *p*-Hydroxybenzaldehyde. 4-Formylphenol. $C_7H_6O_2$; mol wt 122.12. C 68.85%, H 4.95%, O 26.20%. Widely distributed in plants in very small amounts. Obtained as a byproduct from the Reimer-Tiemann reaction for salicylaldehyde from phenol: L. F. Fieser, M. Fieser, *Organic Chemistry* (Reinhold, New York, 3rd ed., 1956) p 681; *cf.* Reimer, Tiemann, *Ber.* **9**, 824 (1876); Herzfeld, Tiemann, *Ber.* **10**, 64 (1877).

Needles from water. Slight, agreeable, aromatic odor, mp 116°. Sublimes at atmospheric pressure without decomposition. Reacts like a weak monobasic acid: Ka at 25° = 2.2×10^{-8}. Dipole moment: 4.19. Sparingly sol in cold water: 1.38 g/100 ml H_2O at 30.5°; more sol in hot water. Freely sol in alc, ether. Slightly sol in benzene: 3.68 g/100 ml C_6H_6 at 65°.

4857. *p*-Hydroxybenzoic Acid. $C_7H_6O_3$; mol wt 138.12. C 60.87%, H 4.38%, O 34.75%. Prepn from *p*-bromophenol: Gilinan, Arntzen, *J. Am. Chem. Soc.* **69**, 1537 (1947); from *p*-hydroxybenzaldehyde: Pearl, *J. Org. Chem.* **12**, 85 (1947); from phenol + potassium ethyl carbonate: Jones, *Chem. & Ind. (London)* **1958**, 228; from potassium phenolate + CO_2: **Brit. pat. 942,418** (1963 to Inventa). A metabolic product of *Penicillium patulum:* Tanenbaum, Bassett, *Biochim. Biophys. Acta* **28**, 21 (1958).

Crystals, mp 213-214°. d 1.46. Soluble in about 125 parts water, freely in alcohol, slightly in chloroform; sol in ether, acetone. Practically insol in carbon disulfide. Ferric chloride does not color its aq soln.

USE: In org. syntheses; intermediate for dyes, fungicides.

4858. α-Hydroxybenzylphosphinic Acid. $C_7H_9O_3P$; mol wt 172.12. C 48.85%, H 5.27%, O 27.89%, P 18.00%. Prepd from benzaldehyde (3 moles) and H_3PO_2 (1 mole): Ville, *Ann. Chim. (Paris)* [6] **23**, 305 (1891); *Beilstein* **7**, 232; Viout, *J. Rech. Cent. Nat. Rech. Sci.* no. **28**, 15-31 (1954), *C.A.* **50**, 7078d (1956).

Leaflets from butanol, mp 110°. Freely sol in water, alcohol; also sol in benzene. Practically insol in anhydr ether.

Sodium salt, *Phos, Phoselit, Phosilite.* Very freely sol in water. The pH is on the acid side, but aq solns are suitable for injection.

THERAP CAT: Sodium salt as nutrient.

4859. 4′-Hydroxybutyranilide. *N-(4-Hydroxyphenyl)-butanamide;* *p*-hydroxybutyranilide; *N*-butyroyl-*p*-aminophenol; Suconox-4. $C_{10}H_{13}NO_2$; mol wt 179.22. C 67.02%, H 7.31%, N 7.82%, O 17.85%. Prepn: Kuhn *et al., Z. Physiol. Chem.* **247**, 197 (1937); Fierz-David, Kuster, *Helv. Chim. Acta* **22**, 82 (1939); Rohmann, Friedrich, *Arch. Pharm.* **278**, 456 (1940).

Needles from water, mp 139-140°. Slightly sol in cold water, more sol in hot water, sol in alcohol.

USE: As antioxidant: Young, Cottle, **U.S. pat. 2,654,722** (1953 to Standard Oil).

4860. γ-Hydroxybutyrate. *4-Hydroxybutanoic acid; γ-hydroxybutyric acid;* 4-hydroxybutyrate; GHB. $C_4H_8O_3$; mol wt 104.11. C 46.15%, H 7.75%, O 46.11%. $HOCH_2CH_2CH_2COOH$. Naturally occurring metabolite of GABA; thought to function as a neurotransmitter or neuromodulator. The highest conc in humans is found in fetal cerebellum and adult hypothalamus. Prepn of salts: A. Saytzeff, *Ann.* **171**, 258 (1874); C. S. Marvel, E. R. Birkhimer, *J. Am. Chem. Soc.* **51**, 260 (1929). Clinical pharmacokinetics: S. D. Ferrara *et al., Brit. J. Clin. Pharmacol.* **34**, 231 (1992). GC/MS determn in biological fluids: K. M. Gibson *et al., Biomed. Environ. Mass Spectrom.* **19**, 89 (1990). Acute toxicity: B. Bruguerolle *et al., Thérapie* **32**, 375 (1977). Clinical studies: L. Scrima *et al., Sleep* **13**, 479 (1990); L. Gallimberti *et al., Neuropsychopharmacology* **9**, 77 (1993). Review of biochemistry and pharmacology: G. Tunnicliff, *Gen. Pharmacol.* **23**, 1027-1034 (1992). Review in treatment of alcoholism: G. Biggio *et al.,* in *Adv. Biochem. Psychopharmacol.* vol. **47** (Raven Press, New York, 1992) pp 281-288. *Review:* M. Mamelak, *Neurosci. Biobehav. Rev.* **13**, 187-198 (1989). Brief overview of public health issues: *J. Am. Med. Assoc.* **265**, 447-448 (1991).

Sodium salt, $C_4H_8NaO_3$, γ-OH, sodium oxybate, sodium γ-oxybutyrate, *Somatomax PM, Wy-3478, NSC-84223, Somsanit, Gamma-OH.* Crystals from alcohol. LD_{50} in male, female rats (mg/kg): 2,000, 1,650 i.p. (Bruguerolle).

THERAP CAT: Anesthetic (intravenous). In treatment of narcolepsy; in treatment of alcoholism.

4861. β-Hydroxybutyric Acid. *3-Hydroxybutanoic acid.* $C_4H_8O_3$; mol wt 104.11. C 46.15%, H 7.75%, O 46.11%. $CH_3CHOHCH_2COOH$.

d-Form, prepd by the action of *Aspergillus griseus* on the *dl*-form: McKenzie, Harden, *J. Chem. Soc.* **83**, 430 (1903). Crystals. $[\alpha]_D^{10}$ +24.3° (c = 2.226). Sol in water, alcohol, ether.

l-Form, found in the urine of diabetics (as much as 30 g per day). Isoln: Fischer, Scheibler, *Ber.* **42**, 1221 (1909); Shaffer, Marriott, *J. Biol. Chem.* **16**, 268 (1913). Hygroscopic, monoclinic crystals, mp 45.5-48°. $[\alpha]_D^{25}$ −24.5° (c = 5). K at 22° = 3.86×10^{-5}. Freely sol in water, alcohol, ether. Sparingly sol in benzene. On distn it dec into crotonic acid and water.

dl-Form, prepd from acetoacetic ester by the action of sodium amalgam: Wislicenus, *Ann.* **149**, 207 (1869); Marian, *Biochem. Z.* **150**, 283 (1924); by the oxidation of aldol: Wurtz, *Compt. Rend.* **76**, 1167 (1873); from crotonic acid by heating with dil acid: Wacker, **Ger. pat. 441,003;** *Frdl.* **15,** 135; by heating crotonitrile with KOH soln: Bruylants, *Bull. Soc. Chim. Belg.* **31,** 182 (1922). Hygroscopic syrup. Volatile with steam. Sol in water, alcohol, ether. On distn it dec into crotonic acid and water.

4862. 3-Hydroxycamphor. *3-Hydroxy-1,7,7-trimethyl-bicyclo[2.2.1]heptan-2-one;* oxycamphor; Oxaphor. $C_{10}H_{16}$-O_2; mol wt 168.24. C 71.39%, H 9.59%, O 19.02%. Prepn: Lucius, Brüning, **Ger. pat. 91,718** (1897); Manasse, *Ber.* **30,** 659 (1897); Lapworth, Chapman, *J. Chem. Soc.* **79,** 377 (1901).

Needles from benzene + petr ether, mp 205-206°. Sol in 50 parts water, more in hot water; very sol in alcohol, chloroform, ether.

THERAP CAT: Antipruritic (topical).

4863. Hydroxychloroquine. *2-[[4-[(7-Chloro-4-quino-linyl)amino]pentyl]ethylamino]ethanol; 7-chloro-4-[4-[ethyl-(2-hydroxyethyl)amino]-1-methylbutylamino]quinoline;* 7-chloro-4-[4-(N-ethyl-N-β-hydroxyethylamino)-1-methyl-butylamino]quinoline; 7-chloro-4-[5-(N-ethyl-N-2-hydroxy-ethylamino)-2-pentyl]aminoquinoline; oxychloroquine; oxi-chlorochine. $C_{18}H_{26}ClN_3O$; mol wt 335.88. C 64.37%, H 7.80%, Cl 10.56%, N 12.51%, O 4.76%. Prepd by reacting a mixture of 4,7-dichloroquinoline, phenol and N'-ethyl-N'-β-hydroxyethyl-1,4-pentadiamine at 125-130°: Surrey, Hammer, *J. Am. Chem. Soc.* **72,** 1814 (1950); Surrey, **U.S. pat. 2,546,658** (1951 to Sterling Drug). Use in combination with cyclophosphamide and azathioprine, *q.q.v.* in the treatment of rheumatoid arthritis: D. J. McCarty, G. F. Carrera, *J. Am. Med. Assoc.* **248,** 1718 (1982). Reassessment in the treatment of rheumatoid arthritis: *Am. J. Med.* **75,** no. 1A, 1-56 (1983). Series of articles on clinical use: *ibid.* **85,** Suppl. 4A, 1-71 (1988).

Crystals from ethylene dichloride and Skellysolve B; mp 89-91°.

Diphosphate, $C_{18}H_{26}ClN_3O.2H_3PO_4$. Recrystallized from ethanol, mp 168-170° (dec).

Sulfate, $C_{18}H_{26}ClN_3O.H_2SO_4$, *Ercoquin, Plaquenil Sulfate, Quensyl.* White crystalline powder; odorless but has a bitter taste. pH of aq solns about 4.5. Exists in two forms, the usual form mp ~240°, the other mp ~198°. Freely sol in water. Practically insol in alcohol, chloroform, ether.

THERAP CAT: Antimalarial; antirheumatic; lupus erythematosus suppressant.

4864. 1α-Hydroxycholecalciferol. *(1α,3β,5Z,7E)-9,10-Secocholesta-5,7,10(19)-triene-1,3-diol;* 1α-hydroxyvitamin D_3; 1α-OH-CC; alfacalcidol; Alfarol; Alpha D_3; EinsAlpha, Etalpha; One-Alpha; Vetalpha. $C_{27}H_{44}O_2$; mol wt 400.65. C 80.94%, H 11.07%, O 7.99%. Synthetic analog of calci-triol, *q.v.,* the hormonal form of vitamin D_3, which shows identical potency with respect to stimulation of intestinal calcium absorption and bone mineral mobilization. This common activity is thought to be due either to the presence in both of a 1α hydroxy group or more likely, to the conversion of the 1α-OH-CC *in vivo* to 1α,25-dihydroxycholecalciferol: Haussler *et al., Proc. Nat. Acad. Sci. USA* **70,** 2248 (1973). Synthesis from cholesterol: Holick *et al., Science* **180,** 190 (1973); Barton *et al., J. Am. Chem. Soc.* **95,** 2748

(1973); Fürst *et al., Helv. Chim. Acta* **56,** 1708 (1973); M. Morisaki *et al., Chem. Pharm. Bull.* **23,** 3272 (1975); T. Sato *et al., ibid.* **26,** 2933 (1979). Total synthesis: R. G. Harrison *et al., Tetrahedron Letters* **1973,** 3649; *eidem, J. Chem. Soc. Perkin Trans. I* **1974,** 2654; P. J. Kocienski, B. Lythgoe, *ibid.* **1980,** 1400. Synthesis of the 1β-epimer from the 1α-form and biological properties: H. E. Paaren *et al., Chem. Commun.* **1977,** 890.

mp 134-136° (Harrison); also reported as mp 138-139.5° (Fürst). $[\alpha]_D^{25}$ +28° (ether). uv max (ether): 264 nm (ϵ 18000) (Harrison); also reported as ϵ 20200 (Barton).

1β-Epimer, *1β-hydroxycholecalciferol, 1β-hydroxyvitamin D_3.*

THERAP CAT: Vitamin D source.

THERAP CAT (VET): Vitamin D source.

4865. 24-Hydroxycholesterol. *(3β,24S)-Cholest-5-ene-3,24-diol.* $C_{27}H_{46}O_2$; mol wt 402.66. C 80.54%, H 11.51%, O 7.95%. Two stereoisomers exist: *cholest-5-ene-3β,24β-diol* and *cholest-5-ene-3β,24α-diol.* Isoln of the naturally occur-ring 24β-epimer, from horse brain: Ercoli *et al., Boll. Soc. Ital. Biol. Sper.* **29,** 494 (1953), *C.A.* **49,** 4744i (1955); from beef spinal cord: Fieser *et al., J. Org. Chem.* **22,** 1380 (1957). Prepn and separation of isomers: Ercoli, Ruggieri, *Gazz. Chim. Ital.* **83,** 720 (1953); *eidem, J. Am. Chem. Soc.* **75,** 3284 (1953). Configurations at C-24: Klyne, Stokes, *J. Chem. Soc.* **1954,** 1979.

24β-Epimer, *cerebrostenediol, cerebrosterol.* Needles from acetone, mp 175-176°. $[\alpha]_D^{20}$ −48.2° (c = 1.06 in $CHCl_3$).

24β-Epimer dibenzoate, $C_{41}H_{54}O_4$, crystals from ether + acetone, mp 182-183°. α_D −19° (c = 1.20 in $CHCl_3$). Absorption max in chloroform: 5.85, 6.2, 6.86, 7.8 μ.

24α-Epimer, crystals, mp 182-183°. $[\alpha]_D^{20}$ −26.8° (c = 0.672 in $CHCl_3$).

24α-Epimer dibenzoate, crystals, mp 141-142°. $[\alpha]_D^{26}$ −11.8° (c = 1 in $CHCl_3$).

4866. 25-Hydroxycholesterol. *(3β)-Cholest-5-ene-3,25-diol;* Δ^5-cholestene-3β,25-diol. $C_{27}H_{46}O_2$; mol wt 402.66. C 80.54%, H 11.51%, O 7.95%. An important intermediate in the synthesis of 25-hydroxycholecalciferol, *q.v.* Synthesis: A. Ryer *et al., J. Am. Chem. Soc.* **72,** 4247 (1950); W. G. Dauben, H. L. Bradlow, *ibid.* 4248. Isoln from autoxida-tion products of cholesterol, *q.v.:* L. F. Fieser *et al., J. Org. Chem.* **22,** 1380 (1957). Prepn from pregnenolone, *q.v.:* T. A. Narwid, M. R. Uskokovic, **U.S. pat. 3,856,780** (1974). Alternate syntheses: M. Morisaki *et al., Chem. Pharm. Bull.* **21,** 457 (1973); A. Rotman, Y. Mazur, *Chem. Commun.* **1974,** 15; J. J. Partridge *et al., Helv. Chim. Acta* **57,** 764 (1974); T. A. Narwid *et al., ibid.* 771; W. G. Salmond *et al., Tetrahedron Letters* **1977,** 987, 1237, 1695; K. Ochi *et al.,*

Chem. Pharm. Bull. **27**, 252 (1979); M. Riediker, J. Schwartz, Tetrahedron Letters **22**, 4655 (1981).

Colorless needles from methanol, mp 178-180°. $[\alpha]_D^{25}$ −39.0° (c = 1.05 in chloroform).

20(S)-isomer, crystals from chloroform, mp 189.5-190.5°. $[\alpha]_D^{25}$ −41.50°(c = 0.9278 in chloroform).

4867. Hydroxycodeinone. 5α-7,8-Didehydro-4,5α-epoxy-14-hydroxy-3-methoxy-17-methylmorphinan-6-one; 14-hydroxycodeinone. $C_{18}H_{19}NO_4$; mol wt 313.35. C 68.99%, H 6.11%, N 4.47%, O 20.42%. Preparation from thebaine: Freund, Speyer, J. Prakt. Chem. **94**, 135 (1916). From codeine: Merck, **Ger.** pat. **411,530**; Frdl. **15**, 1516; K. W. Bentley, The Chemistry of the Morphine Alkaloids (Oxford, 1954). Improved synthesis: F. M. Hauser et al., J. Med. Chem. **17**, 1117 (1974). Isoln from Papaver bracteatum Lindl.: H. G. Theuns et al., Phytochemistry **16**, 753 (1977).

Plates from 96% alcohol + few drops chloroform, dec 275°. Freely sol in CHCl₃, methyl Cellosolve, petr ether, ethyl acetate; slightly sol in alcohol. Practically insol in water, ether; also insol in aq alkaline solns.

Hydrochloride monohydrate, $C_{18}H_{19}NO_4 \cdot HCl \cdot H_2O$, rods from water, dec 285-286°. $[\alpha]_D^{20}$ −150° (c = 2.5).

Note: This is a controlled substance (opiate) listed in the U.S. Code of Federal Regulations, Title 21 Part 1308.12 (1995).

4868. Hydroxydione Sodium. 21-(3-Carboxy-1-oxopropoxy)-5β-pregnane-3,20-dione sodium salt; 21-hydroxypregnane-3,20-dione sodium hemisuccinate; hydroxydione succinate; Presuren; Viadril. $C_{25}H_{35}NaO_6$; mol wt 454.54. C 66.06%, H 7.76%, Na 5.06%, O 21.12%. Prepd by palladium reduction of deoxycorticosterone followed by treatment with succinic anhydride and formation of the sodium salt: Laubach, **U.S.** pat. **2,708,651** (1955 to Pfizer); Laubach et al., Science **122**, 78 (1955).

Lyophilized, fluffy white powder. Dec 193-203°. uv max: 280 nm (ε 93.2). Sol in water, in mildly alkaline buffer solns, acetone, chloroform. pH of 2% aq soln 8.5-9.8.

Free acid, mp 195-197°. $[\alpha]_D^{20}$ +95° in chloroform.

THERAP CAT: Anesthetic (intravenous).

4869. p-Hydroxyephedrine. 4-Hydroxy-α-[1-(methylamino)ethyl]benzenemethanol; p-hydroxy-α-[1-(methylami-

no)ethyl]benzyl alcohol; 1-(4-hydroxyphenyl)-2-methylaminopropanol; α-(1-methylaminoethyl)-p-hydroxybenzyl alcohol; oxyephedrin; Carnigen; Edornat; Methylsympatol; Suprifen. $C_{10}H_{15}NO_2$; mol wt 181.23. C 66.27%, H 8.34%, N 7.73%, O 17.66%. Prepn: Stolz, **U.S.** pat. **1,878,021** (1932 to Winthrop). Prepn of hydrochlorides of d- and l-forms: Takamatsu, Minaki, J. Pharm. Soc. Japan **76**, 1230 (1956), C.A. **51**, 4305e (1957).

Cryst powder, mp 152-154°. Sparingly sol in water, alcohol, ether; readily sol in NaOH soln and dil acids.

Hydrochloride, $C_{10}H_{15}NO_2 \cdot HCl$, crystals, mp 209-211°. Sol in 3 parts water, 5 parts glycerol, 10 parts 90% alcohol. Sparingly sol in abs alcohol, acetone.

d-Form hydrochloride, crystals, dec 223-224°. $[\alpha]_D^{22}$ +33.8°.

l-Form hydrochloride, crystals, dec 222-223°. $[\alpha]_D^{22}$ −35.5°.

THERAP CAT: Adrenergic.

4870. N-Hydroxyethylpromethazine Chloride. N-(2-Hydroxyethyl)-N,N,α-trimethyl-10H-phenothiazine-10-ethanaminium chloride; (2-hydroxyethyl)dimethyl(1-methyl-2-phenothiazin-10-ylethyl)ammonium chloride; N,N-dimethyl-N-[α-methyl-β-phenothiazin-10-yl)ethyl]-N-(β-hydroxyethyl)ammonium chloride; [1-(10-phenothiazinylmethyl)ethyl]-2-hydroxyethyldimethylammonium chloride; Aprobit. $C_{19}H_{25}ClN_2OS$; mol wt 364.94. C 62.53%, H 6.90%, Cl 9.71%, N 7.68%, O 4.38%, S 8.79%. Prepd from N,N-dimethyl-N-[α-methyl-β-phenothiazin-10-yl)ethyl]amine and glycol chlorohydrin in methyl ethyl ketone: Carlsson, Karlson, **Brit.** pat. **881,379** (1961 to Aktiebolaget Recip).

Crystals from isopropanol, mp 233°.
THERAP CAT: Antihistaminic.

4871. Hydroxyglutamic Acid. β-Hydroxyglutamic acid; α-amino-β-hydroxyglutaric acid. $C_5H_9NO_5$; mol wt 163.13. C 36.81%, H 5.56%, N 8.59%, O 49.04%. An amino acid classified as nonessential with respect to its growth effect in rats. "Not among the naturally occurring hydrolytic products of proteins"—Schmidt, Chemistry of the Amino Acids and Proteins (Springfield, Ill., 2nd ed., 1944) p 1090. Synthesis from ethyl α-isonitrosoacetone dicarboxylate: Harington, Randall, Biochem. J. **25**, 1917 (1931); Levene, Schormüller, J. Biol. Chem. **106**, 595 (1934).

L-Form, prisms from water. Dec 135° (if completely anhydrous; hydrated crystals soften at 100-105°). $[\alpha]_D^{20}$ +1.2° (c = 2); $[\alpha]_D^{20}$ +17.6° (c = 2 in 6N HCl). pK₁' 2.09; pK₂' 4.18; pK₃' 9.20. Freely sol in water, acetic acid. Insol in alc, ether.

DL-Form, orthorhombic prisms, needles from water. Dec 198°. Very sol in hot water.

4872. 8-Hydroxy-7-iodo-5-quinolinesulfonic Acid. 7-Iodo-8-hydroxyquinoline-5-sulfonic acid; m-iodo-o-hydroxyquinolineanasulfonic acid; Ferron; Loretin. $C_9H_6INO_4S$; mol wt 351.12. C 30.79%, H 1.72%, I 36.14%, N 3.99%, O 18.23%, S 9.13%. Obtained from the potassium salt of 8-hydroxy-5-quinolinesulfonic acid by the action of KI, bleaching powder, and HCl: Claus, *Arch. Pharm.* **231**, 704 (1893); **Ger. pat. 72,942** (1893); by the action of KI and Ca(OCl)$_2$ on 8-hydroxy-5-quinolinesulfonic acid: Botton, *C.A.* **52**, 18407e (1958).

OH
I
N
SO$_3$H

Sulfur yellow, almost odorless and tasteless, crystalline powder. mp 260-270° (dec). One gram dissolves in 500 ml cold, 170 ml boiling water; slightly sol in alcohol. Practically insol in ether or oils.

Compd with chloroquine (2:1), **chloquinate, chlochinate, cloquinate, Resotren.** Prepn: Koenig, Andersag, **U.S. pat. 2,650,224** (1953 to Schenley Industries).

Mixture of 80% of 8-hydroxy-7-iodo-5-quinolinesulfonic acid and 20% sodium bicarbonate, **chiniofon, quiniofon, Quinoxyl, Sefona, Yatren, Yochinol.**

USE: As colorimetric reagent for ferric ion.

THERAP CAT: Antiamebic; antiseptic.

4873. 4-Hydroxyisophthalic Acid. *4-Hydroxy-1,3-benzenedicarboxylic acid.* $C_8H_6O_5$; mol wt 182.13. C 52.76%, H 3.32%, O 43.92%. Prepd by high pressure carbonation of potassium phenate (Kolbe-Schmitt reaction): Baine *et al., J. Org. Chem.* **19**, 510 (1954) or by boiling salicylic acid with carbon tetrachloride in alkaline medium in the presence of Cu powder: **Ger. pat. 258,887** (1913 to Zeltner & Landau); *Chem. Zentr.* **1913**, I, 1641; *Frdl.* **11**, 210. Major constituent of the brown dust residue from sublimation of salicylic acid: Hunt *et al., Chem. & Ind. (London)* **1955**, 417.

HOOC
COOH
OH

Branched needles from water, platelets from dil alc. Dec 314-315°. One gram dissolves in 3 liters of water at 24°, in 160 ml at 100°. Freely sol in alcohol, ether.

Dimethyl ester, crystals, mp 97.5°.

Diethyl ester, crystals, mp 54-55°.

Note: 4-Hydroxyisophthalic acid has been proposed as an improvement over aspirin: Chesher *et al., Nature* **175**, 206 (1955).

4874. Hydroxylamine. H_3NO; mol wt 33.03. H 9.15%, N 42.41%, O 48.44%. NH$_2$OH. Prepn as the hydrochloride: W. L. Semon, *Org. Syn.* **coll. vol. I**, 318 (1932). Prepn: Hurd, *Inorg. Syn.* **1**, 87 (1939); Benson *et al., J. Am. Chem. Soc.* **78**, 4202 (1956). Crystal structure: Meyers, Lipscomb, *Acta Cryst.* **5**, 583 (1955). Toxicity data: Riemann, *Acta Pharmacol. Toxicol.* **6**, 285 (1950); R. P. Smith, W. R. Layne, *J. Pharmacol. Exp. Ther.* **165**, 30 (1969). Mutagenic action: Phillips, Brown in *Progr. Nucl. Acid Res. Mol. Biol.* **7**, 349-368 (1967). *Reviews:* Mason, "Hydroxylamine" in *Mellor's* vol. **VIII**, supplement 2, *Nitrogen* (part 2), 115-157 (1967); Jones in *Comprehensive Inorganic Chemistry* vol. **2**, J. C. Bailar, Jr. *et al.,* Eds. (Pergamon Press, Oxford, 1973) pp 265-276.

Unstable, large white flakes or needles, mp 33°, bp$_{22}$ 58°. d$_4^0$ 1.2255; d$_4^{40}$ 1.204. K at 20° = 1.07×10^{-8}. Very sol in water. Very sol in liq ammonia and methanol. The soly in the higher alcohols decreases with increasing mol wt. Sparingly sol in ether, benzene, carbon disulfide, chloroform. Very hygroscopic. Dec by hot water. Undergoes rapid decompn at room temps esp in the presence of atm moisture and CO$_2$. Detonates in test tube heated with flame. LD$_{50}$ i.p. in mice: 1.83 mmol/kg (Smith, Layne).

Hydrochloride, $H_3NO.HCl$, *oxammonium hydrochloride.* Monoclinic columnar crystals; slowly dec when moist. d$_{17}$ 1.67. mp about 151°. One gram dissolves in about 1 ml water (83 g in 100 ml water at 17°); 19 ml alcohol; 8 ml methanol. Sol in glycerol, propylene glycol. Insol in cold ether. pH of 0.2 molar aq soln 3.2. Keep well closed. LD$_{50}$ orally in mice: 408 mg/kg (Riemann).

Sulfate, $(H_3NO)_2.H_2SO_4$, *oxammonium sulfate.* Crystals, mp about 170°. Freely sol in water.

Caution: Skin irritant. May cause methemoglobinemia, sulfhemoglobinemia, cyanosis, convulsions, hypotension and coma: *Clinical Toxicology of Commercial Products,* R. E. Gosselin *et al.,* Eds. (Williams & Wilkins, Baltimore, 5th ed., 1984) Section II, p 117.

USE: As reducing agent in photography; in synthetic and analytical chemistry; to purify aldehydes and ketones. As antioxidant for fatty acids and soaps. As dehairing agent for hides.

4875. Hydroxylupanine. *[2S-(2α,7β,7aβ,14β,14aα)]-Dodecahydro-2-hydroxy-7,14-methano-2H,11H-dipyrido[1,2-a:1′,2′-e][1,5]diazocin-11-one;* octalupine; oxylupanine; 13-hydroxylupanine. $C_{15}H_{24}N_2O_2$; mol wt 264.37. C 68.15%, H 9.15%, N 10.60%, O 12.10%. From seed of *Lupinus perennis* L., *L. angustifolius* L. and *L. polyphyllus* Lindl., *Leguminosae:* Rink, Schäfer, *Arch. Pharm.* **287**, 290 (1954); Winterfeld, Pies, *ibid.* **290**, 537 (1957); Bohlmann, Winterfeldt, *Ber.* **93**, 1956 (1960). Structure: Galinovsky *et al., Monatsh.* **81**, 77 (1950). Biosynthesis: Schuette *et al., Arch. Pharm.* **296**, 438 (1963).

H
N
N
OH
O
H

Crystals from dry acetone, mp 169-170°. $[\alpha]_D^{20}$ +45.6° (c = 1.49 in ethanol); $[\alpha]_D^{20}$ +65.1° (c = 1.09 in water). Sol in water, alc, chloroform; slightly sol in benzene, ether.

Methiodide, $C_{15}H_{24}N_2O_2.CH_3I$, needles from methanol + ether, mp 254-255°.

4876. Hydroxymercurichlorophenols. Prepared from *o*-chlorophenol and mercuric oxide or mercuric salts: F. C. Whitmore, *Organic Compounds of Mercury* (Chemical Catalog Co., New York, 1921) p 269; **D.R.P. 234,851** (1910 to Bayer) in *Frdl.* **10**, 1272.

HO
Cl
(HgOH)$_n$ n = 1, 2

p-Form (n = 1), $C_6H_5ClHgO_2$, *(3-chloro-4-hydroxyphenyl)hydroxymercury.* Major ingredient of *Semesan.*

Sulfate, *Uspulun.* See *Beilstein,* 2nd suppl., vol. **16**, 683.

USE: Seed disinfectant.

4877. 1-(Hydroxymethyl)-5,5-dimethylhydantoin. *1-(Hydroxymethyl)-5,5-dimethyl-2,4-imidazolidinedione;* monomethyloldimethylhydantoin; MDMH; methylol dimethylhydantoin. $C_6H_{10}N_2O_3$; mol wt 158.16. C 45.57%, H 6.37%, N 17.71%, O 30.35%. Prepd by treating 5,5-dimethylhydantoin with formaldehyde: Mackey, **U.S. pat. 2,762,708** (1956 to GAF).

OH
H$_3$C
H$_3$C
N
O
N
H
O

Crystals, mp 100°. Freely sol in water, methanol, ethanol, acetone; slightly sol in ethyl acetate. Practically insol in ether, trichloroethylene, carbon tetrachloride.

USE: As preservative in cosmetic prepns. The compound liberates formaldehyde steadily at a very slow rate in the presence of water at pH 6. The available formaldehyde is about 19% (w/w), and 0.1-10% is usuallly mixed with the cosmetic product. When heated in the dry state MDMH forms a water-soluble *dimethylhydantoin formaldehyde resin*, compatible with gelatin, polyvinyl acetate, ethyl cellulose. The resin is used in hair lacquers.

4878.   17-Hydroxy-16-methylene-Δ⁶-progesterone. *17-Hydroxy-16-methylenepregna-4,6-diene-3,20-dione;* 6-dehydro-17α-hydroxy-16-methyleneprogesterone. $C_{22}H_{28}O_3$; mol wt 340.46. C 77.61%, H 8.29%, O 14.10%. Prepn: Syhora, Mazac, *Coll. Czech. Chem. Commun.* **29,** 2351 (1964). Prepn of the 17-acetate: **Brit. pat. 901,293** (1962 to Upjohn); Syhora, Mazac, *loc. cit.;* **Brit. pat. 963,427** (1964 to E. Merck).

Crystals from methanol, mp 196.5-197°. $[\alpha]_D^{20}$ −72.5°. uv max: 283 nm (log ε 4.51).

17-Acetate, $C_{24}H_{30}O_4$, *17-acetoxy-16-methylenepregna-4,6-diene-3,20-dione, 17α-acetoxy-6-dehydro-16-methyleneprogesterone, Superlutin.* Crystals from ethyl acetate, mp 233-234°. $[\alpha]_D^{20}$ −132°. uv max: 283 nm (log ε 4.50).

THERAP CAT: Progestogen (oral).

4879.   5-(Hydroxymethyl)-2-furaldehyde. *5-(Hydroxymethyl)-2-furancarboxaldehyde;* 5-(hydroxymethyl)-2-furancarbonal; 5-(hydroxymethyl)-2-furfural; HMF; 5-hydroxymethyl-2-formylfuran. $C_6H_6O_3$; mol wt 126.11. C 57.14%, H 4.80%, O 38.06%. Prepn from the fructose portion of the sugar molecule in 57% yield: Haworth, Jones, *J. Chem. Soc.* **1944,** 667; Haworth, Wiggins, **Brit.** pats. **591,858** (1947) and **600,871** (1948). Improved process: Garber, Jones, **U.S.** pat. **2,929,823** (1960 to Merck & Co.). Purification: Jones, Lange, **U.S.** pat. **2,994,645** (1961 to Merck & Co.). Outline of process using cornstarch, glucose and sucrose as raw materials: Medwick, *Chem. Eng. News* (Sept. 11, 1961) 75. Prepn from molasses: Jones, Lange, **U.S.** pat. **3,066,-150** (1962 to Merck & Co.); Hales *et al.,* **U.S.** pat. **3,071,599** (1963 to Atlas Chem.).

Needles from ether + petr ether, mp 31.5°. Avoid contact with eyes. Produces harmless yellow stains on skin. Odor of chamomile flowers. Very slightly volatile with steam (as compared with furfural). d_4^{25} 1.2062. $bp_{0.02}$ 110°. n_D^{18} 1.5627. uv max: 283 nm. Heat of combustion: 665 kcal/mol. Freely sol in water, methanol, ethanol, acetone, ethyl acetate, dimethylformamide. Sol in ether, benzene, chloroform. Less sol in carbon tetrachloride. Sparingly sol in petr ether. *Keep protected from light and air.*

USE: In the synthesis of dialdehydes, glycols, ethers, aminoalcohols, acetals. Aq acid catalyzes ring opening.

4880.   N-(Hydroxymethyl)nicotinamide. *N-(Hydroxymethyl)-3-pyridinecarboxamide;* 3-pyridinecarboxylic acid hydroxymethylamide; pyridine-3-carboxylic acid *N*-methylolamide; Bilamid(e); Felosan; Nikoform; Choligen. $C_7H_8N_2O_2$; mol wt 152.15. C 55.26%, H 5.30%, N 18.41%, O 21.03%. Prepn: Graf, *J. Prakt. Chem.* **138,** 292 (1933); Chechelska, Urbanski, *Rocz. Chem.* **27,** 396 (1953), *C.A.* **49,** 1033f (1955).

Crystals, mp 141-142°. Freely sol in hot alcohol and water; sparingly sol in cold alcohol and water.

Hydrochloride, $C_7H_8N_2O_2 \cdot HCl$, crystals, mp > 120° (dec).

THERAP CAT: Cholagogue.

4881.   1-Hydroxy-2-naphthoic Acid. 1-Naphthol-2-carboxylic acid. $C_{11}H_8O_3$; mol wt 188.18. C 70.21%, H 4.28%, O 25.51%. Obtained by the action of CO_2 on the sodium salt of α-naphthol under pressure at 120-140°.

White to reddish crystals. mp 191-192°. Almost insol in cold water; freely sol in alcohol, benzene, ether, alkalies.

4882.   3-Hydroxy-2-naphthoic Acid. *3-Hydroxy-2-naphthalenecarboxylic acid;* 3-naphthol-2-carboxylic acid. $C_{11}H_8O_3$; mol wt 188.18. C 70.21%, H 4.28%, O 25.51%. Obtained by the action of CO_2 on the sodium salt of β-naphthol under pressure at 280-290°: Schmitt, Burkard, *Ber.* **20,** 2702 (1887); Ger. pat. **50,341; Frdl. 2,** 133; **Ger.** pat. **436,034;** *Chem. Zentr.* **1926, I,** 1717; *Frdl.* **15,** 295; **Ger.** pat. **436,524;** *Chem. Zentr.* **1927, I,** 182; *Frdl.* **15,** 298; Schwenk, *Chem. Ztg.* **53,** 335 (1929).

Very pale yellow crystals, mp 222-223°. Practically insol in cold water. Slightly sol in hot water. Sol in benzene, chloroform. Freely sol in alc, ether. Sol in alkaline solns.

Magnesium salt, *Regacholyl.*

USE: The water-sol sodium salt has been used to solubilize riboflavin: Arnold *et al., Poultry Sci.* **31,** 350 (1952).

4883.   Hydroxypethidine. *4-(3-Hydroxyphenyl)-1-methyl-4-piperidinecarboxylic acid ethyl ester; 4-(m-hydroxyphenyl)-1-methylisonipecotic acid ethyl ester;* ethyl 4-(*m*-hydroxyphenyl)-1-methylisonipecotate; demidone; oxipethidine; oxypetidin; Hoechst 10446; Win-771; Bemidone; Biphenal. $C_{15}H_{21}NO_3$; mol wt 263.34. C 68.42%, H 8.04%, N 5.32%, O 18.23%. Prepn: Morrison, Rinderknecht, *J. Chem. Soc.* **1950,** 1467; Eisleb, **Ger.** pat. **752,755** (1952 to I. G. Farben), *C.A.* **52,** 7361e (1958).

Crystals from ethanol, mp 110°.

Hydrochloride, $C_{15}H_{21}NO_3 \cdot HCl$, crystals, mp 173-174°. Soluble in water; slightly sol in alc.

Note: This is a controlled substance (opiate) listed in the U.S. Code of Federal Regulations, Title 21, Part 1308.11 (1995).

THERAP CAT: Analgesic (narcotic).

4884.   Hydroxyphenamate. *2-Phenyl-1,2-butanediol 1-carbamate;* carbamic acid β-ethyl-β-hydroxyphenethyl ester; β-ethyl-β-hydroxyphenethyl carbamic acid ester; β-ethyl-β-hydroxyphenethyl carbamate; 2-hydroxy-2-phenylbutyl carbamate; AL-0361; Listica. $C_{11}H_{15}NO_3$; mol wt 209.25. C 63.14%, H 7.23%, N 6.69%, O 22.94%. Prepd from β-ethyl-β-hydroxyphenethyl alcohol and ethyl chloroformate followed by reaction with ammonia: Sifferd, Braitberg, **U.S. pat. 3,066,164** (1962 to Armour-Pharm.). Pharmacology and toxicology: Bastian, Clements, *Dis. Nerv. Sys.* **22**, 9 (1961).

Crystals, mp 55-56.5°. Soly in water at 25°: 2.5% w/v. LD_{50} orally in mice: 830 mg/kg (Bastian, Clements).

THERAP CAT: Anxiolytic.

4885. N-(4-Hydroxyphenyl)glycine. *p*-Hydroxyphenylaminoacetic acid; *p*-hydroxyanilinoacetic acid; photoglycine; Glycin; Iconyl; Monazol. $C_8H_9NO_3$; mol wt 167.16. C 57.48%, H 5.43%, N 8.38%, O 28.71%. Prepd from *p*-aminophenol and chloracetic acid: Vater, *J. Prakt. Chem.* **29**, 291 (1884); Meldola *et al.*, *J. Chem. Soc.* **111**, 552 (1917); Galatis, *Helv. Chim. Acta* **4**, 576 (1921).

Shiny leaflets from water, browns at 200°, begins to melt at 220°, completely melted at 245-247° (decompn). Sparingly sol in water, alcohol, acetone, ether, chloroform, ethyl acetate, benzene, glacial acetic acid. Sol in alkalies and mineral acids. Freely sol in warm 20% hydrochloric acid.

USE: Photographic developer. In determination of iron; detection and determination of phosphorus and silicon. Acid indicator in bacteriology.

4886. 17α-Hydroxyprogesterone. *17-Hydroxypregn-4-ene-3,20-dione;* 4-pregnen-17α-ol-3,20-dione; Gestageno; Prodox. $C_{21}H_{30}O_3$; mol wt 330.47. C 76.33%, H 9.15%, O 14.52%. Isoln from adrenal glands: Pfiffner, North, *J. Biol. Chem.* **132**, 459 (1940); **139**, 855 (1941); von Euw, Reichstein, *Helv. Chim. Acta* **24**, 879 (1941). Prepn: Julian *et al.*, **U.S. pat. 2,648,662** (1953 to Glidden); Ringold *et al.;* Stork *et al.*, **U.S. pats. 2,802,839** and **2,805,203** (both 1957 to Syntex); Chemerda *et al.;* Cutler, Chemerda; Dulaney, McAleer; **U.S. pats. 2,777,843; 2,786,856/7; 2,813,060** (all 1957 to Merck & Co.); Cutler *et al.*, *J. Org. Chem.* **24**, 1629 (1959); Pederson, **U.S. pat. 3,000,883** (1961 to Upjohn).

Rhombic or hexagonal leaflets from acetone or alcohol, mp 222-223° (rapid heating). With slow heating the substance undergoes molecular rearrangement accompanied by partial resolidification and becomes completely molten only at 276°. $[\alpha]_D^{17}$ +105.6° (c = 1.0417 in chloroform).

Acetate, $C_{23}H_{32}O_4$, *17α-acetoxyprogesterone.* Crystals from chloroform + methanol, mp 239-240°. uv max: 240 nm (log ϵ 4.33).

Caproate, $C_{27}H_{40}O_4$, *17α-hydroxyprogesterone hexanoate,* **Delalutin, Hyproval P.A., Lentogest, Pharlon, Proge, Proluton Depot, Teralutil.** Prepn: Kaspar *et al.*, **U.S. pat. 2,753,-360** (1956 to Schering AG). Comprehensive description: K. Florey, *Anal. Profiles Drug Subs.* **4** 209-224 (1975). Dense needles from isopropyl ether or methanol, mp 119-121°. $[\alpha]_D^{20}$ +61° (c = 1 in chloroform). Soly (mg/ml): sesame oil 25-29; levulinic acid butyl ester 350-400.

THERAP CAT: Progestogen.

THERAP CAT (VET): Estrus regulator.

4887. Hydroxyproline. *trans-4-Hydroxy-L-proline;* Hyp; L$_s$-hydroxyproline; (−)-(2S,4R)-4-hydroxyproline; 4-hydroxy-2-pyrrolidinecarboxylic acid. $C_5H_9NO_3$; mol wt 131.13. C 45.80%, H 6.92%, N 10.68%, O 36.60%. Amino acid analog of proline, *q.v.*; non-essential for human development. Shows *cis-trans* isomerism. Constituent of collagen, *q.v.* Isoln from gelatin: E. Fischer, *Ber.* **35**, 2660 (1902). Synthesis: H. Leuchs, *Ber.* **38**, 1937 (1905); and structure determn: H. Leuchs, J. F. Brewster, *Ber.* **46**, 986 (1913). Stereochemistry: Hudson, Neuberger, *J. Org. Chem.* **15**, 24 (1950). Early chemistry and biochemistry: *Amino Acids and Proteins,* D. M. Greenberg, Ed. (Charles C. Thomas, Springfield, IL, 1951) 950 pp., *passim;* J. P. Greenstein, M. Winitz, *Chemistry of the Amino Acids* vols **1-3** (John Wiley and Sons, Inc., New York, 1961) pp. 2018-2042, *passim.* Flow sheets of four different syntheses: *Chem. & Eng. News* **40**, 40 (Nov. 12, 1962). One pot synthesis: S. G. Ramaswamy, E. Adams, *J. Org. Chem.* **42**, 3440 (1977). Proof of occurence in plants: D. Ashford, A. Neuberger, *Trends Biochem. Sci.* **5**, 245 (1980). Effects of *cis-trans* isomerism on protein structure: N. Panasik, Jr. *et al.*, *Int. J. Pept. Protein Res.* **44**, 262 (1994). *Review:* R. Kuttan, A. N. Radhakrishnan, *Adv. Enzymol.* **37**, 273-348 (1973). Review of metabolism: E. Adams, L. Frank, *Ann. Rev. Biochem.* **49**, 1005-1061 (1980).

Rhombs or needles from water, mp 274°. $[\alpha]_D$ −76.5° (c = 2.5 in water). pK_1' 1.82; pK_2' 9.65. Soly in water at 0°: 288.6 g/l; at 25°: 361.1 g/l; at 50°: 451.8 g/l; at 65°: 516.7 g/l. Very slightly sol in alcohol. Insol in ether.

cis-Form, *allohydroxyproline,* (2S,4S)-4-hydroxyproline. mp 238-241°. $[\alpha]_D^{18}$ −58.1° (c = 5.2 in water).

4888. Hydroxypropyl Cellulose. *Cellulose 2-hydroxypropyl ether;* oxypropylated cellulose; Klucel; Lacrisert. Non-ionic water soluble ether of cellulose, *q.v.* that produces solns having a wide range of viscosity (200-2500 cp). Prepn: **Neth. pat. Appl. 6,401,036;** E. D. Klug, **U.S. pats. 3,278,520, 3,278,521** (1964, 1966, 1966 all to Hercules). Use in the treatment of dry eye syndrome: T. P. Werblin *et al.*, *Ophthalmology* **88**, 78 (1981); P. Huguet *et al.*, *Bull. Soc. Ophthalmol. Fr.* **81**, 1173 (1981). Review of chemistry, physical properties and uses: E. D. Klug in *Encyclopedia of Polymer Science and Technology* vol. **15** (Interscience, New York, 1971) pp 307-314; A. J. Desmarais, *Industrial Gums,* R. L. Whistler, Ed. (Academic Press, New York, 2nd ed., 1973) pp 649-672.

Off-white powder, softens at 130°. Sol in many polar organic solvents. Ppts from water at 40-45°. Thermoplastic.

USE: As emulsifier, stabilizer, whipping aid, protective colloid, film former or thickener in foods; as binder in ceramics and glazes; in hair and cosmetic prepns; in vacuum-formed containers and blow-molded bottles; as suspending agent in PVC polymerization. Pharmaceutic aid (tablet coating agent).

THERAP CAT: Topical protectant.

4889. Hydroxypropyl Methylcellulose. *Cellulose 2-hydroxypropyl methyl ether;* hypromellose; HPMC; Gonak; Goniosol; Lacril; Methocel; Ultra Tears. Non-ionic water soluble ether of methylcellulose, *q.v.* that produces solns having a wide range of viscosity (400-15,000 cp). Prepn: A. B. Savage, **U.S. pat. 2,949,252** (1960 to Dow). Review of chemistry, physical properties and use: *idem, Encyclopedia of Polymer Science and Technology* vol. **3** (Interscience, New York, 1965); pp 496-511; G. K. Greminger, A. B. Savage, *Industrial Gums,* R. L. Whistler, Ed. (Academic Press, New York, 1973) pp 619-647.

Powder. Dissolves slowly in cold water. Insol in hot water. Sol in most polar organics. Has thermogelling properties. Has higher salt tolerance and is more sol than methylcellulose.

USE: As emulsifier, film former, protective colloid, stabilizer, suspending agent, or thickener in foods. Pharmaceutic aid (suspending agent; tablet excipient; demulcent; viscosity increasing agent); ophthalmic lubricant. In adhesives, asphalt emulsions, caulking compounds, tile mortars, plastic mixes, cements, paints. As sticker for agricultural sprays and dusts.

4890. 8-Hydroxyquinoline. *8-Quinolinol;* oxyquinoline; hydroxybenzopyridine; oxybenzopyridine; phenopyridine; oxychinolin; oxine. C_9H_7NO; mol wt 145.16. C 74.47%, H 4.86%, N 9.65%, O 11.02%. Prepn: Z. H. Skraup, *Monatsh.* **1**, 316 (1880); **3**, 536 (1882); R. H. F. Manske *et al., Can. J. Res.* **27F**, 359 (1949). Toxicity study: Bernstein *et al., Toxicol. Appl. Pharmacol.* **5**, 599 (1963). *Review:* J. P. Phillips, *Chem. Rev.* **56**, 271-297 (1956). Book: R. G. W. Hollingshead, *Oxine and Its Derivatives,* **I-IV** (Butterworth, London, 1954/56).

White crystals or crystalline powder. mp 76°. bp ~267°. Almost insol in water, ether. Freely sol in alc, acetone, chloroform, benzene, aq mineral acids. LD_{50} i.p. in mice: 48 mg/kg (Bernstein).

Sulfate, $(C_9H_7NO)_2.H_2SO_4$, *Chinosol.* Pale yellow, cryst powder; slight saffron odor; burning taste. mp 175-178°. Freely sol in water; sol in about 100 parts glycerol, slightly in alcohol. Insol in ether.

Aluminum sulfate, $C_{27}H_{24}AlN_3O_{15}S_3$, *Nyxolan, Aloxyn.*

USE: As fungistat; chelating agent in determn of trace metal ions. Aluminum sulfate as antiperspirant, deodorant.

THERAP CAT: Antiseptic. Aluminum sulfate as anthelmintic (Nematodes).

4891. 8-Hydroxy-5-quinolinesulfonic Acid. $C_9H_7NO_4S$; mol wt 225.22. C 48.00%, H 3.13%, N 6.22%, O 28.41%, S 14.24%. Prepn: K. Matsumura, *J. Am. Chem. Soc.* **49**, 810 (1927); N. K. Chawla, M. M. Jones, *Inorg. Chem.* **3**, 1549 (1964).

Pale yellow, needle-like crystals or cryst powder; odorless. mp 322-324°. Freely sol in water, slightly in organic solvents.

USE: In determn of trace metal ions.

4892. Hydroxystilbamidine. *4-[2-[4-(Aminoiminomethyl)phenyl]ethenyl]-3-hydroxybenzenecarboximidamide;* 2-hydroxy-4,4'-stilbenedicarboxamidine; 2-hydroxy-4,4'-diamidinostilbene; 2-hydroxy-4,4'-diguanylstilbene; 2-hydroxystilbamide. $C_{16}H_{16}N_4O$; mol wt 280.33. C 68.55%, H 5.75%, N 19.99%, O 5.71%. Prepn: J. N. Ashley, J. O. Harris, *J. Chem. Soc.* **1946**, 567; A. J. Ewins *et al.,* **Brit. pat. 574,486;** A. J. Ewins, **U.S. pat. 2,510,047** (1946, 1950 both to May & Baker). Organ and tissue distribution in animals: I. Snapper *et al., Cancer* **4**, 1246 (1951). Pharmacology and antiprotozoal activity: I. Snapper *et al., Trans. N.Y. Acad. Sci.* **14**, 269 (1952). Probe for studying nucleic acid conformation: B. Festy, *C.R. Acad. Sci. Ser. D* **266**, 1433 (1968); B. Festy, M. Daune, *Biochemstry* **12**, 4827 (1973); B. Festy *et al., Biochim. Biophys. Acta* **407**, 24 (1975). Crystal structure: C. Courseille *et al., C.R. Acad. Sci. Ser. C* **274**, 1921 (1972). Use as a fluorochrome for selective staining of nuclei: L. B. Murgatroyd, *Histochemstry* **74**, 107 (1982). *Review:* B. Festy in *Antibiotics* **vol. 5**, pt. 2, F. E. Hahn, Ed. (Springer-Verlag, New York, 1979) pp 223-235.

Yellow microcrystals from nitrobenzene, mp 235°. LD_{50} in mice (mg/g): 0.027 i.v.; 0.14 s.c. (Ewins, 1950).

Isethionate, $C_{20}H_{28}N_4O_9S_2$, yellow crystals, discolored by light, mp 286° (dec). Freely sol in water. Soly in alcohol ~1.0 g/100 ml. Practically insol in ether. pH of a 1% aq soln 3.3 to 5.3. Solns show strong yellow fluorescence under uv light. Solns should be freshly prepared. Although the hydroxy compd is more stable in soln, such solns should not be stored and must not be used if cloudy.

THERAP CAT: Antiprotozoal (Leishmania).

4893. Hydroxystreptomycin. Reticulin (the antibiotic). $C_{21}H_{39}N_7O_{13}$; mol wt 597.58. C 42.21%, H 6.58%, N 16.41%, O 34.81%. Antibiotic substance produced by *Streptomyces reticuli:* Hosoya *et al., Japan. J. Exp. Med.* **20**, 327 (1949), *C.A.* **45**, 3459i (1951); by *S. griseocarneus* (from a Japanese soil): Stodola *et al., J. Am. Chem. Soc.* **73**, 2290 (1951); Benedict, Stodola, **U.S. pat. 2,617,755** (1952 to U.S. Secy. Agr.); by *Streptomyces* NA 232-M1: Grundy *et al., Antibiot. & Chemother.* **1**, 309 (1951); by *S. subrutilus:* Arai *et al., J. Antibiot.* **17A**, 23 (1964). Identity of reticulin (the antibiotic) and hydroxystreptomycin: Hosoya *et al., Japan. J. Exp. Med.* **22**, 303 (1952), *CA.* **48**, 3477a (1954). Toxicity: Ambrose, *Proc. Soc. Exp. Biol. Med.* **76**, 466 (1951).

Trihydrochloride, $C_{21}H_{42}Cl_3N_7O_{13}$. The physical characteristics approx those of streptomycin. The specific rotation in water is 91° under conditions which give 86.1° for streptomycin trihydrochloride. Hydroxystreptomycin trihydrochloride, when assayed against *Bacillus subtilis,* was found to be equiv to 784 μg of streptomycin base/mg. The corresponding value of streptomycin is 842 μg/mg. LD_{50} s.c. in mice: 865 mg/kg (Ambrose).

4894. Hydroxytetracaine. *4-Butylamino-2-hydroxybenzoic acid 2-dimethylaminoethyl ester; p-butylaminosalicylic acid 2-dimethylaminoethyl ester;* 2-dimethylaminoethyl *p*-butylaminosalicylate; hydroxamethocaine; Rhenocain; Salicain. $C_{15}H_{24}N_2O_3$; mol wt 280.37. C 64.26%, H 8.63%, N 9.99%, O 17.12%. Prepn: **Brit. pats. 736,960** (1955) and **760,003** (1956 to Rheinpreussen AG); Grimme, Schmitz, *Ber.* **84**, 734 (1951).

Hydrochloride, $C_{15}H_{24}N_2O_3.HCl$, crystals from water, mp 157°. Soly in water at 20°: about 4%.

Hemihydrate, prisms from ligroin, mp 48°.

THERAP CAT: Anesthetic (local).

4895. 5-Hydroxytryptophan. 5-HTP. $C_{11}H_{12}N_2O_3$; mol wt 220.23. C 59.99%, H 5.49%, N 12.72%, O 21.79%. Precursor of serotonin. Synthesis from 5-benzyloxyindole: Ek, Witkop, *J. Am. Chem. Soc.* **76**, 5579 (1954); Shaw, Morris, *Biochem. Prepns.* **9**, 92 (1962); from 5-benzyloxy-tryptophan: Frangatos, Chubb, *Can. J. Chem.* **37**, 1374 (1959); Frangatos, **Can.** pat. **619,472** (1961 to Frank W. Horner); Ash, **Brit.** pat. **845,034** (1960 to May & Baker); from tryptophan: Renson *et al., Biochem. Biophys. Res. Commun.* **6**, 20 (1961). Prepn of L- and D-forms: A. J. Morris, M. D. Armstrong, *J. Org. Chem.* **22**, 306 (1957). Crystal and molecular structure of DL-form: Wakahara *et al., Tetrahedron Letters* **1970**, 3003. Use of L-5HTP in treatment of myoclonus: M. H. Van Woert, D. Rosenbaum, *Adv. Neurol.* **26**, 107 (1979); L. J. Thal *et al., Ann. Neurol.* **7**, 570 (1980). Clinical trial in migraine: F. Titus *et al., Eur. Neurol.* **25**, 327 (1986). *Review:* M. H. Van Woert, *Orphan Drugs,* F. E. Karch, Ed. (Marcel Dekker, New York, 1982) pp 13-31. Review of antidepressant efficacy: W. F. Byerley *et al., J. Clin. Psychopharmacol.* **7**, 127-137 (1987).

L-form

DL-Form, *Prétonine, Quietim.* Minute rods or needles from ethanol, dec 298-300°. uv max (H_2O at pH 6.0): 278 nm. Soly in water at 5°: 1.0 g/100 ml; at 100°: 5.5 g/100 ml. Soly in 50% boiling alc: 2.5 g/100 ml. Aq solns are stable at low pH.

L-Form, *oxitriptan, L-5HTP, Cincofarm, Levothym, Lévotonine, Oxyfan, Serotonyl, Telesol, Triptene, Tript-Oh.* Crystals, $[\alpha]_D^{20}$ −32.5° (H_2O); $[\alpha]_D^{20}$ +16.0° (4*N* HCl).

D-Form, crystals, $[\alpha]_D^{20}$ +32.2° (H_2O).

THERAP CAT: Antidepressant; antimigraine; in treatment of myoclonus.

4896. Hydroxyurea. Hydroxycarbamide; Hydrea; Litalir. $CH_4N_2O_2$; mol wt 76.06. C 15.79%, H 5.30%, N 36.83%, O 42.07%. $H_2NCONHOH$. Prepn from hydroxylamine HCl and KCN: Hantzsch, *Ann.* **299**, 99 (1898). Alternate route: Graham, **U.S.** pat. **2,705,727** (1955 to du Pont).

Needles from alc, mp 133-136°. Freely sol in water, hot alcohol.

THERAP CAT: Antineoplastic.

4897. Hydroxyzine. 2-[2-[4-[(4-Chlorophenyl)phenyl-methyl]-1-piperazinyl]ethoxy]ethanol; 1-(p-chloro-α-phenyl-benzyl)-4-(2-hydroxyethoxyethyl)piperazine; 1-(p-chlorodi-phenylmethyl)-4-[2-(2-hydroxyethoxy)ethyl]piperazine; N-(4-chlorobenzhydryl)-N'-(hydroxyethyloxyethyl)piperazine; 1-(p-chlorobenzhydryl)-4-[2-(2-hydroxyethoxy)ethyl]dieth-ylenediamine; UCB-4492; Tran-Q; Tranquizine. $C_{21}H_{27}$-ClN_2O_2; mol wt 374.91. C 67.28%, H 7.26%, Cl 9.46%, N 7.47%, O 8.54%. H_1 receptor antagonist. Outline of commercial prepn: *Chem. Week* 79(5), 70 (Aug. 4, 1956); Morren, **U.S.** pat. **2,899,436** (1959 to UCB). Pharmacology and metabolism: Cannizaro, *Boll. Chim. Farm.* **104**, 39 (1965); Close *et al., Ind. Chim. Belge* **33**, 94 (1968); *eidem, Proc. Eur. Soc. Study Drug Toxicity* **9**, 144 (1968); S F Pong, C. L. Huang, *J. Pharm. Sci.* **63**, 1527 (1974). Pharmacokinetics and antihistaminic activity: F. E. R. Simons *et al., J. Allergy Clin. Immunol.* **73**, 69 (1984); S. Ting *et al., ibid.* **75**, 63 (1985). Clinical trials of efficacy in allergic rhinitis: L. Wong *et al., ibid.* **67**, 223 (1981); in urticaria, R. P. Harvey *et al., ibid.* **68**, 262 (1981); as anti-emetic: R. McKenzie *et al., Anesth. Analg.* **60**, 783 (1981); as pre-surgical sedative: G. Wallace, L. J. Mindlin, *ibid.* **63**, 571 (1984). Toxicity data: E. I. Goldenthal, *Toxicol. Appl. Pharmacol.* **18**, 185 (1971). Comprehensive description: J. Tsau, N. DeAngelis, *Anal. Profiles Drug Subs.* **7**, 319-341 (1978).

Dihydrochloride, $C_{21}H_{27}ClN_2O_2.2HCl$, *Alamon, Atarax, Aterax, Durrax, Orgatrax, Quiess, Vistaril Parenteral.* Crystals, mp 193°. Bitter taste. Soly in mg/ml: water < 700; chloroform 60; acetone 2; ether < 0.1. Solns are unstable to intense uv light. LD_{50} in rats (mg/kg): 126 i.p.; 950 orally (Goldenthal).

Pamoate, $C_{44}H_{43}ClN_2O_8$, *Equipose, Masmoran, Paxistil, Vistaril Pamoate.* Crystals. Practically insol in water.

THERAP CAT: Anxiolytic. Antihistaminic.

THERAP CAT (VET): Has been used as a tranquilizer.

4898. Hyenanchin. [1aS-(1aα,1bβ,2β,5β,6α,6aβ,7β,-7aα,8R*)]-Hexahydro-1b,6,8-trihydroxy-6a-methyl-8-(1-methylethenyl)spiro[2,5-methano-7H-oxireno[3,4]cyclopent-[1,2-d]oxepin-7,2'-oxiran)-3(2H)-one; hyaenanchin; hyenanchin; mellitoxin. $C_{15}H_{18}O_7$; mol wt 310.30. C 58.06%, H 5.85%, O 36.09%. Isolated from fruit of *Hyaenanche globosa* Lamb., *Euphorbiaceae:* Henry, *J. Chem. Soc.* **117**, 1619 (1920). Structure and identity with mellitoxin: Jommi *et al., Chim. Ind. (Milan)* **46**, 549 (1964), *C.A.* **61**, 5697 (1964).

Crystals, mp 225-235°. Soly in water at 15°: 1.18%; more sol in hot water; sparingly sol in alcohol, acetone, ethyl acetate. $[\alpha]_D^{15}$ +14.7° (water).

4899. Hygrine. (R)-1-(1-Methyl-2-pyrrolidinyl)-2-propanone; 2-acetonyl-1-methylpyrrolidine; N-methyl-2-acetonylpyrrolidine. $C_8H_{15}NO$; mol wt 141.21. C 68.04%, H 10.71%, N 9.92%, O 11.33%. Occurs in leaves of *Erythroxylon coca* Lam., *Erythroxylaceae* of diverse origin: Liebermann, *Ber.* **22**, 677 (1889). Synthesis: Galinovsky *et al., Monatsh.* **82**, 551 (1951); Lukes *et al., Coll. Czech. Chem. Commun.* **24**, 2433 (1959); Leonard, Cook, *J. Am. Chem. Soc.* **81**, 5627 (1959). Enzymatic synthesis: Tuppy, Faltaous, *Monatsh.* **91**, 167 (1960). Stereochemistry: Galinovsky *et al., ibid.* **84**, 798 (1953). Absolute configuration: Lukes *et al., Coll. Czech. Chem. Commun.* **25**, 483 (1960).

Liquid. bp_{11} 76.5°; bp_{14} 81°. n_D^{20} 1.4555. Sol in alcohol, chloroform, dil acids; slightly sol in water.

Oxime, $C_8H_{16}N_2O$, crystals from ether, mp 123-124°.

4900. Hygromycin. (E)-5-Deoxy-5-[[3-[4-[(6-deoxy-β-D-arabino-hexofuranos-5-ulos-1-yl)oxy]-3-hydroxyphenyl]-2-methyl-1-oxo-2-propenyl]amino]-1,2-O-methylene-D-neo-ino-sitol; homomycin; hygromycin A; 1703-18B; ST-4331. C_{23}-$H_{29}NO_{12}$; mol wt 511.48. C 54.01%, H 5.71%, N 2.74%, O 37.54%. Antibiotic substance produced by *Streptomyces hygroscopicus* (Jensen) Waksman & Henrici, from forest soil near Indianapolis, Ind: R. L. Mann *et al., Antibiot. & Chemother.* **3**, 1279 (1953). Produced also by *Streptomyces noboritoensis* from soil collected in Kanawaga Prefecture.

Identity of homomycin and hygromycin: K. Isono *et al., J. Antibiot.* **10A**, 21 (1957). Structure: *eidem, ibid.* 160; R. L. Mann, D. O. Woolf, *J. Am. Chem. Soc.* **79**, 120 (1957). Abs config: K. Kakinuma, Y. Sakagami, *Agr. Biol. Chem.* **42**, 279 (1978). Synthesis of the sugar component: M. Nakajima, S. Takahashi, *ibid.* **31**, 1082 (1967). Total synthesis: N. Chida *et al., Chem. Commun.* **1989**, 436. Peptide synthesis inhibitor: M. Guerrero, J. Modolell, *Eur. J. Biochem.* **107**, 409 (1980).

Amorphous white powder. mp 105-109°, dec above 160°. Also reported as mp 110-112° (dec) (Chida). Weakly acidic. pKa 8.9. $[\alpha]_D^{25}$ −126° (c = 1 in water). Freely sol in water, alc. Practically insol in the less polar solvents. uv max (dil HCl): 272, 214 nm ($E_{1cm}^{1\%}$ 306, 416).

4901. Hygromycin B. *O-6-Amino-6-deoxy-L-glycero-D-galacto-heptopyranosylidene-(1→2-3)-O-β-D-talopyranosyl-(1→5)-2-deoxy-N³-methyl-D-streptamine;* Hygromix. $C_{20}H_{37}$-N_3O_{13}; mol wt 527.53. C 45.54%, H 7.07%, N 7.97%, O 39.43%. Antibiotic substance produced by *Streptomyces hygroscopicus* (Jensen) Waksman & Henrici, from soil samples from Nebraska and Indiana: Mann, Bromer, *J. Am. Chem. Soc.* **80**, 2714 (1958). Prodn: McGuire, Mann, U.S. pat. **3,018,220** (1962 to Lilly). Structure: Neuss *et al., Helv. Chim. Acta* **53**, 2314 (1970). Studies on mode of action: A. Gonzalez *et al., Biochim. Biophys. Acta* **521**, 459 (1978).

Amorphous powder, dec 160-180°. Weakly basic. pKa 7.1, 8.8. $[\alpha]_D^{20}$ +20.2° (c = 1 in water). Freely sol in water, methanol and ethanol. Practically insol in less polar solvents. Absorption spectrum: Mann, Bromer, *loc. cit.*

THERAP CAT (VET): Anthelmintic (swine and chicken).

4902. Hygrophylline. *(1α,14α)-1,2-Dihydro-12,14-dihydroxysenecionan-11,16-dione.* $C_{18}H_{27}NO_6$; mol wt 353.42. C 61.17%, H 7.70%, N 3.96%, O 27.16%. Pyrrolizidine alkaloid isolated from *Senecio hygrophyllus* Dyer and Sm., *Compositae:* Richardson, Warren, *J. Chem. Soc.* **1943**, 452. Structure and stereochemistry: Schlosser, Warren, *ibid.* **1965**, 5707. NMR spectrum: S. E. Drewes *et al., J. Chem. Soc. Perkin Trans. I* **1981**, 287. X-ray crystallography: M. F. Mackay *et al., Acta Crystallogr.* **C41**, 395 (1985).

Prisms from acetone, mp 173-174°. $[\alpha]_D^{20}$ −67.3° (c = 2.9 in ethanol).

4903. Hymecromone. *7-Hydroxy-4-methyl-2H-1-benzo-pyran-2-one;* 7-hydroxy-4-methylcoumarin; 7-hydroxy-4-methyl-2-oxo-3-chromene; 4-methylumbelliferone; β-methylumbelliferone; imecromone; 4-MU; Bilcolic; Biliton H;

Cantabilin; Cantabiline; Cholonerton; Cholspasmin; Cumarote C; Himecol; Medilla; Mendiaxon. $C_{10}H_8O_3$; mol wt 176.17. C 68.18%, H 4.58%, O 27.25%. Prepd from resorcinol and ethyl acetoacetate: Pechmann, Duisberg, *Ber.* **16**, 2119 (1883); Russell, Frye, *Org. Syn.* **21**, 23 (1941); Woods, Sapp, *J. Org. Chem.* **27**, 3703 (1962). Use in determn of nitric acid: F. J. Welcher, *Organic Analytical Reagents* vol. 1 (Van Nostrand, New York, 1947) pp 214-215.

Crystals from alc, mp 194-195° (Woods, Sapp*)*, formerly reported as 185-186° (Pechmann, Duisberg*)*. uv max (methanol): 221, 251, 322.5 nm. Blue fluorescence in alcohol + water. Sol in methanol, glacial acetic acid; slightly sol in ether, chloroform. Practically insol in cold water.

USE: Analytical reagent.

THERAP CAT: Choleretic; antispasmodic.

4904. Hymecromone *O,O*-Diethyl Phosphorothioate. *7-Hydroxy-4-methylcoumarin O-ester with O,O-diethyl phosphorothioate; O,O-diethyl O-(4-methyl-7-coumarinyl) phosphorothioate; O,O-diethyl O-(4-methyl-7-coumarinyl) thiophosphate; O,O-diethyl O-(4-methylumbelliferone) phosphorothioate; 4-methylumbelliferone O,O-diethyl phosphorothioate;* E-838; Potasan. $C_{14}H_{17}O_5PS$; mol wt 328.33. C 51.22%, H 5.22%, O 24.37%, P 9.43%, S 9.77%. Prepn: Schrader, Kükenthal, U.S. pat. **2,583,744** (1952 to Bayer). *Review:* G. Schrader, *Die Entwicklung neuer insektizider Phosphorsäure Ester* (Verlag Chemie, 1963) p 187.

Long needles from petr ether. Weak aromatic odor. mp 38°. $bp_{1.0}$ 210° (dec). d_4^{38} 1.260 (liq); n_D^{37} 1.5685 (liq). Very sparingly sol in water. Slightly sol in petr ether. Sol in most other organic solvents. Aq solns of emulsions adjusted to pH 7 to 8 show a blue fluorescence. LD_{50} i.p. in rats: 15 mg/kg.

Caution: Acts as cholinesterase inhibitor *in vivo;* symptoms similar to Parathion.

USE: Systemic insecticide, especially effective against the Colorado beetle.

4905. Hymexazol. *5-Methyl-3(2H)-isoxazolone;* 3-hydroxy-5-methylisoxazole; F-319; RTY-319; Tachigaren. $C_4H_5NO_2$; mol wt 99.09. C 48.49%, H 5.09%, N 14.14%, O 32.29%. Prepn: **Neth. pat. Appl. 6,511,925;** I. Iwai *et al.,* U.S. pat. **3,544,584** (1966, 1970 both to Sankyo); I. Iwai, N. Nakamura, *Chem. Pharm. Bull* **14**, 1277 (1966). Metabolism in plants: S. Kamimura *et al., Phytopathology* **64**, 1273 (1974). Properties and analytical methods: T. Nakamura *et al., Anal. Methods Pest. Plant Growth Regul.* **10**, 215 (1978). GC determn in plants and soils: *idem et al., Ann. Rep. Sankyo Res. Lab.* **28**, 130 (1976). Plant growth promoting activity: Y. Ota, *Japan. Agric. Res. Quart.* **9**, 1 (1975). Mode of fungicidal action: T. Nakanishi, H. D. Sisler, *J. Pest. Sci.* **8**, 173 (1983). Field studies: P. A. Payne, G. E. Williams, *Crop Prot.* **9**, 371 (1990). Series of articles on chemistry, biological activity and metabolism: *Ann. Rep. Sankyo Res. Lab.* **25**, 1-51 (1973). Brief review: R. Ishitsuka, *Japan Pest. Info.* **18**, 27-30 (1974).

White needles from *n*-hexane, mp 84-85°. Also reported as mp 86-87° (Nakamura, 1983). Vapor pressure (25°) 1.01 × 10⁻³ Torr. Soly at 25° (g/100 ml): ~8.5 in water. Very sol in methanol, ethanol, isopropanol, acetone, methyl isobutyl ketone, tetrahydrofuran, dioxane, DMF, ethylene glycol, chloroform. Fairly sol in diethyl ether, benzene, xylene, trichloroethylene. Slightly sol in *n*-hexane, carbon disulfide. LD$_{50}$ in male, female mice, rats (mg/kg): 2148, 1968, 4678, 3909 orally; 1297, 1167, 1924, 1884 s.c.; 445, 514, > 1000, > 1000 i.v.; in rats, rabbits (mg/kg): > 10000, > 2000 dermally (Nakamura, 1978).

USE: Agricultural fungicide and plant growth regulator.

4906. Hyodeoxycholic Acid. (3α,5β,6α)-3,6-Dihydroxycholan-24-oic acid; 3α,6α-dihydroxy-5β-cholanic acid; α-hyodeoxycholic acid; hyodesoxycholic acid. C$_{24}$H$_{40}$O$_4$; mol wt 392.58. C 73.43%, H 10.27%, O 16.30%. Isoln from pig bile: Wieland, Gumlich, *Z. Physiol. Chem.* **215**, 18 (1933); Trickey, U.S. pat. **2,547,726** (1951 to U.S. Rubber); Fogle, U.S. pat. **2,745,849** (1956 to Armour); Buckley, Ziegler, U.S. pat. **2,758,120** (1956 to Canada Packers); Liebig, U.S. pat. **3,006,927** (1961 to Riedel-de Haen). Configuration: Moffett, Hoehn, *J. Am. Chem. Soc.* **69**, 1995 (1947).

Crystals from ethyl acetate, mp 196-197°. [α]$_D^{20}$ +8° (alc). Moderately sol in ethanol, glacial acetic acid. Less sol in ether, acetone, ethyl acetate, benzene.

4907. Hyoscyamine. [3(S)-endo]-α-(Hydroxymethyl)-benzeneacetic acid 8-methyl-8-azabicyclo[3.2.1]oct-3-yl ester; 1αH,5αH-tropan-3α-ol(−)-tropate (ester); 3α-tropanyl S-(−)-tropate; l-tropic acid ester with tropine; l-tropine tropate; daturine; duboisine; l-hyoscyamine; Cystospaz; Levsin. C$_{17}$H$_{23}$NO$_3$; mol wt 289.37. C 70.56%, H 8.01%, N 4.84%, O 16.59%. From *Hyoscyamus niger* L., *Atropa belladonna* L., *Datura stramonium* L., and other *Solanaceae*: Ladenburg, *Ann.* **206**, 274 (1881). Identity with duboisine and daturine: Beckurts, *Apoth. Ztg.* **27**, 683 (1912). Obtained by resolution of atropine: Werner, Miltenberger, *Ann.* **631**, 163 (1960). Prepd from (−)-acetyltropoyl chloride and atropine hydrochloride: Fodor *et al.*, *Acta Chim. Acad. Sci. Hung.* **28**(4), 409 (1961), *C.A.* **61**, 1903g (1964). Configuration: Fodor, Csepreghy, *Tetrahedron Letters* no. 7, 16 (1959).

Silky, tetragonal needles from evaporating alc. *Keep well closed and protect from light and heat;* easily racemized. mp 108.5°. [α]$_D^{20}$ −21.0° (alc). K at 19° is 1.9×10⁻¹². One gram dissolves in 281 ml water (pH 9.5), 69 ml ether, 150 ml benzene, 1 ml chloroform. Freely sol in alcohol, dil

acids. Absorption spectra: Dobbie, Fox, *J. Chem. Soc.* **103**, 1194, 1195 (1913).

Hydrobromide, C$_{17}$H$_{23}$NO$_3$.HBr, deliquescent crystals, mp 152°. pH 5.4 (1 in 100). Levorotatory. Very sol in water. One gram dissolves in 3 ml alcohol, 1.2 ml chloroform, 2260 ml ether.

Hydrochloride, C$_{17}$H$_{23}$NO$_3$.HCl. Crystals. *Poisonous!* mp 149-151°. Freely sol in water, alcohol.

Methyl bromide, C$_{17}$H$_{23}$NO$_3$.CH$_3$Br, *N-methylhyoscyaminium bromide*. Crystals, mp 210-212°. Freely sol in water, dil alc; slightly sol in abs alc.

Sulfate dihydrate, (C$_{17}$H$_{23}$NO$_3$)$_2$.H$_2$SO$_4$.2H$_2$O, *Egacene, Egazil Duretter, Peptard*. Needles from alcohol, mp 206° when dry. [α]$_D^{15}$ −29° (c = 2). pH 5.3 (1 in 100). One gram dissolves in 0.5 ml water, about 5.0 ml alcohol. Very slightly sol in chloroform, ether.

THERAP CAT: Anticholinergic.

4908. Hyoscyamus. Henbane; hog's bean; insane root; poison tobacco; black henbane. Dried leaves, with or without the tops, of *Hyoscyamus niger* L., *Solanaceae*, yielding 0.040% hyoscyamus alkaloids (chiefly hyoscyamine and scopolamine). The leaves and seeds have been used as smooth muscle relaxants. The alkaloid content of the seeds is less than that of the leaves. Egyptian henbane (*Hyoscyamus muticus* L., *Solanaceae*) contains about 0.5% alkaloids. *Habit.* Europe, Africa, Asia, naturalized in U.S.; cultivated in England. *Constit.* Leaves: scopolamine, hyoscyamine, hyoscipicrin, choline.

4909. Hypalon®. Elastomer made by substituting chlorine and sulfonyl chloride groups into polyethylene. Prepd from polyethylene of average mol wt of about 20,000; contains approx one chlorine atom for each seven carbon atoms and one sulfonyl chloride group for every ninety carbon atoms. Three types of Hypalon are available: 20, 30, and 40 with sp gr of 1.12, 1.28, and 1.18, respectively. Prepn: Reed, Horn, U.S. pat. **2,046,090** (1933). *Reviews:* Ullmann's *Encyklopädie der Technischen Chemie* **vol. 9**, 342 (1957); Keeley, "Hypalon Synthetic Rubber" in *Introduction to Rubber Technology*, Morton, Ed. (Reinhold, New York, 1959) p 349.

4910. Hypaphorine. (S)-α-Carboxy-N,N,N-trimethyl-1H-indole-3-ethanaminium inner salt; 1-trimethylammonio-3-(3-indolyl)propionate. C$_{14}$H$_{18}$N$_2$O$_2$; mol wt 246.31. C 68.27%, H 7.37%, N 11.37%, O 12.99%. The trimethylbetaine of tryptophan. Poisonous alkaloid from seeds of *Erythrina americana* Mill., *E. sandwicensis* Degener, *E. cristagalli* L., and *Abrus precatorius* Linn., *Leguminosae*: Folkers, Koniuszy, *J. Am. Chem. Soc.* **61**, 1232 (1939); **62**, 1677 (1940); Deulofeu *et al.*, *J. Chem. Soc.* **1939**, 1841; Tung, Lao, *C.A.* **55**, 17770i (1961). Structure and synthesis: von Romburgh, Barger, *J. Chem. Soc.* **99**, 2068 (1911).

Crystals from dil alc, dec 237°; after purification through the nitrate, dec 255°. [α]$_D^{27}$ +113° (c = 0.52). Sol in 12 parts water, slightly in alc. Almost insol in other usual solvents.

Hydrochloride, C$_{14}$H$_{18}$N$_2$O$_2$.HCl, crystals from water, dec 231-232°, [α]$_D^{32}$ +90° (c = 0.5). Moderately sol in water.

Nitrate, C$_{14}$H$_{18}$N$_7$O$_2$.HNO$_3$, dec 224°, [α]$_D^{20}$ +95°. Sol in 170 parts water.

4911. Hypericin. 1,3,4,6,8,13-Hexahydroxy-10,11-dimethylphenanthro[1,10,9,8-opqra]perylene-7,14-dione; 4,5,7,-4',5',7'-hexahydroxy-2,2'-dimethylnaphthodianthrone; hypericum red; Cyclo-Werrol. C$_{30}$H$_{16}$O$_8$; mol wt 504.45. C 71.43%, H 3.20%, O 25.37%. Occurs in *Hypericum* spp. (St. John's wort). Isoln from *Hypericum perforatum* L., *Hypericaceae*: Brockmann *et al.*, *Ann.* **553**, 1 (1942). Synthesis from bromoemodin trimethylether: Brockmann, Muxfeld, *Naturwiss.* **40**, 411 (1953). Also obtained by irradiation of oxypenicilliopsin produced by the fungus *Penicilliopsis cla-*

variaeformis Solms-Laubach: Brockmann, Neef, *ibid.* **38**, 47 (1951). Structure: Brockmann *et al., ibid.* **37**, 540 (1950); Brockmann, Sanne, *ibid.* **40**, 509 (1953). Total synthesis: Brockmann, Kluge, U.S. pat. **2,707,704** (1955 to Schenley); Brockmann *et al., Ber.* **90**, 2302 (1957); Brockmann, Eggers, *ibid.* **91**, 547 (1958).

Solvated blue-black needles from pyridine + methanolic HCl, dec 320°. Freely sol in pyridine and other organic bases yielding cherry-red solns with red fluorescence. Almost insol in most other organic solvents. Sol in alkaline aq solns; below pH 11.5 solns are red, above pH 11.5 they are green with red fluorescence. Absorption and fluorescence spectra: Scheibe, Schöntag, *Ber.* **75**, 2019 (1942); Brockmann, Neef, *loc. cit.*

THERAP CAT: Antidepressant.

4912. Hypochlorous Acid. ClHO; mol wt 52.46. Cl 67.58%, H 1.92%, O 30.50%. HClO. Known in aq soln only. Formed by the action of water on chlorine: Cady, *Inorg. Syn.* **5**, 160 (1957). A 25% soln of hypochlorous acid is obtained when chlorine hydrate and mercuric oxide are distilled at low pressure according to the eq $2Cl_2.6H_2O + HgO \rightarrow 2HClO + HgCl_2 + 11H_2O$: Goldschmidt, *Ber.* **52**, 753 (1919). *Review:* J. A. Wojtowicz, "Chlorine Monoxide, Hypochlorous Acid, and Hypochlorites", in Kirk-Othmer *Encyclopedia of Chemical Technology* **vol. 5** (Wiley-Interscience, New York, 3rd ed., 1979) pp 580-611.

Aq soln (25%), greenish yellow liquid. May be stored for a few days if kept at $-20°$. Dec slowly to Cl_2, O_2 and $HClO_4$. Very weak acid. K at $25° = 3.2 \times 10^{-8}$. Strong oxidizing agent. When a 25% soln is evapd in a vacuum, chlorine monoxide, Cl_2O, is given off. Cl_2O in soln with HClO may be removed by shaking with carbon tetrachloride, but will form again. *Protect from light.*

Note: An aqueous soln of the potassium salt is known as *Javelle water* or *Eau de Javelle* (also used for sodium hypochlorite, *q.v.*).

USE: Disinfectant.

4913. Hypoglycine A. *[S-(R*,S*)]-α-Amino-2-methyl-enecyclopropanepropanoic acid; 2-methylenecyclopropane-alanine;* 2-amino-4,5-methylenehex-5-enoic acid; α-amino-β-(2-methylenecyclopropyl)propionic acid; hypoglycin; hypoglycin A. $C_7H_{11}NO_2$; mol wt 141.17. C 59.56%, H 7.85%, N 9.92%, O 22.67%. Hypoglycemic principle from the akee plant, *Blighia sapida* Kon., *Sapindaceae:* Hassall *et al., Nature* **173**, 356 (1954); Hassall, Reyle, *Biochem. J.* **60**, 334 (1955). Structure: de Ropp *et al., J. Am. Chem. Soc.* **80**, 1004 (1958); Renner *et al., Helv. Chim. Acta* **41**, 589 (1958); Ellington *et al., J. Chem. Soc.* **1959**, 80. Synthesis: Carbon *et al., J. Am. Chem. Soc.* **80**, 1002 (1958); Black, Landor, *Tetrahedron Letters* **1963**, 1065.

Yellow plates from methanol + water, mp 280-284°. $[\alpha]_D^{32}$ +9.2°.

Methyl ester hydrochloride, $C_8H_{13}NO_2.HCl$, needles from methanol + ether, mp 151-152°. $[\alpha]_D^{22} +36°$ (c = 2.0).

4914. Hypoglycine B. *N-L-γ-Glutamyl-3-(2-methylene-cyclopropyl)alanine;* γ-L-glutamyl-α-amino-β-(2-methylene-cyclopropyl)propionic acid dipeptide; γ-L-glutamylhypoglycine. $C_{12}H_{18}N_2O_5$; mol wt 270.29. C 53.33%, H 6.71%, N

10.36%, O 29.60%. Isoln *see* Hypoglycine A. Structure: Jöhl, Stoll, *Helv. Chim. Acta* **42**, 156 (1959); Hassall, John, *J. Chem. Soc.* **1960**, 4112. Synthesis: Jöhl, Stoll, *Helv. Chim. Acta* **42**, 716 (1959).

Yellow needles from acetone + water, mp 194-195°, 200-206°. $[\alpha]_D^{32}$ +9.6 (c = 1.12). Neutralization equivalent 175. N-2,4-Dinitrophenylhydrazone, $C_{18}H_{20}N_4O_9$, needles from methanol + water, mp 170-172°. $[\alpha]_D^{19} -84°$ (c = 0.6 in 4% aq NaHCO₃). uv max: 359 nm (ε 16,600).

Phenylhydantoin derivative, $C_{19}H_{21}N_3O_4S$, needles from alc + water, mp 186-188°. uv max: 268-269 nm (ε 16,000).

4915. Hypophosphoric Acid. $H_4O_6P_2$; mol wt 161.98. H 2.49%, O 59.27%, P 38.24%. Prepn from its salts: Salzer, *Ann.* **187**, 322 (1877); **211**, 1 (1882). *Review:* Ohashi, "Lower Oxo Acids of Phosphorus and Their Salts" in *Topics in Phosphorus Chemistry,* **Vol. 1**, M. Grayson, E. J. Griffith, Eds. (Interscience, New York, 1964) pp 113-187.

Orthorhombic plates, mp 70° (easily forms a dihydrate, mp 55°), usually available only in aq soln. The aq acid is colorless and odorless; dec on concn at atm pressure. Readily forms crystallizable normal and acid sodium salts.

USE: Acid sodium salts in baking powders.

4916. Hypophosphorous Acid. H_3O_2P; mol wt 66.00. H 4.58%, O 48.49%, P 46.93%. Conveniently prepd by treating NaH_2PO_2 with an ion-exchange resin: Klement, *Z. Anorg. Allgem. Chem.* **260**, 267 (1949). *Review:* Ohashi, "Lower Oxo Acids of Phosphorus and Their Salts" in *Topics in Phosphorus Chemistry,* **Vol. 1**, M. Grayson, E. J. Griffith, Eds. (Interscience, New York, 1964) pp 113-187.

The water-free acid forms deliquesc crystals; supercools to a colorless, odorless, oily liquid. d 1.493. mp 26.5°. Dec by heat into H_3PO_4 and spontaneously flammable PH_3. Miscible with water, alcohol, ether. Oxidized by hot H_2SO_4; SO_2 and S formed. It is a powerful reducing agent. Marketed in aq solns of various concns, *e.g.,* 50%, d 1.274; 30-32%, d 1.13; 10%, d 1.04. $K_1 = 8 \times 10^{-2}$.

4917. Hypoxanthine. *1,7-Dihydro-6H-purin-6-one; purin-6(1H)-one;* sarcine; sarkin. $C_5H_4N_4O$; mol wt 136.11. C 44.12%, H 2.96%, N 41.16%, O 11.75%. Desmotropic forms: purin-6-ol; 1H-purin-6-ol; 3H-purin-6-ol; 9H-purin-6-ol; purin-6(3H)-one; 9H-purin-6(1H)-one. Formed in the animal body during the breakdown of nucleic acids; formation from adenosine continues after death. Also widely distributed in the vegetable kingdom. Numerous syntheses, *e.g.,* from 2,6,8-trichloropurine: Fischer, *Ber.* **30**, 2226 (1897); by oxidation of adenine: Krüger, *Z. Physiol. Chem.* **18**, 445 (1894). By reduction of uric acid: Sundwik, *ibid.* **76**, 486 (1912); by condensing ethyl cyanoacetate and thiourea in the presence of Na-ethoxide: Traube, *Ann.* **331**, 64 (1904); from mercapto-4-hydroxy-6-aminopyrimidine: Taylor, Cheng. *J. Org. Chem.* **25**, 148 (1960). *Reviews:* Levene, Bass, *Nucleic Acids,* A.C.S. Monograph Series no. 56 (New York 1931); Chargaff, Davidson, *Nucleic Acids,* **vols. I & II** (Academic Press, New York, 1955).

Small octahedra, needles from water, dec 150° without melting. Sol in 1400 parts water, 70 parts boiling water; sol in dil acids, alkalies. Combines with one equivalent of acid or with two equivalents of base. pK_b (25°): 8.7. Absorption spectrum: Dhere, *Compt. Rend.* **141,** 720.

4918. Hypusine. *(R)-N⁶-(4-Amino-2-hydroxybutyl)-L-lysine;* (2S,9R)-2,11-diamino-9-hydroxy-7-azaundecanoic acid; N^6-(4-amino-2-hydroxybutyl)-2,6-diaminohexanoic acid. $C_{10}H_{23}N_3O_3$; mol wt 233.31. C 51.48%, H 9.94%, N 18.01%, O 20.57%. Naturally occurring L-amino acid formed by the post-translational modification of lysine. Name coined to indicate the condensation product of **hydroxyputrescine with lysine.** Isoln: T. Shiba *et al., Biochim. Biophys. Acta* **244,** 523 (1971); from archaebacteria: H. Schümann, F. Klink, *System. Appl. Microbiol.* **11,** 103 (1989). In eukaryotes occurs only at a single position in one protein, eukaryotic protein synthesis initiation factor 5A, (eIF-5A, formerly eIF-4D): M. H. Park *et al., Proc. Nat. Acad. Sci. U.S.A.* **78,** 2869 (1981); H. L. Cooper *et al., ibid.* **80,** 1854 (1983). Biosynthesis: M. H. Park *et al., J. Biol. Chem.* **257,** 7217 (1982); M. H. Park *et al., ibid.* **259,** 12123 (1984). Synthesis and stereochemistry: T. Shiba *et al., Bull. Chem. Soc. Japan.* **55,** 899 (1982). Total synthesis: C. M. Tice, B. Ganem, *J. Org. Chem.* **48,** 5043, 5048 (1983); R. J. Bergeron *et al., ibid.* **58,** 6804 (1993). Biological significance in eukaryotes: J. W. B. Hershey *et al., Biochim. Biophys. Acta* **1050,** 160 (1990). HPLC determn in eukaryotic samples: S. Beninati *et al., Anal. Biochem.* **184,** 16 (1990). Chromatographic determn in archaebacteria: D. Bartig, F. Klink, *J. Chromatog.* **606,** 43 (1992). Review of biological significance: M. H. Park *et al., Trends Biochem. Sci.* **18,** 475-479 (1993).

Dihydrochloride, $C_{10}H_{23}N_3O_3 \cdot 2HCl$, white crystals, mp 239-241° (dec). $[\alpha]_D^{23}$ +8.3° (c = 0.96 in 6M HCl).

I

4919. Ibafloxacin. *9-Fluoro-6,7-dihydro-5,8-dimethyl-1-oxo-1H,5H-benzo[ij]quinolizine-2-carboxylic acid;* R-835; S-25930. $C_{15}H_{14}FNO_3$; mol wt 275.28. C 65.45%, H 5.13%, F 6.90%, N 5.09%, O 17.44%. Tricyclic fluorinated quinolone-type antibacterial; related to flumequine, *q.v.* Prepn: R. M. Stern, **Eur. pat. Appl.** 109,284; *idem*, **U.S. pat.** **4,472,405** (both 1984 to Riker Labs.). Antimicrobial spectrum: L. J. V. Piddock *et al.*, *Eur. J. Clin. Microbiol.* **5**, 303 (1986); K. V. I. Rolston *et al.*, *Eur. J. Clin. Microbiol. Infect. Dis.* **7**, 681 (1988). Synthesis, absolute configuration and antibacterial activity of enantiomers: J. F. Gerster *et al.*, *J. Med. Chem.* **30**, 839 (1987).

Off-white solid, mp 269-272°.
S-Form, colorless crystals from DMF, mp 274-276°. $[\alpha]_D$ −107.4° (c = 0.1109 g/10mL in dichloroethane).
R-Form, colorless crystals from DMF, mp 274-276°. $[\alpha]_D$ +106.8° (c = 0.1165 g/10mL in dichloroethane).
THERAP CAT (VET): Antibacterial.

4920. Ibogaine. *12-Methoxyibogamine.* $C_{20}H_{26}N_2O$; mol wt 310.44. C 77.38%, H 8.44%, N 9.02%, O 5.15%. Indole alkaloid of the *iboga* group. Isoln from root (1.27%), rootbark (2 to 6%), stems (1.95%) and leaves (0.35%) of the shrub *Tabernanthe iboga* Baill., *Apocynaceae*, found in Africa: Dybowski, Landrin, *Compt. Rend.* **133**, 748 (1901); Haller, Heckel, *ibid.* 850, 1236; from other *Apocynaceae*: H. Achenbach, B. Raffelsberger, *Z. Naturforsch.* **35B**, 219, 885 (1980); N. Ghorbel *et al.*, *J. Nat. Prod.* **44**, 717 (1981); T. Mulamba *et al.*, *ibid.* 184; B. Richard *et al.*, *ibid.* **46**, 283 (1983). Purification: Schlittler *et al.*, *Helv. Chim. Acta* **36**, 1341 (1953). Revised extraction procedure: Dickel *et al.*, *J. Am. Chem. Soc.* **80**, 123 (1958). Review of early isolation work: Lebeau, Janot, *Traité de Pharmacie Chimique* **vol. 4** (Masson et Cie., Paris, 1956) pp 2982-2988. Structure: Bartlett *et al.*, *J. Am. Chem. Soc.* **80**, 126 (1958). Mass spectrum: Biemann, Friedmann-Spiteller, *ibid.* **83**, 4805 (1961). Synthesis: Büchi *et al.*, *ibid.* **88**, 3099 (1966); Rosenmund *et al.*, *Ber.* **108**, 1871 (1975). Derivs: Taylor, **U.S. pat. 2,877,229** (1959 to Ciba). Absolute configuration: K. Blàha *et al.*, *Tetrahedron Letters* **1972**, 2763. Interatomic distances similar to those of serotonin: J. M. Kelley, R. H. Adamson, *Pharmacology* **10**, 28 (1973). NMR spectrum: E. Wenkert *et al.*, *Helv. Chim. Acta* **59**, 2437 (1976). Determn in biological fluids: E. Bertol *et al.*, *J. Chromatog.* **117**, 239 (1976). Iboga extracts said to be used by African natives while stalking game, to enable them to remain motionless for as long as 2 days while retaining mental alertness. Neuropharmacological studies: Schneider, Sigg, *Ann. N.Y. Acad. Sci.* **66**, 765 (1957); S. Gershon, W. J. Lang, *Arch. Int. Pharmacodyn. Ther.* **135**, 31 (1962). Cardiovascular effects: J. A. Schneider, R. K. Rinehart, *ibid.* **110**, 92 (1957). Serotonergic properties: R. S. Sloviter *et al.*, *J. Pharmacol. Exp. Ther.* **214**, 231 (1980). Experimental use in treatment of heroin addiction: H. S. Lotsof, **U.S. pat. 4,499,096** (1985). *Reviews:* W. I. Taylor, "The Iboga and Voacanga Alkaloids" in *The Alkaloids, Chemistry and Physiology* **Vol. 8**, R. H. F. Manske, Ed. (Academic Press, New York, 1965) p 203-235, *idem, ibid.* **Vol. 11** (1968), pp 79-98.

Prismatic needles from abs ethanol, mp 152-153°. Sublimes$_{0.01}$ 150°. $[\alpha]_D^{20}$ −53° (in 95% ethanol). pKa 8.1 in 80% methylcellosolve. uv max (methanol): 226, 298 nm (log ϵ 4.39, 3.93). Sol in ethanol, ether, chloroform, acetone, benzene. Practically insol in water.
Hydrochloride, $C_{20}H_{26}N_2O.HCl$, crystals. Dec 299-300°. $[\alpha]_D^{25}$ −63° (ethanol); $[\alpha]_D^{25}$ −49° (H2O). Soluble in water, methanol, ethanol. Slightly sol in acetone, chloroform. Practically insol in ether.
Note: This is a controlled substance (hallucinogen) listed in the U.S. Code of Federal Regulations, Title 21 Part 1308.11 (1995).

4921. Ibopamine. *2-Methylpropanoic acid 4-[2-(methylamino)ethyl]-1,2-phenylene ester;* 4-[2-(methylamino)ethyl]-*o*-phenylene diisobutyrate; *N*-methyldopamine diisobutyric ester; 3,4-di-*o*-isobutyryl epinine. $C_{17}H_{25}NO_4$; mol wt 307.39. C 66.43%, H 8.20%, N 4.56%, O 20.82%. Inotropic agent with dopaminergic and adrenergic agonist activities. Prepn: C. Casagrande, G. Ferrari, **Ger. pat.** 2,734,678; *eidem*, **U.S. pat.** **4,218,470** (1978, 1980 both to Simes). Pharmacology: G. F. Melloni *et al.*, *Curr. Ther. Res.* **25**, 406 (1979); *eidem, ibid.* **26**, 466 (1979). Diuretic effect in chronic renal failure: S. Stefoni *et al.*, *Brit. J. Clin. Pharmacol.* **11**, 69 (1981); in ascitic liver disease: G. F. Melloni *et al., ibid.* **12**, 813 (1981). Acute hemodynamic effects in congestive heart failure: P. Ghirardi *et al., ibid.* **19**, 613 (1985). α- and β-adrenergic activity: A. J. Nichols, R. R. Ruffolo, Jr., *J. Pharmacol. Exp. Ther.* **242**, 455 (1987). Series of articles on synthesis, pharmacology, clinical efficacy: *Arzneimittel-Forsch.* **36**, 285-408 (1986). Post-marketing surveillance study: D. Sher, V. Ferrari, *ibid.* **37**, 869 (1987). Series of articles on clinical use in congestive heart failure: *Cardiology* **77**, Suppl. 5, 1-95 (1990). Review of pharmacodynamics, pharmacokinetics and therapeutic efficacy: J. M. Henwood, P. A. Todd, *Drugs* **36**, 11-31 (1988).

Hydrochloride, $C_{17}H_{25}NO_4.HCl$, *SB-7505, Inopamil, Scandine*. Crystals from ethyl acetate, mp 132°.
THERAP CAT: Cardiotonic.

4922. Ibotenic Acid. *α-Amino-2,3-dihydro-3-oxo-5-isoxazoleacetic acid; α-amino-3-hydroxy-5-isoxazoleacetic acid;* amino-(3-hydroxy-5-isoxazolyl)acetic acid. $C_5H_6N_2O_4$; mol wt 158.11. C 37.98%, H 3.82%, N 17.72%, O 40.48%. Fly-killing and narcosis-potentiating amino acid structurally similar to kainic acid, *q.v.*, extracted from poisonous mushroom species. Isoln from *Amanita pantherina* (DC.) Fr., and *A. muscaria* (L.) Fr., *Agaricaceae*: Takemoto *et al., J. Pharm. Soc. Japan* **84**, 1233 (1964); Eugster *et al., Tetrahedron Letters* **1965**, 1813. Structure: Takemoto *et al., J. Pharm. Soc. Japan* **84**, 1186, 1232 (1964). Syntheses: Gagneux *et al., Tetrahedron Letters* **1965**, 2081; Sirakawa *et al., Chem. Pharm. Bull.* **14**, 89 (1966); Kishida *et al., ibid.* **14**, 92 (1966); **15**, 1025 (1967). Improved synthesis: Nakamura, *ibid.* **19**, 46 (1971). Industrial pats: **Belg. pat. 665,249**, *C.A.* **65**, 2266e (1966); Gagneux *et al.*, **U.S. pat. 3,459,862** (1965, 1969, both to Geigy); Kishida *et al.*, **Japan. pats. 15,975-('68)** and **25,780('69)** (both to Sankyo), *C.A.* **70**, 77944p (1969); **72**, 13054g (1970). Pharmacology: Theobald *et al.*,

Arzneimittel-Forsch. **18,** 311 (1968); Johnston *et al., Biochem. Pharmacol.* **17,** 2488 (1968). Exhibits potent neuroexcitatory activity: *eidem, Nature* **248,** 804 (1974). Chemistry review: Eugster, *Fortschr. Chem. Org. Naturst.* **27,** 261-321 (1969); Catalfomo, Eugster, *Bull. Narcotics* **22,** 33-41 (1970). Excitatory and possible sedative actions on spinal neurons: D. R. Curtis *et al., J. Physiol.* **291,** 19 (1979); in cerebral cortex: E. Puil, *Can. J. Physiol. Pharmacol.* **59,** 1025 (1981). Use as experimental neurotoxic agent: A. Contestabile *et al., Experientia* **40,** 524 (1984).

Crystals from water or methanol, mp 151-152° (anhydrous); mp 144-146° (monohydrate). LD$_{50}$ in mice, rats (mg/kg): 15, 42 i.v.; 38, 129 orally (Theobald).

USE: Neurobiological tool.

4923. Ibudilast. *2-Methyl-1-[2-(1-methylethyl)pyrazolo-[1,5-a]pyridin-3-yl]-1-propanone;* 3-isobutyryl-2-isopropyl-pyrazolo[1,5-a]pyridine; KC-404; Ketas. C$_{14}$H$_{18}$N$_2$O; mol wt 230.31. C 73.01%, H 7.88%, N 12.16%, O 6.95%. Leukotriene D$_4$ antagonist. Prepn: T. Irikura *et al.,* **Ger. pat. 2,315,801;** *eidem,* **U.S. pat. 3,850,941** (1973, 1974 both to Kyorin). Pharmacology and antiallergic activity: K. Nishino *et al., Japan. J. Pharmacol.* **33,** 267 (1983); H. Nagai *et al., ibid.* 1215. *In vitro* cerebral vasodilating activity: M. Ohashi *et al., Arch. Int. Pharmacodyn.* **280,** 216 (1986); *in vivo* activity: W. M. Armstead *et al., J. Pharmacol. Exp. Ther.* **244,** 138 (1988). Bronchodilating activity in animals: S. Mue *et al., Arch. Int. Pharmacodyn.* **283,** 153 (1986). Antiplatelet activity in animals: M. Ohashi *et al., ibid.* 321; M. Ohashi *et al., Gen. Pharmacol.* **17,** 385 (1986).

Crystals from hexane, mp 53.5-54°. Slightly sol in water, freely sol in organic solvents. LD$_{50}$ i.v. in mice: 260 mg/kg (Irikura, 1973).

THERAP CAT: Antiallergic; antiasthmatic; vasodilator (cerebral).

4924. Ibufenac. *4-(2-Methylpropyl)benzeneacetic acid; (p-isobutylphenyl)acetic acid;* p-isobutyl-α-toluic acid; Dytransin; Ibunac. C$_{12}$H$_{16}$O$_2$; mol wt 192.26. C 74.97%, H 8.39%, O 16.64%. Prepn: J. S. Nicholson, S. S. Adams, **Brit. pat. 971,700;** *eidem,* **U.S. pat. 3,385,887** (1964, 1968 both to Boots Pure Drug). Pharmacology: S. S. Adams *et al., J. Pharm. Pharmacol.* **20,** 305 (1968). Toxicity data: S. S. Adams *et al., Nature* **200,** 271 (1962).

Crystals, mp 85-87°. Very slightly sol in water. Freely sol in many organic solvents. LD$_{50}$ orally in mice: 1.8 g/kg (Adams).

THERAP CAT: Analgesic, anti-inflammatory.

4925. Ibuprofen. *α-Methyl-4-(2-methylpropyl)benzeneacetic acid; p-isobutylhydratropic acid;* (±)-2-(4-isobutylphenyl)propionic acid; RD-13621; Adran; Advil; Anco; Amibufen; Anflagen; Apsifen; Artril 300; Bluton; Brufen; Brufort; Buburone; Butylenin; Dansida; Dentigoa; Dolgin; Dolgirid; Dolgit; Dolocyl; Dolo-Dolgit; Ebufac; Emodin; Epobron; Femadon; Fenbid; Gynofug; Haltran; Ibu-Attri-

tin; Ibumetin; Ibuprocin; Ibutad; Ibutid; Ibutop; Inabrin; Inoven; Lamidon; Lebrufen; Liptan; Lobufen; Medipren; Motricit; Motrin; Mynosedin; Napacetin; Nobfen; Nobgen; Novogent N; Nuprin; Nurofen; Opturem; Pediaprofen; Proflex; Prontalgin; Recidol; Roidenin; Seclodin; Suspren; Tabalon; Trendar; Urem. C$_{13}$H$_{18}$O$_2$; mol wt 206.28. C 75.69%, H 8.80%, O 15.51%. Prepn: Nicholson, Adams, **Brit. pat. 971,700;** *eidem,* **U.S. pats. 3,228,831** and **3,385,-886** (1964, 1966, 1968 all to Boots Pure Drug); T. Shiori, N. Kawai, *J. Org. Chem.* **43,** 2936 (1978); J. T. Pinhey, B. A. Rowe, *Tetrahedron Letters* **21,** 965 (1980). Pharmacology: Adams *et al., Arch. Pharmacodyn. Ther.* **178,** 115 (1969). Metabolism and toxicity: Adams *et al., Toxicol. Appl. Pharmacol.* **15,** 310 (1969). Toxicity: G. Orzalesi *et al., Arzneimittel-Forsch.* **27,** 1006 (1977). Series of articles on pharmacology, pharmacokinetics, adverse effects, and clinical efficacy: *Am. J. Med.* **77**(1A), 1-125 (1984). *Review:* L. Cavallini, G. Lucchetti, *Gazz. Med. Ital.* **134,** 7 (1975).

Colorless, crystalline stable solid, mp 75-77°. Relatively insol in water. Readily sol in most organic solvents. LD$_{50}$ in male mice, rats (mg/kg): 495, 626 i.p.; 1255, 1050 orally (Orzalesi).

Lysine salt, C$_{13}$H$_{18}$O$_2$·C$_6$H$_{14}$N$_2$O$_2$, Solprofen.

S-Form L-lysine salt monohydrate, C$_{13}$H$_{18}$O$_2$·C$_6$H$_{14}$N$_2$O$_2$·H$_2$O, dexibuprofen lysine, MK-223.

Methylglucamine salt, C$_{13}$H$_{18}$O$_2$·C$_7$H$_{17}$NO$_5$, Artrene.

THERAP CAT: Anti-inflammatory.

THERAP CAT (VET): Anti-inflammatory.

4926. Ibuproxam. *N-Hydroxy-α-methyl-4-(2-methylpropyl)benzeneacetamide; dl-2-(4-isobutylphenyl)propionohydroxamic acid;* G-277; Ibudros. C$_{13}$H$_{19}$NO$_2$; mol wt 221.30. C 70.56%, H 8.65%, N 6.33%, O 14.46%. Hydroxylamine deriv of ibuprofen, *q.v.,* to which it is converted *in vivo.* Prepn: G. Orzalesi, R. Selleri, **Ger. pat. 2,400,531;** *eidem,* **U.S. pat. 4,082,707** (1974, 1978 both to Manetti & Roberts). Metabolism in rats: G. Orzalesi *et al., Arzneimittel-Forsch.* **27,** 1012 (1977); in humans: *eidem, ibid.* **30,** 1607 (1980). Pharmacological study: *eidem, ibid.* **27,** 1006 (1977). Thermal decomposition: S. Chimichi *et al., J. Pharm. Sci.* **69,** 521 (1980). Physico-chemical properties: M. Mannelli *et al., Boll. Chim. Farm.* **119,** 203 (1980).

Crystals from acetone/petr ether, mp 119-121°. Sol in methanol, ethanol, acetone, ethyl ether. Practically insol in water and petr ether. LD$_{50}$ in mice, rats (g/kg): >2, >3 orally (Orzalesi, Selleri, 1978).

THERAP CAT: Anti-inflammatory.

4927. Ibutilide. (±)-N-[4-[4-(Ethylheptylamino)-1-hydroxybutyl]phenyl]methanesulfonamide. C$_{20}$H$_{36}$N$_2$O$_3$S; mol wt 384.58. C 62.46%, H 9.44%, N 7.28%, O 12.48%, S 8.34%. Prepn: J. B. Hester, **Eur. pat. Appl. 164,865;** *idem,* **U.S. pat. 5,155,268** (1985, 1992 both to Upjohn); J. B. Hester *et al., J. Med. Chem.* **34,** 308 (1991). Electrophysiologic effects: L. V. Buchanan *et al., J. Cardiovasc. Pharmacol.* **22,** 10, 540 (1993). Effect on facilitated defibrillation: R. C. Wesley, Jr. *et al., Am. J. Physiol.* **264,** H1269 (1993). Mechanism of action studies: K. S. Lee, *J. Pharmacol. Exp. Ther.* **262,** 99 (1992); G. S. Friedrichs *et al., ibid.* **266,** 1348 (1993).

Fumarate, $(C_{20}H_{36}N_2O_3S)_2 \cdot C_4H_4O_4$, *U-70226E.* Crystals from acetone, mp 117-119°. uv max (95% ethanol): 228, 267 nm (ε 16670, 894).

THERAP CAT: Antiarrhythmic (class III).

4928. Iceland Moss. *Cetraria islandica* (L.) Ach., *Parmeliaceae*, a lichen growing in all northern countries. Exported from Iceland, Norway, and Sweden. The gum from the powdered plant appears to be a hemicellulose contg uronic acid, galactose, mannose, and glucose; *cf.* Mantell, *The Water-Soluble Gums*, New York, 1947.

About 60% of dried Iceland moss dissolves when boiled with water contg a little sodium bicarbonate. The soln forms a jelly when cold.

USE: Manuf sea biscuits which are somewhat more resistant to weevil infestation than when wheat flour alone is used. In foods for convalescents. Manuf sizing agents for rayon; hair-setting lotions, other cosmetics.

4929. Ichthammol. Ammonium bituminosulfonate; ammonium ichthosulfonate; ammonium sulfobituminate; ammonium sulfoichthyolate; bitumol; bituminol; ichthammonium; ammonium bithiolicum; ichthosulfol; Ichthyol; Hirathiol; Ichden; Ichtammon; Ichthadone; Ichthymall; Ichthysalle; Ichthalum; Ichthium; Ichtopur; Ichthosan; Ichthynat; Ichthyopon; Lithol; Petrosulpho; Perichthol; Piscarol; Pisciol; Saurol; Subitol; Sulfogenol; Thilaven; Thiolin; Thiozin; Trasulphane; Tumenol; Leukochthol; Ichthosauran; Amsubit; Bitulan. Obtained by sulfation and ammoniation of a distillate from mineral deposits (bituminous schists) originally found near Seefeld, Tyrol. Contains satd and unsatd hydrocarbons, nitrogenous bases, acids, and several thiophene derivs. Analysis shows at least 2.5% NH_3 and at least 10% S. Also contains traces of some 20 minerals and *"zoomelanoidic"* acids. Method of prepn: Schröter, **Ger.** pat. **35,216** (1885); Helmers, **Ger.** pat. **76,128** (1892). Similar deposits occur in Asia east of Lake Baikal where the oil is known as *stone oil, barakshin, Asil;* sold in India for remedial purposes as *saladjidi:* Gerbrein, *Photo-Journal* (Montreal, 1969, July 2-9) p 19. *Review:* Wernicke, *Chem. Ztg.* **60**, 85-87 (1936).

Pale yellow or (usually) brownish-black, thick, viscous liquid. Bituminous odor. Miscible with water, glycerol, propylene glycol, fats, oils, carbowaxes, lanolin. Partially sol in alcohol, ether.

An injectable form is marketed as *Adnexol.*

THERAP CAT: Anti-infective (topical).

THERAP CAT (VET): Demulcent, emollient, antiseptic.

4930. Ichthyopterin. *2-Amino-6-(1,2-dihydroxypropyl)-4,7(1H,8H)-pteridinedione;* 6-(1,2-dihydroxypropyl)isoxanthopterine; fluorescyanine. $C_9H_{11}N_5O_4$; mol wt 253.22. C 42.69%, H 4.38%, N 27.66%, O 25.27%. Found as a chromoprotein in skin and scales of fish. Isoln from minnows: Huttel, Springling, *Ann.* **554**, 69 (1943); from carp: Polonovski *et al., Helv. Chim. Acta* **29**, 1328 (1946). Structure: Kauffmann, *Ann.* **625**, 133 (1959).

Rosettes of thin needles. Thermostable. uv max (0.1*N* NaOH): 343, 276, 258, 220 nm; (0.1*N* HCl): 343, 289, 212 nm. Maximum fluorescence (blue) between 265 and 390 nm at a pH between 5 and 8. Sol in water (0.3 mg/ml), hot ethanol, methanol, and pyridine. Practically insol in organic solvents not miscible with water.

4931. Idarubicin. *(7S-cis)-9-Acetyl-7-[(3-amino-2,3,6-trideoxy-α-L-lyxo-hexopyranosyl)oxy]-7,8,9,10-tetrahydro-6,9,-11-trihydroxy-5,12-naphthacenedione;* (1S,3S)-3-acetyl-1,2,-3,4,6,11-hexahydro-3,5,12-trihydroxy-6,11-dioxo-1-naphthacenyl-3-amino-2,3,6-trideoxy-α-L-*lyxo*-hexopyranoside; 4-demethoxydaunomycin; 4-demethoxydaunorubicin; DMDR; IMI-30; NSC-256439. $C_{26}H_{27}NO_9$; mol wt 497.50. C 62.77%, H 5.47%, N 2.82%, O 28.94%. Orally active anthracycline; analog of daunorubicin, *q.v.* Prepn: B. Patelli *et al.* **Ger.** pat. **2,525,633;** *eidem*, **U.S.** pat. **4,046,878** (1976, 1977 both to Soc. Farmac. Ital.); and antitumor activity: F. Arcamone *et al., Cancer Treat. Rep.* **60**, 829 (1976). Total synthesis for larger scale preparation: M. J. Broadhurst *et al., Chem. Commun.* **1982**, 158. Synthesis of optically pure isomers: Y. Kimura *et al., Bull. Chem. Soc. Japan.* **59**, 423 (1986). Metabolism and biodistribution in rats: G. Zini *et al., Cancer Chemother. Pharmacol.* **16**, 107 (1986). HPLC determn in plasma: S. S. N. De Graaf *et al., J. Chromatog.* **491**, 501 (1989). Clinical pharmacokinetics: H. C. Gillies *et al., Brit. J. Clin. Pharmacol.* **23**, 303 (1987). Clinical evaluation of cardiac toxicity: F. Villani *et al., Eur. J. Cancer Clin. Oncol.* **25**, 13 (1989). Reviews of pharmacology and antitumor efficacy: A. M. Casazza, *Cancer Treat. Rep.* **63**, 835-844 (1979); F. Ganzina *et al., Invest. New Drugs* **4**, 85-105 (1986). Symposium on clinical experience in acute leukemias: *Sem. Oncol.* **17**, Suppl. 2, 1-36 (1989).

Hydrochloride, $C_{26}H_{27}NO_9 \cdot HCl$, *Idamycin, Zavedos.* Orange crystalline powder, mp 183-185° (Arcamone); also reported as mp 172-174° (Broadhurst). $[\alpha]_D^{20}$ +205° (c = 0.1 in methanol) (Arcamone); also reported as $[\alpha]_D^{20}$ +188° (c = 0.10 in methanol) (Kimura).

THERAP CAT: Antineoplastic.

4932. Idebenone. *2-(10-Hydroxydecyl)-5,6-dimethoxy-3-methyl-2,5-cyclohexadiene-1,4-dione;* 6-(10-hydroxydecyl)-2,-3-dimethoxy-5-methyl-1,4-benzoquinone; 2,3-dimethoxy-5-methyl-6-(10'-hydroxydecyl)-1,4-benzoquinone; 6-(10-hydroxydecyl)ubiquinone; CV-2619; Avan; Daruma; Mnesis. $C_{19}H_{30}O_5$; mol wt 338.44. C 67.43%, H 8.93%, O 23.64%. Ubiquinone derivative with protective effects against cerebral ischemia. Prepn: H. Morimoto *et al.*, **Ger.** pat. **2,519,-730;** *eidem*, **U.S.** pat. **4,271,083** (1975, 1981 both to Takeda); K. Okamoto *et al., Chem. Pharm. Bull.* **30**, 2797 (1982); C.-A. Yu, L. Yu, *Biochemistry* **21**, 4096 (1982). Effect on ischemia-induced amnesia in rats: N. Yamazaki *et al., Japan. J. Pharmacol.* **36**, 349 (1984). Metabolism in animals: T. Kobayashi *et al., J. Pharmacobio.-Dyn.* **8**, 448 (1985). Disposition: H. Torii *et al., ibid.* 457. Pharmacokinetics and tolerance in humans: M. F. Barkworth *et al., Arzneimittel-Forsch.* **35**, 1704 (1985). Series of articles on pharmacology and clinical studies: *Arch. Gerontol. Geriatr.* **8**, 193-366 (1989). Review of chemistry, toxicology and pharmacology: I. Zs-Nagy, *Arch. Gerontol. Geriatr.* **11**, 177-186 (1990).

Orange needles from ligroin, mp 46-50° (Morimoto); also reported as crystals from hexane + ethyl acetate, mp 52-53° (Okamoto). Sol in organic solvents. Practically insol in water.

THERAP CAT: Nootropic.

4933. Idose. $C_6H_{12}O_6$; mol wt 180.16. C 40.00%, H 6.71%, O 53.28%. Prepn of D-idose by reduction of D-idonolactone: Fischer, Fay, *Ber.* **28**, 1975 (1895); from D-galactose: Sorking, Reichstein, *Helv. Chim. Acta* **28**, 1 (1945); from tri-O-acetyl-1,6-anhydro-β-D-idopyranose: *eidem, ibid.* 662. Prepn of L-idose by reduction of L-idonolactone: Fischer, Fay, *loc. cit.*; by hydrolysis of 1,2-O-isopropylidene-L-idofuranose: Meyer, Reichstein, *Helv. Chim. Acta* **29**, 152 (1946); von Vargha, *Ber.* **87**, 1351 (1954). Improved synthesis: M. Blanc-Muesser, J. Defaye, *Synthesis* **1977**, 568. Structure: S. F. Dyke, *The Carbohydrates* (Interscience, New York, 1960) p 45. Conformation: Reeves, *J. Am. Chem. Soc.* **72**, 1499 (1950). *Review:* R. L. Whistler, M. L. Wolfrom, *Methods in Carbohydrate Chemistry* (Academic Press, New York, 1962) pp 140-145.

β-D-Idose

D-Form: Syrup. $[\alpha]_D^{13}$ +15.8° (c = 2.3).
Phenylosazone, $C_{18}H_{22}N_4O_4$, yellow needles from alc, mp 168-169°.
L-Form: Syrup. $[\alpha]_D^{20}$ −17.4° (c = 3.6).
1,2-O-Isopropylidene-L-idofuranose, plates from ethyl acetate, mp 113-114°. $[\alpha]_D^{26}$ −20° (c = 2.7 in methanol).

4934. Idoxuridine. *2'-Deoxy-5-iodouridine;* 1-(2-deoxy-β-D-ribofuranosyl)-5-iodouracil; 5-iodo-2'-deoxyuridine; IDU; IDUR; IUDR; Dendrid; Emanil; Herpes-Gel; Herplex; Idexur; Idoxene; Idulea; Iduridin; Kerecid; Ophthalmadine; Stoxil; Virudox; mol wt 354.10. C 30.53%, H 3.13%, I 35.84%, N 7.91%, O 22.59%. Prepn: Prusoff, *Biochim. Biophys. Acta* **32**, 295 (1959); Cheong *et al., J. Biol. Chem.* **235**, 1441 (1960); Chang, Welch, *J. Med. Chem.* **6**, 428 (1963); Amiard, Torelli, Fr. pat. **1,336,866** (1963 to Roussel-UCLAF), *C.A.* **60**, 3082g (1964), corresponds to Brit. pat. **1,024,156**; Prystas, Sorm, *Coll. Czech. Chem. Commun.* **29**, 121 (1964). Crystal and molecular structure: Camerman, Trotter, *Acta Cryst.* **18**, 203 (1965). *Review:* W. H. Prusoff *et al.* in *Antibiotics* vol. **5**(pt. 2), F. E. Hahn, Ed. (Springer-Verlag, New York, 1979) pp 236-261.

Crystals from water, triclinic, dec 160° (Prusoff; Chang, Welch), 190-195° (Cheong *et al.*), 240° (Amiard, Torelli), over 175° (Prystas, Sorm). uv max (water): 288 nm (log ε 3.87). $[\alpha]_D^{25}$ +7.4° (c = 0.108 in water); $[\alpha]_D^{20}$ +29° (N soda). Following properties, *see* Ravin, Gulesich, *J. Am. Pharm. Assoc.* [NS] **4**, 122 (1964). pKa 8.25. pH of 0.1% aq soln, about 6. Soly at 25° in mg/ml: 2.0 in water; 2.0 in 0.2N HCl; 74.0 in 0.2N NaOH; 4.4 in methanol; 2.6 in alc; 0.014 in ether; 0.003 in chloroform; 1.6 in acetone; 1.8 in ethyl acetate; 5.7 in dioxane. LD$_{50}$ i.p. in mice: 1800 mg/kg.

α-Anomer, *1-(2-deoxy-α-D-erythro-pentofuranosyl)-5-iodouracil, 1-(2-deoxy-α-D-ribofuranosyl)-5-iodouracil, α-2'-deoxy-5-iodouridine.* Prepn: Prystas, Sorm, *loc. cit.* Crystals

from water, dec 170°. $[\alpha]_D^{25}$ +21.8° (c = 0.170). uv max (water): 288 nm (log ε 3.88).

THERAP CAT: Antiviral.

4935. Idrocilamide. *N-(2-Hydroxyethyl)-3-phenyl-2-propenamide; N-(2-hydroxyethyl)cinnamamide;* LCB-29; Brolitène; Srilane. $C_{11}H_{13}NO_2$; mol wt 191.23. C 69.09%, H 6.85%, N 7.32%, O 16.73%. Prepn: M. Bayssat *et al.,* Ger. pat. **2,015,447** corresp to U.S. pat. **3,659,014** (1970, 1972 to LIPHA); *Chim. Ther.* **8**, 202 (1973). Pharmacology: Grand *et al., Eur. J. Med. Chem.* **9**, 205 (1974). Metabolism: Belleville *et al., Therapie* **29**, 829 (1974).

White crystals from ethyl acetate or acetone, mp 100-102°. Soluble in alcohol; slightly sol in water. LD$_{50}$ orally in mice, rats: > 2950, > 3000 mg/kg (Grand).

THERAP CAT: Muscle relaxant (skeletal).

4936. Ifenprodil. *α-(4-Hydroxyphenyl)-β-methyl-4-(phenylmethyl)-1-piperidineethanol; 4-benzyl-α-(p-hydroxyphenyl)-β-methyl-1-piperidineethanol;* 2-(4-benzylpiperidino)-1-(4-hydroxyphenyl)-1-propanol; 1-methyl-2-hydroxy-2-(4-hydroxyphenyl)ethyl-1-(4-benzylpiperidine); RC-61-91. $C_{21}H_{27}NO_2$; mol wt 325.45. C 77.50%, H 8.36%, N 4.30%, O 9.83%. Prepn and pharmacology: M. Carron *et al.,* Fr. pat. M5733; U.S. pat. **3,509,164** (1968, 1970 to Robert et Carrière); C. Carron *et al., Arzneimittel-Forsch.* **21**, 1992 (1971).

mp 114°.
Neutral tartrate, $(C_{21}H_{27}NO_2)_2 \cdot C_4H_6O_6$, *Cerocral, Dilvax, Vadilex.* Crystals from methanol, mp 178-180°. Sol in alc, water. Very slightly sol in acetone, chloroform. Practically insol in ether. LD$_{50}$ in male Swiss mice (mg/kg): 17 i.v.; 120 i.p.; 275 orally (Carron, 1970).

THERAP CAT: Vasodilator (cerebral and peripheral).

4937. Ifosfamide. *N,3-Bis(2-chloroethyl)tetrahydro-2H-1,3,2-oxazaphosphorin-2-amine 2-oxide; 3-(2-chloroethyl)-2-[(2-chloroethyl)amino]tetrahydro-2H-1,3,2-oxazaphosphorin-2-oxide;* iphosphamid(e); isoendoxan; isophosphamide; A-4942; Asta Z-4942; MJF-9325; NSC-109724; Z-4942; Holoxan; Ifex; Ifomide; Mitoxana. $C_7H_{15}Cl_2N_2O_2P$; mol wt 261.09. C 32.20%, H 5.79%, Cl 27.16%, N 10.73%, O 12.26%, P 11.86%. Cytostatic agent, related structurally to cyclophosphamide, *q.v.* Prepn: Fr. pat. **1,530,962** (1968 to Asta), *C.A.* **71**, 49998m (1969); H. Arnold *et al.,* U.S. pat. **3,732,340** (1973 to Asta). Chemical properties: H. Arnold, *Proc. 5th Int. Congr. Chemother. Vienna* (Verhandlungen, Vienna, 1967) **2**, pp 751-754. Pharmacology: N. Brock, *ibid.* pp 155-161. Molecular structure and conformation: H. A. Brassfield *et al., J. Am. Chem. Soc.* **97**, 4143 (1975). Mass spectrometry: M. Przybylski *et al., Biomed. Mass Spectrom.* **4**, 209 (1977). Mechanism of action: S. Tomita *et al., Chemother. (Tokyo)* **25**, 3014 (1977). Metabolism: A. Takamizawa *et al., Chem. Pharm. Bull.* **25**, 2900 (1977); *eidem, J. Med. Chem.* **17**, 1237 (1974). Toxicity studies: R. Marcy *et al., IRCS Med. Sci. Libr. Compend.* **5**, 427, 478 (1977). Mutagenicity studies: D. Wald, *J. Mutat. Res.* **56**, 319 (1978); G. R. Mohn, J. Ellenberger, *ibid.* **32**, 331 (1976). Clinical studies: P. J. Creaven *et al., Cancer Treat. Rep.* **60**, 445, 451 (1976); J. Schnitker *et al., Arzneimittel-Forsch.* **26**, 1793 (1976). Symposium on clinical efficacy and comparison with cyclophosphamide: *Cancer Chemother. Pharmacol.* **18**, Suppl. 2, S1-S58 (1986). Reviews of pharmacology, toxic effects and clinical activity: M. Zalupski, L. H. Baker, *J. Nat. Cancer Inst.* **80**, 556-566 (1988); S. E. Schoenike, W. J. Dana, *Clin. Pharm.* **9**, 179-191 (1990).

Crystals from anhyd ether, mp 39-41°. LD_{50} in rats (mg/kg): 160 i.p. (Arnold, 1973); also reported as 150 i.p. (Brock).

THERAP CAT: Antineoplastic.

4938. Ignatia. Ignatius bean; St. Ignatius' bean. Dried, ripe seed of *Strychnos ignatii* Berg., *Loganiaceae*. *Habit.* Philippine Islands, naturalized in Vietnam, Cambodia. *Constit.* 2-3% strychnine, about 1% brucine, igasuric acid, loganin.

THERAP CAT: Bitter tonic.

4939. Illudins. Anti-tumor antibiotic substances produced by the poisonous basidiomycetes *Clitocybe illudens* (now called *Omphalotus illudens*): Anchel *et al.*, *Proc. Nat. Acad. Sci. USA* **36**, 300 (1950); **38**, 927 (1952); and Lampteromyces japonicus: Nakanishi *et al.*, *Nature* **197**, 292 (1963); Endo *et al.*, *Chem. Commun.* **1970**, 309. Structure: McMorris, Anchel, *J. Am. Chem. Soc.* **87**, 1594 (1965); Matsumoto *et al.*, *Tetrahedron* **21**, 2671 (1965); Tada *et al.*, *Chem. Pharm. Bull.* **12**, 853 (1964). Stereochemistry: Nakanishi *et al.*, *ibid.* 856; *Tetrahedron* **21**, 1231 (1965); Matsumoto *et al.*, *Bull. Chem. Soc. Japan* **37**, 1716 (1964). Abs config of illudin S: Harada, Nakanishi, *Chem. Commun.* **1970**, 310. Total synthesis of illudin M: Matsumoto *et al.*, *J. Am. Chem. Soc.* **90**, 3280 (1968); *Tetrahedron Letters* **1970**, 1171; of illudin S: *eidem, ibid.* **1971**, 2049. *In vitro* antitumor activity in human cancer cells: M. J. Kelner *et al.*, *J. Nat. Cancer Inst.* **82**, 1562 (1990).

Illudin M   R = H
Illudin S   R = OH

Illudin M. $C_{15}H_{20}O_3$. *(3'S-trans)-2',3'-Dihydroxy-2',2',4',6'-tetramethylspiro[cyclopropane-1,5'-[5H]inden]-7'(6'H)-one.* Rods, mp 120-122.5° (Matsumoto); from ethanol-water, mp 128-130° (McMorris). uv max (ethanol): 228, 318 nm (ε 13900, 3600).

Monoacetate, $C_{17}H_{22}O_4$, mp 75-76° from petr ether.

Illudin S. $C_{15}H_{20}O_4$. *[2'S-(2'α,3'β,6'α)]-2',3'-Dihydroxy-3',-6'-dihydroxy-2'-(hydroxymethyl)-2',4',6'-trimethylspiro[cyclopropane-1,5'-[5H]inden]-7'(6'H)-one, lampterol, lunamycin (obsolete).* Needles from acetone, mp 124-126°. uv max (ethanol): 233, 319 nm (ε 13200, 3600). Sol in polar organic solvents.

Diacetate, $C_{19}H_{24}O_5$, crystals from petr ether, mp 99-100°. uv max (ethanol): 227, 313 nm (ε 12900, 3400).

4940. Iloprost. *5-[Hexahydro-5-hydroxy-4-(3-hydroxy-4-methyl-1-octen-6-ynyl)-2(1H)-pentalenylidene]pentanoic acid;* 5-{(E)-(1S,5S,6R,7R)-7-hydroxy-6-[(E)-(3S,4RS)-3-hydroxy-4-methyl-1-octenyl]bicyclo[3.3.0]octan-3-ylidene}-pentanoic acid; (E)-(3aS,4R,5R,6aS)-hexahydro-5-hydroxy-4-[(E)-(3S,4RS)-3-hydroxy-4-methyl-1-octen-6-ynyl]-Δ²⁽¹ᴴ⁾ᐟ⁶-pentalenevaleric acid; ciloprost; ZK-36374. $C_{22}H_{32}O_4$, mol wt 360.49. C 73.30%, H 8.95%, O 17.75%. Prostacyclin analog; 1:1 mixture of 16α- and 16β-methyl diastereomers. Prepn: W. Skuballa *et al.*, Ger. pat. **2,845,770**; *eidem*, U.S. pat. **4,692,464** (1980, 1987 both to Schering AG). Pharmacology: K. Schrör *et al.*, *Arch. Pharmacol.* **316**, 252 (1981). Clinical pharmacokinetics: W. Krause, T. Krais, *Eur. J. Clin. Pharmacol.* **30**, 61 (1986). Mechanism of action: E. Tremoli *et al.*, *Thromb. Res.* **Suppl. XI**, 33 (1990). Clinical study in limb ischemia: J. Dormandy, *Therapie* **46**, 319

(1991); in Raynaud phenomenon: F. M. Wigley *et al.*, *Ann. Intern. Med.* **120**, 199 (1994). Review of synthesis: W. Skuballa, H. Vorbrüggen, *Adv. Prostaglandin Thromboxane Leukotriene Res.* **11**, 299-305 (1983); of pharmacology and therapeutic potential: S. M. Grant, K. L., Goa, *Drugs* **43**, 889-924 (1992).

Colorless oil.

Tromethamine, $C_{22}H_{32}O_4 \cdot C_4H_{11}NO_3$, *Endoprost, Ilomedin.*

THERAP CAT: Antithrombotic; vasodilator (peripheral).

4941. Imazamethabenz. *2-[4,5-Dihydro-4-methyl-4-(1-methylethyl)-5-oxo-1H-imidazol-2-yl]-4(and 5)-methylbenzoic acid methyl ester;* imazamethabenz methyl; imazethabenz; AC-222293; AC-293; CL-222293; Assert; Dagger. $C_{16}H_{20}$-N_2O_3; mol wt 288.35. C 66.65%, H 6.99%, N 9.72%, O 16.65%. Selective, post-emergence imidazolinone herbicide; mixture of *methyl 2-(4-isopropyl-4-methyl-5-oxo-2-imidazolin-2-yl)-p-toluate* and *methyl 6-(4-isopropyl-4-methyl-5-oxo-2-imidazolin-2-yl)-m-toluate* (approx 3:2). Prepn of isomeric mixture: M. Los, U.S. pat. **4,188,487** (1980 to American Cyanamid). Activity, physical properties and toxicity: D. L. Shaner *et al.*, *Proc. Brit. Crop Prot. Conf.-Weeds* **1982**, 25; K. Hedlund, L. Andersson, *Weeds Weed Control* **28**, 1 (1987). Activity of component isomers: D. L. Shaner *et al.*, *Proc. Brit. Crop Prot. Conf.-Weeds* **1982**, 333. Mechanism of action: J. B. Pillmoor, J. C. Caseley, *Pestic. Biochem. Physiol.* **27**, 340 (1987). Persistence and mobility in soil: R. Allen, J. C. Caseley, *Proc. Brit. Crop Prot. Conf.-Weeds* **1987**, 569. Field studies: K. Kirkland, N. E. Shafer, *ibid.* **1982**, 33; A. A. Hudson, S. C. E. Townsend, *ibid.* **1985**, 923; in food crops: S. D. Miller, H. P. Alley, *Weed Technol.* **1**, 29 (1987).

m - form   p - form

Off-white fine powder with a tendency to form easily friable aggregates; slight musty odor. Softening begins at 108-117°, melting starts at 113-122° and is completed at 144-153°. Soly (g/100 ml) at 25°: acetone 18.2, DMSO 23.8, distilled water 0.13 (p-isomer), 0.22 (m-isomer), n-heptane 0.04, isopropyl alcohol 14.4, methanol 24.4, methylene chloride 30.0, toluene 3.9. Soly (g/100 g) at 25°: xylene < 5, DMF 30. Partition coefficient in n-octanol/water: 35 (p-isomer), 66 (m-isomer). LD_{50} in rats, rabbits (mg/kg): > 5000, 4500 orally, > 2000, > 2000 dermally (Hedlund, Anderson).

USE: Herbicide.

4942. Imazapyr. *2-[4,5-Dihydro-4-methyl-4-(1-methylethyl)-5-oxo-1H-imidazol-2-yl]-3-pyridinecarboxylic acid;* 2-(4-isopropyl-4-methyl-5-oxo-2-imidazolin-2-yl)nicotinic acid. $C_{13}H_{15}N_3O_3$; mol wt 261.28. C 59.76%, H 5.79%, N 16.08%, O 18.37%. Inhibitor of acetolactate synthase, a key enzyme in branched chain amino acid biosynthesis. Prepn: M. Los, Eur. pat. **Appl. 41,623**; *idem*, U.S. pat. **4,798,619** (1981, 1989 both to Am. Cyanamid). Description: R. Paxman *et al.*, *Proc. 38th N. Z. Weed Pest Control Conf.*, 73

(1985). Mode of action: D. Shaner *et al., Proc. Brit. Crop Prot. Conf. - Weeds* **1985**, 147. Determn in soil: O. Nováková, *Chromatographia* **39**, 62 (1994). Persistence in soil: S. Vizantinopoulos, P. Lolos, *Bull. Environ. Contam. Toxicol.* **52**, 404 (1994). Field trials in sod: K. A. Griffin *et al., Crop Sci.* **34**, 202 (1994).

Crystals from acetone + hexane, mp 170-172.5°. pK_1 1.9; pK_2 3.6. Soly in water (pH 7): ~1.0 g/l. LD_{50} orally in rats: >5000 mg/kg; dermally in rabbits: >2000 mg/kg (Paxman).

Isopropylamine salt, $C_{13}H_{15}N_3O_3 \cdot C_3H_9N$, *AC-252925, CL-252925, Arsenal, Chopper.* mp 160-180° (dec).

USE: Herbicide.

4943. Imazaquin. *2-[4,5-Dihydro-4-methyl-4-(1-methylethyl)-5-oxo-1H-imidazol-2-yl]-3-quinolinecarboxylic acid;* 2-(5-isopropyl-5-methyl-4-oxo-2-imidazolin-2-yl)-3-quinolinecarboxylic acid; AC-252214. $C_{17}H_{17}N_3O_3$; mol wt 311.34. C 65.58%, H 5.50%, N 13.50%, O 15.42%. Pre- and postemergence imidazolinone herbicide especially for use in soybean crops. Prepn: M. Los, **Eur. pat. Appl. 41,623** (1981 to Am. Cyanamid), *C.A.* **96**, 199687q (1982). Alternate prepn: D. R. Maulding, R. F. Doehner, Jr., U.S. pat. **4,459,408** (1984 to Am. Cyanamid). Inhibition of branched-chain amino acid biosynthesis: D. L. Shaner *et al., Plant Physiol.* **76**, 545 (1984). Metabolism by plants: D. L. Shaner, P. A. Robson, *Weed. Sci.* **33**, 469 (1985). Persistence in soil: G. Basham *et al., ibid.* **35**, 576 (1987). Herbicidal activity: M. W. Beale *et al., Proc. Ann. Meet. Northeast. Weed Sci. Soc.* **38**, 36 (1984); W. F. Congleton *et al., Weed Technol.* **1**, 186 (1987).

Crystals from hexane + ethyl acetate, mp 219-222° (dec). Slightly sol in some organic solvents. Soly in water at 25°: 60-120 ppm. LD_{50} orally in rats: 5000 mg/kg; LC_{50} (96 hr) in rainbow trout: 100 mg/l (Congleton).

Ammonium salt, $C_{17}H_{20}N_4O_3$, *Scepter.* Sol in water.

USE: Herbicide.

4944. Imazethapyr. *2-[4,5-Dihydro-4-methyl-4-(1-methylethyl)-5-oxo-1H-imidazol-2-yl]-5-ethyl-3-pyridinecarboxylic acid;* (±)-5-ethyl-2-(4-isopropyl-4-methyl-5-oxo-1H-imidazolin-2-yl)nicotinic acid; AC-263499; CL-263499. $C_{15}H_{19}N_3O_3$; mol wt 289.33. C 62.27%, H 6.62%, N 14.52%, O 16.59%. Selective imidazolinone herbicide. Inhibitor of acetolactate synthase, a key enzyme in the biosynthesis of branched chain amino acids. Prepn: M. Los, **Eur. pat. Appl. 41,623;** *idem,* U.S. pat. **4,798,619** (1981, 1989 both to Am. Cyanamid). Comprehensive description: T. R. Peoples *et al., Proc. Brit. Crop Prot. Conf. - Weeds* **1985**, 99-106. Determn in soil: O. Nováková, *Chromatographia* **39** 62 (1994). Metabolism in corn: N. M. Mallipudi *et al., J. Agric. Food Chem.* **42**, 1213 (1994).

White to off-white crystalline solid, mp 172-175°. pK_1 2.1; pK_2 3.9. Soly in water (25°): 1415 ppm. LD_{50} orally in rats and mice: >5000 mg/kg; dermally in rabbits: >2000 mg/kg (Peoples).

Ammonium salt, $C_{15}H_{22}N_4O_3$, *Pivot, Pursuit.*

USE: Herbicide.

4945. Imiclopazine. *1-[2-[4-[3-(2-Chloro-10H-phenothiazin-10-yl)propyl]-1-piperazinyl]ethyl]-3-methyl-2-imidazolidinone;* 3-chloro-10-[γ-[N'-[β-(1-methyl-2-oxoimidazolidin-3-yl)ethyl]-N-piperazinyl]propyl]phenothiazine; chlorimipiphenine. $C_{25}H_{32}ClN_5OS$; mol wt 486.08. C 61.77%, H 6.64%, Cl 7.29%, N 14.41%, O 3.29%, S 6.60%. Prepn: **Belg. pat. 668,972** (1965 to Asta), *C.A.* **65**, 8934f (1966).

Free base, $bp_{0.01}$ 260°.

Dihydrochloride, $C_{25}H_{32}ClN_5OS \cdot 2HCl$, *P-4241, Ponsital.* Crystals from dioxane or aq acetone.

THERAP CAT: Antipsychotic.

4946. Imidacloprid. *1-[(6-Chloro-3-pyridinyl)methyl]-4,5-dihydro-N-nitro-1H-imidazol-2-amine;* 1-[(6-chloro-3-pyridinyl)methyl]-N-nitro-2-imidazolidinimine; 1-(6-chloro-3-pyridylmethyl)-N-nitroimidazolidin-2-ylideneamine; 1-(2-chloro-5-pyridylmethyl)-2-(nitroimino)imidazolidine; BAY NTN 33893; Admire; Confidor; Gaucho; Merit; Premier. $C_9H_{10}ClN_5O_2$; mol wt 255.66. C 42.28%, H 3.94%, Cl 13.87%, N 27.39%, O 12.52%. Chloronicotinyl insecticide; targets the nicotinic acetylcholine receptor. Prepn: K. Shiokawa *et al.,* **Eur. pat. Appl. 192,060;** *eidem,* U.S. pat. **4,742,060** (1986, 1988 both to Nihon Tokushu Noyaku Seizo KK). Properties and bioactivity: A. Elbert *et al., Brighton Crop Prot. Conf. - Pests Dis.* **1990**, 21. Field trials: H. E. Schmeer *et al., ibid.* 29. *Review:* W. Leicht, *Pestic. Outlook* **4**, 17-21 (1993).

Crystals, mp 143.8°; second crystalline modification, mp 136.4°. Vapor pressure (20°): 2.0×10^{-9} mbar. Soly in water (20°): 0.51 g/l. LD_{50} in rats (mg/kg): ~450 orally; >5000 dermally (Elbert).

USE: Insecticide.

4947. Imidapril. *[4S-[3[R*(R*)],4R*]]-3-[2-[[1-(Ethoxycarbonyl)-3-phenylpropyl]amino]-1-oxopropyl]-1-methyl-2-oxo-4-imidazolidinecarboxylic acid;* (4S)-3-[(2S)-2-[N-[(1S)-1-(ethoxycarbony)l-3-phenylpropyl]amino]propionyl]-1-methyl-2-oxoimidazolidine-4-carboxylic acid. $C_{20}H_{27}N_3O_5$; mol wt 405.45. C 59.25%, H 6.71%, N 10.36%, O 23.68%. Angiotensin converting enzyme (ACE) inhibitor. Prepn: N. Yoneda *et al.,* **Eur. pat. Appl. 95,163;** *eidem,* U.S. pat. **4,508,727** (1983, 1985 both to Tanabe Seiyaku); K. Hayashi *et al., J. Med. Chem.* **32**, 289 (1989). Series of articles on pharmacology, pharmacokinetics and metabolism in animals: *Arzneimittel-Forsch.* **42**, 446-512 (1992). HPLC determn in plasma and urine: K. Tagawa *et al., J. Chromatog.* **617**, 95 (1993). Clinical pharmacology: P. Démolis *et al., Fundam. Clin. Pharmacol.* **8**, 80 (1994). Clinical trial in hypertension: M. J. Vandenburg *et al., Brit. J. Clin. Pharmacol.* **37**, 265 (1994).

Colorless crystals from ethyl acetate + *n*-hexane, mp 139-140°. $[\alpha]_D^{20}$ −71.7° (c = 0.5 in ethanol).

Monohydrochloride, $C_{20}H_{27}N_3O_6 \cdot HCl$, *TA-6366, Novaloc, Tanapril.* Colorless crystals, mp 214-216° (dec). $[\alpha]_D^{20}$ −64.1 (c = 0.5 in ethanol).

Diacid, $C_{18}H_{23}N_3O_6$, *imidaprilat.* Crystals, mp 239-241° (dec). $[\alpha]_D^{19}$ −88.4° (c = 1 in 5% $NaHCO_3$).

THERAP CAT: Antihypertensive.

4948. Imidazole. Glyoxaline; 1,3-diazole; iminazole; miazole; pyrro[b]monazole; 1,3-diaza-2,4-cyclopentadiene. $C_3H_4N_2$; mol wt 68.08. C 52.93%, H 5.92%, N 41.15%. Prepd by the action of ammonia on glyoxal: Debus, *Ann. 107*, 204 (1858); from glyoxal, ammonia, and formaldehyde: Radziszewski, *Ber.* **15**, 1493 (1882); Behrend, Schmitz, *Ann. 277*, 338 (1893); vapor phase synthesis from formamide and ethylenediamine in presence of a dehydrogenation catalyst: Green, U.S. pat. **3,255,200** (1966 to Air Products and Chemicals). Crystal structure: B. M. Craven *et al.*, *Acta Crystallogr.* **33B**, 2585 (1977). Acute toxicity: Nishie *et al.*, *Toxicol. Appl. Pharmacol.* **14**, 301 (1969). *Review:* Pyman, *J. Soc. Dyers Colourists* **36**, 107 (1920). *Monograph:* K. Hofmann, *Imidazole and Its Derivatives* (Interscience, New York, 1953). Review of imidazole chemistry: Grimmett, *Advan. Heterocyclic Chem.* **12**, 103-183 (1970).

Stout prisms from benzene. mp 90-91°. bp_{760} 257°; bp_{20} 165-168°; bp_{12} 138.2°. Weak base. pK (25°): 6.92. Absorption spectrum: Rosanov, *J. Russian Physiol. Chem. Soc.* **48**, 1241 (1916); *Chem. Zentr.* **1923**, III, 1080. Freely sol in water, alcohol, ether, chloroform, pyridine; slightly sol in benzene; very sparingly sol in petr ether. LD_{50} in mice (mg/kg): 610 i.p.; 1880 orally (Nishie).

4949. Imidazole Salicylate. *2-Hydroxybenzoic acid compd with 1H-imidazole (1:1); mono(2-hydroxybenzoate)-1H-imidazole;* salizolo; ITF-182; Flogozen; Selezen. $C_{10}H_{10}$-N_2O_3; mol wt 206.20. C 58.25%, H 4.89%, N 13.59%, O 23.28%. Prepn from equimolar amounts of salicylic acid with imidazole: M. Brissemoret, *Bull. Soc. Chim. France* Ser. 3, **35**, 316 (1906). Use as anti-inflammatory agent: Belg. pat. **889,704** corresp to G. Sportoletti, U.S. pat. **4,329,340** (1981, 1982 both to Italfarmaco). Pharmacology: P. G. Pagella *et al.*, *Arzneimittel-Forsch.* **33**, 716 (1983). Penetration of inflamed sites: *eidem, ibid.* **34**, 208 (1984). Pharmacokinetics: H. P. Kuemmerle *et al.*, *Int. J. Clin. Pharmacol. Ther. Toxicol.* **22**, 521 (1984). Series of clinical studies: *Boll. Chim. Farm.* **122**, 37S-63S (1983). *Review:* R. Fantozzi, *Drugs Exptl. Clin. Res.* **10**, 853-856 (1984).

Crystals from methanol-ether, mp 123-124°. uv max: 300 nm ($E_{1cm}^{1\%}$ 182.5). Soly in water > 100 mg/cc. LD_{50} in male, female rats, mice (mg/kg): 763, 724, 595, 685 s.c.; 422, 434, 462, 435 i.v.; 1211, 1430, 1034, 1091 orally (Pagella).

USE: In heat sensitive copying materials: Belg. pat. **766,-987** corresp to R. C. Desjariais, U.S. pat. **3,694,247** (1971, 1972 both to Scott Graphics).

THERAP CAT: Anti-inflammatory; antipyretic; analgesic.

4950. 2-Imidazolidinone. 2-Imidazolidone; ethylene urea. $C_3H_6N_2O$; mol wt 86.09. C 41.85%, H 7.02%, N

32.54%, O 18.58%. Prepd from ethylenediamine and carbon dioxide under the influence of heat and pressure: Mulvaney, Evans, *Ind. Eng. Chem.* **40**, 393 (1948); from ethylenediamine and urea: Schweitzer, *J. Org. Chem.* **15**, 471 (1950). Systematic survey and bibliography: Klaus Hofmann, *Imidazole,* Part I (Interscience, New York, 1953).

Needles from chloroform, mp 131°. Very sol in water and in hot alc; difficultly sol in ether.

USE: Manuf high polymers, finishing agents for textiles and leather. In the formulation of plasticizers, lacquers, and adhesives. Insecticide: Simkover, U.S. pat. **3,242,044** (1966 to Shell).

4951. Imidocarb. *N,N'-Bis[3-(4,5-dihydro-1H-imidazol-2-yl)phenyl]urea; 3,3'-di-2-imidazolin-2-ylcarbanilide.* C_{19}-$H_{20}N_6O$; mol wt 348.41. C 65.50%, H 5.79%, N 24.12%, O 4.59%. Prepn: Brit. pat. **1,007,334**; R. Fischer, R. Hirt, U.S. pat. **3,338,917** (1965, 1967 both to Wander). Babesicidal effect in mice and rats: G. Schmidt *et al.*, *Res. Vet. Sci.* **10**, 530 (1969); E. Beveridge, *ibid.* 534. Effect on exptl anaplasmosis in calves: T. O. Roby, *ibid.* **13**, 519 (1972). Comparison of the dipropionate and tetracycline, *q.v.*, in canine ehrlichiosis: J. E. Price, T. T. Dolan, *Vet. Rec.* **107**, 275 (1980). Efficacy in *Babesia felis* infection: F. T. Potgieter, *J. S. Afr. Vet. Assoc.* **52**, 289 (1981). Effect on *B. ovis* infection in sheep: S. A. Michael, A. H. El Refaii, *Trop. Animal Health Prod.* **14**, 1 (1982), *C.A.* **96**, 192991 (1982).

Dihydrochloride, $C_{19}H_{20}N_6O \cdot 2HCl$, *imizocarb, 4A65.* Solid, mp 350° (dec). LD_{50} in mice, rats (mg/kg): 107, 150 s.c. (Beveridge).

Dipropionate, $C_{25}H_{32}N_6O_7$, *imizol, Imizad Equine Injection.*

THERAP CAT (VET): Antiprotozoal (Babesia).

4952. Iminodiacetic Acid. *N-(Carboxymethyl)glycine;* iminodiethanoic acid; diglycine; IDA. $C_4H_7NO_4$; mol wt 133.10. C 36.10%, H 5.30%, N 10.52%, O 48.08%. HOOC-CH_2NHCH_2COOH. Obtained from nitrilotriacetic acid, $N(CH_2COOH)_3$, by HCl-hydrolysis in a bomb tube: Schwarzenbach *et al.*, *Helv. Chim. Acta* **28**, 1133 (1945); by oxygenation in presence of palladium/carbon catalyst: Tetenbaum, Stone, *Chem. Commun.* **1970**, 1699. Iminodiacetic acid and nitrilotriacetic acid are formed upon boiling chloroacetic acid with concd aq ammonia: Heintz, *Ann.* **149**, 88 (1869). *See also* Martell, Bersworth, *J. Org. Chem.* **15**, 46 (1950).

Orthorhombic crystals, dec 247.5° (commercial grade, mp 220-250°). pKa_1 2.98; pKa_2 9.89. Forms salts with acids and bases. Soly in water at 5°: 2.43 g/100 ml. Practically insol in acetone, methanol, ether, benzene, carbon tetrachloride, heptane.

Hydrochloride, $C_4H_8ClNO_4$, crystals, dec 238-239°. Concd aq solns yield the free acid when adjusted to pH 2 with NaOH.

Sodium salt monohydrate, $C_4H_6NNaO_4 \cdot H_2O$, freely sol in water.

Forms complexes with Mg, Ca, Ba.

USE: Has been suggested as intermediate in the manuf of surface active agents, complex salts, chelating agents.

4953. 4,4'-Iminodicyclohexanecarboxylic Acid. 4,4'-Dicarboxydicyclohexylamine. $C_{14}H_{23}NO_4$; mol wt 269.34. C 62.43%, H 8.61%, N 5.20%, O 23.76%. Prepd by hydrogenation of *p*-aminobenzoic acid: Ferber, Brückner, *Ber.* **76**, 1019 (1943); Chinoporos, *Anal. Chem.* **34**, 437 (1962).

Minute needles from dil ethanol, dec 188-190°. Sol in abs alcohol.

USE: Detection of ferric ions: Chinoporos, *loc. cit.*

4954. Imipenem. *[5R-[5α,6α(R*)]]-6-(1-Hydroxyethyl)-3-[[2-[(iminomethyl)amino]ethyl]thio]-7-oxo-1-azabicyclo-[3.2.0]hept-2-ene-2-carboxylic acid monohydrate;* N-form-imidoylthienamycin monohydrate; imipemide; MK-787. $C_{12}H_{17}N_3O_4S.H_2O$; mol wt 317.37. C 45.42%, H 6.03%, N 13.24%, O 25.21%, S 10.10%. Extremely broad-spectrum semi-synthetic antibiotic; first stable derivative of thienamy-cin, *q.v.* Prepn: W. J. Leanza *et al., J. Med. Chem.* **22,** 1435 (1979); T. W. Miller, **Eur.** pat. **Appl. 6639** (1980 to Merck & Co.), *C.A.* **93,** 155845y (1980); B. G. Christensen *et al.,* U.S. pat. **4,194,047** (1980 to Merck & Co.). Totally syn-thetic prepn without formation of thienamycin: I. Shinkai *et al., Tetrahedron Letters* **23,** 4903 (1982). HPLC determn in serum: C. M. Myers, J. L. Blumer, *Antimicrob. Ag. Chemother.* **26,** 78 (1984). Series of articles on *in vitro* activ-ity, pharmacokinetics, clinical efficacy of combination with cilastatin sodium, *q.v.,* a renal dehydropeptidase I inhibitor: *J. Antimicrob. Chemother.* **12,** Suppl. D, 1-155 (1983); *Rev. Infect. Dis.* **7,** Suppl. 3, S389-S536 (1985); *Am. J. Med.* **78,** Suppl. 6A, 1-167 (1985); *Infection* **14,** Suppl. 2, S111-S180 (1986). Comprehensive description: E. R. Oberholtzer in *Analytical Profiles of Drug Substances* **vol. 17,** K. Florey, Ed. (Academic Press, New York, 1988) pp 73-114.

Crystals from water-ethanol. $[α]_D^{25}$ +86.8° (c = 0.05 in 0.1*M* phosphate, pH 7). pKa$_1$ ~3.2, pKa$_2$ ~9.9. uv max (water): 299 nm (ε 9670, 98% NH_2OH ext). Soly (mg/ml): water 10, methanol 5, ethanol 0.2, acetone <0.1, dimethyl-formamide <0.1, dimethylsulfoxide 0.3.

Combination with cilastatin sodium, *Imipem, Primaxin, Tenacid, Tienam, Tracix, Zienam.*

THERAP CAT: Antibacterial.

4955. Imipramine. *10,11-Dihydro-N,N-dimethyl-5H-di-benz[b,f]azepine-5-propanamine; 5-(3-dimethylaminopropyl)-10,11-dihydro-5H-dibenz[b,f]azepine;* N-(γ-dimethylamino-propyl)iminodibenzyl; imizin; G-22355. $C_{19}H_{24}N_2$; mol wt 280.41. C 81.38%, H 8.63%, N 9.99%. Tricyclic antidepres-sant. Prepn: Haefliger, Schindler, U.S. pat. **2,554,736** (1951 to Geigy); *eidem, Helv. Chim. Acta* **37,** 472 (1954). Reviews of pharmacology: Crismon, *Psychopharmacol. Bull.* **4,** 151 pp (Oct. 1967); Glassman, Perel, *Arch. Gen. Psychiat.* **28,** 649 (1973). Toxicity data: A. Tobe *et al., Arzneimittel-Forsch.* **31,** 1278 (1981). Comparative clinical trials in depression: R. S. Lipman *et al., Arch. Gen. Psy-chiatry* **43,** 68 (1986); in anxiety: R.J. Kahn *et al., ibid.* 79. Comprehensive description: D. N. Kender, R. E. Schiess-wohl, *Anal. Profiles Drug Subs.* **14,** 37-75 (1985).

Free base, bp$_{0.1}$ 160°.

Hydrochloride, $C_{19}H_{24}N_2.HCl$, *Berkomine, Chrytemin, Deprinol, Efuranol, Feinalmin, Imavate, Imidol, Imilanyle,* *Imiprin, Iramil, Janimine, Melipramine, Presamine, Pryleu-gan, Tofranil.* Crystals from acetone, mp 174-175°. Ac-quires a yellow to reddish discoloration under the influence of light. Freely sol in water, less sol in alcohol, sparingly sol in acetone. LD$_{50}$ in mice, rats (mg/kg): 400, 490 orally; 110, 90 i.p. (Tobe).

Pamoate, *Tofranil-PM.*

THERAP CAT: Antidepressant.

4956. Imipramine *N***-Oxide.** *10,11-Dihydro-N,N-di-methyl-5H-dibenz[b,f]azepine-5-propanamine N-oxide; 5-[3-(dimethylamino)propyl]-10,11-dihydro-5H-dibenz[b,f]azepine N-oxide;* N-(γ-dimethylaminopropyl)iminodibenzyl N-oxide; IPNO. $C_{19}H_{24}N_2O$; mol wt 296.41. C 76.99%, H 8.16%, N 9.45%, O 5.40%. Identification as a metabolite of imipra-mine, *q.v.:* V. Fishman, H. Goldenberg, *Proc. Soc. Exp. Biol. Med.* **110,** 187 (1962). Prepn: **Fr.** pat. **M2508;** H. Dyrsting, J. B. Pedersen, U.S. pat. **3,574,852** (1964, 1971, both to Dumex). Pharmacology: W. Theobald *et al., Med. Pharmacol. Exp.* **15,** 187 (1966). Metabolism: M. H. Bick-el, M. Baggiolini, *Biochem. Pharmacol.* **15,** 1155 (1966); R. Minder *et al., Arch. Pharmakol.* **268,** 334 (1971). GLC determn in biological samples: H. J. Weder, M. H. Bickel, *J. Chromatog.* **37,** 181 (1968); HPLC determn: E. Koyama *et al., Therap. Drug. Monitor.* **15,** 224 (1993). Clinical trial: W. Rapp *et al., Acta. Psychiat. Scand.* **49,** 77 (1973). Clini-cal pharmacokinetics: A. Nagy, T. Hansen, *Acta Pharma-col. Toxicol.* **42,** 58 (1978).

White needle-shaped crystals, mp 120-123° (dec). Sol in methanol, ether, acetone and benzene. Strongly hygro-scopic.

Monohydrate, mp 75-79°.

Hydrochloride, $C_{19}H_{24}N_2O.HCl$, *Imiprex.* Colorless cry-stalline powder, mp 167-174°; also reported as mp 153-155° (dec). Very bitter taste. An aq. soln is moderately acidic. Sol in 75 parts water, in 12 parts 95% ethanol, 4.5 parts chloroform. Nearly insol in ether. LD$_{50}$ in rats, mice (mg/kg): 90, 150 i.p. (Dyrsting, Pedersen).

THERAP CAT: Antidepressant.

4957. Imiquimod. *1-(2-Methylpropyl)-1H-imidazo[4,5-c]quinolin-4-amine;* 4-amino-1-isobutyl-1H-imidazo[4,5-c]-quinoline; R-837; S-26308. $C_{14}H_{16}N_4$; mol wt 240.31. C 69.97%, H 6.71%, N 23.31%. Prepn: J. F. Gerster, **Eur.** pat. **Appl. 145,340** (1985 to Riker). Pharmacology: M. Chen *et al., Antimicrob. Ag. Chemother.* **32,** 678 (1988); D. I. Bernstein *et al., Vaccine* **13,** 72 (1995). Mechanism of action study: K. Megyeri *et al., Mol. Cell. Biol.* **15,** 2207 (1995). Clinical trial in cancer patients: P. L. Witt *et al., Cancer Res.* **53,** 5176 (1993).

Crystals from DMF, mp 292-294°.

THERAP CAT: Antiviral; immunomodulator.

4958. Immunoglobulins. *Immune globulins;* gamma globulins; γ-globulins; Ig's; Cytotect; Endobulin; Gama-stan; Gamimune; Gammagard; Gammagee; Gammabulin; Gammar; Immuglobin; Intraglobin; Kabiglobulin; Sando-globulin; Venilon. A heterogeneous group of proteins syn-thesized by lymphocytes and plasma cells and found in human and animal blood serum, urine, spinal fluid, lymph nodes, spleen, and other body fluids and tissues. They function as antibodies by recognizing determinants on the surfaces of foreign substances (antigens) and binding to them. Immunoglobulins are divided into classes arising from differences in structure and in biological properties; major classes are designated as *IgG, IgM, IgA, IgD, IgE.* IgG, IgM, and IgA constitute about 80%, 10% and < 10%, resp, of the total γ-globulins. Normal serum concentrations of IgD and IgE are low, but IgE levels are elevated in allergic persons. All immunoglobulins have a general structure consisting of two heavy chains and two light chains of varying lengths held together by disulfide bonds. Heavy chains have mol weights of approx 55,000 and are differentiated into the γ, α, δ, ε, or μ types; light chains have mol weights of approx 20,000 and are of the K (*k*) or L (λ) type. IgM and IgA contain an additional J chain, *cf.* S. P. Hauptman, T. B. Tomas, *J. Biol. Chem.* **250**, 3891 (1975). System of nomenclature: Rowe, *Nature* **228**, 509 (1970). Variations in amino acid sequence occur in the N-terminal half of the chains, while the C-terminal half is constant: Hieschmann, Craig, *Proc. Nat. Acad. Sci. USA* **53**, 1403 (1965). These variations, due to genetic factors and antigen specificity, are infinite in normal Ig's, making structural determinations by usual methods of protein chemistry difficult. First complete amino acid sequence of an IgG molecule: G. E. Edelman *et al., ibid.* **63**, 78 (1969). Review of structural studies: Cohen, Milstein, *Nature* **214**, 449 (1967); Porter in *MTP Int. Rev. Sci.: Biochem. Ser. One* vol. **10**, R. R. Porter, Ed. (Univ. Park Press, Baltimore, 1973) pp 159-197; R. S. Nezlin, Y. K. Sykulev, *Mol. Immunol.* **19**, 347-356 (1982). Review of therapeutic uses: E. R. Stiehm, *Pediatrics* **63**, 301-319 (1979). Comprehensive review: Finger, Seeliger, in *Research in Immunochemistry and Immunobiology,* vol. **1**, J. B. G. Kwapinski, Ed. (University Park Press, Baltimore, 1972) pp 3-70; J. N. Goldman, M. B. Goldman, *J. Am. Med. Assoc.* **251**, 774-787 (1984). Books: R. Grubb, *The Genetic Markers of Human Immunoglobulins* (Springer-Verlag, New York, 1970) 152 pp; L. E. Glynn, M. W. Steward, *Structure and Functions of Antibodies* (Wiley-Interscience, New York, 1981) 231 pp; *Pathology of Immunoglobulins: Diagnostic and Clinical Aspects* vol. **2**, S. E. Ritzmann, Ed. (G. R. Liss, New York, 1982) 396 pp.

Monoclonal antibodies. Identical antibodies (immunoglobulins) produced against a single antigenic determinant by a single cell's progeny, or clone. When cells from myelomas (malignant tumors of the immune system) are fused with lymphocytes obtained from animals that have been immunized with a particular antigen, the resulting hybrid cells, termed *hybridomas,* possess the "immortality" of the myeloma cells and the lymphocytes' ability to produce specific antibodies, *cf.* G. Kohler, C. Milstein, *Nature* **256**, 495 (1975). Monoclonal antibodies produced by this method are now widely used in research. Exptl use in detection of metastatic tumor: P. J. Moldofsky *et al., N. Engl. J. Med.* **311**, 106 (1984). *Reviews:* A. S. Tung, *Ann. Rep. Med. Chem.* **16**, 243-256 (1981); *Monoclonal Hybridoma Antibodies: Techniques and Applications,* J. G. R. Hurrell, Ed. (CRC Press, Boca Raton, 1982) 240 pp.

USE: Monoclonal antibodies as tools in biological research and clinical medicine.

THERAP CAT: Passive immunizing agent.

4959. Imolamine. *4-[2-(Diethylamino)ethyl]-5-imino-3-phenyl-Δ²-1,2,4-oxadiazoline;* 3-phenyl-4-diethylaminoethyl-5-imino-1,2,4-oxadiazoline. $C_{14}H_{20}N_4O$; mol wt 260.34. C 64.59%, H 7.74%, N 21.52%, O 6.15%. Prepn: Aron-Samuel, Sterne, **Fr. pat. M2023** (1963), *C.A.* **60**, 2952f (1964). Pharmacology: Molino, Belluardo, *Gazz. Int. Med. Chir.* **71**, 2100 (1966), *C.A.* **66**, 74830e (1967).

bp$_{0.2}$ 165°.

Hydrochloride, $C_{14}H_{20}N_4O \cdot HCl$, *LA-1211, Angolon, Irrigor.* Crystals, mp 154-155°.

THERAP CAT: Antianginal.

4960. Imperatorin. *9-[(3-Methyl-2-butenyl)oxy]-7H-furo[3,2-g][1]benzopyran-7-one; 6-hydroxy-7-(3-methyl-2-butenyloxy)-5-benzofuranacrylic acid δ-lactone;* 8-isoamylenoxypsoralen; marmelosin; ammidin; pentosalen. $C_{16}H_{14}O_4$; mol wt 270.28. C 71.10%, H 5.22%, O 23.68%. From roots of *Imperatoria osthruthium* L., *Umbelliferae:* Späth, Holzen, *Ber.* **66**, 1137 (1933); from seeds of *Angelica archangelica* L.: Späth, Vierhapper, *ibid.* **71**, 1667 (1938); from fruit of *Pastinaca sativa* L., *Umbelliferae:* Soine *et al., J. Am. Pharm. Assoc.* **45**, 426 (1956). Identity with marmelosin: Späth *et al., Ber.* **70**, 1021 (1937). Identity with ammidin: Fahmy, Abu-Shady, *Quart. J. Pharm. Pharmacol.* **21**, 499 (1948). Partial synthesis: Späth, Holzen, *Ber.* **68**, 1123 (1935). Synthesis: Späth, Vierhapper, *ibid.* **70**, 248 (1937).

Prisms from ether, long fine needles from hot water, mp 102°. Practically insol in cold water. Very sparingly sol in boiling water; freely sol in chloroform; sol in benzene, alcohol, ether, petr ether, alkali hydroxides. uv max: 302, 265, 250 nm (log ε 3.95, 4.00, 4.24).

4961. Imperialine. *3,20-Dihydroxycevan-6-one;* raddeamine; sipeimine. $C_{27}H_{43}NO_3$; mol wt 429.64. C 75.48%, H 10.09%, N 3.26%, O 11.17%. Isolation from bulb of *Fritillaria imperialis* L., *Liliaceae:* Fragner, *Ber.* **21**, 3284 (1888); Boit, *ibid.* **87**, 472 (1954); from *Petilium eduardi:* Shakirov *et al., C.A.* **60**, 13280h (1964). Identity with sipeimine and raddeamine: Chu *et al., C.A.* **53**, 7503e (1959); Nurridinov, Yunusov, *C.A.* **57**, 15165h (1962). Structure: Nuriddinov *et al., Khim. Prir. Soedin* **3**, 316 (1967); *C.A.* **68**, 69168g (1968). Prepn of the β-D-glucopyranoside: Shakirov *et al., C.A.* **63**, 1858e (1965). Isolation and structure of the β-D-glucopyranoside: *eidem, C.A.* **63**, 3007f (1965). Absolute configuration: S. Itô *et al., Tetrahedron Letters* **1976**, 3161.

Prisms from methanol, mp 267°. $[\alpha]_D^{22} - 38.5°$ (c = 1.5 in chloroform).

β-D-Glucopyranoside, $C_{33}H_{53}NO_8$, *edpetiline.* Crystals from methanol, mp 272-276°. $[\alpha]_D - 57.89°$ (c = 0.449 in methanol).

4962. Improsulfan. *3,3'-Iminobis-1-propanol dimethanesulfonate (ester); N,N*-bis(3-methylsulfonyloxypropyl)amine; bis(3-mesyloxypropyl)amine. $C_8H_{19}NO_6S_2$; mol wt 289.37. C 33.21%, H 6.62%, N 4.84%, O 33.17%, S 22.16%. Alkylating agent related to busulfan, *q.v.* Prepn and antitumor activity of the hydrochloride: M. M. El-Merzabani, Y. Sakurai, *Gann.* **56**, 589 (1965), *C.A.* **64**, 8793b (1965); Y. Sakurai, **Japan. pat.** 67 26,616 (1967), *C.A.* **69**, 51596f (1968). Prepn of the tosylate: M. Nakanishi *et al.,* **Ger. pat.**

2,059,377 (1971 to Yoshitomi), *C.A.* **75**, 63127r (1972). Absorption, distribution, excretion, metabolism: Y. Kato *et al., Yakugaku Zasshi* **94**, 1107 (1974), *C.A.* **82**, 38441a (1975). Pharmacology: N. Brock, J. Kuhlmann, *Arzneimittel-Forsch.* **24**, 1139 (1974); H. Imamura *et al., Oyo Yakuri* **9**, 861 (1975), *C.A.* **84**, 12492j (1976). Antitumor effect: T. Okumoto *et al., Chem. Pharm. Bull.* **25**, 3003 (1977); T. Iwaguchi, H. Kitagawa, *Gann.* **69**, 123 (1968), *C.A.* **89**, 157282 (1978). Clinical study: S. J. Altman *et al., Cancer* **35**, 1145 (1975). Toxicity study: N. Rakieten *et al., U.S. NTIS Rep.* **PB-196468** (1970) 17 pp, *C.A.* **75**, 47359n (1971). Mutagenicity study: W. F. Benedict *et al., Cancer Res.* **37**, 2209 (1977).

Hydrochloride, $C_8H_{19}NO_6S_2.HCl$, *Yoshi 864, NSC-102627.* Crystals, mp 94-95°. LD_{50} i.v. in rats: 75 mg/kg (Brock, Kuhlmann).

Tosylate, $C_8H_{19}NO_6S_2.C_7H_8O_3S$, *864T, NSC-140117, Protecton.* White crystals, mp 113-118°. Faint, characteristic odor and bitter taste. Sol in water, methanol, acetonitrile. Slightly sol in chloroform. LD_{50} in male Wistar rats (mg/kg): 137 orally; 73 i.v.; 131 i.p.; 147 s.c., *Japan. Med. Gaz.* **17**(3), 5 (1980).

THERAP CAT: Antineoplastic.

4963. Inaperisone. *(±)-1-(4-Ethylphenyl)-2-methyl-3-(1-pyrrolidinyl)-1-propanone;* *(±)-4'-ethyl-2-methyl-3-(1-pyrrolidinyl)propiophenone.* $C_{16}H_{23}NO$; mol wt 245.36. C 78.32%, H 9.45%, N 5.71%, O 6.52%. Prepn: **Japan.** Kokai **84** 33,276; Y. Itho *et al.,* U.S. pat. **4,638,009** (1984, 1987 both to Hokuriku). Pharmacology: K. Morikawa *et al., Arzneimittel-Forsch.* **37**, 331 (1987); *idem et al., Eur. J. Pharmacol.* **213**, 409 (1992). Pharmacokinetics in rats: O. Nagata *et al., Drug Metab. Dispos.* **18**, 902 (1990). GC/MS determn in human serum: *idem et al., J. Chromatog.* **496**, 456 (1989). Series of articles on toxicology: *Yakuri to Chiryo* **20**, 1027-1115 (1992). Comparative clinical evaluation: K. Hiraga *et al., Rinsho Yakuri* **23**, 635 (1992), *C.A.* **118**, 247431y (1993).

Hydrochloride, $C_{16}H_{23}NO.HCl$, *HSR-770, HY-770, Riran.* Colorless needles from methyl ethyl ketone (isolated as ¼ hydrate), mp 152-153°. Also reported as a white to yellowish white crystalline powder. Bitter taste. Very sol in water and methanol. LD_{50} orally in mice: 425 mg/kg (Itho).

THERAP CAT: Muscle relaxant (skeletal).

4964. Indaconitine. *16-Ethyl-1,6,19-trimethoxy-4-(methoxymethyl)aconitane-3,8,10,11-tetrol 8-acetate 10-benzoate;* acetylbenzoylpseudaconine. $C_{34}H_{47}NO_{10}$; mol wt 629.75. C 64.85%, H 7.52%, N 2.22%, O 25.41%. From root of *Aconitum chasmanthum* Stapf. *Ranunculaceae:* Dunstan, Andrews, *J. Chem. Soc.* **87**, 1620 (1905). Structure: Gilman, Marion, *Can. J. Chem.* **42**, 2700 (1964).

Needles from ether, mp 202-203°. $[\alpha]_D^{21}$ +18.28° (c = 2.18 in alc). Indaconitine and its salts are highly toxic: Cash,

Dunstan, *Proc. Roy. Soc. (London)* **B176**, 468 (1905). Sol in acetone, chloroform, alcohol, ether; practically insol in water, petr ether.

Hydrobromide, prisms from water, mp 183-187°; mp 217-218° from alcohol + ether. $[\alpha]_D$ −17.27° (c = 2.991).

Hydrochloride trihydrate, needles from alcohol + ether. Anhydr salt mp 166-171°. Very hygroscopic. $[\alpha]_D^{20}$ −15.83° (c = 1.921).

4965. Indalpine. *3-[2-(4-Piperidinyl)ethyl]-1H-indole;* 4-[2-(3-indolyl)ethyl]piperidine; LM-5008; Upstene. $C_{15}H_{20}N_2$; mol wt 228.34. C 78.90%, H 8.83%, N 12.27%. Selective serotonin uptake inhibitor. Prepn: A. A. Champseix *et al.,* Ger. pat. **2,618,152;** *eidem,* U.S. pat. **4,064,255** (both 1977 to Mar-Pha Soc. Etud. Exploit. Marques). HPLC determn: C. Jozefczak *et al., J. Chromatog.* **230**, 87 (1982). Spectrofluorimetric determn and clinical pharmacology: A. Iliopoulou *et al., Pharmatherapeutica* **2**, 613 (1981). Pharmacology: G. LeFur *et al., Life Sci.* **23**, 1959 (1978); C. Gueremy *et al., J. Med. Chem.* **23**, 1306 (1980); R. Ashkenazi *et al., Brit. J. Pharmacol.* **79**, 765, 915 (1983). Clinical study: A. Wauters, *Acta Psychiat. Belg.* **83**, 69 (1983).

Monohydrochloride, $C_{15}H_{21}ClN_2$, mp 167°.
THERAP CAT: Antidepressant.

4966. Indan. *2,3-Dihydro-1H-indene;* hydrindene. C_9H_{10}; mol wt 118.18. C 91.47%, H 8.53%. Occurs in coal tar: Krämer, Spilker, *Ber.* **29**, 552 (1896). Prepn: *eidem, ibid.* **33**, 2257 (1900); Grosse *et al., Ind. Eng. Chem.* **38**, 1041 (1946); Carpenter, Easter, *J. Org. Chem.* **19**, 87 (1954); Juday, *ibid.* **22**, 532 (1957); Hunter, Aldridge, U.S. pat. **3,082,267** (1963 to Esso); Bestmann *et al., Ann.* **718**, 33 (1968).

Liquid. d_4^{20} 0.9639; d_4^{50} 0.9378. mp −51.4°. bp_{762} 176.5°; bp_{12} 61-64°. n_D^{20} 1.5383. uv max (isooctane): 272, 265, 258 nm (log ϵ 3.18, 3.08, 2.90). Insol in water. Sol in organic solvents.

1-Methylindan, $C_{10}H_{12}$. Liq, d_4^{20} 0.9402. bp_{760} 186-190°; bp_{10} 60°. n_D^{20} 1.5222.

4967. Indanazoline. *N-(2,3-Dihydro-1H-inden-4-yl)-4,-5-dihydro-1H-imidazol-2-amine;* N-(2-imidazolin-2-yl)-N-(4-indanyl)amine. $C_{12}H_{15}N_3$; mol wt 201.27. C 71.61%, H 7.51%, N 20.88%. Prepn: H. J. May, A. Berg, Ger. pat. **2,136,325;** *eidem,* U.S. pat. **3,882,229** (1973, 1975 both to Nordmark); H. J. May, *Arzneimittel-Forsch.* **30**, 1733 (1980). Toxicity data: W. Worstmann *et al., ibid.* 1760. Series of articles on galenical development, pharmacology, pharmacokinetics and toxicity: *ibid.* 1738-1787.

Crystals from petr ether, mp 109-113°.
Hydrochloride, $C_{12}H_{15}N_3.HCl$, *E-VA-16, Farial.* Crystals from isopropanol, mp 182-184°. LD_{50} in male, female mice

---

Consult the Name Index before using this section.

(mg/kg): 179, 233 orally; 22.3, 26.9 i.v.; in male, female rats: 481, 542 orally; 16.3, 17.6 i.v. (Worstmann).

THERAP CAT: Decongestant (nasal); vasoconstrictor.

4968. Indanthrene®. *6,15-Dihydro-5,9,14,18-anthra-zinetetrone; N,N'-dihydro-1,2,1',2'-anthraquinonazine.* $C_{28}H_{14}N_2O_4$; mol wt 442.43. C 76.01%, H 3.19%, N 6.33%, O 14.47%. Vat dye discovered by René Bohn (1901). Prepn: Fierz-David, Blangey, *Grundlegende Operationen der Far-benchemie* (Vienna, 5th ed., 1943) pp 304-305; Thielert, Bauman, U.S. pat. 2,693,469 (1954 to Bayer); Sutter, Fiori-ni, U.S. pat. 2,831,860 (1958 to Ciba); Kastner, U.S. pat. 3,138,612 (1964 to Allied Chem.). Structure: Weinstein, Merrit, *J. Am. Chem. Soc.* 81, 3759 (1959).

Blue powder, dec 470-500°. uv max on cellophane film: 278 nm. Practically insol in organic solvents. Sol in concd H_2SO_4, in dil alkali solns. *Indanthrene Blue R*, the usual commercial grade, is extremely stable to light and heat, but sensitive to chlorine. A purer grade, which is not as sensitive to chlorine, is sold as *Indanthrene Brilliant Blue FF*.

USE: Mainly to dye cotton.

4969. Indapamide. *3-(Aminosulfonyl)-4-chloro-N-(2,3-dihydro-2-methyl-1H-indol-1-yl)benzamide; 4-chloro-N-(2-methyl-1-indolinyl)-3-sulfamoylbenzamide; N-(3-sulfamyl-4-chlorobenzamido)-2-methylindoline; S-1520; SE-1520; Baja-ten; Damide; Fludex; Indaflex; Indamol; Ipamix; Lozol; Natrilix; Noranat; Tandix; Veroxil.* $C_{16}H_{16}ClN_3O_3S$; mol wt 365.84. C 52.53%, H 4.41%, Cl 9.69%, N 11.49%, O 13.12%, S 8.77%. Prepn: Beregi *et al.,* **Fr. pat. 2,003,311;** *eidem,* U.S. pat. 3,565,911 (1969, 1971, both to Sci. Union et Cie, Soc. Franc. Recherche Med.). Pharmacology: Leary *et al., Curr. Ther. Res. Clin. Exp.* 15, 571 (1973); D. B. Campbell, R. A. Moore, *Postgrad. Med. J., Suppl.* 57, 7 (1981). Acute toxicity data: J. Kyncl *et al., Arzneimittel-Forsch.* 25, 1491 (1975). Symposium on pharmacology and clinical efficacy: *Am. J. Med.* 84, Suppl. 1B, 1-111 (1988); *Am. J. Cardiol.* 65, Suppl. H, 1H-80H (1990).

Crystals from isopropanol/water, mp 160-162°. LD_{50} in rats, mice, guinea pigs (mg/kg): 393-421, 410-564, 347-416 i.p.; 394-440, 577-635, 272-358 i.v.; > 3000 all species orally (Kyncl).

Hemihydrate, $C_{16}H_{16}ClN_3O_3S \cdot \frac{1}{2}H_2O$, *Pressural.*

THERAP CAT: Antihypertensive; diuretic.

4970. 1H-Indazole. *Isoindazole; benzopyrazole.* $C_7H_6N_2$; mol wt 118.14. C 71.17%, H 5.12%, N 23.71%. Prepn: Stephenson, *Org. Syn.,* coll. vol. III, 475 (1955); Ainsworth, *Org. Syn.* 39, 27 (1959); Huisgen, Bast, *ibid.* 42, 69 (1962).

Needles from hot water, mp 146.5°. bp_{743} 267-270°. Sol in hot water, alcohol, ether.

4971. Indecainide. *9-[3-[(1-Methylethyl)amino]propyl]-9H-fluorene-9-carboxamide;* 9-carbamoyl-9-(3-isopropylami-nopropyl)fluorene; 9-[3-(isopropylamino)propyl]-9-(amino-carboyl)fluorene; ricainide. $C_{20}H_{24}N_2O$; mol wt 308.42. C 77.89%, H 7.84%, N 9.08%, O 5.19%. Prepn: W. B. Lace-field, R. L. Simon, U.S. pats. 4,197,313 and 4,452,745 (1980, 1984 both to Lilly). Pharmacokinetics in animals: T. L. Lindstrom, G. W. Whitaker, *Drug Metab. Dispos.* 12, 683 (1984). Metabolism: *eidem, ibid.* 691. LC determn in biological fluids: K. Z. Farid *et al., J. Chromatog.* 337, 329 (1985). Toxicological studies: G. E. Sandusky, Jr., D. B. Meyers, *Fundam. Appl. Toxicol.* 5, 175 (1985). Clinical evaluation in cardiac arrhythmias: P. F. Nestico *et al., Am. J. Cardiol.* 59, 1332 (1987); P. J. Podrid *et al., ibid.* 61, 764 (1988).

Crystals from Skelly B, mp 94-95°.

Hydrochloride, $C_{20}H_{24}N_2O \cdot HCl$, *LY-135837, Decabid.* Crystals from chloroform, mp 216.5-217°; also reported as crystals from fresh ethanol and diethyl ether, mp 203-204°. LD_{50} orally in male, female mice, rats: 100, 96, 103, 82 mg/kg (Sandusky, Meyers).

THERAP CAT: Antiarrhythmic (cardiac depressant).

4972. Indeloxazine Hydrochloride. *2-[(1H-Inden-7-yl-oxy)methyl]morpholine hydrochloride;* (±)-2-[(7-indenyloxy)-methyl]morpholine hydrochloride; YM-08054-1; Elen; Noin. $C_{14}H_{18}ClNO_2$; mol wt 267.76. C 62.80%, H 6.78%, Cl 13.24%, N 5.23%, O 11.95%. Serotonin uptake inhibitor. Prepn: M. Masuo *et al.,* **Ger. pat 2,601,703;** *eidem,* U.S. pat. 4,109,088 (1976, 1978 both to Yamanouchi). Synthesis and resolution of isomers: T. Kojima *et al., Chem. Pharm. Bull.* 33, 3766 (1985). Pharmacology and toxicity: S. Tachikawa *et al., Arch. Int. Pharmacodyn.* 238, 81 (1979). Inhibition of synaptosomal uptake of serotonin and nor-adrenaline: M. Harada, H. Maeno, *Biochem. Pharmacol.* 28, 2645 (1979). GLC determn in human plasma: A. G. Hayes, T. Chang, *J. Chromatog.* 272, 176 (1983).

Polymorphic; pale yellow needles from methanol, mp 169-170°; colorless, acicular crystals from acetone, mp 155-156°. LD_{50} in mice (mg/kg): 47 i.v. (Tachikawa).

(+)-Form, crystals from ethanol, mp 112-113°. $[\alpha]_D^{21}$ +4.9° (c = 5 in methanol).

(−)-Form, crystals from isopropanol, mp 142-142.5°. $[\alpha]_D^{20}$ −4.9° (c = 5 in methanol).

THERAP CAT: Antidepressant; nootropic.

4973. Indene. *Indonaphthene.* C_9H_8; mol wt 116.16. C 93.06%, H 6.94%. Found in the tars from coal, lignite, and crude petr. Isoln from coal tar: Weissgerber, *Ber.* 42, 569 (1909); *Brennstoff-Chemie* 5, 208 (1924); Weissgerber, Seidler, *Ber.* 60, 2088 (1927). Prepn from acetylene over activated charcoal at 625°: Zelinsky, *Ber.* 57, 264 (1924). Manuf from tetrahydronaphthalene by passing over SiO_2-Al_2O_3 catalyst at 670°: Brit. pat. 578,083 (1946). Prepn from o-$BrCH_2C_6H_4CH_2CH_2Br$ and triphenylphosphine: Bestmann *et al., Ann.* 718, 33 (1968).

Liquid. d_4^4 1.0081; d_4^{20} 0.9968; d_4^{50} 0.9692. mp $-1.8°$. bp_{760} 181.6°; bp_{400} 157.8°; bp_{200} 135.6°; bp_{100} 114.7°; bp_{60} 100.8°; bp_{40} 90.7°; bp_{20} 73.9°; bp_{10} 58.5°; bp_5 44.3°; $bp_{1.0}$ 16.4°. $n_D^{18.5}$ 1.5773. Absorption spectrum: Morton, de Gouveia, *J. Chem. Soc.* **1934**, 911. Insol in water. Miscible with most organic solvents. Polymerizes and oxidizes on standing. Concd H_2SO_4 forms metaindene, $(C_9H_8)_{16-22}$.

4974. Indenolol. *1-[1H-Inden-4(or 7)-yloxy]-3-[(1-methylethyl)amino]-2-propanol;* (±)-1-[inden-4(or 7)-yloxy]-3-(isopropylamino)-2-propanol; YB-2; Sch-28316Z. $C_{15}H_{21}$-NO_2; mol wt 247.34. C 72.84%, H 8.56%, N 5.66%, O 12.94%. Non-selective β-adrenergic blocker. Prepn: M. Murakami *et al.,* *Ger. pat.* **1,955,229;** *eidem,* U.S. pat. **4,045,482** (1970, 1977 both to Yamanouchi); K. Murase *et al., Yakugaku Zasshi* **92,** 1358 (1972), *C.A.* **78,** 71723 (1973). Indenolol is a tautomeric mixture of the 7- and 4-indenyloxy isomers in a 2:1 ratio, respectively. Unambiguous synthesis of the two isomers: *eidem, Chem. Pharm. Bull.* **24,** 552 (1976). β-Blocking and cardiovascular properties: T. Takenaka, S. Tachikawa, *Arzneimittel-Forsch.* **22,** 1864 (1972). General pharmacology: M. Takeda *et al., Oyo Yakuri* **7,** 469 (1973), *C.A.* **80,** 44015 (1974). Comparative pharmacological study with other β-blockers: W. Bartsch *et al., Arzneimittel-Forsch.* **27,** 1022 (1977). Dose-response studies: F. E. Okupa *et al., Clin. Pharmacol. Ther.* **29,** 434 (1981). Pharmacokinetics: R. Sega *et al., J. Clin. Pharmacol.* **25,** 337 (1985). Clinical evaluation in hypertension: B. Trimarco *et al., ibid.* 328; L. Poggesi *et al., Clin. Pharmacol. Ther.* **41,** 344 (1987).

Crystals from *n*-hexane/ether, mp 88-89°. Hydrochloride, $C_{15}H_{21}NO_2 \cdot HCl$, *Pulsan, Securpres.* Crystals from ethanol/ether, mp 147-148°. LD_{50} in mice: 26 mg/kg i.v. (Bartsch).

THERAP CAT: Antihypertensive, antiarrhythmic, antianginal.

4975. Indican (Metabolic Indican). *1H-Indol-3-ol hydrogen sulfate ester;* indol-3-yl sulfate; 3-indoxylsulfuric acid. $C_8H_7NO_4S$; mol wt 213.21. C 45.07%, H 3.31%, N 6.57%, O 30.02%, S 15.04%. Unstable as the free acid. Occurs in urine of mammals, also in blood plasma. Prepn of potassium salt from potassium indoxyl and potassium bisulfate: Baeyer, *Ber.* **14,** 1745 (1881); from chlorosulfonic acid and *N*-acetylindoxyl in pyridine: Schwenk, Jolles, *Biochem. Z.* **69,** 467 (1915); by persulfate oxidation of indole: Boyland *et al., Biochem. J.* **62,** 546 (1956).

Potassium salt, $C_8H_6KNO_4S$, *indol-3-yl potassium sulfate, urinary indican, potassium indoxyl sulfate.* Light brown plates from aq alc, dec 179-180° with sublimation. Very sol in water. Practically insol in cold alcohol.

4976. Indican (Plant Indican). *1H-Indol-3-yl-β-D-glucopyranoside;* 3-(β-glucosido)indole; indoxyl-β-D-glucoside. $C_{14}H_{17}NO_6$; mol wt 295.29. C 56.94%, H 5.80%, N 4.74%, O 32.51%. Isolated from *Indigofera* spp., *Leguminosae* and from *Polygonum tinctorium* Ait., *Polygonaceae.* Extraction

procedure for the glucoside: Hoogewerff, Meulen, *Rec. Trav. Chim.* **19,** 166 (1900); Perkin, Bloxam, *J. Chem. Soc.* **91,** 1715 (1907); Perkin, Thomas, *ibid.* **95,** 793 (1909). About 3% indican is found in 4-day old air-dried leaves of *Indigofera tinctoria.* Synthesis: Robertson, *J. Chem. Soc.* **1927,** 1937; Robertson, Waters, *ibid.* **1933,** 30; Freudenberg *et al., Ber.* **85,** 641 (1952).

Trihydrate, orthorhombic spear-shaped needles, mp 57-58°. $[\alpha]_{546}^{19}$ $-66°$. After heating *in vacuo* (oil pump) over H_2SO_4 for 48 hrs the hemihydrate is obtained, mp 101°. Robertson obtained the anhydrous compound by heating in the steam oven for 3 hrs, at 110° for 1 hr, and at 160° for 5 minutes. Perkin and Bloxam obtained anhyd indican by treating the trihydrate with boiling alcohol and boiling benzene. Anhydr indican decomposes at 178-180°. $[\alpha]_D^{15}$ $-66°$ (c = 2); $[\alpha]_{546}^{19}$ $-78°$ (c = 0.6). Sol in water, methanol, alcohol, acetone; slightly sol in ether, benzene, chloroform, ethyl acetate, carbon disulfide. Anhydr indican is sparingly sol in anhydr alcohol.

4977. Indigo. *2-(1,3-Dihydro-3-oxo-2H-indol-2-ylidene)-1,2-dihydro-3H-indol-3-one;* [Δ²,²'-biindoline]-3,3'-dione; Δ²,²'-bipseudoindoxyl; indigotin; indigo blue; D & C Blue No. 6; C.I. Pigment Blue 66; C.I. Vat Blue 1; C.I. 73000. $C_{16}H_{10}N_2O_2$; mol wt 262.27. C 73.27%, H 3.84%, N 10.68%, O 12.20%. Probably the oldest known coloring matter. Originally obtained from various species of *Indigofera, Leguminosae,* indigenous to Bengal, Java, Guatemala, in which it occurs as a glucoside. First synthesis: Baeyer, *Ber.* **11,** 1296 (1878); **12,** 456 (1879). First commercial prepn: Heumann, *Ber.* **23,** 3043, 3431 (1890). Newer prepns: Harley-Mason, *J. Chem. Soc.* **1950,** 2907; Ziegler, Kappe, *Monatsh.* **96,** 889 (1965); J. Gosteli, *Helv. Chim. Acta* **60,** 1980 (1977). Indigo occurs in *cis* and *trans* forms, in solid state it is in the *trans* form: Posner, *Ber.* **56,** 31 (1923); **57,** 1313 (1924); **59,** 1799 (1926). Reviews on indigo and indigoid dyes: Lubs, *The Chemistry of Synthetic Dyes,* ACS Monograph Series no. **127** (Reinhold, New York, 1955) pp 551-576; Schweizer, *Künstliche Organische Farbstoffe und ihre Zwischenprodukte* (Springer Verlag, 1964) pp 320-334; *Colour Index* vol. 4 (3rd ed., 1971) p 4595.

Dark-blue powder with coppery luster. Sublimes at about 300°; dec 390°. Practically insol in water, alcohol, ether, and dil acids. Dissolves in nonpolar solvents with red and in polar solvents with blue color. With fuming H_2SO_4 it forms a sol sulfonic acid.

USE: As textile dye. Approved by FDA for use in sutures.

4978. Indigo Carmine. *2-(1,3-Dihydro-3-oxo-5-sulfo-2H-indol-2-ylidene)-2,3-dihydro-3-oxo-1H-indole-5-sulfonic acid disodium salt;* 3,3'-dioxo-[Δ²,²'-biindoline]-5,5'-disulfonic acid disodium salt; disodium 5,5'-indigotin disulfonate; sodium indigotin disulfonate; soluble indigo blue; indigotine; Acid Blue 74; C.I. Acid Blue 74; C.I. Food Blue 1; FD & C Blue No. 2; C.I. 73015. $C_{16}H_8N_2Na_2O_8S_2$; mol wt 466.36. C 41.21%, H 1.73%, N 6.01%, Na 9.86%, O 27.45%, S 13.75%. Synthesis and structure determination: Vorlander, Schubart, *Ber.* **34,** 1860 (1901). Prepn and properties: Matthews, *Color Trade J.* **6,** 96 (1920). *See also Colour Index* vol. 4 (3rd ed., 1971) p 4597.

Dark-blue powder with coppery luster. Sensitive to light. Its solns have a blue or bluish-purple color. One gram dissolves in about 100 ml water at 25°. Slightly sol in alcohol. Practically insol in most other organic solvents. It is also marketed as a paste with water, the dye contents varying according to specification or requirements of the user. It almost always contains sodium chloride or sulfate used for "salting" it out. Indigo carmine is very sensitive to oxidizing agents. The color is readily discharged by nitric acid, chlorates, etc. The color of the aq soln fades on standing.

USE: As dye; in a functional kidney test; in coloring nylon surgical sutures. As a reagent for the detection of nitrate, chlorate, and in testing milk. Approved by the FDA for use in food and ingested drugs: *Fed. Regist.* **48**, 5252 (1982).

4979. Indinavir. *[1(1S,2R),5(S)]-2,3,5-Trideoxy-N-(2,3-dihydro-2-hydroxy-1H-inden-1-yl)-5-[2-[[(1,1-dimethylethyl)-amino]carbonyl]-4-(3-pyridinylmethyl)-1-piperazinyl]-2-(phenylmethyl)-D-erythro-pentonamide;* (αR,γS,2S)-α-benzyl-2-(*tert*-butylcarbamoyl)-γ-hydroxy-N-[(1S,2R)-2-hydroxy-1-indanyl]-4-(3-pyridylmethyl)-1-piperazinevaleramide; *N*-(2(*R*)-hydroxy-1(*S*)-indanyl)-2(*R*)-(phenylmethyl)-4(*S*)-hydroxy-5-[1-[4-(3-pyridylmethyl)-2(*S*)-(*N-tert*-butylcarbamoyl)piperazinyl]]pentanamide. $C_{36}H_{47}N_5O_4$; mol wt 613.80. C 70.45%, H 7.72%, N 11.41%, O 10.43%. Member of the novel hydroxyaminopentane amide class of HIV-1 protease inhibitors. Prepn: J. P. Vacca *et al.,* **Eur.** pat. **Appl.** 541,168; *eidem,* U.S. pat. **5,413,999** (1993, 1995 both to Merck & Co.); B. D. Dorsey *et al., J. Med. Chem.* **37**, 3443 (1994). Diastereoselective synthesis: D. Askin *et al., Tetrahedron Letters* **35**, 673 (1994); P. E. Maligres *et al., ibid.* **36**, 2195 (1995). Crystal structure of HIV-II protease complex: Z. Chen *et al., J. Biol. Chem.* **269**, 26344 (1994). Antiviral activity and pharmacokinetics: J. P. Vacca *et al., Proc. Nat. Acad. Sci. USA* **91**, 4096 (1994). HPLC determn in plasma and urine: E. Woolf *et al., J. Chromatog. A.* **692**, 45 (1995). Metabolism in humans: S. K. Balani *et al., Drug Metab. Dispos.* **23**, 266 (1995).

Occurs as monohydrate, crystals from wet ethyl or isopropyl acetate, loss of water below 100° followed by recrystallization of anhydrous form with mp 153-154°; second anhydrous form, mp 167.5-168°. Soly in water (mg/ml): 0.015 (unbuffered), >1.5 (pH 4.0). $[\alpha]_D^{22}$ +24.1° (c = 0.0133 in chloroform).

Sulfate, $C_{36}H_{47}N_5O_4 \cdot H_2SO_4$, *MK-639, Crixivan.* Occurs as the monoethanolate, crystals from absolute ethanol, softens at 135°, mp 150-153° (dec). Converts to hydrate upon loss of ethanol and exposure to moist air.

THERAP CAT: Antiviral.

4980. Indium. In; at. wt 114.82; at. no. 49; valence 3, 2, 1. Natural isotopes: 115 (95.77%); 113 (4.23%); [115]In is a β⁻ emitter, $T_{1/2}$ 6 × 10¹⁴ years. Artificial radioactive isotopes: 107-112; 114; 116-124. Occurrence in the earth's crust: 1 × 10⁻⁵%. Discovered in sphalerite ore by Reich and Richter in 1863. Generally found in zinc blendes. Monograph: M. T. Ludwick, *Indium* (Indium Corp. of America, Utica, N.Y., 1950). *Review:* Wade, Banister in

Comprehensive Inorganic Chemistry vol. **1**, J. C. Bailar, Jr. *et al.,* Eds. (Pergamon Press, Oxford, 1973) pp 997-1000, 1065-1117; E. F. Milner, C. E. T. White in Kirk-Othmer *Encyclopedia of Chemical Technology* vol. **13** (Wiley-Interscience, New York, 3rd ed., 1981) pp 207-212 .

Soft, white metal with bluish tinge. Emits a "tin cry" on bending. Ductile, malleable, softer than lead, leaves a mark on paper. Quite stable in air. Crystallizes and is diamagnetic. d^{20} 7.3. mp 155°. bp 2000°. Sp heat: 0.0568 cal/g/°C. Hardness (Mohs') = 1.2. Unaffected by water; attacked by mineral acids. Very resistant to alkalies.

Caution: Indium salts are relatively nontoxic when administered orally; highly toxic when administered subcutaneously or intravenously. Experimental animal poisoning has produced injury to blood, heart, liver, kidneys: E. Browning, *Toxicity of Industrial Metals* (Appleton-Century-Crofts, New York, 2nd ed., 1969) pp 164-168.

USE: In bearing alloys; as a thin film on moving surfaces made from other metals. In dental alloys. In semiconductor research. In nuclear reactor control rods (in the form of an Ag-In-Cd alloy).

4981. Indium Antimonide. InSb; mol wt 236.58. In 48.53%, Sb 51.47%. Prepd by melting together stoichiometric amounts of indium and antimony in evacuated ampuls or in zone-refining apparatus: Kleppa, *J. Am. Chem. Soc.* **77**, 897 (1955); Harmon, *J. Electrochem. Soc.* **103**, 128 (1956). *Reviews:* Welker, Weiss in *Solid State Physics,* vol. **3** (Academic Press, New York, 1956); Minden, *Sylvania Technologist* **11**, no. 1, 13-25 (Jan. 1958); Hulme, Mullin, *Solid State Electronics* **5** (Pergamon Press, 1962) 211-247.

Crystals (zinc blende structure). mp 535°. d at mp 5.74 (solid); 6.48 (liq). Dielectric constant = 15.9. Energy gap at 25° = 0.18 ev. Hole mobility 1250 cm²/volt-sec. Electron mobility approx 80,000 cm²/volt-sec.

USE: In semiconductor electronics. Grown p-n junctions have been made by doping a melt with an acceptor impurity such as zinc or cadmium, and dipping in an n-type crystal. Rate-grown junctions have also been made. Broad-area surface junctions have been produced by out-diffusing antimony in vacuum from the surface of an n-type crystal, producing a p-n junction just inside the surface. Also has photoconductive, photoelectromagnetic, and magnetoresistive properties. Useful as an infrared detector and filter, and in Hall effect devices.

4982. Indium Arsenide. AsIn; mol wt 189.74. As 39.49%, In 60.51%. InAs. Prepd by fusion of the elements in an evacuated, sealed tube: Gans *et al., Compt. Rend.* **237**, 310 (1953); Talley, Enright, *Phys. Rev.* **94**, 1931 (1954); Harmon *et al., ibid.* **104**, 1562 (1956); Minden, *Sylvania Technologist* **11**, no. 1, 17 (Jan. 1958).

Metallic appearance. Small single crystals have been grown by a modification of the Czochralski technique: Gremmelmaier, *Z. Naturforsch.* **11A**, 463 (1956). mp 943°. Hardly attacked by mineral acids. Energy gap: 0.35 ev. Electron mobility: approx 33,000 cm²/volt-sec. Hole mobility: 460 cm²/volt-sec. Dielectric constant 11.7.

USE: In semiconductor electronics.

4983. Indium Gallium Aluminum Phosphide. InGaAlP. Crystal lattice matched to GaAs substrates for production of visible (670-690 nm) light lasers. Conforms to the general formula $In_{0.5}(Ga_{(1-x)}Al_x)_{0.5}P$. Prepn by molecular beam epitaxy: H. Asahi *et al., J. Appl. Phys.* **53**, 4928 (1982); M. J. Hafich *et al., J. Vac. Sci. Technol. B.* **10**, 969 (1992). Refractive indices: H. Tanaka *et al., J. Appl. Phys.* **59**, 985 (1986). Photoluminescence studies: S. Naritsuka *et al., J. Electron. Mater.* **20**, 687 (1991). Schottky barrier energy: A. Nanda *et al., Appl Phys. Letters* **61**, 81 (1992). Thermal analysis: G. Hatakoshi *et al., Trans. Inst. Electron., Inf. Commun. Eng.* **E71**, 315 (1988); K. Itaya *et al., IEEE J. Quantum Electron.* **29**, 2068 (1993). Use in lasers: G. Hatakoshi *et al., Japan. J. Appl. Phys.* **31**, 501 (1992); J. Rennie *et al., IEEE J. Quantum Electron.* **29**, 1857 (1993).

USE: Laser diode arrays for high-density optical systems and information processing systems such as barcode readers.

4984. Indium Oxide. Indium sesquioxide. In_2O_3; mol wt 277.63. In 82.71%, O 17.29%.

White to pale-yellow powder. d 7.18. Volatilizes at 850°. Insol in water. Sol in hot mineral acids.

USE: In glass manufacture.

4985. Indium Phosphide. InP; mol wt 145.79. In 78.75%, P 21.25%. Prepd from white phosphorus and indium iodide at 400°: Thiel, Koelsch, *Z. Anorg. Chem.* **66**, 319 (1910); from phosphorus vapor and heated indium metal: Jandelli, *Gazz. Chim. Ital.* **71**, 58 (1941). Synthesis in zone melting furnace at 1010° from a non-stoichiometric melt: Minden, *Sylvania Technologist* **11**, no. 1, 18 (Jan. 1958).

Brittle mass with metallic appearance, not easily attacked by mineral acids. mp 1070°. Dielectric constant: 10.8. Energy gap: 1.3 ev at 25°. Electron mobility: approx 4600 cm²/volt-sec. Hole mobility: approx 150 cm²/volt-sec. Solid solns of InP can cover the energy gap continuously from 0.3 to 1.3 ev. Rectification has been observed in InP although it is more characteristic of a Schottky type barrier than the minority carrier injection phenomenon observed in germanium.

USE: In electronics for research on semiconductors.

4986. Indium Selenide. InSe; mol wt 193.78. In 59.25%, Se 40.75%. Prepd from the elements: Klemm *et al.*, *Z. Anorg. Allgem. Chem.* **219**, 45 (1934); Schubert *et al.*, *Naturwiss.* **41**, 448 (1954).

Black crystals, rhombohedric structure, mp 660°.

USE: In semiconductor research.

4987. Indium Sulfate. In₂O₁₂S₃; mol wt 517.83. In 44.35%, O 37.08%, S 18.58%. In₂(SO₄)₃.

White, hygroscopic powder. d 3.44. Sol in water. *Keep well closed.* MLD in rabbits: 1.8 g/kg orally; 0.67 mg/kg i.v., McCord *et al.*, *J. Ind. Hyg. Toxicol.* **24**, 243 (1942).

4988. Indium Telluride. In₂Te₃; mol wt 612.44. In 37.50%, Te 62.50%. Prepn of black brittle crystals by double furnace technique: Inuzuka, Sugaike, *Proc. Japan. Acad.* **30**, 383 (1954), *C.A.* **49**, 2922e (1955).

Pycnometric d 5.78; x-ray d 5.798. mp 667°: Klemm, Ulrich, *Z. Anorg. Allgem. Chem.* **219**, 45 (1934), *C.A.* **29**, 1730 (1935). Activation energies: 0.94 ev at 300-600°K, 2.4 ev at 600-900°K.

USE: In semiconductor research.

4989. Indium Trichloride. Indium chloride. Cl₃In; mol wt 221.18. Cl 48.09%, In 51.91%. InCl₃.

Yellowish, deliquesc crystals. d 4.0. mp 586°; sublimes at 500°. Freely sol in water. *Keep tightly closed.* MLD s.c. in rats: 10.2 mg/kg; MLD i.v. in rabbits: 0.64 mg/kg, McCord *et al.*, *J. Ind. Hyg. Toxicol.* **24**, 243 (1942).

USE: In electroplating using a soln of the salt with dextrose and NaCN. This soln is stable, though it turns dark on standing and deposits a mud which, however, contains no indium.

4990. Indium Trifluoride. Indium fluoride; indic fluoride. F₃In; mol wt 171.81. F 33.17%, In 66.83%. InF₃. Prepd by heating In₂O₃ or (NH₄)₃InF₆ in a current of fluorine: Hannebohn, Klemm, *Z. Anorg. Allgem. Chem.* **229**, 342 (1936); Ensslin, Dreyer, *ibid.* **249**, 119 (1942); Kwasnik in *Handbook of Preparative Inorganic Chemistry*, vol. 1, G. Brauer, Ed. (Academic Press, New York, 2nd ed., 1963) pp 228-229; from In and HF: Brewer *et al.*, *J. Inorg. Nucl. Chem.* **9**, 56 (1959); Muetterties, Castle, *ibid.* **18**, 148 (1961).

Colorless substance, d 4.39. mp 1170°. bp > 1200°. Soly in water (25°) 0.040 g/100 ml. Freely sol in dil acids. Stable in hot and cold water.

Trihydrate, crystals. Soly in water (22°) 8.49 g/100 ml, indicating a complex.

4991. Indobufen. (±)-4-(1,3-Dihydro-1-oxo-2H-isoindol-2-yl)-α-ethylbenzeneacetic acid; (±)-2-[p-(1-oxo-2-isoindolinyl)phenyl]butyric acid; 1-oxo-2-[p-[(α-ethyl)carboxymethyl]phenyl]isoindoline; 2-[4-(1-carboxypropyl)phenyl]-1-isoindolinone; K-3920; Ibustrin. C₁₈H₁₇NO₃; mol wt 295.34. C 73.20%, H 5.80%, N 4.74%, O 16.25%. Platelet aggregation inhibitor with anti-inflammatory and analgesic activity. Prepn: R. W. J. Carney, G. de Stevens, **Ger. pat.** **2,034,240** (1971 to Ciba), *C.A.* **74**, 12547p (1971); P. N. Giraldi *et al.*, **Ger. pat.** **2,154,525** (1972 to Carlo Erba), *C.A.*

77, 88292v (1972); G. Nannini *et al.*, *Arzneimittel-Forsch.* **23**, 1090 (1973). New process: K. Noda *et al.*, **Eur. pat. Appl.** **47,674** corresp to U.S. pat. **4,400,520** (1982, 1983 to Hisamitsu). Use as anti-inflammatory agent: R. W. J. Carney, G. de Stevens, U.S. pat. **3,997,669** (1976 to Ciba-Geigy). Use as platelet aggregation inhibitor: P. N. Giraldi *et al.*, **Belg. pat.** **818,033** corresp to U.S. pat. **4,010,274** (1974, 1977 both to Carlo Erba). Pharmacokinetics and bioavailability: L. M. Fuccella *et al.*, *Eur. J. Clin. Pharmacol.* **15**, 323 (1979); V. Tamassia *et al.*, *ibid.* 329. Pharmacology: M. Bergameaschi *et al.*, *Pharmacol. Res. Commun.* **16**, 979 (1984). HPLC determn in plasma: E. Wahlin-Boll *et al.*, *Eur. J. Clin. Pharmacol.* **20**, 375 (1981). Clinical trial in cerebrovascular disease: P. Carrieri *et al.*, *Pharmatherapeutica* **3**, 410 (1983). Clinical comparison with aspirin: G. Orefice *et al.*, *Acta Neurol.* (Napoli) **6**, 97 (1984).

Crystals from ethanol, mp 182-184°.

THERAP CAT: Antithrombotic.

4992. Indocyanine Green. 2-[7-[1,3-Dihydro-1,1-dimethyl-3-(4-sulfobutyl)-2H-benz[e]indol-2-ylidene]-1,3,5-heptatrienyl]-1,1-dimethyl-3-(4-sulfobutyl)-1H-benz[e]indolium inner salt, sodium salt; anhydro-3,3,3',3'-tetramethyl-1,1'-bis(4-sulfobutyl)-4,5,4',5'-dibenzoindotricarbocyanine hydroxide inner salt sodium salt; Fox green; Cardio-Green. C₄₃H₄₇N₂NaO₆S₂; mol wt 774.98. C 66.64%, H 6.11%, N 3.61%, Na 2.97%, O 12.39%, S 8.28%. Prepn: Heseltine, Brooker, U.S. pat. **2,895,955** (1959 to Kodak).

A tricarbocyanine type of dye with infrared absorbing properties; peak absorption at about 800 nm. Has little or no absorption in the visible. Commercial product contains moisture and about 5% sodium iodide as contaminant.

USE: In infrared photography; in prepn of Wratten filters.

THERAP CAT: Diagnostic aid (blood volume determination, cardiac output, hepatic function).

4993. Indole. 2,3-Benzopyrrole. C₈H₇N; mol wt 117.15. C 82.02%, H 6.02%, N 11.96%. Obtained from the 240-260° fraction from coal tar: Weissgerber, *Ber.* **43**, 3520 (1910); from feces: Bergeim, *J. Biol. Chem.* **32**, 17 (1917). Prepn from o-formotoluide: Tyson, *Org. Syn.* coll. vol. III, 479 (1955); by dehydrocyclizing ortho alkyl anilines: Erner *et al.*, U.S. pat. **2,953,575** (1960 to Houdry Process); from N-(2-tolyl)-N'-methyl-N'-phenylformamidine: Lorenz *et al.*, *J. Org. Chem.* **30**, 2531 (1965). Toxicity data: Smyth *et al.*, *Am. Ind. Hyg. Assoc. J.* **23**, 95 (1962). Comprehensive review: Van Order, Lindwall, *Chem. Rev.* **30**, 69 (1942); D. W. Bannister in Kirk-Othmer *Encyclopedia of Chemical Technology* vol. 13 (Wiley-Interscience, New York, 3rd ed., 1981) pp 213-222.

Leaflets, mp 52°. bp₇₆₂ 253°; bp₂₈ 128-133°. Intense fecal odor. Volatile with steam. Soluble in hot water, hot alcohol, ether, benzene. LD₅₀ orally in rats: 1 g/kg (Smyth).

USE: In highly dil solns the odor is pleasant, hence indole has been used in perfumery.

4994. Indoleacetic Acid. *1H-Indole-3-acetic acid;* heteroauxin; IAA. C₁₀H₉NO₂; mol wt 175.19. C 68.56%, H

5.18%, N 8.00%, O 18.27%. Plant hormone; recognized as the principal auxin of higher plants. Prepd by the reaction of indole with potassium glycolate at 250°: Johnson, Crosby, *J. Org. Chem.* **28**, 1246 (1963). From indole and chloroacetic acid: Shagalov *et al., U.S. pat.* **3,320,281** (1967). *Reviews:* Leopold in *The Hormones,* vol. IV, G. Pincus *et al.,* Eds. (Academic Press, New York, 1964) pp 1-66; Thimann in *The Physiology of Plant Growth and Development,* M. B. Wilkins, Ed. (McGraw-Hill, New York, 1969) pp 1-45.

Leaflets or cryst powder from water, mp 168-170°. pK 4.75. Sparingly sol in water or chloroform; freely sol in alc; sol in acetone, ether.

USE: Plant growth regulator.

4995. Indolebutyric Acid. *1H-Indole-3-butanoic acid; indole-3-butyric acid;* 4-(3-indolyl)butyric acid. $C_{12}H_{13}NO_2$; mol wt 203.24. C 70.92%, H 6.45%, N 6.89%, O 15.74%. Prepn: Jackson, Manske, *J. Am. Chem. Soc.* **52**, 5029 (1930); by heating indole, γ-butyrolactone, and sodium hydroxide, followed by acidification of the product: Fritz, *U.S. pat.* **3,051,723** (1962 to Union Carbide); by decarboxylation of 2-carboxyindole-3-butyric acid: Bowman, Islip, *Chem. & Ind. (London)* **1971**, 154. Toxicity studies: Anderson *et al., Proc. Soc. Exp. Biol. Med.* **34**, 138 (1936); Pesonen, *Acta Endocrinol.* **5**, 409 (1950). Hypoglycemic effect in rats: Mirsky *et al., Endocrinology* **59**, 715 (1956).

White or slightly yellow crystals. Slight characteristic odor. mp 123-125°. Practically insol in water, chloroform. Sol in alc, ether, acetone. LD_{50} i.p. in mice: 100 mg/kg (Anderson).

USE: Indolebutyric acid suitably diluted (*Hormodin, Seradix*) is used for promoting and accelerating root formation of plant cuttings.

4996. Indolmycin. *[S-(R*,S*)]-5-[1-(1H-Indol-3-yl)eth-yl]-2-(methylamino)-4(5H)-oxazolone; 5-(1-indol-3-ylethyl)-2-(methylamino)-2-oxazolin-4-one;* 2-methylamino-5α-(β-indolyl)ethyl-2-oxazolin-4-one; PA-155A. $C_{14}H_{15}N_3O_2$; mol wt 257.29. C 65.36%, H 5.88%, N 16.33%, O 12.44%. Antibiotic substance produced by *Streptomyces albus:* Rao, *Antibiot. & Chemother.* **10**, 312 (1960); Marsh *et al., ibid.* 316; **Brit. pat. 862,685** (1961 to Pfizer). Structure: M. Schach von Wittenau, H. Els, *J. Am. Chem. Soc.* **83**, 4678 (1961). Total synthesis: *eidem, ibid.* **85**, 3425 (1963); Preobrazhenskaya *et al., Tetrahedron* **24**, 6131 (1968); T. Takeda, T. Mukaiyama, *Chem. Letters* **1980**, 163. Abs config: T. H. Chan, R. K. Hill, *J. Org. Chem.* **35**, 3519 (1970). Biosynthetic studies: Hornemann *et al., Chem. Commun.* **1969**, 245; *eidem, J. Am. Chem. Soc.* **93**, 3028 (1971).

Long rectangular prisms from methanol or ethyl acetate, mp 209-210°. $[\alpha]_D^{25}$ −214° (c = 2 in methanol). uv max: 218 nm ($E_{1cm}^{1\%}$ 1960). Weakly basic, stable to heat. Slightly sol in water, benzene, ether; moderately in lower alcohols, acetone.

4997. 3-Indolylacetone. *Indol-3-yl-2-propanone.* $C_{11}H_{11}NO$; mol wt 173.21. C 76.28%, H 6.40%, N 8.09%, O

9.24%. Prepn: Brown *et al., J. Chem. Soc.* **1952**, 3172; Williamson, *ibid.* **1962**, 2834; Morris *et al., Brit. pat.* **974,-895** (1964 to Parke, Davis).

Brownish rhombs from benzene or needles from aq methanol. mp 115-117.5°. uv max (ethanol): 221, 280, 289 nm (ε 35100, 6400, 5300).

4998. Indomethacin. *1-(4-Chlorobenzoyl)-5-methoxy-2-methyl-1H-indole-3-acetic acid;* 1-(p-chlorobenzoyl)-5-methoxy-2-methyl-3-indolylacetic acid; Amuno; Argun; Artracin; Artrinovo; Bonidon; Catlep; Chibro-Amuno; Chrono-Indocid; Confortid; Dolcidium; Durametacin; Elmetacin; Idomethine; Imbrilon; Inacid; Indacin; Indocid; Indocin; Indomed; Indomee; Indomethine; Indomethine; Indomod; Indo-Phlogont; Indoptic; Indoptol; Indorektal; Indo-Tablinen; Indoxen; Inflazon; Infrocin; Inteban SP; Lausit; Mezolin; Mikametan; Mobilan; Rheumacin LA; Tannex; Vonum. $C_{19}H_{16}ClNO_4$; mol wt 357.79. C 63.78%, H 4.51%, Cl 9.91%, N 3.91%, O 17.89%. Prepn: T. Y. Shen *et al., J. Am. Chem. Soc.* **85**, 488 (1963); T. Y. Shen, *U.S. pat.* **3,161,654** (1964 to Merck & Co.). Alternate process: **Belg. pat. 679,678** (1966 to Sumitomo). Pharmacology: Winter *et al., J. Pharmacol. Exp. Ther.* **141**, 369 (1963). Metabolic studies: Yesair *et al., Biochem. Pharmacol.* **19**, 1579 (1970). Indomethacin blocks prostaglandin biosynthesis: see *Prostaglandin Synthetase Inhibitors—Their Effects on Physiological Functions and Pathological States,* H. J. Robinson, J. R. Vane, Eds. (Raven Press, New York, 1974) 395 pp. Toxicity: C. D. Klaassen, *Toxicol. Appl. Pharmacol.* **38**, 127 (1976). *Review:* T. Y. Shen, C. A. Winter in *Advances in Drug Research* vol. 12, A. B. Simmons, Ed. (Academic Press, New York, 1977) pp 89-245; *Semin. Arthritis Rheum.* **12**, Suppl. 1, 77-151 (1982).

Crystals exhibiting polymorphism, mp for one form ~155°, for the other ~162°. uv max (ethanol): 230, 260, 319 nm (ε 20800, 16200, 6290). pKa 4.5. Sol in ethanol, ether, acetone, castor oil. Practically insol in water. Stable in neutral or slightly acidic media; dec by strong alkali. LD_{50} i.p. in rats: 13 mg/kg (Klaassen).

Sodium salt trihydrate, $C_{19}H_{15}ClNNaO_4.3H_2O$, *Osmosin.* Pale yellow crystalline powder. pH of 1% soln: 8.4. Very sol in methanol; sol in water, ethanol. Very slightly sol in chloroform, acetone.

Meglumine salt, *Liometacen.*

THERAP CAT: Anti-inflammatory, antipyretic, analgesic.

4999. Indoprofen. *4-(1,3-Dihydro-1-oxo-2H-isoindol-2-yl)-α-methylbenzeneacetic acid; p-(1-oxo-2-isoindolinyl)hydratropic acid;* 2-[4-(1-carboxyethyl)phenyl]-1-isoindolinone; 1-oxo-2-[p-[(α-methyl)carboxymethyl]phenyl]isoindoline; 2-[4-(1-oxo-2-isoindolinyl)phenyl]propionic acid; IPP; K-4277; Bor-Ind; Flosin; Flosint; Isindone; Praxis; Reumofene. $C_{17}H_{15}NO_3$; mol wt 281.31. C 72.58%, H 5.37%, N 4.98%, O 17.06%. Nonsteroidal anti-inflammatory agent. Prepn: **Belg. pat. 753,600;** R. W. J. Carney, G. de Stevens, *U.S. pat.* **4,316,850** (1970, 1982 both to Ciba); P. N. Giraldi *et al., Ger. pat.* **2,154,525** (1972 to Carlo Erba), *C.A.* **77**, 88292v (1972); G. Nannini *et al., Arzneimittel-Forsch.* **23**, 1090 (1973). Synthesis and resolution of enantiomers: F. Buzzetti *et al., Ger. pat.* **2,258,088** (1974 to Carlo Erba), *C.A.* **81**, 13382y (1974); T. Kametani *et al., J. Heterocycl. Chem.* **15**, 369 (1978). Abs config: S. De Munari *et al., Tetrahedron Letters* **21**, 2273 (1980). Pharmacology: A.

Buttinoni *et al., Arzneimittel-Forsch.* **23,** 1100 (1973). Absorption, excretion and metabolism: G. C. Goldaniga *et al., ibid.* **24,** 1603 (1974); L. F. Chasseaud *et al., ibid.* 1606. GLC determn of enantiomers in plasma: G. P. Tosolini *et al., J. Pharm. Sci.* **63,** 1072 (1974). Biological activity and toxicity data: A. Buttinoni *et al., J. Pharm. Pharmacol.* **35,** 603 (1983).

Colorless scales from ethanol, mp 213-214°. LD_{50} in rats (mg/kg): 58.66 i.v.; 60.83 orally (Buttinoni).

(+)-Form, *dexindoprofen, Nedius.* Colorless scales from isopropanol, mp 205-207°. $[\alpha]_D^{20}$ +48° (c = 0.05 in DMSO). LD_{50} in rats (mg/kg): 31.98 i.v.; 33.75 orally (Buttinoni).

(−)-Form, colorless scales from isopropanol, mp 205-207°. $[\alpha]_D^{20}$ −48° (c = 0.05 in DMSO). LD_{50} in rats (mg/kg): 555.39 i.v.; 538.02 orally (Buttinoni).

THERAP CAT: Analgesic; anti-inflammatory.

5000. Indoramin. *N-[1-[2-(1H-Indol-3-yl)ethyl]-4-piperidinyl]benzamide;* 3-[2-(4-benzamidopiperidino)ethyl]indole; Wy-21901. $C_{22}H_{25}N_3O$; mol wt 347.46. C 76.05%, H 7.25%, N 12.09%, O 4.60%. α_1-Adrenergic blocking agent with antihypertensive and bronchodilating activity. Prepn: J. L. Archibald, J. L. Jackson, **S. Afr. pat. 68 03204;** *eidem,* U.S. pat. **3,527,761** (1969, 1970 both to Wyeth); J. L. Archibald *et al., J. Med. Chem.* **14,** 1054 (1971). Pharmacological study: R. B. Royds *et al., Clin. Pharmacol. Ther.* **13,** 380 (1972). Pharmacokinetics: G. H. Draffan *et al., Brit. J. Clin. Pharmacol.* **3,** 489 (1976). Antihypertensive activity: G. S. Stokes *et al., Clin. Pharmacol. Ther.* **25,** 783 (1979). Cardiovascular effects in man: A. J. Coleman *et al., J. Int. Med. Res.* **7,** 511 (1979). Effects on respiration in guinea pigs: C. Hamer, D. M. Temple, *Agents Actions* **10,** 399 (1980). Determn of therapeutic concentrations: A. J. Swaisland, *Analyst* **106,** 717 (1981). Symposium on pharmacology and clinical studies: *Brit. J. Clin. Pharmacol.* **12,** Suppl. 1, 1S-140S (1981). Review of pharmacology, therapeutic efficacy: B. Holmes, E. M. Sorkin, *Drugs* **31,** 467-499 (1986).

Crystals from ethanol, mp 208-210°.

Hydrochloride, $C_{22}H_{25}N_3O \cdot HCl$, *Baratol, Doralese, Vidora, Wydora, Wypres, Wypresin.* Crystals, mp 230-232°. Recryst from isopropanol gives different cryst modification, mp 258-260°.

THERAP CAT: Antihypertensive.

5001. Indospicine. *(S)-2,7-Diamino-7-iminoheptanoic acid;* L-6-*amidinonorleucine;* L-2-amino-6-amidinohexanoic acid; L-α-amino-ε-amidinocaproic acid. $C_7H_{15}N_3O_2$; mol wt 173.22. C 48.54%, H 8.73%, N 24.26%, O 18.47%. $H_2N-C=NH(CH_2)_4CH(NH_2)COOH$. Teratogenic and hepatotoxic factor found in extracts of *Indigofera spicata* Forsk (*Indogofera endecaphylla* Jacq.), *Leguminosae.* Isoln: Hegarty, Pound, *Nature* **217,** 354 (1968). Isoln, structure and biological studies: *eidem, Aust. J. Biol. Sci.* **23,** 831 (1970). Total synthesis: Culvenor *et al., Aust. J. Chem.* **24,** 371 (1971). *Review:* Hegarty: *Australas. J. Dermatol.* **14,** 35-38 (1973).

Monohydrochloride monohydrate, $C_7H_{15}N_3O_2 \cdot HCl \cdot H_2O$, needle-like crystals from aq ethanol, mp 131-134°. $[\alpha]_D^{22}$ +18° (c = 1.1 in 5N HCl).

5002. Infasurf. CLSE; CLL. Chloroform/methanol extract of calf lung lavage fluid. Consists of 35 mg/ml phospholipid, 55-70% of which is dipalmitoylphosphatidylcholine, and of low mol wt surfactant proteins. Also contains 0.9 mg/ml fatty acids and 1.8 mg/ml neutral lipids. *See also:* Beractant, Poractant Alfa. Prepn: R. H. Notter *et al., Pediatr. Res.* **16,** 515 (1982). Physiologic effects in preterm lambs: J. J. Cummings *et al., Am. Rev. Respir. Dis.* **145,** 999 (1992). Surface tension: E. M. Scarpelli *et al., Am. J. Perinatol.* **9,** 414 (1992). Separation of hydrophobic constituents: S. B. Hall *et al., J. Lipid Res.* **35,** 1386 (1994). Clinical trial in neonates: M. S. Kwong *et al., Pediatrics* **76,** 585 (1985). Efficacy of prophylactic administration to neonates: J. Kattwinkel *et al., ibid.* **92,** 90 (1993).

THERAP. CAT: Pulmonary surfactant. In treatment of respiratory distress syndrome.

5003. Infusorial Earth. Siliceous earth; diatomaceous earth; fossil flour; kieselguhr; Celite; Super-Cel. Siliceous frustules and fragments of various species of diatoms. *See also* Silicon Dioxide.

White to light gray to pale buff powder. Insol in water, acids, or dil alkalies. Capable of taking up and holding about four times its wt of water.

USE: Clarifying agent. Largely used as an absorbent for liquids and for dispersing fluid extracts in powder form; also in cataplasms and as constituent of and excipient for pill masses. Clarifying oils, varnishes; filtering liquids; manuf heat insulators, fire brick, and fire- and acid-proof packing materials; filler for paper, paints; absorbent dynamite; in metal polishes, dentifrices, nail polishes; in chromatography.

5004. Inhibins. Gonadal polypeptide hormones produced by Sertoli cells in males and granulosa cells in females, which selectively inhibits the secretion of FSH, *q.v.* Dimer, mol wt 32,000 Da, isolated in two isoforms, inhibin A and inhibin B, which have the same α subunit linked to similar but distinct β subunits. Dimers of the β subunits have been isolated which stimulate the secretion of FSH, *see* activins. Name "inhibin" coined upon identification as a water-soluble, testicular factor with pituitary inhibitory activity: D. R. McCullagh, *Science* **76,** 19 (1932). Specific inhibition of FSH: H. F. Klinefelter *et al., J. Clin. Endocrinol.* **2,** 615 (1942); B. P. Setchell, F. Jacks, *J. Endocrinol.* **62,** 675 (1974). Identification in ovarian follicular fluid: F. H. de Jong, R. M. Sharpe, *Nature* **263,** 71 (1976); N. B. Schwartz, C. P. Channing, *Proc. Nat. Acad. Sci. USA* **74,** 5721 (1977). Localization of production to Sertoli cells: A. Steinberger, E. Steinberger, *Endocrinology* **99,** 918 (1976); to granulosa cells: G. F. Erickson, A. J. W. Hsueh, *ibid.* **103,** 1960 (1978). Immunohistochemical confirmation of production sites: P. Cuevas *et al., Biochem. Biophys. Res. Commun.* **142,** 23 (1987). Identification of inhibin-like proteins from human seminal plasma: N. G. Seidah *et al., FEBS Letters* **175,** 349 (1984); C. H. Li *et al., Proc. Nat. Acad. Sci. USA* **82,** 4041 (1985). Isolation of inhibins and higher mol wt precursors: K. Miyamoto *et al., Biochem. Biophys. Res. Commun.* **129,** 396 (1985). Purification and characterization of inhibins A and B: J. Rivier *et al., ibid.* **133,** 120 (1985); N. Ling *et al., Proc. Nat. Acad. Sci. USA* **82,** 7217 (1985). Each subunit is synthesized as a larger precursor molecule, then processed to the mature form. Complete amino acid sequence of mature form α, β_A and β_B subunits: A. J. Mason *et al., Nature* **318,** 659 (1985); A. J. Mason *et al., Biochem. Biophys. Res. Commun.* **135,** 957 (1986). Demonstration of negative feedback mechanism between secretion of FSH and inhibin: S.-Y. Ying *et al., Proc. Nat. Acad. Sci. USA* **84,** 4631 (1987). Proposed nomenclature: H. G. Burger, *J. Endocrinol.* **117,** 159 (1988). Approach to nonradiometric assays: R. Schwall *et al., Prog. Clin. Biol. Res.* **285,** 205 (1988). Review of early bioassays: B. Hudson *et al., J. Reprod. Fert., Suppl.* **26,** 17-29 (1979). Reviews of purification and physiology: C. P. Channing *et al., Proc. Soc. Exp. Biol. Med.* **178,** 339-361 (1985); F. H. de Jong, D. M. Robertson, *Mol. Cell. Endocrinol.* **42,** 95-103 (1985). Review of possible use as contraceptive: A. R. Sheth, S. B. Moodbidri, *Adv. Contracept.* **2,** 131-139 (1986); P. Franchimont, *Male Contraception: Advances and Future Prospects,* G. I. Zatuchni *et al.,* Eds. (Harper & Row, Philadelphia,

1986) pp 408-418. *Reviews:* F. H. de Jong, *Oxford Rev. Reprod. Biol.* **9**, 1-53 (1987); N. Ling *et al., Vitam. Horm.* (New York) **44**, 1-46 (1988).

5005. Inosine. Hypoxanthine riboside; 9-β-D-ribofuranosylhypoxanthine; hypoxanthosine; Inosie; Oxiamine; Ribonosine; Trophicardyl. $C_{10}H_{12}N_4O_5$; mol wt 268.23. C 44.78%, H 4.51%, N 20.89%, O 29.82%. In meat and meat extracts, in sugar beets. Prepd from adenosine by incubation with purified adenosine deaminase from intestine: Kalckar, *J. Biol. Chem.* **167**, 445 (1947); also by the action of sodium nitrite and acetic acid on adenosine: Levene, Jacobs, *Ber.* **43**, 3161 (1910); by the use of barium nitrite and H_2SO_4: Reiff *et al.,* U.S. pat. 3,049,536 (1962 to Zellstoff-Fabrik Waldhof). Fermentation method: Motozaki *et al.,* U.S. pat. 3,111,459 (1963 to Ajinomoto). Structure: Bredereck, *Ber.* **66**, 198 (1933); *Z. Physiol. Chem.* **223**, 61 (1934); Gulland, Holiday, *J. Chem. Soc.* **1936**, 765.

Dihydrate, long rectangular plates from water, mp 90°. Anhydrous needles from 80% alc, dec 218° (rapid heating). $[\alpha]_D^{18}$ −49.2° (c = 0.9 in H_2O). $[\alpha]_{white}^{20}$ −73° (0.5 g + 2 ml *N* NaOH + 3 ml H_2O). 100 ml of the satd water soln at 20° contain 1.6 g inosine. Absorption spectrum: Kalckar, *loc. cit.* uv max (pH 6.0): 248.5 nm (ε 12200). Boiling with 0.1*N* H_2SO_4 yields hypoxanthin and D-ribose.

THERAP CAT: Activates cellular functions.

5006. Inosine Pranobex. *Inosine mono[4-(acetylamino)benzoate] (salt) compd with 1-(dimethylamino)-2-propanol (1:3);* inosine:dimethylaminoisopropanol acetamidobenzoate (1:3); inosiplex; methisoprinol; NP-113; NPT-10381; Aviral; Delimmun; Imunoviral; Isoprinosin; Isoprinosina; Isoprinosine; Isoviral; Modimmunal; Pranosina; Pranosine; Viruxan. $C_{52}H_{78}N_{10}O_{17}$; mol wt 1115.25. C 56.00%, H 7.05%, N 12.56%, O 24.39%. Immunostimulant complex formed from the *p*-acetamidobenzoate salt of dimethylaminoisopropanol and inosine in a 3:1 molar ratio. Prepn: P. Gordon, *Ger. pat.* 1,965,431; *idem,* U.S. pat. 3,646,007 (1971, 1972 both to Newport Pharm.). Antiviral activity: E. R. Brown, P. Gordon, *Can. J. Microbiol.* **18**, 1463 (1972); R. L. Muldoon *et al., Antimicrob. Ag. Chemother.* **2**, 224 (1972). Stimulatory effect on T-cell function: L. Binderup, *Int. J. Immunopharmacol.* **7**, 93 (1985). Pharmacology and therapeutic potential: D. M. Campoli-Richards *et al., Drugs* **32**, 383 (1986). Clinical immunopharmacology: A. J. Glasky, J. F. Gordon, *Cancer Detect. Prev. Suppl.* **1**, 597 (1987). Clinical trial in subacute sclerosing panencephalitis (SSPE): C. E. Jones *et al., Lancet* **1**, 1034 (1982); G. Gascon *et al., Brain Devel.* **15**, 346 (1993). Clinical trial in pre-AIDS patients: C. Pedersen *et al., N. Engl. J. Med.* **322**, 1757 (1990). Review of efficacy in HIV infection: C. De Simone *et al., Int. J. Immunopharmacol.* **13**, Suppl. 1, 19-27 (1991).

Neutral water-soluble solid. LD_{50} in mice and rats (mg/kg): >4000 orally and i.p. (Gordon).

THERAP CAT: Immunomodulator; antiviral.

5007. Inosinic Acid. *5′-Inosinic acid;* 5-inosinic acid; muscle inosinic acid; t-inosinic acid; hypoxanthine riboside-

5-phosphoric acid; IMP. $C_{10}H_{13}N_4O_8P$; mol wt 348.21. C 34.49%, H 3.76%, N 16.09%, O 36.76%, P 8.90%. Prepn from meat extract: Levene, Bass, *Nucleic Acids* (New York, 1931) p 229; from dried sardines: Yoshida, Kageyama, **Japan.** pat. 732('56) (to Ajinomoto), *C.A.* **51**, 3870b (1957). Structure: Levene, Bass, *op. cit.,* pp 187-192; Bredereck, *Ber.* **66**, 198 (1933); Levene, Tipson, *J. Biol. Chem.* **111**, 313 (1935). Also prepd from muscle by enzymatic deamination of muscle adenylic acid: Ostern, *Biochem. Z.* **254**, 65 (1932); by hydrolysis of inosine triphosphate: Kleinzeller, *Biochem. J.* **36**, 729 (1942). Studies on the enzymatic synthesis: Greenberg, *J. Biol. Chem.* **190**, 611 (1951); Korn *et al., ibid.* **217**, 875 (1955). Microbial fermentation method using mutant strains of *Micrococcus glutamicus:* Kinoshita *et al.,* U.S. pat. 3,232,844 (1966 to Kyowa).

Syrup, solidifies to a glass when dried over H_2SO_4. Agreeable sour taste. pK_1 = 2.4; pK_2 = 6.4. Absorption spectrum: Kalckar, *J. Biol. Chem.* **167**, 445 (1947). Freely sol in water, in formic acid; very sparingly sol in alcohol, ether. On boiling with acid hydrolyzes to 1 mol H_3PO_4, 1 mol hypoxanthine, 1 mol D-ribose.

Disodium salt dihydrate, $C_{10}H_{11}N_4Na_2O_8P.2H_2O$, barely sol in alcohol, ether, acetone; soly in water at 20° about 13 g/100 ml. Kawasaki, *New Food Ind. (Tokyo)* **3**, no. 1, 17 (1961).

Barium salt, $C_{10}H_{11}BaN_4O_8P$. Hemipentadecahydrate, lustrous leaflets. Becomes anhydr at 100° *in vacuo.* $[\alpha]_D^{20}$ −18.5° (0.3 g of anhydr Ba salt in 10 ml of 2.5% HCl).

USE: Its salts as flavor intensifiers, like sodium glutamate. Examples of mixtures of sodium inosinate and sodium glutamate or other salts: Toi *et al.,* U.S. pat. 3,109,741 (1963 to Ajinomoto).

5008. Inositol. *myo-Inositol; meso*-inositol; *i*-inositol; hexahydroxycyclohexane; cyclohexanehexol; cyclohexitol; meat sugar; inosite; mesoinosite; phaseomannite; dambose; nucite; bios I; rat antispectacled eye factor; mouse antialopecia factor. $C_6H_{12}O_6$; mol wt 180.16. C 40.00%, H 6.71%, O 53.28%. Widely distributed in plants and animals. Growth factor for animals and microorganisms. Isoln from heart muscle: Scherer, *Ann.* **73**, 322 (1850); from liver: Woolley, *J. Biol. Chem.* **139**, 29 (1941). Synthesis: Wieland, Wishart, *Ber.* **47**, 2082 (1914); Anderson, Wallis, *J. Am. Chem. Soc.* **70**, 2931 (1948). Obtained commercially from corn steep liquor, since inositol is present as phytic acid in corn: Bartow, Walker, *Ind. Eng. Chem.* **30**, 300 (1938); U.S. pat. 2,112,553 (1938); Hoglan, Bartow, *J. Am. Chem. Soc.* **62**, 2397 (1940); Elkin, Meadows, U.S. pat. 2,414,365 (1947); **Brit.** pat. 601,273 (1948 to Corn Prod. Refining). Nine possible stereoisomers: Seven are optically inactive or *meso.* Two optically active forms, the racemic form, and several *cis,trans*-isomers occur naturally. The prevalent natural form is *cis*-1,2,3,5-*trans*-4,6-cyclohexanehexol which is described here. *Reviews:* R. Beckmann, *m-Inosit* (Editio Cantor, Aulendorf, 1953); several authors in *The Vitamins,* vol. 2, W. H. Sebrell, Jr., R. S. Harris, Eds. (Academic Press, New York, 1954) pp 321-386; *ibid.* vol. 3 (2nd ed., 1971) pp 340-415.

Anhydr, non-hygroscopic crystals from water or acetic acid above 80°. Sweet taste. d 1.752. mp 225-227°. Optically inactive. Soly in water at 25°: 14 g/100 ml soln; at 60°: 28 g/100 ml soln. Slightly sol in alc. Practically insol in ether and other common organic solvents. Aq solns are neutral to litmus.

Dihydrate, efflorescent crystals from water below 50°. d 1.524. mp 218°. Becomes anhydr at 100°.

Monophosphate, $C_6H_{13}O_9P$. Prepn: Posternak, Posternak, *Helv. Chim. Acta* **12**, 1165 (1929); McCormick, Carter, *Biochem. Prepn.* **2**, 65 (1952). Crystals from water + alcohol, dec 195-197°. Titrates as a dibasic acid. Freely soluble in water (1 g dissolves in 3 ml H_2O). Practically insol in abs ethanol, ether. Remarkably resistant to hydrolysis by boiling with strong alkali. May be hydrolyzed by boiling with 6N HCl for 14 hrs.

THERAP CAT: Vitamin B complex; lipotropic.

5009. Inositol Niacinate. *myo-Inositol hexa-3-pyridinecarboxylate;* hexanicotinoyl inositol; hexanicotinyl *cis*-1,2,3,-5-*trans*-4,6-cyclohexane; inositol hexanicotinate; *meso*-inositol hexanicotinate; Dilcit; Dilexpal; Mesotal; Esantene; Hämovannid; Hexanicit; Hexopal; Linodil; Mesonex; Palohex. $C_{42}H_{30}N_6O_{12}$; mol wt 810.73. C 62.22%, H 3.73%, N 10.37%, O 23.68%. Prepn: Badgett, Woodward, *J. Am. Chem. Soc.* **69**, 2907 (1947).

Crystals, mp 254.3-254.9°. Practically insol in water. Sol in dil acids.

THERAP CAT: Vasodilator (peripheral).

5010. Insularine. $C_{38}H_{40}N_2O_6$; mol wt 620.75. C 73.53%, H 6.50%, N 4.51%, O 15.46%. From root of *Cissampelos insularis* Makino and *C. ochiaiana* Yamamoto, *Menispermaceae*. Isoln: Kondo, Yano, *J. Pharm. Soc. Japan* **47**, 815 (1927); Kondo, Tomita, *Arch. Pharm.* **274**, 76 (1936). Structure: Tomita, Kikuchi, *J. Pharm. Soc. Japan* **77**, 997 (1957).

Amorphous powder, $[\alpha]_D^7$ +28°, mp 160°. Readily sol in ether. Absorption spectrum: Ochai, *J. Pharm. Soc. Japan* **49**, 425 (1929).

5011. Insulin. Polypeptide hormone produced by pancreatic beta cells that regulates carbohydrate homeostasis. Converted by proteolysis from the single chain proinsulin, *q.v.*, to the active dimer composed of 51 amino acid residues; mol wt ~6000. Regulates carbohydrate and lipid metabolism, and influences protein synthesis. Insulin was the first protein for which the chemical structure and mol wt were determined. Also the first commercial health care product produced by recombinant DNA technology.

Because of its solubility at physiological pH, insulin is rapidly absorbed after subcutaneous injection. Various complexes with protamine and/or zinc have been prepd to improve drug delivery. In addition to biological source (human, porcine or bovine), insulin formulations for therapeutic use are classified according to onset and duration of action. Isoln: F. G. Banting, C. H. Best, *J. Lab. Clin. Med.* **7**, 251 (1921-22). Crystallization: Abel, *Proc. Nat. Acad. Sci. USA* **12**, 132 (1926). Purification and properties: J. Lens, *Biochim. Biophys. Acta* **2**, 76 (1948). Complete amino acid sequence of bovine insulin: F. Sanger, H. Tuppy, *Biochem. J.* **49**, 463, 481 (1951); F. Sanger, E. O. P. Thompson, *ibid.* **53**, 353, 366 (1953). Identification of 2 chain structure: A. P. Ryle *et al., ibid.* **60**, 541 (1955). Review of structure determination: F. Sanger, *Science* **129**, 1340 (1959). Structure of human insulin: D. Nichol, L. F. Smith, *Nature* **187**, 483 (1960). Crystal structure: D. C. Hodgkin, *Verh. Schweiz. Naturforsch. Ges.* **150**, 93 (1970). Synthesis of human insulin: P. G. Katsoyannis *et al., J. Am. Chem. Soc.* **88**, 164, 166 (1966); by the enzymatic modification of porcine insulin: M. A. Ruttenberg, *Science* **177**, 623 (1972). Review of synthetic insulins: P. G. Katsoyannis, *Recent Progr. Horm. Res.* **23**, 505-563 (1967). Synthesis of human insulin gene: H. M. Hsiung *et al., Nucleic Acids Res.* **6**, 1371 (1979); **7**, 2199 (1979); **8**, 5753 (1980); S. A. Narang *et al., Nucleic Acids Symp. Ser.* **7**, 377 (1980). Review of the development and production of human insulin by recombinant DNA technology: I. S. Johnson, *Science* **219**, 632-637 (1983). Molecular basis of insulin action: M. P. Czech, *Ann. Rev. Biochem.* **46**, 359 (1977). Review of biosynthesis: D. F. Steiner *et al., Recent Progr. Horm. Res* **25**, 207-282 (1969). Review of the structure and function of the insulin receptor: J. Lee, P. F. Pilch, *Am. J. Physiol.* **266**, C319-C334 (1994). Symposium on the physiological regulation of insulin secretion and the pathogenesis of diabetes: *Diabetologica* **37**, Suppl. 2, S1-S187 (1994). Review of bioactivity, pharmacokinetics and therapeutic efficacy of human insulin: R. N. Brogden, R. C. Heel, *Drugs* **34**, 350-371 (1987). Review of insulin formulations and therapy: J. A. Galloway, R. E. Chance, *Horm. Metab. Res.* **26**, 591-598 (1994). History: M. Bliss, *The Discovery of Insulin* (Univ. Chicago Press, Chicago, 1982) 304 pp.

HUMAN INSULIN

Crystals, hexagonal system, usually obtained as flat rhombohedra and contg 0.4% Zn. Readily sol in dil acids and alkalies. Isoelectric point 5.30 to 5.35.

Bovine insulin, *Hypurin*.

Porcine insulin, *Iletin II, Velosulin*. Differs from human insulin by a single amino acid substitution.

Recombinant human insulin, *biosynthetic human insulin, insulin (prb), Huminsulin, Humulin, Humulina*. Human insulin prepd by recombinant DNA technology.

Semi-synthetic human insulin, *insulin (emp), Biohulin, Novolin, Orgasuline*. Human insulin prepd by enzymatic modification of porcine insulin.

Zinc insulin. Crystalline prepn of insulin containing 0.45-0.9% zinc. Formulated as suspensions in physiological saline; size of the particles determines the duration of action. Formulations are designated as prompt (or semilente), lente and extended (or ultralente).

Protamine zinc insulin, PZI insulin. Suspensions of insulin modified by the addition of zinc chloride and protamine sulfate. White or almost white suspension, pH 7.1-7.4. Onset of action occurs from 4-6 hrs after s.c. injection; duration of action is 36 hrs.

Isophane insulin, NPH insulin, neutral protein Hagedorn insulin. Crystallized prepn of protamine, zinc and insulin. Prepn: H. C. Hagedorn *et al., J. Am. Med. Assoc.* **106**, 177 (1936). *Review:* P. Felig, *ibid.* **251**, 393-396 (1984). White suspensions of rod-shaped crystals ~30 nm in length, pH 7.1-7.4. Onset of action is 3-4 hrs following s.c. injection; duration of action is 18-28 hrs.

Insulin ^{131}I, radio-iodinated insulin. Prepn: Burrows *et al., J. Clin. Invest.* **36**, 393 (1957); Grodsky *et al., Arch. Biochem. Biophys.* **81**, 264 (1959).

USE: Insulin ^{131}I used in the study of insulin binding factors from insulin resistant sera.

THERAP CAT: Antidiabetic.

5012. Insulinase. An enzyme that hydrolyzes insulin and is prepd from hog pancreas: Brink, Lewis, U.S. pat. **2,957,809** (1960 to Merck & Co.). May be obtained from commercial pancreatin or trypsin. Even the purified crystals contain large amounts of elastase. *Review:* Thomas, *Postgrad. Med. J. Suppl.* **49**, 940 (1973).

5013. Integrelin. Fibrinogen receptor antagonist. Specifically inhibits fibrinogen binding to the platelet integrin GPIIb-IIIa which prevents platelet aggregation and subsequent clot formation. Synthetic, disulfide-linked cyclic heptapeptide based on *barbourin*, a specific GPIIb-IIIa antagonist isolated from venom of the southeastern pigmy rattlesnake, *Sistrurus m. barbouri.* Contains the Lys-Gly-Asp (KGD) amino acid sequence, a modification of the Arg-Gly-Asp (RGD) recognition site which appears to confer specificity: R. M. Scarborough *et al., J. Biol. Chem.* **266**, 9359 (1991). Design and structure-activity study of integrin antagonists: *idem et al., ibid.* **268**, 1066 (1993). Pharmacology: F. A. Nicolini *et al., Circulation* **89**, 1802 (1994); K. Uthoff *et al., ibid.* **90**, pt 2, II-269 (1994). Clinical trial during coronary intervention: J. E. Tcheng *et al., ibid.* **91**, 2151 (1995).

THERAP CAT: Antithrombotic.

5014. Integrins. Family of transmembrane glycoproteins involved in cellular adhesion and signal transduction. Name derived from their ability to "integrate" activities of the extracellular matrix (ECM) and the cytoskeleton. Integrins exhibit widespread evolutionary distribution: identified in mammals and other vertebrates, insects, and yeast; homologues have been identified in plants. At least 20 have been identified in mammals; composed of $\alpha\beta$ heterodimers selected from among 16 α and 8 β subunits. The α subunits vary in size from 120-200 kDa with ~1000 amino acid residues and usually consist of a heavy and light chain joined by a disulfide bond; β subunits generally range from 90-120 kDa with ~800 amino acids. Three major subfamilies have been characterized. Most integrins bind to the Arg-Gly-Asp (RGD) amino acid sequence found on components of the ECM, such as fibronectin, *q.v.* Others bind to cell membrane proteins, such as the intercellular cell adhesion molecules (ICAMs), or to soluble ligands, such as fibrinogen, *q.v.* Several recognize more than one ligand. Integrins anchor cells to the ECM or to adjacent cells, regulate cell spreading and motility, and transduce extracellular stimuli into a variety of intracellular signals. Implicated in a variety of physiological processes including embryological development, wound healing, immune functions, thrombosis, and metastasis. Identification of membrane glycoproteins involved in cell adhesion: D. E. Wylie *et al., J. Cell Biol.* **80**, 385 (1979). Identification of the transmembrane link between the ECM and the cytoskeleton: J. W. Tamkun *et al., Cell* **46**, 271 (1986). Description of integrin family: R. O. Hynes, *ibid.* **48**, 549 (1987); C. A. Buck, A. F. Horwitz, *Ann. Rev. Cell Biol.* **3**, 179-205 (1987). Role of RGD sequence in cell adhesion: E. Ruoslahti, M. D. Pierschbacher, *Science* **238**, 491 (1987). Review of integrin structure and ligand binding: A. Sonnenberg, *Curr. Top. Microbiol. Immunol.* **184**, 7-35 (1993); D. S. Tuckwell, M. J. Hum-

phries, *Crit. Rev. Oncol. Hematol.* **15**, 149-171 (1993). Review of pharmacology: D. Cox *et al., Med. Res. Rev.* **14**, 195-228 (1994). Review of role in signal transduction: A. Richardson, J. T. Parsons, *BioEssays* **17**, 229-236 (1995); E. A. Clark, J. S. Brugge, *Science* **268**, 233-239 (1995).

β_1-*Integrins, VLA antigens, VLA integrins, very late activation antigens.* Widely distributed on various cell types. Share a common β_1 chain complexed with various α chains. Bind to ECM components such as fibronectin, collagen, laminin. Ca^{2+}/Mg^{2+}-dependent. Review: L. G. M. Baldini, L. M. Cro, *Leukemia Lymphoma* **12**, 197-203 (1994).

β_2-*Integrins, leukocyte integrins, Leu-Cam proteins, leukocyte adhesion molecules.* Contain the β_2 chain; found on leukocytes. Bind to ICAMs, complement 3, and fibrinogen. Play an important role in regulating the immune system.

β_3-*Integrins, cytoadhesins.* Contain the β_3 subunit; found on platelets, megakaryocytes and some melanoma cells. Ligands include fibrinogen, fibronectin, von Willebrand factor, vitronectin, and thrombospondin. Includes $\alpha_{IIb}\beta_3$, also known as *platelet glycoprotein IIb/IIIa (GPIIb-IIIa)*, a fibrinogen receptor involved in platelet aggregation. Review: M. H. Ginsberg *et al., Thromb. Haemostasis* **70**, 87-93 (1993).

5015. Interferon. IFN. A family of species-specific vertebrate proteins that confer non-specific resistance to a broad range of viral infections, affect cell proliferation and modulate immune responses. Discovered by A. Isaacs and J. Lindenmann, *Proc. Roy. Soc.* **B147**, 258 (1957) while studying viral interference. Originally produced by the interaction of inactivated influenza virus with chick chorioallantoic membranes; subsequently found to be inducible by viable virus in a variety of cells: D. C. Burke, A. Isaacs, *Brit. J. Exp. Pathol.* **39**, 452 (1958); in human leukocytes: I. Gresser, *Proc. Soc. Exp. Biol. Med.* **108**, 799 (1961). Host cell specificity and physical properties: D. A. J. Tyrrell, *Nature* **184**, 452 (1959); T. C. Merigan, *Science* **145**, 811 (1964). Production of acid-labile interferon by mitogen-stimulated human leukocytes: E. F. Wheelock, *ibid.* **149**, 310 (1966). Three major interferons, alpha, beta and gamma, *q.q.v.*, have been identified based on antigenic and physicochemical properties, the nature of the inducer, and the cellular source from which they are derived, *cf. Nature* **286**, 110 (1980). Known collectively as type I interferon, IFNs-α and -β are structurally related, are stable at pH 2, and compete for the same cell surface receptor. IFN-γ, also known as type II interferon, is structurally unrelated to type I IFNs, is acid labile, and has a different receptor. Receptor binding study: A. A. Branca, C. Baglioni, *Nature* **294**, 768 (1981). *Reviews:* S. Baron, F. Dianzoni, Eds., *Tex. Rep. Biol. Med.* **35**, 1-573 (1977); *Ann. N.Y. Acad. Sci.* **350**, entitled "Regulatory Function of Interferons", J. Vilcek *et al.,* Eds. (1980) 643 pp ; S. Baron *et al.,* Eds., *Tex. Rep. Biol. Med.* **41**, 1-715 (1982). Review of immunobiology and clinical significance: E. R. Stiem *et al., Ann. Intern. Med.* **96**, 80-93 (1982); of pharmacology and toxicology: G. J. Mannering, L. B. Deloria, *Ann. Rev. Pharmacol. Toxicol.* **26**, 455-515 (1986); of IFN-gene family: C. Weissmann, H. Weber, *Prog. Nucl. Acid Res. Mol. Biol.* **33**, 251-300 (1986). Symposium on antiviral activity of natural and recombinant IFNs: *Antiviral Res.* **5**, Suppl. 1, 131-257 (1985). Review of structure, function and nomenclature: K. C. Zoon, *Interferon* **9**, 1-12 (1987). Books: *Interferons and Interferon Inducers,* N. Finter, Ed. (North Holland Publ. Co., Amsterdam, 1973) 598 pp; W. E. Stewart, *The Interferon System* (Springer, New York, 1979) 421 pp; *Interferons,* T. C. Merigan, R. M. Friedman, Eds. (Academic Press, New York, 1982) 481 pp; *Interferon,* K. Munk, H. Kirchner, Eds. (S. Karger, New York, 1982) 233 pp; H. Strander, *Interferon Treatment of Human Neoplasia* (Academic Press, New York, 1986) 265 pp.

5016. Interferon-α. Alfa-interferon; alpha-interferon; IFN-α; LeIF; leukocyte interferon; lymphoblastoid interferon. Family of highly homologous species-specific proteins that inhibits viral replication and cellular proliferation and modulates the immune response. One of the type I interferons. Produced by peripheral blood leukocytes or lymphoblastoid cells upon exposure to live or inactivated virus, double-stranded RNA, or bacterial products. Multiple subtypes have been identified which contain 165-166 amino acids; mol wt approximately 18,000-20,000 daltons. Production by virus stimulated human leukocytes: I. Gresser, *Proc. Soc. Exp. Biol. Med.* **108**, 799 (1961). Purification: C. B. Anfinsen *et al., Proc. Nat. Acad. Sci. USA* **71**, 3139 (1974); K. Berg *et al., J. Immunol.* **114**, 640 (1975). Comparison with interferon-β, *q.v.:* R. L. Cavalieri *et al., Proc. Nat. Acad. Sci. USA* **74**, 3287 (1977). Amino acid analysis: M. Rubenstein *et al., ibid.* **76**, 640 (1979). Partial sequence of human leukocyte IFN: W. P. Levy *et al., ibid.* **77**, 5102 (1980); of human lymphoblastoid IFN: K. C. Zoon *et al., Science* **207**, 527 (1980). Production by recombinant DNA technology: S. Nagata *et al., Nature* **284**, 316 (1980); D. V. Goeddel *et al., ibid.* **287**, 411 (1980). Comparison of structures and biological activities of HuIFN-α subtypes: M. Streuli *et al., Science* **209**, 1343 (1980); D. V. Goeddel *et al., Nature* **290**, 20 (1981). Receptor binding study: A. A. Branca, C. Baglioni, *ibid.* **294**, 768 (1981). Series of articles on therapeutic use: *Eur. J. Hematol.* **45**, Suppl., 1-39 (1990). Symposia on clinical antineoplastic activity: *Sem. Oncol.* **12**, Suppl. 5, 1-34 (1985); *ibid.* **13**, Suppl. 2, 1-101 (1986); *Invest. New Drugs* **5**, Suppl, S1-S77 (1987). Review of clinical efficacy in viral infections: M. Ho, *Ann. Rev. Med.* **38**, 51-59 (1987); in Kaposi's sarcoma: S. E. Krown, *Sem. Oncol.* **14**, Suppl. 3, 27-33 (1987); in hairy cell leukemia: B. D. Cheson, A. Martin, *Ann. Int. Med.* **106**, 871-878 (1987); in multiple sclerosis: R. L. Knobler, *Neurology* **38**, Suppl. 2, 58-61 (1988).

Interferon alfa-2a, $C_{860}H_{1353}N_{227}O_{255}S_9$, *IFN-αA, Ro-22-8181, Canferon, Roferon-A.*

Interferon alfa-2b, $C_{860}H_{1353}N_{229}O_{255}S_9$, *IFN-α2, Sch-30500, YM-14090, Cibian, Introna, Intron A, Viraferon.*

Interferon alfa-2c, Berofor Alpha 2.

Interferon alfa-n1, NSC-339140, Sumiferon, Wellferon. Mixture of natural α-interferons.

Interferon alfa-n3, Alferon. Mixture of natural α-interferons.

THERAP CAT: Antiviral; antineoplastic.

5017. Interferon-β. Beta-interferon; fibroblast interferon; FIF; IFN-β; IFN-β1; Feron; Fiblaferon; Frone; Naferon. Cytokine with antiviral, antiproliferative and immunomodulatory activity produced by fibroblasts in response to stimulation by live or inactivated virus or by certain synthetic polynucleotides. One of the type I interferons. Glycoprotein containing 166 amino acids; mol wt ~ 20,000 daltons. High yield production by human fibroblast cell cultures: E. A. Havell, J. Vilcek, *Antimicrob. Ag. Chemother.* **2**, 476 (1972). Partial purification: C. B. Anfinsen *et al., Proc. Nat. Acad. Sci USA* **71**, 3139 (1974); K. Berg *et al., J. Immunol.* **114**, 640 (1975). Purification and initial characterization: E. Knight, Jr., *Proc. Nat. Acad. Sci. USA* **73**, 520 (1976). Comparison with interferon-α, *q.v.:* R. L. Cavalieri *et al., ibid.* **74**, 3287 (1977). Review of production, purification and potential applications: W. A. Carter, J. S. Horoszewicz, *Pharmacol. Ther.* **8**, 359-377 (1980). Amino acid analysis, partial sequence: E. Knight, Jr., *et al., Science* **207**, 525 (1980); S. Stein *et al., Proc. Nat. Acad. Sci. USA* **77**, 5716 (1980). Amino acid sequence: T. Taniguchi *et al., Gene* **10**, 11 (1980). Production by recombinant DNA technology: T. Taniguchi *et al., Proc. Nat. Acad. Sci. USA* **77**, 5230 (1980); R. Derynck *et al., Nature* **285**, 542 (1980); D. V. Goeddel *et al., Nucleic Acids Res.* **8**, 4057 (1980). Comparative antiproliferative activity of natural IFNs-α and -β: E. C. Borden *et al., Cancer Res.* **42**, 4948 (1982). Clinical evaluation of recombinant HuIFN-β in colorectal cancer: P. K. Lillis *et al., Cancer Treat. Rep.* **71**, 965 (1987); P. L. Triozzi *et al., ibid.* 983; of natural form in herpes simplex infections: M. Glezerman *et al., Lancet* **1**, 150 (1988). *Note:* A second factor produced by fibroblasts under conditions for IFN-β production has been identified: P. B. Sehgal, A. D. Sagar, *Nature* **288**, 95 (1980). Originally designated *IFN-β2,* subsequently shown to be identical to the human *β-cell differentiation factor* or *BSF-2* and to the *hybridoma growth factor* or *HGF:* P. B. Sehgal *et al., Science* **235**, 731 (1987). Referred to in the literature as *interleukin-6:* J. Van Damme *et al., Eur. J. Biochem.* **168**, 543 (1987); P. Poupart *et al., EMBO J.* **6**, 1219 (1987).

Interferon Beta-1b, IFN-beta*ser*, *Betaseron.* Non-glycosylated polypeptide produced in *E. coli*; consists of 165 amino acid residues, 18,500 Da. Synthetic mutein having a serine substituted for the cysteine residue at position 17 of the native molecule. Prepn: D. F. Mark *et al., Proc. Nat. Acad. Sci. USA* **81**, 5662 (1984). Brief review of clinical use and therapeutic potential: D. E. Goodkin, *Lancet* **344**, 1057-1060 (1994).

THERAP CAT: Antineoplastic; antiviral; immunomodulator.

5018. Interferon-γ. Gamma-interferon; IFN-γ; immune IFN; ImIFN; type II interferon. Lymphokine produced by antigen or mitogen stimulated T-cells with antiviral, antiproliferative, and immunomodulatory activity. Natural IFN-γ is composed of 143 amino acids. Mol wt ranges from ~16,000 to 25,000 daltons depending on the degree of glycosylation. Forms non-covalent dimers of approx mol wt 40,000 to 50,000 daltons. IFN-γ is structurally unrelated to interferons-α and -β, *q.q.v.*, is acid-labile and has a different cell surface receptor. Discovery of interferon produced by mitogen stimulated human leukocytes: E. F. Wheelock, *Science* **149**, 310 (1965). Characterization of the immune interferon response: J. A. Green *et al., ibid.* **164**, 1415 (1969). Large-scale production by cell culture: M. P. Langford *et al., Infect. Immun.* **26**, 36 (1979). Purification and characterization: Y. K. Yip *et al., Proc. Nat. Acad. Sci. USA* **79**, 1820 (1982). Cloning and expression of a 146 amino acid recombinant HuIFN-γ: P. W. Gray *et al., Nature* **295**, 503 (1982). Structure of HuIFN-γ gene: P. W. Gray, D. V. Goeddel, *ibid.* **298**, 859 (1982). Complete amino acid sequence of natural IFN-γ and comparison with recombinant form: E. Rinderknecht *et al., J. Biol. Chem.* **259**, 6790 (1984). Comparison with type I interferons of antiviral and antiproliferative effects: B. Y. Rubin, S. L. Gupta, *Proc. Nat. Acad. Sci. USA* **77**, 5928 (1980). Receptor binding study: A. A. Branca, C. Baglioni, *Nature* **294**, 768 (1981). Enhancement of macrophage activation and proposed identity with *macrophage activating factor* (*MAF*): L. F. Nathan *et al., J. Exp. Med.* **158**, 670 (1983); J. Le *et al., J. Immunol.* **131**, 2821 (1983). Pharmacokinetics and toxicology in cancer patients: R. Kurzrock *et al., Oncology* **42**, Suppl. 1, 41 (1985). Clinical evaluation in cancer: S. Vadhan-Raj *et al., Cancer Treat. Rep.* **70**, 609 (1986); in rheumatoid arthritis: E. M. Lemmel *et al., Rheumatol. Int.* **8**, 87 (1988). Review of biological activities: G. R. Adolf, *Oncology* **42**, Suppl. 1, 33-40 (1985); E. M. Bonnem, R. K. Oldham, *J. Biol. Resp. Mod.* **6**, 275-301 (1987). Review of IFN-γ induction and interrelationship with other cytokines: H. M. Johnson *et al., Ann. N.Y. Acad. Sci.* **524**, 208-217 (1988).

Interferon Gamma-1b, $C_{734}H_{1166}N_{204}O_{216}S_5$, *N2-L-methionyl-1-139-interferon* γ (*human lymphocyte protein moiety reduced*)*, Immukin.*

Cys-Tyr-Cys-interferon γ, *N2-[N-(N-L-cysteinyl-L-tyrosyl)-L-cysteinyl]-interferon* γ (*human lymphocyte protein moiety reduced*)*, S-6810, Actimmune, Gammaferon, Immuneron, Polyferon.*

THERAP CAT: Antiviral, antineoplastic, immunomodulator.

5019. Interleukin-1. IL-1; lymphocyte activating factor; LAF; B-cell activating factor; BAF; T-cell replacing factor; TRF; endogenous pyrogen; leukocytic endogenous mediator; mononuclear cell factor. Mol wt 12,000-18,000. Immunoenhancing, pyrogenic, polypeptide factor produced in blood and in a variety of tissues by mononuclear phagocytes responding to antigens or inflammatory agents. Responsible for a wide variety of bioactivities. Elicits a non-antigen specific amplification of cellular and humoral immune responses. Induces production of interleukin-2 (IL-2), collagenase and prostaglandins, *q.q.v.* Initial identification of factor inducing T-cell proliferation: I. Gery *et al., J. Exp. Med.* **136**, 128, 143 (1972); of factor enhancing antibody production: D. D. Wood, S. L. Gaul, *J. Immunol.* **113**, 925 (1974). Isoln and preliminary chemical characterization: I. Gery, R. E. Handschumacher, *Cell. Immunol.* **11**, 162 (1974); D. D. Wood, *J. Immunol.* **123**, 2395 (1979). Definition and nomenclature: L. A. Aarden *et al., ibid.* 2928. Identity with endogenous pyrogen: L. J. Rosenwasser, C. A. Dinarello, *Cell. Immunol.* **63**, 134 (1981). Purification of murine IL-1: S. B. Mizel, D. Mizel, *J. Immunol.* **126**, 834 (1981); of human IL-1: J. A. Schmidt, *J. Exp. Med.* **160**, 772 (1984). Exists in several biochemically distinct forms exhibiting charge heterogeneity and differing in amino acid sequence. Produced as ~30,000 mol wt precursor which is

subsequently converted to low mol wt form. Cloning and expression of cDNA for murine precursor IL-1: P. T. Lomedico et al., *Nature* **312**, 458 (1984); for human precursor IL-1: P. E. Auron et al., *Proc. Natl. Acad. Sci. USA* **81**, 7907 (1984); for 2 distinct forms of human IL-1 (IL-1α and IL-1β): C. J. March et al., *Nature* **315**, 641 (1985). Amino acid sequence of dominant species corresponding to IL-1β: P. Cameron et al., *J. Exp. Med.* **162**, 790 (1985). Crystallization of recombinant human IL-1β: D. B. Carter et al., *Proteins: Struct. Funct. Genet.* **3**, 121 (1988). Characterization of membrane-associated IL-1: E. A. Kurt-Jones et al., *Proc. Natl. Acad. Sci. USA* **82**, 1204 (1985). Induction of IL-2 production by IL-1: K. A. Smith et al., *J. Exp. Med.* **151**, 1551 (1980); S. Gillis, S. B. Mizel, *Proc. Natl. Acad. Sci. USA* **78**, 1122 (1981). Effect on B-cells in antibody production: D. D. Wood, *J. Immunol.* **123**, 2400 (1979); P. Lipsky, *Contemp. Topics Mol. Immunol.* **10**, 195 (1985). Effect in inflammatory response: S. B. Mizel et al., *Proc. Natl. Acad. Sci. USA* **78**, 2474 (1981). Role in the pathogenesis of the acute-phase response: C. A. Dinarello, *N. Engl. J. Med.* **311**, 1413 (1984). Comparison of bioactivity of 4 forms of human IL-1: D. D. Wood et al., *J. Immunol.* **134**, 895 (1985). Review of inter-relationship with lymphokines: N. M. Kouttab et al., *Clin. Chem.* **30**, 1539-1545 (1984). Reviews: J. J. Oppenheim et al., *Fed. Proc.* **41**, 257-262 (1982); S. B. Mizel, *Immunol. Rev.* **63**, 51-72 (1982); K. Bendtzen, *Allergy* **38**, 219-226 (1983); C. A. Dinarello, *Rev. Infect. Dis.* **6**, 51-95 (1984).
Isoelectric point (human): 6.8-7.3 (dominant species); 5.3-5.8 (minor species). Isoelectric point (murine): 4.5-5.5. Sensitive to trypsin, chymotrypsin, sodium dodecyl sulfate. Insensitive to 2-mercaptoethanol, neuraminidase, sodium periodate, iodacetamide. Stable at pH 3-10; stable at 56° for 60 min; unstable after 2 min at 100°.

5020. Interleukin-2. T-Cell Growth Factor; TCGF; Thymocyte Stimulating Factor; Costimulator; Lymphocyte Mitogenic Factor; IL-2. Mol wt ~15,000 (human); ~30,000 (murine). Soluble immunoenhancing glycoprotein (lymphokine) produced by T-lymphocytes following activation by antigens or mitogens in the presence of IL-1, q.v. Induces T-cell growth and proliferation; enhances natural killer cell activity. Potentiates the release of γ-interferon, B-cell growth factor, and B-cell differentiation factor. Restores T-cell function in immunodeficient disease states. Identification of soluble mitogenic factor produced by T-cells in response to antigen: R. S. Geha, E. Merler, *Cell. Immunol.* **10**, 86 (1974). Identification of T-cell growth promoting factor produced by mitogen stimulated lymphocytes: D. A. Morgan et al., *Science* **193**, 1007 (1976). Definition and nomenclature: L. A. Aarden et al., *J. Immunol.* **123**, 2928 (1979). Induction of IL-2 production by IL-1: K. A. Smith et al., *J. Exp. Med.* **151**, 1551 (1980). Review of early studies: J. Watson, D. Mochizuki, *Immunol. Rev.* **51**, 257 (1980). Purification and partial amino acid sequence: R. J. Robb et al., *Proc. Nat. Acad. Sci. USA* **80**, 5990 (1983). Production of human IL-2 by cultured malignant T-cells: S. Gillis, **Japan. Kokai 82 175,120** coresp to U.S. pat. **4,401,756** (1982, 1983 to Immunex); by T-cell hybridoma: idem, **Japan. Kokai 82 181,698** coresp to U.S. pat. **4,473,642** (1982, 1984 to Immunex). Expression in monkey cells of cloned cDNA for human IL-2 and predicted amino acid sequence: T. Taniguchi et al., *Nature* **302**, 305 (1983); microbial expression of human IL-2: eidem, **Eur. pat. Appl. 142,268** (1985 to Ajinomoto; Japanese Foundation for Cancer Research), *C.A.* **103**, 49160m (1985). Isolation of human IL-2 gene: S. Mita et al., *Biochem. Biophys. Res. Commun.* **117**, 114 (1983). Synthesis of IL-2 peptides: A. Altman et al., *Proc. Nat. Acad. Sci. USA* **81**, 2176 (1984). Biological activity of recombinant IL-2: S. A. Rosenberg et al., *Science* **223**, 1412 (1984). Structure of murine IL-2 and comparison with human IL-2: N. Kashima et al., *Nature* **313**, 402 (1985). Purification and cloning of the T-cell surface receptor for IL-2: W. J. Leonard et al., *Nature* **311**, 626 (1984). Studies on the mechanism of IL-2 signaling: M. A. Tigges et al., *Science* **243**, 781 (1989). Potential therapeutic use: A. W. Boylston, B. M. Vose, *Clin. Immunol. Allergy* **3**, 229 (1983). Clinical trial in AIDS: J. D. Lifson et al., *Lancet* **1**, 698 (1984); in metastatic cancer: S. A. Rosenberg et al., *N. Engl. J. Med.* **313**, 1485 (1985); R.

L. Kradin et al., *Lancet* **1**, 577 (1989). Review of clinical studies: D. R. Parkinson, *Semin. Oncol.* **15**, Suppl. 6, 10-26 (1988). Reviews: K. Bendtzen, *Allergy* **38**, 219-226 (1983); J. J. Ferrar et al., *Ann. Rep. Med. Chem.* **19**, 191-200 (1984); N. M. Kouttab et al., *Clin. Chem.* **30**, 1539-1545 (1984). Book: *Contemporary Topics in Molecular Immunology*, F. P. Inman et al., Ed., Vol. 10, "The Interleukins", S. Gillis, F. P. Inman, Eds. (Plenum Press, New York, 1985).
Isoelectric point: 6.5-6.8 (human); 3.9-5.0 (mouse). Sensitive to trypsin, chymotrypsin, sodium dodecyl sulfate. Stable in 2-mercaptoethanol, urea (2-4 mol/liter), neuraminidase. Stable at pH 3-10; at 56°C for 60 min. Unstable after 30 min. at 70°C.
Recombinant human interleukin-2, *aldesleukin, Leuferon-2, Proleukin.*
THERAP CAT: Antineoplastic; immunomodulator.

5021. Interleukin-3. Multipotent colony stimulating factor; IL-3; multi-CSF. Hematopoietic growth factor that promotes the survival, differentiation and proliferation of multipotent hematopoietic stem cells and of committed progenitor cells of the megakaryocyte, granulocyte-macrophage, erythroid, eosinophil, basophil and mast cell lineages. Enhances thrombopoiesis, phagocytosis, antibody-mediated cellular cytotoxicity and eosinophil metabolism. Species-specific, variably glycosylated peptide; human: 14-28 kDa, 133 amino acids; murine: 22-32 kDa, 140 amino acids. Produced primarily by activated T-lymphocytes; also by natural killer cells and mast cells. Undetectable in normal serum, postulated to be involved in the immune response. Purification from murine myelomonocytic leukemia cells (WEHI-3): J. N. Ihre et al., *J. Immunol.* **129**, 2431 (1982). Cloning of murine IL-3: M. C. Fung et al., *Nature* **307**, 233 (1984); T. Yokota et al., *Proc. Nat. Acad. Sci. USA* **81**, 1070 (1984); of human: Y.-C. Yang et al., *Cell* **47**, 3 (1986). Bioactivity: A. G. Leary et al., *Blood* **70**, 1343 (1987); Y. Sonoda et al., *Proc. Nat. Acad. Sci. USA* **85**, 4360 (1988). Review of molecular and cellular biology: C. F. Morris et al., *Immunol. Ser.* **49**, 177-214 (1990); of structure and function: A. Lindemann, R. Mertelsmann, *Cancer Invest.* **11**, 609-623 (1993). Review of clinical studies with recombinant human IL-3: A. Ganser, ibid. 212-218; E. G. E. de Vries et al., *Stem Cells* **11**, 72-80 (1993); in combination therapy with GM-CSF, q.v.: A. Ganser et al., ibid. 465-473.
Fusion protein with GM-CSF, *PIXY321, GM-CSF/IL-3 fusion protein.* Hybrid molecule consisting of the active domains of human GM-CSF and IL-3 coupled by a flexible amino acid sequence. Construction and expression in yeast: B. M. Curtis et al., *Proc. Nat. Acad. Sci. USA* **88**, 5809 (1991). Evaluation in cancer patients: S. Vadhan-Raj et al., *J. Clin. Oncol.* **12**, 715 (1994). Review: idem, *Stem Cells* **12**, 253-261 (1994).

5022. Interleukin-1 Receptor Antagonist. IL-1ra; interleukin-1 inhibitor protein; IRAP. Endogenous inhibitor of interleukin-1, q.v. Modulates the normal, acute inflammatory response to infection and injury. First known naturally occurring protein that functions as a specific receptor anatagonist of a cytokine. Variably glycosylated, 152 amino acid protein; 3 isoforms have been isolated. mol wt ~22 kDa (glycosylated); ~17 kDa (nonglycosylated). Structurally homologous with IL-1 and produced by the same cells. Binds to IL-1 receptors but does not initiate any biological response. Isoln of an IL-1 inhibitor produced by human monocytes: W. P. Arend et al., *J. Immunol.* **134**, 3868 (1985). Isoln from human urine: P. Seckinger et al., ibid. **139**, 1541 (1987). Purification: C. H. Hannum et al., *Nature* **343**, 336 (1990). Cloning and expression of human IL-1ra: S. P. Eisenberg et al., ibid. 341. Therapeutic potential in IL-1-mediated disease: R. C. Thompson et al., *Int. J. Immunopharmacol.* **14**, 475 (1992); C. A. Dinarello, *Immunol. Today* **14**, 260 (1993). Review: W. P. Arend, *Adv. Immunol.* **54**, 167-227 (1993).
Anakinra, N²-L-methionyl-interleukin 1 receptor antagonist (human isoform x reduced), Antril. Nonglycosylated, recombinant human IL-1ra expressed in *E. coli*; mol wt 17 kDa. Prepn: C. H. Hannum et al., **Eur. pat. Appl. 343,684;** eidem, U.S. pat. **5,075,222** (1989, 1991 both to Synergen). Biological properties: W. P. Arend et al., *J. Clin. Invest.* **85**,

1694 (1990). Clinical trial in sepsis syndrome: C. J. Fisher, Jr. *et al., Crit. Care Med.* **22**, 12 (1994).

THERAP CAT: Anti-inflammatory. In treatment of sepsis syndrome.

5023. Inula. Elecampane; scabwort; elfwort; horseheal. Dried rhizome and roots of *Inula helenium* L., *Compositae. Habit.* Central Asia, Europe; naturalized in U.S. *Constit.* Inulin, volatile oil, alantol, helenin, alantic acid, acrid resin.

5024. Inulin. Dahlin; alantin; alant starch. Mol wt approx 5000. Polysaccharide of *Compositae* which partially or completely replaces starch as a reserve food. Isoln from dahlia tubers: McDonald, "Polyfructosans and Difructose Anhydrides" in *Advan. Carbohyd. Chem.* vol. **2**, 254 (1946); from Jerusalem artichoke tubers: Bacon, Edelman, *Biochem. J.* **48**, 114 (1951). Structure: E. G. V. Percival, *Structural Carbohydrate Chemistry* (J. Garnet Miller, London, 2nd ed., 1962) p 274.

R = CH₂OH
n = approx 35

Spherical crystals from water. Hygroscopic in moist air. $[\alpha]_D^{20}$ −40° (c = 2) for the anhydr. Sol in hot water; slightly sol in cold water and organic solvents. Yields D-fructose and D-glucose upon acid hydrolysis.

Acetate, fine powder from methanol. $[\alpha]_D^{20}$ −34° (c = 1.5 in chloroform); $[\alpha]_D^{20}$ −43° (c = 1.8 in acetic acid).

Trimethylinulin, powder from hot water + acetone, mp 140°. $[\alpha]_D^{20}$ −55° (c = 1.03 in chloroform); $[\alpha]_D^{20}$ −54° (c = 1.09 in benzene).

THERAP CAT: Diagnostic aid (renal function).

5025. Invertase. Invertin; saccharase; sucrase. Enzymes, obtained primarily from yeast as well as other sources, which catalyze the hydrolysis of sucrose into fructose and glucose. Since sucrose is both a β-fructofuranoside and an α-glucoside, it is split by two different types of enzymes, *β-fructofuranosidases* or *β-h-fructosidases* and certain *α-glucosidases* or *α-n-glucosidoinvertases*, which attack the sucrose molecule from the fructose and glucose end, respectively. β-Fructofuranosidase, generally obtained from yeast, is characterized by its ability to hydrolyze raffinose, while α-glucosidase is inactive toward raffinose. *Reviews:* K. Myrbäck, "Invertases" in *The Enzymes*, vol. **4**, P. D. Boyer *et al.*, Eds. (Academic Press, New York, 1960) pp 379-396; Lampen, "Yeast and Neurospora Invertases", *ibid.*, vol. **5** (3rd ed., 1971) pp 291-305; Cochrane, *Soc. Chem. Ind. (London) Monograph* no. **11**, 25-31 (1959) (Pub. 1961); Meister, *Wallerstein Lab. Commun.* **28**, 7 (1965).

USE: For preparation of invert sugar from sucrose; as analytical reagent for sucrose.

5026. Invert Sugar. Nulomoline; Calorose; Invesol; Insubeta; Travert. A mixture of about 50% glucose (dextrose) and 50% fructose (levulose) obtained by hydrolysis of sucrose. Hydrolysis of the sucrose may be carried out with acids or enzymes. Invert sugar is slightly levorotatory, reduces Fehling's soln and can be fermented. Honey is mostly invert sugar. Due to the levulose, it is somewhat sweeter than sucrose. The commercial product is obtained by inversion of a 96% cane sugar soln. The inversion is

carried out at pH 3 to 4 by means of invertase and dil HCl. The acid is usually neutralized with sodium carbonate to pH 6.5. At this point, the dextrose crystallizes and the entire mass is beaten into a creamy, plastic product.

USE: In food products, in confectionery. As a humectant, like glycerol, to hold moisture and to prevent drying out. In brewing.

THERAP CAT: Nutrient (parenteral).

5027. Iobenguane. *[(3-Iodophenyl)methyl]guanidine;* *m*-iodobenzylguanidine; MIBG. $C_8H_{10}IN_3$; mol wt 275.09. C 34.93%, H 3.66%, I 46.13%, N 15.27%. Norepinephrine analog with specific affinity for tissues of sympathetic nervous system and related tumors; prepd as ^{123}I and ^{131}I labeled forms. Prepn and imaging studies: D. M. Wieland *et al., J. Nucl. Med.* **21**, 349 (1980); *eidem,* U.S. pat. **4,584,187** (1986). Improved synthesis: P. A. P. M. van Doremalen, A. G. M. Janssen, *J. Radioanal. Nucl. Chem. Lett.* **96**, 97 (1985). Metabolism in man: T. J. Mangner *et al., J. Nucl. Med.* **27**, 37 (1986). HPLC determn in serum and urine: D. Schwabe *et al., J. Chromatog. Biomed. Appl.* **487**, 177 (1989). Radiopharmacokinetics: S. Ertl *et al., Nucl. Med. Commun.* **8**, 643 (1987). Clinical evaluation of myocardial imaging: D. Fagret *et al., Eur. J Nucl. Med.* **15**, 624 (1989). Diagnostic use in pheochromocytoma: B. Shapiro *et al., J. Nucl. Med.* **26**, 576 (1985); therapeutic use: M. Krempf *et al., J. Clin. Endocrinol. Metab.* **72**, 455 (1991). Symposia on therapeutic and diagnostic use in neuroblastoma: *Advances in Neuroblastoma Research* **2**, A. E. Evans *et al.*, Eds. (Alan R. Liss, Inc., New York, 1988) p 643-726; *Med. Pediatr. Oncol.* **15**, 157-228 (1987). Review of pharmacology: J. C. Sisson, D. M. Weiland, *Am. J. Physiol. Imaging* **1**, 96-103 (1986); of biodistribution and clinical studies: A. R. Wafelman *et al., Eur. J Nucl. Med.* **21**, 545-559 (1994).

Sulfate, $(C_8H_{10}IN_3)_2.H_2SO_4$, colorless crystals from water + ethanol, mp 166-167°.

THERAP CAT: Radiolabeled forms as antineoplastic; diagnostic aid (radioactive imaging agent).

5028. Iobenzamic Acid. *N-(3-Amino-2,4,6-triiodobenzoyl)-N-phenyl-β-alanine;* β-[N-(3-amino-2,4,6-triiodobenzoyl)phenylamino]propionic acid; isobenzamic acid; Bilibyk; Osbil. $C_{16}H_{13}I_3N_2O_3$; mol wt 662.00. C 29.03%, H 1.98%, I 57.51%, N 4.23%, O 7.25%. Prepn: **Brit.** pat. **870,321** (1961 to Oesterreichische Stickstoffwerke). Toxicity: Lindner *et al., Arzneimittel-Forsch.* **11**, 384 (1961).

Crystals, mp 133-134.5°. Sol in acetone, dioxane, DMF; sparingly sol in cold ethanol, ether, benzene, toluene. Practically insol in water. LD_{50} orally in mice: 2.87 g/kg (Lindner).

Methyl ester, $C_{17}H_{15}I_3N_2O_3$, crystals, mp 156-157°.

THERAP CAT: Diagnostic aid (radiopaque medium—cholecystographic).

5029. Iocarmic Acid. *3,3'-[(1,6-Dioxo-1,6-hexanediyl)diimino]bis[2,4,6-triiodo-5-[(methylamino)carbonyl]benzoic acid]; 5,5'-(adipoyldiimino)bis[2,4,6-triiodo-N-methylisophthalamic acid];* Myelotrast. $C_{24}H_{20}I_6N_4O_8$; mol wt 1253.87. C 22.99%, H 1.61%, I 60.73%, N 4.47%, O 10.21%. Prepn: **Brit.** pat. **1,033,695**; U.S. pat. **3,290,366** (both 1966 to Mallinckrodt); G. B. Hoey *et al., J. Med. Chem.* **9**, 964 (1966). Clinical studies: Kunze, Schiefer, *Deut. Med. Wochenschr.* **97**, 245 (1972). Diagnostic use in myelography: P. Ahlgren, *Acta Radiol. Diag.* **13**, 753 (1972); in hysterosalpingography: H. E. Schutte, *Diag. Imaging* **51**, 277 (1982). Spinal toxicity in rats: I. O. Skalpe, A. Torvik, *Invest. Radiol.* **10**, 154 (1975).

Crystals from dimethylformamide, mp 302° (dec).

Di-*N*-methylglucamine salt, $C_{38}H_{54}I_6N_6O_{18}$, *dimeglumine iocarmate*, *LM-280, Dimeray, Dimer-X, Dirax.* Soly in water (25°): 65 g/100 ml.

THERAP CAT: Diagnostic aid (radiopaque medium).

5030. Iocetamic Acid. *3-[Acetyl-(3-amino-2,4,6-triiodophenyl)amino]-2-methylpropanoic acid; N-acetyl-N-(3-amino-2,4,6-triiodophenyl)-2-methyl-β-alanine; N-acetyl-N-(2,4,6-triiodo-3-aminophenyl)-β-aminoisobutyric acid; DRC-1201; MP-620; Cholebrine; Cholimil.* $C_{12}H_{13}I_3N_2O_3$; mol wt 613.96. C 23.48%, H 2.13%, I 62.01%, N 4.56%, O 7.82%. Prepn: **Neth.** pat. **Appls. 6,515,305; 6,607,275** (both 1967 to Dagra), *C.A.* **67**, 108422m (1967); **69**, 10263b (1968); Korver, *Rec. Trav. Chim.* **87**, 308 (1968). Pharmacology and toxicology: J. M. Janbroers *et al., Toxicol. Appl. Pharmacol.* **14**, 232, 246 (1969). Metabolic studies: Neleman, *Pharm. Weekblad.* **102**, 1039 (1967). Clinical studies: Hekster, *Radiol. Clin. Biol.* **37**, 338 (1968). Use in cholecystography: B. Goldberg, *Radiol. Clin.* **46**, 42 (1977).

White to light cream-colored powder, mp 224-225° (**Neth.** pat. **Appl. 6,515,305**). Also reported as mp range 191-212° (Korver). Practically insol in water. Very slightly sol in ether, ethanol, benzene; slightly sol in acetone, chloroform (Janbroers). LD$_{50}$ in rats (g/kg): 7.1 orally; 0.70 i.v. (Janbroers). Consists of 2 isomers having mps 232° and 200-201° and pKa's of 4.25 and 4.0 respectively. The lower melting compd is approx two times as soluble.

THERAP CAT: Diagnostic aid (radiopaque medium).

5031. Iodamide. *3-(Acetylamino)-5-[(acetylamino)-methyl]-2,4,6-triiodobenzoic acid; α,5-diacetamido-2,4,6-triiodo-m-toluic acid; 3-acetamido-5-(acetamidomethyl)-2,4,6-triiodobenzoic acid; ametriodinic acid; SH-926.* $C_{12}H_{11}I_3N_2O_4$; mol wt 627.94. C 22.95%, H 1.77%, I 60.63%, N 4.46%, O 10.19%. Prepn: Felder, Pitre, **Fr.** pat. **1,382,277** and **U.S.** pat. **3,360,436** (1964 and 1967 to Eprova); Felder *et al., Helv. Chim. Acta* **48**, 259 (1965). Pharmacology and toxicology: F. Bonati *et al., Arzneimittel-Forsch.* **15**, 222 (1965); Z. B. Zsebök, L. Szlavy, *ibid.* **17**, 1380 (1967); *eidem, Int. Z. Klin. Pharmakol. Ther. Toxikol.* **3**, 157 (1970). Clinical trials: Grothuesmann, *Arzneimittel-Forsch.* **15**, 233 (1965). Comprehensive description: D. Pitré in *Analytical Profiles of Drug Substances* **vol. 15**, K. Florey, Ed. (Academic Press, New York, 1986) pp 337-365.

Crystals from acetic acid, mp 255-257°. Soly in water: 0.3 g/100 ml (22°).

N-Methyl-D-glucamine salt, $C_{19}H_{28}I_3N_3O_9$, *iodamide meglumine, Isteropac E.R., Jodomiron, Opacist E.R., Renovue-65, Renovue-DIP, Uromiro.* Sparingly sol in methanol, slightly sol in water, ethanol. Practically insol in ether, chloroform. LD$_{50}$ in mice, rats, rabbits (g/kg): 9.0, 11.4, 13.2 i.v.; in rats, guinea pigs (g/kg): 17.9, 15.0 i.p. (Bonati).

USE: Diagnostic aid (radiopaque medium).

5032. Iodic Acid. HIO_3; mol wt 175.91. H 0.57%, I 72.14%, O 27.29%. Prepd by the oxidation of iodine with nitric acid, perchloric acid, or hydrogen peroxide: Baxter, Tilley, *Z. Anorg. Allgem. Chem.* **61**, 295 (1909); Lamb *et al., J. Am. Chem. Soc.* **42**, 1643 (1920); Bray, Caulkins, *ibid.* **53**, 44 (1931). Electrolytic method: Willard, Ralston, *Trans. Am. Electrochem. Soc.* **62**, 239 (1932).

Orthorhombic crystals. Darkens upon exposure to light. d$_4^0$ 4.629. mp 110° (dec). Decompn to $HIO_3.I_2O_5$ starts at 70° and decompn to I_2O_5 is complete at 220°. Not hygroscopic, yet very freely sol in water: 269 g/100 ml H_2O at 20°; 295 g/100 ml H_2O at 40°. Density of aq solns (w/w) at 18°: 1% = 1.0071; 10% = 1.0900; 20% = 1.1969; 40% = 1.4640; 75% = 2.4710. Strong electrolyte, especially when dil. Sol in nitric acid, dil alcohol. Insol in abs alc, ether, chloroform. *Keep well closed and protected from light.*

THERAP CAT: Astringent, disinfectant.

5033. Iodinated Glycerol. Iodopropylidene glycerol; Organidin. $C_6H_{11}IO_3$; mol wt 258.06. C 27.93%, H 4.30%, I 49.18%, O 18.60%. Isomeric mixture of 67-75% of *2-(1-iodoethyl)-1,3-dioxolane-4-methanol* or *2,3-(2-iodopropylidenedioxy)propanol* and 33-25% of *2-(2-iodoethyl)-1,3-dioxolane-4-methanol* or *2,3-(3-iodopropylidenedioxy)propanol.* Prepn: Manchey *et al., U.S.* pat. **2,872,378** (1959 to The Denver Chem. Manuf.).

Pale yellow liquid. Pungent, bitter aftertaste. d 1.797. n$_D$ 1.547. Sol in ether, chloroform, isobutyl alc, methyl acetate, ethyl acetate, methyl formate, tetrahydrofuran.

THERAP CAT: Expectorant.

5034. Iodine. I; at. wt 126.9045; at. no. 53; valences 1 to 7 (usually monovalent); elemental state: I_2. A halogen. Abundance in igneous rocks: 3×10^{-5}% by wt; in seawater 5×10^{-8}% by wt. Natural isotope: 127 (100%); isotopes range in mass number from 117 to 139; radioactive tracer elements: 124, 125, 128, 131, 132. Discovered in 1811 by Courtois. Classed among the rarer elements. Extracted from Chilean nitrate-bearing earth (caliche) and from seaweed. Prepn of ultra-pure iodine for research purposes according to the equation $2KI + CuSO_4.5H_2O \rightarrow CuI + K_2SO_4 + \frac{1}{2}I_2 + 5H_2O$: Schmeisser in *Handbook of Preparative Inorganic Chemistry*, Vol. 1, G. Brauer, Ed. (Academic Press, New York, 2nd ed., 1963) p 275. *Reviews: MTP Int. Rev. Sci.: Inorg. Chem., Ser. One,* Vol. 3, V. Gutmann, Ed. (Butterworths, London, 1972); Downs, Adams, "Chlorine, Bromine, Iodine and Astatine" in *Comprehensive Inorganic Chemistry* vol. 2, J. C. Bailar, Jr. *et al.,* Eds. (Pergamon Press, Oxford, 1973) p 1107-1594; C. J. Mazac in Kirk-Othmer *Encyclopedia of Chemical Technology* vol. 13 (Wiley-Interscience, New York, 3rd ed., 1981) pp 649-677.

Bluish-black scales or plates; diatomic; metallic luster; characteristic odor; sharp, acrid taste; violet corrosive vapor. mp 113.60°. bp 185.24°. d (solid, 25°) 4.93. Vapor pressure (solid): 0.030 mm (0°); 0.305 mm (25°); 2.154 mm (50°); 26.78 mm (90°). Heat capacity at constant pressure (25°) 13.011 cal/mole/°C: Shirley, Giauque, *J. Am. Chem. Soc.* **81**, 4778 (1959). Total soly in water (25°) 0.0013 moles/l with negligible formation of HOI (6.4×10^{-6} moles/l); freely sol in aq solns of HI or iodides. Soly in organic solvents (g I_2/100 g soln, 25°): benzene 14.09; CS_2 16.47; ethanol 21.43; ethyl ether 25.20; cyclohexane 2.719; CCl_4 (35°) 2.603: Hildebrand, Jenks, *J. Am. Chem Soc.* **42**,

2180 (1920); Hildebrand *et al., ibid.* **72**, 1017 (1950); sol in chloroform, glacial acetic acid, glycerol oils. Solutions of iodine in aq solns of inorganic iodides are brown or deep brown, depending on the concn of the iodine. Solvents contg nitrogen atoms, such as pyridine, amines, or quinoline dissolve iodine to form brown solns. Chloroform, carbon tetrachloride, carbon disulfide, and especially phosphorus trichloride give violet solns. The violet color is also given by fluorinated amines. The soly in water is also increased by alkali bromides, but decreased by sulfates and nitrates. Less reactive than bromine; $E°$ (aq) $\frac{1}{2}I_2/I^-$ 0.535 V dissociation energy (25°): 36.115 kcal. For reactions, complexes, and formation of monatomic iodine: *see* Chlorine. Iodine stains may be removed with sodium thiosulfate soln or ammoniated alc. *Incompat:* alkaloids, starch, tannins.

Tamed Iodine, *see* Iodophor.

Caution: Potential symptoms of overexposure are irritation of eyes and nose; lacrimation; headache; tight chest; skin burns, rash; cutaneous hypersensitivity. *See NIOSH Pocket Guide to Chemical Hazards* (DHHS/NIOSH 90-117, 1990) p 128. Ingestion of large quantities causes abdominal pain, vomiting and diarrhea due to the highly corrosive action of iodine on the GI tract. *See Clinical Toxicology of Commercial Products,* R. E. Gosselin *et al.,* Eds. (Williams & Wilkins, Baltimore, 5th ed., 1984) Sect. III, pp 213-214.

USE: Manuf iodine compds, germicides, antiseptics. To reduce friction of hard surfaces, including stainless steel and glass. Catalyst in the alkylation and condensation of aromatic amines; in sulfations and sulfonations. Artificial isotopes of iodine are used in biological, biochemical and chemical structure research. Important reagent in analytical chemistry.

THERAP CAT: Antihyperthyroid; Anti-infective (topical).

THERAP CAT (VET): Internally for goiter, hypothyroidism, in iodine deficiency. Topically as antiseptic, disinfectant, counterirritant and to promote absorption.

5035. Iodine Colloidal. Elemental iodine in a colloidal state, stabilized with a suitable "protecting" colloid. Prepn: Chandler, Miller, *J. Phys. Chem.* **31**, 1091 (1927). This iodine is of a brick-red color and was supplied as a powder contg 25% iodine or in the form of an aq suspension contg 16% elemental iodine.

THERAP CAT (VET): Has been used as an anthelmintic in poultry.

5036. Iodine Heptafluoride. F_7I; mol wt 259.89. F 51.17%, I 48.83%. IF_7. Prepd by passing fluorine through liq IF_5 at 90°, then heating the vapors to 270°: Ruff, Keim, *Z. Anorg. Allgem. Chem.* **193**, 176 (1930); Kwasnik in *Handbook of Preparative Inorganic Chemistry,* **vol. 1**, G. Brauer, Ed. (Academic Press, New York, 2nd ed., 1963) pp 160-161. May also be prepd from fluorine and dried PdI_2 or KI to minimize formation of the impurity, IOF_5: Bartlett, Levchuk, *Proc. Chem. Soc.* **1963**, 342; Selig *et al., J. Phys. Chem.* **71**, 2739 (1967). *Reviews:* Kemmitt, Sharp, *Advan. Fluorine Chem.* **4**, 246-247 (1965); Stein, "Physical and Chemical Properties of Halogen Fluorides" in *Halogen Chemistry,* **Vol. 1**, V. Gutmann, Ed. (Academic Press, New York, 1967) pp 133-224; Meinert, *Z. Chem.* **7**, 41-57 (1967).

Colorless gas. Crystals when solid. Moldy, acrid odor. *Attacks glass and quartz.* d (liq; 6°) 2.8. mp 6.45°. Sublimes at 4.77°. Easily forms a supercooled liquid which boils at +4.5°. Trouton constant 26.4. Water dissolves the gas without violence, some decompn. Readily absorbed by NaOH soln.

Caution: Highly irritating to skin, mucous membranes. *See also* Fluorine.

USE: Fluorinating agent.

5037. Iodine Monobromide. BrI; mol wt 206.81. Br 38.64%, I 61.36%. IBr. Prepd from the elements, excess Br is removed by heating to 50° in a current of CO_2: Bornemann, *Ann.* **189**, 202 (1877); Terwogt, *Z. Anorg. Allgem. Chem.* **47**, 203 (1905); *Gmelin's, Iodine* (8th ed.) **8**, p 631 (1933); Schmeisser in *Handbook of Preparative Inorganic Chemistry* vol. 1, G. Brauer, Ed. (Academic Press, New York, 2nd ed., 1963) p 291-292. Crystal structure: Swink, Carpenter, *Acta Cryst.* **24B**, 429 (1968).

Brownish-black crystals or very hard solid. d 4.416. mp 40°. bp 116° (decompn). Sol in water, alcohol, ether, car-

bon disulfide, glacial acetic acid. May be stored in brown glass-stoppered bottles. *Keep in a cool place.*

Caution: Vapors are corrosive to the eyes and mucous membranes.

5038. Iodine Monochloride. Wijs' chloride. ClI; mol wt 162.36. Cl 21.84%, I 78.16%. ICl. Prepd from the elements: Cornog, Karges, *Inorg. Syn.* **1**, 165 (1939); Buckles, Bader, *ibid.* **9**, 130 (1967).

Black crystals or reddish-brown liquid. The crystals occur in two modifications: α-form (stable), black needles (ruby red by transmitted light), mp 27.2°; β-form (labile), black platelets (brownish red by transmitted light), mp 13.9°. bp 97° (decompn). d_4^{29} 3.10. Sol in water, alc, ether, CS_2, acetic acid. May be stored in brown glass-stoppered bottles. Not hygroscopic, but forms I_2O_5 in the presence of air.

Caution: Attacks the skin, forming dark, painful patches (effective countermeasure: immediate washing with 20% HCl).

USE: In Wijs' soln (iodine monochloride in glacial acetic acid), used to determine iodine values of fats and oils.

THERAP CAT: Anti-infective (topical).

5039. Iodine Pentafluoride. F_5I; mol wt 221.90. F 42.81%, I 57.19%. IF_5. Prepd by passing fluorine over iodine with cooling: Gore, *Phil. Mag.* [4] **41**, 309 (1871); Moissan, *Compt. Rend.* **135**, 563 (1902); *Bull. Soc. Chim. France* [3], **29**, 6 (1903); Ruff, Keim, *Z. Anorg. Allgem. Chem.* **201**, 245 (1931); Ruff, Braida, *ibid.* **220**, 43 (1934); Kwasnik in *Handbook of Preparative Inorganic Chemistry,* **vol. 1**, G. Brauer, Ed. (Academic Press, New York, 2nd ed., 1963) pp 159-160. *Reviews:* Kemmitt, Sharp, *Advan. Fluorine Chem.* **4**, 247-248 (1965); Stein, "Physical and Chemical Properties of Halogen Fluorides" in *Halogen Chemistry,* **Vol. 1**, V. Gutmann, Ed. (Academic Press, New York, 1967) pp 133-224; Meinert, *Z. Chem.* **7**, 41-57 (1967).

Liquid. mp 9.43°. bp 100.5°. d^{25} 3.19. Thermodynamic data: Osborne *et al., J. Phys. Chem.* **54**, 3790 (1971). *Fumes in air!* Attacks glass, especially when hot. Violent reaction with water. Organic compds carbonize on contact, sometimes with conflagration. Instant reaction with S, P, Si, Bi, W, As, usually with incandescence.

USE: Mild fluorinating agent.

5040. Iodine Pentoxide. Iodic anhydride. I_2O_5; mol wt 333.81. I 76.03%, O 23.97%. Prepd by thermal dehydration of iodic acid at 195°: Lamb *et al., J. Am. Chem. Soc.* **42**, 1644 (1920). Crystal structure: Selte, Kjekshus, *Acta Chem. Scand.* **24**, 1912 (1970).

Needle shaped, hygroscopic crystals (an impure product may have a pink cast). d^{25} 5.08. Decompn into I_2 and O_2 begins at 275° and proceeds rapidly at 350°. Very sol in water (187.4 g/100 ml at 13°) forming HIO_3. Insol in abs alc, ether, chloroform, carbon disulfide. Dissolves in nitric acid from which it crystallizes as I_2O_5 if the HNO_3 concn is > 50%. Strong oxidizing agent, may detonate when triturated with oxidizable matter. *Keep well closed and protected from light.*

USE: Oxidizes carbon monoxide to carbon dioxide with the formn of iodine (Ditte's reaction, which proceeds rapidly at 65° or above, but slowly at room temp). This reaction is used in gas analysis and for removing carbon monoxide from the air (in respirators preferably in presence of some H_2SO_4).

5041. Iodine Trichloride. Cl_3I; mol wt 233.26. Cl 45.60%, I 54.40%. ICl_3. Prepd by adding finely powdered iodine to an excess of liq chlorine, which is then boiled away: Thomas, Depuis, *Compt. Rend.* **143**, 282 (1906); Booth, Morris, *Inorg. Syn.* **1**, 167 (1939). Alternate process: Birk, *Angew. Chem.* **41**, 751 (1928); *Z. Anorg. Allgem. Chem.* **172**, 399 (1928); Wilke-Dörfurt, Wolff, *ibid.* **185**, 333 (1930).

Long yellow or brownish needles or fluffy powder. Pungent, irritating odor. Corrosive to human skin. *Must be stored in well closed amber glass bottles, preferably in refrigerator.* d^{-4} 3.203. mp ~33°. Volatile at room temp. mp 101° at 16 atm (pressure of its satd vapor). A pressure of 1 atm is reached at 64°. Complete decompn into ICl and Cl_2 at 77°. Generally used as a 20 to 35% soln in concd HCl.

Caution: Concd solns are strongly irritating.

USE: Chlorinating and oxidizing agent.
THERAP CAT: Anti-infective (topical).

5042. Iodinin. *1,6-Phenazinediol 5,10-dioxide;* 1,5-dihydroxyphenazine *N,N'*-dioxide. $C_{12}H_8N_2O_4$; mol wt 244.21. C 59.02%, H 3.30%, N 11.47%, O 26.21%. Antibiotic pigment from *Chromobacterium iodinum:* Clemo, McIlwain, *J. Chem. Soc.* **1938**, 479; Hegedüs in *Emil Barell Jubilee Volume* (1946), p 388; from *Waksmania aerata* and *Pseudomonas iodinum:* Gerber, Lechevalier, *Biochemistry* 3, 598 (1964). Structure: Clemo, Daglish, *J. Chem. Soc.* **1950**, 1481. Synthesis: Matsumura, Takeda, *Nippon Kagaku Zasshi* 81, 515 (1960), *C.A.* **56**, 470a (1962). Biosynthetic studies on incorporation of shikimic acid, *q.v.*, into iodinin: U. Hollstein *et al.*, *Tetrahedron Letters* **1978**, 2987; R. B. Herbert *et al.*, *J. Chem. Soc. Perkin Trans. I* **1979**, 2411; T. Etherington *et al.*, *ibid.* 2416.

Purple crystals with coppery sheen from chloroform, dec 236°. pK 12.5. Stable in acid, unstable in alkali. Sol in benzene, toluene, xylene, carbon disulfide, chloroform, ethyl acetate. Slightly sol in hot alc. Practically insol in cold alc, ether, acetic acid, petr ether, amyl alcohol. Insol in water. Sol in concd sulfuric acid and in glacial acetic acid with red color. Sol in NaOH solns giving brilliantly blue solns which deposit green crystals of the unstable sodium deriv.

5043. Iodipamide. *3,3'-[(1,6-Dioxo-1,6-hexanediyl)di-imino]bis[2,4,6-triiodobenzoic acid];* *3,3'-(adipoyldiimino)-bis[2,4,6-triiodobenzoic acid];* N,N'-adipylbis(3-amino-2,4,6-triiodobenzoic acid); adipic acid di(3-carboxy-2,4,6-triiodo-anilide); adipiodone; Cholografin; Cholospect. $C_{20}H_{14}I_6N_2O_6$; mol wt 1139.77. C 21.08%, H 1.24%, I 66.81%, N 2.46%, O 8.42%. Prepn: Priewe, Rutkowski, U.S. pat. **2,776,241** (1957 to Schering AG); Kotler-Brajtburg *et al.*, *Roczniki Chem.* 36, 763 (1962), *C.A.* 58, 5568a (1963). Purification: Cassebaum, Drux, East Ger. pat. 33,738 (1964), *C.A.* 63, 11441b (1965). Pharmacology: Fischer, Varga, *Acta Physiol.* 38, 135 (1970). Metabolic studies: Kiyono *et al.*, *Radioisotopes* 20, 78 (1971). Pharmacology and toxicity study: F. J. Rosenberg *et al.*, *Invest. Radiol.* 15, S142 (1980). Comprehensive description: H. H. Lerner in *Analytical Profiles of Drug Substances* vol. 3, K. Florey, Ed. (Academic Press, New York, 1974) pp 333-363.

Crystals, dec 306-308°. $n_D^{21.5}$ 1.3294 (c = 0.445 in methanol). Soly at 20°: methanol 0.8%; ethanol 0.3%; acetone 0.2%; ether 0.1%. Practically insol in water, benzene. LD_{50} in mice, rats (mg/kg): 2380 ±290, 4430 ±310 i.v. (Rosenberg).

Disodium salt, $C_{20}H_{12}I_6N_2Na_2O_6$. Prepd by dissolving the free acid in a dil aq soln of NaOH and buffering to pH 6.5-7.7. Used as 20% soln.

Bis[*N*-methylglucamine] salt, $C_{34}H_{48}I_6N_4O_{16}$, *iodipamide meglumine, Biligrafin, Cholegrafin, Endocistobil, Endografin, Intrabilix, Transbilix*. More sol in water than the disodium salt. Used as 40% soln.

THERAP CAT: Diagnostic aid (radiopaque medium—cholecystographic).
THERAP CAT (VET): Diagnostic aid (radiopaque medium).

5044. Iodixanol. *5,5'-[(2-Hydroxy-1,3-propanediyl)bis-(acetylimino)]bis[N,N'-bis(2,3-dihydroxypropyl)-2,4,6-triiodo-1,3-benzenedicarboxamide];* 5,5'-[(2-hydroxytrimethylene)-bis(acetylimino)]bis[N,N'-bis(2,3-dihydroxypropyl)-2,4,6-triiodoisophthalamide]; 1,3-bis(acetylamino)-N,N'-bis[3,5-bis(2,3-dihydroxypropylaminocarbonyl)-2,4,6-triiodophen-yl]-2-hydroxypropane; 2-5410-3A; Acupaque; Visipaque. $C_{35}H_{44}I_6N_6O_{15}$; mol wt 1550.19. C 27.12%, H 2.86%, I 49.12%, N 5.42%, O 15.48%. Nonionic, dimeric x-ray contrast medium. Prepn: P. E. Hansen *et al.*, *Eur. pat.* **Appl.** **108,638** (1984 to Nyegaard), *C.A.* **101**, 151599g (1984). HPLC determn in plasma: H. Nomura *et al.*, *J. Chromatog. Biomed. Appl.* 572, 333 (1991). Clinical pharmacokinetics: M. G. Svaland *et al.*, *Invest. Radiol.* 27, 130 (1992). Clinical studies in cerebral arteriography: Y. Palmers *et al.*, *Eur. J. Radiol.* 17, 203 (1993); in cardiac angiography: J. A. Hill *et al.*, *Am. J. Cardiol.* 74, 57 (1994); in urography: R. M. Conroy *et al.*, *Clin. Radiol.* 49, 337 (1994). Use as density gradient medium: T. Ford *et al.*, *Anal. Biochem.* 220, 360 (1994); J. Graham *et al.*, *ibid.* 367.

mp 240-250°. Soluble in water. Viscosity (37°) 11.1 mPa; osmolality 290 mOsm/kg H_2O; pH 7.2-7.6 (c = 320 mgI/ml). 50% aqueous soln (w/v): viscosity 8.3 mPa; n 1.4128 ±0.0003; d 1.266 ±0.003; osmolality 200 ±5 mOsm. LD_{50} i.v. in mice: 21 gI/kg (Palmers).

THERAP CAT: Diagnostic aid (radiopaque medium).

5045. Iodized Oil. Lipiodol. An iodine addition product of vegetable oils, contg 38-42% organically combined iodine.

Thick, viscous, oily liq; alliaceous odor; oleaginous taste; dec on exposure to air and light, becoming brown. *Keep in small, nearly full containers, protected from light.*

THERAP CAT: Diagnostic aid (radiopaque medium).
THERAP CAT (VET): X-ray contrast medium.

5046. Iodoacetic Acid. $C_2H_3IO_2$; mol wt 185.95. C 12.92%, H 1.63%, I 68.25%, O 17.21%. CH_2ICOOH. Prepd by treating chloroacetic acid in acetone soln with NaI.

Colorless or white crystals. mp 82-83°. Sol in water, alc; very slightly sol in ether.

5047. Iodoalphionic Acid. *4-Hydroxy-3,5-diiodo-α-phenylbenzenepropanoic acid;* β-(4-hydroxy-3,5-diiodophen-yl)-α-phenylpropionic acid; 3,5-diiodo-α-phenylphloretic acid; Coletrast; Priodax; Pheniodol; Dikol; Iodobil; Jodobil; Biliognost; Tenicid; Biliselectan. $C_{15}H_{12}I_2O_3$; mol wt 494.07. C 36.47%, H 2.45%, I 51.37%, O 9.71%. Prepd by iodinating β-(4-hydroxyphenyl)-α-phenylpropionic acid with iodine in an aq soln of alkali iodide in the presence of NH_3 or with iodine in acetic anhydride as the iodinating agent, *cf.* **Brit.** pat. 559,024 (1944), *C.A.* 40, 1883 (1946). Prepn of optically active forms: Tullar, Hoppe, U.S. pat. **2,552,696** (1951 to Sterling Drug). Toxicity data: J. O. Hoppe, S. Archer, *Am. J. Roentgenol. Rad. Ther.* 69, 630 (1953).

dl-Form, faintly yellowish powder. dec 157-162°. Crystals from acetic acid, mp 163.5-163.8°. Insol in water. Sol in alcohol, ether, alkali carbonate and in alkali hydroxide solns; slightly sol in benzene and chloroform. Forms a water-sol disodium salt. LD_{50} in mice (g/kg): 3.8 orally (Hoppe, Archer).

l-Form, hydrated crystals, mp 80-85°, $[\alpha]_D^{25}$ −59° (c = 1.5 in 95% alcohol).

l-Form disodium salt, mp > 230°, $[\alpha]_D^{26}$ −56.6° (c = 1.5 in water).

d-Form, hydrated crystals, mp 80-85°, $[\alpha]_D^{25}$ +59° (c = 1.5 in 95% alcohol).

d-Form disodium salt, mp > 230°, $[\alpha]_D^{26}$ +56.6° (c = 1.5 in water).

THERAP CAT: Diagnostic aid (radiopaque medium—cholecystographic).

5048. *p*-Iodoaniline. C_6H_6NI; mol wt 219.02. C 32.90%, H 2.76%, N 6.40%, I 57.94%.

White crystals. mp 67-68°. Slightly sol in water, freely in alcohol, chloroform, ether.

5049. *o*-Iodoanisole. C_7H_7IO; mol wt 234.04. C 35.92%, H 3.01%, I 54.22%, O 6.84%. Prepd by diazotizing *o*-anisidine and decomposing the diazonium salt with KI.

Yellow liquid; darkens on exposure to air. d^{20} 1.80. bp_{730} 238-240°. Insol in water; miscible with alcohol, chloroform, ether. *Protect from light.*

5050. Iodobenzene. C_6H_5I; mol wt 204.01. C 35.32%, H 2.47%, I 62.20%. Obtained by diazotizing aniline and then treating with an aq soln of KI, or by the action of HNO_3 on a mixture of C_6H_6 and iodine.

Colorless liquid; rapidly becomes yellow; characteristic odor. d_4^{15} 1.8384. bp 188-189°. mp −30°. n_D^{18} 1.621. Insol in water; miscible with alcohol, chloroform, ether. *Protect from light.*

5051. *o*-Iodobenzoic Acid. $C_7H_5IO_2$; mol wt 248.02. C 33.90%, H 2.03%, I 51.17%, O 12.90%. C_6H_4ICOOH. Prepd by diazotizing *o*-aminobenzoic acid and treating the resulting diazonium compound with aq KI.

White needles. d 2.25. mp 162°. Sparingly sol in water; sol in alcohol, ether.

5052. Iodochlorhydroxyquin. *5-Chloro-7-iodo-8-quinolinol;* 5-chloro-8-hydroxy-7-iodoquinoline; chloroiodoquin; iodochlorohydroxyquinoline; iodochloroxyquinoline; Amebil; Alchloquin; Amoenol; Bactol; Barquinol; Budoform; Chinoform; Clioquinol; Cliquinol; Eczecidin; Enteroquinol; Entero-Septol; Entero-Vioform; Enterozol; Entrokin; Hi-Enterol; Iodoenterol; Nioform; Quinambicide; Rometin; Vioform; Vioformio. C_9H_5ClINO; mol wt 305.50. C 35.38%, H 1.65%, Cl 11.60%, I 41.54%, N 4.58%, O 5.24%. Prepn: Ger. pat. **117,767** (1899 to Ciba); U.S. pat. **641,491** (1900); A. Das, S. L. Mukherji, *J. Org. Chem.* **22**, 1111 (1957). Toxicity study: Davis *et al.*, *Am. J. Trop. Med.* **24**, 29 (1944). Comprehensive description: G. Padmanabhan *et al.* in *Analytical Profiles of Drug Substances* vol. **18**, K. Florey, Ed. (Academic Press, New York, 1989) pp 57-90.

Brownish-yellow, bulky powder. Practically odorless. Dec ~178-179°. uv max (water/conc. HCl): 266 nm ($A_{1cm}^{1\%}$ 990); (0.1N methanolic NaOH): 269 nm ($A_{1cm}^{1\%}$ 1120); (ethanol): 255 nm ($A_{1cm}^{1\%}$ 1570). One part dissolves in 43 parts of boiling alc, 128 parts chloroform, 17 parts of boiling ethyl acetate, 170 parts of cold acetic acid, 13 parts of boiling acetic acid. Almost insol in water, cold alc, ether. Soly at room temp (mg/ml): water < 0.01, methanol 1.9, ethanol 1.3, ether 5.4, chloroform 14.9, 0.1N NaOH 17.3, 0.1N HCl 0.02, acetonitrile between 1.7-2.5, tetrahydrofuran between 20-100, ethyl acetate between 6.7-10, carbon disulfide between 10-20, dimethyl sulfoxide between 20-100, dimethyl acetamide between 20-100. Darkens upon exposure to light. LD_{50} orally in cats: 400 mg/kg (Davis).

Caution: Has been linked with the occurrence of subacute myelo-optic neuropathy (S.M.O.N. syndrome) in Japan, T. Tsubaki *et al.*, *Lancet* **1**, 696 (1971); *see also ibid.* **1**, 534 (1977); J. Tateishi in *Drug-Induced Sufferings*, T. Soda, Ed. (Excerpta Medica, New York, 1980) pp 464-472.

THERAP CAT: Anti-infective (topical); antiamebic.

THERAP CAT (VET): Has been used as an intestinal anti-infective.

5053. Iodofenphos. *Phosphorothioic acid O-(2,5-dichloro-4-iodophenyl) O,O-dimethyl ester; O-*(2,5-dichloro-4-iodophenyl) *O,O*-dimethyl phosphorothioate; C-9491; Nuvanol N. $C_8H_8Cl_2IO_3PS$; mol wt 413.00. C 23.27%, H 1.95%, Cl 17.17%, I 30.73%, O 11.62%, P 7.50%, S 7.76%. Organophosphorous insecticide with stomach and contact activity. Prepn: E. Beriger *et al.*, S. Afr. pat. **65 06,107;** *eidem*, U.S. pat. **3,468,984** (1965, 1969 both to Ciba). Metabolism and anti-cholinesterase activity: F. R. Johannsen, C. O. Knowles, *J. Econ. Entomol.* **63**, 693 (1970). GLC determn in cattle urine and tissues: M. C. Ivey, D. D. Oehler, *J. Agr. Food Chem.* **24**, 1049 (1976). Crystal and molecular structure: R. G. Baughman, P.-J. Yu, *ibid.* **30**, 293 (1982). Use to control flies and ticks in cattle: R. O. Drummond *et al.*, *J. Econ. Entomol.* **65**, 1354 (1972). *Review:* B. C. Haddow, T. G. Marks, *Proc. 5th Brit. Insect. Fungic. Conf., 1969* **2**, 531 (1970).

Pale yellow oil, solidifies to crystals, mp 72-73° (Beriger). Also reported as white crystals, mp 75° (Haddow). Sol in acetone, xylene. Slightly sol in alcohol. Soly in water: < 2 ppm. LD_{50} in rats, female mice (mg/kg): 2100, > 10000, orally (Haddow, Marks).

USE: Insecticide.

THERAP CAT (VET): Ectoparasiticide.

5054. Iodoform. *Triiodomethane.* CHI_3; mol wt 393.73. C 3.05%, H 0.26%, I 96.69%. Prepn from acetone, sodium hypochlorite, potassium iodide, and sodium hydroxide: Glass, *Quart. J. Pharm. Pharmacol.* **8**, 351 (1935); from chloroform + methyl iodide: Soroos, Hinkamp, *J. Am. Chem. Soc.* **67**, 1642 (1945); by electrolysis: Glasstone, *Ind. Chem.* **7**, 315 (1931). Description of the iodoform reaction: Seelye, Turney, *J. Chem. Ed.* **36**, 572 (1959). Toxicity study: Kutob, Plaa, *Toxicol. Appl. Pharmacol.* **4**, 354 (1962).

Yellow powder or crystals; mp ~120°; dec at high temp with evolution of iodine. Unctuous touch; characteristic, disagreeable odor. Volatile with steam. d 4.1. Very slightly sol in water. One gram dissolves in 60 ml cold alcohol, 16 ml boiling alcohol, 10 ml chloroform, 7.5 ml ether, 80 ml glycerol, 3 ml carbon disulfide, 34 ml olive oil; freely sol in benzene, acetone, slightly sol in petr ether. LD_{50} s.c. in mice: 1.6 mmoles/kg (Kutob, Plaa).

Incompat: Mercuric oxide, calomel, silver nitrate, tannin, balsam Peru directly mixed.

THERAP CAT: Anti-infective (topical).

THERAP CAT (VET): Antiseptic, disinfectant for superficial lesions and in the female reproductive tract.

5055. *o*-Iodohippurate Sodium. *N-(2-Iodobenzoyl)glycine monosodium salt;* sodium *o*-iodohippurate; Hippodin; Jodairol. $C_9H_7INNaO_3$; mol wt 327.05. C 33.05%, H

2.16%, I 38.80%, N 4.28%, Na 7.03%, O 14.68%. "Contains not less than 38.5% nor more than 39% iodine, when calcd to the dried substance." Prepd by the condensation of *o*-iodobenzoic acid with glycine, and treatment with NaOH soln: A. P. Sachs, U.S. pat. **2,135,474** (1938 to Zonite Prods.). Diagnostic use in post-transplant renography: H. Huland, K. Bischoff, *J. Urology* **129**, 925 (1983); J.-I. D. Jorgensen, J. Ladefoged, *Dan. Med. Bull.* **34**, 50 (1987).

Dihydrate, crystals. Freely sol in water; sol in alcohol and in dil solns of alkalies. An aq soln is neutral or faintly alkaline to litmus.

[131]I labelled form, *Hippuran I 131, Hipputope.*
[123]I labelled form, *Nephroflow.*

USE: Diagnostic aid (radiopaque medium); labelled form as diagnostic aid (radioactive imaging agent).

5056. *o*-Iodophenol. C_6H_5IO; mol wt 220.01. C 32.76%, H 2.29%, I 57.68%, O 7.27%. Prepd by the action of iodine on *o*-chloromercuriphenol in chloroform: Whitmore, Hanson, *Org. Syn.* **4**, 37 (1925).

Plates from ligroin, mp 43°. d^{80} 1.8757. bp_{160} 186-187°. Sol in alcohol, ether, chloroform, carbon disulfide, benzene; fairly sol in hot water.

5057. *p*-Iodophenol. C_6H_5IO; mol wt 220.01. C 32.76%, H 2.29%, I 57.68%, O 7.27%. Obtained by action of iodine in alkaline KI upon phenol; or by diazotization of *p*-aminophenol and subsequent replacement of the diazonium group by iodine.

White or reddish crystals; characteristic odor. mp 93-94°. Slightly sol in water, freely in alc, ether. *Protect from light.*

Note: The designation "phenol iodide", or more properly "iodized phenol", is applied to a soln of 1 part iodine in 4 parts phenol.

5058. Iodophors. Tamed Iodine. The term iodophor may be applied to any product in which surface active agents (such as nonoxynol, *q.v.*) or polymers (such as polyvinylpyrrolidone, *cf.* povidone-iodine) act as carriers and solubilizing agents for iodine. Other examples: *Biopal CVL-10; Dermevan; Idonyx; Iobac; Ioprep; Iosan; Kleenodyne; Rhudane; Showersan; Wescodyne; Westamine X.* An iodophor usually enhances the bactericidal activity of iodine, reduces vapor pressure and odor. Staining is almost nonexistent and wide dilution with water is possible. Examples of prepn: Shelanski, Winicov, U.S. pat. **2,710,277** (1955 to West Labs); Scheib *et al.*, U.S. pat. **2,977,315** (1961 to Lazarus Labs). Germicidal activity: C. A. Lawrence *et al.*, *J. Am. Pharm. Assoc.* **46**, 500 (1957). Reviews of manufacture, properties and use: D. R. Batey, *Aust. J. Dairy Technol.* **31**, 5-7 (1976); A. Leitmanova *et al.*, *Pharmazie* **42**, 809-813 (1987).

USE: Germicide.

THERAP CAT: Antiseptic; disinfectant.
THERAP CAT (VET): Antiseptic; disinfectant.

5059. Iodophthalein Sodium. Soluble iodophthalein; tetraiodophenolphthalein sodium; T.I.P.P.S.; tetraiodophthalein sodium; tetiothalein sodium; Iodeikon; Cholepulvis; Keraphen; Shadocol; Bilitrast; Iodognost; Stipolac; Tetraiode; Foriod; Iodtetragnost; Antinosin; Cholumbrin; Iodorayoral; Opacin; Photobiline; Piliophen; Videophel; Radiotetrane; Nosophene sodium; Iodophene sodium. $C_{20}H_8I_4Na_2O_4$; mol wt 865.88. C 27.74%, H 0.93%, I 58.62%,

Na 5.31%, O 7.39%. Isomeric with phentetiothalein sodium, *q.v.* Prepd by the action of iodine (in KI soln) on phenolphthalein in aq alk soln: Classen, Löb, *Ber.* **28**, 1603 (1895); **Ger.** pats. **85,930** (1894); **88,390** (1895). Also obtained by treating an alk soln of phenolphthalein with iodine monochloride: Kalle, **Ger.** pat. **143,596** (1900). Prepn of free acid: Orndorff, Mahood, *J. Am. Chem. Soc.* **40**, 937 (1918). Toxicity data: J. O. Hoppe *et al.*, *J. Med. Chem.* **13**, 997 (1970). Metabolism in rabbits: D. Pitre, L. Fumagalli, *Farmaco Ed. Sci.* **32**, 76 (1977).

Trihydrate, pale-blue to violet crystals. Saline and astringent taste. Somewhat hygroscopic. Gradually decomp on exposure to air due to absorption of CO_2 and becomes incompletely sol. One g dissolves in about 7 ml water, forming a clear, deep-blue dichroic soln. Slightly sol in alc. *Keep well closed.* LD_{50} in mice (mg/kg): 360 i.v., 3800 orally (Hoppe).

Free acid, $C_{20}H_{10}I_4O_4$, *iodophthalein, tetraiodophthalein, Iodophene, Nosophen.* Light-yellow, odorless, tasteless powder; dec at about 200°. Practically insol in water. Slightly sol in alcohol; sol in alkalies, chloroform, ether.

THERAP CAT: Diagnostic aid (radiopaque medium—cholecystographic).

5060. Iodopsin. One of 3 visual pigments found in retinal cone cells. Absorption maximum approx 560 nm. Biological activity is similar to that of rhodopsin, *q.v.* Composed of the chromophore, 11-*cis* retinal, *q.v.*, bound to photopsin, the specific protein component of cone pigments. (*See* Opsins). Isoln from chicken retinas: G. Wald, *Nature* **140**, 545 (1937). Prepn from 11-*cis* retinal and opsin: G. Wald *et al.*, *J. Gen. Physiol.* **38**, 623 (1955). Methods for purification, prepn, and assay: R. Hubbard *et al.*, "Methodology of Vitamin A and Visual Pigments", in *Methods in Enzymology* Vol. **18**, D. B. McCormick, L. D. Wright, Eds. (Academic Press, New York, 1971) pp 615-653. Exposure to light initiates the conversion of iodopsin through a series of distinct intermediates to yield opsin and *trans*-retinal. Photochemistry: R. Hubbard, A. Kropf, *Nature* **183**, 448 (1959); T. Yoshizawa, G. Wald, *ibid.* **214**, 566 (1967). Studies on the iodopsin binding site: H. Matsumoto *et al.*, *Biochim. Biophys. Acta* **404**, 300 (1975). Three distinct pigments are responsible for color vision: W. B. Marks *et al.*, *Science* **143**, 1181 (1964); P. K. Brown, G. Wald, *ibid.* **144**, 45 (1964). The variation in the absorption maximum of the 3 pigments is regulated by differences in the apoprotein portion of the molecule: R. Hubbard, L. Sperling, *Exp. Eye Res.* **17**, 581 (1973); B. Honig *et al.*, *J. Am. Chem. Soc.* **101**, 7084 (1979). *Reviews:* G. Wald, *Science* **162**, 230-239 (1968); H. Matsumoto, T. Yoshizawa, *Methods in Enzymology* Vol. **81**, L. Packer, Ed. (Academic Press, New York, 1982) pp 154-160. General review of the visual process: P. S. Zurer, *Chem. & Eng. News* **61**, 24-35 (Nov. 28, 1983).

5061. Iodopyracet. 3,5-Diiodo-4-oxo-1(4H)-pyridineacetic acid compound with 2,2'-iminobis[ethanol] (1:1); bis-(hydroxyethyl)ammonium 3,5-diiodo-4-pyridone-*N*-acetate; diethanolamine 3,5-diiodo-4-pyridone-*N*-acetate; 3,5-diiodo-4-pyridone-*N*-acetic acid diethanolamine salt; diodone; RP-3203; Diatrast; Diodrast; Iopyracil; Nosylan; Neo-Methiodal; Neo-Skiodan; Neo-Tenebryl; Nosydrast; Oparenol; Per-Abrodil; Per-Radiographol; Pyelosil; Pylumbrin; Savac; Umbradil; Uriodone; Vasiodone; Xumbradil. $C_{11}H_{16}I_2N_2O_5$; mol wt 510.07. C 25.90%, H 3.16%, I 49.76%, N 5.49%, O 15.68%. Prepn: **Fr.** pat. **728,634** (1931 to I. G. Farbenhind AG); J. Reitmann, U.S. pat. **1,993,039** (1935 to Winthrop). Pharmacology, toxicology and radiographic studies: R. St. A. Heathcote, R. A. Gardner, *Brit. J. Radiol.* **6**, 304

(1933). HPLC determn in biological fluids: P. Hekman, C. A. M. Van Ginneken, *J. Chromatog.* **182**, 492 (1980).

Odorless powder, dec 155-157°. Soly: about 36% in water; about 12% in methanol, much less sol in ice-cold methanol. Practically insol in acetone, ether, chloroform. pH of aq solns: 5 to 8, depending on concn. 3,5-Diiodo-4-pyridone-*N*-acetic acid separated by pptn with dil HCl and dried at 100°, mp 245-249°, and contains not less than 61.5% and not more than 63.5% iodine: *N.F.* **XI.**

Iodopyracet Injection is a 35% (w/v) soln of the diethanolamine salt of 3,5-diiodo-4-pyridone-*N*-acetic acid. Contains about 22% I (w/v). d_4^{25} 1.185. Neutral to litmus.

Iodopyracet Concentrated Solution; Diodrast Concentrated Solution is a 70% soln (w/v) of the diethanolamine salt of 3,5-diiodo-4-pyridone-*N*-acetic acid. Contains about 44% I (w/v). Neutral to litmus.

Iodopyracet Compound Solution; Diodrast Compound Solution is a soln contg approx 40.5% (w/v) of the diethanolamine salt and approx 9.5% (w/v) of the diethylamine salt of 3,5-diiodo-4-pyridone-*N*-acetic acid. Contains about 25% I. Neutral to litmus. d_4^{25} 1.270.

THERAP CAT: Diagnostic aid (radiopaque medium—urographic).

THERAP CAT (VET): Diagnostic aid (radiopaque medium).

5062. Iodopyrrole. *2,3,4,5-Tetraiodo-1H-pyrrole;* Iodol. C_4HI_4N; mol wt 570.68. C 8.42%, H 0.18%, I 88.95%, N 2.45%. Prepd by iodination of pyrrole: Potts, *J. Chem. Soc.* **1953**, 3711; Treibs, Kolm, *Ann.* **614**, 176 (1958). Isoln from shale-oil naphtha: Janssen *et al., J. Am. Chem. Soc.* **73**, 4040 (1951).

Needles from ethanol, dec 162-164°, also reported as dec 150-160° and 140-150°. One gram dissolves in 4900 ml water, 9 ml alcohol, 1.5 ml ether, 105 ml chloroform, 155 ml glycerol; sol in fixed oils.

THERAP CAT: Antiseptic (local).

THERAP CAT (VET): Antiseptic, disinfectant for superficial lesions and in the female reproductive tract.

5063. Iodoquinol. *5,7-Diiodo-8-quinolinol;* diiodohydroxyquin; diiodo-oxyquinoline; 5,7-diiodo-8-hydroxyquinoline; SS-578; Diodoquin; Disoquin; Floraquin; Dyodin; Dinoleine; Searlequin; Diodoxylin; Rafamebin; Ioquin; Direxiode; Stanquinate; Yodoxin; Zoaquin; Enterosept; Embequin. $C_9H_5I_2NO$; mol wt 396.95. C 27.23%, H 1.27%, I 63.94%, N 3.53%, O 4.03%. Prepd by the action of iodine monochloride on 8-hydroxyquinoline: Papesch, Burtner, *J. Am. Chem. Soc.* **58**, 1314 (1936); by the action of KIO_3 on 8-hydroxyquinoline: Zeifman, *C.A.* **34**, 3745. Electrolytic prepn: Brown, Berkowitz, *Trans. Electrochem. Soc.* **75**, 385 (1939). *See also* Claus, **Ger.** pat. **78,880;** Passek, **Ger.** pat. **411,050;** Matsumura, *C.A.* **21,** 1461 (1927); Pirrone, Cherubino, *C.A.* **28**, 3073 (1934).

Crystals from xylene. The medicinal grade is a yellowish-brown powder. mp 200-215° (extensive decompn). Almost insol in water. Sparingly sol in alcohol, ether, and acetone; sol in hot pyridine and in hot dioxane.

THERAP CAT: Antiamebic.

5064. N-Iodosuccinimide. Succiniodimide; NIS. $C_4H_4INO_2$; mol wt 224.99. C 21.35%, H 1.79%, I 56.41%, N 6.23%, O 14.22%. Prepd by iodination of silver succinimide: Djerassi, Lenk, *J. Am. Chem. Soc.* **75**, 3493 (1953).

Needles from dioxane + carbon tetrachloride. mp 200-201°. Sol in acetone, methanol; moderately sol in dioxane. Practically insol in CCl_4, ether. Dec in water. *Protect from light and avoid contact with skin and mucous membranes.*

USE: Iodination of ketones and aldehydes.

5065. 3-Iodotyrosine. Monoiodotyrosine. $C_9H_{10}INO_3$; mol wt 307.09. C 35.20%, H 3.28%, I 41.33%, N 4.56%, O 15.63%. Prepn of L-form: Hillman, Hillman-Elies, *Z. Physiol. Chem.* **305**, 177 (1956); Pitt-Rivers, *Chem. & Ind. (London)* **1956**, 21. Prepn of L- and DL-form: Harington, Rivers, *Biochem. J.* **38**, 320 (1944).

L-Form, crystals from water, dec 204-206°. $[\alpha]_D^{20}$ —4.4° (c = 5 in 1N HCl). Dissolves in 15 parts boiling water.

DL-Form monohydrate, crystals from water, dec 200-201°.

5066. Iofetamine ^{123}I. (±)-*4-(Iodo-^{123}I)-α-methyl-N-(1-methylethyl)benzeneethanamine;* (±)-*p-iodo-^{123}I-N-isopropyl-α-methylphenethylamine;* [^{123}I](±)-*N-isopropyl-p-iodoamphetamine.* $C_{12}H_{18}^{123}IN.$ Lipid soluble radioactive brain imaging agent. Prepn: R. M. Baldwin *et al.,* **Eur.** pat. **Appl. 11,858** (1980 to Hoffmann-La Roche); *eidem,* **U.S.** pat. **4,360,511** (1982 to Medi-Physics). Detailed synthesis: L. Carlsen, K. Andresen, *Eur. J. Nucl. Med.* **7**, 280 (1982). Modified synthesis for improved yields: A. Najafi, *J. Labelled Compd. Radiopharm.* **24**, 1167 (1987). Preparative HPLC purification: J. Mertens *et al., Nucl. Med. Commun.* **5**, 705 (1984). Brain uptake and localization: H. S. Winchell *et al., J. Nucl. Med.* **21**, 940 (1980); H. S. Winchell *et al., ibid.* 947. Distribution kinetics in animals: P. Som *et al., Int. J. Nucl. Med. Biol.* **12**, 185 (1985); in humans: B. L. Holman *et al., J. Nucl. Med.* **25**, 25 (1984). Diagnostic use for cerebral function in Alzheimer's disease: K. A. Johnson *et al., Arch. Neurol.* **45**, 392 (1988). Reviews of radiopharmaceutical production and diagnostic use with single photon emission computed tomography (SPECT) imaging: M. B. Cohen *et al., Appl. Radiat. Isot.* **37**, 749-763 (1986); of diagnostic applications in stroke: C. H. Park *et al., RadioGraphics* **8**, 305-326 (1988); of metabolism and kinetics: R. M. Baldwin, J.-L. Wu, *J. Nucl. Med.* **29**, 122-124 (1988).

$bp_{0.05}$ 91-92°. Also reported as $bp_{0.6}$ 98-99° (Najafi). Hydrochloride, $C_{12}H_{18}^{123}IN.HCl,$ *IMP, ^{123}I labeled IMP, ^{123}I-M123, Perfusamine, Spectamine.* Crystals from ether, mp 156-158°.

THERAP CAT: Diagnostic aid (radioactive imaging agent).

5067. Ioglycamic Acid. *3,3'-[Oxybis[(1-oxo-2,1-ethanediyl)imino]]bis[2,4,6-triiodobenzoic acid]; 3,3'-[oxybis(methylenecarbonylimino)]bis[2,4,6-triiodobenzoic acid]; 3,3'-(diglycoloyldiimino)bis(2,4,6-triiodobenzoic acid);* diglycolic acid bis(2,4,6-triiodo-3-carboxanilide); N,N-(oxydiacetyl)-bis[3-amino-2,4,6-triiodobenzoic acid]. $C_{18}H_{10}I_6N_2O_7$; mol wt 1127.71. C 19.17%, H 0.89%, I 67.52%, N 2.48%, O 9.93%. Prepn: Priewe et al., Ber. **87**, 651 (1954); Priewe, Rutkowski, U.S. pats. **2,776,241**; **2,853,424** (1957, 1958 both to Schering AG).

Occurs in three cryst modifications: Neudert, Röpke, Helv. Chim. Acta **41**, 855 (1958). mp with baking at 222°, sintering at 227°, splitting of iodine at 245°, dec 281°.

Meglumine salt, $C_{18}H_{10}I_6N_2O_7 \cdot 2(C_7H_{17}NO_5)$, dimeglumine ioglycamide, Biligram. Water soluble.

THERAP CAT: Diagnostic aid (radiopaque medium—cholecystographic).

5068. Iohexol. *5-[Acetyl(2,3-dihydroxypropyl)amino]-N,N'-bis(2,3-dihydroxypropyl)-2,4,6-triiodo-1,3-benzenedicarboxamide; N,N'-bis(2,3-dihydroxypropyl)-5-[N-(2,3-dihydroxypropyl)acetamido]-2,4,6-triiodoisophthalamide;* Win-39424; Compd 545; Omnipaque; Exypaque. $C_{19}H_{26}I_3N_3O_9$; mol wt 821.14. C 27.79%, H 3.19%, I 46.36%, N 5.12%, O 17.54%. Nonionic radio-contrast medium. Prepn: V. Nordal, H. Holtermann, Ger. pat. **2,726,196** corresp to U.S. pat. **4,250,113** (1977, 1981 both to Nyegaard). Pharmacology and toxicology: Acta Radiol. (Stockholm) Suppl. **362**, 1-134 (1980). Acute toxicity: S. Salvesen, ibid. 73. Fibrillatory potential in dogs: G. L. Wolf et al., Invest. Radiol. **16**, 320 (1981). Comparative clinical studies in coronary angiography: G. B. J. Mancini et al., Am. J. Cardiol. **51**, 1218 (1983); I. D. Sullivan et al., Br. Heart J. **51**, 643 (1984); M. A. Bettmann et al., Radiology **153**, 583 (1984). Review: T. Almén, Acta Radiol. (Stockholm) Suppl. **366**, 9-19 (1983).

Crystals from butanol, mp 174-180°. Sol in water. Stable in aqueous solutions. Viscosity (cP): 6.2 at 37°; 12.6 at 20° (c = 200 mg Iodine/ml). LD_{50} in male, female rats, mice (g Iodine/kg): 15.0, 12.3, 24.3, 25.1 i.v. (Salvesen).

THERAP CAT: Diagnostic aid (radiopaque medium).

5069. Iomeglamic Acid. *5-[(3-Amino-2,4,6-triiodophenyl)methylamino]-5-oxopentanoic acid; 3'-amino-2',4',6'-triiodo-N-methylglutaranilic acid; N-methyl-N-(3-amino-2,4,6-triiodophenyl)glutaramic acid; N-methyl-N-(2,4,6-triiodo-3-aminophenyl)glutaramidic acid;* RG-270; Falignost. $C_{12}H_{13}I_3N_2O_3$; mol wt 613.96. C 23.48%, H 2.13%, I 62.01%, N 4.56%, O 7.82%. Prepn: Cassebaum, Dierbach, East Ger. pat. **67,209** (1969), C.A. **72**, 66654j (1970). Biotransformation: Pfeifer et al., Pharmazie **27**, 403 (1972). Pharmacology: H. Bekker et al., ibid. 411. Clinical evaluation: Barke, Zbl. Pharm. Pharmakother. Laboratoriumsdiagn. **110**, 1117 (1971).

Yellow to brown crystals from acetic acid, mp 169°. Freely sol in DMF; slightly sol in ethanol, chloroform. Practically insol in water. LD_{50} i.v. in mice: 500 mg/kg (Bekker).

THERAP CAT: Diagnostic aid (radiopaque medium—cholecystographic).

5070. Ionone. Irisone. $C_{13}H_{20}O$; mol wt 192.30. C 81.20%, H 10.48%, O 8.32%. Mixture in which the major constituents are α-ionone, *(E)-4-(2,6,6-trimethyl-2-cyclohexen-1-yl)-3-buten-2-one,* and β-ionone, *(E)-4-(2,6,6-trimethyl-1-cyclohexen-1-yl)-3-buten-2-one.* The latter is a key intermediate in the synthesis of vitamin A. Isoln from the volatile oil of *Boronia megastigma* Nees., Rutaceae: Naves, Parry, Helv. Chim. Acta **30**, 419 (1946). Prepn by condensing citral with acetone: Tremann, Schmidt, Ber. **33**, 3703 (1900); Hibbert, Cannon, J. Am. Chem. Soc. **46**, 119 (1924); Krishna, Joshi, J. Org. Chem. **22**, 224 (1957). Prepn from isobutylene and vinyl methyl ketone: Pasedach, Seefelder, Ger. pat. **1,000,374** (1957 to BASF), C.A. **54**, 1595g (1960). Prepn by cyclization of pseudoionone with H_2SO_4: Kimel et al., J. Org. Chem. **22**, 1611 (1957); Kaiser, Kimel, U.S. pat. **2,877,271** (1959 to Hoffmann-La Roche). Review: Naves, J. Soc. Cosmet. Chem. **22**, 439 (1971).

α-Form   β-Form

Liquid. Odor reminiscent of cedar wood; in very dilute alcoholic soln it resembles the odor of violets. d_{25}^{25} 0.933-0.967. n_D^{20} 1.503-1.508. bp_{12} 126-128°. uv max: α-ionone 228.5 nm (ε 14300); β-ionone 293.5 nm (ε 8700). Miscible with abs alc. Soluble in 2 to 3 parts of 70% alcohol, ether, chloroform, benzene. Very slightly sol in water.

Caution: May cause allergic reactions, Shultheiss, Dermatol. Wochenschr. **135**, 629 (1967).

USE: In perfumery.

5071. Iopamidol. *(S)-N,N'-Bis[2-hydroxy-1-(hydroxymethyl)ethyl]-5-[(2-hydroxy-1-oxopropyl)amino]-2,4,6-triiodo-1,3-benzenedicarboxamide; (S)-N,N'-bis[2-hydroxy-1-(hydroxymethyl)ethyl]-2,4,6-triiodo-5-lactamidoisophthalamide; 5-(α-hydroxypropionylamino)-2,4,6-triiodoisophthalic acid di(1,3-dihydroxyisopropylamide); iomapidol;* B-15000; SQ-13396; Iopamiro; Iopamiron; Isovue; Jopamiro; Niopam; Solutrast. $C_{17}H_{22}I_3N_3O_8$; mol wt 777.09. C 26.28%, H 2.85%, I 48.99%, N 5.41%, O 16.47%. Non-ionic radiocontrast medium. Prepn: E. Felder, D. Pitre, Ger. pat. **2,547,789**; eidem, U.S. pat. **4,001,323** (1976, 1977 both to Savac). Physicochemical properties, preclinical studies: E. Felder et al., Farmaco Ed. Sci. **32**, 835 (1977). Determn of optical purity: eidem, Farmaco Ed. Prat. **32**, 3 (1982). In vitro effects: B. Schulze, H. K. Beyer, Arzneimittel-Forsch. **31**, 1067 (1981). Development, initial testing: M. Sovak et al., Radiology **142**, 115 (1982). Clinical studies in angiography: A. J. Molyneux, P. W. Sheldon, Brit. J. Radiol. **55**, 117 (1982); G. H. Whitehouse, S. L. Snowdon, Clin. Radiol. **33**, 231 (1982); in myelography: B. Drayer et al., Am. J. Neuroradiol. **3**, 59 (1982). Comprehensive description: E. Felder et al., in Analytical Profiles of Drug Substances vol. 17, K. Florey, Ed. (Academic Press, New York, 1988) pp 115-154.

White, odorless cryst powder. Dec at about 300° without melting. $[\alpha]_D^{20}$ -2.01° (c = 10 in water). pKa (25°) 10.70. Very sol in water, methanol. Practically insol in chloroform. Miscible with boiling ethanol. LD$_{50}$ in mice, rats, rabbits, dogs (g/kg): 44.5, 28.2, 19.6, 34.7 i.v. (Felder). Also reported as: LD$_{50}$ in mice (mg iodine/kg body wt): 21,800 i.v.; 20,000 i.p.; 1500 intracerebral (Felder, Pitre).

THERAP CAT: Diagnostic aid (radiopaque medium).

5072. Iopanoic Acid. *3-Amino-α-ethyl-2,4,6-triiodobenzenepropanoic acid; 3-amino-α-ethyl-2,4,6-triiodohydrocinnamic acid;* iodopanoic acid; 3-(3-amino-2,4,6-triiodophenyl)-2-ethylpropanoic acid; 3-(3-amino-2,4,6-triiodophenyl)-2-ethylpropionic acid; Cistobil; Colepax; Telepaque; Teletrast. $C_{11}H_{12}I_3NO_2$; mol wt 570.94. C 23.14%, H 2.12%, I 66.68%, N 2.45%, O 5.60%. Oral radiocontrast medium. Prepn: S. Archer, U.S. pat. **2,705,726** (1955 to Sterling Drug). Optical resolution: Pitré, Boveri, *J. Med. Chem.* **11**, 406 (1968). Toxicity data: P. Tirone, G. Rosati, *Farmaco Ed. Prat.* **31**, 397 (1976). Review of pharmacology and clinical efficacy in oral cholecystography: R. N. Berk *et al., N. Engl. J. Med.* **290**, 204 (1974). Comprehensive description: D. Pitre in *Analytical Profiles of Drug Substances* vol. **14**, K. Florey, Ed. (Academic Press, New York, 1985) pp 181-206.

dl-Form, cream-colored solid, mp 155.2-157°. pKa 4.8. Insol in water. Sol in dil alkali, in 95% alc, in other organic solvents. LD$_{50}$ in mice, rats (mg/kg): 285, 320 i.v.; 1540, 2870 orally (Tirone).

l-Form, crystals, mp 162-163°. $[\alpha]_D^{20}$ -5.2° ± 0.1° (c = 2 in ethanol).

d-Form, crystals, mp 162°. $[\alpha]_D^{20}$ +5.1 ± 0.1° (c = 2 in ethanol).

Sodium salt, *Bilijodon-Natrium.*

THERAP CAT: Diagnostic aid (radiopaque medium—cholecystographic).

5073. Iopentol. *5-[Acetyl(2-hydroxy-3-methoxypropyl)-amino]-N,N′-bis(2,3-dihydroxypropyl)-2,4,6-triiodo-1,3-benzenedicarboxamide; N,N′*-bis(2,3-dihydroxypropyl)-5-[*N*-(2-hydroxy-3-methoxypropyl)acetamido]-2,4,6-triiodoisophthalamide; Imagopaque. $C_{20}H_{28}I_3N_3O_9$; mol wt 835.17. C 28.76%, H 3.38%, I 45.59%, N 5.03%, O 17.24%. Nonionic, water soluble vascular contrast medium. Prepn: K.

Wille, Eur. pat. **Appl. 105,752** (1984 to Nyegaard A/S), *C.A.* **101**, 72456e (1984). Synthesis and preparative LC purification: W. Skjold, A. Berg, *J. Chromatog.* **366**, 299 (1986). Effects on blood-brain barrier in animals: A. A. Michelet, *Acta Radiol.* **28**, 329 (1987).

Osmolality (300 mg I/ml): 0.64 mol/kg H_2O. Viscosity at 20° (300 mg I/ml): 13.2 mPa. Partition coefficient in 1-octanol/water: 0.007.

THERAP CAT: Diagnostic aid (radiopaque medium).

5074. Iophendylate. *4-Iodo-iota-methylbenzenedecanoic acid ethyl ester;* ethyl 10-(*p*-iodophenyl)undecylate; ethyl 10-(*p*-iodophenyl)hendecanoate. $C_{19}H_{29}IO_2$; mol wt 416.34. C 54.81%, H 7.02%, I 30.48%, O 7.69%. Chief component of *Ethiodan, Mulsopaque, Myodil, Neurotrast, Pantopaque;* may also contain other isomers including 11-(*p*-iodophenyl)undecylate. Prepn: W. H. Strain *et al., J. Am. Chem. Soc.* **64**, 1436 (1942); *eidem,* U.S. pat. **2,348,231** (1944 to Noned Corp. and Kodak); W. Baker *et al., J. Soc. Chem. Ind.* **63**, 223 (1944). Pharmacology and toxicity: T. B. Steinhausen *et al., Radiology* **43**, 230 (1944). Clinical evaluation of diagnostic use in myelography: G. H. Ramsey *et al., ibid.* 236. Clinical comparison with metrizamide, *q.v.,* in diagnostic radiology: S. A. Keiffer *et al., ibid.* **129**, 695 (1978). Evaluation of use in MR imaging of the spine: I. F. Braun *et al., Am. J. Neuroradiol.* **7**, 997 (1986); A. K. Anand *et al., Comput. Radiol.* **11**, 165 (1987).

Viscous liquid. Darkens slowly in air. d$_{20}^{20}$ 1.240-1.263. n_D^{25} 1.5230-1.5280. Saponif equiv 395-420. Very slightly sol in water. Freely sol in alcohol, benzene, chloroform, ether. LD$_{50}$ in mice, rats (g/kg): 4.6, 19 i.p. (Steinhausen).

THERAP CAT: Diagnostic aid (radiopaque medium).

THERAP CAT (VET): Diagnostic aid (radiopaque medium).

5075. Iophenoxic Acid. *α-Ethyl-3-hydroxy-2,4,6-triiodobenzenepropanoic acid; α-ethyl-3-hydroxy-2,4,6-triiodohydrocinnamic acid;* α-(2,4,6-triiodo-3-hydroxybenzyl)butyric acid; α-ethyl-β-(3-hydroxy-2,4,6-triiodophenyl)propionic acid; α-ethyl-β-(2,4,6-triiodo-3-hydroxyphenyl)propionic acid; triiodoethionic acid; Teridax. $C_{11}H_{11}I_3O_3$; mol wt 571.92. C 23.10%, H 1.94%, I 66.57%, O 8.39%. Prepn: Papa *et al., J. Am. Chem. Soc.* **75**, 1107 (1953); **Brit.** pat. **726,987** (1955 to Sterling Drug). Toxicity data: J. O. Hoppe *et al., J. Med. Chem.* **13**, 997 (1970).

Crystals from benzene + petr ether, mp 143-144°. LD$_{50}$ i.v. in mice: 374 mg/kg (Hoppe).

THERAP CAT: Diagnostic aid (radiopaque medium—cholecystographic).

5076. Iopromide. *N,N′-Bis(2,3-dihydroxypropyl)-2,4,6-triiodo-5-[(methoxyacetyl)amino]-N-methyl 1,3-benzenedicarboxamide; N,N′*-bis(2,3-dihydroxypropyl)-2,4,6-triiodo-5-(2-methoxyacetamido)-N-methylisophthalamide; 5-methoxyacetylamino-2,4,6-triiodoisophthalic acid [(2,3-dihydroxy-N-methylpropyl)-(2,3-dihydroxypropyl)]diamide; Ultravist. $C_{18}H_{24}I_3N_3O_8$; mol wt 791.12. C 27.33%, H 3.06%, I 48.12%, N 5.31%, O 16.18%. Nonionic, injectable radiocontrast medium. Prepn: U. Speck *et al.,* **Ger.** pat. **2,909,-439**; *eidem,* U.S. pat. **4,364,921** (1980, 1982 both to Schering AG). Physicochemical properties and pharmacological profile: W. Muetzel, U. Speck, *Am. J. Neuroradiol.* **4**, 350 (1983); P. Dawson *et al., Acta Radiol. Diagn.* **25**, 253 (1984). Clinical experience: K.-J. Wolf *et al., ibid.* **24**, 55 (1983).

Colorless solid. Stable in aqueous solutions. Viscosity (cP): 4.8 at 37°; 10.2 at 20° (c = 300 mg iodine/ml). LD_{50} in mice, rats (g iodine/kg body weight): 16.5, 11.4 i.v. (Muetzel).

THERAP CAT: Diagnostic aid (radiopaque medium).

5077. Iopronic Acid. *2-[[2-[3-(Acetylamino)-2,4,6-tri-iodophenoxy]ethoxy]methyl]butanoic acid;* (±)-2-[[2-[3-(3-acetamido-2,4,6-triiodophenoxy)ethoxy]methyl]butyric acid; 3-[2-(3-acetamido-2,4,6-triiodophenoxy)ethoxy]-2-ethyl-propionic acid; B-11420; SQ-21983; Bilimiro; Bilimiron; Oravue; Videobil. $C_{15}H_{18}I_3NO_5$; mol wt 673.03. C 26.77%, H 2.70%, I 56.57%, N 2.08%, O 11.89%. Oral cholecysto-graphic agent. Prepn: E. Felder, D. Pitre, **Ger.** pat. **2,128,-902;** *eidem,* **U.S.** pat. **3,842,124** (1972, 1974 both to Bracco); E. Felder *et al., Farmaco Ed. Sci.* **31,** 349 (1976). Pharmacology, toxicology: P. Tirone, G. Rosati, *Farmaco Ed. Prat.* **31,** 397, 437 (1976). Series of articles on metabolism: *ibid.* 516-546; 755. Intestinal absorption: J. R. Amberg *et al., Invest. Radiol.* **15,** S136 (1980). Effect on bile flow and composition: J. L. Barnhart *et al., ibid.* S124. Toxicity: P. Tirone, G. Rosati, *Farmaco Ed. Prat.* **31,** 397 (1976).

Crystals from 50% ethanol, mp 130°. LD_{50} in mice, rats, dogs (mg/kg): 1950, 5650, >3000 orally; 1090, 1000, 835 i.v. (Tirone).

THERAP CAT: Diagnostic aid (radiopaque medium—cholecystographic).

5078. Iopydol. *1-(2,3-Dihydroxypropyl)-3,5-diiodo-4(1H)-pyridinone.* $C_8H_9I_2NO_3$; mol wt 420.97. C 22.83%, H 2.15%, I 60.29%, N 3.33%, O 11.40%. Prepn: Reitmann, **Ger.** pat. **579,224** (1933 to I.G. Farbenind.), *C.A.* **27,** 4880 (1933).

Solid, mp 161°.

THERAP CAT: Diagnostic aid (radiopaque medium—bronchographic).

5079. Iopydone. *3,5-Diiodo-4(1H)-pyridinone.* $C_5H_3I_2$-NO; mol wt 346.89. C 17.31%, H 0.87%, I 73.17%, N 4.04%, O 4.61%. Prepn: Dohrn, Diedrich, *Ann.* **494,** 284 (1932).

Crystals, dec 321°. Practically insol in ordinary solvents. Sol in caustic alkalies.

THERAP CAT: Diagnostic aid (radiopaque medium—bronchographic).

5080. Iothalamic Acid. *3-(Acetylamino)-2,4,6-triiodo-5-[(methylamino)carbonyl]benzoic acid;* 5-acetamido-2,4,6-tri-iodo-N-methylisophthalamic acid. $C_{11}H_9I_3N_2O_4$; mol wt 613.92. C 21.52%, H 1.48%, I 62.01%, N 4.56%, O 10.42%. Synthesis: Hoey *et al., J. Med. Chem.* **6,** 24 (1963).

Crystals, dec about 285°.

N-Methylglucamine salt, $C_{18}H_{26}I_3N_3O_9$, *meglumine iothalamate, Conray, Contrix "28", Cysto-Conray.*

Sodium salt, $C_{11}H_8I_3N_2NaO_4$, *sodium iothalamate, Angio-Conray, Angio-Contrix "48", Conray-400, Medio-Contrix "38".*

[131]I labeled sodium salt, *Glofil-131.*

THERAP CAT: Diagnostic aid (radiopaque medium).

5081. Iothion. *1,3-Diiodo-2-propanol;* iopropane; 1,3-diiodo-2-hydroxypropane; α-diiodohydrin; 1,3-diiodoiso-propyl alcohol. $C_3H_6I_2O$; mol wt 311.89. C 11.55%, H 1.94%, I 81.38%, O 5.13%. $CH_2ICH(OH)CH_2I$. Prepn from dichlorohydrin and KI: Claus, *Ann.* **168,** 24 (1873); from glycerol, HCl and NaI: Lusignani, *Boll. Chim. Farm.* **78,** 557 (1939), *C.A.* **34,** 2322 (1940).

Yellowish, heavy, oily liquid. d 2.4-2.5. One gram dissolves in 80 ml water, 20 ml glycerol; freely soluble in alcohol, benzene, $CHCl_3$, ether, carbon disulfide, oils; dec by alkalies.

5082. Iotrolan. *5,5'-[(1,3-Dioxo-1,3-propanediyl)bis-(methylimino)]bis[N,N'-bis[2,3-dihydroxy-1-(hydroxymethyl)-propyl]-2,4,6-triiodo-1,3-benzenedicarboxamide];* 5,5'-[malon-ylbis(methylimino)]bis[N,N'-bis[2,3-dihydroxy-1-(hydroxy-methyl)propyl]-2,4,6-triiodoisophthalamide]; iotrol; DL-3117; SH-437; ZK-39482; Isovist. $C_{37}H_{48}I_6N_6O_{18}$; mol wt 1626.24. C 27.33%, H 2.98%, I 46.82%, N 5.17%, O 17.71%. Nonionic, isotonic contrast medium. Prepn: M. Sovak, R. Ranganathan, **Eur.** pat. **Appl. 33,426;** *eidem,* **U.S.** pat. **4,341,756** (1981, 1982 both to Univ. Cal., Berkeley). Neurotoxicity study: M. Sovak *et al., Radiology* **142,** 115 (1982). Neuropharmacology: M. Sovak *et al., Am. J. Neuroradiol.* **4,** 319 (1983); and cerebral distribution: J.-C. Castel *et al., Neuroradiol.* **29,** 206 (1987). Arthrographic evaluation in animals: J. Guerra *et al., Invest. Radiol.* **19,** 228 (1983). Cardiovascular effects in animals: P. Lanzer *et al., ibid.* **20,** 746 (1985). Clinical study of patient tolerance and diagnostic adequacy for intrathecal use: B. Hammer, E. Deisenhammer, *Neuroradiol.* **27,** 337 (1985); B. Hoffmann *et al., ibid.* **29,** 380 (1987); M. D. Malnor *et al., Radiology* **158,** 845 (1986).

Colorless glassy solid. Soly at 20° (mgI/ml): water >400. LD_{50} in mice, rats (gI/kg): >26, 12.7 i.v. (Sovak, 1982).

THERAP CAT: Diagnostic aid (radiopaque medium).

5083. Ioversol. *N,N'-Bis(2,3-dihydroxypropyl)-5-[(hydr-oxyacetyl)(2-hydroxyethyl)amino]-2,4,6-triiodo-1,3-benzenedi-*

Consult the Name Index before using this section.

carboxamide; *N,N'*-bis(2,3-dihydroxypropyl)-5-[*N*-(2-hydroxyethyl)glycolamido]-2,4,6-triiodoisophthalamide; MP-328; Optiray. $C_{18}H_{24}I_3N_3O_9$; mol wt 807.12. C 26.79%, H 3.00%, I 47.17%, N 5.21%, O 17.84%. Nonionic, low osmolality, radiographic contrast agent. Prepn: Y. Lin, **Eur. pat. Appl. 83,964;** *idem,* **U.S. pat. 4,396,598** (both 1983 to Mallinckrodt). Clinical pharmacokinetics: R. A. Wilkins *et al., Invest. Radiol.* **24,** 781 (1989). Comparative clinical study in intravenous urography: A. J. Kaufman *et al., Urol. Radiol.* **12,** 56 (1990). Toxicity studies: W. H. Ralston *et al., Invest. Radiol.* **24,** Suppl. 1, S2 (1989). Symposium on pharmacology, toxicology and diagnostic use in contrast enhanced computed tomographic scanning and angiography: *ibid.* S1-S76.

mp 186-198°. Soly in water > 125% wt/vol. Octanol + water partition coefficient: 4×10^{-4} (c = 0.01 mgI/ml). Osmolality: 702 mOsm/kg H_2O (c = 320 mgI/ml), Viscosity (cps): 11.6 at 20°; 9.9 at 25°; 5.8 at 37° (c = 320 mgI/kg). d^{37} 1.371 (c = 320 mgI/ml). LD_{50} in mice, rats, rabbits, dogs (gI/kg): 17, 15, > 25, > 12 i.v.; in male, female rats (mgI/kg): 1000, > 1200 intracisternally (Ralston).

THERAP CAT: Diagnostic aid (radiopaque medium).

5084. Ioxaglic Acid. *3-[[[[3-(Acetylmethylamino)-2,4,6-triiodo-5-[(methylamino)carbonyl]benzoyl]amino]acetyl]-amino]-5-[[(2-hydroxyethyl)amino]carbonyl]-2,4,6-triiodobenzoic acid; N-(2-hydroxyethyl)-2,4,6-triiodo-5-[2-[2,4,6-triiodo-3-(N-methylacetamido)-5-(methylcarbamoyl)benzamido]acetamido]isophthalamic acid;* P-286. $C_{24}H_{21}I_6N_5O_8$; mol wt 1268.89. C 22.72%, H 1.67%, I 60.01%, N 5.52%, O 10.09%. Ionic contrast medium of low osmolality. Prepn: G. Tilly *et al.,* **Ger. pat. 2,523,567;** *idem,* **U.S. pat. 4,014,986** (1975, 1977 both to Guerbet). Effect on peripheral arterial blood flow in dogs: R. M. Steiner *et al., Clin. Radiol.* **31,** 621 (1980). Effect on thyroid function: R. G. Grainger, G. W. Pennington, *Brit. J. Radiol.* **54,** 768 (1981). Clinical study in femoral angiography: S. Suzuki *et al., Acta Radiol.* **23,** 87 (1982); in cardiac radiology: R. L. Feldman, *Am. J. Cardiol.* **61,** 1334 (1988). Toxicity study: M. Sovak *et al., Invest. Radiol.* **16,** 438 (1981). Mutation study: W. Hadnagy *et al., Mutat. Res.* **104,** 249 (1982).

Mixture of *ioxaglate meglumine* and *ioxaglate sodium,* MP-302, Hexabrix.

THERAP CAT: Diagnostic aid (radiopaque medium).

5085. Ioxilan. *5-[Acetyl(2,3-dihydroxypropyl)amino]-N-(2,3-dihydroxypropyl)-N'-(2-hydroxyethyl)-2,4,6-triiodo-1,3-benzenedicarboxamide; N-(2,3-dihydroxypropyl)-5-[N-(2,3-dihydroxypropyl)acetamido]-N'-(2-hydroxyethyl)-2,4,6-triiodoisophthalamide; ioxitol;* Oxilan. $C_{18}H_{24}I_3N_3O_8$; mol wt 791.12. C 27.33%, H 3.06%, I 48.12%, N 5.31%, O 16.18%. Nonionic iodinated contrast medium. Prepd not claimed: M. Sovak, R. Ranganathan, **PCT Int. pat. Appl. 00,757;** *eidem,* **U.S. pat. 5,035,877** (1987, 1991 both to Cook Imaging). HPLC analysis of isomers: S. J. Foster, M. Sovak, *Invest. Radiol.* **23,** Suppl. 1, S106 (1988). Pharmacology:

R. W. Katzberg *et al., ibid.* **25,** 46 (1990). Clinical evaluation: M. R. Callantine *et al., ibid.* Suppl. 1, S107; C. L. McIntosh, R. Reed, *ibid.* **29,** Suppl. 2, S40 (1994). Brief review of physical properties: M. Sovak, *ibid.* **23,** Suppl. 1, S79-S83 (1988).

Sol in water. Osmolality (mOsm/kg H_2O): 690 (c = 350 mg I/ml); 570 (c = 300 mg I/ml). Viscosity (37°) 4.6-4.7 cps (c = 300 mg I/ml).

THERAP CAT: Diagnostic aid (radiopaque medium).

5086. Ipecac. Ipecacuanha. Dried rhizome and roots of *Cephaelis ipecacuanha* (Brot.) A. Rich., *Rubiaceae,* known as Rio or Brazilian Ipecac; or of *Cephaelis acuminata* Karsten, known as Cartagena, Nicaragua or Panama Ipecac. *Habit.* Brazil to Bolivia; cultivated in India and Malaysia. *Constit.* Emetine, cephaeline, emetamine, psychotrine, methyl psychotrine, protoemetine and resin. Cartagena ipecac contains ~2.2-2.5% total alkaloids, with twice as much cephaeline as emetine. Determn of major alkaloids in ipecacuanha root: A. G. Davidson, S. M. Hassan, *J. Pharm. Biomed. Anal.* **2,** 441 (1984); in pharmaceutical preparations by HPLC: D. A. Elvidge *et al., J. Chromatog.* **463,** 107 (1989). Review of distribution and chemistry of ipecac alkaloids: M.-M. Janot in *The Alkaloids,* vol. 3, R. H. F. Manske, H. L. Holmes, Eds. (Academic Press, New York, 1953) pp 363-394; A. Brossi, S. Teitel, *ibid.* vol. 13, 189-212 (1971). Review of use in acute poisoning: W. D. King, *Clin. Toxicol.* **17,** 353-358 (1980).

THERAP CAT: Emetic; expectorant.

THERAP CAT (VET): Emetic; expectorant.

5087. Ipodate. *3-[[(Dimethylamino)methylene]amino]-2,4,6-triiodobenzenepropanoic acid; 3-[(dimethylaminomethylene)amino]-2,4,6-triiodohydrocinnamic acid; 2,4,6-triiodo-3-[(dimethylaminomethylene)amino]hydrocinnamic acid; β-(3-dimethylaminomethyleneamino-2,4,6-triiodophenyl)propionic acid.* $C_{12}H_{13}I_3N_2O_2$; mol wt 597.96. C 24.10%, H 2.19%, I 63.67%, N 4.68%, O 5.35%. Prepn: Priewe, Poljak, *Ber.* **93,** 2347 (1960). Toxicity data: J. O. Hoppe *et al., J. Med. Chem.* **13,** 997 (1970).

Crystals, mp 168-169° (dec 225°). Practically insol in water. Very sol in methanol, ethanol, chloroform, acetone, dil sulfuric acid. Estimated pK 5-5.5.

Sodium salt, $C_{12}H_{12}I_3N_2NaO_2$, *sodium ipodate, Biloptin, Oragrafin.* Bitter leaflets from water + acetone, mp 303-304° (decompn with evolution of iodine). Very freely sol in water; freely sol in methanol, ethanol. Practically insol in acetone, ether. Soly in DMF and DMSO about 33 g/100 ml; in dimethylacetamide about 66 g/100 ml. LD_{50} in mice (mg/kg): 290 i.v.; 2570 orally (Hoppe).

Calcium salt, $C_{24}H_{24}CaI_6N_4O_4$, *calcium ipodate, Solu-Biloptin, Oragrafin Calcium.* Crystals, mp 298-302°. Soly in water at 20°: 0.1%. Sol in chloroform, dimethylformamide, hot propylene glycol.

Ethyl ester, $C_{14}H_{17}I_3N_2O_2$, *SH-617L, Myelographin.*

THERAP CAT: Diagnostic aid (radiopaque medium—cholecystographic).

5088. Ipomea. Mexican scammony (root); Orizaba jalap root. Dried root of *Ipomoea orizabensis* Ledenois,

Convolvulaceae. Active constituent is the resin. Yields not less than 15% total ipomea resins. Different from *Ipomoea violacea* var. *Pearly Gates* Hort., *Convolvulaceae* and *Ipomoea rubrocoerulea* var. *praecox, morning-glory, ololiuqui,* which contain ergot alkaloids. Occurrence of lysergic acid derivatives and of ergolines in Ipomea: A. Hofmann, H. Tscherter, *Experientia* **16**, 414 (1960); D. Stauffacher *et al., Helv. Chim. Acta* **48**, 1379 (1965).

THERAP CAT: Cathartic.

5089. Ipratropium Bromide. *(endo,syn)-(±)-3-(3-Hydroxy-1-oxo-2-phenylpropoxy)-8-methyl-8-(1-methylethyl)-8-azoniabicyclo[3.2.1]octane bromide; 3α-hydroxy-8-isopropyl-1αH,5αH-tropanium bromide (±)-tropate;* 8-isopropylnoratropine methobromide; *N*-isopropylnoratropinium bromomethylate; Sch-1000; Atem; Atrovent; Bitrop; Itrop; Narilet; Rinatec. $C_{20}H_{30}BrNO_3$; mol wt 412.37. C 58.25%, H 7.33%, Br 19.38%, N 3.40%, O 11.64%. Anticholinergic. Prepn: K. Zeile *et al.,* **S. Afr.** pat. **67 07,766;** *eidem,* **U.S.** pat. **3,505,337** (1968, 1970, both to Boehringer, Ing.); W. Schulz *et al., Arzneimittel-Forsch.* **26**, 960 (1976). Chemistry and pharmacokinetics: W. Deckers, *Postgrad. Med. J.* **51**, Suppl. 7, 76 (1975). Pharmacology and toxicology: A. Engelhardt, H. Klupp, *ibid.* 82. Series of articles on pharmacology, toxicology, pharmacokinetics and clinical studies: *Arzneimittel-Forsch.* **26**, 974-985, 989-1020 (1976). Toxicity data: L. Sarafana *et al., ibid.* 985. *Review:* R. Bauer *et al.,* in *Pharmacological and Biochemical Properties of Drug Substances* vol. 2, M. E. Goldberg, Ed. (Am. Pharm. Assoc., Washington, DC, 1979) pp 489-515. Symposium on pharmacology, toxicology and clinical efficacy: *Am. J. Med.* **81**(5A), 1-102 (1986). Review of clinical toxicology: J. D. Truwit, *Crit. Care Clinics* **7**, 639-657 (1991).

White crystals from *n*-propanol, mp 230-232°. Freely sol in water and lower alcohols. Insol in ether, chloroform and fluorohydrocarbons. Fairly stable in neutral and acid solns; rapidly hydrolyzed in alkaline soln. LD_{50} in male, female mice (mg/kg): 1001, 1083 orally; 12.29, 14.97 i.v.; 300, 340 s.c. LD_{50} in male, female rats (mg/kg): 1663, 1779 orally; 15.89, 15.70 i.v. (Sarafana).

THERAP CAT: Bronchodilator; antiarrhythmic.

5090. Ipriflavone. *7-(1-Methylethoxy)-3-phenyl-4H-1-benzopyran-4-one;* 7-isopropoxy-3-phenyl-4*H*-1-benzopyran-4-one; 7-isopropoxy-3-phenylchromone; 7-isopropoxyisoflavone; FL-113; TC-80; Iprosten; Osten; Osteofix; Yambolap. $C_{18}H_{16}O_3$; mol wt 280.32. C 77.12%, H 5.75%, O 17.12%. Isoflavone derivative with anti-anginal and anti-osteopenic activity. Prepn: L. Feuer *et al.,* **Ger.** pat. **2,125,-245** (1971 to Chinoin), *C.A.* **76**, 72407e (1972). Use as anabolic in animals: *eidem,* **U.S.** pat. **3,833,730;** in humans; *eidem,* **U.S.** pat. **3,949,085** (1974, 1976 both to Chinoin). Cardiovascular properties in stable angina: V. Grubich *et al., Lancet* **1**, 211 (1979). Cardiological effects in animals: L. Feuer *et al., Arzneimittel-Forsch.* **31**, 953 (1981). Metabolism and disposition in rats: K. Yoshida *et al., Radioisotopes* **34**, 612, 618 (1985). Effect in rats on estrogen-stimulated calcitonin secretion: I. Yamazaki, *Life Sci.* **38**, 757 (1986); I. Yamazaki, M. Kinoshita, *ibid.* 1535; on glucocorticoid-induced osteoporosis: I. Yamazaki *et al., ibid.* 951.

Crystals from acetone, mp 115-117°.

THERAP CAT: Calcium regulator.

5091. Iprindole. *6,7,8,9,10,11-Hexahydro-N,N-dimethyl-5H-cyclooct[b]indole-5-propanamine; 5-[3-(dimethylamino)propyl]-6,7,8,9,10,11-hexahydro-5H-cyclooct[b]indole;* 1-(3-dimethylaminopropyl)-2,3-hexamethyleneindole; pramindole (obsolete); Tertran. $C_{19}H_{28}N_2$; mol wt 284.44. C 80.23%, H 9.92%, N 9.85%. Prepn: Rice, Freed, **Belg.** pat. **623,933** corresp to **U.S.** pat. **3,282,942** (1963, 1966, both to Am. Home Prods.); Rice *et al., J. Med. Chem.* **7**, 313 (1964). Pharmacology: Gluckman, Baum, *Psychopharmacologia* **15**, 169 (1969); Baum *et al., Eur. J. Pharmacol.* **13**, 287 (1971). Mechanism of action: Miller *et al., Experientia* **26**, 863 (1970); Lemberger *et al., Biochem. Pharmacol.* **19**, 3021 (1970). Freeman, Sulser, *J. Pharmacol. Exp. Ther.* **183**, 307 (1972).

Hydrochloride, $C_{19}H_{28}N_2 \cdot HCl$, *Wy-3263, Prondol, Galatur.* Crystals from methanol + acetone, mp 146-147°.

THERAP CAT: Antidepressant.

5092. Iproclozide. *4-(Chlorophenoxy)acetic acid 2-(1-methylethyl)hydrazide;* Sursum. $C_{11}H_{15}ClN_2O_2$; mol wt 242.71. C 54.44%, H 6.23%, Cl 14.61%, N 11.54%, O 13.18%. Monoamine oxidase inhibitor. Prepn: Libermann, Denis, *Bull. Soc. Chim. France* **1961**, 1952. GC determn in urine: R. M. DeSagher *et al., Anal. Chem.* **47**, 1144 (1976).

Crystals, mp 93-94°.

THERAP CAT: Antidepressant.

5093. Iprodione. *3-(3,5-Dichlorophenyl)-N-(1-methylethyl)-2,4-dioxo-1-imidazolidinecarboxamide;* glycophene; promidione; RP-26019; ROP-500F; NRC-910; LFA-2043; FA-2071; Rovral; CHIPCO-26019. $C_{13}H_{13}Cl_2N_3O_3$; mol wt 330.17. C 47.29%, H 3.97%, Cl 21.48%, N 12.73%, O 14.54%. Prepn: M. Sauli, **Ger.** pat. **2,149,923** corresp to **U.S.** pat. **3,755,350** (1972, 1973 to Rhône-Poulenc). Activity: L. Lacroix *et al., Phytiatr.-Phytopharm.* **23**, 165 (1974). Structural rearrangement in ethanolic soln: B. K. Cooke *et al., Pestic. Sci.* **10**, 393 (1979). Thermal degradn: J. Gomez *et al., J. Agr. Food Chem.* **30**, 180 (1982). *Review:* L. Lacroix *et al., Anal. Methods Pestic. Plant Growth Regul.* **11**, 247-261 (1980).

Colorless crystals, mp ~136°. Vapor press. at 20°: <10^{-6} mm Hg. Soly in water at 20°: 13 mg/l. Soly at 20° (g/l): ethanol 25; methanol 25; acetone 300; dichloromethane 500; DMF 500. LD_{50} orally in mice, rats: 3500, 4000 mg/kg, *RTECS* Vol. I, R. J. Lewis, R. L. Tatken, Eds. (1980) p 929.

USE: Fungicide.

5094. Iproniazid. *4-Pyridinecarboxylic acid 2-(1-methylethyl)hydrazide; isonicotinic acid 2-isopropylhydrazide;* 1-isonicotinoyl-2-isopropylhydrazine; 1-isonicotinyl-2-isopropylhydrazine. $C_9H_{13}N_3O$; mol wt 179.22. C 60.32%, H 7.31%, N 23.45%, O 8.93%. Monoamine oxidase inhibitor. Prepd by treating isoniazid with isopropyl bromide in abs alcohol in the presence of sodium: McMillan *et al., J. Am. Pharm. Assoc., Sci. Ed.* **42**, 457 (1953). Alternate synthesis: Fox, Gibas, *J. Org. Chem.* **18**, 994 (1953). Comprehensive description: F. Belal, H. Abdel-Aliem in *Analytical Profiles of Drug Substances* vol. **20**, K. Florey, Ed. (Academic Press, New York, 1991) pp 337-368.

Needles from benzene + ligroin, mp 112.5-113.5°. Freely sol in water, alcohol; pH of aq solns about 6.7.

Dihydrochloride, $C_9H_{13}N_3O.2HCl$, rhomboid crystals from isopropanol, dec at 227-228°. Freely sol in water; aq solns are very acidic.

Phosphate, $C_9H_{13}N_3O.II_3PO_4$, *Marsilid.* White to slightly yellowish powder, melting range 175-184°. pH of a 5% soln: 2.7-3.6. uv spectrum: J. Kracmar, J. Kracmarova, *Pharmazie* **34**, 27 (1979). uv max (methanol): 265 nm ($A_{1cm}^{1\%}$ 166); (water): 264 nm ($A_{1cm}^{1\%}$ 176); (0.1N HCl): 267 nm ($A_{1cm}^{1\%}$ 179); (0.1N KOH): 244, 272, 308 nm ($A_{1cm}^{1\%}$ 113, 121, 135). Soly at 25° (mg/ml): water ~188; methanol ~21; ethanol (96%) ~9; chloroform ~0.6; diethyl ether <0.1; hexane <0.1.

THERAP CAT: Antidepressant.

5095. Ipronidazole. *1-Methyl-2-(1-methylethyl)-5-nitro-1H-imidazole;* 2-isopropyl-1-methyl-5-nitroimidazole; Ro-7-1554; Ipropran. $C_7H_{11}N_3O_2$; mol wt 169.18. C 49.70%, H 6.55%, N 24.84%, O 18.91%. Prepn: **Brit.** pat. **1,119,636;** M. Hoffer, M. Mitrovic, **U.S.** pat. **3,634,446** (1968, 1972 both to Hoffmann-La Roche); K. Butler, **S. Afr.** pat. **66 07,466** (1968 to Pfizer), *C.A.* **71**, 3384e (1969). Structure-activity relationship: K. Butler *et al., J. Med. Chem.* **10**, 891 (1967). Activity studies as an antihistomonal agent: M. Mitrovic *et al., Antimicrob. Ag. Chemother.* **1968**, 445; M. Mitrovic, E. G. Schildknecht, *Poultry Sci.* **49**, 86 (1970); as a growth promotant: W. L. Marusich *et al., ibid.* 98. Toxicity studies: *eidem, ibid.* 92.

White plates, mp 60°. LD_{50} orally in poults: 640 ± 25 mg/kg (Marusich, p 92).

Hydrochloride, mp 177-182°. Water-soluble.

THERAP CAT (VET): Antiprotozoal (Histomonas).

5096. Ipsapirone. *2-[4-[4-(2-Pyrimidinyl)-1-piperazin-yl]butyl]-1,2-benzisothiazolin-3(2H)-one 1,1-dioxide;* isapirone. $C_{19}H_{23}N_5O_3S$; mol wt 401.49. C 56.84%, H 5.77%, N 17.44%, O 11.96%, S 7.99%. Nonbenzodiazepine anxiolytic; 5-hydroxytryptamine (5-HT_1) receptor agonist. Prepn: W. Dompert *et al.,* **Ger.** pat. **3,321,969;** P.-R. Seidel *et al.,* **U.S.** pat. **4,818,756** (1984, 1989 both to Troponwerke). Pharmacology and serotonin (5-HT_1) receptor binding study: J. Traber *et al., Brain Res. Bull.* **12**, 741 (1984). 5-HT_1 specific binding activity: W. U. Dompert *et al., Arch. Pharmacol.* **328**, 467 (1985); T. Glaser, J. Traber, *ibid.* **329**, 211 (1985). Behavioral effects in mice: R. J. DeSouza *et al., Brit. J. Pharmacol.* **89**, 377 (1986). Neurophysiological effects in comparison with buspirone, *q.v.:* M. J. Rowan, R. Anwyl, *Eur. J. Pharmacol.* **132**, 93 (1987). HPLC determn in biological samples: G. Bianchi, S. Caccia, *J. Chromatog.* **431**, 477 (1988).

Crystals from isopropanol, mp 137-138°.

Hydrochloride, $C_{19}H_{23}N_5O_3S.HCl$, *Bay q 7821, TVX Q 7821.* Crystals from isopropanol, mp 221-222°.

THERAP CAT: Anxiolytic.

5097. Irbesartan. *2-Butyl-3-[[2'-(1H-tetrazol-5-yl)[1,1'-biphenyl]-4-yl]methyl]-1,3-diazaspiro[4.4]non-1-en-4-one;* 2-butyl-3-[p-(o-1H-tetrazol-5-ylphenyl)benzyl]-1,3-diazaspiro-[4.4]non-1-en-4-one; 2-n-butyl-4-spirocyclopentane-1-[(2'-(tetrazol-5-yl)biphenyl-4-yl)methyl]-2-imidazolin-5-one; BMS-186295; SR-47436. $C_{25}H_{28}N_6O$; mol wt 428.54. C 70.07%, H 6.59%, N 19.61%, O 3.73%. Angiotensin II type 1 (AII_1)-receptor antagonist. Prepn: C. Bernhart *et al.,* PCT Int. pat. Appl. **91 14,679;** *eidem,* U.S. pat. **5,270,317** (1991, 1993 both to Sanofi); *eidem et al., J. Med. Chem.* **36**, 3371 (1993). Pharmacology: C. Cazaubon *et al., J. Pharmacol. Exp. Ther.* **265**, 826 (1993). Clinical evaluation in hypertension: A. H. van den Meiracker *et al., Hypertension* **25**, 22 (1995).

Crystals from 96% ethanol, mp 180-181°.

THERAP CAT: Antihypertensive.

5098. Iridium. Ir; at. wt 192.22; at. no. 77; valences 1, 3, 4; also 2, 5, 6. Two naturally occurring isotopes: 191 (38.5%); 193 (61.5%); artificial, radioactive isotopes: 182-190; 191; 192; 194-198. Occurrence in earth's crust about 0.001 ppm. Discovered in 1804 by Tennant. Occurs in nature in the metallic state, usually as a natural alloy with osmium (osmiridium); found in small quantities alloyed with native platinum (platinum mineral) or with native gold. Recovery and purification from osmiridium: Deville, Debray, *Ann. Chim. Phys.* **61**, 84 (1861); from the platinum mineral: Wichers, *J. Res. Nat. Bur. Stand.* **10**, 819 (1933). Reviews of prepn, properties and chemistry of iridium and other platinum metals: Gilchrist, *Chem. Rev.* **32**, 277-372 (1943); W. P. Griffith, *The Chemistry of the Rare Platinum Metals* (John Wiley, New York, 1967) pp 1-41, 227-312; Livingstone in *Comprehensive Inorganic · Chemistry* **vol. 3**, J. C. Bailar Jr. *et al.*, Eds. (Pergamon Press, Oxford, 1973) pp 1163-1189, 1254-1274.

Silver-white, very hard metal; face-centered cubic lattice. mp 2450°; bp ~4500°. d_4^{20} 22.65; highest sp gr of all elements. Specific heat 0.0307 cal/g/°C. Mohs' hardness 6.5. Pure iridium is not attacked by any acids including aqua regia; only slightly by fused (non-oxidizing) alkalies. It is superficially oxidized on heating in the air. Is attacked by fluorine and chlorine at a red heat; attacked by potassium sulfate or by a mixture of potassium hydroxide and nitrate on fusion; attacked by lead, zinc or tin. The powdered metal is oxidized by air or oxygen at a red heat to the dioxide, IrO_2, but on further heating the dioxide dissociates into its constituents.

USE: In manufacturing crucibles; in hardening platinum; in making nibs for fountain-pen points.

5099. Iridium Hexafluoride. F_6Ir; mol wt 306.21. F 37.23%, Ir 62.77%. IrF_6. Prepd by direct fluorination of iridium: Ruff, Fischer, *Z. Anorg. Allgem. Chem.* **179**, 161 (1929); Robinson, Westland, *J. Chem. Soc.* **1956**, 4481. Crystal structure: Siegel, Northrop, *Inorg. Chem.* **5**, 2187 (1966).

Golden yellow, cubic crystals. d (calc) 4.82. Very hygroscopic. mp 44.4°. bp 53°. Volatilizes on slow heating. Reduced to IrF_4 by halogens. Decomposes in aqueous soln. Does not react with dry glass below 150°. May be stored in evacuated quartz ampuls.

5100. Iridium Sesquioxide. Ir_2O_3; mol wt 432.43. Ir 88.90%, O 11.10%.

Blue-black powder. Decomposes at about 1000° into metal and oxygen. Insol in water; slowly dissolved by boiling HCl. Oxidized to IrO_2 by HNO_3.

5101. Iridium Trichloride. Cl_3Ir; mol wt 298.58. Cl 35.62%, Ir 64.38%. $IrCl_3$. Prepn: Dillamore, Edwards, *J. Inorg. Nucl. Chem.* **31**, 2427 (1969). Several known modifications. Crystal structure: Brodersen *et al.*, *Naturwiss.* **52**, 205 (1966); Babel, Deigner, *Z. Anorg. Allgem. Chem.* **339**, 57 (1965).

α-Form; brown, monoclinic crystals. β-Form; red, orthorhombic crystals. Olive-green modification also prepd: Dillamore, Edwards, *loc. cit.* Decomposes at 763°. Insol in water, acids, alkalies.

USE: Catalyst for chlorination and polymerization reactions.

5102. Iridomyrmecin. *[4S-(4α,4aβ,7β,7aβ)]-Hexahydro-4,7-dimethylcyclopenta[c]pyran-3(1H)-one; 2-(hydroxymethyl)-α,3-dimethylcyclopentaneacetic acid δ-lactone;* Iridomyrmexin; Iridomirmecina. $C_{10}H_{16}O_2$; mol wt 168.24. C 71.39%, H 9.59%, O 19.02%. Isoln from the Argentine ant *Iridomyrmex humilis* Mayr., *Dolichoderinae:* Pavan, *Ricerca Sci.* **19**, 1011 (1949); *Chim. Ind. (Milan)* **37**, 625 (1955); Fusco *et al.*, *ibid.* **251**, 958; Cavill *et al.*, *Austr. J. Chem.* **9**, 288 (1956); **10**, 352 (1957). Synthesis: Korte *et al.*, *Tetrahedron* **6**, 201 (1959); Sisido *et al.*, *J. Org. Chem.* **29**, 3361 (1964); R. S. Matthews, J. K. Whitesell, *ibid.* **40**, 3312 (1975); Y. Yamada *et al.*, *Chem. Letters* **1978**, 1405; P. Callant *et al.*, *Tetrahedron* **37**, 2085 (1981). Stereochemistry: Korte *et al.*, *Ber.* **94**, 1952 (1961); McConnell *et al.*, *Tetrahedron Letters* **1962**, 445; Büchi, Manning, *Tetrahedron* **18**, 1049 (1962). Activity as bactericide and insecticide: M. Pavan, *Z. Hyg. Infektion-Krankh.* **134**, 136 (1952), *C.A.* **46**, 9796f (1952).

Prisms from petr ether. Aromatic odor, resembling catnip. Salty taste. mp 60-61°. Sublimes$_{0.01}$ 50-55°. bp$_{1.5}$ 104-

108°. $[\alpha]_D^{20}$ +210° (c = 4 in ethanol); $[\alpha]_D^{17}$ +205° (c = 0.223 in carbon tetrachloride). n_D^{65} 1.4607. Sparingly sol in water: 0.2 g/100 ml H_2O (25°). Sol in fats and fat solvents. Freely sol in ether.

USE: Insecticide.

THERAP CAT: Antibacterial.

5103. Irigenin. *5,7-Dihydroxy-3-(3-hydroxy-4,5-dimethoxyphenyl)-6-methoxy-4H-1-benzopyran-4-one; 3′,5,7-trihydroxy-4′,5′,6-trimethoxyisoflavone.* $C_{18}H_{16}O_8$; mol wt 360.32. C 60.00%, H 4.48%, O 35.52%. From rhizome of *Iris florentina* L., *Iridaceae:* de Laire, Tiemann, *Ber.* **26**, 2010 (1893). Structure: Baker, *J. Chem. Soc.* **1928**, 1022. Synthesis: Baker *et al.*, *Tetrahedron Letters* no. **5**, 6 (1960); Farkas, Várady, *Ber.* **93**, 2685 (1960); Baker *et al.*, *J. Chem. Soc. (C)* **1970**, 1219.

Yellow plates or needles from dil alc, mp 185°. uv max (abs ethanol): 267 nm. Sol in warm alcohol, benzene, chloroform; practically insol in water, ether, petr ether.

Triacetate, $C_{24}H_{22}O_{11}$, prisms from dil acetic acid, mp 127-128°.

7-Glucoside, $C_{24}H_{26}O_{13}$, *iridin,* 7-(glucosyloxy)-3′,5-di-hydroxy-4′,5′,6-trimethoxyisoflavone. Monohydrate, needles from dil alc, mp 208°. When anhydrous, mp 217°. One gram of anhydrous iridin dissolves in 500 ml water, 43 ml acetone. Sol in hot alcohol; practically insol in ether, ethyl acetate, benzene, chloroform.

Note: There also is an oleoresin called *iridin* (Extractum Iridis) from *Iris versicolor* (blue flag) by extraction with 60% alc. Another substance named *iridine* is a protamine from the sperm of rainbow trout: Felix, Mager, *Z. Physiol. Chem.* **249**, 124 (1937).

5104. Irinotecan. *[1,4′-Bipiperidine]-1′-carboxylic acid (S)-4,11-diethyl-3,4,12,14-tetrahydro-4-hydroxy-3,14-dioxo-1H-pyrano[3′,4′:6,7]indolizino[1,2-b]quinolin-9-yl ester;* 7-ethyl-10-[4-(1-piperidino)-1-piperidino]carbonyloxycamptothecin; (+)-7-ethyl-10-hydroxycamptothecine 10-[1,4′-bi-piperidine]-1′-carboxylate; (+)-(4S)-4,11-diethyl-4-hydroxy-9-[(4-piperidinopiperidino)carbonyloxy]-1H-pyrano[3′,4′:6,-7]indolizino[1,2-b]quinoline-3,14-(4H,12H)-dione. $C_{33}H_{38}$-N_4O_6; mol wt 586.69. C 67.56%, H 6.53%, N 9.55%, O 16.36%. DNA topoisomerase inhibitor; semisynthetic derivative of camptothecin, *q.v.* Prepn: **Japan.** Kokai **85 19,790**; T. Miyasaka *et al.*, U.S. pat. **4,604,463** (1985, 1986 both to Yakult Honsha); S. Sawada *et al.*, *Chem. Pharm. Bull.* **39**, 1446 (1991). Antitumor activity and toxicity data: T. Kunimoto *et al.*, *Cancer Res.* **47**, 5944 (1987). De-esterified *in vivo* to its active metabolite, *7-ethyl-10-hydroxycamptothecin (SN-38)*: N. Kaneda *et al.*, *ibid.* **50**, 1715 (1990). Mechanism of action study: Y. Kawato *et al.*, *ibid.* **51**, 4187 (1991). Clinical evaluation: Y. Shimada *et al.*, *J. Clin. Oncol.* **11**, 909 (1993). Review of irinotecan and other

Consult the Name Index before using this section.

camptothecin analogs: W. J. Slichenmyer *et al.*, *J. Nat. Cancer Inst.* **85**, 271-291 (1993).

Pale yellow powder, mp 222-223°.

Hydrochloride trihydrate, $C_{33}H_{38}N_4O_6 \cdot HCl.3H_2O$, *CPT-11*, *DQ-2805*, *Campto*. Slightly pale yellow needles or crystalline powder from water, mp 256.5°. $[\alpha]_D^{20}$ +67.7° (c = 1 in water). uv max (ethanol): 221, 254, 359, 372 nm (ϵ 53800, 36600, 26200, 25300). pH of 2% aqueous soln = 4. LD_{50} in mice (mg/kg): 177.5 i.p.; 765.3 orally (Kunimoto).

THERAP CAT: Antineoplastic.

5105. Irisolone. *7-(4-Hydroxyphenyl)-9-methoxy-8H-1,3-dioxolo[4,5-g][1]benzopyran-8-one;* "*Sosan*"; 4'-hydroxy-5-methoxy-6,7-methylenedioxyisoflavone. $C_{17}H_{12}O_6$; mol wt 312.28. C 65.39%, H 3.87%, O 30.74%. From rhizomes of *Iris nepalensis* D. Don, *Iridaceae*. Isoln and structure: Gopinath *et al.*, *Tetrahedron* **16**, 201 (1961). Synthesis: Fukui, Matsumato, *Bull. Chem. Soc. Japan* **38**, 887 (1965).

Pale yellow crystals from ethyl acetate, mp 269-270°. uv max: 270 nm (log ϵ 4.62), inflection 330 nm (log ϵ 3.97). Methyl ether, $C_{18}H_{14}O_6$, plates, mp 184-185°.

5106. Iron. Fe; at. no. 26; at. wt 55.847; valences 2, 3; seldom 1, 4, 6. Four naturally occurring isotopes: 54 (5.82%), 56 (91.66%), 57 (2.19%); 58 (0.33%); artificial, radioactive isotopes: 52; 53; 55; 59-61. Second most abundant metal in earth's crust after aluminum: about 5%. The earth's core is believed to consist mainly of iron. Important ores include hematite (Fe_2O_3), magnetite (Fe_3O_4), limonite [FeO(OH).nH_2O] and *siderite* ($FeCO_3$). Study of iron and its compds by Mössbauer spectroscopy: Danon, "^{57}Fe: Metal, Alloys and Inorganic Compounds" in *Chemical Applications of Mössbauer Spectroscopy*, V. I. Goldanskii, R. H. Herber, Eds. (Academic Press, New York, 1968) p 159-313. Ions involved in oxygen transport, electron transport, nitrogen fixation and a number of other biological processes: Nielands, "Evolution of Biological Iron Binding Centers" in *Struct. Bonding* **11**, 145-170 (1972). Review of biology, pharmacology and toxicity of iron compounds: several authors, *Clin. Toxicol.* **4**, 525-642 (1971). Comprehensive reviews: Feldmann, Schenck in *Ullmanns Encyklopädie der Technischen Chemie* vol. 6 (München-Berlin, 1955) pp 261-407; Nicholls in *Comprehensive Inorganic Chemistry* vol. 3, J. C. Bailar, Jr. *et al.*, Eds. (Pergamon Press, Oxford, 1973) pp 979-1051; W. A. Knepper in Kirk-Othmer *Encyclopedia of Chemical Technology* vol. 13 (Wiley-Interscience, New York, 3rd ed., 1981) pp 735-753.

Silvery-white or gray, soft, ductile, malleable, somewhat magnetic metal. Holds magnetism only after hardening (as alloy steel, *e.g.*, Alnico). Supplied as ingots, powder, wire, sheets, etc. Takes a bright polish; can be rolled, hammered, bent, particularly when red hot. Stable in dry air but readily oxidizes in moist air, forming "rust" (chiefly oxide, hydrated). In powder form it is black to gray. Commercial iron usually contains some C, P, Si, S and Mn. d pure 7.86; cast 7.76; wrought 7.25-7.78; steel 7.6-7.78. mp pure 1535°; cast 1000-1300°; wrought 1500°; steel 1300°. bp 3000°. Electrical resistivity (20°): 9.71 microhm-cm. Readily attacked by dil mineral acids and attacked or dissolved by organic acids; not appreciably attacked by cold concd H_2SO_4 or HNO_3, but is attacked by the hot acids.

USE: Alloyed with C, Mn, Cr, Ni, and other elements to form steels. ^{55}Fe and ^{59}Fe used in tracer studies; the former in biological studies.

5107. Irone. 6-Methylionone. $C_{14}H_{22}O$; mol wt 206.33. C 81.50%, H 10.75%, O 7.75%. The fragrant principle of violets, best isolated from the rhizomes of iris or from orris oil: F. Tiemann, P. Kruger, *Ber.* **26**, 2675 (1893); Ruzicka *et al.*, *Helv. Chim. Acta* **16**, 1143 (1933); Naves, Mazuyer, *Les Parfums Naturels* (Paris, 1939). Irone, as isolated, is an isomeric mixture of α-, β-, and γ-irones, *q.v.*,

with each having the same molecular formula. Synthesis of isomeric mixture: Eschinazi, U.S. pat. **3,019,265** (1962 to Givaudan).

Light yellow, slightly viscous liquid. Very stable, does not discolor. In dil alcoholic soln the characteristic odor of violets is best appreciated.

USE: In perfumery for orris and violet compositions.

5108. α-Irone. *4-(2,5,6,6-Tetramethyl-2-cyclohexen-1-yl)-3-butene-2-one.* Main perfume ingredient of violets. Both the *d*-(1,5)-*cis*- and *d*-(1,5)-*trans*-forms occur in nature. Isoln from irone mixture and synthesis: Naves, *Helv. Chim. Acta* **31**, 893, 1103 (1948); Ruzicka *et al.*, *ibid.* **31**, 257 (1948). Synthesis: Barton, Mousseron-Canet, *J. Chem. Soc.* **1960**, 271. Industrial preparation: Kaiser, Kimel, U.S. pat. **2,877,271** (1959 to Hoffmann-La Roche); Eschenmoser *et al.*, U.S. pats. **3,413,351; 3,470,241** (1968, 1969 to Firmenich & Cie). Nomenclature in early papers: Naves, *Helv. Chim. Acta* **32**, 969 (1949). Stereochemistry: Tribolet, Schinz, *ibid.* **37**, 2184 (1954); Naves, Ardizio, *Bull. Soc. Chim. France* **1955**, 1479; V. Rautenstrauch, G. Ohloff, *Helv. Chim. Acta* **54**, 1776 (1971); V. Rautenstrauch *et al.*, *ibid.* **67**, 325 (1984). Stereoselective synthesis of *cis*- and *trans*-α-irones: S. Torii *et al.*, *J. Org. Chem.* **45**, 16 (1980); of (±)-*cis*-α-irone: C. Nussbaumer, G. Fráter, *ibid.* **52**, 2096 (1987).

(+)-cis-α-Irone   (+)-trans-α-Irone

All forms are viscous liquids with characteristic odors. *dl*-*cis*-α-Irone, d_4^{20} 0.9360; n_D^{20} 1.50098. uv max (ethanol): 227 nm (ϵ 15400), Naves, Bachmann, *Helv. Chim. Acta* **32**, 394 (1949).

d-*cis*-α-Irone, $[\alpha]_D^{20}$ +109° (CH_2Cl_2).

dl-*trans*-α-Irone, d_4^{20} 0.9347; n_D^{20} 1.50119. uv max (ethanol): 229 (ϵ 15450).

d-*trans*-α-Irone, $[\alpha]_D^{20}$ +420° (CH_2Cl_2).

5109. β-Irone. *4-(2,5,6,6-Tetramethyl-1-cyclohexen-1-yl)-3-buten-2-one.* Frequent byproduct in synthesis of α-irone. Stereoselective synthesis of (±)-form: S. Torii *et al.*, *J. Org. Chem.* **45**, 16 (1980). For general refs *see* α-irone.

(+)-β-Irone

Oil with odor similar to that of β-ionone. $bp_{0.1}$ 85-90°; $bp_{0.7}$ 99-104°; bp_{11} 125°. $[\alpha]_D^{20}$ +59° (CH_2Cl_2). d_4^{21} 0.9434; n_D^{21} 1.5178; n_D^{25} 1.5162. uv max: 295 nm (log ϵ 4.05).

5110. γ-Irone. *4-(2,2,3-Trimethyl-6-methylenecyclohexyl)-3-buten-2-one.* Principal fragrant constituent of natural iris oil. Isoln from irone mixture: Ruzicka *et al.*, *Helv. Chim. Acta* **30**, 1807 (1947); Ruzicka *et al.*, *ibid.* **31**, 257 (1948). Synthesis of *dl*-γ-irone: Favre, Schinz, *ibid.* **41**, 1368 (1958). Synthesis of (±)-*cis*- and *trans*-γ-irone: F. Leyendecker, M.-T. Comte, *Tetrahedron* **43**, 85 (1987); T. Kawanobe *et al.*, *Agr. Biol. Chem.* **51**, 791 (1987). Stereoselective synthesis of (±)-*cis*-γ-irone: C. Nussbaumer, G. Fráter, *Helv. Chim. Acta* **71**, 619 (1988). Stereochemistry: *See* α-irone.

(+)-cis-γ-Irone

Oil. bp$_{0.06}$ 85-88°. bp$_2$ 114-116°. d$_4^{15}$ 0.939; n$_D^{15}$ 1.505. [α]$_D^{20}$ +2° (methylene chloride). uv max: 230 nm (log ε 4.2). USE: In perfumery.

5111. Iron Pentacarbonyl. Iron carbonyl; pentacarbonyliron. C$_5$FeO$_5$; mol wt 195.90. C 30.66%, Fe 28.51%, O 40.84%. Fe(CO)$_5$. Prepn from CO and Fe ore: Wallis, Townshend, U.S. pat. **2,378,053** (1945 to International Nickel); from CO and Fe or FeSO$_4$.7H$_2$O: Reppe *et al.*, *Ann.* **582**, 116 (1953); from CO and Fe amalgams: Ettmayer, Jangg, *Monatsh.* **92**, 834 (1961); from CO and steel turnings: Shipman, **Brit.** pat. **897,204** (1962 to ICI). Convenient lab prepn of small quantities of Fe(CO)$_5$ from Fe-(CO)$_4$I$_2$: Hieber, Lagally, *Z. Anorg. Allgem. Chem* **245**, 295 (1940). Thermodynamic data: Cotton *et al.*, *J. Am. Chem. Soc.* **81**, 800 (1959); Leadbetter, Spice, *Can. J. Chem.* **37**, 1923 (1959). Toxicity: Sunderman *et al.*, *Arch. Ind. Health* **19**, 11 (1959). *Reviews:* Cable, Sheline, *Chem. Rev.* **56**, 1 (1956); Wender *et al.*, *The Chemistry and Catalytic Properties of Cobalt and Iron Carbonyls* (U.S. Govt. Printing Office, Washington, 1962) 83 pp; H. Alper, "Organic Syntheses with Iron Pentacarbonyl" in *Organic Syntheses via Metal Carbonyls* vol. 2, I. Wender, P. Pino, Eds. (John Wiley, New York, 1977) pp 545-593. Review of *iron tetracarbonyl*, the photochemically produced intermediate: M. Poliakoff, *Chem. Soc. Rev.* **7**, 527-540 (1978).

Colorless to yellow, oily liquid. *Pyrophoric in air;* burns to Fe$_2$O$_3$. Dec by light to Fe$_2$(CO)$_9$ and CO. mp −20°. bp 103°. d$_4^{20}$ 1.46-1.52; n$_D^{22}$ 1.453. Critical temp 285-288°; critical pressure 29.6 atm. Flash pt −15°. Heat capacity at constant pressure (14°) 56.9 cal/mole/°C. Latent heat of fusion 3161 cal/mol; latent heat of vaporization 9.6 kcal/-mole. Heat of combustion −386.9 kcal/mole; heat of formation [Fe(CO)$_5$(liq)] −182.6 kcal/mole. Practically insol in water; readily sol in most organic solvents including ether, benzene, petr ether, acetone, ethyl acetate, carbon tetrachloride, carbon disulfide; slightly sol in alcohol; practically insol in liquid ammonia. *Protect from light and air.* LD$_{50}$ in mice, rats: 2.19, 0.91 mg/l inhalation, exposure for 30 min. (Sunderman).

Caution: Lung irritant. Affects central nervous system; causes liver and kidney damage: Gage, *Brit. J. Ind. Med.* **27**, 1 (1970).

USE: To make finely divided iron, so-called carbonyl iron, which is used in the manuf of powdered iron cores for high frequency coils used in the radio and television industry; as antiknock agent in motor fuels; as catalyst and reagent in organic reactions.

5112. Iron Sorbitex. *Glucitol iron complex, compd with citric acid;* iron sorbitol; iron sorbitol citrate; Jectofer. Sterile soln (pH 7.2-7.9) of a complex of iron, sorbitol, and citric acid, stabilized with dextrin in a dil soln of sorbitol. Contains 50 ± 2 mg/ml of elemental iron. Average mol wt <5000. Prepn: Lindvall, Andersson, *Brit. J. Pharmacol. Chemother.* **17**, 358 (1961).

The prepn is hypertonic, stable in serum and rapidly absorbed from muscle, does not produce hemolysis, affects coagulation only at very high concns.

THERAP CAT: Hematinic.

5113. Irsogladine. *6-(2,5-Dichlorophenyl)-1,3,5-triazine-2,4-diamine;* 2,4-diamino-6-(2,5-dichlorophenyl)-*s*-triazine; dicloguamine. C$_9$H$_7$Cl$_2$N$_5$; mol wt 256.09. C 42.21%, H 2.76%, Cl 27.69%, N 27.35%. Gastric cytoprotectant with benzoguanamine skeleton. Prepn: H. Murai *et al.*, **Ger.** pat. **2,506,814**; *eidem*, U.S. pat. **3,966,728** (1975, 1976 both to Nippon Shinyaku). Series of articles on pharmacology: F. Ueda *et al.*, *Arzneimittel-Forsch.* **34**, 474-491 (1984). Tissue distribution studies: T. Ando *et al.*, *ibid.* **36**, 1221

(1984). Metabolism: M. Sugiyama *et al.*, *ibid.* 1229. Phase I study: M. Nakashima *et al.*, *ibid.* **34**, 492 (1984). Mechanism of antiulcer action study: F. Ueda *et al.*, *Oyo Yakuri* **33**, 157 (1987), *C.A.* **107**, 17604a (1987).

Colorless crystals from dioxane, mp 268-269°.

Maleate, C$_9$H$_7$Cl$_2$N$_5$.C$_4$H$_4$O$_4$, *MN-1695, Gaslon.* Crystals from dioxane, mp 205° (dec). LD$_{50}$ in male, female mice, rats (mg/kg): 6035, 5697, 3898, 2917 orally; 2841, 3216, 1600, 1524 s.c.; 775, 1006, 558, 545 i.p. (Ueda, 1984 p 474).

THERAP CAT: Antiulcerative.

5114. Isanic Acid. *17-Octadecene-9,11-diynoic acid;* bolecic acid; erythrogenic acid. C$_{18}$H$_{26}$O$_2$; mol wt 274.40. C 78.79%, H 9.55%, O 11.66%. CH$_2$=CH(CH$_2$)$_4$C≡C-C≡C(CH$_2$)$_7$COOH. Found in vegetable oils obtained from tropical and equatorial regions. Isoln from boleko oil: De Vries, *Oléagineux* **12**, 427 (1957); U.S. pat. **2,789,993** (1957 to UCB). Structure: Steger, Van Loon, *Rec. Trav. Chim.* **59**, 1156 (1940); Gunstone, Sealy, *J. Chem. Soc.* **1963**, 5772; Morris, *ibid.* 5779. Synthesis: Black, Weedon, *ibid.* **1953**, 1785.

Crysts from petr ether, mp 39.5°. d$_4^{45}$ 0.9309; d$_4^{60}$ 0.91966; d$_4^{78}$ 0.9095; n$_D^{50}$ 1.49148; n$_D^{78}$ 1.4860. Turns pink on exposure to air and light. Tendency to polymerize and explode, especially when heated above 250°. Sol in acetone, ethanol, isopropanol. Moderately sol in petr ether.

5115. Isatide. *1,1′,3,3′-Tetrahydro-3,3′-dihydroxy-[3,3′-bi-2H-indole]-2,2′-dione; 3,3′-dihydroxy-[3,3′-biindoline]-2,2′-dione;* 3,3′-dihydroxy-3,3′-dihydroisoindigo; isatyde. C$_{16}$H$_{12}$N$_2$O$_4$; mol wt 296.28. C 64.86%, H 4.08%, N 9.45%, O 21.60%. Prepn by dimerization reduction of isatin with ammonium hydrosulfide: Laurent, *Ann. Chim.* **3**, 372 (1841); by piperidine-catalyzed condensation of isatin and dioxindole: Heller, *Ber.* **37**, 943 (1904); Wahl, Hansen, *Compt. Rend.* **178**, 393 (1924); from isatin and acenaphthene in sunlight: Oliveri-Mandala, Deleo, *Gazz. Chim. Ital.* **79**, 337 (1949). Structure: Lefèvre, *Bull. Soc. Chim. France* **19**, 113 (1916); Sumpter, *J. Am. Chem. Soc.* **54**, 2917 (1932); Bergmann, *ibid.* **77**, 1549 (1955).

Prisms from alcohol or ether, dec 245°. Difficultly sol in alcohol and ether; practically insol in water.

5116. Isatin. *Indole-2,3-dione;* 2,3-indolinedione; 2,3-diketoindoline. C$_8$H$_5$NO$_2$; mol wt 147.13. C 65.31%, H 3.43%, N 9.52%, O 21.75%. May be obtained by oxidation of indigo or of oxygenated indoles such as indoxyl, oxindole, or dioxindole: Erdmann, *J. Prakt. Chem.* [1] **24**, 1 (1841); Laurent, *ibid.* **25**, 430 (1842); **Ger.** pat. **229,815** in *Frdl.* **10**, 353 (1910); **Japan.** pat. **152,932** (1942 to ICI); *C.A.* **44**, 1544d (1950). Synthesis: Sandmeyer, *Helv. Chim. Acta* **2**, 234 (1919); **Brit.** pat. **128,122** in *C.A.* **13**, 2375 (1919); Marvel, Hiers, *Org. Syn. coll. vol.* I (2nd ed., 1941) p 327; Wibaut, Gerling, *Rec. Trav. Chim.* **50**, 41 (1931); Neunhoeffer, Lehmann, *Ber.* **94**, 2960 (1961); Ziegler *et al.*, *Monatsh.* **94**, 453 (1963). May be isolated from the urine of rabbits that are fed *o*-nitrophenylglyoxylic acid: Bohm, *Z. Physiol. Chem.* **265**, 210 (1940). Pharmacology: Singh, *Indian Vet. J.* **48**, 672 (1971). *Reviews:* Heller, *Ueber Isatin, Isatyd, Dioxindol und Indophenin* (F. Enke, Stuttgart, 1931); Sumpter, *Chem. Rev.* **34**, 393 (1944). Discussion of chemistry of isatin: Morton, *Chemistry of Heterocyclic Compounds* (New York, 1946) pp 126-132.

Orange-colored monoclinic prisms. mp 203.5° (partial sublimation). Absorption spectrum: Hartley, Dobbie, *J. Chem. Soc.* **75**, 647, 656. Freely sol in boiling alcohol; sol in ether and in boiling water with reddish-brown color; sol in alkali hydroxide solns with a violet color becoming yellow on standing. The alc soln imparts a persistent, disagreeable odor to the human skin. Extremely weak base, forms a crystalline perchlorate, $C_8H_5NO_2 \cdot HClO_4 \cdot 2H_2O$.
USE: Manuf vat dyes. In analytical chemistry as a reagent for cuprous ions, mercaptans, thiophene, and indican.

5117. Isatropic Acid. *1,2,3,4-Tetrahydro-1-phenyl-1,4-naphthalenedicarboxylic acid;* 1-phenyltetrahydro-1,4-naphthalenedicarboxylic acid; 1-phenyltetralin-1,4-dicarboxylic acid; diatropic acid; α-isatropic acid; *trans-isatropic acid.* $C_{18}H_{16}O_4$; mol wt 296.32. C 72.96%, H 5.44%, O 21.60%. Prepn: Staudinger, Ruzicka, *Ann.* **380**, 296 (1911); Smith, Forsberg, *C.A.* **42**, 7283 (1948). Structure: Voigtlander, Engelbert, *Ann.* **625**, 196 (1959).

dl-Form, crystals from chloroform + petr ether, mp 238.5-239°. Sparingly sol in hot water; moderately sol in alcohol, acetic acid; practically insol in ether, benzene, ligroin, carbon disulfide.
d-Form, crystals, sinters at 234°, dec at 230°. $[\alpha]_D^{20}$ +9.44° (1 g/15 ml abs alc). K at 18° = 1.20×10^{-4}. Soly in water 0.0047%.

5118. Isaxonine. *N-(1-Methylethyl)-2-pyrimidinamine;* 2-(isopropylamino)pyrimidine. $C_7H_{11}N_3$; mol wt 137.18. C 61.29%, H 8.08%, N 30.63%. Neurotropic agent that promotes neurite outgrowth. Prepn: D. J. Brown, J. S. Harper, *J. Chem. Soc.* **1965**, 5542. Prepn of pharmacologically active salts: A. Esanu, **Belg.** pat. **788,648**; *idem,* U.S. pat. **3,984,414**; *idem,* **Ger.** pat. **2,707,095**; *idem,* U.S. pat. **4,073,-895** (1973, 1976, 1977, 1978 all to Soc. d'Etudes Prod. Chim.). Nerve growth promoting action: A. Hugelin *et al., Compt. Rend. Ser. D* **285**, 1339 (1977); *idem, Experientia* **35**, 626 (1979). Enhancement of muscle re-innervation in man: A. Sebille, A. Hugelin, *Brit. J. Pharmacol.* **9**, 275 (1980). Series of articles on *in vitro* effects, pharmacology, clinical studies: *Nouv. Presse Med.* **11**, 1189-1280 (1982).

mp 27-28°. bp_{12} 92-93°.
Phosphate, $C_7H_{11}N_3 \cdot H_3PO_4$, *Nerfactor.* White crystalline powder, mp 125°. LD_{50} in mice (g/kg): 2.83 orally; 1.53 i.p. (Esanu, 1978).
THERAP CAT: In treatment of peripheral neuropathies.

5119. Isazofos. *Phosphorothioic acid O-[5-chloro-1-(1-methylethyl)-1H-1,2,4-triazol-3-yl] O,O-diethyl ester;* O-(5-chloro-1-isopropyl-1H-1,2,4-triazol-3-yl) O,O-diethyl phosphorothioate; isazophos; CGA-12223; Brace; Miral; Triumph. $C_9H_{17}ClN_3O_3PS$; mol wt 313.74. C 34.45%, H

5.46%, Cl 11.30%, N 13.39%, O 15.30%, P 9.87%, S 10.22%. Organophosphate pesticide. Prepn: D. Dawes, B. Böhner, **Ger.** pat. **2,260,015**; *eidem,* U.S. pat. **3,867,396** (1973, 1975 both to Ciba-Geigy); B. Böhner *et al., Kem.-Kemi* **1**, 585 (1974). Physical properties and nematocidal activity: D. Dawes *et al., Meded. Fac. Landbouwwet., Rijksuniv. Gent* **39**, 727 (1974). Persistence in turfgrass thatch and soil: H. D. Niemczyk, H. R. Krueger, *J. Econ. Entomol.* **80**, 950 (1987).

Amber liquid, $bp_{0.001}$ 100°; bp_{760} 170°. n_D^{20} 1.4867. d^{20} 1.22. Vapor pressure at 20°: 1.3×10^{-4} mm Hg. Soly (20°): 150 ppm in water. Miscible with methanol, chloroform, benzene, hexane. LD_{50} in rats (mg/kg): 60 orally; 250-700 dermally (Dawes).
Caution: Cholinesterase inhibitor.
USE: Nematocide; turf insecticide.

5120. Isbogrel. *(E)-7-Phenyl-7-(3-pyridinyl)-6-heptenoic acid;* CV-4151. $C_{18}H_{19}NO_2$, mol wt 281.35. C 76.84%, H 6.81%, N 4.98%, O 11.37%. Thromboxane A_2 (TXA_2) synthetase inhibitor. Prepn: S. Terao, K. Nishikawa, **Eur.** pat. **Appl. 98,690**; *eidem,* U.S. pat. **4,518,602** (1984, 1985 both to Takeda); and inhibition of TXA_2: K. Kato *et al., J. Med. Chem.* **28**, 287 (1985). Pharmacology: Z. Terashita *et al., Thromb. Res.* **41**, 223 (1986). Antithrombotic effect in experimental thrombi: T. Matsubara *et al., Am. Heart J.* **118**, 837 (1989). Clinical pharmacokinetics: T. Kuzuya *et al., Cardiovasc. Drugs Ther.* **2**, 693 (1988). Clinical trial: R. Hattori *et al., Japan. Circulat. J.* **55**, 324 (1991).

mp 114-115°.
THERAP CAT: Antithrombotic.

5121. Isepamicin. *(S)-O-6-Amino-6-deoxy-α-D-glucopyranosyl-(1→4)-O-[3-deoxy-4-C-methyl-3-(methylamino)-β-L-arabinopyranosyl-(1→6)]-N¹-(3-amino-2-hydroxy-1-oxopropyl)-2-deoxy-D-streptamine;* 1-N-[(S)-3-amino-2-hydroxypropionyl]gentamicin B; HAPA-B; Sch-21420. $C_{22}H_{43}N_5-O_{12}$; mol wt 569.61. C 46.39%, H 7.61%, N 12.30%, O 33.71%. Aminoglycoside antibiotic; semisynthetic derivative of gentamicin B. Prepn: M. J. Weinstein *et al.,* **Belg.** pat. **818,431**; J. J. Wright, P. J. L. Daniels, **Belg.** pat. **824,-657** (both 1975 to Sherico); J. J. Wright *et al.,* U.S. pat. **4,002,742** (1977 to Schering Corp.); T. L. Nagabhushan *et al., J. Antibiot.* **31**, 681 (1978). Antibacterial spectrum *in vitro* and *in vivo:* G. H. Miller *et al., ibid.* 688. Comparison with amikacin of nephrotoxic potential: L. I. Rankin *et al., Antimicrob. Ag. Chemother.* **16**, 491 (1979). Series of articles on antibacterial activity, toxicology and clinical efficacy: *Chemotherapy (Tokyo)* **35**, Suppl 5, 1-610 (1985); on toxicology: *Japan. J. Antibiot.* **39**, 3164-3328 (1986), *C.A.* **107**, 32666r-32673r (1987). Acute toxicity: T. Morino *et al., ibid.* 3164, *C.A.* **107**, 32666s (1987).

Sulfate, *Exacin, Isepacin*. $[\alpha]_D^{26}$ 110.9° (c = 1 in water as the disulfate hydrate). LD$_{50}$ in male, female mice, rats (mg/kg): 234, 236, 489, 476 i.v.; 3312, 3320, 3451, 3392 s.c.; > 5000 orally both species (Morino).

THERAP CAT: Antibacterial.

5122. Isethionic Acid. *2-Hydroxyethanesulfonic acid;* hydroxyethylsulfonic acid. $C_2H_6O_4S$; mol wt 126.13. C 19.04%, H 4.79%, O 50.74%, S 25.42%. HOCH$_2$CH$_2$SO$_3$H. Obtained by heating ethylene, chlorosulfonic acid, and water: Klason, *J. Prakt. Chem.* [2] **19**, 234 (1879); from ether and SO$_3$: Hubner, *Ann.* **223**, 198 (1884); from ethylenesulfite + NaHCO$_3$: Smith, U.S. pat. **2,899,461** (1959 to Dow).

Colorless, syrupy, strongly acid liquid. Kept over H$_2$SO$_4$, it solidifies to a very hygroscopic, cryst mass. Miscible with water, alc. Forms readily cryst salts with organic bases.

Caution: Irritating to skin, mucous membranes.

5123. Isinglass. Ichthyocolla; fish glue. The inner membrane of the swimming bladder of *Acipenser huso* and other species of sturgeon and of hake. *Constit.* Chiefly glutin.

Thin, white or yellowish, semi-transparent, pearly iridescent, horny sheets. Sol in hot water, hot dil alcohol. In cold water, softens and becomes adhesive.

USE: As nutrient and food instead of gelatin; as adhesive, in "court plaster"; for clarifying (fining) wine and beer; as protective colloid in manuf various chemicals; in glass and porcelain cements; for lustering and stiffening silks and other fabrics; with pyroxylin for waterproofing fabrics.

5124. Isoaminile. α-[2-(Dimethylamino)propyl]-α-(1-methylethyl)benzeneacetonitrile; 4-(dimethylamino)-2-isopropyl-2-phenylvaleronitrile; 3-cyano-5-dimethylamino-3-phenyl-2-methylhexane; α-(β-dimethylaminopropyl)-α-isopropylphenylacetonitrile; Aprecon; Dimyril; Nullatuss. $C_{16}H_{24}N_2$; mol wt 244.38. C 78.64%, H 9.90%, N 11.46%. Prepn: **Brit.** pat. **822,695**; Stühmer, Funke, U.S. pat. **2,934,557** (1959, 1960, both to Kali-Chemie).

Liquid, bp$_3$ 138-146°.

Citrate, $C_{16}H_{24}N_2 \cdot C_6H_8O_7$, *Perocan*. Crystals from alcohol + ether, mp 63-64°.

Cyclamate, $C_{22}H_{37}N_3O_3S$, *Mucalan, Peracon*. Prepn: Dickinson, U.S. pat. **3,074,996** (1963 to Abbott).

Tartrate, $C_{16}H_{24}N_2 \cdot C_4H_6O_6$, crystals, mp 64°.

THERAP CAT: Antitussive.

5125. Isoamyl Acetate. Amylacetic ester. $C_7H_{14}O_2$; mol wt 130.19. C 64.58%, H 10.84%, O 24.58%. CH$_3$COO-CH$_2$CH$_2$CH(CH$_3$)$_2$. The technical product is also known as *pear oil* or *banana oil*.

Colorless, neutral liq; pear-like odor and taste. d$_4^{15}$ 0.876. Pure isoamyl acetate bp 142°. n$_D^{21}$ 1.400; the ordinary grade of commerce boils between 120-145°. Flash pt, closed cup: 92° F (33° C); open cup: 100° F (38° C). Sol in 400 parts water; miscible with alcohol, ether, ethyl acetate, amyl alcohol. Soly of water in isoamyl acetate (25°) 1.6% by volume.

Caution: Potential symptoms of overexposure are irritation of eyes, nose and throat; narcosis; dermatitis. *See NIOSH Pocket Guide to Chemical Hazards* (DHHS/NIOSH 90-117, 1990) p 130.

USE: In alcohol solution as a pear flavor in mineral waters and syrups; as solvent for old oil colors, for tannins, nitrocellulose, lacquers, celluloid, and camphor; swelling bath sponges; covering unpleasant odors, perfuming shoe polish; manuf artificial silk, leather or pearls, photographic films, celluloid cements, waterproof varnishes, bronzing liquids, and metallic paints; dyeing and finishing textiles. A special grade of the amyl acetate has been used for burning in the Hefner lamp serving as a photometric standard.

5126. Isoamylamine. *3-Methyl-1-butanamine; isopentylamine;* isobutylcarbylamine; 3-methylbutylamine. C_5H_{13}-N; mol wt 87.16. C 68.90%, H 15.03%, N 16.07%. (CH$_3$)$_2$-CHCH$_2$CH$_2$NH$_2$. Prepn from acetamide + isoamylbromide: Erickson, *Ber.* **59B**, 2665 (1926); from α-methylhydroxylamine + isoamylmagnesium chloride: Sheverdina, Kocheshkov, *J. Gen. Chem. USSR* **8**, 1825 (1938). Isoln from *Clostridium sordelli:* Prevot, Thouvenot, *Ann. Inst. Pasteur* **103**, 925 (1962).

Colorless liq; strong ammonia odor. d^{18} 0.751. bp 95°. n$_D^{18}$ 1.4096. Miscible with water, alcohol, chloroform, ether.

Hydrochloride, $C_5H_{13}N \cdot HCl$, crystals from acetone, mp 215°. One gram dissolves in 0.5 ml water, 20 ml chloroform; sol in alcohol.

Caution: Irritating to skin, mucous membranes.

5127. Isoamyl Benzoate. $C_{12}H_{16}O_2$; mol wt 192.26. C 74.97%, H 8.39%, O 16.64%. C$_6$H$_5$COOCH$_2$CH$_2$CH(CH$_3$)$_2$.
Colorless liquid, d$_4^{15}$ 0.993. bp 260-262°. Insol in water; miscible with alcohol.

USE: In perfumery and cosmetics.

5128. Isoamyl Bromide. *1-Bromo-3-methylbutane.* C_5-H$_{11}$Br; mol wt 151.05. C 39.76%, H 7.34%, Br 52.90%. (CH$_3$)$_2$CHCH$_2$CH$_2$Br. Prepd from isoamyl alcohol, bromine and phosphorus or by refluxing isoamyl alcohol with HBr in presence of H$_2$SO$_4$: Kamm, Marvel, *Org. Syn.* **vol. 1,** p 4 (1921).

Colorless liquid. d$_4^{15}$ 1.210. bp 120-121°. mp −112°. n$_D^{15}$ 1.4433. Slightly sol in water (0.02 g/100 ml at 16.5°); miscible with alcohol, ether.

USE: In organic synthesis.

5129. Isoamyl Butyrate. $C_9H_{18}O_2$; mol wt 158.24. C 68.31%, H 11.47%, O 20.22%. CH$_3$(CH$_2$)$_2$COOCH$_2$CH$_2$-CH(CH$_3$)$_2$.
Colorless liq; aromatic pear-like odor. d$_{15}^{19}$ 0.866. bp 179°. Slightly sol in water; miscible with alcohol, ether.

USE: Manuf artficial rum and fruit essences.

5130. Isoamyl Chloride. *1-Chloro-3-methylbutane.* C_5-H$_{11}$Cl; mol wt 106.60. C 56.34%, H 10.40%, Cl 33.26%. (CH$_3$)$_2$CHCH$_2$CH$_2$Cl. Prepd by refluxing isoamyl alcohol and HCl in presence of ZnCl$_2$; also by direct chlorination of isopentane.

Liquid. d^0 0.893. bp ∼100°. n$_D^{20}$ 1.4103. Slightly sol in water; miscible with alcohol, ether.

5131. Isoamyl Cyanide. *4-Methylpentanenitrile;* isocapronitrile. $C_6H_{11}N$; mol wt 97.16. C 74.17%, H 11.41%, N 14.42%. (CH$_3$)$_2$CHCH$_2$CH$_2$CN. Prepd from KCN and isoamyl chloride in boiling alcohol.

Liquid, very disagreeable odor. d$_4^{20}$ 0.806. bp 155-156°. mp −51°. n$_D^{20}$ 1.406. Insol in water; miscible with alc, ether.

5132. Isoamyl Ether. *1,1'-Oxybis[3-methylbutane];* isoamyl oxide; diisoamyl ether. $C_{10}H_{22}O$; mol wt 158.28. C 75.88%, H 14.01%, O 10.11%. [(CH$_3$)$_2$CHCH$_2$CH$_2$]$_2$O.

Colorless liquid; pleasant, fruity odor. d$_4^{12}$ 0.783. bp 172°. n$_D^{20}$ 1.408. Insol in water; miscible with alc, chloroform, ether.

USE: Solvent in Grignard reaction; also as solvent of odorous principles; manuf lacquers; regenerating rubber.

5133. Isoamyl Formate. $C_6H_{12}O_2$; mol wt 116.16. C 62.04%, H 10.41%, O 27.55%. $HCOOCH_2CH_2CH(CH_3)_2$.

Colorless liquid; fruity odor. d^{20} 0.877. bp 123-124°. n_D^{20} 1.391. Sol in 300 parts water; miscible with alc, ether. LD_{50} orally in rats: 9840 mg/kg, P. M. Jenner et al., Food Cosmet. Toxicol. **2**, 327 (1964).

USE: Artificial fruit syrups.

5134. Isoamyl Iodide. 1-Iodo-3-methylbutane. $C_5H_{11}I$; mol wt 198.05. C 30.32%, H 5.60%, I 64.08%. $(CH_3)_2CH$-CH_2CH_2I. Obtained from isoamyl alcohol, iodine, and phosphorus, or by distillation of the alcohol with HI.

Liquid; quickly becomes brown on exposure to air and light. d_{15}^{18} 1.515. bp 147°. Slightly sol in water; miscible with alc, ether. Keep well closed and protected from light.

5135. Isoamyl Isovalerate. Isoamyl valerianate; amyl valerate; amyl isovalerate; apple oil. $C_{10}H_{20}O_2$; mol wt 172.27. C 69.72%, H 11.70%, O 18.58%. $(CH_3)_2CHCH_2$-$COOCH_2CH_2CH(CH_3)_2$.

Colorless liquid; apple-like odor. d_4^{19} 0.858. bp 191-194°. n_D^{19} 1.413. Very slightly sol in water; miscible with alc, ether.

USE: As apple essence for flavoring liqueurs and candy.

5136. Isoamyl Nitrate. 3-Methyl-1-butanol nitrate. C_5-$H_{11}NO_3$; mol wt 133.15. C 45.10%, H 8.33%, N 10.52%, O 36.05%. $(CH_3)_2CHCH_2CH_2NO_3$.

Colorless liquid. d_4^{22} 0.996. bp 147-148°. n_D^{22} 1.4122. Slightly sol in water; miscible with alcohol, ether.

Caution: Do not confuse with isoamyl nitrite, q.v.

5137. Isoamyl Nitrite. Isopentyl nitrite. $C_5H_{11}NO_2$; mol wt 117.15. C 51.26%, H 9.46%, N 11.96%, O 27.31%. $(CH_3)_2CHCH_2CH_2ONO$. Prepd by nitrosation of isopentyl alcohol: Bevillard, Choucroun, Bull. Soc. Chim. France **1957**, 337.

Yellowish, transparent, flammable liquid; penetrating fragrant, somewhat fruity odor. Unstable and dec on exposure to air and light. The pure nitrite has d_{25}^{25} 0.875. bp 97-99°, but volatilizes readily at much lower temps. n_D^{21} 1.3871. Very slightly sol in water; miscible with alc, chloroform, ether. Forms an explosive mixture with air or oxygen. Keep in tightly closed containers protected from light and in cool place. Incompat. Alcohol, antipyrine, caustic alkalies, alkaline carbonates, potassium iodide, bromides, ferrous salts.

5138. Isoamyl Phthalate. 1,2-Benzenedicarboxylic acid bis(3-methylbutyl) ester; diisoamyl phthalate; amyl phthalate. $C_{18}H_{26}O_4$; mol wt 306.40. C 70.56%, H 8.55%, O 20.89%. Colorless, practically odorless liquid. d 1.028. bp_{40} 225°. Insol in water; sol in organic solvents.

USE: Plasticizer for nitrocellulose and resin lacquers; preventing foam in manuf of glue; in rubber cements.

5139. Isoamyl Salicylate. $C_{12}H_{16}O_3$; mol wt 208.26. C 69.21%, H 7.74%, O 23.05%. $C_6H_4(OH)COOC_5H_{11}$.

Colorless liquid; pleasant odor. d_{15}^{19} 1.048. bp 274-278°. Flash pt 132°. n_D^{20} 1.506. Almost insol in water, miscible with alcohol, chloroform, ether.

USE: In perfumery and soaps.

5140. Isoapo-β-erythroidine. $C_{15}H_{15}NO_2$; mol wt 241.29. C 74.67%, H 6.27%, N 5.80%, O 13.26%. Separation from apo-β-erythroidine: Sauvage et al., Science **109**, 627 (1949); Sauvage, Boekelheide, J. Am. Chem. Soc. **72**, 2062 (1950); Koniuszy, Folkers, ibid. **73**, 333 (1951). Synthesis and structure: Blake et al., ibid. **87**, 1397 (1965).

Crystals, mp 146-148°. uv max (ethanol): 379, 288, 253 nm (ε 6500, 10,800, 16,800).

5141. Isoascorbic Acid. D-erythro-Hex-2-enonic acid γ-lactone; D-araboascorbic acid; erythorbic acid; isovitamin C; saccharosonic acid; glucosaccharonic acid; D-erythro-3-ketohexonic acid lactone; D-erythro-3-oxohexonic acid lactone; erycorbin; Mercate "5"; Neo-Cebicure. $C_6H_8O_6$; mol wt 176.13. C 40.92%, H 4.58%, O 54.50%. Epimer of L-ascorbic acid. Has one-twentieth of the vitamin C activity of L-ascorbic acid. Prepd by treating methyl 2-keto-D-gluconate with sodium methoxide: Maurer, Schiedt, Ber. **66**, 1054 (1933); **67**, 1239 (1934); Ohle et al., ibid. **67**, 324 (1934); Baird et al., J. Chem. Soc. **1934**, 63; Reichstein et al., Helv. Chim. Acta **17**, 516 (1934). Synthesis from sucrose: Heimann, Reiff, Pharm. Zentralh. **93**, 97 (1954). Production by Penicillium spp: Takahashi et al., Nature **188**, 411 (1960); U.S. pat. **3,052,609** (1962 to Sankyo); Bull. Agr. Chem. Soc. Japan **24**, 533 (1960), C.A. **55**, 1788 (1961). Conformation: Matsui et al., Agr. Biol. Chem. **27**, 185 (1963).

Shiny granular crystals from water or dioxane. Dec 174°. $[\alpha]_D^{16.5}$ −17° (c = 1.8 in 0.01N HCl); $[\alpha]_D^{20}$ −16.6° (H_2O). Soluble in water, alc, pyridine. Moderately sol in acetone. Slightly sol in glycerol.

Sodium salt, sodium erythorbate, Mercate "20", Neo-Cebitate. Crystals, sol in water. pH of aq solns of the sodium salt between 5 and 6. A 10% soln, made from commercial grade, may have a pH of 7.2 to 7.9. The free acid is more sol in water (40 g/100 ml H_2O) than the sodium salt (16 g/100 ml H_2O).

USE: Antioxidant and antimicrobial agent for foods: Kadin, Osadca, J. Agr. Food Chem. **7**, 358 (1959).

5142. Isobenzan. 1,3,4,5,6,7,8,8-Octachloro-1,3,3a,4,7,-7a-hexahydro-4,7-methanoisobenzofuran; 1,3,4,5,6,7,10,10-octachloro-4,7-endomethylene-4,7,8,9-tetrahydrophthalan; SD-4402; R-6700; Omtan; Telodrin. $C_9H_4Cl_8O$; mol wt 411.75. C 26.25%, H 0.98%, Cl 68.88%, O 3.89%. Prepn: Brit. pat. **772,212** (1957 to Ruhrchemie); Korte, Stiasni, Ann. **656**, 140 (1962); Feichtinger, Linden, Chem. & Ind. (London) **1965**, 1938. Toxicity study: Tauberger, Arch. Exp. Pathol. Pharmakol. **232**, 227 (1957).

Crystals from heptane, mp 120-122°. Sol in acetone, benzene, toluene, ether, xylene, heavy aromatic naphtha. LD_{50} i.v. in rats: 1.8 mg/kg (Tauberger).

USE: Insecticide.

5143. Isoborneol. exo-1,7,7-Trimethylbicyclo[2.2.1]heptan-2-ol; exo-2-bornanol; exo-2-camphanol. $C_{10}H_{18}O$; mol wt 154.25. C 77.87%, H 11.76%, O 10.37%. Prepn of dl-form: Pickard, Littlebury, J. Chem. Soc. **91**, 1973 (1907); Truett, Moulton, J. Am. Chem. Soc. **73**, 5913 (1951); Ziegler, Brit. pat. **803,178** (1958). Prepn of the l-form by reduction of d-camphor with lithium aluminum hydride: Trevoy, Brown, J. Am. Chem. Soc. **71**, 1675 (1949). Separation of isoborneol from its endo-isomer, borneol, via the p-nitrobenzoate deriv: Truett, Moulton, loc. cit. Resolution of the dl-form: Pickard, Littlebury, loc. cit.; Kenyon, Priston, ibid. **127**, 1472 (1925). Configuration (isoborneol = exo-form; borneol = endo-form): Toivonen et al., Acta Chem. Scand. **3**, 991 (1949). Review: J. L. Simonsen, The Terpenes vol. II (University Press, Cambridge, 2nd ed., 1949) pp 365-367; A. R. Pinder, The Chemistry of the Terpenes (Chapman & Hall, London, 1960) pp 22-24, 101, 103, 105-107, 111.

l-Isoborneol

dl-Form, crystals from petr ether. Sublimes on heating, mp 212° (in a sealed tube). Practically insol in water. Readily sol in alcohol, ether, chloroform.

d-Form, crystals from petr ether, mp 214°. Approx $[\alpha]_D$ +34.3° in alc soln: Picard, Littlebury, *loc. cit.*

l-Form, crystals from petr ether, mp 214°. Approx $[\alpha]_D$ −34.3° in alc soln.

5144. Isobornyl Thiocyanoacetate. *Thiocyanatoacetic acid exo-1,7,7-trimethylbicyclo[2.2.1]hept-2-yl ester;* terpinyl thiocyanoacetate; Thanite. $C_{13}H_{19}NO_2S$; mol wt 253.37. C 61.63%, H 7.56%, N 5.53%, O 12.63%, S 12.66%. *See also* Borneol. May be prepared by treating isoborneol with chloroacetyl chloride and KCNS: J. N. Borglin, U.S. pat. **2,217,611** (1940 to Hercules).

Technical grade, yellow oily liquid with terpene-like odor. Contains 82% or more of isobornyl thiocyanoacetate with other terpenes. $bp_{0.06}$ 95°. Flash pt 82° (180°F). d_4^{25} 1.1465. Acid no. 1.19. n_D^{25} 1.512. Very sol in alcohol, benzene, chloroform, ether. Practically insol in water.

USE: Insecticide, especially in cattle sprays.

5145. Isobutyl Acetate. $C_6H_{12}O_2$; mol wt 116.16. C 62.04%, H 10.41%, O 27.55%. $CH_3COOCH_2CH(CH_3)_2$. Prepn from isobutyl alcohol + acetic acid: **Belg. pat. 505,-023** (1951 to Soc. Belge de l'Azote et des Prod. Chim. du Marly); from methyl isobutyl ketone: White, Emmons, *Tetrahedron* **17,** 31 (1962). Isoln from wood-rotting fungus, *Endoconidiophora coerulescens* Münch.: Birkinshaw, Morgan, *Biochem. J.* **47,** 55 (1950).

Colorless liquid. d_4^{20} 0.871. bp 118°. mp −99°. Flash pt, closed cup: 64° F (18° C). n_D^{19} 1.3907. Sol in 180 parts water, freely in alc.

Caution: Potential symptoms of overexposure are headache; drowsiness; irritation of eyes, upper respiratory system and skin; anesthesia. *See NIOSH Pocket Guide to Chemical Hazards* (DHHS/NIOSH 90-117, 1990) p 130.

USE: Flavoring, solvent.

5146. Isobutyl Alcohol. *2-Methyl-1-propanol;* isopropylcarbinol; 1-hydroxymethylpropane; fermentation butyl alcohol. $C_4H_{10}O$; mol wt 74.12. C 64.82%, H 13.60%, O 21.58%. $(CH_3)_2CHCH_2OH$. Present in fusel oil; also produced by fermentation of carbohydrates: Baraud, Genevois, *Compt. Rend.* **247,** 2479 (1958); Sukhodol, Chatskii, *Spirt. Prom.* **28,** 35 (1962), *C.A.* **57,** 5124e (1962). Prepn: Wender *et al., J. Am. Chem. Soc.* **71,** 4160 (1949); Schreyer, U.S. pat. **2,564,130** (1951 to du Pont); Pistor, U.S. pat. **2,753,366;** Harrer, Rühl, **Ger.** pat. **1,011,865;** Himmler, Schiller, U.S. pat. **2,787,628** (1956, 1957, 1957 all to BASF). Toxicity study: Smyth *et al., Arch. Ind. Hyg. Occup. Med.* **10,** 61 (1954).

Colorless, refractive liq; flammable; odor like that of amyl alcohol, but weaker. d^{15} 0.806. bp 108°. mp −108°. Flash pt, closed cup: 82°F (28° C). n_D^{15} 1.3976. Sol in about 20 parts water; misc with alcohol, ether. LD_{50} orally in rats: 2.46 g/kg (Smyth).

Caution: Potential symptoms of overexposure are irritation of eyes and throat; headache, drowsiness; skin irritation, skin cracking. *See NIOSH Pocket Guide to Chemical Hazards* (DHHS/NIOSH 90-117, 1990) p 132.

USE: Manuf esters for fruit flavoring essences; solvent in paint, varnish removers.

5147. Isobutylamine. *2-Methyl-1-propanamine;* 2-methylpropylamine; 1-amino-2-methylpropane. $C_4H_{11}N$; mol wt 73.14. C 65.69%, H 15.16%, N 19.15%. $(CH_3)_2CH$-CH_2NH_2. Prepn from isobutyl alcohol + NH_3: **Brit.** pat. **847,799** (1960 to Cellulose-Polymères et Dérivés "Cepede"); Shirley, Speranza, U.S. pat. **3,128,311** (1964 to Jefferson Chem.). Isoln from fungi: von Kamienski, *Planta* **50,** 331 (1958). Acute toxicity: K. L. Cheever *et al., Toxicol. Appl. Pharmacol.* **62,** 150 (1982).

Liquid. d_4^{25} 0.724. mp −85°. bp 68-69°. n_D^{17} 1.3988. Miscible with water, alcohol, ether. LD_{50} (14 day) in male, female rats (mg/kg): 224.4, 231.8 orally (Cheever).

Caution: Skin contact can result in erythema, blistering. Inhalation causes headache, dryness of nose and throat.

5148. Isobutyl *p*-Aminobenzoate. *4-Aminobenzoic acid 2-methylpropyl ester;* Cycloform; Isobutyl Kelo-form; Isocaine. $C_{11}H_{15}NO_2$; mol wt 193.25. C 68.37%, H 7.82%, N 7.25%, O 16.56%. Prepn: Adams *et al., J. Am. Chem. Soc.* **48,** 1758 (1926).

Crystals from benzene, mp 65°. Slightly sol in water; sol in alcohol, benzene, ether, acetone, olive oil.

USE: Sunscreen.

THERAP CAT: Anesthetic (topical).

5149. Isobutylbenzene. *(2-Methylpropyl)benzene;* 2-methyl-1-phenylpropane. $C_{10}H_{14}$; mol wt 134.22. C 89.49%, H 10.51%. $C_6H_5CH_2CH(CH_3)_2$. Prepd from bromobenzene, isobutyl iodide, and sodium preferably in the presence of benzene: Wreden, Znatovicz, *Ber.* **9,** 1606 (1876); Schramm, *Monatsh.* **9,** 616 (1888); by passing vapors of dimethylbenzyl carbinol and hydrogen over activated charcoal: Zelinsky, Gawerdowskaya, *Ber.* **61,** 1052 (1928).

Liquid. d_4^{20} 0.8673. bp_{760} 170.5°; bp_{400} 145.2°; bp_{200} 120.7°; bp_{100} 99.0°; bp_{60} 84.1°; bp_{40} 73.2°; bp_{20} 54.7°; bp_{10} 37.3°; bp_5 21.1°; $bp_{1.0}$ −9.8°. n_D^{20} 1.4928. Critical temp 377.1°; crit pressure 23,636 mm Hg.

5150. Isobutyl Bromide. *1-Bromo-2-methylpropane.* C_4H_9Br; mol wt 137.02. C 35.06%, H 6.62%, Br 58.32%. $(CH_3)_2CHCH_2Br$. Obtained from isobutyl alc and PBr_3: Noller, Dinsmore, *Org. Syn.* **coll. vol. II,** 358 (1943).

Colorless liquid. d^{15} 1.272. bp 91.5°. mp −119°. n_D^{15} 1.4391. Slightly sol in water (0.6 g/l); miscible with alcohol, ether.

5151. Isobutyl *n*-Butyrate. $C_8H_{16}O_2$; mol wt 144.21. C 66.63%, H 11.18%, O 22.19%. $CH_3CH_2CH_2COOCH_2$-$CH(CH_3)_2$.

Liquid. d 0.866. bp 157°. n_D^{20} 1.4035. Slightly sol in water; miscible with alcohol, ether.

5152. Isobutyl Carbamate. $C_5H_{11}NO_2$; mol wt 117.15. C 51.26%, H 9.46%, N 11.96%, O 27.31%. $H_2NCOOCH_2$-$CH(CH_3)_2$. Prepd from isobutyl chloroformate and NH_3 in benzene soln.

Crystals, d 0.956. mp 67°; also stated at 61°. bp 206-207°. Insol in water. Sol in alcohol, ether.

5153. Isobutyl Chloride. *1-Chloro-2-methylpropane.* C_4H_9Cl; mol wt 92.57. C 51.90%, H 9.80%, Cl 38.30%. $(CH_3)_2CHCH_2Cl$. Formed by the action of HCl or PCl_5 on isobutyl alcohol.

Liquid. d^{15} 0.883. bp 68-69°. mp −131°. n_D^{15} 1.40096. Insoluble in water. Miscible with alcohol, ether.

5154. Isobutyl Chlorocarbonate. *Carbonochloridic acid 2-methylpropyl ester;* isobutyl chloroformate. C_5H_9-ClO_2; mol wt 136.58. C 43.97%, H 6.64%, Cl 25.96%, O 23.43%. $ClCO_2CH_2CH(CH_3)_2$. Prepd from isobutyl alcohol and phosgene.

Clear liquid; vapors irritate eyes and mucous membranes. d 1.040. bp 130°. Gradually dec by water and alcohol. Miscible with benzene, chloroform, ether.

5155. Isobutylene. *2-Methylpropene;* isobutene. C_4H_8; mol wt 56.11. C 85.63%, H 14.37%. $CH_2=C(CH_3)_2$. Obtained from refinery streams by absorption on 65% H_2SO_4 at about 15°: Packie, Rupp, U.S. pats. **2,424,186, 2,509,885;** Draeger, U.S. pat. **2,456,260;** Steele, Epps Jr., U.S. pat. **2,497,191** (1947, 1950, 1948, 1950, all to Standard Oil); Peters, Gothman; Edwards, Wesselhoft, U.S. pats. **2,962,-537; 3,129,265** (1960, 1964, both to Esso). Separation from a mixed C_4 stream using 50% H_2SO_4: **Brit.** pats. **824,573, 858,645** and **Fr.** pat. **1,337,232** (1959, 1961 and 1963 to Compagnie Francaise de Raffinage); Valet *et al., Hydrocarbon Process. Petrol. Refiner* **41,** No. 5, 119 (1962); Martel, *Chem. Eng.* **72,** No. 7, 66 (1965). Prepn: Verdol, U.S. pat. **3,170,000** (1965 to Sinclair). *Review:* Kennedy, Kirshenbaum, "Isobutylene" in *Vinyl and Diene Monomers* (part 2), E. C. Leonard, Ed. (Wiley-Interscience, New York, 1971) pp 691-756.

Gas. bp_{760} —6.900°; bp_{100} —49.309°; bp_{30} —67.90°; bp_{10} —81.95°; bp_1 —105.06°. d_4^{20} 0.5942; d_4^{25} 0.5879; d_4^{30} 0.5815. Practically insol in water. Very sol in alc, ether, sulfuric acid.

Caution: Simple asphyxiant.

USE: Primarily used to produce diisobutylene, trimers, butyl rubber, and other polymers; also to produce antioxidants for foods, packaging, food supplements, and for plastics: Hatch, *Petrol. Refiner* **39,** No. 6, 207 (1960).

5156. Isobutyl Ether. *1,1'-Oxybis[2-methylpropane];* diisobutyl ether. $C_8H_{18}O$; mol wt 130.23. C 73.78%, H 13.93%, O 12.29%. $(CH_3)_2CHCH_2OCH_2CH(CH_3)_2$.

Colorless liquid, characteristic odor. d^{15} 0.761. bp 122-124°. Insol in water. Miscible with alcohol, ether.

5157. Isobutyl Formate. Tetryl formate. $C_5H_{10}O_2$; mol wt 102.13. C 58.80%, H 9.87%, O 31.33%. $HCOO-CH_2CH(CH_3)_2$.

Liquid. d_4^{20} 0.885. bp 98°. mp —95°. n_D^{20} 1.3858. Sol in 100 parts water; miscible with alc, ether.

5158. Isobutyl Iodide. *1-Iodo-2-methylpropane.* C_4H_9I; mol wt 184.02. C 26.11%, H 4.93%, I 68.96%. $(CH_3)_2CH-CH_2I$. Obtained by distilling isobutyl alcohol and HI, or from isobutyl alcohol, iodine and phosphorus.

Liquid; becomes brown on exposure. d^{20} 1.605. bp 120°. mp 93°. n_D^{20} 1.4960. Insol in water. Miscible with alcohol, ether. *Keep well closed and protected from light.*

5159. Isobutyl Isobutyrate. *2-Methylpropanoic acid 2-methylpropyl ester.* $C_8H_{16}O_2$; mol wt 144.21. C 66.63%, H 11.18%, O 22.19%. $(CH_3)_2CHCOOCH_2CH(CH_3)_2$.

Liquid. d_4^0 0.875. bp 147°. mp —81°. n_D^{20} 1.3999. Insol in water. Miscible with alcohol.

5160. Isobutyl Isovalerate. Isobutyl valerate. $C_9-H_{18}O_2$; mol wt 158.24. C 68.31%, H 11.47%, O 20.22%. $(CH_3)_2CHCH_2COOCH_2CH(CH_3)_2$.

Liquid; ethereal odor. d^{20} 0.853. bp 170-172°. n_D^{20} 1.4064. Insol in water. Miscible with alcohol, ether.

USE: Flavoring and manuf fruit essences.

5161. Isobutyl Mercaptan. *2-Methyl-1-propanethiol.* $C_4H_{10}S$; mol wt 90.19. C 53.27%, H 11.18%, S 35.55%. $(CH_3)_2CHCH_2SH$. May be prepd from isobutyl bromide and KHS in alcohol, *see ref under* other butyl mercaptans.

Mobile liq. Heavy skunk odor. mp —79°. bp 88°. Flammable. d_4^{20} 0.8357. n_D^{20} 1.43859. Slightly sol in water; very sol in alcohol, ether, liquid hydrogen sulfide.

5162. Isobutyl Nitrate. $C_4H_9NO_3$; mol wt 119.12. C 40.33%, H 7.62%, N 11.76%, O 40.29%. $(CH_3)_2CHCH_2O-NO_2$.

Liquid. d^{20} 1.015. bp 123-125°. n_D^{20} 1.4028. Insol in water. Miscible with alcohol, ether.

5163. Isobutyl Nitrite. $C_4H_9NO_2$; mol wt 103.12. C 46.59%, H 8.80%, N 13.58%, O 31.03%. $(CH_3)_2CHCH_2-NO_2$. Prepd from isobutyl alcohol, $NaNO_2$, and dil H_2SO_4.

Colorless liq. d^{22} 0.870. bp 67°. n_D^{22} 1.3715. Slightly sol and gradually dec by water. Miscible with alcohol.

5164. Isobutyl Propionate. $C_7H_{14}O_2$; mol wt 130.19. C 64.58%, H 10.84%, O 24.58%. $CH_3CH_2COOCH_2CH-(CH_3)_2$.

Liquid; agreeable, ethereal odor. d_4^0 0.888. bp 137°. mp —71°. n_D^{20} 1.3975. Insol in water. Miscible with alc.

USE: Manuf fruit essences.

5165. Isobutyl Stearate. $C_{22}H_{44}O_2$; mol wt 340.59. C 77.58%, H 13.02%, O 9.40%. $CH_3(CH_2)_{16}COOCH_2CH-(CH_3)_2$.

Paraffin-like, crystal substance at low temp, mp about 20°.

USE: Waterproof coatings, polishes, face creams, rouges, ointments, soaps, rubber manuf, dye solns, inks, lubricants.

5166. Isobutyl Sulfide. *1,1'-Thiobis[2-methylpropane];* diisobutyl sulfide. $C_8H_{18}S$; mol wt 146.30. C 65.68%, H 12.40%, S 21.92%. $[(CH_3)_2CHCH_2]_2S$. Obtained by heating isobutyl chloride or potassium isobutyl sulfate with K_2S.

Liquid. d_4^{10} 0.836. bp 171-173°. Insol in water. Miscible with alcohol, ether.

5167. Isobutyl Thiocyanate. Isobutyl sulfocyanate. C_5H_9NS; mol wt 115.20. C 52.13%, H 7.87%, N 12.16%, S 27.84%. $(CH_3)_2CHCH_2SCN$. Obtained by distilling potassium isobutyl sulfate with NaCSN.

Liquid, bp 174-176°. Almost insol in water. Miscible with alcohol.

5168. Isobutyl Urethane. *(2-Methylpropyl)carbamic acid ethyl ester.* $C_7H_{15}NO_2$; mol wt 145.20. C 57.90%, H 10.41%, N 9.65%, O 22.04%. $(CH_3)_2CHCH_2NHCOOC_2H_5$. Obtained by shaking isobutylamine with ethyl chloroformate and aq KOH in the cold.

Liquid; apple like odor. d_4^{20} 0.943. Solidif below —65°. bp_{17} 96°. n_D^{20} 1.4288. Insol in water. Sol in alcohol.

5169. Isobutyraldehyde. *2-Methylpropanal;* isobutylaldehyde; isobutyric aldehyde. C_4H_8O; mol wt 72.11. C 66.63%, H 11.18%, O 22.19%. $(CH_3)_2CHCHO$. Manuf by the oxo process from propylene, carbon monoxide, and hydrogen at 130-160° and 1500-3000 psi in the presence of a cobalt catalyst: Roelen, U.S. pat. **2,327,066** (1943); also prepared by air oxidation of isobutyl alcohol: Fossek, *Monatsh.* **2,** 614 (1881); Lipp, *Ann.* **205,** 2 (1880). Toxicity study: Smyth *et al., Arch. Ind. Hyg. Occup. Med.* **10,** 61 (1954). *Review:* P. D. Sherman in Kirk-Othmer *Encyclopedia of Chemical Technology* vol. 4 (Wiley-Interscience, New York, 3rd ed., 1978) pp 376-386.

Flammable liq. Pungent odor. d_4^{20} 0.7938. mp —65.9°. bp_{760} 64°. n_D^{20} 1.3730. Heat of combustion 599.9 kcal. Flash pt (open cup) <20°F. Soly in water at 20°: 11 g/100 ml H_2O. Miscible with ethanol, ether, carbon disulfide, acetone, benzene, toluene, chloroform. Forms an azeotrope with water contg 94% isobutyraldehyde. The azeotrope bp_{760} 59°. Weight per gallon of isobutyraldehyde: 6.55 lbs. Oxidizes slowly on exposure to air, forming isobutyric acid. LD_{50} orally in rats: 3.7 g/kg (Smyth).

USE: In the synthesis of pantothenic acid, valine, leucine, cellulose esters, perfumes, flavors, plasticizers, resins, gasoline additives.

5170. Isobutyric Acid. *2-Methylpropanoic acid;* isopropylformic acid. $C_4H_8O_2$; mol wt 88.11. C 54.53%, H 9.15%, O 36.32%. $(CH_3)_2CHCOOH$. Prepn from 1-nitroisobutane: Lippincott, Hass, *Ind. Eng. Chem.* **31,** 118 (1939); from methallyl chloride: Towle, Hall, U.S. pat. **2,667,508** (1954 to Standard Oil Co. of Indiana); from propylene: Alderson, U.S. pat. **3,020,314** (1962 to du Pont); from 2-methylpropane + CO with HF catalyst: Friedman, Cotton, *J. Org. Chem.* **27,** 481 (1962).

Liquid; pungent odor like that of butyric acid, but not as unpleasant. bp_{760} 152-155°. mp —47°. d^{20} 0.950. n_D^{20} 1.3930. Flash pt, open cup: 170° F (77° C). Sol in 6 parts of water; misc with alcohol, chloroform, ether.

Caution: Mild irritant.

5171. Isobutyronitrile. *2-Methylpropanenitrile;* isopropyl cyanide; 2-cyanopropane. C_4H_7N; mol wt 69.11. C 69.52%, H 10.21%, N 20.27%. $(CH_3)_2CHCN$. Prepn: E. A. Letts, *Ber.* **5,** 669 (1872); A. E. Arbusow, *ibid.* **43,** 2296 (1910); R. E. Kent *et al., Org. Syn.* coll. vol. III, 493 (1955);

A. Ahmad, *Synthesis* **1976**, 418. Acute toxicity and human industrial experience: H. Zeller *et al., Zentralbl. Arbeitsmed. Arbeitsschutz* **19**, 225 (1969), *C.A.* **71**, 128340 (1969). Acute toxicity and metabolism: H. Tanii, K. Hashimoto, *Arch. Toxicol.* **55**, 47 (1984).

Colorless liquid. bp_{740} 99-102°. bp 101-103° (Kent); also reported as bp 103-103.5° (Arbusow), bp 104° (Ahmad), bp 107-108° (Letts). $d_D^{16.25}$ 0.7731. n_D^{25} 1.3713. Log P (*n*-octanol/water): 0.46. LD_{50} (mg/kg): 25 i.p. in mice; 200 orally in rats (Zeller). LD_{50} orally in male mice: 0.3652 mmol/kg (Tanii).

USE: Solvent.

5172. Isocarboxazid. *5-Methyl-3-isoxazolecarboxylic acid 2-benzylhydrazide;* 3-(*N*-benzylhydrazinocarbonyl)-5-methylisoxazole; 1-benzyl-2-(5-methyl-3-isoxazolylcarbonyl)hydrazine; Ro-5-0831; Marplan. $C_{12}H_{13}N_3O_2$; mol wt 231.25. C 62.33%, H 5.67%, N 18.17%, O 13.84%. Monoamine oxidase inhibitor. Prepn: Gardner, Wenis, U.S. pat. **2,908,688** (1959 to Hoffmann-La Roche); *J. Med. Pharm. Chem.* **2**, 133 (1960). Comprehensive description: B. C. Rudy, B. Z. Senkowski in *Analytical Profiles of Drug Substances* vol. 2, K. Florey, Ed. (Academic Press, New York, 1973) pp 295-314.

Crystals from methanol, practically tasteless. mp 105-106°. Very sparingly sol in hot water (0.05%), somewhat more (1 to 2%) in 95% alc, in glycerol, in propylene glycol.

THERAP CAT: Antidepressant.

5173. Isochondrodendrine. $O^7, O^{7'}$-*Didemethylcycleanine;* isobebeerine. $C_{36}H_{38}N_2O_6$; mol wt 594.71. C 72.71%, H 6.44%, N 4.71%, O 16.14%. From *Chondodendron tomentosum* Ruiz and Pav., *Menispermaceae:* Faltis, *Monatsh.* **33**, 873 (1912); Faltis, Neumann, *ibid.* **42**, 311 (1921); Dutcher, *J. Am. Chem. Soc.* **68**, 419 (1946); from the drug *Radix pareirae bravae:* King, *J. Chem. Soc.* **1940**, 737. Structure: Faltis, Dietreich, *Ber.* **67**, 231 (1934); Jeffreys, *J. Chem. Soc.* **1956**, 4451.

Needles from methanol, dec 288°. $[\alpha]_D^{20}$ −29° (c = 1.3 in chloroform). Sol in alcohol, benzene, chloroform.

5174. Isocil. *5-Bromo-6-methyl-3-(1-methylethyl)-2,4-(1H,3H)-pyrimidinedione;* 5-bromo-3-isopropyl-6-methyluracil; Hyvar General Weed Killer. $C_8H_{11}BrN_2O_2$; mol wt 247.09. C 38.89%, H 4.49%, Br 32.34%, N 11.34%, O 12.95%. Prepn: Bucha *et al., Science* **137**, 537 (1962). Toxicity study: G. W. Bailey, J. L. White, *Residue Rev.* **10**, 97 (1965). Superseded by bromacil, *q.v.*

Crystals from 20% aq ethanol, mp 158-159°. LD_{50} orally in rats: 3400 mg/kg (Bailey, White).

USE: Herbicide.

5175. Isocinchomeronic Acid. *2,5-Pyridinedicarboxylic acid.* $C_7H_5NO_4$; mol wt 167.12. C 50.31%, H 3.02%, N 8.38%, O 38.29%. Prepd by treating 5-ethyl-2-picoline (aldehyde-collidine) with suitable oxidizing agents such as HNO_3, $KMnO_4$, or H_2SeO_3: Meyer, Staffen, *Monatsh.* **34**, 517 (1913); Jordan, *Ind. Eng. Chem.* **44**, 332 (1952); Kato, *Bull. Chem. Soc. Japan* **34**, 636 (1961); by oxidation of 5-acetyl-2-methylpyridine: Binns, Swan, *J. Chem. Soc.* **1962**, 2831; from the lutidine fraction of coal tar lutidine: Lukes *et al., Coll. Czech. Chem. Commun.* **26**, 3044 (1961).

Triclinic leaflets or prisms from dilute HCl, mp 254°. Sublimes as nicotinic acid when heated above mp. Practically insol in cold water, alcohol, ether, benzene. Slightly sol in boiling water, boiling alcohol. Appreciably sol in hot, dil aq solns of mineral acids.

Monohydrate, crystals from water or alcohol, dec 238°.
Dihydrazide, crystals, mp 268-269°.

USE: Intermediate in the manuf of nicotinic acid.

5176. Isoconazole. *1-[2-(2,4-Dichlorophenyl)-2-[(2,6-dichlorophenyl)methoxy]ethyl]-1H-imidazole; 1-[2,4-dichloro-β-[(2,6-dichlorobenzyl)oxy]phenethyl]imidazole.* $C_{18}H_{14}Cl_4N_2O$; mol wt 416.13. C 51.95%, H 3.39%, Cl 34.08%, N 6.73%, O 3.84%. Prepn: E. F. Godefroi *et al., J. Med. Chem.* **12**, 784 (1969); *eidem,* **Ger.** pat. **1,940,388**; *eidem,* U.S. pats. **3,717,655** and **3,839,574** (1970, 1973, 1974 all to Janssen). *In vitro* activity: H. J. Kessler, *Arzneimittel-Forsch.* **29**, 1344 (1979). Animal study: H. J. Kessler *et al., ibid.* 1352. Antimicrobial activity in humans: H. Wendt, H. J. Kessler, *ibid.* 846. Bioavailability: U. Täuber, M. Rzadkiewicz, *Mykosen* **22**, 201 (1979).

Nitrate, $C_{18}H_{14}Cl_4N_2O \cdot HNO_3$, *R-15454, Fazol, Gyno-Travogen, Travogen, Travogyn.* Solid, mp 182-183°.

THERAP CAT: Antibacterial; antifungal.

5177. Isocorybulbine. *(13S-trans)-5,8,13,13a-Tetrahydro-3,9,10-trimethoxy-13-methyl-6H-dibenzo[a,g]quinolizin-2-ol; 3,9,10-trimethoxy-13α-methyl-13aβ-berbin-2-ol;* 2-hydroxy-13-methyl-3,9,10-trimethoxyberbine. $C_{21}H_{25}NO_4$; mol wt 355.43. C 70.96%, H 7.09%, N 3.94%, O 18.01%. From tubers of *Corydalis cava* (L.) Schweigg. & Korte *(C. tuberosa* DC., *Fumariaceae):* Gadamer, *Arch. Pharm.* **240**, 19 (1902). Structure: Späth, Dobrowsky, *Ber.* **58**, 1274 (1925). Synthesis: Späth, Holter, *ibid.* **59**, 2800 (1926). Abs configuration: P. W. Jeffs, *Experientia* **21**, 690 (1965); K. Iwasa *et al., J. Org. Chem.* **46**, 4744 (1981).

Leaflets, mp 179-180°; also given as 187-188°. $[\alpha]_D^{15}$ +301° in chloroform. Sol in alcohol, dil acids.

Methiodide, $C_{21}H_{25}NO_4 \cdot CH_3I$, crystals, mp 218-221°.

5178. Isocorydine. *(S)-5,6,6a,7-Tetrahydro-1,2,10-trimethoxy-6-methyl-4H-dibenzo[de,g]quinolin-11-ol; 1,2,10-trimethoxy-6α-aporphin-11-ol;* 11-hydroxy-1,2,10-trimethoxy-aporphine; artabotrine; luteanine. $C_{20}H_{23}NO_4$; mol wt 341.41. C 70.36%, H 6.79%, N 4.10%, O 18.75%. From tubers of *Corydalis cava* (L.) Schweigg. & Korte *(C. tuberosa* DC., *Fumariaceae); Artabotrys suaveolens* Blume, *Anonaceae; Papaver oreophilum* Rupr.; *Phoebe clemensii* Allen *(Lauraceae)* and others. Isoln: Gadamer, *Arch. Pharm.* **249**, 669 (1911); Johns, Lamberton, *Aust. J. Chem.* **20**, 1277 (1967); Pfeifer, Mann, *Pharmazie* **23**, 82 (1968). Structure: Späth, Berger, *Ber.* **64**, 2038 (1931); Gulland *et al., J. Chem. Soc.* **1931**, 2885; Barger, Sargent, *ibid.* **1939**, 991. Identity with artabotrine: Schlittler, Huber, *Helv. Chim. Acta* **35**, 111 (1952); with luteanine: Manske, *Can. J. Res.* **21B**, 13 (1943). Total synthesis: Kametani *et al., Tetrahedron* **27**, 5367 (1971). Pharmacology: Berezhinskaya *et al., Farmakol. Toksikol. (Moscow)* **31**, 44 (1968), *C.A.* **68**, 94521z (1968).

Plates from ethanol or acetone, mp 185°. $[\alpha]_D^{20}$ +195° (chloroform). Sol in ether, chloroform, alcohol, acetone, alkali hydroxide. Practically insol in water, alkali carbonates.

5179. Isocorypalmine. *5,8,13,13a-Tetrahydro-3,9,10-trimethoxy-6H-dibenzo[a,g]quinolizin-2-ol; 3,9,10-trimethoxy-13aα-berbin-2-ol; d-tetrahydrocolumbamine.* $C_{20}H_{23}NO_4$; mol wt 341.41. C 70.36%, H 6.79%, N 4.10%, O 18.75%. From tubers of *Corydalis cava* (L.) Schweigg. & Korte *(C. tuberosa* DC.) and *C. nobilis* Pers., *Fumariaceae:* Gadamer *et al., Arch. Pharm.* **265**, 675 (1927); Manske, *Can. J. Res.* **18B**, 288 (1940). Structure: Späth, Burger, *Ber.* **59**, 1486 (1926). Synthesis of *dl*-form: Govindachari *et al., ibid.* **92**, 1654 (1959). Configuration: Corrodi, Hardegger, *Helv. Chim. Acta* **39**, 889 (1956).

Crystals, mp 239-241°. $[\alpha]_D^{20}$ 30.3° (c = 0.4 in chloroform). Sol in alcohol, ether.

dl-Form, crystals from methanol, mp 216°.

5180. Isocrotonic Acid. *cis-2-Butenoic acid.* $C_4H_6O_2$; mol wt 86.09. C 55.81%, H 7.02%, O 37.17%. CH_3-$CH=CHCOOH$. Prepd by reacting acetaldehyde with malonic acid in the presence of pyridine: Auwers *et al., Ann.* **432**, 46 (1923); from ethyl acetoacetate: Hatch, Nesbitt, *J. Am. Chem. Soc.* **72**, 727 (1950); from but-2-ynoic acid: Allan *et al., J. Chem. Soc.* **1955**, 1862; from $CH_3CH=CH$-$MgBr + CO_2$: Normant, Maitte, *Bull. Soc. Chim. France* **1956**, 1439. Structure: Auwers, Wissebach, *Ber.* **56**, 715 (1923); Plisov, Bogatskii, *Zh. Obshch. Khim.* **27**, 360 (1957), *C.A.* **51**, 15401c (1957).

Liquid. bp$_5$ 55-56°; bp$_{10}$ 62-64°; bp$_{760}$ 168-169°. mp 13.5-14.5°; crystallized from pentane. n_D^{20} 1.4450. uv max (95% ethanol): 205.5 nm (ε 13500).

Methyl ester, $C_5H_8O_2$, liquid. bp 118°. n_D^{20} 1.4175. uv max (95% ethanol): 205.5 nm (ε 14000).

Benzyl ester, $C_{11}H_{12}O_2$, liquid. bp$_{10}$ 121-122°. n_D^{20} 1.5110.

Caution: Can cause severe irritation of skin, mucous membranes.

5181. Isocyanic Acid. Carbimide. CHNO; mol wt 43.03. C 27.92%, H 2.34%, N 32.55%, O 37.19%. $HN=C=O$. Prepn: J. Liebig, F. Wöhler, *Ann.* **39**, 29 (1846); Steyermark, *J. Org. Chem.* **28**, 586 (1963); R. Vorhoeve, L. E. Trimble, *Science* **202**, 525 (1978). Crystal structure: Von Dohlen, Carpenter, *Acta Cryst.* **8**, 646 (1955). Review of prepn and properties: D. J. Belson, A. N. Strachan, *Chem. Soc. Reviews* **11**, 41-56 (1982). *See also* Cyanic Acid.

Store in dil solns of carbon tetrachloride or ether at $-30°$ to slow polymerization. The free acid in the vapor phase or in ether soln gives no indication of being a mixture, and all the evidence supports the iso structure, HNCO, whereas in aq soln cyanic acid HOCN is present: N. V. Sidgwick, *The Chemical Elements and Their Compounds* **vol. I** (Oxford, 1950) p 673.

Caution: Strongly acidic, will blister skin.

5182. Isodurene. *1,2,3,5-Tetramethylbenzene.* $C_{10}H_{14}$; mol wt 134.22. C 89.49%, H 10.51%. Occurs in coal tar. Prepd by the action of dimethyl sulfate on 2,4,6-trimethylphenylmagnesium bromide: Smith, *Org. Syn.* **11**, 66 (1931).

Liquid. d_4^0 0.8961; d_4^{20} 0.8906. mp $-24°$. bp$_{760}$ 197.9°; bp$_{400}$ 173.7°; bp$_{200}$ 149.9°; bp$_{100}$ 128.3°; bp$_{60}$ 115.4°; bp$_{40}$ 105.8°; bp$_{20}$ 91.0°; bp$_{10}$ 77.8°; bp$_5$ 65.8°; bp$_{1.0}$ 40.6°. n_D^{20} 1.5134, n_{He}^{20} 1.51126. Insol in water. Sol in alcohol; very sol in ether.

5183. Isoestradiol. *(8α,17β)-Estra-1,3,5(10)-triene-3,-17-diol;* $\Delta^{1,3,5}$-8-epiestratriene-3,17β-diol; 8α-estradiol; 8-epiestradiol; 8-isoestradiol-17β. $C_{18}H_{24}O_2$; mol wt 272.39. C 79.37%, H 8.88%, O 11.75%. Prepn by high pressure hydrogenation of dihydroequilin: Serini, Logemann, *Ber.* **71**, 186 (1938); by hydrogenation of equilenin with Raney nickel at 2800 psi and 85°: Dauben, Ahramjian, *J. Am. Chem. Soc.* **78**, 633 (1956). Prepn of *dl*-form from *dl*-equilenin and stereochemistry: Johnson *et al., ibid.* **80**, 661 (1958).

d-Form, crystals from dil methanol + chloroform, mp 181°. $[\alpha]_D^{20}$ +18° (16 mg in 2 ml dioxane).

3-Benzoate, $C_{25}H_{28}O_3$, crystals from ethyl acetate, mp 190°. $[\alpha]_D^{20}$ +9.5° (17 mg in 2 ml dioxane).

dl-Form, needles from dil methanol, mp 213.5-214°.

5184. 8-Isoestrone. *(8α)-3-Hydroxyestra-1,3,5(10)-trien-17-one;* $\Delta^{1,3,5}$-8-epiestratrien-3-ol-17-one; 8α-estrone; 8-epiestrone. $C_{18}H_{22}O_2$; mol wt 270.37. C 79.96%, H 8.20%, O 11.84%. Prepn: Serini, Logemann, *Ber.* **71**, 186 (1938); Johnson *et al., J. Am. Chem. Soc.* **80**, 661 (1958); Ananchenko, Torgov, *Tetrahedron Letters* **1963**, 1553.

dl-Form, prisms from methanol, mp 254-255°.

dl-Methyl ether, $C_{19}H_{24}O_2$, blades from methanol, mp 153-155°.

dl-Benzoate, $C_{25}H_{26}O_3$, prisms from methanol, mp 197-198°.

5185. Isoetharine. *4-[1-Hydroxy-2-[(1-methylethyl)-amino]butyl]-1,2-benzenediol; 3,4-dihydroxy-α-[1-(isopropyl-amino)propyl]benzyl alcohol; α-(1-isopropylaminopropyl)-protocatechuyl alcohol; N-isopropylethylnorepinephrine;* 1-(3,4-dihydroxyphenyl)-2-isopropylamino-1-butanol; etyprenaline; isoetarine; Win-3046; Dilabron; Neoisuprel. $C_{13}H_{21}$-NO_3; mol wt 239.31. C 65.25%, H 8.84%, N 5.85%, O 20.06%. Prepn: Bockmühl *et al.,* Ger. pat. **638,650** (1936 to I. G. Farben.), *C.A.* **31,** 3209⁴ (1937). Pharmacology: Lands *et al., J. Pharmacol. Exp. Ther.* **99,** 45 (1950); *eidem, J. Am. Pharm. Assoc.* **47,** 744 (1958).

Hydrochloride, $C_{13}H_{21}NO_3 \cdot HCl$, *Asthmalitan, Numotac.* Crystals from methanol + ether. mp 212-213° (dec).

THERAP CAT: Bronchodilator.

5186. Isoeugenol. *2-Methoxy-4-(1-propenyl)phenol;* 4-hydroxy-3-methoxy-1-propenylbenzene; 4-propenylguaiacol. $C_{10}H_{12}O_2$; mol wt 164.20. C 73.15%, H 7.37%, O 19.49%. Occurs in ylang-ylang and other essential oils. Prepd from eugenol: West, *J. Soc. Chem. Ind.* **59,** 275 (1940); Pal'gi, *Zh. Obshch. Khim.* **28,** 2239 (1958). Stereochemistry: von Auwers, *Ber.* **68,** 1346 (1935); Puxeddu, Rattu, *Gazz. Chim. Ital.* **67,** 647 (1937). Toxicity study: P. M. Jenner *et al., Food Cosmet. Toxicol.* **2,** 327 (1964).

trans-form

Oily liquid; easily becomes somewhat yellow. d_4^{25} 1.080. bp 266°; bp_8 128-130°. mp −10°. n_D^{19} 1.5739. Slightly sol in water; misc with alcohol, ether. LD_{50} orally in rats: 1560 mg/kg (Jenner).

trans-Form, crystals, mp 33°, bp_{12} 140°. d_4^{20} 1.087. n_{He}^{20} 1.5778.

cis-Form, liquid, bp_{11} 133°. d_4^{20} 1.088. n_{He}^{20} 1.5724.

USE: Manuf vanillin.

5187. Isofenphos. *2-[[Ethoxy[(1-methylethyl)amino]-phosphinothioyl]oxy]benzoic acid 1-methylethyl ester; salicylic acid isopropyl ester O-ester with O-ethyl isopropylphosphoramidothioate;* 1-methylethyl 2-[[ethoxy[(1-methylethyl)amino]phosphinothioyl]oxy]benzoate; O-ethyl O-2-isopropoxycarbonylphenyl isopropylphosphoramidothioate; isophenphos; Bay 92114; SRA-12869; Amaze; Oftanol. $C_{15}H_{24}$-NO_4PS; mol wt 345.40. C 52.16%, H 7.00%, N 4.06%, O

18.53%, P 8.97%, S 9.28%. Selective soil insecticide. Prepn: G. Schrader *et al.,* **S. Afr.** pat. **68 7,378,** *C.A.* **72,** 3239g (1970); **Fr.** pat. **1,600,932** corresp to *eidem,* U.S. pat. **3,621,-082** (1969, 1970, 1971 to Bayer). Determn of residues: M. J. Brown, I. H. Williams, *Pestic. Sci.* **7,** 545 (1976).

Oil, $bp_{0.01}$ 120°. d_4^{20} 1.13. Vapor press at 20°: 4×10^{-6} mm Hg. Soly in water at 20°: 23.8 mg/kg. Sol in dichloromethane, cyclohexanone, acetone, alc, ether, benzene. LD_{50} orally in rats, mice: 28, 91.3 mg/kg, *RTECS* Vol. I, R. J. Lewis, R. L. Tatken, Eds. (1980) p 310.

USE: Insecticide.

5188. Isofezolac. *1,3,4-Triphenyl-1H-pyrazole-5-acetic acid;* LM-22102; Sofenac. $C_{23}H_{18}N_2O_2$; mol wt 354.41. C 77.95%, H 5.12%, N 7.90%, O 9.03%. Non-steroidal anti-inflammatory agent; prostaglandin synthetase inhibitor. Prepn: C. Gueremy, C. Renault, **Ger.** pat. **2,312,256;** *eidem,* U.S. pat. **3,984,431** (1973, 1976 to Société Générale de Recherches et d'Applications Scientifiques). Pharmacology: J. Mizoule *et al., Arch. Int. Pharmacodyn.* **238,** 305 (1979). HPLC determn in biological fluids: A. Bannier *et al., J. Chromatog.* **227,** 213 (1982).

Crystals from acetonitrile, mp 200°. LD_{50} in mice, rats (mg/kg): 215, 13 orally (Mizoule).

THERAP CAT: Anti-inflammatory; antipyretic; analgesic.

5189. Isoflavone. *3-Phenyl-4H-1-benzopyran-4-one;* 3-phenylchromone. $C_{15}H_{10}O_2$; mol wt 222.24. C 81.07%, H 4.54%, O 14.40%. Prepn from *o*-hydroxyphenyl benzyl ketone, sodium dust and ethyl formate: Joshi, Venkataraman, *J. Chem. Soc.* **1934,** 513. Structure: Warburton, *Quart. Rev. (London),* **8,** 67 (1954).

Leaflets and needles from petr ether, mp 148°. uv max: 245, 307 nm (log ε 4.41, 3.82).

5190. Isoflupredone. *(11β)-9-Fluoro-11,17,21-trihydroxypregna-1,4-diene-3,20-dione;* 1-dehydro-9α-fluorohydrocortisone; 9-fluoroprednisolone. $C_{21}H_{27}FO_5$; mol wt 378.44. C 66.65%, H 7.19%, F 5.02%, O 21.14%. Prepn: R. Fried *et al., J. Am. Chem. Soc.* **77,** 4181 (1955). Microbiological prepn: A. Nobile *et al., J. Am. Chem. Soc.* **77,** 4184 (1955); E. Vischer *et al., Helv. Chim. Acta* **38,** 1502 (1955); T. H. Stoudt *et al., Arch. Biochem. Biophys.* **59,** 304 (1955). Prepn of the 21-acetate: R. F. Hirschmann *et al., ibid.* 3166; J. A. Hogg *et al., ibid.* 4438; Ch. Meystre *et al., Helv. Chim. Acta* **39,** 734 (1956); A. Wellstein *et al.,* **Ger.** pat. **1,020,329** (1957 to Ciba), *C.A.* **54,** 25562h (1960); **Brit.** pat. **843,214** (1960 to Schering), *C.A.* **55,** 5597b (1961). Absorption spectra: L. L. Smith, W. H. Muller, *J. Org. Chem.* **23,** 960 (1958). Mass spectrometry: P. Toft *et al., Can. J. Pharm. Sci.* **7,** 53 (1972). Urinary excretion in horses: D. I. Chapman *et al., Vet. Rec.* **100,** 447 (1977).

Crystals from acetone, mp 263-266° (dec). $[\alpha]_D^{23}$ +108° (c = 0.611 in ethanol). uv max (ethanol): 240 nm (ϵ 15800) (Vischer), also reported as mp 274-275° (dec), $[\alpha]_D^{23}$ +94° (alcohol)(Fried).

21-acetate, $C_{23}H_{29}FO_6$, *U-6013, Predef.* Crystals from acetone/isopropyl ether or methanol, mp 244-246° (dec). $[\alpha]_D^{23}$ +108° (c = 0.735 in dioxane). uv max (ethanol): 240 (ϵ 16250).

THERAP CAT (VET): Anti-inflammatory.

5191. Isoflurane. *2-Chloro-2-(difluoromethoxy)-1,1,1-trifluoroethane; 1-chloro-2,2,2-trifluoroethyl difluoromethyl ether;* compd 469; Aerrane; Forane; Forene. $C_3H_2ClF_5O$; mol wt 184.49. C 19.53%, H 1.09%, Cl 19.22%, F 51.49%, O 8.67%. $CF_3CHClOCHF_2$. Prepn: Croix, Terrell, **Ger.** pat. **1,814,962** (1969); Terrell, U.S. pats. **3,535,388; 3,535,-425** (both 1970 to Air Reduction); Terrell *et al., J. Med. Chem.* **14,** 517 (1971). Series of articles on pharmacology: *Anesthesiology* **35,** 8-53 (1971); Byles *et al., Can. Anaesth. Soc. J.* **18,** 376-407 (1971). Enantiomeric resolution: J. Meinwald *et al., Science* **251,** 560 (1991).

Clear, colorless liquid having a slight odor. bp 48.5°. Vapor pressure at 25°: 330 mm. sp gr 1.45. n_D^{20} 1.3002. Nonflammable; soda lime stable. Easily miscible with organic liquids including fats and oils.

USE: Solvent and dispersant for fluorinated materials.

THERAP CAT: Anesthetic (inhalation).

THERAP CAT (VET): Anesthetic (inhalation).

5192. Isoflurophate. *Phosphorofluoridic acid bis(1-methylethyl) ester; phosphorofluoridic acid diisopropyl ester;* isopropyl fluophosphate; diisopropyl fluorophosphonate; diisopropyl fluorophosphate; diisopropylphosphorofluoridate; fluostigmine; isofluorphate; DFP; Diflupyl; Dyflos; Floropryl; Fluropryl. $C_6H_{14}FO_3P$; mol wt 184.15. C 39.13%, H 7.66%, F 10.32%, O 26.07%, P 16.82%. Prepd by the action of PCl_3 on isopropanol, chlorinating the resulting intermediate, and converting the diisopropyl chlorophosphate by means of sodium fluoride: Hardy, Kosolapoff, U.S. pat. **2,409,039** (1946); *see also* Edgewood Arsenal, *Chemical Warfare Service TDMR 832* (April, 1944); **Brit.** pat. **601,210** (1948); Lange, von Krueger, *Ber.* **65,** 1598 (1932); Saunders, Stacey, *J. Chem. Soc.* **1948,** 695; B. C. Saunders, *Some Aspects of the Chemistry and Toxic Action of Organic Compounds Containing Phosphorus and Fluorine* (Cambridge, 1957) p 46. Toxicity data: R. G. Horton *et al., J. Pharmacol. Exp. Ther.* **87,** 414 (1946).

Liquid. Forms HF in presence of moisture. d 1.055. mp −82°. bp_5 46°; bp_9 62°; bp_{760} 183° (by extrapolation). Vapor pressure at 20°: 0.579 mm. n_D^{25} 1.3830. Soly in water at 25°: 1.54% w/w (dec; pH about 2.5). Sol in vegetable oils. Not very sol in mineral oils. The anhydr compd or oil solns are stable in glass containers at room temp. LD_{50} in mice (mg/kg): 3.71 s.c.; 36.8 orally (Horton).

Caution: Highly toxic; cholinesterase inactivator. Do not inhale vapors. Avoid contact with skin.

THERAP CAT: Cholinergic (ophthalmic).

THERAP CAT (VET): Has been used as a miotic.

5193. L-Isoglutamine. *(S)-4,5-Diamino-5-oxopentanoic acid;* 4-amino-L-glutaramic acid; glutamic acid α-amide. $C_5H_{10}N_2O_3$; mol wt 146.15. C 41.09%, H 6.90%, N 19.17%, O 32.84%. Prepn by the action of NH_3 on *N*-carbobenzyloxy-L-glutamic acid anhydride: Kozo Narita, *J. Chem. Soc.*

Japan, Pure Chem. Sect. **74,** 832 (1953); from 1-tosylpyroglutamide by reduction with Na in NH_3: Swan, du Vigneaud, *J. Am. Chem. Soc.* **76,** 3110 (1954); from the γ-methyl ester of *N*-tritylglutamic acid: Amiard, Heymes, U.S. pat. **2,927,118** (1960 to UCLAF).

Crystals from acetone, dec 181°. $[\alpha]_D^{21}$ +20.5° (c = 6.1 in H_2O). pK_1' 3.81; pK_2' 7.88. Sol in water. Sparingly sol in organic solvents.

5194. Isoladol. *β-Amino-4-methoxy-α-(4-methoxyphenyl)benzeneethanol; 2-amino-1,2-bis(p-methoxyphenyl)ethanol;* isodianisylethanolamine; di-*p*-methoxyphenylhydroxyethylamine; Evadol. $C_{16}H_{19}NO_3$; mol wt 273.33. C 70.31%, H 7.01%, N 5.12%, O 17.56%. Prepd by the condensation of anisaldehyde and glycine: Read, Campbell, *J. Chem. Soc.* **1930,** 2676; Lespagnol, *Pharmacie Chimique* (Paris, 3rd ed., 1950) p 695.

dl-Form, glistening prisms from alcohol, mp 135.5°. Sol in ethyl acetate.

dl-Form hydrochloride, *LC-2, Evaclin.* Crystals, dec about 200°. Sol in water.

l-Form, needles from abs alcohol, mp 111-112°. $[\alpha]_D^{20}$ −150° (c = 0.952 in abs alcohol).

l-Form hydrochloride, crystals from ethyl acetate, dec 188-190°. $[\alpha]_D^{20}$ −99.0° (c = 0.5805). Sol in water.

THERAP CAT: Analgesic.

5195. Isolan®. *Dimethylcarbamic acid 3-methyl-1-(1-methylethyl)-1H-pyrazol-5-yl ester;* 1-isopropyl-3-methyl-5-pyrazolyl dimethylcarbamate; G-23611. $C_{10}H_{17}N_3O_2$; mol wt 211.26. C 56.85%, H 8.11%, N 19.89%, O 15.15%. Carbamate insecticide: Gerguson, Alexander, *J. Agr. Food Chem.* **1,** 288 (1953); Muller, Spindler, *Experientia* **10,** 91 (1954). Prepd by treating the K salt of the enol form of 1-isopropyl-3-methyl-5-pyrazolone with dimethylcarbamoyl chloride: **Swiss** pats. **279,553** and **282,655; Brit.** pat. **681,-376** (all 1952 to Geigy); Kost, Sagitullin, *Zh. Obshch. Khim.* **33,** 867 (1963), *C.A.* **59,** 8724 (1963). Toxicity studies: T. B. Gaines *et al., Nature* **209,** 88 (1966); T. B. Gaines, *Toxicol. Appl. Pharmacol.* **14,** 515 (1969).

Liquid, $bp_{0.7}$ 103°, $bp_{2.5}$ 117.5-118°. d 1.07. Vapor pressure at 20° = 0.001 mm Hg. Unlike most insecticides, Isolan is more toxic to rats dermally than orally (Gaines, 1966). LD_{50} in male, female rats (mg/kg): 23, 13 orally; 5.6, 6.2 dermally (Gaines, 1969).

USE: Insecticide.

5196. Isoleucine. *L-Isoleucine;* Ile; I; 2-amino-3-methylvaleric acid; α-amino-β-methylvaleric acid; (2S,3S)-2-amino-3-methylpentanoic acid. $C_6H_{13}NO_2$; mol wt 131.17. C 54.94%, H 9.99%, N 10.68%, O 24.39%. An essential amino acid for human development. Isoln from beet-sugar mother liquors: F. Ehrlich, *Ber.* **37,** 1809 (1904). Structure and identification of isomers: *idem, Ber.* **40,** 2538 (1907); J. P. Greenstein *et al., J. Biol. Chem.* **204,** 307 (1953). Early chemistry and biochemistry: *Amino Acids and Proteins,* D. M. Greenberg, Ed. (Charles C. Thomas, Springfield, IL, 1951) 950 pp., *passim*; J. P. Greenstein, M. Winitz, *Chemistry of the Amino Acids* vol 1-3 (John Wiley and Sons, Inc., New York, 1961) pp. 2043-2074, *passim*. TLC determ in serum from patients with Maple Sugar Urine Disease (MSDU): R. J. Allen *et al., Clin. Chem.* **18**, 413 (1972). Metabolism of Ile and isomers in MSDU: U. Wendel *et al., Pediatr. Res.* **25**, 11 (1989); and clinical consequences: T. Mogos *et al., Rev. Roum. Med. Int.* **32**, 57 (1994). Review of biosynthesis: M. Iaccarino *et al., Curr. Topics Cell Regul.* **14**, 29-73 (1978).

Waxy, shiny, rhombic leaflets from alc. Bitter taste. Sublimes 168-170°. Dec 284°. $[\alpha]_D^{20}$ +11.29° (c = 3); +40.61° (c = 4.6 in 6.1N HCl); +11.09° (c = 3.3 in 0.33N NaOH). $[M]_D$ +53.5° (5N HCl); +64.2° (glacial acetic acid). Soly in water (g/l): 37.9 at 0°; 41.2 at 25°; 48.2 at 50°; 60.8 at 75°; 82.6 at 100°. Sparingly sol in hot alc (0.13% w/w at 80°), hot acetic acid. Insol in ether.

DL-Form, *(±)-erythro-2-amino-3-methylpentanoic acid.* Glistening rhombic or monoclinic plates from dil alc. Dec 292°. pK_1' 2.32; pK_2' 9.76. Soly in water (g/l): 18.3 at 0°; 22.3 at 25°; 30.3 at 50°; 46.1 at 75°; 78.0 at 100°.

L-allo-Form, *α-Ile,* L(+)-alloisoleucine, *(+)-threo-2-amino-3-methylpentanoic acid.* Waxy leaflets. Sweet taste. Dec 280°. $[\alpha]_D^{20}$ +14.0° (c = 2); +38.1° (c = 2 in 6N HCl). $[M]_D$ +53.1° (5N HCl); +55.7° (glacial acetic acid). pK_1' 2.27; pK_2' 9.62. One part dissolves in 34.2 parts water at 20°; 0.82 part dissolves in 100 parts 80% alcohol at 20°; 1.97 parts dissolve in 100 parts 80% alcohol; 0.19 part dissolves in 100 parts abs alcohol at 90°.

5197. Isolysergic Acid. *(8α)-9,10-Didehydro-6-methylergoline-8-carboxylic acid.* $C_{16}H_{16}N_2O_2$; mol wt 268.32. C 71.62%, H 6.01%, N 10.44%, O 11.93%. Parent compd of the ergotinine group of alkaloids from ergot. Prepn: Smith, Timmis, *J. Chem. Soc.* **1936**, 1440; Craig *et al., J. Biol. Chem.* **125**, 289 (1938). Differs from lysergic acid by the α-configuration at C-8. Stereochemical studies: Stenlake, *J. Chem. Soc.* **1955**, 1626; Leemann, Fabbri, *Helv. Chim. Acta* **42**, 2696 (1959).

Dihydrate from water, mp 218° (dec). $[\alpha]_D^{20}$ +281° (pyridine). pKa 3.44, pKb 8.61. More sol in water and pyridine than lysergic acid.

Methyl ester, rods from benzene, mp 170-174°. $[\alpha]_D^{20}$ +179° (c = 0.5 in chloroform).

5198. Isomaltol. *1-(3-Hydroxy-2-furanyl)ethanone;* 3-hydroxy-2-furyl methyl ketone. $C_6H_6O_3$; mol wt 126.11. C 57.14%, H 4.80%, O 38.06%. Trace constituent of bread. Obtained by the action of enzymes on starch: Backe, *Compt. Rend.* **151**, 78 (1910). Synthesis from α-lactose and a secondary amine: Hodge, Nelson, U.S. pat. **3,054,805** (1962 to USDA). Structure: Hodge, Nelson, *Cereal Chem.* **38**, 207 (1961); Fischer, Hodge, *J. Org. Chem.* **29**, 776 (1964).

Crystals from water or ether, mp 98-103°. uv max (abs methanol): 280 nm ($E_{1cm}^{1\%}$ 1270). Sol in alcohols, acetone, chloroform, benzene, ethyl acetate, hot water. Practically insol in cold water and pet ether.

Benzoate, $C_{13}H_{10}O_4$, crystals from benzene-toluene, mp 100-101°.

p-Nitrophenylhydrazone, $C_{12}H_{11}N_3O_4$, deep red crystals from nitrobenzene, mp 100-101°.

5199. Isometamidium Chloride. *3-Amino-8-[3-[3-(aminoiminomethyl)phenyl]-1-triazenyl]-5-ethyl-6-phenylphenanthridinium chloride; 8-[3-(m-amidinophenyl)-2-triazeno]-3-amino-5-ethyl-6-phenylphenanthridinium chloride; 7-m-amidinophenyldiazoamino-2-amino-10-ethyl-9-phenylphenanthridinium chloride;* M & B 4180 A; Samorin. $C_{28}H_{26}ClN_7$; mol wt 496.01. C 67.80%, H 5.28%, Cl 7.15%, N 19.77%. Prepn: Wragg *et al., Nature* **182**, 1005 (1958). Structure: Berg, *ibid.* **188**, 1106 (1960).

Red crystals from aq methanol, dec 244-245°.
THERAP CAT (VET): Antitrypanosomal agent.

5200. Isomethadol. *β-[2-(Dimethylamino)-1-methylethyl]-α-ethyl-β-phenylbenzeneethanol; 6-(dimethylamino)-5-methyl-4,4-diphenyl-3-hexanol;* 1-dimethylamino-2-methyl-3,3-diphenyl-4-hexanol; 3-hydroxy-4,4-diphenyl-4-methyl-5-dimethylaminohexane. $C_{21}H_{29}NO$; mol wt 311.47. C 80.98%, H 9.38%, N 4.50%, O 5.14%. Prepn of α-*dl*-form by lithium aluminum hydride reduction of *dl*-isomethadone: May, Mosettig, *J. Org. Chem.* **13**, 663 (1948). Prepn of α-isomers by lithium aluminum hydride reduction of the optical isomers and of β-isomers by sodium-propanol reduction: May, Eddy, *ibid.* **17**, 1210 (1952). Stereochemistry of isomers: Portoghese, Williams, *Tetrahedron Letters* **1966**, 6299.

β-*dl*-Form, prisms from alcohol + water, mp 107-108.5°.
β-*dl*-Form hydrochloride, crystals from alcohol + ether, mp 252-254°.
β-*d*-Form, slim rods from alcohol + water, mp 94-95°; $[\alpha]_D^{20}$ +13.3° (c = 1.65).
β-*d*-Form hydrochloride, thin prisms from alcohol + ether, mp 241-243°; $[\alpha]_D^{20}$ +13.5° (c = 1.78).
β-*l*-Form, mp 93.5-94.5°; $[\alpha]_D^{20}$ −13.8° (c = 1.09).
β-*l*-Form hydrochloride, mp 241-243°; $[\alpha]_D^{20}$ −13.5° (c = 1.56).
α-*dl*-Form, mp 100-102.5°.
α-*dl*-Form hydrochloride, mp 231-233° (sinters at 200°).
α-*d*-Form, mp 124.5-125.5°; $[\alpha]_D^{20}$ +19.1° (c = 0.63).
α-*d*-Form hydrochloride hemihydrate, mp 202-204°; $[\alpha]_D^{20}$ +10.1° (c = 1.89).
α-*l*-Form, prisms from alcohol + water, mp 125-126°; $[\alpha]_D^{20}$ −19.7° (c = 1.43).

α-l-Form hydrochloride hemihydrate, rectangular plates, mp 202-204°; $[\alpha]_D^{20}$ — 9.6° (c = 2.93).

5201. Isomethadone. *6-(Dimethylamino)-5-methyl-4,4-diphenyl-3-hexanone;* 6-dimethylamino-4,4-diphenyl-5-methyl-3-hexanone; 1-dimethylamino-2-methyl-3,3-diphenyl-4-hexanone; isoamidone; Isoadanone. $C_{21}H_{27}NO$; mol wt 309.45. C 81.51%, H 8.79%, N 4.53%, O 5.17%. Prepn of *dl*-form: Easton *et al., J. Am. Chem. Soc.* **70**, 76 (1948); Walton *et al., J. Chem. Soc.* **1949**, 648; Larsen, Tullar, U.S. pat. 2,773,901 (1956 to Sterling Drug). Prepn of *d*- and *l*-forms and resolution of *dl*-form: Larsen *et al., J. Am. Chem. Soc.* **70**, 4194 (1948); Howe, Sletzinger, *ibid.* **71**, 2935 (1949); Larsen, Tullar, *loc. cit.* Configuration of *l*-form, the isomer with greater analgesic activity: Beckett *et al., Chem. & Ind. (London)* **1960**, 1418. Pharmacology: Eddy *et al., J. Pharmacol. Exp. Ther.* **98**, 121 (1950); Winter, Flataker, *ibid.* 305.

dl-Form, slightly yellow, very viscous liq, bp_{12} 215°.
dl-Form hydrobromide, $C_{21}H_{27}NO.HBr$, crystals from water, mp 149-150°.
d-Form, oil, $[\alpha]_D^{25}$ +20.8°.
d-Form hydrochloride, $C_{21}H_{27}NO.HCl$, mp 231-232°. $[\alpha]_D^{25}$ +70°.
d-Form hydrochloride monohydrate, $C_{21}H_{27}NO.HCl.H_2O$, mp 176-177°. $[\alpha]_D^{25}$ +66°.
l-Form, **Liden.** Oily liq, $bp_{0.6}$ 162-165°. $[\alpha]_D^{25}$ — 20° (c = 1.5 in 95% ethanol).
l-Form hydrobromide, crystals, mp 217-218°. $[\alpha]_D^{25}$ —59° (c = 1.5).
l-Form hydrochloride, mp 231-233°. $[\alpha]_D^{25}$ —70° (c = 1.5 in water), —90° (methanol). Sol in water, alcohol. pH of 1% aq soln: 5 to 6.5. Solutions and tablets are stable. LD_{50} s.c. in mice: 21 mg/kg (Winter, Flataker).
l-Form hydrochloride monohydrate, mp 173-174°.
Note: This is a controlled substance (opiate) listed in the U.S. Code of Federal Regulations, Title 21 Part 1308.12 (1995).

THERAP CAT: Analgesic (narcotic).

5202. Isometheptene. *N,6-Dimethyl-5-hepten-2-amine; N,1,5-trimethyl-4-hexenylamine;* 6-methylamino-2-methyl-heptene; 2-methyl-6-methylamino-2-heptene; methylisooctenylamine; methyloctenylamine; Octin; Octon; Octanil. $C_9H_{19}N$; mol wt 141.26. C 76.53%, H 13.56%, N 9.92%. $(CH_3)_2C=CHCH_2CH_2CH(NHCH_3)CH_3$. Prepn: Klavehn, Wolf, U.S. pats. 2,230,753-4 (1941 to E. Bilhuber). Toxicity data: Walton *et al., J. Pharmacol. Exp. Ther.* **92**, 214 (1948).

Colorless, oily liq; characteristic amine odor. Strong base. Volatile with steam. d 0.795. bp 176-178°; bp_7 58-59°. n_D^{15} 1.4472. Practically insol in water. Freely sol in alc, ether, acetone, chloroform.
Hydrochloride, $C_9H_{19}N.HCl$, crystals, very hygroscopic, mp 148-150°. Sol in water, alc. LD_{50} in mice (mg/kg): 17.5 i.v.; 171 s.c. (Walton).
Bitartrate (acid tartrate), $C_{13}H_{25}NO_6$, crystals, mp 78-80° (preliminary sintering). Freely sol in water, alcohol.
Mucate, $C_{24}H_{48}N_2O_8$, bitter crystals, mp 152° (rapid heating). Freely sol in water. Soly in alc about 5 g/100 ml. Almost insol in ether, chloroform. LD_{50} orally in dogs: 148 mg/kg (Walton).

THERAP CAT: Adrenergic.
THERAP CAT (VET): Sympathomimetic. Antispasmodic for gut and urinary tract.

5203. Isoniazid. *4-Pyridinecarboxylic acid hydrazide; isonicotinic acid hydrazide;* isonicotinoylhydrazine; isonicotinylhydrazine; INH; rimitsid; tubazid; RP-5015; FSR-3; Cotinazin; Dinacrin; Ditubin; Hycozid; Iscotin; Isobicina; Isocid; Isolyn; Isonex; Isonizida; Isozid; Laniazid; Mybasan;

Neoteben; Nicizina; Niconyl; Nicotibina; Nydrazid; Pycazide; Pyricidin; Rimifon; Tibinide; Tubilysin; Tyvid. $C_6H_7N_3O$; mol wt 137.14. C 52.55%, H 5.14%, N 30.64%, O 11.67%. Prepn: Meyer, Mally, *Monatsh.* **33**, 400 (1912); Lock, *Pharm. Ind.* **14**, 366 (1952); Urbanski *et al., Rocz. Chem.* **27**, 161 (1953); Gasson, U.S. pat. 2,830,994 (1958 to Distillers). Compositions for combating tuberculosis: H. H. Fox, U.S. pat. 2,596,069 (1952 to Hoffmann-La Roche). Pharmacokinetics: W. W. Weber, D. W. Hein, *Clin. Pharmacokinet.* **4**, 401 (1979). Mechanism of action study: K. Johnsson *et al., J. Am. Chem. Soc.* **117**, 5009 (1995). Toxicity data: E. H. Jenney, C. C. Pfeiffer, *J. Pharmacol. Exp. Ther.* **122**, 110 (1958). Evaluation of carcinogenic risk: *IARC Monographs* **4**, 159 (1974). Comprehensive description: G. A. Brewer, *Anal. Profiles Drug Subs.* **6**, 183-258 (1977). Clinical trial in HIV-positive patients: J. W. Pape *et al., Lancet* **342**, 268 (1993). Review of use in prevention and treatment of TB: American Thoracic Society, *Am. Rev. Respir. Dis.* **134**, 355-363 (1986); T. J. Jordan *et al., ibid.* **144**, 1357-1360 (1991).

Crystals from alc, mp 171.4°. uv max (water): 266 nm ($E_{1cm}^{1\%}$ 378); (0.01N HCl): 265 nm ($E_{1cm}^{1\%}$ ~420). Soly in water at 25°: about 14%; at 40°: about 26%; in alc at 25°: about 2%, in boiling alc: about 10%; in chloroform: about 0.1%. Practically insol in ether, benzene. pH of a 1% aq soln 5.5 to 6.5. Aqueous solns may be sterilized at 120° for 30 min. LD_{50} in mice (mg/kg): 151 i.p., 149 i.v. (Jenney, Pfeiffer).

4-aminosalicylate, $C_{13}H_{14}N_4O_4$, *pasiniazide, GEWO-399, Paraniazide, Dipasic.* An equimolecular salt produced by dissolving the components in hot dil alc, followed by cooling and evapn: Charonnat, Boime, *Compt. Rend.* **236**, 2140 (1953). Prepn: Swiss pat. 303,085 (1955 to Hoffmann-La Roche). Yellow crystals from methanol or ethanol, mp 142-144°. Sparingly sol in water. uv max 272, 303 nm ($E_{1cm}^{1\%}$ 550, 445).

Methanesulfonate, $C_7H_9N_3O_4S$, *4-pyridinecarboxylic acid 2-(sulfomethyl)hydrazide, methaniazide.* Prepn: Logemann, U.S. pat. 2,759,944 (1956 to Carlo Erba). Crystals, dec 187-189°.
Methanesulfonate sodium, $C_7H_8N_3NaO_4S$, *sodium isonicotinylhydrazide methanesulfonate, Neoiscotin.* Yellow crystals from water, dec 164-167°.
Methanesulfonate calcium, $C_{14}H_{16}CaN_6O_8S_2$, *Neo-Tizide.* Crystals, dec 215-220°.

THERAP CAT: Antibacterial (tuberculostatic).
THERAP CAT (VET): Antibacterial (tuberculostatic); anti-actinomycotic agent.

5204. Isonicotinic Acid. *4-Pyridinecarboxylic acid;* γ-pyridinecarboxylic acid; γ-picolinic acid. $C_6H_5NO_2$; mol wt 123.11. C 58.54%, H 4.09%, N 11.38%, O 25.99%. Prepd by permanganate oxidation of γ-picoline: Gilman, Broadbent, *J. Am. Chem. Soc.* **70**, 2757 (1948); Fields, U.S. pat. 2,946,801 (1960 to Standard Oil); from 4-ethylpyridine: Wibaut, Arens, *Rec. Trav. Chim.* **60**, 137 (1941); from citric acid to citrazinic acid to 2,6-dichloroisonicotinic acid: Behrmann, Hofmann, *Ber.* **17**, 2681 (1884); Wibaut, *Rec. Trav. Chim.* **63**, 141 (1944). Prepn of ethyl ester: Burrus, Powell, *J. Am. Chem. Soc.* **67**, 1469 (1945).

Platelets, mp 319°. Sublimes at 260° and 15 mm pressure. K at 25° = 1.1 × 10^{-5}. pH of satd aq soln 3.6. Sparingly sol in cold water (0.52 g/100 ml at 20°), more sol in hot water. Practically insol in benzene, ether, boiling alcohol.

Methyl ester, $C_7H_7NO_2$, liq, slight odor, similar to mint or oil of wintergreen. mp 8.5°. bp_{21} 104°; bp_{760} 209° (slight decompn).

Ethyl ester, liq, ester-like odor. Needles, when cooled by salt-ice mixture, mp 23°. d_4^{15} 1.0091 (liq). bp_5 78.5°; bp_{15} 110°; bp_{760} 220°. Sol in alcohol, ether, chloroform, benzene. Practically insol in water.

5205. Isonicotinic Acid Diethylamide. *N,N-Diethyliso-nicotinamide;* pyridine-4-carboxylic acid diethylamide. C_{10}-$H_{14}N_2O$; mol wt 178.23. C 67.39%, H 7.92%, N 15.72%, O 8.98%. Prepd by the action of phosphorus oxychloride upon a mixture of isonicotinic acid and diethylamine: Carrara, **U.S. pat. 2,858,317** (1958 to Lepetit).

Slightly viscous liq. $bp_{1.0}$ 119-120°. n_D^{20} 1.525. Miscible with water, ether, chloroform, acetone, alcohol.

5206. Isonipecotic Acid. *4-Piperidinecarboxylic acid;* hexahydroisonicotinic acid. $C_6H_{11}NO_2$; mol wt 129.16. C 55.80%, H 8.58%, N 10.84%, O 24.77%. Prepd by reduction of isonicotinic acid in glacial acetic acid in the presence of platinum oxide: Wibaut, *Rec. Trav. Chim.* **63,** 141 (1944); by reduction of isonicotinic acid in H_2O and concd HCl with PtO_2: Sperber *et al.,* **U.S. pat. 2,739,968** (1956 to Schering); Freifelder, **U.S. pat. 3,159,639** (1964 to Abbott).

Needles. Darkens at about 300°; mp 336°.

Hydrochloride, $C_6H_{11}NO_2$·HCl, mp 300° (Wibaut); mp 293° with decompn (Sperber). Sol in ethanol and methanol; freely sol in water.

5207. Isonitrosoacetone. *2-Oxopropanal 1-oxime;* pyruvaldoxime; propanone 1-oxime. $C_3H_5NO_2$; mol wt 87.08. C 41.38%, H 5.79%, N 16.09%, O 36.75%. CH_3CO-CH=NOH. Prepd from acetone by treatment with sodium nitrite in acetic acid at 0°: Küster, *Z. Physiol. Chem.* **155,** 174 (1926); by the action of methyl nitrite on acetone in ether in the presence of HCl: Slater, *J. Chem. Soc.* **117,** 589 (1920).

Leaflets from ether + petr ether or from carbon tetrachloride, mp 69°. Sublimes easily forming shiny needles. Volatile with steam. Freely sol in water, ether. Moderately sol in warm chloroform, benzene, carbon tetrachloride. Practically insol in petr ether. Sol in alkalies forming yellow solns. Faintly acid to litmus. K at 25° = 4.1 × 10⁻⁹.

5208. Isonitrosoacetophenone. *α-Oxobenzeneacetalde-hyde aldoxime;* benzoylformaldoxime. $C_8H_7NO_2$; mol wt 149.15. C 64.42%, H 4.73%, N 9.39%, O 21.45%. C_6H_5CO-CH=NOH.

Plates or prisms, mp 126-128°. Slightly sol in cold water; more sol in hot water; sol in alkalies and alkali carbonates.

USE: As a reagent for the detection of ferrous ions with which it gives a blue color sol in chloroform.

5209. Isonixin. *N-(2,6-Dimethylphenyl)-1,2-dihydro-2-oxo-3-pyridinecarboxamide;* 2-pyridone-3-carboxylic acid 2,6-xylidide; 2-hydroxy-2',6'-nicotinoxylidide; Nixyn. C_{14}-$H_{14}N_2O_2$; mol wt 242.28. C 69.41%, H 5.82%, N 11.56%, O 13.21%. Non-steroidal anti-inflammatory agent. Prepn: J.-A. Canicio Chimeno, **Belg. pat. 820,578;** *idem,* **U.S. pat.**

4,031,105 (1975, 1977 both to Hermes). Prepn, pharmacology and toxicity: R. Cadena *et al., Arzneimittel-Forsch.* **27,** 1457 (1977). Toxicological study: M. T. Mitjavila *et al., ibid.* 1460. Absorption, excretion studies: G. Carrera *et al., ibid.* **29,** 1401 (1979). Tolerance in humans: T. M. Serra *et al., Therapie* **35,** 173 (1980). HPLC determn in plasma: F. González Lopez *et al., Farmaco, Ed. Prat.* **38,** 273 (1983).

White crystalline powder from ethanol, mp 266-267°. Sol in chloroform and strong alkali. Practically insol in acids and water. LD_{50} in male, female mice, male, female rats (mg/kg): 7000, 8000, > 6000, > 6000 orally; all > 2000 i.p. (Cadena).

THERAP CAT: Analgesic, anti-inflammatory.

5210. Isooctane. *2,2,4-Trimethylpentane;* isobutyltrimethylmethane. C_8H_{18}; mol wt 114.23. C 84.12%, H 15.88%. $(CH_3)_3CCH_2CH(CH_3)_2$. Produced by the refining of petroleum.

Highly flammable, mobile liquid. Odor of gasoline. Antiknock octane no. 100. mp −107.45°. bp 99.3°. d_4^{20} 0.69194. n_D^{20} 1.39157. Dipole moment: 0. Flash pt, closed cup: 10°F (−12°C). Practically insol in water; somewhat sol in abs alcohols; sol in benzene, toluene, xylene, chloroform, ether, carbon disulfide, carbon tetrachloride, DMF and oils, except castor oil.

Note: The name isooctane has also been applied to 2-methylheptane.

USE: In determining octane numbers of fuels; in spectrophotometric analysis; as solvent and thinner.

5211. Isooctyl Alcohol. $C_8H_{18}O$; mol wt 130.23. C 73.78%, H 13.93%, O 12.29%. A mixture of closely related isomeric branched-chain primary alcohols: RCH_2OH where R represents a branched heptyl radical. The branching consists mostly of methyl groups located in the 3-, 4-, or 5-positions.

5212. Isopentyl Alcohol. *3-Methyl-1-butanol;* isoamyl alcohol; isobutyl carbinol; primary isoamyl alcohol; fermentation amyl alcohol. $C_5H_{12}O$; mol wt 88.15. C 68.13%, H 13.72%, O 18.15%. $(CH_3)_2CHCH_2CH_2OH$. A major component of commercial amyl alcohol, fusel oil *q.v.,* and *potato-spirit oil.* Prepn from butadiene, CO, H_2O, plus catalyst: Alderson, **U.S. pat. 3,020,314** (1962 to du Pont); from methylbutenes: Brown, Zweifel, *J. Am. Chem. Soc.* **82,** 1504 (1960). Early review of metabolism and toxicity: H. W. Haggard *et al., J. Ind. Hyg. Toxicol.* **27,** 1 (1945). Review of manuf by fractionation of fusel oil and via chlorination of pentanes, and properties: W. L. Faith *et al.,* Eds., *Industrial Chemicals* (John Wiley, New York, 2nd ed., 1957) pp 107-114. Acute toxicity data: Smyth *et al., Am. Ind. Hyg. Assoc. J.* **30,** 470 (1969).

Liquid; characteristic, disagreeable odor; pungent, repulsive taste. bp_{760} 130.5°. mp −117.2°. d_4^{15} 0.813. n_D^{20} 1.4075. Flash pt, closed cup: 114°F (45°C); open cup: 132°F (55°C). Slightly sol in water (2 g/100 ml at 14°); misc with alcohol, ether, benzene, chloroform, petr ether, glacial acetic acid, oils. LD_{50} orally in rats: 7.07 ml/kg (Smyth).

Caution: Potential symptoms of overexposure are irritation of eyes, nose and throat; narcosis; headache, dizziness; dyspnea, nausea, vomiting and diarrhea; skin cracking. *See NIOSH Pocket Guide to Chemical Hazards* (DHHS/NIOSH 90-117, 1990) p 130. *See also Patty's Industrial Hygiene and Toxicology* vol. **2C,** G. D. Clayton, F. E. Clayton, Eds. (Wiley-Interscience, New York, 3rd ed., 1982) pp 4588-4599.

USE: Solvent for fats, resins, alkaloids, etc.; manuf isoamyl (amyl) compds, isovaleric acid, mercury fulminate, pyroxylin, artificial silk, lacquers, smokeless powders; in microscopy; for dehydrating celloidin solns; for determining fat in milk.

Consult the Name Index before using this section.

5213. Isophorone. *3,5,5-Trimethyl-2-cyclohexen-1-one;* α-isophorone; 1,5,5-trimethyl-3-oxocyclohexene; 1,3,3-trimethylcyclohexen-5-one; isoforon; isoacetophorone. $C_9H_{14}O$; mol wt 138.21. C 78.21%, H 10.21%, O 11.58%. Prepn: E. Knoevenagel, *Ann.* **297**, 113 (1897). Synthesis: H. Ueda *et al.*, *Agr. Biol. Chem.* **30**, 1004 (1966); D. S. Torok, W. J. Scott, *Tetrahedron Letters* **34**, 3067 (1993). Colorimetric field test for determn in air: P. Andrew, R. Wood, *Analyst* **95**, 691 (1970). Solid-phase microextraction/GC determn in aq. samples: J.-Y. Horng, S.-D. Huang, *J. Chromatog. A* **678**, 313 (1994). Brief description: P. D. Sherman, Jr., A. J. Papa in Kirk Othmer *Encyclopedia of Chemical Technology* **vol. 13** (Wiley-Interscience, New York, 3rd ed., 1981) pp 918-922. Review of prepn and purification: G. S. Salvapati, M. Janardanarao, *J. Sci. Ind. Res.* **42**, 261-267 (1983). Review of toxicology and human exposure: *Toxicological Profile for Isophorone*, (PB90-180225, 1989) pp 101.

Clear liquid with peppermint-like odor. bp 215.3°. Freezing pt −8.1°. Flash pt (open cup): 184°F (84°C). Autoignition pt: 864°F (462°C). d_4^{20} 0.9229. d_4^{20} 0.9613. n_D^{20} 1.4778. uv max (MeOH): 235.5 nm (ε 14300). Vapor pressure at 20°: 0.3 mm Hg. Soly in water (mg/l): 12000 (20°); 14500 (25°). Sol in ether, acetone alcohol. LD_{50} in male, female rats and male mice (mg/kg): 2700 ±200, 2100 ±200, 2200 ±200 orally (PB90-180225).
USE: Solvent in some printing inks, paints, lacquers and adhesives.

5214. Isophthalic Acid. *1,3-Benzenedicarboxylic acid;* m-phthalic acid. $C_8H_6O_4$; mol wt 166.13. C 57.84%, H 3.64%, O 38.52%. Obtained by the oxidation of m-xylene: Smith, *J. Am. Chem. Soc.* **43**, 1920 (1921); Weisemann, Fragen, *World Petrol. Congr., Proc. 5th, N.Y. 1959* **4**, 197 (1960); Brill, *Ind. Eng. Chem.* **52**, 837 (1960); Bhattacharyya, Ganguly, *J. Indian Chem. Soc.* **38**, 463 (1961); Saffer, Barker, **U.S. pat. 3,089,906** (1963 to Mid-Century Corp.); Hay, **Brit. pat. 951,192** (1964 to General Electric).

Cryst powder, mp 345-348°; sublimes without decompn. Sol in 8000 parts cold water, 460 boiling water; freely sol in alcohol, glacial acetic acid; practically insol in benzene, petr ether.

5215. Isophytol. *3,7,11,15-Tetramethyl-1-hexadecen-3-ol;* 2,6,10,14-tetramethylhexadec-15-en-14-ol; 2,6,10-trimethyl-14-vinylpentadecan-14-ol. $C_{20}H_{40}O$; mol wt 296.54. C 81.01%, H 13.60%, O 5.40%. Decompn product of chlorophyll. Synthesis from linalool or citral: Fischer, Löwenberg, *Ann.* **475**, 183 (1929); Karrer *et al.*, *Helv. Chim. Acta* **26**, 1741 (1943); Sarycheva *et al.*, *Zh. Obshch. Khim.* **28**, 647 (1958); Maurit *et al., ibid.* **32**, 2483 (1962); from acetylene: Nazarov *et al., ibid.* **28**, 1444 (1958); Blaha, Weichet, **Czech. pat. 88,887** (1959), *C.A.* **54**, 2167c (1960); from pseudoionone and propargyl alcohol: Sato *et al.*, *J. Org. Chem.* **28**, 45 (1963).

Oily liquid. d_4^{20} 0.8519. n_D^{20} 1.4571. $bp_{0.01}$ 107-110°; $bp_{0.06}$ 125-128°. Practically insol in water; sol in the usual organic solvents.
USE: Prepn of vitamins E and K_1.

5216. Isopilosine. *[3S-[3α(S*),4β]]-Dihydro-3-(hydroxyphenylmethyl)-4-[(1-methyl-1H-imidazol-5-yl)methyl]-2(3H)-furanone;* carpiline; carpidine. $C_{16}H_{18}N_2O_3$; mol wt 286.33. C 67.12%, H 6.34%, N 9.78%, O 16.76%. This compound was originally called pilosine (the *cis*-isomer of isopilosine): H. W. Voigtländer, W. Rosenberg, *Arch. Pharm.* **292**, 579 (1959). Isoln from leaves of *Pilocarpus microphyllus* Stapf, *Rutaceae:* F. L. Pyman, *J. Chem. Soc.* **101**, 2260 (1912). Synthesis and abs config: Link, Bernauer, *Helv. Chim. Acta* **55**, 1053 (1972). Crystal structure: W. E. Oberhaensli, *Cryst. Struct. Commun.* **1**, 203 (1972).

Needles from alcohol, mp 182-182.5°. $[α]_D^{20}$ +37.6° (alcohol).
Pilosine, needles from alcohol, mp 179°. $[α]_D^{20}$ +83.9° (alcohol). uv max (alcohol): 210 nm (log ε 4.10).

5217. Isopimaric Acid. *[1R-(1α,4aβ,4bα,7α,10aα)]-7-Ethenyl-1,2,3,4,4a,4b,5,6,7,8,10,10a-dodecahydro-1,4a,7-trimethyl-1-phenanthrenecarboxylic acid; 13β-methyl-13-vinylpodocarp-7-ene-15-oic acid;* miropinic acid. $C_{20}H_{30}O_2$; mol wt 302.46. C 79.42%, H 10.00%, O 10.58%. Isoln from *Ducrydium biforme* Pilg., *Podocarpaceae:* Hosking, Brandt, *Ber.* **68**, 1311 (1935); from the bled resin of *Podocarpus ferrugineus, Podocarpaceae:* Brandt, Neubauer, *J. Chem. Soc.* **1940**, 683; from *Pinus palustris* Mill., *Pinaceae:* Harris, Sanderson, *J. Am. Chem. Soc.* **70**, 2079, 2081 (1948). Identity of isopimaric and miropinic acids: Brossi, Jeger, *Helv. Chim. Acta* **33**, 722 (1950). Structure: Antkowiak *et al.*, *J. Org. Chem.* **27**, 1930 (1962). Stereochemistry: Ireland, Newbould, *ibid.* **1931**; Antkowiak *et al.*, *Can. J. Chem.* **43**, 1257 (1965); Bose *et al.*, *Indian J. Chem.* **5**, 228 (1967).

Needles from methanol or ethanol, mp 160°. $[α]_D^{16}$ −3.6° (10.4% soln in 1:1 alcohol:chloroform). Freely sol in chloroform, benzene; moderately sol in ether, alcohol; slightly sol in light petroleum. Absorption spectrum: Brossi, Jeger, *loc. cit.*

5218. Isoprene. *2-Methyl-1,3-butadiene.* C_5H_8; mol wt 68.12. C 88.16%, H 11.84%. Isoln from the products of the pyrolysis of natural rubber: Williams, *Trans. Roy. Soc., London* **150**, 241 (1860); Boonstra, van Amerongen, *Ind. Eng. Chem.* **41**, 161 (1949). Structure: Euler, *Ber.* **30**, 1989 (1897). Prepn from turpentine: Tilden, *J. Chem. Soc.* **45**, 410 (1884); Bibb, **U.S. pat. 2,386,537** (1945 to Newport Ind.); from terpenes: Davis *et al.*, *Ind. Eng. Chem.* **38**, 53 (1946); Bourbon, **U.S. pat. 2,547,684** (1951 to Manufacture de Caoutchouc Michelin); by condensation of isobutylene and formaldehyde in acetic acid: Blomquist, Verdol, *J. Am. Chem. Soc.* **77**, 78 (1955); Chaffe *et al.*, **Fr. pat. 1,294,716** (1962 to Inst. Francais du Petrole); by dehydrogenation of isoamylenes from cracked gasolines: **Brit. pat. 875,346** (Appl. Apr. 21, 1960 to Shell Int. Res. Maatschappij); Voge *et al.*, **U.S. pat. 3,110,746** (1963 to Shell); by dimerization of propylene followed by pyrolysis of the resulting 2-methyl-2-pentene: Gorin, Oblad, **U.S. pat. 2,404,056** (1946 to Socony-Vacuum Oil); **Brit. pat. 913,852** (1962 to Goodyear); by dehydrogenation of isopentane: Dempsey, U.S. pat. **2,914,-588** (1959 to Sun Oil); Owen, U.S. pat. **2,982,795** (1961 to Phillips Petroleum); Stevenson, U.S. pat. **3,088,986** (1963 to Air Products & Chem.); by reacting isoamylenes with methanol and pyrolizing and dehydrogenating the resulting tertiary ethers: Verdol, Walker, U.S. pat. **2,972,645** (1961 to Sinclair Refining). Toxicity: Gostinskii, *C.A.* **62,** 15338a (1965). Reviews of isoprene and its polymers: Bean *et al.,* "Isoprene Polymers" in *Encyclopedia of Polymer Science and Technology* vol. **7** (Interscience, New York, 1967) pp 782-855; Bailey, "Isoprene" in *Vinyl and Diene Monomers* (part 2), E. C. Leonard, Ed. (Wiley-Interscience, New York, 1971) pp 997-1148; W. M. Saltman in Kirk-Othmer *Encyclopedia of Chemical Technology* vol. **13** (Wiley-Interscience, New York, 3rd ed., 1981) pp 818-837; of carcinogenic risk: *IARC Monographs* **60,** 215-232 (1994).

Colorless, volatile liquid. Unstable, oxidizable, flammable. bp_{760} 34.067. mp $-145.95°$. d_4^{20} 0.681; d_{20}^{20} 0.6805. n_D^{20} 1.42160. Practically insol in water. Miscible with alcohol or ether. LD_{50} for mice: 144 mg isoprene vapors/l air (Gostinskii).
Caution: Respiratory irritation is a potential symptom of overexposure. May act as a CNS depressant and asphyxiant at high concns. *See Patty's Industrial Hygiene and Toxicology* vol. **2B,** G. D. Clayton, F. E. Clayton, Eds. (John Wiley & Sons, New York, 4th ed., 1994) p 1252-1253.
USE: Manuf "synthetic" natural rubber, butyl rubber. Copolymer in the production of synthetic elastomers.

5219. Isopromethazine. *N,N,β-Trimethyl-10H-phenothiazine-10-ethanamine;* *10-(2-dimethylamino-1-methylethyl)phenothiazine;* isomethazine; Lilly 01526; RP-4460; Isophenergan; Fen-Bridal. $C_{17}H_{20}N_2S$; mol wt 284.43. C 71.79%, H 7.09%, N 9.85%, S 11.27%. Prepn: Charpentier, *Compt. Rend.* **225,** 306 (1947); Charpentier, Ducrot, *ibid.* **232,** 415 (1951); Charpentier *et al., ibid.* 2232; Berg, Ashley, U.S. pat. **2,607,773** (1952 to Rhône-Poulenc); Edge, Wragg, *J. Pharm. Pharmacol.* **5,** 279 (1953).

Hydrochloride, $C_{17}H_{20}N_2S.HCl$, *diprazin.* Crystals, mp 193-194°. More sol than promethazine.
THERAP CAT: Antihistaminic.

5220. Isopropalin. *4-(1-Methylethyl)-2,6-dinitro-N,N-dipropylbenzenamine;* *2,6-dinitro-N,N-dipropylcumidine;* 4-isopropyl-2,6-dinitro-*N,N*-dipropylaniline; EL-179; Paarlan. $C_{15}H_{23}N_3O_4$; mol wt 309.37. C 58.24%, H 7.49%, N 13.58%, O 20.69%. Selective pre-plant herbicide. Prepn: Q. F. Soper, U.S. pat. **3,257,190** (1966 to Lilly). Persistence in soil: T. Golab, W. A. Althaus, *Weed Sci.* **23,** 165 (1975); L. L. Gingerich, R. L. Zimdahl, *ibid.* **24,** 431 (1976); R. R. Romanowski, A. W. Libik, *ibid.* **26,** 258 (1978).

Red-orange liquid. Soly in water at 25°: 0.1 mg/l. Readily sol in organic solvents. Decomp by uv light. LD_{50} orally in rats: 5000 mg/kg, *RTECS* Vol. **I,** R. J. Lewis, R. L. Tatken, Eds. (1980) p 163.
USE: Herbicide.

5221. Isopropamide Iodide. *γ-(Aminocarbonyl)-N-methyl-N,N-bis(1-methylethyl)-γ-phenylbenzenepropanaminium iodide;* (3-carbamoyl-3,3-diphenylpropyl)diisopropylmethylammonium iodide; 2,2-diphenyl-4-diisopropylaminobutyramide methiodide; R-79; Darbid; Priamide; Tyrimide. $C_{23}H_{33}IN_2O$; mol wt 480.43. C 57.50%, H 6.92%, I 26.41%, N 5.83%, O 3.33%. Prepn: P. Janssen *et al., Arch. Int. Pharmacodyn.* **103,** 82 (1955); M. E. Speeter, U.S. pat. **2,-823,233** (1958 to Bristol). Crystal and molecular structure: N. Datta *et al., J. Chem. Soc., Perkin Trans. II* **1977,** 781. Comprehensive description: R. S. Santoro *et al.,* in *Analytical Profiles of Drug Substances* vol. **2,** K. Florey, Ed. (Academic Press, New York, 1973) pp 315-338; A. Post, R. S. Santoro, *ibid.* vol. **12** (1983) pp 721-732.

Crystals or amorphous powder, mp 198-201° (dec) (Janssen); also reported as mp 189.0-191.5° (Santoro). Sensitive to light. Freely sol in boiling water, in methanol, ethanol, chloroform. Practically insol in ether.
Free base, $C_{22}H_{30}N_2O$, mp 84-86°.
Note: Ingredient of *Stelabid* (also contg trifluoperazine, q.v.).
THERAP CAT: Anticholinergic.
THERAP CAT (VET): Anticholinergic.

5222. Isopropenyl Acetate. *1-Propen-2-ol acetate;* 1-propen-2-yl acetate. $C_5H_8O_2$; mol wt 100.12. C 59.98%, H 8.05%, O 31.96%. Prepn from ketene + acetone: Hull, Agett; Hagemeyer, U.S. pats. **2,481,669; 2,476,860** (both 1949 to Eastman Kodak); Young, U.S. pats. **2,461,016; 2,-511,423** (1949, 1950 both to Carbide & Carbon Chem.); Mawer, U.S. pat. **2,684,980** (1954 to I.C.I.); Buttner, Enk, U.S. pat. **2,867,653** (1959 to Wacker-Chemie). Use as reagent for acylation of potential enols: Hagemeyer, Hull, *Ind. Eng. Chem.* **41,** 2920 (1949); Hull, Agett, U.S. pat. **2,482,-066** (1949 to Eastman Kodak); Jeffery, Satchell, *J. Chem. Soc.* **1962,** 1876, 1906. Toxicity study: Smyth *et al., J. Ind. Hyg. Toxicol.* **31,** 60 (1949).

Liquid. bp 97°; bp_{200} 58-60°. n_D^{20} 1.4001. LD_{50} orally in rats: 3.0 g/kg (Smyth).
Caution: Mild irritant. Narcotic in high concns.
USE: Reagent for acylation of potential enols.

5223. 4-(5-Isopropenyl-2-methyl-1-cyclopenten-1-yl)-2-butanone. *4-[2-Methyl-5-(1-methylethenyl)-1-cyclopenten-1-yl]-2-butanone;* 4-(2-methyl-5-isopropenyl-1-cyclopenten-1-yl)-2-butanone; Pentione. $C_{13}H_{20}O$; mol wt 192.30. C 81.20%, H 10.48%, O 8.32%. Prepn: Kimel, Sax, U.S. pat. **2,799,706** (1957 to Hoffmann-La Roche).

Pale yellow liquid. Citrus-like odor with woody background. d_{25}^{25} 0.9218. $bp_{0.5}$ 72-75°. n_D^{25} 1.4800. Freely sol in diethyl phthalate. At least 10% soly in 70% ethanol, in mineral oil and in corn oil.

USE: In perfumery, i.e., in milled soaps, in modified eau de Cologne, in neroli and orange-blossom compositions. Blends well with ionones.

5224. Isopropyl Acetate. $C_5H_{10}O_2$; mol wt 102.13. C 58.80%, H 9.87%, O 31.33%. $CH_3COOCH(CH_3)_2$. Toxicity study: Smyth et al., Arch. Ind. Hyg. Occup. Med. **10**, 61 (1954).

Colorless liquid. d_4^{20} 0.870. bp 89°. Flash pt, open cup: 40°F (4°C); closed cup: 36°F (2°C). Sol in 23 parts water at 27°; miscible with alc, ether. LD_{50} orally in rats: 6.75 g/kg (Smyth).

Caution: Potential symptoms of overexposure are irritation of eyes, nose and skin; dermatitis; narcosis. See NIOSH Pocket Guide to Chemical Hazards (DHHS/NIOSH 90-117, 1990) p 132.

USE: Solvent for cellulose derivatives, plastics, oils and fats; in perfumery.

5225. Isopropyl Acetoacetate. 3-Oxobutanoic acid 1-methylethyl ester. $C_7H_{12}O_3$; mol wt 144.17. C 58.32%, H 8.39%, O 33.29%. $CH_3COCH_2COOC_3H_7$.

Liquid. d_4^{25} 0.957. bp ~205° with decompn. Slightly sol in water, freely in alcohol, ether.

5226. Isopropylacetone. 4-Methyl-2-pentanone; methyl isobutyl ketone; hexone. $C_6H_{12}O$; mol wt 100.16. C 71.95%, H 12.08%, O 15.97%. $CH_3COCH_2CH(CH_3)_2$. Manuf: Faith, Keyes & Clark's Industrial Chemicals, F. A. Lowenheim, M. K. Moran, Eds. (Wiley-Interscience, New York, 4th ed., 1975) pp 543-546. Toxicity data: Smyth et al., Arch. Ind. Hyg. Occup. Med. **4**, 119 (1951).

Colorless liquid; faint, ketonic and camphor odor. d_4^{20} 0.801. bp 117-118°. n_D^{20} 1.396. Flash pt, closed cup: 73°F (23°C). mp −84.7°. Moderately sol in water (1.91%); misc with alcohol, benzene, ether. LD_{50} orally in rats: 2.08 g/kg (Smyth).

Caution: Potential symptoms of overexposure are irritation of eyes and mucous membranes; headache; narcosis, coma; dermatitis. See NIOSH Pocket Guide to Chemical Hazards (DHHS/NIOSH 90-117, 1990) p 124.

USE: Solvent for gums, resins, nitrocellulose, etc.

5227. Isopropyl Alcohol. 2-Propanol; isopropanol; secondary propyl alcohol; dimethyl carbinol; petrohol. C_3H_8O; mol wt 60.10. C 59.96%, H 13.42%, O 26.62%. $CH_3CHOHCH_3$. Manuf from propylene: Faith, Keyes & Clark's Industrial Chemicals, F. A. Lowenheim, M. K. Moran, Eds. (Wiley-Interscience, New York, 4th ed., 1975) pp 496-501. Monograph: L. F. Hatch, Isopropyl Alcohol (McGraw-Hill, New York, 1961) 184 pp. Toxicity: Smyth, Carpenter, J. Ind. Hyg. Toxicol. **30**, 63 (1948).

Flammable liq. Slight odor resembling that of a mixture of ethanol and acetone. Slightly bitter taste (not potable!). mp −88.5°; fp −89.5°. bp_{760} 82.5°; bp_{400} 67.8°; bp_{200} 53.0°; bp_{100} 39.5°; bp_{60} 30.5°; bp_{40} 23.8°, bp_{20} 12.7°; bp_{10} 2.4°; bp_5 −7.0°; $bp_{1.0}$ −26.1°. d_4^{20} 0.78505; d_4^{83} 0.728. Flash pt, closed cup: 11.7°C (53°F). Autoignition temp 455.6°C (852°F). Lower explosive limit in air: 2.5% (v/v). n_D^8 1.3852; n_D^{15} 1.3802; n_D^{20} 1.37723; n_D^{25} 1.3749. Absorption spectrum: Brode, J. Phys. Chem. **30**, 61 (1926). Miscible with water, alcohol, ether, chloroform. Insol in salt solns. May be recovered from aq mixtures by salting out with sodium chloride, sodium sulfate, sodium hydroxide, etc. Forms an azeotrope with water (bp_{760} 80.37°, d_4^0 0.83361) isopropanol 87.7% (w/w). Freezing points of mixtures with water (v/v, per cent by volume): Isopropanol 15% = −3.3°; 25% = −6.7°; 30% = −8.3°; 35% = −11.1°; 40% = −13°; 45% = −17.8°; 50% = −23°; 60% = −32°. LD_{50} orally in rats: 5.8 g/kg (Smyth, Carpenter).

Caution: Potential symptoms of overexposure are mild irritation of eyes, nose and throat; drowsiness, dizziness and headache; dry cracking skin. See NIOSH Pocket Guide to Chemical Hazards (DHHS/NIOSH 90-117, 1990) p 132.

USE: In antifreeze compositions; as solvent for gums, shellac, essential oils; in the extraction of alkaloids; in quick-drying oils; in quick-drying inks; in denaturing ethyl alcohol; in body rubs; hand lotions, after-shave lotions and similar cosmetics. Solvent for creosote, resins, gums; in manuf of acetone, glycerol, isopropyl acetate. Pharmaceutic aid (solvent).

THERAP CAT: Antiseptic.

THERAP CAT (VET): Antiseptic, rubefacient.

5228. Isopropylamine. 2-Propanamine; 2-aminopropane. C_3H_9N; mol wt 59.11. C 60.96%, H 15.35%, N 23.70%. $(CH_3)_2CHNH_2$. Prepn from acetone + NH_3: Skita, Keil, Ber. **61**, 1682 (1928); Norton et al., J. Org. Chem. **19**, 1054 (1954); from acetone oxime: Winans, Adkins, J. Am. Chem. Soc. **55**, 2056 (1933). Toxicity study: H. F. Smyth et al., Arch. Ind. Hyg. Occup. Med. **4**, 119 (1951).

Colorless, flammable liquid; ammonia odor; strong base. d_4^{15} 0.694. mp −101°. bp 33-34°. n_D^{15} 1.3770. Flash pt, open cup: −15°F (−26°C). Miscible with water, alcohol, ether. LD_{50} orally in rats: 820 mg/kg (Smyth).

Caution: Potential symptoms of overexposure are irritation of eyes, nose, throat and skin; pulmonary edema; visual disturbance; skin and eye burns; dermatitis. See NIOSH Pocket Guide to Chemical Hazards (DHHS/NIOSH 90-117, 1990) p 134.

5229. Isopropyl Bromide. 2-Bromopropane. C_3H_7Br; mol wt 122.99. C 29.30%, H 5.74%, Br 64.97%. $CH_3CHBrCH_3$. Obtained by heating isopropyl alcohol with HBr.

Colorless liquid. d_4^{20} 1.31. mp −89°. bp 59-60°. n_D^{20} 1.4251. Slightly sol in water; miscible with alcohol, benzene, chloroform, ether.

5230. Isopropyl Chloride. 2-Chloropropane. C_3H_7Cl; mol wt 78.54. C 45.88%, H 8.98%, Cl 45.14%. $CH_3CHClCH_3$. Prepd by refluxing isopropyl alcohol with concd HCl and $ZnCl_2$.

Liquid. d^{15} 0.868. mp −117°. bp 35-36°. Flash pt, closed cup: −26°F (−32°C). Slightly sol in water; miscible with alcohol, ether.

5231. Isopropyl Ether. 2,2′-Oxybispropane; 2-isopropoxypropane; diisopropyl ether; DIPE. $C_6H_{14}O$; mol wt 102.18. C 70.53%, H 13.81%, O 15.66%. $(CH_3)_2CHOCH(CH_3)_2$. Prepn: E. Erlenmeyer, Ann. **126**, 305 (1863). Physical properties and solubility: H. R. Fife, E. W. Reid, Ind. Eng. Chem. **22**, 513 (1930). Manuf process: M. N. Harandi, H. Owen, U.S. pat. **5,208,387** (1993 to Mobil). GC determn in gasolines: N. G. Johansen, J. High Resolut. Chromatog. Chromatog. Commun. **7**, 487 (1984). Acute toxicity: E. T. Kimura et al., Toxicol. Appl. Pharmacol. **19**, 699 (1971). Comparison with MTBE, ETBE and TAME, q.q.v., as gasoline oxygenate: M. J. McNally et al., Oil Gas J. **90**, 39 (May 25, 1992).

Liquid. bp_{760} 68.27°. fp −85.89°. d_4^{20} 0.72813. n_D^{20} 1.36888. Surface tension in air (23°): 32 dynes/cm. Viscosity (25°): 0.379 cP. Vapor pressure (20°): 158 mm Hg. Specific heat (22-27°): 0.526 cal/g. Flash pt, closed cup: −6.7°F. Latent heat of vaporization (67.5°): 68.2 cal/g. Soly of commercial grade in water (19°): 1.71 wt%. Misc with most organic solvents. Ethanolamines, glycols, glycerol and their monoesters have limited soly; diesters are sol. Readily forms peroxides when exposed to oxygen unless stabilized with inhibitors such as hydroquinone or BHT. LD_{50} in 14 day old, young adult, adult rats (ml/kg): 6.4, 16.5, 16.0 orally (Kimura).

Caution: Potential symptoms of overexposure are irritation of eyes and nose; respiratory discomfort; dermatitis. See NIOSH Pocket Guide to Chemical Hazards (DHHS/NIOSH 90-117, 1990) p 134.

USE: As solvent; fuel additive.

5232. Isopropylidene Glycerol. 2,2-Dimethyl-1,3-dioxolane-4-methanol; 2,2-dimethyl-4-hydroxymethyl-1,3-dioxolane; glycerol dimethylketal; acetone glycerol; Solketal. $C_6H_{12}O_3$; mol wt 132.16. C 54.53%, H 9.15%, O 36.32%. Prepn: Fisher, Ber. **28**, 1167 (1895); Fisher, Pfähler, Ber. **53**, 1607 (1920); Hibbert, Morazain, Can. J. Res. **2**, 38 (1930); Smith, Lindberg, Ber. **64**, 510 (1931); M. M. Maglio, C. A. Burger, J. Am. Chem. Soc. **68**, 529 (1946); M. Renoll, M. S. Newman, Org. Syn. coll. vol. III, 502 (1955); Mikschik, Austrian pat. **180,926** (1955 to Chemomedica Chemikalien), C.A. **49**, 15951 (1955); Williams, Brit. pat. **802,022** (1958 to Peter Spence & Sons); Brit. pat. **819,835** (1959 to Bayer);

Perez *et al.*, **Span. pat. 499,129** (1982 to Calipe), *C.A.* **97**, 55794v (1982). Synthesis of D(+)-form: E. Baer, H. Fischer, *J. Biol. Chem.* **128**, 463 (1939); of L(−)-form: *eidem, J. Am. Chem. Soc.* **61**, 761 (1939). Conversion of D(+)- to L(−)-form: *eidem, ibid.* **67**, 944 (1945).

Practically odorless liq of medium mobility. d_4^{20} 1.064. bp_{10} 82°. n_D^{20} 1.4383. Viscosity at 20° = 11 cp. Practically nonflammable at ordinary storage temps: Flash pt 90°C (194°F). Evaporation no. about 600 (ether = 1). Miscible with water, alcohols, acetals, esters, ethers, aromatic hydrocarbons, chlorinated hydrocarbons, gasolines, petr ether, turpentine, oils. LD_{50} in rats (g/kg): 7 orally; 3 i.p. (Perez). D(+)-Form, $[\alpha]_D$ +13.6°; $[\alpha]_D$ +10.8° (c = 15.19 in benzene). L(−)-Form, $[\alpha]_D$ −13.6°; $[\alpha]_D$ −10.8° (c = 15.19 in benzene).

USE: Versatile solvent and plasticizer. Pharmaceutic aid (solubilizing and suspending agent).

5233. Isopropyl Iodide. *2-Iodopropane.* C_3H_7I; mol wt 169.99. C 21.20%, H 4.15%, I 74.65%. CH_3CHICH_3. Prepared by distilling isopropyl alcohol with HI; or from glycerol, iodine, water, and phosphorus.

Colorless liquid but readily discolors in air and light. d_4^{20} 1.703. mp −90°. bp 89-90°. n_D^{20} 1.5026. Sol in 720 parts water; miscible with alcohol, benzene, chloroform, ether.

5234. Isopropyl Myristate. *Tetradecanoic acid 1-methylethyl ester.* $C_{17}H_{34}O_2$; mol wt 270.46. C 75.50%, H 12.67%, O 11.83%. $CH_3(CH_2)_{12}COOCH(CH_3)_2$. The commercial product usually appears as a mixture of myristate with small amounts of esters of palmitic and other satd fatty acids. Physical properties: Bonhorst *et al.*, *Ind. Eng. Chem.* **40**, 2379 (1948).

Liquid of low viscosity, practically odorless. mp ~+3°. bp_2 140.2°; bp_{20} 192.6°; dec 208°. d^{20} 0.8532; d^{99} 0.7942. n_D^{25} 1.432-1.434. Withstands oxidation and does not readily become rancid. Sol in castor oil, cottonseed oil, acetone, chloroform, ethyl acetate, ethanol, toluene, and mineral oil. Practically insol in water, glycerol, propylene glycol. Dissolves many waxes, cholesterol, lanolin.

USE: In cosmetic and topical medicinal prepns where good absorption through the skin is desired. A jellied isopropyl myristate was marketed as *Estergel*.

5235. Isopropyl Nitrite. *Nitrous acid 1-methylethyl ester;* 2-propanol nitrite; nitrous acid isopropyl ester. $C_3H_7NO_2$; mol wt 89.09. C 40.44%, H 7.92%, N 15.72%, O 35.92%. $(CH_3)_2CHONO$. Prepd by treating isopropyl alcohol with nitrosyl chloride: Kornblum *et al.*, *J. Am. Chem. Soc.* **77**, 5531 (1955); by adding concd sulfuric acid and isopropyl alcohol to sodium nitrite: Levin, Hartung, *Org. Syn.* **24**, 26 (1944).

Pale yellow oil. d_4^0 0.856; d_4^{25} 0.844. bp_{752} 39-39.5°; bp_{745} 39-40°. n_D^{20} 1.3520.

Caution: Can cause vasodilation with fall in blood pressure, tachycardia, headache. Large doses can cause methemoglobinuria with cyanosis. Severe poisoning results in shock which can end fatally.

USE: Jet propellant.

5236. Isoproterenol. *4-[1-Hydroxy-2-[(1-methylethyl)-amino]ethyl]-1,2-benzenediol; 3,4-dihydroxy-α-[(isopropyl-amino)methyl]benzyl alcohol;* α-(isopropylaminomethyl)protocatechuyl alcohol; isoprenaline; isopropylarterenol; 1-(3,4-dihydroxyphenyl)-2-isopropylaminoethanol; isopropylaminomethyl-(3,4-dihydroxyphenyl)carbinol; *N*-isopropyl-β-dihydroxyphenyl-β-hydroxyethylamine; dihydroxyphenylethanolisopropylamine; *N*-isopropylnoradrenaline; epinephrine isopropyl homolog; A-21; Aludrine; Aleudrin; Isuprel; Norisodrine; Asiprenol; Asmalar; Neo-Epinine; Novodrin; Isupren; Neodrenal; Isopropydrin; Assiprenol; Respifral; Bellasthman; Saventrine; Proternol; Isorenin; Vapo-N-Iso; Isonorin. $C_{11}H_{17}NO_3$; mol wt 211.26. C 62.54%, H 8.11%,

N 6.63%, O 22.72%. β-Adrenergic agonist. Catecholamine derivative prepd by the reaction of 3,4-dihydroxy-α-haloacetophenone with an excess of isopropylamine: Scheuing, Thomä, Ger. pat. **723,278** (1942 to Boehringer, Ing.); U.S. pat. **2,308,232** (1943); from guaiacol and chloral hydrate: Beke *et al.*, *Pharm. Zentralh.* **92**, 237 (1953). Resolution: Kerschbaum, Benedikt, *Monatsh.* **83**, 1090 (1952); Beccari *et al.*, *Science* **118**, 249 (1953); Delmar *et al.*, U.S. pat. **2,715,-141** (1955 to Delmar Chem.). Configuration: Pratesi *et al.*, *Farmaco Ed. Sci.* **15**, 3 (1960). Toxicity: E. I. Goldenthal, *Toxicol. Appl. Pharmacol.* **18**, 185 (1971). Effect on gastric acid secretion: M. J. Daly, *Scand. J. Gastroenterol.* **89**, 3 (1984). Use in treatment of primary pulmonary hypertension: D. A. Pietro *et al.*, *N. Engl. J. Med.* **310**, 1032 (1984). Review of pharmacology and comparison with other β-adrenoceptor agonists: V. T. Popa, *J. Asthma* **21**, 183-207 (1984). Comprehensive description: M. Tariq, A. A. Al-Badr, *Anal. Profiles Drug Subs.* **14**, 391-422 (1985).

dl-Form, crystals from alc, mp 155.5°. pKa 8.64. LD_{50} in male, female rats (mg/kg): 3675, 4282 orally (Goldenthal).

dl-Form hydrochloride, $C_{11}H_{17}NO_3.HCl$, *Aerolone, Aerotrol, Euspiran, Isomenyl, Isovon, Mistarel, Suscardia.* Crystals from alc, mp 170-171°. Soly: one gram dissolves in 3 ml water, in 50 ml ethanol (95%). Less sol in abs ethanol. Practically insol in chloroform, ether, benzene. pH of 1% aq soln about 5. Aq solns turn brownish-pink upon prolonged exposure to air or upon addition of alkali. LD_{50} orally in rats: 2221 ±93 mg/kg (Goldenthal).

dl-Form sulfate dihydrate, $(C_{11}H_{17}NO_3)_2.H_2SO_4.2H_2O$, *Aludrin, Isomist, Propal.* Crystals from acetone + methanol, mp 128° (some decomp). One gram dissolves in about 4 ml water. Slightly sol in alc. Practically insol in chloroform, ether, benzene. pH of a 1% aq soln about 5.

l-Form, crystals, mp 164-165°. $[\alpha]_D^{19}$ −45.0° (c = 2 in 2*N* HCl).

l-Form hydrochloride, crystals, dec 162-164°. $[\alpha]_D^{20}$ −50°.

l-Form *d*-bitartrate dihydrate, *Isolevin.* Crystals, mp 80-83° (sinters at 78°). $[\alpha]_D^{19}$ −14.9° (c = 2.31).

THERAP CAT: Bronchodilator.

THERAP CAT (VET): Sympathomimetic. Chiefly as bronchodilator.

5237. Isoproturon. *N,N-Dimethyl-N′-[4-(1-methylethyl)phenyl]urea; 3-p-cumenyl-1,1-dimethylurea; N-(4-isopropylphenyl)-N′,N′-dimethylurea;* I.P.U.; HOE-16410; CGA-18731; Arelon; Graminon; Tolkan. $C_{12}H_{18}N_2O$; mol wt 206.29. C 69.87%, H 8.79%, N 13.58%, O 7.76%. Pre- and post-emergence herbicide for control of annual grasses and broad-leaved weeds. Prepn and herbicidal properties: C. W. Todd, U.S. pat. **2,655,447** (1953 to du Pont); **Belg. pat. 770,928** (1971 to Pepro); A. Thizy *et al.*, U.S. pat. **4,295,877** (1981 to Philagro). Physical properties, herbicidal activity: *eidem, Compt. Rend. Ser. D* **274**, 2053 (1972); J. Rognon *et al.*, *Meded. Fac. Landbouwwetensch. Rijksuniv. Gent* **37**, 663 (1972), *C.A.* **79**, 1260c (1973). Microbial degradation of labelled isoproturon: J.-C. Fournier *et al.*, *Chemosphere* **4**, 207 (1975). Efficacy in winter cereals: R. T. Hewson, *Proc. 12th Brit. Weed Contr. Conf.* 75 (1974); P. Gonzales *et al.*, *Weed Res.* **23**, 39 (1983). HPLC determn in soil: G. Kulshrestha, R. Khazanchi, *J. Chromatog.* **318**, 144 (1985).

White crystals from benzene, mp 158°. Soly in water at 22°: 70 mg/l. LD_{50} in mice, rats (mg/kg): 3350, 3600 orally (Thizy, 1972).

USE: Herbicide.

5238. Isopyrocalciferol. *(3β,9β,22E)-9-Ergosta-5,7,22-trien-3-ol;* 9β-ergosterol. $C_{28}H_{44}O$; mol wt 396.66. C 84.79%, H 11.18%, O 4.03%. Differs sterically from pyrocalciferol at C-10, from ergosterol at C-9. *See ref under* Pyrocalciferol. Stereochemistry: Castells *et al., J. Chem. Soc.* **1959,** 1159. Selenium dehydrogenation gives Diel's hydrocarbon (3'-methyl-1,2-cylopentenophenanthrene).

Prisms from ether-methanol. mp 112-115°. uv max: 262, 280 nm. $[\alpha]_D^{20}$ +332°, $[\alpha]_{546}^{20}$ +415° (c = 1.5 in chloroform). Precipitated by digitonin.

Acetate, $C_{30}H_{46}O_2$, mp 113-115°, $[\alpha]_D^{20}$ +333°, $[\alpha]_{546}^{20}$ +423° (c = 2 in chloroform).

5239. Isoquassin. Picrasmin. $C_{22}H_{28}O_6$; mol wt 388.46. C 68.02%, H 7.27%, O 24.71%. Bitter principle from Jamaica quassia, *Picrasma excelsa* (Sw.) Planch., Simaroubaceae. Isoln: Clark, *J. Am. Chem. Soc.* **60,** 1146 (1938). Identity of isoquassin and picrasmin: Adams, Whaley, *ibid.* **72,** 375 (1950). Isomer of quassin, *q.v.:* Valenta *et al., Tetrahedron* **18,** 1433 (1962).

Crystals from methanol, mp 222-225°. $[\alpha]_D^{20}$ +45.4° (chloroform). uv max: 258 nm (ε 12500).

5240. Isoquercitrin. *2-(3,4-Dihydroxyphenyl)-3-(β-D-glucofuranosyloxy)-5,7-dihydroxy-4H-1-benzopyran-4-one; 3,3',4',5,7-pentahydroxyflavone-3-glucoside;* quercetin-3-glucoside; isotrifoliin; trifoliin. $C_{21}H_{20}O_{12}$; mol wt 464.38. C 54.32%, H 4.34%, O 41.34%. From flowers of *Gossypium herbaceum* L., *Malvaceae:* Perkin, *J. Chem. Soc.* **95,** 2181 (1909); from flowers of *Aesculus hippocastanum* L., *Hippocastanaceae:* Hörhammer, Wagner, *Arch. Pharm.* **290,** 224 (1957); from *Tropaeolum majus* L., *Tropaeolaceae:* Delaveau, *Compt. Rend.* **252,** 1510 (1961); from *Arnica montana* L., *Compositae:* Friedrick, *Naturwiss.* **49,** 541 (1962). Structure: Attree, Perkin, *J. Chem. Soc.* **1927,** 234. Identity with trifoliin and isotrifoliin: Hattori *et al., J. Chem. Soc. Japan* **58,** 844 (1937), *C.A.* **32,** 219c (1938). Synthesis: Ice, Wender, *J. Am. Chem. Soc.* **74,** 4606 (1952); *eidem,* U.S. pat. **2,727,890** (1955 to the USAEC).

Yellow needles from water, mp 225-227°. uv max: 257, 369 nm. Practically insol in cold but sparingly sol in boiling water. Sol in alkaline solns with a deep yellow tint.

5241. Isoquinoline. 2-Benzazine; benzo[c]pyridine. C_9H_7N; mol wt 129.16. C 83.69%, H 5.46%, N 10.84%. Occurs in coal tar: Hoogewerff, van Dorp, *Rec. Trav. Chim.* **4,** 125, 285 (1885); Weissgerber, *Ber.* **47,** 3175 (1914); Harris, Pope, *J. Chem. Soc.* **121,** 1029 (1922). Modern isoln processes: H. J. V. Winkler, *Der Steinkohlenteer und seine Aufarbeitung* (Essen, 1951). Isoln by a freeze-out method: Kjellman, U.S. pat. **2,483,420** (1946 to Koppers). Purifica-

tion: Freiser, Glowacki, *J. Am. Chem. Soc.* **71,** 514 (1949). Toxicity data: Smyth *et al., Arch. Ind. Hyg. Occup. Med.* **4,** 119 (1951). Synthesis: D. L. Boger *et al., Tetrahedron* **37,** 3977 (1981). Toxicity data: Smyth *et al., Arch. Ind. Hyg. Occup. Med.* **4,** 119 (1951). Comprehensive survey of isoquinoline alkaloids: Gensler, *Heterocyclic Compounds* vol. **4,** R. C. Elderfield, Ed. (Wiley, New York, 1952) pp 344-490; T. Kametani, *Total Synthesis of Natural Products* vol. **3,** J. ApSimon, Ed. (Wiley, New York, 1977) pp 1-272.

Liquid. Pungent odor resembling that of a mixture of anise oil and benzaldehyde. Hygroscopic platelets when solid. d_4^{30} 1.09101; d_2^{80} 1.05143. mp 26.48°. n_D^{30} 1.62078. Viscosity at 30°: 3.2528 cp, at 100°: 1.0230 cp, at 200°: 0.4223 cp. bp_{743} 242.2°; bp_{760} 243.25°. Dipole moment: 2.49. More basic than quinoline, pKa (25°): 8.60. Almost insol in water. Miscible with many organic solvents. Sol in dil acids. LD_{50} orally in rats: 360 mg/kg (Smyth).

Sulfate, $C_9H_9NO_4S$, crystals, mp 209-209.5°. Sol in water.

USE: Synthesis of dyes, insecticides, antimalarials, rubber accelerators.

5242. Isorubijervine. *Solanid-5-ene-3β,18-diol.* $C_{27}H_{43}NO_2$; mol wt 413.64. C 78.40%, H 10.48%, N 3.39%, O 7.74%. From various species of *Veratrum, Liliaceae.* Isoln and structure: Jacobs, Craig, *J. Biol. Chem.* **148,** 41 (1943); **159,** 617 (1945); Pelletier, Jacobs, *J. Am. Chem. Soc.* **75,** 4442 (1953). Stereochemistry: Sato, Latham, *ibid.* **78,** 3146 (1956); Höhne *et al., Tetrahedron* **22,** 673 (1966). *Review:* Morgan, Barltrop, *Quart. Revs.* **12,** 34 (1958).

Crystals from alcohol, mp 235-237°. $[\alpha]_D^{25}$ +6.5° (c = 0.97 in abs ethanol). Forms a sparingly sol digitonide.

Hydrobromide, $C_{27}H_{44}BrN_2O$, fine needles from alcohol + ether, dec 290-295°.

5243. Isosafrole. *5-(1-Propenyl)-1,3-benzodioxole; 1,2-(methylenedioxy)-4-propenylbenzene.* $C_{10}H_{10}O_2$; mol wt 162.19. C 74.06%, H 6.21%, O 19.73%. Purification and separation from safrole: Balbiano, *Ber.* **42,** 1505 (1911); Hoering, Baum, *ibid.* 3082. Prepn: Bert, *Compt. Rend.* **213,** 873 (1941); Naves, Ardizio, *Bull. Soc. Chim. France* **1957,** 1053; Fengeas, *ibid.* **1964,** 1892; Cabiddu *et al., Ann. Chim. (Rome)* **52,** 1261 (1962). Review and evaluation of studies of carcinogenic action in laboratory animals: *IARC Monographs* **10,** 231-241 (1976).

trans - Form

trans-Form, liquid, odor of anise. bp_{760} 253°; bp_{100} 179.5°; bp_{20} 135.6°; $bp_{3.4}$ 85-86°. mp 8.2°. d_4^{20} 1.1206. n_D^{20} 1.5782. uv max (96% alc): 305, 267, 259.5 nm (ε 5340; 11600; 12160). Miscible with alc, ether, benzene. Sol in 8 parts of 90% alcohol.

cis-Form, liq. $bp_{3.5}$ 77-79°. mp −21.5°. d_4^{20} 1.1182. n_D^{20} 1.5691. uv max (96% alc): 296.5, 259 nm (ε 4450; 10000).

USE: Manuf heliotropin; to modify oriental perfumes; to strengthen soap perfumes; in small quantities together with methyl salicylate in root beer and sarsaparilla flavors.

5244. Isosorbide. *1,4:3,6-Dianhydro-D-glucitol;* 1,4:3,6-dianhydrosorbitol; AT-101; NSC-40725; Hydronol; Ismotic; Isobide. $C_6H_{10}O_4$; mol wt 146.14. C 49.31%, H 6.90%, O 43.79%. Prepd by acid dehydration of D-glucitol: Haworth, Wiggins, **Brit.** pat. 600,870 (1948). Purification: Hartmann, **U.S.** pat. 3,160,641 (1964 to Atlas Chemical Industries). Review of prepn, structure, properties: Wiggins, *Advan. Carbohyd. Chem.* **5,** 191-228 (1950). Stereochemistry: Cope, Shen, *J. Am. Chem. Soc.* **78,** 3177 (1956). Conformation studies: Hopton, Thomas, *Can. J. Chem.* **47,** 2395 (1969). Pharmacodynamics in man: Nodine *et al., Clin. Pharmacol. Ther.* **14,** 196 (1973).

Crystals, mp 61-64°. $[\alpha]_D$ +44°.

THERAP CAT: Diuretic.

5245. Isosorbide Dinitrate. *1,4:3,6-Dianhydro-D-glucitol dinitrate;* dinitrosorbide; 1,4:3,6-dianhydrosorbitol 2,5-dinitrate; Astridine; Cardio 10; Cardis; Carvanil; Carvasin; Cedocard; Coroviss; Dignonitrat; Dilatrate; Diniket; Disorlon; Duranitrat; EureCor; Flindix; Frandol; Glentonin; IBD; Imtack; Isdin; ISDN; Iso-Bid; Isocard; Isoket; Iso-Mack; Iso-Puren; Isorbid; Isostenase; Isotrate; Langoran; Laserdil; Maycor; Myorexon; Nitorol; Nitrol; Nitrosorbon; Nosim; Rifloc Retard; Rigedal; Risordan; Soni-Slo; Sorbangil; Sorbichew; Sorbidilat; Sorbid SA; Sorbitrate; Sorquad; Vascardin; Vasorbate; Vasotrate. $C_6H_8N_2O_8$; mol wt 236.14. C 30.52%, H 3.41%, N 11.86%, O 54.20%. Coronary vasodilator. Prepn from sorbitol: Goldberg, *Acta Physiol. Scand.* **15,** 173 (1948), *C.A.* **42,** 5564 (1948); Kochergin, Titkova, *C.A.* **54,** 8647h (1960). Absorption and disposition in man: H. Laufen *et al., Arzneimittel-Forsch.* **33,** 980 (1983). Comprehensive description: L. A. Silvieri, N. J. DeAngelis, *Anal. Profiles Drug Subs.* **4,** 225-244 (1975).

Hard colorless crystals, mp 70°. $[\alpha]_D^{20}$ +135° (alc). Sparingly sol in water (1.089 mg/ml, also given as 1.0 g dissolves in 900 ml H_2O). Freely sol in organic solvents, such as acetone, alc, ether.

Isosorbide-5-mononitrate, $C_6H_9NO_6$, *Corangin, Elan, Elantan, Imdur, Ismo, Isomonat, Monicor, Monit, Mono-Cedocard, Monoclair, Monoket, Mono Mack, Monosorb, Olicard, Pentacard.* Metabolite of isosorbide dinitrate.

THERAP CAT: Antianginal.

5246. β-Isosparteine. *[7S-(7α,7aβ,14α,14aβ)]-Dodecahydro-7,14-methano-2H,6H-dipyrido[1,2-a:1′,2′-e][1,5]diazocine;* pusilline. $C_{15}H_{26}N_2$; mol wt 234.38. C 76.87%, H 11.18%, N 11.95%. Isoln from *Lupinus pusillus,* Pursh., *Leguminosae:* Marion, Fenton, *J. Org. Chem.* **13,** 780 (1948). Identity with pusilline: Greenhalgh, Marion, *Can. J. Chem.* **34,** 456 (1956); identity of monohydrochloride with *nonalupine* and *spathulatine:* Marion, *ibid.* **29,** 959 (1951). Synthesis: Bohlmann, *Ber.* **90,** 653 (1957). Absolute configuration: Okuda, Tsuda, *Chem. & Ind. (London)* **1961,** 1115.

Oil, $bp_{0.2}$ 100-110°. $[\alpha]_D^{32}$ −15.3° (c = 2.3 in abs alc). Sol in methanol, ethanol. Practically insol in water.

Hydrochloride monohydrate, $C_{15}H_{26}N_2.HCl.H_2O$, prismatic needles from alcohol or acetone. Melts at 88-90°, resolidifies, and remelts at 235.5-236.5°. $[\alpha]_D$ −1.88° (c = 5.90 in chloroform); −2.32° (c = 6.42 in alc); −2.44° (c = 6.42 in water). Sol in water, benzene, alcohol, chloroform, carbon tetrachloride, acetone, ethyl acetate; slightly sol in petr ether, ether. The aq soln is slightly alkaline to litmus.

Dihydrochloride, $C_{15}H_{26}N_2.2HCl$, clusters of fine needles from ethyl acetate, mp 270-273° (sinters at 262°).

Hydriodide, $C_{15}H_{26}N_2.HI$, prismatic needles from acetone, mp 252.5° (sinters at 250.5°).

Methiodide, $C_{15}H_{26}N_2.CH_3I$, diamond-shaped crystals from ethyl acetate, mp 250-252°.

5247. Isothebaine. *(S)-5,6,6a,7-Tetrahydro-2,11-dimethoxy-6-methyl-4H-dibenzo[de,g]quinolin-1-ol; 2,11-dimethoxy-6aα-aporphin-1-ol;* 1-hydroxy-2,11-dimethoxyaporphine. $C_{19}H_{21}NO_3$; mol wt 311.38. C 73.29%, H 6.80%, N 4.50%, O 15.41%. Occurs as *d*-form in the root and whole plant of *Papaver orientale* L., *Papaveraceae,* collected in late autumn. Isoln: Gadamer, *Arch. Pharm.* **249,** 39 (1911). Structure: Bentley, Blues, *J. Chem. Soc.* **1956,** 1732; Bentley, Dyke, *J. Org. Chem.* **22,** 429 (1957). Synthesis of *dl*-form: Battersby, Brown, *Proc. Chem. Soc. (London)* **1964,** 85; H. Hara *et al., Chem. Pharm. Bull* **29,** 1083 (1981). Synthesis of natural and racemic forms: Battersby *et al., J. Chem. Soc.* **1965,** 4550. Biosynthesis: *eidem, Chem. Commun.* **1965,** 230.

Light-sensitive, rhombic crystals. mp 203-204°. $[\alpha]_D^{18}$ +285° in alc. Sol in alcohol, chloroform, slightly in ether.

5248. Isothipendyl. *N,N,α-Trimethyl-10H-pyrido[3,2-b][1,4]benzothiazine-10-ethanamine; 10-(2-dimethylamino-2-methylethyl)-10H-pyrido[3,2-b][1,4]benzothiazine;* 10-(2-dimethylaminopropyl)-1-azaphenothiazine; *N*-dimethylaminoisopropylthiophenylpyridylamine; D-201. $C_{16}H_{19}N_3S$; mol wt 285.41. C 67.33%, H 6.71%, N 14.72%, S 11.23%. Prepn: H. L. Yale, F. Sowinski, *J. Am. Chem. Soc.* **80,** 1651 (1958). Pharmacology: A. von Schlichtegroll, *Arzneimittel-Forsch.* **8,** 489 (1958). Clinical evaluation: N. W. Klehr *et al., Therapiewoche* **34,** 6050 (1984); E. Milsmann, P. Rohdewald, *Dermatologica* **170,** 230 (1985). Spectroscopic determn: M. H. Abdel-Hay *et al., J. Pharm. Belg.* **45,** 259 (1990). HPLC determn: A. Turcant *et al., Clin. Chem.* **37,** 1210 (1991).

$bp_{0.4mm}$ 171-174°. LD_{50} in mice (mg/kg): 65 ±2 i.p.; 222 ±18 orally (Schlichtegroll).

Hydrochloride, $C_{16}H_{19}N_3S.HCl$, *Andantol, Andanton, Nilergex.* Sol in water.

THERAP CAT: Antihistaminic.

5249. Isovaleraldehyde. *3-Methylbutanal;* isovaleral; isovaleric aldehyde. $C_5H_{10}O$; mol wt 86.13. C 69.72%, H 11.70%, O 18.58%. $(CH_3)_2CHCH_2CHO$. Occurs in orange, lemon, peppermint, eucalyptus and other oils. Made by oxidation of isoamyl alcohol with $Na_2Cr_2O_7$ and H_2SO_4.

Colorless liquid; pungent apple-like odor. d_{20}^{20} 0.785. mp −51°. bp 92-93°. n_D^{20} 1.3902. Sparingly sol in water; miscible with alcohol, ether.

USE: In artificial flavors and perfumes.

5250. Isovaleramide. $C_5H_{11}NO$; mol wt 101.15. C 59.37%, H 10.96%, N 13.85%, O 15.82%. $(CH_3)_2CHCH_2$-$CONH_2$.
Crystals. d 0.965. mp 135-137°. bp 232°. Sol in water, alcohol.

5251. Isovaleric Acid. *3-Methylbutanoic acid;* isovalerianic acid; isopropylacetic acid; delphinic acid. $C_5H_{10}O_2$; mol wt 102.13. C 58.80%, H 9.87%, O 31.33%. $(CH_3)_2CH$-CH_2COOH. Occurs in hop oil, tobacco and several other plants: Schmeltz *et al., J. Assoc. Offic. Agr. Chem.* **46**, 779 (1963). Prepn: Caldwell, U.S. pat. **2,484,486** (1949 to Kodak); Gustak, Sajko, *Arkiv. Kemi.* **24**, 11 (1952), *C.A.* **49**, 163e (1955); L'vov *et al., Zh. Prikl. Khim.* **35**, 700 (1962). "Valeric" acid of commerce is isovaleric acid. Toxicity data: L. Orö, A. Wretlind, *Acta Pharmacol. Toxicol.* **18**, 141 (1961).
Liquid. Acid taste; disagreeable, rancid-cheese odor. d_4^{20} 0.931. mp −37°. bp 175-177°. n_D^{20} 1.4043. Sol in 24 parts water; sol in alcohol, chloroform, ether. *Keep tightly closed.* LD_{50} i.v. in mice: 1120±30 mg/kg (Orö, Wretlind).

USE: In flavors, perfumes, manuf sedatives.

5252. Isovaleryl Chloride. C_5H_9ClO; mol wt 120.58. C 49.81%, H 7.52%, Cl 29.40%, O 13.27%. $(CH_3)_2CH$-CH_2COCl.
Liquid. d_4^{20} 0.985. bp 114-116°. n_D^{20} 1.4156. Dec by water, alcohol.

5253. Isovaleryl Diethylamide. *N,N-Diethyl-3-methylbutyramide;* N,N-diethylisovaleramide; Valyl; Xalyl. C_9H_{19}-NO; mol wt 157.26. C 68.74%, H 12.18%, N 8.91%, O 10.17%. Prepd from isovaleric acid and diethylamine: Liebrecht, Ger. pat. **129,967** (1902); *Frdl.* **6**, 1226; from ethyl isovalerate and lithium diethylamide: Hamell, Levine, *J. Org. Chem.* **15**, 162 (1950); by reduction of *N,N*-diethyl-*β*-methylcrotonamide with lithium aluminum hydride: Snyder, Putnam, *J. Am. Chem. Soc.* **76**, 1893 (1954).

Liquid, aromatic odor, peppery taste. bp 210-212°, bp_{14} 93-95°, bp_5 77-78°. n_D^{20} 1.4422. d_{20} 0.8764. One gram dissolves in about 25 ml water; sol in alcohol, ether.

THERAP CAT: Sedative.

5254. 2-Isovalerylindane-1,3-dione. *2-(3-Methyl-1-oxobutyl)-1H-indene-1,3(2H)-dione;* PMP; Valone. $C_{14}H_{14}O_3$; mol wt 230.26. C 73.03%, H 6.13%, O 20.84%. Prepn: Shapiro *et al., J. Org. Chem.* **25**, 1860 (1960).

Yellow solid from methanol, mp 68-69°. Practically insol in water, sol in most organic solvents.

USE: Insecticide, rodenticide.

5255. Isovaline. DL-2-Amino-2-methylbutyric acid; α-amino-α-methylbutyric acid. $C_5H_{11}NO_2$; mol wt 117.15. C 51.26%, H 9.46%, N 11.96%, O 27.31%. Prepn: Kurono, *Biochem. Z.* **134**, 427 (1922); Levene, Steiger, *J. Biol. Chem.* **76**, 299 (1928); Bucherer, Steiner, *J. Prakt. Chem.* **141**, 5 (1934); Upham, Dermer, *J. Org. Chem.* **22**, 799 (1957). Optical enantiomorphs: Ehrlich, Wendel, *Biochem. Z.* **8**, 438 (1908); Fischer, von Gravenitz, *Ann.* **406**, 5 (1914); Baker *et*

al., J. Am. Chem. Soc. **74**, 4701 (1952); Greenstein *et al., J. Biol. Chem.* **204**, 307 (1953).

Monoclinic prisms, mp 307-308° (closed tube). Sublimes at around 300°. Soly in cold water about 39 g/100 ml; in alc about 6.6 g/100 g alc at 75°. Slightly sol in ether.
L-Form, crystals. $[\alpha]_D^{25}$ +11.13° (c = 5 in H_2O). $[M]_D$ +9.7° (5N HCl); $[M]_D$ +26.3° (glacial acetic acid).
D-Form, long needles from water + acetone. $[\alpha]_D^{25}$ −11.28° (c = 5 in H_2O).

5256. Isoxaben. *N-[3-(1-Ethyl-1-methylpropyl)-5-isoxazolyl]-2,6-dimethoxybenzamide;* benzamizole; EL-107; NA-8318; Flexidor; Gallery. $C_{18}H_{24}N_2O_4$; mol wt 332.40. C 65.04%, H 7.28%, N 8.43%, O 19.25%. Selective, pre-emergent herbicide for broadleaf weeds and annual grasses. Prepn: K. W. Burow, **Brit.** pat. **Appl.** 2,084,140; *see also:* idem, U.S. pat. **4,636,243** (1982, 1987 both to Lilly). Physical properties and herbicidal activity: F. Huggenberger *et al., Proc. Brit. Crop Prot. Conf. - Weeds* **1982**, 47. Mobility and degradation in soil: F. Huggenberger, P. J. Ryan, *ibid.* **1985**, 947. Effect on cellular growth and metabolism: A. Lefebvre *et al., Weed Res.* **27**, 125 (1987). Field trials: F. Huggenberger, F. Gueguen, *Crop Prot.* **6**, 75 (1987); F. O. Colbert, D. H. Ford, *Proc. West. Soc. Weed Sci.* **40**, 155 (1987).

White crystalline solid, mp 176-179°. Soly at 25° (mg/ml): methanol 50-100; ethyl acetate 50-100; acetonitrile 30-50; toluene 4-5; hexane 0.07-0.08; water 0.001-0.002. Vapor pressure at 30°: < 3.9 × 10^{-7} mm Hg. LD_{50} in mice, rats (mg/kg): > 10000 orally both species; LC_{50} in rats by inhalation: > 1.99 mg/l (Huggenberger, 1982).

USE: Herbicide.

5257. Isoxepac. *6,11-Dihydro-11-oxodibenz[b,e]oxepin-2-acetic acid;* oxepinac; HP-549; P-720549; Artil. $C_{16}H_{12}O_4$; mol wt 268.27. C 71.64%, H 4.51%, O 23.86%. Nonsteroidal anti-inflammatory with analgesic and antipyretic activity. Prepn: K. Ueno *et al.,* **Japan. Kokai** 74 124,086 (1974 to Daiichi), *C.A.* **82**, 170735d (1975); G. C. Helsley *et al.,* **Ger.** pat. **2,442,060;** *eidem,* U.S. pat. **4,585,788** (1975, 1986 both to Hoechst); K. Ueno *et al., J. Med. Chem.* **19**, 941 (1976); D. E. Aultz *et al.,* **20**, 66 (1977). Pharmacology: H. B. Lassman *et al., Arch. Int. Pharmacodyn. Ther.* **227**, 142 (1977). Disposition and metabolism in animals and humans: H. P. A. Illing, J. M. Fromson, *Drug Metab. Dispos.* **6**, 510 (1978). HPLC determn in plasma: J. A. Slack, *J. Chromatogr.* **221**, 431 (1980); H. K. L. Hundt, L. W. Brown, *ibid.* **225**, 482 (1981). Clinical studies: S. Sasaki, *Arzneimittel-Forsch.* **28**, 462 (1978); L. B. Svendsen *et al., Scand. J. Rheumatol.* **10**, 186 (1981); J. Scott, E. C. Huskisson, *Rheumatol. Rehabil.* **21**, 48 (1982).

Crystals from ethyl acetate, mp 131-132.5°. LD_{50} orally in rats: 199 mg/kg (Ueno).

THERAP CAT: Anti-inflammatory.

5258. Isoxicam. *4-Hydroxy-2-methyl-N-(5-methyl-3-isoxazolyl)-2H-1,2-benzothiazine-3-carboxamide 1,1-dioxide;* 4-hydroxy-3-(5-methyl-3-isoxazolocarbamyl)-2-methyl-2H-

1,2-benzothiazine 1,1-dioxide; W-8495; Floxicam; Maxicam; Pacyl; Vectren. $C_{14}H_{13}N_3O_5S$; mol wt 335.34. C 50.14%, H 3.91%, N 12.53%, O 23.86%, S 9.56%. Nonsteroidal anti-inflammatory drug with antipyretic and analgesic properties. Prepn: H. Zinnes et al., Ger. pat. **2,208,351**; eidem, U.S. pat. **3,787,324** (1972, 1974 both to Warner-Lambert). Anti-inflammatory properties: G. DiPasquale et al., Agents Actions **5**, 256 (1975); ibid. **6**, 748 (1976). Pharmacological studies: G. DiPasquale, D. Mellace, ibid. **7**, 481 (1977); K. Rainsford, ibid. 573; G. DiPasquale et al., Res. Commun. Chem. Pathol. Pharmacol. **19**, 529 (1978). Pharmacokinetics in man: E. U. Kölle et al., Arzneimittel-Forsch. **33**, 582 (1983). Series of articles on pharmacology, safety and clinical efficacy: Am. J. Med. **79**, Suppl. 4B, 1-42 (1985); Brit J. Clin. Pharmacol. **22**, Suppl. 2, 107S-190S (1986). Review: Semin. Arthritis Rheum. **12**, Suppl. 2, 153-183 (1982).

Crystals from 1,4-dioxane, mp 265-271° (dec). Sodium salt, $C_{14}H_{12}N_3NaO_5S$, crystals from ethanol. mp 270-272° (dec).

THERAP CAT: Anti-inflammatory.

5259. Isoxsuprine. *4-Hydroxy-α-[1-[(1-methyl-2-phenoxyethyl)amino]ethyl]benzenemethanol; p-hydroxy-α-[1-[(1-methyl-2-phenoxyethyl)amino]ethyl]benzyl alcohol; p-hydroxy-N-(1-methyl-2-phenoxyethyl)norephedrine; 1-(p-hydroxyphenyl)-2-(1-methyl-2-phenoxyethylamino)-1-propanol; 2-(3-phenoxy-2-propylamino)-1-(p-hydroxyphenyl)-1-propanol.* $C_{18}H_{23}NO_3$; mol wt 301.39. C 71.73%, H 7.69%, N 4.65%, O 15.93%. Prepn: Moed, van Dijk, Rec. Trav. Chim. **75**, 1215 (1956); **Brit. pats. 832,286** and **832,287** (both 1960 to N. V. Philip Gloeilampenfabrieken); Moed, **U.S. pat. 3,056,836** (1962 to N. Am. Philips). Configuration: Dirkx, Rec. Trav. Chim. **83**, 535 (1964). Toxicity data: Goldenthal, Toxicol. Appl. Pharmacol. **18**, 185 (1971). Clinical evaluation in Raynaud's Phenomenon: H. Wesseling et al., Eur. J. Clin. Pharmacol. **20**, 329 (1981); in intermittent claudication: S. H. Skotnicki et al., Angiology **35**, 685 (1984). Veterinary trial in navicular disease: A. S. Turner, C. M. Tucker, Equine Vet. J. **21**, 338 (1989).

Crystals, mp 102.5-103.5°. Hydrochloride, $C_{18}H_{23}NO_3$·HCl, *Dilavase, Duvadilan, Duviculine, Isolait, Navilox, Suprilent, Vadosilan, Vasodilan, Vasoplex, Vasotran.* Bitter crystals from water, mp 203-204°. Soly in water at 25° about 2.0% w/v; also sol in ethanol. LD$_{50}$ in rats (mg/kg): 1750 orally; 164 i.p. (Goldenthal).

Resinate, *Defencin.*

THERAP CAT: Vasodilator (peripheral).

THERAP CAT (VET): Treatment of navicular disease.

5260. Isradipine. *4-(4-Benzofurazanyl)-1,4-dihydro-2,-6-dimethyl-3,5-pyridinedicarboxylic acid methyl 1-methylethyl ester;* isopropyl 4-(2,1,3-benzoxadiazol-4-yl)-1,4-dihydro-5-methoxycarbonyl-2,6-dimethyl-3-pyridinecarboxylate; 4-(2,1,3-benzoxadiazol-4-yl)-2,6-dimethyl-1,4-dihydro-3-isopropyloxycarbonylpyridine-5-carboxylic acid methyl ester; isrodipine; PN-200-110; Clivoten; DynaCirc; Dynacrine; Esradin; Lomir; Prescal; Rebriden. $C_{19}H_{21}N_3O_5$; mol wt 371.39. C 61.45%, H 5.70%, N 11.31%, O 21.54%. Dihydropyridine calcium channel blocker. Prepn: P. Neumann,

Ger. pat. **2,949,491**; idem, U.S. pat. **4,466,972** (1980, 1984 both to Sandoz). Prepn of enantiomers: A. Vogel, Ger. pat. **3,320,616** (1983 to Sandoz), C.A. **101**, 7162s (1984). Comparative study of in vitro effects on human and canine cerebral arteries: E. Müller-Schweinitzer, P. Neumann, J. Cereb. Blood Flow Metab. **3**, 354 (1983). Effect on α-adrenoceptor mediated vasoconstriction in rats: K. Jie et al., Arch. Int. Pharmacodyn. **278**, 72 (1985). Pharmacokinetics: F. L. S. Tee, J. M. Jaffe, Eur. J. Clin. Pharmacol. **32**, 361 (1987). Clinical evaluation in angina and coronary artery disease: C. E. Handler, E. Sowton, ibid. **27**, 415 (1984); in hypertension: E. B. Nelson et al., Clin. Pharmacol. Ther. **40**, 694 (1986). Comparison of hemodynamic effects of enantiomers: R. P. Hof et al., J. Cardiovasc. Pharmacol. **8**, 221 (1986). Series of articles on pharmacology and clinical use: Am. J. Med. **86**, 1-146 (1989).

mp 168-170°. S(+)-Form, **PN-205-033**. Crystals from ether + hexane, mp 142°. $[\alpha]_D^{20}$ +6.7° (c = 1.5 in ethanol). R(−)-Form, **PN-205-034**. Crystals from ether + hexane, mp 140°. $[\alpha]_D^{20}$ −6.7° (c = 1.67 in ethanol). THERAP CAT: Antihypertensive; antianginal.

5261. Itaconic Acid. *Methylenesuccinic acid;* propylenedicarboxylic acid. $C_5H_6O_4$; mol wt 130.10. C 46.16%, H 4.65%, O 49.19%. Obtained by dry distillation of citric acid and subsequent treatment of the anhydride with water. Produced on a large scale by submerged aerobic fermentation using Aspergillus terreus and low cost carbohydrates from beet or cane: Kane et al., U.S. pat. **2,385,283** (1945 to Pfizer). Synthesis from propargyl chloride, carbon monoxide, nickel carbonyl and water: Chiusoli, U.S. pat. **3,025,-320** (1962 to Montecatini).

Hygroscopic crystals; characteristic odor. d 1.63. mp 162-164° with decompn. Also reported as mp 172° [Kinoshita, Acta Phytochem. (Japan) **5**, 273 (1931)]. One gram dissolves in 12 ml water, 5 ml alcohol; very slightly sol in benzene, chloroform, ether, carbon disulfide, petr ether. *Keep well closed.*

5262. Itraconazole. *4-[4-[4-[4-[[2-(2,4-Dichlorophenyl)-2-(1H-1,2,4-triazol-1-ylmethyl)-1,3-dioxolan-4-yl]methoxy]phenyl]-1-piperazinyl]phenyl]-2,4-dihydro-2-(1-methylpropyl)-3H-1,2,4-triazol-3-one;* (±)-1-sec-butyl-4-[p-[4-[p-[[(2R*,4S*)-2-(2,4-dichlorophenyl)-2-(1H-1,2,4-triazol-1-ylmethyl)-1,3-dioxolan-4-yl]methoxy]phenyl]-1-piperazinyl]phenyl]-Δ²-1,2,4-triazolin-5-one; oriconazole; R-51211; Itrizole; Sporanox; Triasporin. $C_{35}H_{38}Cl_2N_8O_4$; mol wt 705.64. C 59.57%, H 5.43%, Cl 10.05%, N 15.88%, O 9.07%. Orally active antimycotic structurally related to ketoconazole, q.v. Prepn: J. Heeres, L. J. J. Backx, Eur. pat. Appl. **6,711**; eidem, U.S. pat. **4,267,179** (1980, 1981 both to Janssen); J. Heeres et al., J. Med. Chem. **27**, 894 (1984). In vitro activity: A. Espinel-Ingroff et al., Antimicrob. Ag. Chemother. **26**, 5 (1984). HPLC determn in biological samples: R. Woestenborghs et al., J. Chromatog. **413**, 332 (1987). Symposium on pharmacology and clinical efficacy: Rev. Infect. Dis. **9**, Suppl 1, S1-S152 (1987). Toxicity data: H. Van Cauteren et al., ibid. S43. Review of clinical pharmacokinetics: J. Heykants et al., Mycoses **32**, Suppl 1, 67-87 (1989); of clinical efficacy in dermatophytosis: P. De Doncker, G. Cauwenbergh, Brit. J. Clin. Pract. **Suppl. 71**, 118-122

(1990). *Review:* A. M. Sugar, *Curr. Clin. Topics Inf. Dis.* **13**, 74-98 (1993).

Crystals from toluene, mp 166.2°. pKa 3.7. Lipophilic; partition coefficient (*n*-octanol/aq buffer of pH 8.1): 5.66. Practically insol in water and dil acidic solns. LD_{50} (14 day) in mice, rats, dogs (mg/kg): >320, >320, >200 orally (Van Cauteren).

THERAP CAT: Antifungal.

5263. Itramin Tosylate. *2-Aminoethanol nitrate mono-(4-methylbenzenesulfonate); 2-aminoethanol nitrate mono-p-toluenesulfonate;* 2-nitratoethylaminotoluene-*p*-sulfonate; Cardisan; Tostram; Nilatil. $C_9H_{14}N_2O_6S$; mol wt 278.29. C 38.84%, H 5.07%, N 10.07%, O 34.50%, S 11.52%. Prepn: **Swed. pat. 168,308** (1959 to Aktiebolaget Pharmacia), *C.A.* **54**, 24405d (1960).

Crystals from ethanol, mp 132-133°.
THERAP CAT: Vasodilator.

5264. Ivermectin. 22,23-Dihydroabamectin; 22,23-dihydroavermectin B_1; 22,23-dihydro C-076B_1; MK-933; Cardomec; Cardotek-30; Eqvalan; Heartgard 30; Ivomec; Mectizan; Zimecterin. Semi-synthetic deriv of abamectin, *q.v.*, one of the avermectins, *q.v.* Ivermectin contains at least 80% of 22,23-dihydroavermectin B_{1a} and not more than 20% of 22,23-dihydroavermectin B_{1b}. Prepn: **Japan. Kokai 79 61198;** J. C. Chabala, M. H. Fisher, **U.S. pat. 4,199,569** (1979, 1980 both to Merck & Co.); J. C. Chabala *et al., J. Med. Chem.* **23**, 1134 (1980). Approaches to synthesis: H. Mrozik *et al., Tetrahedron Letters* **23**, 2377 (1982); M. E. Jung *et al., ibid.* **28**, 5977 (1987); M. E. Jung, L. J. Street, *Heterocycles* **27**, 45 (1988). Antiparasitic activity: J. R. Egerton *et al., Brit. Vet. J.* **136**, 88 (1980); E. T. Lyons *et al., Am. J. Vet. Res.* **41**, 2069 (1980); L. S. Blair, W. C. Campbell, *ibid.* 2108. Acaricidal efficacy: C. A. Wilkins *et al., Am. J. Vet. Res.* **41**, 2112 (1980); R. O. Drummond *et al., J. Econ. Entomol.* **74**, 432 (1981). Irreversible effects on adult parasites in onchocerciasis: A. P. Plaisier *et al., J. Infect. Dis.* **172**, 204 (1995). Metabolism in animals: S.-H. L. Chiu *et al., Drug Metab. Dispos.* **14**, 590 (1986). Pharmacokinetics in horses and sheep: S. E. Marriner *et al., J. Vet. Pharmacol. Ther.* **10**, 175 (1987). Evaluation in human onchocerciasis (river blindness): M. A. Aziz *et al., Lancet* **2**, 171 (1982); P. T. Soboslay *et al., Trop. Med. Parasitol.* **38**, 8 (1987). Clinical trial in onchocerciasis: A. T. White *et al., J. Infec. Dis.* **156**, 463 (1987). HPLC determn in cattle and sheep tissues: P. C. Tway *et al., J. Agr. Food Chem.* **29**,

1059 (1981); in human plasma and milk: R. Chiou *et al., J. Chromatog.* **416**, 196 (1987). Review of chemistry and biology: W. C. Campbell *et al., Science* **221**, 823-828 (1983). Review of use in treatment of heartworm in dogs: W. C. Campbell, *Sem. Vet. Med. Surg.* **2**, 48-55 (1987). Review of clinical use and pharmacology: W. C. Campbell, G. W. Benz, *J. Vet. Pharmacol. Ther.* **7**, 1-16 (1984); T. B. Barragry, *Can. Vet. J.* **28**, 512-517 (1987); of pharmacology: J. L. Bennett *et al., Parasitol. Today* **4**, 226-228 (1988). *Book: Ivermectin and Abamectin,* W. C. Campbell, Ed. (Springer-Verlag, New York, 1989) 363 pp.

Off-white powder. $[\alpha]_D$ +71.5 ± 3° (c = 0.755 in chloroform). uv max (methanol): 238, 245 nm (ε 27100, 30100). Soly in water: ~4 μg/ml. Virtually insol in satd hydrocarbons such as cyclohexane. Highly sol in methyl ethyl ketone, propylene glycol, polyethylene glycol.

Component B_{1a}, $C_{48}H_{74}O_{14}$, *5-O-demethyl-22,23-dihydroavermectin A_{1a}, 22,23-dihydroavermectin B_{1a}, 22,23-dihydro C-076B_{1a}.* Crystals from ethanol/water, mp 155-157°.

Component B_{1b}, $C_{47}H_{72}O_{14}$, *5-O-demethyl-25-de(1-methylpropyl)-22,23-dihydro-25-(1-methylethyl)avermectin A_{1a}, 22,23-dihydroavermectin B_{1b}.*

THERAP CAT: Anthelmintic (Onchocerca).

THERAP CAT (VET): Anthelmintic, insecticide, acaricide.

5265. Ixbut. A native name for *Euphorbia lancifolia* Schlecht, *Euphorbiaceae.* Used as a galactagogue: Serrano, *Time,* August 1st, **1949**, p 38; F. Rosengarten, Jr., *Botan. Museum Leaflets (Harvard Univ.)* **26**(9-10), 277-309 (Nov.-Dec. 1978).

5266. Ixodin. Hirudin-like anticoagulant from *Ixodes ricinus* L., *Acari* and other blood-sucking ticks: Sabbatani, *Arch. Ital. Biol.* **31**, 37 (1899). Thrombokinase inhibitor. Extraction procedure: F. Markwardt, *Blutgerinnungshemmende Wirkstoffe aus blutsaugenden Tieren* (Jena, 1963) pp 88-92.

White powder. Soluble in 80% ethanol, 90% methanol and 75% acetone. Practically insol in ether. Stable at pH 1.0-12.0. Adsorbed on Amberlite IRC 50 (H-form). Acts like a high-molecular weight polypeptide. Attacked by papain.

J

5267. Jalap. Dried tuberous root of *Exogonium purga* (Hayne) Lindl. *(E. jalapa* Baill., *Ipomoea purga* Hayne), *Convolvulaceae. Habit.* Mexico; culivated in India. *Constit.* 7-12% resin, gum, sugar.

THERAP CAT: Cathartic.

5268. Jambul. Jamboo; Java plum; jumbul. Bark, fruit and seeds of *Syzygium jambolanum* (Lam.) DC. *(Eugenia jambolana* Lam.), *Myrtaceae. Habit.* East Indies. *Constit.* Bark: resin, tannin. Fruit: volatile and fixed oils, resin, tannin. Seeds: resin, fat, gallic acid, albumin.

THERAP CAT: Antidiarrheal.

5269. Janus Green B. *3-(Diethylamino)-7-[[4-(dimethylamino)phenyl]azo]-5-phenylphenazinium chloride;* C.I. 11050. $C_{30}H_{31}ClN_6$; mol wt 511.07. C 70.51%, H 6.11%, Cl 6.94%, N 16.44%. Prepd by diazotizing 3-amino-7-(diethylamino)-5-phenylphenazinium chloride *(N,N*-diethylphenosafranine) and coupling resulting diazo compd with *N,N*-dimethylaniline: *Colour Index* **vol. 4** (3rd ed., 1971) p 4015.

USE: As dye for cotton, wool; as biological stain; in electrodeposition of copper, Brown, Fellows, U.S. pat. **2,882,209** (1959 to Udylite Res. Corp.).

5270. Japan Wax. Vegetable wax; sumach wax; Japan tallow. A fat expressed from mesocarp of the fruit of *Rhus succedanea* L., *Anacardiaceae. Habit.* Japan and China. *Constit.* 10-15% palmitin; stearin, olein; 1% japanic acid and homologs. Brief review: C. S. Letcher in Kirk-Othmer *Encyclopedia of Chemical Technology* **vol. 24** (Wiley-Interscience, New York, 3rd ed., 1984) p 470.

Pale yellow, flat cakes, disks or squares with a greasy feel; somewhat tallow-like, rancid odor and taste. d 0.97-0.98. mp 53.5-55°. Acid number 22-23. Saponification number 217-237. Iodine number 10-15. Insol in water or cold alcohol; sol in benzene, carbon disulfide, petr ether, ether, hot alc, alkalies.

USE: As a substitute for beeswax in wax varnishes or candles, ingredients in plasters, ointments; floor waxes, furniture polish. As a plasticizer in dental impression compounds.

5271. Japonilure. *[R-(Z)]-5-(1-Decenyl)dihydro-2(3H)-furanone;* [(4R,5Z)]-tetradecen-4-olide; *(R,Z)-5-(dec-1-en-yl)oxacyclopentan-2-one;* IN-60. $C_{14}H_{24}O_2$; mol wt 224.34. C 74.95%, H 10.78%, O 14.26%. Pheromone produced by the female Japanese beetle, *Popillia japonica* Newman (Coleoptera: Scarabaeidae); biological activity is inhibited by its antipode. Isoln and synthesis: J. H. Tumlinson *et al., Science* **197**, 789 (1977). Description of attractant activity: J. H. Tumlinson in *Adv. Pest. Sci., 4th Int. Congr. Pest. Chem.* **2,** H. Geissbühler, Ed (Pergamon Press, Oxford, England, 1979) 315-322. Use in beetle traps: T. L. Ladd, Jr., M. G. Klein, *J. Econ. Entomol.* **79,** 84 (1986); A. Martins *et al., Ecol. Bull.* **39,** 101 (1988). Stereospecific synthesis: S. Chattopadhyay *et al., Syn. Commun.* **20,** 1299 (1990); T. Ebata *et al., Biosci. Biotech. Biochem.* **56,** 818 (1992). Isolation from *Anomala* spp. as component of sex pheromone system and field evaluation: W. S. Leal *et al., J. Chem. Ecol.* **20,** 1643 (1994); W. S. Leal *et al., ibid.* 1667.

bp$_{0.1mm}$ 130-134°. $[\alpha]_D^{21}$ −69.0° (c = 1.42 in CHCl$_3$). $[\alpha]_D^{25}$ −70.2° (c = 0.5 in CHCl$_3$).

5272. Jasmolins. Active insecticidal constituents of pyrethrum flowers. Isoln and structure: Godin *et al., J. Econ. Entomol.* **58,** 548 (1965); Godin *et al., J. Chem. Soc. (C)* **1966,** 322. Stereochemistry: Begley *et al., Chem. Commun.* **1972,** 1276.

Jasmolin I   R = CH$_3$
Jasmolin II   R = COOCH$_3$

Jasmolin I, $C_{21}H_{30}O_3$, *[1R-[1α[S*(Z)],3β]]-2,2-dimethyl-3-(2-methyl-1-propenyl)cyclopropanecarboxylic acid 2-methyl-4-oxo-3-(2-pentenyl)-2-cyclopenten-1-yl ester, 4',5'-dihydropyrethrin I.* Liquid. uv max (Spectrosol hexane): 219 nm (ε 21500).
Jasmolin II, $C_{22}H_{30}O_5$, *[1R-[1α[S*(Z)],3β(E)]]-3-(3-methoxy-2-methyl-3-oxo-1-propenyl)-2,2-dimethylcyclopropanecarboxylic acid 2-methyl-4-oxo-3-(2-pentenyl)-2-cyclopenten-1-yl ester, 4',5'-dihydropyrethrin II.* Liquid. uv max (Spectrosol hexane): 229 nm (ε 22900).

5273. Jasmone. *3-Methyl-2-(2-pentenyl)-2-cyclopenten-1-one.* $C_{11}H_{16}O$; mol wt 164.25. C 80.44%, H 9.82%, O 9.74%. Found in the volatile portion of oil from jasmine flowers. Natural jasmone is the *cis*-ketone. Isoln and structure: Ruzicka, Pfeiffer, *Helv. Chim. Acta* **16,** 1208 (1933). Stereochemistry: Crombie, Harper, *J. Chem. Soc.* **1952,** 869. Synthesis of *cis*-jasmone: H. Hunsdiecker, *Ber.* **75,** 447 (1942); Stork, Borch, *J. Am. Chem. Soc.* **86,** 936 (1964); Büchi, Wuest, *J. Org. Chem.* **31,** 977 (1966); Sakan *et al., Chem. Letters* **1973,** 713; P. Bakuzis, M. L. F. Bakuzis, *J. Org. Chem.* **42,** 2362 (1977). Synthesis of dihydrojasmone and *cis*-jasmone: C. S. Subramaniam *et al., J. Chem. Soc. Perkin Trans. I* **1979,** 2346; T. Kato *et al., Chem. Pharm. Bull.* **28,** 349 (1980). Synthesis of *trans*-jasmone: Sisido *et al., J. Org. Chem.* **29,** 2290 (1964). Comprehensive synthetic reviews: R. A. Ellison, *Synthesis* **1973,** 397; T. L. Ho, *Synth. Commun.* **4,** 265 (1974).

cis - Form

cis-Form, oil, odor of jasmine, bp$_{27}$ 146°. n_D^{20} 1.4978. uv max: 235 nm (ε 12,000).
trans-Form, oil, odor of jasmine, bp$_{23}$ 142°. n_D^{20} 1.4974. uv max: 234 nm (ε 12,300).

USE: In perfumery.

5274. Jatrorrhizine. *5,6-Dihydro-3-hydroxy-2,9,10-trimethoxydibenzo[a,g]quinolizinium; 7,8,13,13a-tetradehydro-3-hydroxy-2,9,10-trimethoxyberbinium;* jateorrhizine; neprotin. $[C_{20}H_{20}NO_4]^+$. From root of *Jateorhiza palmata* (DC.) Miers *(J. columba* Miers), *Menispermaceae:* Feist, *Arch. Pharm.* **245,** 586 (1907); from *Berberis asiatica* Roxb. ex DC. and *B. thunbergii* DC., *Berberidaceae:* Chatterjee *et al., J. Indian Chem. Soc.* **31,** 83 (1954); Tomita, Kikuchi, *J. Pharm. Soc. Japan* **76,** 597 (1956); from *Coptis teeta* Wall., *Ranunculaceae:* Chatterjee *et al., J. Indian Chem. Soc.* **29,** 97 (1952); from *Mahonia acanthifolia* Wall., *M. borealis* Takeda and *M. simonsii* Takeda, *Berberidaceae:* Chatterjee, Guha, *J. Am. Pharm. Assoc.* **39,** 577 (1950); Chatterjee *et al., ibid.* **40,** 36 (1951). Structure: Späth, Duschinsky, *Ber.* **58,** 1939 (1925).

Consult the Name Index before using this section.

Iodide, $C_{20}H_{20}INO_4$, reddish-yellow needles, mp 208-210°. Sol in water and alcohol.

5275. Javanicin. *5,8-Dihydroxy-6-methoxy-2-methyl-3-(2-oxopropyl)-1,4-naphthalenedione; 3-acetonyl-5,8-dihydroxy-6-methoxy-2-methyl-1,4-naphthoquinone.* $C_{15}H_{14}O_6$; mol wt 290.27. C 62.07%, H 4.86%, O 33.07%. Antibiotic substance produced by *Fusarium javanicum:* Arnstein, Cook, *J. Chem. Soc.* **1947**, 1021. Prepd by reduction of fusarubin: Ruelius, Gauhe, *Ann.* **569**, 38 (1950). Structure: Birch, Donovan, *Chem. & Ind. (London)* **1954**, 1047; Whalley, *ibid.* **1958**, 131; Hardegger *et al., Helv. Chim. Acta* **47**, 2027 (1964).

Red crystals with a coppery luster from ethanol, decomp 207.5-208°. Absorption max (alc): 303, 305 nm (log ε 3.97, 3.90); in chloroform: 307, 510 nm (log ε 3.99, 3.86).

Diacetyljavanicin, $C_{19}H_{18}O_8$, needles from dil acetone, dec 207-208°. Absorption max (alc): 221, 290, 426 nm (log ε 4.57, 4.17, 3.79).

5276. Jerusalem Artichoke. Topinambur. The subterranean stem tuber of *Helianthus tuberosus* L., *Compositae,* a kind of sunflower native to North America. Used as food by American Indians. Contains up to 20% (wet basis) of polysaccharides usable as sweetening agents: Rubin, U.S. pat. **2,782,123** (1957).

5277. Jervine. *(3β,23β)-17,23-Epoxy-3-hydroxyveratraman-11-one.* $C_{27}H_{39}NO_3$; mol wt 425.61. C 76.20%, H 9.24%, N 3.29%, O 11.28%. Steroidal alkaloid found in *Veratrum grandiflorum* (Maxim.) Loes F., *V. album* L., *V. viride* Sol., *Liliaceae:* Saito, *Bull. Chem. Soc. Japan* **9**, 15 (1934), *C.A.* **28**, 2463 (1934); Poethke, *Arch. Pharm.* **275**, 357, 571 (1937); **276**, 170 (1938); **282**, 56 (1944); Seiferle *et al., J. Econ. Entomol.* **35**, 35 (1942); Jacobs, Craig, *J. Biol. Chem.* **160**, 555 (1945). Structure: Fried *et al., J. Am. Chem. Soc.* **73**, 2970 (1951). *Reviews:* Wintersteiner in Graff, *Essays in Biochemistry* (Wiley, New York, 1956) pp 308-321; *idem, Festschrift Arthur Stoll* (Basel, 1957) pp 166-176; L. F. Fieser, M. Fieser, *Steroids* (Reinhold, New York, 1959) pp 870-877. Stereochemistry: Sicher, Tichy, *Tetrahedron Letters* no. 12, 6 (1959); Augustine, *Chem. & Ind. (London)* **1961**, 1448; Mitsuhashi, Shimizu, *Tetrahedron* **19**, 1027 (1963); Bailey *et al., Tetrahedron Letters* **1963**, 555; Masamune *et al., ibid.* **1965**, 489. Revised stereochemistry: Scott *et al., ibid.* **1967**, 2381; Kupchan, Suffness, *J. Am. Chem. Soc.* **90**, 2730 (1968); Sprague *et al., Tetrahedron* **27**, 4857 (1971). Total synthesis: Kutney *et al., Can. J. Chem.* **53**, 1796 (1975). Comparative toxicity: O. Krayer *et al., J. Pharmacol. Exp. Ther.* **82**, 167 (1944); K. Tanaka, *ibid.* **113**, 89 (1955).

Needles from methanol + water, mp 243.5-244.5° (Saito). $[\alpha]_D^{20}$ −150° (ethanol) (Saito); $[\alpha]_D^{20}$ −167.6° (chloroform) (Poethke). uv max: 250, 360 nm (ε 15000, 60). LD_{50} i.v. in mice: 9.3 mg/kg (Krayer). LD_{50} s.c. in male mice: 29 mg/kg (Tanaka).

Diacetyljervine, $C_{31}H_{43}NO_5$, mp 173-175°. $[\alpha]_D$ −112°. uv max (ethanol): 250, 360 nm (ε 16400, 80).

Hydrochloride, mp 300-302°.

5278. Jesaconitine. *(1α,3α,6α,14α,15α,16β)-20-Ethyl-1,6,16-trimethoxy-4-(methoxymethyl)aconitane-3,8,13,14,15-pentol 8-acetate 14-(4-methoxybenzoate).* $C_{35}H_{49}NO_{12}$; mol wt 675.77. C 62.21%, H 7.31%, N 2.07%, O 28.41%. From *Aconitum fischeri* Reich. and *A. sachalinense* F. Schmidt, *Ranunculaceae:* Makoshi, *Arch. Pharm.* **247**, 243 (1909); Majima, Morio, *Ber.* **57**, 1472 (1924). Structure: Ochiai *et al., J. Pharm. Soc. Japan* **75**, 545 (1955); Keith, Pelletier, *Chem. Commun.* **1967**, 993 (corr. *ibid.* **1968**, 1739); *eidem, J. Org. Chem.* **33**, 2497 (1968).

Amorphous powder, dec 128-131°. Sol in ether, dil acids. Perchlorate, $C_{35}H_{49}NO_{12}\cdot HClO_4$, prisms from alcohol, dec 230-232°. $[\alpha]_D$ −16.7° (methanol).

5279. Jojoba Oil. Oil of jojoba. A liquid wax ester mixture extracted from ground or crushed seeds from *Simmondsia chinensis* and *S. californica* Nutt. *Buxaceae,* desert shrubs native to Arizona, California, and northern Mexico: Greene, Foster, *Bot. Gaz.* **94**, 826 (1933); Green *et al., J. Chem. Soc.* **1936**, 1750; McKinney, Jamieson, *Oil Soap (Chicago)* **13**, 289 (1936). Similar to sperm whale oil, it is composed essentially of C_{20} and C_{22} straight chain monoethylene acids and alcohols in the form of esters: Molaison *et al., J. Am. Oil Chem. Soc.* **36**, 379 (1959); Miwa, *ibid.* **48**, 259 (1971). Solvent effects in extraction: Knoepfler *et al., ibid.* **36**, 644 (1959). Comparison of sulfurized jojoba and sperm whale oils as high pressure lubricants: T. K. Miwa *et al., ibid.* **56**, 765 (1979). Potential chemical utilization studies: Fore *et al., ibid.* **37**, 387 (1960); J. D. Johnson, C. W. Hinman, *Science* **208**, 460 (1980). Possible uses: J. H. Brown, *Mfg. Chem.* **50**(6), 47 (1979). *Reviews:* Knoepfler, Vix, *J. Agr. Food Chem.* **6**, 118 (1958); *Products from Jojoba: A Promising New Crop for Arid Lands,* Committee on Jojoba Utilization, Natl. Res. Council, 1975; *Jojoba: New Crop for Arid Lands, New Raw Material for Industry,* Natl. Res. Council, 1985.

Liquid wax. fp 10.6 to 7°; mp 6.8 to 7°, bp_{757} (under N_2) 398°. Fire point 338°. Pour point 10°. d^{25} 0.8642. n^{25} 1.4648. Iodine no. 81.7. Saponification value 92.2. Acid value 0.32. Avg mol wt of wax esters: 606. Highly stable, and resistant to bacterial degradation; can be stored for years without becoming rancid.

USE: Potentially as lubricant, fuel, chemical feedstock, substitute for sperm whale oil. For other potential uses *see* Fore *et al., loc. cit.,* Knoepfler, Vix, *loc. cit,* Johnson, Hinman, *loc. cit.*

5280. Josamycin. *Leucomycin V 3-acetate 4^B-(3-methylbutanoate); leucomycin A$_3$;* EN-141; Iosalide; Jomybel; Josamina. $C_{42}H_{69}NO_{15}$; mol wt 828.01. C 60.92%, H 8.40%, N 1.69%, O 28.98%. Macrolide antibiotic produced by *Streptomyces narbonensis* var. *josamyceticus* nov. var. Isoln and characterization: T. Osono *et al., J. Antibiot.* **20A,** 174 (1967). Manuf process: H. Umezawa,T. Osono, **Japan. pat. 66 21,759** (1966 to Microbiochemical Res. Found.), *C.A.* **66,** 54258w (1967). Prepn: Y. Oka *et al.,* **Japan. Kokai 76 41,497** (to Yamanouchi), *C.A.* **85,** 121788b (1976). Identification with leucomycin A$_3$: S. Omura *et al., J. Antibiot.* **23,** 511 (1970). Structure: *eidem, ibid.* **27,** 366 (1974). Absolute configuration: A. Ducruix *et al., Chem. Commun.* **1976,** 947. Stereospecific total synthesis: K. Tatsuta *et al., Tetrahedron Letters* **1980,** 2837. Retrosynthetic studies: K. C. Nicolaou *et al., J. Am. Chem. Soc.* **103,** 1222 (1981). A 17-membered aglycone was proposed at one time, *see* T. Osono, H. Umezawa in *Drug Action and Drug Resistance in Bacteria,* **1,** S. Mitsuhashi, Ed. (University Park Press, Baltimore, 1971) pp 41-120; T. Osono *et al., J. Antibiot.* **27,** 366 (1974). Pharmacology: K. Kuriaki *et al., Japan. J. Antibiot.* **22,** 232 (1969). *Review:* T. Osono, H. Umezawa, *loc. cit.*

R =

Colorless needles from benzene, mp 130-133° (after drying under reduced pressure at 100° for 5 hrs). $[\alpha]_D^{25}$ −70° (c = 1 in ethanol). uv max (0.001N HCl): 232 nm (E$_{1cm}^{1\%}$ 325). pKa 7.1 (40% aq methanol). Very sol in methanol, ethanol, acetone, chloroform, ethyl acetate, dioxane, acidic water. Sol in butanol, ether, CCl$_4$, benzene, toluene. Practically insol in water, petr ether, ligroin, *n*-hexane.

Propionate, $C_{45}H_{73}NO_{16}$, *Josacine, Josamy, Josaxin, Wilprafen.*

THERAP CAT: Antibacterial.

THERAP CAT (VET): Antibacterial.

5281. Juglans. Butternut; white walnut; lemon walnut; oil nut. Dried inner root bark of *Juglans cinerea* L., *Juglandaceae,* collected in the autumn. *Habit.* North America. *Constit.* Resinoid juglandin, juglone, juglandic acid, fixed and volatile oils, tannin.

5282. Juglone. *5-Hydroxy-1,4-naphthalenedione; 5-hydroxy-1,4-naphthoquinone;* 8-hydroxy-1,4-naphthoquinone; C.I. 75500; C.I. Natural Brown 7; nucin; regianin. $C_{10}H_6O_3$; mol wt 174.16. C 68.97%, H 3.47%, O 27.56%. Coloring matter occurring in various *Juglandaceae* spp: Brissemoret, Combes, *Compt. Rend.* **141,** 838 (1905). Isoln from walnut shells: Combes, *Bull. Soc. Chim.* **1**(4), 800 (1907). Isoln and description of sedative properties in fish, mammals: B. A. Westfall *et al., Science* **134,** 1617 (1961). Isoln from pecans and identification as inhibitory agent of mycelial growth of *Fusicladium effusum:* P. A. Hedin *et al., J. Agr. Food Chem.* **28,** 340 (1980). Synthesis: Bernthsen, Semper, *Ber.* **20,** 938 (1887); Willstätter, Wheeler, *Ber.* **47,** 2798 (1914); Teuber, Götz, *Ber.* **87,** 1236 (1954); C. Grundmann, *Synthesis* **1977,** 644. Tumor-promoting activity study: B. L. Van Duuren *et al., J. Med. Chem.* **21,** 26 (1978). Use as pH indicator: S. S. Sawhney, B. M. L. Bhatia, *J. Indian Chem. Soc.* **57,** 438 (1980).

Yellow needles from benzene + petr ether, mp 155°. Sublimes. Absorption max (methanol): 420 nm (log ε 3.56). Volatile with steam. Slightly sol in hot water; freely sol in chloroform, benzene; sol in alcohol, ether; sol in aq solns of alkalies giving a purplish-red soln.

5283. Julocrotine. *N-[2,6-Dioxo-1-(2-phenylethyl)-3-piperidinyl]-2-methylbutanamide;* yulocrotine; β-phenylethylimide of *N*-α-methylbutyrylglutamic acid. $C_{18}H_{24}N_2O_3$; mol wt 316.40. C 68.33%, H 7.65%, N 8.85%, O 15.17%. Isoln from root of *Julocroton montevideensis* Klotzsch, *Euphorbiaceae:* Anastasi, *Anales Asoc. Quim. Argentina* **23,** 348 (1925), *C.A.* **20,** 2332 (1926). Structure: Nakano *et al., J. Org. Chem.* **26,** 1184 (1961).

Crystals from ether + petr ether, mp 108-109°. $[\alpha]_D$ −9° (c = 1.24 in chloroform), −50.1° (c = 1.19 in methanol). uv max (ethanol): 252, 258, 264, 268 nm (log ε 2.43, 2.44, 2.30, 2.18). Sol in alc, benzene, chloroform, ether, slightly in petr ether.

5284. Juniper Berries. Dried ripe fruit of *Juniperus communis* L., *Cupressaceae.* The wood and tops are also employed. *Habit.* Northern Europe, Asia, North America. *Constit.* Volatile oil, juniperin, resin, proteins, formic, acetic and malic acids.

USE: In liquor manuf, perfumery and cosmetics: A. Pomini, *Minerva Farm.* **11,** 207 (1962).

THERAP CAT: Diuretic.

5285. Juniper Tar. Oil of cade; empyreumatic oil of juniper; oil of juniper tar; Haarlem oil; Harlem oil; Tilly drops; Holland balsam; silver drops; silver balsam; Kaparlem; Caparlem. Volatile oil from wood of *Juniperus oxycedrus* L., *Cupressaceae. Constit.* Chiefly cadinene. Toxicity data: P. M. Jenner *et al., Food Cosmet. Toxicol.* **2,** 327 (1964). *Review:* D. L. J. Opdyke, *ibid.* **13,** 733-734 (1975).

Dark brown, more or less viscid liq; smoky odor; acrid, slightly aromatic taste. d$_{25}^{25}$ 0.950-1.055. n$_D^{20}$ 1.510-1.530. Very slightly sol in water; sol in 3 vols ether, in chloroform, amyl alcohol, glacial acetic acid, oil turpentine; partly sol in alcohol or petr ether. LD$_{50}$ orally in rats: 8014 mg/kg (Jenner).

USE: As a gin-like flavor; in perfumery.

THERAP CAT: Anti-eczematic (topical).

5286. Justicidins. Lignans from various species of *Justicia, Acanthaceae.* Isoln and structures for justicidins A, B, C, D, E and F have been elucidated. Isoln of A and B: Munakata *et al., Tetrahedron Letters* **1965,** 4167; Ohta *et al., Agr. Biol. Chem.* **33,** 610 (1969); Okigawa *et al., Tetrahedron* **26,** 4301 (1970). Structure of A: Govindachari *et al., Tetrahedron Letters* **1967,** 3517; Horii *et al., Chem. Pharm. Bull.* **19,** 535 (1971). Synthesis of B: Munakata *et al., Tetrahedron Letters* **1967,** 2831; Munakata, Katsura, **Brit. pat. 1,178,341** (1970), *C.A.* **72,** 100365c; T. Momose *et al., Chem. Pharm. Bull.* **26,** 3195 (1978). Isoln and structures of C and D: Ohta, Munakata, *Tetrahedron Letters* **1970,** 923. Synthesis of C: Horii *et al., Chem. Pharm. Bull.* **17,** 1878 (1969). Structure of E: Wada, Munakata, *Tetrahedron Letters* **1970,** 2017. Synthesis of E: Holmes, Stevenson, *J. Org. Chem.* **36,** 3450 (1971). Synthesis of D, E, and F: Z. I. Horii *et al., Chem. Pharm. Bull.* **25,** 1803 (1977).

Justicidin A   R = OCH₃
Justicidin B   R = H

Justicidin A, $C_{22}H_{18}O_7$, *9-(1,3-benzodioxol-5-yl)-4,6,7-tri-methoxynaphtho[2,3-c]furan-1(3H)-one, diphyllin methyl ether.* Crystals, mp 263°. uv max (CHCl₃): 265, 295, 315, 335 nm (log ε 4.35, 4.13, 4.13, 3.33).

Justicidin B, $C_{21}H_{16}O_6$, *9-(1,3-benzodioxol-5-yl)-6,7-di-methoxynaphtho[2,3-c]furan-1(3H)-one, dehydrocollinusin.* Crystals, mp 240°. uv max (CHCl₃): 260, 295, 310, 350 nm (log ε 4.52, 4.13, 4.13, 3.41).

USE: Piscicidal agent.

5287. Juvenile Hormones. JH. Family of hormones, secreted by the corpora allata, controlling the larval metamorphosis of insects; so named since they induce the retention of insects' juvenile characteristics and prevent maturation. Isoln from the abdomen of the male wild silk moth, *Hyalophora cecropia* L.: C. M. Williams, *Nature* **178**, 212 (1956); H. Röller, J. S. Bjerke, *Life Sci.* **4**, 1617 (1965). Structure elucidation of JH I: H. Röller *et al., Angew. Chem. Int. Ed.* **6**, 179 (1967). Racemic synthesis of JH I: K. H. Dahm *et al., J. Am. Chem. Soc.* **89**, 5292 (1967). Toxicity of racemic JH I: J. B. Siddall, M. Slade, *Nature New Biol.* **229**, 158 (1971). Abs config of JH I: K. Nakanishi *et al., Chem. Commun.* **1971**, 1235; A. S. Meyer *et al., Proc. Nat. Acad. Sci. USA* **68**, 2312 (1971). Stereoselective syntheses of JH I: E. J. Corey *et al., ibid.* **90**, 5618 (1968); W. S. Johnson *et al., ibid.* 6225. Enantioselective syntheses of JH I: P. Loew, W. S. Johnson, *ibid.* **93**, 3765 (1971); D. J. Faulkner, M. R. Petersen, *ibid.* 3766. Isoln, structure, and abs config of JH II and III: K. J. Judy *et al., Proc. Nat. Acad. Sci. USA* **70**, 1509 (1973). Isoln and structure of *JH 0*: B. J. Bergot *et al., Science* **210**, 336 (1980); and *iso-JH 0* (MeJH I): *idem et al.* in *Juvenile Hormone Biochemistry*, G. E. Pratt, G. T. Brooks, Eds. (Elsevier-North Holland Biomedical Press, Amsterdam, 1981) pp 33-45. Biosynthesis,

isoln, and structure elucidation of *JH III bisepoxide* (JHB₃): D. S. Richard *et al., Proc. Nat. Acad. Sci. USA* **86**, 1421 (1989). Abs config of JHB₃: A. J. Herlt *et al., Chem. Commun.* **1993**, 1497. Enantioselective synthesis of JH III: K. Mori, H. Mori, *Tetrahedron* **43**, 4097 (1987); of JH I and II: K. Mori, M. Fujiwhara, *ibid.* **44**, 343 (1988); of JHB₃: R. W. Rickards, R. D. Thomas, *Tetrahedron Letters* **34**, 8369 (1993). Study of regulatory role: M. Cusson *et al., Arch. Insect Biochem. Physiol.* **25**, 329 (1994). Book: *The Juvenile Hormones*, L. I. Gilbert, Ed. (Plenum Press, New York, 1976) 572 pp. Review of biosynthesis: D. A. Schooley, F. C. Baker in *Comprehensive Insect Physiology, Biochemistry and Pharmacology* vol. **8**, G. A. Kerkut, L. I. Gilbert, Eds. (Pergamon Press, New York, 1985) pp 363-389. Review of techniques for identification and quantification: F. C. Baker in *Morphogenetic Hormones of Arthropods* Part 1, A. P. Gupta, Ed. (Rutgers Univ. Press, New Brunswick, 1990) pp 389-453. Review of discovery and identification of JHB₃: C.-M. Yin, *Zool. Stud.* **33**, 237-245 (1994). Review of molecular mechanisms of action: G. Jones, *Annu. Rev. Entomol.* **40**, 147-169 (1995).

	R	R'
JH I	CH₂CH₃	CH₂CH₃
JH II	CH₂CH₃	CH₃
JH III	CH₃	CH₃

Juvenile hormone I, $C_{18}H_{30}O_3$, *[2R-[2α(2E,6E),3α]]-7-ethyl-9-(3-ethyl-3-methyloxiranyl)-3-methyl-2,6-nonadienoic acid methyl ester; methyl (2E,6E,10R,11S)-10,11-epoxy-7-ethyl-3,11-dimethyl-2,6-tridecadienoate; C-18 JH.* n_D^{24} 1.4732. $[\alpha]_D^{24}$ +14.9° (c = 0.935 in chloroform). $[\alpha]_D^{22.5}$ +14.5° (c = 0.78 in methanol).

Juvenile hormone II, $C_{17}H_{28}O_3$, *[2R-[2α(2F,6E),3α]]-9-(3-ethyl-3-methyloxiranyl)-3,7-dimethyl-2,6-nonadienoic acid methyl ester; methyl (2E,6E,10R,11S)-10,11-epoxy-3,7,11-trimethyl-2,6-tridecadienoate, C-17 JH.* $n_D^{23.5}$ 1.4774. $[\alpha]_D^{24.5}$ +17.6° (c = 0.590 in methanol).

Juvenile hormone III, $C_{16}H_{26}O_3$, *[R-(E,E)]-9-(3,3-dimethyloxiranyl)-3,7-dimethyl-2,6-nonadienoic acid methyl ester, methyl (2E,6E,10R)-10,11-epoxy-3,7,11-trimethyl-2,6-dodecadienoate, C-16 JH.* Colorless oil. n_D^{24} 1.4736. $[\alpha]_D^{24}$ +6.71° (c = 0.57 in methanol).

K

5288. Kaempferol. *3,5,7-Trihydroxy-2-(4-hydroxyphenyl)-4H-1-benzopyran-4-one; 3,4′,5,7-tetrahydroxyflavone;* nimbecetin; pelargidenolon 1497; populnetin; rhamnolutein; robigenin; swartziol; trifolitin. $C_{15}H_{10}O_6$; mol wt 286.24. C 62.94%, H 3.52%, O 33.54%. Plant flavonoid isolated from *Delphinium consolida* L., *Ranunculaceae:* A. G. Perkin, E. J. Wilkinson, *J. Chem. Soc.* **81,** 585 (1902); from grapefruit *(Citrus paradisi* Macf., *Rutaceae):* Dunlap, Wender, *Anal. Biochem.* **4,** 110 (1962); from stems and seeds of *Cuscuta reflexa* Roxb., *Convolvulaceae:* K. W. Gopinath *et al., J. Sci. Ind. Res.* **21B,** 601 (1962), *C.A.* **58,** 7134e (1963). Identity with swartziol: M. R.-R. Paris, L. Bézanger-Beauquesne, *Compt. Rend.* **242,** 1761 (1956). Structure: S. von Kostanecki, A. Rozycki, *Ber.* **34,** 3721 (1901). Total synthesis: Y.-H. Lu *et al., Yao Hsueh Hsueh Pao* **15,** 477 (1980), *C.A.* **94,** 174808a (1981). Convenient synthesis: M. Ichikawa *et al., Org. Prep. Proc. Int.* **14,** 183 (1982). Mutagenicity studies: J. T. MacGregor, L. Jurd, *Mutat. Res.* **54,** 297 (1978); A. A. Hardigree, J. L. Epler, *ibid.* **58,** 231 (1978). HPLC determination in soybean: J. D. Gaynor *et al., Chromatographia* **25,** 1049 (1988).

Yellow needles, mp 276-278°; also reported as light yellow powder from ethanol-water, mp 278-280° (dec) (Ichikawa). uv max: 265, 365 nm. Slightly sol in water; sol in hot alcohol, ether or alkalies.
3,7-Dirhamnoside, $C_{27}H_{30}O_{14}$, *kaemferitrin, lespedin.* Isoln from the leaves of *Indigo arrecta* Benth., *Leguminosae:* Perkin, *J. Chem. Soc.* **91,** 435 (1907); from leaves of *Trichosanthus cucumeroides* Maxim., *Cucurbitaceae:* Nakaoki, Morita, *J. Pharm. Soc. Japan* **77,** 108 (1957). Identity with lespedin: Hatlori, *Nature* **168,** 788 (1952). Crystals from dil alc, mp 201-203°. Slightly sol in boiling water and cold alcohol.
3-Glucoside, $C_{21}H_{20}O_{11}$, *astragalin.* From *Podophyllum peltatum* L. and *P. emodi* Wall., *Berberidaceae.* Isoln and structure: von Wartburg, Kuhn, *Experientia* **21,** 67 (1965). Crystals from methanol, mp 175-178°. $[\alpha]_D^{20}$ −16.9° (c = 0.45 in methanol).

5289. Kainic Acid. *[2S-(2α,3β,4β)]-2-Carboxy-4-(1-methylethenyl)-3-pyrrolidineacetic acid;* 2-carboxy-3-carboxymethyl-4-isopropenylpyrrolidine; digenic acid; α-kainic acid; L_s-xylo-kainic acid; Digenin; Helminal. $C_{10}H_{15}NO_4$; mol wt 213.23. C 56.33%, H 7.09%, N 6.57%, O 30.01%. Anthelmintic principle from the dried red alga *Digenea simplex* (Wulf.) Ag., *Rhodomelaceae:* Murakami *et al., J. Pharm. Soc. Japan* **73,** 1026 (1953); **Japan.** pat. **4947('54),** *C.A.* **49,** 13604i (1955); Katsuya *et al.,* **Japan.** pat. **1942('64)** (to New-Japan Pharmaceutical Co.). Structure: Watase, Nitta, *Bull. Chem. Soc. Japan* **30,** 889 (1957); Watase *et al., ibid.* **31,** 714 (1958). Eight theoretical stereoisomers, Nitta *et al., Nature* **181,** 761 (1958); **Brit.** pat. **795,750** (1958 to Takeda). Synthesis: Ueno; Tatsuoka *et al.,* **U.S.** pats. **2,902,492** and **2,954,384** (1959, 1960 both to Takeda); W. Oppolzer, H. Andres, *Helv. Chim. Acta* **62,** 2282 (1979). Excitotoxic amino acid used to identify a specific subset of EAA receptors. Consequently the receptors are known as kainate receptors. Neurotoxic activity: J. V. Nadler, *Life Sci.* **24,** 289 (1979). Mechanism of neurotoxicity: E. G. McGeer *et al., Adv. Neurol.* **23,** 577 (1979); J. T. Coyle *et al., ibid.* 593; J. W. Ferkany *et al., Nature* **298,** 757 (1982); J. Garthwaite, G. Garthwaite, *ibid.* **305,** 138 (1983). Induces epileptogenic lesions: J. V. Nadler, *Neurosci. Res. Program Bull.* **19,** 369 (1981). Autoradiographic characterization of binding sites: J. T. Greenamyre *et al., J. Pharmacol. Exp.*

Ther. **233,** 254 (1985). *Book: Kainic Acid as a Tool in Neurobiology,* E. G. McGeer *et al.,* Eds. (Raven Press, New York, 1978). *Review:* J. T. Coyle, *Ciba Found. Symp.* **126,** 186-203 (1987).

Needles, dec 251°. $[\alpha]_D^{24}$ −14.8° (c = 1.01). Intense absorption at 6.05 and 11.2 μ. Sol in water. Insol in ethanol. Stable in boiling aq solns.
USE: Neurobiological tool.
THERAP CAT: Anthelmintic (Nematodes).

5290. Kallidin. N^2-L-*Lysylbradykinin;* kallidin-10; kallidin II. $C_{56}H_{85}N_{17}O_{12}$; mol wt 1188.40. C 56.60%, H 7.21%, N 20.04%, O 16.16%. A hypotensive and smooth muscle-stimulating principle formed by the proteolysis of kininogen by glandular and other kallikreins *(q.v.).* The decapeptide structure is a homologue of bradykinin *(q.v.):* Werle *et al., Z. Physiol. Chem.* **326,** 174 (1961). Synthesis: Nicolaides *et al., Biochem. Biophys. Res. Commun.* **6,** 210 (1961); Pless *et al., Helv. Chim. Acta* **45,** 394 (1962).

Lys–Arg–Pro–Pro–Gly–Phe–Ser–Pro–Phe–Arg

Amorphous precipitate. $[\alpha]_D^{21}$ −57° (c = 1 in *N* acetic acid). R_f value in a butanol/acetic acid/water system (70:10:20) 0.15.
THERAP CAT: Vasodilator.

5291. Kallikrein. Kallidinogenase; Callicrein; Padreatin; Padukrein; Glumorin; Depot-Glumorin; Circuletin; Kalirechin; Onokrein P; Padutin; Prokrein; Promotin. Hypotensive enzyme which releases kinins from plasma proteins. Major sources in the body are blood plasma, glandular tissues, and urine, occurring abundantly in the pancreas, parotid and submaxillary glands, in intestinal wall, in feces, in duodenal juice, and to a lesser degree in kidney. Isoln from mammalian pancreas or urine: Abelous, Bardier, *C.R. Soc. Biol.* **64,** 848 (1908); Webster *et al., Proc. Soc. Exp. Biol. Med.* **93,** 181 (1956). Separation into two components, kallikreins A and B: E. Habermann, *Z. Physiol. Chem.* **328,** 15 (1962); F. Fiedler, E. Werle, *ibid.* **348,** 1087 (1967); C. Kutzbach, G. Schmidt-Kastner, *ibid.* **353,** 1099 (1972). Prepn of high purity material: Werle, Trautschold, **Ger.** pat. **1,102,973** (1960 to Bayer). Review on pig pancreatic kallikrein: F. Fiedler, *Methods Enzymol.* **45B,** 289-303 (1976); on human kallikrein and prekallikrein: R. W. Colman, A. Bagdasarian, *ibid.* 303-322. Plasma kallikrein differs from glandular or urinary kallikrein. The latter two liberate kallidin, *q.v.;* the former releases bradykinin, *q.v.,* both from the common precursor, kininogen. Pharmacology: Franz, Marquardt, *Arzneimittel-Forsch.* **10,** 779 (1960). *Reviews:* Schachter, *Physiol. Rev.* **49,** 509 (1969); Suzuki *et al., Advan. Exp. Med. Biol.* **8,** 15 (1970).
THERAP CAT: Vasodilator.

5292. Kamala. Kamila; kameela; spoonwood. Glands and hairs covering the fruits of *Mallotus philippinensis* (Lam.) Muell. Arg., (also known as *Rottlera tinctoria* Roxb.) *Euphorbiaceae. Habit.* Philippine Islands, India, China, Australia. *Constit.* Rottlerin, isoallorottlerin, flavonoids, resins, wax. Detemn of constituents: Khorana, Motiwala, *Indian J. Pharm.* **11,** 37 (1949); M. Lounasmaa *et al., Planta Med.* **28,** 16 (1975); C. J. Widén, H. S. Puri, *ibid.* **40,** 284 (1980); V. K. Ahluwalia *et al., Indian J. Chem.* **27B,** 238 (1988). Anthelmintic effects: S. S. Gupta *et al., Indian J. Physiol. Pharmacol.* **28,** 63 (1984).
THERAP CAT: Anthelmintic (Cestodes).
THERAP CAT (VET): Has been used as anthelmintic.

5293. Kanamycin. Antibiotic complex produced by *Streptomyces kanamyceticus* Okami & Umezawa from Japanese soil: Umezawa *et al., J. Antibiot.* **10A,** 181 (1957); **U.S.** pat. **2,931,798** (1960). Comprised of three components, kanamycin A, the major component (usually designated as kanamycin) and kanamycins B and C, two minor congeners. Isoln and purification of kanamycins A and B and their salts: Johnson *et al.*, and Johnson, Hardcastle, U.S. pats. **2,936,307** and **2,967,177** (1960, 1961 both to Bristol-Myers). Separation process: Rothrock, Putter, U.S. pat. **3,032,547** (1962 to Merck & Co.). Prepn of kanamycin C: Murase *et al., J. Antibiot.* **14A,** 156 (1961). Studies on kanamycin B: Wakazawa *et al., ibid.* 180, 187. Structure of kanamycin A: Ogawa *et al., ibid.* **11A,** 169 (1958); Cron *et al., J. Am. Chem. Soc.* **80,** 4741 (1958). Structure of kanamycin B: Ito *et al., J. Antibiot.* **17A,** 189 (1964). Structure of kanamycin C: Murase, *ibid.* **14A,** 367 (1961). Abs config of kanamycin A: Hichens, Rinehart, *J. Am. Chem. Soc.* **85,** 1547 (1963); Umezawa *et al., Bull. Chem. Soc. Japan* **39,** 1244 (1966). Crystal structure of kanamycin A: Koyama *et al., Tetrahedron Letters* **1968,** 1875. Monograph: *Ann. N.Y. Acad. Sci.* **vol. 76,** Art. 2, pp 17-408 (1958). Synthesis of kanamycin A: Umezawa *et al., J. Antbiot.* **21,** 367 (1968); Nakajima *et al., Tetrahedron Letters,* **1968,** 623; Umezawa *et al., Bull. Chem. Soc. Japan* **42,** 533 (1969). Synthesis of kanamycin B: *eidem, J. Antibiot.* **21,** 424 (1968); *Bull. Chem. Soc. Japan* **42,** 537 (1969). Chemical conversion of kanamycin B to kanamycin C: S. Toda *et al., J. Antibiot.* **30,** 1002 (1977). Synthesis of kanamycin C: Umezawa *et al., Bull. Chem. Soc. Japan* **41,** 533 (1968); *J. Antibiot.* **21,** 162 (1968). Effects on protein synthesis: Suzuki *et al., ibid.* **23,** 99 (1970). Toxicity data (kanamycin A sulfate): Zel'tser *et al., Antibiotiki* **19,** 552 (1974). Comprehensive description: P. J. Claes *et al.,* in *Analytical Profiles of Drug Substances* **vol. 6,** K. Florey, Ed. (Academic Press, New York, 1977) pp 259-296.

	R	R'
Kanamycin A	NH$_2$	OH
Kanamycin B	NH$_2$	NH$_2$
Kanamycin C	OH	NH$_2$

Kanamycin A, C$_{18}$H$_{36}$N$_4$O$_{11}$, *O-3-amino-3-deoxy-α-D-glucopyranosyl-(1→6)-O-[6-amino-6-deoxy-α-D-glucopyranosyl-(1→4)]-2-deoxy-D-streptamine.* Crystals from methanol + ethanol. $[\alpha]_D^{24}$ +146° (0.1*N* H$_2$SO$_4$). LD$_{50}$ i.v. in mice: 583 mg/kg (Wakazawa).

Kanamycin A sulfate, *Cantrex, Cristalomicina, Enterokanacin, Kamycin, Kamynex, Kanabristol, Kanacedin, Kanamytrex, Kanasig, Kanatrol, Kanicin, Kannasyn, Kantrex, Kantrox, Klebcil, Otokalixin, Resistomycin, Ophtalmokalixan, Kantrexil, Kano, Kanescin, Kanaqua.* (U.S.P. requires that kanamycin sulfate contains not less than 75% kanamycin A on an anhydr basis). Irregular prisms, dec over a wide range above 250°C. Freely sol in water. Practically insol in the common alcohols and nonpolar solvents. LD$_{50}$ in mice: 20.7 g/kg orally; 1450 mg/kg i.p. (Zel'tser).

Kanamycin B, C$_{18}$H$_{37}$N$_5$O$_{10}$, *bekanamycin, aminodeoxykanamycin, NK-1006.* Crystals, mp 178-182° (dec). $[\alpha]_D^{18}$ +130° (c = 0.5 in water). $[\alpha]_D^{21}$ +114° (c = 0.98 in water). Sol in water, formamide; slightly sol in chloroform, isopropyl alcohol. Practically insol in the common alcohols and nonpolar solvents. LD$_{50}$ i.v. in mice: 136 mg/kg (Wakazawa).

Kanamycin B sulfate, *Coltericin, Kanendomycin, Kanendos.* Pharmacokinetics: F. Di Nola *et al., Minerva Med.* **70,** 1803 (1979).

Kanamycin C, C$_{18}$H$_{36}$N$_4$O$_{11}$. Crystals from methanol + ethanol, dec above 270°. $[\alpha]_D^{20}$ +126° (H$_2$O). Sol in water;

slightly sol in formamide. Practically insol in the common alcohols and nonpolar solvents.

THERAP CAT: Antibacterial.

THERAP CAT (VET): Antibacterial.

5294. Kaolin. Bolus alba; China clay; porcelain clay; white bole; argilla. Essentially a hydrated aluminum silicate, approximately H$_2$Al$_2$Si$_2$O$_8$·H$_2$O. Prepared for pharmaceutical and medicinal purposes by levigating with water to remove sand, etc.

White or yellowish-white, earthy mass or white powder; unctuous when moist. Insol in water, cold acids or in alkali hydroxides.

USE: Manuf porcelain, pottery, bricks, Portland cement; ultramarine, color lakes, refractory mortar; plaster material, filler for paper; electric and heat insulators; clarifying liquids; drying and emollient agent.

THERAP CAT: Adsorbent.

THERAP CAT (VET): Topical and G.I. adsorbent. Poultice.

5295. Karanjin. *3-Methoxy-2-phenyl-4H-furo[2,3-h]-1-benzopyran-4-one.* C$_{18}$H$_{12}$O$_4$; mol wt 292.29. C 73.97%, H 4.14%, O 21.90%. From *Pongamia glabra* Vent., *Leguminosae:* Beal, Katti, *J. Am. Pharm. Assoc.* **14,** 1086 (1925); Rao, Rao, *J. Indian Chem. Soc.* **17,** 526 (1940); Bhat *et al., J. Am. Oil Chem. Soc.* **33,** 197 (1956). Structure: Limaye, *Rasayanam* **1,** 1 (1936), *C.A.* **31,** 2206⁹ (1937); Manjunath *et al., Ber.* **72B,** 39 (1939). Synthesis: Seshadri, Venkateswarlu, *Proc. Indian Acad. Sci.* **13A,** 404 (1941); **17A,** 16 (1943); Kawase *et al., Bull. Chem. Soc. Japan* **28,** 273 (1955); Rao, Seshadri, *Proc. Indian Acad. Sci.* **33A,** 168 (1951); Aneja *et al., Tetrahedron* **2,** 203 (1958); Raizada *et al., J. Sci. Ind. Res.* **19B,** 76 (1960).

Needles from methanol, mp 157-158°. Sol in methanol, ethanol, chloroform, benzene, ether, concd H$_2$SO$_4$, HNO$_3$, HOAc, HCl; practically insol in petr ether, dil mineral acids.

5296. Karaya Gum. Gum karaya; kadaya; katilo; kullo; kuteera; sterculia; Indian tragacanth; mucara. The dried exudate of the tree *Sterculia urens* Roxb., *Sterculiaceae,* found in India, especially in the Gujerat region and in the central provinces: Toothaker, *The Soluble Gums* (Philadelphia, 1921); Mantell, *The Water-Soluble Gums* (New York, 1947). Constituents and structure: Hirst, Dunstan, *J. Chem. Soc.* **1953,** 2332. Structure is a partially acetylated polysaccharide containing about 8% acetyl groups and about 37% uronic acid residues. *Reviews:* F. Smith, R. Montgomery, *The Chemistry of Plant Gums and Mucilages* (Reinhold, New York, 1959); Goldstein, Alter, in *Industrial Gums,* R. L. Whistler, Ed. (Academic Press, New York, 2nd ed., 1973) pp 273-287.

Finely ground white powder, faint odor of acetic acid. Acid to litmus. Absorbs water rapidly to form viscous mucilages at low concs. Viscosity decreases on addn of acid or alkali. Color of the soln lightens in acidic media and darkens in alkaline soln due to the presence of tannins. Gum karaya loses viscosity forming ability when stored in the dry state, the loss being greater for a powdered material than for the crude gum. Cold storage inhibits this degradation.

Note: Karaya gum occurring in broken irregular pieces having a somewhat crystalline appearance has been referred to commercially as 'crystal' gum.

USE: As denture adhesive; as binder in paper manuf; as stabilizer, thickener, texturizer, emulsifier in foods; as thickening agent for dyes in textile industry. A substitute for gum tragacanth.

THERAP CAT: Cathartic.

5297. Karsil. *N-(3,4-Dichlorophenyl)-2-methylpentan-amide; 3',4'-dichloro-2-methylvaleranilide;* Niagara 4562. $C_{12}H_{15}Cl_2NO$; mol wt 260.16. C 55.40%, H 5.81%, Cl 27.25%, N 5.38%, O 6.15%. Prepd from 3,4-dichloroaniline and 2-methylvaleryl chloride: Dorschner *et al.*, **Brit.** pat. **869,169** (1961 to FMC).

Crystals, mp 106-107°.

USE: Herbicide.

5298. Kava. Kava-kava; ava-ava; kawa. Dried rhizome and roots of *Piper methysticum* Forst., *Piperaceae.* *Habit.* Polynesia. Most important constituents are: kawain, dihydrokawain, methysticin, dihydromethysticin, and yangonin: Borsche, Lewinsohn, *Ber.* **66**, 1792 (1933) and references to preceding papers therein. Chemical and pharmacological investigation of the kava constituents: Klohs *et al.*, *J. Med. Pharm. Chem.* **1**, 95 (1959); Meyer, Kretzschmar, *Klin. Wochenschr.* **44**, 902 (1966). Review of chemistry, pharmacology and historical sketch: *U.S. Public Health Service Publ. No. 1645,* D. H. Efron, Ed., pp 103-181 (1967). *Note:* Kava is also the popular name for the intoxicating drink prepared from the plant's roots.

5299. Kawain. *[R-(E)]-5,6-Dihydro-4-methoxy-6-(2-phenylethenyl)-2H-pyran-2-one;* 5-hydroxy-3-methoxy-7-phenyl-2,6-heptadienoic acid δ-lactone; 4-methoxy-6-(β-phenylvinyl)-5,6-dihydro-α-pyrone; 4-methoxy-6-styryl-5,6-dihydro-α-pyrone; kavain; gonosan. $C_{14}H_{14}O_3$; mol wt 230.26. C 73.03%, H 6.13%, O 20.84%. From the rhizome and roots of *Piper methysticum* Forst., *Piperaceae* (kava): Borsche, Peitzsch, *Ber.* **63**, 2414 (1930); Hänsel, Beiersdorff, *Naturwiss.* **45**, 573 (1958); *eidem, Arzneimittel-Forsch.* **9**, 581 (1959). Synthesis: Fowler, Henbest, *J. Chem. Soc.* **1950**, 3642; Kostermans, *Nature* **166**, 788 (1950); *idem, Rec. Trav. Chim.* **70**, 79 (1951); Z. H. Israili, E. E. Smissman, *J. Org. Chem.* **41**, 4070 (1976). Abs config: Snatzke, Hänsel, *Tetrahedron Letters* **1968**, 1797. Crystal and molecular structure: A. Yoshino, W. Nowacki, *Z. Kristallogr., Kristallgeom., Kristallphys., Kristallchem.* **136**, 66 (1972), *C.A.* **78**, 63591z (1973).

Rods from methanol + ether, mp 105-106°. bp$_{0.1}$ 195-197°. $[\alpha]_D^{20}$ +105° (abs alc). uv max (methanol): 210, 245, 282 nm (log ε 4.38, 4.44, 2.81). Practically insol in water. Sol in acetone, ether, methanol; slightly sol in hexane.

(±)-Form, needles from methanol, mp 146-147°.

Dihydrokawain, $C_{14}H_{16}O_3$; *marindinin.* Crystals from ether, mp 58-60°. $[\alpha]_D^{24}$ +31° (methanol). uv max (methanol): 236 nm (log ε 4.14). Sol in alc, chloroform; moderately sol in ether. Practically insol in petr ether, water.

5300. Kebuzone. *4-(3-Oxobutyl)-1,2-diphenyl-3,5-pyr-azolidinedione;* 1,2-diphenyl-4-(γ-ketobutyl)-3,5-pyrazolidinedione; 1,2-diphenyl-4-(3'-oxobutyl)-3,5-dioxopyrazolidine; ketophenylbutazone; KPB; Chebutan; Chepirol; Chetazolidin; Chetil; Copirene; Ketason; Ketazone; Pecnon; Phloguron; Recheton. $C_{19}H_{18}N_2O_3$; mol wt 322.36. C 70.79%, H 5.63%, N 8.69%, O 14.89%. Prepn: Deuss *et al.,* U.S. pat. **2,910,481** (1959 to Geigy). Review of pharmacology: Horakova *et al., Pharmacotherapeutica* **1950-1959,**

335-350 (1963), *C.A.* **60**, 6072g (1964). Metabolism: Nemecek *et al., Arzneimittel-Forsch.* **16**, 1339 (1966); Queisnerova, Nemecek, *Cesk. Farm.* **20**, 55 (1971), *C.A.* **75**, 47077u (1971).

Crystals, mp 115.5-116.5° or 127.5-128.5° depending on cryst form.

THERAP CAT: Antirheumatic.

5301. Kefir Fungi. Kefir grains; kefir seeds. A conglomeration of various fungi, including *Dispora caucasica, Schizomycetes* and a species of *Saccharomyces.*

Grayish-yellow lumps, irregular in size; firm, toughly gelatinous consistency, becoming cartilaginous and brittle when dry.

USE: Preparing a nutritious beverage of fermented milk known as kefir.

5302. Keratin. A protein obtained from hair, wool, horn, nails, claws, beaks, scales, membranes of egg shells and nerve tissue. There are two types of keratins—hard keratin of hair, horn, nails, etc., and soft keratin (pseudokeratin) of the epidermis, and whalebone: Rudall, *Advan. Protein Chem.* **7**, 253-290 (1952). The keratins contain all of the common amino acids and differ from other fibrous structural proteins chiefly by their high cystine content. Sequence studies revealed no preferable grouping or periodicity. Amino acid compn of a few keratins: Tristram in H. Neurath, K. Bailey, *The Proteins* vol. 1A (Academic Press, New York, 1953) p 220. Stability of the protein is due to frequent primary valence cross-links (disulfide bonds) and secondary valence cross-links (hydrogen bonds) between neighboring polypeptide chains. Prepn of different forms for different purposes: Grassmann, **Ger.** pats. **673,203** and **682,257** (both 1939). Molecular structure of α-keratin: Fraser *et al., Nature* **203**, 1231 (1964); of β-keratin: M. L. Huggins, *Macromolec.* **13**, 465 (1980). *Reviews:* Ward, Lundgren, *Advan. Protein Chem.* **9**, 243-297 (1954); Crewther *et al., ibid.* **20**, 191 (1965); Bradbury, *ibid.* **27**, 111 (1973).

Characteristic properties ascribed to keratins: *(a)* insolubility in water, including aq solns of salts, hydrotropic substances, and dil acids and bases at temps not much above room temp; *(b)* resistance to proteolytic enzymes; *(c)* resistance to hydrolysis; *(d)* solvolysis by mixtures of substances which break the —S—S— bonds and the hydrogen bonds.

USE: Coating "enteric" pills which are unaffected in the stomach, but dissolved by the alkaline intestinal secretions. Detailed coating instructions: Dale, *Pharm. J.* **129**, 494 (1932). Other uses are in formulations of foam type extinguishers and in the production of protein hydrolyzates.

5303. Keratinase. M-Zyme. A proteolytic enzyme, commercially produced by a strain of *Streptomyces fradiae:* Nickerson, Noval, U.S. pat. **2,988,487** (1961 to Rutgers Res. and Educ. Found.). Can convert about 50% of the dry wt of wool into a water-sol form, permitting hair separation from hides. The amount of enzyme needed to dehair hides is not enough to cause much digestion of the hair itself, which is recovered and used in felt making. The enzyme attacks at the base of the hair shaft, which is particularly sensitive to the action of keratinase.

Soluble in water. Non-dialyzable. Optimum efficiency at pH 8.5-9.5. Becomes inactive by heating to 100° for 5 min. The ability to digest keratin appears to be dependent on the presence of at least trace amounts of metal ions which form chelates.

USE: In the dehairing of hides and skins: Robison, Nickerson, U.S. pat. **2,988,488** (1961 to Mearl). In depilatory compositions for human use: Mattin, Greenstein, U.S. pat. 2,988,485 (1961 to Mearl).

5304. Kermesic Acid. *9,10-Dihydro-3,5,6,8-tetrahydroxy-1-methyl-9,10-dioxo-2-anthracenecarboxylic acid;* C.I. Natural Red 3; C.I. 75460. $C_{16}H_{10}O_8$; mol wt 330.25. C 58.19%, H 3.05%, O 38.76%. Principle constituent of kermes, one of the oldest known insect dyes. Isoln and characterization: O. Dimroth, *Ber.* **43**, 1387 (1910); O. Dimroth, W. Scheurer, *Ann.* **399**, 43 (1913). Structure: O. Dimroth, R. Fick, *Ann.* **411**, 315 (1916). Revised structure: D. D. Gadgil *et al., Tetrahedron Letters* **1968**, 2223. Synthesis: D. W. Cameron *et al., Chem. Commun.* **1978**, 688; *eidem, Aust. J. Chem.* **34**, 2401 (1981).

Dark red rosettes from acetic acid, mp > 320° (dec). Absorption max: 276, 312, 498 nm (log ε 4.52, 4.12, 3.96). Slightly sol in cold water; sol in hot water (yellowish-red soln). Violet-red in conc H_2SO_4, turning blue on addn of boric acid. Violet in aq NaOH.

USE: As brilliant scarlet dye.

5305. Kerosene. Kerosine. A mixture of petroleum hydrocarbons, chiefly of the methane series having from 10 to 16 carbon atoms per molecule. It constitutes the fifth fraction in the distillation of petroleum (after the pctr ethers and before the oils). A typical analysis of the kerosene fraction from Midcontinent crude includes *n*-dodecane, three alkyl derivatives of benzene, naphthalene, 1- and 2-methyl-5,6,7,8-tetrahydronaphthalene. Toxicity study: W. B. Deichmann *et al., Ann. Int. Med.* **21**, 803 (1944). Toxicological studies and recommended treatment of kerosene poisoning: H. W. Gerarde, *Toxicol. Appl. Pharmacol.* **1**, 462 (1959).

Pale yellow or water-white, mobile, oily liquid. Characteristic, not altogether disagreeable odor. d ~0.80. bp 175-325°. Flash pt 150-185°F. (65-85°C). Insol in water. Misc with other petroleum solvents. A water-white, deodorized form of kerosene is marketed under the trade name *Deobase*. Kerosene is deodorized and decolorized by washing with (fuming) sulfuric acid, followed by sodium plumbite soln and sulfur (Doctor sweetening). LD_{50} orally in rabbits: 28 ml/kg (Deichmann).

Caution: Defatting action on skin can lead to irritation, infection. Inhalation of high concns causes headache, drowsiness, coma. Swallowing causes G.I. irritation with vomiting, diarrhea. Aspiration of vomitus causes serious pneumonitis. Induction of vomiting following ingestion is contraindicated.

USE: In kerosene lamps, flares, and stoves; as degreaser and cleaner; Deobase formerly used as a solvent in cosmetics and in fly spray.

5306. Ketamine. (±)-2-(2-Chlorophenyl)-2-(methylamino)cyclohexanone; $C_{13}H_{16}ClNO$; mol wt 237.73. C 65.68%, H 6.78%, Cl 14.91%, N 5.89%, O 6.73%. Prepn: C. L. Stevens, **Belg.** pat. **634,208**; *idem,* U.S. pat. **3,254,124** (1963, 1966 both to Parke, Davis). Isoln of optical isomers: T. W. Hudyma *et al.,* **Ger.** pat. **2,062,620** (1971 to Bristol-Myers), *C.A.* **75**, 118119x (1971). Clinical pharmacology of racemate and enantiomers: P. F. White *et al., Anesthesiology* **52**, 231 (1980). Toxicity: E. J. Goldenthal, *Toxicol. Appl. Pharmacol.* **18**, 185 (1971). Enantioselective HPLC determn in plasma: G. Geisslinger *et al., J. Chromatog.* **568**, 165 (1991). Comprehensive description: W. C. Sass, S. A. Fusari in *Analytical Profiles of Drug Substances* vol. 6, K. Florey, Ed. (Academic Press, New York, 1977) pp 297-322. Review of pharmacology and use in veterinary medicine: M. Wright, *J. Am. Vet. Med. Assoc.* **180**, 1462-1471 (1982). Review of pharmacology and clinical experience: D. L. Reich, G. Silvay, *Can. J. Anaesth.* **36**, 186-197 (1989); in

pediatric procedures: S. M. Green, N. E. Johnson, *Ann. Emerg. Med.* **19**, 1033-1046 (1990).

Crystals from pentane-ether, mp 92-93°. uv max (0.01*N* NaOH in 95% methanol): 301, 276, 268, 261 nm ($A_{1cm}^{1\%}$ 5.0, 7.0, 9.8, 10.5). pKa 7.5. pH of 10% aq soln 3.5.

Hydrochloride, $C_{13}H_{16}ClNO \cdot HCl$, *CI-581, Ketaject, Ketalar, Ketanarkon, Ketanest, Ketaset, Ketavet, Vetalar.* White crystals, mp 262-263°. Soly in water: 20 g/100 ml. LD_{50} in adult mice, rats (mg/kg): 224 ±4, 229 ±5 i.p. (Goldenthal).

THERAP CAT: Anesthetic (intravenous).

THERAP CAT (VET): Anesthetic (intravenous).

5307. Ketanserin. *3-[2-[4-(4-Fluorobenzoyl)-1-piperidinyl]ethyl]-2,4[1H,3H]-quinazolinedione;* R-41468; Ketensin; Serefrex; Taseron. $C_{22}H_{22}FN_3O_3$; mol wt 395.43. C 66.82%, H 5.61%, F 4.80%, N 10.63%, O 12.14%. Specific serotonin ($5HT_2$)-receptor antagonist. Prepn: J. Vandenheck *et al., Eur.* pat. **Appl. 13,612;** *eidem,* U.S. pat. **4,335,-127** (1980, 1982 both to Janssen). X-ray structure: O. M. Peeters *et al., Cryst. Struct. Commun.* **11**, 375 (1982). Receptor binding profile: J. E. Leysen *et al., Life Sci.* **28**, 1015 (1981). Pharmacology: J. M. Van Neuten *et al., J. Pharmacol. Exp. Ther.* **218**, 217 (1981). HPLC determn in plasma: A. T. Kacprowicz *et al., J. Chromatog.* **272**, 417 (1983). Clinical efficacy in intermittent claudication: J. De Cree *et al., Lancet* **2**, 775 (1984); in Raynaud's phenomenon: J. R. Seibold, A. H. M. Jageneau, *Arthritis Rheum.* **27**, 139 (1984); in hypertension: A. Amery *et al., J. Cardiovasc. Pharmacol.* **6**, 182 (1984). Series of articles on pharmacology and clinical studies: *ibid.* **7**, Suppl. 7, S1-S182 (1985); on pharmacokinetics and metabolism: *Arzneimittel-Forsch.* **38**, 775-800 (1988). Review of pharmacology and clinical efficacy in hypertension and vascular disease: R. N. Brogden, E. M. Sorkin, *Drugs* **40**, 903-949 (1990).

Crystals from 4-methyl-2-pentanone, mp 227-235°. Soly (g/100 ml): 0.001 in water; 0.038 in ethanol; 2.34 in DMF. pKa 7.5.

Tartrate, $C_{22}H_{22}FN_3O_3 \cdot C_4H_6O_6$, *R-49945, Ket, Perketan, Serepress, Sufrexal.*

THERAP CAT: Antihypertensive.

5308. Ketazolam. *11-Chloro-8,12b-dihydro-2,8-dimethyl-12b-phenyl-4H-[1,3]oxazino[3,2-d][1,4]benzodiazepine-4,7-(6H)-dione;* U-28774; Anseren; Ansieten; Anxon; Contamex; Loftran; Unakalm. $C_{20}H_{17}ClN_2O_3$; mol wt 368.82. C 65.13%, H 4.65%, Cl 9.61%, N 7.60%, O 13.01%. Prepn: J. Szmuszkovicz, **Ger.** pat. **1,947,226** corresp to U.S. pat. **3,573,282** (1970, 1971 both to Upjohn); U.S. pat. **3,575,965** (1971 to Upjohn). Synthesis and structure: J. Szmuszkovicz *et al., Tetrahedron Letters* **1971**, 3665. Analysis by HPLC: D. J. Weber, *J. Pharm. Sci.* **61**, 1797 (1972). Pharmacology: V. H. Sethy, *Arch. Exp. Pathol. Pharmakol.* **301**, 157 (1978). Clinical studies: L. F. Fabre, *J. Int. Med. Res.* **4**, 50 (1976); L. A. Gottschalk, J. B. Cohn, *Psychopharmacol. Bull.* **14**, 39 (1978).

Colorless prisms from chloroform/ether, mp 182-183.5° (sinters at 170°). uv max (ethanol): 202, 241 nm (ϵ 40600; 18400).

Note: This is a controlled substance (depressant) listed in the U.S. Code of Federal Regulations, Title 21 Part 1308.14 (1995).

THERAP CAT: Anxiolytic.

5309. Ketene. *Ethenone;* carbomethene. C_2H_2O; mol wt 42.04. C 57.14%, H 4.80%, O 38.06%. $CH_2=C=O$. Prepd by the thermal decompn of acetone, diketene or acetic anhydride: Hurd, *Org. Syn.* **coll. vol. I,** 330 (2nd ed., 1941); S. Andreades, H. D. Carlson, *ibid.* **coll. vol. V,** 679 (1973). Structure of ketene dimer as 3-buteno-β-lactone: Blomquist, Baldwin, *J. Am. Chem. Soc.* **70,** 29 (1948); Hurd, Blanchard, *ibid.* **72,** 1461 (1950); Katz, Lipscomb, *J. Org. Chem.* **17,** 515 (1953). Toxicology: J. F. Treon *et al., J. Ind. Hyg. Toxicol.* **31,** 209 (1949). *Review* on the prepn of stable ketenes: R. S. Ward in *The Chemistry of Ketenes, Allenes and Related Compounds,* **Part 1,** S. Patai, Ed. (Wiley, New York, 1980) pp 223-277; on synthetic uses of ketenes: W. T. Brady, *ibid.* pp 279-308.

Gas. Penetrating odor. mp $-150°$. bp $-56°$. Electron diffraction pictures: Beach, Stevenson, *J. Chem. Physics* **6,** 75 (1938). Fairly sol in acetone. All operations with ketene should be carried out in an efficient hood (Andreades, Carlson).

Caution: Potential symptoms of overexposure are irritation of skin, eyes, nose, throat and lungs; pulmonary edema. *See NIOSH Pocket Guide to Chemical Hazards* (DHHS/NIOSH 90-117, 1990) p 134.

USE: For the conversion of higher acids into their anhydrides; for acetylation in the manuf of cellulose acetate and aspirin.

5310. Kethoxal. *3-Ethoxy-1,1-dihydroxy-2-butanone; β-ethoxy-α-ketobutyraldehyde;* 3-ethoxy-2-oxobutyraldehyde hydrate; U-2032. $C_6H_{12}O_4$; mol wt 148.16. C 48.64%, H 8.16%, O 43.20%. Also depicted as $CH_3CH(OC_2H_5)$-COCHO monohydrate. Prepn: Rappen, *J. Prakt. Chem.* **157,** 177 (1941); Tiffany *et al., J. Am. Chem. Soc.* **79,** 1682 (1957). Antiviral activity: H. E. Rennis, *Ann. N.Y. Acad. Sci.* **173/1,** 527 (1970).

Yellow oil. bp_{11} 54-58°. bp_{760} 145°. Initial $n_D^{23.7}$ 1.4348. Miscible with alc; sol in benzene with yellow color; one part dissolves in ten parts of water.

Bis(thiosemicarbazone), $C_8H_{16}N_6OS_2$, *2,2'-[1-(1-ethoxyethyl)-1,2-ethanediylidene]bishydrazinecarbothioamide, gloxazone, BW-356-C-61, NSC-82116, Contrapar.*

THERAP CAT: Antiviral.

THERAP CAT (VET): Bis(thiosemicarbazone) as anaplasmodastat.

5311. Ketipic Acid. *3,4-Dioxohexanedioic acid;* 2,3-diketoadipic acid; oxalodiacetic acid. $C_6H_6O_6$; mol wt 174.11. C 41.39%, H 3.47%, O 55.14%. $HOOCCH_2COCOCH_2$-COOH. Prepn: Fittig *et al., Ann.* **249,** 182 (1888); Stachel, *Arch. Pharm.* **295,** 735 (1932); **296,** 479 (1963). Structure studies; occurs in the solid form as the γ-lactol: Stachel, *Ann.* **689,** 118 (1965).

Dec at 150° forming diacetyl. Sparingly sol in ether. Sol in glacial acetic acid, concd hydrochloric acid. Practically insol in water, alc, chloroform, CS_2, benzene, petr ether.

Diethyl ester, $C_{10}H_{14}O_6$, prisms from alcohol, mp 83°. bp_{30} 230°. Soluble in ether, chloroform.

5312. Ketobemidone. *1-[4-(3-Hydroxyphenyl)-1-methyl-4-piperidyl]-1-propanone;* 4-(m-hydroxyphenyl)-1-methyl-4-piperidyl ethyl ketone. $C_{15}H_{21}NO_2$; mol wt 247.34. C 72.84%, H 8.56%, N 5.66%, O 12.94%. Prepn: Suter, *Abstracts of First National Medicinal Chemistry Symposium of the A.C.S.* (Ann Arbor, 1948) p 20; **Brit.** pats. **591,992** and **609,763** (1947 and 1948 to Ciba); Eisleb, **U.S.** pats. **2,167,-351; 2,242,575** and **2,248,018** (1939 and 1941 to Winthrop); **Ger.** pat. **752,755** (1952 to I. G. Farben). Metabolism in man: U. Bondesson *et al., Drug Metab. Dispos.* **9,** 376 (1981).

Crystals, mp 156-157°.

Hydrochloride, $C_{15}H_{21}NO_2.HCl$, *Hoechst 10720, Win-1539, Cliradon, Cymidon, Ketogan, Ketogin.* Crystals, mp 201-202°. Sol in water; slightly sol in alcohol. Aq solns may be sterilized by boiling for short periods.

Note: This is a controlled substance (opiate) listed in the U.S. Code of Federal Regulations, Title 21 Part 1308.11 (1995).

THERAP CAT: Analgesic (narcotic).

5313. Ketoconazole. *cis-1-Acetyl-4-[4-[[2-(2,4-dichlorophenyl)-2-(1H-imidazol-1-ylmethyl)-1,3-dioxolan-4-yl]methoxy]phenyl]piperazine;* R-41400; Fungarest; Fungoral; Ketoderm; Ketoisdin; Nizoral; Orifungal M; Panfungol. $C_{26}H_{28}$-$Cl_2N_4O_4$; mol wt 531.44. C 58.76%, H 5.31%, N 13.34%, N 10.54%, O 12.04%. Orally active, broad-spectrum antimycotic. Prepn: J. Heeres *et al.,* **Ger.** pat. **2,804,096,** *eidem,* **U.S.** pats. **4,144,346** and **4,223,036** (1978, 1979, 1980, all to Janssen); *eidem, J. Med. Chem.* **22,** 1003 (1979). Pharmacokinetics: E. W. Gascoigne *et al., Clin. Res. Rev.* **1,** 177 (1981); C. Brass *et al., Antimicrob. Ag. Chemother.* **21,** 151 (1982). HPLC determn in human serum: V. L. Pascucci *et al., J. Pharm. Sci.* **72,** 1467 (1983). Series of articles on animal and human studies: *Rev. Infect. Dis.* **2,** 519-692 (1980). Effect on hepatic enzymes *in vitro* and *in vivo:* K. N. Buchi *et al., Biochem. Pharmacol.* **35,** 2845 (1986); J. K. Ritter, M. R. Franklin, *Toxicol. Letters* **36,** 51 (1987). Case reports of hepatic toxicity: J. K. Heiberg, E. Svejgaard, *Brit. Med. J.* **283,** 825 (1981); R. Rollman, L. Loof, *Brit. J. Dermatol.* **108,** 376 (1983). Controlled clinical trials: E. A. Petersen *et al., Ann. Intern. Med.* **93,** 791 (1980); W. T. Hughes *et al., J. Infect. Dis.* **147,** 1060 (1983); H. W. Jolly *et al., Cutis* **31,** 208 (1983). Clinical evaluation as inhibitor of steroid synthesis: N. Sonino, *N. Engl. J. Med.* **317,** 812 (1987). Review of pharmacology and therapeutic efficacy: C. A. Sohn, *Clin. Pharm.* **1,** 217 (1982); R. C. Heel *et al., Drugs* **23,** 1-36 (1982). Series of articles on clinical efficacy and therapeutic experience: *Drugs Exptl. Clin. Res.* **12,** 397-427 (1986).

Crystals from 4-methyl-2-pentanone, mp 146°. LD$_{50}$ in mice, rats, guinea pigs, dogs (mg/kg): 44, 86, 28, 49 i.v.; 702, 227, 202, 780 orally (Heel).

THERAP CAT: Antifungal.

5314. α-Ketoglutaric Acid. *2-Oxopentanedioic acid; 2-oxoglutaric acid;* 2-oxo-1,5-pentanedioic acid. C$_5$H$_6$O$_5$; mol wt 146.10. C 41.11%, H 4.14%, O 54.76%. HOOCCH$_2$-CH$_2$COCOOH. Plays an important role in amino acid metabolism (transamination) *see* Severo Ochoa, "Enzymic Mechanisms in the Citric Acid Cycle" in *Advances in Enzymology* **15**, 183-270 (1954). Prepn: Friedman, Kosower, *Org. Syn. coll. vol.* III, 510 (1955); Bottorff, Moore, *ibid.* **coll. vol.** V, 687 (1973). Microbial synthesis using a strain of *Pseudomonas:* Lockwood *et al.,* U.S. pat. **2,443,919** (1948); Berger, Witt, U.S. pat. **2,841,616** (1958).

Crystals from acetone-benzene, mp 113.5°. Freely sol in water, alcohol. Very sparingly sol in ether.

Diethyl ester, C$_9$H$_{14}$O$_5$, liq, bp$_{23}$ 160°, bp$_{13}$ 144°.

Compound with L(+)-ornithine, C$_{10}$H$_{18}$N$_2$O$_7$, *L(+)-ornithine α-ketoglutarate, Ornicetil.*

USE: Can be converted to L-glutamic acid by *Aeromonas* spp. *see* Good, U.S. pat. **2,933,434** (1960 to International Minerals & Chem. Corp.).

5315. 2-Keto-L-gulonic Acid. *L-xylo-2-Hexulosonic acid;* 2-oxo-L-gulonic acid. C$_6$H$_{10}$O$_7$; mol wt 194.14. C 37.12%, H 5.19%, O 57.69%. Important intermediate in vitamin C manufacture. Prepn from sorbitol: Reichstein, Grüssner, *Helv. Chim. Acta* **17**, 311 (1934); Reichstein, U.S. pat. **2,301,811** (1942); from L-sorbose: Haworth *et al.,* Brit. pat. **443,901** (1936); by microbial conversion: Huang, U.S. pat. **3,043,749** (1962 to Pfizer); Motizuki *et al.,* U.S. pat. **3,234,105** (1966 to Takeda); Z. Y. Yuan *et al., Ann. N.Y. Acad. Sci.* **672**, 628 (1992). Biosynthesis and metabolism in bacteria: S. Makover *et al., Biotechnol. Bioeng.* **17**, 1485 (1975). NMR structure determn: T. C. Crawford *et al., J. Am. Chem. Soc.* **102**, 2220 (1980). HPLC separation in fermentation broth: R. A. Lazarus, J. L. Seymour, *Anal. Biochem.* **157**, 360 (1986). Review of early syntheses: T. C. Crawford, S. A. Crawford, *Adv. Carbohyd. Chem. Biochem.* **37**, 79-159 (1980).

```
        COOH
         |
         C=O
         |
   HO —  CH
         |
   HC — OH
         |
   HO —  CH
         |
        CH₂OH
```

Crystals from water (may be washed with acetone). mp 171° (slight dec). [α]$_D^{18}$ −48° (c = 1). Moderately sol in water. Strong acid. Reduces boiling Fehling's soln rapidly.

Methyl ester, crystals, mp 155-157°. [α]$_D^{18}$ −25° (c = 1 in methanol).

Ethyl ester, crystals, [α]$_D^{20}$ −14.5° (c = 0.63 in abs alc).

5316. Ketoprofen. *3-Benzoyl-α-methylbenzeneacetic acid; m-benzoylhydratropic acid;* 2-(3-benzoylphenyl)propionic acid; RP-19583; Alrheumat; Alrheumun; Capisten; Dexal; Epatec; Fastum; Iso-K; Kefenid; Ketopron; Lertus; Menamin; Meprofen; Orudis; Orugesic; Oruvail; Oscorel; Profenid; Toprec; Toprek. C$_{16}$H$_{14}$O$_3$; mol wt 254.29. C 75.57%, H 5.55%, O 18.88%. Prepn: **Fr.** pat. **M6444**; D. Farge *et al.,* **S. Afr.** pat. **68 00,524**; *eidem,* U.S. pat. **3,641,-127** (1968, 1968, 1972 all to Rhône-Poulenc); G. A. Pinna *et al., Farmaco Ed. Sci.* **35**, 684 (1980). Clinical data: Gyory *et al.,* Brit. Med. J. **4**, 398 (1972); Cathcart *et al., Ann. Rheum. Dis.* **32**, 62 (1973). Pharmacokinetics in man: T. Ishizaki *et al., Eur. J. Clin. Pharmacol.* **18**, 407 (1980). Toxicity data: K. Ueno *et al., J. Med. Chem.* **19**, 941 (1976). Comprehensive description: G. G. Liversidge, *Anal. Profiles Drug Subs.* **10**, 443-471 (1981).

Crystals from 6:20 benzene-petr ether, mp 94°. uv max (methanol): 255 nm (log ε 4.33). Sol in ether, alc, acetone, chloroform, DMF, ethyl acetate. Slightly sol in water. LD$_{50}$ orally in rats: 101 mg/kg (Ueno).

Lysine salt, C$_{16}$H$_{14}$O$_3$·C$_6$H$_{14}$N$_2$O$_2$, *Artrosilene.*

THERAP CAT: Anti-inflammatory; analgesic.

THERAP CAT (VET): Anti-inflammatory; analgesic.

5317. 11-Ketoprogesterone. *Pregn-4-ene-3,11,20-trione;* 11-oxoprogesterone; U-1258; Ketogestin. C$_{21}$H$_{28}$O$_3$; mol wt 328.45. C 76.79%, H 8.59%, O 14.61%. Prepn: Reichstein, Fuchs, *Helv. Chim. Acta* **23**, 688 (1940); from 11α-hydroxyprogesterone: Whiting, Schneider, U.S. pat. **3,019,238** (1962 to Upjohn).

Crystals, mp 171.5-173°. [α]$_D^{25}$ +270° (chloroform). Practically insol in water. Sol in acetone, chloroform.

THERAP CAT (VET): Has been used in ketosis.

5318. Ketorolac. *(±)-5-Benzoyl-2,3-dihydro-1H-pyrrolizine-1-carboxylic acid;* 5-benzoyl-1,2-dihydro-3H-pyrrolo-[1,2-a]pyrrole-1-carboxylic acid; RS-37619. C$_{15}$H$_{13}$NO$_3$; mol wt 255.27. C 70.58%, H 5.13%, N 5.49%, O 18.80%. Prostaglandin biosynthesis inhibitor. Prepn and separation of isomers: **Belg.** pat. **856,681;** J. M. Muchowski, A. F. Kluge, U.S. pat. **4,089,969** (both 1978 to Syntex). Alternate processes: J. M. Muchowski, R. Greenhouse, U.S. pat. **4,347,186** (1982 to Syntex); F. Franco *et al., J. Org. Chem.* **47**, 1682 (1982); J. B. Doherty, U.S. pat. **4,496,741** (1985 to Merck & Co.). Absolute configuration: A. Guzman *et al., J. Med. Chem.* **29**, 589 (1986). Structure-activity relationships: J. M. Muchowski *et al., ibid.* **28**, 1037 (1985). Pharmacology and analgesic, anti-inflammatory profile of ketorolac and its tromethamine salt: W. H. Rooks *et al., Agents Actions* **12**, 684 (1982); *eidem, Drugs Exp. Clin. Res.* **11**, 479 (1985). Clinical comparison with acetaminophen in postoperative pain: H. J. McQuay *et al., Clin. Pharmacol. Ther.* **39**, 89 (1986).

Crystals from ethyl acetate + ether, mp 160-161°. uv max in methanol: 245, 312 nm (ε 7080, 17400). pKa 3.49 ±0.02. LD$_{50}$ orally in mice: ~200 mg/kg (Rooks).

(±)-Form tromethamine salt, C$_{19}$H$_{24}$N$_2$O$_6$, *Acular, Dolac, Lixidol, Tarasyn, Toradol, Toratex.*

(+)-Form, crystals from hexane + ethyl acetate, mp 174° (Guzman); also reported as mp 154-156° (Muchowski, Kluge). [α]$_D$ +173° (c = 1 in methanol).

(−)-Form, crystals from hexane + ethyl acetate, mp 169-170° (Guzman); also reported as mp 153-155° (Muchowski, Kluge). [α]$_D$ −176° (c = 1 in methanol).

THERAP CAT: Analgesic; anti-inflammatory.

5319. Ketotifen. *4,9-Dihydro-4-(1-methyl-4-piperidinylidene)-10H-benzo[4,5]cyclohepta[1,2-b]thiophen-10-one.* C$_{19}$-H$_{19}$NOS; mol wt 309.43. C 73.75%, H 6.19%, N 4.53%, O 5.17%, S 10.36%. Prepn: J. P. Bourquin, **Ger.** pat. **2,111,-**

071 corresp to U.S. pat. **3,682,930** (1971, 1972 both to Sandoz); E. Waldvogel *et al., Helv. Chim. Acta* **59**, 866 (1976). Pharmacology: U. Martin, D. Roemer, *Arzneimittel-Forsch.* **28**, 770 (1978). Clinical studies: B. Wüthrick, P. Radielovic, *Deut. Med. Wochenschr.* **103**, 1865 (1978); H. Gmür, M. Scherrer, *Schweiz. Med. Wochenschr.* **109**, 881 (1979). Review: U. Martin *et al.,* in *Pharmacological and Biochemical Properties of Drug Substances* vol. 3, M. E. Goldberg, Ed. (Am. Pharm. Assoc., Washington, DC, 1981) pp 424-460.

Crystals from ethyl acetate, mp 152-153°.
Fumarate, $C_{19}H_{19}NOS.C_4H_4O_4$, *HC-20511, Allerkif, Totifen, Zaditen, Zasten.* Crystals, mp 192° (dec).
THERAP CAT: Antiasthmatic.

5320. Khat. Chat; quat. Leaves of *Catha edulis* Forsk., *Celastraceae;* widely used in E. Africa and Yemen as an amphetamine-like stimulant. Most important constituent is the alkaloid cathinone, *q.v.;* also norpseudoephedrine, cathidine, cathedulin. Isoln and characterization of constituents of khat: O. Wolfes, *Arch. Pharm.* **268**, 81 (1930); H. Friebel, R. Brilla, *Naturwiss.* **50**, 354 (1963); M. Cais *et al., Tetrahedron* **31**, 2727 (1975); R. L. Baxter *et al., Chem. Commun.* **1976**, 463. Pharmacology of khat extracts: G. A. Alles *et al., J. Med. Pharm. Chem.* **3**, 323 (1961). Reviews: A. Getahun, A. D. Krikorian, *Econ. Bot.* **27**, 353-377, 378-389 (1973); P. Kalix, *Gen. Pharmacol.* **15**, 179-187 (1984); P. Nencini, A. M. Ahmed, *Drug Alc. Depend.* **23**, 19-29 (1989).

5321. Khellin. *4,9-Dimethoxy-7-methyl-5H-furo[3,2-g]-[1]benzopyran-5-one;* 5,8-dimethoxy-2-methyl-4′,5′-furo-6,7-chromone; 5,8-dimethoxy-2-methyl-6,7-furanochromone; 4,-9-dimethoxy-7-methyl-5-oxofuro[3,2-g]-1,2-chromene; 4,9-dimethoxy-7-methyl-5-oxofuro[3,2-g][1]benzopyran; 4,9-dimethoxy-7-methyl-5-oxo-1,8-dioxabenz[f]indene; visammin; Kellin; Kelamin; Kelicor; Gynokhellan; Kelicorin; Keloid; Norkel; Simeskellina; Vasokellina; Visnagalin; Visnagen; Methafrone; Eskel; Amicardine; Viscardan; Corafurone; Cardio-Khellin; Benecardin; Ammivisnagen; Khelfren; Lynamine; Coronin; Ammicardine; Ammipuran; Ammivin. $C_{14}H_{12}O_5$; mol wt 260.25. C 64.61%, H 4.65%, O 30.74%. Found in seeds of *Ammi visnaga* Lam., *Umbelliferae* (toothpick ammi; chellah; khella). Isoln and structure: Späth, Gruber, *Ber.* **71**, 106 (1938). Synthesis: Baxter *et al., J. Chem. Soc.* **1949**, S 30; Gardner *et al., J. Org. Chem.* **15**, 841 (1950); Schönberg, Sina, *J. Am. Chem. Soc.* **72**, 1611, 3396 (1950); M. W. Reed, H. W. Moore, *J. Org. Chem.* **53**, 4166 (1988). Crystal structure: J. P. Beale, *Cryst. Struct. Commun.* **2**, 125 (1973). Comprehensive description: M. A. Hassan, M. U. Zubair in *Analytical Profiles of Drug Substances* vol. **9**, K. Florey, Ed. (Academic Press, New York, 1980) pp 371-396.

Crystals from methanol. Bitter taste. mp 154-155°. $bp_{0.05}$ 180-200°. uv max (alc): 250, 338 nm ($E_{1cm}^{1\%}$ 1600, 200). Soly in g/100 ml at 25°: water 0.025; acetone 3.0; methanol 2.6; isopropanol 1.25; ether 0.5; Skellysolve B 0.15. Much more sol in hot water and hot methanol. Stable when mixed with

the usual tabletting excipients. LD_{50} orally in rats: 80 mg/kg.
THERAP CAT: Vasodilator.

5322. Khellol Glucoside. *7-[(β-D-Glucopyranosyloxy)methyl]-4-methoxy-5H-furo[3,2-g][1]benzopyran-5-one;* 2-hydroxymethyl-5-methoxyfuranochrome glucoside; khellinin. $C_{19}H_{20}O_{10}$; mol wt 408.36. C 55.88%, H 4.94%, O 39.18%. From seeds of *Ammi visnaga* Lam., *Umbelliferae:* Späth, Gruber, *Ber.* **74**, 1549 (1941); Illing, *Arzneimittel-Forsch.* **7**, 497 (1957); from *Eranthis hyemalis* L., *Ranunculaceae:* Egger, *Z. Naturforsch.* **16B**, 697 (1962). Structure: Fabbrini, Franchi, *Ann. Chim. (Rome)* **49**, 894 (1959), *C.A.* **54**, 4558f (1960).

Crystals from ethanol, mp 179°. uv max (ethanol): 250, 325 nm. Soluble in acetic acid, hot ethanol; slightly sol in hot methanol. Practically insol in acetone, ethyl acetate, chloroform, ether, cold alkali.
Tetraacetate, flakes from ethanol, mp 153°. Sol in ethanol, acetone, ethyl acetate. Practically insol in petr ether.

5323. Kiku Oil. Obtained by distillation from leaves and flowers of *Chrysanthemum indicum* L., *Compositae.* Produced in Japan: Perrier, *Bull. Soc. Chim. France* [3] **23**, 216 (1900). *Constit. l-*Camphene, camphor carvone, xanthophyll, coumarin, angelic acid esters. Used as a folk remedy in Japan in a manner comparable with the use of chamomile and mint in Europe, but also against intestinal worms.
Colorless or greenish oil. Odor reminiscent of oil of Eucalyptus. d_4^{15} 0.932. n_D^{18} 1.4931. Acid to moist litmus paper. Ten grams dissolves in 100 grams abs alc at 95°. Almost insol in alc at 70°.

5324. Kinetin. *N-(2-Furanylmethyl)-1H-purin-6-amine;* N^6-*furfuryladenine;* 6-furfurylaminopurine. $C_{10}H_9N_5O$; mol wt 215.21. C 55.81%, H 4.22%, N 32.54%, O 7.43%. A cell division factor found in various plant parts and in yeast. Isoln from autoclaved water slurries of deoxyribonucleic acid: Miller *et al., J. Am. Chem. Soc.* **77**, 1392 (1955). Structure and synthesis from furfurylamine and 6-methylmercaptopurine: Miller *et al., ibid.* **78**, 1375 (1956); U.S. pat. **2,903,455** (1959 to Wisc. Alumni Res. Found.). Physiologically active at very great dilutions, but only in presence of auxin, *see* Indoleacetic Acid. Crystal structure: M. Soriano-Garcia, R. Parthasarathy, *Acta Crystallogr.* **33B**, 2674 (1977). *Review:* Miller, *Ann. Rev. Plant Physiol.* **12**, 395 (1961).

Platelets from abs ethanol, mp 266-267° (sealed tube). Sublimes at 220°. pKa_1 2.7; pKa_2 9.9. uv max (ethanol): 268 nm (ε 18,650). Slightly soluble in cold water, methanol, ethanol. Freely soluble in dil aq HCl or NaOH. Can be extracted from neutral aq solns by shaking with ether.
Note: Kinetin is also a brand of hyaluronidase, *q.v.*
USE: Plant growth regulator. To augment growth of microbial cultures: **Belg. pat. 632,589** (1963 to Hoechst).

5325. Kino. Resin kino; gum kino. Dried juice from trunk of *Pterocarpus marsupium* Roxb., *Leguminosae.* Habit. Western Africa, East India, Ceylon. *Constit.* 70-80% kinotannic acid; kino-red, pyrocatechol, kinoin, gum. Contains not less than 45% alcohol-soluble and not less than 80% water-soluble material.

THERAP CAT: Astringent.

THERAP CAT (VET): Has been used as astringent.

5326. Kinoprene. *(E,E)-3,7,11-Trimethyl-2,4-dodeca-dienoic acid 2-propynyl ester;* 2-propynyl *(E,E)*-3,7,11-trimethyl-2,4-dodecadienoate; ENT-70531; ZR-777; Enstar. $C_{18}H_{28}O_2$; mol wt 276.42. C 78.21%, H 10.21%, O 11.58%. Juvenile hormone mimic. Prepn: C. A. Hendrick, **Ger. pat. 2,246,924** (1973 to Zoecon), *C.A.* **79,** 41975w (1973). Prepd not claimed: *idem,* **U.S. pat. 3,833,635** (1974 to Zoecon). Activity: R. A. Hamlen, *J. Econ. Entomol.* **68,** 223 (1975); A. DeLoof *et al., Entomol. Exp. Appl.* **26,** 301 (1979).

Amber liquid, bp$_{0.04}$ 115-116°. Vapor press at 20°: 7.19 × 10⁻⁶ mm Hg. Soly in water: 5.22 ppm. Sol in most organic solvents. LD$_{50}$ orally in rats: 4900 mg/kg, *RTECS* Vol. I, R. J. Lewis, Sr., R. L. Tatken (1979) p 600.

USE: Insect growth regulator.

5327. Kistrin. *Kistrin (Agkistrodon rhodostoma reduced).* Member of a homologous family of low molecular weight, cysteine-rich proteins known as *disintegrins.* Isolated from various viper venoms; characterized by an Arg-Gly-Asp (RGD) binding site and by inhibition of platelet aggregation by antagonism of glycoprotein IIb-IIIa integrin receptors. A single chain polypeptide, mol wt ∼7318, consists of 68 amino acids with six intramolecular disulfide bonds; no regular secondary structure. Isolated from the venom of the Malayan pit viper *Agkistrodon rhodostoma.* Purification, chemical and biological characterization: M. S. Dennis *et al., Proc. Nat. Acad. Sci. USA* **87,** 2471 (1989); R. A. Lazarus, M. S. Dennis, **PCT Int. pat. Appl. 90 15,072** (1990 to Genentech). Preliminary structural study: M. Adler *et al., Science* **253,** 445 (1991). NMR structural study: M. Adler, G. Wagner, *Biochemistry* **31,** 1031 (1992). Enhancement of coronary arterial thrombolysis in dogs: T. Yasuda *et al., Circulation* **83,** 1038 (1991). Binding interaction with glycoprotein IIb-IIIa: M. S. Dennis *et al., Proteins* **15,** 312 (1993). Brief review: R. J. Gould *et al., Proc. Soc. Exp. Biol. Med.* **195,** 168-171 (1990).

5328. Kitol. *[1α,2α,5α(1E,3E),6β(1E,3E,5E)]-3,6-Dimethyl-5-[2-methyl-4-(2,6,6-trimethyl-1-cyclohexen-1-yl)-1,3-butadienyl]-6-[4-methyl-6-(2,6,6-trimethyl-1-cyclohexen-1-yl)-1,3,5-hexatrienyl]-3-cyclohexene-1,2-dimethanol.* $C_{40}H_{60}O_2$; mol wt 572.92. C 83.86%, H 10.56%, O 5.59%. One of the provitamins A. Obtained from mammalian liver oil: Embree, Shantz, *J. Am. Chem. Soc.* **65,** 910 (1943); Clough *et al., Science* **105,** 436 (1947); Barua, Morton, *Biochem. J.* **45,** 309 (1949); Chatan, Fridenson, *Compt. Rend.* **234,** 1094 (1952). Purification: Tawara, Fukazawa, *C.A.* **47,** 12483e (1953). Conversion to vitamin A: Libermann, Grundland, *Compt. Rend.* **224,** 1033 (1947). Structure: Burger *et al., Chem. Commun.* **1965,** 588; Giannotti *et al., ibid.* **1966,** 28; Giannotti *et al., Bull. Soc. Chim. France* **1966,** 3299; Burger, Garbers, *J. Chem. Soc. Perkin Trans. I* **1973,** 590.

Crystals from methanol, mp 88-90° (Embree); 98-100° (Chatan); 72° (Tawara). [α]$_D$ −2.6° (c = 1.1 in chloroform). Labile in light, air, petr ether. Kitol has no vitamin A activity. uv max: 290 nm (E$^{1\%}_{1cm}$ 586).

Diphenylazobenzoate, exists in two forms, probably geometric isomers, one mp 125-126°, the other 153-155°.

5329. Kodel®. Polyester staple and filament fiber. Prepn from dimethyl terephthalate and 1,4-cyclohexane-dimethanol: Kibler *et al.,* **U.S. pat. 2,901,466** (1959 to Eastman-Kodak). Review and structure: R. W. Moncrieff, *Man-Made Fibres* (John Wiley, New York, 4th ed., 1963) pp 389-393.

Solid, mp 290-295°. Burns slowly. Sp gr 1.22. Fiber has good resistance to acids and alkalies. Treatment of fabric with trichloroethylene and methylene chloride causes it to shrink; boiling water and perchloroethylene induce slight shrinkage. Fabric has good crease resistance.

5330. Kojic Acid. *5-Hydroxy-2-(hydroxymethyl)-4H-pyran-4-one;* 5-hydroxy-2-(hydroxymethyl)-4-pyrone; 2-hydroxymethyl-5-hydroxy-γ-pyrone. $C_6H_6O_4$; mol wt 142.11. C 50.71%, H 4.26%, O 45.03%. Antibiotic substance produced in an aerobic process by a variety of microorganisms from a wide range of carbon sources. Isoln from *Aspergillus oryzae:* Saito, *Bot. Mag. Tokyo* **21,** 249 (1907). Structure: Yabuta, *J. Chem. Soc.* **125,** 575 (1924); Heyns, Vogelsang, *Ber.* **87,** 13 (1954). Synthesis: Stacey, Turton, *J. Chem. Soc.* **1946,** 661; Lichtenthaler, Heidel, *Angew. Chem. Int. Ed.* **8,** 978 (1968). Industrial prepn: **Brit. pat. 826,244** (1959 to Pfizer), *C.A.* **54,** 11372e (1960). Mutagenicity study: C. I. Wei *et al., Toxicol. Letters* **59,** 213 (1991). *Review:* Beélik, *Advan. Carbohyd. Chem.* **11,** 145-183 (1956); Wilson, "Miscellaneous *Aspergillus* Toxins" in *Microbial Toxins,* vol. VI, A. Ciegler *et al.,* Eds. (Academic Press, New York, 1971) pp 235-250.

Prismatic needles from acetone, ethanol + ether or methanol + ethyl acetate, mp 153-154°. pKa 7.90, 8.03. Freely sol in water, ethanol, acetone; sparingly sol in ether, ethyl acetate, chloroform, pyridine. Absorption spectrum: Stacey, Turton.

USE: Converted to maltol and ethyl maltol, flavor-enhancing additives. Food additive to inhibit tyrosinase.

5331. Kola. Cola; Soudan coffee; Bissy nuts; gooroo nuts; guru nuts. Dried cotyledons of *Cola nitida* Schott and Endl. or of other species of *Cola, Sterculiaceae. Habit.* West Africa; naturalized in West Indies, India, Ceylon. *Constit.* About 1.5% caffeine; theobromine, kola-red, kolatin; the glucoside kolanin; kolazyme—an enzyme; kola-tannin, glucose, gum.

THERAP CAT: Analeptic.

5332. Kopsine. *3-Hydroxy-22-oxokopsan-1-carboxylic acid methyl ester.* $C_{22}H_{24}N_2O_4$; mol wt 380.44. C 69.46%, H 6.36%, N 7.36%, O 16.82%. From *Kopsia fructicosa* A.D., *Apocynaceae:* Bhattacharya *et al., J. Am. Chem. Soc.* **71,** 3370 (1949). Structure: Spiteller, *Monatsh.* **93,** 1220 (1962); Govindachari *et al., Helv. Chim. Acta* **45,** 1146 (1962); **46,** 572 (1963); Guggisberg *et al., ibid.* **52,** 76 (1969).

Crystals from alc, dec 217-218°. $[\alpha]_D^{27}$ − 14.3° ± 1° (c = 2 in chloroform). uv max (ethanol): 240, 278, 285-286 nm (log ε 4.08, 3.37, 3.35). Sol in chloroform; sparingly sol in methanol, ethanol, ethyl acetate, benzene, ether. Practically insol in petr ether, water.

Methiodide, $C_{23}H_{27}IN_2O_4$, crystals from methanol, dec 194-196°.

Oxalate, $C_{24}H_{26}N_2O_8$, prisms from alcohol + acetone, dec 154°. Sol in water.

5333. Kosin. From flowers of *Hagenia abyssinica* J. J. Gmel. (*Brayera anthelmintica* Kunth.), *Rosaceae*: Leichsenring, *Arch. Pharm.* **232**, 54 (1894). Occurs as a mixture of α- and β-kosin: Lobeck, *ibid.* **239**, 672 (1901); Hems, Todd, *J. Chem. Soc.* **1937**, 562. Structure of α-kosin: Riedl, *Ber.* **89**, 2600 (1956). Synthesis of α-kosin and probable structure of β-kosin: Orth, Riedl, *Ann.* **663**, 83 (1963).

α-Kosin   R = CH₃   R' = H
β-Kosin   R = H   R' = CH₃

α-Kosin, $C_{25}H_{32}O_8$, 5,5′-*methylenebis[4,6-dihydroxy-2-methoxy-3-methylisobutyrophenone]*. Yellow needles from ethanol, mp 160-160.5°. uv max: 227, 290 nm (ε 30800, 24400). Soluble in alcohol, benzene, chloroform, ether, glacial acetic acid, alkalies.

Tetraacetate, $C_{33}H_{40}O_{12}$, needles from benzene + hexane, mp 124°.

β-Kosin. Yellow prisms from methanol, mp 120°. uv max: 228, 292 nm (ε 30300, 21260).

THERAP CAT: Has been used as anthelmintic (Cestodes).

5334. Krebiozen. A white powder "chemically separated from horses' serum after stimulation of their cell network by the injection of *Actinomyces bovis.*" Claimed to be a lipopolysaccharide. Developed by Stevan Durovic. Used experimentally in the treatment of cancer: Szujewski, *J. Am. Med. Assoc.* **148**, 929 (1952). Quackery Congress Symposium: James F. Holland, "The Krebiozen Story", *J. Am. Med. Assoc.* **200**, 213-218 (1967). *See also*: W. F. Janssen, *Anal. Chem.* **50**, 197A (1978); G. A. Curt, "Unsound Methods of Cancer Treatment" in *Cancer: Principles and Practice of Oncology*, V. T. DeVita, Jr. *et al.*, Eds. (J. B. Lippincott, Philadelphia, 4th ed., 1993) pp 2734-2747.

5335. Krypton. Kr; at. wt 83.80; at. no. 36. Group VIIIA(18), also known as Group 0. A noble gas characterized by an electronic structure in which the outer *p* subshell is entirely filled. Naturally occurring stable isotopes (mass numbers): 78 (0.35%); 80 (2.25%); 82 (11.6%); 83 (11.5%); 84 (57.0%); 86 (17.3%); known artificial radioactive isotopes: 71-77; 79; 81; 85; 87-95; 97. Isoln from the final residues obtained after evaporation of liq air: Ramsay, Travers, *Proc. Roy. Soc.* **63 [A]**, 405 (1898). Concentration in air: 1.14 ppm by vol. Obtained commercially from the atmosphere by distillation-liquefaction process. Prepn: Lepape, *Compt. Rend.* **187**, 231 (1928); Claude, *ibid.* 581. Monograph: *Argon, Helium and the Rare Gases,* vols. 1, 2, G. A. Cook, Ed. (Interscience, New York, 1963) 818 pp. Diagnostic use of ⁸¹ᵐKr in lung ventilation imaging: A.

Zwijnenburg *et al., Clin. Phys. Physiol. Meas.* **9**, 147 (1988). *Reviews*: A. H. Cockett, K. C. Smith, "The Monatomic Gases" in *Comprehensive Inorganic Chemistry* vol. 1, J. C. Bailar, Jr. *et al.*, Eds. (Pergamon Press, Oxford, 1973) pp 139-211; N. Bartlett, F. O. Sladky, "The Chemistry of Krypton, Xenon and Radon" *ibid.* pp 213-249; S.-C. Hwang, W. R. Weltmer, Jr. in Kirk Othmer *Encyclopedia of Chemical Technology* vol. 13 (John Wiley & Sons, 4th ed., 1995) pp 1-38; G. J. Schrobilgen, J. M. Whalen, *ibid.* pp 38-53; *Chemistry of the Elements,* N. N. Greenwood, A. Earnshaw, Eds. (Pergamon Press, New York, 1984) pp 1042-1059.

Colorless, odorless, inert, monatomic gas; will form compds with highly electronegative elements such as O, F, Cl. Condenses to a colorless liquid. Triple pt temp 115.95°K, press 73.15 kPa. Critical temp 209.4°K, critical press 5502 kPa, critical d 908 kg/m³. Gas: d⁰ (101.3 kPa) 3.7493 kg/m³, d (normal bp) 8.6 kg/m³. Liquid: normal bp 119.8°K, d (normal bp) 2415 kg/m³, d (triple pt) 2451 kg/m³, heat of vaporization (normal bp) 9.050 kJ/mol. Solid: d (triple pt) 2826 kg/m³, heat of vaporization (triple pt) 10.77 kJ/mol, heat of fusion (triple pt) 1.64 kJ/mol; solid form exists as face-centered cubic crystals at normal pressure. Excitation potentials: 9.91; 10.03; 10.56; 10.64 ev; ionization potentials: 14.00; 14.66 ev: Brocklehurst, *Quart. Rev.* **22**, 147 (1968). Soly of gas in water (20°): 59.4 cm³/kg water. Slightly soluble in water with formation of a hydrate; ideal formula Kr.5.75H₂O. A hexadeuterate has been prepd. Soluble in liquid oxygen. Emission spectra: T. Jacksier, R. M. Barnes, *Appl. Spectrosc.* **48**, 65 (1994).

Krypton difluoride, F₂Kr. Prepn: Turner, Pimentel, *Science* **140**, 974 (1963). Colorless solid; decomposes rapidly at room temperature. d (calc) 3.24. Vapor pressure (mm Hg): 10 ±1 at − 15.5°; 29 ±2 at 0°; 73 ±3 at 15°.

Caution: Can act as a simple asphyxiant by displacing air. *See: Matheson Gas Data Book* (Matheson Co., Inc., 4th ed., East Rutherford, NJ, 1966) pp 313-314.

USE: As gas filler in special purpose electric bulbs. To initiate and sustain the arc in metal-vapor discharge tubes, such as mercury and sodium vapor lamps. To determine surface areas of fine solids by adsorption techniques.

THERAP CAT: ⁸¹ᵐKr as diagnostic aid (radioactive imaging agent).

5336. Kurchessine. *(3β,20S)-N,N,N′,N′-Tetramethyl-pregn-5-ene-3,20-diamine; 3β,20α-bis(dimethylamino)-5-pregnene.* $C_{25}H_{44}N_2$; mol wt 372.64. C 80.58%, H 11.90%, N 7.52%. A steroidal alkaloid from Kurchi bark. Isoln from *Holarrhena antidysenterica* Wall., *Apocynaceae*: Tschesche, Otto, *Ber.* **95**, 1144 (1962); Labler, Sorm, *Coll. Czech. Chem. Commun.* **28**, 2345 (1963). Synthesis by methylation of kurchamine: Tschesche, Wiensz, *Ber.* **91**, 1504 (1958). Structure: Labler, Sorm, *loc. cit.* Possible identity with *sarcodinine* from *Sarcococca pruniformis* Lindl., *Euphorbiaceae*: Chatterjee *et al., Tetrahedron Letters* **1965**, 67. Related alkaloids, *kurchiline, kurchiphylline, kurchiphyllamine,* from Kurchi leaves: Janot *et al., Bull. Soc. Chim. France* **1964**, 2158.

Needles. mp 140-141° (from acetone). $[\alpha]_D^{20}$ −37° (c = 1.9); $[\alpha]_D^{22}$ −36° (c = 1.112 in CHCl₃); $[\alpha]_D^{22}$ −17° (c = 1.050 in methanol).

5337. Kurcholessine. *(3β,4α,5α,7β)-3-(Dimethylamino)-4-methyl-5-conanine-5,7-diol.* $C_{25}H_{44}N_2O_2$; mol wt 404.64. C 74.21%, H 10.96%, N 6.92%, O 7.91%. A steroidal alkaloid from Kurchi bark. Isoln from *Holarrhena anti-*

dysenterica Wall., *Apocynaceae:* Tschesche, Otto, *Ber.* **95,** 1144 (1962). Structure: Tschesche *et al.*, *Tetrahedron Letters* **1964,** 1659.

Slender needles from ethanol; microcryst powder from acetone. mp 218.5-221.5°. $[\alpha]_D^{20}$ −4° ±2° (CHCl₃), +5° ±2° (c = 0.01, abs ethanol). Very sol in chloroform, methanol, ethanol, pyridine; slightly sol in warm acetone. Practically insol in ethyl acetate, benzene, petr ether, ether, water.

5338. Kyanmethin. *2,6-Dimethyl-4-pyrimidinamine; 4-amino-2,6-dimethylpyrimidine;* 6-amino-2,4-dimethylpyrimidine. $C_6H_9N_3$; mol wt 123.16. C 58.52%, H 7.37%, N 34.12%. Prepn from acetonitrile and potassium methoxide: Ronzio, Cook, *Org. Syn. coll. vol.* **III,** 71 (1955).

Needles from alcohol or scales from benzene, mp 182-183°. One gram dissolves in 0.64 ml water at 18° or in 5.25 parts alcohol at 18°.

5339. Kynurenic Acid. *4-Hydroxy-2-quinolinecarboxylic acid; 4-hydroxyquinaldic acid.* $C_{10}H_7NO_3$; mol wt 189.17. C 63.49%, H 3.73%, N 7.40%, O 25.37%. Found in the urine of some animals as a metabolic product of tryptophan; its excretion is stepped up in avitaminoses B_1, B_2 and B_6. Isoln and syntheses: Späth, *Monatsh.* **42,** 89 (1921); Besthorn, *Ber.* **54,** 1330 (1921). Alternate syntheses: Benassi, *Gazz. Chim. Ital.* **91,** 1097 (1961); Wald, Joullie, *J. Org. Chem.* **31,** 3369 (1966); Jordanides, *Ann.* **729,** 244 (1969).

Yellow needles, mp 282-283°. Soly in water: about 0.9% at 100°. Sol in hot alc. Insol in ether.

Methyl ester, $C_{11}H_9NO_3$, yellow crystals, mp 224°. *Cf.* xanthurenic acid.

USE: In nutrition studies, specifically in vitamin B deficiency diseases.

5340. Kynurenine. *α,2-Diamino-γ-oxobenzenebutanoic acid;* 3-anthraniloylalanine. $C_{10}H_{12}N_2O_3$; mol wt 208.22. C 57.69%, H 5.81%, N 13.45%, O 23.05%. An amino acid produced in the body from tryptophan. Isoln from urine of rabbits that had been fed tryptophan: Matsuoka, Yoshimatsu, *Z. Physiol. Chem.* **143,** 206 (1925); Butenandt *et al., ibid.* **279,** 27 (1943); Heidelberger *et al., J. Biol. Chem.* **179,** 143 (1949). Structure and synthesis: Butenandt *et al., loc. cit.* Laboratory prepn by oxidation of L-tryptophan with a *Pseudomonas* sp.: Hayaishi, Meister, *Biochem. Prepn.* **3,** 108 (1953). From acetyltryptophan: Warnell, Berg, *J. Am. Chem. Soc.* **76,** 1708 (1954); Auerbach, Knox, *Methods Enzymol.* **3,** 620 (1957).

L-Kynurenine hydrate, $3C_{10}H_{12}N_2O_3 \cdot H_2O$, leaflets from water, dec 180-190°. $[\alpha]_D^{20}$ −29° (c = 0.4). Slightly sol in water (more sol than the DL-form). Forms a molecular compound with sucrose, $C_{22}H_{34}N_2O_{14} \cdot H_2O$, rosettes; dec 145-153°, $[\alpha]_D^{21}$ +14.5° (c = 0.7). Soluble in water.

L-Kynurenine sulfate monohydrate, $C_{10}H_{12}N_2O_3 \cdot H_2SO_4 \cdot H_2O$, needles from water + alcohol. Darkens at 165°, dec 195°. $[\alpha]_D^{20}$ +7.3° (c = 1). uv max (pH 7.0): 230, 257, 360 nm (ε 18,900, 7500, 4500). Soluble in water, slightly in alcohol. Kynurenine sulfate requires 3 molecules alkali for neutralization in alcoholic soln.

L-Kynurenine diacetate, $C_{14}H_{16}N_2O_5$, obtained by acetylating L-kynurenine with ketene. Needles, mp 198°.

Anhydro-L-kynurenine monoacetate, $C_{12}H_{12}N_2O_3$, obtained by acetylating L-kynurenine with acetic anhydride in pyridine. Needles from alc, darkens at 215°, dec 237°.

DL-Kynurenine sulfate, $C_{10}H_{12}N_2O_3 \cdot H_2SO_4$, crystals from water + alcohol. Darkens at 166°, dec 194°. Soluble in water. Slightly sol in alcohol.

USE: In biochemical investigations.

L

5341. Labetalol. *2-Hydroxy-5-[1-hydroxy-2-[(1-methyl-3-phenylpropyl)amino]ethyl]benzamide; 5-[1-hydroxy-2-[(1-methyl-3-phenylpropyl)amino]ethyl]salicylamide;* ibidomide. $C_{19}H_{24}N_2O_3$; mol wt 328.41. C 69.49%, H 7.37%, N 8.53%, O 14.62%. A specific competitive antagonist at both α- and β-adrenergic receptor sites. Prepn: L. H. Lunts, D. T. Collin, Ger. pat. 2,032,642; *eidem*, U.S. pat. 4,012,444 (1971, 1977 both to Allen & Hanburys); J. E. Clifton *et al.*, *J. Med. Chem.* **25**, 670 (1982). Pharmacology: J. B. Farmer *et al.*, *Brit. J. Pharmacol.* **45**, 660 (1972). Clinical studies: J. G. Collier *et al.*, *ibid.* **44**, 286 (1972); D. A. Richards *et al.*, *Brit. J. Clin. Pharmacol.* **1**, 505 (1974). Metabolism in animals and man: R. Hopkins *et al.*, *Biochem. Soc. Trans.* **4**, 726 (1976). HPLC determn in serum: T. F. Woodman, B. Johnson, *Ther. Drug Monit.* **3**, 371 (1981). Synthesis of stereoisomers and comparison of cardiovascular properties: *J. Med. Chem.* **25**, 1363 (1982). Series of articles on pharmacology, metabolism and clinical studies: *Brit. J. Clin. Pharmacol.* **3**, Suppl. 3, 681S-824S (1976); *ibid.* **8**, Suppl. 2, 85S-244S (1979); *ibid.* **13**, Suppl. 1, 1S-141S (1982). Toxicity: K. Shimpo *et al.*, *Hokkaido Igaku Zasshi* **53**, 15 (1978), *C.A.* **90**, 66465v (1974). *Review:* R. T. Brittain *et al.* in *Pharmacological and Biochemical Properties of Drug Substances* vol. 2, M. E. Goldberg, Ed. (Am. Pharm. Assoc., Washington, DC, 1979) pp 229-254.

Hydrochloride, $C_{19}H_{24}N_2O_3 \cdot HCl$, *AH-5158A, Sch-15719W, Amipress, Ipolab, Labelol, Labrocol, Normodyne, Presdate, Pressalolo, Trandate.* White crystalline solid from ethanol-ethyl acetate, mp 187-189°. Sol in water, ethanol. Insol in ether, chloroform. LD_{50} in male, female mice, male, female rats (mg/kg): 114, 120, 113, 107 i.p.; 47, 54, 60, 53 i.v.; 1450, 1800, 4550, 4000 orally (Shimpo).

(R,R)-Isomer, *see* Dilevalol.

THERAP CAT: Antihypertensive.

5342. Laccaic Acid. *C.I. Natural Red 25;* lac dye; C.I. 75450. Pigment found in the lac resin produced by the insect *Coccus laccae* (*Laccifer lacca* Kerr) on certain trees in India. Isoln: R. E. Schmidt, *Ber.* **20**, 1285 (1887). Chemistry: O. Dimroth, S. Goldschmidt, *Ann.* **399**, 62 (1913). Originally thought to be one compound, laccaic acid has been separated into four components: A (major), B, C and D. Isoln of A: R. Burwood *et al.*, *J. Chem. Soc. (C)* **1965**, 6067. Structure of A: *eidem*, *Tetrahedron Letters* **1966**, 3059; *eidem*, *J. Chem. Soc. (C)* **1967**, 842; E. D. Pandhare *et al.*, *Indian J. Chem.* **7**, 977 (1969). Isoln and structure of B: *eidem*, *Tetrahedron Letters* **1967**, 2437; N. S. Bhide *et al.*, *Indian J. Chem.* **7**, 987 (1969); of C: A. V. R. Rao *et al.*, *ibid.* 188; of D: A. R. Mehandale *et al.*, *Tetrahedron Letters* **1968**, 2231. Synthesis of D: D. W. Cameron *et al.*, *Chem. Commun.* **1978**, 688; *eidem*, *Aust. J. Chem.* **34**, 2401 (1981).

Laccaic Acid A　R = CH₂CH₂NHCOCH₃
Laccaic Acid B　R = CH₂CH₂OH
Laccaic Acid C　R = CH₂CH(NH₂)COOH

Laccaic Acid D

Laccaic acid A, $C_{26}H_{19}NO_{12}$, *laccaic acid A₁.* Red platelets from methanol, chars at 230°. Absorption max (conc H_2SO_4): 302, 361, 518, 558 nm (log ϵ 4.33, 4.11, 4.32, 4.37).

Laccaic acid B, $C_{24}H_{16}O_{12}$, red needles from methanol.

Laccaic acid C, $C_{25}H_{17}NO_{13}$, dark red needles from methanol, dec > 360°.

Laccaic acid D, $C_{16}H_{10}O_7$, *xanthokermesic acid.* Yellow needles from water, dec > 300°.

USE: As crimson dye.

5343. Lachesine. *N-Ethyl-2-[(hydroxydiphenylacetyl)-oxy]-N,N-dimethylethanaminium chloride; ethyl(2-hydroxyethyl)dimethylammonium chloride benzilate;* E-3. $C_{20}H_{26}-ClNO_3$; mol wt 363.88. C 66.02%, H 7.20%, Cl 9.74%, N 3.85%, O 13.19%. Prepn: Ford-Moore, Ing, *J. Chem. Soc.* **1947**, 55. Muscarinic receptor agonist: F. Roberts, R. P. Stephenson, *Brit. J. Pharmacol.* **57**, 395 (1976); D. A. Brown, S. Fatherazi, *ibid.* **59**, 500P (1977); N. J. M. Birdsall *et al.*, *Mol. Pharmacol.* **14**, 723 (1978).

Crystals from ethanol + acetone, mp 213°. LD_{50} in mice: 0.8 mg/20 g i.p.; 3.2 mg/20 g s.c.; 20.0 mg/20 g orally.

5344. Lacidipine. *(E)-4-[2-[3-(1,1-Dimethylethoxy)-3-oxo-1-propenyl]phenyl]-1,4-dihydro-2,6-dimethyl-3,5-pyridinedicarboxylic acid diethyl ester;* diethyl *(E)*-4-[2-[2-(tert-butoxycarbonyl)vinyl]phenyl]-2,6-dimethyl-1,4-dihydropyridine-3,5-dicarboxylate; 4-[o-[(E)-2-carboxyvinyl]phenyl]-1,-4-dihydro-2,6-dimethyl-3,5-pyridinedicarboxylic acid 4-tert-butyl diethyl ester; GR-43659X; GX-1048; Caldine; Lacipil; Lacirex; Motens. $C_{26}H_{33}NO_6$; mol wt 455.55. C 68.55%, H 7.30%, N 3.07%, O 21.07%. Dihydropyridine calcium channel blocker. Prepn: C. Semeraro *et al.*, Ger. pat. 3,529,997; *eidem*, U.S. pat. 4,801,599 (1986, 1989 both to Glaxo). Clinical pharmacology: M. Safar *et al.*, *Clin. Pharmacol. Ther.* **46**, 94 (1989).

Crystalline solid from ethyl acetate, mp 174-175°.

THERAP CAT: Antihypertensive.

5345. Lacmoid. Resorcin blue; Iris Blue B; Fluorescent Blue; C.I. 51400. Prepn: Traub, Hock, *Ber.* **17**, 2615 (1884); Musso *et al.*, *Angew. Chem.* **73**, 434 (1961). Discussion of structure controversy: H. J. Conn's Biological Stains, R. D. Lillie, Ed. (Williams & Wilkins, Baltimore, 8th ed., 1969) pp 283-284. *See also: Colour Index* vol. 4 (3rd ed., 1971) p 4468.

Dark-violet, lustrous scales or granules. Freely sol in methanol, ethanol, amyl alcohol, glacial acetic acid, acetone, phenol. Sparingly sol in water, ether. Practically insol in chloroform, benzene, petr ether.

USE: As acid-base indicator in 0.2% soln in alcohol. pH: 4.4 red; 6.4 blue. Satisfactory for titrating mineral acids, strong bases, many alkaloids; determining alkalinity and temporary hardness in water analysis. Biological stain; dye for wool, silk. Not adapted for carbonates, weak inorganic and organic acids, weak bases. Lacmoid is more sensitive than litmus, particularly in form of test paper.

5346. Lactaroviolin. *4-Methyl-7-(1-methylethenyl)-1-azulenecarboxaldehyde; 7-isopropenyl-4-methyl-1-azulenecar-*

boxaldehyde; 1-formyl-4-methyl-7-isopropenylazulene. C_{15}-$H_{14}O$; mol wt 210.28. C 85.68%, H 6.71%, O 7.61%. Antibiotic pigment produced by the fungus *Lactarius deliciosus:* Willstaedt, *Ber.* **68**, 333 (1935); **69**, 997 (1936); Benesová *et al., Chem. & Ind. (London)* **1954**, 1202; Sorm *et al.*, ibid. 1511; Heilbron, Schmid, *Helv. Chem. Acta* **37**, 2018 (1954).

Purple crystals from petr ether, mp 58°. Sol in the usual organic solvents. Practically insol in water.

5347. Lactate Dehydrogenase. Lactic dehydrogenase; serum lactic dehydrogenase. Enzyme found in almost all animal tissues, in microorganisms, and in plants. Isoln from heart muscle, rat skeletal muscle, and Jensen sarcoma: Straub, *Biochem. J.* **34**, 483 (1940); Kubowitz, Ott, *Biochem. Z.* **314**, 94 (1943); Meister, *Biochem. Prepn.* **2**, 18 (1952). Catalyzes the equilibrium reaction of pyruvic acid to lactic acid. Plays an important role in the equilibrium of carbohydrate catabolism and anabolism. Used as fuel for aerobic tissues such as the heart. Structure is a tetramer of mol wt about 140,000. Consists of units of mol wt about 35,000. Two types of subunits are distinguishable: M (muscle) type and H (heart) type. Lactate dehydrogenases of heart and muscle are mainly H_4 and M_4; all other possible hybrids have been found in various tissues. Elevations of lactate dehydrogenase activity have been found in myocardial infarction, hepatocellular necrosis, metastatic carcinoma, diabetic ketosis, sickle cell anemia, malignant lymphoma, infectious mononucleosis, and cerebral infarction: Standjord *et al., J. Am. Med. Assoc.* **182**, 1099 (1962). Comprehensive reviews: Everse, Kaplan, *Advan. Enzymol. Relat. Areas Mol. Biol.* **37**, 61 (1973); Holbrook *et al.*, in *The Enzymes,* vol. XI (part A), P. D. Boyer, Ed. (Academic Press, New York, 3rd ed., 1975) pp 191-292.
USE: In the determination of pyruvate (used in conjunction with reduced coenzyme). In the diagnosis of myocardial infarction and leukemia.

5348. D-Lactic Acid. *(R)-2-Hydroxypropanoic acid;* D(−)-lactic acid; levorotatory lactic acid; *l*-lactic acid; D-Milchsäure (German). $C_3H_6O_3$; mol wt 90.08. C 40.00%, H 6.71%, O 53.28%. Obtained by resolution of DL-lactic acid: Purdie, Walker, *J. Chem. Soc.* **61**, 754 (1892); Borsook *et al., J. Biol. Chem.* **102**, 449 (1933). Convenient laboratory prepn from glucose using *Lactobacillus leichmannii:* Brin, *Biochem. Prepn.* **3**, 61 (1953).

Crystals from ether + isopropyl ether, mp 52.8°. $[\alpha]_{546}^{21.5}$ −2.6° (c = 8). pK = 3.83. Sol in water, alcohol, acetone, ether, glycerol. Practically insol in chloroform. Forms salts with many metals. Most of these salts are dextrorotatory.
Zinc D(+)-lactate, $Zn(C_3H_5O_3)_2.2H_2O$, crystals, $[\alpha]_D^{14}$ +8.18° (c = 2.5).

5349. DL-Lactic Acid. *2-Hydroxypropanoic acid;* racemic lactic acid; ordinary lactic acid; α-hydroxypropionic acid; Milchsäure (German); Lactovagan; Tonsillosan (Lösung). $C_3H_6O_3$; mol wt 90.08. C 40.00%, H 6.71%, O 53.28%. Occurs in sour milk as a result of lactic acid bacteria; also found in molasses due to partial conversion of sugars, in apples and other fruits, tomato juice, beer, wines, opium, ergot, foxglove, and several higher plants, especially during germination. Lactic acid is prepd technically by "lactic acid fermentation" of carbohydrates such as glucose, sucrose, lactose with *Bacillus acidi lacti* or related organisms

such as *Lactobacillus delbrueckii, L. bulgaricus* etc. The fermentation is carried out at relatively high temps. Produced commercially by fermentation of whey, cornstarch, potatoes, molasses. Review on the production of lactic acid by fermentation: S. C. Prescott, C. G. Dunn, *Industrial Microbiology* (McGraw-Hill, New York, 3rd ed., 1959) pp 304-331. Chem prepns from acetaldehyde and CO in dil H_2SO_4 at 130-200° and 900 atm: Loder, U.S. pat. **2,265,945** (1938 to du Pont); by hydrolysis of hexoses with NaOH: Lock, U.S. pat. **2,382,889** (1943). Prepn of crystalline lactic acid: Borsook *et al., J. Biol. Chem.* **102**, 449 (1933). Toxicity data: Smyth *et al., J. Ind. Hyg. Toxicol.* **23**, 259 (1941).

Crystals, mp 16.8°. bp_{14-15} 122°; $bp_{0.5-1}$ 82-85°. K at 25° 1.38×10^{-4}. Heat of combustion at constant pressure 3615 cal/kg. Volatile with superheated steam. Sol in water, alc, furfurol; less sol in ether. Practically insol in chloroform, petr ether, carbon disulfide. *Pharm. Incompat:* Oxidizing agents, iodides, HNO_3, albumin. Caustic in concd solns. LD_{50} orally in rats: 3.73 g/kg (Smyth).
Barium salt, $C_6H_{10}BaO_6$, *barium lactate.* Powder. *Poisonous!* Sol in water, dil alcohol.
Copper salt dihydrate, $C_6H_{10}CuO_6.2H_2O$, *cupric lactate.* Green to blue crystals. Readily sol in water. Practically insol in alcohol.
USE: In dyeing baths, as mordant in printing woolen goods, solvent for water-insoluble dyes (alcohol-soluble induline, nigrosine, spirit-blue); reducing chromates in mordanting wool; manuf cheese, confectionery; acidulant in beverages; for acidulating worts in brewing, for removing *Clostridium butyricum* in manuf of yeast; dehairing, plumping, and decalcifying hides; solvent for cellulose formate; flux for soft solder; manuf lactates which are used in food products, in medicine, and as solvents; plasticizer, catalyst in the casting of phenolaldehyde resins.
THERAP CAT: Acidulant.
THERAP CAT (VET): Has been used as a caustic, and in dilute solutions to irrigate tissues; as an intestinal antiseptic and antifermant.

5350. L-Lactic Acid. *(S)-2-Hydroxypropanoic acid;* L(+)-lactic acid; dextrorotatory lactic acid; *d*-lactic acid; sarcolactic acid; paralactic acid; Fleischmilchsäure; L-Milchsäure. $C_3H_6O_3$; mol wt 90.08. C 40.00%, H 6.71%, O 53.28%. Occurs in small quantities in the blood and muscle fluid of man and animals. The lactic acid concn increases in muscle and blood after vigorous activity. L(+)-Lactic acid is also present in liver, kidney, thymus gland, human amniotic fluid, and other organs and body fluids. Obtained by resolution of DL-lactic acid: Purdie, Walker, *J. Chem. Soc.* **61**, 754 (1892); Borsook *et al., J. Biol. Chem.* **102**, 449 (1933). Convenient laboratory prepn from glucose by fermentation by *Lactobacillus delbrueckii:* Brin, *Biochem. Prepn.* **3**, 61 (1953). Prepn from hexoses using *B. dextrolacticus:* Andersen, Greaves, *Ind. Eng. Chem* **34**, 1522 (1942). Monograph: M. Brin, R. H. Dunlop, "Chemistry and Metabolism of L-and D-Lactic Acids", *Ann. N.Y. Acad. Sci.* vol. **119**, art. 3, 851-1165 (1965).

Crystals from acetic acid or chloroform, mp 53°. $[\alpha]_{546.1}^{21-22}$ +2.6° (c = 2.5). pK at 25°, 3.79. Forms salts with many metals. The salts are more sol in water than the salts of the racemic acid. Most of the salts are levorotary.
Zinc L(−)-lactate dihydrate, $Zn(C_3H_5O_3)_2.2H_2O$, prisms. $[\alpha]_D^{25}$ −8.2° (c = 2.5 in water).

5351. Lactic Acid Lactate. *2-Hydroxypropanoic acid 1-carboxyethyl ester;* 2-(lactoyloxy)propanoic acid; 2-(2-hydroxypropanoyloxy)propanoic acid. $C_6H_{10}O_5$; mol wt 162.14.

C 44.45%, H 6.22%, O 49.34%. Prepd by heating lactic acid at 120° for 10 hours: Dietzel, Krug, *Ber.* **58**, 1307 (1925).

Pale yellow, clear, odorless oil. Sol in water and in the usual organic solvents.

Methyl ester, $C_7H_{12}O_5$. Prepn: Claborn, U.S. pat. **2,371,-281** (1945 to the people of the U.S.). $bp_{7.8}$ 107°; n_D^{20} 1.4313. USE: The methyl ester as a solvent or plasticizer.

5352. Lactitol. *4-O-β-D-Galactopyranosyl-D-glucitol;* *β*-galactoside sorbitol; lactit; lactit M; lactite; lactobiosit; lactosit; lactositol; $C_{12}H_{24}O_{11}$; mol wt 344.32. C 41.86%, H 7.03%, O 51.11%. Polyol sweetner; relative sweetness compared to sucrose is 36%. Prepd by hydrogenation of lactose, *q.v.*: M. J. B. Senderens, *Compt. Rend.* **170**, 47 (1920); M. L. Wolfrom *et al.*, *J. Am. Chem. Soc.* **60**, 571 (1938). Pharmacology: D. H. Patil *et al.*, *Brit. J. Nutr.* **57**, 195 (1987). Crystal structure: J. A. Kanters *et al.*, *Acta Crystallogr.* **C46**, 2408 (1990); J. Kivikoski *et al.*, *Carbohyd. Res.* **223**, 45 (1992). Toxicology: E. J. Sinkeldam *et al.*, *J. Am. Coll. Toxicol.* **11**, 165 (1992). Clinical trial in chronic hepatic encephalopathy: O. Riggio *et al.*, *Hepatogastroenterology* **37**, 524 (1990); as a laxative: L. Goovaerts, G. P. Ravelli, *Acta Ther.* **19**, 61 (1993). Review of properties and applications: J. A. van Velthuijsen, *J. Agric. Food Chem.* **27**, 680-686 (1979); of chemistry and use in foods: C. H. den Uyl, *Dev. Sweeteners* **3**, 65-81 (1987).

Crystals from absolute ethanol, mp 146°. $[\alpha]_D^{23}$ +14° (c = 4 in water). Sol in water, dimethyl sulfoxide, *N,N*-dimethylformamide; slightly sol in ethanol, ether. Strongly hygroscopic.

Monohydrate, *Importal, Portolac*. White, sweet, odorless, crystalline solid. Non-hygroscopic. mp 94-97° (van Velthuijsen), water of crystallization evaporates 145°-185°; also reported as mp 120° (den Uyl). $[\alpha]_D^{22}$ +12.3°. Soly at 25° (g/100 g solvent): water 206; ethanol 0.75; ether 0.4; DMSO 233; DMF 39; at 50°: water 512; ethanol 0.88; at 75°: water 917.

Dihydrate, *Lacty*. White, sweet, odorless, crystalline powder. Data for food grade, mp 75°. $[\alpha]_D^{25}$ +13.5° - 15.0°. pH of 10% solution 4.5 - 8.5. 140 g will dissolve in 100 ml water at 25°.

USE: Sweetener in food.

THERAP CAT: Laxative. In treatment of hepatic encephalopathy.

5353. Lactobacillic Acid. *2-Hexylcyclopropanedecanoic acid;* 11,12-methyleneoctadecanoic acid; phytomonic acid. $C_{19}H_{36}O_2$; mol wt 296.49. C 76.97%, H 12.24%, O 10.79%. A lipid constituent of various microorganisms. Isoln from *Lactobacillus arabinosus*: K. Hofmann, R. A. Lucas, *J. Am. Chem. Soc.* **72**, 4328 (1950); from *Agrobacterium tumefaciens* and identity of phytomonic acid with lactobacillic acid: K. Hofmann, F. Tausig, *J. Biol. Chem.* **213**, 425 (1955). Structure: K. Hofmann *et al.*, *J. Am. Chem. Soc.* **80**, 5717 (1958). Abs config: J. F. Tocanne, *Tetrahedron* **28**, 363 (1972).

Crystals from acetone, mp 27.8-28.8°. Soluble in acetone, petr ether.

Methyl ester, $C_{20}H_{38}O_2$, liq, bp_3 187-187.5°. Soluble in many fat solvents.

Amide, $C_{19}H_{37}NO$, *lactobacillamide*. Crystals, mp 79.4-81.5°. Soluble in dimethylformamide.

5354. Lactobionic Acid. *4-O-β-D-Galactopyranosyl-D-gluconic acid;* 4-(β-D-galactosido)-D-gluconic acid. $C_{12}H_{22}O_{12}$; mol wt 358.30. C 40.23%, H 6.19%, O 53.58%. Obtained by oxidation of lactose: Fischer, Meyer, *Ber.* **22**, 362 (1889); Ruff, Ollendorff, *ibid.* **33**, 1806 (1900); Isbell, *J. Res. NBS* **11**, 713 (1933); Margariello, U.S. pat. **2,746,916** (1956 to Nat. Dairy Res. Labs.); Eddy, *Nature* **181**, 904 (1958); Nishizuka *et al.*, *J. Biol. Chem.* **235**, PC13 (1960). Manuf from lactose: Y. Sato *et al.*, **Ger.** pat. **2,038,230** (1971 to Hayashibara Co.), *C.A.* **74**, 142296c (1971). Crystal structure of calcium salt: W. J. Cook, C. E. Bugg, *Acta Crystallogr.* **B29**, 215 (1973). NMR studies: T. Taga *et al.*, *Bull. Chem. Soc. Japan* **51**, 2278 (1978). For therapeutic use *see* Erythromycin Lactobionate.

Syrup. Freely sol in water, slightly sol in methanol, ethanol, glacial acetic acid. Dehydration by distillation with dioxane yields *lactobionic δ-lactone*, $C_{12}H_{20}O_{11}$, non-deliquescent crystals, dec 195-196°. Shows mutarotation. $[\alpha]_D^{20}$ +53.0° initial (c = 8.8) → $[\alpha]_D^{20}$ +22.6° final (240 minutes).

Calcium salt, $C_{24}H_{42}CaO_{24}$, *calcium lactobionate*. Pentahydrate, hairlike needles in brushlike groups. When anhydr, slender needles from small amts of anhydr ethanol. $[\alpha]_D^{20}$ +23.7° (c = 6.28). n_D^{20} 1.4583 (concd syrup just before crystallization). Freely sol in water.

5355. p-Lactophenetide. *N-(4-Ethoxyphenyl)-2-hydroxypropanamide; p-lactophenetidide;* lactyl-p-phenetidin-*N*-(p-ethoxyphenyl)lactamide; Fenolactine; Lactophenin; Phenolactine. $C_{11}H_{15}NO_3$; mol wt 209.25. C 63.14%, H 7.23%, N 6.69%, O 22.94%. Prepn: Shapiro *et al.*, *J. Am. Chem. Soc.* **81**, 6322 (1959).

Slightly bitter crystals from ethyl acetate + hexane, mp 117-118°. One gram dissolves in 330 ml cold, 55 ml boiling water, 8.5 ml alcohol; slightly sol in ether, petr ether.

THERAP CAT: Analgesic, antipyretic.

5356. Lactose. *4-O-β-D-Galactopyranosyl-D-glucose;* 4-(β-D-galactosido)-D-glucose; milk sugar. $C_{12}H_{22}O_{11}$; mol wt 342.30. C 42.11%, H 6.48%, O 51.41%. Present in milk of mammals: human 6.7%; cow's 4.5%. Milk at body temp contains lactose as an equilibrium mixture of 2 parts of α-lactose and 3 parts of β-lactose. By-product of the cheese industry, produced from whey: Davis, *Can. Dairy and Ice Cream J.* **19**, 52 (1940); *Milk Trade Gaz.* **12**, 4 (1941); F. Ullmann, *Encyklopädie der Technischen Chemie,* **VII**, 579 (2nd ed., 1931). Structure and configuration: Zemplén, *Ber.* **59**, 2402 (1926); Levene, Sobotka, *J. Biol. Chem.* **71**, 471 (1926); Levene, Wintersteiner, *ibid.* **75**, 315 (1927); Haworth, Long, *J. Chem. Soc.* **1927**, 544; Hudson, *J. Am. Chem. Soc.* **52**, 1712 (1930); Hassid, Ballou in *The Carbohydrates,* W. Pigman, Ed. (Academic Press, New York, 1957) p 495. Synthesis: Haskins *et al.*, *J. Am. Chem. Soc.* **64**, 1852 (1942). *Reviews:* Whittier, *Chem. Rev.* **2**, 85-125

(1926); *J. Dairy Sci.* **27**, 505-537 (1944); Weisberg, *ibid.* **37**, 1106-1115 (1954); L. A. W. Thelwall, *Dev. Food Carbohyd.* **2**, 275-326 (1980). Comprehensive description: H. G. Brittain *et al.,* in *Analytical Profiles of Drug Substances* **vol 20,** K. Florey Ed. (Academic Press, New York, 1991) pp 369-398.

On hydrolysis with 2% H_2SO_4 or with emulsin lactose yields 1 mol D-glucose and 1 mol D-galactose. Reduces Fehling's soln.

α-Lactose monohydrate, is the usual milk sugar and the lactose of pharmacy. Monoclinic sphenoidal crystals from water. Faintly sweet taste. Stable in air, but readily absorbs odors. d^{20} 1.53. Becomes anhydrous at 120°. mp 201-202° (rapid heating). Shows mutarotation. $[\alpha]_D^{20}$ +92.6° → +83.5° (10 min.) → +69° (50 min) → +52.3° (22 hrs, c = 4.5). The final value is obtained instantly in the presence of a trace of NH_3. U.S.P. requires +54.4° to +55 9° (c − 10). One gram dissolves in 5 ml water, in 2.6 ml boiling water; very slightly sol in alcohol. Insol in chloroform, ether. Ka at 16.5° = 6.0×10^{-13}. d_4^{20} of aq solns calcd for the monohydrate: 5.2% = 1.018; 10.2% = 1.038; 20.0% = 1.078; 30.2% = 1.123; 50.9% = 1.226; 60.8% = 1.281; 69.1% = 1.330.

β-Lactose, $C_{12}H_{22}O_{11}$. Obtained by crystallizing concd solns of α-lactose above 93.5°. Somewhat sweeter than the α-form. $[\alpha]_D^{25}$ +34° (3 min) → +39° (6 min) → +46° (1 hr) → +52.3° (22 hrs). One gram dissolves in 2.2 ml water at 15°, in 1.1 ml boiling water. After a few days crystals of the less sol α-monohydrate appear from satd solns.

USE: Both forms of lactose are employed, with the α-form predominating: as a nutrient in preparing modified milk and food for infants and convalescents (Whittier, "Lactose and Its Utilization," *loc. cit*). In baking mixtures. Pharmaceutic aid (tablet and capsule excipient and diluent). To produce lactic acid fermentation in ensilage and food products. As chromatographic adsorbent in analytical chemistry. In culture media. For many other uses see the comprehensive review by Weisberg "Recent Progress in the Manufacture and Use of Lactose," *loc. cit.*

THERAP CAT (VET): Added to cow's milk for feeding orphan foals.

5357. Lactucarium—"French". Thridace. Inspissated juice of *Lactuca sativa* L., var. *capitata* L., *Compositae*. *Constit.* Lactucin, hyoscyamine, mannite.

Brown pieces or powder; bitter taste; opium-like odor. Partly sol in water, alcohol, ether.

5358. Lactucarium—"German". "Lettuce opium". Dried milk-juice of *Lactuca virosa* L., *Compositae* (wild lettuce). *Constit.* About 0.2% lactucin; about 50% lactucerol; hyoscyamine, lactucic acid, caoutchouc, volatile oil, mannite.

Brown powder or irregular pieces; wax-like when cut; bitter taste. Partly sol in water, alcohol, ether. *Keep dry.*

THERAP CAT: Sedative.

5359. Lactucin. *[3aR-(3aα,4β,9aα,9bβ)]-3,3a,4,5,9a,9b-Hexahydro-4-hydroxy-9-(hydroxymethyl)-6-methyl-3-methyleneazuleno[4,5-b]furan-2,7-dione.* $C_{15}H_{16}O_5$; mol wt 276.29. C 65.21%, H 5.84%, O 28.95%. From various *Lactuca* spp and *Cichorium intybus* L., *Compositae*. Isoln: Schenck, Graf, *Arch. Pharm.* **274**, 537 (1936); **275**, 36 (1937); Schenck *et al.,* ibid. **294**, 17 (1961). Purification: Späth *et al.,* *Monatsh.* **82**, 114 (1951). Structure: Dolejs *et al.,* *Coll. Czech. Chem. Commun.* **23**, 2195 (1958); Barton, Narayanan, *J. Chem. Soc.* **1958**, 963; Michl, Högenauer, *Monatsh.* **89**, 317 (1958). Revised stereochemistry: Bachelor, Itô, *Can. J. Chem.* **51**, 3626 (1973).

Crystals from methanol, sinters at 218°, mp 228-233°. $[\alpha]_D$ +49° (c = 0.90 in methanol), +77.9° (c = 3.44 in pyridine). uv max: 257 nm (ε 14,000). Soluble in water, ethanol, methanol, ethyl acetate, dioxane anisol.

p-Hydroxyphenylacetate hydrate, $C_{23}H_{22}O_7$, *intybin, lactucopicrin.* Identity with intybin: Schmitt, *Bot. Arch.* **40**, 516 (1940), *C.A.* **36**, 5616 (1942). Structure: Michl, Högenauer, *Monatsh.* **91**, 500 (1960). Crystals from water, dec 148-151°. $[\alpha]_D^{17.5}$ +67.3° (pyridine).

5360. Lactulose. *4-O-β-D-Galactopyranosyl-D-fructose;* 4-D-galactopyranosyl-4-D-fructofuranose; 4-*O*-β-D-galactosyl-D-fructose; 4-β-D-galactosido-D-fructose; Bifiteral; Cephulac; Duphalac; Generlac; Lactuflor; Laevilac; Normase. $C_{12}H_{22}O_{11}$; mol wt 342.30. C 42.11%, H 6.48%, O 51.41%. Prepd from lactose: Montgomery, Hudson, *J. Am. Chem. Soc.* **52**, 2101 (1930); Oosten, *Rec. Trav. Chim.* **86**, 673 (1967); K. B. Hicks, F. W. Parrish, *Carbohyd. Res.* **82**, 393 (1980).

Hexagonal clustered plates from methanol, mp 169.0° (1°/min). Sweeter than lactose, but not as sweet as sucrose. Shows mutarotation; constant value after 24 hrs: −51.4° (c = 4 in H_2O at pH 4.8 and 20°). Soly in water (w/w) at 30°: 76.4%; at 60°: 81%; at 90°: > 86%. Acid hydrolysis yields galactose and fructose.

THERAP CAT: Laxative.

5361. Laidlomycin. *16-Deethyl-3-O-demethyl-16-methyl-3-O-(1-oxopropyl)monensin;* $C_{37}H_{62}O_{12}$; mol wt 698.89. C 63.59%, H 8.94%, O 27.47%. Polyether ionophore antibiotic; structurally related to monensin, *q.v.* Isoln from *Streptomyces spp.* and antimicrobial activity: F. Kitame *et al., J. Antibiot.* **27**, 884 (1974). Structure determn: F. Kitame, N. Ishida, *ibid.* **29**, 759 (1976). Prepn of esters: A. F. Kluge, R. D. Clark, *Eur. pat. Appl.* **24,189** (1981 to Syntex). ^{13}C-NMR study and enhancement of biological activities: R. D. Clark *et al., J. Antibiot.* **35**, 1527 (1982). Mechanism of action study: U. Gräfe *et al., J. Basic Microbiol.* **29**, 391 (1989). HPLC determn in fermentation broth: M. Beran *et al., Chromatographia* **31**, 603 (1991). Effect on cattle growth: M. T. Van Koevering *et al., 1991 Animal Science Research Report,* (Agricultural Experiment Station, Oklahoma State University, Stillwater, OK, 1991) 241; M. L. Galyean *et al., J. Anim. Sci.* **70**, 2950 (1992).

Colorless prisms from chloroform-ethyl acetate. mp 151-153°. $[\alpha]_D^{22}$ +51.3° (c=0.2 in $CHCl_3$). Sol in $CHCl_3$, acetone, alcohols. Insol in water, *n*-hexane. LD_{50} in mice (mg/kg): 5 i.p.; 1 i.v.; 2.5 s.c. (Kitame).

Propionate potassium, $C_{40}H_{65}KO_{13}$, *RS-11988*, *Cattlyst.* mp 190-192°.

THERAP CAT (VET): Growth promotant.

5362. Lamifiban. *(S)-[[1-[2-[[4-(Aminoiminomethyl)-benzoyl]amino]-3-(4-hydroxyphenyl)-1-oxopropyl]-4-piperidinyl]oxy]acetic acid;* [[1-[*N*-(*p*-amidinobenzoyl)-L-tyrosyl]-4-piperidinyl]oxy]acetic acid; Ro 44-9883. $C_{24}H_{28}N_4O_6$; mol wt 468.51. C 61.53%, H 6.02%, N 11.96%, O 20.49%. Specific nonpeptide platelet fibrinogen receptor (GPIIb/IIIa) antagonist. Prepn: L. Alig *et al., Eur. pat.* **Appl. 505,868;** *eidem,* U.S. pat. **5,378,712** (1992, 1995 both to Hoffmann-La Roche); L. Alig *et al., J. Med. Chem.* **35,** 4393 (1992). Pharmacology: J.-P. Carteaux *et al., Thromb. Haemostasis* **70,** 817 (1993); Y. Takiguchi *et al., ibid.* **73,** 683 (1995).

Crystals (zwitterionic form) from water, mp above 200° (dec). $[\alpha]_D^{20}$ +29.8° (c = 0.86 in 1*N* HCl).

Trifluoroacetate salt, $C_{24}H_{28}N_4O_6\cdot C_2HF_3O_2$, mp 125-130° (dec). LD_{50} i.v. in mice: 250 mg/kg (Alig, 1995).

THERAP CAT: Antithrombotic.

5363. Laminaran. Laminarin. A polysaccharide found in brown seaweed and occurring principally in the *Laminaria* spp. Linear polymer composed of β-$(1\to3)$-linked glucose residues; may contain small amounts of β-$(1\to6)$ linkages as interresidue linkages or as branch points and 2-3% D-mannitol as end groups. Two forms of laminaran are recognized; they are referred to as soluble and insoluble laminaran: Percival, Ross, *J. Chem. Soc.* **1951,** 720. Structure: Peat *et al., ibid.* **1958,** 724, 729; **1960,** 175; Goldstein *et al., Chem. & Ind. (London)* **1959,** 124; Annan *et al., ibid.* **1962,** 984; Annan *et al., J. Chem. Soc.* **1965,** 885; Maeda, Nisizawa, *Carbohyd. Res.* **7,** 97 (1968). Structure of soluble laminaran from *Eisenia bicyclis:* T. Usui *et al., Agr. Biol. Chem.* **43,** 603 (1979). NMR studies of laminaran: D. Gagnaire, *Org. Magn. Res.* **11,** 344 (1978); H. Friebolin *et al., ibid.* **12,** 216 (1979). Review: W. A. P. Black, E. T. Dewar in *Industrial Gums,* R. L. Whistler, Ed. (Academic Press, New York, 2nd ed., 1973) pp 137-145.

Water-insoluble laminaran, isolated from *L. cloustoni* Edmondst., *Laminariaceae,* is pptd spontaneously from the aq acid extract of the plant. Has lower degree of branching than the sol form. Typical analysis of the dry material: 92.5% polyglucose, 0.4% non-volatile matter; $[\alpha]_D^{16}$ −13.4° (c = 0.9). Amorphous triacetate, $(C_{12}H_{16}O_8)_n$, $[\alpha]_D^{16}$ −63.5° (c = 0.4 in chloroform).

Soluble form, isolated from *L. digitata,* is separated from the acidified extract only after addition of a precipitant such as ethanol. Typical analysis (on dry basis): 91.2% polyglucose, 1.0% non-volatile matter; $[\alpha]_D^{18}$ −11.9° (c = 2.1).

Sulfate, *laminaran hydrogen sulfate.* Laminaran can be sulfated to varying degrees. Highly sulfated products have anticoagulant properties comparable to heparin, while laminarans with few sulfate groups are antilipemic only: Besterman, Evans, *Brit. Med. J.* **1957,** I, 310.

5364. Laminin. Abundant structural component of the basal lamina; critical to the stability of the extracellular matrix and to the adhesion of cells to the basement membrane. Family of heterotrimeric glycoproteins composed of a heavy chain, designated α (also known as A) and 2 light chains, designated β (B1) and γ (B2), which are linked by disulfide bonds to form an asymetrical cross-shaped structure. Eight genetically distinct laminin subunits have been identified: $\alpha1$, $\alpha2$, $\alpha3$, $\beta1$, $\beta2$, $\beta3$, $\gamma1$, and $\gamma2$. Seven different assembly forms (laminins-1 to -7) are known and ap-

pear to be tissue specific and developmentally regulated. Exhibits diverse biological activities. Influences adhesion, growth, morphology and differentiation of a variety of cells via specific receptors including several of the integrin type. Isoln from murine Engelbreth-Holm-Swarm (EHS) tumor: R. Timpl *et al., J. Biol. Chem.* **254,** 9933 (1979). Review of biological activities: H. K. Kleinman *et al., J. Cell. Biochem.* **27,** 317-325 (1985); of role in neural development: V. Nurcombe, *Pharmacol. Ther.* **56,** 247 (1992). Tissue distribution: E. Engvall *et al., Cell Regul.* **1,** 731 (1990). Review of laminin binding proteins and receptors: R. P. Mecham, *Ann. Rev. Cell Biol.* **7,** 71-91 (1991); and role in metastasis: V. Castronovo, *Invas. Metast.* **13,** 1-30 (1993). Structure and function of laminin isoforms: E. Engvall, *Kidney Int.* **43,** 2-6 (1993); K. Tryggvason, *Curr. Opin. Cell Biol.* **5,** 877-882 (1993). Nomenclature: R. E. Burgeson *et al., Matrix Biol.* **14,** 209 (1994). Review of structure: R. Timpl, J. C. Brown, *ibid.* 275-281.

Laminin-1, *EHS-laminin.* Prototype laminin produced by murine EHS tumor. Contains $\alpha1$ (also known as A or Ae), $\beta1$ (B1, B1e) and $\gamma1$ (B2, B2e) subunits.

Laminin-2, *merosin, laminin M.* Variant found in striated muscle, placental trophoblast and Schwann cell basement membranes. Contains $\alpha2$ (also known as M or Am), $\beta1$ and $\gamma1$ subunits. Identification: I. Leivo, E. Engvall, *Proc. Nat. Acad. Sci. USA* **85,** 1544 (1988); K. Ehrig *et al., ibid.* **87,** 3264 (1990).

Laminin-3, *s-laminin, synaptic laminin.* Contains $\alpha1$, $\beta2$ (also known as s or B1s), and $\gamma1$ chains. Identification in neuromuscular junction: D. D. Hunter *et al., Nature* **338,** 229 (1989).

Laminin-5, *kallinin, nicein.* Contains $\alpha3$, $\beta3$, and $\gamma2$ subunits. Isoln from human keratinocytes: P. Rousselle *et al., J. Cell Biol.* **114,** 567 (1991).

5365. Lamivudine. *(2R-cis)-4-Amino-1-[2-(hydroxymethyl)-1,3-oxathiolan-5-yl]-2(1H)-pyrimidinone;* (−)-2'-deoxy-3'-thiacytidine; (−)-1-[(2R,5S)-2-(hydroxymethyl)-1,3-oxathiolan-5-yl]cytosine; 3'-thia-2',3'-dideoxycytidine; 3TC; (−)NGPB-21; (−)-BCH-189; GR-109714X; Epivir. $C_8H_{11}N_3O_3S$; mol wt 229.26. C 41.91%, H 4.84%, N 18.33%, O 20.94%, S 13.99%. Reverse transcriptase inhibitor. Prepn: J. A. V. Coates *et al.,* **PCT Int. pat. Appl. 17,159,** *C.A.* **117,** 111989 (1991). Synthesis of enantiomers: J. W. Beach *et al., J. Org. Chem.* **57,** 2217 (1992); of (−)-enantiomer: D. C. Humber *et al., Tetrahedron Letters* **33,** 4625 (1992). Comparative *in vitro* anti-HIV activity: J. A. V. Coates *et al., Antimicrob. Ag. Chemother.* **36,** 733 (1992). Clinical pharmacokinetics: R. van Leeuwen *et al., AIDS* **6,** 1471 (1992). HPLC determn in urine: D. M. Morris, K. Selinger, *J. Pharmaceut. Biomed. Anal.* **12,** 255 (1994).

White solid from methanol/ethyl acetate. $[\alpha]_D^{21}$ −132° (c = 1.08 in methanol). Also reported as crystals from boiling ethanol. mp 160-162°. $[\alpha]_D^{21}$ −135° (c = 0.38 in methanol).

THERAP CAT: Antiviral.

5366. Lamoparan. Org-10172. Low molecular weight heparinoid derived from porcine intestinal mucosa; mixture of sulfated glycosaminoglycans with mean mol wt 6500 daltons. Prepn: A. L. M. Sanders *et al., Eur. pat.* **Appl. 66,908;** *eidem,* U.S. pat. **4,438,108** (1982, 1984 both to Akzo). Exhibits antithrombotic activity comparable to standard heparin but with diminished hemorrhagic effects: D. G. Meuleman *et al., Thromb. Res.* **27,** 353 (1982); H. ten Cate *et al., ibid.* **38,** 211 (1985). Preliminary pharmacokinetics and tissue distribution in humans: I. D. Bradbrook *et al., Brit. J. Clin. Pharmacol.* **23,** 667 (1987). Clinical evaluation during hemodialysis: C. P. Henny *et al., Lancet* **1**, 890 (1983); in acute thrombotic stroke: A. G. G. Turpie *et al., ibid.* **1**, 523 (1987). White amorphous, slightly hygroscopic powder. $[\alpha]_D^{20}$ +30° to +70°.

THERAP CAT: Antithrombotic.

5367. Lamotrigine. *6-(2,3-Dichlorophenyl)-1,2,4-triazine-3,5-diamine;* 3,5-diamino-6-(2,3-dichlorophenyl)-1,2,4-triazine; LTG; BW-430C; Lamictal. $C_9H_7Cl_2N_5$; mol wt 256.09. C 42.21%, H 2.76%, Cl 27.69%, N 27.35%. Prepn: M. G. Baxter *et al.*, **Eur.** pat. **Appl. 21,121** (1981 to Wellcome Foundation); D. A. Sawyer *et al.*, U.S. pat. **4,-602,017** (1986). Anticonvulsant activity: A. A. Miller *et al., Epilepsia* **27**, 483 (1986). Mechanism of action study: M. J. Leach *et al., ibid.* 490. Evaluation of CNS effects: A. F. Cohen *et al., Brit. J. Clin. Pharmacol.* **20**, 619 (1985). Pharmacokinetics: A. F. Cohen *et al., Clin. Pharmacol. Ther.* **42**, 535 (1987). Series of articles on clinical pharmacology, efficacy and safety: *Epilepsia* **32**, Suppl. 2, S9-S21 (1991). *Review:* K. L. Goa *et al., Drugs* **46**, 152-176 (1993).

Crystals from isopropanol, mp 216-218° (uncorr). LD_{50} in mice, rats (mg/kg): 250, >640 orally (Sawyer).

THERAP CAT: Anticonvulsant.

5368. Lanatosides. Family of four glycosides, A, B, C, D, isolated from various species of *Digitalis* including *D. lanata* Ehrh., *Scrophulariaceae:* Stoll, Kreis, *Helv. Chim. Acta* **16**, 1049 (1933); Ligeti, *Pharmazie* **12**, 433 (1957); **14**, 162 (1959); Angliker *et al., Ann.* **607**, 131 (1957); *D. Lutea* L., *Scrophulariaceae:* Cole, Gisvold, *J. Am. Pharm. Assoc., Sci. Ed.* **47**, 654 (1958). Structure: Tschesche *et al., Ber.* **92**, 2258 (1959); Uskert, *Ann.* **638**, 199 (1960); Kuhn *et al., Helv. Chim. Acta* **45**, 881 (1962). Absorption spectrum: Bell, *J. Am. Pharm. Assoc., Sci. Ed.* **49**, 277 (1960). Cardioactivity of lanatoside C: K.-O. Haustein, *Pharmacology* **11**, 117 (1974). Mitogenicity: L. L. G. Hammarström, C. I. E. Smith, *J. Immunol.* **120**, 694 (1978). HPLC determn: V. Y. Davydov *et al., J. Chromatog.* **248**, 49 (1982); of lanatoside C: F. Orosz *et al., Anal. Biochem.* **156**, 171 (1986).

Lanatoside A	R = digitoxigenin
Lanatoside B	R = gitoxigenin
Lanatoside C	R = digoxigenin
Lanatoside D	R = diginatigenin

Lanatoside A, $C_{49}H_{76}O_{19}$, *(3β,5β)-3-[(O-β-D-glucopyranosyl-(1→4)-O-3-O-acetyl-2,6-dideoxy-β-D-ribo-hexopyranosyl-(1→4)-O-2,6-dideoxy-β-D-ribo-hexopyranosyl-(1→4)-2,6-dideoxy-β-D-ribo-hexopyranosyl)oxy]-14-hydroxycard-20(22)-enolide, digilanide A, Adigal.* Long, flat prisms from methanol, dec 245-248°. $[\alpha]_D^{20}$ +31.6° (0.48 g in 25 ml 95% alc); $[\alpha]_D^{20}$ +23.2° (0.95 g in 25 ml dioxane). Soluble in 20 parts methanol, 40 parts alcohol, 225 parts chloroform, 16,000 parts water.

Lanatoside B, $C_{49}H_{76}O_{20}$, *(3β,5β,16β)-3-[(O-β-D-glucopyranosyl-(1→4)-O-3-O-acetyl-2,6-dideoxy-β-D-ribo-hexopyranosyl-(1→4)-O-2,6-dideoxy-β-D-ribo-hexopyranosyl-(1→4)-2,6-*

dideoxy-β-D-ribo-hexopyranosyl)oxy]-14,16-dihydroxycard-20-(22)-enolide, digilanide B. Long, flat prisms from alcohol, dec 245-248° after drying in high vacuum at 150° $[\alpha]_D^{20}$ +36.7° (0.47 g in 25 ml 95% alc); $[\alpha]_D^{20}$ +31.8° (0.25 g in 14.1 ml dioxane). Soluble in 20 parts methanol, 40 parts alcohol, 550 parts chloroform. Nearly insol in water.

Lanatoside C, $C_{49}H_{76}O_{20}$, *(3β,5β,12β)-3-[(O-β-D-glucopyranosyl-(1→4)-O-3-O-acetyl-2,6-dideoxy-β-D-ribo-hexopyranosyl-(1→4)-O-2,6-dideoxy-β-D-ribo-hexopyranosyl-(1→4)-2,6-dideoxy-β-D-ribo-hexopyranosyl)oxy]-12,14-dihydroxycard-20-(22)-enolide, digilanide C, Allocor, Cedilanid, Ceglunat, Celadigal, Cetosanol, Lanimerck (Suppositories).* Long, flat prisms from alcohol, dec 248-250°, after drying in vacuum at 150°. $[\alpha]_D^{20}$ +33.4° to +33.7° (200 mg dry weight in 10 ml alcohol). One gram dissolves in 20,000 ml methanol, 2000 ml chloroform. Freely sol in pyridine, dioxane. Practically insol in ether, petr ether.

Lanatoside D, $C_{49}H_{76}O_{21}$, *(3β,5β,12β,16β)-3-[(O-β-D-glucopyranosyl-(1→4)-O-3-O-acetyl-2,6-dideoxy-β-D-ribo-hexopyranosyl-(1→4)-O-2,6-dideoxy-β-D-ribo-hexopyranosyl-(1→4)-2,-6-dideoxy-β-D-ribo-hexopyranosyl)oxy]-12,14,16-trihydroxycard-20(22)-enolide, digilanide D.* Needles from methanol and water, dec 242-250°. $[\alpha]_D^{20}$ +40.5° (c = 5.95 in methanol). uv max: 220 nm (log ε 4.16).

THERAP CAT: Cardiotonic.

5369. Lankamycin. Kujimycin B. $C_{42}H_{72}O_{16}$; mol wt 833.02. C 60.56%, H 8.71%, O 30.73%. Macrolide antibiotic produced by *Streptomyces violaceoniger* from soil of Ceylon. Isolation: Gaumann *et al., Helv. Chim. Acta* **43**, 601 (1960). Yields on hydrolysis an aglucone, monoacetyllankolid, $C_{26}H_{48}O_{10}$, and two sugar-like substances, lankavose (D-chalcose, *q.v.*) and acetylarcanose: Keller-Schierlein, Roncari, *ibid.* **45**, 138 (1962). Structure: *eidem, ibid.* **47**, 78 (1964); Egan, Martin, *J. Am. Chem. Soc.* **92**, 4129 (1970). Stereochemistry: Muntwyler, Keller-Schierlein, *Helv. Chim. Acta* **55**, 460 (1972). Identity with kujimycin B: Omura *et al., J. Antibiot.* **22**, 629 (1969).

Crystals from ether + petr ether. Double mp, 147-150° and 181-182°. $[\alpha]_D^{20}$ −94° (c = 1.23 in alc.). uv max: 289 nm (log ε 1.50).

5370. Lanoconazole. *(E)-(±)-α-[4-(2-Chlorophenyl)-1,-3-dithiolan-2-ylidene]-1H-imidazole-1-acetonitrile;* latoconazole; TJN-318; NND-318; Astat. $C_{14}H_{10}ClN_3S_2$; mol wt 319.84. C 52.57%, H 3.15%, Cl 11.08%, N 13.14%, S 20.05%. Prepn: A. Soe *et al.,* **Japan.** Kokai **85 218,387;** *idem et al.,* U.S. pat. **4,636,519** (1985, 1987 both to Nihon Nohyaku). *In vivo* antifungal activity: H. Oka *et al., Arzneimittel-Forsch.* **42**, 345 (1992); Y. Niwano *et al., Anti-*

microb. Ag. Chemother. **38**, 2204 (1994). Toxicity study: P. L. Munt *et al., Oyo Yakuri* **43**, 195 (1992).

Light yellow crystals, mp 141.5°. LD_{50} in male, female mice, rats (mg/kg): 3224, 2715, 993, 652 orally; 2158, 1743, 1655, 2596 i.p.; > 5000 both species s.c. LD_{50} dermally in rats: > 5000 mg/kg (Munt).

THERAP CAT: Antifungal.

5371. Lanolin. Wool fat; oesipos; agnin; alapurin; Agnolin; Lanum; Lanain; Lanalin; Lanesin; Lanichol; Laniol. Lanolin is the "fat-like" secretion of the sebaceous glands of sheep which is deposited onto the wool fibers. Chemically it is a wax rather than a fat, being a complex mixture of esters and polyesters of 33 high-molecular-weight alcohols and 36 fatty acids. The alcohols are of three types: aliphatic alcohols, steroid alcohols, and triterpenoid alcohols; the acids are also of three types: saturated nonhydroxylated acids, unsaturated nonhydroxylated acids and hydroxylated acids. Liquid lanolin is rich in low molecular weight, branched aliphatic acids and alcohols while waxy lanolin is rich in high molecular weight, straight-chain acids and alcohols. Reviews on composition, derivs, modifications and uses: Barnett, *Drug & Cosmet. Ind.* **80**, 744 (1957); **83**, 292 (1958); Leideritz, *Chemiker Ztg.* **83**, 707 (1959); F. Fawaz *et al., Ann. Pharm. Franc.* **31**, 217-226 (1973). Rheological properties of lanolin: F. Puisieux, *Pharm. Acta Helv.* **51**, 289 (1976). Monograph: E. V. Truter, *Wool Wax, Chemistry and Technology* (Interscience, 1956).

Lanolin contains about 25-30% water. It is a yellowish-white unctuous mass; slight odor. Practically insol in water. Sol in chloroform or ether with the separation of the water.

Anhydr lanolin is a yellowish tenacious, semisolid fat; slight odor or practically odorless. mp 38-42°. Practically insol in, but mixes with about twice its weight of water, without separation. Sparingly sol in cold, more in hot alcohol; freely sol in benzene, chloroform, ether, carbon disulfide, acetone, petr ether.

Acetylated lanolin, *Acetadeps, Modulan.* Almost odorless, pale yellow, semi-solid, unctuous mass. mp 36°. Sol in mineral oils, in some vegetable oils. Readily dispersed in oil-in-water emulsions. Will not emulsify to form water-in-oil emulsions.

USE: Pharmaceutic aid (ointment base).

THERAP CAT (VET): Ointment base, hoof dressing.

5372. Lanosterol. *(3β)-Lanosta-8,24-dien-3-ol;* kryptosterol. $C_{30}H_{50}O$; mol wt 426.73. C 84.44%, H 11.81%, O 3.75%. The core steroid from which all others are derived by biological modification. From wool fat of sheep: Windaus, Tschesche, *Z. Physiol. Chem.* **190**, 51 (1930). Identity with kryptosterol: Ruzicka *et al., Helv. Chim. Acta* **28**, 759 (1945). Structure: Voser *et al., ibid.* **35**, 2414 (1952); Barnes *et al., J. Chem. Soc.* **1953**, 571. Stereochemistry: *eidem, ibid.* **1953**, 576. Prepn from isocholesterol: Bloch, Urech, *Biochem. Prepn.* **6**, 32 (1958). Prepn by cyclization of squalene: Cornforth *et al., Ciba Foundation Symposium of Terpenes and Sterols* **1958**, 119; van Tamelen *et al., J. Am. Chem. Soc.* **88**, 4752 (1966). Mechanism of the squalene to lanosterol conversion: *eidem, ibid.* **104**, 6479, 6480 (1982).

Crystals, mp 138-140°. $[\alpha]_D^{20}$ +62.0° (chloroform). Infrared absorption max 6.124; 9.69; 12.22 microns (KBr disks of lanosterol).

Lanosteryl acetate, crystals, mp 131.5-133°, $[\alpha]_D^{20}$ +62.5° (c = 1.12 in chloroform).

5373. Lansoprazole. *2-[[[3-Methyl-4-(2,2,2-trifluoroethoxy)-2-pyridinyl]methyl]sulfinyl]-1H-benzimidazole;* 2-(2-benzimidazolylsulfinylmethyl)-3-methyl-4-(2,2,2-trifluoro-

ethoxy)pyridine; A-65006; AG-1749; Agopton; Lanzor; Ogast; Takepron; Zoton. $C_{16}H_{14}F_3N_3O_2S$; mol wt 369.37. C 52.03%, H 3.82%, F 15.43%, N 11.38%, O 8.66%, S 8.68%. Gastric proton-pump inhibitor. Prepn: A. Nohara, Y. Maki, *Eur.* pat. *Appl.* 174,726; *eidem,* U.S. pat. **4,628,-098** (both 1986 to Takeda). Pharmacology: H. Satoh *et al., J. Pharmacol. Exp. Ther.* **248**, 806 (1989). Mechanism of action study: H. Nagaya *et al., ibid.* **252**, 1289 (1990). Clinical pharmacology and effect on human gastric acid secretion: P. Müller *et al., Aliment. Pharmacol. Ther.* **3**, 193 (1989).

mp 178-182° (dec).

THERAP CAT: Antiulcerative.

5374. Lanthanum. La; at. wt 138.9055; at. no. 57; valence 3. A rare earth metal; member of the lanthanide series. Naturally occurring isotopes (mass numbers): 139 (99.91%); 138 (0.09%), radioactive, $T_{1/2}$ 1.06 × 10^{11} years; known artificial radioactive isotopes: 123-137; 140-149. Estimated abundance in earth's crust: 18-35 ppm. Found in association with cerium and other light lanthanons. Sources of commercial importance are the rare earth minerals monazite and bastnaesite; also found in cerite. Discovery and isoln: Mosander, *Pogg. Ann.* **47**, 207 (1839). Sepn: James, *J. Am. Chem. Soc.* **34**, 757 (1912). Prepn of metal: Mazzi, *Atti X. Congr. Internaz. Chim.* **3**, 604 (1949); Spedding *et al., Ind. Eng. Chem.* **44**, 553 (1952). Toxicity study: Cochran *et al., Arch. Ind. Hyg. Occup. Med.* **1**, 637 (1950). Reviews of prepn, properties and compds: *The Rare Earths,* F. H. Spedding, A. H. Daane, Eds. (Krieger, Huntington, N.Y., 1971, reprint of 1961 ed.) 641 pp; Hulet, Bode, "Separation Chemistry of the Lanthanides and Transplutonium Actinides" in *MTP Int. Rev. Sci.: Inorg. Chem., Ser. One,* **Vol. 7**, K. W. Bagnall, Ed. (University Park Press, Baltimore, 1972) pp 1-45; Vickery, "Scandium, Yttrium, Lanthanum" in *Comprehensive Inorganic Chemistry* vol. 3, J. C. Bailar Jr. *et al.*, Eds. (Pergamon Press, Oxford, 1973) pp 329-353; Moeller, "The Lanthanides" *ibid.* vol. 4, 1-101; F. H. Spedding in Kirk Othmer *Encyclopedia of Chemical Technology* vol. 19 (John Wiley & Sons, New York, 3rd ed., 1982) pp 833-854; *Chemistry of the Elements,* N. N. Greenwood, A. Earnshaw, Eds. (Pergamon Press, New York, 1984) pp 1102-1110, 1423-1449. Brief review of properties: G. T. Seaborg, *Radiochim. Acta* **61**, 115-122 (1993).

White, malleable metal; tarnishes in air. Three crystalline forms: hexagonal α-form, d 6.162, transforms to β-form at 310°; face-centered cubic β-form, d 6.19, transforms to γ-form at 864°; body-centered cubic γ-form, d 5.97, exists at > 864°. mp 920°. bp 3464°. Heat of fusion: 6.201 kJ/mol. Heat of sublimation (25°): 431.0 kJ/mol. E°(aq) La³⁺/La −2.52 V (calc). Very active; dec water slowly in the cold, more readily on heating. Readily attacked by mineral acids; not attacked by cold concd H_2SO_4. Burns in air at about 450° producing a mixture of oxide and nitride; forms the hydride on heating in hydrogen. Forms alloys with several metals.

Oxide, La_2O_3, *lanthana, lanthanum sesquioxide, lanthanum trioxide,* almost white, amorphous powder. d 6.51. mp > 2000°. Insol in water. Sol in dil mineral acids with formation of salts. Absorbs CO_2 from the air.

Hydroxide, $La(OH)_3$, white, amorphous precip, prepd by adding excess of caustic alkali to a lanthanum salt soln. Strongly basic, displaces ammonia from ammonium salts, absorbs CO_2 from air. On dehydration yields $La_2O_3 \cdot H_2O$.

Chloride, $LaCl_3$, heptahydrate, triclinic crystals. Sol in water or alc. On heating in HCl the anhyd salt (mp 852°) is formed. LD_{50} in rats: 4.2 g/kg orally; 350 mg/kg i.p. (Cochran).

Carbonate octahydrate, $C_3La_2O_9$, *artificial lanthanite.* White, crystalline powder. Practically insol in water. Freely sol in dil mineral acids.

Sulfate, $La_2(SO_4)_3$, nonahydrate, hexagonal prisms. Prepd by treating a lanthanum salt with a slight excess of sulfuric acid. Dec at white heat. Is the least sol of the rare earth sulfates; soly in water decreases with increase in temp. Insol in alc. Forms double salts with alkali or ammonium hydroxide. Anhydr salt prepd by heating hydrate. LD_{50} in rats: > 5000 g/kg orally; 275 mg/kg i.p. (Cochran).

Nitrate, $La(NO_3)_3$, hexahydrate, white deliquesc crystals, mp ~40°, at higher temp forms a basic salt. bp 126°. Very sol in water, alc. Forms double salts with bivalent ion nitrates and ammonium nitrates. Keep well closed. LD_{50} in rats: 4.5 g/kg orally; 450 mg/kg i.p. (Cochran).

USE: Oxide in glass to improve optical properties. La^{3+} used in experimental biology as a specific antagonist of calcium: Weiss, Ann. Rev. Pharmacol. **14**, 343 (1974).

5375. Lanthionine. (R)-S-(2-Amino-2-carboxyethyl)-L-cysteine; 3,3'-thiodi-L-alanine; L-lanthionine; [R-(R*,R*)]-bis(2-amino-2-carboxyethyl)sulfide. $C_6H_{12}N_2O_4S$; mol wt 208.24. C 34.61%, H 5.81%, N 13.45%, O 30.73%, S 15.40%. A rare amino acid found in proteins; sulfide analog of cystine. First isolated as artifacts of wool hydrolysates. Isoln: M. J. Horn et al., J. Biol. Chem. **138**, 141 (1941); from chick embryo: N. H. Sloane, K. G. Untch, Biochemistry **5**, 2658 (1966); from silkworm and Japanese oak moth: D. R. Rao et al., ibid. **6**, 1208. Synthesis: V. du Vigneaud, G. B. Brown, J. Biol. Chem. **138**, 151 (1941). Structure studies: I. W. Stapleton, O. A. Weber, Int. J. Peptide Protein Res. **3**, 243 (1971). Crystal structure: G. R. Desiraju, D. R. Rao, Acta Crystallogr. **C46**, 627 (1990).

Elongated hexagonal plates. mp 295-296° (dec). $[\alpha]_D^{25}$ +9.4° (c = 1.4 in 2.4N NaOH); $[\alpha]_D^{22}$ +8.6° (c — 5 in 2.4N NaOH); $[\alpha]_D^{22}$ +6.0° (c = 1 in 1N NaOH).

meso-Form, six-sided plates having a triangular appearance from dil NH_3. Softens at 270°, dec 304°. Stable to alkalies. Sol in dil acids and alkalies. Sparingly sol in water. Insol in alcohol, ether, chloroform, acetone.

DL-Form, elongated hexagonal plates. Chars at 240°, dec 286-292°.

D-Form, elongated hexagonal plates. Darkens at 245°, dec 293-295°. $[\alpha]_D^{21}$ — 8.0° (c = 5 in 2.4N NaOH).

5376. Lapachol. 2-Hydroxy-3-(3-methyl-2-butenyl)-1,4-naphthalenedione; lapachic acid; taiguic acid; tecomin; greenhartin; NSC-11905. $C_{15}H_{14}O_3$; mol wt 242.27. C 74.36%, H 5.82%, O 19.81%. Yellow crystalline material derived from the heartwood of Asian and South American bignoniaceous plants, esp Surinam greenheart, Taigu wood, Lapacho heartwood and Bethabarra wood: Arnoudon, Compt. Rend. **41**, 1152 (1857); Stein, J. Prakt. Chem. **99**, 1 (1866); Paterno, Gazz. Chim. Ital. **9**, 506 (1879); Greene, Hooker, Am. Chem. J. **11**, 267 (1889). Structure: Paterno, Gazz. Chim. Ital. **12**, 337 (1882); Hooker J. Chem. Soc. **61**, 611 (1892); **69**, 1355 (1896); J. Am. Chem. Soc. **58**, 1168 (1936). Synthesis: Fieser, ibid. **49**, 857 (1927); Hooker, ibid. **58**, 1181 (1936); G. R. Pettit, L. E. Houghton, J. Chem. Soc. (C) **1971**, 509. Although it is related structurally to vitamin K, q.v., it does not possess antihemorrhagic activity: H. J. Almquist, A. A. Klose, J. Am. Chem. Soc. **61**, 1923 (1939); L. F. Fieser et al., J. Biol. Chem. **137**, 659 (1941). It is reported to be an inhibitor of respiratory processes: E. G. Ball et al., ibid. **168**, 257 (1947). Lapachol has also exhibited antitumor activity vs Walker 256 carcinoma: K. V. Rao et al., Proc. Am. Assoc. Cancer Res. **8**, 55 (1967). Mass spectrometry: T. A. Elwood et al., Org. Mass Spectrom. **3**, 841 (1970). Chromatographic detection: M. H. Simatupang et al., J. Chromatog. **52**, 180 (1970). Pharmacology: S. M. Sieber et al., Cancer Treat. Rep. **60**, 1127 (1976). Toxicity study: R. K. Morrison et al., Toxicol. Appl. Pharmacol. **17**, 1 (1970). Review of antitumor activity: K. V. Rao, Cancer Chemother. Rep. (pt. 2) **4** (4), 11-17 (1974).

Yellow prisms from alcohol or ether, mp 140°. uv max: 251.5, 278, 331 nm (log ε 4.38, 4.28, 3.43). Soluble in alcohol, chloroform, benzene, acetic acid, slightly sol in ether, hot water. Sol in aq NaOH solns forming a bright red sodium salt. LD_{50} in male, female BALB/c mice (g/kg): 0.487; 0.792 orally (Morrison).

5377. Lappa. Burdock; clotbur; bardana. Dried first-year root of Arctium lappa L., or of Arctium minus Bernh., Compositae. Habit. Europe, Northern Asia; naturalized in N. America. Constit. Volatile oil, bitter principle, inulin, tannin.

THERAP CAT: Dermatologic.

5378. Lappaconitine. (1α,14α,16β)-20-Ethyl-1,14,16-trimethoxyaconitane-4,8,9-triol 4-[2-(acetylamino)benzoate]. $C_{32}H_{44}N_2O_8$; mol wt 584.71. C 65.73%, H 7.58%, N 4.79%, O 21.89%. From tubers and herb of Aconitum septentrionale Kölle, A. orientale Mill., A. excelsum Reichb., A. ranunculaefolium, Ranunculaceae: Schulze, Ulfert, Arch. Pharm. **260**, 230 (1922); Kuzovkov, Massagetov; Platonova et al., Zh. Obshch. Khim. **25**, 178 (1955); **28**, 258 (1958), C.A. **50**, 1852e (1956), C.A. **52**, 12883i (1958), resp.; Mollov et al., Compt. Rend. Acad. Bulgare Sci. **17**, 251 (1964), C.A. **61**, 12324g (1964). Structure: Khaimova et al., Tetrahedron Letters **1964**, 2711. Revised structure: V. A. Tel'nov et al., Khim. Prir. Soedin. **6**, 583 (1970), C.A. **74**, 42527k (1971). Analgesic activity studies in mice and rats: N. Ono, J. Satoh, Arzneimittel-Forsch. **38**, 892 (1988). Toxicology: F. Dybing et al., Acta Pharmacol. Toxicol. **7**, 337 (1951).

Bitter crystals, mp 217-218°. $[\alpha]_D^{18}$ +27° (chloroform). Sol n benzene; slightly sol in alcohol, ether. Practically insol in ater. LD_{50} in mice (mg/kg): 6.9 i.v.; 9.1 i.p.; approx 20 orally (Dybing).

5379. Lapyrium Chloride. 1-[2-Oxo-2-[[2-[(1-oxododecyl)oxy]ethyl]amino]ethyl]pyridinium chloride; 1-[[(2-hydroxyethyl)carbamoyl]methyl]pyridinium chloride laurate (ester); N-(lauroylcolaminoformylmethyl)pyridinium chloride; N-(acylcolaminoformylmethyl)pyridinium chloride; emulsept (obsolete); E-607; Emcol E-607. $C_{21}H_{35}ClN_2O_3$; mol wt 398.97. C 63.22%, H 8.84%, Cl 8.89%, N 7.02%, O 12.03%. Cationic surfactant. Prepn: A. K. Epstein, B. R. Harris, U.S. pat. **2,290,173** (1942). Bactericidal potency: A. K. Epstein et al., Proc. Soc. Exp. Biol. Med. **53**, 238 (1943). Toxicity study: A. E. Vivino, T. Koppanyi, J. Am. Pharm. Assoc., Sci. Ed. **35**, 169 (1946).

Powder.

Stearoyl analog, $C_{27}H_{47}ClN_2O_3$, *quaternium-7, Emcol E-607S.*

USE: Cationic emulsifier; deodorant; detergent-germicide; antistatic agent. Pharmaceutical aid (surfactant).

5380. Lard. Adeps; axungia porci. Purified internal fat from abdomen of the hog.

Soft, white unctuous mass; slight characteristic odor, bland taste. d 0.917. mp 36-42°. Insoluble in water; very slightly sol in alcohol, freely in benzene, chloroform, ether, carbon disulfide, petr ether. Sapon. no. 195-203. Iodine no. 46-70. *Keep cool and in tight containers.*

USE: In manuf ointments, particularly when absorption is desired.

5381. Lard, Benzoinated. Prepd by heating lard with 1% benzoin powder at temps not above 60°. Less likely to become rancid than pure lard.

USE: As vehicle for medicinal agents and in manufacture of ointments.

5382. Lard Oil. The oil expressed from lard at a low temp. *Constit.* Olein, stearin.

Colorless or pale yellow liq. d 0.905-0.915. Solidif −2° to +4°. n_D^{20} 1.470-1.472. Sapon no. 195-197. Iodine no. 56-74. Acid no. 1.5-2.5. Insol in water or cold alc. Freely sol in benzene.

USE: As lubricant, manuf soap, oiling wool, illuminant.

5383. Larkspur. Delphinium; knight's spur; lark's-heel; lark's-claw; staggerweed. Dried ripe seeds of *Delphinium ajacis* L., *Ranunculaceae. Habit.* Central Europe, cultivated in U.S. *Constit.* Calcatripine, volatile oil, gum, resin, fixed oil, gallic and aconitic acids. Identification and toxicity of alkaloids: G. D. Manners *et al., J. Range Manage.* **45**, 63 (1992).

Caution: Toxic. Poisoning from percutaneous absorption may occur.

THERAP CAT: Pediculicide.

5384. Lasalocid A. *[2R-[2α[2S*(3R*,4S*,5S*,7R*),-3S*,5S*],5α,6β]]-6-[7-[5-Ethyl-5-(5-ethyltetrahydro-5-hydroxy-6-methyl-2H-pyran-2-yl)tetrahydro-3-methyl-2-furanyl]-4-hydroxy-3,5-dimethyl-6-oxononyl]-2-hydroxy-3-methylbenzoic acid;* 3-methyl-6-[7-ethyl-4-hydroxy-3,5-dimethyl-6-oxo-7-[5-ethyl-3-methyl-5-(5-ethyl-5-hydroxy-6-methyl-2-tetrahydropyranyl)-2-tetrahydrofuryl]heptyl]salicylic acid; antibiotic X-537A; ionophore X-4537A; Ro-2-2985; X-537A; Bovatec. $C_{34}H_{54}O_8$; mol wt 590.80. C 69.12%, H 9.21%, O 21.66%. Ionophorous (transport-inducing) antibiotic isolated from an unidentified *Streptomyces* from soil samples of Hyde Park, Mass.: Berger *et al., J. Am. Chem. Soc.* **73**, 5295 (1951). Prepn and activity of title compd and derivs: Stempel, Westley, **Ger.** pat. **2,040,998** corresp to U.S. pat. **3,715,372** (1971, 1973 to Hoffmann-La Roche); Westley *et al., J. Med. Chem.* **16**, 397 (1973). Structure: *eidem, Chem. Commun.* **1970**, 71. Crystal and molecular structure studies: Johnson *et al., ibid.* 72; *J. Am. Chem. Soc.* **92**, 4428 (1970). Total synthesis: T. Nakata *et al., ibid.* **100**, 2933 (1978); R. E. Ireland *et al., ibid.* **102**, 1155 (1980). Complete assignment of ^{13}C-NMR spectra of lasalocid A and its sodium salt: H. Seto *et al., J. Antibiot.* **31**, 289 (1978). Effect in coccidiosis: G. M. J. Horton, P. H. G. Stockdale, *Am. J. Vet. Res.* **42**, 433 (1981). Biosynthesis: Westley *et al., Chem. Commun.* **1970**, 1467; C. R. Hutchinson *et al., J. Am. Chem. Soc.* **103**, 5953, 5956 (1981). Mode of action studies: Lin, Kun, *Biochem. Biophys. Res. Commun.* **50**, 820 (1973). Isoln and structure of four homologs, lasalocids B, C, D, E from *Streptomyces lasaliensis:* J. W. Westley *et al., J. Antibiot.* **27**, 744 (1974).

Crystals, mp 110-114°; also reported as mp 100-109° (unsharp). $[\alpha]_D^{25}$ −7.55° (methanol). uv max (50% aq isopropanol): 248, 318 nm (ϵ 6750, 4200). Sol in organic solvents. Insol in water. LD_{50} in mice (mg/kg): 40 i.p. (Berger); also reported as LD_{50} in mice (mg/kg): 146 orally, 64 i.p., J. W. Westley, "Antibiotics (Polyether)" in Kirk-Othmer *Encyclopedia of Chemical Technology,* **Vol. 3** (Interscience, New York, 3rd ed., 1978) p 61.

Sodium salt, $C_{34}H_{53}NaO_8$, *Avatec.* Crystals from benzene-ligroin, mp (open capillary) 191-192° (dec), also reported as mp 168-171°. $[\alpha]_D^{25}$ −30° (c = 1 in methanol). uv max (50% aq isopropanol): 308 nm (ϵ 4100).

THERAP CAT (VET): Coccidiostat (for poultry).

5385. Laserpitin. *[1R-[1α,3aα,4β(Z),6β,8β(Z),8aβ]]-2-Methyl-2-butenoic acid decahydro-1,6-dihydroxy-3a,6-dimethyl-1-(1-methylethyl)-5-oxo-4,8-azulenediyl ester.* $C_{25}H_{38}O_7$; mol wt 450.57. C 66.64%, H 8.50%, O 24.86%. From root of *Laserpitium latifolium* L., *Umbelliferae:* Külz, *Arch. Pharm.* **221**, 161 (1883); Morgenstern, *Monatsh.* **33**, 709 (1912). Structure: M. Holub *et al., Tetrahedron Letters* **1965**, 1441, 2855. Absolute configuration: M. Holub *et al., Coll. Czech. Chem. Commun.* **35**, 3597 (1970).

Crystals, mp 118°. Practically insol in water. Sol in alcohol, chloroform, ether, petr ether, fatty oils.

USE: Used as seasoning and flavoring agent in antiquity (Greece and Rome).

5386. Lasiocarpine. *[1S-[1α(Z),7[S*(R*)],7aα]]-2-Methyl-2-butenoic acid 7-[[2,3-dihydroxy-2-(1-methoxyethyl)-3-methyl-1-oxobutoxy]methyl]-2,3,5,7a-tetrahydro-1H-pyrrolizin-1-yl ester.* $C_{21}H_{33}NO_7$; mol wt 411.50. C 61.30%, H 8.08%, N 3.40%, O 27.22%. Hepatotoxic pyrrolizidine alkaloid isolated from *Heliotropium lasiocarpum* Fish. et C. Mey. *Boraginaceae:* G. Menschikoff, *Ber.* **65**, 974 (1932), *C.A.* **26**, 4818[3] (1932); G. Menschikoff, J. Schdanowitsch, *Ber.* **69**, 1110 (1936), *C.A.* **30**, 5227[9] (1936). Structure: L. J. Drummond, *Nature* **167**, 41 (1951); R. Adams, B. L. Van-Duvren, *J. Am. Chem. Soc.* **76**, 6379 (1954). Review and evaluation of studies of carcinogenicity and toxicity in laboratory animals: *IARC Monographs* **10**, 281-290, 333-342 (1976). Comprehensive reviews: L. Bull *et al., The Pyrrolizidine Alkaloids* (North-Holland, Amsterdam, 1965) 293 pp; F. L. Warren in *The Alkaloids* vol. **12**, R. H. F. Manske, Ed. (Academic Press, New York, 1970) pp 245-331; D. J. Robins, *Fortschr. Chem. Org. Naturst.* **41**, 115-203 (1982).

Colorless leaflets from petr ether. mp 94-95.5°. $[\alpha]_D$ −4° (10% alc). Sol on ether, alc, benzene; difficultly sol in water.

5387. Latanoprost. *[1R-[1α(Z),2β(R*),3α,5α]]-7-[3,5-Dihydroxy-2-(3-hydroxy-5-phenylpentyl)cyclopentyl]-5-heptenoic acid 1-methylethyl ester;* isopropyl (Z)-7-[(1R,2R,3R,5S)-3,5-dihydroxy-2-[(3R)-3-hydroxy-5-phenylpentyl]cyclopentyl]-5-heptenoate; 13,14-dihydro-17-phenyl-18,19,20-trinor-

PGF$_{2\alpha}$-isopropyl ester; PhXA-41. C$_{26}$H$_{40}$O$_5$; mol wt 432.60. C 72.19%, H 9.32%, O 18.49%. Prostaglandin analog; prodrug of active acid metabolite. Prepd but not claimed: J. Ivanics *et al.*, **PCT Int. pat. Appl. 93 00,329** (1993 to Kabi Pharmacia; Chinoin). Prepn: B. Resul *et al.*, *J. Med. Chem.* **36**, 243, 2242 (corr.) (1993). Clinical pharmacology: C. B. Toris *et al.*, *Ophthalmology* **100**, 1297 (1993). Clinical evaluation in glaucoma: P. Rácz *et al.*, *Arch. Ophthalmol.* **111**, 657 (1993); as single therapy and in combination with timolol, *q.v.*: A. H. Rulo *et al.*, *Brit. J. Ophthalmol.* **78**, 899 (1994). Review of preclinical pharmacology: J. Stjernschantz *et al.*, *Adv. Prostaglandin Thromboxane Leukot. Res.* **23**, 513-518 (1995).

Colorless oil. [α]$_D^{20}$ +31.57° (c = 0.91 in acetonitrile).
THERAP CAT: Antiglaucoma.

5388. Latrunculins. Highly potent toxins isolated from the Red Sea sponge *Latrunculia magnifica* Keller. Also found in the Pacific nudibranch *Chromodoris elisabethina* and the Fijian sponge *Spongia mycofijiensis*. When *L. magnifica* is squeezed into an aquarium, toxins are exuded. The fish experience agitation, hemorrhage, loss of balance and death within 4-6 minutes. Isoln and biological activities: I. Néeman *et al.*, *Marine Biol.* **30**, 293 (1975). Latrunculins (LAT) A through D, M and 6,7-epoxy-LAT A have been isolated and characterized. They are the first family of natural products possessing the 2-thiazolidinone moiety. All but latrunculin M are 16- or 14- membered macrolides, infrastructures previously unknown in marine isolates. Structure elucidation of latrunculins A and B: A. Groweiss *et al.*, *J. Org. Chem.* **48**, 3512 (1983); of A through D: Y. Kashman *et al.*, *Tetrahedron* **41**, 1905 (1985). Chemistry of latrunculins A and B; characterization of latrunculin M and 6,7-epoxy-LAT A: D. Blasberger *et al.*, *Ann.* **1989**, 1171. Total asymmetric synthesis of latrunculin A: J. D. White, M. Kawasaki, *J. Am. Chem. Soc.* **112**, 4991 (1990); *eidem*, *J. Org. Chem.* **57**, 5292 (1992); of B: R. Zibuck *et al.*, *J. Am Chem. Soc.* **108**, 2451 (1986); of A, B, C and M: A. B. Smith, III *et al.*, *ibid.* **114**, 2995 (1992). Comparative study with cytochalasin D, *q.v.*, on the effects on morphology, actin organization and cell processes: I. Spector *et al.*, *Cell Motil. Cytoskel.* **13**, 127 (1989).

Latrunculin A   Latrunculin B

Latrunculin A, C$_{22}$H$_{31}$NO$_5$S, [1R-[1R*,4Z,8E,10Z,12S*,-15R*,17R*(R*)]]-4-(17-hydroxy-5,12-dimethyl-3-oxo-2,16-dioxabicyclo[13.3.1]nonadeca-4,8,10-trien-17-yl)-2-thiazolidinone, LAT-A. Foam. [α]$_D^{24}$ +152° (c = 1.2 in chloroform). uv max (methanol): 218 nm (ε 23500).

Latrunculin B, C$_{20}$H$_{29}$NO$_5$S, [1R-[1R*,4Z,8Z,10S*,13R*,-15R*(R*)]]-4-(15-hydroxy-5,10-dimethyl-3-oxo-2,14-dioxabicyclo[11.3.1]heptadeca-4,8-dien-15-yl)-2-thiazolidinone, LAT-B. Crystals. [α]$_D^{24}$ +112° (c = 0.48 in chloroform). uv max (methanol): 212 nm (ε 17200).
USE: In elucidation of molecular mechanisms of motile processes.

5389. Laudanidine. (R)-2-Methoxy-5-[(1,2,3,4-tetrahydro-6,7-dimethoxy-2-methyl-1-isoquinolinyl)methyl]phenol; *l*-laudanine; tritopine. C$_{20}$H$_{25}$NO$_4$; mol wt 343.42. C 69.95%, H 7.34%, N 4.08%, O 18.64%. The *l*-form of laudanine, *q.v.* Traces in opium: Hesse, *Ann.* **282**, 208 (1894); Kauder, *Arch. Pharm.* **228**, 419 (1890). Structure and synthesis: Späth, Bernhauer, *Ber.* **58**, 200 (1925); Späth, Seka, *ibid.* 1272; Späth, Burger, *Monatsh.* **47**, 733 (1926); Frydman *et al.*, *Tetrahedron* **4**, 342 (1958). Configuration: Corrodi, Hardegger, *Helv. Chim. Acta* **39**, 889 (1956).
Prisms from alcohol, mp 185°. [α]$_D^{15}$ −88° (p = 5 in chloroform). [α]$_D^{22}$ −94.7 ± 1.3°. The hydrochloride is more sol than laudanine hydrochloride and makes possible a separation of the two alkaloids.

5390. Laudanine. (±)-2-Methoxy-5-[(1,2,3,4-tetrahydro-6,7-dimethoxy-2-methyl-1-isoquinolinyl)methyl]phenol; dl-laudanidine. C$_{20}$H$_{25}$NO$_4$; mol wt 343.42. C 69.95%, H 7.34%, N 4.08%, O 18.64%. In the alkaline mother liquors from morphine extraction, in amounts corresp to about 0.005% of the opium employed: L. F. Small, R. E. Lutz, *Chemistry of the Opium Alkaloids*, Supplement No. 103, Public Health Reports, Washington (1932) p 34. Structure: Späth, *Monatsh. Chem.* **41**, 297 (1920). Synthesis: Späth, Lang, *ibid.* **42**, 273 (1921); Frydman *et al.*, *Tetrahedron* **4**, 342 (1958).

Orthorhombic prisms from alcohol or chloroform, mp 167°. d$_4^{20}$ 1.26. uv max: 284 nm (log ε 3.78). Sol in benzene, chloroform, hot alcohol; sparingly sol in ether. Practically insol in water.

5391. Laudanosine. (S)-1-[(3,4-Dimethoxyphenyl)methyl]-1,2,3,4-tetrahydro-6,7-dimethoxy-2-methylisoquinoline; 1,-2,3,4-tetrahydro-6,7-dimethoxy-2-methyl-1-veratrylisoquinoline; N-methyltetrahydropapaverine. C$_{21}$H$_{27}$NO$_4$; mol wt 357.45. C 70.56%, H 7.61%, N 3.92%, O 17.90%. Occurs in opium (0.0008%). It is the last alkaloid to be separated from morphine extraction mother liquors; occurs as (+)-form. Synthesis: Pictet, Finkelstein, *Ber.* **42**, 1979 (1909); Frydman *et al.*, *Tetrahedron* **4**, 342 (1958); Elliott, *J. Heterocycl. Chem.* **9**, 853 (1972). Asymmetric synthesis: M. Konda *et al.*, *Chem. Pharm. Bull.* **23**, 1025 (1975); of (R)-(−)-laudanosine: M. Konda *et al.*, *ibid.* **25**, 69 (1977); R. E. Gawley, G. A. Smith, *Tetrahedron Letters* **29**, 301 (1988). Configuration: Leithe, *Ber.* **64**, 2827 (1931); Faltis, Adler, *Arch. Pharm.* **284**, 281 (1951); Corrodi, Hardegger, *Helv. Chim. Acta* **39**, 889 (1956).

Crystals from light petr (30-60°), mp 89°. $[\alpha]_D^{16}$ +106° (c = 1.6 in 97% alc); $[\alpha]_D^{16}$ +130° (chloroform); $[\alpha]_D^{22}$ +52.2 ±1.3° (chloroform). $[\alpha]_D^{20}$ +82.5° (ethanol). Absorption spectrum: Dobbie, Lauder, *J. Chem. Soc.* **83**, 626 (1903). Practically insol in water. Freely sol in alcohol, chloroform, ether, hot petr ether.

(−)-Form, colorless needles from ethanol, mp 83-85°. $[\alpha]_D^{20}$ −84.8° (c = 0.466 in ethanol).

(±)-Form, crystals from dil alc, mp 114-115.5°.

5392. Laudexium Methyl Sulfate. *2,2′-(1,10-Decanediyl)bis[1-[(3,4-dimethoxyphenyl)methyl]-1,2,3,4-tetrahydro-6,-7-dimethoxy-2-methylisoquinolium] bis(methyl sulfate); 2,2′-decamethylenebis[1,2,3,4-tetrahydro-6,7-dimethoxy-2-methyl-1-veratrylisoquinolinium methyl sulfate];* decamethylenebis-[1,2,3,4-tetrahydro-6,7-dimethoxy-1-(3,4-dimethoxybenzyl)-2-methylisoquinolinium methyl sulfate]; decamethylene-α-ω-bis[1-(3′,4′-dimethoxybenzyl)-1,2,3,4-tetrahydro-6,7-dimethoxy-2-methylisoquinolinium methosulfate]; Compd 20; Laudissine; Laudolissin. $C_{54}H_{80}N_2O_{16}S_2$; mol wt 1077.37. C 60.20%, H 7.48%, N 2.60%, O 23.76%, S 5.95%. Prepn: Taylor, *J. Chem. Soc.* **1952**, 142; Eastland *et al.*, *Brit.* pat. **695,298** (1953 to Allen & Hanbury).

Cream colored granules from alcohol + ether, mp 172-174° after darkening at 164-166°.

THERAP CAT: Muscle relaxant (skeletal).

5393. Laureline. *(R)-6,7,7a,8-Tetrahydro-11-methoxy-7-methyl-5H-benzo[g]-1,3-benzodioxolo[6,5,4-de]quinoline; 10-methoxy-1,2-(methylenedioxy)aporphine.* $C_{19}H_{19}NO_3$; mol wt 309.36. C 73.77%, H 6.19%, N 4.53%, O 15.52%. From bark of *Laurelia novae-zelandiae* A. Cunn., *Lauraceae:* Aston, *J. Chem. Soc.* **97**, 1381 (1910). Synthesis: Schlittler, *Helv. Chim. Acta* **15**, 394 (1932); Faltis *et al.*, *Ber.* **77B**, 686 (1945); Gibson *et al.*, *J. Chem. Soc. (C)* **1970**, 2234; Govindachari *et al.*, *Indian J. Chem.* **8**, 475 (1970).

Cubes from petr ether. $[\alpha]_D^{20}$ −99.2° (c = 0.736 in 50% alc). Absorption spectrum: Girardet, *J. Chem. Soc.* **1931**, 2636. Feebly basic. Readily oxidizes in air. Sol in alcohol, ether. Practically insol in water.

Hydrochloride, $C_{19}H_{19}NO_3$·HCl, crystals, mp 280°. $[\alpha]_D^{20}$ −57°. Sparingly sol in water.

(S)-Form, cubes from petr ether, mp 114-115°. $[\alpha]_D^{20}$ +97.9° (c = 1.12 in abs alc); $[\alpha]_D^{20}$ +100° (c = 0.560 in CS_2).

(±)-Form, rough needles from petr ether, mp 115-116°.

Nitrate, $C_{19}H_{19}NO_3$·HNO₃, crystals, mp 238-240°. Sparingly sol in water.

Hydriodide, $C_{19}H_{19}NO_3$·HI, crystals. Practically insol in water.

Oxalate, $(C_{19}H_{19}NO_3)_2$·(COOH)₂; moderately sol in water.

5394. Laurel Oil. Laurel berry oil. Fixed oil from fresh fruit of *Laurus nobilis* L., *Lauraceae. Constit.* Chiefly the lauryl alcohol esters of lauric, stearic, etc., acids and a volatile oil, the so-called laurel camphor.

Greenish, fatty solid; the green color due to presence of some chlorophyll. d ~0.88. mp ~40°. n_D^{25} 1.4783. Sapon no. 198-199. Iodine no. 68-80. Insol in water. Sparingly sol in alcohol; sol in benzene, ether, carbon disulfide.

5395. Laurepukine. *(7R-cis)-6,7,7a,8-Tetrahydro-7-methyl-5H-benzo[g]-1,3-benzodioxolo[6,5,4-de]quinoline-12-ol 7-oxide.* $C_{18}H_{17}NO_4$; mol wt 311.34. C 69.44%, H 5.50%, N 4.50%, O 20.56%. From bark of *Laurelia novae-zelandiae* A. Cunn., *Lauraceae.* Isoln: Aston, *J. Chem. Soc.* **97**, 1381 (1910). Structure: Girardet, *Helv. Chim. Acta* **14**, 504 (1931); Govindachari *et al.*, *Ber.* **91**, 36 (1958). Revised structure: Weiss *et al.*, *Helv. Chim. Acta* **54**, 1342 (1971).

Needles from chloroform + ether, mp 230-231°. $[\alpha]_D$ −222° (c = 0.022 in chloroform). Practically insol in water; sol in benzene, chloroform.

5396. Lauric Acid. *Dodecanoic acid;* laurostearic acid; dodecoic acid. $C_{12}H_{24}O_2$; mol wt 200.32. C 71.95%, H 12.08%, O 15.97%. $CH_3(CH_2)_{10}COOH$. Isoln from coconut oil: Dale, Meara, *J. Sci. Food Agr.* **6**, 162 (1955); Naudet *et al.*, *Bull. Soc. Chim. France* **1959**, 718; from arecanut fat: Pathak, Mathur, *J. Sci. Food Agr.* **5**, 461 (1954); from *Holoptelea integrifolia* seed oil: Badami, *ibid.* **13**, 297 (1962). Synthesis: Ballard *et al.*, U.S. pat. **2,572,238** (1951 to Shell); Langenbeck, Richter, *Ber.* **89**, 202 (1956); Sprowls, U.S. pat. **2,782,214** (1957 to Baker Castor Oil); Zapesochnoya *et al.*, *Zh. Obshch. Khim.* **33**, 2552 (1963). Toxicity study: L. Orö, A. Wretlind, *Acta Pharmacol. Toxicol.* **18**, 141 (1961).

White, crystalline powder; slight odor of bay oil, mp 44°; also reported as 48°. d_5^4 0.869. bp₁₀₀ 225°; bp₂₀ 160-165°. n_D^{82} 1.4183. Insol in water. 1 g dissolves in 1 ml alcohol, 2.5 ml propyl alcohol; freely sol in benzene, ether. LD₅₀ i.v. in mice: 131 ±5.7 mg/kg (Orö, Wretlind).

5397. Laurocapram. *1-Dodecylhexahydro-2H-azepin-2-one;* 1-dodecylazacycloheptan-2-one; N-dodecyl-ε-caprolactam; N-0252; Azone. $C_{18}H_{35}NO$; mol wt 281.48. C 76.81%, H 12.53%, N 4.98%, O 5.68%. Caprolactam derivative used to enhance percutaneous absorption of physiologically active agents. Also exhibits intrinsic anti-inflammatory activity. Prepn: A. P. Swain *et al.*, *J. Org. Chem.* **18**, 1087 (1953). Improved synthesis: V. J. Rajadhyaksha *et al.*, Eur. pat. Appl. **95,096**; *eidem*, U.S. pat. **4,422,970** (both 1983 to Nelson). Use as skin penetrant: V. J. Rajadhyaksha, U.S. pats. **3,989,816**; **4,405,616** (1976, 1983 to Nelson). Anti-inflammatory activity: E. L. Nelson, U.S. pat. **4,310,-525** (1982 to Nelson). Pharmacology: R. B. Stoughton, *Arch. Dermatol.* **118**, 474 (1982). Comprehensive description: *idem*, *Drug Dev. Ind. Pharm.* **9**, 725 (1983). Topical antiviral activity *in vivo*: M. F. Leonard *et al.*, *Chemotherapy (Basel)* **33**, 151 (1987).

Clear, colorless liquid, mp −7°. bp₅₀ᵤ 160°. d 0.91; n 1.4701. Insol in water. Freely sol in most organic solvents. LD₅₀ rats, mice (g/kg): 8 i.v., i.p. (Stoughton).

USE: Pharmaceutic aid (excipient).

5398. Lauroguadine. *N,N′′′-[4-(Dodecyloxy)-1,3-phenylene]bisguanidine; 1,1′-[4-(dodecyloxy)-m-phenylene]diguanidine;* 2,4-diguanidinophenyl lauryl ether; 2,4-diguanidino-1-dodecyloxybenzene; 2,4-diguanidino-1-lauryloxybenzene; 2,4-diguanidinophenyl dodecyl ether; dodecyl 2,4-diguanidinophenyl ether. $C_{20}H_{36}N_6O$; mol wt 376.55. C 63.80%, H 9.64%, N 22.32%, O 4.25%. Prepn of the dihydrochloride: Pasini, *Farmaco Ed. Sci.* **8**, 646 (1953); *idem, Rend. Sci.*

Farmitalia **1**, 405 (1954); **Brit. pat 730,394** (1955 to Farmitalia).

Dihydrochloride monohydrate, $C_{20}H_{36}N_6O.2HCl.H_2O$, *P-7, Farmidril.* Crystals from water + ethanol. Sinters from 135-210°, dec 250°. Soly (g/100 ml): 0.2 in cold water; ~10 in boiling water; 84.2 in methanol (25°). Practically insol in acetone, benzene. Aq solns are neutral to litmus.

THERAP CAT: Antiprotozoal (Trichomonas).

5399. Laurolinium Acetate. *4-Amino-1-dodecyl-2-methylquinolinium acetate; 4-amino-1-dodecyquinaldinium acetate;* 1-dodecyl-4-aminoquinaldinium acetate; Laurodin. $C_{24}H_{38}N_2O_2$; mol wt 386.58. C 74.57%, H 9.91%, N 7.25%, O 8.28%. Prepn: D. Caldwell *et al., J. Pharm. Pharmacol.* **13**, 554 (1961); D. Caldwell, L. R. Rowe, U.S. pat. **2,997,-476** (1961 to Allen & Hanburys). Toxicity data: W. A. Cox, P. F. D'Arcy, *J. Pharm. Pharmacol.* **15**, 129 (1963).

Crystals from acetone, mp 170-171°. Freely sol in water. About 1 g dissolves in 2 ml of water at 20°. LD_{50} in mice (mg/kg): 131 ±36.2 orally; 30.2 ±5.6 s.c.; 6.0 ±04 i.v.; 2.3 ±0.2 i.p. (Cox, D'Arcy).

THERAP CAT: Antiseptic.

5400. Laurotetanine. *(S)-5,6,6a,7-Tetrahydro-1,2,10-trimethoxy-4H-dibenzo[de,g]quinolin-9-ol; 1,2,10-trimethoxy-6aα-noraporphin-9-ol;* Litsoeine. $C_{19}H_{21}NO_4$; mol wt 327.38. C 69.71%, H 6.47%, N 4.28%, O 19.55%. From the bark of *Litsea citrata* Blume (*Tetranthera citrata* (Blume) Nees), *Lauraceae* and allied plants. Isoln: Greshoff, *Ber.* **23**, 3537 (1890); Filippo, *Arch. Pharm.* **236**, 601 (1898). Structure: Barger *et al., Ber.* **66**, 450 (1933). Synthesis: Kikkawa, *C.A.* **53**, 17163i (1959).

Monohydrate, needles, mp 125°. $[\alpha]_D^{25}$ +98.5°. Practically insol in water. Freely sol in alcohol, chloroform, ethyl acetate, slightly in ether.

5401. 3-O-Lauroylpyridoxol Diacetate. *Dodecanoic acid 4,5-bis[(acetyloxy)methyl]-2-methyl-3-pyridinyl ester; lauric acid ester with pyridoxol diacetate (ester);* 5-lauroyloxy-6-methyl-3,4-pyridinedimethanol diacetate; 3-lauroyloxy-2-picoline-4,5-dimethanol diacetate; 2-methyl-3-lauroyloxy-4,5-diacetoxymethylpyridine; Epixine; Rosamit. $C_{24}H_{37}NO_6$; mol wt 435.56. C 66.18%, H 8.56%, N 3.22%, O 22.04%. Prepn: **Belg. pat. 640,827** (1964 to Soc. Belge Azote Prod. Chim. Marly), *C.A.* **63**, 587h (1965).

Crystals, mp 44°. Practically insol in water; sol in ether, chloroform, ethanol, ethylene dichloride.

THERAP CAT: Antiseborrheic.

5402. Lauryl Bromide. *1-Bromododecane;* dodecyl bromide. $C_{12}H_{25}Br$; mol wt 249.23. C 57.83%, H 10.11%, Br 32.06%. $CH_3(CH_2)_{10}CH_2Br$. Prepd by the action of hydrobromic acid on primary *n*-lauryl alcohol in the presence of sulfuric acid: Kamm, Marvel, *Org. Syn.* **1**, 7 (1921). Liquid. bp_{45} 175-180°. Insol in water. Sol in alc, ether.

5403. Lavender. Garden lavender; true lavender. Flowers of *Lavandula officinalis* Chaix (*L. vera* DC.), *Labiatae. Habit.* Mediterranean region. *Constit.* Volatile oil.

USE: For fumigating; in perfumery; to keep moths from clothes; manuf oil lavender. Pharmaceutic aid (perfume).

5404. Lawesson's Reagent. *2,4-Bis(4-methoxyphenyl)-1,3,2,4-dithiadiphosphetane 2,4-disulfide;* anisyldithiophosphinic anhydride. $C_{14}H_{14}O_2P_2S_4$; mol wt 404.48. C 41.57%, H 3.49%, O 7.91%, P 15.32%, S 31.71%. Thionating agent for conversion of carbonyls to thiocarbonyls. Prepn: II. Z. Lecher *et al., J. Am. Chem. Soc.* **78**, 5018 (1956); and delineation of use: B. S. Pedersen *et al., Bull. Soc. Chim. Belg.* **87**, 223 (1978); S. Scheibye *et al., ibid.,* 229. Mechanistic study: T. B. Rauchfuss, G. A. Zank, *Tetrahedron Letters* **27**, 3445 (1986). Use in thionations: A. Z.-Q. Khan, J. Sandström, *J. Chem. Soc. Perkins Trans. I* **1988**, 2085; S. Araki *et al., Bull. Chem. Soc. Japan* **61**, 2977 (1988); as reducing agent for sulfoxides: H. Bartsch, T. Erker, *Tetrahedron Letters* **33**, 199 (1992). Review of thionation reactions: M. P. Cava, M. I. Levinson, *Tetrahedron* **41**, 5061-5087 (1985).

mp 229°, should be stored under anhydrous conditions.

USE: Thiation reagent.

5405. Lawrencium. Lr; formerly Lw; at. no. 103; valence 3. Man-made radioactive element. No stable nuclides; known isotopes (mass numbers): 253-262. Longest-lived known isotope: 262 ($T_{1/2}$ 216 minutes, rel. at. mass 262.11). Prepn of first isotope ^{257}Lr ($T_{1/2}$ 0.65 sec, α-emitter): A. Ghiorso *et al., Phys. Rev. Lett.* **6**, 473 (1961). Prepn of isotope originally assigned mass number 257, later changed to 258 ($T_{1/2}$ 4.3 seconds, α-emitter) by bombardment of Cf with boron ions: Eskola *et al., Phys. Rev. C* **4**, 632 (1971). Prepn of isotopes including ^{256}Lr ($T_{1/2}$ 28 seconds) by irradiating ^{243}Am with ^{18}O ions: Donets *et al., At. Energ. (USSR)* **19**, 109 (1965), *C.A.* **64**, 1542c (1966). *Reviews*: C. Keller, *The Chemistry of the Transuranium Elements* (Verlag Chemie, Weinheim, English Ed., 1971) pp 609-612; Silva, "Trans-Curium Elements" in *MFP Int. Rev. Sci.: Inorg. Chem., Ser. One* vol. 8, A. G. Maddock, Ed. (University Park Press, Baltimore, 1972) pp 71-105; Ghiorso, *Handb. Exp. Pharmakol.* **36**, 691-715 (1973); Taylor, *ibid.* 717-738; R. J. Silva in *The Chemistry of the Actinide Elements* vol. 2, J. J. Katz *et al.,* Eds. (Chapman and Hall, New York, 1986) pp 1099-1103; *The Elements Beyond Uranium*, G. T. Seaborg, W. D. Loveland, Eds. (John Wiley & Sons, Inc., New York, 1990) p 49-51.

Caution: Radiation hazard; handling requires special equipment and shielding facilities (Katz *et al., loc cit.,* vol. 2, p. 1128).

5406. Lawsone. *2-Hydroxy-1,4-naphthalenedione; 2-hydroxy-1,4-naphthoquinone.* $C_{10}H_6O_3$; mol wt 174.16. C 68.97%, H 3.47%, O 27.56%. From leaves of *Lawsonia inermis* L. and *L. alba* Lam., *Lythraceae:* Latif, *Indian J. Agr. Sci.* **29**, No. 2-3, 147 (1959), *C.A.* **55**, 14828g (1961). Synthesis: Fieser, *J. Am. Chem. Soc.* **70**, 3165 (1948); Jain, Seshadri, *Proc. Indian Acad. Sci.* **35A**, 233 (1952); Eistert, Müller, *Ber.* **92**, 2071 (1959).

Yellow prisms from acetic acid, dec 195-196°.

THERAP CAT: Ultraviolet screen.

5407. Lazabemide. *N-(2-Aminoethyl)-5-chloro-2-pyridinecarboxamide; N-(2-aminoethyl)-5-chloropicolinamide.* $C_8H_{10}ClN_3O$; mol wt 199.64. C 48.13%, H 5.05%, Cl 17.76%, N 21.05%, O 8.01%. Member of a novel class of potent, reversible, and extremely selective monoamine oxidase type B (MAO-B) inhibitors. Prepn: R. Imhof, E. Kyburz, *Brit. pat.* **Appl. 2,163,746;** *eidem,* *U.S. pat.* **4,764,-522** (1986, 1988 both to Hoffmann-LaRoche). HPLC determn in plasma: R. Wyss, W. Philipp, *J. Chromatog.* **507**, 187 (1990). Neurochemical profile: W. E. Haefely *et al., Advan. Neurol.* **53**, 505 (1990). Enzyme binding studies: J. Saura *et al., J. Neurosci.* **12**, 1977 (1992). Clinical trial: Parkinson Study Group, *Ann. Neurol.* **33**, 350 (1993).

Monohydrochloride, $C_8H_{10}ClN_3O.HCl$, *Ro-19-6327.* Crystals from methanol/ether, mp 193-195°. Partition coefficient (*n*-octanol/water) ~0.1. pKa 8.9. LD_{50} orally in mice: 1000-2000 mg/kg (Imhof, Kyburz, 1986).

THERAP CAT: Antiparkinsonian.

5408. Lazurite. Lapis lazuli; lasurite. Composition: $(Na,Ca)_4(AlSiO_4)_3(SO_4,S,Cl)$. E. S. Dana, *A System of Mineralogy* (John Wiley, New York, 6th ed., 1901) pp 432-433; C. S. Hurlbut, Jr., *Dana's Manual of Mineralogy* (John Wiley, New York, 17th ed., 1959) p 503.

Blue, blue-violet or greenish-blue, translucent, cubic or dodecahedral crystals. d 2.4. Dec by HCl with pptn of SiO_2 and evolution of H_2S.

USE: In manuf of vases, ornamental furniture, mosaics; in paints, jewelry.

5409. LBF. *Lactobacillus bulgaricus* factor. Growth factor occurring in products derived from both animal and plant sources and in culture filtrate of certain microorganisms: Williams *et al., J. Biol. Chem.* **177**, 933 (1949); Vitucci *et al., Arch. Biochem. Biophys.* **34**, 409 (1951); Peters *et al., J. Am. Chem. Soc.* **75**, 1688 (1953). Contains pantetheine *q.v.* which is oxidized during purification to the disulfide, pantethine *q.v.* Natural occurrence of several different forms of LBF each being a mixed disulfide of pantetheine: Rasmussen *et al., Proc. Soc. Exp. Biol. Med.* **73**, 658 (1950); Brown, Snell, *J. Biol. Chem.* **198**, 375 (1952). Coenzyme A digested with intestinal phosphatase shows 2-4 LBF-active components: Long, Williams, *J. Bacteriol.* **61**, 195 (1951). *Review:* Snell, Brown, "Pantethine and Related Forms of the Lactobacillus Bulgaricus Factor (LBF)" in *Advan. Enzymol.* **14**, 49-71 (1953).

5410. Lead. Pb; at. wt 207.2; at. no. 82; valence 2, 4. Four naturally occurring isotopes: 204 (1.40%); 206 (25.2%); 207 (21.7%); 208 (51.7%); artificial, radioactive isotopes: 195-203; 205; 209-214. One of the metals known to the ancient world. Extent of occurrence in earth's crust about 15 g/ton, also expressed as 0.002% (depth of crust: 16 km).

Occurs chiefly as sulfide in galena, other minerals include anglesite ($PbSO_4$), *cerussite* ($PbCO_3$), *mimetite* [$PbCl_2.3Pb_3(AsO_4)_2$] and *pyromorphite* [$PbCl_2.3Pb_3(PO_4)_2$]. Recovery from ore and purification: Heuser, *Metall.* **9**, 675 (1955), *C.A.* **49**, 14609 (1955); Ziegfeld, *Eng. Mining J.* **153**, 82 (1952), *C.A.* **46**, 2975 (1952). Prepn of high purity lead: Piontelli, Fagnani, *Chim. Ind. (Milan)* **34**, 629 (1952), *C.A.* **47**, 12062 (1953); Giesen, *Technik (Berlin)* **2**, 393 (1947), *C.A.* **42**, 852 (1948); Hughes *J. Electrochem. Soc.* **101**, 267 (1954); Baralis, Marone, *Met. Ital.* **59**, 494 (1967), *C.A.* **67**, 119613a (1967). Reviews of lead, its alloys and compds: W. Hofmann, *Lead and Lead Alloys, Properties and Technology* (Springer, New York, Eng. Ed., 1970) 551 pp; Abel in *Comprehensive Inorganic Chemistry* vol. 2, J. C. Bailar, Jr. *et al.,* Eds. (Pergamon Press, Oxford, 1973) pp 105-146; H. E. Howe in Kirk-Othmer *Encyclopedia of Chemical Technology* vol. **14** (Wiley-Interscience, New York, 3rd ed., 1981) pp 98-139. Review of carcinogenicity studies of lead and lead compds: *IARC Monographs* **23**, 325-415 (1980).

Bluish-white, silvery, gray metal. Highly lustrous when freshly cut, tarnishes upon exposure to air. Very soft and malleable, easily melted, cast, rolled, and extruded. Cubic crystal structure. mp 327.4°; bp 1740°. d_4^{20} 11.34; d (at mp) 10.65: Schneider *et al., Naturwiss.* **41**, 326 (1954). Heat of vaporization (1740°) 206 cal/g. Heat capacity (20°): 0.031 cal/g/°C. Resistivity (μ-ohm-cm) at 20°: 20.65; at 100°: 27.02; at 320°: 54.76; at 330°: 96.74. Vapor pressure at 1000°: 1.77 mm Hg. E° (aq) Pb/Pb²⁺ +0.126 v. Coefficient of linear expansion (0-100°) 29×10^{-6}, (20-300°) 31.3×10^{-6}, ($-183°$ to $+14°$) 27×10^{-6}; thermal conductivity varies from 0.083 at 50° to 0.077 at 225°: Francl, Kingery, *J. Am. Ceram. Soc.* **37**, 80 (1954); viscosity of molten lead (327.4°) 3.2 centipoises, (400°) 2.32 cp, (600°) 1.54 cp, (800°) 1.23 cp. Heat capacity and heat of fusion study: Douglas, Dever, *J. Am. Chem. Soc.* **76**, 4824 (1954); hardness 1 on Mohs' scale; Brinell hardness (high purity Pb) 4.0: McLellan, *Am. Mineralogist* **30**, 635 (1945). Reacts with hot concd nitric acid, with boiling concd hydrochloric or sulfuric acid. Attacked by pure water, weak organic acids in the presence of oxygen. Resistant to tap water, hydrofluoric acid, brine, solvents.

Caution: *Acute Toxicity:* most common in young children with history of pica; anorexia, vomiting, malaise, convulsions due to increased intracranial pressure. May leave permanent brain damage. Blood lead increased above 0.05 mg%. *Chronic Toxicity:* children show weight loss, weakness, anemia. Lead poisoning in adults is usually occupational due mainly to inhalaton of lead dust or fumes. Wristdrop and colic rarely occur. More often there are vague G.I. and CNS complaints. Pb content of blood > 0.05 mg% and of urine > 0.08 mg per liter support a diagnosis of Pb poisoning. Provocative excretion test using Edathamil may be helpful in confirming excess Pb absorption. Review of toxicity: *Clinical Toxicology of Commercial Products,* R. E. Gosselin *et al.,* Eds. (Williams & Wilkins, Baltimore, 4th ed., 1976) Section III, pp 194-202; *Lead Toxicity,* R. L. Singhal, J. A. Thomas, Eds. (Urban & Schwarzenberg, Baltimore, 1980) 514 pp.

USE: Construction material for tank linings, piping, and other equipment handling corrosive gases and liqs used in the manuf of sulfuric acid, petr refining, halogenation, sulfonation, extraction, condensation; for x-ray and atomic radiation protection; manuf of tetraethyllead, pigments for paints, and other organic and inorganic lead compds; bearing metal and alloys; storage batteries; in ceramics, plastics, and electronic devices; in building construction; in solder and other lead alloys; in the metallurgy of steel and other metals. *Review* of uses, corrosion metallurgy: Mullarkey, *Ind. Eng. Chem.* **49**, 1607 (1957).

5411. Lead Acetate. Neutral lead acetate; normal lead acetate; sugar of lead; salt of Saturn. $C_4H_6O_4Pb$; mol wt 325.29. C 14.77%, H 1.86%, O 19.67%, Pb 63.70%. Pb-$(CH_3COO)_2$. Toxicity data: W. R. Bradley, W. G. Fredrick, *Ind. Med.* **10**, *Ind. Hyg. Sect.* **2**, 15 (1941). Review of safety assessment in hair coloring: A. J. Cohen, F. J. C. Roe, *Food Chem. Toxicol.* **29**, 485-507 (1991).

Occurs as trihydrate, colorless crystals or white granules or powder; slight acetic odor; slowly effloresces. *Poisonous!* Takes up CO_2 from air and becomes incompletely sol. d

2.55. mp 75° when rapidly heated; at a little above 100° it begins to lose acetic acid; dec completely above 200°. One gram dissolves in 1.6 ml water, 0.5 ml boiling water, 30 ml alcohol; freely sol in glycerol. Aq solns of lead acetate dissolve lead monoxide. pH of 5% aq soln at 25° = 5.5-6.5. *Keep well closed.* LD_{50} i.p. in rats: 15 mg Pb/100g (Bradley, Fredrick).

Note: This substance may reasonably be anticipated to be a carcinogen: *Seventh Annual Report on Carcinogens* (PB95-109781, 1994) p 238.

USE: Mordant in cotton dyes; lead coating for metals; drier in paints, varnishes and pigment inks; colorant in hair dyes. Weighting silks; manuf lead salts, chrome-yellow; also for various analytical procedures, e.g. detection of sulfide, determination of CrO_3, MoO_3.

THERAP CAT: Astringent.

THERAP CAT (VET): Astringent and sedative (usually in lotions) for bruises and superficial inflammation. Has been used internally in diarrheas.

5412. Lead Antimonate(V). Naples yellow. Approx $Pb_3(SbO_4)_2$.

Orange-yellow powder. Insol in water, dil acids.

USE: As pigment in oil painting, staining glass, crockery and porcelain.

5413. Lead Arsenate. Approx $PbHAsO_4$. Occurs in nature as the mineral *schultenite*.

White, heavy powder. *Poisonous!* d 5.79. At ~280° loses H_2O and is converted into pyroarsenate. Insol in water; sol in HNO_3, caustic alkalies. LD_{50} in rats, rabbits: approx 825, 125 mg/kg orally, Voigt *et al.*, *J. Am. Pharm. Assoc.* 37, 122 (1948).

USE: As constituent of various insecticides for larvae of gypsy moth, boll weevil, etc.

THERAP CAT (VET): Has been used as a teniacide; insecticide.

5414. Lead Arsenite. Approx $Pb(AsO_2)_2$.

White powder. *Poisonous!* d 5.85. Insol in water; sol in dil HNO_3.

USE: As insecticide like the arsenate.

5415. Lead Azide. N_6Pb; mol wt 291.24. N 28.86%, Pb 71.14%. $Pb(N_3)_2$. Prepd from sodium azide and lead nitrate: Schenk in *Handbook of Preparative Inorganic Chemistry* vol. 1, G. Brauer, Ed. (Academic Press, New York, 2nd ed., 1963) p 763. Most complete description. B. T. Fedoroff *et al.*, *Encyclopedia of Explosives and Related Items* vol. 1 (Picatinny Arsenal, Dover, N.J., 1960) pp A545-A587.

Needles or white powder. Explodes at 350° or on percussion. Heat of formation (25°): +110.5 kcal/mol. Soly in water: 0.023% at 18°; 0.09% at 70°. Freely sol in acetic acid. Insol in NH_4OH.

USE: As primer in explosives. Generally used in the form of dextrinated lead azide.

5416. Lead Borate. Approx $Pb(BO_2)_2 \cdot H_2O$.

White powder. *Poisonous!* Insoluble in water; soluble in dil HNO_3.

USE: Drier for varnishes and paints; with other metals (e.g., Ag) in galvanoplasty for production of conducting coatings on glass, pottery, porcelain, and chinaware.

5417. Lead Bromate. Br_2O_6Pb; mol wt 463.00. Br 34.52%, O 20.73%, Pb 44.75%. $Pb(BrO_3)_2$.

Monohydrate, colorless crystals. *Poisonous!* d 5.53. Dec at 180°. Slightly sol in cold water, freely sol in hot water.

Note: Pure lead bromate is not dangerous, but when made by pptng lead acetate with an alkali bromate, it may detonate or explode on heating, striking or rubbing because some acetate is occluded.

5418. Lead Bromide. Br_2Pb; mol wt 367.01. Br 43.54%, Pb 56.46%. $PbBr_2$.

White, cryst powder. *Poisonous!* d 6.66. mp 373°. On solidifying forms a horn-like mass. Sol in about 200 parts cold water, 20 parts boiling water; insol in alcohol.

5419. Lead Butyrate. *Butyric acid lead salt.* $C_8H_{14}O_4$-Pb; mol wt 381.40. C 25.19%, H 3.70%, O 16.78%, Pb 54.33%. $Pb(C_4H_7O_2)_2$.

Colorless scales or viscid mass. *Poisonous!* mp ~90°. Insol in water; sol in dil HNO_3.

5420. Lead Chlorate. Cl_2O_6Pb; mol wt 374.10. Cl 18.95%, O 25.66%, Pb 55.39%. $Pb(ClO_3)_2$.

Colorless, deliquesc crystals. *Poisonous!* d 3.9. Dec at 230°. Sol in 0.7 part water, freely in alcohol.

5421. Lead Chloride. Cl_2Pb; mol wt 278.11. Cl 25.50%, Pb 74.50%. $PbCl_2$. Occurs in nature as the mineral *cotunnite.*

White, cryst powder. *Poisonous!* d 5.85. mp 501°. bp 950°. Sol in 93 parts cold water, 30 parts boiling water; readily sol in soln of NH_4Cl, NH_4NO_3, alkali hydroxides; slowly in glycerol. MLD orally in guinea pigs: 1.5-2.0 g/kg, *Handbook of Toxicology* vol. 1, W. S. Spector, Ed. (Saunders, Philadelphia, 1956) pp 176-177.

USE: Manuf Pattison's white lead, Verona Yellow, Turner's Patent Yellow, lead oxychloride; as solder and flux.

5422. Lead Chromate(VI). Chrome yellow; Cologne yellow; King's yellow; Leipzig yellow; Paris yellow; C.I. Pigment Yellow 34; C.I. 77600. CrO_4Pb; mol wt 323.19. Cr 16.09%, O 19.80%, Pb 64.11%. $PbCrO_4$. Occurs in nature as the minerals *crocoite, phoenicochroite*. Ref: *Colour Index* vol. 4 (3rd ed., 1971) p 4677.

Yellow or orange-yellow powder. d 6.3. mp 844°. It is one of the most insol salts (0.2 mg/l H_2O). Insol in acetic acid; sol in solns of fixed alkali hydroxides, in dil HNO_3. LD_{75} i.p. in guinea pigs: 156 mg/kg, *Handbook of Toxicology* vol. 1, W. S. Spector, Ed. (Saunders, Philadelphia, 1956) pp 176-177.

USE: Pigment in oil and water colors; printing fabrics, decorating china and porcelain; in chemical analysis of organic substances; in traffic paints.

Note: Basic lead chromates of various shades of color from brown-yellow to red are used as pigments.

5423. Lead Chromate(VI) Oxide. *Chromic acid lead-(2+) salt (1:2);* basic lead chromate; red lead chromate; chrome red; chromium lead oxide; Persian red; Austrian cinnabar. $CrPb_2O_5$; mol wt 546.39. Cr 9.52%, Pb 75.84%, O 14.64%. $PbCrO_4 \cdot PbO$. *See: Colour Index* vol. 4, (3rd ed., 1971) p 4677.

Red powder. Insol in water.

USE: As pigment.

5424. Lead Dioxide. Lead oxide brown; lead peroxide; lead superoxide. O_2Pb; mol wt 239.20. O 13.38%, Pb 86.62%. Occurs in nature as the mineral *plattnerite*. Lab prepn from lead acetate and calcium hypochlorite: Newell, Maxson, *Inorg. Syn.* 1, 45 (1939); by hydrolysis of lead acetate: Kuhn, Hammer, *Ber.* 83, 413 (1950).

Dark-brown powder; evolves oxygen when heated, first forming Pb_3O_4, at high temp PbO. d 9.38. Insol in water; sol in HCl with evolution of Cl; in dil HNO_3 in presence of H_2O_2, oxalic acid, or other reducers; sol in alkali iodide solns with liberation of iodine; soluble in hot caustic alkali solns. LD_{50} i.p. in guinea pigs: 200 mg/kg, *Handbook of Toxicology* vol. 1, W. S. Spector, Ed. (Saunders, Philadelphia, 1956) pp 176-177.

USE: Electrodes in batteries; oxidizing agent in manuf dyes; as discharge in dyeing with indigo; manuf rubber substitutes; with amorphous phosphorus as ignition surface for matches; pyrotechny; manuf pigments; in anal. chemistry.

5425. Lead Fluoride. Lead difluoride; plumbous fluoride. F_2Pb; mol wt 245.20. F 15.50%, Pb 84.50%. PbF_2. Prepd by treating lead carbonate or hydroxide with hydrogen fluoride and evaporating the soln; by mixing solns of potassium fluoride and lead acetate; by precipitation from the soln of a lead salt by HF; by the action of fluorine on lead: Ruff, *Die Chemie des Fluors* (Springer, Berlin, 1920) p 33; Eméleus in *Fluorine Chemistry* vol. 1, J. H. Simons, Ed. (Academic Press, New York, 1950) p 51; Kwasnik in *Handbook of Preparative Inorganic Chemistry* vol. 1, G. Brauer, Ed. (Academic Press, New York, 1963) pp 218-219. *Review:* Kemmitt, Sharp, *Advan. Fluorine Chem.* 4, 188 (1965).

White to colorless crystals. *Poisonous!* Dimorphous: orthorhombic, converted to cubic above 316°. d (orthorhombic) 8.445; d (cubic) 7.750. mp 824°. bp 1293°. Soly in water (g/100 ml): 0.057 (0°); 0.065 (20°). The soly increases in the presence of HNO_3 or nitrates.

5426. Lead Formate. $C_2H_2O_4Pb$; mol wt 297.24. C 8.08%, H 0.68%, O 21.53%, Pb 69.71%. $Pb(CHO_2)_2$.
White, lustrous prisms or needles. *Poisonous!* d 4.63. Dec at 190°. Sol in 65 parts cold water, 6 parts boiling water; insol in alcohol.

5427. Lead Hexafluorosilicate. Lead fluosilicate; lead silicofluoride. F_6PbSi; mol wt 349.28. F 32.64%, Pb 59.32%, Si 8.04%. $PbSiF_6$.
Dihydrate, colorless crystals. Sol in water. *Poisonous!*
USE: In refining lead by electrolytic methods.

5428. Lead Hydroxide. Lead oxide hydrate; basic lead hydroxide. $H_2O_4Pb_3$; mol wt 687.61. H 0.29%, O 9.31%, Pb 90.40%. $3PbO.H_2O$. Prepn and structure: Oswald *et al., Helv. Chim. Acta* **51**, 1389 (1968).
White powder; absorbs CO_2 from the air. *Poisonous!* d 7.41. Insol in H_2O; sol in dil acids, fixed alkali hydroxides.

5429. Lead Hypophosphite. $H_4O_4P_2Pb$; mol wt 337.18. H 1.20%, O 18.98%, P 18.37%, Pb 61.45%. $Pb(H_2PO_2)_2$.
Hygroscopic, cryst powder; dec at elevated temps. *Poisonous!* Slightly sol in cold water; more in hot water; insol in alcohol. *Keep well closed and protected from light.*

5430. Lead Iodide. I_2Pb; mol wt 461.01. I 55.06%, Pb 44.94%. PbI_2.
Bright yellow, heavy odorless powder. *Poisonous!* d 6.16. mp 402°. One gram dissolves in 1350 ml cold, 230 ml boiling water; sol in concd solns of alkali iodides; freely sol in soln of sodium thiosulfate; sol in 200 parts cold 90 parts hot aniline; insol in alcohol or cold HCl. *Protect from light.*
USE: Bronzing, gold pencils, mosaic gold, printing, photography.

5431. Lead Lactate. $C_6H_{10}O_6Pb$; mol wt 385.34. C 18.70%, H 2.62%, O 24.91%, Pb 53.77%. $Pb(C_3H_5O_3)_2$.
White, heavy, cryst powder. *Poisonous!* Slowly absorbs CO_2 from the air, becoming partly insol. Sol in water, hot alcohol. *Keep well closed.*

5432. Lead Molybdate(VI). MoO_4Pb; mol wt 367.14. Mo 26.13%, O 17.43%, Pb 56.44%. $PbMoO_4$. Occurs as the mineral *wulfenite*.
Yellow powder. Insol in water; sol in HNO_3 or NaOH when freshly pptd.
USE: In pigments.

5433. Lead Monoxide. Lead oxide yellow; plumbous oxide; litharge; massicot; lead protoxide. OPb; mol wt 223.20. O 7.17%, Pb 92.83%. PbO. Prepn of high purity PbO: Kwestroo, Huizing, *J. Inorg. Nucl. Chem.* **27**, 1591 (1965); Kwestroo *et al., ibid.* **29**, 39 (1967). Mfg processes: Faith, Keyes & Clark's *Industrial Chemicals,* F. A. Lowenheim, M. K. Moran, Eds (Wiley-Interscience, New York, 4th ed., 1975) pp 509-513. Toxicity study: W. R. Bradley, W. G. Fredrick, *Ind. Med.* **10**, *Ind. Hyg. Sect.* **2**, 15 (1941).
Exists in two forms: red to reddish-yellow, tetragonal crystals, stable at ordinary temp; yellow, orthorhombic crystals, stable above 489°: Petersen, *J. Am. Chem. Soc.* **63**, 2617 (1941). *Poisonous!* At 300-450° in the air, it converts slowly into Pb_3O_4, but at higher temp reverts to PbO. d 9.53. mp 888°. Insol in water, alcohol. Sol in acetic acid, dil HNO_3, in warm solns of fixed alkali hydroxides. LD_{50} i.p. in rats: 40 mg Pb/100g (Bradley, Fredrick).
Caution: Avoid breathing dust. Wear dust mask approved by U.S. Bureau of Mines for this purpose. Wash thoroughly before eating or smoking. Keep away from feed or food products.
USE: In ointments, plasters; preparing soln of lead subacetate. Glazing pottery; glass flux for painting on porcelain and glass; lead glass; varnishes; with glycerol as metal cement; producing iridescent colors on brass and bronze; coloring sulfur-containing substances, *e.g.,* hair, nails, wool, horn; manuf artificial tortoise shell and horn; pigment for rubber; manuf boiled linseed oil; in assay of gold and silver ores.

5434. Lead Nitrate. N_2O_6Pb; mol wt 331.21. N 8.46%, O 28.98%, Pb 62.56%. $Pb(NO_3)_2$.
White or colorless translucent crystals. *Poisonous!* d 4.53. One g dissolves in 2 ml cold, 0.75 ml boiling water, in 2500 ml abs alcohol, 75 ml abs methanol; insol in concd HNO_3. The aq soln is slightly acid. pH of 20% aq soln at 25° = 3.0-4.0. USE: Manuf matches and special explosives; as mordant in dyeing and printing on textiles; mordant for staining horn, mother-of-pearl; oxidizer in dye industry; sensitizer in photography; process engraving.
THERAP CAT (VET): Has been used as a caustic in equine canker.

5435. Lead Oleate. *Oleic acid lead salt.* Approx $Pb(C_{18}H_{33}O_2)_2$.
Granular, wax-like mass. Insol in water; sol, when fresh, in alc, benzene, ether, oil turpentine. LD orally in guinea pigs: 8.0 g/kg, *Handbook of Toxicology* vol. 1, W. S. Spector, Ed. (Saunders, Philadelphia, 1956) pp 176-177.
USE: In varnishes; in extreme pressure lubricants.

5436. Lead Oxalate. C_2O_4Pb; mol wt 295.22. C 8.14%, O 21.68%, Pb 70.19%.
White, heavy powder. Dec at 300°. d 5.28. *Poisonous!* Insol in water; sol in dil HNO_3, fixed alkali hydroxides; sparingly sol in acetic acid.

5437. Lead Phosphate. $O_8P_2Pb_3$; mol wt 811.54. O 15.77%, P 7.63%, Pb 76.59%. $Pb_3(PO_4)_2$.
White powder. d 6.9. mp 1014°. *Poisonous!* Insol in water, alcohol. Sol in HNO_3, fixed alkali hydroxides.
Note: This substance may reasonably be anticipated to be a carcinogen: *Seventh Annual Report on Carcinogens* (PB 95-109781, 1994) p 238.
USE: Stabilizer for styrene and casein plastics; in special glasses.

5438. Lead Selenate. O_4PbSe; mol wt 350.16. O 18.28%, Pb 59.17%, Se 22.55%. $PbSeO_4$. Prepd by adding a soln of lead nitrate to sodium selenate: Lenher, Kao, *J. Am. Chem. Soc.* **47**, 1521 (1925).
Orthorhombic crystals. d_4^{20} 6.37. Dec by heat. Sol in concd acids; insol in water.

5439. Lead Selenite. O_3PbSe; mol wt 334.16. O 14.36%, Pb 62.01%, Se 23.63%. $PbSeO_3$. Prepd by adding selenious acid or an alkali selenite to a soln of lead chloride or nitrate: Berzelius, cited in *Mellor's* vol. X, 833 (1930).
Powder. mp ~500°. Melts to a yellow liquid. Dec at a bright-red heat giving off selenium oxide. Very sparingly sol in water; difficultly dec by boiling sulfuric acid.

5440. Lead Sesquioxide. *Lead trioxide; plumbous plumbate.* O_3Pb_2; mol wt 462.40. O 10.38%, Pb 89.62%. Pb_2O_3.
Reddish-yellow powder; converted at 370° in air to Pb_3O_4, dec at ~530° to PbO. Insol in water; dec by concd HCl or H_2SO_4 with the liberation of Cl or oxygen, respectively.

5441. Lead Sodium Thiosulfate. Lead sodium hyposulfite. $Na_4O_9PbS_6$; mol wt 635.55. Na 14.47%, O 22.66%, Pb 32.60%, S 30.27%. $Na_4Pb(S_2O_3)_3$.
White, small, heavy crystals. *Poisonous!* Sparingly sol in water, more in thiosulfate solns.
USE: Manuf matches.

5442. Lead Stearate. *Stearic acid lead salt.* Approx $Pb(C_{18}H_{35}O_2)_2$.
White powder, mp ~125°. *Poisonous!* Insol in water; sol in hot alcohol.
USE: In extreme pressure lubricants; as drier in varnishes.

5443. Lead Subacetate. Lead monosubacetate; monobasic lead acetate. $C_4H_{10}O_8Pb_3$; mol wt 807.72. C 5.95%, H 1.25%, O 15.85%, Pb 76.96%. $Pb(C_2H_3O_2)_2.2Pb(OH)_2$.
White, heavy powder. *Poisonous!* Sol in 16 parts cold, 4 parts boiling water with alkaline reaction. On exposure to air absorbs CO_2 and becomes incompletely sol. *Keep well closed.*
USE: In sugar analysis to remove coloring matters, etc., from solns before polarizing; for clarifying and decolorizing other solns of organic substances.

5444. Lead Sulfate. O_4PbS; mol wt 303.26. O 21.10%, Pb 68.32%, S 10.57%. $PbSO_4$. Occurs as the minerals: *anglesite, lanarkite*.

White, heavy, cryst powder. *Poisonous!* d 6.2. mp 1170°. Sol in about 2225 parts water; more soluble in dil HCl or HNO_3, less in dil H_2SO_4; sol in NaOH, ammonium acetate or tartrate soln; sol in concd hydriodic acid; insol in alcohol. LD_{75} i.p. in guinea pigs: 290 mg/kg, *Handbook of Toxicology* vol. 1, W. S. Spector, Ed. (Saunders, Philadelphia, 1956) pp 178-179.

USE: Instead of white lead as pigment; with zinc in galvanic batteries; manuf minium, in lithography; preparing rapidly drying oil varnishes; weighting fabrics.

5445. Lead Sulfide. PbS; mol wt 239.27. Pb 86.60%, S 13.40%. Occurs as the mineral *galena*. Toxicity study: W. R. Bradley, W. G. Fredrick, *Ind. Med.* **10**, *Ind. Hyg. Sect.* **2**, 15 (1941).

Black powder. Insol in water. Sol in HNO_3, hot, dil HCl. LD_{50} i.p. in rats: 160 mg Pb/100 g (Bradley, Fredrick).

USE: Glazing earthenware.

5446. Lead Telluride. PbTe; mol wt 334.80. Pb 61.89%, Te 38.11%. Found in nature as the mineral *altaite*. Prepd from lead nitrate, sodium carbonate and powdered tellurium: Montignie, *Bull. Soc. Chim. France* **1947**, 750. Prepn of single crystals by heating stoichiometric quantities of the elements in a graphite cup or fused quartz tube: Brady, *J. Electrochem. Soc.* **101**, 466 (1954).

Silver-gray cubic crystals. d_4^{20} 8.16. mp 905°. Most of the crystal is *p*-type, the *n*-type material being present in the surface layer. Energy gap 0.27 ev. Electron mobility 2240 cm^2/volt-sec. Hole mobility 860 cm^2/volt-sec. Resistivity 0.005 ohm-cm (*p*-type), 0.00090 ohm-cm (*n*-type). Not attacked by hydrochloric, hydrofluoric, perchloric and acetic acids or their mixtures; not attacked by solns of 30% potassium hydroxide or of alkali metal sulfides. Dil nitric acid turns the surface black, while concd nitric acid produces lighter gray surface and turns the black surface to gray. Hot concd sulfuric acid produces a reddish-violet surface.

USE: In photoconductor cells; in semiconductor research.

5447. Lead Tetraacetate. $C_8H_{12}O_8Pb$; mol wt 443.38. C 21.67%, H 2.73%, O 28.87%, Pb 46.73%. $Pb(CH_3COO)_4$. Prepd from Pb_3O_4 and glacial acetic acid preferably in the presence of some acetic anhydride: Dimroth, Schweizer, *Ber.* **56**, 1375 (1923); Bailar, *Inorg. Syn.* **1**, 47 (1939); Baudler in *Handbook of Preparative Inorganic Chemistry* vol. 1, G. Brauer, Ed. (Academic Press, New York, 2nd ed., 1963) p 767. Prepn by electrolysis: Fioshin, Gus'kov, *Dokl. Akad. Nauk SSSR* **112**, 303 (1957), *C.A.* **51**, 16146 (1957); Sataev *et al.*, *Khim. Prom. (Moscow)* **46**, 892 (1970), *C.A.* **74**, 49005x (1971). Reviews of prepn and use as oxidizing agent: Criegee "Oxidations with Lead Tetraacetate" in *Oxidation in Organic Chemistry*, Part A, K. B. Wiberg, Ed. (Academic Press, New York, 1965) pp 277-366; Zyka, *Pure Appl. Chem.* **13**, 569-581 (1966).

Colorless monoclinic prisms from glacial acetic acid. Turns pink easily. Unstable in air. Hydrolyzed by water with the formation of brown lead dioxide and acetic acid. *Avoid contact with skin.* d_4^{17} 2.228. mp 175-180°. Sol in hot glacial acetic acid, benzene, chloroform, tetrachloroethane, nitrobenzene. Dissolves in concd halogen acids with the formation of haloplumbic acids, H_2PbX_6. The dry material can be stored in sealed, evacuated ampuls.

USE: Selective oxidizing agent in organic syntheses: Criegee, *Angew. Chem.* **53**, 321 (1940); *Newer Methods of Preparative Organic Chemistry* (Interscience, N. Y., 1948) pp 1-17.

5448. Lead Tetrafluoride. Plumbic fluoride. F_4Pb; mol wt 283.19. F 26.83%, Pb 73.17%. PbF_4. Prepd by passing fluorine diluted with CO_2 or N_2 over PbF_2 at 300°: v. Wartenberg, *Z. Anorg. Allgem. Chem.* **244**, 339 (1940). *Review:* Kemmitt, Sharp, *Advan. Fluorine Chem.* **4**, 187 (1965).

White, tetragonal crystals, d 6.7. mp ~600°. Readily hydrolyzes and turns brown (forms PbO_2) in the presence of moisture.

USE: Has been proposed as a fluorinating agent for hydrocarbons.

5449. Lead Tetroxide. Lead oxide red; red lead; minium; lead orthoplumbate; mineral orange; mineral red; Paris red; Saturn red; C.I. Pigment Red 105; C.I. 77578. O_4Pb_3; mol wt 685.60. O 9.33%, Pb 90.67%. Pb_3O_4. The article of commerce contains about 90% Pb_3O_4; the remainder being chiefly lead monoxide. Prepn: M. Baudler in *Handbook of Preparative Inorganic Chemistry* vol. 1, G. Brauer, Ed. (Academic Press, New York, 1963) pp 755-757. Structure: S. T. Gross, *J. Am. Chem. Soc.* **65**, 1107 (1943). *Review: Mellor's* vol. 7 (1930) pp 672-680.

Bright-red, heavy powder. *Poisonous!* Dec at ~500° with evolution of oxygen. d 9.1. Insol in water or alcohol; sol in excess glacial acetic acid, in hot HCl with evolution of Cl, in dil HNO_3 in presence of H_2O_2. LD_{50} i.p. in guinea pigs: 220 mg/kg, *Handbook of Toxicology* vol. 1, W. S. Spector, Ed. (Saunders, Philadelphia, 1956) pp 176-177.

USE: Plasters and ointments; manuf colorless glass; glaze for faience; flux for porcelain painting, protective paint for iron and steel; oil-color for ship paints, varnishes; coloring rubber; cement for glass, gas and steam pipes; storage batteries; pencils for writing on glass; manuf lead peroxide, matches.

5450. Lead Thiocyanate. Lead sulfocyanate. C_2N_2-PbS_2; mol wt 323.37. C 7.43%, N 8.66%, Pb 64.08%, S 19.83%. $Pb(SCN)_2$. Prepn: Gardner, Weinberger, *Inorg. Syn.* **1**, 85 (1939).

White, odorless powder. d 3.82. *Poisonous!* Sol in ~200 parts cold, 50 parts boiling water; also sol in alkali hydroxide and thiocyanate solns.

USE: Reverse dyeing with aniline black; manufacture of safety matches and cartridges.

5451. Lead Vanadate(V). Lead metavanadate. O_6-PbV_2; mol wt 405.08. O 23.70%, Pb 51.15%, V 25.15%. $Pb(VO_3)_2$.

Yellow powder. Insol in water. Dec by HNO_3.

USE: Manuf other vanadium compds; as pigment.

5452. Lecithin. Phosphatidylcholine; Lecithol; Vitellin; Kelecin; Granulestin. Phosphatide found in all living organisms (plants and animals). Significant constituent of nervous tissue and brain substance. A mixture of the diglycerides of stearic, palmitic, and oleic acids, linked to the choline ester of phosphoric acid. Commercial grades contain 2.2% P. Isoln from eggs: Sinclair, *Can. J. Res.* **26B**, 777 (1948). Product of commerce is predominantly soybean lecithin obtained as a by-product in the manuf of soybean oil: Stanley in K. S. Markley, *Soybeans* vol. II (Interscience, New York, 1951) pp 593-647. Soybean lecithin contains palmitic acid 11.7%, stearic 4.0%, palmitoleic 8.6%, oleic 9.8%, linoleic 55.0%, linolenic 4.0%, C_{20} to C_{22} acids (includes arachidonic) 5.5%. Synthesis of a mixed acid α-lecithin: de Haas, van Deenen, *Tetrahedron Letters* **1960** (no. 9), 1. Synthetic L-α-(distearoyl)lecithin is identical with hydrogenated egg yolk lecithin and L-α-(dipalmitoyl)lecithin is identical with colfosceril palmitate, *q.v.*, a natural phosphatide of brain, lung, and spleen. Commercial grades of natural lecithin are reported to contain a potent vasodepressor substance: McQuarrie, Andersen, U.S. pat. **2,931,818** (1960 to Cutter Labs.). Comprehensive monograph: G. B. Ansell, J. N. Hawthorne, *Phospholipids* (Elsevier, New York, 1964) 439 pp; A. Wendel, "Lecithin" in Kirk-Othmer *Encyclopedia of Chemical Technology* vol. 15 (Wiley-Interscience, New York, 4th ed., 1995) pp 192-210.

R = fatty acids

Waxy mass when the acid value is about 20. Pourable, thick fluid when the acid value is around 30. Color is nearly white when freshly made, but rapidly becomes yellow to brown in air. d_4^{24} 1.0305. Iodine value 95; saponification value 196. Insoluble but swells up in water and in NaCl soln forming a colloidal suspension. Soluble in about 12 parts cold, abs alcohol; sol in chloroform, ether, petr ether, in mineral oils and fatty acids; sparingly sol in benzene. Insol in acetone; practically insol in cold vegetable and animal oils.

USE: Edible and digestible surfactant and emulsifier of natural origin. Used in margarine, chocolate and in the food industry in general. In pharmaceuticals and cosmetics. Many other industrial uses, e.g. treating leather and textiles.

THERAP CAT: Lipotropic.

5453. Lectins. Agglutinins; affinitins; phytoagglutinins; phasins; protectins. A group of proteins, widely distributed in nature, that have the ability to agglutinate erythrocytes and many other types of cells. Although their existence has been known since 1899, when Stillmark isolated a hemagglutinin from castor beans, the term "lectin" (from the Latin *legere*, to choose) was first introduced by W. C. Boyd and E. Slapleigh in *Science* **119**, 419 (1954). It is now used to designate "a sugar-binding protein or glycoprotein of non-immune origin which agglutinates cells and/or precipitates glycoconjugates", I. J. Goldstein *et al., Nature* **285**, 66 (1980). Lectins are found primarily in seeds of plants, but also occur in roots, leaves and bark. In addition, they are present in invertebrates such as clams, snails, and horseshoe crabs, and in several vertebrate species. The term *phytohemagglutinin* is used to refer to plant lectins. Important members of the lectin family include concanavalin A, abrin, ricin, *q.q.v.*, as well as *soybean agglutinin* or *SBA* and *wheat germ agglutinin* or *WGA*. Lectins vary considerably in chemical and physical properties; only a limited number have been purified. Mol wts of 17,000 to 400,000 have been reported and most lectins have been found to contain Mn^{2+} and Ca^{2+}. Nearly all lectins can be inhibited by free oligo- or monosaccharides of appropriate specificity. Although their physiological functions in plants or in other organisms are unknown, lectins exhibit a variety of unusual biological properties. Some are specific in their reactions with human blood groups; some induce mitosis in lymphocytes. WGA from wheat germ lipase has been shown to agglutinate mouse tumor cells more readily than cells from normal tissue: J. C. Aub *et al., Proc. Nat. Acad. Sci. USA* **50**, 613 (1963); M. M. Burger, A. R. Goldberg, *ibid.* **57**, 359 (1967). Soybean agglutinin and concanavalin A have been shown to agglutinate cell lines transformed by viral or chemical carcinogens: M. Inbar, L. Sachs, *Nature* **223**, 710 (1969); *eidem, Proc. Nat. Acad. Sci. USA* **63**, 1418 (1969); B. A. Sela *et al., J. Membrane Biol.* **3**, 267 (1970). Some plant lectins mimic the direct effects of insulin on nuclear envelope phosphorylation: F. Purrello *et al., Science* **221**, 462 (1983). Soybean agglutinin has also been used in bone marrow transplants in patients with severe combined immunodeficiency: Y. Reisner *et al., Blood* **61**, 341 (1983). *Reviews:* N. Sharon, H. Lis, *Science* **177**, 949-955 (1972); *eidem, Ann. Rev. Biochem.* **42**, 541-574 (1973); L. Sequeira, *Ann. Rev. Phytopathol.* **16**, 453-481 (1978). Book: *Lectins: Biology, Biochemistry, Clinical Biochemistry* vol. **1**, T. C. Bog-Hansen, Ed. (de Gruyter, New York, 1981) 414 pp.

USE: As tools for studying cell surface properties; in cancer research.

5454. Ledol. *[1aR-(1aα,4α,4aβ,7α,7aβ,7bα)]-Decahydro-1,1,4,7-tetramethyl-1H-cycloprop[e]azulen-4-ol;* "Ledum camphor". $C_{15}H_{26}O$; mol wt 222.37. C 81.02%, H 11.79%, O 7.19%. Occurs in the essential oil from leaves of *Ledum palustre* L.: Grassmann, *Repert. Pharm.* **38**, 53 (1931); Hjelt, *Ber.* **28**, 3087 (1895); from *L. groenlandicum* Veder; *L. columbianum* Piper, *Ericaceae:* Cain, Lynn, *J. Am.*

Pharm. Assoc. **23**, 666 (1934); Penfold, *J. Proc. Roy. Soc. N.S. Wales* **59**, 206 (1925). Structure: G. Büchi *et al., Tetrahedron Letters* **1959** (no. 6), 14. Stereochemistry: L. Dolejs, F. Sorm, *Tetrahedron Letters* **1959** (no. 17), 1; revised stereochemistry: G. Büchi *et al., J. Am. Chem. Soc.* **91**, 6473 (1969).

Needles from alc, mp 104-105°. Sublimes easily, even below the mp. bp_{760} 292°. n_D^{110} 1.4667. $[\alpha]_D^{20}$ +28° (c = 10 in chloroform). Practically insol in water. Sol in alc (about 10% w/v). Soluble in other organic solvents.

Chromate, $C_{30}H_{50}O_4Cr$, ruby-red prisms, mp 92°. $[\alpha]_{671}^{20}$ +30° (c = 2 in chloroform).

5455. Lefetamine. *(R)-N,N-Dimethyl-α-phenylbenzeneethanamine; (−)-N,N-dimethyl-1,2-diphenylethylamine; (−)-N,N-dimethyl-α-phenylphenethylamine.* $C_{16}H_{19}N$; mol wt 225.33. C 85.29%, H 8.50%, N 6.22%. Centrally acting analgesic with stereochemical resemblance to morphine, *q.v.* Prepn and activity: K. Ogiu *et al., J. Pharm. Soc. Japan* **80**, 283 (1960); *eidem*, **Japan.** pat. **23,087**('61) (to Res. Found. Practical Life). Absolute configuration: M. Nakazaki, *Chem. & Ind. (London)* **1962**, 1577. NMR data indicates 22% eclipsed conformation which appears to enhance stereoselectivity for morphine receptors: T. Sasaki *et al., J. Med. Chem.* **9**, 847 (1966). Pharmacology: H. Nakamura, M. Shimizu, *Arch. Ind. Pharmacodyn. Ther.* **221**, 105 (1976). Clinical study: J. P. Famaey, T. L. Peeters, *Brux. Med.* **56**, 21 (1976). Analgesic activity: M. Nozaki *et al., Life Sci.* **33**, Suppl. 1, 431 (1983). Opiate agonist activity: M. Graziella de Montis *et al., Pharmacol. Res. Commun.* **17**, 471 (1985).

bp_6 142-147°. $[\alpha]_D^{20}$ −124.2° (ethanol).

Hydrochloride, $C_{16}H_{19}N.HCl$, *Santenol, SPA.* mp 218-219°. $[\alpha]_D^{20}$ −91.7° (water).

THERAP CAT: Analgesic.

5456. Leghemoglobin. *Legoglobin.* Hemoglobin-like red pigment present in the root nodules of leguminous plants. Isolation from soya beans: Keilin, Wang, *Nature* **155**, 227 (1945); Appleby, *Biochim. Biophys. Acta* **60**, 226 (1962). Mol wt is approx one-fourth that of hemoglobin: Ehrenberg, Ellfolk, *Acta Chem. Scand.* **17**, S343 (1963). Resolved into four components on DEAE-cellulose column: Ellfolk, *ibid.* **14**, 609 (1960). Suggested to act as an oxidoreduction catalyst in the symbiotic nitrogen fixation: *idem, ibid.* **15**, 975 (1961). Primary structure of soybean leghemoglobin: Ellfolk, Sievers, *ibid.* **25**, 3532 (1971).

5457. Leiopyrrole. *N,N-Diethyl-2-[2-(2-methyl-5-phenyl-1H-pyrrol-1-yl)phenoxy]ethanamine; 1-[o-(2-diethylaminoethoxy)phenyl]-2-methyl-5-phenylpyrrole; 2-methyl-1-(2-β-diethylaminoethoxyphenyl)-5-phenylpyrrole; DV-714.* $C_{23}H_{28}N_2O$; mol wt 348.49. C 79.27%, H 8.10%, N 8.04%, O 4.59%. Preparation: Buu-Hoi *et al., J. Med. Pharm. Chem.* **1**, 23 (1959).

Base, bp_{13} 232°. n_D^{20} 1.6025.

Hydrochloride, $C_{23}H_{28}N_2O \cdot HCl$, **Leioplegil.** Needles from carbon tetrachloride, mp 138°. Readily sol in water; aq soln turns yellow on exposure to air or light.

THERAP CAT: Antispasmodic.

5458. Lemon Peel. Outer rind of fresh ripe fruit of *Citrus limonum* (L.) Risso (*C. medica* var. *limon* L.), *Rutaceae. Habit.* Northern India; cultivated in California, West Indies, Italy, Spain. *Constit.* Volatile oil, hesperidin, bitter extractive.

USE: As a flavor in medicines; also in beverages, confectionery, and cooking.

5459. Lenacil. *3-Cyclohexyl-6,7-dihydro-1H-cyclopentapyrimidine-2,4-(3H,5H)-dione;* 3-cyclohexyl-5,6-trimethyleneuracil; 3-cyclohexyl-1,5,6,7-tetrahydro-2H-cyclopentapyrimidine-2,4(3H)-dione; Du Pont 634; Venzar. $C_{13}H_{18}N_2O_2$; mol wt 234.30. C 66.64%, H 7.74%, N 11.96%, O 13.66%. Prepn: S. Senda, H. Fujimura, **Japan.** pat. **62 4892** (1962), *C.A.* **59,** 642g (1963); and herbicidal activity: E. J. Soboczenski, **U.S.** pat. **3,235,360** (1966 duPont Nemours). Synthesis: S. Senda *et al., Chem. Pharm. Bull.* **21,** 1894 (1973). Mode of action: C. E. Hoffmann, *Pestic. Chem., Proc. 2nd Int. Congr. Pestic.* **5,** A. S. Tahori, Ed., (Gordon and Breach Science Publishers, New York, 1972) pp 65-85. Field evaluation: J. G. Hilton, W. E. Bray, *Proc. 12th Brit. Weed Contr. Conf.* 485 (1974). GC determn in soils: D. J. Caverly, R. C. Denney, *Analyst* **102,** 576 (1977). Soil persistence: M. Tena *et al., Weed Res.* **23,** 245 (1982). Field tolerance: D. V. Clay, *Test Agrochem. Cultiv.* **11,** 80 (1990).

Crystals from methanol or dioxane, mp > 290° (Senda, Fujimura). Also reported as light grey crystals from DMF, mp 310°-313° (Soboczenski). Technical product properties: odorless, white to tan crystalline solid. d^{20} 1.32 kg/l. Vapor pressure (25°): 0.2 μPa. Soly in water 6 mg/l; in xylene: 2g/l. LD_{50} (80% wettable powder) orally in rats: > 11,000 mg/kg. LC_{50} 96 hr in bluegill sunfish; rainbow trout: 100-1000; 135 mg/l. LC_{50} 8 day in bobwhite quail, Peking duck (mg/kg): 2300, > 5620. (DuPont technical data sheet).

USE: Herbicide.

5460. Lenampicillin. *[2S-[2α,5α,6β(S*)]]-6-[(Aminophenylacetyl)amino]-3,3-dimethyl-7-oxo-4-thia-1-azabicyclo[3.2.0]heptane-2-carboxylic acid (5-methyl-2-oxo-1,3-dioxol-4-yl)methyl ester;* ampicillin (5-methyl-2-oxo-1,3-dioxolen-4-yl)methyl ester; 6-[D(−)-α-aminophenylacetamido]penicillanic acid (5-methyl-2-oxo-1,3-dioxol-4-yl)methyl ester. $C_{21}H_{23}N_3O_7S$; mol wt 461.50. C 54.66%, H 5.02%, N 9.11%, O 24.27%, S 6.95%. Orally active ampicillin prodrug. Prepn: F. Sakamoto *et al.,* **Eur.** pat. **Appl. 39,086;** *eidem,* **U.S.** pat. **4,342,693** (1981, 1982 both to Kanebo); F. Sakamoto *et al., Chem. Pharm. Bull.* **32,** 2241 (1984); S. Ikeda *et al., ibid.* 4316. Metabolism in man, dogs, rats: N. Awata *et al., Japan. J. Antibiot.* **38,** 1776, 1785, *C.A.* **104,** 161479u, 122550r (1985). Human pharmacokinetics: A. Saito, M. Nakashima, *Antimicrob. Ag. Chemother.* **29,** 948 (1986). Series of articles on antibacterial activity, mutagenicity, pharmacology, clinical trials: *Chemotherapy* (Tokyo)

32, Suppl. 8, pp 1-772 (1984). Acute toxicity data: F. Ogino *et al., ibid.* 31.

Hydrochloride, $C_{21}H_{23}N_3O_7S \cdot HCl$, **KB-1585, KBT-1585, Takacillin, Varacillin.** Crystals from isopropanol-ethyl acetate, mp 145° (dec). LD_{50} in male, female rats, male, female mice (mg/kg): approx 10000, approx 10000, 8294, 8492 orally; 4362, 4471, 3576, 4284 s.c.; 876, 838, 711, 775 i.v.; in dogs (mg/kg): > 300 orally (Ogino).

THERAP CAT: Antibacterial.

5461. Lenthionine. *1,2,3,5,6-Pentathiepane.* $C_2H_4S_5$; mol wt 188.38. C 12.75%, H 2.14%, S 85.11%. Odorous principle from the edible mushroom *Shiitake Lentinus edodes* (Berk.) Sing. Isoln and synthesis: Morita, Kobayashi, *Tetrahedron Letters* **1966,** 573. Simple, efficient synthesis: I. Still, G. W. Kutney, *ibid.* **22,** 1939 (1981).

Crystals from methylene chloride, mp 60-61°.

5462. Lentinan. LC-33. $(C_6H_{10}O_5)_n$; mol wt 400,000-800,000. Neutral polysaccharide isolated from the edible mushroom, *Lentinus edodes* (Berk.) Sing. Primary structure is a β-1,3-D-glucan having 2 β-1,6-glucopyranoside branchings for every 5 β-1,3 linear linkages. Isoln and antitumor activity: G. Chihara *et al., Nature* **222,** 637 (1969); *eidem, Cancer Res.* **30,** 2776 (1970). Immunostimulant activity: Y. Y. Maeda, G. Chihara, *Nature* **229,** 634 (1971). Prepn of stable aqueous solution: M. Fujii *et al.,* **Neth.** pat. **Appl. 7,601,114** corresp to **U.S.** pat. **4,207,312** (1976, 1980 both to Ajinomoto; Morishita). Structural study: T. Sasaki *et al., Carbohydrate Res.* **47,** 99 (1976). Preclinical evaluation and toxicity studies: G. Chihara, *Adv. Exp. Med. Biol.* **166,** 189 (1983). Mechanism of action studies: H. Miyakoshi, T. Aoki, *Int. J. Immunopharmacol.* **6,** 365, 373 (1984). Clinical trial in gastric and colorectal cancer: T. Taguchi *et al., Adv. Exp. Med. Biol.* **166,** 181 (1983); K. Okuyama *et al., Cancer* **55,** 2498 (1985). *Review:* G. Chihara, *Int. J. Tissue Res.* **4,** 207-225 (1982).

White powder, dec 250°. Sol in aqueous alkali, formic acid. Slightly sol in hot water, DMSO. Insol in cold water, alcohol, ether, chloroform, pyridine, hexamethylphosphoramide. Stable against sulfuric and hydrochloric acids. $[\alpha]_D^{20}$ +13.5-14.5° (in 2% NaOH); +19.5-21.5° (in 10% NaOH).

THERAP CAT: Antineoplastic; immunomodulator.

5463. Leonurine. *4-Hydroxy-3,5-dimethoxybenzoic acid 4-[(aminoiminomethyl)amino]butyl ester;* 4-hydroxy-3,5-dimethoxybenzoic acid δ-guanidinobutyl ester; syringic acid δ-guanidinobutyl ester; [4-(4-hydroxy-3,5-dimethoxybenzoyloxy)butyl]guanidine; 4-guanidino-1-butanol syringate. $C_{14}H_{21}N_3O_5$; mol wt 311.34. C 54.01%, H 6.80%, N 13.50%, O 25.69%. Isoln from leaves of *Leonurus sibiricus* L., *Labiatae:* Kubota, Nakajima, *C.A.* **25,** 771 (1931). Structure: Goto *et al., Tetrahedron Letters* **1962,** 545. Revised structure: Kishi *et al., ibid.* **1968,** 637. Structure and synthesis: Sugiura *et al., Tetrahedron* **25,** 5155 (1969). Alternate synthesis: K. F. Cheng *et al., Experientia* **35,** 571 (1979). Biological effect: H. W. Yeung *et al., Planta Med.* **31,** 51 (1977).

Hydrochloride monohydrate, $C_{14}H_{21}N_3O_5 \cdot HCl \cdot H_2O$, mp 193-194°. pKa 7.9 in water.

5464. Lepidine. *4-Methylquinoline;* cincholepidine. $C_{10}H_9N$; mol wt 143.19. C 83.88%, H 6.34%, N 9.78%. Prepn from 2-chlorolepidine: Neumann *et al., Org. Syn.* **26,** 45 (1946); from aniline hydrochloride: Campbell, U.S. pat. **2,451,611** (1948 to du Pont); Bach, Rast, *J. Prakt. Chem.* **17,** 63 (1962); from ethylacetanilide: Ardashev, Minkin, *Zh. Obshch. Khim.* **28,** 1578 (1958).

Colorless, oily liq; quinoline odor; turns reddish-brown in light. bp 261-263°; bp$_{14-15}$ 126-127°; bp$_{6-7}$ 115-120°; bp$_{1.5-2}$ 90-95°. d$_4^{20}$ 1.0826. n$_D^{20}$ 1.6190. mp 0°. Slightly sol in water; miscible with alc, benzene, ether. *Protect from light.*

5465. Leptandra. Culver's root; black root. Dried rhizome and roots of *Veronicastrum virginicum* (L.) Farw. (*Leptandra virginica* (L.) Nutt.), *Scrophulariaceae. Habit.* North America. *Constit.* A sterol; esters of cinnamic, methoxycinnamic and fatty acids; resin saponin, tannin, sugars.
THERAP CAT: Cathartic.

5466. Leptin. OB protein. Peptide hormone produced by adipose tissue that is thought to act on receptors in the brain to regulate body weight and fat deposition. Name derived from the Greek word "leptos" meaning "thin." Protein product of the *obese (ob)* gene, mutation of which causes marked obesity and diabetes in mice. Secreted by adipocytes; mol wt 16 kDa. Identification of weight-regulating substance in blood of normal mice: D. L. Coleman, *Diabetologia* **9,** 294 (1973). Cloning of mouse and human *ob* genes: Y. Zhang *et al., Nature* **372,** 425 (1994). Biological activity as regulator of body weight in mice: M. A. Pelleymounter *et al., Science* **269,** 540 (1995); J. L. Halaas *et al., ibid.* 543. Identification of brain as target tissue: L. A. Campfield *et al., ibid.* 546. Increased *ob* gene expression in obese humans: R. V. Considine *et al., J. Clin. Invest.* **95,** 2986 (1995).

5467. Leptodactyline. *3-Hydroxy-N,N,N-trimethylbenzeneethanaminium; (m-hydroxyphenethyl)trimethylammonium.* $[C_{11}H_{18}NO]^+$. Occurs in skin of the amphibian species of the genus *Leptodactylus:* its probable precursor is *m*-tyrosine. Isoln as chloride: Erspamer, *Arch. Biochem. Biophys.* **82,** 431 (1959). Prepd as chloride: Buck *et al., J. Am. Chem. Soc.* **60,** 1789 (1938).

Chloride, $C_{11}H_{18}ClNO$, nodules of small spindles from alcohol + ether, mp 220°. Soluble in water, alcohol, HCl. Practically insol in ether, ethyl acetate.
THERAP CAT: Neuromuscular blocker.

5468. Leptophos. *Phenylphosphonothioic acid O-(4-bromo-2,5-dichlorophenyl) O-methyl ester; O-(4-bromo-2,5-dichlorophenyl) O-methyl phenylphosphonothioate; O-methyl O-2,5-dichloro-4-bromophenyl phenylthiophospho-*

nate; VCS-506; Abar; Phosvel. $C_{13}H_{10}BrCl_2O_2PS$; mol wt 412.07. C 37.89%, H 2.45%, Br 19.39%, Cl 17.21%, O 7.77%, P 7.52%, S 7.78%. Prepn and use as an insecticide: Richter, Hanna; U.S. pats. **3,459,836, 3,551,563, 3,577,482** (1969, 1970, 1971, all to Velsicol). Metabolism in mice and on cotton plants: Holmstead *et al., Arch. Environ. Contam. Toxicol.* **1,** 133 (1973). Toxicological studies in mammals: Kamel *et al., J. Eur. Toxicol.* **6,** 70 (1973); Abou-Donia *et al., Experientia* **30,** 63 (1974); T. B. Gaines, R. E. Linder, *Fundam. Appl. Toxicol.* **7,** 299 (1986).

Tan waxy solid. mp range 55-67°. Sp gr at 25° = 1.53. Solubility at 25° (gm/100 ml): benzene, 133; xylene, 73; acetone, 62; cyclohexane,14; heptane, 7; isopropanol, 2.4; water, 0.03 ppm. Stable to acids; hydrolyzed slowly under strongly alkaline conditions. LD$_{50}$ in adult male, female rats (mg/kg): 19, 20 orally (Gaines, Linder).
USE: Insecticide.

5469. Lercanidipine. *1,4-Dihydro-2,6-dimethyl-4-(3-nitrophenyl)-3,5-pyridinedicarboxylic acid 2-[(3,3-diphenylpropyl)methylamino]-1,1-dimethylethyl methyl ester;* methyl 1,1,N-trimethyl-N-(3,3-diphenylpropyl)-2-aminoethyl 1,4-dihydro-2,6-dimethyl-4-(3-nitrophenyl)pyridine-3,5-dicarboxylate; methyl 1,1-dimethyl-2-[N-(3,3-diphenylpropyl)-N-methylamino]ethyl 2,6-dimethyl-4-(3-nitrophenyl)-1,4-dihydropyridine-3,5-dicarboxylate; masnidipine. $C_{36}H_{41}N_3O_6$; mol wt 611.74. C 70.68%, H 6.76%, N 6.87%, O 15.69%. Dihydropyridine calcium channel blocker. Prepn: D. Nardi *et al., Eur. pat. Appl.* **153,016;** *eidem,* U.S. pat. **4,-705,797** (1985, 1987 both to Recordati). Pharmacology: G. Bianchi *et al., Pharmacol. Res.* **21,** 193 (1989). Clinical evaluation in hypertension: E. Rimoldi *et al., Acta Therap.* **20,** 23 (1994).

Hydrochloride, $C_{36}H_{41}N_3O_6 \cdot HCl$, *Rec-15-2375, R-75.* Prepd as the hemihydrate, mp 119-123°. LD$_{50}$ in mice (mg/kg): 83 i.p.; 657 orally (Nardi).
THERAP CAT: Antihypertensive.

5470. Lesopitron. *2-[4-[4-(4-Chloro-1H-pyrazol-1-yl)-butyl]-1-piperazinyl]pyrimidine.* $C_{15}H_{21}ClN_6$; mol wt 320.82. C 56.16%, H 6.60%, Cl 11.05%, N 26.20%. Selective serotonin 5HT$_{1A}$-receptor agonist. Prepn: A. Colombo Pinol *et al., Eur. pat. Appl.* **382,637;** *eidem,* U.S. pat. **5,128,-343** (1990, 1992 both to Esteve). Receptor binding studies and pharmacology: B. Costall *et al., J. Pharmacol. Exp. Ther.* **262,** 90 (1992); M. Ballarin *et al., Brit. J. Pharmacol.* **113,** 425 (1994). Gastroprotective effects: A. J. Farré *et al., J. Pharmacol. Exp. Ther.* **272,** 832 (1995).

Consult the Name Index before using this section.

Dihydrochloride, $C_{15}H_{21}ClN_6 \cdot 2HCl$, **E-4424.** Crystals, mp 194-197.5°.

THERAP CAT: Anxiolytic.

5471. Lethane® 60. *Dodecanoic acid 2-thiocyanatoethyl ester; lauric acid 2-thiocyanatoethyl ester;* lauric acid ester with 2-hydroxyethyl thiocyanate; 2-thiocyanoethyl dodecanoate; 2-thiocyanoethyl laurate. $C_{15}H_{27}NO_2S$; mol wt 285.45. C 63.12%, H 9.53%, N 4.91%, O 11.21%, S 11.23%. $CH_3(CH_2)_{10}COOCH_2CH_2SCN$. Prepn: Hester, U.S. pat. **2,220,521** (1941 to Rohm & Haas). Toxicity study: E. E. Kenaga, *Bull. Entomol. Soc. Amer.* **12,** 161 (1966).

Liquid, $bp_{0.1}$ 160-190°. LD_{50} in rats, rabbits, dogs (mg/kg): 500, 200, 250 orally (Kenaga).

Caution: Moderately irritating to skin and mucous membranes, and, in high concns, narcotic.

USE: Insecticide.

5472. Lethane® 384. *Thiocyanic acid 2-(2-butoxyethoxy)ethyl ester;* 2-butoxy-2'-thiocyanodiethyl ether; 2-[2-(butoxy)ethoxy]ethyl ester of thiocyanic acid; butyl carbitol rhodanate; 1-butoxy-α-(2-thiocyanoethoxy)ethane. $C_9H_{17}NO_2S$; mol wt 203.31. C 53.17%, H 8.43%, N 6.89%, O 15.74%, S 15.77%. $C_4H_9OCH_2CH_2OCH_2CH_2SCN$. Prepn: Bruson, U.S. pat. **2,372,809** (1945 to Rohm & Haas).

Liquid, $bp_{0.25}$ 120-125°. Commercial product, brownish oil, is a mixture with petroleum distillate which constitutes 47% by weight or 50% by volume. Practically insol in water. Miscible with hydrocarbons and most organic solvents. LD_{50} in rats: 90 mg/kg, *RTECS Vol. II,* R. J. Lewis, R. L. Tatken, Eds. (1980) p 712.

Caution: Effects similar to Lethane 60 but may be more severe. Absorbed through intact skin.

USE: Insecticide.

5473. Letosteine. *2-[2-[(2-Ethoxy-2-oxoethyl)thio]ethyl]-4-thiazolidinecarboxylic acid;* 2-[2-[(carboxymethyl)-thio]ethyl]-4-thiazolidinecarboxylic acid 2-ethyl ester; Viscotiol. $C_{10}H_{17}NO_4S_2$; mol wt 279.38. C 42.99%, H 6.13%, N 5.01%, O 22.91%, S 22.96%. Cyclic cysteine derivative. Prepn: M. X. Chodkiewicz, Ger. pat. **2,410,307;** *eidem,* U.S. pat. **4,032,534** (1974, 1979 to Ferlux-Chemie). Effect on viscoelasticity of mucus: E. Puchelle *et al., Eur. J. Respir. Dis.* **61** (Suppl 110), 195 (1980). In treatment of bronchitis: V. Macquet *et al., Lille Med.* **24,** 735 (1979); P. Freour, J. Vergeret, *Bordeaux Med.* **12,** 1553 (1979); P. Freour, ibid. **13,** 1431 (1980); P. Meroni *et al., Clin. Ter.* **105,** 109 (1983). Comparison with sulfur-containing compounds in bronchitis: D. Rodde, *C.R. Ther. Pharmacol. Clin.* **1**(5), 31 (1983).

Crystals from ethanol, mp 142°.

THERAP CAT: Mucolytic.

5474. Letrozole. *4,4'-(1H-1,2,4-Triazol-1-ylmethylene)-bisbenzonitrile;* 1-[bis(4-cyanophenyl)methyl]-1,2,4-triazole; 4-[1-(4-cyanophenyl)-1-(1,2,4-triazol-1-yl)methyl]benzonitrile; CGS-20267. $C_{17}H_{11}N_5$; mol wt 285.31. C 71.57%, H 3.89%, N 24.55%. Nonsteroidal aromatase inhibitor; structurally related to fadrozole, *q.v.* Prepn: R. M. Bowman *et al.,* **Eur** pat. **Appl. 236,940;** *eidem,* U.S. pat. **4,978,672** (1987, 1990 both to Ciba-Geigy). Pharmacology: A. S. Bhatnagar *et al., J. Steroid Biochem. Mol. Biol.* **37,** 1021 (1990). Clinical suppression of estrogen biosynthesis: L. M. Demers *et al.,* ibid. **44,** 687 (1993). EIA and HPLC determn in biological fluids: C. U. Pfister *et al., J. Pharm. Sci.* **83,** 520 (1994). Clinical evaluation in breast cancer: A. Lipton *et al., Cancer* **75,** 2132 (1995).

mp 181-183°.

THERAP CAT: Antineoplastic.

5475. Leucine. *L-Leucine;* Leu; L; 2-amino-4-methylvaleric acid; α-aminoisocaproic acid; (S)-2-amino-4-methylpentanoic acid. $C_6H_{13}NO_2$; mol wt 131.17. C 54.94%, H 9.99%, N 10.68%, O 24.39%. An essential amino acid for human development. Discovery of leucine is attributed to Proust in 1819 who reported its separation from fermented milk curds. Early chemistry and biochemistry: *Amino Acids and Proteins,* D. M. Greenberg, Ed. (Charles C. Thomas, Springfield, IL, 1951) 950 pp., *passim;* J. P. Greenstein, M. Winitz, *Chemistry of the Amino Acids* **vols 1-3** (John Wiley and Sons, Inc., New York, 1961) pp. 2075-2096, *passim.* Chromatographic determn of specific activity in biological fluids: E. P. Donahue *et al., J. Chromatog.* **571,** 29 (1991); separation of isotopes: P. Q. Baumann *et al.,* ibid. **573,** 11 (1992). Metabolism in humans: K. J. Motil *et al., Metabolism* **30,** 783 (1981); in children receiving parenteral nutrition: O. Goulet et al., *Am. J. Physiol.* **265,** E540 (1993). Kinetics modelling in humans: C. Cobelli *et al.,* ibid. **261,** E539 (1991).

White glistening hexagonal plates from aq alc. d_{18} 1.293. Sublimes at 145-148°. Dec 293-295° (rapid heating, sealed tube). $[M]_D$ +21.0° (5N HCl); $[M]_D$ +29.5° (glacial acetic acid). $[\alpha]_D^{25}$ −10.8° (c = 2.2); $[\alpha]_D^{26}$ +15.1° in 6N HCl (38 mols HCl per mol leucine); $[\alpha]_D^{20}$ +7.6° in 3N NaOH (30 mols NaOH per mol leucine). pKa (25°): 9.6; Kb = 2.3 × 10^{-2}. R_f value 0.79. Soly in water (g/l): 22.7 (0°); 24.26 (25°); 28.87 (50°); 38.23 (75°); 56.38 (100°); in 99% alcohol: 0.72; in acetic acid: 10.9; insol in ether.

DL-Form. Leaflets fom water. Sweet taste. Dec 332° (also reported as 290°). Sublimes. pK_1 2.36; pK_2 9.60. Soly in water (g/l): 7.97 (0°); 9.91 (25°); 14.06 (50°); 22.76 (75°); 42.06 (100°); in 90% alcohol: 1.3; insol in ether.

Hydrochloride, $C_6H_{13}NO_2 \cdot HCl$, crystals. Freely sol in water.

USE: Nutrient.

5476. Leucinocaine Mesylate. *2-(Diethylamino)-4-methyl-1-pentanol 4-aminobenzoate (ester) monomethanesulfonate (salt);* 2-diethylamino-4-methylpentyl p-aminobenzoate methanesulfonate; p-aminobenzoic acid β-diethylaminohexyl ester methanesulfonate; p-aminobenzoic acid N,N-diethylleucinol ester methanesulfonate; p-aminobenzoyl-N-1-diethylamino-1-isobutylethanol methanesulfonate; 2-methyl-4-diethylaminopentan-5-ol p-aminobenzoate; Panthesin. $C_{18}H_{32}N_2O_5S$; mol wt 388.53. C 55.65%, H 8.30%, N 7.21%, O 20.59%, S 8.25%. Prepn: Lebeau, Janot, *Traité de Pharmacie Chimique* **vol. III** (Masson et Cie, Paris, 1955-1956) p 1296. Prepn of hydrochloride: Karrer, U.S. pat. **1,555,217** and see also Ger. pat. **464,484** (1925, 1928 to

Chem. Fabrik Flora). Efficacy and toxicity in animals: R. N. Bieter *et al., J. Pharmacol. Exp. Ther.* **57**, 221 (1936).

Crystals, mp 171°. Sol in 3 parts water, alcohol. Practically insol in ether, petr ether, ethyl acetate, acetone, benzene. MLD in rabbits (mg/kg): 20 i.v.; 10.8 intraspinally (Bieter).

THERAP CAT: Anesthetic (local).

5477. Leucocyanidin. *2-(3,4-Dihydroxyphenyl)-3,4-di-hydro-2H-1-benzopyran-3,4,5,7-tetrol; 3,3',4,4',5,7-flavan-hexol; 3,3',4,4',5,7-hexahydroxyflavane; leucocyanidol;* Flavan; Hamaméliode P; Résivit. $C_{15}H_{14}O_7$; mol wt 306.27. C 58.83%, H 4.61%, O 36.57%. From petals of Asiatic cotton flowers (*Gossypium* spp): Stephens, *Arch. Biochem. Biophys.* **18**, 449 (1948); from *Butea frondosa* Koen. ex Roxb., *Leguminosae:* Ganguly, Seshadri, *Tetrahedron* **6**, 21 (1959). Prepn from quercetin: Bauer *et al., Chem. & Ind. (London)* **1954,** 433; from taxifolin: Freudenberg, Weinges, *Ann.* **613**, 61 (1958). Metabolism: Claveau, Masquelier, *Can. J. Pharm. Sci.* **1**, 74 (1966). *See also* Bioflavonoids.

Monohydrate, crystals from ethyl acetate + petr ether, mp above 355°. uv max (ethanol): 285 nm. Soluble in water, alcohol, acetone. Practically insol in ether, chloroform, petr ether.

THERAP CAT: Capillary protectant.

5478. Leucodrin. *[5S-[5α(S*),8α(R*),9β]]-8-(1,2-Di-hydroxyethyl)-9-hydroxy-4-(4-hydroxyphenyl)-1,7-dioxaspiro-[4.4]nonane-2,6-dione.* $C_{15}H_{16}O_8$; mol wt 324.29. C 55.56%, H 4.97%, O 39.47%. From leaves of *Leucadendron concinnum* R. Br., *Proteaceae:* Hesse, *Ann.* **290**, 314 (1896); Rapson, *J. Chem. Soc.* **1938,** 282. Structure: Rapson, *ibid.* **1940,** 1271; Perold, Pachler, *Proc. Chem. Soc.* **1964,** 62; *eidem, J. Chem. Soc. (C)* **1966,** 1918.

Prisms from water, mp 212-212.5°. $[\alpha]_D^{17}$ −19.2° (c = 3.5 in 40% alc). Slightly sol in water, alcohol.

5479. Leucoglycodrin. *8-(1,2-Dihydroxyethyl)-4-[4-(D-glucopyranosyloxy)phenyl]-9-hydroxy-1,7-dioxaspiro[4.4]non-ane-2,6-dione.* $C_{21}H_{26}O_{13}$; mol wt 486.43. C 51.85%, H 5.39%, O 42.76%. From leaves of *Leucadendron concinnum* R. Br., *Proteaceae:* Merck, *Merck's Jahresber.* **1895**, 3. Structure: Murray, Bradshaw, *Tetrahedron Letters* **1966,** 3773.

Hemihydrate, $C_{21}H_{26}O_{13} \cdot \frac{1}{2}H_2O$, amorphous solid, mp 220-222°. uv max: 201, 226, 276, 280-284 nm (ε 19880, 9240, 2100, 1960-1820).

5480. Leucomycins. Kitasamycin; C-637; Ayermicina; Sineptina; Stereomycine; Syneptine. Macrolide antibiotic complex similar to carbomycin, *q.v.* and erythromycin, *q.v.*, produced by *Streptomyces kitasatoensis* Hata: T. Hata *et al., J. Antibiot.* **6A**, 87 (1953); Sano, *ibid.* **7A**, 93 (1954). Early work performed on a mixture of at least six components, A_1, A_2, B_1-B_4, of which A_1 was the most biologically active: J. Abe *et al., J. Chem. Soc. Japan, Pure Chem. Sect.* **81**, 969 (1960); T. Watanabe *et al., Bull. Chem. Soc. Japan* **33**, 1100, 1104 (1960). Leucomycin isolated from an improved strain of *S. kitasatoensis* is a mixture of at least eight biologically active components, A_1 and new substances A_3-A_9: Hata *et al.,* U.S. pat. **3,535,309** (1970 to Kitasato Institute and Toyo Jozo). Leucomycins A_1, A_3-A_9, U and V are substituted variations of the structure represented below. Structure of A_1: T. Watanabe *et al., Angew. Chem.* **76**, 792 (1964); T. Hata *et al., Chem. Pharm. Bull. (Japan)* **15**, 358 (1967); of A_3: S. Omura *et al., Tetrahedron Letters* **1967**, 609, 1267; of A_4 through A_9: *eidem, J. Antibiot.* **20A**, 234 (1967); **21**, 272 (1968). Conformation: *eidem, Tetrahedron* **28**, 2839 (1972). Revised configuration at C-9: L. A. Freiberg *et al., J. Org. Chem.* **39**, 2474 (1974). Liquid chromatography of A_1, A_{3-9}, U, V: S. Omura *et al., J. Antibiot.* **26**, 795 (1973). Antibacterial and antimycoplasmal activity: Iwata, Akiba, *ibid.* **15A**, 258 (1962); Omura *et al., ibid.* **21**, 532 (1968); **25**, 105 (1972). *In vitro* and preliminary clinical studies: B. C. Stratford, S. Dixson, *Med. J. Australia* **1**, 1029 (1974). Antimicrobial transformation of A_5 to V: K. Singh, S. Rakhit, *ibid.* **32**, 78 (1979). Biosynthesis: C. Kitao *et al., ibid.* **32**, 1055 (1979); S. Omura *et al., ibid.* **36**, 611 (1983). *Reviews:* Desvignes, *Ann. Pharm. Franc.* **21**, 569 (1963); Toju, Omura, "Chemical and Biological Studies on Leucomycins (Kitasamycins)" in *Drug Action and Drug Resistance in Bacteria,* vol. I, S. Mitsuhashi, Ed. (University Park Press, Baltimore, 1971) pp 267-291; Keller-Schierlein in *Fortschr. Chem. Org. Naturst.* **30**, 313-460 (1973); S. Omura, A. Nakagawa, *J. Antibiot.* **28**, 401-433 (1975).

Leucomycin A₁

Leucomycin A_1, A_2, B_1-B_4 complex (Sano, *loc. cit.*). Powder from acetone; stable at room temp; dec 125°. $[\alpha]_D^{20}$ −67.1 (ethanol). uv max (ethanol): 232, 285 nm ($E_{1cm}^{1\%}$ 228, 8.6). Freely sol in most organic solvents. Insol in petr ether. Slightly sol in water.

Leucomycin A_3-A_9 complex (Hata *et al.,* U.S. pat. **3,535,-309**). White powder; mp 128-145°. uv max (methanol): 231 nm ($E_{1cm}^{1\%}$ 353). $[\alpha]_D^{25}$ −53° (CHCl₃). Sol in methanol, ethanol, ethyl acetate, butyl acetate, acetone, benzene, and chloroform; difficultly sol in water. Insol in petr ether. LD₅₀ in mice: > 650 mg/kg i.v.; > 1000 mg/kg orally.

Leucomycin A_1, $C_{40}H_{67}NO_{14}$. $[\alpha]_D^{25}$ −66.0° (CHCl₃). uv max (methanol): 232 nm ($E_{1cm}^{1\%}$ 400). pKa' (50% ethanol): 6.69.

Triacetylleucomycin A_1, $C_{46}H_{73}NO_{17}$, crystals, mp 125-126°. $[\alpha]_D^{25}$ −82.5° (c = 1.3 in CHCl₃).

Leucomycin A_3, see Josamycin.

THERAP CAT: Antibacterial.

5481. Leucopterin. *2-Amino-5,8-dihydro-4,6,7(1H)-pteridinetrione; 2-amino-4,6,7-pteridinetriol; 2-amino-4,6,7-trihydroxypteridine; 2-amino-4,6,7-trihydroxypyrimido[4,5-b]pyrazine.* $C_6H_5N_5O_3$; mol wt 195.14. C 36.93%, H 2.58%, N 35.89%, O 24.60%. A colorless substance found in the wings of butterflies (especially of white-winged butterflies). May be obtained from xanthopterin by dehydrogenation: Wieland, Purrmann, *Ann.* **544**, 172 (1940). Synthesis by fusing 2,4,5-triamino-6-hydroxypyrimidine with an excess of oxalic acid: Purrmann, *ibid.* **188**. *See also* ref *under* Xanthopterin and review on pterins by Purrmann, *Fortschr. Chem. Org. Naturst.* **4**, 64 (1945).

Fine colorless crystals forming yellow Na and Ag salts, and a sparingly sol NH_4 salt. Soluble in alkaline solns with blue fluorescence.

5482. Leukotrienes. LTs. A family of endogenous metabolites of arachidonic acid via the lipoxygenase pathway, chemically related to the prostaglandins and thromboxanes, *q.q.v.* The name leukotrienes was applied because of their origin in leukocytes and their conjugated triene structures. [For a description of the nomenclature of individual leukotrienes *see* B. Samuelsson, S. Hammarström, *Prostaglandins* **19**, 645 (1980)]. Members of the group are potent bronchoconstrictors that play an important pathophysiological role in immediate hypersensitivity reactions; they have been proposed as mediators of the inflammatory process and some are potent chemotactic agents. Initial studies of novel arachidonate metabolites in rabbit polymorphonuclear leukocytes: P. Borgeat *et al., J. Biol. Chem.* **251**, 7816 (1976); **252**, 8772 (1977). Subsequent studies and structure of a dihydroxyeicosatetraenoic acid (leukotriene B, LTB_4): P. Borgeat, B. Samuelsson, *ibid.* **254**, 2643 (1979). LTB_4 is formed from an unstable intermediate oxido-eicosatetraenoic acid, leukotriene A or LTA_4: *eidem, Proc. Nat. Acad. Sci. USA* **76**, 3213 (1979). Stereochemistry and enzymatic conversion of LTA_4 to LTB_4: O. Radmark *et al., Biochem. Biophys. Res. Commun.* **92**, 954 (1980). Synthesis of the four optical isomers of LTA_4: J. Rokach *et al., Tetrahedron Letters* **22**, 2759, 2763 (1981). Total synthesis of LTB_4: E. J. Corey *et al., J. Am. Chem. Soc.* **102**, 7984 (1980); Y. Guindon *et al., Tetrahedron Letters* **23**, 739 (1982). Formation of the 20-hydroxy and 20-carboxy metabolites of LTB_4: G. Hansson *et al., FEBS Letters* **130**, 107 (1981); total synthesis: R. Zamboni, J. Rokach, *Tetrahedron Letters* **23**, 4751 (1982). Earlier studies had described a "slow-reacting substance of anaphylaxis" (SRS-A or SRS) released from guinea pig and cat lung by cobra venom and in guinea pig lung after anaphylactic shock, *cf.* W. S. Feldberg, C. H. Kellaway, *J. Physiol.* **94**, 187 (1938); C. H. Kellaway, E. R. Trethewie, *Quart. J. Exp. Physiol.* **30**, 121 (1940). The relationship between SRS and members of the leukotriene family was established following publication of the uv spectrum of purified SRS-A, which showed the presence of the conjugated triene, *cf.* H. R. Morris *et al., FEBS Letters* **87**, 203 (1978). Purification and structure of LTC_4, an SRS from mouse mastocytoma cells: R. C. Murphy *et al., Proc. Nat. Acad. Sci. USA* **76**, 4275 (1979); S. Hammarström *et al., Biochem. Biophys. Res. Commun.* **91**, 1266 (1979). Total synthesis of LTC_4: E. J. Corey *et al., J. Am. Chem. Soc.* **102**, 1436, 3663 (1980); J. Rokach *et al., Tetrahedron Letters* **21**, 1485 (1980). Identity of synthetic LTC_4 with SRS from mouse mastocytoma cells: S. Hammarström *et al., Biochem. Biophys. Res. Commun.* **92**, 946 (1980). Structure of the SRS from rat basophil leukemia cells (RBL-1) and identification as a leukotriene (LTD_4): H. R. Morris *et al., Prostaglandins* **19**, 185 (1980). It was subsequently proposed that LTC_4 was an intermediate in the biosynthesis of LTD_4. Identity of SRS-A released in sensitized guinea pig lung perfusates and LTD_4: *eidem, Nature* **285**, 104 (1980). Assignment of stereochemistry: *eidem, Prostaglandins* **20**, 601 (1980). Detection of LTA_4 as an intermediate in the biosynthesis of LTC_4 and LTD_4: S. Hammarström, B. Samuelsson, *FEBS Letters* **122**, 83 (1980). The sulfone of LTC_4, which has also been proposed as a natural product, *cf.* H. Ohnishi *et al., Prostaglandins* **20**, 655 (1980), has been found to be as potent as LTC_4: T. Jones *et al., ibid.* **24**, 279 (1982). Synthesis: Y. Girard *et al., Tetrahedron Letters* **23**, 1023 (1982). Discovery of LTF_4: M. E. Anderson *et al., Proc. Nat. Acad. Sci. USA* **79**, 1088 (1982). Synthesis: F. Ellis *et al., Tetrahedron Letters* **23**, 3735 (1982). It is now known that SRS-A is made up of varying amounts of cys-

teine-containing members of the leukotrienes, *i.e.* leukotrienes C_4, D_4, and E_4. These three LTs are generally found to be 100 to 1000 times more potent, on a molar basis, than histamine or prostaglandins in their effects on pulmonary airways. Review of chemistry and structure elucidation: D. A. Clark, A. Marfat, *Ann. Rep. Med. Chem.* **17**, 291-300 (1982). Comprehensive review of synthesis of leukotrienes and other lipoxygenase-derived products: J. G. Atkinson, J. Rokach, in *Handbook of Eicosanoids: Prostaglandins and Related Lipids* IB, A. L. Willis, Ed. (CRC Press, Boca Raton, 1987) pp 175-263. General reviews: P. Sirois, P. Borgeat, *Int. J. Immunopharmacol.* **2**, 281-293 (1980); P. Borgeat, P. Sirois, *J. Med. Chem.* **24**, 121-126 (1981); B. Samuelsson, *Int. Arch. Allergy Appl. Immunol.* **66**, Suppl. 1, 98-106 (1981); L. S. Wolfe, *J. Neurochem.* **38**, 1-14 (1982); J. L. Marx, *Science* **215**, 1380-1384 (1982); B. Samuelsson, S. Hammarström, *Vitam. Horm.* **39**, 1-30 (1982). Books: *SRS-A and Leukotrienes*, P. J. Piper, Ed. (Research Studies Press, London, 1981) 279 pp; *Advances in Prostaglandin, Thromboxane, and Leukotriene Research*, vol. 9, B. Samuelsson, R. Paoletti, Eds. (Raven Press, New York, 1982) 341 pp.

Leukotriene A_4, $C_{20}H_{30}O_3$, *[2S-[2α,3β(1E,3E,5Z,8Z)]]-3-(1,3,5,8-tetradecatetraenyl)oxiranebutanoic acid, leukotriene A, LTA_4*. Unstable. Characterized as its methyl ester: mp 28-32°. $[\alpha]_D^{25}$ −27° (hexane). uv max (methanol): 270, 278, 290 nm (ε 43900, 56700, 43100), *cf.* J. Rokach *et al., Tetrahedron Letters* **22**, 2759 (1981); I. Ernest *et al., ibid.* **23**, 167 (1982).

Leukotriene B_4, $C_{20}H_{32}O_4$, *[S-[R*,S*-(E,Z,E,Z)]]-5,12-dihydroxy-6,8,10,14-eicosatetraenoic acid, leukotriene B, LTB_4*. uv max (methanol): 260, 270.5, 281 nm (ε 38000, 50000, 39000).

Leukotriene C_4, $C_{30}H_{47}N_3O_9S$, *[R-[R*,S*-(E,E,Z,Z)]]-N-[S-[1-(4-carboxy-1-hydroxybutyl)-2,4,6,9-pentadecatetraenyl]-N-L-γ-glutamyl-L-cysteinyl]glycine, leukotriene C, leukotriene C_1, LTC_4*. uv max (methanol): 270, 280, 290 nm (ε 32000,

40000, 31000). Can be stored for several days without appreciable decomposition in frozen (−20°) pH 6.8 phosphate buffer under argon or as the tripotassium salt frozen in water. Biological activity destroyed after incubation with soybean lipoxygenase and uv max shifts to 308 nm.

Leukotriene D₄, C₂₅H₄₀N₂O₆S, [R-[R,S*-(E,E,Z,Z)]]-N-[S-[1-(4-carboxy-1-hydroxybutyl)-2,4,6,9-pentadecatetraenyl]-L-cysteinyl]glycine, leukotriene D, LTD₄.* uv max (methanol): 270, 280, 290 nm (ε 31000, 40000, 31000). Storage, destruction of biological activity, uv shift are the same as for LTC₄.

Leukotriene E₄, C₂₃H₃₇NO₅S, [5S-[5R,6S*(S*),7E,9E,11Z,14Z]]-6-[(2-amino-2-carboxyethyl)thio]-5-hydroxy-7,9,11,14-eicosatetraenoic acid, leukotriene E, LTE₄.* uv max (ethanol) of methyl ester: 269, 280, 291 nm (ε 28200, 35200, 28900).

5483. Leupeptins. Class of modified tripeptide protease inhibitors produced by various species of *Actinomycetes*. Two major components, *leupeptin Ac-LL* and *leupeptin Pr-LL*, consisting of L-leucyl-L-leucyl-DL-argininal modified at the amino terminal by acetyl- or propionyl-, respectively, have been isolated. Various minor analogs, in which valine or isoleucine replaces either or both leucines, have also been found. Isolation from *Actinomycetes*: T. Aoyagi *et al., J. Antibiot.* **22**, 283 (1969); S.-I. Kondo *et al., Chem. Pharm. Bull.* **17**, 1896 (1969). Structure and synthesis of leupeptins Ac-LL and Pr-LL: K. Kawamura *et al., ibid.* 1902; K. Maeda *et al., J. Antibiot.* **24**, 402 (1971). Improved purification of leupeptin Ac-LL: M. C. Y. Ning, R. J. Beynon, *Int. J. Biochem.* **18**, 813 (1986). Inhibition of proteases: T. Aoyagi *et al., J. Antibiot.* **22**, 558 (1969); and synthesis of analogs: G. Borin *et al., Z. Physiol. Chem.* **362**, 1435 (1981). Effects on protein degradation in normal and diseased muscle: P. Libby, A. L. Goldberg, *Science* **199**, 534 (1978); I. Nonaka *et al., Acta Neuropathol.* **58**, 279 (1982); R. P. Hummel, III *et al., J. Surg. Res.* **45**, 140 (1988). HPLC determn of leupeptin Ac-LL in serum and muscle: M. Kai *et al., J. Chromatog.* **345**, 259 (1985).

Note: In some sources leupeptin refers only to leupeptin Ac-LL.

USE: Enzyme inhibitor in biological preparations.

5484. Leuprolide. *6-D-Leucine-9-(N-ethyl-L-prolinamide)-10-deglycinamideluteinizing hormone-releasing factor (pig);* leuprorelin; (D-Leu⁶)-des-Gly¹⁰-LH-RH-ethylamide. C₅₉H₈₄N₁₆O₁₂; mol wt 1209.42. C 58.59%, H 7.00%, N 18.53%, O 15.87%. Synthetic nonapeptide agonist analog of LH-RH, *q.v.* Prepn: M. Fujino *et al., Ger. pat.* **2,446,005** (1975 to Takeda), *C.A.* **83**, 10895y (1975); R. L. Gendrich *et al., U.S. pat.* **4,005,063** (1977 to Abbott). Synthesis: J. A. Vilchez-Martinez *et al., Biochem. Biophys. Res. Commun.* **59**, 1226 (1974); M. Fujino *et al., ibid.* **60**, 406 (1974). Comparison of biological activity with natural LH-RH: D. H. Coy *et al., ibid.* **67**, 576 (1975). Pharmacokinetics: L. T. Sennello *et al., J. Pharm. Sci.* **75**, 158 (1986). Clinical efficacy in prostatic carcinoma: M. B. Garnick *et al., New Engl. J. Med.* **311**, 1281 (1984); in benign prostatic hypertrophy: L. M. Eri, K. J. Tveter, *J. Urol.* **150**, 359 (1993). Clinical trial in endometriosis: J. M. Wheeler *et al., Am. J. Obstet. Gynecol.* **167**, 1367 (1992).

5-oxoPro–His–Trp–Ser–Tyr–D-Leu–Leu–Arg–ProNHC₂H₅

Fluffy solid. [α]²⁵_D −31.7° (c = 1 in 1% acetic acid).

Monoacetate (salt), C₅₉H₈₄N₁₆O₁₂·C₂H₄O₂, *leuprolide acetate, Abbott 43818, A-43818, TAP-144, Carcinil, Enantone, Leuplin, Lucrin, Lupron, Prostap.*

THERAP CAT: Antineoplastic (hormonal); LH-RH agonist.

5485. Levallorphan. *17-(2-Propenyl)morphinan-3-ol; l,N-allyl-3-hydroxymorphinan; l-3-hydroxy-N-allylmorphinan.* C₁₉H₂₅NO; mol wt 283.41. C 80.52%, H 8.89%, N 4.94%, O 5.65%. Prepn from 3-hydroxy-N-methylmorphinan: Schnider, Grüssner, *Helv. Chim. Acta* **34**, 2211 (1951). Alternate route: Hellerbach *et al., ibid.* **39**, 429 (1956). Comprehensive description: B. C. Rudy, B. Z. Senkowski in *Analytical Profiles of Drug Substances* **vol. 2**, K. Florey, Ed. (Academic Press, New York, 1973) pp 339-361.

Crystals from dilute ethanol, mp 180-182°. [α]²⁰_D −88.9° (c = 3 in methanol).

Tartrate, C₁₉H₂₅NO.C₄H₆O₆, *Lorfan.* Crystals from ethanol, mp 176-177°. [α]¹⁶_D −39°. Soluble in water.

THERAP CAT: Narcotic antagonist.

THERAP CAT (VET): Narcotic antagonist.

5486. Levamisole. *(S)-2,3,5,6-Tetrahydro-6-phenylimidazo[2,1-b]thiazole;* (−)-6-phenyl-2,3,5,6-tetrahydroimidazo[2,1-b]thiazole; Levovermax; Totalon. C₁₁H₁₂N₂S; mol wt 204.30. C 64.67%, H 5.92%, N 13.71%, S 15.70%. Biological response modifier with anthelmintic activity. Prepn of racemate: A. H. M. Raeymaekers *et al., U.S. pat.* **3,274,209** (1966 to Janssen); *eidem, J. Med. Chem.* **9**, 545 (1966). Pharmacology: D. Thienpoint *et al., Nature* **209**, 1084 (1966). Prepn and abs config of isomers: A. H. M. Raeymaekers *et al., Tetrahedron Letters* **1967**, 1467. *See also:* M. W. Bullock *et al., J. Med. Chem.* **11**, 169 (1968); U.S. pat. **3,565,907** (1971 to Am. Cyanamid); Dewar *et al., U.S. pat.* **3,579,530** (1971 to ICI). Stereospecific inhibition of succinic dehydrogenase: H. Van den Bossche, P. A. J. Janssen, *Biochem. Pharmacol.* **18**, 35 (1969). Exptl effect on IL-1 production: E. S. Kimball, *Ann. N.Y. Acad. Sci.* **685**, 259 (1993). Review of pharmacology: H. Schneiden, *Int. J. Immunopharmacol.* **3**, 9 (1981); of clinical use in parasitic infections: M. J. Miller, *Drugs* **20**, 122-130 (1980). Determn in animal tissues by ELISA: J. J. Silverlight, R. Jackman: *Analyst* **119**, 2705 (1994); by LC-MS: A. Cannavan *et al., ibid.* **120**, 331 (1995). Clinical trial with fluorouracil in colon cancer: C. G. Moertel *et al., Ann. Int. Med.* **122**, 321 (1995). Review of immunopharmacology and clinical trials in cancer treatment: H. C. Stevenson *et al., J. Clin. Oncol.* **9**, 2052-2066 (1991); M. De Brabander *et al., Anticancer Res.* **12**, 177-188 (1992); W. K. Amery, J. P. Bruynseels, *Int. J. Immunopharmacol.* **14**, 481-486 (1992).

mp 60-61.5°. [α]²⁵_D −85.1° (c = 10 in chloroform).

Hydrochloride, C₁₁H₁₂N₂S.HCl, *R-12564, Ascaridil, Decaris, Ergamisol, Levacide, Levadin, Levasole, Meglum, Nemicide, Nilverm, Ripercol, Solaskil, Spartakon, Tramisol.* mp 227-229°. [α]²⁰_D −124° ± 2° (c = 0.9, water). Sol in water. Stable in acid media.

DL-Form, *tetramisole, tetramizole.* Crystals, mp 87-89°.

DL-Form hydrochloride, *Bayer 9051, McN-JR-8299, R-8299.* Crystals, mp 264-265°. Sol in water (21 g/100 ml at 20°), methanol, propylene glycol; sparingly sol in ethanol. Slightly sol in chloroform, hexane, acetone. LD₅₀ in mice, rats (mg/kg): 22, 24 i.v.; 84, 130 s.c.; 210, 480 orally (Thienpoint).

D-(+)-Form, *dexamisole.* mp 60-61.5°. [α]²⁵_D +85.1° (c = 10 in chloroform).

D-(+)-Form hydrochloride, *R-12563.* mp 227-227.5°. [α]²⁰_D +125° (c = 0.7 in water).

THERAP CAT: Anthelmintic (nematodes); immunomodulator.

THERAP CAT (VET): Anthelmintic (nematodes).

5487. Levcromakalim. *(3S-trans)-3,4-Dihydro-3-hydroxy-2,2-dimethyl-4-(2-oxo-1-pyrrolidinyl)-2H-1-benzopyran-6-carbonitrile;* (3S,4R)-3-hydroxy-2,2-dimethyl-4-(2-oxo-1-pyrrolidinyl)-6-chromancarbonitrile; (−)-6-cyano-3,4-dihydro-2,2-dimethyl-trans-4-(2-oxo-1-pyrrolidinyl)-2H-benzo[b]pyran-3-ol; lemakalim; (−)-cromakalim; BRL-38227. C₁₆H₁₈N₂O₃; mol wt 286.33. C 67.12%, H 6.34%, N 9.78%, O 16.76%. Potassium channel opener. Prepn of racemate:

J. M. Evans *et al.*, **Eur.** pat. **Appl. 76,075;** *eidem*, U.S. pat. **4,446,113** (1983, 1984 both to Beecham); of isomers: E. Faruk, **Eur.** pat. **Appl. 120,428** (1984 to Beecham); V. A. Ashwood *et al.*, *J. Med. Chem.* **29**, 2194 (1986). Mechanism of action and hemodynamic effects: R. P. Hof *et al.*, *Circ. Res.* **62**, 679 (1988). Pharmacology: J. C. Clapham *et al.*, *Arzneimittel-Forsch.* **41**, 385 (1991). Clinical evaluation in hypertension: S. Suzuki *et al.*, *ibid.* **45**, 859 (1995).

Crystals from ethyl acetate, mp 242-244°. $[\alpha]_D^{26}$ −52.2° (c = 1 in chloroform).

(+)-Form, crystals from ethyl acetate, mp 243-245°. $[\alpha]_D^{26}$ +53.5° (c = 1 in chloroform).

(±)-Form, *cromakalim, BRL-34915.* Crystals from ethyl acetate, mp 230-231°.

THERAP CAT: Antihypertensive.

5488. Levobunolol. *(S)-5-[3-[(1,1-Dimethylethyl)amino]-2-hydroxypropoxy]-3,4-dihydro-1(2H)-naphthalenone; (−)-5-[3-(tert-butylamino)-2-hydroxypropoxy]-3,4-dihydro-1(2H)-naphthalenone; l-bunolol; W-6421A.* $C_{17}H_{25}NO_3$; mol wt 291.39. C 70.07%, H 8.65%, N 4.81%, O 16.47%. Nonselective β-adrenoceptor antagonist. Prepn of racemate: C. F. Schwender *et al.*, *J. Med. Chem.* **13**, 684 (1970); J. Shavel, Jr., S. Farber, **Ger.** pat. **1,948,144;** *eidem*, U.S. pat. **3,641,152** (1970, 1972 both to Warner-Lambert). Resolution of enantiomers: J. Shavel, Jr., C. F. Schwender, **Ger.** pat. **2,046,043;** *eidem*, U.S. pat. **3,649,691** (1971, 1972 both to Warner-Lambert); R. D. Dennis *et al.*, U.S. pat. **4,463,176** (1984 to Bristol Myers, Mead Johnson). Cardiovascular pharmacology, anti-arrhythmic activity of racemate and isomers: R. D. Robson, H. R. Kaplan, *J. Pharmacol. Exp. Ther.* **175**, 157, 168 (1970). Adrenoceptor blocking activity: H. R. Kaplan *et al.*, *Eur. J. Pharmacol.* **16**, 237 (1971). Metabolism in humans: F. J. DiCarlo *et al.*, *Clin. Pharmacol. Ther.* **22**, 858 (1977); F.-J. Leinweber *et al.*, *Pharmacology* **16**, 70 (1978). Corneal permeability: R. D. Schoenwald, H.-S. Huang, *J. Pharm. Sci.* **72**, 1266, 1272 (1983). HPLC determn in biological fluids: H. Hengy, E. U. Kolle, *J. Chromatog.* **338**, 444 (1985); D. D. S. Tang-Liu *et al.*, *J. Liq. Chromatog.* **9**, 2237 (1986). Clinical evaluation in hypertension: E. Arce-Gomez *et al.*, *Curr. Ther. Res.* **19**, 386 (1976); comparison with timolol, *q.v.*, in glaucoma and ocular hypertension: A. Cinotti *et al.*, *Am. J. Ophthalmol.* **99**, 11 (1985); D. Long *et al.*, *ibid.* 18. *Review:* H. R. Kaplan, "Levobunolol" in *Pharmacology of Antihypertensive Drugs*, A. Scriabine, Ed. (Raven Press, New York, 1980) pp 317-323. Brief review of long-term treatment of glaucoma: G. D. Novack, *Gen. Pharmacol.* **17**, 373-377 (1986).

LD$_{50}$ in male, female rats, mice (mg/kg): 700, 800, 1530, 1220 orally; 25, 28, 78, 84 i.v.; LD$_{50}$ in male, female hamsters, dogs (mg/kg): 435, 500, 100, 100 orally (Kaplan, 1980).

Hydrochloride, $C_{17}H_{25}NO_3 \cdot HCl$, *W-7000A, Betagan, Vistagan, Gotensin.* Crystals from methanol-ether, mp 209-

211°. $[\alpha]_{589}^{24}$ −19.6±0.7° (c = 2.90 in methanol). uv max (NaOH): 221, 253, 310 nm (ε 24700, 9000, 2400).

THERAP CAT: Antiglaucoma

5489. Levocabastine. *[3S-[1(cis),3α,4β]]-1-[4-Cyano-4-(4-fluorophenyl)cyclohexyl]-3-methyl-4-phenyl-4-piperidinecarboxylic acid; (−)-trans-1-[cis-4-cyano-4-(p-fluorophenyl)cyclohexyl]-3-methyl-4-phenylisonipecotic acid; (−)-cabastine.* $C_{26}H_{29}FN_2O_2$; mol wt 420.53. C 74.26%, H 6.95%, F 4.52%, N 6.66%, O 7.61%. Histamine H$_1$-receptor antagonist. Prepn: R. Stokbroekx *et al.*, **Eur.** pat. **Appl. 34,415;** *eidem*, U.S. pat. **4,369,184** (1981, 1983 both to Janssen); and pharmacology: R. A. Stokbroekx *et al.*, *Drug Dev. Res.* **8**, 87 (1986). Pharmacology: K. Tasaka *et al.*, *Arzneimittel-Forsch.* **40**, 1295 (1990). Clinical pharmacology: N. Rombaut *et al.*, *Ann. Allergy* **67**, 75 (1991). Clinical trial in children with allergic conjunctivitis: M. Rimas *et al.*, *Allergy* **45**, 18 (1990); in allergic rhinitis: M. Schata *et al.*, *J. Allergy Clin. Immunol.* **87**, 873 (1991). Review of pharmacology and therapeutic use: K. L. Dechant, K. L. Goa, *Drugs*, **41**, 202-224 (1991).

Hydrochloride, $C_{26}H_{29}FN_2O_2 \cdot HCl$, *R-50547, Levophta, Livostin.*

THERAP CAT: Antihistaminic.

5490. Levodopa. *3-Hydroxy-L-tyrosine; (−)-3-(3,4-dihydroxyphenyl)-L-alanine; L-dopa; β-(3,4-dihydroxyphenyl)-L-alanine; (−)-2-amino-3-(3,4-dihydroxyphenyl)propanoic acid; Bendopa; Deadopa; Dopaflex; Dopal; Dopaidan; Dopalina; Dopar; Doparkine; Doparl; Dopasol; Dopaston; Dopastral; Cidandopa; Doprin; Eldopal; Eldopar; Eldopatec; Eurodopa; Laradopa; Maipedopa; Larodopa; Ledopa; Parda; Levopa; Veldopa (formerly Weldopa).* $C_9H_{11}NO_4$; mol wt 197.19. C 54.82%, H 5.62%, N 7.10%, O 32.45%. Naturally occurring form of dopa, *q.v.*, the biological precursor of the catecholamines. Prepn from *l*-3-nitrotyrosin: Wasser, Lewandowski, *Helv. Chim. Acta* **4**, 657 (1921); from 3-(3,4-methylenedioxyphenyl)-L-alanine: Yamada *et al.*, *Chem. Pharm. Bull.* **10**, 693 (1962); from L-tyrosine: Vorbrüggen, Krolikiewicz, *Ber.* **105**, 1168 (1972); Bretschneider *et al.*, *Helv. Chim. Acta* **56**, 2857 (1973); from *Vicia faba* beans: Wysong, U.S. pat. **3,253,023** (1966 to Dow Chem.); by fermentation of L-tyrosine: Sih *et al.*, *J. Am. Chem. Soc.* **91**, 6204 (1969); Florent, Renaut, **Ger. Offen. 2,102,793** (1971 to Rhône-Poulenc), *C.A.* **75**, 108505f (1971). Sepn from racemate: Vogler, Baumgartner, *Helv. Chim. Acta* **35**, 1776 (1952); **Neth.** pat. **Appl. 6,514,950** corresp to U.S. pat. **3,405,159** (1966 and 1968 to Merck & Co.). Molecular conformation: Becker *et al.*, *Biochem. Biophys. Res. Commun.* **41**, 444 (1970). Metabolism studies: Shaw *et al.*, *J. Biol. Chem.* **226**, 255 (1957); Calne *et al.*, *Brit. J. Pharmacol.* **37**, 57 (1969). Acute toxicity: *Rx Bulletin* **1**, 16 (November, 1970). Hemodynamic effects in congestive heart failure: S. I. Rajfer *et al.*, *N. Engl. J. Med.* **310**, 1357 (1984). Series of articles on clinical efficacy in Parkinson's disease: *Advan. Neurol.* **45**, 457-510 (1986). Reviews on L-dopa and parkinsonism: Barbeau, *Can. Med. Assoc. J.* **101**, 791 (1969); Pletscher *et al.*, *Schweiz. Med. Wochenschr.* **100**, 797 (1970); Calne, Sandler, *Nature* **226**, 21 (1970); *L-Dopa and Parkinsonism*, A. Barbeau, Ed. (F. A. Davis, Philadelphia, 1970). Comprehensive description: R. Gomez *et al.*, *Anal. Profiles Drug Subs.* **5**, 189-223 (1976).

Colorless to white, odorless and tasteless crystals or crystalline powder. Needles from water, mp 276-278° (dec)

(Yamada); also reported as mp 284-286° (Wysong). $[\alpha]_D^{13}$ —13.1° (c = 5.12 in 1N HCl). uv max (0.001N HCl): 220.5, 280 nm (log ε 3.79, 3.42). Readily sol in dil HCl and formic acid. Soly in water: 66 mg/40 ml. Practically insol in ethanol, benzene, chloroform and ethyl acetate. In the presence of moisture, L-dopa is rapidly oxidized by atmospheric oxygen and darkens. LD_{50} in mice, rats, rabbits (mg/kg): 3650, 4000, 609 orally (*Rx Bulletin*).

THERAP CAT: Antiparkinsonian.

5491. Levomepate. *[3(S)-endo]-α-(Hydroxymethyl)-α-methylbenzeneacetic acid 8-methyl-8-azabicyclo[3.2.1]oct-3-yl ester; 1αH,5αH-tropan-3α-ol (—)-2-methyl-2-phenylhydracrylate (ester); (—)-2-methyl-2-phenylhydracrylic acid 3α-tropanyl ester; (—)-α-methylhyoscyamine; 3α-tropanyl (—)-2-methyl-2-phenylhydracrylate; tropine (—)-α-methylnitropate; Dispan.* $C_{18}H_{25}NO_3$; mol wt 303.40. C 71.26%, H 8.31%, N 4.62%, O 15.82%. Prepd from tropine and (—)-β-acetoxy-α-methyl-α-phenylpropionyl chloride: Vecchi *et al.*, **Brit. pat. 874,015** (1959 to Lepetit); Melone *et al.*, *J. Org. Chem.* **25**, 859 (1960). Pharmacology: Bianchi *et al.*, *Toxicol. Appl. Pharmacol.* **10**, 424 (1967).

Hydrochloride, $C_{18}H_{25}NO_3 \cdot HCl$, crystals from ethyl acetate, mp 210-212°. $[\alpha]_D^{20}$ —6.8°.

THERAP CAT: Anticholinergic.

5492. Levomethadyl Acetate. *[S-(R*,R*)]-β-[2-(Dimethylamino)propyl]-α-ethyl-β-phenylbenzeneethanol acetate (ester); (—)-6-(dimethylamino)-4,4-diphenyl-3-heptanol acetate (ester); α-l-acetylmethadol; levo-α-acetylmethadol; LAAM.* $C_{23}H_{31}NO_2$; mol wt 353.50. C 78.15%, H 8.84%, N 3.96%, O 9.05%. Longest-acting enantiomer of methadyl acetate, *q.v.* Duration of action due to active metabolites. Prepn: A. Pohland *et al.*, *J. Am. Chem. Soc.* **71**, 460 (1949). Synthesis of metabolites: F. I. Carroll *et al.*, *J. Org. Chem.* **41**, 3521 (1976). Metabolism in rats: G. L. Henderson *et al.*, *Drug Metab. Dispos.* **5**, 321 (1977); M. Man *et al.*, *ibid.* **8**, 55 (1980); in man: B. S. Finkle *et al.*, *J. Anal. Toxicol.* **6**, 100 (1982). Determn of LAAM and metabolites by HPLC: C.-H. Kiang *et al.*, *J. Chromatog.* **222**, 81 (1981); by GLC: K. Verebey *et al.*, *ibid.* **343**, 339 (1985). Analgesic activity in mice: N. B. Eddy *et al.*, *J. Org. Chem.* **17**, 321 (1952); in man: A. S. Keats, H. K. Beecher, *J. Pharmacol. Exp. Ther.* **105**, 210 (1952). Pharmacology in monkeys: S. J. Mule, A. L. Misra, *Ann. N.Y. Acad. Sci.* **311**, 199 (1978). Comparison with methadone in the treatment of heroin addiction: T. J. Crowley *et al.*, *Psychopharmacology* **86**, 458 (1985). Review of use in treatment of narcotic addiction: J. D. Blaine *et al.*, *Ann. N.Y. Acad. Sci.* **362**, 101-115 (1981).

Hydrochloride, $C_{23}H_{31}NO_2 \cdot HCl$, *ORLAAM.* Crystals from ethanol-ether, mp 215°. $[\alpha]_D^{25}$ —60° (c = 0.2). Sol in water. LD_{50} in mice (mg/kg): 110.0 s.c., 172.8 orally (Eddy).

THERAP CAT: In treatment of narcotic addiction.

5493. Levophacetoperane. *α-Phenyl-2-piperidinemethanol acetate; threo-1-acetoxy-1-phenyl-1-(2-piperidyl)meth-* ane; acetic acid α-phenyl-2-piperidylmethyl ester; 1-phenyl-1-(2'-piperidyl)-1-acetoxymethane; phacetoperane; RP-8228. $C_{14}H_{19}NO_2$; mol wt 233.31. C 72.07%, H 8.21%, N 6.00%, O 13.72%. Prepn: Jacob, Joseph, U.S. pat. **2,928,835** (1960 to Rhône-Poulenc).

Hydrochloride, $C_{14}H_{19}NO_2 \cdot HCl$, *Lidepran.* Crystals from acetone + ether, mp 229-230°. Levorotatory.

THERAP CAT: Antidepressant; anorexic.

5494. Levopimaric Acid. *[1R-(1α,4aβ,4bα,10aα)]-1,2,-3,4,4a,4b,5,9,10,10a-Decahydro-1,4a-dimethyl-7-(1-methylethyl)-1-phenanthrenecarboxylic acid; 13-isopropylpodocarpa-8(14),12-dien-15-oic acid;* $\Delta^{6,8(14)}$*-abietadienoic acid; l-pimaric acid; β-pimaric acid; l-sapietic acid.* $C_{20}H_{30}O_2$; mol wt 302.46. C 79.42%, H 10.00%, O 10.58%. Isolation from American pine oleoresin: Palkin, Harris, *J. Am. Chem. Soc.* **55**, 3677 (1933); from French galipot, from *Pinus maritima:* Ruzicka, Bacon, *Helv. Chim. Acta* **20**, 1542 (1937); from *Pinus palustris:* Harris, Sanderson, *J. Am. Chem. Soc.* **70**, 334, 3671 (1948). Structure: Ruzicka, Kaufmann, *Helv. Chim. Acta* **23**, 1346 (1940); *cf.* Arbuzov, *Chem. Zentr.* **1942**, II, 893. Stereochemistry: Schuller, Lawrence, *J. Am. Chem. Soc.* **83**, 2563 (1961); Burgstahler *et al.*, *ibid.* **83**, 4660 (1961); Weiss *et al.*, *Chem. & Ind. (London)* **1962**, 1286; Dauben, Coates, *J. Org. Chem.* **28**, 1698 (1963). Conformation: Burgstahler *et al.*, *Chem. Commun.* **1971**, 121; Weiss *et al.*, *ibid.* **1972**, 17.

Orthorhombic crystals, mp 150°; $[\alpha]_D^{20}$ —280.4° (c = 0.7 in alcohol); $[\alpha]_D^{14}$ —266.6° (c = 0.4 in chloroform). Absorption max 273 nm: Kraft, *Ann.* **520**, 133 (1935). Practically insol in water, sol in most organic solvents. Forms a crystalline ammonium salt.

Methyl ester, $C_{21}H_{32}O_2$, crystals from methanol, mp 64°. $[\alpha]_D^{20}$ —268° (c = 1 in ether).

Molecular compd with quinone, $C_{26}H_{34}O_4$, yellow prisms from methanol, mp 214°. $[\alpha]_D^{20}$ —148° (c = 0.7 in chloroform).

5495. Levopropoxyphene. *α-[2-(Dimethylamino)-1-methylethyl]-α-phenylbenzeneethanol propanoate(ester); α-l-4-dimethylamino-3-methyl-1,2-diphenyl-2-butanol propionate; α-l-4-dimethylamino-1,2-diphenyl-3-methyl-2-butanol propionate; l-propoxyphene.* $C_{22}H_{29}NO_2$; mol wt 339.48. C 77.84%, H 8.61%, N 4.13%, O 9.43%. Prepn: Pohland, Sullivan, *J. Am. Chem. Soc.* **77**, 3400 (1955). Stereoselective synthesis: Pohland *et al.*, *J. Org. Chem.* **28**, 2483 (1963). Toxicity data: E. I. Goldenthal, *Toxicol. Appl. Pharmacol.* **18**, 185 (1971). *See also* Propoxyphene.

Crystals from petr ether, mp 75-76°. $[\alpha]_D^{25}$ -68.2° (c = 0.6 in chloroform).

Hydrochloride, $C_{22}H_{29}NO_2 \cdot HCl$, crystals from methanol + ethyl acetate, mp 163-164°. $[\alpha]_D^{25}$ -60.1° (c = 0.7).

2-Naphthalenesulfonate, $C_{22}H_{29}NO_2 \cdot C_{10}H_8SO_3$, *levopropoxyphene napsylate, Novrad, Letusin, Contratuss.* LD_{50} orally in female rats: 1455 ±77 mg/kg (Goldenthal).

THERAP CAT: Antitussive.

5496. Levorphanol. *17-Methylmorphinan-3-ol;* (−)-3-hydroxy-*N*-methylmorphinan; levorphan; lemoran; Ro-1-5431. $C_{17}H_{23}NO$; mol wt 257.38. C 79.33%, H 9.01%, N 5.44%, O 6.22%. Orally active synthetic morphine analog. Prepn of racemate from 2-methyl-1-benzyl-1,2,3,4,5,6,7,8-octahydroisoquinoline: Grewe, *Naturwiss.* **33**, 333 (1946); *Angew. Chem.* **A59**, 198 (1947); Grewe, Mondon, *Ber.* **81**, 279 (1948); Swiss pat. **280,674** (1952 to Hoffmann-La Roche), *C.A.* **47**, 7554 (1953). Prepn of isomers: Schnider, Grüssner, *Helv. Chim. Acta* **34**, 2211 (1951); Vogler, U.S. pat. **2,744,112** (1956 to Hoffmann-La Roche). Absolute configuration: Corrodi *et al., Helv. Chim. Acta* **42**, 212 (1959). Analgesic activity and toxicity data: L. O. Randall, G. Lehmann, *J. Pharmacol. Exp. Ther.* **99**, 163 (1950). HPLC determn in plasma: R. Lucek, R. Dixon, *J. Chromatog.* **341**, 239 (1985). Clinical pharmacokinetics: R. Dixon *et al., Res. Commun. Chem. Pathol. Pharmacol.* **41**, 3 (1983).

Crystals, mp 198-199°. $[\alpha]_D^{20}$ -56° (c = 3 in absolute alcohol).

Tartrate dihydrate, $C_{17}H_{23}NO \cdot C_4H_6O_6 \cdot 2H_2O$, *Ro-1-5431/7, Dromoran, Levo-Dromoran.* Crystals, mp 113-115° (when anhydrous, mp 206-208°). $[\alpha]_D^{20}$ -14° (c = 3 in water). pH of a 0.2% aq soln 3.4 to 4.0. One gram dissolves in 45 ml water, in 110 g alcohol, in 50 g ether.

dl-Form, *racemorphan, methorphinan.* Crystals from anisole and dil alcohol, mp 251-253°.

dl-Form hydrobromide, $C_{17}H_{23}NO \cdot HBr$, *Nu-2206.* Crystals, mp 193-195°. Sol in water; sparingly sol in alcohol. Practically insol in ether. LD_{50} i.v. in mice: 41 mg/kg (Randall, Lehmann).

d-Form, *dextrorphan, Ro-1-6794.* Crystals, mp 198-199°. $[\alpha]_D^{20}$ +56.3° (c = 3 in abs alcohol).

d-Form tartrate monohydrate, $C_{17}H_{23}NO \cdot C_4H_6O_6 \cdot H_2O$, crystals, mp 183-185°. $[\alpha]_D^{20}$ +34.6° (c = 3 in water). Sol in water.

Note: Levorphanol, racemorphan and their salts are controlled substances (opiates) listed in the U.S. Code of Federal Regulations, Title 21 Part 1308.12 (1995).

THERAP CAT: Analgesic (narcotic).

5497. Levothyroxine Sodium. *O-(4-Hydroxy-3,5-diiodophenyl)-3,5-diiodo-L-tyrosine monosodium salt; L-thyroxine sodium salt;* sodium levothyroxine; L-3,3',5,5'-tetraiodothyronine sodium salt; Eltroxin; Euthyrox; Laevoxin; Letter; Levaxin; Levothroid; Levothyrox; Oroxine; Synthroid Sodium; Thyroxevan. $C_{15}H_{10}I_4NNaO_4$; mol wt 798.86. C 22.55%, H 1.26%, I 63.54%, N 1.75%, Na 2.88%, O 8.01%. The sodium salt of the amino acid L-thyroxine, *q.v.*, an ac-

tive physiological principle obtained from the thyroid gland of domesticated animals used for food by man, or prepd synthetically. Prepn: Chalmers *et al., J. Chem. Soc.* **1949**, 3424. Solubility in phosphate buffer solns: Evert, *J. Phys. Chem.* **64**, 478 (1960). Clinical pharmacology in hypothyroidism: P. W. Ladenson *et al., Am. J. Med.* **73**, 467 (1982). Bioavailability and metabolism in hypothyroidism: L. H. Fish *et al., N. Engl. J. Med.* **316**, 764 (1987). Comprehensive description: A. Post, R. J. Warren, *Anal. Profiles Drug Subs.* **5**, 225-281 (1976).

Pentahydrate, triclinic crystals or cream-colored powder. Odorless and tasteless. Somewhat hygroscopic. d_4^{20} 2.381. $[\alpha]_D^{20}$ -4.4° (c = 3 in 70% ethanol). Soly at 25° in water: about 15 mg/100 ml. (Higher solys have been reported). Sol in mineral acids and in solns of alkali hydroxides and carbonates. More sol in alcohol. Very slightly sol in chloroform, ether. pH of a satd water soln: 8.35 to 9.35.

THERAP CAT: Treatment of hypothyroidism.

THERAP CAT (VET): In thyroid deficiency (dogs). Has been used in obesity, renal insufficiency, chronic skin conditions. Has also been used in lowered fertility in bulls; and to stimulate lactation.

5498. Levulinic Acid. *4-Oxopentanoic acid;* laevulinic acid; β-acetylpropionic acid. $C_5H_8O_3$; mol wt 116.12. C 51.72%, H 6.94%, O 41.34%. $CH_3COCH_2CH_2COOH$. Laboratory procedure from starch or cane sugar by boiling with HCl: McKenzie, *Org. Syn.* coll. vol. I, 335 (1941). Produced commercially from low grade cellulose. By-product of furfural manuf. Extensive review: Leonard, *Ind. Eng. Chem.* **48**, 1331 (1956).

Plates or leaflets (commercial product is yellow), mp 33-35°. bp 245-246°. d 1.1447. n_D^{16} 1.442. Freely sol in water, alcohol, ether; essentially insol in aliphatic hydrocarbons. *Protect from light.*

Phenylhydrazone, $C_{11}H_{14}N_2O_2$, *Antithermin.* Leaflets, mp 108°. Slightly sol in cold water, more sol in hot water; freely sol in alcohol, chloroform, ether, dil acids.

USE: In organic syntheses; in the manuf of nylon, synthetic rubbers, plastics, and medicinals.

5499. LH. *Luteinizing hormone;* ICSH; interstitial cell stimulating hormone. Mol wt about 30,000. A glycoprotein gonadotrop(h)ic hormone found in the anterior lobe of the pituitary gland. First isolated from sheep pituitaries: Li *et al., Endocrinology* **27**, 803 (1940); *Science* **92**, 355 (1940); *J. Am. Chem. Soc.* **64**, 367 (1942); from pig pituitaries: Shedlovsky *et al., Science* **92**, 178 (1940); Chow *et al., Endocrinology* **30**, 650 (1942). Stimulates the synthesis of progesterone in the ovaries. Together with FSH stimulates the release of estrogen from Graafian follicles. Also induces the process of ovulation in which the mature ovum is extruded from the follicle and, following this, the cells which hitherto surrounded it are converted under the influence of LH into lutein cells (corpus luteum). In the male, stimulates the interstitial cells of testes to secrete testosterone. The glycoprotein structure contains two dissimilar subunits designated as α and β, *cf.* HCG, FSH, TSH. Each of the dissociated subunits shows little of the biological activity associated with the native hormone. Complete amino acid sequence of ovine LH subunits: Papkoff *et al., J. Am. Chem. Soc.* **93**, 1531 (1971); Liu *et al., J. Biol. Chem.* **247**, 4351, 4365 (1972); of human LH subunits: H. T. Keutmann *et al., Endocrin. Res. Commun.* **5**, 57 (1978); *eidem, Biochem. Biophys. Res. Commun.* **90**, 842 (1979). Reviews: Ward *et al., Recent Progr. Horm. Res.* **29**, 533 (1973); Papkoff *et al., ibid.* 563; J. G. Pierce, T. F. Parsons, *Ann. Rev. Biochem.* **50**, 465-495 (1981).

White powder. Isoelectric point 4.6 (sheep), 7.45 (pig). Soluble in water. Chymotrypsin, trypsin and pepsin destroy the gonadotrop(h)ic action of LH. Picrolonic, flavianic, picric, and trichloroacetic acids precipitate LH with retention of its physiological activity.

THERAP CAT: Gonadotropic hormone.

5500. LH-RH. *Luteinizing hormone-releasing factor;* LH-RF; luteinizing hormone-releasing hormone; LRF; LRH; gonadorelin; gonadotropin-releasing factor; gonadoliberin; luliberin; LH-RH/FSH-RH; Fertagyl; Fertiral; Kryptocur; Relefact LH-RH. $C_{55}H_{75}N_{17}O_{13}$; mol wt 1182.31. C 55.87%, H 6.39%, N 20.14%, O 17.59%. Neurohumoral hormone produced in the hypothalamus which stimulates the secretion of the pituitary hormones, LH (luteinizing hormone) and FSH (follicle-stimulating hormone) *q.q.v.,* which in turn produce changes resulting in the induction of ovulation. Isoln from porcine hypothalamic extracts: A. V. Schally *et al., Biochem. Biophys. Res. Commun.* **43**, 393 (1971). Structure: Matsuo *et al., ibid.* 1334; Baba *et al., ibid.* **44**, 459 (1971); Burgus *et al., C.R. Acad. Sci., Ser. D.* **273**, 1611 (1971). Solid phase synthesis: Monahan *et al., ibid.* 508; Monahan, Rivier, *Biochem. Biophys. Res. Commun.* **48**, 1100 (1972); Coy *et al., Methods Enzymol.* **37**, 416 (1975); D. H. Rich, S. K. Gurwara, *Tetrahedron Letters* **1975**, 301. Industrial prepn: Geiger *et al., Ger.* pat. **2,213,737** (1973 to Hoechst). Confirmation of biological activity: A. V. Schally *et al., Science* **173**, 1036 (1971). Pharmacokinetics: T. W. Redding, *J. Clin. Endocrinol. Metab.* **37**, 626 (1973). Reviews of physiology and implications in fertility control: A. V. Schally *et al., Fert. Steril.* **22**, 703 (1971); R. Guillemin, *Contraception* **6**, 1 (1972). Review of LH-RH and other hypothalamic hormones: A. V. Schally *et al., Science* **179**, 341 (1973); A. V. Schally, *ibid.* **202**, 18-28 (1978); of LH-RH and somatostatin, *q.v.:* S. M. McCann, *Ann. Rev. Pharmacol. Toxicol.* **22**, 491-515 (1982). General reviews: J. Sandow, *Clin. Endocrinol.* **18**, 571-592 (1983); S. M. McCann, *J. Endocrinol. Invest.* **6**, 243-251 (1983).

5-oxoPro–His–Trp–Ser–Tyr–Gly–Leu–Arg–Pro–GlyNH₂

$[\alpha]_D^{25}$ −50° (1% acetic acid). Chymotrypsin, papain, subtilisin, thermolysin destroy the pituitary hormone releasing action of LH-RH.

Acetate (salt) hydrate, *gonadorelin acetate, Cystorelin, Hypocrine, Lutrelef, Lutrepulse.* $C_{55}H_{75}N_{17}O_{13}$·$xC_2H_4O_2$·yH_2O, as the diacetate tetrahydrate or a mixture of monoacetate and diacetate hydrates.

Hydrochloride, *AY-24031, HRF, Factrel.*

THERAP CAT: Gonad-stimulating principle.

THERAP CAT (VET): Gonad-stimulating principle.

5501. Liatris. Deer's tongue; vanilla plant. Leaves of *Trilisa odoratissima* (Walt.) Cass. (*Liatris odoratissima* (Walt.) Willd.), *Compositae.* Habit. U.S., Virginia to Florida and Louisiana. *Constit.* Volatile oil, coumarin.

USE: In perfumery and for perfuming smoking, chewing and snuff tobacco.

5502. Licheniformins. Antibiotic substances produced by *Bacillus licheniformis.* Isoln: Callow *et al., Brit. J. Exp. Pathol.* **28**, 418 (1947). Separation of licheniformins A, B, and C: Callow, Work, *Biochem. J.* **51**, 558 (1952). Mol wt as measured by sedimentation: licheniformin A, 4400; licheniformin B, 3800; and licheniformin C, 4800: Ogston, *ibid.* **51**, 569 (1952).

All three constituents in the form of hydrochlorides are white, amorphous, slightly hygroscopic powders, melting with decompn at indefinite temperatures.

5503. Lichenin. Moss starch. $C_6H_{10}O_5$; mol wt 162.14. C 44.45%, H 6.22%, O 49.34%. Polyglucan from *Cetraria islandica* (L.) Ach., *Parmeliaceae* (Iceland Moss). Isoln: Peat *et al., J. Chem. Soc.* **1957**, 3916. Linear polysaccharide structure composed of regular sequences of two (1→4)- and one (1→3)-β-D-glucopyranosyl residues: Perlin, Suzuki, *Can. J. Chem.* **40**, 50 (1962); Fleming, Manners, *Biochem. J.* **100**, 4, 24 (1966). Exhibits antineoplastic activity: Shibata, **Japan.** pat. 17,147('71), *C.A.* **75**, 67474z (1971); Tadahiro *et al., Chem. Pharm. Bull.* **20**, 2445 (1972).

NMR spectrum: D. Gagnaire, M. Vincedon, *Bull. Soc. Chim. France* **1977**, 479.

White powder. Sol in boiling water, in HCl. $[\alpha]_D$ +18.4°.

5504. Lidamidine. *N-(2,6-Dimethylphenyl)-N′-[imino-(methylamino)methyl]urea;* 1-(2,6-dimethylphenyl)-3-methylamidinourea. $C_{11}H_{16}N_4O$; mol wt 220.27. C 59.98%, H 7.32%, N 25.44%, O 7.26%. Amidinourea with antisecretory, antimotility properties. Prepn as hydrochloride: **Belg.** pat. **844,832** (1977 to Rorer), *C.A.* **88**, 22432m (1977). *See also:* J. Diamond, G. H. Douglas, **U.S.** pat. **4,147,804** (1979 to Rorer). Prepn, structure activity relationship: G. H. Douglas *et al., Arzneimittel-Forsch.* **28**, 1435 (1978). Physical-chemical properties: J. J. Zalipsky *et al., ibid.* 1441. Effect on α_2-adrenergic receptors and electrolyte absorption: T. Durbin *et al., Gastroenterology* **82**, 1352 (1982). Series of articles on pharmacology, metabolism, pharmacokinetics: *Arzneimittel-Forsch..* **28**, 1448-1480 (1978). Toxicity: B. J. Chou *et al., ibid.* 1471. Clinical comparison with loperamide, *q.v.,* in acute diarrhea: G. Gasbarrini *et al., ibid.* **36**, 1843 (1986). Brief review: G. Friedman, *Am. J. Gastroenterol.* **80**, 143 (1985).

Hydrochloride, $C_{11}H_{16}N_4O$·HCl, *WHR-1142A, Lidarral, Smodin.* White powder, mp 194-197°. uv max (H_2O): 262, 271 nm (ε 626, 524). Soly at 25° (mg/ml): water 153.55, methanol 297.94, ethanol 88.55, chloroform 4.62, hexane 0.01. LD_{50} in male mice, male, female rats (mg/kg): 260, 267, 160 orally; in mice (mg/kg): 56 i.v. (Chou).

THERAP CAT: Antiperistaltic; antidiarrheal.

5505. Lidocaine. *2-(Diethylamino)-N-(2,6-dimethylphenyl)acetamide; 2-diethylamino-2′,6′-acetoxylidide;* ω-diethylamino-2,6-dimethylacetanilide; lignocaine; Cuivasil; Duncaine; Leostesin; Lidothesin; Rucaina; Xylocaine; Xylocitin; Xylotox. $C_{14}H_{22}N_2O$; mol wt 234.34. C 71.76%, H 9.46%, N 11.95%, O 6.83%. Long-acting, membrane stabilizing agent against ventricular arrhythmia. Originally developed as local anesthetic. Prepn: N. M. Löfgren, B. J. Lundqvist, **U.S.** pat. **2,441,498** (1948 to Astra); A. D. H. Self, A. P. T. Easson, **Brit.** pat. **706,409** (1954 to May & Baker); I. P. S. Hardie, E. S. Stern, **Brit.** pat. **758,224** (1956 to J. F. Macfarlane & Co.); Zhuravlev, Nikolaev, *Zh. Obshch. Khim.* **30**, 1155 (1960). Toxicity studies: E. R. Smith, B. R. Duce, *J. Pharmacol. Exp. Ther.* **179**, 580 (1971); G. H. Kronberg *et al., J. Med. Chem.* **16**, 739 (1973). Review of pharmacokinetics: N. L. Benowitz, W. Meister, *Clin. Pharmacokinet.* **3**, 177 (1978). Review of action as local anesthetic: Löfgren, *Studies on Local Anesthetics: Xylocaine, A New Synthetic Drug* (Hoeggstroms, Stockholm, 1948); Cooper, *Pharm. J.* **171**, 68 (1953). Reviews of anti-arrhythmic agents: J. L. Anderson *et al., Drugs* **15**, 271 (1978); L. H. Opie, *Lancet* **1**, 861 (1980); E. Carmeliet, *Ann. N.Y. Acad. Sci.* **427**, 1 (1984). Comprehensive description: K. Groningsson *et al., Anal. Profiles Drug Subs.* **14**, 207-243 (1985); M. F. Powell, *ibid.* **15**, 761-779 (1986).

Needles from benzene or alcohol, mp 68-69°. bp_4 180-182°; bp_2 159-160°. Insol in water. Sol in alcohol, ether, benzene, chloroform, oils.

Hydrochloride monohydrate, $C_{14}H_{22}N_2O$·HCl·H_2O, *Lidesthesin, Lignavet, Odontalg, Sedagul, Xylocard, Xyloneural.* Crystals, mp 77-78°; anhydrous, mp 127-129°. Very sol in water, alcohol; sol in chloroform. Insol in ether. pH

Consult the Name Index before using this section.

of 0.5% aq soln: 4.0-5.5. LD_{50} in mice (mg/kg): 292 orally (Smith, Duce); 105 i.p.; 19.5 i.v. (Kronberg).

THERAP CAT: Anesthetic (local); antiarrhythmic (class IB).

THERAP CAT (VET): Anesthetic (local).

5506. Lidofenin. *N-(Carboxymethyl)-N-[2-[(2,6-dimeth-ylphenyl)amino]-2-oxoethyl]glycine;* *N-(2,6-dimethylacetani-lide)iminodiacetic acid;* *N-(2,6-dimethylphenylcarbamoyl-methyl)iminodiacetic acid;* [[(2,6-xylylcarbamoyl)methyl]-imino]diacetic acid; HIDA. $C_{14}H_{18}N_2O_5$; mol wt 294.31. C 57.14%, H 6.16%, N 9.52%, O 27.18%. Iminodiacetic acid analog. Prepn of compd and ^{99m}Tc-complex: M. D. Loberg *et al.,* **Belg. pat. 840,556;** *eidem,* **U.S. pat. 4,017,596** (1976, 1977 both to Research Corp.); and imaging studies: M. D. Loberg *et al., J. Nucl. Med.* **17,** 633 (1976). Structural studies: M. D. Loberg, A. T. Fields, *Int. J. Appl. Radiat. Isotop.* **29,** 167 (1978). Chromatographic determn of radiochemical purity: A. M. Zimmer *et al., Eur. J. Nucl. Med.* **7,** 88 (1982). Biodistribution and clinical kinetics: J. Ryan *et al., J. Nucl. Med.* **18,** 995 (1977). Clinical studies of diagnostic use as hepatobiliary imaging agent: H. S. Weissmann *et al., Am. J. Radiol.* **132,** 523 (1979); R. W. Nicholson *et al., Brit. J. Radiol.* **53,** 878 (1980).

Crystals from water, mp 201-203°. LD_{50} i.v. in male mice: 168 mg/kg (Ryan).

^{99m}Tc-complex, *^{99m}Tc-HIDA, TechneScan HIDA.*

THERAP CAT. ^{99m}Tc-complex as diagnostic aid (radioactive imaging agent).

5507. Lidoflazine. *4-[4,4-Bis(4-fluorophenyl)butyl]-N-(2,6-dimethylphenyl)-1-piperazineacetamide;* 4-[4,4-bis(p-fluorophenyl)butyl]-1-piperazineaceto-2',6'-xylidide; 1-[4,4-di(4-fluorophenyl)butyl]-4-[(2,6-dimethylanilinocarbonyl)-methyl]piperazine; McN-JR-7094; R-7904; Angex; Clinium; Klinium; Ordiflazine; Corflazine. $C_{30}H_{35}F_2N_3O$; mol wt 491.62. C 73.29%, H 7.18%, F 7.73%, N 8.55%, O 3.25%. Calcium blocking agent. Prepn: **Neth. pat. Appl. 6,507,312** corresp to H. K. F. Hermans, W. K. Schaper, **U.S. pat. 3,267,104** (1965, 1966, both to Janssen). Crystal structure: G. Germain *et al., Acta Crystallogr.* **33B,** 1971 (1977). Tissue specificity of calcium-blocking properties: J. M. Van Neuten, D. Wellens, *Arch. Int. Pharmacodyn. Ther.* **242,** 329 (1979). Cardioprotective effects: W. Daenen, W. Flameng, *Angiology* **32,** 543 (1981). Effects in angina: F. L. Gobel *et al., Circulation* **65,** 1 Pt. 2, 127 (1982); W. Shapiro *et al., ibid.* **143.**

Crystals, mp 159-161°. Almost insol in water (<0.01%), very sol in chloroform (>50%), but much less sol in other common organic solvents: W. K. Schaper *et al., J. Pharmacol. Exp. Ther.* **152,** 265 (1966).

THERAP CAT: Vasodilator (coronary).

5508. Light Green SF Yellowish. *N-Ethyl-N-[4-[[4-[ethyl[(3-sulfophenyl)methyl]amino]phenyl](4-sulfophenyl)-methylene]-2,5-cyclohexadien-1-ylidene]-3-sulfobenzenemeth-anaminium inner salt, disodium salt; C.I. Acid Green 5;* ethyl[4-[p-[ethyl(m-sulfobenzyl)amino]-α-(p-sulfophenyl)-benzylidene]-2,5-cyclohexadien-1-ylidene](m-sulfobenzyl)-ammonium hydroxide inner salt disodium salt; C.I. 42095; FD & C Green No. 2; Lissamine Green SF. $C_{37}H_{34}N_2Na_2$-

O_9S_3; mol wt 792.86. C 56.05%, H 4.32%, N 3.53%, Na 5.80%, O 18.16%, S 12.13%. Prepn and refs: *Colour Index* **vol. 4** (3rd ed., 1971) p 4385. Biological use: H. J. Conn's *Biological Stains,* R. D. Lillie, Ed. (Williams & Wilkins, Baltimore, 9th ed., 1977) pp 257-258, 582. Toxicity data: F. C. Lu, A. Lavalle, *Can. Pharm. J.* **97,** 30 (1964). Chronic toxicity study: W. H. Hansen *et al., Food Cosmet. Toxicol.* **4,** 389 (1966).

A reddish-brown powder. Sol in water to a green soln which turns yellowish-brown with HCl and then gradually fades. Addition of NaOH almost completely decolorizes the soln yielding a dull violet ppt. LD_{50} orally in rats: >2 g/kg (Lu, Lavalle).

Caution: Delisted by FDA in 1966 for use in foods, drugs and cosmetics.

USE: As dye, biological stain.

5509. Lignans. Plant products of low molecular weight formed primarily by the oxidative coupling of p-hydroxyphenylpropene units in which the two units may be linked by an oxygen bridge. The monomeric precursor units are cinnamic acid, cinnamyl alcohol, propenylbenzene and allylbenzene. The term lignan or *Haworth lignan* is applied to compounds derived by coupling acid and/or alcohol while the compounds derived by coupling propenyl and/or allyl derivatives are called *neolignans:* O. R. Gottlieb, *Fortschr. Chem. Org. Naturst.* **35,** 1-72 (1978). Lignans occur widely and have been obtained from roots, heartwood, foliage, fruit and resinous exudates of plants. Lignans are optically active compounds. They represent the dimer stage intermediate between monomeric propylphenol units and lignin. Naturally occurring trimers and tetramers have not been reported. Occurrence of lignans, enterolactone, *q.v.,* and enterodiol, in man and animal species: S. R. Stitch *et al., Nature* **287,** 238 (1980); K. D. R. Setchell, *ibid.* **740.** Synthesis of first lignans found in man and animals: G. Cooley *et al., Tetrahedron Letters* **22,** 349 (1981).

5510. Lignin. The most abundant natural aromatic organic polymer found in all vascular plants. Lignin together with cellulose, *q.v.,* and hemicellulose, *q.v.,* are the major cell wall components of the fibers of all wood and grass species. Lignin is composed of coniferyl, p-coumaryl and sinapyl alcohols in varying ratios in different plant species. Monographs: F. E. Brauns, D. A. Brauns, *The Chemistry of Lignin,* Supplement Volume covering the literature 1949-1958 (Academic Press, New York, 1960) 804 pp; I. A. Pearl, *The Chemistry of Lignin* (Marcel Dekker, New York, 1967) 360 pp. Structural aspects and applications: H. Veeramani, G. A. Wani, *Chem. Ind. Dev.* **11,** 13-25 (1977). Chemistry and structure: C. A. Reddy, L. Forney, *Dev. Ind. Microbiol.* **19,** 27-34 (1978). *Reviews:* Nord, Schubert, *Sci. Am.* **199,** no. 4, 104-113 (Oct. 1958); I. A. Pearl, "Lignin as a Raw Material for the Production of Pure Chemicals," *J. Chem. Ed.* **35,** 502 (1958); D. W. Goheen, C. H. Hoyt in Kirk-Othmer *Encyclopedia of Chemical Technology* **vol. 14,** (Wiley-Interscience, New York, 3rd ed., 1981) pp 294-312.

USE: Source of vanillin, syringic aldehyde, dimethyl sulfoxide. Extender for phenolic plastics, to strengthen rubber (esp for shoe soles), as oil mud additive, to stabilize asphalt emulsions, to precipitate proteins.

5511. Lignoceric Acid. *Tetracosanoic acid.* $C_{24}H_{48}O_2$; mol wt 368.64. C 78.20%, H 13.12%, O 8.68%. $C_{23}H_{47}$-

COOH. Obtained from beechwood tar or by the distillation of rotten oak wood: Sullivan, *Ind. Eng. Chem.* **8**, 1027 (1916). Most natural fats contain small amounts (0.2-1%). The seed fat of the Indian tree *Adenanthera pavonina* is said to contain 25%. Synthesis: Fieser, Szmuszkovicz, *J. Am. Chem. Soc.* **70**, 3352 (1948).

Crystals, mp 84.15°. n_D^{100} 1.4287. Soly in 91.53% ethanol: 0.182 g/100 ml. Neutralization value 152.2.

Methyl ester, $C_{25}H_{50}O_2$, platelets, mp 58-59.8°.

5512. Lignum Vitae. The wood of *Guaiacum officinale* L., or *G. sanctum* L., Zygophyllaceae. *See also* Guaiac.

5513. Ligroin. V.M.&P. naphtha; varnish makers' and painters' naphtha; refined solvent naphtha; solvent naphtha; Benzoline; Canadol. Term that has been applied to petroleum fractions of the same nature as described for petroleum benzin, *q.v.*, but of higher density, higher boiling range and higher flash pt. Defined by ASTM prior to 1950 as synonymous with petroleum benzin and petroleum ether: ASTM standard specification **D 288-49**, 865-867 (1949).

Refined solvent, mobile, flammable liquid. $d_{15.6}^{15.6}$ 0.850 to 0.870. Distillation range at 760 mm: percentage recovered at 130° = not more than 5; percentage recovered at 145° = not less than 90. End point (dry point) = not above 155°. Technical benzin (high boiling petr ether) usually has a $d_{15.6}^{15.6}$ 0.730-0.750 and bp_{760} 90-120°.

5514. Limaprost. *(2E,11α,13E,15S,17S)-11,15-Dihydroxy-17,20-dimethyl-9-oxoprosta-2,13-dien-1-oic acid;* 17S,20-dimethyl-*trans*-2,3-didehydro-PGE₁; 9-oxo-11α,15α-dihydroxy-17S,20-dimethylprosta-*trans*-2,*trans*-13-dienoic acid; 17S-methyl-ω-homo-*trans*-Δ²-PGE₁; ONO-1206; OP-1206; Opalmon; Prorenal. $C_{22}H_{36}O_5$; mol wt 380.52. C 69.44%, H 9.54%, O 21.02%. Derivative of prostaglandin E₁, *q.v.* Prepn: M. Hayashi *et al.*, Ger. pat. 3,002,677 (1980 to Ono); *eidem*, U.S. pat. 4,294,849 (1981 to Warner-Lambert). Cardiovascular pharmacology in animals: T. Tsuboi *et al.*, *Arch. Int. Pharmacodyn.* **247**, 89 (1980). Effect on smooth muscle *in vitro*: P. G. Adaikan, S. M. M. Karim, *Prostaglandins Med.* **6**, 449 (1981). Coronary vasodilating effects in primates: S. R. Kottegoda *et al.*, *Prostaglandins Leukotrienes Med.* **8**, 343 (1982). Clinical hemodynamics in chronic lung disease: T. Ishizaki *et al.*, *Chest* **85**, 382 (1984).

White crystals, mp 97-100°.

THERAP CAT: Antianginal.

5515. Limestone. Natural calcium carbonate; agricultural limestone; Agstone; lithographic stone; Solnhofen stone. A term originally applied only to minerals consisting largely of $CaCO_3$, such as Portland stone, dolomite, marble, and chalk, now used indiscriminately to designate technical and agricultural grades of calcium carbonate. *Review:* R. S. Boynton in Kirk-Othmer *Encyclopedia of Chemical Technology* vol. **14** (Wiley-Interscience, New York, 3rd ed., 1981) pp 343-382.

5516. Lime Sulfurated Solution. Vleminckx's soln; Vleminckx's lotion; calcium oxysulfide solution. Made by boiling 16.5 parts freshly slaked lime with 25 parts sublimed sulfur and water to make 100 ml; the resulting soln contg calcium polysulfide and thiosulfate.

Brown, clear liquid.

THERAP CAT: Antiseptic (topical); scabicide.

THERAP CAT (VET): Insecticide, scabicide, has been used in boils and fistulas.

5517. Limettin. *5,7-Dimethoxy-2H-1-benzopyran-2-one; 5,7-dimethoxycoumarin;* citropten. $C_{11}H_{10}O_4$; mol wt 206.20. C 64.07%, H 4.89%, O 31.04%. From rind of fruit of *Citrus lima* Lunan (*C. limetta* Auth.), Rutaceae (lime): Tilden, Beck, *J. Chem. Soc.* **57**, 323 (1890); from W. Indian lime oil: Caldwell, Jones, *ibid.* **1945**, 570; from citrus oils:

Stanley, Vannier, U.S. pat. 2,889,337 (1959 to U.S.D.A.). Synthesis: Schmidt, *Arch. Pharm.* **242**, 288 (1904); Heyes, Robertson, *J. Chem. Soc.* **1936**, 1831.

Needles from methanol, mp 147-148°. uv max (alcohol): 222, 247, 250.5, 324 nm (log ε 4.03, 3.84, 3.84, 4.18). Almost insol in boiling water, ether, petr ether. Freely sol in alcohol, chloroform, acetone.

5518. Limonene. *1-Methyl-4-(1-methylethenyl)cyclohexene; p-mentha-1,8-diene;* cinene; cajeputene; kautschin. $C_{10}H_{16}$; mol wt 136.24. C 88.16%, H 11.84%. Occurs in various ethereal oils, particularly in oils of lemon, orange, caraway, dill and bergamot. Isoln of *d*-limonene from mandarin peel oil *(Citrus reticulata* Blanco, *Rutaceae):* Kugler, Kováts, *Helv. Chim. Acta* **46**, 1480 (1963). *Review:* J. L. Simonsen, *The Terpenes* vol. I (University Press, Cambridge, 2nd ed., 1947) pp 143-165.

dl-Form, *inactive limonene, dipentene.* Liquid. Pleasant lemon-like odor. bp_{763} 175.5-176.5°. $d_4^{20.85}$ 0.8402. n_D 1.4744. Practically insol in water. Miscible with alcohol. With dry HCl or HBr it forms monohalides, and with aq HCl or HBr, the dihalide.

d-Form, liquid. bp_{763} 175.5-176°. d_4^{21} 0.8402. n_D^{21} 1.4743. $[α]_D^{19.5}$ +123.8°.

l-Form, liquid. bp_{763} 175.5-176.5°. $d_4^{20.5}$ 0.8407. n_D^{21} 1.474. $[α]_D^{19.5}$ −101.3°.

Caution: Skin irritant, sensitizer.

USE: Solvent, manuf resins; wetting and dispersing agent.

5519. Limonin. *Limonoic acid di-δ-lactone; 8-(3-furyl)-decahydro-2,2,4a,8a-tetramethyl-11H,13H-oxireno[d]pyrano-[4',3':3,3a]isobenzofuro[5,4-f][2]benzopyran-4,6,13(2H,5aH)-trione;* limonoic acid 3,19:16,17-dilactone. $C_{26}H_{30}O_8$; mol wt 470.52. C 66.37%, H 6.43%, O 27.20%. Tetracyclic triterpenoid; bitter principle of citrus fruits. Isoln: Bernays, *Ann.* **40**, 317 (1841). Improved isoln and purification: P. G. Pifferi *et al.*, *Ital. J. Food Sci.* **5**, 269 (1993). Structure: A. Melera *et al.*, *Helv. Chim. Acta* **40**, 1420 (1957). Stereochemistry: D. Arigoni *et al.*, *Experientia* **16**, 41 (1960); S. Arnott *et al., ibid.* 49. Partial synthesis: C. Lüthy *et al., Helv. Chim. Acta* **57**, 1060 (1974). HPLC determn in citrus juice: W. W. Widmer, *J. Agric. Food Chem.* **39**, 1472 (1991). Review of chemistry: V. P. Maier *et al., ACS Symposium Series* **143**, 63-82 (1980); and removal of juice bitterness: R. F. H. Dekker, *Aust. J. Biotechnol.* **2**, 65 (1988).

Bitter crystals from methylene chloride + isopropanol or acetic acid, mp 298°. $[α]_D$ −128° (c = 1.21 in acetone). uv max: 207, 285 nm (ϵ 7000, 38). Slightly sol in water, ether; sol in alcohol, glacial acetic acid.

5520. Linalool. *3,7-Dimethyl-1,6-octadien-3-ol;* 2,6-dimethyl-2,7-octadien-6-ol; linalol. $C_{10}H_{18}O$; mol wt 154.25. C 77.87%, H 11.76%, O 10.37%. $(CH_3)_2C$=CHCH$_2$CH$_2$C-$(CH_3)(OH)CH$=CH$_2$. Chief constituent of linaloe oil; also occurs in oils of Ceylon cinnamon, sassafras, orange flower, bergamot, *Artemisia balchanorum,* ylang ylang, etc.: Tiemann, *Ber.* **31**, 808 (1898); Walbaum, Stephan, *ibid.* **33**, 2305 (1900); Hesse, Zeitschel, *J. Prakt. Chem.* **66**, 493 (1902); Rafanova *et al.,* U.S.S.R. pat. **103,725** (1956), *C.A.* **51**, 3656c (1957); Naves, *Helv. Chim. Acta* **42**, 1692 (1959). Presence in essential oils: *idem, Compt. Rend.* **251**, 900 (1960). Absolute configuration: Prelog, Watanabe, *Ann.* **603**, 1 (1957). Synthesis of *dl*-linalool: Ruzicka, Fornasir, *Helv. Chim. Acta* **2**, 182 (1919); Surmatis, U.S. pat. **2,848,-502** (1958 to Hoffmann-La Roche); Nair, Pandit, *Tetrahedron Letters* **1966**, 5097. *Review:* J. L. Simonsen, *The Terpenes* vol. I (University Press, Cambridge, 2nd ed., 1947), pp 57-68.

l-Form, *licareol.* Colorless liq. bp$_{760}$ 198°; bp$_{25}$ 98-98.3°; bp$_{14}$ 86-87°. d^{20} 0.8622. n_D^{22} 1.4604. [α]$_D^{20}$ −20.1°. Practically insol in water. Miscible with alcohol, ether.

d-Form, *coriandrol.* bp$_{760}$ 198-200°; bp$_{26}$ 114-114.5°; bp$_{15.5}$ 93-94°; bp$_{12}$ 86°. d$_4^{20}$ 0.8733. n_D^{20} 1.4673. [α]$_D^{20}$ +19.3°. Soluble in 10 vol 50% alc, 4 vol 60% alc.

dl-Form, bp$_{720}$ 194-197°; bp$_{14}$ 89-91°. d^{15} 0.865.

USE: In perfumery instead of bergamot or French lavender oil since it has an odor similar to these oils.

5521. Linalyl Acetate. *3,7-Dimethyl-1,6-octadien-3-yl acetate;* Bergamol. $C_{12}H_{20}O_2$; mol wt 196.29. C 73.43%, H 10.27%, O 16.30%. CH$_3$COOC$_{10}$H$_{17}$. Most valuable constituent of bergamot and lavender oils, also found in many other volatile oils.

Liquid; bergamot odor. d$_4^{20}$ 0.895. bp 220°. n_D^{20} 1.4460. Insol in water. Miscible with alcohol, ether.

USE: In perfumery.

5522. Linamarin. *2-(β-D-Glucopyranosyloxy)-2-methylpropanenitrile;* phaseolunatin. $C_{10}H_{17}NO_6$; mol wt 247.25. C 48.58%, H 6.93%, N 5.67%, O 38.83%. From the seed skins or embryos of flax: Jorissen, Hairs, *Bull. Acad. Roy. Sci. Belg.* [3] **21**, 529 (1891); André *et al., Compt. Rend.* **231**, 590 (1950); Lüdtke, *Biochem. Z.* **323**, 428 (1953). Synthesis: Fischer, Anger, *Ber.* **52**, 854 (1919). Biosynthesis in white clover: Butler, Butler, *Nature* **187**, 780 (1960).

Bitter needles, mp 142-143°. [α]$_D^{18}$ −29°. Freely sol in water, cold alcohol, hot acetone; slightly in hot ethyl acetate, ether, benzene, chloroform. Practically insol in petr ether. Evolves HCN with linseed meal but not with emulsin.

Tetraacetate, $C_{18}H_{25}NO_{10}$, needles from alcohol, mp 140-141°. [α]$_D^{14}$ −10.8° (acetone). Sol in acetone, ethyl acetate, chloroform, glacial acetic acid, benzene, warm methanol and ethanol. Practically insol in petr ether.

5523. Linarin. *7-[[6-O-(6-Deoxy-α-L-mannopyranosyl)-β-D-glucopyranosyl]oxy]-5-hydroxy-2-(4-methoxyphenyl)-4H-1-benzopyran-4-one;* acacetin-β-rutinoside; linarigenin-glucoside; 5,7-dihydroxy-4'-methoxyflavone-D-glucosido-L-rhamnoside; buddleoflavonoloside. $C_{28}H_{32}O_{14}$; mol wt 592.55. C 56.76%, H 5.44%, O 37.80%. From the flowers of *Linaria vulgaris* Mill., Scrophulariaceae: Merz, Wu, *Arch. Pharm.* **274**, 126 (1936); from *Cirsium oleraceum* Scop., Compositae: Wagner *et al., ibid.* **293**, 1053 (1960). Structure: Baker *et al., J. Chem. Soc.* **1951**, 691. Synthesis: Zemplén, Bognàr, *Ber.* **74**, 1818 (1941).

Monohydrate, needles from methanol, mp 268-270°. [α]$_D^{26}$ −100° (0.07 g in 10 ml glacial acetic acid); [α]$_D^{24}$ −87° (0.05 g in pyridine). Practically insol in water and the usual organic solvents. Sol in nitrobenzene, phenol, aniline, pyridine, concd acids and alkalies. The water of crystn cannot be removed at 100° *in vacuo* over P$_2$O$_5$ (Merz); may be removed at 138° in high vacuum (Zemplén). Hydrolysis gives 5,7-dihydroxy-4'-methoxyflavone, D-glucose, and L-rhamnose.

5524. Linatine. *1-[(4-Amino-4-carboxy-1-oxobutyl)amino]-D-proline; N-(D-2-carboxy-1-pyrrolidinyl)-L-glutamine;* 1-[N-(γ-L-glutamyl)amino]-D-proline. $C_{10}H_{17}N_3O_5$; mol wt 259.26. C 46.33%, H 6.61%, N 16.21%, O 30.86%. A vitamin B$_6$ antagonist. The first reported naturally occurring hydrazino acid having antibacterial properties. Isoln from flaxseed (*Linum usitatissimum*), characterization, and synthesis: Klosterman *et al., Biochemistry* **6**, 170 (1967); Lamoureux, *Diss. Abstr. B* **28**, 4908 (1968). Bacterial inhibition: Parsons *et al., Antimicrob. Ag. Chemother.* **1967**, 415.

Amorphous powder. [α]$_D^{25}$ +46.4° (c = 2.8 in water). Attempts to obtain well-defined crystals have failed; melts with dec over a wide temperature range. Very sol in water. Practically insol in anhydr organic solvents.

L-Isomer, amorphous solid. [α]$_D^{24}$ −34.6° (c = 2 in water).

5525. Lincomycin. *(2S-trans)-Methyl 6,8-dideoxy-6-[[(1-methyl-4-propyl-2-pyrrolidinyl)carbonyl]amino]-1-thio-D-erythro-α-D-galacto-octopyranoside;* lincolnensin; U-10149; NSC-70731; Lincolcina. $C_{18}H_{34}N_2O_6S$; mol wt 406.54. C 53.18%, H 8.43%, N 6.89%, O 23.61%, S 7.89%. Antibiotic produced by *Streptomyces lincolnensis* var. *lincolnensis.* Isoln: Mason *et al., Antimicrob. Ag. Chemother.* **1962**, 555; Herr, Bergy, *ibid.* 560; Bergy, Herr and Bergy *et al.,* U.S. pats. **3,086,912** and **3,155,580** (1963, 1964, both to Upjohn). Structure: Hoeksema *et al., J. Am. Chem. Soc.* **86**, 4223 (1964). Prepn of derivs: Argoudelis *et al.,* U.S. pat. **3,380,-992** (1968 to Upjohn). Synthesis: Magerlein, *Tetrahedron Letters* **1970**, 33. Synthesis of carbohydrate moiety: I. Hoppe, U. Schollkopf, *Ann.* **1980**, 1474. Stereoselective synthesis: S. Knapp, P. J. Kukkola, *J. Org. Chem.* **55**, 1632 (1990). Prepn of hydrochloride hemihydrate: Neth. pat. Appl. **6,409,689** (1965 to Upjohn); *C.A.* **63**, 5458f (1965). Clinical study: F. Puleo *et al., Gazz. Med. Ital.* **138**, 401 (1979). Mechanism of action: F. N. Chang in *Antibiotics* vol. 5(pt. 1), F. E. Hahn, Ed. (Springer-Verlag, New York,

1979) pp 127-134. Toxicity data: Gray *et al., Toxicol. Appl. Pharmacol.* **6**, 476 (1964).

Free base, pKa' 7.6. More stable in salt form. Sol in methanol, lower alcohols, ethyl acetate, acetone, chloroform. Slightly sol in water (Bergy).

Hydrochloride hemihydrate, $C_{18}H_{34}N_2O_6S.HCl.\frac{1}{2}H_2O$, *Frademicina, Lincocin, Mycivin, Waynecomycin.* Formerly obtained as needle-like crystals of low specific gravity from aq soln by rapid addition of acetone at low temps; now obtained as crystals of higher specific gravity, with cubic crystal structure and greater soly in HCl, by slow addition of acetone (Upjohn pat. 1965). mp 145-147°. $[\alpha]_D^{25}$ +137° (water). Freely sol in water, methanol, ethanol; sparingly sol in most organic solvents other than hydrocarbons. LD_{50} in mice, rats (g/kg): 1 i.p.; 4 orally (Gray).

Hydrochloride monohydrate, *Albiotic, Cillimycin, Lincomix.*

THERAP CAT: Antibacterial.

THERAP CAT (VET): Antibacterial.

5526. Lindane. *(1α,2α,3β,4α,5α,6β)-1,2,3,4,5,6-Hexachlorocyclohexane;* γ-HCH; γ-benzene hexachloride; gamma benzene hexachloride; gamma hexachlor; ENT-7796; Aparasin; Aphtiria; Esoderm; γ-BHC; Gammalin; Gamene; Gammexane; Gexane; Jacutin; Kwell; Lindafor; Lindatox; Lorexane; Quellada; Streunex; Tri-6; Viton. $C_6H_6Cl_6$; mol wt 290.83. C 24.78%, H 2.08%, Cl 73.14%. Lindane, the gamma isomer, is the active isomer among the eight well-described stereoisomers of hexachlorocyclohexane. Prepns sold for pharmaceutical or medicinal purposes now contain at least 99% pure γ-isomer. Prepn: T. Hardie, U.S. pat. **2,218,148** (1940 to ICI); R. E. Slade, *Chem. & Ind. (London)* **1945**, 314; F. A. Gunther, *ibid.* **1946**, 399. Metabolism: R. Engst *et al., Residue Rev.* **68**, 59 (1977); **72**, 71 (1979). Acute toxicity: T. B. Gaines, *Toxicol. Appl. Pharmacol.* **14**, 515 (1969). Reviews of carcinogenicity studies: M. D. Reuber, *Environ. Res.* **19**, 460-481 (1979); S. D. Vesselinovitch, F. W. Carlborg, *Toxicol. Pathol.* **11**, 12 (1983). Review of isomers: I. Hornstein, *Science* **121**, 206 (1955); S. H. Safe in Kirk-Othmer *Encyclopedia of Chemical Technology* vol. 6 (John Wiley & Sons, New York, 4th ed., 1993) pp 135-139.

Crystals, mp 112.5°. Slight musty odor. Vapor pressure at 20°: 9.4 × 10^{-6} mm Hg. n_D^{20} 1.644. Soly in g/100 g at 20°: acetone 43.5, benzene 28.9, $CHCl_3$ 24.0, ether 20.8, ethanol 6.4. Insol in water. LD_{50} in male, female rats (mg/kg): 88, 91 orally (Gaines).

Caution: Poisoning may occur by ingestion, inhalation, or percutaneous absorption. *Acute:* Dizziness, headache, nausea, vomiting, diarrhea, tremors, weakness, convulsions, dyspnea, cyanosis, circulatory collapse. *Chronic:* Hepatic damage has occurred in exptl animals. Topical use may cause local sensitivity reactions. Vapors may irritate eyes, nose, throat. See: *Clinical Toxicology of Commercial Products,* R. E. Gosselin *et al.,* Eds. (Williams & Wilkins, Baltimore, 5th ed., 1984) Section III, pp 239-241. Lindane and other hexachlorocyclohexane isomers may reasonably be anticipated to be carcinogens: *Seventh Annual Report on Carcinogens* (PB95-109781, 1994) p 241.

USE: Insecticide.

THERAP CAT: Pediculicide; scabicide.

THERAP CAT (VET): Ectoparasiticide.

5527. Lindlar Catalyst. $Pd-Pb-CaCO_3$. Prepn: Lindlar, *Helv. Chim. Acta* **35**, 446 (1952); Lindlar, Dubuis cited by Fieser, Fieser, *Reagents for Organic Synthesis* (New York, 1967) p 566.

USE: In selective hydrogenation of triple bonds to *cis*-double bonds.

5528. Lineatin. *[1R-(1α,2β,5α,7β)]-3,3,7-Trimethyl-4,9-dioxatricyclo[3.3.1.0^{2,7}]nonane;* 3,3,7-trimethyl-2,9-dioxatri-

cyclo[3.3.1.0^{4,7}]nonane; 4,6,6-lineatin. $C_{10}H_{16}O_2$; mol wt 168.24. C 71.39%, H 9.59%, O 19.02%. Isoln of the unique tricyclic aggregation pheromone from ambrosia beetles, *Trypodendron lineatum* (Olivier): J. G. MacConnell *et al., J. Chem. Ecol.* **3**, 549 (1977). Synthesis of (±)-form: K. Mori, M. Sasaki, *Tetrahedron Letters* **1979**, 1329; K. N. Slessor *et al., J. Org. Chem.* **45**, 2290 (1980); K. Mori *et al., Tetrahedron Letters* **23**, 1921 (1982); L. Skattebol, Y. Stenstrom, *ibid.* **24**, 3021 (1983); B. D. Johnston *et al., J. Org. Chem.* **50**, 114 (1985). Synthesis of racemate and optical isomers: K. Mori, M. Sasaki, *Tetrahedron* **36**, 2197 (1980). Short stereoselective synthesis: I. Aljancic-Solaja *et al., Helv. Chim. Acta* **70**, 1302 (1987). Comparative activity of the isomers: J. H. Borden *et al., Can. Entomol.* **112**, 107 (1980).

Oil, bp_{10} 70°. $[\alpha]_D^{24}$ +66.3° (c = 3.1 in $CHCl_3$).

USE: Insect sex attractant.

5529. Linoleic Acid. *(Z,Z)-9,12-Octadecadienoic acid;* 9,12-linoleic acid; linolic acid. $C_{18}H_{32}O_2$; mol wt 280.45. C 77.09%, H 11.50%, O 11.41%. An essential fatty acid, *q.v.* Major constituent of many vegetable oils, *e.g.,* cottonseed, soybean, peanut, corn, sunflower seed, safflower, poppy seed, linseed, and perilla oils, where it occurs as a glyceride. Characteristic ingredient of semi-drying oils. Isoln: Swern, Parker, *J. Am. Oil Chem. Soc.* **30**, 5 (1953); Parker *et al., Biochem. Prepn.* **4**, 86 (1955); McCutcheon, *Org. Syn.* **coll. vol. III**, 526 (1955). Summary of work on structure: T. P. Hilditch, *The Chemical Constitution of Natural Fats* (Chapman & Hall, London, 2nd ed. 1956). Synthesis: Raphael, Sondheimer, *J. Chem. Soc.* **1950**, 2102; Gensler, Thomas, *J. Am. Chem. Soc.* **73**, 4601 (1951); Walborsky *et al., ibid.* 2590; Nigam, Weedon, *J. Chem. Soc.* **1956**, 4052; Osbond, Wickens, *Chem. & Ind. (London)* **1959**, 1288. Review of physiological role in mammals: H. S. Hansen, *Trends Biochem. Sci.* **11**, 263 (1986).

Colorless oil. Easily oxidized by air, cannot be distilled without decompn. Storage in ester form is recommended. d_4^{18} 0.9038; d_4^{22} 0.9007. mp −12°. $bp_{1.4}$ 202°; bp_{16} 230°. $n_D^{11.5}$ 1.4715; n_D^{20} 1.4699; $n_D^{21.5}$ 1.4683; n_D^{50} 1.4588. Iodine value: 181.1. Thiocyanogen value 96.7. Freely sol in ether. Sol in abs alc. One ml dissolves in 10 ml petr ether. Miscible with dimethylformamide, fat solvents, oils.

Aluminum salt, $Al(C_{18}H_{31}O_2)_3$. Yellow lumps or powder; linseed oil odor. Practically insol in water. Sol in oils, fixed alkali hydroxides.

Methyl ester, *see* Methyl Linoleate.

Ethyl ester, *see* Ethyl Linoleate.

Cyclohexylamide, $C_{24}H_{43}NO$, *linolexamide, N-cyclohexyl-linoleamide, Clinolamide.*

USE: Manuf paints, coatings, emulsifiers, vitamins. Aluminum salt used to manuf lacquers.

THERAP CAT: Nutrient (essential fatty acid).

THERAP CAT (VET): Dietary supplement.

5530. Linolenic Acid. *(Z,Z,Z)-9,12,15-Octadecatrienoic acid; α-linolenic acid.* $C_{18}H_{30}O_2$; mol wt 278.44. C 77.65%, H 10.86%, O 11.49%. An essential fatty acid, *q.v.* Occurs as the glyceride in most drying oils. Synthesis: Nigam, Weedon, *J. Chem. Soc.* **1956**, 4049; Osbond, Wickens, *Chem. & Ind. (London)* **1959**, 1288. Biosynthetic studies: C. G. Kannangara *et al., Biochem. Biophys. Res. Commun.* **52**, 648 (1973); B. S. Jacobson *et al., ibid.* 1190; C. J. Bedord *et al., Arch. Biochem. Biophys.* **185**, 15 (1978). Effects on lipid metabolism in rat tissue: M. L. Garg *et al., Lipids* **23**, 847 (1988). Review of dietary linolenic acid in mammals: J. Tinoco *et al., ibid.* **14**, 166-171 (1979); in man: N. Zöllner, *Prog. Lipid Res.* **25**, 177-180 (1986).

Colorless liquid. d_4^{18} 0.914. bp_1 230-232°. Insol in water. Sol in organic solvents.

THERAP CAT: Nutrient (essential fatty acid).

5531. γ-Linolenic Acid. *(Z,Z,Z)-6,9,12-Octadecatrienoic acid; cis-6,cis-9,cis-12-octadecatrienoic acid; gamolenic acid; GLA; Viacutan.* $C_{18}H_{30}O_2$; mol wt 278.44. C 77.65%, H 10.86%, O 11.49%. Polyunsaturated fatty acid produced in the body as the Δ^6-desaturase metabolite of linoleic acid, *q.v.* Converted to dihomo-γ-linolenic acid, a biosynthetic precursor of monoenoic prostaglandins such as PGE_1. Present to varying extents in the fatty acid fraction of evening primrose oil (7-10%), in borage oil (18-26%), in black currant oil (15-20%) and in oils from different fungal sources (6-24%). Isoln from evening primrose oil, *q.v.*: A. Heiduschka, K. Luft, *Arch. Pharm.* **257**, 33 (1919). Proposed structure: Eibner *et al., Chem. Umschau.* **34**, 312 (1927). Confirmation of structure: J. P. Riley, *J. Chem. Soc.* **1949**, 2728. Discussion of occurrence, esp. in fungi: R. Shaw, *Biochim. Biophys. Acta* **98**, 230 (1965). Synthesis: J. M. Osbond *et al., J. Chem. Soc.* **1961**, 2779; J. M. Osbond, *ibid.* 5270. Metabolism studies: J. F. Mead, D. R. Howton, *J. Biol. Chem.* **229**, 575 (1957); K. J. Stone *et al., Lipids* **14**, 174 (1979). Effect of source on essential fatty acid and prostanoid metabolite formation: D. K. Jenkins *et al., Med. Sci. Res.* **16**, 525 (1988).

Hexabromide deriv, $C_{18}H_{30}Br_6O_2$, crystals from ethyl methyl ketone, mp 201-202°.

USE: Nutrient.

THERAP CAT: In treatment of atopic eczema.

THERAP CAT (VET): Dermatological.

5532. Linseed. Flaxseed; linum. Dried ripe seeds of *Linum usitatissimum* L., *Linaceae.* Source of linseed oil. *Constit.* 30-40% oil, about 6% mucilage, about 25% proteins and linamarin.

USE: Emollient.

THERAP CAT (VET): Poultice (crushed seeds), demulcent (boiled in water).

5533. Linseed Oil. A drying oil obtained by expression of linseed. *Constit.* Glycerides of linolenic, linoleic, oleic, stearic, palmitic and myristic acids. *Ref:* T. P. Hilditch, *The Chemical Constitution of Natural Fats* (London, 3rd ed., 1956) p 175 sqq; E. W. Eckey, *Vegetable Fats and Oils* (New York, 1954) pp 535-547.

Yellowish liquid, peculiar odor, bland taste. Exposed to air it gradually thickens, becomes darker, and acquires a more pronounced odor and taste. d 0.925-0.935. n_D^{40} 1.4725-1.4750. Does not congeal above −20°. Sapon no.: 187-195. Iodine no. not below 170. Unsaponifiable matter not over 1.5%. Slightly sol in alcohol, miscible with chloroform, ether, petr ether, carbon disulfide, oil turpentine.

USE: In varnishes, paints, putty, oilcloths, linoleum, printing inks, artificial rubber, tracing cloth, tanning and enameling leather; applied to paper and fabrics to render them waterproof and tough. Emollient.

THERAP CAT (VET): Laxative.

5534. Linuron. *N'-(3,4-Dichlorophenyl)-N-methoxy-N-methylurea;* methoxydiuron; Du Pont Herbicide 326; HOE-2810; Afalon; Linurex; Lorox. $C_9H_{10}Cl_2N_2O_2$; mol wt 249.10. C 43.40%, H 4.05%, Cl 28.47%, N 11.25%, O 12.85%. Selective pre- and post-emergence herbicide. Preparation: Sherer, Heller, U.S. pat. **2,960,534** (1960 to Hoechst). Degradn in soil: G. F. Kempson-Jones, R. J. Hance, *Pestic. Sci.* **10**, 449 (1979). Toxicity study: G. W. Bailey, J. L. White, *Residue Rev.* **10**, 97 (1965).

mp 93-94°. Vapor pressure at 24°: 1.5×10^{-5} mm Hg. Soly in water: 75 ppm. Partially sol in acetone, alcohol, benzene, toluene, xylene. LD_{50} orally in rats: 1500 mg/kg (Bailey, White).

USE: Herbicide.

5535. Liothyronine. *O-(4-Hydroxy-3-iodophenyl)-3,5-diiodo-L-tyrosine; L-3-[4-(4-hydroxy-3-iodophenoxy)-3,5-diiodophenyl]alanine;* 4-(3-iodo-4-hydroxyphenoxy)-3,5-diiodophenylalanine; 3,5,3'-triiodothyronine; T-3. $C_{15}H_{12}I_3$-NO_4; mol wt 650.98. C 27.68%, H 1.86%, I 58.48%, N 2.15%, O 9.83%. Amino acid found in human plasma and thyroid gland, similar to thyroxine, *q.v.* Gross, Pitt-Rivers, *Lancet* **1952**, I, 439. Formation from diiodothyronine: Roche *et al., Biochem. Biophys. Acta* **11**, 215 (1953). Isoln from thyroid gland and synthesis: Gross, Pitt-Rivers, *Biochem. J.* **53**, 645 (1953); Roche *et al., Bull. Soc. Chim. France* **4**, 462 (1957); Plati, Wenner, U.S. pat. **2,784,222** (1957 to Hoffmann-La Roche); Pitt-Rivers, Gross, U.S. pat. **2,823,164** (1958 to Nat. Res. Dev. Corp.); Razdan, Wetherill, U.S. pat. **2,993,928** (1961 to Glaxo). Elevated levels of T-3 in victims of sudden infant death syndrome: G. Kocsard-Varo, *Med. J. Aust.* **2**, 789 (1973); M. A. Chacon, J. T. Tildon, *J. Pediatr.* **99**, 758 (1981). Monograph: Pitt-Rivers, R. Tata, *The Thyroid Hormones* (Pergamon Press, 1959).

Crystals, dec 236-237°. $[\alpha]_D^{29.5}$ +21.5° (c = 4.75 in a mixture of 1 part N HCl + 2 parts ethanol). Possesses 5 times the activity of L-thyroxine (goiter prevention in rats). Insol in water, alc, propylene glycol. Sol in dil alkalies with the formation of a brownish, water-soluble, sodium salt.

Sodium salt, $C_{15}H_{11}I_3NNaO_4$, *liothyronine sodium, sodium L-triiodothyronine, Cyomel, Cytobin, Cytomine, Cytomel, Cynomel, Tertroxin, Triostat, Triothyrone.*

Hydrochloride, $C_{15}H_{12}I_3NO_4$.HCl, *Thybon.* Long birefringent needles, decomp 202-203°. $[\alpha]_D^{29.5}$ +21.5° (c = 4.75 in a mixture of 1 vol N HCl and 2 vols ethanol).

DL-Triiodothyronine hydrochloride, *Trionine.*

THERAP CAT: Thyroid hormone.

THERAP CAT (VET): Thyroid hormone.

5536. Lipase. *Triacylglycerol lipase.* An enzyme (or more exactly a group of enzymes) belonging to the esterases. Hydrolyzes fat (present in ester form, such as glycerides) yielding fatty acids and glycerol. Catalyzes digestion. Widely distributed in the plant world, also in molds, bacteria, milk and milk products, and in animal tissues, especially in the pancreas. Isoln from castor beans: H. Gibian in *Ullmann's Encyklopädie der technischen Chemie,* 3rd ed., vol. 7, 406-407 (1956). Purification of pancreatic lipase: Marchis-Mouren *et al., Arch. Biochem. Biophys.* **83**, 309 (1959). Review of milk lipases: Chandan, Shahani, *J. Dairy Sci.* **47**, 471 (1964). Comprehensive reviews: Wills, *Advan. Lipid Res.* **3**, 197-240 (1965); Desnuelle in *The Enzymes* vol. 7, P. D. Boyer, Ed. (Academic Press, New York, 3rd ed., 1972) pp 575-616.

The optimum temp for enzyme action is between 35° and 37° at pH 5-6. Lipase contains sulfhydryl groups and is inactivated by substances that inhibit such compds. It is activated by substances that keep SH groups in the reduced state, such as glutathione, cysteine, and ascorbic acid. The addition of acid activates lipase preparations. Castor-oil lipase is activated by sulfuric, oxalic, formic, acetic and butyric acids. Acetic, salicylic and hydrochloric acids increase the action of lipase derived from various organs of the pig. Caprylic and caproic acids increase the action of lipase derived from certain mold fungi. Almost all organic solvents decrease lipase activity, petr ether being an exception.

USE: To split fats without damaging sensitive constituents, such as vitamins or unsaturated fatty acids. In food processing for flavor improvement; in detergents for the improvement of cleaning action. For review of industrial applications of microbial lipases, see Seitz, *J. Am. Oil Chem. Soc.* **51**, 12 (1974).

THERAP CAT: Digestive enzyme.

5537. Lipoprotein Lipase. *Diacylglycerol lipase;* clearing factor. A specific lipase which preferentially hydrolyzes triglycerides in the presence of a lipoprotein complex, forming glycerol and unesterified fatty acids: Korn, *Methods Enzymol.* **5**, 542 (1962). The normal substrate is turbid lipemic or chylomicron-containing plasma. This substance is clarified by the enzyme, hence the term clearing factor. First detected in the plasma of animals injected intravenously with heparin: Hahn, *Science* **98**, 19 (1943); Robinson, French, *Phamacol. Rev.* **12**, 241 (1960). Prepn from post heparin human plasma: Baskys, *Diss. Abstr.* **20**, 1146 (1959). It has since been prepd from several tissues of normal, untreated animals. Prepn from chicken adipose tissue: Korn, Quigley, *J. Biol. Chem.* **226**, 833 (1957); from human heart: Schnatz *et al., Am. J. Physiol.* **205**, 401 (1963). Role in the metabolism of triglycerides by adipose tissues: Rodbell, Scow in K. Rodahl, *Fat as a Tissue* (McGraw-Hill, New York, 1964) pp 110-126. *Reviews:* Jensen, *Progr. Chem. Fats Other Lipids* **11**, 347 (1971); Desnuelle in *The Enzymes*, vol. **7**, P. D. Boyer, Ed. (Academic Press, New York, 3rd ed., 1972) pp 606-609.

5538. Liposomes. Liposom Forte. Structures (smectic mesomorphs) formed spontaneously by polar lipids in aqueous media at temperatures allowing the fatty acyl chains to be fluid. Lipid molecules are arranged in concentric bilayers separated from each other by aqueous compartments. Within each bilayer, the molecules are arranged so that their hydrophobic tails point away from the aqueous compartment and their hydrophilic heads point toward it. Liposomes are usually spherical but shape can vary. The largest are fractions of a millimeter; unilamellar liposomes (vesicles) can be as small as several microns. Liposomes are able to entrap solutes and can interact with cellular structures. They can carry into cells biologically active materials which do not normally penetrate the plasma membrane. Encapsulation in liposomes protects these materials from contact with tissue and organ cultures, from inactivation within a living organism's circulatory system or, for pesticides and fertilizers, from removal by rain or irrigation. Liposomes may interact with cells by stable adsorption onto the cell surface, endocytosis (the intact liposome is delivered to the lysosomal apparatus), fusion with the plasma membrane (liposome content is released into the cytoplasm) and lipid exchange between liposome and cell surface without cell association of liposome contents. Liposomes are used as models in biological membrane research, as experimental targeted drug delivery systems and as carriers for other biologically active materials. *See also* Phospholipids. Geometry and thermodynamics: J. N. Israelachvili *et al., Biochim. Biophys. Acta* **470**, 185 (1977). Effect on brain metabolism: G. Toffano *et al.,* "Phospholipid Liposomes Effect on Brain Dopamine Metabolism and on the Secretion of GH" in *Neuroendocrinology: Biological and Clinical Aspects,* A. Polleri, R. M. Macleod, Eds. (Academic Press, New York, 1979) pp 345-349. Interaction with nucleic acids: B. Brosius *et al., J. Biomolec. Struct. Dyn.* **1**, 1535 (1984). Synthesis and use as carriers for biologically active materials: D. P. Papahadjopoulos, F. C. Szoka, Jr., **Belg. pat. 874,408** corresp to **U.S. pat. 4,235,871** (1979, 1980); for pesticides: *eidem,* **U.S. pat 4,241,046** (1980). Inhibition of erythrocytic malaria by liposomes containing neutral glycolipids: C. R. Alving *et al., Science* **205**, 1142 (1979). In ophthalmology: H. E. Schaeffer, D. L. Krohn, *Invest. Ophthalmol. Vis. Sci.* **22**, 220 (1982). In delivery of spermatozoal antigen for fertility control: L. Mettler, A. B. Czuppon, **Ger. pat. 3,301,951** (1984 to Nattermann).

In transfer of nucleic acids: I. J. Mettler, **Brit.** pat. **Appl. 2,140,822** (1984 to Stauffer). In gene transfer (preproinsulin I): C. Nicolau, *Biochem. Soc. Trans.* **12**, 349 (1984). Rigid, non-phospholipid liposomes as drug delivery system in hepatoma ascites tumor in mice: K. R. Patel *et al., Biochim. Biophys. Acta* **797**, 20 (1984). Review of early studies: D. A. Tyrrell *et al., ibid.* **457**, 259-302 (1976); of pharmacology of phospholipid liposomes: G. Toffano, A. Bruni, *Pharm. Res. Comm.* **12**, 829-845 (1980). Books: *Liposomes in Biological Systems,* G. Gregoriadis, A. C. Allison, Eds. (Wiley, New York, 1980) 412 pp; *Liposomes: From Physical Structure to Therapeutic Applications,* C. G. Knight, Ed. (Elsevier, Amsterdam, 1981) pp 323-348; *Liposome Technology,* G. Gregoriadis, Ed. (CRC Press, Boca Raton, 1984) 3 vols.

Immunosome. Immunogenic structures consisting of liposomes attached to antigens. Liposomes as immunological adjuvant: A. C. Allison, G. Gregoriadis, *Nature* **252**, 252 (1974). Immunosomes in influenza virus vaccine: L. Thibodeau *et al., Eur.* pat. **Appl. 47,480** (1982 to Inst. Armand Frappier); rabies vaccine: P. Perrin *et al., Ann. Virol.* **135E**, 183 (1984); vaccine against polysaccharide encapsulated bacteria: H. Snippe *et al., Eur.* pat. **Appl. 97,407** (1984 to Rijksuniv. Utrecht). Antibodies to hepatitis B surface antigen elicited: A. R. Neurath *et al., J. Gen. Virol.* **65**, 1009 (1984).

5539. Lipotropic Hormone. *Lipotropin; pituitary lipotropic hormone; lipid-mobilizing hormone;* adipokinetic hormone; lipolytic hormone; lipotrophin; LPH. Hypophyseal hormone which stimulates the release of fatty acids from adipose tissue. *β-LPH* is a single chain polypeptide containing 89-93 amino acid residues and has been proposed as the biosynthetic precursor of β-MSH and β-endorphin, *q. v., see* M. Chretien *et al., Can. J. Biochem.* **57**, 1111 (1979). *γ-LPH* consists of 58 amino acids and is identical to the sequence of the first 58 residues of β-LPH. Both contain sequences common to ACTH, *q. v.,* and β-melanotropin, *q. v.* Sequences differ slightly among mammalian species. Isoln and proposed structure of ovine β-LPH: C. H. Li, *Nature* **201**, 924 (1964); Y. Birk, C. H. Li, *J. Biol. Chem.* **239**, 1048 (1964); C. H. Li *et al., Nature* **208**, 1093 (1965); C. H. Li, *Arch. Biol. Med. Exp.* **5**, 55 (1968); revised structure: Chretien *et al., Int. J. Peptide Protein Res.* **4**, 263 (1972). Isoln and characterization of porcine β-LPH: Gilardeau, Chretien, *Can. J. Biochem.* **48**, 1017 (1970). Amino acid sequence of porcine γ-LPH: Graf *et al., Acta Biochim. Biophys.* **5**, 305 (1970). Isoln, characterization of bovine β-LPH: P. Lohmar, C. H. Li, *Biochem. Biophys. Res. Commun.* **77**, 1088 (1977); of rat β-LPH: M. Rubinstein *et al., Proc. Nat. Acad. Sci. USA* **74**, 3052 (1977); of whale β-LPH: H. Kawauchi *et al., Int. J. Peptide Protein Res.* **15**, 171 (1980); of turkey β-LPH: W. C. Chang *et al., ibid.* **261**. Initial isoln, partial structure of human β-LPH: G. Cseh *et al., FEBS Letters* **2**, 42 (1968); *eidem, ibid.* **21**, 344(1972). Isoln, characterization, amino acid sequence of human β-LPH: C. H. Li, D. Chung, *Int. J. Peptide Protein Res.* **17**, 131 (1981). *Review:* D. T. Krieger *et al., Recent Progr. Horm. Res.* **36**, 277-344 (1980).

5540. Lisinopril. *(S)-1-[N²-(1-Carboxy-3-phenylpropyl)-L-lysyl]-L-proline dihydrate;* MK-521; Acerbon; Alapril; Carace; Coric; Novatec; Prinil; Prinivil; Tensopril; Vivatec; Zestril. $C_{21}H_{31}N_3O_5 \cdot 2H_2O$; mol wt 441.52. C 57.13%, H 7.99%, N 9.52%, O 25.37%. Orally active angiotensin-converting enzyme (ACE) inhibitor. Prepn: A. A. Patchett *et al., Eur.* pat. **Appl. 12,401;** E. E. Harris *et al.,* **U.S. pat. 4,374,829;** A. A. Patchett, M. T. Wu, **U.S. pat. 4,555,502** (1980, 1983, 1985, all to Merck & Co.). Prepn and activity in rats, dogs: A. A. Patchett *et al., Nature* **288**, 280 (1980); M. T. Wu *et al., J. Pharm. Sci.* **74**, 352 (1985). Hemodynamic effects in heart failure in dogs: C. S. Sweet *et al., J. Cardiovasc. Pharmacol.* **8**, Suppl. 1, S15 (1986). Activity in normotensive man: J. Biollaz *et al., Clin. Pharmacol. Ther.* **29**, 665 (1981); *eidem, Brit. J. Clin. Pharmacol.* **14**, 363 (1982); G. P. Hodsman *et al., ibid.* **17**, 233 (1984). Human pharmacodynamics: J. A. Millar *et al., ibid.* **14**, 347 (1982). Metabolism in man: E. H. Ulm *et al., ibid.* 357. Radioimmunoassay: P. J. Worland, B. Jarrott, *J. Pharm. Sci.* **75**, 512 (1986). Preliminary clinical trials in hypertension: H. H. Rotmensch *et al., Am. J. Cardiol.* **53**, 116 (1984); in congestive heart failure: K. Dickstein *et al.*, *Am. Heart J.* **112**, 11 (1986). Series of articles in hypertension and in congestive heart failure: *Am. J. Med.* **85**, Suppl. 3B, 1-59 (1988). Comprehensive description: D. P. Ip *et al.*, *Anal. Profiles Drug Subs. Excip.* **21**, 233-276 (1992).

White to off-white crystalline, odorless powder. uv max (0.1N NaOH): ~246, ~254, ~258, ~261, ~267 nm ($A_{1cm}^{1\%}$ ~4.0, ~4.5, ~5.1, ~5.1, ~3.7); (0.1N HCl): ~246, ~253, ~258, ~264, ~267 nm ($A_{1cm}^{1\%}$ ~3.2, ~3.9, ~4.5, ~3.0, ~2.8). $[\alpha]_{405nm}^{25}$ ~ -120° (c = 1 in 0.25M pH 6.4 zinc acetate); $[\alpha]_{436nm}^{25}$ ~ -96° (c = 1 in 0.25M pH 6.4 zinc acetate). Partition coefficient at room temp (0.1M pH 7 phosphate buffer/*n*-octanol): 10.2 ±0.5. Soly (mg/ml): water 97; methanol 14; ethanol <0.1; acetone <0.1; acetonitrile <0.1; chloroform <0.1; *N,N*-dimethylformamide <0.1. pKa_1 (25°) 2.5; pKa_2 4.0; pKa_3 6.7; pKa_4 10.1.
Mixture with hydrochlorothiazide, *Prinzide*.
THERAP CAT: Antihypertensive.

5541. Lisuride. *N'-[(8α)-9,10-didehydro-6-methylergolin-8-yl]-N,N-diethylurea*; 9-(3,3-diethylureido)-4,6,6a,7,8,9-hexahydro-7-methylindolo[4,3-*f,g*]quinoline; 1,1-diethyl-3-(D-6-methylisoergolen-8-yl)urea; *N*-(D-6-methyl-8-isoergolenyl)-*N,N*-diethylurea; methylergol carbamide; lysuride. $C_{20}H_{26}N_4O$; mol wt 338.45. C 70.98%, H 7.74%, N 16.55%, O 4.73%. Dopamine D_2-receptor agonist. Prepn: V. Zikan, M. Semonsky, *Coll. Czech. Chem. Commun.* **25**, 1922 (1960); *eidem, Pharmazie* **23**, 147 (1968). Pharmacology and toxicity: Z. Votava, I. Lamplova, *Physiol. Bohemoslov.* **12**, 37 (1963), *C.A.* **59**, 9221d (1963).

Crystals from benzene, mp 186°. $[\alpha]_D^{20}$ +313° (c = 0.60 in pyridine).
Maleate, $C_{20}H_{26}N_4O \cdot C_4H_4O_4$, *Apodel, Cuvalit, Dopergin, Eunal, Lysenyl, Revanil.* Prisms from ethanol, mp 200° (dec). $[\alpha]_D^{20}$ +288° (c = 0.5 in methanol). uv max (methanol): 313 nm. LD_{50} i.v. in mice: 14.4 mg/kg (Votava, Lamplova).
THERAP CAT: Antimigraine; prolactin inhibitor; antiparkinsonian.

5542. Lita®. Protein-based fat mimetic consisting of water-dispersible microparticles of zein, *q.v.*, a hydrophobic prolamine derived from corn. Prepn: L. E. Stark, A. T. Gross, U.S. pat. **5,021,248** (1991 to Enzytech); *eidem*, U.S. pat. **5,145,702** (1992 to Opta Food Ingred.). Brief description: R. Iyengar, A. Gross in *Biotechnology and Food Ingredients*, I. Goldberg, R. Williams, Eds. (Van Nostrand Reinhold, New York, 1991) pp 287-313; M. S. Miller in *Protein Functionality in Food Systems*, N. S. Hettiarachchy, G. R. Zeigler, Eds. (Marcel Dekker, New York, 1994) pp 435-465.

Hydrophobic, smooth spheroidal particles. Size distribution: 0.3-3.0 μm. Insol in water. Sol in ethanol. Colloidal suspensions stable to 95°.
USE: Fat substitute for food products.

5543. Lithium. Li; at. wt 6.941; at. no. 3; valence 1. Group IA(1). Alkali metal. Occurrence in earth's crust: 20 ppm. Natural isotopes: 7 (92.58%); 6 (7.42%); artificial radioactive isotopes: 5, 8, 9; all are unstable ($T_{1/2}$ < 1 sec). Discovered as salt in 1817 by Arfvedson: *Ann. Chim. Phys.* [2] **10**, 82 (1819); metal prepared independently by Davy and Brandé in 1818. Occurs in a number of minerals: *spodumene* ($LiAlS_2O_6$), *lepidolite* [$K(Li,Al)_3(Si,Al)_4O_{10}(F,OH)_2$], *petalite* ($LiAlSi_4O_{10}$), *amblygonite* (AlPO$_4$.LiF), and *triphylite* ($LiFePO_4$). Also recovered from natural brines. Prepn of the metal by electrochemical processes: Guntz, *Compt. Rend.* **117**, 732 (1893); Ruff, Johannsen, *Z. Electrochem.* **12**, 186 (1906); by reduction of the oxide with magnesium or aluminum: Warren, *Chem. News* **74**, 6 (1896); Hanson, U.S. pat. **2,028,390** (1936). Reviews of biology, pharmacology and toxicity of lithium ion: Schou, "Lithium in Psychiatry—A Review" in *Psychopharmacology, A Review of Progress 1957-1967*, D. H. Efron, Ed. (Public Health Service Publication No. 1836, 1968) pp 701-718; Doig *et al.*, *J. Chem. Ed.* **50**, 343-345 (1973); Samuel, Gottesfeld, *Endeavour* **32**, 122-128 (1973); Saran, Gaind, *Clin. Toxicol.* **6**, 257-269 (1973); Schou, *Ann. Rev. Pharmacol. Toxicol.* **16**, 231-243 (1976). *Reviews:* Hart, Beumel, "Lithium and its Compounds" in *Comprehensive Inorganic Chemistry* vol. 1, J. C. Bailar, Jr. *et al.*, Eds. (Pergamon Press, Oxford, 1973) pp 331-367; R. Bach, J. R. Wasson in Kirk Othmer *Encyclopedia of Chemical Technology* vol. 14 (Wiley-Interscience, New York, 3rd ed., 1981) pp 448-476; *Chemistry of the Elements*, N. N. Greenwood, A. Earnshaw, Eds. (Pergamon Press, New York, 1984) pp 75-116; *Patty's Industrial Hygiene and Toxicology* vol. 2C, G. D. Clayton, F. E. Clayton, Eds. (John Wiley & Sons, Inc., New York, 4th ed., 1994) pp 2087-2097. Review of pharmacology: N. J. Birch, J. D. Phillips, *Adv. Inorg. Chem.* **36**, 49-75 (1991); and clinical use: T. H. Price, G. R. Heninger, *N. Engl. J. Med.* **331**, 591-598 (1994).
Silvery-white metal; body-centered cubic structure; becomes yellowish on exposure to moist air. mp 180.54°. bp 1347°. Also reported as bp 1336 ± 5°: Hartman, Schneider, *Z. Anorg. Chem.* **180**, 275 (1929). d^{20} 0.534. Hardest of the alkali metals. Heat capacity at constant pressure (25°): 5.892 cal/mole deg. *See:* Douglass *et al.*, *J. Am. Chem. Soc.* **77**, 2144 (1955). E° (aqueous) Li/Li$^+$ 3.045 V. Reacts with H_2O forming the hydroxide and H_2. Reacts violently with inorganic aicds; reacts slowly with cold H_2SO_4. Sol in liq ammonia forming a blue soln. Does not react with oxygen at room temp; forms Li_2O when heated to 100° or higher. Emits characteristic crimson color (670.8 nm) in flame. *Flammable. Keep under mineral oil or other liquid free from oxygen or water.* Li metal ignites in air near its mp; presents fire and explosion risk when exposed to water, N, acids or oxidizing reagents.
Caution: Potential symptoms of mild overexposure to the Li ion by ingestion of lithium salts are impaired concentration, lethargy, irritability, muscle weakness; moderate overexposure may cause disorientation, confusion, drowsiness; severe overexposure may cause impaired consciousness with progression to coma, convulsions, and impaired renal function (Price, Heninger). *See also: Clinical Toxicology of Commercial Products*, R. E. Gosselin *et al.*, Eds. (Williams & Wilkins, Baltimore, 5th ed., 1984) Section III, pp 241-245.
USE: In prodn of high-strength, low-density aluminum alloys for the aircraft industry; extremely tough, low-density alloys with aluminum and magnesium used for armour plate and aerospace components. In polymerization catalysts for the polyolefin plastics industry; manuf of high-strength glass and glass-ceramics. As anode in electrochemical cells and batteries; as chemical intermediate in organic syntheses. Lithium stearate as thickener and gelling agent to transform oils into lubricating greases.

5544. Lithium Acetate. Quilonorm; Quilonum. $C_2H_3LiO_2$; mol wt 65.99. C 36.40%, H 4.58%, Li 10.52%, O 48.49%. Prepn: *Gmelin's, Lithium* (8th Ed.) **20**, pp 230-232 (1927). Prepn and solubilities: N. V. Sidgewick, J. A. H.

R. Gentle, *J. Chem. Soc.* **121**, 1837 (1922). Crystal structure of anhydrous: C. Saunderson, R. B. Ferguson, *Acta Crystallogr.* **14**, 321 (1961). Pharmacokinetics in psychiatric patients: J.-L. Evrard *et al., Acta Psychiatr. Scand.* **58**, 67 (1978).

mp 286° (slight dec). Changes to dihydrate at 56.5°.

Dihydrate, rhombic crystals. Begins to melt at 49° (Gmelin). Congruent mp 57.8° (Sidgwick, Gentle). Freely sol in water or alcohol. Its aq soln is practically neutral.

THERAP CAT: Antimanic.

5545. Lithium Acetylsalicylate. Hydropyrin; Litmopyrine; Tyllithin. $C_9H_7LiO_4$; mol wt 186.09. C 58.09%, H 3.79%, Li 3.73%, O 34.39%. $CH_3COOC_6H_4COOLi$.

White, slightly hygroscopic powder; dec in moist air. Sol in 1 part water, 4 parts alcohol.

5546. Lithium Amide. Lithamide. H_2LiN; mol wt 22.96. H 8.78%, Li 30.23%, N 61.00%. $LiNH_2$. Prepd from Li + NH_3: Titherley, *J. Chem. Soc.* **65**, 517 (1894); Ruff, Geisel *Ber.* **39**, 840 (1906); **44**, 505 (1911); Campbell *et al., Proc. Indiana Acad. Sci.* **50**, 123 (1940); Schenk in *Handbook of Preparative Inorganic Chemistry* vol. 1, G. Brauer, Ed. (Academic Press, New York, 2nd ed., 1963) p 463. *See also* Inorg. Syn. **2**, 135. Crystal structure: Juza, Opp, *Z. Anorg. Allgem. Chem.* **266**, 313 (1951).

Tetragonal crystal structure, d 1.18. mp 380-400°. Sublimes in NH_3 current. When heated to 450°, NH_3 is given off and Li_2NH forms, which dec ~750-800°. Heat of formation: 42 kcal/mole at 18° and 760 mm Hg. Stable in air-tight containers at room temp. Dec by water to form LiOH + NH_3, slowly when in lumps, faster as the particle size decreases. Insol in anhydr ether, benzene, toluene.

USE: In Claisen condensations, alkylation of nitriles and ketones, synthesis of ethynyl compds, acetylenic carbinols.

5547. Lithium Benzoate. $C_7H_5LiO_2$; mol wt 128.06. C 65.66%, H 3.94%, Li 5.42%, O 24.99%.

White, cryst powder. One gram dissolves in 3 ml water, 13 ml alcohol, 10 ml boiling alcohol. Soly in water is increased by sodium benzoate. The aq soln is slightly alkaline to litmus; pH ~8.

USE: Has been used as a lubricant for compressing tablets.

5548. Lithium Bitartrate. Tartarlithine. $C_4H_5LiO_6$; mol wt 156.02. C 30.79%, H 3.23%, Li 4.45%, O 61.53%.

Monohydrate, white cryst powder, sol in water.

5549. Lithium Borate. Lithium biborate; lithium tetraborate. $B_4Li_2O_7$; mol wt 169.12. B 25.57%, Li 8.21%, O 66.22%. $Li_2B_4O_7$.

Pentahydrate, white, cryst powder. Slightly sol in water. Practically insol in alcohol.

USE: In enamels.

5550. Lithium Borohydride. *Lithium tetrahydroborate.* BH_4Li; mol wt 21.78. B 49.63%, H 18.51%, Li 31.86%. $LiBH_4$. Prepd by the action of diborane on ethyllithium; by the action of aluminum borohydride on ethyllithium: Schlesinger, Brown, *J. Am. Chem. Soc.* **62**, 3429-35 (1940). Review of lithium and other metal tetrahydroborates: James, Wallbridge, *Prog. Inorg. Chem.* **11**, 99-231 (1970).

Orthorhombic crystals, mp 268°; dec 380°. d 0.66. Stable under ordinary conditions, but dec in moist air. Reacts with hydrogen chloride to form hydrogen, diborane, and lithium chloride; forms lithium boromethoxide and hydrogen with methyl alcohol. Sol in water at pH > 7. Aq solns dec slowly if the pH is > 7, and rather vigorously when acidified. Sol in ether, tetrahydrofuran, aliphatic amines.

USE: Strong reducing agent. Used in the reduction of compds contg ketonic, aldehydic, or ester carbonyls and a nitrile group, where reduction of the carbonyl, but not of the nitrile group, is wanted. In the determination of free carboxyl groups in peptides and proteins; after esterification and acetylation, only the ester groups, and none of the peptide bonds are reduced.

5551. Lithium Bromide. BrLi; mol wt 86.85. Br 92.01%, Li 7.99%. LiBr. Usually contains water and 85-90% LiBr.

White, very deliquesc, slightly bitter, granular powder. mp 547° when anhydr. Sol in 0.6 part water, 0.4 part boiling water; freely sol in alcohol, glycol; sol in ether, amyl alcohol. The aq soln is neutral or only slightly alkaline to litmus. *Keep well closed.*

Caution: Large doses may cause CNS depression. Chronic absorption may cause skin eruptions and CNS disturbances due to bromide. May also cause disturbed blood electrolyte balance.

USE: As humectant, in air-conditioning systems.

THERAP CAT: Sedative, hypnotic.

5552. Lithium Carbonate. Camcolit; Carbolith; Carbolithium; Ceglution; Eskalith; Hypnorex; Limas; Liskonum; Lithane; Lithicarb; Lithobid; Lithonate; Lithotabs; Phasal; Plenur; Priadel; Quilonorm-retard; Quilonum-retard; Téralithe. CLi_2O_3; mol wt 73.89. C 16.25%, Li 18.79%, O 64.96%. Li_2CO_3. Purification: Caley, Elving, *Inorg. Syn.* **1**, 1 (1939). Initial report of effect in mental illness: J. F. J. Cade, *Med. J. Aust.* **2**, 349 (1949). Early review of biology and pharmacology of lithium: M. Schou, *Pharmacol. Rev.* **9**, 17-58 (1957); of toxicity: *idem, Acta Pharmacol. Toxicol.* **15**, 70-84 (1958). Thermal behavior: A. Reisman, *J. Am. Chem. Soc.* **80**, 3558 (1958). Acute toxicity data: H. F. Smyth *et al., Am. Ind. Hyg. Assoc. J.* **30**, 470 (1969). Toxicological studies in mammals: Gralla, McIlhenny, *Toxicol. Appl. Pharmacol.* **21**, 428 (1972). Pharmacokinetics: J. S. Ku *et al., Int. J. Clin. Pharmacol. Ther. Toxicol.* **25**, 648 (1987). Clinical studies: J. Rybakowski *et al., Int. Pharmacopsychiatry* **15**, 86 (1980); B. F. Kjellman *et al., Acta Psychiatr. Scand.* **62**, 32 (1980). Review of development of use: G. Chouinard, *Union Med. Can.* **109**, 221-224, 226, 304 (1980). Comprehensive description: H. C. Stober, *Anal. Profiles Drug Subs.* **15**, 367-391 (1986).

White, light, alkaline powder. d 2.11. mp 720° ±1°. One gram dissolves in 78 ml cold, 140 ml boiling water; dissolved by dil acids. Practically insol in alcohol. LD_{50} orally in rats: 0.71 g/kg (Smyth).

USE: In the production of glazes on ceramic and electrical porcelain.

THERAP CAT: Antimanic.

5553. Lithium Chloride. ClLi; mol wt 42.39. Cl 83.63%, Li 16.37%. LiCl. Prepn: Pray, *Inorg. Syn.* **5**, 153 (1957). Review of prepn and properties: Oliver in *Mellor's* vol. **II**, supplement II, *The Alkali Metals* (part 1) 179-215 (1961).

Deliquesc, cubic crystals, granules or cryst powder; sharp saline taste. mp 613° when anhydr; bp 1360°; d 2.07. One gram dissolves in 1.3 ml cold, 0.8 ml boiling water; sol in alcohol, acetone, amyl alcohol, pyridine. The aq soln is neutral or slightly alkaline. *Keep well closed.* LD_{50} i.p. in mice: 1.06 g/kg, *Handbook of Toxicology* vol. **1**, W. S. Spector, Ed. (Saunders, Philadelphia, 1956) pp 178-179.

Caution: Prolonged absorption may cause disturbed electrolyte balance, impaired renal function, CNS disturbances.

USE: Manuf mineral waters; in pyrotechnics; soldering aluminum; in refrigerating machines.

THERAP CAT: Antidepressant.

5554. Lithium Chromate(VI). $CrLi_2O_4$; mol wt 129.88. Cr 40.04%, Li 10.69%, O 49.28%. Li_2CrO_4.

The anhydr salt is appreciably sol in methanol and ethanol. Eutectic with water below −60°. *Keep well closed.*

Dihydrate, yellow, deliquesc, cryst powder. Becomes anhydr at 74.6°. Very sol in water.

USE: Corrosion inhibitor for water-cooled atomic reactors. Soln as low temp heat transfer medium.

5555. Lithium Citrate. *2-Hydroxy-1,2,3-propanetricarboxylic acid trilithium salt; citric acid trilithium salt;* Litarex; Lithonate S. $C_6H_5Li_3O_7$; mol wt 209.92. C 34.33%, H 2.40%, Li 9.92%, O 53.35%.

Tetrahydrate, white granules or cryst powder; feebly alkaline taste; deliquesces on exposure to moist air; loses all of its water at 105°. Sol in 1.5 parts water, slightly in alc. pH ~8. *Keep well closed.*

USE: Formerly in soft drinks.
THERAP CAT: Antidepressant.

5556. Lithium Dichromate(VI). Lithium bichromate. $Cr_2Li_2O_7$; mol wt 229.87. Cr 45.24%, Li 6.04%, O 48.72%. $Li_2Cr_2O_7$.
Dihydrate, yellowish-red, hygroscopic, cryst powder. Sol in less than 1 part water. *Keep well closed.*

5557. Lithium Fluoride. FLi; mol wt 25.94. F 73.24%, Li 26.76%. LiF. Prepd from LiOH and HF or by dissolving Li_2CO_3 in excess HF, evaporating to dryness, and heating to red heat: v. Wartenberg, Schultz, *Z. Elektrochem.* **27**, 568 (1921); Kwasnik in *Handbook of Preparative Inorganic Chemistry* vol. 1, G. Brauer, Ed. (Academic Press, New York, 2nd ed., 1963) p 235.
Cubic crystals (NaCl lattice) or white fluffy powder. d^{20} 2.640. mp 848°; bp 1681°. Volatilizes at 1100-1200°. Soly in water (25°): 0.13 g/100 ml. Soluble in acids. With hydrofluoric acid it forms lithium bifluoride, $LiHF_2$. With lithium hydroxide it forms a double salt LiF.LiOH, mp 462°. LD orally in guinea pigs: 200 mg/kg.
USE: As flux for soldering and welding aluminum, in the manuf of vitreous enamels and glazes. Lithium fluoride prisms are used in infrared spectrophotometers.

5558. Lithium Formate. $CHLiO_2$; mol wt 51.96. C 23.12%, H 1.94%, Li 13.36%, O 61.59%. HCOOLi.
Monohydrate, colorless or white crystals. d 1.46. Sol in 3 parts water. The aq soln is practically neutral.

5559. Lithium Hydride. HLi; mol wt 7.95. H 12.68%, Li 87.32%. LiH. Prepd by the direct combination of hydrogen and lithium: Gibb, *Trans. Electrochem. Soc.* **93**, 198-211 (1948). Review of prepn and properties: Truter in *Mellor's* vol. II, supplement II, *The Alkali Metals* (part 1) 131-145 (1961).
Cubic crystals, darkens rapidly on exposure to light, the commercial product is usually gray. mp 680°. d 0.76-0.77. No solvent known. Rapidly dec in water to form lithium hydroxide and hydrogen. Reacts with the lower alcohols, carboxylic acids, chlorine, and ammonia at 400° to liberate hydrogen.
Caution: Potential symptoms of overexposure are burns eyes and skin; burns mouth and esophagus (if ingested); nausea; muscle twitches; mental confusion; blurred vision. *See NIOSH Pocket Guide to Chemical Hazards* (DHHS/ NIOSH 90-117, 1990) p 136.
USE: Reducing agent; condensing agent with ketones and acid esters; desiccant; in hydrogen generators; 1 g in water liberates ~2.8 liters of hydrogen at STP.

5560. Lithium Hydroxide. Lithium hydrate. HLiO; mol wt 23.95. H 4.21%, Li 28.98%, O 66.81%. LiOH.
Prepn: Barnes, *J. Chem. Soc.* **1931**, 2605; Cohen, *Inorg. Syn.* **5**, 3 (1957); Bravo, *ibid.* **7**, 1 (1963). Review of prepn, properties and uses: Oliver in *Mellor's* vol. II, supplement II, *The Alkali Metals* (part 1) 159-171 (1961).
Granular, free-flowing powder; acrid, strongly alkaline. Readily absorbs CO_2 and water from air. d 2.54. mp 471°. Sol in water, slightly sol in alcohol. *Keep tightly closed.*
Monohydrate, small monoclinic crystals. d_4^{20} 1.51. Heat of formation -188.9 kcal/mol at 25°. Heat of soln -0.87 kcal/mol at 25°. Soly in water (w/w) at 0°: 10.7%; at 20°: 10.9%; at 100°: 14.8%. Slightly sol in alcohol. pH of a 1.0N soln about 14.
Caution: Strongly alkaline and hence caustic. Very irritating to skin. Systemic toxicity similar to other lithium compds. *See Lithium.*
USE: In photographic developers; in alkaline storage batteries; in prepn of other lithium salts where use of carbonate is not practical; as catalyst in the production of alkyd resins, in esterifications; in the production of lithium soaps, sulfonates, greases.

5561. Lithium Iodide. ILi; mol wt 133.85. I 94.81%, Li 5.19%. LiI.
Trihydrate, white, deliquesc granules or fused masses; becomes yellow on exposure to air, due to liberation of iodine. mp 73°; mp 446° when anhydr. Sol in about 0.5 part water or alcohol, freely in amyl alcohol or acetone. The aq soln is neutral or slightly alkaline. *Keep tightly closed and protected from light.*
USE: In photography.

5562. Lithium Nitrate. $LiNO_3$; mol wt 68.95. Li 10.07%, N 20.32%, O 69.62%.
Colorless, deliquesc granules. d 2.38. mp ~255°. Sol in ~2 parts water; sol in alcohol. The aq soln is neutral. *Keep well closed.*

5563. Lithium Oxalate. $C_2Li_2O_4$; mol wt 101.90. C 23.57%, Li 13.62%, O 62.80%. $Li_2C_2O_4$.
White crystals. d 2.12. Sol in 15 parts water.

5564. Lithium Oxide. Li_2O; mol wt 29.88. Li 46.46%, O 53.54%. Prepd from lithium peroxide: Cohen, *Inorg. Syn.* **5**, 5 (1957); from lithium hydroxide: Bravo, *ibid.* **7**, 3 (1963). Review of prepn and properties: Oliver in *Mellor's* vol. II, supplement II, *The Alkali Metals* (part 1) 146-158 (1961).
Finely divided powder or crusty material. d_4^4 2.013. mp 1570°: van Arkel *et al., Can. J. Chem.* **31**, 1009 (1953); 1427° (1700°K): Brewer, Margrave, *J. Phys. Chem.* **59**, 421 (1955). Readily absorbs carbon dioxide and water from the atm. At elevated temp attacks glass, silica, many metals.

5565. Lithium Perchlorate. $ClLiO_4$; mol wt 106.39. Cl 33.32%, Li 6.52%, O 60.15%. $LiClO_4$.
Small crystals. d_4^{25} 2.43. mp 236°. Decompn starts at ~400° and becomes rapid at 430° yielding lithium chloride and oxygen. Heat of formation: -99.94 kcal/mol at 25°. Soly in water (w/w) at 0°: 29.9%; at 25°: 37.5%; at 100°: 71.5%. Appreciably sol in alcohol, acetone, ether, ethyl acetate.
Caution: May be irritating on contact with skin, mucous membranes.
USE: Oxidizing agent.

5566. Lithium Selenate. *Selenic acid dilithium salt.* Li_2O_4Se; mol wt 156.84. Li 8.85%, O 40.80%, Se 50.34%. Prepd by roasting lithium selenite in air or by roasting lithium carbonate with selenium or selenium oxide: Lenher, Wechter, *J. Am. Chem. Soc.* **47**, 1522 (1925).
Monohydrate, monoclinic crystals. d 2.565. Stable in air. Readily sol in water.

5567. Lithium Selenite. *Selenious acid dilithium salt.* Li_2O_3Se; mol wt 140.84. Li 9.86%, O 34.08%, Se 56.06%. Prepd from a soln of selenious acid in lithium hydroxide at 60°: Nilson, *Bull. Soc. Chim.* [2] **21**, 253 (1874); *ibid.* [2] **23**, 262 (1875).
Monohydrate, acicular crystals. Hygroscopic. More sol in cold water than in hot water.

5568. Lithium Silicate. Lithium metasilicate. Li_2O_3Si; mol wt 89.97. Li 15.43%, O 53.35%, Si 31.22%. Li_2SiO_3. Prepd by fusing Li_2CO_3 with SiO_2: Schwarz, Sturm, *Ber.* **47**, 1737 (1914).
Orthorhombic needles. d_4^{25} 2.52. mp 1201°. Heat of formation (solid): -434.9 kcal/mol. Heat of formation (liq) -374.6 kcal/mol. Latent heat of fusion (1177°) = 7.24 kcal/mol, also reported as -80.2 cal/g. Insol in cold water, dec by boiling water, dilute hydrochloric acid.
USE: To calibrate thermoelements.

5569. Lithium Sulfate. Lithiophor; Lithium-Duriles. Li_2O_4S; mol wt 109.95. Li 12.63%, O 58.21%, S 29.17%. Li_2SO_4.
Monohydrate, colorless crystals; loses the water at 130°. d 2.06. Sol in 2.6 parts water. Almost insol in alcohol. The aq soln is neutral.
THERAP CAT: Antidepressant.

5570. Lithium Tetracyanoplatinate(II). Platinous lithium cyanide; lithium platinocyanide. $C_4Li_2N_4Pt$; mol wt 313.03. C 15.35%, Li 4.43%, N 17.90%, Pt 62.32%. $Li_2Pt(CN)_4$.
Pentahydrate, greenish-yellow crystals. Slightly sol in water.
USE: In x-ray photography.

5571. Lithocholic Acid. *(3α,5β)-3-Hydroxycholan-24-oic acid; 3α-hydroxycholanic acid; 17β-(1-methyl-3-carboxypropyl)etiocholan-3α-ol.* $C_{24}H_{40}O_3$; mol wt 376.58. C

76.55%, H 10.71%, O 12.75%. Found in ox bile, human bile, rabbit bile, and in ox and pig gallstones. Isoln: Fischer, *Z. Physiol. Chem.* **73**, 234 (1911). Characterization: Wieland, Weyland, *ibid.* **110**, 123 (1920). Prepn from cholic or from desoxycholic acid: Hoehn, Mason, *J. Am. Chem. Soc.* **62**, 569 (1940); Sarel, Yanuka, *J. Org. Chem.* **24**, 2018 (1959). Crystal and molecular structure by x-ray diffraction: S. K. Arora *et al., Acta Crystallogr.* **32B**, 415 (1976).

Hexagonal leaflets from alcohol, prisms from acetic acid, mp 184-186°. $[\alpha]_D^{20}$ +33.7° (c = 1.5 in abs ethanol); $[\alpha]_D^{19}$ +23.3° (Wieland); $[\alpha]_D^{20}$ +32.1° (Fischer). Freely sol in hot alc. More sol in ether than cholic or desoxycholic acid. Sol in about 10 times its weight of ethyl acetate. Slightly sol in glacial acetic acid (about 0.2 g in 3 ml). More sol in benzene than desoxycholic acid. Insol in petr ether, gasoline, ligroin, water.

Methyl ester, $C_{25}H_{42}O_3$, crystallizes with ½ mol methanol, mp 125-127°.

Ethyl ester, $C_{26}H_{44}O_3$, crystals, mp 92-93°.

Benzyl ester, $C_{31}H_{46}O_3$, crystals, mp 145-148°.

Acetyllithocholic acid, $C_{26}H_{42}O_4$, crystals, mp 169°.

Acetyllithocholic acid methyl ester, $C_{27}H_{44}O_4$, flat needles from pentane, mp 123-130°.

Acetyllithocholic acid ethyl ester, $C_{28}H_{46}O_4$, crystals, mp 90-91°.

5572. Lithopone. *C.I. Pigment White 5;* Griffith's zinc white. A white pigment consisting of a mixture of zinc sulfide, barium sulfate and some zinc oxide. Several grades of lithopone are commercially available. The zinc sulfide content varies from 26 to 60%. The commercial importance of lithopone has decreased since the introduction of titanium pigments.

USE: In water and oil paints to provide thixotropy, improve gloss and flow.

5573. Litmocidin. Antibiotic substance produced by *Proactinomyces cyaneusantibioticus.* Isoln: Gause, *J. Bacteriol.* **51**, 649 (1946); Brazhnikova, *ibid.* 655; Abou-Zeid, El-Gammal, *Z. Allg. Mikrobiol.* **11**, 5 (1971). Belongs to the class of pigments and shows same qualitative reactions as anthocyanidine: Brazhnikova, *C.A.* **41**, 5576h (1947). Approx mol wt of 398-418: Paskhina, *Biokhimiya* **21**, 448 (1956).

5574. Litmus. Lacmus; tournesol; turnsole; lacca musica; lacca coerulea. Mol wt about 3300. Blue coloring matter from various species of lichens, particularly *Variolaria, Lecanora,* and *Roccella. Habit.* Scandinavia, shores of Mediterranean, Azores, California, East India, Madagascar. *Constit.* Chiefly azolitmin and erythrolitmin combined with alkalies: lecanoric acid, orcein, erythrolein. Manuf almost exclusively in Holland. Structure studies: Beecken *et al., Angew. Chem.* **73**, 665 (1961). Contains in small amounts α,β,γ-amino and hydroxyorcein.

Blue powder, lumps or cubes. Partly soluble in water or alcohol.

USE: As acid-base indicator; pH: 4.5 red, 8.3 blue. For preparing litmus papers; in microscopy to color culture media for diagnostic purposes. Has been used for coloring beverages.

5575. Liver Extract. An extract made from the livers of mammals. Upon ingestion or injection in a suitable dosage form it increases the number of red blood corpuscles in the blood of persons afflicted with pernicious anemia. Contains folic acid and vitamin B_{12} activity.

Some commercial products are: *Intraheptol; Pernaemon; Desiver; Anahaemin; Campolon; Campovit; Cromaton; Cur-*

ethyl; Examen; Ficalon; Hepalon; Hepatopron; Hormantoxone; Hoban; Neo-Hepatex; Pernaemyl; Pernexin; Perniciosan; Plexan; Reticulogen; Ripason; Sykoton; Tenelon; Hepol.

Note: The definition given here comprises all types of liver extracts and does not differentiate between oral, injectable crude and injectable refined forms, or between solids and solns.

THERAP CAT: Hematinic; folic acid and vitamin B_{12} source.

5576. Livetins. Major water-soluble proteins found in egg yolk: Shepard, Hottle, *J. Biol. Chem.* **179**, 349 (1949). Prepn and mol wts of α- and β-livetin: Martin *et al., Can. J. Biochem. Physiol.* **35**, 241 (1957). Prepn and mol wt of γ-livetin: Martin, Cook, *ibid.* **36**, 153 (1958). Separation of four fractions, α_1-, α_2-, β-, and γ-livetins, and preparation of β- and γ-livetins: Öberdorfer, *Z. Physiol. Chem.* **331**, 280 (1963). Protein constituents of the livetin fraction of egg yolk were identified as follows: α-livetin identified as serum albumin; β-livetin as an α_2-glycoprotein with a sedimentation coeff of 2.95; and γ-livetin as serum γ-globulin: Williams, *Biochem. J.* **83**, 346 (1962). Heterogeneity of livetin proteins in egg yolk: S. F. Hui, R. H. Common, *Can. J. Biochem.* **44**, 1357 (1966); W. M. McIndoe, J. Culbert, *Int. J. Biochem.* **10**, 659 (1979).

5577. Lobelanidine. *[2α(R*),6α(S*)]-1-Methyl-α,α′-diphenyl-2,6-piperidinediethanol; α,α′-[(1-methyl-2,6-piperidinediyl)dimethylene]dibenzyl alcohol.* $C_{22}H_{29}NO_2$; mol wt 339.48. C 77.84%, H 8.61%, N 4.13%, O 9.43%. From herb of *Lobelia inflata* L., *Lobeliaceae:* Wieland *et al., Ann.* **444**, 40 (1925). Synthesis: Wieland, Drishaus, *ibid.* **473**, 102 (1929). Comprehensive review of H. Wieland's work on Lobelia alkaloids with complete references is given by Schöpf, *Naturwiss.* **30**, 359 (1942).

Scales from alc, mp 150°. Distills unchanged *in vacuo.* Freely sol in benzene, chloroform, acetone; slighty sol in ether, petr ether; almost insol in water.

Hydrochloride, $C_{22}H_{29}NO_2$·HCl, needles from alcohol, mp 135-138°.

Hydrobromide, $C_{22}H_{29}NO_2$·HBr, crystals, mp 189°.

5578. Lobelanine. *cis-2,2′-(1-Methyl-2,6-piperidinediyl)bis[1-phenylethanone]; 2,2″-(1-methyl-2,6-piperidinediyl)diacetophenone.* $C_{22}H_{25}NO_2$; mol wt 335.45. C 78.77%, H 7.51%, N 4.18%, O 9.54%. After lobeline, the most abundant alkaloid of *Lobelia inflata* L., *Lobeliaceae.* Isoln: Wieland *et al., Ann.* **444**, 40 (1925). Synthesis: Schöpf, Lehmann, *ibid.* **518**, 1 (1935); Parker *et al., J. Chem. Soc.* **1959**, 2433.

Rosettes of needles from petr ether or ether, mp 99°. Freely sol in alcohol, acetone, benzene, chloroform; slightly sol in water, ether.

Hydrochloride, $C_{22}H_{25}NO_2$·HCl, crystals from dil alcohol, dec 188°. Sol in chloroform; slightly sol in abs alcohol, cold water.

Hydrobromide, $C_{22}H_{25}NO_2$·HBr, dec 188°.

Nitrate, $C_{22}H_{25}NO_2$·HNO$_3$, mp 153-154°.

5579. Lobelia. Indian tobacco; wild tobacco; emetic herb; asthma weed; bladder pod; vomit wort. Dried leaves and tops of *Lobelia inflata* L., *Lobeliaceae.* (The seeds are also used.) *Habit*. Canada, U.S. *Constit*. Leaves and tops: chiefly lobeline, also lobelidine, lobelanine, lobelanidine, and other alkaloids. Seeds: lobeline, fixed oil.

THERAP CAT: Expectorant.

THERAP CAT (VET): Has been used as an expectorant.

5580. Lobeline. *[2R-[2α,6α(S*)]]-2-[6-(2-Hydroxy-2-phenylethyl)-1-methyl-2-piperidinyl]-1-phenylethanone; 2-[6-(β-hydroxyphenethyl)-1-methyl-2-piperidyl]acetophenone;* α-lobeline; inflatine. $C_{22}H_{27}NO_2$; mol wt 337.46. C 78.30%, H 8.06%, N 4.15%, O 9.48%. From herb and seeds of *Lobelia inflata* L., *Lobeliaceae* (Indian tobacco): H. Wieland, *Ber*. **54**, 1784 (1921). Structure: H. Wieland, O. Dragendorff, *Ann*. **473**, 83 (1929). Synthesis: H. Wieland, I. Drishaus, *ibid*. 102; G. Schering, L. Winterhalder, *ibid*. 126. Absolute configuration: Schöpf, Müller, *Ann*. **687**, 241 (1965). Pharmacology: Cambar *et al.*, *Arch. Int. Pharmacodyn. Ther*. **177**, 1 (1969); Korczyn *et al.*, *ibid*. **182**, 370 (1969); Mansuri *et al.*, *Arzneimittel-Forsch*. **23**, 1721 (1973). Comprehensive description: F. J. Muhtadi, *Anal. Profiles Drug Subs*. **19**, 261-313 (1990).

Needles from alc, ether, benzene, mp 130-131°. $[\alpha]_D^{15}$ −43° (alc). Very slightly sol in water or in petr ether; sol in hot alcohol, or in chloroform, benzene, ether.

Hydrochloride, $C_{22}H_{27}NO_2$·HCl, *Zoolobelin, Lobron*. Rosettes of slender needles from alcohol, mp 178-180°. $[\alpha]_D^{20}$ −43° (c = 2). uv max (methanol) 245, 280 nm (log ε 4.08, 3.05). One gram dissolves in 40 ml water, 12 ml alcohol; very sol in chloroform; very slightly sol in ether. Water solns are slightly acid to litmus. A 1% soln in water has a pH of 4.0-6.0.

Sulfate, $(C_{22}H_{27}NO_2)_2$·H_2SO_4, *Bantron, Unilobin, Lobeton, Lobidan, Toban*. Hygroscopic. Crystals from alc. $[\alpha]_D^{20}$ −25° (c = 2). Soluble in about 30 parts water, slightly in alcohol.

(±)-Form, *lobelidine*. Prisms, mp 110°.

THERAP CAT: Respiratory stimulant.

THERAP CAT (VET): Has been used as a respiratory stimulant, ruminatoric.

5581. Lobenzarit. *2-[(2-Carboxyphenyl)amino]-4-chlorobenzoic acid;* N-*(2-carboxyphenyl)-4-chloroanthranilic acid; 4-chloro-2,2'-iminodibenzoic acid;* CCA. $C_{14}H_{10}ClNO_4$; mol wt 291.69. C 57.65%, H 3.46%, Cl 12.15%, N 4.80%, O 21.94%. Anti-arthritic agent with an immunomodulating mechanism of action. Prepn: M. Tanemura *et al.*, **Ger. pat.** 2,526,092; *eidem*, **U.S. pat.** 4,092,426 (1976, 1978 both to Chugai). CCA lacks anti-inflammatory activity in carrageenin paw edema assay or cotton pellet granuloma test, but suppresses secondary inflammation in adjuvant arthritic rats: Y. Ohsugi *et al.*, *J. Pharm. Pharmacol*. **29**, 636 (1977). Prevention of autoimmune kidney disease in mice: *eidem*, *ibid*. **30**, 126 (1978). Immunoregulating activity: *eidem*, *Int. J. Immunopharmacol*. **2**, 224 (1980). Multicenter clinical trial in rheumatoid arthritis: Y. Shiokawa *et al.*, *J. Rheumatol*. **11**, 615 (1984).

Crystals, mp > 306°. LD_{50} in male, female rats (mg/kg): 2100, 2600 orally (Tanemura).

Disodium salt, $C_{14}H_8ClNNa_2O_4$, *Carfenil*. Solid, mp 388° (dec).

THERAP CAT: Antirheumatic.

5582. Lochnericine. *(5α,6α,7α,12β,19α)-2,3-Didehydro-6,7-epoxyaspidospermidine-3-carboxylic acid methyl ester.* $C_{21}H_{24}N_2O_3$; mol wt 352.43. C 71.57%, H 6.86%, N 7.95%, O 13.62%. Indole alkaloid. From *Vinca rosea* Linn. (*Catharanthus roseus* G. Don.), *Apocynaceae*: Gorman *et al.*, *J. Am. Pharm. Assoc*. **48**, 256 (1959); Svoboda *et al.*, *ibid*. **48**, 659 (1959); Moza, Trojánek, *Coll. Czech. Chem. Commun*. **28**, 1419 (1963). Structure: Moza *et al.*, *Tetrahedron Letters* **1964**, 2561.

Crystals from methanol, dec 190-193°. $[\alpha]_D^{27}$ −432° (chloroform). pKa in 66% DMF: 4.2. uv max (ethanol): 227, 299, 328 nm (log ε 4.10, 4.15, 4.32).

5583. Lochneridine. *(20α)-2,16-Didehydro-20-hydroxycuran-17-oic acid methyl ester.* $C_{20}H_{24}N_2O_3$; mol wt 340.42. C 70.57%, H 7.11%, N 8.23%, O 14.10%. From *Vinca rosea* Linn., *Apocynaceae*: Svoboda *et al.*, *J. Pharm. Sci*. **50**, 409 (1961). Structure: Nakagawa *et al.*, *Chem. & Ind. (London)* **1962**, 1986.

Prisms from methanol, dec 211-214°. $[\alpha]_D^{26}$ +607.5° (chloroform). pKa 5.5 in 66% DMF. uv max (ethanol): 230, 293, 328 nm (log ε 4.04, 3.94, 4.07).

5584. Locust Bean Gum. Carob flour; Johannisbrotmehl; Arobon. Mol wt about 310,000. The ground kernel endosperms of tree pods of *Ceratonia siliqua* L., *Leguminosae* (St. John's bread). Consists of proteins such as albumins, globulins, prolamins, gluteline; carbohydrates such as reducing sugars, sucrose, dextrins, pentosans; ash; fat; crude fiber; moisture. *Refs*: Plaut *et al.*, *Bull. Res. Council Israel* **3**, No. 11, 129 (1953), *C.A.* **48**, 5397 (1954); Griffiths, *Mfg. Chemist* **20**, 321 (1949); Coit, *Econ. Bot*. **5**, 82 (1951). Use as coffee, chocolate, cocoa substitute, extender: W. A. Meer, *Manuf. Confectioner* **59**, 41 (1979). Caffeine, theobromine content: W. J. Craig, T. T. Nguyen, *J. Food Sci*. **49**, 302 (1984). *Review*: F. Rol in *Industrial Gums*, R. L. Whistler, Ed. (Academic Press, New York, 2nd ed., 1973) pp 323-337.

Yellow-green color, is odorless and tasteless, but acquires a leguminous taste when boiled in water.

USE: Stabilizer, thickener, and binder in foods and cosmetics. Coffee, chocolate, cocoa substitute. Sizing and finishing agent in textiles. As fiber bonding in paper manuf. Drilling mud additive.

THERAP CAT: Adsorbent-demulcent.

5585. Lodoxamide. *2,2'-[(2-Chloro-5-cyano-1,3-phenylene)diimino]bis(2-oxoacetic acid);* N,N'-*(2-chloro-5-cyano-m-phenylene)dioxamic acid.* $C_{11}H_6ClN_3O_6$; mol wt 311.64. C 42.40%, H 1.94%, Cl 11.38%, N 13.48%, O 30.80%. Mast cell stabilizer. Prepn: C. M. Hall, J. B. Wright, **Ger. pat.** 2,362,409; *eidem*, **U.S. pat.** 3,993,679 (1974, 1976 both to Upjohn); J. B. Wright *et al.*, *J. Med. Chem*. **21**, 930 (1978). HPLC determn in human plasma: I. L. Honigberg *et al.*, *J. Chromatog*. **276**, 213 (1983). Mechanism of action study:

M. K. Church, J. Hiroi, *Brit. J. Pharmacol.* **90**, 421 (1987). Clinical evaluation in asthma: R. L. Case *et al., J. Am. Med. Assoc.* **247**, 661 (1982); in allergic eye disease: G. T. Fahy *et al., Eur. J. Ophthalmol.* **2**, 144 (1992).

mp 212° (dec). uv max (0.1*N* NaOH) 239.5 (ε 23800).

Diethyl ester, $C_{15}H_{14}ClN_3O_6$, *diethyl N,N'-(2-chloro-5-cyano-m-phenylene)dioxamate, lodoxamide ethyl, U-42718.* Crystals from ethanol, mp 177-179°.

Tromethamine salt, $C_{11}H_6ClN_3O_6.2C_4H_{11}NO_3$, *lodoxamide tromethamine, lodoxamide trometamol, U-42585E, Alomide.*

THERAP CAT: Antiallergic; topically in allergic conjunctivitis.

5586. Lofentanil. *cis-(—)-3-Methyl-4-[(1-oxopropyl)-phenylamino]-1-(2-phenylethyl)-4-piperidinecarboxylic acid methyl ester;* (—)-*cis*-3-methyl-1-phenethyl-4-(*N*-phenylpropionamido)isonipecotic acid methyl ester. $C_{25}H_{32}N_2O_3$; mol wt 408.54. C 73.50%, H 7.89%, N 6.86%, O 11.75%. Prepn: P. A. J. Janssen, G. H. P. Van Daele, Ger. pat. **2,610,228** corresp to U.S. pat. **3,998,834** (both 1976 to Janssen). Receptor affinity and pharmacological potency: K. D. Stahl *et al., Eur. J. Pharmacol.* **46**, 199 (1977). Structural study: C. De Ranter *et al., Arch. Int. Physiol. Biochim.* **87**, 1031 (1979). Antinociceptive effects in cats: A. S. Tung, T. L. Yaksh, *Pain* **12**, 343 (1982). Opiate receptor binding studies of ^{3}H-lofentanil: W. Gommeren, J. E. Leysen, *Arch. Int. Pharmacodyn. Ther.* **258**, 171 (1982); B. Ilien *et al., ibid.* 313; P. M. Laduron, P. Janssen, *Life Sci.* **31**, 457 (1982).

Oxalate, $C_{27}H_{34}N_2O_7$, *R-34995.* Solid, mp 177°.
THERAP CAT: Analgesic (narcotic).

5587. Lofepramine. *1-(4-Chlorophenyl)-2-[[3-(10,11-dihydro-5H-dibenz[b,f]azepin-5-yl)propyl]methylamino]ethanone; 4'-chloro-2-[[3-(10,11-dihydro-5H-dibenz[b,f]azepin-5-yl)propyl]methylamino]acetophenone; N-methyl-N-(4-chlorobenzoylmethyl)-3-(10,11-dihydro-5H-dibenzo[b,f]azepin-5-yl)propylamine; lopramine.* $C_{26}H_{27}ClN_2O$; mol wt 418.97. C 74.54%, H 6.50%, Cl 8.46%, N 6.69%, O 3.82%. Psychotropic drug related to imipramine, *q.v.* Prepn: E. Eriksoo *et al.,* **Brit.** pat. **1,177,525**; *eidem,* U.S. pat. **3,637,660** (1970, 1972 both to AB Leo). Chemistry and pharmacology: E. Eriksoo, O. Rohte, *Arzneimittel-Forsch.* **20**, 1561 (1970). Absorption and metabolism: J. R. Tulic *et al., Acta Pharmacol. Toxicol.* **32**, 304 (1973). Distribution and excretion: G. Plym Forshell, *Xenobiotica* **5**, 73 (1975). Pharmacokinetics: G. Plym Forshell *et al., Eur. J. Clin. Pharmacol.* **9**, 291 (1976). Clinical study: S. Wright, L. Herrmann, *Arzneimittel-Forsch.* **26**, 1167 (1976).

Crystals from methanol or acetone, mp 104-106°. Easily oxidized by air and other oxidizing agents to desipramine and *p*-chlorobenzoic acid.

Hydrochloride, $C_{26}H_{27}ClN_2O.HCl$, *Leo 640, Amplit, Gamanil, Gamonil, Timelit, Tymelyt.* Crystals from butanone, mp 152-154°. Sol in methanol, ethanol, chloroform. Practically insol in water. LD_{50} in mice, rats (mg/kg): >2500, >1000 orally; 920, >1000 i.p.; >1000, >1000 s.c. (Eriksoo, Rohte).

THERAP CAT: Antidepressant.

5588. Lofexidine. *2-[1-(2,6-Dichlorophenoxy)ethyl]-4,-5-dihydro-1H-imidazole; 2-[1-(2,6-dichlorophenoxy)ethyl]-2-imidazoline.* $C_{11}H_{12}Cl_2N_2O$; mol wt 259.13. C 50.99%, H 4.67%, Cl 27.36%, N 10.81%, O 6.17%. Vasoactive agent related structurally to clonidine, *q.v.* Prepn of the HCl salt: H. Baganz, H. J. May, **S. Afr.** pat. **68 00,850** corresp to U.S. pat. **3,966,757** (1968, 1976 both to Nordmark); of the free base: *eidem,* **Ger.** pat. **1,935,479** (1971 to Nordmark), *C.A.* **74**, 87979 (1971). Pharmacological studies: J. Velly, *J. Pharmacol.* **8**, 351 (1977); B. Jarrot *et al., Biochem. Pharmacol.* **28**, 141 (1979). NMR data and cardiovascular effects: P. B. M. Timmermans, P. A. Van Zwieten, *Eur. J. Med. Chem.* **15**, 323 (1980). Hypotensive and sedative properties: P. Birch *et al., Brit. J. Pharmacol.* **68**, 107 (1980). Effects in hypertension: N. D. Vlachakis *et al., Fed. Proc.* **39**, 4844 (1980). Series of articles on pharmacology, toxicology, clinical studies: *Arzneimittel-Forsch.* **32**, 915-993 (1982). Toxicity studies: T. H. Tsai *et al., ibid.* 955.

Crystals, mp 126-128°.
Hydrochloride, $C_{11}H_{12}Cl_2N_2O.HCl$, *MDL-14042A, Ba-168, Britlofex, Lofetensin, Loxacor.* Crystals from ethanol/-ether or 2-propanol, mp 221-223° (**U.S.** patent); also reported as mp 230-232° (**Ger.** patent). Very sol in water, ethanol. Slightly sol in 2-propanol. Practically insol in ether. LD_{50} in mice, rats, dogs (mg/kg): between 74-147 orally (all species); between 8-18 i.v. (all species) (Tsai).

THERAP CAT: Antihypertensive.

5589. Loflucarban. *N-(3,5-Dichlorophenyl)-N'-(4-fluorophenyl)thiourea; 3,5-dichloro-4'-fluorothiocarbanilide; Fluonilid.* $C_{13}H_9Cl_2FN_2S$; mol wt 315.20. C 49.54%, H 2.88%, Cl 22.50%, F 6.03%, N 8.89%, S 10.17%. Prepd from *p*-fluorophenyl isothiocyanate and 3,5-dichloroaniline or from 3,5-dichlorophenyl isothiocyanate and *p*-fluoroaniline: **Belg.** pat. **613,154** (1962 to Madan), *C.A.* **58**, 474f (1963).

Crystals from ethanol, mp 148°. Soluble in ethyl oleate, isopropyl myristate.

THERAP CAT: Antifungal.

5590. Loganin. *1-(β-D-Glucopyranosyloxy)-1,4a,5,6,7,-7a-hexahydro-6-hydroxy-7-methylcyclopenta[c]pyran-4-carboxylic acid methyl ester;* 7-hydroxy-6-desoxyverbenalin. $C_{17}H_{26}O_{10}$; mol wt 390.39. C 52.30%, H 6.71%, O 40.98%.

Key intermediate in the biosynthesis of indole alkaloids. First isolated from the seeds but chiefly from the pulp of the fruit of *Strychnos nux-vomica* L., *Loganiaceae:* Dunstan, Short, *Pharm. J.* **14**, 1025 (1883); Merz, Krebs, *Arch. Pharm.* **275**, 217 (1937); Merz, Lehmann, *ibid.* **290**, 543 (1957). Structure: Sheth *et al., Tetrahedron Letters* **1961**, 394; Büchi, Manning, *Tetrahedron* **18**, 1049 (1962). Crystal structure: Lentz, Rossmann, *Chem. Commun.* **1969**, 1269; P. G. Jones *et al., Acta Crystallogr.* **B36**, 481 (1980). Abs config: Inouye *et al., Tetrahedron* **26**, 3905 (1970). Total synthesis: Büchi *et al., J. Am. Chem. Soc.* **92**, 2165 (1970); Partridge *et al., ibid.* **95**, 532 (1973); Büchi *et al., ibid.* 540; B.-W. Au-Yeung, I. Fleming, *Chem. Commun.* **1977**, 81; I. Fleming, B.-W. Au-Yeung, *Tetrahedron* **37**, Suppl. 9, 13 (1981); K. Hiroi *et al., Chem. Letters* **1981**, 559. Biosynthetic studies: Battersby, "Biosynthesis II—Terpenoid Indole Alkaloids", in *The Alkaloids* vol. 1, The Chemical Society (Burlington House, London, 1971) pp 31-47.

Crystals, mp 222-223°. $[\alpha]_D^{20} - 82.1°$ (water). Freely sol in water; less sol in 96% alcohol; sparingly in abs alcohol. Practically insol in ether, petr ether, ligroin, ethyl acetate, acetone, chloroform.

5591. Loline. *[2R-(2α,3α,3aβ,4α,6aβ)]-Hexahydro-N-methyl-2,4-methano-4H-furo[3,2-b]pyrrol-3-amine;* festucine; methyl-*N*-depropionyldecorticasine. $C_8H_{14}N_2O$; mol wt 154.21. C 62.31%, H 9.15%, N 18.17%, O 10.37%. One of a group of pyrrolizidine alkaloids whose common feature is a unique oxygen bridge. Isoln from seeds of *Lolium cuneatum* Nevski, *Gramineae:* S. Y. Yunusov, S. T. Akramov, *J. Gen. Chem. USSR (Engl. Transl.)* **25**, 1765 (1955); from tall fescue hay, *Festuca arundinaceae* Schreb., *Gramineae:* Yates, Tookey, *Aust. J. Chem.* **18**, 53 (1965). Identity with festucine: Aasen, Culvenor, *ibid.* **22**, 2021 (1969). Identity with methyl-*N*-depropionyldecorticasine: Ribas *et al., An. Quim.* **64**, 516 (1968), *C.A.* **69**, 109764c (1968). Abs config: R. B. Bates, S. R. Morehead, *Tetrahedron Letters* **1972**, 1629. Synthesis of bisquaternary salts: N. P. Abdullaev *et al., Khim. Prir. Soedin.* **3**, 371 (1977), *C.A.* **88**, 23216 (1978). Synthetic and structural studies: S. R. Wilson *et al., J. Org. Chem.* **46**, 3887 (1981).

Dihydrochloride, $C_8H_{14}N_2O.2HCl$, needles from abs ethanol, dec 237-242°. $[\alpha]_D^{15}$ +4.6° (c = 4.37 in water). pKa 8.25 and 2.5-3.0.

5592. Lomefloxacin. *1-Ethyl-6,8-difluoro-1,4-dihydro-7-(3-methyl-1-piperazinyl)-4-oxo-3-quinolinecarboxylic acid.* $C_{17}H_{19}F_2N_3O_3$; mol wt 351.35. C 58.11%, H 5.45%, F 10.81%, N 11.96%, O 13.66%. Fluorinated quinolone antibacterial. Prepn: Y. Itoh *et al.,* **Ger. pat.** 3,433,924; *eidem,* U.S. pat. 4,528,287 (both 1985 to Hokuriku Pharm.). *In vitro* and *in vivo* activity: T. Hirose *et al., Antimicrob. Ag. Chemother.* **31**, 854 (1987). Comparative antibacterial spectrum *in vitro:* R. Wise *et al., ibid.* **32**, 617 (1988). HPLC determn in urine and bile and preliminary pharmacokinetics in rats: A. Saito *et al., ibid.* 156. Supplement on antibacterial spectrum, pharmacokinetics and clinical efficacy: *Chemotherapy (Japan)* **36**, Suppl. 2, 1-1418 (1988).

Colorless needles from ethanol, mp 239-240.5°. LD_{50} in mice (mg/kg): 245.6 i.v.; >4000 orally (Itoh).

Monohydrochloride, $C_{17}H_{19}F_2N_3O_3.HCl$, *NY-198, SC-47111, Bareon, Chimono, Lomebact, Maxaquin, Uniquin.* Colorless needles from water, mp 290-300° (dec).

THERAP CAT: Antibacterial.

5593. Lomerizine. *1-[Bis(4-fluorophenyl)methyl]-4-[(2,3,4-trimethoxyphenyl)methyl]piperazine;* 1-[bis(*p*-fluorophenyl)methyl]-4-(2,3,4-trimethoxybenzyl)piperazine; 1-(2,3,4-trimethoxybenzyl)-4-[bis(4-fluorophenyl)methyl]piperazine. $C_{27}H_{30}F_2N_2O_3$; mol wt 468.54. C 69.21%, H 6.45%, F 8.11%, N 5.98%, O 10.24%. Diphenylpiperazine calcium channel blocker; selective cerebral vasodilator. Prepn: H. Ohtaka *et al.,* **Eur. pat. Appl.** 159,566; *eidem,* U.S. pat. 4,663,325 (1985, 1987 both to Kanebo); *eidem et al., Chem. Pharm. Bull.* **35**, 3270 (1987). Structure-activity study: H. Ohtaka, G. Tsukamoto, *ibid.* 4117. HPLC determn in plasma and brain tissue: H. Waki, S. Ando, *J. Chromatog.* **494**, 408 (1989). Pharmacology: T. Kanazawa *et al., J. Cardiovasc. Pharmacol.* **16**, 430 (1990). Protective effect in cerebral ischemia: H. Hara *et al., Arch. Int. Pharmacodyn.* **304**, 206 (1990); V. Danielisova, M. Chavko, *Neurochem. Res.* **19**, 1503 (1994).

Dihydrochloride, $C_{27}H_{30}F_2N_2O_3.2HCl$, *KB-2796.* Colorless crystals from acetonitrile, mp 214-218° (dec). Also reported as mp 204-207 (dec). LD_{50} orally in mice: 300 mg/kg (Ohtaka).

THERAP CAT: Antimigraine.

5594. Lomustine. *N-(2-Chloroethyl)-N'-cyclohexyl-N-nitrosourea;* 1-(2-chloroethyl)-3-cyclohexyl-1-nitrosourea; NSC-79037; RB-1509; Belustine; CCNU; Cecenu; CeeNU; CiNU. $C_9H_{16}ClN_3O_2$; mol wt 233.70. C 46.26%, H 6.90%, Cl 15.17%, N 17.98%, O 13.69%. Chloroethylnitrosourea derivative with antitumor activity. Similar to carmustine, chlorozotocin, nimustine, ranimustine, *q.q.v.* Synthesis: T. P. Johnston *et al., J. Med. Chem.* **9**, 892 (1966). Clinical review: S. K. Carter, J. W. Newman, *Cancer Chemother. Rep. Part 3* **1**, 136 (1968). Physiological disposition: V. T. Oliverio *et al., Cancer Res.* **30**, 1330 (1970). Toxicology and pharmacology: G. R. Thompson, R. E. Larson, *Toxicol. Appl. Pharmacol.* **21**, 405 (1972); V. T. Oliverio, *Cancer Chemother. Rep. Part 3* **4**, 13 (1973). Comprehensive description: F. J. Al-Shammary, *Anal. Profiles Drug Subs.* **19**, 315-340 (1990).

Yellow powder, mp 90°. Freely sol in chloroform; sol in ethanol, acetone. Soly in water, 0.1*N* NaOH, 0.1*N* HCl or 10% ethanol: <0.05 mg/ml; in absolute ethanol: 70

mg/ml. The bulk drug should be stored in a deep freeze and protected from moisture. LD_{50} in male mice (mg/kg): 51 orally; 56 i.p.; 61 s.c. (Thompson, Larson).

Note: This substance may reasonably be anticipated to be a carcinogen: *Seventh Annual Report on Carcinogens* (PB95-109781, 1994) p 126.

THERAP CAT: Antineoplastic.

5595. Lonapalene. *6-Chloro-2,3-dimethoxy-1,4-naph-thalenediol diacetate;* 6-chloro-1,4-diacetoxy-2,3-dimethoxy-naphthalene; RS-43179. $C_{16}H_{15}ClO_6$; mol wt 338.74. C 56.73%, H 4.46%, Cl 10.47%, O 28.34%. Selective 5-lipoxygenase inhibitor. Prepn: G. H. Jones *et al.,* **Eur. pat. Appl. 107,512;** *eidem,* U.S. pat. **4,466,981** (both 1984 to Syntex); *eidem, J. Med. Chem.* **29,** 1504 (1986); S. T. Perri, H. W. Moore, *Tetrahedron Letters* **28,** 4507 (1987). Clinical pharmacology: A. K. Black *et al., J. Invest. Dermatol.* **95,** 50 (1990). Percutaneous absorption and metabolism: P. A. Lehman *et al., Pharm Res.* **9,** 1145 (1992). HPLC determn in pharmaceutical prepns: R. A. Kenley *et al., Drug Dev. Ind. Pharm.* **11,** 1781 (1985); of metabolites in urine: C.-H. Kiang, C. Lee, *J. Chromatog.* **565,** 339 (1991). Clinical evaluation in psoriasis: A. Lassus, S. Forsstrom, *Brit. J. Dermatol.* **113,** 103 (1985).

Crystals from ether-petroleum ether, mp 93-94° (Jones); also reported as mp 87.5-89.5° (Perri, Moore). Partition coefficient (*n*-octanol/water): 3.26.

THERAP CAT: Antipsoriatic.

5596. Lonazolac. *3-(4-Chlorophenyl)-1-phenyl-1H-pyr-azole-4-acetic acid.* $C_{17}H_{13}ClN_2O_2$; mol wt 312.76. C 65.29%, H 4.19%, Cl 11.34%, N 8.96%, O 10.23%. Prepn: R. A. Newberry, **Brit. pat. 1,373,212** (1974 to Wyeth), *C.A.* **82,** 72987t (1975); G. Rainer, U.S. pat. **4,146,721** (1979 to Byk Gulden); G. Rainer *et al., Arzneimittel-Forsch.* **31,** 649 (1981). Pharmacology: R. Riedel, *ibid.* 655. Clinical studies: G. Lonauer *et al., Z. Rheumatol.* **40,** 161 (1981); W. Siegmeth, P. Placheta, *Wien Klin. Wochenschr.* **94,** 145 (1982).

Crystals from ethanol/water, mp 150-151°. uv max (methanol): 281 nm (ε 24800); (0.1*N* NaOH): 281 nm (ε 23700). pKa 4.3. LD_{50} in male mice, rats (mg/kg): 195, 165 i.v. (Riedel).

Calcium salt, $C_{34}H_{24}CaCl_2N_4O_4$, *Argun L, Irritren.* Solid, melts between 270-290° (dec). uv max (0.1*N* NaOH): 280 nm (ε 46400). LD_{50} in male, female mice, male, female rats (mg/kg): 670, 845, 730, 1000 orally (Riedel).

THERAP CAT: Anti-inflammatory.

5597. Longifolene. *[1S-(1α,3aβ,4α,8aβ)]-Decahydro-4,8,8-trimethyl-9-methylene-1,4-methanoazulene;* junipene; kuromatsuene. $C_{15}H_{24}$; mol wt 204.36. C 88.16%, H

11.84%. Tricyclic sesquiterpene present to the extent of 5-10% in Indian turpentine oil which is produced commercially from the Himalayan pine, *Pinus longifolia* Roxb., *Pinaceae:* J. L. Simonsen, *J. Chem. Soc.* **117,** 570 (1920). Structure: R. H. Moffett, D. Rogers, *Chem. & Ind. (London)* **1953,** 916; P. Naffa, G. Ourisson, *ibid.* 917. Abs config: G. Ourisson, *Bull. Soc. Chim. France* **1955,** 895. Identity of longifolene with junipene and kuromatsuene: S. Akiyoshi *et al., Tetrahedron* **9,** 237 (1960). Total synthesis of naturally occurring (+)-form: E. J. Corey *et al., J. Am. Chem. Soc.* **86,** 478 (1964); of the (±)-form: *eidem, ibid.* **83,** 1251 (1961); J. E. McMurray, S. J. Isser, *ibid.* **94,** 7132 (1972); R. A. Volkmann *et al., ibid.* **97,** 4777 (1975). Review of chemistry of longifolene and its derivs: S. Dev, *Accounts Chem. Res.* **14,** 82-88 (1981); *idem, Fortschr. Chem. Org. Naturst.* **40,** 50-104 (1981).

Viscous oil, bp_{706} 254-256°, bp_{15} 126-127°. $[α]_D^{18}$ +42.73. d_4^{18} 0.9319; n_D^{20} 1.5040. Insol in water; sol in benzene.

Borane deriv, $C_{30}H_{51}B$, *dilongifolylborane.* Heavy, snow-white, shiny plates, mp 160-161° (sealed, evacuated capillary). Strongly dimeric. Sparingly sol in common organic solvents. Prepn and use in hydroboration: P. K. Jadhav, H. C. Brown, *J. Org. Chem.* **46,** 2988 (1981).

USE: Borane deriv as a chiral hydroborating agent.

5598. Lonidamine. *1-[(2,4-Dichlorophenyl)methyl]-1H-indazole-3-carboxylic acid;* 1-(2,4-dichlorobenzyl)-1H-indazole-3-carboxylic acid; diclondazolic acid; DICA; AF-1890; Doridamina. $C_{15}H_{10}Cl_2N_2O_2$; mol wt 321.16. C 56.10%, H 3.14%, Cl 22.08%, N 8.72%, O 9.96%. Antitumor agent with antispermatogenic activity; inhibits cellular respiration in neoplastic cells. Prepn: G. Palazzo, B. Silvestrini, **Ger. pat. 2,310,031;** *eidem,* U.S. pat. **3,895,026** (1973, 1975 both to Angelini Francesco); G. Corsi, G. Palazzo, *J. Med. Chem.* **19,** 778 (1976). Antispermatogenic activity: T. J. Lobl, *Arch. Androl.* **2,** 353 (1979). *In vitro* and *in vivo* antitumor activity: B. Silvestrini *et al., Brit. J. Cancer* **47,** 221 (1983). Mechanism of action in neoplastic cells: A. Floridi *et al., Exp. Mol. Pathol.* **42,** 293 (1985). Pharmacology: V. Cioli *et al., Arzneimittel-Forsch.* **34,** 455 (1984). HPLC determn in plasma and urine: R. Leclaire *et al., J. Chromatog.* **277,** 427 (1983). Toxicological study: R. Heywood *et al., Chemotherapy (Basel)* **27,** Suppl. 2, 91 (1981). Series of articles on mechanism of action, pharmacokinetics and clinical applications in spermatogenesis and tumors: *ibid.* 5-120; on pharmacology and clinical evaluation in cancers: *Oncology* **41,** Suppl. 1, 1-121 (1984). Symposium on clinical evaluations in cancers: *Semin. Oncol.* **18,** Suppl 4, 1-72 (1991).

Crystals from ethanol, mp 207°. Sol in methanol, acetic acid. LD_{50} in mice, rats (mg/kg): 900, 1700 orally; 435, 525 i.p. (Heywood).

THERAP CAT: Antineoplastic.

5599. Lonomycins. Polyether antibiotics active against gram-positive bacteria, esp. coccidia. The major component is lonomycin A (initially named lonomycin). Lonomycin B and C are minor congeners. Isoln from *Streptomyces ribosidificus* strain TM-481 as sodium salt: M. Shibata *et al.,* **Japan. Kokai 75 49,495** (1975 to Taisho Pharm.), *C.A.* **83,** 191339p (1975); S. Omura *et al., J. Antibiot.* **29,** 15 (1976). Identity of A with antibiotic DE-3936; fermentation, isolation, physicochemical properties and biological activity: M. Ohshima *et al., ibid.* 354. Structure of A: N. Otake, M. Koenuma, *Tetrahedron Letters* **1975**, 4147; C. Riche, C. Pascard-Billy, *Chem. Commun.* **1975**, 951; of B, C: H. Seto *et al., J. Antibiot.* **31**, 929 (1978); T. Mizutani *et al., ibid.* **33**, 1224 (1980). ^{13}C-NMR spectrum of A: H. Seto *et al., ibid.* 979. Ionophorous activity of A: M. Mitani, N. Otake, *ibid.* **31**, 750 (1978). Cardiovascular effects of A: K. Kaneko *et al., Res. Commun. Chem. Pathol. Pharmacol.* **67**, 17 (1990); K. Tsuchida *et al., Arch. Int. Pharmacodyn.* **314**, 25 (1991). Total synthesis of A: D. A. Evans *et al., J. Am. Chem. Soc.* **117**, 3448 (1995).

Lonomycin A

Sodium salts are sol in lower alcohols, acetone, ethyl acetate, benzene, chloroform, ether. Slightly sol in *n*-hexane, petr ether. Insol in water.

Lonomycin A, $C_{44}H_{76}O_{14}$, *antibiotic DE-3936, emericid.* Colorless prisms from benzene-petr ether, mp 109-114°. $[\alpha]_D^{20}$ +66.6° (c = 1 in $CHCl_3$); $[\alpha]_D^{23}$ +57.5° (c = 0.40 in CH_2Cl_2). pKa' (66% acetone) 5.9. Sol in most organic solvents. Insol in water.

Lonomycin A sodium salt, $C_{44}H_{75}NaO_{14}$, colorless prisms from benzene-petr ether, mp 173-176 (dec) (Ohshima); also reported as mp 188-189° (Omura). $[\alpha]_D^{25}$ +49.8° (c = 1 in CH_3OH), $[\alpha]_D^{25}$ +67.0° (c = 1 in $CHCl_3$) (Ohshima); also reported as $[\alpha]_D^{25}$ +47° (c = 1 in CH_3OH) (Omura). LD_{50} in mice (mg/kg): 45.8 orally; 13.0 i.p.; 37.5 s.c. (Ohshima).

Lonomycin B sodium salt, $C_{44}H_{75}NaO_{14}$, white prisms from *n*-hexane-benzene, mp 181-182°. $[\alpha]_D$ +48.2° (c = 0.5 in CH_3OH).

Lonomycin C sodium salt, $C_{43}H_{73}NaO_{14}$, white prisms from *n*-hexane-benzene, mp 186-187°. $[\alpha]_D$ +49.2° (c = 0.5 in CH_3OH).

5600. Looplure. *(Z)-7-Dodecen-1-ol acetate; cis-7-dodecenyl acetate; cis-1-acetoxy-7-dodecene;* ENT-33266. $C_{14}H_{26}O_2$; mol wt 226.36. C 74.29%, H 11.58%, O 14.14%. Sex pheromone of the female cabbage looper moth, *Trichoplusia ni* (Hübner). Isoln, identification and synthesis: R. S. Berger, *Ann. Entomol. Soc. Am.* **59**, 767 (1966). Synthesis and activity of isomers: N. Green *et al., J. Med. Chem.* **10**, 533 (1967); H. H. Toba *et al., J. Econ. Entomol.* **63**, 1048 (1970). Alternate syntheses: W. Seidel *et al., Ber.* **110**, 3544 (1977); H. J. Bestmann *et al., Ber.* **112**, 1923 (1979); M. Horiike, C. Hirano, *Agr. Biol. Chem.* **44**, 2229 (1980); D. Basavaiah *et al., J. Org. Chem.* **47**, 1792 (1982). Toxicity study: M. Beroza *et al., Toxicol. Appl. Pharmacol.* **31**, 421 (1975).

Oil, bp$_{0.05}$ 98-100°. n_D^{25} 1.4420. LD_{50} orally in rats: > 13,430 mg/kg (Beroza).

USE: Insect sex attractant.

5601. Loperamide. *4-(4-Chlorophenyl)-4-hydroxy-N,N-dimethyl-α,α-diphenyl-1-piperidinebutanamide; 4-(p-chlorophenyl)-4-hydroxy-N,N-dimethyl-α,α-diphenyl-1-piperidinebutyramide.* $C_{29}H_{33}ClN_2O_2$; mol wt 477.05. C 73.02%, H 6.97%, Cl 7.43%, N 5.87%, O 6.71%. Prepn: Janssen *et al., Fr.* **2,100,711;** *eidem,* U.S. pat. **3,714,159** (1972, 1973 to Janssen); Stokbroekx *et al., J. Med. Chem.* **16**, 782 (1973). Series of articles on pharmacology, toxicology, metabolism, and clinical studies: *Arzneimittel-Forsch.* **24**, 1640-1665 (1974). Toxicity studies: C. J. E. Niemegeers *et al., ibid.* 1633, 1636. *Review:* D. A. Shriver *et al.,* in *Pharmacological and Biochemical Properties of Drug Substances* vol. 3, M. E. Goldberg, Ed. (Am. Pharm. Assoc., Washington, DC,

1981) pp 461-476. Comprehensive description: J. Van Rompay, J. E. Carter in *Analytical Profiles of Drug Substances* vol. 19, K. Florey, Ed. (Academic Press, New York, 1990) pp 341-365.

Hydrochloride, $C_{29}H_{33}ClN_2O_2 \cdot HCl$, *PJ-185, R-18553, Arret, Blox, Brek, Dissenten, Fortasec, Imodium, Imosec, Imossel, Lopemid, Lopemin, Loperyl, Suprasec, Tebloc.* Crystals from isopropanol, mp 222-223°. uv max (0.1*N* HCl/2-propanol, 10/90 v/v): 253, 259, 265, 273 nm (ε 532, 648, 581, 233). Soly (g/100ml): water (pH 1.7) 0.14; citrate-phosphate (pH 6.1) 0.008; citrate-phosphate (pH 7.9) <0.001; methanol 28.6; ethanol 5.37; 2-propanol 1.11; dichloromethane 35.1; acetone 0.20; ethyl acetate 0.035; diethyl ether <0.001; hexane <0.001; toluene 0.001; *N,N*-dimethylformamide 10.3; tetrahydrofuran 0.32; 4-methyl-2-pentanone 0.020; propylene glycol 5.64; polyethylene glycol 400 1.40; dimethylsulfoxide 20.5; 2-butanone 0.18. pKa 8.66. Practically insol in H_2O at physiological pH (0.002%). Stable, can be stored for several years under normal conditions; not hygroscopic; not affected by light. LD_{50} in mice (mg/kg): 75 s.c.; 28 i.p.; 105 orally; in rats (mg/kg): 185 orally (Niemegeers).

THERAP CAT: Antidiarrheal.

5602. Lophophorine. *(S)-6,7,8,9-Tetrahydro-4-methoxy-8,9-dimethyl-1,3-dioxolo[4,5-h]isoquinoline; N-methylanhalonine.* $C_{13}H_{17}NO_3$; mol wt 235.28. C 66.36%, H 7.28%, N 5.95%, O 20.40%. In *Lophophora williamsii* (Lemaire) Coult., Cactaceae. Extraction procedure: Kauder, *Arch. Pharm.* **237**, 191 (1899). Structure: Späth, Gangl, *Monatsh.* **44**, 104, 106 (1923). Synthesis: Späth, Kesztler, *Ber.* **68**, 1667 (1935).

Oily liquid. bp$_{0.02}$ 140-145° (air bath temp). $[\alpha]_D^{25}$ −47° (c = 1 in chloroform). Sol in ether, chloroform.

Hydrochloride, $C_{13}H_{17}NO_3 \cdot HCl$, needles from alcohol. $[\alpha]_D^{17}$ −9.5° (c = 1). Sol in water, alcohol.

5603. Lophotoxin. *[1R-(1R*,2S*,4S*,10R*,12R*,-14R*,15R*)]-2-(Acetyloxy)-12-methyl-4-(1-methylethenyl)-17-oxo-11,16,18,19-tetraoxapentacyclo[12.2.2.1^{6,9}.0^{1,15}.0^{10,12}]-nonadeca-6,8-diene-7-carboxaldehyde;* LTX. $C_{22}H_{24}O_8$; mol wt 416.43. C 63.45%, H 5.81%, O 30.74%. A novel neuromuscular toxin isolated from several spp of Pacific gorgonians (sea fans and whips) of the genus *Lophogorgia*. Although it belongs to the cembrene class of diterpenoids, the ability of LTX to cause irreversible postsynaptic blockade at the neuromuscular junction resembles the mode of action of the protein snake venom α-bungarotoxin (*cf.* bungarotoxins). Isoln and structure: W. Fenical *et al., Science* **212**, 1512 (1981). Molecular mode of action: J. R. Smythies, *Med. Hypotheses* **7**, 1457 (1981), *C.A.* **96**, 137527 (1982). Toxicity study: P. Culver, R. S. Jacobs, *Toxicon* **19**, 825 (1981). Description of the cytotoxic, ichthytoxic, and antibacterial activity of alcoholic extracts of gorgonians: R. S. Jacobs *et al., Fed. Proc.* **40**, 26 (1981); V. J. Paul, W. Fenical, *J. Org. Chem.* **45**, 3401 (1981).

White needles, mp 164-166°. $[\alpha]_D^{27}$ +14.2° (c = 1.7 in chloroform). LD$_{50}$ in mice (mg/kg): 8.9 s.c. (Culver, Jacobs).

5604. Loprazolam. *(Z)-6-(2-Chlorophenyl)-2,4-dihydro-2-[(4-methyl-1-piperazinyl)methylene]-8-nitro-1H-imidazo[1,2-a][1,4]benzodiazepin-1-one.* $C_{23}H_{21}ClN_6O_3$; mol wt 464.91. C 59.42%, H 4.55%, Cl 7.63%, N 18.08%, O 10.32%. Annelated 1,4-benzodiazepine deriv. Prepn: J. B. Taylor, D. R. Harrison, **Ger. pat.** 2,605,652; *eidem,* **U.S. pat.** 4,044,142 (1976, 1977 both to Roussel-UCLAF). CNS activity: J. R. Agar *et al., J. Med. Chem.* **20**, 1035 (1977). Pharmacological profile: T. G. Johns *et al., Arch. Int. Pharmacodyn. Ther.* **240**, 53 (1979). Effects on sleep in normal volunteers: I. Hindmarch, C. A. Clyde, *Drugs Exp. Clin. Res.* **2**, 61 (1980). Review of pharmacology and efficacy in insomnia: B. G. Clark *et al., Drugs* **31**, 500-516 (1986).

Crystals from chloroform/ether, mp 214-215°. LD$_{50}$ in mice: > 1000 mg/kg orally (Agar).

Methanesulfonate, $C_{23}H_{21}ClN_6O_3 \cdot CH_3SO_3H$, *loprazolam mesylate, HR-158, RU-31158, Dormonoct, Havlane, Somnovit, Sonin.* Crystals from methylene chloride, mp 205-210°.

Note: This is a controlled substance (depressant) listed in the U.S. Code of Federal Regulations, Title 21 Part 1308.14 (1995).

THERAP CAT: Sedative, hypnotic.

5605. Loprinone. *1,2-Dihydro-5-imidazo[1,2-a]pyridin-6-yl-6-methyl-2-oxo-3-pyridinecarbonitrile;* 1,2-dihydro-5-imidazo[1,2-a]pyridin-6-yl-6-methyl-2-oxonicotinonitrile; olprinone. $C_{14}H_{10}N_4O$; mol wt 250.26. C 67.19%, H 4.03%, N 22.39%, O 6.39%. Inotropic agent which inhibits cAMP phosphodiesterase. Prepn: M. Yamanaka *et al.,* **Japan. Kokai 86 218,589;** *eidem,* **U.S. pat.** 4,751,227 (1986, 1988 both to Eisai). Synthesis: M. Yamanaka *et al., J. Label. Compd. Radiopharmaceut.* **31**, 125 (1992). Pharmacology: M. Tajimi *et al., Arch. Pharmacol.* **344**, 602 (1991). Pharmacokinetics in dogs: Y. Uemura *et al., J. Pharm. Pharmacol.* **45**, 1077 (1993). Review of pharmacology, toxicology and clinical evaluation: M. Endoh, *Cardiovasc. Drug Rev.* **11**, 432-450 (1993).

Crystals from N,N-dimethylformamide, mp > 300°.

Hydrochloride monohydrate, $C_{14}H_{10}N_4O \cdot HCl \cdot H_2O$, *E-1020.* White to yellowish-white odorless crystalline powder, mp > 300°. Sol in water (pH 3-5) up to 3 mg/ml. LD$_{50}$ in

male, female rats, mice (mg/kg): 7,804, > 10,000, > 10,000, > 10,000 orally; 176, 240, 242, 269 i.v.; 2,133, 2,890, 3,898, 4,479 s.c. (Endoh).

THERAP CAT: Cardiotonic.

5606. Loracarbef. *[6R-[6α,7β(R*)]]-7-[(Aminophenylacetyl)amino]-3-chloro-8-oxo-1-azabicyclo[4.2.0]oct-2-ene-2-carboxylic acid monohydrate;* (6R,7S)-3-chloro-7-[(R)-phenylglycinamido]-8-oxo-1-azabicyclo[4.2.0]oct-2-ene-2-carboxylic acid monohydrate; carbacefaclor; KT-3777; LY-163892 monohydrate; Lorabid. $C_{16}H_{16}ClN_3O_4 \cdot H_2O$; mol wt 367.79. C 52.25%, H 4.93%, Cl 9.64%, N 11.43%, O 21.75%. Carbacephem analog of cefaclor, *q.v.* Prepn: T. Hirata *et al.,* **Eur. pat. Appl.** 14,476; *eidem,* **U.S. pat.** 4,708,956 (1980, 1987 both to Kyowa Hakko); I. Matsukuma *et al., Chem. Pharm. Bull.* **37**, 1239 (1989). Prepn of the crystalline monohydrate: C. E. Pasini, **Eur. pat. Appl.** 311,366 (1989 to Lilly). Enantioselective synthesis: C. C. Bodurow *et al., Tetrahedron Letters* **30**, 2321 (1989). Antibacterial spectrum: K. Sato *et al., J. Antibiot.* **42**, 1844 (1989). β-Lactamase stability: R. N. Jones, A. L. Barry, *Eur. J. Clin. Microbiol.* **6**, 570 (1987). Pharmacology in animals: T. Shetler *et al., Arzneimittel-Forsch.* **43**, 60 (1993). Pharmacokinetics in children: J. D. Nelson *et al., Antimicrob. Ag. Chemother.* **32**, 1738 (1988).

mp 205-215° (dec) (Matsukuma). $[\alpha]_D^{21}$ +34.0 (c = 0.35 in water).

THERAP CAT: Antibacterial.

5607. Lorajmine. *(17R,21α)-Ajmalan-17,21-diol 17-(chloroacetate);* 17-monochloroacetylajmaline; 17-chloroacetylajmaline; MCAA. $C_{22}H_{27}ClN_2O_3$; mol wt 402.92. C 65.58%, H 6.75%, Cl 8.80%, N 6.95%, O 11.91%. Semisynthetic ajmaline derivative. Prepn from ajmaline: **Belg. pat.** 622,395 (1962 to Inverni Della Beffa), *C.A.* **59**, 8816d (1963); A. Bonati, A. Bocchia, *Farmaco Ed. Sci.* **18**, 84 (1963). Improved prepn: A. Bonati, **Ger. pat.** 2,023,949; *eidem,* **U.S. pat.** 3,741,972 (1970, 1973 both to Inverni Della Beffa). In vitro and in vivo anti-arrhythmic activity: C. Capra, *Farmaco Ed. Sci.* **19**, 865 (1964); cardiac action: Y. Katano *et al., Arzneimittel-Forsch.* **23**, 483 (1973); cardiovascular effects and toxicity: C. Capra *et al., Farmaco Ed. Prat.* **35**, 49 (1980). Determination in plasma by TLC fluorescence: L. J. Dombrowski *et al., J. Pharm. Sci.* **64**, 643 (1975). GC-MS determination in blood: St. D. Clemans *et al., Arzneimittel-Forsch.* **27**, 1128 (1977). Comparison with other ajmaline esters: M. Salmona *et al., J. Pharm. Sci.* **64**, 1561 (1975); P. Jaillon *et al., Arch. Int. Pharmacodyn.* **224**, 310 (1976). Clinical studies: M. Cafiero *et al., Minerva Cardioangiol.* **32**, 59 (1984). Book: G. G. Gensini, Ed., *Concepts on the Mechanism and Treatment of Arrhythmias* (Futura Publishing Co., Mount Kisco, N.Y., 1974) 281 pp.

Crystals, mp 232-238°. $[\alpha]_D^{20}$ +27.5° (c = 1 in chloroform).

Hydrochloride, $C_{22}H_{27}ClN_2O_3 \cdot HCl$, **Win-11831, Nevergor, Ritmos Elle.** Crystals from methyl ethyl ketone, mp 243-246°. $[\alpha]_D^{20}$ +29° (c = 1 in ethanol). LD_{50} in mice, rats (mg/kg): 176, 139 i.p.; 370, 480 orally (Capra).

THERAP CAT: Antiarrhythmic (cardiac depressant).

5608. Loratadine. *4-(8-Chloro-5,6-dihydro-11H-benzo-[5,6]cyclohepta[1,2-b]pyridin-11-ylidene)-1-piperidinecarboxylic acid ethyl ester;* 11-[N-(ethoxycarbonyl)-4-piperidylidene]-8-chloro-6,11-dihydro-5H-benzo[5,6]cyclohepta[1,2-b]-pyridine; Sch-29851; Claritin; Clarityn; Lisino. $C_{22}H_{23}ClN_2O_2$; mol wt 382.89. C 69.01%, H 6.05%, Cl 9.26%, N 7.32%, O 8.36%. Nonsedating-type histamine H_1-receptor antagonist. Prepn: F. J. Villani, U.S. pat. **4,282,233** (1981 to Schering); F. J. Villani *et al., Arzneimittel-Forsch.* **36**, 1311 (1986). Pharmacology: A. Barnett *et al., Agents Actions* **14**, 590 (1984). Evaluation of CNS effects in humans: C. M. Bradley, A. N. Nicholson, *Eur. J. Clin. Pharmacol.* **32**, 419 (1987). Clinical pharmacokinetics: E. Radwanski *et al., J. Clin. Pharmacol.* **27**, 530 (1987). Clinical trial in allergic rhinitis: R. J. Dockhorn *et al., Ann. Allergy* **58**, 407 (1987); in comparison with terfenadine, *q.v.*: F. Horak *et al., Arzneimittel-Forsch.* **38**, 124 (1988); G. Bruttmann *et al., J. Allergy Clin. Immunol.* **83**, 411 (1989). Review of pharmacology and therapeutic efficacy: M. Haria *et al., Drugs* **48**, 617-637 (1994).

Crystals from acetonitrile, mp 134-136°.
THERAP CAT: Antihistaminic.

5609. Lorazepam. *7-Chloro-5-(2-chlorophenyl)-1,3-dihydro-3-hydroxy-2H-1,4-benzodiazepin-2-one;* Wy-4036; Ativan; Emotival; Lorax; Lorsilan; Pro Dorm; Psicopax; Punktyl; Quait; Securit; Sedatival; Sedazin; Somagerol; Tavor; Temesta; Wypax. $C_{15}H_{10}Cl_2N_2O_2$; mol wt 321.16. C 56.10%, H 3.14%, Cl 22.08%, N 8.72%, O 9.96%. Prepn and activity studies: Childress, Gluckman, *J. Pharm. Sci.* **53**, 577 (1964); Bell, **Belg.** pat. **621,819,** *C.A.* **60**, 2992h (1964); **Brit.** pat. **1,057,492; U.S.** pat. **3,296,249** (1963, 1967, 1967 all to Am. Home Prod.); *idem, J. Med. Chem.* **11**, 457 (1968). Determn by HPLC: I. Jane, A. McKinnon, *J. Chromatog.* **323**, 191 (1985). Psychopharmacological studies: Stein, Berger, *Science* **166**, 253 (1969); Elliot *et al., Clin. Pharmacol. Ther.* **12**, 468 (1971). Series of articles on pharmacology, metabolism, clinical studies: *Arzneimittel-Forsch.* **21**, 1047-1065, 1073-1102 (1971). Toxicity study: G. Owen *et al., ibid.* 1065. Comprehensive description: J. G. Rutgers, C. M. Shearer, *Anal. Profiles Drug Subs.* **9**, 397-426 (1980). Review of clinical pharmacology and therapeutic use: B. Ameer, D. J. Greenblatt, *Drugs* **21**, 161-200 (1981).

Crystals, mp 166-168°. uv max (methanol): 229 nm, (1N NaOH): 233 nm, (1N HCl): 237 nm. Soly (mg/ml): water 0.08, chloroform 3, alcohol 14, propylene glycol 16, ethyl acetate 30. pK_1 13; pK_2 11.5. LD_{50} in mice, rats (mg/kg): 3178, > 5000 orally (Owen).

Note: This is a controlled substance (depressant) listed in the U.S. Code of Federal Regulations, Title 21 Part 1308.14 (1995).

THERAP CAT: Anxiolytic.

5610. Lorcainide. *N-(4-Chlorophenyl)-N-[1-(1-methyl-ethyl)-4-piperidinyl]benzeneacetamide;* 4'-chloro-N-(1-iso-propyl-4-piperidyl)-2-phenylacetanilide. $C_{22}H_{27}ClN_2O$; mol wt 370.92. C 71.24%, H 7.34%, Cl 9.56%, N 7.55%, O 4.31%. Prepn: S. Sanczuk, H. K. F. Hermans, **Ger.** pat. **2,642,856** (1977 to Janssen), *C.A.* **87**, 53094 (1977); *eidem,* **U.S.** pat. **4,126,689** (1978 to Janssen) Determn of lorcainide and its metabolites by GLC: R. Woestenborghs *et al., J. Chromatog.* **164**, 169 (1979). Disposition and anti-arrhythmic effects: U. Klotz *et al., Int. J. Clin. Pharmacol. Biopharm.* **17**, 152 (1979). Pharmacokinetics and tissue distribution: U. Klotz, E. Golbs, *Arzneimittel-Forsch.* **30**, 619 (1980). Clinical study: P. Somani, S. di Giorgi, *Chest* **78**, 658 (1980). Toxicology: G. J. Barton, R. Marsboom, "A Brief Summary of Animal Toxicology Studies on Lorcainide", in *Prognosis and Pharmacotherapy of Life-Threatening Arrhythmias,* Jähnchen *et al.,* Eds. (Royal Soc. Med. London, 1981) pp 23-25. Review of pharmacology and efficacy: C. E. Eriksson, R. N. Brogden, *Drugs* **27**, 279-300 (1984). Symposium: *Am. J. Cardiol.* **54**, 1B-54B (1984).

LD_{50} in male mice, male and female rats (mg/kg): 18.8, 19.3, 18.6 i.v.; 483, 395, 435 orally (Barton, Marsboom).

Hydrochloride, $C_{22}H_{27}ClN_2O \cdot HCl$, **R-15889, Ro-13-1042, Lopantrol, Lorivox, Remivox.** Crystals from 2-propanone and 2-propanol, mp 263°.

THERAP CAT: Antiarrhythmic (class IC).

5611. Lormetazepam. *7-Chloro-5-(2-chlorophenyl)-1,3-dihydro-3-hydroxy-1-methyl-2H-1,4-benzodiazepin-2-one;* N-methyllorazepam; Wy-4082; Ergocalm; Loramet; Noctamid. $C_{16}H_{12}Cl_2N_2O_2$; mol wt 335.19. C 57.33%, H 3.61%, Cl 21.15%, N 8.36%, O 9.55%. Analog of lorazepam, *q.v.* Prepn: S. C. Bell, **Belg.** pat. **621,819,** *C.A.* **60**, 2993b (1964); *idem,* **U.S.** pat. **3,296,249** (1963, 1967 both to Am. Home Products); *idem, J. Med. Chem.* **11**, 457 (1968); A. Nudelman *et al., J. Pharm. Sci.* **63**, 1886 (1974). Pharmacokinetics and biotransformation in humans: M. Hümpel *et al., Eur. J. Drug Metab. Pharmacokinet.* **4**, 237 (1979). Hypnotic effect: A. Doenicke *et al., Anaesthesist* **28**, 578 (1979). Comparative study: H. Ott *et al., ibid.* 29. Absorption, distribution, excretion of ^{14}C-lormetazepam: R. Girkin *et al., Xenobiotica* **10**, 401 (1980). Radioimmunologic study: M. Hümpel *et al., Clin. Pharmacol. Ther.* **28**, 673 (1980).

Crystals from ethanol/THF, mp 205-207°.

Note: This is a controlled substance (depressant) listed in the U.S. Code of Federal Regulations, Title 21 Part 1308.14 (1995).

THERAP CAT: Sedative, hypnotic.

5612. Lornoxicam. *6-Chloro-4-hydroxy-2-methyl-N-2-pyridinyl-2H-thieno[2,3-e]-1,2-thiazine-3-carboxamide 1,1-dioxide;* 6-chloro-4-hydroxy-2-methyl-3-(2-pyridylcarbamo-yl)-2H-thieno[2,3-e]-1,2-thiazine-1,1-dioxide; chlortenoxicam; Ro-13-9297; TS-110. $C_{13}H_{10}ClN_3O_4S_2$; mol wt 371.82. C 41.99%, H 2.71%, Cl 9.53%, N 11.30%, O 17.21%, S 17.25%. Cyclo-oxygenase inhibitor; structurally similar to tenoxicam, *q.v.* Prepn: R. Pfister *et al.*, Ger. pat. **2,838,851**; *eidem*, U.S. pat **4,180,662** (both 1979 to Hoffmann-La Roche). Clinical pharmacokinetics: S. I. Ankier *et al.*, *Postgrad. Med. J.* **64**, 752 (1988). Symposium on pharmacology and clinical experience: *ibid.* **66**, Suppl. 4, S1-S50 (1990). Overview of pharmacology and safety assessment: T. P. Pruss *et al.*, *ibid.* S18.

Orange to yellow crystals, mp 225-230° (dec). pKa_2 4.7. uv max: 371 nm. Partition coefficient (n-octanol/pH 7.4 buffer): 1.8. LD_{50} orally in mice, rats, rabbits, dogs, monkeys: > 10 mg/kg (Pruss).

THERAP CAT: Anti-inflammatory; analgesic.

5613. Losartan. *2-Butyl-4-chloro-1-[[2'-(1H-tetrazol-5-yl)[1,1'-biphenyl]-4-yl]methyl]-1H-imidazole-5-methanol;* 2-n-butyl-4-chloro-5-hydroxymethyl-1-[[2'-(1H-tetrazol-5-yl)-biphenyl-4-yl]methyl]imidazole; 2-butyl-4-chloro-1-[p-(o-1H-tetrazol-5-ylphenyl)benzyl]imidazole-5-methanol. $C_{22}H_{23}ClN_6O$, mol wt 422.92. C 62.48%, H 5.48%, Cl 8.38%, N 19.87%, O 3.78%. Non-peptide angiotensin II receptor antagonist. Prepn: D. J. Carini, J. J. V. Duncia, **Eur. pat. Appl. 253,310**; D. J. Carini *et al*, U.S. pat. **5,138,069** (1988, 1992 both to Du Pont); D. J. Carini *et al.*, *J. Med. Chem.* **34**, 2525 (1991). Angiotensin receptor binding study: A. T. Chiu *et al.*, *J. Pharmacol. Exp. Ther.* **252**, 711 (1990); receptor subtype specificity: A. T. Chiu *et al.*, *Biochem. Biophys. Res. Commun.* **172**, 1195 (1990). Pharmacology of active metabolite, *EXP-3174*: P. C. Wong *et al.*, *J. Pharmacol. Exp. Ther.* **255**, 211 (1990). Comparison of antihypertensive activity with EXP-3174: P. C. Wong *et al.*, *Eur. J. Pharma-*

col. **202**, 323 (1991). Clinical experience in hypertension: M. A. Weber, *Am. J. Hypertension* **5**, 247S, (1992). Review of *in vivo* pharmacology: P. C. Wong *et al.*, *ibid.* **4**, 288S-298S (1991).

Light yellow solid, mp 183.5-184.5°. pKa: 5-6.

Monopotassium salt, $C_{22}H_{22}ClKN_6O$, *Du Pont 753, DUP-753, MK-954, Cozaar.*

Monopotassium salt mixture with hydrochlorthiazide, *Hyzaar.*

THERAP CAT: Antihypertensive.

5614. Loteprednol Etabonate. *(11β,17α)-17-[(Ethoxy-carbonyl)oxy]-11-hydroxy-3-oxoandrosta-1,4-diene-17-carbox-ylic acid chloromethyl ester;* chloromethyl 17α-ethoxycar-bonyloxy-11β-hydroxyandrosta-1,4-diene-3-one-17β-carbox-ylate; 17α-ethoxycarbonyloxy-Δ'-cortienic acid chloromethyl ester; CDDD-5604; HGP-1; P-5604; Lenoxin. $C_{24}H_{31}ClO_7$; mol wt 466.96. C 61.73%, H 6.69%, Cl 7.59%, O 23.98%. Ophthalmic corticosteroid. Prepn: N. S. Bodor, **Belg. pat. 889,563** (1981 to Otsuka); *idem*, U.S. pat. **4,996,-335** (1991). Physicochemical properties: M. Alberth *et al.*, *J. Biopharm. Sci.* **2**, 115 (1991). HPLC determn in plasma and urine: G. Hochhaus *et al.*, *J. Pharm. Sci.* **81**, 1210 (1992). Metabolism and transdermal permeability: N. Bodor *et al.*, *Pharm. Res.* **9**, 1275 (1992). Evaluation of effect on intraocular pressure: J. D. Bartlett *et al.*, *J. Ocul. Pharmacol.* **9**, 157 (1993).

Crystals from THF + hexane, mp 220.5-223.5°. Soly at 25° (mg/ml): 0.0005 in water; 0.037 in 50% propylene gly-col + water. Lipophilicity (log K): 3.04.

THERAP CAT: Anti-inflammatory (topical).

5615. Lotrifen. *2-(4-Chlorophenyl)-(1,2,4)triazolo[5,1-a]isoquinoline;* 2-(p-chlorophenyl)-s-triazolo[5,1-a]isoquino-line; L-12717; DL-717-IT; Canocenta; Privaprol. $C_{16}H_{10}ClN_3$; mol wt 279.73. C 68.70%, H 3.60%, Cl 12.67%, N 15.02%. Non-hormonal antifertility agent. Prepn: **Belg. pat. 815,498**; A. Omodei-Salé *et al.*, U.S. pat. **4,075,341** (1974, 1978 both to Lepetit). Pharmacokinetics: G. Galli-Zani *et al.*, *J. Pharmacobio.-Dyn.* **5**, 55 (1981). Pregnancy-terminating effect in dogs: G. Galliani, A. Omodei-Salé, *J. Small Anim. Pract.* **23**, 295 (1982). Effect on subsequent fertility: G. Galliani *et al.*, *IRCS Med. Sci.* **12**, 433, 435 (1984). *Review:* A. Assandri *et al.*, *Rev. Drug Metab. Drug Interact.* **4**, 237 (1982).

Crystals mp 238-240°.

THERAP CAT (VET): Abortifacient.

5616. Lovastatin. *[1S-[1α(R*),3α,7β,8β(2S*,4S*),-8aβ]]-2-Methylbutanoic acid 1,2,3,7,8,8a-hexahydro-3,7-di-methyl-8-[2-(tetrahydro-4-hydroxy-6-oxo-2H-pyran-2-yl)eth-yl]-1-naphthalenyl ester;* (1S,3R,7S,8S,8aR)-1,2,3,7,8,8a-hexahydro-3,7-dimethyl-8-[2-[(2R,4R)-tetrahydro-4-hydr-oxy-6-oxo-2H-pyran-2-yl]ethyl]-1-naphthalenyl (S)-2-meth-ylbutyrate; 1,2,6,7,8,8a-hexahydro-β,δ-dihydroxy-2,6-di-methyl-8-(2-methyl-1-oxobutoxy)-1-naphthaleneheptanoic acid δ-lactone; 2β,6α-dimethyl-8α-(2-methyl-1-oxobutoxy)-mevinic acid lactone; mevinolin; 6α-methylcompactin;

monacolin K; MK-803; Lovalip; Mevacor; Mevinacor; Mevlor; Sivlor. $C_{24}H_{36}O_5$; mol wt 404.55. C 71.26%, H 8.97%, O 19.77%. Fungal metabolite; potent inhibitor of HMG-CoA reductase, the rate controlling enzyme in cholesterol biosynthesis. Isoln from *Monascus ruber:* A. Endo, *J. Antibiot.* **32**, 852 (1979); from *Aspergillus terreus:* R. L. Monaghan *et al.*, U.S. pat. 4,231,938 (1980 to Merck & Co.). Structure and biochemical properties: A. W. Alberts *et al.*, *Proc. Nat. Acad. Sci. USA* **77**, 3957 (1980). Synthesis: M. Hirama, M. Iwashita, *Tetrahedron Letters* **24**, 1811 (1983); D. L. J. Clive *et al.*, *J. Am. Chem. Soc.* **110**, 6914 (1988). Biosynthesis: M. D. Greenspan, J. B. Yudkovitz, *J. Bacteriol.* **162**, 704 (1985); R. N. Moore *et al.*, *J. Am. Chem. Soc.* **107**, 3694 (1985). HPLC determn in plasma and bile: R. J. Stubbs *et al.*, *J. Chromatog.* **383**, 438 (1986). Stimulation of receptor-mediated clearance of low density lipoproteins: D. W. Bilheimer *et al.*, *Proc. Nat. Acad. Sci. USA* **80**, 4124 (1983). Effects on lipoprotein metabolism: S. M. Grundy, G. L. Vega, *J. Lipid Res.* **26**, 1464 (1985). Multicenter clinical comparison with gemfibrozil, *q.v.:* M. J. Tikkanen *et al.*, *Am. J. Cardiol.* **62**, 35J (1988). Review of syntheses: T. Rosen, C. H. Heathcock, *Tetrahedron* **42**, 4909-4951 (1986). Review of clinical experience: J. A. Tobert, *Circulation* **76**, 534-538 (1987); *idem, Am. J. Cardiol.* **62**, 28J-34J (1988). Comprehensive description: G. S. Brenner *et al.* in *Analytical Profiles of Drug Substances and Excipients* **vol. 21**, H. G. Brittain, Ed. (Academic Press, San Diego, 1992) pp 277-305.

White crystals, mp (under N_2): 174.5°. $[\alpha]_D^{25}$ +323° (c = 0.5 g in 100 ml acetonitrile). uv max: 231, 238, 247 nm (A$^{1\%}$ 532, 621, 418). Soly at room temp (mg/ml): acetone 47, acetonitrile 28, *n*-butanol 7, *i*-butanol 14, chloroform 350, *N,N*-dimethylformamide 90, ethanol 16, methanol 28, *n*-octanol 2, *n*-propanol 11, *i*-propanol 20, water 0.4 × 10⁻³. LD₅₀ orally in mice: > 1000 mg/kg (Endo).
THERAP CAT: Antihypercholesterolemic.

5617. Loxapine. *2-Chloro-11-(4-methyl-1-piperazinyl)-dibenz[b,f][1,4]oxazepine;* oxilapine; CL-62362; S-805; SUM-3170. $C_{18}H_{18}ClN_3O$; mol wt 327.81. C 65.95%, H 5.53%, Cl 10.81%, N 12.82%, O 4.88%. Prepn: Neth. pat. Appl. 6,406,089 corresp to Schmutz *et al.*, U.S. pat. 3,546,226 (1964, 1970 both to Wander); *eidem Helv. Chim. Acta* **50**, 245 (1967); Coppola, U.S. pat. 3,412,193 (1968 to Am. Cyanamid). Crystal structure: D. B. Cosulich, F. M. Lovell, *Acta Crystallogr.* **33B**, 1147 (1977). Pharmacology: Schmutz *et al.*, *Chim. Ther.* **2**, 424 (1967); Latimer, *J. Pharmacol. Exp. Ther.* **166**, 151 (1969). Toxicity data: Stille *et al.*, *Arzneimittel-Forsch.* **15**, 841 (1965). Toxicity studies: Mineshita *et al.*, *Oyo Yakuri* **4**, 293 (1970), *C.A.* **76**, 81145v (1972). Review of pharmacology and therapeutic efficacy: R. C. Heel *et al.*, *Drugs* **15**, 198-217 (1978).

Pale yellowish crystals from petr ether, mp 109-110°. LD₅₀ orally in mice: 65 mg/kg (Stille).
Hydrochloride, $C_{18}H_{18}ClN_3O$·HCl, *Loxitane C.*
Succinate, $C_{18}H_{18}ClN_3O$·$C_4H_6O_4$, *CL-71563, Loxapac, Loxitane.*
THERAP CAT: Anxiolytic.

5618. Loxiglumide. *(±)-4-[(3,4-Dichlorobenzoyl)-amino]-5-[(3-methoxypropyl)pentylamino]-5-oxopentanoic acid;* (±)-4-(3,4-dichlorobenzamido)-N-(3-methoxypropyl)-N-pentylglutaramic acid; CR-1505. $C_{21}H_{30}Cl_2N_2O_5$; mol wt 461.39. C 54.67%, H 6.55%, Cl 15.37%, N 6.07%, O 17.34%. Cholecystokinin A (CCK-A) antagonist. Prepn: F. Makovec *et al.*, PCT Int. pat. Appl. 87 03,869; *eidem*, U.S. pat. 4,769,389 (1987, 1988 both to Rotta). Pharmacology and receptor binding: I. Setnikar *et al.*, *Arzneimittel-Forsch.* **37**, 703 (1987). Pharmacokinetics: *idem et al., ibid.* **38**, 716 (1988). Effect on bilio-pancreatic secretion: W. E. Schmidt *et al.*, *Digestion* **46** Suppl. 2, 232 (1990). Clinical evaluation in irritable bowel syndrome: P. A. Cann *et al.*, *Ann. N.Y. Acad. Sci.* **713**, 449 (1994); in nonulcer dyspepsia: A. S. B. Chua *et al. ibid.* 451.

Crystals from acetone, mp 113-115°. pKa ~5. Soly in water: 0.01%.
THERAP CAT: Gastroprokinetic.

5619. Loxoprofen. *α-Methyl-4-[(2-oxocyclopentyl)methyl]benzeneacetic acid;* (±)-p-[(2-oxocyclopentyl)methyl]hydratropic acid. $C_{15}H_{18}O_3$; mol wt 246.31. C 73.15%, H 7.37%, O 19.49%. Non-steroidal anti-inflammatory prodrug; the active metabolite is the *trans*-cyclohydroxypentane. Prepn of racemate: A. Terada *et al.*, Ger. pat. 2,814,556; *eidem*, U.S. pat. 4,161,538 (1978, 1979 both to Sankyo); of enantiomers: A. Terada, E. Misaka, Eur. pat. Appl. 55,588; *eidem*, U.S. pat. 4,400,534 (1982, 1983 both to Sankyo). Use as topical anti-inflammatory: A. Terada *et al.*, Belg. pat. 889,149 (1981 to Sankyo), *C.A.* **96**, 199326w (1982). Metabolism: Y. Tanaka *et al.*, *Chem. Pharm. Bull.* **31**, 3656 (1983); *eidem, ibid.* **32**, 1040 (1984). Structure of metabolites: S. Naruto *et al., ibid.* 258. Optical inversion of (2R)- to (2S)-isomer after oral administration to rats: H. Nagashima *et al., ibid.* 251. Inhibition of *in vivo* and *in vitro* prostaglandin synthesis: K. Matsuda *et al.*, *Biochem. Pharmacol.* **33**, 2473 (1984).

Colorless oil. bp₀.₃ 190-195°. mp 108.5-111°.
Sodium salt, $C_{15}H_{17}NaO_3$, *CS-600, Loxonin.*
THERAP CAT: Anti-inflammatory; analgesic.

5620. Lucanthone Hydrochloride. *1-[[2-(Diethylamino)ethyl]amino]-4-methyl-9H-thioxanthen-9-one hydrochloride;* 1-(2-diethylaminoethylamino)-4-methylthiaxanthone hydrochloride; MS-752; RP-3735; Miracil D; Nilodin; Miracol; Tixantone. $C_{20}H_{25}ClN_2OS$; mol wt 376.95. C 63.73%, H 6.68%, Cl 9.41%, N 7.43%, O 4.24%, S 8.51%. Prepd by the reaction of 1-chloro-4-methylthiaxanthone with *asym*-diethylethylenediamine: Mauss, *Naturwiss.* **33**, 253 (1946); *idem, Ber.* **81**, 19 (1948); Sharp, *J. Chem. Soc.* **1951**, 2961; Archer, Suter, *J. Am. Chem. Soc.* **74**, 4296 (1952). Review of mode of action: Weinstein, Hirschberg, *Progr. Mol. Subcell. Biol.* **2**, 232 (1971). *Review:* Hirschberg in *Antibiotics* **vol. 3,** J. W. Corcoran, F. E. Hahn, Eds. (Springer-Verlag, New York, 1975) pp 274-303.

Yellow crystals from alcohol, mp 195-196°. Freely sol in water. Aq soln (orange) is neutral. Slightly sol in alcohol. Free base, mp 64-65°. Sol in the usual organic solvents.

THERAP CAT: Anthelmintic (Schistosoma).

5621. Lucensomycin. Antibiotic FI 1163; FI-1163; Etruscomicina; Etruscomycin. $C_{36}H_{53}NO_{13}$; mol wt 707.82. C 61.09%, H 7.55%, N 1.98%, O 29.39%. Polyene antifungal antibiotic isolated from cultures of *Streptomyces lucensis:* Arcamone *et al., Giorn. Microbiol.* **4,** 119 (1957); Arcamone, Perego, *Ann. Chim. (Rome)* **49,** 345 (1959); Marini, Pennella, *Proc. Symp. Antibiotics Prague* (May 1959) p 148; Arcamone *et al.,* U.S. pat. **3,170,837** (1965 to Farmitalia). Structure: Guadiano *et al., Tetrahedron Letters* **1966,** 3559, 3567; *Gazz. Chim. Ital.* **96,** 1470 (1966); *Chim. Ind. (Milan)* **48,** 1327 (1966). Revised structure: R. C. Pandey, K. L. Rinehart, *J. Antibiot.* **29,** 1035 (1976).

Crystalline powder. $[\alpha]_D^{20}$ +296° (pyridine), +50° (methanolic 0.1*N* HCl). uv max: 218, 278, 290, 303, 318 nm ($E_{1cm}^{1\%}$ 300, 370, 780, 1170, 1098). Practically insol in water, anhydr alcohol, non-polar solvents. Sol in pyridine, dimethylformamide. Unstable beyond pH 6-8, and to heat, light, or air. LD_{50} orally in mice: 1263 mg/kg.

THERAP CAT: Antifungal.

5622. Luciferin. A generic term referring to a substrate which, upon oxidation by the enzyme luciferase, produces bioluminescence. Luciferins isolated from different species may vary greatly in structure, although in many cases identical structures have been found in widely diverse animals. The most widely studied luciferins are those isolated from the sea pansey, *Renilla reniformis,* the ostracod, *Cypridina hilgendorfii,* the limpet, *Latia neritoides,* and the firefly, *Photinus pyralis* (see separate entry, Firefly Luciferin). Structure of *Renilla reniformis* luciferin: K. Hori *et al., Proc. Nat. Acad. Sci. USA* **74,** 4285 (1977). *Review:* M. J. Cormier *et al., Fortschr. Chem. Org. Naturst.* **30,** 1-60 (1973).

5623. Lucifer Yellow CH. *6-Amino-2-[(hydrazinocarbonyl)amino]-2,3-dihydro-1,3-dioxo-1H-benz[de]isoquinoline-5,8-disulfonic acid dilithium salt.* $C_{13}H_9Li_2N_5O_9S_2$; mol wt 457.26. C 34.15%, H 1.98%, Li 3.04%, N 15.32%, O 31.49%, S 14.03%. Highly fluorescent dye that reveals functional connection between cells by its movement from cell to cell, termed "dye-coupling": W. W. Stewart, *Cell* **14,** 741 (1978). Prepn: *idem, J. Am. Chem. Soc.* **103,** 7615 (1981). As tracer for electron microscopy: A. R. Maranto, *Science* **217,** 953 (1982). Selective uptake by retinal cells: P. V. Sarthy *et al., J. Comp. Neurol.* **206,** 371 (1982). *Review:* W. W. Stewart, *Nature* **292,** 17-21 (1981).

Fluffy orange hygroscopic powder. Absorption max (water): 280, 428 nm (ε 24200, 11900). Sol in water.

USE: As intracellular marker in biological systems.

5624. Lufenuron. *N-[[[2,5-Dichloro-4-(1,1,2,3,3,3-hexafluoropropoxy)phenyl]amino]carbonyl]-2,6-difluorobenzamide;* 1-[2,5-dichloro-4-(1,1,2,3,3,3-hexafluoropropoxy)-phenyl]-3-(2,6-difluorobenzoyl)urea; fluphenacur; CGA-184699; Program. $C_{17}H_8Cl_2F_8N_2O_3$; mol wt 511.15. C 39.95%, H 1.58%, Cl 13.87%, F 29.73%, N 5.48%, O 9.39%. Chitin synthesis inhibitor. Prepn: J. Drabek, M. Böger, Eur. pat. **Appl. 179,022;** *eidem,* U.S. pat. **4,798,837** (1986, 1989 both to Ciba-Geigy). Effect on fleas following oral administration in cats: W. F. Hink *et al., J. Med. Entomol.* **28,** 424 (1991). HPLC determn: M. Zakson *et al., Pestic. Sci.* **35,** 117 (1992). Review of properties and field trials in crops: F. Buholzer *et al., Med. Fac. Landbouww.* **57,** 781-790 (1992).

Colorless, odorless, crystalline solid, mp 174.1°. Vapor pressure (25°): < 4 × 10⁻⁶ Pa. Soly (20°): water < 0.1 ppm; acetone 40%; cyclohexanone 35%; xylene 5%; methanol 4.5%. Partition coefficient (*n*-octanol/water): 5.12. LD_{50} in rats (mg/kg): > 2000 orally; LC_{50} in rats (mg/m³): > 2350 (Buholzer).

THERAP CAT (VET): Ectoparasiticide.

5625. Lumazine. *2,4(1H,3H)-Pteridinedione.* $C_6H_4N_4O_2$; mol wt 164.12. C 43.91%, H 2.46%, N 34.14%, O 19.50%. Prepn from 4,5-diamino-2,6-dihydoxypyrimidine and glyoxal sodium bisulfate in the presence of ammonia: Weijlard *et al., J. Am. Chem. Soc.* **67,** 804 (1945); U.S. pat. **2,479,442** (1950 to Merck & Co.); from 5,6-diaminouracil hydrosulfite + ethylene glycol disulfonic acid disodium salt + HCl: Dallacker, Steiner, *Ann.* **660,** 98 (1962).

Yellow to orange needles from water, mp 348-349°. Soly in water at 25° = 0.125%. Sol in acetic acid. Aq solns have a bluish-green fluorescence when neutral, a green fluorescence when alkaline, and a blue fluorescence when acid.

5626. Lumichrome. *7,8-Dimethylbenzo[g]pteridine-2,4-(1H,3H)-dione; 7,8-dimethylalloxazine;* 6,7-dimethylalloxazine. $C_{12}H_{10}N_4O_2$; mol wt 242.24. C 59.50%, H 4.16%, N 23.13%, O 13.21%. Irradiation product of riboflavine. Isoln: Karrer *et al., Helv. Chim. Acta* **17,** 1010 (1934). Prepn: Tishler, Wellman, U.S. pat. **2,417,143** (1947 to Merck & Co.); Cresswell, Wood, *J. Chem. Soc.* **1960,** 4768;

Bardos *et al.*, U.S. pat. 3,057,865 (1962 to Armour Pharm. Co.); Seng, Ley, *Angew. Chem.* **84**, 1061 (1972).

Pale yellow crystals from pyridine or pyridine + alcohol, mp > 300°. Sparingly sol in methanol and 90% hot ethanol, water and chloroform. The aqueous, alcoholic and chloroformic solns fluoresce blue. Absorption spectrum: Karrer *et al.*, *loc. cit.*

5627. Lumiflavine. *7,8,10-Trimethylbenzo[g]pteridine-2,4(3H,10H)-dione;* 7,8,10-trimethylisoalloxazine. $C_{13}H_{12}N_4O_2$; mol wt 256.26. C 60.93%, H 4.72%, N 21.86%, O 12.49%. Irradiation product of riboflavine. Isoln: Warburg, Christian, *Biochem. Z.* **266**, 377 (1933). Structure: Kuhn, Rudy, *Ber.* **67**, 1298 (1934). Prepn: Tishler, Wellman, U.S. pat. 2,417,143 (1947 to Merck & Co.); Birch, Moye, *J. Chem. Soc.* **1958**, 2622; Yoneda *et al.*, *Chem. Pharm. Bull.* **20**, 1832 (1972).

Orange crystals from 12% acetic acid, mp 320°. Purified by sublimation in vacuum, mp 330°. Freely sol in chloroform; very sparingly sol in water and most organic solvents. Aq and chloroformic solns fluoresce green. uv max: 269, 355, 445 nm (ε 38,800, 11,700, 11,800) in 0.1*N* NaOH and 264, 373, 440 nm (ε 34,700, 11,400, 10,400) in 0.1*N* HCl.

5628. Luminol. *5-Amino-2,3-dihydro-1,4-phthalazinedione;* o-aminophthalylhydrazide; 3-aminophthalic hydrazide; o-aminophthaloyl hydrazide. $C_8H_7N_3O_2$; mol wt 177.16. C 54.24%, H 3.98%, N 23.72%, O 18.06%. Prepn from 3-nitrophthalic acid: Huntress *et al.*, *J. Am. Chem. Soc.* **56**, 241 (1934); Redemann, Redemann, *Org. Syn.* **29**, 78, 8 (1949); L. Fieser, *Organic Experiments* (Boston, 1964) pp 240-242. Chemiluminescent: Weber, *Ber.* **75**, 565 (1942); Weber *et al.*, *ibid.* **76**, 366 (1943).

Crystals, mp 319-320°. Oxidation of luminol is accompanied by a striking emission of light.
Note: Not to be confused with Luminal.
USE: Detection of copper, iron, peroxides, cyanides.

5629. Lumisterol. *(3β,9β,10α,22E)-Ergosta-5,7,22-trien-3-ol.* $C_{28}H_{44}O$; mol wt 396.66. C 84.79%, H 11.18%, O 4.03%. Differs from ergosterol by spatial arrangement involving the methyl group at C-10. Prepd by ultraviolet irradiation of a benzene-alcohol soln of ergosterol: Askew *et al.*, *Proc. Roy. Soc. (London)* **B109**, 488 (1932). Stereochemistry: Castells *et al.*, *J. Chem. Soc.* **1959**, 1159. Dehydrogenation with selenium gives Diel's hydrocarbon, 3′-methyl-1,2-cyclopentenophenanthrene.

Needles from acetone-methanol, mp 118°. $[\alpha]_D^{19}$ +191.5°, $[\alpha]_{546}^{19}$ +235.4° (c = 2 in acetone). uv max: 265, 280 nm: Windaus *et al.*, *Ann.* **493**, 259 (1932); Heilbron *et al.*, *J. Chem. Soc.* **1937**, 411. Practically insol in water. Sol in organic solvents. Forms a monomolecular compd with calciferol, mp 122°.
Acetate, $C_{30}H_{46}O_2$, mp 100°. $[\alpha]_D^{19}$ +130.5°, $[\alpha]_{546}^{19}$ +163° (c = 1.8 in acetone).

5630. Lunacridine. *(R)-3-(2-Hydroxy-3-methylbutyl)-4,8-dimethoxy-1-methyl-2(1H)-quinolinone.* $C_{17}H_{23}NO_4$; mol wt 305.37. C 66.86%, H 7.59%, N 4.59%, O 20.96%. From bark of *Lunasia costulata* Miq., and *L. quercifolia* K. Schum. et Lauterb., *Rutaceae.* (+)-Form is naturally occurring. Isoln: Boorsma, *Bull. Inst. Bot. Buitenzorg* **21**, 8 (1904). Structure: Goodwin, Horning, *J. Am. Chem. Soc.* **81**, 1908 (1959). Synthesis: Clarke, Grundon, *J. Chem. Soc.* **1964**, 438. Synthesis and stereochemistry: R. M. Bowman *et al.*, *J. Chem. Soc. Perkin Trans. I* **1973**, 1051. Short synthesis of racemate: M. Ramesh, P. Shanmugam, *Ind. J. Chem.* **24D**, 767 (1985).

Crystals from methanol/water, mp 86-87°. $[\alpha]_{589}^{25}$ +28.1°; $[\alpha]_{436}^{25}$ +76.5° (c − 0.935 in ethanol). uv spectra: Goodwin, Horning, *loc. cit.* Very slightly sol in water; freely sol in alcohol, benzene, chloroform, ether, ethyl acetate, carbon disulfide, petr ether.
Hydroperchlorate, $C_{17}H_{24}ClNO_8$, shiny flakes from methanol/ether, mp 146-148°; remelts at 193-195°. $[\alpha]_{589}^{25}$ +22.3°; $[\alpha]_{436}^{25}$ +60.7° (c = 0.750 in ethanol).
(±)-Form, prisms from petr ether, mp 72-74°. uv max (methanol): 239, 258, 285, 292, 333 nm (ε 24000, 26300, 8300, 7800, 3100).

5631. Lunacrine. *(R)-3,9-Dihydro-8-methoxy-9-methyl-2-(1-methylethyl)furo[2,3-b]quinolin-4(2H)-one.* $C_{16}H_{19}NO_3$; mol wt 273.33. C 70.31%, H 7.01%, N 5.12%, O 17.56%. From bark of *Lunasia costulata* Miq., and *L. amara* Blanco, *Rutaceae.* Isoln: Boorsma, *Bull. Inst. Bot. Buitenzorg* **6**, 15 (1900). Structure: Goodwin, Horning, *J. Am. Chem. Soc.* **81**, 1908 (1959). Synthesis: Clarke, Grundon, *J. Chem. Soc.* **1964**, 438. Stereochemistry: Bowman *et al.*, *J. Chem. Soc. Perkin Trans. I* **1973**, 1051.

Fine needles from ethyl acetate, mp 117-119°. $[\alpha]_{589}^{25}$ −50.4°; $[\alpha]_{436}^{25}$ −116° (c = 0.806 in ethanol). Practically insol in water. Freely sol in alcohol, benzene, chloroform, ether, carbon disulfide, ethyl acetate; slightly sol in petr ether.

dl-Form, prisms from ethyl acetate + pentane, mp 146-148°. $[\alpha]_{589}^{25}$ +2.3°; $[\alpha]_{436}^{25}$ +4.2° (c = 0.864 in ethanol).

5632. Lunasine. *(R)-2,3-Dihydro-4,8-dimethoxy-9-methyl-2-(1-methylethyl)furo[2,3-b]quinolinium.* $[C_{17}H_{22}\text{-}NO_3]^+$. From bark of *Lunasia costulata* Miq., and *L. quercifolia* K. Schum. et Lauterb., *Rutaceae*. Isoln: Boorsma, *Bull. Inst. Bot. Buitenzorg* **6**, 14 (1900). Structure: Price, *Aust. J. Chem.* **12**, 458 (1959). Stereochemistry: Bowman *et al., J. Chem. Soc. Perkin Trans I* **1973**, 1051.

(−)-form perchlorate, $C_{17}H_{22}NO_3.ClO_4$, microcrystalline powder from methanol + ether, mp 195-196°. $[\alpha]_D^{20}$ −29.3° (c = 0.46 in methanol).

5633. Lunine. *(S)-7,10-Dihydro-10-methyl-8-(1-methylethyl)-1,3-dioxolo[4,5-h]furo[2,3-b]quinolin-6(8H)-one.* $C_{16}H_{17}NO_4$; mol wt 287.32. C 66.89%, H 5.96%, N 4.88%, O 22.27%. From leaves of *Lunasia costulata* Miq., and *L. amara* Blanco, *Rutaceae*: Boorsma, *Bull. Inst. Bot. Buitenzorg* **21**, 8 (1904); Goodwin *et al., J. Am. Chem. Soc.* **81**, 6209 (1959). Structure: Goodwin *et al., ibid.* **81**, 3065 (1959).

Prisms from ethyl acetate, mp 227-230°; or mixture of prisms and needles from methanol + water, mp 224-226°. $[\alpha]_{589}^{25}$ −38.5°; $[\alpha]_{436}^{25}$ −87.4° (c = 0.926 in chloroform). uv max (ethanol): 222, 247, 314, 325 nm (log ε 4.31, 4.60, 4.02, 4.01). Practically insol in water. Freely sol in alcohol, benzene, chloroform, slightly in ether, ethyl acetate.

5634. Lunularic Acid. *2-Hydroxy-6-[2-(4-hydroxyphenyl)ethyl]benzoic acid; 6-(p-hydroxyphenethyl)salicylic acid.* $C_{15}H_{14}O_4$; mol wt 258.27. C 69.76%, H 5.46%, O 24.78%. Dihydrostilbene growth inhibitor isolated from liverworts and algae. Isoln from *Lunularia cruciata* (L.) Dum., *Lunulariaceae*, and characterization: I. F. M. Valio *et al., Nature* **223**, 1178 (1969); I. F. M. Valio, W. W. Schwab, *J. Exp. Bot.* **21**, 138 (1970); from more than 70 species of liverwort: J. Gorham, *Phytochemistry* **16**, 249 (1977); from cultures of *Marchantia polymorphia*: S. Abe, Y. Ohta, *ibid.* **22**, 1917 (1983). Synthesis: Y. Arai *et al., Tetrahedron Letters* **1972**, 1615; S. Haneck, K. Schreiber, *East Ger.* pat. **126,866** (1977 to Akad. Wissenschaft. D.D.R.), *C.A.* **88**, 152249t (1978). Effect on liverwort growth: J. Gorham, *Phytochemistry* **17**, 99 (1978). Metabolism by *L. cruciata*: R. J. Pryce, *ibid.* **11**, 1355 (1972).

Pale yellow crystals from methanol/water, mp 192°. uv max in neutral ethanol: 280, 287, 308 nm (ε 3300, 3600, 4200); in weakly alkaline ethanol: 300 nm (ε 6600).
USE: Growth inhibitor for lower plants.

5635. Lupanine. *(7α,7aα,14α,14aβ)-Dodecahydro-7,14-methano-2H,11H-dipyrido[1,2-a:1′,2′-e][1,5]diazocin-11-one; dodecahydro-7,14-methano-4H,6H-dipyrido[1,2-a:1′,2′-e][1,-*

5]diazocin-4-one. $C_{15}H_{24}N_2O$; mol wt 248.37. C 72.54%, H 9.74%, N 11.28%, O 6.44%. Racemic lupanine is found in white lupins, *d*-lupanine is found in blue lupins, *l*-lupanine has been prepd from the natural racemic form. Structure: Davis, *Arch. Pharm.* **235**, 199, 218, 229 (1897); Clemo, Leitch, *J. Chem. Soc.* **1928**, 1811; Clemo *et al., ibid.* **1931**, 429; **1933**, 644; Ing, *ibid.* **1933**, 504; Karrer *et al., Helv. Chim. Acta* **11**, 1062 (1928), **13**, 1292 (1930); Winterfeld, Holschneider, *Ber.* **64**, 137, 2415 (1931), **66**, 1751 (1933); *Ann.* **499**, 109 (1932); Winterfeld, Kneuer, *Ber.* **64**, 150 (1931); Winterfeld, Hoffman, *Arch. Pharm.* **275**, 5, 65 (1937). Abs config of *l*-form: Okuda *et al., Chem. & Ind. (London)* **1961**, 1116. Crystal structure of *dl*-form: H. Doucerain *et al., Acta Crystallogr.* **B32**, 3213 (1976). Biosynthetic study of *d*-form: W. M. Golebewski, I. D. Spenser, *J. Am. Chem. Soc.* **106**, 7925 (1984). Further ref. *under* Sparteine.

dl-Form, orthorhombic prisms from acetone, mp 98-99°. $bp_{1.0}$ 185-195°. Sol in water, alcohol, ether, chloroform. Insol in petr ether. *See:* Soldaini, *Arch. Pharm.* **231**, 321 (1893); Couch, *J. Am. Chem. Soc.* **56**, 1423 (1934).
Dihydrochloride, $C_{15}H_{24}N_2O.2HCl$, deliquescent prisms, mp 185° (dec).
Hydrochloride dihydrate, $C_{15}H_{24}N_2O.HCl.2H_2O$, crystals, mp 177-178°. mp 250-252° (dry).
d-Form, *2-oxosparteine.* Syrup crystallizing difficultly in hygroscopic needles, mp 40-44°. bp_3 190-193°. n_D^{24} 1.5444. $[\alpha]_D^{25}$ +84° (c = 4.8 in alc). Freely sol in water, alcohol, chloroform, ether. Sol in petr ether.
Hydrochloride dihydrate, $C_{15}H_{24}N_2O.HCl.2H_2O$, rhombic crystals from water, mp 127° (dry).
l-Form, *hydrorhombinine.* Viscous oil, $bp_{1.0}$ 186-188°. $[\alpha]_D$ about −61° in acetone.
α-Isolupanine. *cis-cis*-Isomer of lupanine: Marion, Leonard, *Can. J. Chem.* **29**, 355 (1951). Isoln of *d*-form from *Lupinus sericeus* Pursh., *L. perennis* L. and *L. augustifolius*, *Leguminosae*: Marion *et al., ibid.* **31**, 181 (1953); Rink, Schäfer, *Arch. Pharm.* **287**, 290 (1954); Winterfeld, Pies, *ibid.* **290**, 537 (1957). Needles from petr ether, mp 75-76°. $[\alpha]_D^{26}$ +39° (c = 0.77). $[\alpha]_D^{26}$ +65.9° (c = 3.4 in abs ethanol).

5636. Lupeol. *(3β)-Lup-20(29)-en-3-ol;* monogynol B; β-viscol; fagarasterol. $C_{30}H_{50}O$; mol wt 426.73. C 84.44%, H 11.81%, O 3.75%. Abundant plant triterpene. Occurs in the skin of lupin seeds, in chicle, in the latex of fig trees and of rubber plants. Was detected in cocoons of *Bombyx mori*. Isoln: Cohen, *Rec. Trav. Chim.* **28**, 368 (1909); Ruzicka *et al., Helv. Chim. Acta* **20**, 1567 (1937). Identity with β-viscol: Meyer, Jeger, *ibid.* **31**, 1868 (1948). Identity with monogynol B: Chatterji *et al., J. Sci. Ind. Res.* **18B**, 262 (1959). Structure: Ruzicka *et al., Helv. Chim. Acta* **28**, 942 (1945); Ames *et al., J. Chem. Soc.* **1951**, 450. Structure of hydrochloride: Halsall *et al., ibid.* **1952**, 2862. Configuration: Barton, Holmes, *ibid.* **1952**, 78; Djerassi *et al., J. Am. Chem. Soc.* **77**, 5330 (1955). NMR studies: Buckley *et al.,*

Chem. & Ind. (London) **1971**, 298. Total synthesis: Stork *et al., J. Am. Chem. Soc.* **93**, 4945 (1971).

Needles from alcohol or acetone, mp 215°. $[\alpha]_D^{20}$ +27.2° (c = 4.8 in chloroform). Freely sol in ether, benzene, petr ether, warm alcohol. Practically insol in water, dil acid and alkalies.

Acetate, $C_{32}H_{52}O_2$, needles from acetone, mp 218°. $[\alpha]_D^{20}$ +47.3° (c = 2 in chloroform).

Benzoate, $C_{37}H_{54}O_2$, prisms from acetone, mp 273-274°. $[\alpha]_D^{20}$ +61° (0.78 g in 25 ml chloroform).

Hydrochloride, $C_{30}H_{50}O$.HCl, needles from ethanol, mp 211-212°. $[\alpha]$ —31° (c = 1.1 in chloroform).

5637. Lupinine. *[1R-trans]-Octahydro-2H-quinolizine-1-methanol.* *l*-lupinine; (—)-lupinine. $C_{10}H_{19}NO$; mol wt 169.27. C 70.96%, H 11.31%, N 8.27%, O 9.45%. Naturally occurring *l*-form isolated from seeds and herb of *Lupinus luteus* L. and other *L.* species, *Leguminosae* also found in *Anabasis aphylla* L., *Chenopodiaceae.* Extraction procedure: J. F. Couch, *J. Am. Chem. Soc.* **56**, 2434 (1934). Structure and synthesis: R. W. Willstätter, E. Fourneau, *Ber.* **35**, 1910 (1902); P. Karrer *et al., Helv. Chim. Acta* **11**, 1062 (1928); K. Winterfeld, F. W. Holschneider, *Ber.* **64B**, 137, 692 (1931); K. Winterfeld, *ibid.* 692. Synthesis of racemic lupinine: F. W. Holschneider, K. Winterfeld, *Arch. Pharm.* **277**, 192 (1939); G. C. Gerrans *et al., Tetrahedron Letters* **1975**, 4171; T. Iwashita *et al., J. Org. Chem.* **47**, 230 (1982). Synthesis of racemic lupinine and *epi-lupinine:* G. R. Clemo *et al., J. Chem. Soc.* **1937**, 965; H. Takahata *et al., Chem. Pharm. Bull.* **34**, 4523 (1986); of *l-epi-lupinine:* M. L. Bremner, S. M. Weinreb, *Tetrahedron Letters* **24**, 261 (1983). Absolute configuration of ( )-form: R. C. Cookson, *Chem. & Ind. (London)* **1953**, 339. Crystal structure: A. Koziol *et al., Acta Crystallogr.* **B34**, 3491 (1978). Bio synthesis: Soucek, Schutte, *Angew. Chem.* **74**, 901 (1962); W. M. Golebiewski, I. D. Spenser, *J. Am. Chem. Soc.* **106**, 1441 (1984).

Stout, orthorhombic prisms from acetone, mp 68.5-69.2°. bp_4 160-164°; bp_{755} 269-270°. $[\alpha]_D^{26}$ —25.9° (c = 3 in water); $[\alpha]_D^{28}$ —21° (c = 9.5 in alcohol). Sol in water, alcohol, chloroform, ether. It is a strong base.

l-Form hydrochloride, $C_{10}H_{20}ClNO$, orthorhombic prisms, mp 208-213°. $[\alpha]_D$ —14°.

dl-Form, crystals from acetone, mp 58.5-59.5°.

5638. Luprostiol. *[1S-[1α(Z),2β(R*),3α,5α]]-7-[2-[[3-(3-Chlorophenoxy)-2-hydroxypropyl]thio]-3,5-dihydroxycyclopentyl]-5-heptenoic acid;* 9α,11α,15-trihydroxy-16-*m*-chlorophenoxy-13-thia-17,18,19,20-tetranor-5-prostenoic acid; prostianol. EMD-34946; Pronilin; Prosolvin; Reprodin. $C_{21}H_{29}ClO_6S$; mol wt 444.98. C 56.68%, H 6.57%, Cl 7.97%, O 21.57%, S 7.21%. Analog of prostaglandin $F_{2\alpha}$, *q.v.* Prepn: J. Kraemer *et al., Ger. pat.* **2,513,371** (1976 to E. Merck), *C.A.* **86**, 55069k (1977); *eidem, U.S. pat.* **4,309,-441** (1982 to E. Merck). Luteolytic study in cattle: H. De Vries, H. Feenstra, *J. Vet. Pharmacol. Ther.* **2**, 223 (1979). Estrus cycle regulation: D. Schams, H. Karg, *Theriogenology* **17**, 499 (1982), *C.A.* **97**, 50430p (1982).

THERAP CAT (VET): Luteolytic.

5639. Lupulin. Glandular trichomes separated from strobiles of *Humulus lupulus* L., *Moraceae* (hops). *Habit.*

Europe, Asia, North America; widely cultivated. *Constit.* Lupamaric acid, humulol, about 3% volatile oil; resin, choline, wax (myricin), tannin, asparagin. Not less than 60% of lupulin is sol in ether.

THERAP CAT: Aromatic bitter, sedative.

THERAP CAT (VET): Formerly in nymphomania.

5640. Lupulon. *3,5-Dihydroxy-2,6,6-tris(3-methyl-2-butenyl)-4-(3-methyl-1-oxobutyl)-2,4-cyclohexadien-1-one; β-bitter acid; β-lupulic acid.* $C_{26}H_{38}O_4$; mol wt 414.59. C 75.32%, H 9.24%, O 15.44%. Antimicrobial constituent of hops. Isoln from commercial hops: Bungener, *Bull Soc. Chim. France* [2] **45**, 487 (1886); Barth, Lintner, *Ber.* **31**, 2022 (1898); Wöllmer, *Ber.* **49**, 780 (1916); Lewis, *et al., J. Clin. Invest.* **28**, 916 (1949). Structure: Wöllmer, *Ber.* **58**, 672 (1925); Wieland, *ibid.* **102**, 2012; Govaert, Verzele, *Bull. Soc. Chim. Belg.* **58**, 432 (1949); Riedl, *Ber.* **85**, 692 (1952).

Prisms from 90% methanol, mp 92-94°. Bitter taste esp in alc soln. Turns yellow and amorphous within a few days with development of an odor. Perfectly stable *in vacuo* even at 60°. Slightly acid reaction. Monobasic acid. Optically inactive. Soluble in methanol, ethanol, petr ether, hexane, isooctane. Slightly sol in neutral or acidic aq solns. Forms a sodium salt which is readily sol in water. The addition of 0.1% ascorbic acid exerts a marked protective action on the bacteriostatic activity of lupulon steamed or autoclaved at a concentration of 4 ppm in phosphate buffers at pH 6.5 and 8.5. LD_{50} orally in rats: 1.8 g/kg.

5641. Luteolin. *2-(3,4-Dihydroxyphenyl)-5,7-dihydroxy-4H-1-benzopyran-4-one; 3′,4′,5,7-tetrahydroxyflavone;* digitoflavone; cyanidenon 1470. $C_{15}H_{10}O_6$; mol wt 286.24. C 62.94%, H 3.52%, O 33.54%. Found in many plants in glycosidic combination, *e.g.,* as the arabinoside: Perkin, *J. Chem. Soc.* **69**, 800 (1896); Fleischer, *Ber.* **32**, 1186 (1899); Perkin, Horsfall, *J. Chem. Soc.* **77**, 1315 (1900); Hayashi, Inoue, *Acta Phytochim. (Japan)* **15**, 53 (1949); *C.A.* **43**, 8450 (1949). Identity with digitoflavone: Kiliani, Mayer, *Ber.* **34**, 3577 (1901). Synthesis: Hutchins, Wheeler, *J. Chem. Soc.* **1939**, 91. See also Bioflavonoids.

Monohydrate, yellow needles from alc, dec 328-330°. Sublimes in high vacuum. Sparingly sol in water. Sol in alkalies forming yellow solns.

5-Glucoside, $C_{21}H_{20}O_{11}$, galuteolin. From seeds of *Galega officinalis* L., *Leguminosae:* Barger, White, *Biochem. J.* **17**, 836 (1923). Structure: Nakamura, Hukuti, *J. Pharm. Soc. Japan* **60**, 449 (1940); *C.A.* **34**, 7910³ (1940); Nordström, Swain, *J. Chem. Soc.* **1953**, 2764. Yellow needles from hot dil alc, dec 280°. Practically insol in water. Slightly sol in abs alcohol; sol in hot dil alcohol.

7-Glucoside, $C_{21}H_{20}O_{11}$, cynaroside. From *Achillea millefolium* L., *Compositae:* Hörhammer *et al., Acta Chim. Acad. Sci. Hung.* **40**, 463 (1964). Yellow needles from alc, mp 254-256°. uv max (CH_3OH): 350, 255 nm (log ε 4.30, 4.27).

5642. Lutetium. Lutecium. Lu; at. wt 174.967; at. no. 71; valence 3. A lanthanide; belongs to the yttrium group of rare earth metals. Naturally occurring isotopes (mass numbers): 175 (97.41%); 176 (2.59%), radioactive, $T_{1/2}$ 3.59 × 10^10 years, β^- emitter. Known artificial radioactive isotopes: 151; 153-174; 177-181. Abundance in earth's crust: 0.75-0.8 ppm. Occurs in xenotime, gadolinite and other rare earth minerals. Discovered independently: Urbain, *Compt. Rend.* **145**, 759 (1907); **146**, 406 (1908); **152**, 141 (1911); and called *cassiopeium*: von Welsbach, *Sitzungsber. Akad. Wiss. Wien* **1907**, 468; *idem, Monatsh.* **29**, 181 (1908); **30**, 695 (1909). Isoln in form of salt: Prandtl, *Z. Anorg. Chem.* **238**, 321 (1938); Marsh, *J. Chem. Soc.* **1952**, 4804. Sepn by ion exchange: Spedding *et al., J. Am. Chem. Soc.* **76**, 2557 (1954). Toxicity study: Haley, *J. Pharm. Sci.* **54**, 663 (1965). Reviews of prepn, properties and compds: *The Rare Earths,* F. H. Spedding, A. H. Daane, Eds. (Krieger, Huntington, N.Y., 1971, reprint of 1961 ed.) 641 pp; Hulet, Bode, "Separation Chemistry of the Lanthanides and Transplutonium Actinides" in *MTP Int. Rev. Sci: Inorg. Chem., Ser. One* vol. 7, K. W. Bagnall, Ed. (University Park Press, Baltimore, 1972) pp 1-45; Moeller, "The Lanthanides" in *Comprehensive Inorganic Chemistry* vol. 4, J. C. Bailar, Jr. *et al.,* Eds. (Pergamon Press, Oxford, 1973) pp 1-101; F. H. Spedding in Kirk Othmer *Encyclopedia of Chemical Technology* vol. **19** (John Wiley & Sons, New York, 3rd ed., 1982) pp 833-854; *Chemistry of the Elements,* N. N. Greenwood, A. Earnshaw, Eds. (Pergamon Press, New York, 1984) pp 1423-1449. Brief review of properties: G. T. Seaborg, *Radiochim. Acta* **61**, 115-122 (1993).

Silvery-white metal. Hexagonal close-packed crystals, d 9.8404. mp 1663°. bp 3402°. Heat of fusion: 18.65 kJ/mol. Heat of sublimation (25°): 427.6 kJ/mol.

Oxide, Lu_2O_3, cubic crystals.

Chloride, $LuCl_3$, colorless crystals. Sublimes above 750°. mp 892° ±2°; sol in water. LD_{50} in mice: 315 mg/kg i.p.; 7.1 g/kg orally (Haley).

Sulfate, $Lu_2(SO_4)_3$, octahydrate, soly in water (g/100 g): 42.27 (20°); 16.93 (40°).

5643. 2,6-Lutidine. *2,6-Dimethylpyridine; α,α'-lutidin.* C_7H_9N; mol wt 107.16. C 78.46%, H 8.47%, N 13.07%. Isolated from the basic fraction of coal tar: Heap *et al., J. Am. Chem. Soc.* **43**, 1936 (1921); from a bone oil fraction: Ladenburg, Roth, *Ber.* **18**, 51 (1885). Synthesis from ethyl acetoacetate, formaldehyde, and ammonia: Singer, McElvain, *Org. Syn. coll.* **vol. II**, 214 (1943).

Oily liquid. Odor of pyridine + peppermint. d_4^{20} 0.9252. mp −5.8°. bp_{760} 144°; bp_7 79°. n_D^{20} 1.49797. Solubility in water at 45.3° = 27.2% (w/w); at 48.1° = 18.1%; at 57.5° = 12.1%; at 74.5° = 9.5%. Also sol in alcohol, ether. Miscible with dimethylformamide and tetrahydrofuran.

5644. Lututrin. Lutrexin. A uterus-relaxing factor obtained from the corpus luteum of sow ovaries and standardized for potency in terms of units of activity on the guinea pig uterus. *Ref: J. Am. Med. Assoc.* **156**, 1252 (1954).

THERAP CAT: Uterine relaxant.

5645. Lyapolate Sodium. *Ethenesulfonic acid homopolymer sodium salt;* sodium lyapolate; sodium apolate; sodium polyethylene sulfonate; polyethylene sodium sulfonate; PES; Peson. $[CH_2CH(SO_3Na)]_n$.

THERAP CAT: Anticoagulant.

5646. Lycoctonine. *(1α,6β,14α,16β)-20-Ethyl-4-(hydroxymethyl)-1,6,14,16-tetramethoxyaconitane-7,8-diol;* royline. $C_{25}H_{41}NO_7$; mol wt 467.60. C 64.22%, H 8.84%, N 3.00%, O 23.95%. Isomeric with delsoline. Originally isolated from *Aconitum lycoctonum* L., *Ranunculaceae:* Hubschmann, *Schweiz. Wochschr. Pharm.* **3**, 269 (1865). Widely distributed in *Aconitum* and *Delphinium* spp., *Ranunculaceae.* Structure: Edwards *et al., Can. J. Chem.* **34**, 1315 (1956); Anet *et al., ibid.* **35**, 400 (1957). Identity with royline: Edwards, Rodger, *ibid.* **37**, 1187 (1959). Stereochemistry: Przybylska, Marion, *ibid.* 1843. Revised con-

figuration: S. W. Pelletier *et al., J. Am. Chem. Soc.* **103**, 6536 (1981).

Crystals. Bitter taste. mp 143°. $[α]_D^{20}$ +53° (ethanol). Slightly sol in water, ether, benzene; freely sol in alcohol, chloroform.

Hydrochloride, mp 152°.

N-Acetylanthranilic acid ester, see Ajacine.

N-Succinylanthranilic acid ester, $C_{36}H_{48}N_2O_{10}$, *lycaconitine.* Isoln: Schulze, Bierling, *Arch. Pharm.* **251**, 8 (1913). White, amorphous powder, mp 111-114°. $[α]_D$ +42°; also reported as +31.5°. Insol in water. Sol in alcohol, chloroform, ether.

5647. Lycodine. [4a*R*-(4aα,5α,10bα,12*R**)]-2,3,4,4a,-5,6-Hexahydro-12-methyl-1*H*-5,10b-propano-1,7-phenanthroline. $C_{16}H_{22}N_2$; mol wt 242.36. C 79.29%, H 9.15%, N 11.56%. Isoln from *Lycopodium annotinum* L., *Lycopodiaceae:* F. A. L. Anet, C. R. Eves, *Can. J. Chem.* **36**, 902 (1958). Structure: W. A. Ayer, G. G. Iverach, *ibid.* **38**, 1823 (1960); F. A. L. Anet, M. V. Rao, *Tetrahedron Letters* **1960**, 9. Synthesis of (±)-form: E. Kleinman, C. H. Heathcock, *ibid.* **1979**, 4125; C. H. Heathcock *et al., J. Am. Chem. Soc.* **104**, 1054 (1982).

Cryst powder, mp 118-119°. $[α]_D$ −10° (c = 1.01 in ethanol). pKa's 3.97, 8.08.

5648. Lycofawcine. *(5β,8R,15S)-15-Methyllycopodane-5,8,12-triol 5-acetate;* base L. $C_{18}H_{29}NO_4$; mol wt 323.43. C 66.84%, H 9.04%, N 4.33%, O 19.79%. From *Lycopodium fawcettii* Lloyd and Underwood, *Lycopodiaceae:* Burnell *et al., Can. J. Chem.* **41**, 3091 (1963). Structure: Ayer *et al., ibid.* **43**, 328 (1965).

Liquid. pKa 9.7.

Perchlorate, crystals, dec 290-294°.

Methiodide, $C_{19}H_{32}INO_4$, crystals from acetone, mp 281-282°.

5649. Lycomarasmine. *N-[2-[(2-Amino-2-oxoethyl)-amino]-2-carboxyethyl]-L-aspartic acid;* Welkstoff. $C_9H_{15}N_3O_7$; mol wt 277.23. C 38.99%, H 5.45%, N 15.16%, O 40.40%. Antibiotic peptide (tomato-wilting agent) produced by the fungus *Fusarium lycopersici.* Isoln: Plattner, Clauson-Kaas, *Helv. Chim. Acta* **28**, 188 (1945). Structure: *eidem, Experientia* **1**, 195 (1945); Hardegger *et al., Helv. Chim. Acta* **46**, 60 (1963).

Crystals, dec 227-229°. $[\alpha]_D^{20}$ −42° to −48° (aq soln at pH 7). Acid reaction. Sparingly sol in water; freely sol in dil acids or alkalies.

5650. Lycopene. ψ,ψ-*Carotene; (all-trans)-lycopene.* $C_{40}H_{56}$; mol wt 536.88. C 89.49%, H 10.51%. Carotenoid occurring in ripe fruit, especially in tomatoes. One kg of fresh ripe tomatoes yields 0.02 g lycopene. Isoln procedure: Willstätter, Escher, *Z. Physiol. Chem.* **64**, 47 (1910). Chromatographic sepn from other carotenoids: Winterstein, *ibid.* **215**, 52 (1933); Winterstein, Stein, *ibid.* **220**, 250 (1933). Structure: Willstätter, Escher, *loc. cit.*; Karrer and collaborators: *Helv. Chim. Acta* **11**, 751, 1201 (1928); **12**, 285 (1929); **13**, 1084 (1930); **14**, 435 (1931); Kuhn, Grundmann, *Ber.* **65**, 898, 1880 (1932). Synthesis: Karrer *et al., Helv. Chim. Acta* **33**, 1349 (1950); Isler *et al., ibid.* **39**, 463 (1956). Commercial prepn: Kabbe *et al.,* **Ger. pat.** **1,168,890** (1964 to Bayer).

Long, deep red needles from carbon disulfide + ethanol, from methylene chloride + methanol, mp 172-173°. Absorption max (*trans*-form): 446, 472, 505 nm ($E_{1cm}^{1\%}$ 2250, 3450, 3150). 15,15′-*cis*-Form: mp ~105° then solidifies again; absorption max: 361, 444, 470, 502 nm ($E_{1cm}^{1\%}$ 1110, 1280, 1660, 1280). One gram dissolves in 50 ml carbon disulfide, in 3 l boiling ether, in 12 l boiling petr ether, in 14 l hexane at 0°. Sol in chloroform, benzene. Almost insol in methanol, ethanol.

5651. Lycophyll. ψ,ψ-*Carotene-16,16′-diol; (all-trans)-lycopene-16,16′-diol.* $C_{40}H_{56}O_2$; mol wt 568.88. C 84.45%, H 9.92%, O 5.62%. Carotenoid pigment, isoln from *Solanum dulcamara* L. and *Lycopersicum esculentum* Mill., Solanaceae and structure: Zechmeister, Cholnoky, *Ber.* **69**, 422 (1936). Revised structure: Cholnoky, Szabolcs, *Tetrahedron Letters* **1968**, 1931. Stereochemistry: Kelly *et al., Acta Chem. Scand.* **25**, 1607 (1971). Synthesis: Kjosen, Liaaenjensen, *ibid.* 1500.

Purple leaflets from benzene + methanol, needles from benzene + petr ether, mp 179°. Absorption max (benzene): 521, 487, 456 nm. Freely sol in carbon disulfide, less in benzene, ethanol, very slightly sol in petr ether. Dipalmitate, purple needles from benzene + methanol, mp 76°.

5652. Lycopodine. *(15R)-15-Methyllycopodan-5-one.* $C_{16}H_{25}NO$; mol wt 247.38. C 77.68%, H 10.19%, N 5.66%, O 6.47%. From herb of *Lycopodium complanatum* L., Lycopodiaceae: Bödeker, *Ann.* **208**, 363 (1881). Structure: Harrison *et al., Can. J. Chem.* **39**, 2086 (1961). Stereochemistry: Anet, *Tetrahedron Letters* **1960**, no. 20, 13. Synthesis of (±)-form: G. Stork *et al., J. Am. Chem. Soc.* **90**, 1647 (1968); W. A. Ayer *et al., ibid.* 1648; C. H. Heathcock *et al., ibid.* **100**, 8036 (1978); **104**, 1054 (1982); D. Schu-

mann *et al., Ann.* **1982**, 1700. Review of synthetic studies: Stork, *Pure Appl. Chem.* **17**, 383 (1968).

Bitter prisms, mp 114-115°; mp (racemate) 130-131°. $[\alpha]_D$ −24° (alc). Sol in water, alc, benzene, chloroform, ether.

5653. Lycopodium. Club-moss spores; Lycopodium seed (spores); vegetable sulfur. Spores of *Lycopodium clavatum* L., Lycopodiaceae. The spores of *Lycopodium annotinum* L., and of *L. anceps* Wall., Lycopodiaceae can also be used. The spores contain a substance called *selagnine*. *Habit.* North America, Europe, Asia; cultivated in Russia.

The spores are a fine yellowish powder which is highly flammable. Odorless and tasteless. Unctuous to the touch and easily sticking to the fingers. Lycopodium powder is very mobile and when poured on a flat surface should form an even layer, without visible lumps or dimples. When observed in chloral hydrate soln it is seen that the powder consists of unicellular lycopod spores, about 30 μ in diameter, in the shape of triangular pyramids with a convex base and rounded angles; a three-radical suture runs from the top of the pyramid along its facets. After warming and crushing the spores between glass slides, they burst along the suture and yield drops of oil, assuming a red color with alkalies. Adulteration usually consists of the admixture of flour (detected microscopically with iodine soln which stains the starch grains of the flour violet). Other admixtures may be pine pollen and sawdust. When mounted with chloral hydrate, pine pollen is larger than lycopodium, it is oval and has two lateral flying sacs, filled with air, and appearing black at the beginning of the observation. Sawdust is easily detected by the phloroglucinol test.

USE: As covering for pills or suppositories; in explosives, pyrotechnics, flashlight powders; as "dry parting compound" in foundry work for ornamental and nameplate castings. Review: Appel, *Scientific Monthly* **78**, 268 (1954).

THERAP CAT: Adsorbent.

5654. Lycopus. Bugleweed; sweet bugle; water bugle. Whole plant of *Lycopus virginicus* L., Labiatae. *Habit.* N. America. *Constit.* Volatile oil, resin, tannin, glucoside.

5655. Lycoramine. *1,2-Dihydrogalanthamine; 3,4-dihydrogalanthamine.* $C_{17}H_{23}NO_3$; mol wt 289.37. C 70.56%, H 8.01%, N 4.84%, O 16.59%. One of the minor alkaloids of *Lycoris radiata* Herb., Amaryllidaceae: Kondo, Ishiwata, *Ber.* **70B**, 2427 (1937). From *Narcissus* spp.: Boit *et al., ibid.* **90**, 725 2197 (1957). Tentative structure: Kobayashi, Uyeo, *Chem. & Ind. (London)* **1956**, 177. Structure elucidation: William, Rogers, *Proc. Chem. Soc.* **1964**, 357. Synthesis of the naturally occurring (−)-form: Barton, Kirby, *J. Chem. Soc.* **1962**, 806; of the (±)-form: N. Hazama *et al., J. Chem. Soc. (C)* **1968**, 2947; Y. Misaka *et al., ibid.* **2954**; A. G. Schultz *et al., J. Am. Chem. Soc.* **99**, 8065 (1977); S. F. Martin, P. J. Garrison, *J. Org. Chem.* **46**, 3567 (1981); **47**, 1513 (1982).

Plates from acetone, mp 121°. $[\alpha]_D^{27}$ −98° (alcohol). Freely sol in water; sol in alc, acetone. Its salts hydrolyze easily.

5656. Lycorine. *(1α,2β)-3,12-Didehydro-9,10-[methyl-enebis(oxy)]galanthan-1,2-diol; 3,3a-didehydrolycoran-1α,2β-diol;* 2,4,5,7,12b,12c-hexahydro-1*H*-[1,3]dioxolo[4,5-*j*]pyr-rolo[3,2,1-*de*]phenanthridine-1,2-diol; amarylline; belama-rine; narcissine; galanthidine. $C_{16}H_{17}NO_4$; mol wt 287.32. C 66.89%, H 5.96%, N 4.88%, O 22.27%. Alkaloid isolated from the bulbs of *Lycoris radiata* L., *Narcissus pseudonar-cissus* L., *N. tazetta* L., from *Buphane disticha* Herb., in *Crinum* spp., *Amaryllis belladonna* L., *Clivia miniata* Regel and other *Amaryllidaceae*. Extraction procedure: Cook, Loudon in Manske-Holmes, *Alkaloids* **vol. II** (Academic Press, 1952) p 336. Identity of lycorine and galanthidine: Proskurnina, *Doklady Akad. Nauk SSSR* **90,** 565 (1953), *C.A.* **49,** 12500c (1955). Structure: Takeda *et al., J. Am. Chem. Soc.* **80,** 2562 (1958). Stereochemistry: Nakagawa, Uyeo, *J. Chem. Soc.* **1959,** 3736; K. Kotera *et al., Tetrahed-ron Letters* **1966,** 2009. Crystal structure: R. Roques, *Acta Crystallogr.* **30B,** 296 (1974). Attempted synthesis: Dyke *et al., Tetrahedron* **29,** 213 (1973). Synthesis: Y. Tsuda *et al., Chem. Commun.* **1975,** 933; *eidem, J. Chem. Soc. Perkin Trans. I,* **1979,** 1358; T. Sano *et al., Heterocycles* **14,** 1097 (1980); S. F. Martin, C. Tu, *J. Org. Chem.* **46,** 3763 (1981). Biosynthesis: Archer *et al., Proc. Chem. Soc.* **1963,** 168; Fugati, Mazza, *Chem. Commun.* **1972,** 936. *Review:* W. C. Wildman in *The Alkaloids,* R. H. F. Manske, Ed., vol. XI (Academic Press, New York, 1968) pp 307-400.

Stout prisms from alcohol, mp 275-280° (dec). $[\alpha]_D^{16}$ −129° (c = 0.16 in 98% alc). Alkaline reaction to litmus; salts hydrolyze easily. Sparingly sol in alcohol, chloroform, petr ether. Sol in dilute acids. Practically insol in water, alkalies.

Hydrochloride, $C_{16}H_{17}NO_4 \cdot HCl$, long needles from water, mp 217° (dec with slight preliminary sintering). $[\alpha]_D^{20}$ +43°.

Hydrochloride monohydrate, elongated prisms from water, mp 206°. Soluble in 20 parts water.

Methiodide, $C_{16}H_{17}NO_4 \cdot CH_3I$, two forms exist which may be stereoisomeric about the nitrogen atom. α- Form: poly-hedra from alc, dec 247°. $[\alpha]_D^{20}$ −46° (c = 1.52). Freely sol in water; sparingly sol in hot alcohol. β-Form: prisms contg 1H_2O from water, dec 198°, after recryst from alco-hol, dec 281°. $[\alpha]_D^{20}$ +123° (c = 2.44). Freely sol in water and hot alcohol.

5657. Lycoxanthin. *ψ,ψ-Carotene-16-ol; (all-trans)-ly-copen-16-ol.* $C_{40}H_{56}O$; mol wt 552.88. C 86.90%, H 10.21%, O 2.89%. Carotenoid isolated from *Solanum dulcamara* L., *Lycopersicum esculentum* Mill., *Solanaceae; Tamus commu-nis* L., *Dioscoreaceae.* Isoln by chromatography and struc-ture: Zechmeister, Cholnoky, *Ber.* **69,** 422 (1936). Struc-ture: Karrer *et al., Helv. Chim. Acta* **13,** 268, 1084 (1930); **14,** 614, 843 (1931); Winterstein, *Angew. Chem.* **72,** 902 (1960). Revised structure: Cholnoky, Szabolcs, *Tetrahed-ron Letters* **1968,** 1931. Stereochemistry: Kelly *et al., Acta Chem. Scand.* **25,** 1607 (1971). Total synthesis: Kjosen, Liaaen-jensen, *ibid.* 1500.

Purple needles from carbon disulfide. Reddish-brown round or acicular cryst aggregates from benzene + petr ether, mp 168°. Absorption max (acetone): 448, 474 ($E_{1cm}^{1\%}$

3080), 505 nm. Sol in carbon disulfide and benzene. Mod-erately sol in petr ether. Sparingly sol in alc.

Monoacetate, deep purple needles from benzene+ meth-anol, mp 137°. Freely sol in carbon disulfide; sparingly sol in alcohol, petr ether.

5658. Lymecycline. *[4S-(4α,4aα,5aα,6β,12aα)]-N⁶-[[[[4-(Dimethylamino)-1,4,4a,5,5a,6,11,12a-octahydro-3,6,-10,12,12a-pentahydroxy-6-methyl-1,11-dioxo-2-naphthacen-yl]carbonyl]amino]methyl]-L-lysine;* tetracyclinemethylene lysine; *N*-lysinomethyltetracycline; Armyl; Ciclolysal; Mucomycin; Tetralisal; Tetramyl; Tetralysal. $C_{29}H_{38}N_4O_{10}$; mol wt 602.64. C 57.80%, H 6.36%, N 9.30%, O 26.55%. Semi-synthetic antibiotic related to tetracycline, *q.v.* Prepn: Blackwood *et al.,* U.S. pat. **3,042,716** (1962 to Pfizer); Lau-ria, Logemann, **Ger.** pat. **1,134,071** (1962 to Carlo Erba), *C.A.* **58,** 492 (1963); Tubaro, Raffaldoni, *Boll. Chim. Farm.* **100,** 9 (1961).

Sodium salt, $C_{29}H_{37}N_4NaO_{10}$. uv max (CH_3OH): 376 nm.

THERAP CAT: Antibacterial.

5659. Lynestrenol. *(17α)-19-Norpregn-4-en-20-yn-17-ol;* 17α-ethinyl-17β-hydroxyestr-4-ene; 17α-ethynylestr-4-en-17β-ol; 3-desoxynorlutin; ethinylestrenol; Exluton(a); Exlu-tena; Orgametril; Orgametil. $C_{20}H_{28}O$; mol wt 284.44. C 84.45%, H 9.92%, O 5.62%. Prepn: de Winter *et al., Chem. & Ind. (London)* **1959,** 905. Metabolism: Kamyab *et al., Biochem. J.* **103,** 14P (1967); *J. Endocrinol.* **42,** 337 (1968). Pharmacokinetics and metabolism: H. Kuhl *et al., Contra-ception* **26,** 303 (1982). Clinical trial of use in normophasic oral contraceptives: N. Dombrowicz *et al., ibid.* **22,** 537 (1980). Review of carcinogenicity studies: *IARC Mono-graphs* **21,** 407-415 (1979).

Solid, mp 158-160°. $[\alpha]_D$ −13° (chloroform).

Mixture with ethinyl estradiol, Anacyclin, Fysionorm, Minilyn, Noracyclin, Ovanon, Ovoresta, Yermonil.

Mixture with mestranol, Lyndiol.

THERAP CAT: Progestogen. In combination with estrogen as oral contraceptive.

5660. Lyovac® Antivenin. Black widow spider anti-venin. Prepd from the serum of horses immunized against the venom of the New World spider *Latrodectus mactans* Fabr., *Araneae.*

One unit of antivenin neutralizes one average mouse lethal dose of black widow spider venom when the antivenin and the venom are injected simultaneously in mice under suita-ble conditions.

THERAP CAT: Antivenin.

5661. Lypressin. *8-L-Lysinevasopressin;* Schweine-Vasopressin; Phe³-Lys⁸-oxytocin; Diapid; Postacton; Synto-pressin. $C_{46}H_{65}N_{13}O_{12}S_2$; mol wt 1056.23. C 52.31%, H 6.20%, N 17.24%, O 18.18%, S 6.07%. Purification after isolation from hog pituitary glands: Ward, du Vigneaud, *J. Biol. Chem.* **222,** 951 (1956). Structure: du Vigneaud *et al., J. Am. Chem. Soc.* **75,** 4880 (1953); Bartlett *et al., ibid.* **78,** 2905 (1956). Synthesis: du Vigneaud *et al., loc. cit.;* Bartlett *et al., loc. cit.;* Bodanszky *et al., ibid.* **82,** 3195 (1960); Bois-sonnas, Huguenin, *Helv. Chim. Acta* **43,** 182 (1960); Zaoral,

Coll. Czech. Chem. Commun. **30**, 1853 (1965); Meienhofer, Sano, *J. Am. Chem. Soc.* **90**, 2996 (1968); Meienhofer, Trzeciak, *Proc. Nat. Acad. Sci. USA* **68**, 1006 (1971). Configuration: Schally, Barret, *J. Am. Chem. Soc.* **87**, 2497 (1965). *Review:* E. Schröder, K. Lübke, *The Peptides* vol. **II** (Academic Press, New York, 1966) pp 336-350.

Cys–Tyr–Phe–Gln–Asn–Cys–Pro–Lys–GlyNH₂

THERAP CAT: Antidiuretic; vasopressor.

5662. Lysalbinic Acid. A colloidal substance prepd by the action of caustic alkalies on albumin, usually egg albumin, athough it also may be obtained from serum albumin or from casein (caseolysalbinic acid): Paal, *Ber.* **35**, 2195 (1902); **Ger.** pats. **129,031** (1902) and **132,322** (1902). Composition: C 52.9%, H 7.0%, N 14.9%, S 1.2%.

Usually isolated as the Na salt, a brittle yellowish mass, readily sol in water, forming a colloidal soln which foams. It is precipitated by alc, and it may be purified by dialysis.

USE: As a protective colloid for metal sols, especially gold, silver, and mercury sols. "Colloidal gold" is prepd by adding a soln of gold chloride to a soln of lysalbinic acid. The intensely red hydrosol is precipitated with alcohol and dried. The precipitate forms hard, bronze-colored granules which may be stored for a long time, yielding a red, colloidal gold soln upon addn of water. Analysis of dried granules shows about 30% Au. *See* Paal, *Ber.* **35**, 2236-2244 (1902), and **Ger.** pats. **170,433** and **180,730** (1906 and 1907 to Kalle).

5663. Lysergamide. *9,10-Didehydro-6-methylergoline-8β-carboxamide;* lysergic acid amide; ergine. $C_{16}H_{17}N_3O$; mol wt 267.33. C 71.89%, H 6.41%, N 15.72%, O 5.98%. Isoln from *Rivea corymbosa* (L.) and from *Ipomoea tricolor* Cav., *Convolvulaceae:* Hofmann, Tscherter, *Experientia* **16**, 414 (1964). Prepn from lysergic acid hydrazide: Ainsworth, **U.S.** pat. **2,756,235** (1956 to Lilly); from lysergic acid and phosgene-dimethylformamide complex: Patelli, Bernardi, **U.S.** pat. **3,141,887** (1964 to Farmitalia). Microbiological production: Rutschmann, Kobel, **U.S.** pat. **3,219,545** (1965 to Sandoz).

Prisms from methanol, dec 242°. $[\alpha]^{20}_{5461}$ +15° (c = 0.5 in pyridine).

Methanesulfonate, $C_{16}H_{17}N_3O.CH_3SO_3H$, prisms from methanol + acetone, dec 232°.

Note: This is a controlled substance (depressant) listed in the U.S. Code of Federal Regulations, Title 21 Part 1308.13 (1995).

5664. Lysergic Acid. *9,10-Didehydro-6-methylergoline-8-carboxylic acid.* $C_{16}H_{16}N_2O_2$; mol wt 268.32. C 71.62%, H 6.01%, N 10.44%, O 11.93%. Lysergic acid and isolysergic acid are the main cleavage products formed on alkaline hydrolysis of the alkaloids which are characteristic of ergot. Jacobs, Craig *et al., J. Biol. Chem.* **104**, 547 (1934); **125**, 289 (1938); **130**, 399 (1939); **145**, 487 (1942); *J. Org. Chem.* **10**, 76 (1945). High-yield production by *Claviceps paspali:* Arcamone *et al., Proc. Roy. Soc. (London), Ser. B,* **155**, 26 (1961). Total synthesis: Kornfeld *et al., J. Am. Chem. Soc.* **76**, 5256 (1954); **78**, 3087 (1956); M. Julia *et al., Tetrahedron Letters* **1969**, 1569; V. W. Armstrong *et al., ibid.* **1976**, 4311; W. Oppolzer *et al., Helv. Chim. Acta* **64**, 478 (1981); R. Ramage *et al., Tetrahedron* **37**, Suppl. 9, 157 (1981); J. Rebek, D. F. Tai, *Tetrahedron Letters* **24**, 859 (1983). Stereochemistry: Stoll *et al., Helv. Chim. Acta* **37**, 2039 (1954); Stenlake, *J. Chem. Soc.* **1955**, 1626; Leeman, Fabbri, *Helv.*

Chim. Acta **42**, 2696 (1959). Absolute configuration: Stadler, Hofmann, *ibid.* **45**, 2005 (1962).

Hexagonal scales, plates with one or two moles H_2O from water, mp 240° (dec). $[\alpha]^{20}_D$ +40° (c = 0.5 in pyridine). Behaves as an acid and base, pKa 3.44, pKb 7.68. Moderately sol in pyridine. Sparingly sol in water and in neutral organic solvents; sol in NaOH, NH_4OH, Na_2CO_3, and HCl solns. Slightly sol in dil H_2SO_4.

Methyl ester, thin leaflets from benzene, mp 168°.

Note: This is a controlled substance (depressant) listed in the U.S. Code of Federal Regulations, Title 21 Part 1308.13 (1995).

5665. Lysergide. *9,10-Didehydro-N,N-diethyl-6-methylergoline-8β-carboxamide; N,N-diethyl-D-lysergamide;* D-lysergic acid diethylamide; LSD; LSD-25; lysergsäure diethylamid. $C_{20}H_{25}N_3O$; mol wt 323.44. C 74.27%, H 7.79%, N 12.99%, O 4.95%. Microbial formation by *Claviceps paspali* over the hydroxyethylamide: Arcamone *et al., Proc. Roy. Soc. (London)* **155B**, 26 (1961). Partial synthesis: Stoll, Hofmann, *Helv. Chim. Acta* **26**, 944 (1943); **38**, 421 (1955). Industrial prepn: Pioch; Garbrecht, **U.S.** pats. **2,736,728**; **2,774,763** (both 1956 to Lilly); Patelli, Bernardi, **U.S.** pat. **3,141,887** (1964 to Farmitalia). Isotope-labeled LSD: Stoll *et al., Helv. Chim. Acta* **37**, 820 (1954). Toxicity data: E. Rothlin, *Ann. N.Y. Acad. Sci.* **66**, 668 (1957). *Review:* Hofer, *Clin. Pharmacol. Ther.* **6**, 183 (1965). Book: *The Use of LSD in Psychotherapy and Alcoholism*, H. A. Abramson, Ed. (Bobbs-Merrill, Indianapolis, 1967) 697 pp.

Pointed prisms from benzene, mp 80-85°. $[\alpha]^{20}_D$ +17° (c = 0.5 in pyridine). uv max (ethanol): 311 nm ($E^{1\%}_{1cm}$ 257). LD_{50} in mice, rats, rabbits (mg/kg): 46, 16.5, 0.3 i.v. (Rothlin).

D-Tartrate, $C_{46}H_{64}N_6O_{10}$, solvated, elongated prisms from methanol, mp 198-200°. $[\alpha]^{20}_D$ +30°. Soluble in water.

Note: This is a controlled substance (hallucinogen) listed in the U.S. Code of Federal Regulations, Title 21 Part 1308.11 (1995).

USE: In biochemical research as antagonist to serotonin. Has been used experimentally as adjunct in study and treatment of mental disorders.

5666. Lysidine. *4,5-Dihydro-2-methyl-1H-imidazole; 2-methyl-2-imidazoline;* methylglyoxalidine. $C_4H_8N_2$; mol wt 84.12. C 57.11%, H 9.59%, N 33.30%. Prepn: Chitwood, Reid, *J. Am. Chem. Soc.* **57**, 2424 (1935); King, McMillan, *ibid.* **68**, 1774 (1946); Kyrides, **U.S.** pats. **2,392,326** and **2,404,299** (both 1946 to Monsanto); Ahrens, **U.S.** pat. **2,813,862** (1957 to Organon).

Needles from benzene, mp 105°. bp 198-200°. Soluble in water, alc, chloroform; less sol in benzene, carbon tetrachloride, petr ether; practically insol in ether. *Keep well closed. Incompat:* Acids, metallic salts, alkaloids.

5667. Lysine. L-*Lysine;* Lys; K; (S)-2,6-diaminohexanoic acid; α,ε-diaminocaproic acid. $C_6H_{14}N_2O_2$; mol wt 146.19. C 49.30%, H 9.65%, N 19.16%, O 21.89%. An essential amino acid for human development; probably the most limited in the food chain. Identified in a hydrolysate of casein by Drechsel in 1889. Named "lysatine" from the Greek for "loosing", as it produced urea on treatment with barium hydroxide; changed to "lysine" in 1891. Verification of structure: E. Fischer, F. Weigert, *Ber.* **35**, 3772 (1902). Early chemistry and biochemistry: *Amino Acids and Proteins,* D. M. Greenberg, Ed. (Charles C. Thomas, Springfield, IL, 1951) 950 pp, *passim;* J. P. Greenstein, M. Winitz, *Chemistry of the Amino Acids* **vols. 1-3** (John Wiley & Sons, Inc., New York, 1961) pp 2097-2124, *passim.* Review of supplementation of cereals: C. Feldberg, C. P. Hetzel, *Food Technol.* **12**, 496 (1958); of wheat proteins: S. B. Vaghefi *et al., Am. J. Clin. Nutr.* **27**, 1231-1246 (1974). Review of metabolism in mammals: F. C. Fellows, M. H. R. Lewis, *Biochem. J.* **136**, 329-334 (1973); of metabolic errors: N. A. J. Carson, *Clinics Endocrinol. Metabol.* **3**, 71-86 (1974). Brief review of production by fermentation: *Nutr. Rev.* **43**, 88-90 (1985). Review of synthesis: G. Galili, *Plant Cell* **7**, 899-906 (1995).

Needles from water, hexagonal plates from dil alcohol. Darkens at 210°; dec 224.5°. $[\alpha]_D^{20}$ +14.6° (c = 6.5); $[\alpha]_D^{13}$ +25.9° (c = 2 in 6.0N HCl). pK₁ 2.20; pK₂ 8.90; pK₃ 10.28 at 38°. Very freely sol in water. Insol in common neutral solvents.

Dihydrochloride, $C_6H_{14}N_2O_2.2HCl$, crystals from ethanol + ether, mp 193°. $[\alpha]_D^{20}$ +15.3° (c = 2).

Monohydrochloride, $C_6H_{14}N_2O_2.HCl$, *Darvyl, Enisyl, Lyamine.* Crystals from dil ethanol. mp 263-264° when anhydrous. $[\alpha]_D^{25}$ +14.6° (c = 2 in 0.6N HCl).

Monoorotate, $C_{11}H_{18}N_4O_6$, *Lysortine.* Prepn: **Brit.** pat. **922,361** (1962 to A.E.C. Soc. Chim. Org. Biol.). Crystals, dec 315°.

USE: Enrichment of cereals and feeds.

5668. Lysine Acetylsalicylate. DL-*Lysine mono[2-(acetyloxy)benzoate]; lysine monosalicylate acetate;* aspirin lysine salt; LAS; Aspidol; Aspisol; Delgesic; Flectadol; Lysal; Quinvet; Venopirin; Vetalgina. $C_{15}H_{22}N_2O_6$; mol wt 326.35. C 55.21%, H 6.79%, N 8.58%, O 29.42%. Water soluble, injectable aspirin derivative. Prepn: **Fr.** pat. **1,295,304** (1962 to Equilibre Biologique), *C.A.* **58**, 1536f (1963). Prepn of D,L isomers and racemate: **Fr.** pat. **2,115,060** (1972 to Metabio), *C.A.* **78**, 75877k (1972). Comparison with aspirin: S. Rampon *et al., Rhumatologie* **24**, 141 (1972). Pharmacokinetics: H. von Ross *et al., Klin. Wochenschr.* **56**, 1119 (1978); F. Gentit, *Gaz. Med. France* **86**, 4539 (1979). Veterinary use as analgesic: P. Richez *et al., J. Vet. Pharmacol. Therap.* **2**, 231 (1979); *eidem, ibid.* **3**, 121 (1980). Analgesic activity in post-operative pain: K. Korttila *et al., Br. J. Anaesth.* **52**, 613 (1980); in articular pain: C. Diaz *et al., Clin. Ther.* **4**, 121 (1981). Use in treatment of premature labor: F. Wolff *et al., Arch. Gynecol.* **233**, 15 (1982). Comparison with aspirin in production of gastromucosal damage: J.-F. Bretagne *et al., Gastroenterol. Clin. Biol.* **8**, 28 (1984).

Crystals from ethanol, mp 154-156°. Sol in water; slightly sol in ethanol. Insol in methanol, acetone, ether.

THERAP CAT: Analgesic; antipyretic; anti-inflammatory.

THERAP CAT (VET): Analgesic; antipyretic; anti-inflammatory.

5669. L-Lysine L-Glutamate. L-Glutamic acid L-lysine salt. $C_{11}H_{23}N_3O_6$; mol wt 293.32. C 45.04%, H 7.90%, N 14.33%, O 32.73%. $H_2N(CH_2)_4CH(NH_2)COOH.HOOC-(CH_2)_2CH(NH_2)COOH.$ Prepd from DL-lysine and L-glutamic acid as an intermediate in the resolution of DL-lysine: Emmick, Rogers, U.S. pats. **2,556,907; 2,657,230** (1951, 1953 both to du Pont).

Monohydrate, crystals, $[\alpha]^{22}$ +3.73 (12% in water).

USE: As a flavor and nutritive additive to food: **Brit.** pat. **882,163** (1961 to Stamicarbon N.V.).

5670. Lysostaphin. Antibiotic protein complex produced by *Staphylococcus staphyloliticus* with highly specific lytic activity against other *Staphylococcus* species. Contains three enzymes: a hexosaminidase, an amidase, and the major component, an endopeptidase which cleaves the polyglycine cross-linkages in the staphylococcal cell wall. Isoln and antibacterial spectrum: Schindler, Schuhardt, *Proc. Nat. Acad. Sci. USA* **51**, 414 (1964). Prepn by fermentation: *eidem,* U.S. pat. **3,278,378** (1966); Zygmunt, Browder, U.S. pat. **3,398,056** (1968 to Mead Johnson). The endopeptidase is a single polypeptide chain of mol wt 25,000: H. R. Trayer, C. E. Buckley, *J. Biol. Chem.* **245**, 4842 (1970). Identification of active principle: H. P. Browder *et al., Biochem. Biophys. Res. Commun.* **19**, 383 (1965). Characterization of three enzyme components: T. Wadstrom, O. Vesterberg, *Acta Pathol. Microbiol. Scand.* **79**, 248 (1971); O.-J. Iversen, A. Grov, *Eur. J. Biochem.* **38**, 293 (1973). Clinical studies to eradicate *Staph. aureus* from nasal carriers: R. L. Harris *et al., Antimicrob. Ag. Chemother.* **1967**, 110; K. E. Quickel *et al., Appl. Microbiol.* **1971**, 446. Use as identification method for *Staphylococcus* sp.: K. H. Schleifer, W. E. Kloos, *J. Clin. Microbiol.* **1**, 337 (1975); P. J. Severance *et al., ibid.* **11**, 724 (1980); disk modification: B. Poutrel, J.-P. Caffin, *ibid.* **13**, 1023 (1982). *Review:* Zygmunt, Tavormina, *Progress in Drug Research* **Vol. 16**, E. Jucker, Ed. (Birkhauser Verlag, Basel, 1972) pp 309-333.

Isoelectric pt: pH 10.4-11.4. uv max: 278 nm. Sedimentation coeff 2.35 S. Destroyed by pepsin or trypsin. Inhibited by Hg^{2+}, Cu^{2+}, Zn^{2+}. LD_{50} (7-day) in mice, rats (mg/kg): 820, 530 i.v. (Zygmunt).

5671. Lysozyme. Muramidase; N-acetylmuramide glycanohydrolase; N-acetylmuramyl hydrolase; globulin G_1. Mol wt about 14,400 ± 100. Mucolytic enzyme with antibiotic properties, first discovered by A. Fleming: *Proc. Roy. Soc. London* **93B**, 306 (1922). Found in tears, nasal mucus, milk, saliva, blood serum, in a great number of tissues and secretions of different animals, vertebrates and invertebrates, in egg white, in some molds, and in the latex of different plants. Isoln from egg white: Alderton *et al., J. Biol. Chem.* **157**, 43 (1945); Alderton, Fevold, *ibid.* **164**, 1 (1946); *Biochem. Prepns* **1**, 67 (1949); Sophianopoulos *et al., J. Biol. Chem.* **237**, 1107 (1962); from *Ficus* latex: Meyer *et al., ibid.* **163**, 733 (1946). Structure consists of a single polypeptide chain of 129 amino acid subunits of 20 different kinds cross-linked by four disulfide bridges. Chromatographic studies: Goncalves *et al., Arch. Biochem. Biophys.* **60**, 171 (1956); King, Craig, *J. Am. Chem. Soc.* **80**, 3366 (1958). Complete amino acid sequence of egg white lysozyme: Jolles *et al., Biochem. Biophys. Acta* **78**, 668 (1963); Canfield, *J. Biol. Chem.* **238**, 2698 (1963). Synthetic studies of egg white lysozyme: L. E. Barston *et al.* in *Chemistry and Biology of Peptides, Proc. 3rd Am. Peptide Symp.,* J. Meienhofer, Ed. (Ann Arbor Science Publishers, Inc., Michigan, 1972) pp 231-233; J. J. Sharp *et al., J. Am. Chem. Soc.* **95**, 6097 (1973). Three dimensional structure: Blake *et al.,*

Nature **206**, 757 (1965); North, *Science J.* **2**(11), 55 (1966); Phillips, *Sci. Am.* **215**, 78 (Nov. 1966); Blake *et al., Proc. Roy. Soc. London* **167B**, 365 (1967). Primary chemical structure and tentative complete sequence of human milk lysozyme: Jolles, Jolles, *Helv. Chim. Acta* **52**, 2671 (1969). Complete primary structure of human milk lysozyme and comparison with lysozymes of various origins: *eidem, ibid.* **54**, 2668 (1971). Dissolves bacterial cell wall mucopolysaccharides by hydrolyzing the β-(1→4) linkages between *N*-acetyl-D-muramic acid and 2-acetylamino-2-deoxy-D-glucose residues. Also acts on chitin. *Reviews:* Salton, *Bacteriol. Rev.* **21**, 82 (1957); Acker, Hartsell, *Sci. Am.* **202**, 132 (June 1960); Jolles, "Lysozyme" in P. D. Boyer, H. Lardy, K. Myrback, *The Enzymes*, **vol. 4**, (Academic Press, New York, 2nd ed., 1960) pp 431-445; *idem, Angew. Chem.* **76**, 20 (1964); Raftery, Dahlquist, *Fortschr. Chem. Org. Naturst.* **27**, 340 (1969); Hamaguchi, Hayashi, "Lysozyme" in *Proteins, Structure and Function,* **vol. 1**, M. Funatsu *et al.,* Eds. (Kodansha, Tokyo, Wiley, New York, 1972) pp 85-222. Book: *Lysozyme,* E. F. Osserman *et al.,* Eds. (Academic Press, New York, 1974).

Crystals. Isoelectric pt: pH 10.5-11.0. Fairly stable in acid soln. Not affected by heat up to 55°: Cotterill, Winter, *Poultry Sci.* **33**, 1185 (1954).

Hydrochloride, **Acdeam, Antalzyme, Immunozima, Lanzyme, Leftose, Likinozym, Lisozima, Murazyme, Neutase, Neuzyme, Toyolysom-DS.**

THERAP CAT: Mucolytic enzyme. Antiviral.

5672. Lyxoflavine. *5-Deoxy-5-(3,4-dihydro-7,8-dimethyl-2,4-dioxobenzo[g]pteridin-10(2H)-yl)-L-arabinitol;* 6,7-dimethyl-9-(1-L-lyxityl)isoalloxazine; L-lyxoflavine. $C_{17}H_{20}N_4O_6$; mol wt 376.37. C 54.25%, H 5.36%, N 14.89%, O 25.51%. The L-lyxose analog of riboflavine. Isoln from human myocardium: Pallares, Garza, *Arch. Biochem. Biophys.* **22**, 63 (1949); Gardner, *ibid.* **34**, 98 (1951). Synthesis: Wenis, U.S. pat. 2,734,054 (1956 to Hoffmann-La Roche); Folkers, U.S. pat. 2,760,865 (1956 to Merck & Co.).

Yellow to orange needles from water, dec 283-284°. $[\alpha]_D^{23}$ −49° (c = 0.26 in 0.05N NaOH). Sparingly sol in water,

even less sol in alcohol. Freely sol with decompn in alkaline aq solns.

USE: Growth-promoting agent in feedstuffs.

5673. D-Lyxose. *α-D-Lyxose.* $C_5H_{10}O_5$; mol wt 150.13. C 40.00%, H 6.71%, O 53.28%. Prepd by the oxidation of calcium D-galactonate: Clark, *J. Biol. Chem.* **31**, 605 (1917); Fletcher *et al., J. Am. Chem. Soc.* **72**, 4546 (1950); by γ-irradiation of lactose: Adachi, *J. Dairy Sci.* **45**, 1427 (1962). Structure of β-D-lyxose: Hordvik, *Acta Chem. Scand.* **15**, 1781 (1961).

α-D-Lyxose

Hygroscopic monoclinic prisms from ethanol + ether, mp 106-107°. Sweet taste. d^{20} 1.545. Shows mutarotation: $[\alpha]_D^{20}$ +5.5° → −14.0° (c = 0.82 in water). Freely sol in water. One part dissolves in 38 parts abs alc at 17°; 100 ml of 90% alc satd at 20° contain 7.9 g.

M

5674. Mabuterol. *4-Amino-3-chloro-α-[[(1,1-dimethyl-ethyl)amino]methyl]-5-(trifluoromethyl)benzenemethanol; 4-amino-α-[(tert-butylamino)methyl]-3-chloro-5-(trifluoro-methyl)benzyl alcohol; 1-(4'-amino-3'-chloro-5'-trifluoro-methylphenyl)-2-tert-butylaminoethanol; ambuterol.* $C_{13}H_{18}ClF_3N_2O$; mol wt 310.75. C 50.25%, H 5.84%, Cl 11.41%, F 18.34%, N 9.01%, O 5.15%. Orally active β_2-adrenergic agonist related to clenbuterol, *q.v.* Prepn and resolution of isomers: **Belg. pat. 808,743** (1974 to Thomae); G. Engelhardt *et al.*, **U.S. pat. 4,119,710** (1978 to Boehringer Ing.); *eidem, Arzneimittel-Forsch.* **34**, 1612 (1984). Series of articles on pharmacology, pharmacokinetics, toxicology and clinical studies: *ibid.* 1625-1700. Toxicity data: K. Amemiya *et al., ibid.*, 1680. Teratological study: A. M. Hoberman *et al., J. Am. Coll. Toxicol.* **4**, 91 (1985). Determn in human urine and plasma by enzyme immunoassay: I. Yamamoto *et al., J. Immunoassay* **6**, 261 (1985).

dl-Form hydrochloride, $C_{13}H_{18}ClF_3N_2O \cdot HCl$, *KF-868, PB-868Cl, Broncholin.* Crystals from ethyl acetate + ether, mp 205-206°. Fairly sol in water. LD_{50} in male, female mice, male, female rats (mg/kg): 41.5, 51.1, 26.4, 28.1 i.v.; 60.3, 60.0, 76.3, 78.3 i.p.; 113.0, 125.7, 117.2, 123.1 s.c.; 220.8, 199.9, 319.3, 305.6 orally (Amemiya).

d-Form hydrochloride, mp >194° (slow dec). $[\alpha]_{364}^{20}$ +154.9° (c = 1 in methanol).

l-Form hydrochloride, mp >194° (slow dec). $[\alpha]_{364}^{20}$ −154.8° (c = 1 in methanol).

THERAP CAT: Bronchodilator; antiasthmatic.

5675. Macassar Oil. Kusum oil; Kon oil; Paka oil. From the nut kernels of *Schleichera trijuga* Willd., *Sapindaceae* (Ceylon oak). *Habit.* India, Burma, Ceylon, Java. *Constit.* Glycerides of oleic acid 60%, arachidic 20-25%, palmitic 5-8% stearic 2-6%; free acetic acid 1-2%. Unsaponifiable matter 1.5-3% (this may include cyanide-containing glycosides which are removed by careful refining). *Reviews:* Sen-Gupta, *J. Soc. Chem. Ind. (London)* **39**, 88T (1920); Dhingra *et al., ibid.* **48**, 281T (1929).

Yellowish-white oil. Pleasant odor. Melting range: initial 21°, complete transparency 31°. d_{13}^{99} 0.860. n_D^{21} 1.46757, n_D^{27} 1.46655; $n_D^{31.5}$ 1.4646; n_D^{45} 1.4636. Acid value 10-70; saponification value 220-230; iodine value 48-58; Reichert-Meissl value 16. Miscible with ether, chloroform, other vegetable oils, petrolatum, lanolin.

USE: In hair oil formulations.

5676. Maclurin. *(3,4-Dihydroxyphenyl)(2,4,6-trihydroxyphenyl)methanone; 2,3',4,4',6-pentahydroxybenzophenone;* morintannic acid; moritannic acid; laguncurin; kino-yellow; C.I. Natural Yellow 11; C.I. 75240. $C_{13}H_{10}O_6$; mol wt 262.22. C 59.55%, H 3.84%, O 36.61%. From wood of *Chlorophora tinctoria* (L.) Gaud. *(Morus tinctoria* L., *Maclura tinctoria* (L.) D. Don), *Moraceae* (old fustic): Haley, Bassin, *J. Am. Pharm. Assoc.* **40**, 111 (1951); Laidlow, Smith, *Chem. & Ind. (London)* **1959**, 1604; from *Morus alba* Linn., *Moraceae:* Spada *et al., Gazz. Chim. Ital.* **86**, 46 (1956). Identity with laguncurin and kino-yellow: Nierenstein, *Quart. J. Pharm. Pharmacol.* **16**, 11 (1943). *See also Colour Index* vol. 4 (3rd ed., 1971) p 4627.

Yellow needles from ethanol, mp 222-222.5°. Sol in 190 parts water; freely sol in alcohol or ether.

USE: Dyeing fabrics.

5677. Macromerine. *α-[(Dimethylamino)methyl]-3,4-dimethoxybenzenemethanol; α-[(dimethylamino)methyl]veratryl alcohol;* 3,4-dimethoxy-α-[(dimethylamino)methyl]-benzyl alcohol; dimethylaminomethyl 3,4-dimethoxyphenyl carbinol. $C_{12}H_{19}NO_3$; mol wt 225.29. C 63.98%, H 8.50%, N 6.22%, O 21.31%. Prepn of *dl*-form: La Manna *et al., Farmaco Ed. Sci.* **15**, 9 (1960); Chapman *et al., Proc. Roy. Soc. London, Ser. B,* **163**, (990), 116 (1965); Hodgkins *et al., Tetrahedron Letters* **1967**, 1321; of *d*- and *l*-forms: La Manna *et al., loc. cit.* Isoln of *l*-form from cactus, *Coryphantha macromeris* (Engelm.) Lem., Britton and Rose, *Cactaceae:* Hodgkins *et al., loc. cit.*

dl-Form, crystals, mp 46-47°.

l-Form, crystals, mp 66-67.5°. $[\alpha]_D^{25}$ −147.01° (c = 0.0390 g/ml chloroform), −42.61° (c = 0.0200 g/ml abs alcohol).

d-Form, crystals, mp 60-62°.

dl-Form hydrochloride, $C_{12}H_{19}NO_3 \cdot HCl$, crystals from ethanol, mp 163-164°.

l-Form hydrochloride, crystals, mp 178-179°. $[\alpha]_D^{18}$ −41.3° (c = 2.04 in 50% ethanol).

d-Form hydrochloride, crystals from ethanol, mp 178-179°. $[\alpha]_D^{18}$ +51.5° (c = 2.2 in 50% ethanol).

Note: Both the *dl*- and *l*-forms are physiologically active, and caused hallucinogenic reactions when tested in animals (Hodgkins *et al., loc. cit.*).

5678. Macrophage Colony-Stimulating Factor. *Colony-stimulating factor 1;* M-CSF; CSF-1. Species specific hematopoietic growth factor which stimulates the development of committed progenitor cells to monocytes/macrophages. Also potentiates phagocytic activity and monocyte-mediated tumor cell cytotoxicity. Heavily glycosylated homodimer; produced in response to specific activating agents by a variety of cell types including monocytes, fibroblasts and endothelial cells. In humans, several molecular species have been identified which are produced via alternative splicing mechanisms from a single-copy gene. The primary transcripts include a 90 kDa glycoprotein, which is the principle form detected in serum and urine, and a fully bioactive, membrane-associated 40-50 kDa dimer. Murine M-CSF is similar to the human long form. Isoln from human urine: E. R. Stanley *et al., Fed. Proc.* **34**, 2272 (1975); from cultured human pancreatic carcinoma cells: M. Wu *et al., J. Biol. Chem.* **254**, 6226 (1979). Purification of murine M-CSF: E. R. Stanley, P. M. Heard, *ibid.* **252**, 4305 (1977). Characterization of human M-CSF: S. K. Das *et al., Blood* **58**, 630 (1981); K. Motoyoshi *et al., ibid.* **60**, 1378 (1982). Cloning of short form (human clone pcCSF-17): E. S. Kawasaki *et al., Science* **230**, 291 (1985); of long form (human clone p3ACSF-69): G. G. Wong *et al., ibid.* **235**, 1504 (1987). Review of molecular biology: E. S. Kawasaki, M. B. Ladner, *Immunol. Ser.* **49**, 155-176 (1990); of bioactivity and clinical studies: D. H. Munn, N.-K. V. Cheung, *Sem. Oncol.* **19**, 395-407 (1992).

Mirimostim, 1-214-colony-stimulating factor 1 (human clone p3ACSF-69 protein moiety reduced) homodimer, Costilate, Leukoprol.

4-221-Colony-stimulating factor 1 (human clone p3ACSF-69 protein moiety reduced). Nonglycosylated, 218 amino acid polypeptide produced in *E. coli* by recombinant DNA technology. Prepn and renaturation into bioactive homodimer: R. Halenbeck *et al., BioTechnology* **7**, 710 (1989). Clinical evaluation in invasive fungal infections: J. Nemunaitis *et al., Blood* **82**, 1422 (1993). Immunomodulatory effects and pharmacokinetics: R. M. Bukowski *et al.*, *J. Clin. Oncol.* **12**, 97 (1994).

THERAP CAT: Immunomodulator.

5679. Macusines. Isoln of A and B from *Strychnos toxifera* Schomb., *Loganiaceae*: A. R. Battersby *et al.*, *J. Chem. Soc.* **1960**, 1848. Structure of A: A. T. McPhail *et al.*, *Proc. Chem. Soc.* **1961**, 223; of B: A. R. Battersby, D. A. Yeowell, *ibid.* 17. Config of A: M.-M. Janot *et al.*, *Bull. Soc. Chim. France* **1962**, 1079. Crystal and molecular structure of A: A. T. McPhail *et al.*, *J. Chem. Soc.* **1963**, 1832. Sepn of C and abs config of A, B and C: A. R. Battersby, D. A. Yeowell, *ibid.* **1964**, 4419.

	R_1	R_2
Macusine A	CH_2OH	$COOCH_3$
Macusine B	H	CH_2OH
Macusine C	$COOCH_3$	CH_2OH

Macusine A, $[C_{22}H_{27}N_2O_3]^+$, *(16R)-17-hydroxy-16-(methoxycarbonyl)-4-methylsarpaganium.* Chloride, prisms from ethanol + ether, mp 252°. $[\alpha]_D^{25}$ 57.5° (c — 1.45).

Macusine B, $[C_{20}H_{25}N_2O]^+$, *(4α)-17-hydroxy-4-methylsarpaganium.* Chloride, prisms from ethanol + ether, dec 248-249°. $[\alpha]_D^{22}$ +15.6° (c = 1.21). uv max (water): 222, 273, 280, 291 nm (log ϵ 4.61, 3.84, 3.82, 3.74).

Macusine C, $[C_{22}H_{27}N_2O_3]^+$, the C-16 epimer of A. Chloride, crystals from ethanol + ether, dec 264-265°. $[\alpha]_D^{24}$ —60.8° (c = 2.13). uv max (water): 222, 272, 278, 289 nm (log ϵ 4.63, 3.88, 3.87, 3.76).

5680. Maddrell's Salt. Polymeric sodium metaphosphate. Upon heating NaH_2PO_4 or $NaNH_4HPO_4$ to 250° and then slowly to 350°, first $Na_2H_2P_2O_7$ and then the insol Maddrell's salt is obtained, which is stable to about 500°; at 505° it changes to sodium trimetaphosphate, and above 607° to sodium hexametaphosphate. According to Maddrell, *Ann.* **61**, 63 (1847) the salt may be prepd by heating two parts $NaNO_3$ with one part H_3PO_4. An improved procedure is described by v. Knorre, *Z. Anorg. Chem.* **24**, 397 (1900). Structure: Jost, *Acta Cryst.* **16**, 428 (1963). Reviews: *see* Sodium Metaphosphate.

White powder. Practically insol in water and in aq solns of pyrophosphates and hexametaphosphates. Sol in mineral acids. Does not form complexes with Fe(II), Fe(III), U(IV) salts.

USE: In dentifrices, polishing agents, abrasive detergents.

5681. MADU. *2'-Deoxy-5-(methylamino)uridine; 5-(methylamino)-2'-deoxyuridine.* $C_{10}H_{15}N_3O_5$; mol wt 257.25. C 46.69%, H 5.88%, N 16.33%, O 31.10%. Prepn: Visser *et al.*, *Biochim. Biophys. Acta* **76**, 463 (1963). Active *in vitro* against herpes infections: Nemes, Hilleman, *Proc. Soc. Exp. Biol. Med.* **119**, 515 (1965); Shen, U.S. pat. **3,322,-627** (1967 to Merck & Co.).

Crystals from ethanol + water, mp 180-182°.
THERAP CAT: Antiviral.

5682. Maduramicin. *(3R,4S,5S,6R,7S,22S)-23,27-Didemethoxy-2,6,22-tridemethyl-11-O-demethyl-22-[(2,6-dideoxy-3,4-di-O-methyl-β-L-arabino-hexopyranosyl)oxy]-6-methoxylonomycin A monoammonium salt;* antibiotic X-14868A ammonium salt; CL-273703; Cygro. $C_{47}H_{83}NO_{17}$; mol wt 934.17. C 60.43%, H 8.96%, N 1.50%, O 29.12%. Polyether antibiotic chemically related to the lonomycins, *q.v.* Isoln as sodium salt from *Nocardia* sp X-14868 and biological activity: C.-M. Liu *et al.*, U.S. pat. **4,278,663** (1981 to Hoffmann-La Roche); as free acid hydrate from *Actinomadura yumaense* sp nov.: D. P. Labeda *et al.*, U.S. pat. **4,407,946** (1983 to Am. Cyanamid). Fermentation and properties: C.-M. Liu *et al.*, *J. Antibiot.* **36**, 343 (1983). ^{13}C-NMR spectrum: S. Rajan, *ibid.* **37**, 1495 (1984). Biosynthetic studies: H.-R. Tsou *et al.*, *ibid.* 1651; H.-R. Tsou *et al.*, *ibid.* **40**, 94 (1987). Antimalarial activity: L. Oronsky, U.S. pat. **4,496,549** (1985 to Am. Cyanamid). Nematocidal activity: I. B. Wood, U.S. pat. **4,510,134** (1985 to Am. Cyanamid).

Sodium salt, $C_{47}H_{79}NaO_{17}$, crystals from ethyl acetate + n-hexane, mp 193-195°. $[\alpha]_D$ +40.6° (chloroform). $[\alpha]_D$ +23.8° (methanol).

Note: Not to be confused with *maduramycin* isolated from *Actinomadura rubra*, $C_{28}H_{22}O_{10}$: W. F. Fleck *et al.*, *16th Interscience Conference on Antimicrob. Ag. Chemother.*, Chicago, 1976, *Abstracts of Papers*, no. 51.

THERAP CAT (VET): Coccidiostat.

5683. Mafenide. *4-(Aminomethyl)benzenesulfonamide; α-amino-p-toluenesulfonamide;* p-(aminomethyl)benzenesulfonamide; 4-homosulfanilamide; maphenide; Marfanil; Mesudrin; Mesudin; Sulfamylon; Homosulfamine; Ambamide; Neofamid; Septicid; Emilene; Homonal; Paramenyl. C_7H_{10}-N_2O_2S; mol wt 186.23. C 45.15%, H 5.41%, N 15.04%, O 17.18%, S 17.22%. Prepn: Klarer, U.S. pat. **2,288,531** (1942 to Winthrop); Bergeim, Braker, *J. Am. Chem. Soc.* **66**, 1459 (1944); Kusami *et al.*, *J. Pharm. Soc. Japan* **64**, no. 9a, 51 (1944); Komokina *et al.*, *J. Appl. Chem. (USSR)* **21**, 681 (1948); Angyal, Jenkin, *Aust. J. Sci. Res.* **3A**, 461 (1950). Toxicity study: T. W. Skulan, J. O. Hoppe, *Life Sci.* **5**, 2279 (1966).

Crystals from alc, mp 151-152°. Sol in dil alkali, acid.
Hydrochloride, $C_7H_{10}N_2O_2S.HCl$, crystals from 95% alc, mp 256°. Neutral in soln. LD_{50} in rats, mice (mg/kg): 1170, 900 i.v. (Skulan, Hoppe).
Acetate, $C_9H_{14}N_2O_4S$, *Mafatate, Mefamide.* LD_{50} in rats, mice (mg/kg): 2040, 1580 i.v. (Skulan, Hoppe).
Propionate, $C_7H_{10}N_2O_2S.C_3H_6O_2$, *Sulfomyl.* Crystals, mp 158°. Readily sol in water.
THERAP CAT: Antibacterial.

5684. Magainins. A class of potent antimicrobial peptides isolated from the glanular gland of clawed toad *Xenopus laevis.* Name coined by Zasloff from the Hebrew word for "shield". Consists of magainins I and II, both containing 23 amino acid residues and differing only at the 10 (I: Gly, II: Lys) and 22 (I: Lys, II: Asn) positions. Isolation and characterization: M. Zasloff, *Proc. Nat. Acad. Sci. USA* **84**, 5449 (1987). Identification of post-secretory degradation product: T. W. Schwartz *et al., Nature* **329**, 494 (1987). Brief overview: J. Alper, *Chem. Week* (November 4, 1987) p 66; M. Cannon, *Nature* **328**, 478 (1987).

Gly–Ile–Gly–Lys–Phe–Leu–His–Ser–Ala–Gly–Lys–Phe
|
Ser–Lys–Met–Ile–Glu–Gly–Val–Phe–Ala–Lys–Gly

Magainin I

5685. Magaldrate. *Aluminum magnesium hydroxide (AlMg(OH)₇) monohydrate;* magnesium aluminate hydrate; monalium hydrate; AY-5710; Dynese; Riopan; Ripon. Formerly treated as [Mg(OH)₄]₄[(HO)₄Al(OH)(HO)Al(OH)₄].2H₂O. Byk Gulden company information states that based on ^{27}Al broadline nuclear resonance spectra, radiographic investigation and chemical analytical finds, the chemical formula for Riopan powder is: $Mg_{1.5}^+ xAl(OH)_6(SO_4)_x(H_2O)_{0.5-x}$ with 0.325 < x < 0.462. Prepd by mixing together and allowing to react at 0° to 50° an alkali aluminate soln (contg 3-5 mols of Na₂O or K₂O per mol Al₂O₃) with a magnesium salt soln in such proportions that 1 g of Al combines with 0.9 to 3.0 g of Mg, and washing the pptd product: Hallmann, U.S. pat. **2,923,660** (1960 to Byk-Gulden Lomberg).

White, odorless, crystalline powder which contains the equivs of not < 28.0% and not > 39.0% MgO and not < 17.0% and not > 25.0% Al₂O₃.

THERAP CAT: Antacid.

5686. Magenta I. *4-[(4-Aminophenyl)(4-imino-2,5-cyclohexadien-1-ylidene)methyl]-2-methylbenzenamine monohydrochloride; C.I. Basic Violet 14;* α⁴-(p-aminophenyl)-α⁴-(4-imino-2,5-cyclohexadien-1-ylidene)-2,4-xylidine hydrochloride; 2-methyl-4,4'-[(4-imino-2,5-cyclohexadien-1-ylidene)methylene]dianiline hydrochloride; fuchsine; magenta; rosaniline hydrochloride; C.I. 42510. C₂₀H₂₀ClN₃; mol wt 337.85. C 71.10%, H 5.97%, Cl 10.49%, N 12.44%. Prepn: Fischer, Fischer, *Ann.* **194**, 276, 290 (1878); *Ber.* **13**, 2204 (1880); J. T. Scalan, *J. Am. Chem. Soc.* **57**, 887 (1935). Evaluation of carcinogenicity studies: *IARC Monograph* **4**, 57 (1974). *See also: Colour Index* vol. 4 (3rd ed., 1971) p 4389.

Metallic green, lustrous crystals, dec above 200°. Absorption max (ethanol): 543 nm (ε 93000). 2.65 parts dissolve in 1000 parts water. Practically insol in ether. Sol in alcohol with a carmine red color.

USE: As a dye or in manuf of other dyes.

THERAP CAT: Antifungal.

5687. Magnesium. Mg; at. wt 24.305; at. no. 12; valence 2. An alkaline earth metal. One of the most common elements in the earth's crust: 2.1% by weight. Isotopes: 24 (78.70%); 25 (10.13%); 26 (11.17%). Found naturally only in the form of its compounds in magnesite, carnallite, *dolomite* [CaMg(CO₃)₂], epsomite, kieserite, and many other minerals; found in sea-water; in animal and vegetable kingdom. First obtained in metallic form by Davy in 1808 by electrolysis of a mixture of magnesia and mercuric oxide. Methods of prepn: Deville, Caron, etc, cited in *Gmelin's,*

Magnesium (8th ed.) **27A**, 121 (1937). Prepd industrially by reduction of MgO-contg materials. Examples of large-scale processes: Weiss, U.S. pat. **3,264,097** (1966 to Ver. Aluminium-Werke). Review of magnesium and its compounds: Goodenough, Stenger, "Magnesium, Calcium, Strontium, Barium and Radium" in *Comprehensive Inorganic Chemistry,* vol. 1, J. C. Bailar, Jr. *et al.,* Eds. (Pergamon Press, Oxford, 1973) pp 591-664; L. F. Lockwood *et al.,* in Kirk-Othmer *Encyclopedia of Chemical Technology* vol. 14 (Wiley-Interscience, New York, 3rd ed., 1981) pp 570-615.

Silvery-white metal; hexagonal close-packed structure. Slowly oxidizes in moist air. Available as bars, ribbons, wire and powder. mp 651°. bp 1100°. d²⁰ 1.738. Sp heat (20°) 0.245 cal/g. Heat of fusion 88 cal/g. Electrical resistivity 4.46 μohm-cm. E° (aq) Mg²⁺/Mg −2.37 V. Reacts very slowly with water at ordinary temp, less slowly at 100°. Reacts readily with dil acids with liberation of hydrogen; reacts with aq solns of ammonium salts, forming a double salt. Reduces carbon monoxide, carbon dioxide, sulfur dioxide, nitric oxide, and nitrous oxide at a red heat. Burns in air; continues to burn in a current of steam. Combines directly with nitrogen, sulfur, the halogens, phosphorus, and arsenic. Reacts with methyl alcohol at 200° giving magnesium methylate.

Caution: Particles embedded in skin can produce gaseous blebs with a protracted course. Inhalation of the dust is irritating; fumes can cause metal fume fever.

USE: As a constituent of light alloys; for manuf precision instruments, optical mirrors; in pyrotechnics; in metallurgy as deoxidizing and desulfurizing agent; instead of zinc in dry batteries; for flash bulbs and flares, alumino-thermics, ignition of thermite mixture, intense signal lights; for Grignard reagents; in the recovery of titanium.

5688. Magnesium Acetate. Cromosan. C₄H₆MgO₄; mol wt 142.39. C 33.74%, H 4.25%, Mg 17.07%, O 44.94%. Mg(C₂H₃O₂)₂.

Tetrahydrate, colorless or white, deliquesc crystals. d 1.45. mp about 80°. Very sol in water or alcohol. The aq soln is neutral or slightly acid. *Keep well closed.* LD₅₀ i.v. in mice: 18 mg/kg.

5689. Magnesium Acetylsalicylate. *2-(Acetyloxy)benzoic acid magnesium salt;* magnesium aspirin; Apyron; Fyracyl; Magisal; Magnespirin; Novacetyl. C₁₈H₁₄MgO₈; mol wt 382.61. C 56.51%, H 3.69%, Mg 6.35%, O 33.45%. (CH₃COOC₆H₄COO)₂Mg.

White, nonhygroscopic, almost tasteless and odorless powder. Freely soluble in water; less sol in alcohol.

THERAP CAT: Has been used as analgesic, antipyretic.

5690. Magnesium Amide. Magnesium diamide. H₄MgN₂; mol wt 56.35. H 7.15%, Mg 43.13%, N 49.71%. Mg(NH₂)₂. Prepd by the action of ammonia on an ether soln of magnesium diethyl or on magnesium activated with iodine (at 400°): Terentiew, *Z. Anorg. Chem.* **162**, 351 (1927); Schlenk, Jr., *Ber.* **64**, 738 (1931). Prepd from metal and liquid ammonia at high pressure: Juza, Jacobs, *Angew. Chem., Int. Ed.* **5**, 247 (1966); *eidem, Z. Anorg. Allgem. Chem.* **370**, 245 (1969).

White powder or crystals. d₄²⁵ 1.39. Catches fire in air. Dec on heating. Reacts violently with water, evolving ammonia.

USE: Polymerization catalyst.

5691. Magnesium Benzoate. C₁₄H₁₀MgO₄; mol wt 266.54. C 63.09%, H 3.78%, Mg 9.12%, O 24.01%. Mg(C₇H₅O₂)₂.

Trihydrate, white, odorless powder. mp about 200°. Sol in 20 parts water, freely in hot water, alcohol. The aq soln is neutral or slightly acid.

5692. Magnesium Bisulfate. Magnesium hydrogen sulfate; acid magnesium sulfate. H₂MgO₈S₂; mol wt 218.45. H 0.92%, Mg 11.13%, O 58.59%, S 29.36%. Mg(HSO₄)₂.

White powder. Sol in water.

5693. Magnesium Borate. "Antifungin". B₂MgO₄; mol wt 109.92. B 19.67%, Mg 22.11%, O 58.22%. Mg(BO₂)₂. Various magnesium borate minerals occur in nature. These include: *ascharite, camsellite, inderite, kotoite, kurnakovite, paternoite, pinnoite, szaibelyite.*

Octahydrate, white powder. Slightly sol in water.
USE: Antiseptic; fungicide.

5694. Magnesium Bromate. Br_2MgO_6; mol wt 280.11. Br 57.05%, Mg 8.68%, O 34.27%. $Mg(BrO_3)_2$.
Hexahydrate, colorless or white crystals; loses all water at about 200°; dec at higher temp. Sol in 1.5 parts water.

5695. Magnesium Bromide. Br_2Mg; mol wt 184.11. Br 86.80%, Mg 13.20%. $MgBr_2$.
Hexahydrate, colorless, very deliquesc crystals or white granules; bitter taste. mp about 165° with decompn. Sol in 0.3 part water; sol in alcohol. The aq soln is neutral. *Keep well closed.*
USE: In organic syntheses.
THERAP CAT: Sedative, anticonvulsant.

5696. Magnesium Carbonate Hydroxide. *Carbonic acid, magnesium salt (1:1), mixt. with magnesium hydroxide, hydrate;* Marinco C. Usually a basic carbonate, approx $(MgCO_3)_4.Mg(OH)_2.5H_2O$; approx mol wt 485; contains 40-42% MgO. Prepn of various hydrates: Pond, Heneghan, U.S. pat. **3,169,826** (1965 to Merck & Co.). Magnesium carbonate minerals include *magnesite* and *lansfordite.*
White, odorless, bulky powder or light friable masses; at about 700° is converted into MgO. Sol in about 3300 parts CO_2-free water; more sol in water contg CO_2; sol in dil acids with effervescence. Insol in alcohol. Imparts slight alkaline reaction to water.
Hydrotalcite, Altacite, Hi-Ti, Nacid, Talcid, an aluminum magnesium hydroxide carbonate hydrate of the general formula $Al_2O_3.6MgO.CO_2.12H_2O$. Prepn: Kumuru *et al.,* U.S. pat. **3,539,306** (1970 to Kyowa).
USE: Fireproofing, heat insulating; preparing effervescent magnesium citrate; clarifying liqs by filtration; in tooth and face powders, in polishing compds; manuf mineral waters, pigments, paper; filler for rubber.
THERAP CAT: Magnesium carbonate hydroxide as antacid, cathartic; hydrotalcite as antacid.
THERAP CAT (VET): Laxative.

5697. Magnesium Chlorate. Cl_2MgO_6; mol wt 191.21. Cl 37.08%, Mg 12.71%, O 50.21%. $Mg(ClO_3)_2$.
Hexahydrate, white, very deliquesc crystals or cryst powder; bitter taste; d 1.80. mp ~35°. Sol in 0.9 part water; slightly sol in alcohol. *Keep well closed.* LD_{50} orally in rats: 5.25 g/kg, Ulrich, *J. Pharmacol. Exp. Ther.* **35,** 1 (1929).

5698. Magnesium Chloride. Magnogene. Cl_2Mg; mol wt 95.21. Cl 74.47%, Mg 25.53%. $MgCl_2$. Prepd from magnesium ammonium chloride hexahydrate in the presence of HCl: Bryce-Smith, *Inorg. Syn.* **6,** 9 (1960). Toxicity study: H. F. Smyth *et al., Am. Ind. Hyg. Assoc. J.* **30,** 470 (1969).
Soft, highly deliquescent leaflets. mp 712° (rapid heating). d 2.41, also reported as d 2.325. Slow heating releases chlorine at 300°. Can be distilled in a stream of hydrogen. Attacks fused silica when melted. Sol in water, with the evolution of much heat, giving a clear soln. Easily forms alcoholates and etherates.
Hexahydrate, deliquescent crystals. d 1.56. At 100° loses $2H_2O$ (17.7%); at 110° begins to lose some HCl. By strong ignition is converted into oxychloride. mp, when rapidly heated, ~118° with decompn. One gram dissolves in 0.6 ml water, 0.3 ml boiling water, 2 ml alcohol. Its aq soln is neutral. *Keep well closed.* LD_{50} orally in rats: 8.1 g/kg (Smyth).
USE: Fireproofing wood; in disinfectants; various magnesia cements; fire extinguishers; dressing cotton fabrics; in floor-sweeping compds; carbonizing wool; manuf parchment paper or artificial leather; as addition to casein glue; as a reagent in analytical chemistry.
THERAP CAT: Cathartic.
THERAP CAT (VET): Has been used in bovine hypomagnesemia.

5699. Magnesium Citrate. *2-Hydroxy-1,2,3-propanetricarboxylic acid magnesium salt (2:3);* Citramag. $C_{12}H_{10}Mg_3O_{14}$; mol wt 451.12. C 31.95%, H 2.23%, Mg 16.16%, O 49.65%. $Mg_3(C_6H_5O_7)_2$.

Tetradecahydrate, white, odorless, crystalline powder or granules. Slightly sol in water, the soly is increased by citric acid or alkali citrates; sol in dil acids.
THERAP CAT: Cathartic.

5700. Magnesium Citrate, Dibasic. *2-Hydroxy-1,2,3-propanetricarboxylic acid magnesium salt (1:1);* magnesium citrate sol; acid magnesium citrate; Citresia. $C_6H_6MgO_7$; mol wt 214.41. C 33.61%, H 2.82%, Mg 11.34%, O 52.23%. Prepn of stable, hydrated compd: Davenport, Costa, U.S. pat. **2,260,004** (1941 to Pfizer).
Pentahydrate, white or slightly yellow, odorless, granules or powder. Sol in ~5 parts water, becoming less sol with age or by loss of water; insol in alcohol.

5701. Magnesium Fluoride. Afluon; Sellaite. F_2Mg; mol wt 62.30. F 60.99%, Mg 39.01%. MgF_2. Prepd from $MgCO_3$ + HF: Klemm *et al., Z. Anorg. Allgem. Chem.* **176,** 13 (1928).
Colorless substance (rutile-type lattice). Slight violet fluorescence. d 3.148. Mohs' hardness: 6. mp 1248°. bp 2260°. Very sparingly sol in water 87 mg/l (18°). Slightly sol in dil acids, especially nitric acid. May be stored in glass bottles. LD orally in guinea pigs: 1.0 g/kg, *Handbook of Toxicology,* vol. 1, W. S. Spector, Ed. (Saunders, Philadelphia, 1956) pp 180-181.
USE: In the ceramics and glass industry.

5702. Magnesium Formate. $C_2H_2MgO_4$; mol wt 114.34. C 21.01%, H 1.76%, Mg 21.26%, O 55.97%. $Mg(HCOO)_2$.
Dihydrate, crystals or granules. Sol in water; insol in alcohol. Its aq soln is practically neutral.

5703. Magnesium Germanide. $GeMg_2$; mol wt 121.22. Ge 59.90%, Mg 40.10%. Mg_2Ge. Prepd by fusion of the components: Winkler, *Helv. Phys. Acta* **28,** 633 (1955).
Crystals, mp 1115°.
USE: In semiconductor research.

5704. Magnesium Hexafluorosilicate. Magnesium silicofluoride; magnesium fluosilicate. F_6MgSi; mol wt 166.38. F 68.51%, Mg 14.61%, Si 16.88%. $MgSiF_6$.
Hexahydrate, white, efflorescent, odorless crystals. d 1.788. At ~120° loses SiF_4. Soluble in water; insol in alcohol. pH of 1% aq soln: 3.1. LD orally in guinea pigs: 200 mg/kg, *Handbook of Toxicology,* vol. 1, W. S. Spector, Ed. (Saunders, Philadelphia, 1956) pp 180-181.
USE: For mothproofing of textile fabrics.

5705. Magnesium Hydride. H_2Mg; mol wt 26.32. H 7.66%, Mg 92.34%. Prepd by thermal decompn of diethylmagnesium at 200° in high vacuum: Wiberg, Bauer, *Ber.* **85,** 593 (1952); from the elements: Ellinger *et al., J. Am. Chem. Soc.* **77,** 2647 (1955).
White, nonvolatile mass or tetragonal crystals. d 1.45. Dec at 280° in high vacuum. Strong reducing agent. Ignites spontaneously on contact with air, forming MgO and H_2O. On contact with water a violent evolution of hydrogen takes place. On contact with methanol, magnesium alcoholate and hydrogen are formed. Forms double hydrides with boron hydride and aluminum hydride.

5706. Magnesium Hydroxide. Magnesium hydrate; Marinco H. H_2MgO_2; mol wt 58.32. H 3.46%, Mg 41.68%, O 54.87%. $Mg(OH)_2$. Occurs in nature as the mineral *brucite.* Prepn from magnesium chloride or sulfate and NaOH: Perlard, Waldron, U.S. pat. **3,127,241** (1964 to Dow). Industrial method for producing magnesium hydroxide having variable particle sizes: Chisholm, U.S. pat. **3,232,708** (1966 to FMC).
Amorphous powder. Practically insol in water (1:80,000). Sol in dil acids. Imparts slight alkaline reaction to water. pH of aq slurry: 9.5-10.5. Absorbs CO_2 in the presence of water. *Keep well closed.*
Suspension of 30% magnesium hydroxide in water, *Hydro-Magma.*
THERAP CAT: Antacid; cathartic.
THERAP CAT (VET): Laxative.

5707. Magnesium Iodide. I_2Mg; mol wt 278.11. I 91.26%, Mg 8.74%. MgI_2.

Octahydrate, white, deliquesc powder; readily discolors in air and light. Very sol in water; sol in alcohol. The aq soln is neutral or slightly alkaline. *Keep well closed and protected from light.*

5708. Magnesium Lactate. *2-Hydroxypropanoic acid magnesium salt.* $C_6H_{10}MgO_6$; mol wt 202.45. C 35.60%, H 4.98%, Mg 12.01%, O 47.42%. $(CH_3CHOHCOO)_2Mg$.

Trihydrate, white cryst, very bitter powder. One gram dissolves in 25 ml cold water, 3.5 ml boiling water; slightly sol in alcohol. The aq soln is slightly acid.

THERAP CAT: Cathartic.

5709. Magnesium Mandelate. Magdelate. $C_{16}H_{14}$-MgO_6; mol wt 326.59. C 58.84%, H 4.32%, Mg 7.44%, O 29.39%. $Mg[C_6H_5CH(OH)COO]_2$.

White, odorless powder. Stable in air. Slightly sol in cold water; sol in about 250 parts boiling water; insol in alcohol. The aq soln is practically neutral. Dec by acids forming free mandelic acid.

THERAP CAT: Urinary antiseptic.

5710. Magnesium Nitrate. MgN_2O_6; mol wt 148.31. Mg 16.39%, N 18.89%, O 64.72%. $Mg(NO_3)_2$. Hydrated form occurs in nature as the mineral *nitromagnesite.*

Hexahydrate, colorless, clear, deliquesc crystals. d 1.464. mp ~95°. Sol in 0.8 part water; freely sol in alcohol. The aq soln is neutral. *Keep well closed.*

USE: In pyrotechnics; in the concentration of nitric acid.

5711. Magnesium Oleate. *9-Octadecenoic acid magnesium salt.* $C_{36}H_{66}MgO_4$; mol wt 587.22. C 73.63%, H 11.33%, Mg 4.14%, O 10.90%. $Mg(C_{18}H_{33}O_2)_2$. Commercial form usually contains some stearate and palmitate.

Yellowish powder or mass. Insol in water; partly sol in alcohol, ether, petr ether.

USE: As an addition to naphtha to prevent spontaneous ignition in cleaning establishments.

5712. Magnesium Oxalate. *Ethanedioic acid magnesium salt.* C_2MgO_4; mol wt 112.32. C 21.39%, Mg 21.64%, O 56.98%. MgC_2O_4.

Dihydrate, white powder. Sol in about 1500 parts water, in dil mineral acid; insol in alcohol.

5713. Magnesium Oxide. Magnesia; calcined magnesia; magnesia usta; Magcal; Maglite. MgO; mol wt 40.30. Mg 60.30%, O 39.70%. Occurs in nature as the mineral *periclase.* Commercial prepn from magnesite ores: Adams, U.S. pat. **3,320,029** (1967 to Northwest Magnesite).

White, very fine, odorless powder. Available in a very bulky form termed "Light" or in a dense form termed "Heavy." Takes up CO_2 and H_2O from the air, the light form more readily than the heavy. mp 2800°. Highly reflective in visible and near uv region. Combines with water to form magnesium hydroxide. Very slightly sol in pure water, soly increased by CO_2; sol in dil acids; insol in alcohol. Imparts a slight alkaline reaction to water. pH of satd aq soln 10.3. *Keep well closed.*

Caution: Potential symptoms of overexposure to fumes are irritation of eyes and nose; metal fume fever (coughing, chest pain and flu-like fever). *See NIOSH Pocket Guide to Chemical Hazards* (DHHS/NIOSH 90-117, 1990) p 138.

USE: Manuf refractory crucibles, fire bricks, magnesia cements and boiler scale compounds, "powdered" oils, casein glue. Reflector in optical instruments; white color standard. Insulator at low temp.

THERAP CAT: Antacid.

THERAP CAT (VET): Antacid, laxative, in hypomagnesemia.

5714. Magnesium Perborate. B_2MgO_6; mol wt 141.92. B 15.23%, Mg 17.13%, O 67.64%. $Mg(BO_3)_2$.

Heptahydrate, white, bulky powder; gradually dec with loss of active oxygen. Sparingly sol in water with partial decompn; sol in dil acids.

USE: As of sodium perborate; with pptd chalk in tooth powders.

5715. Magnesium Perchlorate. Anhydrone; Dehydrite. Cl_2MgO_8; mol wt 223.21. Cl 31.77%, Mg 10.89%, O 57.34%. $Mg(ClO_4)_2$.

White, very hygroscopic, granular or flaky powder; dec above 250°. Dissolves in water with evolution of a consider-

able amount of heat. Crystallizes from water with $6H_2O$. *Keep tightly closed.*

Caution: Can cause irritation of skin, mucous membranes.

USE: As a drying agent for gases. The article of commerce may contain an amount of water equivalent to a dihydrate, but even the trihydrate is said to be effective for drying gases.

5716. Magnesium Permanganate. $MgMn_2O_8$; mol wt 262.18. Mg 9.27%, Mn 41.91%, O 48.82%. $Mg(MnO_4)_2$. As polymerization catalyst: Pengilly *et al.,* U.S. pat. **3,576,-790** (1971 to Goodyear). In purification of benzene: Ingwalson *et al.,* U.S. pat. **3,478,092** (1969 to Velsicol).

Bluish-black, deliquesc crystals. Freely sol in water. *Keep well closed.*

USE: Catalyst.

5717. Magnesium Peroxide. Magnesium dioxide; Magnesium Perhydrol; Magnesium Superoxol. MgO_2; mol wt 56.30. Mg 43.17%, O 56.83%. The article of commerce contains 15-25% MgO_2, the balance being magnesium hydroxide. For medicinal use the 25% article is preferred.

White, tasteless, odorless powder. Insol in water and gradually dec by it with liberation of oxygen; sol in dil acids, forming hydrogen peroxide; when strongly heated loses all the peroxide oxygen. *Keep well closed.*

THERAP CAT: Antacid, anti-infective.

5718. Magnesium Phosphate, Dibasic. Magnesium hydrogen phosphate; secondary magnesium phosphate. $HMgO_4P$; mol wt 120.28. H 0.84%, Mg 20.21%, O 53.21%, P 25.75%. $MgHPO_4$. Trihydrate occurs in nature as the minerals *newberyite, phosphor-roesslerite.*

Trihydrate, white, cryst powder. d 2.13. Slightly sol in water; sol in dil acids.

THERAP CAT: Cathartic.

5719. Magnesium Phosphate, Monobasic. Magnesium biphosphate; acid magnesium phosphate; primary magnesium phosphate. $H_4MgO_8P_2$; mol wt 218.28. H 1.85%, Mg 11.13%, O 58.64%, P 28.38%. $Mg(H_2PO_4)_2$.

Trihydrate, white powder. Soluble in water.

USE: In fireproofing wood.

5720. Magnesium Phosphate, Tribasic. "Neutral" magnesium phosphate; tertiary magnesium phosphate; trimagnesium phosphate. $Mg_3O_8P_2$; mol wt 262.86. Mg 27.74%, O 48.69%, P 23.57%. $Mg_3(PO_4)_2$. Octahydrate occurs in nature as the mineral *bobierrite.*

Pentahydrate, white, cryst powder. Loses last mol of water at ~400°. Insol in water; sol in dil mineral acids.

THERAP CAT: Antacid.

5721. Magnesium Potassium Selenate. Potassium magnesium selenate. $K_2MgO_8Se_2$; mol wt 388.42. K 20.13%, Mg 6.26%, O 32.95%, Se 40.66%. $K_2Mg(SeO_4)_2$. Prepd from a soln of aq magnesium and potassium selenates: Tutton, cited in *Mellor's* **10,** 864 (1930).

Hexahydrate, monoclinic prisms. d_4^{20} 2.365. Freely sol in water.

5722. Magnesium Pyrophosphate. $Mg_2O_7P_2$; mol wt 222.55. Mg 21.84%, O 50.32%, P 27.83%. $Mg_2P_2O_7$.

Trihydrate, white powder; loses its water at 100°. d 2.56. Insol in water. Sol in dil mineral acids.

5723. Magnesium Salicylate. *2-Hydroxybenzoic acid magnesium salt;* Analate; Lorisal; Magan; Mobidin; Triact. $C_{14}H_{10}MgO_6$; mol wt 298.53. C 56.33%, H 3.38%, Mg 8.14%, O 32.16%. $Mg[C_6H_4(OH)COO]_2$.

Tetrahydrate, white, odorless, efflorescent, cryst powder. Sol in 13 parts water; sol in alc. The aq soln is slightly acid.

THERAP CAT: Anti-infective (intestinal).

5724. Magnesium Selenate. MgO_4Se; mol wt 167.26. Mg 14.53%, O 38.26%, Se 47.21%. $MgSeO_4$. Prepn: Mitscherlich, *Pogg. Ann.* **9,** 623 (1827); Berzelius, *ibid.* **32,** 11 (1834); Huff, McCrosky, *J. Am. Chem. Soc.* **51,** 1457 (1929).

Hexahydrate, monoclinic crystals. d 1.928. Stable in air. Sol in water.

5725. Magnesium Selenide. MgSe; mol wt 103.27. Mg 23.54%, Se 76.46%. Prepd by the action of hydrogen selenide on anhyd magnesium chloride at a red heat and by the action of selenium vapor carried in a current of nitrogen on powdered magnesium: Fonzes-Diacon, *Contribution a l'Etude des Séléniures Métalliques* (Montpellier, 1901); by dropping selenium into molten magnesium: Liddell, *Chem. Met. Eng.* **25**, 102, 263, 453 (1921).

Light brown powder. d 4.21. Unstable in air. Dec in water.

5726. Magnesium Selenite. MgO_3Se; mol wt 151.26. Mg 16.07%, O 31.73%, Se 52.20%. $MgSeO_3$. Hexahydrate prepd by adding sodium selenite to a soln of magnesium chloride: Nilson, cited in *Mellor's* **10**, 826 (1930); by treating a soln of magnesium chloride or sulfate with selenious acid and adding sodium carbonate to start precipitation: Boutzoureano, *Recherches sur les Sélénites* (Paris, 1889).

Orthorhombic crystals. Insol in water; sol in dil acids. Forms monoclinic prisms of the dihydrate when heated in a sealed tube to 150°. When heated to 100°, it loses 5 mols of water leaving the monohydrate.

5727. Magnesium Silicates. Several varieties of magnesium silicate are known. *See also* asbestos, talc.

Magnesium metasilicate, $MgSiO_3$. Occurs in nature as the minerals *clinoenstatite, enstatite, protoenstatite*. White monoclinic crystals, dec at 1557°. d_4^{25} 3.192. Practically insol in water; very slightly sol in HF.

Magnesium orthosilicate, Mg_2SiO_4. Occurs in nature as the mineral *forsterite*. White orthorhombic crystals. mp 1910°; d 3.21. Practically insol in water.

Magnesium trisilicate, $Mg_2Si_3O_8$, *magnesium mesotrisilicate, Magnosil, Petimin, Trisomin.* Occurs in nature as the minerals *meerschaum, parasepiolite, sepiolite.* Prepd from sodium silicate and magnesium sulfate: Glass, *Quart. J. Pharm. Pharmacol.* **9**, 445 (1936); Uyeda, U.S. pat. **3,272,-594** (1966 to Merck & Co.). Odorless, tasteless, slightly hygroscopic powder. Usually contains some water of hydration. Practically insol in water.

Magnesium trisilicate pentahydrate, *Sellagen*.

Serpentine (mineral), $Mg_3O_7Si_2$. Occurs in nature as the dihydrate, $(OH)_4MgSi_2O_5$; exists in two forms: *antigorite*, a platy variety, and *chrysotile*, a fibrous variety. The latter is the most common form of asbestos, *q.v.*

USE: An activated magnesium silicate, *Florisil*, a hard, porous, granular substance, is used in vitamin analysis, chromatography, antibiotic processing: Simons, U.S. pat. **2,393,625** (1946 to Floridin).

THERAP CAT: Trisilicate as antacid.

THERAP CAT (VET): Trisilicate as antacid; gastric sedative.

5728. Magnesium Silicide. Mg_2Si; mol wt 76.70. Mg 63.38%, Si 36.62%. Prepd by heating finely powdered Mg and Si in a proportion of 20 to 6: Gire, *Compt. Rend.* **196**, 1404 (1933), *C.A.* **27**, 3678 (1933); Winkler, *Helv. Phys. Acta* **28**, 633 (1955), *C.A.* **50**, 6908c (1956). Prepn and conversion to silanes: Emeleus, Maddock, *J. Chem. Soc.* **1946**, 1131.

Slate-blue, cubic crystals. d_4^{20} ~2.0. mp 1085°. Dec at 550° *in vacuo* yielding Mg_3Si_2. Heat of formation: 18.5 cal/mol. Dec by water, HCl.

USE: In semiconductor research. Has been used to build Mg-Si rectifiers.

5729. Magnesium Stannide. Mg_2Sn; mol wt 167.32. Mg 29.05%, Sn 70.95%. Prepn by fusion of a stoichiometric mixture of the elements: Winkler, *Helv. Phys. Acta* **28**, 633 (1955); Korenblit, Kolesnikov, *Zh. Tekh. Fiz.* **26**, 941 (1956), *C.A.* **50**, 11475b (1956). Magnetic susceptibility and electrical conductivity measurements: Boltaks, *Doklady Akad. Nauk SSSR* **64**, 487, 653, (1949), *C.A.* **43**, 4528f, 4533f (1949). Electrical conductivity, magnetoresistance: Frederikse *et al.*, *Phys. Rev.* **103**, 67 (1956).

Bluish-white, metallic substance. Calcium fluoride crystal structure. mp 778°. Resistivity (25°): 42,000 microhm-cm; thermoelectric power 90 microvolts per degree for 15-100°. Energy of activation: 22,100 cal/mole. Soluble in water, dilute HCl.

USE: In semiconductor research.

5730. Magnesium Stearate. *Octadecanoic acid magnesium salt.* $C_{36}H_{70}MgO_4$; mol wt 591.25. C 73.13%, H 11.93%, Mg 4.11%, O 10.82%. $Mg(C_{18}H_{35}O_2)_2$. The commercial preparation also contains palmitate.

White powder. Insoluble in water; dec by dil acids.

USE: In baby dusting powders; as tablet lubricant.

5731. Magnesium Sulfate. MgO_4S; mol wt 120.37. Mg 20.19%, O 53.17%, S 26.64%. $MgSO_4$. Monohydrate occurs in nature as the mineral *kieserite*.

Trihydrate, odorless crystals. Prepn: Bennett, U.S. pat. **3,297,413** (1967 to Dow).

Heptahydrate, *bitter salts, epsom salts, Mg 5-Sulfat.* Occurs in nature as the mineral *epsomite*. Efflorescent crystals or powder; bitter, saline, cooling taste. d 1.67. Soly in water (g/100 ml): at 20° = 71; at 40° = 91. Slightly sol in alcohol. Its aq soln is neutral. pH 6-7. On exposure to dry air at ordinary temp it loses ~$1H_2O$; at 70-80° loses $4H_2O$; at 100° loses $5H_2O$; at 120° loses $6H_2O$, rapidly reabsorbing water when exposed to moist air; loses the last mol of water at ~250°: Gmelin's, *Magnesium* (8th ed) **27**, 210-211, 223-226 (1939). *Dried magnesium sulfate* is prepd by heating the heptahydrate until ~25% of its weight is lost. Used in *Morison's paste.* Keep well closed.

Caution: Parenteral use or use in presence of renal insufficiency may lead to magnesium intoxication.

USE: Weighting cotton and silk; increasing the bleaching action of chlorinated lime; manuf mother-of-pearl and frosted papers; fire-proofing fabrics; dyeing and printing calicos; in fertilizers; explosives; matches; mineral water; tanning leather.

THERAP CAT: Heptahydrate as anticonvulsant; cathartic.

THERAP CAT (VET): Purgative, in general anesthesia, in hypomagnesemia; externally (in strong solns) in local inflammations, infected wounds.

5732. Magnesium Sulfite. MgO_3S; mol wt 104.37. Mg 23.29%, O 45.99%, S 30.72%. $MgSO_3$. Occurs as tri- and hexahydrates.

Hexahydrate, colorless crystals or white, cryst powder; gradually oxidizes to sulfate on exposure to air; loses all its water at 200°; dec at higher temp. Sol in ~150 parts water, slightly more in boiling water.

USE: A soln of magnesium sulfite in sulfurous acid (magnesium bisulfite) is used in the manuf of paper pulp.

5733. Magnesium Thiocyanate. Magnesium sulfocyanate. $C_2MgN_2S_2$; mol wt 140.47. C 17.10%, Mg 17.30%, N 19.94%, S 45.65%. $Mg(SCN)_2$.

Tetrahydrate, colorless or white deliquesc crystals. Freely sol in water or alcohol. *Keep well closed.*

5734. Magnesium Thiosulfate. Magnesium hyposulfite; Magnosulf; Antichoc Hipmag. MgO_3S_2; mol wt 136.44. Mg 17.81%, O 35.18%, S 47.01%. MgS_2O_3.

Hexahydrate, colorless or white crystals; loses $3H_2O$ at 170°. d 1.82. Sol in 2 parts water; insol in alcohol. Its aq soln is neutral.

5735. Magneson. *4-[(4-Nitrophenyl)azo]-1,3-benzenediol; 4-(p-nitrophenylazo)resorcinol; 2,4-dihydroxy-4'-nitroazobenzene.* $C_{12}H_9N_3O_4$; mol wt 259.22. C 55.60%, H 3.50%, N 16.21%, O 24.69%. *Ref:* Suitzu, Okuma, *J. Soc. Chem. Ind. (Japan)* **29**, 132 (1926); Ruigh, *J. Am. Chem. Soc.* **51**, 1456 (1929); Engel, *ibid.* **52**, 1812 (1930).

Brownish-red powder. Practically insol in water; sol in dil aq NaOH.

USE: For the detection of magnesium with which it yields a bright blue color in alkaline soln. Also used to determine molybdenum with which it forms a red-violet complex: Nikitina, Andrianova, *C.A.* **75**, 136789v (1971).

5736. Magnoflorine. *(S)-5,6,6a,7-Tetrahydro-1,11-dihydroxy-2,10-dimethoxy-6,6-dimethyl-4H-dibenzo[de,g]quinolinium;* thalictrine. $[C_{20}H_{24}NO_4]^+$. From *Magnolia grandiflora* L., *Magnoliaceae*. Isoln and structure: Nakano, *Chem. Pharm. Bull.* **2**, 329 (1954). From *Cocculus trilobus*

D.C., *Menispermaceae:* Nakano, *ibid.* **4,** 69 (1956); from *Thalictrum thunbergii* D.C., *Ranunculaceae:* Fujita, Tomimatsu, *ibid.* **6,** 107 (1958); from *Aristolochia clematitis* L., *Aristolochiaceae:* Pailer, Pruckmayr, *Monatsh.* **90,** 145 (1959). Identity with thalictrine: Gopinath *et al., J. Sci. & Ind. Res. (India)* **18B** 444 (1959). Synthesis: Tomita, Kikkawa, *J. Pharm. Soc. Japan* **77,** 195 (1957).

Iodide, $C_{20}H_{24}INO_4$, crystals from methanol + acetone, dec 248-249°. $[\alpha]_D^{15}$ 220.1° (methanol). uv max: 270, 310 nm (log ϵ 3.75, 3.59).

5737. Magnoline. *1,2,3,4-Tetrahydro-1-[[4-[2-hydroxy-5-[(1,2,3,4-tetrahydro-7-hydroxy-6-methoxy-2-methyl-1-isoquinolinyl)methyl]phenoxy]phenyl]methyl]-6-methoxy-2-methyl-7-isoquinolinol.* $C_{36}H_{40}N_2O_6$; mol wt 596.72. C 72.46%, H 6.76%, N 4.69%, O 16.09%. From bark of *Magnolia macrophylla* Michx. and *M. fuscata* Andr., *Magnoliaceae:* Lloyd, *Am. J. Pharm.* **1891,** 438; Proskurnina, Orékhov, *Bull. Soc. Chim. France* **5,** 1357 (1938). Structure: *eidem, C.A.* **35,** 2520⁹ (1941).

Crystals, mp 178-179°. $[\alpha]_D$ −9.6° (pyridine). Slightly sol in alcohol, acetone, chloroform; practically insol in water, benzene, ether.

Trimethyl ether, $C_{39}H_{46}N_2O_6$, crystals, mp 109-110°.

5738. Magnolol. *5,5'-Di-2-propenyl-[1,1'-biphenyl]-2,2'-diol;* 5,5'-diallyl-2,2'-biphenyldiol; 2,2'-bichavicol; dehydrodichavicol; 2,2'-dihydroxy-5,5'-diallylbiphenyl. $C_{18}H_{18}O_2$; mol wt 266.34. C 81.17%, H 6.81%, O 12.01%. Bioactive constituent of *Magnoliae Cortex,* the bark of *Magnolia officinalis,* Rehd. et Wils., *Magnoliaceae,* known in Chinese traditional medicine as *houpo,* or of *M. obovata,* Thunb., called *wakoboku* in Japanese. Isoln: Y. Sugii, *Yakugaku Zasshi* **50,** 183 (1930), *C.A.* **24,** 3505 (1930). Synthesis: H. Erdtman, J. Runeberg, *Acta Chem. Scand.* **11,** 1060 (1957); J. Runeberg, *ibid.* **12,** 188 (1958). Isoln from seeds of *M. grandiflora* L.: F. S. El-Feraly, W.-S. Li, *Lloydia* **41,** 442 (1978). CNS depressant effects: K. Watanabe *et al., Japan. J. Pharmacol.* **25,** 605 (1975); *eidem, Planta Med.* **49,** 103 (1983). Anti-inflammatory and analgesic effects: J.-P. Wang *et al., Arch. Pharmacol.* **346,** 707 (1992). HPLC determn: T.-H. Tsai, C.-F. Chen, *J. Chromatog.* **598,** 143 (1992).

mp 101.5-102°. uv max: 293 nm (log ϵ 3.90). Sol in ethanol.

5739. Malachite Green. *N-[4-[[4-(Dimethylamino)-phenyl]phenylmethylene]-2,5-cyclohexadien-1-ylidene]-N-*

methylmethanaminium chloride; bis[p-(dimethylamino)phenyl]phenylmethylium chloride; C.I. Basic Green 4; C.I. 42000; Aniline Green; China Green; Victoria Green B; Victoria Green WB; New Victoria Green Extra O; New Victoria Green Extra I; New Victoria Green Extra II; Diamond Green B; Diamond Green Bx; Diamond Green P Extra; Solid Green O; Light Green N; Benzal Green; Benzaldehyde Green; Fast Green. $C_{23}H_{25}ClN_2$; mol wt 364.92. C 75.70%, H 6.91%, Cl 9.72%, N 7.68%. Triphenylmethane dye with fungicidal and limited antiseptic activity. Prepn: *Colour Index* **Vol. 4** (3rd ed., 1971) p 4380. For dyeing it is prepd as a double salt with $ZnCl_2$. Conformational changes in excited state: S. Saikan, J. Sei, *J. Chem. Phys.* **79,** 4154 (1983). Use in electron microscopy: R. G. Pourcho *et al., Stain Technol.* **53,** 29 (1978). As stain in determn of prostaglandin E and F compounds: E. J. Singh *et al., J. Chromatog.* **105,** 195 (1975). In determn of inorganic phosphate: W. Hohenwallner, E. Wimmer, *Clin. Chim. Acta* **45,** 169 (1973); S. G. Carter, D. W. Karl, *J. Biochem. Biophys. Methods* **7,** 7 (1982). Antiseptic activity: J. E. Madden *et al., Surg. Forum* **22,** 63 (1971). Toxicology: N. Brock, A. Erhardt, *Arzneimittel-Forsch.* **1,** 5 (1951); T. D. Bills, L. L. Marking, *Investigations in Fish Control* **75,** 6 (1977); S. Clemmensen *et al., Arch. Toxicol.* **56,** 43 (1984). Reviews of use as fungicide in fish culture: N. C. Nelson, *U.S. NTIS Reps.* PB-235450 (1974) 79 pp, *C.A.* **82,** 150006p (1975), PB-235451 (1974) 33 pp, *C.A.* **83,** 1547j (1975); of use as biological stain: R. D. Lillie, *H. J. Conn's Biological Stains* (Williams & Wilkins Co., Baltimore, 9th ed., 1977) 248-250, 579-580.

Green crystals with metallic luster. Very sol in water; sol in alcohol, methanol, amyl alcohol. Water solns are blue-green, absorption max 616.9 nm; yellow below pH 2. LD_{50} in mice (mg/kg): 80 orally; 4.2 i.p. (Brock, Erhardt). LC_{50} in various fish species (mg/l): 0.0305-0.383 (Bills, Marking).

Note: The term malachite green applies to the oxalate as well as the chloride.

USE: For directly dyeing silk, wool, jute and leather; dyeing cotton after mordanting. Biological stain. Clinical reagent (inorganic phosphate assay). As spot test reagent for detecting sulfurous acid and cerium. As acid-base indicator: pH: 0.0 yellow, 2.0 green; 11.6 blue-green, 14 colorless.

THERAP CAT (VET): Fungicide and parasiticide in fish.

5740. Malathion. *[(Dimethoxyphosphinothioyl)thio]butanedioic acid diethyl ester;* diethyl mercaptosuccinate *S*-ester with *O,O*-dimethyl phosphorothioate; *S*-(1,2-dicarbethoxyethyl) *O,O*-dimethyldithiophosphate; insecticide no. 4049; carbofos; malathon (obsolete); mercaptothion; phosphothion; ENT-17034; Cythion; Derbac-M; Malamar 50; Malaspray; Organoderm; Prioderm; Suleo-M. $C_{10}H_{19}O_6PS_2$; mol wt 330.36. C 36.36%, H 5.80%, O 29.06%, P 9.38%, S 19.41%. Cholinesterase inhibitor. Prepn: Johnson *et al., J. Econ. Entomol.* **45,** 279 (1952); Cassaday, U.S. pat. **2,578,-652** (1951 to Am. Cyanamid). Purification: Usui, U.S. pat. **2,962,521** (1960 to Sumitomo). Treatment of *Pediculus humanus* (head lice) infestation: D. Taplin *et al., J. Am. Med. Assoc.* **247,** 3103 (1982). Toxicity data: T. B. Gaines, *Toxicol. Appl. Pharmacol.* **14,** 515 (1969). Review of distribution, transport and fate in the environment: M. S. Mulla *et al., Residue Rev.* **81,** 1-159 (1981); of carcinogenic risk: *IARC Monographs* **30,** 103-129 (1983).

Deep brown to yellow liq, mp 2.9°. bp$_{0.7}$ 156-157°. Characteristic odor. d$_4^{25}$ 1.23. n$_D^{25}$ 1.4985. Vapor pressure at 30°: 4×10^{-5} mm Hg. Slightly sol in water (145 ppm). Misc with many organic solvents including alcohols, esters, ketones, ethers, aromatic and alkylated aromatic hydrocarbons and vegetable oils. Limited soly in certain paraffin hydrocarbons. Petroleum ether is sol to about 35% in malathion. Hydrolyzed at pH > 7.0 or < 5.0. Stable in an aq soln buffered to pH 5.26. LD$_{50}$ in female, male rats (mg/kg): 1000, 1375 orally (Gaines).

Caution: Potential symptoms of overexposure are miosis, aching eyes, blurred vision and lacrimation; eye and skin irritation; salivation; anorexia, nausea, vomiting, abdominal cramps, diarrhea, giddiness, confusion and ataxia; rhinorrhea, headache; tightening of chest, wheezing and laryngeal spasms. *See NIOSH Pocket Guide to Chemical Hazards* (DHHS/NIOSH 90-117, 1990) p 138.

USE: Insecticide.

THERAP CAT: Pediculicide.

THERAP CAT (VET): Ectoparasiticide.

5741. Maleamic Acid. *(Z)-4-Amino-4-oxo-2-butenoic acid;* maleic acid monoamide. C$_4$H$_5$NO$_3$; mol wt 115.09. C 41.75%, H 4.38%, N 12.17%, O 41.71%. Prepd by passing ammonia into a mixture of maleic anhydride and an inert solvent such as xylene, dioxane, etc.: Robinson, Humburger, U.S. pat. **2,459,964** (1949 to Beck, Koller).

Crystals from alcohol, mp 172-173°. Also reported as mp 178-180° (dec). Very sol in water, hot alcohol. Practically insol in ether, chloroform, benzene.

5742. Maleanilic Acid. *(Z)-4-Oxo-4-(phenylamino)-2-butenoic acid;* N-phenylmaleamic acid. C$_{10}$H$_9$NO$_3$; mol wt 191.19. C 62.82%, H 4.74%, N 7.33%, O 25.11%. Use of esters as fungicides: Ligett *et al.*, U.S. pat. **2,885,319** (1959 to Pittsburgh Coke & Chem.).

Monoclinic yellow crystals, dec 192°. Apparent density 1.418 g/ml at 30°. Soly in g/100 ml soln at 27°: acetonitrile 0.2; acetone 0.3; benzene < 0.1; butyl Cellosolve 1.1; carbon tetrachloride < 0.1; chloroform < 0.1; dimethylformamide 12.9; dioxane 0.9; ethanol 0.2; ether < 0.1; methanol 0.3; toluene < 0.1; water < 0.1. Shows hydrolysis in aq soln. Dissociates to maleic anhydride and aniline when heated above 100° *in vacuo*.

USE: Esters as fungicides.

5743. Maleic Acid. *(Z)-Butenedioic acid;* toxilic acid; *cis*-1,2-ethylenedicarboxylic acid. C$_4$H$_4$O$_4$; mol wt 116.07. C 41.39%, H 3.47%, O 55.14%. Prepd by the catalytic oxidation of benzene over vanadium pentoxide: Bhattacharyya, Venkataraman, *J. Appl. Chem. (London)* **8**, 728 (1958); Saffer, Olenberg, Fr. pat. **1,321,416** (1963 to Scientific Design), *C.A.* **59**, 11265d (1963). Crystal structure: Shahat, *Acta Cryst.* **5**, 763 (1952). *Review:* W. D. Robinson, R. A. Mount in Kirk-Othmer *Encyclopedia of Chemical*

Technology **vol. 14** (Wiley-Interscience, New York, 3rd ed., 1981) pp 770-793.

White crystals from water, mp 138-139°; from alcohol and benzene, mp 130-131°. Faint, acidulous odor; characteristic repulsive, astringent taste. d 1.59. Is converted in part into the much higher-melting fumaric acid (mp 287°) when heated to a temp slightly above the melting point. Freely sol in water or alcohol. Sol in acetone, glacial acetic acid. Slightly sol in ether. Practically insol in benzene.

Caution: Strong irritant.

USE: Manuf artificial resins; to retard rancidity of fats and oils in 1:10,000 (these are said to keep 3 times longer than those without the acid); dyeing and finishing wool, cotton, and silk; preparing the maleate salts of antihistamines and similar drugs.

5744. Maleic Anhydride. *2,5-Furandione;* cis-butenedioic anhydride; toxilic anhydride. C$_4$H$_2$O$_3$; mol wt 98.06. C 49.00%, H 2.06%, O 48.95%. May be prepd by sublimation of maleic acid and P$_2$O$_5$ under reduced pressure: Kempf, *J. Prakt. Chem.* [2] **78**, 239 (1908). Commercial production by catalytic vapor-phase oxidation of benzene or other suitable hydrocarbons: Weiss, Downs, *Ind. Eng. Chem.* **12**, 228 (1920); U.S. pat. **1,318,633** (1920 to Barrett Co.). Many other syntheses. Review of commercial methods of manufacture: Ashcroft, Clifford, *Chem. Prods.* **24**, 11 (1961), *C.A.* **55**, 9724d (1961); Faith, Keyes & Clark's *Industrial Chemicals*, F. A. Lowenheim, M. K. Moran, Eds. (Wiley-Interscience, New York, 4th ed., 1975) pp 514-518. *Review:* W. D. Robinson, R. A. Mount in Kirk-Othmer *Encyclopedia of Chemical Technology* **vol. 14** (Wiley-Interscience, New York, 3rd ed., 1981) pp 770-793. Book: B. C. Trivedi, B. M. Culbertson, *Maleic Anhydride* (Plenum, New York, 1982) 872 pp.

Orthorhombic needles from chloroform; also readily by sublimation. Commercial grades are furnished in fused form, as briquettes. d 1.48. mp 52.8°. bp$_{760}$ 202.0°; bp$_{400}$ 179.5°; bp$_{200}$ 155.9°; bp$_{100}$ 135.8°; bp$_{60}$ 122.0°; bp$_{40}$ 111.8°; bp$_{20}$ 95.0°; bp$_{10}$ 78.7°; bp$_5$ 63.4°. Specific heat: 0.285 (solid); 0.396 (liq). Sol in water, forming maleic acid. Soly at 25° (g/100 g): acetone 227; ethyl acetate 112; chloroform 52.5; benzene 50; toluene 23.4; *o*-xylene 19.4; carbon tetrachloride 0.60; ligroin 0.25. Sol in dioxane. Sol in alc with ester formation.

Caution: Potential symptoms of overexposure are conjunctivitis; photophobia, double vision; nasal and upper respiratory irritation; bronchial asthma; dermatitis. *See NIOSH Pocket Guide to Chemical Hazards* (DHHS/NIOSH 90-117, 1990) p 138.

USE: In Diels-Alder syntheses (as a dienophile), manuf alkyd-type of resins, dye intermediates, pharmaceuticals, agricultural chemicals (maleic hydrazide, malathion), in copolymerization reactions.

5745. Maleic Hydrazide. *1,2-Dihydro-3,6-pyridazinedione;* maleic acid hydrazide; MH; Fazor; Malazide; Regulox. C$_4$H$_4$N$_2$O$_2$; mol wt 112.09. C 42.86%, H 3.60%, N 24.99%, O 28.55%. Prepd by treating maleic anhydride with hydrazine hydrate in alcohol: Arndt *et al.*, *C.A.* **43**, 579 (1949); *see also* Curtius, Foerstinger, *J. Prakt. Chem.* [2] **51**, 391 (1895). From maleic acid and a hydrazine salt of a strong inorganic acid: Harris, Schoene, U.S. pat. **2,575,954** (1951 to U.S. Rubber). Alternate prepn from hydrazine sulfate and maleic anhydride in aqueous NaOH: Amatsu, Karasawa, *C.A.* **51**, 18014c (1957); from hydrazine hydrate and maleic anhydride in glacial acetic acid: Feuer *et al.*, *J. Am. Chem. Soc.* **80**, 3790 (1958). Has the ability to inhibit growth of plants without killing them: Schoene, Hoffmann, *Science* **109**, 588 (1949). Toxicity data: R. Ben-Dyke *et al.*,

World Rev. Pest Contr. **9**, 119 (1970). *Review:* Massey, *Mfg. Chem.* **26**, 197-200 (1955). Review and chromatographic studies: Fishbein, *Chromatography of Environmental Hazards* (Elsevier, New York, 1972) pp 161-166. Review of toxicology: R. Ponnampalam *et al., Regul. Toxicol. Pharmacol.* **3**, 38-47 (1983).

Crystals from water, dec 260°. Also reported as mp > 300° (Feuery). Slightly sol in hot alcohol, more sol in hot water. LD_{50} orally in rats: 3800-6800 mg/kg; dermally in rabbits: > 4000 mg/kg (Ben-Dyke).

USE: Experimentally in horticulture and agriculture. To control suckering of tobacco. In the synthesis of pyridazine.

5746. Maleuric Acid. *(Z)-4-[(Aminocarbonyl)amino]-4-oxo-2-butenoic acid; N-carbamoylmaleamic acid;* maleylurea. $C_5H_6N_2O_4$; mol wt 158.11. C 37.98%, H 3.82%, N 17.72%, O 40.48%. Prepd from maleic anhydride and urea: Dunlap, Phelps, *Am. Chem. J.* **19**, 492 (1897); Batt *et al., J. Am. Chem. Soc.* **76**, 3663 (1954).

Crystals from hot water, dec 158.5°. uv max: 235 nm (ϵ 8720). Sol in hot acetic acid. Practically insol in cold water, cold acetic acid, acetone, ligroin, chloroform, alc, ether.

Methyl ester, crystals, mp 113-114°. Sol in hot water, methanol, ethanol, acetone, dioxane.

Butyl ester, crystals, mp 95-98°. Sol in ethanol, acetone, benzene, chloroform.

5747. Malic Acid. *Hydroxybutanedioic acid;* hydroxysuccinic acid. $C_4H_6O_5$; mol wt 134.09. C 35.83%, H 4.51%, O 59.66%. The naturally occurring isomer is the L-form which has been found in apples and many other fruits and plants. Prepn of D- and DL-forms, and resolution of racemic mixture: McKenzie *et al., J. Chem. Soc.* **123**, 2875 (1923). Solubilities: Descamps, *Bull. Soc. Chim. Belg.* **49**, 91 (1940). Microbial production of L-form: Kitahara; Abe *et al.,* U.S. pats. **2,972,566; 3,063,910** (1961, 1962, both to Kyowa). Configuration: J. A. Mills, W. Klyne in *Progress in Stereochemistry* vol. 1, W. Klyne, Ed. (Academic Press, New York, 1954) pp 182-183; E. L. Eliel, *Stereochemistry of Carbon Compounds* (McGraw-Hill, New York, 1962) pp 97-98; Cymerman-Craig, Roy, *Tetrahedron* **21**, 1847 (1965). Review of uses: *Mfg Chemist Aerosol News* **35**, 56 (December, 1964). *Review:* S. E. Berger, "Hydroxy Dicarboxylic Acids" in Kirk-Othmer *Encyclopedia of Chemical Technology* vol **13** (Wiley-Interscience, New York, 3rd ed., 1981) pp 103-110.

L-Malic Acid

DL-Form, crystals, mp 131-132°. Soly in g/100 g solvent at 20°: methanol 82.70, diethyl ether 0.84, ethanol 45.53, acetone 17.75, dioxane 22.70, water 55.8. Practically insol in benzene.

D-(+)-Form, crystals, mp 101°.

L-(−)-Form, *apple acid.* Crystals from acetone, or acetone + CHCl₃, mp 100°. Dec ~140°. $[\alpha]_D$ −2.3° (c = 8.5). Soly in g/100 g solvent at 20°: methanol 197.22, diethyl ether 2.70, ethanol 86.60, acetone 60.66, dioxane 74.35, water 36.35. Practically insol in benzene.

USE: Intermediate in chemical synthesis. Chelating and buffering agent. Flavoring agent, flavor enhancer and acidulant in foods.

5748. Mallow. Common mallow; high mallow; cheeseflower. Leaves of *Malva sylvestris* L., and *M. rotundifolia* L., *Malvaceae. Habit.* Europe, Asia, naturalized in U.S. *Constit.* Pectin, tannin, coloring matter.

5749. Malonic Acid. *Propanedioic acid;* methanedicarboxylic acid. $C_3H_4O_4$; mol wt 104.06. C 34.63%, H 3.87%, O 61.50%. $HOOCCH_2COOH$. Prepd from malic acid: Dessaignes, *Ann.* **107**, 251 (1858). Made by the interaction of monochloroacetic acid and NaCN followed by hydrolysis of the resulting cyanoacetic acid: Weiner, *Org. Syn.* coll. vol. II, 376 (1943). Prepn from diethyl malonate: Britton, Monroe, U.S. pat. **2,373,011** (1945 to Dow); from ligneous wastes: Grangaard, U.S. pat. **2,928,868** (1960 to Kimberly-Clark); from sodium acetate: Normant, Angelo, *Bull. Soc. Chim. France* **1962**, 810. *Review:* D. W. Hughes in Kirk-Othmer *Encyclopedia of Chemical Technology* vol. **14** (Wiley-Interscience, New York, 3rd ed., 1981) pp 794-810.

Small crystals, mp ~135° with decompn; sublimes *in vacuo*. d 1.63. One gram dissolves in 0.65 ml water, about 2 ml alcohol, 1.1 ml methanol, 3 ml propyl alcohol, 13 ml ether, 7 ml pyridine.

Diethyl Ester, *see* Ethyl Malonate.

Caution: Strong irritant.

USE: In manuf of barbiturates.

5750. Malononitrile. *Propanedinitrile;* methylene cyanide; dicyanomethane; cyanoacetonitrile. $C_3H_2N_2$; mol wt 66.06. C 54.54%, H 3.05%, N 42.40%. $CH_2(CN)_2$. Prepn: Henri, *Compt. Rend.* **102**, 1394 (1886); B. B. Corson *et al., Org. Syn.* coll. vol. **2**, 379 (1943); A. R. Surrey, *J. Am. Chem. Soc.* **65**, 2471 (1943); *idem,* U.S. pat. **2,389,217** (1945 to Winthrop); *idem, Org. Syn.* coll. vol. **3**, 535 (1955). Toxicity study: G. R. N. Jones, M. S. Israel, *Nature* **228**, 1315 (1970). *Reviews:* F. Freeman, *Chem. Rev.* **69**, 591-624 (1969); A. J. Fatiadi, *Synthesis* **1978**, 165-204, 241-282.

Colorless solid, mp 32°. bp₇₆₀ 218-219°, bp₂₀ 109°. d_4^{20} 1.1910; n_D^{34} 1.4146. Very sol in alc, ether; sol in water, acetone, benzene. LD_{50} i.p. in mice: 12.9 mg/kg (Jones, Israel).

USE: In organic synthesis.

5751. Malotilate. *1,3-Dithiol-2-ylidenepropanedioic acid bis(1-methylethyl) ester;* diisopropyl 1,3-dithiol-2-ylidenemalonate; NKK-105; Hepation; Kantec. $C_{12}H_{16}O_4S_2$; mol wt 288.39. C 49.98%, H 5.59%, O 22.19%, S 22.24%. Prepn from diisopropyl malonate: K. Taninaka *et al.,* Belg. pat. **834,631** corresp to U.S. pat. **4,035,387** (1976, 1977 both to Nihon Nohyaku); from the corresponding ketene mercaptide: H. Matsui *et al.,* U.S. pat. **4,327,223** (1982 to Nihon Nohyaku). Effect on CCl₄-induced liver injury in rats: Y. Imaizumi *et al., Japan. J. Pharmacol.* **31**, 15 (1981). Enhancement of rat liver protein synthesis: *eidem, ibid.* **32**, 369 (1982). Pharmacokinetics and pharmacodynamics: M. Buhrer *et al., Eur. J. Clin. Pharmacol.* **30**, 407 (1986). Clinical evaluation in liver cirrhosis: S. Takase *et al., Gastroenterol. Japan* **23**, 639 (1988).

Pale yellow crystals, mp 60.5°. Sol in benzene, cyclohexane, *n*-hexane, ether.

THERAP CAT: Hepatoprotectant.

5752. Maltol. *3-Hydroxy-2-methyl-4H-pyran-4-one;* 3-hydroxy-2-methyl-4-pyrone; 3-hydroxy-2-methyl-γ-pyrone; larixinic acid; Palatone; Veltol. $C_6H_6O_3$; mol wt 126.11. C 57.14%, H 4.80%, O 38.06%. Found in the bark of young larch trees (*Larix decidua* Mill.), in pine needles (*Abies alba* Mill., *Pinaceae),* in chicory, in wood tars and oils, in roasted malt. Isoln from these sources and structure: Kiliani, Bazlen, *Ber.* **27**, 3115 (1894); Feuerstein, *Ber.* **34**, 1804 (1901); Erdmann, Schaefer, *Ber.* **43**, 2398 (1910); Reichstein, Beitter, *Ber.* **63**, 824 (1930), *cf.* Peratoner, Tamburello, *Chem. Zentr.* **76**, II 680 (1905). Also obtained by alkaline hydrolysis of streptomycin salts: Schenck, Spielman, *J. Am. Chem. Soc.* **67**, 2276 (1945). Synthesis: Spielman, Freifelder, *ibid.* **69**, 2908 (1947); Chawla, McGonigal, *J. Org. Chem.* **39**, 3281 (1974). Novel high-yield synthesis: T. M. Brennan *et al., Tetrahedron Letters* **1978**, 331; P. D. Weeks *et al., J. Org. Chem.* **45**, 1109 (1980). History and comparison with isomaltol: Hodge, Nelson, *Cereal Chem.* **38**, 207 (1961).

Monoclinic prisms from chloroform, orthorhombic bipyramidal crystals + monoclinic prisms from 50% methanol, mp 161-162°. Fragrant, caramel-like odor. Begins to sublime at 93°. Volatile with steam. uv max (0.1*N* HCl): 274 nm (E_m 8400); (0.1*N* NaOH): 317 nm (E_m 7300). pH of 0.5% aq soln 5.3. One gram dissolves in 85 ml water. Freely sol in hot water, chloroform; sol in alcohol; sparingly sol in benzene, ether, petr ether; sol in aq alkali hydroxides giving yellow solns.

USE: Flavoring agent, to impart "freshly baked" odor and flavor to bread and cakes.

5753. Maltose. *4-O-α-D-Glucopyranosyl-D-glucose;* malt sugar; maltobiose; 4-(α-D-glucosido)-D-glucose; Maltos; Martos-10. $C_{12}H_{22}O_{11}$; mol wt 342.30. C 42.11%, H 6.48%, O 51.41%. Monohydrate obtained in about 80% yield by enzymatic (diastase) degradation of starch: Gore, U.S. pat. **1,657,079** (1928). Structure: Haworth, Peat, *J. Chem. Soc.* **1926**, 3094; Hassid, Ballou in W. Pigman, *The Carbohydrates* (Academic Press, New York, 1957) p 498. Crystal and molecular structure: F. Takusagawa, R. A. Jacobson, *Acta Crystallogr.* **B34**, 213 (1978). *Review:* E. Tarelli, *Dev. Food Carbohydr.* **2**, 187-227 (1980); R. Khan, *Carbohydr. Chem. Biochem.* **39**, 213-278 (1981).

Monohydrate, crystals from water or dil alc, mp 102-103°. Does not lose its water of crystn by drying at room temp *in vacuo* over H_2SO_4 or P_2O_5. About one-third as sweet as sucrose. *Cf.* Isbell, Pigman, *J. Res. NBS* **18**, 141 (1937). Shows mutarotation. $[\alpha]_D^{20}$ +111.7° → +130.4° (c = 4). pKa (21°): 12.05. Sol in water, slightly sol in alcohol. Practically insol in ether.

USE: Nutrient, sweetener, in culture media, in prepd bee food. Parenteral supplement of sugar for diabetics. Fermentable intermediate in brewing. Stabilizer for polysulfides. In pharmaceutical dispensing.

5754. Malvidin Chloride. *3,5,7-Trihydroxy-2-(4-hydroxy-3,5-dimethoxyphenyl)-1-benzopyrylium chloride; 3,4',5,7-tetrahydroxy-3',5'-dimethoxyflavylium chloride;* enidin; primulidin; syringidin; 3,4',5,7-tetrahydroxy-3',5'-dimethoxy-2-phenylbenzopyrylium chloride. $C_{17}H_{15}ClO_7$; mol wt 366.75. C 55.67%, H 4.12%, Cl 9.67%, O 30.54%. Found as the diglucoside (malvin) in wild malve (*Primula viscosa* All., *Primulaceae*) and as the monoglucoside in blue grapes: Willstätter, Mieg, *Ann.* **408**, 122 (1915); Karrer, Widmer, *Helv. Chim. Acta* **10**, 5 (1927). Structure: Anderson, Nabenhauer, *J. Am. Chem. Soc.* **48**, 2997 (1926). Synthesis: Bradley, Robinson, *J. Chem. Soc.* **1928**, 1541.

Prisms or rhombic tablets, appearing red by transmitted light with a green luster when dry or with a steel-blue luster when in contact with solvent. Usually obtained as the mono- or dihydrate. The anhydr salt is very hygroscopic. Not melted at 300°. Sol in abs alc, giving a violet-red soln. Sparingly sol in water. Also sol in amyl alc. In methanol the substance is first sol with purple color, then begins to separate as red crystals which are violet by transmitted light.

3-β-Glucoside tetrahydrate, $C_{23}H_{25}ClO_{12}$·4H$_2$O, *3-(glucosyloxy)-4',5,7-trihydroxy-3',5'-dimethoxyflavylium chloride,* enin, cyclamin. Synthesis: Levy *et al., J. Chem. Soc.* **1931**, 2701. Dark prisms with a green metallic shine, from alcohol + HCl. Sparingly sol in water, alcohol, glycerol. Practically insol in benzene, chloroform, ether.

3,5-Diglucoside, $C_{29}H_{35}ClO_{17}$, *3,5-bis(glucosyloxy)-4',7-dihydroxy-3',5'-dimethoxyflavylium chloride,* malvin, malvoside. Synthesis: Robinson, Todd, *ibid* **1932**, 2299. Reddish-brown prisms or needles with a green shine from dil HCl, dec 165°.

3-Galactoside hemihendecahydrate, $C_{23}H_{25}ClO_{12}$·5½H$_2$O, *primulin.* Bronze metallic needles or prisms from methanol + HCl. Bluish-violet by transmitted light. Very sol in methanol, ethanol, cold water, forming deep red solns; slightly sol in acetone. Practically insol in ethyl acetate.

5755. Manaca. Manacan; camganiba; geratacaca; vegetable mercury. Dried root of *Brunfelsia hopeana* Benth. (*Franciscea uniflora* Pohl), *Solanaceae*. Habit. Brazil. *Constit.* The alkaloid manacine, a second principle probably identical with gelsemic acid, and a fluorescent substance probably identical with esculetin.

THERAP CAT: Formerly as antisyphilitic.

5756. Mancozeb. *[[1,2-Ethanediylbis(carbamodithioato)](2—)]manganese mixt. with [[1,2-ethanediylbis(carbamodithioato)](2—)]zinc; ethylenebis(dithiocarbamic acid) manganese zinc complex;* manzeb; manganese ethylenebis-(dithiocarbamate) (polymeric) complex with zinc salt; zinc manganese ethylenebisdithiocarbamate; Dithane M-45; Manzate; Manzin; Nemispor; Penncozeb; Vondozeb. Polymeric salt of ethylenebisdithiocarbamic acid containing approx 20% manganese and 2.5% zinc; related to maneb and zineb, *q.q.v.* Displays activity against a wide range of foliage fungal diseases. Prepn: **New Zealand** pat. **131,543**; C. B. Lyon *et al.,* U.S. pat. **3,379,610** (1965, 1968 both to Rohm & Haas). Metabolic fate in plants, animals: W. R. Lyman, "Metabolic fate of Dithane M-45," in *Pestic. Terminal Residues,* Invited Pap. Int. Symp., A. S. Tahori, Ed. (Butterworth, New York, 1971) pp 243-256. Residue studies: *WHO Pestic. Residues Ser.* **vol. 4**, (WHO, FAO, 1975) 451; W. H. Newsome, *J. Agric. Food Chem.* **27**, 1188 (1979). HPLC determn: K. Gustafsson *et al., ibid.* **29**, 729 (1981). Demonstrates low phytotoxicity: L. H. Cornford, *Proc. 36th N.Z. Weed and Pest Control Conf.,* 258 (1983). In potato blight: H. W. Platt, *Can. J. Plant Pathol.* **5**, 38 (1983). *In vitro* activity against *Alternaria alternata:* V. A. Bourbos *et al., Brit. Crop Prot. Council Monograph* No. 31

x : y   10:1

(Fungic. Crop Prot., Vol. 2), 461 (1985). Aquatic toxicology: C. J. Van Leeuwen *et al., Aquat. Toxicol.* **7**, 145 (1985). Insol in water.

USE: Fungicide.

5757. Mandelic Acid. α-*Hydroxybenzeneacetic acid; dl*-mandelic acid; racemic mandelic acid; α-hydroxy-α-toluic acid; α-hydroxyphenylacetic acid; phenylhydroxyacetic acid; phenylglycolic acid; amygdalic acid; amygdalinic acid; para-mandelic acid; Uromaline. $C_8H_8O_3$; mol wt 152.15. C 63.15%, H 5.30%, O 31.55%. $C_6H_5CH(OH)COOH$. Prepd by the action of warm, dil alkali upon dichloroacetophenone: *Org. Syn.* **23**, 48 (1943); by hydrolysis of mandelonitrile (prepd from benzaldehyde and hydrogen cyanide or from benzaldehyde, sodium bisulfite, and sodium cyanide): *Org. Syn.* **coll. vol. I**, 336 (1941); *ibid.* **coll. vol. III**, 538 (1955); L. F. Fieser, *Organic Experiments* (Boston, 1964) p 109. May be prepd by boiling amygdalin with HCl.

Orthorhombic plates from water, mp 119°. Darkens and dec on prolonged exposure to light. d 1.30. Can be distilled rapidly *in vacuo* at 2 mm without much decompn. Acid to litmus. pK (25°) 3.37. One gram dissolves in 6.3 ml water, 1 ml alc; freely sol in ether, isopropyl alcohol.

Calcium salt, $C_{16}H_{14}CaO_6$, *calcium mandelate, Camdelate.* Contains 88.88% mandelic acid. White powder. One gram dissolves in 80 ml boiling water. Slightly sol in cold water. Insol in alc. The aq soln is slightly acid.

Sodium salt, $C_8H_7NaO_3$, *sodium mandelate.* Cryst powder; slight aromatic odor. Very sol in water; sol in alcohol; the aq soln is neutral or slightly alkaline to litmus.

THERAP CAT: Antiseptic (urinary).

5758. Mandelic Acid Isoamyl Ester. α-*Hydroxybenzeneacetic acid 3-methylbutyl ester; mandelic acid isopentyl ester;* isoamyl mandelate; Atractyl; Spasmol; Mandaverm; Vermiparin; Spasmostenyl. $C_{13}H_{18}O_3$; mol wt 222.28. C 70.24%, H 8.16%, O 21.59%. Prepd by esterification of DL-mandelic acid with an excess of isoamyl alcohol in the presence of concd H_2SO_4: Rona *et al., Biochem. Z.* **181**, 50 (1927). Improved procedure: Brock *et al., Arzneimittel-Forsch.* **2**, 166 (1952).

Oily liquid. bp$_{11}$ 172° (Rona); bp$_{12}$ 155° (Brock). Practically insol in water. Miscible with fat solvents. Easily hydrolyzed by esterases.

Caution: The pure, undiluted liquid is a skin irritant.

THERAP CAT: Adjunct in the treatment of gastric and intestinal spasms.

THERAP CAT (VET): Has been used as an anthelmintic.

5759. Mandelonitrile. α-*Hydroxybenzeneacetonitrile;* mandelic acid nitrile; benzaldehyde cyanohydrin. C_8H_7NO; mol wt 133.15. C 72.17%, H 5.30%, N 10.52%, O 12.02%. $C_6H_5CH(OH)CN$. Isoln of *d*-form from peach flower buds: Jones, Enzie, *Science* **134**, 284 (1961). Prepn of *d*-form by hydrolysis of amygdalin: Auld, *J. Chem. Soc.* **95**, 927 (1909); Smith, *Ber.* **64**, 427 (1931). Resolution of *dl*-form: Feist, *Arch. Pharm.* **247**, 226 (1909). Synthesis of *l*-form using cotton fibers: Bredig, Gerstner, *Biochem. Z.* **250**, 414 (1932). Asymmetric synthesis: Krieble, Wieland, *J. Am. Chem. Soc.* **43**, 164 (1921); Prelog, Wilhelm, *Helv. Chim. Acta* **37**, 1634 (1954); Tsuboyama, *Bull. Chem. Soc. Japan* **35**, 1004 (1962). It has been suggested that mandelonitrile may be responsible for the alleged anticancer activity of Laetrile (*q.v.*): see Culliton, *Science* **182**, 1000 (1973).

dl-Form, yellow, oily liquid, mp −10°. Decomp at 170°. d 1.115-1.120. Almost insol in water. Freely sol in alcohol, chloroform, ether. LD s.c. in rabbits: 6 mg/kg.

d-Form, $[\alpha]_D^{25}$ +43.75° (c = 5.006 in benzene).

USE: Preparing bitter almond water, by mixing 11 g with 500 g alc and 1489 g water; the mixture contains 0.1% HCN.

5760. Mandelonitrile Glucoside. α-(β-D-*Glucopyranosyloxy)benzeneacetonitrile.* $C_{14}H_{17}NO_6$; mol wt 295.29. C

56.94%, H 5.80%, N 4.74%, O 32.51%. Prepn of *d*-form by action of yeast on amygdalin: Fischer, *Ber.* **28**, 1508 (1895); Auld, *J. Chem. Soc.* **93**, 1276 (1908). Isoln of *d*-form from *Prunus serotina,* Ehrk, *P. macrophylla* Sieb et Zucc., *Rosaceae:* Power, Moore, *ibid.* **95**, 243 (1909); **97**, 1099 (1910); Kariyone, Matsushima, *J. Pharm. Soc. Japan* no. **514**, 1061 (1924); from *Eucalyptus corynocalyx* F.v.M., *Myrtaceae:* Finnemore *et al., J. Proc. Roy. Soc. N.S. Wales* **69**, 209 (1936); from *Pteridium aquilinum* (L.) Kühn, *Polypodiaceae:* Kofod, Eyjolfssen, *Tetrahedron Letters* **1966**, 1289. Isoln of *dl*-form from *Prunus laurocerasus* L. and *Cotoneaster microphylla* Wall., *Rosaceae:* Winkler, Simon, *Ann.* **31**, 263 (1839); Hérissey, *J. Pharm. Chim.* **24**, 537 (1906). Isoln of *l*-form from *Sambucus nigra* L., *Caprifoliaceae:* Guignart, *Compt. Rend.* **141**, 16 (1905); Bourquelot, Danjot, *ibid.* 598 (1905). Prulaurasin is *dl*-mandelonitrile glucoside and sambunigrin is *l*-mandelonitrile glucoside: Caldwell, Courtauld, *J. Chem. Soc.* **91**, 671 (1907). Biosynthesis of *d*-mandelonitrile glucoside: Mentzer, Favre-Bonvin, *Compt. Rend.* **253**, 1072 (1961). Abs config of isomers: U. Schwarzmaier, *Ber.* **109**, 3250 (1976).

d-Form, *prunasin.* Needles from chloroform, mp 147-148°. $[\alpha]_D^{14}$ −29.94°. Soluble in water, alcohol, acetone. Converted by alkalies to prulaurasin.

dl-Form, *prulaurasin.* Slightly bitter needles from ethyl acetate + ether, mp 123-125°. $[\alpha]_D$ −54° (water). Soluble in water, alcohol. Practically insol in ether. On hydrolysis yields *dl*-mandelic acid.

l-Form, *sambunigrin.* Bitter needles from hot ethyl acetate, mp 151-152°. $[\alpha]_D^{18}$ −75.1°. Soluble in water, alcohol, ethyl acetate. On hydrolysis yields *l*-mandelic acid.

5761. Maneb. [[1,2-*Ethanediylbis[carbamodithioato]]-(2−)]manganese; [ethylenebis(dithiocarbamato)]manganese;* ethylenebis[dithiocarbamic acid] manganous salt; manganous ethylenebis[dithiocarbamate]; Dithane M-22; Manex; Trimangol. $(C_4H_6MnN_2S_4)_n$. Polymeric salt of ethylenebisdithiocarbamic acid; related to zineb and mancozeb, *q.q.v.* Prepd from aq nabam, *q.v.,* by neutralizing with acetic acid and adding MnCl$_2$ soln: A. L. Flenner, U.S. pat. **2,504,404** (1950 to Du Pont), see also W. F. Hester, U.S. pat. **2,317,-765** (1943 to Rohm & Haas). Fungicidal activity is due to degradation products, principally *ethylenethiuram monosulfide:* R. A. Ludwig *et al., Can. J. Bot.* **33**, 42 (1955). *In vitro* activity against *Alternaria alternata:* V. A. Bourbos *et al., Brit. Crop. Prot. Council Monograph* **No. 31** (Fungic. Crop Prot., Vol. 2), 461 (1985). Decomposition of maneb to ethylenebisdithiocarbamate and ethylene thiourea, *q.v.:* W. J. Trotter, J. Pardue, *J. Food Safety* **4**, 59 (1982). HPLC determn: K. Gustafsson *et al., J. Agr. Food Chem.* **29**, 729 (1981). Toxicological studies in rats: D. W. R. Bleyl, H. Seidler, *Nährung* **29**, 421 (1985); in aquatic organisms: C. J. Van Leeuwen *et al., Aquat. Toxicol.* **7**, 145 (1985).

Yellow powder. Crystals from alcohol. Moderately sol in water. Sol in chloroform, pyridine.

USE: Agricultural fungicide.

5762. Manganese. Mn; at. wt 54.9380; at. no. 25; valences 2, 4, 7; 1, 3, 5, 6 rare. One stable isotope: 55; artificial radioactive isotopes 49-54; 56-58. Widely-distributed, abundant element; constitutes 0.085% of earth's crust. Occurs in the minerals pyrolusite, hausmannite, manganite, *manganosite* (MnO), *braunite* ($3Mn_2O_3.MnSiO_3$), and in several others; occurs in minute quantities in water, plants and animals. First isolated by Gahn in 1774. Prepn of the metal: John *et al.*, cited by Mellor, *A Comprehensive Treatise on Inorganic and Theoretical Chemistry,* **12**, 163 (1932); A. H. Sully, *Manganese* (Academic Press, New York, 1955) 305 pp. Review of physical properties: Meaden, *Met. Rev.* **13**, 97-114 (1968). Review of manganese and its compds: Kemmitt in *Comprehensive Inorganic Chemistry,* vol. **3**, J. C. Bailar Jr. *et al.,* Eds. (Pergamon Press, Oxford, 1973) pp 771-876; L. R. Matricardi, J. H. Downing in Kirk-Othmer *Encyclopedia of Chemical Technology* vol. **14**, (Wiley-Interscience, New York, 3rd ed., 1981) pp 824-843.

Steel gray, lustrous, hard, brittle metal. Exists in four allotropic forms: α-form, body-centered cubic, stable below 710° (approx), d^{20} 7.47; β-form, cubic, stable in the range 710-1079°, d^{20} 7.26; γ-form or electrolytic Mn, face-centered cubic, stable in the range 1079-1143°C, d^{1100} 6.37; δ-form, body-centered cubic, stable from 1143° to mp, d^{1143} 6.28. γ-Mn when stabilized at room temp is face-centered tetragonal, d^{20} 7.21. mp 1244°. bp 2095°. Sp heat 0.115 cal/g/°C; latent heat of fusion: 3.5 kcal/g-atom. Mohs' hardness 5.0. Superficially oxidized on exposure to air. Burns with an intense white light when heated in air. Dec water slowly in the cold, rapidly on heating; pure electrolytic manganese is not attacked by water at ordinary temp; slightly attacked by steam. Reacts with dil mineral acids with evolution of hydrogen and formation of divalent manganous salts. Reacts with aq solns of sodium or potassium bicarbonate. When heated in nitrogen above 2000° burns to form a nitride. Converted by fluorine into the di- and trifluoride, by chlorine into the dichloride. In form of powder reduces most metallic oxides on heating. On heating, reacts directly with carbon, phosphorus, antimony or arsenic.

Caution: Potential symptoms of overexposure are parkinsonism; asthenia, insomnia and mental confusion; metal fume fever (dry throat, coughing, tight chest, dyspnea, rales and flu-like fever); lower back pain; vomiting; malaise; fatigue. *See NIOSH Pocket Guide to Chemical Hazards* (DHHS/NIOSH 90-117, 1990) p 138.

USE: In manuf of steel; for rock crushers, railway points and crossings, wagon buffers; as a constituent of several alloys, *e.g.,* ferromanganese, copper manganese, Manganin.

5763. Manganese Acetate. $C_4H_6MnO_4$; mol wt 173.03. C 27.77%, H 3.50%, Mn 31.75%, O 36.99%. $Mn(CH_3COO)_2$.
Tetrahydrate, pale red, transparent monoclinic crystals. d 1.59. Sol in water or alcohol. LD_{50} orally in rats: 3.73 g/kg, Smyth *et al., Am. Ind. Hyg. Assoc. J.* **30**, 470 (1969).
USE: As mordant in dyeing; manuf bister; drier for paints and varnishes (1-1.5:1000).

5764. Manganese Borate. Approx $MnB_4O_7.8H_2O$.
Brownish-white powder. Insol in water or alcohol. Dec on long contact with water; sol in dil acids.
USE: In drying varnishes and oils; as drier for linseed oil, 1:500; also in leather industry.

5765. Manganese Bromide. Manganese dibromide. Br_2Mn; mol wt 214.75. Br 74.42%, Mn 25.58%. $MnBr_2$.
Tetrahydrate, rose-red, slightly deliquesc crystals. mp 64° with some decompn. Sol in 0.5 parts water; sol in alcohol. The aq soln is slightly acid. *Keep well closed.*

5766. Manganese Carbonate. $CMnO_3$; mol wt 114.95. C 10.45%, Mn 47.79%, O 41.76%. $MnCO_3$. (Usually contains some H_2O). Occurs in nature as the mineral *rhodochrosite.*
Pink to almost white powder when freshly pptd, but gradually becomes light brown in the air. Rhombohedral, calcite structure. d 3.1. Insol in water or alcohol. Sol in dil acid. *Keep well closed.*
USE: As pigment—"manganese white"; drier for varnishes; in feeds.

5767. Manganese Carbonyl. *Decacarbonyldimanganese.* $C_{10}Mn_2O_{10}$; mol wt 389.98. C 30.80%, Mn 28.17%, O 41.03%. $Mn_2(CO)_{10}$. First prepd by reduction of MnI_2 with a Grignard reagent under CO pressure: Hurd *et al., J. Am. Chem. Soc.* **71**, 1899 (1949). Improved prepns: Brimm *et*

al., ibid. **76**, 3831 (1954); Closson, *ibid.* **80**, 6167 (1958); Podall *et al., ibid.* **82**, 1325 (1960); Calderazzo, *Inorg. Chem.* **4**, 293 (1965). Crystal structure: Dahl, Rundle, *Acta Cryst.* **16**, 419 (1963); molecular structure: Almenningen *et al., Acta Chem. Scand.* **23**, 685 (1969). *Reviews: Organic Syntheses via Metal Carbonyls* **vol. 1**, I. Wender, P. Pino, Eds. (Interscience, New York, 1968) *passim;* Anisimov *et al., Usp. Khim.* **37**, 380-408 (1968); *Russ. Chem. Rev.* **37**, 184-197 (1968) (English translation).
Golden yellow, monoclinic crystals. d^{25} 1.75. mp 154-155°. Stable under an atm of CO. In absence of CO, begins to dec at 110°. Sol in organic solvents. Insol in water. Less stable to air, heat and light in soln.
USE: Catalyst; antiknock additive.

5768. Manganese Chloride. Manganous chloride; manganese dichloride. Cl_2Mn; mol wt 125.84. Cl 56.34%, Mn 43.66%. $MnCl_2$.
Tetrahydrate, reddish, slightly deliquesc, monoclinic crystals. d 2.01. mp 58°. Sol in 0.7 part water; sol in alcohol. Insol in ether. pH of 0.2 molar aq soln 5.5. *Keep well closed.* LD s.c. in mice: 180-250 mg/kg; i.v. in dogs: 201.6 mg/kg, *Handbook of Toxicology,* Vol. **1**, W. S. Spector, Ed. (Saunders, Philadelphia, 1956) pp 182-183.
USE: In dyeing (manganese bister); disinfecting; purifying natural gas; linseed oil drier; in electric batteries.

5769. Manganese Difluoride. Manganous fluoride; manganese fluoride. F_2Mn; mol wt 92.93. F 40.89%, Mn 59.11%. MnF_2. Prepd from manganese carbonate and hydrogen fluoride: Moissan, Venturi, *Compt. Rend.* **130**, 1158 (1900); Kwasnik in *Handbook of Preparative Inorganic Chemistry,* Vol. **1**, G. Brauer, Ed. (Academic Press, New York, 2nd ed., 1963) pp 262-263.
Pink, quadratic prisms (tetragonal structure, rutile type) or reddish powder. *Poisonous!* d 3.98. mp 856°. Soly in water (g/100 ml): 0.66 (40°); 0.44 (60°); 0.48 (100°). Insol in alc. Sol in dil hydrofluoric acid, concd hydrochloric or nitric acid. Lowest published lethal dose in guinea pigs: 200 mg/kg orally; 700 mg/kg s.c., *Toxic Substances List,* H. E Christensen, Ed. (1973) p 574. Tetrahydrate, obtained by dissolving manganese carbonate in hydrofluoric acid, evaporating, and drying *in vacuo.*

5770. Manganese Dioxide. Manganese binoxide; manganese peroxide; manganese superoxide; black manganese oxide. MnO_2; mol wt 86.94. Mn 63.19%, O 36.81%. Occurs in nature as the mineral *pyrolusite,* or made artificially (pptd). The native product is heavy, steel-gray when in lumps, black when powdered; the pptd product is a brownish-black, fine powder. Both usually contain some Mn_3O_4 and some water. When ignited evolves oxygen, leaving Mn_3O_4. Lab prepn: Moore *et al., J. Am. Chem. Soc.* **72**, 856 (1950); Covington *et al., Trans. Faraday Soc.* **58**, 1975 (1962). Review of use as reagent: J. S. Pizey, *Synthetic Reagents* vol. 2 (John Wiley, New York, 1974) pp 143-174.
Tetragonal crystals (rutile structure). Insol in water, nitric or cold sulfuric acid; slowly dissolves in cold HCl with evolution of Cl_2; in presence of hydrogen peroxide or oxalic acid it dissolves in dil H_2SO_4 or HNO_3. *Caution: Manganese dioxide is a strong oxidizer, hence it should not be heated or rubbed with organic matter or other oxidizable substances, e.g., sulfur, sulfides, phosphides, hypophosphites, etc.* LD i.v. in rabbits: 45 mg/kg, *Handbook of Toxicology,* **Vol. 1**, W. S. Spector, Ed. (Saunders, Philadelphia, 1956) pp 182-183.
USE: The mineral is the source of manganese and all its compds; largely used in manuf manganese steel; oxidizer; in alkaline batteries (dry cells); for making amethyst glass, decolorizing glass; painting on porcelain, faience and majolica. The ppt is used in electrotechnics, pigments, browning gun barrels, drier for paints and varnishes, printing and dyeing textiles.

5771. Manganese Hypophosphite. $H_4MnO_4P_2$; mol wt 184.91. H 2.18%, Mn 29.71%, O 34.61%, P 33.50%. $Mn(H_2PO_2)_2$.
Monohydrate, pink, odorless, almost tasteless crystals or powder. When heated evolves spontaneously flammable phosphine. One gram dissolves in 6.5 ml water, 6 ml boiling water. Insol in alcohol.

5772. Manganese Iodide. Manganese diiodide. I_2Mn; mol wt 308.75. I 82.21%, Mn 17.79%. MnI_2.

Tetrahydrate, rose-red crystals; rapidly becomes brown on exposure to air and light, due to liberation of iodine. Very sol in water with gradual decompn; sol in alcohol. The aq soln is slightly acid. *Keep tightly closed and protected from light.*

5773. Manganese Nitrate. MnN_2O_6; mol wt 178.95. Mn 30.70%, N 15.65%, O 53.64%. $Mn(NO_3)_2$. Prepn: Dehnicke, Strähle, *Ber.* **97**, 1502 (1964). *Review: Gmelin's, Manganese* (8th ed.) **56**, part C3, 267-281 (1975).

Colorless; hygroscopic. Sol in water, dioxane, tetrahydrofuran, acetonitrile.

Tetrahydrate, pink, deliquesc cryst masses at temp below 20°. d 2.129. mp 37.1°. Very sol in water, sol in alcohol. The aq soln is slightly acid.

Hexahydrate, rose-colored, deliquesc, monoclinic needles (colorless after several crystns). d 1.8. mp 25.0°. Freely sol in water, alcohol.

USE: Intermediate in manuf of reagent grade MnO_2; in prepn of porcelain colorants.

5774. Manganese Oleate. Approx $Mn(C_{18}H_{33}O_2)_2$.

Brown, granular mass. Insol in water. Sol in ether, oleic acid; slightly sol in alcohol.

USE: As a drier in varnishes.

5775. Manganese Oxalate. C_2MnO_4; mol wt 142.96. C 16.80%, Mn 38.43%, O 44.77%. MnC_2O_4.

Dihydrate, white cryst powder; dec at 150°. Slightly sol in water; sol in acids.

5776. Manganese Oxide. Manganomanganic oxide. Mn_3O_4; mol wt 228.81. Mn 72.03%, O 27.97%. Occurs in nature as the mineral *hausmannite*. Prepn: Moore *et al., J. Am. Chem. Soc.* **72**, 856 (1950).

Brownish-black powder. d 4.7. Insol in water. Sol in HCl with evolution of chlorine.

5777. Manganese Phosphate, Dibasic. $HMnO_4P$; mol wt 150.92. H 0.67%, Mn 36.40%, O 42.41%, P 20.52%. $MnHPO_4$. Trihydrate. Exists in two modifications: gray form prepd by decompn of $MnCO_3$ by phosphoric acid; pink form prepd by reaction of $Mn_3(PO_4)_2$ with phosphoric acid (P_2O_5 concn 0.6-28%): Goloshchapov *et al., Russ. J. Inorg. Chem.* **11**, 504 (1966).

Pink form, usually contains some tribasic phosphate. Very slightly sol in water; sol in dil acids. Gray form, sol only in hot concd HCl.

5778. Manganese Pyrophosphate. MnO_7P_2; mol wt 228.88. Mn 24.00%, O 48.93%, P 27.07%. MnP_2O_7.

Trihydrate, white or nearly white powder. Insol in water. Sol in excess of alkali pyrophosphate, in acids.

5779. Manganese Selenide. MnSe; mol wt 133.90. Mn 41.03%, Se 58.97%. Prepd by the action of manganese salts on alkali selenides: Berzelius *et al.,* cited by Mellor, *A Comprehensive Treatise on Inorganic and Theoretical Chemistry* **10**, 798 (1930); by the action of a manganese salt on a soln of hydrogen selenide out of contact with air: Moser, Atynsky, *Monatsh.* **45**, 235 (1925).

Gray-black cubic crystals. d^{15} 5.59. Completely oxidized when heated to redness in oxygen. Insol in cold water. Dec in hot water and dil acids.

5780. Manganese Sesquioxide. Mn_2O_3; mol wt 157.87. Mn 69.60%, O 30.40%. Occurs in nature as the mineral *manganite*, $Mn_2O_3 \cdot H_2O$. Prepn: Moore *et al., J. Am. Chem. Soc.* **72**, 856 (1950).

Black, fine powder. d 4.50. Insol in water. Sol in HCl with evolution of chlorine.

5781. Manganese Silicate. Approx $MnSiO_3$. Occurs in nature as the minerals *rhodonite, manganjustite, tephroite*.

Red crystals or yellowish-red powder. The pptd article is a yellowish-red powder. d of the mineral 3.48. Insol in water.

USE: As color for special glass; producing red glazes on pottery.

5782. Manganese Sulfate. MnO_4S; mol wt 151.00. Mn 36.38%, O 42.38%, S 21.24%. $MnSO_4$. Forms several hydrates. The article of commerce is usually a mixture of the tetra- and pentahydrates.

Monohydrate, pale red, slightly efflorescent crystals; loses all water at 400-450°. Sol in about 1 part cold, 0.6 part boiling water; insol in alcohol.

USE: In dyeing; for red glazes on porcelain; boiling oils for varnishes; in fertilizers for vines, tobacco; in feeds.

THERAP CAT (VET): Nutritional factor (essential trace element in all animals); prevention of perosis in poultry.

5783. Manganese Sulfide. Manganese monosulfide. MnS; mol wt 87.00. Mn 63.14%, S 36.86%. Occurs in nature as the mineral *alabandite* or *manganblende*. Prepn: Mehmed, Haraldsen, *Z. Anorg. Chem.* **235**, 194 (1938).

Pink, green or brown-green powder. Three cryst modifications: α-form, green cubic crystals; β-form, red cubic crystals; γ-form, red hexagonal crystals. Practically insol in water; sol in dil acids. In moist condition it readily oxidizes in air to the sulfate.

5784. Manganese Trifluoride. Manganic fluoride. F_3Mn; mol wt 111.93. F 50.92%, Mn 49.08%. MnF_3. Prepd by fluorination of manganese iodide according to the equation $2MnI_2 + 13F_2 \rightarrow 2MnF_3 + 4IF_5$: Moissan, *Compt. Rend.* **130**, 622 (1900); v. Wartenberg, *Z. Anorg. Allgem. Chem.* **244**, 346 (1940); Kwasnik in *Handbook of Preparative Inorganic Chemistry,* vol. 1, G. Brauer, Ed. (Academic Press, New York, 2nd ed., 1963) pp 263-264.

Red mass; monoclinic crystals. d 3.54. Stable to 600°. Hydrolyzed by water. May be stored in glass ampuls.

USE: Fluorinating agent in organic chemistry.

5785. Mangostin. *1,3,6-Trihydroxy-7-methoxy-2,8-bis-(3-methyl-2-butenyl)-9H-xanthen-9-one;* 1,3,6-trihydroxy-7-methoxy-2,8-di(3-methyl-2-butenyl)xanthone. $C_{24}H_{26}O_6$; mol wt 410.47. C 70.23%, H 6.38%, O 23.39%. From various parts of the mangosteen tree (*Garcinia mangostana L., Guttiferae*): Schmid, *Ann.* **93**, 83 (1855); Dragendorff, ibid. **482**, 280 (1930). Structure: Yates, Stout, *J. Am. Chem. Soc.* **80**, 1691 (1958); Scheinmann, *Chem. Commun.* **1967**, 1015; Stout *et al., ibid.* **1968**, 211.

Yellow crystals from benzene, mp 181.6-182.6°. uv max (ethanol): 243, 259, 318, 351 nm (log ε 4.54, 4.44, 4.38, 3.86). Practically insol in water; sol in alcohol, ether, acetone, chloroform, ethyl acetate.

3,6-Dimethylmangostin, $C_{26}H_{30}O_6$, pale yellow needles from ethanol, mp 123.3-123.8°. uv max (ethanol): 245, 262, 314, 350 nm (log ε 4.50, 4.53, 4.36, 3.81).

5786. Manidipine. *1,4-Dihydro-2,6-dimethyl-4-(3-nitrophenyl)-3,5-pyridinedicarboxylic acid 2-[4-(diphenylmethyl)-1-piperazinyl]ethyl ester;* 2-[4-(diphenylmethyl)-1-piperazinyl]ethyl methyl (±)-1,4-dihydro-2,6-dimethyl-4-(*m*-nitrophenyl)-3,5-pyridinecarboxylate; franidipine. $C_{35}H_{38}N_4O_6$; mol wt 610.71. C 68.84%, H 6.27%, N 9.17%, O 15.72%. Dihydropyridine calcium channel blocker. Prepn: K. Meguro, A. Nagaoka, **Eur. pat. Appl. 94,159**; eidem, **U.S. pat. 4,892,875** (1983, 1990, both to Takeda); K. Meguro *et al., Chem. Pharm. Bull.* **33**, 3787 (1985). Synthesis and bioactivity of optical isomers: M. Kajino *et al., ibid.* **37**, 2225 (1989). Acute toxicity study: S. Chiba *et al., Yakuri to Chiryo* **17**, Suppl. 4, 961 (1989), *C.A.* **111**, 146499b (1989). Determn of enantiomers in human serum: M. Yamaguchi *et al., J. Chromatog.* **575**, 123 (1992). Clinical evaluation in hypertension: K. Mizuno *et al., Curr. Ther. Res.* **52**, 248 (1992).

Light yellow crystals from isopropyl ether + hexane, mp 125-128°.

Dihydrochloride, $C_{35}H_{38}N_4O_6\cdot 2HCl$, *CV-4093, Calslot.* Two crystalline forms. α-form: yellow crystals, mp 157-163°. β-form: light yellow, fine crystals, mp 174-180°. Also prepd as the monohydrate, mp 167-170°. LD_{50} in male, female mice, male, female rats (mg/kg): 387, 340, 222, 199 s.c.; 62.2, 68.0, 66.5, 48.8 i.p.; 190, 171, 247, 156 orally (Chiba).

THERAP CAT: Antihypertensive.

5787. Manna. Dried exudation of *Fraxinus ornus* L., *Oleaceae. Habit.* Mediterranean Basin, Asia Minor, Spain. *Constit.* 40-60% mannitol; 10-16% mannotetrose; 6-16% mannotriose; glucose, mucilage, fraxin.

One gram dissolves in 5 ml water, 150 ml 90% alcohol.

THERAP CAT: Cathartic.

5788. Mannitol. D-Mannitol; mannite; manna sugar; cordycepic acid; Diosmol; Manicol; Mannidex; Osmitrol; Osmosal; Resectisol. $C_6H_{14}O_6$; mol wt 182.17. C 39.56%, H 7.75%, O 52.70%. Widespread in plants and plant exudates; obtained from manna and seaweeds: *The Carbohydrates,* W. Pigman, Ed. (Academic Press, New York, 1957) pp 249-250. Forms a stable, equimolar compound with H_2O_2: S. Tanatar, *J. Russ. Phys. Chem. Soc.* **40,** 376, C.A. **3,** 883 (1909). Prepd by electrolytic reduction of glucose: Creighton, *Can. Chem. Process Inds.* **26,** 690 (1942), *C.A.* **37,** 1088⁵ (1943); Wolfrom *et al., J. Am. Chem. Soc.* **68,** 578 (1946). Prepn from seaweed: Sorensen, Kristensen, U.S. pat. **2,516,350** (1950); by hydrogenation of invert sugar, monosaccharides, and sucrose: Kasehagen, and Kasehagen, Luskin, U.S. pats. **2,642,462, 2,749,371,** and **2,759,024** (1953, 1956, and 1956, all to Atlas Powder). Review of prepn: Pigman, *loc. cit.* Novel synthesis: M. Makkee *et al., Chem. Commun.* **1980,** 930.

Orthorhombic needles from alc, mp 166-168°. Sweetish taste. d^{20} 1.52. $bp_{3.5}$ 290-295°. Is inactive or very slightly levorotatory in distilled water. Forms sodium mannitoborate on addition of borax giving a greater rotation: $[\alpha]_D^{20}$ +23° → 24° after 1 hr in a soln of 10 g mannitol +12.8 g borax + sufficient H_2O to make 100 ml. One gram dissolves in ~5.5 ml water, 83 ml alcohol; more sol in hot water. Insol in ether. Sol in pyridine, aniline, aq solns of alkalies. One gram dissolves in 18 ml glycerol (d 1.24). Soly tables: Creighton, Klauder, *J. Franklin Inst.* **195,** 687 (1923). pKa (18°): 13.50.

USE: Used with boric acid in the manuf of dry electrolytic condensers for radio applications; in making artificial resins and plasticizers; in pharmacy as excipient and diluent for solids and liqs; in analytical chemistry for boron determinations; in the manuf of mannitol hexanitrate. Used in the food industry as anticaking and free-flow agent, flavoring agent, lubricant and release agent, stabilizer and thickener and nutritive sweetener.

THERAP CAT: Diuretic. Diagnostic aid (renal function).

5789. Mannitol Hexanitrate. Mannitol nitrate; nitromannite; nitromannitol; Maxitate; Medemanol; Dilangil; Moloid; Mannitrin; Nitranitol; Manexin. $C_6H_8N_6O_{18}$; mol wt 452.16. C 15.94%, H 1.78%, N 18.59%, O 63.69%. Made by nitration of mannitol: Fleury *et al., Mem. Poudres* **31,** 107 (1949).

Long needles in regular clusters from alc. mp 106-108°. Soluble in alc, in ether; insol in water. Explodes on percussion. Its stability at ordinary temps is such that it may be used commercially, but it is distinctly less stable than nitroglycerol at 75°. Its employ for pharmaceutical prepns is only in admixture with carbohydrate substances in dilutions corresponding to 1 part of mannitol hexanitrate to 9 or more parts of carbohydrate. In such dilutions mannitol hexanitrate is considered nonexplosive.

THERAP CAT: Vasodilator.

5790. Mannomustine. *1,6-Bis(2-chloroethylamino)-1,6-dideoxy-D-mannitol dihydrochloride;* 1,6-dideoxy-1,6-di(2-chloroethylamino)-D-mannitol dihydrochloride; 1,6-di(2-chloroethylamino)-1,6-dideoxy-D-mannitol dihydrochloride; mannitol nitrogen mustard; BCM; Degranol. $C_{10}H_{24}Cl_4N_2$-O_4; mol wt 378.12. C 31.76%, H 6.40%, Cl 37.50%, N 7.41%, O 16.93%. Prepn: Vargha *et al., J. Chem. Soc.* **1957,** 805. Toxicity: Scherf *et al., Arzneimittel-Forsch.* **20,** 1467 (1970).

Crystals from 80% ethanol, dec 239-241°. $[\alpha]_D^{20}$ +18.46° (c = 1.81 in H_2O). (Free base, crystals, dec 278°). Sol in water; slightly sol in ethanol. Aq solns are mildly acidic and stable at room temps. Claimed to be considerably less toxic than mechlorethamine. LD_{50} i.v. in rats: 56 mg/kg (Scherf).

THERAP CAT: Antineoplastic.

5791. D-Mannose. Seminose; carubinose. $C_6H_{12}O_6$; mol wt 180.16. C 40.00%, H 6.71%, O 53.28%. Prepn of α-form by treating ivory nut shavings with H_2SO_4: Isbell, *J. Res. Nat. Bur. Stand.* **26,** 47 (1941); Isbell in *Methods in Carbohydrate Chemistry,* R. L. Whistler, M. L. Wolfrom, Eds. (Academic Press, New York, 1962) pp 145-147. Prepn and stability of α- and β-forms: Reeves, *J. Am. Chem. Soc.* **72,** 1499 (1950); J. Sowden in *The Carbohydrates,* W. Pigman, Ed. (Academic Press, New York, 1957) pp 94-95.

α-D-Mannose

α-Form, crystals from methanol, mp 133°. $[\alpha]_D$ +29.3° → +14.2° (water).

β-Form, orthorhombic, bisphenoidal needles from alcohol or acetic acid, dec 132°. Sweet taste with bitter aftertaste. d^{20} 1.54. Shows mutarotation. $[\alpha]_D^{20}$ −17.0° → +14.2° (c = 4). One gram dissolves in 0.4 ml water, 120 ml methanol, 250 ml abs ethanol, 3.5 ml pyridine. pKa (18°): 11.98. Reduces Fehling's soln; is fermented by yeast.

Phenylhydrazone, $C_{12}H_{18}N_2O_5$, crystals from dil ethanol, mp 199-200°. $[\alpha]_D^{20}$ +26.3° → +33.8° (pyridine).

$CaCl_2$-addition compd tetrahydrate, $C_6H_{12}O_6\cdot CaCl_2\cdot 4H_2O$, mp 101-102°. $[\alpha]_D^{20}$ −31.3° → +6.0° (c = 9).

5792. Maprotiline. *N-Methyl-9,10-ethanoanthracene-9(10H)-propanamine;* 9-(γ-methylaminopropyl)-9,10-dihydro-9,10-ethanoanthracene; 1-(3-methylaminopropyl)dibenzo[*b,e*]bicyclo[2.2.2]octadiene. $C_{20}H_{23}N$; mol wt 277.41. C 86.59%, H 8.36%, N 5.05%. Prepn: P. Schmidt *et al.*, U.S. pat. **3,399,201** (1968 to Ciba). Synthesis, NMR, mass spectra: M. Wilhelm, P. Schmidt, *Helv. Chim. Acta* **52**, 1385 (1969). Toxicity: R. Hess *et al., Boll. Chim. Farm.* **112**, 782 (1973), *C.A.* **81**, 33479p (1974). Review of pharmacology and therapeutic efficacy: R. M. Pinder *et al., Drugs* **13**, 321 (1977). Comprehensive description: S. K. Suh, J. B. Smith in *Analytical Profiles of Drug Substances* vol. **15**, K. Florey, Ed. (Academic Press, New York, 1986) pp 393-426.

mp 92-94°.

Hydrochloride, $C_{20}H_{23}N \cdot HCl$, *Ba-34276, Deprilept, Ludiomil, Psymion.* Crystals from isopropanol, mp 230-232°. Freely sol in methanol and chloroform, slightly sol in water. Practically insol in isooctane. LD_{50} in mice, rats (mg/kg): ~750, ~900 orally (Hess).

THERAP CAT: Antidepressant.

5793. Margaric Acid. Heptadecanoic acid. $C_{17}H_{34}O_2$; mol wt 270.46. C 75.50%, H 12.67%, O 11.83%. CH_3-$(CH_2)_{15}COOH$. Prepn: Kaufmann, Stamm, *Ber.* **91**, 2121 (1958); Bhattacharyya *et al., Chem. & Ind. (London)* **1959**, 1352; Hünig, Ledle, *Ber.* **93**, 913 (1960). Metabolism: Boyer, Scheig, *Lipids* **4**, 615 (1969). Toxicity study: L. Orö, A. Wretlind, *Acta Pharmacol. Toxicol.* **18**, 141 (1961).

Crystals from alcohol, mp 61°. d 0.853. bp_{100} 227°. n_D^{60} 1.4342. Insol in water. Freely sol in ether; slightly sol in alcohol. LD_{50} i.v. in mice: 360.3 mg/kg (Orö, Wretlind).

5794. Marrubiin. *[2aS-(2aα,5aβ,6α,7α,8aα,8bα)]-6-[2-(3-Furanyl)ethyl]decahydro-6-hydroxy-2a,5a,7-trimethyl-2H-naphtho[1,8-bc]furan-2-one; 15,16-epoxy-6β,9-dihydroxy-8βH-labda-13(16),14-dien-19-oic acid* γ*-lactone;* 5-[2-(3-furyl)ethyl]decahydro-5,8-dihydroxy-1,4a,6-trimethyl-1-naphthoic acid γ-lactone. $C_{20}H_{28}O_4$; mol wt 332.44. C 72.26%, H 8.49%, O 19.25%. Diterpene lactone principle isolated from white horehound, *Marrubium vulgare* (Tourn.) L., *Labiatae:* Harms, *Arch. Pharm.* **83**, 144 (1842); Ludwig, Kromayer, *ibid.* **158**, 257 (1861); Nicholas, *J. Pharm. Sci.* **53**, 895 (1964). Alternate view that marrubiin is an artefact generated from *premarrubiin* during the isoln: Henderson, McCrindle, *J. Chem. Soc. (C)* **1969**, 2014. Structure and stereochemistry: Cocker *et al., J. Chem. Soc.* **1953**, 2540; *Chem. & Ind. (London)* **1954**, 1561; **1955**, 1484; Fulke, McCrindle, *ibid.* **1965**, 647. Total stereochemistry: Wheeler *et al., Tetrahedron* **23**, 3909 (1967); Appleton *et al., J. Chem. Soc. (C)* **1967**, 1943. Synthesis: Mangoni *et al., Tetrahedron* **28**, 611 (1972).

Crystals from alc, mp 160°. $[\alpha]_D^{20}$ +35.8° (c = 3.1 in chloroform); $[\alpha]_D^{24}$ +45° (acetone). uv max: 208, 212, 216 nm (log ε 3.75, 3.75, 3.70). One gram dissolves in 60 ml alc. Freely sol in chloroform, acetone, hot alc, pyridine; sparingly sol in ether, benzene. Practically insol in water.

5795. Mastic. Balsam tree; pistachia galls; mastiche; mastix; lentisk; Mastisol. Concrete resinous exudation from *Pistacia lentiscus* L., *Anacardiaceae. Habit.* Mediterranean Islands, especially Chios. *Constit.* Volatile oil (about 2%); masticinic, masticonic acids; masticoresene.

Pale yellow or greenish-yellow, globular, elongated or pear-shaped tears; slightly balsamic odor and terebene taste. Almost insol in water. Nearly completely sol in alcohol; 1 g dissolves in 0.5 ml chloroform, 0.5 ml ether; partially sol in oil turpentine.

USE: In tooth cements, plasters, lacquers, chewing gums, and incense; also for retouching negatives.

5796. Maté. Paraguay tea; yerba maté; St. Bartholomew's tea; Jesuit's tea. Leaves of *Ilex paraguensis* St. Hil., *Aquifoliaceae*, grown in Brazil, Uruguay, Argentina, and Paraguay. *Review:* Cheney, *Econ. Botany* **1**, 243 (1947). For constituents *see:* Siesto, *Quaderni Nutr.* **18**, 35 (1959), *C.A.* **57**, 13133d (1962); Roberts, *Chem. & Ind. (London)* **1956**, 985; Paula, *Rev. Brasil. Quim.* (Sao Paulo) **42**, 202 (1956), *C.A.* **51**, 3167i (1957). Among its constituents are caffeine, other purines, tannins or tannoid substances, such as chlorogenic acid and derivatives.

5797. Matico. Dried leaves of *Piper elongatum* Vahl. (*P. angustifolium* R. & P.), *Piperaceae. Habit.* Peru, Bolivia, Brazil, Mexico, Cuba. *Constit.* 1-3.5% volatile oil, maticin (a bitter principle); artanthic acid, tannin, mucilage, resin.

THERAP CAT: Astringent.

5798. Matricarin. *[3S-(3α,3aα,4α,9aα,9bβ)]-4-(Acetyloxy)-3,3a,4,5,9a,9b-hexahydro-3,6,9-trimethylazuleno[4,5-b]furan-2,7-dione.* $C_{17}H_{20}O_5$; mol wt 304.34. C 67.09%, H 6.62%, O 26.29%. Isoln from *Matricaria chamomilla* L. and *Artemisia tilesii* Ledeb, *Compositae:* Z. Cekan *et al., Coll. Czech. Chem. Commun.* **24**, 1554 (1959). Structure: W. Herz, K. Ueda, *J. Am. Chem. Soc.* **83**, 1139 (1961). Molecular conformation: M. Martinez V. *et al., J. Nat. Prod.* **51**, 221 (1988).

Plates from benzene + petr ether and acetone + petr ether, mp 190-191°. $[\alpha]_D^{23}$ +23.5° (c = 0.65 in chloroform). uv max: 255 nm (ε 15,100).

Desacetoxymatricarin, $C_{15}H_{18}O_3$. Isoln from *Artemisia* spp.: M. Holub, V. Herout, *Coll. Czech. Chem. Commun.* **27**, 2980 (1962); H. H. A. Linde, M. S. Ragab, *Helv. Chim. Acta* **50**, 1961 (1967). Synthesis and stereochemistry: E. H. White *et al., Tetrahedron* **25**, 2099 (1969). Crystals from ether, mp 205-205.5°, also reported as mp 202-203° (Holub). $[\alpha]_D^{24}$ +52.5° (c = 1 in chloroform).

Desacetylmatricarin monohydrate, $C_{15}H_{18}O_4 \cdot H_2O$, crystals from benzene + acetone, mp 123-125°, resolidifies and mp 143-146°.

Tetrahydromatricarin, $C_{17}H_{24}O_5$, needles from benzene + petr ether, mp 175-178°. $[\alpha]_D^{22}$ +32.3° (c = 1.92 in chloroform).

5799. Matrine. *Matridin-15-one;* sophocarpidine. $C_{15}H_{24}N_2O$; mol wt 248.37. C 72.54%, H 9.74%, N 11.28%, O 6.44%. Occurs naturally as the (+)-form in the Chinese drug *Kuh Seng* or the Japanese *Shinkyogan*, the dried roots of *Sophora angustifolia* Sieb. & Zucc., *S. flavescens* Ait. and other *Sophora* spp., *Leguminosae.* Isomeric with lupanine, *q.v.* Isoln: Kondo, *Arch. Pharm.* **266**, 1 (1928); Winterfeld, Kneuer, *Ber.* **64**, 150 (1931); Orechov, Proskurnina, *Ber.* **67**, 77 (1934); Briggs, Ricketts, *J. Chem. Soc.* **1937**, 1795. Identity with sophocarpidine: Orechov *et al., Ber.* **68**, 429 (1935). Alternate method of isoln: F. Bohlmann *et al., Ber.* **91**, 2189 (1958). Structure: Tsuda, Murakami, *Ber.* **69**, 429 (1936). Configuration: Bohlmann *et al., Ber.* **91**, 2176

(1958). Absolute configuration: Okuda *et al., Chem. Pharm. Bull.* **14**, 314 (1966); Cervinka, *Z. Chem.* **7**, 190 (1967). Synthetic studies: Tsuda *et al., J. Org. Chem.* **21**, 1481 (1956); **23**, 1179 (1958). Synthesis of *matridine (deoxymatrine)*: Mandell, Singh, *J. Am. Chem. Soc.* **83**, 1766 (1961). Total synthesis of (±)-matrine: Mandell *et al., ibid.* **87**, 5234 (1965); J. Chen *et al., Chem. Commun.* **12**, 905 (1986). Synthesis of (+)-matrine: Okuda *et al., Chem. Pharm. Bull.* **14**, 275 (1966). Biosynthesis: Schütte *et al., Arch. Pharm.* **295**, 34 (1962).

Has been obtained in four forms: α-Form: needles or flat prisms, mp 76°. β-Form: orthorhombic prisms, mp 87°, $[\alpha]_D^{20}$ +38° (alc). γ-Form: liquid, bp₆ 223°, d₄²⁰ 1.088, n_D^{85} 1.5287. δ-Form: prisms, mp 84°. Sol in water, benzene, chloroform, ether, carbon disulfide; slightly sol in petr ether.

5800. Maytansine. Maitansine; NSC-153858. $C_{34}H_{46}$-ClN_3O_{10}; mol wt 692.21. C 59.00%, H 6.70%, Cl 5.12%, N 6.07%, O 23.11%. Anti-leukemic ansa macrolide, member of a class of compounds that includes the rifamycins, streptovaricins etc. First ansa compound isolated from a plant rather than produced by a microorganism. Isoln from *Maytenus ovatus* Loes., *Celastraceae* and structure elucidation: S. M. Kupchan *et al., J. Am. Chem. Soc.* **94**, 1354 (1972); S. M. Kupchan, **Ger.** pat. **2,241,418** (1974); S. M. Kupchan *et al.,* **U.S.** pat. **3,896,111** (1975 to Research Corp.). Crystal structure and absolute configuration of the (3-bromopropyl) ether: R. F. Bryan *et al., J. Chem. Soc. Perkin Trans. II* **1973**, 897. Synthetic studies: A. I. Meyers *et al., Tetrahedron Letters* **1974**, 717; **1975**, 1745, 1749; E. J. Corey, M. G. Bock, *ibid.* **1975**, 2643; E. J. Corey *et al., ibid.* **1978**, 1051; A. I. Meyers, J. P. Hudspeth, *ibid.* **22**, 3925 (1981). Total synthesis of (±)-form: A. I. Meyers *et al., J. Am. Chem. Soc.* **102**, 6597 (1980); of naturally occurring (−)-form: E. J. Corey *et al., ibid.* 6613. Cytotoxic action: M. K. Wolpert-Defilippes, *Biochem. Pharmacol.* **24**, 751 (1975). Clinical studies in carcinoma, melanoma: J. A. Neidhart *et al., Cancer Treat. Rep.* **64**, 675 (1980); D. L. Ahmann *et al., ibid.* 721. Toxicity: G. M. Mugera, J. M. Ward, *ibid.* **61**, 1333 (1977).

mp 171-172°. $[\alpha]_D^{26}$ −145° (c = 0.055 in chloroform). uv max (ethanol): 233, 254, 282, 290 nm (ε 29800, 27200, 5690, 5520). LD₅₀ in rats (mg/kg): 0.48 s.c. (Mugera, Ward).

5801. Mazindol. *5-(4-Chlorophenyl)-2,5-dihydro-3H-imidazo[2,1-a]isoindol-5-ol;* 5-(4-chlorophenyl)-2,3-dihydro-5-hydroxy-5H-imidazo[2,1-a]isoindole; SaH-42548; Magrilon; Mazildene; Sanorex; Terenac; Teronac. $C_{16}H_{13}ClN_2O$; mol wt 284.74. C 67.49%, H 4.60%, Cl 12.45%, N 9.84%, O 5.62%. Prepn: W. J. Houlihan, **Ger.** pat. **1,814,540,** *C.A.* **71**, 81368s (1969); W. J. Houlihan, M. K. Eberle, **Ger.** pat.

1,930,488; *eidem,* **U.S.** pat. **3,597,445** (1969, 1970, 1970 all to Sandoz); T. S. Sulkowski, **U.S.** pat. **3,763,178** (1973 to American Home Prod.). Clinical trials as appetite depressant: C. Sirtori *et al., Am. J. Med. Sci.* **261**, 341 (1971); A. J. Hadler, *J. Clin. Pharmacol.* **12**, 453 (1972). Effect on human prolactin and growth hormone responses: D. A. Thompson *et al., Metabolism* **30**, 1015 (1981). Clinical evaluation in Duchenne muscular dystrophy: P. J. Collipp *et al., J. Med. Genet.* **21**, 254 (1984).

White crystalline solid from ethanol, mp 215-217° (Sulkowski); also reported as crystals from acetone-hexane, mp 198-199° (Houlihan, Eberle). uv max (95% ethanol): 223, 268.5, 272 nm (ε 19000, 4400, 4400). Sol in ethanol. Insol in water.

Note: This is a controlled substance (stimulant) listed in the U.S. Code of Federal Regulations, Title 21 Part 1308.14 (1995).

THERAP CAT: Anorexic; CNS stimulant.

5802. Mazipredone. *(11β)-11,17-Dihydroxy-21-(4-methyl-1-piperazinyl)pregna-1,4-diene-3,20-dione;* 11β,17α-dihydroxy-3,20-dioxo-21-(4-methyl-1-piperazinyl)pregna-1,4-diene; 11β,17-dihydroxy-21-(4-methyl-1-piperazinyl)-Δ¹-progesterone. $C_{26}H_{38}N_2O_4$; mol wt 442.60. C 70.56%, H 8.65%, N 6.33%, O 14.46%. Prepn: Tuba *et al.,* **Hung.** pat. **150,350** (1963 to Gedeon Richter), *C.A.* **60**, 3057h (1964).

Crystals from tetrahydrofuran + ligroin, dec 199°.

Hydrochloride, $C_{26}H_{38}N_2O_4$.HCl, *Depersolone.* Crystals, dec 246°. Sol in water.

THERAP CAT: Anti-inflammatory.

5803. MCPA. *(4-Chloro-2-methylphenoxy)acetic acid;* (4-chloro-o-toloxy)acetic acid; 2-methyl-4-chlorophenoxyacetic acid; MCP; Agritox; Agroxone; Cornox; Methoxone. $C_9H_9ClO_3$; mol wt 200.62. C 53.88%, H 4.52%, Cl 17.67%, O 23.92%. Prepn: Synerholm, Zimmerman, *Contrib. Boyce Thompson Inst.* **14**, 91 (1945); Templeman, Sexton, *Proc. Roy. Soc.* **133B**, 300 (1946); Foster, **Brit.** pats. **573,479** and **573,510** (both 1945 to ICI); Skeeters, **U.S.** pat. **2,740,810** (1956 to Diamond Alkali). Toxicity study: Rowe, Hymas, *Am. J. Vet. Res.* **15**, 622 (1954). Review of carcinogenic risk: *IARC Monographs* **30**, 255-269 (1983).

Plates from benzene, mp 120°. Practically insol in water. LD₅₀ orally in rats: 700 mg/kg (Rowe, Hymas).

Sodium salt, $C_9H_8ClNaO_3$, *Chiptox.* Very sol in water.

USE: Herbicide.

5804. MDA. *α-Methyl-1,3-benzodioxole-5-ethanamine;* 3,4-methylenedioxyamphetamine; 3,4-methylenedioxy-α-methyl-β-phenylethylamine; 3,4-methylenedioxyphenylisopropylamine; "Love"; EA-1299. $C_{10}H_{13}NO_2$; mol wt 179.22. C 67.02%, H 7.31%, N 7.82%, O 17.85%. Entactogen; synthetic amphetamine derivative with stimulant and hallucinogenic properties. Metabolite of MDMA, *q.v.* Prepn: C. Mannich, W. Jacobsohn, *Ber.* **43**, 189 (1910); U. Braun *et al., J. Pharm. Sci.* **69**, 192 (1980). Pharmacology: L. E. Markert, D. C. S. Roberts, *Pharmacol. Biochem. Behav.* **39**, 569 (1991); K. M. Hegadoren *et al., Psychopharmacology* **118**, 295 (1995). HPLC determn in whole blood: R. E. Michel *et al., J. Neurosci. Methods* **50**, 61 (1993). GC/MS determn in hair: P. Kintz *et al., J. Chromatog. B.* **670**, 162 (1995). Acute toxicity studies: P. N. Thiessen, D. A. Cook, *Clin. Toxicol.* **6**, 193 (1973); H. F. Hardman *et al., Toxicol. Appl. Pharmacol.* **25**, 299 (1973). Review of toxicity, mechanism of action, and psychoactive effects: R. P. Climko *et al., Int. J. Psychiatry Med.* **16**, 359-372 (1986-1987).

Almost colorless oil. bp_{22} 157°. $bp_{0.2}$ 80-90°.
Hydrochloride, $C_{10}H_{13}NO_2$·HCl. Crystals from isopropanol/ether, mp 187-188° (Braun); also reported as mp 180-181° (Mannich, Jacobsohn). Easily sol in water, alcohol. LD_{50} in mice, rats, guinea pigs (mg/kg): 68, 27, 28 i.p. (Hardman).
Note: This is a controlled substance (hallucinogen) listed in the U.S. Code of Federal Regulations, Title 21 Part 1308.11 (1995).

5805. MDE. *(±)-N-Ethyl-α-methyl-1,3-benzodioxole-5-ethanamine; N-ethyl-3,4-methylenedioxyamphetamine; N-ethyl-3,4-methylenedioxyphenylisopropylamine; 3,4-methylenedioxyethamphetamine; MDEA; "Eve".* $C_{12}H_{17}NO_2$; mol wt 207.27. C 69.54%, H 8.27%, N 6.76%, O 15.44%. Entactogen; N-ethyl derivative of MDA, *q.v.* Prepn: U. Braun *et al., J. Pharm. Sci.* **69**, 192 (1980); F. T. Noggle, Jr. *et al., J. Assoc. Off. Anal. Chem.* **70**, 981 (1987). Pharmacology: M. Johnson *et al., Biochem. Pharmacol.* **38**, 4333 (1989); K. M. Hegadoren *et al., Psychopharmacology* **118**, 295 (1995). Determn in human urine using immunoassays and GC/MS: G. W. Kunsman *et al., J. Anal. Toxicol.* **14**, 149 (1990). HPLC determn in tablets: M. Longo *et al., J. Liq. Chromatog.* **17**, 649 (1994). Impact on public health and abuse potential: G. P. Dowling *et al., J. Am. Med. Assoc.* **257**, 1615 (1987); R. O. Bost, *J. Forensic Sci.* **33**, 576 (1988); B. Tehan *et al., Anaesthesia* **48**, 507 (1993). Psychological effects in humans: L. Hermle *et al., Neuropsychopharmacology* **8**, 171 (1993). Neuroendocrine and cardiovascular effects in humans: E. Gouzoulis *et al., ibid.* 187.

Viscous, colorless oil. $bp_{0.2}$ 85-95°.
Hydrochloride, fine white granular solid from ethanol-ether, mp 197-198° (Noggle). Also reported as white solid from isopropanol-ether, mp 201-202° (Braun). LD_{50} i.p. in mice: 102 mg/kg (Noggle).
Note: This is a controlled substance (hallucinogen) listed in the U.S. code of Federal Regulations, Title 21 Part 1308.11 (1995).

5806. MDMA. *N,α-Dimethyl-1,3-benzodioxole-5-ethanamine;* 3,4-methylenedioxymethamphetamine; N-methyl-3,4-methylenedioxyphenylisopropylamine; "Ecstasy"; "E"; "XTC"; "Adam". $C_{11}H_{15}NO_2$; mol wt 193.25. C 68.37%, H 7.82%, N 7.25%, O 16.56%. Entactogen; N-methyl derivative of MDA, *q.v.* (S)-Form possesses more potent CNS activity. Prepn: **Ger. pat.** 274,350 (1914 to E. Merck), *C.A.* **8**, 3350 (1914); U. Braun *et al., J. Pharm. Sci.* **69**, 192 (1980). Spectral and chromatographic analyses: K. Bailey *et al., J. Assoc. Offic. Anal. Chem.* **58**, 62 (1975). GC/MS determn in urine: B. K. Gan *et al., J. Forensic Sci.* **36**, 1331 (1991). HPLC determn in whole blood: R. E. Michel *et al., J. Neurosci. Methods* **50**, 61 (1993). Acute toxicity study: H. F. Hardman *et al., Toxicol. Appl. Pharmacol.* **25**, 299

(1973). Series of articles on history, pharmacology, and neurotoxicity: *Ann. N.Y. Acad. Sci.* **600**, 601-715 (1990). Review of biochemical actions: M. Rattray, *Essays Biochem.* **26**, 77-87 (1991); of pharmacology, toxicology, and human experience: T. D. Steele *et al., Addiction* **89**, 539-551 (1994); A. R. Green *et al., Psychopharmacology* **119**, 247-260 (1995).

Oil, $bp_{0.4}$ 100-110°.
Hydrochloride, $C_{11}H_{15}NO_2$·HCl, crystals from isopropanol/n-hexane, mp 147-148° (Bailey). Crystals from isopropanol/ether, mp 152-153° (Braun). uv max (ethanol): 286 nm (ε 3843). LD_{50} in mice, rats, guinea pigs (mg/kg): 97, 49, 98 i.p. (Hardman).
Note: This is a controlled substance (hallucinogen) listed in the U.S. Code of Federal Regulations, Title 21 Part 1308.11 (1995).

5807. Mebendazole. *(5-Benzoyl-1H-benzimidazol-2-yl)-carbamic acid methyl ester; 5-benzoyl-2-benzimidazolecarbamic acid methyl ester;* methyl 5-benzoyl-2-benzimidazolecarbamate; R-17635; Bantenol; Equivurm Plus; Lomper; Mebenvet; Noverme; Ovitelmin; Pantelmin; Telmin; Vermicidin; Vermirax; Vermox. $C_{16}H_{13}N_3O_3$; mol wt 295.30. C 65.08%, H 4.44%, N 14.23%, O 16.25%. Prepn: J. L. H. Van Gelder *et al.,* **Ger. pat.** 2,029,637; *eidem,* **U.S. pat.** 3,657,267 (1971, 1972 to Janssen). Activity studies: Walker, Knight, *Vet. Rec.* **90**, 58 (1972). Clinical studies: Brugmans *et al., J. Am. Med. Assoc.* **217**, 313 (1971); Callear, Neave, *Brit. Vet. J.* **127**, xli (1971); *eidem, ibid.* **129**, 79 (1973). Comprehensive description: A. A. Al-Badr, M. Tariq, in *Analytical Profiles of Drug Substances* vol. 16, K. Florey, Ed. (Academic Press, New York, 1987) pp 291-326.

Crystals from acetic acid and methanol, mp 288.5°. Soluble in formic acid. Practically insol in water, ethanol, ether, chloroform. LD_{50} orally: > 80 mg/kg in sheep; > 40 mg/kg in mice, rats and chickens (Van Gelder).
THERAP CAT: Anthelmintic (Nematodes).
THERAP CAT (VET): Anthelmintic.

5808. Mebeverine. *3,4-Dimethoxybenzoic acid 4-[ethyl-[2-(4-methoxyphenyl)-1-methylethyl]amino]butyl ester; veratric acid 4-[ethyl(p-methoxy-α-methylphenethyl)amino]butyl ester;* 3,4-dimethoxybenzoic acid 4-[ethyl(p-methoxy-α-methylphenethyl)amino]butyl ester; 4-[ethyl(p-methoxy-α-methylphenethyl)amino]butyl 3,4-dimethoxybenzoate; 4-[N-[2-(p-methoxyphenyl)-1-methylethyl]-N-ethylamino]butyl 3,4-dimethoxybenzoate. $C_{25}H_{35}NO_5$; mol wt 429.56. C 69.90%, H 8.21%, N 3.26%, O 18.62%. Smooth muscle relaxant. Prepn: **Belg. pat.** 609,490, *C.A.* **59**, 517b (1963) and T. Kralt *et al.,* **Ger. pat.** 1,126,889 corresp to **U.S. pat.** 3,265,577 (1962, 1962, 1966 to N. V. Philips). Pharmacology: G. Bertaccini *et al., Farmaco Ed. Sci.* **30**, 823 (1975).

Hydrochloride, $C_{25}H_{35}NO_5$·HCl, *Colofac, Duspatalin, Duspatal.* Crystals from ethyl methyl ketone, mp 105-107° (Ger. patent); also reported as mp 129-131° (Belg. patent).

THERAP CAT: Antispasmodic.

5809. Mebhydroline. *2,3,4,5-Tetrahydro-2-methyl-5-(phenylmethyl)-1H-pyrido[4,3-b]indole; 5-benzyl-2,3,4,5-tetrahydro-2-methyl-1H-pyrido[4,3-b]indole;* 3-methyl-9-benzyl-1,2,3,4-tetrahydro-γ-carboline; 5-benzyl-1,2,3,4-tetrahydro-2-methyl-γ-carboline; N-methyl-9-benzyltetrahydro-γ-carboline; Incidal. $C_{19}H_{20}N_2$; mol wt 276.38. C 82.57%, H 7.29%, N 10.14%. Prepn: Hörlein, *Ber.* **87**, 463 (1954); **Brit.** pat. **721,171** (1954 to Bayer); Hörlein, **U.S.** pat. **2,786,059** (1957 to Schenley). Pharmacology of Diazoline: Kharkevich, *Farmakol. Toksikol.* **20**(no. 6), 46 (1957).

Minute crystals, mp 95°. bp_1 207-215°. Practically insol in water. Freely sol in methanol, ethanol, acetone, chloroform. Slightly sol in ether.

1,5-Naphthalenedisulfonate salt, $(C_{19}H_{20}N_2)_2$·$C_{10}H_8O_6S_2$, *Diazoline, Fabahistin, Omeril.* White powder, dec 280°. Practically insol in water. Sparingly sol in hot glacial acetic acid. Sol in hot formamide, giving a yellow soln.

THERAP CAT: Antihistaminic.

5810. Mebiquine. *8-[(Dihydroxybismutho)oxy]-6-methylquinoline; dihydroxy(6-methyl-8-quinolinolato)bismuth;* 6-methyl-8-bismuthohydroxyquinoline; DV1; Diarétyl. $C_{10}H_{10}BiNO_3$; mol wt 401.17. C 29.94%, H 2.51%, Bi 52.09%, N 3.49%, O 11.96%. Prepn by the addition of 6-methyl-8-hydroxyquinoline to a solution of bismuth nitrate: Riviere *et al.,* **Ger.** pat. **1,811,581**; *eidem,* **U.S.** pat. **3,591,591** (1969, 1971 to Ugine Kuhlmann). Parasiticidal properties: R. Cavier *et al., Ann. Pharm. Franc.* **31**, 273 (1973).

Yellow powder, insol in water, alc, weak alkaline solns. Sol in strong acids. LD_{50} orally in mice: > 10,000 mg/kg (Cavier).

THERAP CAT: Antidiarrheal.

5811. MeBmt. *[2S-(2R*,3S*,4S*,6E)]-3-Hydroxy-4-methyl-2-(methylamino)-6-octenoic acid;* (4R)-4-[(E)-2-butenyl]-4,N-dimethyl-L-threonine. $C_{10}H_{19}NO_3$; mol wt 201.27. C 59.68%, H 9.52%, N 6.96%, O 23.85%. Characteristic C_9 amino acid of cyclosporins, *q.v.* Identification in cyclosporin a: A. Ruegger *et al., Helv. Chim. Acta* **59**, 1075 (1976). Syntheses: R. M. Wenger, *ibid.* **66**, 2308 (1983); D. A. Evans, A. E. Weber, *J. Am. Chem. Soc.* **108**, 6757 (1986); U. Schmidt, W. Siegel, *Tetrahedron Letters* **28**, 2849 (1987). Appears to be necessary for full biological activity of cyclosporin: R. M. Wenger, *Angew. Chem. Int. Eng. Ed.* **24**, 77 (1985); D. H. Rich *et al., J. Med. Chem.* **29**, 978 (1986).

Crystals from ethanol, mp 240-241° (Wenger). $[\alpha]_D^{20}$ +13.5° (c = 0.50 in water, pH 7). Also reported as mp 242-243° (Evans, Weber). $[\alpha]_D$ +17° (c = 0.51 in 0.4N aq HCl).

5812. Mebrofenin. *N-[2-[(3-Bromo-2,4,6-trimethylphenyl)amino]-2-oxoethyl]-N-(carboxymethyl)glycine;* [[[(3-bromomesityl)carbamoyl]methyl]imino]diacetic acid; N-(3-bromo-2,4,6-trimethylacetanilide)iminodiacetic acid; trimethylbromo-IDA; Choletec. $C_{15}H_{19}BrN_2O_5$; mol wt 387.23. C 46.53%, H 4.95%, Br 20.63%, N 7.23%, O 20.66%. Prepn: A. D. Nunn, M. D. Loberg, **Belg.** pat. **891,534**; *eidem,* **U.S.** pat. **4,418,208** (1982, 1983 both to Squibb). Physicochemical properties of ^{99m}technetium complex used for hepatobiliary tract imaging: A. D. Nunn *et al., J. Nucl. Med.* **24**, 423 (1983). Pharmacokinetics in rats: A. R. Fritzberg *et al., J. Pharm. Sci.* **73**, 1861 (1984). Evaluation as cholescintigraphic imaging agent in animals: A. van Aswegen *et al., Nucl. Med. Biol.* **13**, 509 (1986); J. Kapuscinski *et al., Nucl. Med.* **25**, 188 (1986); in humans: W. C. Klingensmith *et al., Radiology* **146**, 181 (1983). Toxicity study: M. Jiang *et al., Zhonghua Heyixue Zazhi* **4**, 214 (1984), *C.A.* **102**, 42241j (1985).

Crystals from ethanol, mp 198-200° LD_{50} in mice, rats (mg/kg): 213.8, 226.4 i.v. (Jiang).

THERAP CAT: ^{99m}Tc complex as diagnostic aid (radioactive imaging agent).

5813. Mebutamate. *2-Methyl-2-(1-methylpropyl)-1,3-propanediol dicarbamate;* 2-sec-butyl-2-methyl-1,3-propanediol dicarbamate; carbamic acid 2-sec-butyl-2-methyltrimethylene ester; 2-sec-butyl-2-methyltrimethylenecarbamate; 2-methyl-2-sec-butyl-1,3-propanediol dicarbamate; 2,2-dicarbamoyloxymethyl-3-methylpentane; dicamoylmethane; W-583; Capla; Butatensin; Carbuten; Dormate; Mebutina; Prean; Sigmafon; Vallene; Mega; No-Press; Axiten; Ipotensivo. $C_{10}H_{20}N_2O_4$; mol wt 232.28. C 51.71%, H 8.68%, N 12.06%, O 27.55%. Prepn: Berger, Ludwig, **U.S.** pat. **2,878,280** (1959 to Carter Prod.). Mechanism of action: Kletzkin, *Arch. Int. Pharmacodyn. Ther.* **164**, 71 (1966). Metabolism: Edelson, Douglas, *ibid.* **173**, 182 (1968).

Crystals, mp 77-79°. Soly in water: ~0.1%. Sol in most organic solvents. On heating or boiling with acid or alkali, hydrolyzes to the alcohol, ammonia and carbon dioxide.

Note: This is a controlled substance (depressant) listed in the U.S. Code of Federal Regulations, Title 21 Part 1308.14 (1995).

THERAP CAT: Antihypertensive.

5814. Mecamylamine. *N,2,3,3-Tetramethylbicyclo[2.2.1]heptan-2-amine; N,2,3,3-tetramethyl-2-norbornanamine;* N,2,3,3-tetramethyl-2-norcamphanamine; 3-methylaminoisocamphane; 2-methylaminoisocamphane; 2-methylamino-2,3,3-trimethylnorbornane; N-methyl-2-isocamphanamine; 3β-methylamino-2,2,3-trimethylbicyclo[2.2.1]heptane; mecamine; $C_{11}H_{21}N$; mol wt 167.29. C 78.98%, H 12.65%, N 8.37%. Prepn from 3-formamidoisocamphane: Stein *et al., J. Am. Chem. Soc.* **78**, 1514 (1956); Stone *et al., J. Med. Pharm. Chem.* **5**, 665 (1962); Pfister, Stein, **U.S.** pat. **2,831,027** (1958 to Merck & Co.). From 3-aminoisocamphane: **Brit.** pat. **856,862** (1961 to Lepetit).

Oily liquid. bp$_{4.0}$ 72°. n_D^{25} 1.4881. Slightly sol in water. Hydrochloride, C$_{11}$H$_{21}$N.HCl, *Inversine, Mevasine.* Crystals, dec 245.5-246.5°. Bittersweet. Soly in water (g/100 ml): 21.2; in alc: 8.2; in glycerol: 10.4; in isopropanol: 2.1. pH of 1% aq soln 6.0-7.5. Solns can be sterilized by autoclaving.

THERAP CAT: Antihypertensive.

5815. Mechlorethamine. *2-Chloro-N-(2-chloroethyl)-N-methylethanamine; 2,2'-dichloro-N-methyldiethylamine; N-methyl-2,2'-dichlorodiethylamine; di(chloroethyl)methylamine; methylbis(β-chloroethyl)amine; methyldi(2-chloroethyl)amine; chlormethine; nitrogen mustard; stickstofflost; MBA; HN2.* C$_5$H$_{11}$Cl$_2$N; mol wt 156.05. C 38.48%, H 7.10%, Cl 45.44%, N 8.98%. CH$_3$N(CH$_2$CH$_2$Cl)$_2$. A nitrogen mustard prepd by action of thionyl chloride on 2,2'-(methylimino)diethanol in trichloroethylene: Prelog, Stepan, *Coll. Czech. Chem. Commun.* **7**, 93 (1935); Hanby, Rydon, *J. Chem. Soc.* **1947**, 513; Abrams *et al., J. Soc. Chem. Ind. (London)* **68**, 280 (1949); T. W. Doyle, D. M. Vyas in Kirk-Othmer, *Encyclopedia of Chemical Technology* vol. 5 (John Wiley & Sons, New York, 4th ed., 1993) pp 876-879, 900-901. Toxicity data: W. P. Anslow *et al., J. Pharmacol. Exp. Ther.* **91**, 224 (1947).

Mobile liquid. Faint odor of herring. *Vesicant, necrotizing irritant. Never use without appropriate gas mask.* Volatility at 25° = 3.581 mg/l. d$_4^{25}$ 1.118. mp −60°. bp$_{18}$ 87°; bp$_{10}$ 75°; bp$_5$ 64°; bp$_2$ 59°. Very slightly sol in water. Miscible with DMF, CS$_2$, CCl$_4$, many other organic solvents and oils. The undiluted liq dec on standing and forms polymeric quaternary ammonium salts which are insol in the free base.

Hydrochloride, C$_5$H$_{11}$Cl$_2$N.HCl, *Caryolysine, Cloramin, Dichloren, Embichen, Embikhine, Erasol, Mustargen hydrochloride, Mustine hydrochloride, Nitrogranulogen.* Hygroscopic leaflets from acetone or chloroform, mp 109-111°. Very sol in water; sol in alcohol. Initial pH of a 2% aq soln: 3.0-4.0. Dry crystals are stable at temps up to 40°. LD$_{50}$ in rats (mg/kg): 1.1 i.v.; 1.9 s.c. (Anslow).

Caution: Direct contact may cause severe skin and eye irritation; injury to deeper ocular structures, particularly the iris and lens. *See* Patty's *Industrial Hygiene and Toxicology* vol. 2E, G. D. Clayton, F. E. Clayton, Eds. (John Wiley & Sons, Inc., New York, 4th ed., 1994) pp 3297-3299. Mechlorethamine hydrochloride may reasonably be anticipated to be a carcinogen: *Seventh Annual Report on Carcinogens* (PB95-109781, 1994) p 273.

N-Oxide hydrochloride, *see* separate entry.

USE: Base as a gas warfare agent.

THERAP CAT: Hydrochloride as an antineoplastic.

THERAP CAT (VET): Hydrochloride has been used as an antineoplastic.

5816. Mechlorethamine Oxide Hydrochloride. *2-Chloro-N-(2-chloroethyl)-N-methylethanamine N-oxide hydrochloride; 2,2'-dichloro-N-methyldiethylamine N-oxide hydrochloride; N-methyl-2,2'-dichlorodiethylamine N-oxide hydrochloride; methylbis(β-chloroethyl)amine N-oxide hydrochloride; methyldi(2-chloroethyl)amine N-oxide hydrochloride; Nitromin; Mitomen; Mustron.* C$_5$H$_{12}$Cl$_3$NO; mol wt 208.51. C 28.80%, H 5.80%, Cl 51.01%, N 6.72%, O 7.67%. Prepd by treating 2,2'-dichloro-*N*-methyldiethylamine with hydrogen peroxide and acetic anhydride in ether, followed by shaking with 10% hydrochloric acid: Aiko *et al., J. Pharm. Soc. Japan* **72**, 1297 (1952, English

Text), *C.A.* **47**, 1289 (1953). *Cf.* Stahmann, Bergmann, *J. Org. Chem.* **11**, 586 (1946).

Prisms from acetone, mp 109-110°. Sol in water.

THERAP CAT: Antineoplastic.

5817. Meclizine. *1-[(4-Chlorophenyl)phenylmethyl]-4-[(3-methylphenyl)methyl]piperazine; 1-(p-chloro-α-phenylbenzyl)-4-(m-methylbenzyl)piperazine; 1-(p-chlorobenzhydryl)-4-(m-methylbenzyl)piperazine; 1-(p-chlorobenzhydryl)-4-(m-methylbenzyl)diethylenediamine; meclozine; parachloramine.* C$_{25}$H$_{27}$ClN$_2$; mol wt 390.96. C 76.81%, H 6.96%, Cl 9.07%, N 7.17%. Synthesis: Morren, **Belg.** pat. **502,889** (1951); **Brit.** pat. **705,979** (1954), **U.S.** pat. **2,709,169** (1955 to U.C.B.). Outline of reactions: Grivsky, *Ind. Chim. Belge* **17**, 735 (1952). Clinical evaluation in vertigo: B. Cohen, J. M. B. Vianney de Jong, *Arch. Neurol.* **27**, 129 (1972).

bp$_2$ 230°.

Dihydrochloride monohydrate, C$_{25}$H$_{27}$ClN$_2$.2HCl.H$_2$O, *UCB-5062, Ancolan, Antivert, Bonamine, Bonine, Calmonal, Diadril, Histametizine, Navicalm, Neo-Istafene, Peremesin, Postafene, Sabari, Sea-Legs, Veritab.* Freely sol in chloroform, pyridine; slightly sol in dil acids, alc. Practically insol in water (0.1 gm/100 ml), ether.

THERAP CAT: Antiemetic.

THERAP CAT (VET): Antiemetic.

5818. Meclocycline. *[4S-(4α,4aα,5α,5aα,12aα)]-7-Chloro-4-(dimethylamino)-1,4,4a,5,5a,6,11,12a-octahydro-3,-5,10,12,12a-pentahydroxy-6-methylene-1,11-dioxo-2-naphthacenecarboxamide; 7-chloro-6-methylene-5-hydroxytetracycline; GS-2989; NSC-78502.* C$_{22}$H$_{21}$ClN$_2$O$_8$; mol wt 476.87. C 55.41%, H 4.44%, Cl 7.43%, N 5.87%, O 26.84%. Semisynthetic antibiotic derived from tetracycline, *q.v.* Prepn: R. K. Blackwood *et al.,* **U.S.** pat. **2,984,686** (1961 to Pfizer); *eidem, J. Am. Chem. Soc.* **83**, 2773 (1961); of the free base and 5-sulfosalicylate salt: *eidem, ibid.* **85**, 3943 (1963). *In vitro* study: L. Lucca, G. Vittadini, *G. Ital. Chemioter.* **26**, 203 (1979). ^{13}C-NMR study: E. Mazzola *et al., J. Pharm. Sci.* **69**, 229 (1980). Toxicity study: F. Bernardi *et al., G. Ital. Chemioter.* **17**, 276 (1970).

uv max (methanol, 0.01N HCl): 245, 347 nm (log ϵ 4.34, 410). LD$_{50}$ in mice (mg/kg): > 5000 orally, 425 i.p. (Bernardi).

5-Sulfosalicylate, C$_{29}$H$_{27}$ClN$_2$O$_{14}$S, *Meclan, Mecloderm, Meclosorb, Meclutin, Traumatociclina.* uv max (methanol, 0.01N HCl): 239, 268, 346 nm (log ϵ 4.46, 4.07, 4.11).

THERAP CAT: Antibacterial.

5819. Meclofenamic Acid. *2-[(2,6-Dichloro-3-methylphenyl)amino]benzoic acid; N-(2,6-dichloro-m-tolyl)anthranilic acid; meclophenamic acid; CI-583; INF-4668; Arquel.* C$_{14}$H$_{11}$Cl$_2$NO$_2$; mol wt 296.15. C 56.78%, H 3.74%, Cl 23.94%, N 4.73%, O 10.80%. Prepn: Scherrer, Short, **Ger.** pat. **1,149,015,** *C.A.* **61**, 1801d (1964) and **U.S.** pat. **3,313,-848** (1963, 1967 to Parke, Davis); Juby *et al., J. Med Chem.* **11**, 111 (1968). Pharmacology and toxicology: Winder *et al., J. Pharmacol. Exp. Ther.* **148**, 422 (1965); Winder, *Ann. Phys. Med.* **1966**, 7; Kaump, *ibid.* 16. Series of articles on structure-activity relationships, mechanism of action, clinical studies: *Arzneimittel-Forsch.* **33**, 619-680 (1983). Effect on prostaglandin E receptor binding: M. C. P. Rees *et al.*, *Lancet* **2**, 541 (1988).

White crystals from acetone-water, mp 257-259°; also reported as mp 248-250°. Soly (mg/ml): water 0.03; 0.1*N* NaOH 28. pH of satd aq soln: ~6.9.

Sodium salt monohydrate, $C_{14}H_{10}Cl_2NNaO_2.H_2O$, *Lenidolor, Meclodol, Meclomen, Movens.* mp 289-291°. Soly in water: 15 mg/ml (slightly turbid). pH 8.7.

THERAP CAT: Anti-inflammatory; antipyretic.

THERAP CAT (VET): Anti-inflammatory.

5820. Meclofenoxate. *(4-Chlorophenoxy)acetic acid 2-(dimethylamino)ethyl ester;* dimethylaminoethyl *p*-chlorophenoxyacetate; centrophenoxine; meclofenoxane; acephen; ANP-235; Analux; Cetrexin; Proseryl. $C_{12}H_{16}ClNO_3$; mol wt 257.72. C 55.93%, H 6.26%, Cl 13.76%, N 5.43%, O 18.62%. Prepn: Rumpf, Thuillier, **Fr.** pat. **M398** (1962 to Centre Nat'l. Recherche Sci.), *C.A.* **57**, 16768e (1962); Thuillier *et al.*, *Compt. Rend.* **249**, 2081 (1959); Thuillier, Rumpf, *Bull. Soc. Chim. France* **1960**, 1786. Pharmacology: Petkov, *C.A.* **65**, 20717g (1966); Liberman, *Farmakol. Toksikol.* **30**, 409 (1967). Auxin activity: Conti, *Boll. Chim. Farm.* **107**, 325 (1968).

Hydrochloride, $C_{12}H_{16}ClNO_3.HCl$, *Cellative, Clocete, Lucidril, Methoxynal, Proserout, Brenal, Marucotol, Helfergin.* Crystals from isopropyl alcohol or acetone, mp 135-139°. Sol in cold water. Sparingly sol in cold isopropyl alcohol, acetone. Practically insol in benzene, ether, chloroform. A 5% soln in water has pH 6. LD_{50} in mice (mg/kg): 330 i.v.; 1750 orally; 845 i.p. (Rumpf, Thuillier).

USE: Plant growth regulator.

THERAP CAT: Cerebral stimulant.

5821. Mecloqualone. *3-(2-Chlorophenyl)-2-methyl-4(3H)-quinazolinone;* 2-methyl-3-(2-chlorophenyl)-4-quinazolone; Nubarene. $C_{15}H_{11}ClN_2O$; mol wt 270.72. C 66.55%, H 4.10%, Cl 13.10%, N 10.35%, O 5.91%. Prepd from acetylanthranilic acid and *o*-chloroaniline in the presence of $POCl_3$: Jackman *et al.*, *J. Pharm. Pharmacol.* **12**, 529 (1960); Closa, *J. Prakt. Chem.* [4] **14**, 84 (1961). Metabolism: Dubnick, Towne, *Toxicol. Appl. Pharmacol.* **22**, 82 (1972); Daenens, Van Bovan, *Arzneimittel-Forsch* **24**, 195 (1974).

Crystals mp 126-128°.

Hydrochloride, $C_{15}H_{11}ClN_2O.HCl$, crystals, mp 239-241°.

Note: This is a controlled substance (depressant) listed in the U.S. Code of Federal Regulations, Title 21 Part 1308.11 (1995).

THERAP CAT: Sedative; hypnotic.

5822. Mecloralurea. *N-Methyl-N'-(2,2,2-trichloro-1-hydroxyethyl)urea;* trichloroethylolmethylurea; Heraldium. $C_4H_7Cl_3N_2O_2$; mol wt 221.47. C 21.69%, H 3.19%, Cl 48.02%, N 12.65%, O 14.45%. Prepn from chloral hydrate and *N*-methylurea: Chattaway, James, *J. Chem. Soc* **1934**, 109; **Fr.** pat. **M5341; Brit.** pat. **1,129,437** corresp to Cas-

taigne, **U.S.** pat. **3,510,557** (1967, 1968 and 1970 to Centre d'Etudes l'Ind. Pharm.).

White crystalline powder from boiling water, mp 135-140°. Partly sol in water, acetic acid, ether; insol in alcohol, pyridine. Practically insol in hydrocarbons and chloroform.

THERAP CAT: Anxiolytic.

5823. Mecloxamine. *2-[1-(4-Chlorophenyl)-1-phenylethoxy]-N,N-dimethyl-1-propanamine; 2-[(p-chloro-α-methyl-α-phenylbenzyl)oxy]-N,N-dimethylpropylamine; β-dimethylaminoisopropyl 4-chloro-α-methylbenzhydryl ether.* $C_{19}H_{24}ClNO$; mol wt 317.86. C 71.80%, H 7.61%, Cl 11.15%, N 4.41%, O 5.03%. Prepn: **Brit.** pat. **875,060** (1960 to Astra Werke).

$bp_{0.6}$ 154-160°.

Citrate, $C_{19}H_{24}ClNO.C_6H_8O_7$, component of *Melidorm.* Crystals, mp 120-124°.

THERAP CAT: Anticholinergic; sedative; hypnotic.

5824. Meconic Acid. *3-Hydroxy-4-oxo-4H-pyran-2,6-dicarboxylic acid;* oxychelidonic acid. $C_7H_4O_7$; mol wt 200.10. C 42.02%, H 2.01%, O 55.97%. From opium which contains 4 to 6%. Prepn: Thoms, Pietrulla, *Ber. Pharm. Ges.* **31**, 4 (1921); Wibaut, Kleinpool, *Rec. Trav. Chim.* **66**, 24 (1947).

Monohydrate, prisms from concd aq solns. One gram dissolves in 4 ml hot water, 50 ml methanol, 50 ml ethyl acetate, 100 ml acetone; freely sol in ethanol, benzene.

Trihydrate, orthorhombic pyramidal prisms from dil aq solns. Becomes anhyd when heated to 100-102° for 20 min. Dec with evolution of CO_2 when heated to 120° or when boiled with water. The slight brown tint often characteristic of meconic acid prepns is due to traces of iron.

5825. Meconin. *6,7-Dimethoxy-1(3H)-isobenzofuranone; 6,7-dimethoxyphthalide;* opianyl; meconinic acid lactone. $C_{10}H_{10}O_4$; mol wt 194.19. C 61.85%, H 5.19%, O 32.96%. Isolated from opium by Dublanc in 1832. Occurs also in the root of *Hydrastis canadensis* L., Ranunculaceae: Freund, *Ber.* **22**, 456, 459 (1889). Is formed in oxidation of noscapine with HNO_3. Synthesis from *o*-veratric acid: Edwards *et al.*, *J. Chem. Soc.* **1925**, 195; Wilson *et al.*, *J. Org. Chem.* **16**, 792 (1951). From opianic acid: Brown, Newbold, *J. Chem. Soc.* **1952**, 4878.

White, optically inactive needles; sharp bitter taste. mp 102-103°. uv max: 213, 308 nm (ε 25000, 3800). Sol in 700 parts cold, 22 parts boiling water; sol in alcohol, benzene, chloroform, ether, glacial acetic acid; slowly sol in alkalies with formation of alkali salt of meconinic acid, (CH_2OH)-

$(CH_3O)_2C_6H_2COOH$. The acid itself is unstable, rapidly changing to lactone.

5826. Mecoprop. (±)-2-(4-Chloro-2-methylphenoxy)-propanoic acid; (±)-2-[(4-chloro-o-tolyl)oxy]propionic acid; mechlorprop; MCPP; CMPP; RD-4593; Astix CMPP; Iso-Cornox; Compitox; Compitox Plus; Proponex-Plus. $C_{10}H_{11}$-ClO_3; mol wt 214.65. C 55.96%, H 5.17%, Cl 16.52%, O 22.36%. Prepn: M. E. Synerholm, P. W. Zimmerman, *Contrib. Boyce Thompson Inst.* **14**, 91 (1945). Studies on plant growth regulation: C. H. Fawcett *et al., Ann. Appl. Biol.* **40**, 231 (1953); and comparison of enantiomers: M. Matell, *Kungl. Lantbruks-Hogsk. Ann.* **20**, 207 (1953); B. Aberg, *ibid.* 241. GLC determn: H. G. Higson, D. Butler, *Analyst* **85**, 657 (1960). Crystal structure: G. Smith *et al., Acta Crystallogr.* **B36**, 992 (1980). Herbicidal activity: G. B. Lush, *Proc. 3rd Brit. Weed Contr. Conf.* 625 (1956); E. L. Leafe, *ibid.* 633; B. Wallgren, *Weeds Weed Contr.* **24th** Swedish *Weed Conf.* 30 (1983); of (+)-enantiomer: J. Toll, *Weeds Weed Contr.* **28th** Swedish *Weed Conf.* 100 (1987). Degradation in soils: L. Lindholm *et al., Acta Agr. Scand.* **32**, 429 (1982); A. E. Smith, *Bull. Environ. Contam. Toxicol.* **34**, 656 (1985). Toxicological studies: M. R. Gurd *et al., Food Cosmet. Toxicol.* **3**, 883 (1965); H. G. Verschuuren *et al., Toxicology* **3**, 349 (1975); R. Roll, G. Matthiaschk, *Arzneimittel-Forsch.* **33**, 1479 (1983). EC-GLC determn in tissues and biological fluids: J. De Beer *et al., Vet. Hum. Toxicol.* **21**, Suppl., 172 (1979). HPLC resolution of enantiomers: B. Blessington *et al., J. Chromatog.* **396**, 177 (1987).

Solid, mp 93-94°. LD$_{50}$ in rats (mg/kg): 1210 orally, 402 i.p. (Verschuuren).

(+)-Form, *Mecoprop-P, Duplosan KV*. Solid, mp 95-96°. $[\alpha]_D^{25}$ +19° (alcohol).

Sodium salt, $C_{10}H_{11}ClNaO_3$. LD$_{50}$ i.p. in rats, mice: 500, 600 mg/kg; orally in mice: 650 mg/kg (Gurd).

Diethylamine salt, $C_{14}H_{22}ClNO_3$, *Mecopar*. LD$_{50}$ in rats, mice (mg/kg): 1060 ±120, 600 ±35 orally; 350, 400 i.p. (Gurd).

Potassium salt, $C_{10}H_{11}ClKO_3$, *Mecomec, Hedonal MCPP*.
USE: Herbicide.

5827. Mecrylate. 2-Cyano-2-propenoic acid methyl ester; 2-cyanoacrylic acid methyl ester; methyl 2-cyanoacrylate; methyl α-cyanoacrylate; AD/here; Coapt. $C_5H_5NO_2$; mol wt 111.10. C 54.05%, H 4.54%, N 12.61%, O 28.80%. $CH_2=C(C\equiv N)COOCH_3$. Prepn: McKeever, U.S. pat. **2,-912,454** (1959 to Rohm & Haas); McKeever, Raterink, U.S. pat. **2,926,188** (1960 to Rohm & Haas).

Liquid, bp$_{1.8}$ 47-49°. n_D^{25} 1.443.

USE: Manuf of polymers and adhesives, *see* U.S. pats. **2,776,232** and **2,794,788** (1957 to Eastman Kodak). Surgical aid (tissue adhesive).

5828. Mecysteine Hydrochloride. L-Cysteine methyl ester hydrochloride; methyl cysteine hydrochloride; methyl β-mercaptoalanine hydrochloride; methyl α-amino-β-mercaptopropionate hydrochloride; LJ-48; Acdrile; Visclair. $C_4H_{10}ClNO_2S$; mol wt 171.65. C 27.99%, H 5.87%, Cl 20.65%, N 8.16%, O 18.64%, S 18.68%. $HSCH_2CH(NH_2)$-$COOCH_3$.HCl. Prepn: Bergmann, Michalis, *Ber.* **63**, 987 (1930); Žervas, Theodoropoulos, *J. Am. Chem. Soc.* **78**, 1359 (1956).

Crystals from methanol, mp 140-141°. $[\alpha]_D^{20}$ −2.9° (methanol).

THERAP CAT: Mucolytic.

5829. Medazepam. 7-Chloro-2,3-dihydro-1-methyl-5-phenyl-1H-1,4-benzodiazepine; Ansilan; Diepin; Medazepol; Megasedan; Narsis; Nobrium; Psiquium; Resmit; Rudotel; Tranquilax. $C_{16}H_{15}ClN_2$; mol wt 270.76. C 70.98%, H 5.58%, Cl 13.09%, N 10.35%. Prepn: L. H. Sternbach *et al., J. Org. Chem.* **28**, 2456 (1963); G. A. Archer *et al.,* **Belg.** pat. **620,773**, *C.A.* **59**, 10095b (1963); E. Reeder, L. H. Sternbach, U.S. pat. **3,243,427** (1963, 1966 both to Hoff-

mann-La Roche); S. Inaba *et al., Chem. Pharm. Bull.* **20**, 1628 (1972); M. Mihalic *et al., J. Heterocycl. Chem.* **14**, 941 (1977). Pharmacology: L. O. Randall *et al., Arch. Int. Pharmacodyn. Ther.* **185**, 135 (1970). Crystal structure: G. Gilli *et al., Acta Crystallogr.* **B34**, 3793 (1978).

Colorless prismatic crystals from ether + petr ether, mp 95-97°. LD$_{50}$ in mice (mg/kg): 360 i.p., 1070 orally (Randall).

Hydrochloride, $C_{16}H_{15}ClN_2$.HCl, orange-red crystalline powder. Freely sol in water, alcohol.

Note: This is a controlled substance (depressant) listed in the U.S. Code of Federal Regulations, Title 21 Part 1308.14 (1995).

THERAP CAT: Anxiolytic.

5830. Medetomidine. 4-[1-(2,3-Dimethylphenyl)ethyl]-1H-imidazole; (±)-4-(α,2,3-trimethylbenzyl)imidazole; 4-[(α-methyl)-2,3-dimethylbenzyl]imidazole. $C_{13}H_{16}N_2$; mol wt 200.28. C 77.96%, H 8.05%, N 13.99%. α_2-Adrenergic agonist. Prepn: A. J. Karjalainen *et al.,* **Brit.** pat. **Appl. 2,101,114**; A. J. Karjalainen, K. O. A. Kurkela, U.S. pat. **4,544,664** (1983, 1985 both to Farmos). Receptor binding study: R. Virtanen *et al., Eur. J. Pharmacol.* **150**, 9 (1988). Sedative and cardiovascular effects in humans: M. Scheinin *et al., Brit. J. Clin. Pharmacol.* **24**, 443 (1987). Veterinary evaluation in cats: D. Stenberg *et al., J. Vet. Pharmacol. Ther.* **10**, 319 (1987).

Hydrochloride, $C_{13}H_{16}CN_2$.HCl, *MPV-785, Domitor*.

d-Form, *dexmedetomidine, (S)-medetomidine, MPV-1440*. Pharmacokinetics: K. T. Kivisto *et al., Eur. J. Clin. Pharmacol.* **46**, 345 (1994). Clinical evaluation as surgical premedicant: M. Virkkila *et al., Anaesthesia* **49**, 853 (1994).

THERAP CAT (VET): Sedative; analgesic.

5831. Medibazine. 1-(1,3-Benzodioxol-5-ylmethyl)-4-(diphenylmethyl)piperazine; 1-(diphenylmethyl)-4-piperonyl-piperazine; 1-benzhydryl-4-piperonylpiperazine. $C_{25}H_{26}N_2$-O_2; mol wt 386.49. C 77.69%, H 6.78%, N 7.25%, O 8.28%. Prepn: **Belg.** pat. **616,371**; Regnier *et al.,* U.S. pat. **3,119,-826** (1962, 1964, both to Science Union). Pharmacology: Laubie *et al., Arch. Int. Pharmacodyn. Ther.* **151**, 313 (1964); Laubie, Schmitt, *ibid.* **155**, 1 (1965).

Dihydrochloride, $C_{25}H_{26}N_2O_2$.2HCl, *Vialibran*. Solid, mp 288°.

THERAP CAT: Vasodilator (coronary); bronchodilator.

5832. Medicagol. 3-Hydroxy-6H-[1,3]dioxolo[5,6]ben-zofuro[3,2-c][1]benzopyran-6-one; 7-hydroxy-11,12-(methyl-enedioxy)coumestan; 7-hydroxy-5',6'-methylenedioxybenzofurano(3',2':3,4)coumarin. $C_{16}H_8O_6$; mol wt 296.24. C 64.87%, H 2.72%, O 32.41%. Occurs in alfalfa having viral leafspot infections. Synthesis: Livingston *et al.*, *J. Org. Chem.* **30**, 2353 (1965); Jurd, *J. Pharm. Sci.* **54**, 1221 (1965); Fukui *et al.*, *Experientia* **24**, 536 (1968).

mp 326-327°. uv max (ethanol): 245, 270, 297, 310, 348 nm (log ε 4.29, 3.91, 3.85, 4.03, 4.46).

5833. Medicarpin. *(6aR-cis)-6a,11a-Dihydro-9-methoxy-6H-benzofuro[3,2-c][1]benzopyran-3-ol;* (−)-3-hydroxy-9-methoxypterocarpan; demethylhomopterocarpin. $C_{16}H_{14}$-O_4; mol wt 270.28. C 71.10%, H 5.22%, O 23.68%. Antifungal phytoalexin produced by leguminous species. Isoln from the heartwood of *Swartzia madagascariensis* Desv., *Caesalpinioideae:* S. H. Harper *et al.*, *Chem. & Ind. (London)* **1965**, 562; from alfalfa, *Medicago sativa* L., *Leguminosae:* D. G. Smith *et al.*, *Physiol. Plant Pathol.* **1**, 41, (1971); from red clover, *Trifolium pratense* L., *Leguminosae:* V. J. Higgins, D. G. Smith, *Phytopathology* **62**, 235 (1972). ^{13}C-NMR: A. A. Chalmers *et al.*, *Tetrahedron* **33**, 1735 (1977). HPLC: J. Koster *et al.*, *J. Chromatog.* **270**, 392 (1983). Synthesis of racemic form: W. Cocker *et al.*, *J. Chem. Soc. (C)* **1965**, 1034. Biosynthetic studies: P. M. Dewick, *Chem. Commun.* **1975**, 656; S. W. Banks *et al.*, *ibid.* **1982**, 157; H. A. M. Al-Ani, P. M. Dewick, *J. Chem. Soc. Perkin Trans. I* **1984**, 2831. Antifungal properties: L. J. Duczek, V. J. Higgins, *Can. J. Bot.* **54**, 2620 (1976); H. D. Van Etten, *Phytochemistry* **15**, 655 (1976); A. O. Latunde-Dada, J. A. Lucas, *Physiol. Plant Pathol.* **26**, 31 (1985).

Prisms from benzene, mp 127.5-128.5°. $[\alpha]_D^{22}$ −226° (chloroform). uv max: 207, 282, 287, 310 nm (log ε 4.86, 3.97, 4.01, 3.38).

(±)-Form, prisms from ethyl acetate/light petroleum, mp 194-195°.

5834. Medifoxamine. *N,N-Dimethyl-2,2-diphenoxyethanamine; (dimethylamino)acetaldehyde diphenyl acetal; N,N'-dimethyl-2,2-diphenoxyethylamine.* $C_{16}H_{19}NO_2$; mol wt 257.33. C 74.68%, H 7.44%, N 5.44%, O 12.43%. Prepn: Fr. pat. **M5498** (1967 to Gerda), *C.A.* **72**, 12358x (1970); M. A. Brunet *et al.*, *Bull. Soc. Chim. France* **6**, 2000 (1967). Pharmacology and toxicity: A. Vagne *et al.*, *Therapie* **26**, 553 (1971). Clinical pharmacology: M. A. Randhawa *et al.*, *Human Psychopharmacol.* **3**, 195 (1988). Comparative clinical trial with clomipramine, *q.v.*, in depression: H. Scharbach *et al.*, *Psychol. Med.* **18**, 1485 (1986).

Fumarate, $C_{16}H_{19}NO_2 \cdot C_4H_4O_4$, *LG-152, Clédial, Gerdaxyl.* Crystals from 95% ethanol, mp 128.5°. LD_{50} orally in rats: 750 mg/kg (Vagne).
THERAP CAT: Antidepressant.

5835. Medmain. *3-Ethyl-N,N,2-trimethyl-1H-indol-5-amine; 5-dimethylamino-3-ethyl-2-methylindole;* 3-ethyl-2-methyl-5-dimethylaminoindole; 2-methyl-3-ethyl-5-dimethylaminoindole; antiserotonin. $C_{13}H_{18}N_2$; mol wt 202.30. C 77.18%, H 8.97%, N 13.85%. Prepn: Shaw, *J. Am. Chem. Soc.* **76**, 1384 (1954).

Crystals, mp 100-102°.
THERAP CAT: Serotonin inhibitor.

5836. Medrogestone. *6,17-Dimethylpregna-4,6-diene-3,20-dione;* 6,17α-dimethyl-6-dehydroprogesterone; AY-62022; Colpro; Colprone; Prothil. $C_{23}H_{32}O_2$; mol wt 340.51. C 81.13%, H 9.47%, O 9.40%. Prepn: Deghenghi, Gaudry, *J. Am. Chem. Soc.* **83**, 4668 (1961); Deghenghi *et al.*, *Tetrahedron* **19**, 289 (1963); *J. Med. Chem.* **6**, 301 (1963); Deghenghi, U.S. pats. **3,133,913, 3,210,387**; Morand, Deghenghi, U.S. pat. **3,170,936** (1964, 1965, and 1965, all to Am. Home Prod.). HPLC determn in serum: W. T. Robinson, L. Cosyns, *Arzneimittel-Forsch.* **29**, 882 (1979). Comparative clinical evaluation in benign prostatic hyperplasia: D. F. Paulson, R. D. Kane, *J. Urol.* **113**, 811 (1975); in endometrial carcinoma: K. C. Podratz *et al.*, *Obstet. Gynecol.* **66**, 106 (1985).

Crystals from ether, mp 144-146°. $[\alpha]_D^{23}$ +79° (c = 1 in chloroform). uv max: 288 nm (ε 25000).
THERAP CAT: Progestogen.

5837. Medronic Acid. *Methylenebisphosphonic acid; methylenediphosphonic acid;* methanebisphosphonic acid; methanediphosphonic acid; MDP. $CH_6O_6P_2$; mol wt 176.00. C 6.82%, H 3.44%, O 54.54%, P 35.20%. Prepn: G. Schwarzenbach, J. Zurc, *Monatsh.* **81**, 202 (1950), *C.A.* **44**, 8205f (1950); V. A. Ginsburg, J. Gen. *Chem. USSR* **28**, 710 (1958), *C.A.* **52**, 17091g (1958); J. A. Cade, *J. Chem. Soc.* **1959**, 2266. Physical properties: K. Moedritzer, R. R. Irani, *J. Inorg. Nucl. Chem.* **22**, 297 (1961). Prepn of ^{99m}Tc-complexes: G. Subramanian, J. G. McAfee, U.S. pat. **4,032,625** (1977 to Research Corp.); J. Kroesbergen *et al.*, *Int. J. Nucl. Med. Biol.* **12**, 419 (1986). HPLC separation of ^{99m}Tc-complexes: D. J. Hoch, T. C. Pinkerton, *Appl. Radiat. Isotop.* **37**, 593 (1986). Biodistribution, toxicity and clinical evaluation for skeletal imaging: G. Subramanian *et al.*, *J. Nucl. Med.* **16**, 744 (1975). Comparative biodistribution studies: G. Subramanian *et al.*, *Radiology* **149**, 823 (1983). Clinical pharmacokinetics: L. Rosenthall *et al.*, *Clin. Nucl. Med.* **2**, 232 (1977). Clinical studies for bone imaging: U. Büll *et al.*, *Brit. J. Radiol.* **50**, 629 (1977); for neuroblastoma imaging: I. Garty *et al.*, *Clin. Nucl. Med.* **14**, 515 (1989).

Crystals from acetic acid, mp 199-200°. LD_{50} in mice, rabbits (mg/kg): 45-50 i.v. (Subramanian, 1975).

THERAP CAT: ^{99m}Tc complex as diagnostic aid (radioactive imaging agent).

5838. Medroxyprogesterone. *(6α)-17-Hydroxy-6-methylpregn-4-ene-3,20-dione;* 17α-hydroxy-6α-methylprogesterone; 6α-methyl-17α-hydroxyprogesterone; 6α-methyl-4-pregnen-17α-ol-3,20-dione. $C_{22}H_{32}O_3$; mol wt 344.49. C 76.70%, H 9.36%, O 13.93%. Orally active progestogen; formerly used in combinations as oral contraceptive. Prepn: J. C. Babcock *et al., J. Am. Chem. Soc.* **80,** 2904 (1958); **Brit.** pat. **866,381;** G. B. Spero, U.S. pat. **3,377,364** (1961, 1968 both to Upjohn); P. Ruggieri, C. Ferrari, U.S. pat. **3,043,832** (1962 to Ormonoterapia Richter); B. Camerino *et al.,* U.S. pat. **3,061,616** (1962 to Farmitalia). HPLC determn in plasma: J. Read, G. Mould, *J. Chromatog.* **341,** 437 (1985). Comparative clinical trial with norethisterone, *q.v.,* as injectable contraceptive: H. K. Toppozada *et al., Contraception* **28,** 1 (1983). Clinical trial in advanced breast cancer: M. Izuo *et al., Cancer* **56,** 2576 (1985). Review of pharmacology and clinical uses: I. S. Fraser, E. Weisberg, *Med. J. Aust.* **1,** Suppl. 1, 3-19 (1981).

Crystals from chloroform, mp 220-223.5°. $[\alpha]_D^{25}$ +75° (in chloroform). uv max (ethanol): 241 nm (ε 16000).

17-Acetate, $C_{24}H_{34}O_4$, *17α-acetoxy-6α-methylprogesterone, 6-α-methyl-17α-acetoxyprogesterone, MAP, Amen, Clinovir, Curretab, Cycrin, Depo-Clinovir, Depo-Provera, Farlutal, Gestapuran, G-Farlutal, Hysron, Lutoral, Nadigest, Nidaxin, Oragest, Perlutex, Prodasone, Provera, Sodelut G, Veramix.* Crystals from methanol, mp 207-209°. $[\alpha]_D$ +61° (in chloroform). uv max (ethanol): 240 nm (ε 15900).

Mixture of acetate with ethinyl estradiol, *Provest.*

THERAP CAT: Progestogen; contraceptive (injectable); antineoplastic (hormonal).

THERAP CAT (VET): Progestogen; estrus regulator.

5839. Medrylamine. *2-[(4-Methoxyphenyl)phenylmethoxy]-N,N-dimethylethanamine; 2-(p-methoxy-α-phenylbenzyloxy)-N,N-dimethylethylamine;* 4-methoxy-Benadryl; *p*-methoxybenzhydryl β-dimethylaminoethylether; β-(p-methoxybenzhydryloxy)ethyldimethylamine; phenyl-*p*-methoxyphenylcarbinyl dimethylaminoethyl ether; Histaphen(e); Postafen Salve. $C_{18}H_{23}NO_2$; mol wt 285.39. C 75.76%, H 8.12%, N 4.91%, O 11.21%. Prepn: Morren, U.S. pat. **2,668,856** (1954 to U.C.B.).

Hydrochloride, $C_{18}H_{23}NO_2$·HCl, crystals, mp 141°.
THERAP CAT: Antihistaminic.

5840. Medrysone. *(6α,11β)-11-Hydroxy-6-methylpregn-4-ene-3,20-dione;* 11β-hydroxy-6α-methylprogesterone; 6α-methyl-11β-hydroxyprogesterone; hydroxymesterone; U-8471; HMS; Medrocort; Ophtocortin; Spectamedryn. $C_{22}H_{32}O_3$; mol wt 344.49. C 76.70%, H 9.36%, O 13.93%. Prepn and use as an intermediate: Sebek *et al.,* Spero, Thompson, U.S. pats. **2,864,837** and **2,968,655** (1958 and

1961, both to Upjohn). In treatment of ocular inflammation: Bedrossian, *Arch. Ophthalmol.* **81,** 184 (1969).

Crystals, mp 155-158°. $[\alpha]_D$ +189° (in chloroform).
THERAP CAT: Glucocorticoid.

5841. Medullipin. *Medullipin I;* antihypertensive neutral renomedullary lipid; ANRL. Antihypertensive lipid prohormone produced by the renomedullary interstitial cells of the renal papilla under the control of the renal artery perfusion pressure. Converted to the active form, **medullipin II,** by the cytochrome P-450 dependent enzyme system of the liver. Medullipin II acts as as vasodilator that suppresses sympathetic activity, causes diuresis-natriuresis, and has a suppressive action on the CNS. Constitutes a feedback control of the renin-angiotensin system. Extraction from renomedullary interstitial cells: E. E. Muirhead *et al., Lab. Invest.* **35,** 162 (1977); from renal venous effluent: E. E. Muirhead *et al., J. Lab. Clin. Med.* **99,** 64 (1982). Biological activity: E. E. Muirhead *et al., Hypertension* **5,** Suppl I, I-112 (1983); G. Karlström *et al., Acta Physiol. Scand.* **137,** 521 (1989). Activation by the liver: E. E. Muirhead *et al., Trans. Assoc. Am. Physicians* **101,** 226 (1988). Effect in spontaneously hypertensive rats: E. E. Muirhead *et al., Hypertension* **17,** 1092 (1991). Review of medullipin system of blood pressure control: E. E. Muirhead, *Am J. Hypertens.* **4,** 556S-568S (1991).

5842. Mefenamic Acid. *2-[(2,3-Dimethylphenyl)amino]benzoic acid; N-(2,3-xylyl)anthranilic acid;* CI-473; INF-3355; Bafhameritin-M; Bonabol; Coslan; Lysalgo; Mefenacid; Namphen; Parkemed; Ponalar; Ponstan; Ponstel; Ponstil; Ponstyl; Pontal; Tanston; Vialidon. $C_{15}H_{15}NO_2$; mol wt 241.29. C 74.67%, H 6.27%, N 5.80%, O 13.26%. Prepn: **Belg.** pat. **605,302** (1961 to Parke, Davis). Pharmacology: C. V. Winder *et al., J. Pharmacol. Exp. Ther.* **138,** 405 (1962); Mokhort, Korkhova, *Farmakol. Toksikol. (Kiev)* **1968,** 85, *C.A.* **71,** 29080c (1969); C. V. Winder *et al., Arthritis Rheum.* **12,** 472 (1969); and toxicology: U. Jahn, R. W. Adrian, *Arzneimittel-Forsch.* **19,** 36 (1969). Crystal structure: J. F. McConnell, F. Z. Company, *Cryst. Struct. Commun.* **5,** 861 (1976).

Crystals, mp 230-231° (effervescence). pKa 4.2. uv max (0.1N NaOH): 285, 340 nm. Soly in H_2O, pH 7.1 (g/100 ml): 0.0041 (25°); 0.008 (37°). Soluble in solns of alkali hydroxides; sparingly sol in ether, chloroform; slightly sol in ethanol. LD_{50} orally in mice, rats: 630, 790 mg/kg (Jahn, Adrian).

Sodium salt, $C_{15}H_{14}NNaO_2$, white powder. Soly in H_2O: > 5 g/100 ml. LD_{50} in mice (mg/kg): 600 orally; 150 i.p. (Mokhort, Korkhova).

THERAP CAT: Anti-inflammatory.

5843. Mefenorex. *N-(3-Chloropropyl)-α-methylbenzeneethanamine; N-(3-chloropropyl)-α-methylphenethylamine;* 1-phenyl-2-(3-chloropropylamino)propane. $C_{12}H_{18}$ClN; mol wt 211.73. C 68.07%, H 8.57%, Cl 16.74%, N 6.62%. Prepn: Beschke *et al.,* **Ger.** pat. **1,210,873** (1966 to Hoffmann-La Roche), *C.A.* **64,** 19486c (1966). Clinical pharmacology: P. Netter *et al., Arzneimittel-Forsch.* **28,** 1310 (1978). GC-MS determn in urine: A. Franceschini *et* *al., J. Chromatog.* **541,** 109 (1991). Pharmacokinetics and metabolism: S. Rendic *et al., Eur. J. Drug Metab. Pharmacokinet.* **19,** 107 (1994). *Review:* J. Engel *et al., Drug Alcohol Depend.* **17,** 229-234 (1986).

Hydrochloride, $C_{12}H_{18}ClN.HCl$, *Ro-4-5282, Incital, Pondinil, Pondinol, Rondimen.* Solid, mp 128-130°.

THERAP CAT: Anorexic.

5844. Mefexamide. *N-[2-(Diethylamino)ethyl]-2-(4-methoxyphenoxy)acetamide;* (*p*-methoxyphenoxy)acetic acid *N*-[2-(diethylamino)ethyl]amide; mexephenamide; NP-297; Mefexadyne; Timodyne. $C_{15}H_{24}N_2O_3$; mol wt 280.37. C 64.26%, H 8.63%, N 9.99%, O 17.12%. Prepn: Thuillier, Rumpf, *Bull. Soc. Chim. France* **1960,** 1786. Pharmacology: Thuillier, *Arzneimittel-Forsch.* **14,** 556 (1964). Clinical study in psychiatric patients: V. Faust *et al., Schweiz. Rundsch. Med. Prax.* **61,** 1177 (1972).

Hydrochloride, $C_{15}H_{25}ClN_2O_3$, mp 112°.

THERAP CAT: CNS stimulant.

5845. Mefloquine Hydrochloride. (*R*,S**)-(±)-α-2-*Piperidinyl-2,8-bis(trifluoromethyl)-4-quinolinemethanol monohydrochloride;* DL-*erythro*-α-2-piperidyl-2,8-bis(trifluoromethyl)-4-quinolinemethanol monohydrochloride; WR-142490; Ro-21-5998; Lariam. $C_{17}H_{17}ClF_6N_2O$; mol wt 414.78. C 49.23%, H 4.13%, Cl 8.55%, F 27.48%, N 6.75%, O 3.86%. Prepn: C. J. Ohnmacht *et al., J. Med. Chem.* **14,** 926 (1971). *See also:* G. Grethe, T. Mitt, *Ger.* pat. **2,806,-909,** *C.A.* **90,** 22838q (1979); H. Bömches, B. Hardegger, *U.S.* pat. **4,507,482** (1979, 1985 both to Hoffmann-La Roche). Prepn, activity, and absolute configuration of isomers: F. I. Carroll, J. T. Blackwell, *J. Med. Chem.* **17,** 210 (1974). Mechanism of action studies: M. W. Davidson *et al., J. Med. Chem.* **20,** 1117 (1977); R. E. Brown *et al., Life Sci.* **25,** 1857 (1979). Photochemistry: G. A. Epling, U. C. Yoon, *Chem. Letters* **1982** (2), 211. Pharmacokinetics: D. E. Schwartz *et al., Chemotherapy (Basel)* **28,** 70 (1982). HPLC determn in blood and plasma: I. M. Kapetanovic *et al., J. Chromatog.* **277,** 209 (1983). Clinical studies: G. M. Trenholme *et al., Science* **190,** 792 (1975); F. Tin *et al., Bull. WHO* **60,** 913 (1982); J. M. Kofe Ekue *et al., ibid.* **61,** 713 (1983); J.-M. de Souza *ibid.* 809, 815. Comprehensive description: P. Lim in *Analytical Profiles of Drug Substances* vol. 14, K. Florey, Ed. (Academic Press, New York, 1985) pp 157-180.

Odorless, white powder, mp 259-260° (dec). Sol in ethanol, ethyl acetate; slightly sol in water.

Combination with sulfadoxine and pyrimethamine, *Fansimef.*

THERAP CAT: Antimalarial.

5846. Mefluidide. *N-[2,4-Dimethyl-5-[[(trifluoromethyl)sulfonyl]amino]phenyl]acetamide;* 5-acetamido-2,4-di-

methyltrifluoromethanesulfonanilide; methafluoridamid; MBR-12325; VEL-3973; Vistar; Embark. $C_{11}H_{13}F_3N_2O_3S$; mol wt 310.30. C 42.58%, H 4.22%, F 18.37%, N 9.03%, O 15.47%, S 10.33%. Prepn: T. L. Fridlinger, *Ger.* pat. **2,-406,475** corresp to *U.S.* pat. **3,894,073** (1974, 1975 to 3M). Metabolism: G. W. Ivie, *J. Agr. Food Chem.* **28,** 1286 (1980).

Cryst solid, mp 183-185°. Vapor press at 25°: < 10⁻⁴ mm Hg. pKa 4.6. Soly at 25° (g/l): water 0.18; benzene 0.31; dichloromethane 2.1; 1-octanol 17; methanol 310; acetone 350. Aq solns decomp in uv light. Mildly corrosive to metals.

USE: Plant growth regulator; herbicide.

5847. Mefruside. *4-Chloro-N¹-methyl-N¹-[(tetrahydro-2-methyl-2-furanyl)methyl]-1,3-benzenedisulfonamide; 4-chloro-N¹-methyl-N¹-(tetrahydro-2-methylfurfuryl)-m-benzenedisulfonamide;* 2-[(4-chloro-N¹-methyl-3-sulfamoylbenzenesulfonamido)methyl]-2-methyltetrahydrofuran; *N*-(4-chloro-3-sulfamoylbenzenesulfonyl)-*N*-methyl-2-furfurylamine; B-1500; Baycaron. $C_{13}H_{19}ClN_2O_5S_2$; mol wt 382.89. C 40.78%, H 5.00%, Cl 9.26%, N 7.32%, O 20.89%, S 16.75%. Prepn: H. Horstmann *et al., Brit.* pat. **1,031,916** corresp to *U.S.* pat. **3,356,692** (1966, 1967 to Bayer AG); of *dl-, d-,* and *l*-forms, *eidem, Arzneimittel-Forsch.* **17,** 653 (1967).

dl-Form, crystals, mp 149-150°.
d-Form, crystals, mp 146°. [α]²⁰₅₇₈ +5.4° (c = 2.026 in methanol).
l-Form, crystals, mp 146°. [α]²⁰₅₇₈ −5.5° (c = 2.100 in methanol). More active as a diuretic than the *d*-form.

THERAP CAT: Diuretic.

5848. Megacins. Antibiotic proteins produced by strains of *Bacillus megaterium,* which are highly specific in their antibacterial activity against strains of homologous species. Isoln of megacin A from *B. megaterium:* Ivanovics, Alföldi, *Nature* **174,** 465 (1954); Holland, *Biochem. J.* **78,** 641 (1961). Destroys the cytoplasmic membrane of sensitive bacteria: Nagy, *Acta Microbiol. Acad. Sci. Hung.* **6,** 337 (1959). Isoln of megacin C from *B. megaterium:* Holland, Roberts, *J. Gen. Microbiol.* **35,** 271 (1964). Has a mode of action quite different from megacin A and is more reminiscent of colicins, *q.v. Review:* Holland in *Antibiotics* vol. 1, D. Gottlieb, P. Shaw, Eds. (Springer-Verlag, New York, 1967) pp 688-695.

5849. Megestrol Acetate. *17-Hydroxy-6-methylpregna-4,6-diene-3,20-dione acetate;* 17α-acetoxy-6-methylpregna-4,-6-diene-3,20-dione; 6-dehydro-6-methyl-17α-acetoxyprogesterone; 6-methyl-Δ⁴,⁶-pregnadien-17α-ol-3,20-dione acetate; Maygace; Megace; Megestat; Megestil; Nia; Niagestin; Ovaban. $C_{24}H_{32}O_4$; mol wt 384.52. C 74.97%, H 8.39%, O 16.64%. Orally active progestogen; formerly used in combinations as oral contraceptive. Prepn: Ringold *et al., J. Am. Chem. Soc.* **81,** 3712 (1959); Dodson, Sollman, *U.S.* pat. **2,891,079** (1959 to Searle); Kirk *et al., Brit.* pat. **870,-286,** *C.A.* **56,** 10248i (1962), and *U.S.* pat. **3,356,573** (1967 to Brit. Drug Houses); Cross, *U.S.* pat. **3,400,137** (1968 to Syntex). Biological effectiveness studies: Chang, Kincl, *Steroids* **12,** 689 (1968). Metabolic studies: Cooper, Kellie, *ibid.* **11,** 133 (1968). Soly and diffusion studies: Sundaram, Kincl, *ibid.* **12,** 517 (1968). Comparative clinical trial with tamoxifen in metastatic breast cancer: J. C. Allegra *et al., Semin. Oncol.* **12,** Suppl. 6, 61 (1985). Evaluation in prevention of hot flashes: C. L. Loprinzi *et al., N. Engl. J. Med.*

331, 347 (1994). Review of carcinogenicity studies: *IARC Monographs* 21, 431-439 (1979).

Crystals from methanol, mp 214-216°. Soly at 37°: in water (2 μg/ml); in plasma (24 μg/ml). $[\alpha]_D^{24}$ +5° (chloroform). uv max (ethanol): 287 nm (log ϵ 4.40).

Mixture with ethinyl estradiol, *Co-Ervonum, Kombiquens, Noval, Nuvacon, Planovin, Tri-Ervonum, Volidan, Weradys.*

Mixture with mestranol, *Delpregnin.*

THERAP CAT: Progestogen. Palliative treatment of breast and endometrial carcinoma.

THERAP CAT (VET): Progestogen. Estrus regulator.

5850. Meglumine Acetrizoate. *1-Deoxy-1-(methylamino)-D-glucitol 3-(acetylamino)-2,4,6-triiodobenzoate (salt); 3-acetamido-2,4,6-triiodobenzoic acid methylglucamine salt;* 3-acetylamino-2,4,6-triiodobenzoate of methylglucamine; methylglucamine 3-acetylamino-2,4,6-triiodobenzoate; methylglucamine acetrizoate; Fortombrine M; Fortoshade M; Jodozoat Meglumin; Vasurix. $C_{16}H_{23}I_3N_2O_8$; mol wt 752.08. C 25.55%, H 3.08%, I 50.62%, N 3.72%, O 17.02%. $C_9H_6I_3NO_3 \cdot C_7H_{17}NO_5$.

Minute crystals. Freely sol in water. Aq solns have a pH of ~7.0. The iodine content of the dry powder is reported as 50.67%.

THERAP CAT: Diagnostic aid (radiopaque medium).

5851. Meglumine Diatrizoate. *1-Deoxy-1-(methylamino)-D-glucitol 3,5-bis(acetylamino)-2,4,6-triiodobenzoate salt;* 3,5-diacetamido-2,4,6-triiodobenzoic acid methylglucamine salt; urografic acid methylglucamine salt; diatrizoate meglumine; diatrizoate methylglucamine; methylglucamine diatrizoate; meglumine amidotrizoate; Angiografin; Cardiografin; Cystografin; Hypaque Cysto; Hypaque Meglumine; Renografin; Reno M; Urovist. $C_{18}H_{26}I_3N_3O_9$; mol wt 809.13. C 26.72%, H 3.24%, I 47.05%, N 5.19%, O 17.80%. $C_{11}H_9I_3N_2O_4 \cdot C_7H_{17}NO_5$. Distribution and metabolism: H. Langecker *et al., Arch. Exp. Pathol. Pharmakol.* 222, 584 (1954). Clinical evaluation of diagnostic use in contrast enhanced computer tomography: J. S. Morrow, *Curr. Ther. Res.* 27, 229 (1980).

Rhombic needles, slightly sweet taste. mp 189-193° (dec). Soly in water at 20°: 89 g/100 ml.

Mixture with sodium diatrizoate, *MD-76, Gastrografin, Renovist, Urografin.*

THERAP CAT: Diagnostic aid (radiopaque medium).

THERAP CAT (VET): X-ray contrast medium.

5852. Meglutol. *3-Hydroxy-3-methylpentanedioic acid; 3-hydroxy-3-methylglutaric acid;* dicrotalic acid; medroglutaric acid; HMG; HMGA; CB-337; Lipoglutaren; Mevalon. $C_6H_{10}O_5$; mol wt 162.14. C 44.45%, H 6.22%, O 49.34%. Hypolipidemic agent that decreases the rate of cholesterol synthesis. Prepn via Reformatsky reaction between ethyl acetoacetate and ethyl bromoacetate: R. Adams, B. L. Van Duuren, *J. Am. Chem. Soc.* 75, 2377 (1953). Synthesis via oxidation of diallylmethylcarbinol with ozone and hydrogen peroxide: H. J. Klosterman, F. Smith, *ibid.* 76, 1229 (1954); J. L. Rabinowitz, *Biochem. Prepns.* 6, 25 (1958). *Caution:* attempts to increase batch size of the ozonolysis resulted in an explosion, *cf. Chem. & Eng. News* 51(6), 29 (1973). Improved synthesis: A. Yavrouian *et al., Synthesis* 1981, 791. Hypocholesteremic properties: Z. H. Beg, M. Siddiqi, *Experientia* 23, 380 (1967); M. Siddiqi, Z. H. Beg, U.S. pat. 3,629,449 (1971); Z. H. Beg, P. J. Lupien, *Biochim. Biophys. Acta* 260, 439 (1972); A. N. K. Yusufi, M. Siddiqi, *Atherosclerosis* 20, 517 (1974); C. D. Padova *et al., Life Sci.* 30, 1907 (1982). Biosynthetic studies: L. W. White, H. Rudney, *Biochemistry* 9, 2713 (1970); L. Hagenfeldt, K. Hell-

strom, *Life Sci.* 9, Pt. 2, 991 (1970). Organ distribution: L. L. Savoie, P. J. Lupien, *Can. J. Physiol. Pharmacol.* 53, 638 (1975). Effectiveness in familial hypercholesteremia: P. J. Lupien *et al., J. Clin. Pharmacol.* 19, 120 (1979). GC study in patients with organic acidurias: K. Tanaka, D. J. Hine, *J. Chromatog.* 239, 301 (1982). Toxicological study: L. L. Savoie, P. J. Lupien, *Arzneimittel-Forsch.* 25, 1284 (1975).

Cryst from ether/petr ether, mp 108-109°. Sol in water. Stable when stored dry. LD_{50} in mice (g/kg): 7.33 orally; 3.23 i.p. (Savoie, Lupien).

THERAP CAT: Antihyperlipoproteinemic.

5853. Melamine. *1,3,5-Triazine-2,4,6-triamine;* 2,4,6-triamino-s-triazine; cyanurotriamide. $C_3H_6N_6$; mol wt 126.12. C 28.57%, H 4.80%, N 66.63%. Usually prepd by heating dicyanodiamide, $H_2NC(=NH)NHC\equiv N$, under pressure: Mackay, U.S. pat. 2,737,513 (1956 to Am. Cyanamid). Alternate methods starting with urea: Mackey, U.S. pat. 2,760,961 (1956 to Am. Cyanamid); Pomot *et al.,* U.S. pat. 3,111,519 (1963 to Office Natl. Ind. de l'Azote). X-ray and neutron crystal structure: J. N. Varghese *et al., Acta Crystallogr.* 33B, 2102 (1977). Review of mfg processes: Faith, Keyes & Clark's *Industrial Chemicals,* F. A. Lowenheim, M. K. Moran, Eds. (Wiley-Interscience, New York, 4th ed., 1975) pp 519-523.

Monoclinic prisms. mp <250°. Sublimes. d^{250} 1.573. Slightly sol in water; very slightly sol in hot alc; insol in ether.

USE: Forms synthetic resins with formaldehyde.

5854. Melanins. Pigments responsible for the dark color of skin, hair, feathers, fur, insect cuticle, soil; found also in fungi, bacteria, and pathological human urine where it is an indication of melanotic tumors. Structures are highly irregular polymers produced in the form of granules which may be bound to protein material. *Allomelanins* are found in the plant kingdom and are produced from nitrogen-free precursors. *See:* aspergillin, humic acid. *Eumelanins* and *phaeomelanins* are found in the animal kingdom. Eumelanins are black or brown, insoluble, nitrogenous pigments produced by the oxidative polymerization of 5,6-dihydroxyindoles derived enzymatically from tyrosine via dopa. One of the best characterized is sepiomelanin, *q.v.* Phaeomelanins are sulfur-containing, alkali-soluble, yellow to reddish-brown pigments produced by oxidative polymerization of cysteinyldopas via 1,4-benzothiazine intermediates. Biosynthesis from tyrosine: H. S. Raper, *Physiol. Rev.* 8, 245 (1928); H. S. Mason, *J. Biol. Chem.* 172, 83 (1948). *In vitro* prepn from dopa: L. E. Arnow, *Science* 87, 308 (1938). Series of articles on structure and biosynthesis: M. Piattelli, R. A. Nicolaus, *Tetrahedron* 15, 66 (1961); M. Piattelli *et al., ibid.* 18, 941 (1962); *eidem, ibid.* 19, 2061 (1963); R. A. Nicolaus *et al., ibid.* 20, 1163 (1964). NMR-study: G. A. Duff *et al., Biochemistry* 27, 7112 (1988). Determn of eumelanin metabolites in human urine: S. Pavel, W. van der Slik, *J. Chromatog.* 375, 392 (1986). Review of biosynthesis: J. M. Pawelek, A. M. Körner, *Am. Scientist* 70, 136-145 (1982); of photochemistry and photobiology: M. R. Chedekel, *Photochem. Photobiol.* 35, 881-885 (1982); M. R. Chedekel, L. Zeise, *Lipids* 23, 587-591 (1988). *Reviews:* R. A. Nicolaus, *Melanins* (Hermann, Paris, 1968); G. Prota, *Med. Res. Rev.* 8, 525-556 (1988).

5855. Melanostatin. Melanocyte-stimulating hormone release-inhibiting factor; MIF; MRIH; MSH release-inhibiting hormone; melanotropin inhibiting factor. An inhibiting factor which mediates hypothalamic control of MSH, *q.v.,* a pituitary hormone: A. J. Kastin, A. V. Schally, *Gen. Comp.*

Endocrinol. **7**, 452 (1966); A. V. Schally, A. J. Kastin, *Endocrinology* **79**, 768 (1966). Although several substances have MIF activity, the exact nature of MSH release inhibition is not known, *cf.* A. J. Kastin *et al., Yale J. Biol. Med.* **46**, 617 (1973); *eidem, Fed. Proc.* **39**, 2931 (1980). The first factor with melanocyte-release inhibiting activity to be isolated was the side-chain tripeptide of oxytocin, *q.v.,* now referred to as MIF-I: M. E. Celis *et al., Proc. Nat. Acad. Sci. USA* **68**, 1428 (1971). Isoln of bovine MIF-I and identity with the tripeptide (Pro-Leu-GlyNH₂): R. M. G. Nair *et al., Biochem. Biophys. Res. Commun.* **43**, 1376 (1971). Synthesis: H. Irie *et al., Chem. Letters* **1980**, 705. Extrapituitary effects of MIF-I are known, particularly potentiation of DOPA-induced behavioral changes: N. P. Plotnikoff *et al., Life Sci.* **10**, 1279 (1971); *eidem, Neuroendocrinology* **14**, 271 (1974). Clinical studies in Parkinson's disease: A. J. Kastin, A. Barbeau, *Can. Med. Assoc. J.* **107**, 1079 (1972); A. Barbeau, *Lancet* **2**, 683 (1975); V. E. Schneider *et al., Arzneimittel-Forsch.* **28**, 1296 (1978). *Reviews:* A. V. Schally *et al., Recent Progr. Horm. Res.* **24**, 497-581 (1968); A. J. Kastin *et al., Life Sci.* **25**, 401-414 (1979); *eidem* in *Polypeptide Hormones*, R. F. Beers, E. G. Bassett, Eds. (Raven Press, New York, 1980) pp 223-224; *eidem, Fed. Proc.* **39**, 2931-2936 (1980).

MIF-I, $C_{13}H_{24}N_4O_3$, L-prolyl-L-leucylglycinamide, melanostatin I (ox), MSH release-inhibiting factor I.

5856. Melarsoprol. *2-[4-[(4,6-Diamino-1,3,5-triazin-2-yl)amino]phenyl]-1,3,2-dithiarsolane-4-methanol; p-[(4,6-diamino-s-triazin-2-yl)amino]dithiobenzenearsonous acid 3-hydroxypropylene ester; 2-p-(4,6-diamino-s-triazin-2-yl)amino)phenyl-4-hydroxymethyl-1,3,2-dithiarsoline; 2-(4-melamin-2-ylphenyl)-4-hydroxymethyl-1,3-dithia-2-arsolane; Mel B; Arsobal.* $C_{12}H_{15}AsN_6OS_2$; mol wt 398.34. C 36.18%, H 3.80%, As 18.81%, N 21.10%, O 4.02%, S 16.10%. Prepn: Friedheim, U.S. pats. 2,659,723 (1953) and 2,772,303 (1956).

Practically insol in water, cold ethanol, methanol. Sol in propylene glycol.

THERAP CAT: Antiprotozoal (Trypanosoma).

5857. Melatonin. *N-[2-(5-Methoxy-1H-indol-3-yl)ethyl]acetamide; N-acetyl-5-methoxytryptamine; Regulin.* $C_{13}H_{16}N_2O_2$; mol wt 232.28. C 67.22%, H 6.94%, N 12.06%, O 13.78%. A hormone of the pineal gland, also produced by extra-pineal tissues, that lightens skin color in amphibians by reversing the darkening effect of MSH, *q.v.* Melatonin has been postulated as the mediator of photic-induced antigonadotrophic activity in photoperiodic mammals and has also been shown to be involved in thermoregulation in some ectotherms and in affecting locomotor activity rhythms in sparrows. Isoln from the pineal glands of beef cattle: Lerner *et al., J. Am. Chem. Soc.* **80**, 2587 (1958); Wurtman *et al., Science* **141**, 277 (1963). Structure: Lerner *et al., J. Am. Chem. Soc.* **81**, 6084 (1959). Crystal and molecular structure: A. Wakahara, *Chem. Lett.* **1972**, 1139. Synthesis from 5-methoxyindole as starting material by two different routes: Szmuszkovicz *et al., J. Org. Chem.* **25**, 857 (1960). Biochemical role of melatonin: *Chem. & Eng. News* **45**, 40 (May 1, 1967). Pharmacological studies: Barchas *et al., Nature* **214**, 919 (1967). Identification of antigonadal action sites in mouse brain: J. D. Glass, G. R. Lynch, *Science* **214**, 821 (1981). Binding studies in human hypothalamus: S. M. Reppert *et al., Science* **242**, 78 (1988). Efficacy in control of estrus in red deer: G. W. Asher, *Animal Reprod. Sci.* **22**, 145 (1990). *Reviews:* M. K. Vaughn, *Int. J. Rev. Physiol.* **24**, 41-95 (1981); D. C.Klein *et al., Life Sci.* **28**, 1975-1998 (1981). Book: *Advan. Biosci.* **vol. 29**, N. Birau, W. Schlott, Eds. (Pergamon Press, New York, 1981) 420 pp. Review of etiological role in clinical disease: A. Miles, D. Philbrick, *CRC Crit. Rev. Clin. Lab. Sci.* **25**, 231-253 (1987); in psychiatric disorders: *eidem, Biol. Psychiat.* **23**, 405-425 (1988).

Pale yellow leaflets from benzene, mp 116-118°. uv max: 223, 278 nm (ε 27550, 6300).

THERAP CAT (VET): Control of estrus.

5858. Meldrum's Acid. *2,2-Dimethyl-1,3-dioxane-4,6-dione; malonic acid cyclic isopropylidene ester;* isopropylidene malonate; 2,2-dimethyl-4,6-diketo-1,3-dioxane. $C_6H_8O_4$; mol wt 144.13. C 50.00%, H 5.59%, O 44.40%. Prepd from malonic acid and acetone in the presence of acetic anhydride: Meldrum, *J. Chem. Soc.* **93**, 598 (1908). Prepn and structure: Davidson, Bernhard, *J. Am. Chem. Soc.* **70**, 3426 (1948); P. Schuster *et al., Monatsh.* **95**, 53 (1964); G. H. Bihlmayer *et al., ibid.* **98**, 564 (1967). *Review:* H. McNab, *Chem. Soc. Rev.* **7**, 345-358 (1978).

Crystals from acetone + water, dec 94-95°. pKa 5.1.

USE: In organic synthesis as a substitute for acyclic malonic esters and generally as a C_3O_2 synthon.

5859. Melengestrol. *17α-Hydroxy-6-methyl-16-methylenepregna-4,6-diene-3,20-dione; 6-methyl-16-methylene-17α-hydroxy-Δ⁶-progesterone.* $C_{23}H_{30}O_3$; mol wt 354.49. C 77.93%, H 8.53%, O 13.54%. Prepn of the acetate: Kirk *et al.;* Petrow, **Brit.** pat. 886,619 and U.S. pat. 3,117,966 (1962, 1964, both to Brit. Drug Houses).

Acetate, $C_{25}H_{32}O_4$, *17α-acetoxy-6-methyl-16-methylene-4,6-pregnadiene-3,20-dione, MGA.* Crystals, mp 224-226°. $[\alpha]_D^{23}$ −127° (c = 0.31 in chloroform). uv max (ethanol): 287 nm (log ε 4.35).

THERAP CAT: Antineoplastic; progestogen.
THERAP CAT (VET): Progestogen.

5860. Melezitose. *O-α-D-Glucopyranosyl-(1 → 3)-β-D-fructofuranosyl-α-D-glucopyranoside.* $C_{18}H_{32}O_{16}$; mol wt 504.44. C 42.86%, H 6.39%, O 50.75%. Trisaccharide built from 2 mols glucose and 1 mol fructose. Acid hydrolysis yields at first glucose and turanose. Occurs in a manna that forms upon the Douglas fir, jack pine *(Pinus virginiana* Mill., *Pinaceae)* and other trees. During periods of drought, bees collect it from the manna in sufficient quantity to change the character of the honey, making it unsuitable for maintaining the bees, but providing a source of melezitose: Hudson, Sherwood, *J. Am. Chem. Soc.* **42**, 116 (1920). Procedure: Bates, *NBS Circular* **C440**, p 472 (1942). Structure: Hehre, Carlson, *Arch. Biochem. Biophys.* **36**, 158 (1952); Hehre, *Advan. Carbohyd. Chem.* **8**, 277 (1953). *Review:* E. B. Rathbone, *Dev. Food Carbohyd.* **2**, 145-185 (1980).

Consult the Name Index before using this section.

Dihydrate, crystals from water. The water of crystn is given up at 110°. When anhydr, mp 153-154°. $[\alpha]_D^{20}$ +88° (c = 4). Sol in water, very sparingly sol in alc. Not fermented by baker's yeast; does not reduce Fehling's soln.

5861. Melibiose. 6-O-α-D-Galactopyranosyl-D-glucose; 6-(α-D-galactosido)-D-glucose. $C_{12}H_{22}O_{11}$; mol wt 342.30. C 42.11%, H 6.48%, O 51.41%. Prepd from raffinose by fermentation with top yeast which removes the fructose: Hudson, Harding, J. Am. Chem. Soc. **37**, 2734 (1915); Fletcher, Diehl, ibid. **74**, 5774 (1952). Structure: Haworth, Leitch, J. Chem. Soc. **113**, 188 (1918); Charlton et al., ibid. **1926**, 99; Charlton et al., ibid. **1927**, 1527; Haworth et al., ibid. 3146; Levene, Jorpes, J. Biol. Chem. **86**, 403 (1930). Synthesis: Helferich, Bredereck, Ann. **465**, 166 (1928).

Dihydrate, monoclinic crystals from water or dil alcohol. mp 84-85°. Shows mutarotation. $[\alpha]_D^{20}$ +111.7° → 129.5° (c = 4). One gram dissolves in 0.4 ml water, 8.5 ml methanol, 220 ml abs alcohol. Dilute acids hydrolyze melibiose to D-glucose and D-galactose. Also split by emulsin and by bottom yeast. Reduces Fehling's soln. 3.5 g melibiose dihydrate are about as sweet as 1.0 g sucrose.

5862. Melilot. Sweet clover; yellow Melilot; yellow sweet clover. Dried leaves and flowering tops of Melilotus officinalis (L.) Lam., Leguminosae. Habit. Europe, Asia, natural to some extent in U.S. Constit. Coumarin, resin, volatile oil.
An extract is marketed as Esberiven.

5863. Melilotoside. (E)-3-[2-(β-D-Glucopyranosyloxy)-phenyl]-2-propenoic acid. $C_{15}H_{18}O_8$; mol wt 326.30. C 55.21%, H 5.56%, O 39.23%. From flowers of Melilotus altissimus Thuill. and M. arvensis Wallr., Leguminosae: Charaux, Bull. Soc. Chim. Biol. **7**, 1056 (1925). Synthesis: Shinoda, Imaida, J. Pharm. Soc. Japan **54**, 107 (1934), C.A. **31**, 100⁵ (1937).

Monohydrate, yellowish needles; slightly bitter, acidulous, astringent taste. mp 240-241° with decompn. $[\alpha]_D$ −60.9° (50% alcohol). Sol in water, alcohol; slightly sol in acetone, ethyl acetate.

5864. Melinamide. (Z,Z)-N-(1-Phenylethyl)-9,12-octa-decadienamide; N-(α-methylbenzyl)linoleamide; MBLA; AC-223; Artes. $C_{26}H_{41}NO$; mol wt 383.62. C 81.41%, H 10.77%, N 3.65%, O 4.17%. Linoleic acid derivative. Prepn and use in lowering blood cholesterol levels: Fr. pat. **1,476,-596**; T. Seki et al., U.S. pat. **3,621,043** (1967, 1971 both to Sumitomo). Physical properties and effect on cholesterol levels in rabbits: K. Toki et al., J. Atheroscler. Res. **7**, 708 (1967). Comparative activity of melinamide and its optical

isomers: H. Fukushima, K. Nakatani, ibid. **9**, 65 (1969); H. Fukushima et al., ibid. **10**, 403 (1969). Mechanism of action studies: D. Kritchevsky, S. A. Tepper, Arzneimittel-Forsch. **21**, 1024 (1971); D. Kritchevsky et al., Lipids **12**, 16 (1977); K. Natori et al., Japan. J. Pharmacol. **42**, 517 (1986). Metabolism: A. Hirohashi et al., Xenobiotica **6**, 329 (1976).

Oil. mp <4°. $bp_{0.03}$ 200-215°; $bp_{0.07}$ 200-204°. n_D^{23} 1.5050; n_D^{30} 1.4863. Saponification value 0.9. Iodine value 127.0.
THERAP CAT: Antihyperlipoproteinemic.

5865. Melinonine A. (19α)-16,17-Didehydro-16-(meth-oxycarbonyl)-4,19-dimethyloxayohimbanium. $[C_{22}H_{27}N_2-O_3]^+$. Calabash-curare alkaloid from Strychnos melinoniana Baillon, Loganiaceae. Isoln and structure: Schlittler, Hohl, Helv. Chim. Acta **35**, 29 (1952); Bächli et al., ibid. **40**, 1167 (1957).

Chloride, needles from alcohol + acetone, mp 260-261°. $[\alpha]_D^{20}$ −107° (c = 0.65 in water). uv max (50% ethanol): 225 nm (log ε 4.68).
Iodide, mass spectrum: Hesse et al., Helv. Chim. Acta **48**, 674 (1965).

5866. Melitracen. 3-(10,10-Dimethyl-9(10H)-anthra-cenylidene)-N,N-dimethyl-1-propanamine; N,N,10,10-tetra-methyl-Δ⁹⁽¹⁰ᴴ⁾,ᵞ-anthracenepropylamine; 9,10-dihydro-10,10-dimethyl-9-(3-dimethylaminopropylidene)anthracene; 9-[3-(dimethylamino)propylidene]-10,10-dimethyl-9,10-dihydro-anthracene; N,N-dimethyl-3-(10,10-dimethyl-9(10H)-anth-rylidene)propylamine. $C_{21}H_{25}N$; mol wt 291.44. C 86.55%, H 8.65%, N 4.81%. Prepn of the hydrochloride: Holm, Acta Chem. Scand. **17**, 2437 (1963); idem, Brit. pat. **939,856** corresp to U.S. pat. **3,177,209** (1963, 1965, both to Kefalas A/S). Crystal structure: J. Lopez de Lerma et al., Acta Crystallogr. **B35**, 1739 (1979). Toxicity data: P. V. Petersen et al., Acta Pharmacol. Toxicol. **24**, 121 (1966).

Hydrochloride, $C_{21}H_{25}N.HCl$, U-24973A, Melixeran, Trausabun, Dixeran. Crystals from acetone, mp 245-248°. LD_{50} i.v. in mice: 52 mg/kg (Petersen).
THERAP CAT: Antidepressant.

5867. Melittin. Melittin I; mellitin; Forapin. $C_{131}H_{229}-N_{39}O_{31}$; mol wt 2846.50. C 55.28%, H 8.11%, N 19.19%, O 17.42%. Strongly basic polypeptide, the principal component of the venom of the honey bee, Apis mellifica (mellifera), comprising 40-50% of the dried venom. A "direct" hemolysin, one of the hemolytic principles present in the venom, the other being "indirect" acting phospholipase A. Sepn and isoln: Neumann et al., Naturwiss. **39**, 286 (1952); Neumann, Haberman, Arch. Exp. Pathol. Pharmakol. **222**,

367 (1954); Habermann, Reiz, *Biochem. Z.* **341**, 451 (1965). Melittin is the first polypeptide whose biological effects can be understood on the basis of its primary structure. Fluci-dation of structure and correlation with activity: E. Haber-mann, J. Jentsch, *Z. Physiol. Chem.* **348**, 37 (1967). Confor-mation studies: R. Bazzo *et al., Eur. J. Biochem.* **173**, 139 (1988). About 10% of the melittin is thought to be for-mylated at the *N*-terminus: Kreil, Kreil-Kiss, *Biochem. Bio-phys. Res. Commun.* **27**, 275 (1967). Isoln and structure of *N$^\alpha$-formyl melittin*: Lübke *et al., Experientia* **27**, 765 (1971). Synthesis of melittin and related peptides: Lübke, Schröder, *Peptides*, H. C. Beyerman, A. van der Linde, W. M. van den Brink, Eds. (North-Holland Publishing Company, Amster-dam, 1967) pp 271-279; Dorman, Markley, *J. Med. Chem.* **14**, 5 (1970); Schröder *et al., Experientia* **27**, 764 (1971). Solid phase synthesis and purification: M. T. Tosteson *et al., Biochemistry* **26**, 6627 (1987). Review of biochemistry and pharmacology: Habermann, *Science* **177**, 314 (1972).


```
        1
Gly–Ile–Gly–Ala–Val–Leu–Lys–Val–Leu–Thr–Thr–Gly–Leu–Pro
                                                         |
             26
NH₂Gln–Gln–Arg–Lys–Arg–Lys–Ile–Trp–Ser–Ile–Leu–Ala
```

Cream white, water soluble powder. $[\alpha]_D^{21}$ $-89.52°$ (c = 0.409).

THERAP CAT: Antirheumatic.

5868. Mellitic Acid. *Benzenehexacarboxylic acid;* mellic acid. $C_{12}H_6O_{12}$; mol wt 342.17. C 42.12%, H 1.77%, O 56.11%. $C_6(COOH)_6$. Preparation from carbonaceous material: M. Kiebler, U.S. pat. **2,461,740** (1949 to Carnegie Inst. of Tech.); Germain *et al., Bull. Soc. Chim. France* **1962**, 779; from tetrahalophthalic acid: Brussel, Ony, *ibid.* **1951**, 565; Juettner, U.S. pat. **3,067,246** (1962).

Crystals. mp 286-288° in sealed tube with decompn. Freely sol in water or alcohol; sol in boiling concd H_2SO_4 without decompn.

5869. Meloxicam. *4-Hydroxy-2-methyl-N-(5-methyl-2-thiazolyl)-2H-1,2-benzothiazine-3-carboxamide 1,1-dioxide;* Metacam; $C_{14}H_{13}N_3O_4S_2$; mol wt 351.41. C 47.85%, H 3.73%, N 11.96%, O 18.21%, S 18.25%. Prepn: G. Trum-mlitz *et al.,* **Ger.** pat. **2,756,113** (1979 to Thomae); *eidem,* U.S. pat. **4,233,299** (1980 to Boehringer Ingelheim). Phar-macology: G. Engelhardt, G. Trummlitz, *Drugs Exp. Clin. Res.* **16**, 53 (1990). Tissue distribution in rats: U. Busch, G. Engelhardt, *ibid.* 49. Pharmacodynamics and pharmaco-kinetics in horses: P. Lees *et al., Brit. Vet. J.* **147**, 97 (1991). Physicochemical properties: R.-S. Tsai *et al., Helv. Chim. Acta* **76**, 842 (1993). Clinical efficacy in dogs: A. J. Hen-derson *et al., Prakt. Tierarzt.* **75**, 179 (1994).

Crystals from ethylene chloride, mp 254° (dec). pKa: 4.08 in water; 4.24 ± 0.01 in water/ethanol (1:1); 4.63 ± 0.03 in water/ethanol (1:4). Log P_{oct} 3.02. LD_{50} orally in mice: 470 mg/kg (Trummlitz, 1980).

THERAP CAT (VET): Anti-inflammatory.

5870. Melperone. *1-(4-Fluorophenyl)-4-(4-methyl-1-piperidinyl)-1-butanone; 4'-fluoro-4-(4-methylpiperidino)-butyrophenone;* γ-(4-methylpiperidino)-*p*-fluorobutyrophen-one; methylperone; flubuperone. $C_{16}H_{22}FNO$; mol wt 263.36. C 72.97%, H 8.42%, F 7.21%, N 5.32%, O 6.08%. Neuroleptic agent related structurally to haloperidol, *q.v.* Prepn: **Belg.** pat. **651,144;** S. E. H. Hernestam *et al.,* U.S. pat. **3,816,433** (1964, 1974 both to Ferrosan). Distribution of ^{14}C melperone: N. Einer-Jensen, E. Hansson, *Acta Phar-macol. Toxicol.* **23**, 65 (1965). Pharmacological and toxico-logical studies: J. A. Christensen *et al., ibid.* 109; R. Hey-wood, A. K. Palmer, *Farmaco Ed. Prat.* **29**, 586 (1974). Dopamine-receptor binding in relation to clinical effect: I.

Creese *et al., Science* **192**, 481 (1976). Sedative and sleep-inducing properties: R. Kretzschmer *et al., Arzneimittel-Forsch.* **26**, 1073 (1976). Clinical studies in anxiety: W. J. Poeldinger, *Therapiewoche* **30**, 4862 (1980); L. F. Fabre, M. J. Napoliello, *Curr. Ther. Res.* **30**, 427 (1981).

Liquid, $bp_{0.1}$ 120-125°.

Hydrochloride, $C_{16}H_{22}FNO.HCl$, *FG-5111, Buronil, Eunerpan.* Crystals, mp 209-211°. LD_{50} in rats, mice (mg/kg): 330, 230 orally; 40, 35 i.v. (Christensen).

THERAP CAT: Antipsychotic.

5871. Melphalan. *4-[Bis(2-chloroethyl)amino]-L-phen-ylalanine; p-di(2-chloroethyl)amino-L-phenylalanine;* L-phenylalanine mustard; alanine nitrogen mustard; L-PAM; melfalan; L-sarcolysine; NSC-8806; CB-3025; Alkeran; Sar-coclorin. $C_{13}H_{18}Cl_2N_2O_2$; mol wt 305.20. C 51.16%, H 5.94%, Cl 23.23%, N 9.18%, O 10.48%. Syntheses: Bergel, Stock, *J. Chem. Soc.* **1954**, 2409; **1955**, 1223; *eidem,* U.S. pats. **3,032,584; 3,032,585** (both 1962 to NRDC); Larionov, *Lancet* **2**, 169 (1955). Toxicity: W. C. J. Ross, *Biochem. Pharmacol.* **13**, 969 (1964). Neurotoxicity study: M. G. Donelli *et al., J. Pharm. Pharmacol.* **18**, 760 (1966). Muta-tion study: J. McCann *et al., Proc. Nat. Acad. Sci. USA* **72**, 5135 (1975). Biliary excretion in rats: K. H. Byington *et al., Biochem. Pharmacol.* **29**, 2518 (1980). Review of carcin-ogenicity studies: *IARC Monographs* **9**, 167-180 (1975). Re-view: R. L. Furner, R. K. Brown, *Cancer Treat. Rep.* **64**, 559-574 (1980).

Needles from methanol (monosolvate), mp 182-183° (dec). $[\alpha]_D^{25}$ +7.5° (c = 1.33 in 1.0N HCl); $[\alpha]_D^{25}$ $-31.5°$ (c = 0.67 in methanol). Soluble in ethanol, propylene glycol. Practically insol in water. LD_{50} i.p. in rats: 14.7 μmol/kg (Ross).

D-Form, *D-sarcolysine, medphalan, CB-3026, NSC-35051.* Needles from methanol (monosolvate), mp 181.5-182° (dec). $[\alpha]_D^{21}$ $-7.5°$ (c = 1.26 in 1.0N HCl).

DL-Form, *merphalan, sarcolysine.* Tiny needles from methanol, mp 180-181°.

Note: This substance has been listed as a known car-cinogen: *Seventh Annual Report on Carcinogens* (PB95-109781, 1994) p 60.

THERAP CAT: Antineoplastic.

5872. Memantine. *3,5-Dimethyltricyclo[3.3.1.1^{3,7}]dec-an-1-amine; 3,5-dimethyl-1-adamantanamine;* 1-amino-3,5-dimethyladamantane; DMAA; D-145. $C_{12}H_{21}N$; mol wt 179.31. C 80.38%, H 11.80%, N 7.81%. Deriv of adaman-tine, *q.v.*, with anti-parkinson activity. Prepn of the hydro-chloride: K. Gerzon *et al., J. Med. Chem.* **6**, 760 (1963); of the free base and hydrochloride: J. Mills, E. Krumkalns, U.S. pat. **3,391,142** (1968 to Lilly). GC and mass spec studies of memantine metabolites: W. Wesemann *et al., Arzneimittel-Forsch.* **27**, 1471 (1977). Effects in parkinsoni-an patients: P.-A. Fischer *et al., ibid.* 1487. Series of arti-cles on distribution, effects on neurobiological processes, clinical studies in control of micturition and limb muscle mobility: *ibid.* **32**, 1236-1276 (1982). Clinical studies as antispasmodic agent: H. Rohde, *Fortschr. Med.* **100**, 2023 (1982). Pharmacodynamics and pharmacokinetics: W. Wesemann *et al., Arzneimittel-Forsch.* **33**, 1122 (1983).

Oil, n_D^{25} 1.4941.

Hydrochloride, $C_{12}H_{21}N.HCl$, *Akatinol*. Cryst from alcohol/ether, mp 258° (Mills, Krumkalns); also reported as mp 290-295° (Gerzon).

THERAP CAT: Muscle relaxant (skeletal).

5873. Menadiol. *2-Methyl-1,4-naphthalenediol;* 2-methyl-1,4-naphthohydroquinone; 2-methyl-1,4-naphthoquinol; dihydrovitamin K_3. $C_{11}H_{10}O_2$; mol wt 174.20. C 75.84%, H 5.79%, O 18.37%. Synthetic naphthoquinol derivative having physiological properties of Vitamin K, *q.v.*; starting material for prepn of menaquinones, *q.v.* Prepn: K. Fries, W. Lohmann, *Ber.* **54**, 2912 (1921); D. W. MacCorquodale *et al., J. Biol. Chem.* **131**, 357 (1939); C. D. Snyder, H. Rapoport, *J. Am. Chem. Soc.* **96**, 8046 (1974). Crystal structure: J. Gaultier, C. Hauw, *Acta Crystallogr.* **25B**, 51 (1969). Bioactivity and tissue distribution: M. J. Thierry, J. W. Suttie, *J. Nutr.* **97**, 512 (1969).

White needles from dil alcohol, mp 168-170° (MacCorquodale); also reported as mp 181° (Gaultier, Hauw). Slightly sol in benzene, chloroform; easily sol in acetone, alcohol.

Diacetate, $C_{15}H_{14}O_4$, *1,4-diacetoxy-2-methylnaphthalene, acetomenaphthone, vitamin K_4, Kapilin, Prokayvit Oral, Vitavel K.* Prepn: Sah *et al., Ber.* **73**, 762 (1940); *Rec. Trav. Chim.* **59**, 461 (1940). Crystals, mp 112-114°. Practically insol in water. Slightly sol in cold alc; sol in 3.3 parts boiling alc, in acetic acid.

Dibutyrate, $C_{19}H_{22}O_4$, *Karanum.* Prepn: von Werder, *Ger.* pat. 734,220 (1943 to E. Merck). Crystals, mp ~53°. Practically insol in water. Sol in alcohol, benzene, oils and fats.

Sodium diphosphate, $C_{11}H_8Na_4O_8P_2$, *2-methyl-1,4-naphthalenediol diphosphoric acid ester tetrasodium salt, Kappadione, Synkavit, Synkayvite.* Prepn: Fieser, Fry, *J. Am. Chem. Soc.* **62**, 228 (1940). Prepd as the hexahydrate, white to pinkish powder. Salty taste. Very sol in water. Practically insol in methanol, ethanol, ether, acetone.

THERAP CAT: Vitamin (prothrombogenic).

5874. Menadione. *2-Methyl-1,4-naphthalenedione;* 2-*methyl-1,4-naphthoquinone;* menaphthone; Vitamin $K_{2(0)}$; Vitamin K_3; Kanone; Kappaxin; Kayklot; Kayquinone; Klottone; Kolklot; Thyloquinone. $C_{11}H_8O_2$; mol wt 172.18. C 76.73%, H 4.68%, O 18.58%. Synthetic naphthoquinone derivative having physiologic properties of vitamin K. Prepn: Fieser, *J. Biol. Chem.* **133**, 391 (1940). Toxicity study: Molitor, Robinson, *Proc. Soc. Exp. Biol. Med.* **43**, 725 (1940). Alkylated *in vivo* to the bioactive metabolite, menaquinone-4, *q.v.*: C. Martius, H. O. Esser, *Biochem. Z.* **331**, 1 (1958). Metabolism and tissue distribution: W. V. Taggart, J. T. Matschiner, *Biochemistry* **8**, 1141 (1969). Im-

proved synthesis: W. Adam *et al., Angew. Chem. Int. Ed. Engl.* **33**, 2475 (1994).

Bright yellow crystals. Very faint acrid odor. mp 105-107°. Stable in air; dec by sunlight. Insol in water. One gram dissolves in about 60 ml alcohol, in 10 ml benzene, in 50 ml vegetable oils; moderately sol in chloroform and in carbon tetrachloride. The alcoholic soln is neutral to litmus. Solutions may be heated to 120° without dec. Destroyed by alkalies and reducing agents. *Keep protected from light.* LD_{50} orally in mice: ~0.5 g/kg (Molitor, Robinson).

Sodium bisulfite, $C_{11}H_9NaO_5S$, *1,2,3,4-tetrahydro-2-methyl-1,4-dioxo-2-naphthalenesulfonic acid sodium salt, Abor K, Hykinone, Izokappa, Klotogen, K-Thrombin.* Prepn: Moore, *J. Am. Chem. Soc.* **63**, 2049 (1941). Structure: Carmack *et al., ibid.* **72**, 844 (1950). Occurs as the trihydrate. White, hygroscopic crystals. Discolors and may turn purple under the influence of light. One gram dissolves in ~2 ml water. Slightly sol in alcohol. Almost insol in ether, benzene.

Dimethylpyrimidinol bisulfite, $C_{17}H_{18}N_2O_6S$, *Hetrazeen.* Prepn: J. B. Nanninga *et al.,* U.S. pat. 3,328,169 (1967 to Heterochemical). Efficacy in swine diets: R. W. Seerley *et al., J. Anim. Sci.* **42**, 599 (1976). Cryst powder, mp 215-217°. Soly in water ~1 g/100 ml. Slightly sol in alcohol. Insol in ether, benzene.

THERAP CAT: Vitamin (prothrombogenic).

THERAP CAT (VET): Vitamin (prothrombogenic); antidote to bishydroxycoumarin poisoning, including sweet clover poisoning.

5875. Menadoxime. *[[(3-Methyl-4-oxo-1(4H)-naphthalenylidene)amino]oxy]acetic acid ammonium salt; menadione carboxymethoxime ammonium salt;* menaphthone carboxymethoxime ammonium salt; carboxymethylmenadione monoxime ammonium salt; Kapilon injectable. $C_{13}H_{14}N_2O_4$; mol wt 262.27. C 59.54%, H 5.38%, N 10.68%, O 24.40%. Prepn: Holland, **Brit.** pat. 621,934 (1949 to Glaxo).

Crystals from alc. Sol in water. Aq solns are neutral. May be sterilized by autoclaving and stored in ampuls if protected from light. Shows high antihemorrhagic activity.

Free acid, $C_{13}H_{11}NO_4$, yellow platelets from alc, mp 162-163°. Forms water-sol salts.

THERAP CAT: Vitamin (prothrombogenic).

5876. Menaquinones. Vitamin K_2; 2-methyl-3-*all-trans*-polyprenyl-1,4-naphthoquinones. Group of prenylated napthoquinone derivatives having the physiological activity of vitamin K, *q.v.* Nomenclature is based on the number of isoprene residues comprising the side chain. Compds of varying chain length are produced by bacteria, particularly by normal intestinal flora, and may serve as a source of vitamin K for humans. Birds and animals are capable of alkylating menadione, *q.v.*, to produce menaquinone 4. Originally isolated from putrefied fish meal: R. W. McKee *et al., J. Biol. Chem.* **131**, 327 (1939). Identification of menaquinone 6: S. B. Binkley *et al., ibid.* **133**, 721 (1940). Isoln of menaquinone 7 from cultures of *Bacillus brevis:* M. Tishler, W. L. Sampson, *Proc. Soc. Exp. Biol. Med.* **68**, 136 (1948). Structures and syntheses of series of menaquinones: O. Isler *et al., Helv. Chim. Acta* **41**, 786 (1958). Review of isoln, props, synthesis: O. Isler, O. Wiss, *Vitam. Horm.* **17**, 53-90 (1959); H. Mayer, O. Isler, *Methods Enzymol.* **18C**, 469-547 (1971); of biosynthesis: D. R. Threlfall, *Vitam. Horm.* **29**, 153-200 (1971); R. Bentley, R. Meganathan, *Microbiol. Rev.* **46**, 241-280 (1982). Menaquinones with side chains of up to 15 repeating units have been described: T. Sakano *et al., Cehm. Pharm. Bull.* **34**, 4322 (1986). HPLC determn in bone: S. J. Hodges *et al., J. Bone Mineral Res.* **8**, 1005 (1993). Metabolism and distribution: H. H. W. Thijssen, M. J. Drittij-Reijnders, *Brit. J. Nutr.* **72**, 415 (1994).

Absorption and bioavailability in humans: J. M. Conly *et al., Am. J. Gastroenterol.* **89**, 915 (1994).

Menaquinone 4

Menaquinone 4, $C_{31}H_{40}O_2$, *(E,E,E)-2-Methyl-3-(3,7,11,15-tetramethyl-2,6,10,14-hexadecatetraenyl)-1,4-naphthalenedione; menatetrenone; MK 4; Vitamin $K_{2(20)}$; Kaytwo.* Yellow crystals from alcohol, mp 35°. uv max: 248 nm ($E_{1cm}^{1\%}$ 439).

Menaquinone 6, $C_{41}H_{56}O_2$, *(all-E)-2-(3,7,11,15,19,23-Hexamethyl-2,6,10,14,18,22-tetracosahexaenyl)-3-methyl-1,4-naphthalenedione; 2-difarnesyl-3-methyl-1,4-naphthoquinone; farnoquinone; MK 6; Vitamin $K_{2(30)}$.* Yellow crystals from acetone + alcohol or petr ether, mp 50°. uv max (petr ether): 243, 248, 261, 270, 325-328 nm ($E_{1cm}^{1\%}$ 304, 320, 290, 292, 53).

Menaquinone 7, $C_{46}H_{64}O_2$, *(all-E)-2-(3,7,11,15,19,23,27-Heptamethyl-2,6,10,14,18,22,26-octacosaheptaenyl)-3-methyl-1,4-naphthalenedione; Vitamin $K_{2(35)}$.* Light yellow microcrystalline plates from petr ether, mp 54°. bp$_{0.0002}$ 200° (some dec). uv max (petr ether): 243, 248, 261, 270, 325-328 nm ($E_{1cm}^{1\%}$ 278, 195, 266, 267, 48). Slightly less sol than vitamin K_1 in the same organic solvents.

THERAP CAT: Vitamin (prothrombogenic).

5877. Menazon. *S-[(4,6-Diamino-1,3,5-triazin-2-yl)methyl]phosphorodithioic acid O,O-dimethyl ester; S-[(4,6-diamino-s-triazin-2-yl)methyl] O,O-dimethyl phosphorodithioate; O,O-dimethyl S-[(4,6-diamino-s-triazin-2-yl)methyl] phosphorodithioate; 2-dimethoxyphosphinothioylthiomethyl-4,6-diamino-s-triazine; ENT-25760; PP-175; Saphicol; Saphizon-DP; Sayfos.* $C_6H_{12}N_5O_2PS_2$; mol wt 281.30. C 25.62%, H 4.30%, N 24.90%, O 11.38%, P 11.01%, S 22.80%. Prepn: Calderbank *et al.*, **Ger. pat. 1,118,789** corresp to U.S. pat. **3,169,964** (1961, 1965 both to I.C.I.). Activity: *eidem, Chem. & Ind. (London)* **1961**, 630. Toxicity: T. B. Gaines, *Toxicol. Appl. Pharmacol.* **14**, 515 (1969).

Crystals, dec 160-162°, mp 164-166°. Vapor press. at 25°: 1×10^6 mm Hg. Low solubility in water and organic solvents. LD_{50} orally in male, female rats: 1020, 1450 mg/kg (Gaines).

Caution: Cholinesterase inhibitor.

USE: Insecticide; acaricide.

5878. Menbutone. *4-Methoxy-γ-oxo-1-naphthalenebutanoic acid; 3-(4-methoxy-1-naphthoyl)propionic acid; β-(1-methoxy-4-naphthoyl)propionic acid; γ-oxo-4-methoxy-1-naphthalenebutyric acid; Ictéryl.* $C_{15}H_{14}O_4$; mol wt 258.27. C 69.76%, H 5.46%, O 24.78%. Prepn: Ruzicka, Waldman, *Helv. Chim. Acta* **15**, 907 (1932); Fieser, Hershberg, *J. Am. Chem. Soc.* **58**, 2314 (1936); Burtner, U.S. pat. **2,623,065** (1952 to Searle).

Crystals, mp 172-173°.

Magnesium salt, $C_{30}H_{26}MgO_8$, *Hepalande.*

THERAP CAT: Choleretic.

5879. Mendelevium. Md; formerly Mv; at. no. 101; valence 3, 2. Man-made radioactive element. No stable nuclides; known isotopes (mass numbers): 247-252, 254-259. Longest-lived known isotope 258 ($T_{1/2}$ 55 days, α-emitter, rel. at. mass 258.10). Prepn of isotope ^{256}Md ($T_{1/2}$ 1.27 hrs; decays by electron capture to ^{256}Fm) by bombarding ^{253}Es with helium ions: A. Ghiorso *et al., Phys. Rev.* **98**, 1518 (1955). Prepn of isotope ^{258}Md by bombardment of ^{255}Es with ^{4}He ions: Fields *et al., Nucl. Phys. A* **154**, 407 (1970). *Reviews:* C. Keller, *The Chemistry of the Trans-uranium Elements* (Verlag Chemie, Weinheim, English Ed., 1971) pp 595-600; Silva, "Trans-Curium Elements" in *MTP Int. Rev. Sci.: Inorg. Chem. Ser. One* vol. 8, A. G. Maddock, Ed. (University Park Press, Baltimore, 1972) pp 71-105; *Comprehensive Inorganic Chemistry* vol. 5, J. C. Bailar, Jr. *et al.*, Eds. (Pergamon Press, Oxford, 1973) *passim; Handb. Exp. Pharmakol.* **36**, 689-738 (1973); R. J. Silva in *The Chemistry of the Actinide Elements* vol. 2, J. J. Katz *et al.*, Eds. (Chapman and Hall, New York, 1986) pp 1092-1095; *The Elements Beyond Uranium*, G. T. Seaborg, W. D. Loveland, Eds. (John Wiley & Sons, Inc., New York, 1990) p 38-46.

Caution: Radiation hazard; handling requires special equipment and shielding facilities (Katz *et al., loc cit.*, vol. 2, p. 1128).

5880. Menhaden Oil. Pogy oil; mossbunker oil. Obtained along the East Coast of North America from the menhaden fish, *Brevoortia tyrannis,* somewhat larger than a herring. The oil contains (in the form of glycerides) about 6% myristic acid, 16% palmitic acid, 30% linoleic acid, 19% C_{20}- and 11% C_{22}-acids (highly unsaturated).

Reddish oil. Characteristic, distasteful fishy odor and taste. d 0.925-0.933. mp 38.5-47.2°. n_D^{20} 1.480. Sapon no. 191-200. Iodine no. 139-180. Acid no. 3.0-11.6. Sol in ether, benzene, petr ether, naphtha, kerosene, CS_2.

USE: Substitute for linseed oil. In leather dressing formulations. Hydrogenated menhaden oil can be used as a substitute for tallow in soap-making.

5881. Menogaril. *[2R-(2α,3β,4α,5β,6α,11α,13α)]-4-(Dimethylamino)-3,4,5,6,11,12,13,14-octahydro-3,5,8,10,13-pentahydroxy-11-methoxy-6,13-dimethyl-2,6-epoxy-2H-naphthaceno[1,2-b]oxocin-9,16-dione; 7-con-O-methylnogarol; 7-OMEN; 7(R)-O-methylnogarol; menogarol; TUT-7; NSC-269148; U-52047; Tomoscar.* $C_{28}H_{31}NO_{10}$; mol wt 541.55. C 62.10%, H 5.77%, N 2.59%, O 29.54%. Semi-synthetic derivative of nogalamycin, *q.v.* Prepn: P. F. Wiley, D. J. Houser, U.S. pat. **4,086,245** (1978 to Upjohn). Total synthesis: M. Kawasaki *et al., Tetrahedron* **44**, 5727 (1988). Antitumor activity: G. L. Neil *et al., Cancer Treat. Rep.* **63**, 1971 (1979). HPLC determn in plasma: J. E. Brown *et al., J. Chromatog.* **226**, 521 (1981); J. P. McGovren *et al., Anal. Chem.* **56**, 1587 (1984). Pharmacokinetics and metabolism: M. J. Egorin *et al., Cancer Res.* **46**, 1513 (1986). Mechanism of action: K. Wierzba *et al., Japan J. Cancer Res.* **81**, 842 (1990). Clinical studies in non-small cell lung cancer: R. B. Vance *et al., Investig. New Drugs* **9**, 73 (1991); in non-Hodgkin's lymphoma: J. Skillings *et al., ibid.*, 79. Brief review of antitumor, biochemical, and toxicological studies: J. P. McGovren *et al., ibid.* **2**, 359-367 (1984).

Orange crystals, mp 247-249° (dec). $[\alpha]_D^{20}$ +857° (c = 0.112 in $CHCl_3$). Also reported as mp 250-254° (dec). $[\alpha]_D^{20}$ +867° (c = 0.045 in $CHCl_3$); $[\alpha]_D$ +958° (c = 0.163 in

CHCl₃). uv max (ethanol): 235, 251, 257, 290, 479 nm (ε 41200, 25500, 24150, 10500, 15530).

THERAP CAT: Antineoplastic.

5882. Menthol. *(1α,2β,5α)-5-Methyl-2-(1-methylethyl)-cyclohexanol; 3-p-menthanol; l-menthol; hexahydrothymol;* peppermint camphor. $C_{10}H_{20}O$; mol wt 156.27. C 76.86%, H 12.90%, O 10.24%. Obtained from peppermint oil or other mint oils, or prepd synthetically by hydrogenation of thymol. Chromatographic sepn from mint oils: Chang, U.S. pat. **2,760,993** (1956 to Iowa State College). Toxicity data: P. M. Jenner *et al., Food Cosmet. Toxicol.* **2**, 327 (1964).

Crystals or granules; peppermint taste and odor. d 0.890. mp 41–43°. bp 212°. n_D^{25} 1.458. $[α]_D^{18}$ − 50° (10% alc soln). Slightly sol in water; very sol in alcohol, chloroform, ether, petr ether; freely sol in glacial acetic acid, liq petrolatum. LD_{50} orally in rats: 3180 mg/kg (Jenner).

Incompat: Butylchloral hydrate, camphor, phenol, chloral hydrate, Exalgine, betanaphthol, resorcinol or thymol in triturations; potassium permanganate, chromium trioxide, pyrogallol.

USE: In liqueurs, confectionery, perfumery, cigarettes, cough drops, and nasal inhalers.

THERAP CAT: Antipruritic (topical).

THERAP CAT (VET): Has been used as a mild local anesthetic, antiseptic; internally as a carminative and gastric sedative.

5883. l-Menthone. *(2S-trans)-5-Methyl-2-(1-methylethyl)cyclohexanone; (1R,4S)-(−)-p-menthan-3-one;* 1-methyl-4-isopropylcyclohexan-3-one. $C_{10}H_{18}O$; mol wt 154.25. C 77.87%, H 11.76%, O 10.37%. Of the four optically active isomers of methone, the one occuring most frequently in nature. Found in various volatile oils, such as pennyroyal, peppermint, geranium: Simonsen, *The Terpenes* vol. I (University Press, Cambridge, 2nd ed., 1947) pp 314-327. Prepd by chromic acid oxidation of l-menthol: Hussey, Baker, *J. Org. Chem.* **25**, 1434 (1960); Brown, Garg, *J. Am. Chem. Soc.* **83**, 2952 (1961).

Bitter liq; slight peppermint odor. bp 207°. bp₄₁ 116-119°. mp −6°. d_4^{20} 0.895. n_D^{20} 1.4505; n_D^{23} 1.4490. $[α]_D^{20}$ −24.8°; $[α]_D^{27}$ −28.9°. Slightly sol in water; sol in organic solvents.

Apinol obtained by dry distln of wood of *Pinus palustris* Mill. (*P. australis* Michx.), *Pinaceae* is chiefly l-menthone: *J. Pharm. Chim.* **18**, 139, 177, 208 (1918), *C.A.* **13**, 56⁹ (1919). Amber-colored oil, bp ∼182.2°, d 0.946.

USE: In perfume and flavor compositions.

5884. Menthyl Acetate. *5-Methyl-2-(1-methylethyl)cyclohexanol acetate.* $C_{12}H_{22}O_2$; mol wt 198.31. C 72.68%, H 11.18%, O 16.14%. $CH_3COOC_{10}H_{19}$. Present in peppermint oil.

Colorless liquid, characteristic odor. d_4^{20} 0.919. bp 227°. n_D^{20} 1.4468. $[α]_D^{20}$ −79.42°. Slightly sol in water; miscible with alcohol, ether.

USE: In perfumery; emphasizes floral notes, especially that of rose, used in toilet waters having a lavender odor. Has

been suggested for flavoring extracts having caraway or mint flavors.

5885. Menthyl Borate. *5-Methyl-2-(1-methylethyl)cyclohexanol monoester with boric acid.* Estoral. $C_{30}H_{57}BO_3$; mol wt 476.59. C 75.61%, H 12.05%, B 2.27%, O 10.07%. $(C_{10}H_{19})_3BO_3$.

White, tasteless cryst powder; faint menthol odor. Insol in water or alcohol; freely sol in chloroform, ether. Dec into its constituents when in soln.

5886. Menthyl Salicylate. *2-Hydroxybenzoic acid 5-methyl-2-(1-methylethyl)cyclohexyl ester.* $C_{17}H_{24}O_3$; mol wt 276.38. C 73.88%, H 8.75%, O 17.37%. $HOC_6H_4COO-C_{10}H_{19}$.

Clear, yellowish, syrupy liquid; odorless or slight fruity odor. d_{25}^{25} 1.045. Insol in water; miscible with most organic solvents. *Keep well closed and protected from light.*

USE: As a "sun screen" to filter out ultraviolet light in prepns for preventing sunburn.

5887. Menthyl Valerate. *(1α,2β,5α)-3-Methylbutanoic acid 5-methyl-2-(1-methylethyl)cyclohexyl ester; isovaleric acid p-menth-3-yl ester;* Validol. $C_{15}H_{28}O_2$; mol wt 240.39. C 74.95%, H 11.74%, O 13.31%. Contains about 30% free menthol.

Liquid; menthol and valerian odor; cooling, faintly bitter taste. d 0.906-0.908. Insol in water. Freely sol in alcohol, chloroform, ether, oils; dec by alkalies.

THERAP CAT: Sedative.

5888. Menyanthes. Buck bean; bog bean; marsh trefoil; water shamrock. Dried leaves or roots of *Menyanthes trifoliata* L., *Gentianaceae.* Habit. Europe, Asia, N. America. *Constit.* Menyanthin.

5889. Meobentine. *N-[(4-Methoxyphenyl)methyl]-N',-N''-dimethylguanidine;* 1-(p-methoxybenzyl)-2,3-dimethylguanidine. $C_{11}H_{17}N_3O$; mol wt 207.28. C 63.74%, H 8.27%, N 20.27%, O 7.72%. Antidysrhythmic, antifibrillatory deriv of guanidine, *q.v.* Prepn of the sulfate: R. A. Maxwell, E. Walton, **Ger.** pat. **2,030,693** corresp to U.S. pat. **3,949,089** (1971, 1976 both to Burroughs Wellcome). Antidysrhythmic effects and tissue concentration: K. B. Touw *et al., Pharmacologist* **19**, 268 (1977); K. B. Touw, *Diss. Abstr. B* **39**, 5340 (1979). Pharmacokinetics by radioimmunoassay: J. W. A. Findlay *et al., Pharmacologist* **21**, 337 (1979). Pharmacological study: W. B. Wastila *et al., J. Pharm. Pharmacol.* **33**, 594 (1981).

Sulfate, $(C_{11}H_{17}N_3O)_2 \cdot H_2SO_4$, *Rythmatine.* Cryst, mp 273-274°.

THERAP CAT: Antiarrhythmic (class III).

5890. Meparfynol. *3-Methyl-1-pentyn-3-ol;* ethyl ethynyl methyl carbinol; 2-ethynyl-2-butanol; methylparafynol; methylpentynol; Allotropal; Anti-Stress; Apridol; Atemorin; Atempol; Dalgol; Dorison; Dormalest; Dormidin; Dormigen; Dormiphen; Dormison; Dormosan; Formison; Hesofen; Hexofen; Imnudorm; Oblivon; Pentadorm; Perlopal; Riposon; Seral; Somnesin. $C_6H_{10}O$; mol wt 98.14. C 73.43%, H 10.27%, O 16.30%. Prepd from methyl ethyl ketone and sodium acetylide in liquid ammonia or by a Grignard reaction: **Ger.** pat. **285,770** (1913 to Bayer); also **Ger.** pats. **289,800** and **291,185** *see Frdl.* **12**, 55, 56, 57; Sung Wouseng, *Ann. Chim.* [10] **1**, 343 (1924); Rupe, Vonaesch, *Ann.* **442**, 80 (1925); Carothers, Coffman, *J. Am. Chem. Soc.* **54**, 4071 (1932); Campbell *et al., ibid.* **60**, 2882 (1938); Campbell, Campbell, *Proc. Indiana Acad. Sci.* **50**, 123-127 (1940), *C.A.* **35**, 5471; Smith, U.S. pat. **2,385,547** (1945 to Commercial Solvents); A. W. Johnson, *The Chemistry of the Acetylenic Compounds* vol. I (Edward Arnold & Co., London, 1946) p 278; Hurd, McPhee, *J. Am. Chem. Soc.* **69**, 239 (1947); P. Piganiol, *Acetylene Homologs and Derivatives*

(Brooklyn, 1950); Papa *et al.*, *Arch. Biochem. Biophys.* **33**, 482 (1951); Thiele, Martinez, *Ciencia (Mex.)* **15**, 70 (1955). An optically active form has been described by Hickmann, Kenyon, *J. Chem. Soc.* **1955**, 2051. Toxicity data: Margolin *et al.*, *Science* **114**, 384 (1951).

Mobile liquid. Acrid odor, burning taste. d_4^{20} 0.8688. d_{20}^{20} 0.8721; 7.28 lbs/U.S. gal. bp_{760} 121-122°; bp_{37} 50°; $bp_{6.5}$ 20°. mp −30.6°. n_D^{20} 1.4318. Flash pt 101.3°F. Surface tension at 25°: 23.8 dynes/cm. 5% aq soln: 34.1 dynes/cm. Soly in water at 25°: 12.8 g/100 ml. Sol in ether. Miscible with acetone, benzene, carbon tetrachloride, Cellosolve, cyclohexanone, diethylene glycol, ethyl acetate, kerosene, methyl ethyl ketone, mineral spirits, ethanolamine, neatsfoot oil, petr ether, soybean oil, Stoddard solvent. LD_{50} orally in mice, rats, guinea pigs: 600-900 mg/kg (Margolin).

Carbamate, $C_7H_{11}NO_2$, *N-Oblivon, Oblivon C, Trusono.* Prepn: McLamore *et al.*, *J. Org. Chem.* **20**, 1379 (1955); from 3-methyl-1-pentyn-3-ol, trichloroacetic acid, KOCN: McCrea *et al.*, *Brit. pat.* 761,817 (1956 to British Schering). Acute toxicity: Soehring *et al.*, *Arzneimittel-Forsch.* **5**, 161 (1955). Crystals from ether + petr ether. mp 55.8-57° (McLamore); from cyclohexane, mp 53.5-55° (McCrea). bp_{16} 120-121°; $bp_{0.01}$ 95°. Solubility in water: 1.6 g/100 ml. LD_{50} (4 hr) s.c. in mice: 0.56 g/kg (Soehring).

THERAP CAT: Hypnotic; sedative.

5891. Mepartricin. *Partricin methyl ester;* methylpartricin; 3N-634; SPA-S-160; Ipertrofan; Orofungin; Tricandil; Tricanginc. Methyl ester of the heptaene macrolide antibiotic complex, partricin, *q.v.* Prepn: T. Bruzzese *et al.*, *Experientia* **28**, 1515 (1972); T. Bruzzese, R. Ferrari, *Ger. pat.* 2,154,436 (1972 to SPA); *eidem*, *U.S. pat.* 3,780,173 (1973). Alternate prepn: R. C. Pandey, *J. Antibiot.* **30**, 158 (1977); R. C. Tweit *et al.*, *ibid.* **35**, 997 (1982). Structure of mepartricins A and B: J. Golik *et al.*, *ibid.* **33**, 904 (1980). Antimycotic activity: W. Ritzerfield, *Farmaco Ed. Sci.* **27**, 235 (1972); activity vs *Trichomonas vaginalis*: G. Pucci, S. Ripa, *Farmaco Ed. Prat.* **28**, 293 (1973). Hemolytic activity studies: B. Cybulska *et al.*, *Biochem. Pharmacol.* **33**, 41 (1984). Stability data: S. Mizuba, K. Lee, *Dev. Ind. Microbiol.* **16**, 380 (1975). Use of mepartricin and its complexes in treatment of benign prostatic hypertrophy: T. Bruzzese, L. Ferrari, *U.S. pat.* 4,237,117 (1980 to SPA). Clinical studies: M. De Bernardi *et al.*, *Curr. Ther. Res.* **43**, 1159 (1988); T. Lotti *et al.*, *ibid.* **44**, 402 (1988).

Mepartricin A   R' = CH₃
Mepartricin B   R' = H

Deep yellow crystalline material. uv max (ethanol): 401, 378, 359, 340 nm. Slightly sol in water, aq alkali, ether, petr ether, benzene. Sol in acetone, alcohols, pyridine, DMF, DMSO. LD_{50} in mice: > 2 g/kg orally; 200 mg/kg i.p. (Bruzzese, 1972).

Mepartricin A, $C_{60}H_{88}N_2O_{19}$, *40-demethyl-3,7-dideoxo-3,7-dihydroxy-N⁴⁷-methyl-5-oxocandicidin D methyl ester cyclic 15,19-hemiacetal, gedamycin methyl ester.* Lemon yellow powder from ether, mp 145-149° (dec). uv max (methanol): 400, 377, 357, 339, 287, 240, 234, 204 nm (ε 79326, 92454, 68094, 51685, 14199, 24612, 26505, 16092).

Mepartricin B, $C_{59}H_{86}N_2O_{19}$. Lemon yellow powder from ether, mp 154-158° (dec). uv max (methanol): 402, 379, 359, 340, 285, 233, 204 nm (ε 81101, 94729, 64171, 41558, 16196, 23835, 21696).

Complex with sodium lauryl sulfate, *SPA-S-222, Montricin.*

THERAP CAT: Antifungal; antiprotozoal (Trichomonas). In treatment of benign prostatic hypertrophy.

5892. Mepazine. *10-[(1-Methyl-3-piperidinyl)methyl]-10H-phenothiazine;* (*N*-methyl-3-piperidyl)methylphenothiazine; mepasin; pecazine; MPMP; P-391; III-2318; Paxital; Lacumin; Pacatal; Nothiazine; Pacatol. $C_{19}H_{22}N_2S$; mol wt 310.46. C 73.51%, H 7.14%, N 9.02%, S 10.33%. Prepn: Schuler, *U.S. pat.* 2,784,185 (1952 to Promonta).

bp_4 230-235°.

Hydrochloride monohydrate, $C_{19}H_{23}ClN_2S.H_2O$, crystals, mp 180-181°. Photosensitive. Protect from light. Slightly bitter taste. Very slightly sol in water; freely sol in abs ethanol; sol in chloroform. Practically insol in ether, benzene.

THERAP CAT: Antipsychotic.

THERAP CAT (VET): Tranquilizer.

5893. Mepenzolate Bromide. *3-[(Hydroxydiphenylacetyl)oxy]-1,1-dimethylpiperidinium bromide; N-methyl-3-piperidyl benzilate methyl bromide;* N-methyl-3-piperidyl diphenylglycolate methobromide; Cantil; Cantril; Gastropidil; Trancolon. $C_{21}H_{26}BrNO_3$; mol wt 420.35. C 60.01%, H 6.23%, Br 19.01%, N 3.33%, O 11.42%. Prepn: Biel, *U.S. pat.* 2,918,408 (1959 to Lakeside Labs.). Prepn of base: Biel *et al.*, *J. Am. Chem. Soc.* **77**, 2250 (1955). Crystal structure: J. M. Leger *et al.*, *Acta Crystallogr.* **B35**, 886 (1979). Toxicity data: E. I. Goldenthal, *Toxicol. Appl. Pharmacol.* **18**, 185 (1971).

Crystals, mp 228-229° (dec). LD_{50} orally in rats: 742±47 mg/kg (Goldenthal).

Free base, $bp_{0.03}$ 175-176°.

Note: N-Methyl-3-piperidyl benzilate is a controlled substance (hallucinogen) listed in the U.S. Code of Federal Regulations, Title 21, Part 1308.11 (1995).

THERAP CAT: Anticholinergic.

5894. Meperidine. *1-Methyl-4-phenyl-4-piperidinecarboxylic acid ethyl ester; 1-methyl-4-phenylisonipecotic acid ethyl ester;* N-methyl-4-phenyl-4-carbethoxypiperidine; ethyl 1-methyl-4-phenylpiperidine-4-carboxylate; isonipecaine; pethidine. $C_{15}H_{21}NO_2$; mol wt 247.34. C 72.84%, H 8.56%, N 5.66%, O 12.94%. Prepn: Eisleb, *U.S. pat.* 2,167,351 (1939 to Winthrop); Smissman, Hite, *J. Am. Chem. Soc.* **81**, 1201 (1959). Pharmacokinetics: L. E. Mather, P. J. Meffin, *Clin. Pharmacokinet.* **3**, 352 (1978). Metabolism: S. Y. Yeh *et al.*, *J. Pharm. Sci.* **70**, 867 (1981). Toxicity data: O. W. Barlow, J. R. Lewis, *J. Pharmacol. Exp. Ther.* **103**, 147 (1951). Comprehensive description: N. P. Fish, N. J. DeAngelis in *Anal. Profiles Drug Subs.* **1**, 175-205 (1972).

Hydrochloride, C₁₅H₂₁NO₂.HCl, *Algil, Alodan, Central-gin, Demerol hydrochloride, Dispadol, Dolantin, Dolestine, Dolosal, Mefedina.* Minute crystals, mp 186-189°. Slightly bitter taste. Stable to air. Just acid to litmus (1:10 soln). Sol in water, acetone, ethyl acetate; slightly sol in alcohol, isopropanol. Insol in benzene, ether. Aq solns may be sterilized by boiling for short periods without dec. LD₅₀ orally in rats: 170 mg/kg (Barlow, Lewis).

Note: This is a controlled substance (opiate) listed in the U.S. Code of Federal Regulations, Title 21 Part 1308.12 (1995).

THERAP CAT: Analgesic (narcotic).

THERAP CAT (VET): Analgesic (narcotic); sedative; anesthetic.

5895. Mephenesin. *3-(2-Methylphenoxy)-1,2-propanediol; 3-(o-tolyloxy)-1,2-propanediol;* 1,2-dihydroxy-3-(2-methylphenoxy)propane; α-(o-tolyl)glyceryl ether; glyceryl o-tolyl ether; o-cresyl glycerol ether; cresoxypropanediol; cresoxydiol; α:β-dihydroxy-γ-(2-methylphenoxy)propane; BDH-312; Atensin; Avosyl; Avoxyl; Curythan; Daserol; Decontractyl; Dioloxol; Glyotol; Glykresin; Kinavosyl; Lissephen; Memphenesin; Mepherol; Mephesin; Mephson; Mervaldin; Myanesin; Myanol; Myodetensine; Myolysin; Myopan; Myoserol; Myoten; Oranixon; Prolax; Relaxar; Relaxil; Renarcol; Rhex; Sansdolor; Sinan; Spasmolyn; Stilalgin; Thoxidil; Tolansin; Tolcil; Tolhart; Tolosate; Toloxyn; Tolserol; Tolseron; Tolulexin; Tolulox; Tolyspaz; Walconesin. C₁₀H₁₄O₃; mol wt 182.22. C 65.92%, H 7.74%, O 26.34%. Prepn: P. Morch, *Arch. Pharm. Chem.* **54**, 327 (1947), *C.A.* **42**, 2058 (1948); **Brit. pat.** 589,821 (1947 to Carroll and Boake Roberts); of the carbamate: W. A. Lott, E. Pribyl, **U.S. pat. 2,609,386** (1952 to Squibb). Pharmacology: F. M. Berger, W. Bradley, *Brit. J. Pharmacol.* **1**, 265 (1946); and toxicity data: P. E. Dresel, I. H. Slater, *Proc. Soc. Exp. Biol. Med.* **79**, 286 (1952); A. P. Roszkowski, *J. Pharmacol. Exp. Ther* **129**, 75 (1960).

Crystals. Bitter taste. Produces numbness of the tongue. mp 70-71°. uv max (0.005% aq soln): 270 nm (E 0.395). Freely sol in alcohol, propylene glycol, chloroform. At 20° one part dissolves in 85 parts water, in 11 parts ether. Urea and its derivatives, particularly urethan, increase the water soly. One part dissolves in 60 parts of 5% urethan soln, in 40 parts of 10%, in 4.5 parts of 25%. Aq solns are stable, can be sterilized by heating, and are compatible and freely miscible with solns of sodium chloride, glucose, and derivatives of barbituric and thiobarbituric acids. pH of satd aq soln ∼6. LD₅₀ in mice, rats, hamsters (mg/kg): 471, 283, 322 i.p.; 990, 945, 821 orally (Roszkowski). LD₅₀ in mice (mM/kg): 2.83 i.p.; 10.53 orally (Dresel, Slater).

Carbamate, C₁₁H₁₅NO₄, *Tolseram.* Crystals from water, mp 93°. (Also a hemihydrate, mp 80-84°.) uv max (ethanol): 271, 277 nm (A₁cm 1% 72.7, 64.1). Has a lower water-solubility and a higher oil-solubility than mephenesin. Soly in water about 0.3%, in chloroform about 2.0%. Freely sol in alcohol. LD₅₀ in mice (mM/kg): 2.77 i.p.; 7.67 orally (Dresel, Slater).

THERAP CAT: Muscle relaxant (skeletal).

THERAP CAT (VET): Muscle relaxant.

5896. Mephenoxalone. *5-[(2-Methoxyphenoxy)methyl]-2-oxazolidinone;* 5-(o-methoxyphenoxymethyl)-2-oxazolidinone; metoxadone; methoxydon(e); methoxadone; AHR-233; OM-518; Lenetran; Trepidone; Ekilan; Dorsiflex; Control-Om; Tranpoise; Dorsilon; Placidex; Riself; Xerene. C₁₁H₁₃-NO₄; mol wt 223.23. C 59.19%, H 5.87%, N 6.27%, O 28.67%. Prepn: Lunsford *et al., J. Am. Chem. Soc.* **82**, 1166 (1960); Lunsford, **U.S. pat. 2,895,960** (1959 to A. H. Robins). Toxicity data: E. I. Goldenthal, *Toxicol. Appl. Pharmacol.* **18**, 185 (1971).

Crystals from 95% ethanol, mp 143-145°. Water insol. LD₅₀ orally in rats: 3820 ±17 mg/kg (Goldenthal).

THERAP CAT: Anxiolytic; muscle relaxant (skeletal).

5897. Mephentermine. *N,α,α-Trimethylbenzeneethanamine; N,α,α-trimethylphenethylamine;* 2-methylamino-2-methyl-1-phenylpropane; N-methyl-ω-phenyl-tert-butylamine. C₁₁H₁₇N; mol wt 163.26. C 80.93%, H 10.50%, N 8.58%. α-Adrenergic agonist. Prepd by hydrogenating 1-chloro-2-methylamino-2-methyl-1-phenylpropane hydrochloride in methanol in the presence of palladium barium carbonate: Bruce, Szabo, **U.S. pat. 2,597,445** (1952 to Wyeth). Alternate process: Abel *et al.,* **U.S. pat. 2,590,079** (1952 to Wyeth).

Liquid. Fishy amine odor. Alkaline reaction. Freely sol in alc. Sol in ether. Practically insol in water. LD₅₀ i.p. in mice: 100-110 mg/kg, *Handbook of Toxicology* **vol. 1**, W. S. Spector, Ed. (Saunders, Philadelphia, 1956) pp 184-185.

Sulfate dihydrate, (C₁₁H₁₇N)₂.H₂SO₄.2H₂O, *Wyamine, Mephine.* Crystals. One gram dissolves in 20 ml water, about 150 ml ethanol (95%). Practically insol in chloroform. pH of aq soln about 6.

THERAP CAT: Antihypotensive.

5898. Mephenytoin. *5-Ethyl-3-methyl-5-phenyl-2,4-imidazolidinedione;* 5-ethyl-3-methyl-5-phenylhydantoin; 3-methyl-5,5-phenylethylhydantoin; "methyl hydantoin"; phenylethylmethylhydantoin; 3-ethylnirvanol; methoin; Insulton; Mesontoin; Mesantoin; Phenantoin; Sedantoinal; Gerot-Epilan; Sacerno. C₁₂H₁₄N₂O₂; mol wt 218.26. C 66.04%, H 6.47%, N 12.84%, O 14.66%. Prepd from 5,5-ethylphenylhydantoin by the action of 1 mol of dimethylsulfate: **Fr. pat.** 769,667; **Swiss pat.** 166,004 (both 1934 to Sandoz). By treating phenylmethylureidoacetonitrile with HCl: **Swiss pat.** 179,692 (1935 to Sandoz).

Crystals, mp 136-137°. Insol in water. Forms a water-soluble sodium salt which has an alkaline reaction. LD₁₀₀ i.p. in rats: 270 mg/kg.

THERAP CAT: Anticonvulsant.

5899. Mephobarbital. *5-Ethyl-1-methyl-5-phenyl-2,4,6-(1H,3H,5H)-pyrimidinetrione; 5-ethyl-1-methyl-5-phenylbarbituric acid;* 5-phenyl-5-ethyl-3-methylbarbituric acid; N-methylethylphenylbarbituric acid; methylphenobarbital;

Phemiton; Prominal; Mebaral; Isonal. $C_{13}H_{14}N_2O_3$; mol wt 246.27. C 63.40%, H 5.73%, N 11.38%, O 19.49%. Prepn: Taub, Kropp, **Ger. pat. 537,366** (1929 to I. G. Farben.).

White, tasteless crystals. mp 176°. Slightly sol in cold, freely in hot water or in alcohol.

Caution: May be habit forming. This is a controlled substance (depressant) listed in the U.S. Code of Federal Regulations, Title 21 Parts 329.1 and 1308.14 (1995).

THERAP CAT: Anticonvulsant; sedative; hypnotic.

THERAP CAT (VET): Anticonvulsant; sedative; hypnotic.

5900. Mephosfolan. *(4-Methyl-1,3-dithiolan-2-ylidene)-phosphoramidic acid diethyl ester; phosphonodithioimidocarbonic acid cyclic propylene P,P-diethyl ester;* 2-diethoxyphosphinylimino-4-methyl-1,3-dithiolane; cyclic propylene (diethoxyphosphinyl)dithioimidocarbonate; EI-47470; ENT-25991; Cytrolane. $C_8H_{16}NO_3PS_2$; mol wt 269.33. C 35.68%, H 5.99%, N 5.20%, O 17.82%, P 11.50%, S 23.81%. Insecticidal contact and stomach poison with systemic activity: D. L. Bull *et al., J. Econ. Entomol.* **57**, 112 (1964). Prepn: J. B. Lovell, **U.S. pat. 3,197,365** (1965 to Am. Cyanamid). Metabolism: J. Zulalian, R. C. Blinn, *J. Agr. Food Chem.* **25**, 1033 (1977). Photodegradation: C. C. Ku *et al., ibid.* **27**, 1046 (1979).

Yellow to amber liquid, $bp_{0.001}$ 120°. n_D^{26} 1.5354. Soly in water at 25°: 57 g/kg. Sol in acetone, ethanol, benzene, 1,2-dichloroethane. Stable at neutral pH; hydrolyzed by acid or alkali.

USE: Insecticide, acaricide.

5901. Mepindolol. *1-[(1-Methylethyl)amino]-3-[(2-methyl-1H-indol-4-yl)oxy]-2-propanol; 1-(isopropylamino)-3-[(2-methylindol-4-yl)oxy]-2-propanol.* $C_{15}H_{22}N_2O_2$; mol wt 262.35. C 68.67%, H 8.45%, N 10.68%, O 12.20%. β-Adrenergic blocker; the 2-methyl analog of pindolol, *q.v.* Prepn: F. Troxler, **Swiss pats. 469,002 and 472,404** (both 1969 to Sandoz); F. Seeman *et al., Helv. Chim. Acta* **54**, 2411 (1971). Pharmacokinetics: R. Gugler *et al., Arzneimittel-Forsch.* **25**, 1067 (1975). Pharmacodynamics: J. Bonelli *et al., Eur. J. Clin. Pharmacol.* **15**, 1 (1979). Clinical studies: H. M. Beumer *et al., Int. J. Clin. Pharmacol. Biopharm.* **16**, 249 (1978); M. Sukerman *et al., Curr. Ther. Res.* **25**, 384 (1979).

Crystals from ethyl acetate, mp 100-102° (Seeman). Also reported as 95-97° (Troxler).

Sulfate salt, $(C_{15}H_{22}N_2O_2)_2 \cdot H_2SO_4$, *SH-E-222, Betagon, Corindolan, Mepicor.*

THERAP CAT: Antihypertensive; antianginal.

5902. Mepiprazole. *1-(3-Chlorophenyl)-4-[2-(5-methyl-1H-pyrazol-3-yl)ethyl]piperazine;* 3-[2-(N′-m-chlorophenyl-piperazino)ethyl]-5-methylpyrazole; EMD-16923. $C_{16}H_{21}$-ClN_4; mol wt 304.82. C 63.05%, H 6.94%, Cl 11.63%, N

18.38%. Psychotropic agent with CNS-depressant effects. Prepn: K. Volker *et al.,* **Brit. pat. 1,124,710** corresp to U.S. pat. 3,491,097 (1968, 1970 both to E. Merck); *eidem, Eur. J. Med. Chem.* **10**, 154, 162 (1975). Pharmacological study: M. A. Ruch-Monachon *et al., Arch. Int. Pharmacodyn. Ther.* **219**, 326 (1976). Effects on brain biogenic amine metabolism: C. Seyfried *et al., Arzneimittel-Forsch.* **26**, 1088 (1976). EEG effects in rabbits: S. Watanabe *et al., ibid.* **29**, 274 (1979).

Crystals, mp 106°.

Dihydrochloride, $C_{16}H_{21}ClN_4 \cdot 2HCl$, *H-4007, Psigodal.* Cryst from ethanol, mp 234°.

THERAP CAT: Tranquilizer.

5903. Mepiquat Chloride. *1,1-Dimethylpiperidinium chloride;* BAS-083; BAS-85559X; Pix. $C_7H_{16}ClN$; mol wt 149.66. C 56.18%, H 10.78%, Cl 23.69%, N 9.36%. Prepn: B. Zeeh *et al.,* **Ger. pat. 2,207,575** corresp to U.S. pat. 3,905,798 (1973, 1975 to BASF).

mp > 350°. LD_{50} orally in rats: 1420 mg/kg, *RTECS* Vol. II, R. J. Lewis, R. L. Tatken, Eds. (1980) p 390.

USE: Plant growth regulator.

5904. Mepitiostane. *(2α,3α,5α,17β)-2,3-Epithio-17-[(1-methoxycyclopentyl)oxy]androstane; cyclopentanone 2α,3α-epithio-5α-androstan-17β-yl methyl acetal;* 10364-S; Thiodelone; Thioderon. $C_{25}H_{40}O_2S$; mol wt 404.66. C 74.20%, H 9.96%, O 7.91%, S 7.92%. Orally active deriv of epitiostanol, *q.v.* Prepn: T. Komeno, **S. Afr. pat. 68 00,565** corresp to U.S. pat. 3,567,713 (1968, 1971 both to Shionogi); T. Komeno *et al., Shionogi Kenkyusho Nempo* **19**, 3 (1969), *C.A.* **72**, 28554u (1970). Inhibitory effect on mammary tumors in rats: O. Takatani, S. Kumaoka, *Gann* **68**, 337 (1977), *C.A.* **87**, 112187u (1977). Clinical study in advanced breast cancer: K. Inoue *et al., Cancer Treat. Rep.* **62**, 743 (1978). Toxicity studies: *Oyo Yakuri* **16**, 739-812 (1978), *C.A.* **90**, 146152f, 146153g (1979).

Cryst, mp 98-101°. $[\alpha]_D^{20}$ +22.5 ±0.5° (c = 1 in chloroform).

THERAP CAT: Antineoplastic.

5905. Mepivacaine. *N-(2,6-Dimethylphenyl)-1-methyl-2-piperidinecarboxamide; 1-methyl-2′,6′-pipecoloxylidide; dl-N-methylpipecolic acid 2,6-dimethylanilide; dl-N-methylhexahydropicolinic acid 2,6-dimethylanilide.* $C_{15}H_{22}N_2O$; mol wt 246.35. C 73.13%, H 9.00%, N 11.37%, O 6.49%. Prepn: Ekenstam *et al., Acta Chem. Scand.* **11**, 1183 (1957);

U.S. pat. **2,799,679** (1957 to A. B. Bofors). Prepn of other salts: Rinderknecht, *Helv. Chim. Acta* **42**, 1324 (1959); **Brit. pat. 826,668** (1960 to Crookes Labs.). Resolution of isomers: Tullar, *J. Med. Chem.* **14**, 891 (1971); Friberger, Aberg, *Acta Pharm. Suecica* **8**, 361 (1971). Pharmacology: Helmy *et al.*, *J. Egypt. Med. Assoc.* **50**, 688 (1967). Metabolism: Reynolds, *Brit. J. Anaesth.* **43**, 33 (1971). Toxicity data: G. Aberg, *Acta Pharmacol. Toxicol.* **31**, 273 (1972).

Crystals from ether, mp 150-151°.

Hydrochloride, $C_{15}H_{22}N_2O \cdot HCl$, *Carbocaina, Carbocaine hydrochloride, Chlorocain, Meaverin, Mepicaton, Mepident, Mepivastesin, Optocain, Scandicain.* mp 262-264°. Sol in water. LD_{50} in mice, rats (mg/kg): 280, 500 s.c. (Aberg). (+)-Form, *dexivacaine.*

THERAP CAT: Anesthetic (local).

THERAP CAT (VET): Anesthetic (local).

5906. Mepixanox. *3-Methoxy-4-(1-piperidinylmethyl)-9H-xanthen-9-one;* 3-methoxy-4-piperidinylmethyl-9-oxo-10-oxa-9,10-dihydroanthracene; mepixanthone; Pimexone. $C_{20}H_{21}NO_3$; mol wt 323.39. C 74.28%, H 6.55%, N 4.33%, O 14.84%. Prepn: M. Sparaci, **S. Afr. pat. 69 02,150** (1969 to Mondi), *C.A.* **72**, 111,300d (1970). Alternate process: D. Milani, U.S. pat. **3,646,030** (1972 to Mondi). CNS activity: P. Da Re *et al.*, *J. Med. Chem.* **13**, 527 (1970). Pharmacology: R. Guira *et al.*, *Minerva Pneumolog.* **21**, 317 (1982). Pharmacokinetics in man: G. Grossi *et al.*, *Curr. Ther. Res.* **38**, 141 (1985). HPLC determn in serum: G. Grossi *et al.*, *J. Chromatog.* **309**, 214 (1984). Clinical comparison with doxapram, *q.v.*, in chronic bronchopulmonary disease: M. Parziale *et al.*, *G. Ital. Mal. Torace* **38**, 323 (1984). Clinical evaluations in chronic obstructive lung disease: D. Olivieri *et al.*, *Arch. Monaldi Tisiol. Mal. Appar. Respir.* **35**, 117 (1980); L. Bertoli *et al.*, *Minerva Pneumolog.* **23**, 323 (1984).

White crystalline powder from ethyl acetate, mp 159-160°. LD_{50} i.p. in mice: 70.73 mg/kg (Sparaci).

Hydrochloride, $C_{20}H_{22}ClNO_3$, white crystalline solid from ethanol, mp > 200° (dec).

THERAP CAT: Respiratory stimulant.

5907. Meprednisone. *(16β)-17,21-Dihydroxy-16-methylpregna-1,4-diene-3,11,20-trione;* 16β-methylprednisone; Betapred; Betapar; Deltacortene Beta; Betalone; Deltisona B. $C_{22}H_{28}O_5$; mol wt 372.46. C 70.94%, H 7.58%, O 21.48%. Prepn: E. P. Oliveto *et al.*, *J. Am. Chem. Soc.* **80**, 4428 (1958); D. Taub *et al.*, *ibid.* 4435; D. Taub *et al.*, *ibid.* **82**, 4012 (1960); G. G. Nathansohn *et al.*, *Experientia* **17**, 448 (1961); **Brit. pat. 901,092** (1962 to Scherico); R. Rausser, E. P. Oliveto, U.S. pat. **3,164,618** (1965 to Schering).

Crystals, mp 200-205°. $[\alpha]_D$ +200° (dioxane). uv max (methanol): 239 nm ($E_{1cm}^{1\%}$ 416).

21-Acetate, $C_{24}H_{30}O_6$, mp 232-235°. $[\alpha]_D$ +210° (dioxane). uv max (methanol): 238 nm ($E_{1cm}^{1\%}$ 358).

THERAP CAT: Glucocorticoid.

5908. Meprobamate. *2-Methyl-2-propyl-1,3-propanediol dicarbamate; carbamic acid 2-methyl-2-propyltrimethylene ester;* 2,2-di(carbamoyloxymethyl)pentane; 2-methyl-2-propyltrimethylene carbamate; procalmadiol; procalmidol; Amosene; Andaxin; Aneural; Artolon; Atraxin; Ayeramate; Bamo 400; Biobamat; Calmiren; Cirpon; Cyrpon; Ecuanil; Equanil; Fas-Cile 200; Gadexyl; Holbamate; Kesso-Bamate; Klort; Mar-Bate; Mepavlon; Meposed; Meprin; Meprindon; Meproban; Meprocompren; Meprol; Meprospan; Meprotabs; Meproten; Meprotil; Meptran; Mesmar; Miltaun; Miltown; Morbam; My-trans; Nervonus; Oasil; Panediol; Perequil; Pertranquil; Placidon; Probamyl; Promate; Quaname; Quanil; Reostral; Restenil; Setran; Sowell; Tamate; Trankvilan; Tranlisant; Tranquilan; Tranquiline; Urbilat. $C_9H_{18}N_2O_4$; mol wt 218.25. C 49.53%, H 8.31%, N 12.84%, O 29.32%. Prepn: Ludwig, Piech, *J. Am. Chem. Soc.* **73**, 5779 (1951); Berger, Ludwig, U.S. pat. **2,724,720** (1955 to Carter Prod.). Synthesis of 2-methyl-2-propyl-1,3-propanediol: Fries, Mönkemeyer, **Swiss pat. 373,026** (1963 to Chemische Werke Hüls). Pharmacology: F. M. Berger, *J. Pharmacol. Exp. Ther.* **112**, 413 (1954). Comprehensive description: C. Shearer, P. Rulon, *Anal. Profiles Drug Subs.* **1**, 207-232 (1972); C. Shearer, *ibid.* **11**, 587-591 (1982).

Crystals from hot water, mp 104-106°. Characteristic bitter taste. Soly in water at 20°: 0.34% (w/w), at 37°: 0.79% (w/w). Easily forms supersatd solns with hot water. Freely sol in most organic solvents. Aq solns are neutral. The substance is stable in dilute acid and alkali and is not broken down in gastric or intestinal juices. LD_{50} i.p. in mice: 800 mg/kg (Berger).

Note: This is a controlled substance (depressant) listed in the U.S. Code of Federal Regulations, Title 21 Part 1308.14 (1995).

THERAP CAT: Anxiolytic.

5909. Meprylcaine. *2-Methyl-2-(propylamino)-1-propanol benzoate (ester);* β-(n-propylamino)-β,β-dimethylethyl benzoate; 2-methyl-2-propylaminopropyl benzoate; Oracaine. $C_{14}H_{21}NO_2$; mol wt 235.33. C 71.46%, H 8.99%, N 5.95%, O 13.60%. Prepn: Reasenberg, U.S. pat. **2,767,207** (1956 to Mizzy).

Oil. Practically insol in water. Sol in alcohol, ether, acetone, oils.

Hydrochloride, $C_{14}H_{21}NO_2 \cdot HCl$, crystals, mp 150-151°. Sol in water, alcohol. pH of 1:50 aq soln: 5.7.

THERAP CAT: Anesthetic (local).

5910. Meptazinol. *3-(3-Ethylhexahydro-1-methyl-1H-azepin-3-yl)phenol;* 1-methyl-3-ethyl-3-(*m*-hydroxyphenyl)-hexahydro-1*H*-azepine. $C_{15}H_{23}NO$; mol wt 233.35. C 77.21%, H 9.93%, N 6.00%, O 6.86%. Mixed opioid agonist-antagonist. Prepn of the hydrobromide: J. F. Cavalla, A. C. White, Ger. pat. **1,941,534** corresp to U.S. pat. **3,729,-465** (1970, 1973 both to Wyeth); of the free base: *eidem,* U.S. pat. **4,197,241** (1980 to Wyeth). Properties: P. G. Goode, A. C. White, *Brit. J. Pharmacol.* **43,** 462 (1971). HPLC determn in plasma: T. Frost, *Analyst* **106,** 999 (1981). Pharmacokinetics in man: G. Davies, *Eur. J. Clin. Pharmacol.* **23,** 535 (1982). Human pharmacology and abuse potential: R. E. Johnson, D. R. Jasinski, *Clin. Pharmacol. Ther.* **41,** 426 (1987). Clinical studies: R. K. Price, A. N. Latham, *Curr. Med. Res. Opin.* **8,** 54 (1982); C. E. Parker, A. F. Langrick, *J. Int. Med. Res.* **10,** 408 (1982). Symposium on clinical studies: *Postgrad. Med. J.* **59,** Suppl. 1, 1-94 (1983). Review of pharmacology and clinical efficacy: B. Holmes, A. Ward, *Drugs* **30,** 285-312 (1985).

Cryst from acetonitrile, mp 127.5-133°. Hydrochloride, $C_{15}H_{23}NO \cdot HCl$, *Wy-22811, Meptid.*

THERAP CAT: Analgesic (narcotic).

5911. Mequitazine. *10-(1-Azabicyclo[2.2.2]oct-3-yl-methyl)-10H-phenothiazine;* 10-(3-quinuclidinylmethyl)phenothiazine; LM-209; Butix; Metaplexan; Mircol; Primalan; Zesulan. $C_{20}H_{22}N_2S$; mol wt 322.47. C 74.49%, H 6.88%, N 8.69%, S 9.94%. Prepn: G. Gueremy *et al.,* Ger. pat. **2,009,-555** corresp to U.S. pat. **3,987,042** (1970, 1976 to Sogeras). Absorption, distribution, and excretion: A. Uzan *et al., Xenobiotica* **6,** 633 (1976). Biotransformation: *eidem, ibid.* 649. Pharmacologic study: *eidem, J. Pharm. Pharmacol.* **31,** 701 (1979). Clinical studies: J. Blamoutier, *Curr. Med. Res. Opin.* **5,** 366 (1978); P. Laugier, M. Orusco, *ibid.* 371.

Crystals from acetonitrile, mp 130-131°.

THERAP CAT: Antihistaminic.

5912. Meralein Sodium. *(3',6'-Dihydroxy-2',7'-diiodo-spiro[3H-2,1-benzoxanthiole-3,9'-[9H]xanthen]-4'-yl)hydroxymercury S,S-dioxide monosodium salt;* o-[6-hydroxy-5-(hydroxymercuri)-2,7-diiodo-3-oxo-3*H*-xanthen-9-yl]benzenesulfonic acid sodium salt; 2,7-diiodo-4-hydroxymercuriresorcinsulfonphthalein monosodium salt; monohydroxymercuri-diiodoresorcinsulfonphthalein sodium salt; sodium meralein; Merodicein. $C_{19}H_9HgI_2NaO_7S$; mol wt 858.73. C 26.58%, H 1.06%, Hg 23.36%, I 29.56%, Na 2.68%, O 13.04%, S 3.73%. Prepn: Dunning, Farinholt, *J. Am. Chem. Soc.* **51,** 804 (1929). Pharmacology and toxicology: Macht, Cook, *J. Pharmacol. Exp. Ther.* **43,** 571 (1931).

Green scales, turn dark red on pulverizing. Sol in water; aq soln slightly fluorescent.

THERAP CAT: Anti-infective (topical).

5913. Meralluride. *[3-[[[(3-Carboxylato-1-oxopropyl)-amino]carbonyl]amino]-2-methoxypropyl]hydroxymercurate-(1—) sodium, mixture with 3,7-dihydro-1,3-dimethyl-1H-purine-2,6-dione;* N-[[2-methoxy-3-[(1,2,3,6-tetrahydro-1,3-dimethyl-2,6-dioxopurin-7-yl)mercuri]propyl]carbamoyl]-succinamic acid; Dilurgen; Mercardan; Mercuhydrin; Mercuretin. $C_{16}H_{23}HgN_6NaO_8$; mol wt 650.97. C 29.52%, H 3.56%, Hg 30.81%, N 12.91%, Na 3.53%, O 19.66%. Structure: Pearson, Sigal, *J. Org. Chem.* **15,** 1055 (1950). Toxicity data: E. I. Goldenthal, *Toxicol. Appl. Pharmacol.* **18,** 185 (1971).

White to slightly yellow powder, slowly dec on exposure to light. Saturated soln is acid to litmus. Slightly sol in water; sol in hot water, glacial acetic acid, and solns of alkali hydroxides. Almost insol in alc, chloroform, ether. LD_{50} s.c. in rats: 28±7 mg/kg (Goldenthal).

THERAP CAT: Diuretic.

THERAP CAT (VET): Diuretic.

5914. Merbromin. *(2'7'-Dibromo-3',6'-dihydroxy-3-oxospiro[isobenzofuran-1(3H),9'-[9H]xanthen]-4'-yl)hydroxymercury disodium salt; [2,7-dibromo-9-(o-carboxyphenyl)-6-hydroxy-3-oxo-3H-xanthen-4-yl]hydroxymercury disodium salt;* mercurochrome; dibromohydroxymercurifluorescein disodium salt; no. 220 sol; Mercurochrome-220 Soluble; Chromargyre; Planochrome; Flavurol; D.O.M.F.; Mercurophage; Mercurocol; Gallochrome; Gynochrome; Mercurome; Asceptichrome; Mercuranine. $C_{20}H_8Br_2HgNa_2O_6$; mol wt 750.66. C 32.00%, H 1.07%, Br 21.29%, Hg 26.72%, Na 6.13%, O 12.79%. Prepd by treating dibromofluorescein with mercuric acetate and sodium hydroxide: White, *J. Am. Chem. Soc.* **42,** 2355 (1920); also prepd by the action of mercuric acetate on dibromofluorescein sodium: Rymill, Corran, *Quart. J. Pharm. Pharmacol.* **7,** 543 (1934), see also U.S. pat. **1,535,003** (1925).

Trihydrate, iridescent green scales or granules. Freely sol in water, giving a carmine-red soln. Very dil solns (1:2000) possess a yellow-green fluorescence. pH of 0.5% soln: 8.8. One gram dissolves in 50 grams of 94% alcohol, in 8.1 grams methanol: Denoel, *J. Pharm. Belg.* **22,** 423 (1940); **23,** 75 (1941). Practically insol in alcohol, acetone, chloroform, ether. *Incompat.* Acids, most alkaloidal salts and most local anesthetics. Colors the skin carmine-red. The stains may be removed by washing first with permanganate soln and then with oxalic acid soln.

THERAP CAT: Antibacterial.

THERAP CAT (VET): Antiseptic.

5915. Mercamphamide. *3-[[3-(Hydroxymercuri)-2-methoxypropyl]carbamoyl]-1,2,2-trimethylcyclopentanecarboxylic acid;* N-[3-(hydroxymercuri)-2-methoxypropyl]camphoramic acid; N-(γ-hydroxymercuri-β-methoxy)propylcamphoramic acid. $C_{14}H_{25}HgNO_5$; mol wt 487.95. C 34.46%, H 5.16%, Hg 41.11%, N 2.87%, O 16.39%. Prepn: Molnar, U.S. pat. 2,117,901 (1938); Werner, Scholz, *J. Am. Chem. Soc.* **76**, 2453 (1954).

Sodium salt, $C_{14}H_{24}HgNNaO_5$. Prepd as a mixture of 20% of the free acid plus 80% of the sodium salt. Mixture contains about 40% Hg and is described as a white, bitter powder; very slightly sol in water or ether; sol in alcohol and dil solns of sodium hydroxide. An aq suspension is practically neutral. pH about 8.

Mixture of sodium salt with theophylline, *mercurophyllin, Novurit.*

THERAP CAT: Diuretic.

5916. 2-Mercaptobenzothiazole. *2(3H)-Benzothiazolethione;* 2-benzothiazolethiol; MBT; Captax; Dermacid; Mertax; Thiotax. $C_7H_5NS_2$; mol wt 167.26. C 50.27%, H 3.01%, N 8.37%, S 38.34%. Prepd industrially by reacting aniline, carbon disulfide, and sulfur at elevated press. and temps: Kelly, U.S. pat. 1,631,871. Purification by treatment with a per-acid salt in alkaline medium: Weyker, Ebel, U.S. pat. 2,730,528 (1956 to Am. Cyanamid). Improved process: Szlatinay, U.S. pat. 3,031,073 (1962 to Monsanto).

Pale yellow, monoclinic needles or leaflets. Disagreeable odor. d 1.42. mp 180.2-181.7° (the technical product mp 170-175°). Practically insol in water. Soly at 25° (g/100 ml) in alcohol: 2.0; ether 1.0; acetone 10.0; benzene 1.0; carbon tetrachloride <0.2; naphtha <0.5. Moderately sol in glacial acetic acid. Sol in alkalies and alkali carbonate solns.

Zinc salt, *Bantex.* Light yellow powder, d_4^{25} 1.70.

Sodium salt, *Nuodex 84.*

USE: Rubber vulcanization accelerator. Salts used as fungicide.

5917. 2-Mercaptoethanol. *β-Mercaptoethanol;* 2-hydroxy-1-ethanethiol; 2-hydroxyethyl mercaptan; monothioethyleneglycol; thioglycol. C_2H_6OS; mol wt 78.14. C 30.74%, H 7.74%, O 20.48%, S 41.04%. $SHCH_2CH_2OH$. Prepn: Woodward, *J. Chem. Soc.* **1948**, 1892; Peppel, Signaigo; Jones, U.S. pats. 2,402,665; 3,493,192 (1946, 1968 both to du Pont). Properties: Bennett, *J. Chem. Soc.* **121**, 2139 (1922). IR: Thompson, *J. Am. Chem. Soc.* **61**, 1398 (1939). Toxicology: K. White *et al., J. Pharm. Sci.* **62**, 237 (1973).

Liquid with very strong, disagreeable odor. bp_{742} 157-158° (dec). d_4^{20} 1.1143. n_D^{20} 1.4996. The pure liquid is miscible with water, alc, ether and benzene. LD_{50} in mice (mg/kg): 322.0 i.p.; 344.8 orally (White).

Caution: Irritating to eyes, nose and skin.

USE: In organic synthesis; as biochemical research tool.

5918. Mercaptomerin Sodium. *(T-4)-[3-[[(3-Carboxylato-2,2,3-trimethylcyclopentyl)carbonyl]amino]-2-methoxypropyl][mercaptoacetato(2→)-O,S]mercurate(2→) disodium;* N-(γ-carboxymethylmercaptomercuri-β-methoxy)propylcamphoramic acid disodium salt; Diucardyn sodium; Thiomerin

sodium. $C_{16}H_{25}HgNNa_2O_6S$; mol wt 606.01. C 31.71%, H 4.16%, Hg 33.10%, N 2.31%, Na 7.59%, O 15.84%, S 5.29%. Prepn: Lehman, U.S. pat. 2,576,349 (1951 to Wyeth); Wendt, Bruce, *J. Org. Chem.* **23**, 1448 (1958); Wendt, U.S. pat. 2,834,795 (1958 to Am. Home Prod.).

Hygroscopic white powder. Dec 150-155°. Freely soluble in water. Soluble in alcohol. Practically insoluble in ether, benzene, chloroform.

THERAP CAT: Diuretic.

THERAP CAT (VET): Diuretic.

5919. 6-Mercaptopurine. *1,7-Dihydro-6H-purine-6-thione; purine-6-thiol;* 6MP; Leukerin; Mercaleukin; Puri-Nethol; Purinethol. $C_5H_4N_4S$; mol wt 152.18. C 39.46%, H 2.65%, N 36.82%, S 21.07%. Prepd from hypoxanthine and phosphorus pentasulfide: G. B. Elion *et al., J. Am. Chem. Soc.* **74**, 411 (1952); G. H. Hitchings, G. B. Elion, U.S. pat. 2,697,709 (1954 to Burroughs Wellcome). Improved procedure: A. G. Beaman, R. K. Robins, *J. Am. Chem. Soc.* **83**, 4042 (1961). Prepn from 7-aminothiazolo[5,4-d]pyrimidine: G. H. Hitchings, G. B. Elion, U.S. pat. 2,721,866 (1955 to Burroughs Wellcome). Metabolized in the body to 6-thiouric acid (6-mercapto-2,8-purinediol): G. B. Elion *et al., J. Am. Chem. Soc.* **81**, 3042 (1959). Pharmacology: H. Froberg, M. S. Schencking, *Arch. Toxicol.* **32**, 1 (1974). HPLC determn in body fluids: J. L. Rudy *et al., Ann. Clin. Biochem.* **25**, 504 (1988). Clinical trial with methotrexate in acute lymphoblastic leukemia: S. Koizumi *et al., Cancer* **61**, 1292 (1988). Comprehensive description: S. A. Benezra, P. R. B. Foss in *Analytical Profiles of Drug Substances* vol. 7, K. Florey, Ed. (Academic Press, New York, 1978) pp 343-357. Review in treatment of Crohn's disease and ulcerative colitis: D. H. Present, *Gastroenterol. Clin. North Am.* **18**, 57-71 (1989); in treatment of leukemia: B. A. Kamen, *Semin. in Hematol.* **28**, 12-14 (1991).

Monohydrate, yellow prisms from water. Becomes anhydrous at 140°, dec 313-314°. uv max (0.1N NaOH): 230, 312 nm (ε 14000, 19600); (0.1N HCl): 222, 327 nm (ε 9240, 21300); (methanol): 216, 329 nm (ε 8940, 19300). pKa_1 7.77, pKa_2 11.17. Insol in water, acetone, ether. Sol in hot ethanol, in alkaline solns with slow decompn. LD_{50} in mice, hamsters (mg/kg): 157, 364 i.p. (Froberg, Schencking).

THERAP CAT: Antineoplastic; immunosuppressant.

THERAP CAT (VET): Antineoplastic.

5920. Mercufenol Chloride. *o-(Chloromercuri)phenol;* o-hydroxyphenylmercuric chloride. C_6H_5ClHgO; mol wt 329.15. C 21.89%, H 1.53%, Cl 10.77%, Hg 60.94%, O 4.86%. Prepd from phenol and mercuric acetate: Whitmore, Hansen, *Org. Syn. coll. vol.* I, 161 (1941); Kaplan, Mellick, U.S. pat. 2,502,382 (1950 to Edwal Labs.).

Feathery crystals, mp 150-152°. Slightly sol in cold water, moderately sol in boiling water; freely sol in alcohol and hot benzene, sparingly sol in chloroform. *Poisonous!*

USE: Disinfectant.

5921. Mercumallylic Acid. *[3-(3-Carboxy-2-oxo-2H-1-benzopyran-8-yl)-2-methoxypropyl]hydroxymercurate(1→) hydrogen; [3-[3-(hydroxymercuri)-2-methoxypropyl]salicylid-*

ene]malonic acid δ-lactone; 8-[3-(hydroxymercuri)-2-meth-oxypropyl]-2-oxo-2H-1-benzopyran-3-carboxylic acid; 8-(2-methoxy-3-hydroxymercuripropyl)coumarin-3-carboxylic acid; 8-[3-(hydroxymercuri)-2-methoxypropyl]-3-carboxy-coumarin; 8-[3-(hydroxymercuri)-2-methoxypropyl]couma-rin-3-carboxylic acid. $C_{14}H_{14}HgO_6$; mol wt 478.85. C 35.12%, H 2.95%, Hg 41.89%, O 20.05%. Prepn: A. Schlesinger et al., U.S. pat. 2,667,442 (1954 to Endo); L. H. Werner, C. R. Scholz, J. Am. Chem. Soc. 76, 2453 (1954). Toxicology of mercumatalin sodium: H. Blumberg et al., J. Pharmacol. Exp. Ther. 105, 336 (1952).

Powder, bitter taste. mp 155-160° (Schlesinger); mp 197° (Werner, Scholz). One gram dissolves in about 4.2 parts of 1N NaOH. Slightly sol in acetic acid. Very slightly sol in water, alcohol, chloroform. Practically insol in ether.

Sodium salt compd with theophylline, $C_{21}H_{21}HgN_4NaO_8$, *mercumatilin sodium, Cumertilin Sodium*. Dry, light yellow powder. Very sol in water. LD_{50} in rats (mg Hg/kg): 9.8 i.v.; 238 orally (Blumberg).

THERAP CAT: Diuretic.

5922. Mercuric Acetate. $C_4H_6HgO_4$; mol wt 318.68. C 15.08%, H 1.90%, Hg 62.94%, O 20.08%. $Hg(CH_3COO)_2$. Prepd from HgO and acetic acid: Wagenknecht, Juza in *Handbook of Preparative Inorganic Chemistry*, vol. 2, G. Brauer, Ed. (Academic Press, New York, 2nd ed., 1965) pp 1120-1121.

Crystals or cryst powder; slight acetic odor; sensitive to light. *Poisonous!* d 3.28. mp 178-180° (overheating results in decompn). One g dissolves in 2.5 ml cold, 1 ml boiling water; sol in alcohol. *Keep well closed and protected from light.* Aq solns decomp on standing, yielding a yellow ppt.

USE: Chiefly for mercuration of organic compounds; for the absorption of ethylene.

5923. Mercuric Arsenate. $AsHHgO_4$; mol wt 340.52. As 22.00%, H 0.30%, Hg 58.91%, O 18.79%. $HgHAsO_4$. Yellow powder. *Poisonous!* Insoluble in water. Sol in HCl or HNO_3.

5924. Mercuric Benzoate. $C_{14}H_{10}HgO_4$; mol wt 442.82. C 37.97%, H 2.28%, Hg 45.30%, O 14.45%. $Hg(C_7H_5O_2)_2$.

Monohydrate, odorless, cryst powder; sensitive to light. *Poisonous!* Sol in 90 parts cold, 40 parts boiling water; freely sol in NaCl soln; slightly sol in alcohol. When boiled with water or alc it hydrolyzes to a basic salt and free benzoic acid. *Protect from light.*

THERAP CAT: Formerly as antisyphilitic.

5925. Mercuric Bromide. Br_2Hg; mol wt 360.40. Br 44.34%, Hg 55.66%. $HgBr_2$. Usually prepd from the elements: Jander, Brodersen, Z. Anorg. Chem. 261, 261 (1950).

White crystals or cryst powder; sensitive to light. *Poisonous!* d 6.05. mp 237°; sublimes at higher temp. Sol in about 200 parts cold, 25 parts boiling water; freely sol in hot alcohol, in methanol, HCl, HBr, alkali bromide solns; slightly sol in chloroform. *Protect from light.*

5926. Mercuric Chloride. Mercury bichloride; corrosive sublimate; mercury perchloride; corrosive mercury chloride. Cl_2Hg; mol wt 271.50. Cl 26.12%, Hg 73.88%. $HgCl_2$.

Crystals or white granules or powder. *Violent poison! May be fatal if swallowed.* d 5.4. mp 277°; volatilizes unchanged at about 300°; also slightly volatile at ordinary temp; appreciably so at 100°. One gram dissolves in 13.5 ml water (soly is increased by HCl or alkali chlorides), in 2.1 ml boiling water, 3.8 ml alc, 1.6 ml boiling alc, 200 ml benzene, 22 ml ether, 12 ml glycerol, 40 ml acetic acid; also sol in methanol, acetone, ethyl acetate; slightly sol in carbon di-

sulfide, pyridine. pH about 4.7. Also reported as 3.2 for 0.2 molar aq soln. Coagulates albumin; produces with NaOH a yellow ppt (difference from calomel, which turns black). *Incompat:* Formates, sulfites, hypophosphites, phosphates, sulfides, albumin, gelatin, alkalies, alkaloid salts, ammonia, lime water, antimony and arsenic, bromides, borax, carbonates; reduced iron; copper, iron, lead, silver salts; infusions of cinchona, columbo, oak bark or senna; tannic acid; vegetable astringents.

Caution: Highly toxic. Corrosive to mucous membranes. Ingestion may cause severe nausea, vomiting, hematemesis, abdominal pain, diarrhea, melena, renal damage, prostration. 1 or 2 g is frequently fatal. Poisoning and death also have occurred from intrauterine douches and application of alcoholic soln to large areas of skin. Do not breathe dust. Keep away from feed or food products. Wash hands before eating or smoking. *Antidote:* Dimercaprol (BAL). Review of clinical toxicology of mercury and its compounds: H. B. Gerstner, J. E. Huff, J. Toxicol. Environ. Health 2, 491-526 (1977).

USE: Preserving (kyanizing) wood and anatomical specimens; also embalming; disinfecting; browning and etching steel and iron; intensifier in photography; white reserve in fabric printing; tanning leather; electroplating aluminum; depolarizer for dry batteries; freeing gold from lead; magic photograms; mordant for rabbit and beaver furs; staining wood and vegetable ivory pink; manuf of ink for mercurography; treating seed potatoes; manuf other mercury compds. As an important reagent in anal. chemistry.

THERAP CAT: Topical antiseptic, disinfectant.

THERAP CAT (VET): Caustic, antiseptic, disinfectant.

5927. Mercuric Chloride, Ammoniated. *Mercury amide chloride;* aminomercuric chloride; mercury ammonium chloride; ammoniated mercury; white precipitate; white mercuric precipitate. ClH_2HgN; mol wt 252.07. Cl 14.06%, H 0.80%, Hg 79.58%, N 5.56%. $HgNH_2Cl$. Prepn: Sen, Z. Anorg. Allgem. Chem. 33, 197 (1903); Wagenknecht, Juza in *Handbook of Preparative Inorganic Chemistry*, vol 2, G. Brauer, Ed. (Academic Press, New York, 2nd ed., 1965) p 1114.

Odorless powder; d 5.38; earthy, styptic, metallic taste. *Poisonous!* Volatile without melting at dull red heat. Insol in water or alcohol. Sol in warm HCl, HNO_3, acetic acid, in cold soln ammonium carbonate or sodium thiosulfate. *Incompat:* Not to be used in conjunction with sulfur or iodine.

Caution: May produce allergic dermatitis; prolonged use may cause local pigmentation of skin or eyelids. Absorption following vigorous dermal application has been reported with resulting ptyalism. Oral ingestion causes epigastric pain, nausea, purging. *Antidote:* Dimercaprol (BAL).

Note: French "White Precipitate" (Precipité Blanc) is calomel. Do not confuse the two when French prescriptions are filled.

THERAP CAT: Anti-infective (topical).

THERAP CAT (VET): Anti-infective (topical).

5928. Mercuric Cyanide. Cianurina. C_2HgN_2; mol wt 252.63. C 9.51%, Hg 79.40%, N 11.09%. $Hg(CN)_2$. Obtained by evaporating a soln of HgO in aq HCN: Biltz, Z. Anorg. Allgem. Chem. 170, 161 (1928).

Colorless, odorless, tetragonal crystals or white powder; darkens on exposure to light. *Violent poison!* d 3.996. Dec at 320°. One gram dissolves in 13 ml water, 3 ml boiling water, 13 ml alcohol, 4 ml methanol; slowly sol in glycerol; slightly sol in ether. *Protect from light.*

THERAP CAT: Topical antiseptic.

THERAP CAT (VET): Has been used as a topical antiseptic.

5929. Mercuric Dichromate(VI). Mercury bichromate. Cr_2HgO_7; mol wt 416.58. Cr 24.96%, Hg 48.15%, O 26.88%. $HgCr_2O_7$.

Red, heavy, cryst powder. Practically insol in water; sol in HCl or HNO_3. *Poisonous!*

5930. Mercuric Fluoride. Mercury difluoride. F_2Hg; mol wt 238.59. F 15.93%, Hg 84.07%. HgF_2. Prepd from $HgCl_2$ and F_2: Henne, Midgley, J. Am. Chem. Soc. 58, 886 (1936).

White powder or cubic crystals. Very sensitive to moisture. d^{15} 8.95. mp 645°. bp > 650°. Turns yellow and hydrolyzes in the prolonged presence of water.

Dihydrate, obtained when mercuric oxide is dissolved in excess of 50% HF.

USE: In the fluorination of organic compds.

5931. Mercuric Iodate. HgI$_2$O$_6$; mol wt 550.40. Hg 36.44%, I 46.11%, O 17.44%. Hg(IO$_3$)$_2$.

White powder. *Poisonous!* On warming, dec at 175° to give mercuric iodide and oxygen. Soly in water (20°) 0.002 g/100 g; soly increased by NaCl or KI.

5932. Mercuric Iodide, Red. Mercury biniodide. HgI$_2$; mol wt 454.40. Hg 44.14%, I 55.86%. Scarlet-red, heavy, odorless, almost tasteless powder; sensitive to light; at 130° becomes yellow, and then red on cooling. *Poisonous!* d 6.28. mp 259°. bp ~350° and sublimes. Soly in water (25°) 0.006 g/100 g. 1 g dissolves in 115 ml alcohol, 20 ml boiling alcohol, about 120 ml ether, about 60 ml acetone, 910 ml chloroform, 75 ml ethyl acetate, 260 ml carbon disulfide, readily in alkali iodides, HgCl$_2$, Na$_2$S$_2$O$_3$; sol in 230 ml olive oil, 50 ml castor oil. *Protect from light.*

USE: In anal. chemistry for preparation of Nessler's Reagent (alkaline mercuric potassium iodide solution).

THERAP CAT: Antiseptic (topical).

THERAP CAT (VET): Has been used as a topical antiseptic, counterirritant, vesicant.

5933. Mercuric Nitrate. Mercury pernitrate. HgN$_2$O$_6$; mol wt 324.60. Hg 61.80%, N 8.63%, O 29.57%. Hg(NO$_3$)$_2$.

Hydrate, white or slightly yellow, deliquesc, cryst powder; odor of nitric acid. *Poisonous!* d 4.3. Sol in a small amount of water; with much water or on boiling with water, an insol basic salt is formed; sol in dil acids. *Keep well closed and protect from light.*

USE: Manufacture of felt; mercury fulminate; destroying phylloxera.

5934. Mercuric Oleate. *9-Octadecenoic acid mercury salt;* oleate of mercury. C$_{36}$H$_{66}$HgO$_4$; mol wt 763.51. C 56.63%, H 8.71%, Hg 26.27%, O 8.38%. Hg(C$_{17}$H$_{33}$COO)$_2$. Commercial product prepd by dissolving yellow mercuric oxide in oleic acid; contains 24-26% HgO and about 10% excess oleic acid.

Yellowish-brown, somewhat transparent, ointment-like mass; odor of oleic acid. *Poisonous!* On keeping, the mercury is partly reduced to mercurous state or metal. Practically insol in water. Slightly sol in alcohol or ether, freely in fixed oils. *Protect from light.*

THERAP CAT: Has been used as ectoparasiticide.

5935. Mercuric Oxide, Red. Red precipitate. HgO; mol wt 216.59. Hg 92.61%, O 7.39%. Contains 99-99.5% HgO.

Bright red or orange-red, heavy, odorless, crystalline powder or scales; orthorhombic structure; yellow when finely powdered. *Poisonous!* Dec on exposure to light into mercury and oxygen; at 400° becomes almost black, but red again on cooling; at 500° dec into mercury and oxygen. d 11.14. Practically insol in water; sol in dil HCl or HNO$_3$ or in solns of alkali cyanides or iodides, slowly in solns of alkali bromides; insol in alcohol. *Protect from light.*

USE: In marine bottom paints, diluting pigments for painting on porcelain, with graphite as depolarizer in dry batteries. In Kjeldahl nitrogen determination; and as reagent for citric acid, thiophene, glucose, aldehyde, urea, acetone. As reagent and catalyst in organic reactions: J. S. Pizey, *Synthetic Reagents,* **vol. 1** (John Wiley, New York, 1974) pp 295-319.

THERAP CAT: Antiseptic (topical).

THERAP CAT (VET): Has been used as a topical antiseptic.

5936. Mercuric Oxide, Yellow. Yellow precipitate. HgO; mol wt 216.59. Hg 92.61%, O 7.39%. Contains 99-99.5% HgO.

Yellow or orange-yellow, heavy, odorless powder, orthorhombic structure. *Poisonous!* Becomes red on heating and yellow again on cooling; more finely divided and more reactive than red mercuric oxide. Other physical properties and

solubilities: *see* Mercuric Oxide, Red. *Protect from light.* *Incompat:* Reducing agents.

USE: Similar to that of the red oxide; in the mfr of organic mercurials. In anal. chemistry for determining Zn or HCN; detecting acetic acid in formic acid, CO in gas mixtures.

THERAP CAT: Topical antiseptic (ophthalmic).

THERAP CAT (VET): Has been used as a topical antiseptic, fungicide, in chronic skin conditions, conjunctivitis, corneal ulcers.

5937. Mercuric Oxycyanide. *Mercury cyanide oxide.* C$_2$Hg$_2$N$_2$O; mol wt 469.21. C 5.12%, Hg 85.50%, N 5.97%, O 3.41%. HgO.Hg(CN)$_2$. For reasons explained below, the article of commerce contains about 33% mercuric oxycyanide and about 67% mercuric cyanide.

White, orthorhombic crystals or cryst powder d 4.44. One gram dissolves in 80 ml cold water, more sol in hot water. *Violent poison! It explodes when touched with a flame or by percussion;* hence for commerce it is made with an excess of mercuric cyanide which eliminates the danger of explosion.

THERAP CAT: Antiseptic (topical).

5938. Mercuric Salicylate. Mercury subsalicylate. C$_7$H$_4$HgO$_3$; mol wt 336.70. C 24.97%, H 1.20%, Hg 59.58%, O 14.26%. Approx C$_7$H$_4$O$_3$Hg. Article of commerce contains 54-58% Hg. Prepn: C. H. Rogers *et al., Inorganic Pharmaceutical Chemistry* (Lea & Febriger, Philadelphia, 4th ed., 1948) pp 409-411.

White or slightly yellowish or pinkish, odorless powder. *Poisonous!* Insoluble in water or alcohol. Sol in warm solns of alkali halides, fixed alkali hydroxides and carbonates. *Incompat:* Alkali iodides.

THERAP CAT: Topical antiseptic.

5939. Mercuric Sodium p-Phenolsulfonate. Mercury and sodium phenolsulfonate; mercuriphenoldisulfonate sodium; Hermophényl. C$_{12}$H$_8$HgNa$_2$O$_8$S$_2$; mol wt 590.89. C 24.39%, H 1.36%, Hg 33.95%, Na 7.78%, O 21.66%, S 10.85%. Prepd by the action of the sodium salt of p-phenolsulfonic acid on yellow mercuric oxide: Lumière, Chevrottier, *Compt. Rend.* **132,** 145 (1901); Lumière, Perrin, *Pharmacie* **1,** 102 (1943). The commercial product is a mixture of the monosulfonate (80%) and the disulfonate.

Na$^+$ $^-$O$_3$S—⟨⟩—O—Hg—O—⟨⟩—SO$_3^-$ Na$^+$

White powder. One gram dissolves in 5 ml water. The aq soln does not coagulate albuminous matter.

Caution: Ingestion or percutaneous absorption may cause mercury poisoning.

USE: As germicide in soaps and lotions. Usual concn in soap about 1:100.

THERAP CAT: Antiseptic (local).

5940. Mercuric Stearate. *Octadecanoic acid mercury salt.* C$_{36}$H$_{70}$HgO$_4$; mol wt 767.54. C 56.34%, H 9.19%, Hg 26.13%, O 8.34%. Hg(C$_{17}$H$_{35}$COO)$_2$.

Yellowish, granular powder. Toxic because of its solubility in lipids. Practically insol in water or alcohol. Sparingly sol in fatty oils.

5941. Mercuric Subsulfate. *Mercury oxide sulfate;* mercury oxonium sulfate; Turpeth mineral; basic mercuric sulfate. Hg$_3$O$_6$S; mol wt 729.83. Hg 82.45%, O 13.15%, S 4.39%. HgSO$_4$.2HgO. Proved to be mercury oxonium sulfate in x-ray studies of its structure. The polymeric oxonium cations (Hg$_3$O$_2$)$_x$ form 2-dimensional infinite layers. The sulfate ions lie inside the loops of the wide-meshed cation lattice.

Lemon-yellow, heavy, odorless powder. *Poisonous!* Practically insol in water. Sol in acids.

5942. Mercuric Succinimide. Mercuric imidosuccinate. C$_8$H$_8$HgN$_2$O$_4$; mol wt 396.75. C 24.22%, H 2.03%, Hg 50.56%, N 7.06%, O 16.13%.

White, small crystals or cryst powder; odorless and stable in air; affected by light. *Poisonous!* One gram dissolves in 20 ml water, 5 ml boiling water, 300 ml alcohol. Practically insol in ether. *Protect from light.* On adding NaOH to its aq soln a yellowish-white ppt forms which is reduced to metallic Hg by heat; heated with zinc dust, pyrrole evolves which colors a pine splinter moistened with HCl red. The aq soln is not pptd by albumin.

THERAP CAT: Antibacterial.

5943. Mercuric Sulfate. Mercury bisulfate. HgO_4S; mol wt 296.65. Hg 67.62%, O 21.57%, S 10.81%. $HgSO_4$.

White, odorless granules or cryst powder. *Poisonous!* d 6.47. Dec by water into a yellow insol, basic sulfate and free H_2SO_4; sol in HCl, hot dil H_2SO_4, concd soln of NaCl. *Protect from light.*

USE: Electrolyte for primary batteries; with NaCl for extracting gold and silver from roasted pyrites; as a reagent for wine coloring, barbital, and cystine.

5944. Mercuric Sulfide, Black. Ethiops mineral. HgS; mol wt 232.66. Hg 86.22%, S 13.78%. Occurs as a mineral in California. Prepd by passing hydrogen sulfide through a soln of mercuric chloride in hydrochloric acid.

Black or grayish-black, heavy, odorless, tasteless, amorphous powder. Also occurs as black, cubic crystals (β-form). Transition temp (red to black) 386°. Black form can exist indefinitely in metastable state at room temp. Insol in water, alcohol, dil mineral acids.

USE: As pigment for horn, rubber, etc.

5945. Mercuric Sulfide, Red. Vermilion; Chinese red; C.I. Pigment Red 106; C.I. 77766. HgS; mol wt 232.66. Hg 86.22%, S 13.78%. Occurs in nature as the mineral *cinnabar.* Prepd from mercuric acetate, ammonium thiocyanate, glacial acetic acid and hydrogen sulfide: Newell *et al., Inorg. Syn.* **1**, 19 (1939). *See also Colour Index* vol. 4 (3rd ed., 1971) p 4682.

Bright scarlet-red powder, lumps, hexagonal crystals (α-form); blackens on exposure to light, particularly in presence of H_2O or alkali hydroxides. At about 250° becomes brownish, at higher temp black, but red again on cooling. When ignited in air it dec into metal and sulfur, the latter burning to SO_2. Practically insol in water; not attacked by HNO_3 or cold HCl; dec by hot concd H_2SO_4; sol in aqua regia with separation of S, in warm hydriodic acid with evolution of H_2S. *Protect from light.*

USE: For coloring plastics, sealing wax, and with $FeSO_4$ for marking linen; manuf fancy colored papers; as pigment.

THERAP CAT: Antibacterial.

5946. Mercuric Thiocyanate. Mercuric sulfocyanate; mercuric sulfocyanide. $C_2HgN_2S_2$; mol wt 316.76. C 7.58%, Hg 63.33%, N 8.84%, S 20.25%. $Hg(SCN)_2$. Prepd from $Hg(NO_3)_2$ + KSCN: Peters, *Z. Anorg. Allgem. Chem.* **77**, 157 (1912).

Odorless powder. When crystalline, usually in radially arranged needles. *Poisonous!* When heated it swells up to many times its original vol, decomposing finally into mercury, nitrogen, etc., at about 165°. Slightly sol in cold water (0.069 g/100 ml H_2O at 25°), more sol in boiling water with decompn; sol in dil HCl, in solns of alkali cyanides, chlorides. *Protect from light.*

USE: For Pharaoh's serpents (fireworks); intensifier in photography.

5947. Mercurophen. Sodium 4-(hydroxymercuri)-2-nitrophenolate; sodium hydroxymercuri-*o*-nitrophenolate; 4-(hydroxymercuri)-2-nitrophenol sodium salt. $C_6H_4HgN-NaO_4$; mol wt 377.68. C 19.08%, H 1.07%, Hg 53.11%, N 3.71%, Na 6.09%, O 16.94%. Prepn from *o*-nitrophenol and $Hg(OCOCH_3)_2$: Schamberg *et al.,* U.S. pat. **1,390,972** (1922). Structure: Malcolm, *J. Bacteriol.* **22**, 403 (1931).

Brick-red, odorless powder. *Poisonous!* Sol in hot water. Solns are deep amber in color.

THERAP CAT: Antiseptic, disinfectant.

THERAP CAT (VET): Antiseptic, disinfectant.

5948. Mercurous Acetate. $C_4H_6Hg_2O_4$; mol wt 519.27. C 9.25%, H 1.16%, Hg 77.26%, O 12.32%. $Hg_2(CH_3COO)_2$. Prepd from $Hg_2(NO_3)_2$ + CH_3COONa: Wagenknecht, Juza, in *Handbook of Preparative Inorganic Chemistry* vol. 2, G. Brauer, Ed. (Academic Press, New York, 2nd ed., 1965) p 1120.

Lustrous leaflets or cryst powder. Darkens on exposure to light. Sol in about 100 parts water; sol in dil acetic acid. Practically insol in alcohol, ether. Aq solns dec quickly under the influence of light and heat. *Keep well closed and protected from light.*

THERAP CAT: Antibacterial.

5949. Mercurous Bromide. Br_2Hg_2; mol wt 560.99. Br 28.49%, Hg 71.51%. Hg_2Br_2.

White, odorless powder; darkens on exposure to light. d 7.3. Sublimes at approx 390° (dec). Insol in water, alcohol, ether; dec by hot HCl or alkali bromides. *Protect from light.*

5950. Mercurous Chlorate. $Cl_2Hg_2O_6$; mol wt 568.08. Cl 12.48%, Hg 70.62%, O 16.90%. $Hg_2(ClO_3)_2$.

White crystals; dec at about 250° to form oxygen, mercuric oxide and mercuric chloride. d 6.41. Sparingly sol in water; hydrolyzed in hot water to form basic salts.

5951. Mercurous Chloride. Calomel; mild mercury chloride; mercury monochloride; mercury protochloride; mercury subchloride; precipité blanc; Calogreen. Cl_2Hg_2; mol wt 472.09. Cl 15.02%, Hg 84.98%. Hg_2Cl_2.

White, odorless, tasteless, heavy powder; slowly dec by sunlight into mercuric chloride and metallic mercury; sublimes at 400-500° without melting. d 7.15. Practically insol in water (0.00020 g/100 ml H_2O at 25°). HCl or alkali and alkaline earth chlorides increase soly in water. Insol in alcohol, ether. Dec by solns of alkali iodides, bromides or cyanides into the mercuric salt and metallic mecury; solns of alkali chlorides act similarly but slowly. It is blackened by ammonia, caustic alkali and alkaline earth solns. *Protect from light.* Incompat: Bromides, iodides, alkali chlorides, sulfates, sulfites, carbonates, hydroxides, lime water, acacia, ammonia, golden antimony sulfide, cocaine, cyanides, copper salts, hydrogen peroxide, iodine, iodoform, lead salts, silver salts, soap, sulfides.

Caution: Excessive doses may cause mercury poisoning. *Antidote:* Dimercaprol (BAL). If laxation from oral mercurous chloride should not occur, saline laxative must be administered to prevent possibility of mercury poisoning, cf. *Clinical Toxicology of Commercial Products,* R. E. Gosselin *et al.,* Eds. (Williams & Wilkins, Baltimore, 4th ed., 1976) Section II, p 95.

USE: Dark green Bengal lights; calomel paper, mixed with gold in painting on porcelain; for calomel electrodes; as fungicide; in agriculture to control root maggots on cabbage and onions.

THERAP CAT: Cathartic, diuretic, antiseptic, antisyphilitic.

THERAP CAT (VET): Has been used as cathartic, and locally as an antiseptic and desiccant.

5952. Mercurous Fluoride. F_2Hg_2; mol wt 439.18. F 8.65%, Hg 91.35%. Hg_2F_2. Prepd from Hg_2CO_3 + HF: Henne, Renoll, *J. Am. Chem. Soc.* **60**, 1060 (1938); also by addition of sodium fluoride to mercurous nitrate soln. Probably formed in the reaction between fluorine and mercury at room temp: Eméleus in *Fluorine Chemistry* vol. 1, J. H. Simons, Ed. (New York, 1950) p 39. The prepn should be carried out in the dark.

Small, yellow cubic crystals, d_4^{15} 8.73. mp 570° (dec). Sublimes at ~240°. The crystals blacken in light or with ammonia fumes. Hydrolyzed in water to form mercury, mercuric oxide and hydrofluoric acid.

5953. Mercurous Iodide. Yellow mercury iodide; mercury protoiodide. Hg_2I_2; mol wt 654.99. Hg 61.25%, I 38.75%.

Bright-yellow, amorphous, heavy, odorless powder. Darkens or becomes greenish on exposure to light, HgI_2 and metallic mercury being formed. d 7.70. mp 290° when rapidly heated with partial decompn into Hg and HgI_2. Insol in water, alcohol or ether; sol in solns of mercurous or mercuric nitrates; cold ammonia, its solns or alkali iodide dec it into mercury and mercuric iodide. *Protect from light.* "Green" mercury iodide is made from metallic mercury and iodine, the green color being due to presence of some uncombined mercury. *Incompat:* Soluble iodides.

Caution: Never prescribe mercury iodide yellow (or green) with a sol iodide, because a highly *poisonous* mercuric iodide is formed!

THERAP CAT: Antibacterial.

5954. Mercurous Nitrate. Mercury protonitrate. Hg_2-N_2O_6; mol wt 525.19. Hg 76.39%, N 5.33%, O 18.28%. $Hg_2(NO_3)_2$. Normally exists as dihydrate. Prepn of anhydr salt: Potts, Allred, *Inorg. Chem.* **5**, 1066 (1966).

Dihydrate, colorless crystals, usually with slight odor of HNO_3. mp ~70° with decompn. *Poisonous!* d 4.78. Sol in 13 parts water contg 1% HNO_3; with water alone a basic salt is formed. Blackened by ammonia, caustic alkali, and alkaline earth solns. *Keep well closed and protected from light.*

USE: Fire gilding, blackening brass.

5955. Mercurous Oxide. Mercury oxide black. No evidence that Hg_2O has been isolated. X-ray data shows that solid formed by addition of sodium hydroxide to mercurous nitrate is an intimate mixture of mercury and mercuric oxide: Fricke, Ackermann, *Z. Anorg. Allgem. Chem.* **211**, 233 (1933).

Black or brownish-black powder. d 9.8. Insol in water. Sol in HNO_3. HCl converts it into calomel. *Protect from light.*

5956. Mercurous Sulfate. Hg_2O_4S; mol wt 497.24. Hg 80.68%, O 12.87%, S 6.45%. Hg_2SO_4.

White to slightly yellow cryst powder. Becomes gray on exposure to light with production of mercury and mercuric sulfate. d 7.56. Slightly sol in water (0.06 g/100g at 25°); sol in dil HNO_3.

USE: For making electric batteries; with zinc sulfate in the standard Clark cell and with cadmium sulfate in the standard Weston cell.

5957. Mercury. Hydrargyrum; liquid silver; quicksilver. Hg; at. wt 200.59; at. no. 80; valences 1, 2. Group 2b element. Abundance in earth's crust 0.5 ppm. Natural isotopes: 202 (29.80%); 200 (23.13%); 199 (16.84%); 201 (13.22%); 198 (10.02%); 204 (6.85%); 196 (0.146%); known isotopes range in mass number from 189 to 206. Obtained by roasting cinnabar (mercuric sulfide). General reviews: Roberts, *Advan. Inorg. Chem. Radiochem.* **11** (Academic Press, New York, 1968) pp 309-339; Aylett, "Group IIB" in *Comprehensive Inorganic Chemistry* vol. 3 (Pergamon Press, Oxford, 1973) pp 187-328; H. J. Drake in Kirk-Othmer *Encyclopedia of Chemical Technology* vol. 15 (Wiley-Interscience, New York, 3rd ed., 1981) pp 143-156. Review of clinical toxicology: H. B. Gerstner, J. E. Huff, *J. Toxicol. Environ. Health* **2**, 491-526 (1977).

Silver-white, heavy, mobile, liquid metal; slightly volatile at ordinary temp; solid mercury is a tin-white, ductile, malleable mass which may be cut with a knife. mp −38.87°; bp 356.72°; d^{25} 13.534. Heat capacity at constant pressure (25°) 6.687 cal/mole deg. Vapor pressure (25°): 2×10^{-3} mm; heat of vaporization (25°): 14.652 kcal/mole: Busey, Giauque, *J. Am. Chem. Soc.* **75**, 806 (1953). Surface tension (25°): 484 dynes/cm; electrical resistivity (20°): 95.76 μohm cm. When pure does not tarnish on exposure to air at ordinary temp, but when heated to near the boiling point slowly oxidizes to HgO. Forms alloys with most metals except iron and combines with sulfur at ordinary temperatures. E^0 (aq) Hg/Hg^{2+} −0.854 V; E^0 (aq) 2 Hg/Hg_2^{2+} −0.789 V. Soly in water (25°): 0.28 μmoles/l; data on soly in organic solvents: Spencer, Voigt, *J. Phys. Chem.* **72**, 464 (1968). Reacts with HNO_3 and hot, concd H_2SO_4; does not

react with dil HCl, cold H_2SO_4, or alkalies. Reacts with ammonia solns in air to form Hg_2NOH, *Millon's base.* Mercury salts when heated with Na_2CO_3 yield metallic Hg and are reduced to metal by H_2O_2 in the presence of alkali hydroxide. Cu, Fe, Zn and many other metals ppt metallic Hg from neutral or slightly acid solns of mercury salts. Soluble ionized *mercuric* salts give a yellow ppt of HgO with NaOH and a red ppt of HgI_2 with alkali iodide. *Mercurous* salts give a black ppt with alkali hydroxides and a white ppt of calomel with HCl or sol chlorides. They are slowly dec by sunlight. *Poisonous!*

Caution: Readily absorbed via respiratory tract (elemental mercury vapor, mercury compd dusts), intact skin, and G.I. tract, although occasional incidental swallowing of metallic mercury is without harm. Spilled and heated elemental mercury is particularly hazardous. *Acute:* sol salts have violent corrosive effects on skin and mucous membranes; severe nausea, vomiting, abdominal pain, bloody diarrhea; kidney damage; death usually within 10 days. *Chronic:* inflammation of mouth and gums, excessive salivation, loosening of teeth; kidney damage; muscle tremors, jerky gait, spasms of extremities; personality changes, depression, irritability, nervousness. Phenyl and alkyl mercurials can cause skin burns and be absorbed by the skin. Burning sensation is delayed several hours and thus gives no warning. Alkyls have affinity for brain tissue and may cause permanent damage. Phenyls are no more toxic than inorganic Hg. *Antidote:* Dimercaprol (BAL). *See* E. Browning, *Toxicity of Industrial Metals* (Appleton-Century Crofts, New York, 2nd ed., 1969) pp 226-242.

USE: In barometers, thermometers, hydrometers, pyrometers; in mercury arc lamps producing ultraviolet rays; in switches, fluorescent lamps; in mercury boilers; manuf all mercury salts, mirrors; as catalyst in oxidation of organic compds; extracting gold and silver from ores; making amalgams, electric rectifiers, mercury fulminate; also in dentistry; in determining N by Kjeldahl method, for Millon's reagent; as cathode in electrolysis, electroanalysis, and many other uses. Also in pharmaceuticals, agricultural chemicals, anti-fouling paints.

5958. Mercury Mass. Blue pill; blue mass. Contains 32-34% metallic Hg. Bluish-gray mass. The rest is honey, licorice, althea, glycerol, and some mercury oleate.

Caution: Can cause systemic mercury poisoning.

THERAP CAT: Cathartic.

THERAP CAT (VET): Has been used as a laxative.

5959. Merisoprol Hg 197. *Hydroxy(2-hydroxypropyl)-mercury-^{197}Hg; 1-(hydroxymercuri-^{197}Hg)-2-propanol;* Merprane. $C_3H_8HgO_2$.

THERAP CAT: Diagnostic aid (renal function).

5960. Meropenem. *[4R-[3(3S*,5S*),4α,5β,6β(R*)]]-3-[[5-[(Dimethylamino)carbonyl]-3-pyrrolidinyl]thio]-6-(1-hydroxyethyl)-4-methyl-7-oxo-1-azabicyclo[3.2.0]hept-2-ene-2-carboxylic acid trihydrate; (1R,5S,6S)-2-[(3S,5S)-5-(dimethylaminocarbonyl)pyrrolidin-3-ylthio]-6-[(R)-1-hydroxyethyl]-1-methylcarbapen-2-em-3-carboxylic acid trihydrate;* ICI-194660; SM-7338; Merrem; Meronem. $C_{17}H_{25}N_3O_5$·S·$3H_2O$; mol wt 437.51. C 46.67%, H 7.14%, N 9.60%, O 29.26%, S 7.33%. Carbapenem. Prepn: M. Sunagawa *et al.,* **Eur.** pat. **Appl.** 126,587; M. Sunagawa, U.S. pat. 4,943,569 (1984, 1990 both to Sumitomo). Structure-activity study: M. Sunagawa *et al., J. Antibiot.* **43**, 519 (1990). Crystal structure: K. Yanagi *et al., Acta Crystallogr.* **C48**, 1737 (1992). Pharmacokinetics: R. Wise *et al., Antimicrob. Ag. Chemother.* **34**, 1515 (1990). Series of articles on antimicrobial activity, metabolism: *J. Antimicrob. Chemother.*

24, Suppl. A, 1-320 (1989); and clinical performance: *ibid.* **36,** Suppl. A, 1-223 (1995).

THERAP CAT: Antibacterial.

5961. Merphyrin. *[7,12-Bis(1-hydroxyethyl)-3,8,13,17-tetramethyl-21H,23H-porphine-2,18-dipropanoato(4—)-N²¹,-N²²,N²³,N²⁴]mercurate(2—) disodium;* mercuri-hematoporphyrin disodium salt; M. H.; hematoporphyrinmercury disodium salt. $C_{34}H_{34}HgN_4Na_2O_6$; mol wt 841.24. C 48.54%, H 4.07%, Hg 23.84%, N 6.66%, Na 5.47%, O 11.41%. Prepd by action of mercuric acetate upon hematoporphyrin and treating the product with sodium methoxide: **Japan.** pat. **1273('59); Brit.** pat. **824,705** (1959 to Daiichi), *C.A.* **54,** 9972 (1960); Obika *et al.,* **U.S.** pat. **3,004,985** (1961 to Daiichi).

Dark brown powder. Sol in water (a 1:1000 soln is slightly alkaline). Aq solns are affected by light. Slightly sol in methanol. Practically insol in acetone, benzene, chloroform.

5962. Mersalyl. *[3-[[2-(Carboxylatomethoxy)benzoyl]-amino]-2-methoxypropyl]hydroxymercurate(1—) sodium; o-[[3-(hydroxymercuri)-2-methoxypropyl]carbamoyl]phenoxyacetic acid sodium salt;* sodium *o-[(3-hydroxymercuri-2-methoxypropyl)carbamoyl]phenoxyacetate; N-(γ-hydroxymercuri-β-methoxypropyl)salicylamide-O-acetic acid sodium salt;* mercuramide; Mercusal; Mersalin; Salyrgan. $C_{13}H_{16}$-HgNNaO₆; mol wt 505.85. C 30.87%, H 3.19%, Hg 39.65%, N 2.77%, Na 4.54%, O 18.98%. Prepn: Diels, Beccard, *Ber.* **39,** 4125 (1906); Bockmühl, Schwarz, **Ger.** pat. **423,031** (1925 to Hoechst); *Frdl.* **15,** 1609; B.I.O.S. *Final Report* no. **766,** p 132; Foye *et al., J. Am. Pharm. Assoc.* **41,** 273 (1952); Werner, Scholz, **U.S.** pat. **2,705,716** (1955 to Ciba). Toxicity data: E. B. Robbins, K. K. Chen, *J. Am. Pharm. Assoc.* **40,** 249 (1951).

Bitter crystals. *Poison! Somewhat deliquescent.* Gradually dec by light. One gram dissolves in about 1 ml water, in about 3 ml ethanol (95%), in 2 ml abs methanol. Practically insol in ether, chloroform. Aq solns are alkaline to litmus. Solns of mersalyl contg sodium chloride or other salts may become toxic unless some substance, such as theophylline which inhibits the decompn of the mercurial complex, is present. LD_{50} i.v. in mice, rats: 72.6, 17.7 mg/kg (Robbins, Chen).

Note: Neptal and *Diursal* also contain theophylline; *cf* U.S. pat. **2,213,457** (1940).

THERAP CAT: Diuretic.
THERAP CAT (VET): Diuretic.

5963. Mesaconic Acid. *(E)-2-Methyl-2-butenedioic acid;* methylfumaric acid; *trans*-1-propene-1,2-dicarboxylic acid. $C_5H_6O_4$; mol wt 130.10. C 46.16%, H 4.65%, O 49.19%. Prepd by heating citraconic anhydride with nitric acid: Shriner *et al., Org. Syn.* **11,** 74 (1931); from ethyl itaconate: Jones *et al., J. Chem. Soc.* **1954,** 1865; from *N,N*-diethyl-5-diethylamino-2,3-dihydro-3-furamide: Sauer *et al., J. Am. Chem. Soc.* **81,** 693 (1959). Obtained almost instantly by the action of sunlight on citraconic acid in ether + chloroform in the presence of a little bromine: Fittig, Langworthy, *Ber.* **26,** 46 (1893); *Ann.* **304,** 119, 149 (1899); by action of β-radiation on citraconic acid + bromine: Lavigne, Levine, **U.S.** pat. **2,979,445** (1961 to California Res. Corp.).

Orthorhombic needles from alcohol, monoclinic tablets from ethyl acetate, mp 204-205°. d 1.466. Sublimes. bp 250° (decompn). K_1 at 25° = 8.5 × 10⁻⁴; K_2 = 15 × 10⁻⁶. 100 g water dissolves 2.7 g at 18° and 117.9 g at the boiling point; 100 g 90% alc dissolves 30.6 g at 17° and 95.7 g at the boiling point. Sol in ether; sparingly sol in chloroform, carbon disulfide and ligroin.

Dimethyl ester, $C_7H_{10}O_4$, liq, bp 206°. d_4^{21} vac 1.12011. n_D^{20} 1.45575. Sol in 122 parts water at 15°.

Diethyl ester, $C_9H_{14}O_4$, liq, bp 229°. d_4^{20} vac 1.04675. n_4^{20} 1.44936.

5964. Mesalamine. *5-Amino-2-hydroxybenzoic acid;* 5-aminosalicylic acid; 5-amino-2-hydroxybenzene-1-carboxylic acid; *m*-aminosalicylic acid; fisalamine; mesalazine; 5-ASA; Asacol; Asacolitin; Claversal; Lixacol; Mesasal; Pentasa; Rowasa; Salofalk. $C_7H_7NO_3$; mol wt 153.14. C 54.90%, H 4.61%, N 9.15%, O 31.34%. Prepn by reduction of *m*-nitrobenzoic acid with Zn dust and HCl: H. Weil *et al., Ber.* **55B,** 2664 (1922); by electrolytic reduction: Le Guyader, Peltier, *Compt. Rend.* **253,** 2544 (1961). Use in manufacture of dyes: **Brit.** pat. **751,386** (1956 to J. R. Geigy). Identification as active metabolite of sulfasalazine, *q.v.:* A. K. Azad Khan *et al., Lancet* **2,** 892 (1977); P. A. M. Van Hees *et al., Gut* **21,** 632 (1980). HPLC determn in serum: E. Brendel *et al., J. Chromatog.* **385,** 299 (1987). Bioavailability, plasma level and excretion: S. N. Rasmussen *et al., Gastroenterology* **83,** 1062 (1982). Clinical evaluation in Crohn's disease: *eidem, ibid.* **85,** 1350 (1983). Clinical trials in ulcerative colitis: M. Campieri *et al., Lancet* **2,** 270 (1981); L. S. Friedman *et al., Am. J. Gastroenterol.* **81,** 412 (1986); in comparison with olsalazine, *q.v.:* S. S. Rao *et al., Scand. J. Gastroenterol.* **22,** 332 (1987). Brief review: G. Friedman, *Am. J. Gastroenterol.* **81,** 141-144 (1986).

White to pinkish crystals, dec ~280°. Slightly sol in cold water, alc; more sol in hot water; sol in HCl.

USE: In manuf of light-sensitive paper, azo and sulfur dyes.

THERAP CAT: Anti-inflammatory (gastrointestinal). Treatment of ulcerative colitis.

5965. Mescaline. *3,4,5-Trimethoxybenzeneethanamine; 3,4,5-trimethoxyphenethylamine;* mezcaline. $C_{11}H_{17}NO_3$; mol wt 211.26. C 62.54%, H 8.11%, N 6.63%, O 22.72%. Psychotomimetic alkaloid isolated from *peyote* (mescal buttons), the flowering heads of *Lophophora williamsii* (Lemaire) Coult., *Cactaceae.* Isoln: A. Heffter, *Ber.* **29,** 221 (1896). Structure and synthesis: E. Späth, *Monatsh.* **40,** 129 (1919); K. H. Slotta, H. Heller, *Ber.* **63,** 3029 (1930); E. Späth, F. Becke, *Monatsh.* **66,** 327 (1935); M. U. Tsao, *J.*

Am. Chem. Soc. **73**, 5495 (1951); K. Banholzer *et al.*, *Helv. Chim. Acta* **35**, 1577 (1952). Novel synthesis: M. N. Aboul-Enein, A. I. Eid, *Acta Pharm. Suec.* **16**, 267 (1979). MS determn: S. P. Jindal, T. Lutz, *Eur. J. Mass Spectrom. Biochem. Med. Environ. Res.* **2**, 117 (1982). Pharmacokinetics in rabbits: C. Van Peteghem *et al.*, *Eur. J. Drug Metab. Pharmacokinet.* **7**, 1 (1982). Mode of action study: M. E. Trulson *et al.*, *Eur. J. Pharmacol.* **96**, 151 (1983). Use in evaluating serotonin S_2 antagonists: C. J. E. Niemegeers *et al.*, *Drug Dev. Res.* **3**, 123 (1983). Evaluation of use with chlorpromazine, *q.v.*, in various psychoses: H. C. B. Denber, S. Merlis: *J. Nerv. Ment. Dis.* **122**, 463 (1955). Toxicity data: L. B. Speck, *J. Pharmacol. Exp. Ther.* **119**, 78 (1957); H. F. Hardman *et al.*, *Toxicol. Appl. Pharmacol.* **25**, 299 (1973). *Reviews:* A. R. Patel, *Progress in Drug Research* **vol. 11**, E. Jucker, Ed. (Birkhäuser Verlag, Basel, 1968) pp 11-47; G. J. Kapadia, M. B. E. Fayez, *J. Pharm. Sci.* **59**, 1699-1727 (1970).

Crystals, mp 35-36°. bp_{12} 180°. Moderately sol in water; sol in alcohol, chloroform, benzene. Practically insol in ether, petr ether. Takes up CO_2 from the air and forms a crystalline carbonate. LD_{50} i.p. in rats: 370 mg/kg (Speck).
Hydrochloride, $C_{11}H_{17}NO_3$.HCl, needles, mp 181°. Sol in water, alcohol. LD_{50} in mice, rats, guinea pigs (mg/kg): 212, 132, 328 i.p. (Hardman).
Sulfate dihydrate, $(C_{11}H_{17}NO_3)_2.H_2SO_4.2H_2O$, prisms, mp 183-186°. Sol in hot water, methanol; sparingly sol in cold water, ethanol.
Acid sulfate, $C_{11}H_{17}NO_3.H_2SO_4$, crystals, mp 158°.
N-Benzoylmescaline, needles from aq alc, mp 121°. Very sol in alcohol, ether.
N-Methylmescaline, occurs naturally, bp 130-140°.
N-Acetylmescaline, occurs naturally, mp 94°.
Caution: This is a controlled substance (hallucinogen) listed in the U.S. Code of Federal Regulations, Title 21 Part 1308.11 (1995).

5966. Mesembrine. *(3aS-cis)-3a-(3,4-Dimethoxyphenyl)octahydro-1-methyl-6H-indol-6-one;* 3a-(3,4-dimethoxyphenyl)tetrahydro-1-methyl-6(3a*H*)-indolinone. $C_{17}H_{23}$-NO_3; mol wt 289.37. C 70.56%, H 8.01%, N 4.84%, O 16.59%. Alkaloid used in preparing *Channa*, a drug of Southwest Africa. Occurs naturally as the (−)-form. From *Sceletium expansum* L., *S. tortuosum* L., Aizoaceae: Hartwick, Zwicky, *Apoth. Ztg.* **29**, 925 (1914); Rimington *et al.*, *J. Vet. Sci. Animal Ind.* **9**, 187 (1938), *C.A.* **32**, 4279⁹ (1938). Structure: Popelak *et al.*, *Naturwiss.* **47**, 156 (1960). Configuration: P. W. Jeffs *et al.*, *J. Am. Chem. Soc.* **91**, 3831 (1969). Synthesis of (±) form: Shamma, Rodriguez, *Tetrahedron Letters* **1965**, 4847; O. Hoshino *et al.*, *Heterocycl.* **10**, 61 (1978); of (±)-form and *trans* isomer: Oh-Ishi, Kugita, *Chem. Pharm. Bull.* **18**, 299 (1970). Synthesis of (+)-form: Yamada, Otani, *Tetrahedron Letters* **1971**, 1133; *eidem, Chem. Pharm. Bull.* **21**, 2130 (1973). Stereoselective synthesis of (±)-form: Wijnberg, Speckamp, *Tetrahedron Letters* **1975**, 3963; *eidem, Tetrahedron* **34**, 2579 (1978); S. F. Martin *et al.*, *J. Org. Chem.* **44**, 3391 (1979); S. Takano *et al.*, *Chem Letters* **1981**, 1385. Enantioselective synthesis of

natural mesembrine: *eidem, Tetrahedron Letters* **22**, 4479 (1981). Biosynthesis: Jeffs *et al.*, *J. Am. Chem. Soc.* **93**, 3752 (1971); *eidem, Chem. Commun.* **1977**, 60. Review of mesembrine alkaloids: A. Popelak, G. Lettenbauer in *The Alkaloids*, R. H. F. Manske, Ed., **vol. IX** (Academic Press, New York, 1967) pp 467-481; R. V. Stevens in *The Total Synthesis of Natural Products* **vol. 3**, J. ApSimon, Ed. (Wiley, New York, 1977) pp 443-453.
Pale yellow oil. $bp_{0.3}$ 186-190°. $[\alpha]_D^{20}$ −55.4° (CH_3OH). Freely sol in alcohol, chloroform, acetone; slightly sol in ether. Practically insol in benzene, petr ether, alkalies.
Hydrochloride, $C_{17}H_{23}NO_3$.HCl, mp 205-206°. $[\alpha]_D^{20}$ −8.4° (CH_3OH).
(+)-Form. (Partially optically active). Pale yellow oil. $[\alpha]_D^{20}$ +16.1° (c = 1.32 in CH_3OH).
(+)-Form hydrochloride, $C_{17}H_{23}NO_3$.HCl, crystals from 2-propanol, mp 206.5-207.5°. $[\alpha]_D^{20}$ +7.3° (c = 0.465 in CH_3OH).
(±)-Form. Colorless oil. $bp_{0.07}$ 178°.
(±)-Form hydrochloride, $C_{17}H_{23}NO_3$.HCl, mp 179-181°.

5967. Mesitylene. *1,3,5-Trimethylbenzene; sym*-trimethylbenzene. C_9H_{12}; mol wt 120.19. C 89.94%, H 10.06%. Occurs in coal tar and in petroleum crudes; prepd by dehydrating acetone with H_2SO_4: Adams, Hufferd, *Org. Syn.* **2**, 41 (1922).

Liquid; peculiar odor. d_4^{20} 0.8637. mp −44.8°. n_D^{18} 1.49541. bp_{760} 164.7°; bp_{100} 98.9°; bp_{20} 61°; bp_{10} 47.4°; $bp_{1.0}$ 9.6°. Practically insol in water (100 g H_2O dissolve 0.002 g). Miscible with alcohol, ether, benzene.

5968. Mesityl Oxide. *4-Methyl-3-penten-2-one;* isopropylideneacetone. $C_6H_{10}O$; mol wt 98.14. C 73.43%, H 10.27%, O 16.30%. $(CH_3)_2C$=CHCOCH₃. Made by distilling diacetone alcohol with a small amount of iodine: Conant, Tuttle, *Org. Syn.* **1**, 53 (1921). Condensation of acetone to mesityl oxide using sulfonated polystyrene-divinylbenzene resin as ion exchange catalyst: Klein, Banchero, *Ind. Eng. Chem.* **48**, 1278 (1956). Believed to be a mixture of two isomers. Toxicity data: *Handbook of Toxicology vol. 1*, W. S. Spector, Ed. (Saunders, Philadelphia, 1956) pp 342-343.
Colorless, oily liq; honey-like odor. d_4^{15} 0.8592. bp_{760} 130°; bp_{100} 72.1°; bp_{20} 26°; $bp_{1.0}$ −8.7°. mp −41.5°; also reported as mp −59°. Can be made to crystallize at low temp in petr ether. n_D^{22} 1.4425. Absorption spectrum: Morton, *J. Chem. Soc.* **1926**, 719. Sol in about 30 parts water; miscible with most organic liqs. Flash pt: 87°F (30.6°C). LC for rats in air: 2500 ppm (Spector).
Caution: Potential symptoms of overexposure are irritation of eyes, skin and mucous membranes; narcosis, coma. *See NIOSH Pocket Guide to Chemical Hazards* (DHHS/NIOSH 90-117, 1990) p 140.
USE: Solvent for nitrocellulose, many gums and resins, particularly vinyl resins. In lacquers, varnishes and enamels. In making methyl isobutyl ketone.

5969. Mesna. *2-Mercaptoethanesulfonic acid sodium salt;* sodium mercaptoethanesulfonate; D-7093; UCB-3983; Mesnex; Mistabron; Mistabronco; Mucofluid; Uromitexan. $C_2H_5NaO_3S_2$; mol wt 164.18. C 14.63%, H 3.07%, Na 14.00%, O 29.23%, S 39.06%. $[HSCH_2CH_2SO_3]^-Na^+$. Sulfhydryl donor used to reduce the urotoxic effects of antineoplastic alkylating agents; also has mucolytic activity. Prepn: I. M. Lipovich, *J. Appl. Chem. USSR* **18**, 718 (1945); C. H. Schramm *et al.*, *J. Am. Chem. Soc.* **77**, 6231 (1955). Synthesis and properties: V. E. Petrun'kin, *C.A.* **51**, 5693a (1957); *ibid.* **54**, 24379c (1960). Use as mucolytic: Neth. pat. **Appl. 6,605,816**; H. Morren, U.S. pat. **3,567,835** (1966, 1971 both to U.C.B.); as uroprotective agent: N. Brock, Ger. pat. **2,756,018**; *idem*, U.S. pat. **4,220,660** (1979, 1980 both to Asta-Werke AG). Pharmacology, toxicity, and uroprotective effects in animals: N. Brock *et al.*, *Eur. J.*

Cancer Clin. Oncol. **18**, 1377 (1982). HPLC determn in plasma and urine: C. A. James, H. J. Rogers, *J. Chromatog.* **382**, 394 (1986). Clinical pharmacokinetics: C. A. James *et al., Brit. J. Clin. Pharmacol.* **23**, 561 (1987). Clinical study of mucolytic effects: M. Tekeres *et al., Clin. Ther.* **4**, 56 (1981). Symposium on pharmacology, toxicity and clinical uroprotective efficacy: *Cancer Treat. Rev.* **10**, Suppl. A, 1-192 (1983). *Reviews:* I. C. Shaw, M. I. Graham, *ibid.* **14** 67-86 (1987); S. E. Schoenike, W. J. Dana, *Clin. Pharm.* **9**, 179-191 (1990).

Freely sol in water, sparingly sol in organic solvents. LD_{50} in male, female mice, male, female rats (mg/kg): 1887, 2048, 2098, 1683 i.v.; 2005, 2098, 1529, 1251 i.p.; 6102, > 7200, 4440, 4679 orally (Brock).

THERAP CAT: Mucolytic; detoxifying agent.

5970. Mesoridazine. *10-[2-(1-Methyl-2-piperidinyl)eth-yl]-2-(methylsulfinyl)-10H-phenothiazine;* thioridazine-2-sulfoxide; TPS-23. $C_{21}H_{26}N_2OS_2$; mol wt 386.58. C 65.25%, H 6.78%, N 7.25%, O 4.14%, S 16.59%. Dopamine receptor blocking agent; analog of thioridazine. Prepn: Renz *et al.,* U.S. pat. **3,084,161** (1963 to Sandoz). Pharmacology and toxicology: Loew *et al., Boll. Chim. Farm.* **106**, 332-371 (1967). Effects on dopaminergic function in comparison with sulforidazine and thioridazine, *q.q.v.:* D. M. Niedzwiecki *et al., J. Pharmacol. Exp. Ther.* **228**, 636 (1984); C. D. Kilts *et al., ibid.* **231**, 334 (1984). GLC determn in plasma: E. C. Dinovo *et al., J. Pharm. Sci.* **65**, 667 (1976). Clinical evaluation in sleep disorders: K. Adam *et al., Brit. J. Clin. Pharmacol.* **3**, 157 (1976); as antipsychotic: R. Axelsson, *Curr. Ther. Res.* **21**, 587 (1977). Toxicity studies: S. Maruyama *et al., Niigata Igakkai Zasshi* **81**, 611 (1967), *C.A.* **68**, 76856h (1968). Brief description: *J. Am. Med. Assoc.* **216**, 313 (1971).

Oily product.

Benzenesulfonate, $C_{27}H_{32}N_2O_4S_3$, *mesoridazine besylate, NC-123, Lidanar, Lidanil, Serentil.* LD_{50} in mice (mg/kg): 33 i.v.; 611 s.c.; 346 orally (Maruyama).

Tartrate, $C_{25}H_{32}N_2O_7S_2$, crystals from ethyl acetate, mp 115-120°.

THERAP CAT: Antipsychotic.

5971. Mesoxalic Acid. *Oxopropanedioic acid;* keto-malonic acid; oxomalonic acid. $C_3H_2O_5$; mol wt 118.05. C 30.52%, H 1.71%, O 67.77%. HOOCCOCOOH. Occurs in *Medicago sativa* L., *Leguminosae;* has been found in beet molasses. Prepd by boiling a soln of alloxan and lead acetate: Deichsel, *J. Prakt. Chem.* [1] **93**, 194 (1864). By electrolysis of *d*-tartaric acid in alkaline soln: *Chem. Zentr.* **1922**, III, 871. Laboratory prepn from dibromomalonic acid: Conrad, Reinbach, *Ber.* **35**, 1819 (1902); from malonic ester and N_2O_3: Curtiss, *Am. Chem. J.* **35**, 477 (1906).

Monohydrate, $C_3H_2O_5 \cdot H_2O$, *dihydroxymalonic acid.* Begins to melt at 113-114° and is clear at 121°. Very sol in water; sol in alc, ether.

Diethyl ester, $C_7H_{10}O_5$, *ethyl oxomalonate.* Liquid. bp_{19} 105-107°. $d_4^{15.6}$ 1.1419. $n_D^{15.6}$ 1.419. Prepn: A. W. Dox, *Org. Syn.* **coll. vol. I**, 266 (2nd ed., 1941). Diethyl ester of dihydroxymalonic acid ($C_7H_{12}O_6$) is obtained from the crude ester mixture by fractional distn. Crystals, mp 57°. Sol in water, alcohol.

5972. Mesquite Gum. Sonora; Prosopis gum. Collected from small thorny trees abundant in the arid regions of the Western United States and as far south as Chile: *Prosopis juliflora* (Swartz) DC., *P. dulcis* Kunth., *P. horrida* Kunth., *P. inermis* H.B.K., *P. glandulosa* Torr., *P. pubescens*

Benth., *P. spicigera* L., and other species of *Prosopis, Leguminosae.* Mesquite gum resembles acacia (gum arabic) in its physical and chemical characteristics. Review of structure work: F. Smith, R. Montgomery, *The Chemistry of Plant Gums and Mucilages* (Reinhold, New York, 1959) pp 175, 288-291.

USE: Substitute for acacia. Potential source of L-arabinose and D-glucuronic acid, *cf.* C. L. Mantell, *The Water-Soluble Gums* (Reinhold, New York, 1947) pp 72-73.

5973. Mestanolone. *(5α,17β)-17-Hydroxy-17-methylan-drostan-3-one;* 17β-hydroxy-17α-methyl-3-androstanone; 17α-methylandrostan-17β-ol-3-one; 17α-methylandrostan-3-on-17β-ol; Androstalone. $C_{20}H_{32}O_2$; mol wt 304.47. C 78.90%, H 10.59%, O 10.51%. Prepd by the oxidation of 17-methyl-3,17-androstanediol: Ruzicka *et al., Helv. Chim. Acta* **18**, 1487 (1935); Swiss. pat. **208,080** (1940 to Ciba).

Crystals from ethyl acetate, mp 192-193°. Insol in water. Sol in acetone, alcohol, ether, ethyl acetate.

THERAP CAT: Androgen.

5974. Mesterolone. *(1α,5α,17β)-17-Hydroxy-1-methyl-androstan-3-one;* 1α-methyl-5α-androstan-17β-ol-3-one; 1α-methyl-5α-dihydrotestosterone; Androviron; Proviron; Mestoranum. $C_{20}H_{32}O_2$; mol wt 304.47. C 78.90%, H 10.59%, O 10.51%. Prepn of acetate: R. Wiechert, Ger. pat. **1,122,-944** corresp to U.S. pat. **3,361,773** (1962, 1968 to Schering, AG).

Crystals from ethyl acetate, mp 203.5-205.0°. $[\alpha]_D^{20}$ +17.6° (c = 0.875 in $CHCl_3$).

Acetate, $C_{22}H_{34}O_3$, *17β-acetoxy-1α-methyl-5α-androstan-3-one.* Crystals, mp 169-170°. $[\alpha]_D^{25}$ +16.5° (c = 0.88 in $CHCl_3$).

Note: This is a controlled substance (anabolic steroid) in the U.S. Code of Federal Regulations, Title 21 Part 1308.13, as defined in Part 1308.02 (1995).

THERAP CAT: Androgen.

5975. Mestilbol. *(E)-4-[1-Ethyl-2-(4-methoxyphenyl)-1-butenyl]phenol; (E)-α,α'-diethyl-4'-methoxy-4-stilbenol;* diethylstilbestrol monomethyl ether; 3-p-hydroxyphenyl-4-p-methoxyphenyl-3-hexene; monomestrol. $C_{19}H_{22}O_2$; mol wt 282.38. C 80.82%, H 7.85%, O 11.33%. Prepn: Reid, Wilson, *J. Am. Chem. Soc.* **64**, 1625 (1942); U.S. pat. **2,385,-468** (1945). The monoether is sepd from the diether by its greater soly in 0.4N alcoholic KOH. Other syntheses: Ger. pat. **708,202** (1941); Wiles, Biggerstaff, *J. Am. Chem. Soc.* **67**, 789 (1945). *See also* Dimestrol.

Needles from benzene + petr ether, mp 116-117.5°; leaflets from 70% alc, mp 114°; v. Pallos, *Arch. Gynäkol.* **170**, 355, 385 (1940), reports mp 120-121°. Distills at 185-195° at

0.3 mm Hg. Is generally more sol than the dimethyl ether of diethylstilbestrol. Practically insol in water. Sol in alcohol, dil aq or alcoholic solns of alkali hydroxides, and in vegetable oils; freely sol in acetone, ether.

5976. Mestranol. *(17α)-3-Methoxy-19-norpregna-1,3,5-(10)-trien-20-yn-17-ol;* 17α-ethynyl-3-methoxy-estratrien-17β-ol; 17α-ethynylestradiol 3-methyl ether; Menophase; Norquen; Ovastol. $C_{21}H_{26}O_2$; mol wt 310.44. C 81.25%, H 8.44%, O 10.31%. Orally active estrogenic steroid. Prepn: F. B. Colton, **U.S. pat. 2,666,769** (1954 to Searle); *J. Am. Chem. Soc.* **79,** 1123 (1957). Comprehensive description: H. A. El-Obeid, A. A. Al-Badr, *Anal. Profiles Drug Subs.* **11,** 375-406 (1982). Clinical pharmacokinetics: J. W. Goldzieher *et al., Contraception* **21,** 17 (1980). Effect on carbohydrate metabolism: W. N. Spellacy *et al., Metabolism* **31,** 106 (1982). Randomized, double-blind clinical trials: S. Koetsawang *et al., Contraception* **25,** 231 (1982); A. Sheth *et al., ibid.* 243. Evaluation of carcinogenic risk: *IARC Monographs* **21,** 257 (1979).

Crystals from methanol or acetone, mp 150-151°. uv max (methanol): 279, 287.5 nm ($E_{1cm}^{1\%}$ 82, 14.4). Sol in ethanol, ether, chloroform, dioxane, acetone. Slightly sol in methanol. Practically insol in water.
Note: Also used in combination with chlormadinone acetate, ethynodiol, lynestranol, norethindrone or norethynodrel, *q.q.v.* Has been used in combination with megestrol acetate, *q.v.* This substance may reasonably be anticipated to be a carcinogen: *Seventh Annual Report on Carcinogens* (PB95-109781, 1994) p 201.
THERAP CAT: Estrogen; in combination with progestogen as oral contraceptive.

5977. Mesulfen. *2,7-Dimethylthianthrene;* 2,6-dimethylthianthrene; 2,6-dimethyldiphenylene disulfide; mesulphen; Mitigal; Odylen; Sudermo; Peligal; Neosulfine. $C_{14}H_{12}S_2$; mol wt 244.38. C 68.81%, H 4.95%, S 26.24%. Prepn: Cohen, Skirrow, *J. Chem. Soc.* **75,** 890 (1899); Barber, Smiles, *ibid.* **1928,** 1149; Rumpf, *Bull. Soc. Chim. France* **7,** 632 (1940).

Needles from acetic acid, ethyl acetate, or alcohol, mp 123°. bp_3 184°; bp_{14} 228-231°. Freely sol in acetone, chloroform, ether, petr ether; moderately sol in abs alcohol, ethyl acetate; practically insol in water.
THERAP CAT: Scabicide, antipruritic.
THERAP CAT (VET): Scabicide, antipruritic.

5978. Metabutoxycaine. *3-Amino-2-butoxybenzoic acid 2-(diethylamino)ethyl ester;* 2-butoxy-3-aminobenzoic acid β-diethylaminoethyl ester; β-diethylaminoethyl 2-butoxy-3-aminobenzoate. $C_{17}H_{28}N_2O_3$; mol wt 308.42. C 66.20%, H 9.15%, N 9.08%, O 15.56%. Prepn: **Brit. pat. 728,527** (1955 to Novocol Chem.), *C.A.* **50,** 5750 (1956).

Hydrochloride, $C_{17}H_{28}N_2O_3 \cdot HCl$, *Primacaine.* Crystals, mp 117-119°. uv max (water): 313 nm ($A_{1cm}^{1\%}$ 6.70). Freely sol in water, ethanol. Moderately sol in acetone, chloroform. Very slightly sol in ether. pH of a 2% aq soln about 5.6.
THERAP CAT: Anesthetic (local).

5979. Metachrome Yellow. *2-Hydroxy-5-[(3-nitrophenyl)azo]benzoic acid monosodium salt; C.I. Mordant Yellow 1;* C.I. 14025; alizarine yellow GG; salicyl yellow; sodium *m*-nitrobenzeneazosalicylate. $C_{13}H_8N_3NaO_5$; mol wt 309.21. C 50.50%, H 2.61%, N 13.59%, Na 7.43%, O 25.87%. Prepd from diazotized *m*-nitroaniline and salicylic acid in alkaline medium: R. Nietzki, **U.S. pat. 424,019;** *Colour Index* vol. 4 (3rd ed., 1971) p 4058.

Yellow powder. Slightly sol in cold water; more sol in hot water.
USE: Biological stain and acid-base indicator. pH: 10.2 colorless, 12.0 yellow.

5980. Metaclazepam. *7-Bromo-5-(2-chlorophenyl)-2,3-dihydro-2-(methoxymethyl)-1-methyl-1H-1,4-benzodiazepine;* brometazepam; metuclazepam; Ka-2547; KC-2547. $C_{18}H_{18}BrClN_2O$; mol wt 393.71. C 54.91%, H 4.61%, Br 20.30%, Cl 9.00%, N 7.12%, O 4.06%. Benzodiazepine deriv with alkoxymethyl group replacing carbonyl in parent compd. Prepn: W. Milkowski *et al.,* **Ger. pat. 2,221,558;** *eidem,* **U.S. pat. 4,098,786** (1973, 1978 both to Kali-Chemie). Effect on psychomotor performance and memory in volunteers: Z. Subhan, I. Hindmarch, *Drugs Exp. Clin. Res.* **9,** 567 (1983). Interaction with ethanol: H. R. Musch, M. Ruhland, *Arzneimittel-Forsch.* **32,** 567 (1982); H. J. Mallach *et al., Blut-alkohol* **20,** 196 (1983); V. Schmidt *et al., Med. Welt* **35,** 32 (1984). Clinical studies: G. Laakman *et al., Arzneimittel-Forsch.* **30,** 1233 (1980).

Hydrochloride, $C_{18}H_{18}BrClN_2O \cdot HCl$, *Talis.* Crystals from ethanol, mp 193-196°. LD_{50} orally in NMRI white mice: 1578 mg/kg (Milkowski).
THERAP CAT: Anxiolytic.

5981. [2.2]Metacyclophane. *Tricyclo[9.3.1.1^{4,8}]hexadeca-1(15),4,6,8(16),11,13-hexaene;* di-*m*-xylylene; *m*-dixylylene. $C_{16}H_{16}$; mol wt 208.30. C 92.26%, H 7.74%. Prepd by the action of phenyllithium on *m*-xylene dibromide: Allinger *et al., J. Am. Chem. Soc.* **83,** 1974 (1961); *eidem, J. Org. Chem.* **32,** 2272 (1967).

Orthorhombic prisms from ether, mp 132.5°. bp_{760} 290°; bp_{12} 170°. Sparingly sol in alc. More sol in ether, benzene.

5982. Metalaxyl. *N-(2,6-Dimethylphenyl)-N-(methoxyacetyl)-DL-alanine methyl ester;* methyl *N*-(2-methoxyacetyl)-*N*-(2,6-xylyl)-DL-alaninate; metaxanin; CGA-48988; Apron; Ridomil; Subdue. $C_{15}H_{21}NO_4$; mol wt 279.34. C 64.50%, H

7.58%, N 5.01%, O 22.91%. Acylalanine fungicide systemically active against phytopathogens of the order *Peronosporales.* Prepn: A. Hubele, **Ger.** pat. 2,515,091 corresp to **U.S.** pat. 4,151,299 (1975, 1979 to Ciba-Geigy). Activity: A. Kerkenaar, A. K. Sijpesteijn, *Pestic. Biochem. Physiol.* **15,** 71 (1981). In potato blight: H. W. Platt, *Can. J. Plant Pathol.* **5,** 38 (1983). Enzyme-linked immunosorbent assay of residues in foods: W. H. Newsome, *J. Agr. Food Chem.* **33,** 528 (1985). Comprehensive description: P. A. Urech, F. J. Schwinn, *Phytiatr.-Phytopharm.* **27,** 239-247 (1978).

Whitish crystals, mp 71-72°. Vapor press at 20°: 2.2 × 10⁻⁶ mm Hg. Soly in water at 20°: 7.1 g/l. Readily sol in most organic solvents. LD_{50} in rats (mg/kg): 669 orally (Urech).

USE: Agricultural fungicide.

5983. Metaldehyde. Metacetaldehyde. A polymer of acetaldehyde. $(C_2H_4O)_n$. Prepd by polymerization of acetaldehyde in the presence of HCl, H_2SO_4 at low temp: Kekule, Zincke, *Ann.* **162,** 125 (1872); in the presence of pyridine and HBr: Wilder, **U.S.** pat. 2,426,961 (1947 to Publicker Ind.).

Prisms, mp 246° in sealed tube. Sublimes at about 112°, but dec with partial regeneration of aldehyde above 80°. Practically insol in water. Sol in benzene, chloroform; sparingly sol in alcohol, ether. LD_{50} orally in rats: 630 mg/kg, *Toxic Substances List,* H. E. Christensen, Ed. (1973) p 587.

Caution: Ingestion may cause severe abdominal pain, nausea, vomiting, diarrhea, marked rise in body temp, convulsions, coma.

USE: In compressed form as a fuel instead of alcohol; slug and snail poison.

5984. Metamfepramone. *2-(Dimethylamino)-1-phenyl-propanone; 2-(dimethylamino)propiophenone;* benzoyl-α-(dimethylamino)ethane; α-(dimethylamino)propiophenone; metamfepyramone; *N*-methylephedrone; dimepropion. $C_{11}H_{15}NO$; mol wt 177.25. C 74.54%, H 8.53%, N 7.90%, O 9.03%. Prepn from 2-bromopropiophenone and dimethylamine: Thomson, Stevens, *J. Chem. Soc.* **1932,** 1932; from *N,N*-dimethyl-2-(dimethylamino)propionamide and phenylmagnesium bromide: Eidebenz, *Arch. Pharm.* **280,** 49 (1942); from α-acetylbenzyl alcohol and dimethylamine: Iwao *et al., J. Pharm. Soc. Japan* **74,** 551 (1954). Prepn of *d*-form: Freudenberg, Nikolai, *Ann.* **510,** 223 (1934). Resolution of *dl*-form: Takamatsu, *J. Pharm. Soc. Japan* **76,** 1219 (1956).

dl-Form, bp₁₃ 126°.
dl-Form hydrochloride, $C_{11}H_{15}NO.HCl$, *Effilone.* Crystals, mp 202-204° (Iwao) also reported as dec 201-202° (Eidebenz).
l-Form, $[\alpha]_D^{26}$ —60°.
l-Form hydrochloride, prisms, mp 197-199°. $[\alpha]_D^{26}$ —52.5°.
THERAP CAT: Anorexic.

5985. Metamitron. *4-Amino-3-methyl-6-phenyl-1,2,4-triazin-5(4H)-one;* BAY DRW 1139; Goltix. $C_{10}H_{10}N_4O$; mol wt 202.22. C 59.40%, H 4.98%, N 27.71%, O 7.91%. Prepn: K. Dickore *et al.,* **Ger.** pat. 2,107,757; *eidem,* **U.S.** pat. 3,847,914 (1972, 1974 both to Bayer); W. Draber *et al., Ann.* **1976,** 2206. Chemistry and mode of action: *idem et*

al., Proc. 8th Int. Plant Prot. Congress **1,** 203 (1975). Comprehensive description: H. Lembrich, *Pflanzenschutz-Nachr.* **31,** 197 (1978). Degradation in soil: R. Allen, A. Walker, *Pestic. Sci.* **18,** 95 (1987). HPLC determn in surface water: R. B. Geerdink, *J. Chromatog.* **543,** 244 (1991).

Crystals from isopropanol/water, mp 169°. uv max (methanol): 312 nm (log ε 4.06). Vapor pressure (20-70°): < 1 × 10⁻⁴ mbar. Soly (g/100g): water 0.18; isopropanol 0-1; methylene chloride 1-5; toluene 0-1; ligroin (80-110°) 0-1. Stable in acid media; rapidly decomp at pH > 10. LD_{50} orally in male, female rats, male, female mice: 3343, 1832, 1450, 1463 mg/kg; dermally in rats: > 500 mg/kg. LC_{50} (4 hr) in rats, mice, hamsters (mg/m³): > 331, > 206, > 206 by inhalation (Lembrich).

USE: Herbicide.

5986. Metampicillin. *[2S-[2α,5α,6β(S*)]]-3,3-Dimethyl-6-[[(methyleneamino)phenylacetyl]amino]-7-oxo-4-thia-1-azabicyclo[3.2.0]heptane-2-carboxylic acid;* D-6-[α-(methyleneamino)phenylacetamido]penicillanic acid; methampicillin; Bonopen; Fedacilina; Micinovo; Pravacilin; Ruticina; Suvipen; Viderpen. $C_{17}H_{19}N_3O_4S$; mol wt 361.42. C 56.50%, H 5.30%, N 11.63%, O 17.71%, S 8.87%. Semisynthetic antibiotic related to penicillin. Prepn: **Belg.** pat. 661,232, *C.A.* **65,** 3884e (1966); Gradnik, **Brit.** pat. 1,081,-093, *C.A.* **68,** 114595g (1968), (1965 and 1967, both to E.R.A.S.M.E.). Synthesis: Gradnik *et al., Farmaco, Ed. Sci.* **26,** 20 (1971). Antibacterial activity studies: Sutherland *et al., Chemotherapy* **17,** 145 (1972). Pharmacokinetics: Fleischmann *et al., Farmaco, Ed. Prat.* **26,** 106 (1971). Clinical studies: Farina, *Minerva Med.* **60,** 1999 (1969); Ginocchi, *ibid.* 2003; Cardinale, Arrotta, *ibid.* 2011.

Sodium salt, $C_{17}H_{18}N_3NaO_4S$, *Ocelina, Magnipen, Venzoquimpe.*
THERAP CAT: Antibacterial.

5987. Metanephrine. *4-Hydroxy-3-methoxy-α-(methylaminomethyl)benzenemethanol;* α-(methylaminomethyl)vanillyl alcohol; 3-*O*-methylepinephrine; 3-*O*-methyladrenaline; 1-(4-hydroxy-3-methoxyphenyl)-2-methylaminoethanol. $C_{10}H_{15}NO_3$; mol wt 197.23. C 60.90%, H 7.67%, N 7.10%, O 24.34%. A naturally occurring derivative of epinephrine, found in the urine and in certain tissues. Various methods of prepn: Külz, Hornung, **Ger.** pat. 682,394 (1939); *Chem. Zentr.* **1940, I,** 1078, *C.A.* **36,** 3011 (1942); Axelrod *et al., J. Biol. Chem.* **233,** 697 (1958); Heacock, Hutzinger, *Chem. & Ind. (London)* **1961,** 595.

dl-Form hydrochloride, $C_{10}H_{15}NO_3.HCl$, prisms from ethanol + ether (also reported as crystals from dil acetone). Dec 175°. uv max (ethanol): 231, 280 nm (ε 7600, 3100).

5988. Metanilic Acid. *3-Aminobenzenesulfonic acid;* *m*-sulfanilic acid; aniline-*m*-sulfonic acid. $C_6H_7NO_3S$; mol wt 173.19. C 41.61%, H 4.07%, N 8.09%, O 27.71%, S 18.51%. Usually obtained by reduction of 3-nitrobenzenesulfonic acid: Fierz-David, Blangey, *Fundamental Processes of Dye Chemistry* (Interscience, New York, 1949) pp 120-123; A. I. Vogel, *Practical Organic Chemistry* (Longmans, London, 3rd ed., 1959) p 589. Large-scale process: *FIAT Final Rept.* **1313** (I), 187-191 (1948). The industrial reduction with iron filings and dil acid gives up to 90% yields, in the lab it seldom exceeds 55%. Better yields with small amounts are claimed for a hydrazine-Raney nickel reduction (about 75%): Gialdi *et al.*, *Farmaco Ed. Sci.* **14**, 765 (1959) or by using WS_2 (about 94%): Ehrmann, **Fr.** pat. **1,336,648** (1963 to BASF), *C.A.* **60**, 2846d (1964).

Anhydrous, orthorhombic needles from water. d 1.69. Very slow crystn yields triclinic prisms of the sesquihydrate. Crystallographic data for both forms: Hall, Maslen, *Acta Cryst.* **18**, 301-306 (1965). Photomicrograph of the sesquihydrate: *Helv. Chim. Acta* **12** (1929), facing page 666. Both forms dec on heating without melting. K at 25° = 2.00 × 10^{-4}. Soly of the sesquihydrate in water (16.8°): 2.37% (w/w). The soly of the anhyd form in water is given as 0.79% (w/w) at 0° and as 6.50% (w/w) at 85°. Sparingly sol in methanol, ethanol.

Sodium salt, $H_2NC_6H_4SO_3Na$, minute crystals from water, dec 302-304°.

USE: The sodium salt in the manuf of azo dyes. In the synthesis of certain sulfa drugs.

5989. Metanil Yellow. *3-[[4-(Phenylamino)phenyl]azo]benzenesulfonic acid monosodium salt; C.I. Acid Yellow 36;* *m*-[(*p*-anilinophenyl)azo]benzenesulfonic acid sodium salt; sodium salt of metanilylazodiphenylamine; Ext. D & C Yellow No. 1; C.I. 13065; Tropaeolin G. $C_{18}H_{14}N_3NaO_3S$; mol wt 375.38. C 57.59%, H 3.76%, N 11.19%, Na 6.12%, O 12.79%, S 8.54%. Prepn: Welcher, *Organic Analytical Reagents*, **vol. 4** (Van Nostrand, 1948) p 516; *Colour Index* **vol. 4** (3rd ed., 1971) p 4045.

Brownish-yellow powder. Sol in water, alc; moderately sol in benzene, ether; slightly sol in acetone.

Note: The provisional approval for external use in drugs and cosmetics was terminated by the FDA, see *FDA Consumer,* February, 1979, p 5.

USE: As indicator in 0.1% soln, of which 2 drops are required for 10 ml liquid. pH: 1.2 red to 2.3 yellow.

5990. Metaphanine. *(8β,10β)-8,10-Epoxy-8-hydroxy-3,4-dimethoxy-17-methylhasubanan-7-one.* $C_{19}H_{23}NO_5$; mol wt 345.40. C 66.07%, H 6.71%, N 4.06%, O 23.16%. A member of the hasubanan alkaloids; also possesses an intramolecular hemiketal ring. Isolated from stems of *Stephania japonica* Miers, *Menispermaceae* from which the alkaloids **stephanine** and protostephanine, *q.v.* are also obtained: Kondo, Sanada, *J. Pharm. Soc. Japan* **514**, 5 (1924); *Yakugaku Zasshi* **44**, 5, 1034 (1924); **48**, 177, 930 (1927); Kondo, Watanabe, *ibid.* **58**, 268 (1938). Structure and stereochemistry: Tomita *et al.*, *Tetrahedron Letters* **1964**, 3605; *Chem. Pharm. Bull.* **13**, 695 (1965). Synthesis of *dl*-form: Ibuka *et al.*, *Tetrahedron Letters* **1972**, 1393; *eidem, Chem. Pharm. Bull.* **22**, 907 (1974).

Needles, mp 232°. Also reported as colorless prisms, mp 205-206° for the *dl*-form (Ibuka). pKa 6.03. Sol in water, alcohol.

Dihydrometaphanine, $C_{19}H_{25}NO_5$, cryst, mp 211°. $[\alpha]_D^{20}$ +72° (in chloroform). pKa 6.76.

5991. Metapramine. *10,11-Dihydro-N,5-dimethyl-5H-dibenz[b,f]azepin-10-amine; 10,11-dihydro-5-methyl-10-(methylamino)-5H-dibenz[b,f]azepine;* RP-19560; Timaxel. $C_{16}H_{18}N_2$; mol wt 238.33. C 80.63%, H 7.61%, N 11.75%. Psychotropic agent, related structurally to imipramine, *q.v.* Prepn: J. C. Fouche, C. G. A. Gueremy, **S. Afr.** pat. **68 00345** corresp to **U.S.** pat. **3,622,565** (1968, 1971 both to Rhone-Poulenc). Determn in plasma by GC: A. R. Viala *et al., Anal. Chem.* **49**, 2354 (1977). HPLC method: J. P. Sommadossi *et al., J. Chromatog.* **228**, 205 (1982). Pharmacological and clinical effects: P. Dick, *Encephale* **4**, 41 (1978). Clinical studies: L. F. Gayral *et al., ibid.* 365; E. J. Caille, J.-P. Brun, *Psychol. Med.* **13**, 1879 (1981). Biotransformation in animals and man: B. Decouvelaere *et al., Therapie* **37**, 249 (1982).

Hydrochloride, $C_{16}H_{19}ClN_2$, crystals from isopropanol + ether, mp 238-240°. Injectable formulation. The fumarate is used for tablet formulations.

THERAP CAT: Antidepressant.

5992. Metaproterenol. *5-[1-Hydroxy-2-[(1-methylethyl)amino]ethyl]-1,3-benzenediol; 3,5-dihydroxy-α-[(isopropylamino)methyl]benzyl alcohol;* 1-(3,5-dihydroxyphenyl)-2-isopropylaminoethanol; 1-(3,5-dihydroxyphenyl)-1-hydroxy-2-isopropylaminoethane; orciprenaline. $C_{11}H_{17}NO_3$; mol wt 211.26. C 62.54%, H 8.11%, N 6.63%, O 22.72%. Prepn and resolution of *d*- and *l*-forms: **Belg.** pat. **611,502**; O. Thoma, K. Zeile, **U.S.** pat. **3,341,594** (1961, 1967 both to Boehringer, Ing.). Pharmacology: H. H. Pelz, *Am. J. Med. Sci.* **253**, 321 (1967). Metabolism: K. Tatsumi *et al., Yakugaku Zasshi* **90**, 639 (1970); *ibid.* **91**, 680 (1971). Toxicity: E. I. Goldenthal, *Toxicol. Appl. Pharmacol.* **18**, 185 (1971). Review of clinical toxicology: J. D. Truwit, *Crit. Care Clinics* **7**, 639-657 (1991).

Crystals, mp 100°.

Sulfate, $(C_{11}H_{17}NO_3)_2 \cdot H_2SO_4$, *TH-152, Alotec, Alupent, Metaprel, Novasmasol.* Crystals from 90% ethanol, mp 202-203°. LD_{50} in rats (mg/kg): 42 orally (Goldenthal).

Hydrochloride, $C_{11}H_{17}NO_3 \cdot HCl$, crystals from methanol + ether, mp 147°.

d-Form hydrochloride, crystals, mp 212-213°. [α]$_D$ +45.2° (c = 5 in methanol).

l-Form hydrochloride, crystals, mp 212°. [α]$_D$ −45°.

THERAP CAT: Bronchodilator.

5993. Metaraminol. *[R-(R*,S*)]-α-(1-Aminoethyl)-3-hydroxybenzenemethanol; (−)-α-(1-aminoethyl)-m-hydroxybenzyl alcohol; m-hydroxynorephedrine; m-hydroxypropadrine; m-hydroxyphenylpropanolamine; m-hydroxy-α-(1-aminoethyl)benzyl alcohol; 2-amino-1-(m-hydroxyphenyl)-1-propanol; 1-(m-hydroxyphenyl)-2-amino-1-propanol; α-(m-hydroxyphenyl)-β-aminopropanol; metaradrine; Pressonex.* $C_9H_{13}NO_2$; mol wt 167.21. C 64.65%, H 7.84%, N 8.38%, O 19.14%. Prepn from m-hydroxyisonitrosopropiophenone: **Brit. pat. 353,361** (1930 to I. G. Farben); *l*-form: **Brit. pat. 396,951** (1932 to I. G. Farben); from *l*-m-hydroxyphenylacetylcarbinol: **Swiss pat. 162,367** (1931 to I. G. Farben); *see also* **U.S. pats. 1,948,162** and **1,951,302**; Hartung, **U.S. pat. 1,995,709** (1935 to Sharp & Dohme). Pharmacology: A. Cession-Fossion, *Arch. Int. Pharmacodyn. Ther.* **172,** 421 (1968). Toxicity data: O. H. Siegmund *et al., J. Pharmacol. Exp. Ther.* **92,** 207 (1948).

Bitartrate (hydrogen L-tartrate), $C_{13}H_{19}NO_8$, *Aramine, Icoral B, Pressorol.* Crystals, mp 176-177°. Freely sol in water. pH of 1% aq soln about 3.5.

Hydrochloride, $C_9H_{13}NO_2$.HCl, hygroscopic crystals, [α]$_D^{20}$ −19.75°. Freely sol in water. LD$_{50}$ i.p. in mice: 440 mg/kg (Siegmund).

Oxalate dihydrate, $C_9H_{13}NO_2.C_2H_2O_4.2H_2O$, crystals, mp 190°. [α]$_D^{20}$ −21.66°. Sol in water.

THERAP CAT: Adrenergic.

5994. Metaxalone. *5-(3,5-Dimethylphenoxymethyl)-2-oxazolidinone; AHR-438; Skelaxin.* $C_{12}H_{15}NO_3$; mol wt 221.26. C 65.14%, H 6.83%, N 6.33%, O 21.69%. Prepn: Lunsford *et al., J. Am. Chem. Soc.* **82,** 1166 (1960); Lunsford, **U.S. pat. 3,062,827** (1962 to A. H. Robins).

Crystals from ethyl acetate, mp 121.5-123°.

THERAP CAT: Muscle relaxant (skeletal).

5995. Metazocine. *1,2,3,4,5,6-Hexahydro-3,6,11-trimethyl-2,6-methano-3-benzazocin-8-ol; 2'-hydroxy-2,5,9-trimethyl-6,7-benzomorphan; methobenzmorphan.* $C_{15}H_{21}$-NO; mol wt 231.34. C 77.88%, H 9.15%, N 6.05%, O 6.92%. Prepd from 3,4-lutidine methiodide: May, Fry, *J. Org. Chem.* **22,** 1366 (1957); May, Ager, *ibid.* **24,** 1432 (1959).

Plates from dil methanol, mp 232-235°.

Hydrochloride monohydrate, $C_{15}H_{21}NO.HCl.H_2O$, rods from abs alcohol + ether, mp 194-196°.

Note: This is a controlled substance (opiate) listed in the U.S. Code of Federal Regulations, Title 21 Part 1308.12 (1995).

THERAP CAT: Analgesic (narcotic).

5996. Metcaraphen. *1-(3',4'-Dimethylphenyl)-1-cyclopentanecarboxylic acid 2-diethylaminoethyl ester; 2-diethylaminoethyl 1-(3',4'-dimethylphenyl)-1-cyclopentanecarboxylate; Netrin.* $C_{20}H_{31}NO_2$; mol wt 317.47. C 75.67%, H 9.84%, N 4.41%, O 10.08%. Prepd by treating the parent acid chloride with 2-diethylaminoethanol: Martin, Häfliger, **U.S. pat. 2,404,588** (1946 to Geigy).

Liquid. Faint amine odor, bp$_{0.05}$ 126-128°.

Hydrochloride, $C_{20}H_{31}NO_2$.HCl, crystals from alcohol + ethyl acetate. Sol in water.

THERAP CAT: Anticholinergic.

5997. Meteloidine. *[1α,3β(E),5α,6α,7α]-2-Methyl-2-butenoic acid 6,7-dihydroxy-8-methyl-8-azabicyclo[3.2.1]oct-3-yl ester; (E)-1αH,5αH-tropane-3α,6β,7β-triol 3-(2-methylcrotonate);* 3-(3,6,7-tropanetriol) tiglate; 6,7-dihydroxytropinetiglic acid ester; 6,7-dihydroxy-3-tigloyloxytropane. C_{13}-$H_{21}NO_4$; mol wt 255.31. C 61.16%, H 8.29%, N 5.49%, O 25.07%. From leaves of *Datura meteloides* DC., *Solanaceae:* Pyman, Reynolds, *J. Chem. Soc.* **93,** 2077 (1908); King, *ibid.* **115,** 487 (1919); from *D. ferox* L.: Evans, Wellendorf, *ibid.* **1959,** 1406. Stereochemistry: Heusner, *Z. Naturforsch.* **9b,** 683 (1954). Synthesis: Zeile, Heusner, *Arch. Pharm.* **292,** 238 (1959).

Flat needles from benzene, mp 141-142°. uv max: 217 nm (ε 12,200). Freely sol in alcohol, chloroform, acetone; sparingly sol in water, ether, benzene.

Hydrobromide dihydrate, $C_{13}H_{22}BrNO_4.2H_2O$, chisel-shaped needles, when anhydr, mp 250°.

5998. Metepa. *1,1',1''-Phosphinylidynetris[2-methylaziridine]; tris[2-methyl-1-aziridinyl]phosphine oxide; tris[1-methylethylene]phosphoric triamide; methyl aphoxide; methapoxide; MAPO.* $C_9H_{18}N_3OP$; mol wt 215.24. C 50.22%, H 8.43%, N 19.52%, O 7.43%, P 14.39%. Prepd by treatment of 2-methylethyleneimine with POCl$_3$ in alk medium: Parke *et al.,* **U.S. pat. 2,606,902** (1952 to Am. Cyanamid). Toxicity study: Gaines, *Toxicol. Appl. Pharmacol.* **14,** 515 (1969).

Liquid, bp 90-92° (0.15-0.3 mm). LD$_{50}$ in male, female rats (mg/kg): 136, 213 orally (Gaines).

USE: Chemosterilant; in creaseproofing and flameproofing textiles.

5999. Metergoline. *[[(8β)-1,6-Dimethylergolin-8-yl]methyl]carbamic acid phenylmethyl ester;* D-8β-[(carbobenzoxyamino)methyl]-1,6-dimethyl-10α-ergoline; D-*N*-carbobenzoxydihydro-1-methyllysergamine I; D-8β-[(carboxyamino)methyl]-1,6-dimethylergoline I benzyl ester; D-*N*-carboxydihydro-1-methyllysergamine I benzyl ester; D-[(4,6,-6a,7,8,9,10,10a-octahydro-4,7-dimethyl-10aα-indolo[4,3-*fg*]-quinolin-9β-yl)methyl]carbamic acid benzyl ester; methergoline; Liserdol; Contralac. $C_{25}H_{29}N_3O_2$; mol wt 403.52. C 74.41%, H 7.24%, N 10.41%, O 7.93%. Prepn: Bernardi *et al.*, *Gazz. Chim. Ital.* **94**, 936 (1964); Camerino *et al.*, U.S. pat. **3,238,211** (1966 to Farmitalia). Metabolic studies: Arcamone *et al.*, *Boll. Chim. Farm.* **110**, 704 (1971). Mode of action study: L. Krulich *et al.*, *Endocrinology* **108**, 1115 (1981). Clinical antiprolactin activity: F. Scapin *et al.*, *Eur. J. Clin. Pharmacol.* **22**, 181 (1982); A. Caballero *et al.*, *J. Reprod. Med.* **32**, 115 (1987).

Crystals from benzene + ether, mp 146-149°. $[\alpha]_D^{28}$ −7 ±2°. uv max: 291 nm ($E_{1cm}^{1\%}$ 165). Very sol in pyridine; sol in alc, acetone, chloroform. Practically insol in benzene, ether, water. LD_{50} orally in mice: 430 mg/kg.

THERAP CAT: Prolactin inhibitor.
THERAP CAT (VET): Prolactin inhibitor.

6000. Metescufylline. *[(7-Hydroxy-4-methyl-2-oxo-2H-1-benzopyran-6-yl)oxy]acetic acid compd with 7-[2-(diethylamino)ethyl]-3,7-dihydro-1,3-dimethyl-1H-purine-2,6-dione (1:1);* 7-[2-(diethylamino)ethyl]theophylline compd with [(7-hydroxy-4-methyl-2-oxo-2H-1-benzopyran-6-yl)oxy]acetic acid (1:1); etamiphylline methesculetol; methescufylline; theophylline methesculetol; Veinartan; Venarterin. $C_{25}H_{31}N_5O_8$; mol wt 529.55. C 56.70%, H 5.90%, N 13.23%, O 24.17%. Prepn: P. Chabrier *et al.*, Fr. pat. **M1234** (1962 to Lab. Dausse), *C.A.* **57**, 16769f (1962).

Crystals, mp 124°. Slightly sol in cold water; freely sol in hot water. Practically insoluble in acetone. LD_{50} i.v. in mice: 260 mg/kg (Chabrier).

THERAP CAT: Capillary protectant.

6001. Metformin. *N,N-Dimethylimidodicarbonimidic diamide; 1,1-dimethylbiguanide; N,N-dimethyldiguanide; N'-dimethylguanylguanidine;* DMGG; LA-6023. $C_4H_{11}N_5$; mol wt 129.17. C 37.20%, H 8.58%, N 54.22%. Oral hypoglycemic agent. Prepn: Werner, Bell, *J. Chem. Soc.* **121**, 1790 (1922); Shapiro *et al.*, *J. Am. Chem. Soc.* **81**, 3728 (1959). Use as antidiabetic: J. J. Sterne, U.S. pat. **3,174,-901** (1965 to Jan Marcel Didier Aron-Samuel). Toxicity: *Rx Bulletin* **3**, 25 (1972). Pharmacokinetics in man: G. T. Tucker *et al.*, *Brit. J. Clin. Pharmacol.* **12**, 235 (1981). Review of pharmacology: L. S. Hermann, *Diabete Metab.* **5**, 233-245 (1979). Efficacy in NIDDM: R. A. DeFronzo *et al.*, *N. Engl. J. Med.* **333**, 541 (1995). Metabolic effects and mechanism of action study: M. Stumvoll *et al.*, *ibid.* 550.

Hydrochloride, $C_4H_{11}N_5 \cdot HCl$, *Diabetosan, Diabex, Glucophage, Metiguanide.* Prisms from water, mp 232° (Werner, Bell); crystals from propanol, mp 218-220° (uncorr) (Shapiro). Sol in water, 95% alcohol. Practically insol in ether, chloroform. LD_{50} in rats (mg/kg): 1000 orally, 300 s.c. (*Rx Bulletin*).

p-Chlorophenoxyacetate (salt), $C_4H_{11}N_5 \cdot C_8H_7ClO_3$, *Glucinan.*

Embonate, $(C_4H_{11}N_5)_2 \cdot C_{23}H_{16}O_6$, *metformin pamoate, Stagid.*

THERAP CAT: Antidiabetic.

6002. Methabenzthiazuron. *N-2-Benzothiazolyl-N,N'-dimethylurea; 1-(2-benzothiazolyl)-1,3-dimethylurea;* metabenzthiazuron; MBU; Bayer 5633; Bayer 74283; Tribunil. $C_{10}H_{11}N_3OS$; mol wt 221.28. C 54.28%, H 5.01%, N 18.99%, O 7.23%, S 14.49%. Derivative of urea. Prepn and use as pre-emergence herbicide: N. E. Searle, U.S. pat. **2,756,135** (1956 to du Pont). Use as pre- and post-emergence herbicide in wheat and barley: H. Hack *et al.*, Brit. pat. **1,085,430** (1967 to Bayer). Herbicidal properties: H. Hack, *Pflanzenschutz-Nachr.* **22**, 331 (1969). Toxicity studies: G. Kimmerle, E. Löser, *ibid.* 351. Use in winter cereals: D. C. Clark *et al.*, *Proc. 12th Brit. Weed Control Conf.* 163 (1974). Mode of action: G. F. Collet, *Weed Res.* **9**, 340 (1969). Long-term effect on soil: P. L. Huge, *Pflanzenschutz-Nachr.* **34**, 97 (1981). Brief review: P. Lours, *Def. Veg.* **24**, 91 (1970).

White crystals from benzene, mp 119-120.5°. Soly in water at 20°: 59 ppm. Sol in organic solvents. Vapor pressure at 20°: < 10^{-6} mm Hg. LD_{50} in mice (mg/kg): >1000 orally; in male, female rats (mg/kg): >2500, >2500 orally; 540, 315 i.p. (Kimmerle, Löser).

USE: Selective herbicide.

6003. Methacholine Chloride. *2-(Acetyloxy)-N,N,N-trimethyl-1-propanaminium chloride;* acetyl-β-methylcholine chloride; *O*-acetyl-β-methylcholine chloride; (2-hydroxypropyl)trimethylammonium chloride acetate; (2-acetoxypropyl)trimethylammonium chloride; trimethyl-β-acetoxypropylammonium chloride; Amechol; Provocholine. C_8H_{18}-$ClNO_2$; mol wt 195.69. C 49.10%, H 9.27%, Cl 18.12%, N 7.16%, O 16.35%. Parasympathomimetic bronchoconstrictor. Prepn: R. T. Major, J. K. Cline, U.S. pat. **2,040,146** (1936 to Merck & Co.). Mechanism of ganglionic blockade in cats: R. L. Volle, *J. Pharmacol. Exp. Ther.* **158**, 66 (1967). Clinical diagnostic efficacy in bronchial asthma: S. L. Spector, R. S. Farr, *J. Allergy Clin. Immunol.* **56**, 308 (1975); J. G. Easton, I. Hirata, *Ann. Allergy* **50**, 171 (1983).

White, hygroscopic needles from ether, mp 172-173°. Slight odor of dead fish. Freely sol in water, alcohol, chloroform. Insol in ether. Aq solns are neutral to litmus. Should not be handled in very moist atmosphere. Bromide is less hygroscopic.

Antidote: Atropine.

THERAP CAT: Cholinergic. Diagnostic aid (bronchial asthma).

6004. Methacrifos. *(E)-3-[(Dimethoxyphosphinothioyl)oxy]-2-methyl-2-propenoic acid methyl ester; 3-hydroxy-2-*

methylacrylic acid methyl ester, O-ester with O,O-dimethyl phosphothioate; methyl (E)-3-(dimethoxyphosphinothioyloxy)-2-methylacrylate; CGA-20168; Damfin. $C_7H_{13}O_5PS$; mol wt 240.22. C 35.00%, H 5.45%, O 33.30%, P 12.89%, S 13.35%. Organophosphorus insecticide effective against arthropod pests in stored grains. Prepn: E. Beriger, L. Pinter, **S. Afr. pat. 67 04,184** corresp to U.S. pat. **3,594,454** (1967, 1971 both to Ciba); eidem, **Belg. pat. 766,000** corresp to U.S. pat. **3,923,932** (1971, 1975 both to Ciba-Geigy). GLC determn of residues in stored grain: J. Desmarchelier et al., Pestic. Sci. **8**, 473 (1977). Efficacy and long-term stability: R. L. Kirkpatrick et al., J. Econ. Entomol. **75**, 277 (1982). Comparative field trial in stored sorghum: M. Bongston et al., Pestic. Sci. **14**, 385 (1983). Comprehensive description: R. Wyniger et al., Proc. Brit. Crop Prot. Conf.—Pests Dis. **1977**, 1033.

Oil, $bp_{0.01}$ 90°. Vapor pressure (20°): 8.3×10^{-4} mm Hg. Soly in water (20°): ~400 ppm. Readily sol in many organic solvents. Highly toxic to rainbow trout. LD_{50} in rats (mg/kg): 678 orally; >3100 dermally (Wyniger).

USE: Grain protectant, insecticide, acaricide.

6005. Methacrylic Acid. 2-Methyl-2-propenoic acid; α-methylacrylic acid. $C_4H_6O_2$; mol wt 86.09. C 55.81%, H 7.02%, O 37.17%. Occurs in oil from Roman chamomile. Prepd by dehydration of α-hydroxyisobutyric acid: Crawford, **U.S. pat. 2,143,941** (1939 to I.C.I.); by hypochlorite oxidation of methyl α-alkylvinyl ketone: Meitzner, **U.S. pat. 2,192,142** (1940 to Rohm & Haas); by hydrolysis of acetone cyanohydrin: Crawford, **Brit. pat. 405,699** (1932 to I.C.I.); by oxidation of methacrolein: Bauer, **U.S. pat. 2,153,406** (1939 to Rohm & Haas). Toxicity study: W. Deichmann, J. Ind. Hyg. Toxicol. **23**, 343 (1941). Review: J. W. Nemec, L. S. Kirch in Kirk-Othmer Encyclopedia of Chemical Technology vol. **15** (Wiley-Interscience, New York, 3rd ed., 1981) pp 346-376.

Long prisms, mp 16°, forming a corrosive liquid. Acrid, repulsive odor. d_4^{20} 1.0153. bp_{760} 163°; bp_{30} 81°; bp_{12} 63°. n_D^{20} 1.43143. Flash pt, open cup: 170°F (76°C). Sol in warm water; miscible with alc, ether. Polymerizes easily, especially on heating or in the presence of traces of HCl. The polymer forms a ceramic-looking mass, sol in abs alc, from which it is precipitated by ether.

Methyl ester, **methyl methacrylate,** polymerizes easily, forming a clear plastic known as **Lucite, Plexiglas, Perspex.** Sol in methyl ethyl ketone, tetrahydrofuran, esters, aromatic and chlorinated hydrocarbons. LD_{50} orally in rats: 8.4 g/kg (Deichmann). Review of carcinogenic risk: IARC Monographs **60**, 445-474 (1994).

Caution: Potential symptoms of overexposure are irritation of eyes, skin and mucous membranes; may cause dermatitis, eye and skin burns. See NIOSH Pocket Guide to Chemical Hazards (DHHS/NIOSH 94-117, 1994) p 194, 214.

USE: In the manuf of methacrylate resins and plastics.

6006. Methacrylonitrile. 2-Methyl-2-propenenitrile; isopropenylnitrile; α-methylacrylonitrile; isopropene cyanide. C_4H_5N; mol wt 67.09. C 71.61%, H 7.51%, N 20.88%. $CH_2{=}C(CH_3)C{\equiv}N$. Prepd by vapor-phase catalytic oxidation of methallylamine: Peters et al., Ind. Eng. Chem. **40**, 2046 (1948); **U.S. pat. 2,375,016** (1945 to Shell); by the dehydration of methacrylamide: Kung, **U.S. pat. 2,373,190** (1945 to Goodrich); from isopropylene oxide and ammonia: Spillane, Kayser, **U.S. pat. 2,557,703** (1951 to Allied Chem.). Toxicity study: Smyth et al., Am. Ind. Hyg. Assoc. J. **23**, 95 (1962).

Liquid. d_4^{20} 0.8001; d_4^{30} 0.7896. mp −35.8°. bp_{760} 90.3°. n_D^{20} 1.4007; n_D^{30} 1.3954. Flash pt, open cup: 55°F (13°C). Viscosity at 20°, 0.392 cp. Surface tension at 20°, 24.4 dynes/cm. Soly in water at 20° = 2.57 wt-%, at 50° = 2.69 wt-%. Soly of water in methacrylonitrile at 20° = 1.62 wt-%, at 50° = 2.83 wt-%. Misc with acetone, octane, toluene at 20-25°. LD_{50} orally in rats: 0.25 ml/kg (Smyth).

Caution: Lacrimator, insidious poison, delayed skin reaction.

USE: In prepn of homopolymers and copolymers; as an intermediate in the prepn of acids, amides, amines, esters, nitriles.

6007. Methacycline. [4S-(4α,4aα,5α,5aα,12aα)]-4-Dimethylamino-1,4,4a,5,5a,6,11,12a-octahydro-3,5,10,12,12a-pentahydroxy-6-methylene-1,11-dioxo-2-naphthacenecarboxamide; 6-methyleneoxytetracycline; 6-methylene-5-hydroxytetracycline; metacycline; Bialatan. $C_{22}H_{22}N_2O_8$; mol wt 442.43. C 59.73%, H 5.01%, N 6.33%, O 28.93%. Broad spectrum, semi-synthetic antibiotic related to tetracycline, q.v. Prepn from oxytetracycline, q.v.: R. K. Blackwood et al., **U.S. pat. 2,984,686** (1961 to Pfizer); eidem, J. Am. Chem. Soc. **83**, 2773 (1961); **85**, 3943 (1963). Solubility data: J. R. Marsh, P. J. Weiss, J. Assoc. Offic. Anal. Chem. **50**, 257 (1967). Toxicity data: E. I. Goldenthal, Toxicol. Appl. Pharmacol. **18**, 185 (1971). Comparative clinical study with ampicillin, q.v., in chronic bronchitis: S. Chodosh et al., Chest **69**, 587 (1976).

Hydrochloride, $C_{22}H_{22}N_2O_8 \cdot HCl$, **Adriamicina, Ciclobiotic, Germiciclin, Metadomus, Metilenbiotic, Londomycin, Optimycin, Physiomycine, Rindex, Rondomycin.** Obtained as crystals containing 0.5 mol water and 0.5 mol methanol from methanol + acetone + concd HCl + ether. Yellow, crystalline powder, dec ~205°. Bitter taste. Soluble in water, sparingly sol in alcohol. Practically insol in ether, chloroform. uv max (methanol + 0.01N HCl): 253, 345 nm (log ε 4.37, 4.19). LD_{50} in rats, mice (mg/kg): 252, 288 i.p. (Goldenthal).

THERAP CAT: Antibacterial.

6008. Methadone Hydrochloride. 6-Dimethylamino-4,4-diphenyl-3-heptanone hydrochloride; 1,1-diphenyl-1-(2-dimethylaminopropyl)-2-butanone hydrochloride; 4,4-diphenyl-6-dimethylamino-3-heptanone hydrochloride. $C_{21}H_{28}ClNO$; mol wt 345.91. C 72.92%, H 8.16%, Cl 10.25%, N 4.05%, O 4.63%. Prepn: Eisleb, Office of Publication Board, Department of Commerce, Report no. **PB-981**, 96A; Schultz et al., J. Am. Chem. Soc. **69**, 2454 (1947); Easton et al., ibid. **2941** (1947). Resolution: Larsen et al., ibid. **70**, 4194 (1948); Howe, Sletzinger, ibid. **71**, 2935 (1949); Brode, Hill, J. Org. Chem. **13**, 191 (1948); Howe, Tishler, **U.S. pat. 2,644,010** (1953 to Merck & Co.); Zaugg, **U.S. pat. 2,983,-757** (1961 to Abbott). Pharmacokinetics in patients with chronic pain: C. E. Inturrisi et al., Clin. Pharmacol. Ther. **41**, 392 (1987). Toxicity data: Finnegan et al., J. Pharmacol. Exp. Ther. **92**, 269 (1948); Winter, Flataker, ibid. **98**, 305 (1950). Review: Eddy, J. Am. Pharm. Assoc., Pract. Pharm. Ed. **8**, 536 (1947). Comprehensive description: R.

H. Bishara, *Anal. Profiles Drug Subs.* **3**, 365-439 (1974). Review of use in opioid dependence: E. C. Senay, *Int. J. Addict.* **20**, 803-821 (1985).

dl-Form, *AN-148, Hoechst 10820, Algidon, Algolysin, Amidon, Butalgin, Depridol, Dolophine, Fenadone, Heptadon, Heptanon, Ketalgin, Mecodin, Mephenon, Miadone, Moheptan, Phenadone, Physeptone, Tussol.* Platelets from alcohol + ether, mp 235°. Bitter taste. uv max: 292 nm. Soly (g/100 ml): water 12; alcohol 8; isopropanol 2.4. Practically insol in ether, glycerol. The pH of a 1% aq soln: 4.5-5.6. The free base (mp 78°) is pptd from solns of pH > 6. Aq solns can be autoclaved at 120° for one hour without loss of potency. LD_{50} orally in rats: 95 mg/kg (Finnegan).

l-Form, *Levadone, L-Polamidon.* Crystals, mp 241°. $[\alpha]_D^{20}$ $-145°$ (c = 2.5); $[\alpha]_D^{20}$ $-169°$ (c = 2.1 in alc). LD_{50} s.c. in rats: 44 mg/kg (Winter, Flataker).

Note: This is a controlled substance (opiate) listed in the U.S. Code of Federal Regulations, Title 21 Part 1308.12 (1995).

THERAP CAT: Analgesic (narcotic).

6009. Methadyl Acetate. β-*[2-(Dimethylamino)propyl]-α-ethyl-β-phenylbenzeneethanol acetate (ester); 6-(dimethylamino)-4,4-diphenyl-3-heptanol acetate (ester); O-acetyl-6-dimethylamino-4,4-diphenyl-3-heptanol; 3-acetoxy-6-dimethylamino-4,4-diphenylheptane; 5-acetoxy-2-dimethylamino-4,4-diphenylheptane; acetylmethadol; acemethadone; amidolacetate; race-acetylmethadol.* $C_{23}H_{31}NO_2$; mol wt 353.50. C 78.15%, H 8.84%, N 3.96%, O 9.05%. Congener of methadone, *q.v.* Prepn of β-*dl*-form: M. Bockmühl, G. Ehrhart, *Ann.* **561**, 52 (1948); of α-*dl*-form: M. E. Speeter *et al., J. Am. Chem. Soc.* **71**, 57 (1949); R. L. Clark, U.S. pats. **2,565,592; 2,668,814** (1951, 1954 to Merck & Co.); M. E. Speeter, U.S. pat. **2,649,445** (1953 to Bristol); of α-d and α-l-forms: A. Pohland *et al., J. Am. Chem. Soc.* **71**, 460 (1949). The α-*dl*-form is more active and less toxic than the β-*dl*-form. The α-l-form (levomethadyl acetate, *q.v.*) is less active than the α-d-form but is longer acting. Metabolism of the α-*dl*-form: R. E. McMahon *et al., J. Pharmacol. Exp. Ther.* **149**, 436 (1965). Analgesic, depressant activity of α-*dl*-form in mice, rats: N. B. Eddy *et al., ibid.* **98**, 121 (1950); of enantiomers: N. B. Eddy *et al., J. Org. Chem.* **17**, 321 (1952). Comparison of methadone with α-*dl*-form in the treatment of heroin addiction: J. H. Jaffe *et al., J. Am. Med. Assoc.* **211**, 1834 (1970).

α-*dl*-Form hydrochloride, $C_{23}H_{32}ClNO_2$, crystals from ethyl acetate, mp 213-214°. Sol in water. pH 4-5. LD_{50} in mice (mg/kg): 61.0 s.c., 118.3 orally (Eddy, 1952).

α-*dl*-Form hydrobromide, $C_{23}H_{32}BrNO_2$, crystals, mp 193-194.5°. Sol in water.

α-*d*-Form, *alphacetylmethadol.*

α-*d*-Form hydrochloride, crystals from ethanol-ether, mp 215°. $[\alpha]_D^{25}$ +61.2° (c = 0.2). Sol in water. LD_{50} in mice (mg/kg): 72.2 s.c., 130.4 orally (Eddy, 1952).

α-*l*-Form, *See* levomethadyl acetate.

β-*l*-Form, *betacetylmethadol.*

β-*dl*-Form hydrochloride, crystals, mp 215-217°. Sol in water. LD_{50} in mice (mg/kg): 42.0 s.c., 80.2 orally (Eddy, 1952).

Note: This is a controlled substance (opiate) listed in the U.S. Code of Federal Regulations, Title 21 Part 1308.11 (1995).

6010. Methafurylene. *N-(2-Furanylmethyl)-N',N'-dimethyl-N-2-pyridinyl-1,2-ethanediamine; 2-[(2-dimethylaminoethyl)furfurylamino]pyridine; N,N-dimethyl-N'-(2-pyridyl)-N'-furfurylethylenediamine; N'-(2-furylmethyl)-N'-*

(2-pyridyl)-*N,N-dimethylethylenediamine.* $C_{14}H_{19}N_3O$; mol wt 245.32. C 68.54%, H 7.81%, N 17.13%, O 6.52%. Prepn: Vaughan, Anderson, *J. Am. Chem. Soc.* **70**, 2607 (1948); Kyrides, Zienty, *ibid.* **71**, 1122 (1949); Horclois, U.S. pat **2,502,151** (1950 to Rhône-Poulenc).

Light yellow oil, $bp_{0.2}$ 117.5-118°.

Hydrochloride, $C_{14}H_{19}N_3O.HCl$, needles from ethyl acetate, mp 117-119°.

Dihydrogen citrate, $C_{14}H_{19}N_3O.C_6H_8O_7$, crystals from methanol-ether or from methyl ethyl ketone, mp 95-97°.

Fumarate, $C_{14}H_{19}N_3O.C_4H_4O_4$, *F-151, Foralamin.*

THERAP CAT: Antihistaminic.

6011. Methallatal. *5-Ethyldihydro-5-(2-methyl-2-propenyl)-2-thioxo-4,6(1H,5H)-pyrimidinedione; 5-ethyl-5-(2-methylallyl)-2-thiobarbituric acid; ethyl-β-methylallylthiobarbituric acid; V-12; Mosidal.* $C_{10}H_{14}N_2O_2S$; mol wt 226.30. C 53.08%, H 6.24%, N 12.38%, O 14.14%, S 14.17%. Prepn: Volwiler, Tabern, U.S. pat. **2,153,729** (1939 to Abbott).

Crystals, mp 160-161°. Insol in water. Forms an alkaline sodium salt which is sol in water, alcohol.

Note: This is a controlled substance (depressant) listed in the U.S. Code of Federal Regulations, Title 21, Part 1308.13 (1995).

THERAP CAT: Antiemetic.

6012. Methallenestril. *β-Ethyl-6-methoxy-α,α-dimethyl-2-naphthalenepropanoic acid; β-(6-methoxy-2-naphthyl)-α,α-dimethylvaleric acid; 3-(6-methoxy-2-naphthyl)-2,2-dimethylpentanoic acid; Vallestril.* $C_{18}H_{22}O_3$; mol wt 286.37. C 75.50%, H 7.74%, O 16.76%. Prepn: A. Horeau, J. Jacques, *Compt. Rend.* **224**, 862 (1947); J. Jacques, A. Horeau, *Bull. Soc. Chim. France* **1948**, 711; R. Gay, A. Horeau, *ibid.* **1955**, 955.

Crystals from dil methanol, mp 139-140°. Sol in ether, vegetable oils.

THERAP CAT: Estrogen.

6013. Methallibure. *N-Methyl-N'-(1-methyl-2-propenyl)-1,2-hydrazinedicarbothioamide; 1-methyl-6-(1-methylallyl)-2,5-dithiobiurea; N-methylthiocarbamoyl-N'-[(1-methylallyl)thiocarbamoyl]hydrazine; metallibure; ICI-33828; Aimax.* $C_7H_{14}N_4S_2$; mol wt 218.35. C 38.51%, H 6.46%, N 25.66%, S 29.37%. Prepn: Paget *et al.,* **Brit.** pat. **878,177** (1961 to I.C.I.). Inhibition of pituitary gonadotropic function in animals: Paget *et al., Nature* **192**, 1191 (1961). Prevention of fetus nidation when given orally to rats: Harper, *J. Reprod. Fert.* **7**, 211 (1964).

Crystals, dec 198-200°.

THERAP CAT (VET): Anterior pituitary activator (for swine).

6014. Methamidophos. *Phosphoramidothioic acid O,S-dimethyl ester;* O,S-dimethyl phosphoramidothioate; Bayer 71628; ENT-27396; Ortho 9006; SRA-5172; Monitor; Tamaron. $C_2H_8NO_2PS$; mol wt 141.13. C 17.02%, H 5.71%, N 9.92%, O 22.67%, P 21.95%, S 22.72%. Prepn: W. Lorenz *et al.,* Belg. pat. **666,143** (1965 to Bayer), *C.A.* **65,** 16864f (1966); **Neth. pat. Appl. 6,602,588** corresp to P. S. Magee, U.S. pat. **3,309,266** (both 1967 to Chevron). Activity: I. Hammann, *Pflanzenschutz-Nachr.* **23,** 140 (1970). Acute toxicity: T. B. Gaines, R. E. Linder, *Fundam. Appl. Toxicol.* **7,** 299 (1986). Review of properties and metabolism: A. M. A. Khasawinah *et al., Pestic. Biochem. Physiol.* **9,** 211-221 (1978).

Crystals from ether, mp 54°. n_D^{40} 1.5092; $d^{44.5}$ 1.31. Vapor press. at 30°; 3×10^{-4} mm Hg. Readily sol in water, ethanol. LD_{50} in adult male, female rats (mg/kg): 25, 27 orally (Gaines, Linder).

USE: Insecticide; acaricide.

6015. Methamphetamine. *(S)-N,α-Dimethylbenzene-ethanamine;* (S)-(+)-N,α-dimethylphenethylamine; d-N-methylamphetamine; d-deoxyephedrine; d-desoxyephedrine; 1-phenyl-2-methylaminopropane; d-phenylisopropylmethyl-amine; methyl-β-phenylisopropylamine; Norodin. $C_{10}H_{15}N$; mol wt 149.24. C 80.48%, H 10.13%, N 9.39%. Central stimulant. Can be prepd by reducing ephedrine or pseudo-ephedrine: Emde, *Helv. Chim. Acta* **12,** 365 (1929). Prepn by reducing the condensation product of benzyl methyl ketone and methylamine: A. Ogata, *J. Pharm. Soc. Japan* **451,** 751 (1919), *C.A.* **14,** 745 (1920). Synthesis from D-phenylalanine: D. B. Repke *et al., J. Pharm. Sci.* **67,** 1167 (1978). Stereochemistry-pharmacology aspects: Patil *et al., J. Pharmacol. Exp. Ther.* **155,** 1, 13 (1967). Toxicity data: A. M. Lands *et al., J. Pharmacol. Exp. Ther.* **89,** 382 (1947). Review of clinical trials in bulimia: H. G. Pope, Jr., J. I. Hudson, *J. Clin. Psychiatry* **47,** 339 (1986). Review of pharmacology and abuse potential: A. K. Cho, *Science* **249,** 631-634 (1990).

Hydrochloride, $C_{10}H_{15}N.HCl$, *"speed", "meth", "ice", Amphedroxyn, Desfedrin, Desoxyfed, Desoxyn, Destim, Doxephrin, Drinalfa, Efroxine, Gerobit, Hiropon, Isophen, Madrine, Methampex, Methedrine, Methylisomyn, Pervitin, Semoxydrine, Soxysympamine, Syndrox, Tonedron.* Crystals, mp 170-175°. Bitter taste. $[\alpha]_D^{25}$ +14 to +20°. Sol in water, alcohol, chloroform. Practically insol in ether. A 1% aq soln is neutral or slightly acid to litmus. LD_{50} i.p. in mice: 70 mg/kg (Lands).

Note: This is a controlled substance (stimulant) listed in the U.S. Code of Federal Regulations, Title 21 Part 1308.12 (1995).

THERAP CAT: Anorexic. In attention deficit disorder with hyperactivity.

THERAP CAT (VET): Sympathomimetic, CNS stimulant.

6016. Metham Sodium. *Methylcarbamodithioic acid sodium salt; methyldithiocarbamic acid sodium salt;* sodium methyldithiocarbamate; N-methylaminodithioformic acid sodium salt; sodium N-methylaminodithioformate; N-meth-

ylaminomethanethionothiolic acid sodium salt; sodium N-methylaminomethanethionothiolate; sodium metham; metam sodium; sodium metam; SMDC; carbathione; trimaton; VPM. $C_2H_4NNaS_2$; mol wt 129.18. C 18.60%, H 3.12%, N 10.84%, Na 17.80%, S 49.64%. $CH_3NHCSSNa$. Prepd from methylamine, carbon disulfide, and NaOH: Compin, *Bull. Soc. Chim. France* [4] **27,** 464 (1920); from methyl iso-thiocyanate and NaSH: Iliceto, D'Angeli, *Gazz. Chim. Ital.* **89,** 1950 (1959). Use as soil fumigant: S. C. Dorman, A. B. Lindquist, U.S. pat. **2,766,554** (1956 to Stauffer). Conversion to methyl isothiocyanate, *q.v.,* in soil: N. J. Turner, M. E. Corden, *Phytopathology* **53,** 1388 (1963); J. H. Smelt, M. Leistra, *Pestic. Sc.* **5,** 401 (1974).

Dihydrate, *N-869, Maposol, Sistan, Vapam.* Crystals, anhydr at 130°. Unpleasant odor, similar to that of carbon disulfide. Non-flammable. Soly in water at 20° = 72.2 g/100 ml H_2O. Moderately sol in alc. Sparingly sol in other solvents. Concd aq solns are stable. Dec when in dil soln and in the presence of acids and heavy metal salts. LD_{50} orally in mice, rats: 285, 820 mg/kg, *RTECS* **vol. I,** R. J. Lewis, R. L. Tatken, Eds. (1980) p 451.

Caution: Irritating to skin, mucous membranes.

USE: Soil fumigant to control weeds and weed seeds, nematodes, fungi, and soil insects.

6017. Methandriol. *(3β,17β)-17-Methylandrost-5-ene-3,17-diol;* methylandrostenediol; MAD; mestenediol; Masdiol; Metocryst; Metildiolo; Androdiol; Metidione; Nabadial; Neosteron; Diolandrone; Stenediol; Protandren; Neostene; Crestabolic; Diolostene; Metendiol; Metandiol; Methandiol; Methanabol; Methostan; Neutrormone; Neutrosteron; Androteston-M; Megabion (Japanese); Notandron. $C_{20}H_{32}O_2$; mol wt 304.47. C 78.90%, H 10.59%, O 10.51%. Prepd by the action of methylmagnesium iodide on 3β-hydroxy-5-androsten-17-one, also called 5,6-dehydroandrosterone: Ruzicka *et al., Helv. Chim. Acta* **18,** 1487 (1935); Miescher, Klarer, *ibid.* **22,** 962 (1939). Absorption spectrum in H_2SO_4: Bernstein, Lenhard, *J. Org. Chem.* **18,** 1153 (1953). NMR: Hampel, Kraemer, *Tetrahedron* **22,** 1601 (1966).

Crystals from ethyl acetate, mp 205.5-206.5°. $[\alpha]_D^{20}$ −73° (alc). Insol in water. Slightly sol in some organic solvents.

Diacetate, $C_{24}H_{36}O_4$, crystals from hexane, mp 145-146°. $[\alpha]_D^{21}$ −59° (c = 0.984 in alc).

Dipropionate, $C_{26}H_{40}O_4$, *Probolin.*

Note: This is a controlled substance (anabolic steroid) in the U.S. Code of Federal Regulations, Title 21 Part 1308.13, as defined in Part 1308.02 (1995).

THERAP CAT: Anabolic.

6018. Methandrostenolone. *(17β)-17-Hydroxy-17-methylandrosta-1,4-dien-3-one;* 17α-methyl-17β-hydroxyandrosta-1,4-dien-3-one; 1-dehydro-17α-methyltestosterone; methandienone; Danabol; Nerobol; Nabolin; Stenolon; Dianabol. $C_{20}H_{28}O_2$; mol wt 300.44. C 79.96%, H 9.39%, O 10.65%. Anabolic steroid. Prepn by microbial dehydrogenation of 17α-methyltestosterone: Vischer *et al., Helv. Chim. Acta* **38,** 1502 (1955); by reduction of 17α-methyltestosterone with selenium dioxide: Meystre *et al., ibid.* **39,**

734 (1956); Wettstein *et al.,* U.S. pat. **2,900,398** (1959 to Ciba).

Crystals from acetone + ether. mp 163-164°. $[\alpha]_D^{26}$ 0° (c = 1.15 in chloroform). uv max: 245 nm (ϵ 15600).

Note: This is a controlled substance (anabolic steroid) in the U.S. Code of Federal Regulations, Title 21 Part 1308.13, as defined in Part 1308.02 (1995).

THERAP CAT: Androgen.

THERAP CAT (VET): Anabolic.

6019. Methane. Marsh gas; methyl hydride. CH_4; mol wt 16.04. C 74.87%, H 25.13%. Widely distributed in nature. American natural gas is about 85% methane. The earth's atm contains 0.00022% by vol. Major constituent of the atm of the outer planets (Jupiter, Saturn, Uranus, Neptune), exact figures in *Landolt-Börnstein,* **vol. III** (Springer, 6th ed., 1952) p 59; G. P. Kuiper, *The Atmospheres of the Earth and the Planets* (University of Chicago Press, 1949). Pure carbon combines directly with pure hydrogen at temperatures above 1100° forming methane. Above 1500° amount of methane formed increases with temperature: Pring, *J. Chem. Soc.* **97,** 498 (1910). Can be prepd from sodium acetate and sodium hydroxide, or from aluminum carbide and water: Matthews, *J. Am. Chem. Soc.* **21,** 647 (1899); Carroll, *J. Phys. Chem.* **22,** 148 (1918). Prepd commercially from natural gas or by fermentation of cellulose and sewage sludge: Cost, U.S. pat. **2,583,090** (1952 to Elliott Co.); Le Paige, de Dommartin, Fr. pat. **994,032** (1951), *C.A.* **51,** 10836i (1957); Oswald, Golueke, *Mech. Eng.* **86,** 40 (1964).

Colorless, odorless, non-poisonous, flammable gas. Burns with a pale, faintly luminous flame. d_4^0 0.554 (air = 1) or 0.7168 g/liter. mp −182.6°. bp −161.4°. Crit temp −82.25°; crit pressure 45.8 atm. Heat of combustion 978 Btu/cu ft at 25° (a kilogram of CH_4 yields 13,300 kcal). Forms exposive mixtures with air, the loudest explosions occur when one vol of methane is mixed with 10 vols of air (or 2 vols of oxygen). Air contg less than 5.53% methane no longer explodes. Air contg more than 14% methane burns without noise. Autoignition temp 650°. Soly in water at 17°: 3.5 ml/100 ml H_2O. Sol in alc, ether, other organic solvents.

Caution: Simple asphyxiant.

USE: Constituent of illuminating and cooking gas, in the manuf of hydrogen, hydrogen cyanide, ammonia, acetylene, formaldehyde, in organic syntheses.

6020. Methanearsonic Acid. *Methylarsonic acid;* methylarsinic acid; monomethylarsinic acid. CH_5AsO_3; mol wt 139.97. C 8.58%, H 3.60%, As 53.53%, O 34.29%. CH_3-$AsO(OH)_2$. Prepd from sodium arsenite and methyl iodide: Quick, Adams, *J. Am. Chem. Soc.* **44,** 809 (1922). The disodium salt is easily prepd by treating sodium arsenite with dimethyl sulfate at 85°: Uhlinger, Cook, *Ind. Eng. Chem.* **11,** 105 (1919). Other routes are by the reaction of methyl chloride with sodium arsonate under pressure: Miller *et al.,* U.S. pat. **2,442,372** (1948); by the reaction of dimethyl sulfate with a solution of arsenic trioxide in sodium hydroxide: Schwerdle, U.S. pat. **2,889,347** (1959). Acute toxicity: T. B. Gaines, R. E. Linder, *Fundam. Appl. Toxicol.* **7,** 299 (1986).

Monoclinic, spear-shaped plates from abs alcohol. Pleasant acid taste. mp 161°. Strong dibasic acid. Freely sol in water; sol in alcohol.

Monosodium salt, CH_4AsNaO_3, *monosodium methanearsonate, MSMA, Ansar 6.6, Bueno, Daconate, Weed Hoe.* LD_{50} in adult male, female rats (mg/kg): 1105, 1059 orally (Gaines, Linder).

Disodium salt, $CH_3AsNa_2O_3$, *disodium monomethanearsonate, DSMA, Ansar 8100, Arrhenal, Arsinyl, Clout, Crab-E-Rad, Dal-E-Rad, Sodar.* Hydrated crystals contg $5H_2O$ or $6H_2O$. One gram dissolves in about one ml water; slightly sol in alcohol. LD_{50} in adult male, female rats (mg/kg): 928, 821 orally (Gaines, Linder).

USE: Herbicide.

6021. Methanesulfonic Acid. Methylsulfonic acid. CH_4O_3S; mol wt 96.11. C 12.50%, H 4.20%, O 49.94%, S 33.36%. CH_3SO_2OH. Prepd from sulfur trioxide and methane: Snyder, Grosse, U.S. pat. **2,493,038** (1950 to Houdry Process); by oxidation of dimethyl disulfide: Johnson,

Wolff, U.S. pat. **2,697,722** (1954 to Standard Oil of Indiana); Proell *et al., Ind. Eng. Chem.* **40,** 1129 (1948). Other prepns and chemistry: Suter, *The Organic Chemistry of Sulfur* (Wiley, New York, 1944).

Solid. d_4^{18} 1.4812. mp 20°. bp_{10} 167°; bp_1 122°. Soly at 26-28° in wt %: hexane, 0; methylcyclopentane, 0; benzene, 1.50; toluene, 0.38; *o*-chlorotoluene, 0.23; ethyl disulfide, 0.47. Thermally stable at moderately elevated temps. Not hydrolyzed by boiling water or hot aq alkali. Corrosive to iron, steel, brass, copper, lead.

Ethyl ester, *see* Ethyl Methanesulfonate.

Methyl ester, *see* Methyl Methanesulfonate.

Caution: Strong irritant.

USE: As catalyst in polymerization, alkylation and esterification reactions; as a solvent.

6022. Methanesulfonyl Chloride. CH_3ClO_2S; mol wt 114.55. C 10.49%, H 2.64%, Cl 30.95%, O 27.93%, S 27.99%. CH_3SO_2Cl. Prepd from methanesulfonic acid and thionyl chloride: Hearst, Noller, *Org. Syn.,* coll. vol. IV, 571 (1963).

Liquid. d_4^{18} 1.4805. bp_{730} 161°; bp_{18} 62°. n_D^{23} 1.451. Practically insol in water; sol in alcohol, ether.

6023. Methanethiol. Methyl mercaptan; mercaptomethane; thiomethyl alcohol; methyl sulfhydrate. CH_4S; mol wt 48.11. C 24.97%, H 8.38%, S 66.65%. CH_3SH. Occurs in "sour" gas of W. Texas, in coal tar, and in petroleum distillates. Isolated from roots of *Raphanus sativus.* Produced in the intestinal tract by the action of anaerobic bacteria on albumin. Evolved from *Penicillium brevicaule* bread cultures containing disulfides. Prepn from sodium methyl sulfate and KHS: Klason, *Ber.* **20,** 3409 (1887); Arndt, *ibid.* **54,** 2236 (1921); catalytically from methanol and hydrogen sulfide: Kramer, Reid, *J. Am. Chem. Soc.* **43,** 880 (1921); from methyl chloride and sodium hydrosulfide: Scott *et al., Ind. Eng. Chem.* **47,** 876 (1955). Toxicity data: *Handbook of Toxicology* **vol. 1,** W. S. Spector, Ed. (Saunders, Philadelphia, 1956) pp 344-345. Review on occurrence, preparation, properties and reactions: E. E. Reid, *Organic Chemistry of Bivalent Sulfur* **vol. I** (Chemical Publishing Co., New York, 1958) pp 15-261.

Flammable gas; odor of rotten cabbage. mp −123°. bp_{760} 5.95°; d_4^{20} 0.8665; d_4^{25} 0.9600. Critical temp 196.8°. Critical pressure 71.4 atm. Heat capacity (solid at 14.97-146.57°K): 0.773-17.47 cal/deg/mole; (liq at 154.16-271.06°K): 21.27-21.13 cal/deg/mole, Russell *et al., J. Am. Chem. Soc.* **64,** 165 (1942). Azeotrope with isobutane (14.9% methanethiol) bp −13.00°. Soly in water at 20°: 23.30 g/l. Forms a cryst hydrate. LC for rats in air: 10000 ppm (Spector).

Sodium salt heminonahydrate, $CH_3SNa.4\frac{1}{2}H_2O$, needles. Freely sol in water, methanol. Practically insol in ether.

Copper salt, CH_3SCu, pale yellow crystals. Practically insol in water, ethanol, ether, benzene.

Caution: Potential symptoms of overexposure to methanethiol are narcosis; cyanosis; convulsions; pulmonary irritation. *See NIOSH Pocket Guide to Chemical Hazards* (DHHS/NIOSH 90-117, 1990) p 154.

USE: Intermediate in manuf of jet fuels, pesticides, fungicides, plastics; synthesis of methionine.

6024. Methanol. Methyl alcohol; carbinol; wood spirit; wood alcohol. CH_4O; mol wt 32.04. C 37.48%, H 12.58%, O 49.93%. CH_3OH. Originally obtained by the destructive distillation of wood, now usually manuf from hydrogen and carbon monoxide or carbon dioxide, also by oxidation of hydrocarbons. *Review:* Faith, Keyes & Clark's *Industrial Chemicals,* F. A. Lowenheim, M. K. Moran, Eds. (Wiley-Interscience, New York, 4th ed., 1975) pp 524-529; L. E. Wade *et al.,* in Kirk-Othmer *Encyclopedia of Chemical Technology* **vol. 15** (Wiley-Interscience, New York, 3rd ed., 1981) pp 398-415. Review of metabolism and toxicology: J. Liesivuori, H. Savolainen, *Pharmacol. Toxicol.* **69,** 157-163 (1991).

Flammable, poisonous, mobile liq. Slight alcoholic odor when pure; crude material may have a repulsive, pungent odor. Burns with a non-luminous, bluish flame. d_4^0 0.8100; d_4^{15} 0.7960; d_4^{20} 0.7915; d_4^{25} 0.7866. mp −97.8°. bp_{760} 64.7°; bp_{400} 49.9°; bp_{200} 34.8°; bp_{100} 21.2°; bp_{60} 12.1°; bp_{40} +5.0°; bp_{20} −6.0°; bp_{10} −16.2°; bp_5 −25.3°; $bp_{1.0}$ −44.0°; n_D^{15}

1.33066; n_D^{20} 1.3292. Vapor density: 1.11 (air = 1). Flash pt, closed cup: 54°F (12°C). Ignition temp 470°C (878°F). Explosive limits (%-vol in air): 6.0 to 36.5. Crit temp 240.0°; crit pressure 78.5 atm. Specific heat at 20-25° = 0.595 to 0.605. Dipole moment 1.69. Miscible with water, ethanol, ether, benzene, ketones, and most other organic solvents. Forms azeotropes with many compds. *Density, freezing and boiling point data* of *methanol-water mixtures:* 10% methanol by vol (d_4^{25}, fp, bp): 0.9836, −5°, 92.8°; 20% methanol: 0.9695, −12°, 87.8°; 30% methanol: 0.9572, −21°, 84.0°; 40% methanol: 0.9423, −33°, 80.9°; 50% methanol: 0.9259, −47°, 78.3°; 60% methanol: 0.9082, −57°, 75.9°. Methanol usually is a better solvent than ethanol, dissolves many inorganic salts, *e.g.*, sodium iodide 43%, calcium chloride 22%, ammonium nitrate 14%, copper sulfate 13%, silver nitrate 4%, ammonium chloride 3.2%, sodium chloride 1.4%.

Caution: Poisoning may occur from ingestion, inhalation or percutaneous absorption. *Acute Effects:* Headache, fatigue, nausea, visual impairment or complete blindness (may be permanent), acidosis, convulsions, mydriasis, circulatory collapse, respiratory failure, death. Death from ingestion of less than 30 ml has been reported. Usual fatal dose 100-250 ml. *Chronic:* Visual impairment, *cf.* Patty's *Industrial Hygiene and Toxicology* vol. 2C, G. D. Clayton, F. E. Clayton, Eds. (Wiley-Interscience, New York, 3rd ed., 1982) pp 4528-4541.

USE: Industrial solvent. Raw material for making formaldehyde and methyl esters of organic and inorganic acids. Antifreeze for automotive radiators and air brakes; ingredient of gasoline and diesel oil antifreezes. Octane booster in gasoline. As fuel for picnic stoves and soldering torches. Extractant for animal and vegetable oils. To denature ethanol. Softening agent for pyroxylin plastics. Solvent and solvent adjuvant for polymers. Solvent in the manuf of cholesterol, streptomycin, vitamins, hormones, and other pharmaceuticals.

6025. Methantheline Bromide. *N,N-Diethyl-N-methyl-2-[(9H-xanthen-9-ylcarbonyl)oxy]ethanaminium bromide; diethyl(2-hydroxyethyl)methylammonium bromide xanthene-9-carboxylate;* β-diethylaminoethyl 9-xanthenecarboxylate methobromide; MTB-51; SC-2910; Banthine Bromide; Avagal; Uldumont; Vagantin; Metaxan; Methanide; Xanteline; Gastron; Gastrosedan; Methanthine Bromide; Vagamin; Metanyl; Doladene; Asabaine. $C_{21}H_{26}BrNO_3$; mol wt 420.35. C 60.01%, H 6.23%, Br 19.01%, N 3.33%, O 11.42%. Prepd from 9-xanthenecarboxylic acid and 2-diethylaminoethanol: J. W. Cusic, R. A. Robinson, *J. Org. Chem.* **16**, 1921 (1951); *eidem,* U.S. pat. **2,659,732** (1952 to Searle). Pharmacology: W. E. Hambourger *et al., J. Pharmacol. Exp. Ther.* **99**, 245 (1950).

Crystals from isopropanol, mp 175-176°. Bitter taste. Very slightly hygroscopic. Freely sol in water, alcohol. Practically insol in ether. pH (2% aq soln): 5.0-5.5. Aq solns tend to hydrolyze after a few days. The corresponding chloride is very hygroscopic. uv max (alc): 246, 282 nm ($E_{1cm}^{1\%}$ 135, 69). LD$_{50}$ i.p. in mice: 76 mg/kg (Hambourger).

THERAP CAT: Anticholinergic.

THERAP CAT (VET): Anticholinergic; antispasmodic; antisecretory agent.

6026. Methaphenilene. *N,N-Dimethyl-N'-phenyl-N'-(2-thienylmethyl)-1,2-ethanediamine; N,N-dimethyl-N'-phenyl-N'-(2-thenyl)ethylenediamine;* 00836; RP-2740; W-50 base; Diatrin base. $C_{15}H_{20}N_2S$; mol wt 260.40. C 69.19%, H 7.74%, N 10.76%, S 12.31%. Prepd by a sodamide condensation of *N,N*-dimethyl-*N'*-phenylethylenediamine with 2-thenyl chloride in dry toluene: Leonard, Solmssen, *J. Am. Chem. Soc.* **70**, 2066 (1948).

Dark yellow oil. bp$_7$ 183-185°. n_D^{25} 1.5902.

Hydrochloride, $C_{15}H_{20}N_2S\cdot HCl$, *Diatrin, Enstamine, Nilhistin.* Crystals from ethanol, mp 186-187°. Freely soluble in water. Soluble in alc. LD$_{50}$ i.p. in mice: 117 mg/kg.

THERAP CAT: Antihistaminic.

THERAP CAT (VET): Antihistaminic.

6027. Methapyrilene. *N,N-Dimethyl-N'-2-pyridinyl-N'-(2-thienylmethyl)-1,2-ethanediamine; 2-[(2-dimethylaminoethyl)-2-thenylamino]pyridine; N,N-dimethyl-N'-(2-pyridyl)-N'-(2-thenyl)ethylenediamine; N,N-dimethyl-N'-(α-pyridyl)-N'-(2-methylthienyl)ethylenediamine;* thenylpyramine; AH-42; Thenylene; Pyrathyn; Semikon; Thionylan; Histadyl (formerly); Restryl; Rest-On; Sleepwell; Paradormalene; Pyrinistab; Pyrinistol; Lullamin. $C_{14}H_{19}N_3S$; mol wt 261.39. C 64.33%, H 7.33%, N 16.08%, S 12.27%. Prepd by heating a 2-thenyl halide with an alkali metal salt of *N,N*-dimethyl-*N'*-(2-pyridyl)ethylenediamine: Kyrides, U.S. pat. **2,581,-868** (1952 to Monsanto). Alternate syntheses: Weston, *J. Am. Chem. Soc.* **69**, 980 (1947); Clapp *et al., ibid.* 1549. Toxicity data: H. M. Lee *et al., Proc. Soc. Exp. Biol. Med.* **80**, 458 (1952). Carcinogenicity study: W. Lijinsky *et al., Science* **209**, 817 (1980). Study of mechanism of hepatocarcinogenicity: K. L. Steinmetz *et al., Carcinogenesis* **9**, 959 (1988). Clinical pharmacokinetics: E. P. Calandre *et al., Clin. Pharmacol. Ther.* **29**, 527 (1981). HPLC determn in feed and sleep aid tablets: B. Shaikh, M. R. Hallmark, *J. Assoc. Off. Anal. Chem.* **64**, 889 (1981); in feed, urine and wastewater: H. C. Thompson, Jr., C. L. Holder, *J. Chromatog.* **283**, 251 (1984).

Liquid. bp$_{0.45}$ 125-135°; bp$_3$ 173-175°. n_D^{25} 1.5842 (also reported as 1.5835). LD$_{50}$ in mice, guinea pigs (mg/kg): 182.2 ±12.8, 374.9 ±34.5 orally; in mice (mg/kg): 19.85 ±0.69 i.v. (Lee).

Hydrochloride, $C_{14}H_{19}N_3S\cdot HCl$, bitter crystals, mp 162°. uv max: 238 nm ($E_{1cm}^{1\%}$ 623); min: 272 nm. One gram dissolves in about 0.5 ml water, in 5 ml alcohol, in 3 ml chloroform. Practically insol in ether, benzene.

Fumarate, $(C_{14}H_{19}N_3S)_2\cdot 3C_4H_4O_4$. Prepn: Meyer, **Brit.** pat. **694,805** (1953 to Monsanto). Crystals, mp 135-136°.

THERAP CAT: Antihistaminic.

THERAP CAT (VET): Antihistaminic.

6028. Methaqualone. *2-Methyl-3-(2-methylphenyl)-4(3H)-quinazolinone; 2-methyl-3-o-tolyl-4(3H)-quinazolinone;* 3,4-dihydro-2-methyl-4-oxo-3-o-tolylquinazoline; metolquizolone; MAOA; MTQ; ortonal; QZ-2; RIC-272; Rorer 148; TR-495; Cateudyl; Citexal; Dormigoa; Dormogen; Dormutil; Dorsedin; Fadormir; Holodorm; Hyminal; Hypcol; Hyptor; Ipnofil; Melsomin; Mequin; Mollinox; Motolon; Nobedorm; Noctilene; Normi-Nox; Omnyl; Optinoxan; Parminal; Parest; Quaalude; Roulone; Rouqualone; Sindesvel; Somnafac; Sonal; Somberol; Somnomed; Soverin; Torinal; Tuazol; Tuazolone. $C_{16}H_{14}N_2O$; mol wt 250.30. C 76.78%, H 5.64%, N 11.19%, O 6.39%. Prepn: Kacker, Zaheer, *J. Indian Chem. Soc.* **28**, 344 (1951); Lab. Toraude, **Brit.** pat. **843,073** (1960); Klosa, *J. Prakt. Chem.* **14**, 84 (1961); **20**, 283 (1963). Improved synthesis: M. S. Manhas *et al., Synthesis* **5**, 309 (1977). Estimation in biol. materials: Akagi *et al., Chem. Pharm. Bull.* **11**, 62 (1963). Metabolism: R. Bonnichsen *et al., Clin. Chim. Acta* **60,** 67 (1975); W.G. Stillwell *et al., Drug Metab. Dispos.* **3,** 287 (1975). Toxicity data: E. I. Goldenthal, *Toxicol. Appl. Pharmacol.* **18,** 185 (1971). Comprehensive description: D. M. Patel *et al., Anal. Profiles Drug Subs.* **4,** 245-267 (1975).

Crystals, mp 120°; also given as mp 114-116°. uv max (ethanol): 225, 263, 304, 316 nm; (0.01N HCl): about 234, 269 nm. Sol in ethanol, ether, chloroform. Practically insol in water. LD$_{50}$ orally in rats: 255 mg/kg (Goldenthal).

Hydrochloride, $C_{16}H_{14}N_2O.HCl$, *Melsedin, Mequelon, Metadorm, Methased, Optimil, Paxidorm, Revonal, Riporest, Sedaquin, Sleepinal, Somnium, Toquilone, Toraflon.* Crystals, mp 255-265°. Sol in ether, ethanol. Practically insol in water.

Note: This is a controlled substance (depressant) listed in the U.S. Code of Federal Regulations, Title 21 Part 1308.11 (1995).

THERAP CAT: Sedative, hypnotic.

6029. Metharbital. *5,5-Diethyl-1-methyl-2,4,6(1H,3H,- 5H)-pyrimidinetrione; 5,5-diethyl-1-methylbarbituric acid;* Gemonil. $C_9H_{14}N_2O_3$; mol wt 198.22. C 54.53%, H 7.12%, N 14.13%, O 24.21%. Prepn: Halpern, Jones, *J. Am. Pharm. Assoc.* **38,** 352 (1949); Snyder, Link, *J. Am. Chem. Soc.* **75,** 1881 (1953).

Crystals from benzene + petr ether, mp 155°. Solubility (g/100 ml soln) in water 0.12; in alcohol 4.3; in ether 2.6. pH of satd aq solns: 5.6-5.7. pK 8.45: Fox, Shugar, *Bull. Soc. Chim. Belg.* **61,** 44 (1952).

Caution: May be habit forming. This is a controlled substance (depressant) listed in the U.S. Code of Federal Regulations, Title 21 Parts 329.1 and 1308.13 (1995).

THERAP CAT: Anticonvulsant.

6030. Methargen. *3,3′-Methylenebis(2-naphthalenesulfonic acid) disilver(1+) salt; 2,2′-dinaphthylmethane-3,3′-disulfonic acid disilver salt;* silver methylenebis(2-naphthyl-3-sulfonate); methylenedinaphthalenesulfonic acid disilver salt; silver dinaphthylmethane disulfonate; Viacutan. $C_{21}H_{14}Ag_2O_6S_2$; mol wt 642.21. C 39.28%, H 2.20%, Ag 33.59%, O 14.95%, S 9.99%. Prepn of similar compds: Craven, Pritchard, U.S. pat. **2,539,728** (1951 to Ward, Blenkinsop).

THERAP CAT: Topical antiseptic.

6031. Methazolamide. *N-[5-(Aminosulfonyl)-3-methyl-1,3,4-thiadiazol-2(3H)-ylidene]acetamide; N-(4-methyl-2-sulfamoyl-Δ²-1,3,4-thiadiazolin-5-ylidene)acetamide;* 5-acetylamino-4-methyl-Δ²-1,3,4-thiadiazoline-2-sulfonamide; Neptazane. $C_5H_8N_4O_3S_2$; mol wt 236.28. C 25.42%, H 3.41%, N 23.71%, O 20.31%, S 27.14%. Carbonic anhydrase inhibitor. Prepn: Young *et al., J. Am. Chem. Soc.* **78,** 4649

(1956); U.S. pat. **2,783,241** (1957 to Am. Cyanamid); Pala, *Farmaco Ed. Sci.* **13,** 650 (1958).

Crystals from water, mp 213-214°. pKa 7.30. uv max (95% ethanol): 254 nm (log ε 3.66); (0.1N NaOH): 247 nm (log ε 3.61).

THERAP CAT: Diuretic.

6032. Methazole. *2-(3,4-Dichlorophenyl)-4-methyl-1,- 2,4-oxadiazolidine-3,5-dione;* oxydiazol; VCS-438; Paxilon; Probe. $C_9H_6Cl_2N_2O_3$; mol wt 261.06. C 41.41%, H 2.32%, Cl 27.16%, N 10.73%, O 18.39%. Pre- and post-emergence herbicide. Prepn: J. Krenzer, **S. Afr. pat. 67 01,543;** *idem,* **U.S. pat. 3,437,664** (1968, 1969 to Velsicol). Metabolism: H. W. Dorough *et al., Drug Metab. Dispos.* **2,** 129 (1974). Persistence in the soil: W. Bond, H. A. Roberts, *Weed Res.* **16,** 23 (1976); A. Walker, *Pestic. Sci.* **9,** 326 (1978). Mechanism of phytotoxicity: J. Verity *et al., Weed Res.* **21,** 243, 307, 317 (1981). Acute toxicity: T. B. Gaines, R. E. Linder, *Fundam. Appl. Toxicol.* **7,** 299 (1986).

Light tan solid, mp 123-124°. Soly at 25°: water 1.5 ppm; xylene 55 g/l. LD$_{50}$ in adult male, female rats (mg/kg): 777, 925 orally (Gaines, Linder).

USE: Herbicide.

6033. Methcathinone. *2-(Methylamino)-1-phenyl-1-propanone;* α-methylaminopropiophenone; monomethylpropion; ephedrone; Jeff; mulka; cat; cosmos; Jee cocktail. $C_{10}H_{13}NO$; mol wt 163.22. C 73.59%, H 8.03%, N 8.58%, O 9.80%. Oxidation product of ephedrine possessing amphetamine-like central stimulant properties. Prepn: J. F. Hyde *et al., J. Am. Chem. Soc.* **50,** 2287 (1928); K. Y. Zhingel *et al., J. Forensic Sci.* **36,** 915 (1991). Prepn of enantiomers and chromatographic analysis: J. DeRuiter *et al., J. Chromatog. Sci.* **32,** 552 (1994). Comparative pharmacology: R. Young, R. A. Glennon, *Pharmacol. Biochem. Behav.* **45,** 229 (1993). Pharmacology of enantiomers: R. A. Glennon *et al., ibid.* **50,** 601 (1995). Absorption and metabolism: S. L. Markantonis *et al., J. Pharm. Pharmacol.* **38,** 515 (1986). GLC/MS determn in urine: *eidem, Chim. Chron.* **14,** 211 (1985). Impact on public health: T. S. Emerson, J. E. Cisek, *Ann. Emerg. Med.* **22,** 1897 (1993).

Hydrochloride, $C_{10}H_{13}NO.HCl$, crystals from alcohol-acetone, mp 176-177°.

l-Form hydrochloride, $C_{10}H_{13}NO.HCl$. Prepn: Y. J. L'Italien, M. C. Rebstock, U.S. pat. **2,802,865** (1957 to Parke, Davis). Crystals from ethanol-ether or chloroform-petr ether, mp 182-183°. Pungent taste. [α]$_D^{25}$ −53° (c = 1 in water). Sol in water. LD$_{50}$ in mice, rats (mg/kg): 106, 90 iv; 233, 86 sc; 342, 99 orally (Emerson).

Note: This is a controlled substance (stimulant) listed in the U.S. Code of Federal Regulations, Title 21 Part 1308.11 (1995).

6034. Methdilazine. *10-[(1-Methyl-3-pyrrolidinyl)-methyl]-10H-phenothiazine.* $C_{18}H_{20}N_2S$; mol wt 296.44. C 72.93%, H 6.80%, N 9.45%, S 10.82%. Prepn: Feldkamp,

Wu, U.S. pat. **2,945,855** (1960 to Mead Johnson). Toxicity data: E. I. Goldenthal, *Toxicol. Appl. Pharmacol.* **18**, 185 (1971).

Crystals, mp 87-88°.

Hydrochloride, $C_{18}H_{20}N_2S.HCl$, *Dilosyn, Disyncran, Tacaryl.* Crystals from isopropyl alc, mp 187.5-189°. LD_{50} orally in rats: 320 mg/kg (Goldenthal).

THERAP CAT: Antipruritic.

6035. Methemoglobin. Hemiglobin; ferrihemoglobin; met Hb. Oxidation product of the normal blood pigment, hemoglobin, in which the iron is present in the ferric state. Methemoglobin may be formed through the direct action of oxidants, through the coaction of hydrogen donors and atmospheric oxygen, or through the autoxidation of hemoglobin which occurs to a small extent in normal blood. Methemoglobin does not have the capacity to combine with molecular oxygen. *Reviews:* Bodansky, *Pharmacol. Rev.* **3**, 144 (1951); Jaffé in *The Red Blood Cell*, Bishop, Surgenor, Eds. (Academic Press, 1964) pp 397-422.

Cleaved by acids and bases to yield globin and ferriprotoporphyrin.

6036. Methenamine. *1,3,5,7-Tetraazatricyclo[3.3.1.-1³,⁷]decane; hexamethylenetetramine;* HMT; HMTA; hexamine; 1,3,5,7-tetraazaadamantane; hexamethylenamine; Aminoform; Ammoform; Cystamin; Cystogen; Formin; Uritone; Urotropin. $C_6H_{12}N_4$; mol wt 140.19. C 51.41%, H 8.63%, N 39.97%. From formaldehyde and ammonia: Meissner, Schwiedessen, U.S. pat. **2,762,800** (1956); *see also* U.S. pat. **2,762,799;** Faith, Keyes & Clark's *Industrial Chemicals*, F. A. Lowenheim, M. K. Moran, Eds. (Wiley-Interscience, New York, 4th ed., 1975) pp 445-448. Review of chemistry: E. M. Smolin, L. Rapoport, *s-Triazines and Derivatives* (Interscience, New York, 1959) Chapter X, pp 545-596. Comprehensive review: J. F. Walker, *Formaldehyde* (Reinhold, New York, 3rd ed., 1964) Chapter 19, pp 511-551. Review of oral toxicity: P. Restani, C. L. Galli, *Crit. Rev. Toxicol.* **21**, 315-328 (1991).

Crystals or granules or powder; odorless. Sublimes at about 263° without melting and with partial decompn; somewhat volatile at lower temp. In contact with a flame it readily burns with smokeless flame. One gram dissolves in 1.5 ml water, 12.5 ml alc, 320 ml ether, 10 ml chloroform. pH of 0.2 molar aq sol 8.4.

USE: In adhesives, coatings, and sealing compounds; in the chemical detection of metals; in the preservation of hides; as cross-linking agent for hardening phenol-formaldehyde resin and vulcanizing rubber; as corrosion inhibitor for steel; as dye fixative; as fuel tablets for camping stoves; as stabilizer for lubricating and insulating oils; with sodium phenate and NaOH as absorber of poisonous gases; the parent substance for the manufacture of explosive compounds (*see* Cyclonite); antimicrobial food additive.

THERAP CAT: Antibacterial (urinary).
THERAP CAT (VET): Urinary antiseptic.

6037. Methenamine Allyl Iodide. Allyl iodide hexamine; allyl iodide hexamethylenetetramine; allyliodourotropine. $C_9H_{17}IN_4$; mol wt 308.17. C 35.08%, H 5.56%, I 41.18%, N 18.18%. $C_6H_{12}N_4.C_3H_5I$. Prepd from allyl iodide and methenamine: F. J. Welcher, *Org. Anal. Reagents*

(Van Nostrand, New York, 1st ed., 1947) p 121; Hurd, Evans, *Ind. Eng. Chem., Anal. Ed.* **5**, 16 (1933).

Cryst substance. mp ~148° with decompn. Very sol in water. Practically insol in chloroform, ether.

USE: Reagent for detection and determination of cadmium, with which it forms a characteristic cryst ppt in neutral or slightly acid soln.

6038. Methenamine Anhydromethylenecitrate. Hexamethylenetetramine anhydromethylenecitrate; β-(hydroxymethoxy)tricarballylic acid γ-lactone hexamethylenetetramine salt; Helmitol; Formanol; Citramin; Uropurgol; Urotropin New. $C_{13}H_{20}N_4O_7$; mol wt 344.32. C 45.35%, H 5.85%, N 16.27%, O 32.53%. $C_7H_8O_7.(CH_2)_6N_4$. Prepn: Settimj, *Ann. Chim. Applicata* **39**, 393 (1949), *C.A.* **46**, 1965 (1952).

White, cryst powder. mp ~175° with decompn. Sol in 10 parts water; very slightly sol in alcohol, ether. Dec by acids or alkalies with liberation of formaldehyde. *Incompat:* Acids, alkalies, hot water.

THERAP CAT: Antibacterial (urinary).

6039. Methenamine Hippurate. Hexamethylenetetramine hippurate; Haiprex; Hippramine; Hiprex; Urex; Urotractan; Viapta. $C_{15}H_{21}N_5O_3$; mol wt 319.36. C 56.41%, H 6.63%, N 21.93%, O 15.03%. $C_6H_{12}N_4.C_9H_9NO_3$. Prepd by refluxing one mole hexamethylenetetramine with one mole hippuric acid in methanol: Galat, U.S. pat. **3,004,026** (1959).

Crystals, mp 105-110°. Freely sol in water and alcohol.

THERAP CAT: Antibacterial (urinary).

6040. Methenamine Mandelate. Mandelamine; hexydaline; hexamethylenetetramine phenylglycolate; hexamethylenamine mandelate; Mandacon; Purerin; Uro-Cedulamin; Mandamina; Mandurin; Reflux; Renelate; Mandoz; Cedulamin; Mantropine; Uromandelin; Uronamin. $C_{14}H_{20}N_4O_3$; mol wt 292.34. C 57.52%, H 6.90%, N 19.17%, O 16.42%. $C_6H_5CHOHCOOH.(CH_2)_6N_4$.

Crystals or white, cryst powder; odorless or practically so, but on long keeping may develop a slight odor. mp 128-130°. Solns have a pH about 4. Slightly sol in acetone, ether; sol in alcohol, chloroform; very sol in water.

THERAP CAT: Antibacterial (urinary).

6041. Methenamine Salicylate. Saliformin; Formin salicylate. $C_{13}H_{18}N_4O_3$; mol wt 278.31. C 56.10%, H 6.52%, N 20.13%, O 17.25%. $(CH_2)_6N_4.C_6H_4(OH)COOH$.

White, cryst powder; acidulous taste. Freely sol in water or alcohol.

6042. Methenamine Sulfosalicylate. Hexalet; Hexal; Sulfhexet. $C_{13}H_{18}N_4O_6S$; mol wt 358.38. C 43.57%, H 5.06%, N 15.63%, O 26.79%, S 8.95%.

Monohydrate, odorless, cryst powder. mp ~190° with decompn. Sol in 8 parts water; slightly sol in alcohol.

THERAP CAT: Antibacterial (urinary).

6043. Methenamine Tetraiodine. Hexamethylenetetramine tetraiodide; methenamine tetraiodide; Siomine; Iodoformine; Mirion. $C_6H_{12}I_4N_4$; mol wt 647.81. C 11.12%, H 1.87%, I 78.36%, N 8.65%. $(CH_2)_6N_4I_4$. Prepd by adding potassium mercuric iodide to an aq soln of methenamine: U.S. pat. **1,226,394** (1917).

Reddish powder. Slight odor and taste. Deflagrates at 138°. Almost insol in water. Slightly sol in alcohol, chloroform, ether, carbon disulfide. Sol in acetone, aq solns of sodium or potassium iodides, sodium thiosulfate and in dil HCl. Decompn is liable to occur when in aq soln.

THERAP CAT: Antiseptic, iodide source.
THERAP CAT (VET): Antiseptic, iodine therapy.

6044. Methenolone. *(5α,17β)-17-Hydroxy-1-methylandrost-1-en-3-one;* 1-methyl-Δ¹-androsten-17β-ol-3-one; méténolone. $C_{20}H_{30}O_2$; mol wt 302.46. C 79.42%, H 10.00%, O 10.58%. Prepn: Wiechert, Kaspar, *Ber.* **93**, 1710 (1960); Popper, Ger. pat. **1,023,764** (1958 to Schering, AG).

Crystals from isopropyl ether, mp 149.5-152°. mp 160-161° (Popper). $[\alpha]_D$ +58.9°.

17-Acetate, $C_{22}H_{32}O_3$, *Primobolan Tablets, Primonabol.* Crystals from isopropyl ether, mp 138-139°. uv max (methanol): 240 nm (ϵ 13300). Sol in methanol, ether, chloroform.

17-Enanthate, $C_{27}H_{42}O_3$, *methenolone enanthate, 17β-heptanoyloxy-1-methyl-5α-androst-1-en-3-one, Primobolan-Depot, Primonabol Depot.* Used as repository form.

Note: This is a controlled substance (anabolic steroid) in the U.S. Code of Federal Regulations, Title 21 Part 1308.13, as defined in Part 1308.02 (1995).

THERAP CAT: Anabolic.

6045. Methestrol. *4,4'-(1,2-Diethyl-1,2-ethanediyl)bis-[2-methylphenol]; 4,4'-(1,2-diethylethylene)di-o-cresol;* 3,4-bis(3-methyl-4-hydroxyphenyl)hexane; dimethylhexestrol; promethestrol; γ-promethestrol. $C_{20}H_{26}O_2$; mol wt 298.43. C 80.50%, H 8.78%, O 10.72%. Prepd by hydrolysis of the dipropionate: Niederl *et al., J. Am. Chem. Soc.* **70,** 508 (1948). Chromatographic identification and determn: R. W. Roos, *J. Pharm. Sci.* **63,** 594 (1974). Pharmacologic study: P. H. Jellinck, A. M. Newcombe, *Biochem. Pharmacol.* **29,** 3031 (1980). Pharmaco-therapeutic study: C. B. Hammond, W. S. Maxson, *Fertil. Steril.* **37,** 5 (1982).

Crystals from dil acetic acid, mp 145°.

Dipropionate, $C_{26}H_{34}O_4$, *Meprane Dipropionate.* Cryst, mp 115°. Freely sol in ether, ethyl acetate, benzene; slightly sol in alc. Practically insol in water, dil acids.

THERAP CAT: Estrogen.

6046. Methetoin. *5-Ethyl-1-methyl-5-phenyl-2,4-imidazolidinedione; 5-ethyl-1-methyl-5-phenylhydantoin;* 1-methylnirvanol; Deltoin. $C_{12}H_{14}N_2O_2$; mol wt 218.26. C 66.04%, H 6.47%, N 12.84%, O 14.66%. Prepn: **Ger.** pat. **611,057** (1935 to Sandoz); Long *et al., J. Am. Chem. Soc.* **70,** 900 (1948).

Crystals, mp 210°.

THERAP CAT: Anticonvulsant.

6047. Methicillin Sodium. *[2S-(2α,5α,6β)]-6-[(2,6-Dimethoxybenzoyl)amino]-3,3-dimethyl-7-oxo-4-thia-1-azabicyclo[3.2.0]heptane-2-carboxylic acid monosodium salt;* 6-(2,6-dimethoxybenzamido)penicillanic acid sodium salt; sodium 2,6-dimethoxyphenylpenicillin; 2,6-dimethoxyphenylpenicillin sodium salt; sodium 6-(2,6-dimethoxybenzamido)penicillinate; 2,6-dimethoxybenzoylpenin sodium salt; dimethoxyphenecillin sodium; sodium methicillin; BRL-1241; X-1497; Azapen; Belfacillin; Celpillina; Celbenin; Cinopenil; Dimocillin; Flabelline; Penistaph; Staphcillin. $C_{17}H_{19}N_2NaO_6S$; mol wt 402.40. C 50.74%, H

4.76%, N 6.96%, Na 5.71%, O 23.86%, S 7.97%. Semisynthetic antibiotic related to penicillin. Prepn starting with 6-aminopenicillanic acid: Doyle *et al., J. Chem. Soc.* **1962,** 1457; Doyle *et al.,* U.S. pat. **2,951,839** (1960); Glombitza, *Ann.* **673,** 166 (1964). Physical-chemical properties: Cotta-Ramusino, Intonti, *Farmaco, Ed. Prat.* **16,** 227 (1961).

Monohydrate, crystals from acetone, mp 196-197° (dec). $[\alpha]_D^{20}$ +230° (c = 5); +225° (c = 1). uv max: 281 nm ($E_{1cm}^{1\%}$ 55); min 264 nm. Solubilities (mg/ml) at 20°: water > 300; ethanol 40; ether < 0.03; acetone 0.35; chloroform 0.06; iso-octane < 0.03.

THERAP CAT: Antibacterial.

THERAP CAT (VET): Antimicrobial.

6048. Methidathion. *Phosphorodithioic acid S-[(5-methoxy-2-oxo-1,3,4-thiadiazol-3(2H)-yl)methyl] O,O-dimethyl ester; phosphorodithioic acid O,O-dimethyl ester S-ester with 4-(mercaptomethyl)-2-methoxy-Δ²-1,3,4-thiadiazolin-5-one;* dithiophosphoric acid *O,O'-dimethyl-S-[(5-methoxy-1,3,4-thiadiazol-2(3H)-one-3-yl)methyl] ester; O,O'-dimethyl-S-[(2-methoxy-1,3,4-thiadiazol-5(4H)-one-4-yl)methyl]* dithiophosphate; GS-13005; Supracide; Ultracid. $C_6H_{11}N_2O_4PS_3$; mol wt 302.34. C 23.84%, H 3.67%, N 9.27%, O 21.17%, P 10.24%, S 31.82%. Synthesis and degradation products: Rüfenacht, *Helv. Chim. Acta* **51,** 518 (1968); Fr. pat. **1,335,755** (1963 to Geigy), *C.A.* **60,** 1764g (1964). Metabolism in rats: Esser *et al., Helv. Chim. Acta* **51,** 513 (1968). Acute toxicity: T. B. Gaines, R. E. Linder, *Fundam. Appl. Toxicol.* **7,** 299 (1986). Carcinogenicity studies: J. A. Quest *et al., Regul. Toxicol. Pharmacol.* **12,** 117 (1990).

Crystals from methanol, mp 39-40°. Soly in water < 1%. Readily sol in benzene, acetone, methanol, xylene and other org solvents. LD_{50} in adult male, female rats (mg/kg): 31, 32 orally (Gaines, Linder).

USE: Insecticide, acaricide.

6049. Methimazole. *1,3-Dihydro-1-methyl-2H-imidazole-2-thione; 1-methylimidazole-2-thiol;* 1-methyl-2-mercaptoimidazole; mercazolyl; thiamazole; Basolan; Danantizol; Favistan; Frentirox; Mercazole; Metazolo; Tapazole; Thacapzol; Thycapsol; Strumazol. $C_4H_6N_2S$; mol wt 114.17. C 42.08%, H 5.30%, N 24.54%, S 28.09%. Prepd by treating aminoacetaldehyde diethyl acetal with methyl isothiocyanate: Wohl, Marckwald, *Ber.* **22,** 1354 (1889); from thiocyanic acid and N-substituted amino acetals: Jones *et al., J. Am. Chem. Soc.* **71,** 4000 (1949). Metabolism: D. S. Sitar, D. P. Thornhill, *J. Pharmacol. Exp. Ther.* **184,** 432 (1973). Comprehensive description: H. Y. Aboul-Enein, A. A. Al-Badr in *Analytical Profiles of Drug Substances* vol. 8, K. Florey, Ed. (Academic Press, New York, 1979) pp 351-370. Review of pharmacology and clinical experience: D. S. Cooper, *N. Engl. J. Med.* **311,** 1353-1362 (1984).

Leaflets from alc, mp 146-148°. bp 280° (some decompn). uv max (0.1N H_2SO_4): 211, 251.5 nm ($E_{1cm}^{1\%}$ 593, 1528).

Freely sol in water. Sol in alcohol, chloroform. Sparingly sol in ether, petr ether, benzene.

Methiodide, *Jomezol*.

USE: In cyanide-free silver electroplating.

THERAP CAT: Antihyperthyroid.

6050. Methiocarb. *3,5-Dimethyl-4-(methylthio)phenol methylcarbamate; methylcarbamic acid 4-(methylthio)-3,5-xylyl ester;* 4-(methylthio)-3,5-xylyl methylcarbamate; 4-methylthio-3,5-dimethylphenyl *N*-methylcarbamate; mercaptodimethur; metmercapturon; Bayer 37344; H-321; Draza; Mesurol. $C_{11}H_{15}NO_2S$; mol wt 225.31. C 58.64%, H 6.71%, N 6.22%, O 14.20%, S 14.23%. Prepn: E. Schegk *et al.,* **Brit.** pat. **912,895;** *eidem,* **U.S.** pat. **3,313,684** (1962, 1967 both to Bayer); E. E. Gilbert, J. A. Otto, **U.S.** pat. **3,358,012** (1967 to Allied). Molluscicidal activity: H. H. Crowell, *J. Econ. Entomol.* **60,** 1048 (1967). Bird repellent properties: E.W. Schafer, R. B. Brunton, *J. Wildl. Manage.* **35,** 569 (1971). Toxicity study: T. B. Gaines, *Toxicol. Appl. Pharmacol.* **14,** 515 (1969).

White crystalline powder, mp 121.5°. Insol in water. Sol in organic solvents. Unstable in alk media. LD_{50} in male, female rats (mg/kg): 70, 60 orally (Gaines).

USE: Insecticide; molluscicide; bird repellent.

6051. Methiodal Sodium. *Iodomethanesulfonic acid sodium salt;* sodium iodomethanesulfonate; Skiodan; Abrodil; Radiographol; Segosin; Diagnorenol. CH_2INaO_3S; mol wt 243.99. C 4.92%, H 0.83%, I 52.01%, Na 9.42%, O 19.67%, S 13.14%. CH_2ISO_3Na. Prepd by the action of sodium sulfite on methylene iodide at 70° in water-alcohol soln: Ossenbeck *et al.,* **U.S.** pat. **1,842,626** (1932 to Winthrop); from iodoform + sodium sulfite: Allardt, **U.S.** pat. **1,867,793** (1932 to Schering-Kahlbaum AG).

Crystals. Slightly saline taste followed by sweetish aftertaste. Freely sol in water (70 g/100 ml); slightly sol in alcohol (2.5 g/100 ml), benzene, ether, acetone.

THERAP CAT: Diagnostic aid (radiopaque medium—urographic).

6052. Methionic Acid. *Methanedisulfonic acid.* $CH_4O_6S_2$; mol wt 176.17. C 6.82%, H 2.29%, O 54.49%, S 36.40%. $CH_2(SO_3H)_2$. Prepn from methane + sulfur trioxide: Snyder, Grosse, **U.S.** pat. **2,493,038** (1950 to Houdry Process); by H_2SO_4 oxidation of acetic acid: Schwab, Neuwirth, *Ber.* **90,** 567 (1957); from $MeSO_3H$ + SO_3: Crowder, Gilbert, **U.S.** pat. **2,842,589** (1958 to Allied Chem.). Prepn of aluminum salt: Christian, Jenkins, *J. Am. Pharm. Assoc.* **39,** 633 (1950); **U.S.** pat. **2,504,107** (1950 to Purdue Res. Found.).

Crystals, mp 96-100°.

Aluminum salt, $C_3H_6Al_2O_{18}S_6$, crystals from water + alcohol. Hygroscopic. Soly in water at 27°: 69 w/v. pH of 5% aq soln = 3.5.

USE: Antiperspirant.

THERAP CAT: Aluminum salt as topical astringent.

6053. Methionine. *L-Methionine;* Met; M; 2-amino-4-(methylthio)butyric acid; *α*-amino-*γ*-methylmercaptobutyric acid; (*S*)-2-amino-4-(methylthio)butanoic acid; *γ*-methylthio-*α*-aminobutyric acid; Acimethin. $C_5H_{11}NO_2S$; mol wt 149.21. C 40.25%, H 7.43%, N 9.39%, O 21.44%, S 21.49%. Essential amino acid for human development. Universal translation start signal although usually missing from mature protein. Isoln from casein: J. H. Mueller, *Proc. Soc. Exp. Biol. Med.* **19,** 161 (1922). Early chemistry and biochemistry: *Amino Acids and Proteins,* D. M. Greenberg, Ed. (Charles C. Thomas, Springfield, IL, 1951) 950 pp., *passim;* J. P. Greenstein, M. Winitz, *Chemistry of the Amino Acids* vols. **1-3** (John Wiley and Sons, Inc., New York, 1961) pp. 2125-2155, *passim.* Determn and distribution in non-protein fractions: J. Giovanelli, S. H. Mudd, *J. Biochem. Bio-*

phys. Methods **11,** 1 (1985). GC-MS determn in biological fluids: S. P. Stabler *et al., Anal. Biochem.* **162,** 185 (1987). Evaluation as tracer in cancer imaging in mice: R. Kubota, *J. Nucl. Med.* **36,** 484 (1995). Clinical evaluation in acetaminophen overdose: A. N. Hamlyn *et al., J. Int. Med. Res.* **9,** 226 (1981). Clinical use as radiolabel in hyperparathyroidism: P. Hellman *et al., Surgery* **116,** 974 (1994). Review of metabolism and clinical significance in man: L. D. Fleisher, G. E. Gaull, *Clin. Endocrinol. Metab.* **3,** 37-55 (1974); and in carcinogenesis: T. L. Gatton-Umphress, *Hosp. Pract.* **28,** 83-90 (1993). Review of toxicity: N. J. Benevenga, *J. Agric. Food. Chem.* **22,** 2-9 (1974). Review of biosynthesis: I. G. Old *et al., Prog. Biophys. Mol. Biol.* **56,** 145-185 (1991). Review as translation start signal: T. Meinnel *et al., Biochimie* **75,** 1061-1075 (1993).

Minute hexagonal plates from dil alc, mp 280-282° (dec, sealed capillary). $[\alpha]_D^{25}$ −8.11° (c = 0.8). $[\alpha]_D^{20}$ +23.40° (c = 5.0 in 3*N* HCl). Sol in water, but the crystals are somewhat water-repellent at first. Sol in warm, dil alcohol. Insol in abs alcohol, ether, petr ether, benzene, acetone.

D-Form. Converted by deamination, followed by transamination with resultant inversion to the L-form. Comparative study with L-form of metabolism in plants: M. Pokorny *et al., Phytochemistry* **9,** 2175 (1970). Evaluation in parenteral nutrition: K J. Printen *et al., Am. J. Clin. Nutr.* **32,** 1200 (1979). *Review:* L. D. Stegink, D Amino Acids in *Clin. Nutr. Update: Amino Acids,* H. L. Greene *et al.,* Eds. (American Medical Association, Chicago, IL, 1977) pp 198-206. $[\alpha]_D^{25}$ +8.12° (c = 0.8). $[\alpha]_D^{25}$ −21.18° (c − 0.8 in 0.2*N* HCl).

DL-Form, *racemethionine, Amurex, Banthionine, Dyprin, Lobamine, Metione, Pedameth, Urimeth.* Platelets from alc, mp 281° (decompn). d 1.340. pK_1 2.28; pK_2 9.21. pH of 1% aq soln 5.6-6.1. R_f value 0.77. Soly in water (g/l) at 0°: 18.18; at 25°: 33.81; at 50°: 60.70; at 75°: 105.2; at 100°: 176.0. Sol in dil acids, alkalies. Very slightly sol in 95% alcohol. Insol in ether.

THERAP CAT: Hepatoprotectant; antidote (acetaminophen poisoning); urinary acidifier.

THERAP CAT (VET): Nutritional supplement; urinary acidifier.

6054. Methionine Hydroxy Analog. *2-Hydroxy-4-(methylthio)-butanoic acid; 2-hydroxy-4-(methylthio)butyric acid;* 2-hydroxy-4-(methylthio)butyric acid; Alimet. $C_5H_{10}O_3S$; mol wt 150.20. C 39.98%, H 6.71%, O 31.96%, S 21.35%. $CH_3SCH_2CH_2CH(OH)COOH$. Prepn (not claimed) and use as poultry feed additive: E. S. Blake, R. J. Wineman, **U.S.** pat. **2,745,745** (1956 to Monsanto). HPLC determn in feeds: A. Baudichau *et al., J. Sci. Food Agr.* **38,** 1 (1987). Use as a feed additive for livestock: A. Papas *et al., J. Nutr.* **104,** 653 (1974); A. K. Clark, A. H. Rakes, *J. Dairy Sci.* **65,** 1493 (1982); D. H. Reifsnyder *et al., J. Nutr.* **114,** 1705 (1984). Efficacy of calcium salt vs free acid: K. P. Boebel, D. H. Baker, *Poultry Sci.* **61,** 1167 (1982).

Calcium salt, $C_{10}H_{18}CaO_6S_2$, *MHA.*

USE: Dietary supplement in livestock.

6055. Methioprim. *4-Amino-2-methylthio-5-pyrimidinemethanol;* 4-amino-2-methylmercapto-5-pyrimidinemethanol; 4-amino-5-hydroxymethyl-2-methylthiopyrimidine. $C_6H_9N_3OS$; mol wt 171.22. C 42.09%, H 5.30%, N 24.54%, O 9.34%, S 18.73%. Synthesis starting with ethyl ethoxyethylenecyanoacetate: Ulbricht, Price, *J. Org. Chem.* **21,** 567 (1956).

Needle-like prisms from benzene, mp 126-127°.
USE: Tumor antagonist in mice.

6056. Methiotriazamine. *1,6-Dihydro-6,6-dimethyl-1-[4-(methylthio)phenyl]-1,3,5-triazine-2,4-diamine; 4,6-diamino-1,2-dihydro-2,2-dimethyl-1-[p-(methylthio)phenyl]-s-triazine;* 4,6-diamino-1-(4-methylmercaptophenyl)-1,2-dihydro-2,2-dimethyl-1,3,5-triazine. $C_{12}H_{17}N_5S$; mol wt 263.37. C 54.73%, H 6.51%, N 26.59%, S 12.18%. Prepn: Schalit, Cutler, *J. Org. Chem.* **24**, 573 (1959).

White prisms, mp 149-152°.
Hydrochloride, needles from water, mp 204-207°.
Note: Ingredient of the coccidiostat *Trithiadol* which contains bithional 50%, methiotriazamine 10%, and calcium sulfate.

6057. Methisazone. *2-(1,2-Dihydro-1-methyl-2-oxo-3H-indol-3-ylidene)hydrazinecarbothioamide; 1-methylindole-2,3-dione 3-thiosemicarbazone;* N-methylisatin 3-thiosemicarbazone; BW-33-T-57; Marboran; Viruzona. $C_{10}H_{10}N_4OS$; mol wt 234.28. C 51.27%, H 4.30%, N 23.91%, O 6.83%, S 13.69%. Prepn: Bauer, Sadler, *Brit. J. Pharmacol.* **15**, 101 (1960); **Brit. pat.** 975,357 (1964 to Wellcome Found.).

Crystals from butanol, mp 245°.
THERAP CAT: Antiviral.

6058. Methitural. *Dihydro-5-(1-methylbutyl)-5-[2-(methylthio)ethyl]-2-thioxo-4,6(1H,5H)-pyrimidinedione; 5-(1-methylbutyl)-5-[2-(methylthio)ethyl]-2-thiobarbituric acid;* 5-(2-methylthioethyl)-5-(2-pentyl)-2-thiobarbituric acid. $C_{12}H_{20}N_2O_2S_2$; mol wt 288.44. C 49.97%, H 6.99%, N 9.71%, O 11.09%, S 22.23%. Prepn: Zima, Von Werder, **U.S. pat.** 2,802,827 (1957 to E. Merck).

Sodium salt, $C_{12}H_{19}N_2NaO_2S_2$, *methioturiate, AM-109, Sch-3132, Neraval, Thiogenal.* Very hygroscopic, yellow crystals. Slight odor of mercaptans. Freely sol in water. pH of a 10% aq soln ~9.5. Water solns are unstable as evidenced by a deepening of color and formation of a cloudy precipitate upon autoclaving and exposure to light. The addition of sodium carbonate has a stabilizing effect and prevents precipitation for about 24 hrs: Irwin *et al., J. Pharmacol. Exp. Ther.* **116**, 317 (1956).
Caution: May be habit forming. This is a controlled substance (depressant) listed in the U.S. Code of Federal Regulations, Title 21 Parts 329.1 and 1308.13 (1995).
THERAP CAT: Sedative, hypnotic.

6059. Methixene. *1-Methyl-3-(9H-thioxanthen-9-ylmethyl)piperidine;* 9-(N-methyl-3-piperidylmethyl)thioxanthene. $C_{20}H_{23}NS$; mol wt 309.48. C 77.62%, H 7.49%, N 4.53%, S 10.36%. Smooth muscle relaxant. Prepn: Caviezel *et al., Pharm. Acta Helv.* **33**, 447 (1958); Schmutz, **U.S. pat.** 2,905,590 (1959 to Wander).

Slightly yellow viscous liquid, $bp_{0.07}$ 171-175°.
Hydrochloride monohydrate, $C_{20}H_{23}NS \cdot HCl \cdot H_2O$, *Tremoquil, Methixart, Trest, Tremonil, Tremaril, Tremarit, Cholinfall, Methyloxan.* Flakes from ether, mp 215-217°. uv max (dil HCl): 268 nm (ε 10250). Sol in water, alcohol, chloroform.
THERAP CAT: Anticholinergic.

6060. Methocarbamol. *3-(2-Methoxyphenoxy)-1,2-propanediol 1-carbamate;* 3-(o-methoxyphenoxy)-2-hydroxypropyl 1-carbamate; 2-hydroxy-3-(o-methoxyphenoxy)propyl 1-carbamate; guaiacol glyceryl ether carbamate; AHR-85; Neuraxin; Miolaxene; Lumirelax; Etroflex; Delaxin; Robamol; Traumacut; Tresortil; Relestrid; Robaxin. $C_{11}H_{15}NO_5$; mol wt 241.24. C 54.77%, H 6.27%, N 5.81%, O 33.16%. Prepn from 3-(o-methoxyphenoxy)-2-hydroxypropyl chlorocarbonate: Murphey, **U.S. pat.** 2,770,649 (1956 to A. H. Robins).

Crystals from benzene, mp 92-94°. Soly in water at 20°: 2.5 g/100 ml. Sol in alcohol, propylene glycol.
THERAP CAT: Muscle relaxant (skeletal).
THERAP CAT (VET): Muscle relaxant (skeletal).

6061. Methohexital Sodium. *1-Methyl-5-(1-methyl-2-pentynyl)-5-(2-propenyl)-2,4,6(1H,3H,5H)-pyrimidinetrione sodium salt; 5-allyl-1-methyl-5-(1-methyl-2-pentynyl) sodium salt;* α-dl-1-methyl-5-(1-methyl-2-pentynyl)-5-allylbarbituric acid sodium salt; α-dl-1-methyl-5-(1-methyl-2-pentynyl)barbituric acid sodium salt; methohexitone sodium; Brevital; Brevital Sodium; Brevimytal Sodium; Brietal Sodium. $C_{14}H_{17}N_2NaO_3$; mol wt 284.29. C 59.15%, H 6.03%, N 9.85%, Na 8.09%, O 16.88%. Prepn: Doran, **U.S. pat** 2,872,448 (1959 to Lilly).

Minute crystals. Soluble in water.
Caution: May be habit forming. This is a controlled substance (depressant) listed in the U.S. Code of Federal Regulations, Title 21 Parts 329.1 and 1308.14 (1995).
THERAP CAT: Anesthetic (intravenous).
THERAP CAT (VET): Ultra-short acting anesthetic.

6062. Methomyl. *N-[[(Methylamino)carbonyl]oxy]ethanimidothioic acid methyl ester; N-[(methylcarbamoyl)oxy]thioacetimidic acid methyl ester; S-methyl N-[(methylcarbamoyl)oxy]thioacetimidate; methyl O-(methylcarbamoyl)thiolacetohydroxamate; Insecticide 1179; Lannate; Nudrin. $C_5H_{10}N_2O_2S$; mol wt 162.21. C 37.02%, H 6.21%, N 17.27%, O 19.73%, S 19.77%. Prepn: A. G. Jelinek, **S. Afr. pat.** 68 00,093; *idem,* **U.S. pat.** 3,506,698 (1968, 1970 both to du Pont). Metabolism studies: J. Harvey *et al., J. Agr. Food Chem.* **21**, 769, 775, 781 (1973). Crystal structure: Sim, Waite, *J. Chem. Soc. (B)* **1971**, 752. Acetylcholinesterase activity: Pickering, Pickering, *Arch. Toxicol.* **27**, 292 (1971). Toxicity studies: A. M. Kaplan, H. Sherman, *Toxicol. Appl. Pharmacol.* **40**, 1 (1977).

Crystals, mp 78-79°. d_4^{24} 1.2946. Soly at 25° (w/w): water 5.8; methanol 100; ethanol 42; isopropanol 22; acetone 73. LD_{50} orally in male rats: 17 mg/kg (Kaplan, Sherman).

USE: Insecticide.

6063. Methoprene. *(E,E)-11-Methoxy-3,7,11-trimethyl-2,4-dodecadienoic acid 1-methylethyl ester;* isopropyl *(2E,-4E)-11-methoxy-3,7,11-trimethyl-2,4-dodecadienoate;* ZR-515; Altosid; Precor. $C_{19}H_{34}O_3$; mol wt 310.48. C 73.50%, H 11.04%, O 15.46%. Juvenile hormone mimic. Prepn: **Belg.** pat. **778,242;** *see also* C. A. Henrick, **U.S.** pat. **3,818,-047** (1972, 1974 both to Zoecon); *idem et al., J. Org. Chem.* **40**, 1 (1975); L. Novák *et al., Ann.* **1982**, 1173. Effect on mustard aphid: P. J. Rup, R. K. Gill, *Insect Sci. Applic.* **14**, 173 (1993). Field trial in a saltwater marsh against mosquitoes: V. L. Kramer *et al., J. Am. Mosq. Control Assoc.* **9**, 127 (1993). Mechanism of action study: E. M. Berger *et al., Dev. Biol.* **151**, 410 (1992). HPLC determn in tobacco: S. S. Yang, *Chromatographia* **33**, 309 (1992). GC determn in aquatic microcosms: D. H. Ross *et al., J. Am. Mosq. Control Assoc.* **10**, 202 (1994). Impact on the fathead minnow: *idem et al., ibid.* 211. Review of properties, metabolism, toxicology, and use: J. B. Siddall, *Environ. Health Perspect.* **14**, 119-126 (1976). Review of analytical methods: L. L. Dunham, W. W. Miller in *Analytical Methods for Pesticides and Plant Growth Regulators* **vol. X**, G. Zweig, Ed. (Academic Press, New York, 1978) pp 95-109.

Amber liquid, $bp_{0.06}$ 135-136°. d^{20} 0.9261 g/ml. Vapor pressure at 25°: 2.37×10^{-5} mm Hg. Soly in water at 25°: 1.39 ppm. Sol in org solvents. Temperature stable up to at least 42° for 24 months. Stable in water, org solvents, and in the presence of dil alkali and acid. LD_{50} orally in rats: > 34500 mg/kg (Siddall).

USE: Insecticide; insect growth regulator.
THERAP CAT (VET): Ectoparasiticide.

6064. Methopterin. *N-[4-[[(2-Amino-1,4-dihydro-4-oxo-6-pteridinyl)methyl]methylamino]benzoyl]-L-glutamic acid; N-[4-[[(2-amino-4-hydroxy-6-pteridinyl)methyl]methylamino]benzoyl]glutamic acid;* 10-methylpteroylglutamic acid; N^{10}-methylfolic acid. $C_{20}H_{21}N_7O_6$; mol wt 455.43. C 52.75%, H 4.65%, N 21.53%, O 21.08%. Prepn: Cosulich, Smith, *J. Am. Chem. Soc.* **70**, 1922 (1948); Cosulich, U.S. pat. **2,563,707** (1951 to Am. Cyanamid).

Monohydrate, yellow spherulites. uv max (0.1N NaOH): 255, 302, 368 nm ($\epsilon \times 10^{-3}$ 26, 27, 9); (0.1N HCl): 307 nm ($\epsilon \times 10^{-3}$ 25).

6065. Methotrexate. *N-[4-[[(2,4-Diamino-6-pteridinyl)methyl]methylamino]benzoyl]-L-glutamic acid;* 4-amino-N^{10}-methylpteroylglutamic acid; 4-amino-10-methylfolic acid; methylaminopterin; amethopterin; MTX; CL-14377;

Emtexate; Ledertrexate; A-Methopterin; Rheumatrex. C_{20}-$H_{22}N_8O_5$; mol wt 454.45. C 52.86%, H 4.88%, N 24.66%, O 17.60%. A folic acid antagonist. Prepn: D. R. Seeger *et al., J. Am. Chem. Soc.* **71**, 1753 (1949); J. M. Smith, D. B. Cosulich, **U.S.** pat. **2,512,572** (1950 to American Cyanamid). Metabolism: M. V. Freeman, *J. Pharmacol. Exp. Ther.* **122**, 154 (1958); E. S. Henderson *et al., Cancer Res.* **25**, 1008, 1018 (1965). Toxicity studies: P. T. Condit *et al., Cancer* **13**, 222-249 (1960); *ibid.* **23**, 126 (1969). Pharmacokinetic models: K. B. Bischoff *et al., J. Pharm. Sci.* **59**, 149 (1970); *eidem, ibid.* **60**, 1128 (1971). Comparative study in treatment of acute lymphocytic leukemia in children: A. I. Freemon *et al., N. Engl. J. Med.* **308**, 477 (1983). Short-term efficacy in rheumatoid arthritis: M. E. Weinblatt *et al., ibid.* **312**, 818 (1985). Toxicity data: H. R. Scherf *et al., Arzneimittel-Forsch.* **20**, 1467 (1970). Comprehensive description: A. R. Chamberlin *et al.,* in *Analytical Profiles of Drug Substances* **vol. 5**, K. Florey, Ed. (Academic Press, New York, 1976) pp 283-306. Review of metabolism and pharmacokinetics: W. E. Evans, *Appl. Pharmacokinet.* **1980**, 518-548. Review of clinical pharmacology: J. R. Bertino, *Cancer Chemother.* **3**, 359-375 (1981); J. Jolivet *et al., N. Engl. J. Med.* **309**, 1094-1104 (1983). Symposium on clinical experience in rheumatoid arthritis: *J. Rheumatol.* **12**, Suppl. 12, 1-44 (1985).

Monohydrate, yellow crystals from dil HCl, dec 185-204° (bath preheated to 160°). uv max (0.1N HCl): 244, 307 nm; (0.1N NaOH): 257, 302, 370 nm. Soluble in alkaline solns with decompn. LD_{50} i.v. in rats: 14 mg/kg (Scherf). Disodium salt, $C_{20}H_{20}N_8Na_2O_5$, *Folex, Mexate.*
THERAP CAT: Antineoplastic; antirheumatic.

6066. Methotrimeprazine. *(R)-2-Methoxy-N,N,β-tri-methyl-10H-phenothiazine-10-propanamine; 10-(3-dimethyl-amino-2-methylpropyl)-2-methoxyphenothiazine;* 2-methoxy-10-(2-methyl-3-dimethylaminopropyl)phenothiazine; 3-methoxy-10-(3-dimethylamino-2-methylpropyl)phenothiazine; levomepromazine; 2-methoxytrimeprazine; levomeprazine; RP-7044; Sinogan-Debil; Tisercin; Neozine; Nirvan; Nozinan; Levoprome. $C_{19}H_{24}N_2OS$; mol wt 328.48. C 69.47%, H 7.36%, N 8.53%, O 4.87%, S 9.76%. Prepn: Courvoisier *et al., C.R. Soc. Biol.* **151**, 1378 (1957); Jacob, Robert, **U.S.** pat. **2,837,518** (1958 to Rhône-Poulenc).

Maleate, $C_{19}H_{24}N_2OS \cdot C_4H_4O_4$, *Minozinan, Milezin, Neuractil, Neurocil, Sofmin, Veractil.* Crystals, darkened by light. Dec about 190°. Sparingly sol in water (0.3% at 20°) and in ethanol (0.4%). pH of a 0.3% aq soln is 4.3. The free base is levorotatory: $[\alpha]_D^{20} - 17°$ (c = 5 in chloroform).
THERAP CAT: Analgesic.

6067. Methoxamine. *α-(1-Aminoethyl)-2,5-dimethoxy-benzenemethanol; α-(1-aminoethyl)-2,5-dimethoxybenzyl alcohol;* 2-amino-1-(2,5-dimethoxyphenyl)-1-propanol; *β*-hydroxy-*β*-(2,5-dimethoxyphenyl)isopropylamine; *β*-(2,5-dimethoxyphenyl)-*β*-hydroxyisopropylamine; 2,5-dimethoxy-norephedrine; $C_{11}H_{17}NO_3$; mol wt 211.26. C 62.54%, H 8.11%, N 6.63%, O 22.72%. α_1-Adrenergic agonist. Prepn: Baltzly *et al.,* **U.S.** pat. **2,359,707** (1944 to Burroughs Wellcome). Metabolism: A. Klutch, M. Bordun, *J. Med. Chem.* **10**, 860 (1967). Clinical pharmacology: N. T. Smith, C.

Whitcher, *Anesthesiology* **28**, 735 (1967); P. D. Snashall *et al., Clin. Sci. Mol. Med.* **54**, 283 (1978). HPLC determn in plasma: I. A. Al-Meshal *et al., J. Liq. Chromatog.* **12**, 1589 (1989). Therapeutic use: P. M. C. Wright *et al., Anesth. Analg.* **75**, 56 (1992); L. Cabanes *et al., N. Engl. J. Med.* **326**, 1661 (1992). Comprehensive description: A. M. Al-Obaid, M. M. El-Domiaty, *Anal. Profiles Drug Subs.* **20**, 399-431 (1991).

Hydrochloride, $C_{11}H_{17}NO_3 \cdot HCl$, *Vasoxine, Vasoxyl, Vasylox.* Crystals, mp 212-216°. pKa (25°C) 9.2. Very sol in water: One gram dissolves in 2.5 ml water, in 12 ml ethanol. Practically insol in ether, benzene, chloroform. pH of a 2% aq soln between 4.5 and 5.5.

THERAP CAT: Antihypotensive.

6068. Methoxsalen. *9-Methoxy-7H-furo[3,2-g][1]benzopyran-7-one; 6-hydroxy-7-methoxy-5-benzofuranacrylic acid δ-lactone;* 9-methoxypsoralen; 8-methoxy-4′,5′:6,7-furocoumarin; 8-methoxy[furano-3′,2′:6,7-coumarin]; ammoidin; xanthotoxin; 8-methoxypsoralen; 8-MOP; 8-MP; Meladinine; Meloxine; Oxsoralen. $C_{12}H_8O_4$; mol wt 216.19. C 66.67%, H 3.73%, O 29.60%. Naturally occurring analog of psoralen, *q.v.*, found in spp. of *Leguminosae, Umbelliferae,* and *Rutaceae.* Isolation: Priess, *Ber. Deut. Pharm. Ges.* **21**, 227 (1911); Thoms, *Ber.* **44**, 3325 (1911); **45**, 3705 (1912); Jois *et al., J. Indian Chem. Soc.* **10**, 41 (1933); Späth *et al., Ber.* **73**, 1361 (1933); Schonberg, Sina, *Nature* **160**, 468 (1947); **161**, 481 (1948); *J. Am. Chem. Soc.* **72**, 4826 (1950). Synthesis: Späth, Pailer, *Ber.* **69**, 767 (1936); Lagercrantz, *Acta Chem. Scand.* **10**, 647 (1956); Stanley, Vannier, U.S. pat. **2,889,337** (1959 to USDA); P. Nore, E. Honkanen, *J. Heterocycl. Chem.* **17**, 985 (1980). Use in treatment of psoriasis and mycosis fungoides: J. A. Parrish *et al., Int. J. Dermatol.* **19**, 379 (1980). Acute toxicity data: Hakim *et al., J. Pharmacol. Exp. Ther.* **131**, 394 (1961). Phototoxicity study: A. Kornhauser *et al., Science* **217**, 733 (1982). Comprehensive description: M. A. Loutfy, M. A. Hassan, *Anal. Profiles Drug Subs.* **9**, 427-454 (1980). Review of use in photochemotherapy: T. F. Anderson, J. J. Voorhees, *Ann. Rev. Pharmacol. Toxicol.* **20**, 235-258 (1980); *Acta Derm. Venereol. Suppl.* **106**, 9-42 (1982).

Silky needles from hot water or benzene + petr ether, long rhombic prisms from alcohol + ether. mp 148°. Odorless. Bitter taste followed by tingling sensation. pH 5.5. uv max: 219, 249, 300 nm (log ε 4.32, 4.35, 4.06). Practically insol in cold water. Sparingly sol in boiling water, liq petrolatum, ether. Sol in boiling alcohol, acetone, acetic acid, vegetable fixed oils, propylene glycol, benzene. Freely sol in chloroform. Sol in aq alkalies with ring cleavage, but is reconstituted upon neutralization. LD_{50} i.p. in rats: 470 ±30 mg/kg (Hakim).

THERAP CAT: Pigmentation agent.

6069. Methoxyamine. *O-Methylhydroxylamine;* methoxylamine; α-methylhydroxylamine; hydroxylamine methyl ether. CH_5NO; mol wt 47.06. C 25.52%, H 10.71%, N 29.77%, O 34.00%. CH_3ONH_2. Prepd by treating hydroxylamine disulfonic acid with methyl sulfate: Goldfarb, *J. Am. Chem. Soc.* **67**, 1852 (1945). May also be prepd from hydroxyurethan: A. H. Blatt, *J. Am. Chem. Soc.* **61**, 3494 (1939).

Free base. Mobile liquid. Fishy, amine odor. *Highly poisonous!* bp_{760} 49-50°. Miscible with water, alcohol, ether.

Hydrochloride, $CH_5ON \cdot HCl$, nacreous scales from alcohol+ ether, mp 149-151°. Sol in water, alcohol.

Caution: Strong irritant.

USE: Analytical reagent for aldehydes and ketones.

6070. Methoxychlor. *1,1′-(2,2,2-Trichloroethylidene)-bis[4-methoxybenzene];* 1,1,1-trichloro-2,2-bis(p-methoxyphenyl)ethane; 2,2-di-p-anisyl-1,1,1-trichloroethane; DMDT; methoxy-DDT; Marlate. $C_{16}H_{15}Cl_3O_2$; mol wt 345.65. C 55.60%, H 4.37%, Cl 30.77%, O 9.26%. Commercial prepn by the condensation of anisole with chloral in the presence of sulfuric acid: Schneller, Smith, *Ind. Eng. Chem.* **41**, 1027 (1949). Activity: P. Läuger *et al., Helv. Chim. Acta* **27**, 892 (1944). Toxicity study: H. C. Hodge *et al., J. Pharmacol. Exp. Ther.* **99**, 140 (1950).

Dimorphic crystals, mp 78-78.2° or 86-88°. Practically insol in water. Soluble in alc. The solubilities are approx those of DDT, *q.v.* LD_{50} orally in rats: 5.0 g/kg (Hodge).

Caution: Potential symptoms of overexposure in animals are trembling, convulsions; kidney and liver damage. Potential occupational carcinogen. See *NIOSH Pocket Guide to Chemical Hazards* (DHHS/NIOSH 90-117, 1990) p 140. See also *Clinical Toxicology of Commercial Products,* R. E. Gosselin *et al.,* Eds. (Williams & Wilkins, Baltimore, 5th ed., 1984) Section II, p 284.

USE: Insecticide.

THERAP CAT (VET): Ectoparasiticide.

6071. Methoxyethylbenzeneboronic Acid. *(R)-[2-(1-Methoxyethyl)phenyl]boronic acid.* $C_9H_{13}BO_3$; mol wt 180.01. C 60.05%, H 7.28%, B 6.01%, O 26.66%. Synthesis of (+)-form and use: K. Burgess, A. M. Porte, *Agnew. Chem. Int. Ed. Engl.* **33**, 1182 (1994). Chemoenzymatic stereosynthesis of both enantiomers and use: S. M. Resnick *et al., J. Org. Chem.* **60**, 3546 (1995).

R-form

White solid from *n*-hexane, mp 72°-74°. $[Ga]_D^{25}$ +27 (c = 0.5 in CH_2Cl_2) (Burgess); $[\alpha]_D$+27.2 (c = 4.6 in CH_2Cl_2) (Resnick).

(−)-*S*-Form, $[\alpha]_D$−28.7 (c = 2.3 in CH_2Cl_2).

USE: NMR shift reagent for enantiomeric assay of diols.

6072. Methoxyflurane. *2,2-Dichloro-1,1-difluoro-1-methoxyethane; 2,2-dichloro-1,1-difluoroethyl methyl ether;* 1,1-difluoro-2,2-dichloroethyl methyl ether; DA-759; Metofane; Penthrane; Pentrane. $C_3H_4Cl_2F_2O$; mol wt 164.97. C 21.84%, H 2.44%, Cl 42.98%, F 23.03%, O 9.70%. $CH_3OCF_2CHCl_2$. Prepn and properties: Miller *et al., J. Am. Chem. Soc.* **70**, 431 (1948); Park *et al., ibid.* **73**, 861 (1951); Brit. pat. **754,976** (1956 to Standard Tele. & Cable).

Liquid, bp 105°; bp_{100} 51°. mp −35°. n_D^{20} 1.3861; n_D^{25} 1.3839. d_4^{20} 1.4262 (Miller); d_4^{20} 1.4226 (Park).

Note: The medicinal grade contains butylated hydroxytoluene as antioxidant.

THERAP CAT: Anesthetic (inhalation).

THERAP CAT (VET): Anesthetic.

6073. 10-Methoxyharmalan. *4,9-Dihydro-6-methoxy-1-methyl-3H-pyrido[3,4-b]indole;* 1-methyl-6-methoxy-3,4-dihydro-2-carboline; 3,4-dihydromethoxyharman. $C_{13}H_{14}N_2$O; mol wt 214.27. C 72.87%, H 6.59%, N 13.07%, O 7.47%. Prepn from 5-methoxytryptamine: Späth, Lederer, *Ber.* **63**, 2102 (1930); Petrova *et al., Zh. Obshch. Khim.* **33**, 1333 (1963), *C.A.* **59**, 10149b (1963). Weak serotonin inhibitor:

M. M. Airaksinen *et al., Acta Pharmacol. Toxicol.* **46**, 308 (1980).

Crystals, mp 208-209°. R_f 0.74 in 4:1:5 butanol, acetic acid, water.

6074. 3-Methoxy-4-hydroxyphenylglycol. *1-(4-Hydroxy-3-methoxyphenyl)-1,2-ethanediol;* (4-hydroxy-3-methoxyphenyl)ethylene glycol; vanylglycol; MHPG. $C_9H_{12}O_4$; mol wt 184.19. C 58.69%, H 6.57%, O 34.75%. Major metabolite of norepinephrine, *q.v.* Isoln from rat urine: J. Axelrod *et al., Biochim. Biophys. Acta* **36**, 576 (1959); from cat liver and kidney: M. Goldstein, S. B. Gertner, *Nature* **187**, 147 (1960). Metabolism studies on urinary MHPG levels in patients with affective disorders: K. Greenspan *et al., J. Psychiat. Res.* **7**, 171 (1970); J. W. Maas *et al., Biochem. Pharmacol.* **1974**, Suppl., pt. 2, 907; J. J. Schildkraut *et al., Biol. Markers Psychiatry Neurol., Proc. Conf.* **1981**, E. Usdin, I. Hanin, Eds. (Pergamon, Oxford, 1982) pp 23-33. Studies on MHPG levels in cerebrospinal fluid (CSF): E. Garelis *et al., Brain Res.* **79**, 1 (1974); P. Frattini *et al., Clin. Chim. Acta* **125**, 97 (1982); in CSF of patients with CNS disorders: T. N. Chase *et al., J. Neurochem.* **21**, 581 (1973); M. G. Ziegler *et al., Am. J. Psychiatry* **134**, 565 (1977). Direct method to estimate rate of prodn of MHPG by human brain *in vivo:* J. W. Maas *et al., Science* **205**, 1025 (1979). Review of origin and distribution in body fluids: E. M. DeMet, A. E. Halaris, *Biochem. Pharmacol.* **28**, 3043-3050 (1979). Use as biochemical marker to identify mental disorders: T. H. Maugh, *Science* **214**, 39 (1981).

USE: As biochemical tool.

6075. 2-(Methoxymethyl)-5-nitrofuran. 5-Nitro-2-furfurylmethyl ether; Furbenal; Furaspor. $C_6H_7NO_4$; mol wt 157.13. C 45.87%, H 4.49%, N 8.91%, O 40.73%. Prepd by nitration of furfuryl methyl ether in acetic anhydride: Gilman, Burtner, *Iowa State Coll. J. Sci.* **6**, 389 (1932), *C.A.* **27**, 288 (1933).

Light yellow, oily liquid. d_{20}^{20} 1.283; bp_3 104-105°. bp_4 114-117°; n_D^{20} 1.5325-1.5343. Miscible with ethanol. Soly in water 11 g/l; peanut oil 170 g/l; Skellysolve B 24 g/l. Unstable to alkali and direct sunlight.

THERAP CAT: Antifungal.

6076. 2-Methoxynaphthalene. Methyl β-naphthyl ether; nerolin "old"; Yara yara. $C_{11}H_{10}O$; mol wt 158.20. C 83.52%, H 6.37%, O 10.11%. Prepn: Hiers, Hager, *Org. Syn. coll. vol. I*, 59 (1941).

Leaflets from ether, mp 72°. bp 272°. Practically insol in water. Sparingly sol in alc; sol in ether, carbon disulfide, benzene.

6077. Methoxyphenamine. *2-Methoxy-N,α-dimethylbenzeneethanamime; o-methoxy-N,α-dimethylphenethylamine; β-(o-methoxyphenyl)isopropylmethylamine;* α-(2-methoxyphenyl)-β-methylaminopropane; $C_{11}H_{17}NO$; mol wt 179.26. C 73.70%, H 9.56%, N 7.81%, O 8.93%. β-Adrenergic agonist. Prepn: Woodruff *et al., J. Am. Chem. Soc.* **62**, 922 (1940); Heinzelman, *ibid.* **75**, 921 (1953); Morishita, **Japan.** pat. **61 2921** (1961 to Nippon Shinyaku), *C.A.* **55**, 24677f (1961).

Oil, bp_2 97-99°.

Hydrochloride, $C_{11}H_{17}NO.HCl$, *Proasma, Orthoxine.* Bitter crystals from ether + alcohol, mp 129-131°. Freely sol in water, alcohol, chloroform. Slightly sol in ether, benzene. pH 5.3-5.7 (5% aq soln).

THERAP CAT: Bronchodilator.

6078. 3-(o-Methoxyphenyl)-2-phenylacrylic Acid. α-*[(2-Methoxyphenyl)methylene]benzeneacetic acid; β-(o-meth-oxyphenyl)-α-phenylacrylic acid;* α-phenyl-o-methoxycinnamic acid. $C_{16}H_{14}O_3$; mol wt 254.29. C 75.57%, H 5.55%, O 18.88%. Prepn: Buckles *et al., J. Am. Chem. Soc.* **73**, 4972 (1951); Crawford, Moore, *J. Chem. Soc.* **1955**, 3445; Alexander, Barthel, *J. Org. Chem.* **23**, 389 (1958).

cis - Form

trans-Form, plates from benzene, mp 184°.
cis-Form, prisms from benzene, mp 131.5°.
Magnesium salt, $C_{32}H_{26}MgO_6$, *AN-1022, Bilcrine.*
THERAP CAT: Choleretic.

6079. N-(p-Methoxyphenyl)-p-phenylenediamine. *N-(4-Methoxyphenyl)-1,4-benzenediamine;* 4-amino-4'-methoxydiphenylamine; 4-methoxy-4'-aminodiphenylamine; Variamine Blue base. $C_{13}H_{14}N_2O$; mol wt 214.27. C 72.87%, H 6.59%, N 13.07%, O 7.47%. Prepd from the corresp nitroso compd: Willstätter, Kubli, *Ber.* **42**, 4139 (1909).

Needles from ligroin, mp 102°. bp_{12} 238°. Freely sol in alcohol, ether, benzene. Slightly sol in water, petr ether.

Sulfate, $(C_{13}H_{14}N_2O)_2.H_2SO_4$, crystals, sol in water. Addition of ferric chloride produces intensely blue solns.

6080. Methoxypromazine. *2-Methoxy-N,N-dimethyl-10H-phenothiazine-10-propanamine; 10-(3-dimethylaminopropyl)-2-methoxyphenothiazine;* 3-methoxy-10-(3'-dimethylaminopropyl)phenothiazine; 2-methoxy-10-(3'-dimethylaminopropyl)phenothiazine; methopromazine; RP-4632. $C_{18}H_{22}N_2OS$; mol wt 314.45. C 68.75%, H 7.05%, N 8.91%, O 5.09%, S 10.20%. Prepd by condensing 3-methoxyphenothiazine with 1-dimethylamino-3-chloropropane in the presence of sodamide: Charpentier *et al., Compt. Rend.* **235**, 59 (1952). Alternate route: **Brit.** pat. **789,276** (1958 to May & Baker).

Crystals, mp 44-48°.

Maleate, $C_{18}H_{22}N_2OS.C_4H_4O_4$, *Mopazine, Tentone, Vetomazin.* Crystals, mp 141-145°. Should be protected from light during storage. Soly at 25° (g/100 ml) in water: 0.3; in methanol: 4; in chloroform: 12.5; in dimethylformamide: 100. Very slightly sol in ethanol, benzene. Practically insol in ether. pH of satd aq soln about 5.

Oxalate, $C_{18}H_{22}N_2OS.C_2H_2O_4$, crystals, dec 178-179°.

THERAP CAT: Antipsychotic.

6081. 6-Methoxy-α-tetralone. *3,4-Dihydro-6-methoxy-1(2H)-naphthalenone;* 6-methoxy-3,4-dihydro-1(2H)-naphthalenone; 1-keto-6-methoxy-1,2,3,4-tetrahydronaphthalene; 5-keto-2-methoxy-5,6,7,8-tetrahydronaphthalene. $C_{11}H_{12}O_2$; mol wt 176.22. C 74.98%, H 6.86%, O 18.16%. Prepn: Papa, *J. Am. Chem. Soc.* **71**, 3246 (1949); Ananchenko *et al., Tetrahedron* **18**, 1355 (1962).

Crystals from methanol or ligroin, mp 80°. bp_1 135-139°, bp_{11} 171°.

USE: In the synthesis of derivatives of estrane and 19-norsteroids (Ananchenko *et al., loc. cit.*).

6082. 1-Methoxy-3-(trimethylsilyloxy)-1,3-butadiene. *(E)-[(3-Methoxy-1-methylene-2-propenyl)oxy]trimethylsilane;* Danishefsky's diene. $C_8H_{16}O_2Si$; mol wt 172.30. C 55.77%, H 9.36%, O 18.57%, Si 16.30%. A silyl enol ether used in Diels-Alder reactions; highly reactive and nucleophilic diene. Prepn and ${}^1H$-NMR: S. Danishefsky, T. Kitahana, *J. Am. Chem. Soc.* **96**, 7807 (1974); S. Danishefsky *et al., Org. Syn.* **61**, 147 (1983). In synthesis of vernolepin, *q.v.: eidem, J. Am. Chem. Soc.* **99**, 6066 (1977); of pentenolactone: *eidem, ibid.* **100**, 6536 (1978). Regiospecific addition to juglone, *q.v.:* R. K. Boeckman *et al., ibid.* 7098. *Review:* S. Danishefsky, *Accounts Chem. Res.* **14**, 400 (1981).

Oil. bp_{23} 78-81°.

USE: In organic syntheses.

6083. 5-Methoxytryptamine. *5-Methoxy-1H-indole-3-ethanamine;* 3-(2-aminoethyl)-5-methoxyindole; meksamin; mexamine. $C_{11}H_{14}N_2O$; mol wt 190.25. C 69.45%, H 7.42%, N 14.72%, O 8.41%. Prepn: Supniewski *et al., C.A.* **55**, 15458 (1961). Proposed as potentiator for hypnotics, sedatives. Claimed to be more active than serotonin: Mashkovsky, Arutyunyan, *Farmakol. Toksikol.* **26** (no. 1), 10 (1963).

Crystals from ethanol, mp 121-122°.

Hydrochloride, $C_{11}H_{14}N_2O.HCl$, crystals, dec 248°.

6084. Methscopolamine Bromide. *[7(S)-(1α,2β,4β,5α,-7β)]-7-(3-Hydroxy-1-oxo-2-phenylpropoxy)-9,9-dimethyl-3-oxa-9-azoniatricyclo[3.3.1.0²,⁴]nonane bromide; 6β,7β-epoxy-3α-hydroxy-8-methyl-1αH,5αH-tropanium bromide tropate (ester);* N-methylscopolammonium bromide; hyoscine methyl bromide; scopolamine methobromide; scopolamine methyl bromide; epoxymethamine bromide; Diopal; Holopon; Mescopil; Neo-Avagal; Pamine; Restropin. $C_{18}H_{24}BrNO_4$; mol wt 398.30. C 54.28%, H 6.07%, Br 20.06%, N 3.52%, O 16.07%. Prepd by the action of methyl bromide on scopolamine base: Visscher, U.S. pat. **2,753,288** (1956 to Upjohn). Treatment of duodenal ulcer by transdermally ad-

ministered methscopolamine bromide: R. P. Walt *et al., Brit. Med. J.* **284**, 1736 (1982).

Crystals from ethanol, dec 214-217°. Freely sol in water, in dil ethanol. Slightly sol in abs ethanol.

THERAP CAT: Anticholinergic.

6085. Methsuximide. *1,3-Dimethyl-3-phenyl-2,5-pyrrolidinedione; N,2-dimethyl-2-phenylsuccinimide;* 1,3-dimethyl-3-phenyl-2,5-dioxopyrrolidine; mesuximide; Celontin; Petinutin. $C_{12}H_{13}NO_2$; mol wt 203.24. C 70.92%, H 6.45%, N 6.89%, O 15.74%. Prepd by the action of methylamine on α-methyl-α-phenylsuccinic acid: Miller, Long, *J. Am. Chem. Soc.* **73**, 4895 (1951); *eidem,* U.S. pat. **2,643,257** (1953 to Parke, Davis).

Crystals from dil alc, mp 52-53°. $bp_{0.1}$ 121-122°. Freely sol in methanol, alc. LD_{50} orally in mice: 1.55 g/kg.

THERAP CAT: Anticonvulsant.

6086. Methyclothiazide. *6-Chloro-3-(chloromethyl)-3,-4-dihydro-2-methyl-2H-1,2,4-benzothiadiazine-7-sulfonamide 1,1-dioxide; 6-chloro-3-chloromethyl-3,4-dihydro-2-methyl-7-sulfamoyl-1,2,4-benzothiadiazine 1,1-dioxide;* 6-chloro-3-chloromethyl-2-methyl-7-sulfamyl-3,4-dihydro-1,2,4-benzothiadiazine 1,1-dioxide; Aquatensen; Duretic; Enduron; Enduronum; Naturon. $C_9H_{11}Cl_2N_3O_4S_2$; mol wt 360.24. C 30.01%, H 3.08%, Cl 19.68%, N 11.66%, O 17.77%, S 17.80%. Prepn: Close *et al., J. Am. Chem. Soc.* **82**, 1132 (1960). Description: Ford, *Curr. Ther. Res.* **2**, 422 (1960). Comprehensive description: J. A. Raihle in *Analytical Profiles of Drug Substances* vol. 5, K. Florey, Ed. (Academic Press, New York, 1976) pp 307-326.

Crystals from alcohol + water, mp 225°. uv max (methanol): 226, 267, 311 nm (ϵ 39300, 21250, 3300). pKa 9.4. Very sol in acetone, pyridine; slightly sol in methanol, ethanol. Almost insol in water, chloroform, benzene.

THERAP CAT: Diuretic; antihypertensive.

6087. Methyl Abietate. Abalyn. $C_{21}H_{32}O_2$; mol wt 316.48. C 79.70%, H 10.19%, O 10.11%. $C_{19}H_{29}COOCH_3$. The article of commerce is a mixture of the methyl esters of the rosin acids.

Colorless to yellow, almost odorless, thick liquid. d_{20}^{20} 1.040. bp 360-365° with decompn. n_D^{20} 1.530. Flash pt 180-218°. Insol in water. Miscible with usual organic solvents, also with aliphatic hydrocarbons. Dissolves ester gums, rosin, many synthetic resins as well as ethyl cellulose, rubber, etc.

USE: As a solvent for ester gums, rosin, many synthetic resins, ethyl cellulose, rubber, etc.; in the manuf of varnish resins; as ingredient in adhesives.

6088. N-Methylacetanilide. *N-Methyl-N-phenylacetamide;* acetomethylanilide; Exalgin. $C_9H_{11}NO$; mol wt 149.19. C 72.46%, H 7.43%, N 9.39%, O 10.72%. Made by the action of acetic anhydride on methylaniline.

Orthorhombic rods from alcohol, plates from ether or petr ether. mp 102-104°. bp_{712} 253°. One gram dissolves in 60 ml water, 2 ml boiling water, 2 ml alcohol, 1.5 ml chloroform, 10 ml ether.

6089. Methyl Acetate. $C_3H_6O_2$; mol wt 74.08. C 48.64%, H 8.16%, O 43.20%. CH_3COOCH_3.
Colorless liquid; pleasant odor. mp −98°. bp 56.9°. d_4^{20} 0.9342; d_4^{25} 0.9279. n_D^{20} 1.3614: Mumford, Phillips, *J. Chem. Soc.* **1950**, 75. Flash pt, closed cup: 14°F (−10°C). Sol in water; miscible with alcohol, ether.
Caution: Potential symptoms of overexposure are irritation of nose and throat; headache, drowsiness; optic atrophy. Direct contact may cause skin and eye irritation. *See NIOSH Pocket Guide to Chemical Hazards* (DHHS/NIOSH 90-117, 1990) p 142.
USE: Solvent for nitrocellulose, acetylcellulose, and many resins and oils; manuf artificial leather.

6090. Methyl Acetoacetate. *3-Oxobutanoic acid methyl ester.* $C_5H_8O_3$; mol wt 116.12. C 51.72%, H 6.94%, O 41.34%. $CH_3COCH_2COOCH_3$. Toxicity data: H. F. Smyth, C. P. Carpenter, *J. Ind. Hyg. Toxicol.* **30,** 63 (1948).
Liquid. d 1.078-1.080. bp 169-171° (slight dec). mp −80°. n_D^{20} 1.418. Flash pt 82°. Sol in 2 parts water; miscible with alc, ether; gives a deep red color with $FeCl_3$. LD_{50} orally in rats: 3.0 g/kg (Smyth, Carpenter).

6091. Methyl Acetylsalicylate. *2-(Acetyloxy)benzoic acid methyl ester;* acetylsalicylic acid methyl ester; methylaspirin; Methylrodin; Methylrhodine. $C_{10}H_{10}O_4$; mol wt 194.19. C 61.85%, H 5.19%, O 32.96%. Prepd from acetylsalicylic acid and diazomethane: Herzig, Tichatschek, *Ber.* **39,** 1559 (1906); from methyl salicylate and acetic anhydride: Erdmann, *J. Prakt. Chem.* [2] **56,** 154 (1897); Thorp, U.S. pat. **1,255,950** (1918).

Plates from petr ether. mp 51-52°. bp_9 134-136°. Very sparingly sol in water; sol in alc, ether, chloroform, glycerol, propylene glycol, oils.
USE: Fixative for perfumes.

6092. Methyl Acrylate. *2-Propenoic acid methyl ester;* acrylic acid methyl ester. $C_4H_6O_2$; mol wt 86.09. C 55.81%, H 7.02%, O 37.17%. $CH_2=CHCOOCH_3$. Convenient prepn from ethylene chlorohydrin: Ger. pat. **571,123** (1928 to Rohm & Haas). For direct syntheses from acetylene and carbon monoxide see W. Reppe, *Chemie & Technik der Acetylen-Druck-Reaktionen* (Weinheim, 2nd ed., 1952). Toxicity study: H. F. Smyth, C. P. Carpenter, *J. Ind. Hyg. Toxicol.* **30,** 63 (1948).
Liquid. Acrid odor. *Lacrimator.* d_4^{20} 0.9561; d_{20}^{20} 0.9574; d_4^0 0.9702; d_4^{-5} 0.9868; d_4^{-10} 0.9929. Weighs 8.0 lbs/gal. mp −76.5°. bp_{608} 70°; bp_{428} 60°; bp_{298} 50°; bp_{200} 40°; bp_{88} 20°; bp_{54} 10°; $bp_{41.5}$ 5°; bp_{32} 0°; $bp_{24.5}$ −5°; $bp_{18.5}$ −10°. n_D^{20} 1.401. Sp heat at −60° = 0.444 cal/g/°C; heat of vaporization 8.25 kcal/mol; heat of combustion 502.88 kcal/mol. Soly in

water at 20° = 6 g/100 ml, at 40° = 5 g/100 ml. Soly of water in methyl acrylate at 20° = 1.8 ml/100 g. Sol in alc, ether. Azeotropes. 9.5% water = bp 73°; 49.0% methanol = bp 61°. Easily polymerizes on standing. The polymerization process can be speeded up by heat, light, and peroxides. If pure, the monomer can be stored below +10° without incurring polymerization. LD_{50} orally in rats: 0.3 g/kg (Smyth, Carpenter).
Polymer: Transparent, elastic substance. Practically no odor. Little adhesive power. Resists the usual solvents.
Caution: Potential symptoms of overexposure to the monomer are irritation of eyes, upper respiratory system and skin. *See NIOSH Pocket Guide to Chemical Hazards* (DHHS/NIOSH 90-117, 1990) p 142. *See also Clinical Toxicology of Commercial Products,* R. E. Gosselin *et al.,* Eds. (Williams & Wilkins, Baltimore, 5th ed., 1984) Section II, p 409.
USE: The monomer in manuf leather finish resins, textile and paper coatings, and plastic films. Produces the hardest resin of the acrylate ester series.

6093. Methylal. *Dimethoxymethane;* formal; formaldehyde dimethyl acetal. $C_3H_8O_2$; mol wt 76.10. C 47.35%, H 10.60%, O 42.05%. $CH_2(OCH_3)_2$. Prepn by catalytic vapor-phase oxidation of methanol in the presence of small amts of HCl; from methanol and formaldehyde: Frevel, Hedelund, U.S. pats. **2,663,742** and **2,691,684** (1953, 1954, both to Dow); from paraformaldehyde + methanol in the presence of $CaCl_2$ + HCl: R. Rambaud, D. Besserre, *Bull. Soc. Chim. France* **1955**, 45.
Colorless, clear, volatile, flammable liq; chloroform odor; pungent taste. bp_{760} 41.6°; bp_{754} 41.5°. mp −105°. d_4^{14} 0.8669; d_4^{20} 0.8593. n_D^{18} 1.3589. Flash pt, closed cup: 0°F (−10°C). Sol in 3 parts water, miscible with alcohol, ether, oils.
Caution: Potential symptoms of overexposure are mild irritation of eyes and upper respiratory system; anesthesia, irritation of skin. *See NIOSH Pocket Guide to Chemical Hazards* (DHHS/NIOSH 90-117, 1990) p 144.
USE: In perfumery; manuf artificial resins; reaction medium for Grignard and Reppe reactions.

6094. Methyl Allyl Trisulfide. *Methyl-2-propenyl trisulfide; allyl methyl trisulfide;* methyl 2-propenyl trisulfane; MATS. $C_4H_8S_3$; mol wt 152.31. C 31.54%, H 5.29%, S 63.16%. $CH_3SSSCH_2CH=CH_2$. Component of garlic oil which inhibits platelet aggregation. Synthesis, identification by GC: D. M. Oaks *et al., Anal. Chem.* **36,** 1560 (1964). Isoln from garlic oil, effect on platelet-rich plasma: T. Ariga *et al., Lancet* **1,** 150 (1981). Improved synthesis: A. W. Mott, G. Barany, *Synthesis* **1984**, 657. Use as antithrombotic agent: **Japan. Kokai 82 209,218** (1982 to Alcon), *C.A.* **98,** 95692n (1983).
Oil, $bp_{0.05}$ 28-30°.

6095. Methylamine. *Methanamine;* monomethylamine; aminomethane. CH_5N; mol wt 31.06. C 38.67%, H 16.23%, N 45.10%. CH_3NH_2. Occurs in herring brine, in urine of dogs after eating meat, in certain plants such as *Mentha aquatica*, in crude methanol together with di- and trimethylamine. Made by heating methyl alcohol, ammonium chloride, and zinc chloride to ~300°; by heating ammonium chloride and formaldehyde: Marvel, Jenkins, *Org. Syn.* **3,** 67 (1923); from methanol + ammonia: Smith, U.S. pat. **2,456,599** (1948 to Comm. Solvents).
Flammable gas at ordinary temp and pressure. Fuming liq when cooled in ice and salt mixture. $d_4^{-10.8}$ 0.699. mp −93.5°. bp_{760} −6.3°; bp_{400} −19.7°; bp_{200} −32.4°; bp_{100} −43.7°; bp_{10} −73.8°. Flash pt 32.5°F (0°C). Stronger base than ammonia: pK_b (25°): 3.35. Absorption spectrum: Bielecki, Henri, *Compt. Rend.* **156,** 1861 (1913). One vol of water at 12.5° dissolves 1154 vols of the gas, and 959 vols at 25°. 100 ml of a benzene soln satd at 25° contain 10.5 g methylamine. Sol in alc; miscible with ether. Good solvent for many organic substances. Marketed in liquefied form or as a 33% aq soln. LD s.c. in mice: 2.5 g/kg.
Hydrochloride, deliquescent tetragonal tablets from alcohol, mp 227-228° with sublimation. bp_{15} 225-230°. Sol in water, in abs alcohol. 100 g of boiling abs alcohol dissolve 23.01 g. Insol in chloroform, acetone, ether, ethyl acetate. *Keep well closed.*

Caution: Potential symptoms of overexposure are irritation of eyes and respiratory system; coughing; skin and mucous membrane burns; dermatitis; conjunctivitis. *See NIOSH Pocket Guide to Chemical Hazards* (DHHS/NIOSH 90-117, 1990) p 144.

USE: Methylamine is used in tanning. Methylamine and methylamine hydrochloride are used in organic synthesis for introducing the methylamino group.

6096. 2-Methylaminoethanol. Methyl(β-hydroxyethyl)amine; methylethylolamine. C_3H_9NO; mol wt 75.11. C 47.97%, H 12.08%, N 18.65%, O 21.30%. CH_3NHCH_2-CH_2OH. The *in vivo* precursor of choline. Has been isolated from a mutant of *Neurospora crassa* which has lost its ability to synthesize choline: Horowitz, *J. Biol. Chem.* **162**, 413 (1946). Prepd by mixing ethylene oxide with concd methylamine soln with external cooling: Knorr, Matthes, *Ber.* **31**, 1069 (1898); Lowe *et al.*, **Brit.** pat. 763,434 (1956 to Oxirane Ltd.); Nikolaev *et al.*, *Zh. Obshch. Khim.* **33**, 391 (1963).

Viscous liquid. Fishy odor. d^{20} 0.937. bp_{760} 155-156°; bp_{12} 64-65°. n_D^{20} 1.4385. Miscible with water, alcohol, ether. Corrosive to skin, cork, metals. Strong base. Forms a deliquescent salt with HCl. LD_{50} orally in rats: 2.34 g/kg, *Toxic Substances List*, H. E. Christensen, Ed. (1974) p 339.

Caution: Irritating to skin, eyes, mucous membranes.

6097. *p*-Methylaminophenol Sulfate. Monomethyl-*p*-aminophenol sulfate; *p*-hydroxymethylaniline sulfate; Photol; Verol; Rhodol; Armol; Elon; Genol; Graphol; Photo-Rex; Pictol; Planetol; Metol. $C_{14}H_{20}N_2O_6S$; mol wt 344.39. C 48.83%, H 5.85%, N 8.13%, O 27.87%, S 9.31%. $(HOC_6H_4NHCH_3)_2 \cdot H_2SO_4$.

Crystals. Discolors in air. mp ~260° with decompn. Sol in 20 parts cold, 6 parts boiling water; slightly sol in alc; insol in ether. *Keep well closed and protected from light.*

USE: Photographic developer; dyeing furs.

6098. Methylaniline. *N-Methylbenzenamine;* monomethylaniline. C_7H_9N; mol wt 107.16. C 78.46%, H 8.47%, N 13.07%. $C_6H_5NHCH_3$. Made by heating aniline chloride and methyl alcohol under pressure.

Colorless or slightly yellow liquid; becomes brown on exposure to air. d_4^{20} 0.989. mp $-57°$. bp 194-196°. $n_D^{21.2}$ 1.5702. Slightly sol in water; sol in alc, ether. LD orally in rabbits: 280 mg/kg.

Caution: Potential symptoms of overexposure are weakness, dizziness and headache; dyspnea, cyanosis; methemoglobinemia; pulmonary edema; liver and kidney damage. *See NIOSH Pocket Guide to Chemical Hazards* (DHHS/NIOSH 90-117, 1990) p 156.

6099. Methyl Anthranilate. *2-Aminobenzoic acid methyl ester;* methyl 2-aminobenzoate; neroli oil (artificial). $C_8H_9NO_2$; mol wt 151.17. C 63.56%, H 6.00%, N 9.27%, O 21.17%. Occurs in neroli, ylang-ylang, bergamot, jasmine, other essential oils and in grape juice; also obtained synthetically by esterifying anthranilic acid with CH_3OH in presence of HCl.

Crystals. d 1.168. mp 24-25°. bp_{15} 135.5°. Slightly sol in water; freely sol in alcohol or ether. LD_{50} orally in rats, mice: 2910, 3900 mg/kg, P. M. Jenner *et al.*, *Food Cosmet. Toxicol.* **2**, 327 (1964).

USE: As perfume for ointments; manuf synthetic perfumes.

6100. 2-Methylanthraquinone. *2-Methyl-9,10-anthracenedione;* β-methylanthraquinone. $C_{15}H_{10}O_2$; mol wt 222.24. C 81.07%, H 4.54%, O 14.40%. Occurs in teakwood. Prepd by oxidation of 2-methylanthracene; by formation from phthalic anhydride and toluene: Fieser, *Org. Syn.* **4**, 43 (1925); from 6- and 7-methylanthraquinone-1-carboxylic acids with powdered copper in quinoline: Fieser,

Martin, *J. Am. Chem. Soc.* **58**, 1443 (1936); from 1,4-naphthoquinone and isoprene: Carothers, Berchet, *ibid.* **55**, 2813 (1933).

Needles from alcohol, mp 177°. Sublimes. Very sol in benzene, toluene, xylene; sol in alcohol, ether, glacial acetic acid, concd H_2SO_4; insol in water.

6101. Methylarbutin. *4-Methoxyphenyl-β-D-glucopyranoside;* methylarbutoside. $C_{13}H_{18}O_7$; mol wt 286.28. C 54.54%, H 6.34%, O 39.12%. Occurs together with arbutin in leaves of *Arctostaphylos uva-ursi* (L.) Spreng. and allied *Ericaceae*. Prepn and physical properties: Lindpainter, *Arch. Pharm.* **277**, 398 (1939).

Monohydrate, bitter needles, mp 158-160° and again at 175°. $[\alpha]_D$ $-63°$ (water). Soluble in hot water or alcohol; slightly sol in ether.

6102. 3-Methylarsacetin. *[4-(Acetylamino)-3-methylphenyl]arsonic acid; N-acetyl-3-methylarsanilic acid;* 3-methyl-4-acetaminophenylarsanilic acid; 3-methyl-4-acetylaminophenylarsonate; Orsudan. $C_9H_{12}AsNO_4$; mol wt 273.12. C 39.58%, H 4.43%, As 27.43%, N 5.13%, O 23.43%. Prepn: Benda, Kahn, *Ber.* **41**, 1672 (1908).

Needles or rods from water, dec 306°. Sol in alkaline hydroxide and carbonate; slightly sol in hot water, methyl alcohol; very slightly sol in ethyl alcohol and other organic solvents; practically insol in dil acid.

Sodium salt, white, cryst powder. Sol in 4 parts cold water, 2.5 parts hot water.

THERAP CAT: Antimalarial.

6103. Methylbenzethonium Chloride. *N,N-Dimethyl-N-[2-[2-[methyl-4-(1,1,3,3-tetramethylbutyl)phenoxy]ethoxy]-ethyl]benzenemethanaminium chloride;* benzyldimethyl[2-[2-(p-1,1,3,3-tetramethylbutylcresoxy)ethoxy]ethyl]ammonium chloride; [2-[2-(p-octylcresoxy)ethoxy]ethyl]dimethylbenzylammonium chloride; benzyldimethyl[2-[2-[4-(1,1,3,3-tetramethylbutyl)tolyloxy]ethoxy]ethyl]ammonium chloride; Diaparene chloride; Hyamine 10X. $C_{28}H_{44}ClNO_2$; mol wt 462.12. C 72.78%, H 9.60%, Cl 7.67%, N 3.03%, O 6.92%.

Consult the Name Index before using this section.

Monohydrate, crystals. Bitter taste. mp 161-163°. Freely sol in water, alcohol, Cellosolve, chloroform, hot benzene.

THERAP CAT: Anti-infective (topical).

6104. Methyl Benzoate. *Benzoic acid methyl ester;* essence of Niobe; oil of Niobe. $C_8H_8O_2$; mol wt 136.15. C 70.57%, H 5.92%, O 23.50%. $C_6H_5COOCH_3$. Toxicity data: Smyth *et al., Arch. Ind. Hyg. Occup. Med.* **10**, 61 (1954).

Colorless, transparent liquid; pleasant odor. d_4^{15} 1.094. mp ~ −15°. bp 198-200°. n_D^{15} 1.5205. Flash pt, closed cup: 181°F (82°C). Insol in water; misc with alcohol, ether, methanol. LD_{50} orally in rats: 3.43 g/kg (Smyth).

USE: In perfumes (Peau d'Espagne).

6105. Methyl Benzoylsalicylate. *2-(Benzoyloxy)benzoic acid methyl ester;* benzosalin. $C_{15}H_{12}O_4$; mol wt 256.26. C 70.31%, H 4.72%, O 24.97%.

Crystals. mp 85°. bp 385°. Insol in water. One gram dissolves in 35 ml alcohol; sol in benzene, chloroform, ether.

6106. α-Methylbenzylamine. *α-Methylbenzenemethanamine;* α-phenylethylamine; 1-phenylethylamine; α-aminoethylbenzene. $C_8H_{11}N$; mol wt 121.18. C 79.29%, H 9.15%, N 11.56%. Prepn from acetophenone and ammonium formate: A. W. Ingersoll *et al., J. Am. Chem. Soc.* **58**, 1808 (1936); by reduction of acetophenone in liquid ammonia: J. C. Robinson, H. R. Snyder, *Org. Syn.* **coll. vol. III**, 717 (1955). Stereospecific synthesis of *R*-(+)-form: H. Takahashi *et al., Chem. Pharm. Bull.* **29**, 3387 (1981). Resolution of *(dl)*-methylbenzylamine: W. Theikeeker, H.-G. Winkler, *Ber.* **87**, 690 (1954). Use of *(R)*-(+)-methylbenzylamine as resolving agent: E. J. Corey, J. Mann, *J. Am. Chem. Soc.* **95**, 6832 (1973). Use as chiral intermediate in synthesis of α-methyl α-amino nitriles: K. Weinges *et al., Ber.* **110**, 2098 (1977); of β-amino acids: M. Furukawa *et al., Chem. Pharm. Bull.* **26**, 260 (1978). Toxicity: H. F. Smyth *et al., Arch. Ind. Hyg. Occup. Med.* **4**, 119 (1951).

dl-Form, liquid. Aromatic odor. Absorbs CO_2 from air. d_4^{15} 0.9395. bp_{18} 80-81°. Strong base. Soly in water at 20° about 4.2%. Misc with alcohol, ether. LD_{50} orally in rats: 0.94 g/kg (Smyth).

(+)-Form, liquid, bp 184-186°. d_4^{22} 0.950. $[\alpha]_D^{22}$ +40.3° (neat).

(−)-Form, liquid, bp_{12} 73°. d_4^{22} 0.950. $[\alpha]_D^{22}$ −40.3° (neat).

USE: As resolving agent; chiral intermediate.

6107. Methyl Blue. *[[4-[Bis[4-[(sulfophenyl)amino]phenyl]methylene]-2,5-cyclohexadien-1-ylidene]amino]benzenesulfonic acid disodium salt;* sodium triphenyl-*p*-rosanilinetrisulfate; brilliant cotton blue; Helvetia blue; C.I. Acid Blue 93; C.I. 42780. $C_{37}H_{27}N_3Na_2O_9S_3$; mol wt 799.81. C 55.56%, H 3.40%, N 5.25%, Na 5.75%, O 18.00%, S 12.03%. *Do not confuse with methylene blue.* Prepn: *Colour Index* vol. 4 (3rd ed., 1971) p 4403.

Dark blue powder. Sol in water. Absorption max about 607 nm.

USE: As a coloring; dye for cotton and silk; biological stain.

THERAP CAT: Antiseptic.

6108. Methyl Bromide. *Bromomethane;* monobromomethane; Embafume. CH_3Br; mol wt 94.94. C 12.65%, H 3.19%, Br 84.16%. Prepd industrially by the action of hydrobromic acid on methanol. Several modifications of the process, *e.g.,* sulfuric acid is added to sodium bromide and methanol, methyl bromide being removed by distillation. Acute toxicity: D. D. Irish *et al., J. Ind. Hyg. Toxicol.* **22**, 218 (1940). Atmospheric emission and potential ozone depletion from biomass burning: S. Manö, M. O. Andreae, *Science* **263**, 1255 (1994).

Colorless gas. *Poisonous!* Usually odorless; sweetish, chloroform-like odor at high concns. Burning taste. Nonflammable in air, but burns in oxygen. mp −93.66°; bp 3.56°: Egan, Kemp, *J. Am. Chem. Soc.* **60**, 2097 (1938). Vapor press. (20°): 1420 mm Hg. d_4^0 1.730. d_{gas}^{20} 3.974 g/l; n_D^{-20} 1.4432. Viscosity at 0°: 0.397 cp. Spec. heat at −96.6°: 0.165 cal/g/°C; at −13.0°: 1.97 cal/g/°C; at 25°: 0.107 cal/g/°C. Crit temp 194°. Soly in water (20°, 748 mm): 1.75 g/100 g of soln. Forms a cryst hydrate, $CH_3Br.20H_2O$, below 4°. Freely sol in alcohol, chloroform, ether, carbon disulfide, carbon tetrachloride, benzene. LC for rats in air (6 hrs): 514 ppm (Irish).

Caution: Potential symptoms of overexposure by inhalation are headache; visual disturbance; vertigo; nausea, vomiting, anorexia, abdominal pain; malaise, profound weakness, slurring of speech, staggering gait; hand tremor; convulsions; mental confusion; dyspnea; pulmonary edema; coma; death from respiratory or circulatory collapse. Onset of symptoms may be delayed. Direct contact may cause eye irritation, skin irritation or vesiculation. Potential occupational carcinogen. *See NIOSH Pocket Guide to Chemical Hazards* (DHHS/NIOSH 90-117, 1990) p 146; *Clinical Toxicology of Commercial Products,* R. E. Gosselin *et al.,* Eds. (Williams & Wilkins, Baltimore, 5th ed., 1984) Section III, pp 280-284.

USE: In ionization chambers. For degreasing wool. Extracting oils from nuts, seeds, flowers. Insect fumigant for mills, warehouses, vaults, ships, freight cars, also as soil fumigant. Because of its toxicity, there has been considerable concern against distributing this gas in small fire extinguishers.

6109. 2-Methyl-1-butanol. Active amyl alcohol; *dl-sec*-butyl carbinol. $C_5H_{12}O$; mol wt 88.15. C 68.13%, H 13.72%, O 18.15%. $CH_3CH_2CH(CH_3)CH_2OH$. One of the major components of fusel oil, *q.v.* Prepn: Hawthorne, *J. Org. Chem.* **23**, 1788 (1958); **Brit. pat. 883,375** (1961 to Continental Oil). Isoln of (−)-form by fractional distillation of fusel oil: Milburn, Truter, *J. Chem. Soc.* **1954**, 3344. Prepn of (+)-form: Carnmalm, *Chem. & Ind. (London)* **1956**, 1093. Toxicity study: H. F. Smyth *et al., Am. Ind. Hyg. Assoc. J.* **23**, 95 (1962). Review of manuf by fractionation of fusel oil and *via* chlorination of pentanes, and properties: *Industrial Chemicals,* W. L. Faith *et al.,* Eds. (John Wiley, New York, 2nd ed., 1957) pp 107-114.

Liquid, bp 128°. d_4^{20} 0.816. n_D^{20} 1.4104. Flash pt, open cup: 122°F (50°C). Slightly sol in water (3.6 g/100 g at 30°); misc with alcohol, ether. LD_{50} orally in rats: 4.92 ml/kg (Smyth).

(−)-Form, $[\alpha]_D^{20}$ −4.75°.

(+)-Form, d_{20}^{20} 0.826. n_D^{20} 1.411.

6110. 3-Methyl-2-butanol. *dl-sec*-Isoamyl alcohol; *sec*-isopentyl alcohol; isopropyl methyl carbinol. $C_5H_{12}O$; mol wt 88.15. C 68.13%, H 13.72%, O 18.15%. $(CH_3)_2CHCH(OH)CH_3$. Prepn: Brown, Subba Rao, *J. Am. Chem. Soc.* **81**, 6423 (1959); Cook, *J. Org. Chem.* **27**, 3873 (1962). Review of manuf by fractionation of fusel oil and *via* chlorination of pentanes, and properties: *Industrial Chemicals,* W. L. Faith *et al.,* Eds. (John Wiley, New York, 2nd ed., 1957) pp 107-114.

Liquid, bp 113-114°. bp_{742} 109.5-110.5°. d^{19} 0.819. n_D^{20} 1.4091. Flash pt, closed cup: 103°F (39°C); open cup: 95°F (35°C). Slightly sol in water (2.8 g/100 g at 30°); miscible with alcohol, ether.

Caution: Potential symptoms of overexposure are irritation of eyes, nose and throat; narcosis; headache, dizziness;

---

dyspnea, nausea, vomiting and diarrhea; skin cracking. *See NIOSH Pocket Guide to Chemical Hazards* (DHHS/NIOSH 90-117, 1990) p 130.

6111. Methyl *tert*-Butyl Ether. *2-Methoxy-2-methylpropane; tert-butyl methyl ether;* MTBE. $C_5H_{12}O$; mol wt 88.15. C 68.13%, H 13.72%, O 18.15%. $(CH_3)_3C(OCH_3)$. Prepn: L. Henry, *Rec. Trav. Chim.* **23**, 324 (1904); from methanol and *t*-butyl alcohol: J. F. Norris, G. W. Rigby, *J. Am. Chem. Soc.* **54**, 2088 (1932); from methanol and isobutylene: K. R. Edlund, T. W. Evans, U.S. pat. **1,968,601** (1934 to Shell); *eidem, Ind. & Eng. Chem.* **28**, 1184 (1936); R. D. Morin, A. E. Bearse, *ibid.* **43**, 1596 (1951); from *t*-butyl alcohol and diazomethane: M. Neeman *et al., Tetrahedron* **6**, 36 (1959). Use as chromatographic eluent: C. J. Little *et al., J. Chromatog.* **169**, 381 (1979). Experimental use to dissolve cholesterol gallstones *in vivo*: M. J. Allen *et al., Gastroenterology* **88**, 122 (1985); *eidem, N. Engl. J. Med.* **312**, 217 (1985). Method for therapeutic use in dissolving cholesterol calculi: J. L. Thistle, M. J. Allen, U.S. pat. **4,758,596** (1988 to Research Corp.). Acute toxicity: D. F. Marsh, C. D. Leake, *Anesthesiology* **11**, 455 (1950). Reviews of toxicity: R. von Burg, *J. Appl. Toxicol.* **12**, 73-74 (1992); E. Reese, R. D. Kimbrough, *Environ. Health Perspect.* **101**, Suppl. 6, 115-131 (1993); M. G. Costantini, *ibid.* 151-160. Liquid, bp 55.2°. mp −109°. d_4^{20} 0.7404, n_D^{20} 1.3689. Vapor pressure at 25°: 245 mm Hg. Flash pt −28°C. Ignition temp: 224°C. Soly in water: 4.8 g/100 g. Soly of water in methyl *t*-butyl ether: 1.5 g/100 g. Unstable in acid soln. LC_{50} in mice (15 min): 1.6 mmol/liter of atmosphere (Marsh).

USE: Octane booster in gasoline. Chromatographic eluent esp in HPLC.

THERAP CAT: Cholelitholytic agent.

6112. Methyl Butyl Ketone. *2-Hexanone.* $C_6H_{12}O$; mol wt 100.16. C 71.95%, H 12.08%, O 15.97%. $CH_3CO-C_4H_9$. Toxicity study: H. F. Smyth *et al., Arch. Ind. Hyg. Occup. Med.* **10**, 61 (1954). Review of human and animal toxicity: P. M. J. Bos *et al., Am. J. Ind. Med.* **20**, 175-194 (1991).

Colorless liquid. d 0.830. bp 127°. Slightly soluble in water; sol in alcohol, ether. LD_{50} orally in rats: 2.59 g/kg (Smyth).

Caution: Potential symptoms of overexposure are irritation of eyes and nose; peripheral neuropathy; weakness, paresthesia; dermatitis; headache; drowsiness. *See NIOSH Pocket Guide to Chemical Hazards* (DHHS/ NIOSH 90-117, 1990) p 122.

6113. 2-Methyl-3-butyn-2-ol. 2-Hydroxy-2-methyl-3-butyne. C_5H_8O; mol wt 84.12. C 71.39%, H 9.59%, O 19.02%. Toxicity data: W. Keil *et al., Arzneimittel-Forsch.* **4**, 477 (1954); K. Soehring *et al., ibid.* **5**, 161 (1955).

Liquid. d_{20}^{20} 0.8672. 7.24 lbs/gal. mp +2.6°. bp_{760} 104-105°; bp_{80} 52°; bp_{12} 20°. Flash pt 77°F (25°C). n_D^{20} 1.4211. Surface tension at 25° = 23.8 dynes/cm; 5% aq soln = 41.7 dynes/cm. Miscible with water, acetone, benzene, carbon tetrachloride, Cellosolve, cyclohexanone, diethylene glycol, ethyl acetate, kerosine, methyl ethyl ketone, mineral spirits, monoethanolamine, neatsfoot oil, petr ether, soybean oil, Stoddard solvent. Azeotrope with water, bp 90.7°, contains 28.4% H_2O. LD_{50} in mice (mg/kg): 1800 orally (Keil); 2340 s.c. (Soehring).

6114. Methyl Butyrate. *Butanoic acid methyl ester.* $C_5H_{10}O_2$; mol wt 102.13. C 58.80%, H 9.87%, O 31.33%. $CH_3(CH_2)_2COOCH_3$.

Colorless liquid. $d_{}^{20}$ 0.898. mp about −95°. bp 102°. Flash pt 14°. n_D^{20} 1.3879. Sol in about 60 parts water; miscible with alcohol, ether.

USE: Manuf artificial rum and fruit essences.

6115. Methyl Carbamate. Urethylane; methylurethane. $C_2H_5NO_2$; mol wt 75.07. C 32.00%, H 6.71%, N

18.66%, O 42.63%. $H_2NCOOCH_3$. Prepn from silver or mercuricyanate with H_2S and methanol: Birkenbach, Kolb, *Ber.* **68**, 901 (1935). From urea and methanol: Ger. pat. **753,127** (1940 to I.G. Farbenind.).

White crystals, mp 52-54°. bp 177°. Freely sublimes even at room temp. Freely sol in water, alcohol.

6116. Methyl Carbitol®. *2-(2-Methoxyethoxy)ethanol;* diethylene glycol monomethyl ether; methyl digol. $C_5H_{12}O_3$; mol wt 120.15. C 49.98%, H 10.07%, O 39.95%. $CH_3O-CH_2CH_2OCH_2CH_2OH$. Toxicity data: Smyth *et al., J. Ind. Hyg. Toxicol.* **23**, 259 (1941).

Liquid. d_4^{20} 1.035. mp $< -84°$. bp 193°. n_D^{27} 1.4264. Flash pt, open cup: 200°F (93°C). Miscible with water, alc, glycerol, ether, acetone, DMF. LD_{50} orally in rats: 9.21 g/kg (Smyth).

USE: Used in the same way as 2-ethoxyethanol where a solvent with a higher boiling point is required.

6117. Methyl Carbonate. *Carbonic acid dimethyl ester.* $C_3H_6O_3$; mol wt 90.08. C 40.00%, H 6.71%, O 53.28%. $CO(OCH_3)_2$. Made from phosgene and methyl alcohol.

Liquid. $d_{}^{17}$ 1.065. mp 0.5°. bp 90-91°; also stated as 89.7°. n_D^{20} 1.3687. Insol in water. Miscible with alc, ether.

6118. Methyl Cellosolve®. *2-Methoxyethanol;* ethylene glycol monomethyl ether. $C_3H_8O_2$; mol wt 76.10. C 47.35%, H 10.60%, O 42.05%. $HOCH_2CH_2OCH_3$. Prepn from ethylene oxide + methanol: Finch, Hagemeyer, U.S. pat. **2,748,171** (1956 to Kodak); from ethylene glycol + diazomethane: Hesse, Majumdar, *Ber.* **93**, 1129 (1960).

Liquid. *Poisonous!* bp_{760} 124.43°; bp_{20} 34-41°. d_4^{20} 0.9663. Flash pt 115°F. n_D^{20} 1.4028. Miscible with water, alcohol ether, glycerol, acetone, dimethylformamide. LD_{50} orally in rats: 2.46 g/kg; lethal concn for mice in air: 1500 ppm.

Caution: Potential symptoms of overexposure are headache, drowsiness and weakness; ataxia, tremor and somnolence; anemic pallor, irritation of eyes. Potential occupational carcinogen. *See NIOSH Pocket Guide to Chemical Hazards* (DHHS/NIOSH 90-117, 1990) p 146. Effects of excessive exposure centered primarily on brain, blood and kidneys. Readily absorbed through skin. *See* Patty's *Industrial Hygiene and Toxicology* **vol. 2C**, G. D. Clayton, F. E. Clayton, Eds. (Wiley-Interscience, New York, 3rd ed., 1982) pp 3911-3919.

USE: Solvent for low-viscosity cellulose acetate, natural resins, some synthetic resins and some alcohol-soluble dyes; in dyeing leather, sealing moistureproof cellophane; in nail polishes, quick-drying varnishes and enamels, wood stains. In modified Karl Fischer reagent: Peters, Jungnickel, *Anal. Chem.* **27**, 450 (1955).

6119. Methyl Cellosolve® Acetate. *2-Methoxyethanol acetate;* ethylene glycol monomethyl ether acetate. $C_5H_{10}O_3$; mol wt 118.13. C 50.84%, H 8.53%, O 40.63%. $CH_3OCH_2-CH_2OOCCH_3$. Toxicity data: H. F. Smyth *et al., J. Ind. Hyg. Toxicol.* **23**, 259 (1941). Brief review: E. S. Brown *et al.,* "Glycols (Ethylene and Propylene)" in Kirk-Othmer *Encyclopedia of Chemical Technology* **vol. 11** (Wiley-Interscience, New York, 3rd ed., 1980) pp 942-946.

Colorless liquid. d_{20}^{20} 1.0067. bp 145°. mp −65.1°. n_D^{20} 1.4019. Flash pt (open cup): 55.6°. Miscible with water, most organic solvents, oils; dissolves gums, resins. LD_{50} orally in rats: 3.4 g/kg (Smyth).

Caution: Potential symptoms of overexposure are kidney damage; brain damage; eye irritation. *See NIOSH Pocket Guide to Chemical Hazards* (DHHS/NIOSH 90-117, 1990) p 146. *See also* Patty's *Industrial Hygiene and Toxicology* **vol. 2C**, G. D. Clayton, F. E. Clayton, Eds. (Wiley-Interscience, New York, 3rd ed., 1982) pp 4022-4024.

USE: Industrial solvent.

6120. Methylcellulose. *Cellulose methyl ether;* Methocel MC; Cellothyl; Syncelose; Bagolax; Celevac; Cellucon; Cologel; Cellumeth; Hydrolose; Nicel; Tylose. Prepd from wood pulp or chemical cotton by treatment with alkali and methylation of the alkali cellulose with methyl chloride. *Review and bibliography:* Ott, *Cellulose and Cellulose Derivatives* (Wiley-Interscience, New York, 2nd ed., 1954/55); Greminger, Savage, in *Industrial Gums*, R. L. Whistler, Ed. (Academic Press, New York, 2nd ed., 1973) pp 619-647.

White granules. Odorless, tasteless. Sol in cold water. Insol in hot water. An aq soln is best prepd by dispersing the granules in hot (but not boiling) water with stirring and chilling to +5°. The soln is then stable at room temp. Presence of inorganic salts increases the viscosity. The soly is dependent upon degree of substitution. Commercial methylcellulose has a methoxyl content of 29% (degree of substitution about 1.8). Clear films may be cast from aq soln.

USE: As a substitute for water-soluble gums; to render paper greaseproof, in adhesives, as thickening agent in cosmetics, as protective colloid in emulsions, as binder and stabilizer in foods. As bulk producer in the formulation of dietetic foods. Pharmaceutic aid (suspending agent).

THERAP CAT (VET): Laxative.

6121. Methyl Chloride. *Chloromethane.* CH_3Cl; mol wt 50.49. C 23.79%, H 5.99%, Cl 70.22%. Review of mfg processes: Faith, Keyes & Clark's *Industrial Chemicals*, F. A. Lowenheim, M. K. Moran, Eds. (Wiley-Interscience, New York, 4th ed., 1975) pp 530-538.

Colorless gas; compresses to a colorless liq of ethereal odor and sweet taste. mp −97°; bp −23.7°; n_D (liq at −23.7°) 1.3712; McGovern, *Ind. Eng. Chem.* **35**, 1230 (1943). Slightly sol in water; misc with chloroform, ether, glacial acetic acid; sol in alcohol. Soly at 20° (ml/100 ml): water 303; benzene 4723; carbon tetrachloride 3756; glacial acetic acid 3679; ethanol 3740. Explosive limits in air 8.1 to 17.2% v/v. Lethal concn for mice in air: 3146 ppm, *RTECS* Vol. II, R. J. Lewis, R. L. Tatken, Eds. (1980) p 57.

Caution: Potential symptoms of overexposure are dizziness, nausea and vomiting; visual disturbance; staggering; slurred speech; convulsions, coma; liver and kidney damage; frostbite. Direct contact may cause skin irritation. Potential occupational carcinogen. *See NIOSH Pocket Guide to Chemical Hazards* (DHHS/NIOSH 90-117, 1990) p 146. *See also* Patty's *Industrial Hygiene and Toxicology* **vol. 2B**, G. D. Clayton, F. E. Clayton, Eds. (Wiley-Interscience, New York, 3rd ed., 1981) pp 3436-3442.

USE: As a refrigerant.

THERAP CAT: Anesthetic (local).

6122. Methyl Chloroacetate. *Chloroacetic acid methyl ester.* $C_3H_5ClO_2$; mol wt 108.52. C 33.20%, H 4.64%, Cl 32.67%, O 29.49%. $CH_2ClCOOCH_3$.

Colorless liquid. d_{20}^{20} 1.238. mp 33°. bp 130-132°. Insol in water; miscible with alcohol, ether.

USE: As solvent.

6123. Methyl Chlorocarbonate. *Carbonochloridic acid methyl ester;* methyl chloroformate. $C_2H_3ClO_2$; mol wt 94.50. C 25.42%, H 3.20%, Cl 37.52%, O 33.86%. $ClCO_2$-CH_3. Made from phosgene and methyl alcohol.

Clear liquid. d_4^{20} 1.223. bp 71°. Slightly sol in water and gradually dec by it; miscible with alcohol, benzene, chloroform, ether.

Caution: Vapors strongly irritating to eyes.

6124. 3-Methylcholanthrene. *1,2-Dihydro-3-methyl-benz[j]aceanthrylene;* 20-methylcholanthrene; 3-MECA; 3-MC. $C_{21}H_{16}$; mol wt 268.36. C 93.99%, H 6.01%. From desoxycholic acid: Wieland, Schlichting, *Z. Physiol. Chem.* **150**, 267 (1925); Wieland, Wiedersheim, *ibid.* **186**, 229 (1930); Wieland, Dane, *ibid.* **219**, 240 (1933). From cholic acid: Cook, Haslewood, *J. Chem. Soc.* **1934**, 428; Fieser, Seligman; *J. Am. Chem. Soc.* **58**, 2482 (1936); *cf.* Bachmann, *J. Org. Chem.* **1**, 347 (1936). Methylcholanthrene has been produced also by an unusual pyrolytic degradation of cholesterol derivatives. Total synthesis: Buchta, Güllich, *Angew. Chem.* **70**, 190 (1958); P. W. Tang, C. A. Maggiulli,

J. Org. Chem. **46**, 3429 (1981); S. A. Jacobs, R. G. Harvey, *Tetrahedron Letters* **22**, 1093 (1981). *Review:* E. Clar, *Polycyclic Hydrocarbons* (Academic Press, 1964) 2 vols.

Pale yellow, slender prisms from benzene + ether, mp 179-180°. bp_{80} 280°; d^{20} 1.28. Absorption spectrum: Fieser, Hershberg, *J. Am. Chem. Soc.* **60**, 940 (1938). Sol in benzene, xylene, toluene; slightly sol in amyl alcohol. Insol in water.

USE: Exptlly in cancer research.

6125. Methylcobalamin. *Cobinamide Co-methyl deriv. dihydrogen phosphate (ester) inner salt 3'-ester with 5,6-dimethyl-1-α-D-ribofuranosyl-1H-benzimidazole;* methyl-5,6-dimethylbenzimidazolylcobalamin; mecobalamin; methyl vitamin B_{12}; MeCbl; Algobaz; Hitocobamin-M; Methycobal; Methylcobaz. $C_{63}H_{91}CoN_{13}O_{14}P$; mol wt 1344.40. C 56.28%, H 6.82%, Co 4.38%, N 13.54%, O 16.66%, P 2.30%. One of the biologically active forms of vitamin B_{12}, *q.v.*; differing only by the substitution of a methyl for the cyano group. Coenzyme required in the biosynthesis of methionine. Identification as requirement for methionine biosynthesis by *E. coli:* R. L. Kisliuk, D. D. Woods, *Biochem. J.* **75**, 467 (1960); J. R. Guest *et al.*, *Nature* **195**, 340 (1962). Prepn: O. Müller, G. Müller, *Biochem. Z.* **336**, 299 (1962); D. Dolphin, *Methods Enzymol.* **18**, Pt. C, 34 (1971); D. Autissier *et al.*, *Bull. Soc. Chim. France* **1980**, part 2, 192. Interrelationship with folate metabolism: B. Shane, E. L. R. Stokstad, *Ann. Rev. Nutr.* **5**, 115-141 (1985). Discussion of mechanism of action: W. S. Beck, *Am. J. Hematol.* **34**, 83-89 (1990). *Review:* "Vitamin B_{12}" in *Vitamins*, W. Friedrich, Ed. (de Gruyter, Berlin, 1988) pp 837-928.

Bright red crystals from water/acetone. uv max (pH 7): 522, 342, 266 nm (ε 9357, 14416, 19897); (0.1N HCl): 462, 304, 261 nm (ε 9599, 22855, 24737).

THERAP CAT: Vitamin (hematopoietic).

6126. Methylconiine. *1-Methyl-2-propylpiperidine.* C_9-$H_{19}N$; mol wt 141.26. C 76.53%, H 13.56%, N 9.92%. The *d*-form is stated to occur in hemlock in small quantities. The *l*-form can be isolated from residues left in the prepn of coniine by crystn of the hydrobromides. Isoln: Ahrens, *Ber.* **35**, 1330 (1902); Hess, Eichel, *ibid.* **50**, 1386 (1917). Prepns: Leonard, Barthel, *J. Am. Chem. Soc.* **71**, 3098 (1949); Gaumeton, Glacet, *Bull. Soc. Chim. France* **1959**, 1501; Lukes, Cerny, *Coll. Czech. Chem. Commun.* **24**, 1287 (1959); Cervinka, *ibid.* **26**, 673 (1961).

dl-Form, $bp_{10.5}$ 56.6°.
d-Form, oily liq, bp 173-174°. d^{24} 0.8318. $[\alpha]_D^{24}$ +81°. n_D^{13} 1.4538. Slightly sol in cold water; sol in organic solvents.
l-Form, $[\alpha]_D^{24}$ −84°. Has essentially the same characteristics as the *d*-form.

6127. 3'-Methyl-1,2-cyclopentenophenanthrene. *16,17-Dihydro-17-methyl-15H-cyclopenta[a]phenanthrene;* Diels' hydrocarbon. $C_{18}H_{16}$; mol wt 232.33. C 93.06%, H 6.94%. Prepd from cholesterol, cholic acid, or suitable sapogenins: Diels, Rickert, *Ber.* **68**, 267 (1935); starting with 2-acetyl-phenanthrene: Gamble *et al.*, *J. Chem. Soc.* **1935**, 443, 644.

Crystals from acetic acid, mp 126-127°. Absorption spectrum: Hillemann, *Ber.* **69**, 2610 (1936).

6128. 5-Methylcytosine. *4-Amino-5-methyl-2(1H)-pyrimidinone.* $C_5H_7N_3O$; mol wt 125.13. C 47.99%, H 5.64%, N 33.58%, O 12.79%. Occurs in a nucleic acid obtained from tubercle bacillus. Isoln: Johnson, Coghill, *J. Am. Chem. Soc.* **47**, 2838 (1925). Synthesis from ethyl bromide addition product of thiourea and from ethyl formyl propionate: Wheeler, Johnson, *Am. Chem. J.* **29**, 492 (1903).

Prisms from water, may contain ½ mol H_2O. mp 270° (effervescence). One gram dissolves in 29 ml water.

6129. Methyl Demeton. *Phosphorothioic acid O-[2-(ethylthio)ethyl] O,O-dimethyl ester mixt with S-[2-(ethylthio)ethyl] O,O-dimethyl phosphorothioate;* demeton-methyl; methyl-mercaptophos; methyl systox; Bayer 21/116; Meta-Systox. $C_6H_{15}O_3PS_2$; mol wt 230.29. C 31.29%, H 6.57%, O 20.84%, P 13.45%, S 27.85%. Isomeric mixture consisting of demeton-O-methyl and demeton-S-methyl (O(and S)-[2-(ethylthio)ethyl] O,O-dimethyl phosphorothioate). *Ref:* Henglein, Schrader, *Z. Naturforsch.* **10b**, 12 (1955). Prepn: **Brit.** pat. **814,332** (1959 to Bayer).

Commercial product is a light yellow liquid. Hydrolyzed by alkali. LD_{50} orally in rats: 65 mg/kg.
Demeton-O-methyl, colorless oil, $bp_{0.15}$ 74°. d_4^{20} 1.190. n_D^{20} 1.5063. Soly in water at room temp: 330 ppm. Soluble in organic solvents. LD_{50} orally in rats: 180 mg/kg.
Demeton-S-methyl, pale yellow oil, $bp_{0.15}$ 89°. d_4^{20} 1.207. n_D^{20} 1.5065. Soly in water at room temp: 3,300 ppm. Sol in organic solvents. LD_{50} orally in rats: 40 mg/kg.
Caution: Cholinesterase inhibitor. Absorbed readily through skin. For symptoms *see* Parathion.
USE: Insecticide.

6130. p-Methyldiphenhydramine. *N,N-Dimethyl-2-[(4-methylphenyl)phenylmethoxy]ethanamine; N,N-dimethyl-2-[(p-methyl-α-phenylbenzyl)oxy]ethylamine; β-dimethylamino-ethyl 4-methylbenzhydryl ether; p-methylbenzhydryl 2-dimethylaminoethyl ether.* Neo-Benodine; Toladryl. $C_{18}H_{23}NO$; mol wt 269.39. C 80.26%, H 8.61%, N 5.20%, O 5.94%. Prepn: Rieveschl, **U.S.** pat. **2,527,962** (1950 to Parke, Davis); **Brit.** pat. **683,483** (1952 to Fabriken Brocades-Stheeman & Pharmacia).

Hydrochloride, $C_{18}H_{23}NO\cdot HCl$, crystals from acetone, mp 150-152° (free base $bp_{0.1}$ 143°).
THERAP CAT: Antihistaminic.

6131. Methyldiphenylamine. *N-Methyl-N-phenylbenzenamine;* diphenylmethylamine. $C_{13}H_{13}N$; mol wt 183.25. C 85.21%, H 7.15%, N 7.64%. $(C_6H_5)_2NCH_3$. Prepd by heating diphenylamine, methyl alcohol and HCl.
Liquid. d_4^{20} 1.0476. mp −7.6°. bp 296-297°; also given as 282-290°, 293°. n_D^{20} 1.6193. Insol in water; sol in alc, ether.
USE: Manuf dyes; as reagent similar to diphenylamine.

6132. Methyldopa. *3-Hydroxy-α-methyl-L-tyrosine; L-3-(3,4-dihydroxyphenyl)-2-methylalanine;* L-α-methyl-3,4-dihydroxyphenylalanine; L-2-amino-2-methyl-3-(3,4-dihydroxyphenyl)propionic acid; α-methyldopa; AMD; MK-351; Aldomet; Aldometil; Aldomine; Dopamet; Dopegyt; Elanpres; Equibar; Lederdopa; Medomet; Medopa; Medopren; Meth-

oplain; Sembrina; Presinol. $C_{10}H_{13}NO_4$; mol wt 211.22. C 56.87%, H 6.20%, N 6.63%, O 30.30%. Exists as the sesquihydrate. Prepn: Pfister, Stein, U.S. pat. **2,868,818** (1959 to Merck & Co.). Resolution: Jones *et al.*, U.S. pat. **3,158,-648** (1964 to Merck & Co.); *cf.* Slates *et al.*, *J. Org. Chem.* **29**, 1424 (1964). Resolution and configuration: Tristram, *ibid.* 2053. Synthesis from asymmetric intermediates: Reinhold *et al.*, *J. Org. Chem.* **33**, 1209 (1968). Prepn of the ethyl ester hydrochloride: **Fr.** pat. **M2153** (1963 to Merck & Co.); of pharmaceutical dosage forms: Marcus, **U.S.** pat. **3,230,143** (1966 to Merck & Co.). Mode of action: Finch, Haeusler, *Brit. J. Pharmacol.* **45**, 167p (1972); Day *et al.*, *ibid.* 168p. Review of pharmacology: A. Scriabine, Ed. in *Pharmacology of Antihypertensive Drugs* (Raven, New York, 1980) pp 43-54.

L-Form sesquihydrate, crystals from water. Minute anhydr crystals from methanol, considerably hygroscopic, dec ~300°. $[\alpha]_D^{23}$ −4.0° ± 0.5° (c = 1 in 0.1N HCl). uv max: 281 nm (ε 2780). Soly in water at 25°: ~10 mg/ml. pH of satd aq soln about 5.0. (Soly of D-form in water at 25°: ~10 mg/ml, of DL-form: ~18 mg/ml.) Practically insol in the common organic solvents. Sol in dil mineral acids.
Ethyl ester hydrochloride, $C_{12}H_{18}ClNO_4$, *methyldopate hydrochloride.* After drying, mol wt 275.7 (1 mg = 0.77 mg of the free acid). Soluble in water to the extent of 10-300 mg/ml at pH 2-7. May be stabilized with sodium bisulfite and/or disodium EDTA.
THERAP CAT: Antihypertensive.

6133. 3-O-Methyldopa. *3-Methoxy-L-tyrosine;* 3-(4-hydroxy-3-methoxyphenyl)-L-alanine; L-3-methoxy-4-hydroxyphenylalanine; OM-dopa; L-3-MTO; OMD. $C_{10}H_{13}NO_4$; mol wt 211.22. C 56.87%, H 6.20%, N 6.63%, O 30.30%. A major metabolite of L-dopa in man and animals, which has a considerably longer biological half-life than L-dopa: Pletscher *et al.*, *Brain Res.* **4**, 106 (1967); Bartholini, Pletscher, *J. Pharmacol. Exp. Ther.* **161**, 14 (1968); Kuruma *et al.*, *Eur. J. Pharmacol.* **10**, 189 (1970). Proposed as a precursor of dopamine *via* its partial demethylation in the organism: Bartholini *et al.*, *Nature* **230**, 533 (1971); Chalmers *et al.*, *Brit. J. Pharmacol.* **43**, 455p (1971); Bartholini *et al.*, *Life Sci.* **14**, 323 (1974). *See also* Carlsson, Waldeck, *Arch. Pharmacol.* **272**, 441 (1972); Bartholini, Pletscher, *ibid.* **274**, 404 (1972). Metabolic pathways: Bartholini *et al.*, *J. Pharmacol. Exp. Ther.* **183**, 65 (1972). Parkinsonism treatment studies: Calne *et al.*, *Clin. Pharmacol. Ther.* **14**, 386 (1973).

6134. Methyleneaminoacetonitrile. α-Hydroformamine cyanide. $C_3H_4N_2$; mol wt 68.08. C 52.93%, H 5.92%, N 41.15%. $CH_2=NCH_2CN$. The actual molecular formula is $C_9H_{12}N_6$: Johnson, Rinehart, *J. Am. Chem. Soc.* **46**, 768, 1653 (1924). Prepd by the action of formaldehyde on a mixture of ammonium chloride, potassium cyanide and acetic acid: Adams, Langley, *Org. Syn.* **coll. vol. I** (2nd ed., 1941) p 355.
Trimeric form, orthorhombic crystals from alc or acetone, mp 129°. Sol in hot water, alc; slightly sol in benzene.

6135. Methylene Azure. Azure I. A somewhat variable mixture obtained by oxidation of methylene blue, contg primarily azure A and azure B, *q.v.* Prepn: Bernthsen, *Ann.* **230**, 169 (1885); Kehrmann, *Ber.* **39**, 1804 (1906). *Review:* H. J. Conn's *Biological Stains*, R. D. Lillie, Ed. (Williams & Wilkins, Baltimore, 9th ed., 1977) pp 493-502.
Green glistening crystals. Forms blue soln in water; very sparingly sol in alcohol with a reddish brown fluorescence.
Mixture with methylene blue, *azure II, azure blue II, methylene azure II.* Deep-green powder. Sol in water with

blue color, less sol in alcohol, slightly sol in chloroform. Insol in ether.

Mixture with methylene blue and eosin, *azure II eosin.* Chief ingredient of Giemsa stain. Green powder. Slightly sol in water; sol in alcohol, methanol, glycerol.

USE: Biological stain.

6136. 4,4'-Methylenebis[2-chloroaniline]. *4,4'-Methylenebis[2-chlorobenzenamine];* 4,4'-diamino-3,3'-dichlorodiphenylmethane; di-(4-amino-3-chlorophenyl)methane; methylenebis(o-chloroaniline); MOCA; MBOCA; DACPM. $C_{13}H_{12}Cl_2N_2$; mol wt 267.16. C 58.45%, H 4.53%, Cl 26.54%, N 10.49%. Prepn: Finger, *J. Prakt. Chem.* [2] **79,** 493 (1909); Mayer, *Ber.* **47,** 1161 (1914). Acute toxicity: G. Ya. Kel'man *et al., Kauch. Rezina* **26**(9), 28 (1967), *C.A.* **67,** 120013m (1967). Review of carcinogenicity studies: *IARC Monographs* vol. **4,** 65-71 (1974); of toxicology and human exposure: *Toxicological Profile for 4,4'-Methylenebis(2-chloroaniline)* (PB95-100186, 1994) 146 p.

Flakes from alcohol, mp 110°. Slightly sol in water; sol in dil acids, ether, alcohol. LD_{50} orally in mice: 880 mg/kg (Kel'man).

Note: This substance may reasonably be anticipated to be a carcinogen: *Seventh Annual Report on Carcinogens* (PB95-109781, 1994) p 246.

USE: Curing agent for polyurethane and epoxy resins. Research tool for studying carcinogens.

6137. Methylene Blue. *3,7-Bis(dimethylamino)phenothiazin-5-ium chloride; C.I. Basic Blue 9;* methylthioninium chloride; tetramethylthionine chloride; 3,7-bis(dimethylamino)phenazathionium chloride; Swiss blue; C.I. 52015; solvent blue 8; Urolene Blue. $C_{16}H_{18}ClN_3S$; mol wt 319.86. C 60.08%, H 5.67%, Cl 11.08%, N 13.14%, S 10.03%. May crystallize with 3, 4 and 5 mols H_2O. First prepd by Caro in 1876. Now usually prepd from dimethylaniline and thiosulfuric acid: H. E. Fierz-David, L. Blangey, *Fundamental Processes of Dye Chemistry* (Interscience, New York, 1949) p 311. *See also Colour Index* vol. **4** (3rd ed., 1971) p 4470.

Trihydrate, dark green, odorless crystals with bronze luster or cryst powder. Absorption max: 668, 609 nm. One gram dissolves in ~25 ml water, ~65 ml alcohol; sol in chloroform. Insol in ether. In aq soln it is decolorized by Zn dust and dil H_2SO_4, but color is restored on exposure to air and more rapidly upon addn of NH_4OH. Forms double salts with many inorganic salts. *Incompat:* Caustic alkali, dichromates, alkali iodides, reducing agents.

USE: Stain in bacteriology; reagent for several chemicals; in mixed indicators; as oxidation-reduction indicator.

THERAP CAT: Antimethemoglobinemic; antidote (cyanide).

THERAP CAT (VET): Antiseptic; disinfectant; antidote (cyanide and nitrate).

6138. Methylene Bromide. *Dibromomethane.* CH_2Br_2; mol wt 173.83. C 6.91%, H 1.16%, Br 91.93%. Prepn: Hartman, Dreger, *Org. Syn. coll.* vol. **I,** 357 (1941). Manuf along with bromochloromethane, from dichloromethane: Lake, Asadorian, U.S. pat. **2,553,518** (1951 to Dow).

Liquid, bp 97°. mp −52.7°. d_4^{20} 2.4956. n_D^{20} 1.5419. Soly (g/1000 g water): 11.70 (15°); 11.93 (30°). Miscible with alcohol, ether, acetone.

6139. α-Methylene Butyrolactone. *4,5-Dihydro-3-methylene-2(3H)-furanone.* $C_5H_6O_2$; mol wt 98.10. C 61.22%, H 6.16%, O 32.62%. Isoln from aq extracts of

Erythronium americanum Ker., *Liliaceae* (dogtooth violet): Cavallito, Haskell, *J. Am. Chem. Soc.* **68,** 2332 (1946).

Liquid. Polymerizes easily when heated above 70°. bp_2 57-60°. Sol in water. pH 4 (150 mg/ml). The lactone is present in all parts of the plant and can be extracted easily with water, but not with ethanol, ether, ethyl acetate or chloroform.

Caution: Irritating to the skin.

6140. Methylene Chloride. *Dichloromethane;* methylene dichloride; methylene bichloride. CH_2Cl_2; mol wt 84.93. C 14.14%, H 2.37%, Cl 83.48%. Prepn by chlorination of methane: Lukes *et al.,* U.S. pat. **2,792,435** (1957 to Diamond Alkali); Pitt, Bender, U.S. pat. **2,979,541** (1961 to Stauffer); Burks, Obrecht, U.S. pat. **3,126,419** (1964 to Stauffer). Review of mfg processes: Faith, Keyes & Clark's *Industrial Chemicals,* F. A. Lowenheim, M. K. Moran, Eds. (Wiley-Interscience, New York, 4th ed., 1975) pp 530-538. Toxicity data: E. T. Kimura *et al., Toxicol. Appl. Pharmacol.* **19,** 699 (1971).

Colorless liquid; vapor is not flammable and when mixed with air is not explosive. Soluble in ~50 parts water; miscible with alc, ether, DMF. bp_{760} 39.75°. mp −95°. d_4^0 1.36174; d_4^{15} 1.33479; d_4^{20} 1.3255; d_4^{30} 1.30777. n_D^{20} 1.4244. LD_{50} orally in young adult rats: 1.6 ml/kg (Kimura).

Caution: Potential symptoms of overexposure are fatigue, weakness, sleepiness, lightheadedness; numbness or tingle of limbs; nausea; irritation of eyes and skin. *See NIOSH Pocket Guide to Chemical Hazards* (DHHS/NIOSH 90-117, 1990) p 150. This substance may reasonably be anticipated to be a carcinogen: *Seventh Annual Report on Carcinogens* (PB95-109781, 1994) p 161.

USE: Solvent in paint removers, for cellulose acetate; degreasing and cleaning fluids; as solvent in food processing. Pharmaceutic aid (solvent). Aerosol propellant; insecticide.

6141. Methylenedigallic Acid. *2,2'-Methylenebis[3,4,5-trihydroxybenzoic acid].* $C_{15}H_{12}O_{10}$; mol wt 352.25. C 51.15%, H 3.43%, O 45.42%. Made by heating a mixture of gallic acid, formaldehyde and dil HCl on the water bath.

Crystals. Insol in cold, sparingly in hot water; sol in alcohol, alkalies.

6142. 5,5'-Methylenedisalicylic Acid. *3,3'-Methylenebis[6-hydroxybenzoic acid];* 4,4'-dihydroxydiphenylmethane-3,3'-dicarboxylic acid. $C_{15}H_{12}O_6$; mol wt 288.26. C 62.50%, H 4.20%, O 33.30%. Prepd from salicylic acid and formaldehyde in the presence of sulfuric acid: Clemmensen, Heitman, *J. Am. Chem. Soc.* **33,** 737 (1911).

Wedge-like cryst from acetone + benzene. Bitter taste. Dec 238° (higher-melting material may be impure). Turns red at 180° and starts giving off CO_2. Freely sol in methanol, ethanol, ether, acetone, glacial acetic acid. Very slightly sol in hot water. Practically insol in benzene, chloroform, carbon disulfide, petr ether.

Bacitracin salt, see Bacitracin Methylenedisalicylic Acid.

Diacetyl deriv, $C_{15}H_{10}O_6(CH_3CO)_2$, white powder, mp 142°, practically insol in water, sol in acetone, ethanol.

6143. Methylene Iodide. *Diiodomethane.* CH_2I_2; mol wt 267.84. C 4.48%, H 0.75%, I 94.76%. Prepd by the reduction of iodoform with sodium arsenite: Adams, Marvel, *Org. Syn.* vol. 1, p 57 (1921). Also by heating iodoform with sodium acetate in alcohol: Bagnara, *Eng. Mining J. Press* **116**, 51 (1923).

Very heavy, highly refractive liq. Darkens on exposure to light, air, and moisture. d_4^{20} 3.32537. Solidifies in leaflets at 5.2° or in thin needles at 5.7°. Usually cooling to 0° is necessary to start crystn. mp 6.0°. bp_{760} 181°; bp_{70} 107°; bp_{11} 68°. $n_D^{10.5}$ 1.7559; n_D^{15} 1.7425. Viscosity in centipoises: 3.35 (10°); 2.80 (20°); 2.39 (30°). Absorption spectrum: Lowry, Sass, *J. Chem. Soc.* **1926**, 624; Stepanov, *Acta Physicochim. URSS* **20**, 174 (1945). Sol in about 70 parts water; miscible with alcohol, propanol, isopropanol, hexane, cyclohexane, ether, chloroform, benzene. CH_2I_2 dissolves sulfur and phosphorus (more than 1:1 at 25°).

USE: In separating mixtures of minerals. In determining the specific gravity of minerals and other substances. In the manufacture of x-ray contrast media.

6144. Methylenomycins. Members of a family of cyclopentenoid antibiotics related structurally to sarkomycins, *q.v.*, and having *in vitro* activity vs gram-positive and gram-negative organisms. Isoln from *Streptomyces violaceoruber*, physical, chemical, biological properties: M. Arai *et al.*, **Japan. Kokai 73 19796** (1973 to Sankyo), *C.A.* **78**, 157861 (1973); T. Haneishi *et al.*, *J. Antibiot.* **27**, 386 (1974). Structures of methylenomycins A and B: *eidem, ibid.* 393. Crystal and molecular structure of (±)-A: B. H. Toder, A. B. Smith, *J. Cryst. Mol. Struct.* **8**, 1 (1979). Stereospecific total synthesis of (±)-A: R. M. Scarborough *et al.*, *J. Am. Chem. Soc.* **99**, 7085 (1977); *eidem, ibid.* **102**, 3904 (1980); and absolute configuration: K. Sakai *et al.*, *Tetrahedron Letters* **1979**, 2365. Stereospecific total synthesis and absolute configuration of (+)-A: J. Jernow *et al.*, *J. Org. Chem.* **44**, 4210 (1979). Revised structure and total synthesis of B: *eidem, ibid.* 4212. Concise synthesis of B: M. Mikolajczyk, R. Zurawinski, *Synlett.* **8**, 575 (1991). Prepn of analogs of A and structure-activity correlations: T. Haneishi *et al., J. Antibiot.* **27**, 400 (1974). Toxicity: *eidem, ibid.* 386. Methylenomycin A is the first example of an antibiotic in which all information required for synthesis is carried by a plasmid, SCP1: L. F. Wright, D. A. Hopwood, *J. Gen. Microbiol.* **95**, 96 (1976). Review of biosynthesis: U. Hornemann, D. A. Hopwood, *Antibiotics* vol. IV, J. W. Corcoran, Ed. (Springer-Verlag, New York, 1981) pp 123-131. General review: A. Terehara *et al., Heterocycles* **13**, 353-371 (1979).

Methylenomycin A   Methylenomycin B

Methylenomycin A, $C_9H_{10}O_4$, *[1S-(1α,2α,5α)]-1,5-dimethyl-3-methylene-4-oxo-6-oxabicyclo[3.1.0]hexane-2-carboxylic acid.* Colorless crystals from chloroform/carbon tetrachloride, mp 115° (dec). mp of the (±)-form: 88.5-89°; after subl (70-75°, 0.025 mm Hg), 107.5-108°. $[\alpha]_D^{20}$ +42.3° (c = 1 in chloroform). uv max (methanol): 224 nm (ε 6300). Sol in benzene, chloroform, ethyl acetate, acetone, methanol, water. Slightly sol in *n*-hexane, CCl_4. pKa′ 3.65. LD_{50} in mice (mg/kg): 1500 orally, 75 i.p. (Haneishi).

Methylenomycin B, $C_8H_{10}O$, *2,3-dimethyl-5-methylene-2-cyclopenten-1-one.* Neutral colorless oil. uv max (methanol): 240 nm (ε 7650). Sol in ether, benzene, chloroform, ethyl acetate, acetone, alcohols. Slightly sol in *n*-hexane, petr ether. LD_{50} in mice (mg/kg): 260 orally, 245 i.p. (Haneishi).

6145. N-Methylephedrine. α-[1-(Dimethylamino)ethyl]benzenemethanol; erythro-α-[1-(dimethylamino)ethyl]benzyl alcohol; 2-dimethylamino-1-phenylpropanol; N,N-dimeth-

ylnorephedrine. $C_{11}H_{17}NO$; mol wt 179.26. C 73.70%, H 9.56%, N 7.81%, O 8.93%. Isoln of *l*-form from *Ephedra distachya* L. (*E. vulgaris* Rich.), and allied *Gnetaceae*: Smith, *J. Chem. Soc.* **1927**, 2056; Wolfes, *Arch. Pharm.* **268**, 327 (1930). Prepn of *l*-form: Smith, *loc. cit.*; of *d*-, *l*-, and *dl*-forms, and resolution of racemic mixture: Nagai, Kanao, *Ann.* **470**, 157 (1929); of *dl*-form: Pfanz, Müller, *Arch. Pharm.* **288**, 11 (1955).

dl-Form, crystals from petr ether or methanol, mp 63.5-64.5°. Readily sol in the usual solvents.

dl-Form hydrochloride, **Metheph.** Crystals from acetone, mp 207-208°.

d-Form, crystals, mp 87-87.5°. $[\alpha]_D^{20}$ +29.2° (c = 4 in methanol).

d-Form hydrochloride, crystals from ethyl acetate, mp 192°. $[\alpha]_D^{20}$ +30.1°.

l-Form, crystals from petr ether, mp 87-88°. $[\alpha]_D$ −29.5° (c = 4.5 in methanol).

l-Form hydrochloride, crystals from ethyl acetate or alcohol, mp 192°. $[\alpha]_D^{20}$ −29.8° (c = 4.6). Readily sol in water; less sol in alcohol; sparingly sol in acetone.

β-Camphorsulfonate, *N-methylephedrine camsylate, Tybraine.*

THERAP CAT: Analeptic.

6146. N-Methylepinephrine. *4-[2-(Dimethylamino)-1-hydroxyethyl]-1,2-benzenediol; α-[(dimethylamino)methyl]-3,4-dihydroxybenzyl alcohol; α-(3,4-dihydroxyphenyl)-2-dimethylaminoethanol; α-(3,4-dihydroxyphenyl)-α-hydroxy-β-dimethylaminoethane; dimethylaminomethyl-(3,4-dihydroxyphenyl) carbinol; α-(dimethylaminomethyl)protocatechuyl alcohol; N-methyladrenaline.* $C_{10}H_{15}NO_3$; mol wt 197.23. C 60.90%, H 7.67%, N 7.10%, O 24.34%. Prepn and resolution of racemic mixture: Manna, Campiglio, *Farmaco Ed. Sci.* **14**, 317 (1959). Configuration: Manna, Ghislandi, *ibid.* **19**, 377 (1964).

DL-Form, **Methadren(e).** Crystals from alcohol + ethyl acetate, mp 142-143°.

D(−)-Form, crystals from ethyl acetate, mp 149-150°. $[\alpha]_D^{18}$ −65.1° (c = 1.41 in 0.5N HCl).

L(+)-Form, crystals, mp 149-150°. $[\alpha]_D^{18}$ +62.3° (c = 1.4).

THERAP CAT: Adrenergic.

6147. Methylergonovine. *[8β(S)]-9,10-Didehydro-N-[1-(hydroxymethyl)propyl]-6-methylergoline-8-carboxamide; N-[α-(hydroxymethyl)propyl]-D-lysergamide; D-lysergic acid (+)-butanolamide-(2); d-lysergic acid-dl-hydroxybutylamide-2; methylergometrine; methylergobasine.* $C_{20}H_{25}N_3O_2$; mol wt 339.44. C 70.77%, H 7.42%, N 12.38%, O 9.43%. A homolog of ergonovine contg one more CH_2 group. Prepd by condensing isolysergic acid azide with d-2-amino-1-butanol and subjecting the reaction product to an isomerization treatment: Stoll, Hofmann, U.S. pat. **2,265,-207** (1941 to Sandoz).

Shiny crystals from benzene, mp 172° (some decompn). $[\alpha]_D^{20}$ −45° (c = 0.4 in pyridine). Sparingly sol in water. Freely sol in alcohol, acetone.

Maleate, $C_{20}H_{25}N_3O_2 \cdot C_4H_4O_4$, *Basofortina, Methergin, Methergine, Metenarin, Methylergobrevin, Ryegonovin, Spametrin-M*. White to pinkish-tan microcryst powder; odorless; bitter taste. Slightly sol in water, alcohol; very slightly sol in chloroform, ether.

Tartrate, $(C_{20}H_{25}N_3O_2)_2 \cdot C_4H_6O_6$, clusters of needles, usually with 2 mols of methanol of crystn. Easily sol in water, alcohol, very slightly sol in ether, chloroform. pH of aq solns 5.0-5.8.

THERAP CAT: Oxytocic.

6148. Methyl Ether. *Oxybismethane;* dimethyl ether. C_2H_6O; mol wt 46.07. C 52.14%, H 13.13%, O 34.73%. CH_3OCH_3. Prepn: J. Dumas, E. Peligot, *Ann.* **15**, 1 (1835); and thermal properties: R. M. Kennedy *et al., J. Am. Chem. Soc.* **63**, 2267 (1941). Atmospheric chemistry: S. M. Japar *et al., Int. J. Chem. Kinet.* **22**, 1257 (1990). Review: H. Höver in *Ullmann's Encyclopedia of Industrial Chemistry* vol. **A8** (VCH, Weinheim, 1987) pp 541-544; of use in diesel engines: A. M. Rouhi, *Chem. Eng. News* **73**, 37-39 (May 29, 1995).

Colorless, flammable gas with a slight ethereal odor. mp −141.50°. bp −24.82°. d^{25} (1 atm) 1.91855 g/l. Heat capacity at 210.51°K: 23.88 cal/deg/mole. Heat of vaporization at bp: 5141.0 cal/mole. Heat of fusion: 1179.8 cal/mole. Flash pt −41°. One vol water takes up 37 vols gas.

USE: Aerosol propellant; alternative diesel fuel; chemical intermediate.

6149. Methyl Ethyl Ketone. *2-Butanone;* ethyl methyl ketone; MEK; 2-oxobutane. C_4H_8O; mol wt 72.11. C 66.63%, H 11.18%, O 22.19%. $CH_3COCH_2CH_3$. Prepn from ethyl 2-methylacetoacetate: J. Schramm, *Ann.* **398**, 242 (1913). Manuf by dehydration of 2-butanol and by catalytic oxidation of *n*-butenes: A. J. Papa, P. D. Sherman, Jr. in Kirk-Othmer *Encyclopedia of Chemical Technology* vol. **13** (Wiley-Interscience, New York, 3rd ed., 1981) pp 903-907. Toxicity: H. F. Smyth *et al., Am. Ind. Hyg. Assoc. J.* **23**, 95 (1962).

Flammable liquid; acetone-like odor. d_4^{20} 0.805. mp −86°. bp 79.6°. Flash pt, closed cup: 21°F (−6°C). n_D^{15} 1.3814. Sol in ∼4 parts water (27.5%); less sol at higher temp; miscible with alcohol, ether, benzene. Constant boiling mixture with water, bp 73.4°, contains 88.7% methyl ethyl ketone. Soly of water in methyl ethyl ketone: 12.5% at 25°. LD_{50} orally in rats: 6.86 ml/kg (Smyth).

Caution: Potential symptoms of overexposure are irritation of eyes and nose; headache; dizziness; vomiting. *See NIOSH Pocket Guide to Chemical Hazards* (DHHS/NIOSH 90-117, 1990) p 48.

USE: As solvent; in the surface coating industry; manuf smokeless powder; colorless synthetic resins.

6150. α-Methylfentanyl. *N-[1-(1-Methyl-2-phenylethyl)-4-piperidinyl]-N-phenylpropanamide; N-[1-(α-methylphenethyl)-4-piperidyl]propionanilide; 1-(1-methyl-2-phenylethyl)-4-(N-propanilido)piperidine.* $C_{23}H_{30}N_2O$; mol wt 350.50. C 78.82%, H 8.63%, N 7.99%, O 4.56%. Potent derivative of fentanyl, *q.v.* Prepn: P. A. J. Janssen, *Fr. pat.* M2430; idem, *U.S. pat.* **3,164,600** (1964, 1965 both to Janssen). This substance has erroneously been referred to as *"China White"*, the street term for very pure Southeast Asian heroin. Initial identification of "China White" as α-methyl-

fentanyl: T. C. Kram *et al., Anal. Chem.* **53**, 1379 A (1981). Molecular structure determn using tandem mass spectrometry: M. T. Cheng *et al., ibid.* **54**, 2204 (1982). Identification and quantification in tissue: T. J. Gillespie *et al., J. Anal. Toxicol.* **6**, 139 (1982). Confirmation of identity of "China White": S. Suzuki *et al., Chem. Pharm. Bull.* **34**, 1340 (1986). Immunoassay for detection of use in racehorses: J. McDonald *et al., Res. Commun. Chem. Pathol. Pharmacol.* **57**, 389 (1987).

Hydrochloride, $C_{23}H_{30}N_2O \cdot HCl$, crystals from isopropanol, mp 272.8-273.6°.

Note: This is a controlled substance (opiate) listed in the U.S. Code of Federal Regulations, Title 21 Part 1308.11 (1995).

6151. Methyl Fluorosulfonate. *Fluorosulfuric acid methyl ester,* methyl fluorosulfate, methyl fluosulfonate, Magic Methyl. CH_3FO_3S; mol wt 114.10. C 10.53%, H 2.65%, F 16.65%, O 42.07%, S 28.10%. CH_3OSO_2F. Prepn from dimethyl ether and fluosulfonic acid: J. Meyer, G. Schramm, *Z. Anorg. Allgem. Chem.* **206**, 24 (1932); from dimethyl sulfate and fluosulfonic acid: R. W. Alder, *Chem. & Ind. (London)* **1973**, 983. Electrochemical prepn: J. P. Coleman, D. Pletcher, *Tetrahedron Letters* **1974**, 147. Powerful methylating agent: M. G. Ahmed *et al., Chem. Commun.* **1968**, 1533. Human toxicity: D. M. W. vanden Ham, D. van der Meer, *Chem. & Eng. News* **54**, 5 (Aug. 30, 1976); eidem, *Chem. & Ind. (London)* **1976**, 782. Toxicity study: M. Hite *et al., Am. Ind. Hyg. Assoc. J.* **40**, 600 (1979).

Volatile liq, bp 92-94°, mp −95°. *Severe Poison!* d 1.412. n_D^{20} 1.3326. Good solvent for most organic compounds. Proton NMR absorption at tau 5.88. LD_{50} orally in mice: < 112 mg/kg; LC_{50} 1 hr for rats: 5-6 ppm (Hite).

Caution: Extremely toxic to humans. (vanden Ham, van der Meer). Exposure can cause fatal pulmonary edema.

USE: In organic synthesis as methylating agent.

6152. N-Methylformamide. *N-Monomethylformamide;* NMF; MMF; NSC-3051. C_2H_5NO; mol wt 59.07. C 40.67%, H 8.53%, N 23.71%, O 27.09%. $HCONHCH_3$. Prepn: A. Gautier, *Ann.* **151**, 239 (1869); G. F. D'Alelio, E. E. Reid, *J. Am. Chem. Soc.* **59**, 109 (1937); J. A. Marsella, G. P. Pez, *J. Mol. Catal.* **35**, 65 (1986); J. J. Cappon *et al., Recl. Trav. Chim. Pays-Bas* **113**, 318 (1994). Dielectric spectrum: J. Barthel *et al., Chem. Phys. Letters* **167**, 62 (1990). Toxicology: S. P. Langdon *et al., Toxicology* **34**, 173 (1985). Review of metabolism, toxicity, and pharmacology: G. L. Kennedy, Jr., *Crit. Rev. Toxicol.* **17**, 129-182 (1986). Review of antitumor activity and clinical evaluation: K. Clagett-Carr *et al., J. Clin. Oncol.* **6**, 906-918 (1988).

mp −5.4°. bp 180-185°. bp_{90} 131°. d_4^{25} 0.9961. Sol in acetone, alcohol, water. LD_{50} in mice (mg/kg): 2300 i.p.; 2600 orally; 1580 i.v.; 2700 i.m. (Langdon).

USE: Solvent.

6153. Methyl Formate. *Formic acid methyl ester.* $C_2H_4O_2$; mol wt 60.05. C 40.00%, H 6.71%, O 53.28%. $HCOOCH_3$.

Colorless flammable liquid, agreeable odor. d_{15}^{15} 0.987. bp 31.5°. n_D^{20} 1.3440. Flash pt, closed cup: −2°F (−19°C). mp ∼ −100°. Sol in about 3.3 parts water; miscible with alcohol.

Caution: Potential symptoms of overexposure are eye and nose irritation; chest oppression, dyspnea; visual disturbance; CNS depression. *See NIOSH Pocket Guide to Chemical Hazards* (DHHS/NIOSH 90-117, 1990) p 150. *See also* Patty's *Industrial Hygiene and Toxicology* vol. **2A**, G. D.

Clayton, F. E. Clayton, Eds. (Wiley-Interscience, New York, 3rd ed., 1981) p 2263.

USE: Fumigant and larvicide for tobacco and food crops. Fire hazard is avoided by use with CO_2.

6154. *N*-Methylglucamine. *1-Deoxy-1-(methylamino)-D-glucitol; N-methyl-D-glucamine; meglumine.* $C_7H_{17}NO_5$; mol wt 195.22. C 43.07%, H 8.78%, N 7.18%, O 40.98%. Prepd from D-glucose and methylamine: Karrer, Herkenrath, *Helv. Chim. Acta* **20**, 83 (1937). Efficacy and toxicity in treatment of leishmaniasis: R. G. Muller *et al., Arch. Inst. Cardiol. Mex.* **52**, 155 (1982); P. Bouree *et al., Pathol. Biol.* **33**, 607 (1985). Antileishmanial activity in dogs: W. L. Chapman *et al., Am. J. Vet. Res.* **45**, 1028 (1984).

$$CH_2NHCH_3$$
$$|$$
$$HC-OH$$
$$|$$
$$HO-CH$$
$$|$$
$$HC-OH$$
$$|$$
$$HC-OH$$
$$|$$
$$CH_2OH$$

Crystals from methanol, mp 128-129°. Does not polymerize or dehydrate unless heated above 150° for prolonged periods. $[\alpha]_D^{18}$ −18.5° (Karrer); $[\alpha]_D^{20}$ −23° (Rhône-Poulenc data sheet). Soly (g/100 ml): water at 25°: ~100; alcohol at 25°: 1.2; alcohol at 70°: 21. pH of 1% aq soln: 10.5. Forms salts with acids and complexes with metals. Salts with alkyl aryl sulfonic acids act as detergents.

Antimonate, $C_7H_{17}NO_5 \cdot HSbO_3$, *RP-2168, Glucantim, Glucantime, Protostib.* Powder. Soly in water about 35% ww. Practically insol in alcohol, ether, chloroform. pH of aq solns 6-7.

USE: In the synthesis of surface active agents, pharmaceuticals, dyes.

THERAP CAT: Antimonate as antiprotozoal (Leishmania).

6155. *N*-Methyl-α-L-glucosamine. *2-(Methylamino)-2-deoxy-α-L-glucopyranose.* $C_7H_{15}NO_5$; mol wt 193.20. C 43.52%, H 7.83%, N 7.25%, O 41.41%. Together with streptose forms the streptobiosamine moiety of streptomycin: Kuehl *et al., J. Am. Chem. Soc.* **69**, 3032 (1947). Prepn from D-glucose by *Streptomyces griseus:* Silverman, Rieder, *J. Biol. Chem.* **235**, 1251 (1960); from the antibiotic, bluensomycin: Bannister, Argoudelis, *J. Am. Chem. Soc.* **85**, 34 (1963). *Review:* Lemieux, Wolfrom, *Advan. Carbohyd. Chem.* **3**, 337 (1948).

$$R = CH_2OH$$
$$R' = CH_3$$

Glass. $[\alpha]_D^{25}$ −62° (c = 1 in methanol).

Hydrochloride, $C_7H_{15}NO_5 \cdot HCl$, needles from ethanol, mp 160-163°. Freely sol in water. Shows mutarotation. $[\alpha]_D^{25}$ −103° → −88° (c = 0.6).

N-acetyl deriv, mp 165-166°. $[\alpha]_D^{25}$ −51° (c = 0.4).

Pentaacetyl deriv, $C_{17}H_{25}NO_{10}$, mp 160.5-161.5°. $[\alpha]_D^{25}$ −100° (c = 0.7 in chloroform).

6156. α-Methylglucoside. *Methyl-α-D-glucopyranoside.* $C_7H_{14}O_6$; mol wt 194.18. C 43.30%, H 7.27%, O 49.44%. Prepd by refluxing finely powdered glucose with methanol-HCl: Fischer, *Ber.* **26**, 2405; **27**, 2987; **28**, 1151 (1895); Helferich, Schäfer, *Org. Syn.* **coll. vol. I**, (2nd ed.), 1941) 364. Enzymatic synthesis by means of α-glucosidase from yeast: Bourquelot *et al., Compt. Rend.* **156**, 491 (1913). Prepd industrially by reacting glucose with methanol in the presence of a cation exchange material: *Chem. Eng. News* **33**, 4592 (1955). Monograph: G. N. Bollenback, *Methyl Glucoside, Preparation, Physical Constants, Derivatives* (Academic Press, New York, 1958).

Orthorhombic bisphenoidal crystals, d_4^{30} 1.46. mp 168°. $bp_{0.2}$ 200°. $[\alpha]_D^{20}$ +158.9° (p = 10). pKa (25°): 13.71. Soly at 17° in water 63% (w/w); in 80% alcohol 7.3%; in 90% alcohol 1.6%. Practically insol in ether. Soly also reported at 20° as 108 g/100 g H_2O and as 5.2 g/100 g methanol.

USE: Manuf reconstituted and upgraded drying oils; tall oil esters and varnishes; fatty acid esters; plasticizers; nonionic surface active agents; fast and hard drying alkyd resins, and so-called plasticizing alkyds. In making glucoside hydroxypropyl ethers for polyurethan foam production.

6157. Methyl Green. *4-[[4-(Dimethylamino)phenyl][4-(dimethylimino)-2,5-cyclohexadien-1-ylidene]methyl]-N-ethyl-N,N-dimethylbenzenaminium bromide chloride.* C.I. 42590; ethyl green. $C_{27}H_{35}BrClN_3$; mol wt 516.95. C 62.73%, H 6.82%, Br 15.46%, Cl 6.86%, N 8.13%. Usually sold as the double salt with $ZnCl_2$. Prepn: *Colour Index* **vol. 4** (3rd ed., 1971) p 4394.

Green powder. Sol in water. Yellow soln in conc H_2SO_4, turning green on dilution.

Note: The term methyl green also applies to the trimethyl analog.

USE: Dyeing and printing textiles; as biological stain.

6158. Methylguanidine. $C_2H_7N_3$; mol wt 73.10. C 32.86%, H 9.65%, N 57.48%. Isoln by Brieger in 1888 from decomposing horsemeat; described as a product of putrefaction: I. Greenwald, *J. Am. Chem. Soc.* **41**, 1109 (1919). Prepn: E. Erlenmeyer, *Ber.* **3**, 896 (1870); E. A. Werner, *J. Bell, J. Chem. Soc.* **121**, 1790 (1922); R. Phillips, H. T. Clarke, *J. Am. Chem. Soc.* **45**, 1755 (1923); E. Philippi, K. Morsch, *Ber.* **60**, 2120 (1927); T. L. Davis, E. N. Rosenquist, *J. Am. Chem. Soc.* **59**, 2112 (1937).

Colorless, deliquesc, strongly alkaline mass. Very sol in water; sol in alcohol. Reduces permanganate. MLD s.c. in rats: 250 mg/kg.

6159. Methylhexaneamine. *4-Methyl-2-hexanamine; 2-amino-4-methylhexane; 1,3-dimethylamylamine; Forthane.* $C_7H_{17}N$; mol wt 115.22. C 72.97%, H 14.87%, N 12.16%. Prepn: U.S. pats. **2,350,318**; **2,386,273** (1944, 1945, both to Lilly).

Liquid, amine odor. d 0.7620-0.7655. n_D^{25} 1.4150-1.4175. bp_{760} 130-135°. Very slightly sol in water. Freely sol in alc, chloroform, ether, dil acids. LD_{50} i.p. in mice: 185 mg/kg.

THERAP CAT: Adrenergic.

6160. Methylhydrazine. Monomethylhydrazine; MMH. CH_6N_2; mol wt 46.07. C 26.07%, H 13.13%, N 60.80%. CH_3NHNH_2. Early prepns: Beilstein 4, 546; 1st suppl., 560. Prepn of sulfate: Hatt, Org. Syn. coll. vol. I, 395 (1943); Audrieth, Diamond, J. Am. Chem. Soc. 76, 4869 (1954). Manuf and properties: Knight, Hydrocarbon Process. Petrol. Refiner 41, 179 (1962). Toxicity and metabolism: Witkin, Arch. Ind. Health 13, 34 (1956); Dost et al., Biochem. Pharmacol. 15, 1325 (1966); Gregory et al., Clin. Toxicol. 4, 435 (1971); Magee et al. in Proceedings of the Fifth International Congress on Pharmacology, San Francisco, 1972, vol. 2, T. A. Loomis, Ed. (Karger, New York, 1973) pp 140-149.

Clear liquid, odor characteristic of short chain, organic amines. d^{25} 0.874. mp −52.4°. bp 87.5°. Heat capacity (25°): 32.25 cal/mole/°C. Ignition temp: 196°. Flash pt (Cleveland open cup): 70°. Flammability limits in air (%-vol): 2.5 to 97 ± 2%. Miscible with water, hydrazine, low mol wt monohydric alcohols. Sol in hydrocarbons. Mildly alkaline base. Strong reducing agent. Ignites spontaneously on contact with strong oxidizing agents such as fluorine, chlorine trifluoride, nitrogen tetroxide, fuming nitric acid. LD_{50} orally in mice, rats: 33.0, 32.5 mg/kg (Witkin); orally in rats: 70.7 mg/kg (Gregory).

Sulfate, $CH_3NHNH_2.H_2SO_4$, white plates from 80% ethanol. mp 141-142°.

Caution: Potential symptoms of overexposure to methylhydrazine are irritation of eyes, vomiting, diarrhea, respiratory system irritation, tremors and ataxia; anoxia, cyanosis; convulsions. Potential occupational carcinogen. See NIOSH Pocket Guide to Chemical Hazards (DHHS/NIOSH 90-117, 1990) p 152.

USE: In rocket fuel; intermediate in chemical syntheses.

6161. Methyl Iodide. Iodomethane. CH_3I; mol wt 141.94. C 8.46%, H 2.13%, I 89.41%. Prepd from methyl alcohol, iodine and red phosphorus: King, Org. Syn. coll. vol. II, 399 (1943); from potassium iodide and methyl sulfate: Hartman, ibid. 404. Review of toxicity and carcinogenic risk: IARC Monographs 41, 213-227 (1986); ibid. Suppl. 7, p 47, 52 (1987); Evaluation of the Potential Carcinogenicity of Methyl Iodide (PB93-196343, 1988) 17 pp.

Colorless, transparent liquid; turns brown on exposure to light. Poisonous! d_4^{20} 2.28. mp −66.5°. bp 42.5°; dec at 270°. n_D^{21} 1.5293. Sol in about 50 parts water; miscible with alcohol, ether. Protect from light. LD_{50} orally in rats: 76 mg/kg; LD_{50} s.c. in mice: 0.78 mmoles/kg (IARC, 1986).

Caution: Potential symptoms of overexposure are nausea, vomiting; vertigo, ataxia; slurred speech, drowsiness; dermatitis; direct contact may cause irritation of eyes, skin and respiratory system. See NIOSH Pocket Guide to Chemical Hazards (DHHS/NIOSH 94-116, 1994) p 210. See also Patty's Industrial Hygiene and Toxicology vol. 2E, G. D. Clayton, F. E. Clayton, Eds. (John Wiley & Sons, Inc., New York, 4th ed., 1994) pp 4030-4034; Clinical Toxicology of Commercial Products, R. E. Gosselin et al., Eds. (Williams & Wilkins, Baltimore, 5th ed., 1984) Section II, pp 158-159. This substance has been listed as reasonably anticipated to be a carcinogen: Fourth Annual Report on Carcinogens (NTP 85-002, 1985) p 132; delisted due to IARC reevaluation as unclassifiable as to its human carcinogenicity: Fifth Annual Report on Carcinogens (NTP 89-239, 1989) p 340.

USE: Methylating agent; in microscopy because of its high refractive index; as imbedding material for examining diatoms; in testing for pyridine. Light sensitive etching agent for electronic circuits; component in fire extinguishers.

6162. Methyl Isobutyrate. 2-Methylpropanoic acid methyl ester. $C_5H_{10}O_2$; mol wt 102.13. C 58.80%, H 9.87%, O 31.33%. $(CH_3)_2CHCOOCH_3$.

Colorless, mobile liquid. d_D^{20} 0.891. mp −84° to −85°. bp 93°. n_D^{20} 1.3840. Slightly sol in water; miscible with alcohol, ether.

6163. Methyl Isocyanate. Isocyanatomethane; isocyanic acid methyl ester; MIC. C_2H_3NO; mol wt 57.05. C

42.11%, H 5.30%, N 24.55%, O 28.04%. $CH_3N=C=O$. Intermediate in the manufacture of insecticides and herbicides including carbaryl, q.v. Prepn via Curtius rearrangement: J. W. Boehmer, Rec. Trav. Chim. 55, 379 (1933); by heating N,N-diphenyl-N'-methylurea: W. Siefken, Ann. 562, 75 (1949); by phosgenation of bis(trimethylsilyl)methylamine: V. F. Mironov et al., Zh. Obshch. Khim. 39, 2598 (1969). Used in prepn of α-aryl-β-methylureas: J. W. Boehmer, loc. cit.; of semicarbazides: Ch. C. P. Pacilly, Rec. Trav. Chim. 55, 101 (1936); in conversion of aldoximes to nitriles: J. A. Albright, M. L. Alexander, Org. Prep. Proced. Int. 4, 215 (1972). Toxicity of vapor to rats, mice, humans; physical properties: G. Kimmerle, A. Eben, Arch. Toxicol. 20, 235 (1960). Uptake and distribution studies in animals: J. S. Ferguson et al., Toxicol. Appl. Pharmacol. 94, 104 (1988). Acute oral toxicity: E. H. Vernot et al., ibid. 42, 417 (1977).

Liquid, bp 39-40°. d^{20} 0.96. Vapor press. at 4.2°: 200 torr; at 13.5°: 300 torr; at 20.6°: 400 torr; at 31.2°: 600 torr. LD_{50} in male rats (mg/kg): 140 single oral dose (Vernot). LC_{50} in rats (4 hours exposure to vapor): 5 ppm (Kimmerle, Eben).

Note: An industrial accident during the manufacture of carbaryl in Bhopal, India on December 3, 1984 resulted in the leakage of an unknown amount of methyl isocyanate into the air. Over 2,000 people died and an estimated 200,000 were exposed to the vapor: Chem. & Eng. News 62, 6 (Dec. 10, 1984); ibid. 63, 14 (Feb. 11, 1985). Series of articles on follow-up studies on survivors: Indian J. Exp. Biol. 26, 149-176, 201-204 (1988). Review of human toxicity: P. S. Mehta et al., J. Am. Med. Assoc. 264, 2781-2787 (1990).

Caution: Potential symptoms of overexposure are irritation of eyes, nose and throat; coughing, secretions, chest pain and dyspnea; asthma; eye and skin injury. See NIOSH Pocket Guide to Chemical Hazards (DHHS/NIOSH 90-117, 1990) p 152. Highly volatile. Exposure to 2 ppm for 1-5 min produced tears and irritation of the nose and throat (Kimmerle, Eben).

USE: In organic synthesis; in manufacture of carbamate pesticides.

6164. Methyl Isopropyl Ketone. 3-Methyl-2-butanone; 2-acetylpropane. $C_5H_{10}O$; mol wt 86.13. C 69.72%, H 11.70%, O 18.58%. $(CH_3)_2CHCOCH_3$. Prepn: E. Frankland, B. F. Duppa, Ann. 138, 328 (1866). Large scale prepn: T. W. Mears et al., J. Res. Nat. Bur. Stand. 44, 299 (1950). Miscibility profile: W. M. Jackson, J. S. Drury, Ind. Eng. Chem. 51, 1491 (1959). Thermodynamic properties: J. L. Hales et al., Trans. Faraday Soc. 63, 1876 (1967). Acute toxicity: H. Tanii et al., Toxicol. Letters 30, 13 (1986).

Colorless, flammable liquid with an acetone-like odor. bp_{760} 94.2°. fp −94.4°. d^{20} 0.8100; d^{25} 0.8061. n_D^{20} 1.3887; n_D^{25} 1.3861. Log P (n-octanol/water): 0.56. LD_{50} orally in male mice: 29.86 mmol/kg (Tanii).

Caution: Direct contact may cause skin and eye irritation. See Patty's Industrial Hygiene and Toxicology vol 2C, G. D. Clayton, F. E. Clayton, Eds. (Wiley-Interscience, New York, 4th ed., 1994) pp 1776-1779.

USE: Solvent.

6165. Methyl Isothiocyanate. Isothiocyanatomethane; methyl mustard oil; Trapex. C_2H_3NS; mol wt 73.12. C 32.85%, H 4.14%, N 19.16%, S 43.85%. $CH_3N=CS$. Obtained by the action of CS_2 on methylamine.

Crystals, mp 35-36°. bp 119°. n_D^{37} 1.5258. Slightly sol in water; freely sol in alcohol, ether. LD_{50} orally in rats: 305 mg/kg, Toxic Substances List, H. E. Christensen, Ed. (1974) p 477.

Caution: Highly irritating.

USE: Pesticide.

6166. Methyl Isovalerate. 3-Methylbutanoic acid methyl ester. $C_6H_{12}O_2$; mol wt 116.16. C 62.04%, H 10.41%, O 27.55%. $(CH_3)_2CHCH_2COOCH_3$.

Liquid; valerian odor. d_4^{20} 0.881. bp 116-117°. Slightly sol in water; miscible with alcohol, ether.

6167. Methyl Lactate. 2-Hydroxypropanoic acid methyl ester. $C_4H_8O_3$; mol wt 104.11. C 46.15%, H 7.75%, O

46.11%. $CH_3CH(OH)COOCH_3$. Prepd by heating 1 mole lactic acid condensation polymer with 2.5-5 moles of methanol and a small quantity of H_2SO_4 at 100° for 1-4 hours in a heavy-walled bottle: Rehberg, *Org. Syn.* coll. vol. III, 47-48 (1955).

Colorless, transparent liquid. d^{19} 1.09. bp 144-145°. n_D^{16} 1.4156. Soluble in alcohol, ether; dec by water.

Caution: Mild irritant.

USE: Cellulose acetate solvent.

6168. Methyl Linoleate. *9,12-Octadecadienoic acid methyl ester.* $C_{19}H_{34}O_2$; mol wt 294.48. C 77.50%, H 11.64%, O 10.87%. $CH_3(CH_2)_4CH=CHCH_2CH=CH-(CH_2)_7COOCH_3$. Prepn from safflower-seed oil: Parker *et al., Biochem. Prepn.* **4,** 88 (1955).

Colorless oil. d_4^{18} 0.8886. mp −35°. bp_{16} 212°; bp_4 192°. n_D^{25} 1.4593. Iodine value 172.4. More stable to air oxidation than linoleic acid. Miscible with dimethylformamide, fat solvents, oils.

USE: In the vitamin industry.

6169. Methyl Malonate. *Propanedioic acid dimethyl ester;* dimethyl malonate. $C_5H_8O_4$; mol wt 132.12. C 45.46%, H 6.10%, O 48.44%. $CH_3OOCCH_2COOCH_3$.

Liquid. d_4^{20} 1.154. mp −62°. bp 180-181°. n_D^{17} 1.4149. Slightly sol in water; miscible with alcohol, ether, oils.

6170. Methyl Methanesulfonate. *Methanesulfonic acid methyl ester;* methyl methanesulfonic acid; *as*-dimethyl sulfite; methyl mesylate; MMS. $C_2H_6O_3S$; mol wt 110.13. C 21.81%, H 5.49%, O 43.58%, S 29.12%. $CH_3SO_2OCH_3$. Prepn: A. Arbusow, P. Pischtschimuka, *Chem. Zentr.* II, 685 (1909); W. E. Bissenger *et al., J. Am. Chem. Soc.* **70,** 3940 (1948). Metabolism: E. A. Barnsley, *Biochem. J.* **106,** 18p (1968). Mutagenicity studies: U. H. Ehling *et al., Mutat. Res.* **5,** 417 (1968); J. Moutschen, *ibid.* **8,** 581 (1969). Review of carcinogenicity studies: *IARC Monographs* 7, 253-260 (1974). Review of comparative mutagenicity of MMS and ethyl methanesulfonate, *q.v.:* S. Kondo, *Environ. Sci. Res.* **24,** 743-785 (1981).

Liquid, bp_{753} 203°, $bp_{0.6}$ 59°. d_4^{20} 1.2943; n_D^{20} 1.4140. Soly in water about 1:5, in DMF about 1:1. Slightly sol in nonpolar solvents.

Note: This substance may reasonably be anticipated to be a carcinogen: *Seventh Annual Report on Carcinogens* (PB95, 109781, 1994) p 253.

USE: Exptlly as mutagen, teratogen, brain carcinogen.

6171. 3-Methyl-6,7-methylenedioxy-1-piperonylisoquinoline. *5-(1,3-Benzodioxol-5-ylmethyl)-7-methyl-1,3-dioxolo[4,5-g]isoquinoline.* $C_{19}H_{15}NO_4$; mol wt 321.33. C 71.02%, H 4.71%, N 4.36%, O 19.92%. Papaverine homolog made by synthesis: *Merck's Jahresbericht* **44,** 15 (1930).

Crystals from methanol, mp 141°. Almost insol in water; sol in benzene, chloroform, in hot alcohol, hot methanol, hot ether. A weak base.

Hydrochloride monohydrate, crystals from water, mp about 158°. Anhydr from abs alcohol, dec 253°. One gram dissolves in 100 ml water at 25°; freely sol in hot water; moderately sol in 95% alcohol. Aq solns are acid to litmus.

Note: This drug was formerly known as *Eupaverin, Syntaverin.*

6172. Methyl N-Methylnipecotate. *1-Methyl-3-piperidinecarboxylic acid methyl ester; N-methylnipecotic acid methyl ester;* dihydroarecoline. $C_8H_{15}NO_2$; mol wt 157.21. C 61.12%, H 9.62%, N 8.91%, O 20.35%. Prepn: Stoll *et al., Helv. Chim. Acta* **33,** 375 (1950); Sperber *et al., J. Am.*

Chem. Soc. **81,** 704 (1959). Electron distribution and dissociation constants: E. Mutschler *et al., Arzneimittel-Forsch.* **17,** 833 (1967). Pharmacologic studies: O. Nieschulz, *ibid.* **22,** 804 (1972). Structure-conformation activity study: G. Lambrecht, E. Mutschler, *ibid.* **24,** 1725 (1974).

Oil, bp_{12} 83-84°. n_D^{25} 1.4520. pKa (20°): 8.66. Methobromide, $C_9H_{18}BrNO_2$, *Adipsin.*

THERAP CAT: Sialagogue.

6173. N-Methylmyosmine. *3-(4,5-Dihydro-1-methyl-1H-pyrrol-2-yl)pyridine; N-methyl-2-(3-pyridyl)-2-pyrroline; 4,5-dihydro-β-nicotyrine.* $C_{10}H_{12}N_2$; mol wt 160.22. C 74.97%, H 7.55%, N 17.48%. Preparation: Korte, Schulze-Steinen, *Ber.* **95,** 2444 (1962).

Light yellow oil, $bp_{0.1}$ 60°. uv max (methanol): 212, 243 nm (log ε 3.70, 3.92).

6174. Methyl Nicotinate. *3-Pyridinecarboxylic acid methyl ester;* nicotinic acid methyl ester; Midalgol. $C_7H_7NO_2$; mol wt 137.14. C 61.31%, H 5.14%, N 10.21%, O 23.33%. Obtained by passing HCl gas into a hot methanol soln of nicotinic acid: Engler, *Ber.* **27,** 1787 (1894); by treatment with tetramethylammonium hydroxide in methanol: Prelog, Piantanida, *Z. Physiol. Chem.* **244,** 56 (1936); by heating a mixture of sulfuric acid, selenium and quinoline, followed by refluxing with methanol: Kaufman, *J. Am. Chem. Soc.* **67,** 497 (1945); from nicotinyl chloride hydrochloride and methanol in pyridine: Charonnat *et al., Bull. Soc. Chim. France* **1948,** 1014.

Crystals, mp 39°. bp_{760} 209°; bp_3 70-72°. Soluble in water, alcohol, benzene.

THERAP CAT: Rubefacient.

6175. Methyl Nitrate. *Nitric acid methyl ester.* CH_3NO_3; mol wt 77.04. C 15.59%, H 3.93%, N 18.18%, O 62.30%. CH_3ONO_2. Prepn from methanol and nitric acid in the presence of sulfuric acid: Black, Babers, *Org. Syn.* coll. vol. II, 412 (1943).

Liquid. Solid at −83°. bp_{760} 64.6° (explodes). d_4^{20} 1.2075. n_D^{20} 1.3748. Slightly sol in water. Sol in alcohol, ether.

USE: Has been used as rocket propellant. Does not need external oxygen for combustion.

6176. Methyl p-Nitrobenzenesulfonate. *p-Nitrobenzenesulfonic acid methyl ester.* $C_7H_7NO_5S$; mol wt 217.20. C 38.71%, H 3.25%, N 6.45%, O 36.83%, S 14.76%. Reagent for selective methylation of cysteine residues in chemical modification of proteins: Nakagawa, Bender, *J. Am. Chem. Soc.* **91,** 1566 (1969). Procedure: Heinrikson, *Biochem. Biophys. Res. Commun.* **41,** 967 (1970). Prepd by the general method of reacting p-nitrobenzenesulfonyl chloride with alcohols. See Morgan, Cretcher, *J. Am. Chem. Soc.* **70,** 375 (1948).

White to cream crystals, mp 93-95°.
USE: Methylating agent.

6177. N-Methyl-N'-nitro-N-nitrosoguanidine. N-Methyl-N-nitroso-N'-nitroguanidine; MNNG. $C_2H_5N_5O_3$; mol wt 147.09. C 16.33%, H 3.43%, N 47.61%, O 32.63%. Prepn: A. F. McKay, G. F. Wright, J. Am. Chem. Soc. **69**, 3028 (1947). Reviews of carcinogenicity and mutagenicity studies: IARC Monographs **4**, 183-195 (1974); U. Sinha, B. B. Chattoo, J. Sci. Ind. Res. **3**, 499-505 (1975).

Yellow crystals from methanol, mp 118° (dec). Reacts with aq KOH to form diazomethane: A. F. McKay, J. Am. Chem. Soc. **70**, 1974 (1948). Reacts at acid pH to give methylnitroguanidine.
Note: This substance may reasonably be anticipated to be a carcinogen: Seventh Annual Report on Carcinogens (PB95-109781, 1994) p 255.
USE: Exptlly as carcinogen and mutagen. Formerly in prepn of diazomethane.

6178. Methyl Nonyl Ketone. 2-Undecanone; 2-hendecanone. $C_{11}H_{22}O$; mol wt 170.30. C 77.58%, H 13.02%, O 9.40%. $CH_3(CH_2)_8COCH_3$. Primary constituent of rue oils distilled mainly from Ruta montana L. and R. graveolens. Prepn: E. v. Gorup-Besanez, F. Grimm, Ann. **157**, 275 (1871); V. Fiandanese et al., Tetrahedron Letters **25**, 4805 (1984); T. Aoyama, T. Shioiri, Synthesis **1988**, 228. LC determn in formulations: R. J. Bushway, J. Assoc. Offic. Anal. Chem. **73**, 743 (1990). Physical properties: E. Guenther, D. Althausen, The Essential Oils vol. 2 (D. Van Nostrand, New York, 1949) p 377. Brief review: D. L. J. Opdyke, Food Cosmet. Toxicol. **13**, 869-870 (1975).
Oily liquid. Strong odor. mp 12.1°. bp$_{761}$ 231.5-232.5°. bp$_7$ 99°. d$_4^{20}$ 0.8260-0.8263. n$_D^{30}$ 1.42527. LD$_{50}$ dermally in rabbits: > 5 g/kg; LD$_{50}$ orally in rats, mice: > 5, 3.88 g/kg (Opdyke).
USE: In the compounding of some synthetic essential oils. As fragrance additive in soaps, detergents, creams, lotions, and perfume. As dog and cat repellant.

6179. Methylol Riboflavine. Hyflavin. Mixture of methylol (CH$_2$OH) derivatives of riboflavine formed by the action of formaldehyde on riboflavine in weakly alkaline soln. The number of methylol groups in the ribityl moiety varies from 1 to 3. Prepn: Schoen, Gordon, Arch. Biochem. **22**, 149 (1949); U.S. pat. **2,587,533** (1952 to Endo Prod.).

X = H or CH$_2$OH

Orange to yellow, hygroscopic powder. May have a slight odor of formaldehyde. Sol in water. Practically insol in alcohol, benzene, chloroform, ether. Dextrorotatory. The pH of a 10% aq soln is between 6.7 and 7.9. The dry powder is unstable and loses its biological activity in the course of several months with the liberation of formaldehyde and the partial formation of products practically insol in water.
THERAP CAT: Vitamin (enzyme co-factor).

6180. Methyl Orange. 4-[[(4-Dimethylamino)phenyl]azo]benzenesulfonic acid sodium salt; sodium p-dimethylaminoazobenzenesulfonate; helianthine B; C.I. Acid Orange 52; C.I. 13025; Orange III; Gold Orange; Tropaeolin D. $C_{14}H_{14}N_3NaO_3S$; mol wt 327.34. C 51.37%, H 4.31%, N 12.84%, Na 7.02%, O 14.66%, S 9.80%. Prepn from sulfanilic acid sodium nitrite + dimethylaniline: L. Gattermann, Die Praxis des organischen Chemikers (de Gruyter, Berlin, 40th ed., 1961) pp 260-261. See also Colour Index vol. 4 (3rd ed., 1971) p 4043.

Orange-yellow powder or cryst scales. Sol in 500 parts water; more sol in hot water. Practically insol in alcohol.
USE: As indicator in 0.1% aq soln. pH: 3.1 red, 4.4 yellow. Employed for titrating most mineral acids, strong bases, estimating alkalinity of waters; useless for organic acids. In dyeing and printing of textiles.

6181. Methyl Oxalate. Ethanedioic acid dimethyl ester; dimethyl oxalate. $C_4H_6O_4$; mol wt 118.09. C 40.68%, H 5.12%, O 54.19%. $CH_3OOCCOOCH_3$.
Colorless crystals. d^{54} 1.148. mp 54°. bp 163-164°. n$_D^{82.1}$ 1.379. Sol in 17 parts water, in alcohol, ether.

6182. Methylparaben. 4-Hydroxybenzoic acid methyl ester; methyl p-hydroxybenzoate; Nipagin M; Tegosept M; Methyl Chemosept; Methyl Parasept. $C_8H_8O_3$; mol wt 152.15. C 63.15%, H 5.30%, O 31.55%. Prepn: Ladenburg, Fitz, Ann. **141**, 247 (1867); Zbarskii, C.A. **33**, 9312³ (1939). Identification in the vaginal secretions of female dogs in estrus: M. Goodwin et al., Science **203**, 559 (1979).

White needles, mp 131°. bp 270-280° (dec). One gram dissolves in 400 ml water, 40 ml warm oil, about 70 ml warm glycerol; freely sol in alcohol, acetone, ether. The soly in water is also given as 0.25% w/w at 20°, and as 0.30% w/w at 25°.
USE: As preservative in foods, beverages and cosmetics.

6183. Methyl Parathion. Phosphorothioic acid O,O-dimethyl O-(4-nitrophenyl) ester; O,O-dimethyl O-p-nitrophenyl phosphorothioate; O,O-dimethyl O-p-nitrophenyl thiophosphate; dimethyl parathion; parathion-methyl; metaphos; E-601; ENT-17292; Dalf (obsolete); Folidol-M; Metacide; Nitrox 80; Penncap M. $C_8H_{10}NO_5PS$; mol wt 263.21. C 36.51%, H 3.83%, N 5.32%, O 30.39%, P 11.77%, S 12.18%. Cholinesterase inhibitor. Prepn: Fletcher et al., J. Am. Chem. Soc. **72**, 2461 (1950). Manuf: Faith, Keyes & Clark's Industrial Chemicals, F. A. Lowenheim, M. K. Moran, Eds. (Wiley-Interscience, New York, 4th ed., 1975) pp 552-555. Toxicity data: T. B. Gaines, Toxicol. Appl. Pharmacol. **14**, 515 (1969). Review of carcinogenic risk: IARC Monographs **30**, 131-152 (1983).

Crystals from cold methanol, mp 37-38°. n_D^{25} 1.5367. d_4^{20} 1.358. Soly in water 50 ppm; sol in most organic solvents. LD_{50} in male, female rats (mg/kg): 14, 24 orally; 67, 67 dermally (Gaines).

USE: Insecticide.

6184. 5-Methyl-5-(3-phenanthryl)hydantoin. *5-Methyl-5-(3-phenanthrenyl)-2,4-imidazolidinedione;* 5-(3'-phenanthryl)-5-methylhydantoin. $C_{18}H_{14}N_2O_2$; mol wt 290.32. C 74.47%, H 4.86%, N 9.65%, O 11.02%. Prepn: Nitz *et al., Arzneimittel-Forsch.* **5,** 337 (1955); **Brit. pat. 774,394** (1957 to Cassella Farbw.).

Crystals from ethanol, mp 236-237°.
Sodium salt, *Bagrosin-Natrium.*
THERAP CAT: Anticonvulsant.

6185. *N*-Methylphenazonium Methosulfate. *5-Methyl-phenazinium methyl sulfate;* phenazine methosulfate. $C_{14}H_{14}N_2O_4S$; mol wt 306.34. C 54.89%, H 4.61%, N 9.14%, O 20.89%, S 10.47%. Prepn: Kehrmann, *Ber.* **46,** 341 (1913); Hillemann, *ibid.* **71,** 34 (1938); Dickens, McIlwain, *Biochem. J.* **32,** 1615 (1938). Mutagenicity study: S. Venitt, C. Crofton-Sleigh, *Mutat Res.* **68,** 107 (1979).

Flat yellow to brown parallelepipeds from alc, mp 155-157° (Hillemann); mp 167° (Dickens). Oxidation-reduction potential at 30° and pH 7: E_h = +0.080 v.

USE: As an electron carrier in place of the flavine enzyme of Warburg in the hexosemonophosphate system: Dickens, *loc. cit.* In the prepn of succinic dehydrogenase: Green *et al., J. Biol. Chem.* **217,** 551 (1955).

6186. Methylphenidate. *α-Phenyl-2-piperidineacetic acid methyl ester;* methyl phenidylacetate; methyl α-phenyl-α-(2-piperidyl)acetate; methylphenidan. $C_{14}H_{19}NO_2$; mol wt 233.31. C 72.07%, H 8.21%, N 6.00%, O 13.72%. Prepn: L. Panizzon, *Helv. Chim. Acta* **27,** 1748 (1944); M. Hartmann, L. Panizzon, **U.S. pat. 2,507,631** (1950 to Ciba). Sepn of isomers: R. Rometsch, **U.S. pat. 2,957,880** (1960 to Ciba). Toxicity data: E. N. Greenblatt, A. C. Osterberg, *J. Pharmacol. Exp. Ther.* **131,** 115 (1961). Comprehensive description: G. R. Padmanabhan, *Anal. Profiles Drug Subs.* **10,** 473-497 (1981). Pharmacokinetics: N. R. Srinivas *et al., Pharm. Res.* **10,** 14 (1993). Clinical efficacy in attention deficit-hyperactivity disorder: W. E. Pelham, Jr. *et al., J. Consult. Clin. Psychol.* **61,** 506 (1993); R. G. Klein, *Encephale* **19,** 89 (1993).

$bp_{0.6mm}$ 135-137°. Sol in alcohol, ethyl acetate, ether. Practically insol in water, petr ether.

Hydrochloride, $C_{14}H_{19}NO_2$.HCl, *Ciba 4311b, Centedrin, Ritalin hydrochloride.* Crystals, mp 224-226°. pKa 8.9. Sol in water, alc, chloroform. A 5% aq soln is neutral to litmus. LD_{50} orally in mice: 190 mg/kg (Greenblatt, Osterberg).

Note: This is a controlled substance (stimulant) listed in the U.S. Code of Federal Regulations, Title 21 Part 1308.12 (1995).

THERAP CAT: CNS stimulant.

6187. 3-Methyl-5-phenylhydantoin. *3-Methyl-5-phenyl-2,4-imidazolidinedione;* Norantoin; Nuvarone. $C_{10}H_{10}N_2O_2$; mol wt 190.20. C 63.15%, H 5.30%, N 14.73%, O 16.82%. Prepn: Pinner, *Ber.* **21,** 2325 (1888); Kjaer, *Acta Chem. Scand.* **4,** 892 (1950); Klosa, *Arch. Pharm.* **285,** 274 (1952).

Fine needles from aq ethanol, mp 162-163°.
THERAP CAT: Anticonvulsant.

6188. 3'-Methylphthalanilic Acid. *2-[[(3-Methylphenyl)amino]carbonyl]benzoic acid; N-m-*tolylphthalamic acid; Duraset; Duraset 20W; Tomaset. $C_{15}H_{13}NO_3$; mol wt 255.27. C 70.58%, H 5.13%, N 5.49%, O 18.80%. Preparation: Smith, Hoffmann, **U.S. pat. 2,556,665** (1951 to U.S. Rubber).

Crystals, mp 149-151°. Slightly sol in water; sol in polar solvents with rapid decompn to the imide. LD_{50} orally in rats: 5230 mg/kg.

USE: As a fruit set, *i.e.* to prevent premature drop of apples, pears, cherries, peaches, and of certain vegetable crops, such as tomatoes and peppers. Antishock treatment for plants.

Caution: On ingestion is hydrolyzed to produce *m*-toluidine. This may cause kidney injury, anemia, methemoglobinemia.

6189. Methylprednisolone. *(6α,11β)-11,17,21-Trihydr-oxy-6-methylpregna-1,4-diene-3,20-dione;* 1-dehydro-6α-methylhydrocortisone; Δ¹-6α-methylhydrocortisone; 6α-methyl-11β,17α,21-triol-1,4-pregnadiene-3,20-dione; Medrate; Medrol; Medrone; Metastab; Metrisone; Promacortine; Suprametil; Urbason. $C_{22}H_{30}O_5$; mol wt 374.48. C 70.56%, H 8.07%, O 21.36%. Prepn: Spero *et al., J. Am. Chem. Soc.* **78,** 6213 (1956); Fried, *ibid.* **81,** 1235 (1959); Sebek, Spero, **U.S. pat. 2,897,218** (1959 to Upjohn); Gould, **U.S. pat. 3,053,832** (1962 to Schering). Review of clinical toxicology: J. D. Truwit, *Crit. Care Clinics* **7,** 639-657 (1991).

Crystals, mp 228-237°. $[\alpha]_D^{20}$ +83° (dioxane). uv max (95% ethanol): 243 nm (α_M 14875).

21-Acetate, $C_{24}H_{32}O_6$, *Depo-Medrate, Depo-Medrol, Depo-Medrone, Mepred, Vetacortyl.* Crystals, mp 205-208°. $[\alpha]_D^{20}$ +101° (dioxane). uv max (95% ethanol): 243 nm (α_M 14825). Practically insol in water.

21-Phosphate disodium salt, $C_{22}H_{29}Na_2O_8P$, *Medrol Stabisol.*

21-Succinate sodium salt, $C_{26}H_{33}NaO_8$, *Urbason-Solubile, Solu-Medrol.* Prepn: Sebek, Spero, *loc. cit.* Sol in water; forms buffered soln suitable for i.v. injection.

Aceponate, $C_{27}H_{36}O_7$, *Advantan.*

THERAP CAT: Glucocorticoid.

THERAP CAT (VET): Glucocorticoid.

6190. p-(2-Methylpropenyl)phenol Acetate. *p*-Acetoxyphenylisobutene; *p*-isobutenylphenol acetate; AO-12; Isotyl. $C_{12}H_{14}O_2$; mol wt 190.24. C 75.76%, H 7.42%, O 16.82%. Prepn: v. Braun, *Ann.* **472,** 71 (1929).

Liquid, odor of anise, bp_{15} 148°.

6191. Methyl Propionate. *Propanoic acid methyl ester.* $C_4H_8O_2$; mol wt 88.11. C 54.53%, H 9.15%, O 36.32%. $CH_3CH_2COOCH_3$.

Colorless liquid. d_4^{20} 0.915. mp −87°. bp 79.7°. Flash pt −2°. n_D^{19} 1.3769. Sol in 16 parts water; miscible with alcohol, ether.

USE: In organic synthesis.

6192. Methyl Propyl Ether. *1-Methoxypropane;* Neothyl. $C_4H_{10}O$; mol wt 74.12. C 64.82%, H 13.60%, O 21.58%. $CH_3CH_2CH_2OCH_3$. Prepd by passing a mixture of methyl and propyl alcohol through a hot layer of β-naphthalenesulfonic or benzenesulfonic acids: Krafft, *Ber.* **26,** 2832 (1893); *Ger. pat.* **69,115;** *Frdl.* III, 11. Laboratory prepn from sodium propylate and methyl iodide: Chancel, *Ann.* **151,** 305 (1869); Michael, Wilson, *Ber.* **39,** 2573 (1906); Bennett, Phillip, *J. Chem. Soc.* **1928,** 1931.

Mobile, flammable liquid. d_4^0 0.7494; d_4^{13} 0.7356. bp_{761} 38.8°. $n_D^{14.3}$ 1.36019. Soly in water (25°): 5 ml/100 ml H_2O. Miscible with alc, ether. Absorbs some water. Oil/water coefficient 10 ± 1. Anesthetic index 2.5. Methyl propyl ether used for anesthesia usually contains 0.002 diphenylamine as stabilizer.

THERAP CAT: Anesthetic (inhalation).

6193. Methyl Propyl Ketone. *2-Pentanone;* ethyl acetone. $C_5H_{10}O$; mol wt 86.13. C 69.72%, H 11.70%, O 18.58%. $CH_3CH_2CH_2COCH_3$. Toxicity study: H. F. Smyth *et al., Am. Ind. Hyg. Assoc. J.* **23,** 95 (1962).

Colorless liquid. d_4^{20} 0.809. mp −78°. bp 102°. n_D^{20} 1.3895. Almost insol in water. Misc with alcohol, ether. LD_{50} orally in rats: 3.73 g/kg (Smyth).

Caution: Potential symptoms of overexposure are irritation of eyes and mucous membranes; headache; dermatitis; narcosis, coma. *See NIOSH Pocket Guide to Chemical Hazards* (DHHS/NIOSH 90-117, 1990) p 174.

USE: Solvent.

6194. 5-Methylpyrazole-3-carboxylic Acid. $C_5H_6-N_2O_2$; mol wt 126.11. C 47.62%, H 4.80%, N 22.21%, O 25.37%. Prepn: Knorr, MacDonald, *Ann.* **279,** 219 (1894);

Lehninger, *J. Am. Chem. Soc.* **64,** 2507 (1942). Hypoglycemic metabolite of 3,5-dimethylpyrazole: Smith *et al., J. Med. Chem.* **8,** 350 (1965).

Crystals from water, mp 236-237°.

Methyl ester, $C_6H_8N_2O_2$, crystals, mp 75-78°.

Ethyl ester, $C_7H_{10}N_2O_2$, crystals, mp 83°.

Caution: A powerful hypoglycemic agent in rats (characteristic of pyrazoles).

6195. Methyl Pyridyl Ketone. *1-(3-Pyridinyl)ethanone; 3-acetylpyridine;* β-acetylpyridine; methyl 3-pyridyl ketone; methyl β-pyridyl ketone. C_7H_7NO; mol wt 121.14. C 69.41%, H 5.82%, N 11.56%, O 13.21%. A nicotinic acid antagonist. Prepd in dry distillation of Ca-nicotinate with Ca-acetate. The ketone is isolated through the phenylhydrazone: Engler, Kiby, *Ber.* **22,** 597 (1889).

Liquid, bp 220°. Freely sol in acids.

Ketoxime, $NC_5H_4C(CH_3)=NOH$, crystals from alcohol or benzene, mp 112°.

Ketoxime hydrochloride, crystals from alcohol, mp 204°.

Phenylhydrazone, $NC_5H_4C(CH_3)=NNHC_6H_5$, yellow needles from alcohol, mp 137°.

6196. Methyl 4-Pyridyl Ketone Thiosemicarbazone. *2-[1-(4-Pyridinyl)ethylidene]hydrazinecarbothioamide.* $C_8-H_{10}N_4S$; mol wt 194.26. C 49.46%, H 5.19%, N 28.84%, S 16.51%. Prepn: Schäfer *et al.,* U.S. pat. **2,744,905** (1956 to Schenley).

Crystals from methanol, dec 229-231°.

Hydrochloride, $C_8H_{10}N_4S.HCl$, *Depreton.*

THERAP CAT: Antihypertensive.

6197. 1-Methylpyrrolidone. *1-Methyl-2-pyrrolidinone; N-methyl-α-pyrrolidinone; N-methyl-γ-butyrolactone; NMP; 1-methylazacyclopentan-2-one; MP; M-Pyrol.* C_5H_9-NO; mol wt 99.13. C 60.58%, H 9.15%, N 14.13%, O 16.14%. Dipolar aprotic solvent. Commercially prepd by condensation of butyrolactone with methylamine. Prepn: J. Tafel, O. Wassmuth, *Ber.* **40,** 2831 (1907); E. Späth, J. Lintner, *ibid.* **69,** 2727 (1936). Use in extraction of unsaturated/aromatic compounds: E. Mueller, G. Hoehfeld, *Proc. 8th World Petrol. Cong.* **4** (Applied Sci. Publishers, London, 1971) pp 213-219; H. Klein, H. M. Weitz, *Hydrocarbon Process.* **47,** 135 (November, 1968); A. Sequeira *et al., ibid.* **58,** 155 (September, 1979). Physical properties and solubility: *Industrial Solvents Handbook* (3rd Ed.) 580-582 (1985). Reference guide to properties and use: *M-Pyrol®, N-Methyl-2-Pyrrolidone Handbook,* ISP (subsidiary of GAF Corp.), 1988. Acute toxicity: W. Bartsch *et al., Arzneimittel-Forsch.* **26,** 1581 (1976). *Review:* E. V. Hort, L. R. Anderson in Kirk-Othmer *Encyclopedia of Chemical Technology* vol. **19** (Wiley-Interscience, New York, 3rd ed., 1982) pp 514-520.

Clear liquid. Fp −24.4°. bp$_{760}$ 202°. n$_D^{25}$ 1.4690. d$_4^{25}$ 1.027. Flash point, closed cup: 199°F; open cup: 204°F. Viscosity at 25° = 1.65 cp. Surface tension at 25°: 40.7 dynes/cm. Dipole moment at 25°: 4.09 D. Dielectric constant at 25°: 32.2. Vapor pressure at 23.2°, 65°: 0.33, 5 mm Hg. Heat of combustion: 719 kcal/mol; of vaporization: 127.3 kcal/kg. Misc with water, alcohol, ether, acetone, ethyl acetate, chloroform, benzene. LD$_{50}$ in mice, rats (ml/kg): 3.5, 2.2 i.v.; 4.3. 2.4 i.p.; 7.5, 3.8 orally (Bartsch).

USE: Industrial solvent primarily used in the extraction of aromatics from lube oils. Other applications include recovery and purification of acetylenes, olefins and diolefins, gas purification, and aromatics extraction from feedstocks. Also, used as polymer solvent.

6198. *N*-Methylpyrroline. *N-Methyl-2,5-dihydropyrrole.* C$_5$H$_9$N; mol wt 83.13. C 72.24%, H 10.91%, N 16.85%. In leaves of *Nicotiana tabacum* L. and *Atropa belladonna* L. By reducing *N*-methylpyrrole with zinc granules in dil acetic or hydrochloric acid.

Almost colorless liquid. Unpleasant, ammonia-like odor. Fumes in air. bp 79-80°. Strong base. Miscible with water; sol in alcohol, ether, chloroform.

6199. Methyl Red. *2-[[4-(Dimethylamino)phenyl]azo]benzoic acid; C.I. Acid Red 2;* C.I. 13020. C$_{15}$H$_{15}$N$_3$O$_2$; mol wt 269.30. C 66.90%, H 5.61%, N 15.60%, O 11.88%. Prepn by diazotization of anthranilic acid and coupling with dimethylaniline: Clarke, Kirner, *Org. Syn.* **2,** 47 (1922). *See also Colour Index* vol. **4** (3rd ed., 1971) p 4043.

Glistening, violet crystals from toluene. mp 181-182°. pKa$_1$ 2.5; pKa$_2$ 9.5; pKb 4.8. Almost insol in water; sol in alcohol and in acetic acid.

USE: As indicator in 0.1% alcoholic soln; pH: 4.4 red, 6.2 yellow. Used for titrating NH$_3$, weak organic bases, *e.g.*, alkaloids; not suitable for organic acids, except oxalic and picric acid. Methyl red is easily reduced, thereby losing its color, and readings should be made promptly. It is gradually being replaced by sulfonphthalein indicators, such as bromcresol green, which are more stable and exhibit a sharper change in color.

6200. Methyl Salicylate. *2-Hydroxybenzoic acid methyl ester;* wintergreen oil; betula oil; sweet birch oil; teaberry oil. C$_8$H$_8$O$_3$; mol wt 152.15. C 63.15%, H 5.30%, O 31.55%. Present in leaves of *Gaultheria procumbens* L., *Ericaceae*, in the bark of *Betula lenta* L., *Betulaceae*, but mostly prepd by esterification of salicylic acid with methanol. The product of commerce is about 99% pure.

Colorless, yellowish or reddish, oily liq; odor and taste of gaultheria. mp −8.6°. bp 220-224°. d$_{25}^{25}$ 1.184. d of the natural ester is ∼1.180. n$_D^{20}$ 1.535-1.538. Flash pt, closed

cup: 210°F (99°C). Slightly sol in water: one gram in about 1500 ml; sol in chloroform, ether. Misc with alcohol, glacial acetic acid. LD$_{50}$ orally in rats: 887 mg/kg, P. M. Jenner *et al.*, *Food Cosmet. Toxicol.* **2,** 327 (1964).

Caution: Ingestion of relatively small amounts may cause severe poisoning and death (average lethal dose: 10 ml in children, 30 ml in adults). *Symptoms of poisoning:* nausea, vomiting, acidosis, pulmonary edema, pneumonia, convulsions, death, *cf.*, *Clinical Toxicology of Commercial Products,* R. E. Gosselin *et al.*, Eds. (Williams & Wilkins, Baltimore, 4th ed., 1976) Section III, pp 295-303.

USE: In perfumery; for flavoring candies, etc.

THERAP CAT: Counterirritant.

THERAP CAT (VET): Counterirritant.

6201. Methyl Silicone Resins. Tympanol. Polymeric methyl silicon oxides, probably of a cross-linked siloxane structure: Rochow, Gilliam, *J. Am. Chem. Soc.* **63,** 798 (1941). Prepd by hydrolyzing dimethyldichlorosilane or its esters followed by oxidation with air and a catalyst to the desired CH$_3$/Si ratio: Hyde, DeLong, *ibid.* 1194; by hydrolysis of dimethyldichlorosilane mixed with methyltrichlorosilane or silicon tetrachloride followed by cocondensation of the products: Rochow, Gilliam, *loc. cit.;* by partially methylating silicon tetrachloride to the desired CH$_3$/Si ratio and hydrolyzing the reaction mixture: U.S. pat. **2,258,218** (1941).

Characterized by excellent thermal stability and good resistance to oxidation. Oxidize slowly in air at 300° to silica. Can be heated to 550° *in vacuo* or to 500° in hydrogen without dec or melting.

CH$_3$/Si = 1.2. Solid. d 1.20. n$_D^{20}$ 1.425. Hardens in 2 hours at 100°.

CH$_3$/Si = 1.3. Solid. d 1.15. n$_D^{20}$ 1.422. Hardens in 1 .5 hours at 120°.

CH$_3$/Si = 1.4. Solid. d 1.08. n$_D^{20}$ 1.421. Hardens in 4 hours at 141°.

CH$_3$/Si = 1.5. Solid. d 1.06. n$_D^{20}$ 1.418. Hardens in 24 hours at 100°.

CH$_3$/Si < 1.2. Sticky syrups. Harden at room temp.

CH$_3$/Si > 1.5. Oily liquids. Set to a soft gel after several days or weeks at 200°.

USE: In electrical insulation; in heat-resistant paints and varnishes; in protective and decorative finishes.

6202. 4'-(Methylsulfamoyl)sulfanilanilide. *4-Amino-N-[4-[(methylamino)sulfonyl]phenyl]benzenesulfonamide; N^1-methyl-N^4-sulfanilylsulfanilamide; DB87; Diseptal B; Neo-Uliron.* C$_{13}$H$_{15}$N$_3$O$_4$S$_2$; mol wt 341.41. C 45.73%, H 4.43%, N 12.31%, O 18.74%, S 18.78%. Prepn: Fr. pat. **817,034** (1937 to I. G. Farbenind.), *C.A.* **32,** 1715 (1938).

Crystals, mp 141°. Slightly sol in water; more sol in alcohol or acetone; freely sol in aq solns of sodium carbonate and bicarbonate.

THERAP CAT: Antibacterial.

6203. Methyl Sulfate. *Sulfuric acid monomethyl ester;* methylsulfuric acid; methyl hydrogen sulfate; monomethyl sulfate; acid methyl sulfate; "Methylsäure". CH$_4$O$_4$S; mol wt 112.11. C 10.71%, H 3.60%, O 57.09%, S 28.60%. CH$_3$-OSO$_2$OH. Prepn from methanol and methyl chlorosulfonate: Levaillant, Simon, *Compt. Rend.* **169**, 855 (1919); from methanol and sulfuric acid: Dumas, Peligot, *Ann.* **15**, 40 (1835); from sulfur trioxide and methanol below 0°: Chamberlain *et al.*, *BIOS Final Rept.* no. **1482**, pp 6 and 8 (1946). *Review:* C. M. Suter, *The Organic Chemistry of Sulfur* (John Wiley, 1944) pp 18-23.

Oily liquid; does not solidify at −30°. d_4^{15} 1.45-1.47. Freely sol in water, less sol in alcohol. Miscible with ether.

Barium salt, $C_2H_6BaO_8S_2$, *barium methyl sulfate.* Dihydrate, efflorescent crystals. Sol in water, alc.

Calcium salt, $C_2H_6CaO_8S_2$, *calcium methyl sulfate, calcium sulfomethylate.* Crystals. Soluble in water.

See also Dimethyl Sulfate.

USE: In sulfonation; as solvent in bromination of indigo dyes.

6204. Methyl Sulfide. *Thiobismethane.* C_2H_6S; mol wt 62.14. C 38.66%, H 9.73%, S 51.61%. $(CH_3)_2S$. Obtained by distilling potassium methyl sulfate with concn aq K_2S.

Liquid; disagreeable odor. d_4^{21} 0.846. mp −83°. bp 36.2°; also stated as 38°. Insol in water; sol in alcohol, ether.

USE: As solvent for anhydr mineral salts.

6205. N-Methyltaurine. *2-Methylaminoethanesulfonic acid; β-methylaminoethane-α-sulfonic acid.* $C_3H_9NO_3S$; mol wt 139.18. C 25.89%, H 6.52%, N 10.06%, O 34.49%, S 23.04%. $CH_3NHCH_2CH_2SO_3H$. Prepn: E. Dittrich, *J. Prakt. Chem.* **18**, 63 (1878); J. W. Schick, E. F. Degering, *Ind. Eng. Chem.* **39**, 906 (1947).

Prisms, mp 241-242°. Freely sol in water. Insol in alcohol, ether.

Sodium salt, *N-methylaminoethane sodium sulfonate, sodium N-methyltaurate.* Usually sold as an aq soln contg 35% of N-methyltaurine sodium salt. d_4^{25} 1.21. At its freezing point (−28°) the soln becomes a suspension of crystals. Also available as a 65% soln.

USE: Intermediate in the manuf of surface-active agents.

6206. 17-Methyltestosterone. *(17β)-17-Hydroxy-17-methylandrost-4-en-3-one;* 17α-methyl-Δ⁴-androsten-17β-ol-3-one; Android; Glosso-Stérandryl; Metandren; Neohombreol M; Orchisterone-M; Oreton Methyl; Perandren (lozenges); Testred. $C_{20}H_{30}O_2$; mol wt 302.46. C 79.42%, H 10.00%, O 10.58%. Prepn from 17-methyl-Δ⁵,⁶-androstene-3,17-diol: Oppenauer, *Rec. Trav. Chim.* **56**, 137 (1937); U.S. pat. **2,384,335** (1945 to Alien Property Custodian); from 17-methyl-Δ⁴-androsten-17-ol with CrO_3: Miescher, Wettstein, U.S. pats. **2,374,370; 2,374,369** (both 1945 to Ciba); from androstenedione 3-enol ethyl ether: Miescher, U.S. pat. **2,386,331** (1945 to Ciba); from dehydroandrosterone: Julian *et al.*, U.S. pat. **2,435,013** (1945 to Glidden).

Crystals from hexane. mp 161-166°. Stable in air. $[\alpha]_D^{25}$ +69° to +75° (dioxane). Soluble in alcohol, methanol, ether and in other organic solvents. Sparingly sol in vegetable oils. Practically insol in water.

Note: This is a controlled substance (anabolic steroid) in the U.S. Code of Federal Regulations, Title 21 Part 1308.13, as defined in Part 1308.02 (1995).

THERAP CAT: Androgen.

THERAP CAT (VET): Has been used as an androgenic agent.

6207. 17α-Methyltestosterone 3-Cyclopentyl Enol Ether. *(17β)-3-(Cyclopentyloxy)-17-methylandrosta-3,5-dien-17-ol;* RP-12222; Pandrocine; Penmestrol. $C_{25}H_{38}O_2$; mol wt 370.58. C 81.03%, H 10.34%, O 8.63%. Prepn: Ercoli, Gardi, *J. Am. Chem. Soc.* **82**, 746 (1960); Ercoli, U.S. pat. **3,019,241** (1962).

Crystals, mp 148-152°. $[\alpha]_D$ −150° (in dioxane).

THERAP CAT: Androgen.

6208. 4-Methyl-5-thiazoleethanol. *5-Hydroxyethyl-4-methylthiazole;* 5-(2-hydroxyethyl)-4-methylthiazole; 4-methyl-5-(β-hydroxyethyl)thiazole. C_6H_9NOS; mol wt 143.21. C 50.32%, H 6.33%, N 9.78%, O 11.17%, S 22.39%. The thiazole moiety of thiamine (vitamin B_1). Several syntheses, *e.g.*, by condensing thioformamide with bromoacetopropanol: Buchman, *J. Am. Chem. Soc.* **58**, 1803 (1936); or with γ,γ-dichloro-γ,γ-diacetodipropyl ether: Stein, Stevens, **Fr. pat. 945,198** and **Brit. pat. 641,426** (1947 and 1950 to Merck & Co.); Londergan, Schmitz, U.S. pat. **2,654,760** (1953 to du Pont).

Oily, viscous liquid. d_4^{24} 1.196. Characteristic disagreeable odor of thiazole compds, becoming somewhat pleasant at extreme dilutions and imparting a nut-like flavor. bp_7 135°; bp_3 123-124°, bp_1 103°. Sol in alcohol, ether, benzene, chloroform. Very sol in water.

Hydrochloride, $C_6H_9NOS.HCl$, hygroscopic crystals, sol in water, alcohol.

USE: Intermediate in the synthesis of vitamin B_1.

THERAP CAT: Sedative, hypnotic.

6209. Methyl Thiocyanate. Methyl sulfocyanate. C_2H_3NS; mol wt 73.12. C 32.85%, H 4.14%, N 19.16%, S 43.85%. $CH_3SC≡N$. Prepd from dimethyl sulfate and aq barium thiocyanate.

Colorless liquid; onion odor. d^{20} 1.068. mp −51°. bp 130-133°. n_D^{20} 1.4697. Very slightly sol in water; miscible with alcohol, ether. LD orally in cats: 8.5 mg/kg.

6210. Methylthiouracil. *2,3-Dihydro-6-methyl-2-thioxo-4(1H)-pyrimidinone;* 6-methyl-2-thiouracil; 4-methyl-2-thiouracil; MTU; Alkiron; Antibason; Basecil; Basethyrin; Methiacil; Methicil; Methiocil; Muracil; Prostrumyl; Strumacil; Thimecil; Thyreostat I. $C_5H_6N_2OS$; mol wt 142.18. C 42.24%, H 4.25%, N 19.70%, O 11.25%, S 22.55%. Prepn: List, *Ann.* **236**, 1 (1886).

Crystals, bitter taste, dec 326-331°. Sublimes readily. Very slightly sol in cold water, ether. One part dissolves in about 150 parts boiling water. Slightly sol in alcohol, acetone. Practically insol in benzene, chloroform. Freely sol in aq solns of ammonia and alkali hydroxides. A satd aq soln is neutral or slightly acid to litmus. MLD orally in rabbits: 2.5 g/kg.

THERAP CAT: Thyroid inhibitor.

THERAP CAT (VET): Antithyroid substance to promote growth and finishing.

6211. Methyltrienolone. *(17β)-17-Hydroxy-17-methylestra-4,9,11-trien-3-one;* 17β-hydroxy-17-methyl-19-norandrosta-4,9,11-trien-3-one; 17α-methyl-4,9,11-estratrien-17β-ol-3-one; R-1881. $C_{19}H_{24}O_2$; mol wt 284.40. C 80.24%, H 8.51%, O 11.25%. Prepn: Velluz *et al.*, *Compt. Rend.* **257,**

569 (1963); Neth. pat. appl. **6,401,555** (1964 to Roussel-UCLAF), *C.A.* **62**, 10498d (1965).

Crystals from diisopropyl ether, mp 170°. $[\alpha]_D^{20}$ −58.7° (c = 0.5 in ethanol).

THERAP CAT: Anabolic.

6212. α-Methyl-*m*-tyrosine. *3-Hydroxy-α-methylphenylalanine;* α-methyl-3-(*m*-hydroxyphenyl)alanine; α-MMT. $C_{10}H_{13}NO_3$; mol wt 195.22. C 61.53%, H 6.71%, N 7.17%, O 24.59%. Prepn of DL-form: Stein *et al., J. Am. Chem. Soc.* **77**, 700 (1955); Pfister, Stein, U.S. pat. **2,868,818** (1959 to Merck & Co.).

Crystals from methanol, dec 296-297°.

THERAP CAT: Inhibitor of catecholamine synthesis.

6213. 6-Methyluracil. *6-Methyl-2,4(1H,3H)-pyrimidinedione;* 4-methyluracil. $C_5H_6N_2O_2$; mol wt 126.11. C 47.62%, H 4.80%, N 22.21%, O 25.37%. Prepn: Boese, U.S. pat. **2,138,756** (1938 to Carbide and Carbon Chem.); Gleason, U.S. pat. **2,174,239** (1939 to Standard Oil). Synthesis from urea and ethyl acetoacetate: J. J. Donleavy, M. A. Kise, *Org. Syn. coll.* **vol. II**, 422 (1943); by isomerization of 2-hydroxy-4-methoxy-6-methylpyrimidine: McOmie *et al., J. Chem. Soc.* **1957**, 1830; from urea and diketene: Khromov-Borisov, Karlinskaya, *J. Gen. Chem. USSR* **26**, 1728 (1956); **U.S.S.R.** pat. **101,690** (1955), *C.A.* **51**, 12989f (1957).

Crystals from glacial acetic acid, dec > 300°. uv max: 277 nm (log ε 3.83). pH of aq soln 13.

6214. Methyl Vinyl Ketone. *3-Buten-2-one;* Δ³-2-butenone; methylene acetone; acetyl ethylene; δ-oxo-α-butylene. C_4H_6O; mol wt 70.09. C 68.55%, H 8.63%, O 22.83%. $CH_3COCH=CH_2$. Prepn by condensation of acetone and formaldehyde to 3-ketobutanol and dehydration to methyl vinyl ketone: Merling, Köhler, *J. Soc. Chem. Ind.* **29**, 1037 (1910); White, Haward, *J. Chem. Soc.* **1943**, 25. Prepn from vinylacetylene: Conaway, U.S. pat. **1,967,225** (1934 to du Pont). For review and polymerization characteristics *see* "Vinyl Ketone Polymers" in *Encyclopedia of Polymer Science and Technology* **vol. 14** (Interscience, New York, 1971) pp 617-636.

Liquid with pungent odor. bp_{760} 81.4°. n_D^{20} 1.4086. d_4^{20} 0.8636; d_4^{25} 0.8407. Easily soluble in water, methanol, ethanol, ether, acetone, glacial acetic acid. Slightly sol in hydrocarbons. Forms a binary azeotrope with water, bp_{760} 75° (12% water). uv spectrum and electric moments: Rogers, *J. Am. Chem. Soc.* **69**, 2544 (1947). Polymerizes on standing. LD_{50} in mice and rats: 35 mg/kg, *C.A.* **72**, 124809b (1970).

Caution: Readily absorbed through skin causing general poisoning of the organism. Irritating to mucous membranes and respiratory tract.

USE: Alkylating agent; commercial starting material for plastics; as intermediate in the synthesis of steroids and vitamin A.

6215. Methymycin. $C_{25}H_{43}NO_7$; mol wt 469.62. C 63.94%, H 9.23%, N 2.98%, O 23.85%. Antibiotic substance produced by a streptomycete from soil near Oswego, N.Y. Has a macrolide structure (12-membered lactone ring, *compare* Picromycin). Isoln and antibacterial activity: Donin *et al., Antibiot. Ann.* **1**, 179 (1953-4). Production using *Streptomyces venezuelae* cultures: Dutcher *et al.*, U.S. pat. **2,916,-483** (1959 to Olin Mathieson). Structure: C. Djerassi, J. A. Zderic, *J. Am. Chem. Soc.* **78**, 6390 (1956). Absolute configuration: Rickards, Smith, *Tetrahedron Letters* **1970**, 1025; Manwaring *et al., ibid.* 1029. Biosynthesis: Birch *et al., J. Chem. Soc.* **1964**, 5274. Synthesis: Masamune *et al., J. Am. Chem. Soc.* **97**, 3512 (1975). Possible mechanism of action: Wilhelm *et al., Antimicrob. Ag. Chemother.* **1967**, 236.

Prisms from abs ethanol, mp 195.5-197°. $[\alpha]_D^{22}$ +61° (c = 0.7 in methanol), +74° (c = 1.1 in chloroform). Basic substance, pKb' 5.7. uv max (methanol): 223, 322 nm (ε 10500, 47). Very slightly sol in water, hexane. Sol in methanol, acetone, chloroform, dil acids. Moderately sol in ethanol, ether.

Sulfate, $(C_{25}H_{43}NO_7)_2.H_2SO_4$, crystals from methanol + acetone. Sol in water, methanol.

Acid sulfate, $C_{25}H_{43}NO_7.H_2SO_4$, plates from acetone. Sol in water, acetone.

6216. Methyprylon. *3,3-Diethyl-5-methyl-2,4-piperidinedione;* 2,4-dioxo-3,3-diethyl-5-methylpiperidine; 3,3-diethyl-2,4-dioxo-5-methylpiperidine; Noctan; Dimerin; Noludar. $C_{10}H_{17}NO_2$; mol wt 183.25. C 65.54%, H 9.35%, N 7.64%, O 17.46%. Prepn: Frick, Lutz, U.S. pat. **2,680,116** (1954 to Hoffmann-La Roche). *cf.* **Ger.** pat. **930,206** (1953). Comprehensive description: B. C. Rudy, B. Z. Senkowski, *Anal. Profiles Drug Subs.* **2**, 363-382 (1973).

Crystals. Bitter taste. mp 74-77°. uv max (alcohol): 295 nm ($A_{1cm}^{1\%}$ 2.0). Sol in water, alcohol, benzene, chloroform.

Caution: May be habit forming. This is a controlled substance (depressant) listed in the U.S. Code of Federal Regulations, Title 21 Part 1308.13 (1995).

THERAP CAT: Sedative, hypnotic.

6217. Methysergide. *[8β(S)]-9,10-Didehydro-N-[1-(hydroxymethyl)propyl]-1,6-dimethylergoline-8-carboxamide; N-[1-(hydroxymethyl)propyl]-1-methyl-D-lysergamide;* 1-methylmethylergonovine; 1-methyl-*d*-lysergic acid butanolamide; 1-methyl-*d*-lysergic acid (+)-1-hydroxy-2-butylamide; UML-491. $C_{21}H_{27}N_3O_2$; mol wt 353.46. C 71.36%, H 7.70%, N 11.89%, O 9.05%. Serotonin receptor antagonist. Prepn: **Brit.** pat. **854,569** (1960); Hofmann, Troxler, U.S. pat. **3,218,324** (1965 to Sandoz). Comparative pharmacology: Z. Votava *et al., Arzneimittel-Forsch.* **16**, 220 (1966); P. N. Chambers, P. B. Marshall, *J. Pharm. Pharmacol.* **19**, 65 (1967). Mechanism of action: D. A. Curran *et al., Res. Clin. Stud. Headache* **1**, 74 (1967); J. E. Hardebo *et al., Neurology* **28**, 64 (1978); S. W. J. Lamberts, R. M. MacLeod, *Endocrinology* **103**, 287 (1978).

Crystals, mp 194-196°. $[\alpha]_D^{20}$ −45° (c = 0.5 in pyridine). Tartrate, $C_{46}H_{60}N_6O_{10}$, crystals, sparingly sol in water. Dimaleate, dec ~165°. Sol in methanol, less sol in water (1:250). Practically insol in abs ethanol.

Hydrogen maleate, $C_{21}H_{27}N_3O_2 \cdot C_4H_4O_4$, *methysergide maleate, Deseril, Désernil, Sansert.*

THERAP CAT: Antimigraine.

6218. Methysticin. *[R-(E)]-6-[2-(1,3-Benzodioxol-5-yl)-ethenyl]-5,6-dihydro-4-methoxy-2H-pyran-2-one; 5-hydroxy-3-methoxy-7-[3,4-(methylenedioxy)phenyl]-2,6-heptadienoic acid δ-lactone;* 4-methoxy-6-[β-(3',4'-methylenedioxyphenyl)vinyl]-5,6-dihydro-α-pyrone; 6-(3',4'-methylenedioxystyr-yl)-4-methoxy-5,6-dihydro-2H-pyran-2-one; kavahin; kavatin. $C_{15}H_{14}O_5$; mol wt 274.27. C 65.69%, H 5.14%, O 29.17%. One of the active components of the intoxicating beverage Kava from the Pacific Islands. From root of *Piper methysticum* Forst., *Piperaceae* (kava). Pomeranz, *Monatsh.* **10**, 783 (1889); Hänsel, Beiersdorff, *Arzneimittel-Forsch.* **9**, 581 (1955). Structure: Borsche *et al., Ber.* **60**, 2113 (1927). Synthesis of racemate: Klohs *et al., J. Org. Chem.* **24**, 1829 (1959). Absolute config: Snatzke, Hänsel, *Tetrahedron Letters* **1968**, 1797. Crystal structure: P. Engel, W. Nowacki, *Z. Kristallographie.* **136**, 437 (1972). ^{13}C NMR: A. Banerji *et al., Org. Magnet. Reson.* **13**, 345 (1980). HPLC determn: R. M. Smith *et al., J. Chromatog.* **283**, 303 (1984). Anticonvulsant effects: Kretzschmar, Meyer, *Arch. Int. Pharmacodyn. Ther.* **177**, 261 (1969). Neuroprotective activity: C. Backhauss, J. Krieglstein, *Eur. J. Pharmacol.* **215**, 265 (1992).

Crystals from methanol, mp 132-134°. uv max (alcohol): 226, 267, 306 nm (log ε 4.40, 4.14, 3.93). Practically insol in water. Sol in alcohol, ether, acetone.

(±)-Dihydromethysticin, $C_{15}H_{16}O_5$, prisms from methanol, mp 118°. uv max (methanol): 232, 288 nm (log ε 4.18, 3.56).

6219. Metiazinic Acid. *10-Methyl-10H-phenothiazine-2-acetic acid;* (10-methyl-2-phenothiazinyl)acetic acid; N-methyl-3-phenothiazinylacetic acid; methiazic acid; methiazinic acid; metiazic acid; RP-16091; Soridermal; Soripal. $C_{15}H_{13}NO_2S$; mol wt 271.34. C 66.40%, H 4.83%, N 5.16%, O 11.79%, S 11.82%. Prepn: **Fr. pat. M4163;** Farge *et al.,* **Neth. pat. Appl. 6,614,516; U.S. pat. 3,424,748** (1966, 1967, 1969 all to Rhône-Poulenc); Messer *et al., Compt. Rend. Ser. C* **265** (14), 758 (1967). Series of articles on prepn, pharmacology, toxicology and metabolism: *Arzneimittel-Forsch.* **19**, 1193-1221 (1969). Toxicity: L. Joulou *et al., ibid.* 1198.

Crystals from benzene, mp 146°. Sol in acetone, ether, chloroform. Forms a water-sol sodium salt. uv max (0.1N NaOH): 253, 305 nm; min 280 nm. LD_{50} in mice, rats (mg/kg): 800, ~500 orally (Joulou).

THERAP CAT: Anti-inflammatory.

6220. Meticrane. *3,4-Dihydro-6-methyl-2H-1-benzo-thiopyran-7-sulfonamide 1,1-dioxide; 6-methylthiochroman-7-sulfonamide 1,1-dioxide;* 6-methyl-7-sulfamoylthiochro-man 1,1-dioxide; Arresten; Fontilix. $C_{10}H_{13}NO_4S_2$; mol wt 275.35. C 43.62%, H 4.76%, N 5.09%, O 23.24%, S 23.29%. Prepn and use as diuretic: Boissier, Malen, **Fr. pats. M2790** and **1,365,504** (1962, 1963 to Soc. Ind. Fabric. Antibiot.).

Crystals from methyl Cellosolve, mp 236-237°.

THERAP CAT: Diuretic.

6221. Metipranolol. *4-[2-Hydroxy-3-[(1-methylethyl)-amino]propoxy]-2,3,6-trimethylphenol 1-acetate;* 1-(4-hydr-oxy-2,3,5-trimethylphenoxy)-3-(isopropylamino)-2-propanol 4-acetate; methypranol; trimepranol; VUFB-6453; Betamet; Betanol; Disorat; Glauline; Glausyn; Turoptin. $C_{17}H_{27}NO_4$; mol wt 309.41. C 65.99%, H 8.80%, N 4.53%, O 20.68%. β-Adrenergic blocker. Prepn: L. Blaha *et al., Czech.* pat. **128,471** (1968), *C.A.* **71**, 3129a (1969). Pharmacology: W. Bartsch *et al., Arzneimittel-Forsch.* **27**, 1022 (1977). Plasma levels and pharmacodynamics in man: O. Mayer *et al., Int. J. Clin. Pharmacol., Ther. Toxicol.* **18**, 113 (1980). Metabolic effects: R. P. Faupel, R. Gotzen, *Med. Klin.* **74**, 929 (1979). Comparative effect in hypertension: K. Hayduk *et al., Therapiewoche* **29**, 7528, 7530 (1979). Review of pharmacology and clinical trials in glaucoma and ocular hypertension: P. E. Battershill, E. M. Sorkin, *Drugs* **36**, 601-615 (1988). Comprehensive description: J. Dohnal *et al.* in *Analytical Profiles of Drug Substances* vol. 19, K. Florey, Ed. (Academic Press, New York, 1990) pp 367-396.

Crystals from cyclohexane, mp 105-107° (Blaha), also reported as 108.5-110.5° (Florey). uv max (methanol): 278, 274 nm ($A_{1cm}^{1\%}$ 51.3 50.5). Freely sol in 95% ethanol, chloroform, benzene; slightly sol in ether. Practically insol in water. LD_{50} in mice (mg/kg): 31 i.v. (Bartsch).

Hydrochloride, $C_{17}H_{27}NO_4 \cdot HCl$, *Betamann, Optipranolol.* Sol in water.

THERAP CAT: Antihypertensive; antiglaucoma agent.

6222. Metitepine. *1-[10,11-Dihydro-8-(methylthio)di-benzo[b,f]thiepin-10-yl]-4-methylpiperazine;* 8-methylthio-10-(4-methylpiperazino)-10,11-dihydrodibenzo[b,f]thiepine; methiothepin; methiothepin; Ro-8-6837. $C_{20}H_{24}N_2S_2$; mol wt 356.56. C 67.37%, H 6.78%, N 7.86%, S 17.99%. Serotonin (5-HT$_2$) receptor antagonist; also exhibits affinity for 5-HT$_1$-receptors. Prepn: **Neth. pat. Appl. 6,608,618;** M. Protiva *et al., U.S. pat. 3,379,729* (1966, 1968 both to SPOFA); and pharmacology: K. Pelz *et al., Coll. Czech. Chem. Commun.* **33**, 1895 (1968); J. O. Jilek *et al., ibid.* **39**, 3338 (1974). Receptor-blocking study: M.-A. Monachon *et* al., Arch. Pharmacol. **274**, 192 (1972). Use in classification of 5-HT receptors: P. B. Bradley et al., Neuropharmacology **25**, 563 (1986); E. J. Mylecharane, Clin. Exp. Pharmacol. Physiol. **16**, 517 (1989).

Crystals from ethanol, mp 88-89°.

Maleate, $C_{20}H_{24}N_2S_2 \cdot C_4H_4O_4$, crystals from ethanol, mp 171-173°. LD_{50} in mice (mg/kg): 51 i.v.; 94 orally (Jilek).

USE: Biochemical tool in serotonin receptor binding studies.

6223. Metizoline. 4,5-Dihydro-2-[(2-methylbenzo[b]-thien-3-yl)methyl]-1H-imidazole; 2-[(2-methylbenzo[b]thien-3-yl)methyl]-2-imidazoline; 2-methyl-3-(Δ^2-imidazolinyl-methyl)benzo[b]thiophene; benazoline (obsolete). $C_{13}H_{14}$-N_2S; mol wt 230.33. C 67.79%, H 6.13%, N 12.16%, S 13.92%. Prepn: **Fr. pats. M1614; 1,355,049** (1963, 1964, both to E. Merck), C.A. **58**, 12574e (1963); **61**, 4146e (1964). Pharmacology and toxicity: G. F. Rosati, M. G. Poletto, Farmaco Ed. Prat. **21**, 204 (1966).

Crystals, mp 156-157°.

Hydrochloride, $C_{13}H_{14}N_2S \cdot HCl$, **H-1032, Elsyl, Eunasin.** Crystalline powder, mp 244-246°. Slightly sol in water; very sol in ethanol, methanol. Practically insol in chloroform. LD_{50} in mice (mg/kg): 49 i.p.; 9.1 i.v.; 155 orally; in rats (mg/kg): 74 orally (Rosati, Poletto).

THERAP CAT: Adrenergic (vasoconstrictor); nasal decongestant.

6224. Metobromuron. N'-(4-Bromophenyl)-N-methoxy-N-methylurea; 3-(p-bromophenyl)-1-methoxy-1-methylurea; Ciba 3126; Pattonex; Patoran. $C_9H_{11}BrN_2O_2$; mol wt 259.10. C 41.72%, H 4.28%, Br 30.84%, N 10.81%, O 12.35%. Prepd by bromination of N-phenyl-N'-methoxy-N'-methylurea: H. Martin et al., **Swiss pat. 405,821**; eidem, **U.S. pat. 3,288,851** (both 1966 to Ciba). Toxicity study: S. Novakova, S. Dinoeva, Egeszsegtudomany **17**, 233 (1973), C.A. **80**, 56258j (1974).

Crystals from cyclohexane, mp 95-96°. Vapor pressure at 20°: 3×10^{-6} mm Hg. Soly in water at 20° = 320 ppm. Sol in methanol, ethanol, acetone, chloroform. LD_{50} orally in rats: 3875 mg/kg (Novakova, Dinoeva).

USE: Herbicide.

6225. Metochalcone. 1-(2,4-Dimethoxyphenyl)-3-(4-methoxyphenyl)-2-propen-1-one; 2',4,4'-trimethoxychalcone; 3'-(4-methoxyphenyl)-1-(2,4-dimethoxyphenyl)-2-propen-1-one; CB-1314; Lesidrin; Vesidril; Vesidryl. $C_{18}H_{18}O_4$; mol wt 298.34. C 72.47%, H 6.08%, O 21.45%. Prepn from p-methoxyacetophenone and 2,4-dimethoxybenzaldehyde: Freudenberg et al., Ber. **90**, 957 (1957).

Crystals from aq methanol, mp 97°.

THERAP CAT: Choleretic; diuretic.

6226. Metoclopramide. 4-Amino-5-chloro-N-[(2-diethylamino)ethyl]-2-methoxybenzamide; 4-amino-5-chloro-N-[2-(diethylamino)ethyl]-o-anisamide; 4-amino-5-chloro-2-methoxy-N-(β-diethylaminoethyl)benzamide; DEL-1267; Elieten; Eucil; Imperan; Reliveran. $C_{14}H_{22}ClN_3O_2$; mol wt 299.80. C 56.09%, H 7.40%, Cl 11.83%, N 14.02%, O 10.67%. Substituted benzamide with neuroleptic activity. Prepn: **Belg. pat. 620,543**; Thominet, **U.S. pat. 3,177,252** (1962, 1965 to Soc. d'Etudes Sci. et Ind. de l'Ile-de-France); R. Pakula et al., Arch. Pharm. **313**, 297 (1980). Colorimetric determn in plasma and solubility: G. Pitel, T. Luce, Ann. Pharm. Franc. **23**, 673 (1965). Dopamine D_2 receptor antagonist (adenylate cyclase independent receptors): J. W. Kebabian, D. B. Calne, Nature **277**, 93 (1979). Pharmacology: M. A. Smith, F. J. Salter, Drug Intell. Clin. Pharm. **14**, 169 (1980). Pharmacokinetics: H. Vergin et al., Arzneimittel-Forsch. **33**, 458 (1983). Review of efficacy in psychiatric disorders: E. D. Peselow, M. Stanley, Adv. Biochem. Psychopharmacol. **35**, 163-194 (1982); of pharmacology and clinical use: R. A. Harrington et al., Drugs **25**, 451-494 (1983); ibid., Suppl. 1, 1-88. Comprehensive description: D. Pitré, R. Stradi, Anal. Profiles Drug Subs. **16**, 327-361 (1987).

mp 146.5-148°. Soly at 25° (g/100 ml): water 0.02; 95% ethanol 2.30; abs ethanol 1.90; benzene 0.10; chloroform 6.60.

Dihydrochloride monohydrate, $C_{14}H_{22}ClN_3O_2 \cdot 2HCl \cdot H_2O$, **Emetid, Gastronerton, Primperan.** Crystals, dec 145°. Soly at 25° (g/100 ml): water 48; ethanol (95%) 9; abs ethanol 6; benzene 0.10; chloroform 0.10. Stable in acids solns. Unstable in strongly alkaline solns.

Monohydrochloride monohydrate, $C_{14}H_{22}ClN_3O_2 \cdot HCl \cdot$-$H_2O$, **AHR-3070-C, Cerucal, Clopromate, Duraclamid, Emperal, Gastrese, Gastrobid, Gastromax, Gastrosil, Gastro-Tablinen, Gastrotem, Gastro-Timelets, Maxeran, Maxolon, Meclopran, Metamide, Metoclol, Metocobil, Metramid, Moriperan, Mygdalon, Parmid, Paspertin, Peraprin, Plasil, Pramiel, Reglan.** White crystalline powder, mp 182.5-184°. Sol in water.

THERAP CAT: Antiemetic.

6227. Metocurine Iodide. 6,6',7',12'-Tetramethoxy-2,2,-2',2'-tetramethyltubocuraranium diiodide; (+)-O,O'-dimethylchondrocurarine diiodide; d-tubocurarine iodide dimethyl ether; dimethyl tubocurarine iodide; tubocurarine dimethyl ether iodide; Metubine Iodide. $C_{40}H_{48}I_2N_2O_6$; mol wt 906.64. C 52.99%, H 5.34%, I 27.99%, N 3.09%, O 10.59%. Prepd by methylation of *d*-tubocurarine with methyl iodide.

Crystals, dec 257-267°. $[\alpha]_D^{22}$ +148 to +158° (c = 0.25). uv max: 280 nm. ($E_{1cm}^{1\%}$ 74). Slightly soluble in water (about 300 ml/100 ml), in dil HCl, in dil NaOH. Very slightly sol in alcohol; practically insol in benzene, chloroform, ether.

Trihydrate, large prisms from water, appreciably sol in methanol.

THERAP CAT: Muscle relaxant (skeletal).

6228. Metofenazate. *3,4,5-Trimethoxybenzoic acid 2-[4-[3-(2-chloro-10H-phenothiazin-10-yl)propyl]-1-piperazinyl]ethyl ester;* 2-[4-[3-(2-chlorophenothiazin-10-yl)propyl]-1-piperazinyl]ethyl 3,4,5-trimethoxybenzoate; *N*-[β-(3,4,5-trimethoxybenzoyloxy)ethyl]-*N'*-[γ-(3-chloro-10-phenothiazinyl)propyl]piperazine; methophenazine; perphenazine 3,4,5-trimethoxybenzoate. $C_{31}H_{36}ClN_3O_5S$; mol wt 598.16. C 62.25%, H 6.07%, Cl 5.93%, N 7.02%, O 13.37%, S 5.36%. Prepn: *Magy. Kém. Lapja* **17**, 169 (1962), *C.A.* **57**, 11314d (1962); Toldy *et al.*, *Acta Chim. Acad. Sci. Hung.* **42**, 351 (1964).

Crystals from ethanol, mp 102-107°.
Difumarate, $C_{31}H_{36}ClN_3O_5S.2C_4H_4O_4$, *Frenolon*.
THERAP CAT: Antipsychotic.

6229. Metofoline. *1-(p-Chlorophenethyl)-1,2,3,4-tetrahydro-6,7-dimethoxy-2-methylisoquinoline;* 1-(p-chlorophenethyl)-2-methyl-6,7-dimethoxy-1,2,3,4-tetrahydroisoquinoline; methopholine; Ro-4-1778/1. $C_{20}H_{24}ClNO_2$; mol wt 345.87. C 69.45%, H 6.99%, Cl 10.25%, N 4.05%, O 9.25%. Prepn: Brossi *et al.*, *Helv. Chim. Acta* **43**, 1459 (1960); Besendorf *et al.*, U.S. pats. **3,067,203; 3,146,266** (1962, 1964, both to Hoffmann-La Roche); Brossi *et al.*, in *Analgetics*, de Stevens, Ed. (Med. Chem. Ser. vol. 5, 1965) p 281. Metabolism: Sciorelli, *Experientia* **23**, 934 (1967).

Colorless leaflets from aqueous methanol, mp 110-111°. LD_{50} s.c. in mice: 180 mg/kg.
Hydrochloride, crystals, mp 105-106°. uv max (ethanol): 283, 287 nm.
THERAP CAT: Analgesic.

6230. Metolachlor. *2-Chloro-N-(2-ethyl-6-methylphenyl)-N-(2-methoxy-1-methylethyl)acetamide;* 2-chloro-6'-ethyl-*N*-(2-methoxy-1-methylethyl)acet-*o*-toluidide; α-chloro-2'-ethyl-6'-methyl-*N*-(1-methyl-2-methoxyethyl)acetanilide; metelilachlor; CGA-24705; Dual; Pennant. $C_{15}H_{22}ClNO_2$; mol wt 283.80. C 63.48%, H 7.81%, Cl 12.49%, N 4.94%, O 11.28%. Selective pre-emergence herbicide. Prepn: C. Vogel, R. Aebi, Ger. pat. **2,328,340;** *eidem*, U.S. pat. **3,937,730** (1973, 1976 to Ciba-Geigy). Metabolism: L. L. McGahen, J. M. Tiedje, *J. Agr. Food Chem.* **26**, 414 (1978).

Clear, odorless liquid. $bp_{0.001}$ 100°. n_D^{20} 1.5301. Vapor pressure at 20°: 1.3×10^{-5} mm Hg. Soly in water at 20°: 530 ppm. Sol in most organic solvents. LD_{50} orally in rats: 2780 mg/kg, *RTECS* **vol. 1**, R. J. Lewis, Sr., R. L. Tatken, Eds. (1979) p 50.

USE: Herbicide.

6231. Metolazone. *7-Chloro-1,2,3,4-tetrahydro-2-methyl-3-(2-methylphenyl)-4-oxo-6-quinazolinesulfonamide;* 7-chloro-1,2,3,4-tetrahydro-2-methyl-4-oxo-3-o-tolyl-6-quinazolinesulfonamide; 2-methyl-3-o-tolyl-6-sulfamyl-7-chloro-1,2,3,4-tetrahydro-4-quinazolinone; SR-720-22; Diulo; Metenix; Mykrox; Oldren; Xuret; Zaroxolyn. $C_{16}H_{16}ClN_3O_3S$; mol wt 365.84. C 52.53%, H 4.41%, Cl 9.69%, N 11.49%, O 13.12%, S 8.77%. Prepn: B. V. Shetty, U.S. pat. **3,360,518** (1967 to Wallace & Tiernan); B. V. Shetty *et al.*, *J. Med. Chem.* **13**, 886 (1970). Pharmacology: E. Belair *et al.*, *Arch. Int. Pharmacodyn. Ther.* **177**, 71 (1969); E. J. Belair, *Res. Comm. Chem. Pathol. Pharmacol.* **2**, 98 (1971). HPLC determn in urine: D. Farthing *et al.*, *J. Chromatog.* **534**, 228 (1990). Clinical evaluations as diuretic in chronic renal disease: R. Schoonees *et al.*, *N.Y. State J. Med.* **71**, 566 (1971); in hypertension: C. L. Curry *et al.*, *Clin. Ther.* **9**, 47 (1986); in congestive heart failure: A. Kiyingi *et al.*, *Lancet* **335**, 29 (1990).

Crystals from ethanol or butanol, mp 252-254°. LD_{50} in mice (mg/kg): > 5000 orally; > 1500 i.p. (Shetty, 1967).
THERAP CAT: Diuretic; antihypertensive.

6232. Metomidate. *1-(1-Phenylethyl)-1H-imidazole-5-carboxylic acid methyl ester; 1-(α-methylbenzyl)imidazole-5-carboxylic acid methyl ester;* methyl 1-(α-methylbenzyl)imidazole-5-carboxylate; methoxymol; methomidate. $C_{13}H_{14}N_2O_2$; mol wt 230.27. C 67.81%, H 6.13%, N 12.17%, O 13.90%. Injectable hypnotic related to etomidate, *q.v.* Prepn: **Belg.** pat. **662,474**; E. F. Godefroi, C. A. M. van der Eijcken, U.S. pat. **3,354,173** (1965, 1967 both to Janssen); E. F. Godefroi *et al.*, *J. Med. Chem.* **8**, 220 (1965). Anaesthetic effect of combination with azaperone, *q.v.*: J. F. F. Callear, J. F. E. Van Gestel, *Vet. Rec.* **92**, 284 (1973); of combination with fentanyl, *q.v.*: W. Erhardt, *J. Small Animal Pract.* **19**, 401 (1978); C. J. Green, *Lab. Animal* **15**, 171 (1981).

Hydrochloride, $C_{13}H_{14}N_2O_2$.HCl, *R-7315, Hypnodil*. Crystals from methanol-ether, mp 173-174°. LD_{50} i.v. in rats: 50 mg/kg (Godefroi).
THERAP CAT (VET): Hypnotic.

6233. Metopimazine. *1-[3-[2-(Methylsulfonyl)-10H-phenothiazin-10-yl]propyl]-4-piperidinecarboxamide; 1-[3-[2-*

(methylsulfonyl)phenothiazin-10-yl]propyl]isonipecotamide; 10-[3-(4-carbamoylpiperidino)propyl]-2-(methanesulfonyl)-phenothiazine; 1-[3-[2-(methylsulfonyl)phenothiazin-10-yl]propyl]-4-piperidinecarboxamide; EXP-999; RP-9965; Vogalene. $C_{22}H_{27}N_3O_3S_2$; mol wt 445.61. C 59.30%, H 6.11%, N 9.43%, O 10.77%, S 14.39%. Prepn: Jacob, Robert, **Ger. pat. 1,092,476** (to Rhône-Poulenc), C.A. **56**, 8723h (1962). Toxicity study: E. I. Goldenthal, Toxicol. Appl. Pharmacol. **18**, 185 (1971).

Solid, mp 170-171°. LD_{50} in male rats (mg/kg): 976 orally, 1080 s.c. (Goldenthal).
THERAP CAT: Antiemetic.

6234. Metopon. (5α)-4,5-Epoxy-3-hydroxy-5,17-dimethylmorphinan-6-one; methyldihydromorphinone. $C_{18}H_{21}NO_3$; mol wt 299.37. C 72.22%, H 7.07%, N 4.68%, O 16.03%. Prepn from thebaine: Small et al., J. Am. Chem. Soc. **58**, 1457 (1936). From dihydrocodeinone enol acetate: Small et al., J. Org. Chem. **3**, 204 (1938). Structure: Stork, Bauer, J. Am. Chem. Soc. **75**, 4373 (1953).

Needles from alc; sinters 235°, mp 243-245° (evac tube). $[\alpha]_D^{24}$ −141° (c = 1 in alc). Slightly sol in organic solvents. Hydrochloride, $C_{18}H_{21}NO_3 \cdot HCl$, crystals from alc, dec 315-318° (evac tube). $[\alpha]_D^{24}$ −105° (c = 1). Freely sol in water. Sparingly sol in alc. Slightly sol in chloroform. Very slightly sol in ether. Insol in benzene. A 1% aq soln has a pH of ∼5.0.
Note: This is a controlled substance (opiate) listed in the U.S. Code of Federal Regulations, Title 21 Part 1308.12 (1995).
THERAP CAT: Analgesic (narcotic).

6235. Metoprolol. 1-[4-(2-Methoxyethyl)phenoxy]-3-[(1-methylethyl)amino]-2-propanol; (±)-1-(isopropylamino)-3-[p-(β-methoxyethyl)phenoxy]-2-propanol; CGP-2175; H-93/26. $C_{15}H_{25}NO_3$; mol wt 267.37. C 67.38%, H 9.42%, N 5.24%, O 17.95%. β_1-Adrenergic blocker lacking intrinsic sympathomimetic activity. Prepn: A. E. Brandstrom et al., **Ger. pat. 2,106,209**; eidem, U.S. pat. **3,873,600** (1971, 1975 to AB Hässle). Pharmacology: B. Ablad et al., Life Sci. **12**, 107 (1973); Johansson, Eur. J. Pharmacol. **24**, 194 (1973). Effect on mortality in acute myocardial infarction: A. Hjalmarson et al., Lancet **2**, 823 (1981). Clinical trials in acute myocardial infarction: L. Rydén et al., N. Engl. J. Med. **308**, 614 (1983); Drugs **29**, Suppl. 1, 2-8 (1985). Toxicology: N.-O. Bodin et al., Acta Pharmacol. Toxicol. **36**, Suppl. V, 96 (1975). Review of pharmacology, therapeutic efficacy: P. Benfield et al., Drugs **31**, 376-429 (1986); B. Ablad et al., in Pharmacology of Antihypertensive Drugs, A. Scriabine, Ed. (Raven, New York, 1980) pp 247-262. Comprehensive description: J. R. Luch, Anal. Profiles Drug Subs. **12**, 325-356 (1983).

Succinate, $(C_{15}H_{25}NO_3)_2 \cdot C_4H_6O_4$, **Toprol-XL.**
Tartrate, $(C_{15}H_{25}NO_3)_2 \cdot C_4H_6O_6$, **Beloc, Betaloc, Lopressor, Lopresor, Prelis, Seloken, Selopral, Selo-Zok.** Soly (mg/ml) at 25°: water > 1000; methanol > 500; chloroform 496; acetone 1.1; acetonitrile 0.89; hexane 0.001. uv max (water): 223 nm (ε 23400). LD_{50} in female mice, male rats (mg/kg): 118, ∼90 i.v.; 2090, 3090 orally (Bodin).
THERAP CAT: Antihypertensive; antianginal; antiarrhythmic (class II).

6236. Metoquinone. $C_{20}H_{24}N_2O_4$; mol wt 356.42. C 67.40%, H 6.79%, N 7.86%, O 17.96%. $[C_6H_4(OH)NH-CH_3]_2 \cdot C_6H_4(OH)_2$. A salt-like compound of 1 mol hydroquinone and 2 mols p-monomethylaminophenol.
White crystals. mp ∼135° with decompn and blackening. Sol in 100 parts water; less sol in alcohol, acetone; very slightly sol in benzene, chloroform, ether.
USE: As a photographic developer.

6237. Metoserpate. (3β,16β,17α,18α,20α)-11,17,18-Trimethoxyyohimban-16-carboxylic acid methyl ester; O-methyl-18-epireserpic acid methyl ester; 18-epi-O-methylreserpic acid methyl ester; methyl 18-epi-O-methylreserpate; methyl O-methyl-18-epireserpate. $C_{24}H_{32}N_2O_5$; mol wt 428.53. C 67.27%, H 7.53%, N 6.54%, O 18.67%. Prepn: Robison et al., Experientia **17**, 14 (1961), J. Am. Chem. Soc. **83**, 2694 (1961); Robison, Lucas; and Ziegler, U.S. pats. **3,120,532; 3,126,390** and **3,151,117** (1964) and **3,264,303** (1966) (all to Ciba).

Crystals from benzene + cyclohexane, dec 248.5-251.5°, also reported as 239-241°. $[\alpha]_D^{26}$ −37.5° (CHCl₃).
Hydrochloride, $C_{24}H_{32}N_2O_5 \cdot HCl$, **Su-9064, Avicalm, Pacitran (obsolete).** Solid, dec 240-242°. Sol in water.
THERAP CAT (VET): Sedative.

6238. Metralindole. 2,4,5,6-Tetrahydro-9-methoxy-4-methyl-1H-3,4,6a-triazafluoranthene; 3-methyl-8-methoxy-3H-1,2,5,6-tetrahydropyrazino[1,2,3-ab]-β-carboline. $C_{15}H_{17}N_3O$; mol wt 255.32. C 70.56%, H 6.71%, N 16.46%, O 6.27%. Heterocyclic β-carboline derivative used as an antidepressant. Prepn: R. G. Glushkov et al., **Ger. pat. 2,357,-320**; eidem, U.S. pat. **4,088,647** (1974, 1978 both to All Union Khim.-Farm Instit. USSR); eidem, Khim.-Farm. Zh. **16**, 1054 (1982), C.A. **98**, 53831b (1983). Psychotropic activity, toxicity: M. D. Mashkovsky et al., **Ger. pat. 2,356,-091**; eidem, U.S. pat. **3,959,470** (1974, 1976 both to All Union Khim.-Farm. Instit. USSR). Pharmacology: N. I. Andreeva, M. D. Mashkovsky, Farmakol. Toxicol. **43**, 133 (1980). Effect on the EEG of the brain cortex: L. F. Roshchina, ibid. **349**. Effect on adrenergic neurotransmission: L. V. Panasiuk et al., ibid. **45**, 13 (1982). Clinical studies: M. D. Mashkovsky, N. I. Andreeva, Zh. Nevropatol. Psikhiatr. **84**, 410 (1984).

Crystals, mp 164-165°.

Hydrochloride, $C_{15}H_{17}N_3O \cdot HCl$, *Incazan*. Off-white odorless powder with bitter taste. mp 305-308°. Sol in water. LD_{50} orally in mice: 445 mg/kg (Mashkovsky, 1976).

THERAP CAT: Antidepressant.

6239. Metribuzin. *4-Amino-6-(1,1-dimethylethyl)-3-(methylthio)-1,2,4-triazin-5(4H)-one; 4-amino-6-tert-butyl-3-(methylthio)-as-triazin-5(4H)-one;* Bay 94337; Lexone; Sencor; Sencoral. $C_8H_{14}N_4OS$; mol wt 214.29. C 44.84%, H 6.59%, N 26.15%, O 7.47%, S 14.96%. Pre- and post-emergence triazone herbicide. Prepn: **Belg.** pat. **697,083;** K. Westphal *et al.,* U.S. pat. **3,671,523** (1967, 1972 both to Bayer AG). Physicochemical properties, toxicology and herbicidal activity: L. Eue, *Pflanzenschutz-Nachr.* **25,** 175 (1972). Toxicity data: E. Loeser, G. Kimmerle, *ibid.* 186. HPLC determn in plant tissues: C. E. Parker, G. H. Degen, *J. Liq. Chromatog.* **6,** 725 (1983).

White crystalline solid, mp 125-126.5°. d_4^{20} 1.28. Vapor pressure at 20°: $< 10^{-5}$ mm Hg. Sol in methanol, ethanol. Soly in water: 1200 ppm. LD_{50} orally in rats: 2200 mg/kg; LC_{50} in rainbow trout: > 10 ppm (Loeser, Kimmerle).

USE: Herbicide.

6240. Metrizamide. *2-[[3-(Acetylamino)-5-(acetyl-methylamino)-2,4,6-triiodobenzoyl]amino]-2-deoxy-D-glucose; 2-[3-acetamido-2,4,6-triiodo-5-(N-methylacetamido)benz-amido]-2-deoxy-D-glucose;* Win 39103; Amipaque. $C_{18}H_{22}I_3N_3O_8$; mol wt 789.10. C 27.40%, H 2.81%, I 48.25%, N 5.33%, O 16.22%. X-ray contrast agent related to metrizoic acid, *q.v.* Prepn: H. O. Torsten *et al.,* **Ger.** pat. **2,031,724;** *eidem,* U.S. pat. **3,701,-771** (1971, 1972 both to Nyegaard). Metabolism: J. G. Johansen, S. Kolmannskog, *Invest. Radiol.* **13,** 93 (1978). Pharmacology: T. W. Morris *et al., ibid.* 74; G. F. Dibona, *Proc. Soc. Exp. Biol. Med.* **157,** 453 (1978). Series of articles on clinical use: *Ann. Radiol.* **22,** 195-206 (1979). Toxicity studies: S. Salvesen, *Radiology* **123,** 241 (1977); M. Sovak *et al., ibid.* 242; P. Aspelin, *Invest. Radiol.* **11,** 309 (1976). *Review:* L. Hol *et al.,* in *Pharmacological and Biochemical Properties of Drug Substances* vol. 1, M. E. Goldberg, Ed. (Am. Pharm. Assoc., Washington, DC, 1977) pp 387-412.

White crystals from isopropyl alcohol, mp 230° (dec), after drying in vacuo at room temp and 70°. $[\alpha]_D^{20}$ +18.0° (c = 10% in 0.1N HCl; equilibrated with respect to mutarota-

tion). Very sol in water (50% w/v) at room temp. LD_{50} i.v. in mice: 15 g/kg (Torsten); 18.6 g/kg (Salveson); 11.5 g/kg (Sovak); 17.3 g/kg (Aspelin).

THERAP CAT: Diagnostic aid (radiopaque medium).

6241. Metrizoic Acid. *3-(Acetylamino)-5-(acetylmeth-ylamino)-2,4,6-triiodobenzoic acid; 3-acetamido-2,4,6-triiodo-5-(N-methylacetamido)benzoic acid;* N-methyl-3,5-diacet-amido-2,4,6-triiodobenzoic acid. $C_{12}H_{11}I_3N_2O_4$; mol wt 627.94. C 22.95%, H 1.77%, I 60.63%, N 4.46%, O 10.19%. Prepn: Pitré, Fumigalli, *Farmaco Ed. Sci.* **17,** 340 (1962); **Brit.** pat. **973,881** (1964 to Nyegaard), *C.A.* **62,** 16139g (1965). Series of articles on use in separation of leukocytes from blood and bone marrow: A. Boyum, *Scand. J. Clin. Lab. Invest.* **21,** Suppl. 97, 7-106 (1968).

Crystals, mp 281-282°.

Sodium salt, *metrizoate sodium, Isopaque, Triosil.*

THERAP CAT: Diagnostic aid (radiopaque medium).

6242. Metronidazole. *2-Methyl-5-nitroimidazole-1-ethanol;* 1-(2-hydroxyethyl)-2-methyl-5-nitroimidazole; 1-(β-ethylol)-2-methyl-5-nitro-3-azapyrrole; Bayer 5630; RP-8823; Arilin; Clont; Deflamon; Flyzol; Flagyl; Fossyol; Ginelavir; Klion; MetroGel; Metrolag; Metrolyl; Metrotop; Orvagil; Rathimed; Sanatrichom; Trichazol; Tricocet; Tri-chocide; Tricho Cordes; Tricho-Gynaedron; Trivazol; Vagi-len; Vagimid; Zadstat. $C_6H_9N_3O_3$; mol wt 171.16. C 42.11%, H 5.30%, N 24.55%, O 28.04%. Prepn: Jacob *et al.,* U.S. pat. **2,944,061** (1960 to Rhône-Poulenc); Cossar *et al., Arzneimittel-Forsch.* **16,** 23 (1966). Activity studies: Bock, *ibid.* **11,** 587 (1961). Metabolism: Ings *et al., Biochem. Pharmacol.* **15,** 515 (1966). Comprehensive description: L. L. Wearley, G. D. Anthony, *Anal. Profiles Drug Subs.* **5,** 327-344 (1976). Review of activity, pharmacokinetics and use in anaerobic infection: R. N. Brogden *et al., Drugs* **16,** 387-417 (1978). Symposium: *Surgery* **93,** 123-234 (1983).

Cream-colored crystals, mp 158-160°. Soly at 20° (g/100 ml): water 1.0; ethanol 0.5; ether <0.05; chloroform <0.05. Sparingly sol in DMF. Sol in dil acids. pH of satd aq soln: 5.8.

Hydrochloride, $C_6H_9N_3O_3 \cdot HCl$, *SC-326421, Flagyl I.V.*

Note: Metronidazole may reasonably be anticipated to be a carcinogen: *Seventh Annual Report on Carcinogens* (PB95-109781, 1994) p 257.

THERAP CAT: Antiprotozoal (Trichomonas).

THERAP CAT (VET): Antiprotozoal (Trichomonas); anti-amebic; antibacterial.

6243. Metron S. *6-Methyl-N-(1-methylethyl)-2-heptan-amine; N-isopropyl-1,5-dimethylhexylamine;* 2-isopropyl-amino-6-methylheptane; Metron (Japanese). $C_{11}H_{25}N$; mol wt 171.33. C 77.12%, H 14.71%, N 8.18%. Prepn: Ota *et al., Yakugaku Zasshi* **80,** 1153 (1960), *C.A.* **55,** 3825d (1961).

Oil, bp$_{23}$ 84-85°.
THERAP CAT: Antihistaminic.

6244. Metsulfuron Methyl. *2-[[[[(4-Methoxy-6-methyl-1,3,5-triazin-2-yl)amino]carbonyl]amino]sulfonyl]benzoic acid methyl ester;* N-[(4-methoxy-6-methyl-1,3,5-triazin-2-yl)aminocarbonyl]-2-methoxycarbonylbenzenesulfonamide; DPX-T6376; Ally; Allie. C$_{14}$H$_{15}$N$_5$O$_6$S; mol wt 381.37. C 44.09%, H 3.96%, N 18.36%, O 25.17%, S 8.41%. Sulfonyl urea used as pre- and postemergence herbicide. Prepn and herbicidal activity: G. Levitt, Eur. pat. Appl. **7687** (1980 to Du Pont); *idem,* U.S. pat. **4,383,113** (1983 to Du Pont). Comparison with other herbicides in cereals: A. Aamisepp, *Weeds Weed Control* **26**, 26 (1985). Efficacy in control of gorse: A. I. Popay *et al., Proc. 38th N.Z. Weed Pest Control Conf.* **1985**, 94. Review of properties, herbicidal activity: P. Lefebvre, *Def. Veg.* **38**, 331 (1984).

White crystals, mp 163-166°. Vapor pressure at 25°: 5.8×10^{-5} mm Hg. Soly in water: 0.27 g/l at pH 4.59; 9.5 g/l at pH 6.11. LD$_{50}$ in male and female rats, in male and female rabbits (mg/kg): > 5000, > 2000 orally. LC$_{50}$ (96 hour) in American perch, rainbow trout: 150 ppm. LD$_{50}$ in wild ducks: 2510 mg/kg (Lefebvre).
USE: Herbicide.

6245. Meturedepa. *[Bis(2,2-dimethyl-1-aziridinyl)phosphinyl]carbamic acid ethyl ester;* ethyl [bis(2,2-dimethyl-1-aziridinyl)phosphinyl]carbamate; ethyl N-[bis(2,2-dimethylethylenimido)phosphoro]carbamate; AB-132; Turloc. C$_{11}$H$_{22}$N$_3$O$_3$P; mol wt 275.29. C 47.99%, H 8.06%, N 15.26%, O 17.44%, P 11.25%. Prepn: **Brit.** pat. **911,764** (1962 to Armour), *C.A.* **60**, 1701h (1964).

Crystals, mp 119-121°.
THERAP CAT: Antineoplastic.

6246. Metyrapone. *2-Methyl-1,2-di-3-pyridyl-1-propanone;* 2-methyl-1,2-bis(3-pyridyl)-1-propanone; methopyrapone; mepyrapone; metopyrone; methbipyranone; Su-4885; Metopirone; Metroprione. C$_{14}$H$_{14}$N$_2$O; mol wt 226.28. C 74.31%, H 6.24%, N 12.38%, O 7.07%. Prepn: Chart *et al., Experientia* **14**, 151 (1958); Bencze, Allen, *J. Am. Chem. Soc.* **81**, 4015 (1959); U.S. pat. **2,923,710** (1960 to Ciba).

Crystals from ether and pentane, mp 50-51°.
THERAP CAT: Diagnostic aid (pituitary function).

6247. Metyridine. *2-(β-Methoxyethyl)pyridine;* methyridine; Dekelmin; Promintic. C$_8$H$_{11}$NO; mol wt 137.18. C 70.04%, H 8.08%, N 10.21%, O 11.66%. Prepn: Arnall, Greenhalgh, **Brit.** pat. **889,748** (1962 to ICI and Midland Tar Distillers); U.S. pat. **3,223,710** (1965).

Sweet smelling liquid, bp$_{17}$ 94-96°. d^{20} 0.988. n$_D^{20}$ 1.4975. pKa = 5.5. Very sol in water and common solvents.
Sulfate, *Mintic.*
Hydrochloride, mp 104-105°.
THERAP CAT (VET): Anthelmintic.

6248. Metyrosine. *α-Methyl-L-tyrosine;* α-methyl-p-tyrosine; α-methyltyrosine; 4-hydroxy-α-methylphenylalanine; α-methyl-3-(p-hydroxyphenyl)alanine; metirosine; L-α-MT; α-MPT; MK-781; Demser. C$_{10}$H$_{13}$NO$_3$; mol wt 195.22. C 61.53%, H 6.71%, N 7.17%, O 24.59%. An inhibitor of the first and rate-limiting reaction in catecholamine biosynthesis, the hydroxylation of tyrosine to dopa. Prepn: **Neth.** pat. **6,607,757** (1966 to Merck & Co.), *C.A.* **67**, 91108p (1967). Prepn of DL-form: Stein *et al., J. Am. Chem. Soc.* **77**, 700 (1955); Potts, *J. Chem. Soc.* **1955**, 1632; Pfister, Stein, U.S. pat. **2,868,818** (1959 to Merck & Co.); Saari, *J. Org. Chem.* **32**, 4074 (1967). Metabolism and biochemical and pharmacologic effects in man: Engelman *et al., J. Clin. Invest.* **47**, 568, 577 (1968). Review of pharmacology and clinical use: R. N. Brogden *et al., Drugs* **21**, 81-89 (1981).

Crystals, mp 310-315°.
DL-Form, crystals from water, dec 320° (Stein *et al., loc. cit.),* also reported as dec 330-332° (Potts, *loc. cit.).* Soly in water at room temp: 0.57 mg/ml.
THERAP CAT: Tyrosine hydroxylase inhibitor; as antihypertensive in pheochromocytoma.

6249. Mevaldic Acid. *3-Hydroxy-3-methyl-5-oxopentanoic acid;* 3-hydroxy-3-methylglutaraldehydic acid. C$_6$H$_{10}$O$_4$; mol wt 146.14. C 49.31%, H 6.90%, O 43.79%. Prepn by acid hydrolysis of 3-hydroxy-3-methyl-5,5-dimethoxypentanoic acid: Shunk *et al., J. Am. Chem. Soc.* **79**, 3294 (1957). Exists in equilibrium with *β,δ-dihydroxy-β-methyl-δ-valerolactone.*

Unstable and reactive compd. Isolated in soln only.

6250. Mevalonic Acid. *3,5-Dihydroxy-3-methylpentanoic acid; 3,5-dihydroxy-3-methylvaleric acid;* β,δ-dihydroxy-β-methylvaleric acid; hiochic acid. C$_6$H$_{12}$O$_4$; mol wt 148.16. C 48.64%, H 8.16%, O 43.20%. Precursor in the biosynthesis of cholesterol. Occurs in equilibrium with the δ-lactone. Isoln from distillers' solns: Wright *et al., J. Am. Chem. Soc.* **78**, 5273 (1956). Synthesis: Wolf *et al., ibid.* **79**, 1486 (1957); Hoffmann *et al., ibid.* 2316; Eggerer *et al., Ann.* **608**, 71 (1957); U.S. pats. **2,915,398; 2,915,531-3; 2,915,551** (1959 to Merck & Co.); Shunk *et al.,* U.S. pat. **3,014,963** (1961 to Merck & Co.); Hulcher, Hosick, U.S. pat. **3,119,842** (1964); F. C. Huang *et al., J. Am. Chem. Soc.* **97**, 4144 (1975). Resolution: Shunk *et al., ibid.* **79**, 3294 (1957); U.S. pat. **2,945,059** (1960 to Merck & Co.). Synthesis of the δ-lactone: R. H. Cornforth *et al., Tetrahedron* **18**, 1351 (1962); E. L. Eliel, K. Soai, *Tetrahedron Letters* **22**, 2859 (1981); A. Banerji, G. P. Kalena, *Syn. Commun.* **12**, 225 (1982). Review of organic and biochemistry: Wagner, Folkers, *Endeavour* **20**, 177-187 (Oct. 1961). Prepn of 5-phosphate: Robinson, Wittreich, U.S. pat. **3,014,057** (1961 to Merck & Co.). Review of role of mevalonic acid in sterol biosynthesis: G. J. Schroepfer, Jr., *Ann. Rev. Biochem.* **50**, 585-621 (1981); E. Caspi, *Tetrahedron* **42**, 3-50 (1986).

Oily liquid. Very sol in water, but also sol in organic solvents, especially polar organic solvents.

δ-Lactone, $C_6H_{10}O_3$, *mevalolactone, mevalonic lactone, β-hydroxy-β-methyl-δ-valerolactone.* Hygroscopic crystals, mp 28°.

Benzhydrylamide, $C_{19}H_{23}NO_3$, crystals from benzene + petr ether, mp 96-97°. $[\alpha]_D^{20}$ −2.0° (c = 2 in ethanol).

6251. Mevastatin. *[1S-[1α(R*),7β,8β(2S*,4S*),8aβ]]-2-Methylbutanoic acid 1,2,3,7,8,8a-hexahydro-7-methyl-8-[2-(tetrahydro-4-hydroxy-6-oxo-2H-pyran-2-yl)ethyl]-1-naphthalenyl ester;* 7-[1,2,6,7,8,8a-hexahydro-2-methyl-8-(methylbutyryloxy)naphthyl]-3-hydroxyheptan-5-olide; 2β-methyl-8α-(2-methyl-1-oxobutyl)mevinic acid lactone; compactin; 6-demethylmevinolin; CS-500; ML-236 B. $C_{23}H_{34}O_5$; mol wt 390.52. C 70.74%, H 8.78%, O 20.48%. Fungal metabolite which is a potent inhibitor of HMG-CoA reductase, the rate controlling enzyme in cholesterol biosynthesis. Isoln from *Penicillium citrinum:* A. Endo *et al.*, Ger. pat. **2,524,355** corresp to U.S. pat. **3,983,140** (1975, 1976 to Sankyo). Isoln from *P. brevicompactum*, crystal and molecular structure: A. G. Brown *et al.*, *J. Chem. Soc. Perkin Trans I* **1976**, 1165. Inhibition of HMG-CoA reductase activity: A. Endo *et al.*, *FEBS Letters* **72**, 323 (1976); M. S. Brown *et al.*, *J. Biol. Chem.* **253**, 1121 (1978). Therapeutic effects in primary hypercholesterolemia: A. Yamamoto *et al.*, *Atherosclerosis* **35**, 259 (1980). Total synthesis: N. Y. Wang *et al.*, *J. Am. Chem. Soc.* **103**, 6538 (1981); M. Hirama, M. Uei, *ibid.* **104**, 4251 (1982); N. N. Girotra, N. L. Wendler, *Tetrahedron Letters* **23**, 5501 (1982); C.-T. Hsu *et al.*, *J. Am. Chem. Soc.* **105**, 593 (1983); P. A. Grieco *et al.*, *ibid.* 1403; D. L. J. Clive *et al.*, *J. Am. Chem. Soc.* **110**, 6914 (1988). Review of syntheses: T. Rosen, C. H. Heathcock, *Tetrahedron* **42**, 4909-4951 (1986). Review of mevastatin and related compounds: A. Endo, *J. Med. Chem.* **28**, 401-405 (1985).

Crystals from aq ethanol, mp 152°. $[\alpha]_D^{22}$ +283° (c = 0.48 in acetone). uv max: 230, 237, 246 nm (log ε 4.28, 4.30, 4.11).

6252. Mevinphos. *3-[(Dimethoxyphosphinyl)oxy]-2-butenoic acid methyl ester;* 3-hydroxycrotonic acid methyl ester dimethyl phosphate; 1-methoxycarbonyl-1-propen-2-yl dimethyl phosphate; methyl 3-(dimethoxyphosphinyloxy)crotonate; O,O-dimethyl 1-carbomethoxy-1-propen-2-yl phosphate; 2-carbomethoxy-1-methylvinyl dimethyl phosphate; ENT-22374; OS-2046; Phosdrin. $C_7H_{13}O_6P$; mol wt 224.15. C 37.51%, H 5.85%, O 42.83%, P 13.82%. Cholinesterase inhibitor. Prepn: A. R. Stiles, U.S. pat. **2,685,552** (1954 to Shell). Activity: R. A. Corey *et al.*, *J. Econ. Entomol.* **46**, 386 (1953). Configuration of isomers: Fukuto *et al.*, *J. Org. Chem.* **26**, 4620 (1961). Metabolism and degradn: K. I. Beynon *et al.*, *Residue Rev.* **47**, 55 (1973). Toxicity study: T. B. Gaines, *Toxicol. Appl. Pharmacol.* **14**, 515 (1969).

(E) - isomer

Commercial product is a yellow liq mixture of *(E)-* (also referred to as *cis-*) and *(Z)-* (also referred to as *trans-*) isomers, the *(E)*-isomer being about 100 times more potent in insecticidal activity than the *(Z)*-isomer: Casida *et al.*, *J. Agr. Food Chem.* **4**, 236 (1956). d_4^{20} 1.25. n_D^{20} 1.4494. bp_1 106-107.5°. Misc with water, acetone, benzene, carbon tetrachloride, chloroform, ethyl and isopropyl alcohols, toluene and xylene. One gram dissolves in 20 ml carbon disulfide and 20 ml kerosene. Practically insol in hexane. LD_{50} in female, male rats (mg/kg): 3.7, 6.1 orally; 4.2, 4.7 dermally (Gaines).

Caution: Potential symptoms of overexposure are miosis; rhinorrhea; headache; tight chest, wheezing, laryngeal spasm, salivation and cyanosis; anorexia, nausea, vomiting, abdominal cramps and diarrhea; paralysis; ataxia, convulsions; low blood pressure, cardiac irregularities; irritation of skin and eyes. *See NIOSH Pocket Guide to Chemical Hazards* (DHHS/NIOSH 90-117, 1990) p 180.

USE: Insecticide.

6253. Mexacarbate. *4-(Dimethylamino)-3,5-dimethylphenol methylcarbamate (ester); methylcarbamic acid 4-(dimethylamino)-3,5-xylyl ester;* 4-dimethylamino-3,5-xylyl methylcarbamate; Zectran. $C_{12}H_{18}N_2O_2$; mol wt 222.29. C 64.84%, H 8.16%, N 12.60%, O 14.40%. Outline of prepn starting with 3,5-xylenol. Marquardt in *Analyt. Methods for Pesticides* **2** (Academic Press, New York, 1964) pp 581-596. Metabolic studies: Roberts *et al.*, *J. Agr. Food Chem.* **17**, 107 (1969). Toxicity studies in insects: El-Aziz *et al.*, *J. Econ. Entomol.* **62**, 318 (1969); in lab animals: T. B. Gaines, *Toxicol. Appl. Pharmacol.* **14**, 515 (1969); in birds, other wildlife: Tucker, Crabtree, *ibid.* 1307; Tucker, Haegel, *Toxicol. Appl. Pharmacol.* **20**, 57 (1971).

Crystals, mp 85°. Vapor pressure <0.1 mm at 139°. Practically insol in water (0.01% at 25°). Freely sol in acetone (162.3% w/w), acetonitrile (142% w/w), methylene chloride (120.6% w/w), ethanol (116.0% w/w), benzene (102.0% w/w). Sparingly sol in petroleum solvents (1.05% w/w). LD_{50} in male, female rats (mg/kg): 37, 25 orally (Gaines).

USE: Pesticide for use in the control of snails and slugs.

6254. Mexazolam. *10-Chloro-11b-(2-chlorophenyl)-2,3,7,11b-tetrahydro-3-methyloxazolo[3,2-d][1,4]benzodiazepin-6(5H)-one;* 10-chloro-11b-(o-chlorophenyl)-2,3,7,11b-tetrahydro-3-methyloxazolo[3,2-d][1,4]benzodiazepin-6(5H)-one; CS-386; Melex. $C_{18}H_{16}Cl_2N_2O_2$; mol wt 363.24. C 59.52%, H 4.44%, Cl 19.52%, N 7.71%, O 8.81%. Prepn from 2-amino-2',5-dichlorobenzophenone: R. Tachikawa *et al.*, Ger. pat. **1,812,252**; *eidem*, U.S. pat. **3,772,371** (1969, 1973 to Sankyo). Synthesis via Schiff base, physical properties: T. Miyadera *et al.*, *J. Med. Chem.* **14**, 520 (1971). Acid-base equilibrium: M. Ikeda, T. Nagai, *Chem. Pharm. Bull.* **30**, 3810 (1982). Effect on socially induced suppression in monkeys: T. Kamioka *et al.*, *Psychopharmacology (Berlin)* **52**, 17 (1977); on gastric contraction in cats: N. Iwata *et al.*, *J. Pharmacobio-Dyn.* **3**, 413 (1980). Toxicity studies: H. Masuda *et al.*, *Sankyo Kenkyusho Nempo* **30**, 175 (1978), *C.A.* **90**, 180217d.

Crystals, mp 172-175°. pKa 6.69. LD$_{50}$ in male, female mice; male, female rats (mg/kg): 4687, 4571, 810, 4500 orally; > 6000, > 6000, > 4000, > 4000 i.p. or s.c. (Masuda).

THERAP CAT: Anxiolytic.

6255. Mexenone. *(2-Hydroxy-4-methoxyphenyl)(4-methylphenyl)methanone; 2-hydroxy-4-methoxy-4'-methylbenzophenone;* benzophenone-10; Uvistat. C$_{15}$H$_{14}$O$_3$; mol wt 242.27. C 74.36%, H 5.82%, O 19.81%. Prepn: Hardy, Forster, U.S. pat. 2,773,903 (1956 to Am. Cyanamid).

THERAP CAT: Ultraviolet screen.

6256. Mexicain. A cysteine proteinase obtained from the fruit of *Pileus mexicanus* probably identical with *Leucopremna mexicana* (A.DC.) Standl., *Caricaceae:* Castaneda-Agullo *et al., Science* **96,** 365 (1942); *J. Biol. Chem.* **159,** 751 (1945). Immunological studies: Estrada-Parra *et al., Rev. Latinoamer. Microbiol. Parasitol.* **1969,** 145, *C.A.* **72,** 1744u (1970). Mol wt about 31,100; lysine and alanine believed to be the *N*- and *C*-terminal amino acids resp: Soriano *et al., Rev. Latinoamer. Quim.* **1975,** 143, *C.A.* **84,** 56489t (1976).
Similar to but not identical with papain. Mexicain presents some advantage over papain, since it is more stable in solns and does not require cysteine. Isoelectric pt 9.12.

6257. Mexiletine. *1-(2,6-Dimethylphenoxy)-2-propanamine;* 1-(2,6-xylyloxy)-2-propylamine; 1-(2',6'-dimethylphenoxy)-2-aminopropane; 1-methyl-2-(2,6-xylyloxy)ethylamine. C$_{11}$H$_{17}$NO; mol wt 179.26. C 73.70%, H 9.56%, N 7.81%, O 8.93%. Prepn: **Fr.** pat. **1,551,055;** H. Köppe *et al.,* U.S. pat. **3,659,019** (1968, 1972 to Boehringer, Ing.); *eidem,* **S. Afr.** pat. **69 03,772** (1970 to Boehringer, Ing.), *C.A.* **73,** 120307j (1970). Pharmacology: B. N. Singh, E. M. Vaughan Williams, *Brit. J. Pharmacol.* **44,** 1 (1972); J. D. Allen *et al., ibid.* **45,** 561 (1972). Characterization of metabolites by mass spectral and ^{13}C-NMR: K. N. Scott *et al., Drug Metab. Dispos.* **1,** 506 (1973). Toxicological study: A. Kast *et al., Iyakuhin Kenkyu* **13,** 922 (1982), *C.A.* **97,** 192878q (1982). Review of pharmacology and therapeutic efficacy: C. Y. C. Chew *et al., Drugs* **17,** 161-181 (1979). Comprehensive description: M. A. Abounassif *et al.* in *Analytical Profiles of Drug Substances* **vol. 20,** K. Florey, Ed. (Academic Press, New York, 1991) pp 433-474.

Hydrochloride, C$_{11}$H$_{17}$NO.HCl, *Ko-1173, Katen, Mexitil, Ritalmex.* White to off-white crystals from ethanol-ether, mp 203-205°. pKa: 9.0. LD$_{50}$ in male, female rats, mice, rabbits (mg/kg): 350, 400, 310, 400, 180, 160 orally; in male, female rats, mice (mg/kg): 27, 30, 43, 50 i.v. (Kast).
THERAP CAT: Antiarrhythmic (class IB).

6258. Mezereum. Mezereon; olive spurge; dwarf bay; magell; paradise plant; spurge flax; wild pepper. Dried bark

(also seed, though not official) from aerial portions of *Daphne mezereum* L., *D. gnidium* L., and *D. laureola* L., *Thymelaeaceae. Habit.* New England, Canada, mountainous Europe, Siberia. *Constit.* Bark: Mezerein and acrid resin, daphnin, umbelliferone, fixed oil. Seed: Fixed oil, acrid resin.
THERAP CAT: Vesicant.

6259. Mezlocillin. *[2S-[2α,5α,6β(S*)]]-3,3-Dimethyl-6-[[[[[3-(methylsulfonyl)-2-oxo-1-imidazolidinyl]carbonyl]-amino]phenylacetyl]amino]-7-oxo-4-thia-1-azabicyclo[3.2.0]-heptane-2-carboxylic acid;* 6R-[2-[3-(methylsulfonyl)-2-oxo-1-imidazolidine carboxamido]-2-phenylacetamido]penicillanic acid; Bay f 1353. C$_{21}$H$_{25}$N$_5$O$_8$S$_2$; mol wt 539.59. C 46.74%, H 4.67%, N 12.98%, O 23.72%, S 11.89%. Semisynthetic, broad-spectrum antibiotic related to penicillin and azlocillin, q.q.v. Prepn: W. Schroeck *et al.,* **Ger.** pat. **2,318,-955** (1974 to Bayer), *C.A.* **82,** 31313b (1975). *In vitro* study: G. P. Bodey, T. Pan, *Antimicrob. Ag. Chemother.* **11,** 74 (1977). Pharmacokinetics: H. Lode *et al., Infection* **5,** 163 (1977); S. J. Pancoast, H. C. Neu, *Clin. Pharmacol. Ther.* **24,** 108 (1978). Clinical pharmacology: B. F. Issell *et al., Antimicrob. Ag. Chemother.* **13,** 180 (1978). Tissue concentration and efficacy: E. Helwing *et al., Med. Klin.* **74,** 112 (1979). Series of articles on antibacterial activity, pharmacology, and clinical trials: *Arzneimittel-Forsch.* **29,** 1915-2032 (1979); *Infection* **10,** Suppl. 3, S121-S266 (1982); *J. Antimicrob. Chemother.* **11,** Suppl. C, 1-108 (1983).

Sodium salt monohydrate, C$_{21}$H$_{24}$N$_5$NaO$_8$S$_2$.H$_2$O, *Baycipen, Baypen, Mezlin.* Pale yellow cryst; sol in water, methanol, DMF. Insol in acetone and ethanol.
THERAP CAT: Antibacterial.

6260. Mianserin. *1,2,3,4,10,14b-Hexahydro-2-methyl-dibenzo[c,f]pyrazino[1,2-a]azepine;* 2-methyl-1,2,3,4,10,14b-hexahydro-2H-pyrazino[1,2-f]morphanthridine. C$_{18}$H$_{20}$N$_2$; mol wt 264.37. C 81.78%, H 7.63%, N 10.60%. Serotonin receptor antagonist. Prepn: **Neth.** pat. **Appl. 6,603,256;** van der Burg, Delobelle, U.S. pat. **3,534,041** (1967, 1970, both to Organon); van der Burg *et al., J. Med. Chem.* **13,** 35 (1970). Pharmacology: Saxena *et al., Eur. J. Pharmacol.* **13,** 295 (1971); Vargaftig *et al., ibid.* **16,** 336 (1971); Van Riezen, *Arch. Int. Pharmacodyn. Ther.* **198,** 256 (1972). Molecular and crystal structure: van Rij, Feil, *Tetrahedron* **29,** 1891 (1973). Review of pharmacology and therapeutic efficacy in depressive illness: R. N. Brogden *et al., Drugs* **16,** 273-301 (1978). 5HT-receptor binding study: B. S. Alexander, M. D. Wood, *J. Pharm. Pharmacol.* **39,** 664 (1987). *Review:* H. van Riezen *et al.,* in *Pharmacological and Biochemical Properties of Drug Substances* **vol. 3,** M. E. Goldberg, Ed. (Am. Pharm. Assoc., Washington, DC, 1981) pp 56-93; *Brit. J. Clin. Pharmacol.* **15,** Suppl. 2, S141-S342 (1983).

Hydrochloride, C$_{18}$H$_{20}$N$_2$.HCl, *GB-94, Org-GB-94, Athymil, Bolvidon, Lantanon, Norval, Tetramide, Tolvin, Tolvon.*

mp 282-284°. LD$_{50}$ in male, female mice (mg/kg): 365, 390 orally; 32.5, 31.0 i.v. (van Riezen).

THERAP CAT: Antidepressant.

6261. Mibefradil. *(1S-cis)-Methoxyacetic acid 2-[2-[[3-(1H-benzimidazol-2-yl)propyl]methylamino]ethyl]-6-fluoro-1,2,3,4-tetrahydro-1-(1-methylethyl)-2-naphthalenyl ester;* (1S,2S)-2-[2-[[3-(2-benzimidazolyl)propyl]methylamino]ethyl]-6-fluoro-1,2,3,4-tetrahydro-1-isopropyl-2-naphthyl methoxyacetate. C$_{29}$H$_{38}$FN$_3$O$_3$; mol wt 495.64. C 70.28%, H 7.73%, F 3.83%, N 8.48%, O 9.68%. Calcium channel blocker. Prepn: Q. Branca *et al.,* **Eur.** pat. **Appl. 268,148;** *eidem,* **U.S.** pat. **4,808,605** (1988, 1989 both to Hoffmann-La Roche). Pharmacology: F. Hefti *et al., Arzneimittel-Forsch.* **40,** 417 (1990). Calcium channel selectivity: G. Mehrke *et al., J. Pharmacol. Exp. Ther.* **271,** 1483 (1994). Metabolism in rats: H. R. Wiltshire *et al., Xenobiotica* **22,** 837 (1992). Clinical evaluation in stable angina pectoris: M. C. M. Portegies *et al., J. Cardiovasc. Pharmacol.* **18,** 746 (1991); in hypertension: R. Schmitt *et al., Clin. Pharmacol. Ther.* **52,** 314 (1992). Review of pharmacology, toxicology and clinical evaluation: J.-P. Clozel *et al., Cardiovasc. Drug Rev.* **9,** 4-17 (1991).

Dihydrochloride, *Ro-40-5967, Posicor.* Bitter tasting, odorless, white crystalline powder, mp 128°. Chemically stable, light insensitive and water soluble. pKa: 4.8; 5.5.

THERAP CAT: Antihypertensive.

6262. Mibolerone. *(7α,17β)-17-Hydroxy-7,17-dimethyl-estr-4-en-3-one;* 7α,17α-dimethyl-19-nortestosterone; U-10997; Cheque; Matenon. C$_{20}$H$_{30}$O$_2$; mol wt 302.46. C 79.42%, H 10.00%, O 10.58%. Synthetic anabolic steroid related to testosterone, *q.v.* Prepn: **Belg.** pat. **610,385** corresp to J. C. Babcock, J. A. Campbell, **U.S.** pat. **3,341,-557** (1962, 1967, both to Upjohn). Biological properties: J. A. Campbell *et al., Steroids* **1,** 317 (1963). Dissolution and solubility rates: W. E. Hamlin *et al., J. Pharm. Sci.* **54,** 1651 (1965). Pharmacodynamics: J. H. Sokolowski, C. W. Kasson, *Am. J. Vet. Res.* **39,** 837 (1978). Efficacy evaluation in dogs: J. H. Sokolowski, *J. Am. Vet. Med. Assoc.* **173,** 983 (1978).

Cryst solid. Soly in deionized water: 0.0454 mg/ml at 37°.

Note: This is a controlled substance (anabolic steroid) in the U.S. Code of Federal Regulations, Title 21 Part 1308.13, as defined in Part 1308.02 (1995).

THERAP CAT (VET): Anabolic; androgen.

6263. Miboplatin. *[SP-4-3-(1R-cis)]-[1,1-Cyclobutane-dicarboxylato(2—)](2-pyrrolidinemethanamine-N^α,N^1)plati-num;* (—)-cis-[(R)-2-(aminomethyl)pyrrolidine](1,1-cyclo-butanedicarboxylato)platinum; (—)-(R)-1,1-cyclobutane-dicarboxylato(2-aminomethylpyrrolidine)platinum(II). C$_{11}$-H$_{18}$N$_2$O$_4$Pt; mol wt 437.36. C 30.21%, H 4.15%, N 6.41%, O 14.63%, Pt 44.60%. Cytotoxic platinum complex. Prepn: M. Honda *et al.,* **Eur.** pat. **Appl. 176,005;** *eidem,* **U.S.** pat. **4,822,892** (1986, 1989 both to Chugai); K. Morikawa *et al., J. Pharm. Sci.* **80,** 837 (1991). Comparative antitumor and nephrotoxic activities of isomers: T. Matumoto *et al., Brit. J. Cancer* **64,** 41 (1991). Mechanism of action studies: M.

Itoh *et al., Chem. Pharm. Bull.* **37,** 2825 (1989); M. Akaboshi *et al., Japan. J. Cancer Res.* **85,** 106 (1994). Clinical evaluation: K. Tamura *et al., Cancer* **66,** 2059 (1990). Acute toxicity: M. Suzuki *et al., Yakuri to Chiryo* **20,** S743 (1992), *C.A.* **117,** 124096 (1992).

Monohydrate, *DWA-2114R, Neoplat.* Colorless crystals from water, mp 248-254°. [α]$_D^{20}$ —40.1° (c = 0.4 in water). LD$_{50}$ in male, female mice, rats (mg/kg): 150, 155, 107, 135 i.v. (Suzuki).

THERAP CAT: Antineoplastic.

6264. Michler's Base. *4,4'-Methylenebis(N,N-dimethyl-benzenamine); 4,4'-methylenebis(N,N-dimethylaniline);* bis(p-dimethylaminophenyl)methane; tetra-base; Arnold's base. C$_{17}$H$_{22}$N$_2$; mol wt 254.38. C 80.27%, H 8.72%, N 11.01%. Prepd by heating dimethylaniline with 40% formaldehyde and concd HCl: Mekel, **Ger.** pat. **1,026,322** (1958 to BASF); Hey, Sanderson, *J. Chem. Soc.* **1960,** 3203; from dimethylaniline + diacetylperoxide: Horner *et al., Ann.* **626,** 1 (1959); from dimethylaniline + tert-butylperbenzoate: Sosnovsky, Yang, *J. Org. Chem.* **25,** 899 (1960).

Lustrous leaflets, mp 90-91°. Sublimes without decompn. bp 390°; bp$_{0.1}$ 155-157°. Insol in water. Sol in benzene, ether, carbon disulfide, acids; slightly sol in cold alcohol; more sol in hot alcohol.

Note: This substance may reasonably be anticipated to be a carcinogen: *Seventh Annual Report on Carcinogens* (PB95-109781, 1994) p 248.

USE: Has been used in manufacture of dyes. As reagent for lead: *Deut. med. Wochschr.* **52,** 1855 (1926).

6265. Michler's Ketone. *Bis[4-(dimethylamino)phen-yl]methanone; 4,4'-bis(dimethylamino)benzophenone;* tetramethyldiaminobenzophenone. C$_{17}$H$_{20}$N$_2$O; mol wt 268.36. C 76.09%, H 7.51%, N 10.44%, O 5.96%. Prepn from dimethylaniline + phosgene: Michler, *Ber.* **9,** 716 (1876); Michler, Dupertius, *ibid.* 1900; from dimethylaniline, AlCl$_3$ and CCl$_4$: Fierz, Koechlin, *Helv. Chim. Acta* **1,** 218 (1918).

White to greenish leaflets, mp 172°. bp > 360° with decompn. Practically insol in water. Sol in alcohol, warm benzene; very slightly sol in ether.

Note: This substance may reasonably be anticipated to be a carcinogen: *Seventh Annual Report on Carcinogens* (PB95-109781, 1994) p 259.

USE: In manuf of dyes and pigments.

6266. Miconazole. *1-[2-(2,4-Dichlorophenyl)-2-[(2,4-dichlorophenyl)methoxy]ethyl]-1H-imidazole; 1-[2,4-dichloro-β-[(2,4-dichlorobenzyl)oxy]phenethyl]imidazole.* C$_{18}$H$_{14}$Cl$_4$-N$_2$O; mol wt 416.13. C 51.95%, H 3.39%, Cl 34.08%, N 6.73%, O 3.84%. Prepn: E. F. Godefroi *et al., J. Med. Chem.* **12,** 784 (1969); E. F. Godefroi, J. Heeres, **Ger.** pat. **1,940,388;** *eidem,* **U.S.** pat. **3,717,655** (1970, 1973 to Janssen). Clinical evaluation: Brugmans *et al.*, *Arch. Dermatol.* **102**, 428 (1970); Godts *et al.*, *Arzneimittel-Forsch.* **21**, 256 (1971). *Review:* P. Janssen, W. Van Bever, in *Pharmacological and Biochemical Properties of Drug Substances* **vol. 2**, M. E. Goldberg, Ed. (Am. Pharm. Assoc., Washington, DC, 1979) pp 333-354; R. C. Heel *et al.*, *Drugs* **19**, 7-30 (1980).

Nitrate, $C_{18}H_{14}Cl_4N_2O \cdot HNO_3$, *R-14889, Aflorix, Albistat, Andergin, Brentan, Conoderm, Conofite, Daktar, Daktarin, Deralbine, Dermonistat, Epi-Monistat, Florid, Fungiderm, Fungisdin, Gyno-Daktarin, Gyno-Monistat, Micatin, Miconal Ecobi, Micotef, Monistat, Prilagin, Vodol.* Crystals, mp 170.5° (Godefroi, Heeres, 1970); 184-185° (Godefroi*).
(+)-Form nitrate, mp 135.3°. $[\alpha]_D^{20}$ +59° (methanol).
(−)-Form nitrate, mp 135°. $[\alpha]_D^{20}$ −58° (methanol).
THERAP CAT: Antifungal (topical).
THERAP CAT (VET): Antifungal (topical).

6267. Micranthine. *6′,7-Epoxy-6-methoxy-2-methyloxyacanthan-12′-ol.* $C_{34}H_{32}N_2O_5$; mol wt 548.64. C 74.43%, H 5.88%, N 5.11%, O 14.58%. Minor alkaloid from bark of *Daphnandra micrantha* (Tul.) Benth., *Monimiaceae.* Isoln: Pyman, *J. Chem. Soc.* **105**, 1679 (1914). Structure: Bick, Todd, *ibid.* **1950**, 1606. Revised structure and stereochemistry: Bick *et al.*, *Tetrahedron Letters* **1972**, 33. Biogenesis: I. Ralph, C. Bick, *Heterocycles* **16**, 2105 (1981).

Needles from ethyl acetate or methanol, mp 194-196°. $[\alpha]_D^{22}$ −231° (c = 0.7 in chloroform). Sol in chloroform, ethanol; sparingly sol in methanol, ethyl acetate, benzene; practically insol in ether.

6268. Micrococcin P. Micrococcin. Antibiotic substance produced by a species of *Micrococcus:* cf. *Brit. J. Exp. Pathol.* **29**, 473 (1948). Isoln of micrococcin P from *B. pumilus:* Heatley, Doery, *Biochem. J.* **50**, 247 (1951). Probable identity of micrococcin and micrococcin P: Abraham *et al.*, *Nature* **178**, 44 (1956). Structure studies: Brookes *et al.*, *J. Chem. Soc.* **1957**, 689; Brookes *et al.*, *ibid.* **1960**, 916; Dean *et al.*, *ibid.* **1961**, 3394; James, Watson, *ibid. C,* **1966**, 1361. The antibiotic is a mixture of two components, P_1 (major) and P_2. Proposed total structure of P_1: J. Walker *et al.*, *Chem. Commun.* **1977**, 706. Revised structure of P_1 and structure of P_2: B. W. Bycroft, M. S. Gowland, *ibid.* **1978**, 256. Mechanism of action: E. Cundliffe, J. Thompson, *Eur. J. Biochem.* **118**, 47 (1981). *Review:* Pestka in *Antibiotics* **vol. 3**, J. W. Corcoran, F. E. Hahn, Eds. (Springer-Verlag, New York, 1975) pp 480-486.

Micrococcin P_1

Needles from ethanol, dec 222-228°. $[\alpha]_D^{21}$ +116° (c = 0.10 in 90% ethanol). uv max (ethanol): 345 nm. Sol in ethanol, chloroform, acetone, glacial acetic acid, pyridine. Insol in ether, benzene, amyl acetate, glycerol. Very slightly sol in water.

6269. Micronomicin. *O-2-Amino-2,3,4,6-tetradeoxy-6-(methylamino)-α-D-erythro-hexopyranosyl-(1→4)-O-[3-deoxy-4-C-methyl-3-(methylamino)-β-L-arabinopyranosyl-(1→6)]-2-deoxy-D-streptamine;* 6′-N-methylgentamicin C_{1a}; gentamicin C_{2b}; sagamicin (formerly); antibiotic KW-1062; KW-1062; XK-62-2. $C_{20}H_{41}N_5O_7$; mol wt 463.58. C 51.82%, H 8.91%, N 15.11%, O 24.16%. Gentamicin C complex antibiotic produced by *Micromonospora sagamiensis* var. *nonreducans:* T. Nara *et al.*, Ger. pat. **2,326,781** (1973 to Kyowa), *C.A.* **80**, 58389b (1974); *eidem*, U.S. pat. **4,045,298** (1977 to Abbott). Isoln, physicochemical and antibacterial properties: R. Okachi *et al.*, *J. Antibiot.* **27**, 793 (1974). Structure: R. S. Egan *et al.*, *ibid.* **28**, 29 (1975). Identity with gentamicin C_{2b}: P. J. L. Daniels *et al.*, *ibid.* **35**. Pharmacology: T. Hashimoto *et al.*, *Japan. J. Antibiot.* **30**, 362 (1977), *C.A.* **89**, 17113z (1978). Toxicity studies: T. Hara *et al.*, *ibid.* 386-449, *C.A.* **89**, 209205c-8f (1978). Biosynthesis: H. Kase *et al.*, *J. Antibiot.* **35**, 1 (1982); *eidem*, *Agr. Biol. Chem.* **46**, 515 (1982). Series of articles on pharmacology, metabolism and clinical studies: *Chemotherapy (Tokyo)* **25**, 1801-2287 (1977). *See also* Gentamicin.

White amorphous powder, mp 260° (dec). $[\alpha]_D^{20}$ +116° (c = 1 in water). Sol in water, methanol. Insol in chloroform, ethyl acetate, benzene, petr ether. LD_{50} i.v. in mice: 93 mg/kg (Okachi).
Sulfate, $(C_{20}H_{41}N_5O_7)_2 \cdot 5H_2SO_4$, *6′-N-methylgentamicin C_{1a} hemipentasulfate, Sagamicin, Santemycin.*
THERAP CAT: Antibacterial.

6270. Midazolam. *8-Chloro-6-(2-fluorophenyl)-1-methyl-4H-imidazo[1,5-a][1,4]benzodiazepine.* $C_{18}H_{13}ClFN_3$; mol wt 325.77. C 66.36%, H 4.02%, Cl 10.88%, F 5.83%, N 12.90%. Short-acting deriv of diazepam, *q.v.* Prepn: R. I. Fryer, A. Walser, Ger. pat. **2,540,522** (1976 to Hoffmann-La Roche), *C.A.* **85**, 21497 (1976); A. Walser *et al.*, *J. Org.*

Chem. **43**, 936 (1978). Determn in blood and urine: C. V. Puglisi *et al., J. Chromatog.* **145**, 81 (1978). Clinical, electroencephalographic, pharmacokinetic studies: C. R. Brown *et al., Anesthesiology* **50**, 467 (1979). Pharmacological study in dogs: D. J. Jones *et al., ibid.* **51**, 430 (1979). Bioassay: F. H. Sarnquist *et al., ibid.* **52**, 149 (1980). Toxicity data: L. Pieri *et al., Arzneimittel-Forsch.* **31**, 2180 (1981). Series of articles on pharmacology, metabolism, pharmacokinetics, clinical experience: *ibid.* 2177-2288; *Brit. J. Clin. Pharmacol.* **16**, Suppl. 1, 1S-199S (1983). Review of pharmacology and therapeutic use: J. W. Dundee *et al., Drugs* **28**, 519-543 (1984).

Colorless crystals from ether/methylene chloride/hexane, mp 158-160°. uv max (2-propanol): 220 nm (ε 30000). Sol in water.

Maleate, $C_{18}H_{13}ClFN_3 \cdot C_4H_4O_4$, *Ro-21-3981/001, Dormicum, Flormidal.* Crystals from ethanol/ether, mp 114-117° (solvated). LD_{50} in male mice (mg/kg): 760 orally; 86 i.v. (Pieri).

Hydrochloride, $C_{18}H_{13}ClFN_3 \cdot HCl$, *Ro-21-3981/003, Hypnovel, Versed.*

Note: This is a controlled substance (depressant) listed in the U.S. Code of Federal Regulations, Title 21 Part 1308.14 (1995).

THERAP CAT: Anesthetic (intravenous).

6271. Midecamycins. Macrolide antibiotic complex produced by *Streptomyces mycarofaciens* nov. sp. Isoln and characterization of the major component, midecamycin A_1: T. Tsuruoka *et al.*, **Japan.** pat. **28834('71)**; *eidem*, **U.S.** pat. **3,761,588** (1973, both to Meiji); *eidem, J. Antibiot.* **24**, 319, 452 (1971). Structure: I. Shigeharu *et al., ibid.* 460. Isoln, properties, structures of the minor components, midecamycins A_2, A_3, A_4: T. Tsuruoka *et al., ibid.* 476, 526. Enzymatic interconversion of A_1 and A_3: Y. Matsuhashi *et al., ibid.* **32**, 777 (1979). Metabolism: T. Shomura *et al., Chem. Pharm. Bull.* **22**, 2427 (1974). Pharmacokinetics: K. Fukaya *et al., Chemotherapy (Tokyo)* **21**, 692 (1973). Pharmacology: K. Mashimo *et al., ibid.* 729. Antibacterial activity: T. Watanabe *et al., ibid.* **25**, 1624 (1977). Toxicity and teratogenicity study in mice: M. Moriguchi *et al., Japan. J. Antibiot.* **15**, 187, 193 (1972), *C.A.* **78**, 11652a, 11653b (1973).

Midecamycin A₁

Midecamycin A_1, $C_{41}H_{67}NO_{15}$, leucomycin V 3,4^B-dipropanoate, espinomycin A, mydecamycin, turimycin P3, SF-837, YL-704B1, Aboren, Medemycin, Midecin, Momicine, Myoxam, Normicina, Rubimycin. Colorless needles from benzene, mp 155-156° (after drying at 80° for 8 hrs in vacuo). $[\alpha]_D^{23}$ −67° (c = 1 in ethanol). uv max (ethanol): 232 nm ($E_{1cm}^{1\%}$ 325). Weakly basic; pKa' 6.9 in 50% aq ethanol. Sol in methanol, ethanol, acetone, chloroform,

ethyl acetate, benzene, ethyl ether, acidic water. Practically insol in *n*-hexane, petr ether, water.

Midecamycin A_3, $C_{41}H_{65}NO_{15}$, 9-deoxy-9-oxoleucomycin V 3,4^B-dipropanoate, antibiotic SF-837A3. White powder, mp 122-125°. $[\alpha]_D^{22}$ −44° (c = 1 in ethanol). uv max (ethanol): 280 nm ($E_{1cm}^{1\%}$ 295). pKa' 7.0 in 50% aq ethanol. Solubilities similar to midecamycin A_1.

THERAP CAT: Antibacterial.

6272. Midodrine. *2-Amino-N-[2-(2,5-dimethoxyphenyl)-2-hydroxyethyl]acetamide; 1-(2',5'-dimethoxyphenyl)-2-glycinamidoethanol.* $C_{12}H_{18}N_2O_4$; mol wt 254.29. C 56.68%, H 7.13%, N 11.02%, O 25.17%. Prepn: K. Wismayr *et al.*, **Austrian** pat. **241,435**; *eidem*, **U.S.** pat. **3,340,298** (1965, 1967 both to OSSW). Hemodynamic effects: G. Hitzenberger *et al., Int. J. Clin. Pharmacol. Ther. Toxicol.* **7**, 323 (1973). Pharmacodynamic actions: H. Pittner *et al., Arzneimittel-Forsch.* **26**, 2145 (1976). Plasma levels: N. Kolassa *et al., Arch. Int. Pharmacodyn. Ther.* **238**, 96 (1979). Use in ejaculation disorders: D. Jonas *et al., Eur. Urol.* **5**, 184 (1979); in circulatory disorders: H. Sazovsky, H. Pittner, *Fortschr. Med.* **97**, 733 (1979). Synthesis of potential metabolites: T. Kappe, W. Witoszynaky, *Arch. Pharm.* **308**, 339 (1975).

Hydrochloride, $C_{12}H_{18}N_2O_4 \cdot HCl$, *ST-1085, Alphamine, Amatine, Gutron, Hipertan, Metligine, Midamine.* Crystals, mp 192-193°.

THERAP CAT: α-Adrenergic (vasoconstrictor). Antihypotensive.

6273. Mifepristone. *(11β,17β)-11-[4-(Dimethylamino)-phenyl]-17-hydroxy-17-(1-propynyl)estra-4,9-dien-3-one; 11β-[4-(N,N-dimethylamino)phenyl]-17α-(prop-1-ynyl)-Δ^{4,9}-estradiene-17β-ol-3-one; RU-486; RU-38486; Mifegyne.* $C_{29}H_{35}NO_2$; mol wt 429.60. C 81.08%, H 8.21%, N 3.26%, O 7.45%. Progesterone receptor antagonist with partial agonist activity. Prepn: J. G. Teutsch *et al.*, **Eur.** pat. **Appl.** **57,115**; *eidem*, **U.S.** pat. **4,386,085** (1982, 1983 both to Roussel-UCLAF). Pharmacology: W. Herrmann *et al., Compt. Rend. Ser. III* **294**, 933 (1982). Pituitary and adrenal responses in primates: D. L. Healy *et al., J. Clin. Endocrinol. Metab.* **57**, 863 (1983). Mechanism of action study: M. Rauch *et al., Eur. J. Biochem.* **148**, 213 (1985). Clinical study as abortifacient: B. Couzinet *et al., N. Engl. J. Med.* **315**, 1565 (1986); as postcoital contraceptive: A. Glasier *et al., ibid.* **327**, 1041 (1992). Review of mechanism of action and clinical applications: E. E. Baulieu, *Science* **245**, 1351-1357 (1989). *Reviews:* I. M. Spitz, C. W. Bardin, *N. Engl. J. Med.* **329**, 404-412 (1993); R. N. Brogden *et al., Drugs* **45**, 384-409 (1993).

mp 150°. $[\alpha]_D^{20}$ +138.5° (c = 0.5 in chloroform).
THERAP CAT: Abortifacient.

6274. Miglitol. *1,5-Dideoxy-1,5-[(2-hydroxyethyl)imino]-D-glucitol; N-(2-hydroxyethyl)moranoline; (2R,3R,4R,-5S)-1-(2-hydroxyethyl)-2-(hydroxymethyl)-3,4,5-piperidinetriol; N-(β-hydroxyethyl)-1-deoxynojirimycin; Bay m 1099.* $C_8H_{17}NO_5$; mol wt 207.23. C 46.37%, H 8.27%, N 6.76%, O 38.60%. α-Glucosidase inhibitor; hydroxyethyl derivative of 1-deoxynojirimycin, *q.v.* Prepn: B. Junge *et al.*, **Ger.** pat.

2,758,025; *eidem,* U.S. pat. **4,639,436** (1979, 1987 both to Bayer AG). Inhibition of α-glucosidase: B. Lembcke *et al., Digestion* **31,** 120 (1985); and hypoglycemic activity: Y. Yoshikuni *et al., J. Pharmacobio-Dyn.* **11,** 356 (1988). Clinical pharmacology: F. P. Kennedy *et al., Clin. Exp. Pharmacol. Physiol.* **14,** 633 (1987). Clinical evaluations in non-insulin dependent diabetes: N. Katsilambros *et al., Arzneimittel-Forsch.* **36,** 1136 (1986); A. R. Scott, R. B. Tattersall, *Diabetic Med.* **5,** 42 (1988); in insulin dependent diabetes: J. Gerard *et al., Int. J. Clin. Pharmacol. Ther. Toxicol.* **25,** 483 (1987); F. P. Kennedy, J. E. Gerich, *Clin. Pharmacol. Ther.* **42,** 455 (1987).

Crystals from ethanol, mp 114°.

THERAP CAT: Antidiabetic.

6275. Mikamycin. Antibiotic complex isolated from *Streptomyces mitakaensis* found in the soil at Mitaka City, Japan. Belongs to the streptogramin family of antibiotics. Originally considered a single substance, it was later found to contain several active components, the main ones being designated mikamycin A and B. Isoln and characterization: Arai, Nakamura *et al., J. Antibiot.* **9A,** 193 (1956); Arai, Karasawa *et al., ibid.* **11A,** 14 (1958); Arai, Okabe *et al., ibid.* 21; Okabe, *ibid.* **12A,** 86 (1959). Both components exhibit synergistic action *in vitro* and *in vivo:* Tanaka *et al., ibid.* **12A,** 290 (1959); Watanabe, *ibid.* **13A,** 62 (1960). Production of mikamycin A and B: Fr. pat. **1,349,946;** Watanabe, U.S. pat. **3,137,640** (both 1964 to Kanegafuchi). Nomenclature: P. Crooy, R. De Neys, *J. Antibiot.* **25,** 371 (1972). *Review:* N. Tanaka in *Antibiotics* **vol. 3,** J. W. Corcoran, F. E. Hahn, Eds. (Springer-Verlag, New York, 1975) pp 487-497.

Mikamycin B

Mikamycin A, *see* Virginiamycin M₁.

Mikamycin B, $C_{45}H_{54}N_8O_{10}$, *4-[4-(dimethylamino)-N-methyl-L-phenylalanine]virginiamycin* S_1, *mikamycin I_A-ostreogrycin B, streptogramin B, pristinamycin I_A, vernamycin B_α.* Isoln and properties: Watanabe *et al., J. Antibiot.* **12A,** 112 (1959); **13A,** 57 (1960). Structure: *ibid.* **13A,** 291, 293 (1960); **14A,** 14 (1961); Cox *et al., Chem. Commun.* **1970,** 1623. Monohydrate, platelets from methanol, mp 160°. Anhydr form obtained by drying at 150° under vacuum for 16 hrs, dec 262-263°. uv max (methanol): 209, 260, 305 nm ($E_{1cm}^{1\%}$ 605, 220, 101). $[\alpha]_D^{20}$ −60.3° (c = 1.0 in methanol). Sol in methanol, ethanol, acetone, butyl acetate, benzene, chloroform. Practically insol in water, petr ether, hexane. Stable at neutral and acid pH, unstable at alkaline pH.

THERAP CAT: Antibacterial.

6276. Milbemectin. E-187; Koromite, Milbeknock. Mixture of milbemycins, *q.v,* in a 3:7 ratio of milbemycin A_3 and milbemycin A_4. Isolated from fermentation broth of *Streptomyces hygroscopicus* subsp. *aureolacrimosus* SANK-60576; structurally related to the avermectins, q.v.: A. Aoki

et al., **Ger.** pat. **2,329,486;** *eidem,* U.S. pat. **3,950,360** (1973, 1976 both to Sankyo); Y. Takiguchi *et al., J. Antibiot.* **33,** 1120 (1980). Synthesis of A_3: M. Hirama *et al., J. Am. Chem. Soc.* **113,** 1830 (1991); S. V. Ley *et al., Tetrahedron Letters* **34,** 7479 (1993). Metabolism in rats: S. Sadakane *et al., J. Pest. Sci.* **17,** 147 (1992). Review: A. Aoki *et al., ibid.* **19,** 245-247 (1994).

Milbemycin A_3 R = CH₃
Milbemycin A_4 R = CH₂CH₃

Stable to heat; relatively unstable under direct sunlight. Relatively unstable to acid; unstable to alkali. LD_{50} in male, female rats, male, female mice (mg/kg): 762, 456, 324, 313 orally. LD_{50} in rats: > 5000 dermally. LD_{50} by inhalation in male, female rats: 1.90, 2.80 mg/l (Aoki, 1994).

Milbemycin A_3, $C_{31}H_{44}O_7$, *(6R,25R)-5-O-demethyl-28-deoxy-6,28-epoxy-25-methylmilbemycin, milbemycin α_1, antibiotic B-41A3.* White crystalline powder, mp 212°-215°. $[\alpha]_D^{20}$ +106° (c = 0.25 in acetone) (Aoki). Also reported as mp 192°-194°. $[\alpha]_D^{20}$ +106° (acetone) (Hirama). uv max (ethanol): 238, 244 nm (ε 27800, 30500). Soly in water (20°): 7.2 ppm; methanol 64.8 g/l; ethanol 41.9 g/l; acetone 66.1 g/l; n-hexane 1.4 g/l; benzene 143.1 g/l; ethyl acetate 69.5 g/l. Vapor pressure (20°): $<1 \times 10^{10}$ mm Hg.

Milbemycin A_4, $C_{32}H_{46}O_7$, *milbemycin α_3, antibiotic B-41A4.* White crystalline powder, mp 212°-215°. $[\alpha]_D^{20}$ +106° (c = 0.25 in acetone). uv max (ethanol): 238, 244 nm (ε 27800, 30500). Soly in water (20°): 0.88 ppm; methanol 458.8 g/l; ethanol 234.0 g/l; acetone 365.3 g/l; n-hexane 6.5 g/l; benzene 524.2 g/l; ethyl acetate 320.4 g/l. Vapor pressure (20°): $<1 \times 10^{10}$ mm Hg.

USE: Acaracide.

6277. Milbemycin Oxime. CGA-179246; Interceptor; Milbemax. Mixture of milbemycin oximes containing ~80% *milbemycin A_4 5-oxime* ($C_{32}H_{45}NO_7$) and 20% *milbemycin A_3 5-oxime* ($C_{31}H_{43}NO_7$). Prepn: J. Ide *et al.,* **Eur.** pat. **Appl. 110,667;** *eidem,* U.S. pat. **4,547,520** (1984, 1985 both to Sankyo). Anthelmintic activity in experimentally induced infections in dogs: D. D. Bowman *et al., Am. J. Vet. Res.* **49,** 1986 (1988); D. D. Bowman *et al., ibid.* **52,** 64 (1991); in cats: B. L. Blagburn *et al., The OVMA Newsletter* (February, 1990), 1; in naturally acquired infections in dogs: D. D. Bowman *et al., Am. J. Vet. Res.* **51,** 487 (1990). Phys-

Component A_4 R = CH₂CH₃
Component A_3 R = CH₃

ical data and toxicity: Ciba-Geigy Material Data Safety Sheet.

Practically odorless white to pale yellow powder, mp 169.6-177.4°. pH of aqueous solution: 6.3. LD_{50} orally in mice: 1000-1500 mg/kg (Ciba-Geigy).

THERAP CAT (VET): Anthelmintic.

6278. Milbemycins. Antibiotic B-41. A family of novel macrolide antibiotics with insecticidal and acaricidal activity; structurally related to avermectin, *q.v.*, aglycones. Isoln from *Streptomyces hygroscopicus* subsp. *aureolacrimosus* and properties: A. Aoki *et al.*, Ger. pat. **2,329,486**; *eidem*, U.S. pat. **3,950,360** (1973, 1976 both to Sankyo); of α_1-α_{10}, β_1-β_3, D-H: Y. Takiguchi *et al.*, *J. Antibiot.* **33**, 1120 (1980). Isoln and properties of D-H: Y. Takiguchi *et al.*, *J. Antibiot.* **36**, 502 (1983); of J and K: M. Ono *et al.*, *ibid.* 509. Structures of milbemycins α_1-α_{10}, β_1: M. Kurabayashi *et al.*, *18th Symp. Chem. Natural Products* (Kyoto, 1974) nos. 309-316; of β_1-β_3: H. Mishima *et al.*, *Tetrahedron Letters* **1975**, 711; of D-H, J, K: H. Mishima *et al.*, *J. Antibiot.* **36**, 980 (1983). Total synthesis of (±)-milbemycin β_3: A. B. Smith *et al.*, *J. Am. Chem. Soc.* **104**, 4015 (1982); of the (+)-form: D. R. Williams *et al.*, *ibid.* 4708. Prepn of milbemycins from avermectins: H. Mrozik *et al.*, *Tetrahedron Letters* **1983**, 5333. Total synthesis studies of D: M. T. Crimmins *et al.*, *ibid.* **28**, 3651 (1987). Prophylactic use of milbemycin D against *Dirofilaria immitis* infection in dogs: M. Tagawa *et al.*, *Japan. J. Vet. Sci.* **47**, 787 (1985). Toxicity data: N. Matsunuma *et al.*, *Sankyo Kenkyusho Nempo* **35**, 71 (1983), *C.A.* **101**, 32870d (1984). Review of syntheses: T. Blizzard *et al.*, in *Recent Prog. Chem. Synth. Antibiotics*, G. Lukacs, M. Ohno, Eds (Springer-Verlag, Berlin, 1990) pp 66-102. Review of structure-activity relationships: W. L. Shoop *et al.*, *Vet. Parasitol.* **59**, 139-156 (1995).

Milbemycin D

Sol in *n*-hexane, benzene, acetone, ethanol, methanol, chloroform; very slightly sol in water.

Milbemycin α_1, see milbemectin.

Milbemycin β_1, $C_{32}H_{48}O_7$, *25α-methylmilbemycin B, antibiotic B-41A1*. Amorphous solid. $[\alpha]_D^{20}$ +160° (c = 0.25 in acetone). uv max (ethanol): 245 nm (ϵ 26500).

Milbemycin D, $C_{33}H_{48}O_7$, *(6R,25R)-5-O-demethyl-28-deoxy-6,28-epoxy-25-(1-methylethyl)milbemycin B, antibiotic B-41D*. Colorless needles from hexane-ethyl acetate (20:1), mp 186-188°. $[\alpha]_D^{27}$ +107° (c = 0.25 in acetone). uv max (ethanol): 244 nm (31000).

THERAP CAT (VET): Antiparasitic.

6279. Mildiomycin. *(S)-4-Amino-1-[4-[(2-amino-3-hydroxy-1-oxopropyl)amino]-9-[(aminoiminomethyl)amino]-6-C-carboxy-2,3,4,7,9-pentadeoxy-α-L-talo-non-2-enopyranosyl]-5-(hydroxymethyl)-2(1H)-pyrimidinone.* $C_{19}H_{30}N_8O_9$; mol wt 514.50. C 44.36%, H 5.88%, N 21.78%, O 27.99%. A nucleoside antibiotic with anti-mildew activity and low toxicity in mammals and fish. Isoln from *Streptoverticillium rimofaciens* B-98891 and characterization: S. Harada, T. Kishi, *J. Antibiot.* **31**, 519 (1978). Taxonomy and fermentation study: T. Iwasa *et al.*, *ibid.* 511. Structure: S. Harada *et al.*, *J. Am. Chem. Soc.* **100**, 4895 (1978); *Tetrahedron* **37**, 1317 (1981). Anti-mildew spectrum: K. Suetomi, T. Kusaka, *Nippon Noyaku Gakkaishi* **4**, 349 (1979), *C.A.* **92**, 17012c (1980).

Hygroscopic solid, mp > 300° (monohydrate). $[\alpha]_D^{20}$ +100° (c = 0.5 in water); +78.5° (c = 0.5 in 0.1N HCl). pKa': 2.8, 4.3, 7.2, > 12. uv max (pH 7 and 0.1N NaOH): 271 nm ($E_{1cm}^{1\%}$ 164); (0.1N HCl): 280 nm ($E_{1cm}^{1\%}$ 247). LD_{50} in rats, mice: ~500-1000 mg/kg i.v.; ~2.5-5.0 g/kg orally (Harada, Kishi).

6280. Milk. Cow's milk. Compn and physical properties (a study of samples representing 8 million quarts of milk from 8 cities): Dahlberg, *Proc. Ann. Conv. Milk Ind. Foundation, Lab. Sec.* **46**, 44 (1953). Review: Patton, *Sci. Am.* **221** (1), 58 (July 1969).

White liquid. d 1.032. mp −0.54°. pH 6.62. Fat 3.82%; total solids 12.43%; protein 3.25%; lactose 4.64%; ash 0.73%. Vitamins and minerals in 100 g: Vitamin A 152 I.U.; riboflavin 0.156 mg; thiamine 0.043 mg; ascorbic acid 1.14 mg; vitamin B_{12} 0.56 γ; vitamin D 2.4 I.U.; vitamin E 0.1 mg; nicotinic acid 0.085 mg; pantothenic acid 0.35 mg; pyridoxine 0.048 mg; biotin 0.0035 mg; folic acid 0.2 γ; inositol 13 mg; choline 13 mg; essential fatty acids 99 mg; Ca 113 mg; P 90 mg; Fe 0.032 mg. Caloric value: 65 cal/100 g.

6281. Milnacipran. *cis-(±)-2-(Aminomethyl)-N,N-diethyl-1-phenylcyclopropanecarboxamide;* 1-phenyl-1-(diethylaminocarbonyl)-2-(aminomethyl)cyclopropane; midalcipran. $C_{15}H_{22}N_2O$; mol wt 246.35. C 73.13%, H 9.00%, N 11.37%, O 6.49%. Serotonin and norepinephrine uptake inhibitor. Prepn: G. Mouzin *et al.*, Eur. pat. Appl. **68,999**; *eidem*, U.S. pat. **4,478,836** (1983, 1984 both to Pierre Fabre); B. Bonnaud *et al.*, *J. Med. Chem.* **30**, 318 (1987). Mechanism of action study: C. Moret *et al.*, *Neuropharmacology* **24**, 1211 (1985). Pharmacology: A. Stenger *et al.*, *Psychopharmacology* **91**, 147 (1987). Clinical inhibition of monoamine uptake: C. Palmier *et al.*, *Eur. J. Clin. Pharmacol.* **37**, 235 (1989). Clinical comparison with amitriptyline, *q.v.*, in depression: M. Ansseau *et al.*, *Psychopharmacology* **98**, 163 (1989).

Hydrochloride, $C_{15}H_{22}N_2O \cdot HCl$, *F-2207*. Crystals from ethanol-ethyl ether, mp 179-181°. LD_{50} orally in mice: 237 mg/kg (Bonnaud).

THERAP CAT: Antidepressant.

6282. Milorganite®. Activated sewage sludge prepd by microbial treatment of sewage, useful as fertilizer and as a source of vitamin B_{12}: Schendel, Johnson, *J. Agr. Food. Chem.* **2**, 23 (1954). Isoln of B_{12}-active product: Miner, Wolnak, U.S. pat. **2,646,386** (1953 to Sewerage Commission of Milwaukee); U.S. pat. **2,941,933** (1960 to UCLAF).

Caution: The dust can cause irritation of eyes, respiratory tract.

6283. Miloxacin. *5,8-Dihydro-5-methoxy-8-oxo-1,3-dioxolo[4,5-g]quinoline-7-carboxylic acid;* 6,7-methylenedioxy-1-methoxy-4-oxo-1,4-dihydroquinoline-3-carboxylic acid; antibiotic AB 206; AB-206; Fuldazin. $C_{12}H_9NO_6$; mol wt 263.21. C 54.76%, H 3.45%, N 5.32%, O 36.47%. Quinolone antibacterial; analog of oxolinic acid, *q.v.* Prepn: T. Nakagome *et al.,* **Ger.** pat. **2,134,451;** *eidem,* **U.S.** pat **3,799,930** (1972, 1974 both to Sumitomo). Mode of action: T. Nagate *et al., Antimicrob. Ag. Chemother.* **17,** 763 (1980). Series of articles on antibacterial activity, absorption and excretion in animals, determn of metabolites by HPLC: *ibid.* **18,** 37-49 (1980). Reproduction studies in rats: T. Yamada *et al., Oyo Yakuri* **19,** 651, 815, 833 (1980), *C.A.* **93,** 197897x, 215570s, 215571t (1980).

Colorless prisms from DMF, mp 264° (dec).
THERAP CAT: Antibacterial.

6284. Milrinone. *1,6-Dihydro-2-methyl-6-oxo-(3,4'-bipyridine)-5-carbonitrile;* 1,2-dihydro-6-methyl-2-oxo-5-(4-pyridinyl)nicotinonitrile; Win-47203; Corotrope. $C_{12}H_9N_3$-O; mol wt 211.22. C 68.24%, H 4.29%, N 19.89%, O 7.57%. Selective phosphodiesterase inhibitor with vasodilating and positive inotropic activity. Prepn: G. Y. Lesher *et al.,* **Belg.** pat. **886,336;** G. Y. Lesher, R. E. Philion, **U.S.** pat. **4,313,-951** (1981, 1982 both to Sterling). Improved process: B. Singh, **U.S.** pat. **4,413,127** (1983 to Sterling); *idem, Heterocycles* **23,** 1479 (1985). Pharmacological profile: A. A. Alousi *et al., J. Cardiovasc. Pharmacol.* **5,** 792, 804 (1983). HPLC determn in body fluids: J. Edelson *et al., J. Chromatog.* **276,** 456 (1983). Clinical studies in congestive heart failure: C. S. Maskin *et al., Circulation* **67,** 1065 (1983); L. S. Sinoway *et al., J. Am. Coll. Cardiol.* **2,** 327 (1983); D. S. Baim *et al., N. Engl. J. Med.* **309,** 748 (1983).

Crystals from DMF + water or from ethanol, mp > 300°. Lactate, $C_{12}H_9N_3O.xC_3H_6O_3$, *Primacor.*
THERAP CAT: Cardiotonic.

6285. Miltefosine. *2-[[(Hexadecyloxy)hydroxyphosphinyl]oxy]-N,N,N-trimethylethanaminium inner salt; choline phosphate hexadecyl ester, hydroxide, inner salt;* hexadecyl 2-(N,N,N-trimethylamino)ethyl phosphate; *n*-hexadecylphosphorylcholine; hexadecylphosphocholine; HPC; D-18506; Miltex. $C_{21}H_{46}NO_4P$; mol wt 407.57. C 61.89%, H 11.38%, N 3.44%, O 15.70%, P 7.60%. Anticancer agent which targets the cell membrane. Synthesis: U. Kaatze *et al., Chem. Phys. Lipids* **27,** 263 (1980); W. J. Hansen *et al., Lipids* **17,** 453 (1982). See also: H. Eibl, **Eur.** pat. **Appl. 225,608;** *idem,* **U.S.** pat. **4,837,023** (1987, 1989 both to Max-Planck Ges. Wissenschaft.). Toxicity and anticancer properties: C. Muschiol *et al., Lipids* **22,** 930 (1987). Structure-activity study: M. R. Berger *et al., Cancer Treat. Rev.* **17,** 143 (1990). HPTLC determn in plasma: I. Rustenbeck, S. Lenzen, *J. Chromatog.* **525,** 85 (1990). Clinical evaluation in topical treatment of cutaneous breast cancer metastases: C. Unger *et al., Cancer Treat. Rev.* **17,** 243 (1990). *Review:* H. Eibl, C. Unger, *ibid.,* 233-242.

mp 232-234° (dec); discoloration at 225°. LD_{50} in rats (mg/kg): 246 orally (Muschiol).
THERAP CAT: Antineoplastic.

6286. Mimosine. *α-Amino-3-hydroxy-4-oxo-1(4H)-pyridinepropanoic acid;* 3-hydroxy-4-oxo-1(4H)-pyridinealanine; β-[N-(3-hydroxy-4-pyridone)]-α-aminopropionic acid; leucenol; leucenine; leucaenine; leucaenol. $C_8H_{10}N_2O_4$; mol wt 198.18. C 48.49%, H 5.09%, N 14.14%, O 32.29%. A naturally occurring amino acid found in large quantities in the seeds and foliage of the legume genera *Mimosa* and *Leucena.* Isoln from *Leucena glauca* (Willd.) Benth., *Leguminosae:* Renz, *Z. Physiol. Chem.* **244,** 153 (1936); Mascré, *Compt. Rend.* **204,** 890 (1937); improved large-scale isoln: N. K. Hart *et al., Heterocycles* **7,** 265 (1977). Structure: Adams, Jones, *J. Am. Chem. Soc.* **69,** 1803 (1947). Synthesis: Adams, Johnson, *ibid.* **71,** 705 (1949). Identity of mimosine and leucenol: Kleipool, Wibaut, *Rec. Trav. Chim.* **69,** 37 (1950); Wibaut, Schuhmacher, *ibid.* **71,** 1017 (1952). Crystal structure: A. Mostad *et al., Acta Chem. Scand.* **27,** 164 (1973). Shown to cause inhibition of hair growth and loss of hair in mice: Crounse *et al., Nature* **194,** 694 (1962). Explanation of the toxicity in animals: J. F. Thompson *et al., Ann. Rev. Biochem.* **38,** 137 (1969).

dl-Form, crystals from water, mp 235-236°. Slightly sol in water, much less sol in methanol and ethanol. Practically insol in the higher alcohols, in dioxane, ethyl acetate, ether, benzene, chloroform, glacial acetic acid, pyridine, Cellosolve. Sol in dil acids or bases and may be recovered from these solns by adjusting the pH so that it is just acid to bromcresol green. uv max: 282 nm (log ϵ 4.23).
dl-Hemihydrate, crystals, darkens at 215-226°, mp 227-228°.
l-Form, crystals from water, mp 225°. $[\alpha]_D^{22} -20°$.
Hydrochloride, $C_8H_{11}ClN_2O_4$, dec 175°, sol in water.
Hydrobromide, $C_8H_{11}BrN_2O_4$, dec 179.5°, sol in water.
USE: Depilatory agent, Hegarty *et al., Aust. J. Agr. Res.* **15,** 153 (1964), *C.A.* **60,** 16405e (1964).

6287. Minaprine. *N-(4-Methyl-6-phenyl-3-pyridazinyl)-4-morpholineethanamine; 4-[2-[(4-methyl-6-phenyl-3-pyridazinyl)amino]ethyl]morpholine.* $C_{17}H_{22}N_4O$; mol wt 298.39. C 68.43%, H 7.43%, N 18.78%, O 5.36%. Prepn: H. Laborit, **Ger.** pat. **2,229,215;** *idem,* **U.S.** pat. **4,169,158** (1973, 1979 to Centre Etudes Exper. Clin. Physio-Biologie); C.-G. Wermuth, A. Exinger, *Agressologie* **13,** 285 (1972). Pharmacology: H. Laborit *et al., ibid.* 291. Metabolism of ^{14}C-minaprine: A. G. Rico *et al., J. Pharmacol.* **9,** 170 (1978). Pharmacokinetics: J. P. Jeanniot *et al., ibid.* 169. Clinical evaluation: A. Garcia-Maffla, M. H. de Garcia, *Pharmatherapeutica* **2,** 265 (1979). Pharmacological evaluation in depression: K. Bizière *et al., Arzneimittel-Forsch.* **32,** 824 (1982). Toxicologic study: A. G. Mazure *et al., Agressologie* **13,** 319 (1972).

Fine buff-colored needles from isopropanol, mp 122°. Insol in water. Slightly sol in cold ethanol, sol in hot ethanol, chloroform.
Dihydrochloride, $C_{17}H_{22}N_4O.2HCl$, *Agr-1240, CB-30038, Brantur, Cantor.* Crystals from abs ethanol, mp 182°.
THERAP CAT: Antidepressant.

6288. Mineral Spirits. *Petroleum spirits;* white spirits; turpentine substitutes. Name applied to various types of hydrocarbon solvents, primarily petroleum distillates, which have flash points above 100°F (38°C) and distillation ranges between 300°F (149°C) and 415°F (213°C). *See: A.S.T.M. Standard Specifications* **D 235-83,** 71-73 (1983).

Type I, *Stoddard solvent, Texsolve S, Varsol 1.* Regular mineral spirits. Clear liquid. Flash pt (min): 38°C (100°F). $d_{15.6}^{15.6}$ 0.754-0.820. Distillation range: initial bp (min), 149°C (300°F); 50% recovered (max), 182°C (360°F); dry pt (max), 208°C (407°F).

Type II. High flash point mineral spirits. Flash pt (min): 60°C (140°F). $d_{15.6}^{15.6}$ 0.768-0.820. Distillation range: initial bp (min), 177°C (350°F); 50% recovered (max), 196°C (385°F); dry pt (max), 211°C (412°F).

Type III. Odorless mineral spirits. Flash pt (min): 38°C (100°F). $d_{15.6}^{15.6}$ 0.775 (max). Distillation range: initial bp (min), 149°C (300°F); 50% recovered (max), 196°C (385°F); dry pt (max), 213°C (415°F).

Type IV, *Texsolve S-2, Varsol 3.* Low dry point mineral spirits. Flash pt (min): 38°C (100°F). $d_{15.6}^{15.6}$ 0.754-0.800. Distillation range: initial bp (min), 149°C (300°F); 50% recovered (max), 174°C (345°F); dry pt (max), 185°C (365°F).

USE: Solvent; paint thinner. In the coatings and dry cleaning industries.

6289. Minocycline. *[4S-(4α,4aα,5aα,12aα)]-4,7-Bis(dimethylamino)-1,4,4a,5,5a,6,11,12a-octahydro-3,10,12,12a-tetrahydroxy-1,11-dioxo-2-naphthacenecarboxamide;* 7-dimethylamino-6-demethyl-6-deoxytetracycline; Minocyn. $C_{23}H_{27}N_3O_7$; mol wt 457.48. C 60.39%, H 5.95%, N 9.19%, O 24.48%. Semi-synthetic antibiotic effective against tetracycline-resistant staphylococci. Prepn: Boothe, Petisi, **U.S. pats. 3,148,212** and **3,226,436** (1964 and 1965 to Am. Cyanamid). Synthesis: Martell, Boothe, *J. Med. Chem.* **10,** 44 (1967); Church *et al., J. Org. Chem.* **36,** 723 (1971); Bernardi *et al., Farmaco Ed. Sci.* **30,** 736 (1975). Activity data: Kradolfer *et al., Antimicrob. Ag. Chemother.* **1966,** 359. Metabolism: Kelly, Kanegis, *Toxicol. Appl. Pharmacol.* **11,** 171 (1967). Toxicity studies: Noble *et al., ibid.* **128.** Clinical evaluation: Frisk, Tunevall, *Antimicrob. Ag. Chemother.* **1968,** 335; Cappel, Klastersky, *Curr. Ther. Res.* **13,** 227 (1971). Comprehensive description: V. Zbinovsky, G. P. Chrekian in *Analytical Profiles of Drug Substances* vol. 6, K. Florey, Ed. (Academic Press, New York, 1977) pp 323-339.

Bright yellow-orange amorphous solid. $[\alpha]_D^{25}$ − 166° (c = 0.524). uv max (0.1N HCl): 352, 263 nm (log ε 4.16, 4.23); (0.1N NaOH): 380, 243 nm (log ε 4.30, 4.38).

Hydrochloride, $C_{23}H_{27}N_3O_7$·HCl, *Minocin, Klinomycin, Minomycin, Vectrin.* Yellow crystalline powder. Slightly hygroscopic; sensitive to light and to surface oxidation.

THERAP CAT: Antibacterial.

6290. Minoxidil. *6-(1-Piperidinyl)-2,4-pyrimidinediamine 3-oxide; 6-amino-1,2-dihydro-1-hydroxy-2-imino-4-piperidinopyrimidine;* 2,3-dihydro-3-hydroxy-2-imino-6-(1-piperidinyl)-4-pyrimidinamine; 2,4-diamino-6-piperidinopyrimidine 3-oxide; 6-piperidino-2,4-diaminopyrimidine 3-oxide; PDP; U-10858; Alopexil; Alostil; Loniten; Lonolox; Minoximen; Normoxidil; Pierminox; Prexidil; Regaine; Rogaine; Tricoxidil. $C_9H_{15}N_5O$; mol wt 209.25. C 51.66%, H 7.23%, N 33.47%, O 7.65%. Prepn: **Neth. pat. Appl. 6,615,385;** W. C. Anthony *et al.,* **U.S. pat. 3,382,247** (1967, 1968 both to Upjohn); J. M. McCall *et al., J. Org. Chem.* **40,** 3304 (1975). *See also* W. C. Anthony, **U.S. pat. 3,644,-364** (1972 to Upjohn). Metabolism: R. C. Thomas *et al., J. Pharm. Sci.* **64,** 1360 (1975). Pharmacology and pharmacokinetics: D. T. Lowenthal *et al., J. Clin. Pharmacol.* **18,** 500 (1978). Percutaneous absorption and excretion: T. J. Franz, *Arch. Dermatol.* **121,** 203 (1985). Clinical studies: O. Andersson, R. Sivertsson, *Acta Med. Scand.* **205,** 213 (1979); M. Moser, *Advan. Cardiol.* **26,** 38 (1979). Clinical trial in early male pattern baldness: E. A. Olsen *et al., J.*

Am. Acad. Dermatol. **13,** 185 (1985). Toxicology: R. G. Carlson, E. S. Feenstra, *Toxicol. Appl. Pharmacol.* **39,** 1 (1977). Review of pharmacology and therapeutic use: V. M. Campese, *Drugs* **22,** 257-278 (1981). Review of topical application in baldness: E. Novak *et al., Int. J. Dermatol.* **24,** 82 (1985). Comprehensive description: D. K. J. Gorecki in *Analytical Profiles of Drug Substances* vol. 17, K. Florey, Ed. (Academic Press, New York, 1988) 185-219.

Crystals from methanol-acetonitrile, mp 248°, dec 259-261° (Anthony, 1972). pKa 4.61. uv max (ethanol): 230, 261, 285 nm (ε 35210, 11210, 11790); (0.01N H_2SO_4): 232, 280 nm (ε 26350, 23850); (0.01N KOH): 231, 261.5, 285 nm (ε 36100, 11400, 12040). Soly (mg/ml): propylene glycol 75, methanol 44, ethanol 29, 2-propanol 6.7, dimethylsulfoxide 6.5, water 2.2, chloroform 0.5, acetone < 0.5, ethylacetate < 0.5, diethyl ether < 0.5, benzene < 0.5, acetonitrile < 0.5. LD_{50} in rats, mice (mg/kg): 49, 51 i.v. (Carlson, Feenstra).

THERAP CAT: Antihypertensive. Antialopecia agent.

6291. Miokamycin. *Leucomycin V 3^B,9-diacetate 3,4^B-dipropanoate;* 9,3''-diacetylmidecamycin; MOM; ponsinomycin; Miocamycin. $C_{45}H_{71}NO_{17}$; mol wt 898.06. C 60.19%, H 7.97%, N 1.56%, O 30.29%. Macrolide antibiotic deriv of midecamycin A_1. Prepn: S. Inoue *et al.,* **Japan. Kokai 74 124087** (1974 to Meiji), *C.A.* **83,** 10735w (1975); S. Omoto *et al., J. Antibiot.* **29,** 536 (1976); T. Nakamura *et al., Chem. Letters* **1978,** 1293. Physico-pharmaceutical properties of crystalline and non-crystalline miokamycin: T. Sato *et al., Chem. Pharm. Bull.* **29,** 2675 (1981). *In vitro* and *in vivo* antibacterial activity: K. Kawaharajo *et al., J. Antibiot.* **34,** 436 (1981). Metabolism: T. Shomura *et al., Chem. Pharm. Bull.* **29,** 2413 (1981). General pharmacological studies: U. Shibata *et al., Japan. J. Antibiot.* **34,** 734 (1981), *C.A.* **96,** 62631x (1982). Laboratory and clinical pediatric studies: Y. Toyonaga *et al., ibid.* **35,** 1475 (1982), *C.A.* **98,** 216y (1983). Mutagenicity study: F. Hirano, U. Takeda, *J. Antibiot.* **34,** 443 (1981).

Tasteless crystals from isopropanol, mp ∼220° (with coloration). $[\alpha]_D^{25}$ − 53° (c = 1.0 in chloroform). $[\alpha]_D^{20}$ −74° (c = 1 in methanol). uv max (methanol): 231 nm ($E_{1cm}^{1\%}$ 342). Sol in methanol, acetone, chloroform. Very slightly sol in water. The non-crystalline form converts gradually to the crystalline form in aqueous suspension. In 0.2% hydroxypropylmethylcellulose soln, the non-crystalline solid was at least 10 times more sol than the crystalline form.

THERAP CAT: Antibacterial.

6292. Mipafox. *N,N'-Bis(1-methylethyl)phosphorodiamidic fluoride; N,N'-diisopropylphosphorodiamidic fluoride;* bis(isopropylamino)fluorophosphine oxide; Isopestox; Pestox XV. $C_6H_{16}FN_2OP$; mol wt 182.18. C 39.56%, H 8.85%, F 10.43%, N 15.38%, O 8.78%, P 17.00%. Prepn: Pound *et al.,* **Brit. pat. 688,787** (1953 to Fisons); Heath, *J. Chem. Soc.* **1956,** 3796.

Crystals, mp 65°.
Caution: Cholinesterase inhibitor.
USE: Insecticide.

6293. Miraculin. Taste-modifying protein; Miralin. A basic glycoprotein of probable mol wt 44,000. Responsible for the taste-changing properties of *Miracle fruit (agbayun)*, the red berries of *Synsepalum dulcificum* (Schum.) Daniell, *Sapotaceae* [alternate name: *Richardella dulcifica* (Schum. and Thonn) Baehni], a shrub indigenous to tropical W. Africa. Sour materials taste sweet after the tasteless mucilaginous pulp of the fruit has been applied to the tongue. First description: Daniell, *Pharm. J.* **11**, 445 (1852). Isolation studies: Inglett *et al., J. Agr. Food Chem.* **13**, 284 (1965); Kurihara, Beidler, *Science* **161**, 1241 (1968). Isoln and purif of miraculin: Brouwer *et al., Nature* **220**, 373 (1968); **Neth. pat. Appl. 6,911,954** corresp to Brouwer *et al.,* **U.S. pat. 3,682,880** (1970, 1972 to Lever Bros.). Max sweetening effect occurs when $4 \times 10^{-7} M$ (in 0.02 M citric acid) is held in the mouth for 3 mins. The protein probably binds to receptors of the taste buds and modifies their function. Mechanism of action studies: Kurihara, Beidler, *Nature*, **222**, 1176 (1969). *Reviews:* Henning, *Pharm. Weekbl.* **106**, 271 (1971); Cagan, *Science* **181**, 32 (1973).

USE: Sweetening agent.

6294. Mirex. *1,1a,2,2,3,3a,4,5,5,5a,5b,6-Dodecachloro-octahydro-1,3,4-metheno-1H-cyclobuta[cd]pentalene;* perchloropentacyclo[5.2.1.0^{2,6}.0^{3,9}.0^{5,8}]decane; hexachloropentadiene dimer; CG-1283; ENT-025719; Dechlorane. $C_{10}Cl_{12}$; mol wt 545.54. C 22.02%, Cl 77.98%. Prepn: Prins, *Rec. Trav. Chim.* **65**, 455 (1946); Newcomer, McBee, *J. Am. Chem. Soc.* **71**, 952 (1949); Gilbert, **U.S. pat. 2,702,305** (1955 to Allied Chem.); Johnson, **U.S. pat. 2,724,730** (1955 to Hooker Electrochem.). Structure: McBee *et al., J. Am. Chem. Soc.* **78**, 1511 (1956). Degradation in the environment to Kepone and related compds: D. A. Carlson *et al., Science* **194**, 939 (1976). Carcinogenicity study: B. M. Ullard *et al., J. Nat. Cancer Inst.* **58**, 133 (1977). Toxicity data: T. B. Gaines, *Toxicol. Appl. Pharmacol.* **14**, 515 (1969). *Reviews:* Ungnade, McBee, *Chem. Rev.* **58**, 249-320 (1958); Alley, *J. Environ. Qual.* **2**, 52-61 (1973); E. M. Waters *et al., Environ. Res.* **14**, 212-222 (1977); K. Kaiser, *Environ. Sci. Technol.* **12**, 520-528 (1978).

Snow-white odorless crystals from benzene, dec 485°. Supports combustion. Practically insol in water. Soly at room temp: dioxane 15.3%; xylene 14.3%; benzene 12.2%; carbon tetrachloride 7.2%; methyl ethyl ketone 5.6%. Practically non-corrosive to metals. LD_{50} orally in female rats (corn oil suspension): 600 mg/kg (Gaines).

Note: This substance may reasonably be anticipated to be a carcinogen: *Seventh Annual Report on Carcinogens* (PB95-109781, 1994) p 260.

USE: Fire retardant for plastics, rubber, paint, paper, electrical goods. Formerly as insecticide for fire ants.

6295. Mirtazepine. *1,2,3,4,10,14b-Hexahydro-2-methyl-pyrazino[2,1-a]pyrido[2,3-c][2]benzazepine;* 2-methyl-1,2,3,-4,10,14b-hexahydrobenzo[c]pyrazino[1,2-a]pyrido[3,2-f]-azepine; 6-azamianserin; mepirzepine; Org-3770; Remeron. $C_{17}H_{19}N_3$; mol wt 265.36. C 76.95%, H 7.22%, N 15.84%. α_2-Adrenergic blocker; analog of mianserin, *q.v.* Prepn: W.

J. Van der Burg, **Ger. pat. 2,614,406**; *idem,* **U.S. pat. 4,062,-848** (1976, 1977 both to AKZO). GC determn in plasma: J. E. Paanakker, H. J. M. Van Hal, *J. Chromatog.* **417**, 203 (1987). Pharmacology and mechanism of action study: T. De Boer *et al., Neuropharmacology* **27**, 399 (1988). Clinical comparison with amitriptyline, *q.v.:* W. T. Smith *et al., Psychopharmacol. Bull.* **26**, 191 (1990).

Crystals from petr ether, mp 114-116°.
THERAP CAT: Antidepressant.

6296. Misch Metal. Mixed rare earth metal. Prepd by electrolysis of fused rare earth chlorides. Misch metal contains 94-99% rare earth metals plus traces of aluminum, calcium, carbon, silicon, and iron. A typical composition is 50-60% cerium, 25-30% lanthanum, 15-17% neodymium, 4-6% praseodymium.

USE: In the manuf of flints for gas and cigarette lighters. In metallurgy: decreases microporosity and improves stress resistance, thermomalleability and heat exchange in refined steel.

6297. Misoprostol. *(11α,13E)-11,16-Dihydroxy-16-methyl-9-oxoprost-13-en-1-oic acid methyl ester;* (±)-methyl-(1R,2R,3R)-3-hydroxy-2-[(4RS)-4-hydroxy-4-methyl-1-octenyl]-5-oxocyclopentaneheptanoate; (±)-15-deoxy-(16RS)-16-hydroxy-16-methyl-PGE$_1$ methyl ester; SC-29333; Cytotec. $C_{22}H_{38}O_5$; mol wt 382.54. C 69.08%, H 10.01%, O 20.91%. Cytoprotective prostaglandin PGE$_1$ analog. Double racemate comprised of the (+)- and (−)-enantiomers of the 16R- and 16S-forms. The pharmacologically active form is the (11R,16S)-enantiomer. Prepn: P. W. Collins, R. Pappo, **Belg. pat. 827,127**; *eidem,* **U.S. pat. 3,965,143** (1975, 1976 both to Searle); P. W. Collins *et al., Tetrahedron Letters* **48**, 4217 (1975). Prepn, activity, NMR data: P. Collins *et al., J. Med. Chem.* **20**, 1152 (1977). HPLC resolution of enantiomers: D. A. Roston, R. Wijayaratne, *Anal. Chem.* **60**, 948 (1988). Mechanism of gastric secretory inhibition: D. G. Colton *et al., Arch. Int. Pharmacodyn. Ther.* **236**, 86 (1978). Symposium on pharmacology and clinical efficacy: *Dig. Dis. Sci.* **30**, Suppl. 11, 114S-205S (1985). Toxicity data: F. N. Kotsonis *et al., ibid.* 142S. Clinical trial in prevention of NSAID-induced ulcer: D. Y. Graham *et al., Ann. Int. Med.* **119**, 257 (1993); to induce labor: H. Fletcher *et al., Obstet. Gynecol.* **83**, 244 (1994).

(11R, 16S)-Form

Light yellow oil. Sol in water. LD_{50} in rats, mice (mg/kg): 40-62, 70-160 i.p.; 81-100, 27-138 orally (Kotsonis).

THERAP CAT: Antiulcerative.

6298. Mitobronitol. *1,6-Dibromo-1,6-dideoxy-D-mannitol;* 1,6-dideoxy-1,6-dibromo-D-mannitol; dibromomannitol; DBM; Myebrol; Myelobromol. $C_6H_{12}Br_2O_4$; mol wt 307.97. C 23.40%, H 3.93%, Br 51.89%, O 20.78%. Cytostatic agent; the first dibromohexitol introduced into clinical use. Manuf: **Brit. pat. 959,407** (1964 to Chinoin). Pharmacobiochemical studies: J. Szabo *et al., Neoplasma* **20**, 13 (1973). Clinical studies: D. F. Chiuten *et al., Cancer* **47**,

442 (1981). Comparative study with busulfan, *q.v.*: R. T. Silver *et al.*, *Cancer* **60**, 1442 (1987).

Crystals from methanol + dichloroethane, mp 176-178°.
THERAP CAT: Antineoplastic.

6299. Mitoguazone. *2,2'-(1-Methyl-1,2-ethanediylidene)bis[hydrazinecarboximidamide]; 1,1'-[(methylethanediylidene)dinitrilo]diguanidine;* methylglyoxal bis(guanylhydrazone); pyruvaldehyde bis(amidinohydrazone). $C_5H_{12}N_8$; mol wt 184.20. C 32.60%, H 6.57%, N 60.83%. Prepn: Baiocchi *et al.*, *J. Med. Chem.* **6**, 431 (1963); Oliverio, Denham, *J. Pharm. Sci.* **52**, 202 (1963). Experimentally effective against myelogenous leukemia: Freireich *et al.*, *Cancer Chemother. Repts.* **16**, 183 (1962).

Hemihydrate, crystals from isopropyl alcohol, dec 225°. uv max at pH 1: 283 nm (ε 38,400), at pH 11: 325 nm (ε 33,500).
Dihydrochloride, $C_5H_{12}N_8 \cdot 2HCl$, *NSC-32946*, *Methyl-GAG*. Crystals from acetone, dec 256-257°.
Diacetate, *NSC-30689*.
THERAP CAT: Antineoplastic.

6300. Mitolactol. *1,6-Dibromo-1,6-dideoxygalactitol;* 1,6-dibromo-1,6-dideoxydulcitol; dibromodulcitol; dibromodulcit; DBD; NSC-104800; Elobromol; Mitolac. $C_6H_{12}Br_2O_4$; mol wt 307.97. C 23.40%, H 3.93%, Br 51.89%, O 20.78%. Diastereoisomer of mitobronitol, *q.v.*, with myelosuppressive effects. Prepn and anticancer activity: **Neth. pat. Appl. 6,600,395**; P. Horváth-Lengyel *et al.*, U.S. pat. **3,993,781** (1966, 1976 both to Chinoin); B. Kellner *et al.*, *Nature* **213**, 402 (1967). Structure-activity correlations: L. Institoris *et al.*, *Arzneimittel-Forsch.* **17**, 145 (1967). Pharmacological study: T. H. Corbett *et al.*, *Cancer* **40**, Suppl. 5, 2660 (1977). Metabolism, pharmacokinetics: I. P. Horváth *et al.*, *Eur. J. Cancer* **15**, 337 (1979). Clinical studies: W. Medina, J. M. Kirkwood, *Cancer Treat. Rep.* **66**, 195 (1982); D. C. Tormey *et al.*, *Am. J. Clin. Oncol.* **5**, 33 (1982).

Crystals from methanol, aq alc, or aq acetone, mp 187-188° (dec). LD_{50} in rats (mg/kg): 1400 orally; 470 i.p. (Kellner).
THERAP CAT: Antineoplastic.

6301. Mitomycins. A group of antitumor antibiotics produced by *Streptomyces caespitosus* (*griseovinaceseus*). Isoln of mitomycins A and B: T. Hata *et al.*, *J. Antibiot.* **9A**, 141 (1956). Isoln of C: S. Wakaki *et al.*, *Antibiot. Chemother.* **8**, 228 (1958); **Brit. pat. 830,874**; T. Hata *et al.*, U.S. pat. **3,660,578** (1960, 1972 both to Kyowa and Kitasato Kenkyusho). Production of the complex: T. Yamamoto, H. Umezawa, **Japan. pat. 2898** (1956), *C.A.* **51**, 9100 (1957). Improved method: A. Gourevitch *et al.*, U.S. pat. **3,042,582**

(1962 to Bristol-Myers). Isoln from *S. verticillatus*, constituents: D. V. Lefemine *et al.*, *J. Am. Chem. Soc.* **84**, 3184 (1962). Structures: J. S. Webb *et al.*, *ibid.* 3185, 3187. Total synthesis of mitomycins A and C: T. Fukuyama *et al.*, *Tetrahedron Letters* **1977**, 4295. Toxicity: S. Kinoshita *et al.*, *J. Med. Chem.* **14**, 13 (1971). Review and evaluation of studies of carcinogenic action of mitomycin C in laboratory animals: *IARC Monographs* **10**, 171-179 (1976). Review of syntheses: Y. Kishi, *J. Nat. Prod.* **42**, 549-568 (1979). General review: R. W. Franck, *Fortschr. Chem. Org. Naturst.* **38**, 1-41 (1979).

Mitomycin C

Mitomycin A, $C_{16}H_{19}N_3O_6$. The 6-methoxy analog of mitomycin C. X-ray crystallographic study: A. Tulinsky, *J. Am. Chem. Soc.* **84**, 3188 (1962); A. Tulinsky, J. H. van den Hende, *ibid.* **89**, 2905 (1967). Red-violet crystals from acetone + carbon tetrachloride, dec 159-161°. Absorption max (water): 215, 318, 530 nm ($E_{1cm}^{1\%}$ 234, 122, 118.8). Sol in water, benzene, toluene, trichloroethylene, nitrobenzene and many organic solvents. Practically insol in xylene, carbon tetrachloride, carbon disulfide, petr ether, ligroin, cyclohexane. LD_{50} i.v. in mice: 2 mg/kg (Hata, 1956; Kinoshita).
Mitomycin B, $C_{16}H_{19}N_3O_6$. Molecular structure: R. Yahashi, I. Matsubara, *J. Antibiot.* **29**, 104 (1976); *ibid.* **31**(6), correction (1978). Violet crystals from acetone + carbon tetrachloride, dec 182-184°. Absorption max (water): 220, 320, 550 nm ($E_{1cm}^{1\%}$ 117.5, 55.0, 9.9). Sol in water and many organic solvents. Practically insol in xylene, carbon tetrachloride, carbon disulfide, petr ether, ligroin, cyclohexane, benzene, toluene, trichloroethylene, nitrobenzene. LD_{50} i.v. in mice: 10 mg/kg (Hata, 1956); also reported as 3 mg/kg (Kinoshita).
Mitomycin C, $C_{15}H_{18}N_4O_5$, *[1aS-(1aα,8β,8aα,8bα)]-6-amino-8-[[(aminocarbonyl)oxy]methyl]-1,1a,2,8,8a,8b-hexahydro-8a-methoxy-5-methylazirino[2',3':3,4]pyrrolo[1,2-a]indole-4,7-dione*, *MMC*, *Ametycine*, *Mitocin-C*, *Mutamycin*. Chemistry and structure: C. L. Stevens *et al.*, *J. Med. Chem.* **8**, 1 (1965). Crystal and molecular structure: K. Ogawa *et al.*, *Bull. Chem. Soc. Japan* **52**, 2334 (1979). Revised absolute configuration: K. Shirahata, N. Hirayama, *J. Am. Chem. Soc.* **105**, 7199 (1983). Mechanism of action: C. Rodigheriero, *Farmaco Ed. Sci.* **33**, 651 (1978). Comprehensive description: J. H. Beijnen *et al.*, *Anal. Profiles Drug Subs.* **16**, 361-401 (1986). Blue-violet crystals, does not melt below 360°. Absorption max (methanol): 216, 360, 560 nm ($E_{1cm}^{1\%}$ 742, 742, 0.06). Sol in water, methanol, acetone, butyl acetate and cyclohexanone; slightly sol in benzene, carbon tetrachloride, ether. Practically insol in petr ether. LD_{50} i.v. in mice: 5 mg/kg (Wakaki); also reported as 9 mg/kg (Kinoshita). Mitomycin C exhibits delayed toxicity to mice; injected mice die from 2 days to 2 weeks after injection.
N-Methylmitomycin C, see Porfiromycin.
THERAP CAT: Mitomycin C as antineoplastic.

6302. Mitotane. *1-Chloro-2-[2,2-dichloro-1-(4-chlorophenyl)ethyl]benzene; 1,1-dichloro-2-(o-chlorophenyl)-2-(p-chlorophenyl)ethane;* 2-(2-chlorophenyl)-2-(4-chlorophenyl)-1,1-dichloroethane; 2,4'-dichlorodiphenyldichloroethane; 2,-2-bis(2-chlorophenyl-4-chlorophenyl)-1,1-dichloroethane; *o,p'*-DDD; CB-313; Lysodren. $C_{14}H_{10}Cl_4$; mol wt 320.04. C 52.54%, H 3.15%, Cl 44.31%. Constituent of commercial DDD which contains about 10% of this *o,p'*-isomer. Prepn from 2,2-dichloro-1-(*o*-chlorophenyl)ethanol with chlorobenzene in presence of H_2SO_4: Haller *et al.*, *J. Am. Chem.*

Soc. **67**, 1600 (1945). Isoln from technical grade DDD: Cueto, Brown, *Endocrinology* **62**, 326 (1958).

Crystals from pentane or methanol, mp 76-78°. Sol in ethanol, isooctane, carbon tetrachloride.

THERAP CAT: Antineoplastic.

6303. Mitoxantrone. *1,4-Dihydroxy-5,8-bis[[2-[(2-hydroxyethyl)amino]ethyl]amino]-9,10-anthracenedione;* 1,4-dihydroxy-5,8-bis[[2-[(2-hydroxyethyl)amino]ethyl]amino]-9,10-anthraquinone; DHAQ; NSC-279836. $C_{22}H_{28}N_4O_6$; mol wt 444.49. C 59.45%, H 6.35%, N 12.60%, O 21.60%. Cytostatic deriv of anthraquinone, *q.v.* Prepn: R. K. Y. Zee-Cheng, C. C. Cheng, *J. Med. Chem.* **21**, 291 (1978); K. C. Murdock *et al., ibid.* **22**, 1024 (1979); *eidem*, Ger. pat. **2,835,661,** *C.A.* **91**, 56706 (1979) and K. C. Murdock, F. E. Durr, U.S. pat. **4,197,249** (1979, 1980 both to Am. Cyanamid). Activity vs exptl tumors in mice: R. E. Wallace *et al., Cancer Res.* **39**, 1570 (1979). Pharmacological study: K. Lu, T. L. Loo, *ibid.* **40**, 1427 (1980). HPLC studies: R. F. Taylor, L. A. Gaudio, *J. Chromatog.* **187**, 212 (1980); D. L. Reynolds *et al., ibid.* **222**, 225 (1981). Plasma levels in humans: *eidem, Int. J. Pharm.* **9**, 67 (1981). Clinical studies: W. R. Wynert *et al., Cancer Treat. Rep.* **66**, 1303 (1982); J. A. Stewart *et al., ibid.* 1327; J. D. Cowan *et al., ibid.* 1779. Safety assessment: B. M. Henderson *et al., ibid.* 1139; B. M. Sparano *et al., ibid.* 1145. Cardiotoxicity study: G. Zbinden, A. K. Beilstein, *Toxicol. Letters* **11**, 289 (1982). Mutagenicity study: A. Nishio *et al., Mutat. Res.* **101**, 77 (1982). Reviews: F. Traganos, *Pharmacol. Ther.* **22**, 199-214 (1983); M. McDonald *et al., Drugs Exp. Clin. Res.* **10**, 745-752 (1984). Comprehensive description: J. H. Beijnen *et al., Anal. Profiles Drug Subs.* **17**, 221-258 (1988).

Crystals from ethanol/hexane, mp 160-162°. Abs max (ethanol): 244, 279, 525, 620, 660 nm (log ε 4.64, 4.31, 3.70, 4.37, 4.38). Sparingly sol in water, slightly sol in methanol. Practically insol in acetonitrile, chloroform, acetone.

Dihydrochloride, $C_{22}H_{28}N_4O_6$.2HCl, DHAD, DHAD, NSC-301739, Novantrone. Hygroscopic blue-black solid from water/ethanol, mp 203-205°. Abs max (water): 241, 273, 608, 658 nm (ε 41000, 12000, 19200, 20900). Sparingly sol in water, slightly sol in methanol. Practically insol in acetonitrile, chloroform, acetone.

THERAP CAT: Antineoplastic.

6304. Mitragynine. *(16E,20β)-16,17-Didehydro-9,17-dimethoxycorynan-16-carboxylic acid methyl ester;* (E)-16,17-didehydro-9,17-dimethoxy-17,18-seco-20α-yohimban-16-carboxylic acid methyl ester; 9-methoxycorynantheidine. $C_{23}H_{30}$-N_2O_4; mol wt 398.50. C 69.32%, H 7.59%, N 7.03%, O 16.06%. Major alkaloid of *Mitragyna speciosa* Korth., *Rubiaceae:* Field, *J. Chem. Soc.* **119**, 887 (1921); Ing, Raison, *ibid.* **1939**, 986. Structural studies: Hendrickson, *Chem. & Ind. (London)* **1961**, 713; Joshi *et al., ibid.* **1963**, 573. Revised structure: Zacharias *et al., Acta Cryst.* **18**, 1039 (1965). Pharmacology: Macko *et al., Arch. Int. Pharmacodyn. Ther.* **198**, 145 (1972).

White, amorphous powder, mp 102-106°. bp_5 230-240°. $[\alpha]_D$ +39° (chloroform). uv max: 226, 292 nm (ε 41,150, 6600). Soluble in alcohol, chloroform, acetic acid.

6305. Mivacurium Chloride. *[R-[R*,R*-(E)]]-2,2'-(1,8-Dioxo-4-octene-1,8-diyl)bis(oxy-3,1-propanediyl)bis(1,2,3,4-tetrahydro-6,7-dimethoxy-2-methyl-1-[(3,4,5-trimethoxyphenyl)methyl]isoquinolinium) dichloride;* (R)-1,2,3,4-tetrahydro-2-(3-hydroxypropyl)-6,7-dimethoxy-2-methyl-1-(3,4,5-trimethoxybenzyl)isoquinolinium chloride, (E)-4-octenedioate (2:1); BW-B1090U; Mivacron. $C_{58}H_{80}Cl_2N_2O_{14}$; mol wt 1100.18. C 63.32%, H 7.33%, Cl 6.44%, N 2.55%, O 20.36%. Non-depolarizing neuromuscular blocking agent. Prepn: R. A. Swaringen, Jr. *et al.,* East Ger. pat. **235,638;** *eidem,* U.S. pat. **4,761,418** (1986, 1988 both to Burroughs Wellcome). Clinical pharmacology: H. H. Ali *et al., Brit. J. Anaesth.* **61**, 541 (1988); J. J. Savarese *et al., Anesthesiology* **68**, 723 (1988). *In vitro* metabolism: D. R. Cook *et al., Anesth. Analg.* **68**, 452 (1989). Clinical evaluation in adults: S. Weber *et al., ibid.* **67**, 495 (1988); R. P. From *et al., Brit. J. Anaesth.* **64**, 193 (1990); in children: J. K. Alifimoff, N. G. Goudsouzian, *ibid.* **63**, 520 (1989).

Amorphous solid. $[\alpha]_D^{20}$ −62.7° (c = 1.9 in water).
THERAP CAT: Muscle relaxant (skeletal).

6306. Mizoribine. *5-Hydroxy-1-β-D-ribofuranosyl-1H-imidazole-4-carboxamide;* 4-carbamoyl-1-β-D-ribofuranosyl-imidazolium-5-olate; HE-69; Bredinin. $C_9H_{13}N_3O_6$; mol wt 259.22. C 41.70%, H 5.05%, N 16.21%, O 37.03%. Nucleoside antibiotic produced by *Eupenicillium brefedianum* with cytotoxic and immunosuppressive activity. Isoln: K. Mizuno *et al.,* Belg. pat. **799,805;** *eidem,* U.S. pat. **3,888,843** (1973, 1975 to Toyo Jozo); *eidem, J. Antibiot.* **27**, 775 (1974). Synthesis: M. Hayashi *et al., Chem. Pharm. Bull.* **23**, 245 (1975); K. Fukukuwa *et al., ibid.* **32**, 1644 (1984). Prepn of the aglycone: T. Atsumi *et al.,* Japan. Kokai **76 88,965** (1976 to Sumitomo), *C.A.* **86**, 106582g (1977). HPLC determn in human serum: K. Takada *et al., J. Chromatog.* **222**, 156 (1981). Anti-arthritic activity: H. Iwata *et al., Experientia* **33**, 502 (1977). Cytotoxic effect and comparison with aglycone: K. Sakaguchi *et al., J. Antibiot.* **28**, 798 (1975). Antitumor spectrum of aglycone: N. Yoshida *et al., Cancer Res.* **43**, 5851 (1983). Pharmacokinetics: K. Takada, *Eur. J. Pharmacol.* **24**, 457 (1983). Clinical trials in renal transplantation: T. Inou *et al., Transplantation Proc.*

12, 526 (1980); *eidem, ibid.* **13**, 315 (1981); A. Tajima *et al., Transplantation* **38**, 116 (1984).

Crystals from methanol, mp > 200° (dec). $[\alpha]_D^{27}$ −35° (c = 0.8 in H_2O). pKa 6.75. uv max (H_2O): 245, 279 nm (E 250, 580). Sol in water. Slightly sol in methanol, ethanol. Insol in most organic solvents. LD_{50} in mice (g/kg): > 1.5 i.v., > 2.4 i.p. (Mizuno, 1975).

Aglycone, $C_4H_5N_3O_2$, *5-hydroxy-1H-imidazole-4-carboxamide; 4-carbamoyl-5-hydroxyimidazole; 4-carbamoylimidazolium-5-olate, SM-108.*

THERAP CAT: Immunosuppressant.

6307. MMT. *Tricarbonyl[(1,2,3,4,5-η)-1-methyl-2,4-cyclopentadien-1-yl]manganese;* methylcyclopentadienylmanganese tricarbonyl; MCMT; tricarbonyl(η^5-methylcyclopentadienyl)manganese(I); TCMn; AK-33X. $C_9H_7MnO_3$; mol wt 218.09. C 49.57%, H 3.24%, Mn 25.19%, O 22.01%. Organometallic antiknock compound used to increase the octane quality of fuel. Prepn: J. E. Brown *et al.,* U.S. pat. **2,818,417** (1957 to Ethyl Corp.). Evaluation of antiknock properties: R. J. Riggs *et al., Proc. 37th Natural Gas Assoc. Am.* **1958**, 51. Effect on auto emissions: W. R. Pierson *et al., J. Air Pollut. Control. Assoc.* **28**, 692 (1978). Use as dopant in Mn deposition: K. Hirabayashi, H. Kozawaguchi, *Japan. J. Appl. Phys.* **25**, 711 (1986). Pyrolysis mechanism in deposition: S. Wen-bin *et al., J. Cryst. Growth* **113**, 1 (1991); I. M. T. Davidson *et al., J. Mater. Chem.* **4**, 13 (1994). GC determn in gasoline: W. A. Aue *et al., Anal. Chem.* **62**, 2453 (1990); and in air: V. S. Gaind *et al., Analyst* **117**, 161 (1992). Review as octane enhancer: J. D. Bailie, *Oil Gas J.* **74**, 69-72 (1976); D. P. Hollrah, A. M. Burns, *ibid.* **89**, 86-90 (1991). Review of toxicology: J. Stara *et al., Proc. 4th Ann. Conf. Environment. Toxicol.* AD-781 031, 251-270 (1973); P. J. Abbott, *Sci. Total Environ.* **67**, 247-255 (1987); and environmental assessment: D. R. Lynam *et al., ibid.* **93**, 107-114 (1990.)

Yellow orange liquid, fp −0.75°. n_D^{20} 1.5873. d_{20}^4 1.3942. Vapor pressure ranges from 8 mm at 100° to 360.6 mm at 200°. Readily sol in hydrocarbons and the usual organic solvents including hexane, alchohols, ether, acetone, ethylene glycol, lubricating oils, and hydrocarbon fuels, such as gasoline and diesel fuel. LD_{50} orally in mice, guinea pig, rabbit: 352; 905; 95 mg/kg; orally in male, female rats: 175 ±33; 89 ±14; LC_{50} inhalation in rat: 0.22 mg/l (Stara).

USE: Antiknock fuel additive; dopant for Mn.

6308. Mocimycin. Antibiotic MYC 8003; delvomycin; kirromycin; MYC-8003. $C_{43}H_{60}N_2O_{12}$; mol wt 796.96. C 64.81%, H 7.59%, N 3.52%, O 24.09%. Antibiotic produced by *Streptomyces ramocissimus*. Isoln and properties: C. Vos, J. Den Admirant, Ger. pat. **2,140,674;** *eidem,* U.S. pats. **3,927,211** and **3,923,981** (1972, 1975, 1975 all to Gist-Brocades). Structural studies: C. Vos, P. E. J. Verwiel, *Tetrahedron Letters* **1973**, 2823, 5173. Total structure and identity with kirromycin: H. Maehr *et al., J. Am. Chem. Soc.* **95**, 8449 (1973). Absolute stereochemistry: *eidem, ibid.* **96**, 4034 (1974). Conversion to aurodox: H. Maehr *et al., J. Antibiot.* **32**, 361 (1979). Biological studies: H. Wolf *et al., Proc. Nat. Acad. Sci. USA* **71**, 4910 (1974); J. A. M. Van De Klundeit *et al., FEBS Letters* **81**, 303 (1977). Review of chemistry: H. Maehr *et al., Can. J. Chem.* **58**, 501-526 (1980). Review of properties and actions: A. Parmeggiani, G. Sander, *Topics in Antibiotic Chemistry* vol. 5, P. G. Sammes, Ed. (Halsted Press, New York, 1980) pp 159-221.

Yellow solid. Weakly acidic. $[\alpha]_D^{22}$ −60° (c = 1 in methanol). uv max (methanol/water): 233, 276, 286, 327 nm. Sol in chloroform, methyl isobutyl ketone, butyl acetate, ethyl acetate, acetone, methanol, alkaline soln. Slightly sol in CCl_4, benzene. Insol in diethyl ether, petr ether, water, acid soln. Stable for 4 hrs in 50% aq methanol at pH 3-12. The solid antibiotic stored at 25° and 37° and low relative humidity shows no activity loss for 5 mos. Stable for 3 mos at 25°, 100% relative humidity; for 2 mos at 37°, 100% relative humidity. LD_{50} i.p. in mice: > 1000 mg/kg (Vos, Den Admirant, 1972).

1-Methyl deriv, see Aurodox.

5,6-Dihydromocimycin, $C_{43}H_{62}N_2O_{12}$. Isoln from *S. ramocissimus* and properties: H. Jongsma *et al.,* Ger. pat. **2,658,977** (1977 to Gist-Brocades), *C.A.* **87**, 135078y (1977); see also C. Vos *et al.,* U.S. pat. **4,062,948** (1977 to Gist-Brocades). Pale yellow solid, dec begins at 123°. $[\alpha]_D^{20}$ 85° (1% methanolic soln). uv max (methanol/water): 233.5, 267, 291, 333 nm (ε 63000, 23000, 19000, 18000). Solubility similar to mocimycin in most solvents.

USE: Animal growth promotant.

6309. Moclobemide. *4-Chloro-N-[2-(4-morpholinyl)-ethyl]benzamide;* p-chloro-N-(2-morpholinoethyl)benzamide; Ro-11-1163; Aurorix; Manerix; Moclamine. $C_{13}H_{17}ClN_2O_2$; mol wt 268.74. C 58.10%, H 6.38%, Cl 13.19%, N 10.42%, O 11.91%. Reversible monoamine oxidase A (MAO-A) inhibitor. Prepn: W. Burkard, P.-C. Wyss, Ger. pat. **2,706,179;** U.S. pat. **4,210,754** (1977, 1980 both to Hoffmann-La Roche). Gas chromatography: K. P. Maguire *et al., J. Chromatog.* **278**, 429 (1983). Human pharmacokinetics: F. A. Wiesel *et al., Eur. J. Clin. Pharmacol.* **28**, 89 (1985). Clinical study: J. K. Larsen *et al., Acta Psychiat. Scand.* **70**, 254 (1984). Review of pharmacology: W. P. Burkard *et al., J. Pharmacol. Exp. Ther.* **248**, 391-399 (1989); of neurochemical profile: M. DaPrada *et al., ibid.* 400-414.

Crystals from isopropanol, mp 137°. LD_{50} in rats (mg/kg): 707 orally (Burkard, Wyss).

Hydrochloride, $C_{13}H_{18}Cl_2N_2O_2$, crystals from isopropanol, mp 208°.

THERAP CAT: Antidepressant.

6310. Modacrylic Fibers. A generic term for acrylic fibers composed of a long chain synthetic polymer containing 35-85% of acrylonitrile units such as *Dynel* and *Verel.* See: R. W. Moncrieff, *Man-Made Fibres* (Wiley, New York, 4th ed., 1963) p 9. *Review:* Kennedy in *Encyclopedia of Polymer Science and Technology* vol. 8, N. M. Bikales, Ed. (Interscience, New York, 1968) pp 812-839.

USE: In pile fabrics, carpeting, paint rollers.

6311. Modafinil. *2-[(Diphenylmethyl)sulfinyl]acetamide;* 2-(benzhydrylsulfinyl)acetamide; CRL-40476. $C_{15}H_{15}NO_2S$; mol wt 273.36. C 65.91%, H 5.53%, N 5.12%, O 11.71%, S 11.73%. α_1-Adrenergic agonist. Prepn: L. Lafon, Ger. pat. **2,809,625;** *eidem,* U.S. pat. **4,177,290** (1978, 1979 both to Lab. Lafon). Effects on vigilance: F. Goldenberg, J. S. Weil, *Sleep (Stuttgart)* **8**, 343 (1988); on normal sleep: B. Saletu *et al., Int. J. Clin. Pharmacol. Res.* **9**, 183 (1989). Clinical evaluation in hypersomnia and narcolepsy: H. Bastuji, M. Jouvet, *Prog. Neuro-Psychopharmacol. Biol. Psychiatry* **12**, 695 (1988).

Crystals from methanol, mp 164-166°.
THERAP CAT: CNS Stimulant.

6312. Mofebutazone. *4-Butyl-1-phenyl-3,5-pyrazolidinedione;* 4-butyl-1-phenyl-3,5-dioxopyrazolidine; 2-phenyl-3,5-dihydroxy-4-butylpyrazolidine; monophenylbutazone; Arcomonol Tablets; Mobutazon; Mobuzon; Mofesal; Monazan; Monobutyl; Monorheumetten; Reumatox. C_{13}-$H_{16}N_2O_2$; mol wt 232.28. C 67.22%, H 6.94%, N 12.06%, O 13.78%. Prepn: Büchi *et al., Helv. Chim. Acta* **36**, 75 (1953); **Brit. pat.** **839,057** (1960 to Comm. Farm. Milanese). Toxicity data: Schoetensack, *Arch. Exp. Pathol. Pharmakol.* **233**, 365 (1958).

Crystals from ethanol + water, mp 102-103°. uv max (ethanol): 240, 275 nm ($E_{1cm}^{1\%}$ 443, 245). LD_{50} i.v. in mice: 600 mg/kg (Schoetensack).
THERAP CAT: Anti-inflammatory.

6313. Mofegiline. *(E)-4-Fluoro-β-(fluoromethylene)-benzenebutanamine;* (E)-2-(fluoromethylene)-4-(*p*-fluorophenyl)butylamine; (E)-2-(4-fluorophenethyl)-3-fluoroallylamine. $C_{11}H_{13}F_2N$; mol wt 197.23. C 66.99%, H 6.64%, F 19.27%, N 7.10%. Irreversible monoamine oxidase-B (MAO-B) inhibitor. Prepn: I. A. McDonald, M. G. Palfreyman, **Eur. pat. Appl.** **295,604** (1988 to Merrell Dow). GC-MS determn in plasma and urine: B. D. Dulery *et al., J. Chromatog.* **571**, 241 (1991). Pharmacokinetics: *eidem et al., Arzneimittel-Forsch.* **43**, 297 (1993). Metabolism: J. Dow *et al., Drug Metab. Dispos.* **22**, 738 (1994). Clinical pharmacology: C. Hinze *et al., J. Neural Transm. Suppl.* **41**, 371-375 (1994). Clinical evaluation in Parkinson's disease: V. V. Myllylä *et al., Adv. Neurol.* **60**, 676 (1993).

Hydrochloride, $C_{11}H_{13}F_2N$·HCl, *MDL-72974A*. Colorless plates, mp 131°.
THERAP CAT: Antiparkinsonian.

6314. Mofezolac. *3,4-Bis(4-methoxyphenyl)-5-isoxazoleacetic acid;* 3,4-di(*p*-methoxyphenyl)-5-isoxazoleacetic acid; N-22. $C_{19}H_{17}NO_5$; mol wt 339.35. C 67.25%, H 5.05%, N 4.13%, O 23.57%. Prostaglandin biosynthetase inhibitor. Prepn: R. G. Micetich *et al.,* **Eur. pat. Appl.** **26,928;** *eidem,* **U.S. pat.** **4,327,222** (1981, 1982 both to CDC Life Sci.). Physico-chemical properties: T. Ushio *et al., Iyakuhin Kenkyu* **22**, 892 (1991). HPLC determn in plasma and urine: T. Marunaka, M. Maniwa, *J. Chromatog.* **422**, 227 (1987). Pharmacology: K. Tanaka *et al., Yakuri to*

Chiryo **18**, 3347 (1990), *C.A.* **114**, 135911 (1991). Acute toxicity: K. Satoh *et al., J. Toxicol. Sci.* **15**, Suppl 2, 1 (1990).

White crystalline powder; slight characteristic odor. mp 147.5°. pKa' ∼3.3. uv max (methanol): 236 nm (ϵ = 1.83 × 10⁴). Soly at 20° (w/v%): DMF 68.7; chloroform 19.5; ethyl acetate 16.5; acetone 7.32; methanol 3.93; anhydr ethanol 3.72; ether 2.90; hexane 2.90 × 10⁻⁵; water 4.85 × 10⁻³. Partition coefficient (chloroform/water): 16.6 (pH 6); 1.87 (pH 7); 0.178 (pH 8). LD_{50} in male, female mice, male, female rats (mg/kg): 1528, 1740, 920, 887 orally; 275, 321, 378, 342 i.p.; 612, 545, 572, 510 s.c. (Satoh).
THERAP CAT: Anti-inflammatory; analgesic.

6315. Molindone. *3-Ethyl-1,5,6,7-tetrahydro-2-methyl-5-(4-morpholinylmethyl)-4H-indol-4-one;* 3-ethyl-6,7-dihydro-2-methyl-5-(morpholinomethyl)indol-4(5H)-one. C_{16}-$H_{24}N_2O_2$; mol wt 276.38. C 69.53%, H 8.75%, N 10.14%, O 11.58%. Prepn: **Belg. pat.** **670,798** (1966 to Endo), *C.A.* **65**, 7148f (1966). Pharmacology: Sugerman, Herrmann, *Clin. Pharmacol. Ther.* **8**, 261 (1967); Claghorn, *Curr. Ther. Res.* **11**, 524 (1969); Guerrero-Figueroa *et al., ibid.* **15**, 508 (1973). Toxicity study: E. I. Goldenthal, *Toxicol. Appl. Pharmacol.* **18**, 185 (1971).

Crystals, mp 180-181°.
Hydrochloride, *EN-1733A, Lidone, Moban.* LD_{50} orally in rats: 261 mg/kg (Goldenthal).
THERAP CAT: Antipsychotic.

6316. Molsidomine. *N-(Ethoxycarbonyl)-3-(4-morpholinyl)sydnone imine;* N-carboxy-3-morpholinosydnonimine ethyl ester; morsydomine; SIN-10; Corvaton; Corvasal; Molsidolat; Morial; Motazomin. $C_9H_{14}N_4O_4$; mol wt 242.23. C 44.63%, H 5.83%, N 23.13%, O 26.42%. Coronary vasodilator; member of a class of non-benzene aromatic, heterocyclic and mesoionic type of compounds previously unknown in the pharmaceutical industry. Developmental work on sydnone imines: Brookes, Walker, *J. Chem. Soc.* **1957**, 4409. Prepn: K. Masuda *et al.,* **Japan. pat.** **6,265**('70); *eidem,* **U.S. pat.** **3,769,283** (1970, 1973 both to Takeda); *eidem, Chem. Pharm. Bull.* **19**, 72 (1971). Stability studies: Asahi *et al., ibid.* **19**, 1079 (1971). Pharmacological studies: Kikuchi *et al., Japan. J. Pharmacol.* **20**, 102, 187, 253 (1970); Hashimoto *et al., Arzneimittel-Forsch.* **21**, 1329 (1971); T. Nakaguchi *et al., Toho Igakkai Zasshi* **17**, 26 (1970). Metabolism: S. Tanayama *et al., Xenobiotica* **4**, 175 (1974). Pharmacokinetics: R. Bergstrand *et al., Eur. J. Clin. Pharmacol.* **27**, 203 (1984). Clinical trial in angina pectoris: P. A. Majid *et al., N. Engl. J. Med.* **302**, 1 (1980); in hypertension: J. Melei *et al., Eur. J. Clin. Pharmacol.* **18**, 231 (1980). Synergism with penbutolol in angina pectoris: A. E. Balestrini *et al., ibid.* **27**, 1 (1984). Review of pharmacology: R.-E. Nitz, V. B. Fiedler, *Pharmacotherapy* **7**, 28 (1987). *Review: Japan. Med. Gaz.* **8**(9), 10 (1971); E. Bassenge, W. R. Kukovetz in *New Drugs Annual: Cardiovascular Drugs,* **Vol. 2**, A. Scriabine, Ed. (Raven Press, New York, 1984) pp 177-191.

Colorless crystals or white cryst powder, practically tasteless and odorless, mp 140-141° (toluene). Freely sol in $CHCl_3$. Sol in dil HCl, ethanol, ethyl acetate, methanol; sparingly sol in water, acetone, benzene. Very slightly sol in ether, petr ether. pK (100°) 3.0 ± 0.1. Most stable in aq solns pH 5-7; least stable in very alkaline solns. uv max ($CHCl_3$): 326 nm. Sensitive to light of $\lambda < 320$ mμ. LD_{50} in male, female mice, male, female rats (mg/kg): 780, 750, 1380, 1350 s.c.; 860, 800, 830, 760 i.v.; 700, 760, 1250, 1250 i.p.; 830, 840, 1050, 1200 orally (Nakaguchi).

THERAP CAT: Antianginal.

6317. Molybdenum. Mo; at. wt 95.94; at. no. 42; valences 2,3,4,5,6. Naturally occurring isotopes: 98 (23.75%); 96 (16.5%); 95 (15.7%); 92 (15.86%); 94 (9.12%); 100 (9.62%); 97 (9.45%); artificial radioactive isotopes: 88-91; 93; 99; 101-105. Its most important ores are molybdenite, MoS_2, and wulfenite, $PbMoO_4$. Occurrence in the earth's crust: 1-1.5 ppm. Discovered in 1778 by Scheele; isolated in 1782 by Hjelm. Methods of preparation: L. Northcott, *Molybdenum* (Academic Press, New York, 1956) 222 pp; Hein, Herzog, in *Handbook of Preparative Inorganic Chemistry* vol. 2, G. Brauer, Ed. (Academic Press, New York, 2nd ed., 1965) pp 1401-1402. Important trace element; participates in biochemical redox reactions such as N_2-fixation: Spence, *Coord. Chem. Rev.* **4**, 475 (1969). Physical properties: Worthing, *Phys. Rev.* [2] **25**, 846 (1925); D. R. Stoll, G. C. Sinke, *Thermodynamic Properties of the Elements*, Advances in Chemistry Series **18**, (American Chemical Society, Washington, 1956) pp 23, 130-131. Review of molybdenum and its compds: Rollinson, "Chromium, Molybdenum and Tungsten" in *Comprehensive Inorganic Chemistry* vol. 3, J. C. Bailar Jr. *et al.*, Eds. (Pergamon Press, Oxford, 1973) pp 622-623, 700-742; R. Q. Barr in Kirk-Othmer *Encyclopedia of Chemical Technology* vol. 15 (Wiley-Interscience, New York, 3rd ed., 1981) pp 670-682. Biochemical review: *Bioinorganic Chemistry* II, K. N. Raymond, Ed. (A.C.S., Washington, 1977) pp 353-430. Symposium on the chemistry and uses of molybdenum and its cmpds: *Polyhedron* **5**, 1-606 (1986).

Dark-gray or black powder with metallic luster or coherent mass of silver-white color; body-centered cubic structure. mp 2622° (Worthing). bp ~4825°. d 10.28. Spec heat 5.68 cal/g-atom/deg; heat of fusion: 6.6 kcal/g-atom; heat of vaporization: 142 kcal/g-atom (Stoll, Sinke). Fairly stable at ordinary temp; oxidized to the trioxide at a red heat; slowly oxidized by steam. Not attacked by water, by dil acids or by concd hydrochloric acid. Practically insol in alkali hydroxides or fused alkalies. Reacts with nitric acid, hot concd sulfuric acid, fused potassium chlorate or nitrate. Attacked by fluorine at ordinary temp, by chlorine or bromine at a red heat.

Caution: Limited data suggest low order of toxicity. See E. Browning, *Toxicity of Industrial Metals* (Appleton-Century-Crofts, New York, 1969) pp 243-248.

USE: In the form of ferromolybdenum for manufg special steels for tools, boiler plate, rifle barrels, propeller shafts; electrical contacts, spark plugs, x-ray tubes, filaments, screens and grids for radio tubes; in the production of tungsten; glass-to-metal seals; nonferrous alloys; in colloidal form as lubricant additive.

6318. Molybdenum Disulfide. MoS_2; mol wt 160.07. Mo 59.94%, S 40.06%. Occurs as the mineral *molybdenite*, which is the principal source of molybdenum. Lab prepn: Bell, Herfert, *J. Am. Chem. Soc.* **79**, 3351 (1957).

Lead-gray, lustrous powder; the artificially prepd sulfide is black and lustrous. d_{15}^{15} 5.06; mp 2375°. Begins to sublime at 450°. Insol in water or dil acids.

USE: Dry lubricant and lubricant additive. Hydrogenation catalyst.

6319. Molybdenum Hexafluoride. F_6Mo; mol wt 209.93. F 54.30%, Mo 45.70%. MoF_6. Prepd by direct fluorination of powdered molybdenum: Ruff, Ascher, *Z. Anorg. Allgem. Chem.* **196**, 418 (1931); from MoO_3 and SF_4: Oppengard *et al.*, *J. Am. Chem. Soc.* **82**, 3825 (1960).

Volatile, white, cubic crystals. Very hygroscopic. d_{liq}^{19} 2.543. mp 17.5°. bp 35.0°. Hydrolyzed by water. Forms blue-white clouds in moist air. Soly in anhydr HF: 1.5 moles/1000 g HF, Frlec, Hyman, *Inorg. Chem.* **6**, 1596 (1967). Should be stored in quartz ampuls.

6320. Molybdenum Sesquioxide. Mo_2O_3; mol wt 239.88. Mo 79.99%, O 20.01%.

Grayish-black powder. Very slightly sol in acids.

Combination with ferrous sulfate *Mol-Iron (obsolete)*.

THERAP CAT: Combination with ferrous sulfate as hematinic.

6321. Molybdenum Trioxide. Molybdic anhydride. MoO_3; mol wt 143.94. Mo 66.65%, O 33.35%. Prepn from ammonium molybdate: Schumb, Hartford, *J. Am. Chem. Soc.* **56**, 2613 (1934). Toxicity studies: L. T. Fairhall *et al.*, "The Toxicity of Molybdenum," *U.S. Public Health Service Bulletin No. 293*, Washington DC (1945) 36 pp.

White or slightly yellow to slightly bluish powder or granules. d_4^{26} 4.696. mp 795°. Melts to dark-yellow liquid which solidifies to a yellowish-white cryst mass; sublimes at higher temp. bp 1155°. Sol in water (28°) 0.490 g/liter. Sol in concd mineral acids, in solns of alkali hydroxides, ammonia or potassium bitartrate; after strong ignition it is very slightly sol in acids.

USE: Chiefly as a reagent for chemical analysis.

6322. Molybdic(VI) Acid. H_2MoO_4; mol wt 161.95. H 1.24%, Mo 59.24%, O 39.52%. Prepd as the monohydrate from ammonium molybdate and nitric acid: Hein, Herzog in *Handbook of Preparative Inorganic Chemistry*, vol. 2, G. Brauer, Ed. (Academic Press, New York, 2nd ed., 1965) pp 1412-1413.

Monohydrate, white powder. d 3.1. Sparingly soluble in cold, more in hot water; soluble in NH_4OH, H_2SO_4, fixed alkalies.

6323. Molybdic Acid, 85%. Consists largely of an ammonium molybdate. Contains 84-86% MoO_3. It is the most widely used form of "molybdic acid".

White or slightly yellow powder. Insol or partly sol in water; sol in ammonia or solns of fixed alkali hydroxides.

USE: Principally for the determination of phosphorus or phosphate and lead; also in glazes for ceramics.

6324. Mometasone Furoate. *(11β,16α)-9,21-Dichloro-17-[(2-furanylcarbonyl)oxy]-11-hydroxy-16-methylpregna-1,4-diene-3,20-dione;* 9,21-dichloro-11β,17-dihydroxy-16α-methylpregna-1,4-diene-3,20-dione 17-(2-furoate); Sch-32088; Elocon. $C_{27}H_{30}Cl_2O_6$; mol wt 521.44. C 62.19%, H 5.80%, Cl 13.60%, O 18.41%. Topical corticosteroid. Prepn: E. L. Shapiro, **Eur.** pat. **Appl. 57,401;** *idem*, **U.S.** pat. **4,472,393** (1982, 1984 both to Schering Corp.); E. L. Shapiro *et al.*, *J. Med. Chem.* **30**, 1581 (1987). Clinical trial in psoriasis: R. S. Medansky *et al.*, *Semin. Dermatol.* **6**, 94 (1987).

Crystals from aq methanol, mp 218-220°. $[\alpha]_D^{26}$ +58.3° (dioxane). uv max (methanol) 247 nm (ϵ 26300).

THERAP CAT: Anti-inflammatory (topical).

6325. Monacetin. *1,2,3-Propanetriol monoacetate;* acetin; monoacetin; glyceryl monoacetate. $C_5H_{10}O_4$; mol wt 134.13. C 44.77%, H 7.51%, O 47.71%. $C_3H_7O_2 \cdot OOCCH_3$. Obtained along with the di- and triacetates by heating glycerol with glacial acetic acid.

Colorless, very hygroscopic liquid; characteristic odor. The commercial product is pale yellow. d_4^{20} 1.206. bp_{17} ~158°; bp_3 129-131°. Soluble in water, alcohol, slightly in ether; insol in benzene. LD_{50} s.c. in rats: 6.6 g/kg.

USE: In manuf of smokeless powder and dynamite; as a solvent for basic dyes; in tanning leather.

6326. Monarda. American horsemint; wild bergamot. Leaves of *Monarda punctata* L., *Labiatae. Habit.* New York to Florida, west to Texas and Wisconsin. *Constit.* Volatile oils: thymol, carvacrol, neryl formate, geranyl formate, cineole, γ-terpinene, α- and β-pinene. Gas chromatographic analysis of the *oil of monarda*: R. W. Scora, *J. Chromatog.* **19**, 601 (1965).

USE: Source for thymol.

THERAP CAT: Aromatic stimulant, carminative.

6327. Monardein Chloride. Monardaein chloride; pelargonidin-3-(6-*p*-coumaroyl)glucosido-5-glucoside chloride. $C_{36}H_{37}ClO_{17}$; mol wt 777.13. C 55.64%, H 4.80%, Cl 4.56%, O 35.00%. Coloring matter of *Monarda didyma* L. ("golden balm"), *Labiatae:* Karrer, Widmer, *Helv. Chim. Acta* **10**, 67 (1927); Karrer, *ibid.* **11**, 837 (1928). Structure: Birkofer *et al., Z. Naturforsch.* **20b**, 424 (1965).

Red powder. $[\alpha]_D$ −374.5° (0.1% HCl). Absorption max (methanolic HCl): 286, 313, 507 nm. Sol in water and methanol. The aq soln is scarlet in color, changed to violet by Na_2CO_3 and to light brown by excess NaOH.

6328. Monellins. Intensely sweet principle from the fruit of the tropical plant *Dioscoreophyllum cumminsii* Diels, *Menispermaceae* ("Serendipity Berry"). First believed to consist of a single polypeptide chain of approx 91 amino acids, mol wt about 10,700. Preliminary data and extraction: Inglett, May, *J. Food Sci.* **34**, 408 (1969); Van der Wel, *FEBS Letters* **21**, 88 (1971); Van der Wel, Loeve, *ibid.* **29**, 181 (1973). Purification: Morris, Cagan, *Biochim. Biophys. Acta* **261**, 114 (1972). Characterized by scientists at the Univ. of Pennsylvania's Monell Chemical Senses Center as the first protein to elicit a sweet taste in man, a "chemo-stimulatory protein". *See* Morris *et al., J. Biol. Chem.* **248**, 534 (1973); Cagan, *Science* **181**, 32 (1973). Monellin is composed of two nonidentical subunits: Z. Bohak, S. Li, *Biochim. Biophys. Acta* **427**, 153 (1976); the presence of both subunits is necessary for the sweet taste to occur: B. Jirgenson, *ibid.* **446**, 255 (1976). Structure of subunit I: G. Hudson, K. Biemann, *Biochem. Biophys. Res. Commun.* **71**, 212 (1976). Complete amino acid sequence of subunits A and B consisting of 44 and 50 amino acids, resp.: G. Frank, H. Zuber, *Z. Physiol. Chem.* **357**, 585 (1976). Crystallographic studies: A. Wlodawer, K. Hodgson, *Proc. Nat. Acad. Sci. USA* **72**, 398 (1975); G. E. Tomlinson, S. H. Kim, *J. Biol. Chem.* **256**, 12476 (1981).

Water soluble. Approximately 3000 times sweeter than sucrose on a weight basis. Essentially free from carbohydrate: <5 μg/mg protein. uv max (0.01M sodium phosphate buffer): 277 nm.

USE: Potential low-calorie sweetener.

6329. Monensin. *2-[5-Ethyltetrahydro-5-[tetrahydro-3-methyl-5-[tetrahydro-6-hydroxy-6-(hydroxymethyl)-3,5-dimethyl-2H-pyran-2-yl]-2-furyl]-2-furyl]-9-hydroxy-β-methoxy-α,γ,2,8-tetramethyl-1,6-dioxaspiro[4.5]decane-7-butyric acid;* monensic acid (obsolete); A-3823A. $C_{36}H_{62}O_{11}$; mol wt 670.88. C 64.45%, H 9.31%, O 26.23%. Polyether anti-

biotic. Major factor in antibiotic complex isolated from *Streptomyces cinnamonensis.* Discovery and isolation: Haney, Hoehn, *Antimicrob. Ag. Chemother.* **1967**, 349. Production: Haney, Hoehn, U.S. pat. **3,501,568** (1970 to Lilly). Structure: Agtarap *et al., J. Am. Chem. Soc.* **89**, 5737 (1967). Crystal structure studies: Lutz *et al., Helv. Chim. Acta* **53**, 1732 (1970); *ibid.* **54**, 1103 (1971). Fermentation studies: Stark *et al., Antimicrob. Ag. Chemother.* **1967**, 353. Chemistry: Agtarap, Chamberlin, *ibid.* 359. Stereocontrolled total synthesis: T. Fukuyama *et al., J. Am. Chem. Soc.* **101**, 262 (1979); D. B. Collum *et al., ibid.* **102**, 2117, 2118, 2120 (1980). ^{13}C-NMR study: J. A. Robinson, D. L. Turner, *Chem. Commun.* **1982**, 148. Biosynthesis: Day *et al., Antimicrob. Ag. Chemother.* **4**, 410 (1973). *Review:* Stark, "Monensin, A New Biologically Active Compound Produced by a Fermentation Process", in *Fermentation Advances,* Pap. Int. Ferment. Symp., 3rd, **1968**, D. Perlman, Ed. (Academic Press, New York, 1969) pp 517-540.

Crystals, mp 103-105° (monohydrate). $[\alpha]_D$ +47.7°. pKa 6.6 (in 66% DMF). Very stable under alkaline conditions. Slightly sol in water; more sol in hydrocarbons; very sol in other organic solvents. LD_{50} of monensin complex in mice, chicks (mg/kg): 43.8 ± 5.2, 284 ± 47 orally (Haney, Hoehn).

Sodium salt, $C_{36}H_{61}NaO_{11}$, **Coban, Romensin, Rumensin.** mp 267-269°. $[\alpha]_D$ +57.3° (methanol). Slightly sol in water; more sol in hydrocarbons; very sol in other organic solvents.

THERAP CAT (VET): Antiprotozoal, antibacterial, antifungal. Coccidiostat in chickens. Feed additive.

6330. Monoamine Oxidase. *Amine oxidase;* adrenaline oxidase; tyraminase; MAO. Since its first description by Hare in 1928 as an enzyme which catalyzed the oxidative deamination of tyramine, it has been found to be widely distributed in animals. This enzyme, which catalyzes the oxidative deamination of a variety of biogenic amines, such as serotonin, norepinephrine, epinephrine, tyramine and dopamine, deaminates compounds in which the amine group is attached to the terminal carbon group. Enzymes other than the classical amine oxidase which catalyze the oxidative deamination of amines, have been found in animals, plants and bacteria. In contrast to the classical amine oxidase, they are inhibited by carbonyl reagents and do not act on *N*-substituted amines. Enzymes of this group include: oxidases in microorganisms, *spermine oxidase, mescaline oxidase* (of rabbit liver), *plant amine oxidase,* and *histaminase, q.v.* Structural studies: Achee, *Biochemistry* **7**, 4329 (1968). Multiple forms of rat brain monoamine oxidase: M. B. H. Youdim *et al., Nature* **223**, 626 (1969). *Review:* H. Blaschko "Amine Oxidase" in *The Enzymes,* vol. **8**, P. D. Boyer *et al.,* Eds. (Academic Press, New York, 2nd ed., 1963) pp 337-351. Book: *Monoamine Oxidase: Structure, Function and Altered Functions,* T. P. Singer *et al.,* Eds. (Academic Press, New York, 1980) 557 pp.

6331. Monobenzone. *4-(Phenylmethoxy)phenol; p-(benzyloxy)phenol;* hydroquinone monobenzyl ether; *p*-quinone benzyl ether; *p*-hydroxyphenyl benzyl ether; benzyl hydroquinone; monobenzyl hydroquinone; Benoquin; Depigman; Pigmex; Benzoquin; Agerite. $C_{13}H_{12}O_2$; mol wt 200.24. C 77.98%, H 6.04%, O 15.98%. Prepd from hydroquinone and benzyl bromide in alcoholic KOH: Schiff, Pellizzari, *Ann.* **221**, 370 (1883).

Lustrous leaflets from water, mp 122.5°. Practically insol in cold water. Soly in boiling water ~1.0 g/100 ml. Sol in alcohol, ether, benzene. LD_{50} i.p. in rats: 4.5 g/kg.

THERAP CAT: Depigmentor.

6332. Monobutyrin. *Butanoic acid 2,3-dihydroxypropyl ester;* 1-butyrylglycerol; α-monobutyrin; glycerol α-mono-butyrate; glyceryl 3-monobutyrate. $C_7H_{14}O_4$; mol wt 162.19. C 51.84%, H 8.70%, O 39.46%. Naturally occurring lipid that stimulates angiogenesis. Secreted by differentiating adipocytes; stimulates blood vessel formation and causes vasodilation. The *(R)*-isomer is bioactive. Prepn of optical isomers: E. Abderhalden, E. Eichwald, *Chem. Ber.* **48**, 1847 (1915); of L-form: E. Baer, H. O. L. Fischer, *J. Am. Chem. Soc.* **67**, 2031 (1945); S. Gronowitz *et al., Chem. Phys. Lipids* **14**, 174 (1975). Purification from adipocytes and identification of angiogenic activity: D. E. Dobson *et al., Cell* **61**, 223 (1990). Biosynthesis: W. O. Wilkison *et al., J. Biol. Chem.* **266**, 16886 (1991); W. O. Wilkison, B. M. Spiegelman, *ibid.* **268**, 2844 (1993). Effect on retinal circulation and probable role in pathology of diabetes: Y.-D. C. Halvorsen *et al., J. Clin. Invest.* **92**, 2872 (1993).

(R)-Form, *3-butyryl-sn-glycerol, L-α-butanoyl glycerol.* Cryst from pure ether at −70°. n_D^{20} 1.4493. sp wt 0.8600. $[\alpha]_D^{18}$ −0.63°; $[\alpha]_D^{25}$ −7.40° (c = 2.04 in pyridine).

6333. Monocrotaline. *(13α,14α)-14,19-Dihydro-12,13-dihydroxy-20-norcrotolanan-11,15-dione;* crotaline; MCT; NSC-28693; NCI-C56462. $C_{16}H_{23}NO_6$; mol wt 325.36. C 59.07%, H 7.13%, N 4.30%, O 29.50%. Toxic pyrrolizidine alkaloid isolated from *Crotalaria spp..* Isoln: R. Adams, E. F. Rogers, *J. Am. Chem. Soc.* **61**, 2815 (1939); R. B. Tinker, W. M. Lauter, *Econ. Bot.* **10**, 254 (1956); C. C. J. Culvenor, L. W. Smith, *Aust. J. Chem.* **16**, 239 (1963). Structure: R. Adams *et al., J. Am. Chem. Soc.* **74**, 5612 (1952). Stereochemistry: D. J. Robins, D. H. G. Crout, *J. Chem. Soc. (C)* **1969**, 1386. Crystal structure: H. Stoeckli-Evans, *Acta Crystallogr.* **B35**, 231 (1979). Stereoselective synthesis: H. Niwa *et al., Tetrahedron* **48**, 10531 (1992). ELISA measurement: M. A. Bober *et al., Toxicon* **27**, 1059 (1989). Toxicology: P. M. Newberne *et al., Toxicol. Appl. Pharmacol.* **18**, 387 (1971); R. A. Roth *et al., ibid.* **60**, 193 (1981). Review of carcinogenicity studies: *IARC Monographs* **10**, 291-302, 333-342 (1976). Comprehensive reviews: L. Bull *et al., The Pyrrolizidine Alkaloids* (North-Holland, Amsterdam, 1968) 293 pp; D. J. Robins, *Fortschr. Chem. Org. Naturst.* **41**, 115-203 (1982). Review of pulmonary toxicity: D. W. Wilson *et al., Crit. Rev. Toxicol.* **22**, 307-325 (1992).

White prisms from abs ethanol, mp 197-198° (dec). $[\alpha]_D^{26}$ −54.7° (c = 5.054 in chloroform). Also reported as color-less crystals from ethanol, mp 187-190° (dec). $[\alpha]_D^{12}$ −55.0° (c = 0.16 in CHCl_3). uv max (96% ethanol): 217 nm (log ε 3.32), Simánek *et al., Coll. Czech. Chem. Commun.* **34**, 1832 (1969). LD_{50} orally in rats: 71 mg/kg (Newberne).

Hydrochloride, $C_{16}H_{23}NO_6 \cdot HCl$, mp 184° (dec). $[\alpha]_D^{28}$ −38.4° (c = 5.2 in water).

Methiodide, $C_{16}H_{23}NO_6 \cdot CH_3I$, mp 205° (dec). $[\alpha]_D^{28}$ +23.4° (c = 3.1 in methanol).

USE: For inducing pulmonary diseases in rats.

6334. Monocrotophos. *(E)-Phosphoric acid dimethyl [1-methyl-3-(methylamino)-3-oxo-1-propenyl] ester; (E)-phosphoric acid dimethyl ester, ester with 3-hydroxy-N-methylcrotonamide;* 3-(dimethoxyphosphinyloxy)-N-methyl-cis-crotonamide; dimethyl 2-methylcarbamoyl-1-methylvinyl phosphate; dimethyl phosphate of 3-hydroxy-N-methyl-cis-crotonamide; C-1414; ENT-27129; SD-9129; Azodrin; Monocron; Nuvacron. $C_7H_{14}NO_5P$; mol wt 223.17. C 37.67%, H 6.32%, N 6.28%, O 35.85%, P 13.88%. General prepn: Whetstone, Stiles, U.S. pat. **2,802,855;** see also Ward, Morales, U.S. pat. **3,400,177** (1957, 1968 both to Shell). Metabolite of dicrotophos, *q.v.:* Menzer, Casida, *J. Agr. Food Chem.* **13**, 102 (1965). Toxicity study: T. B. Gaines, *Toxicol. Appl. Pharmacol.* **14**, 515 (1969).

Crystals, mp 54-55° Commercial product is a reddish-brown solid, mp 25-30°. Vapor pressure at 20°: 7 × 10^6 mm Hg. Misc with water; sol in acetone, ethanol. Practically insol in diesel oils, kerosine. LD_{50} in male, female rats (mg/kg): 17, 20 orally; 126, 112 dermally (Gaines).

USE: Insecticide.

6335. Monoctanoin. Monooctanoin; Capmul 8210; Moctanin. Cholesterol solvent used to dissolve gallstones by direct biliary infusion. Semisynthetic mixture of mono- and di-glycerides of octanoic and decanoic acids of which glycerol 1-octanoate is the major component. Description of prepn from coconut oil and use in treatment of gallstones: V. K. Babayan, U.S. pat. **4,205,086** (1980 to Stokeley-Van Camp). Physical properties and clinical application: J. L. Thistle *et al., Gastroenterology* **78**, 1016 (1980). Factors affecting cholesterol dissolution rate: J. B. Bogardus, *J. Pharm. Sci.* **73**, 906 (1984). Clinical trial: K. R. Palmer, A. F. Hofmann, *Gut* **27**, 196 (1986). Reviews of cholelitholytic agents: G. D. Bell, *Pharmacol. Ther.* **23**, 79-108 (1983); R. J. Fitzgibbons *et al., Surgery Annual* **21**, 237-262 (1989).

Glycerol 1-octanoate, $C_{11}H_{22}O_4$, *octanoic acid 2,3-dihydroxypropyl ester, caprylic acid α-monoglyceride, α-monocaprylin.* Prepn: L. Hartman, *J. Chem. Soc.* **1959**, 4134. Crystals from light petroleum, mp 39.5-40.5°.

THERAP CAT: Cholelitholytic agent.

6336. Monorden. *8-Chloro-1a,14,15,15a-tetrahydro-9,-11-dihydroxy-14-methyl-6H-oxireno[e][2]benzoxacyclotetradecin-6,12(7H)-dione;* 5-chloro-6-(7,8-epoxy-10-hydroxy-2-oxo-3,5-undecadienyl)-β-resorcylic acid μ-lactone; radicicol. $C_{18}H_{17}ClO_6$; mol wt 364.78. C 59.27%, H 4.70%, Cl 9.72%, O 26.32%. Antibiotic substance from *Monosporium bonorden:* Delmotte, Delmotte-Plaquée, *Nature* **171**, 344 (1953). Identity with radicicol: Mirrington *et al., Tetrahedron Letters* **1964**, 365. Structure: McCapra, Scott, *ibid.* **1964**, 869; Mirrington *et al., Aust. J. Chem.* **19**, 1265 (1966). Isoln: **Neth. pat. Appl. 6,506,173** (1965 to Sandoz).

Crystals from chloroform, alcohol or benzene, mp 193.5°. $[\alpha]_D^{20}$ +203° (chloroform). uv max (ethanol): 264, 272 nm (ϵ 13200, 13100).

Diacetate, $C_{22}H_{21}ClO_8$, crystals, mp 189°. uv max: 275 nm (ϵ 13000).

6337. Monosodium Glutamate. *Glutamic acid monosodium salt monohydrate;* monosodium L-glutamate monohydrate; sodium glutamate monohydrate; Chinese seasoning; MSG; RL-50; Accent; Ajinomoto; Glutacyl; Glutavene; Vetsin. $C_5H_8NNaO_4\cdot H_2O$; mol wt 187.13. C 32.09%, H 5.39%, N 7.49%, Na 12.29%, O 42.75%. HOOCCH(NH$_2$)-CH$_2$CH$_2$COONa.H$_2$O. The monosodium salt of the naturally occurring L-form of glutamic acid. Manufactured by fermentation of carbohydrate sources such as sugar beet molasses. Production by hydrolysis of vegetable proteins (*see also* Glutamic Acid): Ikeda, Suzuki, **Brit. pat. 9440,** *C.A.* **5,** 836 (1910); U.S. pat. **1,015,891,** *C.A.* **6,** 717 (1912); from Steffens waste from beet-sugar molasses by acid hydrolysis: Ikeda, U.S. pat. **1,721,820,** *C.A.* **23,** 4591 (1929); by the action of *Micrococcus glutamicus* upon a carbohydrate and subsequent partial neutralization: Kinoshita *et al.,* U.S. pats. **3,002,889** and **3,003,925** (both 1961 to Kyowa Ferment. Ind.). Prepn of cryst Na-glutamate: Shildneck, U.S. pat. **2,306,646,** *C.A.* **37,** 3107 (1943). As a rule, sugar beet products are used in the U.S. and in Europe; other carbohydrate sources such as sugar cane and tapioca are often used in the Orient. Mfg methods: Faith, Keyes, Clark, *Industrial Chemicals* (Wiley, New York, 2nd ed., 1957) p 522. *Review:* T. Yoshida, "L-Monosodium Glutamate (MSG)" in Kirk-Othmer *Encyclopedia of Chemical Technology,* **Vol. 2** (Wiley, New York, 3rd ed., 1978) pp 410-421. Book: *Glutamic Acid: Advances in Biochemistry and Physiology,* L. J. Filer *et al.,* Eds. (Raven Press, New York, 1979) 416 pp.

White, practically odorless, free flowing crystals or crystalline powder. Forms rhombic prisms when crystallized from water. Below $-8°$, it crystallizes as a pentahydrate, which, after filtration and exposure to air, loses water of crystallization and becomes the monohydrate. Commercially preferable crystals are grown in the presence of amino acid contaminants. The optimum concentration is from 0.2 to 0.9% in normally salted food. Slightly levorotatory in water, but dextrorotatory in acid solns (the free L-acid is dextrorotatory). $[\alpha]_D^{25}$ +24.2° to +25.5° (c = 8.0 in 1.0N HCl). pH of 0.2% soln = 7.0. Very sol in water; sparingly sol in alcohol.

USE: Flavor enhancer; swine food additive; with sugar, to improve palatability of bitter drugs.

THERAP CAT: Treatment of hepatic coma.

6338. Monotropein. *[1S-(1α,4aα,7β,7aα)]-1-(β-D-Glucopyranosyloxy)-1,4a,7,7a-tetrahydro-7-hydroxy-7-(hydroxymethyl)cyclopenta[c]pyran-4-carboxylic acid.* $C_{16}H_{22}O_{11}$; mol wt 390.34. C 49.23%, H 5.68%, O 45.09%. From herb of

Monotropa hypopitys L., *Ericaceae:* Bridel, *Compt. Rend.* **176,** 1742 (1923); *Bull. Soc. Chim. Biol.* **5,** 722 (1923). Structure: Inouye *et al., Tetrahedron Letters* **1963,** 1031; Bobbitt *et al., Chem. & Ind. (London)* **1964,** 931. Stereochemistry: Norio *et al., Tetrahedron Letters* **1967,** 2367. GC-mass spec studies: H. Inouye *et al., J. Chromatog.* **118,** 201 (1976). Purgative activity: *eidem, Planta Med.* **25,** 285 (1976).

Monohydrate, needles from water, dec 170-173°. $[\alpha]_D^{18}$ $-127.7°$ (water). uv max: 235 nm (log ϵ 3.98). Soluble in water, alcohol. Practically insol in ethyl acetate.

Pentaacetate, $C_{26}H_{32}O_{16}$, crystals, mp 173-174.5°. $[\alpha]_D^{18}$ $-82.5°$ (ethanol).

6339. Montan Wax. Lignite wax. Obtained by extraction from lignite. Asphalt and resin content and physical properties vary with source of lignite. Brief review: C. S. Letcher in Kirk-Othmer *Encyclopedia of Chemical Technology* **vol. 24** (Wiley-Interscience, New York, 3rd ed., 1984) pp 471-472.

Dark-brown lumps; white or nearly white when bleached. mp 80-86°. Saponification number 88-112. Insol in water. Sol in benzene, chloroform, carbon tetrachloride, hot petr ether; incompletely sol in hot ether or boiling alcohol.

USE: Electric cable insulators; in polishes instead of carnauba wax; manuf candles, waterproof paints and varnishes.

6340. Montelukast. *[R-(E)]-1-[[[1-[3-[2-(7-Chloro-2-quinolinyl)ethenyl]phenyl]-3-[2-(1-hydroxy-1-methylethyl)-phenyl]propyl]thio]methyl]cyclopropaneacetic acid.* $C_{35}H_{36}$-$ClNO_3S$; mol wt 586.19. C 71.71%, H 6.19%, Cl 6.05%, N 2.39%, O 8.19%, S 5.47%. Selective leukotriene D$_4$ receptor antagonist. Prepn: M. L. Belley *et al.,* **Eur. pat. Appl.** **480,717** (1992 to Merck Frosst); M. Labelle *et al., Bioorg. Med. Chem. Lett.* **5,** 283 (1995). Pharmacological profile: T. R. Jones *et al., Can. J. Physiol. Pharmacol.* **73,** 191 (1995). LC determn in human plasma: R. D. Amin *et al., J. Pharm. Biomed. Anal.* **13,** 155 (1995).

Monosodium salt, $C_{35}H_{35}ClNNaO_3S$, *MK-476, Singulair.* THERAP CAT: Antiasthmatic.

6341. Montmorillonite. A clay forming the principal constituent of bentonite and fuller's earth. Approximate formula: $R^+_{0.33}(Al,Mg)_2Si_4O_{10}(OH)_2\cdot nH_2O$ where R^+, in natural material, includes one or more of the cations Na$^+$, K$^+$, Mg^{2+}, and Ca^{2+}, and possibly others. *Caution:* Inhalation of the dust can cause respiratory irritation.

USE: In industrial chromatographic technique, *e.g.,* U.S. pat. **2,626,888** and Brit. pat. **697,060** (1953 to Merck & Co.), describing the elution of vitamin B$_{12}$-active substances adsorbed on montmorillonite adsorbents. Widely used in the petroleum industry. Also as catalyst carrier.

6342. Monuron. *N'-(4-Chlorophenyl)-N,N-dimethylurea;* 1,1-dimethyl-3-(p-chlorophenyl)urea; CMU; Karmex Monuron Herbicide; Telvar. $C_9H_{11}ClN_2O$; mol wt 198.65. C 54.42%, H 5.58%, Cl 17.85%, N 14.10%, O 8.05%. Prepd by reacting p-chlorophenyl isocyanate with dimethylamine: Bucha, Todd, *Science* **114,** 493 (1951); *see also* U.S. pats. **2,655,444-2,655,447.** Toxicity study: G. W. Bailey, J. L. White, *Residue Rev.* **10,** 97 (1965). *Review:* McCall, *Agricultural Chemicals* **7,** 40 (1952).

Consult the Name Index before using this section.

Thin rectangular prisms from methanol, mp 170.5-171.5° (the commercial product melts at 176-177°). Vapor pressure at 25°: 5×10^{-7} mm; at 100°: 178×10^{-5} mm. Slight odor. Stable toward oxygen and moisture under ordinary conditions at neutral pH; elevated temps and more acid or alkaline conditions appreciably raise rate of hydrolysis. Very slightly sol in water and in no. 3 Diesel oil: About 230 ppm at 25°. pH of satd aq soln 6.26. Moderately sol in methanol, ethanol, acetone. Practically insol in hydrocarbon solvents. LD_{50} orally in rats: 3700 mg/kg (Bailey, White).

Trichloroacetate, $C_9H_{11}ClN_2O \cdot C_2HCl_3O_2$, *monuron TCA, Urox.* mp 78-81°. Prepn: Gilbert *et al.,* U.S. pat. 2,782,112 (1957 to Allied Chem.). Effective in control of both weeds and grasses.

Caution: Anemia and methemoglobinemia have been produced in experimental animals.

USE: Herbicide.

6343. Moperone. *1-(4-Fluorophenyl)-4-[4-hydroxy-4-(4-methylphenyl)-1-piperidinyl]-1-butanone; p-fluoro-4-(4'-hydroxy-4'-p-tolylpiperidino)butyrophenone; p-fluoro-4-(4'-hydroxy-4'-p-methylphenylpiperidino)butyrophenone; 1-(3'-p-fluorobenzoylpropyl)-4-hydroxy-4-p-tolylpiperidine; ω-(4-hydroxy-4-p-tolylpiperidino)-p-fluorobutyrophenone; methylperidol;* R-1658. $C_{22}H_{26}FNO_2$, mol wt 355.45. C 74.34%, H 7.37%, F 5.34%, N 3.94%, O 9.00%. Prepn: Janssen *et al., J. Med. Pharm. Chem.* 1, 281 (1959); Janssen, **Brit.** pat. 881,893 (1961).

Crystals, mp 118-119.5°. uv max: 246.5 nm (ε 12200).
Hydrochloride, $C_{22}H_{26}FNO_2 \cdot HCl$, *Luvatren.* Crystals from isopropyl ether, mp 216-218°.

THERAP CAT: Antipsychotic.

6344. Mopidamol. *2,2',2'',2'''-[[4-(1-Piperidinyl)pyrimido[5,4-d]pyrimidine-2,6-diyl]dinitrilo]tetrakisethanol; 2,2',2'',2'''-[(4-piperidinopyrimido[5,4-d]-pyrimidine-2,6-diyl)dinitrilo]tetraethanol; 2,6-bis(diethanolamino)-4-piperidinopyrimido[5,4-d]pyrimidine;* RA-233; *Rapenton.* $C_{19}H_{31}N_7O_4$; mol wt 421.50. C 54.14%, H 7.41%, N 23.26%, O 15.18%. An inhibitor of platelet aggregation with antimetastatic properties. Prepn: **Brit.** pat. 1,051,218 corresp to J. Roch, H. Scheffler, U.S. pat. 3,322,755 (1966, 1967 both to Boehringer, Ing.). Effect on tumor cell metastasis: J. L. Ambrus *et al., J. Med.* 9, 183 (1978); on inhibition of adenosine uptake into platelets: J. P. Lips *et al., Biochem. Pharmacol.* 29, 43 (1980). Therapeutic use: R. L. Clark, *Cancer* 43, 790 (1979); I. L. Bonta *et al., Agents Actions (Suppl.)* 4, 278 (1979).

Deep yellow crystals, mp 157-158°.
THERAP CAT: Antineoplastic.

6345. Moprolol. *1-(2-Methoxyphenoxy)-3-[(1-methylethyl)amino]-2-propanol; 1-(isopropylamino)-3-(o-methoxyphenoxy)-2-propanol.* $C_{13}H_{21}NO_3$; mol wt 239.31. C 65.25%, H 8.84%, N 5.85%, O 20.06%. β-Adrenergic blocker. Prepn: C. D. Lunsford *et al., J. Am. Chem. Soc.* 82, 1166 (1960); **Neth.** pat. Appl. 301,580 corresp to A. F. Crowther, L. H. Smith, U.S. pat. 3,501,769 (1965, 1970 both to ICI); A. F. Crowther *et al., J. Med. Chem.* 12, 638 (1969). Pharmacology: G. Croce *et al., Arzneimittel-Forsch.* 20, 1074 (1970). Anti-anginal and hemodynamic effects: M. Radice *et al., ibid.* 28, 2160 (1978). Resolution of enantiomers: G. Ferrari, V. Vecchietti, **Eur.** pat. Appl. 15,418; *eidem,* U.S. pat. 4,683,245 (1980, 1987 both to Simes). Synthesis of (−)-form: K. Kan *et al., Agr. Biol. Chem.* 49, 207 (1985). Bioavailability of (−)-form in dog: A. Marzo, J. P. Desager, *Curr. Ther. Res.* 30, 442 (1981). Pharmacokinetics of racemate and (−)-form: C. Harvengt, J. P. Desager, *Int. J. Clin. Pharmacol. Ther. Toxicol.* 20, 57 (1982). Clinical comparison of antihypertensive activity of enantiomers: P. Ghirardi, F. Grosso, *J. Cardiovasc. Pharmacol.* 2, 471 (1980). Clinical studies of (−)-form in hypertension: G. Vrebos *et al., Curr. Ther. Res.* 29, 654 (1981); H. Hoeffkes, *ibid.* 30, 88 (1981).

Crystals from cyclohexane, mp 82-83°.
Hydrochloride, $C_{13}H_{21}NO_3 \cdot HCl$, *SD-1601, Omeral.* Crystals, mp 110-112°.

l-Form, *levomoprolol.* Crystals from ethyl acetate, mp 78-80°. [α] −5.5 ±0.2 (c = 5.0 in ethanol).

l-Form hydrochloride, *Levotensin.* Crystals from ethyl acetate/95% ethanol, mp 121-123°. $[\alpha]_D^{25}$ −16.3° (c = 5.0 in ethanol).

THERAP CAT: Antihypertensive.

6346. MOPS. *4-Morpholinepropanesulfonic acid; 3-(N-morpholino)propanesulfonic acid; N-(3-sulfopropyl)morpholine.* $C_7H_{15}NO_4S$; mol wt 209.27. C 40.18%, H 7.22%, N 6.69%, O 30.58%, S 15.32%. One of the zwitterionic amino acids known as "Good" buffers; active in the pH range of 6.5-8.0. Prepn: C. F. H. Allen *et al., Anal. Chem.* 37, 156 (1965); N. E. Good, S. Izawa, "Hydrogen Ion Buffers," in *Methods of Enzymology* vol. 24, A. San Pietro, Ed. (Academic Press, New York, 1972) pp 53-65. Temperature effects on pK: M. Sankar, R. G. Bates, *Anal. Chem.* 50, 1922 (1978); and pH: R. N. Roy *et al., Cryo-Letters* 6, 139 (1985). Effects on "Lowry" protein determn: H. M. Himmel, W. Heller, *J. Clin. Chem. Clin. Biochem.* 25, 909 (1987). Uses as a biological buffer: E. P. Paques *et al., Eur. J. Biochem.* 107, 447 (1980); L. A. Sonna *et al., J. Biol. Chem.* 263, 6625 (1988).

White powder, mp 283.5-284.5°. pKa 7.15. pKa_2 (25°): 7.184; (37°): 7.041. ΔpKa/°C −0.013. Is not completely stable when autoclaved in the presence of glucose.

USE: Biological buffer.

6347. Moquizone. *2,3-Dihydro-1-[(4-morpholinyl)acetyl]-3-phenyl-4(1H)-quinazolinone;* Rec-14-0127. $C_{20}H_{21}N_3O_3$; mol wt 351.41. C 68.36%, H 6.02%, N 11.96%, O 13.66%. Prepn: Bonola *et al., J. Med. Chem.* 11, 1136 (1968); *eidem,* **S. Afr.** pat. 68 03,948, *C.A.* 71, 81406c (1969) and **Swiss** pat. 474,524 corresp to U.S. pat. 3,637,681 (1968, 1969, 1972, all three to SECEPH). Pharmacology: Setnikar *et al., Arzneimittel-Forsch.* 22, 1894 (1972). Toxicology: Setnikar, De Fina, *Toxicol. Appl. Pharmacol.* 16,

571 (1970). Review of pharmacological and clinical aspects: Grossi, Spada, *Clin. Ter.* **50**, 203 (1969).

Crystals from isopropanol, mp 135-137°. Crystals from ethyl acetate or benzene-petr ether, mp 128-130°.

Hydrochloride, $C_{20}H_{21}N_3O_3$.HCl, *Peristil.* Crystals from ethanol-ether, mp 210-214°. Solubility in water (25°): 12.2 g/100 ml. LD_{50} in mice, rats: 1155, 2135 mg/kg orally; 237, 146 mg/kg i.v.

THERAP CAT: Choleretic.

6348. Morantel. *(E)-1,4,5,6-Tetrahydro-1-methyl-2-[2-(3-methyl-2-thienyl)ethenyl]pyrimidine.* $C_{12}H_{16}N_2S$; mol wt 220.34. C 65.41%, H 7.32%, N 12.71%, S 14.55%. Prepn: Austin *et al.*, **Brit. pat. 1,120,587** (1968 to Pfizer).

Crystals from chloroform + benzene, mp 239-241°.
Citrate monohydrate, $C_{12}H_{16}N_2S.C_6H_8O_7.H_2O$, *Exhelm.*
Tartrate, $C_{12}H_{16}N_2S.C_4H_6O_6$, *CP-12009-18, Paratect, Suiminth.*

THERAP CAT (VET): Anthelmintic.

6349. Morazone. *1,2-Dihydro-1,5-dimethyl-4-[(3-methyl-2-phenyl-4-morpholinyl)methyl]-2-phenyl-3H-pyrazol-3-one; 4-[(3-methyl-2-phenylmorpholino)methyl]antipyrine; 1-phenyl-2,3-dimethyl-4-(2'-phenyl-3'-methylmorpholinomethyl)-5-pyrazolone;* R-445; Tarugan. $C_{23}H_{27}N_3O_2$; mol wt 377.49. C 73.18%, H 7.21%, N 11.13%, O 8.48%. Prepn: Hengen *et al., Arzneimittel-Forsch.* **8**, 421 (1958); Siemer, Doppstadt, **U.S. pats. 2,943,022; 3,005,818** (1960, 1961 both to Ravensberg Chem.). Metabolism: Cartoni *et al., J. Chromatog.* **84**, 419 (1973).

Crystals, mp 149-150°. Freely sol in chloroform; sol in methanol, acetone; slightly sol in ether.

Hydrochloride, $C_{23}H_{27}N_3O_2$.HCl, crystals from methanol + acetone, mp 171-172° (dec). Sol in water.

THERAP CAT: Analgesic; anti-inflammatory; antipyretic.

6350. Morclofone. *(4-Chlorophenyl)[3,5-dimethoxy-4-[2-(4-morpholinyl)ethoxy]phenyl]methanone; 4'-chloro-3,5-dimethoxy-4-(2-morpholinoethoxy)benzophenone;* dimeclophenone; K-3712; Medicil; Nitux; Plausitin. $C_{21}H_{24}ClNO_5$; mol wt 405.88. C 62.14%, H 5.96%, Cl 8.73%, N 3.45%, O 19.71%. Prepn: F. Lauria *et al.,* **Ger. pat. 2,016,707** corresp to **U.S. pat. 3,708,482** (1971, 1973 both to Carlo Erba). Antitussive activity in humans: E. Bosisio *et al., Farmaco Ed. Prat.* **26**, 356 (1971).

Cryst from cyclohexane, mp 91-92°. LD_{50} orally in mice: 552 mg/kg (Lauria).

THERAP CAT: Antitussive.

6351. Moricizine. *[10-[3-(4-Morpholinyl)-1-oxopropyl]-10H-phenothiazin-2-yl]carbamic acid ethyl ester; 10-(3-morpholinopropionyl)phenothiazine-2-carbamic acid ethyl ester;* ethmosine; ethmozin; moracizine; EN-313. $C_{22}H_{25}N_3O_4S$; mol wt 427.52. C 61.81%, H 5.89%, N 9.83%, O 14.97%, S 7.50%. Sodium channel blocker. Prepn: A. Gritsenko *et al.,* **Ger. pat. 2,014,201** (1971 to Acad. Med. Sci., USSR), *C.A.* **76**, 34273a (1972); *eidem*, **U.S. pat. 3,740,395** (1973). Pharmacology and toxicity: N. V. Kaverina *et al., Russian Pharmacol. Toxicol.* **35**, 74 (1972). Mechanism of action study: T. Yamane *et al., Brit. J. Pharmacol.* **108**, 812 (1993). HPLC determn in plasma: C. C. Whitney *et al., J. Pharm. Sci.* **70**, 462 (1981). Symposium on pharmacology, pharmacokinetics, clinical safety and efficacy: *Am. J. Cardiol.* **65**, 1D-71D (1990). *Review:* C. A. Clyne *et al., N. Engl. J. Med.* **327**, 255 (1992).

Crystals, mp 156-157°. pKa 6.4.

Hydrochloride, $C_{22}H_{25}N_3O_4S$.HCl, *Ethmozine.* Crystals from dichloroethane, mp 189° (dec). Sol in water, alcohol. LD_{50} in mice, rats (mg/kg): 36, 12 i.v.; in mice (mg/kg): 131 i.p. (Kaverina).

THERAP CAT: Antiarrhythmic (class I).

6352. Morin. *2-(2,4-Dihydroxyphenyl)-3,5,7-trihydroxy-4H-1-benzopyran-4-one; 2',3,4',5,7-pentahydroxyflavone;* 2'-hydroxypelargidenolon 1522; C.I. Natural Yellow 8; C.I. Natural Yellow 11; C.I. 75660. $C_{15}H_{10}O_7$; mol wt 302.24. C 59.61%, H 3.33%, O 37.06%. In wood of old fustic *(Chlorophora tinctoria* (L.) Gaud., *Moraceae),* also called Cuba wood, or yellow Brazil wood. The wood of the Osage orange tree also contains morin and maclurin. Isoln: Perkin, Pate, *J. Chem. Soc.* **67**, 649 (1895); Rolland, *Teintex* **3**, 460 (1938). Synthesis by condensing phloracetophenone dimethyl ether with 2,4-dimethoxybenzaldehyde: von Kostanecki *et al., Ber.* **39**, 625, 4014, 4022 (1906). Comprehensive list of refs in *Colour Index* vol. **4** (3rd ed., 1971) p 4636.

Crystallizes with 1 or 2 mols water. Anhyd needles from abs alcohol, dec 285-290°. One gram dissolves in 4 liters water at 20°, in 1060 ml boiling water. Freely sol in alcohol; slighly sol in ether, acetic acid. Sol in aq alkaline solns with intense yellow color which turns brown on exposure to air. Absorpion spectrum: Grinbaumowna, Marchlewski, *Biochem. Z.* **290**, 261 (1937).

Calico yellow, obtained by the action of bisulfite on fustic extract consists mainly of the bisulfite compound of morin,

$C_{15}H_{10}O_7$ + $NaHSO_3$, and is used in calico printing. The calico yellow of commerce is a good source of morin for the laboratory. The bisulfite is removed with HCl.

USE: As a spot test reagent for Al, Be, Zn, Ga, In, and Sc salts: Feigl, *Spot Tests* (New York, 1954); Beck, *Mikrochim. Acta* **2**, 287 (1937). As luminescence indicator: Kocsis, Zádor, *Z Anal. Chem.* **124**, 42-45 (1942). As a textile dye, morin dyes wool mordanted with chromium: olive yellow; mordanted with aluminum: yellow; mordanted with tin: lemon yellow; mordanted with iron: deep olive brown.

6353. Morindin. *1,5-Dihydroxy-2-methyl-6-[(6-O-β-D-xylopyranosyl-β-D-glucopyranosyl)oxy]-9,10-anthracenedione; 2-[[6-O-(6-deoxy-α-L-mannopyranosyl)-β-D-glucopyranosyl]-oxy]-1,5-dihydroxy-6-methyl-9,10-anthracenedione.* $C_{27}H_{30}$-O_{14}; mol wt 578.53. C 56.06%, H 5.23%, O 38.72%. From bark of *Coprosma australis* Forst. (*C. grandifolia* Hook.), *Rubiaceae:* Briggs, Dacre, *J. Chem. Soc.* **1948**, 564. Structure: Briggs, Le Quesne, *ibid.* **1963**, 3471. On acid hydrolysis yields the aglucone *morindone.* Synthesis and structure proof: B. Vermes *et al., Phytochemistry* **19**, 119 (1980).

Yellow needles from glacial acetic acid, mp 169-171°. $[\alpha]_D^{20}$ −90.0° (c = 0.054 in dioxane). Absorption max (ethanol): 230, 261, 448 nm (log ε 4.15, 3.93, 3.57). Sol in cold water but on boiling forms a water insol form, β-*morindin.* Sol in dioxane, pyridine, acetone, methanol; slightly sol in ethanol, glacial acetic acid. Practically insol in ether, chloroform, benzene, petr ether.

6354. Moroxydine. *N-(Aminoiminomethyl)-4-morpholinecarboximidamide; 4-morpholinecarboximidoylguanidine; N^1,N^1-anhydrobis(β-hydroxyethyl)biguanide; abitilguanide; abitylguanide; ABOB; Virusmin.* $C_6H_{13}N_5O$; mol wt 171.20. C 42.09%, H 7.65%, N 40.91%, O 9.35%. Prepn: **Brit. pat. 776,176** (1957 to AB Kabi). Proposed as influenza suppressant: Melander, *Antibiot. & Chemother.* **10**, 39 (1960).

Hydrochloride, $C_6H_{13}N_5O·HCl$, *Flumidin, Virustat.* Crystals. The acetate, maleate, and nitrate have also been prepd.

THERAP CAT: Antiviral.

6355. Morphazinamide. *N-(4-Morpholinylmethyl)pyrazinecarboxamide; N-morpholinomethylpyrazinamide; morfazinamide; morinamide; B-2310.* $C_{10}H_{14}N_4O_2$; mol wt 222.25. C 54.04%, H 6.35%, N 25.21%, O 14.40%. Prepn from pyrazinamide, morpholine, and formalin: Felder, Tiepolo, **Ger. pat. 1,129,492** (1962 to Bracco).

Crystals, mp 118.5-119.5°. uv max (ethanol): 269, 317 nm (log ε 3.95, 2.77). One gram dissolves in 3 ml water, 30 ml ethanol, 30 ml benzene, 2.5 ml chloroform.

Hydrochloride, $C_{10}H_{14}N_4O_2·HCl$, *B-2311, Piazofolina, Piazolin.* Crystals, mp 196°. One gram dissolves in 2 ml water, 350 ml ethanol, 2000 ml chloroform.

THERAP CAT: Antibacterial (tuberculostatic).

6356. Morphenol. *Phenanthro[4,5-bcd]furan-3-ol;* 3-hydroxy-4,5-oxidophenanthrene; 3-hydroxy-4,5-phenanthrylene oxide. $C_{14}H_8O_2$; mol wt 208.22. C 80.76%, H 3.87%, O 15.37%. First described by Vongerichten, *Ber.* **30**, 2439 (1897). Prepn from morphine: Mosettig, Meitzner, *J. Am. Chem. Soc.* **56**, 2738 (1934). Approach to synthesis: Dendy *et al., J. Chem. Soc.* **1963**, 4040.

Needles from benzene, mp 145°. Sol in methanol, ethanol, ether. Sol in concd H_2SO_4 giving a yellow soln with green fluorescence; also sol in aq NaOH giving a yellow soln with blue fluorescence.

Methyl ether, $C_{15}H_{10}O_2$, mp 68°.

6357. Morpheridine. *1-[2-(Morpholinyl)ethyl]-4-phenyl-4-piperidinecarboxylic acid ethyl ester; 1-(2-morpholinoethyl)-4-phenylisonipecotic acid ethyl ester;* ethyl 1-(2'-morpholinoethyl)-4-phenylpiperidine-4-carboxylate; morpholinoethyl norpethidine. $C_{20}H_{30}N_2O_3$; mol wt 346.47. C 69.33%, H 8.73%, N 8.09%, O 13.85%. Prepn: Anderson *et al., J. Chem. Soc.* **1956**, 4088; Stern, Anderson, **U.S. pat. 2,795,-581** (1957 to J. F. Macfarlan). HPLC determn: I. Jane, A. McKinnon, *J. Chromatog.* **323**, 191 (1956). Analgesic activity in rats: R. A. Millar, R. P. Stephenson, *Brit. J. Pharmacol.* **11**, 27 (1956). Pharmacology: A. F. Green, N. B. Ward, *ibid.* 32.

Liquid. $bp_{0.5}$ 188-192°. n_D^{18} 1.5276.

Dihydrochloride, $C_{20}H_{30}N_2O_3·2HCl$, *TA-1.* Crystals from dil ethanol, mp 264-266° (dec). LD_{50} in mice (mg/kg): 45 i.v. (Green, Ward), 118 i.p. (Millar, Stephenson).

Note: This is a controlled substance (opiate) listed in the U.S. Code of Federal Regulations, Title 21 Part 1308.11 (1995).

6358. Morphinan. [$4aR(4a\alpha,10\alpha,10a\alpha)$]-1,3,4,9,10,-10a-Hexahydro-2H-10,4a-(iminoethano)phenanthrene; 1,2,-3,9,10,10a-hexahydro-10,4a(4H)-iminoethanophenanthrene. $C_{16}H_{21}N$; mol wt 227.35. C 84.53%, H 9.31%, N 6.16%. Parent substance of the morphine alkaloids, codeine, thebaine, etc. Prepd from 1-benzyl-1,2,3,4,5,6,7,8-octahydro-2-methylisoquinoline: Grewe, Mondon, *Ber.* **81**, 279 (1948); from β-cyclohexenyl-N-phenacetylethylamine: Grewe *et al., Ber.* **84**, 527 (1951).

Oily liq, bp$_{0.05}$ 115°.
Hydrochloride, crystals from acetone + ether, dec 229°.
Sulfate, crystals from alcohol + ether, dec 195°.
N-Methylmorphinan, crystals, mp 61°.

6359. Morphine. *(5α,6α)-7,8-Didehydro-4,5-epoxy-17-methylmorphinan-3,6-diol;* morphium; morphia; Dolcontin; Duromorph; Morphina; Nepenthe. $C_{17}H_{19}NO_3$; mol wt 285.34. C 71.56%, H 6.71%, N 4.91%, O 16.82%. Principal alkaloid of opium which contains 9-14% anhydr morphine. Occurs naturally as the (−)-form. Also found in normal brain, blood and liver. Extraction procedures: Achor, Geiling, *Anal. Chem.* **26**, 1061 (1954); Leete, *J. Am. Chem. Soc.* **81**, 3950 (1959). Structure: Knorr, *Ber.* **22**, 1113 (1889); Knorr, Hörlein, *Ber.* **40**, 2032, 3341, 4889 (1907); Gulland, Robinson, *J. Chem. Soc.* **123**, 980 (1925). Total synthesis: Gates, Tschudi, *J. Am. Chem. Soc.* **74**, 1109 (1952); **78**, 1380 (1956); Ginsburg *et al., J. Am. Chem. Soc.* **1951**, 936; **1953**, 1524, 2664; **1954**, 3052. Synthesis of (+)-form: I. Iijima *et al., J. Org. Chem.* **43**, 1462 (1978); of (−)-form: E. J. Bijsterveld, H. J. Sinnige, *Rec. Trav. Chim.* **95**, 24 (1976); H. C. Beyerman *et al., ibid.* **97**, 127 (1978). Biogenesis: Leete, *J. Am. Chem. Soc.* **81**, 3948 (1959). Configuration: G. Stork in *The Alkaloids*, vol. II, Manske, Holmes, Eds. (Academic Press, New York, 1952) pp 171-189; Bick, *Nature* **169**, 755 (1952); Rapoport, Lavigne, *J. Am. Chem. Soc.* **75**, 5329 (1953); Bose, *Chem. & Ind. (London)* **1954**, 130; Mackay, Hodgkin, *J. Chem. Soc.* **1955**, 3261; Kalvoda *et al., Helv. Chim. Acta* **38**, 1847 (1955). Infrared and polarographic data: Seagers *et al., J. Am. Pharm. Assoc., Sci. Ed.* **41**, 640 (1952). Toxicity data: M. E. Buchwald, G. S. Eadie, *J. Pharm. Exp. Ther.* **71**, 197 (1941). GC-MS determn in biological fluids: R. Wasels, F. Belleville, *J. Chromatog A* **674**, 225 (1994). Review: K. W. Bentley, *The Chemistry of the Morphine Alkaloids* (Oxford, 1954) 433 pp. Comprehensive description: F. J. Muhtadi, *Anal. Profiles Drug Subs.* **17**, 259-366 (1988). Review of pharmacology of endogenous morphine: S. Benyhe, *Life Sci.* **55**, 969-979 (1994).

Short, orthorhombic, columnar prisms from anisole, dec 254°, also a metastable phase, mp 197°. High melting form sublimes at 190-200° (0.2 mm pressure at 2 mm distance). Monohydrate, $C_{17}H_{19}NO_3$·H_2O, orthorhombic, sphenoidal prisms, needles from methanol. Dec 254-256° with rapid heating. Darkens on exposure to light. Loses water at 130°. d$_4^{20}$ 1.32. [α]$_D^{20}$ −132° (methanol). pKb at 20° = 6.13; pKa 9.85. pH of satd soln, 8.5. uv max in acid: 285 nm; in alkali: 298 nm. One gram dissolves in about 5000 ml water, 1100 ml boiling water, 210 ml alcohol, 98 ml boiling alc, 1220 ml chloroform, 6250 ml ether, 114 ml amyl alc, 10 ml boiling methanol, 525 ml ethyl acetate. Freely sol in solns of fixed alkali and alkaline earth hydroxides, in phenol, cresols; moderately sol in mixtures of chloroform with alcohols; slightly sol in ammonia, benzene.
Acetate trihydrate, $C_{19}H_{23}NO_5$·$3H_2O$, yellowish-white powder; slight acetic odor. [α]$_D^{15}$ −77° (water). One gram dissolves in 2.25 ml water, 2 ml boiling water, 22 ml alc, 2 ml alc at 60°, 4.5 ml glycerol, 4.75 ml chloroform. Practically insol in ether. *Keep well closed and protected from light.*
Mucate, $C_{23}H_{29}NO_{11}$, *galactaric acid compd with (5α,6α)-7,8-didehydro-4,5-epoxy-17-methylmorphinan-3,6-diol (1:1), morphine hyperduric.*
Tartrate trihydrate, $(C_{17}H_{19}NO_3)_2$·$C_4H_6O_6$·$3H_2O$, crystalline powder. Sol in 11 parts water; slightly sol in alcohol. Practically insol in chloroform, ether, carbon disulfide.
6-Methyl ether, $C_{18}H_{21}NO_3$, *heterocodeine.*
Caution: May be habit forming. This is a controlled substance (opiate) listed in the U.S. Code of Federal Regulations, Title 21 Parts 329.1 and 1308.12 (1995).

THERAP CAT: Analgesic (narcotic).
THERAP CAT (VET): Analgesic (narcotic), preanesthetic, antitussive, antiperistaltic.

6360. Morphine Hydrochloride. Epimor. $C_{17}H_{20}Cl$-NO_3; mol wt 321.80. C 63.45%, H 6.26%, Cl 11.02%, N 4.35%, O 14.92%. Toxicity data: M. E. Buchwald, G. S. Eadie, *J. Pharm. Exp. Ther.* **71**, 197 (1941).
Trihydrate, white flakes or crystalline powder; bitter taste. Loses its water of crystn at about 100° and usually becomes yellowish. mp about 200° (dec). [α]$_D^{25}$ −113.5° (c = 2.2 in H_2O, anhydr basis). One gram dissolves in 17.5 ml water, 0.5 ml boiling water, 52 ml alc, 6 ml alc at 60°; slowly sol in glycerol. Insol in chloroform, ether. pH about 5. *Protect from light.* LD$_{50}$ in mice (mg/kg): 226-318 i.v. (Buchwald, Eadie).
Caution: May be habit forming. This is a controlled substance (opium derivative) listed in the U.S. Code of Federal Regulations, Title 21 Parts 329.1 and 1308.12 (1995).
THERAP CAT: Analgesic (narcotic).
THERAP CAT (VET): Analgesic (narcotic), preanesthetic, antitussive, antiperistaltic. Contraindicated in cats; unreliable in horses.

6361. Morphine Sulfate. *(5α,6α)-7,8-Didehydro-4,5-epoxy-17-methylmorphinan-3,6-diol sulfate salt (2:1);* Kapanol; Moscontin; MS Contin; MST Continus; Oblioser; Oramorph; Relipain; Roxanol. $C_{34}H_{40}N_2O_{10}S$; mol wt 668.77. C 61.06%, H 6.03%, N 4.19%, O 23.92%, S 4.79%.
Pentahydrate, *MST 10 Mundipharma, MST 30 Mundipharma.* White, fine odorless crystals or powder, or cubical masses. Loses some H_2O at ordinary temp; about $3H_2O$ (7.1%) at 100°, the remainder at 130°. Discolors on exposure to light. mp ∼250° with decompn when anhydr. [α]$_D^{25}$ −108.7° (c = 4 in H_2O, anhydr basis). One gram dissolves in 15.5 ml water at 25°, 0.7 ml water at 80°, 565 ml alcohol, 240 ml alcohol at 60°. Insol in chloroform, ether. pH about 4.8. *Keep well closed and protected from light.* Even ampuled aq solns of morphine sulfate will turn brown on storage. No loss of analgesic potency and no increase in toxicity has ever been demonstrated for such discolored solns: *J. Am. Med. Assoc.* **155**, 28 (1954); *J. Pharm. Pharmacol.* **2**, 673 (1950).
Caution: May be habit forming. This is a controlled substance (opium derivative) listed in the U.S. Code of Federal Regulations, Title 21 Parts 329.1 and 1308.12 (1995).
THERAP CAT: Analgesic (narcotic).
THERAP CAT (VET): Analgesic (narcotic), sedative, preanesthetic, gastric sedative. Contraindicated in cats; unreliable in horses.

6362. Morpholine. Tetrahydro-2H-1,4-oxazine; tetrahydro-1,4-oxazine; diethylene oximide; diethylene imidoxide. C_4H_9NO; mol wt 87.12. C 55.15%, H 10.41%, N 16.08%, O 18.36%. Prepd by dehydrating diethanolamine: Knorr, *Ann.* **301**, 1 (1898); Jones, Burns, *J. Am. Chem. Soc.* **47**, 2966 (1925); Hampton, Pollard, *ibid.* **58**, 2338 (1936). Toxicity study: H. F. Smyth *et al., Arch. Ind. Hyg. Occup. Med.* **10**, 61 (1954). Review of morpholine and derivatives: A. L. Wilson, *Ind. Eng. Chem.* **27**, 867-871 (1935). *Monograph:* Morpholine Technical Bulletin, Jefferson Chemical Co. (New York, 1953).

Mobile, hygroscopic liquid. Characteristic amine odor. mp −4.9°. bp$_{760}$ 128.9°; bp$_5$ 20.0°. d$_4^{20}$ 1.007. n_D^{20} 1.4540. Volatile with steam. Does not form an azeotrope with water. Flash pt, open cup: 100°F (38°C). Surface tension at 20° = 37.5 dynes/cm. Viscosity at 20° = 2.23 cp. Dipole moment 1.58. Strong base, pKb 5.6. Immiscible with concd NaOH solns. Miscible with water with evolution of some heat, with acetone, benzene, ether, castor oil, methanol, ethanol, ethylene glycol, 2-hexanone, linseed oil, turpentine, pine oil. Will dissolve 109% dimethylamine; 34% trimethylamine; 33% methylamine; > 5% naphtha; < 1%

paraffin oil; < 5% sulfur. LD_{50} orally in female rats: 1.05 g/kg (Smyth).

Hydrochloride, $C_4H_9NO.HCl$, crystals, dec 175-176°. Sol in water.

Salicylate, $C_{11}H_{15}NO_4$, *Retarcyl, Deposal.* Crystals from ethanol, mp 110-111°. Soluble in water, methanol, ethanol, ethyl acetate, acetone, benzene, chloroform. Practically insol in toluene, xylene, petr ether, ether, carbon tetrachloride.

N-Methylmorpholine, $C_5H_{11}NO$, mobile liquid, characteristic ammonia odor. bp_{764} 116-117°; bp_{760} 115.4°. d_4^{20} 0.9168. n_D^{20} 1.4332. Miscible with water, alcohol, ether.

Caution: Potential symptoms of overexposure are visual disturbance; nose irritation; coughing; respiratory irritation; eye and skin irritation; liver and kidney damage. *See NIOSH Pocket Guide to Chemical Hazards* (DHHS/NIOSH 90-117, 1990) p 156.

USE: Cheap solvent for resins, waxes, casein, dyes. Morpholine fatty acid salts are used as surface-active agents and emulsifiers. Other morpholine compds are used as corrosion inhibitors, antioxidants, plasticizers, viscosity improvers, insecticides, fungicides, herbicides, local anesthetics and antiseptics.

THERAP CAT: Salicylate as analgesic; antipyretic; anti-inflammatory.

6363. 7-Morpholinomethyltheophylline. *3,7-Dihydro-1,3-dimethyl-7-(4-morpholinylmethyl)-1H-purine-2,6-dione;* 1,3-dimethyl-7-morpholinomethylxanthine; Xanturil. $C_{12}H_{17}N_5O_3$; mol wt 279.30. C 51.60%, H 6.13%, N 25.07%, O 17.19%. Prepn: Burckhalter, Dill, *J. Org. Chem.* **24**, 562 (1959).

Crystals from ethanol, mp 177°.
THERAP CAT: Diuretic.

6364. Morphothebaine. *5,6,6a,7-Tetrahydro-10-methoxy-6-methyl-4H-dibenzo[de,g]quinoline-2,11-diol; 10-methoxyaporphine-2,11-diol;* 2,11-dihydroxy-10-methoxyaporphine. $C_{18}H_{19}NO_3$; mol wt 297.35. C 72.71%, H 6.44%, N 4.71%, O 16.14%. Prepd by heating thebaine with HCl. Formation from bulbocapnine: Ayer, Taylor, *J. Chem. Soc.* **1956,** 472. *Review:* Small, Lutz, "Chemistry of the Opium Alkaloids", *U.S. Public Health Reports,* Suppl. No. 103, Washington (1932); R. H. F. Manske, H. L. Holmes, *The Alkaloids,* vol. II (Academic Press, New York, 1952) pp 112-116.

Rhombic crystals, dec 197°. $[\alpha]_D^{15}$ − 130° (c = 0.88 in alc). Sol in alcohol, methanol, benzene, dil alkalies. Absorption spectra: Farmilo, *Bull. Narcotics U.N. Dept. Social Affairs* **6**, No. 3/4, 18 (1954).

Hydrobromide, $C_{18}H_{19}NO_3.HBr$, crystals from water, mp 270-275°. Also an acid hydrobromide, $(B.HBr)_2.HBr$.

Hydrochloride, $C_{18}H_{19}NO_3.HCl$, needles from water, mp 256-260°. Also an acid hydrochloride, $(B.HCl)_2.HCl$.

Absorption spectrum: Csokán, *Z. Anal. Chem.* **124**, 344 (1942).

Hydriodide, $C_{18}H_{19}NO_3.HI$, clusters from dil alc, mp 243-244° (with preliminary softening at 237°).

Methiodide, $C_{18}H_{19}NO_3.CH_3I$, crystals from acetic acid, mp 221-222°.

6365. Mosapramine. *(±)-1'-[3-(3-Chloro-10,11-dihydro-5H-dibenz[b,f]azepin-5-yl)propyl]hexahydrospiro[imidazo[1,2-a]pyridine-3(2H),4'-piperidin]-2-one;* 1'-[3-(3-chloro-10,11-dihydro-5H-dibenz[b,f]azepin-5-yl)propyl]-1,2,3,5,6,-7,8,8a-octahydro-2-oxoimidazo[1,2-a]-pyridine-3-spiro-4'-piperidine; 3-chloro-5-[3-(2-oxo-1,2,3,5,6,7,8,8a-octahydroimidazo[1,2-a]-pyridine-3-spiro-4'-piperidino)propyl]-10,11-dihydro-5H-dibenz[b,f]azepine; clospipramine. $C_{28}H_{35}ClN_4O$; mol wt 479.06. C 70.20%, H 7.36%, Cl 7.40%, N 11.70%, O 3.34%. Dopamine receptor antagonist with adrenoceptor blocking activity; metabolite of clocapramine, *q.v.* Prepn: C. Tashiro, I. Horii, U.S. pat. **4,337,260** (1982 to Yoshitomi); C. Tashiro *et al., Yakugaku Zasshi* **109**, 93 (1989), *C.A.* **112**, 35749j (1990). X-ray crystallographic analysis: I. Ueda, C. Tashiro, *Acta Crystallogr.* **C40**, 422 (1984). Receptor binding studies: M. Setoguchi *et al., Eur. J. Pharmacol.* **112**, 313 (1985). Pharmacokinetics and metabolism: J. Ishigooka *et al., Psychopharmacology* **97**, 303 (1989). Toxicity data: H. Horizoe *et al., Oyo Yakuri* **37**, 529 (1989); *C.A.* **111**, 167174g (1989). HPLC determn in plasma: K. Hikida *et al., Anal. Sci.* **6**, 367 (1990).

Dihydrochloride, $C_{28}H_{35}ClN_4O.2HCl$, *Y-516, Cremin.* White crystals, mp 271°. LD_{50} in male and female mice (mg/kg): 1008, 1293 orally; 74, 116 i.p.; 1147, 1264 s.c. (Horizoe).

THERAP CAT: Antipsychotic.

6366. Motilin. A gastrointestinal peptide hormone whose existence was postulated during experimental chemical stimulation of canine duodenum: J. C. Brown *et al., Gastroenterology* **50**, 333 (1966). Motilin stimulates gastric antrum, duodenum, and colon; it has been reported to be about 50 times more powerful than acetylcholine in its effects on sensitive tissue: U. Strunz *et al., ibid.* **68**, 1485 (1975). In man it is found primarily in the upper small intestine. Although its effects are not known completely, it plays a role in increasing gastric motility and stimulating pepsin output and may also be important in regulating the interdigestive myoelectric complex. *Porcine motilin,* which contains 22 amino acids, has been isolated from mucosa of hog small intestine: J. C. Brown *et al., Can. J. Physiol. Pharmacol.* **49**, 399 (1971). Purification, amino acid composition, *C*-terminal residue: *eidem, Gastroenterology* **62**, 401 (1972). Amino acid sequence: *eidem, Can. J. Biochem.* **51**, 533 (1973); revised sequence: H. Schubert, J. C. Brown, *ibid.* **52**, 7 (1974). First synthetic work: E. Wünsch *et al., Z. Naturforsch.* **28C**, 235 (1973). Synthesis of the complete docosapeptide corresponding to porcine motilin: H. Yajima *et al., Chem. Commun.* **1975,** 159. Alternate syntheses: Y. Kai *et al., Chem. Pharm. Bull.* **23**, 2346 (1975); S. Yamada *et al., J. Am. Chem. Soc.* **97**, 7174 (1975); E. Izeboud, H. C. Beyerman, *Rec. Trav. Chim.* **99**, 124 (1980). Solid phase synthesis: N. Ikota *et al., Chem. Pharm. Bull.* **28**, 3347 (1980); D. H. Coy *et al., Peptides* **3**, 137 (1982). Radioimmunoassay: J. R. Dryburgh, J. C. Brown, *Gastroenterology* **68**, 1169 (1975). Identification of motilin endocrine cells: J. M. Polak *et al., Gut* **16**, 225 (1975). Rodent and canine motilin differ from porcine; human motilin has not yet been purified, although its immunologic properties suggest strong similarity to porcine motilin. The presence of a larger motilin has also been detected immunologically, see J. M. Polak, A. M. J. Buchan, *Gastroenterology* **76**, 1065 (1979). *Reviews:* N. D. Christofides *et al.,* "Physiology of Motilin" in *Gut Hormones*, S. R. Bloom, Ed. (Churchill Livingstone, Edinburgh, 1978) pp 343-350; S. R. Bloom, J. M. Polak, *Clin. Endocrinol. Metab.* **8**, 401-411 (1979); V. Mutt, *Vitam. Horm.* **39**, 231-426 (1982).

$$\overset{1}{\text{Phe}}\text{–Val–Pro–Ile–Phe–Thr–Tyr–Gly–Glu–Leu–Gln–Arg}$$
$$\text{HO–}\overset{22}{\text{Gln}}\text{–Gly–Lys–Asn–Arg–Glu–Lys–Glu–Gln–Met}$$

Porcine Motilin

6367. Motretinide. *(all E)-N-Ethyl-9-(4-methoxy-2,3,6-trimethylphenyl)-3,7-dimethyl-2,4,6,8-nonatetraenamide;* 9-(4-methoxy-2,3,6-trimethylphenyl)-3,7-dimethylnona-2,4,-6,8-tetraen-1-oic acid ethyl amide; Ro-11-1430; Tasmaderm. $C_{23}H_{31}NO_2$; mol wt 353.50. C 78.15%, H 8.84%, N 3.96%, O 9.05%. Aromatic analog of retinoic acid, *q.v.* Prepn: W. Bollag *et al.,* **Ger. pat. 2,414,619** corresp to **U.S. pat. 4,105,-681** (1974, 1978, both to Hoffmann-La Roche). Description of properties: *idem, Chemotherapy* **21**, 236 (1975). ^{13}C-NMR study: G. Englert, *Helv. Chim. Acta* **58**, 2367 (1975). Pharmacology: L. J. Wilkoff *et al., Cancer Res.* **36**, 964 (1976); S. S. Shapiro *et al., ibid.* 3702. Inhibitory effect on radiation-induced neoplasms: L. Harisiades *et al., Nature* **274**, 486 (1978). Use in treatment of acne: K. Nordin *et al., Dermatologica* **162**, 104 (1981).

Crystals from ethanol, mp 179-180°.
THERAP CAT: Antiacne.

6368. Moveltipril. *[R-(R*,S*)]-1-[3-[[2-[(Cyclohexyl-carbonyl)amino]-1-oxopropyl]thio]-2-methyl-1-oxopropyl]-L-proline;* N-[3-(N-cyclohexanecarbonyl-D-alanylthio)-2-meth-ylpropanoyl]-L-proline; (−)-N-[(S)-[3-(N-cyclohexylcarbon-yl-D-alanyl)thio]-2-methylpropionyl]-L-proline; altiopril. $C_{19}H_{30}N_2O_5S$; mol wt 398.52. C 57.26%, H 7.59%, N 7.03%, O 20.07%, S 8.05%. Angiotensin-converting enzyme (ACE) inhibitor; structural analog of captopril. Prepn: S. Tanaka *et al.,* **Belg. pat. 893,553** (1982 to Chugai), *C.A.* **98**, 215995n (1983). Pharmacology: K. Sakai *et al., Tohoku J. Exp. Med.* **152**, 363 (1987). Hemodynamic effects in dogs: K. Noguchi *et al., Japan. J. Pharmacol.* **40**, 373 (1986). Antihypertensive activity: J. Aono *et al., Arch. Int. Pharma-codyn.* **292**, 203 (1988). Tissue specific ACE inhibitory activity: K. Sakai *et al., ibid.* **294**, 228 (1988). Determn by radioimmunoassay: Y. Hinohara *et al., J. Pharmacobio-Dyn.* **11**, 411 (1988).

Solid, mp 113-116°. $[\alpha]_D$ 14.2° (c = 1.05 in methanol). Calcium salt, $C_{38}H_{58}CaN_4O_{10}S_2$, *MC-838, Lowpres.* Practically white powder, mp ∼190°. Bitter taste. $[\alpha]_D^{20}$ −48 to −52° (c = 1 in methanol). Freely sol in water, methanol; sol in ethanol, chloroform. Practically insol in acetone, ethyl acetate. 10% aq soln has pH 5.5-6.5. Stable as aq soln and powder at room temp. LD_{50} in male, female mice, male, female rats (g/kg): all > 10.0 orally; 2.1, 2.3, 1.3, 1.3

i.p.; 3.0, 3.8, 3.4, 3.9 s.c.; and in male, female dogs (g/kg): > 6.0, > 6.0 orally (Sakai).
THERAP CAT: Antihypertensive.

6369. Moxalactam. *7-[[Carboxy(4-hydroxyphenyl)acet-yl]amino]-7-methoxy-3-[[(1-methyl-1H-tetrazol-5-yl)thio]-methyl]-8-oxo-5-oxa-1-azabicyclo[4.2.0]oct-2-ene-2-carboxylic acid;* N-[(6R,7R)-2-carboxy-7-methoxy-3-[[(1-methyl-1H-tetrazol-5-yl)thio]methyl]-8-oxo-5-oxa-1-azabicyclo[4.2.0]-oct-2-en-7-yl]-2-(p-hydroxyphenyl)malonamic acid; 7β-[2-carboxy-2-(4-hydroxyphenyl)acetamido]-7α-methoxy-3-[[(1-methyl-1H-tetrazol-5-yl)thio]methyl]-1-oxa-1-dethia-3-cephem-4-carboxylic acid; lamoxactam; latamoxef. $C_{20}H_{20}N_6O_9S$; mol wt 520.48. C 46.15%, H 3.87%, N 16.15%, O 27.67%, S 6.16%. Oxa-substituted third generation cephalosporin antibiotic (oxacephalosporin). Prepn: M. Narisada, W. Nagata, **Ger. pat. 2,713,370;** *eidem,* **U.S. pat. 4,138,486** (1977, 1979 both to Shionogi); M. Narisada *et al., J. Med. Chem.* **22**, 757 (1979). Laboratory evaluation: T. Yoshida *et al., Antimicrob. Ag. Chemother.* **17**, 302 (1980). Mechanism of action study: Y. Komatsu, T. Nishikawa, *ibid.* 316. Pharmacokinetics: R. Wise *et al., ibid.* **18**, 369 (1980). Comparative *in vitro* activity: T. O. Kurtz *et al., ibid.* 645. Series of articles on pharmacology, teratology, toxicity studies: *Chemotherapy (Tokyo)* **28**, Suppl. 7, 1002-1235 (1980). Clinical evaluation in neonates and infants: U. B. Schaad *et al., J. Pediatr.* **98**, 129 (1981). *Review: Rev. Infect. Dis.* **4**, Suppl., S489-S726 (1982).

Colorless powder, mp 117-122° (dec). $[\alpha]_D^{25}$ −15.3 ±2.6° (c = 0.216 in methanol). uv max (methanol): 276 nm (ε 10200).
Disodium salt, $C_{20}H_{18}N_6Na_2O_9S$, *LY-12735, S-6059, Festa-moxin, Moxalactam, Moxam, Shiomarin.* $[\alpha]_D^{22}$ −45° (water). uv max (water): 270 nm (ε 12000).
THERAP CAT: Antibacterial.

6370. Moxastine. *2-(1,1-Diphenylethoxy)-N,N-dimeth-ylethanamine;* α-methylbenzhydryl 2-dimethylaminoethyl ether; α-methyldiphenhydramine; mephenhydramine; Spofa 325; Alfadryl. $C_{18}H_{23}NO$; mol wt 269.39. C 80.26%, H 8.61%, N 5.20%, O 5.94%. Prepn: Protiva *et al., Chem. Listy* **43**, 257 (1949); **Czech. pat. 86,516** (1957), *C.A.* **53**, 1261 (1959); Z. Vejdelek *et al., Coll. Czech. Chem. Commun.* **49**, 2649 (1984).

bp$_{0.15}$ 129-136°.
Hydrochloride, $C_{18}H_{23}NO\cdot HCl$, mp 168°.
THERAP CAT: Antihistaminic.

6371. Moxaverine. *3-Ethyl-6,7-dimethoxy-1-(phenyl-methyl)isoquinoline;* 1-benzyl-3-ethyl-6,7-dimethoxyisoquin-oline. $C_{20}H_{21}NO_2$; mol wt 307.39. C 78.15%, H 6.89%, N 4.56%, O 10.41%. Prepn of hydrochloride: **Fr. pat. 1,362,-765** (1964 to Orgamol, SA), *C.A.* **62**, 536f (1965), corresp to **Brit. pat. 1,030,022.**

Crystals, mp 78-79°.

Hydrochloride, $C_{20}H_{21}NO_2 \cdot HCl$, *Eupaverin, Eupaverina, Kollateral, Sorbosan.* Crystals, mp 214°; also reported as dec 208-210°. Sol in hot water, hot alc, and many other organic solvents; very sparingly sol in cold water.

Note: The name Eupaverin was formerly used to designate 3-methyl-6,7-methylenedioxy-1-piperonylisoquinoline hydrochloride.

THERAP CAT: Antispasmodic.

6372. Moxestrol. *(11β,17α)-11-Methoxy-19-nor-17-pregna-1,3,5(10)-trien-20-yne-3,17-diol;* 3,17-dihydroxy-11β-methoxy-19-nor-17α-pregna-1,3,5-trien-20-yne; 17α-ethynyl-11β-methoxyestra-1,3,5(10)-triene-3,17β-diol; 11β-methoxy-17α-ethynyl-Δ$^{1,3,5(10)}$-estratriene-3,17β-diol; 11β-methoxy-17α-ethynylestradiol; R-2858; Surestryl. $C_{21}H_{26}O_3$; mol wt 326.44. C 77.27%, H 8.03%, O 14.70%. Prepn: Bertin, Pierdet, **S. Afr.** pat. **67 03,381** corresp to **U.S.** pat. **3,579,-545** and **Fr.** pat. **M6182,** *C.A.* **72,** 90739m (1970), (1968, 1971 and 1968, all to Roussel-UCLAF); Azadian-Boulanger, Bertin, *Chim. Ther.* **8,** 451 (1973). Activity studies: Raynaud, *Steroids* **21,** 249 (1973); Raynaud *et al., Mol. Pharmacol.* **9,** 520 (1973).

Crystals, mp 280°. $[\alpha]_D^{20}$ +29° (c = 0.6 in ethanol). uv max (ethanol): 280 nm ($E_{1cm}^{1\%}$ 58.4).

THERAP CAT: Estrogen.

6373. Moxidectin. *[6R,23E, 25S(E)]-5-O-Demethyl-28-deoxy-25-(1,3-dimethyl-1-butenyl)-6,28-epoxy-23-(methoxyimino)milbemycin B;* 23-methyloxime LL-F282249α; (2aE,4E,-5'R,6R,6'S,8E,11R,13S,15S,17aR,20R,20aR,20bS)-6'-[(E)-1,3-dimethyl-1-butenyl]-5',6,6',7,10,11,14,15,17a,20,20a,-20b-dodecahydro-20,20b-dihydroxy-5',6,8,19-tetramethyl-spiro[11,15-methano-2H,13H,17H-furo[4,3,2-pq]benzo-dioxacyclooctadecin-13,2'-[2H]pyran]-4',17(3'H)-dione 4'-(E)-(O-methyloxime); CL-301423; Cydectin; Moxidec; Vetdectin. $C_{37}H_{53}NO_8$, mol wt 639.83. C 69.46%, H 8.35%, N 2.19%, O 20.00%. Semisynthetic macrolide antibiotic; 23-methyloxime analog of nemadectin, *q.v.* Prepn: B. M. Bain

et al., **Eur.** pat. **Appl. 237,339** (1987 to Glaxo); G. Asato, D. J. France, **Eur.** pat. **Appl. 259,779** (1988 to Am. Cyanamid). Efficacy in cattle against lice: J. D. Webb *et al., J. Econ. Entomol.* **84,** 1266 (1991); against gastrointestinal nematodes: G. L. Zimmerman *et al., Am. J. Vet. Res.* **53,** 1409 (1992); J. C. Williams *et al., Vet. Record* **131,** 345 (1992); against psoroptic mange: J. F. Lonneux, B. Losson, *Vet. Parasitol.* **45,** 147 (1992). Environmental effects on dung-burying beetles: G. T. Fincher, G. T. Wang, *Southwest. Entomol.* **17,** 303 (1992).

THERAP CAT (VET): Anthelmintic, antiparasitic.

6374. Moxisylyte. *4-[2-(Dimethylamino)ethoxy]-2-methyl-5-(1-methylethyl)phenol acetate (ester);* 5-(2-dimethyl-aminoethoxy)carvacrol acetate; 6-acetoxythymol 2-(dimeth-ylamino)ethyl ether; (6-acetoxythymoxy)ethyldimethyl-amine; 4-(2-dimethylaminoethoxy)-5-isopropyl-2-methyl-phenyl acetate; 4-(2-dimethylaminoethoxy)-2-methyl-5-iso-propylphenyl acetate; thymoxamine. $C_{16}H_{25}NO_3$; mol wt 279.38. C 68.79%, H 9.02%, N 5.01%, O 17.18%. α-Adren-ergic blocker. Prepn: Pahlicke, **Ger.** pat. **905,738** (1954 to Diwag), *C.A.* **52,** 16294i (1958); **Brit.** pat. **745,070** (1956 to Veritas Drug), *C.A.* **51,** 471i (1957); A. Buzas *et al., Bull. Soc. Chim. France* **1959,** 839. Pharmacology: K. Greeff, H. J. Schümann, *Arzneimittel-Forsch.* **3,** 341 (1953); J. Mercier *et al., Therapie* **26,** 785 (1971). Toxicology: J. Roquebert, J. Canellas, *ibid.* 775. Clinical-pharmacological studies: A. Wasilewski, *Arzneimittel-Forsch.* **21,** 1183 (1971). α-Adren-ergic antagonist activity: A. T. Birmingham, J. Szolcsanyi, *J. Pharm. Pharmacol.* **17,** 449 (1965); G. M. Drew, *Eur. J. Pharmacol.* **36,** 313 (1976). Metabolism in rats: C. Feniou *et al., J. Pharm. Pharmacol.* **32,** 104 (1980). Series of articles on metabolism: K.-O. Vollmer *et al., Eur. J. Drug Metab. Pharmacokinet.* **10,** 61, 71, 139 (1985). Pharmacokinetics: F. Nielsen-Kudsk *et al., Acta Pharmacol. Toxicol.* **47,** 11 (1980). GLC determn of metabolites in plasma: H. Hengy *et al., Eur. J. Drug Metab. Pharmacokinet.* **10,** 203 (1985). HPLC determn: P. Duchene *et al., J. Chromatog.* **424,** 205 (1988). Clinical trial in Raynaud's disease and chilblains: G. V. Jaffe, J. J. Grimshaw, *Brit. J. Clin. Prac.* **34,** 343 (1980). Clinical study in miosis: F. Grehn *et al., Graefe's Arch. Clin. Exp. Ophthalmol.* **224,** 174 (1986); F. Grehn, *Am. J. Ophthalmol.* **103,** 709 (1987). *Review:* M. Wand, W. M. Grant, *Survey Ophthalmol.* **25,** 75-84 (1980).

Hydrochloride, $C_{16}H_{25}NO_3 \cdot HCl$, *Arlitene, Moxyl, Opilon, Uroalpha, Vasoklin.* Shiny beige needles from ethyl acetate/methanol, mp 208-210°. LD_{50} in mice, rats (mg/kg): 265 ±19, 740 ±51 orally; 200 ±15, 190 ±19 s.c. (Roquebert, Canellas).

THERAP CAT: Vasodilator (peripheral).

6375. Moxonidine. *4-Chloro-N-(4,5-dihydro-1H-imid-azol-2-yl)-6-methoxy-2-methyl-5-pyrimidinamine;* 4-chloro-6-methoxy-2-methyl-5-(2-imidazolin-2-yl)aminopyrimidine; BDF-5895; Cynt; Physiotens. $C_9H_{12}ClN_5O$; mol wt 241.68. C 44.73%, H 5.00%, Cl 14.67%, N 28.98%, O 6.62%. α$_2$-Adrenoceptor agonist. Prepn: **Neth.** pat. **Appl. 7,908,192;** W. Stenzel *et al.,* **U.S.** pat. **4,323,570** (1980, 1982 both to Beiersdorf AG). Series of articles on pharmacology and binding studies: *Arzneimittel-Forsch.* **38,** 1426-1445 (1988). Pharmacokinetics: D. Trenk *et al., J. Clin. Pharmacol.* **27,** 988 (1987); W. Kirch *et al., ibid.* **30,** 1088 (1990). Clinical comparison with clonidine, *q.v.:* V. Plänitz, *ibid.* **27,** 46 (1987).

mp 217-219° (dec).

Hydrochloride, $C_9H_{12}ClN_5O.xHCl$. Crystals from isopropanol/ether, mp 189°.

THERAP CAT: Antihypertensive.

6376. MPTP. *1,2,3,6-Tetrahydro-1-methyl-4-phenylpyridine;* 1-methyl-4-phenyl-1,2,3,6-tetrahydropyridine. $C_{12}H_{15}N$; mol wt 173.26. C 83.19%, H 8.73%, N 8.08%. Piperidine derivative which causes irreversible symptoms of parkinsonism in humans, monkeys. Prepn as hydrochloride by Grignard reaction: A. Ziering *et al., J. Org. Chem.* **12**, 894 (1947). Alternative prepn: C. J. Schmidle, R. C. Mansfield, *J. Am. Chem. Soc.* **78**, 425 (1956). Identification as impurity in "synthetic heroin" and effect on drug users: J. W. Langston *et al., Science* **219**, 979 (1983). Selective destruction of dopaminergic neurons in primates: R. S. Burns *et al., Proc. Nat. Acad. Sci. USA* **80**, 4546 (1983). *In vitro* metabolism by rat brain monoamine oxidase to 1-methyl-4-phenylpyridinium ion *(MPP⁺):* K. Chiba *et al., Biochem. Biophys. Res. Commun.* **120**, 574 (1984). Studies on mechanism of neurotoxicity: J. W. Langston *et al., Science* **225**, 1480 (1984); S. P. Markey *et al., Nature* **311**, 464 (1984). Binding studies in rat brain: C. M. Wieczorek *et al., Eur. J. Pharmacol.* **98**, 453 (1984); B. Parsons, T. C. Rainbow, *ibid.* **102**, 375 (1984); in rat, human brain: J. A. Javitch *et al., Proc. Nat. Acad. Sci. USA* **81**, 4591 (1984). Comparison of idiopathic and MPTP-induced parkinsonism in humans: R. S. Burns *et al., N. Engl. J. Med.* **312**, 1418 (1985). *Review:* T. P. Singer *et al., Trends Biochem. Sci.* **12**, 266-270 (1987); L. M. Sayre, *Toxicol. Lett.* **48**, 121-149 (1989).

Crystals from heptane, mp 40-42°. $bp_{0.8}$ 85-90°.

6377. MSH. *Melanotropin;* melanophore-affecting hormone; melanocyte-stimulating hormone; melanophore hormone; melanophore dilating hormone; melanophore expanding hormone; melanophore-stimulating hormone; melanotropic hormone; chromatophorotropic hormone; melanosome-dispersing hormone; pigmentation hormone. First known as *Intermedin(e)* or *B hormone,* a pituitary factor causing color changes in fish and amphibia: B. Zondek, H. Kron, *Klin. Wochenschr.* **11**, 849 (1932). Subsequently, the hormone was identified as two linear peptides designated as α- and β-MSH. γ-MSH, a third peptide has also been identified but has no significant melanocyte stimulating activity. The peptides are secreted by the pars intermedia of the pituitary gland. The brain may be a secondary site of synthesis. α-MSH derives from ACTH and β-MSH from β-lipotrophic hormone, *q.q.v.* These MSH precursors derive from two sections of pro-opiomelanocortin, *q.v.* A third portion of pro-opiomelanocortin may serve as the precursor of γ-MSH. Biosynthesis: B. G. Jenks *et al., J. Endocrinol.* **98**, 19 (1983); G. J. M. Martens *et al., Gen. Comp. Endocrinol.* **49**, 73 (1983). Distribution and biosynthesis in the brain: D. F. Swaab *et al.,* "The Distribution of MSH and ACTH in the Rat and Human Brain and its Relation to Pituitary Stores," in *Endogenous Peptides and Learning and Memory Processes,* J. L. Martinez, Jr. *et al.,* Eds. (Academic Press, New York, 1981) pp 7-36. The mechanism of control of MSH release from the pituitary is still being investigated. There is evidence for inhibitory control by the brain via dopaminergic neurons, inhibitory control by the peptide melanotropin (MIF-1), *q.v.* and stimulatory control by β-adrenergic agents and the peptide CRF, *q.v.:* G. Schmitt *et al., Neuroendocrinology* **33**, 306 (1981); P. G. Smelik *et al.,* "The Role of Catecholamines in the Control of the Secretion of Pro-Opiocortin-Derived Peptides from the Anterior and Intermediate Lobes and its Implications in the Response to

Stress," in *The Anterior Pituitary Gland,* A. S. Bhatnagar, Ed. (Raven Press, New York, 1983) pp 113-125; L. Proulx-Ferland *et al., J. Steroid Biochem.* **19**, 439 (1983). Although the role of MSH in adaptive color change in lower vertebrates in well known, its physiological significance in mammals appears to be extrapigmentary. Effects on reproduction: M. E. Celis, M. Volosin, *Prog. Clin. Biol. Res.* **87**, 113 (1982). Reviews of behavioral and neurochemical effects: P. C. Datta, M. G. King, *Neurosci. Biobehav. Rev.* **6**, 297-310 (1982); B. E. Beckwith, C. A. Sandman, *Peptides* **3**, 411-420 (1982). General reviews: Li, *Advan. Protein Chem.* **12**, 270-295 (1957); Novales, *Neuroendocrinology* **2**, 241 (1967); R. Schwyzer, *Proc. Roy. Soc. London [Biol.]* **210**, 5-20 (1980); F. L. Strand, C. M. Smith, *Pharmacol. Ther.* **11**, 509-533 (1980). Book: A. J. Thody, *The MSH Peptides* (Academic Press, New York, 1980) 162 pp.

α-*Melanotropin.* Contains 13 amino acids. Mammalian α-MSH is the most potent melanocyte-stimulating peptide known. Mechanism of action: T. K. Sawyer *et al., Am. Zool.* **23**, 1983. Isoln from hog pituitary gland: T. H. Lee, A. B. Lerner, *J. Biol. Chem.* **221**, 943 (1956); from pig pituitary gland: Harris, Lerner, *Nature* **179**, 1346 (1957); Harris, *Biochem. J.* **71**, 451 (1959); from horse pituitary gland: Dixon, Li, *J. Am. Chem. Soc.* **82**, 4568 (1960). Structure: Lee *et al., J. Biol. Chem.* **236**, 1390 (1961). Synthesis: Schwyzer *et al., Helv. Chim. Acta* **46**, 870 (1963). Synthesis and biological activity of dogfish α-MSH: A. Eberle *et al., Helv. Chim. Acta* **61**, 2360 (1978). Activity as centrally administered antipyretic agent: M. T. Murphy *et al., Science* **221**, 192 (1983). Cardiovascular effects: D. J. Diz, D. M. Jacobowitz, *Brain Res.* **270**, 265 (1983). Review of the α-melanotropinergic system and its role in the CNS: T. L. O'Donohue, D. M. Jacobowitz in *Polypeptide Hormones,* R. F. Beers, E. G. Bassett, Eds. (Raven Press, New York, 1980) pp 203-222. $[\alpha]_D^{25}$ −58.5° ± 2.5° (c = 0.38 in 10% acetic acid).

β-*Melanotropin.* Contains 18-22 amino acids differing slightly from one species to another: Lee *et al., J. Biol. Chem.* **236**, 1390 (1961). Isoln from hog pituitary gland: T. H. Lee, A. B. Lerner, *loc. cit.;* from pig pituitary gland: Geschwind *et al., J. Am. Chem. Soc.* **78**, 4494 (1956); **79**, 620 (1957); Harris, Roos, *Biochem. J.* **71**, 434 (1959); from bovine pituitary gland: Geschwind *et al., J. Am. Chem. Soc.* **79**, 6394 (1957); from human pituitary gland: Dixon, *Biochim. Biophys. Acta* **37**, 38 (1960); Harris, *Nature* **184**, 167 (1959). Synthesis of derivatives: Hofmann, *Ann. N.Y. Acad. Sci.* **88**, art. 3, p 689 (1960). Synthesis of bovine β-melanotropin: Schwyzer *et al., Helv. Chim. Acta* **46**, 1975 (1963); S. Lemaire *et al., J. Med. Chem.* **19**, 373 (1976). Solid-phase synthesis of human and monkey β-melanotropins: Wang *et al., Int. J. Peptide Protein Res.* **5**, 33 (1973); of camel β_{c1}-MSH: Li *et al., Biochemistry* **14**, 953 (1975); of camel β_{c2}-MSH: S. Lemaire *et al., loc. cit.;* of dogfish β-MSH: H. Yajima *et al., Chem. Pharm. Bull.* **26**, 571 (1978); of equine β-MSH: J. Izdebski *et al., Int. J. Peptide Res.* **19**, 327 (1982). It is probable that the human pituitary does not produce β-MSH. The substance reported as human β-MSH in early studies seems to have been an artifact formed by enzymatic degradation of β-LPH during the extraction procedure: K. Tanaka *et al., J. Clin. Invest.* **62**, 94 (1978); A. J. Thody, *loc. cit.*

γ-*Melanotropin.* Contains 12 amino acids. Discovered during nucleotide sequencing of cloned DNA for bovine pro-opiomelanocortin: S. Nakanishi *et al., Nature* **278**, 423 (1979). May be an antagonist of β-endorphin or a partial agonist/antagonist of ACTH. Synthesis and biological activity: N. Ling *et al., Life Sci.* **25**, 1773 (1979); K. Okamoto *et al., Chem. Pharm. Bull.* **28**, 2839 (1980); W. A. Bijl *et al., Rec. Trav. Chim.* **100**, 123 (1981). Radioimmunoassay: T. Shibasaki *et al., Life Sci.* **26**, 1781 (1980). Behavioral profile in rats: J. M. Van Ree *et al., ibid.* **28**, 2875 (1981). Comparison with α-MSH: T. L. O'Donohue *et al., Peptdes* **2**, 101 (1981). *Review:* J. M. Van Ree *et al., Ciba Found. Symp.* **81**, 263-276 (1981). $[\alpha]_D^{20}$ −38.0° (c 0.5 in 10% acetic acid). $[\alpha]_D^{23}$ −33.4° (c = 1.0 in 1% acetic acid).

6378. Mucins. High molecular weight glycoproteins, major constituents of saliva, gastric juice, intestinal juice, and other secretions: F. Haurowitz, *Chemistry and Biology of Proteins* (Academic Press, New York, 1950) p 199. Mu-

Consult the Name Index before using this section.

cins are capable of forming viscous solutions and thereby act as lubricants or protectants in cavities of the body or on body surfaces. For prepn (*e.g.* from snails) and their biochemical role *see* P. A. Levene, *The Hexosamines and Mucoproteins* (Longmans, Green, London, 1925). Ovine submaxillary mucin consists of a single polypeptide chain with about 800 disaccharide units attached, each of which is *N*-acetylneuraminyl(2→6) *N*-acetylgalactosamine. Carbohydrate accounts for about 45% of the molecular weight. Structure studies: Pigman, Tanaka, *148th Am. Chem. Soc. Meet.* (Chicago, Aug.-Sept., 1964), Abstracts of Papers, p 11C; Ozeki, Yosizawa, *Arch. Biochem. Biophys.* **142**, 177 (1971); Huser *et al., Z. Physiol. Chem.* **354**, 749 (1973). Glycoproteins in bovine cervical mucus: F. A. Meyer *et al., Adv. Exp. Med. Biol.* **89**, 239 (1977). Review on physical and chemical properties, biosynthesis and function of mucin: several authors in *Mucus in Health and Disease,* M. Elstein, D. V. Parke, Eds. (Plenum Press, New York, 1977) pp 171-311.

Mucins are obtained as greenish-gray or yellow powders forming very viscous solns in water; generally sol in dil alkalies. Insol in acetic acid. Acidic substances with an isoelectric point between pH 3 and 5.

USE: Demulcent; adsorbent.

6379. Mucochloric Acid. *2,3-Dichloro-4-oxo-2-butenoic acid; dichloromalealdehydic acid;* α,β*-dichloro-β-formylacrylic acid; 2,3-dichloromaleic aldehyde acid.* $C_4H_2Cl_2O_3$; mol wt 168.96. C 28.43%, H 1.19%, Cl 41.96%, O 28.41%. Prepn from β,γ-dichloropyromucic acid and bromine water or dil nitric acid: Hill, Jackson, *Am. Chem. J.* **12**, 43 (1890); by heating furfural with manganese dioxide and hydrochloric acid: Simonis, *Ber.* **32**, 2085 (1899); Beattie *et al., J. Chem. Soc.* **1932**, 264; Mowry, *J. Am. Chem. Soc.* **72**, 2535 (1950); by chlorination of butyne-1,4-diol: Dury, *Angew. Chem.* **72**, 864 (1960).

Monoclinic prisms from ether and ligroin, mp 127°. Slightly sol in cold water; sol in hot water; hot benzene, alc. LD_{50} in rats: 0.5-1.0 g/kg orally; 10-25 mg/kg i.p., Fassett in *Industrial Hygiene and Toxicology* **vol. II**, F. A. Patty, Ed. (Interscience, New York, 2nd ed., 1962) p 1977.

Caution: Strong irritant to skin and eyes; a potent skin sensitizer.

6380. Mucochloric Anhydride. *5,5'-Oxybis[3,4-dichloro-2(5H)-furanone]; bis[3,4-dichloro-2(5)-furanonyl] ether;* GC-2466. $C_8H_2Cl_4O_5$; mol wt 319.91. C 30.04%, H 0.63%, Cl 44.33%, O 25.01%. Prepd by refluxing mucochloric acid with benzenesulfonic acid in a mixture of benzene and dioxane: Mowry, *J. Am. Chem. Soc.* **72**, 2535 (1950).

Crystals from benzene + dioxane, α-isomer (racemate): mp 141-143°, β-isomer (*meso* form): mp 180°. Substantially insol in water. Sol in many organic solvents, such as acetone, xylene, cyclohexanone, methylnaphthalenes.

USE: Fungicide. *Ref:* Gilbert, U.S. pat. **2,861,919** (1958 to Allied Chem.).

6381. Muconic Acid. *2,4-Hexadienedioic acid;* 1,3-butadiene-1,4-dicarboxylic acid. $C_6H_6O_4$; mol wt 142.11. C 50.71%, H 4.26%, O 45.03%. HOOCCH=CHCH=CH-COOH. Prepd by oxidation of phenol and peracetic acid: Boeseken, Engelberts, *C.A.* **26**, 2970 (1932); by treatment of ethyl 1,4-dibromoadipate with alcoholic potassium hydroxide: Guha, Sankaran, *Org. Syn.* **26**, 57 (1946); by isomerization of 3-hydroxy-4-carbomethoxybut-1-ene-1-carboxylic acid lactone: Elvidge *et al., J. Chem. Soc.* **1950**, 2235; by

carbonylation of acetylene: Tsuji *et al., J. Am. Chem. Soc.* **86**, 2095 (1964). Configuration and separation of isomers: Boeseken, Kerkhoven, *Rec. Trav. Chim.* **51**, 964 (1932); Elvidge *et al., J. Chem. Soc.* **1953**, 708.

trans,trans-Form, prisms from water, mp 301°. uv max (0.1*N* NaOH): 251, 259, 264 nm (ε 25600, 29100, 25600). One gram dissolves in about 5 liters water at 15°. Quite sol in hot alcohol and glacial acetic acid.

Dimethyl ester, $C_8H_{10}O_4$, crystals from alcohol, mp 159°. Insol in ether.

Diethyl ester, $C_{10}H_{14}O_4$, needles from alcohol, mp 64°; d_4^{100} 0.9829. n_{He}^{99} 1.4675.

cis,cis-Form, prisms from ethanol, mp 194-195°. uv max (0.1*N* NaOH): 251, 258, 264 nm (ε 15600, 17000, 15300). Freely sol in boiling water; sparingly sol in ether.

Dimethyl ester, $C_8H_{10}O_4$, needles from petr ether, mp 75°. Soluble in ether.

Diethyl ester, $C_{10}H_{14}O_4$, minute crystals from alcohol at −25°; mp 13°. Changes to the *trans-trans*-form on storage at room temp.

cis,trans-Form, needles from hot water, mp 190-191°. uv max (0.1*N* NaOH): 251, 259, 265 nm (ε 23400, 25600, 23400).

Dimethyl ester, $C_8H_{10}O_4$, needles from aqueous methanol, mp 75°.

6382. Muira Puama. Potency wood. Wood of *Liriosma ovata* Miers, *Oleaceae,* or according to Rebourgeon, *Acanthea virilis* Wehmer, *Acanthaceae.* Habit. Brazil. Constit. Aromatic resin, muirapuaminc, fat. Studies of components: Auterhoff, Pankow, *Arch. Pharm. (Weinheim)* **301**, 481 (1968); **302**, 209 (1969). *Reviews:* Gaebler, *Deut. Apoth.* **22**, 94 (1970); Steinmetz, *Quart. J. Crude Drug Res.* **11**, 1787 (1971).

6383. Mupirocin. *[2S-[2α(E),3β,4β,5α[2R*,3R*(1R*,-2R*)]]]-9-[[3-Methyl-1-oxo-4-[tetrahydro-3,4-dihydroxy-5-[[3-(2-hydroxy-1-methylpropyl)oxiranyl]methyl]-2H-pyran-2-yl]-2-butenyl]oxy]nonanoic acid; pseudomonic acid A; trans-pseudomonic acid;* BRL-4910A; Bactoderm; Bactroban; Turixin. $C_{26}H_{44}O_9$; mol wt 500.63. C 62.38%, H 8.86%, O 28.76%. Major component of the pseudomonic acids, *q.v.,* an antibiotic complex produced by *Pseudomonas fluorescens* NCIB 10586. Isoln and characterization: A. T. Fuller *et al., Nature* **234**, 416 (1971); K. D. Barrow, G. Mellows, Ger. pat. **2,227,739;** *eidem,* U.S. pat. **3,977,943;** *eidem,* U.S. pat **4,071,536** (1973, 1976, 1978 all to Beecham). Purification: P. J. O'Hanlon *et al.,* Ger. pat. **2,842,358;** *eidem,* U.S. pat. **4,222,942** (1979, 1980 both to Beecham). Structure: E. B. Chain, G. Mellows, *Chem. Commun.* **1974**, 847; *eidem, J. Chem. Soc. Perkin Trans. I* **1977**, 294. Absolute configuration: R. G. Alexander *et al., ibid.* **1978**, 561. Prepn from methyl pseudomonate C: *eidem, Tetrahedron Letters* **22**, 2059 (1981). Total syntheses of (±)-form: B. B. Snider, G. B. Phillips, *J. Am. Chem. Soc.* **104**, 1113 (1982); B. B. Snider *et al., J. Org. Chem.* **48**, 3003 (1983). Biosynthesis: T. C. Feline *et al., J. Chem. Soc. Perkin Trans. I* **1977**, 309. Effect of pH on chemical stability and antibacterial activity: J. P. Clayton *et al., ibid.* **1979**, 838. Inhibition of bacterial protein synthesis: J. Hughes, G. Mellows, *J. Antibiot.* **31**, 330 (1978); of isoleucyl-tRNA synthetase: *eidem, Biochem. J.* **176**, 305 (1978); *eidem, ibid.* **191**, 209 (1980). *In vitro* antibacterial spectrum: R. Sutherland *et al., Antimicrob. Ag. Chemother.* **27**, 495 (1985); M. W. Casewell, R. L. R. Hill, *J. Antimicrob. Chemother.* **15**, 523 (1985). Antimycoplasmal activity *in vitro:* R. M. Banks *et al., J. Antibiot.* **41**, 609 (1988). Clinical evaluations: G. D. Reilly, R. C. Spencer, *ibid.* **13**, 295 (1984); M. W. Casewell, R. L. R. Hill, *ibid.* **17**, 365 (1986). *Reviews:* A. Ward, D. M. Campoli-Richards, *Drugs* **32**, 425-444 (1986); M. W. Casewell, R. L. R. Hill, *J. Antimicrob. Chemother.* **19**, 1-5 (1987).

Crystals from ether, mp 77-78°. $[\alpha]_D^{20}$ −19.3° (c = 1 in methanol). uv max (ethanol): 222 nm (ε 14500).

THERAP CAT: Topical antibacterial.

THERAP CAT (VET): Topical antibacterial.

6384. Muramic Acid. *(R)-2-Amino-3-O-(1-carboxyethyl)-2-deoxy-D-glucose;* 3-O-α-carboxyethyl-D-glucosamine. $C_9H_{17}NO_7$; mol wt 251.24. C 43.03%, H 6.82%, N 5.58%, O 44.58%. Amino sugar found (as the *N*-acetyl derivative) in *peptidoglycan*, the main skeletal component of the bacterial cell wall. Discovery: J. T. Park, *J. Biol. Chem.* **194**, 885 (1952). Isoln from spores of *Bacillus megatherium*: R. E. Strange, F. A. Dark, *Nature* **177**, 186 (1956). Identification and synthesis: R. E. Strange, L. H. Kent, *Biochem. J.* **71**, 333 (1959). Stereospecific synthesis: Y. Matsushima, J. T. Park, *J. Org. Chem.* **27**, 3581 (1962); *eidem, Biochem. Prepn.* **10**, 109 (1963); T. Osawa, R. W. Jeanloz, *J. Org. Chem.* **30**, 448 (1965). Review of peptidoglycan structure: H. J. Rogers, *Ann. N.Y. Acad. Sci.* **235**, 29-51 (1974). Use to determine bacterial levels in mammalian tissues: J. Gilbart *et al., J. Microbiol. Methods* **5**, 271 (1986); in airborne dust: A. Fox *et al., Appl. Environ. Microbiol.* **59**, 4354 (1993).

Crystals from water, mp 152-154° (dec); $[\alpha]_D^{25}$ +103° (c = 0.26 in water) (Matsushima, Park). Also reported as crystals from 90% ethanol, mp 160-162° (dec); $[\alpha]_D^{22}$ +146° (6 minutes) → +116° (31 hrs) (c = 0.57 in water) (Osawa, Jeanloz).

N-Acetylmuramic acid, $C_{11}H_{19}NO_8$, *(R)-2-(acetylamino)-3-O-(1-carboxyethyl)-2-deoxy-D-glucose.* Crystals from ethyl acetate + methanol, mp 119-121°. $[\alpha]_D^{20}$ +56° (10 minutes) → +40° (24 hrs) (c = 0.68 in water).

USE: As chemical marker for the detection of bacterial contamination.

6385. Muramyl Dipeptide. *N²-[N-(N-Acetylmuramoyl)-L-alanyl]-D-α-glutamine;* *N*-acetylmuramyl-L-alanyl-D-isoglutamine; 2-acetamido-2-deoxy-3-O-(D-2-propionyl-L-alanyl-D-isoglutamine)-D-glucopyranose; MDP. $C_{19}H_{32}N_4O_{11}$; mol wt 492.48. C 46.34%, H 6.55%, N 11.38%, O 35.74%. Synthetic immunoadjuvant corresponding to the smallest immunologically active glycopeptide subunit of the bacterial cell wall. Prepn: A. Adam *et al.,* **Ger.** pat. **2,450,355;** *eidem,* **U.S.** pat. **4,235,771** (1975, 1980 both to Agen. Nat. Valor. Recher.); C. Merser *et al., Biochem. Biophys. Res. Commun.* **66**, 1316 (1975); P. Lefrancier *et al., Int. J. Pep. Prot. Res.* **9**, 249 (1977). Structure-activity relationships: F. Ellouz *et al., Biochem. Biophys. Res. Commun.* **59**, 1317 (1974); A. Adam *et al., ibid.* **72**, 339 (1976). Review of immunoregulating activity: L. Chedid *et al., Prog. Allergy* **25**, 63-105 (1978). Tissue distribution in mice: M. Parant *et al., Int. J. Immunopharmacol.* **1**, 35 (1979). Potentiation of

antitumor immunity: S. Sone *et al., J. Biol. Resp. Mod.* **3**, 185 (1984); A. E. Eggers, *ibid.* **7**, 229 (1988).

Crystals from methanol-acetone-ether. $[\alpha]_D^{25}$ +44° (acetic acid).

USE: Immunological adjuvant.

6386. Murexide. *5-[(Hexahydro-2,4,6-trioxo-5-pyrimidinyl)imino]-2,4,6(1H,3H,5H)-pyrimidinetrione monoammonium salt; 5,5'-nitrilodibarbituric acid monoammonium salt;* acid ammonium purpurate; ammonium purpurate. $C_8H_8N_6O_6$; mol wt 284.19. C 33.81%, H 2.84%, N 29.57%, O 33.78%. Prepn from alloxan + NH_3: Hartley, *J. Chem. Soc.* **87**, 1791 (1905); Schwartz, Handritschk, **East Ger.** pat. **17,-589** (1959), *C.A.* **55**, 3630a (1961); from alloxantin + ammonium acetate: Davidson, *J. Am. Chem. Soc.* **58**, 1821 (1936). Structure: Hartley, *J. Chem. Soc.* **87**, 1796 (1905); Winslow, *J. Am. Chem. Soc.* **61**, 2089 (1939); Schreiber, *Mitt. Deut. Pharm. Ges.* **28**, 20 (1958).

Purple-red crystals with green metallic luster. Absorption max in water: 520 nm. Sparingly sol in cold water, more in hot water; practically insol in alcohol, ether. The H_2O soln is deep purple and the aq NaOH soln is deep blue.

USE: Indicator for complexometric titrations.

6387. Murexine. *2-[[3-(1H-Imidazol-4-yl)-1-oxo-2-propenyl]oxy]-N,N,N-trimethylethanaminium;* β-(4-imidazolyl)-acrylcholine; urocanylcholine. $[C_{11}H_{18}N_3O_2]^+$. Neuromuscular blocker found, often in very large quantities, in the median zone of the hypobranchial body of *Murex trunculus* and of other related species of mollusks: Erspamer, Dordoni, *Ric. Sci. Ricostr.* **16**, 1114 (1946); *Arch. Int. Pharmacodyn.* **74**, 263 (1947); Erspamer, *ibid.* **76**, 308 (1948). Isoln: *idem, Experientia* **4**, 226 (1948); M. Roseghini *et al., Eur. J. Biochem.* **12**, 468 (1970); J. E. Blankenship *et al., Comp. Biochem. Physiol.* **51C**, 129 (1975). Synthesis: Pasini *et al., Ann.* **578**, 6 (1952); Pasini, Coda, **U.S.** pat. **2,956,061** (1960 to Società Farmaceutici). Structure: V. Erspamer, O. Benati, *Science* **117**, 161 (1953). Effect on vertebrate and invertebrate muscles: J. C. de Freitas, *Comp. Biochem. Physiol.* **56C**, 57 (1977).

The base is instantly hydrolyzed by water; unstable both in acid or alkaline media.

Chloride, extremely hygroscopic crystals. uv max (pH 4.5): 280-282 nm.

Chloride hydrochloride, hygroscopic microcrystalline powder, mp 219-221° (dec); shows the same uv max as the chloride.

6388. Muscalure. *(Z)-9-Tricosene;* cis-tricos-9-ene. $C_{23}H_{46}$; mol wt 322.62. C 85.63%, H 14.37%. Sex pheromone of the female common house fly, *Musca domestica* L. Isoln and synthesis: Carlson *et al., Science* **174**, 76 (1971). Alternate syntheses: Eiter, *Naturwiss.* **59**, 468 (1972); Cargill, Rosenblum, *J. Org. Chem.* **37**, 3971 (1972); Gribble *et al., Chem. Commun.* **1973**, 735; Ho, Wong, *Can. J. Chem.* **52**, 1923 (1974); K. Abe *et al., Bull. Chem. Soc. Japan* **50**, 2792 (1977); V. N. Odinokov *et al., Tetrahedron Letters* **23**, 1371 (1982).

bp$_{0.1}$ 157-158°. n_D^{26} 1.4517.

USE: Insect attractant.

6389. Muscarine. *[2S-(2α,4β,5α)]-Tetrahydro-4-hy-droxy-N,N,N,5-tetramethyl-2-furanmethanaminium.* [C$_9$H$_{20}$-NO$_2$]$^+$. Alkaloid from the red variety of *Amanita muscaria* (L.) Pers., *Agaricaceae*, the fly fungus, a poisonous mushroom. Also found in some other fungi: *Inocybe patouillardi; I. fastigiata, I. umbrina; I. rimosa*. Isoln procedure for the naturally occurring L-(+)-form: Kuehl *et al., J. Am. Chem. Soc.* **77**, 6663 (1955); Eugster, *Helv. Chim. Acta* **39**, 1002 (1956). Structure and synthesis of racemate: Kögl *et al., Rec. Trav. Chim.* **76**, 109 (1957); Kögl *et al., Experientia* **13**, 137, 138 (1957); Cox *et al., Helv. Chim. Acta* **41**, 229 (1958). Alternate syntheses: **Brit.** pat. 828,395 (1960 to Hoffmann-La Roche); Matsumoto *et al., Tetrahedron* **25**, 5889 (1969); W. C. Still, J. A. Schneider, *J. Org. Chem.* **45**, 3375 (1980). Synthesis of muscarine: Whiting *et al., Can. J. Chem.* **50**, 3322 (1972); A. M. Mubarak, D. M. Brown, *Tetrahedron Letters* **21**, 2453 (1980); *eidem, J. Chem. Soc. Perkin Trans I* **1982**, 809; S. Pochet, T. Huynhdinh, *J. Chem. Org. Chem.* **47**, 193 (1982). Chemistry and pharmacology: Waser, *Pharmacol. Rev.* **13**, 465-515 (1961). Toxicity study: P. J. Fraser, *Brit. J. Pharmacol.* **12**, 47 (1957). *Reviews:* C. H. Eugster, *Advances in Organic Chemistry* **2**, 427-455 (1960); S. Wilkinson, *Quarterly Revs. (London)* **15**, 153-171 (1961). Configurational relationship in the muscarine series: Bollinger, Eugster, *Helv. Chim. Acta* **54**, 2704 (1971).

Chloride, C$_9$H$_{20}$ClNO$_2$, stout prisms from ethanol + acetone, mp 180-181°. Extremely hygroscopic. [α]$_D^{25}$ +8.1° (c = 3.5 in ethanol). Very sol in water, ethanol. Slightly sol in chloroform, ether, acetone. Aq solns are stable. LD$_{50}$ i.v. in mice: 0.23 mg/kg (Fraser).

Caution: Potential symptoms of toxicity following ingestion are profuse sweating, increased salivation, visual disturbances, nausea, vomiting, abdominal colic, diarrhea, headache, and bronchospasm. Very high doses may produce lacrimation, incontinence, bradycardia, hypotension and shock. *See Clinical Toxicology of Commercial Products,* R. E. Gosselin *et al.,* Eds. (Williams & Wilkins, Baltimore, 5th ed., 1984) Section II, p 247.

THERAP CAT: Cholinergic.

6390. Muscazone. *α-Amino-2,3-dihydro-2-oxo-5-oxazoleacetic acid; α-amino-2-oxo-4-oxazoline-5-acetic acid.* C$_5$H$_6$N$_2$O$_4$; mol wt 158.11. C 37.98%, H 3.82%, N 17.72%, O 40.48%. From *Amanita muscaria* (L.) Fr., *Agaricaceae:* Eugster *et al., Tetrahedron Letters* **1965**, 1813. Structure: Fritz *et al., ibid.* **1965**, 2075; Reiner, Eugster, *Helv. Chim. Acta* **50**, 128 (1967). Synthesis: Göth *et al., ibid.* 137.

Crystals, decomp above 190°. uv max at pH 2-7: 212 nm (ε 8700); at pH 12: 220 nm (ε 7500).

6391. Muscimol. *5-(Aminomethyl)-3(2H)-isoxazolone; 5-(aminomethyl)-3-isoxazolol;* 5-aminomethyl-3-hydroxy-isoxazole; 3-hydroxy-5-aminomethylisoxazole; agarin; pantherine. C$_4$H$_6$N$_2$O$_2$; mol wt 114.10. C 42.11%, H 5.30%, N 24.55%, O 28.04%. Potent CNS depressant and GABA agonist isolated from *Amanita muscaria* (L.) Fr., *Agaricaceae.* Structural identity with the insecticidal substances, pantherine and agarin: Onda *et al., Chem. Pharm. Bull.* **12**, 751 (1964); Eugster *et al., Tetrahedron Letters* **1965**, 1813; Bowden, Drysdale, *ibid.* 727. Synthesis: Gagneux *et al., ibid.* 2077; Göth *et al., Helv. Chim. Acta* **50**, 137 (1967); Bowden *et al., J. Chem. Soc. (C)* **1968**, 172. Improved syntheses: Nakamura, *Chem. Pharm. Bull.* **19**, 46 (1971); P.

Krogsgaard-Larsen, S. B. Christensen, *Acta Chem. Scand. Ser. B* **B30**, 281 (1976); A. Barco *et al., J. Chem. Res. (S)* **1979**, 176; B. E. McCarry, M. Savard, *Tetrahedron Letters* **22**, 5153 (1981); V. Jager, M. Frey, *Ann.* **1982**, 817. Conformational structure: L. Brehm *et al., ibid.* **26**, 1298 (1972). Industrial patents: Hafliger, Gagneux, U.S. pats. 3,242,190, 3,397,209 (1966, 1968 both to Geigy). Pharmacology: Theobald *et al., Arzneimittel-Forsch.* **18**, 311 (1968). Review of pharmacology: F. V. DeFeudis, *Neurochem. Res.* **5**, 1047-1068 (1980).

Crystals, mp 175° (dec). LD$_{50}$ in mice (mg/kg): 3.8 s.c., 2.5 i.p.; in rats (mg/kg): 4.5 i.v., 45 orally (Theobald).

USE: As a molecular probe to study GABA receptors.

6392. Muscone. *3-Methylcyclopentadecanone;* muskone; methylexaltone. C$_{16}$H$_{30}$O; mol wt 238.41. C 80.61%, H 12.68%, O 6.71%. The odorous principle of musk. Characterization of naturally occurring (−)-form: Ruzicka, *Helv. Chim. Acta* **9**, 715 (1926). Structure: *ibid.* 1008. Synthesis of (±)-form: Ziegler, Weber, *Ann.* **512**, 164 (1934); Ruzicka, Stoll, *Helv. Chim. Acta* **17**, 1308 (1934); Hunsdiecker, *Ber.* **75B**, 1197 (1942). Review of syntheses: S. Abe *et al., Cosmetics and Perfumery* **88**, 67 (1973). Recent syntheses: Stork, Macdonald, *J. Am. Chem. Soc.* **97**, 1264 (1975); H. G. Fliri *et al., Monatsh. Chem.* **110**, 245 (1979); C. Fehr *et al., Helv. Chim. Acta* **62**, 2655 (1979); K. H. Schulte-Elte *et al., ibid.* 2673; G. Cantoni *et al., J. Org. Chem.* **45**, 1906 (1980). Chiral synthesis of (R)-(−)- and (S)-(+)-forms: Q. Branca, A. Fischli, *Helv. Chim. Acta* **60**, 925 (1977).

(−)-Form, oily liquid. Musk odor. bp 328°; bp$_{0.5}$ 130°. d$_4^{17}$ 0.9221. n_D^{17} 1.4802. [α]$_D^{17}$ −13°. Very slightly sol in water; miscible with alc.

(±)-Form, oily liquid. bp$_{0.01}$ 90°. n_D^{26} 1.4767. uv max (ethanol): 285 nm (log ε 1.54).

6393. Musks. Fixatives for perfumes having a characteristic persistent musky aroma. Obtained from plant and animal sources and by chemical synthesis. Natural sources of musk include the musk glands of the male musk deer, *Moschus moschiferus*, the civet cat, *Viverra civetta*, the Louisiana muskrat, *Abelmoschus moschatus*, and angelica roots, *Angelica archangelica*. Musk odor is exhibited by several categories of chemical compounds. Natural musks are macrocyclic ketones or lactones having approximately fifteen carbons in their ring structures. Synthetic musks are of greater industrial importance and include nitro and non-nitro benzenes, indans, and tetralins; derivatives of hydrindacene, isochroman, naphthindan, and coumarin. Purification of natural and synthetic musks: Thomas, Stephens, U.S. pat. 3,415,813 (1968 to Pfizer). Review of macrocyclic musks: Berends, *Am. Perfum. Cosmet.* **80**, 35 (1965). Review of chemical studies of synthetic musks: T. F. Wood, "Chemistry of the Aromatic Musks" (Givaudan Corp., company literature).

USE: In perfumery.

6394. Mustard, Black. Brown mustard; red mustard. Dried ripe seeds of *Brassica nigra* (L.) Koch, or of *B. juncea* (L.) Cosson, and of varieties of these species *(Cruciferae). Habit.* Europe, Asia, naturalized in U.S. *Constit.* Sinigrin (potassium myronate), myrosin, sinapine sulfocyanate, fixed oil; erucic, behenic, and sinapolic acids.

USE: Source of volatile oil of mustard.

THERAP CAT: Emetic; counterirritant.

THERAP CAT (VET): Counterirritant. Has been used as emetic, carminative.

6395. Mustard Gas. *1,1′-Thiobis[2-chloroethane]; bis-(2-chloroethyl)sulfide; β,β′-*dichloroethyl sulfide; 2,2′-dichlorodiethyl sulfide; bis(β-chloroethyl)sulfide; 1-chloro-2-(β-chloroethylthio)ethane; sulfur mustard; yellow cross liquid; Kampfstoff "Lost"; Yperite. $C_4H_8Cl_2S$; mol wt 159.08. C 30.20%, H 5.07%, Cl 44.57%, S 20.16%. $(ClCH_2-CH_2)_2S$. War gas prepd by treating ethylene with sulfur chloride (Levinstein process): Mann, Pope, *J. Chem. Soc.* **121,** 594 (1922); by treating β,β′-dihydroxyethyl sulfide with HCl gas (German process): Meyer, *Ber.* **19,** 3260 (1886); *Ann.* **240,** 310 (1887); Gomberg, *J. Am. Chem. Soc.* **41,** 1427 (1919). Reactions and derivatives: Helfrich, Reid, *ibid.* **42,** 1208 (1920). Toxicity: Anslow *et al., J. Pharmacol. Exp. Ther.* **93,** 1 (1948). Review of carcinogenicity studies: *IARC Monographs* **9,** 181-192 (1975).
Oily liquid. *Deadly vesicant.* Weak, sweet, agreeable odor. On cooling it forms prisms, mp 13-14°. d^{13} 1.338 (solid); d_4^{20} 1.2741 (liq). bp_{760} 215-217°; bp_{10} 98°. Volatile with steam. n_D^{20} 1.53125. Very sparingly sol in water; sol in fat solvents, other common organic solvents. High lipid soly. Vapor pressure at 0° = 0.025 mm; at 30° = 0.090 mm. Hydrolyzed by alkalies. Recommended neutralizing agent and inactivator: Bleaching powder, sodium hypochlorite. LD_{50} in rats, mice (mg/kg): 3.3, 8.6 i.v. (Anslow).
Caution: Overexposure may cause severe irritation and tissue damage to eyes, skin, and respiratory and GI tracts; symptoms may be delayed in onset. This substance has been listed as a known carcinogen: *Seventh Annual Report on Carcinogens* (PB95-109781, 1994) p 63.
USE: In chemical warfare.

6396. Mustard, White. Yellow mustard. Dried ripe seeds of *Brassica alba* (L.) Boiss. *(Sinapis alba* L.), *Cruciferae. Habit.* Europe, Asia, adventitious in U.S. *Constit. and Use:* As of mustard, black.
THERAP CAT: Emetic; counterirritant.
THERAP CAT (VET)*: See* Mustard, Black.

6397. Muzolimine. *5-Amino-2-[1-(3,4-dichlorophenyl)-ethyl]-2,4-dihydro-3H-pyrazol-3-one;* 3-amino-1-(3,4-dichloro-α-methylbenzyl)-2-pyrazolin-5-one; Bay g 2821; Edrul. $C_{11}H_{11}Cl_2N_3O$; mol wt 272.13. C 48.55%, H 4.07%, Cl 26.06%, N 15.44%, O 5.88%. High-ceiling loop diuretic. Prepn: E. Möller *et al.,* **Ger.** pat. **2,319,278** corresp to **U.S.** pat. **4,018,890** (1974, 1977 both to Bayer); *eidem, Experientia* **33,** 382 (1977). *In vitro* effects: H. J. Kramer, *Pharmatherapeutica* **1,** 353 (1977). Mechanism of action: M. Mussche, N. Lameire, *Curr. Med. Res. Opin.* **4,** 462 (1977). Pharmacology: K. Meng *et al., ibid.* 555. Pharmacokinetics: O. Broers *et al., Eur. J. Clin. Pharmacol.* **15,** 105 (1979); *eidem, Curr. Med. Res. Opin.* **6,** 431 (1980). Use in advanced renal disease: A. D. Canton *et al., Brit. Med. J.* **282,** 595 (1981); P. Schmidt *et al., Eur. J. Clin. Pharmacol.* **20,** 23 (1981). Toxicological studies: D. Lorke, P. Mürmann, *Curr. Med. Res. Opin.* **4,** 716 (1977). Symposium on pharmacology, pharmacodynamics, toxicology, clinical studies: *Clin. Nephrol.* **19,** Suppl. 1, S1-S117 (1983).

Cryst from methanol, mp 127-129°. LD_{50} in mice, rats, dogs, rabbits (mg/kg): 1794, 1559, 2000, 1250 orally (Lorke, Mürmann).
THERAP CAT: Diuretic; antihypertensive.

6398. Mycaminose. *3,6-Dideoxy-3-(dimethylamino)-D-glucose.* $C_8H_{17}NO_4$; mol wt 191.23. C 50.25%, H 8.96%, N 7.32%, O 33.47%. Part of the carbomycin molecule. Synthesis and stereochemistry: Richardson, *Proc. Chem. Soc.* **1961,** 430; Foster *et al., Chem. & Ind. (London)* **1962,** 142. Stereochemistry: Grisebach, Hofheinz, *Angew. Chem.* **74,** 499 (1962).

Hydrochloride monohydrate, crystals from moist isopropanol, mp 116-118°. $[α]_D$ +31° (equilib value, c = 0.96).

6399. Mycarose. *2,6-Dideoxy-3-C-methyl-L-ribo-hexose;* 2,6-didesoxy-3-C-methyl-L-*ribo*-hexose; 2-(4,5-*cis*)-trihydroxy-4,6-dimethyltetrahydropyran. $C_7H_{14}O_4$; mol wt 162.19. C 51.84%, H 8.70%, O 39.46%. Constituent of the macrolides carbomycin A and B, spiramycin A, B and C and tylosin: Woodward, *Angew. Chem.* **69,** 50 (1957); Paul, Tchelitcheff, *Bull. Soc. Chim. France* **1957,** 443; **1960,** 150; Hamill *et al., Antibiot. & Chemother.* **11,** 328 (1961). Part of carbomycin molecule: Regna *et al., J. Am. Chem. Soc.* **75,** 4625 (1953). Stereochemistry: Grisebach, Hofheinz, *Angew. Chem.* **74,** 499 (1962); Foster *et al., Proc. Chem. Soc.* **1962,** 254; Hofheinz *et al., Tetrahedron* **18,** 1265 (1962). Synthesis of DL-mycarose: Korte *et al., ibid.* 1257; Grisebach *et al., Ber.* **96,** 1823 (1963). Synthesis of L-mycarose: Lemal *et al., Tetrahedron* **18,** 1275 (1962); Koto *et al., Bull. Chem. Soc. Japan* **45,** 532 (1972). Synthesis of D-mycarose: Flaherty *et al., J. Chem. Soc. (C)* **1966,** 398. Isoln from aureolic acid: Berlin *et al., Khim. Prir. Soedin.* **8,** 535 (1972), *C.A.* **78,** 4432a (1973).

DL-Form, crystals from acetone + petr ether, mp 110-111°.
L-Form, needles from boiling acetone + chloroform, mp 128.5-130.5°. $[α]_D^{25}$ −31.1° (c = 4). Sol in water. Exhibits only end absorption in the ultraviolet.
3-*O*-Methylmycarose, $C_8H_{16}O_4$, *Cladinose.* Constituent of erythromycin A and B. Liquid, $bp_{0.25}$ at bath temp 120-132°. $[α]_D^{25}$ −23.1° (c = 2.6 in water). Sol in water, alcohol, acetone, ether, benzene, chloroform, carbon tetrachloride; slightly sol in petr ether. Dec by strong acids.

6400. Mycelianamide. *[S-(?,E)]-3-[[4-[(3,7-Dimethyl-2,6-octadienyl)oxy]phenyl]methylene]-1,4-dihydroxy-6-methyl-2,5-piperazinedione.* $C_{22}H_{28}N_2O_5$; mol wt 400.47. C 65.98%, H 7.05%, N 7.00%, O 19.98%. Antibiotic substance found in the mycelium of *Penicillium griseofulvum:* Anslow, Raistrick, *Biochem. J.* **25,** 39 (1931); Oxford *et al., ibid.* **29,** 1102 (1935); **33,** 240 (1939); Oxford, Raistrick, *ibid.* **42,** 323 (1948). Structure: Birch *et al., J. Chem. Soc.* **1956,** 3717; Bates, Schauble, *Tetrahedron Letters* **1963,** 1683. Configuration: Gallina *et al., Gazz. Chim. Ital.* **94,** 1301 (1964). Partial synthesis: Brown, Meehan, *Aust. J. Chem.* **21,** 1581 (1968). Total synthesis: N. Shinmon *et al., Chem. Commun.* **1980,** 1020.

Crystals from ethyl acetate, dec 170-172°. $[α]_{546}^{19}$ −217°; $[α]_{579}^{19}$ −182°. Weak acid. Freely soluble in acetone, dioxane; sparingly sol in other organic solvents; sol in aq sodium carbonate solution, but not in sodium bicarbonate soln. Quickly destroyed by acids and alkalies.

6401. Mycetins. Antibiotic substances produced by *Actinomyces violaceus:* Fainshmidt, Koreniako, *Biokhimiya* **9,** 147 (1944); Krassilnikov, Koreniako, *Microbiologiya* **14,** 80 (1945). Chromatographic and structure studies: Yakubov *et al.*, *ibid.* **31**, 526 (1962); Blinov *et al.*, *J. Chromatog.* **8**, 522 (1962); Yakubov *et al.*, *Antibiotiki* **10**, 771 (1965).

6402. Myclobutanil. *α-Butyl-α-(4-chlorophenyl)-1H-1,2,4-triazole-1-propanenitrile;* 2-(4-chlorophenyl)-2-(1*H*-1,2,4-triazol-1-ylmethyl)hexanenitrile; RH-3866; Nova; Rally; Systhane. $C_{15}H_{17}ClN_4$; mol wt 288.78. C 62.39%, H 5.93%, Cl 12.28%, N 19.40%. Broad spectrum systemic triazole fungicide; ergosterol-biosynthesis inhibitor. Prepn: T. T. Fujimoto, **Eur. pat. Appl.** 145,294 (1985 to Rohm & Haas), *C.A.* **103**, 160519z (1985). Physical properties and field trials: C. Orpin *et al.*, *Proc. Brit. Crop Prot. Conf.—Pests Dis.* **1986**, 55; A. Perrot, *Défense des Végétaux* **40**, 9 (1986). Antifungal activity: J. A. Quinn *et al.*, *Pestic. Sci.* **17**, 357 (1986).

Light yellow crystals, mp 63-68°. $bp_{1.0}$ 202-208°. Vapor pressure at 25°: 1.6×10^{-6} Torr. Soly at 25°: water 142 ppm. Sol in common organic solvents such as ketones, esters, alcohols, aromatic hydrocarbons. Insol in aliphatic hydrocarbons. LD_{50} in male, female rats (mg/kg): 1600, 2229 orally; LD_{50} in rabbits (mg/kg): 7500 dermally (Orpin).

USE: Agricultural fungicide.

6403. Mycobacidin. *4-Oxo-2-thiazolidinehexanoic acid;* ε-[2-(4-thiazolidone)]hexanoic acid; 4-thiazolidone-2-caproic acid; 2-(5-carboxypentyl)-4-thiazolidone; cinnamonin; actithiazic acid; acidomycin. $C_9H_{15}NO_3S$; mol wt 217.29. C 49.75%, H 6.96%, N 6.45%, O 22.09%, S 14.76%. Antibiotic substance produced by *Streptomyces lavendulae*, *Streptomyces virginiae* and several other *Streptomyces* spp. Isoln: Tejera *et al.*, *Antibiot. & Chemother.* **2**, 333 (1952). Production by fermentation: Grundy *et al.*, **U.S. pat.** 2,678,929 (1954 to Abbott). Synthesis: McLamore *et al.*, *J. Am. Chem. Soc.* **75**, 105 (1953); Nienburg, Friedrichsen, **Ger. pat.** 931,651 (1955 to BASF). Toxicity studies: Miyake *et al.*, *Pharm. Bull.* **1**, 84, 89 (1953). *Review:* Caltrider, *Antibiotics*, **vol. I**, D. Gottlieb, P. D. Shaw, Eds. (Springer-Verlag, New York, 1967) pp 666-668.

l-Form, needles from water, ethanol or ethyl acetate, mp 139-140°. $[\alpha]_D^{25}$ −54° (c = 1 in methanol); $[\alpha]_D^{25}$ −60° (c = 1 in ethanol). pK 5.1. Soluble in water, acetone, alcohol, ethylene dichloride, glacial acetic acid. Aq solns are stable over a wide pH range at room temp. Rapidly loses optical activity in dil alkali to give the racemate. Has *in vitro* activity against *Mycobacterium tuberculosis*, *M. ranae* and *M. phlei*. LD_{50} in mice (g/kg): 3.5 i.v., 2 s.c. (Miyake). Methyl ester, $C_{10}H_{17}NO_3S$, needles from ether + hexane, mp 53-54°. $[\alpha]_D^{25}$ −50.9° (methanol).

d-Form, crystals, mp 138-139°. $[\alpha]_D^{25}$ +57° (methanol).

dl-Form, crystals from water, mp 123°.

6404. Mycobacillin. $C_{65}H_{85}N_{13}O_{30}$; mol wt 1528.46. C 51.08%, H 5.61%, N 11.91%, O 31.40%. Antifungal polypeptide antibiotic isolated from culture filtrates of *Bacillus subtilis*: Majumdar, Bose, *Nature* **181**, 134 (1958). Structural studies: *eidem, Biochem. J.* **74**, 596 (1960); *Arch. Biochem. Biophys.* **90**, 154 (1960). Amino acid configuration: Banerjee, Bose, *Nature* **200**, 471 (1963). Nature of the peptide linkages: Sengupta *et al.*, *Biochem. J.* **121**, 839 (1971). *Review:* Banerjee *et al.*, *Antibiotics* **vol. II**, D. Gottlieb, P. D. Shaw, Eds. (Springer-Verlag, New York, 1967) pp 271-275, 445-446.

Ala–D-α-Asp–Pro–D-α-Asp–D-γ-Glu–Tyr
            | |
          α-Asp
            |
D-α-Asp–D-γ-Glu–Leu–D-α-Asp–Ser–Tyr

Needles, mp 235-240°. uv max 277 nm in a solution of *n*-butanol-water-acetic acid-95% ethanol (10:10:2:2).

6405. Mycobactins. Growth factors for *Mycobacterium paratuberculosis (M. johnei)*, the organism responsible for Johne's disease in cattle. Mycobactins are structurally complex *siderophores* (microbial iron chelators). At least nine mycobactins, designated with the letters A, F, H, M, N, P, R, S, and T, have been isolated from various non-pathogenic species of *Mycobacterium*. Earlier studies were actually on *mycobactin P; see* isoln from *M. phlei:* Francis *et al., Nature* **163**, 365 (1949); *eidem, Biochem. J.* **55**, 596 (1953); structure: Snow, *ibid.* **94**, 160 (1965). Separation and identification of mycobactins: White, Snow, *ibid.* **108**, 593 (1968); **111**, 785 (1969). Chemical and biological properties: *eidem, ibid.* **115**, 1031 (1969). On saponification, mycobactin splits into *mycobactic acid* and *cobactin.* Synthetic studies: Black *et al., Aust. J. Chem.* **25**, 2155 (1972). Synthesis of mycobactin S2: P. J. Maurer, M. J. Miller, *J. Am. Chem. Soc.* **105**, 240 (1983). Review of early studies: Rose, Snow, "Mycobactin: A Growth Factor for Acid-Fast Bacilli" in G. E. W. Wolstenholme, M. P. Cameron, C. M. O'Connor, *Ciba Foundation Symposium on Exptl. Tuberculosis Bacillus and Host* **1955**, 41-54. Comprehensive review: Snow, *Bacteriol. Rev.* **34**, 99 (1970).

R^2, R^3, R^5 = H or CH_3
R^1, R^4 = CH_3, CH_2CH_3 or long alkyl chain

Able to chelate with metals, showing strong selectivity toward iron. Isolated as octahedral ferric complexes in the form of red-brown glasses; purified to metal-free mycobactins as white microcrystalline powders. Characterizations, esp mass spectral studies, carried out on the aluminum complexes. Mycobactin solids have definite melting points, are stable to air and heat up to about 100°, and show apple-green fluorescence in uv light. Characteristic uv max (methanol): 250, 311 nm for mycobactins having a methyl substituent on the benzene ring; 243, 249, 304 nm for mycobactins without the methyl substituent. Very sol in chloroform; fairly sol in ethanol; less sol in other alcoholic solvents. Slightly sol in benzene, ether, aliphatic hydrocarbons. Stable to acids; easily broken down under alk. conditions.

USE: In the taxonomy of mycobacteria; in development of iron chelators for clinical use.

6406. Mycolic Acids. High mol wt α-branched, β-hydroxy fatty acids, components of the cell envelopes of all *Mycobacteria*. All known mycolic acids have the basic structure $R^2CH(OH)CHR^1COOH$ where R^1 is a C_{20} to C_{24} linear alkane and R^2 is a more complex structure of 30 to 60 carbon atoms that may contain various numbers of carbon-carbon double bonds and/or cyclopropane rings, methyl branches or oxygen functions such as C=O, $CH_3OCH=$, COOH. The structure of mycolic acids varies by families and species. Three principal categories are known: (1) *corynomycolic acids* ranging from C_{28} to C_{40} found mostly in *Corynebacteria;* (2) *nocardic acids,* also called *nocardomycolic acids,* ranging from C_{40} to C_{60}, produced by strains of *Nocardia;* and (3) mycobacterial mycolic acids ranging from C_{60} to C_{90}. First obtained from a human strain of *Mycobacterium tuberculosis* and studied in an impure form called "unsaponifiable wax": R. J. Anderson, *J. Biol. Chem.* **85**, 351 (1929). Isoln of the first representative of this group of acids from the human tubercle bacillus and derivation of the name "mycolic acids": F. H. Stodola *et al., ibid.* **126**, 505 (1938). Historical reviews: R. J. Anderson, *Fortschr. Chem. Org. Naturst.* **3**, 145-202 (1939); J. Asselineau, E. Lederer, *ibid.* **10**, 170-273 (1953); J. Asselineau, *The Bacterial Lipids* (Hermann-Holden Day, Paris-New York, 1962) pp 82-148; E. Lederer, *Pure and Appl. Chem.* **25**, 135 (1971). Major mycolic acids of *Mycobacterium smegmatis:* M. Y. H. Wong *et al., J. Biol. Chem.* **254**, 5734 (1979); M. Y. H. Wong, G. R. Gray, *ibid.* 5741. Total synthesis of mycolic acids from *M. smegmatis:* H. C. Huang *et al., J. Org. Chem.* **47**, 4018 (1982). Evaluation as immunotherapeutic agents in the treatment of cancer in animal models: M. V. Pimm *et al., Int. J. Cancer* **24**, 780 (1979). Clinical evaluation in humans: G. J. Vosika, *Cancer* **44**, 495 (1979). Mycolic acids are acid fast. They are always isolated in the form of more or less complex mixtures. They occur in nature esterified with carbohydrates: with arabinose in the cell wall, with trehalose in cord factor, *q.v.*

6407. Mycomycin. *3,5,7,8-Tridecatetraene-10,12-diynoic acid;* $C_{13}H_{10}O_2$; mol wt 198.22. C 78.77%, H 5.08%, O 16.14%. HC≡C—C≡CCH=C=CHCH=CHCH=CH-CH₂COOH. Antibiotic substance produced by *Nocardia acidophilus:* Johnson, Burdon, *J. Bacteriol.* **54**, 281 (1947); Johnson, *Soc. Am. Bacteriologists,* 49th General Meeting (May, 1949), *Abstracts of Papers,* p 68; Celmer, Solomons, 121st *A.C.S.* Meeting (Milwaukee, 1952), *Abstracts of Papers,* p 17-J; *J. Am. Chem. Soc.* **74**, 1870, 2245 (1952); **75**, 1372 (1953). Synthesis: Bohlmann, Sucrow, *Angew. Chem.* **76**, 611 (1964).

Needles from methylene chloride at −40°. Very thermolabile. Complete retention of activity is possible only by storage at −40° or lower. mp +75° (deflagrates). $[\alpha]_D^{25}$ −130° (c = 0.4 in ethanol). uv max (ether): 267, 281 nm (ε 61,000, 67,000). Sol in alcohol, ether, amyl acetate, methylene chloride. Forms a water-sol sodium salt. The acid itself is sparingly sol in water and is relatively stable in extremely dil aq solns (approx 0.005 mg/ml), half-life inversely related to concn. Undergoes an unusual rearrangement in normal aq KOH at 27°, involving an allene to acetylene isomerization accompanied by migration of the existing acetylenic bonds. The rearranged acid, *isomycomycin,* has been assigned the structure *3,5-tridecadiene-7,9,11-triynoic acid,* CH₃—C≡C—C≡C—C≡C—CH=CHCH=CH-CH₂COOH.

6408. Mycophenolic Acid. *(E)-6-(1,3-Dihydro-4-hydroxy-6-methoxy-7-methyl-3-oxo-5-isobenzofuranyl)-4-methyl-4-hexenoic acid; 6-(4-hydroxy-6-methoxy-7-methyl-3-oxo-5-phthalanyl)-4-methyl-4-hexenoic acid;* Melbex. $C_{17}H_{20}O_6$; mol wt 320.34. C 63.74%, H 6.29%, O 29.97%. Antibiotic substance produced by *Penicillium brevi-compactum; P. stoloniferum* and related spp. Isoln: Oxford; Raistrick, *Biochem. J.* **26**, 1902 (1932); **27**, 1176, 1473 (1933); Clutterbuck *et al., ibid.* **26**, 1441 (1932); Florey *et al., Lancet* I, 46 (1946); Williams *et al., Antimicrob. Ag. Chemother.* **1968**, 229. Structure: Birkinshaw *et al., Biochem. J.* **50**, 630 (1952); Logan, Newbold, *J. Chem. Soc.* **1957**, 1946. Total syntheses: Birch, Wright, *Aust. J. Chem.* **22**, 2635 (1969); *eidem, Chem. Commun.* **1969**, 788; Canonica *et al., Tetrahedron Letters* **1971**, 2689. Biosyntheses: Canonica *et al., Chem. Commun.* **1970**, 1357; **1971**, 257; Bedford *et al., ibid.* **1971**, 323; Canonica *et al., J. Chem. Soc. Perkins Trans. I* **1972**, 2639. Industrial patents: Borrow *et al.;* Carter; Jefferys *et al., Brit. pats.* 1,157,099; 1,157,100; 1,158,387 (all 1969 to ICI). Activity studies: Williams *et al., J. Antibiot.* **21**, 463 (1968); Ando *et al., ibid.* 649; Suzuki *et al., ibid.* **22**, 297 (1969); Carter *et al., Nature* **223**, 848 (1969). Metabolic studies: Noto *et al., J. Antibiot.* **22**, 165 (1969); Nery, Nice, *J. Pharm. Pharmacol.* **23**, 842 (1971). *Review:* Wilson,

"Miscellaneous *Penicillium* Toxins," in *Microbial Toxins* **vol. VI,** A. Ciegler *et al.,* Eds. (Academic Press, New York, 1971) pp 460-470.

Needles from hot water, mp 141°. Weak dibasic acid. Stable. Almost insol in cold water. Freely sol in alcohol; moderately sol in ether, chloroform; sparingly sol in benzene, toluene. LD_{50} in mice, rats (mg/kg): 2500, 700 orally; 550, 450 i.v. (Wilson).

THERAP CAT: Antineoplastic.

6409. Mycosamine. *3-Amino-3,6-dideoxymannose;* 3-amino-3-deoxyrhamnose; 3,6-dideoxy-3-amino-D-mannopyranose; 3-amino-3,6-dideoxy-D-aldohexose. $C_6H_{13}NO_4$; mol wt 163.17. C 44.17%, H 8.03%, N 8.58%, O 39.22%. An amino sugar which represents the nitrogen-contg moiety of the antifungal antibiotics nystatin, *q.v.,* and amphotericin B, *q.v.* Isoln and structure: Dutcher *et al., Antibiot. Ann.* **1956-1957**, 866. Walters *et al., J. Am. Chem. Soc.* **79**, 5076 (1957); Dutcher *et al., J. Org. Chem.* **28**, 995 (1963). Stereochemistry: von Saltza *et al., J. Am. Chem. Soc.* **83**, 2875 (1961); *eidem, J. Org. Chem.* **28**, 999 (1963).

N-Acetylmycosamine, $C_8H_{15}NO_5$, needles from methanol + acetone, mp 195-197°. $[\alpha]_D^{22}$ −46° (ethanol).
Triacetylmycosamine, $C_{12}H_{19}NO_7$, prisms from acetone or ethanol, mp 185-187°. $[\alpha]_D^{22}$ +85° (ethanol).
Tetraacetylmycosamine, $C_{14}H_{21}NO_8$, needles from benzene, mp 159-161°. $[\alpha]_D^{23}$ +39.3° (ethanol).
Hydrochloride, $C_6H_{13}NO_4$·HCl, prismatic rods from ethanol + ether, mp 162°. $[\alpha]_D^{24}$ −11.5°.

6410. Mylabris. Chinese cantharides; Chinese blistering flies. Dried insect, *Mylabris sidae (Phalerata)* Pallas, *Coleoptera. Habit.* China and East India. *Constit.* Cantharidin, said to be present in larger proportions than in cantharides (1-1.2%).
Black, cylindrical body, rounded above, flattish below; smaller than cantharides; black wing cases marked with a spot at point of insertion and has three tawny bands.
Caution: Highly toxic! *See* Cantharides.
USE: Source for cantharidin.

6411. Myoral. *Calcium [mercaptoacetato(2−)-O,S]aurate(1−) (1:2); [(carboxymethyl)thio]gold calcium salt;* calcium aurothioglycolate; mercaptoacetic acid calcium salt gold derivative. $C_4H_4Au_2CaO_4S_2$; mol wt 614.22. C 7.82%, H 0.66%, Au 64.14%, Ca 6.53%, O 10.42%, S 10.44%. (Au-SCH₂COO)₂Ca. Prepn: Denko, Anderson, *J. Am. Chem. Soc.* **67**, 2241 (1945). Description: Lespagnol, *Pharmacie Chimique* (Paris, 3rd ed., 1950) p 941.
Reddish-brown powder. Stable to air. Practically insol in water. Employed as a suspension in oil.
THERAP CAT: Antirheumatic.

6412. Myosin. Mol wt approx 500,000. The main protein in muscle, which, by hydrolyzing ATP, provides the energy necessary for muscular contraction. Individual molecules are rod-shaped and consist of two globular heads attached flexibly to a tail of approx 1500Å length. All muscle myosins have a similar subunit structure composed of two heavy chains and two distinct pairs of light chains. One of the light chains is involved in a Ca^{2+}-dependent regulatory process in the vertebrate skeletal muscle system. Biological prepn and review: A. G. Szent-Györgyi, *Chemistry of Muscular Contraction* (Academic Press, New York, 2nd ed., 1951) pp 38-57, 146-148. Other reviews: *idem,*

Advan. Enzymol. Relat. Subj. Biochem. **16**, 313-360 (1955); A. G. Szent-Györgyi *et al.*, in *Sulfur in Proteins*, R. Benesch *et al.*, Eds. (Academic Press, New York, 1959) pp 291-295; J. Gergely, *Biochemistry of Muscle Contraction* (Little, Brown, Boston, 1964) pp 3-115; several authors in *Contractile Proteins and Muscle*, K. Laki, Ed. (Dekker, New York, 1971); several authors in *Methods Enzymol.* **vol. 85**, Part B, D. W. Frederiksen, L. W. Cunningham, Eds. (Academic Press, New York, 1982) p 55-130. Review on structure and function of myosin: W. F. Harrington, "Contractile Proteins of Muscle" in *The Proteins*, **vol. 4**, H. Neurath, R. L. Hill, Eds. (Academic Press, New York, 1979) p 245-409. Polymorphism of mammalian myosins and distribution of isoforms in different muscle fibers reviewed by A. G. Weeds in *Plasticity of Muscle*, D. Pette, Ed. (DeGruyter, Berlin, 1980) pp 55-68. Role of myosin light chains in calcium regulation: Kendrick-Jones, *Nature* **249**, 631 (1974). Tertiary crystal structure of globular heads: I. Rayment *et al.*, *Science* **261**, 50 (1993). See also Tropomyosin.

6413. β-Myrcene. *7-Methyl-3-methylene-1,6-octadiene;* 2-methyl-6-methylene-2,7-octadiene. $C_{10}H_{16}$; mol wt 136.24. C 88.16%, H 11.84%. Found in oil of bay, verbena, hop, and others. Isoln: Power, Kleber, *Pharm. Rundschau* **13**, 60 (1895); *see also* Ruzicka, Stoll, *Helv. Chim. Acta* **7**, 272 (1924); Goulding, Roberts, *J. Chem. Soc.* **105**, 2614 (1914); Booker *et al.*, *ibid.* **1940**, 1453; Kugler, Kováts, *Helv. Chim. Acta* **46**, 1480 (1963). Obtained by pyrolysis of β-pinene: Goldblatt, Palkin, *J. Am. Chem. Soc.* **63**, 3517 (1941); Rummelsburg, U.S. pat. **2,444,790** (1948 to Hercules Powder); Houlihan *et al.*, *J. Am. Chem. Soc.* **81**, 4692 (1959). Separation of isomers: Ohloff *et al.*, *Ann.* **675**, 83 (1964). Synthesis: O. P. Vig *et al.*, *Indian J. Chem.* **7**, 450 (1969); **11**, 104 (1973); **13**, 1244 (1975); T. Mandai *et al.*, *Tetrahedron Letters* **22**, 763 (1981). Formation from isoprene: K. Takabe *et al.*, *Synthesis* **1977**, 307.

CH₂ structure

Oil. Pleasant odor. d_4^{20} 0.794. n_D^{20} 1.4709. uv max (ethanol): 226 nm (ε 16,100). Practically insol in water. Sol in alcohol, chloroform, ether, glacial acetic acid.

α-Myrcene. 2-Methyl-6-methylene-1,7-octadiene. bp_{10} 44°. n_D^{25} 1.4661. d_{25}^{25} 0.7959. uv max (isooctane): 224.5 nm (ε 18,600). Not found in nature. Prepn: Mitzner *et al.*, *J. Org. Chem.* **30**, 646 (1965); Vig *et al.*, *J. Indian Chem. Soc.* **50**, 329 (1973).

USE: β-Myrcene as an intermediate in the manuf of perfume chemicals.

6414. Myricetin. *3,5,7-Trihydroxy-2-(3,4,5-trihydroxyphenyl)-4H-1-benzopyran-4-one; 3,3′,4′,5,5′,7-hexahydroxyflavone;* cannabiscetin; delphidenolon 1575. $C_{15}H_{10}O_8$; mol wt 318.24. C 56.61%, H 3.17%, O 40.22%. From the bark of *Myrica nagi* Thunb., *Myriaceae:* Perkin, Hummel, *J. Chem. Soc.* **69**, 1287 (1896). Structure: Perkin, *ibid.* **81**, 203 (1902). Identity with cannabiscetin: Seshadri, Venkateswarlu, *Proc. Indian Acad. Sci.* **23A**, 296 (1946); *C.A.* **40**, 6447 (1946). Synthesis: Kalff, Robinson, *J. Chem. Soc.* **127**, 181 (1925); Rao *et al.*, *J. Sci. Ind. Res.* **8B**, No. 6, 113 (1949). Occurrence in *Hamamelidaceae* and *Anacardiaceae:*

Myricetin structure

Reznek, Egger, *Z. Naturforsch.* **15b**, 247 (1960). Metabolism: Smith, Griffiths, *Biochem. J.* **118**, 53p (1970).

Yellow needles from dil alc, mp 357°. uv max (ethanol): 375, 255 nm. Sparingly sol in boiling water; sol in alcohol. Practically insol in chloroform, acetic acid.

Hexaacetate, $C_{27}H_{22}O_{14}$, crystals, mp 213°.

Hexaethyl ether, $C_{27}H_{34}O_8$, needles from alcohol, mp 149-151°.

3-Rhamnoside, $C_{21}H_{20}O_{12}$, myricitrin. Structure: Hattori, Hayashi, *Acta Phytochim.* **5**, 213 (1931), *C.A.* **26**, 990⁸ (1932). Pale yellow leaflets from water, mp 199-200°. uv max (alc): 262, 352 nm. Sparingly sol in water, abs alcohol.

6415. Myristica. Nutmeg; nux moschata; nuces (semen) nucistae. Dried ripe seed of *Myristica fragrans* Houtt., *Myristicaceae*, deprived of its seed coat and with or without a thin coating of lime. *Habit.* Southern Asia, Moluccas, cultivated in many tropical countries. *Constit.* 25-35% fixed oil and 5-15% volatile oils. The myristicin fraction of the latter, together with elemicin (>25% content) is supposedly responsible for the purported hallucinogenic properties of nutmeg seed: Shulgin, *Nature* **197**, 4865 (1963); Weil, *Econ. Bot.* **19**, 194 (1965). Review of botany, history, chemistry and psychopharmacology: *Public Health Service Publ. No. 1645*, D. H. Efron, Ed., 183-229 (1967).

Caution: Ingestion of large quantities causes drowsiness, stupor, death.

USE: As a spice in food; as source of myristica oil.

THERAP CAT: Carminative.

6416. Myristic Acid. *Tetradecanoic acid.* $C_{14}H_{28}O_2$; mol wt 228.38. C 73.63%, H 12.36%, O 14.01%. CH_3-$(CH_2)_{12}COOH$. Occurs in nutmeg butter *(Myristica fragrans* Houtt.) to the extent of 70-80%; predominates in the fats of the *Myristicaceae;* in palm seed fats it may comprise 20% of the total fatty acids; in milk fats between 8-12% of the total acids. Occurs in most animal and vegetable fats; has been found in considerable amounts (up to 15%) in sperm whale oil, *see* Markley, *Fatty Acids* (New York, 1947). Prepn: G. D. Beal, *Org. Syn.* **coll. vol. I**, 379 (2nd ed., 1941). Prepn from tall-oil fatty acids: Segessmann, Molnar, U.S. pat. **2,481,356** (1949); from 9-ketotetradecanoic acid: Ames *et al.*, *J. Chem. Soc.* **1950**, 174; by electrolysis of methyl hydrogen adipate + decanoic acid: Greaves *et al.*, *ibid.* **1950**, 3326; by Maurer oxidation of myristyl alc: Langenbeck, Richter, *Ber.* **89**, 202 (1956); from cetanol: Selwitz, U.S. pat. **2,969,380** (1961 to Gulf). Wide-line NMR spectrum: A. V. Bailey, R. A. Pittman, *J. Am. Oil Chem. Soc.* **48**, 775 (1971). Toxicity study: L. Orö, A. Wretlind, *Acta Pharmacol. Toxicol.* **18**, 141 (1961).

Crystals from methanol, mp 58.5°. bp_{100} 250.5°; bp_{16} 199°; bp_4 184°. d_4^{54} 0.8622. d_4^{70} 0.8528. d_4^{90} 0.8394. n_D^{60} 1.4305; n_D^{70} 1.4273. Soluble in abs alcohol, methanol, ether, petr ether, benzene, chloroform. Neutralization value: 245.68. Absorption spectrum: Markley, *op. cit.* LD_{50} i.v. in mice: 432.6 mg/kg (Orö, Wretlind).

Ethyl ester, $C_{16}H_{32}O_2$, *ethyl myristate.* Liquid, mp 12°; bp 295°; bp_{30} 195°. d_4^{20} 0.856. Insol in water. Sol in alcohol; slightly sol in ether.

USE: As ingredient in soaps and shaving creams; in lubricants; in coatings for anodized aluminum.

6417. Myristicin. *4-Methoxy-6-(2-propenyl)-1,3-benzodioxole; 5-allyl-1-methoxy-2,3-(methylenedioxy)benzene.* $C_{11}H_{12}O_3$; mol wt 192.21. C 68.74%, H 6.29%, O 24.97%. Aromatic ether extracted from nutmeg, parsley, carrots. Isoln from nutmeg, *Myristica fragrans* Houtt., *Myristicaceae* and characterization: F. B. Power, A. H. Salway, *J. Chem. Soc.* **91**, 2037 (1907); A. T. Shulgin, *Nature* **197**, 379 (1963); from carrots, *Daucus carota* L., *Umbelliferae:* S. G. Yates, R. E. England, *J. Agr. Food Chem.* **30**, 317 (1982); from parsley, *Petroselinium hortense* Hoffman, *Umbelliferae:* M. Balogh *et al.*, *Herba Hung.* **17**, 39 (1978), *C.A.* **91**, 2543e (1979). Synthesis: V. M. Trikojus, D. E. White, *Nature* **144**, 1016 (1939); F. Dallacker, R. Sluysmans, *Monatsh.* **100**, 560 (1969). HPLC sepn: L. W. Wulf *et al.*, *J. Chromatog.* **161**, 271 (1978); A. W. Archer, *ibid.* **438**, 117 (1988). GC-MS study: G. M. Sammy, W. W. Nawar, *Chem. Ind.* **1968**, 1279. MS-MS study of nutmeg extract: D. V. Davis, R. G. Cooks, *J. Agr. Food Chem.* **30**, 495 (1982). Possible psychotropic properties: A. T. Shulgin, *Nature* **210,** 380 (1966).

Colorless oil, bp$_{40}$ 173°. n_D^{20} 1.54032. d$_{20}^{20}$ 1.1437.

6418. Myristyl Alcohol. *1-Tetradecanol;* tetradecyl alcohol. C$_{14}$H$_{30}$O; mol wt 214.39. C 78.43%, H 14.10%, O 7.46%. CH$_3$(CH$_2$)$_{12}$CH$_2$OH. Prepn by sodium reduction of fatty acid esters: Hansley, *Ind. Eng. Chem.* **39,** 55 (1947); by LiAlH$_4$ reduction of fatty acids: Watanabe, *Bull. Chem. Soc. Japan* **34,** 398 (1961); from acetaldehyde + dimethyl-amine: Langenbeck *et al., J. Prakt. Chem.* **8,** 112 (1959).

White crystals, mp 38°. d 0.824. bp$_{15}$ 167°. Practically insol in water. Sol in ether, slightly in alcohol.

USE: As emollient for cold creams, etc., also for making the sulfated alcohol whose sodium salt is applicable as a "wetter" in textiles.

6419. Myristyltrimethylammonium Bromide. *N,N,N-Trimethyl-1-tetradecanaminium bromide; trimethyltetradecyl-ammonium bromide;* tetradonium bromide; Morpan T; Mytab. C$_{17}$H$_{38}$BrN; mol wt 336.40. C 60.70%, H 11.39%, Br 23.75%, N 4.16%. [CH$_3$(CH$_2$)$_{13}$N(CH$_3$)$_3$]$^+$Br$^-$. Cationic germicidal detergent. Prepn and antibacterial activity: R. S. Shelton *et al., J. Am. Chem. Soc.* **68,** 753 (1946). Toxicity and pharmacology: B. Isomaa, K. Bjondahl, *Acta Pharmacol. Toxicol.* **47,** 17 (1980).

White powder, mp 245-250°. Sol in 5 parts water. *Corrosive!* LD$_{50}$ in mice, rats (mg/kg): 12.0, 15.0 i.v. (Issomaa, Bjondahl).

USE: Disinfectant; deodorant; laboratory reagent.

6420. Myrophine. *7,8-Didehydro-4,5-epoxy-17-methyl-3-(phenylmethoxy)morphinan-6-ol tetradecanoate(ester);* 3-benzylmorphine 6-myristate; benzylmorphine myristic acid ester; benzylmorphinyl myristate; myristyl benzylmorphine; 3-benzyloxy-6-myristoyloxy-*N*-methyl-4,5-epoxy-7-mor-phinene; 3-benzyloxy-6-myristoyloxymorphine; myrocodine; myrophinium; Leucodinine; Myricodine; Peronine myristate. C$_{38}$H$_{51}$NO$_4$; mol wt 585.83. C 77.91%, H 8.77%, N 2.39%, O 10.92%. Prepn: Jeanson, U.S. pat. **2,802,828** (1957 to Laborec).

Solid, mp 41°.
Hydrochloride, C$_{38}$H$_{51}$NO$_4$.HCl, solid, mp 198-199°.
Note: This is a controlled substance (opium derivative) listed in the U.S. Code of Federal Regulations, Title 21 Part 1308.11 (1995).
THERAP CAT: Analgesic (narcotic).

6421. Myrrh. Gum-resin myrrh. From *Commiphora abyssinica* (Berg) Eng. or from other species of *Commiphora, Burseraceae.* Yields not less than 30% alcohol-soluble extract. *Habit.* Nubia, Somaliland, Arabia. *Constit.* 20-25% resin, 57-61% gum, 7-17% volatile oil and a bitter principle.
THERAP CAT: Carminative; astringent.

6422. Myrtecaine. *2-[2-(6,6-Dimethylbicyclo[3.1.1]-hept-2-en-2-yl)ethoxy]-N,N-diethylethanamine; 2-[2-(6,6-dimethyl-2-norpinen-2-yl)ethoxy]triethylamine;* 2-homomyr-tenyloxy-1-(diethylamino)ethane; homomyrtenyl β-(diethyl-amino)ethyl ether; Nopoxamine. C$_{17}$H$_{31}$NO; mol wt 265.44.

C 76.92%, H 11.77%, N 5.28%, O 6.03%. Prepn: Goudin, **Brit. pat. 861,900** (1961).

Liquid, bp$_{2-3}$ 135-140°. n_D^{20} 1.477.
THERAP CAT: Anesthetic (local).

6423. Myrtol. Gelomyrtol. The fraction of the volatile oil from *Myrtus communis* L., *Myrtaceae* distilling between 166-180° and consisting chiefly of eucalyptol and *dextro*-pinene with a small quantity of an undefined camphor.

Colorless or slightly yellow, strongly dextrorotatory liquid (due to *dextro*-pinene); pleasant odor resembling that of turpentine and eucalyptol. d ~0.895. [α]$_D^{20}$ +10°. n_D^{20} ~1.465. Freely sol in alcohol, ether.

6424. Mytatrienediol. *3-Methoxy-16-methyl-1,3,5(10)-estratriene-16β,17β-diol;* SC-6924; Anvene; Manvene. C$_{20}$-H$_{28}$O$_3$; mol wt 316.44. C 75.91%, H 8.92%, O 15.17%. Prepn: Tyner, U.S. pat. **2,949,476** (1960 to Searle).

Crystals from benzene + acetone, mp 179-181°. [α]$_D^{20}$ +71° in dioxane.
THERAP CAT: Estrogen; antilipemic; hypocalciuric.

6425. Myxin. *6-Methoxy-1-phenazinol 5,10-dioxide;* 1-hydroxy-6-methoxyphenazine 5,10-dioxide; 3C antibiotic. C$_{13}$H$_{10}$N$_2$O$_4$; mol wt 258.23. C 60.47%, H 3.90%, N 10.85%, O 24.78%. Potent broad-spectrum antibiotic from a *Sorangium* species (strain 3C), a soil-borne myxobacter: Peterson *et al., Can. J. Microbiol.* **12,** 221 (1966). Proposed structure: Edwards, Gillespie, *Tetrahedron Letters* **1966,** 4867. Revised structure and synthesis: Weigele, Leimgruber, *ibid.* **1967,** 715; Sigg, Toth, *Helv. Chim. Acta* **50,** 716 (1967). Activity: Grunberg *et al., Chemotherapia* **12,** 272 (1967). Crystal structure: Hanson, *Acta Cryst., Sect. B* **24,** 1084 (1968). *Review:* Edwards, Gillespie, *Proc. Int. Sym. Drug Res.* **1967,** 236; Leimgruber, Grunberg, *ibid.* 240; Baker, Vezina, *ibid.* 242.

Red needles from acetone, mp 120-130°; also given as mp 149° (dec) (Sigg, Toth). Strongly exothermic at 149°. May explode on drying with heat: Rachlin, *Chem. Eng. News* **45,** 32 (Sept. 4, 1967). Absorption max (0.1*N* HCl): 283, 340, 505 mn (ε 97,000; 5400; 6500). Stable in water after 6 hours at 37° in 0.05*N* acetate buffer at pH 5.0, in 0.05*N* phosphate at pH 6.0 and 7.0, or in 0.005*N* tris-HCl at pH 8.0 and 9.0. Stable at 70° except in phosphate buffer. Acetate, methanol, or aq tris buffer solns remain stable indefinitely at 4°. LD$_{50}$ i.p. in mice: ~40 mg/kg. Capable of inhibiting a wide range of gram-positive and gram-negative bacteria, various fungi and yeasts.

Copper(II) complex, $C_{26}H_{18}CuN_4O_8$, *cuprimyxin, Ro-7-4488/1, UNITOP.* Prepn: Leimgruber *et al.,* **Ger.** pat. **1,931,466** corresp to **U.S.** pat. **3,586,674** (1970, 1971 to Hoffmann-La Roche). Activity studies: Maestrone *et al.,* *Am. J. Vet. Res.* **33,** 185 (1972). Dark green, fine crystals. Absorption max (DMSO): 287, 300, 356, 408, 610 nm (ϵ 68500, 63200, 10000, 10400, 9500).

THERAP CAT (VET): Antibacterial, antifungal.

N

6426. Nabam. *1,2-Ethanediylbiscarbamodithioic acid disodium salt; ethylenebis[dithiocarbamic acid] disodium salt;* disodium ethylenebis[dithiocarbamate]; Dithane D-14; Parzate. $C_4H_6N_2Na_2S_4$; mol wt 256.35. C 18.74%, H 2.36%, N 10.93%, Na 17.94%, S 50.03%. Prepd by treating ethylenediamine with CS_2 in the presence of NaOH: Hester, U.S. pat. **2,317,765** (1943 to Rohm & Haas); Flenner, U.S. pat. **2,504,404** (1950 to du Pont).

Hexahydrate, crystals from alc. Moderately sol in water. Forms a continuous film on plant surfaces, which is said to become insol in water. LD_{50} orally in rats: 395 mg/kg.
Caution: Irritating to skin, mucous membranes, and, in high concns, narcotic. In presence of alc can cause violent vomiting.
USE: Agricultural fungicide.

6427. Nabilone. *trans-(±)-3-(1,1-Dimethylheptyl)-6,-6a,7,8,10,10a-hexahydro-1-hydroxy-6,6-dimethyl-9H-dibenzo-[b,d]pyran-9-one;* LY-109514; Cesamet. $C_{24}H_{36}O_3$; mol wt 372.55. C 77.38%, H 9.74%, O 12.88%. Synthetic cannabinoid with antiemetic, antiglaucoma, and CNS activity. Prepn: R. A. Archer, **Ger.** pat. **2,451,934,** *idem,* U.S. pat. **3,968,125** (1975, 1976 both to Lilly); R. A. Archer *et al., J. Org. Chem.* **42,** 2277 (1977). Pharmacology: P. Stark, R. A. Archer, *Pharmacologist* **17,** 210 (1975); R. M. Orzelek-O'Neil *et al., Toxicol. Appl. Pharmacol.* **54,** 493 (1980). Physiological disposition: A. Rubin *et al., Clin. Pharmacol. Ther.* **22,** 85 (1977). Pharmacokinetics: H. R. Sullivan *et al., Biomed. Mass Spectrom.* **5,** 296 (1978). Behavioral effects: P. Stark, P. B. Dews, *J. Pharmacol. Exp. Ther.* **214,** 124 (1980). Clinical studies as antiemetic in cancer patients: N. Steele *et al., Cancer Treat. Rep.* **64,** 219 (1980); C. J. Williams *et al., Cancer Clin. Trials* **3,** 363 (1980). Clinical evaluation of anxiolytic effects: L. F. Fabre *et al., J. Clin. Pharmacol.* **21,** Suppl 8-9, 377S (1981); R. M. Glass *et al., ibid.* 383S. Review of pharmacology and efficacy as antiemetic: A. Ward, B. Holmes, *Drugs* **30,** 127-144 (1985). Toxicity study in neonatal rats: C. L. Moss *et al., Toxicol. Appl. Pharmacol.* **48,** A120 (1979). Comprehensive description: R. W. Souter, *Anal. Profiles Drug Subs.* **10,** 499-512 (1981).

White crystals from ethyl acetate/hexane, mp 159-160°. uv max (ethanol): 207, 280 nm (ε 47000, 250). pKa in 66% DMF: 13.5.
Note: This is a controlled substance (hallucinogen) listed in the U.S. Code of Federal Regulations, Title 21 Part 1308.12 (1995).
THERAP CAT: Antiemetic.

6428. Nabumetone. *4-(6-Methoxy-2-naphthalenyl)-2-butanone;* 4-(6-methoxy-2-naphthyl)-butan-2-one; BRL-14777; Arthaxan; Balmox; Consolan; Nabuser; Relafen; Relifen; Relifex. $C_{15}H_{16}O_2$; mol wt 228.29. C 78.92%, H 7.06%, O 14.02%. Nonacidic, lipophillic prodrug; metabolized *in vivo* to 6-methoxy-2-naphthylacetic acid which inhibits prostaglandin synthesis. Prepn: A. W. Lake, C. J. Rose, **Ger.** pat. **2,442,305;** *eidem,* U.S. pat. **4,061,779** (1975,

1977 both to Beecham); A. C. Goudie *et al., J. Med. Chem.* **21,** 1260 (1978). Pharmacology: E. A. Boyle *et al., J. Pharm. Pharmacol.* **34,** 562 (1982); B. Nunn, P. D. Chamberlain, *ibid.* 576. Metabolism: R. E. Haddock *et al., Xenobiotica* **14,** 327 (1984). HPLC determn in plasma: J. E. Ray, R. O. Day, *J. Chromatog.* **336,** 234 (1984). Symposia on pharmacology, pharmacokinetics and clinical efficacy: *Int. Congr. Symp. Ser. - Roy. Soc. Med.* **69,** 1-215 (1985); *J. Rheumatol.* **19,** Suppl. 36, 1-91 (1992). *Review:* S. L. Dahl, *Ann. Pharmacother.* **27,** 456-463 (1993).

Crystals from ethanol, mp 80°.
THERAP CAT: Anti-inflammatory; analgesic.

6429. NAD. *Adenosine 5'-(trihydrogen diphosphate) P'→5'-ester with 3-(aminocarbonyl)-1-β-D-ribofuranosylpyridinium inner salt; 3-carbamoyl-1-β-D-ribofuranosylpyridinium hydroxide 5'→5'-ester with adenosine 5'-(trihydrogen pyrophosphate) inner salt;* nicotinamide-adenine dinucleotide; diphosphopyridine nucleotide; DPN; cozymase; coenzyme I; Co I; codehydrogenase; β-NAD; NAD^+; nadide. $C_{21}H_{27}N_7O_{14}P_2$; mol wt 663.43. C 38.02%, H 4.10%, N 14.78%, O 33.76%, P 9.34%. One of the biologically active forms of nicotinic acid, *q.v.* Occurs in living cells primarily in the oxidized state. Serves as a coenzyme of the dehydrogenases, especially in the dehydrogenation of primary and secondary alcohols. NAD usually acts as a hydrogen acceptor, forming NADH which then serves as a hydrogen donor in the respiratory chain. Isoln from erythrocytes: O. Warburg, W. Christian, *Biochem. Z.* **287,** 291 (1936); from rabbit muscle: S. Ochoa, *ibid.* **292,** 68 (1937); from yeast: H. v. Euler, F. Schlenk, *Z. Physiol. Chem.* **246,** 64 (1937); G. A. LePage, *J. Biol. Chem.* **168,** 623 (1947); *Biochem. Prepn.* **1,** 28 (1949). Manuf by fermentation: **Brit.** pat. **1,190,079** (1970 to Kyowa), *C.A.* **73,** 54631g (1970). Synthetic approach: N. A. Hughes *et al., J. Chem. Soc.* **1957,** 3733. Nomenclature: M. Dixon, *Science* **132,** 1548 (1960). Occurs in 2 forms, α-NAD and β-NAD, distinguished by the configuration of the ribosyl nicotinamide linkage. Only the β-anomer is bioactive. Stereochemistry: R. U. Lumieux, J. W. Lown, *Can. J. Chem.* **41,** 889 (1963). NMR studies: N. J. Oppenheimer *et al., Proc. Nat. Acad. Sci. USA* **68,** 3200 (1971); R. H. Sarma, R. J. Mynott, *Chem. Commun.* **1972,** 977; M. Blumenstein, M. A. Raftery, *Biochemistry* **11,** 1643 (1972). Biosynthesis: *Nutr. Rev.* **30,** 139 (1972); J. W. Foster, A. G. Moat, *Microbiol. Rev.* **44,** 83 (1980). Combined bioluminescent assay of NAD and NADH: M. T. Karp, P. I. Vuorinen, *Methods Enzymol.* **122,** 147 (1986). *Review:* "Niacin: Nicotinic Acid, Nicotinamide, NAD(P)" in *Vitamins,* W. Friedrich, Ed. (Walter de Gruyter, Berlin, 1988) pp 473-542.

Very hygroscopic white powder. Freely sol in water without residue. A 1% soln has a pH of ~2. Aq solns are stable for ~1 week. $[\alpha]_D^{20}$ −31.5° (c = 1.2 in water). uv max: 260 nm (ε 18100).
Reduced form, $C_{21}H_{29}N_7O_{14}P_2$, *NADH, DPNH, 1,4-dihydronicotinamide adenine dinucleotide.* uv max: 260, 340 nm (ε 15000, 6290).

6430. Nadifloxacin. (±)-*9-Fluoro-6,7-dihydro-8-(4-hydroxy-1-piperidinyl)-5-methyl-1-oxo-1H,5H-benzo[ij]quinolizine-2-carboxylic acid;* jinofloxacin; OPC-7251. $C_{19}H_{21}$-FN_2O_4; mol wt 360.39. C 63.32%, H 5.87%, F 5.27%, N 7.77%, O 17.76%. Fluorinated quinolone antibacterial. Prepn: H. Ishikawa *et al.,* **Belg. pat. 891,046;** *eidem,* U.S. pat. **4,399,134** (1982, 1983 both to Otsuka); *eidem, Chem. Pharm. Bull.* **37,** 2103 (1989). Toxicity data: K. Hashimoto *et al., Iyakuhin Kenkyu* **21,** 671 (1990), *C.A.* **114,** 156625r (1991). *In vitro* antibacterial activity: K. Vogt *et al., Eur. J. Clin. Microbiol. Infect. Dis.* **11,** 943 (1992). HPLC determn: M. Koike *et al., J. Chromatog.* **526,** 235 (1990). Clinical trial in treatment of acne: I. Kurokawa *et al., J. Am. Acad. Dermatol.* **25,** 674 (1991).

Colorless prisms from EtOH-H_2O, mp 245°-247° (dec). LD_{50} male, female mice and rats (mg/kg): 376.5, 420.6, 225.7, 240.5 i.v. (Hashimoto).

THERAP CAT: Antibacterial (topical).

6431. Nadolol. *5-[3-[(1,1-Dimethylethyl)amino]-2-hydroxypropoxy]-1,2,3,4-tetrahydro-2,3-naphthalenediol;* 1-(*tert*-butylamino)-3-[(5,6,7,8-tetrahydro-*cis*-6,7-dihydroxy-1-naphthyl)oxy]-2-propanol; (2*R*,3*S*)-5-[3-(*tert*-butylamino)-2-hydroxypropoxy]-1,2,3,4-tetrahydronaphthalene-2,3-diol; 2,3-*cis*-1,2,3,4-tetrahydroxy-5-[2-hydroxy-3-(*tert*-butylamino)-propoxy]-2,3-naphthalenediol; SQ-11725; Anabet; Corgard; Solgol. $C_{17}H_{27}NO_4$; mol wt 309.41. C 65.99%, H 8.80%, N 4.53%, O 20.68%. β-Adrenergic blocker. Prepn: F. P. Hauck *et al.,* **Ger. pat. 2,258,995;** *eidem,* **U.S. pat. 3,935,267** (1973, 1976, both to Squibb). Resolution of isomers: F. P. Hauck, J. E. Sundeen, **Ger. pat. 2,421,549** (1974 to Squibb), *C.A.* **82,** 57481e (1975). Metabolism: J. Dreyfuss *et al., J. Clin. Pharmacol.* **17,** 300 (1977). Toxicology: P. L. Sibley *et al., Toxicol. Appl. Pharmacol.* **44,** 379 (1978). Review of pharmacology: R. C. Heel *et al., Drugs* **1,** 1-23 (1980); M. J. Antonaccio, D. B. Evans, in *Pharmacology of Antihypertensive Drugs,* A. Scriabine, Ed. (Raven Press, New York, 1980) pp 295-301. Comprehensive description: L. Slusarek, K. Florey, *Anal. Profiles Drug Subs.* **9,** 455-485 (1980). Book: *International Experience with Nadolol, a Long-Acting β-Blocking Agent,* F. Gross, Ed. (Grune & Stratton, New York, 1981) 229 pp.

Crystalline powder, mp 124-136°. uv max (methanol): 270, 278 nm ($E_{1cm}^{1\%}$ 37.5, 39.1). pKa 9.67. Freely soluble in alc, propylene glycol. Slightly soluble in chloroform. Insol in acetone, benzene, ether, hexane. LD_{50} in mice, rats (mg/kg): 4500, 5300 orally (Antonaccio, Evans).

THERAP CAT: Antihypertensive, antianginal.

6432. Nadoxolol. *N,3-Dihydroxy-4-(1-naphthalenyloxy)butanimidamide;* 3-hydroxy-4-(1-naphthyloxy)butyramidoxime; 4-(α-naphthyloxy)-3-hydroxybutyramidoxime. $C_{14}H_{16}N_2O_3$; mol wt 260.29. C 64.60%, H 6.20%, N 10.76%, O 18.44%. β-Adrenergic blocker. Prepn: Lafon, **Ger. pat. 2,132,113** corresp to **U.S. pat. 3,819,702** (1972, 1974 to Orsymonde). Pharmacology: Duteil *et al., Therapie* **28,** 703 (1973). Clinical studies: Tricot *et al., ibid.* 721.

Hydrochloride, $C_{14}H_{16}N_2O_3 \cdot HCl$, **LL-1530,** **Bradyl.** White crystalline powder from ethyl ether-methanol, mp 188°. Sol in water, methanol, ethanol. Insol in ethyl ether. LD_{50} in mice (mg/kg): 180 i.v.; 1000 orally (Lafon).

THERAP CAT: Antiarrhythmic.

6433. NADP. *Adenosine 5'-(trihydrogen diphosphate) 2'-(dihydrogen phosphate) P'→5'-ester with 3-(aminocarbonyl)-1-β-D-ribofuranosylpyridinium inner salt; 3-carbamoyl-1-β-D-ribofuranosylpyridinium hydroxide 5'→5'-ester with adenosine 2'-(dihydrogen phosphate) 5'-(trihydrogen pyrophosphate) inner salt;* nicotinamide-adenine dinucleotide phosphate; triphosphopyridine nucleotide; TPN; coenzyme II; Co II; phosphocozymase; codehydrase II; codehydrogenase II. $C_{21}H_{28}N_7O_{17}P_3$; mol wt 743.41. C 33.93%, H 3.80%, N 13.19%, O 36.59%, P 12.50%. One of the biologically active forms of nicotinic acid, *q.v.* Differs from NAD, *q.v.,* by an additional phosphate group at the 2'-position of the adenosine moiety. Serves as a coenzyme of hydrogenases and dehydrogenases. Present in living cells primarily in the reduced form (NADPH) and is involved in synthetic reactions. Isoln from horse blood: O. Warburg *et al., Biochem. Z.* **282,** 157 (1935); from hog liver by adsorption on charcoal: G. A. LePage, G. C. Mueller, *J. Biol. Chem.* **180,** 975 (1949); from sheep liver by extraction with hot water: A. Kornberg, B. L. Horecker, *Biochem. Prepn.* **3,** 24 (1953); from human red blood cells by anion-exchange chromatography: G. R. Bartlett, *J. Biol. Chem.* **234,** 449, 459 (1959). Chemical and enzymatic synthesis: D. R. Walt *et al., J. Am. Chem. Soc.* **106,** 234 (1984). Nomenclature: M. Dixon, *Science* **132,** 1548 (1960). Review of synthesis and metabolism: Y. Nishizuka in *Method. Chim.* vol. **11,** F. Korte, Ed. (Academic Press, New York, 1977) pp 84-87. *Review:* "Niacin: Nicotinic Acid, Nicotinamide, NAD(P)" in *Vitamins,* W. Friedrich, Ed. (Walter de Gruyter, Berlin, 1988) pp 473-542.

Grayish-white powder. pKa_1 3.9; pKa_2 6.1. uv max: 260 nm (ε 18000). Sol in water, methanol. Much less sol in ethanol. Practically insol in ether, ethyl acetate.

Reduced form, $C_{21}H_{30}N_7O_{17}P_3$, **NADPH, NADPH₂, TPNH, dihydrocodehydrogenase II.** uv max: 260, 340 nm (ε 18000, 6220).

6434. Nadroparin. Tedegliparin. Low molecular weight fraction of heparin, *q.v.,* prepd from porcine mucosal heparin by selective ethanol precipitation. Mean mol wt 4500 daltons. Prepn: J.-C. Lormeau *et al.,* **Ger. pat. 2,944,-792;** *eidem,* U.S. pats. **4,486,420; 4,692,435** (1980, 1984, 1987 all to Choay); J. Choay *et al., Thromb. Res.* **18,** 573 (1980). Comparison with unfractionated heparin: G. F. Gensini *et al., Haemostasis* **14,** 466 (1984); M. D. Vandenbroek, B. Bordes, *J. Pharm. Clin.* **5,** 35 (1986); A. Doutremepuich *et al., Thromb. Res.* **43,** 691 (1986). Pharmacokinetics and bioavailability in humans: J. Harenberg *et al., ibid.* **44,** 549 (1986). Clinical trial in prophylaxis of postoperative thromboembolism: V. V. Kakkar, *Nouv. Rev. Fr. Hematol.* **26,** 277 (1984); V. V. Kakkar, W. J. G. Murray,

Brit. J. Surg. **72**, 786 (1985); in acute ischemic stroke: R. Kay *et al., N. Engl. J. Med.* **333**, 1588 (1995).

Calcium salt, *CY-216, Fraxiparine.*

THERAP CAT: Antithrombotic.

6435. Naepaine. *2-(Pentylamino)ethanol 4-aminobenzoate (ester); p*-aminobenzoic acid 2-*n*-pentylaminoethyl ester; *p*-aminobenzoic acid 2-*n*-amylaminoethyl ester; 2-*n*-amylaminoethyl *p*-aminobenzoate; 2-*n*-pentylaminoethyl *p*-aminobenzoate; Amylsine. $C_{14}H_{22}N_2O_2$; mol wt 250.34. C 67.17%, H 8.86%, N 11.19%, O 12.78%. Prepn: Goldberg, Whitmore, *J. Am. Chem. Soc.* **59**, 2280 (1937). Prepn of dimorphic cryst forms: Kreider, Menotti, *ibid.* **64**, 1227 (1942).

Solid, mp 66°.

Hydrochloride, $C_{14}H_{23}ClN_2O_2$, dimorphic crystals. Rectangular plates, mp 153.5°; rods, mp 176°. Bitter taste followed by a sense of numbness. Sol in water; sparingly sol in alcohol; insol in ether, chloroform, benzene.

THERAP CAT: Hydrochloride as local anesthetic.

THERAP CAT (VET): Topical anesthetic (ocular).

6436. Nafamostat. *4-[(Aminoiminomethyl)amino]-benzoic acid 6-(aminoiminomethyl)-2-naphthalenyl ester; 6-amidino-2-naphthyl-4-guanidinobenzoate;* nafamstat. C_{19}-$H_{17}N_5O_2$; mol wt 347.38. C 65.70%, H 4.93%, N 20.16%, O 9.21%. Non-peptide protease inhibitor. Prepn: S. Fujii *et al., Eur.* pat. Appl. **48,433;** *eidem,* U.S. pat. **4,454,338** (1982, 1984 both to Torii & Co.); T. Aoyama *et al., Chem. Pharm. Bull.* **33**, 1458 (1985). Inhibitory effects on trypsin, thrombin, kallikrein, plasmin and complement-mediated hemolysis: S. Fujii, Y. Hitomi, *Biochim. Biophys. Acta* **661**, 342 (1981); T. Aoyama *et al., Japan. J. Pharmacol.* **35**, 203 (1984). Effect in experimental acute pancreatitis in rats: M. Iwaki *et al., ibid.* **41**, 155 (1986). Spectrofluorometric determn in biological material: T. Aoyama *et al., Chem. Pharm. Bull.* **33**, 2142 (1985). Clinical anti-complement activity: Y. Miyamoto *et al., Trans. Am. Soc. Artif. Intern. Organs* **31**, 508 (1985).

Dimethanesulfonate, $C_{19}H_{17}N_5O_2 \cdot 2CH_3SO_3H$, *nafamostat mesylate, FUT-175, Futhan.* Crystals from ethyl ether, mp 217-220° (Fujii). Also reported as colorless powder from water, mp 260° (dec) (Aoyama).

THERAP CAT: Enzyme inhibitor (protease).

6437. Nafarelin. *6-[3-(2-Naphthalenyl)-D-alanine]luteinizing hormone-releasing factor(pig);* 5-OXO-L-prolyl-L-histidyl-L-tryptophyl-L-seryl-L-tyrosyl-3-(2-naphthyl)-D-alanyl-L-leucyl-L-arginyl-L-prolylglycinamide; [6-[3-(2-naphthyl)-D-alanine]] LHRH; D-nal(2)^6-LHRH; NAG. $C_{66}H_{83}N_{17}O_{13}$; mol wt 1322.49. C 59.94%, H 6.33%, N 18.00%, O 15.73%. Synthetic peptide agonist analog of LH-RH, *q.v.* Prepn: J. J. Nestor *et al.,* U.S. pat. **4,234,571** (1980 to Syntex); *eidem, J. Med. Chem.* **25**, 795 (1982). Suppression of luteal and placental function in baboons: B. H. Vickery *et al., Fertil. Steril.* **36**, 664 (1981). Inhibition of ovulation in humans: J. A. Gudmundsson *et al., Contraception* **30**, 107 (1984). Radioimmunoassay in plasma and serum: C. Nerenberg *et al., Anal. Biochem.* **141**, 10 (1984). Kinetics and tissue disposition in animals: N. I. Chu *et al., Drug Metab. Dispos.* **13**, 560 (1985). Clinical evaluation in benign prostatic hyperplasia: C. A. Peters, P. C. Walsh, *N. Engl. J. Med.* **317,**

599 (1987); in endometriosis: M. R. Henzl *et al., ibid.* **318,** 485 (1988). Review of clinical studies: P. G. Hoffman *et al., J. Androl.* **8,** Suppl., S17-S22 (1987).

5-oxoPro—His—Trp—Ser—Tyr—N---C---C—Leu—Arg—Pro—GlyNH₂

Monoacetate, $C_{66}H_{83}N_{17}O_{13} \cdot C_2H_4O_2$, mp 188-190°. $[\alpha]_D^{25}$ −27.4° (c = 0.9 in acetic acid).

Acetate hydrate, $C_{66}H_{83}N_{17}O_{13} \cdot xC_2H_4O_2 \cdot yH_2O$, *RS-94991-298, Nacenyl, Nasanyl, Synarel, Synrelina.*

THERAP CAT: Treatment of endometriosis.

6438. Nafcillin Sodium. *[2S-(2α,5α,6β)]-6-[[(2-Ethoxy-1-naphthalenyl)carbonyl]amino]-3,3-dimethyl-7-oxo-4-thia-1-azabicyclo[3.2.0]heptane-2-carboxylic acid monosodium salt;* sodium 6-(2-ethoxy-1-naphthamido)penicillanate; 6-(2-ethoxy-1-naphthamido)penicillin sodium salt; Nafcil; Naftopen; Unipen. $C_{21}H_{21}N_2NaO_5S$; mol wt 436.46. C 57.79%, H 4.85%, N 6.42%, Na 5.27%, O 18.33%, S 7.35%. Semisynthetic antibiotic related to penicillin. Prepn: Doyle *et al.,* **Brit.** pat. **880,400** (1961 to Beecham).

THERAP CAT: Antibacterial.

THERAP CAT (VET): Antibacterial.

6439. Nafiverine. *α-Methyl-1-naphthaleneacetic acid 1,4-piperazinediyldi-2,1-ethanediyl ester;* 1,4-piperazinediethanol bis(α-methyl-1-naphthaleneacetate); *N,N'*-bis[[2-(1-naphthyl)propionyloxy]ethyl]piperazine; *N,N'*-di[α-(1-naphthyl)propionyloxy-2-ethyl]piperazine; Naftidan. $C_{34}H_{38}N_2$-O_4; mol wt 538.69. C 75.81%, H 7.11%, N 5.20%, O 11.88%. Prepn: Pala, **Brit.** pat. **1,016,968** (1966 to de Angeli). Isoln and characterization of isomers: Pala, Montegani, *Farmaco Ed. Sci.* **22**, 3 (1967).

Dihydrochloride, $C_{34}H_{40}Cl_2N_2O_4$, crystals, mp 220-221°.

THERAP CAT: Antispasmodic.

6440. Nafronyl. *Tetrahydro-α-(1-naphthalenylmethyl)-2-furanpropanoic acid 2-(diethylamino)ethyl ester; tetrahydro-α-(1-naphthylmethyl)-2-furanpropionic acid 2-(diethylamino)ethyl ester;* 3-(1-naphthyl)-2-tetrahydrofurfurylpropionic acid 2-(diethylamino)ethyl ester; α-tetrahydrofurfuryl-1-naphthalenepropionic acid 2-(diethylamino)ethyl ester; *N*-diethylaminoethyl β-(1-naphthyl)-β-tetrahydrofuryl isobutyrate; naftidrofuryl; Dubimax; Gevatran; Tridus. C_{24}-$H_{33}NO_3$; mol wt 383.53. C 75.16%, H 8.67%, N 3.65%, O 12.51%. Prepn: Szarvasi, Bayssat, **Fr.** pat. **1,363,948** corresp to **U.S.** pat. **3,334,096** (1964, 1967 both to LIPHA);

Szarvasi *et al., Compt. Rend.* **260**, 3095 (1965); *eidem, Bull. Soc. Chim. France* **1966**, 1838. Activity studies: Fontaine *et al., Compt. Rend. Ser. D* **262**, 719 (1966). Metabolism and toxicology: Fontaine *et al., Chim. Ther.* **4**, 44 (1969); Fontaine *et al., J. Eur. Toxicol.* **11**, 40 (1969). Acute toxicity: P. Bessin *et al., Eur. J. Med. Chem.* **10**, 291 (1975). Clinical studies: C. A. Clyne *et al., Brit. J. Surg.* **67**, 347 (1980); R. M. Greenhalgh, *ibid.* **68**, 265 (1981).

bp$_{0.5}$ 190°. d$_4^{31}$ 1.0465. n$_D^{29}$ 1.5513. LD$_{50}$ in mice (mg/kg): 365 orally (Bessin).

Acid oxalate, C$_{24}$H$_{33}$NO$_3$.C$_2$H$_2$O$_4$, *EU-1806, LS-121, Citoxid, Di-Actane, Dusodril, Praxilene.* Crystals from ethyl acetate, mp 110-111°. Slightly hygroscopic, sol in water.

THERAP CAT: Vasodilator.

6441. Naftalofos. *2-[(Diethoxyphosphinyl)oxy]-1H-benz[de]isoquinoline-1,3(2H)-dione; N-hydroxynaphthalimide diethyl phosphate; O,O-diethyl O-naphthaloximide phosphate; naphthalophos; phthalophos; Bay 9002; Bayer 25820; ENT-25567; S-940; Maretin; Rametin.* C$_{16}$H$_{16}$NO$_6$P; mol wt 349.28. C 55.02%, H 4.62%, N 4.01%, O 27.48%, P 8.87%. Prepn: Lorenz, Wegler, **Ger.** pat. **962,608** and **Austrian** pat. **193,894** (both 1957 to Bayer).

Minute crystals. The commercial product has a brown to tan color. mp 174-179°. Practically insol in water. Sparingly sol in the usual organic solvents. Fairly sol in methylene chloride.

THERAP CAT (VET): Anthelmintic.

6442. Naftifine. *(E)-N-Methyl-N-(3-phenyl-2-propenyl)-1-naphthalenemethanamine; (E)-N-cinnamyl-N-methyl-1-naphthalenemethylamine; N-methyl-N-(1-naphthylmethyl)-3-phenylpropen-1-amine; naftifungin.* C$_{21}$H$_{21}$N; mol wt 287.40. C 87.76%, H 7.36%, N 4.87%. Antimycotic allylamine. Prepn: D. Berney, **Belg.** pat. **853,976;** *idem,* **U.S.** pat. **4,282,251** (1977, 1981 both to Sandoz); A. Blade Fond, E. Mendoza-Villela, **Span.** pat. **504,432** (1982 to Lab. Frumtost-Prem S.A.), *C.A.* **99**, 38173r (1983); H. Loibner *et al., Tetrahedron Letters* **25**, 2535 (1984). In vitro activity against dermatophytes and *Candida* spp: A. Georgopoulos *et al., Antimicrob. Ag. Chemother.* **19**, 386 (1981). Morphological changes induced in *Trichophyton mentagrophytes:* J. C. Meingassner *et al., J. Invest. Dermatol.* **77**, 444 (1981). Specific inhibitor of squalene epoxidase, a key enzyme in fungal ergosterol biosynthesis: F. Paltauf *et al., Biochim. Biophys. Acta* **712**, 268 (1982); N. S. Ryder, *Antimicrob. Ag. Chemother.* **25**, 483 (1984). HPLC determn in human plasma and urine: F. Schatz, H. Haberl, *Arzneimittel-Forsch.* **36**, 1850 (1986). Clinical trials in dermatophytosis: U. Ganzinger *et al., Clin. Trials J.* **19**, 342 (1982).

Colorless, viscous oil, bp$_{0.015\ torr}$ 162-167°.

Hydrochloride, C$_{21}$H$_{21}$N.HCl, *AW-105-843, SN-105-843, Exoderil, Naftin.* mp 177° (from propanol).

THERAP CAT: Antifungal (topical).

6443. Naftopidil. *4-(2-Methoxyphenyl)-α-[(1-naphthalenyloxy)methyl]-1-piperazineethanol; RS-1-[4-(2-methoxyphenyl)-1-piperazinyl]-3-(1-naphthoxy)-2-propanol; 1-(2-methoxyphenyl)-4-[3-(naphth-1-yloxy)-2-hydroxypropyl]-piperazine; KT-611; Avishot; Flivas.* C$_{24}$H$_{28}$N$_2$O$_3$; mol wt 392.50. C 73.44%, H 7.19%, N 7.14%, O 12.23%. α$_1$-Adrenergic blocker and serotonin (5HT$_{1A}$) receptor agonist. Prepn: E. C. Witte *et al.,* **Ger.** pat. **2,408,804;** *eidem,* **U.S.** pat. **3,997,666** (1975, 1976 both to Boehringer Mann.). Clinical pharmacodynamics: R. Kirsten *et al., Eur. J. Clin. Pharmacol.* **46**, 271 (1994). Clinical pharmacokinetics: M. J. G. Farthing *et al., Postgrad. Med. J.* **70**, 363 (1994). HPLC determn in human plasma: G. Niebch *et al., J. Chromatog.* **534**, 247 (1990). Clinical evaluation in BPH: K. Yasuda *et al., Prostate* **25**, 46 (1994). Review of pharmacology and clinical experience: H. M. Himmel, *Cardiovasc. Drug Rev.* **12**, 32-47 (1994).

Crystals from isopropanol, mp 125-126°; also reported as colorless crystals, mp 125-129°. Insol in water. Partition coefficient (octanol/water): 75. LD$_{50}$ in mice, rats (g/kg): 1.3, 6.4 orally (Himmel).

Dihydrochloride, C$_{24}$H$_{28}$N$_2$O$_3$.2HCl, crystals from methanol/ethanol (1:2), mp 212-213°.

THERAP CAT: Antihypertensive; α-blocker in treatment of symptomatic benign prostate hypertrophy.

6444. Nalbuphine. *(5α,6α)-17-(Cyclobutylmethyl)-4,5-epoxymorphinan-3,6,14-triol; N-cyclobutylmethyl-14-hydroxydihydronormorphine.* C$_{21}$H$_{27}$NO$_4$; mol wt 357.45. C 70.56%, H 7.61%, N 3.92%, O 17.90%. Mixed opioid agonist-antagonist. Prepn: **Brit.** pat. **1,119,270** corresp to I. J. Pachter, Z. Matossian, **U.S.** pat. **3,393,197** (both 1968 to Endo). Pharmacology: S. G. Holtzman, *Probl. Drug Depend.* **1976**, 374; H. E. Shannon, S. G. Holtzman, *J. Pharmacol. Exp. Ther.* **201**, 55 (1977); G. J. Schaefer, S. G. Holtzman, *ibid.* **205**, 291 (1978). Plasma determn: S. H. Weinstein *et al., J. Pharm. Sci.* **67**, 547 (1978). Clinical study: T. Tammisto, I. Tigerstedt, *Acta Anaesthesiol. Scand.* **21**, 390 (1977). Review of pharmacology and therapeutic efficacy: J. K. Errick, R. C. Heel, *Drugs* **26**, 191-211 (1983).

Cryst from methanol-acetone, mp 230.5°.

Hydrochloride, $C_{21}H_{27}NO_4 \cdot HCl$, *EN-2234A, Nubain.*
THERAP CAT: Analgesic (narcotic).

6445. Naled. *Phosphoric acid 1,2-dibromo-2,2-dichloroethyl dimethyl ester;* dimethyl 1,2-dibromo-2,2-dichloroethyl phosphate; bromchlophos; ENT-24988; OMS-75; RE-4355; Bromex; Dibrom; Ortho-Dibrom. $C_4H_7Br_2Cl_2O_4P$; mol wt 380.78. C 12.62%, H 1.85%, Br 41.97%, Cl 18.62%, O 16.81%, P 8.13%. Cholinesterase inhibitor. Prepn: **Brit. pat.** 855,157; Ospenson, Kohn, U.S. pat. 2,971,882 (1960, 1961 both to California Res. Corp.). Toxicity study: T. B. Gaines, *Toxicol. Appl. Pharmacol.* **14**, 515 (1969).

Usually obtained as a liquid, slightly pungent odor, bp$_{0.5}$ 110°. d_4^{25} 1.96. Has been crystallized, mp 26.5-27.5°. Vapor pressure at 20° about 2×10^{-3} mm Hg. Practically insol in water. Completely hydrolyzed by water within 48 hrs. Freely sol in aromatic and chlorinated hydrocarbons, ketones, alcohols. Sparingly sol in petroleum solvents and mineral oils. LD$_{50}$ in male rats (mg/kg): 250 orally; 800 dermally (Gaines).
Caution: Potential symptoms of overexposure are miosis, lacrimation; headache; chest tightness, wheezing and laryngeal spasm; salivation; cyanosis; anorexia, nausea, vomiting, abdominal cramps and diarrhea; weakness, twitching and paralysis; giddiness, ataxia and convulsions; low blood pressure; cardiac irregularities; irritation of skin and eyes. *See NIOSH Pocket Guide to Chemical Hazards* (DHHS/NIOSH 90-117, 1990) p 96.
USE: Insecticide; acaricide.

6446. Nalidixic Acid. *1-Ethyl-1,4-dihydro-7-methyl-4-oxo-1,8-naphthyridine-3-carboxylic acid;* 3-carboxy-1-ethyl-7-methyl-1,8-naphthyridin-4-one; 1-ethyl-7-methyl-1,8-naphthyridin-4-one-3-carboxylic acid; Win-18320; Betaxina; Dixiben; Eucistin; Innoxalon; Nalidicron; Nalitucsan; Narigix; NegGram; Negram; Nevigramon; Nicelate; Nogram; Poleon; Specifin; Uriben; Uriclar; Uralgin; Urodixin; Uroman; Uroneg; Uropan; Wintomylon. $C_{12}H_{12}N_2O_3$; mol wt 232.24. C 62.06%, H 5.21%, N 12.06%, O 20.67%. Prepn: Lesher *et al., J. Med. Pharm. Chem.* **5**, 1063 (1962); S. Afr. pat. **R61/3073**; G. Y. Lesher, M. D. Gruett, **Belg. pat. 612,-258**; *eidem,* U.S. pat. **3,149,104** (1962, 1962, 1964 all to Sterling Drug). Synthesis of intermediates: Lappin, *J. Am. Chem. Soc.* **70**, 3348 (1948). Mechanism of action studies: Bourguignon *et al., Antimicrob. Ag. Chemother.* **4**, 479 (1973); Goss, Cook, *Antibiotics,* **vol. 3**, J. W. Corcoran, F. E. Hahn, Eds. (Springer-Verlag, New York, 1975) pp 174-196; A. M. Pedrini, *ibid.* **vol. 5**(pt. 1), F. E. Hahn, Ed. (1979) pp 154-175; H. T. Wright *et al., Science* **213**, 455 (1981). Comprehensive description: P. E. Grubb, *Anal. Profiles Drug Subs.* **8**, 371-397 (1979).

Pale buff, crystalline powder, mp 229-230°. Soly at 23° (mg/ml): chloroform 35; toluene 1.6; methanol 1.3; ethanol 0.9; water 0.1; ether 0.1. LD$_{50}$ in mice (mg/kg): 3300 orally; 500 s.c.; 176 i.v. (Lesher *et al.*).
THERAP CAT: Antibacterial.
THERAP CAT (VET): Antibacterial; has been used in urinary tract infections.

6447. Nalmefene. *(5α)-17-(Cyclopropylmethyl)-4,5-epoxy-6-methylenemorphinan-3,14-diol;* 6-desoxy-6-methylenenaltrexone; nalmetrene; ORF-11676. $C_{21}H_{25}NO_3$; mol wt 339.43. C 74.31%, H 7.42%, N 4.13%, O 14.14%. Struc-

tural analog of naltrexone, *q.v.,* with opiate antagonist activity. Prepn: J. Fishman, U.S. pat. **3,814,768** (1974 to M. J. Lewenstein); E. F. Hahn *et al., J. Med. Chem.* **18**, 259 (1975). Improved synthesis: P. C. Meltzer, J. W. Coe, **Eur. pat. Appl. 140,367**; *eidem,* U.S. pat. **4,535,157** (both 1985 to Key Pharm.). Comparative evaluation of efficacy in mice: R. D. Heilman *et al., Res. Commun. Chem. Pathol. Pharmacol.* **13**, 635 (1976). HPLC determn in human plasma: J. Hsiao, R. Dixon, *ibid.* **42**, 449 (1983). Receptor binding activity in rat brain: M. E. Michel *et al., Meth. Find. Exp. Clin. Pharmacol.* **7**, 175 (1985). Clinical evaluation of narcotic antagonism: T. J. Gal, C. A. DiFazio, *Anesthesiology* **64**, 175 (1986). Pharmacokinetics in humans: R. Dixon *et al., Clin. Pharmacol. Ther.* **39**, 49 (1986); *eidem, J. Clin. Pharmacol.* **27**, 233 (1987). Use as antagonist to opioid-induced immobilization in large animals: T. J. Kreeger *et al., J. Wildl. Dis.* **23**, 619 (1987).

Crystals from ethyl acetate, mp 188-190°.
THERAP CAT: Narcotic antagonist.
THERAP CAT (VET): Narcotic antagonist.

6448. Nalorphine. *(5α,6α)-7,8-Didehydro-4,5-epoxy-17-(2-propenyl)morphinan-3,6-diol;* N-allylnormorphine; allorphine; antorphine; NANM. $C_{19}H_{21}NO_3$; mol wt 311.38. C 73.29%, H 6.80%, N 4.50%, O 15.41%. Prepd from normorphine: McCawley *et al., J. Am. Chem. Soc.* **63**, 314 (1941); amended procedure: Hart, McCawley, *J. Pharmacol.* **82**, 339 (1944); Weijlard, Erickson, *J. Am. Chem. Soc.* **64**, 869 (1942); U.S. pat. **2,364,833** (1944); Weijlard, U.S. pat. **2,-891,954** (1959 to Merck & Co.). Toxicity data: E. I. Goldenthal, *Toxicol. Appl. Pharmacol.* **18**, 185 (1971). Comprehensive description: M. U. Zubair *et al., Anal. Profiles Drug Subs.* **18**, 195-219 (1989).

Crystals from ether, mp 208-209° (not 92-93° as first given by McCawley). $[\alpha]_D^{25}$ $-155.3°$ (c = 3 in methanol). Sparingly sol in water, ether. Sol in alc, acetone, chloroform, dil alkalies. uv max (in acid): 285 nm; (in alkali): 298 nm.
Hydrobromide, $C_{19}H_{21}NO_3 \cdot HBr$, *Lethidrone Norfin.* Crystals from alc, dec 258-259°. Sol in water.
Hydrochloride, $C_{19}H_{21}NO_3 \cdot HCl$, *Nalline.* Crystals from alc, mp 260-263°. Sol in water. Moderately sol in alc. uv max (water): 285 nm. IR and polarographic data: Seagers *et al., J. Am. Pharm. Assoc., Sci. Ed.* **41**, 640 (1952). pH of 0.5% aq soln: 5.0. LD$_{50}$ s.c. in rats: 1460 mg/kg (Goldenthal).
Note: This is a controlled substance listed in the U.S. Code of Federal Regulations, Title 21 Part 1308.13 (1995).
THERAP CAT: Narcotic antagonist.
THERAP CAT (VET): Narcotic antagonist.

6449. Naloxone. *(5α)-4,5-Epoxy-3,14-dihydroxy-17-(2-propenyl)morphinan-6-one;* 17-allyl-4,5α-epoxy-3,14-dihydroxymorphinan-6-one; (−)-N-allyl-14-hydroxynordihydromorphinone; N-allylnoroxymorphone. $C_{19}H_{21}NO_4$; mol wt 327.38. C 69.71%, H 6.47%, N 4.28%, O 19.55%. Specific opioid antagonist. Prepn: **Brit.** pat. **939,287** (1963 to Sankyo); M. J. Lewenstein, J. Fishman, U.S. pat. **3,254,088** (1966); R. A. Olofson *et al., Tetrahedron Letters* **1977**, 1567.

Clinical pharmacology and abuse potential: D. R. Jasinski *et al., J. Pharmacol. Exp. Ther.* **157**, 420 (1967). Metabolism: J. M. Fujimoto, *Proc. Soc. Exp. Biol. Med.* **133**, 317 (1970). Synthesis of (+)-form and stereospecific activity: I. Iijima *et al., J. Med. Chem.* **21**, 398 (1978). Clinical trial in Alzheimer's disease: B. Reisberg *et al., N. Engl. J. Med.* **308**, 721 (1983). Comprehensive description: M. M. A. Hasson *et al., Anal. Profiles Drug Subs.* **14**, 453-489 (1985). Review of pharmacology and therapeutic uses: L. F. Mc-Nicholas, W. R. Martin, *Drugs* **27**, 81-93 (1984); of clinical experience in Alzheimer's disease: C. Stowe, M. L. Gora, *Ann. Pharmacother.* **27**, 447-448 (1993); in drug overdose: J. M. Chamberlain, B. L. Klein, *Am. J. Emerg. Med.* **12**, 650-660 (1994).

Crystals from ethyl acetate, mp 184° (Lewenstein), 177-178° (Sankyo Co.). $[\alpha]_D^{20}$ −194.5° (c = 0.93 in CHCl$_3$). Sol in chloroform. Practically insol in petr ether.

Hydrochloride, C$_{19}$H$_{21}$NO$_4$.HCl, *EN-15304, Nalone, Narcan, Narcanti.* Crystals from ethanol + ether, mp 200-205°. Sol in water, alcohol. Practically insol in ether.

THERAP CAT: Narcotic antagonist.

THERAP CAT (VET): Narcotic antagonist.

6450. Naltrexone. *(5α)-17-(Cyclopropylmethyl)-4,5-epoxy-3,14-dihydroxymorphinan-6-one; N-cyclopropylmethyl-14-hydroxydihydromorphinone; UM-792.* C$_{20}$H$_{23}$NO$_4$; mol wt 341.41. C 70.36%, H 6.79%, N 4.10%, O 18.75%. Congener of naloxone, *q.v.*, with opiate-blocking activity. Prepn: H. Blumberg *et al.,* U.S. pat. **3,332,950** (1967 to Endo). Plasma kinetics in dogs, monkeys: R. H. Reuning *et al., J. Pharm. Sci.* **68**, 411 (1979). Metabolism study: E. J. Cone, *Tetrahedron Letters* **1973**, 2607. Biosynthesis, isoln, identification of major human metabolite: E. J. Cone *et al., J. Pharm. Sci.* **64**, 618 (1975). Physiological and psychological effects of single doses in man: E. R. Gritz *et al., Clin. Pharmacol. Ther.* **19**, 773 (1976). Disposition, metabolism, effects after acute and chronic administration: K. Verebey *et al., ibid.* **20**, 315 (1976). Urinary excretion profile: H. E. Dayton, C. E. Inturrisi, *Drug Metab. Dispos.* **4**, 474 (1976). Enhanced analgesic effect of stress after chronic administration: S. Amir, Z. Amit, *Eur. J. Pharmacol.* **59**, 137 (1979). Effects on brain opiate receptor supersensitivity: R. S. Zukin *et al., Brain Res.* **245**, 285 (1982); A. Tempel *et al., Life Sci.* **31**, 1402 (1982). Operant analysis of effects on human self-administration of heroin: N. K. Mello *et al., J. Pharmacol. Exp. Ther.* **216**, 45 (1981). Clinical study in chronic schizophrenia: M. Ragheb *et al., Int. Pharmacopsychiatry* **15**, 1 (1980); in heroin addiction: B. A. Judson *et al., Drug Alcohol Dependence* **7**, 325 (1981). Preliminary evaluation in infantile autism: *Psychopharmacol. Bull.* **24**, 135 (1988). Toxicity data: R. P. Maickel *et al., Ann. N.Y. Acad. Sci.* **281**, 321 (1976). *Review:* M. S. Gold *et al., Med. Res. Rev.* **2**, 211-246 (1982).

Crystals from acetone, mp 168-170°. LD$_{50}$ in mice (mg/kg): 586 s.c. (Maickel).

Hydrochloride, C$_{20}$H$_{23}$NO$_4$.HCl, *EN-1639A, Antaxone, Celupan, Nalorex, Trexan.* Crystals from methanol, mp 274-276°.

THERAP CAT: Narcotic antagonist.

6451. Nandinine. *(S)-5,8,13,13a-Tetrahydro-10-methoxy-6H-benzo[g]-1,3-benzdioxolo[5,6-a]quinolizin-9-ol; 10-methoxy-2,3-(methylenedioxy)berbin-9-ol; 5,6,13,13a-tetrahydro-9-hydroxy-10-methoxy-2,3-(methylenedioxy)-8H-dibenzo[a,g]quinolizine; tetrahydroberberrubine.* C$_{19}$H$_{19}$NO$_4$; mol wt 325.36. C 70.14%, H 5.89%, N 4.30%, O 19.67%. From root bark of *Nandina domestica* Thunb., *Berberidaceae:* Eijkman, *Ber.* **17**, 1884 (1884); Kitasato, *J. Pharm. Soc. Japan* **522**, 1 (1925). Structure: *idem, Acta Phytochim. (Japan)* **3**, 175 (1927), *C.A.* **22**, 1779 (1928); Späth, Leithe, *Ber.* **63**, 3007 (1930). Configuration: Corrodi, Hardegger, *Helv. Chim. Acta* **39**, 889 (1956). Synthesis: Kametani *et al., J. Chem. Soc. (C)* **1969**, 2036; *eidem, J. Chem. Soc. Perkin I* **1977**, 1151.

dl-Form, colorless needles from ethanol, mp 183-185°.
l-Form, crystals from ether. mp 195-196°. $[\alpha]_D^{15}$ −304° (c = 0.513 in chloroform).
d-Form, crystals from chloroform + ether, mp 195-196°.

6452. Nandrolone. *(17β)-17-Hydroxyestr-4-en-3-one; 17β-hydroxy-4-estren-3-one; 4-estren-17β-ol-3-one; 17β-hydroxy-19-nor-4-androsten-3-one; 19-nortestosterone.* C$_{18}$H$_{26}$O$_2$; mol wt 274.40. C 78.79%, H 9.55%, O 11.66%. Prepd from alkyl ethers of estradiol: Birch, *Quart. Rev.* **4**, 69 (1950); Wilds, Nelson, *J. Am. Chem. Soc.* **75**, 5366 (1953). Alternate syntheses: H. Ueberwasser *et al., Helv. Chim. Acta* **46**, 344 (1963); I. Shimizu *et al., Tetrahedron Letters* **21**, 487 (1980).

Dimorphic crystals, mp 112° and 124°. $[\alpha]_D^{22}$ +55° (c = 0.93 in chloroform). uv max (ethanol): 241 nm (ε 17000). Sol in alcohol, ether, chloroform.

Benzoate, C$_{25}$H$_{30}$O$_3$, Needles from 75% alc, mp 174-175°. $[\alpha]_D^{20}$ +104.5° (alc).

Cyclohexanecarboxylate, C$_{25}$H$_{36}$O$_3$, *19-nortestosterone hexahydrobenzoate, Norlongandron, Nor-Durandron.* Small, elongated prisms from petr ether, mp 88-89°, $[\alpha]_D^{20}$ +50° (c = 0.5 in chloroform).

Cyclohexanepropionate, C$_{27}$H$_{40}$O$_3$, *19-nortestosterone cyclohexylpropionate, Sanabolicum.*

Decanoate, C$_{28}$H$_{44}$O$_3$, *19-nortestosterone decanoate, Deca-Durabolin, Deca-Durabol, Deca-Hybolin, Hybolin Decanoate, Retabolil.* White to yellow crystals, mp 32-35°. Practically insol in water. Freely sol in ethanol, ether, acetone, chloroform, oils.

Dodecanoate, C$_{30}$H$_{48}$O$_3$, *nandrolone laurate, Laurabolin.*

Furylpropionate, C$_{25}$H$_{32}$O$_4$, *19-nortestosterone furylpropionate, NFP, Demelon.*

p-Hexyloxyphenylpropionate, C$_{33}$H$_{46}$O$_4$, *19-nortestosterone-3-(p-hexyloxyphenyl)propionate, Anador, Anadur.* Crystals, mp 53-55°. $[\alpha]_D$ +45° (c = 1.0 in dioxane).

Phenpropionate, C$_{27}$H$_{34}$O$_3$, *19-nortestosterone β-phenylpropionate, Activin, Durabolin, Durabol, Strabolene, Super-*

anabolon, Nandrolin. Crystals, mp 95-96°.   [α]$_D$ +58° (chloroform).

Propionate, C$_{21}$H$_{30}$O$_3$, *19-nortestosterone propionate, Norybol-19, Nortesto.* Crystals from aqueous methanol or isopropyl ether, mp 55-60°.   [α]$_D^{23.5}$ +58° in chloroform (Rao). uv max: 240 nm (ε 16650).

Note: This is a controlled substance (anabolic steroid) in the U.S. Code of Federal Regulations, Title 21 Part 1308.13, as defined in Part 1308.02 (1995).

THERAP CAT: Anabolic.

THERAP CAT (VET): Phenpropionate as anabolic steroid.

6453. Napalm. A coprecipitated aluminum soap from naphthenic acids and the fatty acids of coconut oil developed early in 1942 (Fieser, Harris, Hershberg, Morgana, Novello, Putnam) for prepn of gasoline gels for incendiary munitions: U.S. pats. **2,606,107** (1952); Herron, U.S. pat. **2,684,339** (1954 to Safety Fuel & Chem. Corp.). The name was derived from the *naph*thenic and *palm*itic acids which are its major constituents. Structure and mfg problems: *Chem. & Eng. News* **32**, 2690 (1954). Historical account: L. F. Fieser, *The Scientific Method* (Reinhold, New York, 1964); Bruce, "Chemical Warfare—Flame" in Kirk-Othmer *Encyclopedia of Chemical Technology* (Interscience, New York, 1964) p 888.

USE: Gasoline thickener. In chemical warfare (fire bombs, flame throwers, flame land mines).

6454. Napelline. *(1α,12α,15β)-21-Ethyl-4-methyl-16-methylene-7,20-cycloveatchane-1,12-15-triol;* luciculine. C$_{22}$H$_{33}$NO$_3$; mol wt 359.51. C 73.50%, H 9.25%, N 3.90%, O 13.35%. Isoln from *Aconitum napellus* L., *Ranunculaceae:* Freudenberg, Rogers, *J. Am. Chem. Soc.* **59**, 2572 (1937); Jacobs, Craig, *J. Biol. Chem.* **143**, 611 (1942). The product obtained by these isoln procedures is a mixture of at least three compounds. Structure: Wiesner, Valenta, *Fortschr. Chem. Org. Naturst.* **16**, 53 (1958). Related stereochemical studies: Okamoto *et al., Chem. Pharm. Bull.* **13**, 1270 (1963). Total synthesis of racemic napelline: Wiesner *et al., Can. J. Chem.* **52**, 2353, 2355 (1974); S. P. Sethi *et al., ibid.* **58**, 1889 (1980).

Rectangular plates from ether + petr ether.

Hydrochloride, C$_{22}$H$_{34}$ClNO$_3$, solvated crystals, dec 200-222°.   [α]$_D^{22}$ −93.9° (c = 5). Sol in water.

6455. Naphazoline. *4,5-Dihydro-2-(1-naphthalenylmethyl)-1H-imidazole; 2-(1-naphthylmethyl)imidazoline.* C$_{14}$H$_{14}$N$_2$; mol wt 210.28. C 79.97%, H 6.71%, N 13.32%. α-Adrenergic agonist. Prepd by reacting the acetic acid anhydride of naphthoimidic acid with ethylenediamine: A. Sonn, U.S. pat. **2,161,938** (1939 to Soc. Chem. Ind. Basle); by reacting naphthylthioacetamide with ethylenediamine: Dan. pat. **62,889** (1944 to Lovens Kemiske), *C.A.* **40**, 4398 (1946). Toxicity data: J. Gylfe *et al., Fed. Proc.* **9**, 280 (1950). Comprehensive description: G. M. Wall, *Anal. Profiles Drug Subs. Excip.* **21**, 307-344 (1992).

Hydrochloride, C$_{14}$H$_{14}$N$_2$·HCl, *Ak-Con, Albalon, Clera, Coldan, Iridina Due, Naphcon, Niazol, Opcon, Privine hydro-*

chloride, Rhinantin, Rhinoperd, Sanorin, Sanorin-Spofa, Strictylon. Bitter crystals, mp 255-260°. uv max (ethanol): 223, 270, 280, 287, 291 nm (E$_{1cm}^{1\%}$ 3622, 239, 286, 196, 198). pKa (25°C) 10.35 ±0.02, (35°C) 10.13 ±0.02, (45°C) 9.92 ±0.03. pKa (25°C) 10.35 ±0.02, (35°C) 10.13 ±0.02, (45°C) 9.92 ±0.03. Freely sol in water (40 g dissolve in 100 ml) and in alcohol. Slightly sol in chloroform. Insol in benzene, ether. A 1% aq soln has a pH of ∼6.2. LD$_{50}$ s.c. in rats: 385 mg/kg (Gylfe).

THERAP CAT: Adrenergic (vasoconstrictor); decongestant.

6456. Naphthacene. Tetracene; 2,3-benzanthracene; rubene; chrysogen. C$_{18}$H$_{12}$; mol wt 228.29. C 94.70%, H 5.30%. Occurs in coal tar. Contaminates commercial anthracene to which it imparts a yellow color. Isoln: Cook *et al., Proc. Roy. Soc. London* **B111**, 455 (1932). Isoln from crude anthracene by chromatography: Winterstein *et al., Z. Physiol. Chem.* **230**, 159 (1934). Synthesis by condensing succinic acid and phthalic anhydride in the presence of sodium acetate: Gabriel, Michael, *Ber.* **10**, 1559, 2207 (1877); **11**, 1682 (1878); Roser, *Ber.* **17**, 2744 (1884); Nathanson, *Ber.* **26**, 2582 (1893); Gabriel, Leupold, *Ber.* **31**, 1159, 1272 (1898); Wanag, *Ber.* **70**, 274 (1937); from 1-naphthol and phthalic anhydride: Deichler, Weizmann, *Ber.* **36**, 547, 719 (1903); Ger. pat. **298,345** (1916); from 1,5-dihydroxynaphthalene and phthalic anhydride: Bentley *et al., J. Chem. Soc.* **91**, 411, 1588 (1907); from tetralin and phthalic anhydride: Schroeter, *Ber.* **54**, 2242 (1921); Ger. pat. **346,673** (1918); *cf.* Fieser, *J. Am. Chem. Soc.* **53**, 2329 (1931). Other syntheses: Coulson, *J. Chem. Soc.* **1935**, 77; Weizmann *et al., ibid.* **1939**, 398.

Orange leaflets from xylene. d 1.35. Sublimes *in vacuo.* mp 341° (open capillary tube), mp 357° (copper block). Absorption spectrum: Clar, *Ber.* **69**, 607 (1936). Fluorescence maxima: Krishman, Seshan, *Z. Krist.* **89**, 538 (1934). Difficultly sol in most solvents. Solns show slight green fluorescence in daylight. Does not form a picrate.

6457. Naphthalene. *Naphthalin; naphthene; tar camphor.* C$_{10}$H$_8$; mol wt 128.17. C 93.71%, H 6.29%. Most abundant single constituent of coal tar. Dry coal tar contains about 11%. Crystallizes from the middle or "carbolic oil" fraction of the distilled tar. Purified by hot pressing, which may be followed by washing with H$_2$SO$_4$, NaOH, and water, then by fractional distillation or by sublimation. Manuf: Faith, Keyes & Clark's *Industrial Chemicals,* F. A. Lowenheim, M. K. Moran, Eds. (Wiley-Interscience, New York, 4th ed., 1975) pp 556-562. *Review:* R. M. Gaydos in Kirk-Othmer *Encyclopedia of Chemical Technology* **vol. 15** (Wiley-Interscience, New York, 3rd ed., 1981) pp 698-719.

Monoclinic prismatic plates from ether or by sublimation; also sold as white scales, powder, balls, or cakes, mp 80.2°. Odor of moth balls. Volatilizes appreciably at room temp. d$_4^{20}$ 1.162. d$_4^{100}$ 0.9628. Sublimes appreciably at temps above mp; volatile with steam. bp$_{760}$ 217.9°; bp$_{400}$ 193.2°; bp$_{200}$ 167.7°; bp$_{100}$ 145.5°; bp$_{60}$ 130.2°; bp$_{40}$ 119.3°; bp$_{20}$ 101.7°; bp$_{10}$ 85.8°. Flash pt, open cup 174°F (79°C); closed cup 190°F (88°C). Autoignition temp 1053°F (567°C). n$_D^{100}$ 1.58212. Purple fluorescence in Hg light (petr ether soln). Ultraviolet absorption: Several characteristic bands between 217.5 and 320 nm in hexane. Insol in water. One gram dissolves in 13 ml methanol or ethanol, in 3.5 ml benzene or toluene, in 8 ml olive oil or turpentine, in 2 ml chloroform or carbon tetrachloride, in 1.2 ml carbon disulfide. Very sol in ether, hydronaphthalenes, in fixed and volatile oils.

Caution: Poisoning may occur by ingestion of large doses, inhalation, or skin absorption. *Symptoms and signs:* nausea, vomiting, headache, diaphoresis, hematuria, hemolytic anemia, fever, hepatic necrosis, convulsions, coma,

Clinical Toxicology of Commercial Products, R. E. Gosselin *et al.,* Eds. (Williams & Wilkins, Baltimore, 4th ed., 1976) Section III, pp 242-246.

USE: Manuf phthalic and anthranilic acids which are used in making indigo, indanthrene, and triphenylmethane dyes. Manuf of hydroxyl (naphthols), amino (naphthylamines), sulfonic acid and similar compds used in the dye industries. Manuf of synthetic resins, celluloid, lampblack, smokeless powder. Manuf of hydronaphthalenes (Tetralin, Decalin) which are used as solvents, in lubricants, and in motor fuels. The use of naphthalene as a moth repellent and insecticide is decreasing due to the introduction of chlorinated compds such as *p*-dichlorobenzene.

THERAP CAT: Has been used as antiseptic (topical and intestinal); anthelmintic (Cestodes).

THERAP CAT (VET): Has been used in dusting powders, as an insecticide and internally as an intestinal antiseptic and vermicide.

6458. 1-Naphthaleneacetic Acid. α-Naphthaleneacetic acid; naphthylacetic acid; NAA; Fruitone-N; Phyomone; Planofix; Tre-Hold. $C_{12}H_{10}O_2$; mol wt 186.21. C 77.40%, H 5.41%, O 17.18%. Prepn from naphthalene + chloroacetic acid: Ogata, Ishiguro, *J. Am. Chem. Soc.* **72**, 4302 (1950); Southwick *et al., ibid.* **76**, 754 (1954); U.S. pat. **2,655,531** (1953 to FMC); from naphthylacetonitrile: Wenner, U.S. pat. **2,489,348** (1949 to Hoffmann-La Roche); *J. Org. Chem.* **15**, 548 (1950). Activity: F. E. Gardiner *et al., Science* **90**, 208 (1939). Crystal structure: S. S. Rajan, *Acta Crystallogr.* **B34**, 998 (1978). Toxicity study: G. W. Bailey, J. L. White, *Residue Rev.* **10**, 97 (1965).

Needles from water, mp 134.5-135.5°. Sol in about 30 parts alcohol; freely sol in acetone, ether, chloroform. Soly in water at 17°: 0.38 g/l. LD_{50} orally in rats: 1000 mg/kg (Bailey, White).

USE: Plant growth regulator.

6459. 1,8-Naphthalenediamine. 1,8-Diaminonaphthalene. $C_{10}H_{10}N_2$; mol wt 158.20. C 75.92%, H 6.37%, N 17.71%. Prepd by reducing 1,8-dinitronaphthalene with phosphorus triiodide: Meyer, Müller, *Ber.* **30**, 775 (1897).

Crystals from dil alc, mp 66.5°. bp_{12} 205°; $n_D^{99.4}$ 1.6828; $d_4^{99.4}$ 1.1265. Sublimable. Turns brown on standing. Soluble in alcohol or ether; slightly sol in water or chloroform.

Dihydrochloride, $C_{10}H_{12}Cl_2N_2$, leaflets, mp 280°.

USE: Antioxidant for lubricating oils. Detection of selenium and nitrites.

6460. 1,6-Naphthalenedisulfonic Acid. Ewer-Pick acid. $C_{10}H_8O_6S_2$; mol wt 288.30. C 41.66%, H 2.80%, O 33.30%, S 22.24%. Prepn: Fierz-David, Hasler, *Helv. Chim. Acta* **6**, 1134 (1923).

Crystals. Very sol in water; sol in alcohol; practically insol in ether.

6461. 2,6-Naphthalenedisulfonic Acid. Ebert-Merz β-acid. $C_{10}H_8O_6S_2$; mol wt 288.30. C 41.66%, H 2.80%, O 33.30%, S 22.24%. Prepn: Fierz-David, Richter, *Helv. Chim. Acta* **28**, 257 (1945).

Deliquesc crystals. Very sol in water, alcohol; practically insol in ether.

6462. 2,7-Naphthalenedisulfonic Acid. Ebert-Merz α-acid. $C_{10}H_8O_6S_2$; mol wt 288.30. C 41.66%, H 2.80%, O 33.30%, S 22.24%. Prepn from naphthalene and concd H_2SO_4: Fierz-David, Blangey, *Fundamental Processes of Dye Chemistry* (Interscience, New York, 1949) p 209.

Very deliquesc crystals. Very sol in water, alcohol; practically insol in ether.

USE: In dye chemistry.

6463. 1-Naphthalenesulfonic Acid. α-Naphthalenesulfonic acid. $C_{10}H_8O_3S$; mol wt 208.24. C 57.68%, H 3.87%, O 23.05%, S 15.40%. Made by sulfonating naphthalene with H_2SO_4 at 0°.

Crystals. mp 90° (dihydrate). Freely sol in water or alcohol, slightly in ether.

USE: Manuf α-naphthol. The sodium salt is used for rendering phenols sol in water.

6464. 2-Naphthalenesulfonic Acid. β-Naphthalenesulfonic acid. $C_{10}H_8O_3S$; mol wt 208.24. C 57.68%, H 3.87%, O 23.05%, S 15.40%. Made by sulfonating naphthalene with H_2SO_4 at 160°.

Monohydrate, white to slightly brownish, cryst leaflets; very hygroscopic. mp 124-125°. mp 91° when anhydr. Freely sol in water.

USE: Manuf β-naphthol and intermediates.

6465. 1-Naphthalenethiol. α-Thionaphthol; 1-thionaphthol; 1-mercaptonaphthalene; 1-naphthyl mercaptan. $C_{10}H_8S$; mol wt 160.24. C 74.96%, H 5.03%, S 20.01%. Prepd by catalytic sulfurization of naphthalene with S, S_2Cl_2, etc., followed by hydrogenation: Lazier *et al.;* Signaigo, U.S. pats. **2,402,645; 2,402,686** (both 1946 to du Pont).

Liquid, strong mercaptan odor. Solidif on cooling. d_4^0 1.1729; d_4^{20} 1.607; d_4^{23} 1.1549. bp_{760} 285°; $bp_{10.3}$ 144.8°; bp_2 138-140°. n_D^{20} 1.6802. Sol in ethanol, ether; sparingly sol in aq alkalies. Volatile with steam.

6466. 2-Naphthalenethiol. β-Thionaphthol; 2-thionaphthol; 2-mercaptonapthalene; 2-naphthyl mercaptan. $C_{10}H_8S$; mol wt 160.24. C 74.96%, H 5.03%, S 20.01%. Prepd by catalytic hydrogenation of a sulfonic acid derivative of naphthalene: Lazier, Signaigo, U.S. pat. **2,402,641**

(1946 to du Pont); by reduction of naphthalenesulfonyl chloride with zinc: Holt, U.S. pat. **2,216,840** (1940 to du Pont).

HOOC COOH structure (2-naphthalenethiol, SH)

Crystals from ethanol, disagreeable odor. mp 81°. bp$_{760}$ 286°; bp$_{10.3}$ 146.3°. Very sol in ethanol, ether, petr ether; sparingly sol in water. Slightly volatile with steam.

6467. Naphthalic Acid. *1,8-Naphthalenedicarboxylic acid.* C$_{12}$H$_8$O$_4$; mol wt 216.19. C 66.67%, H 3.73%, O 29.60%. Made by the oxidation of acenaphthene with chromic acid mixture, etc.: Graebe, Gfeller, *Ber.* **25,** 652 (1895); Ogilvie, Wilder, U.S. pat. **2,379,032** (1945 to Allied Chem.); Karishin, Fedorenko, *Zhur. Priklad. Khim.* **29,** 955 (1956), *C.A.* **50,** 14677b (1956).

Crystals, mp 270°. Practically insol in water; freely sol in warm alcohol, slightly in ether.

6468. 1-Naphthoic Acid. *1-Naphthalenecarboxylic acid;* α-naphthoic acid. C$_{11}$H$_8$O$_2$; mol wt 172.18. C 76.73%, H 4.68%, O 18.58%. Prepn from α-bromonaphthalene by Grignard reaction: Gilman *et al., Org. Syn.* **coll. vol. II,** 425 (1943); from 4-aminonaphthalenesulfonic acid by Sandmeyer reaction: Bassilios, *Bull. Soc. Chim. France* **1950,** 757; from β-nitronaphthalene by von Richter reaction: Bunnett, Rauhut, *J. Org. Chem.* **21,** 934, 944 (1956); by oxidation of 1-methylnaphthalene: Aries, U.S. pat. **2,930,802** (1960); Barker, U.S. pat. **2,963,508** (1960 to Mid-Century).

Crystals from hot toluene, mp 160.5-162°. bp 300°. uv max (ethanol): 293 nm (log ε$_m$ 3.9). Slightly sol in hot water; freely sol in hot alcohol, ether.
Caution: Moderately irritating to skin, mucous membranes.

6469. 2-Naphthoic Acid. *2-Naphthalenecarboxylic acid;* β-naphthoic acid; isonaphthoic acid. C$_{11}$H$_8$O$_2$; mol wt 172.18. C 76.73%, H 4.68%, O 18.58%. Prepn from β-acetylnaphthalene + NaOCl: Newman, Holmes, *Org. Syn.* **coll. vol. II,** 428 (1943); from naphthalene + octylsodium: Morton *et al., J. Am. Chem. Soc.* **64,** 2250 (1942); from 2-aminonaphthalenesulfonic acid by Sandmeyer reaction: Wahl, Bassilios, *Bull. Soc. Chim. France* **1947,** 482; by carboxylation of naphthalene: Prichard, U.S. pat. **2,729,673** (1956 to du Pont).

Crystals from 95% alcohol, mp 184-185°. bp > 300°. uv max (ethanol): 235, 280, 335 nm (log ε$_m$ 4.7, 3.8, 3.1). Slightly sol in hot water; sol in alcohol, ether.
Caution: Moderately irritating to skin, mucous membranes.

6470. 1-Naphthol. *1-Naphthalenol;* α-naphthol; alphanaphthol; α-hydroxynaphthalene. C$_{10}$H$_8$O; mol wt 144.17. C 83.31%, H 5.59%, O 11.10%. Prepd by fusing the sodium salt of α-naphthalenesulfonic acid with NaOH: Tyrer, U.S. pats. **2,407,044, 2,407,055** and **2,451,996** (1946 and 1948);

by oxidation of naphthalene: Loeb, U.S. pat. **3,033,903** (1962 to Union Carbide). Toxicity data: H. F. Smyth *et al., Am. Ind. Hyg. Assoc. J.* **23,** 95 (1962).

Prisms, mp 96°. Phenolic odor; disagreeable, burning taste. bp 288°; bp$_{40}$ 184°. d$_4^{98.7}$ 1.0954. uv max: 297, 310, 324 nm. Sublimable; volatile with steam. Darkens in light; reduces ammoniacal silver nitrate. Slightly sol in water, freely in alcohol, benzene, chloroform, ether, alkali hydroxide solns. *Protect from light.* LD$_{50}$ orally in rats: 2.59 g/kg (Smyth).
Caution: See 2-Naphthol.
USE: Manuf dyes, intermediates, synthetic perfumes; also in microscopy.

6471. 2-Naphthol. *2-Naphthalenol;* β-naphthol; betanaphthol; β-hydroxynaphthalene; isonaphthol; C.I. Azoic Coupling Component 1; C.I. Developer 5; C.I. 37500. C$_{10}$H$_8$O; mol wt 144.17. C 83.31%, H 5.59%, O 11.10%. Prepn from sodium naphthalene-2-sulfonate: Schoeffel, Barton, U.S. pat. **2,760,992** (1956 to Sterling); Stevens, Harris, U.S. pat. **2,831,895** (1958 to Dow); Fr. pat. **1,326,175** (1963 to Ciba); by oxidation of naphthalene: Simons, U.S. pat. **2,530,369** (1950 to Phillips Petroleum); from 2-bromonaphthalene + *tert*-butyl hydroperoxide: Lawesson, Yang, *J. Am. Chem. Soc.* **81,** 4230 (1959).

Crystals, mp 121-123°. bp 285-286°. d 1.22. Flash pt 161°. Slight phenolic odor. Darkens with age on exposure to light. Sublimes when heated, distillable *in vacuo;* volatile with vapors of alcohol or water; reduces ammoniacal silver nitrate. uv max (95% ethanol): 226, 265, 275, 286, 320, 331 nm (ε 91,194, 3911, 4559, 3301, 1861, 2163). One gram dissolves in 1000 ml water, 80 ml boiling water, 0.8 ml alcohol, 17 ml chloroform, 1.3 ml ether; sol in glycerol, olive oil, solns of alkali hydroxides. *Protect from light. Incompat.:* Antipyrine, camphor, phenol, ferric salts, menthol, potassium permanganate and other oxidizing agents, urethane.
Sodium salt, C$_{10}$H$_7$NaO, *sodium β-naphtholate, sodium naphthol, Microcidin.* Grayish-white powder; becomes reddish or brownish on exposure to light and air. Soluble in 3 parts water. *Keep well closed and protected from light.*
Caution: Potential symptoms of overexposure to 2-naphthol include crampy abdominal pain, nausea, vomiting and sometimes convulsions. Intestinal or percutaneous absorption may lead to severe nephritis, liver injury and acute hemolytic anemia. Lens opacities and retinal changes may occur. See *Clinical Toxicology of Commercial Products,* R. E. Gosselin *et al.,* Eds. (Williams & Wilkins, Baltimore, 5th ed., 1984) Section II, p 188.
USE: Manuf medicinal organics, dyes, perfumes; the largest single use is probably in making antioxidants for the synthetic rubber industry.
THERAP CAT: Formerly as anthelmintic (Nematodes).
THERAP CAT (VET): Has been used as antiseptic, anthelmintic and counter-irritant in alopecia.

6472. 1-Naphthol-8-amino-3,6-di-sulfonic Acid. *4-Amino-5-hydroxy-2,7-naphthalenedisulfonic acid;* 8-amino-1-naphthol-3,6-disulfonic acid; H acid. C$_{10}$H$_9$NO$_7$S$_2$; mol wt 319.32. C 37.61%, H 2.84%, N 4.39%, O 35.07%, S 20.08%. Prepn: Willard, *Color Trade J.* **15,** 40 (1924); Mow, U.S. pat. **2,272,272** (1942 to Allied Chem.); Hayashi, Yamaguchi *J. Sci.* **2,** 67 (1951), *C.A.* **49,** 2390f (1955); Roos *et al.,* U.S. pat. **2,875,243** (1959 to Bayer).

White crystals or gray powder. Slightly sol in water, alcohol, ether; sol in alkalies.
USE: Prepn of azo dyes.

6473. 1-Naphthol-4,8-disulfonic Acid. *4-Hydroxy-1,5-naphthalenedisulfonic acid;* α-naphtholdisulfonic acid S; Schöllkopf's acid. $C_{10}H_8O_7S_2$; mol wt 304.30. C 39.47%, H 2.65%, O 36.80%, S 21.08%. Prepd by sulfonation of 1-naphthol-8-sulfonic acid sultone: I. G. Farbenindustrie, **PB 74197**, frames 828-829.

Crystals. Soluble in water. The sodium salt is very soluble in water.

6474. 2-Naphthol-3,6-disulfonic Acid. *3-Hydroxy-2,7-naphthalenedisulfonic acid;* R acid. $C_{10}H_8O_7S_2$; mol wt 304.30. C 39.47%, H 2.65%, O 36.80%, S 21.08%. Prepn from β-naphthol and concd H_2SO_4: Fierz-David, Blangey, *Fundamental Processes of Dye Chemistry* (Interscience, New York, 1949) p 197.

Deliquesc needles. Very sol in water, alcohol. Practically insol in ether. The sodium salt is known as *R salt.*
USE: In dye chemistry.

6475. 2-Naphthol-6,8-disulfonic Acid. *7-Hydroxy-1,3-naphthalenedisulfonic acid;* 2-hydroxynaphthalene-6,8-disulfonic acid; G acid. $C_{10}H_8O_7S_2$; mol wt 304.30. C 39.47%, H 2.65%, O 36.80%, S 21.08%. Prepd by heating β-naphthol with concd H_2SO_4, 2-naphthol-3,6-disulfonic acid (R acid) being obtained as a byproduct: Fierz-David, Blangey, *Fundamental Processes of Dye Chemistry* (Interscience, New York, 1949) p 197; *see also* Forster, Keyworth, *J. Soc. Chem. Ind. (London)* **46**, 27T (1927).

Sodium salt, *G salt.* Platelets or prisms. Freely sol in water, sol in dil alcohol.
Barium salt octahydrate, $C_{10}H_6BaO_7S_2.8H_2O$, minute prisms. Freely sol in water. Practically insol in alcohol, even when much dil with water.
USE: In the manuf of azo dyes, *see* Crossley, Resenvelt, *Ind. Eng. Chem.* **16**, 271 (1924).

6476. α-Naphtholphthalein. *3,3-Bis(4-hydroxynaphthalenyl)-1(3H)-isobenzofuranone;* 3,3-bis(4-hydroxy-1-naphthyl)phthalide; di-p-α-naphtholphthalide; p-α-naphtholphthalein. $C_{28}H_{18}O_4$; mol wt 418.45. C 80.37%, H 4.34%, O 15.29%. Prepn: Werner, *J. Chem. Soc.* **113**, 20 (1918). Commercial prepn contains a small amount of the isomeric 3,3-bis(2-hydroxy-1-naphthyl)phthalide.

Colorless powder when pure; usually grayish-red. mp 253-255°. Practically insol in water. Sol in alcohol.
USE: As indicator in 0.1% or 0.04% soln in alc. pH 7.3 colorless to reddish; 8.7 greenish to blue. Particularly adapted for weak acids in strong alcoholic soln.

6477. 1-Naphthol-2-sulfonic Acid. *1-Hydroxy-2-naphthalenesulfonic acid;* Baum's acid; Schaeffer's α-acid. $C_{10}H_8O_4S$; mol wt 224.24. C 53.56%, H 3.60%, O 28.54%, S 14.30%. Prepd by sulfonation of α-naphthol: Hodgson, Hathaway, *J. Soc. Dyers Colourists* **63**, 109 (1947).

Deliquesc crystals. Slightly sol in cold water; sol in boiling water. Practically insol in ether.

6478. 1-Naphthol-4-sulfonic Acid. *4-Hydroxy-1-naphthalenesulfonic acid;* Nevile and Winther's acid. $C_{10}H_8O_4S$; mol wt 224.24. C 53.56%, H 3.60%, O 28.54%, S 14.30%. Prepn from 1,4-diazonaphthalenesulfonic acid + sulfuric acid: Erdmann, *Ann.* **247**, 341 (1888); by sulfonation of α-naphthylcarbonate: Reverdin, *Ber.* **27**, 3458 (1894); from chlorosulfonic acid + α-naphthol: Gebauer-Fülnegg, *Monatsh.* **49**, 195 (1928); Baddiley *et al.*, U.S. pat. **1,452,481** (1923); from sodium naphthionate + SO_2: Binns, Lurie, U.S. pat. **1,880,701** (1933 to Virginia Smelting).

Crystals, dec 170°. K_{OH} 3×10^{-9}. Very sol in water. Salt solns give a blue color with ferric chloride.
USE: Preparation of azo dyes.

6479. 2-Naphthol-6-sulfonic Acid. *6-Hydroxy-2-naphthalenesulfonic acid;* Schaeffer's β-acid. $C_{10}H_8O_4S$; mol wt 224.24. C 53.56%, H 3.60%, O 28.54%, S 14.30%. Prepn from β-naphthol and concd H_2SO_4: Fierz-David, Blangey, *Fundamental Processes of Dye Chemistry* (Interscience, New York, 1949) p 194.

Leaflets. mp 125°. Very sol in water, alcohol. Practically insol in ether.
Sodium salt, $C_{10}H_7NaO_4S$, *Schaeffer's salt.* Light yellow to pink powder. Freely sol in water; slightly sol in alcohol.
USE: In dye chemistry.

6480. Naphthol Yellow S. *8-Hydroxy-5,7-dinitro-2-naphthalenesulfonic acid disodium salt;* Citronin A; Sulfur Yellow S; Acid Yellow S; FD & C Yellow no. 1; Ext. D & C

Yellow no. 7; C.I. Acid Yellow 1; C.I. 10316. $C_{10}H_4N_2Na_2$-O_8S; mol wt 358.20. C 33.53%, H 1.13%, N 7.82%, Na 12.84%, O 35.73%, S 8.95%. Prepn: *Colour Index*, **vol. 4** (3rd ed., 1971) p 4004.

Greenish yellow powder, sol in water to a yellow soln. The free acid is known as flavianic acid, *q.v.*
Caution: Approved by FDA for external use only.
USE: Dye for wool, silk.

6481. 1,2-Naphthoquinone. *1,2-Naphthalenedione;* β-naphthoquinone. $C_{10}H_6O_2$; mol wt 158.16. C 75.94%, H 3.82%, O 20.23%. Prepd by oxidation of 1-amino-2-naphthol with ferric chloride: Fieser, *Org. Syn.* **17,** 68 (1937); by oxidation of naphthalene: Milas, U.S. pat. **2,395,638** (1946 to Research Corp.); by bacterial dissimilation of naphthalene: Murphy, Stone, *Can. J. Microbiol.* **1,** 579 (1955); from tetrachloro-o-benzoquinone: Horner, Dürckheimer, *Z. Naturforsch.* **14b,** 741 (1959). Structure: Hodgson, Hathaway, *Trans. Faraday Soc.* **41,** 115 (1945). Spectroscopic properties: Oliver *et al., Tetrahedron* **24,** 4067 (1968).

Golden yellow needles, dec 145-147°. uv max (abs alc): 250, 340, 405 nm (log ε 4.35, 3.40, 3.40). Sol in alc, benzene, ether, 5% NaOH, 5% NaHCO$_3$, concd H$_2$SO$_4$ with green color. Practically insol in water.
2-Semicarbazone, $C_{11}H_9N_3O_2$, *naftazone, Haemostop Injection, Mediaven, Karbinon.*
USE: As reagent for resorcinol and thalline.
THERAP CAT: 2-Semicarbazone as hemostatic.

6482. 1,4-Naphthoquinone. *1,4-Naphthalenedione;* α-naphthoquinone; 1,4-dihydro-1,4-diketonaphthalene. C_{10}-H_6O_2; mol wt 158.16. C 75.94%, H 3.82%, O 20.23%. Prepn: Fieser, *Org. Syn.* **coll. vol. I,** 383 (2nd ed., 1941); Braude, Fawcett, *ibid.,* **coll. vol. IV,** 698 (1963). Substituted 1,4-naphthoquinones occur in nature, *e.g.,* phthiocol, *q.v.,* various pigments, and the vitamins K.

Yellow triclinic needles from alcohol or petr ether. Odor like that of benzoquinone. d 1.422. mp 126°. Begins to sublime below 100°. Sublimation rates: Kempf, *J. Prakt. Chem.* [2] **78,** 236, 257. Easily volatile with steam. uv spectrum: Purvis, *J. Chem. Soc.* **101,** 1318 (1912). Sparingly sol in cold water, slightly in petr ether, freely in hot alcohol, ether, benzene, chloroform, carbon bisulfide, acetic acid; sol in alkali hydroxide solns giving a reddish-brown soln.

6483. Naphthoresorcinol. *1,3-Naphthalenediol;* 1,3-dihydroxynaphthalene. $C_{10}H_8O_2$; mol wt 160.17. C 74.99%, H 5.03%, O 19.98%. Prepn by cyclization of ethyl phenylacetoacetate: Soliman, West, *J. Chem. Soc.* **1944,** 53; by heating 1-amino-3-hydroxy-4-naphthalenesulfonic acid in acidic soln: Meyer, Bloch, *Org. Syn.* **25,** 73 (1945); from 1,3-naphtholsulfonic acid: Kozlov, Odintsov, *J. Appl.*

Chem. (USSR) **17,** 219 (1944); Kozlov *et al., Zh. Prikl. Khim.* **35,** 880 (1962).

Leaflets, mp 124-125°. Freely sol in water, alcohol, ether.
USE: Reagent for sugars, oils, and for glucuronic acid in urine: Forsyth, *Nature* **161,** 239 (1948); Heyns, Kelch, *Z. Anal. Chem.* **139,** 339 (1953).

6484. 2-Naphthoxyacetic Acid. *2-Naphthalenyloxyacetic acid;* β-naphthoxyacetic acid; *O*-(2-naphthyl)glycolic acid. $C_{12}H_{10}O_3$; mol wt 202.21. C 71.28%, H 4.98%, O 23.74%. Prepd by treating β-naphthol with chloroacetic acid in an alkaline medium: Spitzer, *Ber.* **34,** 3192 (1901); Shirley, *Organic Intermediates* (New York, 1951) p 209.

Prisms from hot water or benzene, mp 156°. Moderately sol in hot water. Soluble in alcohol, ether, acetic acid.
Ethyl ester, $C_{14}H_{14}O_3$, leaflets from alcohol, mp 49°. Sol in alcohol, ether.
USE: As plant hormone, to promote growth of roots on clippings, to prevent fruit from falling prematurely; causes stunted growth when used in excess.

6485. 1-Naphthylamine. *1-Naphthalenamine;* 1-aminonaphthalene; α-naphthylamine; naphthalidine. $C_{10}H_9N$; mol wt 143.19. C 83.88%, H 6.34%, N 9.78%. Prepd by reducing α-nitronaphthalene with Fe and HCl: West, *J. Chem. Soc.* **127,** 494 (1925); from 1-naphthalenecarboxylic acid and hydroxylamine: Snyder *et al., J. Am. Chem. Soc.* **75,** 2014 (1953). Review of carcinogenic risk: *IARC Monographs* **4,** 87-96 (1974).

Needles, becoming red on exposure to air, or a reddish, cryst mass; unpleasant odor. Sublimes; volatile with steam. d 1.13. mp 50°. bp 301°. Flash pt 157°. Sol in 590 parts water; freely sol in alc, ether. Reduces warm ammoniacal silver nitrate. *Keep well closed and protected from light.*
Hydrochloride, $C_{10}H_9N·HCl$, cryst powder; becomes bluish on exposure to air and light. Sol in about 27 parts water; sol in alcohol, ether. *Protect from light.*
Sulfate dihydrate, $(C_{10}H_9N)_2·H_2SO_4·2H_2O$, white to yellowish cryst powder. Slightly sol in water or alcohol.
Caution: Potential symptoms of overexposure to 1-naphthylamine are dermatitis; hemorrhagic cystitis; dyspnea, ataxia and methemoglobinemia; hematuria; dysuria. See *NIOSH Pocket Guide to Chemical Hazards* (DHHS/NIOSH 90-117, 1990) p 158. This substance has been listed as a carcinogen by OSHA: *Fed. Reg.* **39,** 3757 (1974).
USE: Manufacturing dyes; toning prints made with cerium salts; the hydrochloride with sulfanilic acid is a reagent for nitrate.

6486. 2-Naphthylamine. *2-Naphthalenamine;* 2-aminonaphthalene; β-naphthylamine. $C_{10}H_9N$; mol wt 143.19. C 83.88%, H 6.34%, N 9.78%. Prepd by heating β-naphthol with ammonium sulfite and NH$_4$OH at 150°: **Ger.** pat. **117,471** (1900); *Frdl.* **6,** 190; from 2-naphthalenecarboxylic acid and hydroxylamine: Snyder *et al., J. Am. Chem. Soc.* **75,** 2014 (1953). Review of carcinogenic risk: *IARC Monographs* **4,** 97-111 (1974).

White to reddish crystals; volatile with steam. d_4^{98} 1.061. mp 111-113°. bp 306°; also stated as 294°. Sol in hot water, alcohol, ether. Reduces warm ammoniacal silver nitrate.

Acetate, $C_{10}H_9N.CH_3COOH$, white to yellowish scales or flakes; slight odor of acetic acid. Very sol in water and alcohol; sol in ether. *Keep well closed and protected from light.*

Caution: Potential symptoms of overexposure to 2-naphthylamine are dermatitis; hemorrhagic cystitis; dyspnea; ataxia; methemoglobinemia, hematuria; dysuria. *See NIOSH Pocket Guide to Chemical Hazards* (DHHS/NIOSH 90-117, 1990) p 158. 2-Naphthylamine has been listed as a known carcinogen: *Seventh Annual Report on Carcinogens* (PB95-109781, 1994) p 64.

USE: Formerly in manuf of dyes, as antioxidant in rubber.

6487. 1-Naphthylamine-2,7-disulfonic Acid. *1-Amino-2,7-naphthalenedisulfonic acid;* Kalle's acid. $C_{10}H_9NO_6S_2$; mol wt 303.32. C 39.60%, H 2.99%, N 4.62%, O 31.65%, S 21.14%. Prepn from 1-naphthylamine-2,4,7-trisulfonic acid: **Ger. pat. 62,634** (1892 to Kalle); *Frdl.* **3**, 431; from 1-naphthylamine-2,5,7-trisulfonic acid: **Ger. pat. 255,724** (1912 to Bayer); *Frdl.* **11**, 217.

Needles.
Sodium salt, prisms from dil alc. Solns display a blue-green fluorescence.
Barium salt, needles. Very slightly sol in hot water.

6488. 1-Naphthylamine-4,6-disulfonic Acid. *4-Amino-1,7-naphthalenedisulfonic acid;* Dahl's acid II. $C_{10}H_9NO_6S_2$; mol wt 303.32. C 39.60%, H 2.99%, N 4.62%, O 31.65%, S 21.14%. Prepn by sulfonation of α-naphthylamine and separation from mixture of sulfonic acids obtained: Erdmann, *Ann.* **275**, 192 (1893).

Crystals. Sol in 6 parts cold water; very sol in hot water; sol in alcohol.

6489. 1-Naphthylamine-4,7-disulfonic Acid. *4-Amino-1,6-naphthalenedisulfonic acid;* Dahl's acid III. $C_{10}H_9NO_6S_2$; mol wt 303.32. C 39.60%, H 2.99%, N 4.62%, O 31.65%, S 21.14%. Prepn: Erdmann, *Ann.* **275**, 192 (1893).

Crystals. Sol in 145 parts cold water, in 20 parts boiling water. Practically insol in alcohol.

6490. 1-Naphthylamine-4-sulfonic Acid. *4-Amino-1-naphthalenesulfonic acid;* naphthionic acid; Piria's acid. $C_{10}H_9NO_3S$; mol wt 223.25. C 53.80%, H 4.06%, N 6.27%, O 21.50%, S 14.36%. Prepd on a large-scale by heating α-

naphthylamine with equimolar amounts of concd H_2SO_4 at 180-200° or by heating α-naphthylamine acid sulfate: Fierz-David, Blangey, *Fundamental Processes of Dye Chemistry* (Interscience, New York, 1949).

Sesquihydrate, shiny needles from water. d_4^{25} 1.673. Dec on heating without melting (the amide mp 206°). K at 25° = 2.1×10^{-3}. One gram dissolves in 3.45 l of water at 10°, in 3.22 l at 20°, in 1.69 l at 50°, in 438.5 ml at 100°. Very sparingly sol in alcohol, ether. Practically insol in acetic acid and acetic anhydride, but becomes sol when some pyridine is added. Sol in dil solns of alkali hydroxides or carbonates with a blue fluorescence.

Sodium salt tetrahydrate, $C_{10}H_8NNaO_3S.4H_2O$, *naphthionine, 101-E.* Ref: Piria, *Ann.* **78**, 41 (1851). Large monoclinic prisms from water, loses 3½ mols water at 80°, anhyd at 130°. Sweet aftertaste. Freely sol in water with a blue fluorescence. Also sol in 95% alc. Practically insol in ether. Sparingly sol in concd aq and alcoholic caustic solns. pH of a 1% aq soln 6.8. Even ampuled aq solns discolor under the influence of light.

Sodium salt glucoside, $C_{16}H_{20}NNaO_8S$, *naphthionine N-glucoside, 101-G.* Ref: Estève et al., *Ann. Pharm. Franc.* **10**, 680 (1952). White powder, bitter taste. Dec 255°. $[\alpha]_D^{23.5}$ −113° (c = 10). Sol in water; slightly sol in alcohol. Practically insol in ether.

USE: The sodium salt is an important dye intermediate in the manuf of Congo Red, Fast Red A, Azo Rubine, and similar azo dyes.

THERAP CAT: Sodium salt as hemostatic.

6491. 1-Naphthylamine-5-sulfonic Acid. *5-Amino-1-naphthalenesulfonic acid;* Laurent's acid. $C_{10}H_9NO_3S$; mol wt 223.25. C 53.80%, H 4.06%, N 6.27%, O 21.50%, S 14.36%. Prepn from 5-chloro-1-naphthalenesulfonic acid + ammonia: Ochler, **Ger. pat. 72,336** (1893); *Frdl.* **3**, 435; from α-naphthylamine + concd sulfuric acid: Erdmann, *Ann.* **275**, 193, 200 (1893); Sixma, *Rec. Trav. Chim.* **73**, 235 (1954); from 2-naphthylamine-8-sulfonic acid: Blangey, *Helv. Chim. Acta* **39**, 977 (1956).

White crystals. Sol in 950 parts cold water; more sol in hot water.

USE: In dye chemistry.

6492. 1-Naphthylamine-8-sulfonic Acid. *8-Amino-1-naphthalenesulfonic acid;* Peri acid. $C_{10}H_9NO_3S$; mol wt 223.25. C 53.80%, H 4.06%, N 6.27%, O 21.50%, S 14.36%. Prepn: Fierz, Weissenbach, *Helv. Chim. Acta* **3**, 310 (1920); Martin, *Z. Farben-Ind.* **20**, 9, 76 (1928), *C.A.* **22**, 2665 (1928); Shebuev et al., **USSR pat. 165,746** (1964 to State Scientific Res. Inst. of Org. Intermediates & Dyes), *C.A.* **62**, 10392c (1965).

Needles. Sol in 4800 parts cold, 240 parts boiling water; freely sol in glacial acetic acid.

Caution: Irritating to skin.

6493. 2-Naphthylamine-1-sulfonic Acid. *2-Amino-1-naphthalenesulfonic acid;* Tobias acid. $C_{10}H_9NO_3S$; mol wt 223.25. C 53.80%, H 4.06%, N 6.27%, O 21.50%, S 14.36%. Prepd by sulfonation of 2-naphthylamine: Tinker, Hansen, U.S. pat. **1,969,189** (1934 to du Pont). Utilization in viral biology study: N. Sakota *et al., J. Ferment. Technol.* **56**, 53 (1978). TLC study: J. Franc, V. Koudelkova, *J. Chromatog.* **170**, 89 (1979).

Anhydr scales from hot water; hydrated needles from cold water. Slightly sol in cold water; more sol in hot water; very slightly sol in alcohol, ether.

Diethylammonium salt, $C_{10}H_9NO_3S.(C_2H_5)_2NH$, mp 180°.

USE: Dyestuff intermediate.

6494. 2-Naphthylamine-5-sulfonic Acid. *6-Amino-1-naphthalenesulfonic acid;* Dahl's acid. $C_{10}H_9NO_3S$; mol wt 223.25. C 53.80%, H 4.06%, N 6.27%, O 21.50%, S 14.36%. Prepn by reduction of 6-nitro-1-naphthalenesulfonic acid: Kappeler, *Ber.* **45**, 633 (1912); by sulfonation of β-naphthylamine and separation from mixt of sulfonic acids obtained: Green, Vakil, *J. Chem. Soc.* **113**, 35 (1918).

Needles from water. Sol in 3025 parts water at 20°. Almost insol in alcohol.

USE: Dyestuff intermediate.

6495. 2-Naphthyl Benzoate. *2-Naphthalenol benzoate;* betanaphthyl benzoate; benzonaphthol; benzoylnaphthol; Lintrin; Haertolan. $C_{17}H_{12}O_2$; mol wt 248.28. C 82.24%, H 4.87%, O 12.89%. $C_6H_5COOC_{10}H_7$.

White, crystalline powder; darkens with age. mp 107-110°. Almost insol in water. Freely sol in hot alcohol, chloroform, glycerol, oils; slightly sol in ether.

USE: Hardening agent for paraffin.

THERAP CAT: Antiseptic (intestinal).

6496. *N-(1-Naphthyl)ethylenediamine. N-1-Naphthalenyl-1,2-ethanediamine;* 1-amino-2-(α-naphthylamino)ethane. $C_{12}H_{14}N_2$; mol wt 186.26. C 77.38%, H 7.58%, N 15.04%. Prepn from α-naphthylamine and bromoethylphthalimide: Bratton, Marshall, *J. Biol. Chem.* **128**, 537 (1939).

Straw-yellow, viscous liquid; bp$_9$ 204°; bp$_{760}$ ~320° (dec); n_D^{25} 1.6648; d_4^{25} 1.114. The soly in water is about 0.2 g in 100 ml at 25°; more sol in hot water. pH of satd aq soln: 10.5. Readily sol in common organic solvents except petr ether.

Dihydrochloride, $C_{12}H_{14}N_2.2HCl$, long hexagonal prisms; mp 188-190°. Readily sol in 95% alcohol, dil hydrochloric acid, hot water; slightly sol in cold water, acetone, abs alc.

USE: Dihydrochloride in determination of sulfanilamide in body fluids; also in determination of potassium, nitrites, and sulfates.

6497. 1-Naphthylisocyanate. *1-Isocyanatonaphthalene.* $C_{11}H_7NO$; mol wt 169.18. C 78.09%, H 4.17%, N 8.28%, O 9.46%. $C_{10}H_7N=CO$.

Colorless liquid; almost odorless at ordinary temp; vapors have a pungent odor characteristic of isocyanates (*e.g.,* phenylisocyanate). d 1.181. bp 270°. Sol in chloroform, ether, petr ether, alcohol. *Keep well closed and protected from light.*

6498. 1-Naphthylisothiocyanate. *1-Isothiocyanato-naphthalene;* ANIT. $C_{11}H_7NS$; mol wt 185.25. C 71.32%, H 3.81%, N 7.56%, S 17.31%. $C_{10}H_7N=C=S$. Prepn: Cymerman-Craig *et al., Org. Syn.* **coll. vol. IV**, 700 (1963); Jochims, *Ber.* **101**, 1746 (1968). Has been used with pyrethrum as insecticide. Toxicity studies: Schwartz, Warren, *U.S. Pub. Health Repts.* **54**, 1426 (1939); Ambrose, Miller, *Fed. Proc.* **2**, 74 (1943); Becker, Plaa, *Toxicol. Appl. Pharmacol.* **7**, 804 (1965); Capizzo, Roberts, *ibid.* **17**, 262 (1970).

White needles; practically odorless and tasteless, mp 58°. Insol in water. Freely sol in ether, benzene, hot alcohol, acetone, carbon tetrachloride, olive oil, petr ether. LD_{50} orally in mice: 245 mg/kg (Becker, Plaa).

Caution: Hepatotoxic; may cause dermatitis.

6499. 2-Naphthyl Lactate. *2-Hydroxypropanoic acid 2-naphthalenyl ester;* lactic acid β-naphthyl ester; Lactol; Lactonaphthol. $C_{13}H_{12}O_3$; mol wt 216.24. C 72.21%, H 5.59%, O 22.20%. $CH_3CHOHCOOC_{10}H_7$. Prepd by heating equimolar amounts of β-naphthol sodium and sodium lactate to 125-130° in the presence of phosphorus oxychloride: P. Lebeau, *Traité de Pharmacie Chimique* **vol. II** (Masson, Paris, 1955/56) p 964.

Crystals from alcohol. Moderately sol in alcohol. Insol in water, ether.

THERAP CAT: Antiseptic (intestinal).

6500. 2-(2-Naphthyloxy)ethanol. *2-(2-Naphthalenyl-oxy)ethanol;* β-naphthoxyethanol; betanaphthoxyethanol; β-hydroxyethyl 2-naphthyl ether; 2-(β-hydroxyethoxy)naphthalene; ethylene glycol mono-2-naphthyl ether; Anavenol. $C_{12}H_{12}O_2$; mol wt 188.23. C 76.57%, H 6.43%, O 17.00%. Prepd by the condensation of β-naphthol with ethylene chlorohydrin: Rindfusz *et al., J. Am. Chem. Soc.* **42**, 157, 164, 165 (1920); Kirner, Richter, *ibid.* **51**, 3409 (1929); by heating β-naphthol with ethylene oxide and sodium ethoxide in alcohol: Boyd, Marle, *J. Chem. Soc.* **105**, 2117 (1914).

Crystals from benzene + petr ether, mp 76.7°. Insoluble in water. One gram dissolves in 4 g of 95% alcohol, in 2 g acetone; also sol in ether, chloroform. uv max (0.004% in chloroform): 273, 328 nm (E ~1.00, ~0.395).

THERAP CAT (VET): As an adjunct to anesthesia, sedative.

6501. 1-Naphthyl Salicylate. *2-Hydroxybenzoic acid 1-naphthalenyl ester;* α-naphthyl salicylate; α-naphthol salicylate; Alphol. $C_{17}H_{12}O_3$; mol wt 264.28. C 77.26%, H 4.58%, O 18.16%.

Crystalline powder. mp 83°. Insol in water. Freely sol in alcohol, ether, oils.

THERAP CAT: Anti-infective; anti-inflammatory.

6502. 2-Naphthyl Salicylate. *2-Hydroxybenzoic acid 2-naphthalenyl ester;* β-naphthol salicylate; Betol; Naphthalol; Naphthosalol; Salinaphthol. $C_{17}H_{12}O_3$; mol wt 264.28. C 77.26%, H 4.58%, O 18.16%. White, odorless, tasteless, crystalline powder. mp 95°. Insol in water or glycerol. Sparingly sol in cold alcohol; freely sol in boiling alcohol, in benzene, ether.

THERAP CAT: Antiseptic.

6503. Napropamide. *N,N-Diethyl-2-(1-naphthalenyl-oxy)propanamide; N,N-diethyl-2-(1-naphthyloxy)propion-amide;* 2-(α-naphthoxy)-*N,N*-diethylpropionamide; R-7465; Devrinol. $C_{17}H_{21}NO_2$; mol wt 271.36. C 75.25%, H 7.80%, N 5.16%, O 11.79%. Prepn: H. Filles *et al.,* U.S. pat. **3,-480,671** (1969 to Stauffer). Characteristics in the soil: C.

H. Wu *et al., Weed Sci.* **23**, 54 (1975). Toxicity study: T. Kawada *et al., Niigata Igakkai Zasshi* **87**, 289 (1973), *C.A.* **80**, 56257h (1974).

Light brown solid from *n*-pentane, mp 63-64° (tech, 69.5°). Solubility in water at 20°: 70 ppm. LD$_{50}$ in mice (g/kg): > 5 orally, > 1 i.p., > 1 s.c. (Kawada).
USE: Herbicide.

6504. Naproxen. *(S)-6-Methoxy-α-methyl-2-naphthaleneacetic acid; d-2-(6-methoxy-2-naphthyl)propionic acid;* MNPA; RS-3540; Bonyl; Diocodal; Dysmenalgit; Equiproxen; Floginax; Laraflex; Laser; Naixan; Napren; Naprium; Naprius; Naprosyn; Naprosyne; Naprux; Naxen; Nycopren; Pranoxen; Prexan; Proxen; Reuxen; Veradol; Xenar. C$_{14}$H$_{14}$O$_3$; mol wt 230.26. C 73.03%, H 6.13%, O 20.84%. Nonsteroidal anti-inflammatory. Prepn: J. H. Fried, I. T. Harrison, **S. Afr.** pat. **67 07,597;** *eidem,* **U.S.** pat. **3,904,682;** *eidem,* **U.S.** pat. **4,009,197** (1968, 1975, 1977 all to Syntex); I. T. Harrison *et al., J. Med. Chem.* **13,** 203 (1970). Pharmacology: Roszkowski *et al., J. Pharmacol. Exp. Ther.* **179,** 114 (1971). Activity may be due to the ability to inhibit prostaglandin biosynthesis. Mode of action studies: Tomlinson *et al., Biochem. Biophys. Res. Commun.* **46,** 552 (1972). Metabolism: Runkel *et al., J. Pharm. Sci.* **61,** 703 (1972). HPLC determn in plasma and serum: P. J. Streete, *J. Chromatog.* **495,** 179 (1989). Stereoselective synthesis: K. T. Wan, M. E. Davis, *Nature* **370,** 449 (1994). Clinical studies: Katona *et al., Clin. Trials J.* **8,** 3 (1972); Runkel, *Chem. Pharm. Bull.* **20,** 1457 (1972). Review: *Arzneimittel-Forsch.* **25,** 278-332 (1975). Review of pharmacology and therapeutic efficacy: R. N. Brogden *et al., Drugs* **18,** 241-277 (1979). Comprehensive description: F. J. Al-Shammary *et al., Anal. Profiles Drug Subs. Excip.* **21,** 345-373 (1992).

Crystals from acetone-hexane, mp 152-154°. [α]$_D$ +66° (in chloroform). Sol in 25 parts ethanol (96%), 20 parts methanol, 15 parts chloroform, 40 parts ether. Practically insol in water. LD$_{50}$ in mice (mg/kg): 435 i.v.; 1234 orally; in rats (mg/kg): 575 i.p.; 534 orally (Roszkowski).
Piperazine salt, (C$_{14}$H$_{14}$O$_3$)$_2$·C$_4$H$_{10}$N$_2$, *piproxen, Numidan.*
Sodium salt, C$_{14}$H$_{13}$NaO$_3$, *RS-3650, Aleve, Anaprox, Apranax, Axer Alfa, Flanax, Gynestrel, Miranax, Primeral, Synflex.* Crystals from acetone, mp 244-246°. [α]$_D$ −11° (in methanol).
THERAP CAT: Anti-inflammatory; analgesic; antipyretic.
THERAP CAT (VET): Anti-inflammatory.

6505. Naptalam. *2-[(1-Naphthalenylamino)carbonyl] benzoic acid; N-1-naphthylphthalamic acid;* α-naphthylphthalamic acid. C$_{18}$H$_{13}$NO$_3$; mol wt 291.31. C 74.22%, H 4.50%, N 4.81%, O 16.48%. Selective pre-emergence herbicide. Prepn: A. E. Smith, O. L. Hoffmann, **U.S.** pat. **2,-556,665** (1951 to U.S. Rubber). Activity: O. L. Hoffmann, A. E. Smith, *Science* **109,** 588 (1949). Mobility in soil: A. E. Smith *et al., J. Agr. Food Chem.* **5,** 748 (1957).

Crystals from ethanol, d$_4^{20}$ 1.40. mp 203° (technical grade, mp 175-180°). Soly: < 0.02 g/100 ml of water. Sol in alkaline solns, but dec above pH 9.5. Slightly sol in ethanol, acetone, benzene. Hydrolyzed by strong acids and bases.
Sodium salt, C$_{18}$H$_{12}$NNaO$_3$, *ACP-332, NPA-3, Alanap.*
USE: Herbicide; as analytical reagent for thorium and zirconium.

6506. Narasin. *(4S)-4-Methylsalinomycin;* narasin A; Compd 79891; Antibiotic A-28086 factor A; C-7819B; Monteban. C$_{43}$H$_{72}$O$_{11}$; mol wt 765.04. C 67.51%, H 9.49%, O 23.00%. Main component of a polyether antibiotic complex produced by *Streptomyces aureofaciens* NRRL 5758 & NRRL 8092. Production: D. H. Berg *et al.,* **Ger.** pat. **2,-525,095** corresp to **U.S.** pat. **4,038,384** (1975, 1977 to Lilly); L. D. Boeck *et al., Dev. Ind. Microbiol.* **18,** 471 (1976). Isoln and characterization: D. H. Berg, R. L. Hamill, *J. Antibiot.* **31,** 1 (1978). Biosynthetic studies using ^{13}C-NMR study: D. E. Dorman *et al., Helv. Chim. Acta* **59,** 2625 (1976). Structure: J. L. Occolowitz *et al., Biomed. Mass. Spectrom.* **3,** 272 (1976); H. Seto *et al., J. Antibiot.* **30,** 530 (1977). Anticoccidial activity: M. D. Ruff *et al., Poultry Sci.* **59,** 2008 (1980). Total synthesis: Y. Kishi *et al., Front. Chem., Plenary Keynote Lect. IUPAC Congr., 28th* **1981,** K. J. Laidler, Ed. (Pergamon, Oxford, 1982) pp 287-304. HPLC determn in animal feeds: M. R. LaPointe, H. Cohen, *J. Assoc. Offic. Anal. Chem.* **71,** 480 (1988).

Crystals from acetone-water, mp 98-100°; resolidif and remelts at 198-200°. uv max (ethanol): 285 nm (ε 58). [α]$_D^{25}$ −54° (c = 0.2 in methanol). pKa 7.9 (80% aq DMF). Sol in alcohols, acetone, DMF, DMSO, benzene, chloroform, ethyl acetate. Insol in water. LD$_{50}$ i.p. in mice: 7.15 mg/kg (Berg).
THERAP CAT (VET): Coccidiostat; growth stimulant.

6507. Narbomycin. *12-Deoxypicromycin.* C$_{28}$H$_{47}$NO$_7$; mol wt 509.68. C 65.98%, H 9.29%, N 2.75%, O 21.97%. Antibiotic substance produced by *Streptomyces narbonensis* from soil near Cannes, France: Corbaz *et al., Helv. Chim. Acta* **38,** 935 (1955). Structure: Prelog *et al., ibid.* **45,** 4 (1962). Stereochemical studies: Rickards, Smith, *Tetrahedron Letters* **1970,** 1025; H. Ogura *et al., J. Am. Chem. Soc.* **97,** 1930 (1975); *eidem, Tetrahedron* **37,** Suppl. 1, 165 (1981). Isoln and structure of the aglycone *narbonolide:* Hori *et al., Chem. Commun.* **1971,** 304. Synthesis of narbonolide: T. Kaiho *et al., J. Org. Chem.* **47,** 1612 (1982).

Crystals from ether + petr ether, mp 113.5-115°. [α]$_D^{20}$ +68.5° (c = 1.35 in chloroform). uv max (abs ethanol):

225, 286 nm (log ϵ 4.06, 2.23). LD_{50} s.c. in mice: 500 mg/kg (Corbaz).

6508. Narceine. *6-[[6-[2-(Dimethylamino)ethyl]-4-methoxy-1,3-benzodioxol-5-yl]acetyl]-2,3-dimethoxybenzoic acid; 6-[[6-[2-(dimethylamino)ethyl]-2-methoxy-3,4-(methylenedioxy)phenyl]acetyl]-o-veratric acid.* $C_{23}H_{27}NO_8$; mol wt 445.47. C 62.01%, H 6.11%, N 3.14%, O 28.73%. Occurs in opium to the extent of 0.1-0.5%. The separation from morphine mother liquors is tedious: Merck, *Chem. Ztg.* **13**, 525 (1889). Prepn from narcotine or gnoscopine: Roser, *Ann.* **247**, 167 (1888); Frankforter, Keller, *Ann. Chem. J.* **22**, 61 (1899); Frerichs, *Arch. Pharm.* **241**, 259 (1903); Hope, Robinson, *J. Chem. Soc.* **105**, 2100 (1914). Structure: Freund, Frankforter, *Ann.* **277**, 20 (1893); Freund, Michaels, *Ann.* **286**, 248 (1895); Freund, *Ber.* **40**, 194 (1907); Freund, Oppenheim, *Ber.* **42**, 1084 (1909); Addinall, Major, *J. Am. Chem. Soc.* **55**, 1202, 2153 (1933).

The anhydr material is very hygroscopic, mp 138°. uv max (ethanol): 270 nm (log ϵ 3.98). Usually the alkaloid is obtained as the trihydrate. Clusters of silky, prismatic needles from water, mp 176°. pKb at 20° = 10.7, Kb = 2×10^{-11}; pKa = 9.3, Ka = 5×10^{-10}. pH of satd soln = 5.8. One gram dissolves in 770 ml water, 220 ml boiling water. Moderately sol in hot alcohol, nearly insol in benzene, chloroform, ether, petr ether; sol in alkali hydroxide solns forming salts, also in dil mineral acids.

Ethylnarceine hydrochloride, $C_{25}H_{32}ClNO_8$, *Narcyl.* Plates from water, mp 208-210°. Slightly sol in cold water, sol in hot water, alcohol, chloroform; insol in ether.

THERAP CAT: Ethylnarceine hydrochloride as analgesic (narcotic); antitussive.

6509. Narcobarbital. *5-(2-Bromo-2-propenyl)-1-methyl-5-(1-methylethyl)-2,4,6(1H,3H,5H)-pyrimidinetrione; 5-(2-bromoallyl)-5-isopropyl-1-methylbarbituric acid; N-methyl-β-bromoallylisopropylbarbituric acid; 5-isopropyl-5-(β-bromoallyl)-N-methylbarbituric acid; enibomal.* $C_{11}H_{15}BrN_2O_3$; mol wt 303.16. C 43.58%, H 4.99%, Br 26.36%, N 9.24%, O 15.83%. Prepd by heating 5-isopropyl-1-methylbarbituric acid with 2,3-dibromo-1-propene in alkaline soln: Boedecker, Gruber, **Ger.** pat. **613,403** (1937 to Riedel de Haen). *See also* **Ger.** pat. **627,380**; **U.S.** pat. **2,080,071.**

Crystals from dilute ethanol, mp 115°. Bitter taste. Very sparingly sol in water. Sol in methanol, ethanol, pyridine, alkaline aq solns.

Sodium salt, $C_{11}H_{14}BrN_2NaO_3$, *Eunarcon, Narcotal.* Alkaline, water-soluble.

Caution: May be habit forming. This is a controlled substance (depressant) listed in the U.S. Code of Federal Regulations, Title 21 Parts 329.1 and 1308.13 (1995).

THERAP CAT: Sedative, hypnotic; anticonvulsant.

6510. Narcotoline. *[S-(R*,S*)]-6,7-Dimethoxy-3-(5,6,-7,8-tetrahydro-4-hydroxy-6-methyl-1,3-dioxolo[4,5-g]isoquinolin-5-yl)-1(3H)-isobenzofuranone; desmethylnarcotine.*

$C_{21}H_{21}NO_7$; mol wt 399.40. C 63.15%, H 5.30%, N 3.51%, O 28.04%. Found in the shell of ripe poppyseed capsules: Wrede, *Arch. Pharmacol.* **184**, 331 (1937); Baumgarten, Christ, *Pharmazie* **5**, 80 (1950); Pfeifer, Weiss, *ibid.* **10**, 658 (1955); from opium prepns: Pfeifer, *Arch. Pharm.* **290**, 209 (1957). Preliminary stereochemical studies: Battersby, Spenser, *J. Chem. Soc.* **1965**, 1087. Revised stereochemistry: Blaha *et al.*, *Coll. Czech. Chem. Commun.* **29**, 2328 (1964); Snatzke *et al.*, *Tetrahedron* **25**, 5059 (1969).

Rectangular rods from dil methanol, mp 202°. $[\alpha]_D^{20}$ −189° (0.1 g in 25 ml chloroform, 20 cm tube). $[\alpha]_D^{20}$ +5.8° (0.065 g in 5 ml 0.1N HCl in 20 cm tube). Very sparingly sol in water. Moderately sol in warm alc and ether. Freely sol in chloroform, in dil acids and in dil aq solns of KOH and NaOH. Sparingly sol in dil aq solns of Na_2CO_3.

6511. Naringenin. *2,3-Dihydro-5,7-dihydroxy-2-(4-hydroxyphenyl)-4H-1-benzopyran-4-one; 4',5,7-trihydroxyflavanone; naringetol; salipurpol; pelargidanon 1602.* $C_{15}H_{12}O_5$; mol wt 272.26. C 66.17%, H 4.44%, O 29.38%. The aglucon of naringin. Prepn by hydrolysis of naringin: Asahina, Inubuse, *Ber.* **61**, 1514 (1928); Haley, Bassin, *J. Am. Pharm. Assoc.* **40**, 111 (1951). From kino of *Eucalyptus maculata* Hook, *Myrtaceae:* Gell *et al.*, *Aust. J. Chem.* **11**, 372 (1958). Synthesis: Rosenmund, Rosenmund, *Ber.* **61**, 2608 (1928); Zemplén, Bognár, *Ber.* **75**, 648 (1942). Absolute configuration: Gaffield, Waiss, *Chem. Commun.* **1968**, 29.

Needles from dilute alc, mp 251°. uv max: 226, 292 nm. Sol in alcohol, ether, benzene; almost insol in water.

Triacetylnaringenin, $C_{15}H_9(CH_3CO)_3O_5$, crystals, mp 55°.

6512. Naringin. *7-[[2-O-(6-Deoxy-α-L-mannopyranosyl)-β-D-glucopyranosyl]oxy]-2,3-dihydro-5-hydroxy-2-(4-hydroxyphenyl)-4H-1-benzopyran-4-one; 4',5,7-trihydroxyflavanone 7-rhamnoglucoside; naringenin-7-rhamnoglucoside; aurantiin.* $C_{27}H_{32}O_{14}$; mol wt 580.54. C 55.86%, H 5.56%, O 38.58%. In the flowers of grapefruit trees *(Citrus paradisi* Macfad, *Rutaceae),* also in fruit and rind. Most abundant in immature fruit; main bitter component of grapefruit juice. Extraction from grapefruit peel: Zoller, *Ind. Eng. Chem.* **10**, 364 (1918); Pulley, von Loesecke, *J. Am. Chem. Soc.* **61**, 175 (1939); **U.S.** pats. **2,421,062; 2,421,063** (both 1947 to California Fruit Growers Exchange). Structure: Asahina, Inubuse, *C.A.* **23**, 3475 (1929); *cf.* also *Ber.* **61**, 1514 (1928); Horowitz, Gentili, *Tetrahedron* **19**, 773 (1963). Solubility data: G. N. Pulley, *Ind. Eng. Chem., Anal. Ed.* **8**, 360 (1936). *Review:* Kesterson, Hendrickson, *Naringin, A Bitter Principle of Grapefruit,* Univ. Florida Agric. Expt. Station, Bulletin no. 511 (Jan. 1953).

When crystallized from water, it contains 6 to 8 mols H_2O. mp $\sim83°$. After drying at 110° to constant weight, it contains 2 mols H_2O, mp 171°. Bitter taste (1:10,000 H_2O can be tasted). $[\alpha]_D^{19}$ — 82° (alcohol). One gram dissolves in 1000 ml water at 40°. At 75° one gram dissolves in 10 ml water. Sol in acetone, alcohol, warm acetic acid.

6513. Natamycin. *Pimaricin; antibiotic A 5283; tennecetin; CL-12625; Mycophyt; Myprozine; Natacyn; Pimafucin; Synogil.* $C_{33}H_{47}NO_{13}$; mol wt 665.74. C 59.54%, H 7.12%, N 2.10%, O 31.24%. Polyene antifungal antibiotic produced by *Streptomyces natalensis* from soil near Pietermaritzburg, South Africa and by *S. chattanoogensis.* Isoln: Struyk *et al., Antibiot. Ann.* **1957-1958,** 878. Prodn: **Brit.** pat. **844,289** (1960 to Koninklijke Nederlandsche Gist en Spiritus-Fabriek); **Brit.** pat. **846,933** (1960 to Am. Cyanamid). Identity with tennecetin: Divekar *et al., Antibiot. & Chemother.* **11,** 377 (1961). Structure: Golding *et al., Tetrahedron Letters* **1966,** 3551; Meyer, *Chem. Commun.* **1968,** 470. Configuration: Gaudiano *et al., Chim. Ind. (Milan)* **48,** 1327 (1966). Revised structure: R. C. Pandey, K. L. Rinehart, *J. Antibiot.* **29,** 1035 (1976). Soly data: Marsh, Weiss, *J. Assoc. Offic. Anal. Chem.* **50,** 457 (1967). Toxicology: G. J. Levinskas *et al., Toxicol. Appl. Pharmacol.* **8,** 97 (1966). Comprehensive description: H. Brik in *Analytical Profiles of Drug Substances* vol. 10, K. Florey, Ed. (Academic Press, New York, 1981) pp 513-561.

Crystals from methanol + water, darkens $\sim200°$, dec 280-300°. $[\alpha]_D^{20}$ +278° (c = 1 in CH_3COOH). Sensitive to light, but otherwise very stable in the dry state. uv max (methanol + 0.1% CH_3COOH): 220, 280, 290, 303, 318 nm (ϵ 21300, 26630, 52930, 83220, 76230). Practically insol in higher alcohols, ether, esters, aromatic or aliphatic hydrocarbons, chlorinated hydrocarbons, ketones, dioxane, cyclohexanol, oils. LD_{50} orally in male, female rats (g/kg): 2.73, 4.67 (Levinskas).

THERAP CAT: Antifungal (topical).

6514. Natural Gas. American natural gas consists of about 85% methane, 9% ethane, 3% propane, 2% nitrogen, 1% butane. Northern Texas gas contains more nitrogen, also sufficient helium to warrant commercial extraction.

Caution: Narcotic in high concns. Incomplete combustion can result in production of carbon monoxide.

USE: Fuel gas, in the manuf of hydrogen, methane, ammonia.

6515. Nauheim Salts (Artificial). Bad Nauheim salts (artificial). The composition used in European pharmaceutical dispensing practice is as follows: three lbs $CaCl_2$·$2H_2O$, two lbs NaCl, 1½ lbs Na_2CO_3·$10H_2O$ and ½ lb

$NaHCO_3$. This mixture is dissolved in a bath of 50 gallons of water. Then one lb of $NaHSO_4$·H_2O is slowly added.

6516. Nealbarbital. *5-(2,2-Dimethylpropyl)-5-(2-propenyl)-2,4,6(1H,3H,5H)-pyrimidinetrione; 5-allyl-5-neopentylbarbituric acid;* nealbarbitone; neallymal; Nevental; Censedal. $C_{12}H_{18}N_2O_3$; mol wt 238.29. C 60.49%, H 7.61%, N 11.76%, O 20.14%. Synthesis: Brandstrom, *Acta Chem. Scand.* **13,** 613, 615, 619 (1959); **Brit.** pat. **797,017** and U.S. pat. **2,899,435** (1958, 1959, both to Pharmacia). Metabolism: J. N. T. Gilbert *et al., J. Pharm. Pharmacol.* **26,** 119 (1974); J. N. T. Gilbert, J. W. Powell, *Eur. J. Drug Metab. Pharmacokinet.* **3,** 195 (1978).

Slightly bitter crystals, mp 155-157°. Practically insol in water, petr ether. Moderately sol in chloroform; freely sol in alcohol, ether, acetone. Also sol in aq alkaline solns.

Note: This is a controlled substance listed in the U.S. Code of Federal Regulations, Title 21 Part 1308.13 (1995).

THERAP CAT: Sedative, hypnotic.

6517. Neamine. *2-Deoxy-4-O-(2,6-diamino-2,6-dideoxy-α-D-glucopyranosyl)-D-streptamine;* neomycin A. $C_{12}H_{26}N_4O_6$; mol wt 322.36. C 44.71%, H 8.13%, N 17.38%, O 29.78%. Component of the antibiotic complex neomycin, *q.v.* Isoln: Dutcher *et al., J. Am. Chem. Soc.* **73,** 1384 (1951). Identity of neomycin A and neamine: Leach, Teeters, *ibid.* **74,** 3187 (1952). Prepn from neomycin B: Peck *et al., U.S. pat.* **2,691,675** (1954 to Merck & Co.). Structure: Carter *et al., J. Am. Chem. Soc.* **83,** 3723 (1961); Hichens, Rinehart, *ibid.* **85,** 1547 (1963). Synthesis: Umezawa *et al., J. Antibiot.* **20A,** 53 (1967); H. Kohno *et al., Agr. Biol. Chem.* **39,** 1091 (1975); A. Harayama *et al., Bull. Chem. Soc. Japan* **52,** 3626 (1979).

Crystals from water or aq alc, dec 225-226°, $[\alpha]_D^{25}$+112.8° (c = 1). IR spectrum: Leach, Teeters, *J. Am. Chem. Soc.* **73,** 2794 (1951).

Hydrochloride, $C_{12}H_{26}N_4O_6$·4HCl, amorphous, dec 250-260°, $[\alpha]_D^{25}$+83° (c = 1).

N-Acetyl deriv, $C_{12}H_{26}N_4O_6$·$(CH_3CO)_4$, crystals from methanol, mp 334-336°. $[\alpha]_D^{25}$ +87° (c = 1).

6518. Neatsfoot Oil. Fixed oil from feet of neat (bovine) cattle.

Pale yellow liquid; peculiar odor. d 0.915. Solidif 0° to —10°. n_D^{20} 1.4695-1.4708. Sapon no. 192-203. Iodine no. 44-73.2.

USE: Waterproofing and softening leather; lubricant; oiling wool, etc.

6519. Nebivolol. *α,α'-[Iminobis(methylene)]bis[6-fluoro-3,4-dihydro-2H-1-benzopyran-2-methanol]; α,α'-(iminodimethylene)bis[6-fluoro-2-chromanmethanol];* dl-nebivolol; narbivolol; R-65824; Nebilet. $C_{22}H_{25}F_2NO_4$; mol wt 405.44. C 65.17%, H 6.22%, F 9.37%, N 3.45%, O 15.78%. β_1-Adrenergic blocker; prepd as racemic mixture of RSSS and SRRR enantiomers. Prepn: G. R. E. Van Lommen *et al.,* **Eur. pat. Appl. 145,067;** *eidem,* **U.S. pat. 4,654,362** (1985, 1987 both to Janssen); *eidem, J. Pharm. Belg.* **45,** 355 (1990). Pharmacology of racemate: A. Van de Water *et al., J. Cardiovasc. Pharmacol.* **11,** 552 (1988); and of enantiomers: A. Van de Water *et al., Eur. J. Pharmacol.* **156,** 95

(1988). Receptor binding study: P. J. Pauwels *et al., Mol. Pharmacol.* **34**, 843 (1988). HPLC fluorescence determn in plasma: R. Woestenborghs *et al., Methodol. Surv. Biochem. Anal.* **18**, 215 (1988). Clinical pharmacology: J. De Crée *et al., Angiology* **41**, 95 (1990).

THERAP CAT: Antihypertensive.

6520. Nebracetam. (±)-*4-(Aminomethyl)-1-(phenyl-methyl)-2-pyrrolidinone;* (±)-4-(aminomethyl)-1-benzyl-2-pyrrolidinone; WEB-1881. $C_{12}H_{16}N_2O$; mol wt 204.27. C 70.56%, H 7.89%, N 13.71%, O 7.83%. Cognition enhancer with cholinergic and cytoprotectant activity. Prepn: K. H. Weber *et al.,* **Ger. pat.** **3,336,024**; *eidem,* **U.S. pat.** **4,833,140** (1985, 1989, both to Boehringer, Ing.). Pharmacology: F. J. Kuhn *et al., Arch. Int. Pharmacodyn. Ther.* **292**, 13 (1988). Toxicology: T. Yabe *et al., Oyo Yakuri* **44**, 417 (1992), *C.A.* **117**, 245471s (1992). Clinical effect on cognitive processes: T. F. Münte *et al., Neuropsychobiology* **19**, 158 (1988). Clinical evaluation in Alzheimer's disease: K. Urakami *et al., Clin. Neuropharmacol.* **16**, 347 (1993).

Fumarate, $(C_{12}H_{16}N_2O)_2 \cdot C_4H_4O_4$, *WEB-1881FU.* Crystals, mp 192-194°. LD_{50} in male, female rats (mg/kg): 257, 257 i.v.; 2884, 3622 orally (Yabe).

THERAP CAT: Nootropic.

6521. Nebularine. *9-β-D-Ribofuranosyl-9H-purine.* $C_{10}H_{12}N_4O_4$; mol wt 252.23. C 47.62%, H 4.80%, N 22.21%, O 25.37%. Isoln from the mushroom *Clitocybe nebularis* (Batsch.) Quel., *Agaricaceae:* Ehrenberg *et al., Svensk Kem. Tidskr.* **58**, 269 (1946); Löfgren *et al., Acta Chem. Scand.* **8**, 670 (1954); from a streptomycete: Isono, Suzuki, *J. Antibiot.* **13A**, 270 (1960). *In vitro* toxicity towards sarcoma 180 cells, mouse embryonic fibroblasts and epithelial cells: J. J. Biescle *et al., Cancer* **8**, 87 (1955). Synthesis: Brown, Weliky, *J. Biol. Chem.* **204**, 1019 (1953); Fox *et al., J. Am. Chem. Soc.* **80**, 1669 (1958); Hashizume, Iwamura, *Tetrahedron Letters* **1966**, 643; *eidem, J. Org. Chem.* **33**, 1796 (1968). Crystal structure: T. Takeda, *Acta Crystallogr.* **30B**, 825 (1974). Alternate syntheses: V. Nair, S. G. Richardson, *Tetrahedron Letters* **1979**, 1181; P. K. Gupta, D. S. Bhakuni, *Indian J. Chem.* **20B**, 534 (1981). Toxicity studies: Truant, D'Amato, *Fed. Proc.* **14**, 391 (1955).

Small rhombohedra from ethyl methyl ketone + methanol, mp 181-182°; needles from methanol, mp 182-183°. $[\alpha]_D^{25}$ −48.6° (H_2O). uv max (0.1N HCl): 262 nm ($E_{1cm}^{1\%}$ 232); (0.1N NaOH): 263 nm ($E_{1cm}^{1\%}$ 361). Considerably sol in water (about 10%). Slightly sol in cold ethanol. Very slightly sol in acetone, ether, chloroform. Aq solns may be sterilized by boiling without decompn. LD_{50} in rats, guinea pigs (mg/kg): 220, 15 s.c. (Truant, D'Amato).

6522. Nebulin. Family of giant, actin-binding proteins, varying in size from 700-900 kDa, found in the sarcomeres of vertebrate skeletal muscles; comprises ~3% of the total myofibrillar protein. Composed of small repeats of ~35 amino acid residues arranged in a 7-fold super-repeating pattern. Integral component of the cytoskeletal lattice of muscle cells (*see also* Titin). Forms inextensible filaments that are attached to the Z-line and span the length of the thin filaments. Thought to act as a "molecular ruler" that stabilizes the actin polymer and regulates thin filament length. Name derived from its histological localization in the N_2-line of the sarcomere, a nebulous striation within the I-band. Identification: K. Wang, C. L. Williamson, *Proc. Nat. Acad. Sci. USA* **77**, 3254 (1980). Distribution and organization in cytoskeletal matrix of striated muscle: K. Wang, *Adv. Exp. Med. Biol.* **170**, 285-305 (1984). Physiological role: M.-J. G. Chen *et al., J. Biol. Chem.* **268**, 20327 (1993). Quantitative determn by gel electrophoresis: H. L. M. Granzier, K. Wang, *Electrophoresis* **14**, 56 (1993). Structural study: M. Pfuhl *et al., EMBO J.* **13**, 1782 (1994). Complete primary structure: S. Labeit, B. Kolmerer, *J. Mol. Biol.* **248**, 308 (1995). *Review:* J. Trinick, *Trends Biochem. Sci.* **19**, 405-409 (1994).

6523. Neburon. *N-Butyl-N'-(3,4-dichlorophenyl)-N-methylurea;* 3-(3,4-dichlorophenyl)-1-methyl-1-*n*-butylurea; Kloben Neburon. $C_{12}H_{16}Cl_2N_2O$; mol wt 275.18. C 52.38%, H 5.86%, Cl 25.77%, N 10.18%, O 5.81%. Selective pre-emergence herbicide. Prepn: Todd, U.S. pats. **2,655,-444-7** (all 1953 to du Pont). Toxicity study: G. W. Bailey, J. L. White, *Residue Rev.* **10**, 97 (1965).

Crystals from dioxane + water, mp 101.5-103°. Soly in water at 24°: 48 ppm; sparingly sol in hydrocarbon solvents. Stable toward oxidation and water at ordinary temperatures. LD_{50} orally in rats: >11000 mg/kg (Bailey, White).

USE: Herbicide.

6524. Nedocromil. *9-Ethyl-6,9-dihydro-4,6-dioxo-10-propyl-4H-pyrano[3,2-g]quinoline-2,8-dicarboxylic acid; 4,6-dioxo-1-ethyl-10-propyl-4H,6H-pyrano[3,2-g]quinoline-2,8-dicarboxylic acid;* FPL-59002. $C_{19}H_{17}NO_7$; mol wt 371.35. C 61.45%, H 4.61%, N 3.77%, O 30.16%. Anti-allergic compound with the ability to stabilize both mucosal and connective tissue mast cells. Prepn and pharmacology: H. Cairns, D. Cox, **Belg. pat.** **866,622**; *eidem,* **U.S. pat.** **4,474,-787** (1978, 1984 both to Fisons); H. Cairns *et al., J. Med. Chem.* **28**, 1832 (1985). Prevention of histamine release from mast cells, inhibition of bronchoconstriction in monkeys: R. P. Eady *et al., Brit. J. Pharmacol.* **85**, 323 (1985). Early clinical trial in bronchial asthma: S. Lal *et al., Thorax* **39**, 809 (1984).

Yellow powder, mp 298-300° (dec). Disodium salt, $C_{19}H_{15}NNa_2O_7$, *FPL-59002KP, Rapitil, Tilade, Tilarin.* Pale yellow powder.

THERAP CAT: Antiallergic; antiasthmatic.

6525. Neem. Nim. Subtropical shade tree, *Azadirachta indica,* A. Juss. *(Melia azadirachta L.), Meliceae,* native to the arid regions of India, Pakistan, parts of Africa. Most important constituent is the limonoid, azadirachtin, *q.v.* Known for centuries as being free of insects, disease, nematodes. All parts of the tree, but esp the seeds, are resistant. The bark, leaves and fruit have been used in traditional medicinal remedies: R. N. Chopra *et al., Indigenous Drugs of India* (U. N. Dhur & Sons, Calcutta, 2nd ed., 1958) pp 360-363. Antifeedant activity of neem extracts against locusts: J. H. Butterworth, E. D. Morgan, *Chem. Commun.* **1968**, 23; against Carolina grasshopper, walkingstick, field cricket: V. E. Adler, E. C. Uebel, *J. Environ. Sci. Health* **A19**, 393 (1984); against bengalgram podborer: P. C. Sundaru Babu, B. Rajasekaran, *Pesticides* **18**, 58 (1984). Series of articles on chemistry and activity of neem extracts: *Natural Pesticides from the Neem Tree,* Proc. 1st Int. Neem Conf., 1980, H. Schmutterer *et al.,* Eds. (German Agency for Technical Cooperation, Eschborn, 1981) 291 pp.

Note: Margosan-O, a neem formulation, has been approved by the EPA (1985) for limited use as a pesticide on non-food crops: *Chem. & Eng. News* **63**, 51 (May 27, 1985).
USE: Oil, crushed seeds as insect repellent, antifeedant.

6526. Neem Oil. Nim oil; margosa oil. A fixed oil expressed from seed-kernels of the Indian neem tree, *Azadirachta indica* A. Juss. (*Melia azadirachta* L.), Meliaceae. The seed-kernels yield about 10% of the oil. *Constit.* Mostly glycerides and limonoids including azadirachtin, nimbin, nimbiol, *q.q.v.,* salannin; about 2% bitters. Seed extracts as feeding deterrents for Japanese beetles: T. L. Ladd, Jr. *et al., J. Econ. Entomol.* **71**, 810 (1978). Antifeedant activity of oil: R. S. Saxena *et al.,* in *Natural Pesticides of the Neem Tree,* Proc. 1st Int. Neem Conf., 1980, H. Schmutterer *et al.,* Eds. (German Agency for Technical Cooperation, Eschborn, 1981) pp 171-204. Has been used as a traditional medicinal remedy. Cases of poisoning have been reported: D. Sinniah, G. Baskaran, *Lancet* **1**, 487 (1981).

Yellow oil. Odor of garlic. Bitter taste. d_4^{15} 0.925. Sol in ether, chloroform. Practically insol in alcohol, water.
USE: Agricultural insect repellent, antifeedant.

6527. Nefazodone. *2-[3-[4-(3-Chlorophenyl)-1-piperazinyl]propyl]-5-ethyl-2,4-dihydro-4-(2-phenoxyethyl)-3H-1,2,4-triazol-3-one;* 2-[3-[4-(3-chlorophenyl)-1-piperazinyl]propyl]-5-ethyl-4-(2-phenoxyethyl)-2*H*-1,2,4-triazol-3(4*H*)-one. $C_{25}H_{32}ClN_5O_2$; mol wt 470.01. C 63.89%, H 6.86%, Cl 7.54%, N 14.90%, O 6.81%. Selective serotonin 5-HT$_2$ receptor antagonist. Prepn: D. L. Temple, Jr., W. G. Lobeck, Jr., U.S. pat. **4,338,317** (1982 to Mead Johnson). Synthesis and x-ray crystal structure: G. D. Madding *et al, J. Heterocycl. Chem.* **22**, 1121 (1985). Pharmacology: A. S. Eison *et al., Psychopharmacol. Bull.* **26**, 311 (1990). HPLC determn in plasma: J. E. Franc *et al., J. Chromatog.* **570**, 129 (1991). Clinical trial in depression: M. F. D'Amico *et al., Psychopharmacol. Bull.* **26**, 147 (1990). *Review:* W. E. Heydorn, *Exp. Opin. Invest. Drugs* **4**, 131-137 (1995).

Crystals from 2-propanol/heptane, mp 83-84°.
Hydrochloride, $C_{25}H_{32}ClN_5O_2$·HCl, *BMY-13754, MJ-13754-1, Dutonin, Serzone.* Crystals from 2-propanol, with slow cooling yields a polymorph, mp 186.0-187.0°; with rapid cooling yields a polymorph, mp 181.0-182.0° (Madding). Also reported as crystals from ethanol, mp 175-177° (Temple).
THERAP CAT: Antidepressant.

6528. Nefiracetam. *N-(2,6-Dimethylphenyl)-2-oxo-1-pyrrolidineacetamide;* 2-(2-oxo-1-pyrrolidinyl)-*N*-(2,6-dimethylphenyl)acetamide; 2-oxo-1-pyrrolidineaceto-2',6'-xylidide; 2-oxo-1-pyrrolidinylacetic acid 2,6-dimethylanilide; DMMPA; DM-9384; DZL-221; Translon. $C_{14}H_{18}N_2O_2$; mol wt 246.31. C 68.27%, H 7.37%, N 11.37%, O 12.99%. Cyclic deriv of γ-aminobutyric acid. Prepn: H. Betzing *et al.,* Ger. pat. **2,924,011**; *eidem,* U.S. pat. **4,341,790** (1980, 1982 both to Nattermann). Cognition enhancing effects in

rats: T. Sakurai *et al., Japan. J. Pharmacol.* **50**, 47 (1989). HPLC determn in serum and urine: Y. Fujimaki *et al., J. Chromatog.* **575**, 261 (1992). Clinical pharmacokinetics: *eidem, Xenobiotica* **23**, 61 (1993). Series of articles on pharmacology and toxicology: *Arzneimittel-Forsch.* **44**, 193-259 (1994).

Crystals from water, mp 153°. LD$_{50}$ in mice (mg/kg): 421 i.v.; 1766 orally (Betzing, 1982).
THERAP CAT: Nootropic.

6529. Nefopam. *3,4,5,6-Tetrahydro-5-methyl-1-phenyl-1H-2,5-benzoxazocine;* 5-methyl-1-phenyl-1,3,4,6-tetrahydro-5*H*-benz[*f*]-2,5-oxazocine. $C_{17}H_{19}NO$; mol wt 253.34. C 80.60%, H 7.56%, N 5.53%, O 6.32%. A cyclized analog of orphenadrine and diphenhydramine, *q.q.v.;* representative of a new class of centrally acting skeletal muscle relaxants, the benzoxazocines. Prepn: Neth. pat. **6,606,390** (1966 to Rexall); M. W. Klohs *et al.,* U.S. pat. **3,830,803** (1974 to Riker). Pharmacology: Bassett *et al., Brit. J. Pharmacol.* **37**, 69 (1969); Klohs *et al., Arzneimittel-Forsch.* **22**, 132 (1972). Review of pharmacology and therapeutic efficacy: R. C. Heel *et al., Drugs* **19**, 249-267 (1980).

Hydrochloride, $C_{17}H_{19}NO$·HCl, *fenazoxine, R-738, Acupan, Ajan.* mp 238-242°. LD$_{50}$ in mice, rats (mg/kg): 119, 178 orally; 44.5, 28 i.v. (Baltes).
THERAP CAT: Analgesic; antidepressant.

6530. Negamycin. *3,6-Diamino-2,3,4,6-tetradeoxy-L-threo-hexonic acid 2-(carboxymethyl)-2-methylhydrazide; 3,6-diamino-5-hydroxyhexanoic acid 2-(carboxymethyl)-2-methylhydrazide;* [2-(3,6-diamino-5-hydroxy-1-oxohexyl)-1-methylhydrazino]acetic acid. $C_9H_{20}N_4O_4$; mol wt 248.28. C 43.54%, H 8.12%, N 22.57%, O 25.78%. Hydrazide antibiotic isolated from strains related to *Streptomyces purpeofuscus:* Hamada *et al., J. Antibiot.* **23**, 170 (1970). Prepn: *eidem,* Japan. pat. **71 28,835** (1971 to Microbiochem. Res. Found.), *C.A.* **75**, 139416g (1971). Structure and partial synthesis: Kondo *et al., J. Am. Chem. Soc.* **93**, 6305 (1971). Total synthesis of negamycin and its antipode: Shibahara *et al., ibid.* **94**, 4353 (1972). Synthesis of racemic negamycin: W. Streicher *et al., J. Antibiot.* **31**, 725 (1978); G. Pasquet *et al., Tetrahedron Letters* **21**, 931 (1980); A. Pierdet *et al., Tetrahedron* **36**, 1763 (1980). Stereocontrolled synthesis of (+)-negamycin: Y. F. Wang *et al., J. Am. Chem. Soc.* **104**, 6465 (1982); S. DeBernardo *et al., Tetrahedron Letters* **29**, 4077 (1988). Mechanism of action: Mizuno *et al., J. Antibiot.* **23**, 581 (1970); Y. Uehara *et al., Biochim. Biophys. Acta* **442**, 251 (1976).

Colorless powder, mp 110-120° (dec). $[\alpha]_D^{29}$ +2.5° (c = 2). Amphoteric compound. pKa values after treatment with HCl-methanol: 3.55, 8.10, 9.75. Sol in water. Practically insol in methanol, ethanol, butanol, ethyl acetate, chloroform and benzene. LD_{50} i.v. in mice: 400-500 mg/kg (Hameda).

THERAP CAT: Antibacterial.

6531. Negatol®. *Hydroxymethylbenzenesulfonic acid polymer with formaldehyde;* methylenebis(hydroxytoluenesulfonic acid) polymer; dihydroxydimethyldiphenylmethanedisulfonic acid polymer; Albocresil; Albothyl; Negatan. Colloidal product of relatively high mol wt prepd by reacting sulfonated *m*-cresol and formaldehyde, e.g. the product obtained by Thuau, **Can.** pat. **417,272** (1943 to Lilly), *C.A.* **38,** 1077 (1944). Shown to be a mixture of six components: Pasich, Stawinska, *Farm. Polska* **20,** 829 (1964), *C.A.* **62,** 12976h (1965).

Sol in water, forming colloidal solns. Very acid reaction. pH of 5% w/v soln about 1.0.

THERAP CAT: Antiseptic.

THERAP CAT (VET): Antiseptic.

6532. Nemadectin. *[6R,23S,25S,(E)]-5-O-Demethyl-28-deoxy-25-(1,3-dimethyl-1-butenyl)-6,28-epoxy-23-hydroxy-milbemycin B;* antibiotic S-541A; LL-F-28249α; F-28249α; CL-287088. $C_{36}H_{52}O_8$; mol wt 612.80. C 70.56%, H 8.55%, O 20.89%. Macrocyclic antibiotic produced by *Streptomyces cyaneogriseus* ssp. *noncyanogenus* and *Streptomyces thermoachaensis.* Structurally related to the milbemycins and the avermectins, *q.q.v.* Isoln: I. B. Wood *et al.,* **Eur.** pat. **Appl. 170,006;** I. B. Wood, J. A. Pankavich, **U.S.** pat. **4,869,901** (1986, 1989 both to Am. Cyanamid). Chemical and structural characterization: G. T. Carter *et al., Chem. Commun.* **1987,** 402; G. T. Carter *et al., J. Antibiot.* **41,** 519 (1988). NMR spectroscopy: S. Rajan, G. W. Stockton, *Magn. Reson. Chem.* **27,** 437 (1989). Antiparasitic activity in horses: E. T. Lyons *et al., Am. J. Vet. Res.* **50,** 970 (1989); in dogs: M. E. Doscher *et al., Vet. Parasitol.* **34,** 255 (1989); in sheep: J. A. Pankavich *et al, Vet. Rec.* **130,** 241, (1992).

White fluffy solid from *tert*-butanol. $[\alpha]_D^{26}$ +133 (c = 0.3 in acetone). uv max (methanol): 244 nm, (log ϵ 4.47). Readily sol in common organic solvents. Practically insol in water.

THERAP CAT (VET): Antiparasitic.

6533. Nemonapride. *cis-5-Chloro-2-methoxy-4-(methylamino)-N-[2-methyl-2-(phenylmethyl)-3-pyrrolidinyl]benzamide;* (±)-*cis-N*-(1-benzyl-2-methyl-3-pyrrolidinyl)-5-chloro-4-(methylamino)-*o*-anisamide; *cis-N*-(1-benzyl-2-methylpyrrolidin-3-yl)-5-chloro-2-methoxy-4-(methylamino)benzamide; emonapride; YM-09151-2; Emilace. $C_{21}H_{26}ClN_3O_2$; mol wt 387.91. C 65.02%, H 6.76%, Cl 9.14%, N 10.83%, O 8.25%. Selective dopamine D_2 receptor antagonist. Prepn: M. Takashima *et al.,* **Japan. Kokai 79 14,965;** *eidem,* **U.S.** pat. **4,210,660** (1979, 1980 both to Yamanouchi); S. Iwanami *et al., J. Med. Chem.* **24,** 1224 (1981). Pharmacology: M. Yamamoto *et al., Neuropharmacology* **21,** 945 (1982). D_2 dopamine binding studies: C. W. Grewe *et al., Eur. J. Pharmacol.* **81,** 149 (1982); J. F. Plantjé *et al., ibid.* **105,** 73 (1984). Use as a label for D_2 dopamine receptors: A. S. Unis *et al., Life Sci.* **47,** PL151 (1990).

Colorless crystals from isopropanol, mp 152-153° (Takashima, 1980). Also reported as mp 150° (Iwanami).

THERAP CAT: Antipsychotic.

6534. Neoarsphenamine. *Sulfoxylic acid mono[[[5-[(3-amino-4-hydroxyphenyl)diarsenyl]-2-hydroxyphenyl]amino]methyl] ester monosodium salt; [5-[(3-amino-4-hydroxyphenyl)arseno]-2-hydroxyanilino]methanol sulfoxylate sodium;* arsphenamine methylenesulfoxylic acid sodium salt; 3,3'-diamino-4,4'-dihydroxyarsenobenzenemethylenesulfoxylate sodium; Neosalvarsan; Collunovar; N.A.B.; Neo-Arsoluin; Vetarsenobillon; Novarsenobillon; Arsevan; Novarsan; Novarsenobenzol; Miarsenol. $C_{13}H_{13}As_2N_2NaO_4S$; mol wt 466.16. C 33.50%, H 2.81%, As 32.14%, N 6.01%, Na 4.93%, O 13.73%, S 6.88%. The usual medicinal grade contains a small amount of inert inorganic salts and some solvent. *The National Formulary* requires 19+% As. Prepn from arsphenamine + sodium formaldehydesulfoxylate: Krumwiede, *J. Am. Pharm. Assoc.* **8,** 795 (1919); Heyl, Miller, *ibid.* **11,** 432 (1922); Dohr, **U.S.** pat. **1,549,465** (1925); Kober, **U.S.** pat. **1,564,859** (1926); Kraft *et al.,* **Russian** pat. **158,388** (1962).

Yellow powder; odorless or slight odor. Oxidizes in air, becoming darker and more toxic; higher temps accelerate the oxidation; hence marketed in air-evacuated ampuls or filled with a nonoxidizing gas. Very sol in water; sol in glycerol. Slightly sol in alcohol or acetone. Practically insol in chloroform, ether. Its aq soln is practically neutral, unlike arsphenamine, which is acid. LD_{50} i.v. in rats: 300 mg/kg.

Caution: Neoarsphenamine does not require neutralization. Injections should be made with freshly prepd soln. The soln should not be shaken.

THERAP CAT (VET): Has been used in contagious pleuropneumonia, babesiasis, equine petechial fever, eperythrozoonosis.

6535. Neocembrene. *[R-(E,E,E)]-1,5,9-Trimethyl-12-(1-methylethenyl)-1,5,9-cyclotetradecatriene;* cembrene A; neocembrene A. $C_{20}H_{32}$; mol wt 272.47. C 88.16%, H 11.84%. Termite trail pheromone with all-*trans* configuration isolated from *Nasutitermes exitiosus* (Hill): B. P. Moore, *Nature* **211,** 746 (1966). Structure: A. J. Birch *et al., J. Chem. Soc. Perkin Trans. I* **1972,** 2653; V. D. Patil *et al., Tetrahedron* **29,** 341 (1973). Synthesis: M. Kodama *et al., Tetrahedron Letters* **1975,** 3065; Y. Kitahara *et al., Chem. Letters* **1976,** 219; H. Takayanagi *et al., Chem. Commun.* **1978,** 359. Synthesis of neocembrene and isomers: T. Kato *et al., J. Org. Chem.* **45,** 1126 (1980).

Colorless oil, faint wax-like odor. $bp_{0.8}$ 150-152°. n_D^{30} 1.5102.

6536. Neo-cupferron. *N-Hydroxy-N-nitroso-1-naphthalenamine ammonium salt; N-1-naphthyl-N-nitrosohydroxylamine ammonium derivative;* ammonium α-naphthylnitrosohydroxylamine. $C_{10}H_{11}N_3O_2$; mol wt 205.22. C 58.53%, H 5.40%, N 20.48%, O 15.59%. $C_{10}H_7N(NO)ONH_4$.

White to light buff crystals; sensitive to light. mp 125-126° with decompn. Soluble in water, methanol; insol in ether. The aq soln is unstable.

USE: Reagent for iron and copper. *See also* Cupferron.

6537. Neocuproine. *2,9-Dimethyl-1,10-phenanthroline;* 2,9-dimethyl-*o*-phenanthroline. $C_{14}H_{12}N_2$; mol wt 208.26. C 80.74%, H 5.81%, N 13.45%. Prepn: Case, *J. Am. Chem. Soc.* **70**, 3994 (1948); O'Reilly, Plowman, *Aust. J. Chem.* **13**, 145 (1960).

Hemihydrate, $C_{14}H_{12}N_2 \cdot \frac{1}{2}H_2O$, crystals from water or ligroin, mp 159-160°.

Dihydrate, needles from water. Loses water over P_2O_5 or at 80°.

USE: Clinical reagent (blood glucose assay). In spectrophotometric determination of copper, Nebesar, *Anal. Chem.* **36**, 1961 (1964).

6538. Neodymium. Nd; at. wt 144.24; at. no. 60; valences 2, 3, 4. A lanthanide; belongs to the cerium group of rare earth metals. Naturally occurring isotopes (mass numbers): 142 (27.13%); 143 (12.18%); 144 (23.80%), radioactive, $T_{1/2}$ 2.1 × 10^{15} years, α-emitter; 145 (8.30%); 146 (17.19%); 148 (5.76%); 150 (5.64%). Known artificial radioactive isotopes: 127-141; 147; 149; 151; 152. Abundance in earth's crust: 24-40 ppm. Commercially important sources are the rare earth minerals monazite and bastnaesite; also found in cerite and gadolinite. Discovered by von Welsbach in 1885. Sepn from Pr: F. H. Spedding *et al., J. Am. Chem. Soc.* **69**, 2786 (1947); **76**, 2557 (1954); Kauffman, Blank, *J. Chem. Ed.* **37**, 156 (1960); from other rare earths: Spedding *et al., J. Am. Chem. Soc.* **69**, 2812 (1947). Metal prepd by electrolysis of the chloride: Matignon, *Compt. Rend.* **131**, 891 (1900); Sieverts, Roell, *Z. Anorg. Chem.* **150**, 261 (1926); by reduction of the chloride: Bommer, Hohmann, *ibid.* **248**, 357 (1941). Spectrum: Alberston *et al., Phys. Rev.* **61**, 167 (1942). Reviews of prepn, properties and compds: *The Rare Earths,* F. H. Spedding, A. H. Daane, Eds. (Krieger, Huntington, N.Y., 1971, reprint of 1961 ed.) 641 pp; Hulet, Bode, "Separation Chemistry of the Lanthanides and Transplutonium Actinides" in *MTP Int. Rev. Sci.: Inorg. Chem., Ser. One,* vol. 7, K. W. Bagnall, Ed. (University Park Press, Baltimore, 1972) pp 1-45; Moeller, "The Lanthanides" in *Comprehensive Inorganic Chemistry* vol. 4, J. C. Bailar Jr. *et al.,* Eds. (Pergamon Press, Oxford, 1973) pp 1-101; F. H. Spedding in Kirk Othmer *Encyclopedia of Chemical Technology* vol. 19 (John Wiley & Sons, New York, 3rd ed., 1982) pp 833-854; *Chemistry of the Elements,* N. N. Greenwood, A. Earnshaw, Eds. (Pergamon Press, New York, 1984) pp 1423-1449. Review of toxicology: T. J. Haley, *J. Pharm. Sci.* **54**, 663-670 (1965). Brief review of properties: G. T. Seaborg, *Radiochim. Acta* **61**, 115-122 (1993).

Silver-white metal, becomes yellowish on exposure to air. Crystalline forms: hexagonal α-form, d 7.003, transforms to β-form at 868°; body-centered cubic β-form, d 6.80, exists at > 868°. mp 1021°. bp 3074°. Heat of fusion: 7.134 kJ/mol. Heat of sublimation (25°): 327.6 kJ/mol. E°(aq) Nd^{3+}/Nd −2.44 V (calc). Experimental reduction potentials (referred to a normal calomel electrode): −1.870, −1.960 V: Noddack, Brukl, *Angew. Chem.* **50**, 362 (1937). Reacts slowly with cold water; rapidly on heating.

Oxide, Nd_2O_3, blue powder, hexagonal structure, exhibits a slightly red fluorescence. Prepd by heating the hydroxide, carbonate, nitrate or oxalate. Very stable. Sol in dil acids;

soly in water: 5.7×10^{-6} g-mol/l at 29°. Commercial uses: Pings, *Colo. Sch. Mines, Miner. Ind. Bull.* **12**, no. 2 (1969) 19 pp.

Hydroxide, $Nd(OH)_3$, bluish or pink precipitate; on heating at 300-350° is converted into $2Nd_2O_3 \cdot 3H_2O$, grayish-brown; on further increase in temp is converted into $Nd_2O_3 \cdot H_2O$.

Chloride, $NdCl_3$, large purple prisms. Sol in water, in alcohol. Forms addition compds with ammonia. A hexahydrate is obtained from the aq soln; the hydrate is very sol in water (2.46 parts per 1 part of water); mp 124°. LD_{50} in mice (mg/kg): 4000 s.c., 348.3 i.p.; in male mice (mg/kg): 600 i.p., 5250 orally; in rats (mg/kg): 150-250 i.p.; in guinea pigs (mg/kg): 70 i.v., 139.6 i.p.; in rabbits (mg/kg): 200-250 i.v. (Haley).

Sulfate, $Nd_2(SO_4)_3$, pinkish needles; prepd by heating the oxide with concd sulfuric acid. Heat of formation 57.2 kcal. Sol in water; heat of soln 36.5 kcal. Is dec at 700-800°. Forms several double salts. A penta-, an octa-, and a pentadecahydrate have been prepared.

Nitrate, $Nd(NO_3)_3$, penta- and hexahydrates; prepd by adding the oxide to nitric acid. LD_{50} (hexahydrate) in female mice (mg/kg): 270 i.p.; in female rats (mg/kg): 270 i.p., 2750 orally, 6.4 i.v.; in male rats (mg/kg): 66.8 i.v. (Haley).

USE: Oxide as glass coloring agent; in glass filter plate laminated on color TV tubes to improve contrast and brightness; in lasers. Alloyed with iron and boron to produce powerful permanent magnets.

6539. Neoergosterol. *(3β,22E)-19-Norergosta-5,7,9,22-tetraen-3-ol.* $C_{27}H_{40}O$; mol wt 380.61. C 85.20%, H 10.59%, O 4.20%. Prepd from ergosterol: Inhoffen, *Ann.* **497**, 130 (1932); Windaus, Deppe, *Ber.* **70**, 76 (1937); Mosettig, Scheer, *J. Org. Chem.* **17**, 764 (1952). Stereochemistry: Steele *et al., J. Am. Chem. Soc.* **85**, 1134 (1963).

Needles from methanol, mp 152-154°. $[\alpha]_D^{17}$ −11° (c = 2 in chloroform). Practically insol in water; sol in organic solvents. Is precipitated by digitonin.

Methyl ether, $C_{28}H_{42}O$, mp 94°. $[\alpha]_D^{20}$ −5° (c = 1.2 in chloroform).

6540. Neohesperidin Dihydrochalcone. *1-[4-[[2-O-(6-Deoxy-α-L-mannopyranosyl)-β-D-glucopyranosyl]oxy]-2,6-dihydroxyphenyl]-3-(3-hydroxy-4-methoxyphenyl)-1-propanone; 3,5-dihydroxy-4-(3-hydroxy-4-methoxyhydrocinnamoyl)phenyl-2-O-(6-deoxy-α-L-mannopyranosyl)-β-D-glucopyranoside;* neohesperidin DHC; NHDC; Sukor. $C_{28}H_{36}O_{15}$; mol wt 612.58. C 54.90%, H 5.92%, O 39.18%. Prepn from naringen, a flavanone glycoside occurring naturally in grapefruit: Horowitz, Gentile, U.S. pats. **3,087,821** and **3,375,242** (1963, 1968, both to U.S. Secy. Agr.); Robertson *et al., Ind. Eng. Chem. Prod. Res. Develop.* **13**, 125 (1974). Sweetening effect: Inglett *et al., J. Food Sci.* **34**, 101 (1969). Chromatography: Gentile, Horowitz, *J. Chromatog.* **63**, 467 (1971);

J. F. Fisher, *J. Agr. Food Chem.* **25**, 682 (1977). Subchronic oral toxicity study: B. A. R. Lina *et al., Food Chem. Toxicol.* **28**, 507 (1990). *Review:* P. J. Pratter, *Perfum. Flavor.* **5**, 12-18 (1981).

Crystals from acetone, mp 156-158°. 1000 to 1500 times sweeter than sucrose; 20 times sweeter than saccharin.

USE: Sweetening agent, especially in chewing gum and dentifrices.

6541. Neomethymycin. *(12R)-10-Deoxy-12-hydroxymethymycin.* $C_{25}H_{43}NO_7$; mol wt 469.62. C 63.94%, H 9.23%, N 2.98%, O 23.85%. Macrolide antibiotic found in the mother liquors from methymycin, *q.v.* Differs from methymycin only in the location of one hydroxyl group, which results in marked changes in chemical behavior: Djerassi, Halpern, *J. Am. Chem. Soc.* **79**, 2022, 3926 (1957); *Tetrahedron* **3**, 255 (1958). Production from *Streptomyces venezuelae* cultures: Dutcher *et al.,* U.S. pat. **2,916,486** (1959 to Olin Mathieson). Synthesis of the aglycone, *neomethynolide*: J. Inanago *et al., Chem. Letters* **1981**, 1415.

Crystals from ether + hexane, mp 156-158°. $[\alpha]_D^{25}$ +93°. uv max (ethanol): 227.5 nm (log ϵ 4.10). Solvates easily and when crystallized from dilute acetone yields 1 cm long crystals of the acetone solvate. Soluble in alcohol, acetone, chloroform, benzene, ethyl acetate, ether. Slightly sol in water, dibutyl ether. Practically insol in hexane, aliphatic hydrocarbons. *Compare also* Picromycin.

6542. Neomycin. Mycifradin; Fradiomycin; Neomin; Neolate; Neomas; Pimavecort; Vonamycin Powder V. Antibiotic complex composed of neomycins A, B and C. Produced by *Streptomyces fradiae:* Waksman, Lechevalier, *Science* **109**, 305 (1949); Waksman *et al., J. Clin. Invest.* **28**, 934 (1949); Swart *et al., ibid.* 1045; Waksman, Lechevalier, U.S. pat. **2,799,620** (1957). Purification: Jackson, U.S. pat. **2,848,365** (1958 to Upjohn); Haak, U.S. pat. **3,108,996** (1963 to Upjohn). Recovery: Miller, U.S. pat. **3,005,815** (1961 to Merck & Co.); Moses, U.S. pat. **3,022,228** (1962 to Penick); **Brit.** pat. **945,475** (1964 to O.W.G. Chemie). Characterization: Peck *et al., J. Am. Chem. Soc.* **71**, 2590 (1949); Regna, Murphy, *ibid.* **72**, 1045 (1950); Dutcher *et al., ibid.* **73**, 1384 (1951). Structure of neomycins B and C: K. L. Rinehart *et al., ibid.* **79**, 4567, 4568 (1957); *ibid.* **84**, 3216, 3218 (1962). Abs config of neomycin C: M. Hichens, K. L. Rinehart, *ibid.* **85**, 1547 (1963). Total synthesis of neomycin C: S. Umewaza, Y. Nishimura, *J. Antibiot.* **30**, 189 (1977); S. Umewaza *et al., Bull. Chem. Soc. Japan* **53**, 3259 (1980); of neomycin B: T. Usui, S. Umezawa, *J. Antibiot.* **40**, 1464 (1987). Monographs: S. A. Waksman,

Neomycin (Rutgers Univ. Press, New Brunswick, N. J., 1953) 219 pp; K. L. Rinehart, Jr., *The Neomycins and Related Antibiotics* (Wiley, New York, 1964) 137 pp. Comprehensive description: W. F. Heyes, *Anal. Profiles Drug Subs.* **8**, 399-488 (1979).

Neomycin complex is an amorphous base sol in water, methanol and acidified alcohol. Practically insol in common organic solvents. Solns up to 250 mg/ml H_2O may be prepared.

Neomycin A, see Neamine.

Neomycin B, $C_{23}H_{46}N_6O_{13}$*, antibiotique EF 185, Framycetin, Enterfram, Framygen, Soframycin, Actilin.* Identity of neomycin B and framycetin: K. L. Rinehart *et al., ibid.* **82**, 3938 (1960). Yields on hydrolysis neamine and *neobiosamine B.* Structure of neobiosamine B: K. L. Rinehart *et al., J. Am. Chem. Soc.* **82**, 2970 (1960).

Neomycin B hydrochloride, amorphous white powder. $[\alpha]_D^{20}$ +57° (H_2O). Soly in mg/ml at ~28°: water 15.0; methanol 5.7; ethanol 0.65; isopropanol 0.05; isoamyl alcohol 0.33; cyclohexane 0.06; benzene 0.03. Practically insol in acetone, ether, other organic solvents. For additional soly data *see* Weiss *et al., Antibiot. & Chemother.* **7**, 374 (1957).

Neomycin B sulfate, *Biosol, Bykomycin, Endomixin, Fraquinol, Myacine, Neosulf, Neomix, Neobrettin, Nivemycin, Tuttomycin.* Amorphous white powder. Practically tasteless. $[\alpha]_D^{20}$ +54° (c = 2 in H_2O). Soly in mg/ml at ~28°: water 6.3; methanol 0.225; ethanol 0.095; isopropanol 0.082; isoamyl alcohol 0.247; cyclohexane 0.08; benzene 0.05. Practically insol in acetone, ether, chloroform. Aq solns are fairly stable at pH 2 to 9. Highly purified prepns are very stable to alkali and unstable to acids. Refluxing with barium hydroxide for 18 hrs showed no loss of activity. Boiling with mineral acids yields an aldehyde, characterized as furfural, and an organic base.

Neomycin C, $C_{23}H_{46}N_6O_{13}$. Yields on hydrolysis neamine and *neobiosamine C.* Structure of neobiosamine C: K. L. Rinehart, P. Woo, *J. Am. Chem. Soc.* **80**, 6463 (1958).

THERAP CAT: Antibacterial.

THERAP CAT (VET): Antibacterial.

6543. Neomycin Undecylenate. Neodecyllin. 10-Undecenoate salt of the antibiotic neomycin, *q.v.* Commercial prepn contains the equivalent of not less than 30% neomycin base, calculated on the dried basis. Prepn: van De Griendt, U.S. pat. **3,022,286** (1962 to Penick).

THERAP CAT: Antibacterial; antifungal.

6544. Neon. Ne; at. wt 20.1797; at. no. 10. Group VIIIA(18), also known as Group 0. A noble gas characterized by an electronic structure in which the outer *p* subshell is entirely filled. Naturally occurring stable isotopes (mass numbers): 20 (90.51%); 21 (0.27%); 22 (9.92%); known artificial radioactive isotopes: 16-19; 23-25. Longest-lived known isotope: 24 ($T_{1/2}$ 3.38 min, β-emitter). Abundance in igneous rock of the earth's crust: 7×10^{-5} ppm by wt; concentration in air: 18.2 ppm by vol. Obtained commercially from the atmosphere by distillation-liquefaction process. Discovered in 1898 by Travers and Ramsay: *The Discovery of the Rare Gases*, London, 1928. Monograph: *Argon, Helium and the Rare Gases*, **vols. 1, 2**, G. A. Cook, Ed. (Interscience, New York, 1961) 818 pp. *Reviews:* Cockett, Smith, "The Monatomic Gases" in *Comprehensive Inorganic Chemistry* **vol. 1**, J. C. Bailar, Jr. *et al.*, Eds. (Pergamon Press, Oxford, 1973) pp 139-211; S.-C. Hwang, W. R. Weltmer, Jr. in Kirk Othmer *Encyclopedia of Chemical Technology* **vol. 13** (John Wiley & Sons, 4th ed., 1995) p1-38; *Chemistry of the Elements*, N. N. Greenwood, A. Earnshaw, Eds. (Pergamon Press, New York, 1984) pp 1042-1059. Review of use in inductively coupled plamsa-mass spectrometry: S. F. Durrant, *Fres. J. Anal. Chem.* **347**, 389-392 (1993).

Colorless, odorless, tasteless, monatomic, inert gas; will form compds with highly electronegative elements such as O, F, Cl. Does not condense at the temp of liq air; solid at the temp of liq hydrogen; the solid form exists as face-centered cubic crystals at normal pressure. Soly of gas in water (20°): 10.5 cm³/kg water. Triple pt temp 24.562°K, press 43.37 kPa. Critical temp 44.40°K, critical press 2654 kPa, critical d 483 kg/m³. gas: d⁰ (101.3 kPa) 0.89994 kg/m³, d (normal bp) 9.552 kg/m³. Liquid: normal bp 27.102°K, d (normal bp) 1207 kg/m³, d (triple pt) 1247 kg/m³, heat of vaporization (normal bp) 1.741 kJ/mol. Solid: d (triple pt) 1444 kg/m³, heat of vaporization (triple pt) 2.139 kJ/mol, heat of fusion (triple pt) 0.335 kJ/mol. Emission spectra: T. Jacksier, R. M. Barnes, *Appl. Spectrosc.* **48**, 65 (1994).

Caution: Can act as a simple asphyxiant by displacing air. *See: Matheson Gas Data Book* (Matheson Co., Inc., 4th ed., East Rutherford, NJ, 1966) pp359-361.

USE: Gas in neon light tubes; ingredient of gaseous fillers for antifog devices, warning signals, electrical current detectors, high-voltage indicators for high-tension electric lines, lightning arresters, wave-meter tubes; in Ne-He lasers; in mixtures with He and Ar in Geiger counters. Liquid as cryogen to produce low temps.

6545. Neopentane. *2,2-Dimethylpropane;* tetramethylmethane. C_5H_{12}; mol wt 72.15. C 83.24%, H 16.76%. Found in petr naphtha. Prepn from *tert*-butyl iodide and dimethylzinc: Lwow, *Z. Chemie* **1870**, 520; *cf. J. Am. Chem. Soc.* **54**, 3460 (1932); from butylmagnesium iodide and dimethyl sulfate in ether: Ferrario, Fagetti, *Gaz. Chim.* **38, II**, 663 (1908); from amylene vapor and hydrogen in presence of an electric discharge: *ibid.* **62**, 621 (1932); by passing pentanes over aluminum chloride at 140°, *J. Gen. Chem. USSR* **14**, 343 (1944), *C.A.* **39**, 3783 (1945); from neopentylmagnesium chloride and water: Whitmore *et al., J. Am. Chem. Soc.* **56**, 749 (1934).

Liquid or gas. d_4^0 0.613 (liq). Solidifies to form tetragonal crystals, mp −19.8°. bp 9.5°. Insol in water.

6546. Neopentyl Alcohol. *2,2-Dimethyl-1-propanol;* *tert*-butyl carbinol; neoamyl alcohol; neopentanol. $C_5H_{12}O$; mol wt 88.15. C 68.13%, H 13.72%, O 18.15%. $(CH_3)_3C-CH_2OH$. Prepn: Conant *et al., J. Am. Chem. Soc.* **51**, 1249 (1929). Review of manuf (by fractionation of fusel oil and via chlorination of pentanes) and properties: *Industrial Chemicals,* W. L. Faith *et al.,* Eds. (Wiley & Sons, New York, 2nd ed., 1957) pp 107-114.

Volatile crystals, peppermint odor. bp 114°. mp 53°. d_4^{20} 0.812. Vapor pressure: 16 mm at 20°; 50.5 mm at 46°; 533.5 mm at 100°. Slightly sol in water (~3.5% at 25°); miscible with alcohol, ether.

6547. Neopentyl Glycol. *2,2-Dimethyl-1,3-propanediol;* dimethyltrimethylene glycol. $C_5H_{12}O_2$; mol wt 104.15. C 57.66%, H 11.61%, O 30.72%. Prepd from isobutyraldehyde.

Needles from benzene, mp 127°. bp₇₆₀ 208°. Solubility in water ~65% w/w. Freely sol in alcohol, ether.

USE: In the manuf of plasticizers, polyesters, as modifier of alkyd resins.

6548. Neophyl Chloride. *(2-Chloro-1,1-dimethylethyl)-benzene; 1-chloro-2-methyl-2-phenylpropane; (β-chloro-tert-butyl)benzene.* $C_{10}H_{13}Cl$; mol wt 168.67. C 71.21%, H 7.77%, Cl 21.02%. Prepn (68% yield) from benzene and methallyl chloride in the presence of concd sulfuric acid: Whitmore *et al., J. Am. Chem. Soc.* **65**, 1469 (1943); Smith, Sellas, *Org. Syn.* coll. vol. **IV**, 702 (1963).

Liquid. d_4^{25} 1.0379. bp₇₄₁ 221° (dec); bp₉₀ 111°; bp₃₀ 120°; bp₂₀ 105°; bp₁₈ 104°; bp₁₃ 97°; bp₁₀ 95°; bp₁.₀ 53°. n_D^{20} 1.5249; n_D^{25} 1.5228. Considerably less reactive with sodium than neopentyl chloride. Forms a Grignard reagent readily and in good yield. Salt effect in acetolysis: Fainberg, Winstein, *J. Am. Chem. Soc.* **78**, 2763 (1956).

6549. Neopine. *(5α,6α)-8,14-Didehydro-4,5-epoxy-3-methoxy-17-methylmorphinan-6-ol;* β-codeine. $C_{18}H_{21}NO_3$; mol wt 299.37. C 72.22%, H 7.07%, N 4.68%, O 16.03%. An alkaloid from opium, isomeric with codeine. Prepn from 14-bromocodeinone: Conroy, U.S. pat. **2,797,222** (1957 to Merck & Co.); from thebaine: S. Makleit *et al., Acta Chim. Acad. Sci. Hung.* **94**, 161 (1977).

Long needles from petr ether, mp 127.5°. $[\alpha]_D^{23}$ −28° (c = 7.5 in chloroform). The absorption spectrum seems identical with that of codeine: Dobbie, Lauder, *J. Chem. Soc.* **99**, 34 (1911). Solubilities about the same as those of codeine, *q. v.*

Hydrobromide, $C_{18}H_{21}NO_3 \cdot HBr$, prismatic crystals from water, darkens about 240°, dec 283°. $[\alpha]_D^{23}$ +17° (c = 3.7). Relatively insol in water, making possible a separation from other more sol hydrobromides of various opium alkaloids.

Note: Incorrectly called **hydroxycodeine**, *cf.* L. J. Sargent, U. Weiss, *J. Org. Chem.* **25**, 987 (1960).

6550. Neoprene. Duprene; GR-M; polychloroprene; poly(2-chloro-1,3-butadiene). Mol wt range of 100,000-300,000. An oil-resistant synthetic rubber, with predominantly *trans* configuration made by polymerization of chloroprene. Prepn: W. H. Carothers *et al., J. Am. Chem. Soc.* **53**, 4203 (1931). Review: P. J. Johnson in Kirk-Othmer *Encyclopedia of Chemical Technology* vol. **8** (Wiley-Interscience, New York, 3rd ed., 1979) pp 515-534.

d²⁰ approx 1.23. n_D^{20} approx 1.5512. Brittle point: −35°. Softens *ca.* 80°. Has high tensile strength, resilience and abrasion resistance.

USE: In mechanical and automotive products; as wire and cable jackets, gaskets, roof coatings, binder for fibers.

6551. Neopterin. *2-Amino-6-(1,2,3-trihydroxypropyl)-4(3H)-pteridinone; 1-(2-amino-4-hydroxy-6-pteridinyl)-1,2,-3-propanetriol; 6-(1',2',3'-trihydroxy)pterin;* Crithidia factor. $C_9H_{11}N_5O_4$; mol wt 253.22. C 42.69%, H 4.38%, N 27.66%, O 25.27%. Precursor in the biosynthesis of biopterin, *q.v.* Of the four possible isomers, D-*erythro*, L-*erythro*, D-*threo*, L-*threo*, two have been found in nature: the D-*erythro* form, to which the term neopterin originally referred; first isolated from the pupae of bees: Rembold, Buschmann, *Ann.* **662**, 72 (1963); the L-*threo* form, found to be the growth factor for the protozoan *Crithidia fasciculata* and isolatable from cell-free extracts of *Serratia indica:* Kobashi, Iwai, *Agr. Biol. Chem.* **35**, 47 (1971); **36**, 1685, 1695 (1972). Both natural forms found in human urine: Fukushima, Shiota, *J. Biol. Chem.* **247**, 4549 (1972). Early synthetic studies and structure: Rembold, Buschmann, *Ber.* **96**, 1406 (1963). Synthesis of L-(−)-form: Viscontini, Provenzale, *Helv. Chim. Acta* **51**, 1495 (1968). Synthesis of natural D-neopterin: Viscontini *et al., ibid.* **53**, 1202 (1970).

Consult the Name Index before using this section.

D-erythro-Neopterin

D-*erythro*-Form, $[\alpha]_D^{25}$ +45° ± 3° (c = 0.3 in 0.1N HCl).
L-*erythro*-Form, $[\alpha]_D^{25}$ −44° ± 3° (c = 0.3 in 0.1N HCl).
D-*threo*-Form, $[\alpha]_D^{25}$ −92° ± 3° (c = 0.3 in 0.1N HCl).
L-*threo*-Form, pale yellow, spiny crystals. $[\alpha]_D^{25}$ +97° ± 3° (c = 0.3 in 0.1N HCl). uv max in 0.1N NaOH: 255, 363 nm (ε 20,900, 7050); in 0.1N HCl: 248, 323 nm (ε 11,200, 7880). Strong blue fluorescence in neutral or alkaline soln; weak fluorescence in acidic soln.

6552. Neoquassin. *16-Hydroxy-2,12-dimethoxypicrasa-2,12-diene-1,11-dione; 3a,4,5,6a,7,7a,8,11a,11b,11c-decahydro-2-hydroxy-2,10-dimethoxy-3,8,11a,11c-tetramethylphenanthro[10,1-bc]pyran-1,11-dione.* $C_{22}H_{30}O_6$; mol wt 390.48. C 67.67%, H 7.74%, O 24.58%. Found together with the quassin in the mixture of bitter constituents of the wood of *Quassia amara* L., *Simaroubaceae*, known in commerce as Surinam quassia. Forms quassin on oxidation. Isoln: London *et al.*, *J. Chem. Soc.* **1950**, 3431. Structure: Valenta *et al.*, *Tetrahedron Letters* **1960**(20), 25; Carman, Ward, *ibid.* **1961**, 317; Valenta *et al.*, *Tetrahedron* **15**, 100 (1961).

Very bitter polymorphous crystals. Stable form, thick prisms, mp 227.5-228.5°; unstable form, long narrow plates, mp 213°. $[\alpha]_D^{20}$ +41.0° (c = 4.98 in chloroform). uv max ∼255 nm (ε ∼11,650). Sol in cold acetone, chloroform, pyridine, acetic acid, in warm ethyl acetate, benzene, alcohol. Sparingly sol in ether, petr ether.

6553. Neostigmine. *3-[[(Dimethylamino)carbonyl]oxy]-N,N,N-trimethylbenzenaminium;* (3-dimethylcarbamoxyphenyl) trimethylammonium; synstigmin; proserine. $[C_{12}H_{19}N_2O_2]^+$. Synthesis: Aeschlimann, U.S. pat. **1,905,990** (1933 to Hoffmann-La Roche); Yanagisawa, Japan. pat. **51 3071**, *C.A.* **47**, 4908g (1953). Pharmacology and pharmacokinetics: T. N. Calvey *et al.*, *Brit. J. Clin. Pharmacol.* **7**, 149 (1979). Synergism of neostigmine toxicity by lithium: W. M. Davis, N. S. Hatoum, *Toxicology* **17**, 1 (1980). Toxicity: L. O. Randall, G. Lehmann, *J. Pharmacol. Exp. Ther.* **99**, 16 (1950). Comprehensive description: A. A. Al-Badr, M. Tariq, *Anal. Profiles Drug Subs.* **16**, 403-444 (1987).

Bromide, $C_{12}H_{19}BrN_2O_2$, *Juvastigmin (tabl.), Neoesserin, Neostigmin (tabl.), Normastigmin (tabl.), Prostigmin.* Crystals from alcohol + ether, dec ∼167°. One gram dissolves in ∼1 ml water. Sol in alc.
Methyl sulfate, $C_{13}H_{22}N_2O_6S$, *Intrastigmina, Juvastigmin (amp.), Metastigmin, Neostigmin (inj.), Normastigmin (amp.), Prostigmin (amp.), Stiglyn.* Crystals from alc, mp

142-145°. One gram dissolves in ∼10 ml water, less sol in alcohol. LD_{50} in mice (mg/kg): 0.16 i.v.; 0.42 s.c.; 7.5 orally, (Randall, Lehmann).
THERAP CAT: Cholinergic; miotic; antidote (curare).
THERAP CAT (VET): Cholinergic; antidote (curare).

6554. Neotetrazolium Chloride. *3,3'-[1,1'-Biphenyl]-4,-4'-diylbis[2,5-diphenyl-2H-tetrazolium] dichloride; 3,3'-[4,4'-biphenylene]bis[2,5-diphenyl-2H-tetrazolium] dichloride; 2,-2',5,5'-tetraphenyl-3,3'-(p-diphenylene)ditetrazolium chloride; 2,2'-(p-diphenylene)-bis[3,5-diphenyl]ditetrazolium chloride; neotetrazolium blue; neo-T; TP; NTC.* $C_{38}H_{28}Cl_2$-N_8; mol wt 667.60. C 68.37%, H 4.23%, Cl 10.62%, N 16.78%. Prepn: L. J. Pannone, J. B. Rust, U.S. pat. **2,713,-581** (1955 to Montclair Res. Corp. and Ellis-Foster Co.). Purification: G. R. N. Jones, *Histochem. J.* **1**, 59 (1968). *See also* Triphenyltetrazolium chloride.

Light yellow powder, mp 297° (dec).
USE: Staining of microorganisms and plant and animal tissues. In determn of dehydrogenases in histochemical and cytochemical studies: K. Neumann, G. Koch, *Z. Physiol. Chem.* **295**, 35 (1953). In amylase determn.

6555. Neovitamin A. *5-cis*-Vitamin A. $C_{20}H_{30}O$; mol wt 286.46. C 83.86%, H 10.56%, O 5.59%. A naturally occurring isomer of vitamin A. Isolated in cryst form from soupfin shark liver oil. Also present in cod, dogfish, halibut, and California jewfish liver oils to the extent of ∼35% of the total vitamin A content. Synthetic vitamin A concentrates also contain neovitamin A in the same proportion. The biological potencies of neovitamin A and vitamin A, as measured by the U.S.P. growth method in cats, proved to be identical. Isoln: Robeson, Baxter, *J. Am. Chem. Soc.* **69**, 136 (1947); Robeson, U.S. pat. **2,552,908** (1951 to Eastman Kodak). Discussion of stereoisomerism: Zechmeister, *Chem. Rev.* **34**, 267 (1944); Zechmeister, *Vitam. Horm. (New York)* **7**, 57-81 (1949); Vogel-Knobloch, *Chemie und Technik der Vitamine* vol. **I** (Stuttgart, 3rd ed., 1950) p 102. Additional characteristics: Meunier, Jouannetau, *Bull. Soc. Chim. Biol.* **30**, 260 (1948).
Pale yellow needles from ethyl formate, mp 58-60°. More resistant to atmospheric oxidation than vitamin A. uv max: 328 nm. Neovitamin A and its esters react with maleic anhydride at a much slower rate than vitamin A. This reaction forms a basis of assaying neovitamin A in fish liver oils and concentrates.

6556. Nepetalactone. *5,6,7,7a-Tetrahydro-4,7-dimethyl-cyclopenta[c]pyran-1-(4aH)-one.* $C_{10}H_{14}O_2$; mol wt 166.22. C 72.26%, H 8.49%, O 19.25%. The first fully characterized methylcyclopentane monoterpenoid. Isolated from the volatile oil of catnip produced by *Nepeta cataria* L., *Labiatae*: S. M. McElvain *et al.*, *J. Am. Chem. Soc.* **63**, 1558 (1941); J. Meinwald, *ibid.* **76**, 4571 (1954). First believed to be a single entity, it was later shown to be a mixture of the *cis-trans* and *trans-cis* isomers, the *cis-trans* isomer comprising 70-99% of the oil. Structure: S. M. McElvain, E. J. Eisenbraun, *J. Am. Chem. Soc.* **77**, 1599 (1955). Configuration of the *cis-trans* and *trans-cis* isomers: R. B. Bates *et al.*, *ibid.* **80**, 3420 (1958). Separation and properties: R. B. Bates, C. W. Siegel, *Experientia* **19**, 564 (1963). Biosynthesis: F. E. Regnier *et al.*, *Phytochemistry* **7**, 221 (1968). Metabolism in cats: G. R. Waller *et al.*, *Science* **164**, 1281 (1969). Behavioral and toxicological study: J. W. Harney *et al.*, *Lloydia* **41**, 367 (1978). Structural study of isomeric nepetalactones: E. J. Eisenbraun *et al.*, *J. Org. Chem.* **45**, 3811 (1980).

cis-trans-Nepetalactone

Oil. Odor very attractive to cats. d_4^{25} 1.0663. $bp_{0.05}$ 71-72°. $[\alpha]_D^{23}+3.6°$. n_D^{25} 1.4859. Soluble in ether, carbon tetrachloride. When shaken with 10% NaOH yields *nepetalic acid*, crystals, mp 74°, $[\alpha]_D^{30}$ +47.6° (chloroform).

cis-trans-Form, *2-(2-hydroxy-1-methylethenyl)-5-methylcyclopentanecarboxylic acid delta lactone*. Oil, $[\alpha]_D^{21}$ +3.7° (c = 27 in chloroform). $[\alpha]_D^{27.5}$ +11.11° (chloroform). n_D^{25} 1.4878.

trans-cis-Form, mp 37-39°. $[\alpha]_D^{21}$ −24.4° (c = 6.15 in chloroform) (Eisenbraun); also reported as $[\alpha]_D^{27.5}$ +21.9° (chloroform) (Bates, Siegel). n_D^{25} 1.4878.

6557. Neptunium. Np; at. no. 93; valence 3, 4, 5, 6, 7. First man-made transuranium element; no stable nuclides. Known isotopes (mass numbers): 227-242. Discovery of isotope ^{239}Np (α-emitter, $T_{1/2}$ 2.14 × 10^6 years, longest-lived known isotope, rel. at. mass 237.0482): E. McMillan, P. Abelson, *Phys. Rev.* **57**, 1185 (1940); of isotope ^{237}Np ($T_{1/2}$ 2.35 days, β-decay): A. C. Wahl, G. T. Seaborg, *ibid.* **73**, 940 (1948). Prepn of metal: S. Fried, N. Davidson, *J. Am. Chem. Soc.* **70**, 3539 (1948); L. B. Magnusson, T. J. LaChapelle, *ibid.* 3534. Presence in nature: Seaborg, Perlman, *ibid.* **70**, 1571 (1948). Chemical properties: Seaborg, Wahl, *ibid.* 1128. Reviews: C. Keller, *The Chemistry of the Transactinide Elements* (Verlag Chemie, Weinheim, English Ed., 1971) pp 253-332; W. W. Schulz, G. E. Benedict, *Neptunium-237: Production and Recovery*, AEC Critical Review Series (USAEC, Washington, D.C., 1972) 85 pp; *Comprehensive Inorganic Chemistry* vol. 5, J. C. Bailar, Jr. *et al.*, Eds. (Pergamon Press, Oxford, 1973) *passim*; J. A. Fahey in *The Chemistry of the Actinide Elements* vol. 1, J. J. Katz *et al.*, Eds. (Chapman and Hall, New York, 1986) pp 443-498; G. T. Seaborg in Kirk-Othmer *Encyclopedia of Chemical Technology* vol. 1 (Wiley-Interscience, New York, 4th ed., 1991) pp 412-444.

Silvery metal; develops a thin oxide layer upon exposure to air for short periods. Reacts with air at high temperatures to form NpO$_2$. Exhibits 3 allotropic modifications: orthorhombic α-form, d 20.45, transforms to β-form at 280°; tetragonal β-form, d 19.36, transforms to γ-form at 577°; cubic γ-form transforms to liquid at mp 637°. Extrapolated bp 4174°. Np has been obtained in its five oxidation states in soln; the most stable is the pentavalent state. Tetravalent Np is readily oxidized to the hexavalent state by permanganate in the cold, or by strong oxidizing agents; on electrolytic reduction in an atmosphere of nitrogen, the trivalent form is obtained.

Caution: Radiation hazard; handling requires special equipment and shielding facilities (Katz *et al.*, *loc cit.*, vol. 2, p. 1128).

USE: Source material for prodn of ^{238}U (power source).

6558. Nequinate. *6-Butyl-1,4-dihydro-4-oxo-7-(phenylmethoxy)-3-quinolinecarboxylic acid methyl ester; 7-(benzyloxy)-6-n-butyl-1,4-dihydro-4-oxo-3-quinolinecarboxylic acid methyl ester*; 3-methoxycarbonyl-6-n-butyl-7-benzyloxy-4-oxoquinoline; 7-(benzyloxy)-6-n-butyl-4-hydroxy-3-quinolinecarboxylic acid methyl ester; 7-(benzyloxy)-6-n-butyl-3-methoxycarbonyl-4-quinolone; methyl benzoquate; ICI-

55052; Statyl. C$_{22}$H$_{23}$NO$_4$; mol wt 365.43. C 72.31%, H 6.34%, N 3.83%, O 17.51%. Prepn: **Belg. pat. 677,592** and **Brit. pat. 1,070,223** (1966 to I.C.I.), *C.A.* **68**, 68899j (1968). Coccidiostatic activity: Bowie *et al.*, *Nature* **214**, 1349 (1967). HPLC determn in poultry feedstuffs: G. H. J. Merson *et al.*, *Analyst* **110**, 761 (1985).

Crystals, mp 287-288°.

THERAP CAT (VET): Coccidiostat.

6559. Neriifolin. *(3β,5β)-3-[(6-Deoxy-3-O-methyl-α-L-glucopyranosyl)oxy]-14-hydroxycard-20(22)-enolide.* C$_{30}$H$_{46}$-O$_8$; mol wt 534.69. C 67.39%, H 8.67%, O 23.94%. Cardiac glycoside isolated from *Thevetia neriifolia* Juss.: M. Frèrejacque, *Compt. Rend.* **221**, 645 (1945); from *Cerbera odollam* Gaertn.: S. Rangaswami, E. V. Rao, *J. Sci. Ind. Res.* **16B**, 209 (1957); from *Thevetia thevetioides* seeds: J. L. McLaughlin *et al.*, *J. Econ. Entomol.* **73**, 39 (1980). Pharmacology: K. Mezey, *Arch. Int. Pharmacodyn. Ther.* **84**, 367 (1950); K. K. Chen *et al.*, *ibid.* 81.

Rhombic plates from methanol, mp 218-225°. Also reported as mp 208° (Frèrejacque). $[\alpha]_D^{23}$ − 50.2° (CH$_3$OH). uv max (CH$_3$OH): 217 nm (log ϵ 4.1).

2′-Acetate, C$_{32}$H$_{48}$O$_9$, *cerberin, veneniferin, monoacetylneriifolin*. Identity of cerberin with monoacetylneriifolin: M. Frèrejacque, *Compt. Rend.* **226**, 835 (1948); of cerberin with veneniferin: M. Frèrejacque, Durgeat, *ibid.* **228**, 1310 (1949). Coarse prisms from methanol + water, mp 212-215°. $[\alpha]_D^{19}$ −82° (CHCl$_3$). Sol in alc, CHCl$_3$, acetic acid, ether. Practically insol in water.

THERAP CAT: Cardiotonic.

6560. Nerol. *(Z)-3,7-Dimethyl-2,6-octadien-1-ol; cis-2,6-dimethyl-2,6-octadien-8-ol.* C$_{10}$H$_{18}$O; mol wt 154.25. C 77.87%, H 11.76%, O 10.37%. The *cis*-isomer of geraniol, *q.v.*; found in many essential oils. Readily loses water and cyclizes forming dipentene. Isoln from neroli oil: Hesse, Zeitschel, *J. Prakt. Chem.* **66**, 502 (1902). Structure: Verley, *Bull. Soc. Chim. France* **25**, 68 (1919); J. L. Simonsen, *The Terpenes* vol. I (University Press, Cambridge, 2nd ed., 1947) pp 52-54. Stereochemistry: Burrell *et al.*, *Proc. Chem. Soc.* **1959**, 263; Bates *et al.*, *J. Org. Chem.* **28**, 1086 (1963). Synthesis: Yukawa *et al.*, *Bull. Chem. Soc. Japan* **37**, 158 (1964). Stereochemistry and synthesis: Burrell *et al.*, *J. Chem. Soc. (C)* **1966**, 2144; K. Takabe *et al.*, *Chem. Letters* **1977**, 1025.

Liquid. Odor of sweet rose. bp_{745} 224-225°; bp_{25} 125°. d^{15} 0.8813. Optically inactive. uv max: 189-194 nm (ϵ 18000). Sol in abs alc.

Tetrabromide, C$_{10}$H$_{18}$Br$_4$O, crystals, mp 118°.

Allophanate, $C_{12}H_{20}N_2O_3$, needles from petr ether, mp 84-86°.

USE: Base for manuf of perfumes.

6561. Nerolidol. *3,7,11-Trimethyl-1,6,10-dodecatrien-3-ol;* peruviol. $C_{15}H_{26}O$; mol wt 222.37. C 81.02%, H 11.79%, O 7.19%. Found in essential oils from many flowers such as *Melaleuca viridiflora* Soland., *Myrtaceae* and *Myroxylon pereirae* (Royle) Klotzsch, *Leguminosae:* Naves, *Compt. Rend.* **251**, 900 (1960). Isoln: Hellyer, McKern, *J. Proc. Roy. Soc. N.S. Wales* **89**, (Pt. 4), 188 (1955); Sutherland *et al., Aust. J. Chem.* **13**, 357 (1960). Synthesis: Nazarov *et al., Zh. Obshch. Khim.* **28**, 1444 (1958); Ofner *et al., Helv. Chim. Acta* **42**, 2577 (1959); Shvarts, Petrov, *Zh. Obshch. Khim.* **30**, 3598 (1960). Exists in two stereoisomeric forms: Bates *et al., J. Org. Chem.* **28**, 1086 (1963).

trans-form

trans-Form, liquid. $bp_{0.15}$ 78°. n_D^{25} 1.4792.
cis-Form, liquid. $bp_{0.10}$ 70°. n_D^{25} 1.4775.
cis/trans-Form, liquid. bp_3 122°. n_D^{25} 1.4769. d_4^{25} 0.8720. Sol in abs alc (still clearly sol in 3 parts of 70% alc.)

6562. Nerve Growth Factor. NGF. Polypeptide factor which promotes the growth and survival of peripheral sympathetic and sensory neurons and basal forebrain cholinergic neurons. Prototype of the neurotrophins, *q.v.* Synthesized in limiting amounts by target tissues of NGF-responsive cells; anomalous sources of large amounts of NGF include mouse submaxillary gland, snake venom gland, and guinea pig prostate. Bioactive form, also referred to as *β-NGF*, is a tightly associated dimer of 2 identical polypeptide chains. Mol wt of the dimer ~26,500 Da. Initial identification as a nerve growth stimulating factor from mouse sarcoma: R. Levi-Montalcini, V. Hamburger, *J. Exp. Zool.* **116**, 321 (1951); R. Levi-Montalcini, *Ann. N.Y. Acad. Sci.* **55**, 330 (1952). Identification in snake venom: S. Cohen, *J. Biol. Chem.* **234**, 1129 (1959); in mouse submaxillary gland: *idem, Proc. Nat. Acad. Sci. USA* **46**, 302 (1960); in guinea pig prostate: G. P. Harper *et al., Nature* **279**, 160 (1979). Occurs in mouse submaxillary gland as a complex (known as 7S NGF) composed of α-, β-, and γ-subunits, of which only the β-subunit is bioactive. Characterization of subunits: S. Varon *et al., Biochemistry* **7**, 1296 (1968); A. C. Server, E. M. Shooter, *Adv. Protein Chem.* **31**, 339 (1977). Amino acid sequence of mouse NGF: R. H. Angeletti, R. A. Bradshaw, *Proc. Nat. Acad. Sci. USA* **68**, 2417 (1971). Cloning of mouse NGF: J. Scott *et al., Nature* **302**, 538 (1983); of human: A. Ullrich *et al., ibid.* **303**, 821 (1983). Crystal structure: N. Q. McDonald *et al., ibid.* **354**, 411 (1991). Review of NGF receptor: P. A. Barker, R. A. Murphy, *Mol. Cell Biochem.* **110**, 1-15 (1992). Review of pharmacology and potential role in neurodegenerative disease: F. Hefti, P. A. Lapchak, *Adv. Pharmacol.* **24**, 239-273 (1993); in peripheral nerve regeneration: G. Raivich, G. W. Kreutzberg, *Clin. Neurol. Neurosurg.* **95** Suppl., S84-S88 (1993). *Reviews:* R. Levi-Montalcini, R. H. Angeletti, *Physiol. Rev.* **48**, 534-569 (1968); H. Thoenen, Y. A. Barde, *ibid.* **60**, 1284-1335 (1980); R. Levi-Montalcini, *Science* **237**, 1154-1162 (1987); R. A. Bradshaw *et al., Trends Biochem. Sci.* **18**, 48-52 (1993). *Book:* Neurotrophic Factors, S. E. Loughlin, J. H. Fallon, Eds. (Academic Press, San Diego, 1993) 607 pp.

6563. Netilmicin. *O-3-Deoxy-4-C-methyl-3-(methyl-amino)-β-L-arabinopyranosyl-(1→6)-O-[2,6-diamino-2,3,4,6-tetradeoxy-α-D-glycero-hex-4-enopyranosyl-(1→4)]-2-deoxy-N^1-ethyl-D-streptamine; (2S-cis)-4-O-[3-amino-6-(aminomethyl)-3,4-dihydro-2H-pyran-2-yl]-2-deoxy-6-O-[3-deoxy-4-C-methyl-3-(methylamino)-β-L-arabinopyranosyl]-N^1-ethyl-D-streptamine;* 1-N-ethylsisomicin; Sch-20569. $C_{21}H_{41}N_5O_7$; mol wt 475.59. C 53.04%, H 8.69%, N 14.73%, O 23.55%. Broad spectrum semi-synthetic aminoglycoside antibiotic, related to sisomicin, *q.v.* Prepn: M. J. Weinstein *et al., Ger.* pat. **2,437,160** (1975 to Sherico); J. J. Wright, U.S. pat.

4,029,882 (1977 to Schering); J. J. Wright *et al.*, U.S. pat. **4,002,742** (1977 to Schering). Structure and synthesis: J. J. Wright, *Chem. Commun.* **1976**, 206. Biological activity: G. H. Miller *et al., Antimicrob. Ag. Chemother.* **10**, 827 (1976). Radioimmunoassay: A. Broughton *et al., Clin. Chem.* **24**, 717 (1978). Pharmacology: W. Raab, *Adv. Clin. Pharmacol.* **15**, 91 (1978); I. Trestman *et al., Antimicrob. Ag. Chemother.* **13**, 832 (1978). Metabolism and pharmacokinetics: J. C. Pechere *et al., Clin. Pharmacol. Ther.* **23**, 677 (1978); R. E. Brummett *et al., Arch. Otolaryngol.* **104**, 579 (1978). Clinical studies: J. Klastersky *et al., Antimicrob. Ag. Chemother.* **12**, 503 (1977); A. P. Panwalker *et al., ibid.* **13**, 170 (1978). Toxicity studies: L. Albiero *et al., Arch. Int. Pharmacodyn. Ther.* **233**, 343 (1978); F. C. Lust, *J. Int. Med. Res.* **6**, 286 (1978). *Review:* P. Noone, *Drugs* **27**, 548-578 (1984).

$[\alpha]_D^{26}$ +164° (c = 3 in water). LD_{50} in mice (mg/kg): 40 i.v.; 125 i.p.; 175 s.c. (Miller).
Sulfate, $(C_{21}H_{41}N_5O_7)_2.5H_2SO_4$, Certomycin, Netillin, Netilyn, Netromicine, Netromycin, Nettacin, Vectacin, Zetamicin.
THERAP CAT: Antibacterial.

6564. Netobimin. *2-[[[(Methoxycarbonyl)amino][[2-nitro-5-(propylthio)phenyl]amino]methylene]amino]ethanesulfonic acid;* 2-[[[(methoxycarbonyl)amino][[2-nitro-5-(propylthio)phenyl]imino]methyl]amino]ethanesulfonic acid; methyl [N'-[2-nitro-5-(propylthio)phenyl]-N-(2-sulfoethyl)-amidino]carbamate; N-methoxycarbonyl-N'-[2-nitro-5-(propylthio)phenyl]-N''-2-(ethylsulfonic acid)guanidine; totabin; Sch-32481; Hapadex. $C_{14}H_{20}N_4O_7S_2$; mol wt 420.47. C 39.99%, H 4.79%, N 13.32%, O 26.64%, S 15.25%. Phenylguanidine anthelmintic. Prepn: M. M. Nafissi-Varchei, **Eur.** pat. **Appl.** 50,286; *idem,* U.S. pat. **4,406,893** (1982, 1983 both to Schering). Metabolism in animals: P. Delatour *et al., J. Vet. Pharmacol. Therap.* **9**, 230 (1986). Anthelmintic efficacy in cattle: J. C. Williams *et al., Am. J. Vet. Res.* **46**, 2188 (1985).

mp 215° (dec).
Sodium salt, $C_{14}H_{20}N_4NaO_7S_2$, crystals, mp 150° (dec) (Nafissi-Varchei); also reported as yellow powder, mp 160° (Delatour). uv max (methanol): 225, 347 nm. Sol in water, acetone, methanol, dimethylformamide, DMSO. Insol in ether.
THERAP CAT (VET): Anthelmintic.

6565. Netropsin. *4-[[[(Aminoiminomethyl)amino]acetyl]amino]-N-[5-[[(3-amino-3-iminopropyl)amino]carbonyl]-1-methyl-1H-pyrrol-3-yl]-1-methyl-1H-pyrrole-2-carboxamide; N'-(2-amidinoethyl)-4-(2-guanidinoacetamido)-1,1'-dimethyl-N,4'-bi[pyrrole-2-carboxamide];* sinanomycin; congocidine; T-1384. $C_{18}H_{26}N_{10}O_3$; mol wt 430.47. C 50.22%, H 6.09%, N 32.54%, O 11.15%. Basic oligopeptide antibiotic produced by *Streptomyces netropsis* with wide range of antimicrobial activity. Prepn and antibacterial spectrum: A. C. Finlay *et al., J. Am. Chem. Soc.* **73**, 341 (1951); and trypanocidal activity: C. Cosar *et al., Compt. Rend.* **234**, 1498 (1952). Antiviral activity: F. M. Schabel *et al., Proc. Soc. Exp. Biol. Med.* **83**, 1 (1953). Identity with sinanomycin:

K. Watanabe, *J. Antibiot.* **9A**, 102 (1956). Structural studies: M. Julia, N. Joseph, *Compt. Rend.* **243**, 961 (1956). Structure: C. W. Waller *et al., J. Am. Chem. Soc.* **79**, 1265 (1957). Revised structure: M. Julia, N. Préau-Joseph, *Compt. Rend.* **257**, 1115 (1963); *eidem, Bull. Soc. Chim. France* **1967**, 4348. Specific binding to DNA: Ch. Zimmer *et al., J. Mol. Biol.* **58**, 329 (1971); K. Zakrzewska *et al., Nucleic Acids Res.* **11**, 8825, 8841 (1983). Use in gradient separation of DNA: K. M. Tatti *et al., Anal. Biochem.* **89**, 561 (1978). Selective blocking of DNAase I cleavage: Ch. Zimmer *et al., Nucleic Acids Res.* **8**, 2999 (1980). Review of activity: F. E. Hahn in *Antibiotics* **vol. 3**, J. W. Corcoran, F. E. Hahn, Eds. (Springer-Verlag, New York, 1975) pp 79-100. Review of effect on DNA structure and function: Ch. Zimmer, *Prog. Nucleic Acid Res. Mol. Biol.* **15**, 285-318 (1975).

Disulfate, $C_{18}H_{26}N_{10}O_3 \cdot 2H_2SO_4$, needles, mp 288°. Soly in water $\sim$30 mg/ml at 80°, <0.5 mg/ml at 25°. Practically insol in the common organic solvents.

Dihydrochloride, $C_{18}H_{26}N_{10}O_3 \cdot 2HCl$, prisms, mp 228°. LD_{50} in mice (mg/kg): 17 i.v.; 70 s.c.; > 300 orally (Finlay).

USE: Non-intercalative DNA binding agent.

6566. Neuraminic Acid. Prehemataminic acid. C_9H_{17}-NO_8; mol wt 267.24. C 40.45%, H 6.41%, N 5.24%, O 47.90%. Parent acid of a family of amino sugars contg 9 or more carbon atoms (sialic acids or nonulosaminic acids). May be regarded as the aldol condensation product of pyruvic acid and *N*-acetyl-D-mannosamine. Neuraminic acid has not been isolated and characterized as such, but several *N*-and *O*-substituted derivatives are widely distributed in nature. *Review:* Whelan, *Ann. Repts. Progr. Chem.* **54**, 319 (1958); A. Gottschalk, *The Chemistry and Biology of Sialic Acids and Related Substances* (Cambridge Univ. Press, 1960). *See also* Sialic Acids.

6567. Neurine. *N,N,N-Trimethylethenaminium hydroxide; trimethylvinylammonium hydroxide.* $C_5H_{13}NO$; mol wt 103.16. C 58.21%, H 12.70%, N 13.58%, O 15.51%. $CH_2=CHN(CH_3)_3OH$. Found in egg yolk, brain, bile, in cadavers. Formed during putrefaction by dehydration of choline: Hofmann, *Compt. Rend.* **47**, 559 (1858); Renshaw, Ware, *J. Am. Chem. Soc.* **47**, 2992 (1925). Prepn: Meyer, Hopff, *Ber.* **54**, 2277 (1921); from aq trimethylamine and acetylene: Gardner *et al., J. Chem. Soc.* **1949**, 789. Separation and identification in biological fluids: E. Merlevede *et al., Arch. Int. Pharmacodyn.* **122**, 474 (1959). Stereoelectronic config: *Intern. Kongr. Entomol. Verhandl., 11th, Vienna*, **1960**, No. 3, pp 87-93, *C.A.* **57**, 7749b (1962). Toxicity study: Hunt, *J. Pharmacol. Exp. Ther.* **28**, 267 (1926). Syrupy liq. Fishy odor. *Poisonous!* Forms a crystalline trihydrate. Alkaline reaction. Readily absorbs CO_2 from the air. Sol in water, alcohol. Dec readily, forming trimethylamine. Forms an HCl salt. LD s.c. in mice: 46 mg/kg (Hunt).

6568. Neuropeptide Y. Neuropeptide tyrosine; NPY. Neuroendocrine peptide that stimulates ingestive behavior. Named for its five tyrosine (Y) residues. Member of a structurally related peptide family which includes peptide YY and pancreatic polypeptide. Most abundant known peptide in mammalian brain; widely distributed in the neurons of the central and peripheral nervous system and adrenal medullary cells. Composed of a highly conserved sequence of 36 amino acids. Exhibits wide range of physiological effects including anxiolysis, sedation, vasoconstriction, and the stimulation of food intake. Isoln from porcine brain: K. Tatemoto *et al., Nature* **296**, 659 (1982). Amino acid sequence: *idem, Proc. Nat. Acad. Sci. USA* **79**, 5485 (1982). Distribution in human brain: T. E. Adrian *et al., Nature* **306**, 584 (1983). Role of NPY in the antiobesity effect of the *obese* gene product (*see also* Leptin): T. W. Stephens *et al., Nature* **377**, 530 (1995). Conference proceedings: *Neuropeptide Y: Karolinska Inst. Nobel Conf. Series*, V. Mutt *et al.,* Eds. (Raven Press, N.Y., 1989) 353 p. Book: *The Biology of Neuropeptide Y and Related Peptides*, W. F. Colmers, C. Wahlestedt, Eds. (Humana Press, Totowa, N. J., 1993) 564 p. Review of pharmacology: J. G. Wettstein *et al., Pharmacol. Ther.* **65**, 397-414 (1995); of role as potential therapeutic target: L. Grundemar, R. Hakanson, *Trends Pharmacol. Sci.* **15**, 153-159 (1994).

6569. Neurotensin. A basic tridecapeptide, found in mammalian brain and gut, having a wide variety of hormone-like activities. Neurotensin has been shown to induce hypotension in the rat, to stimulate contraction of guinea pig ileum and rat uterus, and to cause relaxation of rat duodenum. There is also evidence that it acts as a CNS neurotransmitter. Isoln from bovine hypothalamus: R. Carraway, S. Leeman, *J. Biol. Chem.* **248**, 6854 (1973). Amino acid sequence of bovine neurotensin: *eidem, ibid.* **250**, 1907 (1975). Synthesis: *eidem, ibid.* 1912; K. Kitagawa *et al., Chem. Pharm. Bull.* **24**, 2692 (1976); H. Yajima *et al., Int. J. Peptide Protein Res.* **14**, 169 (1979). Isoln, amino acid sequence of chicken neurotensin: R. Carraway, Y. M. Bhatnagar, *Peptides* **1**, 167 (1980). Synthesis: H. Yajima *et al., Chem. Pharm. Bull.* **29**, 2587 (1981). Radioimmunoassay and differential distribution of bovine neurotensin in CNS, small intestine, and stomach: R. Carraway, S. Leeman, *J. Biol. Chem.* **251**, 7035, 7045 (1976). CNS effects: C. B. Nemeroff *et al., Brain Res.* **128**, 485 (1977). Receptor binding in brain: G. R. Uhl *et al., ibid.* **130**, 299 (1977). Pharmacokinetics and effect on gastrointestinal and pituitary hormones in man: A. M. Blackburn *et al., J. Clin. Endocrinol. Metab.* **51**, 1257 (1980). Reviews of biological activity: G. R. Uhl, S. H. Snyder, *Advan. Biochem. Psychopharmacol.* **28**, 87-106 (1981); D. R. Brown, R. J. Miller, *Ann. Rep. Med. Chem.* **17**, 271-280 (1982). Book: *Ann. N.Y. Acad. Sci.* **400**, entitled "Neurotensin, a Brain and Gastrointestinal Peptide", C. B. Nemeroff, A. J. Prange, Eds. (1982) pp 1-444.

pyroGlu–Leu–Tyr–Glu–Asn–Lys–Pro–Arg–Arg–Pro–Tyr–Ile–LeuOH

Bovine Neurotensin

Bovine neurotensin triacetate hexahydrate, $C_{84}H_{133}N_{21}$-$O_{26} \cdot 6H_2O$, *neurotensin(ox) triacetate(salt).* Fluffy white powder. $[\alpha]_D^{23}$ −65.6° (c = 0.5 in water).

6570. Neurotrophins. Family of highly conserved, target-derived, secretory proteins that promote the differentiation, growth, and survival of neurons in the peripheral and central nervous systems. Produced in limiting amounts by innervated tissues to enhance neuronal differentiation and survival. Five closely related factors have been identified: nerve growth factor (NGF), *q.v.*, brain-derived neurotrophic factor (BDNF), neurotrophin-3 (NT-3), NT-4, and NT-5. Neurotrophins differ in their sites of synthesis, developmental patterns of expression, and neuronal targets. Amino acid sequences are highly homologous. Most of the sequence differences are contained in 4 specific regions which are presumed to be responsible for the biological specificities of the factors. Purification of BDNF from pig brain: Y.-A. Barde *et al., EMBO J.* **1**, 549 (1982). Cloning and expression of BDNF: J. Leibrock *et al., Nature* **341**, 149 (1989). Identification of rat NT-3: P. C. Maisonpierre *et al., Science* **247**, 1446 (1990); of human NT-3: Y. Kaisho *et al., FEBS Lett.* **266**, 187 (1990); A. Rosenthal *et al., Neuron* **4**, 767 (1990). Isolation of NT-4 from *Xenopus laevis*: F. Hallböök *et al., ibid.* **6**, 845 (1991). Identification of NT-

5 in rat and human: L. R. Berkemeier *et al., ibid.* **7**, 857 (1991). Proposed identity of NT-5 as the mammalian homolog of Xenopus NT-4: N. Y. Ip *et al., Proc. Nat. Acad. Sci. USA* **89**, 3060 (1992). Comparison of neurotrophins and other neurotrophic factors: H. Thoenen, *Trends Neurosci.* **14**, 165 (1991). Discussion of neurotrophin receptors: S. O. Meakin, E. M. Shooter, *ibid.* **15**, 323 (1992). Review of amino acid sequences and molecular structure: T. Ebendal, *J. Neurosci. Res.* **32**, 461-470 (1992). Review of bioactivity and distribution in the nervous system: G. Vantini, *Psychoneuroendocrinology* **17**, 401-410 (1992); of bioactivity and therapeutic potential in neurologic disease: F. F. Eide *et al., Exp. Neurol.* **121**, 200-214 (1993).

Brain-derived neurotrophic factor, BDNF. 119 amino acid peptide; mol wt ~13.5 kDa. pI ~9.99. Localized primarily, but not exclusively, in CNS. Amino acid sequence identical in human, pig, mouse and rat. Exhibits ~50% sequence homology with NGF.

Neurotrophin-3, NT-3, HDNF, hippocampal-derived neurotrophic factor, nerve growth factor-2, NGF-2, neuronotrophin-3. 119 amino acid peptide; mol wt ~13.5 kDa. pI ~9.5. Produced primarily in peripheral tissues; also produced in the CNS. Amino acid sequence identical in human, mouse and rat. Exhibits ~57% sequence homology with NGF; ~58% homology with BDNF.

6571. Neutral Red. $N^8,N^8,3$-*Trimethyl-2,8-phenazinediamine monohydrochloride; C.I. Basic Red 5;* 3-amino-7-dimethylamino-2-methylphenazine hydrochloride; toluylene red; neutral red chloride; nuclear fast red; aminodimethylaminotoluaminozine hydrochloride; C.I. 50040; Kernechtrot; Michrome no. 226. $C_{15}H_{17}ClN_4$; mol wt 288.78. C 62.39%, H 5.93%, Cl 12.28%, N 19.40%. Prepn from N,N-dimethyl-*p*-nitrosoaniline hydrochloride + toluene-2,4-diamine: Witt, *Ber.* **12**, 931 (1879); **Ger.** pat. **15,272** (1880), *Frdl.* **1**, 274; Bernthsen, Schweitzer, *Ann.* **236**, 332 (1886); Pokorny, *J. Soc. Dyers Colour.* **42**, 347 (1926). Toxicity data: F. Stolarsky, T. J. Haley, *Fed. Proc.* **10**, 337 (1951). Brief review: H. J. Conn's *Biological Stains*, R. D. Lillie, Ed. (Williams & Wilkins, Baltimore, 9th ed., 1977) pp 377-380, 596. *See also: Colour Index* vol. **4** (3rd ed., 1971) p 4446.

Dark green powder. Absorption max (50% alc): 533 nm. pKa 6.7. Soluble in water or alcohol with red color. Soly in water 4.0%, in abs alcohol 1.8%, in Cellosolve 3.75%, in ethylene glycol 3.0%. Practically insol in xylene. LD_{50} in mice, rats, rabbits (mg/kg): 142, 112, 97 i.v. (Stolarsky, Haley).

USE: As indicator (0.1% in 60% alcohol) for alkalinity of water, urea, nitrite, etc. pH: 6.8 red; 8.0 yellow. Also used for preparing neutral-red paper. As biological stain (study of the Golgi apparatus in cells).

6572. Neutral Spirits. Distilled from suitable raw materials, is 95% ethanol (v/v) meaning that it is at least 190 proof when distilled. Used for blending with straight whisky (*see* Whisky) and for making gin, cordials, liqueurs and vodka. It may be made from grain or molasses or by redistillation of other beverages, such as brandy or rum, and is almost colorless and contains no extraneous flavoring.

6573. Nevirapine. *11-Cyclopropyl-5,11-dihydro-4-methyl-6H-dipyrido[3,2-b:2',3'-e][1,4]diazepin-6-one;* BI-RG-587. $C_{15}H_{14}N_4O$; mol wt 266.30. C 67.65%, H 5.30%, N 21.04%, O 6.01%. Non-nucleoside, specific HIV-1 reverse transcriptase inhibitor. Prepn: K. D. Hargrave *et al.*, **Eur. pat. Appl. 429,987** (1991 to Boehringer, Ing; Thomae), *C.A.* **115**, 114559n (1991); K. D. Hargrave *et al., J. Med. Chem.* **34**, 2231 (1991). Specific enzyme inhibition: V. J. Merluzzi *et al., Science* **250**, 1411 (1990). Mechanism of action study: R. A. Koup *et al., J. Infect. Dis.* **163**, 966 (1991). Binding

study: J. C. Wu *et al., Biochemistry* **30**, 2022 (1991). Crystal structure: P. W. Mui *et al., J. Med. Chem.* **35**, 201 (1992). Synergism with zidovudine, *q.v., in vitro:* D. Richman *et al., Antimicrob. Ag. Chemother.* **35**, 305 (1991). Pharmacokinetics: S. H. Cheeseman *et al., ibid.* **37**, 178 (1993). *Reviews:* P. M. Grob *et al., AIDS Res. Hum. Retroviruses.* **8**, 145-152 (1992); V. J. Merluzzi, A. S. Rosenthal, *Adv. Exp. Med. Biol.* **312**, 89-94 (1992).

Crystals from pyridine + water, mp 247-249°. pKa 2.8. Lipophilic; partition coefficient 83. Soly in water ~0.1 mg/ml at neutral pH; highly sol at pH < 3.

THERAP CAT: Antiviral.

6574. Niacinamide. *3-Pyridinecarboxamide;* nicotinamide; nicotinic acid amide; Aminicotin; Benicot; Dipegyl; Nicamindon; Nicobion; Nicotamide; Nicotilamide; Pelmin. $C_6H_6N_2O$; mol wt 122.13. C 59.01%, H 4.95%, N 22.94%, O 13.10%. Precursor of the coenzymes NAD and NADP, *q.q.v.* Isolation: H. v. Euler *et al., Z. Physiol. Chem.* **258**, 212 (1939); P. Karrer, H. Keller, *Helv. Chim. Acta* **22**, 1292 (1939). Prepn from 3-cyanopyridine: E. J. Gasson, D. J. Hadley, U.S. pat. **2,904,552** (1959 to Distillers). Alternately prepd by passing NH_3 gas into molten nicotinic acid: A. Truchan, J. B. Davidson, U.S. pat. **2,993,051** (1961 to Cowles Chem.). Isoln and analysis by liquid chromatography: C. Bernofsky, *Methods Enzymol.* **66**, 23 (1980). Extraction from cereal products: R. B. Roy, J. J. Merten, *J. Assoc. Off. Anal. Chem.* **66**, 291 (1983). Toxicity data: F. G. Brazda, R. A. Coulson, *Proc. Soc. Exp. Biol. Med.* **62**, 19 (1946). *Review:* "Niacin: Nicotinic Acid, Nicotinamide, NAD(P)" in *Vitamins*, W. Friedrich, Ed. (Walter de Gruyter, Berlin, 1988) pp 473-542. Comprehensive description: E. M. Abdel Moety *et al., Anal. Profiles Drug Subs.* **20**, 475-555 (1991).

Needles from benzene, mp 128-131°. Distills at 150-160° at 5×10^{-4} mm Hg. Absorption spectrum: Kuhn, Vetter, *Ber.* **68**, 2374 (1935). uv max: 261 nm ($A_{1cm}^{1\%}$ 451). pK (20°C): 3.3. One gram dissolves in about one ml water, in about 1.5 ml alcohol, in 10 ml glycerol. A 10% w/v soln in water is neutral to litmus. Forms cryst salts with acids. LD_{50} s.c. in rats: 1.68 g/kg (Brazda, Coulson).

Note: The terms *niacin*, *Vitamin B_3*, and *Vitamin PP* have been used to refer to both nicotinamide and nicotinic acid, *q.v.*

THERAP CAT: Vitamin (enzyme cofactor).

THERAP CAT (VET): Vitamin (enzyme cofactor).

6575. Nialamide. *4-Pyridinecarboxylic acid 2-[3-oxo-3-[(phenylmethyl)amino]propyl]hydrazide; isonicotinic acid 2-[2-(benzylcarbamoyl)ethyl]hydrazide;* 1-[2-(benzylcarbamoyl)ethyl]-2-isonicotinoylhydrazide; N-[2-(benzylcarbamoyl)ethylamino]isonicotinamide; N-benzyl-β-(isonicotinoylhydrazino)propionamide; N-isonicotinoyl-N'-[β-(N-benzylcarboxamido)ethyl]hydrazide; Espril; Niamid; Niamidal; Niaquitil; Nuredal; Nyazin. $C_{16}H_{18}N_4O_2$; mol wt 298.34. C 64.41%, H 6.08%, N 18.78%, O 10.73%. Prepn: Bloom, Carnahan, U.S. pats. **2,894,972; 3,040,061** (1959, 1962, both to Pfizer).

Slightly bitter crystals from ethyl acetate, mp 152-153°. Sparingly sol in water; freely sol in acidic solvents. LD_{50} in mice (mg/kg): 590 orally; 435 i.p.; 120 i.v., *Toxic Substances List*, H. E. Christensen, Ed. (1974) p 436.

THERAP CAT: Antidepressant.

6576. Niaprazine. *N-[3-[4-(4-Fluorophenyl)-1-piperazinyl]-1-methylpropyl]-3-pyridinecarboxamide; N-[3-[4-(p-fluorophenyl)-1-piperazinyl]-1-methylpropyl]nicotinamide;* Nopron. $C_{20}H_{25}FN_4O$; mol wt 356.44. C 67.39%, H 7.07%, F 5.33%, N 15.72%, O 4.49%. Prepn: R. Y. Mauvernay, **Ger.** pat. **1,957,371** corresp to U.S. pat. **3,712,893** (1970, 1973 both to CERM). Pharmacology: P. Duchene-Marrullaz *et al.*, *Therapie* **26**, 1203 (1971). Psycholeptic effects in mice: J. Hache, J. Tachon, *J. Pharmacol.* **7**, 469 (1976). Brain catecholamine depletion: P. E. Keane, M. S. Benedetti, *Neuropharmacology* **18**, 595 (1979).

Cryst from ethyl acetate, mp 131°. Photosensitive. LD_{50} in mice (mg/kg): 890 orally, 145 i.v. (Mauvernay).

THERAP CAT: Sedative; hypnotic.

6577. Nicametate. *3-Pyridinecarboxylic acid 2-(diethylamino)ethyl ester; nicotinic acid 2-diethylaminoethyl ester;* diethylaminoethyl 3-pyridinecarboxylate; diethylaminoethyl nicotinate; Eucast. $C_{12}H_{18}N_2O_2$; mol wt 222.29. C 64.84%, H 8.16%, N 12.60%, O 14.40%. Prepd from nicotinoyl chloride and diethylaminoethanol or from nicotinic acid and diethylaminoethyl chloride: Blicke, Jenner, *J. Am. Chem. Soc.* **64**, 1722 (1942). Clinical trial in vascular pathology: P. Langeron, *J. Sci. Med. Lille* **90**, 179 (1972). Use of the citrate in cardiovascular surgery: C. Stankowiak *et al.*, *Lille Med.* **18**, 1039 (1973).

Liquid. bp_{10} 155-157°; bp_2 120-125°.

Hydrochloride, $C_{12}H_{18}N_2O_2$.HCl, crystals from acetone, ethanol or isopropyl alcohol, mp 128-129°.

Citrate monohydrate, $C_{12}H_{18}N_2O_2$.$C_6H_8O_7$.H_2O, *Euclidan, Nutrin, Soclidan.*

THERAP CAT: Vasodilator (cerebral, peripheral).

6578. Nicarbazin. *N,N'-Bis(4-nitrophenyl)urea, compd with 4,6-dimethyl-2(1H)-pyrimidinone (1:1); 4,4'-dinitrocarbanilide compd with 4,6-dimethyl-2-pyrimidinol (1:1);* Nicarb; Nicoxin; Nicrazin. $C_{19}H_{18}N_6O_6$; mol wt 426.39. C 53.52%, H 4.26%, N 19.71%, O 22.51%. Prepn: Cuckler *et al.*, *Science* **122**, 244 (1955); Basso, O'Neill, U.S. pats. **2,731,-382; 2,731,383; 2,731,384** (all 1956 to Merck & Co.). Effect on nutritional encephalopathy in chicks: I. Bartov, P. Budowski, *Poultry Sci.* **58**, 597 (1979).

Crystals, dec 265-275°. uv max (concd H_2SO_4): 298 nm ($A_{1cm}^{1\%}$ 670). Practically insol in water. Complex is slowly dec by trituration with water, or more rapidly by dil aq acids. The dry crystals are strongly electrostatic and present some dry-mixing problems.

THERAP CAT (VET): Coccidiostat.

6579. Nicardipine. *1,4-Dihydro-2,6-dimethyl-4-(3-nitrophenyl)-3,5-pyridinedicarboxylic acid methyl 2-[methyl-(phenylmethyl)amino]ethyl ester.* $C_{26}H_{29}N_3O_6$; mol wt 479.53. C 65.12%, H 6.10%, N 8.76%, O 20.02%. Dihydropyridine calcium channel blocker. Prepn: M. Murakami *et al.*, **Belg.** pat. **811,324**; *eidem*, U.S. pat. **3,985,758** (1974, 1976 both to Yamanouchi); M. Iwanami *et al.*, *Chem. Pharm. Bull.* **27**, 1426 (1979). Vasodilator profile: T. Takenaka *et al.*, *Arzneimittel-Forsch.* **26**, 2172 (1976). Absorption, excretion, metabolism: S. Higuchi *et al.*, *Xenobiotica* **7**, 469 (1977). Clinical pharmacology: T. Seki, T. Takenaka, *Int. J. Clin. Pharmacol. Biopharm.* **15**, 267 (1977). Mechanism of action: K. Satoh *et al.*, *Clin. Exp. Pharmacol. Physiol.* **7**, 249 (1980). Acute toxicity study: Y. Odani, T. Sado, *Oyo Yakuri* **18**, 301 (1979), *C.A.* **92**, 104384u (1980). Symposium: *Brit. J. Clin. Pharmacol.* **20**, Suppl. 1, 1S-208S (1985). Clinical synergy with captopril, *q.v.*: K. A. Conrad *et al.*, *Clin. Pharmacol. Ther.* **42**, 113 (1987). Review of pharmacodynamics, pharmacokinetics and therapeutic efficacy: E. M. Sorkin, S. P. Clissold, *Drugs* **33**, 296-345 (1987).

Hydrochloride, $C_{26}H_{29}N_3O_6$.HCl, RS-69216, YC-93, Barizin, Bionicard, Cardene, Dacarel, Lecibral, Lescodil, Loxen, Nerdipina, Nicant, Nicardal, Nicarpin, Nicapress, Nicodel, Nimicor, Perdipina, Perdipine, Ranvil, Ridene, Rycarden, Rydene, Vasodin, Vasonase. Isolated in two crystalline forms from acetone. α-form: mp 179-181°; β-form: mp 168-170°. The forms also have different IR and x-ray diffraction patterns. LD_{50} in male, female rats (mg/kg): 634, 557 orally; 18.1, 25.0 i.v.; in male, female mice: 634, 650 orally; 20.7, 19.9 i.v. (Odani, Sado).

THERAP CAT: Antianginal; antihypertensive.

6580. Nicergoline. *(8β)-10-Methoxy-1,6-dimethylergoline-8-methanol 5-bromo-3-pyridinecarboxylate (ester);* 1-methyllumilysergol 8-(5-bromonicotinate) 10-methyl ether; 4,6,6a,7,8,9,10,10a-octahydro-10aα-methoxy-4,7-dimethylindolo[4,3-fg]quinoline-9-methanol 5-bromonicotinate; 8β-[(5-bromonicotinoyloxy)methyl]-1,6-dimethyl-10α-methoxyergoline; nicotergoline; nimergoline; MNE; FI-6714; Cergodum; Circo-Maren; Dilasenil; Duracebrol; Ergotop; Ergobel; Memoq; Nicergolent; Sermion; Vasospan. $C_{24}H_{26}$-BrN_3O_3; mol wt 484.39. C 59.51%, H 5.41%, Br 16.50%, N 8.67%, O 9.91%. Prepn: Bernardi *et al.*, U.S. pat. **3,228,-943**; Temperilli, **Ger.** pat. **2,112,273** (1966, 1971, both to Farmitalia); Arcari *et al.*, *Experientia* **28**, 819 (1972). Series of articles on pharmacology, clinical studies, tolerability: *Arzneimittel-Forsch.* **29**, 1213-1316 (1979). Toxicity study: B. W. Neumann, F. Lauschner, *ibid.* 1206. Hemodynamic effects in the dog: *ibid.* **31**, 1693 (1981). Use in acute myocardial infarction with diastolic hypertension: E. Triulzi *et al.*, *Farmaco Ed. Prat.* **36**, 449 (1981).

mp 136-138°. LD$_{50}$ in male mice, rats (mg/kg): 860, 2800 orally; 46, 43 i.v. (Neumann, Lauschner).
THERAP CAT: Vasodilator (cerebral, peripheral).

6581. Niceritrol. *3-Pyridinecarboxylic acid 2,2-bis[[(3-pyridinylcarbonyl)oxy]methyl]-1,3-propanediyl ester; nicotinic acid neopentanetetrayl ester;* pentaerythritol tetranicotinate; 8-AL; Perycit; Bufor. $C_{29}H_{24}N_4O_8$; mol wt 556.53. C 62.59%, H 4.35%, N 10.07%, O 23.00%. Prepn: Fr. pat. M2046 (1963 to Societé d'Etudes et de Recherches Pharmacotechniques), *C.A.* **60**, 10656e (1964); Brit. pats. **1,022,880**; **1,053,689** (1966, 1967 to Aktiebolag Bofors), *C.A.* **64**, 19708d (1966), *C.A.* **66**, 104907e (1967); Kuriyama, Kudo, Japan. pat. **67 2359** (1967 to Yoshitomi), *C.A.* **66**, 115935p (1967). Acute toxicity data: T. Sugawara *et al., Oyo Yakuri* **14**, 741 (1977), *C.A.* **88**, 131005v (1977).

Crystals, mp reported as 160-162° and as 163-164°. LD$_{50}$ in mice, rats (g/kg): >20, >20 orally; >5, >5 s.c.; >5, >5 i.p.; in rabbits (g/kg): >10 orally; >5 i.p. (Sugawara).
THERAP CAT: Antihyperlipoproteinemic.

6582. Nickel. Ni; at. wt 58.6934; at. no. 28; valence 2; seldom 1, 3, 4. Group VIII(10). Naturally occurring isotopes (mass numbers): 58 (68.27%), 60 (26.10%), 61 (1.13%), 62 (3.59%), 64 (0.91%); known artificial radioactive isotopes: 53, 55-57, 59 (longest-lived known isotope, T$_{1/2}$ 7.5 × 10^4 years, decay by electron capture), 63, 65-67. Abundance in earth's crust 99 ppm. Discovered by Cronstedt in 1754: Cronstedt, *Mineralogie* (Stockholm, 1758) p 218. Isoln: Berthier, *Ann. Chim. Phys.* [2] **14**, 52 (1820); **25**, 94 (1824). Occurs free in meteorites. Found in many ores as sulfides, arsenides, antimonides and oxides or silicates; chief sources include chalcopyrite, *q.v.*, pyrrhotite, *pentlandite* [(Fe,Ni)$_9$S$_8$] and *garnierite* [3(Mg,Ni)O.2SiO$_2$.2H$_2$O]; other ores include *niccolite* (NiAs) and *millerite* (NiS). Methods of extraction and purification: Mackiw, *Can J. Chem. Eng.* **46**, 3 (1968); Houot, *Ann. Mines* **1969** (April), 9; Queneau, *J. Metals* **22**, 44-48 (1970). Prepn of high purity nickel: Wise, Schaefer, *Metals Alloys* **16**, 424 (1924); from NiO and H$_2$: Glemser in *Handbook of Preparative Inorganic Chemistry* vol. 2, G. Brauer, Ed. (Academic Press, New York, 2nd ed., 1965) pp 1543-1544; by electrolysis: Vu Quang Kinh, Nardin, *Compt. Rend. Ser. C* **266**, 307 (1968). Comprehensive reviews: *Gmelin's, Nickel* (8th ed.) **57**, 5 vols, about 3500 pp (1965-1967); Nicholls in *Comprehensive Inorganic Chemistry* vol. 3, J. C. Bailar, Jr. *et al.,* Eds. (Pergamon Press, Oxford, 1973) pp 1109-1161; J. K. Tien, T. E. Howson in Kirk-Othmer *Encyclopedia of Chemical Technology* vol. 15 (Wiley-Interscience, New York, 3rd ed., 1981) pp 787-801; *Chemistry of the Elements,* N. N. Greenwood, A.

Earnshaw, Eds. (Pergamon Press, New York, 1984) pp 1328-1363. Book: *Nickel Toxicology,* S. S. Brown, F. W. Sunderman, Eds. (Academic Press, New York, 1980) 193 pp. Review of carcinogenicity studies: *IARC Monographs* **11**, 75-112 (1976); of toxicology and human exposure: *Toxicological Profile for Nickel* (PB-160378, 1988) 111 pp.

Lustrous white, hard, ferromagnetic metal; face-centered cubic crystals. mp 1453°. bp (calc) 2732°. Also reported as bp (calc) 2837° (3110°K): D. R. Stull, G. C. Sinke, *Thermodynamic Properties of the Elements,* Advances in Chemistry Series **18** (A.C.S., Washington, 1956). d 8.908. Heat capacity (25°) 6.23 cal/g-atom/°C. Mohs' hardness 3.8. Latent heat of fusion 73 cal/g. Electrical resistivity (20°): 6.844 μohms-cm. E°(aq) Ni/Ni^{2+} 0.250 V. Stable in air at ordinary temp; burns in oxygen, forming NiO; not affected by water; decomposes steam at a red heat. Slowly attacked by dil hydrochloric or sulfuric acid; readily attacked by nitric acid. Not attacked by fused alkali hydroxides.

Caution: Potential symptoms of overexposure are sensitization dermatitis, allergic asthma, pneumonitis. *See NIOSH Pocket Guide to Chemical Hazards* (DHHS/NIOSH 94-116, 1994) p 224. *See also Patty's Industrial Hygiene and Toxicology* vol. 2C, G. D. Clayton, F. E. Clayton, Eds. (John Wiley & Sons, New York, 4th ed., 1994) 2157-2173. This substance and certain nickel compounds may reasonably be anticipated to be carcinogens: *Seventh Annual Report on Carcinogens* (PB95-109781, 1994) p 262.

USE: Nickel-plating; for various alloys such as new silver, Chinese silver, German silver; for coins, electrotypes, storage batteries; magnets, lightning-rod tips, electrical contacts and electrodes, spark plugs, machinery parts; catalyst for hydrogenation of oils and other organic substances. *See also* Raney nickel. Manuf of Monel metal, stainless steels, heat resistant steels, heat and corrosion resistant alloys, nickel-chrome resistance wire; in alloys for electronic and space applications.

6583. Nickel Acetate. C$_4$H$_6$NiO$_4$; mol wt 176.78. C 27.18%, H 3.42%, Ni 33.20%, O 36.20%. Ni(CH$_3$CO$_2$)$_2$.
Tetrahydrate, green, cryst mass or powder; acetic odor. d 1.744. Sol in 6 parts water, in alcohol. *Keep well closed.*
USE: Catalyst; mordant for textiles.

6584. Nickel Acetylacetonate. *Bis(2,4-pentanedionato-O,O')nickel;* bisacetylacetonatonickel(II); bis(2,4-pentanediono)nickel(II); 2,4-pentanedione nickel complex. C$_{10}$H$_{14}$NiO$_4$; mol wt 256.91. C 46.75%, H 5.49%, Ni 22.85%, O 24.91%. Ni(CH$_3$COCHCOCH$_3$)$_2$. Also Ni(acac)$_2$ or Ni-(AA)$_2$. Prepn from acetylacetone and Ni(OH)$_2$: Gach, *Monatsh.* **21**, 103 (1900); from acetylacetone and NiCl$_2$.-6H$_2$O: Charles, Pawlikowski, *J. Phys. Chem.* **62**, 440 (1958); from 4-diethylamino-3-pentene-2-one and NiCl$_2$.-6H$_2$O: Gash, *Can. J. Chem.* **45**, 2109 (1967). *See also* Fernelius, Bryant, *Inorg. Syn.* **5**, 105 (1957). Exists as a trimer in the solid state: Bullen, *Nature* **177**, 537 (1956); Bullen *et al., Inorg. Chem.* **4**, 456 (1965); as a monomer in the vapor phase: Fackler *et al., J. Phys. Chem.* **72**, 4631 (1972). Structure of dihydrate: Montgomery, Lingafelter, *Acta Cryst.* **17**, 1481 (1964).
Emerald-green orthorhombic crystals. mp 229-230°. bp$_{11}$ 220-235°. d^{17} 1.455. uv max (10^{-4}M in CHCl$_3$): 298, 265 nm (log ε 4.34, 4.44). Sol in water, alcohol, chloroform, benzene. Insol in ether, ligroin.
USE: Catalyst.

6585. Nickel Bromide. Nickel dibromide. Br$_2$Ni; mol wt 218.50. Br 73.14%, Ni 26.86%. NiBr$_2$.
Trihydrate, yellowish-green, very deliquesc crystals; loses its water at about 200°, the anhyd salt is a golden-yellow color and sublimable in absence of air. Sol in one part water, in alcohol. *Keep well closed.*

6586. Nickel Carbonate Hydroxide. CH$_4$Ni$_3$O$_7$; mol wt 304.12. C 3.95%, H 1.33%, Ni 57.90%, O 36.83%. NiCO$_3$.2Ni(OH)$_2$. Tetrahydrate occurs in nature as the mineral *zaratite.*
Tetrahydrate, green, odorless powder. Insol in water. Sol in ammonia and in dil acids with effervescence.
USE: Nickel-plating; catalyst for hardening of fats; in ceramic colors and glazes.

6587. Nickel Carbonyl. Nickel tetracarbonyl. C$_4$-NiO$_4$; mol wt 170.74. C 28.14%, Ni 34.38%, O 37.48%. Ni(CO)$_4$. Intermediate in nickel refining. Made by passing carbon monoxide over finely divided nickel: Mond *et al., J. Chem. Soc.* **57,** 749 (1890); Gilliland, Blanchard, *Inorg. Syn.* **2,** 234 (1946). Use of nickel carbonyl in organic synthesis: G. Wilke *et al., Angew. Chem. Int. Ed.* **5,** 151 (1966); M. F. Semmelhack in *Organic Reactions* vol. 19 (Wiley, New York, 1972) p 115; E. J. Corey, H. A. Kirst, *J. Am. Chem. Soc.* **94,** 667 (1972). Kinetic studies: D. H. Stedman *et al., Science* **208,** 1029 (1980). Toxicity study: Hackett, Sunderman, *Arch. Environ. Health* **14,** 604 (1967). *Review:* Nicholls in *Comprehensive Inorganic Chemistry* vol. 3, J. C. Bailar, Jr. *et al.,* Eds. (Pergamon Press, Oxford, 1973) pp 1115-1119.

Colorless, volatile liquid. *Poisonous!* Oxidizes in the air: explodes at about 60°. d^{17} 1.318. bp 43°. mp −19.3°. Crit temp about 200°. Crit pressure about 30 atm. Sol in about 5000 parts water free from air; sol in alcohol, benzene, chloroform, acetone, carbon tetrachloride. LD$_{50}$ in rats (mg/kg): 39 i.p.; 63 s.c.; 66 i.v. (Hackett, Sunderman).

Caution: Potential symptoms of overexposure are sensitization dermatitis, allergic asthma and pneumonitis. *See NIOSH Pocket Guide to Chemical Hazards* (DHHS/NIOSH 90-117, 1990) p 160. *See also Clinical Toxicology of Commercial Products,* R. E. Gosselin *et al.,* Eds. (Williams & Wilkins, Baltimore, 5th ed., 1984) Section II, p. 145. This substance may reasonably be anticipated to be a carcinogen: Seventh Annual Report on Carcinogens (PB95-109781, 1994) p 262.

USE: In organic synthesis; production of high-purity nickel powder and continuous nickel coatings on steel and other metals.

6588. Nickel Chloride. Nickel dichloride. Cl$_2$Ni; mol wt 129.60. Cl 54.71%, Ni 45.29%. NiCl$_2$.

Hexahydrate, green, deliquesc crystals or cryst powder. Monoclinic. Structure reported to be *trans*-[NiCl$_2$(H$_2$O)$_4$].-2H$_2$O. Mizuno, *J. Phys. Soc. Japan* **16,** 1574 (1960), *C.A.* **55,** 26605g (1961). Sol in about one part water, in alcohol. The anhydr salt is golden-yellow, sublimable in absence of air and readily absorbs NH$_3$. The aq soln is acid; pH about 4. *Keep well closed.* LD i.v. in dogs: 40-80 mg/kg, *Handbook of Toxicology,* vol. **1,** W. S. Spector, Ed. (Saunders, Philadelphia, 1956) pp 212-213.

USE: For nickel-plating cast zinc, manuf sympathetic ink. The anhydr salt as absorbent for NH$_3$ in gas masks.

6589. Nickel Cyanide. C$_2$N$_2$Ni; mol wt 110.73. C 21.69%, N 25.30%, Ni 53.01%. Ni(CN)$_2$. Prepn of yellow-brown anhydr salt: Aynsley, Campbell, *J. Chem. Soc.* **1958,** 1723. (The commercial salt usually contains 20-25% water.)

Tetrahydrate, apple-green powder. *Poison!* Insol in water. Slightly sol in dil acids, freely in alkali cyanides, in ammonia, and in ammonium carbonate.

USE: In nickel-plating.

6590. Nickel Dimethylglyoxime. *Bis[(2,3-butanedione dioximato)(1−)-N,N']nickel; bis(dimethylglyoximato)nickel.* C$_8$H$_{14}$N$_4$NiO$_4$; mol wt 288.92. C 33.26%, H 4.88%, N 19.39%, Ni 20.31%, O 22.15%. Prepn: Banks *et al., J. Am. Chem. Soc.* **77,** 324 (1955); F. J. Welcher, *Organic Analytical Reagents* vol. 3 (Van Nostrand, New York, 1947) pp 165-179; Thabet *et al., Inorg. Nucl. Chem. Letters* **8,** 211 (1972). Structure: Godycki, Rundle, *Acta Cryst.* **6,** 487 (1953); Merritt, *Anal. Chem.* **25,** 718 (1953).

Scarlet-red, cryst powder. Sublimes at 250°. Insol in water, acetic acid, ammonia. Sol in dil mineral acids and appreciably sol in abs alcohol.

USE: As sun-fast pigment in paints, lacquers, cellulose compounds and cosmetics.

6591. Nickel Fluoride. Nickel difluoride; nickelous fluoride. F$_2$Ni; mol wt 96.69. F 39.30%, Ni 60.70%. NiF$_2$. Prepn: Henkel, Klemm, *Z. Anorg. Allgem. Chem.* **222,** 74 (1935); Priest, *Inorg. Syn.* **3,** 173 (1950); Rochow, Kukin, *J. Am. Chem. Soc.* **74,** 1615 (1952); Haendler *et al., ibid.* 3167.

Yellowish to green tetragonal crystals (rutile type). d 4.72. Sublimes in HF stream above 1000°. Slightly sol in water (4 g/100 ml at 25°). Aq solns are dec by boiling. Insol in alcohol, ether. LD$_{50}$ in mice: 130 mg/kg i.v., *Toxic Substances List,* H. E. Christensen, Ed. (1973) p 652.

Caution: Chronic exposure may cause mottling of teeth, changes in bones.

6592. Nickel Formate. C$_2$H$_2$NiO$_4$; mol wt 148.73. C 16.15%, H 1.36%, Ni 39.46%, O 43.03%. Ni(HCOO)$_2$. Prepd by reaction of formic acid with Ni: Johnson, U.S. pat. **2,576,072** (1951 to Harshaw Chemical); with NiCO$_3$: Bircumshaw, Edwards, *J. Chem. Soc.* **1950,** 1800.

Dihydrate, fine, green, monoclinic crystals. Becomes anhydr on careful heating to 130-140°; decomposes at 180-200° yielding Ni, CO, CO$_2$, H$_2$, H$_2$O, CH$_4$. d$^{20.2}$ 2.154. Moderately sol in water. Practically insol in alc, formic acid.

USE: Manuf of Ni; prepn of Ni catalysts for organic reactions, particularly hydrogenation catalysts.

6593. Nickel Hydroxide. *"Green nickel oxide".* H$_2$-NiO$_2$; mol wt 92.71. H 2.17%, Ni 63.31%, O 34.52%. Ni-(OH)$_2$.

Monohydrate, apple-green powder. Decomp above 200° to form NiO and H$_2$O. Insol in water. Sol in dil acids, in ammonia.

6594. Nickel Iodide. Nickel diiodide. I$_2$Ni; mol wt 312.50. I 81.22%, Ni 18.78%. NiI$_2$.

Iron-black color. Sublimes in absence of air. Hexahydrate, bluish-green very deliquesc crystals. Very sol in water or alcohol. *Keep well closed.*

6595. Nickel Monoxide. Nickelous oxide; nickel protoxide. NiO; mol wt 74.69. Ni 78.58%, O 21.42%. Occurs as the mineral *bunsenite.*

Green powder; yellow when hot. Insol in water. Sol in acids.

USE: Painting on porcelain.

6596. Nickel Nitrate. N$_2$NiO$_6$; mol wt 182.70. N 15.33%, Ni 32.12%, O 52.54%. Ni(NO$_3$)$_2$.

Hexahydrate, green, deliquesc crystals. d 2.05. mp 56.7°. bp 137°. Sol in 0.4 part water, in alcohol. The aq soln is acid; pH about 4. *Keep well closed.* LD$_{50}$ orally in rats: 1.62 g/kg, H. F. Smyth *et al., Am. Ind. Hyg. Assoc. J.* **30,** 470 (1969).

USE: Nickel-plating; manuf brown ceramic colors.

6597. Nickel Oxalate. C$_2$NiO$_4$; mol wt 146.71. C 16.37%, Ni 40.01%, O 43.62%. NiC$_2$O$_4$.

Dihydrate, light green powder. Insol in water; sol in mineral acids, in solns of ammonium chloride, nitrate, or sulfate.

6598. Nickel Phosphate. Ni$_3$O$_8$P$_2$; mol wt 366.02. Ni 48.11%, O 34.97%, P 16.92%. Ni$_3$(PO$_4$)$_2$.

Octahydrate, light green powder. Insol in water; sol in acids, ammonia.

USE: On ignition yields "nickel yellow"—a pigment used in oil and water colors.

6599. Nickel Sesquioxide. Nickelic oxide; black nickel oxide. Ni$_2$O$_3$; mol wt 165.39. Ni 70.98%, O 29.02%. Contains a variable quantity of water.

Gray-black powder. Dec at ~600° into NiO and oxygen. Insol in water; very slightly sol in cold acid; dissolved by hot HCl with evolution of Cl, and by hot H$_2$SO$_4$ or HNO$_3$ with evolution of oxygen.

6600. Nickel Sulfate. NiO$_4$S; mol wt 154.76. Ni 37.93%, O 41.35%, S 20.72%. NiSO$_4$.

Hexahydrate, two known phases. α-Form, blue to blue-green tetragonal crystals; transition to β-form at 53.3°. β-Form, green transparent crystals; stable at 40°; becomes blue and opaque at room temp. Sweet astringent taste; somewhat efflorescent. Loses 5H$_2$O at ~100°. Greenish-yellow anhyd salt formed at 280°. Sol in 1.4 parts water; sparingly sol in alcohol, more in methanol. The aq soln is acid; pH ~4.5. LD s.c. in guinea pigs: 62 mg/kg, *Handbook of Toxicology* vol. **1**, W. S. Spector, Ed. (Saunders, Philadelphia, 1956) pp 212-213.

Heptahydrate, *single nickel salts.*

USE: In nickel-plating; as mordant in dyeing and printing fabrics; blackening zinc and brass.

6601. Niclofolan. *5,5'-Dichloro-3,3'-dinitro[1,1'-biphenyl]-2,2'-diol; 4,4'-dichloro-6,6'-dinitro-o,o'-biphenol;* 5,5'-dichloro-2,2'-dihydroxy-3,3'-dinitrobiphenyl; menichlopholan; Bayer 9015; ME-3625; Bilevon-M. $C_{12}H_6Cl_2N_2O_6$; mol wt 345.09. C 41.77%, H 1.75%, Cl 20.55%, N 8.12%, O 27.82%. Anthelmintic activity: Meiser, Federmann, U.S. pat. 3,082,151 (1963 to Bayer); P. J. Lane, J. M. Stewart, *Vet. Rec.* **80**, 702 (1967). Pharmacokinetics in desert sheep: B. H. Ali *et al., J. Vet. Pharmacol. Ther.* **13**, 217 (1990).

THERAP CAT (VET): Anthelmintic (fasciolicide).

6602. Niclosamide. *5-Chloro-N-(2-chloro-4-nitrophenyl)-2-hydroxybenzamide;* 2',5-dichloro-4'-nitrosalicylanilide; 5-chloro-N-(2'-chloro-4'-nitrophenyl)salicylamide; 5-chlorosalicyloyl-(o-chloro-p-nitranilide); N-(2'-chloro-4'-nitrophenyl)-5-chlorosalicylamide; Bayer 2353; Cestocide; Niclocide; Ruby; Trédémine; Yomesan. $C_{13}H_8Cl_2N_2O_4$; mol wt 327.12. C 47.73%, H 2.46%, Cl 21.68%, N 8.56%, O 19.56%. Prepn: **Brit.** pat. 824,345 (1959 to Bayer), *C.A.* **54**, 15822b (1960). *See also:* E. Schraufstätter, R. Gönnert, **U.S.** pat. 3,079,297; R. Strufe *et al.*, **U.S.** pat. 3,113,067 (both 1963 to Bayer); Bekhli *et al., Med. Prom. SSSR* **1965**, 25.

Pale yellow crystals, mp 225-230°. Practically insol in water. Sparingly sol in ethanol, chloroform, ether.

Ethanolamine salt, $C_{13}H_8Cl_2N_2O_4.C_2H_7NO$, *clonitrilide, Bayluscid.* Yellow-brown solid, mp 204°.

USE: The ethanolamine salt as a molluscicide.

THERAP CAT: Anthelmintic (Cestodes).

THERAP CAT (VET): Anthelmintic (Cestodes).

6603. Nicoclonate. *3-Pyridinecarboxylic acid 1-(4-chlorophenyl)-2-methylpropyl ester; nicotinic acid p-chloro-α-isopropylbenzyl ester;* 1-(p-chlorophenyl)isobutyl nicotinate; p-chlorophenylisopropylcarbinol nicotinate; 1-(p-chlorophenyl)-1-(nicotinoyloxy)-2-methylpropane; S-486; Lipidium. $C_{16}H_{16}ClNO_2$; mol wt 289.76. C 66.32%, H 5.57%, Cl 12.24%, N 4.83%, O 11.04%. Prepn: J. Nordmann, H. B. Swierkot, **Fr.** pat. M3454; *eidem*, **U.S.** pat. 3,367,939 (1965, 1968 both to Kuhlmann).

Base, white needles from methanol and water, mp 61-62° (Maquenne block); 55-56.5° (Culatti block). Very soluble in lipids; sol in alcohols, benzene, toluene, ether and acetone. Practically insol in water.

Hydrochloride, $C_{16}H_{17}Cl_2NO_2$, crystals from methanol, mp 124-127°C. LD_{50} i.p. in mice: 2.27 g/kg (Nordmann, Swierkot).

THERAP CAT: Antilipemic.

6604. Nicofibrate. *2-(4-Chlorophenoxy)-2-methylpropanoic acid 3-pyridinylmethyl ester; 2-(p-chlorophenoxy)-2-methylpropionic acid 3-pyridylmethyl ester;* 3-hydroxymethylpyridine p-chlorophenoxyisobutyrate; clofenpyride. $C_{16}H_{16}ClNO_3$; mol wt 305.76. C 62.85%, H 5.27%, Cl 11.59%, N 4.58%, O 15.70%. Prepn: **Neth.** pat. **Appl.** 6,610,738 corresp to Bolhofer, **U.S.** pat. 3,369,025 (1967, 1968, both to Merck & Co.); Nakanishi *et al.*, **Japan.** pat. 14,466('68) (to Yoshitomi); Garzia, **Ger.** pat. 1,948,320 (1970 to Ist. Chem. Ital.). Pharmacology: Marmo *et al., Farmaco Ed. Prat.* **26**, 557 (1971).

mp 48-49°. $bp_{0.4}$ 180°.

Hydrochloride, $C_{16}H_{16}ClNO_3.HCl$, *Arterium V.* Crystals from ethyl acetate, mp 115.5-118.5°.

THERAP CAT: Antihyperlipoproteinemic.

6605. Nicofuranose. *β-D-Fructofuranose 1,3,4,6-tetra-3-pyridinecarboxylate; fructose 1,3,4,6-tetranicotinate;* 1,3,4,6-tetranicotinoylfructofuranose; 1,3,4,6-tetranicotinoyl-D-fructose; Vasperdil; Bradilan. $C_{30}H_{24}N_4O_{10}$; mol wt 600.54. C 60.00%, H 4.03%, N 9.33%, O 26.64%. Prepn: Suter, Kuendig, **Swiss.** pat. 366,523 (1963 to Eprova).

Crystals from ethyl acetate, mp 140-142°. $[\alpha]_D^{18}$ −8.5° (after 4-6 hours in $CHCl_3$).

THERAP CAT: Vasodilator (peripheral).

6606. Nicomol. *3-Pyridinecarboxylic acid (2-hydroxy-1,3-cyclohexanediylidene)tetrakis(methylene) ester; nicotinic acid 1,1,3,3-tetraester with 2-hydroxy-1,1,3,3-cyclohexanetetramethanol;* 2-hydroxycyclohexane-1,1,3,3-tetramethanol tetraester with nicotinic acid; tetranicotinic acid 2-hydroxycyclohexa-1,1,3,3-tetramethyl ester; 2,2,6,6-tetrakis(hydroxymethyl)-1-cyclohexanol 2,2,6,6-tetranicotinate; nicotinic acid 2-hydroxy-1,3-cyclohexanediylidenetetrakismethylene ester; 2,2,6,6-tetrakis(nicotinoyloxymethyl)cyclohexanol; K-31; Cholexamin. $C_{34}H_{32}N_4O_9$; mol wt 640.65. C 63.74%, H 5.03%, N 8.75%, O 22.48%. Prepn: **Belg.** pat. 658,422 corresp to Irikura *et al.*, **U.S.** pat. 3,299,077 (1965 and 1967 to Kyorin). Hypocholesterolemic action studies: Aso *et al., J. Atheroscler. Res.* **10**, 391 (1969). Acute toxicity data: T. Sugawara *et al., Oyo Yakuri* **14**, 741 (1977), *C.A.* **88**, 131005v (1977). *Review: Japan. Med. Gaz.* **8**(8), 9 (1971).

Crystals from dilute acetic acid, mp 177-180°. Practically odorless and tasteless. One gram is sol in 3 ml 1N HCl and 10 ml chloroform. Slightly sol in water, ethanol and ether. LD_{50} in mice, male and female rats (g/kg): > 10, 15, < 10 orally; > 5, > 2.5, > 2.5 i.p. (Sugawara).

THERAP CAT: Anticholesteremic.

6607. Nicomorphine. *(5α,6α)-7,8-Didehydro-4,5-epoxy-17-methylmorphinan-3,6-diol di-3-pyridinecarboxylate (ester);* morphine ester with nicotinic acid; morphine bis(nicotinate); morphine bis(pyridine-3-carboxylate); nicotinic acid morphine ester; morphine dinicotinate; Gewalan; Vilan. $C_{29}H_{25}N_3O_5$; mol wt 495.53. C 70.29%, H 5.09%, N 8.48%, O 16.14%. Prepn: Pongratz, Zirm, *Monatsh.* **88**, 330 (1957); *Brit. pat.* **807,115** (1959 to Lannacher Heilmittel).

Crystals, mp 178-178.5° (corr.). Practically insol in water. Sol in ethanol.

Hydrochloride, $C_{29}H_{25}N_3O_5$·HCl, dec 248°. Sol in water with neutral reaction.

Note: This is a controlled substance (opium derivative) listed in the U.S. Code of Federal Regulations, Title 21 Part 1308.11 (1995).

THERAP CAT: Analgesic (narcotic).

6608. Nicorandil. *N-[2-(Nitrooxy)ethyl]-3-pyridinecarboxamide;* N-(2-hydroxyethyl)nicotinamide nitrate (ester); SG-75; Ikorel; Perisalol; Sigmart. $C_8H_9N_3O_4$; mol wt 211.18. C 45.50%, H 4.30%, N 19.90%, O 30.31%. Nicotinamide derivative; exhibits dual mechanism of action as both nitrovasodilator and potassium channel activator. Prepn: H. Nagano *et al.*, *Ger. pat.* **2,714,713**; *eidem*, *U.S. pat.* **4,200,640** (1977, 1980 both to Chugai). Pharmacology: N. Taira *et al.*, *Clin. Exp. Pharmacol. Physiol.* **6**, 301 (1979). Mechanism of action: F. Yoneyama *et al.*, *Cardiovasc. Drugs Ther.* **4**, 1119 (1990). Symposia on pharmacology and clinical efficacy: *Am. J. Cardiol.* **63**, Suppl, 1J-85J (1989); *J. Cardiovasc. Pharmacol.* **20**, Suppl. 3, S1-S108 (1992). Review of clinical potential in ischemic heart disease: H. Purcell, K. Fox, *Brit. J. Clin. Pract.* **47**, 150-154 (1993).

Colorless needles from ether/ethanol, mp 92-93°. LD_{50} in rats (mg/kg): 1200-1300 orally; 800-1000 i.v. (Nagano).

THERAP CAT: Antianginal.

6609. Nicotelline. *3,2':4',3''-Terpyridine;* 2,4-di(3-pyridyl)pyridine; 2,4-di(β-pyridyl)pyridine. $C_{15}H_{11}N_3$; mol wt 233.27. C 77.23%, H 4.75%, N 18.01%. From tobacco leaf: Pictet, Rotschy, *Ber.* **34**, 696 (1901); Noga, *Fach. Mitt. Tabakregie* **1914**, Nos. 1 and 2. Structure: Kuffner, Faderl, *Monatsh.* **87**, 71 (1956). Synthesis: Thesing, Müller, *Angew. Chem.* **68**, 577 (1956); *Ber.* **90**, 711 (1957).

Prismatic needles from warm dil alcohol, hot water or chloroform + petr ether, mp 147-148°. bp > 300°. Sol in hot water, chloroform, alcohol, benzene; slightly sol in cold water, ether, petr ether.

6610. Nicotinamide Ascorbate. *L-Ascorbic acid mixt. with 3-pyridinecarboxamide;* niacinamide ascorbate; ascorbic acid nicotinamide complex; merpress; nicoscorbine; Nicastubin. $C_{12}H_{14}N_2O_7$; mol wt 298.25. C 48.33%, H 4.73%, N 9.39%, O 37.55%. Prepd from one mole of nicotinamide and one mole of ascorbic acid by warming in alcohol, acetone, water: Milhorat, *Proc. Soc. Exp. Biol. Med.* **55**, 52 (1944); Fox, Opferman, *U.S. pat.* **2,433,688** (1947 to Gelatin Prod.); Bailey *et al.*, *J. Am. Chem. Soc.* **67**, 1184 (1945); Weimer, *J. Org. Chem.* **14**, 22 (1949); Najer, Guépet, *Ann. Pharm. Franc.* **12**, 712 (1954); Runti, *Il Farmaco, Ed. Sci.* **10**, 424 (1955); Genot, *Brit. pat.* **771,317** (1957).

Yellow crystals, mp 141-145°, practically odorless. $[\alpha]_D^{20}$ +27.5° (c = 8 in H_2O). pH of 5% aq soln ~4.0. Soly at 20° in water 40%, abs ethanol 2.4%, methanol 10%. Sparingly sol in acetone. Practically insol in benzene, ether. A 45% to 50% w/w soln can be prepd by dissolving in equal volumes of water and propylene glycol.

THERAP CAT: Vitamin.

6611. Nicotine. *(S)-3-(1-Methyl-2-pyrrolidinyl)pyridine;* 1-methyl-2-(3-pyridyl)pyrrolidine; β-pyridyl-α-N-methylpyrrolidine; Habitrol; Nicabate; Nicoderm; Nicolan; Nicopatch; Nicotell TTS; Nicotinell; Tabazur. $C_{10}H_{14}N_2$; mol wt 162.23. C 74.03%, H 8.70%, N 17.27%. From the dried leaves of *Nicotiana tabacum* and *N. rustica* where it occurs to the extent of 2 to 8%, combined with citric and malic acids. Commercial nicotine is entirely a byproduct of the tobacco industry. Extraction procedure: Gattermann, Wieland, *Laboratory Methods of Organic Chemistry* (New York, 24th ed., 1937); Schwyzer, *Die Fabrikation pharmazeutischer und chemisch-technischer Produkte* (Berlin, 1931). Purification through the zinc chloride double salt: Ratz, *Monatsh.* **26**, 1241 (1905). Structure and synthesis: Pinner, *Ber.* **26**, 294 (1893); Pictet, Rotschy, *Ber.* **37**, 1225 (1904); Craig, *J. Am. Chem. Soc.* **55**, 2854 (1933). Recent synthesis: M. Nakane, C. R. Hutchinson, *J. Org. Chem.* **43**, 3922 (1978). Conformation in soln: T. P. Pitner *et al.*, *J. Am. Chem. Soc.* **100**, 246 (1978). HPLC determn in plasma: M. Harlharan *et al.*, *Clin. Chem.* **34**, 724 (1988). Toxicity data: R. B. Barlow, L. J. McLeod, *Brit. J. Pharmacol.* **35**, 161 (1969). Review and bibliography: Jackson, *Chem. Rev.* **29**, 123 (1941). Review of pharmacology: R. W. Ryall in *Neuropoisons: Their Pathophysiological Actions* vol. 2, L. L. Simpson, D. R. Curtis, Eds. (Plenum, New York, 1974) pp 61-97; N. L. Benowitz, *Ann. Rev. Med.* **37**, 21-32 (1986).

Colorless to pale yellow, oily liq; very hygroscopic; turns brown on exposure to air or light. Acrid burning taste. Develops odor of pyridine. bp_{745} 247° (partial decompn);

bp$_{17}$ 123-125°. Volatile with steam. n_D^{20} 1.5282. d_4^{20} 1.0097. $[\alpha]_D^{20}$ −169°. pK$_1$ (15°) 6.16; pK$_2$ 10.96. pH of 0.05M soln: 10.2. Forms salts with almost any acid and double salts with many metals and acids. Absorption spectrum: Purvis, *J. Chem. Soc.* **97**, 1035 (1910); Dobbie, Fox, *ibid.* **103**, 1194 (1913). Misc with water below 60°; on mixing nicotine with water the volume contracts. Very sol in alc, chloroform, ether, petr ether, kerosene, oils. Distribution of nicotine between water and petroleum oils: Norton, *Ind. Eng. Chem., Ind. Ed.* **32**, 241 (1940). LD$_{50}$ in mice (mg/kg): 0.3 i.v.; 9.5 i.p.; 230 orally (Barlow, McLeod).

Hydrochloride, C$_{10}$H$_{14}$N$_2$.HCl, deliquesc crystals. $[\alpha]_D^{20}$ +104° (p = 10).

Dihydrochloride, C$_{10}$H$_{14}$N$_2$.2HCl, deliquesc crystals, very sol in water and alcohol. Nearly insol in ether.

Sulfate, (C$_{10}$H$_{14}$N$_2$)$_2$.H$_2$SO$_4$, *nicotine neutral sulfate*, six-sided tablets. $[\alpha]_D^{20}$ +88° (p = 70). Sol in water, alcohol.

Bitartrate, C$_{10}$H$_{14}$N$_2$.2C$_4$H$_6$O$_6$, *nicotine tartrate*, dihydrate, crystals. mp 90°. $[\alpha]_D^{20}$ +26° (c = 10). Very sol in water or alcohol.

Zinc chloride double salt monohydrate, C$_{10}$H$_{16}$Cl$_4$N$_2$Zn.-H$_2$O, also with 4H$_2$O. Very sol in water; sparingly sol in abs alcohol and ether.

Salicylate, C$_{17}$H$_{20}$N$_2$O$_3$, *Eudermol*. Six-sided plates, mp 118°. $[\alpha]_D^{20}$ +13° (c = 9). Freely sol in water or alcohol.

Polacrilex, *Nicorette*.

Caution: Nicotine can be absorbed from the alimentary canal, the respiratory tract and intact skin. Percutaneous absorption of the free alkaloid is much more rapid than the acid salts. Local effects due to overexposure are burning sensation in mouth and throat, salivation, nausea, abdominal pain, vomiting and diarrhea. GI reactions are less severe but do occur after cutaneous and respiratory exposure. The major systemic effects are a transient stimulation and subsequent depression or paralysis of the CNS. Symptoms include agitation, headache, sweating, dizziness, auditory and visual disturbances, confusion, weakness and incoordination. At first respirations are deep and rapid, the blood pressure is high and the pulse is slow. CNS excitation may be evidenced by tremors or clonic-tonic convulsions. As depression develops, the pupils dilate, the blood pressure falls, and the pulse becomes rapid and irregular. Faintness, prostration, cyanosis and dyspnea progress to collapse. Death from paralysis of respiratory muscles usually follows shortly after collapse. See: *Clinical Toxicology of Commercial Products*, R. E. Gosselin *et al.*, Eds. (Williams & Wilkins, Baltimore, 5th ed., 1984) Section III, pp 311-314.

USE: Insecticide; fumigant. In the U.S. a 40% soln of nicotine sulfate, *Black Leaf 40*, was the commonly used form. As a contact poison it is most effective as soap, *i.e.*, as the laurate, oleate, or naphthenate. As a stomach poison a combination with bentonite has come into use.

THERAP CAT: Treatment of smoking withdrawal syndrome.

THERAP CAT (VET): Ectoparasiticide. Has been used as an anthelmintic.

6612. Nicotinic Acid. *3-Pyridinecarboxylic acid;* pyridine-β-carboxylic acid; P. P. factor; pellagra preventive factor; antipellagra vitamin; niacin; Akotin; Daskil; Niacor; Nicacid; Nicangin; Nicobid; Nicolar; Niconacid; Nico-Span; Wampocap. C$_6$H$_5$NO$_2$; mol wt 123.11. C 58.54%, H 4.09%, N 11.38%, O 25.99%. Precursor of the coenzymes NAD and NADP, *q.q.v.* Widely distributed in nature; appreciable amounts are found in liver, fish, yeast and cereal grains. Dietary deficiency is associated with pellagra. The term "niacin" has also been applied to nicotinamide, *q.v.*, or to other derivatives exhibiting the biological activity of nicotinic acid. Prepn by oxidation of alkyl β-substituted pyridines: A. Ladenburg, *Ann.* **301**, 152 (1898). Synthesis from pyridine: S. M. McElvain, M. A. Goese, *J. Am. Chem. Soc.* **63**, 2283 (1941). Prepn by oxidation of nicotine: S. M. McElvain, *Org. Syn.* **coll. vol I**, 385 (1941); C. F. Woodward *et al.*, *Ind. Eng. Chem.* **36**, 540, 544 (1944). Toxicity study: F. G. Brazda, R. A. Coulson, *Proc. Soc. Exp. Biol. Med.* **62**, 19 (1946). LC determn in foods: G. W. Chase, Jr. *et al.*, *J. AOAC Int.* **76**, 390 (1993). Review of nutritional requirements, bioavailability and relationship with tryptophan: W. J. Darby *et al.*, *Nutr. Rev.* **33**, 289-297 (1975); of

biosynthesis, metabolism and physiological effects: L. M. Henderson, *Ann. Rev. Nutr.* **3**, 289-307 (1983); of pharmacology and therapeutic uses: J. R. DiPalma, W. S. Thayer, *ibid.* **11**, 169-187 (1991). *Review:* "Niacin: Nicotinic Acid, Nicotinamide, NAD(P)" in *Vitamins*, W. Friedrich, Ed. (Walter de Gruyter, Berlin, 1988) pp 473-542.

Needles from alcohol or water, mp 236.6°. Nonhygroscopic and stable in air. Sublimes without decompn. pKa 4.85. pH of satd aq soln 2.7. uv max: 263 nm, Hünecke, *Ber.* **60**, 1451 (1927). One gram dissolves in 60 ml water. Freely sol in boiling water and in boiling alcohol, in alkali hydroxides and carbonates; sol in propylene glycol. Insol in ether. LD$_{50}$ s.c. in rats: 5 g/kg (Brazda, Coulson).

N-Oxide, *see* Oxiniacic Acid.

Sodium salt sesquihydrate, C$_6$H$_4$NNaO$_2$.1½H$_2$O, *Direktan*. White crystals or crystalline powder; stable in air. One gram dissolves in ~1.4 ml water, in ~60 ml alc, in 10 ml glycerol. Insol in ether. pH of aq soln: ~7.

THERAP CAT: Antihyperlipoproteinemic; vitamin (enzyme cofactor).

THERAP CAT (VET): Vitamin (enzyme cofactor).

6613. Nicotinic Acid Benzyl Ester. *3-Pyridinecarboxylic acid phenylmethyl ester;* pyridine-β-carboxylic acid benzyl ester; benzyl nicotinate; Rubriment; Pycaril; Pykaryl. C$_{13}$H$_{11}$NO$_2$; mol wt 213.24. C 73.23%, H 5.20%, N 6.57%, O 15.01%. Prepn: Brit. pat. **817,103** (1959 to Nordmark-Werke).

Liquid, bp$_{3-4}$ 170°.

THERAP CAT: Rubefacient.

6614. Nicotinyl Alcohol. *3-Pyridinemethanol.* β-pyridylcarbinol; 3-hydroxymethylpyridine; nicotinic alcohol; Nu-2121; Roniacol; Ronicol. C$_6$H$_7$NO; mol wt 109.13. C 66.04%, H 6.47%, N 12.84%, O 14.66%. Prepd by catalytic hydrogenation of 3-pyridinecarboxaldehyde: Panizzon, *Helv. Chim. Acta* **24**, supplemental issue in honor of Gadient Engi, p 26E (1941); by lithium aluminum hydride reduction of ethyl nicotinate: Rosenmund, Zymalkowski, *Ber.* **85**, 156 (1952); Cohen, U.S. pat. **2,520,037** (1950); Mosher, Tessieri, *J. Am. Chem. Soc.* **73**, 4926 (1951); of methyl nicotinate: Bohlmann, *Ber.* **86**, 1423 (1953). From 3-cyanopyridine: Chase, U.S. pat. **2,615,896** (1952 to Hoffmann-La Roche); from 3-aminomethylpyridine: Schläpfer, U.S. pat. **2,547,048** (1951 to Hoffmann-La Roche); from thionicotinic acid S-methyl ester: Ruzicka, Prelog, U.S. pat. **2,509,171** (1950 to Ciba).

Very hygroscopic liquid. bp$_{28}$ 154°; bp$_{16}$ 144-145°; bp$_{12}$ 114°; bp$_{0.1}$ 110°. Freely sol in water, ether. Sparingly sol in petr ether.

d-Tartrate, C$_6$H$_7$NO.C$_4$H$_6$O$_6$, *Roniacol Tartrate, Radecol, Niltuvin*. Crystals, sour taste, mp 147-148°. Sol in ether, freely sol in water, alcohol.

USE: Solubilizer for riboflavin: U.S. pat. **2,458,430** (1949).

THERAP CAT: Vasodilator (peripheral).

6615. Nidroxyzone. *1-(2-Hydroxyethyl)-2-[(5-nitro-2-furanyl)methylene]hydrazinecarboxamide; 5-nitro-2-furaldehyde 2-(2-hydroxyethyl)semicarbazone;* NF-67; Furadroxyl. C$_8$H$_{10}$N$_4$O$_5$; mol wt 242.19. C 39.67%, H 4.16%, N 23.13%,

O 33.03%. Prepn: Stillman, Scott, U.S. pat. **2,416,234** (1947 to Eaton).

Bright orange plates from alc, mp 214-216° (dec). Soly in water 1:2000.

THERAP CAT: Anti-infective.

6616. Nidulin. *(E)-2,4,7-Trichloro-3-hydroxy-8-methoxy-1,9-dimethyl-6-(1-methyl-1-propenyl)-11H-dibenzo[b,e]-[1,4]dioxepin-11-one;* methylustin; ustin methyl ether. $C_{20}H_{17}Cl_3O_5$; mol wt 443.71. C 54.14%, H 3.86%, Cl 23.97%, O 18.03%. Antibiotic substance produced by a strain of *Aspergillus nidulans* (Eidam) Wint.: Dean *et al., Nature* **172,** 344 (1953); Dean *et al., J. Chem. Soc.* **1954,** 1432. Structure: Dean *et al., ibid.* **1960,** 4829; Bycroft *et al., ibid.* **1963,** 5148.

Slender, shiny rods from petr ether, mp 180°. Practically insol in water. Sol in aq solns of sodium hydroxide and sodium bicarbonate. Freely sol in chloroform. Sparingly sol in 95% ethanol and in benzene. Very sparingly sol in hot petr ether. A 4% soln in chloroform shows no optical activity. Completely inhibits *in vitro* the growth of *Mycobacterium tuberculosis* for four weeks at a diln of between 1 in 5000 and 1 in 10,000 when tested in the presence of serum by modified Long's medium and the floating-pellicle method. A 0.1% soln also inhibits the growth of the human parasitic fungi *Trichophyton tonsurans* and *Microsporum audouini,* but shows little or no activity towards a wide range of other microorganisms. Inactive against bacteriophage.

6617. Nifedipine. *1,4-Dihydro-2,6-dimethyl-4-(2-nitrophenyl)-3,5-pyridinedicarboxylic acid dimethyl ester;* 4-(2'-nitrophenyl)-2,6-dimethyl-3,5-dicarbomethoxy-1,4-dihydropyridine; Bay a 1040; Adalat(e); Adapress; Aldipin; Alfadat; Anifed; Aprical; Bonacid; Camont; Chronadalate; Citilat; Coracten; Cordicant; Cordilan; Corotrend; Duranifin; Ecodipin; Hexadilat; Introcar; Kordafen; Nifedicor; Nifedin; Nifelat; Nifensar XL; Orix; Oxcord; Pidilat; Procardia; Sepamit; Tibricol; Zenusin. $C_{17}H_{18}N_2O_6$; mol wt 346.34. C 58.96%, H 5.24%, N 8.09%, O 27.72%. Dihydropyridine calcium channel blocker. Prepn: Bossert, Vater, **S. Afr.** pat. **68 01,482;** *eidem,* **U.S.** pat. **3,485,847** (1968, 1969 both to Bayer). Series of articles on pharmacology, pharmacokinetics, biotransformation, clinical studies: *Arzneimittel-Forsch.* **22,** 1-56, 330-388 (1972). Toxicity data: Vater *et al., ibid.* 1. Comparative study in variant angina: D. D. Waters *et al., Am. J. Cardiol.* **47,** 179 (1981). Antihypertensive effects: K. Maeda *et al., Arzneimittel-Forsch.* **32,** 267 (1982). Controlled clinical trial in treatment of Raynaud's phenomenon: R. J. Rodeheffer *et al., N. Engl. J. Med.* **308,** 880 (1983). HPLC determn in plasma: H. Suzuki *et al., J. Chromatog.* **341,** 341 (1985). Review of mechanism of action: T. H. Swanson, C. L. Green, *Gen. Pharmacol.* **17,** 255-260 (1986). Comprehensive description: S. L. Ali, *Anal. Profiles Drug Subs.* **18,** 221-288 (1989).

Yellow crystals, mp 172-174°. uv max (methanol): 340, 235 nm (ε 5010, 21590); (0.1*N* HCl): 338, 238 nm (ε 5740, 20600); (0.1*N* NaOH): 340, 238 nm (ε 5740, 20510). Soly at 20° (g/L): acetone 250, methylene chloride 160, chloroform 140, ethyl acetate 50, methanol 26, ethanol 17. Very light sensitive in soln. LD_{50} in mice, rats (mg/kg): 494, 1022 orally; 4.2, 15.5 i.v. (Vater).

THERAP CAT: Antianginal; antihypertensive.

6618. Nifenalol. *(±)-α-[[(1-Methylethyl)amino]methyl]-4-nitrobenzenemethanol;* α-[(isopropylamino)methyl]-p-nitrobenzyl alcohol; 1-(4-nitrophenyl)-2-(isopropylamino)-ethanol; isophenethanol; INPEA. $C_{11}H_{16}N_2O_3$; mol wt 224.26. C 58.91%, H 7.19%, N 12.49%, O 21.40%. β-Adrenergic blocker. Prepn: Teotino *et al., Farmaco Ed. Sci.* **17,** 252 (1962); *eidem,* **Brit.** pat. **950,682** (1964 to Selvi), *C.A.* **60,** 13184h (1964). Pharmacology: Somani, Lum, *J. Pharmacol. Exp. Ther.* **147,** 194 (1965); Almirante *et al., Farmaco Ed. Sci.* **21,** 637 (1966). Prepn of the optical isomers: Almirante, Murmann, *J. Med. Chem.* **9,** 650 (1966); *eidem,* **Belg.** pat. **673,273** (1966 to Selvi), *C.A.* **66,** 65278d (1967).

mp 98°.

Hydrochloride, $C_{11}H_{16}N_2O_3$·HCl, *Inpea.* mp 181°.

THERAP CAT: Antianginal; antiarrhythmic.

6619. Nifenazone. *N-(2,3-Dihydro-1,5-dimethyl-3-oxo-2-phenyl-1H-pyrazol-4-yl)-3-pyridinecarboxamide; N-antipyrinylnicotinamide;* 2,3-dimethyl-1-phenyl-4-(3-pyridine-carboxamido)-5-pyrazolone; 2,3-dimethyl-4-nicotinamido-1-phenyl-5-pyrazolone; 4-nicotinamido-1,5-dimethyl-2-phenyl-3-pyrazolone; 4-(N-nicotinoylamino)-1-phenyl-2,3-dimethyl-5-pyrazolone; N-nicotinoylaminoantipyrine; Dolongan; Nicopyron; Nikofezon; Niprazina; Phenicazone; Reupiron; Thylin. $C_{17}H_{16}N_4O_2$; mol wt 308.34. C 66.22%, H 5.23%, N 18.17%, O 10.38%. Prepn: Heid, **Ger.** pat. **897,-407** (1953); Trommsdorff, **Belg.** pat. **635,597** (1963); Pongraz, Zirm, *Monatsh.* **88,** 330 (1957); Zorn, Schmidt, *Pharmazie* **12,** 396 (1957).

Crystals from alc, mp 252-253° (also reported as mp 256-258°). Sparingly sol in water, acetone, ethyl acetate, ether; sol in hot water, alcohol, chloroform, dil acids.

THERAP CAT: Analgesic, antipyretic.

6620. Niflumic Acid. *2-[[3-(Trifluoromethyl)phenyl]-amino]-3-pyridinecarboxylic acid; 2-(α,α,α-trifluoro-m-toluidino)nicotinic acid;* 2-[3-(trifluoromethyl)anilino]nicotinic acid; UP-83; Actol; Forenol; Landruma; Nifluril. $C_{13}H_9F_3N_2O_2$; mol wt 282.22. C 55.33%, H 3.21%, F 20.20%, N 9.93%, O 11.34%. Prepn: **Neth.** pat. **Appl. 6,414,717;** C. Hoffmann, A. Faure, **U.S.** pat. **3,415,834** (1965, 1968 both to Lab. U.P.S.A.); *eidem, Bull. Soc. Chim. France* **1966**, 2316. Pharmacological and metabolic studies: Glasson *et al., Biochem. Pharmacol.* **18**, 633 (1969); Boissier *et al., Therapie* **26**, 211 (1971). Toxicity data: J. R. J. Sorenson, *J. Med. Chem.* **19**, 135 (1976). Determn in human plasma by GLC: G. Houin *et al., J. Chromatog.* **223**, 351 (1981).

Crystals from ethanol, mp 204°. LD$_{50}$ in rats (mg/kg): 370 orally; 155 i.p. (Sorenson).

β-Morpholinoethyl ester, C$_{19}$H$_{20}$F$_3$N$_3$O$_3$, *morniflumate, Flomax, Nifluril (suppositories).* Prepn of the ester HCl: Hoffmann, **Ger. pat. 1,802,777**; *idem,* **U.S. pat. 3,708,481** (1969, 1973 both to Hexachimie).

Phthalidyl ester, *see* Talniflumate.

THERAP CAT: Anti-inflammatory.

6621. Nifuradene. *1-[[(5-Nitro-2-furanyl)methylene]amino]-2-imidazolidinone;* oxafuradene (rescinded); NF-246; NSC-6470; Renafur. C$_8$H$_8$N$_4$O$_4$; mol wt 224.18. C 42.86%, H 3.60%, N 24.99%, O 28.55%. Prepn: Gever, Michels, **U.S. pat. 2,746,960** (1956 to Norwich Pharm.); *eidem, J. Am. Chem. Soc.* **78**, 5349 (1956); Michels, **U.S. pat. 2,776,979** (1957 to Norwich Pharm.). Toxicity data: E. I. Goldenthal, *Toxicol. Appl. Pharmacol.* **18**, 185 (1971).

Lemon-yellow solid from nitromethane, mp 261.5-263° (dec). Turns orange on standing, on washing with ethanol or on heating at 75-85°. uv max (water): 387.5, 273 nm (ε 17550, 13200). Soly in water: 88-109 mg/l. LD$_{50}$ orally in rats: 1681 mg/kg (Goldenthal).

THERAP CAT: Antibacterial.

6622. Nifuraldezone. *Aminooxoacetic acid [(5-nitro-2-furanyl)methylene]hydrazide;* 5-nitro-2-furaldehyde semioxamazone; Furamazone. C$_7$H$_6$N$_4$O$_5$; mol wt 226.15. C 37.18%, H 2.67%, N 24.77%, O 35.37%. Prepn: Stillman, Scott, **U.S. pat. 2,416,238** (1947 to Eaton Labs.).

Yellow cryst powder, mp 270° (dec).

THERAP CAT (VET): Antibacterial.

6623. Nifuratel. *5-[(Methylthio)methyl]-3-[[(5-nitro-2-furanyl)methylene]amino]-2-oxazolidinone; 5-[(methylthio)methyl]-3-[(5-nitrofurfurylidene)amino]-2-oxazolidinone;* 5-(methylmercaptomethyl)-3-(5-nitro-2-furfurylideneamino)-2-oxazolidinone; methylmercadone; Inimur; Macmiror; Magmilor; Omnes; Polmiror; Tydantil. C$_{10}$H$_{11}$N$_3$O$_5$S; mol wt 285.28. C 42.10%, H 3.89%, N 14.73%, O 28.04%, S 11.24%. Prepn: **Belg. pat. 635,608** (1963 to Polichimica SAP), *C.A.* **61**, 16069c (1964), corresp to **Brit. pat. 969,126.**

Crystals from acetic acid, mp 182°.

THERAP CAT: Antibacterial; antifungal; antiprotozoal (Trichomonas).

6624. Nifurfoline. *3-(4-Morpholinylmethyl)-1-[[(5-nitro-2-furanyl)methylene]amino]-2,4-imidazolidinedione; 3-(morpholinomethyl)-1-[(5-nitrofurfurylidene)amino]hydantoin;* Urbac. C$_{13}$H$_{15}$N$_5$O$_6$; mol wt 337.29. C 46.29%, H 4.48%, N 20.76%, O 28.46%. Prepn: J. A. Bofill-Auge, J. M. Espinos-Taya, **Span. pat. 297,087** (1964), *C.A.* **63**, 11572e (1966); **Brit. pat. 1,245,095** (1971 to Esteve), *C.A.* **75**, 121382h (1971).

Yellow cryst, mp 206°; also reported as mp 194-196°, J. Klosa, H. Starke, **E. Ger. pat. 508,615** (1966), *C.A.* **66**, 65501w (1967).

THERAP CAT: Antibacterial.

6625. Nifuroquine. *4-(5-Nitro-2-furanyl)-2-quinolinecarboxylic acid 1-oxide;* 2-carboxy-4-[2'-(5'-nitrofuryl)]quinoline 1-oxide; 4-(5-nitro-2-furyl)quinaldic acid 1-oxide; quinaldofur; Abimasten 100. C$_{14}$H$_8$N$_2$O$_6$; mol wt 300.23. C 56.01%, H 2.69%, N 9.33%, O 31.97%. Prepn: R. R. G. Haber, E. Schoenberger, **S. Afr. pat. 67 03,320** corresp to **U.S. pats. 4,217,456** and **4,224,448** (1967, 1980, 1980). Use in bovine mastitis: R. R. G. Haber *et al.,* **Ger. pat. 2,612,250** corresp to **U.S. pat. 4,070,469** (1977, 1978 both to ABIC). *In vitro* study: G. Ziv *et al., Zentralbl. Veterinaermed. (B)* **23**, 301 (1976). Pharmacology: G. Ziv, A. Saran, *ibid.* 310. Activity under anaerobic conditions: G. M. Maluszynska, L. Bassalik-Chabielska, *Pr. Mater. Zootech.* **23**, 63 (1980), *C.A.* **93**, 143793 (1980).

Yellow cryst powder, mp 190° (dec). Practically insol in water.

THERAP CAT (VET): Antibacterial.

6626. Nifuroxazide. *4-Hydroxybenzoic acid [(5-nitro-2-furanyl)methylene]hydrazide; p-hydroxybenzoic acid (5-nitrofurfurylidene)hydrazide;* 1-(p-hydroxybenzoyl)-2-(5-nitrofurfurylidene)hydrazine; 5-nitro-2-furaldehyde p-hydroxybenzoylhydrazone; RC-27109; Adral; Bacifurane; Diarlidan; Dicoferin; Ercefurol; Ercefuryl; Pentofuryl. C$_{12}$H$_9$N$_3$O$_5$; mol wt 275.22. C 52.37%, H 3.30%, N 15.27%, O 29.07%. Prepn: **Fr. pat. 1,327,840** (1963 to Robert et Carriere), *C.A.* **59**, 12763b (1963); M. C. E. Carron, **Brit. pat. 962,706**; *idem,* **U.S. pat. 3,290,213** (1964, 1966 both to Robert et Carriere). Antiseptic and antibacterial properties: M. C. E. Carron *et al., Ann. Pharm. Franc.* **21**, 287 (1963). *In vitro* study of activity spectrum: A. Thabaut, J. L. Durosoir, *Gaz. Med. Fr.* **85**, 4516 (1978), *C.A.* **90**, 1487 (1979).

Crystals from pyridine, mp 298°. Practically insol in water.

THERAP CAT: Intestinal antiseptic.

6627. Nifuroxime. *5-Nitro-2-furancarboxaldehyde oxime; anti-5-nitro-2-furaldoxime;* Micofur. $C_5H_4N_2O_4$; mol wt 156.10. C 38.47%, H 2.58%, N 17.95%, O 41.00%. Prepd by treating 5-nitrofurfural with hydroxylamine in alcohol: Ikeda, *C.A.* **50**, 10701 (1956). Claimed in Stillman *et al.*, U.S. pat. **2,319,481** (1943 to Norwich Pharmacal).

Tasteless, pale yellow or greenish crystals from ethanol. Darkens on exposure to light. mp 156° (Ikeda); mp 163-164° (Stillman *et al.*). Soly at 25° in water: ~1 g/l, in methanol: 89.0 g/l, in 95% ethanol: 39.0 g/l.

THERAP CAT: Topical anti-infective; antiprotozoal (Trichomonas).

6628. Nifurpirinol. *6-[2-(5-Nitro-2-furanyl)ethenyl]-2-pyridinemethanol; 6-[2-(5-nitro-2-furyl)vinyl]-2-pyridinemethanol;* 6-(hydroxymethyl)-2-[2-(5-nitro-2-furyl)vinyl]-pyridine; furpirinol; furpyrinol; P-7138; Furanace. $C_{12}H_{10}N_2O_4$; mol wt 246.22. C 58.54%, H 4.09%, N 11.38%, O 25.99%. Prepn: Fujita *et al.*, *J. Pharm. Soc. Japan* **86**, 1014 (1966); **Brit.** pat. **1,053,730** (1967 to Dainippon), *C.A.* **66**, 115605f (1967); Murakami, Iwanami, **Japan.** pat. **14,072** ('70) (to Yamanouchi), *C.A.* **73**, 45364v (1970). Antibacterial activity: Shimizu, Takase, *Progr. Antimicrob. Anticancer Chemother., Proc. 6th Int. Congr. Chemother.* Tokyo, 1969 (Univ. Park Press, Baltimore, 1970) **2**, p 388; Amend, Ross, *Prog. Fish Cult.* **32**, 19 (1970); Ross, *ibid.* **34**, 18 (1972); Takase *et al.*, *Chem. Pharm. Bull.* **21**, 144 (1973).

Yellow needles from acetone or methanol, mp 170-171°. LD_{50} orally in eels: 1780 mg/kg.

THERAP CAT: Antibacterial.

THERAP CAT (VET): Antibacterial in fish diseases.

6629. Nifurprazine. *6-[2-(5-Nitro-2-furanyl)ethenyl]-3-pyridazinamine; 3-amino-6-[2-(5-nitro-2-furyl)vinyl]pyridazine;* 1-(5-nitro-2-furyl)-2-(6-amino-3-pyridazyl)ethylene; furenapyridazin; Furenazin. $C_{10}H_8N_4O_3$; mol wt 232.20. C 51.73%, H 3.47%, N 24.13%, O 20.67%. Prepn: **Belg.** pat. **630,163** (1963 to Boehringer, Mann.), *C.A.* **60**, 14516a (1964) corresp to **Brit.** pat. **966,832.**

Hydrochloride, $C_{10}H_8N_4O_3$·HCl, *Carofur.* Crystals, mp 290°.

THERAP CAT: Topical antibacterial.

6630. Nifurtimox. *3-Methyl-N-[(5-nitro-2-furanyl)-methylene]-4-thiomorpholinamine 1,1-dioxide; 4-[(5-nitro-furfurylidene)amino]-3-methylthiomorpholine-1,1-dioxide;* tetrahydro-3-methyl-4-[(5-nitrofurfurylidene)amino]-2H-1,4-thiazine 1,1-dioxide; 1-[(5-nitrofurfurylidene)amino]-2-methyltetrahydro-1,4-thiazine 4,4-dioxide; Bay 2502; Lampit. $C_{10}H_{13}N_3O_5S$; mol wt 287.30. C 41.81%, H 4.56%, N 14.63%, O 27.84%, S 11.16%. Prepn from 5-nitrofurfural and 4-amino-3-methyltetrahydro-1,4-thiazine 1,1-dioxide: Herlinger *et al.*, **Ger.** pat. **1,170,957** corresp to U.S. pat. **3,262,930** (1964 and 1966 to Bayer). Series of articles on pharmacology and clinical findings: *Arzneimittel-Forsch.* **22**, 1563-1642 (1972). Toxicity data: K. Hoffmann, *ibid.* 1590.

Orange-red crystals from dil acetic acid, mp 180-182°. LD_{50} in mice, rats (mg/kg): 3720, 4050 by gavage (Hoffmann).

THERAP CAT: Antiprotozoal (Trypanosoma).

6631. Nifurtoinol. *3-(Hydroxymethyl)-1-[[(5-nitro-2-furanyl)methylene]amino]-2,4-imidazolidinedione; 3-(hydroxymethyl)-1-[(5-nitrofurfurylidene)amino]hydantoin;* 1-[(5-nitrofurfurylidene)amino]-3-(hydroxymethyl)hydantoin; Urfadyn. $C_9H_8N_4O_6$; mol wt 268.19. C 40.31%, H 3.01%, N 20.89%, O 35.79%. Prepn: Michels, **Belg.** pat. **611,941** corresp to U.S. pat. **3,446,802** (1962, 1969 to Norwich Pharmacal); Spencer, Michels, *J. Org. Chem.* **29**, 3416 (1964).

Yellow crystals from aq formaldehyde. When heated on a melting-point block, loses formaldehyde, then further decomposes at same temp as nitrofurantoin. uv max (2% in dimethylformamide): 367.5, 265 nm (ϵ 17900, 12800).

THERAP CAT: Antibacterial.

6632. Nifurzide. *5-Nitro-2-thiophenecarboxylic acid [3-(5-nitro-2-furanyl)-2-propenylidene]hydrazide; N^1-(5-nitro-2-furylacrylidene)-N^2-(5-nitro-2-thenoyl)hydrazine;* 5-nitro-2-thiophenecarboxylic acid [3-(5-nitro-2-furyl)allylidene]hydrazide; Ricridene. $C_{12}H_8N_4O_6S$; mol wt 336.28. C 42.86%, H 2.40%, N 16.66%, O 28.55%, S 9.54%. Bactericidal agent related to nitrofurazone, *q.v.* Prepn: E. Szarvasi, L. Fontaine, **Ger.** pat. **2,200,375**; *eidem*, U.S. pat. **3,847,911** (1972, 1974 both to Lipha); E. Szarvasi *et al.*, *J. Med. Chem.* **16**, 281 (1973). Mode of action study: A. Delsarte *et al.*, *Antimicrob. Ag. Chemother.* **19**, 477 (1981).

trans-form

Yellow crystals from DMF/ether, mp 235-236°. LD_{50} orally in mice: 3200 mg/kg (Szarvasi, Fontaine, 1972).

THERAP CAT: Anti-infective.

6633. Nigericin. Antibiotic K 178; antibiotic X-464; azalomycin M; helixin C; polyetherin A. $C_{40}H_{68}O_{11}$; mol wt 724.97. C 66.27%, H 9.45%, O 24.28%. Polyether antibiotic which affects ion transport and ATPase activity in mitochondria; produced by *Streptomyces hygroscopicus* E-749 and structurally related to monensin, *q.v.* Isoln, characterization, production: R. L. Harned *et al.*, *Antibiot. & Chemother.* **1**, 594 (1951); J. Berger *et al.*, *J. Am. Chem. Soc.* **73**, 5295 (1951); J. Shoji *et al.*, *J. Antibiot.* **21**, 402 (1968). Structure: L. K. Steinrauf *et al.*, *Biochem. Biophys. Res. Commun.* **33**, 29 (1968); T. Kubota *et al.*, *Chem. Commun.* **1968**, 1541; T. Kubota, S. Matsutani, *J. Chem. Soc. (C)* **1970**, 695. Use in coccidiosis: M. Gorman, R. L. Hamill, U.S. pat. **3,555,150** (1971 to Lilly). Effect on calcium uptake and membrane potential in mitochondria: H. Rottenberg, A. Scarpa, *Biochemistry* **13**, 4811 (1974). Stimulation of ATPase activity: H. Sze, *Proc. Nat. Acad. Sci. USA* **77**, 5904 (1980). Approach to synthesis: C. P. Holmes, P. A. Bartlett, *J. Org. Chem.* **54**, 98 (1989).

Colorless needles, mp 183.5-185°. $[\alpha]_D^{24}$ +36.2° (c = 0.842 in CHCl$_3$). Sol in alcohols, acetone, ethyl acetate, chloroform, benzene, ether; slightly sol in satd hydrocarbons. Practically insol in water. LD$_{50}$ in mice (mg/kg): 10-15 i.p. (Shoji); also reported as 2.5 i.p. (Harned).

Sodium salt, C$_{40}$H$_{67}$NaO$_{11}$, crystals, mp 245-255° (dec). Sol in chloroform. Practically insol in water.

6634. Nihydrazone. *Acetic acid [(5-nitro-2-furanyl)methylene]hydrazide; acetic acid 5-(nitrofurfurylidene)hydrazide;* 5-nitro-2-furaldehyde acetylhydrazone; N-acetyl-5-nitro-2-furaldehyde hydrazide; 1-(5-nitro-2-furfurylidene)-2-acetylhydrazine; 1-acetyl-2-(5-nitro-2-furfurylidene)hydrazide; NF-64; HC-064; Furiton; Nidrafur. C$_7$H$_7$N$_3$O$_4$; mol wt 197.15. C 42.65%, H 3.58%, N 21.31%, O 32.46%. Prepn: Stillman, Scott, U.S. pats. **2,416,234** and **2,416,236** (1947 to Eaton Labs).

Yellow crystals from acetic acid + ethanol, dec 230-235°. uv max (water): 253, 364 nm (log ε 4.11, 4.23). Soly in water 1:20,000.

THERAP CAT (VET): Antibacterial, antiprotozoal.

6635. Nikethamide. *N,N-Diethyl-3-pyridinecarboxamide; N,N-diethylnicotinamide;* pyridine-3-carboxylic acid diethylamide; nicotinic acid diethylamide; Anacardone; Astrocar; Carbamidal; Cardamine; Cardiamid; Cardimon; Coracon; Coractiv N; Coramine; Cordiamin; Corediol; Cormed; Cormid; Corvitol; Corvotone; Dynacoryl; Eucoran; Inicardio; Niamine; Nicamide; Nicor; Nicorine; Nikardin; Pyricardyl; Salvacard; Stimulin; Ventramine. C$_{10}$H$_{14}$N$_2$O; mol wt 178.23. C 67.39%, H 7.92%, N 15.72%, O 8.98%. Prepd by the action of thionyl chloride on nicotinic acid, followed by treatment with diethylamine hydrochloride; also formed by heating nicotinic acid or quinolinic acid anhydride with diethylamine: Hartmann, Seiberth, U.S. pat. **1,403,117** (1922); from nicotinic acid + benzenesulfonyldiethyl amide: Oxley *et al., J. Chem. Soc.* **1946**, 763. Toxicity data: E. I. Goldenthal, *Toxicol. Appl. Pharmacol.* **18**, 185 (1971).

Slightly viscous liquid or cryst solid. Faintly bitter taste followed by a faint sensation of warmth. d$_4^{25}$ 1.058-1.066 (liq). mp 24-26°. bp$_{760}$ 296-300° (some decompn); bp$_{10}$ 158-159°; bp$_3$ 128-129°; bp$_{0.4}$ 115°. n$_D^{20}$ 1.525-1.526; n$_D^{21}$ 1.522-1.524. Misc with water, ether, chloroform, acetone, alc. Usually marketed as a 25% w/v aq soln, pH 6.0-6.5. Incompatible with Na$_2$CO$_3$ solns which cause precipitation; tannates produce an amorph ppt. LD$_{50}$ i.p. in rats: 272 mg/kg (Goldenthal).

THERAP CAT: CNS and respiratory stimulant.
THERAP CAT (VET): Respiratory stimulant.

6636. Nilutamide. *5,5-Dimethyl-3-[4-nitro-3-(trifluoromethyl)phenyl]-2,4-imidazolidinedione;* 1-(3'-trifluoromethyl-4'-nitrophenyl)-4,4-dimethylimidazoline-2,5-dione; RU-23908; Anandron. C$_{12}$H$_{10}$F$_3$N$_3$O$_4$; mol wt 317.22. C 45.44%, H 3.18%, F 17.97%, N 13.25%, O 20.17%. Prepn: J. Perronnet *et al.,* Ger. pat. **2,649,925**; *eidem,* U.S. pat.

4,097,578 (1977, 1978 both to Roussel-UCLAF). Pharmacology: J.-P. Raynaud *et al., J. Steroid Biochem.* **11**, 93 (1979); M. Moguilewsky *et al., ibid.* **24**, 139 (1986). Kinetics, metabolism and review of clinical studies: J.-P. Raynaud *et al., Prog. Clin. Biol. Res.* **185A**, 99-120 (1985). Effect on ocular response: C. Harnois *et al., Brit. J. Ophthalmol.* **70**, 471 (1986). Review of pharmacology, mechanism of action and effect on LHRH: J.-P. Raynaud *et al., Prostate* **5**, 299-311 (1984). Clinical study of combination with LHRH agonist in prostate cancer: F. Labrie *et al., Int. Congr. Ser.-Excerpta Med.* **655**, 450 (1984).

Crystals from ethanol, mp 149°.
THERAP CAT: Antiandrogen.

6637. Nilvadipine. *2-Cyano-1,4-dihydro-6-methyl-4-(3-nitrophenyl)-3,5-pyridinedicarboxylic acid 3-methyl 5-(1-methylethyl) ester;* 5-isopropyl-3-methyl-2-cyano-1,4-dihydro-6-methyl-4-(m-nitrophenyl)-3,5-pyridinedicarboxylate; isopropyl 6-cyano-5-methoxycarbonyl-2-methyl-4-(3-nitrophenyl)-1,4-dihydropyridine-3-carboxylate; nivadipine; nilvadipine; FR-34235; FK-235; SKF-102362; Escor; Nivadil. C$_{19}$H$_{19}$N$_3$O$_6$; mol wt 385.38. C 59.22%, H 4.97%, N 10.90%, O 24.91%. Dihydropyridine calcium channel blocker. Prepn: Y. Sato, Belg. pat. **879,263**; *idem,* U.S. pat. **4,338,322** (1980, 1982 both to Fujisawa). Structural studies: A. Miyamae *et al., Chem. Pharm. Bull.* **34**, 3071 (1986). Determn in plasma and urine: Y. Tokuma *et al., J. Chromatog.* **415**, 156 (1987). Preliminary pharmacokinetics and resolution of enantiomers: *eidem, J. Pharm. Sci.* **76**, 310 (1987). Pharmacokinetics in rabbits: Y. Nezasa *et al., Kankyo Igaku Kenkyusho Nenpo* **38**, 200 (1987), *C.A.* **107**, 89273q (1987). Mode of action: P. A. Molyvdas, N. Sperelakis, *J. Cardiovasc. Pharmacol.* **8**, 449 (1986). Cardiovascular effects: G. J. Gross *et al., Gen. Pharmacol.* **14**, 677 (1983). Clinical evaluation in hypertension: K. Mizuno *et al., Res. Comm. Chem. Pathol. Pharmacol.* **52**, 3 (1986).

Yellow prisms from ethanol, mp 148-150°.
(+)-Form, $[\alpha]_D^{20}$ +222.42° (c = 1 in methanol).
(−)-Form, $[\alpha]_D^{20}$ −219.62° (c = 1 in methanol).
THERAP CAT: Antihypertensive; antianginal.

6638. Nimbin. *(4α,5α,6α,7α,15β,17α)-6-(Acetyloxy)-7,-15:21,23-diepoxy-4,8-dimethyl-1-oxo-18,24-dinor-11,12-secochola-2,13,20,22-tetraene-4,11-dicarboxylic acid dimethyl ester;* 5-(acetyloxy)-2-(3-furanyl)-3,3a,4a,5,5a,6,9,9a,10,10a-decahydro-6-(methoxycarbonyl)-1,6,9a,10a-tetramethyl-9-oxo-2H-cyclopenta[b]naphtho[2,3-d]furan-10-acetic acid. C$_{30}$H$_{36}$O$_9$; mol wt 540.61. C 66.65%, H 6.71%, O 26.64%. Bitter principle from various parts of the neem tree, *Azadirachta indica* A. Juss. *(Melia azadirachta* L.), *Meliaceae:* Siddiqui, *Current Sci. (India)* **11**, 278 (1942). Structure: C. R. Narayanan *et al., Chem. & Ind. (London)* **1964**, 322; C. R. Narayanan, R. V. Pachapurkar, *Tetrahedron Letters* **1965**, 4333. Stereochemistry: *eidem, J. Org. Chem.* **31**, 2691 (1966); Harris *et al., Tetrahedron* **24**, 1517 (1968). Crystal structure of dihydronimbin: C. R. Narayanan *et al., Acta Crystallogr.* **B36**, 486 (1980).

Crystals from methanol, mp 205°. $[\alpha]_D^{24}$ +170° (abs ethanol). uv max (95% ethanol): 210, 330 nm (ε 32700; 66). Practically insol in water. Sol in ether, abs alcohol.

Dihydronimbin, $C_{30}H_{38}O_9$, crystals from methanol, mp 215-216°. $[\alpha]_D$ +167.5° (chloroform). uv max (95% ethanol): 210, 298 nm (ε 16200; 29).

6639. Nimbiol. *(4aS-trans)-2,3,4,4a,10,10a-Hexahydro-6-hydroxy-1,1,4a,7-tetramethyl-9(1H)-phenanthrenone;* 6-hydroxy-7-methyl-9-oxopodocarpane. $C_{18}H_{24}O_2$; mol wt 272.39. C 79.37%, H 8.88%, O 11.75%. (+)-Form found in the bark of the neem tree, *Azadirachta indica* A. Juss. *(Melia azadirachta L.), Meliaceae:* P. Sengupta *et al., Chem. & Ind. (London)* **1958**, 861. Structure: P. Sengupta *et al., Tetrahedron* **10**, 45 (1960). Also obtained by conversion of podocarpic acid: Bible, *ibid.* **11**, 22 (1960); Wenkert *et al., J. Am. Chem. Soc.* **83**, 2320 (1961). Total synthesis of (±)-form: W. L. Meyer *et al., J. Org. Chem.* **40**, 3686 (1975). Synthesis of (±)-methyl ether: Ramachandran, Dutta, *J. Chem. Soc.* **1960**, 4766; Delobelle, Fétizon, *Bull. Soc. Chim. France* **1961**, 1900; Nasipuri, Roy, *J. Indian Chem. Soc.* **40**, 327 (1963).

Crystals from dilute methanol or platelets by high vac sublimation. mp 248-252°. $[\alpha]_D^{25}$ +33° (chloroform). uv max (abs alcohol): 234, 283 nm (log ε 4.13, 4.10).

Methyl ether, $C_{19}H_{26}O_2$, crystals, mp 143°. $[\alpha]_D^{25}$ +43.7° (chloroform). uv max (alcohol): 207, 232, 279 nm (log ε 4.18, 4.15, 4.12).

(±)-Form, white needles from CH_3OH, mp 237.0-237.5°.

(±)-Methyl ether, $C_{19}H_{26}O_2$. Crystals from hexane. mp 112-113°. uv max (alcohol): 207, 232, 279 nm (log ε 4.13, 4.13, 4.11) (Delobelle, Fétizon). Also reported as needles from CH_3OH, mp 117-118° (Nasipuri, Roy).

6640. Nimesulide. *N-(4-Nitro-2-phenoxyphenyl)methanesulfonamide;* 4-nitro-2-phenoxymethanesulfonanilide; R-805; Aulin; Flogovital; Mesulid; Nisulid. $C_{13}H_{12}N_2O_5S$; mol wt 308.31. C 50.64%, H 3.92%, N 9.09%, O 25.95%, S 10.40%. Prostaglandin synthetase and platelet aggregation inhibitor. Prepn and anti-inflammatory activity: **Belg. pat. 801,812;** G. G. I. Moore, J. K. Harrington, **U.S. pat. 3,840,-597** (both 1974 to Riker). Activity, comparison with other non-steroidal anti-inflammatories: K. F. Swingle *et al., Arch. Int. Pharmacodyn. Ther.* **221**, 132 (1976). Mechanism of action: R. L. Vigdahl, R. H. Tukey, *Biochem. Pharmacol.* **26**, 307 (1977). Chromatographic determn in plasma: S. F. Chang *et al., J. Pharm. Sci.* **66**, 1700 (1977). Pharmacology: K. F. Swingle, G. G. I. Moore, *Drugs Exp. Clin. Res.* **10**, 587 (1984). Clinical trials in rheumatic disorders: R. Weissenbach, *J. Int. Med. Res.* **9**, 349 (1981); in osteoarthritis: M. Reiner, *ibid.* **10**, 92 (1982); in acute inflammation: J. M. Pais, F. M. Rosteiro, *ibid.* **11**, 149 (1983); C. Milvio, *ibid.* **12**, 327 (1984); M. E. Nouri, *Clin. Ther.* **6**, 142 (1984).

Light tan crystals from ethanol, mp 143-144.5°. LD_{50} orally in rats: 324 mg/kg (Swingle, Moore).

THERAP CAT: Anti-inflammatory.

THERAP CAT (VET): Anti-inflammatory.

6641. Nimetazepam. *1,3-Dihydro-1-methyl-7-nitro-5-phenyl-2H-1,4-benzodiazepin-2-one;* 1-methyl-5-phenyl-7-nitro-1,3-dihydro-2H-1,4-benzodiazepin-2-one; 1-methyl-nitrazepam; S-1530; Elimin; Hypnon. $C_{16}H_{13}N_3O_3$; mol wt 295.30. C 65.08%, H 4.44%, N 14.23%, O 16.25%. The desmethyl derivative of nitrazepam, *q.v.* Prepn: L. H. Sternbach *et al., J. Med. Chem.* **6**, 261 (1963); E. Reeder, L. H. Sternbach, **U.S. pats. 3,109,843; 3,144,439; 3,141,890; 3,243,427** (1963, 1964, 1964, 1966 all to Hoffmann-La Roche); O. Keller *et al.,* **U.S. pats. 3,121,114; 3,203,990** (1964, 1965 to Hoffmann-La Roche); Sorrentino, **S. Afr. pat. 67 06,791** (1968 to Dumex), *C.A.* **70**, 57916c (1969); Yamamoto *et al.,* **Ger. pat. 1,816,046** (1970 to Sumitomo), *C.A.* **73**, 120690d (1970); Inaba *et al., Chem. Pharm. Bull.* **19**, 722 (1971). Pharmacology: S. Sakai *et al., Arzneimittel-Forsch.* **22**, 534 (1972).

Pale yellow plates from ethanol, mp 156.5-157.5°. uv max (methanol): 259, 308 nm (ε 15800, 9600). LD_{50} in male, female mice, rats (mg/kg): 910, 750, 1150, 970 orally; 970, 840, 970, 980 i.p.; 1500, 1500, 1000, 1000 s.c. (Sakai).

Note: This is a controlled substance (depressant) listed in the U.S. Code of Federal Regulations, Title 21 Part 1308.14 (1995).

THERAP CAT: Anticonvulsant; muscle relaxant (skeletal).

6642. Nimidane. *4-Chloro-N-1,3-dithietan-2-ylidene-2-methylbenzeneamine;* cyclic methylene (4-chloro-o-tolyl)-dithioimidocarbonate; AC-84633; ENT-29106; Abequito. $C_9H_8ClNS_2$; mol wt 229.75. C 47.05%, H 3.51%, Cl 15.43%, N 6.10%, S 27.91%. Prepn: R. W. Addor, S. Kantor, **Ger. pat. 2,305,517** (1973 to Am. Cyanamid), *C.A.* **79**, 115553f (1973); W. W. Brand, M. W. Bullock, **U.S. pat. 3,842,096** (1974 to Am. Cyanamid). Tickicidal activity: N. K. Amaral *et al., J. Econ. Entomol.* **67**, 387 (1974).

White solid, mp 43-46°.

THERAP CAT (VET): Acaricide.

6643. Nimodipine. *1,4-Dihydro-2,6-dimethyl-4-(3-nitrophenyl)-3,5-pyridinedicarboxylic acid 2-methoxyethyl 1-methylethyl ester;* 2-methoxyethyl 1,4-dihydro-5-(isopropoxycarbonyl)-2,6-dimethyl-4-(3-nitrophenyl)-3-pyridinecarboxylate; isopropyl 2-methoxyethyl 1,4-dihydro-2,6-dimethyl-4-(3-nitrophenyl)-3,5-pyridinedicarboxylate; 2,6-dimethyl-4-(3'-nitrophenyl)-1,4-dihydropyridine-3,5-dicarboxylic acid 3-β-methoxyethyl ester 5-isopropyl ester; Bay e 9736; Admon; Nimotop; Periplum. $C_{21}H_{26}N_2O_7$; mol wt 418.45. C 60.28%, H 6.26%, N 6.69%, O 26.76%. Dihydropyridine calcium channel blocker. Prepn: H. Meyer et al., **Ger.** pat. **2,117,571**; eidem, **U.S.** pat. **3,799,934** (1972, 1974 to Bayer). Pharmacology: R. Towart, S. Kazda, Brit. J. Pharmacol. **67**, 409P (1979); K. Tanaka et al., Arzneimittel-Forsch. **30**, 1494 (1980); L. M. Auer, ibid. **31**, 1423 (1981). Prepn of isomers and pharmacological comparison with racemate: R. Towart et al., ibid. **32**, 338 (1982). Use as cerebral vasodilator: H. Meyer et al., **Brit.** pat. **2,018,134**; eidem, **U.S.** pat. **4,406,906** (1979, 1983 to Bayer). Effect on associative learning in aging rabbits: R. A. Deyo et al., Science **243**, 809 (1989). GC and LC determns in biological fluids: G. J. Krol et al., J. Chromatog. **305**, 105 (1984). Clinical trial in prophylaxis of cerebral vasospasm: G. S. Allen et al., N. Engl. J. Med. **308**, 619 (1983). Toxicology: H. Schlüter, Arzneimittel-Forsch. **36**, 1733 (1986). Series of articles on clinical pharmacology and therapeutic use: Am. J. Cardiol. **55**(3), 139B-153B (1985).

Crystals from petr ether/acetic ester, mp 125°. LD_{50} in mice, rats (mg/kg): 3562, 6599 orally; 33, 16 i.v. (Schlüter).
(+)-Form, $[\alpha]_D^{20}$ +7.9° (c = 0.439 in dioxane).
(−)-Form, $[\alpha]_D^{20}$ −7.93° (c = 0.374 in dioxane).
THERAP CAT: Vasodilator (cerebral).

6644. **Nimorazole.** *4-[2-(5-Nitro-1H-imidazol-1-yl)ethyl]morpholine; N-2-morpholinoethyl-5-nitroimidazole; 1-(2-N-morpholinylethyl)-5-nitroimidazole; nitrimidazine; K-1900; Acterol; Esclama; Naxofem; Naxogin; Nulogyl.* $C_9H_{14}N_4O_3$; mol wt 226.24. C 47.78%, H 6.24%, N 24.76%, O 21.22%. Prepn: **Belg.** pat. **667,262**; Giraldi, Mariotti, **U.S.** pat. **3,399,193** (1965, 1968 both to Carlo Erba); **Neth.** pat. Appls. **6,609,552; 6,609,553**; Gal and Carlson et al., **U.S.** pats. **3,458,528** and **3,646,027** (1967, 1967, 1969, 1972, all to Merck & Co.). Synthesis and biological activity studies: Giraldi et al., Arzneimittel-Forsch. **20**, 52 (1970). Pharmacology and toxicology: de Carneri et al., Progr. Antimicrob. Anticancer Chemother., Proc. 6th Int. Congr. Chemother. Tokyo, 1969 (Univ. of Tokyo Press, 1970, Tokyo) **vol. I**, pp 149-154. Clinical results: Emanueli, de Carneri, ibid **vol. II**, pp 369-372; Evans, Catterall, Brit. Med. J. **IV**, 146 (1971). Metabolic studies: Giraldi, Biochem. Pharmacol. **20**, 339 (1971). Acute toxicity data: B. Cavalleri et al., J. Med. Chem. **21**, 781 (1978).

Crystals from water, mp 110-111°. Slightly sol in water at room temp; sol in alcohols, acetone, chloroform. LD_{50} orally in mice: 1530 mg/kg (Cavalleri).
THERAP CAT: Antiprotozoal (Trichomonas).

6645. **Nimustine.** *N'-[(4-Amino-2-methyl-5-pyrimidinyl)methyl]-N-(2-chloroethyl)-N-nitrosourea.* $C_9H_{13}ClN_6O_2$; mol wt 272.69. C 39.64%, H 4.81%, Cl 13.00%, N 30.82%, O 11.73%. Chloroethylnitrosourea derivative with antitumor activity. Similar to carmustine, chlorozotocin, lomustine, ranimustine, q.q.v. Prepn: H. Nakao et al., **Ger.** pat. **2,257,360**; eidem, **U.S.** pat. **4,003,901** (1973, 1977 both to Sankyo); eidem, Yakugaku Zasshi **94**, 1932 (1974), C.A. **82**, 43263y (1975). Effects on macrophage cytostatic activity in rats: N. Saijo et al., Brit. J. Cancer **42**, 162 (1980). Dis-

tribution, excretion, metabolism in mice: M. Tanaka et al., Cancer Treat. Rep. **64**, 575 (1980). Exptl and clinical effect: N. Saijo et al., Cancer Chemother. Pharmacol. **4**, 165 (1980). Acute toxicity: H. Masuda et al., Sankyo Kenkyusho Nempo **29**, 118 (1977), C.A. **88**, 146298s (1978).

Pale yellow crystals from ethanol, dec 125°.
Hydrochloride, $C_9H_{13}ClN_6O_2$·HCl, NSC-245382, ACNU, Nidran. White to light yellow cryst powder. uv max (0.04N HCl): 245 nm ($E_{1cm}^{1\%}$ 480-510). Sol in methanol, slightly sol in abs ethanol, n-butanol. Practically insol in ethyl acetate, ether, chloroform, benzene, n-hexane. Gradually develops greenish yellow color in light; decomposes slowly in humid air. LD_{50} in mice, rats (mg/kg): 62, 46 i.v. (Masuda).
THERAP CAT: Antineoplastic.

6646. **Ninhydrin.** *2,2-Dihydroxy-1H-indene-1,3(2H)-dione; 2,2-dihydroxy-1,3-indanedione; 1,2,3-indantrione monohydrate; triketohydrinden hydrate.* $C_9H_6O_4$; mol wt 178.14. C 60.68%, H 3.39%, O 35.92%. Prepn: Ruhemann, J. Chem. Soc. **97**, 1438 (1910); Teeters, Shriner, J. Am. Chem. Soc. **55**, 3026 (1933); Wanag, Lode, Ber. **71**, 1267 (1938). Improved syntheses: L. F. Fieser, Experiments in Organic Chemistry (Boston, 3rd ed., 1955) p 123; Becker, Russell, J. Org. Chem. **28**, 1896 (1963). Absorption spectrum: Polonovski et al., Bull. Soc. Chim. [5] **6**, 1557 (1939). Poisonous action on bacteria, mice, guinea pigs: O. Loew, Biochem. Z. **69**, 111 (1915). Acute toxicity study: J. C. Breton et al., Compt. Rend. Soc. Biol. **151**, 719 (1957). Monograph: Henri Plagnol, Influence de la Structure des Composés Aminés dans leur Reaction avec la Ninhydrine (Bordeaux, 1962) 156 pp. Reviews: J. C. Breton, Etudes Chimiques et Biologiques sur la Ninhydrine, Réactif des Aminoacides (Imprimerie E. Drouillard, Bordeaux, 1958) 285 pp; McCaldin, Chem. Rev. **60**, 39 (1960); A. Schoenberg, E. Singer, Tetrahedron **34**, 1285-1300 (1978).

Monohydrate, pale yellow prisms from water or alcohol. Reddens at 125°, swells at 139°, dec 241°. Freely sol in water. LD_{50} i.p. in mice: 78 mg/kg (Breton, 1957).
USE: Reagent for the detection of free amino and carboxyl groups in proteins and peptides, yielding a blue color under the proper conditions.

6647. **Ninopterin.** *N-[4-[[1-(2-Amino-1,4-dihydro-4-oxo-6-pteridinyl)ethyl]amino]benzoyl]-L-glutamic acid; 9-methylpteroylglutamic acid; 9-methylfolic acid; Bremfol.* $C_{20}H_{21}N_7O_6$; mol wt 455.43. C 52.75%, H 4.65%, N 21.53%, O 21.08%. Folic acid analog. Prepn: Hultquist et al., J. Am. Chem. Soc. **71**, 619 (1949); Hultquist, Smith, **Brit.** pat. **667,098** (1952 to Am. Cyanamid). Clinical study in neoplastic diseases: J. C. Wright et al., J. Nat. Med. Assn. **43**, 211 (1951). In vitro evaluation as folic acid antagonist: S. Waxman et al., Chemotherapy (Basel) **28**, 402 (1982).

uv max in 0.1N NaOH: 255, 283, 365 nm (ϵ 26, 30, 8 $\times$ 10³); in 0.1N HCl: 297 nm (ϵ 18 $\times$ 10³).

THERAP CAT: Experimental antineoplastic.

6648. Niobium. Columbium. Nb; at. wt 92.9064; at. no. 41; valence 2, 3, 4, 5; usually pentavalent. One naturally occurring isotope: ⁹³Nb; artificial, radioactive isotopes: 88-92; 94-101. Approximately as abundant as nickel. Occurs in nature together with tantalum in the minerals *columbite* [(Fe,Mn)(Nb,Ta)₂O₆], *pyrochlore* (NaCaNb₂O₆F) and tantalite *(q.v.)*. Discovered by Hatchett in 1801, isolated by Blomstrand in 1866, named after Niobe, daughter of Tantalus. Extracted from columbite which is mined largely in Nigeria and Zaire. Less than 10% of niobium-bearing ores come from the U.S.A., Canada, and Norway. Reviews of niobium and its compds: *Technology of Columbium (Niobium)* B. W. Gonser, E. M. Sherwood, Eds. (Wiley, New York, 1958); G. L. Miller, *Tantalum and Niobium* (Academic Press, New York, 1959) 767 pp; Brown, "The Chemistry of Niobium and Tantalum" in *Comprehensive Inorganic Chemistry*, **Vol. 3**, J. C. Bailar, Jr. *et al.,* Eds. (Pergamon Press, Oxford, 1973) pp 553-622; P. H. Payton in Kirk-Othmer *Encyclopedia of Chemical Technology* vol. 15 (Wiley-Interscience, New York, 3rd ed., 1981) pp 820-840.

Steel-gray, lustrous metal. Ductile and malleable when pure. Lattice structure: body-centered cube, lattice constant: 3.294 Å. d 8.57. mp 2468°. bp 4927°. Sp ht: 6.012 cal/g-atom/°C. Heat of sublimation: 170.9 kcal/g-atom; heat of combustion: 2379 cal/g. Coefficient of linear expansion per °C: 7.1 × 10⁻⁶. Electrical resistivity (20°): 13.2 µohm-cm. Temp coefficient of electrical resistivity per °C: 0.00395. Electron work function: 4.01 ev. Ionization potential: 6.77 V. Inert toward HCl, HNO₃ or aqua regia, but attacked by fusion with alkali hydroxides or oxidizing agents.

USE: In ferrous metallurgy: Ferroniobium (produced by silicon reduction of columbite) is used to alloy stainless steels and metals for welding rods. In niobium base alloys for high temps and nuclear reactions. Niobium has some use as a getter in electronic vacuum tubes.

6649. Niobium Pentachloride. Columbium pentachloride. Cl₅Nb; mol wt 270.17. Cl 65.61%, Nb 34.39%. NbCl₅. Prepn: Epperson *et al., Inorg. Syn.* **7**, 163 (1963); *cf.* Rolsten, *J. Am. Chem. Soc.* **80**, 2952 (1958). Review of niobium halides: Fairbrother in *Halogen Chemistry*, **Vol. 3**, V. Gutmann, Ed. (Academic Press, New York, 1967) pp 123-178.

Yellow, very deliquesc, monoclinic crystals; dec in moist air with evolution of HCl. d 2.75. mp 204.7-209.5°; bp ~250°, but begins to sublime at 125°. Sol in HCl, carbon tetrachloride. LD₅₀ in mice: 830 mg/kg orally, *Toxic Substances List*, H. E. Christensen, Ed. (1972) p 368.

6650. Niobium Pentafluoride. Columbium pentafluoride. F₅Nb; mol wt 187.90. F 50.55%, Nb 49.45%. NbF₅. Prepn from NbCl₅ + HF: Ruff, Zedner, *Ber.* **42**, 492 (1909); Ruff, Schiller, *Z. Anorg. Allgem. Chem.* **72**, 329 (1911); Kwasnik in *Handbook of Preparative Inorganic Chemistry*, **Vol. 1**, G. Brauer, Ed. (Academic Press, New York, 2nd ed., 1963) p 254. Prepn from the elements: Fairbrother, Frith, *J. Chem. Soc.* **1951**, 3051; Junkins *et al., J. Am. Chem. Soc.* **74**, 3464 (1952). Review of transition metal pentafluorides: Peacock, *Advan. Fluorine Chem.* **7**, 113-145 (1973).

Strongly refractive, deliquescent, monoclinic crystals, d⁸⁰₄ 2.6955. mp 80.0°; bp 234.9°. Appreciable vapor pressure at 50°. Hydrolyzes in water, alc, caustic solns. Sparingly sol in carbon disulfide, chloroform. More sol in concd H₂SO₄ than TaF₅. Forms complexes with Lewis bases.

6651. Niobium Pentoxide. Columbium pentoxide. Nb₂O₅; mol wt 265.81. Nb 69.90%, O 30.10%. Separation: Münchow, *Chem. Ztg.* **84**, 490, 527 (1960). Review of polymorphism: Schäfer *et al., Angew. Chem. Int. Ed.* **5**, 40-52 (1966).

White, orthorhombic crystals. Becomes yellow on heating. d 4.6. mp 1520°. Insol in water; sol in HF, hot H₂SO₄.

6652. Niobium Potassium Oxypentafluoride. Columbium potassium oxypentafluoride. F₇K₂NbO; mol wt 320.09. F 41.55%, K 24.43%, Nb 29.02%, O 5.00%. K₂Nb-

OF₅. Exists as monohydrate. Prepn: Hoard, Martin, *J. Am. Chem. Soc.* **63**, 11 (1941).

Monohydrate, lustrous monoclinic leaflets; fatty to touch. Freely sol in hot water. Soly in cold water ~1 part in 13.

6653. Nioxime®. *1,2-Cyclohexanedione dioxime.* C₆H₁₀N₂O₂; mol wt 142.16. C 50.69%, H 7.09%, N 19.71%, O 22.51%. Prepn: Hach *et al., Org. Syn.* **32**, 35 (1952); Boyer, Ellzey, *J. Am. Chem. Soc.* **82**, 2525 (1960); Borger *et al., Bull. Soc. Chem. Belges* **73**, 73 (1964); Arthur, **Fr.** pat. **1,339,224** (1963 to du Pont).

Crystals, mp 185-188° (darkening at 170°).
USE: Analytical reagent for nickel and palladium.

6654. Nipecotic Acid. *3-Piperidinecarboxylic acid; hexahydronicotinic acid.* C₆H₁₁NO₂; mol wt 129.16. C 55.80%, H 8.58%, N 10.84%, O 24.77%. Prepd by hydrogenation of nicotinic acid with platinum oxide catalyst: McElvain, Adams, *J. Am. Chem. Soc.* **45**, 2738 (1923); with rhodium on alumina catalyst: Freifelder, *J. Org. Chem.* **28**, 1135 (1963); *idem,* **U.S.** pat. **3,159,639** (1964 to Abbott).

Crystals, mp 261° (dec). Freely sol in water. Practically insol in abs alcohol and ether.
Hydrochloride, C₆H₁₁NO₂.HCl, mp 240-242°. Very sol in water; slightly sol in alc (also reported as freely sol in alc); slightly sol in chloroform. Practically insol in ether, benzene, and acetone.

6655. Nipradilol. *3,4-Dihydro-8-[2-hydroxy-3-[(1-methylethyl)amino]propoxy]-2H-1-benzopyran-3-ol 3-nitrate;* 3,4-dihydro-8-(2-hydroxy-3-isopropylamino)propoxy-3-nitroxy-2H-1-benzopyran; 8-[2-hydroxy-3-(isopropylamino)propoxy]-3-chromanol 3-nitrate; nipradolol; K-351; Hypadil. C₁₅H₂₂N₂O₆; mol wt 326.35. C 55.21%, H 6.79%, N 8.58%, O 29.42%. β-Adrenergic blocker with vasodilating activity. Prepn: M. Shiratsuchi *et al., Eur.* pat. Appl. **42,299;** *eidem,* **U.S.** pat. **4,394,382** (1981, 1983 both to Kowa). Synthesis: M. Shiratsuchi *et al., Chem. Pharm. Bull.* **35**, 632 (1987). Resolution of isomers: M. Shiratsuchi *et al., Eur.* pat. Appl. **154,511;** *eidem,* **U.S.** pat. **4,727,085** (1985, 1988 both to Kowa). Synthesis and activity of isomers: M. Shiratsuchi *et al., Chem. Pharm. Bull.* **35**, 3691 (1987). Pharmacology: Y. Uchida *et al., Arch. Int. Pharmacodyn.* **262**, 132 (1983). Hemodynamic effects in dogs: H. Hisa *et al., ibid.* **271**, 169 (1984). Effects on cardiac function in dogs: M. Sakanashi *et al., ibid.* **274**, 47 (1985); M. Fujii *et al., Japan. Heart J.* **27**, 233 (1986). Pharmacokinetics and metabolism in rats: S. Kabuto *et al., Arzneimittel-Forsch.* **35**, 1674 (1985); H. Kimata *et al., ibid.* 1680. Metabolic fate in dog and man: M. Yoshimura *et al., Chem. Pharm. Bull.* **33**, 3456 (1985). Clinical evaluation in angina: H. Kishida *et al., Japan. Heart J.* **29**, 309 (1988).

Colorless needles, mp 107-116° (Shiratsuchi, 1983); also reported as mp 110-122° (Shiratsuchi, 1987). LD₅₀ i.v. in mice, rats: 74.0, 73.0 mg/kg; orally in mice: 540 mg/kg (Shiratsuchi, 1988).

THERAP CAT: Antianginal; antihypertensive.

6656. Niridazole. *1-(5-Nitro-2-thiazolyl)-2-imidazolidinone;* 1-(5-nitro-2-thiazolyl)-2-oxotetrahydroimidazole; nitrothiamidazol; Ba-32644; Ciba 32644-Ba; Ambilhar. $C_6H_6N_4O_3S$; mol wt 214.20. C 33.64%, H 2.82%, N 26.16%, O 22.41%, S 14.97%. Prepn: **Belg.** pat. **632,989** (1963 to Ciba), *C.A.* **61,** 1873d (1964), corresp to **Brit.** pat. **986,562;** Lambert *et al., Experientia* **20,** 452 (1964).

Yellow crystals from dimethylformamide, mp 260-262°.
THERAP CAT: Anthelmintic (Schistosoma).

6657. Nisin. $C_{143}H_{230}N_{42}O_{37}S_7$; mol wt 3354.12. C 51.21%, H 6.91%, N 17.54%, O 17.65%, S 6.69%. Polypeptide antibiotic structurally similar to subtilin, *q.v.,* but containing no tryptophan. Produced by *Streptococcus lactis:* A. T. R. Mattick, A. Hirsch, *Nature* **154,** 551 (1944); *eidem, Lancet* **250,** 417 (1946); **253,** 5 (1947). Purification and nature of nisin: Berridge *et al., Biochem. J.* **52,** 529 (1952). Production: Hawley, Hall, **U.S.** pat. **2,935,503** (1960 to Aplin & Barrett). Structure contains 34 amino acid residues, eight of which are rarely found in nature, including lanthionine (two alanines bonded to sulfur at the β-carbons) and β-methyllanthionine. Structure: E. Gross, J. L. Morrell, *J. Am. Chem. Soc.* **93,** 4634 (1971). Confirmation of structure of nisin and its major degradation product: M. Barber *et al., Experientia* **44,** 266 (1988). Partial synthesis: K. Fukase *et al., Bull. Chem. Soc. Japan* **59,** 2505 (1986). Total synthesis: K. Fukase *et al., Tetrahedron Letters* **29,** 795 (1988). Biosynthetic study: G. W. Buchman *et al., J. Biol. Chem.* **263,** 16260 (1988). *Review:* A. Hurst, *Adv. Appl. Microbiol.* **27,** 85-123 (1981).

Abu = α-aminobutyric acid
Dha = dehydroalanine
Dhb = dehydrobutyrine

Crystals from ethanol. Soluble in dilute acids. Stable to boiling in acid soln.
USE: In food processing as a preservative, esp. for cheese and canned fruits and vegetables.

6658. Nisoldipine. *1,4-Dihydro-2,6-dimethyl-4-(2-nitrophenyl)-3,5-pyridinedicarboxylic acid methyl 2-methylpropyl ester;* isobutyl methyl 1,4-dihydro-2,6-dimethyl-4-(o-nitrophenyl)-3,5-pyridinedicarboxylate; isobutyl 1,4-dihydro-5-methoxycarbonyl-2,6-dimethyl-4-(2-nitrophenyl)-3-pyridinecarboxylate; 2,6-dimethyl-3-carbomethoxy-4-(2-nitrophenyl)-5-carbisobutoxy-1,4-dihydropyridine; Bay k 5552; Baymycard; Norvasc; Syscor; Zadipina. $C_{20}H_{24}N_2O_6$; mol wt 388.42. C 61.85%, H 6.23%, N 7.21%, O 24.71%. Dihydropyridine calcium channel blocker. Prepn: E. Wehinger *et al.,* **Ger.** pat. **2,549,568;** *eidem,* **U.S.** pat. **4,154,839** (1977, 1979 both to Bayer). Pharmacology: S. Kazda *et al., Arzneimittel-Forsch.* **30,** 2144 (1980). Preferential effect on vascular smooth muscle in humans: A. Vogt *et al., ibid.* 2162. Pharmacodynamics and pharmacokinetics: F. Pasanisi *et*

al., Eur. J. Clin. Pharmacol. **29,** 21 (1985). Series of articles on pharmacokinetics in animals, humans: *Arzneimittel-Forsch.* **38,** 1093-1110 (1988). Comparison of hemodynamic effects in humans with nifedipine, nitrendipine, *q.q.v.:* N. M. G. Debbas *et al., Eur. J. Clin. Pharmacol.* **30,** 393 (1986). Clinical trial in angina: J. Lam *et al., J. Am. Coll. Cardiol.* **6,** 447 (1985).

Crystals from ethanol, mp 151-152°.
THERAP CAT: Antihypertensive; antianginal.

6659. Nitarsone. *4-Nitrophenylarsonic acid; p-nitrobenzenearsonic acid.* $C_6H_6AsNO_5$; mol wt 247.04. C 29.17%, H 2.45%, As 30.33%, N 5.67%, O 32.38%. Prepn: Jacobs *et al., J. Am. Chem. Soc.* **40,** 1580 (1918); Doak, Freedman, **U.S.** pat. **2,653,160** (1953 to the U.S.A. as represented by the Administrator of the Federal Security Agency); Ruddy, Starkey, *Org. Syn.* **coll. vol. III,** 665 (1955).

Pale yellow leaflets from water, dec 298-300°. Very slightly sol in cold water, cold alcohol; readily sol in warm water, warm alcohol.
THERAP CAT (VET): Antihistomonad.

6660. Nithiazide. *N-Ethyl-N'-(5-nitro-2-thiazolyl)urea;* Hepzide. $C_6H_8N_4O_3S$; mol wt 216.22. C 33.33%, H 3.73%, N 25.91%, O 22.20%, S 14.83%. Prepd by the condensation of ethyl isocyanate with 2-amino-5-nitrothiazole in toluene: Cuckler *et al., Proc. Soc. Exp. Biol. Med.* **92,** 483 (1956); O'Neill *et al.,* **U.S.** pat. **2,755,285** (1956 to Merck & Co.).

Crystals, dec 228°. pKa 7.3. Practically insol in water: 3 mg/100 ml.
Sodium salt, orange plates. Soly in water: 8 g/100 ml. Aq solns are alkaline and unstable.
Potassium salt, crystals. Soly in water: 3 g/100 ml. Aq solns are more stable than those of the Na salt.
THERAP CAT (VET): Antiprotozoal.

6661. Nitracrine. *N,N-Dimethyl-N'-(1-nitro-9-acridinyl)-1,3-propanediamine; 9-[[3-(dimethylamino)propyl]-amino]-1-nitroacridine.* $C_{18}H_{20}N_4O_2$; mol wt 324.38. C 66.65%, H 6.21%, N 17.27%, O 9.86%. Deriv of acridine, *q.v.,* with cytostatic and cytotoxic properties. Prepn: **Fr.** pat. **1,458,183** (1966 to Polfa), *C.A.* **68,** 39493s (1968); A. Ledochowski, B. Stefanska, *Roczniki Chem.* **40,** 301 (1966), *C.A.* **65,** 2219b (1966). Pharmacological studies: J. Gieldanowski *et al., Arch. Immunol. Ther. Exp.* **20,** 399 (1972); *eidem, ibid.* 419. Mechanism of action: J. Konopa *et al., Mater. Med. Pol.* **8,** 258 (1976). Cytotoxicity study: I. Szumiel, M. Walicka, *Neoplasma* **27,** 697 (1980). DNA binding activity: L. Szmigiero, M. Gniazdowski, *Arzneimittel-Forsch.* **31,** 1875 (1981). Comprehensive review: M. Gniazdowski *et al.* in *Antibiotics* Vol. V, part 2, F. E. Hahn, Ed. (Springer-Verlag, New York, 1979) pp 275-297.

Crystals from benzene/petr ether, mp 134-135°. Practically insol in water. Sol in most organic solvents. pKa$_1$ 6.45; pKa$_2$ 8.8.

Dihydrochloride monohydrate, $C_{18}H_{20}N_4O_2.2HCl.H_2O$, C-283, Ledakrin. Orange crystals, mp 223-224°. Sol in water, methanol, ethanol, slightly sol in benzene, diethyl ether. Conc water solns are acidic (pH 4). LD$_{50}$ in rats, mice (mg/kg): 1, 0.72 i.v.; 34, 26 i.g. (Gniazdowski).

THERAP CAT: Antineoplastic.

6662. Nitralin. *4-(Methylsulfonyl)-2,6-dinitro-N,N-dipropylbenzenamine; 4-(methylsulfonyl)-2,6-dinitro-N,N-dipropylaniline;* SD-11831; Planavin. $C_{13}H_{19}N_3O_6S$; mol wt 345.38. C 45.21%, H 5.54%, N 12.17%, O 27.79%, S 9.28%. Prepn: S. B. Soloway, K. D. Swahlen, U.S. pats. **3,227,734** and **3,321,292** (1966, 1967, both to Shell). Persistence in the soil: J. H. Miller *et al., Weed Sci.* **23**, 211 (1975).

Golden crystals, mp 150-151°. Vapor pressure at 20°: 9.3 × 10^{-9} mm Hg; at 30°: 3.3 × 10^{-8} mm Hg. Soly at 22°: water, 0.6 ppm; acetone, 36 g/100 ml; DMSO, 33 g/100 ml. Poorly soluble in common hydrocarbon and aromatic solvents and alcohol. LD$_{50}$ orally in rats: > 2 g/kg.

USE: Herbicide.

6663. Nitramide. $H_2N_2O_2$; mol wt 62.03. H 3.25%, N 45.16%, O 51.59%. H_2NNO_2. Prepd from potassium nitrocarbamate: Thiele, Lachmann, *Ber.* **27**, 1909 (1894); *Ann.* **288**, 267 (1895); Marlies, La Mer, *J. Am. Chem. Soc.* **57**, 2008 (1935); Marlies *et al., Inorg. Syn.* **1**, 68 (1939).

Unstable, shiny, white leaflets from ether + petr ether, mp 72-75° (dec). Dipole moment 3.7. Should be freshly prepd when needed. Does not explode under ordinary conditions. *Handle in glass or platinum!* Sol in ether, alcohol, acetone, water. Slightly sol in benzene. Practically insol in petr ether, chloroform.

6664. Nitramine. *N-Methyl-N,2,4,6-tetranitrobenzenamine; N-methyl-N,2,4,6-tetranitroaniline;* picrylmethylnitramine; picrylnitromethylamine; Tetralite; Tetryl. $C_7H_5N_5O_8$; mol wt 287.15. C 29.28%, H 1.76%, N 24.39%, O 44.58%.

Yellow crystals. d 1.57. mp 130-132°; explodes at about 180-190°. Insol in water. Sol in alc, ether, benzene, glacial acetic acid. *Keep soln in the dark.* Causes yellow staining to skin and hair.

Caution: Potential symptoms of overexposure are sensitization dermatitis, itching, erythema, edema on nasal folds, cheeks and neck; keratitis; sneezing; anemia; fatigue; coughing, coryza; irritability; malaise, headache, lassitude and insomnia; nausea, vomiting. *See NIOSH Pocket Guide to Chemical Hazards* (DHHS/NIOSH 90-117, 1990) p 212.

USE: As indicator, 0.1 g in 60 ml alcohol with water to make 100 ml. pH: 10.8 colorless, 13.0 reddish-brown. One

to five drops of soln required for 10 ml liquid. Salt error said to be small. Also used in explosives.

6665. Nitranilic Acid. *2,5-Dihydroxy-3,6-dinitro-2,5-cyclohexadiene-1,4-dione; 2,5-dihydroxy-3,6-dinitro-p-benzoquinone;* 2,5-dihydroxy-3,6-dinitroquinone. $C_6H_2N_2O_8$; mol wt 230.09. C 31.32%, H 0.88%, N 12.17%, O 55.63%. Prepd by treating hydroquinone diacetate with fuming nitric and concd sulfuric acids: Nietzki, Benckiser, *Ber.* **18**, 499 (1885); Nietzki, *Ber.* **43**, 3458 (1910); Meyer, *Ber.* **57**, 326 (1924); Town, *Biochem. J.* **30**, 1833 (1936).

Solvated, golden-yellow plates from a little water + nitric acid. Effloresces at 100°, deflagrates at 170° without melting. Yellow prisms from ethyl acetate. Absorption max: 460 nm. Stable when kept away from moisture. Freely sol in water, alcohol. Insol in ether. Aq solns dec forming HCN and oxalic acid.

6666. Nitrapyrin. *2-Chloro-6-(trichloromethyl)pyridine;* α,α,α,6-tetrachloro-2-picoline; N-Serve. $C_6H_3Cl_4N$; mol wt 230.91. C 31.21%, H 1.31%, Cl 61.41%, N 6.07%. Prepn: Johnston *et al., Brit.* pat. **957,276**, Belg. pat. **624,-800** (1964, 1963 to Dow), *C.A.* **61**, 1841b (1964).

Crystals, mp 62.5-62.9° (CH$_2$Cl$_2$-pentane). bp$_{11}$ 136-137.5°.

USE: Fertilizer additive to control nitrification and prevent loss of soil nitrogen.

6667. Nitrazepam. *1,3-Dihydro-7-nitro-5-phenyl-2H-1,-4-benzodiazepin-2-one;* LA-1; Ro-4-5360; Benzalin; Calsmin; Eatan; Eunoctin; Imeson; Insomin; Ipersed; Mogadan; Mogadon; Nelbon; Neuchlonic; Nitrados; Nitrenpax; Noctesed; Pelson; Radedorm; Remnos; Somnased; Somnibel; Somnite; Sonebon; Surem; Unisomnia. $C_{15}H_{11}N_3O_3$; mol wt 281.27. C 64.05%, H 3.94%, N 14.94%, O 17.06%. Prepn: Sternbach *et al., J. Med. Chem.* **6**, 261 (1963); O. Keller *et al.,* U.S. pat. **3,121,076** (1964 to Hoffmann-La Roche). Crystal structure: G. Gilli *et al., Acta Crystallogr.* **33B**, 2664 (1977). Pharmacokinetic studies: Rieder, *Arzneimittel-Forsch.* **23**, 212 (1973); L. Kangas, *Acta Pharmacol. Toxicol.* **45**, 16 (1979). Toxicity data: Randall *et al., Schweiz. Med. Wochenschr.* **95**, 334 (1965). *Review:* Rieder, Wendt in *Benzodiazepines,* S. Garattini *et al.,* Eds. (Raven Press, New York, 1973) pp 99-127. Comprehensive description: H. Y. Aboul-Enein *et al., Anal. Profiles Drug Subs.* **9**, 487-517 (1980).

Yellow crystals from ethanol, mp 224-226°. uv max (0.1N H_2SO_4): 277.5 nm (E$_{1cm}^{1\%}$ 1500). Sol in alc, acetone, chloroform, ethyl acetate. Practically insol in water, ether, benzene, hexane. LD$_{50}$ orally in rats: 825 ±80 mg/kg (Randall).

Note: This is a controlled substance (depressant) listed in the U.S. Code of Federal Regulations, Title 21 Part 1308.14 (1995).

THERAP CAT: Anticonvulsant; hypnotic.

6668. Nitrefazole. *2-Methyl-4-nitro-1-(4-nitrophenyl)-1H-imidazole;* 1-(4-nitrophenyl)-2-methyl-4-nitroimidazole; EMD-15700; Altimol. $C_{10}H_8N_4O_4$; mol wt 248.20. C 48.39%, H 3.25%, N 22.57%, O 25.78%. Aldehyde dehydrogenase inhibitor. Prepn: R. Klink, L. Hepding, **Brit. pat. 1,133,408** corresp to U.S. pat. **3,491,105** (1968, 1970 to E. Merck). Use as alcohol deterrent: A. Behpour *et al.,* **Ger. pat. 2,645,709** corresp to U.S. pat. **4,182,770** (1978, 1980 to E. Merck). Effect on enzyme activity: C. A. Seyfried *et al., Alcohol. Clin. Exp. Res.* **6,** 313 (1982). Effect on ethanol consumption: D. E. McMillan, *Drug Dev. Res.* **3,** 193 (1983). Clinical studies: R. Dusing *et al., Arzneimittel-Forsch.* **32,** 903 (1982); D. Roscher, W. Poser, *ibid.* 905.

Crystals from acetone, mp 185-187°.

THERAP CAT: Alcohol deterrent.

6669. Nitrendipine. *1,4-Dihydro-2,6-dimethyl-4-(3-nitrophenyl)-3,5-pyridinedicarboxylic acid ethyl methyl ester;* ethyl 1,4-dihydro-5-(acetoxycarbonyl)-2,6-dimethyl-4-(3-nitrophenyl)-3-pyridinecarboxylate; 3-ethyl-5-methyl-1,4-dihydro-2,6-dimethyl-4-(3-nitrophenyl)-3,5-pyridinedicarboxylate; Bay e 5009; Bayotensin; Baypress; Bylotensin; Deiten; Nidrel. $C_{18}H_{20}N_2O_6$; mol wt 360.37. C 59.99%, H 5.59%, N 7.77%, O 26.64%. Dihydropyridine calcium channel blocker. Prepn: H. Meyer *et al.,* **Ger. pat. 2,117,571;** *idem,* **U.S. pat. 3,799,934** (1972, 1974 both to Bayer); *idem, Arzneimittel-Forsch.* **31,** 407 (1981). Series of articles on pharmacology: *ibid.* 2056-2067. Hemodynamic effects: H. O. Ventura *et al., Am. J. Cardiol.* **51,** 783 (1983). Pharmacokinetics: L. Hansson *et al., Hypertension* **5,** Suppl. II, II-25 (1983). HPLC determn in human serum: R. A. Janis *et al., J. Clin. Pharmacol.* **23,** 266 (1983). Double-blind, controlled clinical trial: U. Brugmann *et al., Hertz* **10,** 53 (1985). Symposium on pharmacology and clinical efficacy: *J. Cardiovasc. Pharmacol.* **6** Suppl. 7, S929-S1113 (1984). Toxicology studies: K. Hoffmann, "Toxicological Studies with Nitrendipine" in *Nitrendipine,* A. Scriabine *et al.,* Eds. (Urban & Schwarzenberg, Baltimore, 1984) pp 25-32. Review: A. Scriabine *et al.,* "Nitrendipine" in *New Drugs Annual: Cardiovascular Drugs* **Vol. 2,** A. Scriabine, Ed. (Raven Press, New York, 1984) pp 37-49.

Crystals from ethanol, mp 158°. Relatively insol in water. LD_{50} in mice, rats (mg/kg): 39, 12.6 i.v.; 2540, > 10000 orally (Hoffmann).

THERAP CAT: Antihypertensive.

6670. Nitrenes. Molecular fragments with six electrons on the nitrogen. Nitrenes are nitrogen analogs of carbenes, *q.v.* They are formed thermally or photochemically from hydrazoic acid or organic azides. As a result of hydrogen shifts, nitrenes are capable of isomerization to imines, of hydrogen abstraction from neighboring molecules affording primary amines, and of ring formation by internal dehydrogenation (insertion reactions). Survey: Horner, Christmann, *Angew. Chem. Int. Ed.* **2,** 599 (1963). Books: T. L. Gilchrist, C. W. Rees, *Carbenes, Nitrenes and Arynes*

(Plenum Press, New York, 1969) 131 pp; W. Lwowski, *Nitrenes* (Wiley-Interscience, New York, 1970). *Reviews: idem, Reactive Intermediates* vol. 1, M. Jones, R. A. Moss, Eds. (Wiley, New York, 1978) pp 197-227; vol. 2 (1981) pp 315-334.

6671. Nitric Acid. Aquafortis; Salpetersäure (German). HNO_3; mol wt 63.01. H 1.60%, N 22.23%, O 76.17%. Usually produced by the catalytic oxidation of ammonia. Reviews of industrial processes: F. D. Miles, *Nitric Acid, Manufacture and Uses* (Oxford Univ. Press, 1961, 1963); W. Sommer in *Ullmann's Encykopädie der technischen Chemie* vol. **15,** pp 3-67 (3rd ed., 1964). Purification by distln: Briner *et al., Helv. Chim. Acta* **18,** 376 (1935); Ward *et al., Inorg. Syn.* **3,** 13 (1950); Kaplan, Schechter, *ibid.* **4,** 52 (1953). *Reviews:* Several authors in *Mellor's* Vol. VIII, supplement II, *Nitrogen* (part 2), 278-352 (1967); Jones in *Comprehensive Inorganic Chemistry* vol. 2, J. C. Bailar, Jr. *et al.,* Eds. (Pergamon Press, Oxford, 1973) pp 375-388; D. J. Newman in Kirk-Othmer *Encyclopedia of Chemical Technology* vol. **15** (Wiley-Interscience, New York, 1981) pp 853-871. *See also* Nitric Acid, Anhydrous.

Colorless liq. Fumes in moist air. Characteristic choking odor. d_4^{25} 1.50269. mp —41.59°. Monohydrate, mp —37.68°. Trihydrate, mp —18.47°. Heat of fusion: 2.503 kcal/mole. Heat of infinite dilution (298.1°K): —7971 cal/mole. bp_{760} 83°. With water it forms a negative azeotrope (min vapor press. max bp), so-called constant boiling acid at 68% HNO_3, bp 120.5°, d_4^{20} 1.41. The oxide-free acid, diluted with an equal vol of water, does not discolor drops of 0.1N $KMnO_4$. Colorless 100% acid cannot be stored in the presence of light without formation of NO_2 which produces the discoloration. Stains woolen fabrics and animal tissue a bright yellow. In the presence of traces of oxides it attacks all base metals except Al and special Cr steels which become passivated. Strong, monobasic acid. Oxidizing agent. Reacts violently with alcohol, turpentine, charcoal, organic refuse. *Nitric acid concentrated* is a water soln contg 70-71% HNO_3. Density of aq solns: d_4^{20} 1.0036 (1% HNO_3 w/w); 1.0543 (10%); 1.1150 (20%); 1.1800 (30%); 1.2463 (40%); 1.3100 (50%); 1.3667 (60%); 1.4134 (70%); 1.4521 (80%); 1.4826 (90%); 1.5129 (100%): *International Critical Tables* III, 58 (1928). *See also* Nitric Acid, Fuming.

Caution: Potential symptoms of overexposure are irritation of eyes, mucous membranes and skin; delayed pulmonary edema, pneumonitis and bronchitis; dental erosion. *See NIOSH Pocket Guide to Chemical Hazards* (DHHS/NIOSH 90-117, 1990) p 160. *See also Clinical Toxicology of Commercial Compounds,* R. E. Gosselin *et al.,* Eds. (Williams & Wilkins, Baltimore, 5th ed., 1984) Section III, pp 8-11.

USE: Manufacture of inorganic and organic nitrates and nitro compounds for fertilizers, dye intermediates, explosives and many different organic chemicals. Pharmaceutic aid (acidifier).

THERAP CAT (VET): Cauterizing agent (for warts).

6672. Nitric Acid, Anhydrous. Prepared by distillation of concd nitric acid with concd sulfuric acid; by treating sodium or potassium nitrate with 100% H_2SO_4 and removing HNO_3 by distillation; by fractional crystallization of concd HNO_3. Review of preparation and properties of pure HNO_3: Stern *et al., Chem. Rev.* **60,** 185-207 (1960). *See also* Nitric Acid.

Liquid; decomposes above freezing point to form NO_2, H_2O and O_2. Develops yellow color due to nitrogen oxide formation. mp —41.59°. d^0 (liq) 1.5492. Forms white, monoclinic crystals; $d^{-41.6}$ (solid) 1.895.

6673. Nitric Acid, Fuming. Defined as concentrated nitric acid containing dissolved nitrogen dioxide. May be prepared from concd nitric acid by passing nitrogen dioxide into it or by adding a small amount of organic reducing agent, such as formaldehyde.

Yellow to brownish-red, clear, strongly fuming, very corrosive liq; evolves suffocating, poisonous, yellowish-red fumes of nitrogen dioxide and nitrogen tetroxide. The density increases as the free NO_2 content increases: concd HNO_3 with 7.5% NO_2 added d_4^{20} = 1.526; with 12.7% NO_2 d_4^{20} = 1.544. Fuming nitric acid A.C.S. reagent has a density of approx 1.5 and assays not less than 90% HNO_3 by titration with NaOH using methyl orange indicator. Miscible with water. *Handle with extreme care.*

USE: In those organic reactions where nitric acid acts more as an oxidizing agent than as a source of hydrogen ions.

6674. Nitric Oxide. Mononitrogen monoxide; nitrogen monoxide. NO; mol wt 30.01. N 46.68%, O 53.32%. Free radical found ubiquitously in mammals. Effector molecule synthesized from arginine, *q.v.*, that participates in many diverse biological actions, including neurotransmission, vasodilation, cytotoxicity of macrophages, and inhibition of platelet aggregation. Prepd industrially by passing air through an electric arc (basis of atmospheric nitrogen fixation) or by oxidation of ammonia over platinum gauze. Laboratory prepn: Blanchard, *Inorg. Syn.* **2**, 126 (1946); Schenk in *Handbook of Preparative Inorganic Chemistry* **vol. 1**, G. Brauer, Ed. (Academic Press, New York, 2nd ed., 1963) pp 485-487. *Reviews:* Beattie, "Nitric Oxide" in *Mellor's* **Vol. VIII**, supplement II, *Nitrogen* (part 2) 216-240 (1967); Jones in *Comprehensive Inorganic Chemistry* **vol. 2**, J. C. Bailar, Jr. *et al.*, Eds. (Pergamon Press, Oxford, 1973) pp 323-334. Reviews of physiological role: S. Moncada *et al.*, *Pharmacol. Rev.* **43**, 109-142 (1991); A. R. Butler, D. L. H. Williams, *Chem. Soc. Rev.* **22**, 233-242 (1993); P. L. Feldman *et al.*, *Chem. Eng. News* **71**, 26 (Dec. 20, 1993); of genotoxicity: K. Victorin, *Mut. Res.* **317**, 43-55 (1994).
Colorless gas. Burns only when heated with hydrogen. triple pt −163.6°. bp −151.8°. d$^{-150.2}$ (liq) 1.27. Relative d (gas) 1.036 (air = 1). Absolute d (gas) 1.227 (air = 1). n_D^{25} 1.0002697. Trouton constant 27.1. Contains an odd number of electrons and is paramagnetic. Crit temp −92.9°. Crit press. 64.6 atm. Heat of formation (18°): −21.5 kcal/mole. Heat of vaporization (bp): 3.293 kcal/mole. Solubility in water (ml/100 ml; 1 atm): 7.38 (0°); 4.6 (20°); 2.37 (60°). Combines with oxygen to form NO_2 (a brown gas) and with chlorine and bromine to form the nitrosyl halides, such as NOCl, *see* N. V. Sidgwick, *Chemical Elements and Their Compounds* **vol. I** (Oxford, 1950) p 683.
Caution: Potential symptoms of overexposure are irritation of eyes, nose and throat; drowsiness; unconsciousness. *See NIOSH Pocket Guide to Chemical Hazards* (DHHS/NIOSH 90-117, 1990) p 162. Inhalation may result in delayed pulmonary edema. *See Matheson Gas Data Book* 6th ed. (1980) p 514. Immediately on contact with air, nitric oxide is converted to the highly poisonous nitrogen dioxide. Respiratory protection and adequate ventilation should be used to avoid overexposure. *See Patty's Industrial Hygiene and Toxicology* **vol. 2F**, G. D Clayton, F. E. Clayton, Eds. (John Wiley & Sons, New York, 4th ed., 1994) pp 4566-4591. *See also* Nitrogen Dioxide.

USE: Manuf of nitric acid; in the bleaching of rayon; as stabilizer (to prevent free-radical decompn) for propylene, methyl ether, etc.

6675. Nitrilotriacetic Acid. *N,N-Bis(carboxymethyl)-glycine;* triglycollamic acid; α,α',α''-trimethylaminetricarboxylic acid; tri(carboxymethyl)amine; triglycine; NTA. $C_6H_9NO_6$; mol wt 191.14. C 37.70%, H 4.75%, N 7.33%, O 50.22%. $N(CH_2COOH)_3$. Prepn: Heintz, *Ann.* **122**, 260 (1862); Michaelis, Schubert, *J. Biol. Chem.* **106**, 331 (1934); Martell, Bersworth, *J. Org. Chem.* **15**, 46 (1950); Singer, Weisberg, U.S. pats. **2,855,428** and **3,061,628** (1958 and 1962, both to Hampshire Chem.). IR studies: Nakamoto *et al.*, *J. Am. Chem. Soc.* **84**, 2081 (1962); Chapman *et al.*, *Proc. Chem. Soc.* **1962**, 336. Solubility data: Bird, *J. Soc. Dyers Col.* **56**, 473 (1940). Toxicity data: *Soap Chem. Spec.* **42**, 58 (1966). pK data: Schwarzenbach *et al.*, *Helv. Chim. Acta* **28**, 828 (1945). *Review:* Souchay, Graizon, *Bull. Soc. Chim. France* **1952**, 34.
Prismatic crystals from hot water, mp 230-235° dec (Michaelis, Schubert). mp 241.5° (dec). 1.28 g dissolves in 1 liter of water at 22.5°, pH of satd aq soln is 2.3 (Bird). At 20° pK 3.03, pK$_2$ 3.07, pK$_3$ 10.70.
Sodium salt, $C_6H_6NNa_3O_6$, *NTANa$_3$, Trilon A.* MLD orally in rats > 4,000 mg/kg (*Soap Chem. Spec.*).
Note: Nitrilotriacetic acid may reasonably be anticipated to be a carcinogen: *Seventh Annual Report on Carcinogens* (PB95-109781, 1994) p 269.

USE: Chelating and sequestering agent; builder in synthetic detergents.

6676. Nitrin. *2-Aminobenzaldehyde phenylhydrazone.* $C_{13}H_{13}N_3$; mol wt 211.27. C 73.91%, H 6.20%, N 19.89%. Prepd by refluxing 2-nitrobenzaldehyde with phenylhydrazine: Knöpfer, *Monatsh.* **31**, 97 (1910).

Needles from acetone, mp 227-229° (dec). Sol in acetone. Sparingly sol in cold alcohol, ether, chloroform, benzene. A soln in alcohol or acetone develops a red color with nitrites upon addn of acid.

USE: Detection of nitrites, colibacilli in urine: Pfeiffer, *Münch. Med. Wochenschr.* **92**, 1315 (1950).

6677. Nitroacetanilide. $C_8H_8N_2O_3$; mol wt 180.16. C 53.33%, H 4.48%, N 15.55%, O 26.64%.

m-Nitroacetanilide, N-(3-nitrophenyl)acetamide, leaflets. mp 151-153°. Sparingly sol in hot water; freely sol in chloroform, nitrobenzene. Practically insol in ether.
o-Nitroacetanilide, yellow leaflets, d 1.42. mp 93-94°. Moderately sol in cold water, freely in boiling water, in cold fixed alkali hydroxide solns; sol in chloroform, alcohol, ether.
p-Nitroacetanilide, prisms. mp 214-216°. Almost insol in cold water. Sol in hot water, alcohol, ether; sol in KOH with orange color.

6678. Nitroakridin 3582. *1-Diethylamino-3-[(2,3-dimethoxy-6-nitro-9-acridinyl)amino]-2-propanol;* 5-(γ-diethylamino-β-hydroxypropyl)amino-2-nitro-7,8-dimethoxyacridine; 9-(3-diethylamino-2-hydroxypropylamino)-6,7-dimethoxy-3-nitroacridine; nitroacridine 3582; W-1889; Entozon. $C_{22}H_{28}N_4O_5$; mol wt 428.49. C 61.67%, H 6.59%, N 13.08%, O 18.67%. Nitroacridine dye with antimicrobial activity. Prepn: Bockmühl, Fehrle, U.S. pat. **2,040,070** (1936 to Winthrop); *BIOS rept.* no. **766** (1946); Miller, Wagner, *J. Org. Chem.* **13**, 891 (1948); Steck *et al.*, *J. Am. Chem. Soc.* **79**, 4414 (1957). Exptl use for *in vitro* identification of hypoxic cells: A. C. Begg *et al.*, *Brit. J. Radiol.* **56**, 970 (1983).

Crystals from acetone, mp 168-169°.
Dihydrochloride, $C_{22}H_{28}N_4O_5 \cdot 2HCl$, orange crystals, dec 219-220°.
THERAP CAT: Antiseptic.

6679. *m*-Nitroaniline. *3-Nitrobenzenamine; m*-nitraniline. $C_6H_6N_2O_2$; mol wt 138.13. C 52.17%, H 4.38%, N 20.28%, O 23.17%. Prepn by nitration of aniline: Holleman *et al.*, *Ber.* **44**, 704 (1911); by reduction of *m*-dinitrobenzene: Brady *et al.*, *J. Chem. Soc.* **1929**, 2266; Kubota *et al.*, *J. Pharm. Soc. Japan* **76**, 801 (1956); Kuhn, U.S. pat **2,768,209** (1956 to Ringwood); from *m*-nitrobenzoic acid: Snyder *et al.*, *J. Am. Chem. Soc.* **75**, 2014 (1953).

Yellow crystals from water, mp 114°. Calc bp 306°. d_4^{25} 0.9011. One gram dissolves in 880 ml water, about 20 ml alcohol, 18 ml ether, 11.5 ml methanol. Forms water-sol salts with mineral acids.

Caution: Highly toxic; absorbed through skin. Avoid breathing dust; avoid contact with skin, eyes, clothing. In case of contact, immediately flush skin or eyes with plenty of water for at least 15 minutes; for eyes, get medical attention. Wash clothing before reusing. Acute exposure can cause methemoglobinemia, cyanosis. Chronic exposure may cause liver damage.

USE: Dyestuff intermediate.

6680. *o*-Nitroaniline. *o*-Nitraniline. $C_6H_6N_2O_2$; mol wt 138.13. C 52.17%, H 4.38%, N 20.28%, O 23.17%. Prepn from *o*-dinitrobenzene: Meisenheimer, Hesse, *Ber.* **52**, 1166 (1919); from o-nitroaniline-*p*-sulfonic acid: Ehrenfeld, Puterbaugh, *Org. Syn. coll. vol.* **I**, 388 (1941); from ψ-*o*-dinitrosobenzene: Boyer et al., *J. Am. Chem. Soc.* **77**, 5688 (1955).

Orange-yellow crystals from boiling water, mp 69-71°. Calc bp 284°. d_4^{25} 0.9015. Slightly sol in cold water; sol in hot water, alcohol, chloroform. Forms water soluble salts with mineral acids.

Caution: See *m*-Nitroaniline.

USE: Dyestuff intermediate.

6681. *p*-Nitroaniline. *p*-Nitraniline. $C_6H_6N_2O_2$; mol wt 138.13. C 52.17%, H 4.38%, N 20.28%, O 23.17%. Prepn from acetanilide: Bart, **U.S. pat. 2,406,578** (1946 to Am. Cyanamid); by Schmidt reaction: Stockel, Hall, *Nature* **197**, 787 (1963).

Bright yellow powder, mp 146°. Calc bp 332°. One gram dissolves in 1250 ml water, 45 ml boiling water, 25 ml alcohol, 30 ml ether; sol in benzene, methanol. Forms water soluble salts with mineral acids.

Caution: Potential symptoms of overexposure are cyanosis, ataxia; tachycardia, tachypnea; dyspnea; irritability; vomiting, diarrhea; convulsions; respiratory arrest; anemia; methemoglobinemia. See *NIOSH Pocket Guide to Chemical Hazards* (DHHS/NIOSH 90-117, 1990) p 162.

USE: Dyestuff intermediate.

6682. Nitroanisole. *Methoxynitrobenzene*; nitrophenyl methyl ether. $C_7H_7NO_3$; mol wt 153.14. C 54.90%, H 4.61%, N 9.15%, O 31.34%.

m-*Nitroanisole.* Crystals. d 1.373. mp 38-39°. bp 258°. Volatile with steam. Insol in water; sol in alc.

o-*Nitroanisole.* Colorless to yellowish liquid. d_4^{20} 1.254. mp +9.4°. bp 277°; also stated as 272-273°. n_D^{20} 1.5620. Volatile with steam. Insol in water; sol in alc, ether.

p-*Nitroanisole.* Crystals. d 1.233. mp 54°. bp 260°; also stated as 274°. Insol in water; freely sol in alc, ether, boiling petr ether; slightly sol in cold petr ether.

USE: *o*-Isomer as dye intermediate; in organic syntheses.

6683. 5-Nitrobarbituric Acid. *5-Nitro-2,4,6(1H,3H,-5H)-pyrimidinetrione*; dilituric acid. $C_4H_3N_3O_5$; mol wt 173.09. C 27.76%, H 1.75%, N 24.28%, O 46.22%.

Trihydrate, prisms and leaflets from water. mp 176° with decompn, when anhydr. Soluble in ~1200 parts cold water; more sol in hot water; sol in alcohol, in sodium hydroxide soln. Insol in ether.

USE: As a microreagent for potassium with which it forms a characteristic precipitate.

6684. Nitrobenzaldehyde. $C_7H_5NO_3$; mol wt 151.12. C 55.64%, H 3.33%, N 9.27%, O 31.76%.

m-*Nitrobenzaldehyde.* Yellowish, cryst powder. mp 58°. bp_{23} 164°. Volatile with steam. Almost insol in water; sol in alcohol, chloroform, ether.

o-*Nitrobenzaldehyde.* Light yellow needles. mp 42-44°. bp_{23} 153°. Volatile with steam. Slightly sol in water, freely in alcohol, benzene, ether.

p-*Nitrobenzaldehyde.* White to yellow crystals, mp 106-107°. Sublimes; slightly volatile with steam. Slightly sol in water or ether; sol in alc, benzene, glacial acetic acid.

USE: *o*-Nitrobenzaldehyde as reagent for isopropyl alcohol and acetone.

6685. Nitrobenzene. Nitrobenzol; essence of mirbane; oil of mirbane. $C_6H_5NO_2$; mol wt 123.11. C 58.54%, H 4.09%, N 11.38%, O 25.99%. Produced by treating benzene with a mixture of HNO_3 and H_2SO_4. *Review:* Faith, Keyes & Clark's *Industrial Chemicals*, F. A. Lowenheim, M. K. Moran, Eds. (Wiley-Interscience, New York, 4th ed., 1975) pp 571-574; K. L. Dunlap in Kirk-Othmer *Encyclopedia of Chemical Technology* **vol. 15** (Wiley-Interscience, New York, 3rd ed., 1981) pp 916-932. Review of toxicity: E. Browning, *Toxicity and Metabolism of Industrial Solvents* (Elsevier, New York, 1965) pp 298-303.

Colorless to pale yellow, oily liquid; odor of volatile oil almond. *Poisonous!* Use only with adequate ventilation. d_4^{15} 1.205. mp +6°. bp 210-211°. Flash pt, closed cup: 190°F (88°C). n_D^{20} 1.5529. Volatile with steam. Sol in ~500 parts water; freely sol in alcohol, benzene, ether, oils. LD_{50} orally in rats: 640 mg/kg, *RTECS Vol. 1*, R. J. Lewis, R. L. Tatken, Eds. (1979) p 217.

Caution: Potential symptoms of overexposure are anoxia; irritation of eyes; dermatitis; anemia. See *NIOSH Pocket Guide to Chemical Hazards* (DHHS/NIOSH 90-117, 1990) p 162.

USE: For the manuf of aniline; in soaps, shoe polishes; for refining lubricating oils; manuf pyroxylin compds.

6686. Nitrobenzoic Acid. $C_7H_5NO_4$; mol wt 167.12. C 50.31%, H 3.02%, N 8.38%, O 38.29%. Prepn of *m*-isomer: O. Kamm, J. B. Segur, *Org. Syn. coll. vol. I* (2nd ed., 1964) p 391; of *p*-isomer: O. Kamm, A. O. Matthews, *ibid.* p 392.

m-**Nitrobenzoic acid,** monoclinic leaflets from water. Bitter taste. d 1.494. mp 142°. Melts in hot water. K (25°) = 3.48 × 10⁻⁴. Absorption spectrum: Purvis, *J. Chem. Soc.* **107,** 971 (1915). One gram dissolves in 320 ml water, 3 ml alc, 4 ml ether, 18 ml chloroform, ~2 ml methanol, 2.5 ml acetone; very slightly sol in benzene, carbon disulfide, petr ether.

o-**Nitrobenzoic acid,** yellowish-white, intensely sweet crystals. d 1.58. mp 147-148°. One gram dissolves in 146 ml water, 3 ml alcohol, 220 ml chloroform, 4.5 ml ether, 2.5 ml acetone, 2.5 ml methanol; very slightly sol in benzene, carbon disulfide, petr ether.

p-**Nitrobenzoic acid,** monoclinic leaflets, plates from benzene. d 1.58. mp 242.4°. Sublimes. One gram dissolves in 2380 ml water, 110 ml alcohol, 12 ml methanol, 150 ml chloroform, 45 ml ether, 20 ml acetone; slightly sol in benzene, carbon disulfide; insol in petr ether.

USE: Manuf intermediates; also as a reagent for alkaloids and thorium.

6687. 4-Nitrobenzoyl Chloride. C₇H₄ClNO₃; mol wt 185.57. C 45.31%, H 2.17%, Cl 19.11%, N 7.55%, O 25.87%. NO₂C₆H₄COCl. Prepd by the action of phosphorus pentachloride on *p*-nitrobenzoic acid: Adams, Jenkins, *Org. Syn.* **3,** 75 (1923).

Bright yellow needles from petr ether. Pungent odor. mp 75°. bp₁₀₅ 205°; bp₁₂ 154°. Decomposed by water and alcohol; sol in ether. *Keep well closed.*

6688. *p*-Nitrobenzyl Cyanide. *4-Nitrobenzeneacetonitrile; p-nitrophenylacetonitrile; p*-nitro-α-toluinitrile. C₈H₆-N₂O₂; mol wt 162.15. C 59.26%, H 3.73%, N 17.28%, O 19.73%. NO₂C₆H₄CH₂CN. Prepd by the action of concd nitric acid on benzyl cyanide: Robertson, *Org. Syn.* **2,** 57 (1922).

Elongated prisms from alcohol. mp 117°. Insol in water; sol in alcohol, ether.

USE: In the prepn of *p*-nitrophenylacetic acid.

6689. *o*-Nitrobiphenyl. *2-Nitro-1,1'-biphenyl; o*-nitrodiphenyl; ONB. C₁₂H₉NO₂; mol wt 199.21. C 72.35%, H 4.55%, N 7.03%, O 16.06%. Prepn from diazotized *o*-nitroaniline and benzene by a modified Gomberg reaction: Elks *et al., J. Chem. Soc.* **1940,** 1284. Prepd industrially by direct nitration of biphenyl (yields ~50% para, 35% ortho).

Orthorhombic bipyramidal plates from ethanol. Characteristic sweetish odor. mp 36.7°. d₄²⁵ 1.44. d₁₅.₅⁴⁰ (liq) 1.189. The technical liq weighs ~10 lbs/gal. bp₇₆₀ 325°; bp₃₀ 205°; bp₁₃ 170°; bp₄ 166°. n_D²⁵ (tech liq) 1.613. Flash pt 179° (354°F). uv max: 325 nm. Dipole moment: 3.79. Viscosity at 45°: 12 cps; at 25°: 38 cps (tech liq). Insol in water. Sol in methanol, ethanol, tetrahydrofurfuryl alcohol, acetone, DMF, carbon tetrachloride, perchlorethylene, mineral spirits, turpentine, glacial acetic acid.

USE: Plasticizer for resins, cellulose acetate and nitrate, polystyrenes; fungicide for textiles; wood preservative; dye intermediate.

6690. *p*-Nitrobiphenyl. *4-Nitro-1,1'-biphenyl; 4-nitrobiphenyl; p*-nitrodiphenyl; PNB. C₁₂H₉NO₂; mol wt 199.21. C 72.35%, H 4.55%, N 7.03%, O 16.06%. Prepn: G. Schultz, *Ann.* **174,** 210 (1874); O. Kühling, *Ber.* **28,** 42 (1895); E. Bamberger, *ibid.* 404; R. L. Jenkins *et al., Ind. Eng. Chem.* **22**(1), 31 (1930); K. Dimroth *et al., Ber.* **101,** 2215 (1968). Crystal structure: M. Prasad *et al., J. Indian Chem. Soc.* **13,**

519 (1936), *C.A.* **31,** 1271⁵ (1937). Review of carcinogenic risk: *IARC Monographs* **vol. 4,** 113-117 (1974).

Needles from alcohol, mp 113.7°. bp₇₆₀ 340°; bp₃₀ 223.7-224.1°. Insol in water. Slightly sol in cold alc; more readily sol in hot alc; sol in chloroform, ether. uv spectrum: D. F. DeTor, H. J. Scheifele, *J. Am. Chem. Soc.* **73,** 1442 (1951).

Caution: Potential symptoms of overexposure are headache, lethargy and dizziness; dyspnea; ataxia, weakness; methemoglobinemia; urinary burning; acute hemorrhagic cystitis. *See NIOSH Pocket Guide to Chemical Hazards* (DHHS/NIOSH 90-117, 1990) p 162. This substance has been listed as a carcinogen by OSHA: *Fed. Reg.* **39,** 3757 (1974).

USE: Formerly in prepn of *p*-biphenylamine, *q.v.*

6691. 2-Nitro-1,1-bis(*p*-chlorophenyl)propane. *1,1'-(2-Nitropropylidene)bis[4-chlorobenzene]; 1,1-bis(p-chlorophenyl)-2-nitropropane;* Prolan. C₁₅H₁₃Cl₂NO₂; mol wt 310.18. C 58.08%, H 4.22%, Cl 22.86%, N 4.52%, O 10.32%. Prepn: Hass, Blickenstaff, U.S. pat. **2,516,186** (1950 to Purdue Res. Found.); Neher, Blickenstaff, *Ind. Eng. Chem.* **43,** 2875 (1951); Hodge, *J. Am. Chem. Soc.* **73,** 2341 (1951). Toxicity data: Lehman, *Assoc. Food Drug Officials U.S. Quart. Bull.* **14,** 82 (1951), *C.A.* **45,** 3517h (1951).

Viscous oil, bp₀.₁₆ 180°. Crystals from abs ethanol, mp 80.5-81.5°. LD₅₀ orally in rats: 750 mg/kg (Lehman). Combination with Bulan®, *Dilan.*

USE: Insecticide.

6692. *m*-Nitrocinnamic Acid. *3-(3-Nitrophenyl)-2-propenoic acid.* C₉H₇NO₄; mol wt 193.16. C 55.96%, H 3.65%, N 7.25%, O 33.13%. NO₂C₆H₄CH=CHCOOH. Prepd by heating a mixture of *m*-nitrobenzaldehyde, sodium acetate and acetic anhydride: Thayer, *Org. Syn.* **5,** 83 (1925). From *m*-nitrobenzaldehyde and malonic acid in the presence of glycine in a little water at 100°: Dakin, *J. Biol. Chem.* **7,** 54 (1909-1910).

trans-Form, prepd by the above methods. White needles from benzene or alcohol. mp 200-201°. Can be sublimed or distilled at atmospheric press. with very little decompn. Absorption spectrum: Purvis, *J. Chem. Soc.* **107,** 971 (1906). One gram dissolves in ~100 ml alcohol at 25°.

cis-Form, prepd by ultraviolet irradiation of a soln in water-alcohol-ammonia giving a 22% yield: Wollring, *Ber.* **47,** 112 (1914). Minute needles from toluene + methanol, mp 158°. By reirradiation of a chloroform soln containing a trace of bromine, the more stable *trans*-form is regenerated.

6693. Nitrodan. *3-Methyl-5-[(4-nitrophenyl)azo]-2-thioxo-4-thiazolidinone; 3-methyl-5-[(p-nitrophenyl)azo]rhodanine;* 3-methyl-5-[(p-nitrophenyl)azo]-2-thio-2,4-thiazolidinedione; CTR-6110; Nidanthel; Everfree. C₁₀H₈N₄O₃S₂; mol wt 296.33. C 40.53%, H 2.72%, N 18.91%, O 16.20%, S 21.64%. Prepn: Benghiat, Howard, U.S. pat. **2,952,673** (1960 to Stauffer); **Belg.** pat. **637,282** (1964 to Cooper, Tinsley Labs.).

Yellow powder, mp 267-268°. Sparingly sol in water.
THERAP CAT (VET): Anthelmintic.

6694. Nitroethane. $C_2H_5NO_2$; mol wt 75.07. C 32.00%, H 6.71%, N 18.66%, O 42.63%. $CH_3CH_2NO_2$. Obtained by vapor phase nitration of ethane with HNO_3: Reidel, *Oil Gas J.* **54**, no. 36, 110-114 (1956). Laboratory prepn by treating 1.5 moles sodium nitrite with 1 mole sodium ethyl sulfate at 125-130° in the presence of 0.0625 moles potassium carbonate: Desseigne, Giral, *Mem. Poudres* **34**, 49-53 (1952), *C.A.* **49**, 836 (1955). Similar procedure according to the equation $EtOSO_2OEt + NaNO_2 \rightarrow EtOSO_2ONa + EtNO_2$: McCombie *et al., J. Chem. Soc.* **1944**, 24.

Oily liquid; pleasant odor. d_{25}^{25} 1.041; d_{20}^{20} 1.052. Flash pt, open cup: 106°F (41.11°C). mp ~ −50°. bp 114-115°. Undergoes thermal decompn at 335-382°. Heating value (liq): 7,720 Btu/lb. Lower limit of flammability in air = 4.0% by volume. Viscosity at 25° = 0.661 centipoise. n_D^{20} 1.3917; $n_D^{24.3}$ 1.39007. Soly in water: 4.5 ml/100 ml H_2O at 20°. Miscible with methanol, ethanol, ether. Sol in chloroform, aq solns of alkalies. Has high heat of absorption. Sudden absorption of the anhyd liq or gas on activated carbon or Hopcalite may result in flames: *Chem. & Eng. News* **30**, 2344 (1952). LD_{50} orally in rats: 1100 mg/kg, *Toxic Substances List,* H. E. Christensen, Ed. (1974) p 332.

Caution: Potential symptom of overexposure is dermatitis. See *NIOSH Pocket Guide to Chemical Hazards* (DHHS/NIOSH 90-117, 1990) p 164.

USE: Solvent; in organic syntheses. Experimentally as liq propellant.

6695. Nitrofen. *2,4-Dichloro-1-(4-nitrophenoxy)benzene; 2,4-dichlorophenyl-p-nitrophenyl ether;* nitraphen; nitrophen; nitrofene; FW-925; TOK. $C_{12}H_7Cl_2NO_3$; mol wt 284.10. C 50.73%, H 2.48%, Cl 24.96%, N 4.93%, O 16.89%. Selective pre- and post-emergence herbicide. Prepn: H. F. Wilson, H. M. Dougal, U.S. pat. 3,080,225 (1963 to Rohm & Haas). Toxicity studies: A. M. Ambrose *et al., Toxicol. Appl. Pharmacol.* **19**, 263 (1971). Carcinogenicity studies: J. F. Robens, *Vet. Human Toxicol.* **22**, 328 (1980). Review of carcinogenic risk: *IARC Monographs* **30**, 271-282 (1983).

Crystalline solid, mp 70-71°. Vapor pressure at 40°: 8 × 10^{-6} mm Hg. Soly in water at 22°: 0.7-1.2 ppm. LD_{50} orally in rats: 3.58 g/kg (Ambrose). −
Note: This substance may reasonably be anticipated to be a carcinogen: *Seventh Annual Report on Carcinogens* (PB95-109781, 1994) p 272.

USE: Herbicide.

6696. Nitrofurantoin. *1-[[(5-Nitro-2-furanyl)methylene]amino]-2,4-imidazolidinedione; N-(5-nitro-2-furfurylidene)-1-aminohydantoin;* 1-(5-nitro-2-furfurylideneamino)hydantoin; Berkfurin; Chemiofuran; Cyantin; Cystit; FuaMed; Furachel; Furalan; Furadantin; Furadantine MC; Furadoin; Furantoina; Furobactina; Furophen T-Caps; Ituran; Macrodantin; Parfuran; Trantoin; Urantoin; Urizept; Urodin; Urolong; Uro-Tablinen; Welfurin. $C_8H_6N_4O_5$; mol wt 238.16. C 40.35%, H 2.54%, N 23.52%, O 33.59%. Prepn from 1-aminohydantoin sulfate and 5-nitro-2-furaldehyde diacetate: Hayes, U.S. pat. 2,610,181 (1952 to Eaton Labs.); Swirska *et al., Przem. Chem.* **11** (34), 306 (1955), *C.A.* **52**, 14079b (1958). Alternate route(s): Michels; Gever, O'Keefe, U.S. pats. 2,898,335; 2,927,110 (1959 and 1960, both to Norwich). Comprehensive description: D. E. Cadwallader, H. W. Jun, in *Analytical Profiles of Drug Substances* **vol. 5**, K. Florey, Ed. (Academic Press, New York, 1976) pp 345-373.

Orange-yellow needles from dil acetic acid, dec 270-272°. pKa 7.2. uv max: 370 nm ($E_{1cm}^{1\%}$ 776). Soly (mg/100 ml): water (pH 7) 19.0; 95% ethanol 51.0; acetone 510; DMF 8000; peanut oil 2.1; glycerol 60; polyethylene glycol 1500. Monohydrate, *Macrobid.*
THERAP CAT: Antibacterial.
THERAP CAT (VET): Antibacterial.

6697. Nitrofurazone. *2-[(5-Nitro-2-furanyl)methylene]-hydrazinecarboxamide; 5-nitro-2-furaldehyde semicarbazone;* Amifur; Furacin; Chemofuran; Furesol; Nifuzon; Nitrofural; Nitrozone; Furacinetten; Furacoccid; Furazol W; Mammex; Furaplast; Coxistat; Aldomycin; Nefco; Vabrocid. $C_6H_6N_4O_4$; mol wt 198.14. C 36.37%, H 3.05%, N 28.28%, O 32.30%. Prepd from 2-formyl-5-nitrofuran and semicarbazide hydrochloride: Stillman, Scott, U.S. pat. 2,416,234 (1947 to Eaton Labs.). Alternate route: Gever, O'Keefe, U.S. pat. 2,927,110 (1960 to Norwich Pharmacal). Toxicity: J. Godwin *et al., J. Am. Med. Assoc.* **133**, 299 (1947). Tumorigenic activity: Morris *et al., Cancer Res.* **29**, 2145 (1969); Ertürk *et al., ibid.* **30**, 1409 (1970).

Pale yellow needles, dec 236-240°. Bitter aftertaste. Darkens on prolonged exposure to light. uv max: 260, 375 nm. Very slightly sol in water (1:4200); slightly sol in alcohol (1:590), in propylene glycol (1:350). Sol in alkaline solns with dark orange color. Insol in ether. pH of satd water soln 6.0 to 6.5. LD_{50} in rats (g/kg): 0.59 orally; 3.0 s.c. (Godwin).
THERAP CAT: Anti-infective (topical).
THERAP CAT (VET): Antimicrobial.

6698. Nitrogen. N; at. wt 14.0067; at. no. 7; valences 3, 5; elemental state: N_2. Two naturally occurring isotopes: 14 (99.635%); 15 (0.365%); five short-lived, artificial, radioactive isotopes: 12; 13; 16-18. Discovered in 1772 by Daniel Rutherford and independently by Scheele and Cavendish. Constitutes about 75.5% by weight or 78.06% by volume of the atmosphere; found frequently in volcanic or mine gases, gases from springs and gases occluded in minerals and rocks; an essential constituent of all living organisms; fixed or combined nitrogen is present in many mineral deposits. Prepn from sodium (and alkaline earth) azides by heating the azide: Tiede, *Ber.* **46**, 4100 (1913); **49**, 1745 (1916); Justi, *Ann. Physik* [5] **10**, 985 (1931). Prepd industrially by fractional distln of liquid air; by removal of oxygen by combustion; by reduction of ammonia. Purification of nitrogen furnished in steel cylinders: Kautsky, Thiele, *Z. Anorg. Allgem. Chem.* **152**, 342 (1926); Kendall, *Science* **73**, 395 (1931); Schenk in *Handbook of Preparative Inorganic Chemistry* **vol. 1**, G. Brauer, Ed. (Academic Press, New York, 2nd ed., 1963) pp 458-460. Review of nitrogen and nitrogen compounds: Jones in *Comprehensive Inorganic Chemistry* **vol. 2**, J. C. Bailar, Jr. *et al.,* Eds. (Pergamon Press, Oxford, 1973) pp 147-388; R. W. Schroeder in Kirk-Othmer *Encyclopedia of Chemical Technology* **vol. 15** (Wiley-Interscience, New York, 3rd ed., 1981) pp 932-941. Books: W. L. Jolly, *The Inorganic Chemistry of Nitrogen* (Benjamin, New York, 1964) 124 pp; *Mellor's* **Vol. VIII,** Supplements I, II, *Nitrogen,* part 1 (1964) 619 pp; part 2 (1967) 676 pp; M. Sittig, *Nitrogen in Industry* (Van Nostrand, Princeton, 1965) 278 pp.

Odorless gas; condenses to a liq, bp −195.79° (77.36°K); solidifies to a snow-white mass, mp −210.01° (63.14°K). d^{gas} (0°, 1 atm) 1.25046 g/l. Critical temp: −147.1°; critical press: 33.5 atm; critical density: 0.311 g/cm^3. Sparingly sol in water: 100 volumes of water absorbs 2.4 volumes of gas at 0°, 1.6 volumes at 20°. Soly in water at 50, 75 and 100° from 25 to 1000 atmospheres: Wiebe *et al., J. Am. Chem. Soc.* **55**, 947 (1933). Soly in liq ammonia: Wiebe *et al., ibid.* 975. Soly in alc: one volume of alcohol dissolves 0.1124 volume of nitrogen at 20°. Liquid oxygen at −195.5° absorbs 50.7% of its weight of gaseous nitrogen. Heat of dissociation of the nitrogen molecule (N_2): 225.1 kcal/mole. Combines with oxygen and hydrogen on sparking, forming nitric oxide and ammonia, respectively. Combines directly with lithium, and at a red heat with calcium, strontium, and barium to form nitrides. Forms cyanides when heated with carbon in presence of alkalies or barium oxide.

Caution: In high concns it is a simple asphyxiant.

USE: In manuf of ammonia, nitric acid, nitrates, cyanides, etc.; in manuf explosives; in filling high-temp thermometers, incandescent bulbs; to form an inert atm for preservation of materials, for use in dry boxes or glove bags. Liquid nitrogen in food-freezing processes; in the laboratory as a coolant. Pharmaceutic aid (air displacement).

6699. Nitrogen Chloride. Nitrogen trichloride; chlorine nitride; trichlorine nitride; Agene. Cl_3N; mol wt 120.36. Cl 88.36%, N 11.64%. NCl_3. Prepared by the action of chlorine gas or hypochlorous acid on ammonium salts: Dulong, *Ann. Chim. Phys.* [1] **86**, 37 (1813); Balard, *ibid.* [2] **57**, 258 (1834); Hentschel, *Ber.* **30**, 1792, 2642 (1897); Noyes, *J. Am. Chem. Soc.* **42**, 2173 (1920); by reaction of anhydr ammonia and anhydr chlorine: Noyes, Haw, *ibid.* 2167; industrially by electrolyzing an acidified soln of ammonium chloride: U.S. pat. **2,118,903** (1938), Ger. pat. **641,816** (1937); Brit. pat. **494,188** (1938); for bleaching flour by blowing an air stream containing hydrogen chloride gas through a bed of finely powdered ammonium chloride and potassium hypochlorite distributed on an inert carrier to form a gas mixture of chlorine dioxide and nitrogen chloride: Brit. pat. **597,199** (1948). *Review:* Richards in *Mellor's* Vol. **VIII**, supplement II, *Nitrogen* (part 2), 411-415 (1967).

Yellow, thick, oily liq, d 1.653, pungent odor, evaporates rapidly in air, very unstable. Dec in light, vapor pressure at room temp is ~150 mm Hg. Explodes when heated to 93°, subjected to a flash of direct sunlight or magnesium light, sealed in glass containers at 60° after 13 seconds, frozen in liq air and thawed *in vacuo*, in contact with ozone, nitric oxide, grease, and several organic substances. Insol in water. Dec in water after 24 hours, sol in carbon disulfide, phosphorus trichloride, benzene, carbon tetrachloride, chloroform.

USE: Bleaching of flour (prohibited in the U.S.A.), wastage control of citrus fruit.

6700. Nitrogen Dioxide. NO_2; mol wt 46.01. N 30.45%, O 69.55%. Damaging component of photochemical smog. Prepd industrially from nitric oxide and air. Convenient lab prepn from lead nitrate: Schenk in *Handbook of Preparative Inorganic Chemistry* vol. 1, G. Brauer, Ed. (Academic Press, New York, 1963) pp 488-489. Ultrapure NO_2 from N_2O_5: Hackspill, Besson, *Bull. Soc. Chim. France, Mem.* [5] **16**, 479 (1949). *Reviews:* Beattie, "Nitrogen Dioxide and Dinitrogen Tetroxide" in *Mellor's* Vol. **VIII**, supplement II, *Nitrogen* (part 2) 246-268 (1967); Jones in *Comprehensive Inorganic Chemistry* vol. 2, J. C. Bailar, Jr. *et al.*, Eds. (Pergamon Press, Oxford, 1973) pp 340-356. Symposium on chemistry and toxicology: *Toxicology* **89**, 1-312 (1994).

Reddish-brown gas. Liquid below 21.15°. Irritating odor. *Deadly poison!* The commercial brown liq under pressure is called *nitrogen tetroxide*. Actually this is an equilibrium mixture of NO_2 and the colorless N_2O_4. d_4^{20} (liq) 1.448; d (gas) 1.58 (air = 1); $d_{gas}^{21.3}$ 3.3 g/liter. mp −9.3°. bp 21.15°. Crit temp 158.2°. Crit press. 99.96 atm. Heat of vaporization (bp) 9.110 kcal/mole. Does not burn, but supports the combustion of carbon, phosphorus, sulfur. Sol in concd sulfuric and nitric acids. Dec in water forming nitric acid and nitric oxide, reacts with alkalies to form nitrates and nitrites. Corrosive to steel when wet, but may be stored in steel cylinders when moisture content is 0.1% or less.

Caution: Potential symptoms of overexposure are coughing, mucoid frothy sputum and dyspnea; chest pain; pulmonary edema, cyanosis, tachypnea and tachycardia; eye irritation. *See NIOSH Pocket Guide to Chemical Hazards* (DHHS/NIOSH 90-117, 1990) p 164. One of the most insidious gases. Inflammation of lungs may cause only slight pain or pass unnoticed, but the resulting edema several days later may cause death. 100 ppm is dangerous for even a short exposure, and 200 ppm may be fatal: Y. Henderson, H. W. Haggard, *Noxious Gases,* A.C.S. Monograph Series, no. **35** (Reinhold, New York, 2nd ed., 1943) pp 134-137, 141. *See also Clinical Toxicology of Commercial Products,* R. E. Gosselin *et al.,* Eds. (Williams & Wilkins, Baltimore, 5th ed., 1984) Section III, pp 319-326.

USE: Intermediate in nitric and sulfuric acid production. Used in the nitration of organic compds and explosives, in the manuf of oxidized cellulose compds (hemostatic cotton). Has been used to bleach flour. Proposed as oxidizing agent in rocket propulsion. *See also:* Sisler, "Reactions in Liquid Dinitrogen Tetroxide," *J. Chem. Ed.* **34**, 555 (1957).

6701. Nitrogen Fluoride. Nitrogen trifluoride. F_3N; mol wt 71.00. F 80.27%, N 19.73%. NF_3. Prepd by electrolysis of melted ammonium acid fluoride, NH_4F_2H: Ruff *et al., Z. Anorg. Allgem. Chem.* **172**, 417 (1928); Ruff, *ibid.* **197**, 273 (1931); Ruff, Staub, *ibid.* **198**, 32 (1931); Kwasnik in *Handbook of Preparative Inorganic Chemistry* vol. **1**, G. Brauer, Ed. (Academic Press, New York, 2nd ed., 1963) pp 181-183. Reviews of prepn and chemistry: Hoffman, Neville, *Chem. Rev.* **62**, 1-18 (1962); Kemmitt, Sharp, *Advan. Fluorine Chem.* **4**, 189-190 (1965).

Colorless gas. Moldy odor. mp −208.5°. bp −129°. d (liq at bp) 1.885. Trouton const 19.9. Insoluble in water. Rather inert chemically. Does not attack glass, mercury. Decomposed by electric sparks.

Caution: Prolonged absorption may cause mottling of teeth, skeletal changes; *see* Sodium Fluoride (chronic effects).

6702. Nitrogen Pentoxide. Dinitrogen pentoxide; nitric anhydride. N_2O_5; mol wt 108.01. N 25.94%, O 74.06%. Usually prepd by dehydration of nitric acid by means of phosphorus pentoxide: Gruenhut *et al., Inorg. Syn.* **3**, 78 (1950). *Reviews:* Beattie, "Dinitrogen Pentoxide" in *Mellor's* Vol. **VIII**, supplement II, *Nitrogen* (part 2), 269-277 (1967); Jones in *Comprehensive Inorganic Chemistry* vol. **2**, J. C. Bailar, Jr. *et al.*, Eds. (Pergamon Press, Oxford, 1973) pp 356-360.

Colorless hexagonal crystals. mp 30°. Sublimes at 32.4° but undergoes moderately fast decompn into O_2 and the NO_2/N_2O_4 equilibrium mixture at temps above −10°. d^{15} 2.05. bp_{760} 47.0°; bp_{100} 7.0°; bp_{10} −20°. Dipole moment 1.39. Freely sol in chloroform without appreciable decomposition. Less sol in carbon tetrachloride. Chloroform solns may be stored at −20° for as long as one week without excessive decompn.

USE: As nitrating agent in chloroform soln.

6703. Nitrogen Selenide. *Selenium nitride.* N_4Se_4; mol wt 371.87. N 15.07%, Se 84.93%. Prepd by passing dry ammonia over selenium tetrachloride: Wöhler, *Bull. Soc. Chim.* [1] **1**, 25 (1859); by treating a dil soln of selenium oxychloride in benzene with dry ammonia and washing the precipitate with water and potassium cyanide: Lenher, Wolesensky, *J. Am. Chem. Soc.* **29**, 215 (1907); by the action of dry ammonia on a dilute soln of selenium monochloride in carbon disulfide: van Valkenburg, Bailar, *ibid.* **47**, 2134 (1925); by the action of dry ammonia on diethyl selenite or dimethyl selenite dissolved in benzene and washing with potassium cyanide: Strecker, Schwarzkopf, *Z. Anorg. Chem.* **221**, 193 (1934); by reacting anhydr, liquid ammonia with selenium dioxide: Jander, Doetsch, *Ber.* **93**, 561 (1960). Crystal structure: Baernighausen *et al., Acta Cryst.* **15**, 615 (1962); **21**, 571 (1966).

Orange-red, amorphous powder or monoclinic crystals. d 4.2. Very hygroscopic. *Explosive.* Sparingly sol in carbon disulfide, benzene, acetic acid; insol in water, ether, abs alc.

6704. Nitroglycerin. *1,2,3-Propanetriol trinitrate;* glyceryl trinitrate; glycerol nitric acid triester; nitroglycerol; trinitroglycerol; glonoin; trinitrin; blasting gelatin; blasting oil; S.N.G.; Adesitrin; Angibid; Angiolingual; Anginine; Angorin; Aquo-Trinitrosan; Cardamist; Cordipatch; Coro-Nitro;

Corditrine; Deponit; Diafusor; Discotrine; Gilucor; GTN; Klavikordal; Lenitral; Lentonitrina; Millisrol; Minitran; Myoglycerin; Nitradisc; Nitran; Nitriderm-TTS; Nitro-Bid; Nitrocine; Nitrocontin; Nitroderm TTS; Nitrodisc; Nitro-Dur; Nitrofortin; Nitrogard; Nitro-Gesanit; Nitroglin; Nitroglyn; Nitrolan; Nitrolande; Nitrolar; Nitrolent; Nitrolingual; Nitro Mack; Nitromex; Nitronal; Nitrong; Nitro-PRN; Nitrorectal; Nitroretard; Nitrosigma; Nitrostat; Nitrozell-retard; Nysconitrine; Percutol; Perlinganit; Reminitrol; Suscard; Sustac; Sustonit; Transderm-Nitro; Transiderm-Nitro; Tridil; Trinalgon; Trinitrosan; Vasoglyn. $C_3H_5N_3O_9$; mol wt 227.09. C 15.87%, H 2.22%, N 18.50%, O 63.41%. Prepn: Sobrero, *Ann.* **64**, 398 (1847); Williamson, *Ann.* **92**, 305 (1854). Review of the early literature: J. W. Lawrie, *Glycerol and the Glycols* (New York, 1928). Review of chemistry and biochemistry: F. J. DiCarlo, *Drug Metab. Rev.* **4**, 1-38 (1975). Review of mechanism of action: S. F. Vatner, G. R. Heyndrickx, *Handb. Exp. Pharmakol.* **40**, 131-161 (1975). Review of the first hundred years: J. R. Parratt, *J. Pharm. Pharmacol.* **31**, 801-809 (1979). Comprehensive description: E. F. McNiff *et al.*, *Anal. Profiles Drug Subs.* **9**, 519-541 (1980). Symposium on nitroglycerin therapy, perspectives and mechanisms: *Am. J. Med.* **74**, no. 6B, 1-94 (1983). Review of pharmacology and clinical studies of intravenous administration in heart disease: E. M. Sorkin *et al.*, *Drugs* **27**, 45 (1984).

$$CH_2\!-\!ONO_2$$
$$CH\!-\!ONO_2$$
$$CH_2\!-\!ONO_2$$

Pale yellow, oily liquid. Sweet burning taste. Produces headache on tasting. Explodes on rapid heating or on concussion. Crystallizes in 2 forms: labile form, mp +2.8°; stable form, mp +13.5°. d_{15}^{15} 1.599; d_4^4 1.6144; d_4^{15} 1.6009; d_4^{25} 1.5918. n_D^{15} 1.474. Begins to dec at 50-60°, appreciably volatile at 100°, evolves nitrous yellow vapors at 135°, explodes at 218°. Vapor pressure at 20°: 0.00026 mm; at 93°: 0.31 mm. One gram dissolves in 800 ml water, in 4 g ethanol, in 18 g methanol, in 120 g carbon disulfide. Misc with ether, acetone, glacial acetic acid, ethyl acetate, benzene, nitrobenzene, pyridine, chloroform, ethylene bromide, dichloroethylene. Sparingly sol in petr ether, liq petrolatum, glycerol. Heat of combustion: 1580 cal/g. On explosion harmless gases are produced: $4C_3H_5(ONO_2)_3 \rightarrow 12CO_2 + 10H_2O + 6N_2 + O_2$.

Caution: Acute poisoning, occurring especially in industrial workers: nausea, vomiting, abdominal cramps, headache, mental confusion, delirium, bradypnea, bradycardia, paralysis, convulsions, methemoglobinemia and cyanosis, circulatory collapse, death. *Chronic poisoning:* severe headache, hallucinations, skin rashes. Alcohol aggravates symptoms. Toxic effects may occur by ingestion, inhalation of dust or absorption through intact skin, *cf.* Patty's *Industrial Hygiene and Toxicology* vol. **2C**, G. D. Clayton, F. E. Clayton, Eds. (Wiley-Interscience, New York, 3rd ed., 1982) pp 4186-4188.

USE: Manuf of *dynamite* (75% nitroglycerol, 24.5% diatomaceous earth, 0.5% sodium carbonate).

THERAP CAT: Antianginal; vasodilator (coronary).

THERAP CAT (VET): Has been used for bronchial asthma in dogs.

6705. Nitroguanidine. $CH_4N_4O_2$; mol wt 104.07. C 11.54%, H 3.87%, N 53.84%, O 30.75%. $H_2NC(NH)NH\text{-}NO_2$. Prepd by the action of concd H_2SO_4 on guanidine nitrate: Davis, *Org. Syn.* **7**, 68 (1927).

α-Form, (the usual stable form): Needles, prisms from water, dec 225-250° (depending on speed of heating) giving off NH_3 fumes. Absorption spectrum: Riegel, Buchwald, *J. Am. Chem. Soc.* **51**, 491 (1929). One liter of water dissolves 4.4 g at 25°; 82.5 g at 100°. Slightly soluble in methanol (<0.5%), alcohol; practically insol in ether. Sol in concd acids from which it is precipitated by water. Sol in cold solns of alkalies (not of carbonates) with slow decompn.

β-Form: Prepd by the nitration of guanidine sulfate with fuming HNO_3. Is converted into the α-form by dissolving in concd H_2SO_4 and pouring into ice water: Davis *et al.*, *J. Am. Chem. Soc.* **47**, 1065 (1925).

USE: Explosive of moderate power. Can be exploded only with a detonator. Intermediate in the synthesis of pharmaceuticals.

6706. Nitrohydrochloric Acid. *Aqua regia;* nitromuriatic acid; chloronitrous acid; chloroazotic acid. Made by mixing 18 ml HNO_3, 82 ml HCl. In addition to the free acids, it contains nitrosyl chloride (NOCl) and some free chlorine.

Yellow, fuming, corrosive, volatile liq; suffocating odor; attacks all metals including gold and platinum. Miscible with water. *Keep in well-closed, glass-stoppered bottles in a cool place and protected from light.*

Caution: Undiluted form is a strong irritant, corrosive.

6707. Nitromersol. *[2-Methyl-5-nitrophenolato(2—)-C^6,O^1]mercury; 5-methyl-2-nitro-7-oxa-8-mercurabicyclo-[4.2.0]octa-1,3,5-triene; 3-(hydroxymercuri)-4-nitro-o-cresol inner salt;* Metaphen. $C_7H_5HgNO_3$; mol wt 351.71. C 23.91%, H 1.43%, Hg 57.03%, N 3.98%, O 13.65%. According to the National Formulary this compd is the anhydride of 4-nitro-3-hydroxymercuri-o-cresol. Prepd by treating 4-nitro-o-cresol with mercuric acetate: U.S. pat. **1,544,293**; reissue **17,563** (1925).

Yellow, odorless, tasteless powder or granules. Insol in water; almost insol in acetone, alcohol, ether, aq soln of sodium carbonate. Sol in solns of alkalies and of ammonia by opening the anhydride ring and the formation of a salt. Also sol in boiling glacial acetic acid.

THERAP CAT: Antiseptic.

THERAP CAT (VET): Disinfectant (topical).

6708. Nitromethane. Nitrocarbol. CH_3NO_2; mol wt 61.04. C 19.68%, H 4.95%, N 22.95%, O 52.42%. Prepd by the interaction of sodium nitrite and sodium chloroacetate: Whitmore, *Org. Syn.* **3**, 83 (1923). Monograph: *Am. Ind. Hyg. Assoc. J.* **22**, 518 (1961). Toxicity study: Weatherby, *Arch. Ind. Health* **11**, 102 (1955).

Oily liquid with a moderately strong, somewhat disagreeable odor. d_4^{25} 1.1322. One gallon weighs 9.5 lbs. Flash pt 112°F. mp −29°. bp_{760} 101.2°; bp_{100} 46.6°; bp_{40} 27.5°; bp_{20} 14.1°; bp_{10} +2.8°; bp_5 −7.9°; $bp_{1.0}$ −29°. n_D^{22} 1.38056. Absorption spectrum: Hantzsch, Voigt, *Ber.* **45**, 106 (1912). Slightly sol in water (9.5% by vol at 20°); sol in alc, ether, DMF. Water solns are acid to litmus. pH of 0.01M aq soln 6.12. Forms an explosive sodium salt which bursts into flame on contact with water. Lethal concn for guinea pigs in air: 1000 ppm. LD_{50} orally in mice: 1.44 g/kg (Weatherby).

Caution: Potential symptom of overexposure is dermatitis. *See NIOSH Pocket Guide to Chemical Hazards* (DHHS/NIOSH 90-117, 1990) p 166.

USE: Rocket fuel; solvent for zein. Used in the coating industry.

6709. Nitromide. *3,5-Dinitrobenzamide.* $C_7H_5N_3O_5$; mol wt 211.13. C 39.82%, H 2.39%, N 19.90%, O 37.89%. Prepd from 3,5-dinitrobenzoyl chloride and ammonium acetate: Finan, Fothergill, *J. Chem. Soc.* **1962**, 2824.

Leaflets from water, mp 183°. Slightly sol in cold water, somewhat more sol in hot water.

Note: An ingredient of *Tristat, Unistat, Unistat-3.*

THERAP CAT (VET): Antibacterial, coccidiostat (for poultry).

6710. Nitron. *1,4-Diphenyl-3-(phenylamino)-1H-1,2,4-triazolium inner salt.* $C_{20}H_{16}N_4$; mol wt 312.37. C 76.90%, H 5.16%, N 17.94%. Prepn: E. Merck, **Ger.** pat. **161,235** (1905); *Frdl.* **8**, 1240. Formerly represented as *3,5,6-triphenyl-2,3,5,6-tetraazabicyclo[2.1.1]hex-1-ene:* Busch, *Ber.* **38**, 858, 4054 (1905). Meso-ionic structure: Baker, Ollis, *Quart. Rev.* **11**, 26 (1957); Olah, *J. Inorg. Nucl. Chem.* **16**, 225 (1961). Use in detection of gold: N. Ganchev, A. Dimitrova, *Mikrochim. Acta* **1971**, 476. In determn of nitrate: A. Hulanicki, M. Maj, *Talanta* **22**, 767 (1975); of molybdenum: T. J. Koralewski, G. A. Parker, *Anal. Chim. Acta* **113**, 389 (1980).

Intensely yellow leaflets from alcohol, solvated needles from chloroform, dec ~189°. Practically insol in water; sol in alcohol, benzene, acetone, chloroform, ethyl acetate, dil acids; slightly sol in ether. The alc soln undergoes partial decompn, indicated by a red color, hence should be protected from light. Dipole moment in benzene: 7.2. Solubility of some Nitron salts in 0.01*N* HCl: Nitrate 0.0099%, perchlorate 0.008%, iodide 0.017%, thiocyanate 0.04%, chromate 0.06%, chlorate 0.12%, nitrite 0.19%, bromide 0.61%.

USE: Detection of nitrate, perchlorate, gold, rhenium. Determination of boron, nitrate, perchlorate, rhenium, tungsten, molybdenum.

6711. 1-Nitronaphthalene. α-Nitronaphthalene. $C_{10}H_7NO_2$; mol wt 173.17. C 69.36%, H 4.07%, N 8.09%, O 18.48%. Prepd from napthalene and a mixture of nitric and sulfuric acids at 50°.

Yellow crystals. d 1.331. mp 59-61°. bp 304°. Insol in water; sol in alcohol, freely in chloroform, ether, carbon disulfide. Gives a dark red soln with concd H_2SO_4.

USE: Deblooming petr oils, 2-3 parts suffice for 1000 parts oil; manuf dyes and intermediates.

6712. 1-Nitro-2-naphthol. *1-Nitro-2-naphthalenol;* α-nitro-β-naphthol; 1-nitro-2-hydroxynaphthalene. $C_{10}H_7NO_3$; mol wt 189.17. C 63.49%, H 3.73%, N 7.40%, O 25.37%.

Yellow needles or platelets, mp 103°. Insol in water. Sol in alcohol, ether, glacial acetic acid, in alkali hydroxide solns.

USE: As a reagent for the determination of cobalt with which it gives a precipitate.

6713. 3-Nitropentane. $C_5H_{11}NO_2$; mol wt 117.15. C 51.26%, H 9.46%, N 11.96%, O 27.31%. $CH_3CH_2CH(NO_2)$-CH_2CH_3.

Colorless liquid; fusel oil odor. d_4^0 0.957. bp 153-155°. Insol in water; sol in alcohol, ether.

6714. 5-Nitro-*o*-phenetidine. *2-Ethoxy-5-nitrobenzenamine;* 5-nitro-2-ethoxyaniline; 1-ethoxy-2-amino-4-nitrobenzene; 1-nitro-3-amino-4-phenyl ethyl ether; 4-nitro-2-aminophenetole; Neo-Douxan. $C_8H_{10}N_2O_3$; mol wt 182.18. C 52.74%, H 5.53%, N 15.38%, O 26.35%. Prepn: Verkade *et al., Rec. Trav. Chim.* **65**, 346 (1946). Toxicology: Grupp, Bilger, *Arzneimittel-Forsch.* **1**, 326 (1951).

Brown-yellow crystals from benzene, mp 97.5-98.5°. 950 times as sweet as cane sugar, claimed to have no aftertaste.

USE: Sweetening agent.

6715. Nitrophenide. *Bis(3-nitrophenyl)disulfide;* m,m'-dinitrodiphenyl disulfide; NP; Megasul. $C_{12}H_8N_2O_4S_2$; mol wt 308.34. C 46.74%, H 2.62%, N 9.09%, O 20.76%, S 20.80%. Prepd by the reduction of *m*-nitrobenzenesulfonylchloride with hydriodic acid: Ekbom, *Ber.* **24**, 336 (1891); Foss *et al., J. Am. Chem. Soc.* **60**, 2729 (1938). Anticoccidial activity: Waletzky *et al., Ann. N.Y. Acad. Sci.* **52**, 543 (1949).

Yellow rhomboid crystals, mp 83°. Freely sol in ether, less sol in alcohol. Insol in water.

THERAP CAT (VET): Coccidiostat.

6716. *m*-Nitrophenol. $C_6H_5NO_3$; mol wt 139.11. C 51.80%, H 3.62%, N 10.07%, O 34.50%. Obtained by boiling diazotized *m*-nitroaniline with water and H_2SO_4: Adams, Wilson, *Org. Syn.* **3**, 87 (1923); Bachmann, Rottschaefer, *ibid.* **18**, 88 (1938); Manske, *ibid.* **coll. vol. I**, 404 (2nd ed., 1941).

Monoclinic prisms from ether or dil HCl. d_4^{20} 1.485; d_4^{100} 1.2797. mp 97°. bp_{70} 194°. Dec when distilled at ordinary pressure. K at 18° = 4.6 × 10⁻⁹. Absorption spectrum: Baly *et al., J. Chem. Soc.* **97**, 586 (1910); Marchlewski, Moroz, *Bull. Soc. Chim.* [4] **35**, 476 (1924). An aq soln satd at 40° contains 3.02 g in 100 g soln; at 98.7°, 40.9 g: Sidgwick *et al., J. Chem. Soc.* **107**, 1202 (1915). Soly (g/100 g solvent) in acetone: 169.35 (0.2°); 1305.9 (84°); in alcohol: 116.9 (1°); 1105.25 (85°); in ether: 105.9 (0.2°); 1065.8 (83°); Carrick, *J. Phys. Chem.* **25**, 635 (1921). Sol in hot and dil acids, in caustic solns; insol in petr ether. LD_{50} orally in mice, rats: 1414, 933 mg/kg, K. C. Back *et al., Reclassification of Materials Listed as Transportation Health Hazards* (TSA-20-72-3; PB214-270, 1972).

USE: As indicator in 0.3% soln in 50% alc. pH: 6.8 colorless, 8.6 yellow.

6717. *o*-Nitrophenol. $C_6H_5NO_3$; mol wt 139.11. C 51.80%, H 3.62%, N 10.07%, O 34.50%. The *o*- and *p*-nitrophenols are formed by the nitration of phenol; the *o*-form is separated by steam distillation.

Light yellow needles or prisms; peculiar, aromatic odor. d 1.495. mp 44-45°. bp 214-216°. Volatile with steam. Slightly sol in cold water, freely in hot water, in alcohol, benzene, ether, carbon disulfide, alkali hydroxides. LD_{50} orally in mice, rats: 1.297, 2.828 g/kg, K. C. Back *et al., Reclassification of Materials Listed as Transportation Health Hazards* (TSA-20-72-3; PB214-270, 1972).

USE: Manuf of many important compds; as indicator in 2% alc soln. pH: 5.0 colorless, 7.0 yellow, but the color change is not sharp and cannot be used where CO_2 is present; as reagent for glucose.

6718. *p*-Nitrophenol. $C_6H_5NO_3$; mol wt 139.11. C 51.80%, H 3.62%, N 10.07%, O 34.50%.

Colorless to slightly yellow, odorless crystals; sweetish, then burning taste. d_4^{20} 1.270. mp 113-114°. Sublimes; slightly volatile with steam. Moderately sol in cold water, freely in alcohol, chloroform, ether; also sol in solns of fixed alkali hydroxides and carbonates. LD_{50} orally in mice, rats: 467, 616 mg/kg, K.C. Back *et al.*, *Reclassification of Materials Listed as Transportation Health Hazards* (TSA-20-72-3; PB214-270, 1972).
USE: Manuf of many important compounds; as indicator in 0.1% alcohol soln. pH: 5.6 colorless, 7.6 yellow.

6719. *p*-Nitrophenylacetic Acid. *4-Nitrobenzeneacetic acid; p-nitro-α-toluic acid.* $C_8H_7NO_4$; mol wt 181.15. C 53.04%, H 3.89%, N 7.73%, O 35.33%. $NO_2C_6H_4CH_2$-COOH. Prepd by hydrolysis of *p*-nitrobenzyl cyanide with 50% H_2SO_4: Robertson, *Org. Syn.* **2**, 59 (1922); **coll. vol. I**, 406 (2nd ed., 1941).
Long, pale yellow needles from water. mp 153°. K at 25° 1.04 × 10⁻⁴. Absorption spectrum: Hewitt *et al.*, *J. Chem. Soc.* **101**, 1774 (1912). Sparingly sol in cold water; sol in alcohol, ether, benzene.
Methyl ester, $C_9H_9NO_4$, mp 54°.
Benzyl ester, $C_{15}H_{13}NO_4$, mp 92°.
Nitrile, *see p*-Nitrobenzyl Cyanide.

6720. 4-Nitro-*o*-phenylenediamine. *4-Nitro-1,2-benzenediamine; 1,2-diamino-4-nitrobenzene; 4-nitro-1,2-diaminobenzene; 4-nitro-1,2-phenylenediamine.* $C_6H_7N_3O_2$; mol wt 153.14. C 47.06%, H 4.61%, N 27.44%, O 20.90%. Prepd by the reduction of 2,4-dinitroaniline using H_2S in ammonia water: Kehrmann, *Ber.* **28**, 1707 (1895); Griffin, Peterson, *Org. Syn.* **coll. vol. III**, 242 (1955). Use as reagent for α-keto acids: Hockenhull, Floodgate, *Biochem. J.* **52**, 38 (1952); Taylor, Smith, *Analyst* **80**, 607 (1955). Toxicity study: C. Burnett *et al.*, *J. Toxicol. Environ. Health* **2**, 657 (1977).

Dark red needles from hot water. mp 201°. Sparingly sol in water. Sol in aq solns of hydrochloric acid. LD_{50} in rats (mg/kg): 3720 orally; > 1600 i.p. (Burnett).
USE: Reagent for α-keto acids.

6721. (4-Nitrophenyl)hydrazine. $C_6H_7N_3O_2$; mol wt 153.14. C 47.06%, H 4.61%, N 27.44%, O 20.90%.
Orange-red leaflets or needles. mp ~157° with dec. Slightly sol in cold water; sol in hot water or hot benzene, alcohol, chloroform, ether, ethyl acetate.
USE: As reagent for ketones, aliphatic aldehydes.

6722. *o*-Nitrophenylpropiolic Acid. *3-(2-Nitrophenyl)-2-propynoic acid.* $C_9H_5NO_4$; mol wt 191.14. C 56.55%, H 2.64%, N 7.33%, O 33.48%.

Yellowish to light brown scales or crystals. mp ~157° with decompn and may explode. Moderately sol in cold water, more in hot water or alcohol, very slightly in chloroform. Almost insol in carbon disulfide and petr ether.
USE: As a reagent for alkaloids and glucose.

6723. (4-Nitrophenyl)urea. $C_7H_7N_3O_3$; mol wt 181.15. C 46.41%, H 3.89%, N 23.20%, O 26.50%. Prepn from *N*-substituted *S*-phenylthiocarbamates: Crosby, Niemann, *J. Am. Chem. Soc.* **76**, 4458 (1954).

Prisms from abs ethanol, needles from dil ethanol, mp 238°. Practically insol in cold water. Sol in boiling water, methanol, ethanol, dimethylformamide. Sparingly sol in ether, benzene.

6724. 1-Nitropropane. $C_3H_7NO_2$; mol wt 89.09. C 40.44%, H 7.92%, N 15.72%, O 35.92%. $CH_3CH_2CH_2NO_2$. Prepd by vapor-phase nitration of propane: Hass *et al.*, U.S. pat. **1,967,667** (1934 to Purdue Res. Found.); *Ind. Eng. Chem.* **28**, 339 (1936); from natural gas: Bachman, Pollack, *ibid.* **46**, 713 (1954).
Liquid. d_4^{25} 0.9934. wt 8.4 lb/U.S. gal. mp −108°. bp_{760} 131.6°; bp_{728} 130°; bp_{401} 110°; bp_{94} 70°; $bp_{7.5}$ 20°; bp_4 10°. n_D^{20} 1.4018. Flash pt 34° (93°F). Heat of formation −40.05 kcal/mole at 25°; heat of combustion 481.33 kcal/mole at 25°. Latent heat of vaporization 10.37 kcal/mole at 25°. Slightly sol in water (1.4 ml/100 ml); 0.5 ml water dissolves in 100 ml of 1-nitropropane. Miscible with many organic solvents.
Caution: Potential symptoms of overexposure are eye irritation, headache, nausea, vomiting and diarrhea. See *NIOSH Pocket Guide to Chemical Hazards* (DHHS/NIOSH 90-117, 1990) p 166.
USE: Solvent for cellulose acetate, vinyl resins, lacquers, synthetic rubbers, fats, oils, dyes, other organic materials; as intermediate, propellant.

6725. 2-Nitropropane. $C_3H_7NO_2$; mol wt 89.09. C 40.44%, H 7.92%, N 15.72%, O 35.92%. $CH_3CH(NO_2)CH_3$. Prepn: Hass *et al.*, U.S. pat. **1,967,667** (1934 to Purdue Res. Found.); *Ind. Eng. Chem.* **28**, 339 (1936); from natural gas: Bachman, Pollack, *ibid.* **46**, 713 (1954). Toxicity and carcinogenicity study: T. R. Lewis *et al.*, *J. Environ Pathol. Toxicol.* **2**, 233 (1979). Health hazard alert: *Am. Ind. Hyg. Assoc. J.* **41**, 18 (1980).
Liquid. d_4^{25} 0.9821. wt 8.4 lb/U.S. gal. mp −93°. bp_{760} 120.3°; bp_{564} 110°; bp_{300} 90°; bp_{95} 60°; bp_7 10°. n_D^{20} 1.3944. Flash pt 24° (75°F). Heat of formn −43.78 kcal/mole at 25°; heat of combustion 477.60 kcal/mole at 25°. Latent heat of vaporization 9.88 kcal/mole at 25°. Slightly sol in water (1.7 ml/100 ml); 0.6 ml water dissolves in 100 ml of 2-nitropropane. Misc with many organic solvents.
Caution: Potential symptoms of overexposure are headache, anorexia, nausea, vomiting, diarrhea and irritation of respiratory system. See *NIOSH Pocket Guide to Chemical Hazards* (DHHS/NIOSH 90-117, 1990) p 166. This substance may reasonably be anticipated to be a carcinogen: *Seventh Annual Report on Carcinogens* (PB95-109781, 1994) p 275.
USE: *See* 1-Nitropropane.

6726. 5′-Nitro-2′-propoxyacetanilide. *N-(5-Nitro-2-propoxyphenyl)acetamide; 1-propoxy-2-acetamino-4-nitrobenzene; Falimint.* $C_{11}H_{14}N_2O_4$; mol wt 238.24. C 55.46%, H 5.92%, N 11.76%, O 26.86%. Prepd by reacting 1-chloro-2,4-dinitrobenzene with sodium alkoxide, followed by partial reduction and treatment with acetic anhydride: Profft, *Chem. Tech. (Leipzig)* **5**, 393 (1953), *C.A.* **47**, 12757a (1953); by nitration of 1-propoxy-2-acetylaminobenzene: Verkade *et al.*, *Rec. Trav. Chim.* **66**, 374 (1947); also **Brit. pat. 597,-835** (1948 to N. V. Polak & Schwarz's Essencefabrieken).

Consult the Name Index before using this section.

Crystals from propanol, mp 102.5-103.5°. Flat taste.
THERAP CAT: Antipyretic, analgesic.

6727. 5-Nitro-2-propoxyaniline. *5-Nitro-2-propoxybenzenamine;* 2-amino-4-nitro-1-propoxybenzene; 1-propoxy-2-amino-4-nitrobenzene; P-4000; Ultrasüss. C_9H_{12}-N_2O_3; mol wt 196.21. C 55.09%, H 6.16%, N 14.28%, O 24.46%. A substance which is 4000 times sweeter than sugar, with no bitter aftertaste. Prepn: Verkade *et al., Rec. Trav. Chim.* **65,** 346-360 (1946); **Neth. pat. Appl. 6,514,769** (1967 to N. V. Koffie en Theehandel), *C.A.* **67,** 108430n (1967); A. J. de Koning, *J. Chem. Ed.* **53,** 521 (1976); J. McK. Woollard, *ibid.* **57,** 464 (1980). Used as a sweetener in some European countries but banned in U.S.A. because of possible toxic effects. Toxicity studies: Fitzhugh *et al., J. Am. Pharm. Assoc.* **46,** 583 (1951).

Orange crystals from *n*-propanol + petr ether. mp 47.5-48.5°. Soly in water at 20°: 136 mg/liter. Stable in boiling water and in dil acids.

6728. 5-Nitroquinaldic Acid. *5-Nitro-2-quinolinecarboxylic acid;* 5-nitroquinaldinic acid. $C_{10}H_6N_2O_4$; mol wt 218.17. C 55.05%, H 2.77%, N 12.84%, O 29.33%. Prepn from quinaldic acid: Besthorn, Ibele, *Ber.* **39,** 2333 (1906).

Yellow crystals, mp 275-278° (dec). Slightly sol in water and in most organic solvents. Readily sol in acetic acid.
USE: In the determination of zinc.

6729. 3-Nitrosalicylic Acid. *2-Hydroxy-3-nitrobenzoic acid.* $C_7H_5NO_5$; mol wt 183.12. C 45.91%, H 2.75%, N 7.65%, O 43.69%. The 3- and 5-nitrosalicylic acids are formed by the nitration of salicylic acid with H_2SO_4 and HNO_3.

Monohydrate, yellowish crystals. mp 123°; mp 148° when anhydr. Slightly sol in water; freely sol in alcohol, benzene, chloroform, ether.
USE: In dye chemistry.

6730. 5-Nitrosalicylic Acid. Anilotic acid. $C_7H_5NO_5$; mol wt 183.12. C 45.91%, H 2.75%, N 7.65%, O 43.69%.
Yellowish crystals. d 1.65. mp 228-230°. One gram dissolves in 1475 ml water; more sol in hot water; freely sol in alcohol, ether.

6731. Nitroscanate. *1-Isothiocyanato-4-(4-nitrophenoxy)benzene; isothiocyanic acid p-(p-nitrophenoxy)phenyl ester;* p-(p-nitrophenoxy)phenyl isothiocyanate; 1-(4-isothiocyanatophenoxy)-4-nitrobenzene; 4-nitro-4'-isothiocyanodiphenyl ether; cantrodifene (obsolete); CGA-23654; GS-23654; Lopatol; Scaniver. $C_{13}H_8N_2O_3S$; mol wt 272.28. C 57.35%, H 2.96%, N 10.29%, O 17.63%, S 11.78%. Prepn: **Fr. pat. 1,491,477** (1967 to Agripat); F. Paltauf *et al.,* **U.S. pat. 3,697,555** (1972 to Ciba-Geigy); A. Martvon, K. Antos, *Chem. Zvesti* **23,** 181 (1969). Activity in dogs: M. A. Gemmell, G. Oudemans, *Res. Vet. Sci.* **19,** 217 (1975).

Crystals, mp 107-113° (Martvon, Antos); also reported as 124-125° (Paltauf). Insol in water. Sol in organic solvents.
THERAP CAT (VET): Anthelmintic.

6732. 3-Nitrosobenzamide. NOBA. $C_7H_6N_2O_2$; mol wt 150.14. C 56.00%, H 4.03%, N 18.66%, O 21.31%. Antiretroviral zinc ejecting agent. Prepn: E. Kun *et al.,* **PCT. Int. pat. Appl. 91 08,902** (1991 to Octamer). Tumoricidal activity: W. G. Rice *et al., Proc. Nat. Acad. Sci. USA* **89,** 7703 (1992). Inhibition of HIV-1 infectivity: W. G. Rice *et al., Nature* **361,** 473 (1993); and site of action: *eidem, Proc. Nat. Acad. Sci. USA* **90,** 9721 (1993); of SIV: A. J. Chuang *et al., FEBS Letters* **326,** 140 (1993).

Light yellow solid. Darkens above 135°, softens with polymerization 150-160°; mp 240-250° (dec). In solution the compound is dark green. uv/vis max (absol. ethanol): 218, 304, 750nm (ϵ 1.5 × 10^4, 5.35 × 10^3, 37.6).

6733. N-Nitrosodiethanolamine. *2,2'-(Nitrosoimino)-bisethanol;* 2,2'-nitrosiminodiethanol; di-(2-hydroxyethyl)-nitrosamine; NDELA. $C_4H_{10}N_2O_3$; mol wt 134.14. C 35.82%, H 7.51%, N 20.88%, O 35.78%. $(HOCH_2CH_2)_2N$-NO. Formed by the action of nitrites on di- or triethanolamine. Prepn: E. R. H. Jones, W. Wilson, *J. Chem. Soc.* **1949,** 547; R. Preussmann, *Ber.* **95,** 1571 (1962); W. Lijinsky *et al., J. Nat. Cancer Inst.* **49,** 1239 (1972). Carcinogenicity study: H. Druckery *et al., Z. Krebsforsch.* **69,** 103 (1967). Impurity in cutting fluids: T. Y. Fan *et al., Science* **196,** 70 (1977); in cosmetics: *eidem, Food Cosmet. Toxicol.* **15,** 423 (1977). Mutagenicity study: A. Hesbert *et al., Mutat. Res.* **68,** 207 (1979).

Light yellow oil, $bp_{0.01}$ 125°. n_D^{20} 1.4849.
Note: This substance may be anticipated to be a carcinogen: *Seventh Annual Report on Carcinogens* (PB95-109781, 1994) p 281.

6734. N-Nitrosodiethylamine. *N-Ethyl-N-nitrosoethanamine;* diethylnitrosamine; DEN; DENA; NDEA. C_4H_{10}-N_2O; mol wt 102.14. C 47.04%, H 9.87%, N 27.43%, O 15.66%. $(C_2H_5)_2NNO$. Detected in trace amounts in tobacco smoke: Druckney, Preussman, *Naturwiss.* **49,** 498 (1962) and in various processed foods: Hedler, Marquardt, *Food Cosmet. Toxicol.* **6,** 341 (1968); Friemuth, Glaeser, *Nahrung* **14,** 357 (1970). Formed by the interaction of nitrite with diethylamine and by the action of nitrate-reducing bacteria. Industrial prepn: Reilly, **Ger. pat. 1,085,166** (1960 to du Pont), *C.A.* **56,** 4594h (1962); Levering, Maury, **U.S. pat. 3,090,786** (1963 to Hercules Powder); Minisci, Galli, *Chim. Ind.* (Milan) **46,** 173 (1964). Hepatotoxicity and carcinogenicity studies: Schmaehl *et al., Naturwiss.* **54,** 341 (1967); Grover, Fischer, *Eur. J. Cancer* **7,** 77 (1971); Bader *et al., Arch. Geschwulstforsch.* **37,** 327 (1971). General review: Magee, Barnes, *Adv. Cancer Res.* **10,** 163-246 (1967).

Slightly yellow liq. d_4^{20} 0.9422. bp 175-177°. bp_5 47°. n_D^{20} 1.4388. Sol in water, alc, ether.
Note: This substance may reasonably be anticipated to be a carcinogen: *Seventh Annual Report on Carcinogens* (PB95-109781, 1994) p 282.
USE: Gasoline and lubricant additive; antioxidant; stabilizer in plastics.

6735. N-Nitrosodimethylamine. *N-Methyl-N-nitrosomethanamine;* dimethylnitrosamine; DMN; DMNA. C_2H_6-N_2O; mol wt 74.08. C 32.43%, H 8.16%, N 37.81%, O 21.60%. $(CH_3)_2NNO$. Reportedly found in trace amounts in tobacco smoke condensates: Rhoades, Johnson, *Nature* **236,** 307 (1972); in cured meat products, notably bacon:

Sen *et al., ibid.* **241,** 473 (1973); in smoked and salted fish: Fazio *et al., J. Agr. Food Chem.* **19,** 250 (1971); Fong, Chan, *Nature* **243,** 421 (1973). Formed by the interaction of nitrite with dimethylamine and by the action of nitrate-reducing bacteria. Industrial prepn: **Brit.** pat. **772,331** (1957 to Olin Mathieson), *C.A.* **51,** 14783a (1957); Ioffe, *Zh. Obshch. Khim.* **28,** 1296 (1958); Norris, *J. Am. Chem. Soc.* **81,** 3346 (1959); Campbell, **U.S.** pat. **2,981,752** (1961 to C.S.C.); Datin, Elliott, **U.S.** pat. **3,136,821** (1964 to Allied Chem.). Metabolism and toxicity: Magee, Vandekar; Magee, *Biochem. J.* **70,** 600, 606 (1958); Heath, *ibid.* **85,** 72 (1962). Carcinogenicity studies: Magee, Barnes, *Brit J. Cancer* **10,** 114 (1956); *see also eidem, Adv. Cancer Res.* **10,** 163-246 (1967). Chemistry: Layne *et al., J. Am. Chem. Soc.* **85,** 435, 1816 (1963).

Yellow liquid. bp 151-153°. bp$_{40}$ 67.1°. d$_4^{20}$ 1.0048. n$_D^{20}$ 1.4368. Very sol in water, alcohol, ether. LD$_{50}$ i.p. in rats: 34 mg/kg (Heath).

Caution: Potential symptoms of overexposure are nausea, vomiting, diarrhea and abdominal cramps; headache; fever; enlarged liver, jaundice; reduced function of liver, kidneys and lungs. *See NIOSH Pocket Guide to Chemical Hazards* (DHHS/NIOSH 90-117, 1990) p 168. This substance may reasonably be anticipated to be a carcinogen: *Seventh Annual Report on Carcinogens* (PB95-109781, 1994) p 286.

USE: Formerly in the prodn of rocket fuels; antioxidant; additive for lubricants; softener of copolymers.

6736. *p*-Nitroso-*N,N*-dimethylaniline. *N,N-Dimethyl-4-nitrosobenzenamine;* Accelerine. C$_8$H$_{10}$N$_2$O; mol wt 150.18. C 63.98%, H 6.71%, N 18.65%, O 10.65%. Prepd in the cold from NaNO$_2$ and a soln of dimethylaniline in HCl.

Green plates or leaflets. mp 92.5-93.5°; also stated as 87-88°. Volatile with steam. Insol in water. Sol in alcohol, ether.

USE: Manuf organic compds; accelerator in vulcanizing; in printing fabrics.

6737. *p*-Nitrosodiphenylamine. *4-Nitroso-N-phenylbenzenamine.* C$_{12}$H$_{10}$N$_2$O; mol wt 198.22. C 72.71%, H 5.08%, N 14.13%, O 8.07%. C$_6$H$_5$NHC$_6$H$_4$NO.

Green plates with bluish luster (from benzene) or steel-blue prisms or plates (from ether + water). mp 144-145°. Slightly sol in water or petr ether; freely sol in alcohol, ether, chloroform, benzene. Dissolves in H$_2$SO$_4$ with a red color, which suddenly changes to violet on warming.

Note: This substance was formerly anticipated to be a carcinogen: *Fifth Annual Report on Carcinogens* (NTP 89-239, 1989) p 212; delisted for insufficient evidence of carcinogenicity: *Sixth Annual Report on Carcinogens* (PB92-120666, 1991) p 13.

USE: Accelerator in vulcanizing rubber.

6738. *N*-Nitrosomorpholine. *4-Nitrosomorpholine;* NMOR. C$_4$H$_8$N$_2$O$_2$; mol wt 116.12. C 41.37%, H 6.94%, N 24.12%, O 27.56%. Prepn: L. Knorr, *Ann.* **301,** 1 (1898); F. Chapman, *J. Chem. Soc.* **1949,** 1631; G. Oláh *et al., Ber.* **89,** 2374 (1956). Metabolism: B. W. Stewart, P. N. Magee, *Biochem. J.* **126,** 21P (1972). Carcinogenicity studies: H. Druckrey *et al., Naturwiss.* **48,** 134 (1961); P. Bannasch, H.-A. Müller, *Arzneimittel-Forsch.* **14,** 805 (1964); *IARC Monographs* **17,** 263 (1978).

Yellow crystals, mp 29°. bp$_{25}$ 139-140°; bp$_{747}$ 224-224.5°. Sol in water. LD$_{50}$ orally in rats: 282 mg/kg (Druckrey).

Note: This substance may reasonably be anticipated to be a carcinogen: *Seventh Annual Report on Carcinogens* (PB95-109781, 1994) p 299.

6739. 1-Nitroso-2-naphthol. *1-Nitroso-2-naphthalenol;* nitroso-β-naphthol. C$_{10}$H$_7$NO$_2$; mol wt 173.17. C 69.36%, H 4.07%, N 8.09%, O 18.48%. Prepared by the addition of H$_2$SO$_4$ to a mixture of β-naphthol dissolved in aq NaOH

and NaNO$_2$: C. S. Marvel, P. K. Porter, *Org. Syn.* **coll. vol. I,** 411 (2nd ed., 1941).

Yellowish-brown needles from petr ether, mp 109-110°. Sol in 1000 parts water, 35 parts alcohol; also sol in hot alcohol, benzene, ether, carbon disulfide, caustic alkali solns, glacial acetic acid; slightly sol in cold petr ether.

USE: To prevent gum formation in gasoline; in analytical chemistry in the determination of cobalt (to separate it from nickel).

6740. 4-Nitrosophenol. Quinone oxime; quinone monoxime. C$_6$H$_5$NO$_2$; mol wt 123.11. C 58.54%, H 4.09%, N 11.38%, O 25.99%.

Pale yellow orthorhombic needles, browns at 126° (dec 144°). *Explodes on contact with concentrated acid, alkali, or fire. A technical grade exploded on storage.* Ka at 25° = 3.3 × 10^{-7}. Moderately sol in water. Sol in dil alkalies giving green to brownish green solns. Sol in alc, ether, acetone.

6741. *N*-Nitrosopyrrolidine. *1-Nitrosopyrrolidine;* NPYR; NO-PYR. C$_4$H$_8$N$_2$O; mol wt 100.12. C 47.99%, H 8.05%, N 27.98%, O 15.98%. Cyclic nitrosamine that occurs in food products and tobacco smoke. Prepn: F. C. Peterson, *Ber.* **21,** 290 (1888). Carcinogenicity studies: H. Druckrey *et al., Z. Krebsforsch.* **69,** 103 (1967); M. Greenblatt *et al., J. Nat. Cancer Inst.* **48,** 1687 (1972); *IARC Monographs* **17,** 313 (1978). Mechanism of mutagenicity: L. I. Hecker *et al., Mutat. Res.* **62,** 213 (1979). Metabolism: R. C. Cottrell *et al., Adv. Exp. Med. Biol.* **136B,** 1165 (1982).

Yellow liquid, bp$_{20}$ 104-106°. Sol in water. LD$_{50}$ orally in rats: 900 mg/kg (Druckrey).

Note: This substance may reasonably be anticipated to be a carcinogen: *Seventh Annual Report on Carcinogens* (PB95-109781, 1994) p 305.

6742. Nitroso-R Salt. *3-Hydroxy-4-nitroso-2,7-naphthalenedisulfonic acid disodium salt;* sodium salt of 1-nitroso-2-hydroxynaphthalene-3,6-disulfonic acid. C$_{10}$H$_5$NNa$_2$O$_8$S$_2$; mol wt 377.26. C 31.84%, H 1.34%, N 3.71%, Na 12.19%, O 33.93%, S 17.00%.

Golden-yellow, fan-shaped crystals. Sol in ~40 parts water; more sol in hot water; slightly sol in methyl or ethyl alcohol.

USE: As a reagent for cobalt and potassium.

6743. 2-Nitro-4-sulfobenzoic Acid. $C_7H_5NO_7S$; mol wt 247.19. C 34.01%, H 2.04%, N 5.67%, O 45.31%, S 12.97%. Prepn by sulfonation of *o*-nitrotoluene and oxidation of the resulting 2-nitro-4-toluenesulfonic acid with potassium permanganate: Hart, *Am. Chem. J.* **1**, 352 (1879-80).

Needles from hydrochloric acid. Stable in air under ordinary conditions.

USE: Alkalimetric standard.

6744. Nitrosyl Chloride. ClNO; mol wt 65.46. Cl 54.16%, N 21.40%, O 24.44%. NOCl. Best prepared from nitrosylsulfuric acid and dry HCl: Coleman *et al.*, *Inorg. Syn.* **1**, 55 (1939).

Non-explosive, very corrosive, reddish-yellow gas; liquid at −5.5°; solid at −61.5°. Decomposed by water. Sol in fuming H_2SO_4. Critical temp 167°; crit press. 92.4 atm. The orange color of aqua regia is produced by nitrosyl chloride.

Caution: Intensely irritating to eyes, skin, mucous membranes. Inhalation may cause pulmonary edema, hemorrhage.

6745. Nitrosyl Fluoride. Nitrogen oxyfluoride. FNO; mol wt 49.00. F 38.77%, N 28.58%, O 32.65%. Preparation: Ruff *et al.*, *Z. Anorg. Allgem. Chem.* **208**, 293 (1932); Balz, Mailänder, *ibid.* **217**, 166 (1934); Faloon, Kenna, *J. Am. Chem. Soc.* **73**, 2937 (1951); Kwasnik in *Handbook of Preparative Inorganic Chemistry* **vol. 1**, G. Brauer, Ed. (Academic Press, New York, 2nd ed., 1963) pp 184-185. *Reviews:* Hoffman, Neville, *Chem. Rev.* **62**, 1-18 (1962); Kemmitt, Sharp, *Advan. Fluorine Chem.* **4**, 194-195 (1965); Woolf, *ibid.* **5**, 1-30 (1965); Schmutzler, *Angew. Chem. Int. Ed.* **7**, 440-455 (1968).

Colorless gas. Often bluish because of impurities. Vigorous reaction with glass, corroding action on quartz. May be kept in quartz ampuls if cooled in liq oxygen. mp −132.5°. bp −59.9°. d (liq at bp) 1.326. d (solid) 1.719. Trouton const. 21.3. Reacts with water to form NO, HNO_3 and HF.

Caution: Highly irritating to skin, eyes, mucous membranes. *See also* Fluorine.

USE: Oxidizer in rocket propellants; stabilizing agent for liquid SO_3; fluorinating agent.

6746. Nitrosylsulfuric Acid. *Sulfuric acid monoanhydride with nitrous acid; nitrosyl sulfate;* chamber crystals; nitrososulfuric acid; nitroxylsulfuric acid; nitrosulfonic acid; nitrosyl hydrogen sulfate; nitro acid sulfite; Nitrose. HNO_5S; mol wt 127.08. H 0.79%, N 11.02%, O 62.95%, S 25.23%. Formed as an intermediate in the lead chamber process for sulfuric acid by the reaction of sulfur dioxide, nitrogen trioxide, oxygen, and water: Clément, Désormes, *Ann. Chim. Phys.* [1] **59**, 329 (1806); Lunge, *J. Chem. Soc.* **47**, 470 (1885). Prepd from sulfur trioxide, nitrogen oxides and water: Döbereiner, *Schweigger's Journ.* **8**, 239 (1812); de Claubry, *Ann. Chim. Phys.* [2] **45**, 284 (1832); Kuhlmann, *ibid.* [3] **1**, 116 (1843); from silver acid sulfate and nitrosyl bromide: Berl *et al.*, *Z. Anorg. Allgem. Chem.* **209**, 264 (1932). *See also* U.S. pats. **1,909,557** and **1,909,558**. The formation of crystals of nitrosylsulfuric acid may be observed by igniting a mixture of 1 part sulfur and 2 or 3 parts potassium nitrate under a bell jar.

Prisms, dec 73.5°. In moist air the crystals dec with the formation of sulfuric and nitric acids and above 50° nitric oxide and nitrogen dioxide are evolved. Sol in sulfuric acid, dec in water.

USE: For bleaching cereal milling products.

6747. Nitrosyl Tetrafluoroborate. Nitrosonium tetrafluoroborate; nitrosyl borofluoride; nitrosyl fluoborate. BF_4NO; mol wt 116.81. B 9.26%, F 65.06%, N 11.99%, O 13.70%. $NOBF_4$. Prepd according to the equation: $2 HBF_4 + N_2O_3 \rightarrow 2 NOBF_4 + H_2O$: Wilke-Dörfurt, Balz, *Z. Anorg. Allgem. Chem.* **159**, 219 (1927); Balz, Mailänder, *ibid.* **217**, 162 (1934); H. S. Booth, D. R. Martin, *Boron Trifluoride and Its Derivatives* (New York, 1949) p 133 sqq. Review of tetrafluoroborates: Sharp, *Advan. Fluorine Chem.* **1**, 68-128 (1960).

Birefringent, orthorhombic, hygroscopic platelets. d_4^{25} 2.185. Sublimes at 0.01 mm and 250° without decompn. Decomposed by water. May be stored in glass bottles if absolutely dry.

USE: In the prepn of diazonium fluoborates.

6748. Nitrotoluene. *Methylnitrobenzene.* $C_7H_7NO_2$; mol wt 137.14. C 61.31%, H 5.14%, N 10.21%, O 23.33%. Nitration of toluol by a mixture of HNO_3 and H_2SO_4 yields principally *o*- and *p*-nitrotoluol. Prepn of *m*-nitrotoluene from 3-nitro-4-amino-toluene and $NaNO_2$: Clark, Taylor, *Org. Syn.* **3**, 91 (1923).

m-Nitrotoluene. Liquid. d_4^{15} 1.1630; d_4^{20} 1.1581; d_4^{59} 1.124; d_4^{121} 1.063. Solidifies in an ice and salt cooling mixture; melts at 15.5°. bp_{760} 231.9°; bp_{100} 156.9°; bp_{40} 130.7°; bp_{20} 112.8°; bp_{10} 96.0°; bp_5 81.0°; $bp_{1.0}$ 50.2°. n_D^{30} 1.5426. Absorption spectrum: Marchlewski, Mayer, *Bull. Acad. Polon.* [A] **1929**, 188. Soly in water at 30°: 0.498 g/l. Miscible with alcohol and ether. Sol in benzene.

o-Nitrotoluene. Yellowish liquid at ordinary temp. d_{15}^{19} 1.1622. mp −10°. bp 222°. n_D^{20} 1.5472. Almost insol in water; sol in alcohol, benzene, petr ether.

p-Nitrotoluene. Yellowish crystals. d 1.286. mp 53-54°. bp 238°. Flash pt 106°. Almost insol in water; sol in alcohol, benzene, ether, chloroform, acetone.

USE: Manuf of dyes, toluidines, nitrobenzoic acids, etc.

6749. Nitrourea. *N*-Nitrocarbamide. $CH_3N_3O_3$; mol wt 105.05. C 11.43%, H 2.88%, N 40.00%, O 45.69%. $NH_2CONHNO_2$. Prepd by the action of concd sulfuric acid upon urea nitrate: Thiele, Lachman, *Ann.* **288**, 281 (1895); Ingersoll, Armendt, *Org. Syn.* **5**, 85 (1925). By dropwise addition of HCl to a cooled mixture of silver cyanate and nitramide in water: Davis, Blanchard, *J. Am. Chem. Soc.* **51**, 1794 (1929).

Platelets from alcohol + petr ether dec 158.4-158.8°. K at 20° = 7.0 × 10⁻³. Absorption spectrum: Baly, Desch, *J. Chem. Soc.* **93**, 1753 (1908). Soluble in hot water, but water solns are unstable. Decompn in aq alkaline solns is almost instantaneous. Freely sol in acetone, alcohol, acetic acid. Sparingly sol in petr ether, chloroform, benzene. Stable to oxidizing agents. Can be detonated, but is not sensitive to percussion or heating.

6750. Nitrous Acid. HNO_2; mol wt 47.01. H 2.14%, N 29.79%, O 68.06%. Formed by the action of strong acids on inorganic nitrites. *Review:* Block, "Nitrous Acid, Hyponitrous Acid and their Salts" in *Mellor's* **Vol. VIII**, supplement II, *Nitrogen* (part 2) 353-408 (1967).

Known only in soln (pale blue in color). Weak acid. K (25°): 4.5 × 10⁻⁴. In water it changes quickly into nitric oxide and nitric acid. Forms stable, water-sol nitrites with Li, Na, K, Ca, Sr, Ba, Ag. Does not form salts with weak polyvalent cations like Al or Be. Forms stable esters with alcohols.

6751. Nitrous Oxide. Dinitrogen monoxide; laughing gas; hyponitrous acid anhydride; factitious air. N_2O; mol wt 44.01. N 63.65%, O 36.35%. Constituent of the earth's atm, about 0.00005% by volume: Slobod, Krogh, *J. Am.*

Chem. Soc. **72**, 1175 (1950). Prepd by thermal decompn of ammonium nitrate: E. H. Archibald, *The Preparation of Pure Inorganic Substances* (Wiley, New York, 1932) p 246; Castner, Kirst, U.S. pat. **2,111,276** (1938 to du Pont). Preparation and purification: Schenk in *Handbook of Preparative Inorganic Chemistry* vol. 1, G. Brauer, Ed. (Academic Press, New York, 2nd ed., 1963) pp 484-485. The chief impurity of the commercial product is N_2, although NO_2, N, O_2, and CO_2 may also be present. Teratogenicity study: G. A. Lane *et al.*, *Science* **210**, 899 (1980). *Reviews:* Beattie, "Nitrous Oxide" in *Mellor's* Vol. VIII, suppl II, *Nitrogen* (part 2) 189-215 (1967); Jones in *Comprehensive Inorganic Chemistry* vol. 2, J. C. Bailar, Jr. *et al.*, Eds. (Pergamon Press, Oxford, 1973) pp 316-323.

Colorless gas. *Asphyxiant.* Slightly sweetish odor and taste. Supports combustion. Very stable and rather inert chemically at room temperatures. Dissociation begins above 300° when the gas becomes a strong oxidizing agent. mp −90.81°; bp_{760} −88.46°; Trouton constant 21.4: Hoge, *J. Res. Nat. Bur. Stand.* **34**, 281 (1945). Dipole moment 0.166. d^{-89} (liq) 1.226; d(S.T.P.) 1.967; d(gas) 1.53 (air = 1). Critical temp 36.5°; crit press. 71.7 atm. Heat of vaporization (bp): 3.956 kcal/mole. While in the steel cylinder nitrous oxide is compressed to the form of gas over liq and has a pressure of ∼800 lbs/sq. in. at room temp. At 20° and 2 atm one liter of the gas dissolves in 1.5 liters of water. Freely sol in sulfuric acid. Sol in alcohol, ether, oils.

Caution: Narcotic in high concns. Less irritating than other oxides of nitrogen.

USE: To oxidize organic compds at temps > 300°; to make nitrites from alkali metals at their boiling points; in rocket fuel formulations (with carbon disulfide); in the prepn of whipped cream.

THERAP CAT: Anesthetic (inhalation); analgesic.

6752. Nitrovin. *2-[3-(5-Nitro-2-furanyl)-1-[2-(5-nitro-2-furanyl)ethenyl]-2-propenylidene]hydrazinecarboximidamide; [[3-(5-nitro-2-furyl)-1-[2-(5-nitro-2-furyl)vinyl]allylidene]amino]guanidine; sym-bis(5-nitro-2-furfurylidene)-acetone guanylhydrazone; 1,5-bis(5-nitro-2-furyl)-3-pentadienone guanylhydrazone; 1,5-bis(5-nitro-2-furyl)-3-pentadienone amidinohydrazone; Panazon; Payzone.* $C_{14}H_{12}N_6O_6$; mol wt 360.29. C 46.67%, H 3.36%, N 23.33%, O 26.64%. Prepn: Uota *et al.*, **Japan.** pat. 2673('52) (to Toyama), *C.A.* **48**, 2115h (1954); Uoda, Tanizaki, **Japan.** pat. 4479('64) (to Fukuju Pharm.), *C.A.* **62**, 10412f (1965).

Blackish violet crystals from ethyl alcohol, mp 217° (dec). Hydrochloride, mp 280° (dec).

THERAP CAT (VET): Growth promoter; antibacterial.

6753. Nitroxoline. *5-Nitro-8-quinolinol; 5-nitro-8-hydroxyquinoline; Enterocol; Nibiol; Noxibiol; Uritrol; Urocoli.* $C_9H_6N_2O_3$; mol wt 190.16. C 56.85%, H 3.18%, N 14.73%, O 25.24%. Prepn: Kostanecki, *Ber.* **24**, 154 (1891); Petrow, Sturgeon, *J. Chem. Soc.* **1954**, 570; Pratt, Duke, *J. Am. Chem. Soc.* **82**, 1155 (1960). *In vitro* antibacterial and antifungal activity: A. Desvignes, P. Leguen, *Ann. Pharm. Franc.* **21**, 803 (1963); M. Medic-Saric *et al.*, *Chemotherapy* **26**, 263 (1980). Toxicological study: O. Angelova *et al.*, *Adv. Antimicrob. Antineoplastic Chemother., Proc. 7th Int. Congr. Chemother.* **1**, 507 (1972). Clinical pharmacokinetics: A. Mrhar *et al.*, *Int. J. Clin. Pharmacol. Biopharm.* **17**, 476 (1979). HPLC determn in plasma and urine: R. H. A. Sorel *et al.*, *J. Chromatog.* **222**, 241 (1981). Clinical evaluation in urinary tract infections: M. R. Jacobs *et al.*, *S. Afr. Med. J.* **54**, 959 (1978); B. Cancet, A. Amgar, *Pathol. Biol.* **35**, 879 (1987).

Yellow needles from alcohol or acetic acid, mp 179.5-181.5°. Freely sol in alkali and hot HCl; sparingly sol in alcohol, ether.

Hydrochloride, $C_9H_7ClN_2O_3$, yellow needles from alcohol, mp 258°.

THERAP CAT: Antibacterial.

6754. Nitroxynil. *4-Hydroxy-3-iodo-5-nitrobenzonitrile; Dovenix.* $C_7H_3IN_2O_3$; mol wt 290.02. C 28.99%, H 1.04%, I 43.76%, N 9.66%, O 16.55%. Preparation: **Neth. pat. Appl. 6,516,359** corresp to Collins *et al.*, U.S. pat. **3,331,738** (1966, 1967 both to May & Baker).

Yellow crystals from benzene, mp 137-138°. Sparingly sol in water; moderately sol in most organic solvents.

D-N-Methylglucamine salt, $C_{14}H_{20}IN_3O_8$, *nitroxynil meglumine, 4-hydroxy-3-iodo-5-nitrobenzonitrile compd with D-1-deoxy-1-(methylamino)glucitol (1:1).* Solid, mp 85-90°.

N-Ethylglucamine, $C_{15}H_{22}IN_3O_8$, *nitroxynil eglumine, 4-hydroxy-3-iodo-5-nitrobenzonitrile compd with 1-deoxy-1-(ethylamino)glucitol (1:1), Trodax.* Readily sol in water with a yellow, odorless and substantially neutral soln. Aq soln is very stable but contamination with calcium and certain other salts can result in pptn of an insol salt of nitroxynil.

THERAP CAT (VET): Anthelmintic (fasciolicide).

6755. Nitryl Chloride. *Nitroxyl chloride.* $ClNO_2$; mol wt 81.46. Cl 43.52%, N 17.19%, O 39.28%. NO_2Cl. Conveniently prepd by the addn of chlorosulfonic acid to nitric acid: Dachlauer, **Ger. pat. 509,405** (1929 to I. G. Farben); Kaplan, Schechter, *Inorg. Syn.* **4**, 52 (1953); Collis *et al.*, *J. Chem. Soc.* **1958**, 438.

Corrosive, toxic, colorless gas. Chlorine-like odor. Vapor density (100°): 2.81 g/l. Dec >120°. bp −14.3°. mp −145°. d_{liq}^0 1.37; d_{liq}^{16} 1.33. Even the purest liquid may have a pale yellow color. Solns in polar solvents are always yellow. The gas or liquid may attack organic matter with explosive violence.

Caution: Strong irritant, corrosive.

USE: Nitrating and chlorinating agent in organic syntheses.

6756. Nitryl Fluoride. FNO_2; mol wt 65.00. F 29.23%, N 21.55%, O 49.23%. Credit for original prepn by the spontaneous combustion of nitric oxide in an atm of fluorine is given to Moissan, Lebeau, *Compt. Rend.* **140**, 1573, 1621 (1905); more easily prepd by mixing nitrogen dioxide and fluorine: Ruff *et al.*, *Z. Anorg. Allgem. Chem.* **208**, 298 (1932); Faloon, Kenna, *J. Am. Chem. Soc.* **73**, 2937 (1951). Reviews of prepn and chemistry: Hoffman, Neville, *Chem. Rev.* **62**, 1-18 (1962); Kwasnik in *Handbook of Preparative Inorganic Chemistry* vol. 1, G. Brauer, Ed. (Academic Press, New York, 2nd ed., 1963) pp 186-187; Kemmitt, Sharp, *Advan. Fluorine Chem.* **4**, 195-196 (1965); Woolf, *ibid.* **5**, 1-30 (1965); Schmutzler, *Angew. Chem. Int. Ed.* **7**, 440-455 (1968).

Colorless gas. Pungent odor. Attacks mucous membranes. mp −166.0°. bp −72.4°. d (liq at bp) 1.796. d (solid) 1.924. Trouton const 21.2. May be stored in quartz ampuls if cooled in liq oxygen. Purification can be accomplished by fractional distillation at reduced press. in dry

Consult the Name Index before using this section.

glass or quartz apparatus. Rapidly hydrolyzed in water to form nitric and hydrofluoric acids. Powerful oxidizing agent, with fluorinating powers; slightly weaker than fluorine. Absorbs mercury completely. Spontaneously ignites iodine, selenium, phosphorus (red and white), arsenic, antimony, boron, silicon, thorium, molybdenum. On mild warming attacks lead, bismuth, chromium, manganese, iron, nickel, tungsten, sulfur, charcoal. Does not react readily with hydrogen in the cold. Converts ethanol to ethyl nitrate; benzene to nitrobenzene.

Caution: See Fluorine.

USE: Oxidizer in rocket propellants; fluorinating agent.

6757. Nivalenol. *(3α,4β,7α)-12,13-Epoxy-3,4,7,15-tetrahydroxytrichothec-9-en-8-one; 3α,4β,7α,15-tetrahydroxyscirp-9-en-8-one.* $C_{15}H_{20}O_7$; mol wt 312.32. C 57.69%, H 6.45%, O 35.86%. *Trichothecene mycotoxin* isolated from *Fusarium nivale:* T. Tatsuno *et al., Chem. Pharm. Bull.* **16**, 2519 (1968). Structure: *eidem, Tetrahedron Letters* **1969**, 2823. Toxicology: T. Tatsuno, *Cancer Res.* **28**, 2393 (1968). Implicated as a chemical warfare agent in Southeast Asia with T-2 toxin, *q.v.:* N. Wade, *Science* **214**, 34 (1981); R. T. Rosen, J. D. Rosen, *Biomed. Mass Spectrom.* **9**, 443 (1982).

Crystals, mp 222-223° (dec). $[\alpha]_D^{24}$ +21.54° (c = 1.3 in ethanol). uv max (methanol): 218 nm (ε 6300). Slightly sol in water; sol in polar organic solvents. LD_{50} i.p. in mice: 40 μg/10 g (Tatsuno).

Caution: Potential symptoms of overexposure include fever, nausea, vomiting, diarrhea, leukopenia, bleeding, sepsis; necrotic lesions of skin and mucosa. *See* M. J. Ellenhorn, D. G. Barceloux in *Medical Toxicology: Diagnosis and Treatment of Human Poisoning* (Elsevier, New York, 1988) pp. 1312-1314.

6758. Nizatidine. *N-[2-[[[2-[(Dimethylamino)methyl]-4-thiazolyl]methyl]thio]ethyl]-N'-methyl-2-nitro-1,1-ethenediamine; N-[4-(6-methylamino-7-nitro-2-thia-5-aza-6-heptene-1-yl)-2-thiazolylmethyl]-N,N-dimethylamine;* LY-139037; ZE-101; ZL-101; Axid; Calmaxid; Cronizat; Distaxid; Gastrax; Naxidine; Nizax; Nizaxid; Zanizal. $C_{12}H_{21}N_5O_2S_2$; mol wt 331.46. C 43.48%, H 6.39%, N 21.13%, O 9.65%, S 19.35%. Histamine H_2-receptor antagonist related to ranitidine, *q.v.* Prepn: R. P. Pioch, **Eur. pat. Appl. 49,618;** *idem,* U.S. pat. **4,375,547** (1982, 1983 both to Eli Lilly). General pharmacology in animals: K. Bemis *et al., Arzneimittel-Forsch.* **39**, 240 (1989). Pharmacokinetics and gastric acid suppression in humans: J. T. Callaghan *et al., Clin. Pharmacol. Ther.* **37**, 162 (1985). Study of effect on hepatic drug metabolism in humans: J. W. Secor *et al., Brit. J. Clin. Pharmacol.* **20**, 710 (1985). Disposition and metabolism in humans: M. P. Knadler *et al., Drug. Metab. Dispos.* **14**, 175 (1986). Symposium on pharmacology and clinical studies: *Scand. J. Gastroenterol.* **22**, Suppl. 136, 1-88 (1987). Comprehensive description: T. J. Wozniak in *Analytical Profiles of Drug Substances* vol. **19**, K. Florey, Ed. (Academic Press, New York, 1990) pp 397-427.

Crystals from ethanol-ethyl acetate, mp 130-132°. uv max (methanol): 240, 325 nm (ε 8400, 19600); (water): 260, 314 nm (ε 11820, 15790). pKa_1 2.1; pKa_2 6.8. Partition coefficient (octanol/water): 0.3 (pH 7.4). Freely sol in chloroform; sol in methanol; sparingly sol in water and buffered solns; slightly sol in ethyl acetate, isopropanol. Essentially

insol in benzene, diethyl ether, octanol. Soly (mg/ml): chloroform > 100.0; methanol between 50.0-100.0; water between 10.0-33.3; isopropanol between 3.33-5.0; ethyl acetate between 1.0-2.0; benzene, diethyl ether, octanol <0.5. LD_{50} in mice, rats (mg/kg): 265, > 300 i.v.; 1685, 1680 orally (Pioch).

THERAP CAT: Antiulcerative.

6759. Nizofenone. *(2-Chlorophenyl)[2-[2-[(diethylamino)methyl]-1H-imidazol-1-yl]-5-nitrophenyl]methanone; 2'-chloro-2-[2-[(diethylamino)methyl]imidazol-1-yl]-5-nitrobenzophenone; 1-[2-(2-chlorobenzoyl)-4-nitrophenyl]-2-(diethylaminomethyl)imidazole.* $C_{21}H_{21}ClN_4O_3$; mol wt 412.88. C 61.09%, H 5.13%, Cl 8.59%, N 13.57%, O 11.63%. Imidazole derivative exhibiting protective activity against cerebral anoxia or ischemia. Prepn: M. Nakanishi *et al.,* **Ger. pat. 2,403,416;** *eidem,* U.S. pat. **3,915,981** (1974, 1975 both to Yoshitomi). Antianoxic effect in animal models: H. Yasuda *et al., Arch. Int. Pharmacodyn.* **233**, 136 (1978). Mechanism of action: *eidem, ibid.* **242**, 77 (1979). Multicenter clinical studies: I. Saito *et al., Neurol. Res.* **5**, 29 (1983); T. Ohta *et al., J. Neurosurg.* **64**, 420 (1986). Toxicity studies: H. Horizoe *et al., Oyo Yakuri* **30**, 627 (1985), *C.A.* **104**, 61754m (1986); K. Okumura *et al., ibid.* 633, *C.A.* **104**, 61755 (1986).

Pale yellow crystals from isopropyl ether, mp 75-76°.

Fumarate, $C_{21}H_{21}ClN_4O_3 \cdot C_4H_4O_4$, *midafenone, Y-9179, Ekonal.* Pale yellow crystals from isopropyl ether, mp 157-158°. LD_{50} in male, female mice, male, female rats (mg/kg): 495, 504, 1711, 1580 orally; 62, 70, 63, 65 i.v.; 270, 278, 1830, 1629 s.c. (Horizoe).

THERAP CAT: Nootropic.

6760. NMDA. *N-Methyl-D-aspartic acid;* N-methyl-D-aspartate; N-methyl-D-asparaginsaure (German). $C_5H_9NO_4$; mol wt 147.13. C 40.82%, H 6.17%, N 9.52%, O 43.50%. Excitotoxic amino acid used to identify a specific subset of excitatory amino acid receptors; consequently the receptors are known as NMDA receptors. *See also* quisqualic acid, kainic acid. Prepn: O. Lutz, *Z. Physiol. Chem.* **70**, 256 (1909); O. Lutz, Br. Jirgensons, *Ber.* **63**, 448 (1930). Isoln from the blood shell, *Scapharca broughtonii:* M. Sato *et al., Biochem. J.* **241**, 309 (1987). Identification as a neuroexcitant: D. R. Curtis, J. C. Watkins, *J. Physiol.* **166**, 1 (1963); of neurotoxic properties: J. W. Olney *et al., Life Sci.* **25**, 537 (1979); R. Zaczek *et al., Neurosci. Letters* **24**, 181 (1981); M. V. Sofroniew, R. C. A. Pearson, *Brain Res.* **339**, 186 (1985). Receptor binding studies: C. K. Mitchell, D. A. Redburn, *Neurosci. Letters* **28**, 241 (1982); D. T. Monaghan, C. W. Cotman, *J. Neurosci.* **5**, 2909 (1985); J. M. M. Olson *et al., Neuroscience* **22**, 913 (1987). Review of NMDA and its receptor: J. C. Watkin, R. H. Evans, *Ann. Rev. Pharmacol. Toxicol.* **21**, 165-204 (1981); T. W. Stone, N. R. Burton, *Prog. Neurobiol.* **30**, 333-368 (1988).

6761. NMN. *3-(Aminocarbonyl)-1-(5-O-phosphono-β-D-ribofuranosyl)pyridinium inner salt; 3-carbamoyl-1-β-D-ribofuranosylpyridinium hydroxide 5'-(dihydrogen phosphate) inner salt;* nicotinamide mononucleotide. $C_{11}H_{15}N_2O_8P$; mol wt 334.22. C 39.53%, H 4.52%, N 8.38%, O

38.30%, P 9.27%. Prepd by incubation of NAD with potato pyrophosphatase in a non-phosphate buffer and in the presence of fluoride: Plaut, Plaut, *Arch. Biochem. Biophys.* **48**, 189 (1954); *Biochem. Prepn.* **5**, 55 (1957). Enzymatic synthesis from nicotinamide by human erythrocytes and hemolyzates: Preiss, Handler, *J. Biol. Chem.* **225**, 759 (1957). Prepn by hydrolysis of NAD by a yeast enzyme: Takei, *Agr. Biol. Chem.* **34**, 23 (1970); by cleavage of NAD with crude rattlesnake venom: Apps, *FEBS Letters* **15**, 277 (1971); R. Jeck, C. Wönckhaus, *Methods in Enzymol.* **66**, 62 (1980). NMR studies and conformation of α- and β-nicotinamide nucleotide: N. J. Oppenheimer, N. O. Kaplan, *Biochemistry* **15**, 3981 (1976).

Can be precipitated from slightly acidic aq soln by a large vol of acetone. Freely sol in water, practically insol in acetone. More stable when stored frozen in soln, than when in dry form.

6762. Nobelium. No; at. no. 102; valence 2, 3. Manmade radioactive element. No stable nuclides; known isotopes (mass numbers): 250-259. Longest-lived known isotope: 259 (T$_{1/2}$ 1.00 hours, α-emitter, rel. at. mass 259.1009). Discovery of element 102 first claimed by P. R. Fields *et al., Phys. Rev.* **107**, 1460 (1957); later experiments failed to confirm these results. Prepn of isotope originally assigned mass 254, later identified as ^{252}No (T$_{1/2}$ 2.3 sec, α-emitter) by ^{244}Cm(^{12}C,4n) reaction: A. Ghiorso *et al., Phys. Rev. Letters* **1**, 18 (1958). Prepn of ^{259}No by bombardment of ^{248}Cm with ^{18}O ions: Silva *et al., Nucl. Phys. A* **216**, 97 (1973), *C.A.* **80**, 43040g (1974). *Reviews:* G. T. Seaborg, *J. Chem. Ed.* **36**, 38-44 (1959); A. Ghiorso, T. Sikkeland, *Phys. Today* **20**(9), 25-32 (1967); G. N. Flerov, *U.S. At. Energy Comm.* **JINR-D7-3444** (1967) 36 pp, *C.A.* **68**, 34658q (1968); C. Keller, *The Chemistry of the Transuranium Elements* (Verlag Chemie, Weinheim, English Ed., 1971) pp 601-607; Silva, "Trans-Curium Elements" in *MTP Int. Rev. Sci.: Inorg. Chem., Ser. One* vol. **8**, A. G. Maddock, Ed. (University Park Press, Baltimore, 1972) pp 71-105; *Comprehensive Inorganic Chemistry* vol. **5**, J. C. Bailar, Jr., *et al.,* Eds. (Pergamon Press, Oxford, 1973) *passim;* Ghiorso, *Handb. Exp. Pharmakol.* **36**, 691-715 (1973); Taylor, *ibid.* 717-738 (1973); R. J. Silva in *The Chemistry of the Actinide Elements* vol. **2**, J. J. Katz *et al.,* Eds. (Chapman and Hall, New York, 1986) pp 1096-1099; *The Elements Beyond Uranium,* G. T. Seaborg, W. D. Loveland, Eds. (John Wiley & Sons, Inc., New York, 1990) pp 46-49.

Caution: Radiation hazard; handling requires special equipment and sheilding facilities (Katz *et al., loc cit.,* vol. **2**, p. 1128).

6763. Nocardamin. *1,12,23-Trihydroxy-1,6,12,17,23,28-hexaazacyclotritriacontane-2,5,13,16,24,27-hexone.* C$_{27}$H$_{48}$-N$_6$O$_9$; mol wt 600.71. C 53.99%, H 8.05%, N 13.99%, O 23.97%. Antibiotic substance produced by a *Nocardia* isolated from old bee-honeycombs: Stoll *et al., Schweiz. Z. Pathol. u. Bakteriol.* **14**, 225 (1951); *Helv. Chim. Acta* **34**,

862 (1951). Structure: Keller-Schierlein, Prelog, *ibid.* **44**, 1981 (1961).

Needles from hot methanol, mp 192-195°. Optically inactive. Has no basic properties. Soluble in hot water, methanol. An 0.5% aq soln at room temp has a pH of 4.55.

Triacetate, C$_{33}$H$_{54}$N$_6$O$_{12}$, crystals, mp 115-117°.

6764. Nocardicin(s). Monocyclic β-lactam antibiotics with antimicrobial activity which inhibit bacterial cell wall biosynthesis. Nocardicins A, B, C, D, E, F, G have been identified. All are produced by *Nocardia uniformis* subsp. *tsuyamenensis,* A being the most important component. Prodn: H. Aoki *et al.,* Ger. pat. **2,242,699** (1973 to Fujisawa), *C.A.* **78**, 134496k (1973); U.S. pat. **3,923,977** (1975 to Fujisawa). Isoln and characterization of A: H. Aoki *et al., J. Antibiot.* **29**, 492 (1976); of B: M. Kurita *et al., ibid.* 1243. Nocardicin A has also been produced by *Actinosynnema mirum:* K. Watanabe *et al., ibid.* **36**, 321 (1983). Structural determination of A and B: M. Hashimoto *et al., J. Am. Chem. Soc.* **98**, 3023 (1976), *J. Antibiot.* **29**, 890 (1976). Isoln, characterization, and structures of C, D, E, F, G: J. Hosoda *et al., Agric. Biol. Chem.* **41**, 2013 (1977). Total synthesis of A: G. A. Koppel *et al., J. Am. Chem. Soc.* **100**, 3933 (1978); W. V. Curran *et al., J. Antibiot.* **35**, 329 (1981); of A and D: T. Kamiya *et al., Tetrahedron* **1979**, 323; of A and B: H. P. Isenring, W. Hofheinz, *Tetrahedron* **1983**, 2591. Biosynthetic studies of A: J. Hosada *et al., Agr. Biol. Chem.* **41**, 2007 (1977); C. A. Townsend *et al., J. Am. Chem. Soc.* **105**, 919 (1983). Series of articles on antimicrobial activity of A: *J. Antibiot.* **30**, 917-944 (1977).

Nocardicin A

Nocardicin A, C$_{23}$H$_{24}$N$_4$O$_9$, *[3S-[1(S*),3R*[Z(S*)]]]-3-[[[4-(3-amino-3-carboxypropoxy)phenyl](hydroxyimino)acetyl]-amino]-α-(4-hydroxyphenyl)-2-oxo-1-azetidineacetic acid.* Colorless needles from acidic water, mp 214-216° (dec). [α]$_D^{25}$ −135° (for the sodium salt). uv max (1/15M phosphate buffer): 272 nm (E$_{1cm}^{1\%}$ 310); (0.1N NaOH): 244, 283 nm (E$_{1cm}^{1\%}$ 460, 270). Sol in alkaline solns, slightly sol in methanol. Insol in chloroform, ethyl acetate, ethyl ether. LD$_{50}$ in male mice, rats (mg/kg): > 8000 orally; > 2000 i.v.; 2500, 2600 i.p. (Aoki).

Nocardicin B, the *E*-isomer of nocardicin A. Colorless needles, mp 262-264° (dec). [α]$_D^{25}$ −162° (for the sodium salt). uv max (ethanol/water): 224, 270 nm (ε 24600, 9700); (ethanol/0.1N NaOH): 245, 280 nm (ε 26000, 11100).

6765. Nodakenin. *2-[1-(β-D-Glucopyranosyloxy)-1-methylethyl]-2,3-dihydro-7H-furo[3,2-g][1]benzopyran-7-one;* nodakenetin glucoside. C$_{20}$H$_{24}$O$_9$; mol wt 408.41. C 58.82%, H 5.92%, O 35.26%. From root of *Peucedanum decursivum* Maxim., *Umbelliferae:* Arima, *Bull. Chem. Soc. Japan* **4**, 16, 113 (1929); Späth, Kainrath, *Ber.* **69**, 2062 (1936). Structure: Späth, Tyray, *ibid.* **72**, 2089 (1939).

Thin leaflets from alc, dec 218-219°. From dil alc or water it crystallizes with one mol H$_2$O in form of yellow prisms, melting at 216°. [α]$_D^{30}$ +56.6°. Sol in hot water or alcohol.

Tetraacetate, $C_{28}H_{32}O_{13}$, crystals from methanol, mp 195-196°.

6766. Noformicin. *5-Amino-N-(3-amino-3-iminopropyl)-3,4-dihydro-2H-pyrrole-2-carboxamide; N-(2-amidinoethyl)-5-imino-2-pyrrolidinecarboxamide; 2-[N-(2-amidinoethyl)carbamoyl]-5-iminopyrrolidine; β-(5-imino-2-pyrrolidinecarboxamido)propamidine; noformycin.* $C_8H_{15}N_5O$; mol wt 197.24. C 48.72%, H 7.67%, N 35.51%, O 8.11%. Antiviral agent produced by *Nocardia formica*, culture no. N.R.R.L. 2470: Peck *et al.*, U.S. pat. **2,804,463** (1957 to Merck & Co.); Gray, *Phytopathology* **45**, 281 (1955). Activity studies: Furusawa *et al.*, *Proc. Soc. Exp. Biol. Med.* **116**, 938 (1964); *eidem, Med. Pharmacol. Exp.* **12**, 259 (1965). Synthesis of (+)- and (±)-forms: Diana, *J. Med. Chem.* **16**, 857 (1973).

(+)-Form dihydrochloride, $C_8H_{17}Cl_2N_5O$. The biologically active form. Crystals from methanol, dec 263-264°. $[\alpha]_D^{25}$ +8.8° (methanol). Sol in water. Somewhat sol in organic solvents. pKa 9.4.

(±)-Form dihydrochloride, mp 252-254°.

6767. Nogalamycin. *[2R-(2α,3β,4α,5β,6α,11β,13α,-14α)]-11-[(6-Deoxy-3-C-methyl-2,3,4-tri-O-methyl α-L-mannopyranosyl)oxy]-4-(dimethylamino)-3,4,5,6,9,11,12,13,14,16-decahydro-3,5,8,10,13-pentahydroxy-6,13-dimethyl-9,16-dioxo-2,6-epoxy-2H-naphthaceno[1,2-b]oxocin-14-carboxylic acid methyl ester;* U-15167; NSC-70845. $C_{39}H_{49}NO_{16}$; mol wt 787.82. C 59.46%, H 6.27%, N 1.78%, O 32.49%. Antitumor antibiotic isolated from *Streptomyces nogalater* var. *nogalater:* B. K. Bhuyan, A. Dietz, *Antimicrob. Ag. Chemother.* **1965**, 836; B. K. Bhuyan *et al.*, U.S. pat. **3,183,157** (1965 to Upjohn). Characterization: P. F. Wiley *et al.*, *Tetrahedron Letters* **1968**, 663. Structure: *eidem, J. Am. Chem. Soc.* **99**, 542 (1977). Stereochemistry: *eidem, J. Org. Chem.* **44**, 4030 (1979). Biosynthesis: *eidem, ibid.* **43**, 3457 (1978). Synthesis of L-nogalose, sugar component: L. Valente *et al.*, *Tetrahedron Letters* **1979**, 1153; J. Yoshimura *et al.*, *Tetrahedron Letters* **1979**, 687. Partial stereospecific synthesis: R. P. Joyce *et al.*, *Tetrahedron Letters* **27**, 4885 (1986). Molecular structure, absolute stereochemistry and interactions with DNA: S. K. Arora, *J. Am. Chem. Soc.* **105**, 1328 (1983). Toxicity data: Z. Hadidian, *PB Report* 173334 (1966). Review of pharmacology: B. K. Bhuyan, C. G. Smith, *Handb. Exp. Pharmacol.* **38**(pt. 2), 623-632 (1975).

Orange-red solid from methanol, mp 195-196° (dec). $[\alpha]_D^{25}$ +425° (c = 0.11 in $CHCl_3$). pKa' 7.45 (60% ethanol). uv max (ethanol): 236, 258, 292 nm (ε 52360, 24755, 9890). Sol in methylene chloride, acetone, chloroform, ethyl acetate. Insol in water, methanol, ethanol. LD_{50} in mice (mg/kg): 11.75 i.v., 4.79 i.p. (Hadidian).

THERAP CAT: Antineoplastic.

6768. Nomifensine. *1,2,3,4-Tetrahydro-2-methyl-4-phenyl-8-isoquinolinamine; 8-amino-1,2,3,4-tetrahydro-2-methyl-4-phenylisoquinoline.* $C_{16}H_{18}N_2$; mol wt 238.33. C 80.63%, H 7.61%, N 11.75%. Novel antidepressant distin-

guished from existing tricyclic and tetracyclic antidepressants by its bicyclic structure. Prepn: **Brit. pat. 1,164,192** corresp to G. Ehrhart *et al.*, U.S. pat. **3,577,424** (1969, 1971, both to Farbwerke Hoechst); I. Hoffmann *et al.*, *Arzneimittel-Forsch.* **21**, 1045 (1971). Pharmacology: I. Hoffmann, *ibid.* **23**, 45 (1973); P. Hunt *et al.*, *J. Pharm. Pharmacol.* **26**, 370 (1974); B. Costall *et al.*, *Psychopharmacologia* **41**, 153 (1975). Pharmacokinetics: L. Vereczkey *et al.*, *ibid.* **45**, 225 (1975). Study on abuse potential in humans: C. Spyraki, H. C. Fibiger, *Science* **212**, 1167 (1981). *Review: Brit. J. Clin. Pharmacol.* **4**, Suppl. 2 (1977) pp 1S-248S; *Int. Pharmacopsychiatry* **17**, Suppl. 1 (1982) pp 1-148.

mp 179-181°.

Maleate, $C_{16}H_{18}N_2 \cdot C_4H_4O_4$, HOE-984, Alival, Hostalival, Merital, Neurolene, Psicronizer. Crystals from ethanol, mp 199-201°. LD_{50} in mice, rats (mg/kg): 400, 430 orally; 90, 72 i.v.; in mice (mg/kg): 410 s.c. (Hoffman, 1973).

THERAP CAT: Antidepressant.

6769. Nomilin. *1-(Acetyloxy)-1,2-dihydroobacunoic acid ε-lactone; 1-(3-furyl)decahydro-11-hydroxy-4b,7,7-11a,-13a-pentamethyloxireno[4,4a]-2-benzopyrano[6,5-g]2]benzoxepin-3,5,9(3aH,4bH,6H)-trione acetate.* $C_{28}H_{34}O_9$; mol wt 514.57. C 65.36%, H 6.66%, O 27.98%. Bitter principle from citrus juice, seeds. Isoln: O. H. Emerson, *J. Am. Chem. Soc.* **70**, 545 (1948). Structure elucidation: D. L. Dreyer, *Tetrahedron* **21**, 75 (1965); *idem, ibid.* **24**, 3273 (1968). Distribution in citrus seeds: S. Hasegawa *et al.*, *J. Agr. Food Chem.* **28**, 922 (1980); R. L. Rouseff, S. Nagy, *Phytochemistry* **21**, 85 (1982). Identification in grapefruit juice: R. L. Rouseff, *J. Agr. Food Chem.* **30**, 504 (1982). Reduction of nomilin bitterness by *Arthrobacter globiformis* cells: S. Hasegawa, V. A. Pelton, *ibid.* **31**, 178 (1983); by *Cornebacterium fascians:* S. Hasegawa *et al.*, *ibid.* **32**, 457 (1984); *eidem, J. Food Sci.* **50**, 330 (1985). Antifeedant effect against *Spodoptera frugiperda:* M. A. Altieri *et al.*, *Protect. Ecol.* **6**, 91 (1984). Biosynthetic pathways: S. Hasegawa *et al.*, *Phytochemistry* **23**, 1601 (1984).

Needles from methanol, mp 278-279°. Slightly sol in 2-propanol, ethyl acetate. $[\alpha]_D^{23}$ −95.7°.

6770. Nonactin. *[1R-(1R*,2R*,5R*,7R*,10S*,11S*,-14S*,16S*,19R*,20R*,23R*,25R*,28S*,29S*,32S*,34S*)]-2,-5,11,14,20,23,29,32-Octamethyl-4,13,22,31,37,38,39,40-octaoxapentacyclo[32.2.1.1^{7,10}.1^{16,19}.1^{25,28}]tetracontane-3,12,21,30-tetrone.* $C_{40}H_{64}O_{12}$; mol wt 736.94. C 65.19%, H 8.75%, O 26.05%. Macrotetrolide antibiotic. Produced by several *Streptomyces* spp.: Corbaz *et al.*, *Helv. Chim. Acta* **38**, 1445 (1955); Wallhäusser *et al.*, *Arzneimittel-Forsch.* **14**, 356 (1964). Structure: Dominguez *et al.*, *Helv. Chim. Acta* **45**, 129 (1962). Crystal structure: Dobler, *ibid.* **55**, 1371 (1972). Steroselective syntheses of *nonactic acid*, building block of nonactin: Gerlach, Wetter, *ibid.* **57**, 2306 (1974);

R. E. Ireland, J.-P. Vevert, *J. Org. Chem.* **45**, 4260 (1980). Total synthesis of (+)- and (−)-nonactic acids: *eidem, Can. J. Chem.* **59**, 572 (1981). Synthesis of nonactin: Gombos *et al., Monatsh.* **106**, 1043 (1975); *eidem, Tetrahedron Letters* **1975**, 3391; Gerlach *et al., Helv. Chim. Acta* **58**, 2036 (1975); U. Schmidt *et al., Ber.* **109**, 2628 (1976). Three homologs, *monactin,* $C_{41}H_{66}O_{12}$, *dinactin,* $C_{42}H_{68}O_{12}$, and *trinactin,* $C_{43}H_{70}O_{12}$ are known: Beck *et al., Helv. Chim. Acta* **45**, 620 (1962); Gerlach, Prelog, *Ann.* **669**, 121 (1963). Activity of nonactin and its homologues: Meyers *et al., J. Antibiot.* **18A**, 128 (1965).

Needles from methanol, mp 147-148°. $[\alpha]_D^{20}$ 0° ±2° (c = 1.2 in CHCl$_3$). uv max (ethanol): slight peak at 264 nm (log ε 1.5). Remarkably inert to chemical compds.

6771. 2-Nonenal. *trans*-2-Nonenaldehyde. $C_9H_{16}O$; mol wt 140.23. C 77.09%, H 11.50%, O 11.41%. Widely distributed in nature; found in beer, coffee, cucumbers, watermelon, palm oil, potatoes, carrots etc. Prepn: R. Delaby, S. Guillot-Allègre, *Bull. Soc. Chim. France* **53**, 301 (1933); J. Braun *et al., Ber.* **67**, 269 (1934); J. Ficini, H. Normant, *Bull. Soc. Chim. France* **1964**, 1294; J. Ficini *et al., Tetrahedron Letters* **1977**, 3589; M. Wado *et al., Chem. Letters* **1977**, 345. Insecticidal activity: P. M. Guerin, M. F. Ryan, *Experientia* **36**, 1387 (1980). Cockroach repellant activity: T. H. Maugh, *Science* **218**, 278 (1982).

Liquid, bp$_{16}$ 100-102°. n_D^{18} 1.4403.
USE: As flavoring agent.

6772. Nonoxynol. α-(4-Nonylphenyl)-ω-hydroxypoly-(oxy-1,2-ethanediyl); polyethyleneglycols mono(nonylphenyl) ether; macrogol nonylphenyl ether; nonoxinol; polyoxyethylene(n)nonylphenyl ether; nonylphenyl polyethyleneglycol ether; nonylphenoxypolyethoxyethanol. Nonionic surfactant mixtures prepd by reacting nonylphenol with ethylene oxide. Prepn of polyoxyethylated alkyl phenols: A. Steindorff *et al.,* **U.S. pat. 2,213,477** (1940 to GAF). Average number of ethylene oxide units (*n*) per molecule is indicated by number following nonoxynol (e.g. nonoxynol-15 for n = 15). Trademarks for nonoxynol series include *Conco NI, Dowfax 9N, Igepal CO, Makon, Neutronyx 600's, Nonipol NO, Polytergent B, Renex 600's, Solar NP, Sterox, Surfonic N, T-DET-N, Tergitol NP, Triton N.* Individual members of series indicated by numerical suffixes. Toxicology study: H. F. Smyth Jr., J. C. Calandra, *Toxicol. Appl. Pharmacol.* **14**, 315 (1969). Inactivation of HIV *in vitro* by nonoxynol-9: M. Malkovsky *et al., Lancet* **1**, 645 (1988). Clinical evaluation of nonoxynol-9 in sexually transmitted diseases: W. C. Louv *et al., J. Infect. Dis.* **158**, 518 (1988). *Review:* C. R. Enyeart, "Polyoxyethylene Alkylphenols" in *Nonionic Surfactants,* M. J. Schick, Ed. (Dekker, New York, 1967) pp 44-85.

Nonoxynol-9

Highly stable compounds. Lower adducts (n < 15) are yellow to almost colorless liquids; higher adducts (n > 20) are pale yellow to off-white pastes or waxes. Lower adducts (n < 6) sol in oil; higher ones sol in water. Review of physical properties: Enyeart, *loc. cit.*

Nonoxynol-9, C-Film, Conco NI-90, Dowfax 9N9, Encare, Gynol II, Igepal CO-630, Intercept, Neutronyx 611, Semicid, Staycept, Tergitol TP-9. av. mol wt 617. A component of *Conceptrol, Delfen, Gentersal, Ortho-creme.* Almost colorless liquid. For Igepal CO-630: d_4^{25} 1.06; solidif pt 26 ±2°F; pour point 37±2°F; flash point: 535-555°F; cloud point (1% aq soln) 126-133°F; viscosity (25°) 175-250 cp. Sol in water, ethanol, ethylene glycol, ethylene dichloride, xylene, corn oil. Insoluble in Stoddard solvent, deodorized kerosene, low viscosity white mineral oil.

Nonoxynol-11, Duragel, Duracreme.

USE: Nonionic surfactants used as detergents, emulsifiers, wetting agents, dispersants, stabilizers, intermediates in synthesis of anionic surfactants, defoaming agents. Nonoxynols-9, 11 as spermatocide. Nonoxynols-4, 15, 30 as pharmaceutic aids (surfactants).

6773. *n*-Nonyl Acetate. *Acetic acid n-nonyl ester; n*-nonyl ethanoate; nonanol acetate; Acetate C-9. $C_{11}H_{22}O_2$; mol wt 186.29. C 70.92%, H 11.90%, O 17.18%. CH_3COO-$(CH_2)_8CH_3$.

Liquid. Pungent odor, suggestive of mushrooms, resembles odor of gardenias when dil. d_4^{15} 0.8785 (commercial grade, d_{25}^{25} 0.864-0.868). bp 208-212°. n_D^{20} 1.4328 (commercial grade, n_D^{20} 1.422-1.426). Insoluble in water. Freely sol in abs alcohol and ether; 33 ml dissolve in 100 ml of 80% ethanol.

Note: The nonyl acetate of commerce may be the above compound or any of its isomers such as diisobutylcarbinyl acetate.

USE: In perfumery.

6774. *n*-Nonyl Alcohol. *1-Nonanol;* nonalol. $C_9H_{20}O$; mol wt 144.26. C 74.93%, H 13.97%, O 11.09%. CH_3-$(CH_2)_7CH_2OH$. Occurs in oil of orange. Prepn by reduction of ethyl Δ^1-nonylenate: Harding, Weizmann, *J. Chem. Soc.* **97**, 304 (1910); by reduction of pelargonic aldehyde: Tomecko, Adams, *J. Am. Chem. Soc.* **49**, 529 (1927); by oxo process: Russum, Hengstebeck, **U.S. pat. 2,638,487** (1953 to Standard Oil of Indiana); by reductive cleavage of oleic acid ozonide: Sousa, Bluhm, *J. Org. Chem.* **25**, 108 (1960); from diborane and propylene trimer: Marshall, Smith, **Brit. pat. 879,242** (1961 to I.C.I.).

Colorless to yellowish liquid; odor of citronella oil. bp$_{760}$ 215°; bp$_{15}$ 107.5°; bp$_{7.5}$ 95.6°. d_4^{20} 0.8279. n_D^{20} 1.4338. Practically insol in water. Miscible with alcohol, ether.

USE: In manufacture of artificial lemon oil.

6775. Nonyl Phenol. Approx mol wt 215. A technical grade mixture of monoalkyl phenols, predominantly *para* substituted. The side chains are isomeric branched-alkyl radicals. Manuf of *p*-nonylphenol: Faith, Keyes & Clark's *Industrial Chemicals,* F. A. Lowenheim, M. K. Moran, Eds. (Wiley-Interscience, New York, 4th ed., 1975) pp 575-578.

Pale yellow liquid. Slight characteristic phenolic odor. Comparatively high viscosity. Hydroxyl no. 253. d_4^{20} 0.950. bp 293-297°. n_D^{20} 1.513. Flash pt (open cup) 300°F. Practically insol in water or dil aq NaOH. Sol in benzene, chlorinated solvents, aniline, heptane, aliphatic alcohols, ethylene glycol.

USE: In the prepn of lubricating oil additives, resins, plasticizers, surface active agents.

6776. (*p*-Nonylphenoxy)acetic Acid. $C_{17}H_{26}O_3$; mol wt 278.39. C 73.35%, H 9.41%, O 17.24%. Prepd by the reaction of sodium chloroacetate with nonyl phenol in alkaline medium: Preston, **Brit. pat. 812,938** and Preston, Taylor, **Brit. pat. 831,883** (1959, 1960, to I.C.I.).

Liquid. d_4^{20} 1.010-1.025. Viscosity at 30° = 5200 cp. Flow pt 5°. Practically insol in water. Miscible with benzene, mineral oil, kerosene, petr ether.

USE: Corrosion inhibitor and antifoaming agent in gasoline and cutting oils.

6777. Noprylsulfamide. *1-[[4-(Aminosulfonyl)phenyl]amino]-3-phenyl-1,3-propanedisulfonic acid disodium salt; 1-phenyl-3-p-sulfamoylanilino-1,3-propanedisulfonic acid disodium salt; N⁴-(disodium 1,3-disulfo-3-phenylpropyl)sulfanilamide; disodium 1-phenyl-3-p-sulfamoylanilino-1,3-propanedisulfonate; disodium p-(γ-phenylpropylamino)benzenesulfonamide-α,γ-disulfonate; RP-40; Solucin; Soluseptasine; Soluseptazine; Solusetazine; Sulphasolucin; Sulphasolutin.* $C_{15}H_{16}N_2Na_2O_8S_3$; mol wt 494.48. C 36.44%, H 3.26%, N 5.67%, Na 9.30%, O 25.88%, S 19.45%. Prepd by the action of sodium bisulfite on N⁴-cinnamylidenesulfanilamide (prepd from cinnamic aldehyde and sulfanilamide): Despois, U.S. pat. 2,262,544 (1941 to Rhône-Poulenc).

Crystals. One gram dissolves in 5 ml water. Neutral reaction. Breaks down in the body with the liberation of free sulfanilamide.

THERAP CAT: Antibacterial.

6778. Noracymethadol. *α-Ethyl-β-[2-(methylamino)propyl]-β-phenylbenzeneethanol acetate; α-dl-6-(methylamino) 4,4-diphenyl 3-heptanol acetate; α-dl-3-acetoxy-4,4-diphenyl-6-methylaminoheptane; α-dl-3-acetoxy-6-methylamino-4,4-diphenylheptane; α-dl-4,4-diphenyl-6-methylamino-3-heptanol acetate.* $C_{22}H_{29}NO_2$; mol wt 339.48. C 77.84%, H 8.61%, N 4.13%, O 9.43%. Metabolite of methadyl acetate, q.v. Prepn: A. Pohland, U.S. pat. 3,021,360 (1962 to Lilly). Metabolism: G. L. Henderson et al., *Drug Metab. Dispos.* 5, 321 (1977); M. Man et al., ibid. 8, 55 (1980). HPLC determn: C.-H. Kiang et al., *J. Chromatog.* 222, 81 (1981).

Hydrochloride, $C_{22}H_{29}NO_2 \cdot HCl$, *NIH-7667.* Crystals from acetone + ether, mp ~216-217°.

Note: This is a controlled substance (opiate) listed in the U.S. Code of Federal Regulations, Title 21 Part 1308.11 (1995).

6779. Norbolethone. *(17α)-(±)-13-Ethyl-17-hydroxy-18,19-dinor-pregn-4-en-3-one; dl-13β,17α-diethyl-17β-hydroxygon-4-en-3-one; Wy-3475; Genabol.* $C_{21}H_{32}O_2$; mol wt 316.48. C 79.70%, H 10.19%, O 10.11%. Prepn of dl-, d-, and l-forms: Smith et al., *J. Chem. Soc.* 1964, 4472. Biological activities of dl-, d-, and l-forms: Edgren et al., *Steroids* 2, 731 (1963). Toxicity study: E. I Goldenthal, *Toxicol. Appl. Pharmacol.* 18, 185 (1971).

Crystals from alcohol, mp 144-145°. uv max: 241 nm (ε 16500). LD_{50} orally in mice: > 5010 mg/kg (Goldenthal).

d-Form, crystals from acetone + hexane, mp 175-176°. $[\alpha]_D$ +20.7° (in chloroform).

l-Form, crystals from acetone + hexane, mp 172-175.5°. $[\alpha]_D$ -18.1° (in chloroform).

THERAP CAT: Anabolic.

6780. Norbormide. *3a,4,7,7a-Tetrahydro-5-(hydroxyphenyl-2-pyridinylmethyl)-8-(phenyl-2-pyridinylmethylene)-4,7-methano-1H-isoindole-1,3(2H)-dione; 5-(α-hydroxy-α-2-pyridylbenzyl)-7-(α-2-pyridylbenzylidene)-5-norbornene-2,3-dicarboximide;* McN-1025; Raticate; Shoxin. $C_{33}H_{25}N_3O_3$; mol wt 511.58. C 77.48%, H 4.93%, N 8.21%, O 9.38%. Prepd by condensation of 2-benzoylpyridine and cyclopentadiene followed by reaction of the obtained fulvene with maleimide: Roszkowski et al., *Science* 144, 412 (1964). Stereoisomerism: Mohrbacher et al., *J. Org. Chem.* 31, 2141 (1966); Abrahamsson, Nilsson, ibid. 3631. Structure-activity studies: Poos et al., *J. Med. Chem.* 9, 537 (1966). Toxicity data: Roszkowski, *J. Pharmacol. Exp. Ther.* 149, 288 (1965).

Crystals from methylene chloride + ether, mp 190-198°. uv max (methanol): 250 nm (ε 17500). LD_{50} in rats (mg/kg): 5.3 orally, 0.65 i.v. (Roszkowski). Practically insol in water unless pH is below 4.

USE: Rodenticide.

6781. Norcarane. *Bicyclo[4.1.0]heptane;* 1,2-methylenecyclohexane. C_7H_{12}; mol wt 96.17. C 87.42%, H 12.58%. Prepd by distilling the barium salt of norcaranecarboxylic acid with ZnO + BaO: Ebel et al., *Helv. Chim. Acta* 12, 19 (1929); by the action of methylene iodide and zinc-copper couple on cyclohexene: Smith, Simmons, *Org. Syn.* 41, 72 (1961); Simmons, U.S. pat. 3,074,984 (1963 to du Pont); from cyclohexene + diazomethane + zinc iodide: Applequist, Badad, *J. Org. Chem.* 27, 288 (1962); from methylene chloride + cyclohexyl lithium: Closs, *J. Am. Chem. Soc.* 84, 809 (1962).

Liquid. bp 116-117°. n_D^{25} 1.4546.

6782. Norcholanic Acid. *(5β)-24-Norcholan-23-oic acid.* $C_{23}H_{38}O_2$; mol wt 346.55. C 79.71%, H 11.05%, O 9.23%. Prepn from ethyl cholanate: Wieland et al., *Z. Physiol. Chem.* 161, 80 (1926); from 3,7-diketonorcholanic acid: Windaus, van Schoor, ibid. 173, 312 (1928); from 12-ketonorcholanic acid: Cook, Haslewood, *J. Chem. Soc.* 1934, 428; from 3α,23-dihydroxycholanic acid: Yanuka et al., *Tetrahedron Letters* 1968, 1725.

Needles from acetic acid, mp 177°.
Methyl ester, $C_{24}H_{40}O_2$, needles from methanol, mp 74°.
Ethyl ester, $C_{25}H_{42}O_2$, prisms from alcohol, mp 67°.

6783. Norcodeine. *(5α,6α)-7,8-Didehydro-4,5-epoxy-3-methoxymorphinan-6-ol; N-desmethylcodeine; normorphine 3-methyl ether.* $C_{17}H_{19}NO_3$; mol wt 285.34. C 71.56%, H 6.71%, N 4.91%, O 16.82%. Prepn from acetylcodeine: von Braun, *Ber.* **47**, 2312 (1914); **Ger.** pat. **286,743** (1914); *Chem. Zentr.* **1915**, II, 862; *Frdl.* **12**, 741; *Houben* **4**, 588; *cf.* **Ger.** pat. **289,273** (1914); from codeine N-oxide: Diels, Fischer, *Ber.* **49**, 1723 (1916); from codeine: Speyer, Walther, *Ber.* **63**, 822 (1930).

Plates or needles from acetone or ethyl acetate, mp 185°. Sparingly sol in water, ether. Moderately sol in acetone. Freely sol in hot methanol, ethanol.
Hydrochloride trihydrate, $C_{17}H_{19}NO_3 \cdot HCl \cdot 3H_2O$, needles from water. When anhydr, dec 309°. Sparingly sol in cold water, more sol in hot water. Freely sol in methanol, ethanol. Almost insol in acetone.
Sulfate, $2C_{17}H_{19}NO_3 \cdot H_2SO_4$, crystals, freely sol in water.
Acetate (salt), $C_{17}H_{19}NO_3 \cdot C_2H_4O_2$, crystals, freely sol in water.

6784. Nordazepam. *7-Chloro-1,3-dihydro-5-phenyl-2H-1,4-benzodiazepin-2-one;* desmethyldiazepam; nordiazepam; DMDZ; A-101; Ro-5-2180; Calmday; Madar; Nordaz; Praxadium; Stilny. $C_{15}H_{11}ClN_2O$; mol wt 270.72. C 66.55%, H 4.10%, Cl 13.10%, N 10.35%, O 5.91%. Desmethyl analog and principal metabolite of diazepam, *q.v.* Prepn: L. H. Sternbach, E. Reeder, *J. Org. Chem.* **26**, 4936 (1961); E. Reeder *et al.*, **Ger.** pat. **1,136,709**; E. Reeder, L. Sternbach, **U.S.** pat. **3,051,701** (both 1962 to Hoffmann-La Roche); A. Stempel, **Belg.** pat. **620,020**; *idem*, **U.S.** pat. **3,202,699** (1963, 1965, both to Hoffmann-La Roche); S. C. Bell *et al.*, *J. Org. Chem.* **27**, 562 (1962); A. Stempel, G. W. Landgraf, *ibid.* 4675. Pharmacology: U. Traversa *et al.*, *J. Pharm. Pharmacol.* **29**, 504 (1977); M. Babbini *et al.*, *Pharmacology* **17**, 121 (178). Metabolism: U. Klotz *et al.*, *Brit. J. Clin. Pharmacol.* **7**, 119 (1979). Pharmacokinetics: M. Konishi, *J. Pharm. Sci.* **67**, 1777 (1978). Clinical study: V. Andreoli *et al.*, *Arzneimittel-Forsch.* **27**, 436 (1977). Toxicity study: L. O. Randall *et al.*, *Curr. Ther. Res.* **7**, 590 (1965). Terato-

genicity study: R. P. Miller *et al.*, *Toxicol. Appl. Pharmacol.* **25**, 453 (1973).
White or pale yellow crystalline powder from acetone, mp 216-217°. uv max (chloroform): 313 nm ($E_{1cm}^{1\%}$ 82). Slightly sol in alc, chloroform. Practically insol in water. LD_{50} in mice (mg/kg): 2750 orally, > 400 i.p. (Randall). Also reported as LD_{50} in mice, rats (mg/kg): 1300, > 5200 orally (company communication).
Note: This is a controlled substance (depressant) listed in the U.S. Code of Federal Regulations, Title 21 Part 1308.14 (1995).
THERAP CAT: Anxiolytic.

6785. Nordefrin. *4-(2-Amino-1-hydroxypropyl)-1,2-benzenediol; α-(1-aminoethyl)-3,4-dihydroxybenzyl alcohol;* 3,4-dihydroxynorephedrine; 3,4-dihydroxyphenylpropanolamine; 3,4-dihydroxyphenylaminopropanol; α-(α-aminoethyl)protocatechuyl alcohol; α-methylnoradrenaline; α-methylnorepinephrine; norhomoepinephrine; isoadrenaline; aminopropanolpyrocatechol; homoarterenol; $C_9H_{13}NO_3$; mol wt 183.21. C 59.00%, H 7.15%, N 7.65%, O 26.20%. α-Adrenergic agonist. Prepn: **Ger.** pat. **254,438** (1912 to Bayer); W. H. Hartung *et al.*, *J. Am. Chem. Soc.* **53**, 4149 (1931); M. Bockmühl *et al.*, **U.S.** pat. **1,948,162** (1934 to Winthrop); Bruckner *et al.*, *Ber.* **76B**, 466 (1943). Configuration: Fodor *et al.*, *Monatsh.* **83**, 1146 (1952). Prepn of optical isomers: **Ger.** pat. **269,327** (1913 to Bayer); Bockmühl, Gorr, **Ger.** pat. **639,126** (1936 to I. G. Farbenind.). Pharmacology: F. P. Luduena *et al.*, *J. Dental Res.* **37**, 206 (1958). HPLC determn in plasma: D. A. Jenner *et al.*, *J. Chromatog.* **224**, 507 (1981). Effect on toxicity of local anesthetics: S. E. Taylor, R. L. Dorris, *Anesth. Prog.* **36**, 79 (1989). Review of use in local anesthesia: T. M. Hensley *et al.*, *J. Foot Surg.* **26**, 504-510 (1987).

Hydrochloride, $C_9H_{13}NO_3 \cdot HCl$, *Corbasil, Cobefrin.* Crystals, dec 178-179°. One gram dissolves in about 1.5 ml water, 15 ml alcohol. Practically insol in ether. Solns are neutral and are easily destroyed by traces of alkali.
(−)-Form, *corbadrine, levonordefrin, Neo-Cobefrin.* Crystals, dec 212-215°. $[\alpha]_D^{25}$ − 31.0° (c = 0.5 in 0.01N HCl). uv max (dil HCl): 278 nm. Practically insol in water. Freely sol in aq solns of mineral acids. Very slightly sol in acetone, chloroform, ethanol, ether. LD_{50} in mice (mg/kg): 12.6 i.v. (Luduena).
THERAP CAT: Vasoconstrictor.

6786. Nordihydroguaiaretic Acid. *(R*,S*)-4,4'-(2,3-Dimethyl-1,4-butanediyl)bis[1,2-benzenediol]; meso-4,4'-(2,3-dimethyltetramethylene)dipyrocatechol;* 2,3-bis(3,4-dihydroxybenzyl)butane; β,γ-dimethyl-α,δ-bis(3,4-dihydroxyphenyl)butane; NDGA; masoprocol; CHX-100; Actinex. $C_{18}H_{22}O_4$; mol wt 302.37. C 71.50%, H 7.33%, O 21.17%. Lipoxygenase inhibitor; occurs as the *meso*-form in the resinous exudate of the creosote bush, *Larrea divaricata, Zygophyllaceae (Covillea tridentata)*. Isoln: C. W. Waller, O. Gisvold, *J. Am. Pharm. Assoc.* **34**, 78 (1945). Prepn from guaiaretic acid dimethyl ether: G. Schroeter *et al.*, *Ber.* **51**, 1587 (1918); R. D. Haworth *et al.*, *J. Chem. Soc.* **1934**, 1423. Synthesis: S. V. Lieberman *et al.*, *J. Am. Chem. Soc.* **69**, 1540 (1947); and configuration of naturally occurring form: C. W. Perry *et al.*, *J. Org. Chem.* **37**, 4371 (1972). Use as antioxidant: W. M. Lauer, **U.S.** pat. **2,373,192** (1945 to U. S. Secr'y of Agriculture). Antiproliferative effect on cultured keratinocytes: D. I. Wilkinson, E. K. Orenberg, *Int. J. Dermatol.* **26**, 660 (1987); on cultured glioma cells: D. E. Wilson *et al.*, *J. Neurosurg.* **71**, 551 (1989). Clinical trial in actinic keratoses: E. A. Olsen *et al.*, *J. Am. Acad. Dermatol.* **24**, 738 (1991).

Crystals from acetic acid, mp 185-186°. uv max (methanol): 283, 218 nm (ε 6660, 13400). Sol in ethanol, methanol, ether, concd H_2SO_4; slightly sol in hot water, chloroform; sol in dil alkalies, developing a deep red color. Practically insol in petr ether, benzene, toluene, dil HCl.

USE: As antioxidant for fats and oils in foods.

THERAP CAT: Antineoplastic.

6787. Norea. *N,N-Dimethyl-N'-(octahydro-4,7-methano-1H-inden-5-yl)urea;* *3-(hexahydro-4,7-methanoindan-5-yl)-1,1-dimethylurea;* noruron; Hercules 7531; Herban. $C_{13}H_{22}N_2O$; mol wt 222.33. C 70.23%, H 9.97%, N 12.60%, O 7.20%. Prepn: Diveley, Pombo, U.S. pat. **3,150,179** (1964 to Hercules). Microbial degradn: Murray *et al.*, *Weed Sci.* **17,** 52 (1969).

Crystals from aq ethanol, mp 176-178°. Sol in acetone, cyclohexanone; slightly sol in benzene, toluene. Practically insol in water, hexane. LD_{50} orally in rats, mice: 2000, 4600 mg/kg, *RTECS* Vol. **II,** R. J. Lewis, R. L. Tatken, Eds. (1979) p 682.

USE: Herbicide.

6788. Norepinephrine. *(R)-4-(2-Amino-1-hydroxyethyl)-1,2-benzenediol;* *α-(aminomethyl)-3,4-dihydroxybenzyl alcohol;* 2-amino-1-(3,4-dihydroxyphenyl)ethanol; 1-(3,4-dihydroxyphenyl)-2-aminoethanol; noradrenaline; levarterenol; Adrenor, Levophed. $C_8H_{11}NO_3$; mol wt 169.18. C 56.80%, H 6.55%, N 8.28%, O 28.37%. Demethylated precursor of epinephrine, *q.v.* Occurs in animals and man, and is a sympathomimetic hormone of both adrenal origin and adrenergic orthosympathetic postganglionic origin in man. Physiologic review: Malmejac, *Physiol. Rev.* **44,** 186 (1964). It has also been found in plants, e.g., *Portulaca olerocea* L., *Portulacaceae:* Fing *et al.*, *Nature* **191,** 1108 (1961). Synthesis of *dl*-form: Payne, *Ind. Chemist* **37,** 523 (1961). Historic review of synthesis: Loewe, *Arzneimittel-Forsch.* **4,** 583 (1954). Resolution of *dl*-form: Tullar, *J. Am. Chem. Soc.* **70,** 2067 (1948); *idem,* U.S. pat. **2,774,789** (1956 to Sterling Drug). Configuration: Pratesi *et al., J. Chem. Soc.* **1959,** 4062. Comprehensive description: C. F. Schwender in *Analytical Profiles of Drug Substances* vol. 1, K. Florey, Ed. (Academic Press, New York, 1972) pp 149-173; T. D. Wilson, *ibid.* vol. 11 (1982) pp 555-586.

Microcrystals, dec 216.5-218°. $[\alpha]_D^{25}$ −37.3° (c = 5 in water with 1 equiv HCl).

Hydrochloride, $C_8H_{11}NO_3$·HCl, *Arterenol.* Crystals, mp 145.2-146.4°. $[\alpha]_D^{25}$ −40° (c = 6). Freely sol in water. Solns slowly oxidize under the influence of light and oxygen in a manner comparable to epinephrine hydrochloride.

d-Bitartrate, $C_8H_{11}NO_3$·$C_4H_6O_6$, *levarterenol bitartrate, Aktamin, Binodrenal.* Obtained as the monohydrate, crystals, mp 102-104°. $[\alpha]_D^{25}$ −10.7° (c = 1.6 in H_2O). When anhydr, mp 158-159° (some decomp). Freely sol in water.

dl-Form, crystals, dec 191°. Sparingly sol in water; very slightly sol in alc, ether; readily sol in dilute acids, caustic.

THERAP CAT: Adrenergic (vasopressor); antihypotensive.

THERAP CAT (VET): Sympathomimetic; vasopressor in shock.

6789. Norethandrolone. *(17α)-17-Hydroxy-19-nor-pregn-4-en-3-one;* 17α-ethyl-19-nortestosterone; 17α-ethyl-17-hydroxy-4-norandrosten-3-one; 17α-ethyl-17-hydroxy-19-norandrost-4-en-3-one; Nilevar; Solevar. $C_{20}H_{30}O_2$; mol wt 302.46. C 79.42%, H 10.00%, O 10.58%. Prepn by catalytic hydrogenation of 17α-ethynyl-19-nortestosterone: Colton, U.S. pat. **2,721,871** (1955 to Searle); *J. Am. Chem. Soc.* **79,** 1123 (1957).

Crystals from methanol, mp 140-141°. uv max: 240 nm (ε 16500). Sol in alcohol, benzene, ether, ethyl acetate. Insol in water.

Note: This is a controlled substance (anabolic steroid) in the U.S. Code of Federal Regulations, Title 21 Part 1308.13, as defined in Part 1308.02 (1995).

THERAP CAT: Androgen.

6790. Norethindrone. *(17α)-17-Hydroxy-19-norpregn-4-en-20-yn-3-one;* 19-nor-17α-ethynyltestosterone; 17α-ethynyl-19-nortestosterone; 17α-ethynyl-17β-hydroxy-4-androsten-3-one; 19-nor-17α-ethynylandrosten-17β-ol-3-one; anhydrohydroxynorprogesterone; 19-norethisterone; norpregneninolone; "mini-pill"; Conludag; Menzol; Micronor; Micronovum; Mini-Pe; Norcolut; Noriday; Norluten; Norlutin; Nor-QD; Primolut N; Utovlan. $C_{20}H_{26}O_2$; mol wt 298.43. C 80.50%, H 8.78%, O 10.72%. Prepn from 19-nor-4-androstene-3,17-dione: Djerassi *et al., J. Am. Chem. Soc.* **76,** 4092 (1954); U.S. pat. **2,744,122** (1956 to Syntex); De Ruggieri, U.S. pat. **2,849,462** (1958). Pharmacokinetics: H. Singh *et al., Am. J. Obstet. Gynecol.* **135,** 409 (1979); M. Humpel, *Contraception* **26,** 83 (1982). Double-blind, comparative clinical trial: S. Koetsawang *et al., ibid.* **25,** 231 (1982); A. Sheth *et al., ibid.* 243. Multicenter trial of combination with ethinyl estradiol: M. Toews *et al., Curr. Ther. Res.* **41,** 509 (1987). Review of carcinogenicity studies: *IARC Monographs* **21,** 441-460 (1979). Comprehensive description: A. P. Schroff, E. S. Moyer, *Anal. Profiles Drug Subs.* **4,** 268-293 (1975).

Crystals from ethyl acetate, mp 203-204°. $[\alpha]_D^{20}$ −31.7° (chloroform). First reported as $[\alpha]_D^{20}$ −25° (chloroform). uv max (ethanol): 240 nm (log ε 4.24).

Mixture with ethinyl estradiol, *Binovum, Brevicon, Brevinor, Conceplan, Modicon, Neocon 1/35, Norimin, Norinyl 1+35, Norquentiel, Ortho-Novum 1/35, Ortho-Novum 7/7/7, Ovcon, Ovysmen, Synphase, Tri-Norinyl, Trinovum.*

Mixture with mestranol, *Norinyl-1, Ortho-Novin 1/50, Ortho-Novum 1/50.*

Acetate, $C_{22}H_{28}O_3$, *Aygestin, Milligynon, Norlutate, Primolut-Nor.* Prepn: O. Engelfried *et al.,* U.S. pat. **2,964,537** (1960 to Schering AG). Crystals from methylene chloride + hexane, mp 161-162°. uv max: 240 nm (ε 18690).

Acetate, mixture with ethinyl estradiol, *Etalontin, Primosiston, Anovlar, Gynovlar, Loestrin, Minovlar, Norlestrin.*

Enanthate, $C_{27}H_{38}O_3$, *Noristerat.* Clinical trial as injectable contraceptive: S. K. Banerjee *et al., Contraception* **30,** 561 (1984).

Note: Norethindrone may reasonably be anticipated to be a carcinogen: *Seventh Annual Report on Carcinogens* (PB95-109781, 1994) p 308.

THERAP CAT: Progestogen. Norethindrone and acetate in combination with estrogen as contraceptive (oral). Enanthate as contraceptive (injectable).

6791. Norethynodrel. *(17α)-17-Hydroxy-19-norpregn-5(10)-en-20-yn-3-one;* 17α-ethynyl-17-hydroxy-5(10)-estren-3-one; 13-methyl-17-ethynyl-17-hydroxy-1,2,3,4,6,7,8,9,11,-12,13,14,16,17-tetradecahydro-15*H*-cyclopenta[*a*]phenanthren-3-one. $C_{20}H_{26}O_2$; mol wt 298.43. C 80.50%, H 8.78%, O 10.72%. Prepn: Colton, U.S. pats. **2,691,028, 2,725,389** (1954, 1955 to Searle). Metabolism studies in mice: Freudenthal *et al., Toxicol. Appl. Pharmacol.* **24**, 125 (1973).

Crystals from aq methanol, mp 169-170°. [α]$_D$ +108° (1% chloroform).

Mixture with mestranol, *Conovid E, Enavid, Enovid.*

THERAP CAT: Progestogen. In combination with estrogen as contraceptive (oral).

6792. Norfenefrine. *α-(Aminomethyl)-3-hydroxybenzene methanol; α-(aminomethyl)-m-hydroxybenzyl alcohol;* 1-(*m*-hydroxyphenyl)-2-aminoethanol; *m*-hydroxyphenylethanolamine; norphenylephrine. $C_8H_{11}NO_2$; mol wt 153.18. C 62.73%, H 7.24%, N 9.14%, O 20.89%. Chem synthesis: Sachs, **Fr. pat. 866,569** (1941), *C.A.* **43**, 5043c (1949); Legerlotz, U.S. pat. **2,312,916** (1943 to Ciba); Credner, Neugebauer, *Arzneimittel-Forsch.* **3**, 462 (1953); Bretschneider, Hörmann, *Monatsh.* **84**, 1021 (1953). Synthesis and resolution: D'Amico *et al., Chim. Ind. (Milan)* **38**, 93 (1956). Biosynthesis: Hartman *et al., J. Am. Chem. Soc.* **77**, 816 (1955). Activity studies: Gersmeyer, Albrecht, *Med. Welt* **1966**, 657. Metabolism: Barac, *C. R. Soc. Biol.* **155**, 1598 (1961).

dl-Form hydrochloride, $C_8H_{11}NO_2$·HCl, *Coritat, Depot-Novadral, Energona, Esbuphon, Molycor-R, Novadral, Stagural, Tonolift, Vingsal, Zondel.* Crystals, mp 159-160°. uv max: 274 nm (E$_{1cm}^{1\%}$ 91.21). Freely soluble in water.

THERAP CAT: Adrenergic.

6793. Norfloxacin. *1-Ethyl-6-fluoro-1,4-dihydro-4-oxo-7-(1-piperazinyl)-3-quinolinecarboxylic acid;* AM-715; MK-366; Baccidal; Barazan; Chibroxin(e); Chibroxol; Floxacin; Fulgram; Gonorcin; Lexinor; Noflo; Nolicin; Noracin; Noraxin; Norocin; Noroxin(e); Norxacin; Sebercim; Uroxacin; Utinor; Zoroxin. $C_{16}H_{18}FN_3O_3$; mol wt 319.34. C 60.18%, H 5.68%, F 5.95%, N 13.16%, O 15.03%. Fluorinated quinolone antibacterial. Prepn: T. Irikura, **Belg. pat. 863,429**; *eidem,* U.S. pat. **4,146,719** (1978, 1979 both to Kyorin); M. Pesson, **Ger. pat. 2,840,910**; *eidem,* U.S. pat. **4,292,317** (1979, 1981 both to Roger Bellon/Dainippon); H. Koga *et al., J. Med. Chem.* **23**, 1358 (1980). Comparative antibacterial activity: K. Hirai *et al., Antimicrob. Ag. Chemother.* **19**, 188 (1981); M. Y. Khan *et al., ibid.* 265; H. H. Gadebusch *et al., Infection* **10**, 41 (1982). Series of articles on *in vivo* and *in vitro* activity, pharmacology, pharmacokinetics, metabolism, early clinical studies, toxicology: *Chemotherapy (Tokyo)* **29**, Suppl. 4, 1-1000 (1981). Acute toxicity data: T. Irikura *et al., ibid.* 783. *In vitro* activity vs gentamicin-resistant *P. aeruginosa:* J. Downs *et al., Antimicrob. Ag. Chemother.* **21**, 670 (1982). Treatment of lower urinary tract infection: H. Giamarellou *et al., Eur. J. Clin.*

Microbiol. **2**, 266 (1983); of penicillin-resistant gonorrhea: S. R. Crider *et al., N. Engl. J. Med.* **311**, 137 (1984). Review of pharmacokinetics, antibacterial activity, clinical trials: B. Holmes *et al., Drugs* **30**, 482-513 (1985); R. C. Rowen *et al., Pharmacotherapy* **7**, 92-110 (1987). Symposium on clinical experience: *Scand. J. Infect. Dis.* Suppl. 48, 1-91 (1986). Comprehensive description: C. Mazuel, *Anal. Profiles Drug Subs.* **20**, 557-600 (1991).

White to light-yellow crystalline powder, mp 220-221°. uv max (0.1*N* NaOH): ~274, 325, 336 nm (A$_{1cm}^{1\%}$ ~1109, 437, 425). pKa$_1$ 6.34; pKa$_2$ 8.75. Partition coefficient (octanol/water): 0.46. Soly at 25° (mg/ml): water 0.28; methanol 0.98; ethanol 1.9; acetone 5.1; chloroform 5.5; diethyl ether 0.01; benzene 0.15; ethyl acetate 0.94; octyl alcohol 5.1; glacial acetic acid 340. Solubility in water is pH dependent, increasing sharply at pH < 5 or at pH > 10. Hygroscopic in air, forms a hemihydrate. LD$_{50}$ in mice, rats (mg/kg): > 4000 orally (both species); 1500 s.c. (both species); 470, > 500 i.m.; 220, 270 i.v. (Irikura).

THERAP CAT: Antibacterial.

6794. Norflurazon. *4-Chloro-5-(methylamino)-2-[3-(trifluoromethyl)phenyl]-3(2H)-pyridazinone; 4-chloro-5-(methylamino)-2-(α,α,α-trifluoro-m-tolyl)-3(2H)-pyridazinone;* 1-(3-trifluoromethylphenyl)-4-methylamino-5-chloropyridazone; SAN-9789; Evital; Solicam; Zorial. $C_{12}H_9Cl$-F_3N_3O; mol wt 303.67. C 47.46%, H 2.99%, Cl 11.67%, F 18.77%, N 13.84%, O 5.27%. Selective pre-emergent herbicide which inhibits carotenoid biosynthesis in susceptible species. Prepn: C. Ebner, M. Schuler, **Belg. pat. 712,832**; *eidem,* U.S. pat. **3,644,355** (1968, 1972 both to Sandoz). Absorption and metabolism: R. H. Strang, R. L. Rogers, *J. Agr. Food Chem.* **22**, 1119 (1974). Mechanism of action: G. Sandmann *et al., Pest. Biochem. Physiol.* **14**, 185 (1980). HPLC determn in edible crops: W. M. Draper, J. C. Street, *J. Agr. Food Chem.* **29**, 724 (1981). Comprehensive description: E. Ummel *et al., Proc. Brit. Weed Contr. Conf.* **13**, 313 (1976).

Crystals from alcohol, mp 183-185°. Soly in water (25°): 28 ppm. LD$_{50}$ orally in rats: 9300 mg/kg (Ummel).

USE: Herbicide.

6795. Norgesterone. *(17α)-17-Hydroxy-19-norpregna-5(10),20-dien-3-one;* 17β-hydroxy-17α-vinylestr-5(10)-en-3-one; 17α-vinyl-5(10)-estren-17β-ol-3-one; norvinodrel; vinylestrenolone. $C_{20}H_{28}O_2$; mol wt 300.44. C 79.96%, H 9.39%, O 10.65%. Prepn: Ruggieri, Ferrari, U.S. pat. **2,983,735** (1961 to Richter); *eidem,* U.S. pat. **3,062,713** (1962). Synthesis of racemic form: Hiscock, Whitehurst, *J. Chem. Soc.* **1965**, 5772. Biological properties: Ruggieri *et al., Steroids* **5**, 73 (1965).

Crystals from diethyl ether-hexane, mp 142-143°. $[\alpha]_D$ +161° (chloroform).

Mixture with ethinyl estradiol, *Vestalin*.

THERAP CAT: Progestogen.

6796. Norgestimate. *(17α)-17-(Acetyloxy)-13-ethyl-18,-19-dinorpregn-4-en-20-yn-3-one 3-oxime;* (+)-13-ethyl-17-hydroxy-18,19-dinor-17α-pregn-4-en-20-yn-3-one oxime acetate (ester); 17α-acetoxy-13-ethyl-17-ethynylgon-4-en-3-one oxime; dexnorgestrel acetime; D-138; ORF-10131. $C_{23}H_{31}NO_3$; mol wt 369.50. C 74.76%, H 8.46%, N 3.79%, O 12.99%. Acetate oxime of D-norgestrel, *q.v.* Prepn: A. P. Shroff, **Ger. pat. 2,633,210**; *idem*, **U.S. pat. 4,027,019** (both 1977 to Ortho). HPLC determn: P. A. Lane *et al., J. Pharm. Sci.* **76**, 44 (1987). Pharmacology: D. W. Hahn *et al., Contraception* **16**, 541 (1977); J. Killinger *et al., ibid.* **32**, 311 (1985). Pharmacodynamics: H. S. Weintraub *et al., J. Pharm. Sci.* **67**, 1406 (1978). Metabolism in female monkeys: S. F. Sisenwine *et al., Contraception* **15**, 25 (1977); in women: K. B. Alton *et al., ibid.* **29**, 19 (1984). Clinical trial as a contraceptive in combination with ethinyl estradiol: B. Rubio-Lotvin, R. Gonzales-Ansorena, *Acta Eur. Fertil.* **9**, 1 (1978).

Crystals from methylene chloride, mp 214-218°. $[\alpha]_D^{25}$ +110°.

Mixture with ethinyl estradiol, *Cilest, Ortho Cyclen, Ortho Tri-Cyclen, Ortrel, TriCilest.*

THERAP CAT: Progestogen. In combination with estrogen as oral contraceptive.

6797. Norgestrel. *(17α)-(±)-13-Ethyl-17-hydroxy-18,-19-dinorpregn-4-en-20-yn-3-one;* Wy-3707; Neogest; Ovrette. $C_{21}H_{28}O_2$; mol wt 312.45. C 80.73%, H 9.03%, O 10.24%. The bioactive enantiomer is levorotatory. Prepn: H. Smith, **Belg. pat. 623,844** (1963); *C.A.* **61**, 4427c (1964); G. A. Hughes, H. Smith, **U.S. pat. 3,959,322** (1976 to Herchel Smith); H. Smith *et al., Experientia* **19**, 394 (1963); *eidem, J. Chem. Soc.* **1964**, 4472. Synthesis of (±)-form: M. Rosenberger *et al., Helv. Chim. Acta* **54**, 2857 (1971); of (−)-form: C. Rufer *et al., Ann.* **702**, 141 (1967); H. Baier *et al., Helv. Chim. Acta* **68**, 1054 (1985). Crystal structure determn of (+)-form: N. J. DeAngelis *et al., Acta Crystallogr.* **31B**, 2040 (1975). Comprehensive description: A. M. Sopirak, L. F. Cullen, *Anal. Profiles Drug Subs.* **4**, 294-318 (1975). Metabolism: F. Z. Stanczyk, S. Roy, *Contaception* **42**, 67

(−)-form

(1990). Clinical efficiacy as oral contraceptive: T. R. Dunson *et al., ibid.* **48**, 109 (1993); as contraceptive implant: H. B. Croxatto, *Ann. Med.* **25**, 155 (1993).

Crystals from methanol, mp 205-207°. uv max (ethanol): 241 nm (ε 16700).

Mixture with ethinyl estradiol, *Lo/Ovral, Ovral, Stédiril.*

(+)-Form, *(8α,9β,10α,13α,14β)-13-ethyl-17-hydroxy-18,19-dinorpregn-4-en-20-yn-3-one, dextronorgestrel.* Crystals, mp 238-242°. $[\alpha]_D^{25}$ +40.7° (CHCl$_3$).

(−)-Form, *17-ethynyl-18-methyl-19-nortestosterone, 13β-ethyl-17α-ethynyl-17β-hydroxygon-4-en-3-one, levonorgestrel, D-norgestrel, dexnorgestrel (obsolete), Microlut, Microval, Norgeston, Norplant.* Crystals from methanol, mp 235-237°. $[\alpha]_D^{20}$ −32.4° (c = 0.496 in CHCl$_3$). uv max (methanol): 241 nm (ε 16770).

Mixture of (−)-form with ethinyl estradiol, *Levlen, Logynon, Microgynon, Nordette, Ovran, Ovranette, Tetragynon, Tri-Levlen, Trinordiol, Triphasil.*

THERAP CAT: Progestogen; oral contraceptive; as contraceptive implant.

6798. Norgestrienone. *(17α)-17-Hydroxy-19-norpregna-4,9,11-trien-20-yn-3-one;* 17α-ethynyl-4,9,11-estratrien-17β-ol-3-one; 17α-ethynyl-17β-hydroxy-4,9,11-estratriene; 17α-ethynyl-13β-methyl-$\Delta^{4,9,11}$-gonatriene-17β-ol-3-one; Ogyline. $C_{20}H_{22}O_2$; mol wt 294.39. C 81.60%, H 7.53%, O 10.87%. Prepn: **Neth. pat. Appl. 6,401,555**, corresp to Nominé *et al.,* **U.S. pat. 3,257,278** (1964 and 1966, both to Roussel-UCLAF); Nominé *et al., Compt. Rend.* **260**, 4545 (1965); **Fr. pat. M3060** and **Neth. pat. Appl. 6,517,141** (1965 and 1966, both to Roussel-UCLAF), *C.A.* **63**, 8449b (1965) and **65**, 15470b (1966).

Pale yellow needles from diisopropyl ether, mp 169°. $[\alpha]_D^{20}$ +63° (c = 0.5 in alc). uv max: 342, 238 nm (ε 29100, 5920). Sol in alcohols, ether, acetone, benzene, chloroform; practically insol in water, dil aq acids and alkalies.

THERAP CAT: Progestogen.

6799. Norhyoscyamine. *[3(S)-endo]-α-(Hydroxymethyl)benzeneacetic acid 8-azabicyclo[3.2.1]oct-3-yl ester; 1αH,-5αH-nortropan-3α-ol (−)-tropate;* 1-tropic acid 3α-nortropanyl ester; pseudohyoscyamine; solandrine. $C_{16}H_{21}NO_3$; mol wt 275.35. C 69.79%, H 7.69%, N 5.09%, O 17.43%. Isoln from *Solanaceae* and structure: Carr, Reynolds, *J. Chem. Soc.* **101**, 946 (1912). Synthesis: Fodor *et al., Ber.* **93**, 2681 (1960). Identity with pseudohyoscyamine: Carr, Reynolds, *loc. cit.;* with solandrine: Petrie, *C.A.* **12**, 2345[7] (1918).

Crystals, mp 140.5°. Strongly basic. $[\alpha]_D$ −23° (alcohol). Slightly sol in water, ether; sol in alcohol, chloroform.

6800. Norleucine. *2-Aminohexanoic acid;* α-aminocaproic acid; glycoleucine; caprine. $C_6H_{13}NO_2$; mol wt 131.17. C 54.94%, H 9.99%, N 10.68%, O 24.39%. CH_3-$(CH_2)_3CH(NH_2)COOH$. An amino acid classified as nonessential with respect to its growth effect in rats. Several syntheses; prepn from α-bromo-*n*-caproic acid by action of

25% ammonia at 50-55°: Marvel, du Vigneaud, *Org. Syn.* **4**, 3 (1925).

DL-Form, lustrous leaflets from water. d 1.172. Dec 327°. pK$_1$ 2.39; pK$_2$ 9.76. Soly in water: 11.49 g/l at 25°, 17.27 g/l at 50°, 28.61 g/l at 75°, 52.0 g/l at 100°. Sparingly sol in alcohol: 0.42 g/100 g at 25°. Sol in acids.

L(+)-Form, slightly sweet, shiny leaflets from water. mp 301° (partial decompn). Sublimes partially at 275-280°. [M]$_D$ +32.1° (5N HCl); +47.9° (glacial acetic acid). [α]$_D^{20}$ +21.3° (c = 4.25 in 6N HCl); +6.26° (c = 0.70 in water).

D(−)-Form, bitter, shiny leaflets from water. mp 301° (partial decompn). Sublimes partially at 275-280°. [α]$_D^{20}$ −22.4° (c = 4.69 in 6N HCl); −4.49° (c = 0.96 in water).

6801. Norlevorphanol. *Morphinan-3-ol;* (−)-3-hydroxymorphinan; 1,3,4,9,10,10a-hexahydro-6-hydroxy-2H-10,4a-iminoethanophenanthrene; 1,3,4,9,10,10a-hexahydro-2H-10,4a-iminoethanophenanthren-6-ol; NIH-7539. C$_{16}$H$_{21}$NO; mol wt 243.35. C 78.97%, H 8.70%, N 5.76%, O 6.57%. Prepn: Schnider, Grüssner, *Helv. Chim. Acta* **34**, 2211 (1951); Hellerbach, *et al.*, *ibid.* **39**, 429 (1956); **Brit. pat. 765,920** (1957 to Hoffmann-La Roche).

Crystals from acetone + methanol, mp 270-272°. [α]$_D^{21}$ −42° ±2° (c = 1 in methanol).

Hydrochloride, C$_{16}$H$_{21}$NO.HCl, crystals from water, mp 320°.

Hydrobromide, C$_{16}$H$_{21}$NO.HBr, crystals from water, mp 222-224°.

Note: This is a controlled substance (opiate) listed in the U.S. Code of Federal Regulations, Title 21 Part 1308.11 (1995).

THERAP CAT: Analgesic (narcotic).

6802. Norlobelanine. *2,2′-(2,6-Piperidinediyl)bis[1-phenylethanone];* isolobelanine. C$_{21}$H$_{23}$NO$_2$; mol wt 321.42. C 78.47%, H 7.21%, N 4.36%, O 9.96%. In herb of *Lobelia inflata* L., *Lobeliaceae.* Isoln from crude lobelanidine nitrate: Wieland *et al.*, *Ann.* **473**, 118 (1929). Synthesis: Schöpf, Lehmann, *ibid.* **518**, 1 (1935). *Cis*- and *trans*-isomers: Ebnöther, *Helv. Chim. Acta* **41**, 386 (1958).

Crystals, mp 120-121°. Practically insol in water. Sol in alcohol, benzene, chloroform, ether, dil acids. On methylation it is converted to lobelanine.

trans-Form, prisms from ethyl acetate + ether, mp 88-91°.

trans-Form hydrochloride, C$_{21}$H$_{23}$NO$_2$.HCl, crystals from alcohol + ether, dec 163-168°.

6803. Normetanephrine. *α-(Aminomethyl)-4-hydroxy-3-methoxybenzenemethanol;* α-(aminomethyl)vanillyl alcohol; 4-hydroxy-3-methoxy-α-(aminomethyl)benzyl alcohol; 1-(4-hydroxy-3-methoxyphenyl)-2-aminoethanol; 3-*O*-methylarterenol; 3-*O*-methylnoradrenaline; 3-*O*-methylnorepinephrine. C$_9$H$_{13}$NO$_3$; mol wt 183.21. C 59.00%, H 7.15%, N 7.65%, O 26.20%. A naturally occurring derivative of epinephrine, found together with metanephrine in urine and in certain tissues. Prepn: Fodor *et al.*, *Acta Chim. Acad. Sci. Hung.* **1**, 395 (1951); *C.A.* **49**, 897 (1955); Axelrod *et al.*, *J. Biol. Chem.* **233**, 697 (1958); Heacock, Hutzinger, *Chem. & Ind. (London)* **1961**, 595.

dl-Form, isolated as the hydrochloride, C$_9$H$_{13}$NO$_3$.HCl, prisms from abs ethanol, dec 206-207°. uv max (abs ethanol): 232, 282 nm (ε 7100, 2970).

6804. Normethadone. *6-(Dimethylamino)-4,4-diphenyl-3-hexanone;* 1-dimethylamino-3,3-diphenyl-4-hexanone; 1,1-diphenyl-1-(2-dimethylaminoethyl)-2-butanone; isoamidone I; desmethylmethadone; phenyldimazone; Hoechst 10582. C$_{20}$H$_{25}$NO; mol wt 295.42. C 81.31%, H 8.53%, N 4.74%, O 5.42%. Prepn from 4-bromo-2,2-diphenylbutanenitrile and dimethylamine: Bockmühl, Ehrhart, *Ann.* **561**, 72 (1948); Easton *et al.*, *J. Am. Chem. Soc.* **74**, 5772 (1952). Toxicity data: Eddy *et al.*, *J. Pharmacol. Exp. Ther.* **98**, 121 (1950).

Oily liquid, bp$_3$ 164-167°. Alkaline reaction.

Hydrochloride, C$_{20}$H$_{25}$NO.HCl, *Ticarda.* Crystals from acetone, mp 174-175°. Sol in water, alcohol. pH of 1% aq soln ~5. LD$_{50}$ s.c. in mice: 90 mg/kg (Eddy).

Note: This is a controlled substance (opiate) listed in the U.S. Code of Federal Regulations, Title 21 Part 1308.11 (1995).

THERAP CAT: Antitussive; analgesic (narcotic).

6805. Normethandrone. *(17β)-17-Hydroxy-17-methylestr-4-en-3-one;* 17α-methyl-19-nortestosterone; methylestrenolone; normethandrolone; normetandrone; methylnortestosterone; Orgasteron; Metalutin; Methalutin. C$_{19}$H$_{28}$O$_2$; mol wt 288.43. C 79.12%, H 9.78%, O 11.09%. Prepn: Djerassi *et al.*, U.S. pats. **2,744,122** and **2,774,777** (1956 to Syntex); *J. Am. Chem. Soc.* **76**, 4092 (1956); De Ruggieri, U.S. pat. **2,849,461** (1958).

Crystals from ether-hexane, mp 156-158° (Kofler). [α]$_D$ +33°. uv max (ethanol): 240 nm (log ε 4.23).

THERAP CAT: Androgen.

6806. Normorphine. *(5α,6α)-7,8-Didehydro-4,5-epoxymorphinan-3,6-diol;* desmethylmorphine. C$_{16}$H$_{17}$NO$_3$; mol wt 271.32. C 70.83%, H 6.32%, N 5.16%, O 17.69%. Prepd by hydrolysis of cyanonormorphine: von Braun, *Ber.* **47**, 2312 (1914); Speyer, Walther, *Ber.* **63**, 852 (1930); Weijlard, Erickson, *J. Am. Chem. Soc.* **64**, 869 (1942); Rapoport, Look, U.S. pat. **2,890,221** (1959 to U.S.A.E.C.).

Sesquihydrate, crystals from water, mp 273°; when anhydr, mp 276-277°. Sparingly sol in hot water, alcohol. Insol in ether, chloroform.

Hydrochloride monohydrate, cryst from water, dec 305°.

Note: This is a controlled substance (opium derivative) listed in the U.S. Code of Federal Regulations, Title 21 Part 1308.11 (1995).

THERAP CAT: Analgesic (narcotic).

6807. Nornicotine. *3-(2-Pyrrolidinyl)pyridine;* 2-(3-pyridyl)pyrrolidine. $C_9H_{12}N_2$; mol wt 148.21. C 72.94%, H 8.16%, N 18.90%. Occurs in ordinary tobacco, in other species of *Nicotiana,* and also in *Duboisia hopwoodii* F. Muell., *Solanaceae.* *Review:* Markwood, *U.S. Dept. Agr. Bur. Entomol. Plant Quarantine* E-561 (1942), Supplement by Roark, E-645 (1945). Synthesis: Mizogruchi, *Chem. Pharm. Bull.* **9**, 818 (1961); M. Nakane, C. R. Hutchinson, *J. Org. Chem.* **43**, 3922 (1978).

Hygroscopic, somewhat viscous liquid, develops a slight amine odor, less pungent than that of nicotine. bp 270°; bp_{11} 131°; bp_3 105-107°. Hardly volatile with steam (difference from nicotine which is readily volatile). d_4^{20} 1.0737. $[\alpha]_D^{22}$ −89° (c = 100). n_D^{20} 1.5378. Miscible with water. Very sol in alcohol, chloroform, ether, petr ether, kerosene, oils. Less volatile and less easily oxidized than nicotine. Commercial nicotine prepns sometimes contain several percent nornicotine. *Cf.* Hoskins, Craig, *Ann. Rev. Biochem.* **15**, 539 (1946). LD i.p. in rats: 23.5 mg/kg; LD i.v. in rabbits: 3 mg/kg.

Dipicrate, $C_9H_{12}N_2.2C_6H_3N_3O_7$, yellow crystals, mp 192° (evac tube).

Caution: Approx one-third as toxic as nicotine. Faintness, prostration, muscular weakness, severe nausea, vomiting, diarrhea, collapse with or without convulsions.

USE: Agricultural or horticultural insecticide.

6808. Nornidulin. *2,4,7-Trichloro-3,8-dihydroxy-1,9-dimethyl-6-(1-methyl-1-propenyl)-11H-dibenzo[b,e][1,4]dioxepin-11-one;* Ustin. $C_{19}H_{15}Cl_3O_5$; mol wt 429.68. C 53.11%, H 3.52%, Cl 24.75%, O 18.62%. Antibiotic substance produced by *Aspergillus ustus:* Kurung, *Science* **102**, 11 (1945); Hogeboom, Craig, *J. Biol. Chem.* **162**, 363 (1946); Doering *et al., J. Am. Chem. Soc.* **68**, 725 (1946). Structure: Dean *et al., J. Chem. Soc.* **1960**, 4829; Bycroft *et al., ibid.* **1963**, 5148.

Crystals from ether + cyclohexane, mp 185-187°. uv max: 266 nm (ε 8120).

6809. Norphenazone. *2,4-Dihydro-5-methyl-2-phenyl-3H-pyrazol-3-one;* 3-methyl-1-phenyl-2-pyrazolin-5-one; 1-phenyl-3-methyl-5-pyrazolone; C.I. developer 1; developer Z; norantipyrine; MCI-186. $C_{10}H_{10}N_2O$; mol wt 174.20. C 68.95%, H 5.79%, N 16.08%, O 9.18%. Free radical scavenger; metabolite of antipyrine, *q.v.* Prepn: L. Knorr, *Ann.* **238**, 137 (1887). Structural determn by NMR and IR: R. Jones *et al., Tetrahedron* **19**, 1497 (1963); F.-D. Höppner *et al., J. Prakt. Chem.* **318**, 555 (1976). Crystal structure: F. Bechtel *et al., Cryst. Struct. Comm.* **3**, 469 (1973). HPLC determn in urine: M. A. Mikati *et al., J. Chromatog.* **433**, 305 (1988). Anti-ischemic and radical scavenging actions: T. Watanabe *et al., J. Pharmacol. Exp. Ther.* **268**, 1597 (1994).

White crystals from benzene, mp 129°-130°. Also reported as mp 127° (Knorr). bp_{265} 287°.

THERAP CAT: In treatment of stroke.

6810. Norpipanone. *4,4-Diphenyl-6-(1-piperidinyl) 3 hexanone;* 1-piperidino-3,3-diphenyl-4-hexanone; 6-(1-piperidyl)-4,4-diphenyl-3-hexanone; Hoechst 10495; Hexalgon. $C_{23}H_{29}NO$; mol wt 335.49. C 82.34%, H 8.71%, N 4.18%, O 4.77%. Prepn: Bockmühl, Ehrhart, *Ann.* **561**, 73 (1949); Dupré *et al., J. Chem. Soc.* **1949**, 500.

Hydrochloride, $C_{23}H_{29}NO.HCl$, *Orfenso.* Crystals, mp 181-182°. Sol in water, alcohol.

Hydrobromide, $C_{23}H_{29}NO.HBr$, crystals, mp 192-193°. Sol in water, alcohol.

Note: This is a controlled substance (opiate) listed in the U.S. Code of Federal Regulations, Title 21 Part 1308.11 (1995).

THERAP CAT: Analgesic (narcotic).

6811. Norpseudoephedrine. *(R*,R*)-α-(1-Aminoethyl)-benzenemethanol;* *threo*-2-amino-1-hydroxy-1-phenylpropane; *threo*-1-phenyl-1-hydroxy-2-aminopropane; nor-ψ-ephedrine; pseudonorephedrine; ψ-norephedrine; cathine; katine. $C_9H_{13}NO$; mol wt 151.21. C 71.49%, H 8.67%, N 9.26%, O 10.58%. Occurs naturally as the D-*threo*-form in the leaves of the khat plant, *Catha edulis* Forsk., *Celastraceae,* an evergreen shrub native to southern Arabia and Ethiopia; also found in smaller amounts in the South American tree *Maytenus krukovii* A. C. Smith, *Celastraceae* and in mother liquors from Ma Huang after recovery of ephedrine. Isoln from *Catha edulis* leaves: Wolfes, *Arch. Pharm.* **268**, 81 (1930); from Ma Huang: Smith, *J. Chem. Soc.* **1927**, 2056; **1928**, 51; **1929**, 2755. Physical measurements: Gibson, Levin, *ibid.* **1929**, 2754. Synthesis: Nagai, Kanao, *Ann.* **470**, 157 (1929); Pfanz, Muller, *Arch. Pharm.* **288**, 11, 65 (1955); Pfanz, Wieduwilt, *ibid.* 563; Sicher, Pankova, *Coll. Czech. Chem. Commun.* **20**, 1419 (1955). Toxicity data for hydrochloride: H. Hofmann *et al., Arzneimittel-Forsch.* **5**, 367 (1955).

D-Form, free base, plates from methanol, mp 77.5-78°. Strongly alkaline reaction. $[\alpha]_{546}^{20}$ +37.9° (c = 3 in methanol). Sol in alcohol, chloroform, ether, dil acids.

Hydrochloride, $C_9H_{13}NO.HCl$, *Amorphan, Adiposetten, Exponcit N, Fasupond, Fugoa, Minusin.* Prisms, mp 180-181°. $[\alpha]_D^{20}$ +43.2° (H_2O). Sol in water. pH of aq soln 5.9-6.1. LD_{50} s.c. in mice: 275 mg/kg (Hofmann).

Sulfate, $(C_9H_{13}NO)_2.H_2SO_4$, hexagonal plates, mp 298°. $[\alpha]_{546}^{20}$ +48.7° (c = 1.4 in H_2O). Sol in water.

dl-Form hydrochloride, $C_9H_{13}NO.HCl$, crystals, mp 169-171°.

For other synthetic forms, *see* Phenylpropanolamine Hydrochloride.

USE: In the optical resolution of externally compensated acids: Gibson, Levin, *loc. cit.*

THERAP CAT: Anorexic.

6812. Nortriptyline. *3-(10,11-Dihydro-5H-dibenzo[a,-d]cyclohepten-5-ylidene)-N-methyl-1-propanamine;* 10,11-dihydro-*N*-methyl-5*H*-dibenzo[*a,d*]cycloheptene-$\Delta^{5,\gamma}$-propylamine; 5-(α-methylaminopropylidene)dibenzo[*a,d*]cyclohepta[1,4]diene; 3-(10,11-dihydro-5*H*-dibenzo[*a,d*]cyclohepten-5-ylidene)-*N*-methylpropylamine; 10,11-dihydro-5-(3-methylaminopropylidene)-5*H*-dibenzo[*a,d*][1,4]cycloheptene; desitriptilina; desmethylamitriptyline; Avantyl; Aventyl; Noritren; Ateben; Psychostyl; Sensaval. $C_{19}H_{21}N$; mol wt 263.38. C 86.65%, H 8.04%, N 5.32%. Prepn: Hoffsommer *et al., J. Org. Chem.* 27, 4134 (1962); Neth. pat. Appl. 6,408,512; E. I. Engelhardt, U.S. pat. 3,922,305 (1965, 1975 both to Merck & Co.). Alternative process: N. L. Wendler, U.S. pat. 3,442,949 (1969 to Merck & Co). GC determn in plasma: J. E. Burch, *J. Chromatog.* 308, 165 (1984). Determn in plasma or serum by radioimmunoassay: J. F. Sayegh, *Neurochem. Res.* 11, 193 (1986). Pharmacokinetics: E. H. Rubin *et al., J. Clin. Psychiatry* 46, 418 (1985). Clinical trial in post-stroke depression: J. R. Lipsey *et al., Lancet* 1, 297 (1984). Anticholinergic effect and clinical efficacy in urinary dysfunction: I. Nissenkorn *et al., Eur. Urol.* 12, 109 (1986). Comprehensive description: J. L. Hale in *Analytical Profiles of Drug Substances* vol. 1, K. Florey, Ed. (Academic Press, New York, 1972) pp 233-247.

Hydrochloride, $C_{19}H_{21}N.HCl$, *Acetexa, Allegron, Altilev, Nortrilen, Norzepine, Pamelor, Sensival, Vividyl.* Crystals from ether + ethanol, mp 213-215°. Sol in ethanol, water, chloroform. Practically insol in ether, acetone, benzene. uv max (methanol): 240 nm (ϵ 13900).

THERAP CAT: Antidepressant.

6813. Norvaline. *2-Aminovaleric acid;* α-aminovaleric acid; 2-aminopentanoic acid. $C_5H_{11}NO_2$; mol wt 117.15. C 51.26%, H 9.46%, N 11.96%, O 27.31%. $CH_3(CH_2)_2CH-(NH_2)COOH$. Prepd by treating butyraldehyde ammonia with HCN and HCl: Slimmer, *Ber.* 35, 404 (1902); from 2-acetylvaleric acid ethyl ester: Hamlin, Hartung, *J. Biol. Chem.* 145, 349 (1942); from acetamidomalonic acid diethyl ester: Archer, Albertson, U.S. pat. 2,445,817 (1948 to Winthrop-Stearns); from 1-nitrobutane: Stiles, Finkbeiner, *J. Am. Chem. Soc.* 85, 616 (1963); U.S. pat. 3,055,936 (1962 to Res. Corp.). Prepn of optically active forms: Abderhalden, Kurton, *Fermentf.* 4, 328; *Chem. Zentr.* 1921, III, 296.

DL-Form, minute leaflets from alcohol or water, mp 303° (closed capillary). pK_1' 2.36; pK_2' 9.72. Sublimes without decompn. One gram dissolves in 10 ml water at 18°. Freely sol in hot water; practically insol in alcohol, ether, chloroform, ethyl acetate, petr ether.

L(+)-Form, crystals from dil alc. mp ~305° (closed capillary). $[M]_D$ +29.2° (5N HCl); $[M]_D$ +41.0° (glacial acetic acid). $[\alpha]_D^{20}$ +23.0° (c = 10 in 20% HCl). Freely sol in hot water; insol in alcohol, ether, chloroform, ethyl acetate, petr ether.

D(−)-Form, minute leaflets. mp ~307°. $[\alpha]_D^{20}$ −24.2° (c = 10 in 20% HCl). Freely sol in hot water; insol in alcohol, ether, chloroform, ethyl acetate, petr ether.

6814. Norvinisterone. *(17α)-17-Hydroxy-19-norpregna-4,20-dien-3-one;* 17-hydroxy-17α-vinyl-4-estren-3-one; 17-hydroxy-13-methyl-17α-vinyl-1,2,3,6,7,8,9,10,11,12,13,-14,16,17-tetradecahydro-15*H*-cyclopenta[*a*]phenanthren-3-one; 17α-vinyl-19-nortestosterone; Nor-Progestelea. $C_{20}H_{28}O_2$; mol wt 300.44. C 79.96%, H 9.39%, O 10.65%. Prepn: Colton, U.S. pats. **2,655,518; 2,802,015** (both 1953 to Searle).

Crystals from ethyl acetate + petr ether, mp 169-171°. $[\alpha]_D$ +36°.

THERAP CAT: Progestogen.

6815. Noscapine. *[S-(R*,S*)]-6,7-Dimethoxy-3-(5,6,7,8-tetrahydro-4-methoxy-6-methyl-1,3-dioxolo[4,5-g]isoquinolin-5-yl)-1(3H)-isobenzofuranone; narcotine; l-α-narcotine; l-α-2-methyl-8-methoxy-6,7-methylenedioxy-1-(6,7-dimethoxy-3-phthalidyl)-1,2,3,4-tetrahydroisoquinoline;* narcosine; methoxyhydrastine; opian; opianine; NSC-5366; Coscopin; Coscotabs; Capval; Longatin; Lyobex; Narcompren; Narcotussin; Nectadon; Nicolane; Nipaxon; Noscapalin; Terbenol; Tusscapine; Vadebex. $C_{22}H_{23}NO_7$; mol wt 413.43. C 63.92%, H 5.61%, N 3.39%, O 27.09%. An opium alkaloid, isolated from the plant *Papaver somniferum* L. *Papaveraceae.* Present in amounts up to 11% depending on season and locality. First isoln: Robiquet, *Ann. Chim. Phys.* [2] 5, 275 (1817). Extractable from the water-insoluble residue remaining from the processing of opium for the manufacture of morphine. Racemization to gnoscopine: Rabe, McMillian, *Ann.* 377, 233 (1910); and structural studies: Perkin, Robinson, *J. Chem. Soc.* 99, 775 (1911); Marshall *et al., ibid.* 1934, 1318. Preliminary stereochemical studies: Ohta *et al., Tetrahedron Letters* 1963, 1857; Battersby, Spenser, *J. Chem. Soc.* 1965, 1087. Revised stereochemistry: Blaha *et al., Coll. Czech. Chem. Commun.* 29, 2328 (1964); Snatzke *et al., Tetrahedron* 25, 5059 (1969). Synthesis of racemate: Kerekes, Bognar, *J. Prakt. Chem.* 313, 923 (1971). Biosynthesis: Battersby, Hirst, *Tetrahedron Letters* 1965, 669. Metabolism: N. Tsunoda, Y. Yoshimura, *Xenobiotica* 9, 181 (1979); *eidem, ibid.* 11, 23 (1981). Pharmacokinetics: B. Dhalstroem *et al., Eur. J. Clin. Pharmacol.* 22, 535 (1982). Clinical evaluation as antitussive: D. W. Empey *et al., ibid.* 16, 393 (1979). HPLC determn in serum: K. M. Jensen, *J. Chromatog.* 274, 381 (1983). Comprehensive description: M. A. Al-Yahya, M. M. A. Hassan, *Anal. Profiles Drug Subs.* 11, 407-461 (1982).

Orthorhombic bisphenoidal prisms, tablets from diacetone. Triboluminescent. d 1.395. mp 176°. Sublimes at 150-160° under 11 mm pressure at 2 mm distance. Very weak base forming unstable salts with acids and strong bases. pK 7.8. uv max (ethanol): 209, 291, 309-310 nm (log ϵ 4.86, 3.60, 3.69). Practically insol in vegetable oils. Slightly sol in NH_4OH, hot solns of KOH and NaOH, forming salts. Salts formed with acids are dextrorotatory and unstable in water.

Hydrochloride, $C_{22}H_{23}NO_7 \cdot HCl$, hemihydrate to tetrahydrate, crystals, very sol in water forming basic salts.

Camphorsulfonate, $C_{32}H_{39}NO_{11}S$, *374-JL, Tulisan.* Contains 35.97% camphosulfonic acid. Prepn: Maillard, U.S. pat. 3,108,106 (1963 to Jacques Logeais). Crystals, mp 188-191°. $[\alpha]_D^{33}$ +32.7° (c = 4.56 in water). Freely sol in methanol, ethanol. Slightly sol in ethyl acetate. Practically insol in ether.

dl-Form, *dl-narcotine, gnoscopine.* Long needles from methanol, mp 232° (dec). pK 7.8. Freely sol in carbon disulfide, hot chloroform; sol in ~1500 parts alcohol; sparingly sol in benzene, water.

THERAP CAT: Antitussive.

6816. Nosiheptide. Multhiomycin; RP-9671; Primofax. $C_{51}H_{43}N_{13}O_{12}S_6$; mol wt 1222.38. C 50.11%, H 3.55%, N 14.90%, O 15.71%, S 15.74%. Polythiazole antibiotic produced by *Streptomyces actuosus.* Isoln and characterization: S. Pinnert *et al., Fr.* pat. 1,392,453; *eidem,* U.S. pat. 3,155,581 (1961, 1964 both to Rhône-Poulenc); F. Benazet *et al., Experientia* 36, 414 (1980). NMR determination of mol wt and elemental formula: H. Depaire *et al., Tetrahedron Letters* 1977, 1397, 1401. Structure and configuration: T. Prange *et al., Nature* 265, 189 (1977); C. Pascard *et al., J. Am. Chem. Soc.* 99, 6418 (1977). Biosynthetic study: D. P. Houck *et al., ibid.* 109, 1250 (1987). Identity with multhiomycin: T. Endo, H. Yonehara, *J. Antibiot.* 31, 623 (1978). Mode of action: E. Cundliffe, J. Thompson, *J. Gen. Microbiol.* 126, 185 (1981). *Review:* F. Benazet *et al., Experientia* 36, 414-416 (1980).

Yellow needles, mp 310-320° (dec). $[\alpha]_D^{20}$ +38° (c = 1 in pyridine). uv max (water/DMF): 242, 322 nm ($E_{1cm}^{1\%}$ 525, 229). Sol in chloroform, dioxane, pyridine, DMF, DMSO; slightly sol in methanol, ethanol, ethyl acetate, benzene. Insol in water and petr ether.

THERAP CAT (VET): Antibacterial; growth promotant.

6817. Novembichin. *2-Chloro-N,N-bis(2-chloroethyl)-propanamine hydrochloride;* 2-chloropropyldi(2-chloroethyl)amine hydrochloride; Novoembichin; Embichin 7; Embikhin 7. $C_7H_{15}Cl_4N$; mol wt 255.01. C 32.97%, H 5.93%, Cl 55.61%, N 5.49%. Prepn: Ford-Moore *et al., J. Chem. Soc.* 1946, 819; E. Wilson, M. Tishler, *J. Am. Chem. Soc.* 73, 3635 (1951).

Crystals from chloroform + ethanol, mp 78-79.4°. Free base, colorless oil, bp 114°.

THERAP CAT: Antineoplastic.

6818. Novobiocin. *N-[7-[[3-O-(Aminocarbonyl)-6-deoxy-5-C-methyl-4-O-methyl-β-L-lyxo-hexopyranosyl]oxy]-4-hydroxy-8-methyl-2-oxo-2H-1-benzopyran-3-yl]-4-hydroxy-3-(3-methyl-2-butenyl)benzamide;* crystallinic acid; streptonivicin; PA-93; U-6591; Albamycin (not albomycin); Biotexin; Cardelmycin; Cathocin; Cathomycin; Inamycin; Spheromycin; Vulcamicina; Vulcamycin; Vulkamycin. $C_{31}H_{36}N_2O_{11}$; mol wt 612.63. C 60.78%, H 5.92%, N 4.57%, O 28.73%. Antibiotic substance produced by *Streptomyces spheroides:* Kaczka *et al., J. Am. Chem. Soc.* 77, 6404 (1955); Wolf, U.S. pat. 3,000,873 (1961 to Merck & Co.); Stammer, Miller; Miller; Wallick, U.S. pats. 3,049,475; 3,049,476; 3,049,534 (all 1962 to Merck & Co.). By *Streptomyces niveus:* Hoeksema *et al., J. Am. Chem. Soc.* 77, 6710 (1955); *Antibiot. & Chemother.* 6, 143 (1956); French, U.S. pat. 3,068,221 (1962 to Upjohn). Structure: Shunk *et al., J. Am. Chem. Soc.* 78, 1770 (1956); Hoeksema *et al., ibid.* 2019; Walton *et al., ibid.* 82, 1489 (1960). Conformation: Golding, Richards, *Chem. & Ind. (London)* 1963, 1081. Revised configuration: O. Achmatowicz *et al., Tetrahedron* 32, 1051 (1976). Synthesis: Stammer, U.S. pat. 2,925,411 (1960); Walton, Spencer, U.S. pat. 2,966,484 (1960 to Merck & Co.); Vaterlaus *et al., Helv. Chim. Acta* 47, 390 (1964). Conversion of *Isonovobiocin* to novobiocin: Caron *et al.,* U.S. pat. 2,983,723 (1961 to Upjohn). Antiviral activity: Chang, Weinstein, *Antimicrob. Ag. Chemother.* 1970, 165. Efficacy in canine respiratory infections: B. W. Maxey, *Vet. Med. Small Anim. Clin.* 75, 89 (1980). Mechanism of action studies: Smith, Davis, *J. Bacteriol.* 93, 71 (1967); H. T. Wright *et al., Science* 213, 455 (1981); I. W. Althaus *et al., J. Antibiot.* 41, 373 (1988). *Review:* Brock in *Antibiotics,* vol. 1, R. Gottlieb, P. Shaw, Eds. (Springer-Verlag, New York, 1967) pp 651-665; M. J. Ryan, *ibid.* vol. 5(pt. 1), F. E. Hahn, Ed. (1979) pp 214-234.

Pale yellow orthorhombic crystals from ethanol. *Sensitive to light.* d 1.3448. Dec at 152-156° (a rarer modification dec 174-178°). Acid reaction: pKa_1 4.3; pKa_2 9.1. $[\alpha]_D^{24}$ −63.0° (c = 1 in ethanol). uv max (0.1N NaOH; 0.1N methanolic HCl; pH 7 phosphate buffer): 307; 324; 390 nm ($E_{1cm}^{1\%}$ 600, 390, 350 resp.). Sol in aq soln above pH 7.5. Practically insol in more acidic solns. Sol in acetone, ethyl acetate, amyl acetate, lower alcohols, pyridine. Additional soly data: Weiss *et al., Antibiot. & Chemother.* 7, 374 (1957).

Monosodium salt, $C_{31}H_{35}N_2NaO_{11}$, *Robiocina.* Minute crystals, dec 220°. $[\alpha]_D^{24}$ −38° (c = 2.5 in 95% ethanol); $[\alpha]_D^{24}$ −33° (c = 2.5 in water). Freely sol in water. A 100 mg/ml soln has a pH of 7.5 and a half-life of ~30 days at 25° and several months at 4°. Soly data: Weiss *et al., loc. cit.* Properties: Birlova, Traktenberg, *Antibiotiki* 13, 997 (1968).

THERAP CAT: Antibacterial.
THERAP CAT (VET): Antimicrobial.

6819. Novoldiamine. *N^1,N^1-Diethyl-1,4-pentanediamine;* 1-diethylamino-4-aminopentane; 4-amino-1-diethylaminopentane; 2-amino-5-diethylaminopentane; δ-diethylamino-α-methylbutylamine; δ-diethylaminoisopentylamine. $C_9H_{22}N_2$; mol wt 158.29. C 68.29%, H 14.01%, N 17.70%.

Prepd commercially from 2-diethylaminoethanol and ethyl acetoacetate. 2-Chlorotriethylamine (formed by the action of thionyl chloride on the alcohol) is condensed with the sodium derivative of ethyl acetoacetate to yield an intermediate ester, which is hydrolyzed and decarboxylated to *novol ketone* (5-diethylamino-2-pentanone). This is hydrogenated in the presence of ammonia to yield novoldiamine. Several other prepns, i.e., from 1,4-pentanediol and diethylamine: Kyrides, U.S. pat. **2,365,825** (1944 to Monsanto). Purification procedure: Jones, U.S. pat. **2,400,934** (1946 to Lilly).

Liquid. Amine odor. d_{26}^{20} 0.819. bp_{753} 200-200.5°. n_D^{26} 1.4403. Sol in water, alcohol, ether.

USE: Manuf quinacrine and other antimalarials having the same basic side chain.

6820. Novonal. *2,2-Diethyl-4-pentenamide;* diethyl-allylacetamide; Epinoval. $C_9H_{17}NO$; mol wt 155.24. C 69.63%, H 11.04%, N 9.02%, O 10.31%. Description: Bockmühl, Schaumann, *Deut. Med. Wochenschr.* **54,** 270 (1928).

White powder, mp 75-76°. Sol in 120 parts water; freely sol in alcohol, ether.

THERAP CAT: Hypnotic.

6821. Noxiptilin. *10,11-Dihydro-5H-dibenzo[a,d]cyclo-hepten-5-one O-[2-(dimethylamino)ethyl]oxime;* 2-[[(10,11-dihydro-5H-dibenz[a,d]cyclohepten-5-ylidene)amino]oxy]-N,N-dimethylethylamine; 5-(2-dimethylaminoethyloximino)-5H-dibenzo[a,d]cyclohepta-1,4-diene; 5-[β-(dimethylamino)-ethoxyimino]-10,11-dihydro-5H-dibenzo[a,d]cycloheptene; noxiptyline; dibenzoxin. $C_{19}H_{22}N_2O$; mol wt 294.40. C 77.52%, H 7.53%, N 9.52%, O 5.43%. Prepn: Engelhard *et al.,* **Ger.** pat. **1,198,353** (1965); Schutz, Hoffmeister, **Ger.** pat. **1,225,169;** Schutz *et al.,* U.S. pat. **3,505,321** (1966, 1970 both to Bayer); Wrigley, Leeming, **Brit.** pat. **1,045,911** (1966 to Pfizer). Toxicity study: F. Hoffmeister *et al., Arzneimittel-Forsch.* **19,** 846 (1969). Series of articles on pharmacology, metabolism, clinical data: *ibid.* 458-467; 846-878.

Free base, $bp_{0.05}$ 160-164°.
Hydrochloride, $C_{19}H_{22}N_2O.HCl$, *Bay 1521, Agedal, Nogédal.* Crystals from ethanol + ethyl acetate, mp 185-187°. LD_{50} s.c. in mice: 240 mg/kg (Hoffmeister).

THERAP CAT: Antidepressant.

6822. Noxythiolin. *N-(Hydroxymethyl)-N'-methylthio-urea; 1-(hydroxymethyl)-3-methyl-2-thiourea;* noxytiolin; Noxyflex-S. $C_3H_8N_2OS$; mol wt 120.18. C 29.98%, H 6.71%, N 23.31%, O 13.31%, S 26.68%. Prepn: A. Aebi, E. Hafstetter, **Brit.** pat. **970,414** (1964 to Ed. Geistlich Söhne), *C.A.* **61,** 15981e (1964). Antibacterial activity *in vitro:* D. Horsfield, *Clin. Trials J.* **4,** 625 (1967); J. I. Blenkharn, *J. Pharm. Pharmacol.* **42,** 589 (1990). Inhibition of bacterial adherence *in vitro:* S. P. Gorman *et al., J. Appl. Bacteriol.* **60,** 311 (1986). Clinical evaluation in peritonitis: F. Antos, *Clin. Trials J.* **17,** 159 (1980); in urinary tract sepsis: M. A. Jones, A. Hasan, *Brit. J. Urol.* **62,** 311 (1988).

Crystals. mp 88-90°. Sol in water (10% w/v), ethanol (4% w/v). pH of 2.5% soln: 6.31-7.0. LD_{50} orally in mice: > 3 g/kg (Aebi, Hafstetter).

THERAP CAT: Antiseptic.

6823. Nucleic Acids. Macromolecules (polynucleotides) found in living cells which are of fundamental importance in controlling the reproduction, growth and metabolism of living systems. Nucleic acids occur in close association with proteins to form *nucleoproteins.* First isolated in 1868-1869 by F. Miescher from the nuclei of pus cells and called *nuclein.* The components of nucleic acids are purine and pyrimidine bases (primarily adenine, guanine, cytosine, thymine, uridine), a sugar (D-2-deoxyribose or D-ribose) and phosphoric acid. A purine or pyrimidine base in glycosidic linkage with the sugar forms a *nucleoside* (e.g. adenosine, thymidine, *q.v.*). In naturally occurring *N*-nucleosides, the carbon in the 1'-position of the sugar is attached in β-glycosyl linkage to the nitrogen in the 9 position of the purine or in the 1 position of the pyrimidine. (Some *C*-nucleosides are also known). The phosphate ester of a nucleoside is a *nucleotide* (e.g. 5'-guanylic acid, 3'-cytidylic acid, *q.v.*) which is the smallest repeating unit of nucleic acids. (These terms also apply to synthetic derivatives and analogs of naturally occurring nucleosides and nucleotides). Comprehensive reviews of bases, nucleotides and nucleosides: A. M. Michelson, *The Chemistry of Nucleosides and Nucleotides* (Academic Press, New York, 1963); Ts'o, "Monomeric Units of Nucleic Acids—Bases, Nucleosides and Nucleotides" in *Biological Macromolecules* vol. **4** entitled *Fine Structure of Proteins and Nucleic Acids,* G. D. Fasman, S. N. Timasheff, Eds. (Dekker, New York, 1970) pp 49-190; C. A. Dekker, L. Goodman, "Nucleosides" in *The Carbohydrates* vol. **2A,** W. Pigman *et al.,* Eds. (Academic Press, New York, 1970) pp 1-68; G. R. Pettit, *Synthetic Nucleotides* vol. **1** (Van Nostrand Reinhold, New York, 1972) 252 pp.
There are two main types of nucleic acids: deoxyribonucleic acid (DNA) and ribonucleic acid (RNA). Both types are present in all types of cells, whether of plant or animal origin. See Deoxyribonucleic Acid and Ribonucleic Acid. Monographs: E. Chargaff, J. N. Davidson, *The Nucleic Acids,* 3 vols. (Academic Press, New York, 1955, 1960); V. R. Potter, *Nucleic Acid Outlines* vol. **I** (Burgess Publishing, Minneapolis, 1960); D. O. Jordan, *The Chemistry of Nucleic Acids* (Butterworth, London, 1960); R. F. Steiner, R. F. Beers, *Polynucleotides* (Elsevier, New York, 1961); E. Chargaff, *Essays on Nucleic Acids* (Elsevier, New York, 1963) 211 pp; E. Harbers *et al., Introduction to Nucleic Acids* (Reinhold, New York, 1968) 403 pp; *Organic Chemistry of Nucleic Acids* (Parts A, B) N. K. Kochetkov, E. I. Budovskii, Eds. (Plenum, New York, Eng. ed., 1971, 1972) 639 pp; J. N. Davidson, *The Biochemistry of the Nucleic Acids* (Academic Press, New York, 7th ed., 1972) 396 pp; *Basic Principles in Nucleic Acid Chemistry* vols. **1, 2,** P. O. P. Ts'o, Ed. (Academic Press, New York, 1974) 636, 519 pp; V. A. Bloomfield *et al., Physical Chemistry of Nucleic Acids* (Harper & Row, New York, 1974) 517 pp; *MTP Int. Rev. Sci.: Biochem. Ser. One* vol. **6** entitled *Biochemistry of Nucleic Acids,* K. Burton, Ed. (University Park Press, Baltimore, 1974) 364 pp. Series of volumes on nucleic acids: *Progress in Nucleic Acid Research and Molecular Biology* (Academic Press, New York, 1963-present).

6824. Nucleocidin. *4'-C-Fluoroadenosine 5'-sulfamate;* 9-(4-fluoro-5-O-sulfamoylpentofuranosyl)adenine; 4'-fluoro-5'-O-sulfamoyladenosine. $C_{10}H_{13}FN_6O_6S$; mol wt 364.31. C 32.97%, H 3.60%, F 5.21%, N 23.07%, O 26.35%, S 8.80%. Antitrypanosomal antibiotic produced by *Streptomyces calvus.* The first naturally occurring derivative of a fluoro sugar. Isoln and activity studies: Thomas *et al., Antibiot. Ann.* **1956-1957,** 716; Hewitt *et al., ibid.* 722. Partial structure: Waller *et al., J. Am. Chem. Soc.* **79,** 1011 (1957). Revised formula and structure: Morton *et al., ibid.* **91,** 1535 (1969). Synthesis: Jenkins *et al., ibid.* **93,** 4323 (1971).

As the monohydrate, crystalline, weakly alkaline substance. mp > 190° (dec). uv max (methanol): 259 nm (ε 15000). LD$_{50}$ i.p. in mice: ~0.2 mg/kg (Thomas).

6825. Nudic Acids. Antibiotic substances produced by the basidiomycete *Tricholoma nudum.* Isoln: Heatley in Florey *et al., Antibiotics* vol. I (Oxford, 1949) p 358; *cf.* Wilkins, Harris, *Ann. Appl. Biol.* **31**, 261 (1944).
Nudic acid A, C$_{14}$H$_{20}$O$_3$, large glistening needles or plates from hexane, mp 123.5°. pK between 3.5 and 4.0. Optically inactive in alc. Slightly sol in hot water; sol in the usual organic solvents; freely sol in sodium hydroxide and sodium carbonate solns.
Nudic acid B *see* Diatretyne II.

6826. Nupharidine. *[1R-(1α,4β,5α,7β,9aα)]-4-(3-Fur-anyl)octahydro-1,7-dimethyl-2H-quinolizine 5-oxide.* C$_{15}$H$_{23}$-NO$_2$; mol wt 249.35. C 72.25%, H 9.30%, N 5.62%, O 12.83%. From rhizome of *Nuphar luteum* (L.) Sibth. & Sm. (*Nymphaea lutea* L.), *Nymphaeaceae:* Goris, Crété, *Bull Sci. Pharmacol.* **17**, 13 (1910). Isoln from *N. japonicum* DC., *Nymphaeaceae* and structure: Kotake *et al., Ann.* **606**, 148 (1957). Total synthesis: Kotake *et al., Bull. Chem. Soc. Japan* **35**, 698 (1962). Abs configuration: Kawasaki *et al., ibid.* **41**, 1264 (1968); La Londe *et al., J. Am. Chem. Soc.* **93**, 2501 (1971). Determn of crystal structure, abs config and stereochemistry by x-ray diffraction methods: J. Ohrt *et al., J. Cryst. Mol. Struct.* **3**, 3 (1973).

Crystals, mp 221°. [α]$_D$ +15°. Soluble in alcohol, chloroform, ether, acetone, amyl alcohol, dil acids. The base is tasteless, the salts bitter.

6827. Nutgall. Galla; galls; Aleppo-galls; Turkey-galls; Mecca-galls. Excrescence from the young twigs of *Quercus infectoria* Oliv. and other allied species of *Quercus, Fagaceae. Habit.* Asia Minor (Levant). *Constit.* 50-60% tannic acid, 2-4% gallic acid, ellagic acids, resin.
Chinese nutgall is the excrescence on the leaf or leafstalk of *Rhus semialata* Murr., *Anacardiaceae.* Contains about 70% tannin.
Incompat. Alkalies, alkaloids; salts of Cu, Fe, Pb, Zn; AgNO$_3$; opium in soln.
Caution: Irritating to mucous membranes.
USE: Manuf tannin and ink; dyeing; tanning.
THERAP CAT: Astringent.
THERAP CAT (VET): Has been used as a topical astringent.

6828. Nux Vomica. Quaker buttons; bachelor's buttons; poison nut; dog buttons; vomit nut. Dried, ripe seeds of *Strychnos nux-vomica* L., *Loganiaceae. Habit.* Southern Asia, Northern Australia. *Constit.* 1-1.4% strychnine, about an equal amount of brucine; strychnicine, loganin, caffeotannic (igasuric) acid, proteins. Nux vomica from Saigon contains 1.6-2% strychnine.

Caution: Extremely poisonous.
THERAP CAT: Formerly as bitter tonic.
THERAP CAT (VET): Has been used as a bitter tonic.

6829. Nybomycin. *8-(Hydroxymethyl)-6,11-dimethyl-2H,4H-oxazolo[5,4,3-ij]pyrido[3,2-g]quinoline-4,10(11H)-dione;* 6,11-dimethyl-8-(hydroxymethyl)pyrido[3,2-g]oxa-zolo[5,4,3-ij]quinoline-4,10(2H,11H)-dione. C$_{16}$H$_{14}$N$_2$O$_4$; mol wt 298.30. C 64.42%, H 4.73%, N 9.39%, O 21.45%. Antibiotic substance produced by *Streptomycete* A 717 isolated from Missouri soil: Strelitz *et al., Proc. Nat. Acad. Sci. USA* **41**, 620 (1955); Eble *et al., Antibiot. & Chemother.* **8**, 627 (1958); Brock, Sokolski, *ibid.* 631. Structure: Rinehart, Renfroe, *J. Am. Chem. Soc.* **83**, 3729 (1961). Revised structure: Rinehart *et al., ibid.* **92**, 6994 (1970). Total synthesis of *deoxynybomycin:* Forbis, Rinehart, *ibid.* 6995. Total synthesis of nybomycin: *eidem, J. Antibiot.* **24**, 326 (1971); *eidem, J. Am. Chem. Soc.* **95**, 5003 (1973).

Needles from acetic acid, mp 325-330°. Sublimes at 250° (15 mm). Optically inactive. uv max (ethanol): 266, 285 nm. Soluble in concd acids. Very slightly sol in water, alkalies, and common organic solvents. Shows antiphage and antibacterial properties. LD$_{50}$ i.p. in mice: 650 mg/kg (Brock, Sokolski). ^{13}C NMR spectrum: A. M. Nadzan, K. L. Rinehart, *J. Am. Chem. Soc.* **99**, 4647 (1977).
Acetate, C$_{18}$H$_{16}$N$_2$O$_5$, crystals from chloroform + ethanol, mp 236-237°.
Succinate, C$_{20}$H$_{19}$N$_2$O$_7$, crystals from dimethylformamide. Practically insol in water.

6830. Nylidrin. *4-Hydroxy-α-[1-[(1-methyl-3-phenyl-propyl)amino]ethyl]benzenemethanol; p-hydroxy-α-[1-[(1-methyl-3-phenylpropyl)amino]ethyl]benzyl alcohol; p-hydr-oxy-N-(1-methyl-3-phenylpropyl)norephedrine; buphenine; 1-(p-hydroxyphenyl)-2-(1'-methyl-3'-phenylpropylamino)-1-propanol; phenyl-sec-butyl norsuprifen.* C$_{19}$H$_{25}$NO$_2$; mol wt 299.41. C 76.22%, H 8.42%, N 4.68%, O 10.69%. Prepn: **Fr. pat. 968,273** (1950 to Troponwerke Dinklage); **Brit.** pats. 669,574-5 (1952); *Chem. & Eng. News* **33**, 2896 (1955); Külz, Schöpf, **U.S.** pats. **2,661,372-3** (1953). Pharmacology: T. T. Yen, D. V. Pearson, *Res. Commun. Chem. Pathol. Pharmacol.* **23**, 11 (1979); B. Fichtl, W. Felix, *Eur. J. Pharmacol.* **65**, 333 (1980).

Crystals from methanol, mp 111-112°.
Hydrochloride, C$_{19}$H$_{25}$NO$_2$.HCl, *SKF-1700-A, Arlidin, Bufedon, Buphedrin, Dilatal, Dilatol, Dilatropon, Dilydrin, Opino, Penitardon, Perdilatal, Rudilin, Rydrin, Tocodily-drin, Tocodrin.* Crystals. Sparingly sol in water; slightly sol in alcohol. Practically insol in ether, chloroform, benzene.
THERAP CAT: Vasodilator (peripheral).

6831. Nylon. Polyamide. Generic term used to describe "a manufactured fiber in which fiber-forming substances are any long-chain synthetic polyamide having recurring polyamide groups (—CONH—) as an integral part of the polymer chain". Formed from various combinations of diacids, diamines, and amino acids. May be formed also by addition polymerization. The linear polyamides have achieved the greatest commercial success. Shorthand nomenclature of nylons involves the use of numbers: a single numeral indicating the number of carbon atoms in a monomer, e.g. nylon 6; two numbers indicating a polymer formed from diamines and dibasic acids, the first numeral indicating the number of carbon atoms separating the nitrogen atoms of the diamine, the second indicating the number of straight-chain carbon atoms in the dibasic acid, e.g. nylon 6,6. First produced by E. I. du Pont de Nemours & Co. according to patents of W. H. Carothers. The name *nylon* was dedicated to public domain on Oct. 27, 1938 at the Herald Tribune Forum where the product itself was announced. *Reviews:* R. W. Moncrieff, *Man-made Fibres* (John Wiley, New York, 1963) pp 335-355; Snider, Richardson, "Polyamide Fibers" in *Encyclopedia of Polymer Science and Technology* vol. 10 (Interscience, New York, 1969) pp 347-460; J. H. Saunders, "Polyamides (Fibers)" in Kirk-Othmer *Encyclopedia of Chemical Technolgy* vol. 18 (Wiley-Interscience, New York, 3rd ed., 1982) pp 372-405. *Book: Nylon Plastics,* M. I. Kohan, Ed. (Wiley-Interscience, New York, 1973).

Crystalline solids characterized by low specific gravity, high strength, durability, high flexibility, and high tensile strength. Soluble in phenol, cresols (especially *m*-cresol), xylenol, formic acid. Insoluble in alcohols, esters, ketones, hydrocarbons. Hydrolysis and degradation occur at higher temperatures, esp in the melt. Stable to aqueous alkali. Degrades rapidly in aqueous acids. Undergoes photodegradation.

USE: In production of synthetic fibers for various textile and domestic uses. Surgical aid (nonabsorbable suture).

6832. Nylon 6. *Poly[imino(1-oxo-1,6-hexanediyl)]; poly(iminocarbonylpentamethylene);* Caprolan; Enkalon; Grilon; Kapron; Mirlon; Perlon; Phrilon; Amilan. Linear polymer obtained by polymerization of ε-caprolactam, *q.v.:* Schlack, U.S. pat. **2,241,321** (1941 to I. G. Farbenind.). The importance of this fiber increased with the discovery that caprolactam can be produced by the nitrosation of cyclohexanecarboxylic acid: Muench *et al.,* U.S. pats. **3,022,291** and **3,108,096** (1962, 1963, both to Snia Viscosa). *Review:* R. W. Moncrieff, *Man-Made Fibres* (John Wiley & Sons, New York, 1963) pp 335-355; H. K. Reimschuessel, *J. Polym. Sci., Macromol. Rev.* **12**, 65-139 (1977).

n = approx 200

Softens at 210°. mp 223°. Can withstand a temp of 100° for long periods of time. d_4^{20} 1.14. Moisture regain is about 4%. Swelling is low; if steeped in water and then centrifuged its volume increases by about 13-14%. Immune to microbiological attack. Resistant to most org chemicals, but dissolved by phenol, cresol, and strong acids.

USE: Tire cord; fishing lines; tow ropes; hose manuf; woven fabrics.

6833. Nylon 46. *Poly[imino-1,4-butanediylimino(1,6-dioxo-1,6-hexanediyl)]; poly(tetramethyleneadipamide);* Stanyl. Symmetrical polyamide with higher melting point and greater tensile strength than nylon 66 or nylon 6, *q.v.* Prepn from 1,4-diaminobutane and adipic acid: W. H. Carothers, U.S. pat. **2,130,948** (1938 to Du Pont). Improved process: E. H. J. P. Bour, J. M. M. Warnier, Eur. pat. Appl. **77,106** corresp to U.S. pat. **4,463,166** (1983, 1984 both to Stamicarbon). Account of prepn by melt polymerization, physical properties: R. J. Gaymans *et al., J. Polym. Sci., Polym. Chem. Ed.* **15**, 537 (1977). Brief account: D. O'Sullivan, *Chem. & Eng. News* **62**, 33 (May 21, 1984).

Clear polymer, mol wt 22,000-45,000. mp 283-319°. d 1.20 (solution-cast film). Insol in most solvents. Sol in formic acid. Slightly sol in trifluoroacetic acid.

USE: Industrial fiber; fiber-reinforced rubber products.

6834. Nystatin. Fungicidin; Biofanal; Diastatin; Candex; Candio-Hermal; Mycostatin; Moronal; Nystan; Nystavescent; O-V Statin. Polyene antifungal antibiotic complex containing 3 biologically active components, A_1, A_2, A_3. Produced by *Streptomyces noursei, S. aureus* and other *Streptomyces* spp: Hazen, Brown, *Science* **112**, 423 (1950); *Proc. Soc. Exp. Biol. Med.* **76**, 93 (1951); Raubitscheck *et al., Antibiot. & Chemother.* **2**, 179 (1952); Cohen, Webb, *Arch. Pediatrics* **69**, 414 (1952); Dutcher *et al., Antibiot. Ann.* **1953-1954**, 191; *eidem, Therapy of Fungus Diseases* (Little, Brown, Boston, 1955) p 168. Review of early literature: Brown, Hazen, *Trans. N.Y. Acad. Sci.,* Ser. II, **19** (1956-1957) pp 447-456. Purification: Vandeputte, U.S. pat. **2,-832,719** (1958 to Olin Mathieson); Renella, U.S. pat. **3,517,-100** (1970 to Am. Cyanamid). Chemistry and partial structure: Birch *et al., Tetrahedron Letters* **1964**, 1491; of nystatin A_1 and A_2: Shenin *et al., Antibiotiki* **13**, 387 (1968). Structure of the aglycone: Manwaring *et al., ibid.* **1969**, 5319. Complete structure of A_1: Chong, Rickards, *ibid.* **1970**, 5145; Borowski *et al., ibid.* **1971**, 685. Revised structure: R. C. Pandey, K. L. Rinehart, *J. Antibiot.* **29**, 1035 (1976). Stereochemical study of A_1: J. M. Lancelin *et al., Tetrahedron Letters* **29**, 2827 (1988). Structure of A_3: J. Zielinski *et al., J. Antibiot.* **41**, 1289 (1988). Mechanism of action: R. W. Holz in *Antibiotics* vol. **5**, pt. 2, F. E. Hahn, Ed. (Springer-Verlag, New York, 1979) pp 313-340. Toxicity study: H. Seneca, *Antibiot. Ann.* **1955-1956**, 697. Comprehensive description: G. W. Michel, *Anal. Profiles Drug Subs.* **6**, 341-421 (1977).

Nystatin A₁

Light yellow powder. Gradually decomposes above 160° without melting by 250°. $[\alpha]_D^{25}$ −10° (glacial acetic acid); +21° (pyridine); +12° (DMF); −7° (0.1N HCl in methanol). uv max (ethanol): 290, 307, 322 nm. Exhibits strong reducing properties. Solubilities determined by Weiss *et al., Antibiot. & Chemother.* **7**, 374 (1957) in mg/ml at about 28°: water 4.0; methanol 11.2; ethanol 1.2; carbon tetrachloride 1.23; chloroform 0.48; benzene 0.28; ethylene glycol 8.75. Solns and aq suspensions begin to lose activity soon after prepn. Aq suspensions are stable for 10 minutes on heating to 100° at pH 7.0; also stable in moderately alkaline media, but labile at pH 9 and pH 2. Heat, light, and oxygen accelerate decompn. Activity not diminished by blood or serum. LD$_{50}$ i.p. in mice: ~200 mg/kg (Seneca).

Nystatin A₁, $C_{47}H_{75}NO_{17}$.

THERAP CAT: Antifungal.

THERAP CAT (VET): Antifungal; growth promotant.

O

6835. Obidoxime Chloride. *1,1'-[Oxybis(methylene)]-bis[4-(hydroxyimino)methyl]pyridinium dichloride; 1,1'-(oxy-dimethylene)bis[4-formylpyridinium]dichloride dioxime; N,-N*-dimethyleneoxidebis(pyridinium-4-aldoxime) dichloride; bis(4-hydroxyiminomethylpyridinium-1-methyl) ether dichloride; bis(isonicotinaldoxime 1-methyl) ether dichloride; BH-6; LüH6; Toksobidin; Toxogonin. $C_{14}H_{16}Cl_2N_4O_3$; mol wt 359.21. C 46.81%, H 4.49%, Cl 19.74%, N 15.60%, O 13.36%. Prepn from pyridine aldoxime and α,α'-dichloro-dimethyl ether to Lüttringhaus *et al.,* U.S. pat. **3,137,702** (1963, 1964 to E. Merck); Lüttringhaus, Hagedorn, *Arzneimittel-Forsch.* **14**, 1 (1964). Pharmacology and toxicology: Erdman, Engelhard, *ibid.* 5; Mayer, Michalek, *Biochem. Pharmacol.* **20**, 3029 (1971); Bajgar *et al., Eur. J. Pharmacol.* **19**, 199 (1972). Hydrolysis studies: Christenson, *Acta Pharm. Suecica* **9**, 309 (1972).

Occurs in two interchangeable isomeric forms (*syn* and *anti*). Crystals from HCl contg 70% alcohol, dec 225°. Also reported as syn, mp 235-236°; anti, mp 218-220° [Leitis *et al., C.A.* **71**, 81098d (1969)]. Freely sol in water, stable in 1-10% aq solns. LD_{50} orally in mice: > 2240 mg/kg.
Dibromide, $C_{14}H_{16}Br_2N_4O_3$, dec 202-203°.
THERAP CAT: Cholinesterase reactivator.

6836. Ochratoxins. Toxic metabolites from *Aspergillus ochraceus* Wilh.: Scott, *Mycopathol. Mycol. Appl.* **25**, 213 (1965); *A. sulphureus* and *A. melleus:* Lai *et al., Appl. Microbiol.* **19**, 542 (1970); and *Penicillium viridicatum* Westling: van Walbeek *et al., Can. J. Microbiol.* **15**, 1281 (1969). As these molds occur widely, some toxins have been found as natural contaminants on corn, peanuts, storage grains, cottonseed and other decaying vegetation. Isolns: van Walbeek *et al., ibid.* **14**, 131 (1968); Shotwell *et al., Appl. Microbiol.* **17**, 765 (1969); Nesheim, *J. Assoc. Offic. Anal. Chem.* **52**, 975 (1969); Natori *et al., Chem. Pharm. Bull.* **18**, 2259 (1970). Structure and stereochemistry: van der Merwe *et al., J. Chem. Soc.* **1965**, 7083; *Nature* **205**, 1112 (1965). Synthesis of ochratoxins A, B and ester deriv: Steyn, Holzapfel, *Tetrahedron* **23**, 4449 (1967); Nesheim, *loc. cit.;* Roberts, Woolven, *J. Chem. Soc. (C)* **1970**, 278. Facile synthesis of A: G. A. Kraus, *J. Org. Chem.* **46**, 201 (1981). Toxicity study: Purchase, Theron, *Food Cosmet. Toxicol.* **6**, 479 (1968). Physicochemical data: A. E. Pohland *et al., Pure Appl. Chem.* **54**, 2220 (1982). Comprehensive reviews: Steyn, "Ochratoxin and other Dihydroisocoumarins" in *Microbial Toxins* vol. VI, A. Ciegler, *et al.,* Eds. (Academic Press, New York, 1971) p 179-205; F. S. Chu, *Crit. Rev. Toxicol.* **2**, 499-524 (1974); P. Krogh in *Natural Toxins*, D. Eaker, P. Wadström, Eds. (Pergamon, New York, 1980) pp 673-680; A. E. Pohland *et al., Pure Appl. Chem.* **64**, 1029-1046 (1992).

Ochratoxin A   Cl

Ochratoxin A, $C_{20}H_{18}ClNO_6$, *(R)-N-[(5-chloro-3,4-dihydro-8-hydroxy-3-methyl-1-oxo-1H-2-benzopyran-7-yl)carbonyl]-L-phenylalanine.* The major ochratoxin component. Crystals from xylene, mp 169°. Exhibits green fluorescence. $[\alpha]_D$ −118° (c = 1.1 in CHCl₃). uv max (ethanol): 215, 333 nm (ε 34000; 2400) (van der Merwe). Also frequently reported as mp 90° from benzene (one mole of benzene of crystalliza-

tion). uv max: 213, 332 nm (ε 36800; 6400) (Steyn, Holzapfel). LD_{50} orally in rats: 20-22 mg/kg (Purchase, Theron).
Ochratoxin B, $C_{20}H_{19}NO_6$, *(R)-N-[(3,4-dihydro-8-hydroxy-3-methyl-1-oxo-1H-2-benzopyran-7-yl)carbonyl]-L-phenylala-nine.* The less toxic dechloro deriv of ochratoxin A. Crystals from methanol, mp 221° (van der Merwe); also reported as 208-209° (Nesheim). Exhibits blue fluorescence. $[\alpha]_D$ −35° (c = 0.15 in ethanol). uv max: 218, 318 nm (ε 37200, 6900).
Ochratoxin C, $C_{22}H_{22}ClNO_6$, *(R)-N-[(5-chloro-3,4-dihydro-8-hydroxy-3-methyl-1-oxo-1H-2-benzopyran-7-yl)carbonyl]-L-phenylalanine ethyl ester.* The equally toxic amorphous ethyl ester of ochratoxin A: Steyn, Holzapfel, *J. S. Afr. Chem. Inst.* **20**, 186 (1967), *C.A.* **69**, 2797p (1968). uv max (ethanol): 333 nm (ε 6500).
Note: Ochratoxin A may reasonably be anticipated to be a carcinogen: *Seventh Annual Report on Carcinogens* (PB95-109781, 1994) p 310.
USE: Exptlly as teratogen and carcinogen.

6837. Ocimene. $C_{10}H_{16}$; mol wt 136.24. C 88.16%, H 11.84%. From the leaves of *Ocimum basilicum* L., *Labiatae; Baronia dentigeroides* Cheel, *Rutaceae; Litsea zeylanica* C. & T. Nees, *Lauraceae; Homoranthus flavescens* A. Cunn., *Myrtaceae.* From the fruits of *Evodia rutaecarpa* (Juss.) Hook. f. & Thoms., *Rutaceae.* Exists in two modifications: α-form, *3,7-dimethyl-1,3,7-octatriene;* and β-form, *3,7-dimethyl-1,3,6-octatriene.* Cis and trans refers to the stereochemistry at the double bond between positions 3 and 4. Isoln and structure: van Romburgh, *Proc. Kon. Ned. Acad. Wetensch.* **3**, 454 (1900); Enklaar, *Rec. Trav. Chim.* **26**, 157 (1907); **27**, 422 (1908); **36**, 215 (1917); **45**, 337 (1926). Structure: Sutherland, *J. Am. Chem. Soc.* **74**, 2688 (1952). Separation of isomers: Ohloff *et al., Ann.* **675**, 83 (1964). Synthesis of *trans*-α-form: O. P. Vig *et al., Indian J. Chem.* **7**, 1111 (1969). Stereospecific synthesis: O. P. Vig *et al., ibid.* **15B**, 25 (1977).

trans-β-Ocimene

Oil. Pleasant odor. Mixture of isomers. bp_{70} 100°. d_4^{20} 0.8006. n_D^{20} 1.4862. uv max (methanol): 233 nm (ε 26,200). Practically insol in water. Sol in alcohol, chloroform, ether, glacial acetic acid.
trans-β-Form, d_4^{20} 0.799. n_D^{20} 1.4893. uv max (ethanol): 232 nm (ε 27,600).
cis-β-Form, d_4^{20} 0.799. n_D^{20} 1.4877. uv max (ethanol): 237.5 nm (ε 21,000).
trans-α-Form, d_4^{20} 0.793. n_D^{20} 1.4802. uv max (ethanol): 231 nm (ε 27,300).
cis-α-Form, d_4^{20} 0.794. n_D^{20} 1.4789. uv max (ethanol): 234.5 nm (ε 21,600).

6838. Octabenzone. *[2-Hydroxy-4-(octyloxy)phenyl]-phenylmethanone; 2-hydroxy-4-(octyloxy)benzophenone;* benzophenone-12; Spectra-Sorb UV 531. $C_{21}H_{26}O_3$; mol wt 326.44. C 77.27%, H 8.03%, O 14.70%. Prepn: Armitage *et al.,* U.S. pat. **3,098,842** (1963 to du Pont).

Crystals, mp 45-46°.
USE: To stabilize polyethylene against deterioration by ultraviolet light.
THERAP CAT: Ultraviolet screen.

6839. Octacaine. *3-(Diethylamino)-N-phenylbutan-amide;* 3-diethylaminobutyranilide. $C_{14}H_{22}N_2O$; mol wt 234.34. C 71.76%, H 9.46%, N 11.95%, O 6.83%. Prepn: Hofstetter, Wilder Smith, *Helv. Chim. Acta* **36**, 1698 (1953); Hofstetter, U.S. pat. **2,851,393** (1958 to E. Geistlich Söhne).

Crystals from petr ether, mp 46-47°. bp_1 200°. Easily sol in stoichiometric amount of HCl, in ether, alc, benzene. Hydrochloride, $C_{14}H_{22}N_2O.HCl$, *Amplicain.* mp 132-134°. Sol in water.

THERAP CAT: Anesthetic (local).

6840. Octacosanol. *1-Octacosanol; n*-octacosanol; octacosyl alcohol. $C_{28}H_{58}O$; mol wt 410.77. C 81.87%, H 14.23%, O 3.89%. $CH_3(CH_2)_{26}CH_2OH$. Constituent of vegetable waxes. Isoln from the wax found on green blades of wheat: Pollard *et al., Biochem. J.* **27**, 1889 (1933); from carnauba wax: Koonce, Brown, *Oil & Soap* **21**, 231 (1944). Synthesis starting with behenic acid: Bleyberg, Ulrich, *Ber.* **64**, 2504 (1931); Francis *et al., Proc. Roy. Soc. (London)* A **158**, 691 (1937).

Crystals from much acetone, mp 83.4°. Insol in water. Sol in carbon disulfide, other fat solvents, oils.

6841. Octafluorocyclobutane. Perfluorocyclobutane; Freon C318. C_4F_8; mol wt 200.03. C 24.02%, F 75.98%. Prepd by pyrolysis of chlorodifluoromethane: Downing *et al.,* U.S. pat. **2,384,821** (1945 to Kinetic Chem.); U.S. pats. **2,551,573; 2,615,926** (1951; 1952 to du Pont). Thermodynamic studies: Furukawa *et al., J. Res. Nat. Bur. Stand.* **52**, 11 (1954); Duus, *Ind. Eng. Chem.* **47**, 1445 (1955).

Nonflammable, nontoxic gas. d_{vapor}^{27} 8.2. mp −41.4°. bp −6.04° (also reported bp −5°). Heat of formation at 25° = 352 kcal. Heat of combustion 1359 ± 39 cal/g.

USE: Refrigerant; heat-transfer medium.

6842. 2,2,3,3,4,4,5,5-Octafluoro-1-pentanol. *1H,1H,-5H*-Octafluoropentanol. $C_5H_4F_8O$; mol wt 232.07. C 25.88%, H 1.74%, F 65.49%, O 6.89%. $CHF_2(CF_2)_3CH_2OH$. Prepd by free radical telomerization of tetrafluoroethylene in methanol: Hanford, Joyce, U.S. pat. **2,562,547** and Joyce, U.S. pat. **2,559,625** (both 1951 to du Pont).

Liquid. d_4^{20} 1.6647. bp_{760} 140-141°. n_D^{20} 1.3178. Surface tension at 20° = 27.6 dynes/cm.

USE: To introduce fluoroalkyl groups into an organic molecule: Kramer, Gilbert, U.S. pat. **2,963,526** (1960 to Esso). Proposed intermediate for plastics, surface-active agents, lubricants, elastomers.

6843. Octamethylcyclotetrasiloxane. $C_8H_{24}O_4Si_4$; mol wt 296.62. C 32.39%, H 8.16%, O 21.58%, Si 37.87%. Isolated from the hydrolysis product of dimethyldichlorosilane: Patnode, Wilcock, *J. Am. Chem. Soc.* **68**, 358 (1946).

Oily liquid. mp 17.5°. bp 175°. bp_{20} 74°. d 0.9558. n_D^{20} 1.3968.

USE: Preparation of methyl silicon oils.

6844. Octamethyltrisiloxane. $C_8H_{24}O_2Si_3$; mol wt 236.53. C 40.62%, H 10.23%, O 13.53%, Si 35.62%. Prepn: Patnode, Wilcock, *J. Am. Chem. Soc.* **68**, 358 (1946).

Liquid. bp 153°. d 0.8200. n_D^{20} 1.3848. mp ∼ −80°. Stable. Inert to most chemical reagents and rubber. Maintains about the same viscosity over a wide temp range. Sol in benzene and the lighter hydrocarbons; slightly sol in alcohol and heavy hydrocarbons.

USE: As a basis for silicone oils or fluids designed to withstand extremes of temp; as a foam suppressant in petroleum lubricating oil.

6845. Octamoxin. *(1-Methylheptyl)hydrazine;* 2-hydrazinooctane; octomoxine; Ximaol. $C_8H_{20}N_2$; mol wt 144.26. C 66.61%, H 13.97%, N 19.42%. Monoamine oxidase inhibitor. Prepd by condensation of methyl hexyl ketone and hydrazine hydrate followed by hydrogenation under pressure: Michel-Ber *et al.,* Brit. pat. **899,385** (1962 to Soc. Civile Auguil).

Sulfate, $C_8H_{20}N_2.H_2SO_4$, *Nimaol.* Crystals, mp 78-80°.

THERAP CAT: Antidepressant.

6846. Octamylamine. *6-Methyl-N-(3-methylbutyl)-2-heptanamine; N-isopentyl-1,5-dimethylhexylamine; N*-isoamyl-1,5-dimethylhexylamine; 2-methyl-6-(3-methylbutylamino)heptane; 2-isoamylamino-6-methylheptane; *N*-(1,5-dimethylhexyl)isopentylamine; Octinum D; Octisamyl; Neo-Octon; Octometine. $C_{13}H_{29}N$; mol wt 199.38. C 78.31%, H 14.66%, N 7.03%. Prepd by the condensation of 2-methyl-6-heptanone with isoamylamine: **Swiss** pat. **258,452** (1942 to Knoll).

Oily liquid. Weak aromatic odor. bp_7 100-101°. Strong base.

Hydrochloride, $C_{13}H_{29}N.HCl$, waxy leaflets, mp 121°. Sol in water, alcohol, ether.

THERAP CAT: Anticholinergic, antispasmodic.

6847. Octane. C_8H_{18}; mol wt 114.23. C 84.12%, H 15.88%. $CH_3(CH_2)_6CH_3$. Found in petroleum. Flammable liquid. mp −56.8°. d_4^{20} 0.7028. bp_{760} 125.6°. n_D^{20} 1.39764. Flash pt, open cup: 72°F (22°C). Insol in water. Slightly sol in alcohol; sol in ether; miscible with benzene, petr ether, gasoline. The term "octane rating" is explained under Gasoline.

Caution: Potential symptoms of overexposure are irritation of eyes and nose; drowsiness; dermatitis; chemical pneumonia. *See NIOSH Pocket Guide to Chemical Hazards* (DHHS/NIOSH 90-117, 1990) p 168.

6848. Octanohydroxamic Acid. *N-Hydroxyoctanamide;* caprylohydroxamic acid; Oct HA; Taselin. $C_8H_{17}NO_2$; mol wt 159.23. C 60.35%, H 10.76%, N 8.80%, O 20.10%. $CH_3(CH_2)_6CONHOH$. Synthesis and chemical study: Inoue, Yukawa, *J. Agr. Chem. Soc. Japan.* **16**, 504 (1940); Ichim *et al., Igiena (Bucharest)* **9**, 319 (1960), *C.A.* **55**, 16662a (1961). Activity as a urease inhibitor: Kobashi *et al., Biochim. Biophys. Acta* **65**, 380 (1962); **227**, 429 (1971); as an antimicrobial: Hase *et al., Chem. Pharm. Bull.* **19**, 363 (1971).

White plates from benzene, mp 78.5-79°. Water soluble. Practically insol in petr ether.

THERAP CAT (VET): Antimicrobial; growth promotant.

6849. 1-Octanol. Caprylic alcohol. $C_8H_{18}O$; mol wt 130.23. C 73.78%, H 13.93%, O 12.29%. $CH_3(CH_2)_6-CH_2OH$. Occurs in the form of esters in some essential oils.

Consult the Name Index before using this section.

Prepd from the esterified products of coconut oil, the methyl caprylate being reduced by Na and alcohol.

Colorless liquid; penetrating, aromatic odor. d_4^{20} 0.827. mp $-16°$ to $-17°$. bp 194-195°. n_D^{20} 1.430. Insol in water. Miscible with alcohol, chloroform, ether.

USE: Manuf of perfumes and esters.

6850. 2-Octanol. Secondary caprylic alcohol; methyl hexyl carbinol; hexylmethylcarbinol. $C_8H_{18}O$; mol wt 130.23. C 73.78%, H 13.93%, O 12.29%. $CH_3(CH_2)_5CH$-$(OH)CH_3$. Prepd by heating sodium ricinoleate with caustic soda in a copper vessel and distilling: Adams, Marvel, *Org. Syn.* **1**, 61 (1921); Kenyon, *ibid.* **6**, 68 (1926); Ellis, Reid, *J. Am. Chem. Soc.* **54**, 1678 (1932).

dl-Form, oily, refractive liquid. Aromatic, yet somewhat unpleasant odor, particularly on heating. d_4^{20} 0.8193. mp $-38.6°$. bp_{760} 178.5°; bp_{60} 107.4°; bp_{20} 83.3°; bp_{10} 70.0°; bp_5 57.6°; $bp_{1.0}$ 32.8°. n_D^{20} 1.42025 [Béhal, *Bull. Soc. Chim.* [4] **25**, 482 (1919)]. Flash pt $\sim140°F$ (60°C). Soly in water: 0.096 ml/100 ml. Miscible with alc, ether.

d-Form, bp_{20} 86°. $[\alpha]_D^{17}$ +9.9°. d_4^{20} 0.8216.
l-Form, bp_{20} 86°. $[\alpha]_D$ $-9.9°$.

USE: In the manuf of perfumes, in disinfectant soaps. To prevent foaming. Solvent for fats and waxes.

6851. Octaverine. *6,7-Dimethoxy-1-(3,4,5-triethoxyphenyl)isoquinoline;* 1-(3,4,5-triethoxyphenyl)-6,7-dimethoxyisoquinoline; oktaverine. $C_{23}H_{27}NO_5$; mol wt 397.47. C 69.50%, H 6.85%, N 3.52%, O 20.13%. Prepd by cyclodehydration and subsequent dehydrogenation of *N*-(3,4,5-triethoxybenzoyl)homoveratrylamine: **Fr. pat.** *760,825* (1934 to Asta Chemische Fabrik); cf. Goldberg, Shapero, *J. Pharm. Pharmacol.* **6**, 171 (1954).

Free base. Insol in water.
Hydrochloride, $C_{23}H_{27}NO_5.HCl$, crystals, mp 199-200°. Sparingly sol in water (1:500). Aq solns are acid to litmus.
THERAP CAT: Antispasmodic.

6852. Octenidine. *N,N'-(1,10-Decanediyldi-1(4H)-pyridinyl-4-ylidene)bis[1-octanamine];* 1,10-bis-[4-(octylamino)-1-pyridinyl]decane; Win-41464. $C_{36}H_{62}N_4$; mol wt 550.92. C 78.49%, H 11.34%, N 10.17%. Prepn: D. M. Bailey, **Belg. pat.** *851,807*; *idem*, **U.S. pat.** *4,206,215* (1977, 1980 both to Sterling); D. M. Bailey *et al., J. Med. Chem.* **27**, 1457 (1984). Structure-activity study: A. M. Slee *et al., Antimicrob. Ag. Chemother.* **23**, 531 (1983). Antiplaque activity: A. M. Slee, J. R. O'Connor, *ibid.* 379; microbiocidal activity: D. M. Sedlock, D. M. Bailey, *ibid.* **28**, 786 (1985). Antimycotic effects: M. A. Ghannoum *et al., J. Antimicrob. Chemother.* **25**, 237 (1990). Clinical evaluation as topical disinfectant: B. Christiansen, *Z. Bakteriol. Mikrobiol. Hyg.* **186**, 368 (1988). Clinical trials in prevention of plaque and gingivitis: M. R. Patters *et al., J. Periodont. Res.* **21**, 154 (1986); B. B. Beiswanger *et al., J. Dent. Res.* **69**, 454 (1990).

Dihydrochloride, $C_{36}H_{62}N_2.2HCl$, *Win-41464-2, Neo Kodan, Octeniderm, Octenisept.* Solid from ether, mp 215-217°.

Disaccharin, $C_{36}H_{62}N_2.2C_7H_5NO_3S$, *Win-41464-6.*
THERAP CAT: Antiseptic (topical).

6853. Octhilinone. *2-Octyl-3(2H)-isothiazolone; 2-octyl-4-isothiazolin-3-one;* RH-893; Kathon. $C_{11}H_{19}NOS$; mol

wt 213.34. C 61.93%, H 8.98%, N 6.57%, O 7.50%, S 15.03%. Prepn: S. N. Lewis *et al.,* **Fr. pat.** *1,555,416* corresp to **U.S. pat.** *3,761,488* (1969, 1973 to Rohm & Haas); *eidem, J. Heterocycl. Chem.* **8**, 571 (1971).

Liquid, $bp_{0.01}$ 120°. uv max (methanol): 280 nm (log ϵ 3.88).

USE: Fungicide. Biocide in cooling-tower water, paints, cutting oils, cosmetics and shampoo; for leather preservation.

6854. Octodrine. *6-Methyl-2-heptanamine;* 6-methyl-2-heptylamine; 2-methyl-6-aminoheptane; 6-amino-2-methylheptane; 2-amino-6-methylheptane; α,ϵ-dimethylhexylamine; 1,5-dimethylhexylamine; SKF-51; Vaporpac. C_8H_{19}-N; mol wt 129.25. C 74.35%, H 14.82%, N 10.84%. α-Adrenergic agonist. Prepd from the corresponding saturated ketone: Rohrmann, Shonle, *J. Am. Chem. Soc.* **66**, 1516 (1944). Pharmacology: E. J. Fellows, *J. Pharmacol. Exp. Ther.* **90**, 351 (1947).

dl-Form, viscous liquid, fishy odor, bp 154-156° n_D^{24} 1.4200.

Hydrochloride, $C_8H_{19}N.HCl$, crystals, sol in water. LD_{50} in mice, rats (mg/kg): 59, 41.5 i.p. (Fellows).
Sulfate, $(C_8H_{19}N)_2.H_2SO_4$, crystals, sol in water.
THERAP CAT: Decongestant.

6855. Octogen. *Octahydro-1,3,5,7-tetranitro-1,3,5,7-tetrazocine;* 1,3,5,7-tetranitro-1,3,5,7-tetrazacyclooctane; cyclotetramethylene tetranitramine; HMX. $C_4H_8N_8O_8$; mol wt 296.16. C 16.22%, H 2.72%, N 37.84%, O 43.22%. Prepn: H. Fischer, *Ber.* **82**, 192 (1949); A. F. McKay *et al., Can. J. Res.* **27**, 462 (1949). Determn in explosives by infrared: W. Selig, *U.S. Atomic Energy Commission* UCRL-7873 (part II), 29 (1965); by HPLC: P. Vouros *et al., Anal. Chem.* **49**, 1039 (1977). Mass spec: E. P. Burrows, *Org. Mass Spectrom.* **26**, 1027 (1991). Thermal analysis of phases using x-ray diffraction: M. Herrmann *et al., Z. Kristallogr.* **204**, 121 (1993). ^{14}N-NMR: M. Kony *et al., J. Org. Chem.* **59**, 5623 (1994). Review of syntheses and properties: B. Singh *et al., Def. Sci. J.* **28**, 41-50 (1978); of thermal behaviour: T. L. Boggs, *Prog. Astronaut. Aeronaut.* **90**, 121-175 (1984).

Highly Explosive! White crystalline solid, mp 281°. Nonhygroscopic. Four polymorphic forms: $\alpha,\beta,\gamma,$ and δ forms. Stability at 300°K $\beta > \alpha > \gamma > \delta$. β is stable at room temp; α at 115-156°C; γ at 156°C; δ at ~170-279°C. Densities are: $\beta = 1.903$; $\alpha = 1.87$; $\gamma = 1.82$; $\delta = 1.78$ g/cm³. Practically insol in water.

USE: In solid propellants and explosives.

6856. Octopamine. *α-(Aminomethyl)-4-hydroxybenzenemethanol; α-(aminomethyl)-p-hydroxybenzyl alcohol;* 1-(*p*-hydroxyphenyl)-2-aminoethanol; norsympatol; norsynephrine; *p*-hydroxyphenylethanolamine; WV-569. C_8H_{11}-NO_2; mol wt 153.18. C 62.73%, H 7.24%, N 9.14%, O 20.89%. A biogenic amine that is the phenol analog of noradrenaline (norepinephrine, *q.v.*). It is a neurosecretory product found in several vertebrates and invertebrates.

Formed by β-hydroxylation of tyramine by the enzyme dopamine β-hydroxylase: Pisano *et al.*, *Biochim. Biophys. Acta* **43**, 566 (1960). Identification: Erspamer, *Nature* **169**, 375 (1952). Found in the salivary glands of *Octopus vulgaris*, *O. macropus*, and of *Eledone moschata: idem, Arzneimittel-Forsch.* **2**, 253 (1952); in mammalian nerves: Molinoff, Axelrod, *Science* **164**, 428 (1969); in cockroach nervous system: Nathanson, Greengard, *ibid.* **180**, 308 (1973). Prepd synthetically: Asscher, U.S. pat. 2,585,988 (1952). The natural D(−) form is 3 times more potent than the L(+) form in producing cardiovascular adrenergic responses in anesthetized dogs and cats: Korol, Soffer, *The Pharmacologist* **5**, 247 (1963). Prepn of D- and L-forms: Kappe, Armstrong, *J. Med. Chem.* **7**, 569 (1964). In invertebrate nervous systems octopamine may function as a neurotransmitter: Saavedra *et al.*, *Science* **185**, 364 (1974). Effects on neuromuscular transmission in crustacean muscle: C. A. Breen, H. L. Atwood, *Nature* **303**, 716 (1983).

D(−)-Form, crystals from hot water which change at about 160° to a compd which melts above 250° (dec). $[\alpha]_D^{25}$ −56.0° (0.1N HCl); −37.4° (H$_2$O).
DL-Form hydrochloride, C$_8$H$_{11}$NO$_2$.HCl, *Epirenor, Norden, Norfen, Norphen (ampules).* Crystals, dec 170°. Freely sol in water.
THERAP CAT: Adrenergic.

6857. Octotiamine. *6-(Acetylthio)-8-[[2-[[(4-amino-2-methyl-5-pyrimidinyl)methyl]formylamino]-1-(2-hydroxyethyl)-1-propenyl]dithio]octanoic acid methyl ester; 8-[[2-[N-[(4-amino-2-methyl-5-pyrimidinyl)methyl]formamido]-1-(2-hydroxyethyl)propenyl]dithio]-6-mercaptooctanoic acid methyl ester S(or 6)-acetate; S-(3-acetylthio-7-carbomethoxyheptylthio)thiamine; thiamine 8-(methyl 6-acetyldihydrothioctate) disulfide; Gerostop; Neuvitan; TATD.* C$_{23}$H$_{36}$N$_4$O$_5$S$_3$; mol wt 544.76. C 50.71%, H 6.66%, N 10.28%, O 14.68%, S 17.66%. Prepn: Ohara *et al.*, U.S. pat. 3,098,856 (1963 to Fujisawa).

E-form

Crystals, mp 106-109°. uv max: 234, 277 nm (ε 16200, 5820).
Hydrochloride, C$_{23}$H$_{36}$N$_4$O$_5$S$_3$.HCl, crystals from ether + abs ethanol, mp 134.5-135°. uv max: 233 nm (ε 23000).
THERAP CAT: Vitamin (enzyme cofactor).

6858. Octoxynol. *α-[4-(1,1,3,3,-Tetramethylbutyl)phenyl]-ω-hydroxypoly(oxy-1,2-ethanediyl);* octylphenoxy polyethoxyethanol; polyethylene glycol *p*-isooctylphenyl ether. Prepd by reacting isooctylphenol with ethylene oxide. Average number of ethylene oxide units (n) per molecule is indicated by number following octoxynol. *See also* nonoxynol. Trademarks for series of octoxynols include *Igepal CA, Polytergent G, Triton X.*

n = 5 to 15

Mixture in which *n* ranges from about 5 to 15. Pale yellow, viscous liquid. d$_4^{25}$ 1.0595. n$_D^{25}$ 1.4894. Miscible with water, alcohol, acetone. Sol in benzene, toluene. Insol in petr ether. pH of 5% aq soln: 7-9. *Octoxynol-9,* average

comp. (n = 9): C$_{34}$H$_{62}$O$_{11}$; av mol wt 647. Trademarks of products where n = 9 to 10: *Conco NIX-100, Igepal CA-630, Neutronyx 605, Triton X-100.* Ingredient of *Preceptin.*
USE: Nonionic detergent, emulsifier, dispersing agent. Octoxynol-9 as spermaticide.

6859. Octreotide. *D-Phenylalanyl-L-cysteinyl-L-phenylalanyl-D-tryptophyl-L-lysyl-L-threonyl-N-[2-hydroxy-1-(hydroxymethyl)propyl]-L-cysteinamide cyclic (2→7)-disulfide;* 1,2-dithia-5,8,11,14,17-pentaazacycloeicosane cyclic peptide deriv; SMS-201-995; Longastatin; Sandostatin. C$_{49}$H$_{66}$N$_{10}$-O$_{10}$S$_2$; mol wt 1019.26. C 57.74%, H 6.53%, N 13.74%, O 15.70%, S 6.29%. Long-acting, octapeptide analog of somatostatin, *q.v.* Prepn: W. Bauer, J. Pless, *Eur. pat.* **Appl. 29,579;** *eidem,* U.S. pat. 4,395,403 (1981, 1983 both to Sandoz). Prepn and pharmacology: W. Bauer *et al.*, *Life Sci.* **31**, 1133 (1982). Opiate antagonist properties: R. Maurer *et al.*, *Proc. Nat. Acad. Sci. USA* **79**, 4815 (1982). Inhibitory effect on human gastroenteropancreatic hormone secretion: M. E. Kraenzlin *et al.*, *Experientia* **41**, 738 (1985). Endocrine profile in humans: E. del Pozo *et al.*, *Acta Endocrinol.* **111**, 433 (1986). Clinical evaluation in acromegaly: G. Plewe *et al.*, *Lancet* **2**, 782 (1984); in autonomic neuropathy: R. D. Hoeldtke *et al.*, *ibid.* **2**, 602 (1986). Symposium on chemistry, pharmacology and clinical trials: *Scand. J. Gastroenterol.* **21**, Suppl. 119, 1-274 (1986); on clinical evaluation in gastrointestinal endocrine tumors: *Am. J. Med.* **82**, Suppl. 5B, 1-99 (1987).

D-Phe—Cys—Phe—D-Trp—Lys—Thr—Cys—NH

$[\alpha]_D^{20}$ −42° (c = 0.5 in 95% acetic acid).
THERAP CAT: Gastric antisecretory agent. Treatment of acromegaly.

6860. Octyl Acetate. *Acetic acid 2-ethylhexyl ester; 2-ethylhexyl acetate.* C$_{10}$H$_{20}$O$_2$; mol wt 172.27. C 69.72%, H 11.70%, O 18.58%. CH$_3$(CH$_2$)$_3$CH(C$_2$H$_5$)CH$_2$OOCCH$_3$. Toxicity data: H. F. Smyth, C. P. Carpenter, *J. Ind. Hyg. Toxicol.* **26**, 269 (1944).
Liquid. d$_{20}^{20}$ 0.873. bp 199°. mp ∼ −80°. n$_D^{20}$ 1.4204. Flash pt, open cup: 190°F (88°C); closed cup: 56°F (13°C). Very slightly sol in water; misc with alcohol, oils, and other organic liquids. LD$_{50}$ orally in rats: 3.0 g/kg (Smyth, Carpenter).
USE: Solvent for nitrocellulose, some resins, waxes, and oils.

6861. *n*-Octyl Bromide. *1-Bromooctane.* C$_8$H$_{17}$Br; mol wt 193.13. C 49.75%, H 8.87%, Br 41.37%. CH$_3$(CH$_2$)$_6$-CH$_2$Br. Prepd from hydrobromic acid and *n*-octanol: O. Kamm, C. S. Marvel, *Org. Syn. coll.* **vol. I**, 30 (2nd ed., 1941); *cf.* Norris *et al.*, *J. Am. Chem. Soc.* **38**, 1076 (1916); Whitmore *et al.*, *ibid.* **67**, 2059 (1945); from PBr$_3$ and *n*-octanol: Coulson *et al.*, *J. Chem. Soc.* **1965**, 2364. Toxicity study: H. F. Smyth *et al.*, *Am. Ind. Hyg. Assoc. J.* **30**, 470 (1969).
Colorless liquid. d$_4^{25}$ 1.108. bp 198-200°. mp −55°. n$_D^{25}$ 1.4503. Insol in water. Miscible with alcohol, ether. LD$_{50}$ orally in rats: 4.49 ml/kg (Smyth).
USE: In organic syntheses.

6862. *sec*-Octyl Bromide. *2-Bromooctane.* C$_8$H$_{17}$Br; mol wt 193.13. C 49.75%, H 8.87%, Br 41.37%. CH$_3$CHBr-(CH$_2$)$_5$CH$_3$. Prepd by the action of PBr$_3$ on *sec*-octyl alcohol: Hsueh, Marvel, *J. Am. Chem. Soc.* **50**, 856 (1928); Reynolds, Adkins, *ibid.* **51**, 279 (1929); Shriner, Young, *ibid.* **52**, 3332 (1930); Ellis, Reid, *ibid.* **54**, 1680 (1932); Coulson *et al.*, *J. Chem. Soc.* **1965**, 2364.
dl-Form, d$_4^{25}$ 1.0878. bp$_{14}$ 72°. bp$_6$ 66°. bp$_3$ 61°. n$_D^{25}$ 1.4442. Insol in water, miscible with alcohol, ether.
l-Form, d$_4^{25}$ 1.0982. bp$_{18}$ 73°. bp$_{14}$ 71°. bp$_3$ 60°. n$_D^{25}$ 1.4475. α$_D^{20}$ −44.91°.
d-Form, d$_4^{25}$ 1.0982. bp$_{18}$ 77°. bp$_{14}$ 71°. bp$_3$ 60°. n$_D^{25}$ 1.4475.

6863. sec-Octyl Iodide. *2-Iodooctane.* $C_8H_{17}I$; mol wt 240.13. C 40.02%, H 7.14%, I 52.85%. $CH_3CHI(CH_2)_5$-CH_3. Prepn: Coulson *et al., J. Chem. Soc.* **1965**, 2364.

dl-Form, oily liquid; discolors in light. d_{15}^{18} 1.318. bp about 210° with decompn; also stated as 190°. *Protect from air and light.*

l-Form, α_D^{20} −64.63°.

6864. Octyl Methoxycinnamate. *3-(4-Methoxyphenyl)-2-propenoic acid 2-ethylhexyl ester;* 2-ethylhexyl *p*-methoxy-cinnamate; Parsol MCX; Parsol MOX. $C_{18}H_{26}O_3$; mol wt 290.40. C 74.45%, H 9.02%, O 16.53%. UV-B blocker. Manuf process and purification: P. Schudel *et al.,* **U.S. pat.** 4,713,473 (1987 to Givaudan). Use in sunscreen preparations: D. H. Liem, L. T. H. Hilderink, *Int. J. Cosmet. Sci.* **1**, 341 (1979). Assessment of photostability: R. Aberturas *et al., Boll. Chim. Farm.* **126**, 208 (1987); A. Deflandre, G. Lang, *Int. J. Cosmet. Sci.* **10**, 53 (1988). HPLC determn in cosmetics: L. Gagliardi *et al., J. Chromatog.* **408**, 409 (1987).

E-form

bp at 1 mbar: 185-195°; bp at 0.1 mbar: 140-150°.
THERAP CAT: Ultraviolet screen.

6865. Ofloxacin. (±)-*9-Fluoro-2,3-dihydro-3-methyl-10-(4-methyl-1-piperazinyl)-7-oxo-7H-pyrido[1,2,3-de]-1,4-benzoxazine-6-carboxylic acid;* ofloxacine; DL-8280; HOE-280; Exocin; Flobacin; Floxil; Floxin; Oflocet; Oflocin; Oxaldin; Tarivid; Visiren. $C_{18}H_{20}FN_3O_4$; mol wt 361.37. C 59.83%, H 5.58%, F 5.26%, N 11.63%, O 17.71%. Broad spectrum, fluorinated quinolone antibacterial. Prepn: I. Hayakawa *et al.,* **Eur.** pat. **Appl. 47,005;** *eidem,* **U.S.** pat. 4,382,892 (1982, 1983 both to Daiichi). Total synthesis: H. Egawa *et al., Chem. Pharm. Bull.* **34**, 4098 (1986). Synthesis and activity of optical isomers: S. Atarashi *et al., ibid.* **35**, 1896 (1987). Antibacterial spectrum of racemate: K. Sato *et al., Antimicrob. Ag. Chemother.* **22**, 548 (1982). Pharmacology and clinical efficacy: *Infection* **14**, Suppl. 1, S1-S109 (1986). Review of antibacterial spectrum, pharmacology, and clinical efficacy: J. P. Monk, D. M. Campoli-Richards, *Drugs* **33**, 346-391 (1987). Symposium on pharmacokinetics and therapeutic use: *Scand. J. Inf. Dis.* **Suppl. 68**, 1-69 (1990). Toxicity data: H. Ohno *et al., Chemotherapy (Tokyo)* **32**, Suppl. 1, 1084 (1984); M. Kato *et al., Arzneimittel-Forsch.* **42**, 365 (1992). Series of articles on pharmacology and toxicology of (−)-form: *ibid.,* 368-418.

Colorless needles from ethanol, mp 250-257° (dec). LD_{50} in male, female mice, male, female rats (mg/kg): 5450, 5290, 3590, 3750 orally; 208, 233, 273, 276 i.v.; >10000, >10000, 7070, 9000 s.c. (Ohno).

S-(−)-Form, *levofloxacin, DR-3355, Cravit.* Prepd as the hemihydrate; needles from ethanol + ethyl ether, mp 225-227° (dec). $[\alpha]_D^{23}$ −76.9 (c = 0.385 in 0.5N NaOH). LD_{50} in male, female mice, male, female rats (mg/kg): 1881, 1803, 1478, 1507 orally (Kato).

THERAP CAT: Antibacterial.

6866. Oil of Amber, Rectified. Obtained by the destructive distillation of amber and purified by redistillation. Consists of a mixture of terpenes with resinous, oxygen-containing substances.

Pale yellow to yellowish-brown, volatile oil; penetrating odor; burning acrid taste. d 0.850-0.920. α_D^{20} +22° to +26°. Insol in water. Sol in about 10 vols alcohol; freely sol in chloroform, ether, carbon disulfide, oils.

6867. Oil of Angelica. Volatile oil from root of *Angelica archangelica* L. (*A. officinalis* Moench.), *Umbelliferae.* *Constit.* Phellandrene, valeric acid.

Yellow liquid. d_{15}^{15} 0.857-0.915. α_D^{20} +16° to +32°. n_D^{20} 1.4800. Almost insol in water. Sol in 6 vols 90% alcohol. *Keep well closed, cool, and protected from light.*

USE: In manuf liqueurs.

6868. Oil of Anise. Volatile oil from dried ripe fruit of *Pimpinella anisum* L., *Umbelliferae,* or of *Illicium verum* Hook. fil., *Magnoliaceae* (Chinese star anise). *Constit.* 80-90% anethole; methylchavicol, anisaldehyde.

Colorless or pale yellow, refractive liquid. d_{15}^{25} 0.978-0.988. Solidif not below 15°. α_D +1° to −2°. n_D^{20} 1.553-1.560. Slightly sol in water; sol in about 3 vols alcohol, freely in chloroform, ether. *Keep cool in well-closed and well-filled containers, protected from light.*

USE: In manuf of liqueurs; flavor for candies, cookies, dentifrices. Pharmaceutic aid (flavor).

THERAP CAT: Carminative, expectorant.

THERAP CAT (VET): Has been used as a carminative.

6869. Oil Anise, Japanese. Volatile oil from fruit of *Illicium anisatum* L., *Magnoliaceae* (Japanese star anise). *Constit.* Chiefly anethole; also safrol, eugenol.

Colorless to slightly yellow liquid. d about 1.006. Solidifies at −10° to −15°. α_D^{20} about −8°.

THERAP CAT: Carminative; expectorant.

6870. Oil of Asarum. Oil of Canada snakeroot. Volatile oil from rhizome of *Asarum canadense* L., *Aristolochiaceae.* *Constit.* Terpene (pinene), methyleugenol, borneol, linalool, geraniol.

Yellowish-brown liq; aromatic odor and taste. d_{15}^{15} 0.93-0.96. α_D^{20} −1.4 to −3.5°. Practically insol in water. Sol in 2 vols 70% alc. *Keep well closed, cool and protected from light.*

6871. Oil of Balm. Oil of melissa balm; oil of lemon balm. Volatile oil from leaves and tops of *Melissa officinalis* L., *Labiatae.* Chiefly citral. Composition studies: Hefendehl, *Arch. Pharm.* **303**, 345 (1970).

Yellow to yellowish-green liquid. d_{15}^{15} 0.89-0.925. Practically insol in water. Sol in alcohol. *Keep well closed, cool, and protected from light.*

6872. Oil of Basil. Volatile oil from leaves of *Ocimum basilicum* L., *Labiatae* (sweet basil). *Constit.* Methylchavicol, eucalyptol, linalool, estragol.

Yellowish to greenish liquid; aromatic odor. d_{20}^{20} 0.905-0.930. α_D^{20} −6° to −22°. Almost insol in water. Sol in 2 vols 80% alc; miscible with ether, chloroform. *Keep well closed, cool and protected from light.*

6873. Oil of Bay. Oil of Myrcia. Volatile oil distilled from leaves of *Pimenta (Myrcia) acris* Kostel., *Myrtaceae.* *Constit.* 40-55% eugenol; myrcene, chavicol, methyleugenol, methylchavicol, citral, *l*-phellandrene; total phenols, 50-65% by volume.

Yellow to brownish-yellow liq; pleasant odor; sharp, spicy taste; becomes brown on exposure to air. d_{15}^{25} 0.962-0.990. α_D^{25} −3°. n_D^{20} 1.500-1.520. Insoluble in water. Very sol in alcohol, carbon disulfide, glacial acetic acid.

USE: Pharmaceutic aid (aromatic). Manuf bay rum.

6874. Oil of Bergamot. Volatile oil expressed from rind of fresh fruit of *Citrus aurantium* L., var. *bergamia* Wight & Arn., *Rutaceae.* *Constit.* 36-45% *l*-linalyl acetate, about 6% *l*-linalool; *d*-limonene, dipentene, bergaptene.

Yellowish-green liquid; agreeable odor. d_{15}^{25} 0.875-0.880. α_D^{25} +8° to +24°. n_D^{20} 1.464-1.467. Acid no. 1-4. Almost insol in water. Sol in 0.5 vol 95% alcohol, 2 vols 80% alcohol. *Keep well closed in a cool place, protected from light.*

USE: For masking many disagreeable odors, such as iodoform, naphthalene, etc. In perfumery, hair oils, pomades.

6875. Oil of Bitter Almond. Volatile oil from dried ripe kernels of bitter almonds or from other kernels containing amygdalin, such as apricots, cherries, plums, and especially peaches. Obtained by macerating with water, then steam distilling. *Constit.* Not less than 95% benzaldehyde; 2-4% HCN and phenoxyacetonitrile.

Colorless to yellow, very refractive liq; characteristic odor and taste of benzaldehyde. d_{25}^{25} 1.038-1.060. n_D^{20} 1.5428-1.5439. Slightly sol in water; miscible with alcohol, ether, oils. *Keep cool and protected from light.*

Caution: Hydrogen cyanide *(q.v.)* component responsible for highly toxic properties. *Very poisonous!*

USE: Only the oil *free from HCN* may be used for liqueurs and foods.

THERAP CAT: Formerly as topical antipruritic.

6876. Oil of Bitter Orange. Volatile oil expressed from fresh peel of *Citrus aurantium* L., *Rutaceae. Constit.* About 90% *d*-limonene; citral, decyl aldehyde, methyl anthranilate, linalool, terpineol.

Pale yellow liquid; bitter taste. d_{25}^{25} 0.842-0.848. α_D^{25} +88° to +98°. Very slightly sol in water; miscible with abs alcohol; sol in 4 vols alcohol, in 1 vol glacial acetic acid. *Keep well closed, cool, and protected from light.*

USE: As flavoring; in perfumery.

6877. Oil of Cajeput. Cajuput oil; cajeputi oil. Volatile oil from fresh leaves and twigs of several varieties of *Melaleuca leucadendron* L., and other species of *Melaleuca, Myrtaceae. Constit.* 50-60% eucalyptol (cineol); *l*-pinene, terpineol; valeric, butyric, benzoic and other aldehydes. Toxicity: P. M. Jenner *et al., Food Cosmet. Toxicol.* **2**, 327 (1964).

Colorless or yellowish liquid, agreeable camphor odor, and bitter aromatic taste. d 0.912-0.925. $\alpha_D^{20} < -4°$. n_D^{20} 1.4660-1.4710. Very slightly sol in water; sol in 1 vol 80% alcohol. Misc with alcohol, chloroform, ether, carbon disulfide. *Keep well closed, cool, and protected from light.* LD$_{50}$ orally in rats: 3870 mg/kg (Jenner).

THERAP CAT: Expectorant, counterirritant, scabicide.

THERAP CAT (VET): Rubefacient, topical antimycotic.

6878. Oil of Calamus. Oil of sweet flag. Volatile oil from rhizome of *Acorus calamus* L., *Araceae* grown in North America, Europe and Asia. *Constit.* of Jammu, India variety: β-asarone 75.8%; calamene 3.84%; calamol 3.2%; α-asarone 1.32%; camphene 0.92%; β-pinene 0.56%; asaronaldehyde 0.2%, Vashist, Handa, *Soap Perfum. Cosmet.* **37**, 135 (1964). Toxicity study: J. M. Taylor *et al., Toxicol. Appl. Pharmacol.* **10**, 405 (1967).

Yellow to yellowish-brown viscid liquid; aromatic odor; bitter taste. d_{20}^{20} 0.960-0.970. α_D^{20} +9° to +31°. n_D^{20} 1.507-1.515. Sapon no. 16-20. Very slightly sol in water; miscible with alcohol. LD$_{50}$ orally in rats (Jammu variety): 777 mg/kg (Taylor). *Keep well closed, cool and protected from light.*

USE: In perfumery. Formerly as minor ingredient in bitter flavors such as vermouth and flavored wines.

6879. Oil of Camphor, Rectified. Formosa oil of camphor; Japanese oil of camphor; white oil of camphor; light oil of camphor. Volatile oil from *Cinnamomum camphora* T. Nees & Eberm., *Lauraceae. Constit.* Safrol, acetaldehyde, camphor, terpineol, eugenol, cineol, *d*-pinene, phellandrene, dipentene, cadinene.

Colorless or yellowish liquid; camphor odor. d_{20}^{20} 0.875-0.900. α_D^{25} +9° to +24°. n_D^{20} 1.465-1.470. Insol in water. Sol in chloroform, ether, oils, in about 3 vols alcohol. *Keep well closed, cool, and protected from light.*

USE: As solvent in paint and lacquer industry; in perfuming of soaps and detergents; in technical odor masking.

THERAP CAT: Rubefacient.

6880. Oil of Caraway. Volatile oil distilled from dried ripe fruit of *Carum carvi* L., *Umbelliferae. Constit.* 53-63% carvone (by vol), *d*-limonene. Composition studies: Von Schantz, Ek, *Sci. Pharm.* **39**, 82 (1971). *Review:* Arctander, *Perfume and Flavor Materials of Natural Origin* (S. Arctander, Elizabeth, N.J., 1960) pp 124-125.

Colorless or pale yellow liq; darkens and thickens with age. d_{25}^{25} 0.900-0.910. α_D^{25} +70° to +80°. n_D^{20} 1.485-1.497. Almost insol in water. Sol in 8 vols 80% or in 1 vol 90% alcohol. *Keep well closed, cool, and protected from light.*

USE: In manuf liqueurs and perfuming soaps; flavor for cookies, candies. Pharmaceutic aid (flavor).

THERAP CAT: Carminative.

6881. Oil of Cardamom. One of the oldest known essential oils; distilled from the ripe seeds of *Elettaria cardamomum* Maton, *Zingiberaceae.* Description of distillation processes: Viehoever, Sung, *J. Am. Pharm. Assoc.* **26**, 872 (1937); E. Guenther, *The Essential Oils* vol. **V** (New York, 1952) pp 85-106. *Constit.* Eucalyptol (cineol), sabinene, *d*,α-terpineol and acetate, borneol, limonene, terpinene, 1-terpinen-4-ol and its formate and acetate. Composition studies: Lewis, *Perfum. Essent. Oil Rec.* **57**, 623 (1966).

Colorless or pale yellow volatile oil. Characteristic, aromatic, agreeable odor and taste. Distills with steam. d_4^{25} 0.917-0.947. α_D^{25} +22° to +44°. n_D^{20} 1.4630-1.4660. Insol in water. Miscible with abs ethanol. Sol in ether; one vol dissolves in 7 vols of 70% alcohol. Phenol coefficient 7.5.

USE: For flavoring liqueurs, pharmaceutical syrups, curry sauces, confectionery, baked goods.

THERAP CAT: Carminative.

6882. Oil of Cascarilla. Volatile oil from bark of *Croton eluteria* (L.) Sw., *Euphorbiaceae.* Chromatographic separation of constituents: Claudelafontaine *et al., Bull. Soc. Chim. France* **1973** (part 2), 2866.

Yellow to greenish liquid. d 0.890-0.925. Rotation +2° to +5° in a 100-mm tube at 20°. Very sol in alcohol, ether. *Keep well closed, cool, and protected from light.*

6883. Oil of Cashew Nut Shell. From *Anacardium occidentale* L., *Anacardiaceae* (Southern India). Consists mostly (90%) of anacardic acid, *q.v.* Review and bibliography: Sanyal, Das, *Indian Pharm.* **10**, 272-283 (1955).

Wijs iodine no. over 250. Polymerizes easily when heated with acid.

USE: In insulating varnishes, typewriter rolls, coldsetting cements, floor tile, brake linings; molding compounds from polymers: Harvey, U.S. pat. **2,767,150** (1956 to The Harvel Corp.).

6884. Oil of Cedar Leaf. Cedar leaf oil. Obtained by steam distillation from the fresh leaves of *Thuja occidentalis* L., *Cupressaceae.* Although this oil is known commercially as cedar leaf oil, the tree from whose leaves it is obtained is not a true cedar; it is a coniferous tree popularly known as "arbor vitae," sometimes erroneously called "white cedar." *Habit.* Canada and Northern U.S. *Constit.* α-pinene, *d*-thujone, α-fenchone.

Yellowish volatile oil. d 0.910-0.920. α_D^{25} −10° to −13°. n_D^{20} 1.4560-1.4590. Soly similar to oil of lavender, *q.v.*

Caution: Ingestion of large quantities causes hypotension, bradycardia, tachypnea, convulsions, death.

USE: Substitute for oil of lavender.

THERAP CAT: Counterirritant.

6885. Oil of Cedar Wood. Volatile oil from wood of *Juniperus virginiana* L., *Cupressaceae,* and other species of cedar. *Constit.* Chiefly cedrene (a terpene), and cedral (cedar camphor).

Colorless or slightly yellow, somewhat viscid liquid. d_{20}^{20} 0.940-0.950. α_D^{20} −25° to −46°. n_D^{20} 1.495-1.510. Insol in water. Sol in 10-20 vols 90% alcohol; sol in ether. *Keep well closed, cool, and protected from light.*

USE: In perfumery; as insect repellent; the thickened oil is used in microscopy as a clearing agent and for use with immersion lenses.

6886. Oil of Celery. Volatile oil from celery seed, *Apium graveolens* L., *Umbelliferae. Constit. d*-Limonene, phenols, sedanolide, sedanoic acid.

Colorless liquid; celery odor. d 0.870-0.895. α_D^{20} +67° to +79°. Slightly sol in water; very sol in alcohol. *Keep well closed, cool, and protected from light.*

USE: As flavor for soft drinks, unpleasant medicaments.

6887. Oil of Champaca. Volatile oil from flowers of *Michelia champaca* L., *Magnoliaceae. Constit.* Esters of benzoic acid, benzaldehyde, benzyl alcohol, isoeugenol.

Light yellow or reddish-yellow to brownish liquid. d_{15}^{15} 0.906-0.935. α_D −12° to −52°. Sapon no. ~77. Slightly

sol in water; sol in chloroform, ether; sparingly sol in alcohol. *Keep well closed, cool and protected from light.*
USE: In perfumes.

6888. Oil of Chenopodium. Oil of American wormseed; Chenopodiol; Chenoposan; Chenoposetten. Volatile oil from fresh, aboveground parts of flowering and fruiting plant of *American wormseed, Chenopodium ambrosioides* L. var. *anthelminticum* (L.) Aellen, *Chenopodiaceae* also known as *Mexican tea, Spanish tea, Jerusalem tea, ambrosia. Habit.* Central America, U.S., Canada. Oil contains 60-70% ascaridole, the principal active constituent; *p*-cymene, α-terpinene, *l*-limonene, methadiene.
Colorless or pale yellow liquid; characteristic disagreeable odor and taste. d_{25}^{25} 0.950-0.980. α_D^{20} −4° to −8°. n_D^{20} 1.4723-1.4790. Insoluble in water. Sol in 8 vols 70% alcohol; partly sol in glacial acetic acid. *Keep well closed, cool and protected from light.*
THERAP CAT: Anthelmintic.
THERAP CAT (VET): Anthelmintic.

6889. Oil of Cherry Laurel. Volatile oil from leaves of *Prunus laurocerasus* L., *Rosaceae. Constit.* HCN, benzaldehyde, benzaldehyde cyanhydrin, benzyl alcohol.
Pale yellow liquid; odor and taste similar to oil of bitter almond. *Very poisonous!* d_{20}^{20} 1.054-1.066. Slightly sol in water; sol in 2 vols 70% alcohol, in benzene, chloroform, ether. *Keep well closed, cool and protected from light.*
Caution: Hydrogen cyanide *(q.v.)* component responsible for highly toxic properties.

6890. Oil of Cinnamon. Oil of cassia; oil of Chinese cinnamon. Volatile oil from leaves and twigs of *Cinnamomum cassia* Nees, *Lauraceae. Constit.* 80-90% cinnamaldehyde; cinnamyl acetate, eugenol.
Yellowish or brownish liquid. Darkens and thickens on exposure to air. d_{25}^{25} 1.045-1.063. α_D −1° to +1°. n_D^{20} 1.6020-1.6060. Slightly sol in water; sol in an equal vol of alcohol and of glacial acetic acid. *Keep well closed, cool and protected from light.*
USE: As flavor in foods and in perfumes.
THERAP CAT: Carminative.

6891. Oil of Cinnamon, Ceylon. Volatile oil from bark of Ceylon cinnamon. *Constit.* 50-65% cinnamaldehyde; 4-8% eugenol; phellandrene.
Light yellow liq; gradually becomes reddish; characteristic odor, d 1.000-1.030. α_D 0° to −2°. n_D^{20} 1.565-1.582.
THERAP CAT: Carminative.

6892. Oil of Citronella. Volatile oil from fresh grass of *Cymbopogon (Andropogon) nardus* (L.) Rendle, *Gramineae. Constit.* Ceylon: about 60% geraniol, about 15% citronellal, 10-15% camphene and dipentene, small quantities of linalool, borneol. Java: 25-50% citronellal, 25-45% geraniol.
Almost colorless to pale yellow liq; gradually becomes reddish; pleasant odor. d Ceylon, 0.897-0.912; Java, 0.885-0.900. α_D^{20}: Ceylon, −6° to −14°; Java, −2° to −5°. n_D^{20} Ceylon 1.479-1.485; Java, 1.468-1.473. Slightly sol in water; sol in 10 vols 80% alcohol. *Keep well closed, cool and protected from light.*
USE: As perfume; insect repellent.

6893. Oil of Clove. Clove oil. Volatile oil from dried flower buds of *Eugenia caryophyllata* Thunb. (*Caryophyllus aromaticus* L.), *Myrtaceae. Constit.* 82-87% eugenol, including ∼10% acetyleugenol; caryophyllene, small quantities of furfural, vanillin, methyl amyl ketone.
Colorless to pale yellow liq, becoming darker and thicker with age. d_{25}^{25} 1.038-1.060. α_D^{20} < −1°10'. n_D^{20} 1.530. bp ∼250°. Insol in water; sol in 2 vols 70% alcohol; very sol in stronger alcohol, in ether, glacial acetic acid. *Keep well closed, cool and protected from light.*
USE: In confectionery, toothpowders; also in microscopy.
THERAP CAT: Local anesthetic (toothache); counterirritant; carminative.

6894. Oil of Copaiba. Volatile oil from copaiba balsam, usually Maracaibo. *Constit.* Chiefly caryophyllene and cadinene.
Colorless or pale yellow liquid. d_{15}^{15} 0.895-0.905, increasing with age. bp 250-275°. α_D −7° to −35°. n_D^{20} 1.495-

1.500. Insol in water; very sol in alcohol, ether, carbon disulfide. *Keep well closed, cool and protected from light.*
USE: In perfumery.

6895. Oil of Coriander. Volatile oil from dried ripe fruit of *Coriandrum sativum* L., *Umbelliferae. Constit.* Chiefly *d*-linalool and its acetate.
Colorless or pale yellow liquid. d_{25}^{25} 0.863-0.875. α_D^{25} +8° to +15°. n_D^{20} 1.4620-1.4720. Almost insol in water; sol in 3 vols 70% alcohol; more sol in stronger alcohol; very sol in chloroform, ether, glacial acetic acid. *Keep well closed, cool and protected from light.*
USE: Flavoring in foods and alcoholic beverages.
THERAP CAT: Carminative.

6896. Oil of Cubeb. Volatile oil from unripe fruit of *Piper cubeba* L.f., *Piperaceae. Constit.* Dipentene, cadinene, cubeb camphor, the latter in old oils.
Colorless, pale green or yellowish liquid. d_{25}^{25} 0.905-0.925. α_D^{20} −20° to −40°. n_D^{20} 1.4800-1.5020. Insol in water; sol in ∼10 vols alcohol; miscible with abs alcohol, chloroform. *Keep well closed, cool and protected from light.*
THERAP CAT: Urinary antiseptic.

6897. Oil of Cumin. Volatile oil from fruit of *Cuminum cyminum* L., *Umbelliferae. Constit.* 30-40% cuminaldehyde; *p*-cymene, β-pinene, dipentene.
Colorless to yellow liquid. d_{25}^{25} 0.900-0.935. α_D^{20} +4° to +8°. n_D^{20} 1.4950-1.5090. Almost insol in water; sol in 10 vols 80% alcohol; more sol in stronger alcohol; very sol in chloroform, ether. *Keep well closed, cool and protected from light.*
USE: Flavoring in Indian curry powder.

6898. Oil of Cypress. Volatile oil from leaves and young branches of *Cupressus sempervirens* L., *Pinaceae. Constit.* Furfural, *d*-pinene, *d*-camphene, cymene, *d*-terpineol, *l*-cadinene, sylvestrene, cypress camphor.
Yellowish liquid. d 0.88-0.89. α_D^{20} +4° to +18°. Slightly sol in water; sol in 2-6 vols 90% alcohol. *Keep well closed, cool and protected from light.*
USE: In perfumery.

6899. Oil of Dill. Dill seed oil; dill fruit oil. Volatile oil from dried ripe fruit of *Anethum graveolens* L., *Umbelliferae. Constit.* About 50% carvone *d*-limonene, phellandrene and other terpenes.
Colorless or pale yellow liquid; characteristic odor. d_{15}^{15} 0.900-0.915. α_D^{20} +70° to +80°. n_D^{20} 1.481-1.492. Insol in water; sol in 1 vol 90% alcohol. *Keep well closed, cool and protected from light.*
THERAP CAT: Aromatic carminative.

6900. Oil of Dwarf Pine Needles. Oil of mountain pine; Pinus Montana oil; Pinus pumilio oil. Volatile oil from fresh leaves of *Pinus montana* Mill. (*P. pumilio* Haenke), *Pinaceae. Constit.* *l*-Pinene, *l*-phellandrene, sylvestrene, dipentene, cadinene, 5-7% bornyl acetate.
Colorless or faintly yellow liq; pleasant odor; bitter taste. d_{25}^{25} 0.853-0.869. α_D^{25} −5° to −12°. n_D^{20} 1.4750-1.4800. Insol in water. Sol in 4.5-8 vols 90% alcohol; very sol in chloroform, ether. *Keep well closed, cool and protected from light.*
USE: Pharmaceutic aid (flavor and perfume).
THERAP CAT: Expectorant.

6901. Oil of Eucalyptus. Dinkum oil. Volatile oil from fresh leaves of *Eucalyptus globulus* Labill and of some other species of *Eucalyptus, Myrtaceae.* A dwarf species, called Mallee in Australia, is richest in oil of Eucalyptus. *Constit.* 70-80% cineole (eucalyptol); α-pinene; phellandrene; terpineol; citronellal; geranyl acetate; eudesmol; eudesmyl acetate; piperitone; volatile aldehydes (principally isovaleric): E. Guenther, *The Essential Oils* vol. 4 (van Nostrand, New York, 1950) pp 437-525.
Colorless to pale yellow liquid; characteristic camphoraceous odor; pungent, spicy, cooling taste. d_{25}^{25} 0.905-0.925. Does not solidify below −15.4°. α_D −5° to +5°. n_D^{20} 1.4580-1.4700. Almost insol in water; sol in 5 vols 70% alcohol; miscible with abs alcohol, oils, fats. *Keep well closed, cool and protected from light.*
THERAP CAT: Expectorant, anthelmintic, local antiseptic.

THERAP CAT (VET): Inhalation expectorant; wound dressing.

6902. Oil of Fennel. Volatile oil from the dried fruit of *Foeniculum vulgare* Mill., *Umbelliferae*. Also from the sweet fennel plant (var. *dulce*). *Constit.* 50-60% anethole, ~20% fenchone, pinene, limonene, dipentene, phellandrene.

Colorless or pale yellow liquid; odor and taste of fennel. d_{25}^{25} 0.953-0.973. α_D^{25} +12° to +24°. n_D^{20} 1.5280-1.5380. Slightly sol in water; sol in 1 vol 90% or in 8 vols 80% alcohol; very sol in chloroform, ether. *Keep well closed, cool and protected from light.*

USE: To cover taste of unpleasant medicines.

THERAP CAT: Carminative.

6903. Oil of Fir. Oil of silver pine; oil of silver fir. Volatile oil from needles and young twigs of *Abies alba* Mill. (*A. picea* Lindl., *A. pectinata* DC.), *Pinaceae*. *Constit.* *l*-Pinene, *l*-limonene, *l*-bornyl acetate.

Colorless, clear liquid; balsamic odor; terebinthinate taste. d_{15}^{15} 0.869-0.875. α_D^{D} −20° to −59°. Insol in water. Sol in 5 vols 90% alc, in ether. *Keep well closed, cool and protected from light.* .

USE: Pharmaceutic aid (flavor and perfume).

THERAP CAT: Expectorant.

6904. Oil of Fir—Siberian. Oil of pine; oleum abietis; Siberian pine needle oil. Volatile oil from fresh leaves of *Abies sibirica* Ledeb., *Pinaceae*. *Constit.* About 40% esters calculated as bornyl acetate; pinene, camphene, dipentene, and phellandrene.

Colorless or pale yellow liquid; aromatic odor; pungent taste. d_{15}^{15} 0.905-0.925. α_D^{20} −32° to −45°. n_D^{20} 1.466-1.476. Sol in an equal vol 90% alcohol. *Keep well closed, cool and protected from light.*

USE: Pharmaceutic aid (flavor and perfume).

THERAP CAT: Expectorant.

6905. Oil of Fleabane. Oil of Canada fleabane; oil of erigeron. Volatile oil from fresh flowering herb of *Conyza canadensis* (L.) Cron. (*Erigeron canadensis* L., *Leptilon canadense* (L.)) Britt., *Compositae*. *Constit.* *d*-Limonene, aldehydes.

Pale yellow liquid; becomes darker and thicker with age and on exposure to air; peculiar odor; aromatic, slightly pungent taste. d_{25}^{25} 0.845-0.865. α_D^{20} ~+45°. Slightly sol in water; sol in 1 vol alcohol; very sol in chloroform, ether. *Keep well closed, cool and protected from light.*

6906. Oil of Garlic. Volatile oil from bulb or entire plant *Allium sativum* L., *Liliaceae*. *Constit.* Diallyl disulfide, diallyl trisulfide, methyl allyl trisulfide, methyl allyl disulfide, diallyl sulfide, q.v. Cardiovascular effects: A. Bordia, H. C. Bansal, *Lancet* **2**, 1491 (1973). Prepn and *in vitro* platelet antiaggregation activity: H. Nishimura, T. Ariga, *ACS Symposium Series* **546**, 128 (1994). Brief review: E. Guenther, *The Essential Oils* vol. 6 (Van Nostrand, New York, 1952) pp 67-69. Comprehensive description: N. A. Shaath et al., *Dev. Food Sci.* **37B**, 2025 (1995).

Clear, pale yellow to reddish brown liquid; strong garlic odor. d^{15} 1.046-1.057. d_{15}^{15} 1.055-1.098. n_D^{20} 1.5578-1.5745. Optically inactive.

USE: Flavoring agent in foods.

6907. Oil of Geranium. Oil of pelargonium geranium; oil of rose geranium. Volatile oil from leaves of *Pelargonium odoratissimum* Ait. and allied species, *Geraniaceae*. *Constit.* Geraniol esters, calculated as geranyl tiglate, 21-30% in Algerian or French oil, 25-35% in Bourbon oil; citronellol; some linalool.

Colorless, greenish or brownish liquid. d_{15}^{15} 0.894-0.905. α_D^{20} −7° to −11°. n_D^{20} 1.4650-1.470. Slightly sol in water; sol in 3 vols 10% alcohol; more sol in stronger alcohol; very sol in chloroform, ether. *Keep well closed, cool and protected from light.*

USE: In perfumery; as odorant for tooth and dusting powders, ointments, etc. In manuf of rhodinol, q.v.

6908. Oil of Geranium—East Indian. Turkish geranium oil; palmarosa oil; Indian grass oil; rusa oil. Volatile oil from *Andropogon schoenanthus* L., *Gramineae*, and allied species grown in India (not Turkey). *Constit.* 85-95% geraniol; citronellol, dipentene; it is practically devoid of esters.

Colorless or light yellow liquid; pleasant rose odor. d_{15}^{15} 0.885-0.896. α_D +1° 40′ to −2°. n_D^{20} 1.476-1.4085.

6909. Oil of Ginger. Volatile oil from the rhizome *Zingiber officinale* Roscoe, *Zingiberaceae*. *Constit.* *l*-Zingiberene, *d*-camphene, phellandrene, borneol, cineol, citral.

Yellowish, viscid liquid. d_{15}^{15} 0.875-0.885. α_D −25° to −45°. n_D^{20} 1.4880-1.4950. Almost insol in water; sparingly sol in alcohol; sol in ether, carbon disulfide.

USE: In mouthwashes, ginger beverages, liqueurs, etc.

6910. Oil of Hops. Volatile oil from strobiles of *Humulus lupulus* L., *Moraceae*. *Constit.* 65-70% humulene, terpenes. Gas chromatography and identification of constituents: Buttery et al., *Nature* **200**, 435 (1963). *See also* review by Stevens, *Chem. Rev.* **67**, 19 (1967).

Light yellow to brownish liquid. d_{15}^{15} 0.855-0.880. Practically inactive optically. n_D^{20} 1.4775. Insoluble in water; slightly sol in alcohol; sol in ether. *Keep well closed, cool and protected from light.*

6911. Oil of Hyssop. Volatile oil from *Hyssopus officinalis* L., *Labiatae*. *Constit.* About 50% pinene, small quantities of aromatic alcohol, probably also sesquiterpenes.

Colorless or greenish-yellow liquid; sharp, camphor-like taste. d_{15}^{15} 0.925-0.940. α_D^{20} −17° to −23°. bp ~200°. Almost insol in water; sol in 2 to 4 vols 80% alcohol. *Keep well closed and protected from light.*

6912. Oil of Juniper. Juniperberry oil. Volatile oil from dried ripe fruit (berries) of *Juniperus communis* L., *Cupressaceae*. *Constit.* Pinene, cadinene, camphene, terpineol, juniper camphor. Analysis by GLC of the oil of six western North American species: F. C. Vasec, R. W. Scora, *Am. J. Bot.* **54**, 781 (1967).

Colorless to pale greenish-yellow liquid. Odor and taste as of the berries. d_{25}^{25} 0.854-0.879. $[\alpha]_D^{25}$ 0° to −15°. n_D^{20} 1.4780-1.4840. Almost insol in water. Sol in 4 vols alcohol. Miscible with carbon disulfide, benzene, chloroform, amyl alcohol. *Keep cool in well-closed and well-filled bottles, protected from light.*

USE: Manuf liqueurs; for preserving catgut ligatures. In perfumery.

THERAP CAT: Diuretic.

6913. Oil of Juniper Wood. Steam-distilled oil from wood or branches of *Juniperus communis* L., *Cupressaceae*. It is not Juniper tar.

Colorless to yellowish, turpentine-like liquid; weak juniper odor. Insol in water. Sol in alcohol.

Caution: Not to be used for preserving catgut or silk.

6914. Oil of Lavender. Volatile oil from fresh flowering tops of *Lavandula officinalis* Chaix (*L. vera* DC.), *Labiatae*. *Constit.* 30-40% esters calculated as linalyl acetate; linalool, pinene, limonene, geraniol, some citral.

Colorless or yellow liquid. d 0.875-0.888. α_D^{25} −3° to −10°. n_D^{20} 1.459-1.470. Slightly sol in water; sol in 4 vols 70% alcohol; miscible with abs alcohol, carbon disulfide. *Keep well closed, cool and protected from light.*

USE: As perfume and flavor.

THERAP CAT: Aromatic, carminative.

6915. Oil of Lemon. Cedro oil. Volatile oil expressed from fresh peel of *Citrus limonum* (L.) Risso (*C. medica* var. *limon* L.), *Rutaceae*. *Constit.* About 90% of limonene, terpinene, phellandrene, and pinene combined; 4-6% aldehydes calculated as citral, some citronellal, geranyl acetate, sesquiterpenes.

Pale yellow or greenish-yellow liquid. d_{25}^{25} 0.849-0.855. α_D^{25} +57° to +65.6°. n_D^{20} 1.4742-1.4755. Slightly sol in water; sol in 3 vols alcohol; miscible with carbon disulfide, glacial acetic acid. *Keep cool, in well-filled and well-closed bottles, protected from light.*

USE: For flavoring medicaments; as a flavor in liqueurs, pastry, foods, beverages; also in perfumes.

6916. Oil of Lemon Grass. Indian oil of verbena; Indian melissa oil. Volatile oil from *Cymbopogon (Andropogon) citratus* (DC.) Stapf, and of *C. flexuosus* (Nees) Stapf,

Gramineae. Constit. 75-85% citral; methylheptenone, citronellal, geraniol, limonene, dipentene.

Reddish-yellow or brownish-red liquid; strong odor of verbena. d$_{15}^{15}$ 0.895-0.908 (0.878-0.882 of West Indian oil). α_D — 3°. n$_D^{20}$ 1.483-1.489. Slightly sol in water; sol in 3 vols 70% alcohol; sol in chloroform, ether. *Keep well closed, cool and protected from light.*

USE: As source of citral (synthesis of vitamin A) and in perfumery.

6917. Oil of Levant Wormseed. Volatile oil from the flowers of *Artemisia maritima* var *stechmanniana* Bess. (*A. pauciflora* Weber), *Compositae. Constit.* Largely eucalyptol (cineol).

Pale yellow to yellowish-green liquid. d$_{20}^{20}$ 0.915-0.940. Insol in water; sol in alcohol, ether. *Keep well closed, cool and protected from light.*

THERAP CAT: Anthelmintic.

6918. Oil of Linaloe. Volatile oil distilled from a Mexican wood (*Bursera delpechiana* Poiss. and probably other species of *Bursera, Burseraceae*). *Constit.* Linalool, geraniol, methylheptenone.

Colorless to yellowish liquid; pleasant odor. d$_{15}^{15}$ 0.875-0.890. α_D — 5° to — 12°. n$_D^{20}$ 1.4638. Slightly sol in water; sol in 2 vols 70% alcohol; sol in ether, chloroform. *Keep well closed, cool and protected from light.*

USE: In perfumery.

6919. Oil of Marjoram. Volatile oil from *Origanum marjorana* L., *Labiatae. Constit.* About 40% terpenes, chiefly terpinene; *d*-terpineol.

Yellow or greenish-yellow liquid. d$_{15}^{15}$ 0.888-0.912. α_D^{20} +13° to +18°. Insol in water; sol in 2 vols 80% alcohol; sol in chloroform, ether.

USE: In perfumes and microscopy.

6920. Oil of Mustard, Expressed. Fixed oil expressed from the mustard seeds of *Brassica alba* (L.) Boiss. and *B. nigra* (L.) Koch, *Cruciferae. Constit.* Chiefly the glycerides of oleic acid and other fatty acids, including arachidic.

Straw-colored or brownish-yellow, or greenish-brown liquid. d$_{15}^{15}$ 0.914-0.916. Solidif — 8° to — 16°. n$_D^{40}$ 1.4655-1.4670. Sapon no. 170-174. Iodine no. 92-97. Insol in water; slightly sol in alcohol; miscible with chloroform, ether, petr ether.

USE: Manuf oleomargarine, soap; as lubricant; salad oil.

6921. Oil of Myrtle. Volatile oil from leaves of *Myrtus communis* L., *Myrtaceae. Constit. d*-Pinene, eucalyptol, dipentene, camphor.

Yellow to greenish liquid; fragrant odor. d$_{15}^{15}$ 0.890-0.915. α_D +10° to +30°. Insol in water; sol in alcohol, chloroform, ether.

6922. Oil of Niaouli. Volatile oil from leaves of *Melaleuca viridiflora* (Soland.) Gaertn., *Myrtaceae. Constit.* About 65% cineol, about 30% terpineol, limonene and *d*-pinene combined.

Slightly yellow liquid; aromatic odor; pungent, refreshing peppermint-like taste. d$_{15}^{15}$ 0.908-0.932. Insol in water or glycerol; sol in alcohol, benzene, ether.

THERAP CAT: Anthelmintic.

6923. Oil of Nutmeg, Expressed. Nutmeg butter; oil of mace. Oil expressed from nutmeg (*Myristica fragrans* Houtt., *Myristicaceae*). *Constit.* Chiefly trimyristin; some volatile oil.

Orange-red to reddish-brown, soft solid. mp 45-51°. Odor and taste of nutmeg. d 0.990-0.995. Sapon no. 172-179. Iodine no. 40-52. Acid no. 17-23. Partly sol in cold alcohol, almost completely sol in hot alcohol; freely sol in chloroform, ether. LD$_{50}$ orally in rats: 3640 mg/kg, P. M. Jenner *et al., Food Cosmet. Toxicol.* **2**, 327 (1964).

Caution: Symptoms similar to Oil of Nutmeg, Volatile.

THERAP CAT: Rubefacient.

6924. Oil of Nutmeg, Volatile. Oil of myristica. Steam-distilled oil from dried kernels of ripe seeds of nutmeg (*Myristica fragrans* Houtt., *Myristicaceae*). *Constit.* 60-80% *d*-Camphene, ~8% *d*-pinene; dipentene, *d*-borneol, *l*-terpineol, ~6% geraniol, safrol, ~4% myristicin. The myristicin fraction together with its more than 25% content of

elemicin is supposed to be responsible for the purported hallucinogenic properties of nutmeg seed: Shulgin, *Nature* **197**, 4865 (1963); Weil, *Econ. Botany* **19**, 194-217 (1965). Toxicity study: P. M. Jenner *et al., Food Cosmet. Toxicol.* **2**, 327 (1964).

Colorless or pale yellow liquid; odor and taste of nutmeg. d$_{25}^{25}$ 0.859-0.924. α_D +10° to +30°. n$_D^{20}$ 1.4740-1.4880. Insoluble in water. Sol in 1 vol alcohol, in 3 vols 90% alcohol. *Keep well closed, cool and protected from light.* LD$_{50}$ orally in rats: 2620 mg/kg (Jenner).

Caution: Ingestion of large quantities produces narcosis, delirium, death.

USE: As a flavor.

THERAP CAT: Carminative.

6925. Oil of Orange. Oil sweet orange. Volatile oil expressed from fresh peel of ripe fruit of the orange (*Citrus aurantium* var *sinensis* L., *Rutaceae*). *Constit.* About 90% *d*-limonene; citral, decyl aldehyde, methyl anthranilate, linalool, terpineol.

Yellow to deep orange liquid; characteristic orange taste and odor. d$_{25}^{25}$ 0.842-0.846. α_D^{25} +94° to +99°. n$_D^{20}$ 1.4723-1.4737. Slightly sol in water; sol in 2 vols 90% alcohol, in 1 vol glacial acetic acid; miscible with abs alcohol, carbon disulfide. *Keep well closed, cool and protected from light.*

USE: Chiefly as flavor and perfume. Pharmaceutic aid (flavor).

THERAP CAT: Expectorant.

6926. Oil of Orange Flowers. Oil of neroli. Volatile oil distilled from fresh orange flowers. *Constit.* Limonene, *l*-linalool, geraniol, 7-18% linalyl acetate; some methyl anthranilate, nerol and neroli camphor.

Yellowish, fluorescent liquid; very intense and pleasant odor; becomes brown on exposure to light. d$_{25}^{25}$ 0.86-0.88. [α]$_{25}^{25}$ +1° 30' to +9° 8'. n$_D^{20}$ 1.475. Slightly sol in water; sol in 1.5-2 vols 80% alcohol with fine violet fluorescence. *Keep well closed, cool and protected from light.*

USE: As perfume and flavor.

6927. Oil of Origanum. Oil of wild marjoram. Volatile oil from flowering tops of *Origanum vulgare* L., *Labiatae. Constit.* Carvacrol, terpenes.

Light yellow liquid. d$_{15}^{15}$ 0.870-0.910. α_D ~ —34°. Very slightly sol in water; very sol in alcohol. *Keep well closed, cool and protected from light.*

USE: In perfumery.

6928. Oil of Parsley. Parsley seed oil. Volatile oil from parsley seeds *Petroselinum hortense*, Hoffm. (*P. sativum* Hoffm., *Carum petroselinum* Benth. & Hook.), *Umbelliferae. Constit.* Chiefly apiol; terpene, *l*-pinene(?).

Colorless or yellow, rather viscid liquid. d$_{15}^{15}$ 1.040-1.100. α_D — 5° to — 11°. n$_D^{20}$ 1.510-1.519. Very slightly sol in water; sol in 8 vols 80% alcohol; sol in ether.

USE: Flavoring in meat sauces, seasonings, spice blends, pickles.

6929. Oil of Patchouli. Patchouli oil. Essential oil from leaves of several *Labiatae* species. The commercial oil is obtained from the cultivated species, *Pogostemon cablin* (Blanco) Benth. (*P. patchouly* Pellet. var *suavis* Hook. f.), *Labiatae.* Major constituent is patchouli alcohol (*q. v.*); minor constituents include patchoulene, azulene, eugenol, and several unidentified sesquiterpenes: Pfau, Plattner, *Helv. Chim. Acta* **19**, 874 (1936); Naoko *et al., Bull. Chem. Soc. Japan* **40**, 597 (1967). *Review:* E. Guenther, *The Essential Oils* vol. **III** (Van Nostrand, New York, 1949) pp 552-575.

Yellowish or greenish to dark brown oil, intense and persistent fragrant odor. Can be stored indefinitely. Odor seems to improve with age. d$_{15}^{15}$ 0.975-0.987. [α]$_D^{20}$ —54° to —65.3°. n$_D^{20}$ 1.5099 to 1.5111. Saponif no. 3.3 to 9.3. Ester no. after acetylation: 17.7 to 22.4. Practically insol in water. Soluble in ether.

USE: In perfume formulations to impart a lasting oriental fragrance, in incense, soaps, cosmetics. To scent fine Indian fabrics and shawls.

6930. Oil of Pennyroyal—American. Oil of hedeoma. Volatile oil from leaves and flowering tops of *Hedeoma* *pulegioides* (L.) Pers., *Labiatae. Constit.* Chiefly pulegone; 2 ketones; acetic, formic and isoheptoic acids.

Pale yellow liq; aromatic odor. d_{25}^{25} 0.920-0.935. α_D^{20} +18° to +22°. n_D^{20} 1.482. Slightly sol in water; sol in 3 vols 70% alcohol; very sol in chloroform, ether. *Keep well closed, cool and protected from light.*

THERAP CAT: Aromatic carminative.

6931. Oil of Pennyroyal—European. Oil of pulegium. Volatile oil from *Mentha pulegium* L., *Labiatae. Constit.* About 85% pulegone.

Yellow or greenish-yellow liquid; aromatic mint-like odor; aromatic taste. d_{15}^{15} 0.960. α_D^{20} +14° to +28°. n_D^{20} 1.475-1.496.

6932. Oil of Pepper. Volatile oil from unripe fruit of the black pepper. *Constit.* Chiefly *l*-phellandrene, sesquiterpenes (caryophyllene).

Colorless or yellowish liquid. d_{15}^{15} 0.890-0.900. α_D about −3° to −5°. n_D^{20} 1.4935-1.4977. Insol in water; sol in about 15 vols 90% alcohol.

USE: As condiment.

6933. Oil of Peppermint. Colpermin; Mintec. Steam-distilled, volatile oil from fresh flowering plant *Mentha piperita* L., *Labiatae.* The Japanese oil, also known as oil of Poho, is the liq portion remaining after the separation of menthol from the oil of *Mentha arvensis* L., *Labiatae. Constit.* Not less than 50% total menthol including 5-9% esters calcd as menthyl acetate; menthyl isovalerate, menthone, inactive pinene, *l*-limonene, cadinene, phellandrene, some acetaldehyde, isovaleric aldehyde, amyl alcohol, dimethyl sulfide.

Colorless to pale yellow liquid; strong, penetrating peppermint odor and pungent taste. d_{25}^{25} 0.896-0.908. α^{25} −18° to −32°. n_D^{20} 1.460-1.471. Very slightly sol in water; sol in 4 vols 90% alcohol.

USE: Pharmaceutic aid (flavor). In flavoring liqueurs.

THERAP CAT: Carminative.

THERAP CAT (VET): Has been used as a carminative.

6934. Oil of Pettigrain. Volatile oil from leaves, twigs and unripe fruit of *Citrus vulgaris* Risso (*C. bigaradia* Loisel.), *Rutaceae. Constit.* 40-80% linalyl acetate; geraniol, geranyl acetate, limonene.

Yellow liquid. d_{15}^{15} 0.887-0.900. α_D^{20} +3° 43′ to −1° 22′. n_D^{20} 1.4623. Slightly sol in water; sol in 2 vols 80% alcohol. *Keep well closed, cool and protected from light.*

USE: In perfumes.

6935. Oil of Pimenta. Oil of allspice; oil of pimento. Volatile oil from fruit of *Pimenta officinalis* Lindl., *Myrtaceae. Constit.* 65-75% by vol of phenols calcd as eugenol and a sesquiterpene.

Colorless, yellow or reddish liquid; odor and taste of allspice; becomes darker with age. d_{25}^{25} 1.018-1.048. α^{25} −4°. Very slightly sol in water; sol in glacial acetic acid, in 1 vol 90% alcohol, in 2 vols 70% alcohol. *Keep well closed, cool and protected from light.*

THERAP CAT: Carminative.

6936. Oil of Pine Needles. Oil of Scotch fir; fir-wood oil. Volatile oil from *Pinus sylvestris* L., *Pinaceae. Constit.* Dipentene, pinene, sylvestrene, cadinene, 3-3.5% bornyl acetate.

Yellowish liquid. d_{15}^{15} 0.884-0.886. α_D^{20} +7° 3′ to +10°. Insol in water. Sol in 10 vols 90% alcohol. *Keep well closed, cool and protected from light.*

USE: Pharmaceutic aid (flavor and perfume).

THERAP CAT: Expectorant.

6937. Oil of Rose. Otto of rose; essence of rose; attar of rose. Volatile oil from fresh flowers of *Rosa gallica* L. and *R. damascena* Mill. and varieties of these species *(Rosaceae). Constit.* 70-75% free geraniol and citronellol; small amounts of their esters; terpenes.

Colorless or pale yellow liquid; viscous at 25°; highly fragrant, rose odor. d_{15}^{30} 0.848-0.863. Congeals at 18-22° to a translucent, cryst mass. α_D^{25} −1° to −4°. n_D^{20} 1.457-1.463. Very slightly sol in water, sparingly sol in alcohol; sol in fatty oils, chloroform. *Keep cool in well-closed and well-filled containers and protected from light.*

USE: Largely in perfumery; for flavoring lozenges, ointments, and toilet prepns.

6938. Oil of Rosemary. Volatile oil from fresh flowering tops of *Rosmarinus officinalis* L., *Labiatae. Constit.* Not less than 10% total borneol; not less than 2.5% esters calculated as bornyl acetate; camphor, eucalyptol, pinene, camphene.

Colorless or pale yellow liquid; characteristic rosemary odor; camphoraceous taste. d_{25}^{25} 0.894-0.912. α_D^{25} −5° to +10°. n_D^{20} 1.464-1.476. Almost insol in water; sol in 10 vols 80% alcohol. *Keep well closed, cool and protected from light.*

USE: In liniments and in hair lotions.

THERAP CAT: Carminative, rubefacient.

6939. Oil of Rue. Volatile oil from *Ruta graveolens* L., *Rutaceae. Constit.* About 90% methyl nonyl ketone, methyl anthranilate.

Pale yellow liquid; characteristic, sharp, unpleasant odor, but odor is pleasant on dilution. d_{15}^{15} 0.832-0.845. Solidif +8° to +10°. Optically inactive or slightly dextrorotatory. n_D^{20} 1.430-1.440. Almost insol in water; sol in 3 vols 70% alcohol. *Keep well closed, cool and protected from light.*

Caution: Frequent dermal contact produces erythema, vesication. Ingestion of large quantities causes epigastric pain, nausea, vomiting, confusion, convulsions, death; may cause abortion.

USE: Flavoring agent in food.

6940. Oil of Santal. Sandalwood oil; arheol; East Indian sandalwood oil. Volatile oil from dried heartwood of *Santalum album* L., *Santalaceae. Constit.* At least 90% total alcohols calculated as santalol, about 2-4% santalol as esters.

Pale yellow, somewhat viscid liquid; characteristic sandalwood odor and taste. d_{25}^{25} 0.965-0.980. α^{25} −15° to −20°. n_D^{20} 1.500-1.510. Very slightly sol in water; sol in 5 vols 70% alcohol. *Keep well closed, cool and protected from light.*

Note: West Indian sandalwood oil is derived from *Amyris balsamifera* and is not official. Its rotation is +24° to +29°. It is less sol in 70% alc than the East Indian.

THERAP CAT: Urinary anti-infective.

THERAP CAT (VET): Formerly as a urinary antiseptic.

6941. Oil of Sassafras. Volatile oil from root of *Sassafras albidum* (Nutt.) Nees (*S. variifolium* (Salisb.) Kuntze: *S. officinale* Nees & Eberm.), *Lauraceae. Constit.* About 80% safrol; small amount of eugenol; pinene, phellandrene, sesquiterpene, *d*-camphor.

Yellow to reddish-yellow liquid; characteristic odor and taste of sassafras. d_{25}^{25} 1.065-1.077. α_D^{20} +2° to +4°. n_D^{20} 1.5250-1.5350. Very slightly sol in water; sol in 2 vols 90% alcohol. *Keep well closed, cool and protected from light.*

USE: To mask disagreeable odor of medicaments.

THERAP CAT: Topical anti-infective; pediculicide; carminative.

6942. Oil of Savin. Volatile oil from fresh tops of *Juniperus sabina* L., *Cupressaceae. Constit.* 35-55% sabinol, partly free and partly as sabinyl acetate; cadiene, pinene.

Colorless or pale yellow liquid. d_{15}^{15} 0.903-0.923. α_D +40° to +60°. n_D^{20} 1.470-1.478. Very slightly sol in water; very sol in alcohol. *Keep well closed, cool and protected from light.*

THERAP CAT: Emmenagogue. Anthelmintic.

6943. Oil of Spearmint. Oil of crispmint; oil of curled mint. Volatile oil from the flowering tops of *Mentha spicata* L. (*M. virdis* L.), *Labiatae. Constit.* At least 50% carvone, *l*-limonene, pinene.

Colorless, yellow or greenish-yellow liq; characteristic spearmint odor and taste. d_{25}^{25} 0.917-0.934. α_D^{20} −48° to −59°. n_D^{20} 1.4820-1.4900. Very slightly sol in water; sol in equal vol 80% alcohol. *Keep well closed, cool and protected from light.*

USE: Pharmaceutic aid (flavor).

THERAP CAT: Carminative.

6944. Oil of Spike. Volatile oil from leaves and tops of *Lavandula spica* L. (*L. latifolia* Vill.), *Labiatae. Constit.* About 35% eucalyptol; camphor, linalool, borneol, terpineol, *d*-camphene, sesquiterpene.

Colorless or pale yellow liq; lavender and eucalyptol odor. d_{15}^{15} 0.900-0.920. α_D $-4°$ to $+6°$. n_D^{20} 1.462-1.469. Practically insol in water; sol in 3 vols 70% alcohol, in 6 vols 65% alcohol. *Keep well closed, cool and protected from light.*

6945. Oil of Sweet Almond. Expressed almond oil. Fixed oil from kernels of varieties of *Prunus amygdalus* Stokes (*Amygdalus communis* L.), *Rosaceae. Constit.* Chiefly glyceryl oleate with small amounts of glycerides of linolic, etc., acids. Stated to contain no stearic acid.

Colorless or pale yellow, almost odorless, oily liq; bland taste. d_{25}^{25} 0.910-0.915. It is clear at $-10°$; congeals near $-20°$. n_D^{40} 1.4593-1.4646. Sapon no. 191-200. Iodine no. 93-100. Insol in water. Slightly sol in alcohol; miscible with benzene, chloroform, ether, petr ether. *Keep well closed, cool and protected from light.*

USE: In perfumery, manuf fine soaps; as lubricant for delicate mechanisms such as watches, firearms, etc. Emollient.

THERAP CAT: Cathartic.

6946. Oil of Sweet Bay. Volatile oil of laurel. Volatile oil from leaves of *Laurus nobilis* L., *Lauraceae. Constit.* Eucalyptol, eugenol, methyl chavicol, pinene; also esters of isobutyric and isovaleric acids.

Pale yellow to greenish liquid. d_{15}^{15} 0.92-0.93. α_D^{20} $-15°$ to $-18°$. Insol in water; sol in alcohol. *Keep well closed, cool and protected from light.*

USE: In perfumes.

6947. Oil of Tansy. Volatile oil from leaves and tops of *Tanacetum vulgare* L., *Compositae. Constit.* Thujone, borneol, camphor.

Yellow liq; becomes brown on exposure to air and light. *Poisonous!* d_{15}^{15} 0.925-0.950. α_D $+30°$ to $+45°$. Almost insol in water; sol in alcohol, chloroform, ether. *Keep cool in well-filled and well-closed containers, protected from light.*

6948. Oil of Thyme. Volatile oil distilled from flowering plant *Thymus vulgaris* L., *Labiatae. Constit.* 20-40% by vol of thymol and carvacrol; cymene, pinene, linalool, bornyl acetate.

Colorless to reddish-brown liq; pleasant thymol odor; sharp taste. d_{25}^{25} 0.894-0.930. α_D^{25} $< -4°$. n_D^{20} 1.4830-1.5100. Very slightly sol in water; sol in 2 vols 80% alcohol. *Keep well closed, cool and protected from light.*

Note: Often misnamed "oil of origanum."

THERAP CAT: Rubefacient, counterirritant, antiseptic, carminative.

6949. Oil of Turpentine. "Spirit of turpentine." Volatile oil distilled from the oleoresin obtained from *Pinus palustris* Mill., *Pinaceae* and other species of *Pinus* yielding only terpene oils.

Colorless liq; characteristic odor and taste, both becoming more pronounced and less agreeable on aging or exposure to air. d_{25}^{25} 0.854-0.868. Greater part distills between 154-170°. n_D^{20} 1.4680-1.4780. Rotation is variable. Insol in water; sol in 5 vols alcohol; miscible with benzene, chloroform, ether, carbon disulfide, petr ether and oils.

Caution: Vapors can cause eye irritation, headache, dizziness, nausea. Inhalation and ingestion can cause bladder irritation. *See also* Turpentine.

USE: Solvent for oils, resins, varnishes; vehicle for paints.

THERAP CAT: Rubefacient, counterirritant.

THERAP CAT (VET): Externally as rubefacient and counterirritant; internally as antiseptic, carminative and expectorant. Has been used as anthelmintic.

6950. Oil of Turpentine, Rectified. Turpentine oil freed from unpleasant odor and taste by treatment with sodium hydroxide and distillation.

d_{25}^{25} 0.853-0.862; otherwise has the properties of the ordinary oil. *Keep well closed, cool and protected from light.*

THERAP CAT: Inhalation expectorant.

THERAP CAT (VET): Counterirritant and rubefacient. Antiseptic, carminative and expectorant. Bronchitis, acute gastric tympany and flatulent colic in horses, rumen tympanites in cattle. Formerly as an anthelmintic in horses, cattle and sheep.

6951. Oil of Turpentine, Sulfurated. Haarlem oil; Dutch oil; Dutch drops. Mixture of 1 part sulfurated linseed oil with 3 parts oil of turpentine.

Note: Not to be confused with "Dutch liquid," which is ethylene dichloride.

Caution: Symptoms similar to oil of turpentine.

6952. Oil of Valerian. Volatile oil from rhizome and root of *Valeriana officinalis* L., *Valerianaceae. Constit.* Bornyl esters of acetic, formic and isovaleric acids; *l*-pinene, *l*-camphene, *l*-limonene.

Yellowish-green to brownish liquid. d_{15}^{15} 0.93-0.96. α_D $-8°$ to $-13°$. Sapon no. 100-150. Acid no. 20-50. Slightly sol in water; very sol in alcohol, chloroform, ether. *Keep well closed, cool and protected from light.*

THERAP CAT: Sedative.

6953. Oil of Vetiver. Vetiver oil; oleum Andropogonis muricati; vetyver oil; khas khas oil; khus oil; cus cus oil; vetivert oil; vetiver oil Java; vetiver oil Haiti; vetiver oil Reunion (Bourbon). Distilled from roots of vetiver grass *Vetiveria zizanioides* Stapf., (*Andropogon muricatus* Retz., *Anatherum zizanioides* (L.) Hitch. & Chase, *Gramineae*), grown chiefly in Java, India, Reunion Island and Haiti. The constituents of vetiver oil vary according to the place of origin. Java and Reunion vetiver oils may contain 8-35% (usually 15-27%) sesquiterpene ketones of which α- and β-vetivones, *q.v.* have been isolated. The content of vetivenols (vetiverols, vetiver alcohols) varies from 45 to 65%. Other isolated components are vetivenyl vetivenate, vetivenic acid ($C_{15}H_{22}O_2$), palmitic acid, benzoic acid, and vetivene ($C_{15}H_{24}$). Major constituents of Indian vetiver oil are khusol, khusitol, khusinol. Review of chemical studies: Anh, Fetizon, *Am. Perfum. Cosmet.* **80**, 40 (1965). Biogenetically significant components: Kaiser, Naegeli, *Tetrahedron Letters* **1972**, 2009; Paknikar *et al.*, *ibid.* **1975**, 2973. Isoln and synthesis of *zizanal* and *epizizanal*, two insect repellent constituents, from Javanese vetiver oil: S. C. Jain *et al.*, *Tetrahedron Letters* **23**, 4639 (1982).

Brown to reddish-brown viscous oil. Aromatic to harsh, woody odor which improves on aging. d_{15}^{15} 0.990-1.040. n_D^{20} 1.5200-1.5280. $[\alpha]_D^{20}$ $+15°$ to $+45°$. Ester value after acetylation 110-165. Soluble in 1-3 volumes of 80% alcohol, becoming slightly turbid upon further dilution. Sol in all proportions in most fixed oils, diethyl phthalate, benzyl benzoate, mineral oil (slight turbidity). Practically insol in glycerol and propylene glycol. Fairly stable to dilute acids and weak alkalies; unstable to strong acids and alkalies. Should be stored in a cool place and protected from light. *Refs:* Kretchmar, Pictet, *Chimia* **8**, 123 (1954); Pfau, Plattner, *Helv. Chim. Acta* **22**, 640 (1939); Guenther, *The Essential Oils*, vol. **IV** (Van Nostrand, New York, 1950) p 156.

USE: In soap and perfumery formulations.

6954. Oil of White Cedar. Oil of arbor vitae; oil of thuja. Volatile oil from leaves of *Thuja occidentalis* L., *Cupressaceae. Constit.* *d*-Thujone, *l*-fenchone, *d*-pinene.

Colorless or pale yellow or greenish-yellow liq; camphor odor; bitter taste. d_{15}^{15} 0.915-0.935. α_D $-5°$ to $-14°$. Insol in water. Sol in 4 vols 70% alcohol. *Keep well closed, cool and protected from light.*

USE: Perfume.

THERAP CAT: Counterirritant.

6955. Oil of Wine, "Heavy". An oily liq obtained by distilling alcohol or ether (or both) with sulfuric acid. *Constit.* Etherin and etherol (C_2H_4)$_n$, polymers of ethylene; ethyl sulfate and ethyl sulfovinate.

Colorless to slightly yellow liq; penetrating odor; sharp, bitter taste. d_{15}^{15} 1.095-1.130. Almost insol in water; miscible with alcohol, ether.

USE: For making ethereal oil which is used in making Hoffman's anodyne; sometimes for flavoring brandy.

6956. Oil of Wormwood. Volatile oil from leaves and tops of *Artemisia absinthium* L., *Compositae. Constit.* Thujyl alcohol and acetate; thujone, phellandrene, cadinene; also a blue oil.

Brownish-green liq. d_{15}^{15} 0.925-0.955. n_D^{20} 1.460-1.4741. Very slightly soluble in water; sol in 2 vols 80% alcohol, in ether. *Keep well closed, cool and protected from light.*

USE: In flavoring of vermouth; formerly in absinthe.
THERAP CAT: Anthelmintic, antimalarial.

6957. Oil of Yarrow. Milfoil oil. Volatile oil from leaves and tops of *Achillea millefolium* L., *Compositae*. *Constit.* Cineol.

Blue liquid. d_{20}^{20} 0.905-0.925. Insol in water; very sol in alc, ether. *Keep well closed, cool and protected from light.*
THERAP CAT: Diaphoretic.

6958. Okadaic Acid. *9,10-Deepithio-9,10-didehydroacanthifolicin;* halochondrine A. $C_{44}H_{68}O_{13}$; mol wt 805.02. C 65.65%, H 8.51%, O 25.84%. First ionophoric polyether identified in marine organisms; isolated from marine black sponges, *Halichondria (okadai* or *melanodocia*). Tumor promoting cytotoxin associated with diarrhetic seafood poisoning. Isoln: K. Tachibana *et al., J. Am. Chem. Soc.* **103,** 2469 (1981). Total synthesis: M. Isobe *et al., Tetrahedron* **43,** 4767 (1987). Contractile effects: S. Shibata *et al., J. Pharmacol. Exp. Ther.* **223,** 135 (1982). Inhibition of protein phosphatases: A. Takai *et al., FEBS Letters* **217,** 81 (1987); C. Bialojan, A. Takai, *Biochem. J.* **256,** 283 (1988). Tumor promoting activity: M. Suganuma *et al., Proc. Nat. Acad. Sci. U.S.A.* **85,** 1768 (1988). Review of mechanism of action and use as a probe for cellular regulation: P. Cohen *et al., Trends Biochem. Sci.* **15,** 98-102 (1990); A. Schönthal, *New Biologist* **4,** 16-21 (1992). Review of tumor promoting activity: H. Fujiki, M. Suganuma, *J. Biochem.* **115,** 1-5 (1994).

Crystals from dichloromethane/hexane, mp 171-175°. $[\alpha]_D^{20}$ +21° (c = 0.33 in CHCl$_3$). Also reported as crystals from benzene-CHCl$_3$, mp 164-166°. $[\alpha]_D^{25}$ +25.4° (c = 0.24 in CHCl$_3$). LD$_{50}$ i.p. in mice: 192 µg/kg (Shibata).
USE: Biochemical tool as tumor promoter and probe of cellular regulation.

6959. Olanzapine. *2-Methyl-4-(4-methyl-1-piperazinyl)-10H-thieno[2,3-b][1,5]benzodiazepine;* LY-170053; Lanzac. $C_{17}H_{20}N_4S$; mol wt 312.44. C 65.35%, H 6.45%, N 17.93%, S 10.26%. Serotonin (5-HT$_2$) and dopamine (D$_1$/D$_2$) receptor antagonist with anticholinergic activity. Prepn: J. K. Chakrabarti *et al., Eur. pat. Appl.* **454,436;** *eidem, U.S. pat.* **5,229,382** (1991, 1993 both to Lilly). Comparative pharmacology: N. A. Moore *et al., Curr. Opin. Invest. Drugs* **2,** 281 (1993). HPLC determn in human plasma: J. T. Catlow *et al., J. Chromatog. B* **668,** 85 (1995).

Crystals from acetonitrile, mp 195°.
THERAP CAT: Antipsychotic.

6960. Olaquindox. *N-(2-Hydroxyethyl)-3-methyl-2-quinoxalinecarboxamide 1,4-dioxide;* Bay Va 9391; Bayo-n-ox; Fedan. $C_{12}H_{13}N_3O_4$; mol wt 263.25. C 54.75%, H 4.98%, N 15.96%, O 24.31%. Prepn: *Fr. pat.* **1,594,628** corresp to K. Ley *et al., U.S. pat.* **3,908,008** (1970, 1975 to Bayer). As feed additive for pigs: K. Bronsch *et al., Z. Tierphysiol., Tierernaehr. Futtermittelkd.* **36,** 211 (1976), *C.A.* **84,** 149717f (1976); R. S. Barber *et al., Anim. Feed Sci. Technol.* **4,** 117 (1979). HPLC analysis: G. F. Bories, *J. Chromatog.* **172,** 505 (1979).

Pale yellow crystals, mp 209° (dec). Slightly sol in water; insol in most organic solvents.
THERAP CAT (VET): Growth stimulant.

6961. Old Yellow Enzyme. *Reduced nicotinamide adenine dinucleotide phosphate dehydrogenase; dihydronicotinamide adenine dinucleotide phosphate diaphorase;* NADPH$_2$ diaphorase; OYE. Mol wt 102,000-106,000. Flavoprotein which catalyzes the oxidation of reduced triphosphopyridine nucleotide by oxygen or ferricytochrome c. When the oxidizing agent is molecular oxygen, OYE functions as a true hydrogen carrier, whereas when ferricytochrome c is the oxidizing agent, only electrons are passed on to the cytochrome, the protons being liberated as free hydrogen ions. Shows diaphorase activity as well as low cytochrome c reductase activity. Isoln from dried brewer's bottom yeast: Warburg, Christian, *Biochem. Z.* **266,** 377 (1933); Theorell, Akeson, *Biochem. Prepn.* **6,** 54 (1958). Each molecule contains two flavin mononucleotide groups: Ehrenberg, *Acta Chem. Scand.* **11,** 1257 (1957). *Review:* Akeson *et al., The Enzymes* vol. **7,** P. D. Boyer *et al.,* Eds. (Academic Press, New York, 1963) pp 477-494.

6962. Oleandomycin. PA-105; Amimycin; Landomycin; Romicil. $C_{35}H_{61}NO_{12}$; mol wt 687.87. C 61.11%, H 8.94%, N 2.04%, O 27.91%. Antibiotic substance produced by *Streptomyces antibioticus* no. ATCC 11891: Sobin *et al.;* Ratajak, Nubel, *U.S. pats.* **2,757,123; 2,842,481** (1956, 1958 to Pfizer). Structure: Hochstein *et al., J. Am. Chem. Soc.* **82,** 3225 (1960). Absolute configuration: Celmer, *ibid.* **87,** 1797 (1965); Celmer, Hobbs, *Carbohyd. Res.* **1,** 137 (1965); S. Omura *et al., Tetrahedron Letters* **1975,** 2939. Synthetic study: K. Tatsuta *et al., ibid.* **29,** 3975 (1988). Activity: Hahn, *Antibiotics* **1,** 378, 755 (1967). For a review of macrolide antibiotics see Keller-Schierlein, *Fortschr. Chem. Org. Naturst.* **30,** 313-460 (1973). Toxicity: H. Sous *et al., Arzneimittel-Forsch.* **8,** 386 (1958).

White amorphous powder. uv max (methanol): 286-289 nm. Moderately sol in water. Sol in dil acids. Freely sol in methanol, ethanol, butanol, acetone. Practically insol in hexane, carbon tetrachloride, dibutyl ether.

Hydrochloride, $C_{35}H_{61}NO_{12}$.HCl, long needles from ethyl acetate, mp 134-135°. $[\alpha]_D^{25}$ − 54° (methanol). Freely sol in water. Forms various cryst hydrates. LD$_{50}$ in mice, rats (mg/kg): 8200, > 10000 orally; 600, 400 i.v. (Sous).

Phosphate, $C_{35}H_{61}NO_{12}$.H$_3$PO$_4$, *Matromycin*.
Triacetyl deriv see Troleandomycin.
THERAP CAT: Antibacterial.
THERAP CAT (VET): Antibacterial.

6963. Oleandrin. *(3β,5β,16β)-16-(Acetyloxy)-3-[(2,6-dideoxy-3-O-methyl-α-L-arabino-hexopyranosyl)oxy]-14-hydroxycard-20(22)-enolide;* neriolin; Corrigen; Folinerin.

C$_{32}$H$_{48}$O$_9$; mol wt 576.73. C 66.64%, H 8.39%, O 24.97%. From the leaves of *Nerium oleander* L., *Apocynaceae* (Laurier rose): Tanret, *Compt. Rend.* **194**, 914 (1932); Neumann, *Ber.* **70**, 1547 (1937). Prepn by enzymic hydrolysis of urechitoxin: Hassall, *J. Chem. Soc.* **1951**, 3193. Structure: Tschesche, *Ber.* **70**, 1554 (1937); Krasso *et al., Helv. Chim. Acta* **46**, 1691 (1963).

Crystals from dil methanol, mp 250°. [α]$_D^{25}$ −48.0° (c = 1.3 in methanol). uv max: 220 nm (log ε 4.20). Practically insol in water. Sol in alcohol, chloroform.

Desacetyloleandrin, C$_{30}$H$_{46}$O$_8$, leaflets from alcohol, mp 238-240°. [α]$_D^{18}$ −24.9°.

THERAP CAT: Cardiotonic; diuretic.

6964. Oleanolic Acid. *(3β)-3-Hydroxyolean-12-en-28-oic acid;* oleanol; caryophyllin. C$_{30}$H$_{48}$O$_3$; mol wt 456.71. C 78.90%, H 10.59%, O 10.51%. Occurs in the free state in leaves of *Olea europaea, Oleaceae,* in leaves of *Viscum album* L., *Loranthaceae,* in buds of *Syzygium aromaticum* (L.) Merr. & Perry, *Myrtaceae* (cloves), in *Swertia japonica* (Maxim.) Makino, and in *Centaurium umbellatum* Gilib. (*Erythraea centaurium* (L.) Pers.), *Gentianaceae;* as acetate in birch bark, as glycoside in many saponins. Isoln procedures (from cloves): Winterstein, Stein, *Z. Physiol. Chem.* **202**, 222 (1931); Ruzicka, Hofmann, *Helv. Chim. Acta* **19**, 114 (1936); Picard *et al., J. Chem. Soc.* **1939**, 1047. Structure: Ruzicka *et al., Helv. Chim. Acta* **29**, 210 (1946). Review: J. Simonsen, W. C. T. Ross, *The Terpenes* vol. 5 (University Press, Cambridge, 1957) pp 221-285. *Cf.* α- and β-amyrin.

Fine, solvated needles from alc. After drying, mp 310°. [α]$_D^{20}$ +83.3° (c = 0.6 in chloroform). pK 2.52. Insol in water. Sol in 65 parts ether, 106 parts 95% alcohol, 35 parts boiling 95% alcohol, 118 parts chloroform, 180 parts acetone, 235 parts methanol.

Acetate, C$_{32}$H$_{50}$O$_4$, needles from methanol, mp 268°. [α]$_D^{17}$ +74.5° (c = 0.6 in CHCl$_3$).

Methyl ester, C$_{31}$H$_{50}$O$_3$, mp 201°. [α]$_D^{20}$ +75° (c = 0.6 in CHCl$_3$).

Acetate of methyl ester, C$_{33}$H$_{52}$O$_4$, needles from alcohol, mp 223°. [α]$_D^{20}$ +70° (c = 0.6 in CHCl$_3$).

6965. Oleic Acid. *(Z)-9-Octadecenoic acid.* C$_{18}$H$_{34}$O$_2$; mol wt 282.47. C 76.54%, H 12.13%, O 11.33%. Obtained by the hydrolysis of various animal and vegetable fats and oils. Prepn from olive oil: *Biochem. Prepn.* **2**, 100 (1952). Separation from olive oil by double fractionation via urea

adducts: Rubin, Paisley, *Biochem. Prepn.* **9**, 113 (1962). Stereochemistry: Thieme, *Ann.* **343**, 354 (1905). Synthesis: Robinson, Robinson, *J. Chem. Soc.* **127**, 175 (1925). ^{13}C-NMR studies: W. Stoffel *et al., Z. Physiol. Chem.* **353**, 1962 (1972); J. G. Batchelor *et al., J. Org. Chem.* **39**, 1698 (1974). Toxicity data: L. Orö, A. Wretlind, *Acta Pharmacol. Toxicol.* **18**, 141 (1961). Exptl use of ^{131}I-labelled oleic acid in myocardial imaging: F. J. Bonte *et al., Radiology* **108**, 195 (1973). Review of diagnostic use of ^{3}H-oleic acid in pancreatic function: N. T. Pedersen, *Digestion* **37**, Suppl. 1, 25-34 (1987).

Pure oleic acid is a colorless or nearly colorless liq (above 5-7°). d$_{25}^{25}$ ~0.895. Solidifies to cryst mass, mp 4°. bp$_{100}$ 286°. At atm pressure it dec when heated at 80-100°. n_D^{18} 1.463; n_D^{26} 1.4585. Iodine no. 89.9; acid value 198.6. On exposure to air, especially when impure, it oxidizes and acquires a yellow to brown color and rancid odor. Practically insol in water. Sol in alcohol, benzene, chloroform, ether, fixed and volatile oils. *Keep well closed, protected from light.* LD$_{50}$ i.v. in mice: 230±18 mg/kg (Orö, Wretlind).

Several grades of the acid are available in commerce, varying in color from pale yellow to red-brown and, depending on the amount of saturated acid present, becoming turbid at 8-16°. The acid of commerce usually contains 7-12% saturated acids, e.g., stearic, palmitic; also some linoleic, etc., unsaturated acids.

Methyl ester, C$_{19}$H$_{36}$O$_2$, *methyl oleate.* Prepd by refluxing oleic acid with *p*-toluene sulfonic acid in methanol: Rubin, Paisley, *loc. cit.* Iodine no. 85.6. d$_4^{18}$ 0.879. n_D^{20} 1.4510. bp$_2$ 168-170°. Miscible with anhydr ethanol, ether.

Ethyl ester, C$_{20}$H$_{38}$O$_2$, *ethyl oleate, (Z)-9-octadecenoic acid ethyl ester.* Yellowish, oily liquid. d 0.87. bp 205-208° (some dec). Insol in water. Misc with alcohol, ether.

Barium salt, C$_{36}$H$_{66}$BaO$_4$, *barium oleate.* Yellowish-white, granular masses. *Poisonous!* Practically insol in water. Slightly sol in boiling alcohol.

Sodium salt, C$_{18}$H$_{33}$NaO$_2$, *Eunatrol.* White powder, slight tallow-like odor. Sol in ~10 parts water, ~20 parts alcohol. Generally contains small quantities of the sodium salts of stearic, etc. acids. Alkaline in aq solns due to hydrolysis but not in alcohol solns.

Caution: Mildly irritating to skin, mucous membranes.

USE: Prepn of Turkey red oil, soft soap and other oleates; in polishing compds; waterproofing textiles, oiling wool; manuf driers; thickening lubricating oils. Pharmaceutic aid (solvent). The barium salt in rodent extermination.

THERAP CAT: Diagnostic aid (pancreatic function).

6966. Oleoresin of Aspidium. Oleoresin of male fern. An ether extract of male fern containing not less than 24% crude filicin.

Dark green, thick liq, usually depositing a granular, cryst substance. Bitter, unpleasant taste. d not less than 1.0. Insoluble in water. Partly sol in alcohol or chloroform. Almost completely sol in ether; not less than 85% sol in petr ether. *Shake well before dispensing.*

THERAP CAT: Anthelmintic.

THERAP CAT (VET): Anthelmintic (Cestodes; Trematodes).

6967. Oleuropein. *[2S-(2α,3E,4β)]-3-Ethylidene-2-(β-D-glucopyranosyloxy)-3,4-dihydro-5-(methoxycarbonyl)-2H-pyran-4-acetic acid 2-(3,4-dihydroxyphenyl)ethyl ester.* C$_{25}$H$_{32}$O$_{13}$; mol wt 540.52. C 55.55%, H 5.97%, O 38.48%. Bitter glucoside; first *secoiridoid* to be isolated. Isolation from olives and the leaves and bark of the olive tree, *Olea europaea* L., *Oleaceae* and structural studies: Panizzi *et al., Gazz. Chim. Ital.* **90**, 1449 (1960); Beyerman *et al., Bull. Soc. Chim. France* **1961**, 1821; Shasha, Leibowitz, *J. Org. Chem.* **26**, 1948 (1961). Isoln from the ripe fruits of *Ligustrum lucidum* and *L. japonicum* Thunb, *Oleaceae:* Inouye, Nishioka, *Tetrahedron* **28**, 4231 (1972). Revised structure and stereochemistry: Inouye *et al., Tetrahedron Letters* **1970**, 2459. Pharmacology: Petkov, Manolov, *Arzneimittel-Forsch.* **22**, 1476 (1972). Partial synthesis: A. Bianco *et al., J. Nat. Prod.* **55**, 760 (1992).

O—β-ᴅ-glucopyranose

Minute crystals from ethyl acetate, mp 87-89°. Hygroscopic. $[\alpha]_D^{20}$ —147° (H_2O, alcohol, or acetone). Shows mutarotation $[\alpha]_D^{20}$ —127° after 9 hrs (H_2O). Freely sol in acetone, ethanol, methanol, pyridine, glacial acetic acid, 5% aq NaOH soln. Moderately sol in water, dioxane, butanol, ethyl acetate, butyl acetate. Practically insol in ether, petr ether, chloroform, benzene, carbon tetrachloride.

6968. Oleyl Alcohol. *(Z)-9-Octadecen-1-ol; cis-9-octadecen-1-ol; Ocenol.* $C_{18}H_{36}O$; mol wt 268.48. C 80.53%, H 13.52%, O 5.96%. $CH_3(CH_2)_7CH=CH(CH_2)_7CH_2OH$. Found in fish oils. Usually obtained as a mixture of C_{16} and C_{18} unsaturated alcohols with C_{18} predominating. Prepd from butyl oleate by a Bouveault-Blanc reduction with sodium and butyl alcohol; or from triolein by hydrogenation in the presence of zinc chromite: Noller, Bannerot, *J. Am. Chem. Soc.* **56**, 1563 (1934); Reid *et al., Org. Syn.* **15**, 72 (1935); Adkins, Gillespie, *ibid.* **coll. vol. III,** 671 (1955). Purification by fractional crystallization at —40° from acetone, followed by distillation: Swern *et al., Oil Soap (Chicago)* **21**, 113 (1944); Loev, Dawson, *J. Am. Chem. Soc.* **78**, 1182 (1956).
Oily liquid. Usually pale yellow. Gives off acrid fumes when heated. d_4^{20} 0.850. mp 13-19°. bp$_8$ 195°. bp$_{1.5}$ 182-184°. (Distilling range at 760 mm: 305-370°.) $n_D^{27.5}$ 1.4582. Insol in water; sol in alcohol, ether.
USE: Chiefly in the manuf of its sulfuric esters which are detergents and wetting agents, as an antifoam agent; metal cutting lubricant; in carbon paper, stencil paper, printing ink; as a plasticizer; for softening and lubricating textile fabrics; carrier for medicaments.

6969. Olibanum. Frankincense; gum thus. Gum resin from *Boswellia carterii* Birdwood and other species of *Boswellia (Burseraceae).* **Habit.** Ethiopia, Egypt, Arabia, Somaliland. **Constit.** 3-8% volatile oil (pinene, dipentene, etc.); about 60% resins; 20% gum (polysaccharide fraction); 6-8% bassorin; bitter principle. Discussion of polysaccharides present: F. Smith, R. Montgomery, *The Chemistry of Plant Gum and Mucilages* (Reinhold, New York, 1959) pp 312-313.

6970. Oligomycins. Macrolide antibiotic complex produced by an actinomycete similar to *Streptomyces diastatochromogenes:* Smith *et al., Antibiot. & Chemother.* **4**, 962 (1954); McCoy, Peterson, **U.S. pat. 2,927,057** (1960 to Wisconsin Alumni Res. Foundation). Complex of several closely related compounds: Masamune *et al., J. Am. Chem. Soc.* **80**, 6092 (1958). Separation of oligomycins A, B and C:

Oligomycin A

Marty, McCoy, *Antibiot. & Chemother.* **9**, 286 (1959). Isoln and activity of D from *Streptomyces rutgersensis:* R. Q. Thompson *et al., Antimicrob. Ag. Chemother.* **1961**, 474. Partial structure of B: Prouty *et al., Biochem. Biophys. Res. Commun.* **44**, 619 (1971); total structure of B: Glehn *et al., FEBS Letters* **20**, 267 (1972). Structures of A and C: G. T. Carter, *Diss. Abstr. Int. B.* **37**, 766 (1976); of D: C. Merienne, T. Staron, *Chem. Commun.* **1978**, 318. *Review:* P. D. Shaw in *Antibiotics* vol. 1, D. Gottlieb, P. D. Shaw, Eds. (Springer-Verlag, New York, 1967) pp 585-610.

Oligomycin A, $C_{45}H_{74}O_{11}$. Two crystalline modifications, mp 140-141° and mp 150-151° (hexagonal crystals, melting with rapid decompn). uv max (abs ethanol): 225 nm (ϵ about 20,000). Solys at 25° (g/100 ml soln): water 0.002; ether 28; benzene 6; Skellysolve B 0.02; abs ethanol 25; glacial acetic acid 37.5; acetone 85.

Oligomycin B, $C_{45}H_{72}O_{12}$, *28-oxooligomycin A.* Potent inhibitor of oxidative phosphorylation: Lardy *et al., Arch. Biochem. Biophys.* **78**, 587 (1958).

Oligomycin C, $C_{45}H_{74}O_{10}$, *12-deoxyoligomycin A.*

Oligomycin D, $C_{44}H_{72}O_{11}$, *26-demethyloligomycin A, rutamycin, A-272, RR-32705.* mp 116-119°. $[\alpha]_D^{20}$ —62° (c = 1.36 in $CHCl_3$).

THERAP CAT: Oligomycin D as antifungal.

6971. Olivacine. *1,5-Dimethyl-6H-pyrido[4,3-b]carbazole;* guatambuinine. $C_{17}H_{14}N_2$; mol wt 246.31. C 82.90%, H 5.73%, N 11.37%. Isolated from the bark and stem of *Aspidosperma olivaceum* Müll. Arg., *A. australe* Müll. Arg., *A. longepetiolatum* Kuhlm., *Apocynaceae:* Schmutz, Hunziger, *Pharm. Acta Helv.* **33**, 341 (1958). Identity with guatambuinine: Marini-Bettolo, Carvalho-Ferreira, *Ann. Chim.* **49**, 869 (1959). Structure: Marini-Bettolo, Schmutz, *Helv. Chim. Acta* **42**, 2146 (1959); Ondetti, Deulofeu, *Tetrahedron* **15**, 160 (1961). Synthesis: Schmutz, Wittwer, *Helv. Chim. Acta* **43**, 793 (1960); Wenkert, Dave, *J. Am. Chem. Soc.* **84**, 94 (1962); Mosher *et al., J. Med. Chem.* **9**, 237 (1966); R. Besselievre, H. Husson, *Tetrahedron Letters* **1976**, 1873; J. Bergman, R. Carlsson, *ibid.* **1978**, 4055; J. P. Kutney *et al., Heterocycles* **16**, 1469 (1981).

Fine yellow needles from dilute methanol, yellow prisms from undiluted methanol, mp 317-325°. Fluoresces in dilute alcoholic soln. uv max (ethanol): 224, 238, 276, 287, 292, 314, 329, 375 nm (log ϵ 4.39, 4.33, 4.70, 4.85, 3.66, 3.80, 3.66). Soly in methanol, acetone, chloroform, carbon tetrachloride, carbon disulfide, tetrahydrofuran, dioxane less than 1%.

6972. Olivanic Acids. A family of naturally occurring carbapenem β-lactamase inhibitors with antibacterial activity, produced by *Streptomyces olivaceus:* D. Butterworth, G. N. Rolinson, **Ger. pat. 2,146,400;** *eidem* **U.S. pat. 3,919,415** (1972, 1975 both to Beecham). Isoln: A. G. Brown *et al., J. Antibiot.* **29**, 668 (1976). The family contains seven members, all containing the 7-oxo-1-azabicyclo[3.2.0]hept-2-ene ring system. Compounds *MM 4550, MM 13902,* and *MM 17880* are sulfated and are predominant when *S. olivaceus* is grown in sodium sulfate-containing media. *MM 22380, MM 22381, MM 22382, MM 22383* are non-sulfated hydroxy analogs and are produced in sulfate-free media or by mutants of *S. olivaceus* unable to complete the sulfation process. Detection, properties, fermentation of the sulfated members: D. Butterworth *et al., ibid.* **32**, 287 (1979); isoln, characterization: J. D. Hood *et al., ibid.* 295. Fermentation, isoln, characterization of the non-sulfated members: S. J. Box *et al., ibid.* 1239. Structures of MM 4550, MM 13902: A. G. Brown *et al., Chem. Commun.* **1977**, 523; of MM 17880: D. F. Corbett *et al., ibid.* 953. Structures of MM 22380-3: A. G. Brown *et al., J. Antibiot.* **32**, 963 (1979). Total synthesis of racemic MM 22383: R. J. Ponsford, R. Southgate, *Chem. Commun.* **1980,** 1085. Comparative antibacterial activity *in vitro:* M. J. Basker *et al., J. Antibiot.* **33,** 878 (1980). MM 22380, MM 22382, MM 22381, MM 22383, MM 13902, and MM 17880 are identical to *epithienamycins* A through F, respectively, *cf.* E. O. Stapley *et al., ibid.* **34,** 628 (1981); P. J. Cassidy *et al., ibid.* 637. Synthetic study on epithienamycins: T. Kametani *et al., J. Chem. Soc. Perkin Trans. I* **1981,** 3048.

MM 4550	R = –S–CH = CH–NHCOCH₃	(E)
MM 13902	R = –S–CH = CH–NHCOCH₃	(E)
MM 17880	R = –S–CH₂CH₂NHCOCH₃	

MM 4550, MM 13902, MM 17880 are isolated as disodium salts. Sol in water, methanol, DMF, DMSO. Practically insol in other organic solvents. uv max (water): of MM 4550: 240, 287 nm ($E_{1cm}^{1\%}$ 268); MM 13902: 227, 307 nm ($E_{1cm}^{1\%}$ 356); MM 17880: 298 nm ($E_{1cm}^{1\%}$ 192). The antibiotics are unstable in aq soln outside a narrow pH range; in acids, degradation leads to changes in uv spectra. Addition of hydroxylamine or cysteine to neutral solns results in rapid degradation.

6973. Olive Oil. A fixed oil obtained from ripe olives, the fruit of the cultivated olive tree *Olea europaea* L., *Oleaceae.* Produced almost exclusively in the countries adjoining the Mediterranean Sea, Spain being the largest producer. Whole olives are crushed in edge runner mills and the oil is expressed in open hydraulic presses. *Constit.* Mixed glycerides of oleic acid 83.5%, of palmitic acid 9.4%, of linoleic acid 4.0%, of stearic acid 2.0%, of arachidic acid 0.9%. Minor constitutents are squalene, up to 0.7%, phytosterol and tocopherols about 0.2%. *Reviews and bibliographies:* José M. de Soroa y Pineda, *El aceite de oliva* (Dossat, Madrid, 1944); R. F. Simari, G. B. Martinenghi, *Olivicoltura e Oleificio* (Hoepli, Milano, 1950); P. G. Garoglio, *Technologia de los Aceites Vegetales* (Mendoza, Argentina, 1951); E. W. Eckey, *Vegetable Fats and Oils* (Reinhold, New York, 1954).

Pale yellow or light greenish-yellow oil with a pleasing delicate flavor. Becomes rancid on exposure to air. Begins to get turbid at +5 to +10°, below 0° it forms a whitish, granular mass. Flash pt 437°F (225°C). Ignition temp 650°F (343°C). d_{15}^{15} 0.914-0.919; d_{25}^{25} 0.909-0.915. n_D^{25} 1.466-1.468; n_D^{40} 1.460-1.464. Titer 17-26°. Acid value 0.2-2.8. Saponification value 187-196. Iodine value 79-90. Thiocyanogen value 75-83. Hydroxyl value 4-12. Reichert-Meissl value 0.2-1.0. Unsaponifiable 0.5-1.3%. Slightly sol in alcohol. Miscible with ether, chloroform, carbon disulfide.

USE: As food in salads, with sardines, for cooking and baking. In the manuf of soaps, textile lubricants, sulfonated oils, cosmetics and pharmaceutical preparations. Emollient.

THERAP CAT (VET): Laxative, emollient.

6974. Olivil. *[2S-(2α,3β,4α)]-Tetrahydro-4-hydroxy-2-(4-hydroxy-3-methoxyphenyl)-4-[(4-hydroxy-3-methoxyphenyl)methyl]-3-furanmethanol.* $C_{20}H_{24}O_7$; mol wt 376.41. C 63.82%, H 6.43%, O 29.75%. From gum-resin of *Olea europaea* L., *Oleaceae.* Isoln: Pelletier, *Ann. Chim. Phys.* **51,** 196 (1833). Structure: Traverso, *Gazz. Chim. Ital.* **90,** 792 (1960); Smith, *Tetrahedron Letters* **1963,** 991. Revised

structure: Ayres, Mhasalkar, *ibid.* **1964,** 335; *eidem, J. Chem. Soc.* **1965,** 3586.

Monohydrate, crystals. mp 118-120°; when anhydr, mp 142-143°. $[\alpha]_D^{12}$ −127°. Sol in hot water, alcohol, acetic acid, fatty oils.

6975. Olivomycins. A mixture of antibiotics produced by *Streptomyces olivoreticuli:* Gauze *et al., Antibiotiki (Moscow)* **7,** 34 (1962); Brazhnikova *et al., ibid.* 39. Similar to the chromomycins, *q.v.:* Berlin *et al., Tetrahedron Letters* **1966,** 1643. Prepn: Gauze, **Brit.** pat. **1,152,748; Fr.** pat. **1,554,-600** (1969, both to Sci. Res. Inst. Antibiot.), *C.A.* **71,** 37490b, 128693m (1969). Composed of olivomycins A (major component), B, C, and D. Separation: Berlin *et al., Khim. Prir. Soedin.* **1967,** 331, *C.A.* **68,** 40017w (1968). Structural elucidation of the aglycone, *olivin,* and the carbohydrate moieties: Berlin *et al., Tetrahedron Letters* **1964,** 1323; **1966,** 1431; Berlin *et al., Khim. Prir. Soedin.* **1969,** 567, *C.A.* **73,** 25823r (1970). Stereochemistry: Bakhaeva *et al., Chem. Commun.* **1967,** 10; Berlin *et al., Khim. Prir. Soedin.* **1972,** 519. Revised structure: J. Thiem, B. Meyer, *Tetrahedron* **37,** 551 (1981). Synthetic studies: J. H. Dodd, S. M. Weinreb, *Tetrahedron Letters* **1979,** 3593; J. H. Dodd *et al., J. Org. Chem.* **47,** 4045 (1982). Series of articles on pharmacology: *Antibiotiki (Moscow)* **17** (1972). Toxicity data: M. Slavik, S. K. Carter, *Adv. Pharmacol. Chemother.* **12,** 1 (1975). *Review:* J. D. Skarbek, M. K. Speedie, in *Antitumor Compounds of Natural Origin* vol. 1, A. Aszalos, Ed. (CRC Press, Boca Raton, 1981) pp 191-235.

Olivomycin A

Olivomycin A, $C_{58}H_{84}O_{26}$, *3^D-O-[2,6-dideoxy-3-C-methyl-4-O-(2-methyl-1-oxopropyl)-α-L-arabino-hexopyranosyl]olivomycin D.* Formerly known as olivomycin and variant I. Yellow crystals from ethanol-hexane, mp 160-165°. $[\alpha]_D^{20}$ −36° (c = 0.5 in ethanol). uv max (ethanol): 228, 277, 318, 406 nm (log ε 4.39, 4.67, 3.81, 4.05). Sol in alc, ether, chloroform. Insol in benzene, carbon tetrachloride, petr ether, water. LD_{50} i.v. in mice: 13.75 mg/kg (Slavik, Carter).

THERAP CAT: Antineoplastic.

6976. Olsalazine. *3,3'-Azobis(6-hydroxybenzoic acid);* **C.I. Mordant Yellow 5;** 3,3'-dicarboxy-4,4'-dihydroxyazobenzene; 5,5'-azobis(salicylic acid); azodisal. $C_{14}H_{10}N_2O_6$; mol wt 302.24. C 55.64%, H 3.33%, N 9.27%, O 31.76%. Dimer of mesalamine, *q.v.,* originally used as a mordant dye. Prepn: **Ger.** pat. **278,613;** C. Mettler, **U.S.** pat. **1,157,169** (1914, 1915 both to J. R. Geigy). Improved prepn: K. H. Agback, A. S. Nygren, **Eur.** pat. **Appl. 36,636;** *eidem,* **U.S.** pat. **4,528,367** (1981, 1985 both to Pharmacia AB). Use in inflammatory intestinal disorders: K. H. Agback *et al.,* **Eur.** pat. **Appl. 36,637;** *eidem,* **U.S.** pat. **4,559,330** (1981, 1985 both to Pharmacia AB). Metabolism, distribution in man: C. P. Willoughby *et al., Gut* **23,** 1081 (1982). HPLC determn in serum, urine, feces: R. A. van Hogezand *et al., J. Chromatog.* **305,** 470 (1984). *In vitro* effect on fecal bacteria: H. Sandberg-Gertzen *et al., Scand. J. Gastroenterol.* **20,** 607 (1985). Clinical trials in patients intolerant to sulfasalazine, *q.v.:* H. Sandberg-Gertzen *et al., Gastroenterol.* **90,** 1024 (1986). Comparison with mesalamine in ulcerative colitis: S. S. Rao *et al., Scand. J. Gastroenterol.* **22,** 326 (1987). Brief review: R. A. Levinson, *Am. J. Gastroenterol.* **80,** 203 (1985).

Disodium salt, $C_{14}H_8N_2Na_2O_6$, *azodisal sodium, disodium azodisalicylate, C.I. 14130, Ph-CJ-91B, Dipentum.* Yellow powder. Sol in water, moderately sol in ethanol.

USE: Mordant dye for wool.

THERAP CAT: Treatment of ulcerative colitis.

6977. Omeprazole. *5-Methoxy-2-[[(4-methoxy-3,5-dimethyl-2-pyridinyl)methyl]sulfinyl]-1H-benzimidazole;* H-168/68; Antra; Gastroloc; Losec; Mepral; Mopral; Omepral; Omeprazen; Parizac; Pepticum; Prilosec. $C_{17}H_{19}N_3O_3S$; mol wt 345.42. C 59.11%, H 5.54%, N 12.16%, O 13.90%, S 9.28%. Gastric proton-pump inhibitor. Prepn: U. K. Junggren, S. E. Sjostrand, **Eur. pat. Appl. 5,129;** *eidem,* **U.S. pat. 4,255,431** (1979, 1981 both to AB Hässle). Pharmacology: P. Muller *et al., Arzneimittel-Forsch.* **33,** 1685 (1983). Mechanism of action study: B. Wallmark *et al., Biochim. Biophys. Acta* **778,** 549 (1984). Studies of active metabolite: W. B. Im *et al., J. Biol. Chem.* **260,** 4591 (1985); P. Lindberg *et al., J. Med. Chem.* **29,** 1327 (1986). Chromatographic determn in plasma and urine: P. Lagerstrom, B. Persson, *J. Chromatog.* **309,** 347 (1984). Pharmacokinetics: P. J. Prichard *et al., Gastroenterol.* **88,** 64 (1985). Clinical trial in Zollinger-Ellison syndrome: C. B. H. W. Lamers *et al., N. Engl. J. Med.* **310,** 758 (1984); in duodenal ulcer: K. Lauritsen *et al., ibid.* **312,** 958 (1985); P. J. Prichard *et al., Brit. Med. J.* **290,** 601 (1985). Survey of preclinical data: *Scand. J. Gastroenterol.* **20,** Suppl 108, 1-120 (1985). Toxicological studies: L. Ekman *et al., ibid.* **33,** 53. Preliminary review of pharmacodynamics, pharmacokinetics and therapeutic use: S. P. Clissold, D. M. Campoli-Richards, *Drugs* **32,** 15-47 (1986).

Crystals from acetonitrile, mp 156°. LD_{50} in mice, rats (g/kg): 0.08, > 0.05 i.v.; >4, >4 orally (Ekman).

THERAP CAT: Antiulcerative; in treatment of Zollinger-Ellison syndrome.

6978. Omoconazole. *(Z)-1-[2-[2-(4-Chlorophenoxy)ethoxy]-2-(2,4-dichlorophenyl)-1-methylethenyl]-1H-imidazole; (Z)-1-[2,4-dichloro-β-[2-(p-chlorophenoxy)ethoxy]-α-methylstyryl]imidazole;* CM-8282. $C_{20}H_{17}Cl_3N_2O_2$; mol wt 423.73. C 56.69%, H 4.04%, Cl 25.10%, N 6.61%, O 7.55%. Prepn: L. Zirngibl *et al.,* **Ger. pat. 2,839,388;** *eidem,* **U.S. pat. 4,210,657** (both 1980 to Siegfried); and crystal structure: K. Thiele *et al., Helv. Chim. Acta* **70,** 441 (1987). Stereospecific synthesis: L. Zirngibl, K. Thiele, **U.S. pat. 4,554,356** (1985 to Siegfried). *In vitro* fungistatic activity: M. Mosse *et al., Pathol. Biol.* **34,** 684 (1986).

Crystals from ethyl acetate/hexane (1:4), mp 89-90°.
Nitrate, $C_{20}H_{17}Cl_3N_2O_2 \cdot HNO_3$, *Sgd-12878, Fangorex, Fongarex.* Crystals from ethyl acetate/ethanol, mp 118-120° (Büchi), 122.5° (Mettler).

THERAP CAT: Antifungal (topical).

6979. Ondansetron. *1,2,3,9-Tetrahydro-9-methyl-3-[(2-methyl-1H-imidazol-1-yl)methyl]-4H-carbazol-4-one.* $C_{18}H_{19}$-

N_3O; mol wt 293.37. C 73.70%, H 6.53%, N 14.32%, O 5.45%. Specific serotonin ($5HT_3$) receptor antagonist. Prepn: I. H. Coates *et al.,* **Eur. pat. Appl. 191,562;** *eidem,* **U.S. pat. 4,695,578** (1986, 1987 both to Glaxo). Pharmacology: A. Butler *et al., Brit. J. Pharmacol.* **94,** 397 (1988). Anxiolytic activity: B. J. Jones *et al., ibid.* **93,** 985 (1988). Antiemetic properties: D. Cunningham *et al., Lancet* **1,** 1461 (1987). Clinical trials as antiemetic for cancer chemotherapy patients: M. G. Kris *et al., J. Clin. Oncol.* **6,** 659 (1988); L. Cubeddu *et al., N. Engl. J. Med.* **322,** 810 (1990). Comparative clinical trial with metoclopramide, *q.v.,* in cisplatin-induced emesis: M. Marty *et al., ibid.* 816.

Crystals from methanol, mp 231-232°.
Hydrochloride dihydrate, $C_{18}H_{19}N_3O \cdot HCl \cdot 2H_2O$, *GR-38032F, GR-C507/75, SN-307, Zofran, Zophren.* White crystalline solid from water/isopropanol, mp 178.5-179.5°.
3S-Form, $[\alpha]_D^{25}$ −14° (c = 0.19 in methanol).
3R-Form, $[\alpha]_D^{24}$ +16° (c = 0.34 in methanol).
THERAP CAT: Antiemetic.

6980. Onion Oil. Obtained from crushed onion seeds, the seeds of *Allium cepa* L., *Liliaceae:* Loew, *Ind. Quim.* **10,** 5 (1948); Phadnis *et al., J. Univ. Bombay* **17A,** no. 24, 62 (1948). Contains allylpropyl bisulfide, *S*-(1-propenyl)cysteine sulfoxide and 1-propenylsulfenic acid which is thought to be the lacrimator in onions: Virtanen, Sprare, *Suom. Kemistilehti* **35B,** 28, 29 (1962); *Chem. Ztg.* **86,** 816 (1962).
Pale yellow oil. Strong odor of onions. d_4^{15} 0.9289. Solidifies at −15°. n_D^{25} 1.4730. $[\alpha]_D^{20}$ −1.5°. Sol in ether, chloroform, carbon disulfide.

6981. Ontianil. *N-(4-Chlorophenyl)-2,6-dioxocyclohexanecarbothioamide;* 4'-chloro-2,6-dioxothiocyclohexanecarboxanilide. $C_{13}H_{12}ClNO_2S$; mol wt 281.76. C 55.42%, H 4.29%, Cl 12.58%, N 4.97%, O 11.36%, S 11.38%. Prepn: H. Ruschig *et al.,* **Ger. pat. 2,039,466;** *eidem,* **U.S. pat. 3,746,765** (1972, 1973 both to Hoechst); of sodium, potassium, calcium salts: *eidem,* **Ger. pat. 2,317,579** (1974 to Hoechst), *C.A.* **82,** 64537f (1975).

Crystals from ethanol, mp 113-115°. LD_{50} in rats (mg/kg): 964 s.c.; 229 i.v. (Ruschig, 1973).
Potassium salt, $C_{13}H_{12}ClKNO_2S$, *Terondit.*
THERAP CAT (VET): Antifungal.

6982. Oosporein. *2,2',5,5'-Tetrahydroxy-4,4'-dimethyl-[bi-1,4-cyclohexadien-1-yl]-3,3',6,6'-tetrone; 3,3',6,6'-tetrahydroxy-5,5'-dimethyl-2,2'-bi-p-benzoquinone;* chaetomidin; iso-oosporein. $C_{14}H_{10}O_8$; mol wt 306.23. C 54.91%, H 3.29%, O 41.80%. Fungal pigment isolated from *Oospora colorans* van Beyma.: Kögl, van Wessem, *Rec. Trav. Chim.* **63,** 5 (1944); from *Chaetomium aureum* Chivers and identity with chaetomidin: Lloyd *et al., J. Chem. Soc.* **1955,** 2163; from *Acremonium* spp: Divekar *et al., Can. J. Chem.* **37,** 2097 (1959); from *Beauveria bassiana* (Bals.) Vuill.: Vining *et al., Can. J. Microbiol.* **8,** 931 (1962). Identity with iso-oosporein: Smith, Thomson, *Tetrahedron* **10,** 148 (1960). Synthesis: J. Kalamar *et al., Helv. Chim. Acta* **57,** 2368 (1974). Biosynthesis: E. Steiner *et al., ibid.* 2377.

Bronze plates from aq methanol, mp 290-295°. uv max (ethanol): 216, 287 nm (log ε 3.51, 4.67).

Tetraacetate, $C_{22}H_{28}O_{12}$, yellow needles from methanol, mp 190°. uv max (ethanol): 262 nm (log ε 4.41).

Tetramethyl ether, $C_{18}H_{18}O_8$, orange needles from aq methanol, mp 123°. uv max (ethanol): 285.5, 394 nm (log ε 4.40, 2.98).

6983. Opianic Acid. *6-Formyl-2,3-dimethoxybenzoic acid; 5,6-dimethoxyphthalaldehydic acid.* $C_{10}H_{10}O_5$; mol wt 210.19. C 57.14%, H 4.80%, O 38.06%. Obtained (together with cotarnine) by heating narcotine with dil HNO_3. Prepn: Wilson et al., J. Org. Chem. **16**, 792 (1951); Blair, J. Chem. Soc. **1955**, 708. NMR studies: Buu-Hoï et al., Bull. Soc. Chim. France **1970**, 137.

Needles from water, mp 150°. Sol in 400 parts cold, 60 parts boiling water; sol in alcohol, ether. uv max: 215, 284 nm (ε 20,700, 6500).

6984. Opiniazide. *4-Pyridinecarboxylic acid [(2-carboxy-3,4-dimethoxyphenyl)methylene]hydrazide; 5,6-dimethoxyphthalaldehydic acid isonicotinoylhydrazone; 1-(2-carboxy-3,4-dimethoxybenzylidene)-2-isonicotinoylhydrazine; 2-carboxy-3,4-dimethoxybenzal isonicotinylhydrazone; saluside; saluzide; saliuzid; saluzid.* $C_{16}H_{15}N_3O_5$; mol wt 329.31. C 58.36%, H 4.59%, N 12.76%, O 24.29%. Description: G. N. Pershin, S.A. Vichkanova, C.A. **51**, 10747e (1957); C.A. **52**, 1485-1486 (1958); V. A. Buskina, C.A. **54**, 744f (1960). Bioavailability: A. F. Zaeko, V. G. Perkova, Farmatsiya (Moscow) **28**, 28 (1979), C.A. **92**, 185805g (1980). Toxicity study: R. A. Akhundov, Mater. Nauchn. Konf. Azgosmedinstituta **1975**, 137, C.A. **86**, 50678y (1977).

LD_{50} i.v. in guinea pigs: 1.634 g/kg (Akhundov).
THERAP CAT: Antibacterial (tuberculostatic).

6985. Opipramol. *4-[3-(5H-Dibenz[b,f]azepin-5-yl)propyl]-1-piperazineethanol; 5-[γ-[4-(β-hydroxyethyl)piperazino]propyl]dibenzo[b,f]azepine; N-[3-[4-(2-hydroxyethyl)-piperazino]propyl]iminostilbene; 4-[3-(5H-dibenzo[b,f]azepin-5-yl)propyl]-1-(2-hydroxyethyl)piperazine.* $C_{23}H_{29}N_3O$; mol wt 363.50. C 76.00%, H 8.04%, N 11.56%, O 4.40%. Prepn: Fr. pat. **M209** (1961 to Rhône-Poulenc), C.A. **58**, 3442f (1963), Schindler. Ger. pat. **1,133,729** and Swiss pats. **359,143** and **360,061** (all 1962 to Geigy), C.A. **58**, 10219a, 10218f, 10218h (1963).

Crystals, mp 100-101°.
Dihydrochloride, $C_{23}H_{29}N_3O.2HCl$, *Dinsidon, Ensidon, Insidon, Nisidana.* Crystals from ethanol, mp 210° (Fr. pat., loc. cit.), mp 228-230° (Schindler, loc. cit.). Sol in water, alcohol; sparingly sol in acetone.
THERAP CAT: Antidepressant; antipsychotic.

6986. Opium. Gum opium; crude opium. Air-dried, milky exudation from incised, unripe capsules of *Papaver somniferum* L., or *P. album* Mill., *Papaveraceae.* Habit. of the plant: Asia Minor, Persia, China, Africa, India; cultivated in the Balkan States, Hungary, Southern Russia. In Japan the strain cultivated from the production of opium is called "Ikkanshu." Appearance and sources: Chem. & Eng. News **32**, 2701 (1954). Constit. About 20 alkaloids, constituting about 25% of the opium; meconic acid, some lactic and sulfuric acids, sugar, resinous and waxy-like substances; 12-25% water. Morphine is the most important alkaloid and occurs to the extent of 10-16%, noscapine 4-8%, codeine 0.8-2.5%, papaverine 0.5-2.5%, thebaine 0.5-2%.

Deodorized opium, *"denarcotized" opium,* powdered opium freed from its odor and nauseating substances by treatment with petr ether. Contains 10-10.5% anhydr morphine.

Granulated opium, opium dried at not above 70°, reduced to a 16-50 mesh powder and adjusted with lactose or other inert diluent to contain 10-10.5% anhydr morphine.

Powdered opium, opium dried at a temp not above 70°, finely powdered amd adjusted with lactose or other inert diluent to contain 10-10.5% anhydr morphine.

Caution: May be habit forming. This is a controlled substance (opiate) listed in the U.S. Code of Federal Regulations, Title 21 Parts 329.1 and 1308.12 (1995).

USE: Largely for the manuf of morphine, codeine and other opium alkaloids.

THERAP CAT: Analgesic (narcotic). Hypnotic.

THERAP CAT (VET): Analgesic (narcotic); antidiarrheal; antitussive.

6987. Opromazine. *2-Chloro-N,N-dimethyl-10H-phenothiazine-10-propanamine 5-oxide; 2-chloro-10-[3-(dimethylamino)propyl]phenothiazine 5-oxide; 10-(γ-dimethylaminopropyl)-3-chlorphenothiazine 9-oxide; 5-oxychlorpromazine; chlorpromazine sulfoxide; Secotil.* $C_{17}H_{19}ClN_2OS$; mol wt 334.87. C 60.98%, H 5.72%, Cl 10.59%, N 8.37%, O 4.78%, S 9.58%. Metabolite of chlorpromazine, q.v.: Salzman, Brodie, J. Pharmacol. Exp. Ther. **118**, 46 (1956). Preparation by enzymatic oxidation of chlorpromazine: Gillette, Kamm, ibid. **130**, 262 (1960). Synthesis: Fr. pat. **1,167,653** (1958 to Rhône-Poulenc). Toxicity data: Minami, Yoshimoto, Nara Igaku Zasshi **8**, 50 (1957), C.A. **51**, 16966a (1957).

Crystals, mp 115°. LD_{50} s.c. in mice: 102 mg/kg (Minami, Yoshimoto).

6988. Opsins. Broad class of species-specific proteins which form the basis of the visual pigments and of bacteriorhodopsin, q.v. Structurally integrated into the rods and cones of the retina of the eye or into the photoreceptor membranes of certain bacteria. Each of these cell types produces a genetically specified opsin which has been classified on the basis of cellular source: *scotopsins* (rods), *photopsins* (cones), and *bacteriopsins* (bacteria). Methods for purification, prepn and assay: R. Hubbard *et al., Methods Enzymol.* **18**, 615-653 (1971). Series of articles on photoreceptor biosynthesis: *ibid.* **81**, 763-815 (1982). Review of biosynthetic process: D. S. Papermaster, B. G. Schneider in *Cell Biology of the Eye*, D. McDevitt, Ed. (Academic Press, New York, 1982) p 475. The photoreceptor activity of visual pigments is due to a carotenoid chromophore, retinal or 3-dehydroretinal, *q.q.v.*, bound as a protonated Schiff base to a lysine moiety in the opsin portion of the molecule: A. R. Osenoff, R. Callender, *Biochemistry* **13**, 4243 (1974). Each pigment has unique physicochemical properties. The most significant is the absorption spectrum which is regulated by electrostatic interactions between the chromophore and the charged or dipolar groups on the opsin: R. Hubbard, L. Sperling, *Exp. Eye Res.* **17**, 581 (1973); B. Honig *et al., J. Am. Chem. Soc.* **101**, 7084 (1979). A visual system, a set of pigments spanning the light sensitivity range of a particular species, is generally based on one type of chromophore combined with various opsins. Visual systems based on retinal are the most widespread in nature. The 11-*cis* isomer is utilized by rhodopsin, *q.v.*, the most common pigment of rod cells, and by the corresponding trichromatic cone pigments (*see* Iodopsin). Bacteriorhodopsin is composed of bacterioopsin and *trans* retinal. Visual systems based on 3-dehydroretinal, exemplified by the pigments porphyropsin and cyanopsin, *q.q.v.*, have been found to occur only in certain fish and amphibians. Visual pigments utilizing both types of chromophore have been found to coexist in the retina of some of these species: T. E. Reuter *et al., J. Gen. Physiol.* **58**, 351 (1971). Exposure to light initiates the bleaching of the pigment through a series of distinct intermediates involving the isomerization and ultimate dissociation of the chromophore from the opsin: R. Hubbard, A. Kropf, *Proc. Nat. Acad. Sci. USA* **44**, 140 (1958); B. Honig *et al., ibid.* **76**, 2503 (1979). This process initiates the mechanism of energy transduction and visual excitation: T. G. Ebrey, B. Honig, *Quart. Rev. Biophys.* **8**, 129-184 (1975); B. Honig, *Ann. Rev. Phys. Chem.* **29**, 31-57 (1978); R. Uhl, E. W. Abrahamson, *Chem. Rev.* **81**, 291 (1981). Review of energy transduction in invertebrate photoreceptors: P. Hillman *et al., Physiol. Rev.* **63**, 668-772 (1983); in bacteriorhodopsin: H. V. Westerhoff, Z. Dancshazy, *Trends Biochem. Sci.* **9**, 112 (1984). General reviews: G. Wald, *Science* **162**, 230-239 (1968); D. F. O'Brien, *ibid.* **218**, 961-966 (1982); A. Maeda, T. Yoshizawa, *Photochem. Photobiol.* **35**, 891-898 (1983); P. S. Zurer, *Chem. & Eng. News* **61**, 24-35 (Nov. 28, 1983). *See also: Methods Enzymol.* **88**, 1-836 (1982).

6989. Orange I. *4-[(4-Hydroxy-1-naphthalenyl)azo]-benzenesulfonic acid monosodium salt;* Tropaeolin OOO no. 1; C.I. Acid Orange 20; FD & C Orange I; Ext. D & C Orange 3; C.I. 14600; α-naphthol orange; sodium azo-α-naphtholsulfanilate. C$_{16}$H$_{11}$N$_2$NaO$_4$S; mol wt 350.33. C 54.86%, H 3.16%, N 8.00%, Na 6.56%, O 18.27%, S 9.15%. Prepared by coupling diazotized sulfanilic acid with α-naphthol: *Colour Index* vol. 4 (3rd ed., 1971) p 4065.

Reddish-brown powder. Sol in water to orange-red soln, in alcohol to orange soln. Acids ppt the aq soln. Sodium hydroxide intensifies the red color of the aq soln. pH: 7.6 brownish-yellow; 8.9 purple.
Caution: Delisted by FDA for use in foods, drugs and cosmetics.

6990. Orange II. *4-[(2-Hydroxy-1-naphthalenyl)azo]-benzenesulfonic acid monosodium salt;* C.I. Acid Orange 7; Betanaphthol orange; D & C Orange No. 4; C.I. 15510;

Mandarin G; Tropaeolin OOO no. 2. C$_{16}$H$_{11}$N$_2$NaO$_4$S; mol wt 350.33. C 54.86%, H 3.16%, N 8.00%, Na 6.56%, O 18.27%, S 9.15%. By coupling β-naphthol with diazotized sulfanilic acid in alkaline soln, *cf. Org. Syn.* coll. vol. II, 36 (1943); *Colour Index* vol. 4 (3rd ed., 1971) p 4078. *Review:* H. J. Conn's *Biological Stains,* R. D. Lillie, Ed. (Williams & Wilkins, Baltimore, 9th ed., 1977) pp 112-113, 573.

Pentahydrate, orange needles from water. One gram dissolves in 20 ml water. Sol in alc. Absorption max 4844 Å.
Caution: Delisted by the FDA for internal use in 1968. May be used externally without restriction.
USE: One of the most common dyes. Biological stain. Infrequently used as indicator: Amber to orange pH 7.4-8.6; orange to red pH 10.2-11.8.

6991. Orange B. *4,5-Dihydro-5-oxo-4-[(4-sulfo-1-naphthalenyl)azo]-1-(4-sulfophenyl)-1H-pyrazole-3-carboxylic acid 3-ethyl ester disodium salt; 5-oxo-4-[(4-sulfo-1-naphthyl)azo]-1-(p-sulfophenyl)-2-pyrazoline-3-carboxylic acid 3-ethyl ester disodium salt; 5-hydroxy-4-[(4-sulfo-1-naphthalenyl)azo]-1-(4-sulfophenyl)-1H-pyrazole-3-carboxylic acid 3-ethyl ester disodium salt; 1-(4-sulfophenyl)-3-ethylcarboxyl-4-(4-sulfonaphthylazo)-5-hydroxypyrazole disodium salt;* C.I. Acid Orange 137; C.I. 19235. C$_{22}$H$_{16}$N$_4$Na$_2$O$_9$S$_2$; mol wt 590.50. C 44.75%, H 2.73%, N 9.49%, Na 7.79%, O 24.39%, S 10.86%. Prepn: W. H. Kretlow *et al.,* U.S. pat. **3,285,906** (1966 to Stange Co.). *Review: Fed. Reg.* **43**, 45611 (1978). *See also Colour Index* vol. 4 (3rd ed., 1971) p 4134.

Dull orange crystals. Absorption max (0.04N ammonium acetate): 442 nm. Soly in water at 77°: 220 g/l. Violet soln in conc H$_2$SO$_4$, changing to fuchsia then red on dilution. Red soln in conc HCl. Yellowish soln in 10% NaOH, changing to brownish yellow on dilution.
USE: In coloring sausage and frankfurter casings.

6992. Orange Peel. Rind of the fruit of *Citrus aurantium* L., *Rutaceae.* Habit. N. India; cultivated near Mediterranean Sea, Spain, W. Indies, Florida, California. Review: A. Y. Leung, *Encyclopedia of Common Natural Ingredients* (John Wiley & Sons, New York, 1980) pp 248-252.
Bitter orange peel. Dried rind of unripe fruit. *Constit.* Volatile oil, hesperidine, naringin, aurantiamarin, acrid resin, gum, tannin.
Sweet orange peel. Fresh rind of ripe fruit of *C. aurantium,* var *sinensis* (also known as *C. sinensis*). *Constit.* Volatile oil, hesperidine, fixed oil, resin, gum, tannin. Closely resembles bitter orange peel, but has an orange-yellow color; sweetish, fragrant odor; aromatic and only slightly bitter taste.
USE: Flavoring agent.

6993. Orazamide. *1,2,3,6-Tetrahydro-2,6-dioxo-4-pyrimidinecarboxylic acid compd with 5-amino-1H-imidazole-4-carboxamide (1:1); 5-aminoimidazole-4-carboxamide orotate; orotic acid compd with 5(or 4)-aminoimidazole-4(or 5)-carboxamide (1:1); 4-amino-5-imidazolecarboxamide orotate; AICA orotate;* Aicamin; Aicorat. C$_9$H$_{10}$N$_6$O$_5$; mol wt 282.22. C 38.30%, H 3.57%, N 29.78%, O 28.35%. Prepn: Fr. pat. **1,351,141** (1964); Haraoka, Kamiya, U.S. pat. **3,-**

271,398 (1966); *eidem,* **Japan.** pat. **26,553('64),** *C.A.* **62,** 10450h (1965) (all three to Fujisawa). Pharmacology and toxicology: Tamura, Shibayama, *Yakugaku Kenkyu* **35,** 94 (1963), *C.A.* **64,** 2652h (1966); Hashimoto, *J. Vitaminol. (Kyoto)* **13,** 9, 19 (1967).

Dihydrate, crystals, dec 284-285°. LD_{50} in mice (g/kg): 0.6 i.p.; > 4.0 orally.

THERAP CAT: Hepatoprotectant.

6994. Orcein. A dye first prepared from lichens (cudbear, *q.v.,* or *archil*). Prepn by oxidation of orcinol with H_2O_2 in the presence of ammonia water: Zulkowski, Peters, *Monatsh.* **11,** 227 (1890). Can be separated into 14 dyes by distribution chromatography: Musso, *Ber.* **89,** 1659 (1956). Structure studies: Beecken *et al., Angew. Chem.* **73,** 665 (1961). Study of binding by elastin, *q.v.,* and certain collagens termed *collastin:* H. Puchter, S. N. Meloan, *Histochemistry* **64,** 119 (1979). Brief review: H. J. Conn's *Biological Stains,* R. D. Lillie, Ed. (Williams & Wilkins, Baltimore, 9th ed., 1977) pp 400-403. Review of early use as textile dye and histological stain: H. Puchter, S. N. Meloan, *loc. cit.*

	R_1	R_2	R_3
α-Aminoorcein	H	O	NH_2
α-Hydroxyorcein	H	O	OH
β- and γ-Aminoorcein	orcinol	O	NH_2
β- and γ-Hydroxyorcein	orcinol	O	OH
β- and γ-Aminoorceimine	orcinol	NH	NH_2

Brownish-red microcryst powder. Practically insol in water, benzene, chloroform, ether, carbon disulfide. Sol in alcohol, acetone or acetic acid with red color, in dil aq alkali with bluish-violet color.

β- and γ-Components are *cis-trans* isomers of the same compd. The eight compds depicted are the major components of orcein.

USE: Biological stain.

6995. Orcinol. *5-Methyl-1,3-benzenediol; 5-methylresorcinol;* orcin; 3,5-dihydroxytoluene. $C_7H_8O_2$; mol wt 124.14. C 67.73%, H 6.50%, O 25.78%. Occurs in many species of lichens: Sastry, Rao, *Curr. Sci.* **10,** 437 (1941). Prepn: Anker, Cook, *J. Chem. Soc.* **1945,** 311; Kisteneva, Rozhdestvenskii, *Zhur. Priklad. Khim.* **22,** 1108 (1949); Stevens, U.S. pat. **2,603,662** (1952 to Gulf); Zimmer, **U.S.** pat. **3,028,410** (1962 to Hooker Chem.). Toxicity studies: I. Veldre *et al., Vop. Gig. Tr. Prof. Patol. Est. SSR* **2,** 160 (1970), *C.A.* **74,** 51746h (1971).

Monohydrate, crystals; sweet but unpleasant taste; reddens on exposure to air due to oxidation. mp about 58°; 107° when anhydr. bp 290°; bp_{14-20} 165-170°; bp_5 147°. Freely sol in water, alcohol, ether; less sol in benzene; slightly sol in chloroform or carbon disulfide. *Keep well closed and*

protected from light. LD_{50} in mice, rats, rabbits, guinea pigs (mg/kg): 772, 844, 2400, 1678 orally (Veldre).

USE: As a reagent for pentoses, lignin, beet sugar, saccharoses, arabinose, diastase.

6996. Oripavine. *(5α)-6,7,8,14-Tetrahydro-4,5-epoxy-6-methoxy-17-methylmorphinan-3-ol; O^3-demethylthebaine.* $C_{18}H_{19}NO_3$; mol wt 297.35. C 72.71%, H 6.44%, N 4.71%, O 16.14%. From *Papaver orientale* L., and *P. bracteatum* Lindl., *Papaveraceae:* Junusov *et al., Ber.* **68,** 2158 (1935); Kiselev, Konovalova, *J. Gen. Chem. USSR* **18,** 142 (1948). Identity of *O*-methyl derivative with thebaine: *eidem, Zhur. Obshch. Khim.* **18,** 855 (1948). Biosynthesis: Gross, Dawson, *Biochemistry* **2,** 186 (1963). *Reviews:* H. L. Holmes in *The Alkaloids,* vol. II, Manske, Holmes, Eds. (Academic Press, New York, 1952) p 167; K. W. Bentley, *The Chemistry of the Morphine Alkaloids* (Oxford, 1954) pp 192-196. Synthesis from morphine: Barber, Rapoport, *J. Med. Chem.* **18,** 1074 (1975).

Crystals, mp 200-201°. $[\alpha]_D^{20}$ −211.8°. Hydrochloride, $C_{18}H_{19}NO_3$·HCl, crystals, dec 244-245°. Methiodide, $C_{18}H_{19}NO_3$·CH_3I, crystals, dec 207-208°.

6997. Orlistat. *[2S-[2α(R*),3β]]-N-Formyl-L-leucine 1-[(3-hexyl-4-oxo-2-oxetanyl)methyl]dodecyl ester; N-*formyl-L-leucine ester with (3S,4S)-3-hexyl-4-[(2S)-2-hydroxytridecyl]-2-oxetanone; (−)-tetrahydrolipstatin; orlipastat; Ro-18-0647; Xenical. $C_{29}H_{53}NO_5$; mol wt 495.74. C 70.26%, H 10.78%, N 2.83%, O 16.14%. Pancreatic lipase inhibitor. Isoln from fermentation broth of *Streptomyces toxytricini:* P. Hadvary *et al.,* Eur. pat. **129,748;** eidem, U.S. pat. **4,598,089** (1985, 1986 both to Hoffmann-LaRoche). Structural elucidation: E. Hochuli *et al., J. Antibiot.* **40,** 1086 (1987). Synthesis and absolute configuration: P. Barbier, F. Schneider, *Helv. Chim. Acta* **70,** 196 (1987). Total synthesis: S. Hanessian *et al., J. Org. Chem.* **58,** 7768 (1993). Mechanism of action: B. Borgström, *Biochim. Biophys. Acta* **962,** 308 (1988). Clinical pharmacology: J. B. Hauptman *et al., Am. J. Clin. Nutr.* **55,** 309S (1992). Clinical study: M. L. Drent, E. A. van der Veen, *Int. J. Obesity* **17,** 241 (1993).

Crystals, mp 43°. $[\alpha]_D^{20}$ −32.0° (c = 1 in chloroform).

THERAP CAT: Treatment of obesity.

6998. Orlon®. Polyacrylonitrile; Fiber A. Obtained by polymerizing acrylonitrile. Review of prepn, properties, and uses: R. W. Moncrieff, *Man-Made Fibres* (John Wiley, New York, 1963) pp 446-467.

White fiber. Sticks at 235°. Ironing temps above 160° may cause yellowing. Sp gr 1.17. Flammability similar to that of rayon and cotton. Generally has very good resistance to mineral acids; excellent resistance to common solvents, oils, greases, neutral salts, sunlight; fairly good resistance to weak alkalies but is degraded by strong alkalies. Resists attack by molds, mildew, insects. 100% polyacrylonitrile fibers rarely used commercially due to difficulty in dyeing.

USE: Fiber suitable for outdoor furnishings (awnings, tents, outdoor furniture), indoor furnishings, anode bags in electroplating, knitwear, rugs.

6999. Ormosinine. *21-Ormosanin-20-yl panamine.* $C_{40}H_{66}N_6$; mol wt 631.00. C 76.14%, H 10.54%, N 13.32%. From seed of *Ormosia dasycarpa* Jacks., *Leguminosae:* Hess, Merck, *Ber.* **52,** 1976 (1919); from *O. panamensis* Benth., *Leguminosae:* Lloyd, Horning, *J. Am. Chem. Soc.* **80,** 1506 (1958). Structural studies: Wilson, *Chem. & Ind. (London)* **1965,** 472; *Tetrahedron* **21,** 2561 (1965). Composed of one molecule of *panamine* and one molecule of *ormosanine;* NMR elucidation of structure: N. S. Bhacca *et al., J. Am. Chem. Soc.* **105,** 2538 (1983).

Needles from ethyl acetate, mp 219-220°. $[\alpha]_{436}^{25}$ +16.0°. $[\alpha]_{589}^{25}$ +8.9° (c = 1.29 in chloroform). Sol in chloroform; slightly sol in ether. Practically insol in water, alcohol. Sublimes to panamine.

7000. Ornidazole. *α-(Chloromethyl)-2-methyl-5-nitro-1H-imidazole-1-ethanol;* 1-(3-chloro-2-hydroxypropyl)-2-methyl-5-nitroimidazole; Ro-7-0207; Madelen; Ornidal; Tiberal. $C_7H_{10}ClN_3O_3$; mol wt 219.63. C 38.28%, H 4.59%, Cl 16.14%, N 19.13%, O 21.85%. Prepn: Neth. pat. **Appl. 6,606,853;** M. Hoffer, U.S. pats. **3,435,049** and **3,493,582** (1966, 1969 and 1970, all to Hoffmann-La Roche). Activity studies: E. Grunberg *et al., Proc. Soc. Exp. Biol. Med.* **133,** 490 (1970). Synthesis and antiprotozoal activity: M. Hoffer, E. Grunberg, *J. Med. Chem.* **17,** 1019 (1974). Pharmacokinetics: D. E. Schwartz, F. Jeunet, *Chemotherapy* **22,** 19 (1976).

Crystals from toluene, mp 77-78°. uv max (2-propanol): 288, 312 nm (ε 3720, 9150). pKa: 2.4 ± 0.1. LD_{50} in rats, mice (mg/kg): 1780, 1420 orally (Grunberg). Also reported as LD_{50} in mice (mg/kg): >2000 orally, >2000 i.p. (Hoffer, Grunberg).

THERAP CAT: Anti-infective.

7001. Ornipressin. *8-L-Ornithinevasopressin;* orn(8)-vasopressin; POR-8. $C_{45}H_{63}N_{13}O_{12}S_2$; mol wt 1042.21. C 51.86%, H 6.09%, N 17.47%, O 18.42%, S 6.15%. Synthetic analog of vasopressin, *q.v.,* in which L-ornithine replaces L-arginine. Prepn: R. L. Huguenin, R. A. Boissonnas, *Helv. Chim. Acta* **46,** 1669 (1963); *eidem,* Fr. pat. **1,396,607** corresp to U.S. pat. **3,299,036** (1965, 1967 both to Sandoz). Separation by reversed-phase HPLC: K. Krummen, R. W. Frei, *J. Chromatog.* **132,** 27 (1977). Pharmacologic properties: B. Berde *et al., Experientia* **20,** 42 (1964). Pharmacodynamics: S. Keppens, H. de Wulf, *Biochim. Biophys. Acta* **588,** 63 (1979). Comparison of hemostatic and cardiovascular effects: L. Saarnivaara, P. Leander, *Anaesthetist* **26,**

144 (1977). Clinical study: D. J. Adendorff, D. Davies, *S. Afr. Med. J.* **51,** 131 (1977).

Cys–Tyr–Phe–Gln–Asn–Cys–Pro–Orn–GlyNH₂

THERAP CAT: Vasoconstrictor.

7002. Ornithine. *L-Ornithine; α,δ-diaminovaleric acid;* 2,5-diaminopentanoic acid. $C_5H_{12}N_2O_2$; mol wt 132.16. C 45.44%, H 9.15%, N 21.20%, O 24.21%. $NH_2(CH_2)_3CH(NH_2)COOH$. An amino acid naturally occuring as the L-form, classified as nonessential for the maintenance of growth in rats. Ornithine has not been isolated from proteins except after hydrolysis with alkali. Under these circumstances DL-ornithine is obtained by secondary decompn and racemization from L-arginine. L-Ornithine has, however, been isolated from tyrocidine: Gordon *et al., Biochem. J.* **37,** 313 (1943), and from gramicidin-S: Synge, *ibid.* **39,** 363 (1945); its presence in bacitracin is suspected. Practical laboratory method for the prepn of L-ornithine from arginine by means of arginase from pig livers: Vickery, Cook, *J. Biol. Chem.* **94,** 393 (1931-2). Prepn of DL-ornithine: Kurtz, *ibid.* **122,** 477 (1938); Kline, U.S. pat. **3,028,424** (1962 to Lilly); K. Warning *et al., Ann.* **1978,** 1707.

Microcrystals from alcohol-ether. Softens at 120°, mp 140°. Usually obtained as a syrup. $[\alpha]_D^{25}$ +11.5° (c = 6.5). Aq solns are alkaline. pK_1' 1.94; pK_2' 8.65; pK_3' 10.76. Freely sol in water, alc; sparingly sol in ether.

L-Form monohydrochloride, $C_5H_{12}N_2O_2.HCl$, dec 233°. $[\alpha]_D^{23}$ +11.0° (c = 5.5). Soluble in water.

L-Form dihydrochloride, $C_5H_{12}N_2O_2.2HCl$. $[\alpha]_D^{23}$ +16.7° (c = 5.3). Soluble in water.

L-Form monosulfate, $2C_5H_{12}N_2O_2.H_2SO_4$, dec 234°. $[\alpha]_D^{25}$ +8.4°.

L-Form aspartate, **Hepa-Merz, Ormeta, Orparan.**

DL-Form, crystals from water. Sparingly sol in alcohol.

DL-Form monohydrochloride, dec 233°.

DL-Form monosulfate, dec 234°.

THERAP CAT: Anticholesteremic.

7003. Ornoprostil. *(11α,13E,15S,17S)-11,15-Dihydroxy-17,20-dimethyl-6,9-dioxoprost-13-en-1-oic acid methyl ester;* methyl (−)-(1R,2R,3R)-3-hydroxy-2-[(E)-(3S,5S)-3-hydroxy-5-methyl-1-nonenyl]-ε,5-dioxocyclopentaneheptanoate; 17S,20-dimethyl-6-oxoprostaglandin E_1 methyl ester; 6-oxo-17S,20-dimethyl-PGE₁ methyl ester; ronoprost; ONO-1308; OU-1308; Alloca; Ronok. $C_{23}H_{38}O_6$; mol wt 410.55. C 67.29%, H 9.33%, O 23.38%. Analog of prostaglandin E_1, *q.v.* Prepn: M. Hayashi *et al.,* Ger. pat. **2,840,-032;** *eidem,* U.S. pat. **4,278,688** (1979, 1981 both to Ono). Effect on gastric mucosal lesions in rats: H. Kuwata *et al., Nippon Shokakibyo Gakkai Zasshi* **82,** 1858 (1985), *C.A.* **104,** 15709c (1986); *eidem, Curr. Clin. Pract. Ser.* **36,** 243 (1986); on experimental esophagitis: S. Inoue *et al., Gendai Iyro* **19,** 1348 (1987), *C.A.* **107,** 626m (1987).

THERAP CAT: Antiulcerative.

7004. Orotic Acid. *1,2,3,6-Tetrahydro-2,6-dioxo-4-pyrimidinecarboxylic acid;* uracil-6-carboxylic acid; whey factor; animal galactose factor; Oropur; Orotyl. $C_5H_4N_2O_4$; mol wt 156.10. C 38.47%, H 2.58%, N 17.95%, O 41.00%. A pyrimidine precursor in animal organisms, found in milk: Bachstez, *Ber.* **64,** 2683 (1931); Hilbert, *J. Am. Chem. Soc.* **54,** 2082 (1932); Johnson, Schroeder, *ibid.* 2942. Synthesis from aspartic acid: Nye, Mitchell, *ibid.* **69,** 1382 (1947). Prepn by condensation of urea with monoethyl ester of oxalacetic acid in methanol: Scriabine, U.S. pat. **2,937,175** (1960 to Rhône-Poulenc). Older syntheses from urea and oxalacetic ester: Müller, *J. Prakt. Chem.* **56,** 488 (1897);

Behrend, Struve, *Ann.* **378**, 165 (1910). Microbial process by a pyrimidine-requiring *Micrococcus glutamicus* mutant: Kinoshita *et al.*, U.S. pat. **3,086,917** (1963 to Kyowa). Classed as a vitamin: Moruzzi *et al.*, *Biochem. Z.* **333**, 318 (1960). Relationship to *vitamin B₁₃*: Manna, Hauge, *J. Biol. Chem.* **202**, 91 (1953).

Crystals from water, mp 345-346°.
Monohydrate, *Lactinium, Oroturic.* Crystals, mp 334°. uv max: 282 nm; min: 255 nm. Sol in water ~1.7 mg/ml.
Methyl ester, $C_6H_6N_2O_4$, crystals, mp 249°.
Ethyl ester, $C_7H_8N_2O_4$, crystals, mp 188-189°.
Choline salt, $C_5H_4N_2O_4 \cdot C_5H_{14}NO$, *choline orotate, Cholergol.*
USE: Has been proposed as feed supplement in combination with methionine to aid growth of calves.
THERAP CAT: Uricosuric.

7005. Orotidine. *1,2,3,6-Tetrahydro-2,6-dioxo-3-β-D-ribofuranosyl-4-pyrimidinecarboxylic acid;* 3-β-D-ribofuranosylorotic acid; 6-carboxyuridine. $C_{10}H_{12}N_2O_8$; mol wt 288.21. C 41.67%, H 4.20%, N 9.72%, O 44.41%. An orotic acid riboside obtained from cultures of *Neurospora crassa* mutants: Michelson *et al.*, *Proc. Nat. Acad. Sci. USA* **37**, 396 (1951). Isoln: Mitchell, Michelson, U.S. pat. **2,788,346** (1957 to California Inst. Res. Found.). Structure: Fox *et al.*, *Biochim. Biophys. Acta* **23**, 295 (1957). Synthesis: Curran, Angier, *J. Org. Chem.* **31**, 201 (1966); U.S. pat. **3,282,-919** (1966 to Am. Cyanamid). Synthesis of orotidine 5'-phosphate: Moffatt, *J. Am. Chem. Soc.* **85**, 1118 (1963).

Needles from methanol + benzene. Turned brown near 200° but failed to melt at 400°. uv max (methanol): 268 nm (ε 8900); in 0.1*N* HCl: 267 nm (ε 9570); in 0.1*N* methanolic NaOH: 265 nm (ε 8960). Soluble in hot water, lower aliphatic alcohols and aq solns of such alcohols.
Cyclohexamine salt, $C_{16}H_{25}N_3O_8$, crystals from ethanol + benzene, mp 183-184°. [α]$_D$ +15° (c = 1). Also isolated as the lead salt.
Orotidine 5'-phosphate trisodium salt trihydrate, $C_{10}H_{10}$-$N_2O_{11}PNa_3 \cdot 3H_2O$, uv max (0.1*N* HCl): 267 nm (ε 9430). Soluble in water.
N³-Methylorotidine methyl ester, $C_{12}H_{16}N_2O_8$, stout crystals from isopropanol, mp 135-137°. uv max (methanol): 271 nm (ε 7620).

7006. Oroxylin A. *5,7-Dihydroxy-6-methoxy-2-phenyl-4H-1-benzopyran-4-one; 5,7-dihydroxy-6-methoxyflavone;* oroxylin. $C_{16}H_{12}O_5$; mol wt 284.27. C 67.60%, H 4.25%, O 28.14%. From root bark of *Oroxylum indicum* Vent., *Bignoniaceae:* Naylor, Chaplin, *Pharm. J.* **20**, 257 (1890); Naylor, Dyer, *J. Chem. Soc.* **79**, 954 (1901); Row *et al.*, *Proc. Indian Acad. Sci.* **28A**, 189 (1948). Structure: Shah *et al.*, *J. Chem. Soc.* **1936**, 591; **1938**, 1555. Synthesis: Murti, Seshadri, *Proc. Indian Acad. Sci.* **29A**, 1 (1949); Sarin, Seshadri, *J. Sci. Ind. Res. (India)* **19B**, 117 (1960); Molho, Gerphagnon, *Bull. Soc. Chim. France* **1963**, 607; Varady, *Tetrahedron Letters* **1965**, 4281.

Yellow plates from ethanol, mp 231-232°. Sol in alc, acetone, hot benzene, ether, alkalies, glacial acetic acid; sparingly sol in chloroform; practically insol in water.
Diacetyl oroxylin, $C_{20}H_{16}O_7$, needles from alcohol, mp 131-132°.

7007. Orphenadrine. *N,N-Dimethyl-2-[(2-methylphenyl)phenylmethoxy]ethanamine; N,N-dimethyl-2-(o-methyl-α-phenylbenzyloxy)ethylamine;* o-methyldiphenhydramine; o-monomethyldiphenhydramine; 2-(phenyl-o-tolylmethoxy)-ethyldimethylamine; phenyl-o-tolylmethyl dimethylaminoethyl ether; β-dimethylaminoethyl 2-methylbenzhydryl ether; BS-5930; Biorphen; Brocasipal. $C_{18}H_{23}NO$; mol wt 269.39. C 80.26%, H 8.61%, N 5.20%, O 5.94%. Prepd by the action of 2-methylbenzhydryl chloride on dimethylaminoethanol: Bijlsma *et al.*, *Arzneimittel-Forsch.* **5**, 72 (1955); Harms, Nauta, *J. Med. Pharm. Chem.* **2**, 57 (1960). Covered, but not described in Rieveschl, U.S. pat. **2,567,351** (1951 to Parke, Davis). Resolution of optical isomers: van der Stelt *et al.*, *Arzneimittel-Forsch.* **19**, 2010 (1969). Synthesis and pharmacological study of metabolites: Den Besten *et al.*, *ibid.* **20**, 538 (1970). Prepn of dosage forms: Harms, U.S. pat. **2,991,225** (1961 to Brocades-Stheeman).

Liquid. bp₁₂ 195°.
Hydrochloride, $C_{18}H_{23}NO \cdot HCl$, *Disipal, Mephenamin.* Crystals, mp 156-157°. Sol in water, alcohol, chloroform. Sparingly sol in acetone, benzene. Practically insol in ether. pH of aq soln about 5.5.
Citrate, $C_{18}H_{23}NO \cdot C_6H_8O_7$, *Banflex, Norflex, X-Otag.*
THERAP CAT: Muscle relaxant (skeletal); antihistaminic.

7008. Orris. White flag. Rhizome of *Iris florentina* L., *I. pallida* Lam. or *I. germanica* L., *Iridaceae.* *Habit.* Northern Italy, Germany, France. *Constit.* Iridin, irone, ionone, resin, starch, volatile oil (butter of orris). *Review:* W. A. Poucher, *Perfumes, Cosmetics and Soaps* vol. 1 (Chapman & Hall, London, 1974) pp 287-290.
Oil, *orris root oil. Constit.* About 85% myristic acid; the odorous principle irone; methyl myristate, oleic aldehyde. Yellowish-white to yellow, semisolid fatty substance; violet-like odor. mp 44-50°. Slightly dextrorotatory. Acid no. 213-222. Sapon no. 2-6.
USE: In perfumes and soaps. As dusting powder.

7009. *o*-Orsellinic Acid. *2,4-Dihydroxy-6-methylbenzoic acid; 6-methyl-β-resorcylic acid;* 4,6-dihydroxy-o-toluic acid; 2,4-dihydroxy-6-methylbenzenecarboxylic acid; orcinolcarboxylic acid. $C_8H_8O_4$; mol wt 168.15. C 57.14%, H 4.80%, O 38.06%. Found in conjugated form or in depside form in *Roccella* and *Lecanora* lichens, postulated to arise by autocondensation of acetoacetic acid. Isoln from the fungus *Chaetomium cochliodes:* Mosbach, *Z. Naturforsch.* **14B**, 69 (1959). Synthesis: Sonn, *Ber.* **61**, 926 (1928); Kloss, Clayton, *J. Org. Chem.* **30**, 3566 (1965).

Consult the Name Index before using this section.

Needles from acetone, mp 176° (effervescence). K (25°) = 1.27 × 10⁻⁴. uv max (0.1N HCl): 214, 260, 296 nm; in 0.1N NaOH: 272 nm. Sol in water, alcohol, glycerol. Soly in ether at 20° = 15.7%. Slightly sol in benzene.
Monohydrate, needles from water, mp 186-189°.
Methyl ester, $C_9H_{10}O_4$, crystals, mp 140°.
Ethyl ester, $C_{10}H_{12}O_4$, crystals, mp 132°.

7010. Orthanilic Acid. *2-Aminobenzenesulfonic acid;* o-sulfanilic acid; o-anilinesulfonic acid. $C_6H_7NO_3S$; mol wt 173.19. C 41.61%, H 4.07%, N 8.09%, O 27.71%, S 18.51%. Prepn from o-nitrobenzenesulfonyl chloride: Wertheim, *Org. Syn.*, coll. vol. II, 271 (1943).

Minute hexagonal plates, dec ~325°. Slow crystallization from water below 13.5° may yield a hemihydrate. K (25°) = 3.3 × 10⁻³. Slowly and sparingly sol in water.

7011. Orthocaine. *3-Amino-4-hydroxybenzoic acid methyl ester;* methyl 3-amino-4-hydroxybenzoate; Orthoform; Orthoform New. $C_8H_9NO_3$; mol wt 167.16. C 57.48%, H 5.43%, N 8.38%, O 28.71%. Prepd by dissolving 3-amino-4-hydroxybenzoic acid in methanol, saturating with HCl gas: Einhorn, Pfyl, *Ann.* **311**, 46 (1900); **Ger. pat. 97,333**; *Chem. Zentr.* **1898**, II, 525. Also prepd by reduction of methyl 3-nitro-4-hydroxybenzoate with aluminum amalgam: Auwers, Röhrig, *Ber.* **30**, 991 (1897); with tin and HCl or with stannous chloride and alcoholic HCl: **Ger. pat. 97,334**; *Chem. Zentr.* **1898**, II, 526.

Needles from benzene, mp 143°. When crystallized from chloroform it sometimes assumes an allotropic form, mp 111°, which on melting changes to the normal form and after solidifying always melts at 143°. Odorless and tasteless. Neutral reaction. Almost insol in cold water. Moderately sol in hot water with gradual decompn forming 3-amino-4-hydroxybenzoic acid and methanol. One gram dissolves in 6 ml alcohol, 50 ml ether; readily dissolves in aq NaOH. Forms water sol salts with HCl and HBr.
Note: The name Orthoform was applied originally to the methyl ester of 4-amino-3-hydroxybenzoic acid, mp 121°, which was also used as a local anesthetic. This compd now is designated as Orthoform Old.
THERAP CAT: Anesthetic (topical).
THERAP CAT (VET): Anesthetic (topical).

7012. Orthoformic Acid. *Methanetriol;* trihydroxymethane. Hypothetical compd, illustrated for nomenclature purposes only. The hydrogens of the hydroxyl groups are replaceable with alkyl groups giving rise to orthoformic esters, which are real compds (produced by various syntheses, e.g. from sodium alcoholates and chloroform). Monograph: H. W. Post, *The Chemistry of the Aliphatic Orthoesters* (Reinhold, New York, 1943), 188 pp.

Trimethyl ester, $C_4H_{10}O_3$, *methyl orthoformate, trimethoxymethane.* Prepd from chloroform and methanol in presence of sodium: Sah, Ma, *J. Am. Chem. Soc.* **54**, 2965 (1932). Liquid. d_4^{20} 0.9676; d_4^{25} 0.9623. bp$_{760}$ 100.6°. n_D^{25} 1.3773.
Triethyl ester, $C_7H_{16}O_3$, *ethyl orthoformate, 1,1′,1′-[methylidynetris(oxy)]tris[ethane], triethoxymethane, triethyl orthoformate, Aethone.* Prepd from chloroform and ethanol in presence of sodium: Chu, Shen, *C.A.* **38**, 2930⁷ (1944). Toxicity study: H. F. Smyth *et al., Arch. Ind. Hyg. Occup. Med.* **4**, 119 (1951). Liquid, sweetish odor resembling that of pine needles. d_4^{20} 0.8909; d_4^{25} 0.8858. bp$_{765}$ 143°. mp < −18°. n_D^{25} 1.3900. Slightly sol in water with decompn; misc with alcohol, ether. LD$_{50}$ orally in rats: 7.06 g/kg (Smyth).

7013. Orthovaleric Acid. Hypothetical compd, illustrated for nomenclature purposes only. The hydrogens of the hydroxyl groups are replaceable with alkyl groups giving rise to orthovaleric esters, which are described compds produced by various syntheses. Monograph: H. W. Post, *The Chemistry of the Aliphatic Orthoesters* (Reinhold, New York, 1943) 188 pp.

Trimethyl ester, *methyl orthovalerate.* Prepd from methyl imidovalerate: McElvain *et al., J. Am. Chem. Soc.* **68**, 1922 (1946). Liquid. d_4^{27} 0.9413. bp$_{760}$ 165°. n_D^{24} 1.4090.

7014. Oryzacidin. $C_8H_{13}NO_5$; mol wt 203.19. C 47.29%, H 6.45%, N 6.89%, O 39.37%. Antibiotic substance produced by a strain of *Aspergillus oryzae:* Shimoda, *J. Agr. Chem. Soc. Japan* **25**, 254 (1951), *C.A.* **46**, 10284 (1952); *idem*, **Japan.** pat. 1594('52), *C.A.* **47**, 6097 (1953). Production of oryzacidin by *Streptomyces: idem*, **Japan.** pat. **13,198**('65), *C.A.* **63**, 12284c (1965).
Hygroscopic needles, dec 162-163°. $[\alpha]_D^{13.5}$ −133°. Sol in water, methanol, ethanol, isopropanol, hot acetone. Slightly sol in butanol, ethyl acetate. Practically insol in ether, acetone, chloroform, carbon tetrachloride.
Quinine salt, tabular crystals, dec 210°.
Oryzacidin A, $C_{22}H_{34}N_4O_{12}$. Orange, amorphous powder, mp 231°. Soluble in ether, benzene, carbon tetrachloride, methanol, ethanol, butyl acetate, chloroform, pyridine, dioxane, acetic acid, toluene, xylene. Slightly sol or insol in water, acetone, carbon disulfide, isopropyl ether, petr ether, hexane. pH of ethanolic soln, 6.6. uv max: 243 nm (E$_{1cm}^{1\%}$ 122.5). Stable in direct sunlight.

7015. Oryzalin. *4-(Dipropylamino)-3,5-dinitrobenzenesulfonamide; 3,5-dinitro-N⁴,N⁴-dipropylsulfanilamide;* EL-119; Dirimal; Ryzelan; Surflan. $C_{12}H_{18}N_4O_6S$; mol wt 346.36. C 41.61%, H 5.24%, N 16.18%, O 27.72%, S 9.26%. Selective pre-emergence herbicide. Prepn: Q. F. Soper, **U.S. pat. 3,367,949** (1968 to Lilly). Soil degradation: T. Golab *et al., Pestic. Biochem. Physiol.* **5**, 196 (1975); E. W. Stoller, L. M. Wax, *J. Environ. Qual.* **6**, 124 (1977). *Review:* O. D. Decker, W. S. Johnson, *Anal. Methods Pestic. Plant Growth Regul.* **8**, 433 (1976).

Yellow-orange crystals, mp 137-138° (tech, 141-142°). Soly in water at 25°: 2.5 ppm. Sol in acetone, ethanol, methanol, acetonitrile; slightly sol in benzene. Practically

insol in hexane. Vapor pressure at 30°: $< 1 \times 10^{-7}$ mm Hg. LD_{50} orally in rats: > 10 g/kg (Decker, Johnson).
USE: Herbicide.

7016. γ-Oryzanol. OZ; γ-OZ; γ-orizanol; Caclate; Gammajust 50; Gamma-OZ; Gammariza; Gammatsul; Guntrin; Hi-Z; Maspiron; Oliver; Oryvita; Oryzaal; Thiaminogen. Mixture of ferulic acid esters of sterols (campestrol, stigmasterol, β-sitosterol) and triterpene alcohols (cycloartanol, cycloartenol, 24-methylenecycloartanol, cyclobranol) extracted from rice bran, corn and barley oils. Isoln from rice bran oil: R. Kaneko, T. Tsuchiya, *J. Chem. Soc. Japan, Ind. Chem. Sect.* **57**, 526 (1954); T. Tsuchiya *et al.*, **Japan.** pat. **56 7182** (1956 to Agcy Ind. Sci. Technol.), *C.A.* **52**, 15848b (1958); *eidem,* **Japan.** pat. **57 4895** (1957 to Bureau Ind. Technics), *C.A.* **52**, 5758i (1958); from corn oil: T. Tsuchiya, O. Okubo, **Japan.** pat. **60 2945** (1960 to Bureau Ind. Technics), *C.A.* **54**, 25609g (1960). Extraction from rice bran, corn and barley oils and purification: T. Yamamoto, **Ger.** pat. **1,301,002** (1969 to Toyo Koatsu), *C.A.* **71**, 128704r (1969). Separation of 3 major components designated oryzanols A, B and C: M. Shimizu *et al.*, *Chem. Pharm. Bull.* **5**, 36 (1957). Structure of A: G. Ohta, M. Shimizu, *ibid.* **40**; of C: *eidem, ibid.* **6**, 325 (1958); G. Ohta, *ibid.* **8**, 5, 9 (1960). Oryzanol B was subsequently found to be a mixture of A and C: M. Shimizu, G. Ohta, *ibid.* **108**. Separation and properties of the seven components of γ-oryzanol: T. Endo *et al.*, *Yukagaku* **17**, 344 (1968), *C.A.* **69**, 37319m (1968); *eidem, ibid.* **18**, 63 (1969), *C.A.* **70**, 89042f (1969). General pharmacology: Y. Yamaji *et al.*, *Oyo Yakuri* **25**, 947 (1983), *C.A.* **99**, 151730h (1983). Antioxidant activity in cytochrome model systems: K. Tajima *et al.*, *Biochem. Biophys. Res. Commun.* **115**, 1002 (1983). Effect on lipid metabolism in rats: M. Shinomiya, *Tohoku J. Exp. Med.* **141**, 191 (1983). Antiulcerative effect on gastric lesions in mice: Y. Ichimaru *et al.*, *Nippon Yakurigaku Zasshi* **84**, 537 (1984), *C.A.* **102**, 55922g (1985). Clinical evaluation in chronic gastritis: T. Arai, *Kitakanto Igaku* **30**, 71, 85 (1980); in hyperlipidemia: G. Yoshino *et al.*, *Curr. Ther. Res.* **45**, 543 (1989).

Oryzanol A

White or slightly yellowish, tasteless powder with little or no odor. Crystals from acetone, mp 135-137°. uv max (heptane): 216, 231, 291, 315 nm.
Oryzanol A, $C_{40}H_{58}O_4$, *(3β)-9,19-cyclolanost-24-en-3-ol 3-(4-hydroxy-3-methoxyphenyl)-2-propenoate, cycloartenyl ferulate.* Prepd as the monohydrate, mp 150-151.5°. $[\alpha]_D$ +40° (c = 0.68). uv max (heptane): 231, 290, 315 nm (log ε 4.15, 4.24, 4.34).
Oryzanol C, $C_{41}H_{60}O_4$, *(3β)-24-methylene-9,19-cyclolanostan-3-ol 3-(4-hydroxy-3-methoxyphenyl)-2-propenoate, 24-methylenecycloartanyl ferulate.*
THERAP CAT: Antiulcerative; antihyperlipoproteinemic. Also used in treatment of menopausal syndrome.

7017. Osajin. *5-Hydroxy-3-(4-hydroxyphenyl)-8,8-dimethyl-6-(3-methyl-2-butenyl)-4H,8H-benzo[1,2-b:3,4-b']dipyran-4-one.* $C_{25}H_{24}O_5$; mol wt 404.46. C 74.24%, H 5.98%, O 19.78%. Isoln from osage oranges, *Maclura pomifera* (Raf.) Schneid. (*M. aurantiaca* Nutt.), *Moraceae:* Walter *et al.*, *J. Am. Chem. Soc.* **60**, 574 (1938); Wolfrom, Mahan, *ibid.* **64**, 308 (1942). Structure: Wolfrom *et al.*, *ibid.* **68**, 406 (1946). Synthesis of osajin 4'-methyl ether: Jain *et al.*, *Tetrahedron* **26**, 1977 (1970). Synthesis of osajin: Jain, Sharma, *J. Org. Chem.* **39**, 2215 (1974).

Light yellow crystals from xylene or 95% ethanol, mp 189° (uncorr), 193° (corr). Practically insol in water, petr ether, moderately sol in benzene, alcohol, warm carbon tetrachloride; very sol in chloroform, ether, acetone, pyridine. uv max (abs alc): 274 nm, Horowitz, Jurd, *J. Org. Chem.* **26**, 2446 (1961).

7018. Osalmid. *2-Hydroxy-N-(4-hydroxyphenyl)benzamide; 4'-hydroxysalicylanilide; N-(p-hydroxyphenyl)salicylamide; N-salicoylaminophenol; oksafenamide; oxaphenamide; Driol; Jestmin; Kanochol; Saryuurin; Yoshicol.* $C_{13}H_{11}NO_3$; mol wt 229.24. C 68.11%, H 4.84%, N 6.11%, O 20.94%. Prepn: Weizmann *et al.*, *J. Org. Chem.* **13**, 796 (1948). Description: *Subsidia Medica* **8**, 103 (1956). HPLC determn in plasma: T. Sadanaga *et al.*, *J. Chromatog.* **223**, 243 (1981).

Crystals, mp 179°. Practically insol in cold water, acetic acid. Slightly sol in warm water, benzene, toluene. Freely sol in methanol, ethanol, ether, acetone.
Diacetate, $C_{17}H_{15}NO_5$, needles from alcohol, mp 151°.
THERAP CAT: Choleretic.

7019. Osaterone. *6-Chloro-17-hydroxy-2-oxapregna-4,-6-diene-3,20-dione.* $C_{20}H_{25}ClO_4$; mol wt 364.87. C 65.84%, H 6.91%, Cl 9.72%, O 17.54%. Androgen receptor antagonist. Prepn: K. Shibata *et al.*, **Eur.** pat. **Appl. 193,871;** *eidem,* **U.S.** pat. **4,785,103** (1986, 1988 both to Teikoku Hormone); K. Shibata *et al.*, *Chem. Pharm. Bull.* **40**, 935 (1992). Pharmacology: M. Murakoshi *et al.*, *Acta Pathol. Japan* **42**, 151 (1992); T. Ichikawa *et al.*, *Endocrine J.* **40**, 425 (1993); M. Murakoshi *et al.*, *ibid.* 479.

Colorless prisms from acetone/hexane, mp 218-221°.
Acetate ester, $C_{22}H_{27}ClO_5$, *2-oxachlormadinone acetate,* *TZP-4238.* Colorless prisms from acetone/hexane, mp 253-255°.
THERAP CAT: Treatment of benign prostatic hypertrophy; antiandrogen.

7020. Osmaron B. Composed of benzoates of primary aliphatic fatty amines obtained from palm kernel oil and corn oil as starting materials. Preponderantly dodecylammonium benzoate and tetradecylammonium benzoate.

Total nitrogen 4.15% by analysis; average benzoic acid content 27.6% by analysis.

Yellowish-brown, thick oil. mp -3 to $-4°$. n_D^{40} 1.4855-1.4885. Vapors of Osmaron B turn wet litmus paper blue and turmeric paper brown. Soluble in methanol, ethanol, ethylene glycol, glycerol, acetone, ethyl acetate, carbon disulfide, formic acid, acetic acid, and lactic acid. Forms turbid solns with ether, gasoline, benzene and its homologs, and chlorinated hydrocarbons which become clear upon deposition of the fine precipitate on standing. *Ref:* E. Benk, *Chem. Ztg.* **75**, 351 (1951).

USE: Disinfectant; cationic surface-active agent; in udder ointments used with milking machines.

7021. Osmium. Os; at. wt 190.2; at. no. 76; valences 1-8; most common states 3, 4, 6. Seven naturally occuring isotopes: 184 (0.02%); 186 (1.6%); 187 (1.6%); 188 (13.3%); 189 (16.1%); 190 (26.4%); 192 (41.0%); artificial radioactive isotopes: 181-183; 185; 191; 193-195. Occurrence in earth's crust ∼0.001 ppm. Found in the mineral osmiridium and in all platinum ores. Discovered by Tennant in 1804. Prepn: Berzelius *et al.*, cited by Mellor, *A Comprehensive Treatise on Inorganic and Theoretical Chemistry* **15**, 687 (1936). Reviews of prepn, properties and chemistry of osmium and other platinum metals: Gilchrist, *Chem. Rev.* **32**, 277-372 (1943); Beamish *et al.* in *Rare Metals Handbook*, C. A. Hampel, Ed. (Reinhold, New York, 1956) pp 291-328; Griffith, *Quart. Rev.* **19**, 254-273 (1965); *idem, The Chemistry of The Rarer Platinum Metals* (John Wiley, New York, 1967) pp 1-125; Livingstone in *Comprehensive Inorganic Chemistry*, vol. 3, J. C. Bailar, Jr. *et al.*, Eds. (Pergamon Press, Oxford, 1973) pp 1163-1189, 1209-1233.

Bluish-white, lustrous metal; close-packed hexagonal structure. d_4^{20} 22.61; long believed to be the densest element; x-ray data show it to be slightly less dense than iridium. mp ∼2700°. bp ∼5500°. Sp heat (0°) 0.0309 cal/g/°C. Hardness 7.0 on Mohs' scale. Electrical resistivity (0°) 8.12 μohms-cm. Stable in air in the cold; when finely divided, is slowly oxidized by air even at ordinary temp to form tetroxide. Attacked by fluorine above 100°; by dry chlorine on heating; not attacked by bromine or iodine. Attacked by aqua regia, by oxidizing acids over a long period of time; barely affected by HCl, H_2SO_4. Burns in vapor of phosphorus to form a phosphide, in vapor of sulfur to form a sulfide. Attacked by molten alkali hydrosulfates, by potassium hydroxide and oxidizing agents. Finely divided osmium absorbs a considerable amount of hydrogen.

Osmarins, high-molecular weight polymers of carbohydrate and osmium. Potential use in the treatment of arthritis: *Chem. & Eng. News* **60**, 8 (April 5, 1982).

USE: As alloy with iridium for pen points and fine machine bearings; as catalyst in the synthesis of ammonia; as catalyst in hydrogenation of organic compounds.

7022. Osmium Hexafluoride. F_6Os; mol wt 304.22. F 37.47%, Os 62.53%. OsF_6. Prepd by fluorination of osmium metal: Weinstock, Malm, *J. Am. Chem. Soc.* **80**, 4466 (1958); Hargreaves, Peacock, *Proc. Chem. Soc. (London)* **1959**, 85. Previously thought to be *osmium octafluoride:* Ruff, Tschirch, *Ber.* **46**, 929 (1913).

Pale yellow, volatile solid. *Highly poisonous! Very corrosive to skin!* mp 32.1°; bp 45.9°. Also reported: mp 33.4°; bp 47.5°, Cady, Hargreaves, *J. Chem. Soc.* **1961**, 1563. Hydrolyzed on exposure to moisture; forms white corrosive fumes which soon turn bluish. May be stored in quartz ampuls.

7023. Osmium Tetrachloride. Cl_4Os; mol wt 332.04. Cl 42.71%, Os 57.29%. $OsCl_4$. Prepn: Kolbin *et al.*, *Russ. J. Inorg. Chem.* **8**, 1270 (1963).

Red crystals. Sublimes at 450°. d_4^{20} 4.38. Soluble in water to give a yellow soln which on standing hydrolyzes into osmium oxides and HCl.

7024. Osmium Tetroxide. Osmic acid. O_4Os; mol wt 254.23. O 25.17%, Os 74.83%. OsO_4. Prepd by heating (at 300-400°) finely divided osmium metal in a stream of air or oxygen. Lab prepn: Grube in *Handbook of Preparative Inorganic Chemistry*, vol. 2, G. Brauer, Ed. (Academic Press, New York, 2nd ed., 1965) pp 1603-1604. Use in treatment of arthritis: M. Nissilä *et al.*, *Scand. J. Rheuma-*

tol. **5**, 111 (1977); A. S. Hendricson *et al.*, *Acta Orthop. Scand.* **52**, 17 (1982). Review of chemistry and biochemistry: W. P. Griffith, *Platinum Metal Rev.* **18**, 94-96 (1974); of toxicity: E. Browning, *Toxicity of Industrial Metals* (Appleton-Century-Crofts, New York, 2nd ed., 1969) pp 261-266.

Pale yellow solid; monoclinic crystals. *Vapor poisonous! Safeguards necessary when opening container.* Acrid, chlorine-like odor. Minimum perceptible concn 0.02 mg/liter of air. mp 40.6°. d 5.10 (calc): Ueki *et al.*, *Acta Cryst.* **19**, 157 (1965). bp$_{760}$ 130.0°; bp$_{400}$ 109.3°; bp$_{200}$ 89.5°; bp$_{100}$ 71.5°; bp$_{60}$ 59.4°. Begins to sublime and distil well below the boiling point. Vapor press at 27°: 11 mm. Critical temp 405°; crit press. 170 atm. Sol in benzene. Soly at 25° (g/100 g): water 7.24; carbon tetrachloride 375: Anderson, Yost, *J. Am. Chem. Soc.* **60**, 1822 (1938). Also sol in alc, ether, ammonium hydroxide, phosphorus oxychloride.

Caution: Potential symptoms of overexposure are lacrimation, visual disturbance and conjunctivitis; headache; coughing; dyspnea; dermatitis. *See NIOSH Pocket Guide to Chemical Hazards* (DHHS/NIOSH 90-117, 1990) p 170.

USE: Oxidizing agent, particularly for converting olefins to glycols. Catalyzes chlorate, peroxide, periodate, and other oxidations: P. N. Rylander, *Organic Syntheses with Noble Metal Catalysts* (Academic Press, New York, 1973) pp 121-144. As fixing and staining agent for cell and tissue studies.

7025. Osteocalcin. Bone Gla protein; BGP. Abundant noncollagenous, calcium-binding protein of bone; constitutes ∼1-2% of the total bone matrix. Also found in tooth dentin and cementum. Contains 46-50 amino acid residues (human: 49); mol wt ∼5800 daltons. Amino acid sequence is highly conserved across species. Characterized by 3 residues of the vitamin K-dependent amino acid, γ-carboxyglutamic acid (Gla), *q.v.* Synthesized by osteoblasts and odontoblasts; modulated by calcitriol, *q.v.* Detectable in serum as an indicator of bone metabolism. Identification in chicken bone: P. V. Hauschka *et al.*, *Proc. Nat. Acad. Sci. USA* **72**, 3925 (1975). Characterization of calf BGP: P. A. Price *et al. ibid.* **73**, 1447 (1976). Biosynthesis in bone cell culture: S. K. Nishimoto, P. A. Price, *J. Biol. Chem.* **254**, 437 (1979). RIA determn in plasma: P. A. Price *et al.*, *J. Clin. Invest.* **66**, 878 (1980). Review of isoln and characterization: C. M. Gundberg *et al.*, *Methods Enzymol.* **107B**, 516-544 (1984). Review of hydroxyapatite binding characteristics: P. V. Hauschka, F. H. Wians, Jr. *Anat. Rec.* **224**, 180-188 (1989). Comprehensive review: P. V. Hauschka *et al.*, *Physiol. Rev.* **69**, 990-1047 (1989). Review of diagnostic methods and clinical applications: M. J. Power, P. F. Fottrell, *Crit. Rev. Clin. Lab. Sci.* **28**, 287-335 (1991). Immunohistochemical study and potential role in bone mineralization and resorption: H. I. Roach, *Cell Biol. Int.* **18**, 617-628 (1994).

Very acidic, pI 4.0. Selective affinity for insoluble Ca^{2+} salts; binds tightly to hydroxyapatite.

7026. Osthole. 7-*Methoxy-8-(3-methyl-2-butenyl)-2H-1-benzopyran-2-one*; 7-*methoxy-8-(3-methyl-2-butenyl)coumarin*; 8-(3-methyl-2-butenyl)herniarin. $C_{15}H_{16}O_3$; mol wt 244.29. C 73.75%, H 6.60%, O 19.65%. From rhizome of *Peucedanum ostruthium* (L.) Koch (*Imperatoria ostruthium* L.) Umbelliferae: Herzog, Krohn, *Arch. Pharm.* **247**, 553 (1909); Butenandt, Marten, *Ann.* **495**, 187 (1932); from *Prangos pabularia* Lindl., *Umbelliferae:* Pigulevskii, Kuznetsova, *Doklady Akad. Nauk. SSSR* **61**, 309 (1948), *C.A.* **43**, 3416d (1949); from *Flindersia bennettiana* F. Muell., *Rutaceae:* Galbraith *et al.*, *Aust. J. Chem.* **13**, 427 (1960). Structure: Späth, Pesta, *Ber.* **66**, 754 (1933). Synthesis:

Späth, Holzen, *ibid.* **67**, 264 (1934); Murayama *et al.*, *Chem. Pharm. Bull.* **20**, 741 (1972).

Prisms from ether. mp 83-84°. uv max: 322, 258 nm (ε 8000, 4300). Practically insol in water; sol in aq alkalies, alcohol, chloroform, acetone, boiling petr ether.

7027. Ostruthin. *(E)-6-(3,7-Dimethyl-2,6-octadienyl)-7-hydroxy-2H-1-benzopyran-2-one; (E)-6-(3,7-dimethyl-2,6-octadienyl)-7-hydroxycoumarin;* 6-(3,7-dimethyl-2,6-octadienyl)umbelliferone. $C_{19}H_{22}O_3$; mol wt 298.38. C 76.48%, H 7.43%, O 16.09%. From the root of *Peucedanum ostruthium* (L.) Koch (*Imperatoria ostruthium* L.), Umbelliferae: Butenandt, Marten, *Ann.* **495**, 187 (1932); from *Eriostemon tomentellus, Rutaceae:* Duffield, Jeffries, *Aust. J. Chem.* **16**, 123 (1963). Structure: Späth, Klager, *Ber.* **67**, 859 (1934); Späth, Kainrath, *ibid.* **70**, 2272 (1937).

Crystals from alcohol, mp 117-119°. Practically insol in water, petr ether. Sol in chloroform, ethyl acetate, hot alc.

Acetate, $C_{21}H_{24}O_4$, leaflets from alcohol, mp 81°. Practically insol in water. Sol in alc, ether, benzene, chloroform.

7028. Ostruthol. *[R-(Z)]-2-Methyl-2-butenoic acid 2-hydroxy-2-methyl-1-[[(7-oxo-7H-furo[3,2-g][1]benzopyran-4-yl)oxy]methyl]propyl ester.* $C_{21}H_{22}O_7$; mol wt 386.40. C 65.28%, H 5.74%, O 28.98%. From rhizome of *Peucedanum ostruthium* L. Koch (*Imperatoria ostruthium* L.), Umbelliferae: Herzog, Koch, *Arch. Pharm.* **247**, 553 (1909). Structure: Späth, Christiani, *Ber.* **66**, 1150 (1933). Synthesis: Chatterjee, Dutta, *Sci. Cult.* **34**, 460 (1968).

Silky needles from benzene, mp 136-137°. $[\alpha]_D^{15}$ −18.3° (pyridine). Slightly sol in water or in ether; sol in hot alcohol, benzene, toluene, pyridine.

7029. Otobain. *9α-(1,3-Benzodioxol-5-yl)-6,7,8,9-tetrahydro-7α,8β-dimethylnaphtho[1,2-d]-1,3-dioxole;* 5,6-methylenedioxy-2,3-dihydro-1-methyl-4-(3',4'-methylenedioxyphenyl)-1,2,3,4-tetrahydronaphthalene; otobite. $C_{20}H_{20}O_4$; mol wt 324.38. C 74.06%, H 6.21%, O 19.73%. Isolated from the otoba fat of *Dialeyanthera otoba* (Humb. & Bonpl.) Warb. (*Myristica obota* Humb. & Bonpl.), Myristicaceae: Baughman *et al.*, *J. Am. Chem. Soc.* **43**, 200 (1921). Structure: Gilchrist *et al.*, *J. Chem. Soc.* **1962**, 1780; Bhacca, Stevenson, *J. Org. Chem.* **28**, 1638 (1963). Synthesis: Maclean, Stevenson, *J. Chem. Soc. (C)* **1966**, 1717; McMurry, Kennedy-Skipton, *Tetrahedron Letters* **1966**, 975.

Needles from ethanol, mp 137-138°. $[\alpha]_D$ −40.5° (c = 3.2 in chloroform). uv max (ethanol): 234, 287 nm (ε 9300, 6700). Sol in ether, hot alcohol. Practically insol in water.

8,6'-Dinitrootobain, $C_{20}H_{18}N_2O_8$, yellow needles from chloroform + methanol, mp 234-236°. $[\alpha]_D$ −170° (c = 2.45 in chloroform).

8,6'-Dibromootobain, $C_{20}H_{18}Br_2O_4$, prisms from chloroform + methanol, mp 197-199°. $[\alpha]_D$ +64° (c = 2.7 in chloroform).

Dehydroootobain, $C_{20}H_{16}O_4$, needles from chloroform + methanol, mp 185-187°.

7030. Ouabagenin. *(1β,3β,5β,11α)-1,3,5,11,14,19-Hexahydroxycard-20(22)-enolide;* G-strophanthidin. $C_{23}H_{34}O_8$; mol wt 438.52. C 63.00%, H 7.81%, O 29.19%. Prepn from ouabain with HCl in cold acetone: Mannich, Siewert, *Ber.* **75**, 737 (1942). Review of structure: Reichstein, Reich, *Ann. Rev. Biochem.* **15**, 155 (1946). Proof of this structure: Tamm *et al.*, *Helv. Chim. Acta* **40**, 1469 (1957), *cf. Experientia* **13**, 185 (1957); Turner, Meschino, *J. Am. Chem. Soc.* **80**, 4862 (1958); Volpp, Tamm, *Helv. Chim. Acta* **42**, 1408, 1418 (1959).

Monohydrate, clusters of needles from water, mp 235-238°. One gram dissolves in about 10 ml boiling water. At room temp the soly in water is <1%. Also sol in dil alcohol. Practically insol in abs alcohol, ether, and chloroform. Becomes anhyd at 100° *in vacuo* over P_2O_5. The anhydr compd is hygroscopic, mp 255-256°. $[\alpha]_D^{17}$ +11.3° (c = 1.27).

Tetraacetylouabagenin trihydrate, $C_{31}H_{42}O_{12}.3H_2O$, needles from 10% alcohol, mp 282-285°.

Dihydroouabagenin, $C_{23}H_{36}O_8.CH_3OH$, solvated crystals from methanol + ether contg 1 mol methanol, mp 261°.

7031. Ouabain. *(1β,3β,5β,11α)-3-[(6-Deoxy-α-L-mannopyranosyl)oxy]-1,5,11,14,19-pentahydroxycard-20(22)-enolide;* G-strophanthin; Gratus strophanthin; acocantherin. $C_{29}H_{44}O_{12}$; mol wt 584.66. C 59.58%, H 7.59%, O 32.84%. Obtained from the seeds of *Strophanthus gratus* (Wall. & Hock.) Baill.; also occurs in *Acokanthera ouabaio* Cathel and other *A.* spp, Apocynaceae. Isoln: Schwartze *et al.*, *J. Pharmacol.* **36**, 481 (1929). Hydrolysis yields one mol ouabagenin and one mol rhamnose: Jacobs, Bigelow, *J. Biol. Chem.* **96**, 647 (1932). Toxicity study: Small *et al.*, *Toxicol. Appl. Pharmacol.* **20**, 44 (1971). *See also* Fieser, Fieser,

Steroids (1959, Reinhold, New York; Chapman & Hall, London) pp 768, 772; Reichstein, Reich, *Ann. Rev. Biochem.* **15**, 155 (1946).

Octahydrate, *Purostrophan, Strodival, Strophoperm.* Shiny plates (from water) which give up their water of crystn at 130°. When anhydr dec about 130°. $[\alpha]_D^{25}$ −31° to −32.5° (c = 1 calcd as anhydr). Stable in air, but affected by light. One gram dissolves in about 75 ml water, in 5 ml boiling water, in 100 ml alcohol, in 8 ml boiling alcohol. Also sol in amyl alcohol, dioxane. Slightly sol in ether, chloroform, ethyl acetate. Aq solns are neutral to litmus. LD_{50} i.v. in rats: 14 mg/kg (Small).

THERAP CAT: Cardiotonic.

THERAP CAT (VET): Cardiotonic, diuretic.

7032. Ovalbumin. Egg albumin. The major protein constituent (75%) of egg white from hen's eggs. Mol wt about 45,000. Produced under hormonal control by the bird oviduct. May be isolated and crystallized readily from the filtrate of an acidified mixture of egg white and an equal volume of satd. ammonium sulfate: Sorensen, Hoyrup, *C. R. Trav. Lab. Carlsberg* **12**, 12 (1917). Alternate method: Kekwick, Cannan, *Biochem. J.* **30**, 227 (1936). Can be separated by electrophoresis and chromatography from about ten other minor components including avidin *(q.v.)*, lysozyme *(q.v.)*, conalbumin *(q.v.)*, and ovomucoid. Structure is a complex protein consisting of a single polypeptide chain of about 400 residues (about half of which are hydrophobic), a maximum of two phosphate residues per mole, and an oligosaccharide side chain composed of only mannose and glucosamine residues. Sequences of *N*- and *C*-terminal segments: Narita, Ishii, *J. Biochem.* **52**, 367 (1962); Thompson *et al.*, *Aust. J. Biol. Sci.* **24**, 525 (1971). *Reviews:* R. C. Warner in Neurath-Bailey, *The Proteins,* vol. **II**, part A (New York, 1954) p 443 sqq; Taborsky, *Advan. Protein Chem.* **28**, 34-50 (1974).

Needles or elongated prisms, frequently forming rosettes. The crystals usually contain 2 mols protein and 3 mols H_2SO_4. $[\alpha]_D^{20}$ −30.7°. Coagulation temp 56°. Sol in electrolyte-free water. Combines with salts, acids, and bases. Denaturation can be induced by heating to 56°, by vigorous shaking, by electric current, and by various chemicals, such as acids, ammonia salts, heavy metal salts, and alcohols. All methods produce complete and irreversible denaturation. Isoelectric point: 4.63.

7033. Ovex. *4-Chlorobenzenesulfonic acid 4-chlorophenyl ester;* *p*-chlorophenyl *p*-chlorobenzenesulfonate; chlorfenson; K-6451; Estonmite; Genite 883; Lethalaire G-58; Mitran; Orthotran; Ovochlor; Ovotox; Ovotran. $C_{12}H_8$-Cl_2O_3S; mol wt 303.17. C 47.54%, H 2.66%, Cl 23.39%, O 15.83%, S 10.58%. Prepn: Barna, **Brit.** pat. **747,368** (1956 to Diamond Alkali).

Crystals from benzene, mp 86.5-86.8°. Practically insol in water. Solubilities at 25° in g/100 ml: 95% alcohol 1.4; acetone 130.0; carbon tetrachloride 41.0; cyclohexanone 110.0; ethylene dichloride 110.0; xylene 78.0; petr oils 2.0-2.7. LD_{50} orally in rats: 2000 mg/kg, *RTECS* Vol. I, R. J. Lewis, R. L. Tatken, Eds. (1979) p 220.

Caution: Irritating to skin.

USE: Miticide; control of powdery mildew.

7034. Oxabolone. *(17β)-4,17-Dihydroxyestr-4-en-3-one;* 4-hydroxy-19-nortestosterone; 19-norandrost-4-ene-4,17β-diol-3-one. $C_{18}H_{26}O_3$; mol wt 290.40. C 74.45%, H 9.02%, O 16.53%. Prepn: Camerino *et al.*, **U.S.** pats. **2,999,870** and **3,145,201** (1961, 1964, both to Farmitalia). Prepn of the cypionate: Camerino, Patelli, **Brit.** pat. **879,622** and **U.S.** pat. **3,020,295** (1961, 1962, both to Farmitalia).

Solid, mp 188-190°. uv max: 278 nm (11600).

17-Cyclopentanepropionate, $C_{26}H_{38}O_4$, *17β-(2-cyclopentylpropionyloxy)-4-hydroxyestr-4-en-3-one, oxabolone cypionate, Steranabol long-acting, Steranabol ritardo.* Crystalline powder, mp 158-160°. $[\alpha]_D^{20}$ +30° (c = 1 in chloroform). uv max (alc): 276 nm ($E_{1cm}^{1\%}$ 315). Practically insol in water, hexane. Slightly sol in methanol; sol in chloroform, dioxane, benzene.

THERAP CAT: Anabolic.

7035. Oxaceprol. *trans-1-Acetyl-4-hydroxy-L-proline;* 4-hydroxy-*N*-acetylproline; 1-acetyl-4-hydroxy-2-pyrrolidinecarboxylic acid; CO-61; AHP-200; Jonctum. C_7H_{11}-NO_4; mol wt 173.17. C 48.55%, H 6.40%, N 8.09%, O 36.96%. Derivative of hydroxyproline having anti-inflammatory activity. Prepn: R. L. M. Synge, *Biochem. J.* **33**, 1924 (1939); J. J. Kolb, G. Toennies, *J. Biol. Chem.* **144**, 193 (1942); O. Leonardo *et al.*, **Ger.** pat. **2,301,358** corresp to U.S. pat. **3,860,607** (1973, 1975 to Richardson-Merrell). Elucidation of cyclic conformation and *cis/trans* isomerism about the amide bond: T. Prange *et al.*, *Biochem. Biophys. Res. Comm.* **61**, 104 (1974). Crystal structure: M. Hospital *et al.*, *Biopolymers* **18**, 1141 (1979). In crystals, the acetyl group is in the *trans* conformation and the ring is puckered. The *trans* form is more stable than the *cis* form in solutions: W. A. Thomas, M. K. Williams, *J. Chem. Soc. Chem. Comm.* **1972**, 994. Anti-inflammatory and wound-healing activity in animals: P. Coirre, B. Coirre, **Brit.** pat. **1,246,-141** (1971) related to P. Coirre *et al.*, **U.S.** pats. **3,891,765** and **3,932,638** (1975, 1976 both to Franco Chimie). Clinical studies of use in several rheumatic conditions: P. Grellat, *Rheumatologie* **27**, 223 (1975); in degenerative joint disease, R. Schubotz, L. Hausmann, *Therapiewoche* **27**, 4248 (1977); in treatment of burns, tumors and other wounds (applied locally): Y. Privat, *Gaz. Med. de France* **84**, 618 (1977).

Crystals from acetone, mp 133-134° (Synge); also reported as mp 126-128° (Leonardo). $[\alpha]_D^{20}$ −116.5° (c = 3.2); $[\alpha]_D^{18}$ −119.5° (c = 3.75). Very sol in alcohol. Sol in water, methanol. Insol in ether, chloroform.

Monohydrate, $C_7H_{13}NO_5$, crystals from moist ethyl acetate or acetone, mp 74-76°.

Zinc salt, $C_{14}H_{20}N_2O_8Zn$, mp 120°.

THERAP CAT: Anti-inflammatory; vulnerary.

7036. Oxacillin. *[2S-(2α,5α,6β)]-3,3-Dimethyl-6-[[(5-methyl-3-phenyl-4-isoxazolyl)carbonyl]amino]-7-oxo-4-thia-1-azabicyclo[3.2.0]heptane-2-carboxylic acid;* 5-methyl-3-phenyl-4-isoxazolylpenicillin; 6-(5-methyl-3-phenyl-2-isoxazoline-4-carboxamido)penicillanic acid; oxazocilline. $C_{19}H_{19}$-N_3O_5S; mol wt 401.44. C 56.85%, H 4.77%, N 10.47%, O 19.93%, S 7.99%. Semi-synthetic antibiotic related to penicillin. Prepn: Doyle, Nayler, **U.S.** pat. **2,996,501** (1961); Doyle *et al.*, *Nature* **192**, 1183 (1961). Toxicity: E. I. Goldenthal, *Toxicol. Appl. Pharmacol.* **18**, 185 (1971).

Sodium salt monohydrate, $C_{19}H_{18}N_3NaO_5S\cdot H_2O$, *penicillin P-12, sodium oxacillin, BRL-1400, Bactocill, Bristopen, Cryptocillin, Micropenin, Oxabel, Penstapho, Pentaphocid, Prostaphlin, Resistopen, Stapenor.* Crystals from isopropanol mp 188° (dec). $[\alpha]_D^{20}$ +201° (c = 1 in water). LD_{50} orally in rats: > 8000 mg/kg (Goldenthal).

THERAP CAT: Antibacterial.

THERAP CAT (VET): Antibacterial.

7037. 1,3,4-Oxadiazole. $C_2H_2N_2O$; mol wt 70.05. C 34.29%, H 2.88%, N 39.99%, O 22.84%. Prepn: Ainsworth, *J. Am. Chem. Soc.* **87**, 5800 (1965).

Liquid. bp 150°. Thermally stable.

7038. Oxadiazon. *3-[2,4-Dichloro-5-(1-methylethoxy)-phenyl]-5-(1,1-dimethylethyl)-1,3,4-oxadiazol-2(3H)-one; 2-tert-butyl-4-(2,4-dichloro-5-isopropoxyphenyl)-Δ²-1,3,4-oxadiazolin-5-one; 5-tert butyl-3-(2,4-dichloro-5-isopropoxy-phenyl)-1,3,4-oxadiazolin-2-one; RP-17623; Ronstar.* $C_{15}H_{18}Cl_2N_2O_3$; mol wt 345.23. C 52.19%, H 5.26%, Cl 20.54%, N 8.11%, O 13.90%. Selective pre-emergence herbicide. Prepn: **Brit. pat. 1,110,500** corresp to J. Metivier, R. Boesch, **U.S. pat. 3,385,862** (both 1968 to Rhône Poulenc). Metabolism: A. Guardigli *et al., Arch. Environ. Contam. Toxicol.* **4**, 145 (1976). Persistence in soil: D. Ambrosi *et al., J. Agr. Food Chem.* **25**, 868 (1977).

Crystals, mp 88-90°. Vapor press at 20°: < 1 × 10⁻⁶ mm Hg. Soly in water at 20°: 0.7 mg/l. Soly at 20° (g/l): ethanol 100; acetone 600; benzene 1000. LD_{50} orally in rats: 3500 mg/kg, *RTECS* Vol. II, R. J. Lewis, R. L. Tatken, Eds. (1979) p 182.

USE: Herbicide.

7039. Oxaflozane. *4-(1-Methylethyl)-2-[3-(trifluoro-methyl)phenyl]morpholine; 4-isopropyl-2-(α,α,α-trifluoro-m-tolyl)morpholine;* tetrahydro-4-isopropyl-2-[3-(trifluoro-methyl)phenyl]-2H-oxazine. $C_{14}H_{18}F_3NO$; mol wt 273.30. C 61.53%, H 6.64%, F 20.85%, N 5.13%, O 5.85%. Non-tricyclic antidepressant agent. Prepn: R. Y. Mauvernay *et al.,* **Fr. pat. 1,564,792;** eidem, **U.S. pat. 3,637,680** (1969, 1972 both to CERM); N. Busch *et al., Eur. J. Med. Chem.-Chem. Ther.* **11**, 201 (1976). Improved synthesis: P. M. Weintraub *et al., J. Org. Chem.* **45**, 4989 (1980). Pharmacological profile: J. Hache *et al., Therapie* **29**, 81 (1974). Urinary metabolism: M. Constantin, J. F. Pognat, *Arzneimittel-Forsch.* **29**, 109 (1979). Preliminary clinical study: A. Rascol *et al., Therapie* **29**, 95 (1974).

Oil, $bp_{0.005}$ 52°; bp_3 99°. n_D^{24} 1.4751.

Hydrochloride, $C_{14}H_{18}F_3NO\cdot HCl$, *CERM-1766, Conflictan.* Crystals from ethyl acetate/ethanol, mp 164°. LD_{50} in mice (mg/kg): 365 orally, 90 i.v. (Mauvernay, 1969); also reported as: 420 orally, 80 i.v. (Hache).

THERAP CAT: Antidepressant.

7040. Oxaflumazine. *10-[3-[4-[2-(1,3-Dioxan-2-yl)eth-yl]-1-piperazinyl]propyl]-2-(trifluoromethyl)-10H-phenothiazine.* $C_{26}H_{32}F_3N_3O_2S$; mol wt 507.62. C 61.52%, H 6.35%, F 11.23%, N 8.28%, O 6.30%, S 6.32%. Prepn: Ratouis, Boissier, *Bull. Soc. Chim. France* **1966**, 2963; eidem, **Ger. pat. 1,911,719** (1969 to S.I.F.A.), *C.A.* **72**, 55480y (1970). Pharmacology: Boissier, Dumont, *Thérapie* **26**, 481 (1971). Metabolism: Boissier *et al., Ann. Pharm. Franc.* **30**, 851 (1972). Clinical studies: Deniker *et al., Thérapie* **26**, 227 (1971).

Disuccinate, $C_{26}H_{32}F_3N_3O_2S\cdot 2C_4H_6O_4$, *SD-270-31, Oxaflumine.* Crystals from acetonitrile, mp 136-138°. uv max (methanol): 259, 310 nm (log ε 4.53, 3.58). LD_{50} in mice (mg/kg): 94 i.v.; 175 i.p.; 919 orally.

THERAP CAT: Antipsychotic.

7041. Oxalacetic Acid. *Oxobutanedioic acid;* oxosuccinic acid; ketosuccinic acid. $C_4H_4O_5$; mol wt 132.07. C 36.38%, H 3.05%, O 60.57%. $HOOCCOCH_2COOH$. Prepd by hydrolyzing sodium diethyloxalacetate with concd HCl: Heidelberger, Hurlbert, *J. Am. Chem. Soc.* **72**, 4704 (1950). Prepn of *cis*-enol form: Heidelberger, *Biochem. Prepn.* **3**, 59 (1953).

Diethyl ester, *see* Ethyl Oxalacetate.

trans-enol Form, *hydroxyfumaric acid.* Crystals from acetone + benzene, mp 184°. K at 17° = 2.76 × 10⁻³. Soluble in water, ethanol, ether. Can be converted into the lower-melting *cis*-form, mp 152°, by dissolving the acid in water and re-isolating it as rapidly as possible.

cis-enol Form, *hydroxymaleic acid.* Crystals from ethyl acetate + carbon tetrachloride, mp 152°. K at 17° = 2.505 × 10⁻³. Sol in ethanol, acetone, ethyl acetate. Sparingly sol in ether. Insol in benzene, chloroform, ligroin.

7042. Oxalenediuramidoxime. *N,N''-Bis(aminocarbon-yl)-N',N'''-dihydroxyethanediimidamide.* $C_4H_8N_6O_4$; mol wt 204.15. C 23.53%, H 3.95%, N 41.17%, O 31.35%. Prepn: Zinkeisen, *Ber.* **1889**, 2952.

Needles from dil alcohol. mp 191-192° (dec). Insol in cold water, in ether, chloroform, benzene, petr ether. Very sol in alcohol, acids, alkalies.

USE: As a reagent for the determination of nickel, with which it gives an orange ppt in dil ammoniacal soln.

7043. Oxalic Acid. *Ethanedioic acid.* $C_2H_2O_4$; mol wt 90.04. C 26.68%, H 2.24%, O 71.08%. $(COOH)_2$. Present in many plants and vegetables, notably in those of the *Oxalis* and *Rumex* families, where it occurs in the cell sap of the plant as the potassium or calcium salt. It is a product of the metabolism of many molds. Several species of *Penicillium* and *Aspergillus* convert sugar into calcium oxalate with 90% yields under optimum conditions. Oxalic acid was formerly manuf by fusion of cellulose matter, e.g. sawdust, with NaOH or by oxidation with HNO_3; it is now made by passing carbon monoxide into concd NaOH or by heating sodium formate in the presence of NaOH or Na_2CO_3: Wallace, U.S. pat. 1,602,802 (1926); Beckman, U.S. pat. 2,687,-433 (1951 to Allied Chem.). Efficient laboratory prepn of anhydrous form: H. T. Clarke, A. W. Davis, *Org. Syn.* coll. vol. I, 421 (2nd ed., 1941). Toxicity study: E. H. Vernot *et al., Toxicol. Appl. Pharmacol.* **42,** 417 (1977). *Review:* Wilson, "Miscellaneous *Aspergillus* Toxins" in *Microbial Toxins,* vol. **VI,** A. Ciegler *et al.,* Eds. (Academic Press, New York, 1971) pp 268-273; C. A. Bernales in Kirk-Othmer *Encyclopedia of Chemical Technology* vol. **16** (Wiley-Interscience, New York, 3rd ed., 1981) pp 618-636.

Dihydrate, monoclinic tablets, prisms, granules. *Poisonous!* $d_4^{18.5}$ 1.653. mp 101-102° giving off water of crystn and starting to sublime. pK_1 1.27; pK_2 4.28. pH of $0.1M$ soln 1.3. One gram dissolves in about 7 ml water, 2 ml boiling water; 2.5 ml alcohol, 1.8 ml boiling alcohol, 100 ml ether, 5.5 ml glycerol. Practically insol in benzene, chloroform, petr ether. $d_4^{17.5}$ of aq solns: 1% (w/w) 1.0035; 3% 1.0105; 5% 1.0175; 10% 1.0350; 13% 1.0455. Oxalic acid can be dehydrated by careful drying at 100°, but considerable loss occurs through sublimation; this, moreover, is harmful to the oven. LD_{50} orally in rats (5% solution): 9.5 ml/kg (Vernot).

Anhydr oxalic acid, crystallized from glacial acetic acid, is orthorhombic, the crystals being pyramidal or elongated octahedra. Hygroscopic, mp 189.5° (dec). Sublimes best at 157°. At higher temps dec into CO_2, CO, formic acid, and H_2O. d_4^{17} 1.90. 100 g of aq soln satd at 15° contain 6.71 g, at 20° 8.34 g, at 25° 9.81 g.

Caution: Caustic and corrosive to skin, mucous membranes. Ingestion may cause severe gastroenteritis with vomiting, diarrhea, melena. Renal damage can occur as a result of formation of excessive calcium oxalate. Convulsions, coma, death from cardiovascular collapse may occur. *See: Clinical Toxicology of Commercial Products,* R. E. Gosselin *et al.,* Ed. (Williams & Wilkins, Baltimore, 4th ed., 1976) Section III, pp 260-263.

USE: As analytical reagent; in calico printing and dyeing; for bleaching straw (hats) and leather; removing paint or varnish, rust or ink stains; cleaning wood; manuf oxalates; blue ink; celluloid; intermediates and dyes; in metal polishes; in indigo dyeing; in purifying methanol; for decolorizing crude glycerol; for stabilizing hydrocyanic acid. As a general reducing agent; in ceramics and pigments; in metallurgy as cleanser; in the paper industry; in photography; in process engraving; in the rubber mfg industry; in making glucose from starch; as condensing agent in organic chemistry.

THERAP CAT (VET): In 5% solution with 5% malonic acid as hemostatic agent.

7044. Oxaliplatin. *[SP-4-2-(1R-trans)]-(1,2-Cyclohexanediamine-N,N')[ethanedioato(2—)-O,O']platinum; [(1R,-2R)-1,2-cyclohexanediamine-N,N'][oxalato (2—)-O,O']platinum; oxalato (1R,2R-cyclohexanediamine)platinum(II); Pt-(oxalato)(trans-l-dach); oxalato (trans-l-1,2-diaminocyclohexane)platinum (II); oxalatoplatin; oxalatoplatinum; l-OHP; RP-54780; NSC-266046.* $C_8H_{14}N_2O_4Pt$; mol wt 397.29. C 24.19%, H 3.55%, N 7.05%, O 16.11%, Pt 49.10%. Third generation platinum complex. Prepn and antitumor activity: Y. Kidani *et al., J. Med. Chem.* **21,** 1315 (1978); Y. Kidani *et al., Gann* **71,** 637 (1980). Crystal structure: M. A. Bruck *et al., Inorg. Chim. Acta* **92,** 279 (1984). FAB and LAMMA MS analysis: J. Claereboudt *et al., J.*

Pharmaceut. Biomed. Anal. **7,** 1599 (1989). Pharmacokinetics: R. Kizu *et al., Cancer Chemother. Pharmacol.* **31,** 475 (1993). Clinical evaluation in colorectal cancer: F. Levi *et al., Eur. J. Cancer* **29A,** 1280 (1993). Brief review of antitumor and clinical activity: G. Mathé *et al., Biomed. Pharmacother.* **43,** 237-250 (1989).

Colorless, thin triangular plates with truncated vertices. Sol in water 7.9 mg/ml.

THERAP CAT: Antineoplastic.

7045. Oxalomolybdic Acid. *[Ethanedioato(2—)-O,O']-trioxomolybdate(2—) dihydrogen; [ethanedioato(1—)-O]hydroxydioxomolybdenum; hydrogen trioxooxalatomolybdate-(VI).* $C_2H_2MoO_7$; mol wt 233.97. C 10.27%, H 0.86%, Mo 41.00%, O 47.87%. $H_2[MoO_3(C_2O_4)]$. Usually contains 1 or 2 H_2O.
Colorless crystals, sol in water.
USE: In invisible inks.

7046. Oxalyl Chloride. *Ethanedioyl dichloride.* $C_2Cl_2O_2$; mol wt 126.93. C 18.93%, Cl 55.86%, O 25.21%. $(COCl)_2$. Obtained from oxalic acid and PCl_5.
Colorless, fuming liquid; penetrating odor. *Poisonous!* d_4^{13} 1.488. mp —12°. bp 63-64°. n_D^{13} 1.4340. Violently dec by water, also by alcohol.
Caution: Severely irritating to skin, eyes, respiratory tract.

7047. Oxamarin. *6,7-Bis[2-(diethylamino)ethoxy]-4-methyl-2H-1-benzopyran-2-one; 6,7-bis[2-(diethylamino)ethoxy]-4-methylcoumarin.* $C_{22}H_{34}N_2O_4$; mol wt 390.52. C 67.66%, H 8.78%, N 7.17%, O 16.39%. Deriv of α-pyrone, related structurally to sulmarin, *q.v.* Prepn: E. Massarani, *Farmaco Ed. Sci.* **12,** 691 (1957); G. Cavallini, E. Massarani, U.S. pat. **2,895,963** (1959 to Maggioni); E. Massarani *et al., J. Med. Pharm. Chem.* **3,** 231 (1961). Comparative study in treatment of varicose syndrome: G. Frantoli *et al., Panminerva Med.* **14,** 290 (1972), *C.A.* **78,** 92741 (1973). Transcutaneous absorption: G. Ciaceri, P. Marini, *Boll. Chim. Farm.* **111,** 321 (1972). Cardiovascular, intestinal smooth muscle effects: *eidem,* G. *Ital. Patol. Sci. Affini* **19,** 11 (1972), *C.A.* **81,** 130823 (1974). Anti-inflammatory action: *eidem, ibid.* 1, *C.A.* **81,** 130952 (1974).

Oil, $bp_{0.5}$ 195°.
Dihydrochloride, $C_{22}H_{34}N_2O_4$·2HCl, *MG-652, Idro P_3.* Crystals from ethanol, mp 224-226° (Massarani, 1957); also reported as mp 234-236° (Massarani *et al.,* 1961).
THERAP CAT: Hemostatic.

7048. Oxametacine. *1-(4-Chlorobenzoyl)-N-hydroxy-5-methoxy-2-methyl-1H-indole-3-acetamide; 1-(p-chlorobenzoyl)-5-methoxy-2-methylindole-3-acetohydroxamic acid;* indoxamic acid; Dinulcid; Flogar. $C_{19}H_{17}ClN_2O_4$; mol wt 372.81. C 61.21%, H 4.60%, Cl 9.51%, N 7.51%, O 17.17%. Deriv of indomethacin, *q.v.* Prepn: R. Aries, *Fr.* pat. **1,599,495** (1969), *C.A.* **73,** 3797h (1970); F. De Martiis *et al., Ger.* pat. **2,008,332** corresp to U.S. pat. **3,624,103** (both 1971 to ABC); *eidem, Boll. Chim. Farm.* **114,** 309 (1975). Pharmaco-toxicological evaluation: *eidem, ibid.* 319. Disposition in rats: L. F. Elsom *et al., Arzneimittel-Forsch.* **29,** 1155 (1979). Pharmacokinetics: P. Dittrich *et al., ibid.* **31,** 518 (1981). Effect on prostaglandin biosynthesis: J. S. Franzone *et al., Farmaco. Ed. Sci.* **35**, 498 (1980). Clinical study: J. Polderman, M. Colon, *J. Int. Med. Res.* **8**, 156 (1980).

Cryst from dioxane, mp 181-182° (dec). Sol in most organic solvents at elevated temperatures. In strong alkali it is hydrolyzed quickly to the debenzoylated product. LD_{50} orally in rats: 96 mg/kg (De Martiis, U.S. patent).

THERAP CAT: Anti-inflammatory.

7049. Oxamic Acid. *Aminooxoacetic acid;* oxamidic acid. $C_2H_3NO_3$; mol wt 89.05. C 26.98%, H 3.40%, N 15.73%, O 53.90%. $H_2NCOCOOH$. Prepd by heating oxamide in water with ammonia, followed by neutralization of resulting ammonium salt: Toussaint, *Ann.* **120**, 237 (1861). Cryst powder from water, or prisms from alcohol, dec about 210 or 214°. Sparingly sol in water. Practically insol in abs alcohol, ether.

Ammonium salt, $H_2NCOCOONH_4$, monoclinic prisms, dec about 226°. Very slightly sol in cold water, alcohol.

7050. Oxamide. *Ethanediamide;* oxalamide; oxalic acid diamide; ethanedioic acid diamide. $C_2H_4N_2O_2$; mol wt 88.07. C 27.28%, H 4.58%, N 31.81%, O 36.34%. $H_2NCO-CONH_2$. Prepd from formamide by glow-discharge electrolysis: Brown *et al., J. Org. Chem.* **27**, 3698 (1962). Crystal structure: G. DeWith, S. Harkema, *Acta Crystallogr.* **33B**, 2367 (1977). Metabolized in body to form oxalic acid.

Triclinic needles, dec 350°. d_4^{20} 1.667. Sparingly sol in hot water, alcohol.

7051. Oxamniquine. *1,2,3,4-Tetrahydro-2-[[(1-methylethyl)amino]methyl]-7-nitro-6-quinolinemethanol; 1,2,3,4-tetrahydro-2-[(isopropylamino)methyl]-7-nitro-6-quinolinemethanol;* 6-hydroxymethyl-2-isopropylaminomethyl-7-nitro-1,2,3,4-tetrahydroquinoline; UK-4271; Mansil; Vansil. $C_{14}H_{21}N_3O_3$; mol wt 279.34. C 60.20%, H 7.58%, N 15.04%, O 17.18%. Prepn: H. C. Richards, **S. Afr. pat. 68 03636;** *idem,* **U.S. pat. 3,821,228** (1967, 1974 both to Pfizer). Discovery of schistosomicidal activity: H. C. Richards, R. Foster, *Nature* **222**, 581 (1969). Metabolism: N. M. Woolhouse, B. Kaye, *Parasitology* **15**, 111 (1977). Genetic activity: T. M. Ong, *Mutat. Res.* **55**, 43 (1978). *In vitro* activity: C. J. Chavasse *et al., Ann. Trop. Med. Parasitol.* **72**, 293 (1978). Mutagenicity study: R. P. Batzinger, E. Bueding, *J. Pharmacol. Exp. Ther.* **200**, 1 (1977). Pharmacology: R. Foster, *Rev. Inst. Med. Trop. Sao Paulo* **15**, Suppl. 1, 189 (1973), *C.A.* **83**, 188417g (1975). Clinical study: R. J. Pitchford, M. Lewis, *S. Afr. Med. J.* **53**, 677 (1978). HPLC determn in biological fluids: H. W. Jun, M. A. Radwan, *Anal. Lett.* **18**, 1345 (1985). Comprehensive description: I. Ahmad *et al.,* in *Analytical Profiles of Drug Substances* **vol. 20,** K. Florey Ed. (Academic Press, New York, 1991) pp 601-625.

Pale yellow crystals from isopropanol, mp 147-149°. Sol in acetone, chloroform and methanol. Sol in about 3300 parts of water at 27°C. pH of a 1% soln 8.0-10.0. uv max (methanol) 205.5, 249.5, 389.5 nm ($A_{1cm}^{1\%}$ 486, 695, 62.5). LD_{50} in mice, rabbits (mg/kg): > 2000, > 1000 i.m., 1300, 800 orally (Foster).

THERAP CAT: Anthelmintic (Schistosoma).

7052. Oxamyl. *2-(Dimethylamino)-N-[[(methylamino)carbonyl]oxy]-2-oxoethanimidothioic acid methyl ester; N',N'-dimethyl-N-[(methylcarbamoyl)oxy]-1-thiooxamimidic acid methyl ester; N,N-dimethyl-α-methylcarbamoyloxyimino-α-(methylthio)acetamide; methyl 1-(dimethylcarbamoyl)-N-(methylcarbamoyloxy)thioformimidate; thioxamyl; DPX-1410;* Vydate. $C_7H_{13}N_3O_3S$; mol wt 219.26. C 38.34%, H 5.98%, N 19.16%, O 21.89%, S 14.62%. Prepn: J. B. Buchanan, **S. Afr. pat. 68 03,629;** *eidem,* **U.S. pats. 3,530,-220,** and **3,658,870** (1968, 1970, 1972 all to du Pont). Decompn: J. Harvey, Jr., J. Han, *J. Agr. Food Chem.* **26**, 536 (1978). Metabolism: *eidem, ibid.* 902; J. Harvey, Jr. *et al., ibid.* 529. Toxicity study: M. Fahmy *et al., J. Agr. Food Chem.* **26**, 550 (1978).

Crystalline solid, slight sulfurous odor. mp 100-102°, changing to a different crystalline form, mp 108-110°. Soly in g/100 ml at 25°: water 28; acetone 67; ethanol 33; 2-propanol 11; methanol 144; toluene 1. LD_{50} orally in rats: 5 mg/kg (Fahmy).

Caution: Toxic to wildlife, fish, bees.

USE: Insecticide, nematocide, acaricide.

7053. Oxanamide. *2-Ethyl-3-propyloxiranecarboxamide; 2,3-epoxy-2-ethylhexanamide;* 2-ethyl-3-propylglycidamide; 2-ethyl-3-propyl-2,3-epoxypropionamide; Quiactin. $C_8H_{15}NO_2$; mol wt 157.21. C 61.12%, H 9.62%, N 8.91%, O 20.35%. Prepn: Wheeler *et al., J. Org. Chem.* **25**, 1021 (1960).

Tasteless, odorless, white crystals from petr ether, mp 90-91°. One part is soluble in 95 parts of water at 30°. LD_{50} i.p. in mice: 720 mg/kg; orally in mice, rats: 1220, 1360 mg/kg, *Toxic Substances List,* H. E. Christensen, Ed. (1974) p 398.

THERAP CAT: Anxiolytic.

7054. Oxandrolone. *(5α,17β)-17-Hydroxy-17-methyl-2-oxaandrostan-3-one;* dodecahydro-3-hydroxy-6-(hydroxymethyl)-3,3a,6-trimethyl-1H-benz[e]indene-7-acetic acid δ-lactone; Anavar; Lonavar; Provitar; Vasorome. $C_{19}H_{30}O_3$; mol wt 306.45. C 74.47%, H 9.87%, O 15.66%. Prepn: Pappo, Jung, *Tetrahedron Letters* **1962**, 365; Pappo, **U.S. pat. 3,128,283** (1964 to Searle).

Crystals, mp 235-238°. $[\alpha]_D^{25}$ −23°. Absorption max: 2.87, 5.79 μ.

Note: This is a controlled substance (anabolic steroid) in the U.S. Code of Federal Regulations, Title 21 Part 1308.13, as defined in Part 1308.02 (1995).

THERAP CAT: Androgen.

7055. Oxantel. *(E)-3-[2-(1,4,5,6-Tetrahydro-1-methyl-2-pyrimidinyl)ethenyl]phenol; (E)-m-[2-(1,4,5,6-tetrahydro-1-methyl-2-pyrimidinyl)vinyl]phenol;* 1-methyl-1,4,5,6-tetrahydro-2-[2-(3-hydroxyphenyl)vinyl]pyrimidine; CP-14445. $C_{13}H_{16}N_2O$; mol wt 216.28. C 72.19%, H 7.46%, N 12.95%, O 7.40%. Analog of pyrantel, *q.v.,* with activity vs whipworms (*Trichuris* spp.). Prepn: J. W. Mc Farland, **S. Afr. pat. 68 04589** corresp to **U.S. pats. 3,579,510** and **3,708,584**

(1968, 1971, 1973 all to Pfizer). Synthesis and evaluation in whipworm control: J. W. Mc Farland, H. L. Howes *J. Med. Chem.* **15**, 365 (1972). Evaluation vs *Trichuris* in dogs: H. L. Howes, *Proc. Soc. Exp. Biol. Med.* **139**, 394 (1972). Efficacy vs *T. trichiuris* in humans: E. L. Lee *et al., Am. J. Trop. Med. Hyg.* **25**, 563 (1976); vs *T. suis* in swine: M. Robinson, *Vet. Parasitol.* **5**, 223 (1979). Anthelmintic effects of combination with pyrantel pamoate: B. Sinniah, D. Sinniah, *Ann. Trop. Med. Parasitol.* **75**, 315 (1981).

Hydrochloride, $C_{13}H_{16}N_2O \cdot HCl$, crystals from ethanol, mp 207-208°. uv max (water): 231, 274 nm (ϵ 12700, 20100).

Pamoate, $C_{49}H_{48}N_4O_8$, *oxantel embonate, CP-14445-16, Telopar.*

Mixture with pyrantel pamoate (1:1), *see* Pyrantel.

THERAP CAT: Anthelmintic (Nematodes).

THERAP CAT (VET): Anthelmintic (Nematodes).

7056. Oxapropanium Iodide. *N,N,N-Trimethyl-1,3-di-oxolane-4-methanaminium iodide; (1,3-dioxolan-4-ylmethyl)-trimethylammonium iodide;* 4-(dimethylaminomethyl)-1,3-dioxacyclopentane methiodide; 1-dimethylamino-2,3-dioxamethylenepropane methiodide; vasodilatateur 2249F; 2249-F; Dilvasene. $C_7H_{16}INO_2$; mol wt 273.11. C 30.78%, H 5.90%, I 46.47%, N 5.13%, O 11.72%. Prepd by the reaction of a cyclic acetal of an α-monohalohydrin of glycerol with dimethylamine which on treatment with methyl iodide results in a quaternary ammonium salt: Fourneau, U.S. pat. **2,445,393** (1948 to Rhône-Poulenc).

Crystals, mp 158-160° (the free base bp$_{21}$ 68°). Freely sol in water. Sol in boiling alcohol (about 40% w/v). Sparingly sol in cold alcohol, ether, chloroform, benzene.

THERAP CAT: Cholinergic.

7057. Oxaprozin. *4,5-Diphenyl-2-oxazolepropanoic acid; 4,5-diphenyl-2-oxazolepropionic acid;* β-(4,5-diphenyl-oxazol-2-yl)propionic acid; Wy-21743; Alvo; Daypro; Durapro; Duraprost; Oxapro. $C_{18}H_{15}NO_3$; mol wt 293.32. C 73.71%, H 5.15%, N 4.78%, O 16.36%. First description of anti-inflammatory properties: K. Brown *et al., Nature* **219**, 164 (1968). Prepn: **Fr.** pat. **2,001,036** (1969 to Inst. Farm. Serono), *C.A.* **72**, 66930w (1970); K. Brown, **Brit.** pat. **1,206,403** and **U.S.** pat. **3,578,671** (1970, 1971 both to Wyeth). Biochemical properties: M. W. Whitehouse *et al., Biochem. Pharmacol.* **20**, 2309 (1971). Metabolism: F. W. Janssen *et al., Drug Metab. Dispos.* **6**, 465 (1978). Pharmacology: D. A. Shriver *et al., Toxicol. Appl. Pharmacol.* **42**, 75 (1977); F. Awouters *et al., J. Pharm. Pharmacol.* **30**, 41 (1978). Clinical studies: R. Jamar, J. Dequeker, *Curr. Med. Res. Opin.* **5**, 433 (1978); J. A. Hubsher *et al., J. Int. Med. Res.* **7**, 69 (1979); *eidem, Arthritis Rheum.* **25**, S117 (1982). Protein binding and clearance: C. A. Homon *et al., Agents Actions* **12**, 211 (1982). Symposium on pharmacology and clinical efficacy: *Semin. Arthritis Rheum.* **15**, Suppl. 3, 1-107 (1986).

Crystals from methanol, mp 160.5-161.5°.

THERAP CAT: Anti-inflammatory.

7058. Oxatomide. *1-[3-[4-(Diphenylmethyl)-1-piperazinyl]propyl]-1,3-dihydro-2H-benzimidazol-2-one;* 1-[3-[4-(diphenylmethyl)-1-piperazinyl]propyl]-2-benzimidazolinone; R-35443; Celtect; Cobiona; Dasten; Tinset. $C_{27}H_{30}N_4O$; mol wt 426.56. C 76.03%, H 7.09%, N 13.13%, O 3.75%. Orally active anti-allergic agent, related structurally to cinnarizine, *q.v.*, and having a novel biphasic mode of action. Prepn: J. Vandenberk *et al.,* **Ger.** pat. **2,714,437;** *eidem,* **U.S.** pat. **4,250,176** (1977, 1981 both to Janssen). Inhibition of release and effects of allergic mediators: F. Awouters *et al., Experientia* **33**, 1657 (1977). *In vitro* study of inhibition and stimulation of histamine release: M. K. Church *et al., Agents Actions* **10**, 4 (1980). *In vivo* study: S. Gatti *et al., Brit. J. Dermatol.* **103**, 671 (1980). Clinical studies: S. W. Barham, F. Moran, *Brit. J. Clin. Pract.* **34**, 323 (1980); E. F. Juniper *et al., Clin. Allergy* **11**, 61 (1981); W. Peremans *et al., Dermatologia* **162**, 42 (1981). Review of pharmacological and clinical studies: D. M. Richards *et al., Drugs* **27**, 210 (1984).

White powder, mp 153.6°. LD$_{50}$ in guinea pigs, mice, rats (mg/kg): 320, >2560, >2560 orally; 23, 27, 30 i.v. (Richards).

THERAP CAT: Antiallergic; antiasthmatic.

7059. Oxazepam. *7-Chloro-1,3-dihydro-3-hydroxy-5-phenyl-2H-1,4-benzodiazepin-2-one;* 7-chloro-3-hydroxy-5-phenyl-1,3-dihydro-2H-1,4-benzodiazepin-2-one; Wy-3498; Abboxapam; Adumbran; Aplakil; Azutranquil; Bonare; Durazepam; Enidrel; Hilong; Isodin; Lederpam; Limbial; Nesontil; Noctazepam; Oxa-Puren; Praxiten; Propax; Quilibrex; Rondar; Serax; Serenal; Serenid; Serepax; Seresta; Sigacalm; Sobril; Tazepam; Uskan; Zaxopam. $C_{15}H_{11}ClN_2$-O_2; mol wt 286.72. C 62.84%, H 3.87%, Cl 12.37%, N 9.77%, O 11.16%. Prepn: Bell, Childress, *J. Org. Chem.* **27**, 1691 (1962); E. Reeder, L. Sternbach, **U.S.** pat. **3,109,843;** E. Reeder *et al.,* **Belg.** pat. **629,227;** *eidem,* **U.S.** pat. **3,340,-253** (1963, 1967, 1967, all to Hoffmann-La Roche). Crystal and molecular structure: G. Gilli *et al., Acta Crystallogr.* **B34**, 2826 (1978). Metabolic studies: Sisenwine *et al., Arzneimittel-Forsch.* **22**, 682 (1972); Knowles, Ruelius, *ibid.* 687. Pharmacokinetics: D. J. Greenblatt, *Clin. Pharmacokinet.* **6**, 89 (1981). Toxicity data: E. I. Goldenthal, *Toxicol. Appl. Pharmacol.* **18**, 185 (1971). Comprehensive description: C. M. Shearer, C. R. Pilla, *Anal. Profiles Drug Subs.* **3**, 441-464 (1974). *Review: Acta Psychiatr. Scand.,* **Suppl. 274**, R. Chinnery, A. Sundwall, Eds. (1978) 128 pp.

Crystals from alcohol, mp 205-206°. Sol in alcohol, chloroform, dioxane. Practically insol in water. LD_{50} in mice, rats (mg/kg): > 5010, > 5010 orally (Goldenthal).

Note: This is a controlled substance (depressant) listed in the U.S. Code of Federal Regulations, Title 21 Part 1308.14 (1995).

THERAP CAT: Anxiolytic.

7060. Oxazidione. *2-(4-Morpholinylmethyl)-2-phenyl-1H-indene-1,3(2H)-dione;* 2-phenyl-2-morpholinomethylin-dane-1,3-dione; mofedione; LD-4610; Transidione; Amplidione. $C_{20}H_{19}NO_3$; mol wt 321.38. C 74.75%, H 5.96%, N 4.36%, O 14.94%. Prepd by reacting 2-phenyl-1,3-indan-dione with 4-(ethoxymethyl)morpholine: Giudicelli, Najer, **S. Afr. pat.** **69 02,814** (1969 to Lab. Dausse), *C.A.* **72,** 111159q (1970).

mp 88° from cyclohexane. LD_{50} orally in mice: 235 mg/kg.

Hydrochloride, mp 150°.

THERAP CAT: Anticoagulant.

7061. Oxazolam. *10-Chloro-2,3,7,11b-tetrahydro-2-methyl-11b-phenyloxazolo[3,2-d][1,4]benzodiazepin-6(5H)-one;* 10-chloro-2,3,5,6,7,11b-hexahydro-2-methyl-11b-phen-ylbenzo[6,7]-1,4-diazepino[5,4-b]oxazol-6-one; oxazolazepam; Hializan; Serenal; Tranquit. $C_{18}H_{17}ClN_2O_2$; mol wt 328.80. C 65.75%, H 5.21%, Cl 10.78%, N 8.52%, O 9.73%. Prepn: Tachikawa *et al.*, **Ger. pat.** **1,812,252;** *eidem,* **U.S. pat.** **3,772,371** (1969, 1973 to Sankyo). Synthesis and pharmacology: Miyadera *et al., J. Med. Chem.* **14,** 520 (1971). Metabolism studies: Shindo *et al., Chem. Pharm. Bull.* **19,** 60 (1971); Yasumura *et al., ibid.* 1929.

Crystals, mp 186-188°. Sol in chloroform; slightly sol in ethanol. Practically insol in water.

Note: This is a controlled substance (depressant) listed in the U.S. Code of Federal Regulations, Title 21 Part 1308.14 (1995).

THERAP CAT: Anxiolytic.

7062. Ox Bile Extract. Purified oxgall; sodium choleate; Bi-Ketolan; Bilein; Bilicholan; Cholatol; Crescefel; Desicol; Doxychol; Glycotauro; Panoxolin; Plebilin. One gram = 8 g ox bile. *Constit.* Chiefly the sodium salts of glycocholic and taurocholic acids; also cholesterol, lecithin, glycocol and choline compounds.

Yellowish-green, soft solid; peculiar odor; partly sweet, partly bitter, disagreeable taste. Very sol in water or alc.

THERAP CAT: Choleretic.

THERAP CAT (VET): Has been used as a choleretic.

7063. Oxcarbazepine. *10,11-Dihydro-10-oxo-5H-di-benz[b,f]azepine-5-carboxamide;* oxacarbazepine; GP-47680; Trileptal. $C_{15}H_{12}N_2O_2$; mol wt 252.27. C 71.42%, H 4.79%, N 11.10%, O 12.68%. Ketoderivative of carbamazepine, *q.v.* Prepn: W. Schindler, **Ger. pat.** **2,011,087** (1970 to Geigy); *idem,* **U.S. pat.** **3,642,775** (1972 to Ciba-Geigy). Metabolism: H. Schütz *et al., Xenobiotica* **16,** 769 (1986). Hyponatremic effects: O. A. Nielsen *et al., Epilepsy Res.* **2,** 269 (1988). Determn of oxcarbazepine and main metabolites by GC in plasma: G. E. Von Unruh, W. D. Paar, *J. Chroma-*

tog. **345,** 67 (1985); by HPLC: A. A. Elyas, V. D. Goldberg, *ibid.* **528,** 473 (1990). Clinical evaluation in treatment of epilepsy: M. Dam *et al., Epilepsy Res.* **3,** 70 (1989); in management of trigeminal neuralgia: J. M. Zakrzewska, P. N. Patsalos, *J. Neurol. Neurosurg. Psychiat.* **52,** 472 (1989). Brief review of pharmacology and clinical studies: *Lancet* **2,** 196-198 (1989).

Crystals from ethanol, mp 215-216°.

THERAP CAT: Anticonvulsant.

7064. Oxeladin. *α,α-Diethylbenzeneacetic acid 2-[2-(di-ethylamino)ethoxy]ethyl ester;* 2-ethyl-2-phenylbutyric acid 2-(2-diethylaminoethoxy)ethyl ester; 2-(2-diethylaminoethoxy)-ethyl α,α-diethylphenylacetate; α,α-diethylphenylacetic acid 2-(2-diethylaminoethoxy)ethyl ester. $C_{20}H_{33}NO_3$; mol wt 335.49. C 71.60%, H 9.91%, N 4.18%, O 14.31%. Prepn: V. Petrow *et al., J. Pharm. Pharmacol.* **10,** 40 (1958); *eidem.* **U.S. pat.** **2,885,404** (1959 to Brit. Drug Houses).

Yellow oil with an acrid odor and bitter taste. $bp_{0.5}$ 150°; $bp_{0.1}$ 140°. Soluble in dil HCl, ethanol, acetone, ether, toluene; practically insol in water. Stable in acids, unstable in alkalies. Non-hygroscopic.

Citrate, $C_{20}H_{33}NO_3 \cdot C_6H_8O_7$, *Pectamol, Pectamon, Paxela-dine, Silopentol.* Small needles from ethyl acetate, mp 90-91°. Sol in water.

THERAP CAT: Antitussive.

7065. Oxendolone. *(16β,17β)-16-Ethyl-17-hydroxyestr-4-en-3-one;* 16β-ethyl-19-nortestosterone; TSAA-291; Prostetin. $C_{20}H_{30}O_2$; mol wt 302.46. C 79.42%, H 10.00%, O 10.58%. Prepn: K. Hiraga *et al.,* **Ger. pat.** **2,100,319;** *eidem,* **U.S. pat.** **3,856,829** (1971, 1974 both to Takeda); K. Yoshioka *et al., Chem. Pharm. Bull.* **23,** 3203 (1975). Stereoselective synthesis and NMR study: G. Goto *et al., ibid.* **25,** 1295 (1977). Synthesis and anti-androgenic activity: *eidem, ibid.* **26,** 1718 (1978). Physico-chemical properties and stabilities: K. Itakura *et al., Takeda Kenkyusho Ho* **37,** 297 (1978), *C.A.* **91,** 20879 (1979). Disposition and metabolism: S. Tanayama *et al., Steroids* **33,** 65 (1979). Series of articles on pharmacology, mechanism of action, anti-androgen effects: *Acta Endocrinol.* **92**(Suppl 2), 1-107 (1979).

Crystals from ether, mp 152-153°. $[\alpha]_D$ +41° (c = 1.0 in ethanol). uv max (ethanol): 240 nm (ϵ 15800). Stable against heat, humidity, indoor light. Converted to the 16α- and 17α-epimers in sunlight. LD_{50} in rats, mice (g/kg): > 10 orally; 5-10 i.m. and i.p. (Hiraga, 1971).

THERAP CAT: Anti-androgen; in treatment of benign prostatic hypertrophy.

7066. Oxenin. *11,12-Didehydro-7,10-dihydro-10-hydroxyretinol;* 3,7-dimethyl-9-(2,6,6-trimethyl-1-cyclohexen-1-yl)nona-2,7-dien-4-yne-1,6-diol. $C_{20}H_{30}O_2$; mol wt 302.46. C 79.42%, H 10.00%, O 10.58%. Intermediate in the vitamin A synthesis. Prepn: Isler *et al., Helv. Chim. Acta* **30**, 1911 (1947); Isler, U.S. pat. **2,451,739** (1948 to Hoffmann La Roche); Kardys, U.S. pat. **3,046,310** (1962 to Pfizer).

Very viscous yellow oil. d_4^{27} 0.9984; n_D^{22} 1.5344. Has been crystallized, mp 58.5-59°.

7067. Oxethazaine. *2,2′-[(2-Hydroxyethyl)imino]bis[N-(1,1-dimethyl-2-phenylethyl)-N-methylacetamide];* N,N-bis[N-methyl-N-phenyl-*tert*-butylacetamido]-β-hydroxyethylamine; oxetacaine; oxethazine; Wy-806; Storocain; Topicain. $C_{28}H_{41}N_3O_3$; mol wt 467.65. C 71.91%, H 8.84%, N 8.99%, O 10.26%. Prepn: Seifter *et al.,* U.S. pat. **2,780,646** (1957 to Am. Home Prods.); Salomon, **Israeli** pat. **10,062** (1957), *C.A.* **52**, 15569 (1958). Pharmacology: Axerio Agnessetti, *An. Real. Acad. Farm.* **35**, 183 (1969), *C.A.* **71**, 122082q (1969). Toxicity studies: J. M. Glassman *et al., Toxicol. Appl. Pharmacol.* **5**, 184 (1963). Clinical trial of combination with antacid in ulcer pain: F. T. M. Cordeiro *et al., Curr. Ther. Res.* **19**, 230 (1976); A. Zanni *et al., Curr. Med. Res. Opin.* **10**, 128 (1986). Evaluation as dental anesthetic: P. A. Brennan, J. D. Langdon, *Brit. J. Oral Maxillofac. Surg.* **28**, 26 (1990).

Crystals from benzene + hexane, mp 104-104.5°. Insol in water. Sol in dil acids.
Hydrochloride, $C_{28}H_{41}N_3O_3 \cdot HCl$, *Emoren.* mp 146-147°. Easily sol in water. LD_{50} in mice, rats (calcd as mg base/kg): 399.9, 625.9 orally; 247.2, 502.3 i.m.; 3.6, 1.3 i.v.; LD_{50} in rabbits (mg base/kg): 0.54 i.v. (Glassman).
Combination with aluminum and magnesium hydroxides, *Mutesa, Muthesa, Oxaine M, Tepilta.*
THERAP CAT: Anesthetic (local).

7068. Oxetorone. *3-Benzofuro[3,2-c][1]benzoxepin-6(12H)-ylidene-N,N-dimethyl-1-propanamine;* N,N-dimethylbenzofuro[3,2-c][1]benzoxepin-$\Delta^{6(12H),\gamma}$-propylamine; 6-(3-dimethylaminopropylidene)benzo[b]benzofurano[2,3-e]oxepine. $C_{21}H_{21}NO_2$; mol wt 319.40. C 78.97%, H 6.63%, N 4.39%, O 10.02%. Novel serotonin and histamine antagonist with antimigraine activity. Prepn: F. Binon, M. Descamps, **Ger.** pat. **1,963,205** corresp to U.S. pat. **3,651,051** (1970, 1972 both to Labaz). Metabolism *in vitro*: E. Rossi *et al., J. Chromatog.* **152**, 228 (1978). Use in treatment of migraine: J. J. Dufresne, *Praxis* **67**, 1148 (1978); J. Florence, *ibid.* 1323; in chronic headache: U. Thoden, *Therapiewoche* **30**, 492 (1980).

Fumarate, $C_{25}H_{25}NO_6$, *L-6257, Nocertone, Oxedix.* Cryst from isopropanol, mp 160°.
THERAP CAT: Analgesic (specific in migraine).

7069. Oxfendazole. *[5-(Phenylsulfinyl)-1H-benzimidazol-2-yl]carbamic acid methyl ester;* methyl-5-(phenylsulfinyl)-2-benzimidazolecarbamate; 5-phenylsulfinyl-2-carbomethoxyaminobenzimidazole; RS-8858; Autoworm; Benzelmin; Repidose; Synanthic; Systamex. $C_{15}H_{13}N_3O_3S$; mol wt 315.35. C 57.13%, H 4.16%, N 13.32%, O 15.22%, S 10.17%. Prepn: C. Beard *et al.,* **Ger.** pat. **2,363,351** corresp to U.S. pat. **3,929,821** (1974, 1975, both to Syntex); E. A. Averkin *et al., J. Med. Chem.* **18**, 1164 (1975). Metabolism: J. P. Bell, R. V. Tomlinson, *Fed. Proc.* **35**, 487 (1976). Radioimmunoassay: C. Nerenberg *et al., J. Pharm. Sci.* **67**, 1553 (1978). Efficacy evaluations: J. L. Duncan, J. G. Reid, *Vet. Rec.* **103**, 332 (1978); N. F. Baker *et al., Am. J. Vet. Res.* **39**, 1258 (1978).

Crystals from chloroform-methanol, mp 253° (dec). LD_{50} in dogs, rats, mice: > 1600, > 6400, > 6400 mg/kg (Averkin).
THERAP CAT (VET): Anthelmintic.

7070. Oxibendazole. *(5-Propoxy-1H-benzimidazol-2-yl)carbamic acid methyl ester;* 5-propoxy-2-benzimidazolecarbamic acid methyl ester; 5-propoxy-2-(carbomethoxyamino)-benzimidazole; SKF-30310; Anthelcide EQ; Equitac. $C_{12}H_{15}N_3O_3$; mol wt 249.27. C 57.82%, H 6.07%, N 16.86%, O 19.26%. Prepn: **Brit.** pat. **1,123,317** corresp to P. P. Actor, J. F. Pagano, U.S. pat. **3,574,845** (1968, 1971 both to SK & F). Comparative anthelmintic effects in mice: C. S. Karunakaron, D. A. Denham, *J. Parasitol.* **66**, 929 (1980). Clinical studies in foals and horses: J. H. Drudge *et al., Am. J. Vet. Res.* **42**, 526 (1981); E. T. Lyons *et al., ibid.* 685; D. K. Hass *et al., ibid.* **43**, 534 (1982).

Cryst, mp 230-230.5°.
THERAP CAT (VET): Anthelmintic.

7071. Oxiconazole Nitrate. *(Z)-1-(2,4-Dichlorophenyl)-2-(1H-imidazol-1-yl)ethanone O-[(2,4-dichlorophenyl)methyl]oxime mononitrate;* 2′,4′-dichloro-2-(imidazol-1-yl)-acetophenone O-(2,4-dichlorobenzyl)oxime nitrate; Sgd-301-76; Ro-13-8996; Gyno-Myfungar; Myfungar; Oceral; Oxistat. $C_{18}H_{14}Cl_4N_4O_4$; mol wt 492.14. C 43.93%, H 2.87%, Cl 28.81%, N 11.38%, O 13.00%. Broad spectrum topical antimycotic agent. Prepn: G. Mixich *et al.,* **Ger.** pat. **2,657,578**; *eidem,* U.S. pat. **4,124,767** (1977, 1978 both to Siegfried). Stereospecific synthesis of *(E)-, (Z)-*isomers, ^{1}H NMR study: G. Mixich, K. Thiele, *Arzneimittel-Forsch.* **29**, 1510 (1979). GLC determn in plasma: M. Zell, L. Herzfeld, *J. Chromatog.* **229**, 111 (1982). *In vitro* and *in vivo* antifungal activity: A. Polak, *Arzneimittel-Forsch.* **32**, 17 (1982). Study of skin, nail penetration: G. Stueltgen, E. Bauer, *Mykosen* **25**, 74 (1982). Comparison of antifungal imidazoles: W. H. Beggs, *IRCS Med. Sci.* **11**, 677 (1983).

Crystals from ethanol, mp 137-138°.
THERAP CAT: Antifungal.

7072. Oxidimethiin. *2,3-Dihydro-5,6-dimethyl-1,4-di-thiin 1,1,4,4-tetraoxide;* tetrathiin; UBI-N252; Harvade. C_6-$H_{10}O_4S_2$; mol wt 210.28. C 34.27%, H 4.79%, O 30.44%, S 30.50%. Prepn: A. D. Brewer *et al.,* **Fr.** pat. **2,228,429** corresp to **U.S.** pat. **3,920,438** (1974, 1975 to Uniroyal). Activity: R. W. Neidermyer *et al., Proc. Brit. Weed Control Conf., 12th* **3,** 959 (1974). Crystal structure: S. K. Arora *et al., Acta Crystallogr.* **B34,** 2918 (1978).

Long white needles from water, mp 166-168°.
USE: Plant growth regulator.

7073. Oxidized Cellulose. Absorbable cellulose; cellulosic acid; polyanhydroglucuronic acid; Oxycel; Hemo-Pak. A cellulose of varied carboxyl content retaining the fibrous structure. Prepd by oxidizing cellulose with nitrogen dioxide: Yackel, Kenyon, *J. Am. Chem. Soc.* **64,** 121 (1942).
The degree of oxidation is sufficiently high to make the product sol in dil alk solns. Insol in water or acidic solns.
THERAP CAT: Hemostatic (local).
THERAP CAT (VET): Hemostatic (local).

7074. Oxidronic Acid. *(Hydroxymethylene)bisphosphonic acid; (hydroxymethylene)diphosphonic acid;* HDP; HMDP. $CH_6O_7P_2$; mol wt 192.00. C 6.26%, H 3.15%, O 58.33%, P 32.26%. Bisphosphonate similar to etidronic acid, *q.v.* Prepn: O. T. Quimby, **Fr.** pat. **1,506,840;** *idem,* **U.S.** pat. **3,422,137** (1967, 1969 both to Procter & Gamble); of sodium salt: *idem et al., J. Org. Chem.* **32,** 4111 (1967). Prepn of ^{99m}Tc-complex: J. A. Bevan, **Eur.** pat. **Appl. 7,676;** *idem,* **U.S.** pat. **4,247,534** (1980, 1981 both to Procter & Gamble); and osteotropic properties: T. S. T. Wang *et al., J. Nucl. Med.* **21,** 767 (1980). Biodistribution: J. A. Bevan *et al., ibid* 961. Radiochemical purity: W. Majewski *et al., J. Nucl. Med. Technol.* **11,** 23 (1983). Clinical evaluation as skeletal imaging agent: P. A. Domstad *et al., Radiology* **136,** 209 (1980). Diagnostic use in Paget's disease: A. Evans *et al., Eur. J. Nucl. Med.* **18,** 757 (1991); for localization of bone metastases: F. Tenenbaum *et al., ibid* **20,** 1168 (1993).

Sodium salt, $CH_4Na_2O_7P_2$, *disodium monohydroxy methylene diphosphonate, oxidronate sodium.* Crystals from methanol/water (6/1), mp 297-300°.
^{99m}Tc-complex, *technetium ^{99m}Tc oxidronate, TechneScan HDP; Osteoscan-HDP.*
THERAP CAT: ^{99m}Tc-complex as diagnostic aid (radioactive imaging agent).

7075. Oxiniacic Acid. *3-Pyridinecarboxylic acid 1-oxide; nicotinic acid 1-oxide;* 3-carboxypyridine *N*-oxide. $C_6H_5NO_3$; mol wt 139.11. C 51.80%, H 3.62%, N 10.07%, O 34.50%. Prepn: Clemo, Koenig, *J. Chem. Soc.* **1949,** S231; Taylor, Crovetti, *J. Org. Chem.* **19,** 1633 (1954). Mass spectrum: Bild, Hesse, *Helv. Chim. Acta* **50,** 1885 (1967). Therapeutic effect: Debay, Thery, **Belg.** pat. **618,-968** (1962), *C.A.* **58,** 12375e (1963).

Needles, mp 254-255° (dec). Slightly soluble in cold water, more sol in hot water, hot glacial acetic acid, hot methanol, less sol in ethanol. Insol in light petroleum, benzene, chloroform. uv max (0.1N H_2SO_4): 220, 260 nm (ϵ 22400, 10200).
Ethanolamine salt, $C_6H_5NO_3 \cdot C_2H_7NO$, *ethanolamine oxiniacate, Novacyl.*
THERAP CAT: Antihyperlipoproteinemic.

7076. Oxiracetam. *4-Hydroxy-2-oxo-1-pyrrolidineacetamide;* 2-(4-hydroxypyrrolidin-2-on-1-yl)acetamide; hydroxypiracetam; CT-848; ISF-2522; Neuractiv; Neuromet. C_6-$H_{10}N_2O_3$; mol wt 158.16. C 45.57%, H 6.37%, N 17.71%, O 30.35%. Analog of piracetam, *q.v.,* with psychostimulant activity. Prepn: G. Pifferi, M. Pinza, **Ger.** pat. **2,635,853;** *eidem,* **U.S.** pat. **4,118,396** (1977, 1978 both to I.S.F.); *eidem, Farmaco Ed. Sci.* **32,** 602 (1977). Effect on learning and memory in animals and prepn of enantiomers: S. Banfi *et al., ibid.* **39,** 16 (1984). Effect in animals with cerebral impairment: S. Banfi *et al., Pharmacol. Res. Commun.* **16,** 67 (1984). Pharmacokinetics in humans: E. Perucca *et al., Eur. J. Drug Metabol. Pharmacokinet.* **9,** 267 (1984). Clinical evaluation in organic brain syndrome: B. Saletu *et al., Neuropsychobiology* **13,** 44 (1985).

White, microcrystalline powder from methanol, mp 165-168°.
(R)-Form, crystals from acetone + water, mp 135-136°. $[\alpha]_D$ +36.2° (c = 1.00 in water).
(S)-Form, crystals from acetone + water, mp 135-136°. $[\alpha]_D$ −36.0° (c = 1.00 in water).
THERAP CAT: Nootropic.

7077. Oxitropium Bromide. *[7(S)-(1α,2β,4β,5α,7β)]-9-Ethyl-7-(3-hydroxy-1-oxo-2-phenylpropoxy)-9-methyl-3-oxa-9-azoniatricyclo[3.3.1.0^{2,4}]nonane bromide; (8r)-6β,7β-epoxy-8-ethyl-3α-hydroxy-1αH,5αH-tropanium bromide (−)-tropate; (−)-N-ethylnorscopolamine methobromide;* Ba-253; Ba-253-BR-L; Oxivent; Tersigat; Ventilat. $C_{19}H_{26}BrNO_4$; mol wt 412.32. C 55.35%, H 6.36%, Br 19.38%, N 3.40%, O 15.52%. Anticholinergic bronchodilating agent. Prepn: K. Zeile *et al.,* **U.S.** pat. **3,472,861** (1969 to Boehringer, Ing.). Comparison with fenoterol, *q.v.:* D. Nolte, *Respiration* **36,** 32 (1978); with ipratropium bromide, *q.v.:* J. Lulling *et al., ibid.* **42,** 188 (1981). Efficacy in bronchial asthma: H. M. Beumer *et al., Int. J. Clin. Pharmacol. Ther. Toxicol.* **19,** 168 (1981); in exercise-induced asthma: K. Larsson, *Respiration* **43,** 57 (1982).

Crystals, mp 203-204° (dec). $[\alpha]_D^{21}$ −25° (c = 2.0 in water).

THERAP CAT: Bronchodilator.

7078. Oxolamine. *N,N-Diethyl-3-phenyl-1,2,4-oxadiazole-5-ethanamine; 5-(2-diethylaminoethyl)-3-phenyl-1,2,4-oxadiazole;* 3-phenyl-5-(β-diethylaminoethyl)-1,2,4-oxadiazole; 683-M. $C_{14}H_{19}N_3O$; mol wt 245.32. C 68.54%, H 7.81%, N 17.13%, O 6.52%. Prepn: **Ger. pat. 1,097,998** (1961 to Angelini Francesco), *C.A.* **56,** 11598h (1962). Teratogenicity studies of oxolamine citrate: L. Nilsson, *Arzneimittel-Forsch.* **17,** 781 (1967).

Liquid, $bp_{0.4}$ 127°.
Hydrochloride, $C_{14}H_{19}N_3O.HCl$, crystals, mp 153-154°.
Citrate, $C_{14}H_{19}N_3O.C_6H_8O_7$, *AF-438, Bredon, Broncatar, Perebron, Prilon, Flogobron, Oxarmin.* Crystals, slightly sol in water, alcohol. uv max (aq soln): 239 nm (ε 260), 273, 283 nm.

THERAP CAT: In inflammatory conditions of respiratory tract.

7079. Oxolinic Acid. *5-Ethyl-5,8-dihydro-8-oxo-1,3-dioxolo[4,5-g]quinoline-7-carboxylic acid;* 1-ethyl-1,4-dihydro-6,7-methylenedioxy-4-oxo-3-quinolinecarboxylic acid; 1-ethyl-6,7-methylenedioxy-4-quinolone-3-carboxylic acid; W-4565; Emyrenil; Inoxyl; Nidantin; Ossian; Oxoboi; Pietil; Prodoxol; Urinox; Uritrate; Uro-Alvar; Urotrate; Uroxin; Uroxol; Utibid. $C_{13}H_{11}NO_5$; mol wt 261.23. C 59.77%, H 4.24%, N 5.36%, O 30.62%. Quinolone antibacterial. Prepn: Kaminsky, Meltzer, **U.S. pat. 3,287,458** (1966 to Warner-Lambert); *eidem, J. Med. Chem.* **11,** 160 (1968). Pharmacology: Turner *et al., Antimicrob. Ag. Chemother.* **1967,** 475 sqq. Metabolism studies: DiCarlo *et al., Arch. Biochem. Biophys.* **127,** 503 (1968); Crew *et al., Xenobiotica* **1,** 193 (1971). Use in urinary tract infections: S. Kalowski *et al., Med. J. Aust.* **21,** 345 (1979); E. Anza *et al., Minerva Med.* **70,** 2333 (1979). Mechanism of action study: H. T. Wright *et al., Science* **213,** 455 (1981). Brief review of therapeutic use: R. Glickman *et al., Am. J. Hosp. Pharm.* **36,** 1077-1079 (1979).

Crystals from DMF, mp 314-316° (dec). LD_{50} in mice, rats (mg/kg): >6000, >2000 orally (Turner).

THERAP CAT: Antibacterial.
THERAP CAT (VET): Antibacterial.

7080. Oxomemazine. *N,N,β-Trimethyl-10H-phenothiazine-10-propanamine 5,5-dioxide; 10-(3-dimethylamino-2-methylpropyl)phenothiazine 5,5-dioxide;* 3-(9,9-dioxo-10-phenothiazinyl)-2-methyl-1-dimethylaminopropane; RP-6847; Dysedon. $C_{18}H_{22}N_2O_2S$; mol wt 330.45. C 65.43%, H 6.71%, N 8.48%, O 9.68%, S 9.70%. Prepn: Jacob, Robert, **U.S. pat. 2,972,612** (1961 to Rhône-Poulenc).

Crystals from heptane, mp 115°.
Hydrochloride, $C_{18}H_{22}N_2O_2S.HCl$, *Doxergan, Imakol.* Crystals from ethanol + isopropanol, mp 250°.

7081. Oxonic Acid. *1,4,5,6-Tetrahydro-4,6-dioxo-1,3,5-triazine-2-carboxylic acid; allantoxanic acid;* 5-azaorotic acid; s-triazine-2,4-dione-6-carboxylic acid; 2,4-dioxo-1,2,3,-4-tetrahydro-1,3,5-triazine-6-carboxylic acid. $C_4H_3N_3O_4$; mol wt 157.09. C 30.58%, H 1.92%, N 26.75%, O 40.74%. Formed during alkaline oxidation of uric acid with H_2O_2 or $KMnO_4$. Prepn and structure: Brandenberger, *Helv. Chim. Acta* **37,** 641 (1954); Brandenberger, Brandenberger, *ibid.* **37,** 2207 (1954); Piskala, Gut, *Coll. Czech. Chem. Commun.* **27,** 1572 (1962). Metabolism: Chelbova *et al., Biochem. Pharmacol.* **19,** 2785 (1970); Cihak, Sorm, *ibid.* **21,** 607 (1972).

Free acid is very unstable; can be isolated as the monopotassium salt. uv max (pH 9): 255 nm (ε 6800); (pH 12): 252 nm (ε 4600).
Oxonic acid amide, $C_4H_4N_4O_3$, microcrystalline substance. Practically insol in organic solvents, sparingly sol in boiling water. Does not melt below 350°, darkens above 300°.

7082. Oxophenarsine Hydrochloride. *2-Amino-4-arsenosophenol hydrochloride;* 3-amino-4-hydroxyphenyl arsinoxide hydrochloride; Ehrlich 5; Arseno 39; Arsenoxide; Mapharsen; Mapharside; Mapharsal; Fontarsan; Arsenosan; Oxiarsolan; Oxarsolan. $C_6H_7AsClNO_2$; mol wt 235.50. C 30.60%, H 3.00%, As 31.81%, Cl 15.05%, N 5.95%, O 13.59%. *Ref:* G. W. Raiziss, J. L. Gavron, *Organic Arsenical Compounds* (Chem. Catalog Co., New York, 1923), pp 136-137; Scott, Tullar, **Can. pat. 405,532** (1942 to Parke, Davis).

Hygroscopic, white or nearly white, powder contg ½ mole ethanol. Readily sol in water, methanol, ethanol, in solns of alkali hydroxides and carbonates, in dil mineral acids. Sparingly sol in glacial acetic acid. Practically insol in acetone, ether. The aq soln is acid to litmus or methyl red, but alkaline to Congo red. Solns darken on exposure to air. Usually marketed mixed with buffering agents to render its soln physiologically compatible with human blood.

THERAP CAT: Antiprotozoal (Trypanosoma).
THERAP CAT (VET): Has been used as an antirickettsial.

7083. Oxophenylarsine. Arsenosobenzene; phenyl arsenoxide; phenylarsine oxide; Arzene. C_6H_5AsO; mol wt 168.03. C 42.89%, H 3.00%, As 44.59%, O 9.52%. Prepn: Blicke, Smith, *J. Am. Chem. Soc.* **51**, 3479 (1929); **52**, 2946 (1930).

Crystals, mp 144-146° (also reported as mp 119-120°).
USE: Pesticide.
THERAP CAT (VET): Coccidiostat (poultry).

7084. L-2-Oxo-4-thiazolidinecarboxylic acid. *(R)-2-Oxo-4-thiazolidinecarboxylic acid;* OTCA; OTC; Procysteine. $C_4H_5NO_3S$; mol wt 147.15. C 32.65%, H 3.42%, N 9.52%, O 32.62%, S 21.79%. Cysteine prodrug. Prepn: J. A. Maclaren, *Aust. J. Chem.* **21**, 1891 (1968); T. Kömives, *Org. Prep. Proced. Int.* **21**, 251 (1989). Crystal structure: N. Ramasubbu, R. Parthasarathy, *Int. J. Pep. Protein Res.* **34**, 153 (1989). Precursor for cysteine and use as glutathione biosynthesis stimulant: J. M. Williamson *et al., Proc. Nat. Acad. Sci. USA* **79**, 6246 (1982); in chicks: T. K. Chung *et al., J. Nutr.* **120**, 158 (1990). Toxicity: R. D. White *et al., Acute Toxic Data* **1**, 164 (1992); R. D. White *et al., Toxicol. Letters* **69**, 15 (1993). Clinical evaluation in HIV infection: G. Giorgi *et al., Curr. Ther. Res.* **52**, 461 (1992); in asymptomatic HIV infection: R. C. Kalayjian *et al., J. Acquir. Immune Defic. Syndr.* **7**, 369 (1994). HPLC determn in plasma: L. W. Webb *et al., J. Chromatogr. B.* **654**, 257 (1994).

Colorless cubes, mp 171-173° (dec) (Maclaren). Also reported as white crystals, mp 173-174° (Kömives). pKa (22°): 3.32. $[\alpha]_D^{20}$ −59.4° (c = 2). Readily sol in alcohol. Insol in benzene and ethyl acetate.

7085. Oxotremorine. *1-[4-(1-Pyrrolidinyl)-2-butynyl]-2-pyrrolidinone;* 1-(2-oxo-1-pyrrolidinyl)-4-(1-pyrrolidinyl)-2-butyne. $C_{12}H_{18}N_2O$; mol wt 206.29. C 69.87%, H 8.79%, N 13.58%, O 7.76%. Active metabolite of tremorine, *q.v.:* Cho *et al., Biochem. Biophys. Res. Commun.* **5**, 276 (1961). Synthesis: Bebbington, Shakeshaft, *J. Med. Chem.* **8**, 274 (1965); Nau, U.S. pat. **3,444,185** (1969 to Soc. Civile Auguil). Cholinergic agent, used experimentally as convulsant in study of parkinsonism: Cho *et al., Proc. Int. Pharmacol. Meeting, 2nd, Prague 1963,* **2**, 75 (1964).

Pale yellow liquid. $bp_{0.6}$ 150-155°; $bp_{0.1}$ 124°. n_D^{25} 1.5156. Picrolonate, $C_{22}H_{26}N_6O_6$, crystals from acetone, mp 157-159°.
USE: Pharmacological tool.

7086. Oxprenolol. *1-[(1-Methylethyl)amino]-3-[2-(2-propenyloxy)phenoxy]-2-propanol;* 1-[o-(allyloxy)phenoxy]-3-(isopropylamino)-2-propanol; 1-(isopropylamino)-2-hydroxy-3-[o-(allyloxy)phenoxy]propane. $C_{15}H_{23}NO_3$; mol wt 265.35. C 67.90%, H 8.74%, N 5.28%, O 18.09%. β-Adrenergic blocker. Prepn: **Belg. pat. 669,402** (1966 to Ciba), *C.A.* **65**, 5402d (1966), corresp to **Brit. pat. 1,077,603.** Metabolic studies: Garteiz, *J. Pharmacol. Exp. Ther.* **179**, 354 (1971). Crystal structure of hydrochloride: J. M. Leger *et al., Acta Crystallogr.* **33B**, 2156 (1977). Long-term pre-

vention study in coronary heart disease: S. H. Taylor *et al., N. Engl. J. Med.* **307**, 1293 (1982).

Crystals from hexane, mp 78-80°.
Hydrochloride, $C_{15}H_{23}NO_3 \cdot HCl$, *Ba-39089, Coretal, Laracor, Paritane, Slow-Pren, Trasicor, Trasacor.* Crystals, mp 107-109°.
THERAP CAT: Antihypertensive, antianginal, antiarrhythmic.

7087. Oxyacanthine. *6,6',7-Trimethoxy-2,2'-dimethyloxyacanthan-12'-ol;* vinetine. $C_{37}H_{40}N_2O_6$; mol wt 608.73. C 73.01%, H 6.62%, N 4.60%, O 15.77%. From root of *Berberis vulgaris* L., *Berberidaceae:* Rüdel, *Arch. Pharm.* **229**, 631 (1891); Späth, Kolbe, *Ber.* **58**, 2280 (1925). Structure: v. Bruchhausen *et al., Ann.* **507**, 144 (1933); Fujita, *J. Pharm. Soc. Japan* **72**, 213 (1952), *C.A.* **47**, 6429b (1953).

White, cryst, bitter powder. mp 216-217°. $[\alpha]_D^{20}$ +131.5° (chloroform). Practically insol in water; sol in alcohol, chloroform, ether, dil acids.
Dihydrochloride, $C_{37}H_{40}N_2O_6 \cdot 2HCl$, mp 270-271°. $[\alpha]_D$ +185.5°. Sol in water.

7088. Oxybenzone. *(2-Hydroxy-4-methoxyphenyl)phenylmethanone;* 2-hydroxy-4-methoxybenzophenone; 4-methoxy-2-hydroxybenzophenone; benzophenone-3; MOB; Cyasorb UV 9 (obsolete); Spectra-Sorb UV 9; Uvinul M-40. $C_{14}H_{12}O_3$; mol wt 228.25. C 73.67%, H 5.30%, O 21.03%. Prepn: König, v. Kostanecki, *Ber.* **39**, 4027 (1906); Hardy, Forster, U.S. pat. **2,773,903** (1956 to Am. Cyanamid); Stanley *et al.*, U.S. pats. **2,861,104/5; 3,073,866** (1958, 1958, 1963 all to General Aniline & Film). Toxicity study: H. -J. Lewerenz *et al., Food Cosmet. Toxicol.* **10**, 41 (1972).

Crystals from isopropanol, mp 66°. Readily sol in most organic solvents. LD_{50} orally in rats: > 12.8 g/kg (Lewerenz).
THERAP CAT: Ultraviolet screen.

7089. Oxybutynin Chloride. *α-Cyclohexyl-α-hydroxybenzeneacetic acid 4-(diethylamino)-2-butynyl ester hydrochloride; α-phenylcyclohexaneglycolic acid 4-(diethylamino)-2-butynyl ester hydrochloride;* 4-diethylamino-2-butynyl phenylcyclohexylglycolate hydrochloride; oxibutinina hydrochloride; Cystrin; Ditropan; Dridase; Pollakisu; Tropax. $C_{22}H_{32}ClNO_3$; mol wt 393.95. C 67.07%, H 8.19%, Cl 9.00%, N 3.56%, O 12.18%. Prepn: **Brit. pat. 940,540** (1963 to Mead Johnson). Toxicity: E. I. Goldenthal, *Toxicol. Appl. Pharmacol.* **18**, 185 (1971).

Crystals, mp 129-130°. LD_{50} orally in rats: 1220 mg/kg (Goldenthal).

THERAP CAT: Anticholinergic. Treatment of symptoms in neurogenic bladder.

7090. Oxychlorosene. Monoxychlorosene; Clorpactin; Clorpactin XCB. $C_{20}H_{35}ClO_4S$; mol wt 407.01. C 59.02%, H 8.67%, Cl 8.71%, O 15.72%, S 7.88%. $C_{20}H_{34}O_3S.HOCl$. Described as a buffered organic hypochlorous acid derivative with slightly acid pH. Contains a long chain hydrocarbon (surface-active agent). The chain (14 carbons) may be branched or straight but actually consists of a constant mixture having predominantly the straight chain. The hydrocarbon chain also has a phenyl substituent which in turn holds a sulfonic acid group. The complex corresponds to a formula: $HO_3S—C_6H_4—(C_{14}H_{29}).HOCl$.

Aq solns are unstable and should be freshly prepd; the solid may be stored in properly stoppered bottles.

Sodium salt, *Clorpactin WCS*.

THERAP CAT: Antiseptic.

7091. Oxycinchophen. *3-Hydroxy-2-phenyl-4-quinolinecarboxylic acid; 3-hydroxy-2-phenylcinchoninic acid;* 3-hydroxycinchophen; HPC; Fenidrone; Magnofenyl; Magnophenyl; Oxinofen; Reumalon. $C_{16}H_{11}NO_3$; mol wt 265.27. C 72.45%, H 4.18%, N 5.28%, O 18.09%. Prepn: Berlingozzi, Capuano, *Atti Accad. Lincei* [V] **33**, II, 91 (1924); John, Fränkel, *J. Prakt. Chem.* **133**, 259 (1932); improved procedure: Marshall, Blanchard, *J. Pharmacol.* **95**, 186 (1949); Kreysa, U.S. pat. **2,776,290** (1957 to Chemo Puro).

Minute, deep yellow prisms from alc, dec 206-207°. Sol in acetic acid, alkalies, hot alc, benzene. Sparingly sol in water, ether. Forms an alkaline, water-sol sodium salt.

THERAP CAT: Antidiuretic; uricosuric.

7092. Oxyclozanide. *2,3,5-Trichloro-N-(3,5-dichloro-2-hydroxyphenyl)-6-hydroxybenzamide; 3,3',5,5',6-pentachloro-2'-hydroxysalicylanilide;* 3,3',5,5',6-pentachloro-2,2'-dihydroxybenzanilide; Zanil. $C_{13}H_6Cl_5NO_3$; mol wt 401.46. C 38.89%, H 1.51%, Cl 44.15%, N 3.49%, O 11.96%. Prepn: Neth. pat. Appl. **6,409,325** corresp to Broome *et al.*, U.S. pat. **3,349,090** (1965, 1967, both to I.C.I.).

Crystals, mp 209-211°.

THERAP CAT (VET): Anthelmintic (Trematodes).

7093. Oxycodone. *(5α)-4,5-Epoxy-14-hydroxy-3-methoxy-17-methylmorphinan-6-one;* dihydrohydroxycodeinone; 14-hydroxydihydrocodeinone; Dihydrone. $C_{18}H_{21}NO_4$; mol wt 315.37. C 68.55%, H 6.71%, N 4.44%, O 20.29%. Prepn

by catalytic reduction of hydroxycodeinone, its oxime, or its bromination products or by reduction of hydroxycodeinone with sodium hydrosulfite. Bibliography: Small, Lutz, *"Chemistry of the Opium Alkaloids,"* Suppl. No. 103 to Public Health Reports, Washington (1932); K. W. Bentley, *The Chemistry of the Morphine Alkaloids* (Oxford, 1954). Oxycodone apparently exists in two tautomeric forms which give the same salts: Freund, Speyer, *J. Prakt. Chem.* **94**, 135 (1916).

Long rods from alc, mp 218-220°. Sol in alc, chloroform; nearly insol in ether, water, KOH, NaOH, NH_4OH. Does not reduce ammoniacal silver or Fehling's soln.

Tautomeric form: Strongly refringent scales from alc, mp 219-220°, more sol in alc than the other form.

Hydrochloride, $C_{18}H_{21}NO_4.HCl$, *Dinarkon, Eubine, Eucodal, Eukodal, Eutagen, Oxikon, Oxycon, Pancodine, Tecodin, Tekodin, Thecodine, Thekodin*. Component of *Percodan*. Long rods from water, dec 270-272°. $[\alpha]_D^{20} -125°$ (c = 2.5). One gram dissolves in 10 ml water. Slightly sol in alc.

Pectinate, *Proladone*, used for prolonged action.

Caution: May be habit forming. This is a controlled substance (opiate) listed in the U.S. Code of Federal Regulations, Title 21 Parts 329.1 and 1308.12 (1995).

THERAP CAT: Analgesic (narcotic).

7094. 4,4'-Oxydi-2-butanol. *4,4'-Oxybis[2-butanol];* 3,3'-dihydroxydibutyl ether; bis[3-hydroxybutyl] ether; DHBE; Diskin; Dyskinébyl; Dis-Cinil; Colenormol. $C_8H_{18}O_3$; mol wt 162.23. C 59.23%, H 11.18%, O 29.59%. $CH_3-CH(OH)CH_2CH_2OCH_2CH_2CH(OH)CH_3$. Prepn: Joulty, Fr. pat. **1,267,084** (1960). Action on bile flow: Bornmann, *Arzneimittel-Forsch.* **2**, 122 (1952). Use as choleretic and antispasmodic: Albot, Toulet, *Presse Med.* **67**, 2053 (1959). General pharmacological properties: Fregnan, Porta, *Arzneimittel-Forsch.* **26**, 2116 (1976).

Clear, almost colorless, bitter fluid. Miscible with water.

THERAP CAT: Choleretic.

7095. 10,10'-Oxydiphenoxarsine. *10,10'-Oxybis-10H-phenoxarsine;* bis(phenoxarsin-10-yl)ether; bis(10-phenoxarsyl)oxide; bis(10-phenoxarsinyl)oxide; OBPAVinyzene. $C_{24}H_{16}As_2O_3$; mol wt 502.23. C 57.40%, H 3.21%, As 29.84%, O 9.56%. Prepn: W. L. Lewis *et al.*, *J. Am. Chem. Soc.* **43**, 891 (1921). Synthesis: K. D. Shvetsova-Shilovskaya *et al.*, *J. Gen. Chem. USSR* **31**, 776 (1961). Use in control of plant growth: M. L. Joseph, J. L. Hardy, U.S. pat. **3,069,252** (1962 to Dow Chemical); as bactericide in plastics: C. C. Yeager, U.S. pat. **3,288,674** (1966 to Scientific Chemicals). Toxicity study: D. G. Clark, *Toxicol. Appl. Pharmacol.* **21**, 315 (1972).

Colorless monoclinic prisms, mp 184-185°. dec. 380°. sp. gr. 1.40-1.42. Sol in alcohol, chloroform, methylene chloride. Practically insol in water (5 ppm at 20°) and alkali. LD_{50} in male rats: 35-50 mg/kg (Ventron, company data sheet).

USE: Primarily for fungicidal and bactericidal protection of plastics.

7096. Oxyfedrine. *[R-(R*,S*)]-3-[(2-Hydroxy-1-methyl-2-phenylethyl)amino]-1-(3-methoxyphenyl)-1-propanone; L-3-[(β-hydroxy-α-methylphenethyl)amino]-3'-methoxypropiophenone;* L-(1-hydroxy-1-phenyl-2-propylamino)-1-(*m*-methoxyphenyl)-1-propanone; oxyphedrine. $C_{19}H_{23}NO_3$; mol wt 313.40. C 72.82%, H 7.40%, N 4.47%, O 15.32%. Partial β-adrenergic agonist with coronary vasodilating and positive inotropic effects. Prepn: K. Thiele, **Belg. pat. 630,296;** *idem,* **U.S. pat. 3,225,095** (1963, 1965 both to Degussa). K. Thiele *et al., Arzneimittel-Forsch.* **16,** 1064 (1966). Absolute configuration determined by circular dichroism: J. Engel *et al., Chem. Ztg.* **105,** 85 (1981). TLC determn: Musumarra, *J. Chromatog.* **350,** 151 (1985). Pharmacology: H. Hueller *et al., Pharmazie* **27,** 242 (1972). Mode of action: P. Mentz, W. Forster, *Arzneimittel-Forsch.* **34,** 1739 (1984); N. Sternitzke, *Z. Kardiol.* **73,** 586 (1984). In prevention of experimental myocardial necrosis in rats: S. D. Seth *et al., Arzneimittel-Forsch.* **34,** 678 (1984). Comparison with atenolol, *q.v.,* in angina pectoris: L. Fananapazir, C. Bray, *Brit. J. Clin. Pharmacol.* **20,** 405 (1985).

L-Form hydrochloride, $C_{19}H_{23}NO_3 \cdot HCl$, *D-563, Ildamen, Modacor.* Crystals from methanol, mp 192-194°. LD_{50} in mice (mg/kg): 29 i.v. (Hueller).
DL-Form hydrochloride, mp 173-175°. LD_{50} in mice (mg/kg): 34 i.v. (Hueller).
THERAP CAT: Antianginal. Treatment of coronary insufficiency.

7097. Oxyfluorfen. *2-Chloro-1-(3-ethoxy-4-nitrophenoxy)-4-(trifluoromethyl)benzene;* 2-chloro-α,α,α-trifluoro-*p*-tolyl-3-ethoxy-4-nitrophenyl ether; RH-2915; Goal. $C_{15}H_{11}ClF_3NO_4$; mol wt 361.70. C 49.81%, H 3.07%, Cl 9.80%, F 15.76%, N 3.87%, O 17.69%. Selective pre- and post-emergence herbicide. Prepn: **Neth. pat. Appl. 7,303,590;** H. O. Bayer *et al.,* **U.S. pat. 3,798,276** (1973, 1974 both to Rohm & Haas). Activity: R. Y. Yih, C. Swithenbank, *J. Agr. Food Chem.* **23,** 592 (1975). Metabolism: I. L. Adler *et al., ibid.* **25,** 1339 (1977).

Orange crystalline solid, mp 83-84°. Soly in water: 0.1 ppm. Sol in most organic solvents. LD_{50} orally in male albino rats: > 5000 mg/kg (Yih, Swithenbank).
USE: Herbicide.

7098. Oxygen. O; at. wt 15.9994; at. no. 8; valence 2. Occurs normally as the diatomic gas O_2, also as ozone O_3. Atomic oxygen (O) can be prepd. Three naturally occurring isotopes: 16 (99.759%); 17 (0.037%); 18 (0.204%); artificial radioactive isotopes: 13-15; 19; 20. The most abundant element on earth; makes up 46.6% of earth's crust; 20.95% by vol of dry air. Obtained on a large scale by liquefaction of air. First obtained by Scheele in 1771 and independently by Priestley in 1774. Monograph: M. Ardon, *Oxygen—Elementary Forms and Hydrogen Peroxide* (Benjamin, New York, 1965) 106 pp. Review of oxygen and its compounds: Ebsworth *et al.,* in *Comprehensive Inorganic Chemistry* **vol. 2,** J. C. Bailar, Jr. *et al.,* Eds. (Pergamon Press, Oxford, 1973) pp 685-794; A. H. Taylor in Kirk-Othmer *Encyclopedia of Chemical Technology* **vol. 16** (Wiley-Interscience, New York, 3rd ed., 1981) pp 653-673.
Colorless, odorless, tasteless, neutral gas; supports combustion. d^0 (gas) 1.429 g/l; d^{-183} (liquid) 1.14 g/ml. mp

−218.4°. bp −182.96°. Critical temp −118.95°. Critical press. 50.14 atm. Heat of vaporization (−183°): 50.9 cal/g. Usually marketed under pressure in metal cylinders. One vol gas dissolves in 32 vols water at 20°; in 7 vols alcohol at 20°; also sol in other organic liquids and usually to a greater extent than in water. *Caution:* Avoid smoking, flames, electric sparks—explosion hazard.
USE: In oxyhydrogen or oxyacetylene flame for welding metals and for lighting (calcium light, etc); submarine work by divers, propellant for rockets. In the production of synthesis gas which can be used in the Fischer-Tropsch process for liquid fuels.
THERAP CAT: Medicinal gas to relieve hypoxia; at hyperbaric pressures in cardiac and other surgery, anaerobic infections, carbon monoxide poisoning; in cryotherapy (liq form).
THERAP CAT (VET): In hypoxia and in conjunction with volatile anesthetics.

7099. Oxymesterone. *(17β)-4,17-Dihydroxy-17-methylandrost-4-en-3-one;* 4,17β-dihydroxy-17α-methyl-3-oxoandrost-4-ene; 4-hydroxy-17α-methyltestosterone; 17α-methyl-4-androstene-4,17β-diol-3-one; oxymestrone; Anamidol; Oranabol; Theranabol. $C_{20}H_{30}O_3$; mol wt 318.46. C 75.43%, H 9.50%, O 15.07%. Prepn: **Brit. pat. 848,288;** Camerino *et al.,* **U.S. pat. 3,060,201** (1960, 1962 to Farmitalia).

Crystals, mp 169-171°. $[\alpha]_D^{20}$ +69° (ethanol). uv max (ethanol): 278 nm ($E_{1cm}^{1\%}$ 406). Practically insol in water. Sol in chloroform, acetone, alcohol.
Note: This is a controlled substance (anabolic steroid) in the U.S. Code of Federal Regulations, Title 21 Part 1308.13, as defined in Part 1308.02 (1995).
THERAP CAT: Androgen; anabolic.

7100. Oxymetazoline. *3-[(4,5-Dihydro-1H-imidazol-2-yl)methyl]-6-(1,1-dimethylethyl)-2,4-dimethylphenol; 6-tert-butyl-3-(2-imidazolin-2-ylmethyl)-2,4-dimethylphenol;* 2-(4-*tert*-butyl-2,6-dimethyl-3-hydroxybenzyl)-2-imidazoline; H-990; Navisin; Hazol; Rhinofrenol; Rhinolitan; Sinerol; Nezeril. $C_{16}H_{24}N_2O$; mol wt 260.38. C 73.81%, H 9.29%, N 10.76%, O 6.14%. Prepd from (4-*tert*-butyl-2,6-dimethyl-3-hydroxyphenyl)acetonitrile and ethylenediamine: Fruhstorfer, Mueller-Calgan, **Ger. pat. 1,117,588** (1961 to E. Merck), *C.A.* **57,** 4674a (1962). Toxicity data: Hotovy *et al., Arzneimittel-Forsch.* **11,** 1016 (1961).

Crystals from benzene, mp 181-183°.
Hydrochloride, $C_{16}H_{24}N_2O \cdot HCl$, *Afrazine, Afrin, Iliadin, Nafrine, Nasivin, Oxilin, Sinex.* Crystals, dec 300-303°. Freely sol in water, alcohol. Practically insol in ether, chloroform, benzene. LD_{50} orally in mice: 10 mg/kg (Hotovy).
Note: Ingredient of *Drixin.*
THERAP CAT: Adrenergic (vasoconstrictor); nasal decongestant.

7101. Oxymetholone. *(5α,17β)-17-Hydroxy-2-(hydroxymethylene)-17-methylandrostan-3-one;* 2-hydroxymethylene-17α-methyldihydrotestosterone; 4,5α-dihydro-2-hydroxymethylene-17α-methyltestosterone; 2-hydroxymethylene-

17α-methyl-17β-hydroxy-5α-androstan-3-one; 2-hydroxy-methylene-17α-methylandrostan-17β-ol-3-one; anasterone; Adroyd; Anapolon; Anadrol; Pardroyd; Plenastril; Prot-anabol; Nastenon; Synasteron. $C_{21}H_{32}O_3$; mol wt 332.48. C 75.86%, H 9.70%, O 14.44%. Anabolic steroid. Prepn: Ringold et al., J. Am. Chem. Soc. **81**, 427 (1959); Ringold, Rosenkranz, **Ger.** pat. **1,070,632** (1959 to Syntex).

Crystals from ethyl acetate, mp 178-180°. $[\alpha]_D +38°$. uv max: 285 nm (log ε 3.99).

Enol acetate, $C_{23}H_{34}O_4$, crystals from hexane, mp 144-148°. $[\alpha]_D +27°$ (ethanol). uv max: 255 nm (log ε 4.09).

Enol propionate, $C_{24}H_{36}O_4$, crystals from hexane, mp 135°. $[\alpha]_D +26°$ (ethanol). uv max: 257 nm (log ε 4.11).

Enol benzoate, $C_{28}H_{36}O_4$, crystals from acetone + water, mp 188-190°. $[\alpha]_D \pm0°$. uv max: 230 nm (log ε 4.19).

Note: Oxymetholone may reasonably be anticipated to be a carcinogen: *Seventh Annual Report on Carcinogens* (PB95-109781, 1994) p 313. This is a controlled substance (ana-bolic steroid) in the U.S. Code of Federal Regulations, Title 21 Part 1308.13, as defined in Part 1308.02 (1995).

THERAP CAT: Androgen.

THERAP CAT (VET): Anabolic steroid for small animals.

7102. Oxymethurea. *N,N'-Bis(hydroxymethyl)urea; N,N'*-dihydroxymethylurea; Methural. $C_3H_8N_2O_3$; mol wt 120.11. C 30.00%, H 6.71%, N 23.32%, O 39.96%. (HO-CH$_2$NH)$_2$CO. Prepn: Einhorn, Hamburger, *Ber.* **41**, 26 (1908); Walter, **U.S.** pat. **1,863,426** (1927); U.S. pat. **2,436,-355** (1946 to du Pont).

Crystals from alcohol, mp 137-139°. Very sol in cold water, hot ethyl alcohol, methanol.

USE: In the textile industry in cotton crease- and shrink-proofing, finishing, drying, sizing; in tanning; pesticides; in photographic developers.

THERAP CAT: Antiseptic.

7103. Oxymorphone. *(5α)-4,5-Epoxy-3,14-dihydroxy-17-methylmorphinan-6-one;* dihydrohydroxymorphinone; dihydro-14-hydroxymorphinone; 14-hydroxydihydromor-phinone. $C_{17}H_{19}NO_4$; mol wt 301.34. C 67.76%, H 6.36%, N 4.65%, O 21.24%. Obtained from dihydrohydroxycodei-none by boiling with concd aq hydrobromic acid: Weiss, *J. Am. Chem. Soc.* **77**, 5891 (1955); Lewenstein, Weiss, **U.S.** pat. **2,806,033** (1957).

Crystals from boiling ethanol, ethyl acetate or benzene. mp 248-249° (dec). Levorotatory. Sol in boiling acetone and chloroform; readily sol in aq alkalies; moderately sol in boiling ethanol; sparingly sol in benzene.

Hydrochloride, $C_{17}H_{19}NO_4 \cdot HCl$, *Numorphan.* Crystals.

Note: This is a controlled substance (opiate) listed in the U.S. Code of Federal Regulations, Title 21 Part 1308.12 (1995).

THERAP CAT: Analgesic (narcotic).

THERAP CAT (VET): Analgesic (narcotic).

7104. Oxypendyl. *4-[3-(10H-Pyrido[3,2-b][1,4]benzo-thiazin-10-yl)propyl]-1-piperazineethanol;* 10-[3-[4-(2-hydr-oxyethyl)-1-piperazinyl]propyl]-10H-pyrido[3,2-b][1,4]ben-

zothiazine; 10-[3-(1-hydroxyethyl-4-piperazinyl)propyl]-4-azaphenothiazine; oxipendyl; D-706. $C_{20}H_{26}N_4OS$; mol wt 370.52. C 64.83%, H 7.07%, N 15.12%, O 4.32%, S 8.65%. Prepn: Schuler, Klebe, *Ann.* **653**, 172 (1962).

bp$_1$ 260-265°, bp$_6$ 280-300°.

Dihydrochloride, $C_{20}H_{26}N_4OS \cdot 2HCl$, *Pervetral.* Crystals, mp 218-220°.

THERAP CAT: Antiemetic.

7105. Oxypertine. *5,6-Dimethoxy-2-methyl-3-[2-(4-phenyl-1-piperazinyl)ethyl]-1H-indole;* 1-[2-(5,6-dimethoxy-2-methyl-3-indolyl)ethyl]-4-phenylpiperazine; Win-18501-2; Equipertine; Forit; Integrin (formerly). $C_{23}H_{29}N_3O_2$; mol wt 379.50. C 72.79%, H 7.70%, N 11.07%, O 8.43%. Prepn: Archer et al., J. Am. Chem. Soc. **84**, 1306 (1962).

THERAP CAT: Antidepressant.

7106. Oxyphenbutazone. *4-Butyl-1-(4-hydroxyphenyl)-2-phenyl-3,5-pyrazolidinedione; 4-butyl-2-(p-hydroxyphenyl)-1-phenyl-3,5-pyrazolidinedione;* 1-phenyl-2-(p-hydroxyphen-yl)-3,5-dioxo-4-n-butylpyrazolidine; 1-(p-hydroxyphenyl)-2-phenyl-4-butylpyrazolidine-3,5-dione; p-hydroxyphenyl-butazone; G-27202; Californit; Crovaril; Flogitolo; Flogoril; Frabel; Neo-Farmadol; Oxalid; Rapostan; Tandacote; Tan-dearil; Visubutina. $C_{19}H_{20}N_2O_3$; mol wt 324.38. C 70.35%, H 6.21%, N 8.64%, O 14.80%. Prepn: Hafliger, **U.S.** pat. **2,745,783** (1956 to Geigy); Pfister, Hafliger, *Helv. Chim. Acta* **40**, 395 (1957).

Monohydrate, *Imbun, Phlogistol, Phlogase, Phlogont.* Crystals, mp 96°. Anhyd crystals from ether + petr ether, mp 124-125°. Acidic reaction. Soluble in ethanol, meth-anol, chloroform, benzene, ether. Forms a water-soluble sodium salt.

THERAP CAT: Anti-inflammatory.

7107. Oxyphencyclimine. *α-Cyclohexyl-α-hydroxybenz-eneacetic acid (1,4,5,6-tetrahydro-1-methyl-2-pyrimidinyl)-methyl ester; 1,4,5,6-tetrahydro-1-methyl-2-pyrimidinylmethyl α-phenylcyclohexaneglycolate;* α-phenylcyclohexaneglycolic acid 1-methyl-2-tetrahydroxypyrimidylmethyl ester; 1-meth-yl-1,4,5,6-tetrahydro-2-pyrimidylmethyl α-cyclohexyl-α-phenylglycolate; Antulcus; Caridan; Daricol; Setrol; Vio-Thene; Daricon; Naridan; Zamanil. $C_{20}H_{28}N_2O_3$; mol wt 344.45. C 69.74%, H 8.19%, N 8.13%, O 13.93%. Prepn: **Brit.** pat. **795,758** (1958 to Pfizer); Faust et al., *J. Am. Chem. Soc.* **81**, 2214 (1959).

Consult the Name Index before using this section.

Hydrochloride, $C_{20}H_{28}N_2O_3 \cdot HCl$, crystals, dec 231-232°. Solubility in water: 1.2 g/100 ml.

THERAP CAT: Anticholinergic.

7108. Oxyphenisatin Acetate. *3,3-Bis[4-(acetyloxy)-phenyl]-1,3-dihydro-2H-indol-2-one; 3,3-bis(p-acetoxyphenyl)-oxindole;* acetphenolisatin; endophenolphthalein; diacetyl-diphenolisatin; diacetyldihydroxydiphenylisatin; diacetyl-hydroxyphenylisatin; diacetoxydiphenylisatin; di(acetoxy-phenyl)oxindole; diphesatin; diacetyldioxyphenylisatin; Isacen; Isocrin; Isaphen; Laxo-Isatin; Promassolax; Sanapert; Bydolax; Cirotyl; Lisagal; Contax; Prulet; Purgaceen; Bisatin. $C_{24}H_{19}NO_5$; mol wt 401.42. C 71.81%, H 4.77%, N 3.49%, O 19.93%. Prepd from isatin by treatment with phenol followed by acetylation in the presence of H_2SO_4 at temps below 100°: Baeyer, Lazarus, *Ber.* **18**, 2641 (1885); Hoffmann-La Roche, **Swiss** pat. **100,806; Ger.** pat. **406,210** (1923); **U.S.** pat. **1,624,675** (1927); **Ger.** pat.**482,435** (1929); Mizuno, Turuga, **Japan.** pat. **129,200** (1939).

Tasteless crystals, mp 242°. Practically insol in water, ether, dilute HCl. Slightly sol in alcohol.

THERAP CAT: Cathartic.

7109. Oxyphenonium Bromide. *2-[(Cyclohexylhydr-oxyphenylacetyl)oxy]-N,N-diethyl-N-methylethanaminium bromide; diethyl(2-hydroxyethyl)methylammonium α-phenyl-cyclohexaneglycolate bromide;* α-phenylcyclohexaneglycolic acid ester diethyl(2-hydroxyethyl)methylammonium bromide; diethylaminoethyl α-phenylcyclohexaneglycolate methylbromide; 2-diethylaminoethyl α-cyclohexyl-α-phenyl-glycolate methobromide; cyclohexylhydroxyphenylacetic acid diethylmethylaminoethyl ester bromide; phenylcyclo-hexyloxyacetic acid diethylaminoethyl ester bromomethyl-ate; Ba-5473; C-5473; Antrenyl; Spasmophen. $C_{21}H_{34}Br-NO_3$; mol wt 428.41. C 58.88%, H 8.00%, Br 18.65%, N 3.27%, O 11.20%. Prepn: **Swiss** pat. **259,948** (1949 to Ciba).

Crystals, mp 189-194° from ethyl acetate + alc. Freely sol in water. Sparingly sol in alc. Aq solns are neutral.

THERAP CAT: Anticholinergic.

7110. Oxypinocamphone. *2-Hydroxy-2,6,6-trimethyl-bicyclo[3.1.1]heptan-3-one; 2-hydroxy-3-pinanone;* 2-hydr-oxypinocamphone; Camphostene; Oxypinone. $C_{10}H_{16}O_2$; mol wt 168.24. C 71.39%, H 9.59%, O 19.02%. Prepn of *l*-form: Kuwata, *J. Am. Chem. Soc.* **59**, 2509 (1937); Harispe *et al., Bull. Soc. Chim. France* **1964**, 1035; of *dl*-form: Ku-

wata, *loc. cit.* Stereochemistry: Carlson *et al., Tetrahedron Letters* **1968**, 5941.

dl-Form, crystals, mp 38.5-39.0°.
l-Form, crystals from ether + petr ether, mp 35.5-36.5°. $[\alpha]_D^{25}$ −18.56° (c = 14.44 in alcohol).

THERAP CAT: Respiratory insufficiency.

7111. Oxytetracycline. *[4S-(4α,4aα,5α,5aα,6β,12aα)]-4-(Dimethylamino)-1,4,4a,5,5a,6,11,12a-octahydro-3,5,6,10,-12,12a-hexahydroxy-6-methyl-1,11-dioxo-2-naphthacenecarb-oxamide;* glomycin; riomitsin; hydroxytetracycline. $C_{22}H_{24}N_2O_9$; mol wt 460.44. C 57.39%, H 5.25%, N 6.08%, O 31.27%. Antibiotic substance isolated from the elaboration products of the actinomycete, *Streptomyces rimosus,* grown on a suitable medium: Finlay *et al., Science* **111**, 85 (1950). Isoln: Regna, Solomons, *Ann. N.Y. Acad. Sci.* **53**, 221 (1950); Regna *et al., J. Am. Chem. Soc.* **73**, 4211 (1951). Production from *Streptomyces rimosus:* Sobin *et al.,* **U.S.** pat. **2,516,080** (1950 to Pfizer). Isoln from *S. xanthophaeus:* Brockmann, Musso, *Naturwiss.* **41**, 451 (1954); Brockmann *et al.,* **Ger.** pat. **913,687** (1954 to Bayer), *C.A.* **53**, 4662h (1959). Solubility data: Weiss *et al., Antibiot. & Chemother.* **7**, 374 (1957). Structure: Hochstein *et al., J. Am. Chem. Soc.* **74**, 3708 (1952). Abs config: Dobrynin *et al., Tetra-hedron Letters* **1962**, 901. Stereochemistry: Schach von Wittenau *et al., J. Am. Chem. Soc.* **87**, 134 (1965). Total synthesis of the *dl*-form: H. Muxfeldt *et al., ibid.* **101**, 689 (1979).

Stability: Oxytetracycline crystals show no loss in potency on heating for 4 days at 100°, the hydrochloride crystals show < 5% inactivation after 4 mos at 56°. Aq solns of the hydrochloride at pH 1.0 to 2.5 are stable for at least 30 days at 25°. Solns at pH 3.0 to 9.0 show no detectable loss in potency on storage at 5° for at least 30 days. Half-life in hours of aq oxytetracycline solns at 37°: pH 1.0 = 114; pH 2.5 = 134; pH 4.6 = 45; pH 5.5 = 45; pH 7.0 = 26; pH 8.5 = 33; pH 10.0 = 14.

Dihydrate, *Abbocin, Berkmycen, Clinimycin, Imperacin, Oxatets, Oxydon, Oxymycin, Stecsolin, Stevacin, Terralon-LA, Terramycin, Unimycin.* Needles from water or methanol, dec 181-182°. $[\alpha]_D^{25}$ −196.6° (0.1N HCl); $[\alpha]_D^{25}$ −2.1° (0.1N NaOH); $[\alpha]_D^{25}$ +26.5° (methanol). uv max (pH 4.5 phosphate buffer 0.1M): 249, 276, 353 nm ($E_{1cm}^{1\%}$ 240, 322, 301). Soly in water at 23° at various pH's: pH 1.2 = 31,400 γ/ml, pH 2.0 = 4600 γ/ml, pH 3.0 = 1400 γ/ml, pH 5.0 = 500 γ/ml, pH 6.0 = 700 γ/ml, pH 7.0 = 1100 γ/ml, pH 9.0 = 38,600 γ/ml. Soly in abs ethanol 12,000 γ/ml, in 95% ethanol 200 γ/ml.

Hydrochloride, *Alamycin, Aquacycline, Bio-Mycin, Duphacycline, Engemycin, Geomycin, Gynamousse, Macocyn, Occrycetin, Oxlopar, Oxybiocycline, Oxybiotic, Oxycyclin, Oxy-Dumocyclin, Oxyject, Oxylag, Oxypan, Oxytetracid, Oxy-tetrin, Terrafungine, Terraject, Tetramel, Tetran, Tetra-Tablinen, Toxinal, Vendarcin.* Yellow platelets from water. Very soluble in water (1 g/ml). Concd aq solns at neutral pH hydrolyze on standing and deposit crystals of oxytetracycline. Soly in abs ethanol: 12,000 γ/ml, in 95% ethanol: 33,000 γ/ml.

Disodium salt dihydrate, $C_{22}H_{22}N_2Na_2O_9.2H_2O$, yellow crystals; darkens on standing. Soly in abs alc: 8,000 γ/ml, in methanol: 1500 γ/ml.

USE: To treat lethal yellowing in palm trees.

THERAP CAT: Antibacterial.

THERAP CAT (VET): Antibacterial.

7112. Oxythiamine. *3-[(1,4-Dihydro-2-methyl-4-oxo-5-pyrimidinyl)methyl]-5-(2-hydroxyethyl)-4-methylthiazolium chloride; 5-(2-hydroxyethyl)-3-[(4-hydroxy-2-methyl-5-pyrimidinyl)methyl]-4-methylthiazolium chloride.* $C_{12}H_{16}ClN_3O_2S$; mol wt 301.80. C 47.76%, H 5.34%, Cl 11.75%, N 13.92%, O 10.60%, S 10.63%. Prepn: F. Bergel, A. R. Todd, *J. Chem. Soc.* **1937**, 1504; M. Soodak, L. R. Cerecedo, *J. Am. Chem. Soc.* **66**, 1988 (1944). Improved prepn: H. N. Rydon, *Biochem. J.* **48**, 383 (1951). Thiamine antagonist activity: A. J. Eusebi, L. R. Cerecedo, *Science* **110**, 162 (1949); L. J. Daniel, L. C. Norris, *Proc. Soc. Exp. Biol. Med.* **72**, 165 (1949); and distribution in tissues: C. J. Gubler, D. S. Murdock, *J. Nutr. Sci. Vitaminol.* **28**, 217 (1982). Proposed mechanism of action: S. A. Strumilo *et al., Biomed. Biochim. Acta* **43**, 159 (1984). Determn by HPLC: B. C. Hemming, C. J. Gubler, *J. Liq. Chromatog.* **3**, 1697 (1980).

Hydrochloride, $C_{12}H_{16}ClN_3O_2S.HCl$, flat needles grouped in rosettes, dec 195°. uv max (acid soln): 265, 258, 228, 223 nm; (alkaline soln): 268, 260, 228, 221 nm. Does not give the thiochrome reaction.

Diphosphate, crystals, mp 127-129°.

Monophosphoric acid ester, hygroscopic crystals, dec 185-186°.

Triphosphoric acid ester, minute crystals, dec 245-255°.

7113. Oxythioquinox. *6-Methyl-1,3-dithiolo[4,5-b]-quinoxalin-2-one; dithiocarbonic acid cyclic S,S-(6-methyl-2,3-quinoxalinediyl) ester;* 6-methyl-2,3-quinoxalinedithiol cyclic S,S-dithiocarbonate; 6-methyl-2-oxo-1,3-dithio[4,5-b]-quinoxaline; chinomethionat(e); quinomethionate; Bayer 36205; Forstan; Morestan. $C_{10}H_6N_2OS_2$; mol wt 234.30. C 51.26%, H 2.58%, N 11.96%, O 6.83%, S 27.37%. Prepn and review (in English) of chemical and fungicidal properties and toxicology: F. Grewe, H. Kaspers, *Pflanzenschutz Nachr. Bayer* **18** (1), 1-23 (1965), *C.A.* **64**, 5689h (1966). Toxicity study: T. B. Gaines, *Toxicol. Appl. Pharmacol.* **14**, 515 (1969).

Yellow crystals from benzene, mp 172°. Practically insol in water. Freely sol in DMF. Sol in hot benzene, toluene, dioxane. Slightly sol in methanol, ethanol, acetone. LD_{50} in male, female rats (mg/kg): 1800, 1100 orally (Gaines).

USE: Acaricide; agricultural fungicide.

7114. Oxytocin. Alpha-hypophamine; ocytocin; Intertocine-S; Perlacton; Pitocin; Syntocinon; Oxystin; Partocon; Synpitan; Uteracon. $C_{43}H_{66}N_{12}O_{12}S_2$; mol wt 1007.20. C 51.28%, H 6.60%, N 16.69%, O 19.06%, S 6.37%. The principal uterus-contracting and lactation-stimulating hormone of the posterior pituitary gland. Isoln: Pierce *et al., J. Biol. Chem.* **199**, 929 (1952). Structure and synthesis: Tuppy, Michl, *Monatsh.* **84**, 1011 (1953); Tuppy, *Biochim. Biophys. Acta* **11**, 449 (1953); du Vigneaud *et al., J. Am. Chem. Soc.* **75**, 4879 (1953); **76**, 3115 (1954); Bodanszky, du Vigneaud, *ibid.* **81**, 2504 (1959); Cash *et al., J. Med. Pharm. Chem.* **5**, 413 (1962); Sakakibara *et al., Bull. Chem. Soc. Japan* **38**, 120 (1965). Solid phase synthesis: Bayer, Hagenmaier, *Tetrahedron Letters* **1968**, 2037; Ives, *Can. J. Chem.* **46**, 2318 (1968). Synthesis of D-oxytocin: Flouret, du Vigneaud, *J. Am. Chem. Soc.* **87**, 3775 (1965).

Description of commercial process: Velluz *et al.,* U.S. pats. **2,938,891** and **3,076,797** (1960, 1963, both to Roussel-UCLAF). Radioimmunoassay: T. Chard, *Clin. Biochem. Anal.* **5**, 209 (1977). *Review:* du Vigneaud, *Experientia Suppl.* **II** (14th Intl. Congr. Pure and Appl. Chem.), 9-26 (1955); R. Caldeyro-Barcia, H. Heller, *Proc. of an Intl. Symp. on Oxytocin* (Montevideo 1959) 443 pp; several authors, *Advan. Exp. Med. Biol.* **2**, 53-104 (1968); C. R. W. Edwards in *Hormones in Blood* vol. **2**, C. H. Gray, V. James, Ed. (Academic Press, New York, 3rd ed., 1979) pp 401-421. Review of role in parturition: A.-R. Fuchs, F. Fuchs, *Advan. Exp. Med.* **1980**, 403-428. Comprehensive description: F. Nachtmann *et al., Anal. Profiles Drug Subs.* **10**, 563-600 (1981).

Cys–Tyr–Ile–Gln–Asn–Cys–Pro–Leu–GlyNH₂

White powder. $[\alpha]_D^{22}$ −26.2° (c = 0.53). Sol in water, 1-butanol, 2-butanol.

Citrate, *Pitocin-Buccal.*

THERAP CAT: Oxytocic.

THERAP CAT (VET): Stimulates milk let-down, uterine contraction.

7115. Ozagrel. *(E)-3-[4-(1H-Imidazol-1-ylmethyl)phenyl]-2-propenoic acid; (E)-4-(imidazol-1-ylmethyl)cinnamic acid;* OKY-046. $C_{13}H_{12}N_2O_2$; mol wt 228.25. C 68.41%, H 5.30%, N 12.27%, O 14.02%. Prepn: K. Iizuka *et al.,* **Ger. pat. 2,923,815;** *eidem,* U.S. pat. **4,226,878** (both 1980 to Ono; Kissei). Synthesis and thromboxane synthetase inhibitory activity: K. Iizuka *et al., J. Med. Chem.* **24**, 1139 (1981). Pharmacology: S. Hiraku *et al., Japan. J. Pharmacol.* **41**, 393 (1986). Pulmonary vascular effects: R. Garcia-Szabo *et al., Prostaglandins* **28**, 851 (1984). Metabolism in animals: M. Shimizu *et al., Iyakuhin Kenkyu* **17**, 289 (1986), *C.A.* **105**, 72012q (1986). HPLC determn in biological fluids: *eidem, ibid.* 298, *C.A.* **105**, 72013r (1986). Clinical pharmacology and evaluation in myocardial infarction: T. Ito *et al., Biomed. Biochim. Acta* **43**, S125 (1984). Clinical evaluation in prevention of cerebral vasospasm: S. Suzuki *et al., Acta Neurochir.* **77**, 133 (1985); in coronary artery disease: M. Shikano *et al., Japan. Heart J.* **28**, 663 (1987). Toxicity data: T. Nishigake *et al., Clin. Rep.* **20**, 2671 (1986). Series of articles on pharmacology: *Pharmacometrics* **31**, 527-565 (1986), *C.A.* **105**, 35349-35352 (1986).

Prisms from ethanol-ether, mp 223-224°.

Hydrochloride, $C_{13}H_{12}N_2O_2.HCl$, crystals from ethanol-ether, mp 214-217°.

Sodium salt, $C_{13}H_{11}N_2NaO_2$, Cataclot, Xanbon. LD_{50} in male, female mice, male, female rats (mg/kg): 1940, 1580, 1150, 1300 i.v.; 3800, 3600, 5900, 5700 orally; 2450, 2100, 2300, 2250 s.c. (Nishigake).

THERAP CAT: Antithrombotic; antianginal.

7116. Ozone. Triatomic oxygen. O_3; mol wt 48.00. Found in the atm in varying proportions (about 0.05 ppm at sea level), since it is produced continuously in the outer layers of the atm by the action of solar uv radiation on the oxygen of the air. So-called sterilizing lamps operate on the same principle. In the laboratory ozone is prepd by passing dry air between two plate electrodes connected to an alternating current source of several thousand volts. The reaction is reversible, and after a little ozone has been produced it is dec at the same rate as it is generated. Obtained in pure form by cooling ozonized air to −180° when it separates as a dark blue liquid. *See also* C. E. Thorp, *Bibliography of Ozone Technology* (Armour Res. Found., Chicago). Lab prepn: *Org. Syn.* **coll. vol. III**, 673 (1955). Monograph: *Ozone Chemistry and Technology,* Advances in Chemistry Series no. **21** (A.C.S., Washington D.C., 1959) 465 pp. *Review:* C. Nebel in Kirk-Othmer *Encyclopedia of Chemical Technology* vol. **16** (Wiley-Interscience, New York, 3rd ed., 1981) pp 683-713.

Bluish, explosive gas or blue liquid. Pleasant, characteristic odor in concns of less than 2 ppm. Irritating and injurious in higher concns. Powerful oxidizing agent. d^0 (gas): 2.144 g/l; $d^{-195.4}$ (liq) 1.614 g/ml. mp −193°. bp −111.9°. Critical temp −12.1°. Critical press. 53.8 atm. Heat of formation 34.4 kcal/mole at 25°. Intense absorption band beginning at about 290 nm. Unstable. Solutions contg ozone explode on warming. Prepn of ozone solns in liq oxygen: Cook, *U.S.* pat. **3,008,902** (1961 to Union Carbide). Although the stability of ozone in aq solns decreases as alkalinity rises, this effect is reversed at high concns. For example, the half life of ozone is 2 min in 1N NaOH; it is increased to 83 hrs in 20N NaOH: Heidt, Landi, *J. Chem. Phys.* **41,** 176 (1964).

Caution: Potential symptoms of overexposure are irritation of eyes and mucous membranes; pulmonary edema; chronic respiratory disease. *See NIOSH Pocket Guide to Chemical Hazards* (DHHS/NIOSH 90-117, 1990) p 172.

USE: As disinfectant for air and water by virtue of its oxidizing power. For bleaching waxes, textiles, oils. In organic syntheses. Forms ozonides which are sometimes useful oxidizing compds.

P

7117. Paclitaxel. *[2aR-[2aα,4β,4aβ,6β,9α(αR*,βS*),-11α,12α,12aα,12bα]]-β-(Benzoylamino)-α-hydroxybenzene-propanoic acid 6,12b-bis(acetyloxy)-12-(benzoyloxy)-2a,3,4,-4a,5,6,9,10,11,12,12a,12b-dodecahydro-4,11-dihydroxy-4a,8,-13,13-tetramethyl-5-oxo-7,11-methano-1H-cyclodeca[3,4]benz-[1,2-b]oxet-9-yl ester;* 5β,20-epoxy-1,2α,4,7β,10β,13α-hexa-hydroxytax-11-en-9-one 4,10-diacetate 2-benzoate 13-ester with (2R,3S)-N-benzoyl-3-phenylisoserine; taxol A; NSC-125973; Taxol. $C_{47}H_{51}NO_{14}$; mol wt 853.92. C 66.11%, H 6.02%, N 1.64%, O 26.23%. Antileukemic and antitumor agent first isolated, as the *l*-form, from the bark of the Pacific yew tree, *Taxus brevifolia, Taxaceae;* promotes the assembly of microtubules and inhibits the tubulin disassembly process. Isoln and structure: M. C. Wani *et al., J. Am. Chem. Soc.* **93**, 2325 (1971). *In vitro* promotion of microtubule assembly: P. B. Schiff *et al., Nature* **277**, 665 (1979). Isoln from *Taxus baccala* L. and *in vitro* inhibition of depolymerization of microtubules into tubulin: G. Chauviere *et al., C.R. Acad. Sci. Paris Ser. II* **293**, 501 (1981). Total synthesis of *taxusin*, which contains the entire ring skeleton: R. A. Holton *et al., J. Am. Chem. Soc.* **110**, 6558 (1988). Total stereosynthesis: R. A. Holton *et al., ibid.* **116**, 1597, 1599 (1994); K. C. Nicolaou *et al., Nature* **367**, 630 (1994). Synthesis and anticancer activity of derivs: D. G. I. Kingston *et al., Studies in Organic Chemistry* vol. **26**, entitled "New Trends in Natural Products Chemistry 1986", Attaur-Rahman, P. W. Le Quesne, Eds. (Elsevier, Amsterdam, 1986) pp 219-235. Production by *Taxomyces andreanae,* an endophytic fungus associated with *T. brevifolia:* A. Stierle *et al., Science* **260**, 214 (1993). Use in study of structure and function of microtubules: S. B. Horwitz *et al., Cold Spring Harbor Symp. Quant. Biol.* **46**, 219 (1982). Review of mechanism of action: J. J. Manfredi, S. B. Horwitz, *Pharmacol. Ther.* **25**, 83-125 (1984); S. B. Horwitz *et al., Ann. N.Y. Acad. Sci.* **466**, 733-744 (1986); S. B. Horwitz, *Trends Pharmacol. Sci.* **13**, 134-136 (1992). Symposium on clinical toxicology, pharmacology and efficacy: *Sem. Oncol.* **20**, Suppl. 3, 1-60 (1993). Review in ovarian cancer: C. D. Runowicz *et al., Cancer* **71**, 1591-1596 (1993). Book: *Taxol®: Science and Applications,* M. Suffness, Ed. (CRC Press, Boca Raton, 1995) 426 pp.

Needles from aq methanol, mp 213-216° (dec). $[\alpha]_D^{20}$ −49° (methanol). uv max (methanol): 227, 273 nm (ϵ 29800, 1700).

USE: Tool in study of structure and function of microtubules.

THERAP CAT: Antineoplastic.

7118. Paclobutrazol. *(R*,R*)-(±)-β-[(4-Chlorophenyl)-methyl]-α-(1,1-dimethylethyl)-1H-1,2,4-triazole-1-ethanol;* 1-*tert*-butyl-2-(*p*-chlorobenzyl)-2-(1,2,4-triazol-1-yl)ethanol; (2RS,3RS)-1-(4-chlorophenyl)-4,4-dimethyl-2-(1H-1,2,4-triazol-1-yl)pentan-3-ol; ICI-PP-333; PP-333; Bonzi; Clipper; Cultar; Parlay. $C_{15}H_{20}ClN_3O$; mol wt 293.80. C 61.32%, H 6.86%, Cl 12.07%, N 14.30%, O 5.45%. Plant growth regulator with fungicidal activity. Prepn: B. C. Baldwin *et al.,* **Ger.** pat. 2,734,426; S. Balasubramanyan, M. C. Shephard, **U.S.** pat. 4,243,405 (1978, 1981 both to ICI). Physical properties and biological activity: B. G. Lever *et al., Proc. Brit. Crop. Prot. Conf.-Weeds* **1982**, 3. Resolution and

activity of isomers: B. Sugavanam, *Pestic. Sci.* **15**, 296 (1984). GC determn in plant tissue: E. A. Stahly, D. A. Buchanan, *HortScience* **21**, 534 (1986). Comparison with daminozide, *q.v.,* of effect on apple trees: G. R. Stinchcombe *et al., J. Hort. Sci.* **59**, 323 (1984).

White crystalline solid, mp 165-166°. d 1.22. Vapor pressure at 20°: 1×10^{-6} Pa. Soly: water 35 mg/l, methanol 15%, propylene glycol 5%, acetone 11%, cyclohexanone 18%, methylene dichloride 10%, hexane 1%, xylene 6%.

USE: Plant growth regulator.

7119. Pactamycin. *2-Hydroxy-6-methylbenzoic acid [1S-[1α,2β,3α,3(R*),4α,5β]]-[5-[(3-acetylphenyl)amino]-4-amino-3-[[(dimethylamino)carbonyl]amino]-1,2-dihydroxy-3-(1-hydroxyethyl)-2-methylcyclopentyl]methyl ester;* NSC-52947; U-15800. $C_{28}H_{38}N_4O_8$; mol wt 558.63. C 60.20%, H 6.86%, N 10.03%, O 22.91%. Antitumor antibiotic produced by *Streptomyces pactum* var *pactum.* Discovery and biological properties: Bhuyan *et al., Antimicrob. Ag. Chemother.* **1961**, 184. Isoln and characterization: Argoudelis *et al., ibid.* 191. Manuf: **Brit.** pat. 980,346 (1965 to Upjohn), *C.A.* **62**, 11115f (1965). Structure: Wiley *et al., J. Org. Chem.* **35**, 1420 (1970). Revised structure: D. J. Duchamp, *Am. Crystallogr. Assn.* (Winter Mtg., Albuquerque, 1972) p 23. Mechanism of action: T. A. Beerman *et al., Adv. Enzyme Regul.* **14**, 207 (1976). ^{13}C-NMR study: D. D. Weller *et al., J. Antibiot.* **30**, 997 (1978). Biosynthesis: D. D. Weller, K. L. Rinehart, *J. Am. Chem. Soc.* **100**, 6757 (1978). *Review:* Goldberg in *Antibiotics* vol. **3**, J. W. Corcoran, F. E. Hahn, Eds. (Springer-Verlag, New York, 1975) pp 498-515.

$[\alpha]_D^{25}$ +79° (ethanol) changing to +23° on standing. $[\alpha]$ changes in acetone on standing from +25° to +76° in 24 hours. Amphoteric. Sol in ethanol, chloroform, methylene chloride, benzene, ether, in solns < pH 5 and > 9.5. Insol in Skellysolve B, cyclohexane; insol at isoelectric pt, pH 8.3. Unstable in solution. LD_{50} in mice (mg/kg): 10.7 orally; 15.6 i.v.; in rats (mg/kg): 1.4 i.v. (Bhuyan).

7120. Palitantin. *[2α,3β(1E,3E),5α,6α]-3-(1,3-Heptadienyl)-5,6-dihydroxy-2-(hydroxymethyl)cyclohexanone.* $C_{14}H_{22}O_4$; mol wt 254.33. C 66.12%, H 8.72%, O 25.16%. Metabolic product of *Penicillium palitans* Westling. Isoln: J. H. Birkinshaw, H. Raistrick, *Biochem. J.* **30**, 801 (1936). Derivs and degradation products: J. H. Birkinshaw, *ibid.* **51**, 271 (1952). Structure: K. Bowden *et al., J. Chem. Soc.* **1959**, 1662. Biosynthesis: P. Chaplen, R. Thomas, *Biochem. J.* **77**, 91 (1960); A. J. Birch, M. Kocor, *J. Chem. Soc.* **1960**, 866. Reactivity studies: A. T. Austin, B. Pearson, *Chem. & Ind. (London)* **1966**, 1228. Stereoselective synthesis of (±)-form: A. Ichihara *et al., Tetrahedron Letters* **1977**, 3473; *eidem, Tetrahedron* **36**, 1547 (1980).

Needles from hot water, mp 165°. $[\alpha]^{23}_{5461}$ +4.4° (c = 0.8 in CHCl$_3$). uv max (ethanol): 323 nm (ϵ 34,000). Sol in hot water, alc, chloroform; slightly sol in cold water, ether.

7121. Palladium. Pd; at. wt 106.42; at. no. 46; valences 2, 4. Six naturally occurring isotopes: 102 (1.0%); 104 (11.0%); 105 (22.2%); 106 (27.3%); 108 (26.7%); 110 (11.8%); artificial, radioactive isotopes: 98-101, 103; 107; 109; 111-115. Abundance in earth's crust 0.001-0.01 ppm. Discovered in 1803 by Wollaston. Belongs to the platinum group of metals. Occurs in nature alloyed with platinum or gold and as a selenide; found in nickel sulfide ores; found in the minerals stibiopalladinite, braggite, porpezite. Isoln: Vauquelin et al., cited by Mellor, *A Comprehensive Treatise on Inorganic and Theoretical Chemistry* **15**, 595 (1936). Reviews of prepn, properties and chemistry of palladium and other platinum metals: Gilchrist, *Chem. Rev.* **32**, 277-372 (1943); Beamish et al., in *Rare Metals Handbook*, C. A. Hampel, Ed. (Reinhold, New York, 1956) pp 291-328; Livingstone in *Comprehensive Inorganic Chemistry* vol. 3, J. C. Bailar, Jr. et al., Eds. (Pergamon Press, Oxford, 1973) pp 1163-1189, 1274-1329. Review of uses: E. M. Wise, *Palladium, Recovery, Properties, Uses* (Academic Press, New York, 1968) 187 pp. Use of organic palladium derivatives in synthesis: P. M. Maitlis, *The Organic Chemistry of Palladium* (Academic Press, New York, 1971); J. Tsuji, *Topics in Current Chemistry* **91**, 29 (1980); B. M. Trost, *Tetrahedron* **33**, 2615 (1977); idem, *Accts. Chem. Res.* **13**, 385 (1980).
Silver-white metal, face-centered cubic structure; occurs also as black powder and as spongy masses which can be compressed to a compact mass. mp 1555°; bp 3167°. d$^{20}_4$ 12.02. Hardness on Mohs' scale 4.8, Brinell hardness 61.0. Spec heat 0.0584 cal/g at 0°C. Electrical resistivity at 0° = 10.0 microohms-cm. Appreciably volatile at high temps. At a red heat is converted into the monoxide. Forms dihalides with fluorine or chlorine at a red heat. Reacts with nitric acid, sulfuric acid, a mixture of hydrochloric and chloric acids. Reacts slightly with concd HCl; more readily in the presence of air or free chlorine. Forms a sulfide when heated with sulfur, a phosphide when heated with phosphorus. Absorbs a considerable amount of hydrogen.
USE: In form of gold, silver, and copper alloys in dentistry; for alloy bearings, springs, balance wheels of watches; for mirrors in astronomical instruments; as catalyzer in manuf of sulfuric acid and in other oxidizing processes; in powder form as catalyst in hydrogenation and in ignition of hydrogen or hydrocarbons with oxygen; the spongy form is used in gas analysis for separating hydrogen from mixtures of gases.

7122. Palladium Chloride. Palladous chloride. Cl$_2$Pd; mol wt 177.33. Cl 39.99%, Pd 60.01%. PdCl$_2$. Prepn: Krustinsons, *Z. Elektrochem.* **44**, 537 (1938). Toxicity study: Orestano, *Boll. Soc. Ital. Biol. Sper.* **8**, 1154 (1933). Review of PdCl$_2$ and other halides: J. H. Canterford, R. Colton, *Halides of the Second and Third Row Transition Metals* (John Wiley, New York, 1968) pp 358-389.
Red crystals. mp 678-680°. Dec at high temp to palladium and chlorine. MLD i.v. in rabbits: 0.0186 g/kg (Orestano).
Dihydrate, dark brown crystals. Sol in water, alcohol, acetone. Reduced in soln by hydrogen or CO to metal. *Keep tightly closed.*
USE: In photography, for preparing pictures to be transferred to porcelain; toning solutions; electroplating parts of clocks and watches; manuf indelible ink; for the prepn of the metal for use as a catalyst; PdCl$_2$ paper is used for detecting CO, to find leaks in buried gas pipes. Prepn of palladium catalysts using PdCl$_2$: Mozingo, *Org. Syn.* coll. vol. III, 685 (1955).

7123. Palladium Diacetate. *Acetic acid palladium salt;* bis(acetato)palladium; diacetatopalladium(II); palladium(II)

acetate. C$_4$H$_6$O$_4$Pd; mol wt 224.51. C 21.40%, H 2.69%, O 28.51%, Pd 47.40%. (CH$_3$COO)$_2$Pd. Prepn from palladium sponge or palladium nitrate and glacial acetic acid: Morehouse et al., *Chem. & Ind. (London)* **1964**, 544; Stephenson et al., *J. Chem. Soc.* **1965**, 3632; from hydrated palladium oxide and acetic acid: Hausman et al., *Fr. pat.* **1,403,398** corresp to U.S. pat. **3,318,891** (1965, 1967 to Engelhard). Crystal structure of trimeric palladium acetate: Skapski, Smart, *Chem. Commun.* **1970**, 658.
Orange-brown crystals; mp 205° (dec). Shown to be trimeric by osmometric determn of mol wt in benzene (37°); monomeric by ebullioscopic measurement in benzene (80°). Crystals are trimeric. Sol in chloroform, methylene dichloride, acetone, acetonitrile, diethyl ether; sol with dec in aq HCl, aq KI solutions. Insol in water, aq NaCl, NaNO$_3$, NaOAc solns, petroleum, alcohols. Dec when warmed with alcohols.
USE: Catalyst.

7124. Palladium Nitrate. Palladous nitrate. N$_2$O$_6$Pd; mol wt 230.43. N 12.16%, O 41.66%, Pd 46.18%. Pd-(NO$_3$)$_2$. Prepn from palladium and nitric acid: *Gmelin's, Palladium* (8th ed.) **65**, 269 (1942). Reported to be the dihydrate: Gatehouse et al., *J. Chem. Soc.* **1957**, 4222.
Brown, deliquesc crystals. Sol in water with turbidity; with much water a brown basic salt precipitates; completely sol in dil HNO$_3$. *Keep well closed, protected from light.*
USE: Sepn of Cl$_2$ and I$_2$; catalyst in organic syntheses.

7125. Palladium Oxide. Palladium monoxide; palladous oxide. OPd; mol wt 122.42. O 13.07%, Pd 86.93%. PdO. Prepn: Shriner, Adams, *J. Am. Chem. Soc.* **46**, 1685 (1924).
Black powder. d 8.3. Insol in water, acids; slightly sol in aqua regia; sol in 48% HBr. Dec when strongly heated; also in the presence of H$_2$.
USE: Reduction catalyst in synthesis of organic compds.

7126. Palmatine. *5,6-Dihydro-2,3,9,10-tetramethoxy-dibenzo[a,g]quinolizinium; 7,8,13,13a-tetradehydro-2,3,9,10-tetramethoxyberbinium;* calystigine. [C$_{21}$H$_{22}$NO$_4$]$^+$. Obtained only in form of its salts. First isolated from Calumba root *(Jateorhiza palmata* (DC.) Miers, *Menispermaceae)*. Now found in many other genera. Palmatine and tetrahydropalmatine, q.v. are probably the most widely distributed Berberis alkaloids. Extraction procedure: Feist, Dschu, *Arch. Pharm.* **263**, 301 (1925). Structure: Feist, Sandstede, *ibid.* **256**, 2, 5 (1918); Späth, Lang, *Ber.* **54**, 3064, 3068 (1921); Späth, Böhm, *Ber.* **55**, 2988 (1922); Späth, Meinhard, *Ber.* **75**, 400 (1942). Synthesis: Späth, Quientensky, *Ber.* **58**, 2267 (1925); R. D. Haworth et al., *J. Chem. Soc.* **1927**, 548; Z. Kiparissides et al., *Can. J. Chem.* **58**, 2770 (1980). Identity with calystigine, alkaloid of the Chinese drug Chi-Kuo-Lan: Huang, Chen, *C.A.* **52**, 15827i (1958).

Palmatine forms addn products with acetone and chloroform, as does berberine.
Iodide, (C$_{21}$H$_{22}$NO$_4$)I, yellow needles from water, dec 239°; also a dihydrate. Sparingly sol in hot water and alc.
Nitrate, (C$_{21}$H$_{22}$NO$_4$)NO$_3$, sesqui- or dihydrate, yellow needles, dec 239°. Freely sol in most solvents.
Chloride trihydrate, (C$_{21}$H$_{22}$NO$_4$)Cl.3H$_2$O, yellowish-green needles from water. Freely sol in hot water and alc.
Sulfate pentahydrate, (C$_{21}$H$_{22}$NO$_4$)$_2$SO$_4$.5H$_2$O, yellow needles, mp 250°. Very sol in alc, sol in water.

7127. Palmidrol. *N-(2-Hydroxyethyl)hexadecanamide;* N-(2-hydroxyethyl)palmitamide. C$_{18}$H$_{37}$NO$_2$; mol wt 299.50. C 72.19%, H 12.45%, N 4.68%, O 10.68%. CH$_3$-(CH$_2$)$_{14}$CONHCH$_2$CH$_2$OH. A naturally occurring anti-inflammatory agent. Isolated from soybean lecithin, egg yolk, and peanut meal: Kuehl et al., J. Am. Chem. Soc. **79**, 5577 (1957). Synthesis by refluxing ethanolamine with palmitic acid: Roe et al., ibid. **74**, 3442 (1952).

Crystals from 95% ethanol or cyclohexane, mp 98-99°.

7128. Palmitic Acid. Hexadecanoic acid; hexadecylic acid; cetylic acid. $C_{16}H_{32}O_2$; mol wt 256.43. C 74.94%, H 12.58%, O 12.48%. $CH_3(CH_2)_{14}COOH$. Occurs as the glyceryl ester in many oils and fats. Obtained from palm oil, Japan wax, or Chinese vegetable tallow. Purification: Magne et al., U.S. pat. **2,791,596** (1957 to Secretary of Agriculture). Toxicity study: L. Orö, A. Wretlind, Acta Pharmacol. Toxicol. **18**, 141 (1961).

White, crystalline scales. d_4^{62} 0.853. mp 63-64°. bp$_{15}$ 215°. n_D^{80} 1.4273. Insol in water. Sparingly sol in cold alcohol or in petr ether; freely sol in hot alcohol, in ether, propyl alcohol, chloroform. LD$_{50}$ i.v. in mice: 57±3.4 mg/kg (Orö, Wretlind).

7129. Palm Oil, From Fruit. Palm butter. Obtained from the fruit Elaeis guineensis Jacq., Palmae. Constit. Chiefly palmitin; some stearin, linolein.

Reddish-yellow to dark dirty red, fatty mass; faint odor of violet. d 0.920-0.927. mp 27-42.5°. n_D^{40} 1.453-1.459. Sapon. no. 200-205. Iodine no. 53-57.

USE: In manuf of soap, candles; lubricant.

7130. Palm Oil, From Seed. An oil obtained from seed of Elaeis guineensis Jacq., Palmae.

White to yellowish. d 0.952. mp 26-30°. Sapon. no. ~247. Iodine no. ~15.

USE: Manuf soap; with coconut oil in manuf plant butter; has also been used in liniments and ointment.

7131. Palustric Acid. [1R-(1α,4aβ,10aα)]-1,2,3,4,4a,5,-6,9,10,10a-Decahydro-1,4a-dimethyl-7-(1-methylethyl)-1-phenanthrenecarboxylic acid; 13-isopropylpodocarpa-8,13-dien-15-oic acid. $C_{20}H_{30}O_2$; mol wt 302.46. C 79.42%, H 10.00%, O 10.58%. Isoln from gum rosin: Loeblich et al., J. Am. Chem. Soc. **77**, 2823 (1955); Joye et al., J. Org. Chem. **30**, 654 (1965). Structure: Shuller et al., J. Am. Chem. Soc. **82**, 1734 (1960). Prepn of the 4-epi-form: Tabacik, Poisson, Bull. Soc. Chim. France **1969**, 3264.

Crystals from methanol, mp 162-167°. [α]$_D$ +71.6°. uv max (0.01N NaOH): 265-266 nm.

Methyl ester, $C_{21}H_{32}O_2$, crystals from methanol, mp 25-27°. [α]$_D$ +67.7°. uv max: 265-266 nm.

7132. Palytoxin. Palytoxin (C51-55 hemiacetal); PTX. $C_{129}H_{223}N_3O_{54}$; mol wt 2680.18. C 57.81%, H 8.39%, N 1.57%, O 32.24%. Potent toxin isolated from zoanthid coral of the genus Palythoa that is the most poisonous non-proteinaceous substance known. Isoln from "Limu-make-o-Hana", the Hawaiian name for the highly toxic coelenterate Palythoa toxica: R. E. Moore, P. J. Scheuer, Science **172**, 495 (1971). Structure: R. E. Moore, G. Bartolini, J. Am. Chem. Soc. **103**, 2491 (1981). Structure of palytoxin from P. tuberculosa of Okinawa (differs from palytoxin from P. toxica at two positions): D. Uemura et al., Tetrahedron Letters **22**, 2781 (1981). Proposed absolute configuration of 60 of the 64 chiral centers: R. E. Moore et al., J. Am. Chem. Soc. **104**, 3776 (1982). Structure and stereochemistry: J. K. Cha et al., ibid. 7369. Discussion of the structural elucidation, see Y. Shimizu, Nature **302**, 212 (1983); review: R. E. Moore, Prog. Chem. Org. Nat. Prod. **48**, 82-202 (1985). Synthetic studies: Y. Kishi, Chemica Scripta **27**, 573 (1987). Pharmacological study: P. N. Kaul et al., Proc. West. Pharmacol. Soc. **17**, 294 (1974). Mechanism of action and treat-

ment of palytoxin poisoning: J. A. Vick et al., Toxicol. Appl. Pharmacol. **34**, 214 (1975). Mode of contractile action on vascular smooth muscle: K. Ito et al., Eur. J. Pharmacol. **46**, 9 (1977). Depolarizing action on frog spinal cord: Y. Kudo, S. Shibata, Brit. J. Pharmacol. **71**, 575 (1980). Toxicology and toxicity studies: J. S. Wiles et al., Toxicon **12**, 427 (1974); K. Ito et al., Arch. Int. Pharmacodyn. Ther. **258**, 146 (1982). Brief review of biology: P. J. Scheuer, Accounts Chem. Res. **10**, 33-39 (1977). Review of synthetic studies and conformational analysis: Y. Kishi, Pure Appl. Chem. **61**, 313-324 (1989).

White amorphous hygroscopic solid. No definite mp; chars when heated to 300°. [α]$_D^{25}$ +26° (water). Insol in chloroform, ether, acetone. Sparingly sol in methanol, ethanol. Sol in pyridine, DMSO, water. LD$_{50}$ in mice: 0.45 μg/kg i.v. (Wiles); 50-100 ng/kg i.p. (Kaul).

Caution: Palytoxin is an intense vasoconstrictor. In intact dogs, doses of > 0.06 μg/kg i.v. caused a transient rise in arterial pressure followed by rapid hypotension and resulted in death within 5 minutes (Ito, 1982).

USE: As a physiological tool to evaluate anti-anginal chemotherapeutic agents.

7133. Pamabrom. 8-Bromo-3,7-dihydro-1,3-dimethyl-1H-purine-2,6-dione compd with 2-amino-2-methyl-1-propanol (1:1); 8-bromotheophylline compd with 2-amino-2-methyl-1-propanol (1:1); 2-amino-2-methyl-2-propanol 8-bromotheophyllinate. $C_{11}H_{18}BrN_5O_3$; mol wt 348.20. C 37.94%, H 5.21%, Br 22.95%, N 20.11%, O 13.78%. Prepn from 2-amino-2-methyl-1-propanol and 8-bromotheophylline: J. M. Holbert, I. W. Grote, U.S. pat. **2,711,411** (1955 to Chattanooga Med.). Prepn and diuretic activity: J. M. Holbert et al., J. Am. Pharm. Assoc., Sci. Ed., **44**, 355 (1955).

Fine white powder, dec 300°. Soly in water > 30 g/100 ml at 25°. pH of satd aq soln 8.0-8.5.

Mixture with acetaminophen and pyrilamine maleate, *Midol PMS, Premsyn PMS, Sunril.*

THERAP CAT: Diuretic.

7134. Pamaquine. N^1,N^1-*Diethyl-N⁴-(6-methoxy-8-quin-olinyl)-1,4-pentanediamine; 8-[[4-(diethylamino)-1-methyl-butyl]amino]-6-methoxyquinoline;* Plasmochin; Aminoquin; Praequine; Béprochine; Gamefar; Quipenyl; Plasmoquine. $C_{19}H_{29}N_3O$; mol wt 315.46. C 72.34%, H 9.27%, N 13.32%, O 5.07%. Preparation by reductive condensation of 1-diethylaminopentan-4-one with 6-methoxy-8-aminoquinoline: Elderfield *et al., J. Am. Chem. Soc.* **70**, 40 (1948). *Review:* Cooper, *U.S. Pub. Health Repts.* **64**, 717 (1949).

Dark yellow oil. $bp_{0.3}$ 175-180°; bp_1 182-194°.

Pamoate, $C_{42}H_{45}N_3O_7$, *pamaquine embonate, pamaquine naphthoate.* Prepn: **Brit.** pat. **295,656** (1927 to I. G. Farbenind). Details and list of German pats. in K. H. Slotta, *Grundriss der modernen Arzneistoff-Synthese* (Stuttgart, 1931). Yellow to orange-yellow odorless, almost tasteless powder. Numbs the tongue. Practically insol in water; sol in alcohol, acetone. *Protect from light.*

THERAP CAT: Naphthoate as antimalarial.

7135. Pamidronic Acid. *(3-Amino-1-hydroxypropylid-ene)bisphosphonic acid;* 3-amino-1-hydroxypropane-1,1-diphosphonic acid; ADP; AHPrBP. $C_3H_{11}NO_7P_2$; mol wt 235.07. C 15.33%, H 4.72%, N 5.96%, O 47.64%, P 26.35%. Calcium metabolism regulator; structurally similar to etidronic acid, *q.v.* Prepn: F. Krueger *et al.,* **Ger.** pat. **2,130,-794** (1973 to Benckiser); *C.A.* **78**, 84528z (1973); K.-H. Worms *et al., Z. Anorg. Allgem. Chem.* **457**, 214 (1979). Improved process: H. Blum, K.-H. Worms, U.S. pat. **4,327,039** (1982 to Henkel). Mechanism of inhibition of osteoclastic bone resorption studies: J.-M. Delaisse *et al., Life Sci.* **37**, 2291 (1985); P. M. Boonekamp *et al., Bone Mineral* **1**, 27 (1986); C. W. G. Lowik *et al., Adv. Exp. Med. Biol.* **208**, 275 (1986). Pharmacokinetics in rodents: F. Wingen, D. Schmähl, *Arzneimittel-Forsch.* **37**, 1037 (1987). Clinical evaluations in Paget's disease: W. B. Frijlink *et al., Lancet* **1**, 799 (1979); G. Heynen *et al., Eur. J. Clin. Invest.* **12**, 29 (1982); in osteolytic bone disease: F. J. M. van Breukelen *et al., Lancet* **1**, 803 (1979). Reproduction toxicity study: P. Graepel *et al., Arzneimittel-Forsch.* **42**, 654 (1992). Prepn of the ^{99m}Tc-ADP complex, and use as a bone scanning agent: O. J. Degrossi *et al., J. Nucl. Med.* **26**, 1135 (1985). Review of pharmacology and therapeutic use in resorptive bone disease: A. Fitton, D. McTavish, *Drugs* **41**, 289-318 (1991).

Disodium salt, $C_3H_9NNa_2O_7P_2$, *pamidronate disodium, GCP-23339A, Aminomux, Aredia.*

THERAP CAT: Antipagetic; inhibitor of tumor-induced hypercalcemia. ^{99m}Tc complex as diagnostic aid (radioactive imaging agent).

7136. Pamoic Acid. *4,4'-Methylenebis[3-hydroxy-2-naphthalenecarboxylic acid]; 4,4'-methylenebis(3-hydroxy-2-naphthoic acid); 4,4'-methylenedi(3-hydroxy-2-naphthoic acid); 2,2'-dihydroxy-1,1'-dinaphthylmethane-3,3'-dicarboxylic acid; embonic acid.* $C_{23}H_{16}O_6$; mol wt 388.38. C 71.13%, H 4.15%, O 24.72%. Prepn from 2-hydroxy-3-naphthoic acid + formaldehyde: Strohback, *Ber.* **34**, 4162 (1901); Brass, Sommer, *ibid.* **61**, 993 (1928); Barber, Gaimster, *J. Appl. Chem. (London)* **2**, 565 (1952). Prepn of salts: Puetzer, **U.S.** pat. **2,397,903** (1946 to Vick Chem.); Barber, **U.S.** pat. **2,641,610** (1953 to May & Baker).

Crystals from dil pyridine, dec above 280° without melting. Practically insoluble in water, alcohol, ether, benzene, acetic acid. Sparingly sol in chloroform; sol in nitrobenzene, pyridine.

Methyl ether, $C_{24}H_{18}O_6$, yellow crystals from dil methanol, dec 277-282°.

7137. Pancreatin. Diastase vera; Creon; Pancrease; Pancrex-Vet; Pankrotanon; Panzytrat; Zypanar. A substance from the fresh pancreas of the hog or ox, contg the enzymes amylopsin, trypsin, and steapsin. It converts not less than 25 times its wt of starch into sol carbohydrates and not less than 25 times its wt of casein into proteoses within 5 mins. (equiv to 150 times in 30 mins.). Its greatest activity is exhibited in neutral or faintly alkaline media; mineral acids or excess alkali hydroxides or carbonates render it inactive. Method of production from cow or pig pancreas: Hoek *et al.,* U.S. pat. **3,223,594** (1965 to N. American Philips).

Yellowish, amorphous powder. Partly sol in water. Insol in alcohol. *Incompat.* Alcohol, acids.

Note: Pancreatin is destroyed in the stomach, hence should be given in salol-coated pills.

THERAP CAT: Digestive aid.

THERAP CAT (VET): In replacement therapy in pancreatic enzyme deficiency; in preparation of "predigested" protein foods.

7138. Pancrelipase. Accelerase; Cotazym; Ilozyme; Ku-Zyme HP; Pancrease; Viokase. Porcine pancreatic enzyme concentrate containing principally lipase, with amylase and protease. Clinical pharmacology and comparison with pancreatin, *q.v.:* Y. W. Cho, D. M. Aviado, *J. Clin. Pharmacol.* **21**, 224 (1981). Clinical trial in cystic fibrosis: E. H. Mischler *et al., Am. J. Dis. Child.* **136**, 1060 (1982); in adult pancreatic insufficiency: S. K. Dutta, D. K. Tilley, *J. Clin. Gastroenterol.* **5**, 51 (1983).

THERAP CAT: Enzyme (digestive adjunct).

7139. Pancuronium Bromide. *1,1'-[(2β,3α,5α,16β,17β)-3,17-Bis(acetyloxy)androstane-2,16-diyl]bis[1-methylpiperidinium] dibromide; 1,1'-(3α,17β-dihydroxy-5α-androstan-2β,16β-ylene)bis[1-methylpiperidinium] dibromide diacetate; 3α,17β-diacetoxy-2β,16β-dipiperidino-5α-androstane dimethobromide; 2β,16β-dipiperidino-5α-androstane-3α,17β-diol diacetate dimethobromide;* poncuronium bromide (rescinded USAN); NA-97; Org-NA-97; Mioblock; Pavulon. $C_{35}H_{60}Br_2N_2O_4$; mol wt 732.68. C 57.38%, H 8.25%, Br 21.81%, N 3.82%, O 8.73%. Prepn: W. R. Buckett *et al., Chim. Ther.* **2**, 186 (1967); *eidem, J. Med. Chem.* **16**, 1116 (1973). Structural studies: Savage *et al., J. Chem. Soc. (B)* **1971**, 410. Structure-activity correlation: Waser, *Anaesthesist* **20**, 23 (1971). Pharmacology: W. R. Buckett, I. L. Bonta, *Fed. Proc.* **25**, 718 (1966); W. R. Buckett *et al., Brit.*

J. Pharmacol. Chemother. 32, 671 (1968); I. L. Bonta et al., Eur. J. Pharmacol. 4, 83, 303 (1968). Comparative study of neuromuscular blocking and vagolytic effect: S. L. Son et al., Anesthesiology 55, 12 (1981). Review: Speight, Avery, Drugs 4, 163-226 (1972).

Odorless crystals with bitter taste, mp 215°. One gram dissolves in 30 parts chloroform, one part water (20°). LD_{50} in mice (mg/kg): 0.047 i.v.; 0.152 i.p.; 0.167 s.c.; 21.9 orally; in rats, rabbits: 0.153, 0.016 i.v. (Buckett, 1968).

THERAP CAT: Muscle relaxant (skeletal).

7140. Pangamic Acid. Controversial mixture of compounds erroneously labeled as *vitamin B_{15}*. Allegedly isolated from apricot kernel: E. T. Krebs, Sr. et al., Int. Rec. Med. 164, 18 (1951). Originally named pangamic acid because of its supposed ubiquity in seeds. There is no clear chemical identity for pangamic acid. An historical review by J. C. Micheau et al., in Chim. Ther. 7, 103 (1972) contains structural studies, analyses, syntheses and discussion of various components of pangamic acid. Products sold as pangamic acid in the U.S. vary considerably in their composition. Some are mixtures of calcium gluconate and N,N-dimethylglycine, q.v.; others contain diisopropylamine dichloroacetate, q.v.: FDA Drug Bulletin vol. 8(6), Dec. 1978-Jan. 1979. Ref.: V. Herbert, Am. J. Clin. Nutr. 32, 1534 (1979); V. Herbert, R. Herbert, in Controversies in Nutrition, L. Ellenbogen, Ed. (Churchill-Livingstone, New York, 1981) pp 159-170.

7141. Panipenem. [5R-[3(S*),5α,6α(R*)]]-6-(1-Hydroxyethyl)-3-[[1-(1-iminoethyl)-3-pyrrolidinyl]thio]-7-oxo-1-azabicyclo[3.2.0]hept-2-ene-2-carboxylic acid; (5R,6S)-6-[(R)-1-hydroxyethyl]-2-[(S)-1-acetimidoylpyrrolidin-3-ylthio]-1-carbapen-2-em-3-carboxylic acid; (+)-(5R,6S)-3-[[(S)-1-acetimidoyl-3-pyrrolidinyl]thio]-6-[(R)-1-hydroxyethyl]-7-oxo-1-azabicyclo[3.2.0]hept-2-ene-2-carboxylic acid; CS-533; RS-533. $C_{15}H_{21}N_3O_4S$; mol wt 339.42. C 53.08%, H 6.24%, N 12.38%, O 18.86%, S 9.45%. Carbapenem antibiotic. Prepn: T. Miyadera et al., J. Antibiot. 36, 1034 (1983). Manufacturing process: A. Yoshida, K. Oda, Eur. pat. Appl. 587,436 (1994 to Sankyo). Antibacterial spectrum in vitro: H. C. Neu et al., Antimicrob. Ag. Chemother. 30, 828 (1986). Series of articles on pharmacology and clinical efficacy in combination with betamipron, q.v: Chemotherapy (Japan) 39 Suppl 3, 1-813 (1991). Toxicology: K. Kimura et al., ibid. 140.

Prepd as the hemihydrate; colorless fine prisms, mp 198-200° (dec). uv max (water): 298 nm (ε 10400). Approximate LD_{50} in male, female mice (mg/kg): 1700-2200, 1300-1700 i.v. (Kimura).

Mixture with betamipron, CS-976, Carbenin. THERAP CAT: Antibacterial.

7142. Pankrin. A proteolytic enzyme (proteinase, endopeptidase). Partial purification and characterization: Grant, Robbins, J. Am. Chem. Soc. 78, 5888 (1956). Isoln from hog or beef pancreas: eidem, U.S. pat. 2,871,165 (1959 to Armour).

Insol in water, sol in a 60% aq alcohol soln at 25°. Maintains proteolytic activity in aq solns of pH 4.7 at 25° for 8 hrs. Incapable of clotting blood plasma, but clots milk. The proteolytic activity upon urea-denatured hemoglobin and upon acetyl-1-tyrosine ethyl ester is such that the ratio of its hemoglobin to ATEE activity is at least 150.

7143. Pantetheine. 2,4-Dihydroxy-N-[3-[(2-mercaptoethyl)amino]-3-oxopropyl]-3,3-dimethylbutanamide; 2,4-dihydroxy-N-[2-[(2-mercaptoethyl)carbamoyl]ethyl]-3,3-dimethylbutyramide; α,γ-dihydroxy-β,β-dimethylbutyryl-β-alanyl-β-aminoethanethiol. $C_{11}H_{22}N_2O_4S$; mol wt 278.37. C 47.46%, H 7.97%, N 10.06%, O 22.99%, S 11.52%. Growth factor for Lactobacillus bulgaricus: Williams et al., J. Biol. Chem. 177, 933 (1949). Intermediate in the pathway of coenzyme A (q.v.) formation in mammalian liver and in some microorganisms. Synthesis: Schwyzer, Helv. Chim. Acta 35, 1903 (1952); Baddiley, Thain, J. Chem. Soc. 1952, 800; King et al., J. Am. Chem. Soc. 75, 1290 (1953); Walton et al., ibid. 76, 1146 (1954); Walton, U.S. pats. 2,744,119 and 2,835,704 (1956, 1958, both to Merck & Co.). Reviews: Snell, Brown, Advan. Enzymol. 14, 49 (1953); Snell, Wittle, Methods Enzymol. 3, 918 (1957).

Syrup or glass. $[\alpha]_D^{20}$ +12.9° (c = 4.5 in water). Microbiological activity: 20,000 LBF units/mg. An LBF unit is that amount of the growth factor contained in one mg of Basamine-Busch, a standard yeast extract manuf by Anheuser-Busch, Inc.

Silver mercaptide, $C_{11}H_{21}AgN_2O_4S$, yellow noncryst solid. $[\alpha]_D^{25}$ +8° (c = 4.27 in 0.9N NaCl). Very sol in water. 11,000 LBF units/mg.

Mercuric mercaptide, $C_{22}H_{42}HgN_4O_8S_2 \cdot C_3H_6O$, crystals from acetone, mp 96-98°. $[\alpha]_D^{27}$ +9.6° (c = 4 in water). 15,500 LBF units/mg. uv max: 260-265 nm (ε 1000). Sol in water, methanol.

S-Benzoylpantetheine, $C_{18}H_{26}N_2O_5S$, crystals from ethyl acetate, mp 116°. $[\alpha]_D^{27}$ +31° (ethanol). Sol in water, ethyl acetate, chloroform.

S-Acetylpantetheine, $C_{13}H_{24}N_2O_5S$, thick syrup. $[\alpha]_D^{27}$ +39° (c = 0.8 in ethanol).

7144. Pantethine. N,N'-[Dithiobis[2,1-ethanediylimino-(3-oxo-3,1-propanediyl)]]bis[2,4-dihydroxy-3,3-dimethylbutanamide]; N,N'-[dithiobis(ethyleneiminocarbonylethylene)]-bis(2,4-dihydroxy-3,3-dimethylbutyramide); D-bis(N-pantothenyl-β-aminoethyl) disulfide; Lipodel; Pantetina; Panthecin; Pantomin; Pantosin. $C_{22}H_{42}N_4O_8S_2$; mol wt 554.73. C 47.63%, H 7.63%, N 10.10%, O 23.07%, S 11.56%. [HOCH$_2$C(CH$_3$)$_2$CHOHCONHCH$_2$CH$_2$CONHCH$_2$CH$_2$S]$_2$. Disulfide dimer of pantetheine, q.v. Growth factor for Lactobacillus bulgaricus: Williams et al., J. Biol. Chem. 177, 933 (1949). Formed by oxidation of pantetheine: Brown, Snell, J. Biol. Chem. 198, 375 (1952). Structure: Snell et al., J. Am. Chem. Soc. 72, 5349 (1950). Synthesis: Wieland, Bokelmann, Naturwiss. 38, 384 (1951); Wittle et al., J. Am. Chem. Soc. 75, 1694 (1953); Viscontini et al., Helv. Chim. Acta 37, 375 (1954); Bowman, Cavalla, J. Chem. Soc. 1954, 1171; Shimizu et al., Chem. Pharm. Bull. 13, 180 (1965). Clinical trial in hyperlipoproteinemia: A. Gaddi et al., Atherosclerosis 50, 73 (1984).

Glassy, colorless to light yellow substance. $[\alpha]_D^{27}$ +13.5° (c = 3.75 in water). Freely sol in water; less sol in ethanol.

Practically insol in ether, acetone, ethyl acetate, benzene, chloroform.

THERAP CAT: Antihyperlipoproteinemic.

7145. Pantolactone. *Dihydro-3-hydroxy-4H-dimethyl-2(3H)-furanone;* pantoic acid γ-lactone; pantoyl lactone; pantoic lactone; 2,4-dihydroxy-3,3-dimethylbutyric acid γ-lactone; α-hydroxy-β,β-dimethyl-γ-butyrolactone. $C_6H_{10}O_3$; mol wt 130.14. C 55.37%, H 7.74%, O 36.88%. A degradation product of pantothenic acid from liver: Williams, Major, *Science* **91**, 246 (1940). Important intermediate in the synthesis of pantothenic acid. May be prepd by condensing isobutyraldehyde with formaldehyde yielding α,α-dimethyl-β-hydroxypropionaldehyde which is condensed with hydrocyanic acid in the presence of calcium chloride to form racemic pantolactone. Various modifications of this procedure exist: Glaser, *Monatsh.* **25**, 46 (1904); Stiller *et al., J. Am. Chem. Soc.* **62**, 1785 (1940); Reichstein, Grüssner, *Helv. Chim. Acta* **23**, 650 (1940); Carter, Ney, *J. Am. Chem. Soc.* **63**, 312 (1941). Vast patent literature, *e.g.,* Beckmann *et al.;* Klein, U.S. pats. **2,967,869** and **3,024,250** (1961, 1962, both to Nopco).

D(−)-Form, hygroscopic crystals from benzene + petr ether, mp 92°. $[\alpha]_D^{25}$ −50.7° (c = 2.05 in H_2O). Can be purified by microsublimation.

L(+)-Form, hygroscopic crystals from benzene, mp 91°. $[\alpha]_D^{25}$ +50.1° (c = 2 in H_2O).

DL-Form, hygroscopic rosettes or prisms, mp 80°, bp_{18} 130°. Freely sol in water. Sol in ether, benzene, chloroform, alcohol, carbon disulfide.

7146. Pantoprazole. *5-(Difluoromethoxy)-2-[[(3,4-dimethoxy-2-pyridinyl)methyl]sulfinyl]-1H-benzimidazole;* SKF-96022; BY-1023. $C_{16}H_{15}F_2N_3O_4S$; mol wt 383.38. C 50.13%, H 3.94%, F 9.91%, N 10.96%, O 16.69%, S 8.36%. Gastric proton pump inhibitor. Prepn: B. Kohl *et al., Eur. pat. Appl.* **166,287**; *eidem,* U.S. pat. **4,758,579** (1986, 1988 both to Byk Gulden); *eidem, J. Med. Chem.* **35**, 1049 (1992). Mechanism of action: W. A. Simon *et al., Biochem. Pharmacol.* **39**, 1799 (1990). HPLC determn in plasma and serum: R. Huber *et al., J. Chromatog.* **529**, 389 (1990). Clinical pharmacokinetics: B. Simon *et al., Z. Gastroenterol.* **28**, 443 (1990). Clinical trial in duodenal ulcer: P. Müller *et al., ibid.* **30**, 771 (1992).

Off-white solid, mp 139-140° (dec). pKa_1 3.92; pKa_2 8.19. Sodium salt, $C_{16}H_{14}F_2N_3NaO_4S$. White to off-white solid, dec > 130°. uv max (methanol): 289 (ϵ 1.64 × 10⁴).

THERAP CAT: Antiulcerative.

7147. Pantothenic Acid. *(R)-N-(2,4-Dihydroxy-3,3-dimethyl-1-oxobutyl)-β-alanine;* D(+)-N-(2,4-dihydroxy-3,3-dimethylbutyryl)-β-alanine; chick antidermatitis factor; vitamin B_5. $C_9H_{17}NO_5$; mol wt 219.24. C 49.31%, H 7.82%, N 6.39%, O 36.49%. A member of the B complex vitamins; essential vitamin for the biosynthesis of coenzyme A in mammalian cells. Occurs ubiquitously in all animal and plant tissue. The richest common source is liver, but jelly of the queen bee contains 6 times as much as liver. Rice bran and molasses are other good sources. Isoln from liver: R. J. Williams *et al., J. Am. Chem. Soc.* **60**, 2719 (1938). Synthesis: E. T. Stiller *et al., ibid.* **62**, 1785 (1940); Reichstein, Grüssner, *Helv. Chim. Acta* **23**, 650 (1940); Grüssner *et al., ibid.* 1276. Absolute configuration: Hill, Chan, *Biochem. Biophys. Res. Commun.* **38**, 181 (1970). Only the natural, dextrorotatory form has vitamin activity. Review of chem-

istry, biochemistry and pharmacology: W. Friedrich, *Vitamins* (de Gruyter, Berlin, 1988) pp 809-835; of metabolism: A. G. Tahiliani, C. J. Beinlich, *Vitamins Hormones* **46**, 165-228 (1991).

Unstable, viscous oil. Extremely hygroscopic. Easily destroyed by acids, bases, heat. $[\alpha]_D^{25}$ +37.5°. Freely sol in water, ethyl acetate, dioxane, glacial acetic acid; moderately sol in ether, amyl alcohol. Practically insol in benzene, chloroform.

Sodium salt, $C_9H_{16}NNaO_5$, very hygroscopic crystals. $[\alpha]_D^{25}$ +27.1° (c = 2). Can be stored in sealed ampuls only.

Calcium salt, $C_{18}H_{32}CaN_2O_{10}$, Calpanate, Pantholin. Prepn: Wehrmeister, U.S. pat. **2,780,645** (1957 to Commercial Solvents); Kagan, U.S. pat. **2,845,456** (1958 to Upjohn). Minute needles from CH_3OH. Sweetish taste with slightly bitter aftertaste. Dec 195-196°. Moderately hygroscopic. $[\alpha]_D^{20}$ +28.2° (c = 5). One gram dissolves in 2.8 ml H_2O. Sol in glycerol; slightly sol in alcohol, acetone. pH of 5% aq soln: 7.2-8.0; pH in CO_2-free water: 8.7.

THERAP CAT: Vitamin.

THERAP CAT (VET): Nutritional factor: dietary essential except in horses, ruminants.

7148. Papain. Papayotin; vegetable pepsin; Arbuz; Nematolyt; Summetrin; Tromasin; Velardon; Vermizym. First recognized member of the class of proteolytic enzymes that needs a free sulfhydryl group for activity. Isolated from the latex of the green fruit and leaves of *Carica papaya* L., *Caricaceae.* Initial isolation and crystallization: Balls *et al., Science* **86**, 379 (1937); Balls, Lineweaver, *J. Biol. Chem.* **130**, 669 (1939). Prepn from commercial dried papaya latex and physical properties: Kimmel, Smith, *ibid.* **207**, 515 (1954); *see also ibid.* 533-573; Becker, *Econ. Bot.* **12**, 62 (1958). Purification: Gibian, Bratfisch, U.S. pat. **2,950,227** (1960 to Schering AG); Lesuk, U.S. pat. **3,011,952** (1961 to Sterling Drug); Blumberg *et al., Eur. J. Biochem.* **15**, 97 (1970). The papain molecule consists of one folded polypeptide chain of 212 residues, mol wt ~23,400. Complete amino acid sequence: Drenth *et al., Nature* **218**, 929 (1968); Mitchel *et al., J. Biol. Chem.* **245**, 3485 (1970). Mechanism of action studies: Morihara, *ibid.* **62**, 250 (1967). Use in treatment of contact lenses to prolong wearing time in keratoconic patients with papillary conjunctivitis: D. R. Korb *et al., Arch. Ophthalmol.* **101**, 48 (1983). *Reviews:* Kimmel, Smith in *Advan. Enzymol. Relat. Subj. Biochem.* **19**, 267-334 (1957); Glazer, Smith in *The Enzymes* vol. **III**, P. D. Boyer, Ed. (Academic Press, New York, 3rd ed., 1971) pp 501-537; *see also* Drenth *et al., ibid.* 485-498 and *eidem, Advan. Protein Chem.* **25**, 79-115 (1971) for a comprehensive review of the structural elucidation.

White or grayish-white, slightly hygroscopic powder. uv max: 278 nm ($A_{1cm}^{1\%}$ 25.0). Incompletely sol in water, glycerol. Practically insol in most organic solvents. *Keep well closed.* Potency varies according to process of prepn, etc. with the usual grade digesting ~35 times its wt of lean meat. Best grades render sol 200-300 times their wt of coagulated egg albumin in alkaline media. A temp. range of 60-90° is favorable for the digestive process with 65° the optimum point. Best pH is 5.0, but it functions also in neutral or alkaline media. Activated by reduction (HCN, H_2S etc.) and inactivated by oxidation (H_2O_2, iodoacetate).

Note: The term papain is currently applied to both the crude dried latex and the crystalline proteolytic enzyme.

USE: For tenderizing meats; for clearing beverages; for bating skins.

THERAP CAT: Enzyme (proteolytic). Debriding agent; digestive aid. Has been used to prevent adhesions; as anthelmintic (Nematodes).

7149. Papaveraldine. *(6,7-Dimethoxy-1-isoquinolinyl)-(3,4-dimethoxyphenyl)methanone; 6,7-dimethoxy-1-isoquinolyl 3,4-dimethoxyphenyl ketone;* 6,7-dimethoxy-1-veratroyl-isoquinoline; xanthaline. $C_{20}H_{19}NO_5$; mol wt 353.37. C

67.98%, H 5.42%, N 3.96%, O 22.64%. The old name papaveraldine is retained to avoid confusion. From opium; whether papaveraldine occurs as such in the poppy plant, or is formed during the process of extraction, has not been investigated. Oxidation of papaverine to papaveraldine by SeO_2: Menon, *Proc. Indian Acad. Sci.* **19A**, 21 (1944). For older references *see:* Small, Lutz, "Chemistry of the Opium Alkaloids," *U.S. Public Health Reports,* **Suppl. No. 103,** Washington (1932). Synthesis from Reissert compds: Popp, McEwen, *J. Am. Chem. Soc.* **79**, 3773 (1957).

Crystals from abs ethanol, mp 208-209°. Sol in benzene, chloroform; slightly sol in alcohol, ether, petr ether; nearly insol in water, alkalies or carbonates.

Hydrochloride, $C_{20}H_{19}NO_5$.HCl, yellow crystals from abs alcohol, mp 200°.

7150. Papaveretum. "Concentrated opium"; Omnopon; Pantopon. A mixture of the hydrochlorides of the opium alkaloids in their approximate natural proportions. Contains approx 50% morphine, 3% codeine, 20% noscapine, and 5% papaverine. Exhibits biological action of morphine and other alkaloids present in opium. Clinical evaluation in intravenous analgesia: J. A. Catling *et al., Brit. Med. J.* **281**, 478 (1980).

Yellowish-gray, cryst powder. Freely sol in water.

Note: This is a controlled substance (opium derivative) listed in the U.S. Code of Federal Regulations, Title 21 part 1308.12 (1995).

THERAP CAT: Analgesic (narcotic).

7151. Papaverine. *1-[(3,4-Dimethoxyphenyl)methyl]-6,-7-dimethoxyisoquinoline; 6,7-dimethoxy-1-veratrylisoquinoline.* $C_{20}H_{21}NO_4$; mol wt 339.39. C 70.78%, H 6.24%, N 4.13%, O 18.86%. Smooth muscle relaxant found in opium (0.8-1.0%). Synthesis: Pictet, Gams, *Compt. Rend.* **149**, 210 (1909); *Ber.* **42**, 2943 (1909). Review of commercial syntheses: Goldberg, *Chem. Prod. Chem. News* **17**, 371 (1954). Improved synthetic procedures: Braz, Chizhov, *Soviet Pharmaceutical Research* **3**, 90-93 (New York, 1958). Biosynthetic studies: Battersby, Harper, *J. Chem. Soc.* **1962**, 3526; Brochmann-Hanssen *et al., J. Pharm. Sci.* **60**, 1672 (1971). Pharmacology and toxicology: Preininger in *The Alkaloids* **vol. 15**, R. H. F. Manske, Ed. (Academic Press, New York, 1975) pp 209-223. Toxicity: S. Levis *et al., Arch. Int. Pharmacodyn.* **123**, 264 (1960). Clinical effect on cerebral blood flow: H. L. Karpman, J. J. Sheppard, *Angiology* **26**, 592 (1975). Clinical evaluation in intermittent claudication: Y. Sheino *et al., ibid.* **34**, 257 (1983). Comprehensive description: M. S. Hifnawy, F. J. Muhtadi, *Anal. Profiles Drug Subs.* **17**, 367-447 (1988).

Triboluminescent, orthorhombic prisms from alcohol + ether, mp 147°. Sublimes at 135-140° at 11 mm pressure and 2 mm distance. d_4^{20} 1.337. pK (25°) 8.07. uv max (ethanol): 239, 278-280, 314, 327 nm (log ε 4.83, 3.86, 3.60, 3.67). Almost insol in water. Sol in hot benzene, glacial acetic acid, acetone; slightly sol in chloroform, carbon tetrachloride, petr ether. Optimal pH for storage of papaverine solns: 2.0-2.8.

Hydrochloride, $C_{20}H_{21}NO_4$.HCl, *Artegodan, Cepaverin, Cerebid, Cerespan, Dynovas, Optenyl, Pameion, Panergon, Papital T.R., Pavabid, Pavacap, Pavacen, Pavadel, Pavagen, Pavakey, Pavased, Spasmo-Nit, Therapav, Vasal, Vasospan.* Monoclinic rods from water, mp 220-225°. uv max (ethanol): 249-250, 280-282, 311 nm (log ε 4.69, 3.80, 3.82). One gram dissolves in about 40 ml water. Sol in alcohol and chloroform. Practically insol in ether. pH of 0.05 molar soln 3.9; pH of 2% aq soln 3.3. LD_{50} in mice, rats (mg/kg): 27.5, 20 i.v.; 150, 370 s.c. (Levis).

Nitrite, $C_{20}H_{21}NO_4$.HNO_2, light yellow, crystalline powder. Slightly soluble in water or alcohol; freely sol in chloroform, acetone.

THERAP CAT: Vasodilator (cerebral).

7152. Papaya. Papaw; Carica; melon tree. Fruit of *Carica papaya* L., *Caricaceae. Habit.* Tropical America and Asia, Florida. *Constit.* Papain, the dried and purified latex of the fruit; carpaine; carposide (a glucoside).

USE: Manufacture of papain.

7153. Papuamine. *[4aS-(4aR*,5aS*,10aS*,11aR*,-15aS*,15bR*,16E,18E,19aR*,19bS*)]-2,3,4,4a,5,5a,6,7,8,9,-10,10a,11,11a,12,13,14,15,15a,15b,19a,19b-Docosahydro-1H-diindeno[2,1-f:1′,2′-1][1,5]diazacyclotridecine.* $C_{25}H_{40}N_2$; mol wt 368.61. C 81.46%, H 10.94%, N 7.60%. Alkaloid originally isolated from *Haliclona sp.,* a marine sponge found in Papua, New Guinea. Exhibits antifungal and antimicrobial activity. Isoln and characterization: B. J. Baker *et al., J. Am. Chem. Soc.* **110**, 965 (1988). Total synthesis: R. M. Borzilleri *et al., ibid.* **116**, 9789 (1994).

White solid, mp 167.5-169°. $[\alpha]_D$ −150° (c = 1.5 in methanol).

Dihydrochloride, white solid, mp 230° (dec). $[\alpha]_D$ −140° (c = 1.3 in CH_3OH). uv max (methanol): 241 nm (ε 3000).

7154. Parabanic Acid. *Imidazolidinetrione;* imidazoletrione; oxalylurea. $C_3H_2N_2O_3$; mol wt 114.06. C 31.59%, H 1.77%, N 24.56%, O 42.08%. Prepd by the condensation of urea with diethyl oxalate in a methanol soln of sodium methoxide: Murray, *Org. Syn.* **37**, 71 (1957).

Crystals, mp about 230°; also stated as 243° with decomposition. Sublimes at 100°. Soluble in about 20 parts water, in alcohol. Its salts are unstable.

7155. Paraffin. Paraffin wax; hard paraffin. A mixture of solid hydrocarbons having the general formula $C_nH_{2n}+2$, obtained from petroleum.

Colorless or white, somewhat translucent, odorless mass; greasy feel; burns with a luminous flame. d about 0.90. mp 50-57°; also available with higher and lower melting ranges. Insol in water or alcohol. Sol in benzene, chloroform, ether, carbon disulfide, oils; miscible when melted with wax, spermaceti, fats.

USE: For raising mp of ointments. Manuf paraffin paper and candles (so-called wax paper or candles); for fixing drawings, etc., on muslin; water-proofing wood, cork, paper, leather; manuf varnishes; to render wooden vessels impermeable to water or alcohol; in lubricants; to cover food products; in floor polishes, cosmetics, electrical insulators; for extracting perfumes from flowers. Pharmaceutic aid (stiffening agent).

7156. Paraffin Chlorinated. Chlorcosane; Cereclor. Prepared by chlorinating a liquid paraffin. Contains about 50% Cl.

Light yellow to amber, thick, oily liq; odorless and stable in air. d 1.00-1.07. Insol in water. Slightly sol in alc; miscible with benzene, chloroform, ether, carbon tetrachloride.

USE: As solvent for dichloramine-T, dissolving about 8%.

7157. Paraflutizide. *6-Chloro-3-[(4-Fluorophenyl)-methyl]-3,4-dihydro-2H-1,2,4-benzothiadiazine-7-sulfon-amide 1,1-dioxide;* 1,1-dioxo-3-(*p*-fluorophenylmethyl)-3,4-dihydro-6-chloro-1,2,4-benzothiadiazine-7-sulfonamide; LD-3612. $C_{14}H_{13}ClFN_3O_4S_2$; mol wt 405.86. C 41.43%, H 3.23%, Cl 8.74%, F 4.68%, N 10.35%, O 15.77%, S 15.80%. Prepn: **Belg. pat. 620,829** (1962 to Lab. Dausse), *C.A.* **58**, 12585g (1963), corresponds to **Brit. pat. 961,641; Fr. pat. M1374** (1962 to Lab. Dausse), *C.A.* **58**, 2462c (1963). Synthesis: J. Klosa, *J. Prakt. Chem.* **33**, 298 (1966).

Crystals from 50% alcohol, mp 238-240°.

THERAP CAT: Diuretic.

7158. Paraformaldehyde. Polyoxymethylene; Para form; Formagene. Also erroneously referred to as *Triformol* or as *"trioxymethylene"*. Polymerized formaldehyde. $(CH_2O)_n$. Obtained by concentrating formaldehyde soln. Use in mummifying dental pulp: I. Curson, *Brit. Dent. J.* **121**, 519 (1966); P. Hobson, *ibid.* **128**, 275 (1970).

White, cryst powder, having an odor of formaldehyde. Slowly sol in cold, more readily in hot water, with evolution of formaldehyde; insol in alcohol, ether; sol in fixed alkali hydroxide solns. *Keep tightly closed.*

USE: For disinfecting sickrooms, clothing, linen, and sickroom utensils. Active ingredient of contraceptive creams. Also used as fumigant; in dentistry; in manuf synthetic resins and artificial horn or ivory.

7159. Paraherquamide. (−)-Paraherquamide; [1′R-(1′α,5′aβ,7′β,8′aβ,9′aβ)]-2′,3′,8′a,9′-tetrahydro-1′-hydroxy-1′,4,4,8′,8′,11′-hexamethylspiro[4H,8H-[1,4]dioxepino[2,3-g]indole-8,7′(8′H)-[5H,6H-5a,9a](iminomethano)[1H]cyclopent[f]indolizine]-9,10′(10H)-dione. $C_{28}H_{35}N_3O_5$; mol wt 493.60. C 68.13%, H 7.15%, N 8.51%, O 16.21%. Oxindole alkaloid fungal metabolite. Isolation from *Penicillium paraherquei* and structure: M. Yamazaki *et al.*, *Maikotokishin* **10**, 27 (1980), *C.A.* **95**, 19321p (1981); M. Yamazaki *et al.*, *Tetrahedron Letters* **22**, 135 (1981); from *P. charlesii* and nematocidal activity: J. G. Ondeyka *et al.*, *J. Antibiot.* **43**, 1375 (1990). Absolute stereochemistry: T. A. Blizzard *et al.*, *J. Org. Chem.* **54**, 2657 (1989). Approach to synthesis: R. M. Williams, T. D. Cushing, *Tetrahedron Letters* **31**, 6325 (1990). Anthelmintic activity in sheep: W. L. Shoop *et al.*, *J. Parasitol.* **76**, 349 (1990); in dogs: W. L. Shoop *et al.*, *Vet. Parasitol.* **40**, 339 (1991). Toxicity: *eidem, Am. J.*

Vet. Res. **53**, 2032 (1992). Mode of action: J. M. Schaeffer *et al.*, *Biochem. Pharmacol.* **43**, 679 (1992).

Colorless prisms, mp 244-247° (dec). $[\alpha]_D^{22}$ −28° (c = 0.43 in methanol). uv max (methanol): 226, 260, 290nm (ε 32400, 6100, 1600). Sol in methanol, ethyl acetate, acetone, DMSO. Practically insol in water.

THERAP CAT (VET): Anthelmintic.

7160. Paraldehyde. *2,4,6-Trimethyl-1,3,5-trioxane;* paracetaldehyde; Paral. $C_6H_{12}O_3$; mol wt 132.16. C 54.53%, H 9.15%, O 36.32%. A polymer of acetaldehyde. Prepd by the polymerization of acetaldehyde catalyzed by HCl and H_2SO_4 at medium to high temp: Kekulé, Zincke, *Ann.* **162**, 125 (1872); Baer, Mahan, U.S. pat. **2,864,827** (1958 to Phillips). Toxicity data: Figot *et al.*, *Acta Pharmacol. Toxicol.* **8**, 290 (1952).

Liquid, characteristic aromatic odor and warm, but disagreeable taste. d_{25}^{25} ~0.994. bp ~124°. mp 12°. n_D^{20} 1.4049. Sol in 8 parts water at 25°, in 17 parts boiling water; miscible with alc, chloroform, ether, oils. Gives acetaldehyde on heating with dil HCl or on warming with several drops concd H_2SO_4. LD_{50} orally in rats: 1.65 g/kg (Figot).

Caution: May be habit forming. This is a controlled substance (depressant) listed in the U.S. Code of Federal Regulations, Title 21 Parts 329.1 and 1308.14 (1995).

USE: Manuf organic compounds.

THERAP CAT: Sedative; hypnotic.

7161. Paramethadione. *5-Ethyl-3,5-dimethyl-2,4-oxazolidinedione; 3,5-dimethyl-5-ethyloxazolidine-2,4-dione;* Paradione. $C_7H_{11}NO_3$; mol wt 157.17. C 53.49%, H 7.05%, N 8.91%, O 30.54%. Prepn: Spielman, U.S. pat. **2,575,693** (1951 to Abbott).

Liquid. Fruity, esterlike odor. d_4^{25} 1.1180-1.1240. n_D^{25} 1.449. Slightly sol in water. Freely sol in alcohol, benzene, chloroform, ether.

THERAP CAT: Anticonvulsant.

7162. Paramethasone. *(6α,11β,16α)-6-Fluoro-11,17,21-trihydroxy-16-methylpregna-1,4-diene-3,20-dione;* 6α-fluoro-16α-methylprednisolone; 16α-methyl-6α-fluoropredniso-lone. $C_{22}H_{29}FO_5$; mol wt 392.47. C 67.33%, H 7.45%, F 4.84%, O 20.38%. Prepn: Edwards *et al.*, *J. Am. Chem. Soc.* **82**, 2318 (1960). Toxicity study: E. I. Goldenthal, *Toxicol. Appl. Pharmacol.* **18**, 185 (1971).

21-Acetate, $C_{24}H_{31}FO_6$, *Cortidene, Dilar, Dillar, Haldrate, Haldrone, Metilar, Monocortin, Paramezone, Syntecort, Stemex.* mp 228-241° (dec). $[\alpha]_D$ +85°. uv max (ethanol): 243 nm (log ε 4.16). Sol in ethanol, acetone; slightly sol in water. LD_{50} i.p. in female rats: 392±23 mg/kg (Goldenthal).

Disodium phosphate, *Soludillar.*

Mixture of 21-acetate and disodium phosphate, *Triniol.*
THERAP CAT: Glucocorticoid.

7163. Paranyline. *4-(9H-Fluoren-9-ylidenemethyl)benz-*
enecarboximidamide; α-fluoren-9-ylidene-p-toluamidine; 9-
(p-guanylbenzylidene)fluorene; 9-(p-guanylbenzal)fluorene;
MER-27. $C_{21}H_{16}N_2$; mol wt 296.37. C 85.11%, H 5.44%, N
9.45%. Prepn: Allen *et al., J. Am. Chem. Soc.* **80**, 591
(1958); Van Campen *et al.,* U.S. pat. **2,877,269** (1959 to
Wm. S. Merrell).

Hydrochloride, $C_{21}H_{16}N_2.HCl$, crystals from methanol +
isopropanol, mp 308°.
THERAP CAT: Anti-inflammatory.

7164. Paraoxon. *Phosphoric acid diethyl 4-nitrophenyl*
ester; diethyl *p*-nitrophenyl phosphate; phosphacol; E-600;
Ester 25; Eticol; Fosfakol; Mintacol; Miotisal A; Soluglaucit. $C_{10}H_{14}NO_6P$; mol wt 275.20. C 43.64%, H 5.13%, N
5.09%, O 34.88%, P 11.26%. Prepd by the action of diethyl
chlorophosphate on sodium *p*-nitrophenolate or by nitration
of diethyl phenyl phosphate: Schrader, *B.I.O.S. Final Rept.*
714, 52 (1947); *Angew. Chem.* **62**, 471 (1950); **Swiss** pat.
257,649 (1949); Fagerlind *et al., Svensk Farm. Tid.* **56**, 303,
C.A. **46**, 9259. Physical properties: E. F. Williams, *Ind.
Eng. Chem.* **43**, 950 (1951). Formation from parathion on
citrus foliage and soil surface: R. C. Spear *et al., J. Agric.
Food Chem.* **23**, 808 (1975); W. F. Spencer *et al., Bull. Environ. Contam. Toxicol.* **14**, 265 (1975). Toxicity study: W.
R. Pickering, J. C. Malone, *Biochem. Pharmacol.* **16**, 1183
(1967). Brief review: G. Schrader, *Die Entwicklung neuer
insektizider Phosphorsäure-Ester* (Verlag Chemie, Weinheim,
1963) pp 259-272.

Oily liq. Slight odor. *Poisonous!* $bp_{1.0}$ 169-170°. d_4^{25}
1.2683. n_D^{20} 1.50959. uv max: 274 nm (ϵ 8.9×10^3). Soly in
water (25°): 2400 μg/mL. Freely sol in ether, other organic
solvents. Aq solns are stable up to pH 7. Roughly 300
times more stable to hydrolysis than tetraethyl pyrophosphate. The uncatalyzed reaction with water is very small
and the overall velocity constant is K = 0.52[OH] +
1×10^{-6} min^{-1}. See Coates, *Ann. Appl. Biol.* **36**, 158 (1949).
LD_{50} orally in rats: 1.8 mg/kg (Pickering, Malone).
Caution: Cholinesterase inhibitor. For symptoms *see*
Parathion.
USE: Insecticide.

7165. Paraquat. *1,1′-Dimethyl-4,4′-bipyridinium; N,N′-*
dimethyl-γ,γ′-dipyridylium; methyl viologen (2+). [$C_{12}H_{14}$-
N_2]$^{2+}$. Non-selective contact herbicide. Prepn of dichloride
and bismethyl sulfate derivs: L. Michaelis, E. S. Hill, *J.
Am. Chem. Soc.* **55**, 1481 (1933); R. C. Brian *et al.,* **Brit.**
pat. **813,531** (1959 to ICI). Activity: R. F. Homer *et al., J.
Sci. Food Agr.* **11**, 309 (1960); A. D. Dodge, *Endeavor* **30**,
130 (1971). Degradation: A. Calderbank, P. Slade, *Outlook
Agr.* **5**, 55 (1966); A. Calderbank, T. E. Tomlinson, *ibid.* 252
(1968); A. Calderbank, *ibid.* **6**, 128 (1970). Toxicity studies:
D. G. Clark *et al., Brit. J. Ind. Med.* **23**, 126 (1966); D. M.
Conning *et al., Brit. Med. Bull.* **25**, 245 (1969); R. D. Kimbrough, T. B. Gaines, *Toxicol. Appl. Pharmacol.* **17**, 679
(1970); J. F. Dasta, *Am. J. Hosp. Pharm.* **35**, 1368 (1978).

Controversial use on marijuana plants: R. J. Smith, *Science*
199, 861 (1978). *Review:* A. A. Akhavein, D. L. Linscott,
Residue Rev. **23**, 97-145 (1968); A. Calderbank, P. Slade in
Herbicides: Chemistry, Degradation and Mode of Action, P.
C. Kearney, D. Kaufman, Eds. (Dekker, New York, 2nd
ed., 1976) pp 501-540.

Dichloride, $C_{12}H_{14}Cl_2N_2$, *PP-148, Gramoxone.* Colorless
crystals, mp 300° (dec). Very sol in water, slightly sol in
lower alcohols. Insol in hydrocarbons. Hydrolyzed by
alkali. Inactivated by inert clays and anionic surfactants.
Corrosive to metal. Non-volatile. Normal potential at 30°:
−0.446 volts. LD_{50} orally in rats: 125 mg/kg (Conning).
Bismethyl sulfate, $C_{14}H_{20}N_2O_8S_2$, *Paraquat I, PP-910.*
Yellow solid. LD_{50} orally in male rats: 100 mg/kg (Kimbrough, Gaines).
Caution: Poison! Causes ulceration of digestive tract,
diarrhea, vomiting, renal damage, jaundice. Edema, hemorrhage, fibrosis of lung, and death from anoxia may result
(Dasta).
USE: Herbicide. Dichloride as biological oxidation-reduction indicator.

7166. Parasorbic Acid. *(S)-5,6-Dihydro-6-methyl-2H-*
pyran-2-one; 5-hydroxy-2-hexenoic acid lactone; δ-Δ^{α,β}-hex-
enolactone; 2-hexen-5,1-olide; sorbic oil. $C_6H_8O_2$; mol wt
112.13. C 64.27%, H 7.19%, O 28.54%. The sole constituent of "Vogelbeeröl," an oil obtained by steam distillation of
the acidified juice of the ripe berries of the mountain ash,
Sorbus aucuparia L., Rosaceae: Hofmann, *Ann.* **110**, 129
(1859); Doebner, *Ber.* **27**, 344 (1894); Kuhn, Jerchel, *Ber.*
76, 413 (1943). Structure: *eidem, ibid.* Synthesis: Haynes,
Jones, *J. Chem. Soc.* **1946**, 954; Lamberti *et al., Rec. Trav.
Chim. Pays-Bas* **86**, 504 (1967). Pharmacology and acute
toxicity: H. J. Meyer, R. Kretzschmar, *Arzneimittel-Forsch.*
19, 617 (1969).

Oily liquid, sweet aromatic odor. bp_{14} 104-105°; bp_{22} 119-
123°. n_D^{25} 1.4682. d_4^{18} 1.079. [α]$_D^{18}$ +49.3°; [α]$_D^{19}$ +210° (c = 2
in alc). Soluble in water; freely sol in alcohol, ether. Aq
solns are neutral and turn acid on storage. LD_{50} in mice
(mg/kg): 420 ±6.3 i.p.; 195 ±13.6 i.v. (Meyer, Kretzschmar).

7167. Parathion. *Phosphorothioic acid O,O-diethyl O-*
(4-nitrophenyl) ester; O,O-diethyl O-p-nitrophenyl phosphorothioate; diethyl-*p*-nitrophenyl monothiophosphate;
DNTP; S.N.P.; E-605; AC-3422; ENT-15108; Alkron; Alleron; Aphamite; Etilon; Folidol; Fosferno; Niran; Paraphos;
Rhodiatox; Thiophos. $C_{10}H_{14}NO_5PS$; mol wt 291.26. C
41.24%, H 4.84%, N 4.81%, O 27.47%, P 10.63%, S 11.01%.
Non-systemic contact and stomach insecticide and acaricide; cholinesterase inhibitor. Original prepn: Thurston,
FIAT Report 949 (1946); Coates, Topley, *BIOS Final Report
1808* (1947). See also Fletcher *et al., J. Am. Chem. Soc.* **70**,
3943 (1948). Conversion to toxic oxygen analogs: *See*
Paraoxon. Toxicity study: T. B. Gaines, *Toxicol. Appl.
Pharmacol.* **14**, 515 (1969). *Review:* Hall, *Advances in
Chemistry Series* **1**, 150 (1950). Review of industrial syntheses: Chadwick, Watt, "Thiophosphates" in *Phosphorus
and its Compounds* vol. 2, J. R. Van Wazer, Ed. (Interscience, New York, 1961) pp 1257-1262. Review of distribution, transport and fate in the environment: M. S.
Mulla *et al., Residue Rev.* **81**, 1-159 (1981); of carcinogenic
risk: *IARC Monographs* **30**, 153-181 (1983).

Pale yellow liquid. bp_{760} 375°; $bp_{0.6}$ 157-162°. mp 6°. n_D^{25} 1.5370. d_4^{25} 1.26. Vapor pressure at 20°: 3.78×10^{-5} mm Hg. Surface tension at 25°: 39.2 dynes/cm. Viscosity at 25°: 15.30 cp. Absorption spectra: Williams, *Ind. Eng. Chem.* **43**, 950 (1951). Freely sol in alcohols, esters, ethers, ketones, aromatic hydrocarbons. Practically insol in water (20 ppm), petr ether, kerosene, and the usual spray oils. Incompatible with substances having a pH higher than 7.5. LD_{50} in female, male rats (mg/kg): 3.6, 13 orally; 6.8, 21 dermally (Gaines).

Caution: Potential symptoms of overexposure are miosis; rhinorrhea; headache; tight chest, wheezing, laryngeal spasm, salivation and cyanosis; anorexia, nausea, vomiting, abdominal cramps and diarrhea; sweating; muscle fasiculation, weakness and paralysis; giddiness, confusion and ataxia; convulsions, coma; low blood pressure; cardiac irregularities; skin and eye irritation. *See NIOSH Pocket Guide to Chemical Hazards* (DHHS/NIOSH 90-117, 1990) p 172. *See also Clinical Toxicology of Commercial Products,* R. E. Gosselin *et al.,* Eds. (Williams & Wilkins, Baltimore, 4th ed., 1976) Section III, pp 263-271.

USE: Insecticide; acaricide.

7168. Parathyroid Hormone. *Parathormone;* PTH. Mol wt about 9,500. Regulatory factor in the homeostatic control of calcium and phosphate metabolism, its principal sites of activity being the skeleton, kidneys, and gastrointestinal tract. Prime function is to raise plasma calcium concns. Acts synergistically with vitamin D_3 *(q.v.)* except in the kidneys where the latter causes phosphate retention. Secretion from the parathyroid gland varies inversely with serum Ca^{2+} concentrations, unlike calcitonin *(q.v.)* which is secreted in direct proportion to serum calcium levels. Structure consists of a single-chain polypeptide of 84 amino acid residues. Sequence varies slightly among mammalian species. Sequence of bovine PTH: Niall *et al., Z. Physiol. Chem.* **351**, 1586 (1970); Brewer, Ronan, *Proc. Nat. Acad. Sci. USA* **67**, 1862 (1970). Sequence of porcine PTH: O'Riordan *et al., Proc. Roy. Soc. Med.* **64**, 1263 (1971). Isoln of human PTH from parathyroid adenomas: O'Riordan *et al., Endocrinology* **89**, 234 (1971). Fragment exhibiting full biological activity consists of about 35 amino acid residues from the *N*-terminal: Potts *et al.* in *Parathyroid Hormone and Thyrocalcitonin (Calcitonin),* R. V. Talmage, L. F. Belanger, Eds. (Excerpta Medica, New York, 1968) p 44; *see* in entirety for review and special studies. Synthesis of active bovine fragment: Potts *et al., Proc. Nat. Acad. Sci. USA* **68**, 63 (1971); of human PTH (1-38): S. Funakoshi *et al., Peptide Chem.* **18**, 223 (1980). Reviews of early literature: Potts *et al., Recent Progr. Horm. Res.* **22**, 101 (1966); Arnaud *et al., Ann. Rev. Physiol.* **29**, 349 (1967). *Reviews:* Behrens, Grinnan, *Ann. Rev. Biochem.* **38**, 83 (1969); Auerbach *et al., Recent Progr. Horm. Res.* **28**, 353 (1972); Parsons, Potts, "Physiology and Chemistry of Parathyroid Hormone" in *Clinics in Endocrinology and Metabolism,* I. MacIntyre, Ed. (Saunders, Philadelphia, 1972) pp 33-78. Biosynthetic review: J. F. Habener *et al., Recent Progr. Horm. Res.* **33**, 249 (1977).

Note: Aqueous solns of the active principles of bovine parathyroid gland are in use under the names: *Parathorm, Para-thor-mone, Paroidin.*

THERAP CAT: Blood calcium regulator.

7169. Parbendazole. *(5-Butyl-1H-benzimidazol-2-yl)-carbamic acid methyl ester;* methyl 5-butyl-2-benzimidazolecarbamate; 5-butyl-2-(carbomethoxyamino)benzimidazole; SKF-29044; Helmatac; Verminum; Worm Guard. $C_{13}H_{17}N_3O_2$; mol wt 247.30. C 63.14%, H 6.93%, N 16.99%, O 12.94%. Prepd from 4-butyl-*o*-phenylenediamine and carbomethoxycyanamide: Actor *et al., Nature* **215**, 321 (1967); Brit. pat. **1,123,317** and Stedman; Actor, Pagano, U.S. pats. **3,480,642; 3,574,845** (1968, 1969, 1971 all to SKF). Activ-

ity: Ostmann, Scheidy, *Progr. Antimicrob. Anticancer Chemother., Proc. Int. Congr. Chemother., 6th, Tokyo, 1969,* **1**, 159 (1970). Identification of metabolites: Dunn *et al., J. Med. Chem.* **16**, 996 (1973).

Crystals from aq ethanol, mp 225-227° (dec). uv max (95% ethanol/1N HCl): 282, 288 nm (ϵ 16200, 20000). Practically insol in water. LD_{50} in mice, rats (g/kg): >4 orally, both species (Actor).

THERAP CAT (VET): Anthelmintic.

7170. Pareira. Pareira brava. Dried root of *Chondodendron platiphyllum* (A. St. Hil.) Miers, *C. microphyllum* (Eichl.) Mold., and *C. tomentosum* Ruiz et Pavon, *Menispermaceae. Habit.* Brazil. *Constit.* Bebeerine, chondrodine, fatty acids, tannin.

THERAP CAT: Diuretic, anti-infective (urinary).

7171. Parethoxycaine. *4-Ethoxybenzoic acid 2-(diethylamino)ethyl ester;* β-diethylaminoethyl *p*-ethoxybenzoate; diethoxin. $C_{15}H_{23}NO_3$; mol wt 265.35. C 67.90%, H 8.74%, N 5.28%, O 18.09%. Prepn: Christiansen, Harris, U.S. pat. **2,404,691** (1946 to Squibb).

Hydrochloride, $C_{15}H_{23}NO_3 \cdot HCl$, *Maxicaine.* Crystals, mp 172.5-173.5°.

THERAP CAT: Anesthetic (local).

7172. Pargyline. *N-Methyl-N-2-propynylbenzenemethanamine; N-methyl-N-2-propynylbenzylamine; N*-benzyl-*N*-methyl-2-propynylamine; *N*-methyl-*N*-propargylbenzylamine; MO-911; A-19120; Eudatin; Supirdyl. $C_{11}H_{13}N$; mol wt 159.23. C 82.97%, H 8.23%, N 8.80%. Monoamine oxidase inhibitor. Prepd from propargyl bromide and benzylmethylamine: Brit. pat. **906,245**; Martin, U.S. pat. **3,155,-584** (1962 and 1964, both to Abbott). Activity as a glucuronyl transferase inducer: Yeh, Mitchell, *Experientia* **28**, 298 (1972).

Free base, bp_{11} 96-97°.

Hydrochloride, $C_{11}H_{13}N \cdot HCl$, *Eutonyl.* Crystals from ethanol + ether, mp 154-155°. Readily sol in water. Aq solns are unstable.

THERAP CAT: Antihypertensive.

7173. Paromomycin. *O-2-Amino-2-deoxy-α-D-glucopyranosyl-(1→4)-O-[O-2,6-diamino-2,6-dideoxy-β-L-idopyranosyl-(1→3)-β-D-ribofuranosyl-(1→5)]-2-deoxy-D-streptamine; O-2,6-diamino-2,6-dideoxy-β-L-idopyranosyl-(1→3)-O-β-D-ribofuranosyl-(1→5)-O-[2-amino-2-deoxy-α-D-glucopyranosyl-(1→4)]-2-deoxystreptamine;* paromomycin I; amminosidin; catenulin; crestomycin; estomycin; hydroxymycin; monomycin A; neomycin E; paucimycin; R-400. $C_{23}H_{45}N_5O_{14}$; mol wt 615.64. C 44.87%, H 7.37%, N 11.38%, O 36.38%. Oligosaccharide-type antibiotic isolated from various *Streptomyces.* From *S. rimosus* forma *paromomycinus:* Frohardt *et al.,* U.S. pat. **2,916,485** (1959 to Parke, Davis); from *S. catenulae:* Davisson, Finlay, U.S. pat. **2,895,876** (1959 to Pfizer); from *S. chrestomyceticus:* Canevazzi, Scotti, *Giorn. Microbiol.* **7**, 242 (1959). Arcamone *et al., ibid.* 251. Identity of paromomycin, catenulin, hydroxymycin and aminosidine: Schillings, Schaffner, *Antimicrob.*

Ag. Chemother. **1961,** 274. Structure: Haskell *et al., J. Am. Chem. Soc.* **81,** 3480, 3482 (1959); Rinehart *et al., ibid.* **84,** 3218 (1962); Hichens, Rinehart, ibid. **85,** 1547 (1963). Probable identity with *zygomycin A*: Horii, *J. Antibiot.* **15A,** 187 (1962). Identity with monomycin A: Konstantinova, Brazhnikova, *Antibiotiki* **10**(1), 34 (1965); with neomycin E: Hessler *et al., J. Antibiot.* **23,** 464 (1970). Toxicity data: A. DiMarco, C. Bertazzoli, *Antibiot. & Chemother.* **11,** 2 (1963). Review of antimicrobial activity: G. L. Coffey *et al., ibid.* **9,** 730 (1959). Review of pharmacology: Gasparini, Pignatelli, *Veterinaria (Milan)* **21,** 7 (1972).

Amorphous white powder. $[\alpha]_D^{25}$ +65° ±3°. Soluble in water; moderately sol in methanol; sparingly sol in abs ethanol. LD_{50} in rats, mice (mg/kg): >1625, >2275 orally; >650, 423 s.c.; 156, 90 i.v. (Coffey).

Sulfate, $C_{23}H_{45}N_5O_{14}.H_2SO_4$, *1600 Antibiotic, FI-5853, Aminoxidin, Aminosidine, Farmiglucin, Farminosidin, Gabbromicina, Gabbromycin, Gabbroral, Humagel, Humatin, Pargonyl, Paramicina, Paricina, Sinosid.* $[\alpha]_D^{25}$ +50.5° (c = 1.5 in water pH 6). LD_{50} in mice (mg/kg): ~15,000 orally; 700 s.c.; 110 i.v. (Di Marco, Bertazzoli).

THERAP CAT: Antibacterial; antiamebic.

THERAP CAT (VET): Antiamebic.

7174. Parotin. A salivary gland hormone. It is a protein of globulin nature having an isoelectric point of pH 5.7. Produced by the parotid gland. Structure studies: Ito *et al., Endocrinol. Japan* **12,** 249 (1966); Shimasaki *et al., ibid.* **14,** 11 (1967). Generally acts on the mesenchymal tissues, esp the hard and connective tissues, to promote their development and growth. It also has a protein-anabolic function. Shows hypocalcemic and leukocytosis-promoting activities: Ito *et al., ibid.* **12,** 298 (1966). *Reviews:* Yosoji Ito, *J. Japan. Biochem. Soc.* **25,** 143-164 (1953); *Ann. N. Y. Acad. Sci.* **85,** art. 1, 228-310 (1960).

7175. Paroxetine. *(3S-trans)-3-[(1,3-Benzodioxol-5-yloxy)methyl]-4-(4-fluorophenyl)piperidine;* (−)-*trans*-4-(*p*-fluorophenyl)-3-[[3,4-(methylenedioxy)phenoxy]methyl]-piperidine; FG-7051; BRL-29060; Aropax; Paxil; Seroxat. $C_{19}H_{20}FNO_3$; mol wt 329.37. C 69.29%, H 6.12%, F 5.77%, N 4.25%, O 14.57%. Serotonin (5-HT) uptake inhibitor. Prepn: J. A. Christensen, R. F. Squires, **Ger.** pat. **2,404,-113;** *eidem,* **U.S.** pats. **3,912,743; 4,007,196** (1974, 1975, 1977 all to Ferrosan); of hydrochloride: R. D. B. Barnes *et al.,* **Eur.** pat. **Appl. 223,403;** U.S. pat. **4,721,723** (1987, 1988 both to Beecham). Characterization of serotonin inhibition: J. Buus Lassen, *Eur. J. Pharmacol.* **47,** 351 (1978). Binding to serotonin transporter complex: E. Habert *et al., ibid.* **118,** 107 (1985); in comparison with imipramine, *q.v.*: S. Z. Langer *et al., J. Recept. Res.* **7,** 499 (1987). Pharmacokinetics: J. Lund *et al., Acta Pharmacol. Toxicol.* **44,** 289 (1979); J. Lund *et al., ibid.* **51,** 351 (1982). HPLC determn in plasma: M. A. Brett *et al., J. Chromatog.* **419,** 438 (1987). Clinical pharmacology: S. M. Hassan *et al., Brit. J. Clin. Pharmacol.* **19,** 705 (1985). Comparison with amitriptyline, *q.v.*, in treatment of depression: A. L. Laursen *et al., Acta Psychiat. Scand.* **71,** 249 (1985); R. Battegay *et al., Neuropsychobiology* **13,** 31 (1985). Review of pharmacology and therapeutic use: K. L. Dechant, S. P. Clissold, *Drugs* **41,** 225-253 (1991).

Hydrochloride, $C_{19}H_{20}FNO_3.HCl$, crystal platelets, mp 118°.

Hydrochloride hemihydrate, $C_{19}H_{20}FNO_3.HCl.\frac{1}{2}H_2O$, crystals, mp 129-131°.

Maleate, crystals from ethanol-ether, mp 136-138°. $[\alpha]_D$ −87° (c = 5 in ethanol). LD_{50} in mice (mg/kg): 845 s.c.; 500 orally (Christensen, Squires, 1977).

THERAP CAT: Antidepressant.

7176. Paroxypropione. *1-(4-Hydroxyphenyl)-1-propanone; 4'-hydroxypropiophenone;* p*-hydroxypropiophenone; paraoxypropiophenone;* p*-propionylphenol; ethyl* p*-hydroxyphenyl ketone;* P.O.P.; B-360; H-365; Profenone; Frenantol; Frenohypon; Paroxon; Possipione; Hypostat. C_9H_{10}-O_2; mol wt 150.18. C 71.98%, H 6.71%, O 21.31%. Prepn: Perkin, *J. Chem. Soc.* **55,** 546 (1889); Goldzweig, Kaiser, *J. Prakt. Chem.* [2] **43,** 86 (1891); Cox, *J. Am. Chem. Soc.* **49,** 1028 (1927); Hartung *et al., ibid.* **53,** 4153 (1931); Farinholt, ibid. **55,** 3386 (1933); Miller, Hartung, *Org. Syn.* **coll. vol. II** (1943) p 543. Derivatives: Buu-Hoi, *Rec. Trav. Chim.* **68,** 759 (1949).

Needles or prisms from water. mp 149°. One part dissolves in 2896 parts of water at 15°, in 30 parts at 100°. Freely sol in alcohol or ether.

THERAP CAT: Pituitary gonadotropic hormone inhibitor.

7177. Parsalmide. *5-Amino-N-butyl-2-(2-propynyloxy)-benzamide;* 5-amino-*O*-propynyl-*N*-butylsalicylamide; 2-propargyloxy-5-amino-*N*-butylbenzamide; MY-41-6; Sinovial; Parsal. $C_{14}H_{18}N_2O_2$; mol wt 246.31. C 68.27%, H 7.37%, N 11.37%, O 12.99%. Prepn: B. Gradnik *et al.,* **Ger.** pat. **2,029,991;** *eidem,* **U.S.** pat. **3,739,030** (1970, 1973 both to Soc. Etude Recher. Appl. Sci. Med.). Synthesis and pharmacology: A. Pedrazzoli *et al., Boll. Chim. Farm.* **115,** 125 (1976). Effect on gastric secretion in laboratory animals: G. Bertaccini *et al., Farmaco Ed. Prat.* **34,** 482 (1979). Inhibition of platelet aggregation: R. Fantasia *et al., Arzneimittel-Forsch.* **32,** 1312 (1982). Clinical comparison with phenylbutazone, *q.v.*, in management of joint pain: A. Bajardi, S. Fantato, *Minerva Med.* **67,** 3371 (1976); S. Menci *et al., ibid.* 3403. Series of articles on anti-inflammatory activity, pharmacology and toxicology: *Boll. Chim. Farm.* **115,** 135-230 (1976), *C.A.* **85,** 13981-13986 (1976).

Crystals from ethanol, mp 83-85°. pKa (aq soln, 20°) 4.6 ±0.03. uv max (abs alcohol): 220 ±1, 327 ±2 nm; (water, pH 2): 284 nm (log ϵ 3.39); (water, pH 8.5): 313 nm (log ϵ 3.40). Very soluble in alcohol, chloroform, acetone, ethyl acetate, dioxane; slightly sol in water. Practically insol in petr ether, cyclohexane.

THERAP CAT: Anti-inflammatory; analgesic.

7178. Parsley Seed. Parsley fruit. Dried ripe seed of *Petroselinum hortense* Hoffm. (*P. sativum* Hoffm.; *Carum petroselinum* Benth. & Hook.), *Umbelliferae. Habit.* Russia, France, Germany; cultivated everywhere. *Constit.* Seed: volatile and fixed oils; apiol, apiolin, apiin, tannin.

USE: Source of apiol.

7179. Parthenin. *[3aS-(3aα,6β,6aα,9aβ,9bα)]-3,3a,4,5,-6,6a,9a,9b-Octahydro-6a-hydroxy-6,9a-dimethyl-3-methylene-azuleno[4,5-b]furan-2,9-dione; 1,6β-dihydroxy-4-oxo-10αH-ambrosa-2,11(13)-dien-12-oic acid γ-lactone;* parthenicin. $C_{15}H_{18}O_4$; mol wt 262.31. C 68.69%, H 6.92%, O 24.40%. From herb of *Parthenium hysterophorus* L., *Compositae.* Parthenin is the substance largely responsible for the allergic contact dermatitis caused by *P. hysterophorus.* Isoln: Arny, *Am. J. Pharm.* **69,** 169 (1897). Isoln and structure: Herz *et al., Tetrahedron Letters* **1961,** 82; Herz *et al., J. Am. Chem. Soc.* **84,** 2601 (1962). Abs config: Emerson *et al., Tetrahedron Letters* **1966,** 6151. Total synthesis of (±)-form: P. Kok *et al., Bull. Soc. Chim. Belg.* **87,** 615 (1978); C. H. Heathcock *et al., J. Am. Chem. Soc.* **104,** 6081 (1982).

Crystals from water, mp 163-166°. $[\alpha]_D^{25}$ +7.02° (c = 2.71 in chloroform). uv max: 215, 340 nm (ε 15,100; 22). Practically insol in water. Sol in alcohol, chloroform, ether, ethyl acetate.

7180. Parthenolide. *[1aR-(1aR*,4E,7aS*,10aS*,-10bR*)]-2,3,6,7,7a,8,10a,10b-Octahydro-1a,5-dimethyl-8-methyleneoxireno[9,10]cyclodeca[1,2-b]furan-9(1aH)-one; 4,-5α-epoxy-6β-hydroxy-germacra-1(10),11(13)-dien-12-oic acid γ-lactone.* $C_{15}H_{20}O_3$; mol wt 248.32. C 72.55%, H 8.12%, O 19.33%. Sesquiterpene lactone found in feverfew, *q.v.,* and in other plants. Isolation from *Chrysanthemum parthenium* (L.) Bernh. *Compositae* and characterization: V. Herout *et al., Chem. & Ind. (London)* **1959,** 1069; M. Soucek *et al., Coll. Czech. Chem. Comm.* **26,** 803 (1961); from *Magnolia grandiflora* L., *Magnoliaceae:* F. S. El-Feraly, Y.-M. Chan, *J. Pharm. Sci.* **67,** 347 (1978). Revised structure and spectral analysis: T. R. Govindachari *et al., Tetrahedron* **21,** 1509 (1965). Absolute configuration: A. S. Bawdekar *et al., Tetrahedron Letters* **1966,** 1225. Crystal structure: A. Quick, D. Rogers, *J. Chem. Soc. Perkins Trans. II,* **4,** 465 (1976). HPLC determn: D. Strack *et al., Z. Naturforsch.* **35,** 915 (1980). Cytotoxicity: K.-H. Lee *et al., Cancer Res.* **31,** 1649 (1971); L. A. J. O'Neill *et al., Brit. J. Clin. Pharmacol.* **23,** 81 (1987).

Colorless plates, mp 115-116°. $[\alpha]_D^{20}$ −81.4° (c = 1.04 in chloroform); $[\alpha]_D^{22}$ −71.4° (c = 0.220 in CH_2Cl_2). uv max: 214 nm (log ε 4.22).

7181. Particin. Ayfactin; SPA-S-132. Heptaene macrolide antibiotic complex produced by *Streptomyces aureofaciens* NRRL 3878. Isoln: T. Bruzzese, R. Ferrari, U.S. pat. **3,773,925** (1973 to SPA). Recovery and purification process: S. Magnaghi *et al.,* **Brit.** pat. **1,462,442** (1977 to SPA), *C.A.* **87,** 66588a (1977). *In vitro* activity: W. Ritzerfeld, *Farmaco Ed. Sci.* **27,** 235 (1972); G. A. Meloni *et al., ibid.* **34,** 183 (1979). Use in treatment of benign prostatic

hypertrophy: T. Bruzzese, L. Ferrari, U.S. pat. **4,237,117** (1980 to SPA). Separation, characterization and structure of components A and B: R. C. Tweet *et al., J. Antibiot.* **35,** 997 (1982).

Particin A   R' = CH₃
Particin B   R' = H

Particin A, $C_{59}H_{86}N_2O_{19}.4H_2O$. Greenish-yellow powder, mp > 300° (dec). uv max (75% methanol in DMF): 232, 240, 247, 288, 342, 358, 378, 400 nm (ε 33476, 32237, 22936, 15493, 58282, 76883, 102308, 89280). pKa's (70% aq DMF): 6.07, 8.91.

Particin B, $C_{58}H_{84}N_2O_{19}.2\frac{1}{2}H_2O$. Brownish-yellow powder, mp > 300° (dec). uv max (75% methanol in DMF): 232, 240, 247, 288, 342, 358, 378, 400 nm (ε 34761, 32826, 23174, 20594, 50207, 73392, 100425, 87558). pKa's (70% aq DMF): 6.31, 8.95. $[\alpha]_D^{26}$ +87.2° (c = 0.06 in DMF).

Complex, amphoteric yellow crystals. uv max (ethanol): 401, 379, 359, 341 nm. Sol in DMF, DMSO, dimethyl acetamide, pyridine. Practically insol in water, common organic solvents. LD_{50} in mice (mg/kg): 300 orally; 0.5 i.p. (Bruzzese).

Methyl ester, *see* Mepartricin.

7182. Parvaquone. *2-Cyclohexyl-3-hydroxy-1,4-naph-thalenedione; 2-cyclohexyl-3-hydroxy-1,4-naphthoquinone; 2-hydroxy-3-cyclohexyl-1,4-naphthoquinone;* BW-993C; Clexon. $C_{16}H_{16}O_3$; mol wt 256.30. C 74.98%, H 6.29%, O 18.73%. Prepn: L. F. Fieser, *J. Am. Chem. Soc.* **70,** 3165 (1948); L. F. Fieser, M. T. Leffler, U.S. pat. **2,553,648** (1951 to Research Corp.). Antimalarial activity: L. F. Fieser, A. P. Richardson, *J. Am. Chem. Soc.* **70,** 3156 (1948). Electron transport inhibition: A. L. Tappel, *Biochem. Pharmacol.* **3,** 289 (1960); by uncoupling: J. L. Howland, *Biochim. Biophys. Acta* **131,** 247 (1967). *In vitro* and *in vivo* antiprotozoal activity: A. T. Hudson *et al., Parasitology* **90,** 45 (1985). Bioassay in serum: N. McHardy, J. Mercer, *Kenya Vet.* **8(2),** 9 (1984). Treatment of East Coast fever in cattle: T. T. Dolan *et al., Vet. Parasitol.* **15,** 103 (1984); of *Theileria annulata* in calves: N. McHardy, D. W. Morgan, *Res. Vet. Sci.* **39,** 1 (1985). Field comparison of antitheilarial activity with buparvaquone, *q.v.:* N. McHardy *et al., ibid.* 39.

Bright yellow needles, mp 135-136°.

THERAP CAT (VET): Antiprotozoal (Theileria).

7183. Passiflora. Passion flower; passion vine; May pops. Dried flowering and fruiting tops of *Passiflora incarnata* L., *Passifloraceae. Habit.* Southeastern U.S. *Constit.* Harman.

THERAP CAT: Sedative, analgesic.

7184. Patchouli Alcohol. *[1R-(1α,4β,4aα,6β,8aα)]-Oc-tahydro-4,8a,9,9-tetramethyl-1,6-methanonaphthalen-1(2H)-ol;* patchouli camphor. $C_{15}H_{26}O$; mol wt 222.37. C 81.02%, H 11.79%, O 7.19%. A tricyclic sesquiterpene alcohol isolated from oil of patchouli: Gadamer, Amenomiya, *Arch.*

Pharm. **241,** 39 (1903). Proposed structure: Treibs, *Ann.* **564,** 141 (1949). Revised structure: Dobler *et al., Proc. Chem. Soc. (London)* **1963,** 383. Structural studies: Büchi *et al., J. Am. Chem. Soc.* **78,** 1262 (1956); **83,** 927 (1961); **84,** 3205 (1962); **86,** 4438 (1964). Synthesis of *dl*-form: Danishevsky, Dumas, *Chem. Commun.* **1968,** 1287; Mirrington, Schmalzl, *J. Org. Chem.* **37,** 2871 (1972); K. Yamada *et al., Tetrahedron* **35,** 293 (1979). Stereoselective total synthesis of racemic form: F. Näf, G. Ohloff, *Helv. Chim. Acta* **57,** 1868 (1974); of natural, racemic and (+)-forms: F. Näf *et al., ibid.* **64,** 1387 (1981). *Review:* Walker, *Mfg. Chem. Aerosol News* **39,** no. 7, 27 (1968).

Large crystals (hexagonal-trapezohedral) from the higher boiling fractions of oil of patchouli or from petr ether, mp 56°. mp (racemate) 39-40° (Danishevsky, Dumas); also reported as 46-47° (Mirrington, Schmalzl). bp$_8$ 140°. d$_4^{20}$ 1.0284. [α]$_D^{20}$ −97.4° (c = 24 in chloroform). n$_D^{65}$ 1.5029. Practically insol in water. Sol in alcohol, ether, common organic solvents.

7185. Patulin. *4-Hydroxy-4H-furo[3,2-c]pyran-2(6H)-one;* clavacin; clavatin; claviformin; expansine; mycoin C$_3$; penicidin. C$_7$H$_6$O$_4$; mol wt 154.12. C 54.55%, H 3.92%, O 41.52%. An antibiotic derived from the metabolites of a number of fungi, e.g., *Aspergillus clavatus, A. claviforme, A. giganteus, A. terreus, Penicillium patulum, P. expansum, P. melinii, P. leucopus, P. urticae* and *Gymnoascus* spp. Isoln and antibacterial activity: Birkinshaw *et al., Lancet* **245,** 625 (1943); Birkinshaw, Michael, U.S. pat. **2,417,584** (1947 to Therap. Res. Corp. of Great Britain); *cf.* Waksman *et al., Science* **96,** 202 (1942); Brack, *Helv. Chim. Acta* **30,** 1 (1947); Norstadt, McCalla, *Appl. Microbiol.* **17,** 193 (1969). Structure: Birkinshaw, *loc. cit.;* Bergel *et al., J. Chem. Soc.* **1944,** 415; Woodward, Singh, *J. Am. Chem. Soc.* **71,** 758 (1949); Shemyakin, Khokhlov, *Doklady Akad. Nauk SSSR* **75,** 41 (1950). Synthesis: Woodward, Singh, *J. Am. Chem. Soc.* **72,** 1428 (1950). Biosynthesis: Bu'Lock, Ryan, *Proc. Chem. Soc.* **1958,** 222; Tanenbaum, Bassett, *J. Biol. Chem.* **234,** 1861 (1959). Inhibition of K$^+$ ion uptake in erythrocytes: Kahn, *J. Pharmacol. Exp. Ther.* **121,** 234 (1957). Physicochemical data: A. E. Pohland *et al., Pure Appl. Chem.* **54,** 2219 (1982). Has carcinogenic activity attributable to α,β-unsaturation together with an external conjugated double bond attached to 4 position of the γ-lactone ring: Dickens, *Brit. Med. Bull.* **20,** 96 (1964). Toxicity: R. Kinosita, T. Shikata, "On Toxic Moldy Rice" in *Mycotoxins in Foodstuffs,* G. N. Wogan, Ed. (The M.I.T. Press, Cambridge, 1965) pp 117-119. Review and evaluation of studies of carcinogenic action in laboratory animals: *IARC Monographs* **10,** 205-210 (1976). *Review:* Ciegler *et al.,* "Patulin, Penicillic Acid and Other Carcinogenic Lactones" in *Microbial Toxins,* **vol. VI,** A. Ciegler *et al.,* Eds. (Academic Press, New York, 1971) p 409-414.

Compact prisms or thick plates from ether or chloroform, mp 111.0°; also reported as 105-108° (Pohland). [α]$_D^{21}$ −6.2 (chloroform). uv max: 276.5 nm (Bergel). Sol in water and the common organic solvents except petr ether; very sol in ethyl or amyl acetate. Unstable in alkali with loss of biological activity. LD$_{50}$ s.c. in mice: 10-15 mg/kg (Kinosita, Shikata).

Acetylpatulin, C$_9$H$_8$O$_5$, prisms from 50% alcohol, mp 118-120°.

7186. Pavoninin-5. *(3β,15α,25R)-26-(Acetyloxy)-3-hydroxycholest-5-en-15-yl 2-(acetylamino)-2-deoxy-β-D-gluco-*

pyranoside. C$_{37}$H$_{61}$NO$_9$; mol wt 663.89. C 66.94%, H 9.26%, N 2.11%, O 21.69%. Major component of a group of closely related shark repellents isolated from the defense secretion of the Pacific sole, *Pardachirus pavoninus.* The *pavoninins* are a series of ichthyotoxic and hemolytic steroid amine glycosides. Isoln, structure determn and evaluation of pavoninins 1-6 as shark repellents: K. Tachibana *et al., Science* **226,** 703 (1984). ^{13}C, ^{1}H-NMR studies, chemical correlations and ichthyotoxicity: *eidem, Tetrahedron* **41,** 1027 (1985). Brief review: *Chem. & Eng. News* **62,** 36 (Nov. 19, 1984).

Sol in ethyl acetate. [α]$_D^{29}$ +21° (c = 0.7 in ethanol).

7187. Pazinaclone. *8-[[2-(7-Chloro-1,8-naphthyridin-2-yl)-2,3-dihydro-3-oxo-1H-isoindol-1-yl]acetyl]-1,4-dioxa-8-azaspiro[4.5]decane;* (±)-2-(7-chloro-1,8-naphthyridin-2-yl)-3-[(1,4-dioxa-8-azaspiro[4.5]dec-8-yl)carbonylmethyl]isoindolin-1-one; DN-2327. C$_{25}$H$_{23}$ClN$_4$O$_4$; mol wt 478.93. C 62.70%, H 4.84%, Cl 7.40%, N 11.70%, O 13.36%. Nonbenzodiazepine exhibiting mixed agonist/antagonist effects with benzodiazepine receptors; activity resides primarily in the (S)-enantiomer. Prepn: G. Goto, Y. Saji, *Eur. pat.* Appl. 174,858; *eidem,* U.S. pat. 4,695,572 (1986, 1987 both to Takeda). Pharmacologic profile: T. Wada, N. Fukuda, *Psychopharmacology* **103,** 314 (1991). HPLC determn in human plasma and urine, of racemate and/or its metabolites: Z. Hussein *et al., J. Chromatog.* **613,** 105 (1993); of enantiomers and/or their metabolites: *eidem, ibid.* 113. Clinical pharmacokinetics of racemate and its enantiomers: Z. Hussein *et al., Brit. J. Clin. Pharmacol.* **36,** 357 (1993). Clinical evaluation and abuse liability of racemate: G. K. Mumford *et al., Exp. Clin. Psychopharmacol.* **3,** 39 (1995).

Colorless crystals from methylene chloride/ethyl acetate, mp 238-239°.

THERAP CAT: Anxiolytic.

7188. Pazufloxacin. *(S)-10-(1-Aminocyclopropyl)-9-fluoro-2,3-dihydro-3-methyl-7-oxo-7H-pyrido[1,2,3-de]-1,4-benzoxazine-6-carboxylic acid;* T-3761. C$_{16}$H$_{15}$FN$_2$O$_4$; mol wt 318.30. C 60.37%, H 4.75%, F 5.97%, N 8.80%, O 20.11%. Fluorinated quinolone. Prepn: H. Narita *et al.,* **Ger. pat. 3,913,245;** *eidem,* U.S. pat. **4,990,508** (1989, 1991 both to Toyama). Synthesis: Y. Todo *et al., Chem. Pharm. Bull.* **42,** 2569 (1994). Antimicrobial activity: Y. Fukuoka *et al., Antimicrob. Ag. Chemother.* **37,** 384 (1993).

Crystals, mp 269°-271.5°. $[\alpha]_D^{25}$ −88.0° (c = 0.5 in 0.05N aq NaOH). LD$_{50}$ i.v. in male mice: > 500 mg/kg (Todo). Methanesulfonate, $C_{16}H_{15}FN_2O_4 \cdot CH_3SO_3H$, *pazufloxacin mesylate, T-3762*. Colorless prisms, mp 258°-259°(dec). $[\alpha]_D^{20}$ −64.2° (c = 1 in 1.0N NaOH). Soly (25°) in water > 200 mg/ml.

THERAP CAT: Antibacterial.

7189. Peach Oil, Expressed. Persic oil. Fixed oil from seed of *Prunus persica* (L.) Stokes and allied spp., *Rosaceae.*

Light yellow liquid; closely resembles expressed oil of almond. d_{15}^{15} 0.917-0.921. Solidif not above −15°. n_D^{40} 1.464-1.465. Sapon no. 189-193. Iodine no. 100-110. Acid no. about 8. Sol in chloroform, ether, petr ether; slightly sol in alcohol.

USE: Emollient.

7190. Peanut. Groundnut; earthnut. Ripe, underground pods with seeds of *Arachis hypogaea* L., *Leguminosae. Habit.* Parana River Valley in Paraguay, Brazil and Argentina. Cultivated in the reasonably warm regions of all continents, e.g., through the southern U.S.A. *Composition:* pericarp (shell) 21-29%, episperm (skin) 1.95-3.2%, kernel (+ germ) 71-75%. Botanically the peanut is kin to peas and beans, but its constituents are more like those of true nuts. *Constit.* of kernels (roasted with skin): proteins 26.2%, oil 48.7%, water 1.8%, carbohydrates 20.6%, ash 2.7%. The chief proteins are arachin (25% in oil-free meal) and conarchin (8%). Both are globulins of different soly: Johns, Jones, *J. Biol. Chem.* **28,** 77 (1916). The vitamin content of peanuts is moderate, the largest portion being in the episperm. Trace mineral content of kernels: iron 20 mg/kg; manganese 8.51 mg/kg; copper 6.8 mg/kg; zinc 16 mg/kg. *Reviews:* J. Adam, *Les Plantes a Matiere Grasse,* **vol. 3,** *L'Arachide* (Paris, 1947); N. J. Morris, F. G. Dollear, *Abstract Bibliography of the Chemistry and Technology of Peanuts,* Southern Regional Res. Lab. (New Orleans, 1949); J. G. Woodruff, C. T. Young in Kirk-Othmer *Encyclopedia of Chemical Technology* **vol. 14** (Interscience, New York, 1967) p 122.

USE: In peanut butter, candy; as salted peanuts; for fodder and seeding; crushed for oil. Peanut proteins have been used to produce a fiber, *Sarelon.* The shells are used in the manuf of furfural, xylose, cellulose, plastics, mucilage, also in fertilizers and cattle feed.

7191. Peanut Oil. Arachis oil; groundnut oil; earthnut oil; katchung oil. Prepd by pressing the shelled and skinned seeds of *Arachis hypogaea* L., *Leguminosae.* N. J. Morris, F. G. Dollear, *Abstract Bibliography of the Chemistry and Technology of Peanuts* (Southern Regional Res. Lab., New Orleans, 1949); E. W. Eckey, *Vegetable Fats and Oils* (Reinhold, New York, 1954). *Constits.* of cold pressed oil: glycerides of the following fatty acids: palmitic 8.3%, stearic 3.1%, arachidic 2.4%, behenic 3.1%, lignoceric 1.1%, oleic 56.0%, linoleic 26.0%. Traces of capric and lauric acids have been reported. Unsaponifiable matter 0.8% (includes tocopherols 0.022 to 0.059%, sterols 0.19 to 0.25%, squalene 0.027% and other hydrocarbons).

Greenish-yellow or almost colorless oil. Mild, pleasant odor; bland taste. d_{15}^{15} 0.917-0.921. d_{25}^{25} 0.910-0.915. Clouds at low room temp. Solidifies at ~−5°. A.S.T.M. cloud point +4.5°; A.S.T.M. pour point +1°. Titer 26 to 32°. Flash pt 540°F (283°C); ignition temp 833°F (443°C). n_D^{25} 1.466-1.470; n_D^{40} 1.4605-1.4645. Acid value 0.08-6; saponification value 188-195; iodine value 84-102; thiocyanogen value 67-73; hydroxyl value 2.5-9.5; Reichert-Meissl value 0.2-1.0; Polenske value 0.2-0.7. Very slowly thickens and becomes rancid on prolonged exposure to air. Miscible with ether, petr ether, chloroform, carbon disulfide. Sol in benzene, carbon tetrachloride, oils; very slightly sol in alc.

USE: Edible oil: for salad oil as is; in hydrogenated state as shortening, in mayonnaise, in confections. For the manuf of margarine, soaps, paints. In pharmacy as vehicle for i.m. medication, in the laboratory as heat transfer medium in melting point apparatus. Pharmaceutic aid (solvent).

THERAP CAT (VET): Has been used in control of pasture bloat.

7192. Pebulate. *Butylethylcarbamothioic acid S-propyl ester;* S-propyl butylethylthiocarbamate; propyl ethyl-*n*-butylthiolcarbamate; PEBC; Stauffer 2061; Tillam. $C_{10}H_{21}$NOS; mol wt 203.35. C 59.07%, H 10.41%, N 6.89%, O 7.87%, S 15.77%. Prepn: Campbell, Klingman, U.S. pat. **2,983,747** (1961 to Stauffer). Toxicity data: G. W. Bailey, J. L. White, *Residue Rev.* **10,** 97 (1965).

Liquid, bp$_{20}$ 142°. n_D^{30} 1.4752. d_4^{30} 0.9458. Slightly sol in water; miscible with acetone, benzene, isopropanol, methanol, xylene. LD$_{50}$ orally in rats: 1.12 g/kg (Bailey, White).

USE: Selective herbicide.

7193. Pecilocin. *[R-(E,E,E)]-1-(8-Hydroxy-6-methyl-1-oxo-2,4,6-dodecatrienyl)-2-pyrrolidinone;* Supral; Variotin. $C_{17}H_{25}NO_3$; mol wt 291.39. C 70.07%, H 8.65%, N 4.81%, O 16.47%. Antifungal antibiotic isolated from *Paecilomyces varioti* Bainier var. *antibioticus:* S. Takeuchi *et al., J. Antibiot.* **12A,** 109, 195 (1959); Sumiki *et al.,* **Brit. pat. 866,425** (1961 to Japan. Antibiot. Res. Assoc. and Nippon Kayaku). Structure: S. Takeuchi *et al., J. Antibiot.* **17A,** 267 (1964). Stereochemistry: S. Takeuchi, H. Yonehara, *Tetrahedron Letters* **1966,** 5197. Revised structure and stereochemistry (E,Z,E to E,E,E): *eidem, J. Antibiot.* **22,** 179 (1969). Synthesis of the *dl*-form: A. Ishida, T. Mukaiyama, *Chem. Letters* **1977,** 467; *eidem, Bull. Chem. Soc. Japan* **51,** 2077 (1978).

Neutral oil with ester-like odor. Does not show definite boiling or dec pt. $[\alpha]_D^{28}$ −5.68° (methanol). Freely sol in methanol, ethanol, acetone, ethyl acetate, benzene, ether, chloroform, pyridine, dioxane, acetic acid; slightly sol in water, petr ether, ligroin. uv max (methanol): *ca.* 318, 324 nm (E$_{1cm}^{1\%}$ 1198). Unstable and gradually loses antifungal activity in desiccator, though it is fairly stable in organic solvents. Unstable under alkaline conditions.

Monohydrate, needles from ethyl acetate + petr ether, mp 41.5-42.5°. uv max: 320 nm (ε 46,000).

THERAP CAT: Antifungal.

7194. Pectin. Mol wt 20,000-400,000. Polysaccharide substance present in cell walls of all plant tissues which functions as an intercellular cementing material. One of the richest sources of pectin is lemon or orange rind which contains about 30% of this polysaccharide. Occurs naturally as the partial methyl ester of α-(1 → 4) linked D-polygalacturonate sequences interrupted with (1 → 2)-L-rhamnose residues. Neutral sugars: D-galactose, L-arabinose, D-xylose and L-fucose form side chains on the pectin molecule. Structure studies: D. A. Rees, A. W. Wight, *J. Chem. Soc. B,* **1971,** 1366. Secondary and tertiary structure in solution and in gels: D. A. Rees, E. J. Welsh, *Angew. Chem. Int. Ed.* **16,** 214 (1977). Review and bibliography: Towle, Christensen, in *Industrial Gums,* R. L. Whistler, Ed. (Academic Press, New York, 2nd ed., 1973) p 429-461. Book: Z. I. Kertesz, *The Pectic Substances* (Interscience, New York, 1951).

Occurs as a coarse or fine powder, yellowish-white in color, practically odorless, and with a mucilaginous taste. Almost completely sol in 20 parts water, forming a viscous soln contg negatively charged, very much hydrated particles. Acid to litmus. Insol in alcohol or in diluted alcohol, and in other organic solvents. Dissolves more readily in water, if first moistened with alcohol, glycerol or sugar syrup, or if first mixed with 3 or more parts of sucrose. Stable under mildly acidic conditions; more strongly acidic or basic conditions cause depolymerization.

USE: In the preparation of jellies and similar food products: Owens et al., "Factors Influencing Gelation with Pectin" in Advances in Chemistry Series, Natural Plant Hydrocolloids (A.C.S., Washington, 1954) pp 10-15.

THERAP CAT (VET): Antidiarrheal.

7195. Pectolinarigenin. 5,7-Dihydroxy-6-methoxy-2-(4-methoxyphenyl)-4H-1-benzopyran-4-one; 5,7-dihydroxy-4',6-dimethoxyflavone; 6-methoxyacacetin; 6-hydroxypelargidenon-6,4'-dimethyl ether 1467. $C_{17}H_{14}O_6$; mol wt 314.29. C 64.97%, H 4.49%, O 30.54%. From leaves of Linaria vulgaris Mill., Scrophulariaceae: Schmid, Rumpel, Monatsh. 57, 421 (1931); Merz, Wu, Arch. Pharm. 274, 126 (1936). Structure: Schmid, Rumpel, Monatsh. 60, 8 (1932). Synthesis: Wessely, Moser, ibid. 56, 97 (1930); Zemplén, Farkas, Ber. 76, 937 (1943); Murti, Seshadri, Proc. Indian Acad. Sci. 30A, 78 (1949), C.A. 44, 3987d (1950); Farkas, Strelisky, Tetrahedron Letters 1970, 187.

Yellow needles from methanol, mp 220-223°. uv max (methanol): 275, 335 nm. Sol in alcohol, acetone, ether, ethyl acetate. Practically insol in water, benzene, chloroform, petr ether.

Diacetate, $C_{21}H_{18}O_8$, needles from 96% alcohol, mp 151°.

7-Rutinoside, $C_{29}H_{34}O_{15}$, pectolinarin, neolinarin. From leaves of L. vulgaris Mill., Scrophulariaceae: Klobk, Compt. Rend. 145, 331 (1907); Zemplén et al., Ber. 75, 489 (1942); from Cirsium oleraceum Scop., Compositae: Wagner et al., Arch. Pharm. 293, 1053 (1960). Structure: Zemplén, Bognár, Ber. 74, 1818 (1941). Yellow crystals from methanol, mp 275°. uv max (methanol): 275, 330 nm (log ε 4.256, 4.365).

7196. Pederin. [2R-[2α,2[S*(S*)],5β,6β]]-2,6-anhydro-3,5,7-trideoxy-1-C-[[hydroxy(tetrahydro-2-methoxy-5,6-dimethyl-4-methylene-2H-pyran-2-yl)acetyl]amino]-5,5-dimethyl-1,8,9-tri-O-methyl-D-mannononitol; N-[[6-(2,3-dimethoxypropyl)tetrahydro-4-hydroxy-5,5-dimethyl-2H-pyran-2-yl]methoxymethyl]tetrahydro-2-methoxy-5,6-dimethyl-4-methylene-2H-pyran-2-glycolamide; pederine; paederine. $C_{25}H_{45}$-NO_9; mol wt 503.63. C 59.62%, H 9.01%, N 2.78%, O 28.59%. The toxic principle isolated from blister beetles, Paederus fuscipes Curt.: Pavan, Bo, Physiol. Comparata et Oecol. 3, 307 (1953), C.A. 48, 10217g (1954); Brit. pat. 932,875 (1963 to Farmitalia). Powerful inhibitor of protein biosynthesis and mitosis. Structure: Cardani et al., Gazz. Chim. Ital. 96, 3 (1966). Revised structure: Matsumoto et al., Tetrahedron Letters 1968, 6297. Biosynthesis: Cardani et al., ibid. 1973, 2815. Total synthesis: F. Matsuda et al., Tetrahedron Letters 23, 4043 (1982); 24, 1277 (1983).

Crystals from hexane, benzene + hexane, ether + hexane. mp 112-112.5°. Slightly sol in water, hexane. Sol in methanol, ethanol, carbon disulfide, chloroform, carbon tetrachloride, benzene, and acids. Practically insol in petr ether, NH_4OH, NaOH.

Pseudopederin, $C_{24}H_{43}NO_9$, N-[[6-(2,3-Dimethoxypropyl)-tetrahydro-4-hydroxy-5,5-dimethyl-2H-pyran-2-yl]methoxymethyl]tetrahydro-α,2-dihydroxy-5,6-dimethyl-4-methylene-2H-pyran-2-acetamide, ψ-paederine, ψ-pederine, pseudopaederine. Isoln from blister beetles, Paederus fuscipes: Quilico et al., Chem. Ind. (Milan) 43, 1434 (1961), cf. Cardani et al., Tetrahedron Letters 1965, 2537; Brit. pat. 932,875 (1963 to Farmitalia). Crystals from benzene, mp 133°.

7197. Pefloxacin. 1-Ethyl-6-fluoro-1,4-dihydro-7-(4-methyl-1-piperazinyl)-4-oxo-3-quinolinecarboxylic acid; pefloxacine; EU-5306; 1589RB; AM-725. $C_{17}H_{20}FN_3O_3$; mol wt 333.36. C 61.25%, H 6.05%, F 5.70%, N 12.60%, O 14.40%. Fluorinated quinolone antibacterial; analog of norfloxacin, q.v. Prepn: M. Pesson, Ger. pat. 2,840,910; idem, U.S. pat. 4,292,317 (1979, 1981 to Roger Bellon/Dainippon). Pharmacology and antibacterial spectrum: Y. Goueffon et al., Compt. Rend. Ser. III 292, 37 (1981). Pharmacokinetics: J. Barre et al., J. Pharm. Sci. 73, 1379 (1984). Bioavailability and metabolism: A. Contrepois et al., J. Antimicrob. Chemother. 14, 51 (1984); G. Montay et al., Antimicrob. Ag. Chemother. 25, 463 (1984). HPLC determn in urine and plasma: eidem, J. Chromatog. 272, 359 (1983). Symposium on pharmacokinetics, clinical efficacy and safety: J. Antimicrob. Chemother. 26, Suppl. B, 1-229 (1990).

Crystals from DMF, mp 270-272° (dec). Slightly sol in water; sol in alkaline and acidic solutions. LD_{50} in mice (mg/kg): 225 i.v., 1000 orally; in rats (g/kg): 1.5 i.p., 2.5 orally (Goueffon).

Methanesulfonate dihydrate, $C_{17}H_{20}FN_3O_3 \cdot CH_4O_3S$-$\cdot 2H_2O$, pefloxacin mesylate, 1589mRB, Peflacine, Peflox.

THERAP CAT: Antibacterial.

7198. Pelargonic Acid. Nonanoic acid; nonylic acid; nonoic acid. $C_9H_{18}O_2$; mol wt 158.24. C 68.31%, H 11.47%, O 20.22%. $CH_3(CH_2)_7COOH$. Occurs as an ester in oil of pelargonium: Redtenbacher, Ann. 59, 41, 52, 54 (1846). Prepn from unsaturated hydrocarbons by the oxo process: Hill, U.S. pat. 2,815,355 (1957 to Standard Oil of Indiana); from tall oil unsaturated fatty acids: Maggiolo, U.S. pat. 2,865,937 (1958 to Welsbach); by oxidation of oleic acid: Mackenzie, Morgan, U.S. pat. 2,820,046 (1958 to Celanese); from rice bran oil fatty acid: Mihara et al., U.S. pat. 3,060,211 (1962 to Toya Koatsu Ind.). Purification: Port, Riser, U.S. pat. 2,890,230 (1959 to U.S.D.A.). Toxicity study: L. Orö, A. Wretlind, Acta Pharmacol. Toxicol. 18, 141 (1961).

Colorless, oily liquid at ordinary temp; crystallizes when cooled; characteristic odor. d_4^{20} 0.907. mp 12.5°. bp_{756} 252-253°; bp_{14} 143-145°; $bp_{6.3}$ 132-133°. n_D^{20} 1.4330; n_D^{40} 1.4245. Practically insol in water. Sol in alcohol, chloroform, ether. LD_{50} i.v. in mice: 224±4.6 mg/kg (Orö, Wretlind).

Caution: Strong irritant.

USE: In the production of hydrotropic salts (hydrotropic salts form aq solns which dissolve sparingly sol substances to a greater extent than water); in the manuf of lacquers, plastics.

7199. Pelargonidin. 3,5,7-Trihydroxy-2-(4-hydroxyphenyl)-1-benzopyrylium chloride; 3,4',5,7-tetrahydroxyflavylium chloride; 3,4',5,7-tetrahydroxy-2-phenylbenzopyrylium chloride. $C_{15}H_{11}ClO_5$; mol wt 306.70. C 58.74%, H 3.62%, Cl 11.56%, O 26.08%. The aglucone of pelargonin: Willstätter, Bolton, Ann. 408, 42 (1914). Synthesis: Malkin, Robinson, J. Chem. Soc. 127, 1190 (1925); Robertson et *al.*, **1928**, 1533. Prepn from kaempferol: Mirza, Robinson, *Nature* **166**, 997 (1950); King, White, *J. Chem. Soc.* **1957**, 3901.

Reddish-brown prisms from 2% HCl or from alcoholic HCl. Not melted at 350°. Absorption max (ethanol + 0.01% HCl): 530 nm (ε 32,000). Sol in alcohol, methanol; moderately sol in water; slightly sol in chloroform.

3,5-Diglucoside, $C_{27}H_{31}ClO_{15}$, *3,5-bis(β-D-glucopyranosyl-oxy)-7-hydroxy-2-(4-hydroxyphenyl)-1-benzopyrylium chloride, pelargonin, monardin, salvinin, punicin.* From flowers of *Pelargonium zonale* Ait. var. meteor, *Geraniaceae:* Willstätter, Bolton, *loc. cit.;* from scarlet roses: Harborne, *Experientia* **17**, 72 (1961). Identity with monardin and salvinin: Robinson, Todd, *J. Chem. Soc.* **1932**, 2488. Identity with punicin: Karrer, Widmer, *Helv. Chim. Acta* **10**, 67 (1927). Structure: Leon *et al., J. Chem. Soc.* **1931**, 2672. Red needles with green luster from methanol + HCl, dec 175-180°. $[\alpha]_D$ −291°. Absorption max (methanol + HCl): 269, 505 nm. Sol in water, alc.

3-Glucoside, $C_{21}H_{21}ClO_{10}$, *3-(β-D-glucopyranosyloxy)-5,7-dihydroxy-2-(4-hydroxyphenyl)-1-benzopyrylium chloride, callistephin.* From purple-red aster *(Callistephus chinensis* (L.) Nees, *Compositae):* Willstätter, Burdick, *Ann.* **412**, 149 (1916); from strawberries: Sondheimer, Kertesz, *J. Am. Chem. Soc.* **70**, 3476 (1948). Structure and synthesis: Robertson, Robinson, *J. Chem. Soc.* **1928**, 1460. Dark brownish-red needles with bronze luster. Absorption max (ethanol + HCl): 515 nm (ε 13,000). Sol in water, methanol, ethanol, 0.5-7% aq HCl; moderately sol in 10% HCl.

7200. Pelletierine. *1-(2-Piperidinyl)-2-propanone;* 2-acetonylpiperidine; punicine. $C_8H_{15}NO$; mol wt 141.21. C 68.04%, H 10.71%, N 9.92%, O 11.33%. From the rootbark of the pomegranate tree, *Punica granatum* L., *Punicaceae.* Isoln: Tanret, *Compt. Rend.* **86**, 1270 (1878). Structure: Gilman, Marion, *Bull. Soc. Chim. France* **1961**, 1993; Drillien, Viel, *ibid.* **1963**, 2393. Abs config: H. C. Beyerman *et al., Rec. Trav. Chim.* **86**, 80 (1967). Synthesis: Anet *et al., Nature* **164**, 501 (1949); T. Nagasaka *et al., Heterocycles,* **29**, 155 (1989).

Isopelletierine, (±)-*pelletierine,* slightly colored oily liq; bp 195°; bp_{11} 102-107°. d_4^{20} 0.988. Sol in alc, ether, chloroform. One gram dissolves in 20 ml water. Should be stored under nitrogen.

Hydrochloride, $C_8H_{15}NO\cdot HCl$, crystals from alcohol + ether, mp 145°. Soluble in water and alcohol.

N-methylpelletierine, $C_9H_{17}NO$, Oily liquid. d_4^{20} 0.948. bp_{13} 96-98°. n_D^{20} 1.46737. Sol in water, petr ether.

Note: In the early literature the names pelletierine and isopelletierine were used interchangeably.

THERAP CAT: Anthelmintic (Cestodes).

7201. Pellitorine. *(E,E)-N-(2-Methylpropyl)-2,4-deca-dienamide; E,E-N-isobutyl-2,4-decadienamide.* $C_{14}H_{25}NO$; mol wt 223.36. C 75.28%, H 11.28%, N 6.27%, O 7.16%. Pungent principle isolated from *Anacyclus pyrethrum* DC., *Compositae:* J. M. Gulland, G. U. Hopton, *J. Chem. Soc.* **1930**, 6. Structure: M. Jacobson, *J. Am. Chem. Soc.* **71**, 366 (1949). Synthesis: *idem, ibid.* **75**, 2584 (1953); L. Crombie, *J. Chem. Soc.* **1955**, 1007; J. Tsuji *et al., Tetrahedron Letters* **1977**, 1917; J. Nokami *et al., ibid.* **21**, 4455 (1980); T. Mandai *et al., Chemistry Letters* **1980**, 313. Stereoselective synthesis: R. Bloch, D. Hassangonzales, *Tetrahedron* **42**, 4975

(1986). Insecticidal activity: R. T. Lalonde *et al., J. Chem. Ecol.* **6**, 35 (1980).

Needles from petr ether, mp 90°. uv max (abs ethanol): 258 nm ($E_{1cm}^{1\%}$ 1330). Sol in organic solvents; sparingly sol in water. Practically insol in dil acid or alkalies.

7202. Pellotine. *1,2,3,4-Tetrahydro-6,7-dimethoxy-1,2-dimethyl-8-isoquinolinol; N-methylanhalonidine;* 8-hydroxy-6,7-dimethoxy-1,2-dimethyl-1,2,3,4-tetrahydroisoquinoline. $C_{13}H_{19}NO_3$; mol wt 237.30. C 65.80%, H 8.07%, N 5.90%, O 20.23%. From the mescal buttons (pellote) of *Lophophora williamsii* (Lemaire) Coult., *Cactaceae.* Structure: Späth *et al., Ber.* **65**, 1771 (1932). Synthesis: Brossi *et al., Helv. Chim. Acta* **47**, 2089 (1964); **49**, 403 (1966); Takido *et al., J. Pharm. Sci.* **59**, 271 (1970). Biosynthetic studies: Battersby *et al., Tetrahedron Letters* **1967**, 563; **1968**, 6111.

Plates from petr ether, mp 112°. Alkaline reaction. Freely sol in alc, acetone, ether, chloroform; sparingly sol in water.

Hydrochloride, $C_{13}H_{19}NO_3\cdot HCl$, prisms, freely sol in water; sparingly sol in alcohol.

Hydriodide, $C_{13}H_{19}NO_3\cdot HI$, prisms, mp 130°, sol in water, alcohol, almost insol in ether.

7203. α-Peltatin. *[5R-(5α,5aβ,8aα)]-5,8,8a,9-Tetra-hydro-10-hydroxy-5-(4-hydroxy-3,5-dimethoxyphenyl)furo-[3',4':6,7]naphtho[2,3-d]-1,3-dioxol-6(5aH)-one;* 8-hydroxy-2-hydroxymethyl-6,7-methylenedioxy-4-(4'-hydroxy-3',5'-dimethoxyphenyl)-1,2,3,4-tetrahydronaphthalene-3-carboxylic acid lactone. $C_{21}H_{20}O_8$; mol wt 400.39. C 63.00%, H 5.03%, O 31.97%. Exists as a glucoside in the rhizomes of *Podophyllum peltatum* L., *Berberidaceae:* von Wartburg *et al., Helv. Chim. Acta* **40**, 1331 (1957). Isoln from resin podophyllum: Hartwell, Detty, *J. Am. Chem. Soc.* **72**, 246 (1950).

Prismatic leaflets from abs ethanol. Begins to sinter at 236°. Dec 242-246°. $[\alpha]_D^{20}$ −124.8° (c = 0.5 in chloroform). Soly in water at 20° ∼30 mg/liter. Fairly sol in chloroform, hot ethanol, acetic acid, acetone, dilute caustic; less sol in benzene, ether, carbon tetrachloride, propylene glycol. Practically insol in petr ether.

α-Peltatin-β-D-glucoside, $C_{27}H_{30}O_{13}$, prismatic needles from acetone, mp 168-171°. $[\alpha]_D^{20}$ −128.9° (c = 0.590 in methanol); $[\alpha]_D^{20}$ −174.4° (c = 0.579 in pyridine).

Caution: Irritates the skin.

7204. β-Peltatin. *[5R-(5α,5aβ,8aα)]-5,8,8a,9-Tetrahy-dro-10-hydroxy-5-(3,4,5-trimethoxyphenyl)furo[3',4':6,7]-naphtho[2,3-d]-1,3-dioxol-6(5aH)-one;* 8-hydroxy-2-hydroxy-methyl-6,7-methylenedioxy-4-(3',4',5'-trimethoxyphenyl)-1,2,3,4-tetrahydronaphthalene-3-carboxylic acid lactone;

β-peltatin A. $C_{22}H_{22}O_8$; mol wt 414.41. C 63.76%, H 5.35%, O 30.89%. Exists as a glucoside in the rhizomes of *Podophyllum peltatum* L., *Berberidaceae:* von Wartburg *et al., Helv. Chim. Acta* **40,** 1331 (1957). Isoln from resin podophyllum: Hartwell, Detty, *J. Am. Chem. Soc.* **70,** 2833 (1948); **72,** 246 (1950); from *Hyptis verticillata* Jacq., *Labiatae:* German, *J. Pharm. Sci.* **60,** 649 (1971).

Prisms from abs ethanol. Dec 238-241° (slight sintering begins at 234°). $[\alpha]_D^{20}$ −122.9° (c = 0.578 in chloroform). Somewhat less sol than α-peltatin. Soly in water at 23°: 13 mg/liter. Fairly sol in chloroform, hot ethanol, acetic acid, acetone, dil caustic; less sol in benzene, ether, carbon tetrachloride, propylene glycol. Practically insol in petr ether.

β-Peltatin-β-D-glucoside, $C_{28}H_{32}O_{13}$, white amorph powder from acetone + ether, dec 156-159°. $[\alpha]_D^{20}$ −122.7° (c = 0.587 in methanol); $[\alpha]_D^{20}$ −169.2° (c = 0.556 in pyridine).
Caution: Irritates the skin.

7205. Pemirolast. *9-Methyl-3-(1H-tetrazol-5-yl)-4H-pyrido[1,2-a]pyrimidin-4-one.* $C_{10}H_6N_6O$; mol wt 228.21. C 52.63%, H 3.53%, N 36.83%, O 7.01%. Inhibitor of chemical mediator release from tissue mast cells. Prepn: P. F. Juby, U.S. pat. **4,122,274** (1978 to Bristol-Myers). LC determn in plasma: H. Cheng *et al., J. Pharm. Sci.* **76,** 918 (1987). Immunopharmacology and mechanism of action studies: Y. Yanagihara *et al., Japan. J. Pharmacol.* **48,** 91, 103 (1988). Clinical evaluation in allergic rhinitis: D. G. Tinkelman, R. B. Berkowitz, *Ann. Allergy* **66,** 162 (1991).

Crystals from DMF, mp 310-311° (dec).
Potassium salt, $C_{10}H_7KN_6O$, *BMY-26517, TBX, Pemilaston.* Yellowish-white crystalline powder. Highly soluble in water.
THERAP CAT: Antiallergic.

7206. Pemoline. *2-Amino-5-phenyl-4(5H)-oxazolone;* phenoxazole; phenylisohydantoin; phenylpseudohydantoin; azoxodone; PIO; LA-956; YH-1; Azoksodon; Cylert; Deltamine; Hyton Asa; Kethamed; Nitan; Pioxol; Pondex; Senior; Sigmadyn; Stimul; Tradon; Volital. $C_9H_8N_2O_2$; mol wt 176.17. C 61.36%, H 4.58%, N 15.90%, O 18.16%. Prepn: W. Traube, R. Ascher, *Ber.* **46,** 2077 (1913); L. Schmidt, H. Scheffler, U.S. pat. **2,892,753** (1959 to Boehringer, Ing.). Prepn of magnesium hydroxide mixture, originally thought to be a chelate complex: W. E. Lange *et al., J. Pharm. Sci.* **51,** 477 (1962); B. H. Candon, M. Chessin, U.S. pat. **3,108,045** (1963 to Purdue Frederick). CNS stimulant activity: L. Schmidt, *Arzneimittel-Forsch.* **6,** 423 (1956). Efficacy in minimal brain dysfunction in hyperkinetic children: C. K. Connors *et al., Psychopharmacologia* **26,** 321 (1972); J. G. Page *et al., J. Learn. Dis.* **7,** 498 (1974). Toxicity data: E. W. Schafer, *Toxicol. Appl. Pharmacol.* **21,** 315 (1972). *Review:* A. T. Dren, R. S. Janicki, in *Pharmacological and Biochemical Properties of Drug Substances* vol. **1,** M. E. Goldberg, Ed. (Am. Pharm. Assoc., Washington, DC, 1977) pp 33-65.

Crystals, mp 256-257° (dec). Practically insol in water, ether, acetone, dil hydrochloric acid. Sol in propylene glycol (1%); in hot alcohol. LD_{50} orally in rats: 500 mg/kg (Schafer).
Magnesium hydroxide mixture, *magnesium pemoline, Abbott 30400, Tamilan.*
Note: This is a controlled substance (stimulant) listed in the U.S. Code of Federal Regulations, Title 21 Part 1308.14 (1995).
THERAP CAT: CNS stimulant.

7207. Pempidine. *1,2,2,6,6-Pentamethylpiperidine.* $C_{10}H_{21}N$; mol wt 155.28. C 77.35%, H 13.63%, N 9.02%. Ganglion blocking agent. Prepn by methylation of 2,2,6,6-tetramethylpiperidine: Leonard, Hauck, *J. Am. Chem. Soc.* **79,** 5289 (1957); Hall, *ibid.* 5447. Description: Spinks, Young, *Nature* **181,** 1397 (1958); Lee *et al., ibid.* 1717.

Liquid. pK at 30° = 11.25. Extremely basic reaction for a tertiary amine. bp_{760} 147°. n_D^{21} 1.4550.
p-Toluenesulfonate, $C_{17}H_{19}NO_3S$, crystals from ethyl acetate + ethanol, mp 162-163°.
Tartrate, $C_{10}H_{21}N.C_4H_6O_6$, *M & B 4486, Pempidil, Pempiten, Perolysen, Tenormal, Tensinol, Tensoral.* Crystals, mp 160°. Sol in alcohol, moderately sol in water.
Hydrochloride, $C_{10}H_{21}N.HCl$, crystals, sol in water, alc.
THERAP CAT: Tartrate as antihypertensive.

7208. Penamecillin. *[2S-(2α,5α,6β)]-3,3-Dimethyl-7-oxo-6-[(phenylacetyl)amino]-4-thia-1-azabicyclo[3.2.0]heptane-2-carboxylic acid (acetyloxy)methyl ester; penicillin G hydroxymethyl ester acetate;* acetoxymethyl benzylpenicillinate; benzylpenicillin acetoxymethyl ester; Wy-20788; Havapen. $C_{19}H_{22}N_2O_6S$; mol wt 406.46. C 56.15%, H 5.46%, N 6.89%, O 23.62%, S 7.89%. Semi-synthetic antibiotic related to penicillin. Prepn: Jansen, Russell, *J. Chem. Soc.* **1965,** 2127; *eidem,* **Brit.** pat. **1,003,479** corresp to U.S. pat. **3,250,-679** (1965, 1966 to John Wyeth & Brother).

Crystals from isopropanol + ethanol, mp 106-108°. $[\alpha]_D^{20}$ +154°. Not inactivated by gastric acid.
THERAP CAT: Antibacterial.

7209. Penbutolol. *(S)-1-(2-Cyclopentylphenoxy)-3-[(1,1-dimethylethyl)amino]-2-propanol; (S)-1-(tert-butylamino)-3-(o-cyclopentylphenoxy)-2-propanol;* (−)-1-tert-butylamino-2-hydroxy-3-(2'-cyclopentylphenoxy)propane. $C_{18}H_{29}NO_2$; mol wt 291.43. C 74.18%, H 10.03%, N 4.81%, O 10.98%. β-Adrenergic blocker. Prepn: H. Ruschig *et al.,* **S. Afr.** pat. **68 07,915;** *eidem,* U.S. pat. **3,551,493** (1969, 1970 both to Hoechst). Preliminary chemistry and pharmacology: G. Härtfelder *et al., Arzneimittel-Forsch.* **22,** 930 (1972). Physicochemical and analytical study: P. Hajdu, D. Damm, *ibid.* **29,** 602 (1979). Pharmacology: J. Kaiser *et al., ibid.* **30,** 420 (1980). Action specificity: J. Kaiser *ibid.* 427. Crystallographic study: J. M. Leger *et al.*, *Mol. Pharmacol.* **17**, 339 (1980). Hemodynamic effects: P. Lund-Johansen, *Eur. J. Clin. Pharmacol.* **16**, 149 (1979). Clinical study in hypertension: J. L. Cangiano *et al.*, *J. Clin. Pharmacol.* **19**, 384 (1979). HPLC determn in plasma: R. K. Bhamra *et al.*, *Biomed. Chromatog.* **1**, 140 (1986). Review of pharmacology and therapeutic efficacy: R. C. Heel *et al.*, *Drugs* **22**, 1-25 (1981).

Crystals, mp 68-72°. $[\alpha]_D^{20}$ −11.5° (c = 1 in methanol). pKa 9.3 (1.5 mmol/l in 25% ethanol). Sol in methanol, ethanol, chloroform.

Sulfate, $(C_{18}H_{29}NO_2)_2 \cdot H_2SO_4$, *HOE-893d, HOE-39-893d, Betapressin, Levatol, Paginol.* Crystals, mp 216-218° (dec). $[\alpha]_D^{20}$ −24.6° (c = 1 in methanol).

THERAP CAT: Antihypertensive; antianginal; antiarrhythmic.

7210. Penciclovir. *2-Amino-1,9-dihydro-9-[4-hydroxy-3-(hydroxymethyl)butyl]-6H-purin-6-one;* 9-[4-hydroxy-3-(hydroxymethyl)but-1-yl]guanine; PCV; BRL-39123. $C_{10}H_{15}N_5O_3$; mol wt 253.26. C 47.43%, H 5.97%, N 27.65%, O 18.95%. Carba analog of ganciclovir, *q.v.*, active against several herpes viruses. Prepn: U. K. Pandit *et al.*, *Syn. Commun.* **2**, 345 (1972), R. L. Jarvest, M. R. Harnden, U.S. pat. **5,075,445** (1991 to Beecham). Synthesis: M. R. Harnden *et al.*, *J. Med. Chem.* **30**, 1636 (1987); J. Hannah *et al.*, *J. Heterocyclic Chem.* **26**, 1261 (1989). Crystal and molecular structures: M. R. Harnden *et al.*, *Nucleosides Nucleotides* **9**, 499 (1990). *In vitro* activity of enantiomers in comparison with acyclovir, *q.v.*: G. Abele *et al.*, *Antiviral Chem. Chemother.* **2**, 163 (1991); against herpes simplex viruses: A. Weinberg *et al.*, *Antimicrob. Ag. Chemother.* **36**, 2037 (1992). Clinical pharmacokinetics: S. E. Fowles *et al.*, *Eur. J. Clin. Pharmacol.* **43**, 513 (1992). HPLC determn in plasma and urine: J. R. McMeekin *et al.*, *Anal. Proc.* **29**, 178 (1992). Review of development and antiviral activity: M. R. Harnden, *Drugs Future* **14**, 347-358 (1989).

White crystalline solid from water, (monohydrate), mp 275°-277°; also reported as colorless matted needles, mp 272-275°. uv max (in water): 253 nm (ε 11500). uv max (aq 0.01*N* NaOH): 215, 268 nm (ε 18140, 10710). Sol in water (20°): 1.7 mg/ml, pH 7.

Sodium salt, *BRL-39123A.* Occurs as monohydrate, stable crystalline solid. Sol in water (20°): >200 mg/ml. 30 mg/ml sol has pH 11.

THERAP CAT: Antiviral.

7211. Pendimethalin. *N-(1-Ethylpropyl)-3,4-dimethyl-2,6-dinitrobenzenamine;* *N*-(1-ethylpropyl)-2,6-dinitro-3,4-xylidine; *N*-(1-ethylpropyl)-3,4-dimethyl-2,6-dinitroaniline; penoxalin; AC-92553; Prowl; Herbadox; Stomp. $C_{13}H_{19}N_3O_4$; mol wt 281.31. C 55.51%, H 6.81%, N 14.94%, O 22.75%. Pre-emergence and pre-planting herbicide. Prepn: R. H. Kupelian, Ger. pat. **2,232,263**, *C.A.* **78**, 84010z (1973) and A. W. Lutz, R. E. Diehl, Ger. pat. **2,241,408**, *C.A.* **78**, 135858s (1973) (both 1973 to Am. Cyanamid). Activity: H. A. Roberts, W. Bond, *Proc. Brit. Weed Control Conf., 12th* **1**, 387 (1974). Persistence in soil: A. Walker, W. Bond, *Pestic. Sci.* **8**, 359 (1977).

Orange-yellow cryst solid, mp 56-57°. Vapor press at 25°: 3×10^{-5} mm Hg. Soly in water at 20°: 0.3 mg/l. Sol in most organic solvents.

USE: Herbicide.

7212. Penethamate Hydriodide. *[2S-(2α,5α,6β)]-3,3-Dimethyl-7-oxo-6-[(phenylacetyl)amino]-4-thia-1-azabicyclo-[3.2.0]heptane-2-carboxylic acid 2-(diethylamino)ethyl ester monohydriodide; penicillin G 2-diethylaminoethyl ester hydriodide;* benzylpenicillin β-diethylaminoethyl ester hydriodide; β-diethylaminoethyl benzylpenicillinate hydriodide; ephicillin hydriodide; penethecillin; Alivin; Broncopen; Estopen; Neopenil; Pulmaxil N; Bronchocillin; Pulmo 500; Leocillin. $C_{22}H_{32}IN_3O_4S$; mol wt 561.48. C 47.06%, H 5.74%, I 22.60%, N 7.48%, O 11.40%, S 5.71%. *Refs:* Jensen *et al.*, *Ugeskrift for Laeger* **112**, 1043, 1075 (1950); *Acta Path. Microbiol. Scand.* **28**, 407 (1951); Woodard, *J. Pharm. Pharmacol.* **4**, 1019 (1952); Ebelhare, Ferlauto, *Antibiot. & Chemother.* **3**, 873 (1953); **Span.** pats. **206,653-5** (1953 to Antibiotics S.A.); Villax, **Brit.** pat. **759,603** (1956). Residue studies in ruminants: J. F. M. Nouws, G. Ziv, *Zentralbl. Veterinaermed. A* **25**, 312 (1978); J. F. M. Nouws *et al.*, *Arch. Lebensmittelhyg* **30**, 4 (1979), *C.A.* **91**, 37354p (1979).

Crystals, mp 178-179°. Slightly soluble in water (0.96% at 20°). pH of satd aq soln 4.5-5.2. Theoretical potency 1058 I.U./mg. In dry form the compd has a shelf life of at least 2 years. Aq solns are unstable, the ester is hydrolyzed to free penicillin and diethylaminoethanol, the velocity of the reaction increasing with rise in temperature and pH. If stored in a cool place (below 25°) sterile aq suspensions retain their potency for a max of 7 days and for 30 days in a refrigerator at 4°.

THERAP CAT: Antibacterial.

THERAP CAT (VET): Antibacterial.

7213. Penfluridol. *1-[4,4-Bis(4-fluorophenyl)butyl]-4-[4-chloro-3-(trifluoromethyl)phenyl]-4-piperidinol; 1-[4,4-bis-(p-fluorophenyl)butyl]-4-(4-chloro-α,α,α-trifluoro-m-tolyl)-4-piperidinol;* 1-(4,4-bis(4-fluorophenyl)butyl)-4-hydroxy-4-(3-trifluoromethyl-4-chlorophenyl)piperidine; R-16341; Semap. $C_{28}H_{27}ClF_5NO$; mol wt 523.97. C 64.18%, H 5.19%, Cl 6.77%, F 18.13%, N 2.67%, O 3.05%. Prepn: H. K. F. Hermans, C. J. E. Niemegeers, Ger. pat. **2,040,231**; *eidem*, U.S. pat. **3,575,990** (both 1971 to Janssen); Sindelár *et al.*, *Coll. Czech. Chem. Commun.* **38**, 3879 (1973). Pharmacology and toxicology: Janssen *et al.*, *Eur. J. Pharmacol.* **11**, 139 (1970). Crystal structure: Koch, *Acta Crystallogr.* **29B**, 1538 (1973).

White, microcrystals, mp 105-107°. Slightly sol in water, dil HCl (<0.5 mg/ml). LD_{50} orally in mice (day 7): 86.8 mg/kg (Janssen).

THERAP CAT: Antipsychotic.

7214. Penicillamine. *3-Mercapto-D-valine; β,β-dimeth-ylcysteine; α-amino-β-methyl-β-mercaptobutyric acid;* DMC; β-thiovaline; Cuprenil; Cuprimine; Depamine; De-pen; Mercaptyl; Pendramine; Perdolat; Sufortan; Trolovol. $C_5H_{11}NO_2S$; mol wt 149.21. C 40.25%, H 7.43%, N 9.39%, O 21.44%, S 21.49%. The most characteristic degradn prod-uct of penicillin type antibiotics. Prepn by hydrolysis of penicillins: E. P. Abraham *et al., Nature* **151,** 107 (1943). Review of syntheses of racemic penicillamine and enantio-mers: H. M. Crooks in *The Chemistry of Penicillin,* H. T. Clarke *et al.,* Eds. (Princeton Univ. Press, 1949) pp 455-472; W. M. Weigert *et al., Angew. Chem. Int. Ed.* **14,** 330 (1975). Polymorphism of D-form: J. A. G. Vidler, *J. Pharm. Phar-macol.* **28,** 662 (1976). Pharmacokinetics and metabolism of the D- and L-forms: Ruiz-Torres, *Arzneimittel-Forsch.* **24,** 914 (1974). In treatment of rheumatoid arthritis: P. Davis, *Clin. & Invest. Med.* **7,** 41 (1983); of Wilson's Disease: A. Deiss, *Ann. Int. Med.* **99,** 398 (1983); of progressive systemic sclerosis: B. Kang *et al., J. Allergy Clin. Immun.* **69,** 297 (1982). Evaluation in fibrotic lung disease: R. Chapela *et al., Int. J. Clin. Pharmacol. Ther. Tox.* **24,** 16 (1986). Re-view of assay methods: N. Kucharczyk, S. Shahiniam, *J. Rheumatol.* **8,** Suppl. 7, 28-84 (1981); of metabolism and therapeutic applications: J. C. Crawhall *et al., Biopharm. & Drug Dispos.* **1,** 73-95 (1979); of side effects: R. S. Levy *et al., J. Am. Acad. Dermatol.* **8,** 548-558 (1983). Toxicity data: Veis *et al., Antibiotiki* **14,** 837 (1969). General re-views: I. A. Jaffe in *Pharmacological and Biochemical Prop-erties of Drug Substances* Vol. **2,** M. E. Goldberg, Ed. (Am. Pharm. Assoc., Washington, DC, 1979) pp 465-478; C. C. Chiu, L. T. Grady in *Analytical Profiles of Drug Substances* Vol. **10,** K. Florey, Ed. (Academic Press, New York, 1981) pp 601-637.

mp 202-206° (Weigert *et al.).* $[\alpha]_D^{25}$ $-63°$ (c = 0.1 in pyridine). LD_{50} in rats (mg/kg): > 10000 orally, > 660 i.p. (Jaffe).

Hydrochloride, $C_5H_{11}NO_2S.HCl$, *Distamine, Metalcap-tase.* Hygroscopic crystals, dec 177.5°. $[\alpha]_D^{25}$ $-63°$ (1N NaOH). Freely sol in water, sol in ethanol. Aq solns are comparatively stable at pH 2-4. LD_{50} i.v. in mice: 2289 mg/kg (Veis).

DL-Form, crystals, dec 201°. pK values: carboxyl 1.8; α-amino 7.9; β-thiol 10.5. LD_{50} orally in rats: 365 mg/kg (Jaffe).

DL-Form hydrochloride, crystals, dec 145-148°.

L-Form, crystals, mp 190-194°. $[\alpha]_D^{25}$ $+63°$ (in 1N NaOH). LD_{50} i.p. in rats: 350 mg/kg (Jaffe).

USE: L-Form as exptl inducer of convulsions in rats, *C.A.* **84,** 159420t (1976).

THERAP CAT: Chelating agent (copper) in Wilson's disease. Antirheumatic.

7215. Penicillamine Cysteine Disulfide. *(R)-3-[(2-Amino-2-carboxyethyl)dithio]-D-valine; 3,3-dimethyl-3-3'-di-thiodialanine;* 1,6-diamino-5-5-dimethyl-3,4-dithiahexane-1,6-dicarboxylic acid. $C_8H_{16}N_2O_4S_2$; mol wt 268.36. C 35.81%, H 6.01%, N 10.44%, O 23.85%, S 23.90%. Prepn: Tabachnick *et al., Nature* **174,** 701 (1954); Schöberl *et al., Angew. Chem.* **68,** 213 (1956); Schöberl, Grafje, *Ann.* **617,** 71 (1958); Levine, *Nature* **187,** 940 (1960).

Crystals, mp 195°.

7216. Penicillamine Disulfide. *3,3'-Dithiobisvaline; 3,3'-dithiodivaline;* 3,3,3',3'-tetramethylcystine. $C_{10}H_{20}N_2$-O_4S_2; mol wt 296.41. C 40.52%, H 6.80%, N 9.45%, O 21.59%, S 21.64%. Prepn: Süs, *Ann.* **561,** 31 (1948); Berg *et al., Brit.* pat. **621,915** (1949 to Merck & Co.); Butenandt *et al., Z. Physiol. Chem.* **285,** 238 (1950).

DL-Form, crystals, mp 181-183°.

D-Form, crystals, mp 204-205°. $[\alpha]_D^{23}$ +27° (c = 1.46 in 1N HCl).

L-Form, crystals, mp 207°. $[\alpha]_D^{22}$ $-26°$ (1N HCl).

7217. Penicillanic Acid. *(2S-cis)-3,3-Dimethyl-7-oxo-4-thia-1-azabicyclo[3.2.0]heptane-2-carboxylic acid.* A build-ing block of penicillin, devoid of significant antibacterial activity: Sheehan *et al., J. Am. Chem. Soc.* **75,** 3292 (1953); **81,** 5838 (1959). Separation from its esters and other peni-cillins by gas chromatog: Evard *et al., Nature* **201,** 1124 (1964). *See also* 6-Aminopenicillanic Acid.

7218. Penicillic Acid. *3-Methoxy-5-methyl-4-oxo-2,5-hexadienoic acid;* γ-keto-β-methoxy-δ-methylene-Δ^α-hex-enoic acid. $C_8H_{10}O_4$; mol wt 170.17. C 56.47%, H 5.92%, O 37.61%. Antibiotic substance produced by the following fungi: *Penicillium puberulum, P. cyclopium, P. thomii, P. suaveolens, P. baarnense, Aspergillus ochraceus, A. melleus.* Isoln: Alsberg, Black, *U.S. Dept. Agric. Bur. Plant Industry,* bull. no. **270,** (1913); Birkinshaw *et al., Biochem. J.* **30,** 394 (1936); Oxford *et al., Chem. & Ind. (London)* **20,** 22 (1942); Karow *et al., Arch. Biochem.* **5,** 279 (1944); Burton, *Nature* **165,** 274 (1950); Natori *et al., Chem. Pharm. Bull.* **18,** 2259 (1970). Acid in tautomeric equilibrium with its lactone. Structure: Birkinshaw *et al., loc. cit.* Physical properties: Kovac, Solcaniova, *Tetrahedron* **25,** 3617 (1969). Synthesis: Raphael, *Nature* **160,** 261 (1947); *J. Chem. Soc.* **1948,** 1508; C. L. Yeh *et al., Tetrahedron Letters* **1978,** 3987. Biosyn-thesis: Birch *et al., J. Chem. Soc.* **1958,** 4582; Bentley, Keil, *J. Biol. Chem.* **237,** 867 (1962). Physicochemical data: A. E. Pohland *et al., Pure Appl. Chem.* **54,** 2219 (1982). Has carcinogenic activity attributable to α,β-unsaturation to-gether with an external conjugated double bond attached to the 4 position of the γ-lactone ring: Dickens, *Brit. Med. Bull.* **20,** 96 (1964). Review and evaluation of studies of carcinogenic action in laboratory animals: *IARC Mono-graphs* **10,** 211-216 (1976). Activity studies: Suzuki *et al., Agr. Biol. Chem.* **35,** 287 (1971). *Review:* Ciegler *et al.,* "Patulin, Penicillic Acid and Other Carcinogenic Lactones" in *Microbial Toxins,* vol. VI, A. Ciegler *et al.,* Eds. (Aca-demic Press, New York, 1971) p 414.

Needles from petr ether, mp 83-84°. uv max: about 220 nm. Acid reaction, turns Congo red paper blue. Moderate-ly sol in cold water (2 g/100 ml); freely sol in hot water, alcohol, ether, benzene, chloroform; slightly sol in hot petr ether. Practically insol in pentane-hexane. LD_{50} s.c. in mice: 100 mg/kg.

Monohydrate, large transparent monoclinic or triclinic, rhombic crystals from water, mp 58-64°.

7219. Penicillinase. β-Lactamase; Neutrapen. Mol wt about 50,000. Enzymes found in many bacteria which destroy penicillins and cephalosporins by catalyzing the hydrolysis of the amide bond in the β-lactam ring. Good penicillinase producers are *Bact. coli*, the *Bacillus subtilis-mesentericus* group, *Bacillus anthracis* and *Staphylococci*. Both intra- and extracellular penicillinase are of protein nature. There are probably as many different penicillinases as there are bacteria producing them. Ion-exchange procedures for the purification of penicillinase: Puetzer, Boschetti, U.S. pat. **2,982,696** (1961 to Schenley). Amino acid sequence studies: Ambler, Meadway, *Nature* **222**, 24 (1969). *Reviews:* Chain *et al.* in *Antibiotics* vol. 2, Flory *et al.*, Eds. (Oxford, 1949) p 1090; Rothe, *Pharmazie* **5**, 25 (1950); Citri, Pollock, *Advan. Enzymol.* **28**, 237 (1966); Citri in *The Enzymes*, P. D. Boyer, Ed. (Academic Press, New York, 3rd ed., 1971) pp 23-46.
USE: In culture media to antagonize antibacterial activity of penicillin.
THERAP CAT: Enzyme (penicillin inactivating).

7220. Penicillin G Benethamine. *[2S-(2α,5α,6β)]-3,3-Dimethyl-7-oxo-6-[(phenylacetyl)amino]-4-thia-1-azabicyclo-[3.2.0]heptane-2-carboxylic acid compd with N-(phenylmethyl)benzeneethanamine (1:1);* benzylpenicillinic acid N-benzyl-β-phenylethylamine salt; N-benzyl-2-phenylethylamine salt of benzylpenicillin; benethamine penicillin G; Benapen; Betapen; Benetolin. $C_{31}H_{35}N_3O_4S$; mol wt 545.70. C 68.23%, H 6.46%, N 7.70%, O 11.73%, S 5.88%. Semi-synthetic antibiotic prepd from penicillin G and N-benzylphenethylamine: Jansen, Hems, Brit. pat. **732,559** (1955 to Glaxo). Clinical studies: Nelson *et al.*, *Brit. Med. J.* **1954**, II, 339; Boger *et al.*, *Antibiot. Ann.* **1954-55**, 123.

Crystals, mp 146-147°. Slight, characteristic, amine taste. Potency ~1100 units/mg (as compared to 1670 units/mg of penicillin G sodium). Very slightly sol in water (0.1 w/v at 40°). Solutions of 1000 units/ml approach the limits of aq soly. This degree of soly may be compared with that of procaine penicillin, approx 4000 units/ml and that of benzathine penicillin, approx 200 units/ml.
THERAP CAT: Antibacterial.
THERAP CAT (VET): Antibacterial.

7221. Penicillin G Benzathine. *[2S-(2α,5α,6β)]-3,3-Dimethyl-7-oxo-6-[(phenylacetyl)amino]-4-thia-1-azabicyclo-[3.2.0]heptane-2-carboxylic acid compd with N,N'-bis(phenylmethyl)-1,2-ethanediamine; penicillin G compd with N,N'-dibenzylethylenediamine (2:1);* N,N'-dibenzylethylenediamine bis[benzylpenicillin]; dibenzylethylenediamine dipenicillin G; penicillin G salt of N,N'-dibenzylethylenediamine; benzethacil; benzathine penicillin G; diamine penicillin; Beacillin; Megacillin suspension; Bicillin; Cillenta; Permapen; Duropenin; Dibencillin; DBED Penicillin; Penidural; Tardocillin; Dibencil; Lentopenil; Vicin Neolin; Pen-Di-Ben; Penidure; Moldamin; Extencilline; Longacilina; Longicil; Penadur; Penditan; Cepacilina. $C_{48}H_{56}N_6O_8S_2$; mol wt 909.14. C 63.41%, H 6.21%, N 9.24%, O 14.08%, S 7.05%. Semi-synthetic antibiotic prepd by adding an aq soln of one mole N,N'-dibenzylethylenediamine diacetate to an aq soln of 2 moles sodium penicillin G: Szabo *et al.*, *Antibiot. & Chemother.* **1**, 499 (1951); Szabo, Bruce, U.S. pat. **2,627,491** (1953 to Wyeth). Comprehensive description: F. Kreuzig in *Analytical Profiles of Drug Substances* vol. 11, K. Florey, Ed. (Academic Press, New York, 1982) pp 463-482.
Hydrated, tasteless crystals. mp 123-124°. $[\alpha]_D^{25}$ +206° (c = 0.105 in formamide). Activity not less than 1050 units/mg (theory: 1307 units/mg). Soly (mg/ml): water 0.15 (about 200 units/ml); benzene 0.38; alc 5.2; acetone 1.5; formamide 28.0. pH of satd aq soln about 6. Solubili-

ties determined by Weiss *et al.*, *Antibiot. & Chemother.* **7**, 374 (1957) in mg/ml: water 0.315; methanol 16.9; ethanol 15.4.
THERAP CAT: Antibacterial.
THERAP CAT (VET): Antibacterial.

7222. Penicillin G Benzhydrylamine. *[2S-(2α,5α,6β)]-3,3-Dimethyl-7-oxo-6-[(phenylacetyl)amino]-4-thia-1-azabicyclo[3.2.0]heptane-2-carboxylic acid compd with α-phenylbenzenemethanamine (1:1); 1,1-diphenylmethylamine benzylpenicillinate;* benzylpenicillin benzhydrylamine; benzhydrylamine penicillin; penicillin aminodiphenylmethane salt; benzylpenicillinic acid benzhydrylamine salt; Orencil; Penidryl. $C_{29}H_{31}N_3O_4S$; mol wt 517.65. C 67.29%, H 6.04%, N 8.12%, O 12.36%, S 6.19%. $C_{16}H_{18}N_2O_4S.C_{13}H_{13}N$. Semi-synthetic antibiotic. Prepd from potassium penicillin G and benzhydrylamine HCl in dil acetone: Hagemann *et al.*, *Ann. Pharm. Franc.* **12**, 565 (1954); Goldsmith *et al.*, *Antibiot. & Chemother.* **5**, 648 (1955); Penau, Hagemann, U.S. pat. **2,710,863** (1955 to Labs. Français de Chimiothérapie).
Crystals from dil acetone, mp 159°. $[\alpha]_D^{20}$ +206° (c = 1). Activity 1150 units/mg. Soly in water at 24°: 0.7%. Insol in vegetable oils. The solid is stable for 2 years at 24° and for 2 days at 100°.
THERAP CAT: Antibacterial.

7223. Penicillin G Calcium. *[2S-(2α,5α,6β)]-3,3-Dimethyl-7-oxo-6-[(phenylacetyl)amino]-4-thia-1-azabicyclo-[3.2.0]heptane-2-carboxylic acid calcium salt (2:1);* benzylpenicillin calcium; calcium penicillin G; calcium benzylpenicillinate; benzylpenicillinic acid calcium salt. $C_{32}H_{34}CaN_4O_8S_2$; mol wt 706.85. C 54.38%, H 4.85%, Ca 5.67%, N 7.93%, O 18.11%, S 9.07%. $(C_{16}H_{17}N_2O_4S)_2Ca$. Antibiotic obtained by extraction of an ether soln of the free acid (prepared from the sodium salt) with satd calcium hydroxide soln to pH 6.5, followed by lyophilizing: Clarke *et al.*, *Chemistry of Penicillin* (Princeton, 1949) p 88.
White powder, granules, scales. Freely sol in water, in isotonic sodium chloride soln, in glucose solns. Also sol in alcohol, glycerol, acetone, ethyl acetate, chloroform. Insol in fixed oils, liquid petrolatum. pH of an aq soln contg 20,000 units per ml is 5.0 to 6.5.
THERAP CAT: Antibacterial.
THERAP CAT (VET): Antibacterial.

7224. Penicillin G Hydrabamine. *N,N'-Bis(dehydroabietyl)ethylenediamine dipenicillin G;* benzylpenicillinic acid N,N'-bis(dehydroabietyl)ethylenediamine double salt; hydrabamine penicillin G; Compocillin. $C_{74}H_{100}N_6O_8S_2$; mol wt 1265.78. C 70.22%, H 7.96%, N 6.64%, O 10.11%, S 5.07%. Semi-synthetic antibiotic. Prepn: L. C. Cheney, U.S. pat. **2,812,325** (1957). Pharmacology: D. W. MacCorquodale *et al.*, *Antibiot. Ann.* **1954-1955**, 133; M. H. Lepper, H. W. Spies, *ibid.* 137; A. F. DeRose *et al.*, *Antibiot. & Chemother.* **5**, 315, 324 (1955). Purification: A. F. DeRose, U.S. pat. **2,812,326** (1957 to Abbott).
Tasteless crystals from dil methanol, dec 171-173°. $[\alpha]_D^{25}$ +115.3° (c = 10 in chloroform). Potency 939 units/mg. Soly in water 9 units/ml (bioassay). Solubilities determined by P. J. Weiss *et al.*, *Antibiot. & Chemother.* **7**, 374 (1957) in mg/ml at about 28°: Water 0.075; methanol 7.3; ethanol 5.2; isopropanol 1.7; isoamyl alcohol 3.1; cyclohexane 0.115; benzene 0.60; toluene 0.39; petr ether 0.0; isooctane 0.055; carbon tetrachloride 0.50; ethyl acetate 1.65; isoamyl acetate 1.4; acetone 3.4; methyl ethyl ketone 3.65; ether 0.70.
THERAP CAT: Antibacterial.

7225. Penicillin G Potassium. *[2S-(2α,5α,6β)]-3,3-Dimethyl-7-oxo-6-[(phenylacetyl)amino]-4-thia-1-azabicyclo-[3.2.0]heptane-2-carboxylic acid monopotassium salt;* benzylpenicillin potassium; potassium penicillin G; potassium benzylpenicillinate; benzylpenicillinic acid potassium salt; Notaral; Crystapen; Hipercilina; Pentid; Tabilin; Eskacillin; Forpen; Hylenta; Cosmopen; Falapen; Hyasorb; Cristapen; M-Cillin; Monopen; Megacillin tablets. $C_{16}H_{17}KN_2O_4S$; mol wt 372.49. C 51.59%, H 4.60%, K 10.50%, N 7.52%, O 17.18%, S 8.61%. For general prepn and references *see* Benzylpenicillin Sodium. Rapid synthesis: A. N. Voldeng *et al.*, *J. Heterocycl. Chem.* **16**, 621 (1979). HPLC determn: J. M. Blaha *et al.*, *J. Pharm. Sci.* **64**, 1384 (1975). Comprehensive description: J. Kirschbaum, *Anal. Profiles Drug Subs.* **15**, 427-507 (1987).

Crystalline antibiotic, moderately hygroscopic. Dec 214-217°. $[\alpha]_D^{22}$ +285-310° (c = 0.7). Freely sol in water, in isotonic sodium chloride soln, and in dextrose solns. Sol in methanol, ethanol, glycerol. pH of 3% aq soln 5.0 to 7.5. Solutions stored at refrigerator temps remain stable for several days. They are rapidly inactivated by acids, alkali hydroxides, and by oxidizing agents. One milligram of benzylpenicillin potassium is equivalent to 1595 units. One international or U.S.P. penicillin unit is equivalent to 0.6 micrograms of benzylpenicillin sodium.
THERAP CAT: Antibacterial.
THERAP CAT (VET): Antibacterial.

7226. Penicillin G Procaine. *[2S-(2α,5α,6β)]-3,3-Dimethyl-7-oxo-6-[(phenylacetyl)amino]-4-thia-1-azabicyclo-[3.2.0]heptane-2-carboxylic acid compd with 2-(diethylamino)ethyl 4-aminobenzoate (1:1) monohydrate; penicillin G compd with 2-(diethylamino)ethyl p-aminobenzoate monohydrate;* benzylpenicillin procaine; procaine benzylpenicillinate; procaine penicillin G; Abbocillin-DC; Afsillin; Ampinpenicillin; Aquacillin; Aquasuspen; Avloprocil; Cilicaine; Crysticillin; Despacilina; Depocillin; Distaquaine; Dorsallin "A.R."; Duracillin; Flo-Cillin Aqueous; Hydracillin; Ilcocillin P; Kabipenin; Ledercillin; Lenticillin; Mammacillin; Megapen; Mylipen; Neoproc; Penaquacaine G; Pen-Fifty; Premocillin; Procanodia; Pro-Pen; Wycillin. $C_{29}H_{38}N_4O_6S$.H_2O; mol wt 588.73. C 59.16%, H 6.85%, N 9.52%, O 19.02%, S 5.45%. Semi-synthetic antibiotic. Prepn: N. P. Sullivan *et al., Science* **107**, 169 (1948); C. J. Sullivan *et al., J. Am. Chem. Soc.* **70**, 1287 (1948); H. W. Rhodehamel, Jr., U.S. pat. **2,515,898** (1950 to Eli Lilly). Crystal structure: Rose, *Anal. Chem.* **27**, 1841 (1955). Toxicity: K. Soehring *et al., Arzneimittel-Forsch.* **1**, 28 (1951). Pharmacokinetics in horses: S. M. Stover *et al., Am. J. Vet. Res.* **42**, 629 (1981); in humans: B. T. Goh *et al., Brit. J. Vener. Dis.* **60**, 371 (1984). Review of use in syphilis: M. W. Adler, *Brit. Med. J.* **288**, 551-553 (1984).
Monoclinic hemimorphic crystals from methanol-water, mp 106-110° (with decompn). d 1.255-1.256. Not appreciably affected by air or light. Aq solns are dextrorotatory. The pH of a satd aq soln is between 5 and 7.5. One gram dissolves in 250 ml water, in 30 ml alcohol, in 60 ml chloroform (*U.S.P.*). Soly profile: Weiss *et al., Antibiot. & Chemother.* **7**, 374 (1957). Soly in mg/ml at about 28°: water 6.8; methanol > 20; isopropanol 6.5; benzene 0.075; toluene 1.05; petr ether 0.12; isooctane 0.0; carbon tetrachloride 0.12; ethyl acetate 3.35. Rapidly inactivated by acids, alkali hydroxides, oxidizing agents. Penicillin potency: approx 1000 units/mg. LD_{50} s.c. in mice: 2.3 g/kg (Soehring).
THERAP CAT: Antibacterial.
THERAP CAT (VET): Antibacterial.

7227. Penicillin N. *[2S-[2α,5α,6β(S*)]]-6-[(5-Amino-5-carboxy-1-oxopentyl)amino]-3,3-dimethyl-7-oxo-4-thia-1-aza-bicyclo[3.2.0]heptane-2-carboxylic acid; 6-(D-5-amino-5-carboxyvaleramido)-3,3-dimethyl-7-oxo-4-thia-1-azabicyclo-[3.2.0]heptane-2-carboxylic acid;* (D-4-amino-4-carboxybutyl)penicillinic acid; cephalosporin N; adicillin; Synnematin B. $C_{14}H_{21}N_3O_6S$; mol wt 359.40. C 46.79%, H 5.89%, N 11.69%, O 26.71%, S 8.92%. Antibiotic substance produced by *Cephalosporium* spp found in sewage outpours: Gottshall *et al., Proc. Soc. Exp. Biol. Med.* **76**, 307 (1951); Abraham *et al., Nature* **171**, 343 (1953); **176**, 551 (1955). Produced also by *Paecilomyces persicimus:* Pisano *et al., Antimicrobial Agents Annual* **1960** (Plenum Press, New York, 1961) pp 41, 48; by *Penicillium chrysogenum:* Flynn *et al., J. Am. Chem. Soc.* **84**, 4594 (1962). Structure: Abraham, Newton, *Biochem. J.* **58**, 103 (1954). Production: Miller *et*

al., U.S. pat. **2,831,797** (1958). Purification: Goodall, Sutcliffe, U.S. pat. **2,899,425** (1959 to ICI).

Soluble in water. Dextrorotatory. Inactivated by penicillinase as is penicillin G, but differs from the common penicillin by its antibacterial activity and hydrophilic character. When an aq soln is kept at pH 2.7 and 37° for 2 hrs, there is a loss of antibacterial activity and an increase in dextrorotation. Active against *Sarcina lutea, Proteus vulgaris, Salmonella typhimurium, Diplococcus pneumoniae.* Shows practically no activity against *B. subtilis* and *Staph. aureus.* The toxicity is somewhat less than that of penicillin G, although penicillin N is excreted more slowly.
Barium salt, white powder. $[\alpha]_D^{20}$ +187° (c = 0.6). Freely sol in water, sparingly sol in methanol. Practically insol in ethanol.
THERAP CAT: Antibacterial.

7228. Penicillin O. *[2S-(2α,5α,6β)]-3,3-Dimethyl-7-oxo-6-[[(2-propenylthio)acetyl]amino]-4-thia-1-azabicyclo-[3.2.0]heptane-2-carboxylic acid; [(allylthio)methyl] penicillin;* allylmercaptomethylpenicillin; allylmercaptomethylpenicillinic acid; penicillin AT. $C_{13}H_{18}N_2O_4S_2$; mol wt 330.43. C 47.25%, H 5.49%, N 8.48%, O 19.37%, S 19.41%. Antibiotic produced by *Penicillium chrysogenum.* Biosynthesis of salts: Behrens *et al., J. Biol. Chem.* **175**, 793 (1948); Rhodehamel, Behrens *et al.,* U.S. pats. **2,528,175** and **2,623,876** (1950, 1952, both to Lilly); Ford *et al., Antibiot. & Chemother.* **3**, 1149 (1953); Ford, U.S. pat. **2,647,894** (1953 to Upjohn); Palecková, Slechta, *C.A.* **50**, 17309g (1956).

2-Chloroprocaine salt monohydrate, $C_{26}H_{37}ClN_4O_6S_2$.H_2O, *chloroprocaine penicillin O, penicillin O 2-chloroprocaine, Depo-Cer-O-Cillin Chloroprocaine.* Slender needles from hot water, mp 79-81°. Practically insol in cold water. Stable in dry form at room temp. Aq suspensions are stable at room temp for 1 week, at refrigerator temps for 3 weeks. Calculated activity: 949 units/mg. Solubilities: Weiss *et al., Antibiot. & Chemother.* **7**, 374 (1957).
Potassium salt, $C_{13}H_{17}KN_2O_4S_2$, *potassium penicillin O, penicillin O potassium.* Crystals from acetone. Soluble in water. Stable in dry form at room temp for at least 3 years. Requires no refrigeration when dry. Aq solns may be kept for 3 days at +10° without significant loss of activity. Behrens' prepn assayed 1630 units/mg. Less toxic than benzylpenicillin in exptl animals.
Procaine salt, $C_{26}H_{38}N_4O_6S_2$, crystals from water.
Sodium salt, $C_{13}H_{17}N_2NaO_4S_2$, *Cer-O-Cillin Sodium.* Crystals from acetone.
THERAP CAT: Antibacterial.

7229. Penicillin S Potassium. *[2S-(2α,5α,6β)]-6-[[[(3-Chloro-2-butenyl)thio]acetyl]amino]-3,3-dimethyl-7-oxo-4-thia-1-azabicyclo[3.2.0]heptane-2-carboxylic acid monopotassium salt; γ-chlorocrotylmercaptomethylpenicillin potassium; γ-chlorocrotylmercaptomethylpenicillinic acid potassium salt;* potassium γ-chlorocrotylmercaptomethylpenicillinate; potassium penicillin S. $C_{14}H_{18}ClKN_2O_4S_2$; mol wt 416.99. C 40.33%, H 4.35%, Cl 8.50%, K 9.38%, N 6.72%, O 15.35%, S 15.38%. Obtained by growing a penicillin-producing mold in a culture medium contg γ-chlorocrotyl-mercaptoacetic acid: Ford *et al., J. Am. Chem. Soc.* **70**,

3522 (1948); Ford *et al., Antibiot. & Chemother.* **3**, 1149 (1953).

Crystals from acetone. Disagreeable odor and taste. Sol in water. Bioassay: 1900 units/mg.

7230. Penicillin V. *[2S-(2α,5α,6β)]-3,3-Dimethyl-7-oxo-6-[(phenoxyacetyl)amino]-4-thia-1-azabicyclo[3.2.0]heptane-2-carboxylic acid; 6-phenoxyacetamidopenicillanic acid;* penicillin phenoxymethyl; phenoxymethylpenicillin; phenoxymethylpenicillinic acid; Acipen-V; Distaquaine V; Fenospen; Meropenin; Oracilline; Oratren; V-Cillin. $C_{16}H_{18}N_2O_5S$; mol wt 350.40. C 54.85%, H 5.18%, N 7.99%, O 22.83%, S 9.15%. Obtained by adding 2-phenoxyethanol to the *Penicillium* culture using yeast autolyzate as source of nitrogen: Brandl *et al., Wien. Med. Wochenschr.* **1953**, 602; Brandl, Margreiter, *Oesterr. Chem.-Ztg.* **55**, 11-21 (1954), *C.A.* **48**, 10296 (1954). Purification: Parker *et al., J. Pharm. Pharmacol.* **7**, 683 (1953). Total synthesis: Sheehan, Henery-Logan, *J. Am. Chem. Soc.* **79**, 1262 (1957); **81**, 3089 (1959); **84**, 2983 (1962). Prepn from 6-aminopenicillanic acid: Glambitza, *Ann.* **673**, 166 (1964). Soly data: Weiss *et al., Antibiot. & Chemother.* **7**, 374 (1957). The biologically active form is the dextrorotatory D-form; the DL-form is half as active. L-Penicillin V has little, if any, antibiotic activity. Toxicity data: E. I. Goldenthal, *Toxicol. Appl. Pharmacol.* **18**, 185 (1971). Comprehensive description of the potassium salt: J. M. Dunham in *Anal. Profiles Drug Subs.* **1**, 249-300 (1972); D. H. Sieh, *ibid.* **17**, 677-748 (1988).

Crystals, dec 120-128°. Stable in air up to 37°; relatively stable to acid. uv max: 268, 274 nm (ε 1330, 1100). Soly in water at pH 1.8 (acidified with HCl) = 25 mg/100 ml. Sol in polar organic solvents. Practically insol in vegetable oils and in liq petrolatum.

Potassium salt, $C_{16}H_{17}KN_2O_5S$, *Antibiocin, Apsin VK, Arcasin, Beromycin, Betapen VK, Calciopen, Cliacil, Compocillin VK, Distakaps V-K, Distaquaine V-K, Dowpen VK, D.Q.V.K., Fenoxypen, Icipen, Isocillin, Ispenoral, Ledercillin VK, Megacillin, Orapen V-K, Ospen* (tabl.), *Ospeneff, Penagen, Penavlon V, Pencompren, Pen-Oral, Pen-Vee K, Pfizerpen VK, PVK, SK-Penicillin VK, Stabicilline, Stabillin V-K, Suspen, Uticillin VK, V-Cil-K, V-Cillin K, Veetids, Vepen.* Soluble in water. $[α]_D^{25}$ +223° (c = 0.2). LD_{50} orally in rats: >1040 mg/kg (Goldenthal).

Sodium salt, $C_{16}H_{17}N_2NaO_5S$, soluble in water.

Calcium salt, $C_{32}H_{34}CaN_4O_{10}S_2$, *Calcipen, Penicals, Septocillin.*

THERAP CAT: Antibacterial.

THERAP CAT (VET): Antibacterial.

7231. Penicillin V Benzathine. *[2S-(2α,5α,6β)]-3,3-Dimethyl-7-oxo-6-[(phenoxyacetyl)amino]-4-thia-1-azabicyclo[3.2.0]heptane-2-carboxylic acid compd with N,N'-bis(phenylmethyl)-1,2-ethanediamine (2:1);* penicillin V DBED; benzathine penicillin V; benzathine benzylpenicillin; Bicillin L-A; Falcopen-V; Ospen (syrup); Pen-Vee. Semi-synthetic antibiotic. Prepn: R. Brunner *et al.*, U.S. pat. 2,820,789 (1958 to American Home Products).

Practically white powder having a characteristic odor, mp 105-109°. Solubilities determined by Weiss *et al., Antibiot. &*

Chemother. **7**, 374 (1957) in mg/ml at about 28°: water 0.321; ethanol 14.6.

THERAP CAT: Antibacterial.

7232. Penicillin V Hydrabamine. *N,N'-Bis(dehydroabietyl)ethylenediamine bis(phenoxymethylpenicillin);* hydrabamine phenoxymethyl penicillin; hydrabamine penicillin V; Abbocillin V; Compocillin-V. $C_{74}H_{100}N_6O_{10}S_2$; mol wt 1297.77. C 68.49%, H 7.77%, N 6.48%, O 12.33%, S 4.94%. Semi-synthetic antibiotic. Prepn: DeRose, U.S. pat. 2,812,326 (1956 to Abbott).

Solubilities determined by Weiss *et al., Antibiot. & Chemother.* **7**, 374 (1957) in mg/ml at about 28°: Water 0.05; methanol 11.05; ethanol 5.8; isopropanol 1.75; isoamyl alcohol 6.85; cyclohexane 0.12; benzene 1.4; toluene 1.07; petr ether 0.06; isooctane 0.065; carbon tetrachloride 3.30; ethyl acetate 4.0; isoamyl acetate 4.9; acetone 10.2; methyl ethyl ketone 13.7; ether 0.095; ethylene chloride > 20; dioxane 7.5.

THERAP CAT: Antibacterial.

7233. Penicilloyl Polylysine. Benzylpenicilloyl polylysine; PPL; Cilligen; Pre-Pen; Testarpen. Prepn from polylysine and a penicillenic acid: Parker *et al., J. Exp. Med.* **115**, 803 (1962). Prepn and use as diagnostic aid: M. A. Stahmann, S. S. Wagle, **Brit.** pat. 1,226,773; *eidem,* U.S. pat. 3,979,508 (1971, 1976 both to Kremers-Urban Co.). Intradermal test for penicillin sensitivity: Brown *et al., J. Am. Med. Assoc.* **189**, 599 (1964); Van Arsdale, *ibid.* **191**, 238 (1965); T. J. Sullivan, *J. Allergy Clin. Immunol.* **68**, 171 (1981).

THERAP CAT: Diagnostic aid (penicillin sensitivity).

7234. Penillic Acids. 2,3,7,7a-Tetrahydro-2,2-dimethylimidazo[5,1-*b*]thiazole-3,7-dicarboxylic acid derivatives. Degradation products of penicillins obtained by acid hydrolysis at pH 2. Review and bibliography: Abrahams in *Antibiotics* vol. 2, H. W. Florey *et al.*, Eds. (Oxford University Press, New York, 1949) p 889.

7235. Penimepicycline. *[2S-(2α,5α,6β)]-3,3-Dimethyl-7-oxo-6-[(phenoxyacetyl)amino]-4-thia-1-azabicyclo[3.2.0]-heptane-2-carboxylic acid compd with [4S-(4α,4aα,5aα,6β,-12aα)]-4-(dimethylamino)-1,4,4a,5,5a,6,11,12a-octahydro-3,-6,10,12,12a-pentahydroxy-N-[[4-(2-hydroxyethyl)-1-piperazinyl]methyl]-6-methyl-1,11-dioxo-2-naphthacenecarboxamide (1:1);* N-[4-(β-hydroxyethyl)diethylenediaminomethyl]tetracycline phenoxymethylpenicillinate; mepicycline phenoxymethylpenicillinate; penicillin V compd with mepicycline; penimepiciclina; Criseocil; Geotricyn; Mepenicycline; Olimpen; Penetracyne; Peniltetra; Prestociclina. $C_{45}H_{56}N_6O_{14}S$; mol wt 937.04. C 57.68%, H 6.02%, N 8.97%, O 23.90%, S 3.42%. $C_{16}H_{18}N_2O_5S.C_{29}H_{38}N_4O_9$. Semi-synthetic antibiotic. Prepn: Pedrazzoli *et al., Boll. Chim. Farm.* **98**, 516 (1959), *C.A.* **54**, 3856a (1960); Gradnik, Pedrazzoli, **Brit.** pat. 897,-826 (1962 to E.R.A.S.M.E.), *C.A.* **57**, 15034e (1962). Properties: Gradnik *et al., Pharm. Acta Helv.* **35**, 529 (1960).

Yellowish-white cryst powder, dec above 143° (not corr). Sensitive to light, heat, air. Slightly bitter taste. pH of 2% aq soln, 5.5-5.7. Soly in water at 20° = 1 g/0.7 ml. $[α]_D^{20}$ −50.5° (c = 2 in methanol).

THERAP CAT: Antibacterial.

7236. Pennyroyal. Hedeoma; squaw mint; mosquito plant. Dried leaves and flowering tops of *Hedeoma pulegioides* (L.) Pers., *Labiatae. Habit.* Canada to Florida and west to Nebraska. *Constit.* Volatile oil (1%), bitter principles, tannin.

THERAP CAT: Diaphoretic, emmenagogue.

7237. Pentaborane(9). Pentaboron nonahydride. B_5H_9; mol wt 63.13. B 85.63%, H 14.37%. Prepd from diborane: Stock, Mathing, *Ber.* **69B**, 1456 (1936); Schlesinger,

Burg, *J. Am. Chem. Soc.* **53**, 4321 (1931); **55**, 4009 (1933). Molecular structure determination by rotational spectroscopy: D. Schwoch *et al., Inorg. Chem.* **16**, 3219 (1977). Review of toxicity: *see* Decaborane(14). *Review:* Greenwood in *Comprehensive Inorganic Chemistry* vol. 1, J. C. Bailar, Jr. *et al.,* Eds. (Pergamon Press, Oxford, 1973) pp 792-801.

Liquid. mp −46.6°; bp 60°. d_4^0 0.61; vp 66 mm Hg at 0°. Dec very slowly at 150°. Ignites spontaneously in air. Hydrolyzes in water after long heating. Reacts with ammonia to form a diammoniate.

Caution: Potential symptoms of overexposure are dizziness, headache, drowsiness and lightheadedness; incoordination, tremor and tonic spasm of face, neck, abdomen and limbs; convulsions; irritation of eyes and skin. *See NIOSH Pocket Guide to Chemical Hazards* (DHHS/NIOSH 90-117, 1990) p 172.

7238. Pentaborane(11). Dihydropentaborane(9); pentaboron undecahydride. B_5H_{11}; mol wt 65.14. B 82.98%, H 17.02%. Prepd from diborane: Burg, Schlesinger, *J. Am. Chem. Soc.* **55**, 4009 (1933).

Liquid. mp −123°. bp 63°; bp$_{53}$ 0°. Unstable. When heated or allowed to stand for long periods of time, it produces diborane, tetraborane, hydrogen, pentaborane, decaborane and brown nonvolatile liqs and solids. Ignites spontaneously in air. Hydrolyzes in water to boric acid and hydrogen. Reacts with ammonia to form a tetraammonate.

7239. Pentabromoacetone. *1,1,1,3,3-Pentabromo-2-propanone.* C_3HBr_5O; mol wt 452.56. C 7.96%, H 0.22%, Br 88.28%, O 3.54%. $CBr_3COCHBr_2$. Prepd by the addition of 12 parts bromine to 1 part acetone: Mulder, *Jahresbericht über die Fortschritte der Chemie* 1864, 330.

Orthorhombic needles or prisms from alcohol or ether. Penetrating odor. mp 76° (sublimes above mp). Volatile with steam. Practically insol in water. Freely sol in organic solvents. Forms bromoform under the influence of alkalies.

7240. Pentacene. Benzo[*b*]naphthacene; 2,3,6,7-dibenzoanthracene; *lin*-naphthoanthracene. $C_{22}H_{14}$; mol wt 278.35. C 94.93%, H 5.07%. Synthesis from *m*-xylene or from 4-benzyl-1,3-dimethylbenzene and benzoyl chloride in presence of aluminum chloride or from terephthalyl chloride and *o*-tolylmagnesium bromide: Clar, John, *Ber.* **62**, 940 (1929); **63**, 2967 (1930); **64**, 981 (1931). Synthesis by reduction of 6,13-pentacenequinone: Bruckner *et al., Tetrahedron Letters* **1960**, no. 1, 5; Bruckner, Tomasz, *Acta Chim. Acad. Sci. Hung.* **28**(4), 405 (1961). Structure: Campbell *et al., Acta Cryst.* **14**, 705 (1961).

Deep blue needles with violet luster from hot nitrobenzene. Sublimes in CO_2 stream under reduced pressure at ∼300° (Clar, John, *loc. cit.*). In presence of air dec > 300°. Practically insol in water; sparingly sol in organic solvents.

Note: Anthracene (3 linear rings) is colorless; naphthacene (4 linear rings) is orange; pentacene (5 linear rings) is blue; hexacene (6 linear rings) is green.

7241. Pentachloroethane. Pentalin. C_2HCl_5; mol wt 202.29. C 11.87%, H 0.50%, Cl 87.63%. CCl_3CHCl_2. Toxicity data: G. S. Barsoum, K. Saad, *Quart. J. Pharm. Pharmacol.* **7**, 205 (1934).

Liquid; chloroform-like odor. d_4^{25} 1.6712; bp 161-162°. mp −29°. n_D^{15} 1.5054. Insol in water. Miscible with alcohol, ether. MLD (mg/kg) in dogs: 1750 orally; 100 i.v.; in rabbits: 700 s.c. (Barsoum, Saad).

Caution: Irritant; narcotic.

7242. Pentachlorophenol. Penta; PCP; penchlorol; Santophen 20. C_6HCl_5O; mol wt 266.34. C 27.06%, H 0.38%, Cl 66.56%, O 6.01%. Prepd by the chlorination of phenol in the presence of a catalyst. Toxicity study: T. B. Gaines, *Toxicol. Appl. Pharmacol.* **14**, 515 (1969).

Needle-like crystals, mp 190-191°; bp ∼309-310° (dec). d_4^{22} 1.978. Very pungent odor only when hot. Sublimes in needles. Almost insol in water (8 mg in 100 ml). Freely sol in alc, ether; sol in benzene; slightly sol in cold petr ether. LD$_{50}$ in male, female rats (mg/kg): 146, 175 orally (Gaines).

Sodium salt, **sodium pentachlorophenate, sodium pentachlorophenoxide, Santobrite, Dowicide G.** Sol in water.

Caution: Ingestion causes increase then decrease of respiration, blood pressure, urinary output; fever; increased bowel action; motor weakness; collapse with convulsions and death. Causes lung, liver, kidney damage; contact dermatitis. May be absorbed through skin. More toxic in organic solvents. Dust causes sneezing. *See:* Patty's *Industrial Hygiene and Toxicology,* vol. 2A, G. D. Clayton, F. E. Clayton, Eds. (Wiley-Interscience, New York, 1981) pp 2604-2612.

USE: Insecticide for termite control; pre-harvest defoliant; general herbicide. Has been recommended for use in the preservation of wood, wood products, starches, dextrins, glues.

7243. Pentacynium Bis(methylsulfate). *4-[2-[(5-Cyano-5,5-diphenylpentyl)dimethylammonio]ethyl]-4-methylmorpholinium bis(methylsulfate); N^1-(5-cyano-5,5-diphenylpentyl)-N^1,N^1,N^2-trimethylethylene-1-ammonium-2-morpholinium bis(methylsulfate); ethylene-1-[(5-cyano-5,5-diphenylpentyl)dimethylammonium]-(4-methylmorpholinium) bis(methylsulfate); pentacyone mesylate; Presidal.* $C_{29}H_{45}N_3O_9S_2$; mol wt 643.82. C 54.10%, H 7.04%, N 6.53%, O 22.37%, S 9.96%. Ganglionic blocker. Prepn: Billinghurst, U.S. pat. 2,851,458 (1958 to Burroughs Wellcome).

Crystals from ethanol, mp 173-175°. Sol in water.
THERAP CAT: Antihypertensive.

7244. 3-Pentadecylcatechol. *3-Pentadecyl-1,2-benzenediol; 3-pentadecylpyrocatechol;* tetrahydrourushiol; hydrourushiol; dihydrorhengol; 3-PDC. $C_{21}H_{36}O_2$; mol wt 320.52. C 78.70%, H 11.32%, O 9.98%. Constituent of the irritant oil of poison ivy (*Toxicodendron radicans* (L.) Kuntze) and other *Toxicodendron* spp. (*Anacardiaceae*). Prepn by hydrogenation of extracts from fruits of *Semecarpus heterophylla:* Backer, Haack, *Rec. Trav. Chim.* **57**, 225 (1938). Synthesis from 2,3-dimethoxybenzaldehyde and tetradecyl chloride: Mason, *J. Am. Chem. Soc.* **67**, 1538 (1945); from *o*-veratraldehyde: Backer, Haack, *loc. cit.;* Dawson *et al., J. Am. Chem. Soc.* **68**, 534 (1946); Keil *et al.,* U.S. pat. 2,451,955 (1948); from 2,3-dibenzyloxybenzaldehyde: Loev, Dawson, *J. Org. Chem.* **24**, 980 (1959); from furan derivs: Boehme, *J. Am. Chem. Soc.* **82**, 499 (1960); from catechol: Hanafusa, Yukawa, *Chem. & Ind. (London)* **1961**, 23. Criticism of reported syntheses and synthesis of dimethyl deriv: Byck, Dawson, *J. Org. Chem.* **32**, 1084 (1967). Total synthesis: E. Wenkert *et al., J. Am. Chem. Soc.* **105**, 2021 (1983). Evaluation of diagnostic patch test: M. V. Dahl *et al., Arch. Dermatol.* **120**, 1022 (1984).

Short needles from toluene or petr ether, mp 59-60°. Can be purified by molecular distn. uv max: 277 nm. Sol in alc, ether, benzene, toluene. Sparingly sol in petr ether.

THERAP CAT: Diagnostic aid (contact allergen).

7245. Pentaerythritol. *2,2-Bis(hydroxymethyl)-1,3-propanediol; tetrakis(hydroxymethyl)methane;* tetramethylolmethane; Metab-Auxil; Penetek; Pentek. $C_5H_{12}O_4$; mol wt 136.15. C 44.11%, H 8.88%, O 47.01%. Prepd by treating acetaldehyde with formaldehyde in an aq soln of calcium hydroxide: H. B. J. Schurink, *Org. Syn.* **coll. vol. I**, 425 (2nd ed., 1941); Fieser, Fieser, *Organic Chemistry* (2nd ed, 1950) p 133. Review of mfg processes: P. W. Sherwood, *Petroleum Refiner,* **Nov. 1956**, p 171-179; Faith, Keyes & Clark's *Industrial Chemicals,* F. A. Lowenheim, M. K. Moran, Eds. (Wiley-Interscience, New York, 4th ed., 1975) pp 598-603. Monograph: E. Berlow *et al., The Pentaerythritols,* ACS Monograph Series no. **136** (Reinhold, New York, 1958).

Ditetragonal crystals from dil HCl, mp 260°. One gram dissolves in 18 ml water at 15°. Sol in ethanol, glycerol, ethylene glycol, formamide. Insol in acetone, benzene, paraffin, ether, carbon tetrachloride.

Trinitrate, *see* Pentrinitrol.

USE: In synthetic resins, in paint and varnish industries.

7246. Pentaerythritol Chloral. *1,1'-[2,2-Bis[(2,2,2-trichloro-1-hydroxyethoxy)methyl]-1,3-propanediylbis(oxy)]bis-[2,2,2-trichloroethanol];* Petrichloral; Periclor. $C_{13}H_{16}Cl_{12}O_8$; mol wt 725.70. C 21.52%, H 2.22%, Cl 58.62%, O 17.64%. Prepn: Bruce, U.S. pat. **2,784,237** (1957 to Am. Home Prod.).

Yellow glass (easily powdered), mp 52-54°. Less odor and aftertaste than choral hydrate. Sol in water. Sol to the extent of 30 or more g/100 ml in ethyl acetate, formamide, ethanol, chloroform, benzene; to the extent of 5 g/100 ml in tetrachloroethylene and isopropanol.

Note: This is a controlled substance (depressant) listed in the U.S. Code of Federal Regulations, Title 21 Part 1308.14 (1995).

THERAP CAT: Sedative, hypnotic.

7247. Pentaerythritol Dichlorohydrin. *2,2-Bis(chloromethyl)-1,3-propanediol;* Dispranol. $C_5H_{10}Cl_2O_2$; mol wt 173.04. C 34.71%, H 5.82%, Cl 40.98%, O 18.49%. $(HOCH_2)_2C(CH_2Cl)_2$. Prepn: Fecht, *Ber.* **40**, 3883 (1907); Rapoport, *J. Am. Chem. Soc.* **68**, 341 (1946); Dee, U.S. pat. **2,763,679** (1956 to Heyden Chem.).

Crystals from water or benzene, mp 79-80° (uncorr), 83° (corr), also reported as 65° and 95°, *see* Rapoport, *loc. cit.* bp_{12} 158.5-160°. Sublimes in high vacuum.

THERAP CAT: Tranquilizer; muscle relaxant.

7248. Pentaerythritol Tetraacetate. *2,2-Bis[(acetyloxy)methyl]-1,3-propanediol diacetate;* tetra-*O*-acetylpentaerythritol; pentaerythrityl tetraacetate; Normosterol. $C_{13}H_{20}O_8$; mol wt 304.30. C 51.31%, H 6.62%, O 42.06%.

Prepn: Wolfrom *et al., J. Am. Chem. Soc.* **73**, 874 (1951); Bonner *et al., J. Chem. Soc.* **1960**, 2914.

Crystals, mp 83-84°.

THERAP CAT: Antilipemic.

7249. Pentaerythritol Tetranitrate. *2,2-Bis[(nitrooxy)methyl]-1,3-propanediol dinitrate (ester);* pentaerythrityl tetranitrate; 2,2-bisdihydroxymethyl-1,3-propanediol tetranitrate; PETN; nitropentaerythritol; penthrit; niperyt; Lentrat; Hasethrol; Peritrate; Mycardol; Nitropenton; Pentral 80; Dilcoran-80; Terpate; Pentrite; Perityl; Pentritol; Pentanitrine; Prevangor; Subicard; Pentryate; Vasodiatol; Neo-Corovas; Pentafin; Quintrate; Pergitral; Pentitrate; Metranil; Cardiacap; Angitet; Nitropenta. $C_5H_8N_4O_{12}$; mol wt 316.14. C 19.00%, H 2.55%, N 17.72%, O 60.73%. Prepn: Acken, Vyverberg, U.S. pat. **2,370,437** (1945 to Du Pont). GC/MS analysis in post-explosion residues: T. Tamiri *et al.* in *Adv. Anal. Detect. Explosives,* J. Yinon, Ed. (Kluwer, Netherlands, 1993) pp 323-334.

Tetragonal holohedra from acetone + alcohol, mp 140°. d_4^{20} 1.773. Soluble in acetone. Practically insoluble in water (1.5 γ/ml). Sparingly sol in alcohol, ether. Does not reduce Fehling's soln (difference from erythrityl tetranitrate). *Caution:* Explodes on percussion. More sensitive to shock than TNT. For medicinal purposes it is dil with an inert ingredient, usually lactose, to prevent accidental explosions.

USE: Mainly in the manuf of detonating fuse (Primacord), a waterproof textile filled with powdered PETN.

THERAP CAT: Vasodilator (coronary).

7250. Pentagastrin. *N-[(1,1-Dimethylethoxy)carbonyl]-β-alanyl-L-tryptophyl-L-methionyl-L-α-aspartyl-L-phenylalaninamide; N-carboxy-β-alanyl-L-tryptophyl-L-methionyl-L-aspartylphenyl-L-alaninamide N-tert-butyl ester;* N-(α-carbamoylphenethyl)-3-[2-[2-[3-(carboxyamino)propionamido]-3-indol-3-ylpropionamido]-4-(methylthio)butyramido]succinamic acid *N-tert*-butyl ester; *N*-[*N*-[*N*-[*N*-(*N-tert*-butoxycarbonyl-β-alanyl)-L-tryptophanyl]-L-methionyl]-L-aspartyl]-L-phenylalaninamide; Boc-β-Ala-Trp-Met-Asp-Phe(NH$_2$); AY-6608; ICI-50123; Gastrodiagnost; Peptavlon. $C_{37}H_{49}N_7O_9S$; mol wt 767.90. C 57.87%, H 6.43%, N 12.77%, O 18.75%, S 4.18%. Prepn: Hardy *et al.,* **Belg. pat. 665,591** corresp to U.S. pat. **3,896,103** (1965, 1975 to I.C.I.); Davey *et al., J. Chem. Soc. (C)* **1966**, 555; Sakakibara *et al., Bull. Chem. Soc. Japan* **41**, 438 (1968). Structure-function relationship studies: Morley *et al., Nature* **207**, 1356 (1965). Formulation and pharmacological studies: Wai *et al., J. Pharm. Pharmacol.* **22**, 923 (1970). *Reviews:* Makhlouf, *Fed. Proc.* **27**, 1322 (1968); Sanders, Schimmel, *Am. J. Med.* **49**, 380 (1970).

Fine, colorless needle-shaped crystals from 2-ethoxyethanol + water; mp 229-230° (dec). $[\alpha]_D^{22}$ −28.8° ±0.5° (DMF). uv max (2*N* NH$_4$OH): 280, 289 nm (ε 5340, 4590). Sol in DMF, DMSO. Almost insol in water, ethanol, ether, benzene.

THERAP CAT: Diagnostic aid (gastric secretion stimulant).

7251. Pentagestrone. *3-(Cyclopentyloxy)-17-hydroxypregna-3,5-dien-20-one;* 17α-hydroxyprogesterone 3-cyclopentyl enol ether. $C_{26}H_{38}O_3$; mol wt 398.59. C 78.35%, H

9.61%, O 12.04%. Prepn of acetate: Ercoli, Gardi, *J. Am. Chem. Soc.* **82**, 746 (1960); of free alcohol and acetate: Ercoli, U.S. pat. **3,019,241** (1962).

Solid, mp 184.5-186.5°. [α]_D −115° (dioxane).

Acetate, C₂₈H₄₀O₄, *17α-acetoxyprogesterone 3-cyclopentyl enol ether, Gestovis.* Solid, mp 137-138°, also reported as 157-158°, *see* Ercoli, **Brit.** pat. **893,315** (1962 to Vismara). [α]_D −147° (dioxane).

THERAP CAT: Progestogen.

7252. Pentahomoserine. *5-Hydroxynorvaline;* α-amino-δ-hydroxy-*n*-valeric acid; 2-amino-5-hydroxypentanoic acid. C$_5$H$_{11}$NO$_3$; mol wt 133.15. C 45.10%, H 8.33%, N 10.52%, O 36.05%. Synthesis: Sörensen, *Bull. Soc. Chim. France* **33**, 1052 (1904); Plieninger, *Ber.* **83**, 271 (1950); Gaudry, *Can. J. Chem.* **29**, 544 (1951). Enzymatic resolution: Berlinguet, Gaudry, *J. Biol. Chem.* **198**, 765 (1952).

L-Form, crystals. [M]_D +38.3° (5*N* HCl); [M]_D +46.6° (glacial acetic acid). [α]$_D^{23}$ −28.8° (c = 2 in 6*N* HCl).

D-Form, crystals from dil alcohol. [α]$_D^{23}$ −28.0° (c = 2 in 6*N* HCl).

DL-Form, crystals from dil alcohol, mp 218°; 218-220°; 223-224°, depending on rate of heating.

7253. Pentamethonium Bromide. *N,N,N,N',N',N'-Hexamethyl-1,5-pentanediaminium bromide; pentamethylenebis-[trimethylammonium bromide];* α,ω-bis(trimethylammonium)pentane dibromide; 1,5-dibromohexamethylpentaneammonium; C-5; Penthonium; Lytensium. C$_{11}$H$_{28}$Br$_2$N$_2$; mol wt 348.16. C 37.95%, H 8.11%, Br 45.90%, N 8.05%. Ganglion blocking agent.

Hygroscopic crystals, slight fishy odor, slightly bitter, saline taste. Aq solns are stable, almost neutral; can be sterilized by autoclaving. The compd is almost completely ionized in aq solns. It contains 54.10% of pentamethonium ion.

THERAP CAT: Antihypertensive.

7254. Pentamidine. *4,4'-[1,5-Pentanediylbis(oxy)]bis-benzenecarboximidamide; 4,4'-(pentamethylenedioxy)dibenzamidine;* 4,4'-diamidino-α,ω-diphenoxypentane; C$_{19}$H$_{24}$N$_4$O$_2$; mol wt 340.43. C 67.04%, H 7.11%, N 16.46%, O 9.40%. Prepn: A. J. Ewins, **Brit.** pat. **507,565** (1939); J. N. Ashley *et al., J. Chem. Soc.* **1942**, 103; of isethionate: G. Newbery, A. P. T. Easson, U.S. pat. **2,394,003** (1946 to May & Baker). Trypanocidal activity: E. M. Lourie, W. Yorke, *Ann. Trop. Med. Parasitol.* **33**, 289 (1939). Preliminary pharmacological studies in animals: R. Wien, *ibid.* **37**, 1 (1943). Activity in fibrinolytic systems: J. D. Geratz, *Thromb. Diath. Haemorrh.* **29**, 154 (1973). Pharmacodynamics in men and mice: T. P. Waalkes *et al., Clin. Pharmacol. Ther.* **11**, 505 (1970). *In vitro* activity against *Pneumocystis carinii:* E. L. Pesanti, C. Cox, *Infec. Immun.* **34**, 908 (1981). Uptake and distribution of aerosolized form in animals: R. J. Debs *et al., Am. Rev. Respir. Dis.* **135**, 731 (1987). *In vivo* efficacy of aerosolized form in rats: *eidem, Antimicrob. Ag. Chemother.* **31**, 37 (1987). Determn in plasma, urine and tissues: T. P. Waalkes, V. T. DeVita, *J. Lab. Clin. Med.* **75**, 871 (1970); by HPLC: C. M. Dickinson *et al., J. Chromatog.* **345**, 91 (1985). Preliminary clinical evaluation in *P. carinii* pneumonia: V. T. DeVita *et al., N. Engl.*

J. Med. **280**, 287 (1968). Comparison with sulfamethoxazole-trimethoprim mixture in *P. carinii* pneumonia in AIDS: J. M. Wharton *et al., Ann. Int. Med.* **105**, 37 (1986). Early review of pharmacology, mode of action and clinical applications: E. B. Schoenbach, E. M. Greenspan, *Medicine* **27**, 327-377 (1948). *Review:* S. Drake *et al., Clin. Pharm.* **4**, 507-516 (1985); M. Sands *et al., Rev. Infec. Dis.* **7**, 625-634 (1985); of activity, pharmacokinetics and therapeutic use: K. L. Goa, D. M. Campoli-Richards, *Drugs* **33**, 242-258 (1987).

Crystallizes as colorless plates from water. Dec 186°.

Dihydrochloride, C$_{19}$H$_{24}$N$_4$O$_2$.2HCl, fine needles from dil HCl, mp 232-234°. LD$_{50}$ in mice (mg/g): 0.028 i.v.; 0.064 s.c. (Wein).

Isethionate, C$_{23}$H$_{36}$N$_4$O$_{10}$S$_2$, *M & B 800, RP-2512, Aeropent, Banambax, NebuPent, Pentacarinat, Pentam, Pneumopent.* Hygroscopic, very bitter crystals, mp ~180°. Slight butyric odor. Sol in water (~1 in 10 at 25°, ~1 in 4 at 100°); sol in glycerol, more readily on warming; slightly sol in alcohol. Insol in ether, acetone, chloroform, liq petr. pH of a 5% w/v soln in water: 4.5 to 6.5.

Dimethanesulfonate, C$_{21}$H$_{32}$N$_4$O$_8$S$_2$, *pentamidine mesylate, Lomidine.* White powder.

THERAP CAT: Antiprotozoal (Trypanosoma, Leishmania); antipneumocystis.

THERAP CAT (VET): Antiprotozoal (Babesia, Leishmania).

7255. Pentane. *n*-Pentane. C$_5$H$_{12}$; mol wt 72.15. C 83.24%, H 16.76%. CH$_3$CH$_2$CH$_2$CH$_2$CH$_3$. Occurs in petroleum; it is a constituent of petr ether. Sepn from natural gasoline: Love, *Petrol. Eng.* **12**, no. 10, 130 sqq (1941), *C.A.* **35**, 7162 (1941); from virgin naphthas: Tongberg *et al., Ind. Eng. Chem.* **30**, 166 (1938). Prepd by dehydration and subsequent hydrogenation of 2- and 3-pentanol: Mair, *Bur. Stand. J. Res.* **9**, 457 (1932); from 2-bromopentane by Grignard reaction: Noller, *Org. Syn.* **11**, 84 (1931).

Flammable liq. d$^0_0$ 0.64529; d$^{20}_4$ 0.62638; d$^{30}_4$ 0.6163; mp −129.7°. bp$_{760}$ 36.1°; bp$_{400}$ 18.5°; bp$_{200}$ +1.9°; bp$_{100}$ −12.6°; bp$_{60}$ −22.2°; bp$_{40}$ −29.2°; bp$_{20}$ −40.2°; bp$_{10}$ −50.1°; bp$_5$ −62.5°; bp$_{1.0}$ −76.6°; n$^{20}_D$ 1.35768. Flash pt, closed cup: < −40°F (−40°C). Explosive limits, % by vol in air: lower 1.4; upper 8.0. Autoignition temp +588°F (+309°C). Soly in water at 16°: 0.36 g/l. Miscible with alc, ether, many organic solvents. Lethal concn for mice in air: 128200 ppm, *Handbook of Toxicology* vol. 1, W. S. Spector, Ed. (Saunders, Philadelphia, 1956) pp 346-347.

Caution: Potential symptoms of overexposure are drowsiness; irritation of eyes and nose; dermatitis; chemical pneumonia. *See NIOSH Pocket Guide to Chemical Hazards* (DHHS/NIOSH 90-117, 1990) p 174.

7256. 1,5-Pentanediol. Pentamethylene glycol; 1,5-dihydroxypentane. C$_5$H$_{12}$O$_2$; mol wt 104.15. C 57.66%, H 11.61%, O 30.72%. CH$_2$(CH$_2$)$_3$CH$_2$OH. Prepd by hydrogenolysis of tetrahydrofurfuryl alcohol in the presence of copper chromite: Connor, Adkins, *J. Am. Chem. Soc.* **54**, 4678 (1932); D. Kaufman, W. Reeve, *Org. Syn.* **coll. vol. III**, 693 (1955). Toxicity study: H. F. Smyth *et al., Am. Ind. Hyg. Assoc. J.* **23**, 95 (1962).

Viscous, oily liquid. Bitter taste. d^{20} 0.9941. mp −18°. bp$_{760}$ 239°, bp$_{3.0}$ 120°. n$^{20}_D$ 1.4499. Flash pt 125° (275°F). Miscible with water, methanol, alc, acetone, ethyl acetate. Soly in ether (25°): 11% w/w. Limited soly in benzene, trichloroethylene, methylene chloride, petr ether, heptane. LD$_{50}$ orally in rats: 5.89 g/kg (Smyth).

USE: As plasticizer in cellulose products and adhesives, in brake fluid compositions. Forms esters and polyesters which can be used as plasticizers, emulsifying agents and resin intermediates.

7257. 1-Pentanol. Pentyl alcohol; *n*-amyl alcohol; *n*-butyl carbinol. C$_5$H$_{12}$O; mol wt 88.15. C 68.13%, H 13.72%, O 18.15%. CH$_3$(CH$_2$)$_4$OH. Prepn from 1-pentene:

Brown, Rao, *J. Am. Chem. Soc.* **81**, 6434 (1959); Brown, U.S. pat. **2,925,437** (1960). Toxicity study: P. M. Jenner *et al., Food Cosmet. Toxicol.* **2**, 327 (1964). Review of manuf by fractionation of fusel oil and via chlorination of pentanes, and properties: *Industrial Chemicals*, W. L. Faith *et al.*, Eds. (John Wiley, New York, 2nd ed., 1957) pp 107-114.

Liquid, mild characteristic odor. bp 137.5°. mp −79°. d_4^{20} 0.8146; d_4^{25} 0.8110. n_D^{20} 1.4103: Mumford, Phillips, *J. Chem. Soc.* **1950**, 75. Flash pt, closed cup: 100°F (38°C). Slightly sol in water (2.7 g/100 ml at 22°); misc with alc, ether. LD_{50} orally in rats: 3030 mg/kg (Jenner).

Caution: Irritating to eyes, respiratory passages. Narcotic: E. Browning, *Toxicity and Metabolism of Industrial Solvents* (Elsevier, New York, 1965) pp 356-367.

USE: In organic syntheses; as solvent.

7258. 2-Pentanol. *dl-sec*-Amyl alcohol; methyl propyl carbinol. $C_5H_{12}O$; mol wt 88.15. C 68.13%, H 13.72%, O 18.15%. $CH_3CH_2CH_2CH(OH)CH_3$. Prepn: Brown, Nakagawa, *J. Am. Chem. Soc.* **77**, 3614 (1955); Brown, Wheeler, *ibid.* **78**, 2199 (1956). Sepn of optical isomers: Adembri, *Ann. Chim. (Rome)* **46**, 62 (1956). Review of manuf by fractionation of fusel oil and via chlorination of pentanes, and properties: *Industrial Chemicals*, W. L. Faith *et al.*, Eds. (John Wiley, New York, 2nd ed., 1957) pp 107-114.

Liquid, characteristic odor. bp 119.3°. bp_{745} 118°. d_4^{20} 0.8098. n_D^{25} 1.4041, n_D^{20} 1.406. Flash pt (open cup) 105°F. Slightly sol in water (16.6 g/100 ml at 20°). Miscible with alcohol, ether.

Caution: See 1-Pentanol.

7259. 3-Pentanol. Diethyl carbinol. $C_5H_{12}O$; mol wt 88.15. C 68.13%, H 13.72%, O 18.15%. $CH_3CH_2CH(OH)CH_2CH_3$. Prepn: Brown, Nakagawa, *J. Am. Chem. Soc.* **77**, 3614 (1955). Toxicity study: H. F. Smyth *et al., Arch. Ind. Hyg. Occup. Med.* **10**, 61 (1954). Review of manuf and properties: *Industrial Chemicals*, W. L. Faith *et al.*, Eds. (John Wiley, New York, 2nd ed., 1957) pp 107-114.

Liquid, characteristic odor. bp 115.6°. bp_{738} 113.5-113.7°. d_4^{25} 0.815; n_D^{25} 1.4077, n_D^{20} 1.4097. Slightly sol in water (5.5 g/100 g at 30°); sol in alcohol, ether. LD_{50} orally in rats: 1.87 g/kg (Smyth).

Caution: See 1-Pentanol.

USE: Flotation agent; solvent in organic synthesis.

7260. Pentapiperide. *α-(1-Methylpropyl)benzeneacetic acid 1-methyl-4-piperidinyl ester; 3-methyl-2-phenylvaleric acid 1-methyl-4-piperidyl ester; 2-phenyl-3-methylpentanoic acid 1-methyl-4-piperidyl ester; 1-methyl-4-piperidyl 3-methyl-2-phenylvalerate.* $C_{18}H_{27}NO_2$; mol wt 289.42. C 74.70%, H 9.40%, N 4.84%, O 11.06%. Prepn: **Brit. pat. 781,382** and Martin, Habicht, U.S. pat. **2,987,517** (1957, 1961 to Cilag).

Fumarate, $C_{22}H_{31}NO_6$, *C-4675*. Crystals, mp 91-93°.

Methosulfate, $C_{20}H_{33}NO_6S$, *pentapiperium methylsulfate, pentapiperide methylsulfate, Crylène, Hycholin, Quilene.* Crystals, mp 110-112°.

THERAP CAT: Anticholinergic; antispasmodic.

THERAP CAT (VET): Anticholinergic.

7261. Pentazocine. *(2α,6α,11R*)-1,2,3,4,5,6-Hexahydro-6,11-dimethyl-3-(3-methyl-2-butenyl)-2,6-methano-3-benzazocin-8-ol;* 2-dimethylallyl-5,9-dimethyl-2'-hydroxybenzomorphan; 3-(3-methyl-2-butenyl)-1,2,3,4,5,6-hexahydro-6,11-dimethyl-2,6-methano-3-benzazocin-8-ol; NSC-107430; Win 20228; Fortral; Pentagin; Sosegon. $C_{19}H_{27}NO$; mol wt 285.43. C 79.95%, H 9.53%, N 4.91%, O 5.61%. Mixed opioid agonist-antagonist. Prepn: Archer, **Belg. pat. 611,000** (1962 to Sterling Drug), *C.A.* **58**, 2440b (1963);

Archer *et al., J. Med. Chem.* **7**, 123 (1964). Review of pharmacology: Brogden *et al., Drugs* **5**, 6-91 (1973). Clinical study in post-operative pain: F. Camu, *Eur. J. Clin. Pharmacol.* **19**, 259 (1981). HPLC analysis in blood and plasma: R. D. Anderson *et al., J. Chromatog.* **227**, 239 (1982). Toxicity data: E. I. Goldenthal, *Toxicol. Appl. Pharmacol.* **18**, 185 (1971).

Crystals from methanol + water, mp 145.4-147.2°. LD_{50} s.c. in male rats: 175±36 mg/kg (Goldenthal).

Hydrochloride, $C_{19}H_{27}NO.HCl$, *Fortralin (tabl.), Fortalgesic (tabl.).*

Lactate, $C_{19}H_{27}NO.C_3H_6O_3$, *Liticon, Pentalgina, Talwin (inj.).*

Note: This is a controlled substance listed in the U.S. Code of Federal Regulations, Title 21 Part 1308.14 (1995).

THERAP CAT: Analgesic (narcotic).

THERAP CAT (VET): Analgesic (narcotic).

7262. 1-Pentene. Propylethylene; α-n-amylene. C_5H_{10}; mol wt 70.13. C 85.63%, H 14.37%. $CH_3(CH_2)_2CH=CH_2$. Occurs in coal tar. Prepd from allyl bromide and ethylmagnesium bromide in ether or better in dipropyl ether: Meisenheimer, Casper, *Ber.* **54**, 1663 (1921); Norris, Joubert, *J. Am. Chem. Soc.* **49**, 885 (1927); Kirrmann, *Compt. Rend.* **184**, 1178 (1927).

Liquid. d_4^{20} 0.6429; bp_{760} 30.1°; n_D^{20} 1.3714. Insol in water. Miscible with alcohol, ether, benzene.

7263. 2-Pentene. β-n-Amylene; *sym*-methylethylethylene. C_5H_{10}; mol wt 70.13. C 85.63%, H 14.37%. $CH_3CH_2CH=CHCH_3$. Prepn of mixture of *cis* and *trans* isomers by dehydration of 2-pentanol: Norris, *Org. Syn. coll. vol. I*, (2nd ed., 1941) p 430. Prepn of the geometrical isomers from *cis*- and *trans*-α-methyl-β-ethylacrylic acids: Lucas, Prater, *J. Am. Chem. Soc.* **59**, 1682 (1937). Prepn of the *cis* form from 1-ethyl-2-iodopropyl acid with quinoline, of the *trans* form from 1-ethyl-2-iodobutyric acid with sodium carbonate: Sherrill, Matlak, *ibid.* 2134; prepn of both isomers from *sec*-amyl alcohol with H_2SO_4 and diatomaceous earth at 90-110° for 3 hrs: Lucas *et al., ibid.* **63**, 22 (1941). Absorption spectra: Carr, Stücklen, *ibid.* **59**, 2138 (1937).

cis-Form, liquid. d_4^{20} 0.6503; d_4^{80} 0.5824; d_4^{30} 0.6392; d_4^0 0.6710. mp −180 to −178°. bp_{760} 37.0°. n_D^{20} 1.38130.

trans-Form, liquid. d_4^{20} 0.6482; d_4^{80} 0.5814; d_4^{30} 0.6381; d_4^0 0.6675. mp −136 to −135°. bp_{760} 35.85°. n_D^{20} 1.37921.

7264. 2-Pentenylpenicillin Sodium. *[2S-(2α,5α,6β)]-3,3-Dimethyl-7-oxo-6-[(1-oxo-3-hexenyl)amino]-4-thia-1-azabicyclo[3.2.0]heptane-2-carboxylic acid monosodium salt; Δ-β,γ-pentenylpenicillin sodium; sodium penicillin F; sodium penicillin I; sodium 2-pentenylpenicillinate.* $C_{14}H_{19}N_2NaO_4S$; mol wt 334.37. C 50.29%, H 5.73%, N 8.38%, Na 6.88%, O 19.14%, S 9.59%. 2-Pentenylpenicillin is produced by certain strains of *Penicillium chrysogenum* and *P. notatum*. Separation and purification by partition chromatography, *cf.* monograph by Boon and Carrington in *Chemistry of Penicillin* (Princeton, 1949). Prepn of cryst penicillin F: Wintersteiner, Adler, U.S. pat. **2,485,227** (1949 to Squibb).

Crystals. A trihydrate and a sesquihydrate have been obtained. The anhydr form exists in two cryst modifications. Usually long, fine, blunt-ended needles from water +

butanol. After drying *in vacuo* over P_2O_5 and then at 55-60° in high vacuum: mp 204-205° (with decompn, open capillary in block preheated to 200°). $[\alpha]_D^{25}$ +316° (c = 0.88). pK in water at 5° = 2.83; at 25° = 2.87. Very sol in water, in isotonic sodium chloride soln and in glucose solns; also sol in alcohol, but is inactivated by this solvent, likewise by glycerol and other primary alcohols; insol in benzene, carbon tetrachloride, liquid petrolatum.

7265. Pentetate Calcium Trisodium. *[N,N-Bis[2-[bis-(carboxymethyl)amino]ethyl]glycinato(5—)]calciate(3—)trisodium;* calcium trisodium pentetate; sodium[[[(carboxymethyl)imino]bis(ethylenenitrilo)]tetraacetato]calciate; [[(carboxymethyl)imino]bis(ethylenenitrilo)]tetraacetic acid calcium complex trisodium salt; trisodium calcium diethylenetriaminepentaacetate; pentacin; Calcium Chel 330; Ditripentat; Penthamil. $C_{14}H_{18}CaN_3Na_3O_{10}$; mol wt 497.36. C 33.81%, H 3.65%, Ca 8.06%, N 8.45%, Na 13.87%, O 32.17%. Prepn: Rubin, Dexter, U.S. pat. 3,062,719 (1962 to Geigy). Effects in uranium poisoning in rats: R. Dagirmanjian *et al., J. Pharmacol. Exp. Ther.* **117,** 20 (1956); in mixed fusion products poisoning: K. Kostial *et al., J. Appl. Toxicol.* **3,** 291 (1983).

Solid. Sol in water. Practically insol in alcohol. Used in aq soln. LD_{50} i.p. in rats: 3.8 g/kg (Dagirmanjian).

THERAP CAT: Chelating agent (plutonium and other transuranium elements).

7266. Pentetic Acid. *N,N-Bis[2-[bis(carboxymethyl)-amino]ethyl]glycine; [[(carboxymethyl)imino]bis(ethylenenitrilo)]tetraacetic acid;* pentacarboxymethyl diethylenetriamine; diethylenetriamine pentaacetic acid; DTPA. $C_{14}H_{23}N_3O_{10}$; mol wt 393.35. C 42.75%, H 5.89%, N 10.68%, O 40.67%. Prepn of the acid: Curme *et al.,* U.S. pat. 2,384,-816 (1945 to Carbide and Carbon Chem.); Samoilova, Yashunskii, U.S.S.R. pat. 144,479 (1962), *C.A.* **57,** 12326d (1962); of Na salt: Brit. pats. 601,816/7 (1948 to Carbide and Carbon Chem.). Structure and infrared spectrum: Nakamoto *et al., J. Am. Chem. Soc.* **85,** 309 (1963).

Trisodium calcium salt, *see* Pentetate Calcium Trisodium.
USE: Chelating agent.
THERAP CAT: Chelating agent (iron).

7267. Pentetreotide. *[R-(R*,R*)]-N-[2-[[2-[Bis(carb-oxymethyl)amino]ethyl](carboxymethyl)amino]ethyl]-N-(carboxymethyl)glycyl-D-phenylalanyl-L-cysteinyl-L-phenylala-nyl-D-tryptophyl-L-lysyl-L-threonyl-N-[2-hydroxy-1-(hydroxy-methyl)propyl]-L-cysteinamide cyclic (3→8)disulfide;* [DTPA-D-Phe¹]-octreotide; DTPA-SMS; SDZ-215-811. $C_{65}H_{87}N_{13}O_{19}S_2$; mol wt 1394.59. C 54.26%, H 6.29%, N 13.06%, O 21.80%, S 4.60%. Somatostatin analog. Prepn of compd and chelate, and binding characteristics: W. H. Bakker *et al., Life Sci.* **49,** 1583 (1991). Synthesis, biological activity and somatostatin receptor-positive tumor imaging: R. Albert *et al., Actual. Chim. Thér.* **21,** 111 (1994). Clinical metabolism and scintigraphic studies: E. P. Krenning *et al., J. Nucl. Med.* **33,** 652 (1992). Diagnostic imaging: E. Bombardieri *et al., Eur. J. Cancer* **31A,** 184 (1995); F. Jamar *et al., J. Nucl. Med.* **36,** 542 (1995); G. Kahaly *et al., ibid.* 550. Review of diagnostic use in neuroendocrine tumors: C. A. Hoefnagel, *Eur. J. Nucl. Med.* **21,** 561-581 (1994); mechanism of action: E. P. Krenning *et al., Sem. Oncol.* **21,** Suppl. 13, 6-14 (1994).

¹¹¹In chelate, $C_{63}H_{84}^{111}InN_{13}O_{19}S_2$, *¹¹¹In pentetreotide,* *[¹¹¹In-DTPA-D-Phe¹]-octreotide, MP-1727, OctreoScan.*

THERAP CAT: ¹¹¹In chelate as diagnostic aid (radioactive imaging agent).

7268. Penthienate Bromide. *2-[(Cyclopentylhydroxy-2-thienylacetyl)oxy]-N,N-diethyl-N-methylethanaminium bro-mide; diethyl(2-hydroxyethyl)methylammonium bromide α-cyclopentyl-2-thiopheneglycolate;* diethyl-2-[1-hydroxy-1-cyclopentyl-1-(2-thienyl)acetoxy]ethylmethylammonium bromide; 2-diethylaminoethyl 2-cyclopentyl-2-(2-thienyl)hy-droxyacetate methobromide; 2-diethylaminoethyl-α-cyclo-pentyl-2-thiopheneglycolate methobromide; Win-4369; Monodral bromide. $C_{18}H_{30}BrNO_3S$; mol wt 420.41. C 51.43%, H 7.19%, Br 19.01%, N 3.33%, O 11.42%, S 7.63%. Prepn: Blicke, Tsao, *J. Am. Chem. Soc.* **66,** 1645 (1944); Blicke, U.S. pat. 2,541,634 (1951).

Pale yellow crystals, mp 124.6°. uv max: 238 nm ($A_{1cm}^{1\%}$ 189). One gram dissolves in 5 ml water. Freely sol in alc, chloroform. Slightly sol in acetone. Insol in ether. pH of a 1% aq soln 5.5 to 7.0.

THERAP CAT: Anticholinergic.

7269. Pentifylline. *1-Hexyl-3,7-dihydro-3,7-dimethyl-1H-purine-2,6-dione; 1-hexyltheobromine;* 1-hexyl-3,7-di-methylxanthine; 3,7-dimethyl-1-hexyl-1H,3H-purin-2,6-dione; SK-7. $C_{13}H_{20}N_4O_2$; mol wt 264.33. C 59.07%, H 7.63%, N 21.20%, O 12.11%. Prepn from theobromine and hexyl halide: Eidebenz, Schuh, Ger. pat. 860,217 (1952 to Chem. Werke Albert); Serchi, *Chimica* **40,** 451 (1964); Chkhikvadze *et al.,* U.S.S.R. pat. 202,152 (1967), *C.A.* **69,** 19368x (1968). Pharmacology: Cugurra, Echinard-Garin, *Arch. Int. Pharmacodyn. Ther.* **123,** 481 (1960); Ramos *et al., ibid.* **153,** 430 (1965); Mohler *et al., Arzneimittel-Forsch.* **16,** 1524 (1966). Metabolic studies: Mohler *et al., Arch. Pharm.* **299,** 448 (1966).

Crystals, mp 82-83°.
Principal ingredient of *Cosaldon* (also contains nicotinic acid): Brit. pat. 815,969 (1959 to Chem. Werke Albert).

USE: Stabilizer of vitamin preparations: Nook, Eidebenz, Ger. pat. 1,810,705 (1970 to Chem. Werke Albert).

THERAP CAT: Vasodilator.

7270. Pentigetide. *N²-[1-[N-(N-L-α-Aspartyl-L-seryl)-L-α-aspartyl]-L-prolyl]-L-arginine;* human IgE pentapeptide; HEPP; Pentyde. $C_{22}H_{36}N_8O_{11}$; mol wt 588.58. C 44.90%, H 6.17%, N 19.04%, O 29.90%. Synthetic pentapeptide with antiallergic activity. Structure corresponds to amino acids 320-324 of the epsilon chain of human immunoglobulin E (IgE). Prepn: R. N. Hamburger, Ger. pat. 2,602,443; *idem,* U.S. pat. 4,171,299 (1976, 1979 both to Univ. California). Inhibition of IgE-mediated immediate hypersensitivity re-sponse: *idem, Science* **189,** 389 (1975). Stimulation of peri-toneal macrophages: E. Tzehoval *et al., Proc. Nat. Acad. Sci. USA* **75,** 3400 (1978). Mechanism of action study: R. N. Hamburger, *Immunology* **38,** 781 (1979). Clinical evalu-ations in allergic rhinitis: G. A. Cohen *et al., Ann. Allergy* **52,** 83 (1984); B. M. Prenner, *ibid.* **58,** 332 (1987).

Asp—Ser—Asp—Pro—Arg

$[\alpha]_D^{20}$ −78.6° (c = 1 in water).
THERAP CAT: Antiallergic.

7271. Pentisomide. (±)-α-[2-[Bis(1-methylethyl)amino]ethyl]-α-(2-methylpropyl)-2-pyridineacetamide; 2-[2-(diisopropylamino)ethyl]-4-methyl-2-(2-pyridyl)pentanamide; propisomide; penticainide; CM-7857; ME-3202. $C_{19}H_{33}N_3$O; mol wt 319.49. C 71.43%, H 10.41%, N 13.15%, O 5.01%. Sodium channel blocker; derivative of disopyramide, q.v. Prepn: H. Demarne et al., Eur. pat. **Appl. 27,412;** eidem, U.S. pat. **4,356,177** (1981, 1982 both to C. M. Industries); C. A. Bernhart et al., J. Med. Chem. **26**, 451 (1983). Electrophysiological study: V. Kühlkamp et al., Int. J. Cardiol. **36**, 69 (1992). HPLC determn in plasma, urine, and tissues: T. A. Plomp, M. J. H. Buijs, J. Chromatog. **612**, 123 (1993). Clinical evaluation: S. G. Priori et al., Am. J. Cardiol. **60**, 1068 (1987); S. G. Mangini et al., Eur. Heart J. **12**, 712 (1991). Review: S. Yuan, S. B. Olsson, Cardiovasc. Drug Rev. **11**, 74-93 (1993).

White crystals from diisopropyl ether, mp 108-109°. Sol in water.
THERAP CAT: Antiarrhythmic (class I).

7272. Pentobarbital. 5-Ethyl-5-(1-methylbutyl)-2,4,6-(1H,3H,5H)-pyrimidinetrione; mebubarbital; pentobarbitone; Neodorm. $C_{11}H_{18}N_2O_3$; mol wt 226.28. C 58.39%, H 8.02%, N 12.38%, O 21.21%. Prepn: **Ger. pat. 293,163** (1916 to Bayer), Frdl. **13**, 799 (1923); and hypnotic activity: E. H. Volwiler, D. L. Tabern, J. Am. Chem. Soc. **52**, 1676 (1930). Toxicity data: E. W. Schafer, Toxicol. Appl. Pharmacol. **21**, 315 (1972). HPLC determn in plasma: M. J. Avram, T. C. Krejcie, J. Chromatog. **414**, 484 (1986). Use in clinical management of elevated intracranial pressure: P. S. Woster, K. L. LeBlanc, Clin. Pharm. **9**, 762 (1990); for euthanasia: A. T. Evans et al., J. Am. Vet. Med. Assoc. **203**, 664 (1993).

Crystals from alcohol, mp 129-130°.
Sodium salt, $C_{11}H_{17}N_2NaO_3$, pentobarbital sodium, soluble pentobarbital, Carbrital, Narcoren, Nembutal, Pentone, Praecicalm, Somnopentyl, Sopental. White powder; slightly bitter taste. Dec ~127°. Freely sol in water, alc. Practically insol in benzene, ether. Aq solns are unstable. LD_{50} orally in rats: 118 mg/kg (Schafer).
Calcium salt, $C_{22}H_{34}CaN_4O_6$, pentobarbital calcium, Repocal.
Caution: May be habit forming. This is a controlled substance (depressant) listed in the U.S. Code of Federal Regulations, Title 21 Parts 329.1, 1308.12 and 1308.13 (1995).
THERAP CAT: Sedative, hypnotic.
THERAP CAT (VET): Anesthetic (intravenous); for euthanasia.

7273. 1-Pentol. 3-Methyl-2-penten-4-yn-1-ol; 1-hydroxy-3-methyl-2-penten-4-yne. C_6H_8O; mol wt 96.13. C 74.97%, H 8.39%, O 16.64%. Prepd by allylic rearrangement of methylvinylethynyl carbinol with moderately strong acid at 80°: Oroshnik, J. Am. Chem. Soc. **78**, 2651 (1956).

cis-Isomer [1'-pentol], oily liquid. $bp_{9.4}$ 65°. n_D^{20} 1.4820. uv max (ethanol): 223 nm (ε 11,000).
p-Nitrobenzoate, mp 61-62°.
trans-Isomer [1''-pentol], oily liquid. $bp_{9.4}$ 73°. n_D^{20} 1.4934. uv max (ethanol): 224 nm (ε 13,100).
p-Nitrobenzoate, mp 63-64°.
Caution: Both isomers tend to polymerize and will explode when heated above 120° in a sealed bomb tube.
USE: Intermediate in vitamin A synthesis.

7274. Pentolinium Tartrate. 1,1'-(1,5-Pentanediyl)bis[1-methylpyrrolidinium] salt with [R-(R*,R*)]-2,3-dihydroxybutanedioic acid (1:2); pentamethylene-1,5-bis(1-methylpyrrolidinium) hydrogen tartrate; pentapyrrolidinium bitartrate; pentolonium bitartrate; M & B 2050A; Ansolysen Tartrate; Ansolysen Bitartrate; Pentilium. $C_{23}H_{42}N_2O_{12}$; mol wt 538.59. C 51.29%, H 7.86%, N 5.20%, O 35.65%. General procedure of prepn: Libman et al., J. Chem. Soc. **1952**, 2305. Ganglionic blocking agent.

Crystals, dec 203°. Non-hygroscopic. Acid taste. Freely sol in water or in 25% aq polyvinylpyrrolidone soln. One gram dissolves in 0.4 ml water, in 810 ml ethanol. Insol in ether, chloroform. A 10% aq soln has a pH of about 3.5.
THERAP CAT: Antihypertensive.

7275. Pentorex. α,α,β-Trimethylbenzeneethanamine; α,α,β-trimethylphenethylamine; 1,1-dimethyl-2-phenylpropylamine; DL-2-phenyl-3-methyl-3-butylamine; phenpentermine. $C_{11}H_{17}N$; mol wt 163.26. C 80.93%, H 10.50%, N 8.58%. Prepn: **Fr. pat. M2594** (1964 to Nordmark), C.A. **61**, 16013f (1964). GC determn in plasma and urine: F. T. Delbeke, M. Debackere, J. Chromatog. **273**, 141 (1983).

bp_{20} 109-111°.
Hydrochloride, $C_{11}H_{17}N.HCl$, crystals, mp 164-166°.
Hydrogen D-tartrate, $C_{11}H_{17}N.C_4H_6O_6$, Liprodène, Modatrop. Crystals, mp 160-162°. $[\alpha]_D^{20}$ +13.4° (c = 0.8 in water).
D-Form hydrogen D-tartrate, crystals, mp 167-169°. $[\alpha]_D^{20}$ +17.9°.
L-Form hydrogen D-tartrate, crystals, mp 164-166°. $[\alpha]_D^{20}$ +3.45°.
THERAP CAT: Anorexic.

7276. Pentosan Polysulfate. Xylan hydrogen sulfate; xylan polysulfate; CB-8061. Semi-synthetic sulfated polyanion composed of β-D-xylopyranose residues with properties similar to heparin, q.v. Mol wt ranges from 1500 to 5000. Prepn: **Swiss pat. 293,566** (1953 to Wander), C.A. **49**, 1787h (1955). Anticoagulant activity: B. Paramelle, Therapie **17**, 719 (1962); J.-B., Dureux et al., ibid. **19**, 879 (1964). Pharmacokinetics: R. Taugner et al., Arch. Int. Pharmacodyn. **189**, 250 (1971). Effects on aggregation of human blood platelets: G. Kindness et al., Thromb. Res. **16**, 97 (1979). Comparison of pentosan polysulfate with heparin: C. Soria et al., ibid. **19**, 455 (1980); M. Ryde et al., ibid. **23**, 435 (1981); A.-M. Fischer et al., Thromb. Haemostasis **47**, 104, 109 (1982). Mechanism of antiadherence activity in urinary bladder: C. L. Parsons et al., Science **208**, 605 (1980); eidem, Infect. Immun. **27**, 876 (1980). Clinical trials in interstitial cystitis: A. Fritjofsson *et al.*, *J. Urol.* **138**, 508 (1987); S. G. Mulholland *et al.*, *Urology* **6**, 552 (1990).

Sodium salt, *sodium pentosan polysulfate, sodium xylan polysulfate, Cartrophen, Elmiron, Fibrase, Fibrezym, Hémoclar, SP-54, Thrombocid.* White odorless powder, slightly hygroscopic. $[\alpha]_D^{20}$ −57°. pH of 10% aq soln ∼6.0. n_D^{20} of 10% aq soln: 1.344. Sol in water.

THERAP CAT: In treatment of interstitial cystitis; antithrombotic.

THERAP CAT (VET): Anti-inflammatory.

7277. Pentostatin. *(R)-3-(2-Deoxy-β-D-erythro-pentofuranosyl)-3,6,7,8-tetrahydroimidazo[4,5-d][1,3]diazepin-8-ol;* 2′-deoxycoformycin; DCF; 2′-dCF; CL-67310465; NSC-218321; CI-825; Nipent. $C_{11}H_{16}N_4O_4$; mol wt 268.27. C 49.25%, H 6.01%, N 20.88%, O 23.86%. Adenosine deaminase inhibitor. Isolation from *Streptomyces antibioticus* and structure determn: P. W. K. Woo *et al.*, *J. Heterocycl. Chem.* **11**, 641 (1974). Isolation and purification: A. Ryder *et al.*, **Ger. pat.** 2,517,596; *eidem,* **U.S. pat.** 3,923,785 (both 1975 to Parke-Davis). Total synthesis: E. Chan *et al.*, *J. Org. Chem.* **47**, 3457 (1982). Preliminary biosynthetic study: J. C. Hanvey *et al.*, *Biochemistry* **27**, 5790 (1988). Inhibition of deaminases *in vitro:* T. Rogler-Brown *et al.*, *Biochem. Pharmacol.* **27**, 2289 (1978); C. Frieden *et al.*, *Biochem. Biophys. Res. Commun.* **91**, 278 (1979); *in vivo:* W. Plunkett *et al.*, *Biochem. Pharmacol.* **28**, 201 (1979). Pharmacokinetics in mice: W. R. McConnell *et al.*, *Drug Metab. Dispos.* **7**, 11 (1979). Toxicology study: J. F. Smyth *et al.*, *Cancer Chemother. Pharmacol.* **1**, 49 (1978). Clinical pharmacology: J. F. Smyth *et al.*, *ibid.* **5**, 93 (1980); P. P. Major *et al.*, *Blood* **58**, 91 (1981); F. J. Cummings *et al.*, *Clin. Pharmacol. Ther.* **44**, 501 (1988). Enzymatic determn in biological fluids: M. M. Chassin *et al.*, *Biochem. Pharmacol.* **28**, 1849 (1979); A. E. Staubus *et al.*, *ibid.* **33**, 1633 (1984). Clinical evaluation in hairy cell leukemia: J. B. Johnston *et al.*, *J. Nat. Cancer Inst.* **80**, 765 (1988). Brief review: P. O'Dwyer *et al.*, *Ann. Int. Med.* **108**, 733 (1988).

White crystals from methanol/water, mp 220-225° (Woo), also reported as 204-209.5° with darkening at >150° (Chan). uv max (water, pH 7): 282 nm (ε 8000); (pH 11): 283 nm (ε 7970); (pH 2): 273 nm (ε 7570 initially, 3143 after 6.5 hrs). $[\alpha]_D^{25}$ +76.4° (c = 1 in water); $[\alpha]_D^{23}$ +73.0° (c = 1, pH 7 buffer). pKa 5.2 in water.

THERAP CAT: Antineoplastic.

7278. Pentoxifylline. *3,7-Dihydro-3,7-dimethyl-1-(5-oxohexyl)-1H-purine-2,6-dione; 1-(5-oxohexyl)theobromine;* 1-(5-oxohexyl)-3,7-dimethylxanthine; 3,7-dimethyl-1-(5-oxohexyl)-1H,3H-purin-2,6-dione; oxpentifylline; vazofirin; BL-191; Azupentat; Durapental; Rentylin; Torental; Trental. $C_{13}H_{18}N_4O_3$; mol wt 278.31. C 56.10%, H 6.52%, N 20.13%, O 17.25%. Identity as a metabolite of pentifylline, *q.v.:* Mohler *et al.*, *Arch. Pharm.* **299**, 448 (1966); Mohler *et al.*, *Arzneimittel-Forsch.* **16**, 1524 (1966). Prepn: **Neth. pat. Appl.** 6,511,581; Mohler *et al.*, **Ger. pat.** 1,235,320; *eidem,* **U.S. pat.** 3,422,107 (1966, 1967, 1969 all to Chem. Werke Albert). Chemistry and synthesis data: W. Mohler, A. Söder, *Arzneimittel-Forsch.* **21**, 1159 (1971). Pharmacology: K. Popendiker *et al.*, *ibid.* 1160; I. Boksay *et al.*, *ibid.* 1174;

U. J. Jovanovic, *ibid.* **22**, 994 (1972). Clinical trials: F. Lehrach, R. Müller, *ibid.* **21**, 1171 (1971); L. Wilbert, *ibid.* **22**, 751 (1972). *Review:* R. Müller, F. Lehrach, *Curr. Res. Med. Opin.* **7**, 253-263 (1981); of pharmacodynamics, pharmacokinetics and therapeutic efficacy: A. Ward, S. P. Clissold, *Drugs* **34**, 50-97 (1987).

Bitter tasting, colorless needles from methanol, mp 105°. uv max: 273, 208 nm ($E_{1cm}^{1\%}$ 365, 935). Soly in water: 77 mg/ml at 25°, 191 mg/ml at 37°; in benzene: 11 g/100 ml. LD_{50} orally in mice: 1385 mg/kg (Popendiker).

THERAP CAT: Vasodilator.

7279. Pentoxyl. *5-Hydroxymethyl-6-methyl-2,4(1H,-3H)-pyrimidinedione; 5-hydroxymethyl-6-methyluracil;* 5-hydroxymethyl-4-methyluracil; 4-methyl-5-hydroxymethyluracil. $C_6H_8N_2O_3$; mol wt 156.14. C 46.15%, H 5.16%, N 17.94%, O 30.74%. Prepn: Endicott, Johnson, *J. Am. Chem. Soc.* **63**, 2063 (1941).

Crystals from boiling water, dec 314-315°.

THERAP CAT: Leukopoietic stimulant.

7280. Pentrinitrol. *2,2-Bis[(nitrooxy)methyl]-1,3-propanediol mononitrate (ester); pentaerythritol trinitrate;* W-2197; Petrin. $C_5H_9N_3O_{10}$; mol wt 271.14. C 22.15%, H 3.35%, N 15.50%, O 59.01%. Coronary vasodilator related to pentaerythritol tetranitrate, *q.v.* Prepn (no data given): **Ger. pats.** 638,422-3 (1936 to Westfälisch-Anhaltisch Sprengstoff), *C.A.* **31**, 1212²(1937); N. J. Marans *et al.*, *J. Am. Chem. Soc.* **76**, 1304 (1954); A. T. Camp *et al.*, *ibid* **77**, 751 (1955); J. Simecek, *Coll. Czech. Chem. Commun.* **27**, 362 (1962). Use in coronary insufficiency: F. J. DiCarlo, **U.S. pat.** 3,419,571 (1968 to Warner-Lambert). Electrical moment and molecular rotation: A. R. Lawrence, A. J. Matuszko, *J. Phys. Chem.* **65**, 1903 (1961). Chromatographic determn: I. W. F. Davidson *et al.*, *J. Chromatog.* **57**, 345 (1971). Metabolism: F. J. DiCarlo *et al.*, *Clin. Pharmacol. Ther.* **22**, 309 (1977). Comparative duration of action: J. Vohra *et al.*, *Aust. N. Z. J. Med.* **9**, 554 (1979).

Viscous liq. n_D^{20} 1.4941. d_4^{20} 1.554. Viscosity: 166.8 centistokes at 40°; 77.8 at 50°. Soly in water: 0.705 g/100 ml at 20°; in benzene: 21.40 g/100 ml at 20°. Very sol in ethanol, ether. Forms a cryst hydrate, mp 32°, when washed with water and allowed to stand overnight at 20°; returns to unhydrated form when allowed to stand at 60° for 2 hrs. When used medicinally it is usually diluted with lactose or mannitol to reduce explosive liability, *see* pentaerythritol tetranitrate.

THERAP CAT: Vasodilator (coronary).

7281. Pentryl. *2-[Nitro(2,4,6-trinitrophenyl)amino]-ethanol nitrate (ester); 2-(N,2,4,6-tetranitroanilino)ethanol nitrate; sym*-trinitrophenylnitroaminoethyl nitrate. $C_8H_6N_6O_{11}$; mol wt 362.17. C 26.53%, H 1.67%, N 23.20%, O 48.59%. Prepd by nitration of 2,4-dinitrophenylaminoethanol: Clark, *Ind. Eng. Chem.* **25**, 1385 (1933); Desseigne, *Mém. poudres* **33**, 255 (1951); B. T. Fedoroff *et al.*, *Encyclo-*

pedia of Explosives and Related Items, **vol. I** (Picatinny Arsenal, Dover, New Jersey, 1960) pp A425-429.

Small, cream-colored crystals from chloroform, mp 129° (slight decomp). d 1.82. Explodes when heated to 235°. Heat of combustion: 911.1 kcal/mole; heat of explosion: 372.4 kcal/mole; heat of formation: 43.4 kcal/mole. Soly (w/w) at 25° in benzene: 0.70; carbon tetrachloride: trace; chloroform: 0.07; ethanol: 0.11; ether: 0.16; ethylene dichloride: 0.72; methanol: 0.67; nitroglycerol: freely sol; toluene: 0.63; water: trace.

USE: High explosive. Base charge in detonators.

7282. tert-Pentyl Alcohol. *2-Methyl-2-butanol; tert-amyl alcohol;* dimethyl ethyl carbinol; ethyl dimethyl carbinol; *tert*-pentanol; amylene hydrate. $C_5H_{12}O$; mol wt 88.15. C 68.13%, H 13.72%, O 18.15%. $(CH_3)_2C(OH)$-CH_2CH_3. Prepd from 2-methyl-2-butene alone and mixed with 2-methyl-1-butene: Fenske, Jones, U.S. pat. **2,858,331** (1958 to Esso); Odioso *et al., Ind. Eng. Chem.* **53,** 209 (1961). Review of manuf by fractionation of fusel oil and *via* chlorination of pentanes, and properties: *Industrial Chemicals,* W. L. Faith *et al.,* Eds. (John Wiley, New York, 2nd ed., 1957) pp 107-114. Toxicity: Schaffarzick, Brown, *Science* **116,** 663 (1952).

Volatile liquid; characteristic odor, burning taste. bp_{765} 102.5°. mp −9.0°. d^{20} 0.8084: Costello, Bowden, *Rec. Trav. Chim.* **77,** 36 (1958). n_D^{20} 1.4052. Flash pt, closed cup: 67°F (19°C); open cup: 70°F (21°C). Sol in 8 parts water; miscible with alcohol, ether, benzene, chloroform, glycerol, oils. *Keep tightly closed and protected from light.* LD_{50} orally in rats: 1.0 g/kg (Schaffarzick, Brown).

Caution: Moderately irritating to mucous membranes. Narcotic in high concns. *See also* 1-Pentanol.

THERAP CAT: Hypnotic.

7283. Pentylenetetrazole. *6,7,8,9-Tetrahydro-5H-tetrazolo[1,5-a]azepine;* α,β-cyclopentamethylenetetrazole; 1,5-pentamethylenetetrazole; 6,7,8,9-tetrahydro-5-azepotetrazole; 1,2,3,3a-tetrazacyclohepta-8a,2-cyclopentadiene; 7,8,-9,10-tetrazabicyclo[5.3.0]-8,10-decadiene; pentetrazol; Cardiazol; Cenalene-M; Cenazol; Coranormol; Corazole; Corvasol; Deumacard; Gewazol; Korazol; Metrazol; Phrenazol; Ventrazol. $C_6H_{10}N_4$; mol wt 138.17. C 52.16%, H 7.29%, N 40.55%. Prepd by the addition of cyclohexanone to a benzene or Tetralin soln of hydrazoic acid: Schmidt, *Ber.* **57,** 704 (1924); U.S. pats. **1,564,631** (1925); **1,599,493** (1926); Prochazka, **Czech.** pat. **92,456** (1959), *C.A.* **56,** 4776e (1962). Synthesis from caprolactam: Glushkov, Golovchinskaya, *Zh. Prikl. Khim.* **32,** 920 (1959); *Med. Prom. SSSR* **14,** no. 1, 12 (1960). Toxicity study: E. I. Goldenthal, *Toxicol. Appl. Pharmacol.* **18,** 185 (1971).

Slightly pungent, bitter crystals, mp 57-60°. Freely sol in water, most organic solvents. Aq solns are neutral to litmus. Very stable, not easily attacked by other substances. LD_{50} in rats (mg/kg): 85±2 s.c., 62 i.p. (Goldenthal).

THERAP CAT: CNS stimulant.

THERAP CAT (VET): CNS stimulant; narcotic antagonist.

7284. p-tert-Pentylphenol. *4-(1,1-Dimethylpropyl)phenol; p-tert-*amylphenol; 2-methyl-2-*p*-hydroxyphenylbutane; Pentaphen. $C_{11}H_{16}O$; mol wt 164.25. C 80.44%, H 9.82%, O 9.74%. Prepd by condensation of *tert*-pentanol or 2-methyl-3-butanol with phenol in the presence of aluminum chloride: Huston, Hsieh, *J. Am. Chem. Soc.* **58,** 439 (1936); Huston *et al., ibid.* **67,** 899 (1945). Physical properties: Pardee, Weinrich, *Ind. Eng. Chem.* **36,** 595 (1944). Toxicity

study: H. F. Smyth *et al., Am. Ind. Hyg. Assoc. J.* **23,** 95 (1962).

Crystals, mp 94-95°. bp 262.5°, bp_{740} 248-250°, bp_{15} 138.5°, bp_3 112-120°. d_4^{20} 0.962. Practically insol in water. Sol in alcohol, ether, benzene, chloroform. LD_{50} orally in rats: 3.08 g/kg (Smyth).

USE: In the manuf of oil-soluble resins; has been recommended as a germicide and fumigant; intermediate for organic mercury germicides, for pesticides, for chemicals used in rubber and petroleum industries.

7285. Peonidin. *3,5,7-Trihydroxy-2-(4-hydroxy-3-methoxyphenyl)-1-benzopyrylium chloride; 3,4',5,7-tetrahydroxy-3'-methoxyflavylium chloride;* 3,4',5,7-tetrahydroxy-3'-methoxy-2-phenylbenzopyrylium chloride. $C_{16}H_{13}ClO_6$; mol wt 336.73. C 57.07%, H 3.89%, Cl 10.53%, O 28.51%. The aglucon of peonin: R. Willstätter, T. J. Nolan, *Ann.* **408,** 136 (1915). Structure and synthesis: T. J. Nolan *et al., J. Chem. Soc.* **1926,** 1968; S. Murakami, R. Robinson, *ibid.* **1928,** 1537.

Monohydrate, reddish-brown needles from aq 20% hydrochloric acid soln. Moderately sol in water giving a reddish-brown soln. Sol in alc giving a purplish-red soln.

3,5-Diglucoside, $C_{28}H_{33}ClO_{16}$, *3,5-bis(β-D-glucopyranosyloxy)-7-hydroxy-2-(4-hydroxy-3-methoxyphenyl)-1-benzopyrylium chloride, peonin.* From deep violet-red peonies: R. Willstätter, T. J. Nolan, *loc. cit.* Synthesis: R. Robinson, A. R. Todd, *J. Chem. Soc.* **1932,** 2488. Deep purple needles with water of crystn from dil HCl, mp 165-167° (dec).

7286. Peplomycin. *(S)-N¹-[3-[(1-Phenylethyl)amino]propyl]bleomycinamide;* N^1-[3-[[(S)-α-methylbenzyl]amino]propyl]bleomycinamide; pepleomycin; NK-631. $C_{61}H_{88}N_{18}$-$O_{21}S_2$; mol wt 1473.61. C 49.72%, H 6.02%, N 17.11%, O 22.80%, S 4.35%. Deriv of bleomycin, *q.v.* with cytostatic activity and less pulmonary toxicity than the natural bleomycin mixture. Prepn of the (*R,S*)-form: H. Umezawa *et al.,* U.S. pat. **3,846,400** (1974 to Microbiochem. Res. Found.); of the (*S*)-form: T. Takita *et al.,* **Ger. pat. 2,828,-933;** *eidem,* U.S. pat. **4,195,018** (1979, 1980 both to Nippon Kayaku); W. Tanaka *et al., Heterocycles* **13,** 469 (1979). Biological study of degradation products: K. Takahashi *et al., J. Antibiot.* **32,** 36 (1979). General pharmacology: Y. Ishii *et al., Japan. J. Antibiot.* **31,** 886 (1978), *C.A.* **91,** 215c (1979). Absorption, distribution, excretion, metabolism: H. Takayama *et al., ibid.* 895, *C.A.* **91,** 32580j (1979). Properties and stability: A. Fuji *et al., Iyakuhin Kenkyu* **10,** 197 (1979), *C.A.* **91,** 9408a (1979). Effect in prostatic cancer: T. Niijima, K. Koiso, *Scand. J. Respir. Dis.* **Suppl. 55,** 177 (1980). Relative pulmonary toxicity: B. I. Sikic *et al., Cancer Treat. Rep.* **64,** 659 (1980). Acute toxicity study: K. Ito *et al., Japan. J. Antibiot.* **31,** 719 (1978), *C.A.* **91,** 68461k (1979). Review of clinical studies: S. Oko, *Recent Results Cancer Res.* **74,** 163 (1980).

R =

Sulfate salt, $C_{61}H_{88}N_{18}O_{21}S_2.H_2SO_4$, *Pepleo Injection.* Pale yellow amorphous powder, mp 196-198°. $[\alpha]_{436}^{25}$ −2.0° (c = 1 in water). pKa 2.9, 4.8, 7.4, 9.0. Sol in water, methanol, acetic acid, DMSO, DMF. Slightly sol in dioxane. Insol in ethyl acetate, acetone, ether, benzene. Stable at 37° for 3 months, 50° for 6 weeks, room temp for 30 months in a sealed container. LD_{50} in male rats, mice (mg/kg): 234, 88 s.c.; 208, 85 i.p.; 245, 51 i.v. (Ito).

THERAP CAT: Antineoplastic.

7287. Pepper. Black pepper. Dried, unripe fruit of *Piper nigrum* L., *Piperaceae.* White pepper is the decorticated ripe fruit of black pepper. **Habit.** India, Malabar coast, Philippines, Sumatra, Java, Ceylon, etc. Predominant pungent principle in black pepper is piperine, *q.v.* Other pungent substances present in small quantities are chavicine, piperidine, *q.q.v.* and piperettine. Isoln of constituents: R. Grewe *et al., Ber.* **103**, 3752 (1970). HPLC determn of constituents: M. Verzele *et al., J. Chromatog.* **172**, 493 (1979); M. Rathnawathie, K. A. Buckle, *ibid.* **264**, 316 (1983); D. E. Games, N. J. Alcock, *ibid.* **294**, 269 (1984). Insecticidal properties of black pepper: H. C. F. Su, *J. Econ. Entomol.* **70**, 18 (1977); W. P. Scott, G. H. McKibben, *ibid.* **71**, 343 (1978); H. C. F. Su, R. Horvat, *J. Agr. Food Chem.* **29**, 115 (1981).

USE: Chiefly as a spice.

7288. Peppermint. Brandy mint; lamb mint. Dried leaves and flowering tops of *Mentha piperita* L., *Labiatae.* **Habit.** Asia, Europe, North America; cultivated in gardens. **Constit.** Volatile oil, tannin, resin, gum.

USE: Flavoring agent. Peppermint oil and peppermint water as pharmaceutic aids (flavor).

7289. Pepsin. Puerzym. Mol wt 34,500. Principal digestive enzyme of gastric juice; controls the degradation of proteins to proteoses and peptones. Distinctive among enzymes for having a very low isoelectric point and a very low pH optimum. Hydrolyzes only peptide linkages. The N.F. grade digests not less than 3000 or more than 3500 times its wt of freshly coagulated and disintegrated egg albumin in 2½ hours at 52° in water acidulated with HCl. Isoln from gastric juice of swine and beef: Northrup, *J. Gen. Physiol.* **13**, 739 (1930); **16**, 615 (1933); from salmon and tuna: Norris, Edam, *J. Biol. Chem.* **134**, 443 (1940); **204**, 673 (1953). Amino acid composition: Brand in *Crystalline Enzymes*, J. H. Northrup *et al.* (Columbia Univ. Press, New York, 2nd ed., 1948) p 26. Active site: Herriott, *J. Gen. Physiol.* **45**, Suppl., 57 (1962). *Review:* Bovey, Yanari in *The Enzymes* vol. **4**, P. D. Boyer *et al.*, Eds. (Academic Press, New York, 1960) pp 63-92.

White or yellowish-white translucent scales or granules, or an amorphous slightly hygroscopic powder, or spongy masses. It has a slightly acid or saline taste. $[\alpha]_D^{26}$ −64.5° (water pH 4.6). Isoelectric point < pH 1.0. Freely sol in water with more or less opalescence. Practically insol in alc, chloroform, ether. Very stable to acid. The activity of pepsin in solns is destroyed by heating above 70°, or by alkalies; dry pepsin is not injured by heating to 100°. *Incompat.* Alc, tannin, alkaline substances and salts of heavy metals.

Note: Pepsin of an activity of 4000, 5000, 6000, 10,000, 15,000, and 20,000 is likewise available, and it has also been obtained in pure crystalline form.

THERAP CAT: Enzyme (digestive).

THERAP CAT (VET): Has been used as a digestive aid in deficiency of pepsin secretion.

7290. Pepstatin. *[1S-[1R*,2R*,4[R*[R*(R*)]]]]-N-(3-methyl-1-oxobutyl)-L-valyl-N-[4-[[2-[[1-(2-carboxy-1-hydroxyethyl)-3-methylbutyl]amino]-1-methyl-2-oxoethyl]amino]-2-hydroxy-1-(2-methylpropyl)-4-oxobutyl]-L-valinamide;* Pepstatin A; *N*-isovaleryl-L-valyl-L-valyl-3-hydroxy-6-methyl-γ-aminoheptanoyl-L-alanyl-3-hydroxy-6-methyl-γ-aminoheptanoic acid. $C_{34}H_{63}N_5O_9$; mol wt 685.90. C 59.54%, H 9.26%, N 10.21%, O 20.99%. A pentapeptide pepsin inhibitor, isolated from cultured broths of *Streptomyces testaceus* Hamada *et* Okami and *Streptomyces argenteolus* var. *toyonakensis:* H. Umezawa *et al., J. Antibiot.* **23**, 259 (1970). Prepn: H. Umezawa *et al.,* Ger. pat. **2,028,403** (1971 to Microbiochemical Res. Found.), *C.A.* **74**, 75201c (1971). Structure: H. Morishima *et al., J. Antibiot.* **23**, 263 (1970). Synthesis: H. Morishima *et al., ibid.* **25**, 551 (1972). Biological properties: T. Aoyagi *et al., ibid.* **24**, 687 (1971). *N-n-*Caproyl and *N-iso*-caproyl derivatives, *pepstatin B* and *pepstatin C*, also isolated from pepstatin-producing *Streptomyces*, as minor components of crude preparations: T. Miyano *et al., ibid.* **25**, 489 (1972). Mechanism of pepsin inhibition: S. Kunimoto *et al., ibid.* 251; 27, 413 (1974); J. Marciniszyn *et al., Adv. Exp. Med. Biol.* **95**, 199 (1977); D. H. Rich, E. T. O. Sun, *Biochem. Pharmacol.* **29**, 2205 (1980); of other acid protease inhibition: D. H. Rich *et al., Biochemistry* **24**, 3165 (1985). Tissue distribution in rats: D. A. W. Grant *et al., Biochem. Pharmacol.* **31**, 2302 (1982). Effect on gastric ulcers in man: O. Bonnevie *et al., Gut* **20**, 624 (1979); L. B. Svendsen *et al., Scand. J. Gastroenterol.* **14**, 929 (1979).

Colorless needles, mp 228-229° (dec). $[\alpha]_D^{27}$ −90.3° (c = 0.288 in methanol). Sol in methanol, ethanol, acetic acid DMSO. Practically insol in benzene, chloroform, ether, and water. LD_{50} in mice, rats, rabbits, dogs (mg/kg): 1090, 875, 820, 450 i.p.; all > 2000 orally (Umezawa, 1970).

7291. Peptide T. $C_{35}H_{55}N_9O_{16}$; mol wt 857.87. C 49.00%, H 6.46%, N 14.69%, O 29.84%. Octapeptide segment of the human immunodeficiency virus (HIV) envelope glycoprotein (gp 120); named peptide T because of its high threonine content. Has been reported to block the *in vitro* binding of HIV envelope to human leukocyte receptor CD4. Isolation, neuropharmacology and anti-HIV activity of peptide and analogs: C. B. Pert *et al., Proc. Nat. Acad. Sci. USA* **83**, 9254 (1986). Characterization of active core structure, T[4-8], and chemotactic effects: C. B. Pert, M. R. Ruff, *Clin. Neuropharmacol.* **9**, Suppl. 4, 482 (1986). Chemotactic effects and structural homology with vasoactive intestinal peptide (VIP), *q.v.:* M. R. Ruff *et al., FEBS Letters* **211**, 17 (1987). Competitive binding studies at VIP receptor: T. D. Nguyen, *Peptides* **9**, 425 (1988). Structural homology with thymosin α_1, *q.v.:* T. D. Nguyen, L. A. Scheving, *Biochem. Biophys. Res. Commun.* **145**, 884 (1987). Evaluation of anti-HIV activity: J. Sodroski *et al., Lancet* **1**, 1428 (1987). Clinical evaluation in treatment of AIDS: L. Wetterberg *et al., ibid.* 159. Chromatographic purification of *[D-ala¹] peptide T amide*, a metabolically stable and more potent analog of peptide T: T. R. Burke, M. Knight, *J. Chromatog.* **411**, 431 (1987). Blood to brain transport of the amide in mice: C. M. Barrera *et al., Brain Res. Bull.* **19**, 629 (1987). Brief review of debate over effectiveness of peptide T: D. M. Barnes, *Science* **237**, 128 (1987).

Ala–Ser–Thr–Thr–Thr–Asn–Tyr–Thr

7292. Peptonized Iron. Iron peptonized; Saferon. A compd of iron oxide and peptone, rendered sol by the presence of Na citrate, contg 16-18% Fe. The iron is in nonionic form, hence not detectable by the usual reactions and is assumed to be more readily assimilated. Prepn: *U.S.D.*, 25th ed., p 1800.

Dark brown, lustrous granules or brown powder. Characteristic odor and taste. Affected by light. Freely sol in water, yielding neutral or alkaline solns. Practically insol in alcohol. *Protect from light.*

THERAP CAT: Hematinic.

7293. Peracetic Acid. *Ethaneperoxoic acid;* peroxyacetic acid; acetyl hydroperoxide. $C_2H_4O_3$; mol wt 76.05. C 31.59%, H 5.30%, O 63.11%. CH_3COOOH. Prepd from acetaldehyde and oxygen in the presence of cobalt acetate: **Ger.** pats. **269,937; 272,738** (1914); *Frdl.* **11**, 73; by the auto-oxidation of acetaldehyde: Wallace; Golding, **U.S.** pats. **2,833,813-4** (1958 to du Pont). A 50% soln may be obtained from acetic anhydride, hydrogen peroxide, and sulfuric acid: D'Ans, Frey, *Ber.* **45**, 1848 (1912); Erlenmeyer, *Helv. Chim. Acta* **8**, 795 (1925).

Liquid, acrid odor. Explodes violently on heating to 110°. Freely sol in water, alcohol, ether, H_2SO_4. Stable in dil aq soln. Strong oxidizing agent.

Caution: Strongly irritating to skin and eyes.

7294. Perazine. *10-[3-(4-Methyl-1-piperazinyl)propyl]-10H-phenothiazine;* N-methylpiperazinyl-N'-propylphenothiazine; 10-(γ-methylpiperazinopropyl)phenothiazine; P-725; Taxilan. $C_{20}H_{25}N_3S$; mol wt 339.50. C 70.76%, H 7.42%, N 12.38%, S 9.44%. Prepn: Hromatka *et al., Monatsh.* **88**, 56, 193 (1957); **91**, 107 (1960); Horclois, **Brit.** pat. **780,193** (1957 to Rhône-Poulenc).

Crystals, mp 51-53°. bp$_{0.001}$ 160-170° (air bath temp).
Dihydrochloride, $C_{20}H_{27}Cl_2N_3S$, hygroscopic needles, dec 228-230°.
Dihydrochloride hemihydrate, platelets from ethanol, mp 225-227°.
Dimaleate, $C_{28}H_{33}N_3O_8S$, crystals from water, mp 210°.

THERAP CAT: Antipsychotic.

7295. Perbenzoic Acid. *Benzenecarboperoxoic acid; peroxybenzoic acid;* benzoyl hydroperoxide. $C_7H_6O_3$; mol wt 138.12. C 60.87%, H 4.38%, O 34.75%. C_6H_5COOOH. Prepd from dibenzoyl peroxide by treatment with a soln of sodium methoxide in methanol at 0°: Braun, *Org. Syn.* **13**, 86 (1933); *cf.* Bergmann, Witte, **Ger.** pat. **409,779**; *Chem. Zentr.* **1925**, I, 1911.

Leaflets from benzene. Acrid odor. mp 41-43°. Very volatile. Sublimes in desiccator. bp$_{15}$ 100-110° (partial decomposition). Volatile with steam. Very sparingly sol in water, but turns liquid upon contact with water; sparingly sol in petr ether; freely sol in other organic solvents. Stability of solns with chloroform, carbon tetrachloride, ether, benzene: Prileshajew, *Chem. Zentr.* **1911**, I, 1280; Kolthoff *et al., J. Polymer Sci.* **2**, 199 (1947), *C.A.* **41**, 4960 (1947). Forms an unstable acid sodium salt, and a somewhat more stable neutral sodium salt.

USE: To convert ethylenic compounds into oxides; in analysis of unsatd compounds, to determine the number of double bonds.

7296. Perchloric Acid. $ClHO_4$; mol wt 100.46. Cl 35.29%, H 1.00%, O 63.71%. $HClO_4$. Prepd from potassium perchlorate and sulfuric acid: Schmeisser in *Handbook of Preparative Inorganic Chemistry* vol. **1**, G. Brauer, Ed. (Academic Press, New York, 2nd ed., 1963) pp 318-320. Comprehensive monograph: J. C. Schumacher, *Perchlorates* (Reinhold, New York, 1960).

The anhydr acid is a colorless, volatile, very hygroscopic liquid. d^{22} 1.768; bp$_{11}$ 19°. Dec when distilled at atmospheric pressure, sometimes with explosive violence. mp −112°. Combines vigorously with water with evolution of heat. Undergoes spontaneous and explosive decompn, hence it is marketed only in mixture with water contg 60-70% $HClO_4$, density 1.5 and 1.6, respectively. The aq acid is very caustic and may deflagrate in contact with oxidizable substances. Density of aq solns at 15°: 1% = 1.0050; 10% = 1.0597; 20% = 1.1279; 30% = 1.2067; 40% = 1.2991; 50% = 1.4103; 60% = 1.5389; 70% = 1.6736. Density of aq solns at 25°: 65.0% = 1.597; 70.0% = 1.664; 75.0% = 1.731.

Caution: Corrosive to skin, mucous membranes.

USE: The acid in analytical chemistry as an oxidizer and for separation of potassium from sodium. Its salts for explosives and for plating of metals.

7297. Perchloryl Fluoride. $ClFO_3$; mol wt 102.45. Cl 34.61%, F 18.54%, O 46.85%. *Reviews:* Downs, Adams in *Comprehensive Inorganic Chemistry* vol. **2**, J. C. Bailar, Jr. *et al.,* Eds. (Pergamon Press, Oxford, 1973) pp 1391-1393; Christe, Schack in *Advan. Inorg. Chem. Radiochem.* **18**, 319-398.

Characteristic sweet odor. Usually stored in cylinders as liq under pressure. mp −147.7°. bp −46.7°. d^{20} (liq) 1.434. Critical temp 95.2°. Crit pressure 53 atm. Crit density: 0.637. Heat of vaporization 4.6 kcal/mol. Trouton constant: 20.4. Dipole moment 0.03. Heat of formation of gas at 25° −5.12 kcal/mol. Sp heat of liq: 0.229 cal/g/°C at −40°; 0.290 at +50°. Does not corrode base metals when anhydr. Shows the greatest resistance to electrical breakdown known for any gas. *Handle with care.* Do not bring in direct contact with reducing agents, alcohols, etc.: *Chem. & Eng. News* **37**, 60 (1959). Explosions have occurred.

Caution: Potential symptom of overexposure is respiratory irritation; direct contact may cause skin burns. *See NIOSH Pocket Guide to Chemical Hazards* (DHHS/NIOSH 90-117, 1990) p 176. May cause methemoglobinemia, cyanosis, alveolar edema, bronchopneumonia. *See Patty's Industrial Hygiene and Toxicology* vol. **2F**, G. D. Clayton, F. E. Clayton, Eds. (John Wiley & Sons, New York, 4th ed., 1994) p 4478.

USE: In organic synthesis to introduce fluorine atoms into organic molecules. As oxidizing agent; insulator for high voltage systems.

7298. Perezone. *(R)-2-(1,5-Dimethyl-4-hexenyl)-3-hydroxy-5-methyl-2,5-cyclohexadiene-1,4-dione; 2-(1,5-dimethyl-4-hexenyl)-3-hydroxy-5-methyl-p-benzoquinone;* pipitzahoic acid. $C_{15}H_{20}O_3$; mol wt 248.32. C 72.55%, H 8.12%, O 19.33%. From roots of *Trixis pipitzahuac* Shaffner *(Perezia adnata* Gr., *Compositae):* Weld, *Ann.* **95**, 188 (1855); from roots of *Radix pereziae:* Mylius, *Ber.* **18**, 463, 480, 937 (1885); Anschütz, *Ber.* **18**, 709 (1885); Fichter *et al., Ann.* **395**, 1, 15 (1913). Structure: Archer, Thomson, *Chem. Commun.* **1965**, 354; Bates *et al., Chem. & Ind. (London)* **1965**, 1793.

Yellow plates from water or alcohol, mp 103-104°. [α]$_D^{20}$ −17° (ether).

7299. Perflubron. *1-Bromo-1,1,2,2,3,3,4,4,5,5,6,6,7,7,-8,8,8-heptadecafluorooctane;* bromoperfluorooctane; perfluoroctyl bromide; PFOB; L-1913; Imagent; LiquiVent; Oxygent. C_8BrF_{17}; mol wt 498.96. C 19.26%, Br 16.01%, F 64.73%. $CF_3(CF_2)_6CF_2Br$. Synthetic second generation brominated fluorocarbon. Prepd not claimed: **Fr. pat. 1,512,068;** L. A. Loree, **U.S. pat. 3,456,024** (1968, 1969 both to Dow Corning). Radiopacity and toxicology: D. M. Long *et al., Radiology* **105,** 323 (1972). Biodistribution and determn in plasma and tissue: F. H. Lee *et al., J. Pharm. Sci.* **67,** 1038 (1978). Clinical trial in abdominal imaging: R. F. Mattrey, *Radiology* **191,** 841 (1994). Toxicity studies: A. R. Burgan *et al., Biomater. Artif. Cells Artif. Organs* **16,** 681 (1988). Review of physicochemical properties and therapeutic potential: D. M. Long *et al., ibid.* 411-420; of oxygen delivery characteristics: J. G. Riess, *ibid.* **20,** 183-202 (1992).

Liquid, bp_{760} 141°. d^{25} 1.928. Vapor pressure (37.5°) 14 torr. Surface tension 18.2 dynes/cm. Insol in water. LD_{50} i.v. in female rats: 41 g/kg (Burgan).

THERAP CAT: Diagnostic aid (contrast agent).

7300. Perfluidone. *1,1,1-Trifluoro-N-[2-methyl-4-(phenylsulfonyl)phenyl]methanesulfonamide;* 1,1,1-trifluoro-4'-(phenylsulfonyl)methanesulfono-o-toluidide; 2-methyl-4-phenylsulfonyltrifluoromethanesulfonanilide; MBR-8251; Destun. $C_{14}H_{12}F_3NO_4S_2$; mol wt 379.38. C 44.32%, H 3.19%, F 15.02%, N 3.69%, O 16.87%, S 16.90%. Selective pre-emergence herbicide. Prepn: Moore, Harrington, **Ger. pat. 2,118,190** (1972 to Minnesota Mining & Mfg.), *C.A.* **77,** 19402u (1972). Activity: W. A. Gentner, *Weed Sci.* **21,** 122 (1973).

Cryst solid, mp 142-144°. pKa 2.5. Vapor press. at 25°: $<1 \times 10^{-5}$ mm Hg. Soly in water at 22°: 60 mg/l. Soly (g/l): acetone 750; benzene 11; dichloromethane 162; methanol 595.

USE: Herbicide.

7301. Perforin. Cytolysin; pore-forming protein; PFP; C9-related protein. Complement-like protein produced by cytotoxic T-lymphocytes (CTLs) and natural killer (NK) cells; released from cytoplasmic granules during the cytolytic process. Perforates the membrane of target cells by forming transmembrane pores that perturb membrane permeability and result in cell lysis. Pore-formation is mediated by the Ca^{2+}-dependent polymerization of perforin within the target cell membrane. Also thought to act as a conduit for the endocytosis of *granzymes,* serine proteinases produced by cytotoxic lymphocytes, that initiate nuclear disintegration. Glycoprotein; mol wt 66-70 kDa. Identification of pore-forming tubular complexes produced by NK cells: E. R. Podack, G. Dennert, *Nature* **302,** 442 (1983). Partial purification of protein: E. R. Podack, P. J. Konigsberg, *J. Exp. Med.* **160,** 695 (1984). Cloning and amino acid sequence of human perforin: M. G. Lichtenheld *et al., Nature* **335,** 448 (1988); of murine: D. M. Lowrey *et al., Proc. Nat. Acad. Sci. USA* **86,** 247 (1989). Confirmation of biological activity: D. Kägi *et al., Nature* **369,** 31 (1994). Review of structure and function: E. R. Podack, *Curr. Top. Microbiol. Immunol.* **178,** 175-184 (1992); C-C. Liu *et al., Immunol. Today* **16,** 194-201 (1995).

7302. Performic Acid. *Methaneperoxoic acid; peroxyformic acid;* permethanoic acid; formyl hydroperoxide. CH_2O_3; mol wt 62.03. C 19.36%, H 3.25%, O 77.39%. HCOOOH. A strong oxidizing agent. A 90% soln is obtained when a mixture of 20 g 100% H_2O_2 and 6.5 g H_2SO_4 is allowed to interact for 2 hrs and is then distilled: D'Ans, Kneip, *Ber.* **48,** 1137 (1915); Greenspan, *J. Am. Chem. Soc.* **68,** 907 (1946).

The 90% soln is a colorless liq. *Prone to explode on contact with metals, their oxides, reducing substances, or on distillation.* Has lower vapor pressure than formic acid. Miscible with water, alc, ether. Sol in benzene, chloroform. Solns are unstable, gassing being noticeable after a few hours, and the effective concn showing a definite decline in 2 hrs.

Caution: Irritant.

USE: For oxidation, epoxidation and hydroxylation reactions.

7303. Perfosfamide. *cis-(±)-2-[Bis(2-chloroethyl)amino]tetrahydro-2H-1,3,2-oxazaphosphorin-4-yl hydroperoxide P-oxide;* 2-[bis(2-chloroethyl)amino]-4-hydroperoxytetrahydro-2H-1,3,2-oxazaphosphorine 2-oxide; cis-4-hydroperoxy-cyclophosphamide; 4-HC; NSC-181815; Pergamid. $C_7H_{15}Cl_2N_2O_4P$; mol wt 293.09. C 28.69%, H 5.16%, Cl 24.19%, N 9.56%, O 21.84%, P 10.57%. Prepn: **Neth. pat. Appl. 7,208,895;** A. Takamizawa, **U.S. pat. 3,808,297** (1972, 1974 both to Shionogi); *idem et al., J. Am. Chem. Soc.* **95,** 985 (1973). Structural study: A. Camerman *et al., Acta Cryst.* **B33,** 678 (1977). Cytostatic activity: A. Takamizawa *et al., J. Med. Chem.* **18,** 376 (1975). Clinical use: D. Biggs *et al., Prog. Clin. Biol. Res.* **389,** 9 (1994). Review of use as purging agent in autologous bone marrow transplantation: A. M. Yeager *et al., Am. J. Pediatr. Hematol. Oncol.* **12,** 245-256 (1990); R. J. Jones, *J. Hematother.* **1,** 343-348 (1992).

White crystals from acetone/ether, mp 107-108°. LD_{50} in rats, mice (mg/kg): 115, 235 i.v.; 131, 181 i.p. (Takamizawa, 1975).

THERAP CAT: Antineoplastic; bone marrow purging agent.

7304. Pergolide. *(8β)-8-[(Methylthio)methyl]-6-propylergoline;* D-6-n-propyl-8β-methylmercaptomethylergoline; LY-141B. $C_{19}H_{26}N_2S$; mol wt 314.49. C 72.56%, H 8.33%, N 8.91%, S 10.20%. Dopaminergic agonist that also decreases plasma prolactin concentrations. Prepn: E. C. Kornfeld, N. J. Bach, **U.S. pat. 4,166,182** (1979 to Lilly). Dopaminergic effects in rats: R. W. Fuller *et al., Life Sci.* **24,** 375 (1979); T. T. Yen *et al., ibid.* **25,** 209 (1979). Clinical pharmacology: L. Lemberger, R. E. Crabtree, *Science* **205,** 1151 (1979). Pharmacological evaluation as antiparkinson agent: W. C. Koller, *Neuropharmacology* **19,** 831 (1980). Clinical study in galactorrhea: J. T. Callaghan *et al., Life Sci.* **28,** 95 (1981); in hyperprolactinemia: S. Francks *et al., Lancet* **2,** 659 (1981); in treatment of pituitary tumors secreting prolactin or growth hormone: D. L. Kleinberg *et al., N. Engl. J. Med.* **309,** 704 (1983). Clinical studies in Parkinson's disease: J. Jankovic, J. Orman, *Adv. Neurol.* **45,** 551 (1986); C. W. Olanow, M. J. Alberts, *ibid.* 555. Comprehensive description: D. J. Sprankle, E. C. Jensen in *Analytical Profiles of Drug Substances and Excipients* **vol. 21,** H. G. Brittain, Ed. (Academic Press, San Diego, 1992) pp 375-413.

Solid, mp 206-209°.

Mesylate, $C_{19}H_{26}N_2S.CH_3SO_3H$, *LY-127809, Celance, Permax.* Crystals, mp 225° (dec); also reported as 258-260° (Sprankle, Jensen). uv max (water): 279 nm (ε 6385); (methanol): 280 nm (ε 6980); (dehydrated ethanol): 281 nm (ε 6993). $α_D^{20}$ between −18.0° and −23.0° (c = 10 mg/ml in DMF). pKa (66% DMF) 7.8. Sparingly sol in DMF, methanol; slightly sol in water, 0.01N HCl, chloroform, acetonitrile, dichloromethane, dehydrated ethanol;

very slightly sol in acetone. Practically insol in 0.1*N* NaOH, 0.1*N* HCl, ether. Partition coefficient at 25° (chloroform/water): 6.14 (pH 2.19); 119.6 (pH 4.02).

THERAP CAT: Antiparkinsonian.

7305. Perhexiline. *2-(2,2-Dicyclohexylethyl)piperidine;* 1,1-dicyclohexyl-2-(2-piperidyl)ethane; perhexilene. $C_{19}H_{35}$-N; mol wt 277.49. C 82.24%, H 12.71%, N 5.05%. Calcium blocking agent. Prepn: **Brit. pat. 1,025,578** (1966 to Richardson-Merrell), *C.A.* **65,** 2229f (1966). Pharmacology: Hudak *et al., J. Pharmacol. Exp. Ther.* **173,** 371 (1970). Clinical studies: Winsor, *Clin. Pharmacol. Ther.* **11,** 85 (1970). Toxicity study: Causa, Perri, *Arzneimittel-Forsch.* **21,** 114 (1971). Series of articles: *Postgrad. Med. J. Suppl.* **49,** 8-132 (1973). Book: *Antiarrhythmic Action and the Puzzle of Perhexiline,* E. M. Vaughan Williams, Ed. (Academic Press, New York, 1980) 143 pp.

Maleate, $C_{19}H_{35}N.C_4H_4O_4$, *Pexid.* mp 188.5-191°. LD_{50} in rats, mice (g/kg): > 7, 4.37 orally (Causa, Perri).

Hydrochloride, $C_{19}H_{35}N.HCl$, white crystalline powder, mp 243-245.5° (dec).

THERAP CAT: Vasodilator (coronary); diuretic.

7306. Pericyazine. *10-[3-(4-Hydroxy-1-piperidinyl)-propyl]-10H-phenothiazine-2-carbonitrile;* 2-cyano-10-[3-(4-hydroxypiperidino)propyl]phenothiazine; periciazine; propericiazine; RP-8908; Aolept; Neulactil; Neuleptil. $C_{21}H_{23}$-N_3OS; mol wt 365.50. C 69.01%, H 6.34%, N 11.50%, O 4.38%, S 8.77%. Psychotherapeutic phenothiazine. Prepn: Robert, Jacques, **Fr. pat 1,212,031** (1960 to Rhône-Poulenc). Hypotensive action in dogs: K. P. Singh *et al., Ind. J. Med. Res.* **58,** 1467 (1970). GLC determn in human urine: A. P. De Leenheer, *J. Pharm. Sci.* **63,** 389 (1974). Clinical studies in psychiatric patients: U. Spiegelberg, G. Kleu, *Arzneimittel-Forsch.* **17,** 159 (1967); J. C. Barker, M. Miller, *Brit. J. Psychiat.* **115,** 169 (1969). Toxicity and metabolism: L. Julou *et al., Proc. Eur. Soc. Study Drug Toxic.* **9,** 11 (1968). Toxicity data: Schafer, *Toxicol. Appl. Pharmacol.* **21,** 315 (1972). Use in spectrophotometry: H. Gowda *et al., Anal. Chem.* **55,** 1816 (1983); A. T. Gowda *et al., Anal. Chim. Acta* **154,** 347 (1983).

Crystals, mp 116-117°. uv max: 232.5, 271.5 nm (log ε 4319, 4503). LD_{50} orally in rats: 395 mg/kg (Schafer).

USE: Spectrophotometric reagent for palladium and ruthenium.

THERAP CAT: Antipsychotic.

7307. Perilla Ketone. *1-(3-Furanyl)-4-methyl-1-pentanone; 1-(3-furyl)-4-methyl-1-pentanone;* β-furyl isoamyl ketone. $C_{10}H_{14}O_2$; mol wt 166.22. C 72.26%, H 8.49%, O 19.25%. Potent pulmonary edematogenic agent. Isoln from the mint plant, *Perilla frutescens* Britton, *Labiatae,* and structure: R. Goto, *J. Pharm. Soc. Japan* **57,** 77 (1937). Synthesis: T. Matsuura, *Bull. Chem. Soc. Japan* **30,** 430 (1957); K. Kondo, M. Matsumoto, *Tetrahedron Letters* **1976,** 4363; T. Kitamura *et al., Synth. Commun.* **7,** 521 (1977); K. Inomata *et al., Chem. Letters* **1979,** 709. Potent lung toxin implicated in emphysema of grazing cattle: B. J. Wilson *et al., Science* **197,** 573 (1977).

Colorless oil, bp 196°. n_D^{20} 1.4781; d_{15}^{15} 0.9920. uv max (ethanol): 207, 253 nm (ε 14100, 5800). Sensitive to oxygen; becomes reddish-orange on standing. LD_{50} in female, male mice, male rats (mg/kg): 2.5, 6, 10 i.p. (Wilson).

7308. Perillaldehyde. *4-(1-Methylethenyl)-1-cyclohexene-1-carboxaldehyde; 4-isopropenyl-1-cyclohexene-1-carboxaldehyde.* $C_{10}H_{14}O$; mol wt 150.22. C 79.96%, H 9.39%, O 10.65%. Isoln from *Perilla arguta* Benth., *Labiatae:* Semmler, Zaar, *Ber.* **44,** 52, 815 (1911); from essential oil of *Sium latifolium* L., *Umbelliferae:* Parczewski, *Dissertationes Pharm.* **12,** 223 (1960), *C.A.* **55,** 7765c (1961); from mandarin peel oil *(Citrus reticulata* Blanco, *Rutaceae):* Kugler, Kováts, *Helv. Chim. Acta* **46,** 1480 (1963). Prepn by chromic oxidation of perilla alcohol: Naves, *ibid.* **29,** 553 (1946); Ritter, Ginsburg, *J. Am. Chem. Soc.* **72,** 2381 (1950); Kergomard, Philibert-Bigou, *Bull. Soc. Chim. France* **1958,** 393, 1174; Naves, Grampoloff, *ibid.* **1960,** 37.

d-Form, liquid. bp_{745} 237°; bp_7 98-100°. d_4^{20} 0.953. n_D^{20} 1.5058. $[α]_D^{20}$ +127° (c = 13.1 in carbon tetrachloride).

l-Form, liquid. bp_{10} 104-105°. d_4^{20} 0.9645. n_D^{20} 1.5069. $[α]_D^{20}$ −146°.

Oxime, $C_{10}H_{15}NO$, *l-Perillaldehyde α-syn-oxime, perillartine,* "perilla sugar". Previously referred to as *l-perillaldehyde α-anti-oxime.* Synthesis: Andô *et al., Science (Tokyo)* **17,** 241 (1947), *C.A.* **45,** 1976d (1951). Clarification of structure: Acton *et al., Experientia* **26,** 473 (1970). Needles, mp 102°. uv max (alc): 232 nm (ε 21800). About 2000 times as sweet as sucrose: Furukawa, *Koryo* **No. 11,** 11, 40 (1950), *C.A.* **44,** 6083g (1950). LD_{50} orally in rats: 2500 mg/kg, *RTECS* **Vol. I,** R. J. Lewis, R. L. Tatken, Eds. (1980) p 568.

USE: The oxime is used as sweetening agent in Japan.

7309. Perimethazine. *1-[3-(2-Methoxy-10H-phenothiazin-10-yl)-2-methylpropyl]-4-piperidinol;* 10-methoxy-10-[2-methyl-3-(4-hydroxypiperidino)propyl]phenothiazine; 3-methoxy-10-[3-(4-hydroxypiperidyl)-2-methylpropyl]phenothiazine; perimetazine; RP-9159; AN-1317; Leptryl. C_{22}-$H_{28}N_2O_2S$; mol wt 384.54. C 68.72%, H 7.34%, N 7.28%, O 8.32%, S 8.34%. Prepn: **Brit. pat. 904,210;** Jacob, Robert, **U.S. pat. 3,075,976** (1962, 1963 both to Rhône-Poulenc). Pharmacology: L. Julon *et al., Compt. Rend. Soc. Biol.* **160,** 1852 (1966). Clinical evaluation as antipsychotic: M. Bourgeois, A.-M. Sicart, *Bord. Med.* **3,** 515 (1970).

Crystalline powder from benzene:cyclohexane (15:85), mp 137-138°.

Hydrochloride, $C_{22}H_{28}N_2O_2S.HCl$. LD_{50} in mice (mg/kg): 115 i.v.; 140 i.p.; 330 s.c.; 310 orally (Julon).

THERAP CAT: Antipsychotic.

7310. Perimycin. Aminomycin; fungimycin; WX-2412; NC-1968. Polyene antifungal antibiotic complex produced by *Streptomyces coelicolor* var. *aminophilus* NRRL 2390; Wooldridge, U.S. pat. **2,956,925** (1960); Mohan *et al.*, *Antimicrob. Ag. Chemother.* **1963**, 462; McDaniel *et al.*, U.S. pat. **3,182,004** (1965 to Warner-Lambert). Purification: Borowski *et al.*, *Antimicrob. Ag. Ann.* **1960**, 532. Found to be a mixture of three active components, perimycin A (major), B, and C. Structural studies: C. H. Lee, C. P. Schaffner, *Tetrahedron Letters* **1966**, 5837; *eidem*, *Tetrahedron* **25**, 2229 (1969). Isoln of perimycins A, B, C and structural elucidation of A: P. Kolodziejczyk *et al.*, *Tetrahedron Letters* **1976**, 3603. Description of unique biological properties: E. Borowski, B. Cybulska, *Nature* **213**, 1034 (1967).

Perimycin A

Amorphous, golden-yellow solid. Has no definite mp but dec slowly with darkening upon heating. uv max (methanol): 383 nm ($E_{1cm}^{1\%}$ 1000). Sol in the following solvents in the presence of water: lower alcohols, pyridine, tetrahydrofuran, acetone, dioxane. Sol in warm methanol, DMF, dimethylsulfoxide, and in the lower fatty acids. Practically insol in water, petr ether, ethyl acetate, benzene.

Perimycin A. $C_{59}H_{88}O_{17}N_2$. The first member of the aromatic subgroup of heptaene macrolides for which the complete structure has been elucidated.

THERAP CAT: Antifungal.

7311. Perindopril. *[2S-[1-[R*,(R*)],2α,3aβ,7aβ]]-1-[2-[[1-(Ethoxycarbonyl)butyl]amino]-1-oxopropyl]octahydro-1H-indole-2-carboxylic acid;* (2S,3aS,7aS)-1-[(S)-N-[(S)-1-carboxybutyl]alanyl]hexahydro-2-indolinecarboxylic acid 1-ethyl ester; (2S)-2-[(1S)-1-carbethoxybutylamino]-1-oxopropyl-(2S,3aS,7aS)-perhydroindole-2-carboxylic acid; S-9490; McN-A-2833. $C_{19}H_{32}N_2O_5$; mol wt 368.47. C 61.93%, H 8.75%, N 7.60%, O 21.71%. Angiotensin-converting enzyme (ACE) inhibitor. Hydrolyzed *in vivo* to the active diacid metabolite. Prepn: G. Remond *et al.*, Eur. pat. Appl. **49,-658** (1982 to Sci. Union et Cie. - Soc. Franc. Rech. Med.); M. Vincent *et al.*, U.S. pat. **4,508,729** (1985 to ADIR). Stereoselective synthesis: M. Vincent *et al.*, *Tetrahedron Letters* **23**, 1677 (1982). NMR study: N. Platzer *et al.*, *Magnet. Res. Chem.* **26**, 296 (1988). Hemodynamic effects in humans: K. R. Lees, J. L. Reid, *Brit. J. Clin. Pharmacol.* **23**, 159 (1987). Clinical evaluation in essential hypertension: T. Morgan *et al.*, *J. Cardiovasc. Pharmacol.* **10**, Suppl. 7, S116 (1987). Pharmacokinetics, pharmacodynamics of the diacid: K. R. Lees, J. L. Reid, *ibid.* **10**, 129 (1987). Review of pharmacology and clinical use: P. A. Todd, A. Fitton, *Drugs* **42**, 90-114 (1991).

tert-Butylamine, $C_{19}H_{32}N_2O_5.C_4H_{11}N$, *perindopril erbumine, S-9490-3, McN-A-2833-109, Aceon, Coversum, Coversyl, Procaptan.*

Diacid form, $C_{17}H_{28}N_2O_5$, *perindoprilat, S-9780.*

THERAP CAT: Antihypertensive.

7312. Periodic Acid. H_5IO_6; mol wt 227.94. H 2.21%, I 55.67%, O 42.11%. Prepd by electrolytic oxidation of iodic acid or from barium periodate and nitric acid: Willard, *Inorg. Syn.* **1**, 172 (1939). Chemistry of periodic acid and periodates: H. Siebert, *Fortschr. Chem. Forsch.* **8**, 470 (1967). Periodic acid and periodates in organic and bioorganic chemistry: A. J. Fatiadi, *Synthesis* **1974**, 229. Book: G. Dryhurst, *Periodate Oxidation of Diol and Other Functional Groups* (Pergamon Press, New York, 1970).

Monoclinic, hygroscopic crystals. mp 122°; dec 130-140° forming I_2O_5, H_2O, and O_2. Freely sol in water. A 38% w/w soln had d_4^{17} 1.3875. Sol in alcohol, slightly in ether. Soly in nitric acid (d 1.42) at 26° = 7.82 g/100 ml. When aq solns are evapd at room temp the "ortho" acid H_5IO_6 crystallizes out. If this is heated in a vacuum at 100° and 12 mm it loses water and is converted to the "meta" acid HIO_4, an intermediate $H_4I_2O_9$ being formed at 80°. Periodic acid is a dibasic acid, much weaker than perchloric acid. K_1 at 25° = 2.3 × 10^{-2}; K_2 = 2 × 10^{-6}. Very easily reduced to iodic acid by nitrous or sulfurous acids and even by hydrochloric and sulfuric acids. Oxidizes organic material.

USE: In organic synthesis.

7313. Periodyl. *12-Hydroxy-9,10-diiodo-9-octadecenoic acid;* diiodoricinstearolic acid; ricinstearolic acid diiodide; 8,9-diiodo-11-hydroxy-8-heptadecene-1-carboxylic acid; Diiodyl; Joristen. $C_{18}H_{32}I_2O_3$; mol wt 550.26. C 39.29%, H 5.86%, I 46.13%, O 8.72%. $CH_3(CH_2)_5CHOHCH_2CI=CI-(CH_2)_7COOH$. Prepn: Mühle, *Ber.* **46**, 2091 (1913); Ger. pat. **296,495** (to Riedel).

Tasteless needles from dil alcohol, mp 62°. Practically insol in water, acids; sol in dil alkalies, alcohol, ether, chloroform; slightly sol in benzene.

THERAP CAT: Iodine source.

7314. Periplanones. Sex pheromones of the American cockroach, *Periplaneta americana* L., found primarily in the alimentary tract and excreta of the insect. They act as close proximity sex-excitants and function over relatively short distances compared to long range insect sex-attractants. Isoln: D. R. A. Wharton *et al.*, *Science* **137**, 1062 (1962). Improved isolns, identification and proposed structures of periplanone B: C. J. Persoons *et al.*, *Tetrahedron Letters* **1976**, 2055; E. Talman *et al.*, *Isr. J. Chem.* **17**, 227 (1978); C. J. Persoons *et al.*, *J. Chem. Ecol.* **5**, 221 (1979). Total synthesis and structure of (±) periplanone B: W. C. Still, *J. Am. Chem. Soc.* **101**, 2493 (1979); H. Hauptmann, G. Muhlbauer, *Tetrahedron Letters* **27**, 1315 (1986); of the (−)-form: T. Kihara *et al.*, *ibid.* 1343. Absolute configuration of B: M. A. Adams *et al.*, *J. Am. Chem. Soc.* **101**, 2495 (1979). Structural and stereochemical studies of periplanone A: C. J. Persoons *et al.*, *J. Chem. Ecol.* **8**, 439 (1982); H. Hauptmann, G. Muhlbauer, *Tetrahedron Letters* **27**, 6189 (1986); Y. Shizuri *et al.*, *ibid.* **28**, 1791, 1795 (1987); L. MacDonald *et al.*, *Heterocycles* **25**, 305 (1987). Short synthesis of periplanone A: Y. Shizuri *et al.*, *Tetrahedron Letters* **29**, 1971 (1988). Use in insecticidal formulations: W. J. Bell *et al.*, *Environ. Entomol.* **13**, 448 (1984); W. J. Bell *et al.*, *Pest Control* **54**, 40 (1986).

Periplanone B

Periplanone A, $C_{15}H_{20}O_2$. *[3S-(3R*,6R*,7E,11Z)]-9-methylene-6-(1-methylethyl)-1-oxaspiro[2.9]dodeca-7,11-dien-2-one.* Found primarily in excreta of *P. americana*, it is unstable and gradually rearranges to a biologically inactive compound. uv max (hexane): 220 nm (log ε 4.18).

Consult the Name Index before using this section.

Periplanone B, $C_{15}H_{20}O_3$, *[1R*,2R*,5S*,6E,10R*]-(±)-8-methylene-5-(1-methylethyl)spiro[11-oxabicyclo[8.1.0]undec-6-ene-2,2'-oxiran)-3-one.* More stable and more active than periplanone A, it is found in both the alimentary tract and excreta of *P. americana*. The ratio of A to B in excreta is 1:10. Crystals, mp 48-50°. uv max (hexane): 226 nm.
(−)-Periplanone B, crystals, mp 47-50°. $[\alpha]_D^{22}$ −667° (c = 0.13 in *n*-hexane).

7315. Periplocin. *3β-[(2,6-Dideoxy-4-O-β-D-glucopyranosyl-β-D-ribo-hexopyranosyl)oxy]-5β,14-dihydroxycard-20(22)-enolide;* glucoperiplocymarin; periplocoside. $C_{36}H_{56}O_{13}$; mol wt 696.83. C 62.05%, H 8.10%, O 29.85%. From *Periploca graeca* L., *Asclepiadaceae:* Lehmann, *Arch. Pharm.* **235**, 157 (1897); Jacobs, Hoffmann, *J. Biol. Chem.* **79**, 519 (1928); from *Strophanthus preussii* Engl. and Pax., *Apocynaceae:* Ruppol, Trukovic, *J. Pharm. Belg.* **10**, 221 (1955), *C.A.* **50**, 12089d (1956). Structure: Stoll, Renz, *Helv. Chim. Acta* **22**, 1193 (1939). Acid hydrolysis yields periplogenin, *q.v.*, and periplobiose (cymarose + glucose, $C_{13}H_{24}O_9$, $[\alpha]_D^{20}$ +32°). Enzymatic hydrolysis with strophanthobiase splits off glucose, yielding periplocymarin. The biose is attached to the hydroxyl group at C-3 of the aglycon.
Dihydrate, fine needles from water. Becomes anhydr after drying for one hour in high vacuum at 105°. mp 224° when bath is preheated to 200°. $[\alpha]_D^{20}$ +23° (c = 0.7 in alcohol). One gram dissolves in ~20 ml boiling water, at 25° the soly is ~1:2500. Freely sol in alcohol; almost insol in ether, chloroform. LD s.c. in rabbits: 10 mg/kg, *Handbook of Toxicology*, vol. 1, W. S. Spector, Ed. (Saunders, Philadelphia, 1956) pp 226-227.
Tetraacetylperiplocin, $C_{44}H_{64}O_{17}$, six-sided prisms from alcohol, mp 195°. $[\alpha]_D^{20}$ +20° (c = 0.5 in alcohol). Sol in alcohol, chloroform; very slightly sol in water.

7316. Periplocymarin. *(2,6-Dideoxy-3-O-methyl-β-D-ribo-hexopyranosyl)oxy-5,14-dihydroxy-card-20(22)-enolide.* $C_{30}H_{46}O_8$; mol wt 534.69. C 67.39%, H 8.67%, O 23.94%. By extracting bark and wood of *Periploca graeca* L., *Asclepiadaceae* with 70% alcohol and treating with the enzyme strophanthobiase from seeds of *Strophanthus courmonti* Sacleux, *Apocynaceae:* Jacobs, Hoffmann, *J. Biol. Chem.* **79**, 519 (1928); Katz, Reichstein, *Pharm. Acta Helv.* **19**, 231 (1944); from *S. hypoleucus* Stapf: von Euw, Reichstein, *Helv. Chim. Acta* **33**, 544 (1950); from *S. ledienii* Stein: Lichti *et al., ibid.* **39**, 1914 (1956); from *S. eminii* Asch. et Pax.: Zelnik, Schindler, *ibid.* **40**, 2110 (1957). Structure: von Euw, Reichstein, *ibid.* **31**, 883 (1948). Acid hydrolysis yields one mol periplogenin *(q.v.)* and one mol cymarose *(q.v.).* The cymarose is attached to the hydroxyl group at C-3 of the aglycon.
Lustrous needles from methanol contg approx one mol CH_3OH of crystn. Bitter taste, but not nearly so marked as that of cymarin. Becomes solvent-free at 100° *in vacuo*. Sinters at 138°. mp 148°. $[\alpha]_D^{27}$ +29° (c = 0.94 in 95% alcohol for the anhydr substance). Readily sol in alcohol, chloroform, acetone; less readily in methanol; very sparingly sol in water and practically insol in ether. Readily hydrolyzed in the cold.

7317. Periplogenin. *(3β,5β)-3,5,14-Trihydroxycard-20-(22)-enolide;* desoxostrophanthidin. $C_{23}H_{34}O_5$; mol wt 390.52. C 70.74%, H 8.78%, O 20.48%. The aglycon of periplocin and periplocymarin; accompanies strophanthidin in *Strophanthus eminii* Asch. et Pax., *Apocynaceae.* Isoln:

Lehmann, *Arch. Pharm.* **235**, 157 (1897); Jacobs, Hoffmann, *J. Biol. Chem.* **79**, 519 (1928); Stoll, Renz, *Helv. Chim. Acta* **22**, 1193 (1939); Lardon, *ibid.* **33**, 639 (1950). Structure: Speiser, Reichstein, *Experientia* **3**, 323 (1947); *eidem, Helv. Chim. Acta* **30**, 2143 (1947); **31**, 622 (1948). Synthesis: Deghenghi, Gaudry, *Tetrahedron Letters* **1963**, 2045; Kamano *et al., J. Org. Chem.* **39**, 2319 (1974).
Solvated prisms from methanol which contain an undetermined amount of methanol, sinter at 140°, mp 235°. $[\alpha]_D^{27}$ +31.5° (c = 1.04 in alcohol). Sol in alcohol and chloroform; slightly sol in ether, water (1:2500); practically insol in benzene, petr ether. pH of aq solns ~7. Strong positive Legal test.
Monobenzoate, $C_{30}H_{38}O_6$, stout glistening wedges from 95% alcohol, mp 235°.
Dihydroperiplogenin, $C_{23}H_{36}O_5$, stout prisms from 25% alcohol, mp 204° (slight preliminary softening).
3-O-Acetylperiplogenin, $C_{25}H_{36}O_6$, crystals, mp 242-244°, $[\alpha]_D^{22}$ +46.9° (c = 0.32 in chloroform).

7318. Perisoxal. *α-(5-Phenyl-3-isoxazolyl)-1-piperidineethanol;* 3-(2-piperidino-1-hydroxyethyl)-5-phenylisoxazole. $C_{16}H_{20}N_2O_2$; mol wt 272.35. C 70.56%, H 7.40%, N 10.29%, O 11.75%. Prepn: **Neth.** pat. **Appl. 6,408,694;** H. Kano *et al.* **U.S.** pat. **3,321,475** (1965, 1967 both to Shionogi); *eidem, J. Med. Chem.* **10**, 411 (1967). Series of articles on pharmacological properties: *Oyo Yakuri* **6**, 1285-1338 (1972), *C.A.* **79**, 49177h-49179k (1973). Studies on physical dependence in rats: K. Yamamoto *et al., ibid.* **16**, 871 (1978), *C.A.* **90**, 180250w (1979). Absorption, excretion, distribution, metabolism: J. Watanabe *et al., Chem. Pharm. Bull.* **27**, 1075 (1979). Toxicity study: H. Kano *et al., J. Med. Chem.* **10**, 411 (1967).

Colorless plates from 50% aq ethanol, mp 107-108°.
Citrate, $(C_{16}H_{20}N_2O_2)_2 \cdot C_6H_8O_7$, S-31252, Isoxal. Colorless prisms from ethanol/acetone, mp 143-145°. LD_{50} s.c. in mice: 416 mg/kg (Kano).
THERAP CAT: Anti-inflammatory; analgesic.

7319. Perivine. *4-Demethyl-3-oxovobasan-17-oic acid methyl ester.* $C_{20}H_{22}N_2O_3$; mol wt 338.41. C 70.99%, H 6.55%, N 8.28%, O 14.18%. Indole alkaloid from *Vinca rosea* Linn., *Apocynaceae:* G. Svoboda, *J. Am. Pharm. Assoc.* **47**, 834 (1958); G. Svoboda *et al., ibid.* **48**, 659 (1959). Structure: Gorman, Sweeny, *Tetrahedron Letters* **1964**, 3105. Cytotoxicity: D. G. I. Kingston, *J. Pharm. Sci.* **67**, 272 (1978).

Prisms from methanol, dec 218-221°. Also reported as 180-181° (Svoboda 1959). $[\alpha]_D^{26}$ −121.4° (chloroform). pKa in 66% DMF: 7.5. uv max (ethanol): 314 nm ($E_{1cm}^{1\%}$ 2.67).

7320. Perlapine. *6-(4-Methyl-1-piperazinyl)-11H-dibenz[b,e]azepine; 6-(4-methyl-1-piperazinyl)morphanthridine;* AW-14'2333; HF-2333; MP-11; Hypnodin. $C_{19}H_{21}N_3$; mol wt 291.40. C 78.32%, H 7.26%, N 14.42%. Prepn: **Brit.** pat. **1,006,156** corresp to Schmutz, Hunziker, **U.S.** pat. **3,389,139** (1965, 1968, both to Wander); Hunziker *et al.,*

Helv. Chim. Acta **49**, 1433 (1966). Pharmacology: Take *et al., Takeda Kenkyusho Ho* **29**, 416 (1970); Yokotani *et al., ibid.* 441; Stille *et al., Psychopharmacologia* **28**, 325 (1973).

Yellow, prismatic crystals from acetone-petrol ether, mp 136-138°. LD_{50} orally in mice: 415 mg/kg.

THERAP CAT: Hypnotic.

7321. Permethrin. *3-(2,2-Dichloroethenyl)-2,2-dimethylcyclopropanecarboxylic acid (3-phenoxyphenyl)methyl ester;* 3-(phenoxyphenyl)methyl (±)-*cis,trans*-3-(2,2-dichloroethenyl)-2,2-dimethylcyclopropanecarboxylate; *m*-phenoxybenzyl (±)-*cis,trans*-3-(2,2-dichlorovinyl)-2,2-dimethylcyclopropanecarboxylate; FMC-33297; NIA-33297; NRDC-143; PP-557; SBP-1513; S-3151; Ambush; Corsair; Dragnet; Ectiban; Eksmin; Nix; Pulvex; Pounce; Pynosect; Ridect Pour-On. $C_{21}H_{20}Cl_2O_3$; mol wt 391.29. C 64.46%, H 5.15%, Cl 18.12%, O 12.27%. Synthetic pyrethroid insecticide, more stable to light and at least as active as the natural pyrethrins and with low mammalian toxicity: M. Elliott *et al., Nature* **246**, 169 (1973). Of the four possible isomers, the (1*R,-trans*)- and the (1*R,cis*)-isomers are the two esters primarily responsible for insecticidal activity: P. E. Burt *et al., Pestic. Sci.* **5**, 791 (1974). Prepn of the racemic mixture: T. Mizutani *et al., Ger. pat.* **2,437,882** (1975 to Sumitomo); F. Mori *et al., Ger. pat.* **2,544,150;** *eidem,* U.S. pat. **4,113,968** (1976, 1978, both to Kuraray). Metabolism: M. Elliott *et al., J. Agr. Food Chem.* **24**, 270 (1976); L. C. Gaughan *et al., ibid.* **26**, 613 (1978). Photodecompn: R. L. Holmstead *et al., ibid.* 590. Analysis: H. Swaine, M. J. Tandy, *Anal. Methods Pestic. Plant Growth Regul.* **13**, 103-120 (1984). Field evaluation to control flies and ticks in cows: S. S. Quisenberry, D. R. Strohbehn, *J. Econ. Entomol.* **77**, 422 (1984); R. B. Davey, E. H. Ahrens, *Am. J. Vet. Res.* **45**, 1008 (1984). Clinical trial in pediatric pediculosis: E. Ares Mazas *et al., Int. J. Dermatol.* **24**, 603 (1985). Toxicity studies in mammals: L. Metker *et al., U.S. NTIS, AD REP.* **1977**, AD-AO47284, 70 pp; F. Cantalamessa, *Arch. Toxicol.* **67**, 510 (1993). *Review:* C. N. E. Ruscoe, *Pestic. Sci.* **8**, 236 (1977).

(1R-cis)-form

Technical material is a mixture of ~60% *trans-* and 40% *cis*-isomers: Colorless crystals to a pale yellow viscous liquid, mp ~35°. $bp_{0.05}$ 220°. d^{20} 1.190-1.272. Vapor pressure at 50° < 1 × 10^{-6} mm Hg. Soly in water: < 1 ppm. Sol or miscible with org solvents except ethylene glycol. LD_{50} orally in female rats: 3801 mg/kg (Metker). LD_{50} in 8 day old rats, male adult rats (mg/kg): 340.5, 1500.0 orally (Cantalamessa). Toxic to bees and fish.

Caution: Mild irritant to skin and eyes (Metker).

USE: Insecticide.

THERAP CAT: Pediculicide.

THERAP CAT (VET): Ectoparasiticide.

7322. Pernambuco. Fernambuco; Brazil wood; Nicaragua wood; Lima wood; redwood. Wood of *Caesalpinia echinata* Lam., *Leguminosae.* Habit. Brazil. Constit. Brazilin.

USE: Dyeing red; manuf of a red lake pigment. With alkalies = purplish-red; with acids = yellow.

7323. Perphenazine. *4-[3-(2-Chloro-10H-phenothiazin-10-yl)propyl]-1-piperazineethanol;* 2-chloro-10-[3-[1-(2-hydroxyethyl)-4-piperazinyl]propyl]phenothiazine; 1-(2-hydroxyethyl)-4-[3-(2-chloro-10-phenothiazinyl)propyl]-piperazine; chlorpiprazine; chlorpiprozine; PZC; Sch-3940; Trilafon; Trilifan; Decentan; Fentazin; Perphenan. $C_{21}H_{26}-ClN_3OS$; mol wt 403.98. C 62.44%, H 6.49%, Cl 8.78%, N 10.40%, O 3.96%, S 7.94%. Prepn: Cusic, U.S. pat. **2,766,-235** (1956); Sherlock, Sperber, U.S. pat. **2,860,138** (1958 to Schering). Metabolism: U. Breyer, H. J. Gaertner, *Adv. Biochem. Psychopharmacol.* **9**, 167 (1974); H. J. Gaertner *et al., Drug Metab. Dispos.* **3**, 437 (1975). Crystal structure: J. J. H. McDowell, *Acta Crystallogr.* **B34**, 686 (1978).

Crystals. Sensitive to light. mp 94-100°. $bp_{0.15}$ 214-218°; bp_1 278-281°. Practically insol in water. Soly (mg/ml): ethanol 153; acetone 82. Practically insol in sesame oil.

Dihydrochloride, $C_{21}H_{28}Cl_3N_3OS$, crystals from alcohol, mp 225-226°.

THERAP CAT: Antipsychotic.

THERAP CAT (VET): Tranquilizer, pre-anesthetic agent.

7324. Peruvoside. *(3β,5β)-3-[(6-Deoxy-3-O-methyl-α-L-glucopyranosyl)oxy]-14-hydroxy-19-oxocard-20(22)-enolide;* cannogenin α-L-thevetoside; Encordin. $C_{30}H_{44}O_9$; mol wt 548.67. C 65.67%, H 8.08%, O 26.24%. Cardenolide (cardiac glycoside) found in seeds of the tropical garden plant *Thevetia peruviana* (Pers.) K. Schum. (*Th. neriifolia* Juss.), *Apocynaceae* (named trumpet flower in Florida). Isoln: Rangaswami, Venkata Rao, *J. Sci. Ind. Res.* **17B**, 331 (1958). Structure: Bloch *et al., Helv. Chim. Acta* **43**, 652 (1960). Pharmacological studies: *Arzneimittel-Forsch.* **18**, 1582-1608 (1968).

Spears from methanol + ether, mp 161-164°. $[\alpha]_D^{22}$ −71.7° (c = 1.54 in methanol). One gram dissolves in ~2500 ml water. Freely soluble in chloroform and acetone, sparingly soluble in methanol and ethanol.

THERAP CAT: In cardiac insufficiency.

7325. Perylene. Dibenz[*de,kl*]anthracene; *peri*-dinaphthalene. $C_{20}H_{12}$; mol wt 252.32. C 95.21%, H 4.79%. Occurs in coal tar. Isoln from pitch distillate: Cook *et al., J. Chem. Soc.* **1933**, 395. From 2,2'-dihydroxy-1,1'-dinaphthyl: Zinke *et al., Monatsh.* **64**, 415 (1934). From the reaction of phenanthrene with acrolein in anhydr HF: Weinmayr, U.S. pat. **2,145,905** (1939); *cf. J. Am. Chem. Soc.* **61**, 949 (1939).

Yellow to colorless crystals from toluene. mp 273-274°. Sublimes 350-400°. d 1.35. Absorption spectrum: Clar, *Ber.* **65**, 848 (1932). Freely sol in CS_2, chloroform; moderately sol in benzene; slightly in ether, alcohol, acetone; very sparingly sol in petr ether. Insol in water.

Monopicrate, $C_{26}H_{15}N_3O_7$, dark violet-blue needles, mp 223-225°.

7326. Petrolatum. Petroleum jelly; paraffin jelly; vasoliment; Stanolind; Vaseline; Saxoline; Cosmoline. Purified mixture of semisolid hydrocarbons, chiefly of the methane series of the general formula $C_nH_{2n}+2$. Actually, petrolatum is a colloidal system of nonstraight-chain solid hydrocarbons and high-boiling liq hydrocarbons, in which most of the liq hydrocarbons are held inside the micelles. Detailed historical account including chemistry and modern mfg methods: Schindler, *Drug Cosmet. Ind.* **89**, 36-37, 76, 78-80, 82 (1961).

Yellowish to light amber or white, semisolid, unctuous mass; practically odorless and tasteless. d_{25}^{60} 0.820-0.865. mp 38-54°. n_D^{60} 1.460-1.474. White petrolatum is transparent in thin layers even at 0°. Practically insol in water, glycerol, alcohol. Sol in benzene, chloroform, ether, petr ether, carbon disulfide, oils.

Premium white petrolatum, *Stanolene.*

USE: As ointment base in pharmaceuticals and cosmetics. Lubricating firearms and machinery, leather grease, shoe polish, rust preventives, modeling clays.

7327. Petrolatum, Liquid. Liquid paraffin; mineral oil; white mineral oil; paraffin oil; Clearteck; Drakeol; Hevyteck; Kremol; Kaydol; Alboline; Nujol; Paroleine; Saxol; Adepsine oil; Glymol. A mixture of liquid hydrocarbons from petroleum.

Colorless, oily liquid; practically tasteless and odorless even when warmed. The density of the "light" oil is usually 0.83-0.860; the "heavy" 0.875-0.905. Surface tension at 25° slightly below 35 dynes/cm. Insol in water, alc. Sol in benzene, chloroform, ether, carbon disulfide, petr ether, oils.

USE: Lubricant. Pharmaceutic aid (vehicle, solvent). As formulation aid in foods. In cosmetics as emollient.

THERAP CAT: Cathartic.

THERAP CAT (VET): Laxative, externally as a protectant, lubricant.

7328. Petroleum. Crude oil; mineral oil; rock oil; coal oil; seneca oil. Consists of a mixture of hydrocarbons from C_2H_6 and up—chiefly of the paraffins, cycloparaffins, or of cyclic aromatic hydrocarbons, with small amounts of benzene hydrocarbons, sulfur, and oxygenated compounds. *Occurrence:* U.S., Mexico, Iran, Russia, Roumania, Poland, Dutch East Indies, etc.

Dark yellow to brown or greenish-black, oily liquid. Insol in water and only a small portion of it may dissolve in alcohol; sol in benzene, chloroform, ether.

USE: Source of gasoline, petr ether, liq and solid petrolatum, fuel and lubricating oils, butane, isopropyl alcohol, etc.

7329. Petroleum Benzin. *Naphtha;* benzin; petroleum naphtha. Term that has been applied to low boiling fractions of petroleum, consisting chiefly of hydrocarbons of the methane series—principally pentanes and hexanes.

Clear, colorless, nonfluorescent, *highly flammable*, volatile liq; characteristic odor; does not solidify in the cold. The vapors mixed with air explode if ignited. *Keep tightly closed in a cool place and away from fire.* d 0.625-0.660; bp between 35-80°. Flash pt $\sim$ −40°. Insol in water. Miscible with abs alc, benzene, chloroform, ether, carbon disulfide, carbon tetrachloride, and oils except castor oil.

Caution: Potential symptoms of overexposure are light-headedness, drowsiness; irritation of eyes, nose and skin; dermatitis. *See NIOSH Pocket Guide to Chemical Hazards* (DHHS/NIOSH 90-117, 1990) p 158.

USE: Pharmaceutic aid (solvent).

THERAP CAT: Counterirritant.

7330. Petroselinic Acid. *cis-6-Octadecenoic acid;* petroselic acid; 5-heptadecylene-1-carboxylic acid; Δ^5-octadecylenic acid. $C_{18}H_{34}O_2$; mol wt 282.47. C 76.54%, H 12.13%, O 11.33%. $CH_3(CH_2)_{10}CH=CH(CH_2)_4COOH$. Isoln from parsley seed oil, the oil extracted from dried ripe seed of *Petroselinum hortense* Hoffm., *Umbelliferae:* Fore *et al., J. Am. Oil Chem. Soc.* **37**, 490 (1960).

Leaflets from petr ether, mp 29.5-30.1°. bp_{18} 237-238°; d_4^{40} 0.8700. n_D^{40} 1.4533. Low temp solubilities: Heptane at −10° = 0.50 g/100 g solution; methanol at −20° = 0.48 g/100 g; ethyl acetate at −20° = 0.73 g/100 g; ether at −20° = 3.52 g/100 g. Ozonolysis yields 85% of adipic acid. Neutralization equivalent: 282.45; iodine value 89.87%.

Methyl ester, $C_{19}H_{36}O_2$, liq, d_4^{20} 0.8767; n_D^{20} 1.4501; bp_{10} 208-210°.

Glyceryl triester, $C_{57}H_{104}O_6$, *glyceryl tripetroselinate, tripetroselin.* Solidifies at 16.5°. n_D^{40} 1.4619.

Amide, $C_{18}H_{35}NO$, needles, mp 76°.

7331. Petunidin. 2-(3,4-Dihydroxy-5-methoxyphenyl)-3,5,7-trihydroxy-1-benzopyrylium chloride; *3,3',4',5,7-penta-hydroxy-5'-methoxyflavylium chloride;* petunidol. $C_{16}H_{13}$-ClO_7; mol wt 352.73. C 54.48%, H 3.71%, Cl 10.05%, O 31.75%. The aglucone of petunin. Willstätter, Burdick, *Ann.* **412**, 217 (1917). Synthesis: Bradley *et al., J. Chem. Soc.* **1930**, 793; Robinson, Robinson, *Biochem. J.* **25**, 1687 (1931). Chromatographic separation: Spaeth, Rosenblatt, *Anal. Chem.* **22**, 1321 (1950).

Gray-brown leaflets or prisms from dil HCl.

3,5-Diglucoside, $C_{28}H_{33}ClO_{17}$, *2-(3,4-dihydroxy-5-methoxy-phenyl)-3,5-bis(β-D-glucopyranosyloxy)-7-hydroxy-1-benzo-pyrylium chloride, petunin.* From *Petunia hybrida* Hort., *Solanaceae:* Willstätter, Burdick, *loc. cit.* Synthesis: Bell, Robinson, *J. Chem. Soc.* **1934**, 1604. Violet plates with a coppery luster from dil HCl, mp $\sim$178°. Absorption max (methanolic HCl): 540 nm.

7332. Peucedanin. *3-Methoxy-2-(1-methylethyl)-7H-furo[3,2-g][1]benzopyran-7-one; 6-hydroxy-2-isopropyl-3-methoxy-5-benzofuranacrylic acid δ-lactone;* 4-methoxy-5-isopropylfuro[2,3:6,7]coumarin; oreoselone methyl ether. $C_{15}H_{14}O_4$; mol wt 258.27. C 69.76%, H 5.46%, O 24.78%. Coumarin deriv obtained from rhizome of *Peucedanum officinale* L., *Umbelliferae:* Schlatter, *Ann.* **5**, 201 (1833); Hlasiwetz, Weidel, *ibid.* **174**, 67 (1874); A. Jassoy, P. Haensel, *Arch. Pharm.* **236**, 662 (1898); Popper, *Monatsh.* **19**, 268 (1898); from *Peucedanum morisonii,* Bess., *Umbelliferae:* G. K. Nikonov, A. A. Ivashenko, *Zh. Obshch. Khim.* **33**, 2740 (1963). Fluorescence spectrum: R. H. Goodwin, F. Kavanagh, *Arch. Biochem.* **27**, 182 (1950). Antitumor activity and toxicity studies: E. M. Vermel, S. A. Kruglyak-Syrkina, *Voprosy Onkol.* **5**, 43 (1959), *C.A.* **53**, 19162h (1959). Structure: E. Späth *et al., Ber.* **64**, 2203 (1931); E. Späth, K. Klager, *ibid.* **66**, 749 (1933). Synthesis: H. Schmid, A. Ebnöther, *Helv. Chim. Acta* **34**, 1982 (1951).

Colorless needles from ether/petr ether, mp 84-87°; also reported as mp 95-97° (Schmid, Ebnöther); from ligroin, mp 102.5° (Nikonov, Ivashenko). uv max (methanol): 255, 295, 340 nm (log ϵ 4.40, 4.05, 3.70). Practically insol in water. Freely sol in chloroform, CS_2; sol in hot alcohol, ether, acetic acid; sparingly sol in benzene, petr ether. LD_{50} orally in mice: 315 mg/kg (Vermel, Kruglyak-Syrkina).

7333. Peyonine. *1-[2-(3,4,5-Trimethoxyphenyl)ethyl]-1H-pyrrole-2-carboxylic acid.* $C_{16}H_{19}NO_5$; mol wt 305.33. C 62.94%, H 6.27%, N 4.59%, O 26.20%. A peyote alkaloid, isolated from *Lophophora williamsii* (Lem.) Coult. Isoln and synthesis: Kapadia, Shah, *Lloydia* **30**, 287 (1967). Structure studies: Kapadia, Highet, *J. Pharm. Sci.* **57**, 191 (1968).

Crystals. mp 131-133.5°. uv max (methanol): 261 nm (ϵ 10000).

7334. Pfeiffer's Substance. 4-(Dimethylamino)antipyrine compd with 5,5-diethylbarbituric acid (1:1); corps de Pfeiffer. $C_{21}H_{29}N_5O_4$; mol wt 415.49. C 60.71%, H 7.04%, N 16.86%, O 15.40%. A molecular compd of aminopyrine and barbital, prepd by repeated crystn of stoichiometric amounts of the components from a minimum of hot water: Pfeiffer, *Z. Physiol. Chem.* **146**, 98 (1925).
Silky needles, mp 113-115°. At 115° the melt is turbid, but turns into a clear liquid at 140°. Freely sol in water. The aminopyrine moiety dissolves in benzene, while the barbital moiety precipitates out.

7335. P-Glycoprotein. Glycoprotein-P; P-gp; P-170; permeability glycoprotein. Highly conserved cell-surface glycoprotein of M_r 170,000 daltons; protein product of the mammalian multidrug resistance (*mdr*) gene. Present in selective human tissues where it functions as an energy-dependent efflux pump. Increased expression in mammalian tumor cells is associated with multidrug resistance (MDR) to cancer chemotherapy agents. Identification and isolation from colchicine-resistant Chinese hamster ovary (CHO) cells: R. J. Juliano, V. Ling, *Biochim. Biophys. Acta* **455**, 152 (1976). Purification from plasma membrane of CHO cells: J. R. Riordan, V. Ling, *J. Biol. Chem.* **254**, 12701 (1979). MDR associated with P-glycoprotein in mammalian cell lines: N. Kartner *et al.*, *Science* **221**, 1285 (1983); N. Kartner *et al.*, *Cancer Res.* **43**, 4413 (1983). Deduced amino acid sequence from mouse *mdr* cDNA: P. Gros *et al.*, *Cell* **47**, 371 (1986); from human *mdr*1 cDNA: C.-J. Chen *et al.*, *ibid.* 381. Transfection and expression of human *mdr*1 gene in cell culture: K. Ueda *et al.*, *Proc. Nat. Acad. Sci. USA* **84**, 3004 (1987). Characterization of ATPase activity: H. Hamada, T. Tsuruo, *Cancer Res.* **48**, 4926 (1988). Detection in normal human tissues: F. Thiebaut *et al.*, *Proc. Nat. Acad. Sci. USA* **84**, 7735 (1987); R. N. Hitchins *et al.*, *Eur. J. Cancer Clin. Oncol.* **24**, 449 (1988); and in human tumor cells: D. R. Bell *et al.*, *J. Clin. Oncol.* **3**, 311 (1985); W. S. Dalton *et al.*, *Blood* **73**, 747 (1989). Review of role in MDR in human cancer: I. Pastan, M. Gottesman, *N. Engl. J. Med.* **316**, 1388 (1987); of biochemistry of MDR: J. A. Endicott, V. Ling, *Ann. Rev. Biochem.* **58**, 137-171 (1989); of mechanism of action: S. Ruetz, P. Gros, *Trends Pharm. Sci.* **15**, 260-263 (1994).

7336. Phalloidin. Phalloidine. $C_{35}H_{48}N_8O_{11}S$; mol wt 788.88. C 53.29%, H 6.13%, N 14.20%, O 22.31%, S 4.06%.

Best known of the toxins isolated from the poisonous green fungus *Amanita phalloides* (Fr.) Seer., *Agaricaceae*, known as "the green death cap" or "deadly agaric": Lynen, Wieland, *Ann.* **533**, 93 (1938). Structure: Wieland, Schön, *ibid.* **593**, 157 (1955); Wieland, Schöpf, *ibid.* **626**, 174 (1959); Wieland, Schnabel, *ibid.* **657**, 225 (1962). Differs from amanitin in rapidity of action; at high dose levels, death of mice or rats occurs within 1 or 2 hours. Phalloidin acts by binding actin, *q.v.*, an essential internal structural protein. Ultrastructural pathology: M. A. Russo *et al.*, *Am. J. Pathol.* **109**, 133 (1982). Toxicity study: Vogt, *Arch. Exp. Pathol. Pharmakol.* **190**, 406 (1938). Review of the chemistry and toxicology of the toxins of *Amanita phalloides*: Wieland, Wieland, *Pharmacol. Rev.* **11**, 87-107 (1959); *see also* T. Wieland, *Fortschr. Chem. Org. Naturst.* **25**, 214-250 (1967); T. Wieland, H. Faulstich, *Crit. Rev. Biochem.* **5**, 185-260 (1978).

Hexahydrate, needles from water, mp 280-282°. uv max (water): 295 nm ($E_{1cm}^{1\%}$ 0.597). Soly in water (0°): 0.5%; much more sol in hot water; freely sol in methanol, ethanol, butanol, pyridine. LD_{50} i.m. in albino mice: 3.3 µg/g (Vogt). LD_{50} i.p. in mice: 2 mg/kg (Wieland, Wieland).
Caution: See Amanitin.

7337. Phanquinone. *4,7-Phenanthroline-5,6-dione;* 4,7-phenanthroline-5,6-quinone; 5,6-dioxo-5,6-dihydro-4,7-phenanthroline; phanchinone; phanquone; Ciba 11925; Entobex. $C_{12}H_6N_2O_2$; mol wt 210.19. C 68.57%, H 2.88%, N 13.33%, O 15.22%. Prepd from 5,(6)-methoxy-4,7-phenanthroline: Druey, Schmidt, *Helv. Chim. Acta* **33**, 1080 (1950); **Brit. pat.** 688,802 (1953 to Ciba).

Crystals from methanol, mp 295° (dec). Sparingly sol in water, alcohol; sol in dil mineral acids.
THERAP CAT: Antiamebic.

7338. Phaseolin. *(6bR-cis)-6b,12b-Dihydro-3,3-dimethyl-3H,7H-furo[3,2-c:5,4-f']bis[1]benzopyran-10-ol;* phaseollin. $C_{20}H_{18}O_4$; mol wt 322.36. C 74.52%, H 5.63%, O 19.85%. Antifungal phytoalexin isolated from French bean (*Phaseolus vulgaris* L., *Leguminosae*): I. A. M. Cruickshank, D. R. Perrin, *Life Sci.* **2**, 680 (1963). Structure: D. R. Perrin, *Tetrahedron Letters* **1964**, 29; D. R. Perrin *et al.*, *ibid.* **1972**, 1673. Crystal structure: C. DeMartinis *et al.*, *Tetrahedron* **34**, 1849 (1978). Biosynthesis: S. L. Hess *et al.*, *Phytopathology* **61**, 79 (1971); P. M. Dewick, M. J. Steele, *Phytochemistry* **21**, 1599 (1982). Total synthesis: S. E. N. Mohamed *et al.*, *J. Chem. Soc. Perkin Trans. I* **1987**, 431. Mode of action: F. D. Van Etten, D. F. Bateman, *Phytopathology* **61**, 1363 (1971). Antifungal properties: I. A. M. Cruickshank,

D. R. Perrin, *Phytopathol. Z.* **70**, 209 (1971); M. A. Gordon *et al., Antimicrob. Ag. Chemother.* **17**, 120 (1980).

Crystals, mp 177-178°. [α]₅₇₈ −145°. pKa 9.13. uv max (ethanol): 207, 230, 280, 286 nm (log ε 4.68, 4.40, 3.97, 3.90).

7339. Phasin. Poisonous agglutinin from beans. A polypeptide composed of glutamic acid, aspartic acid, serine, alanine, tyrosine, lysine and arginine: Piekarski, *Dissertationes Pharm.* **9**, 255 (1957), *C.A.* **52**, 6456i (1958). Prepn: Wienhaus, *Biochem. Z.* **18**, 228 (1909), *C.A.* **3**, 2471 (1909); Piekarski, *loc. cit.*
Amorphous powder. Its toxicity and agglutination properties are destroyed by heating.

7340. α-Phellandrene. 2-Methyl-5-(1-methylethyl)-1,3-cyclohexadiene; *p*-mentha-1,5-diene; 5-isopropyl-2-methyl-1,-3-cyclohexadiene; 4-isopropyl-1-methyl-1,5-cyclohexadiene. C₁₀H₁₆; mol wt 136.24. C 88.16%, H 11.84%. Isoln of *l*-form from essential oils of *Eucalyptus dives* Schau. and *E. phellandra* Baker & Smith, *Myrtaceae:* Smith *et al., J. Chem. Soc.* **123**, 1657 (1923). Isoln of *d*-form from oil of bitter fennel (*Foeniculum vulgare* Mill., *Umbelliferae*): Wallach, *Ann.* **336**, 9 (1904). Structure: Semmler, *Ber.* **36**, 1749 (1903). Synthesis of *dl*-form: Read, Storey, *J. Chem. Soc.* **1930**, 2770; B. Singaram, J. Verghese, *Ind. J. Chem.* **14B**, 1003 (1976). Configuration: Burgstahler *et al., J. Am. Chem. Soc.* **83**, 4660 (1961). Circular dichroism: G. Snatzke *et al., Tetrahedron Letters* **1966**, 4551. Review: J. L. Simonsen, *The Terpenes* vol. 1 (Univ. Press, Cambridge, 2nd ed., 1947) pp 193-204; B. Singaram, J. Verghese, *Perfum. Flavor.* **2**, 33-38 (1978).

l-Form, mobile oil. bp₇₅₈ 171-172°; bp₁₆ 58-59°. d₄²⁰ 0.8410. n_D²⁰ 1.4709. [α]_D²⁰ −217°. Practically insol in water; sol in ether.
d-Form, mobile oil. bp₁₆ 66-68°. d₄²⁵ 0.8463. n_D²⁵ 1.4777. [α]_D¹⁶ +86.4°. Practically insol in water; sol in ether.
Caution: Can be irritating to, and absorbed through, skin. Ingestion can cause vomiting, diarrhea.
USE: In fragrances.

7341. β-Phellandrene. 3-Methylene-6-(1-methylethyl)-cyclohexene; *p*-mentha-1(7),2-diene; 3-isopropyl-6-methylene-1-cyclohexene; 4-isopropyl-1-methylene-2-cyclohexene. C₁₀H₁₆; mol wt 136.24. C 88.16%, H 11.84%. Isoln of *d*-form from oil of water fennel (*Phellandrium aquaticum* L., *Umbelliferae*): Berry *et al., J. Chem. Soc.* **1937**, 1448. Isoln of *l*-form from Canada balsam oil: Macbeth *et al., ibid.* **1938**, 119. Structure: Wallach, *Ann.* **343**, 28 (1905). Synthesis of *dl*-form: Deorha, Sareen, *Rec. Trav. Chim.* **82**, 137 (1965); B. Singaram, J. Verghese, *Ind. J. Chem.* **14B**, 1003 (1976).

d-Form, mobile oil. bp₇₆₀ 171-172°; bp₁₁ 57°. d₄²⁰ 0.8520. n_D²⁰ 1.4788. [α]_D²⁰ +65.2°. Practically insol in water; sol in ether.
l-Form, mobile oil. bp₇₅₈ 178-179°; bp₁₂ 53°. d₁₅¹⁵ 0.8497. n_D²⁰ 1.4800. [α]_D²⁰ −51.9°. Practically insol in water, alcohol; sol in ether.

7342. Phenacaine Hydrochloride. *N,N'-Bis(4-ethoxyphenyl)ethanimidamide monohydrochloride; N¹,N²-bis(p-ethoxyphenyl)acetamidine hydrochloride;* Holocaine Hydrochloride. C₁₈H₂₃ClN₂O₂; mol wt 334.85. C 64.57%, H 6.92%, Cl 10.59%, N 8.37%, O 9.56%. Prepn: **Ger.** pats. **79,868; 80,568.**

Monohydrate, faintly bitter crystals producing transient numbness of the tongue. Stable in air. When anhyd, mp 190-192°. One gram dissolves in 50 ml water. Freely sol in boiling water, alcohol, chloroform. Insol in ether. Incompatible with alkalies and their carbonates and the usual alkaloidal reagents. Aq solns are stable and are not dec by boiling.
THERAP CAT: Topical anesthetic (ophthalmic).
THERAP CAT (VET): Ocular anesthetic.

7343. Phenacemide. *N-(Aminocarbonyl)benzeneacetamide; (phenylacetyl)urea;* phenacetylurea, Epiclase, Phacetur; Phenurone; Phetylureum. C₉H₁₀N₂O₂; mol wt 178.19. C 60.66%, H 5.66%, N 15.72%, O 17.96%. C₆H₅CH₂CO-NHCONH₂. Prepd by the action of aq NH₃ on phenacetylurethan: Basterfield, Greig, *Can. J. Res.* **8**, 454 (1933); of phenacetyl chloride on urea: Spielman *et al., J. Am. Chem. Soc.* **70**, 4189 (1948). Toxicity study: K. Nakamura *et al., Arzneimittel-Forsch.* **18**, 524 (1968).
Crystals from alcohol, mp 212-216°. Very slightly sol in water; slightly sol in alcohol, benzene, chloroform, ether. LD₅₀ in mice, rats (mmol/kg): 5.54, >10 orally (Nakamura).
THERAP CAT: Anticonvulsant.

7344. Phenacetin. *N-(4-Ethoxyphenyl)acetamide; p-acetophenetidide;* *p*-ethoxyacetanilide; acetophenetidin; *para*-acetphenetidin; acet-*p*-phenetidin; *p*-acetophenetide. C₁₀H₁₃NO₂; mol wt 179.22. C 67.02%, H 7.31%, N 7.82%, O 17.85%. Prepn: *Beilstein* vol. XIII, 461. Improved process: Eaker, Campbell, U.S. pat. **2,887,513** (1959 to Monsanto). *Monograph:* P. K. Smith, *Acetophenetidin* (Interscience, New York, 1958) 180 pp. Toxicity: Boyd, *Toxicol. Appl. Pharmacol.* **1**, 240 (1959). Review of toxicity and metabolism studies: L. Fishbein, *IARC Sci. Publ.* **40**, 287-310 (1981). Epidemiologic study of renal morbidity and mortality: U. C. Dubach *et al., N. Engl. J. Med.* **308**, 357 (1983). Evaluation of renal effects: D. P. Sandler *et al., ibid.* **320**, 1238 (1989).

Slightly bitter, cryst scales or powder. mp 134-135°. One gram dissolves in 1310 ml cold water, 82 ml boiling water; 15 ml cold alcohol, 2.8 ml boiling alcohol; 14 ml chloroform, 90 ml ether; sol in glycerol. Gives a pasty mass with phenol, chloral hydrate or pyrocatechol, strong acids or alkalies, salicylic acid, oxidizers, iodine, spirit nitrous ether. LD₅₀ orally in rats: 1.65 g/kg (Boyd).
Note: Component of *APC* tablets, also containing aspirin and caffeine. Phenacetin may reasonably be anticipated to be a carcinogen: *Seventh Annual Report on Carcinogens* (PB95-109781, 1994) p 315.

7345. Phenacetolin. Degener's indicator. Reaction product of concd H_2SO_4 and glacial acetic acid on phenol. Yellowish-brown powder. Slightly sol in water; sol in alc.

USE: Has been employed as indicator, particularly for mixtures of alkali hydroxides and carbonates; the yellow color changes to red when hydroxide is neutralized, and again to yellow when carbonate is fully dec by acid; was also used for determining alkalinity of water. The indicator soln is prepd by digesting 1 g with 100 ml warm alcohol and filtering after cooling; 2-3 drops used for 100 ml liquid.

7346. Phenactropinium Chloride. *3-[(Hydroxyphenyl-acetyl)oxy]-8-methyl-8-(2-oxo-2-phenylethyl)-8-azoniabicyclo-[3.2.1]octane chloride; 8-benzoylmethyl-3α-hydroxy-1αH,-5αH-tropanium chloride mandelate;* 8-phenacylhomatropinium chloride; N-phenacyl-O-dl-mandelyltropinium chloride; Trophenium. $C_{24}H_{28}ClNO_4$; mol wt 429.94. C 67.05%, H 6.56%, Cl 8.25%, N 3.26%, O 14.89%. Prepn: Johnston, Spencer, U.S. pat. **2,828,312** (1958 to T. & H. Smith). Ganglionic blocking agent.

Crystals from hot alcohol, mp 195-197°.
THERAP CAT: Antihypertensive.

7347. Phenacylamine. *2-Amino-1-phenylethanone; 2-aminoacetophenone;* α-aminoacetophenone; ω-aminoaceto-phenone. C_8H_9NO; mol wt 135.17. C 71.09%, H 6.71%, N 10.36%, O 11.84%. Prepn from α-phenylethylamine: H. E. Baumgarten, J. M. Petersen, *J. Am. Chem. Soc.* **82**, 459 (1960); *eidem, Org. Syn.* **coll. vol. V**, 909 (1973).

Hydrochloride, C_8H_9NO.HCl, crystals from isopropanol + HCl, dec 188.5°.

7348. Phenadoxone. *6-(4-Morpholinyl)-4,4-diphenyl-3-heptanone;* 6-tetrahydrooxazine-4,4-diphenyl-3-heptanone; heptazone; morphodone. $C_{23}H_{29}NO_2$; mol wt 351.49. C 78.60%, H 8.32%, N 3.98%, O 9.10%. Prepn: M. Bockmühl, G. Ehrhart, *Ann.* **561**, 52 (1948). Comparison with other analgesics: C. A. Winter, L. Flataker, *J. Pharmacol. Exp. Ther.* **98**, 305 (1950). HPLC determn: I. Jane *et al., J. Chromatog.* **323**, 191 (1985).

mp 75-76°.

Hydrochloride, $C_{23}H_{29}NO_2$.HCl, *Hoechst 10600, CB-11, Hepagin, Heptalgin, Heptalin, Heptone, Supralgin.* Crystals, mp 224-225° (some decompn). pKa (25°): 6.7 in 40% alcohol. pH of 1% aq soln 4.1; of 5% aq soln 3.7. The free base begins to precipitate when the pH is raised to 5.5. uv max (abs alcohol): 260, 295 nm. Soly (g/100 ml) in water: 10 at 25°; 80 at the bp; in methanol: 27; in abs ethanol: 24; in chloroform: 50. Slightly sol in acetone. Almost insol in benzene, ethyl acetate. LD_{50} i.v. in mice: 47.5 mg/kg (Winter, Flataker).

Note: This is a controlled substance (opiate) listed in the U.S. Code of Federal Regulations, Title 21 Part 1308.11 (1995).

THERAP CAT: Analgesic (narcotic).

7349. Phenallymal. *5-Phenyl-5-(2-propenyl)-2,4,6(1H,-3H,5H)-pyrimidinetrione; 5-allyl-5-phenylbarbituric acid;* 5-phenyl-5-allylbarbituric acid; phenallymalum; alphenal; prophenal; Alphenate; Allofenyl; Fenallymal. $C_{13}H_{12}N_2O_3$; mol wt 244.25. C 63.93%, H 4.95%, N 11.47%, O 19.65%. Its methods of prepn are similar to those used for phenobarbital: Slotta, *Grundriss der modernen Arzneistoff-Synthese* (Stuttgart, 1931) p 45. Metabolism: D. J. Harvey *et al., Res. Commun. Chem. Pathol. Pharmacol.* **4**, 247 (1972); *eidem, Drug Metab. Dispos.* **5**, 527 (1977).

Crystals, mp 156-157.5°. Bitter taste. Readily sol in alcohol, chloroform. One gram dissolves in 580 ml water, in 10 ml ether, in 500 ml benzene, in 4000 ml carbon tetrachloride, in 17,500 ml petr ether.

Note: This is a controlled substance (depressant) listed in the U.S. Code of Federal Regulations, Title 21 Part 1308.13 (1995).

THERAP CAT: Sedative, hypnotic.

7350. Phenamacide Hydrochloride. *(±)-α-Aminobenzeneacetic acid 3-methylbutyl ester hydrochloride; 2-phenylglycine isopentyl ester hydrochloride;* isopentyl 2-phenylglycinate hydrochloride; phenylaminoacetic acid isoamyl ester hydrochloride; isoamyl phenylaminoacetate hydrochloride; Aklonin. $C_{13}H_{20}ClNO_2$; mol wt 257.76. C 60.58%, H 7.82%, Cl 13.75%, N 5.43%, O 12.41%. Prepn: Klosa, *Arch. Pharm.* **285**, 332 (1952). Pharmacology: Schultz *et al., Pharmazie* **25**, 560 (1970).

Crystals, mp 154°.
THERAP CAT: Antispasmodic.

7351. Phenamet. *N-[[4-[Bis(2-chloroethyl)amino]phenyl]acetyl]-L-methionine ethyl ester.* $C_{19}H_{28}Cl_2N_2O_3S$; mol wt 435.41. C 52.41%, H 6.48%, Cl 16.28%, N 6.43%, O 11.02%, S 7.36%. Prepn: Golubeva *et al., Dokl. Akad. Nauk SSSR* **119**, 83 (1958); Knunyants *et al., ibid.* **132**, 836 (1960).

Crystals from ethyl acetate + petr ether, mp 93-95°.
THERAP CAT: Antineoplastic.

7352. Phenamidine. *4,4'-Oxybisbenzenecarboximid-amide; 4,4'-oxydibenzamidine;* 4,4'-diamidinodiphenyl ether; $C_{14}H_{14}N_4O$; mol wt 254.29. C 66.13%, H 5.55%, N 22.03%, O 6.29%. Prepn: A. J. Ewins *et al.*, **Brit. pat.** 507,565 (1939 to May & Baker); J. N. Ashley *et al., J. Chem. Soc.* **1942,** 103; of isethionate: G. Newbery, A. P. T. Easson, **U.S. pat.** 2,410,796 (1946 to May & Baker). Trypanocidal activity: E. M. Lourie, W. Yorke, *Ann. Trop. Med. Parasitol.* **33,** 289 (1939). Preliminary pharmacology: R. Wien, *ibid.* **37,** 1 (1943). Use in treatment of Babesia infections in dogs: M. D. Ruff *et al., Am. J. Vet. Res.* **34,** 641 (1973); R. Gothe *et al., Kleintierpraxis,* **32,** 97 (1987). Early review of pharmacology, mode of action and clinical applications: E. M. Schoenbach, E. M. Greenspan, *Medicine* **27,** 327-377 (1948).

Irregular plates from water, mp 215-216°.
Isethionate, $C_{14}H_{14}N_4O.2C_2H_6O_4S_2$, *M & B* **736,** *Lomadine.* Crystals, mp 225° (dec). One part is sol in 1.4 parts water; 300 parts alcohol. Practically insol in ether, chloroform.
THERAP CAT (VET): Antiprotozoal (Babesia).

7353. Phenampromid(e). *N-[1-Methyl-2-(1-piperidinyl)-ethyl]-N-phenylpropanamide; N-(1-methyl-2-piperidinoethyl)-propionanilide;* 1-piperidino-2-(N-propionylanilino)propane. $C_{17}H_{26}N_2O$; mol wt 274.41. C 74.41%, H 9.55%, N 10.21%, O 5.83%. Prepn and analgesic activity: W. B. Wright, Jr. *et al., J. Am. Chem. Soc.* **81,** 1518 (1959); W. B. Wright, Jr., H. J. Brabander, **U.S. pat.** 3,016,382 (1962 to Am. Cyanamid). (−)-Phenampromid(e), the more active enantiomer, has the *R*-configuration. Absolute configuration: Portoghese, *J. Med. Chem.* **8,** 147 (1965). Pharmacology: Kikuchi *et al., Nippon Yakurigaku Zasshi* **57,** 585 (1961), *C.A.* **59,** 2082b (1963). Determn by HPLC: I. Jane, A. McKinnon, *J. Chromatog.* **323,** 191 (1985).

(−)-Phenampromid(e)

Liquid, $bp_{0.2}$ 124-128°. n_D^{28} 1.518.
(−)-Hydrochloride, $C_{17}H_{27}ClN_2O$, crystals from alcoholic HCl + ether, mp 201-202°.
Note: This is a controlled substance (opiate) listed in the U.S. Code of Federal Regulations, Title 21 Part 1308.11 (1995).

7354. Phenanthrene. $C_{14}H_{10}$; mol wt 178.23. C 94.34%, H 5.66%. Isomeric with anthracene. Occurs in coal tar. Isoln: Ostermayer, Fittig, *Ber.* **5,** 933 (1872); Glaser, *ibid.* 982. Purification (from contaminating carbazole and anthracene): Clar, *Ber.* **65,** 852 (1932). Formation from toluene, bibenzil, 9-methylfluorene or stilbene by passage through red-hot tube: Graebe, *Ber.* **7,** 48 (1874); *Ann.* **167,** 161 (1879); *Ber.* **37,** 4145 (1904). Also from coumarone and benzene: Kraemer, Spilker, *Ber.* **23,** 85 (1890). Pschorr synthesis from *o*-nitrobenzaldehyde and phenylacetic acid: *Ber.* **29,** 500 (1896). From diphenylethylene: Cook, Hewett, *J. Chem. Soc.* **1933,** 1098. Diene synthesis from 1-vinylnaphthalene and maleic anhydride: Cohen, Warren, *ibid.* **1937,** 1315. From *o*-phenylbenzoic acid: Schönberg, Warren, *Chem. & Ind. (London)* **58,** 199 (1939). By irradiation of stilbene: Mallory *et al., J. Am. Chem. Soc.* **84,** 4361 (1962). Synthesis by double succinoylation of ben-

zene: Rahman *et al., J. Org. Chem.* **28,** 3571 (1963). Structure: Trotter, *Acta Cryst.* **16,** 605 (1963).

Monoclinic plates from alcohol. d^{25} 1.179; mp 100°; bp 340°. Sublimes in high vacuum. Absorption spectrum: Clar, Lombardi, *Ber.* **65,** 1412 (1932); Mayneord, Roe, *Proc. Roy. Soc. London* A **152,** 317 (1935). Practically insol in water; sol in organic solvents, especially in aromatic hydrocarbons. One gram dissolves in 60 ml cold, 10 ml boiling 95% alcohol, 25 ml abs alcohol, 2.4 ml toluene or carbon tetrachloride, 2 ml benzene, 1 ml carbon disulfide, 3.3 ml anhydr ether. Soluble in glacial acetic acid. Solns exhibit a blue fluorescence. Forms molecular compds with picric acid, picryl chloride, dinitrobenzene and similar nitro compounds. LD_{50} orally in mice: 700 mg/kg, *Toxic Substances List,* H. E. Christensen, Ed. (1973) p 708.
Caution: Can cause photosensitization of skin.

7355. Phenanthrenequinone. *9,10-Phenanthrenedione; 9,10-*phenanthraquinone. $C_{14}H_8O_2$; mol wt 208.22. C 80.76%, H 3.87%, O 15.37%. Prepn by oxidation of phenanthrene: Curtis *et al.*, **U.S. pat.** 2,956,065 (1960 to U.S. Steel); Binder, Koch, **Ger. pat.** 1,166,176 (1964).

Orange-red crysts. mp 206-207°. bp ~360°. d 1.405. Sublimes. Practically insol in water; sol in benzene, ether, glacial acetic acid, hot alc. With concd H_2SO_4 gives a dark green color.

7356. *o*-Phenanthroline. *1,10-Phenanthroline;* 4,5-phenanthroline. $C_{12}H_8N_2$; mol wt 180.21. C 79.98%, H 4.47%, N 15.54%. Made by heating *o*-phenylenediamine with glycerol, nitrobenzene and concd H_2SO_4, or in the same manner from 8-aminoquinoline. Prepn: Madeja, *J. Prakt. Chem.* **17,** 104 (1962). Crystal and molecular structure: S. Nishigaki *et al., Acta Crystallogr.* **B34,** 875 (1978). Review of analytical uses: Vydra, Kopanica, *Chemist-Analyst* **52,** 88-94 (1963).

Monohydrate, white, cryst powder. mp 93-94°; anhydr, 117°. Soluble in ~300 parts water, 70 parts benzene; sol in alcohol, acetone.
USE: Forms a complex compound with ferrous ions which is used under the name of *"Ferroin"* as an indicator in oxidation reduction systems, *e.g.,* titration of ferrous salts. Also used in determination of nickel, ruthenium, silver and other metals.

7357. Phenarsazine Chloride. *10-Chloro-5,10-dihydrophenarsazine; 5-*aza-10-arsenaanthracene chloride; 10-chloro-5,10-dihydroarsacridine; diphenylaminechlorarsine; phenazarsine chloride; adamsite; DM. $C_{12}H_9AsClN$; mol wt 277.58. C 51.92%, H 3.27%, As 26.99%, Cl 12.77%, N 5.05%. Prepd by heating diphenylamine with arsenic trichloride: **Ger. pat.** 281,049 (1914 to I. G. Farben.); Wieland, Reinheimer, *Ann.* **423,** 12 (1921); Lewis, Hamilton, *J. Am. Chem. Soc.* **43,** 2222 (1921); Burton, Gibson, *J. Chem. Soc.* **1926,** 450.

Canary-yellow crystals from carbon tetrachloride. *Poisonous.* Dimorphous. The stable form occurs as orthorhombic crystals; d 1.65; mp 195°; bp 410° (decompn). Sublimes readily. Vapor press. at 20° $= 2 \times 10^{-13}$ mm; volatility 0.02 mg/cu m; heat of volatilization 54.8 cal; spec heat 0.268 cal. Practically insol in water. Slightly sol in benzene, xylene, carbon tetrachloride. Corrodes iron, bronze, brass. (The metastable form melts at 186° if monoclinic, and at 182° if triclinic.)
Caution: Irritating to skin and respiratory tract. Causes profuse watery nasal discharge; severe pain in nose, sinuses, chest; sneezing, coughing, nausea, vomiting, marked depression, weakness. Sensory disturbances may occur later.
USE: As war gas, dispersed in air in the form of minute particles. For riots in combination with tear gas (chloroacetophenone). In the formulation of wood-treating solns, against marine borers and similar pests.

7358. Phenatine. *N-(1-Methyl-2-phenylethyl)-3-pyridinecarboxamide; N-(α-methylphenethyl)nicotinamide;* nicotinic acid β-phenylisopropylamide; nicotinoyl-β-phenylisopropylamine; fenatin; perviton. $C_{15}H_{16}N_2O$; mol wt 240.30. C 74.97%, H 6.71%, N 11.66%, O 6.66%. Synthesis from nicotinoyl chloride and $C_6H_5CH_2CH(CH_3)NH_2$: Arbuzov *et al.*, *Sb. Statei Obshch. Khim., Akad. Nauk SSSR* **1**, 714 (1953), *C.A.* **49**, 1072 (1955). Pharmacology: Arbuzov, *Farmakol. Toksikol.* **31**, 373 (1968). Analysis: P. P. Suprun, *Farm. Zh. (Kiev)* **30**, 49 (1975); O. N. Shcherbina *et al.*, *ibid.* **1979**, 49.

Crystals from benzene, mp 99-100°.
Phosphate, $C_{15}H_{16}N_2O.2H_3PO_4$, crystals from alcohol + ether, mp 162°. Sol in water, warm alcohol. Insol in ether. Aq solns may be boiled for 3 hrs without decompn.

7359. Phenazine. Dibenzopyrazine; dibenzoparadiazine; azophenylene. $C_{12}H_8N_2$; mol wt 180.21. C 79.98%, H 4.47%, N 15.54%. Obtained (with other products) by passing aniline vapor through a red-hot tube: Bernthsen, *Ber.* **19**, 3257 (1886); by heating aniline with nitrobenzene and sodium hydroxide to 140°: Wohl, Aue, *Ber.* **34**, 2446 (1901); Wohl, *Ber.* **36**, 4135 (1903); by heating o-phenylenediamine with pyrocatechol in sealed tube: Ris, *Ber.* **19**, 2206 (1886); Hinsberg, Garfunkel, *Ann.* **292**, 258 (1896); upon distilling 2-aminodiphenylamine with lead monoxide: Fischer, Heiler, *Ber.* **26**, 383 (1893); by heating 2-aminodiphenylamine with 2-nitrodiphenylamine in the presence of sodium acetate: Kehrmann, Havas, *Ber.* **46**, 342 (1913); by heating nitrobenzene with barium oxide: Zerewitinoff, Ostromisslensky, *Ber.* **44**, 2402 (1911); by boiling 2,2'-dinitrodiphenylamine with stannous chloride in hydrochloric and acetic acids, followed by oxidation with hydrogen peroxide: Eckert, Steiner, *Monatsh.* **35**, 1154 (1914).

Pale yellow needles from alcohol or by sublimation. Colorless needles from dilute alcohol. mp 171°; bp above 360°. Practically insol in water. One part dissolves in 50 parts alcohol. Moderately sol in ether, benzene; sol in mineral acids giving yellow to red solns.

7360. Phenazocine. *1,2,3,4,5,6-Hexahydro-6,11-dimethyl-3-(2-phenethyl)-2,6-methano-3-benzazocin-8-ol;* 2'-hydroxy-2-(N,β-phenethyl)-5,9-dimethyl-6,7-benzomorphan; 2'-hydroxy-5,9-dimethyl-2-phenethyl-6,7-benzmorphan; phenethylazocine; phenobenzorphan. $C_{22}H_{27}NO$; mol wt 321.46. C 82.20%, H 8.47%, N 4.36%, O 4.98%. Prepn: E. L. May, N. B. Eddy, *J. Org. Chem.* **24**, 294, 1435 (1959); Ager, May, *ibid.* **25**, 984 (1960); Gordon *et al.*, U.S. pat. **2,959,594** (1960 to SK&F).

(±)-Form, rods from methanol, mp 181-182°.
(−)-Form, needles from methanol, mp 159-159.5°. $[\alpha]_D^{20}$ −122° (c = 0.74 in 95% ethanol).
(±)-Hydrobromide, $C_{22}H_{27}NO.HBr$, *NIH-7519, SKF-6574, Narphen, Prinadol, Xenagol.* Rods from acetone or abs alc + ether, mp 166-170°. LD_{50} s.c. in mice: 332 mg/kg (May, Eddy).
(−)-Hydrobromide, crystals, mp 284-287°. $[\alpha]_D^{20}$ −84.1° (c = 1.12 in 95% ethanol). LD_{50} s.c. in mice: 147 mg/kg (May, Eddy).
Note: This is a controlled substance (opiate) listed in the U.S. Code of Federal Regulations, Title 21 Part 1308.12 (1995).
THERAP CAT: Analgesic (narcotic).

7361. Phenazopyridine Hydrochloride. *3-(Phenylazo)-2,6-pyridinediamine monohydrochloride; 2,6-diamino-3-phenylazopyridine hydrochloride;* β-phenylazo-α,α'-diaminopyridine hydrochloride; Pyridium; Pyridiate; Pyridacil; Mallophene; Sedural; Urodine; Phenazodine. $C_{11}H_{12}ClN_5$; mol wt 249.70. C 52.91%, H 4.84%, Cl 14.20%, N 28.05%. The commercial product may contain some β,β'-bis(phenylazo)-α,α'-diaminopyridine and is therefore identified simply as phenylazo-α,α-diaminopyridine monohydrochloride. Prepd by coupling diazotized aniline with α,α-diaminopyridine: Chichibabin, Zeide, *J. Russ. Phys. Chem. Soc.* **46**, 1216 (1914); *Chem. Zentr.* **1915**, I, 1064; Chichibabin, Ossetrowa, *J. Am. Chem. Soc.* **56**, 1711 (1934); Ostromislensky, *ibid.* 1713; Ger. pat. **515,781** (1927 to Boehringer); U.S. pats. **1,680,108; 1,680,109; 1,680,110; 1,680,111** (all 1928 to Pyridium); Shreve *et al.*, *J. Am. Chem. Soc.* **65**, 2241 (1943). Physical characteristics: Collins, *J. Am. Pharm. Assoc.* **20**, 455 (1931). Toxicity data: B. A. Becker, J. G. Swift, *Toxicol. Appl. Pharmacol.* **1**, 42 (1959). Comprehensive description: K. W. Blessel *et al.*, *Anal. Profiles Drug Subs.* **3**, 465-482 (1974).

Brick-red microcrystals, slight violet luster. Slightly bitter taste. (The free base mp 139°.) Slightly sol in cold water, about 1 part in 300; sol in boiling water, about 1 part in 20. Forms supersatd solns easily. Will precipitate out of a 2% soln at 25° after about 2 days, out of a 1% soln only after months. Aq solns may be stabilized by the addn of 10% glucose. One part is sol in about 100 parts glycerol U.S.P. Sol in ethylene and propylene glycols. Slightly sol in alc, lanolin; sol in acetic acid. Insol in acetone, benzene, chloroform, ether, toluene. Aq solns are yellow to brick-red and slightly acid. Pyridium stains may be removed by soaking stained fabric (cotton, wool, rayon, nylon, Dacron) in 0.25% sodium hydrosulfite soln. LD_{50} orally in rats: 403 mg/kg (Becker, Swift).
Note: This substance may reasonably be anticipated to be a carcinogen: *Seventh Annual Report on Carcinogens* (PB95-109781, 1994) p 316.

THERAP CAT: Analgesic (urinary tract).

THERAP CAT (VET): Analgesic (urinary tract).

7362. Phenbutamide. *N-[(Butylamino)carbonyl]ben-zenesulfonamide; 1-butyl-3-(phenylsulfonyl)urea; N-benzene-sulfonyl-N'-butylurea;* fenbutamide; AL-132; R-131; Diaperos; Ipoglon. $C_{11}H_{16}N_2O_3S$; mol wt 256.33. C 51.54%, H 6.29%, N 10.93%, O 18.73%, S 12.51%. $C_6H_5SO_2NH-CONHC_4H_9$. Prepn: Brit. pat. **808,071** (1959 to Hoechst), *C.A.* **53,** 14056c (1959); Hokfelt, Joenssen, *J. Med. Pharm. Chem.* **5,** 231 (1962); Brzozowaki, *Rocz. Chem.* **43,** 1761 (1969). Activity studies: Tonse *et al., Nature* **193,** 891 (1962); Khurana *et al., Indian J. Med. Res.* **55,** 1084 (1967).

Crystals, mp 130-132°.

THERAP CAT: Antidiabetic.

7363. Phencarbamide. *Diphenylcarbamothioic acid S-[2-(diethylamino)ethyl] ester; diphenylthiocarbamic acid S-[2-(diethylamino)ethyl] ester;* S-[2-(diethylamino)ethyl] di-phenylthiocarbamate; fencarbamide. $C_{19}H_{24}N_2OS$; mol wt 328.48. C 69.47%, H 7.36%, N 8.53%, O 4.87%, S 9.76%. Prepd from diphenylcarbamyl chloride and β-diethylamino-ethanethiol: **Brit.** pat. **871,774;** K. H. Risse *et al.,* U.S. pat. **3,228,949** (1961, 1966 both to Bayer). Pharmacology: W. Wirth *et al., Arch. Int. Pharmacodyn.* **151,** 515 (1964); W. Wirth, R. Goesswald, *ibid.* **155,** 393 (1965); B. R. Madan *et al., ibid.* **185,** 53 (1970).

Crystals, mp 48-49°, bp$_{0.01}$ 120-126°. Practically insol in water. Freely sol in methanol, ether, chloroform. Sol in petr ether.

Hydrochloride, $C_{19}H_{24}N_2OS.HCl$, *Ba-1355, Escorpal.* Crystals, mp 180-181°. Sol in water. LD$_{50}$ orally in rats: 410 mg/kg (Wirth, Goesswald).

1,5-Naphthalenedisulfonate, $C_{19}H_{24}N_2OS.C_{10}H_8O_6S_2$, *phencarbamide napadisilate, Gelosedine.* LD$_{50}$ orally in rats: 920 mg/kg (Wirth, Goesswald).

Component of *Spasmo-Dolviran.*

THERAP CAT: Anticholinergic.

7364. Phencyclidine. *1-(1-Phenylcyclohexyl)piperidine;* angel dust; HOG; PCP; CI-395. $C_{17}H_{25}N$; mol wt 243.39. C 83.89%, H 10.35%, N 5.75%. Prepn: **Brit.** pat. **836,083** and Godefroi *et al.,* U.S. pat. **3,097,136** (1960, 1963 to Parke, Davis); V. H. Maddox *et al., J. Med. Chem.* **8,** 230 (1965). Pharmacology: G. Chen *et al., J. Pharmacol. Exp. Ther.* **127,** 241 (1959); J. C. Munch, *Bull. Narcotics* **26,** 9 (1974). Human metabolism: L. K. Wong, K. Biemann, *Biomed. Mass Spectrom.* **2,** 204 (1975). Toxicity: K. Bailey *et al., J. Pharm. Pharmacol.* **28,** 713 (1976). Extensive bibliography: R. L. Balster, R. S. Pross, *J. Psychedelic Drugs* **10,** 1-15 (1978). *Review:* R. E. Garey, *ibid.* **11,** 265-275 (1979). Review of neuropharmacology: K. M. Johnson, S. M. Jones, *Ann. Rev. Pharmacol. Toxicol.* **30,** 707-750 (1990).

Colorless crystals, mp 46-46.5°. bp$_{1.0}$ 135-137°. uv max (0.1N HCl): 252, 257.5, 262, 268.5 nm (E$_{1cm}^{1\%}$ 7.9, 11.2, 13.0, 9.7).

Hydrochloride, $C_{17}H_{25}N.HCl$, *Sernyl, Sernylan.* Crystals from 2-propanol, mp 233-235°. uv max (ethanol): 254, 258, 262.5, 269 nm (E$_{1cm}^{1\%}$ 7.9, 10.8, 12.7, 10.0). LD$_{50}$ orally in mice: 76.5 mg/kg (Bailey).

Hydrobromide, crystals, mp 214-218°.

Caution: This is a controlled substance (depressant) listed in the U.S. Code of Federal Regulations, Title 21 Part 1308.12 (1995). The ethylamine, pyrrolidine and thiophene analogs are listed as hallucinogens, *ibid.,* Title 21 Part 1308.11.

THERAP CAT: Anesthetic (intravenous).

THERAP CAT (VET): Analgesic; anesthetic.

7365. Phendimetrazine. *3,4-Dimethyl-2-phenylmorpho-line; d-2-phenyl-3,4-dimethylmorpholine; 3,4-dimethyl-2-phenyltetrahydro-1,4-oxazine.* $C_{12}H_{17}NO$; mol wt 191.27. C 75.35%, H 8.96%, N 7.32%, O 8.36%. Prepn from ethylene chlorohydrin and *l*-ephedrine: Otto, *Angew. Chem.* **68,** 181 (1956); from 2-phenyl-3,4-dihydroxy-3,4-dimethylmorpho-line: Boehringer *et al.,* **Brit.** pats. **791,416** (1958); **862,198** (1961). Stereochemistry: Dvornik, Schilling, *J. Med. Chem.* **8,** 466 (1965). Toxicity data: Stegen *et al., Toxicol. Appl. Pharmacol.* **2,** 589 (1960).

bp$_8$ 122-124°; bp$_{12}$ 134-135°.

Bitartrate, *Adphen, Bacarate, Dietrol, Neo-Nilorex, Obepar, Phenazine, Plegine, Statobex, Symetra, Trimstat, Trimtabs.*

Hydrochloride, $C_{12}H_{17}NO.HCl$, *Antapentan* mp 191° (Otto), 208° (Boehringer). [α]$_D^{20}$ +35.7°. LD$_{50}$ in rats (mg/kg): 455 orally; 245 i.p. (Stegen).

Pamoate, $C_{47}H_{50}N_2O_9$, *Fringanor.*

Note: This is a controlled substance (stimulant) listed in the U.S. Code of Federal Regulations, Title 21 Part 1308.13 (1995).

THERAP CAT: Anorexic.

7366. Phenelzine. *(2-Phenylethyl)hydrazine;* β-phenyl-ethylhydrazine; phenalzine. $C_8H_{12}N_2$; mol wt 136.20. C 70.55%, H 8.88%, N 20.57%. $C_6H_5CH_2CH_2NHNH_2$. Monoamine oxidase inhibitor. Prepn: Votocek, Leminger, *Coll. Czech. Chem. Commun.* **4,** 271 (1932), *C.A.* **26,** 5294 (1932); Biel *et al., J. Am. Chem. Soc.* **81,** 2805 (1959); Biel, U.S. pat. **3,000,903** (1959 to Lakeside). Toxicity: J. A. Gylys *et al., Ann. N.Y.Acad. Sci.* **107,** 899 (1963). Comprehensive description: R. E. Daly, *Anal. Profiles Drug Subs.* **2,** 383-407 (1973).

Liquid. bp$_{0.1}$ 74°. n_D^{20} 1.5494.

Acid sulfate, $C_8H_{12}N_2.H_2SO_4$, *W-1544a, Nardelzine, Nardil, Stinerval.* White powder, sol in water. LD$_{50}$ orally in mice: 156 mg/kg (Gylys).

Hydrochloride, $C_8H_{12}N_2.HCl$, crystals, mp 174°.

THERAP CAT: Antidepressant.

7367. Phenesterine. *(3β)-Cholest-5-en-3-ol 4-[bis(2-chloroethyl)amino]benzeneacetate; p-[bis(2-chloroethyl)ami-no]phenylacetic acid cholesterol ester;* cholesteryl p-bis(2-chloroethyl)aminophenylacetate; fenesterin. $C_{39}H_{59}Cl_2NO_2$; mol wt 644.81. C 72.65%, H 9.22%, Cl 11.00%, N 2.17%, O 4.96%. Prepd from p-bis(2-chloroethyl)aminophenylacetic acid chloride and cholesterol: Shkodinskaya *et al., Zh. Ob-shch. Khim.* **32,** 959 (1962), *C.A.* **58,** 2502h (1963).

Crystals, mp 90-90.5°. [α]$_D$ −13.7°.

THERAP CAT: Antineoplastic.

7368. Phenetharbital. *5,5-Diethyl-1-phenyl-2,4,6(1H,-3H,5H)-pyrimidinetrione; 5,5-diethyl-1-phenylbarbituric acid;*

5,5-diethyl-2,4,6-trioxo-1-phenylhexahydropyrimidine; 1-phenyl-5,5-diethylbarbituric acid; N-phenylbarbital; phenidiemal; Fedibaretta; Pyrictal. $C_{14}H_{16}N_2O_3$; mol wt 260.29. C 64.60%, H 6.20%, N 10.76%, O 18.44%. Prepd from ethyl diethylmalonate and phenylurea: Fischer, Dilthey, *Ann.* **335**, 334 (1904); Buck, *J. Am. Chem. Soc.* **58**, 1284 (1936).

Small, thick, glittering plates, mp 178°. Freely sol in hot alcohol, alkali.

Note: This is a controlled substance (depressant) listed in the U.S. Code of Federal Regulations, Title 21 Part 1308.13 (1995).

THERAP CAT: Anticonvulsant.

7369. Phenethicillin Potassium. *[2S-(2α,5α,6β)]-3,3-Dimethyl-7-oxo-6-[(1-oxo-2-phenoxypropyl)amino]-4-thia-1-azabicyclo[3.2.0]heptane-2-carboxylic acid monopotassium salt;* (α-phenoxyethyl)penicillin potassium; α-phenoxyethylpenicillinic acid potassium salt; 6-(α-phenoxypropionamido)penicillanic acid potassium salt; penicillin-152; penicillin MV; penicillin-152 potassium; potassium phenethicillin; Alfacillin; Alpen (obsolete); Brocsil; Broxil; Chemipen; Darcil; Dramcillin-S; Maxipen; Optipen; Oralopen; Peniplus; Penorale; Penova; Pensig; Syncillin; Synthecilline; Synthepen. $C_{17}H_{19}KN_2O_5S$; mol wt 402.51. C 50.73%, H 4.76%, K 9.71%, N 6.96%, O 19.87%, S 7.97%. Semi-synthetic antibiotic related to penicillin. Prepn from 6-aminopenicillanic acid: Perron *et al., Antibiot. Ann.* **1959-1960**, 107; *J. Am. Chem. Soc.* **82**, 3934 (1960); Glombitza, *Ann.* **673**, 166 (1964).

dl-Form, crystals from acetone, dec 230-232°. Much less hygroscopic than benzylpenicillin sodium. Freely sol in water.

l-Form, *epiphenethicillin potassium.* Crystals from 20% butanol, dec 238-239°. $[\alpha]_D^{24}$ +218° (c = 0.01 in water).

THERAP CAT: Antibacterial.

7370. Phenethyl Alcohol. *Benzeneethanol; 2-phenylethanol;* β-phenylethyl alcohol; benzyl carbinol; β-hydroxyethylbenzene. $C_8H_{10}O$; mol wt 122.17. C 78.65%, H 8.25%, O 13.10%. Found in a number of natural essential oils, such as rose, carnation, hyacinth, Aleppo pine, orange blossom, geranium Bourbon, neroli and in the essential oil of champaca. Prepd by reduction of ethyl phenylacetate with sodium in abs alcohol: Bouveault, Blanc, *Bull. Soc. Chim.* [3] **31**, 672 (1904); Leonard, *J. Am. Chem. Soc.* **47**, 1778 (1925); by hydrogenation of phenylacetaldehyde in the presence of nickel catalyst: Skita, Ritter, *Ber.* **43**, 3398 (1910); v. Braun, Kochendörfer, *Ber.* **56**, 2176 (1923); Milligan, Reid, *J. Am. Chem. Soc.* **44**, 204 (1922). Isoln from the fungus *Gibberella fujikuroi:* Cross *et al., J. Chem. Soc.* **1963**, 2937. Antibacterial activity: R. M. E. Richards *et al., J. Pharm. Pharmacol.* **21**, 681 (1969). Use as preservative in ophthalmic solutions: R. M. E. Richards, R. J. McBride, *ibid.* **24**, 145 (1972). Toxicity: Jenner *et al., Food Cosmet. Toxicol.* **2**, 327 (1964).

Liquid. Floral odor, rose character. mp −27°. d_{25}^{25} 1.017 to 1.019. bp_{750} 219-221°; bp_{14} 104°; bp_{12} 98-100°. n_D^{20} 1.530 to 1.533. Two ml dissolve in 100 ml water after thorough shaking. One part is clearly sol in 1 part of 50% alc. Miscible with alcohol, ether. LD_{50} orally in rats: 1790 mg/kg (Jenner).

USE: Pharmaceutic aid (antimicrobial). In flavors and perfumery (esp rose perfumes).

7371. Phenethylamine. *Benzeneethanamine;* β-*phenylethylamine;* 1-amino-2-phenylethane; β-aminoethylbenzene; PEA. $C_8H_{11}N$; mol wt 121.18. C 79.29%, H 9.15%, N 11.56%. $C_6H_5CH_2CH_2NH_2$. Endogenous amine related structurally and pharmacologically to amphetamine: P. Mantegazza, M. Riva, *J. Pharm. Pharmacol.* **15**, 472 (1963). Present in oil of bitter almonds. Found in normal human urine (about 30 μg/liter). Prepn: Johnson, Guest, *Am. Chem. J.* **42**, 346 (1909); Robinson, Snyder, *Org. Syn.* coll. vol. III, 720 (1955). Elevated levels in urine due to paranoid chronic schizophrenia: S. G. Potkin *et al., Science* **206**, 470 (1979); due to stress: M. A. Paulos, R. E. Tessel, *ibid.* **215**, 1127 (1982).

Liquid. Fishy odor. Absorbs CO_2 from air. Does not solidify when cooled in an ice-salt mixture. Strong base. d_4^{25} 0.9640; bp 194.5-195°. Sol in water. Freely sol in alc, ether.

Hydrochloride, $C_8H_{11}N.HCl$, orthorhombic bipyramidal platelets from abs alcohol, mp 217°. Freely sol in water (100 parts H_2O will dissolve 80 parts at 15°). Sol in alc. Insol in ether. LD_{50} s.c. in mice: 470 mg/kg (Mantegazza, Riva).

Caution: Skin irritant and possible sensitizer.

7372. *o*-Phenetidine. *2-Ethoxybenzenamine;* 2-aminophenetole. $C_8H_{11}NO$; mol wt 137.18. C 70.04%, H 8.08%, N 10.21%, O 11.66%.

Oily liquid; rapidly becomes brown on exposure to light and air. mp < −20°. bp 228-230°. Insol in water. Sol in alcohol.

Caution: Absorbed through skin; vapor hazardous.

USE: Manuf of dyes.

7373. *p*-Phenetidine. *4-Ethoxybenzenamine;* 4-aminophenetole; 4-ethoxyaniline; *p*-aminophenyl ethyl ether. $C_8H_{11}NO$; mol wt 137.18. C 70.04%, H 8.08%, N 10.21%, O 11.66%. Prepn by reduction of *p*-nitrophenetole: West, *J. Chem. Soc.* **127**, 494 (1925). Prepn from nitrobenzene, ethanol, Mg, and sulfuric acid: Yukawa, *J. Chem. Soc. Japan* **71**, 547 (1950).

Colorless liquid; becomes red to brown on exposure to air and light. d_4^{16} 1.0652; mp about 3°; bp 253-255°. Practically insol in water. Sol in alcohol. *Keep well closed and protected from light.*

Citrate, $C_8H_{11}NO.C_6H_8O_7$, *Citrophen.* Crystalline powder, mp 186-188°. Slightly acid taste. Sol in about 40 parts water.

Caution: Absorbed through skin; vapor hazardous.

USE: Manuf acetophenetidine, phenocoll, dulcin, dyes.

7374. Phenetole. *Ethoxybenzene;* ethyl phenyl ether; phenyl ethyl ether. $C_8H_{10}O$; mol wt 122.17. C 78.65%, H 8.25%, O 13.10%. $C_6H_5OC_2H_5$. Prepd from phenol or its salts by the use of the following ethylating agents: ethyl chloride, Wohl, *Ber.* **39**, 1953 (1906); ethyl bromide, White *et al., J. Am. Chem. Soc.* **46**, 965 (1924); ethyl *p*-toluenesulfonate, Finzi, *Ann. Chim. Applicata* **15**, 41 (1925); diethyl sulfate or triethyl phosphate, Noller, Dutton, *J. Am. Chem. Soc.* **55**, 424 (1933).

Oily liquid. d_4^{20} 0.967; mp −30°; bp 171-173°; n_D^{20} 1.507. Practically insol in water. Freely sol in alcohol, ether. MLD s.c. in rats: 4.0 g/kg.

7375. Pheneturide. *N-(Aminocarbonyl)-α-ethylbenzeneacetamide; (2-phenylbutyryl)urea;* α-phenyl-α-ethylacetylurea; *N-(α-phenylbutyryl)urea;* α-ethyl-α-phenylacetylurea; ethylphenacemide; EPA; PBU; Benuride. $C_{11}H_{14}N_2O_2$; mol wt 206.24. C 64.06%, H 6.84%, N 13.58%, O 15.51%. Prepd by heating α-phenylbutyric acid chloride with urea: Ger. pat. 249,241 (1912 to Bayer); *Chem. Zentr.* **1912**, II, 396; *Frdl.* **10**, 1165. From α-phenylbutyric acid chloride, urea, and antipyrine: Gold-Aubert, *Helv. Chim. Acta* **41**, 1512 (1958); **Swiss** pat. 374,644 (1964 to Labs. Sapos). Pharmacokinetics: R. L. Galeazzi *et al., J. Pharmacokinet. Biopharm.* **7**, 453 (1979). Toxicity: M. J. Orloff *et al., Neurology* **1**, 377 (1951). Clinical studies: J. C. Bowe, *Brit. J. Clin. Pract.* **27**, 174 (1973); F. B. Gibberd *et al., J. Neurol. Neurosurg. Psychiat.* **45**, 1113 (1982).

dl-Form, needles from ethanol, mp 149-150°.

d-Form, needles from ethanol, mp 168-169°. $[\alpha]_D^{17}$ +54.0° (c = 1 in ethanol); $[\alpha]_D^{22}$ +53.8° (c = 1 in acetone); +48.2° (c = 1 in dioxane).

l-Form, crystals from 50% ethanol, mp 162-163°. $[\alpha]_D^{30}$ −51.6° (c = 1 in ethanol).

THERAP CAT: Anticonvulsant.

7376. Phenformin. *N-(2-Phenylethyl)imidodicarbonimidic diamide; 1-phenethylbiguanide;* phenethyldiguanide; *N′-β-phenethylformamidinyliminourea;* fenformin; fenormin; β-PEBG; PEDG. $C_{10}H_{15}N_5$; mol wt 205.26. C 58.52%, H 7.37%, N 34.12%. Prepn: Shapiro *et al., J. Am. Chem. Soc.* **81**, 2220 (1959); Shapiro, Freedman, U.S. pats. **2,961,377, 3,057,780** (1960, 1962 both to U.S.V.). Metabolism: R. Beckmann, *Diabetologia* **3**, 368 (1967). Experimental model of phenformin-induced lactic acidosis in rats: R. Assan *et al., ibid.* **14**, 261 (1978). Toxicity study: G. Proske *et al., Arzneimittel-Forsch.* **12**, 314 (1962). Comprehensive description: J. E. Moody in *Analytical Profiles of Drug Substances* vol. 4, K. Florey, Ed. (Academic Press, New York, 1975) pp 319-332.

Hydrochloride, $C_{10}H_{15}N_5$·HCl, *Azucaps, DBI, DBI-TD, Debeone-DT, Debinyl, Dibein, Dibotin, Dipar, Feguanide, Glucopostin, Insoral, Lentobetic, Meltrol, Normoglucina.* Crystals from isopropanol, mp 175-178°. Sol in water; pH (0.1*M* aq soln): 6.7. LD$_{50}$ in mice (mg/kg): 450 orally, 19 i.v. (Proske).

Caution: Causes lactic acidosis! FDA has declared this substance an imminent hazard to public health: *Fed. Reg.* **43**, 54995 (1978).

THERAP CAT: Antidiabetic.

7377. Phenglutarimide. *3-[2-(Diethylamino)ethyl]-3-phenyl-2,6-piperidinedione; 2-(2-diethylaminoethyl)-2-phenylglutarimide;* α-phenyl-α-(diethylaminoethyl)glutarimide; 3-phenyl-3-(β-diethylaminoethyl)-2,6-dioxopiperidine. $C_{17}H_{24}N_2O_2$; mol wt 288.39. C 70.80%, H 8.39%, N 9.71%, O 11.10%. Prepn: Tagmann *et al., Helv. Chim. Acta* **35**, 1235 (1952); **37**, 185 (1954); Hoffmann, Tagmann, U.S. pat. **2,664,424** (1953 to Ciba).

Crystals from isopropanol, mp 125-127°. Can also be crystallized from a mixture of ethyl acetate and ligroin.

Hydrochloride, $C_{17}H_{24}N_2O_2$·HCl, *Aturban(e), Aturbal.* Crystals from methanol + ethyl acetate, dec 168-172°, also 176-177°. Freely sol in water. Dec by alkali.

THERAP CAT: Anticholinergic.

7378. Phenicarbazide. *2-Phenylhydrazinecarboxamide; 1-phenylsemicarbazide;* 1-carbamyl-2-phenylhydrazine; Cryogenine; Kryogenin. $C_7H_9N_3O$; mol wt 151.17. C 55.62%, H 6.00%, N 27.80%, O 10.58%. $C_6H_5NHNHCONH_2$. Prepd by mixing 50 g phenylhydrazine with 37 g 90% acetic acid and 250 g water, then slowly adding a soln of 45 g potassium cyanate in 150 g water: Wildman, *Ber.* **26**, 2613 (1893).

Leaflets from water, mp 172°. Very sol in hot water, alcohol, methanol, and acetone; difficultly sol in cold water, ether, benzene, ligroin.

THERAP CAT: Antipyretic.

7379. Phenicin. *2,2′-Dihydroxy 4,4′ dimethyl[bi 1,4-cyclohexadien-1-yl]-3,3′,6,6′-tetrone; 3,3′-dihydroxy-5,5′-dimethyl-2,2′-bi-p-benzoquinone;* phoenicin. $C_{14}H_{10}O_6$; mol wt 274.23. C 61.32%, H 3.68%, O 35.01%. Fungal pigment produced by *Penicillium phoeniceum* and *P. rubrum:* Friedheim, *Helv. Chim. Acta* **21**, 1464 (1938); Charollais *et al., Arch. Sci. (Geneva)* **16**, 474 (1963). Synthesis: Posternak *et al., Helv. Chim. Acta* **26**, 2031 (1943); Musso, Beecken, *Ber.* **92**, 1416 (1959); J. Kalamar *et al., Helv. Chim. Acta* **57**, 2368 (1974). Biosynthesis: E. Steiner *et al., ibid.* 2377.

Yellowish-brown crystals from alc, mp 230-231°. uv max (chloroform): 268, 406 nm (log ε 4.52, 3.36). Sparingly sol in water; freely sol in chloroform, acetic acid, hot alcohol. Very acidic solns are yellow, turning red at pH 1.6-3.6 and violet at pH 4.9-6.

7380. Phenindamine. *2,3,4,9-Tetrahydro-2-methyl-9-phenyl-1H-indeno[2,1-c]pyridine;* 2-methyl-9-phenyl-2,3,4,9-tetrahydro-1-pyridindene; 1,2,3,4-tetrahydro-2-methyl-9-phenyl-2-azafluorene; Nu-1504. $C_{19}H_{19}N$; mol wt 261.37. C 87.31%, H 7.33%, N 5.36%. Prepn: J. T. Plati, W. Wenner, U.S. pat. **2,470,108** (1949 to Hoffmann-La Roche).

Crystals, mp 91°, d 1.17.

Hydrogen tartrate, $C_{19}H_{19}N·C_4H_6O_6$, *Pernovin, Thephorin.* Crystals, mp 165-167°. Soly: about 2.5% in water; sparingly sol in propylene glycol. Insol in 95% ethanol, glycerol, ether. LD$_{50}$ orally in rats: 280 mg/kg.

Note: Thephorin was formerly used to designate theobromine sodium formate.

THERAP CAT: Tartrate as antihistaminic.

7381. Phenindione. *2-Phenyl-1H-indene-1,3(2H)-dione; 2-phenyl-1,3(2H)-indandione;* 2-phenyl-1,3-diketohydrindene; PID; Pindione; Bindan; Hedulin; Dindevan; Dineval; Hemolidione; Indon; Indema; Fenilin; Fenhydren; Rectadione; Cronodione; Thrombasal. $C_{15}H_{10}O_2$; mol wt 222.24. C 81.07%, H 4.54%, O 14.40%. By heating phthalide with benzaldehyde and sodium ethylate soln: Dieckmann, *Ber.* **47**, 1439 (1914).

Leaflets from alcohol, mp 149-151°. Practically insol in cold water. Slightly sol in warm water. pH of satd soln at 25° = 4.5. Readily sol in alkaline solns. Soluble in methanol, ethanol, ether, acetone, benzene, chloroform. Solns in alkalies are red, in concd H_2SO_4 blue.

THERAP CAT: Anticoagulant.

7382. Peniprazine. *(1-Methyl-2-phenylethyl)hydrazine; (α-methylphenethyl)hydrazine;* (1-benzylethyl)hydrazine; 1-phenyl-2-hydrazinopropane; β-phenylisopropylhydrazine; PIH. $C_9H_{14}N_2$; mol wt 150.22. C 71.96%, H 9.39%, N 18.65%. $C_6H_5CH_2CH(CH_3)NHNH_2$. Prepn of DL-form: Biel *et al., J. Am. Chem. Soc.* **80**, 1519 (1958); **81**, 2805 (1959); Biel, U.S. pats. **2,978,461** and **3,000,903** (both to Lakeside). Resolution of stereoisomers: Bernstein *et al., J. Am. Chem. Soc.* **81**, 4433 (1959). Synthesis: Biel *et al., ibid.* 4995.

DL-Form, liquid, $bp_{0.5}$ 82-86°. n_D^{20} 1.5401.
DL-Form hydrochloride, $C_9H_{14}N_2$.HCl, *JB-516, Catral, Catron, Catroniazid, Cavodil.* Crystals, mp 124-125°.
D-Form, bp_{10} 135-138°. n_D^{20} 1.5385. $[\alpha]_D^{25}$ +4.5° (c = 5 in methanol).
D-Form hydrochloride, crystals, mp 152-154°. $[\alpha]_D^{25}$ +12.8° (c = 5 in water); also reported as mp 148-149°, $[\alpha]_D^{25}$ +13.8° (c = 1 in water).
L-Form, bp_{10} 135-138°. n_D^{20} 1.5385. $[\alpha]_D^{25}$ −4.5° (c = 5 in methanol).
L-Form hydrochloride, crystals, mp 152-154°. $[\alpha]_D^{25}$ −12.5° (c = 5 in water); also reported as mp 148-149°, $[\alpha]_D^{25}$ −14.0° (c = 1.0 in water).

THERAP CAT: Antihypertensive.

7383. Pheniramine. *N,N-Dimethyl-γ-phenyl-2-pyridinepropanamine; 2-[α-(2-dimethylaminoethyl)benzyl]pyridine;* 1-phenyl-1-(2-pyridyl)-3-dimethylaminopropane; 3-phenyl-3-(2-pyridyl)-N,N-dimethylpropylamine; prophenpyridamine; propheniramine; Avil; Tripoton. $C_{16}H_{20}N_2$; mol wt 240.35. C 79.96%, H 8.39%, N 11.66%. Synthesis: Sperber *et al.,* U.S. pats. **2,567,245** and **2,676,964** (1951, 1954, both to Schering).

Oily liq. Slightly yellow color. Characteristic amine-like odor. d 1.0081; bp_{13} 181°; bp_2 142°; $bp_{0.5}$ 135°; n_D^{25} 1.5519 to 1.5521. Insol in water. Sol in dil acids, alcohol, benzene, chloroform, ether.
Maleate, $C_{16}H_{20}N_2$.$C_4H_4O_4$, *Daneral, Inhiston, Trimeton (formerly).* Crystals from amyl alcohol, faint amine-like odor, mp 107°. Freely sol in water, alcohol. Slightly sol in ether, benzene. pH of 1% aq soln between 4.3 and 4.9.

p-Aminosalicylate, $C_{16}H_{20}N_2$.$C_7H_7NO_3$, small needles from acetone + ethyl acetate, large octahedra from water, dec 142°. One gram dissolves in 10 ml water, freely sol in alcohol. Sparingly sol in ethyl acetate, ether, acetone.

THERAP CAT: Antihistaminic.

7384. Phenmedipham. *3-(Methylphenyl)carbamic acid 3-[(methoxycarbonyl)amino]phenyl ester; m-hydroxycarbanilic acid methyl ester m-methylcarbanilate;* methyl 3-(*m*-tolylcarbamoyloxy)phenylcarbamate; Schering 38584; Betanal. $C_{16}H_{16}N_2O_4$; mol wt 300.31. C 63.99%, H 5.37%, N 9.33%, O 21.31%. Prepn: **Neth. pat. Appl. 6,604,363** (1966 to Schering AG), *C.A.* **66**, 104813w (1967). Persistence in the soil: K. Kossmann, *Weed Res.* **10**, 349 (1970).

Colorless crystals, mp 139-142 (tech, 143-144°). Soly in water at room temp: < 10 ppm. LD_{50} orally in rats: > 8 g/kg.

USE: Herbicide.

7385. Phenmetrazine. *3-Methyl-2-phenylmorpholine;* 3-methyl-2-phenyltetrahydro-2H-1,4-oxazine; 2-phenyl-3-methyltetrahydro-1,4-oxazine; A-66. $C_{11}H_{15}NO$; mol wt 177.25. C 74.54%, H 8.53%, N 7.90%, O 9.03%. Prepn: Thomae, Wick, U.S. pat. **2,835,669** (1958 to Boehringer, Ing.); Siemer, Hengen, U.S. pat. **3,018,222** (1962 to Ravensberg Chem.); Clark, *J. Org. Chem.* **27**, 3251 (1962); Klosa, *J. Prakt. Chem.* **21**, 12 (1963).

Liquid. bp_{12} 138-140°; bp_1 104°.
Hydrochloride, $C_{11}H_{15}NO$.HCl, *Marsin, Neo-Zine, Preludin.* Crystals from ethanol + ether, mp 182°. One gram dissolves in 0.4 ml water, in 2.0 ml 95% alc, in 2.0 ml chloroform. Sparingly sol in ether.
Note: This is a controlled substance (stimulant) listed in the U.S. Code of Federal Regulations, Title 21 Part 1308.12 (1995).

THERAP CAT: Anorexic.

7386. Phenobarbital. *5-Ethyl-5-phenyl-2,4,6(1H,3H,5H)-pyrimidinetrione; 5-ethyl-5-phenylbarbituric acid;* phenylethylmalonylurea; phenobarbitone; Agrypnal; Barbiphenyl; Barbipil; Eskabarb; Gardenal; Luminal; Phenobal. $C_{12}H_{12}N_2O_3$; mol wt 232.24. C 62.06%, H 5.21%, N 12.06%, O 20.67%. Prepn: **Ger. pat. 247,952** (1911 to Bayer); Hoerlein, U.S. pat. **1,025,872** (1912), *Frdl.* **11**, 926; *Chem. Zentr.* **1912, II**, 212; Chamberlain *et al., J. Am. Chem. Soc.* **57**, 352 (1935); Inman, Bitler, U.S. pat. **2,358,072** (1944 to Kay-Fries Chem.); J. T. Pinhey, B. A. Rowe, *Tetrahedron Letters* **21**, 965 (1980). Toxicity data: Schaffarzick, Brown, *Science* **116**, 663 (1952); E. I. Goldenthal, *Toxicol. Appl. Pharmacol.* **18**, 185 (1971). Series of articles on pharmacology and mechanism of antiepileptic action: *Adv. Neurol.* **27**, 473-562 (1980). Clinical evaluation in febrile convulsions: R. W. Newton, *Arch. Dis. Child.* **63**, 1189 (1988). Comprehensive description: M. K. C. Chao *et al., Anal. Profiles Drug Subs.* **7**, 359-399 (1978).

Crystals (3 different phases), mp 174-178°. uv max (pH 10 buffer): 240 nm ($A_{1cm}^{1\%}$ 431); (0.1N NaOH): 256 nm ($A_{1cm}^{1\%}$ 314). Slightly bitter taste. One gram dissolves in about one liter of water, 8 ml alc, 40 ml chloroform, 13 ml ether, about 700 ml benzene. Soluble in alkali hydroxides and carbonates. A satd aq soln is acid to litmus. pK_1 7.3, pK_2 11.8. LD_{50} orally in rats: 162 ±14 mg/kg (Goldenthal).

Sodium salt, $C_{12}H_{11}N_2NaO_3$, *sol phenobarbital, sol phenobarbitone.* Bitter, slightly hygroscopic crystals or white powder. One gram dissolves in about 1 ml water, about 10 ml alcohol. Insol in ether or chloroform. Aq solns are alkaline to litmus and phenolphthalein, pH ~9.3. LD_{50} orally in rats: 660 mg/kg (Schaffarzick, Brown).

Caution: May be habit forming. This is a controlled substance (depressant) listed in the U.S. Code of Federal Regulations, Title 21 Parts 329.1 and 1308.14 (1995).

THERAP CAT: Anticonvulsant; sedative; hypnotic.

THERAP CAT (VET): Anticonvulsant; sedative.

7387. Phenobutiodil. *2-(2,4,6-Triiodophenoxy)butanoic acid; 2-(2,4,6-triiodophenoxy)butyric acid; α-(2,4,6-triiodophenoxy)butyric acid; 1-(2,4,6-triiodophenoxy)propanecarboxylic acid;* 4114 Th; Baygnostil; Vésipaque. $C_{10}H_9I_3O_3$; mol wt 557.89. C 21.53%, H 1.63%, I 68.24%, O 8.60%. Prepn: Redel *et al., Bull. Soc. Chim. France* **1954**, 342; Redel, Maillard, U.S. pat. **2,796,432** (1957 to Chimie et Atomistique). Pharmacology: H. Bekker *et al., Pharmazie* **27**, 411 (1972).

Crystals from ether + petr ether, mp 124-125°. LD_{50} in mice (mg/kg): 1800 orally; 265 i.v. (Bekker).

THERAP CAT: Diagnostic aid (radiopaque medium—cholecystographic).

7388. Phenocoll. *2-Amino-N-(4-ethoxyphenyl)acetamide; 2-amino-p-acetophenetidide; α-amino-p-acetophenetide; aminoacetophenetidine; 4-glycylaminophenol ethyl ether; glycine p-phenetidide;* 4 *glycylaminophenetol, glycocoll-p-phenetidide; phenokoll;* Phenamine. $C_{10}H_{14}N_2O_2$; mol wt 194.23. C 61.84%, H 7.27%, N 14.42%, O 16.47%. Prepn: Majert, Ger. pats. **59,121; 59,874** (1891 to Schering), *Frdl.* **3**, 915, 918; Ger. pat. **346,809** (1921 to Schering), *Frdl.* **13**, 1066; Karrer, Haebler, *Helv. Chim. Acta* **7**, 534 (1924).

Crystals, mp 100.5°.

Monohydrate, crystals, mp 95°.

Hydrochloride, $C_{10}H_{14}N_2O_2$.HCl, cryst powder. Sol in 20 parts water, in alcohol. More sol in hot water; slightly sol in benzene, chloroform, ether. Aq soln is neutral. *Incompat.* Alkali hydroxides or carbonates.

Salicylate, $C_{10}H_{14}N_2O_2.C_7H_6O_3$, *Salocoll.* Fine needles. Sweetish taste. Sol in 200 parts cold, 20 parts hot water. *Incompat.* As for the hydrochloride; with substances incompat with salicylates.

THERAP CAT: Antipyretic, analgesic.

7389. Phenoctide. *N,N-Diethyl-N-[2-[4-(1,1,3,3-tetramethylbutyl)phenoxy]ethyl]benzenemethanaminium chloride; benzyldiethyl-2-[(p-1,1,3,3-tetramethylbutyl)phenoxy]ethylammonium chloride; β-p-tert-octylphenoxyethyldiethylbenzylammonium chloride;* Octaphen. $C_{27}H_{42}ClNO$; mol wt 432.09. C 75.05%, H 9.80%, Cl 8.20%, N 3.24%, O 3.70%. Prepn: Goldberg, Besly, **Brit.** pat. **703,477** (1954 to Ward Blenkinsop); Erekaev, *C.A.* **54**, 12483i (1960).

Crystals from ethyl acetate. mp 112-114° (Goldberg, Besly); mp 95° (Erekaev).

USE: Orthophosphate as lubricant: Semmens, Summers-Smith, **Brit.** pat. **790,056** (1958 to I.C.I.).

THERAP CAT: Anti-infective (topical).

7390. Phenol. Carbolic acid; phenic acid; phenylic acid; phenyl hydroxide; hydroxybenzene; oxybenzene. C_6H_6O; mol wt 94.11. C 76.57%, H 6.43%, O 17.00%. Obtained from coal tar, or made by fusing sodium benzenesulfonate with NaOH, or by heating monochlorobenzene with aq NaOH under high pressure. The crystalline article of commerce contains at least 98% phenol. Review of mfg processes: A. Dierichs, R. Kubicka, *Phenole und Basen, Vorkommen und Gewinnung* (Akademie-Verlag, Berlin, 1958) 472 pp; Faith, Keyes & Clark's *Industrial Chemicals,* F. A. Lowenheim, M. K. Moran, Eds. (Wiley-Interscience, New York, 4th ed., 1975) pp 612-623. Use in treatment of spasticity: D. E. Garland *et al., Clin. Orthop.* **165**, 217 (1982); *eidem, Arch. Phys. Med. Rehab.* **65**, 243 (1984). Review of use in pain relief: K. M. Wood, *Pain* **5**, 205-229 (1978). Review of toxicology: H. Babich, D. L. Davis, *Regul. Toxicol. Pharmacol.* **1**, 90-109 (1981). Toxicity: W. B. Deichmann, S. Witherup, *J. Pharmacol. Exp. Ther.* **80**, 233 (1944). Review: C. Thurman in Kirk-Othmer *Encyclopedia of Chemical Technology* vol. 17 (Wiley-Interscience, New York, 3rd ed., 1982) pp 373-384.

Colorless, acicular crystals or white, crystalline mass. Characteristic odor, somewhat sickeningly sweet and acrid with a sharp and burning taste. *Poisonous and caustic!* Prone to redden on exposure to air and light, hastened by presence of alkalinity. d 1.071. When free from water and cresols it congeals at 41° and melts at 43°. Ultrapure material mp 40.85°. The commercial product contains an impurity which raises the mp. bp 182°. Flash pt, closed cup: 175°F (79°C). n_D^{41} 1.5425. pKa at 25° = 10.0. pH of aq solns ~6.0. It is liquefied by mixing with ~8% water. One gram dissolves in ~15 ml water, 12 ml benzene; very sol in alcohol, chloroform, ether, glycerol, carbon disulfide, petrolatum, volatile and fixed oils, aq alkali hydroxides. Almost insol in petr ether. LD_{50} orally in rats: 530 mg/kg (Deichmann, Witherup). *Keep well closed and protected from light. Do not handle with bare hands.*

Ammonium salt, $C_6H_6O.NH_3$, *ammonium phenate, ammonium carbolate.* White to pink crystalline masses. Sol in water.

Sodium salt, C_6H_5NaO, *phenolate sodium.*

Aqueous solution with phenolate sodium, *Chloraseptic.*

Caution: Phenol is rapidly absorbed through skin, and from the stomach and lungs. Systemic toxicity can result from ingestion of solutions, or from inhalation or cutaneous absorption of solutions or vapor. This is characterized by an initial excitatory phase, rapidly progressing to fulminating CNS depression with coma. Potential symptoms of acute overexposure are steatorous breathing, mucous rales, froth at mouth and nose, frank pulmonary edema; cyanosis; tremor, convulsions, twitching; death due to respiratory failure. Characteristic odor of phenol on breath. Ingestion may cause burning of mouth and throat; white necrotic lesions in mouth, esophagus and stomach; abdominal pain. Direct contact may cause irritation of eyes, nose and throat; skin burns; dermatitis; ochronosis. Chronic exposure may result in vomiting, difficulty swallowing, excess salivation, diarrhea, anorexia, weight loss; headache, fainting, vertigo, mental disturbances; muscle aches and pain, weakness; damage to liver and kidney, dark urine. *See NIOSH Pocket Guide to Chemical Hazards* (DHHS/NIOSH 90-117, 1990) p 176; *Clinical Toxicology of Commercial Products, R. E. Gosselin et al.,* Eds. (Williams & Wilkins, Baltimore, 5th ed., 1984) Section III, p344-348; *Patty's Industrial Hygiene and Toxicology* vol. **2A,** G. D. Clayton, F. E. Clayton, Eds. (Wiley-Interscience, New York, 3rd ed., 1981) p 2567-2584.

USE: As a general disinfectant, either in soln or mixed with slaked lime, etc., for toilets, stables, cesspools, floors, drains, etc.; for the manuf of colorless or light-colored artificial resins, many medical and industrial organic compds and dyes; as a reagent in chemical analysis. Pharmaceutic aid (preservative).

THERAP CAT: Aqueous soln as topical anesthetic; topical antiseptic; topical antipruritic.

THERAP CAT (VET): Antiseptic caustic. Topical anesthetic in pruritic skin conditions. Has been used internally and externally as an antiseptic.

7391. Phenoldisulfonic Acid. *4-Hydroxy-1,3-benzene-disulfonic acid; 4-hydroxy-m-benzenedisulfonic acid;* 1-phenol-2,4-disulfonic acid. $C_6H_6O_7S_2$; mol wt 254.24. C 28.35%, H 2.38%, O 44.05%, S 25.22%. Conveniently prepd by hydrolysis of the dichloride which is obtained by the action of chlorosulfonic acid on phenol at room temp: Pollak *et al., Monatsh.* **46,** 395 (1925). Monograph: E. E. Gilbert, *Sulfonation and Related Reactions* (Interscience, New York, 1965) 529 pp.

Deliquescent needles, vague melting range from 89° to 100°. Decomp above 100°. Freely sol in water, methanol. Practically insol in ether, petr ether.

USE: In the manuf of aminophenoldisulfonic acids which are intermediates in the dye industry.

7392. Phenolphthalein. *3,3-Bis(4-hydroxyphenyl)-1-(3H)-isobenzofuranone;* 3,3-bis(*p*-hydroxyphenyl)phthalide; α-(*p*-hydroxyphenyl)-α-(4-oxo-2,5-cyclohexadien-1-ylidine)-*o*-toluic acid; Chocolax; Darmol. $C_{20}H_{14}O_4$; mol wt 318.33. C 75.46%, H 4.43%, O 20.10%. Prepd by condensing phenol with phthalic anhydride: Baeyer, *Ann.* **202,** 69 (1880); Herzog, *Chem. Ztg.* **51,** 84 (1927); Hubacher, U.S. pat. **2,192,485** (1940 to Ex Lax); Gamrath, U.S. pat. **2,522,939** (1950 to Monsanto). Subchronic toxicity studies: D. D. Dietz *et al., Fund. Appl. Toxicol.* **18,** 48 (1992). Comprehensive description: F. J. Al-Shammary *et al.* in *Analytical Profiles of Drug Substances* vol. **20,** K. Florey Ed. (Academic Press, New York, 1991) pp 627-664.

Minute, triclinic crystals, often twinned. mp 258-262°. d 1.299. Color: White or yellowish-white, *see also* Yellow Phenolphthalein. Sol in alcohol and ether, very slightly sol in chloroform. One gram dissolves in 12 ml alcohol, in ~100 ml ether. Almost insol in water. Sol in dil solns of alkali hydroxides and hot solns of alkali carbonates forming a red soln. pKa (25°C) 9.7. uv max (methanol): 205, 229, 276 nm (ε 27261.147, 14692.144, 2006.369).

USE: A 1% alcoholic soln as an indicator in titrations of mineral and organic acids and most alkalies. Not suitable

for ammonia. Very sensitive to CO_2, and in estimating carbonates the liq must be boiled. Borax can be titrated with phenolphthalein as an indicator only when glycerol is present, because the color gradually fades away as the acid is added. Usable with a few alkaloids. Colorless to pH 8.5; pink to deep-red, >pH 9.

THERAP CAT: Cathartic.

THERAP CAT (VET): Has been used as a laxative.

7393. Phenolphthalein Sodium. *3,3-Bis(4-hydroxy-phenyl)-1(3H)-isobenzofuranone disodium salt.* $C_{20}H_{12}Na_2-O_4$; mol wt 362.29. C 66.31%, H 3.34%, Na 12.69%, O 17.66%. Prepd from an alcoholic soln of 1 mol phenolphthalein by the addition of 2 mols NaOH. Structure: Buu-Hoi, *Bull. Soc. Chim.* [5] **8,** 165 (1941).

Reddish-brown granular mass with coppery luster, or pale red powder. Dec in air. Soluble in water, giving a deep-red soln. Absorption max: 550-555 nm. The unstable enol form is colorless, as is the trisodium salt.

7394. Phenolphthalin. *2-[Bis(4-hydroxyphenyl)methyl]-benzoic acid;* decolorized phenolphthalein; phthalin; 4′,4″-dihydroxytriphenylmethane-2-carboxylic acid. $C_{20}H_{16}O_4$; mol wt 320.34. C 74.99%, H 5.03%, O 19.98%. Made by boiling phenolphthalein with zinc dust in alkaline soln. Called *"Kastle-Meyer reagent"* when in soln. Prepd by dissolving 2 g phenolphthalin + 20 g KOH in a desired amt of doubly distd H_2O and diluting with an equal vol of 95% ethanol. Detailed directions for prepn of test soln and its use: Lecoq, *Bull. Soc. Chim. Belge* **54,** 186-202 (1945), *C.A.* **41,** 2346i (1947).

Colorless crystals, mp 237°. Insol in water; sol in alcohol, ether, aq alkali. The alkaline solns gradually become pink on exposure to air or other oxidizing substances.

USE: As a reagent for oxidases, blood, HCN, peroxides, copper.

7395. Phenolphthalol. *2-[Bis(4-hydroxyphenyl)methyl]-benzenemethanol; o-[bis(p-hydroxyphenyl)methyl]benzyl alcohol;* dihydroxyphenylmethenylbenzyl alcohol; 2-(4,4′-dihydroxybenzhydryl)benzyl alcohol; bis(4-hydroxyphenyl)-(2-hydroxymethylphenyl)methane; Egmol; Regolax. $C_{20}-H_{18}O_3$; mol wt 306.36. C 78.41%, H 5.92%, O 15.67%. Prepn: Hubacher, *J. Am. Chem. Soc.* **74,** 5216 (1952); Schultz, Geller, *Arch. Pharm.* **288,** 234 (1955); Bulcsu, **Ger.** pat. **1,141,293** (1962 to Iromedica), *C.A.* **59,** 1535b (1963).

Crystals from dil alcohol, mp 201-202°.

Monoacetate, $C_{22}H_{20}O_4$, crystals from benzene or chloroform, mp 171-174°. Soluble in dil NaOH.

Triacetate, $C_{26}H_{24}O_6$, crystals from methanol or ethanol, mp 104-106°. Practically insol in dil alkali.

THERAP CAT: Cathartic.

7396. p-Phenolsulfonic Acid. *4-Hydroxybenzenesulfonic acid;* sulfocarbolic acid. $C_6H_6O_4S$; mol wt 174.18. C 41.38%, H 3.47%, O 36.74%, S 18.41%. Commercially available as a 65% soln. Prepn by hydrolysis of p-chloro- or p-bromobenzenesulfonic acid: Zollinger, Roehling, **U.S.** pat. **1,321,271** (1920); by sulfonation of phenol: Davidson, Byrne, **Brit.** pat. **820,659** (1959 to Hardman & Holden). Prepn of ammonium salt: Oxley *et al., J. Chem. Soc.* **1948,** 303.

Deliquescent needles. Miscible with water, alcohol.

Ammonium salt, $C_6H_9NO_4S$, *ammonium p-phenolsulfonate.* Plates from water, dec 270-271°.

Barium salt, $C_{12}H_{10}BaO_8S_2$, *barium p-phenolsulfonate.* Monohydrate, powder. *Poisonous!* Sol in water; slightly sol in alcohol.

Caution: Irritating to skin.

USE: Intermediate in mfg of pharmaceuticals, dyestuffs. In the Ferrostan process of tin plating (U.S. Steel).

7397. Phenolsulfonphthalein. *4,4'-(3H-2,1-Benzoxathiol-3-ylidene)bisphenol S,S-dioxide;* α-hydroxy-α,α-bis(p-hydroxyphenyl)-o-toluenesulfonic γ-sultone; 3,3-bis(p-hydroxyphenyl)-3H-2,1-benzoxathiole 1,1-dioxide; phenol red; P.S.P.; Sulfonphthal. $C_{19}H_{14}O_5S$; mol wt 354.38. C 64.40%, H 3.98%, O 22.57%, S 9.05%. Prepd by the action of o-sulfobenzoic anhydride or of o-sulfobenzoyl chloride on phenol: Kekulé, Barbaglia, *Ber.* **5,** 876 (1872); Kekulé, *Ber.* **6,** 943 (1873); Heumann, Kochlin, *Ber.* **15,** 1118 (1882); **Ger.** pat. **142,116;** *Chem. Zentr.* **II,** 79 (1903); Orndorff, Sherwood, *J. Am. Chem. Soc.* **45,** 486 (1923). Diagnostic use: G. Dunea, P. Freedman, *J. Am. Med. Assoc.* **204,** 159 (1968); R. D. Wilkes *et al., Vet. Med. Small Anim. Clin.* **76,** 289 (1981).

Bright red to dark red crystals. Stable in air. pK = 7.9. One gram dissolves in ~1300 ml water, in ~350 ml alc, in 500 ml acetone. Almost insol in chloroform, ether. Readily sol in aq alkali hydroxides or carbonates with red color, which is discharged by boiling with zinc dust.

USE: As indicator in 0.02-0.05% alcohol soln. pH 6.8 yellow, 8.4 red.

THERAP CAT: Diagnostic aid (renal function).

THERAP CAT (VET): Diagnostic aid (renal function).

7398. Phenoltetrachlorophthalein. *4,5,6,7-Tetrachloro-3,3-bis(4-hydroxyphenyl)-1(3H)-isobenzofuranone;* 3,4,5,6-tetrachlorophenolphthalein. $C_{20}H_{10}Cl_4O_4$; mol wt 456.11. C 52.67%, H 2.21%, Cl 31.09%, O 14.03%. Prepd by condensation of phenol and tetrachlorophthalic acid or its anhydride: W. R. Orndorff, J. A. Black, *Am. Chem. J.* **41,** 349 (1909); Zalkind, Belikova, *Zh. Prikl. Khim. (Leningrad)* **8,** 1210 (1935). Metabolism and hepatic chromosecretion of

phthalein dyes: P. Hykes, J. Jirsa, *Acta Univ. Carol. Med.* **25,** 135 (1979), *C.A.* **95,** 111282z (1981).

White powder, dec > 300°. Almost insol in water, chloroform, benzene. Sol in alcohol, ether, acetone, glacial acetic acid; sol in aq alkali hydroxides or carbonates with deep purple color when concd, violet-red when dil, and bluish when very dil.

Disodium salt, $C_{20}H_8Cl_4Na_2O_4$, *Chlor-Tetragnost.* Dihydrate, violet crystals. Freely sol in water. Dec on exposure to air.

THERAP CAT: Cathartic; disodium salt as diagnostic aid (hepatic function).

7399. Phenomorphan. *17-(2-Phenylethyl)morphinan-3-ol;* 3-hydroxy-N-(2-phenylethyl)morphinan; (−)-1,3,4,9,10,-10a-hexahydro-11-phenethyl-2H-10,4a-iminoethanophenanthren-6-ol; NIH-7274. $C_{24}H_{29}NO$; mol wt 347.50. C 82.95%, H 8.41%, N 4.03%, O 4.60%. Prepn: Grüssner *et al., Helv. Chim. Acta* **40,** 1232 (1957).

Crystals from dimethylformamide, mp 243-245°.

Hydrobromide, $C_{24}H_{30}BrNO$, crystals from dil alc, mp 300-301°. $[\alpha]_D^{20}$ −63.12° (c = 3.27 in alc).

D-Tartrate monohydrate, $C_{28}H_{35}NO_7.H_2O$, crystals from isopropanol, mp 125-126°. $[\alpha]_D^{20}$ −42.75° (c = 0.983 in water).

Methyl bromide, $C_{25}H_{32}BrNO$, crystals from alcohol, mp 239-240°. $[\alpha]_D^{20}$ −42.81° (c = 1.155 in methanol).

Note: This is a controlled substance (opiate) listed in the U.S. Code of Federal Regulations, Title 21 Part 1308.11 (1995).

7400. Phenoperidine. *1-(3-Hydroxy-3-phenylpropyl)-4-phenyl-4-piperidinecarboxylic acid ethyl ester;* 1-(3-hydroxy-3-phenylpropyl)-4-phenylisonipecotic acid ethyl ester; 1-[γ-hydroxy-γ-phenylpropyl]-4-phenyl-4-carbethoxypiperidine; 3-(4-carbethoxy-4-phenylpiperidino)-1-phenyl-1-propanol; 1-phenyl-3-[(4'-phenyl-4'-carbethoxy)piperidino]-1-propanol; phenoperidin. $C_{23}H_{29}NO_3$; mol wt 367.49. C 75.17%, H 7.95%, N 3.81%, O 13.06%. Prepn: Janssen, Eddy, *J. Med. Pharm. Chem.* **2,** 31 (1960); Pohland, **U.S.** pat. **2,951,-080** (1960 to Lilly); Cutler, Fisher, **U.S.** pat. **2,962,501** (1960 to Merck & Co.).

Hydrochloride, $C_{23}H_{29}NO_3.HCl$, *Lealgin, Operidine.* Crystals from ethyl acetate + methanol, or ethanol, mp 200-202°. Soluble in water.

Note: This is a controlled substance (opiate) listed in the U.S. Code of Federal Regulations, Title 21 Part 1308.11 (1995).

THERAP CAT: Analgesic (narcotic).

7401. Phenopyrazone. *1,4-Diphenyl-3,5-pyrazolidine-dione;* 1,4-diphenyl-3,5-dioxopyrazolidine. $C_{15}H_{12}N_2O_2$; mol wt 252.27. C 71.42%, H 4.79%, N 11.10%, O 12.68%. Prepn: Kraft, **U.S. pat. 2,909,465** (1959 to Knoll).

Crystals, mp 233-234°. May be recrystallized from dioxane or ethyl acetate.

2-Diethylaminoethanol salt, component of *Osadrin.*

THERAP CAT: Analgesic, antipyretic.

7402. Phenosafranin. *3,7-Diamino-5-phenylphenazinium chloride;* C.I. 50200; Safranin B Extra. $C_{18}H_{15}ClN_4$; mol wt 322.80. C 66.98%, H 4.68%, Cl 10.98%, N 17.36%. Prepd by dichromate oxidation of 1 mol *p*-phenylenediamine hydrochloride and 2 mols aniline hydrochloride: Witt, *Ber.* **12**, 939 (1879); **19**, 3121 (1886). *See also Colour Index* vol. **4** (3rd ed., 1971) p 4449.

Green, lustrous needles from dil HCl. Freely sol in water, also sol in alc. Aq and alcoholic solns are purplish-red and have a greenish-yellow fluorescence. Absorption max ~530 nm.

USE: Biological stain.

7403. Phenosulfazole. *4-Hydroxy-N-2-thiazolylbenzenesulfonamide; N-(2-thiazolyl)-1-phenol-4-sulfonamide;* Virazene. $C_9H_8N_2O_3S_2$; mol wt 256.31. C 42.18%, H 3.15%, N 10.93%, O 18.73%, S 25.02%. *Refs:* Sanders *et al., Tex. Rep. Biol. Med.* **6**, 385 (1948); *Chem. & Eng. News* **26**, 2710 (1948); Jungeblut, *Proc. Soc. Exp. Biol. Med.* **70**, 371 (1949); LoGrippo *et al., ibid.* 528; Cox *et al., ibid.* 530; Weil, Warren, *ibid.* 534; Francis, Brown, *ibid.* 535.

Crystals, insol in water. Forms a water-sol sodium salt.

THERAP CAT: Formerly in poliomyelitis.

7404. Phenothiazine. *10H-Phenothiazine;* thiodiphenylamine; dibenzothiazine; AFI-Tiazin; Antiverm; Fentiazin; Helmetina; Lethelmin; Nemazine; Orimon; Phénégic; Phenoverm; Phenovis; Phenoxur; Reconox; Souframine; Vermitin. $C_{12}H_9NS$; mol wt 199.28. C 72.33%, H 4.55%, N 7.03%, S 16.09%. Prepd by fusing diphenylamine with sulfur: Bernthsen, *Ber.* **16**, 2896 (1883); *Ann.* **230**, 73 (1885); **Ger.** pat. **25,150** (1883), *Frdl.* **1**, 252. Improved yields with iodine as catalyst: Knoevenagel, *J. Prakt. Chem.* [2] **89**, 11 (1914); Mitchell, Webb, **U.S. pat. 2,415,363** (1947 to Koppers). Purification: Vierling, **U.S. pat. 2,887,482** (1959); Rigby, **U.S. pat. 3,000,887** (1961 to Shell Oil). Crystal structure: J. D. Bell *et al., Chem. Commun.* **1968**, 1656.

Yellow, rhombic leaflets or diamond-shaped plates from toluene or butanol, mp 185.1°. Sublimes at 130° at 1 mm. bp_{760} 371°; bp_{40} 290°. Freely sol in benzene; sol in ether, in hot acetic acid; slightly sol in alcohol and in mineral oils. Practically insol in petr ether, chloroform, water. Readily oxidized by sunlight or when in presence of a finely divided inert carrier, acquiring a greenish-brown tint. This can be prevented by the admixture of 0.3-1.0% methenamine.

USE: Insecticide; manuf pharmaceuticals.

THERAP CAT (VET): Anthelmintic.

7405. Phenothrin. *2,2-Dimethyl-3-(2-methyl-1-propenyl)cyclopropanecarboxylic acid (3-phenoxyphenyl)methyl ester; 2,2-dimethyl-3-(2-methylpropenyl)cyclopropanecarboxylic acid m-phenoxybenzyl ester;* 3-phenoxybenzyl *cis,trans*-chrysanthemate; S-2539; Sumithrin. $C_{23}H_{26}O_3$; mol wt 350.46. C 78.83%, H 7.48%, O 13.70%. Potent synthetic pyrethroid insecticide. Prepn of racemic mixture: N. Itaya *et al.,* **Ger. pat. 1,926,433** corresp to **U.S. pat. 3,666,789** (1969, 1972 to Sumitomo). Comparative activity of isomers: K. Fujimoto *et al., Agr. Biol. Chem.* **37**, 2681 (1973); Y. Okuno *et al.,* **Ger. pat. 2,348,930** corresp to **U.S. pat. 3,934,023** (1973, 1976 to Sumitomo). Analysis: Y. Takimoto *et al., Anal. Methods Pestic. Plant Growth Regul.* **13**, 133-146 (1984).

(1R-trans)-form

The commercial product is a mixture of isomers. Colorless liquid. d_{25}^{25} 1.06; n_D^{25} 1.5483. Sol in acetone, xylene; insol in water.

USE: Insecticide.

7406. Phenoxazine. Phenazoxine. $C_{12}H_9NO$; mol wt 183.21. C 78.67%, H 4.95%, N 7.65%, O 8.73%. Prepn from 2-aminophenol and 2-aminophenol-HCl: Kehrmann, Neil, *Ber.* **47**, 3107 (1903); de Antoni, *Bull. Soc. Chim. France* **1963**, 2871; by heating 2-amino-2'-hydroxydiphenyl ether in a sealed tube at 270-280° for 40 hrs: Cullinane *et al., J. Chem. Soc.* **1934**, 718.

Leaflets from alc, mp 156°. Freely sol in abs methanol, ethanol, ether, $CHCl_3$, benzene. Sparingly sol in petr ether.

10-Acetylphenoxazine, $C_{14}H_{11}NO_2$, prisms, mp 142°. Sparingly sol in hot water; sol in alc; freely sol in glacial acetic acid.

7407. Phenoxyacetic Acid. Phenoxyethanoic acid; *O*-phenylglycolic acid; phenyl ether glycolic acid; Phenylium. $C_8H_8O_3$; mol wt 152.15. C 63.15%, H 5.30%, O 31.55%. Prepd from phenol and monochloroacetic acid: Giacosa, *J. Prakt. Chem.* [2] **19**, 396 (1879); van Alphen, *Rec. Trav. Chim.* **46**, 148 (1927).

Needles from water, mp 98°; bp 285° (some decompn). K at 25° = 7.54 × 10^{-4}. One gram dissolves in ~75 ml water. Freely sol in alcohol, ether, benzene, carbon disulfide, glacial acetic acid.

Caution: Mild irritant.

USE: Fungicide; keratin exfoliative (to relieve and to soften calluses, corns, and other hard skin surfaces; applied as plasters, pads or in liquids).

7408. Phenoxyacetyl Cellulose. *Cellulose phenoxyacetate;* Enzorb-A. Support used to immobilize proteins, enzymes and microsomes by hydrophobic adsorption. The attractive forces between the hydrophobic phenoxyacetate groups of the support and the hydrophobic surface areas of protein molecules arise from the common repulsion of the aqueous medium. Prepn and properties: L. G. Butler, *Arch. Biochem. Biophys.* **171**, 645 (1975). Applications: J. Dixon *et al., Biotechnol. Bioeng.* **21**, 2113 (1979).

Stable in dry form; slowly hydrolyzed in aq media above pH 8. Resistant to enzymatic hydrolysis.

USE: In enzyme immobilization.

7409. Phenoxybenzamine. *N-(2-Chloroethyl)-N-(1-methyl-2-phenoxyethyl)benzenemethanamine; N-(2-chloroethyl)-N-(1-methyl-2-phenoxyethyl)benzylamine; N-phenoxyisopropyl-N-benzyl-β-chloroethylamine;* bensylyt; 688-A. $C_{18}H_{22}ClNO$; mol wt 303.83. C 71.16%, H 7.30%, Cl 11.67%, N 4.61%, O 5.27%. Prepn: Kerwin, Ullyot, U.S. pat. 2,599,000 (1952 to SK&F).

Crystals from petr ether, mp 38-40°. Sol in benzene.
Hydrochloride, $C_{18}H_{22}ClNO.HCl$, *Dibenyline, Dibenzylin, Dibenzyran.* Crystals from alcohol + ether, mp 137.5-140°. Sol in alcohol, propylene glycol. Very sparingly sol in water. Solutions in propylene glycol should be sterilized by filtration.

Note: The hydrochloride may reasonably be anticipated to be a carcinogen: *Seventh Annual Report on Carcinogens* (PB95-109781, 1994) p 318.

THERAP CAT: Antihypertensive.

7410. 2-Phenoxyethanol. 1-Hydroxy-2-phenoxyethane; ethylene glycol monophenyl ether; β-hydroxyethyl phenyl ether; Phenoxethol; Phenoxetol; Phenyl Cellosolve. $C_8H_{10}O_2$; mol wt 138.17. C 69.55%, H 7.30%, O 23.16%. $C_6H_5OCH_2CH_2OH$. Obtained by treating phenol with ethylene oxide in an alkaline medium: Becker, Barthell, *Monatsh.* **77**, 80 (1947); *see also* Roithner, *ibid.* **15**, 674, 678 (1894); Rindfusz, *J. Am. Chem. Soc.* **41**, 669 (1919). Toxicity study: H. F. Smyth *et al., J. Ind. Hyg. Toxicol.* **23**, 259 (1941).

Oily liquid. Faint aromatic odor. Burning taste. d^{20}_{20} 1.1094; d^{22}_4 1.102; mp 14°; bp_{760} 245.2°; bp_{80} 165°; bp_{25} 137°; bp_{20} 128-130°; n^{20}_D 1.534. Flash pt 250°F. Soly in water: 2.67 g/100 ml. Freely sol in alcohol, ether, NaOH solns. LD_{50} orally in rats: 1.26 g/kg (Smyth).

Acetate, $C_{10}H_{12}O_3$, liquid, bp 243°.

USE: Fixative for perfumes, in org synthesis; as bactericide in conjunction with quaternary ammonium compds; as insect repellent.

THERAP CAT: Antiseptic (topical).

7411. Phenoxypropazine. *(1-Methyl-2-phenoxyethyl)-hydrazine;* fenoxypropazine. $C_9H_{14}N_2O$; mol wt 166.22. C 65.03%, H 8.49%, N 16.85%, O 9.63%. Prepn: Drain *et al., J. Med. Chem.* **6**, 63 (1963).

$bp_{0.2}$ 98-102°.

Maleate, $C_9H_{14}N_2O.C_4H_4O_4$, *Drazine.* Prismatic needles from isopropanol, mp 107-108°. Very sol in water; less sol in ethanol, isopropanol.

THERAP CAT: Formerly as monoamine oxidase inhibitor.

7412. Phenprobamate. *Benzenepropanol carbamate; carbamic acid 3-phenylpropyl ester;* 1-carbamoyloxy-3-phenylpropane; γ-phenylpropyl carbamate; proformiphen; MH-532; Extacol; Palmita; Spantol; Ansepron; Gamaquil; Quamaquil. $C_{10}H_{13}NO_2$; mol wt 179.22. C 67.02%, H 7.31%, N 7.82%, O 17.85%. $H_2NCOOCH_2CH_2CH_2C_6H_5$. Prepn: **Brit. pat. 837,718** (1960 to Siegfried AG). Pharmacology: G. Stille, *Arzneimittel-Forsch.* **12**, 340 (1962).

Shiny leaflets from aq ethanol, mp 101-104°. Slightly bitter, somewhat burning taste. Numbs the tongue slightly afterwards. Soluble in abs ethanol, chloroform, propylene glycol, ethylenediamine, dimethylformamide; sparingly sol in ether, 50% ethanol. Practically insol in water. LD_{50} orally in mice: 840 mg/kg (Stille).

THERAP CAT: Anxiolytic, muscle relaxant (skeletal).

7413. Phenprocoumon. *4-Hydroxy-3-(1-phenylpropyl)-2H-1-benzopyran-2-one; 3-(α-ethylbenzyl)-4-hydroxycoumarin;* 3-(1-phenylpropyl)-4-hydroxycoumarin; Falithrom; Marcoumar; Marcumar; Liquamar. $C_{18}H_{16}O_3$; mol wt 280.32. C 77.12%, H 5.75%, O 17.12%. Prepn: Grüssner, Balthasar, U.S. pat. **2,723,276** (1955 to Hoffmann-La Roche); Junek, Ziegler, *Monatsh.* **87**, 218 (1956); Schroeder, Link, *J. Am. Chem. Soc.* **79**, 3291 (1957); U.S. pat. **2,872,-457** (1959 to Wisconsin Alumni Res. Found.); **Brit. pat. 805,748** (1958 to Geigy); L. R. Pohl, *J. Med. Chem.* **18**, 513 (1975). Resolution: Preis *et al.,* U.S. pat. **3,239,529** (1966 to Wisconsin Alumni Res. Found.). Conformation in soln: E. J. Valente *et al., J. Med. Chem.* **21**, 141, 231 (1978).

Crystals or prisms from dil methanol, mp 179-180°.
THERAP CAT: Anticoagulant.

7414. Phensuximide. *1-Methyl-3-phenyl-2,5-pyrrolidinedione; N-methyl-2-phenylsuccinimide; N-methyl-α-phenylsuccinimide;* Lifène; Milontin; Mirontin; Succitimal. $C_{11}H_{11}NO_2$; mol wt 189.21. C 69.83%, H 5.86%, N 7.40%, O 16.91%. Prepd by the action of methylamine on phenylsuccinic acid: Miller, Long, *J. Am. Chem. Soc.* **73**, 4895 (1951); U.S. pat. **2,643,258** (1953 to Parke, Davis). Toxicity study: Chen, Ensor, *J. Lab. Clin. Med.* **41**, 78 (1953).

Fine crystals from hot 95% ethanol, mp 71-73°. Slightly sol in water (about 4.2 mg/ml at 25°). Readily sol in methanol, ethanol. Aq solns are fairly stable at pH 2-8, but hydrolysis sets in under more alkaline conditions. LD_{50} orally in mice: 960 mg/kg (Chen, Ensor).

THERAP CAT: Anticonvulsant.

7415. Phentermine. *α,α-Dimethylbenzeneethanamine; α,α-dimethylphenethylamine;* phenyl-*tert*-butylamine; α-benzylisopropylamine. $C_{10}H_{15}N$; mol wt 149.24. C 80.48%, H 10.13%, N 9.39%. Prepn: Shelton, Van Campen, U.S. pat. **2,408,345** (1946 to Wm. S. Merrell); Abell *et al.,* U.S. pat. **2,590,079** (1952 to Wyeth).

Oily liquid. bp$_{750}$ 205°; bp$_{21}$ 100°.

Hydrochloride, C$_{10}$H$_{15}$N.HCl, *Adipex-P, Fastin, Wilpo.* Crystals, mp 198°.

Ion exchange resin complex, *Duromine, Ionamin, Linyl, Mirapront, Omnibex.*

Note: This is a controlled substance (stimulant) listed in the U.S. Code of Federal Regulations, Title 21 Part 1308.14 (1995).

THERAP CAT: Anorexic.

7416. Phentetiothalein Sodium. *4,5,6,7-Tetraiodophenolphthalein sodium;* 3,3-bis(4-hydroxyphenyl)-4,5,6,7-tetraiodo-1(3H)-isobenzofuranone disodium salt; 3,4,5,6-tetraiodophenolphthalein disodium salt; phenoltetraiodophthalein sodium; Iso-Iodeikon. C$_{20}$H$_8$I$_4$Na$_2$O$_4$; mol wt 865.88. C 27.74%, H 0.93%, I 58.62%, Na 5.31%, O 7.39%. Prepd by the condensation of phenol and tetraiodophthalic acid or its anhydride: Orndorff, Black, *Am. Chem. J.* **41**, 349 (1909); Zalkind, Belikova, *Zh. Prikl. Khim. (Leningrad)* **8**, 1210 (1935). Isomeric with iodophthalein sodium, *q.v.*

Bronze-purple, odorless, slightly hygroscopic granules. Dec on exposure becoming incompletely sol. Sol in water, alcohol. *Keep tightly closed.*

THERAP CAT: Diagnostic aid (radiopaque medium—cholecystography and hepatic function).

7417. Phentolamine. *3-[[(4,5-Dihydro-1H-imidazol-2-yl)methyl](4-methylphenyl)amino]phenol;* 2-[N-(m-hydroxyphenyl)-p-toluidinomethyl]imidazoline; 2-(m-hydroxy-N-p-tolylanilinomethyl)-2-imidazoline; 2-(N'-p-tolyl-N'-m-hydroxyphenylaminomethyl)-2-imidazoline; C-7337. C$_{17}$H$_{19}$N$_3$O; mol wt 281.36. C 72.57%, H 6.81%, N 14.93%, O 5.69%. α-Adrenergic blocker. Prepn: K. Miescher *et al.,* U.S. pat. **2,503,059** (1950 to Ciba); E. Urech *et al., Helv. Chim. Acta* **33**, 1386 (1950). Pharmacology and toxicity: R. Meier *et al., Proc. Soc. Exp. Biol. Med.* **71**, 70 (1949). HPLC determn in biological samples: B. D. Kerger *et al., Anal. Biochem.* **170**, 145 (1988). Diagnostic use: G. Spergel *et al., J. Am. Med. Assoc.* **211**, 266 (1970). Review of pharmacology and therapeutic applications: L. Gould, C. V. R. Reddy, *Am. Heart J.* **92**, 397-402 (1976).

Crystals, mp 174-175°.

Hydrochloride, C$_{17}$H$_{19}$N$_3$O.HCl, bitter crystals, mp 239-240°. One gram dissolves in 50 ml water, in 70 ml alcohol. Very slightly sol in chloroform. Practically insol in acetone, ethyl acetate. pH of 1% aq soln 4.5-5.5. Aq solns cannot be stored. LD$_{50}$ in rats (mg/kg): 75 i.v.; 275 s.c.; 1250 orally (Meier).

Methanesulfonate, C$_{17}$H$_{19}$N$_3$O.CH$_3$SO$_3$H, *phentolamine mesylate, Regitine, Rogitine.* Crystals, mp 177-181°. One gram dissolves in 50 ml water, 23 ml alcohol, 660 ml chlo-

roform. pH of 1% aq soln 4.5-5.5. Aq solns cannot be stored.

THERAP CAT: Antihypertensive; treatment of pheochromocytoma; diagnostic aid (pheochromocytoma).

7418. Phentydrone. *1,2,3,4-Tetrahydro-9-fluorenone;* THF. C$_{13}$H$_{12}$O; mol wt 184.24. C 84.75%, H 6.57%, O 8.68%. Prepn: Beyerman, Veer, U.S. pat. **2,692,898** (1954 to Organon); House *et al., J. Am. Chem. Soc.* **82**, 1457 (1960).

Light yellow needles or prisms from pentane, mp 81-82° (Beyerman); golden yellow prisms from petr ether, mp 41.5-42.5° (House). bp$_{0.05}$ 139-140°. uv max (alcoholic molar soln): 265, 232 nm (log ε 4.48, 3.62); (carbon tetrachloride): 238, 245, 317 nm (ε 39700, 46900, 960). LD$_{50}$ s.c. in mice: 1-1.3 g/kg.

7419. Phenylacetaldehyde. *Benzeneacetaldehyde;* α-toluic aldehyde; α-toluldehyde; Hyacinthin. C$_8$H$_8$O; mol wt 120.15. C 79.97%, H 6.71%, O 13.32%. C$_6$H$_5$CH$_2$CHO. Prepd by oxidizing phenylethyl alcohol with chromic acid. High-yield synthesis from styrene oxide or styrene glycol: G. Paparatto, G. Gregorio, *Tetrahedron Letters* **29**, 1471 (1988).

Oily, colorless liq which polymerizes and grows more viscous on standing. Odor reminiscent of lilac and hyacinth. (Has been crystallized, mp 33-34°.) d$_{25}^{25}$ 1.023-1.030; bp$_{760}$ 195°; bp$_{18}$ 88°; bp$_{10}$ 78°; n$_D^{20}$ 1.524-1.528. Slightly sol in water. Soluble in alcohol, ether. One part is sol in 2 parts of 80% alc forming a clear solution.

USE: In perfumery; intermediate in organic synthesis.

7420. α-Phenylacetamide. *Benzeneacetamide;* α-toluamide. C$_8$H$_9$NO; mol wt 135.17. C 71.09%, H 6.71%, N 10.36%, O 11.84%. The amide of phenylacetic acid, not to be confused with N-phenylacetamide which is acetanilide. Prepd by Willgerodt reaction from acetophenone or from styrene. Review and lab procedures: Carmack, Spielman in *Org. Reactions* **II** (New York, 1946) p 83. Has also been prepd by heating the ammonium salt of phenylacetic acid: Menschutkin, *Ber.* **31**, 1429 (1898). Lab procedure from benzyl cyanide: Wenner, *Org. Syn.* **coll. vol. IV**, 760 (1963).

Bimorphous plates, leaflets. mp 155°. Distils *in vacuo.* Freely sol in alc. Slightly sol in water, ether, benzene.

USE: In manuf of penicillin G.

7421. Phenyl Acetate. *Acetic acid phenyl ester;* acetylphenol. C$_8$H$_8$O$_2$; mol wt 136.15. C 70.57%, H 5.92%, O 23.50%. C$_6$H$_5$OOCCH$_3$. Prepd from phenol and acetyl chloride.

Colorless, mobile, highly refractive liquid; phenolic odor. d^{20} 1.073; bp 195-196°. n$_D^{20}$ 1.5030 (almost the same as glass). Practically insol in water. Miscible with alc, chloroform, ether; sol in glacial acetic acid. LD$_{50}$ orally in rats: 1.63 ml/kg, H. F. Smyth *et al., Am. Ind. Hyg. Assoc. J.* **30**, 470 (1969).

7422. Phenylacetic Acid. *Benzeneacetic acid;* α-toluic acid. C$_8$H$_8$O$_2$; mol wt 136.15. C 70.57%, H 5.92%, O 23.50%. C$_6$H$_5$CH$_2$COOH. Made by refluxing benzyl cyanide with dil H$_2$SO$_4$ or HCl: Adams, Thal, *Org. Syn.* **2**, 63 (1922). Absorption spectrum: Baly, Tryhorn, *J. Chem. Soc.* **107**, 1065 (1915).

Leaflets on distillation *in vacuo;* plates, tablets from petr ether; mp 76.5°. bp_{760} 265.5°; bp_{100} 198.2°; bp_{40} 173.6°; bp_5 127°; $bp_{1.0}$ 97°. d_4^{77} 1.091. pK (25°): 4.25. Slightly sol in cold, freely in hot water. The aq soln satd at 25° is 0.131*N*. Sol in alcohol, ether. Soly at 25° in chloroform (moles/l): 4.422; in carbon tetrachloride: 1.842; in acetylene tetrachloride: 4.513; in trichlorethylene: 3.299; in tetrachlorethylene: 1.558; in pentachloroethane: 3.252.

Methyl ester, $C_9H_{10}O_2$, liquid, bp 215°.

Ethyl ester, *see* Ethyl phenylacetate.

USE: Starting material in manuf synthetic perfumes, condensation products with aldehydes.

7423. Phenylacetone. *1-Phenyl-2-propanone;* benzyl methyl ketone. $C_9H_{10}O$; mol wt 134.18. C 80.56%, H 7.51%, O 11.92%. $C_6H_5CH_2COCH_3$. Prepn from phenylacetic and acetic acids: R. H. Pickard, J. Kenyon, *J. Chem. Soc.* **105**, 1124 (1914); R. M. Herbst, R. H. Manske, *Org. Syn. coll. vol.* II, 389 (1943); from α-phenylacetoacetonitrile: P. L. Julian, J. J. Oliver, *ibid.* 391; from α-methyl-α-phenylethylene oxide: S. Danilow, E. Venus-Danilowa, *Ber.* **60**, 1050 (1927); from diethyl malonate: H. G. Walker, C. R. Hauser, *J. Am. Chem. Soc.* **68**, 1386 (1946). Conformational calculations: M. Hirota *et al., Tetrahedron* **39**, 3091 (1983). Use as prochiral ketone in enantioselective hydrosilylation: H. Brenner *et al., Ber.* **117**, 1330 (1984).

Oil, mp −16° to −15°. bp_{760} 214°; bp_{14} 100-101°. d_4^{20} 1.0157. n_D 1.5174. uv max (ethanol): 258, 283 nm (ε 255, 150).

Note: This is a controlled substance listed in the U.S. Code of Federal Regulations, Title 21 Part 1308.12 (1995).

USE: In organic synthesis; production of benzyl radicals by photolysis.

7424. Phenyl Acetylsalicylate. *2-(Acetyloxy)benzoic acid phenyl ester;* acetylphenylsalicylate; Acetylsalol; Spiroform; Vesipyrin. $C_{15}H_{12}O_4$; mol wt 256.26. C 70.31%, H 4.72%, O 24.97%. $CH_3COOC_6H_4COOC_6H_5$.

White, crystalline powder. mp 97°; bp_{11} 198°. Insol in water. Sol in alcohol, ether.

THERAP CAT: Analgesic, antipyretic, anti-inflammatory.

7425. Phenylalanine. *L-Phenylalanine;* Phe; F; β-phenylalanine; α-aminohydrocinnamic acid; (*S*)-2-amino-3-phenylpropanoic acid; α-amino-β-phenylpropionic acid. C_9H_{11}-NO_2; mol wt 165.19. C 65.44%, H 6.71%, N 8.48%, O 19.37%. Essential amino acid for human development. Originally isolated from the sprouts of lupine: E. Schulze, J. Barbieri, *Ber.* **12**, 1924 (1879). Early chemistry and biochemistry: *Amino Acids and Proteins,* D. M. Greenberg, Ed. (Charles C. Thomas, Springfield, IL, 1951) 950 pp. *passim;* J. P. Greenstein, M. Winitz, *Chemistry of the Amino Acids* **vols 1-3** (John Wiley and Sons, Inc., New York, 1961) pp. 2156-2177, *passim.* Synthesis from L-tyrosine: V. Viswanatha, V. J. Hruby, *J. Org. Chem.* **45**, 2010 (1980). Colorimetric determn in blood: R. S. Campbell *et al., Ann. Clin. Biochem.* **31**, 140 (1994). Effects on food intake: G. H. Anderson, L. A. Leiter, *Appetite* **11**, Suppl. 48 (1988); P. J. Rogers, J. E. Blundell, *Physiol. Behav.* **56**, 247 (1994). Clinical trial in vitiligo: C. Antoniou *et al., Int. J. Dermatol.* **28**, 545 (1989); A. H. Siddiqui *et al., Dermatology* **188**, 215 (1994). Review of metabolism: H. N. Munro, *J. Toxicol. Environ. Health* **2**, 189-206 (1976). Review of role in phenylketonuria: S. P. Bessman, *Nutr. Rev.* **37**, 209-220 (1979); F. Güttler, *Acta Paediatr. Scand.* **73**, 705-716 (1984). Review of microbial production: T. K. Maiti, S. P. Chatterjee, *Hind. Antibiot. Bull.* **32**, 3-26 (1990).

Monoclinic plates, leaflets from warm concd aq solns. Hydrated needles from dil solns. Dec 283°. Sublimes *in vacuo.* $[\alpha]_D^{20}$ −35.1° (c = 1.94). Soly in water (g/l): 19.8 at 0°; 29.6 at 25°; 44.3 at 50°; 66.2 at 75°; 99.0 at 100°. Very slightly sol in methanol, ethanol.

D-Form, *(R)-phenylalanine, Sabiden.* Occurs naturally in microbial products such as tyrocidine, *q.v.:* A. H. Gordon *et al., Biochem. J.* **37**, 313 (1943). May be used as a source for L-form in man. Biotransformation: S. Tokuhisa *et al., Chem. Pharm. Bull.* **29**, 514 (1981). Evaluation of D-form in multiple sclerosis: A. Winter, *Neurol. Orthopaed. J. Med. Surg.* **5**, 39 (1984). Leaflets from water, dec 285°. $[\alpha]_D^{20}$ +35.0° (c = 2.04); $[\alpha]_D^{20}$ +7.1° (c = 3.8 in 18% HCl). One gram dissolves in 35.5 ml water at 16°. Sparingly sol in methanol.

DL-Form, monoclinic leaflets or prisms from water or alcohol, sweetish taste. Dec 271-273°. Sublimes *in vacuo.* pK_1 2.58; pK_2 9.24. Soly in water (g/l): 9.97 at 0°; 14.11 at 25°; 21.87 at 50°; 37.08 at 75°; 68.9 at 100°.

USE: Component of the artificial sweetner aspartame, *q.v.;* nutrient.

7426. Phenyl Aminosalicylate. *4-Amino-2-hydroxybenzoic acid phenyl ester; p-aminosalicylic acid phenyl ester;* phenyl *p*-aminosalicylic acid; *p*-aminosalol; fenamisal; Phenyl PAS; Pheny-PAS-Tebamin; Tebamin; Tebanyl. $C_{13}H_{11}$-NO_3; mol wt 229.24. C 68.11%, H 4.84%, N 6.11%, O 20.94%. Description of properties: Meyer, *Antibiot. Ann.* **1957-1958**, 614. Prepd by Raney nickel reduction of the corresp nitro ester in ethyl acetate: Friere, U.S. pat. **2,604,-488** (1952 to Rhône-Poulenc).

Crystals from isopropanol, mp 153°. Soly in water: 0.7 mg/100 ml; in serum: 12 mg/100 ml. One gram of phenyl PAS is equivalent to 0.67 g PAS.

THERAP CAT: Antibacterial (tuberculostatic).

7427. *N*-Phenylanthranilic Acid. *2-(Phenylamino)benzoic acid;* 2-anilinobenzoic acid; diphenylamine-2-carboxylic acid. $C_{13}H_{11}NO_2$; mol wt 213.24. C 73.23%, H 5.20%, N 6.57%, O 15.01%. Prepd from *o*-chlorobenzoic acid and aniline: Ullmann, Dieterle, *Ann.* **365**, 322 (1907); Ullmann, *Ber.* **36**, 2382 (1907); from 2-iodobenzoic acid and phenylhydroxylamine: Wieland, Roseeu, *Ber.* **48**, 1120 (1915). Use: Adamovich, Zagorulko, *Zavodskaya Lab.* **9**, 465 (1940), *C.A.* **37**, 1346 (1943).

Leaflets from alcohol, dec 183-184°. Sol in hot alcohol. Very slightly sol in hot water, hot benzene, ether.

USE: Detection of vanadium in steel.

7428. 2-Phenyl-1*H*-benzimidazole. *N,N'*-Benzenyl-*o*-phenylenediamine; phenzidole; Gainex. $C_{13}H_{10}N_2$; mol wt 194.24. C 80.39%, H 5.19%, N 14.42%. Prepn from *N*-benzoyl-*o*-phenylenediamine: v. Auwers, Frese, *Ber.* **59**, 548 (1926); from *o*-phenylenediamine and benzonitrile: Hölljes, Wagner, *J. Org. Chem.* **9**, 31 (1944). Anthelmintic in sheep: Forsyth, *Australian Vet. J.* **38**, 398 (1962).

Needles from benzene, plates from water, mp 291°. Sol in methanol, abs ethanol. Sparingly sol in water, benzene, chloroform.

Hydrochloride, $C_{13}H_{10}N_2 \cdot HCl$, needles from alcohol, dec 328°. Freely sol in water.

THERAP CAT (VET): Has been used as an anthelmintic.

7429. Phenyl Benzoate. *Benzoic acid phenyl ester.* $C_{13}H_{10}O_2$; mol wt 198.22. C 78.77%, H 5.08%, O 16.14%. $C_6H_5COOC_6H_5$.

Monoclinic prisms; geranium odor. d 1.235; mp 70°; bp 314°. Insoluble in water. Freely sol in hot alcohol; slightly sol in cold alcohol or ether.

7430. Phenyl Biguanide. *N-Phenylimidodicarbonimidic diamide;* phenyl diguanide; 1-phenylbiguanide; *N*-phenyl-*N'*-guanylguanidine. $C_8H_{11}N_5$; mol wt 177.21. C 54.22%, H 6.26%, N 39.52%. Prepd by reacting aniline hydrochloride and dicyandiamide in water: Cohn, *J. Prakt. Chem.* [2] **84**, 396; from aniline, dicyandiamide in pyridine contg HCl: Jacobs, Jolles, **Brit. pat. 587,907** (1947 to ICI).

Crystals from water or toluene. Sharp, somewhat bitter taste. mp 144-146°. pK_1 10.76, pK_2 2.13. Freely sol in water and in alcohol.

Hydrochloride, $C_8H_{11}N_5 \cdot HCl$, prisms, mp 237°. Sol in about 115 parts of water at 32°.

Nitrate, $C_8H_{11}N_5 \cdot HNO_3$, crystals, mp 208-209°.

7431. Phenylbutazone. *4-Butyl-1,2-diphenyl-3,5-pyrazolidinedione;* 4-butyl-1,2-diphenyl-3,5-dioxopyrazolidine; 3,5-dioxo-1,2-diphenyl-4-*n*-butylpyrazolidine; flexazone; diphebuzol; fenibutazona; G-13871; R-3-ZON; Ambene; Artrizin; Azolid; Bizolin; Butacote; Butadion; Butapirazol; Butadiona; Butatron; Butoz; Butazolidin; Buzon; Ecobutazone; Equipalazone; Exrheudon N; Fenibutol; Intrabutazone; Intrazone; Mepha-Butazon; Phenyzene; Robizone-V; Tevcodyne; Uzone. $C_{19}H_{20}N_2O_2$; mol wt 308.38. C 74.00%, H 6.54%, N 9.08%, O 10.38%. Prepn: Stenzl, **U.S. pat. 2,562,830** (1951 to Geigy); *cf.* **Brit. pat. 812,449** (1959 to Geigy). Review of synthesis: *Ullmann's Encyklopädie der Technischen Chemie,* **13**, 298 (1962). Physical properties and pharmacology: v. Rechenberg, *Phenylbutazone* (Edward Arnold, London, 1962) 197 pp. Acute toxicity: T. B. Gaines, R. E. Linder, *Fundam. Appl. Toxicol.* **7**, 299 (1986). Soly data: Pulver *et al., Schweiz. Med. Wochenschr.* **86**, 1080 (1956). Comprehensive description: S. L. Ali, *Anal. Profiles Drug Subs.* **11**, 483-521 (1982). Review of hematological effects: G. A. Faich, *Pharmacotherapy* **7**, 25 (1987).

Crystals from ethanol, mp 105°. Soly in water at 22.5°: 0.7 mg/ml (also reported as 2.2 mg/ml), pK 4.5 (from uv in water), pK 4.89 (titration in 50% ethanol), pK 5.25 (titration in 80% 2-methoxyethanol). uv max (acid methanol): 239.5 nm (log ε 4.19).

Sodium salt, $C_{19}H_{19}N_2NaO_2$, *GP-26872, Elmedal.* Crystals, freely sol in water. pH of aq solns ~8.2.

Calcium salt, $C_{38}H_{36}CaN_4O_4$, *Ticinil Calcico.* LD_{50} in adult male, female rats (mg/kg): 1311, 647 orally (Gaines, Linder).

Piperazine salt, $C_{19}H_{20}N_2O_2 \cdot C_4H_{10}N_2$, *pyrazinobutazone, pyrasanone, Carudol.* mp 140-141° (solidifies and remelts at ~180°).

2-Amino-2-thiazoline salt, *see* Thiazolinobutazone.

THERAP CAT: Anti-inflammatory.

THERAP CAT (VET): Analgesic; anti-inflammatory.

7432. α-Phenylbutyramide. *α-Ethylbenzeneacetamide;* 2-phenylbutanamide; α-ethyl-α-phenylacetamide; α-phenyl-α-ethylacetamide; TH-4128; Hyposterol; Natol; Phenetamid; Redusterol. $C_{10}H_{13}NO$; mol wt 163.22. C 73.59%, H 8.03%, N 8.58%, O 9.80%. Prepn from phenylethylacetylurea: **Ger. pat. 249,241** (1912 to Bayer); *Chem. Zentr.* **1912,** II, 396; *Frdl.* **10,** 1164.

Practically tasteless crystals, mp 86°; also given as mp 83-84°. Very sparingly sol in cold water. Soly in boiling water: about 1:50; in ethanol at 25°: 1:10. Sol in many other organic solvents.

THERAP CAT: In hypercholesterolemia.

7433. Phenyl Carbonate. *Carbonic acid diphenyl ester;* diphenyl carbonate. $C_{13}H_{10}O_3$; mol wt 214.22. C 72.89%, H 4.71%, O 22.41%. $C_6H_5OCOOC_6H_5$. Prepn: Eckenroth, *Ber.* **27,** 3410 (1894); Bischoff, *ibid.* **35,** 3434 (1902); Gomberg, Snow, *J. Am. Chem. Soc.* **47,** 198 (1925).

Lustrous needles, mp 80-81°. bp 302-306°; bp_{15} 168°. Practically insol in water. Sol in hot alcohol, benzene, ether, glacial acetic acid.

USE: In the molten state as solvent for nitrocellulose.

7434. 2-Phenyl-6-chlorophenol. *3-Chloro[1,1'-biphenyl]-2-ol;* 6-chlororthoxenol. $C_{12}H_9ClO$; mol wt 204.66. C 70.43%, H 4.43%, Cl 17.32%, O 7.82%.

Pale yellow, viscous liquid. Slight, characteristic odor. d_4^{25} about 1.24; mp 6°; bp 317-318° (dec.) n_D^{30} 1.6237. Insoluble in water. Sol in fixed alkali hydroxide solns and in most organic solvents.

USE: Germicide; fungicide.

7435. 4-Phenyl-2-chlorophenol. *3-Chloro-(1,1'-biphenyl)-4-ol.*

Pale yellow crystals. mp ~77°. bp_7 ~160-162°. Soly and use as for 2-phenyl-6-chlorophenol, *q.v.*

7436. α-Phenylcinnamic Acid. *α-(Phenylmethylene)-benzeneacetic acid; 2,3-diphenylacrylic acid;* 2,3-diphenylpropenoic acid; stilbene-α-carboxylic acid. $C_{15}H_{12}O_2$; mol wt 224.26. C 80.34%, H 5.39%, O 14.27%. Prepn of cis-form: Buckles, Hausman, *J. Am. Chem. Soc.* **70,** 415 (1948); Buckles, *J. Chem. Ed.* **27,** 210 (1950); Buckles, Bremer, *Org. Syn. coll. vol. IV,* 777 (1963); B. H. Patwardhan, G. Bagavant, *Indian J. Chem.* **10,** 59 (1972). Prepn of cis- and trans-forms: L. F. Fieser, *Experiments in Organic Chemistry* (Heath, Boston, 3rd ed., 1955) p 182. Isomerization studies: S. V. Kessar *et al., Indian J. Chem.* **20B,** 4 (1981).

cis-form

cis-Form, silky needles from ether + petr ether, mp 174. pKa 6.1 in 60% ethanol. uv max (ethanol): 223, 280 nm (ε 32100, 19500). Sol in hot water, methanol, ethanol, isopropanol, ether, benzene.

trans-Form, stout prisms from ether + petr ether, mp 138-139°. Less stable than the *cis*-form. pKa 4.8 in 60% ethanol. uv max (ethanol): 222, 289 nm (ε 14500, 22500). More sol than the *cis*-form.

7437. m-Phenylenediamine. *1,3-Benzenediamine;* m-diaminobenzene. $C_6H_8N_2$; mol wt 108.14. C 66.64%, H 7.46%, N 25.90%. Prepd by the reduction of *m*-dinitrobenzene: Kuhn, U.S. pat. **2,768,209** (1956 to Ringwood); Faust, *J. Prakt. Chem.* **6**, 14 (1958); Neilson *et al., J. Chem. Soc.* **1962**, 371; Tallee, Peltier, *Compt. Rend.* **259**, 400 (1964). Toxicity study: C. Burnett *et al., J. Toxicol. Environ. Health* **2**, 657 (1977).

White crystals becoming red on exposure to air. d 1.139; mp 62-63°; bp 284-287°; dipole moment 1.79. Fire point: 175°. Sol in water, methanol, ethanol, chloroform, acetone, dimethylformamide, methyl ethyl ketone, dioxane. Slightly sol in ether, carbon tetrachloride, dibutyl phthalate, isopropanol. Very slightly sol in benzene, toluene, xylene, butanol. *Keep well closed and protected from light.* LD_{50} in rats (mg/kg): 650 orally; 283 i.p. (Burnett).

Hydrochloride, white or slightly red, crystalline powder; becomes darker on exposure to air. Freely sol in water; sol in alcohol.

USE: Manuf dyes; rubber curing agents, ion exchange resins, decolorizing resins, formaldehyde condensates, resinous polyamides, block polymers, textile fibers, urethanes, petroleum additives, rubber chemicals, corrosion inhibitors; in photography; as reagent for gold and bromine. The hydrochloride chiefly as a reagent for nitrite.

7438. o-Phenylenediamine. *1,2-Benzenediamine;* o-diaminobenzene. $C_6H_8N_2$; mol wt 108.14. C 66.64%, H 7.46%, N 25.90%. Made by reducing *o*-nitroaniline with Zn and NaOH.

Brownish-yellow crystals. mp 103-104°. bp 256-258°. Slightly sol in water; freely sol in alcohol, chloroform, ether. LD_{50} in rats (mg/kg): 1070 orally; 516 i.p., C. Burnett *et al., J. Toxicol. Environ. Health* **2**, 657 (1977).
USE: Manufacture of dyes.

7439. p-Phenylenediamine. *1,4-Benzenediamine;* p-diaminobenzene; p-aminoaniline; orsin; C.I. 76076; Ursol D. $C_6H_8N_2$; mol wt 108.14. C 66.64%, H 7.46%, N 25.90%. Prepn: A. Rinne, T. Zincke, *Ber.* **7**, 869 (1874); **Ger.** pat. **202,170** (1907 to BASF), *C.A.* **3**, 382 (1909); A. J. Quick, *J. Am. Chem. Soc.* **42**, 1033 (1920); J. F. Norris, E. O. Cummings, *Ind. & Eng. Chem.* **17**, 305 (1925). *See also: Beilstein* **XIII**, 61 (1930). Crystal structure: A. Domenicano *et al., Acta Crystallogr.* **B33**, 1664 (1977). Mutagenicity studies: B. N. Ames *et al., Proc. Nat. Acad. Sci. USA* **72**, 2423 (1975); W. G. H. Blijleven, *Mutat. Res.* **48**, 181 (1977). Toxicity study: C. Burnett *et al., J. Toxicol. Environ. Health* **2**, 657 (1977).

White to slightly red crystals; darkens on exposure to air. mp 145-147°. bp 267°. Sol in 100 parts cold water; sol in alcohol, chloroform, ether. A black color is developed with 3% H_2O_2; brown with 5% $FeCl_3$ soln. *Keep well closed and protected from light.* LD_{50} in rats (mg/kg): 80 orally, 37 i.p. (Burnett).

Hydrochloride, white to slightly reddish crystals. Freely sol in water, slightly in alcohol, ether.

Caution: Potential symptoms of overexposure are irritation of pharynx and larynx; bronchial asthma; sensitization dermatitis. *See NIOSH Pocket Guide to Chemical Hazards* (DHHS/NIOSH 90-117, 1990) p 178.

USE: Dyeing furs; also in photochemical measurements, accelerating vulcanization; manuf azo dyes, etc. The hydrochloride as reagent for blood, H_2S, amyl alcohol; in testing of milk.

7440. Phenylephrine Hydrochloride. *(R)-3-Hydroxy-α-[(methylamino)methyl]benzenemethanol hydrochloride; 1-m-hydroxy-α-[(methylamino)methyl]benzyl alcohol hydrochloride; l-1-(m-hydroxyphenyl)-2-methylaminoethanol hydrochloride; l-α-hydroxy-β-methylamino-3-hydroxy-1-ethylbenzene hydrochloride; m-methylaminoethanolphenol hydrochloride; metaoxedrin; Adrianol; Ak-Dilate; Ak-Nefrin; Isophrin; m-Sympatol; Mezaton; Neophryn; Neo-Synephrine; Pyracort D; Prefrin; Mydfrin.* $C_9H_{14}ClNO_2$; mol wt 203.67. C 53.08%, H 6.93%, Cl 17.41%, N 6.88%, O 15.71%. α-Adrenergic agonist. Prepn from *m*-hydroxy-ω-chloroacetophenone and methylamine: U.S. pats. **1,932,347; 1,954,389** (1934). Synthesis from *m*-benzyloxybenzaldehyde: Bergmann, Sulzbacher, *J. Org. Chem.* **16**, 84 (1951). Toxicity data: M. R. Warren, H. W. Werner, *J. Pharmacol. Exp. Ther.* **86**, 284 (1946). Comprehensive description: C. A. Gaglia in *Analytical Profiles of Drug Substances* **vol. 3**, K. Florey, Ed. (Academic Press, New York, 1974) pp 483-512.

Bitter crystals, mp 140-145°. (free base mp 169-172°: *U.S.P. XVI*). $[\alpha]_D^{25}$ −46.2° to −47.2°. Freely sol in water or alcohol. The aq soln is neutral to litmus paper. LD_{50} in rats (mg/kg): 17 ±1.1 i.p.; 33 ±2.0 s.c. (Warren, Werner).
THERAP CAT: Mydriatic; decongestant.

7441. Phenylethanolamine. *α-(Aminomethyl)benzenemethanol; α-(aminomethyl)benzyl alcohol; β-hydroxyphenethylamine.* $C_8H_{11}NO$; mol wt 137.18. C 70.04%, H 8.08%, N 10.21%, O 11.66%. Prepn: Dornow, Theidel, *Ber.* **88**, 1267 (1955).

Pale yellow crystals, mp 56-57°. bp_{17} 157-160°. Freely sol in water. Gives an alkaline reaction in water and forms salts with acids under mild conditions.

Sulfate, $(C_8H_{11}NO)_2 \cdot H_2SO_4$, *dl-α-phenyl-β-aminoethanol sulfate, Apophedrin.* Prepn: Roth, *Arch. Pharm.* **292**, 76 (1959). Crystals, mp 275-276° (Kofler). Soluble in water.

USE: Free base as a stopping agent during polymerization of styrene-butadiene rubber; in the hardening of waxes. Intermediate in the manuf of pressor amines.
THERAP CAT: Sulfate as topical vasoconstrictor.

7442. Phenyl Ether. *1,1'-Oxybisbenzene;* diphenyl ether; diphenyl oxide. $C_{12}H_{10}O$; mol wt 170.21. C 84.68%, H 5.92%, O 9.40%. $C_6H_5OC_6H_5$. Prepd by heating sodium phenolate with chlorobenzene.

Liquid; characteristic odor. d^{20} 1.075 (liq); mp 28°, and remains liq at lower temp; bp 259°. Flash pt 115°. Insol in water; sol in alcohol, benzene, ether, glacial acetic acid.

Consult the Name Index before using this section.

Caution: Potential symptoms of overexposure are nausea; irritation of eyes, nose and skin. *See NIOSH Pocket Guide to Chemical Hazards* (DHHS/NIOSH 90-117, 1990) p 178.

USE: As heat transfer medium; in perfuming soaps; in organic syntheses.

7443. 5-(α-Phenylethyl)semioxamazide. *Oxo-[(1-phenylethyl)amino]acetic acid hydrazide; 5-(α-methylbenzyl)semioxamazide.* $C_{10}H_{13}N_3O_2$; mol wt 207.23. C 57.96%, H 6.32%, N 20.28%, O 15.44%. Prepd from α-phenylethylamine, ethyl oxalate, and hydrazine: Leonard, Boyer, *J. Org. Chem.* **15,** 42 (1950).

dl-Form, fine needles from ethanol, mp 157°.

d-Form, crystals, mp 167-168°. $[\alpha]_D^{25}$ +102.0° (c = 1.04 in chloroform).

l-Form, crystals, mp 167-168°. $[\alpha]_D^{25}$ − 102.5° (c = 0.625 in chloroform).

USE: The racemic form as reagent for the characterization of aldehydes and ketones. The optically active forms as resolving agents for carbonyl compds possessing asymmetric carbon atoms.

7444. Phenylglyceryl Ether. *3-Phenoxy-1,2-propanediol; Antodyne.* $C_9H_{12}O_3$; mol wt 168.19. C 64.27%, H 7.19%, O 28.54%. $C_6H_5OCH_2CH(OH)CH_2OH$. Prepd from phenol and 3-chloro-1,2-propylene carbonate: Smith, U.S. pat. **2,967,892** (1961 to Dow). Toxicity study: Hine *et al., J. Pharmacol. Exp. Ther.* **97,** 414 (1949).

Crystals, mp 50-52°, $bp_{0.6}$ 129-142°. Sol in water, alcohol. LD_{50} i.p. in mice: 1280 mg/kg (Hine).

7445. N-Phenylglycine. *Anilinoacetic acid.* $C_8H_9NO_2$; mol wt 151.17. C 63.56%, H 6.00%, N 9.27%, O 21.17%. $C_6H_5NHCH_2COOH$. Prepd from aniline and chloroacetic acid: Curtius, *J. Prakt. Chem.* **38,** 436 (1888); Thorpe, Wood, *J. Chem. Soc.* **103,** 1606 (1913).

Crystals, mp 127-128°. pK (25°): 5.42. Moderately sol in water. Less sol in alc. Sparingly sol in ether. Forms water-sol salts with alkali hydroxides.

Ethyl ester, $C_{10}H_{13}NO_2$, leaflets, mp 58°. Sol in alc, ether.

USE: In the Heumann synthesis of indigo.

7446. α-Phenylglycine. *α-Aminobenzeneacetic acid; α-aminophenylacetic acid; α-amino-α-toluic acid; C-phenylglycine.* $C_8H_9NO_2$; mol wt 151.17. C 63.56%, H 6.00%, N 9.27%, O 21.17%. $C_6H_5CH(NH_2)COOH$. Prepd by hydrolysis of α-aminophenylacetonitrile with dil hydrochloric acid: Marvel, Noyes, *J. Am. Chem. Soc.* **42,** 2264 (1920); R. E. Steiger, *Org. Syn. coll. vol.* III, 84 (1955).

dl-Form, lustrous platelets. Sublimes without melting at ~255°. Slightly sol in the usual organic solvents. Sol in alkalies.

Methyl ester, $C_9H_{11}NO_2$, needles from petr ether, mp 32°. Sol in alcohol, ether, benzene.

Ethyl ester, $C_{10}H_{13}NO_2$, liquid, bp 257°; bp_{16} 149°; bp_5 114-115°; n_D^{25} 1.500.

l-Form, needles from dil alc, mp 305-310°. $[\alpha]_D^{20}$ − 157.8° (dil HCl).

d-Form, crystals. $[\alpha]_D^{20}$ +156° (dil HCl).

7447. Phenylhydrazine. *Hydrazinobenzene.* $C_6H_8N_2$; mol wt 108.14. C 66.64%, H 7.46%, N 25.90%. $C_6H_5NH-NH_2$. Prepd by diazotizing aniline with $NaNO_2$ and HCl, then treating the soln with Na_2SO_3 followed by NaOH: Fischer, *Anleitung zur Darstellung organischer Präparate* (Braunschweig, 10th ed., 1922) p 23; Coleman, *Org. Syn.* **2,** 71 (1922), *coll. vol.* I (2nd ed., 1941) p 442. Mechanism of oxidative hemolysis of erythrocytes: B. Goldberg, A. Stern, *Mol. Pharmacol.* **13,** 832 (1977); B. Vilsen, H. Nielsen, *Biochem. Pharmacol.* **33,** 2739 (1984). Review of interaction with hemoglobin: M. D. Shetlar, H. A. O. Hill, *Environ. Health Perspec.* **64,** 265-281 (1985).

Monoclinic prisms or oil. Turns yellow to dark red on exposure to air and light. Faint aromatic odor. d_4^{20} 1.0978;

mp 19.5°; bp_{760} 243.5° (dec); bp_{100} 173.5°; bp_{40} 148.2°; bp_{20} 131.5°; bp_{10} 115.8°; bp_5 101.6°; $bp_{1.0}$ 71.8°; $n_D^{20.3}$ 1.60813. Weak base, pK (15°): 8.79. Miscible with alcohol, ether, chloroform, benzene. Sparingly sol in water, petr ether; sol in dil acids. Forms a hemihydrate, mp 24°. *Keep well closed and protected from light.*

Hydrochloride, $C_6H_8N_2$.HCl. Prepn: Fischer, *Ann.* **190,** 83 (1878); Brunner, Eiermann, *Ber.* **31,** 1406 (1898); Rüetschi, Trümpler, *Helv. Chim. Acta* **36,** 1649 (1953); Hupfer, **Ger. pat. 1,143,825** (1963 to Hoechst). Leaflets from alc, mp 243-246° (slight browning). Sublimes. Freely sol in water; sol in alcohol. Practically insol in ether.

Caution: Potential symptoms of overexposure are skin sensitization, hemolytic anemia, dyspnea and cyanosis; jaundice; kidney damage; vascular thrombosis. Potential occupational carcinogen. *See NIOSH Pocket Guide to Chemical Hazards* (DHHS/NIOSH 90-117, 1990) p 180. *See also* Patty's *Industrial Hygiene and Toxicology* vol. 2A, G. D. Clayton, R. E. Clayton, Eds. (Wiley-Interscience, New York, 3rd ed., 1981) pp 2792-2795, 2804-2805.

USE: Manuf dyes, antipyrine, nitron (a stabilizer for explosives); reagent for sugars, aldehydes, ketones.

THERAP CAT: Hemolytic.

7448. Phenylhydroxylamine. *N-Hydroxybenzenamine; β-phenylhydroxylamine; N-phenylhydroxylamine.* C_6H_7-NO; mol wt 109.13. C 66.04%, H 6.47%, N 12.84%, O 14.66%. C_6H_5NHOH. Prepd by zinc reduction of nitrobenzene in ammonium chloride soln: Kamm, *Org. Syn.* **4,** 57 (1925).

Needles from satd NaCl soln. mp 82°. *Deteriorates on storage and should be used promptly.* Sol in 50 parts cold, in 10 parts hot water. Freely sol in alcohol, ether, carbon disulfide, chloroform, hot benzene, dil mineral acids, acetic acid. Slightly sol in petr ether. The oxalate is more stable.

USE: Manufacture of cupferron.

7449. Phenyl Isocyanate. *Isocyanatobenzene;* carbanil; phenylcarbimide. C_7H_5NO; mol wt 119.12. C 70.58%, H 4.23%, N 11.76%, O 13.43%. $C_6H_5N=C=O$. Prepd by passing carbonyl chloride into a hot soln of aniline in toluene, saturated with HCl: Hardy, *J. Chem. Soc.* **1934,** 2011.

Liquid, with acrid odor. $d_4^{11.6}$ 1.101; d_4^{15} 1.092; $d_4^{19.6}$ 1.0956; $d_4^{25.9}$ 1.08870. bp 158-168°; bp_{18-20} 58.2-59.5°; bp_{13} 55°. $n_D^{19.6}$ 1.53684; $n_D^{25.9}$ 1.53412.

Caution: Irritating to eyes.

7450. Phenyl Isothiocyanate. *Isothiocyanatobenzene;* isothiocyanic acid phenyl ester; phenyl mustard oil; thiocarbanil. C_7H_5NS; mol wt 135.19. C 62.19%, H 3.73%, N 10.36%, S 23.72%. $C_6H_5N=C=S$. Prepd from ammonium phenyldithiocarbamate by the action of lead nitrate: Dains *et al., Org. Syn.* **coll. vol.** I (2nd ed., 1941) p 447.

Liquid. mp −21°. d_4^{15} 1.1288; d_4^{35} 1.1202; d_4^{50} 1.1061. bp_{760} 221°; bp_{33} 117.1°; bp_{12} 95°. Distills with water without dec. $n_D^{23.4}$ 1.64918. Insol in water. Sol in alcohol, ether.

USE: Derivatizing agent for primary, secondary amines. In sequencing of peptides by Edman degradation. In amino acid analyses by HPLC (Pico-Tag).

7451. Phenylmagnesium Chloride. *Chlorophenylmagnesium.* C_6H_5ClMg; mol wt 136.86. C 52.66%, H 3.68%, Cl 25.90%, Mg 17.76%. C_6H_5MgCl. One of the Grignard reagents. Prepd by reacting chlorobenzene with magnesium at reflux temperatures in the presence of catalytic amounts of an organic nitrate: Ramsden, U.S. pat. **2,816,937** (1957 to M. & T. Corp.).

Sol in ether (ethyl ether, other ethers may be used as solvents). A solution which is about 3 molar has an approx strength of 48%, and a d_4^{20} of ~1.15. Reacts with water, steam or acids to produce toxic and flammable vapors with evolution of heat.

USE: In organic synthesis, especially in the production of hydrocarbons, alcohols, ketones, organic acids, amines, silicones, boranes.

7452. N-Phenylmaleimide. *1-Phenyl-1H-pyrrole-2,5-dione.* $C_{10}H_7NO_2$; mol wt 173.17. C 69.36%, H 4.07%, N 8.09%, O 18.48%. Prepn from maleanilic acid: Cava *et al., Org. Syn.* **41,** 93 (1961).

Canary-yellow needles from cyclohexane, mp 89-89.8°.
USE: Dienophile in the Diels-Alder reaction. Usually gives cryst adducts.

7453. Phenylmercuric Acetate. *(Acetato)phenylmercury;* acetoxyphenylmercury; phenylmercury acetate; PMA; PMAC; PMAS; Ceresan Slaked Lime; Gallotox; Liquiphene; Mersolite; Nylmerate; Phix; Riogen; Scutl; Tag Fungicide; Tag HL-331. $C_8H_8HgO_2$; mol wt 336.74. C 28.53%, H 2.39%, Hg 59.57%, O 9.50%. $C_6H_5HgOOCCH_3$. Made by heating benzene with mercuric acetate: Maynard, *J. Am. Chem. Soc.* **46,** 1510 (1925); **Brit.** pat. **325,846** (1928 to I. G. Farbenind.); Renschler, **U.S.** pat. **2,050,018** (1936 to Hamilton Labs.); Grave *et al., J. Am. Pharm. Assoc.* **25,** 752 (1936). *Review:* F. C. Whitmore, *Organic Compounds of Mercury* (Chemical Catalog Co., New York, 1921) pp 175-176.
Small, lustrous prisms from ethanol, mp 149°. Sol in ~600 parts water; sol in alcohol, benzene, acetone. LD_{50} orally in rats: 22 mg/kg, *RTECS* Vol. II, R. J. Lewis, R. L. Tatken, Eds. (1980) p 46.
An ingredient of *Agrosan GN*: Gates, *Ann. Appl. Biol.* **47,** 502 (1959); and of *Ceresan Dry:* Bayer *Pflanzenschutz Compendium* vol. 1 (Bayer, Leverkusen, 1962) pp 24-27.
USE: Herbicide; fungicide.

7454. Phenylmercuric Chloride. *Chlorophenylmercury;* phenylmercury chloride; Stopspot. C_6H_5ClHg; mol wt 313.15. C 23.01%, H 1.61%, Cl 11.32%, Hg 64.06%. C_6H_5-HgCl. Antibacterial activity: H. L. Friedman, *Ann. N.Y. Acad. Sci.* **65,** 461 (1957). Use as fungicide: U. Prota, *Phytopath. Mediter.* **8,** 87 (1970), *C.A.* **72,** 42138t (1970); M. Makes, Z. Lokaj, *Agrochemia* **16,** 181 (1976), *C.A.* **85,** 187602t (1976). Toxicity: M. Umeda *et al., Japan. J. Exp. Med.* **39,** 47 (1969).
White satiny leaflets, mp 250-252°. Sol in ~20,000 parts cold water, in benzene, ether, pyridine; slightly sol in hot alcohol. LD_{50} s.c. in rats: 30 mg/kg (Umeda).
USE: Agricultural fungicide.

7455. Phenylmercuric Nitrate, Basic. *(Nitrato-O)-phenylmercury;* merphenyl nitrate; Phe-Mer-Nite; Phenmerzyl nitrate. $C_{12}H_{11}Hg_2NO_4$; mol wt 634.40. C 22.72%, H 1.75%, Hg 63.24%, N 2.21%, O 10.09%. C_6H_5HgOH.-$C_6H_5HgNO_3$. Made by boiling benzene with mercuric acetate, then treating the resulting acetate with an alkali nitrate: Pyman, Stevenson, *Pharm. J.* **133,** 269 (1934); T. B. Grave *et al., J. Am. Pharm. Assoc.* **25,** 752 (1936); **Brit.** pat. **446,703** (1936 to Schering-Kahlbaum). Toxicology: R. Wien, *Quart. J. Pharm. Pharmacol.* **12,** 212 (1939).
Pearly lustrous scales, mp 187-190° (dec). Phenol coefficient stated to be ~600 and relatively unaffected by organic matter. Soluble in ~1250 parts water; slightly sol in alcohol, moderately in glycerol. Practically insol in other usual organic solvents. LD_{50} in mice (mg/g): 0.045 s.c.; 0.027 i.v. (Wien).
Caution: Local application of the more concd solns may cause irritation.
USE: Pharmaceutic aid (antimicrobial). Antiseptic, germicide, fungicide esp for tree wounds.
THERAP CAT (VET): Antibacterial, antifungal; chiefly used topically, may also be injected.

7456. Phenylmercury Borate. *[Orthoborato(3—)-O]-phenylmercurate(2—) dihydrogen; (dihydrogen borato)phenylmercury;* phenylmercuric borate; Famosept; Merfen. C_6-H_7BHgO_3; mol wt 338.52. C 21.29%, H 2.08%, B 3.19%, Hg 59.25%, O 14.18%. Prepn: Christiansen, **U.S.** pat. **2,196,384** (1950 to Lever Bros.).

Crystalline powder, mp 112-113°. Sol in water, alcohol, glycerol.
Note: Product of commerce may be an equimolar composition of phenylmercuriborate and phenylmercurihydroxide.
THERAP CAT: Antiseptic (topical).

7457. Phenylmethylbarbituric Acid. *5-Methyl-5-phenyl-2,4,6(1H,3H,5H)-pyrimidinetrione; 5-methyl-5-phenylbarbituric acid;* 2,4,6-trioxo-5-methyl-5-phenylhexahydropyrimidine; Rutonal. $C_{11}H_{10}N_2O_3$; mol wt 218.21. C 60.55%, H 4.62%, N 12.84%, O 22.00%. Prepn: **Ger.** pat. **247,952** (1912 to Bayer); Dvornik *et al., C.A.* **50,** 312f (1956).

Slightly bitter crystals, mp 226°. Practically insol in water. Sol in alcohol, ether, aq solns of the alkalies. Forms a water-sol sodium salt.
Note: Not to be confused with mephobarbital, *q.v.*
Caution: May be habit forming. This is a controlled substance (depressant) listed in the U.S. Code of Federal Regulations, Title 21 Parts 329.1 and 1308.13 (1995).
THERAP CAT: Anticonvulsant; sedative, hypnotic.

7458. o-Phenylphenol. *(1,1'-Biphenyl)-2-ol; 2-biphenylol;* orthoxenol; o-hydroxydiphenyl; 2-hydroxydiphenyl; Dowicide 1. $C_{12}H_{10}O$; mol wt 170.21. C 84.68%, H 5.92%, O 9.40%. Prepn from phenyl ether: Lüttringhaus, Sääf, *Ann.* **542,** 241 (1939); from dibenzofuran: Gilman, Esmay, *J. Am. Chem. Soc.* **75,** 2947 (1953); Müller, **U.S.** pat. **2,862,-035** (1958 to Bayer). Purification: Widiger, **U.S.** pat. **3,087,969** (1963 to Dow). Toxicity data: Hodge *et al., J. Pharmacol. Exp. Ther.* **104,** 202 (1952). Review of carcinogenic risk: *IARC* Monographs **30,** 329-344 (1983).

White, flaky crystals. Mild, characteristic odor. mp 55.5-57.5°; bp 280-284°; bp_{15} 152-154°. Practically insol in water. Sol in fixed alkali hydroxide solns and most organic solvents. LD_{50} orally in rats: 2.48 g/kg (Hodge).
Sodium salt tetrahydrate, $C_{12}H_9NaO.4H_2O$, *Dowicide A, Natriphene.* White flakes. Soly in g/100 g of solvent: water 122; acetone 156; methanol 138; propylene glycol 28. Practically insol in petroleum fractions, pine oil. pH of satd water soln at 25° = 12.0-13.5.
Caution: Toxic symptoms similar to phenol, *q.v.* See *Clinical Toxicology of Commercial Products,* R. E. Gosselin *et al.,* Eds. (Williams & Wilkins, Baltimore, 5th ed., 1984) Section II, p 316.
USE: In the rubber industry; agricultural fungicide; disinfectant.

7459. p-Phenylphenol. *[1,1'-Biphenyl]-4-ol; 4-hydroxydiphenyl;* Paraxenol.
mp 164-165°; bp 305-308°. Possesses approx the soly of the ortho compd.
USE: As intermediate in the manuf of resins; also in the rubber industry.

7460. Phenyl Phthalate. *1,2-Benzenedicarboxylic acid diphenyl ester;* diphenyl phthalate. $C_{20}H_{14}O_4$; mol wt 318.33. C 75.46%, H 4.43%, O 20.10%. $C_6H_4(COOC_6H_5)_2$. Prepd from phenol and phthalic anhydride.

White odorless crystals. d^{74} 1.572; mp 70-73°; bp_{14} ~255°. Insol in water; sol in acetone and other ketones, in liquid esters and chlorinated hydrocarbons.

USE: Plasticizer in nitrocellulose lacquers.

7461. Phenylpropanolamine Hydrochloride. *(R*,S*)-(±)-α-(1-Aminoethyl)benzenemethanol hydrochloride; α-(1-aminoethyl)benzyl alcohol hydrochloride; dl-norephedrine hydrochloride; 2-amino-1-phenyl-1-propanol hydrochloride; α-hydroxy-β-aminopropylbenzene hydrochloride; 1-phenyl-2-amino-1-propanol hydrochloride;* mydriatin; Kontexin; Monydrin; Obestat; Propadrine. $C_9H_{14}ClNO$; mol wt 187.67. C 57.60%, H 7.52%, Cl 18.89%, N 7.46%, O 8.53%. Sympathomimetic amine. Prepn: Vanderbilt, Hass, *Ind. Eng. Chem.* **32**, 34 (1940); cf. Nagai, Kanao, *Ann.* **470**, 157 (1929); Kamlet, U.S. pat. **2,151,517** (1939); Hoover, Hass, *J. Org. Chem.* **12**, 506 (1947). Industrial prepn: Jenkins, Hartung, *Chemistry of Organic Medicinal Products* (New York, 1949). Improved process: Wilbert, Sosis, U.S. pat. **3,028,429** (1962 to Nepera Chem.). Prepn of optically active compds: Rusch *et al.*, Ger. pat. **1,014,553** (1957 to Knoll). Separation of enantiomers: S. Kanao, *J. Pharm. Soc. Japan* **48**, 947 (1928); C. Jarowski, W. H. Hartung, *J. Org. Chem.* **8**, 564 (1943); C. Pettersson, H. W. Stuurman, *J. Chromatog. Sci.* **22**, 441 (1984). Crystal structure: A. Podder *et al.*, *Indian J. Phys.* **53A**, 652 (1979). HPLC determn: V. Das Gupta, A. G. Ghanekar, *J. Pharm. Sci.* **66**, 895 (1977); N. Muhammad, J. A. Bodnar, *J. Liquid Chromatog.* **3**, 113 (1980). Clinical studies of decongestant activity: O. K. Haugeto *et al.*, *J. Otolaryngol.* **10**, 359 (1981); M. Bende *et al.*, *Rhinology* **23**, 43 (1985); in weight loss: S. Altschuler *et al.*, *Int. J. Obesity* **6**, 549 (1982); in stress incontinence: E. Fossberg *et al.*, *Urologia Int.* **38**, 293 (1983). Toxicity: E. I. Goldenthal, *Toxicol. Appl. Pharmacol.* **18**, 185 (1971); P. Pentel, *J. Am. Med. Assoc.* **252**, 1898 (1984). Comprehensive description: I. Kanfer *et al.*, in *Analytical Profiles of Drug Substances* vol. 12, K. Florey, Ed. (Academic Press, New York, 1983) pp 357-383.

Crystals, mp 190-194°. Odor resembling that of crude benzoic acid. pKa 9.44±0.04. Freely sol in water, alcohol. Practically insol in ether, chloroform, benzene. The aq soln is neutral to litmus. LD_{50} orally in rats: 1490 mg/kg (Goldenthal).

(±)-Form base, mp 101-101.5°.

(+)-Form hydrochloride, $[\alpha]_D^{25}$ +32° (water). mp 171-172°.

D-*threo*-Form, *see* Norpseudoephedrine.

THERAP CAT: Decongestant. Anorexic.

THERAP CAT (VET): Bronchodilator. Decongestant (nasal).

7462. Phenylpropylmethylamine. *N,β-Dimethylbenzeneethanamine; dl-N,β-dimethylphenethylamine; dl-N-methyl-2-phenylpropylamine; 1-methylamino-2-phenylpropane; phenpromethamine; 1-methylamino-2-methyl-2-phenylethane;* Vonedrine. $C_{10}H_{15}N$; mol wt 149.24. C 80.48%, H 10.13%, N 9.39%. Synthesis starting with the condensation of chlorobenzene with allyl chloride, followed by ammonolysis: Patrick *et al.*, *J. Am. Chem. Soc.* **68**, 1009 (1946).

Volatile liquid. d_4^{25} 0.915-0.925; bp_{760} 205-210°; bp_{15} 95-96°; n_D^{20} 1.5102. Slightly sol in water (1.2 g/100 ml). Freely sol in alcohol, ether, benzene. Aq solns are strongly alkaline; pH of a soln of 2 drops (~0.1 ml) dil with 10 ml H_2O is ~10.5.

Hydrochloride, $C_{10}H_{15}N.HCl$, mp 144-148°.

THERAP CAT: Adrenergic.

7463. 1-Phenyl-3-pyrazolidinone. 1-Phenyl-3-pyrazolidone; Phenidone. $C_9H_{10}N_2O$; mol wt 162.19. C 66.65%, H 6.21%, N 17.27%, O 9.86%. Prepd by heating phenylhydrazine with β-chloropropionic acid: **Ger.** pat. **53,834** (1889); by acid hydrolysis of 3-amino-1-phenylpyrazoline: Kendall, Duffin, **Brit.** pats. **650,911** and **669,591** (1951, 1952, both to Ilford). *Review:* Kendall, *Brit. J. Photogr.* **100**, 56 (1953).

Leaflets or needles from benzene, mp 121°. One gram dissolves in 10 ml boiling water, in 10 ml hot alcohol, in 37.5 ml boiling benzene. Practically insol in ether, petr ether. Freely sol in dil aq solns of acids and alkalies. Pharmacological tests indicate low oral toxicity. No cases of dermatitis have occurred.

USE: Non-staining, high contrast photographic developer. The amount required is about one-fifth to one-tenth that of Metol.

7464. Phenyl Salicylate. *2-Hydroxybenzoic acid phenyl ester;* Salol. $C_{13}H_{10}O_3$; mol wt 214.22. C 72.89%, H 4.71%, O 22.41%. Made by the action of phosphorus oxychloride on a mixture of phenol and salicylic acid.

White, small crystals or crystalline powder; pleasant aromatic odor and taste. d 1.25; mp 41-43°; bp_{12} 173°. One gram dissolves in 6670 ml water, 6 ml alcohol, 1.5 ml benzene, 5 ml amyl alcohol, 10 ml liq paraffin, 4 ml almond oil. Sol in acetone, chloroform, ether, oils; very slightly sol in glycerol. Solubility at 25° in g/100 g: absolute ethanol 53; ethyl acetate 470; methyl ethyl ketone 620; toluene 460; Stoddard solvent 88; water less than 0.1%.

Incompat. Bromine water, ferric salts; camphor, monobromated camphor, phenol, chloral hydrate, thymol or urethan in trituration.

USE: In the manuf of various polymers for the plastics industry, also in lacquers, adhesives, waxes, polishes. In suntan oils and cremes. As light absorber to prevent discoloration of plastics. Has some plasticizer properties.

THERAP CAT: Analgesic, antipyretic, anti-inflammatory.

THERAP CAT (VET): Has been used externally as a disinfectant, internally as an intestinal antiseptic and antipyretic.

7465. 4-Phenylsemicarbazide. *N-Phenylhydrazinecarboxamide;* anilinoformylhydrazine. $C_7H_9N_3O$; mol wt 151.17. C 55.62%, H 6.00%, N 27.80%, O 10.58%. C_6H_5-NHCONHNH$_2$. Prepd by the action of hydrazine hydrate on phenylurea: Curtius, *J. Prakt. Chem.* [2] **58**, 216 (1898); Wheeler, *Org. Syn.* **coll. vol. I** (2nd ed., 1941) p 450. Orthorhombic plates from water, mp 122°. Insol in ether, difficultly sol in hot water. Freely sol in alcohol, chloroform, dil acids and alkalies.

Hydrochloride, $C_7H_9N_3O.HCl$, prisms, mp 215°; freely sol in water and alcohol.

7466. N-Phenylsulfanilic Acid. *4-(Phenylamino)benzenesulfonic acid;* 4-diphenylaminesulfonic acid; p-anilinobenzenesulfonic acid. $C_{12}H_{11}NO_3S$; mol wt 249.29. C 57.82%, H 4.45%, N 5.62%, O 19.25%, S 12.86%. Prepd by acetylation followed by sulfonation of diphenylamine: Merz, Weith, *Ber.* **6**, 1512 (1873); Gnehm, Werdenberg, *Z.*

Angew. Chem. **12**, 1027 (1899); Sarver, Kolthoff, *J. Am. Chem. Soc.* **53**, 1902 (1931).

Leaves. Becomes blue on exposure to light. Soluble in water, in alcohol. Insol in ether. Dec into diphenylamine and sulfuric acid when heated above 200° with water contg hydrochloric acid.

Sodium salt, $C_{12}H_{10}NNaO_3S$, very sol in water.

Potassium salt, $C_{12}H_{10}KNO_3S$, leaves; slightly sol in alcohol, very sol in water.

Barium salt, $(C_{12}H_{10}NO_3S)_2Ba$, leaflets; slightly sol in water. *Poisonous!*

USE: Colorimetric determination of nitrates; oxidation-reduction indicator; detection of oxidizing substances.

7467. Phenyl Sulfide. *1,1'-Thiobis[benzene];* diphenylsulfide. $C_{12}H_{10}S$; mol wt 186.28. C 77.37%, H 5.41%, S 17.21%. $(C_6H_5)_2S$. Willard, Hall, *J. Am. Chem. Soc.* **44**, 2219 (1922). Toxicity study: H. F. Smyth et al., *Am. Ind. Hyg. Assoc. J.* **23**, 95 (1962).

Colorless, almost odorless liquid. d_{15}^{15} 1.118; mp about −40°; bp 295-297°; n_D^{18} 1.6350. Insol in water. Sol in hot alcohol; misc with benzene, ether, carbon disulfide. LD_{50} orally in rats: 0.49 ml/kg (Smyth).

7468. Phenylthiourea. Phenylthiocarbamide. $C_7H_8N_2S$; mol wt 152.22. C 55.23%, H 5.30%, N 18.40%, S 21.07%. $C_6H_5NHCSNH_2$. Prepd by evaporating an aq soln of aniline hydrochloride and ammonium thiocyanate and carefully heating the residue. Toxicity data: Scheline et al., *J. Med. Pharm. Chem.* **4**, 109 (1961).

Bitter or tasteless needles, depending upon heredity of taster. d 1.3. mp 154°. Soluble in 400 parts cold water, 17 parts boiling water; soluble in alcohol. LD_{50} in rats, rabbits (mg/kg): 3, 40 orally (Scheline).

USE: In medical genetics.

7469. Phenyltoloxamine. *N,N-Dimethyl-2-[2-(phenylmethyl)phenoxy]ethanamine; N,N-dimethyl-2-(α-phenyl-o-tolyloxy)ethylamine;* N,N-dimethyl-2-(α-phenyl-o-toloxy)ethylamine; 2-(2-dimethylaminoethoxy)diphenylmethane; 2-benzhydryl β-dimethylaminoethyl ether; 2-benzylphenyl β-dimethylaminoethyl ether; PRN; bistrimin; C-5581H; Antin; Phenoxadrine; Bristamin. $C_{17}H_{21}NO$; mol wt 255.36. C 79.96%, H 8.29%, N 5.49%, O 6.27%. Prepn: Cheney et al., *J. Am. Chem. Soc.* **71**, 60 (1949); Binkley, Cheney, U.S. pat. 2,703,324 (1955 to Bristol Labs.).

Oily liquid, bp 141-144° (at <0.1 mm Hg).

Dihydrogen citrate, $C_{17}H_{21}NO \cdot C_6H_8O_7$, crystals from water or methanol, mp 138-140°. Soluble in water.

Hydrochloride, $C_{17}H_{21}NO \cdot HCl$, crystals from methyl isobutyl ketone, mp 119-121°. Soluble in water.

Phenyltoloxamine adsorbed on a resin is marketed as *Histionex.*

THERAP CAT: Dihydrogen citrate as antihistaminic.

7470. Phenyl Tolyl Ketone. $C_{14}H_{12}O$; mol wt 196.25. C 85.68%, H 6.16%, O 8.15%. *o*-Isomer made from benzene and *o*-toluic acid chloride in the presence of $AlCl_3$. *p*-Isomer made from benzoyl chloride and toluene in presence of $AlCl_3$.

o-Phenyl tolyl ketone, *(2-methylphenyl)phenylmethanone, 2-methylbenzophenone.* Oily liquid. mp below −18°. bp 309-311°. Insol in water. Freely sol in alcohol, oils, most organic solvents.

p-Phenyl tolyl ketone, *(4-methylphenyl)phenylmethanone, 4-methylbenzophenone.* Crystals, mp 59-60°. bp_{720} 311-312°. Insol in water. Sol in alcohol; easily sol in benzene, ether, oils.

USE: As fixative in perfumery.

7471. Phenyltrimethylammonium Iodide. *N,N,N-Trimethylbenzaminium iodide.* $C_9H_{14}IN$; mol wt 263.12. C 41.08%, H 5.36%, I 48.23%, N 5.32%. $C_6H_5(CH_3)_3NI$. *Ref:* Pass, Ward, *Analyst* **58**, 667 (1933).

White, cryst powder. mp 175°. Sol in water or alcohol.

USE: For the detection and determination of cadmium.

7472. Phenylurea. Phenylcarbamide. $C_7H_8N_2O$; mol wt 136.15. C 61.75%, H 5.92%, N 20.57%, O 11.75%. $C_6H_5NHCONH_2$. Obtained with carbanilide as byproduct by refluxing aniline hydrochloride and urea in water: Davis, Blanchard, *Org. Syn.* **3**, 95 (1923).

Monoclinic prisms from water or alcohol. d 1.302; mp 147° (dec); bp 238°. On slow cooling separates in needles several centimeters in length. Soluble in hot water, hot alcohol, ether, ethyl acetate, glacial acetic acid.

7473. Phenylurethan(e). *Phenylcarbamic acid ethyl ester;* ethyl phenylcarbamate; ethyl carbanilate. $C_9H_{11}NO_2$; mol wt 165.19. C 65.44%, H 6.71%, N 8.48%, O 19.37%. $C_6H_5NHCOOC_2H_5$. Made by the action of aniline on ethyl chloroformate.

White, acicular crystals. d 1.106; mp 52-53°; bp 238° with slight decompn; n_D^{20} 1.5376. Slightly sol in water, freely in alcohol, ether; scarcely attacked by boiling for a short time with HCl or NaOH.

7474. Phenyramidol. *α-[(2-Pyridinylamino)methyl]-benzenemethanol;* 2-(β-hydroxyphenethylamino)pyridine; fenyramidol; $C_{13}H_{14}N_2O$; mol wt 214.27. C 72.87%, H 6.59%, N 13.07%, O 7.47%. Prepn from 2-mandelamidopyridine or 2-aminopyridine: Gray et al., *J. Am. Chem. Soc.* **81**, 4347, 4351 (1959).

Crystals from dil methanol, mp 82-85°. uv max (95% ethanol): 243, 303 nm (log ε 4.24, 3.63). pKa 5.85.

Hydrochloride, $C_{13}H_{14}N_2O \cdot HCl$, *IN-511, MJ-505, NSC-17777, Abbolexin, Anabloc, Analexin, Cabral, Miodar.* Crystals from ethanol + ether, mp 140-142°. Soluble in water. Aq solns are slightly acid and stable in ampuls.

Methiodide, $C_{13}H_{14}N_2O \cdot CH_3I$, crystals from ethanol + ether, mp 164-166°.

THERAP CAT: Analgesic; muscle relaxant (skeletal).

7475. Phenytoin. *5,5-Diphenyl-2,4-imidazolidinedione;* diphenylhydantoin; Difhydan; Dihycon; Di-Hydan; Dilantin; Ekko; Hydantin; Hydantol; Lehydan; Lepitoin; Phenhydan; Zentropil. $C_{15}H_{12}N_2O_2$; mol wt 252.27. C 71.42%, H 4.79%, N 11.10%, O 12.68%. Prepn from benzophenone: Henze, U.S. pat. 2,409,754 (1946 to Parke, Davis); of sodium salt from benzil, urea and NaOH: Biltz, *Ber.* **41**, 1391 (1908); **44**, 411 (1911). Pharmacology: Gillis et al., *J. Pharmacol. Exp. Ther.* **179**, 599 (1971). Toxicity of base: G. Stille, I. Brunckow, *Arzneimittel-Forsch.* **4**, 723 (1954); of sodium salt: G. B. Fink, E. A. Swinyard, *J. Pharmacol. Exp. Ther.* **127**, 318 (1959). Reviews: Damato, *Progr. Cardiov. Dis.* **12**, 1-15 (1969); Dreifus, Watanabe, *Am. Heart J.*

80, 709-713 (1970). Review of carcinogenicity studies: *IARC Monographs* 13, 201-225 (1977). Comprehensive description: J. Philip *et al., Anal. Profiles Drug Subs.* 13, 417-445 (1984).

Powder, mp 295-298°. Practically insol in water. One gram dissolves in about 60 ml alcohol, about 30 ml acetone. Sol in alkali hydroxides. LD_{50} in mice (mg/kg): 92 i.v.; 110 s.c. (Stille, Brunckow).

Sodium salt, $C_{15}H_{11}N_2NaO_2$, *phenytoin soluble, sodium 5,5-diphenyl hydantoinate, Antisacer, Danten, Diphantoine, Diphenin, Diphenylan Sodium, Epanutin, Minetoin, Tacosal, Solantyl.* White powder; bitter, soapy taste. Somewhat hygroscopic. Easily dissociated even by weak acids (incl CO_2 absorbed on exposure to air) regenerating phenytoin. One gram dissolves in 10.5 ml alc; in ~66 ml water (aq soln turbid unless pH adjusted to > 11.7, the pH of a satd soln). Insol in ether, chloroform. LD_{50} orally in mice: 490 mg/kg (Fink, Swinyard).

Note: Phenytoin may reasonably be anticipated to be a carcinogen: *Seventh Annual Report on Carcinogens* (PB95-109781, 1994) p 319.

THERAP CAT: Anticonvulsant; antiepileptic.

THERAP CAT (VET): Sodium salt has been recommended in epileptiform seizures in dogs.

7476. Phethenylate Sodium. *5-Phenyl-5-(2-thienyl)-2,-4-imidazolidinedione monosodium salt; 5-phenyl-5-(2-thienyl)hydantoin sodium;* sodium 5-phenyl-5-(2-thienyl)hydantoinate; *Thiantoin(e) Sodium.* $C_{13}H_9N_2NaO_2S$; mol wt 280.28. C 55.71%, H 3.24%, N 9.99%, Na 8.20%, O 11.42%, S 11.44%. Prepn: Spurlock, U.S. pat. **2,366,221** (1942).

Minute, hygroscopic crystals. Sol in water, alc. Aq solns are alkaline and dec easily.

Base, mp 256-257°.

THERAP CAT: Anticonvulsant.

7477. Phillyrin. *[1S-(1α,3aα,4β,6aα)]-4-[4-(3,4-Dimethoxyphenyl)tetrahydro-1H,3H-furo[3,4-c]furan-1-yl]-2-methoxyphenyl-β-D-glucopyranoside;* phyllyrin; phillyroside; forsythin. $C_{27}H_{34}O_{11}$; mol wt 534.56. C 60.67%, H 6.41%, O 32.92%. From bark of *Phillyrea latifolia* L., and allied *Oleaceae:* Campona, *Ann.* 24, 242 (1837); Eijkman, *Rec. Trav. Chim.* 5, 127 (1886); Kramer, *Compt. Rend.* 196, 814 (1933); Sosa, *Bull. Soc. Chim. Biol.* 29, 918 (1947). Structure: Kaku *et al., J. Pharm. Soc. Japan* 59, 248 (1939); M. Chiba *et al., Chem. Pharm. Bull.* 25, 3435 (1977).

α-Form, needles from dil alc, mp 154-155°. $[α]_D$ +48.4° (alcohol).

β-Form, needles, mp 184-185°. $[α]_D$ +48.5° (alcohol).

7478. Phloionic Acid. *(R*,R*)-(±)-9,10-Dihydroxyoctadecanedioic acid;* floionic acid. $C_{18}H_{34}O_6$; mol wt 346.46. C 62.40%, H 9.89%, O 27.71%. Isoln from cork: Guillemonat, Cesaire, *Bull. Soc. Chim. France* **1949**, 792; Ribas, Seoane, *An. Real Soc. Espan. Fis. Quim. (Madrid)* 50B, 963 (1954), *C.A.* 50, 806f (1956); Brown, Rosen, U.S. pat. **2,872,464** (1959 to Crown Cork). Structure: Duhamel, *Bull. Soc. Chim. France* **1965**, 399. Synthesis: Ruzicka *et al., Helv. Chim. Acta* 25, 1086 (1942); Gensler, Schlein, *J. Am. Chem. Soc.* 77, 4846 (1955). Resolution of isomers: Gender, Mahadevan, *ibid.* 78, 169 (1956); Alvarez-Varquez, Ribas-Marques, *An. Quim.* 64, 783 (1968). Synthesis of (±), (+), and (−) forms: McGhie *et al., Chem. & Ind. (London)* **1972**, 536.

Crystals from ethanol + water, mp 126°.

Dimethyl ester, $C_{20}H_{38}O_6$, crystals from ethanol, mp 77.5-78°.

7479. Phloionolic Acid. *(R*,R*)-9,10,18-Trihydroxyoctadecanoic acid;* 9,10,18-trihydroxystearic acid. $C_{18}H_{36}O_5$; mol wt 332.48. C 65.03%, H 10.91%, O 24.06%. Isolation from cork: Seoane, Ribas, *An. Real Soc. Espan. Fis. Quim. (Madrid)* 47B, 61 (1951), *C.A.* 46, 89i (1952); 51, 8651a (1957); Ribas, Seoane, *ibid.* 50B, 963 (1954), *C.A.* 50, 806f (1956); Brown, Rosen, U.S. pat. **2,872,464** (1959 to Crown Cork). Structure: Seoane *et al., Chem. & Ind. (London)* **1957**, 490; *An. Real Soc. Espan. Fis. Quim. (Madrid)* 55B, 839 (1959); Duhamel, *Bull. Soc. Chim. France* **1965**, 399.

Crystals from ethanol + water, mp 101-102°.

Methyl ester, $C_{19}H_{38}O_5$, crystals from benzene + petr ether, mp 80.5-81.5°.

7480. Phloretin. *3-(4-Hydroxyphenyl)-1-(2,4,6-trihydroxyphenyl)-1-propanone; 2',4',6'-trihydroxy-3-(p-hydroxyphenyl)propiophenone;* β-(p-hydroxyphenyl)phloropropiophenone; β-(p-hydroxyphenyl)-2,4,6-trihydroxypropiophenone. $C_{15}H_{14}O_5$; mol wt 274.27. C 65.69%, H 5.14%, O 29.17%. The aglucon of phlorizin, *q.v.* From root bark of apple trees: Rochleder, *J. Prakt. Chem.* 98, 205 (1866). Structure: Seshadri, *Ann. Rev. Biochem.* 20, 495 (1951). Synthesis: Fischer, Nouri, *Ber.* 50, 611 (1917); Rosenmund, Rosenmund, *ibid.* 61B, 2608 (1928); Shinoda *et al., J. Pharm. Soc. Japan* 49, 797 (1929); Johnston, U.S. pat. **2,789,995** (1957 to Union Carbide).

Needles from dil alcohol, dec 262°. Absorption spectrum: Lambrechts, *Arch. Int. Physiol.* 44, Suppl., 1-39 (1937). Freely sol in alc, methanol, acetone; sol in alkalies in hot glacial acetic acid; very sparingly sol in benzene, chloroform. Practically insol in water, ether, the soly being increased by addition of alcohol.

7481. Phloridzin. *1-[2-(β-D-Glucopyranosyloxy)-4,6-dihydroxyphenyl]-3-(4-hydroxyphenyl)-1-propanone;* phlorhizin; phlorizin; phlorrhizen; phloretin-2'-β-glucoside; 4,6-dihydroxy-2-(β-D-glucosido)-β-(p-hydroxyphenyl)propiophenone. $C_{21}H_{24}O_{10}$; mol wt 436.42. C 57.80%, H 5.54%, O 36.66%. A dihydrochalcone occurring in all parts of the apple tree except the mature fruit. Once thought to occur in pear, plum, cherry trees and other *Rosaceae:* A. H. Williams in *Comparative Phytochemistry*, T. Swain, Ed. (Academic Press, New York, 1966) pp 297-307. Isoln from root bark: De Koninck, *Ann.* **15**, 75, 258 (1835); Stass, *Ann.* **30**, 192 (1839); Bridel, Kramer, *Bull. Soc. Chim. Biol.* **15**, 544 (1933). Hydrolysis by dil mineral acids yields phloretin and glucose. Procedure for acid hydrolysis: Wessely, Sturm, *Monatsh.* **53-54**, 557 (1929); Müller, Robertson, *J. Chem. Soc.* **1933**, 1170. The energy of activation required for the hydrolysis is much less than for other glucosides and approaches that for γ-fructosides (sucrose, raffinose). Is hydrolyzed by saccharase at pH 4.45, by enzymes of *Aspergillus niger* and other enzymes found in invertebrates: Kobert, *Pflüger's Arch. Ges. Physiol.* **99**, 116 (1903); Moelwyn-Hughes, *J. Gen. Physiol.* **13**, 807 (1930); *cf. Trans. Faraday Soc.* **25**, 81 (1929). Synthesis: Zemplen, Bognár, *Ber.* **75B**, 1040 (1942); *cf. ibid.* 645 and **76B**, 386 (1943).

β-D-glucose-O

Dihydrate, long needles from water, mp 110°. Sweet, with bitter aftertaste. $[\alpha]_D^{25}$ −52° (0.16 g in 5 ml of 96% alcohol). One gram dissolves in about one liter of water at 22°, in 64 ml at 60°, in 22 ml at 70°. Freely sol in boiling water; in about 4 parts alcohol, in methanol, amyl alcohol, acetone, ethyl acetate, pyridine, aniline, quinoline and other organic bases; in aq alkaline solns and in glacial acetic acid. Practically insol in ether, chloroform, benzene.

USE: Experimentally to produce glycosuria in animals.

7482. Phloroglucinol. *1,3,5-Benzenetriol;* 1,3,5-trihydroxybenzene; phloroglucin; Dilospan S; Spasfon-Lyoc. $C_6H_6O_3$; mol wt 126.11. C 57.14%, H 4.80%, O 38.06%. Prepn by reducing trinitrobenzoic acid or trinitrobenzene: Pascoe, *Chem. Products* **18**, 454 (1955); from TNT: Kastens, Kaplan, *Ind. Eng. Chem.* **42**, 402 (1950), from picryl chloride: Heertjes, *Rec. Trav. Chim.* **78**, 452 (1959); from *m*-isopropylresorcinyl diacetate: Zimmer, U.S. pat. **3,028,-410** (1962 to Hooker Chem.); from tribromobenzene: McKillop *et al., Syn. Commun.* **4**, 35 (1974).

White rhombic crystals from water, mp 218°. Sweet taste. Sublimes with decompn. Discolors in light. Soluble in 100 parts water, 10 parts alcohol, 0.5 part pyridine; sol in ether. *Protect from light.* LD_{50} in mice (mg/kg): 4550 orally; 4050 i.p.; 5520 s.c., *RTECS* **Vol. II**, R. J. Lewis, R. L. Tatken, Eds. (1980) p 304.

USE: In diazo-type printing and textile dyeing. As reagent for pentoses and pentosans, aldehydes, lignin, HCl, methanol, chloral hydrate, turpentine oil, lignified cell tissue, free acid (HCl) in gastric juice; in microscopy, as an excellent decalcifier of bone specimens.

THERAP CAT: Antispasmodic.

7483. Phlorol. *2-Ethylphenol.* $C_8H_{10}O$; mol wt 122.17. C 78.65%, H 8.25%, O 13.10%. Prepn by heating ethylene and phenol with phosphoric acid at 200°: Ipatieff, *J. Am. Chem. Soc.* **60**, 1162 (1939); by heating ethanol + phenol over thoria + alumina at 350°: Hansch, Robertson, *ibid.* **72**, 4810 (1950); from phenol, ethylene chlorohydrin + sodium: Carlton, Bradbury, *ibid.* **78**, 1069 (1956). Purification: Biddiscombe *et al., J. Chem. Soc.* **1963**, 5764.

Colorless liq; phenol odor. Cooling the liq to −30° gives a metastable cryst form, mp about −28°; when kept for 24-48 hrs, this changes to the stable cryst form, mp −3.4°. bp_{760} 204.52°. d_{25} 1.01459. Practically insol in water. Freely sol in alcohol, benzene, glacial acetic acid.

7484. Pholcodine. *(5α,6α)-7,8-Didehydro-4,5-epoxy-17-methyl-3-[2-(4-morpholinyl)ethoxy]morphinan-6-ol;* 3-[2-(4-morpholinyl)ethyl]morphine; tetrahydro-1,4-oxazinylmethylcodeine; 3-(2-morpholinoethyl)morphine; β-morpholinylethylmorphine; homocodeine; Ethnine; Galenphol; Galphol; Memine; Codylin; Pectolin; Weifacodine. $C_{23}H_{30}N_2O_4$; mol wt 398.50. C 69.32%, H 7.59%, N 7.03%, O 16.06%. Prepn: Chabrier *et al.*, U.S. pat. **2,619,485** (1952 to Dausse). Description: *Ann. Pharm. Franc.* **8**, 261 (1950). Toxicity data: B. Kelentey *et al., Arzneimittel-Forsch.* **8**, 325 (1958).

Monohydrate, crystals, mp 91°. Bitter taste. $[\alpha]_D^{20}$ −95.3° (c = 2 in ethanol). Slightly sol in water (2% w/v), ether. Sol in alc (1:3), chloroform, benzene. pH of 2% aq soln 9.5 to 9.8. LD_{50} s.c. in mice: 0.540 g/kg (Kelentey).

Note: This is a controlled substance (opium derivative) listed in the U.S. Code of Federal Regulations, Title 21 Part 1308.11 (1995).

THERAP CAT: Antitussive.

7485. Pholedrine. *4-[2-(Methylamino)propyl]phenol;* *p*-hydroxy-*N*,α-dimethylphenethylamine; β-(*p*-hydroxyphenyl)isopropylmethylamine; α-(*p*-hydroxyphenyl)-β-methylaminopropane; *p*-hydroxy-*N*-methylbenzedrine; Knoll H_{75}. $C_{10}H_{15}NO$; mol wt 165.24. C 72.69%, H 9.15%, N 8.48%, O 9.68%. α-Adrenergic agonist. Prepd from *p*-methoxybenzyl methyl ketone: **Fr. pat. 822,422, Brit. pat. 482,414, Ger. pat. 674,753** (1937, 1938 and 1939 to Knoll); Hildebrandt and Hildebrandt, Freese, **Ger. pats. 665,793 and 767,161** (1938 and 1951 to Knoll). Synthesis: Savitskii, Makhnenko, *J. Gen. Chem. USSR* **10**, 1819 (1940); Buzas, Dufour, *Bull. Soc. Chim. France* **1950**, 139. Toxicity data: Lindner, *Arch. Exp. Pathol. Pharmacol.* **188**, 675 (1938).

Crystals from methanol, mp 162-163°. Acrid, burning taste. Alkaline reactions. Slightly sol in water. Sol in alcohol, ether; readily sol in dil acids.

Sulfate, $(C_{10}H_{15}NO)_2 \cdot H_2SO_4$, Paredrinol, Pulsotyl, Veritol. Crystals, dec 320-323°. Sol in water. LD s.c. in rats: 500 mg/kg (Lindner).

THERAP CAT: Antihypotensive.

THERAP CAT (VET): Circulatory stimulant.

7486. Phorate. *Phosphorodithioic acid O,O-diethyl S-[(ethylthio)methyl] ester;* O,O-diethyl S-(ethylthio)methyl phosphorodithioate; O,O-diethyl S-ethylmercaptomethyl dithiophosphate; AC-3911; EI-3911; ENT-24042; Rampart; Thimet. $C_7H_{17}O_2PS_3$; mol wt 260.38. C 32.29%, H 6.58%, O 12.29%, P 11.90%, S 36.94%. Cholinesterase inhibitor. Prepn: Schrader, Lorenz, U.S. pat. **2,759,010; Brit. pat. 797,307** (1956, 1958 both to Bayer). Metabolism: J. B. Bowman, J. E. Casida, *J. Agr. Food Chem.* **5**, 192 (1957). Persistence in soil: D. L. Suett, *Pestic. Sci.* **6**, 385 (1975).

Toxicity study: T. B. Gaines, *Toxicol. Appl. Pharmacol.* **14**, 515 (1969).

Clear liquid, $bp_{0.1}$ 75-78°, $bp_{0.8}$ 118-120°, $bp_{2.0}$ 125-127°. d_4^{25} 1.156. n_D^{25} 1.5329. Vapor pressure at 20°: 8.4 × 10^{-4} mm Hg. Soly: 50 ppm in water. Misc with xylene, carbon tetrachloride, dioxane, methyl Cellosolve, dibutyl phthalate, vegetable oils. Stable at room temp. Hydrolyzed in the presence of water and alkali. LD_{50} in female, male rats (mg/kg): 1.1, 2.3 orally; 2.5, 6.2 dermally (Gaines).

USE: Insecticide.

7487. Phorbol. *[1aR-(1aα,1bβ,4aβ,7aα,7bα,8α,9β,-9aα)]-1,1a,1b,4,4a,7a,7b,8,9,9a-Decahydro-4a,7b,9,9a-tetrahydroxy-3-(hydroxymethyl)-1,1,6,8-tetramethyl-5H-cyclopropa[3,4]benz[1,2-e]azulen-5-one.* $C_{20}H_{28}O_6$; mol wt 364.44. C 65.92%, H 7.74%, O 26.34%. Parent alcohol of the tumor promoting compounds in croton oil, *q.v.*, the oil expressed from the seeds of *Croton tiglium* L., *Euphorbiaceae*. Phorbol has a structural skeleton based on cyclopropabenzazulene. Isoln: Flaschenträger, Wigner, *Helv. Chim. Acta* **25**, 569 (1942); Kauffmann, Neumann, *Ber.* **92**, 1715 (1959); S. Tseng *et al.*, *J. Org. Chem.* **42**, 3645 (1977). Structure and stereochemistry: Hecker *et al.*, *Tetrahedron Letters* **1967**, 3165; Pettersen, Ferguson, *Chem. Commun.* **1967**, 716. Unlike its diesters, phorbol does not appear to be co-carcinogenic or enhance chemically-induced mutagenesis: C. J. Soper, F. J. Evans, *Cancer Res.* **37**, 2487 (1977). Mechanism of action study on phorbol esters: A. S. Kraft, W. B. Anderson, *Nature* **301**, 621 (1983). Comprehensive review of phorbol and its esters: Hecker, Schmidt, *Fortschr. Chem. Org. Naturst.* **31**, 377-467 (1974); P. M. Blumberg, *Crit. Rev. Toxicol.* **8**, 199-234 (1981).

Anhydr crystals, dec 250-251°. Two forms of solvated crystals from ethyl acetate: mp 162-163° and 233-234°. Solvated crystals from methanol or ethanol; mp 249-250°. $[\alpha]_D^{24}$ +102° (water). $[\alpha]_D^{20}$ +118° (c = 0.4 in dioxane). uv max (ethanol): 235, 334 nm (ε 5200, 70). Quite sol in polar solvents, including water.

12-Myristate 13-acetate diester, $C_{36}H_{56}O_8$, *12-O-tetradecanoylphorbol-13-acetate, TPA, croton oil factor A_1.* uv max (ethanol): 232, 333 nm (ε 5400, 73). Differentiation of human leukemia cells: J. B. Weinberg, *Science* **213**, 655 (1981).

7488. Phorone. *2,6-Dimethyl-2,5-heptadien-4-one;* diisopropylideneacetone. $C_9H_{14}O$; mol wt 138.21. C 78.21%, H 10.21%, O 11.58%. $(CH_3)_2C=CHCOCH=C(CH_3)_2$. Prepn from isobutenyllithium + CO_2: Braude, Timmons, *J. Chem. Soc.* **1950**, 2000; from acetone: Dolgov, Samsonova, *Zh. Obshch. Khim.* **22**, 632 (1952); Joseph, Blumenthal, *J. Org. Chem.* **24**, 1371 (1959); Tsmur, *Zh. Prikl. Khim.* **34**, 1628 (1961); M. Konieczny, G. Sosnovsky, *Z. Naturforsch.* **33B**, 454 (1978).

Yellow liquid or yellowish-green prisms, mp 28°. bp 198-199°; bp_{17} 88°. d_4^{20} 0.885. n_D^{21} 1.4968.

7489. Phosalone. *Phosphorodithioic acid S-[(6-chloro-2-oxo-3(2H)-benzoxazolyl)methyl] O,O-diethyl ester; phosphorodithioic acid, O,O-diethyl ester, S-ester with 6-chloro-3-(mercaptomethyl)-2-benzoxazolinone;* 3-(O,O-diethyldithiophosphorylmethyl)-6-chlorobenzoxazolinone; 6-chloro-3-

(O,O-diethyldithiophosphorylmethyl)benzoxazolone; *S*-(6-chloro-2-oxobenzoxazolin-3-yl)methyl diethyl phosphorothiolothionate; RP-11974; Zolone. $C_{12}H_{15}ClNO_4PS_2$; mol wt 367.81. C 39.19%, H 4.11%, Cl 9.64%, N 3.81%, O 17.40%, P 8.42%, S 17.44%. Prepn: **Brit.** pat. **1,005,372** (1965 to Rhône-Poulenc). Toxicology, metabolism, insecticidal properties: Colinese, Terry, *Chem. & Ind. (London)* **1968**, 1507. Use as molluscicide: Cessac, **Fr.** pat. **1,453,989** (1966 to Rhône-Poulenc).

Crystals, mp 47.5-48°. Sol in ketones, alcohols and most aromatic solvents. Practically insol in water and aliphatic hydrocarbons. LD_{50} orally in mice, rats: 180-205, 120-175 mg/kg.

USE: Insecticide, acaricide.

7490. Phosfolan. *1,3-Dithiolan-2-ylidenephosphoramidic acid diethyl ester; phosphonodithioimidocarbonic acid cyclic ethylene P,P-diethyl ester;* diethyl 1,3-dithiolan-2-ylidenephosphoramidate; 2-(diethoxyphosphinylimino)-1,3-dithiolane; phospholan; AC-47031; CL-47031; EI-47031; ENT-25830; Cyolane; Cylan. $C_7H_{14}NO_3PS_2$; mol wt 255.30. C 32.93%, H 5.53%, N 5.49%, O 18.80%, P 12.13%, S 25.12%. Prepn: R. W. Addor, J. B. Lovell, **Belg.** pat. **618,155**, *C.A.* **59**, 10066h (1963) and R. W. Addor, **U.S.** pat. **3,197,481** (1962 and 1965 to Am. Cyanamid). Activity: D. L. Bull *et al.*, *J. Econ. Entomol.* **57**, 112 (1964); R. L. Ridgway *et al.*, *ibid.* **59**, 149 (1966). Fate in soil: G. Tantawy *et al.*, *Alexandria J. Agr. Res.* **22**, 315 (1974). Metabolism: I. P. Kapoor, R. C. Blinn, *J. Agr. Food Chem.* **25**, 413 (1977).

Solid, mp 36.5°, $bp_{0.001}$ 115-118°. Sol in water, acetone, benzene, ethanol, cyclohexane, toluene. Slightly sol in ether; sparingly sol in hexane. LD_{50} orally in rats: 29 mg/kg, *RTECS* **Vol. I**, R. J. Lewis, R. L. Tatken, Eds. (1979) p 789.

USE: Insecticide.

7491. Phosgene. *Carbonic dichloride;* carbonyl chloride; chloroformyl chloride. CCl_2O; mol wt 98.92. C 12.14%, Cl 71.68%, O 16.17%. $Cl_2C=O$. Prepn from chlorine + carbon monoxide: Whitehouse, **U.S.** pat. **1,231,226** (1917); Peacock, **U.S.** pat. **1,360,312** (1921); Bradner, **U.S.** pat. **1,457,493** (1923); Douthitt, **U.S.** pat. **2,847,470** (1958 to Texas Co.); from carbon monoxide + nitrosyl chloride: Williams, **U.S.** pat. **1,746,506** (1930 to du Pont Ammonia Corp.); from carbon tetrachloride + oleum: Murphy, Reuter, *Aust. Chem. Inst. J. & Proc.* **15**, 144 (1948). Toxicology: S. A. Cucinell, *Arch. Environ. Health* **28**, 272 (1974). Review: E. E. Hardy in Kirk-Othmer *Encyclopedia of Chemical Technology* vol. 17 (Wiley-Interscience, New York, 3rd ed., 1982) pp 416-425.

Colorless, highly toxic gas; suffocating odor; when much diluted with air there is an odor reminiscent of moldy hay. Condenses at ~0° to a clear, colorless, fuming liquid. d_4^0 1.432. mp −118°. bp_{760} 8.2°. Vapor press at 20°: 1215 mm. Slightly sol in water and slowly hydrolyzed by it; freely sol in benzene, toluene, glacial acetic acid and most liq hydrocarbons.

Caution: Insidious poison as it is not irritating immediately, even when fatal concns are inhaled. May cause severe pulmonary edema (may be quickly fatal) or pneumonia. Inhalation of high concns causes choking, constricted feeling in chest, coughing, painful breathing, bloody sputum. Vapors strongly irritating to the eyes, *cf.* Patty's *Industrial*

Hygiene and Toxicology **vol. 2C**, G. D. Clayton, F. E. Clayton, Eds. (Wiley-Interscience, New York, 3rd ed., 1982) pp 4126-4128.

Note: Paper soaked in alcoholic or carbon tetrachloride soln contg 10% of a mixture of equal parts of *p*-dimethylaminobenzaldehyde and colorless diphenylamine, then dried, will turn from yellow to deep orange in the presence of approx the max allowable concn of phosgene, and should always be used where the generation of this gas is possible or suspected.

USE: For the prepn of many organic chemicals; as a war gas.

7492. Phosmet. *Phosphorodithioic acid S-[(1,3-dihydro-1,3-dioxo-2H-isoindol-2-yl)methyl] O,O-dimethyl ester; phosphorodithioic acid O,O-dimethyl ester S-ester with N-(mercaptomethyl)phthalimide; O,O-dimethyl S-phthalimidomethyl phosphorothionate; N-(mercaptomethyl)phthalimide S-(O,O-dimethyl phosphorodithioate); phthalophos (USSR);* ENT-25705; R-1504; Imidan; Prolate. $C_{11}H_{12}NO_4PS_2$; mol wt 317.33. C 41.64%, H 3.81%, N 4.41%, O 20.17%, P 9.76%, S 20.21%. Prepn: Fancher, U.S. pat. **2,767,194** (1956 to Stauffer). Activity: B. A. Butt, J. C. Keller, *J. Econ. Entomol.* **54**, 813 (1961). GC determn in produce: M. A. Luke *et al., J. Assoc. Offic. Anal. Chem.* **64**, 1187 (1981). Toxicity study: T. B. Gaines, *Toxicol. Appl. Pharmacol.* **14**, 515 (1969).

Off-white, crystalline solid, mp 71.9°. Tech product (95-98% pure), mp 66.5-69.5°. Dec below its boiling point. Vapor pressure at 50°: 1×10^{-3} mm Hg. Soly in water at 25°: 25 ppm. LD_{50} in male, female rats (mg/kg): 113, 160 orally (Gaines).

USE: Insecticide, acaricide.

7493. Phosphamidon. *Phosphoric acid 2-chloro-3-(diethylamino)-1-methyl-3-oxo-1-propenyl dimethyl ester; phosphoric acid dimethyl ester, ester with 2-chloro-N,N-diethyl-3-hydroxycrotonamide;* 2-chloro-2-diethylcarbamoyl-1-methylvinyl dimethyl phosphate; Ciba 570; ENT-25515; Dimecron. $C_{10}H_{19}ClNO_5P$; mol wt 299.69. C 40.08%, H 6.39%, Cl 11.83%, N 4.67%, O 26.69%, P 10.34%. Cholinesterase inhibitor. Prepn: Beriger, Sallmann, U.S. pat. **2,908,605** (1959 to Ciba); Anliker *et al., Helv. Chim. Acta* **44**, 1622 (1961). Toxicity data: T. B. Gaines, *Toxicol. Appl. Pharmacol.* **14**, 515 (1969).

Z-isomer

Oil. d_4^{25} 1.2132. n_D^{25} 1.4718. $bp_{1.5}$ 162°; $bp_{0.001}$ 120°. mp −45°. Vapor pressure at 20°: 2.5×10^{-5} mm Hg. Absorption spectra: Anliker *et al., loc. cit.* Misc with water and most organic solvents except saturated hydrocarbons. One gram dissolves in ∼30 g hexane. Stable in neutral or acid media; hydrolyzed by alkali. LD_{50} orally in rats: 24 mg/kg (Gaines).

Caution: See Parathion.

USE: Insecticide.

7494. Phosphine. H_3P; mol wt 34.00. H 8.89%, P 91.11%. PH_3. Formed in small quantity in the putrefaction of organic matter contg phosphorus. Prepd from white phosphorus and aq alkali hydroxide; also by treatment of PH_4I with KOH: Klement in *Handbook of Preparative Inorganic Chemistry* **vol. 1**, G. Brauer, Ed. (Academic Press, New York, 2nd ed., 1963) pp 525-530; by pyrolysis of phosphorous acid: Gokhale, Jolly, *Inorg. Syn.* **9**, 56 (1967); by hydrolysis of a metal phosphide such as calcium phosphide:

Klement, *loc. cit.;* Baudler *et al., Z. Anorg. Allgem. Chem.* **353**, 122 (1967). Toxicology: L. T. Fairhall, Industrial Toxicology (Hafner, New York, 2nd ed., 1969) pp 91-92.

Gas. *Poisonous!* Odor of decaying fish; bp −87.7°. mp −133°. Spontaneously flammable in air if there is a trace of P_2H_4 present; burns with a luminous flame. Slightly sol in water (0.26 vol. at 20°). Combines violently with oxygen and the halogens. Liberates hydrogen and forms the phosphide when passed over heated metal. Forms phosphonium salts when brought in contact with the halogen acids. Lowest published lethal concn for hamsters (inhalation): 8 ppm, *Toxic Substances List*, H. E. Christensen, Ed. (1973) p 737.

Caution: Potential symptoms of overexposure are nausea, vomiting, abdominal pain and diarrhea; thirst; chest pressure, dyspnea; muscle pain, chills; stupor or syncope. *See NIOSH Pocket Guide to Chemical Hazards* (DHHS/NIOSH 90-117, 1990) p 180.

7495. Phosphinothricin. *2-Amino-4-(hydroxymethylphosphinyl)butanoic acid;* (3-amino-3-carboxypropyl)methylphosphinic acid; homoalanin-4-yl(methyl)phosphinic acid; 2-amino-4-methylphosphinobutyric acid; glufosinate. $C_5H_{12}NO_4P$; mol wt 181.13. C 33.16%, H 6.68%, N 7.73%, O 35.33%, P 17.10%. First reported naturally occurring amino acid containing a phosphinic acid group. Glutamine synthetase inhibitor with herbicidal activity. Active component of antibiotic peptides produced by several *Streptomyces spp.* The naturally occurring form is the L-isomer. Isoln from *S. viridochromogenes* and synthesis of DL-form: E. Bayer *et al., Helv. Chim. Acta* **55**, 224 (1972). Prepn by hydrolysis of *SF-1293*, also known as *bialaphos*, an antibiotic produced by *S. hydroscopicus:* Y. Ogawa *et al., Japan. Kokai* **73 85,538** (1973 to Meiji Seika Kaisha), *C.A.* **80**, 60035b (1974); *eidem, Meiji Seika Kenkyo Nempo* **13**, 42 (1973), *C.A.* **81**, 37806r (1974). Alternate synthesis: H. Gross, T. Gnauk, *J. Prakt. Chem.* **318**, 157 (1976). Crystal structure and absolute configuration of L-phosphinothricin: E. F. Paulus, S. Grabley, *Z. Kristallogr.* **160**, 63 (1982). Biosynthetic studies: H. Seto *et al., J. Antibiot.* **35**, 1719 (1982); *eidem, ibid.* **36**, 96 (1983). Asymmetric synthesis of enantiomers: N. Minowa *et al., Tetrahedron Letters* **25**, 1147 (1984). Activity and use as herbicide: W. Rupp *et al.,* **Belg.** pat. **854,753;** U.S. pat. **4,168,963** (1977, 1979 both to Hoechst AG); P. Langelüddeke *et al., Meded. Fac. Landbouww. Rijksuniv. Ghent* **47**, 95 (1982), *C.A.* **98**, 48584v (1982). Comparative phytotoxicity: K. L. Carlson, O. C. Burnside, *Weed Sci.* **32**, 841 (1984). Mechanism of action study: R. R. Bellinger *et al., ibid.* **33**, 779 (1985).

L-Form, $[\alpha]_D^{25}$ +13.4° (c = 1 in water).
D-Form, $[\alpha]_D^{25}$ −12.4° (c = 1 in water).
DL-Form, mp 229-231° (dec).
DL-Form hydrochloride, $C_5H_{12}NO_4P \cdot HCl$, crystals from water + ethanol, mp 195-198°.
DL-Form monoammonium salt, $C_5H_{15}N_2O_4P$, *glufosinate-ammonium, HOE-661, HOE-39866, Basta, Total.* Sol in water. LD_{50} in male, female mice, male, female rats (mg/kg): 431, 416, 2000, 1620 orally (Langelüddeke).

USE: DL-Form monoammonium salt as nonselective post-emergent herbicide.

7496. Phosphocreatine. *N-[Imino(phosphonoamino)-methyl]-N-methylglycine; N-(phosphonoamidino)sarcosine;* creatine phosphate; creatinephosphoric acid; PC. $C_4H_{10}N_3O_5P$; mol wt 211.11. C 22.76%, H 4.77%, N 19.90%, O 37.89%, P 14.67%. The distribution of phosphocreatine (PC) in various tissues of a number of vertebrates has been determined; skeletal muscle contains much more than other tissues. Although PC is characteristic of the vertebrates, it also occurs in certain invertebrates. Isoln from frog muscle: Eggleton, *Biochem. J.* **21**, 190 (1927); from cat muscle: Fiske, Subbarow, *J. Biol. Chem.* **81**, 629 (1929). Synthesis by phosphorylation of creatine: Zeile, Fawaz, *Z. Physiol. Chem.* **256**, 193 (1938); Ennor, Stocken, *Biochem. J.* **43**, 190

(1948). Enzymic determn in muscle: K. Yoshikawa *et al.*, *Anal. Biochem.* **159**, 303 (1986). Clinical trial as cardioprotective: M. L. Semenovsky *et al.*, *J. Thorac. Cardiovasc. Surg.* **94**, 762 (1987); as anti-arrhythmic: M. Y. Ruda *et al.*, *Am. Heart J.* **116**, 393 (1988). *Review:* Ennor, Morrison, *Physiol. Rev.* **38**, 631 (1958).

Sodium salt, $C_4H_8N_3Na_2O_5P$, *Creatergyl, Neoton.* Occurs as the hexahydrate, platelets from water + ethanol. Very sol in water. (Free acid, pKa_2 is 4.6).

Calcium salt tetrahydrate, $C_4H_8CaN_3O_5P.4H_2O$, hygroscopic crystals. Soluble in water; sparingly sol in alcohol.

Note: The term *phosphagen*, originally a synonym for phosphocreatine, has since been used to describe any naturally occurring phosphorylated guanidine compd.

THERAP CAT: Cardioprotective.

7497. Phosphocysteamine. *2-Aminoethanethiol dihydrogen phosphate (ester); phosphorothioic acid S-(2-aminoethyl) ester;* cysteamine-S-phosphate; MEAP. $C_2H_8NO_3PS$, mol wt 157.13. C 15.29%, H 5.13%, N 8.91%, O 30.55%, P 19.71%, S 20.41%. $H_2NCH_2CH_2SPO_3H_2$. Radioprotectant. Prepn of sodium salt: S. Åkerfeldt, *Acta Chem. Scand.* **13**, 1479 (1959); of free acid: *idem, ibid.* **14**, 1980 (1960). Radioprotective effect: B. Hansen, B. Sörbo, *Acta. Radiol.* **56**, 141 (1961); A. Vacek *et al.*, *Folia Biol.* **17**, 340 (1971). Mechanism of action: B. Shapiro *et al.*, *Radiat. Res.* **44**, 421 (1970). Radiobiological and biochemical profile: J. W. Harris, T. L. Phillips, *ibid.* **46**, 362 (1971). Photoreactivity: S. A. Grachev *et al.*, *Radiochem. Radioanal. Letters* **19**, 283 (1974); H. Neumann, M. Sokolovsky, *Biochim. Biophys. Acta.* **381**, 292 (1975). Clinical evaluation in cystinosis: L. A. Smolin *et al.*, *Pediat. Res.* **23**, 616 (1988); D. S. Theodoropoulos *et al.*, *J. Am. Med. Assoc.* **270**, 2200 (1993). Crystals from glacial acetic acid, mp 156°. pKa: 2.2, 5.0, 10.3.

Sodium salt, $C_2H_7NNaO_3PS$, *sodium hydrogen-S-(2-aminoethyl)phosphorothioate, cystafos, cystaphos, WR-638.* White crystalline water soluble powder. LD_{50} i.p. in male mice: 0.93 ±0.03 g/kg (Hansen).

USE: Research radioprotective agent.

THERAP CAT: Treatment of cystinosis.

7498. Phosphomolybdic Acid. *Molybdophosphoric acid;* dodecamolybdophosphoric acid. Formula approx $24MoO_3.P_2O_5.xH_2O$. Prepn: Wu, *J. Biol. Chem.* **43**, 189 (1920); Hastings, Frediani, *Anal. Chem.* **20**, 382 (1948). Formula also reported to be $20MoO_3.P_2O_5.51H_2O$: *U.S.P. XXI*, 1398. Review of phosphomolybdic acids: *Mellor's, Vol. XI*, pp 659-672 (1931).

Bright yellow crystals. Sol in less than 0.4 part water; very sol in alcohol, ether.

USE: Weighting silks; as reagent for alkaloids, uric acid, xanthine, creatinine, some metals, with hematoxylin as nerve stain in microscopy.

7499. Phosphonium Iodide. H_4IP; mol wt 161.91. H 2.49%, I 78.38%, P 19.13%. PH_4I. Prepd by hydrolysis of a mixture of diphosphorus tetraiodide and white phosphorus: Work, *Inorg. Syn.* **2**, 141 (1946). Improved apparatus for its prepn: Beredjick, *ibid.* **6**, 91 (1960).

Large, transparent, colorless crystals (usually cubes). Tetragonal system. Sublimes at room temp. *Store in sealed ampuls in refrigerator!* Vapor pressure: 50 mm at 20°; 760 mm at 62.5°. mp 18.5° under its own vapor pressure. Heat of fusion 12,680 cal/mol. *Caution:* Heat or traces of moisture or alcohol cause decompn into PH_3 and HI. Will detonate if heated rapidly.

USE: In the laboratory prepn of phosphine.

7500. Phosphoric Acid. Orthophosphoric acid. H_3O_4P; mol wt 98.00. H 3.09%, O 65.31%, P 31.61%. H_3PO_4. Obtained commercially from phosphate rock deposits in Florida, Tennessee, and the Western United States. Phosphate rock is essentially tricalcium phosphate and one of the

large scale processes is based on the equation: $Ca_3(PO_4)_2 + 3H_2SO_4 + 6H_2O \rightarrow 2H_3PO_4 + 3(CaSO_4.2H_2O)$. Description of various processes: W. H. Waggaman, *Phosphoric Acid, Phosphates and Phosphatic Fertilizers* (Reinhold, New York, 1952); *Phosphoric Acid,* Vol. 1, parts I, II, A. V. Slack, Ed. (Dekker, New York, 1968) 1159 pp; Faith, Keyes & Clark's *Industrial Chemicals,* F. A. Lowenheim, M. K. Moran, Eds. (Wiley-Interscience, New York, 4th ed., 1975) pp 628-639. Prepn of ultrapure, cryst H_3PO_4: Simon, Schulze, *Z. Anorg. Allgem. Chem.* **242**, 322 (1939); Weber, King, *Inorg. Syn.* **1**, 101 (1939). *Reviews:* J. R. Van Wazer, *Phosphorus and its Compounds,* Vol. 1, *Chemistry* (Interscience, New York, 1958) pp 479-491; R. B. Hudson *et al.*, "Phosphoric Acids and Phosphates" in Kirk-Othmer *Encyclopedia of Chemical Technology,* Vol. 17 (Wiley-Interscience, New York, 3rd ed., 1982) pp 426-472.

Unstable, orthorhombic crystals, mp 42.35°, or clear, syrupy liquid; easily supercooled into a glass. Pleasing acid taste when suitably diluted. An acid containing ~88% H_3PO_4 will frequently crystallize on prolonged cooling; forms hemihydrate, mp 29.32°. Becomes anhyd at 150°, gradually changes to pyrophosphoric acid at ~200°, and changes to metaphosphoric acid when heated above 300°. The hot concd acid attacks porcelain and granite ware. May be stored in suitable stainless steel containers. Tribasic acid: pK_1: 2.15; pK_2: 7.09; pK_3: 12.32. Other reported values of dissociation constants and viscosity data reviewed by Van Wazer. The pH of a $0.1N$ aq soln is 1.5. Heat of formation (crystals): -306.2 kcal/mole. Heat of soln: $+2.79$ kcal/mole. Sol in water, alc; sol in 8 vols of a 3:1 ether: alcohol mixture. Properties of phosphoric acid solns. d^{25} 1.8741 (100% soln); 1.6850 (85% soln); 1.3334 (50% soln); 1.0523 (10% soln). Density measurements: Christensen, Reed, *Ind. Eng. Chem.* **47**, 1277 (1950); Egan, Luff, *ibid.* 1280. $n_D^{17.5}$ 1.34203 (10% soln); 1.35032 (20% soln); 1.35846 (30% soln). Spec heat (21.3°): 0.4359 (88% soln).

Caution: Potential symptoms of overexposure are irritation of upper respiratory tract, eyes and skin; burns skin and eyes; dermatitis. *See NIOSH Pocket Guide to Chemical Hazards* (DHHS/NIOSH 90-117, 1990) p 182.

USE: In the manuf of superphosphates for fertilizers, other phosphate salts, polyphosphates, detergents. Acid catalyst in making ethylene, purifying hydrogen peroxide. As acidulant and flavor, synergistic antioxidant and sequestrant in food. Pharmaceutical aid (solvent). In dental cements; process engraving; rustproofing of metals before painting; coagulating rubber latex; as analytical reagent.

THERAP CAT (VET): Has been used in lead poisoning.

7501. Phosphoric Acid, Meta. *Metaphosphoric acid;* glacial phosphoric acid. $(HPO_3)_n$; H 1.26%, O 60.01%, P 38.73%.

Transparent, glass-like solid or soft silky masses; hygroscopic. Volatilizes at red heat. Very slowly sol in cold water, slowly changing to H_3PO_4, the change is hastened by boiling; sol in alcohol. *Keep tightly closed.* For greater convenience it is also marketed in form of rods made by the addition of sodium phosphate.

USE: In dentistry for making zinc oxyphosphate cement; as reagent in chemical analysis.

7502. Phosphorous Acid. *Phosphonic acid.* H_3O_3P; mol wt 82.00. H 3.69%, O 58.54%, P 37.77%. Prepd by hydrolysis of PCl_3 according to the equation $PCl_3 + 3H_2O \rightarrow H_3PO_3 + 3HCl$, the rather violent reaction can be slowed down by the initial presence of concd HCl: Milobedzki, Friedmann, *Chem. Pol.* **15**, 76 (1917); *Chem. Z.* **1918**, I, 933; Simon, Fehér, *Z. Anorg. Allgem. Chem.* **230**, 298 (1937). Alternate procedure carrying out the reaction in carbon tetrachloride: Voight, Gallais, *Inorg. Syn.* **4**, 55 (1953). *Review:* Ohashi, "Lower Oxo Acids of Phosphorus and Their Salts" in *Topics in Phosphorus Chemistry*, Vol. 1, M. Grayson, E. J. Griffith, Eds. (Interscience, New York, 1964) pp 113-187.

White, very hygroscopic and deliquesc, crystalline mass; garlic-like taste; slowly oxidized by oxygen (air) to H_3PO_4. d_4^{21} 1.65; d_4^{76} liq 1.597. mp ~$73°$; above $180°$ is dec into PH_3 and H_3PO_4. Very sol in water, alcohol. pK_1 1.29; pK_2 6.74. Usually marketed as a 20% aq soln.

7503. Phosphorus. P; at. wt 30.97376; at. no. 15; valences 3, 5. One naturally occurring isotope: ^{31}P; artificial, radioactive isotopes: 28-30; 32-34. Abundance in earth's crust: about 0.12%. Does not occur free in nature; found in the form of phosphates in the minerals *chlorapatite* [$3Ca_2(PO_4)_2.CaCl_2$], *fluorapatite* [$3Ca(PO_4)_2.CaF_2$], vivianite, wavellite and "phosphate rock" or phosphorite; occurs in small quantities in granite rocks; occurs in all fertile soil; an essential constituent of protoplasm, nervous tissue and bones. Discovered in 1669 by Brandt. Prepn: Ullmann, *Enzyklopädie der Technischen Chemie* **8**, 362 (1931); De-Witt, Skolnik, *J. Am. Chem. Soc.* **68**, 2305 (1946); Skolnik *et al., ibid.* 2310. Lab prepn and purification: Klement in *Handbook of Preparative Inorganic Chemistry* vol. 1, G. Brauer, Ed. (Academic Press, New York, 2nd ed., 1963) pp 518-525. *Reviews:* J. R. Van Waser, *Phosphorus and Its Compounds,* 2 vols. (Interscience, New York, 1958, 1961) 2046 pp; Corbridge, "The Structural Chemistry of Phosphorus Compounds" in *Topics in Phosphorus Chemistry,* Vol. **3**, E. J. Griffith, M. Grayson, Eds. (Interscience, New York, 1966) pp 57-394; Toy, "Phosphorus" in *Comprehensive Inorganic Chemistry,* Vol. **2**, J. C. Bailar, Jr. *et al.,* Eds. (Pergamon Press, Oxford, 1973) pp 389-545; J. R. Van Wazer in Kirk-Othmer *Encyclopedia of Chemical Technology* vol. 17 (Wiley-Interscience, New York, 3rd ed., 1982) pp 473-490.

Phosphorus exists in three main allotropic forms: white, black, and red. The same liquid is obtained on melting. *Caution:* Avoid contact with $KClO_3$, $KMnO_4$, peroxides and other oxidizing agents; explosions may result on contact or friction. Phosphorus atoms exist as symmetrical, tetrahedral P_4 molecules in the liquid phase and in the vapor phase below $800°$; molecules dissociate to P_2 above $800°$.

White phosphorus. Colorless or white, transparent, cryst solid; waxy appearance; darkens on exposure to light. Sometimes called *yellow phosphorus;* color due to impurities. Two allotropic modifications: α-form exists at room temp; cubic crystals containing P_4 molecules; d 1.83. β-Form prepd by conversion of α-form at $-79.6°$; hexagonal crystals; d 1.88. mp 44.1° (vapor press. 0.181 mm); bp 280°. Volatile; sublimes *in vacuo* at ordinary temp when exposed to light. When exposed to air in the dark, emits a greenish light and gives off white fumes. Solubilities in water: one part/300,000 parts water; in abs alc: one g/400 ml; in abs ether: one g/102 ml; in $CHCl_3$: one g/40 ml; in benzene: one g/35 ml; in CS_2: one g/0.8 ml. Soly in oils: one gram phosphorus dissolves in 80 ml olive oil, 60 ml oil of turpentine, ~100 ml almond oil. Ignites at about 30° in moist air; the ignition temp is higher when the air is dry. *Caution: handle with forceps. Keep under water.* The fumes and the element itself are poisonous. Combines directly with the halogens to form tri- or pentahalides; combines with sulfur to form sulfides. Reacts with several metals to form phosphides. Yields orthophosphoric acid when treated with nitric acid. Reacts with alkali hydroxides with formation of phosphine and sodium hypophosphite. *Incompat.* Sulfur, iodine, oil of turpentine, potassium chlorate.

Black Phosphorus. Polymorphic. Orthorhombic crystalline form: stable in air; resembles graphite in texture; produced from the white modification under high pressures: Bridgman, *J. Am. Chem. Soc.* **36**, 1344 (1914); Jacobs, *J. Chem. Phys.* **5**, 945 (1937); Krebs, *Inorg. Syn.* **7**, 60 (1963). d 2.691. Does not catch fire spontaneously. Insol in organic solvents. Amorphous form prepd at lower pressures: Jacobs, *loc. cit.* At higher pressure the orthorhombic form undergoes reversible transition to a rhombohedral structure, d 3.56, and a cubic structure, d 3.83: Jamieson, *Science* **139**, 1291 (1963).

Red phosphorus. Polymorphism: Roth *et al., J. Am. Chem. Soc.* **69**, 2881 (1947); Corbridge, *loc. cit.* Crystal structure of one form, *Hittorf's phosphorus*: Thurn, Krebs, *Acta Cryst.* **25B**, 125 (1969). Red to violet powder. The properties of red phosphorus are intermediate between those of the white and black forms. Sublimes at 416°, triple point 589.5° under 43.1 atm. d 2.34. Insol in organic solvents. Sol in phosphorus tribromide. Less active than the white form; reacts only at high temp. Yields the white modification when distilled at 290°. Catches fire when heated in air to about 260° and burns with formation of the pentoxide. Burns when heated in an atmosphere of chlorine.

Caution: Ingestion of even small amounts of white phosphorus may produce severe G.I. irritation, bloody diarrhea, liver damage, skin eruptions, oliguria, circulatory collapse, coma, convulsions, death. The approx fatal dose is 50 to 100 mg. External contact may cause severe burns. Chronic poisoning (from ingestion or inhalation) is characterized by boney necrosis, especially of the mandible, spontaneous fractures, anemia, weight loss. Red phosphorus is relatively nontoxic unless it contains the white form as an impurity: *Clinical Toxicology of Commercial Products,* M. N. Gleason *et al.,* Eds. (Williams & Wilkins, Baltimore, 1969) Section III, pp 192-195.

USE: White phosphorus: manuf rat poisons; for smoke screens, gas analysis. Red phosphorus: pyrotechnics; manuf safety matches; in organic synthesis; manuf phosphoric acid, phosphine, phosphoric anhydride, phosphorus pentachloride, phosphorus trichloride; manuf fertilizers, pesticides, incendiary shells, smoke bombs, tracer bullets.

7504. Phosphorus Hemitriselenide. P_2Se_3; mol wt 298.83. P 20.73%, Se 79.27%. Prepd by warming a mixture of phosphorus and selenium in the correct proportions: Hahn, *J. Prakt. Chem.* [1] **93**, 430 (1864); Muthmann, Clever, *Z. Anorg. Chem.* **13**, 191 (1897); Meyer, *ibid.* **30**, 258 (1902); **61B**, 1807 (1928).

Dark red mass. Dec by heat. Dec in moist air and water. Sol in potassium hydroxide; insol in carbon disulfide.

7505. Phosphorus Oxybromide. *Phosphoryl bromide;* phosphoryl tribromide. Br_3OP; mol wt 286.69. Br 83.62%, O 5.58%, P 10.80%. $POBr_3$. Prepd according to the equation: $3PBr_5 + P_2O_5 \rightarrow 5POBr_3$: Hönigschmid, Hirschbold-Wittner, *Z. Anorg. Allgem. Chem.* **243**, 355 (1940); Johnson, Nunn, *J. Am. Chem. Soc.* **63**, 141 (1941); Booth, Seegmiller, *Inorg. Syn.* **2**, 151 (1946).

Thin plates, faint orange tint, d 2.822. mp 56° (in hot water). bp_{758} 193° (dec). Slowly hydrolyzes in water forming H_3PO_4 and HBr. Sol in ether, benzene, chloroform, carbon disulfide, concd H_2SO_4. *Store in sealed glass ampuls!*

7506. Phosphorus Oxychloride. *Phosphoryl chloride;* phosphorus chloride. Cl_3OP; mol wt 153.33. Cl 69.36%, O 10.43%, P 20.20%. $POCl_3$. Manuf: Faith, Keyes & Clark's *Industrial Chemicals,* F. A. Lowenheim, M. K. Moran, Eds. (Wiley-Interscience, New York, 4th ed., 1975) pp 646-649.

Colorless, clear, strongly fuming liquid; pungent odor. d^{25} 1.645. bp 105.8°; mp 1.25°: J. R. Van Wazer, *Phosphorus and its Compounds—Chemistry,* vol. 1 (Interscience, New York, 1958) p 254. Reacts exothermically with water, alc. *Keep in tightly closed containers.*

Caution: Intensely irritating to skin, eyes, mucous membranes. Inhalation may cause pulmonary edema.

USE: As chlorinating agent, especially to replace oxygen in organic compounds; as solvent in cryoscopy.

7507. Phosphorus Pentabromide. Phosphoric bromide; phosphorus perbromide. Br_5P; mol wt 430.49. Br 92.81%, P 7.19%. PBr_5. Review: Payne, "The Chemistry of Phosphorus Halides" in *Topics in Phosphorus Chemistry,* Vol. 4, M. Grayson, E. J. Griffith, Eds. (Interscience, New York, 1967) pp 85-155.

Yellow, cryst mass. mp > 100° with decompn. Dec by water or alcohol; sol in carbon disulfide or tetrachloride. *Keep tightly closed. Corrosive.*

USE: Brominating agent for converting organic acids to acyl bromides.

7508. Phosphorus Pentachloride. Phosphoric chloride; phosphorus perchloride. Cl_5P; mol wt 208.24. Cl 85.13%, P 14.87%. PCl_5. Prepn: Maxson, *Inorg. Syn.* **1**, 99 (1939).

Review: Payne, "The Chemistry of Phosphorus Halides" in *Topics in Phosphorus Chemistry,* Vol. 4, M. Grayson, E. J. Griffith, Eds. (Interscience, New York, 1967) pp 85-155.

White to pale yellow, fuming, deliquesc, cryst mass; pungent, unpleasant odor; attacks the eyes and mucous membranes. mp 148° under pressure; sublimes at about 100° without melting. bp 160°. Hydrolyzed by water to form phosphoric acid and hydrogen chloride. Reacts with alcohols (ROH) to form the corresponding chloride (RCl). Sol in carbon disulfide or tetrachloride. *Keep in tightly closed containers and handle with caution. Corrosive.*

Caution: Potential symptoms of overexposure are irritation of eyes and respiratory system; bronchitis; dermatitis. *See NIOSH Pocket Guide to Chemical Hazards* (DHHS/NIOSH 90-117, 1990) p 182.

USE: As catalyst in manuf acetylcellulose; for replacing hydroxyl groups by Cl, particularly for converting acids into acid chlorides.

7509. Phosphorus Pentafluoride. F_5P; mol wt 125.97. F 75.41%, P 24.59%. PF_5. Prepd by treating PF_3 with bromine to form PF_3Br which then disproportionates to PF_5 and PBr_5: Moissan, *Compt. Rend.* **100,** 1348; **101,** 1490 (1885); by heating P_2O_5 with CaF_2: Lucas, Ewing, *J. Am. Chem. Soc.* **49,** 1270 (1927); Booth, Bozarth, *ibid.* **55,** 3890 (1933); from PCl_5 and AsF_3: Thorpe, *Ann.* **182,** 201 (1876); *Proc. Roy. Soc.* **25,** 122 (1877); O. Ruff, *Die Chemie des Fluors* (Berlin, 1920), p 29; Kwasnik in *Handbook of Preparative Inorganic Chemistry,* vol. 1, G. Brauer, Ed. (Academic Press, New York, 2nd ed., 1963) pp 190-191; from PCl_3 and CaF_2: Muetterties *et al., J. Inorg. Nucl. Chem.* **16,** 52 (1960); from phosphoryl fluoride, hydrogen fluoride and sulfur trioxide: Wiesboeck, U.S. pats. **3,584,999; 3,592,594** (1971 to U.S. Steel). *Reviews:* Burg in *Fluorine Chemistry,* vol. 1, J. H. Simons, Ed. (Academic Press, New York, 1950) pp 97-98; Kemmitt, Sharp, *Advan. Fluorine Chem.* **4,** 197-198 (1965); Schmutzler, *ibid.* **5,** 32-285 (1965).

Colorless gas. Fumes strongly in air. d (gas) 5.805 g/l. mp −93.8°. bp −84.6°. Trouton constant 21.8. Dipole moment: zero. High thermal stability. Does not attack dry glass even at 250°, but a slight trace of moisture leads to formation of POF_3 and HF. Water hydrolysis ultimately yields phosphoric acid; intermediates are oxyfluorophosphates. Lewis acid; forms complexes with amines, ethers, nitrates, sulfoxides, organic bases. Forms a cryst addn product $PF_5.NO_2$ at −10° which dissociates on warming. May be stored in steel cylinders.

Caution: Intensely irritating to skin, eyes, mucous membranes. Inhalation may cause pulmonary edema.

USE: Catalyst in ionic polymerization reactions.

7510. Phosphorus Pentaselenide. P_2Se_5; mol wt 456.75. P 13.56%, Se 86.44%. Prepd by melting a mixture of phosphorus and selenium in an atm of CO_2 or N_2: Carius, Bogen, *Ann.* **124,** 57 (1862); Kudchadker *et al., Can. J. Chem.* **46,** 1415 (1968).

Amorphous, glass, black-purple solid. Dec in steam and boiling water. Reacts with CCl_4; insol in carbon disulfide.

7511. Phosphorus Pentasulfide. Phosphoric sulfide; thiophosphoric anhydride; phosphorus persulfide. P_2S_5; mol wt 222.28. P 27.87%, S 72.13%. Exists as P_4S_{10}. Conveniently prepd by fusing red phosphorus with sulfur: Stock, Herscovici, *Ber.* **43,** 1223 (1910); Klement in *Handbook of Preparative Inorganic Chemistry,* vol. 1, G. Brauer, Ed. (Academic Press, New York, 2nd ed., 1963) p 567. Manuf: Faith, Keyes & Clark's *Industrial Chemicals,* F. A. Lowenheim, M. K. Moran, Eds. (Wiley-Interscience, New York, 4th ed., 1975) pp 650-653.

Light yellow, triclinic crystals; peculiar odor. d 2.09. mp 286-290°; bp 513-515°. Dec by water forming H_3PO_4 and H_2S; sol in carbon disulfide, in aq soln of alkali hydroxides. *Keep tightly closed.*

Caution: Potential symptoms of overexposure are apnea, coma and convulsions; irritation of eyes, conjunctival pain, lacrimation, photophobia, kerato-conjunctivitis and corneal vesiculation; irritation of respiratory system; dizziness; headache; fatigue; irritability, insomnia; gastrointestinal disturbance. Direct contact may cause skin irritation. *See NIOSH Pocket Guide to Chemical Hazards* (DHHS/NIOSH 90-117, 1990) p 182.

USE: In manuf of lube oil additives and pesticides. Manuf safety matches, ignition compds, and for introducing sulfur into organic compds.

7512. Phosphorus Pentoxide. Phosphoric anhydride; diphosphorus pentoxide. O_5P_2; mol wt 141.94. O 56.36%, P 43.64%. P_2O_5. Exists as P_4O_{10}. Prepd commercially by burning phosphorus in a current of dry air. Purification: Manley, *J. Chem. Soc.* **121,** 331 (1922); de Decker, McGillavry, *Rec. Trav. Chim.* **60,** 153, 413 (1941). Review of phosphorus oxides: J. R. Van Waser, *Phosphorus and its Compounds—Chemistry,* vol. 1 (Interscience, New York, 1958) pp 267-286.

Very deliquescent crystals. *Corrosive.* Several crystalline and amorphous modifications. Commercial form; hexagonal; d 2.30. mp 340°. Sublimation temperature 360°. Nonflammable. Does not support combustion. Heat of formation: −365.83 kcal/mole. Sp heat: 0.170 cal/g/°C. Heat of fusion: 8.2 kcal/mole. Heat of volatilization: 22.7 kcal/mole of P_4O_{10}. Readily absorbs moisture from the air. Exothermic hydrolysis by water to form phosphoric acid. The reaction with alcohol is similar.

Caution: Strong irritant; corrosive to skin, mucous membranes and eyes.

USE: Drying and dehydrating agent. Condensing agent in organic synthesis.

7513. Phosphorus Sulfochloride. *Thiophosphoryl chloride.* Cl_3PS; mol wt 169.40. Cl 62.79%, P 18.28%, S 18.93%. $PSCl_3$. Prepd from P_2S_5 + PCl_5: Martin, Duvall, *Inorg. Syn.* **4,** 73 (1953). Alternate procedure from PCl_3, $AlCl_3$ and S: Moeller *et al., ibid.* 71.

Fuming liq, crystallizes as α-form at −40.8° or as β-form at −36.2°. bp_{760} 125°. n_D^{25} 1.635. Sol in benzene, carbon tetrachloride, carbon disulfide, chloroform. Hydrolyzes slowly in water, rapidly in alkaline solns. In water the hydrolysis products are orthophosphoric acid, hydrochloric acid, and hydrogen sulfide.

Caution: Strong irritant.

7514. Phosphorus Tribromide. Phosphorous bromide. Br_3P; mol wt 270.69. Br 88.56%, P 11.44%. PBr_3. Prepn: Gray, Maxson, *Inorg. Syn.* **2,** 147 (1946). *Review:* Payne, "The Chemistry of Phosphorus Halides" in *Topics in Phosphorus Chemistry,* vol. 4, M. Grayson, E. J. Griffith, Eds. (Interscience, New York, 1967) pp 85-155.

Colorless, fuming liquid; very penetrating odor. d^{15} 2.85. mp −41.5°. bp 173.2°. Vapor pressure: 10 mm (48°). Dissolved and dec by water or alc; sol in acetone, carbon disulfide. *Keep tightly closed. Corrosive.*

7515. Phosphorus Trichloride. Phosphorous chloride. Cl_3P; mol wt 137.33. Cl 77.45%, P 22.55%. PCl_3. Prepd from red phosphorus and dry chlorine in the presence of refluxing PCl_3: Forbes *et al., Inorg. Syn.* **2,** 145 (1946). Manuf: Faith, Keyes & Clark's *Industrial Chemicals,* F. A. Lowenheim, M. K. Moran, Eds. (Wiley-Interscience, New York, 4th ed., 1975) pp 654-657. *Review:* Payne, "Chemistry of Phosphorus Halides" in *Topics in Phosphorus Chemistry,* vol. 4, M. Grayson, E. J. Griffith, Eds. (Interscience, New York, 1967) pp 85-155.

Colorless, clear, fuming liquid. d_4^{21} 1.574. mp −112°. bp 76°. Vapor pressure: 100 mm (21°). Decomposed by water or alc. Sol in benzene, chloroform, ether, carbon disulfide. *Keep in tightly closed containers and handle with caution.*

Caution: Potential symptoms of overexposure are irritation of eyes, nose and throat; pulmonary edema; burns eyes and skin. *See NIOSH Pocket Guide to Chemical Hazards* (DHHS/NIOSH 90-117, 1990) p 184.

USE: As of phosphorus oxychloride; manuf $POCl_3$, PCl_5; producing iridescent metallic deposits.

7516. Phosphorus Trifluoride. F_3P; mol wt 87.97. F 64.79%, P 35.21%. PF_3. Prepd by halogen exchange between PCl_3 and AsF_3: Moissan, *Compt. Rend.* **100,** 272 (1885); Hoffman, *Inorg. Syn.* **4,** 149 (1953); between PCl_3 and ZnF_2: Williams, *ibid.* **5,** 95 (1957); between PCl_3 and CaF_2 or SbF_3: Booth, Bozarth, *J. Am. Chem. Soc.* **61,** 2927 (1939); Muetterties *et al., J. Inorg. Nucl. Chem.* **16,** 52 (1960); between PCl_3 and HF: Kwasnik in *Handbook of Preparative Inorganic Chemistry,* vol. 1, G. Brauer, Ed. (Academic Press, New York, 2nd ed., 1963) pp 189-190.

Reviews: Burg in *Fluorine Chemistry,* vol. **1**, J. H. Simons, Ed. (Academic Press, New York, 1950) pp 98-100; Kemmitt, Sharp, *Advan. Fluorine Chem.* **4**, 198-199 (1965); Schmutzler, *ibid.* **5**, 32-285 (1965).

Colorless gas. Does not fume in air. Does not attack glass except at high temps. *Poisonous!* d (gas) 3.907 g/l. mp −151.30°. bp −101.38°. Critical temp −2.05°; critical press. 42.69 atm. May be stored in steel cylinders or in glass, also in a gasometer over Hg. Slowly hydrolyzed by water. Absorbed by aq bases at a rate increasing with pH, producing a fluophosphite which boiling aq HNO_3 does not convert to phosphate. Anhyd KOH may be used to dry PF_3 with little loss. Dry NH_3 forms a solid addn product. Aq oxidizers such as chromic acid, permanganate, or bromine rapidly destroy PF_3, and alcohols convert it to alkyl phosphite. Phosphides and fluorides are formed upon reaction with hot metals.

Caution: See Phosphorus Pentafluoride.

7517. Phosphorus Trioxide. Diphosphorus trioxide. O_3P_2; mol wt 109.95. O 43.66%, P 56.34%. P_2O_3. Exists as P_4O_6. Prepd by treating PCl_3 with tetramethylammonium sulfite in liq SO_2: Jander *et al., Ber.* **77**, 689 (1944). Alternate prepn from the elements: Thorpe, Tutton, *J. Chem. Soc.* **57**, 545 (1890); Miller, *ibid.* **1928**, 1847; **1929**, 1823; Wolf, Schmager, *Ber.* **62**, 779 (1929). Chemistry: Riess, Van Wazer, *Inorg. Chem.* **5**, 178 (1966).

Transparent monoclinic crystals or colorless liquid. *Very poisonous!* d_4^{21} 2.135. mp 23.8°. bp 173.1° in nitrogen atm. Disproportionates into red P and P_2O_4 when heated above 210°. Sol in benzene, carbon disulfide. When placed in cold water, H_3PO_3 is formed slowly. Hot water produces a violent reaction with the formation of red phosphorus, phosphine, and H_3PO_4.

7518. Phosphorus Triselenide. Tetraphosphorus triselenide. P_4Se_3; mol wt 360.78. P 34.34%, Se 65.66%. Prepn: Meyer, *Z. Anorg. Chem.* **30**, 258 (1902); **61B**, 1807 (1928); Irgolic *et al., Inorg. Chem.* **4**, 1421 (1965).

Orange-red crystals. Irritating odor. mp 245-246°. d 1.31. bp 360-400°. Phosphoresces at 160°. Flammable when heated in air. Dec in moist air. Sol in carbon tetrachloride, carbon disulfide, chloroform, benzene, toluene, acetone, acetylene dichloride, acetylene trichloride.

7519. Phosphorylcholine. *N,N,N-Trimethyl-2-(phosphonooxy)ethanaminium chloride; choline chloride dihydrogen phosphate;* (2-hydroxyethyl)trimethylammonium chloride phosphate; phosphorylcholine chloride; choline phosphate chloride; choline chloride phosphate; choline phosphoric acid ester (chloride). $C_5H_{15}ClNO_4P$; mol wt 219.60. C 27.35%, H 6.88%, Cl 16.14%, N 6.38%, O 29.14%, P 14.10%. $[(CH_3)_3^+NCH_2CH_2OPO_3H_2]Cl^-$. Prepd by phosphorylation of choline chloride with diphenylphosphoryl chloride: Baer, McArthur, *J. Biol. Chem.* **154**, 451 (1944); Baer, *J. Am. Chem. Soc.* **69**, 1253 (1947); *Biochem. Prepn.* **2**, 96 (1952); by heating choline or choline chloride with pyrophosphoric or polyphosphoric acid: Cherbuliez, Rabinowitz, *Helv. Chim. Acta* **42**, 1154 (1959).

Barium salt, $C_5H_{13}BaClNO_4P$, hydrated glistening leaflets from water + alcohol. Must be dried *in vacuo* (0.5 mm) over P_2O_5 at 100° for 16 hrs. Freely sol in water, practically insol in abs ethanol, ether.

Calcium salt, $C_5H_{13}CaClNO_4P$, *Colifos, Epafosforil, Fosfocolina, Isocolin.*

Magnesium salt, $C_5H_{13}ClMgNO_4P$, *Heparexine.*

THERAP CAT: Hepatobiliary dysfunction.

7520. Phosphoserine. *Serine dihydrogen phosphate (ester);* serine phosphate; O-phospho-DL-serine. $C_3H_8NO_6P$; mol wt 185.07. C 19.47%, H 4.36%, N 7.57%, O 51.87%, P 16.74%. Prepn: Neuhaus, Korkes, *Biochem. Prepn.* **6**, 75 (1958); D. M. Theodoropoulos, *J. Chem. Soc.* **1960**, 5257. Other methods of prepn and isoln from protein hydrolysates: Greenstein, Winitz, *Chemistry of the Amino Acids* (Wiley, New York, 1961), *passim.* Crystal structure: G. H. McCallum, *Nature* **184**, 1863 (1959).

Crystals from ethanol + ether, mp 166-167° (dec).

Note: Used in combination with L-glutamine and vitamin B_{12} as a roborant: *Fosforina B_{12}, Vitasprint B_{12}.*

7521. Phosphotungstic Acid. *Tungstophosphoric acid.* Approx composition $24WO_3.2H_3PO_4.48H_2O$. The H_2O content may vary appreciably. Prepn: Bailar, *Inorg. Syn.* **1**, 132 (1939).

White or slightly yellowish-green, slightly efflorescent crystals or cryst powder. Sol in about 0.5 part water. Also sol in alcohol and in ether.

USE: As reagent for alkaloids and many other nitrogen bases, for phenols, albumin, peptone, amino acids, uric acid, urea, blood, carbohydrates; as biological stain.

7522. Phosvitin. A phosphoprotein of mol wt about 40,000. Isoln from egg yolk: Mecham, Olcott, *J. Am. Chem. Soc.* **71**, 3670 (1949); *Biochem. Prepn.* **2**, 15 (1952); Joubert, Cook, *Can. J. Biochem. Physiol.* **36**, 399 (1958); from roe: Mano, Yoshida, *J. Biochem. (Tokyo)* **66**, 105 (1969). Represents about 7% of the total yolk protein; high in phosphorus content and a ratio of serine to all other amino acid residues of 6:4. *In vitro* synthesis in hens liver: Heald, McLachlan, *Biochem. J.* **94**, 32 (1965); Rudack, Wallace, *Biochim. Biophys. Acta* **155**, 299 (1968). Estradiol-induced synthesis in roosters: Veuving, Gruber, *ibid.* **232**, 524, 529 (1971); Jailkhani, Talwar, *Nature New Biol.* **236**, 239 (1972). Amino acid sequence: Belitz, *Angew. Chem.* **76**, 574 (1964); Shainkin, Perlmann, *J. Biol. Chem.* **246**, 2278 (1971). Conformation: Grizzuti, Perlmann, *ibid.* **245**, 2573 (1970). Anticlotting activity studies: Sato *et al., Am. J. Physiol.* **203**, 1170 (1962). *Review:* Taborsky, *Advan. Protein Chem.* **28**, 50-78 (1974).

THERAP CAT: Anticoagulant.

7523. Phoxim. *4-Ethoxy-7-phenyl-3,5-dioxa-6-aza-4-phosphaoct-6-ene-8-nitrile 4-sulfide; phenylglyoxylonitrile oxime O,O-diethyl phosphorothioate; O,O-diethyl O-(α-cyanobenzylideneamino)phosphorothioate;* α-[[(diethoxyphosphinothioyl)oxy]imino]benzeneacetonitrile; Bay 5621; Bay 77488; Baythion; Sebacil; Volaton. $C_{12}H_{15}N_2O_3PS$; mol wt 298.30. C 48.32%, H 5.07%, N 9.39%, O 16.09%, P 10.38%, S 10.75%. Prepn: **Neth.** pat. **Appl.** 6,605,907; W. Lorenz *et al.,* U.S. pat. 3,591,662 (1966, 1971 both to Bayer). Activity: C. R. Harris, *J. Econ. Entomol.* **63**, 782 (1970); D. C. Read, *ibid.* **69**, 429 (1976). Toxicity studies: J. H. Vinopal, T. R. Fukuto, *Pestic. Biochem. Physiol.* **1**, 44 (1971). Degradation in soil: G. Dräger, *Pflanzenschutz-Nachr.* **30**, 28 (1977).

Pale yellow oil, $bp_{0.01}$ 102°. n_D^{20} 1.5405. d_4^{20} 1.176. Sol in alc, ketones, aromatic hydrocarbons. LD_{50} orally in mice: > 2000 mg/kg (Vinopal, Fukuto).

USE: Insecticide.

7524. Phrenosin. *[1S-[1R*(S*),2S*,3E]]-N-[1-[(β-D-Galactopyranosyloxy)methyl]-2-hydroxy-3-heptadecenyl]-2-hydroxytetracosanamide.* $C_{48}H_{93}NO_9$; mol wt 828.27. C 69.61%, H 11.32%, N 1.69%, O 17.39%. Cerebroside easily separated from a commercially available beef spinal cord lipid concentrate or from nerve tissue. A large proportion of its fatty acid content (25%) is 2-hydroxystearic acid, the rest is cerebronic (2-hydroxylignoceric) acid. Isoln: Radin *et al., J. Biol. Chem.* **219**, 977 (1956); Skipski *et al., Arch. Biochem. Biophys.* **82**, 487 (1959). Synthesis: Shapiro, Flowers, *J. Am. Chem. Soc.* **83**, 3327 (1961).

Consult the Name Index before using this section. **Page 1267**

Cryst from pyridine + acetone, methanol + chloroform, toluene + alcohol (1:1) or chloroform + alcohol + water, [α] +4.5° (c = 2 in pyridine). IR spectrum: H. P. Schwarz et al., *Ann. N.Y. Acad. Sci.* **69**, 116 (1957). Sol in hot dioxane, 1,1-dichloro-1-nitroethane, butanol and acetonitrile + chloroform.

7525. Phthalamide. *1,2-Benzenedicarboxamide;* phthalic acid diamide. $C_8H_8N_2O_2$; mol wt 164.16. C 58.53%, H 4.91%, N 17.06%, O 19.49%. Prepd from phthalic anhydride and anhydr ammonia in benzene: Dominikiewicz, *Arch. Chem. Farm.* **3**, 141 (1937).

Minute crystals, mp 228° (commercial grades mp 221-223°). Melts with decomposition to NH_3 and phthalimide. Slightly sol in cold water, methanol. More sol in hot solvents. Heat of combustion 921.7 cal/mol. Boiling of aq or alcoholic solns produces NH_3 and phthalimide. The presence of acids accelerates decompn.

7526. Phthalazine. 2,3-Benzodiazine; benzo[d]pyridazine; β-phenodiazine. $C_8H_6N_2$; mol wt 130.15. C 73.83%, H 4.65%, N 21.52%. Prepd by reduction of 1-chlorophthalazine with hydriodic acid and red phosphorus: Gabriel, Eschenbach, *Ber.* **30**, 3024 (1897); Paul, *Ber.* **32**, 2015 (1899); by redn of 1-hydrazinophthalazine: Armarego, *J. Appl. Chem.* **11**, 70 (1961); from o-phthalaldehyde + hydrazine sulfate: Smith, Otremba, *J. Org. Chem.* **27**, 879 (1962); from 1,2-benzodinitrile: Carter, Cheeseman, *Org. Prep. Proced. Int.* **6**, 67 (1974).

Pale yellow needles from ether, mp 90-91°. bp 315-317° (dec); bp_{29} 189°; bp_{17} 175°. uv max (water): 218, 261, 292, 305 nm (log ε 4.83, 3.53, 3.18, 3.11). Freely sol in water; sol in ethanol, methanol, benzene, ethyl acetate; less sol in ether. Practically insol in ligroin.
Hydrochloride, $C_8H_6N_2 \cdot HCl$, needles from ether, mp 231° (efferv).

7527. Phthalic Acid. *1,2-Benzenedicarboxylic acid.* $C_8H_6O_4$; mol wt 166.13. C 57.84%, H 3.64%, O 38.52%. Manufactured by catalytic oxidation of o-toluic acid and oxidation of xylene: Taylor, Dean, **U.S. pat. 3,064,046** (1962 to I.C.I.); Cier, **U.S. pat. 3,088,974** (1963 to Esso Res. & Eng.). Isoln from the fungus, *Gibberella fujikuroi:* Cross et al., *J. Chem. Soc.* **1963**, 2937. Toxicity studies: C. B. Shaffer et al., *J. Ind. Hyg. Toxicol.* **27**, 130 (1945); A. R. Singh et al., *J. Pharm. Sci.* **61**, 59 (1972).

Crystals, mp ∼230° when rapidly heated, forming phthalic anhydride and water. One gram dissolves in 160 ml water, 10 ml alc, 205 ml ether, 5.3 ml methanol. Practically insol in chloroform. LD_{50} orally in rats: 7.9 g/kg (Shaffer).
Ethyl ester, $C_{12}H_{14}O_4$, *ethyl phthalate, diethyl phthalate, Neantine, Palatinol A.* Colorless, practically odorless, oily liq; bitter disagreeable taste. d_4^{14} 1.232. bp 295°. Flash pt 140°. n_D^{14} 1.5049. Insol in water. Misc with alcohol, ether and many other organic solvents. LD_{50} i.p. in rats: 5.06 ml/kg (Singh).
Caution: Irritating to mucous membranes, and in high concns, narcotic.
USE: Ethyl phthalate used in manuf celluloid; solvent for cellulose acetate in manuf varnishes and dopes; fixative for perfumes; denaturing alc.

7528. Phthalic Anhydride. *1,3-Isobenzofurandione.* $C_8H_4O_3$; mol wt 148.12. C 64.87%, H 2.72%, O 32.41%. Prepd from naphthalene by oxidation with a mixture of $HgSO_4$ and $CuSO_4$ in presence of H_2SO_4; by passing naphthalene and oxygen over a suitable catalyst at 400-500°. Review of mfg processes: Faith, Keyes & Clark's *Industrial Chemicals,* F. A. Lowenheim, M. K. Moran, Eds (Wiley-Interscience, New York, 4th ed., 1975) pp 658-665.

White, lustrous needles. d 1.53; mp 130.8°; bp 295°. Sublimes. Sol in 162 parts water, more in hot water with conversion into phthalic acid, in 125 parts carbon disulfide; sol in alcohol, sparingly in ether.
Caution: Potential symptoms of overexposure are conjunctivitis; nasal ulcer bleeding, upper respiratory irritation; bronchitis, bronchial asthma; dermatitis. *See NIOSH Pocket Guide to Chemical Hazards* (DHHS/NIOSH 90-117, 1990) p 184.
USE: Manuf phthaleins, phthalates, benzoic acid, synthetic indigo, artificial resins (glyptal).

7529. Phthalimide. *1H-Isoindole-1,3(2H)-dione.* C_8H_5-NO_2; mol wt 147.13. C 65.31%, H 3.43%, N 9.52%, O 21.75%. Prepd from phthalic anhydride and NH_4OH or $(NH_4)_2CO_3$: W. A. Noyes, P. K. Porter, *Org. Syn.* coll. vol. I, 457 (2nd ed., 1941). Teratogenicity study: K. Fickentscher et al., *Pharmazie* **31**, 172 (1976).

Monoclinic prisms from water or by sublimation. mp 238°. Has slightly acidic properties, Ka = 5×10^{-9}. Absorption spectrum: Hartley, Hedley, *J. Chem. Soc.* **91**, 317 (1907). Slightly soluble in water. A satd aq soln contains 0.036 g at 25°; 0.07 g at 40°; about 0.4 g at the bp. 100 g of boiling alcohol dissolves 5 g phthalimide. Almost insol in benzene, petr ether; fairly sol in boiling acetic acid; freely sol in aq alkali hydroxides.

7530. Phthalofyne. *1,2-Benzenedicarboxylic acid mono(1-ethyl-1-methyl-2-propynyl) ester; phthalic acid 1-ethyl-1-methyl-2-propynyl ester;* 1-ethyl-1-methyl-2-propynyl acid phthalate; 3-methyl-1-pentyn-3-yl acid phthalate; ftalofyne; NSC-25614; Whipcide. $C_{14}H_{14}O_4$; mol wt 246.26. C 68.28%, H 5.73%, O 25.99%. Prepn: Sugimoto, Okumura, **Japan. pat. 1833('54)** (Tanabe), *C.A.* **49**, 11711e (1955); **Brit. pat. 736,993** (1955 to Schering).

Crystals from benzene or hexane, mp 96-98°. Weak acid. Slightly sol in water. Unstable in strong alkali.

THERAP CAT (VET): Anthelmintic.

7531. Phthaloyl Chloride. *1,2-Benzenedicarbonyl dichloride.* $C_8H_4Cl_2O_2$; mol wt 203.02. C 47.33%, H 1.99%, Cl 34.92%, O 15.76%. Obtained by the action of PCl_5 on phthalic anhydride.

Colorless, oily liquid. d^{20} 1.409. Solidif +12°; mp 15-16°; bp 280-282°; n_D^{20} 1.5692. Dec by water or alcohol; sol in ether. *Keep tightly closed.*

7532. Phthalylsulfacetamide. *2-[[[4-[(Acetylamino)sulfonyl]phenyl]amino]carbonyl]benzoic acid; 4'-(acetylsulfamyl)phthalanilic acid;* phthalylsulfacetimide; *N¹-acetyl-N⁴-phthaloylsulfanilamide; N¹-acetyl-N⁴-phthaloylsulfanilamide;* phthaloylsulfacetamide; ftalicetimida; *N-[p-(o-carboxybenzamido)benzenesulfonyl]acetamide; N-(o-carboxybenzoyl)-sulfacetamide;* Enterocid; Enterosulfamid; Enterosulfon; Talecid; Thalamyd; Rabalan; Sterathal. $C_{16}H_{14}N_2O_6S$; mol wt 362.36. C 53.03%, H 3.89%, N 7.73%, O 26.49%, S 8.85%. Prepn: Basu, *J. Indian Chem. Soc.* **26**, 130 (1949).

Needles from dil alcohol, mp 196°. Soluble in alcohol. Very sparingly sol in water. Forms a water-sol sodium salt.

THERAP CAT: Antibacterial.

THERAP CAT (VET): Antibacterial.

7533. Phthalylsulfathiazole. *2-[[[4-[(2-Thiazolylamino)sulfonyl]phenyl]amino]carbonyl]benzoic acid; 4'-(2-thiazolylsulfamyl)phthalanilic acid;* 2-(N⁴-phthalylaminobenzenesulfonamido)thiazole; 2-(N⁴-phthalylsulfanilamido)thiazole; phthalylsulfonazole; AFI-Ftalyl; Entexidina; Ftalazol; Intestiazol; Sulfathalidine; Sulftalyl; Taleudron; Talidine; Thalazole; Ultratiazol. $C_{17}H_{13}N_3O_5S_2$; mol wt 403.44. C 50.61%, H 3.25%, N 10.42%, O 19.83%, S 15.90%. May be prepd by condensing sulfathiazole with phthalic anhydride: M. L. Moore, U.S. pats. **2,324,013; 2,324,015** (1943 to Sharp & Dohme). Prepn of 8-hydroxyquinoline salt: Ger. pat. **1,008,296** (1957 to Geistlich Söhne). Toxicity data: P. A. Mattis *et al., J. Pharmacol. Exp. Ther.* **81**, 116 (1944).

Crystals. Slightly bitter taste. Darkens on prolonged exposure to light. Effervesces at 244° to 250°. mp 272-277°

(dec) when the melting point bath is preheated to 220-225°. Slightly sol in alcohol; very slightly sol in ether; readily sol in NaOH or KOH soln, ammonia water and concd HCl. Practically insol in chloroform and water. LD_{50} i.p. in mice: 920 mg/kg (Mattis).

8-Hydroxyquinoline salt, *Ilentazol.* Yellow crystals from dil methanol, dec above 220°. Split into its components by dil acids.

THERAP CAT: Antibacterial.

THERAP CAT (VET): Antibacterial.

7534. Phthiocol. *2-Hydroxy-3-methyl-1,4-naphthalenedione; 2-hydroxy-3-methyl-1,4-naphthoquinone.* $C_{11}H_8O_3$; mol wt 188.18. C 70.21%, H 4.28%, O 25.51%. Antibiotic substance produced by *Mycobacterium tuberculosis:* Terni, *Boll. Soc. Ital. Biol. Sper.* **25**, 60 (1949), *C.A.* **45**, 2054f (1951). Possesses some vitamin K activity. Isoln and synthesis: Anderson, Newman, *J. Biol. Chem.* **101**, 773 (1933); **103**, 197, 405 (1933); L. F. Fieser, *ibid.* **133**, 391 (1940); Tarbell *et al., J. Am. Chem. Soc.* **72**, 379 (1950); Burton, Praill, *J. Chem. Soc.* **1952**, 755; Eistert, Müller, *Ber.* **92**, 2071 (1959); H. Kallmayer, *Arch. Pharm.* **307**, 806 (1974); K. Maruyama, S. Arakawa, *J. Org. Chem.* **42**, 3793 (1977).

Yellow prisms from ether-petr ether, mp 173-174°. Sublimes. Volatile with steam. Slightly sol in water; sol in the usual organic solvents except petr ether. Forms deep red water-sol salts. E_0^{alc} 0.299 volts.

Acetate, $C_{13}H_{10}O_4$, pale yellow crystals, mp 101-102°.

7535. Phycobiliproteins. Deeply colored, highly fluorescent photoreceptor pigments found in blue-green, red and cryptomonad algae that contain a linear tetrapyrrole as the prosthetic group. They are composed of a bile pigment or *phycobilin* and an apoprotein. Phycobiliproteins are classified according to uv-vis absorption maxima as *phycocyanins* (blue pigment), *phycoerythrins* (red pigment), and *allophycocyanins* (pale blue pigment). Phycocyanins and phycoerythrins occur as large mol wt aggregates called *phycobilisomes* that are attached to the photosynthetic membranes. They are closely linked to the chlorophyll containing system for efficient energy transfer. *Phytochrome* is a similar biliprotein that functions in plant photomorphogenesis. It is widely distributed in plants but only in trace amounts. It exists in two forms that are interconverted upon alternate exposure to red and far-red light. *Reviews:* H. W. Siegelman *et al., Biochem. Soc. Symp.* **28**, 107-120 (1968); P. O'Carra, C. O'hEocha in *Chemistry and Biochemistry of Plant Pigments* vol. 1, T. W. Goodwin, Ed. (Academic, New York, 2nd ed., 1976) pp 328-376; A. Bennett, H. W. Siegelman in *Porphyrins* vol. 6, D. Dolphin, Ed. (Academic, New York, 1979) pp 493-520.

7536. Phylloquinone. *[R-[R*,R*-(E)]]-2-Methyl-3-(3,-7,11,15-tetramethyl-2-hexadecenyl)-1,4-naphthalenedione;* 2-methyl-3-phytyl-1,4-naphthoquinone; 3-phytylmenadione; phytomenadione; phytonadione; vitamin K_1; AquaMephyton; Konakion; Mephyton; Mono-Kay; Veda-K_1; Veta-K_1. $C_{31}H_{46}O_2$; mol wt 450.71. C 82.61%, H 10.29%, O 7.10%. Photosynthetic electron carrier; occurs widely in green plants, algae, photosynthetic bacteria. Major dietary source of vitamin K, *q.v.* Isoln from alfalfa: H. Dam *et al., Helv. Chim. Acta* **22**, 310 (1939). Structure: D. W. MacCorquodale *et al., J. Biol. Chem.* **131**, 357 (1939); L. F. Fieser, *J. Am. Chem. Soc.* **61**, 3467 (1939). Partial syntheses from menadione and phytol: H. J. Almquist, A. A. Klose, *ibid.* 2557; S. B. Binkley *et al., ibid.* 2558; L. F. Fieser, *ibid.* 2559. Stereochemistry and total synthesis: H. Mayer *et al., Helv. Chim. Acta* **47**, 221 (1964); L. M. Jackman *et al., ibid.* **48**, 1332 (1965). Synthesis using a π-allylic nickel(I) complex: Sato *et al., J. Chem. Soc. Perkin Trans. I* **1973**, 2289. Alternate synthesis: Y. Tachibana, *Chem. Letters* **1977**, 901.

Metabolic studies: M. J. Shearer *et al., Brit. J. Haematol.* **18**, 297 (1970); **22**, 579 (1972). The *cis* isomer is not bioactive: J. T. Matschiner *et al., J. Nutr.* **102**, 625 (1972). Isoln from chloroplasts: E. Interschick-Niebler, H. K. Lichtenthaler, *Z. Naturforsch.* **36C** 276 (1981). Role in photosystem I: K. Brettel *et al., FEBS Letters* **203**, 220 (1986). Conversion *in vivo* to menaquinone-4, *q.v.*: H. H. W. Thijssen, M. J. Drittij-Reijnders, *Brit. J. Nutr.* **72**, 415 (1994). HPLC determn in foods: S. L. Booth *et al., J. Agric. Food Chem.* **42**, 295 (1994). Clinical efficacy in hemorrhagic disease of newborn: P. M. Loughnan, P. N. McDougall, *J. Paediatr. Child Health* **29**, 171 (1993). Comprehensive description: M. M. A. Hassan *et al., Anal. Profiles Drug Subs.* **17**, 449-531 (1988).

Yellow viscous oil. $[\alpha]_D^{25}$ -0.28° (dioxane). n_D^{20} 1.5263. uv max (petr ether): 242, 248, 260, 269, 325 nm ($E_{1cm}^{1\%}$ 396, 419, 383, 387, 68). Insol in water. Sparingly sol in methanol; sol in ethanol, acetone, benzene, petr ether, hexane, dioxane, chloroform, ether, other fat solvents and in vegetable oils. Stable to air and moisture, but dec in sunlight. Unaffected by dil acids, but destroyed by solns of alkali hydroxides and by reducing agents. *Keep well closed and protected from light.*

Dihydro form, *phytonadiol, dihydrovitamin K$_1$,* α-*phyllohydroquinone.* Waxy mass. Freely sol in ether; sparingly sol in petr ether. Insol in water.

Dihydro form sodium diphosphate, $C_{31}H_{48}Na_2O_8P_2$, *phytonadiol sodium diphosphate, Kayhydrin.* mp 138°. Sol in water and methanol.

2,3-Epoxide, $C_{31}H_{46}O_3$, *vitamin K$_1$ epoxide, vitamin K$_1$ oxide.* Prepn: Fieser *et al., J. Am. Chem. Soc.* **61**, 3216 (1939). Colorless oil. uv max (95% alc): 259, 305 nm (log E_M 3.79, 3.31). Insol in water.

THERAP CAT: Vitamin (prothrombogenic).

THERAP CAT (VET): Vitamin (prothrombogenic); antidote for dicoumarol poisoning.

7537. Physalaemin. Physalemin. $C_{58}H_{84}N_{14}O_{16}S$; mol wt 1265.46. C 55.05%, H 6.69%, N 15.50%, O 20.23%, S 2.53%. An undecapeptide belonging to the group of proteins named tachykinins. Found in skin of the amphibian *Physalaemus fuscumaculatus:* Erspamer *et al., Experientia* **18**, 562 (1962). Structure: Erspamer *et al., ibid.* **20**, 489 (1964); Anastasi *et al., Arch. Biochem. Biophys.* **108**, 341 (1964). Synthesis: Bernardi *et al., Experientia* **20**, 490 (1964); Nakamura *et al.,* **Japan. pat. 25,384**('71) (to Dainippon), *C.A.* **75**, 152083r (1971). Solid-phase synthesis: W. Voelter *et al., Tetrahedron* **28**, 5963 (1972). Biological activities similar to the tachykinins eledoisin and substance P, *q.q.v.* Exerts a powerful hypotensive action, stimulates salivary secretion, intestinal contraction, and vasodilation. Occurrence in other *Physalaemus* spp. and pharmacology: G. Bertaccini *et al., Brit. J. Pharmacol.* **25**, 363 (1965); G. Bertaccini, *Pharmacol. Rev.* **28**, 127 (1976). Differentiation of physalaemin and substance P: Geipert *et al., Arch. Pharmacol.* **265**, 225 (1969). Immunoreactivity study in human lung small-cell carcinoma: L. H. Lazarus *et al., Science* **219**, 79 (1983).

5-oxo-Pro–Ala–Asp–Pro–Asn–Lys–Phe–Tyr–Gly–Leu–MetNH$_2$

Trifluoroacetate dihydrate, $C_{58}H_{84}N_{14}O_{16}S \cdot CF_3COOH \cdot 2H_2O$, dec 180°. $[\alpha]_D^{20}$ -56° (c = 0.2 in ethanol). uv max: 278 nm (ε 1780). Slowly loses activity when incubated in blood. Inactivated by liver and kidney homogenates.

Hydrochloride trihydrate, $C_{58}H_{84}N_{14}O_{16}S \cdot HCl \cdot 3H_2O$, dec about 185°. $[\alpha]_D^{25}$ -43° (c = 1 in 95% acetic acid).

7538. Physodic Acid. *3,8-Dihydroxy-11-oxo-1-(2-oxoheptyl)-6-pentyl-11H-dibenzo[b,e][1,4]dioxepin-7-carboxylic acid;* 4,4′,6-trihydroxy-6-(2-oxoheptyl)-2′-pentyl-2,3′-oxydibenzoic acid-ε-lactone; physodalin. $C_{26}H_{30}O_8$; mol wt

470.52. C 66.37%, H 6.43%, O 27.20%. From the lichens *Parmelia physodes* (L.) Ach.: Hesse, *Ber.* **30**, 1983 (1897); Klosa, *Pharm. Ind.* **15**, 46 (1953); and *Cetraria ciliaris* Ach.: Culberson, *Science* **143**, 255 (1964). Structure: Asahina, Nogami, *Ber.* **67**, 805 (1934); **68**, 77 (1935).

Needles from methanol, mp 205°. uv max (95% ethanol): 256 nm (log ε 4.2). Sol in ether, acetone, hot methanol; somewhat sol in cold ethanol, methanol, hot chloroform; practically insol in benzene, petr ether, carbon disulfide.

Diacetate, $C_{30}H_{34}O_{10}$, plates from acetone + carbon disulfide, mp 153-155.5°.

Methyl ester, $C_{27}H_{32}O_8$, prisms from 80% alcohol, mp 156-157°.

7539. Physostigma. Calabar bean; ordeal bean; chop nut; split nut. Dried ripe seed of *Physostigma venenosum* Balf., *Leguminosae.* Poisonous! Habit. West Africa (near mouths of Niger and Old Calabar Rivers); introduced into India and Brazil. *Constit.* 0.15-0.3% alkaloids consisting of physostigmine (eserine), physovenine, eseridine, eseramine, phytosterol.

7540. Physostigmine. *(3aS-cis)-1,2,3,3a,8,8a-Hexahydro-1,3a,8-trimethylpyrrolo[2,3-b]indol-5-ol methylcarbamate (ester);* Cogmine; Eserine. $C_{15}H_{21}N_3O_2$; mol wt 275.35. C 65.43%, H 7.69%, N 15.26%, O 11.62%. From Calabar beans, the seeds of the vine *Physostigma venenosum* Balf., *Leguminosae:* Jobst, Hesse, *Ann.* **129**, 115 (1864); Hesse, *ibid.* **141**, 82 (1867). Extraction procedure: Schwyzer, *Die Fabrikation pharmazeutischer und chemisch-technischer Produkte* (Berlin, 1931) p 338; Chemnitius, *J. Prakt. Chem.* **116**, 59 (1927). Structure: Stedman, Barger, *J. Chem. Soc.* **127**, 247 (1925). Absolute configuration: R. B. Longmore, B. Robinson, *Chem. & Ind. (London)* **1969**, 622. Crystal structure: Petcher, Pauling, *Nature* **241**, 277 (1973). Synthesis: Julian, Pikl, *J. Am. Chem. Soc.* **57**, 755 (1935); Harley-Mason, Jackson, *J. Chem. Soc.* **1954**, 3651; J. Wijnberg, W. N. Speckamp, *Tetrahedron* **34**, 2399 (1978). Cholinesterase inhibitor: Engelhart, Loewi, *Arch. Exp. Pathol. Pharmakol.* **150**, 1 (1930); Matthes, *J. Physiol.* **70**, 338 (1930). Toxicity data: W. T. Lynch, J. M. Coon, *Toxicol. Appl. Pharmacol.* **21**, 53 (1972); R D. Sofia, L. C. Knabloch, *ibid.* **28**, 227 (1974). Improvement of long-term memory: K. L. Davis *et al., Science* **201**, 272 (1978). Memory enhancement in Alzheimer's disease: L. J. Thal *et al., N. Engl. J. Med.* **308**, 720 (1983); *eidem, Ann. Neurol.* **13**, 491 (1983); L. Gustafson *et al., Psychopharmacol.* **93**, 31 (1987). Review: B. Robinson in *The Alkaloids* Vol. XIII, R. H. F. Manske, Ed. (Academic Press, New York, 1971) pp 213-226. Comprehensive description: F. J. Muhtadi, S. S. El-Hawary in *Analytical Profiles of Drug Substances* vol. **18**, K. Florey, Ed. (Academic Press, New York, 1989) pp 289-350.

Orthorhombic sphenoidal prisms or clusters of leaflets from ether or benzene. mp 105-106° (also an unstable, low-melting form, mp 86-87°). $[\alpha]_D^{17}$ -76° (c = 1.3 in chloroform); $[\alpha]_D^{25}$ -120° (benzene). pKa$_1$ 6.12; pKa$_2$ 12.24.

Slightly sol in water; sol in alc, benzene, chloroform, oils. Solid and solns turn red on exposure to heat, light, air, and on contact with traces of metals. Under certain conditions the oxidation may proceed to yield *eserine blue*, $C_{26}H_{31}N_5O_2$. LD_{50} orally in mice: 4.5 mg/kg (Lynch, Coon).

Salicylate, $C_{22}H_{27}N_3O_5$, *Antilirium*. Acicular crystals, mp 185-187°. uv max (methanol): 239, 252, 303 nm (log ε 4.09, 4.04, 3.78). One gram dissolves in 75 ml water at 25° (pH of 0.5% aq soln 5.8); in 16 ml water at 80°; in 16 ml alcohol, 5 ml boiling alcohol, in 6 ml chloroform, 250 ml ether. *Solns should be kept well closed in light-resistant, alkali-free glass containers and used within a week.* Turns red and loses effectiveness on exposure to heat, light, air. LD_{50} i.p. in mice: 0.64 mg/kg (Sofia, Knabloch).

Sulfate, $(C_{15}H_{21}N_3O_2)_2.H_2SO_4$, deliquescent scales, mp 140° (after drying at 100°). One gram dissolves in 0.4 ml alcohol, 4 ml water (pH of 0.05M soln 4.7), 1200 ml ether. The solns are more prone to change color than those of the salicylate.

Sulfite, $(C_{15}H_{21}N_3O_2)_2.H_2SO_3$, white powder, freely sol in water, alcohol. The aq soln is stated to remain colorless for a long time.

THERAP CAT: Cholinergic (anticholinesterase); miotic.

THERAP CAT (VET): Cholinergic; miotic. In atony of gastrointestinal tract.

7541. Physovenine. *(3aS-cis)-3,3a,8,8a-Tetrahydro-3a,-8-dimethyl-2H-furo[2,3-b]indol-5-ol methylcarbamate (ester).* $C_{14}H_{18}N_2O_3$; mol wt 262.31. C 64.11%, H 6.92%, N 10.68%, O 18.30%. From seed of *Physostigma venenosum* Balf., *Leguminosae:* Salway, *J. Chem. Soc.* **99**, 2148 (1911). Structure: Robinson, *ibid.* **1964**, 1503. Synthesis: Longmore, Robinson, *Chem & Ind (London)* **1965**, 1297. Abs config: *eidem, ibid.* **1969**, 622.

Plates from ether, mp 124-125°. $[\alpha]_D^{22.5}$ −92° (ethanol). uv max (ethanol): 252, 310 nm (ε 13200, 3300). Sol in alcohol, benzene, chloroform; slightly sol in ether. Practically insol in water, petr ether.

7542. Phytic Acid. *myo-Inositol hexakis(dihydrogen phosphate);* inositolhexaphosphoric acid; 1,2,3,4,5,6-cyclohexanehexolphosphoric acid; cyclohexanehexyl hexaphosphate; Alkalovert. $C_6H_{18}O_{24}P_6$; mol wt 660.04. C 10.92%, H 2.75%, O 58.18%, P 28.16%. $C_6H_6[OPO(OH)_2]_6$. Major phosphorus compound in plants; particularly abundant in oil seeds, legumes and cereal grains. Forms insoluble complexes with di- and trivalent cations. Prepn: S. Posternak, U.S. pat. **1,313,014** (1919 to Ciba); *idem, Helv. Chim. Acta* **4**, 155 (1921); M. J. Thomas, U.S. pat. **2,718,523** (1955 to Staley Mfg.); D. S. Bolley *et al.,* U.S. pat. **2,732,395** (1956 to Natl. Lead Co.); A. R. Baldwin *et al.,* U.S. pat. **2,815,360** (1957 to Corn Prods. Ref.). HPLC determn in plants: E. Graf, F. R. Dintzis, *Anal. Biochem.* **119**, 413 (1982); in foods and biological samples: B.E. Knuckles *et al., J. Food Sci.* **47**, 1257 (1982). Review of effect of dietary phytate on zinc metabolism: *Nutr. Rev.* **41**, 64-66 (1983); on mineral bioavailability: J. L. Kelsay, *Am. J. Gastoenterol.* **82**, 983-986 (1987).

Syrupy, straw-colored liquid. Dec on heating. Acid reaction; the pH of a 10% aq soln has been reported as 0.86. Miscible with water, 95% alc, glycerol; sol in water contg alcohol-ether mixtures; very slightly sol in abs alc, methanol. Practically insol in anhydr ether, benzene, chloroform.

Calcium iron salt, obtained in the processing of corn. Grayish powder contg 20% Ca, 14% P, and 2% Fe. Sparingly sol in water, dil mineral acids.

Calcium magnesium salt, *phytin.* White, odorless powder. Poor soly in water. Sol in dil acids.

Sodium salt, $C_6H_9Na_9O_{24}P_6$, *sodium phytate*, *Phytat D.B., Rencal.* Soluble in water with neutral reaction.

USE: Complexing agent for the removal of traces of heavy metal ions (also employed as the sodium salt). Starting material in manufacture of inositol. Fermentation nutrient.

THERAP CAT: Sodium salt as hypocalcemic.

7543. Phytochlorin. *18-Carboxy-20-(carboxymethyl)-8-ethenyl-13-ethyl-2,3-dihydro-3,7,12,17-tetramethyl-21H,23H-porphine-2-propanoic acid;* chlorine e_6; phytochlorin(e) e. $C_{34}H_{36}N_4O_6$; mol wt 596.68. C 68.44%, H 6.08%, N 9.39%, O 16.09%. Prepn from broccoli-leaf extract: Wall, U.S. pat. **2,555,583** (1951 to U.S.D.A.). *Review:* H. Fischer, A. Stern, *Die Chemie des Pyrrols* vol. **II**, 2 (Leipzig, 1940) pp 91-94; *The Chlorophylls*, L. P. Vernon, G. R. Seely, Eds. (Academic Press, New York, 1966) pp 9, 12, 71, 90, 95, 98, 142.

Greenish-brown rectangular plates. May contain one mol water. $[\alpha]_D^{20}$ −141° in acetone. Ether solns are olive-green with deep red fluorescence. Sparingly sol in ethanol, ether, acetone. Freely sol in pyridine.

7544. Phytofluene. *7,7',8,8',11,12-Hexahydro-ψ,ψ-carotene;* 5,6,7,8,9,10,10',9',8',7',6',5'-dodecahydrolycopene; all-*trans*-phytofluene. $C_{40}H_{68}$; mol wt 548.98. C 87.52%, H 12.48%. Polyene hydrocarbon widespread in the vegetable kingdom where it has been observed in chlorophyll-free tissues which contain considerable amounts of carotenoid pigments. Extracted from *Diospyros kaki* L.f., *Ebenaceae, Arbutus unedo* L., *Ericaceae, Pyracantha angustifolia* (Franch.) Schneid., *Rosaceae,* and tomatoes: Zechmeister, Sandonal, *J. Am. Chem. Soc.* **68**, 197 (1946); Wallace, Porter, *Arch. Biochem. Biophys.* **36**, 468 (1952); from the basidiomycete, *Dacromyces stillatus:* Goodwin, *Biochem. J.* **53**, 538 (1953). Structure: Porter, Lincoln, *Arch. Biochem. Biophys.* **27**, 390 (1950); Zechmeister, *Experientia* **10**, 1 (1954).

Pale orange, viscous oil which solidifies upon cooling, forming a glassy mass without apparent crystal structure. $bp_{0.0001}$ 140-185° (bath temp). uv max (petr ether): 367, 348, 332 nm. Freely sol in petr ether, ether, benzene. Practically insol in water, methanol, ethanol. Strong green fluorescence in ultraviolet spectra.

7545. Phytol. *[R-[R*,R*-(E)]]-3,7,11,15-Tetramethyl-2-hexadecen-1-ol.* $C_{20}H_{40}O$; mol wt 296.54. C 81.01%, H 13.60%, O 5.40%. Decompn product of chlorophyll: Willstätter, *Ann.* **354**, 205 (1907); **371**, 1 (1909); **378**, 1, 73 (1911); **418**, 121 (1918). Synthesis: Fischer, Löwenberg, *Ann.* **475**, 183 (1929); Karrer, Ringier, *Helv. Chim. Acta* **22**, 610 (1939); Karrer *et al., ibid.* **26**, 1741 (1943); from ethyl levulinate: Lukes, Zobacova, *Chem. Listy* **51**, 330 (1957); from acetone: Sato *et al., J. Org. Chem.* **32**, 177 (1967). Stereochemistry: Burrell *et al., Proc. Chem. Soc.* **1959**, 263. Abs config: Crabbe *et al., ibid.* **1959**, 264. Stereochemistry and synthesis: Burrell *et al., J. Chem. Soc. (C)* **1966**, 2144. Stereoselective total synthesis of natural phytol: T. Fujisawa *et al., Tetrahedron*

Letters **22**, 4823 (1981); M. Schmid *et al., Helv. Chim. Acta* **65**, 684 (1982). *Review:* J. Simonsen, D. H. R. Barton, *The Terpenes,* vol. III (Cambridge University Press, Cambridge, 1952) pp 345-349.

Oily liquid. d_4^{25} 0.8497. n_D^{25} 1.4595. bp_{10} 203-204°; $bp_{0.03}$ 145°. uv max (abs alcohol): 212 nm (log ϵ 3.04), Bader, *Helv. Chim. Acta* **34**, 1632 (1951). Practically insol in water; sol in the usual organic solvents.

USE: Preparation of vitamins E and K_1.

7546. Phytolacca. Poke root; garget; pocan; American nightshade root; scoke. Dried root of *Phytolacca americana* L. (*P. decandra* L.), *Phytolaccaceae* (berries also used). *Habit.* North America; naturalized in Southern Europe. *Constit.* Resin, tannin, about 10% sugar; phytolaccine, phytolaccic acid, asparagine.

THERAP CAT: Antirheumatic; emetic; ectoparasiticide.

7547. Piberaline. *1-(Phenylmethyl)-4-(2-pyridinylcarbonyl)piperazine;* 1-benzyl-4-picolinoylpiperazine; EGYT 475; Trelibet. $C_{17}H_{19}N_3O$; mol wt 281.36. C 72.57%, H 6.81%, N 14.93%, O 5.69%. Prepn: J. Körösi *et al.,* **Belg. pat.** 781,494; *eidem,* U.S. pat. 3,865,828 (1972, 1975 both to EGYT). Alternate prepns: Z. Budai *et al.,* **Ger. pat.** 2,828,-888, *C.A.* **90**, 168643u (1979); *eidem,* Hung. pat. 17,182, *C.A.* **92**, 215463p (1980) (both 1979 to EGYT); *eidem, Acta Chim. Acad. Sci. Hung.* **105**, 241 (1980), *C.A.* **94**, 192275d (1981). HPLC separation: J. Borda *et al., J. Chromatog.* **258**, 271 (1983). Effect on avoidance behavior in rats: G. Telegdy *et al., Arch. Int. Pharmacodyn. Ther.* **266**, 50 (1983). Metabolism in animals and humans: K. Magyar, *Pol. J. Pharmacol. Pharm.* **39**, 107 (1987). Mechanism of action study: K. Tekes *et al., ibid.* 203.

Dihydrochloride, $C_{17}H_{19}N_3O.2HCl$, crystals from ethanol, mp 214-215°.
Maleate, $C_{17}H_{19}N_3O.C_4H_4O_4$, crystals, mp 169-170°.
Fumarate, $C_{17}H_{19}N_3O.C_4H_4O_4$, crystals, mp 165°.
THERAP CAT: Antidepressant.

7548. Picadex. *1-Piperazinecarbodithioic acid;* 1-piperazinecarbodithioic acid betaine; 1-piperazinecarbodithioic acid inner salt; piperazine carbon disulfide complex; piperazine dithiocarbamate; Choisine; Elmifarma; Parvex; Safersan. $C_5H_{10}N_2S_2$; mol wt 162.28. C 37.01%, H 6.21%, N 17.26%, S 39.52%. Prepn: Schmidt, Wichmann, *Ber.* **24**, 3237 (1891); Herz, *ibid.* **30**, 1584 (1897). As anthelmintic: Leiper, Watkins, U.S. pat. 2,814,583 (1957 to Boots Pure Drug).

Melting pt depends on method of determination: in an open-end capillary tube in a heating bath—sublimes 225-228°; in a sealed capillary decomp 235°. Practically insol in ether, benzene, chloroform, water.
THERAP CAT (VET): Anthelmintic.

7549. Picein. *1-[4-(β-D-Glucopyranosyloxy)phenyl]ethanone;* p-(acetylphenyl)-β-D-glucopyranoside; 4'-(β-D-glucopyranosyloxy)acetophenone; p-hydroxyacetophenone-D-glucoside; piceoside; salinigrin; salicinerein; ameliaroside. $C_{14}H_{18}O_7$; mol wt 298.29. C 56.37%, H 6.08%, O 37.55%. In needles and sprouts of *Pinus picea* L., *Picea excelsa* Link., *Picea glehnii* Mast., *Coniferae:* Tanret, *Bull. Soc. Chim.* [3] **11**, 944 (1894); Kariyone *et al., Yakugaku Zasshi* **79**, 394

(1959), *C.A.* **53**, 14096i (1953). In various willow barks, especially in bark of *Salix discolor* Muhl., *S. nigra* Marsh., *Salicaceae:* Jowett, *J. Chem. Soc.* **77**, 707 (1900); Nonomura, *J. Pharm. Soc. Japan* **75**, 80 (1955). In English mistletoe, *Amelanchier vulgaris* Moench., *Rosaceae:* Bridel *et al., Compt. Rend.* **187**, 56 (1928). Synthesis: Mauthner, *J. Prakt. Chem.* [2] **85**, 564 (1912).

Needles or prisms from methanol; mp 195-196°. $[\alpha]_D^{23}$ −88° (c = 1). One gram dissolves in 50 ml water at 15°, in 1 ml boiling water, in ~650 ml abs alc at 15°, in ~40 ml boiling abs alc, in 140 ml ethyl acetate at 15°. Sol in hot glacial acetic acid. Practically insol in ether, chloroform. Hydrolysis with dil mineral acids or with emulsin yields D-glucose and p-hydroxyacetophenone (*Piceol*). Alkaline hydrolysis yields levoglucosan and p-hydroxyacetophenone: Montgomery *et al., J. Org. Chem.* **10**, 194 (1945).

7550. Picene. 3,4-Benzchrysene; 1,2,7,8-dibenzphenanthrene; dibenzo[*a,i*]phenanthrene; β,β-binaphthyleneethene. $C_{22}H_{14}$; mol wt 278.35. C 94.93%, H 5.07%. Found in tar oils from soft coal: Burg, *Ber.* **13**, 1834 (1880); Lang *et al., ibid.* **97**, 494 (1964); in petroleum residues from the cracking process: Meyer, Hofmann, *Monatsh.* **37**, 681 (1916). Synthesis from cholic acid by selenium dehydrogenation: Ruzicka *et al., Helv. Chim. Acta* **17**, 200 (1934); from o-xylylenedicyanide with o-nitrobenzaldehyde by a double Pschorr synthesis: Waldmann, Pitschak, *Ann.* **527**, 183 (1937); by a diene synthesis with tetrahydrodinaphthyl and maleic anhydride: Weidlich, *Ber.* **71**, 1203 (1938); from 9,10-dihydrophenanthrene: Phillips, *J. Am. Chem. Soc.* **75**, 3223 (1953); from ethyl 4,6-dioxoheptane-1,5-dicarboxylate: Nasipuri, *Chem. & Ind. (London)* **1956**, 795.

Fluorescent plates from ethyl acetate, mp 366-367°. bp 518-520°. Absorption spectrum: Mayneord, Roe, *Proc. Roy. Soc. London* **A152**, 319 (1935). Difficultly sol in most solvents. Somewhat sol in boiling benzene, chloroform, glacial acetic acid; more sol in cumene (isopropylbenzene).

7551. Picilorex. *3-(4-Chlorophenyl)-5-cyclopropyl-2-methylpyrrolidine;* UP-507-04. $C_{14}H_{18}ClN$; mol wt 235.76. C 71.33%, H 7.70%, Cl 15.04%, N 5.94%. Prepn: J. M. Teulon, Ger. pat. 2,446,317; *idem,* U.S. pat. 4,005,103 (1975, 1977 both to Hexachemie). Anorexic activity in animals: G. Dumeur *et al., Brit. J. Pharmacol.* **58**, 437 (1976).

Hydrochloride, $C_{14}H_{18}ClN.HCl$, *Roxenan.* Crystals from isopropanol, mp 191°.
THERAP CAT: Anorexic.

7552. Picloram. *4-Amino-3,5,6-trichloro-2-pyridinecarboxylic acid; 4-amino-3,5,6-trichloropicolinic acid;* Tordon. $C_6H_3Cl_3N_2O_2$; mol wt 241.46. C 29.85%, H 1.25%, Cl 44.05%, N 11.60%, O 13.25%. Prepn: H. Johnston, M. S. Tomita, **Belg. pat.** 628,487, *C.A.* **61**, 1838d (1964), and *eidem,* U.S. pat. 3,285,925 (1963, 1966 both to Dow). Activity: J. W. Hamaker *et al., Science* **141**, 363 (1963). Carcinogenicity study: M. D. Reuber, *J. Toxicol. Environ. Health* **7**, 207 (1981).

Crystals, mp 218-219°. Vapor press. at 35°: 6.16 × 10^{-7} mm Hg. pKa 3.6. Soly in (g/l) at 25°: water 0.43; acetone 19.8; 2-propanol 5.5; dichloromethane 0.6. LD$_{50}$ orally in rats, mice, rabbits: 3.75, 1.5, 2.0 g/kg, *RTECS* Vol. II, R. J. Lewis, R. L. Tatken, Eds. (1979) p 305.

USE: Herbicide.

7553. Picloxydine. *N,N'''-Bis[[(4-chlorophenyl)amino]-iminomethyl]-1,4-piperazine dicarboximidamide; N,N''-bis-[(p-chlorophenyl)amidino]-1,4-piperazinedicarboxamidine;* 1,4-bis($N^{d'}$-p-chlorophenylamidinoamidinyl)piperazine; 1,1'-[1,4-piperazinediylbis(imidocarbonyl)]bis[3-(p-chlorophenyl)guanidine]. C$_{20}$H$_{24}$Cl$_2$N$_{10}$; mol wt 475.38. C 50.53%, H 5.09%, Cl 14.92%, N 29.46%. Heterocyclic biguanide with antibacterial activity. Prepn: J. W. James, L. F. Wiggins, **Brit.** pat. **855,017**; *eidem*, **U.S.** pat. **3,101,336** (1960, 1963 both to Aspro-Nicholas). Prepn and antibacterial spectrum: J. W. James *et al., J. Med. Chem.* **11**, 942 (1968). Bactericidal effect in disinfectant formulations: A. M. Gordon, *J. Clin. Pathol.* **22**, 496 (1969). Mode of action study: B. D. Rawal, J. V. Hardy, *Microbios* **14**, 135 (1974). *In vitro* activity vs *Chlamydia trachomatis:* D. Thomas *et al., Pathol. Biol.* **32**, 544 (1984).

Dihydrochloride, C$_{20}$H$_{24}$Cl$_2$N$_{10}$.2HCl, *Vitabact.* Crystals from water, mp 274°. LD$_{50}$ in mice (mg/kg): 150 i.p. (James).

THERAP CAT: Antibacterial (topical).

7554. α-Picoline. *2-Methylpyridine.* C$_6$H$_7$N; mol wt 93.13. C 77.38%, H 7.58%, N 15.04%. Found in coal tar and in bone oil. Synthesis (40-50% yield) from cyclohexylamine with excess ammonia and zinc chloride at 350°: Nordt, *PB Report 704* (1941). Prepn from ethylene-mercuric acetate adduct and ammonia water (70% yield): Gumboldt, Feichtinger, *Z. Naturforsch.* **4b**, 123 (1949). Toxicity study: H. F. Smyth *et al., Arch. Ind. Hyg. Occup. Med.* **4**, 119 (1951). Physical properties: Biddiscombe *et al., J. Chem. Soc.* **1954**, 1957. Review of the various methods of condensing aldehydes, ketones, ethylene, butadiene, etc., with ammonia: F. Brody, P. R. Ruby in A. Weissberger, *The Chemistry of Heterocyclic Compounds,* **vol. 14**, part I (New York, 1960) pp 99-589. Contains 1851 references.

Colorless liquid; strong unpleasant odor. mp −70°. bp 128-129°. d$_4^{15}$ 0.950; n$_D^{20}$ 1.501. Freely sol in water; miscible with alcohol, ether. LD$_{50}$ orally in rats: 1.41 g/kg (Smyth).

Caution: Irritating to respiratory tract.

USE: Solvent; intermediate in the dye and resins industries.

7555. β-Picoline. *3-Methylpyridine.* C$_6$H$_7$N; mol wt 93.13. C 77.38%, H 7.58%, N 15.04%.

Colorless liq; sweetish, not unpleasant odor. d$_4^{15}$ 0.9613; bp 143-144°; n$_D^{20}$ 1.5043. Miscible with water, alcohol, ether.

USE: Solvent; intermediate in the dye and resins industries; in the manuf of insecticides, waterproofing agents, niacin, and niacinamide.

7556. γ-Picoline. *4-Methylpyridine;* 4-picoline. C$_6$H$_7$N; mol wt 93.13. C 77.38%, H 7.58%, N 15.04%. Found in coal tar, in bone oil, in urine of horses. Isoln from technical picolines: Flaschner, *J. Chem. Soc.* **95**, 669 (1909). Toxicity

study: H. F. Smyth *et al., Arch. Ind. Hyg. Occup. Med.* **10**, 61 (1954).

Flammable liquid. Obnoxious, sweetish odor. Turns brown if not very pure. d$_4^{15}$ 0.9571. bp$_{760}$ 145°. n$_D^{17}$ 1.5064. Kb at 25° = 1.1 × 10^{-8}. Sol in water, alcohol, ether. LD$_{50}$ orally in rats: 1.29 g/kg (Smyth).

USE: Manuf isonicotinic acid and derivatives. In waterproofing agents for fabrics. As solvent for resins.

7557. Picolinic Acid. *2-Pyridinecarboxylic acid; o-pyridinecarboxylic acid.* C$_6$H$_5$NO$_2$; mol wt 123.11. C 58.54%, H 4.09%, N 11.38%, O 25.99%. An isomer of nicotinic acid. Prepn: Singer, McElvain, *Org. Syn.* **coll. vol. III**, 740 (1955). *Review:* Oliveto, "Pyridinecarboxylic Acids" in *Pyridine and Its Derivatives,* **Part 3**, E. Klingsberg, Ed. (Interscience, New York, 1962) p 179.

Needles from water, alcohol or benzene, mp 134-136°. pK 5.4. Sublimes. Very sol in glacial acetic acid; practically insol in ether, chloroform, carbon disulfide.

Hydrochloride, crystals from abs ethanol/ether, mp 210-212° (slow heating).

7558. Picoperine. *N-Phenyl-N-[2-(1-piperidinyl)ethyl]-2-pyridinemethanamine; 1-[2-[N-(2-pyridylmethyl)anilino]-ethyl]piperidine; N-(2-pyridylmethyl)-N-phenyl-N-2-(piperidinoethyl)amine; 1-[2-[phenyl(2-pyridylmethyl)amino]ethyl]piperidine; N-(2-picolyl)-N-phenyl-N-(2-piperidinoethyl)-amine; N-(2-piperidinoethyl)-N-(2-pyridylmethyl)aniline; N-phenyl-N-(2-pyridylmethyl)-2-piperidinoethylamine;* picoperidamine; TAT-3. C$_{19}$H$_{25}$N$_3$; mol wt 295.43. C 77.25%, H 8.53%, N 14.22%. Prepn: **Fr.** pat. **1,511,398** and **U.S.** pat. **3,471,501** (1968 and 1969 to Takeda); Miyano *et al., J. Med. Chem.* **13**, 704 (1970). Prepn of salts: Sawa *et al., Takeda Kenkyusho Ho* **29**(2), 275 (1970), *C.A.* **73**, 98758h (1970); Masuda *et al.,* **Ger.** pat. **1,935,172** (1970 to Takeda), *C.A.* **73**, 98818c (1970). Pharmacology: Y. Kasé *et al., Arzneimittel-Forsch.* **19**, 1916 (1969); *eidem, ibid.* **20**, 37 (1970). *Review: Japan. Med. Gaz.* **8**(11), 9 (1971).

Pale yellow liquid, bp$_4$ 195-196°.

Hydrochloride, C$_{19}$H$_{25}$N$_3$.HCl, *Coben.* Water-soluble white crystalline powder, odorless with slightly bitter taste, mp 183-185°. Freely sol in ethanol and chloroform. Slightly sol in acetone, dioxane, benzene. Practically insol in hexane, ether. LD$_{50}$ in mice (mg/kg): 210 s.c.; 85 i.p.; 17 i.v.; 240 orally (Kasé, 1969).

Tripalmitate, C$_{19}$H$_{25}$N$_3$.3C$_{16}$H$_{32}$O$_2$, *Coben P.* White, tasteless, odorless crystalline powder, mp 57-60°. Freely sol in ethanol, chloroform; slightly sol in dioxane, acetone, benzene. Practically insol in hexane, water. LD$_{50}$ orally in mice: 1900 mg/kg (Kasé, 1969).

THERAP CAT: Antitussive.

7559. Picosulfate Sodium. *4,4'-(2-Pyridinylmethylene)-bisphenol bis(hydrogen sulfate) (ester) disodium salt; 4,4'-(2-pyridylmethylene)diphenolbis(hydrogen sulfate) (ester) disodium salt; 4,4'-(2-picolylidene)bis(phenylsulfuric acid) diso-*

dium salt; 2-picolylidenebis(p-phenyl sodium sulfate); disodium 4,4'-disulfoxydiphenyl-(2-pyridyl)methane; picosulfol; sodium picosulfate; Guttalax-Fher; Laxoberal; Laxoberon; Neopax; Pico-Salax. $C_{18}H_{13}NNa_2O_8S_2$; mol wt 481.41. C 44.91%, H 2.72%, N 2.91%, Na 9.55%, O 26.59%, S 13.32%. Prepn: Pala, **Fr.** pat. **M5832** and **U.S.** pats. **3,528,986, 3,558,643** (1968, 1970, 1971 all to Ist. de Angeli); Seeger, Machleidt, **Ger.** pat. **1,904,322** (1970 to Thomae). Synthesis and physical-chemical data: Pala *et al., Helv. Chim. Acta* **51**, 1164 (1968). Pharmacology: Pala *et al., Arch. Int. Pharmacodyn. Ther.* **164**, 356 (1966). Metabolism: Perego *et al., Arzneimittel-Forsch.* **19**, 1889 (1969).

White crystalline solid from ethanol or methanol, mp 272-275° (dec). uv max (H_2O): 218, 262 nm (ϵ 20450, 6075). Readily sol in water; slightly sol in alcohol. Practically insol in most organic solvents.

THERAP CAT: Cathartic.

7560. Picotamide. *4-Methoxy-N,N'-bis(3-pyridinylmethyl)-1,3-benzenedicarboxamide; 4-methoxy-N,N'-bis(3-pyridylmethyl)isophthalamide; N,N'-bis(3-picolyl)-4-methoxyisophthalamide;* G-137. $C_{21}H_{20}N_4O_3$; mol wt 376.41. C 67.01%, H 5.36%, N 14.88%, O 12.75%. Prepn: R. Selleri *et al., Chim. Ther.* **6**, 203 (1971); G. Orzalesi, R. Selleri, **Fr.** pat. **2,100,850**, *C.A.* **77**, 164502f (1972); prepn and use as inhibitor of blood platelet aggregation: *eidem,* **U.S.** pat. **3,973,026** (1972, 1976 both to Soc. Italo-Brit. L. Manetti-H. Roberts). Crystal structure of monohydrate: E. Foresti *et al., Acta Crystallogr. Sect. C* **42**, 220 (1986). Mechanism of action study: M. Berrettini *et al., Boll. Soc. Ital. Biol. Sper.* **59**, 309 (1983). Clinical studies: R. Schmutzler *et al., Age Ageing* **7**, 246 (1978); V. Coto *et al., Minerva Cardioangiol.* **34**, 601 (1986).

Crystalline powder from benzene, mp 124°. LD_{50} i.p. in male mice: 1205 mg/kg (Orzalesi, Selleri, 1976).

Monohydrate, $C_{21}H_{20}N_4O_3 \cdot H_2O$, *Plactidil.*

THERAP CAT: Antithrombotic; fibrinolytic; anticoagulant.

7561. Picramic Acid. *2-Amino-4,6-dinitrophenol;* 4,6-dinitro-2-aminophenol. $C_6H_5N_3O_5$; mol wt 199.12. C 36.19%, H 2.53%, N 21.10%, O 40.17%. Prepd from picric acid, concd NH_4OH, and H_2S followed by acetic acid neutralization of ammonium salt: Egerer, *J. Biol. Chem.* **35**, 565 (1918). Prepn of ammonium salt from picric acid, aqueous NH_3-soln, and ammonium sulfide: Dehn, **U.S.** pat. **1,472,791** (1924).

Dark red needles from alcohol, prisms from chloroform. mp 169-170°. At 22-25°, 0.065 g dissolves in 100 ml H_2O; not much more sol in hot water. Sparingly sol in ether, chloroform; moderately sol in alcohol; sol in benzene, glacial acetic acid, aniline. Flashes at 210°; in contact with

open flame in glass tube or beaker, ignites rapidly and burns relatively fast: Blinov, *Khim. Prom.* **1959**, 419, *C.A.* **55**, 6866d (1961).

Caution: Toxic symptoms similar to 2,4-dinitrophenol, *q.v.*

USE: Manuf of azo dyes; rarely as indicator (yellow with acids, red with alkalies); reagent for albumin.

7562. Picric Acid. *2,4,6-Trinitrophenol;* picronitric acid; carbazotic acid; nitroxanthic acid. $C_6H_3N_3O_7$; mol wt 229.11. C 31.46%, H 1.32%, N 18.34%, O 48.88%. Prepd by sulfonating phenol then treating with nitric acid: Olsen, Goldstein, *Ind. Eng. Chem.* **16**, 66 (1924); by treating benzene with nitric acid and mercuric nitrate: Teeters, Mueller, **U.S.** pat. **2,455,322** (1948 to Allied Chem.); by nitration of 2-*tert*-butyl-4,6-dinitropenol: Ley, Müller, *Ber.* **89**, 1402 (1956). Crystal structure: E. N. Duesler *et al., Cryst. Struct. Commun.* **7**, 449 (1978).

Pale yellow, odorless, intensely bitter crystals. d 1.763. mp 122-123°. Explodes above 300°. One gram dissolves in 78 ml water, 15 ml boiling water, 12 ml alc, 10 ml benzene, 35 ml chloroform, 65 ml ether. *Incompat:* All oxidizable substances, albumin, gelatin, alkaloids. *Keep in a cool place and remote from fire. Explodes when rapidly heated or by percussion! Note:* For safety in transportation, 10-20% water is usually added.

Caution: Local or generalized allergic reactions may occur following topical use. Ingestion or percutaneous absorption may cause nausea, vomiting, diarrhea, abdominal pain, oliguria, anuria, yellow staining of skin (not icterus), pruritus, skin eruptions, stupor, convulsions, death, *cf. Clinical Toxicology of Commercial Products,* R. E. Gosselin *et al.,* Eds. (Williams & Wilkins, Baltimore, 4th ed., 1976) Section II, p 133.

USE: Explosives, matches; in leather industry; electric batteries; etching copper; manuf colored glass; textile mordant; also as reagent.

7563. Picrocrocin. *(R)-4-(β-D-Glucopyranosyloxy)-2,6,-6-trimethyl-1-cyclohexene-1-carboxaldehyde;* saffron-bitter. $C_{16}H_{26}O_7$; mol wt 330.38. C 58.17%, H 7.93%, O 33.90%. From stigmas of *Crocus sativus* L., *Iridaceae.* Isoln: Kayser, *Ber.* **17**, 2228 (1884). Structure: Kuhn, Winterstein, *Ber.* **67**, 344 (1934). Exerts sex-determining influences in the plant organism: Kuhn, *Angew. Chem.* **53**, 1 (1940). Its moieties are glucose and safranal, *q.v.* Abs config: Buchecker, Eugster, *Helv. Chim. Acta* **56**, 1121 (1973). Synthesis: H. Mayer, J.-M. Santer, *Helv. Chim. Acta* **63**, 1463 (1980).

Crystals, mp 154-156°. $[\alpha]_D^{20}$ $-58°$ (c = 0.6). Bitter taste. Alkali unstable. Sol in water, alcohol; slightly sol in chloroform, ether. Practically insol in petr ether, benzene.

7564. Picrolichenic Acid. *6-Hydroxy-2'-methoxy-2,4'-dioxo-4,6'-dipentylspiro[benzofuran-3(2H),1'-[2,5]cyclohexadiene]-5-carboxylic acid;* picrolichenin. $C_{25}H_{30}O_7$; mol wt 442.51. C 67.86%, H 6.83%, O 25.31%. Bitter acidic principle of crustose lichen, *Pertusaria amara* (Ach.) Nyl., *Pertusariaceae.* Isoln: Alms, *Ann. Pharm.* **1**, 61 (1832). Identity of picrolichenin and picrolichenic acid: Zopf, *Ann.* **321**, 38 (1902). Isoln and structure: Wachmeister, *Acta Chem.*

Scand. **12**, 147 (1958). Synthesis: Davidson, Scott, *J. Chem. Soc.* **1961**, 4075.

Prisms from benzene, dec 184-187°; prisms from aqueous acetic acid, decomp 187-190°. Intense bitter taste similar to that of quinine. Optically inactive. uv max: 245, 270-277 nm (ϵ 24,000, 7800). Readily sol in most organic solvents except benzene and light petroleum.

Methyl picrolichenate, $C_{26}H_{32}O_7$, needles from methanol, mp 102-103.5°. Soluble in cold 0.5N NaOH.

Methyl *O*-methylpicrolichenate, $C_{27}H_{34}O_7$, needles from hexane, mp 80-82°. Practically insol in cold 0.1N NaOH.

7565. Picrolonic Acid. *2,4-Dihydro-5-methyl-4-nitro-2-(4-nitrophenyl)-3H-pyrazol-3-one;* 3-methyl-4-nitro-1-(*p*-nitrophenyl)-2-pyrazolin-5-one. $C_{10}H_8N_4O_5$; mol wt 264.20. C 45.46%, H 3.05%, N 21.21%, O 30.28%.

Yellow leaflets, mp 116-117°; dec at 125°. Sparingly sol in water; sol in alcohol.

USE: As reagent for alkaloids, tryptophan, phenylalanine, and for the detection and estimation of calcium.

7566. Picromycin. (*E*)-14-Ethyl-13-hydroxy-3,5,7,9,-13-pentamethyl-6-[[3,4,6-trideoxy-3-(dimethylamino)-β-D-*xylo*-hexopyranoside]oxy]oxacyclotetradec-11-ene-2,4,10-trione; pikromycin; albomycetin; amaromycin. $C_{28}H_{47}NO_8$; mol wt 525.68. C 63.98%, H 9.01%, N 2.66%, O 24.35%. First macrolide antibiotic isolated. Isoln from *Actinomyces* spp by Lindenbein and Bauer, *see* Brockmann, Henkel, *Naturwiss.* **37**, 138 (1950); *Ber.* **84**, 284 (1951); Brockmann, Bohne, U.S. pat. **2,693,433** (1954 to Schenley). Structure: Brockmann, Oster, *Ber.* **90**, 605 (1957); Anliker, Gubler, *Helv. Chim. Acta* **40**, 119, 1768 (1957). Studies in stereochemistry: Djerassi, Halpern, *J. Am. Chem. Soc.* **79**, 3926 (1957); Ogura *et al., ibid.* **97**, 1930 (1975); *eidem, Tetrahedron* **37**, Suppl. 1, 165 (1981). Revised structure: Rickards, Smith, *Chem. Commun.* **1968**, 1049; Muxfeldt *et al., J. Am. Chem. Soc.* **90**, 4748 (1968). Synthesis from narbonolide: Maezawa *et al., J. Antibiot.* **26**, 771 (1973).

Very bitter, rectangular platelets from methanol, mp 169.5-170°. Stable to heat. $[\alpha]_D^{24}$ +8.2° (c = 3.5 in ethanol); $[\alpha]_D^{20}$ −33.5° (c = 2.07 in chloroform); $[\alpha]_D^{24}$ −50.2° (c = 6.3 in chloroform). uv max (ethanol): 225 nm (log ϵ 3.97). Rotary dispersion data: Djerassi, Halpern, *Tetrahedron* **3**, 268 (1958). Very sparingly sol in water, petr ether, carbon disulfide. Soly in ethanol: 3.5 g/100 ml at 20°. Freely sol

in acetone, benzene, chloroform, ethyl acetate, dioxane. Moderately sol in ether, methanol.

7567. Picropodophyllin. *[5R-(5α,5aα,8aα,9α)]-5,8,8a,-9-Tetrahydro-9-hydroxy-5-(3,4,5-trimethoxyphenyl)furo[3',4':-6,7]naphtho[2,3-d]-1,3-dioxol-6(5aH)-one;* picropodophyllinic acid lactone. $C_{22}H_{22}O_8$; mol wt 414.41. C 63.76%, H 5.35%, O 30.89%. Found in resin podophyllum, a dried alcoholic extract of *Podophyllum peltatum* L., Berberidaceae. Isomer of podophyllotoxin *q.v.;* picropodophyllin does not exist in the fresh plant, but is formed from podophyllotoxin β-D-glucoside during the extraction process. Structure: Schrecker, Hartwell, *J. Am. Chem. Soc.* **76**, 752 (1954); *J. Org. Chem.* **21**, 381 (1956); von Wartburg *et al., Helv. Chim. Acta* **40**, 1331 (1957). Isoln from resin podophyllum: Späth *et al., Ber.* **65**, 1545 (1932). Isoln as the glucoside: von Wartburg *et al., loc. cit.* Synthesis: Gensler, Wang, *J. Am. Chem. Soc.* **76**, 5890 (1954); Gensler *et al., ibid.* **82**, 1714 (1960).

Crystals from acetone. Irritates the skin. mp 214°. Higher-melting modification from abs ethanol, mp 228°. $[\alpha]_D^{20}$ +9.4° (c = 0.7 in chloroform). Soly in water at 25° ~100 mg/l. Fairly sol in chloroform, hot ethanol, acetic acid, acetone, dil caustic; less sol in benzene, ether, carbon tetrachloride, propylene glycol. Practically insol in petr ether.

Picropodophyllin-β-D-glucoside, $C_{28}H_{32}O_{13}$, crystals, mp 231-233°; 252-254°. $[\alpha]_D^{20}$ −10.9° (c = 0.549 in pyridine).

7568. Picrorhiza. Dried rhizome of *Picrorhiza kurroa* Royle, Scrophulariaceae. Habit. Himalayas. Constit. Picrorhizin—a glucoside.

7569. Picrotin. *[1aR-(1aα,2aβ,3β,6β,6aβ,8aS*,8bβ,-9S*)]-Hexahydro-2a-hydroxy-9-(1-hydroxy-1-methylethyl)-8b-methyl-3,6-methano-8H-1,5,7-trioxacyclopenta[ij]cycloprop-[a]azulene-4,8(3H)-dione.* $C_{15}H_{18}O_7$; mol wt 310.30. C 58.06%, H 5.85%, O 36.09%. Nontoxic component of picrotoxin. Prepn from picrotoxin: Meyer, Bruges, *Ber.* **31**, 2958 (1898). Structure studies: Bensted *et al., J. Chem. Soc.* **1952**, 1042; Slater, Wilson, *ibid.* 1597; Holker *et al., ibid.* **1958**, 2987. Total synthesis: E. J. Corey, H. L. Pearce, *Tetrahedron Letters* **21**, 1823 (1980). Toxicity data: C. H. Jarboe *et al., J. Med. Chem.* **11**, 729 (1968). *Review:* Porter, *Chem. Rev.* **67**, 441 (1967).

Fine needles from water; thick, shiny prisms from aq alc soln, mp 248-250°. $[\alpha]_D$ −64.7° (c = 2.31 in abs alc). Freely sol in abs alc, acetic acid, boiling water. Slightly sol in cold water. Practically insol in ether, chloroform, benzene. LD_{50} i.p. in mice: 135 mg/kg (Jarboe).

7570. Picrotoxin. Cocculin. $C_{30}H_{34}O_{13}$; mol wt 602.59. C 59.80%, H 5.69%, O 34.52%. Bitter principle isolated from the seed of *Anamirta cocculus* L. Wight & Arn., *Menispermaceae*, also found in *Tinomiscium philip-*

pinense Diels. Extraction procedure: Clark, *J. Am. Chem. Soc.* **57**, 1111 (1935). For a complete chemical bibliography up to April 1949 *see Helv. Chim. Acta* **32**, 1859 (1949). Picrotoxin is a molecular compd of one mole picrotoxinin, *q.v.*, and one mole picrotin, *q.v.*, into which it is readily separated. Crystal and molecular structure: L. Dupont *et al., Acta Crystallogr.* **B32**, 2987 (1976). Toxicity study: I. Setnikar *et al., J. Pharmacol. Exp. Ther.* **128**, 176 (1960).

Intensely bitter shiny rhomboid leaflets, mp 203°. *Very poisonous!* $[\alpha]_D^{16}$ −29.3° (c = 4 in abs ethanol). One gram dissolves in ~350 ml water, in ~5 ml boiling water, in 13.5 ml 95% ethanol, in ~3 ml boiling alcohol. Sparingly sol in ether, chloroform. Readily sol in strong ammonia water, in aq solns of NaOH. Highly toxic to fish. LD_{50} i.p. in mice: 7.2 mg/kg (Setnikar).

THERAP CAT: CNS and respiratory stimulant.

THERAP CAT (VET): CNS stimulant, antidote to barbiturates.

7571. Picrotoxinin. *[1aR-(1aα,2aβ,3β,6β,6aβ,8aS*,-8bβ,9R*)]-Hexahydro-2a-hydroxy-8b-methyl-9-(1-methylethenyl)-3,6-methano-8H-1,5,7-trioxacyclopenta[ij]cycloprop-[a]azulene-4,8(3H)-dione.* $C_{15}H_{16}O_6$; mol wt 292.29. C 61.64%, H 5.52%, O 32.84%. Toxic component of picrotoxin. Prepn from picrotoxin: Horrmann, *Ber.* **45**, 2090 (1912). Structure: Conroy, *J. Am. Chem. Soc.* **73**, 1889 (1951); **79**, 5550 (1957); Craven, *Tetrahedron Letters* no. 19, 21 (1960). Total synthesis: E. J. Corey, H. L. Pearce, *J. Am. Chem. Soc.* **101**, 5841 (1979). Structure-activity relationship: C. H. Jarboe *et al., J. Med. Chem.* **11**, 729 (1968). Review: Porter, *Chem. Rev.* **67**, 441 (1967).

Very bitter large prisms or small crystals contg water, mp 209.5°. $[\alpha]_D^{17}$ +4.4° (c = 4.28 in abs alc), +3.49° (c = 7.57 in acetone). Soluble in hot common organic solvents, in cold alcohol and chloroform. LD_{50} i.p. in mice: 3 mg/kg (Jarboe).

7572. Picryl Chloride. *2-Chloro-1,3,5-trinitrobenzene.* $C_6H_2ClN_3O_6$; mol wt 247.55. C 29.11%, H 0.81%, Cl 14.32%, N 16.97%, O 38.78%.

Almost white needles. d 1.797. mp 83°. Insol in water; freely sol in benzene, hot chloroform, boiling alcohol, slightly in ether or petr ether.

7573. Picumast. *7-[3-[4-[(4-Chlorophenyl)methyl]-1-piperazinyl]propoxy]-3,4-dimethyl-2H-1-benzopyran-2-one;* 7-[3-[4-(p-chlorobenzyl)-1-piperazinyl]propoxy]-3,4-dimethyl-coumarin; 1-(4-chlorobenzyl)-4-[3-(3,4-dimethylcoumarin-7-yloxy)propyl]piperazine. $C_{25}H_{29}ClN_2O_3$; mol wt 440.97. C 68.09%, H 6.63%, Cl 8.04%, N 6.35%, O 10.88%. Mast cell degranulation inhibitor with histamine antagonist activity. Prepn: E. C. Witte *et al.,* **Ger. pat.** 2,123,924; *eidem,* U.S. pat. 3,810,898 (1972, 1974 both to Boehringer, Mann.); E. C. Witte, *Arzneimittel-Forsch.* **39**, 1309 (1989). Mechanism of action studies: O. H. Wilhelms, *Int. Arch. Allergy Appl. Immunol.* **82**, 544, 547 (1987). Series of articles on pharmacology, pharmacokinetics, and clinical trials: *Arzneimittel-Forsch.* **39**, 1307-1376 (1989).

Crystals from isopropanol, mp 112-114° (Witte, 1974); also reported as mp 115-117° (Witte, 1989).

Dihydrochloride, $C_{25}H_{29}ClN_2O_3$·2HCl, *BM-15100, Auteral.* Crystals from aq ethanol, mp 266-268°.

THERAP CAT: Antiallergic.

7574. Pidotimod. *[R-(R*,S*)]-3-[(5-Oxo-2-pyrrolidin-yl)carbonyl]-4-thiazolidinecarboxylic acid;* 3-L-pyroglutam-yl-L-thiazolidine-4-carboxylic acid; (R)-3-[(S)-5-oxoprolyl]-4-thiazolidinecarboxylic acid; PGT/1A; Axil; Onaka; Pigitil; Polimod. $C_9H_{12}N_2O_4S$; mol wt 244.27. C 44.25%, H 4.95%, N 11.47%, O 26.20%, S 13.13%. Peptide-like biological response modifier. Prepn: S. Poli, **Eur. pat. Appl.** 276,-752; *idem,* U.S. pat. 4,839,387 (1988, 1989 both to Poli). Immunostimulant effect in nude mice: A. Vacca *et al., Int. J. Immunother.* **9**, 85 (1993); on human peripheral blood monocytes: M. O. Borghi *et al., ibid.* **10**, 35 (1994). HPLC determn in plasma and urine: G. Coppi, M. Barchielli, *J. Chromatog.* **563**, 385 (1991). Clinical trial in recurrent acute tonsillitis: P. Careddu *et al., Adv. Otorhinolaryngol. (Basel),* **47**, 328 (1992). Toxicity study: G. Coppi *et al., Arzneimittel-Forsch.* **44**, 1448 (1994). Series of articles on chemistry, pharmacology, toxicology and clinical experience: *ibid.,* 1399-1530.

White or slightly ivory, odorless microcrystalline powder, mp 194-198° (dec). Also reported as white crystals from water, mp 192-194° (Poli). $[\alpha]_D^{25}$ −150° (c = 2 in 5N HCl). pKa 3.03. Soly (g/l): water 37.8; methanol 13.8; ethanol 4.4; DMF 72.4. Practically insol in chloroform, hexane. LD_{50} in mice and rats (mg/kg): >4000 i.v.; >4000 i.m.; >8000 i.p.; >8000 orally (Coppi).

THERAP CAT: Immunomodulator.

7575. Pifarnine. *1-(1,3-Benzodioxol-5-ylmethyl)-4-(3,-7,11-trimethyl-2,6,10-dodecatrienyl)piperazine;* 1-piperonyl-4-(3,7,11-trimethyl-2,6,10-dodecatrienyl)piperazine; U-27; Pifazin. $C_{27}H_{40}N_2O_2$; mol wt 424.63. C 76.37%, H 9.49%, N 6.60%, O 7.54%. Non-anticholinergic gastric anti-secretory agent. Prepn: S. Tricerri *et al.,* **Ger. pat.** 2,310,044 corresp to U.S. pat. 3,875,163 (1973, 1975 both to Pierrel); *eidem, Eur. J. Med. Chem.* **9**, 555 (1974). Pifarnine is a mixture of four stereoisomers: *ZZ, EZ, ZE, EE.* Separation does not give compounds with significantly different activity/toxicity ratios from the mixture. Stereochemistry and pharmacology of the four isomers: G. Guadagnini *et al., ibid.* **10**, 585 (1975). Pharmacological study of pifarnine: A. Bianchetti *et al., Arzneimittel-Forsch.* **25**, 580 (1975). Physical-chemical and analytical studies: G. Guadagnini *et al., Pharm. Ind.* **38**, 296 (1976). Absorption, distribution, excretion: M. Riva *et al., Farmaco Ed. Prat.* **34**, 542 (1979). Clinical studies: A. Porro *et al., ibid.* 85; M. Petrillo *et al., Curr. Ther. Res.* **25**, 457 (1979).

(E,E)-isomer

Light yellow viscous liq with a slight odor and bitter taste. uv max (ethanol): 287 nm ($E_{1cm}^{1\%}$ 94.6). n_D^{20} 1.5235. Relative density at 20°: 1.013-1.015. pK_1 4.10; pK_2 3.25. Readily sol in most org solvents. Slightly sol in aq solns of organic acids. Practically insol in alkali, water. LD_{50} in mice, rats (mg/kg): 2175, 2610 orally; 40.6, 33.3 i.v. (Bianchetti). Approx LD_{50} i.p. in mice: 500 mg/kg (Tricerri, 1974).

THERAP CAT: Antiulcerative.

7576. Piketoprofen. *3-Benzoyl-α-methyl-N-(4-methyl-2-pyridinyl)benzeneacetamide; m*-benzoyl-*N*-(4-methyl-2-pyridyl)hydratropamide; 2-(3-benzoylphenyl)-*N*-(4-methyl-2-pyridyl)propionamide; Calmatel (aerosol). $C_{22}H_{20}N_2O_2$; mol wt 344.41. C 76.72%, H 5.85%, N 8.13%, O 9.29%. Derivative of ketoprofen, *q.v.* used topically as cream or aerosol. Prepn: R. G. W. Spickett *et al.*, **Brit.** pat. **1,436,-502** (1976 to Antonio Gallardo, S.A.). Pharmacology and clinical efficacy: E. Tarrus *et al.*, *2nd Eur. Congr. Biopharm. Pharmacokinet.* **1**, 483 (1984).

Oil, sol in methylene chloride, ethanol. Insol in water. Hydrochloride, $C_{22}H_{20}N_2O_2$·HCl, *Calmatel (cream).* mp 180-182°.

THERAP CAT: Anti-inflammatory (topical).

7577. Pildralazine. *6-[(2-Hydroxypropyl)methylamino]-3(2H)-pyridazinone hydrazone;* 3-hydrazino-6-[(2-hydroxypropyl)methylamino]pyridazine; (±)-1-[(6-hydrazino-3-pyridazinyl)methylamino]-2-propanol; propyldazine; propildazine. $C_8H_{15}N_5O$; mol wt 197.24. C 48.72%, H 7.67%, N 35.51%, O 8.11%. Peripheral vasodilator with hypotensive activity; related to hydralazine, *q.v.* Prepn: G. Pifferi, **Ger.** pat. **2,154,245**; *idem*, **U.S.** pat. **3,769,278** (1972, 1973 both to I.S.F.); G. Pifferi *et al.*, *J. Med. Chem.* **18**, 741 (1975). Pharmacology in animals: L. Dorigotti *et al.*, *Pharmacol. Res. Commun.* **8**, 295 (1976); L. Dorigotti *et al.*, *Arzneimittel-Forsch.* **34**, 876 (1984); in humans: L. Terzoli *et al.*, *Boll. Soc. Ital. Cardiol.* **22**, 1053 (1977). GLC determn in plasma: P. Ventura *et al.*, *J. Chromatog.* **161**, 237 (1978). Clinical evaluation in essential hypertension: R. Pellegrini, G. Abbondati, *Farmaco Ed. Prat.* **32**, 19 (1977). Pharmacokinetics and tissue binding: E. Noack *et al.*, *Arzneimittel-Forsch.* **37**, 407 (1987).

Dihydrochloride, $C_8H_{15}N_5O$·2HCl, *ISF-2123, Atensil.* Crystals from ethanol, mp 206-209° (dec). LD_{50} in mice, rats (mg/kg): 357, 355 i.p.; 1170, 1230 orally (Dorigotti, 1976).

THERAP CAT: Antihypertensive.

7578. Pilocarpine. *(3S-cis)-3-Ethyldihydro-4-[(1-methyl-1H-imidazol-5-yl)methyl]-2(3H)-furanone;* Ocusert Pilo. $C_{11}H_{16}N_2O_2$; mol wt 208.26. C 63.44%, H 7.74%, N 13.45%, O 15.36%. Cholinergic principle from *Pilocarpus jaborandi* Holmes, *Rutaceae.* Isoln: Petit, Polanovski, *Bull. Soc. Chim.* [3] **17**, 557, 702 (1897). Structure: Jowett, *J. Chem. Soc.* **77**, 473, 851 (1900); **83**, 438 (1903). Stereoisomeric with isopilocarpine: Polonovski, Polonovski, *Bull. Soc. Chim.* [4] **31**, 1314 (1922). Has the *cis* configuration; isopilocarpine is *trans:* Zav'yalov, *Doklady Akad. Nauk SSSR* **82**, 257 (1952). Absolute configuration: Hill, Barcza, *Tetrahedron* **22**, 2889 (1966). Synthesis: Preobrashenski *et al.*, *Ber.* **66**, 1187 (1933); Samokhvalov, *Med. Prom. SSSR* **11**, no. 2, 10 (1957); DeGraw, *Tetrahedron* **28**, 967 (1972); Link, Bernauer, *Helv. Chim. Acta* **55**, 1053 (1972). Stereoselective synthesis: A. Noordam *et al.*, *Rec. Trav. Chim.* **98**, 467 (1979). *Review:* Langenbeck, *Angew. Chem.* **60**, 297

(1948); van Rossum *et al.*, *Experientia* **16**, 373 (1960). Toxicity studies: Beccari, *Boll. Chim. Farm.* **106**, 8 (1967). Comprehensive description: A. A. Al-Badr, H. Y. Aboul-Enein, *Anal. Profiles Drug Subs.* **12**, 385-432 (1983).

Oil or crystals, mp 34°. bp_5 260° (partial conversion to isopilocarpine). $[\alpha]_D^{18}$ +106° (c = 2). pK_1 (20°) 7.15; pK_2 (20°) 12.57. Sol in water, alcohol, chloroform; sparingly sol in ether, benzene. Almost insol in petr ether.

Hydrochloride, $C_{11}H_{16}N_2O_2$·HCl, *Akarpine, Almocarpine, Isopto Carpine, Pilopine HS, Pilostat, Salagen.* Hygroscopic crystals from alcohol, mp 204-205°. $[\alpha]_D^{18}$ +91° (c = 2). Freely sol in water, alcohol. Practically insol in ether, chloroform. *Keep well closed and protected from light.*

Nitrate, $C_{11}H_{16}N_2O_2$·HNO$_3$, *Chibro Pilocarpine, Licarpin, Pilo, Pilofrin, Pilagan.* mp 173.5-174.0° (dec). *Poisonous!* $[\alpha]_D$ +77° to +83° (c = 10). One gram dissolves in 4 ml water, 75 ml alcohol. Insol in chloroform, ether.

Incompat. Silver nitrate, mercury bichloride, iodides, gold salts, tannin, calomel, KMnO$_4$, alkalies.

Isopilocarpine, β-pilocarpine. Hygroscopic oily liquid or prisms. bp_{10} 261°. $[\alpha]_D^{18}$ +50° (c = 2). pK_1 (18°) 7.17. Miscible with water and alcohol; very sol in chloroform; less sol in benzene, ether. Almost insol in petr ether.

Isopilocarpine hydrochloride hemihydrate, $C_{11}H_{16}N_2O_2$·HCl·½H$_2$O, scales from alcohol + ether, mp 127°; when anhyd, mp 161°. $[\alpha]_D^{18}$ +39° (c = 5). Sol in 0.27 part water; 2.1 parts alcohol.

Isopilocarpine nitrate, $C_{11}H_{16}N_2O_2$·HNO$_3$, prisms from water, scales from alcohol, mp 159°. $[\alpha]_D^{18}$ +39° (c = 2). Sol in 8.4 parts water, in 350 parts abs alcohol.

THERAP CAT: Antiglaucoma agent; miotic.

THERAP CAT (VET): Parasympathomimetic; miotic; gastric secretory stimulant.

7579. Pilocarpus. Jaborandi. Leaves of *Pilocarpus jaborandi* Holmes (Pernambuco jaborandi), or of *P. microphyllus* Stapf (Maranhao jaborandi), *Rutaceae.* *Habit.* Brazil, Paraguay. *Constit.* About 1% alkaloids of which about 0.5% is pilocarpine; pilocarpidine, isopilocarpine, jaborine, jaboridine, jaboric acid, pilocarpic acid.

THERAP CAT: Sudorific; miotic.

7580. Pilocereine. $C_{45}H_{65}N_3O_6$; mol wt 744.03. C 72.64%, H 8.81%, N 5.65%, O 12.90%. From *Pilocereus sargentianus* Orcutt, *Lophocereus schottii* Britt. and Rose, *Cactaceae.* Isoln: Heyl, *Arch. Pharm.* **239**, 451 (1901); Djerassi *et al.*, *J. Am. Chem. Soc.* **75**, 3632 (1953). Structure: *eidem, ibid.* **84**, 3210 (1962). Biosynthesis: Schütte, Seelig, *Ann.* **730**, 186 (1969); O'Donovan *et al.*, *J. Chem. Soc. (C)* **1971**, 2398.

Needles from ethanol, mp 176.5-177°. uv max (abs ethanol): 284 nm (log ε 3.72). Sol in alcohol, benzene, chloroform, ether, petr ether. Practically insol in water.

Methyl ether, $C_{46}H_{67}N_3O_2$, crystals from hexane, mp 103-105°, solidifies and remelts at 153-155°. uv max: 282, 320 nm (log ε 3.68, 2.05).

Ethyl ether, $C_{47}H_{69}N_3O_6$, crystals from hexane, mp 90-95°, solidifies and remelts at 152-153°.

Acetate, $C_{47}H_{67}N_3O_7$, crystals from ether + acetone, mp 186-186.5°. uv max (abs ethanol): 284 nm (log ε 3.84).

7581. Pilsicainide. *N-(2,6-Dimethylphenyl)tetrahydro-1H-pyrrolizine-7a(5H)-acetamide;* tetrahydro-1H-pyrrolizine-7a(5H)-aceto-2',6'-xylidide; *N-(2,6-dimethylphenyl)-8-pyrrolizidineacetamide.* $C_{17}H_{24}N_2O$; mol wt 272.39. C 74.96%, H 8.88%, N 10.28%, O 5.87%. Structural analog of lidocaine, *q.v.* Prepn: S. Miyano *et al., Eur. pat. Appl.* **89,061;** *idem,* U.S. pat. **4,564,624** (1983, 1986 both to Suntory); *idem, J. Med. Chem.* **28,** 714 (1985). Pharmacology: K. Aisaka *et al., Arzneimittel-Forsch.* **35,** 1239 (1985); T. Hidaka *et al., ibid.* 1381. Clinical evaluation in supraventricular tachycardia: T. Terazawa *et al., Am. Heart J.* **121,** 1437 (1991).

Hydrochloride hemihydrate, $C_{17}H_{24}N_2O \cdot HCl \cdot \frac{1}{2}H_2O$, *SUN-1165, Sunrhythm.* Crystals from ethanol-ether, mp 212-214°. LD_{50} in mice (mg/kg): 410 s.c. (Miyano, 1985).

THERAP CAT: Antiarrhythmic.

7582. Pimaric Acid. *[1R-(1α,4aβ,4bα,7β,10aα)]-7-Ethenyl-1,2,3,4,4a,4b,5,6,7,9,10,10a-dodecahydro-1,4a,7-trimethyl-1-phenanthrenecarboxylic acid; 13α-methyl-13-vinyl-podocarp-8(14)-ene-15-oic acid;* dextropimaric acid; *d-*pimaric acid; *α*-pimaric acid. $C_{20}H_{30}O_2$; mol wt 302.46. C 79.42%, H 10.00%, O 10.58%. Isoln from American rosin: Rimbach, *Ber. Deut. Pharm. Ges.* **6,** 61 (1896); from French galipot from *Pinus maritima* Mill, *Pinaceae:* Ruzicka, Balas, *Helv. Chim. Acta* **6,** 677 (1923); Ruzicka *et al., ibid.* **15,** 915 (1932). Structure: Ruzicka, Sternbach, *ibid.* **23,** 124 (1940); Fleck, *J. Am. Chem. Soc.* **62,** 2044 (1940); Harris, Sanderson, *ibid.* **70,** 2081 (1948). Stereochemistry: Wenkert, Chamberlin, *ibid.* **81,** 688 (1959).

Orthorhombic, bisphenoidal crystals from acetone or acetic acid, mp 217-219°; bp_{18} 282°; $[\alpha]_D^{18}$ +74.7° (c = 0.4 in chloroform); $[\alpha]_D^{20}$ +87.3° (in chloroform).

Hydrochloride, $C_{20}H_{31}O_2Cl$, dec 184°. $[\alpha]_D^{20}$ +13.6° (c = 0.5 in alcohol). Can be reconverted by heating with quinoline at 250°.

Quinidine salt, $C_{40}H_{54}N_2O_4$, mp 90°.

Methyl ester, $C_{21}H_{32}O_2$, mp 69°.

7583. Pimeclone. *2-(1-Piperidinylmethyl)cyclohexanone;* 2-(piperidinomethyl)cyclohexanone; Nu-582; NA-66. $C_{12}H_{21}NO$; mol wt 195.30. C 73.80%, H 10.84%, N 7.17%, O 8.19%. Prepn: Mannich, Honig, *Arch. Pharm.* **265,** 598 (1927); Dimroth *et al., Ber.* **73,** 1399 (1940).

Liquid, bp_{14} 118-120°; bp_6 124°.

Hydrochloride, $C_{12}H_{21}NO \cdot HCl$, *Karion, Spiractin.* Crystals from isopropanol, mp 161-165°.

THERAP CAT: Respiratory stimulant.

7584. Pimefylline. *3,7-Dihydro-1,3-dimethyl-7-[2-[(3-pyridinylmethyl)amino]ethyl]-1H-purine-2,6-dione; 7-[2-[(3-pyridylmethyl)amino]ethyl]theophylline;* 7-(β-3'-picolylaminoethyl)theophylline; pimephylline; ES-771. $C_{15}H_{18}N_6O_2$; mol wt 314.35. C 57.31%, H 5.77%, N 26.73%, O 10.18%. Prepn: **Neth. pat. Appl. 6,600,250;** Suter, Zutter, U.S. pat. **3,350,400** (1966, 1967 both to Eprova). Structure determination, properties and chemistry: *eidem, Pharm. Acta Helv.* **48,** 133 (1973). Pharmacology: Ciaceri, Attaguile, *Gazz. Med. Ital.* **132,** 36, 108 (1973). Metabolism: Pitrè, *Farmaco Ed. Prat.* **29,** 46 (1974).

Crystals from isopropyl acetate, mp 111-112°. uv max (water): 270 nm. Readily sol in cold water, chloroform; sol in acetone, ethanol. LD_{50} in mice (mg/kg): 1900 orally, 402 i.v. (Suter, Zutter, 1973).

Nicotinate, $C_{15}H_{18}N_6O_2 \cdot C_6H_5NO_2$, *ES-902, Teonicon.* Colorless, odorless powder from ethanol, mp 159-160°. uv max (water): 267 nm. Very sol in water (40% soln can be prepd); slightly sol in methanol; very slightly sol in acetone, chloroform. LD_{50} in mice (mg/kg): 2530 orally, 470 i.v. (Suter, Zutter, 1973).

THERAP CAT: Vasodilator (coronary).

7585. Pimelic Acid. *Heptanedioic acid;* 1,5-pentanedicarboxylic acid. $C_7H_{12}O_4$; mol wt 160.17. C 52.49%, H 7.55%, O 39.96%. $HOOC(CH_2)_5COOH$. Prepn starting with cyclohexanone and ethyl oxalate: Snyder *et al., Org. Syn. coll. vol. II,* p 531 (1943); by the action of sodium and isoamyl alcohol on salicylic acid: Müller, *ibid.* p 535.

Monoclinic prisms from benzene, mp 105.7-105.8°. Tends to sublime. bp_{100} 272°; bp_{50} 251.5°; bp_{15} 223°; bp_{10} 212°. 2.52 parts dissolve in 100 parts water at 13.5°; 5 parts dissolve in 100 parts water at 20°. Freely sol in alcohol, ether. Practically insol in cold benzene.

Diethyl ester, $C_{11}H_{20}O_4$, liquid, d_4^{20} 0.99448; bp_{748} 254°; bp_{24} 155°; bp_{15} 140°. Sol in alcohol, ether, ethyl acetate.

7586. Pimenta. Allspice; Jamaica pepper; semen Amomi. Dried, nearly ripe fruit of *Pimenta officinalis* Lindl., *Myrtaceae.* *Habit.* East and West Indies, Central and South America. *Constit.* 3-4.5% volatile oil containing about 70% eugenol and small quantities eugenol methyl ether; cineol, *l-α*-phellandrene and caryophyllene; resin, tannin, fixed oil, sugar, gum.

7587. Piminodine. *4-Phenyl-1-[3-(phenylamino)propyl]-4-piperidinecarboxylic acid ethyl ester; 1-(3-anilinopropyl)-4-phenylisonipecotic acid ethyl ester;* ethyl 4-phenyl-1-[(3-phenylamino)propyl]-4-piperidinecarboxylate; ethyl 1-(3-anilinopropyl)-4-phenylisonipecotate. $C_{23}H_{30}N_2O_2$; mol wt 366.50. C 75.38%, H 8.25%, N 7.64%, O 8.73%. Prepn: Elpern *et al., J. Am. Chem. Soc.* **81,** 3784 (1959).

Consult the Name Index before using this section.

Dihydrochloride, $C_{23}H_{30}N_2O_2.2HCl$, crystals, mp 219.4-222.2°.

Ethanesulfonate, $C_{25}H_{36}N_2O_5S$, *NIH-7590, Win-14098, Alvodine, Pimadin.*

Note: This is a controlled substance (opiate) listed in the U.S. Code of Federal Regulations, Title 21 Part 1308.12 (1995).

THERAP CAT: Analgesic (narcotic).

7588. Pimobendan. *4,5-Dihydro-6-[2-(4-methoxyphenyl)-1H-benzimidazol-5-yl]-5-methyl-3(2H)-pyridazinone;* UD-CG 115; UD-CG 115 BS; 2-(4-methoxyphenyl)-5(6)-(5-methyl-3-oxo-4,5-dihydro-2H-6-pyridazinyl)benzimidazole. $C_{19}H_{18}N_4O_2$; mol wt 334.38. C 68.25%, H 5.43%, N 16.76%, O 9.57%. Positive inotropic agent. Prepn: V. Austel *et al.,* **Ger. pat. 2,837,161;** *eidem,* **U.S. pat. 4,361,563** (1980, 1982 both to Thomae). Cardiovascular effects: J. C. A. von Meel, *Arzneimittel-Forsch.* **35,** 284 (1985); P. D. Verdouw *et al., Eur. J. Pharmacol.* **126,** 21 (1986). Comparison of enantiomer activity: K. Fujino *et al., J. Pharmacol. Exp. Ther.* **247,** 519 (1988). Clinical trial in congestive heart failure: F. Hagemeijer, *Am. Heart J.* **122,** 517 (1991); S. D. Katz *et al., ibid.* **123,** 95 (1992).

Hydrochloride, $C_{19}H_{18}N_4O_2.HCl$. Crystals from methanol and ethereal HCl, mp 311°(dec). LD_{50} orally in mice: ~600 mg/kg (Austel).

THERAP CAT: Cardiotonic.

7589. Pimozide. *1-[1-[4,4-Bis(4-fluorophenyl)butyl]-4-piperidinyl]-1,3-dihydro-2H-benzimidazol-2-one; 1-[1-[4,4-Bis(p-fluorophenyl)butyl]-4-piperidyl]-2-benzimidazolinone;* 1-[4,4-di-(4-fluorophenyl)butyl]-4-(2-oxo-1-benzimidazolinyl)piperidine; R-6238; Orap; Opiran. $C_{28}H_{29}F_2N_3O$; mol wt 461.55. C 72.86%, H 6.33%, F 8.23%, N 9.10%, O 3.47%. Prepn: Janssen, **Fr. pat. M3695** (1965 to Janssen), *C.A.* **66,** 115709t (1967). Pharmacological studies: Janssen *et al., Arzneimittel-Forsch.* **18,** 261, 279, 282 (1968). Metabolism in rats: Soudijn, Wijngaarden, *Life Sci.* **8,** 291 (1969). Clinical experience: Poldinger, *Curr. Ther. Res.* **13,** 23 (1971). Pimozide blocks establishment but not expression of amphetamine-produced environment-specific conditioning: R. J. Beninger, B. L. Hahn, *Science* **220,** 1304 (1983). Use in treatment of acute schizophrenia: B. Shopsin, G. Selzer, *Curr. Ther. Res.* **21,** 755 (1977); in treatment of chronic schizophrenia: F. Kline *et al., ibid.* 768. In management of Gilles de la Tourette's syndrome: M. S. Ross, H. Moldofsky, *Lancet* **1,** 103 (1977); A. K. Shapiro, E. Shapiro, *Am. J. Psychiat.* **140,** 1235 (1983); *eidem, J. Am. Acad. Child Psychiat.* **23,** 161 (1984).

Microcrystals, mp 214-218°. Weak base, pKa 7.32. Almost insol in water (<0.01 mg/ml). Very slightly sol in dil aq solns of organic and mineral acids (<5 mg/ml).

THERAP CAT: Antipsychotic.

7590. Pimpinella. Pimpernel; brunet saxifrage; small saxifrage. Root of *Pimpinella saxifraga* L. or *P. magna* L.

Umbelliferae. Habit. Europe; adventitious in U.S. *Constit.* Volatile oil, resin, benzoic acid, pimpinellin.

THERAP CAT: Aromatic carminative.

7591. Pimpinellin. *5,6-Dimethoxy-2H-furo[2,3-h]-1-benzopyran-2-one;* 4-hydroxy-6,7-dimethoxy-5-benzofuranacrylic acid δ-lactone. $C_{13}H_{10}O_5$; mol wt 246.22. C 63.42%, H 4.09%, O 32.49%. Found in *Pimpinella saxifraga* L., *Heracleum spondylium* L., *H. lanatum* Michx., and *H. panaces* L., *Umbelliferae.* Isoln: Heut, *Arch. Pharm.* **236,** 162 (1898); Herzog, Hancu, *ibid.* **246,** 402 (1908); Wessely, Kallab, *Monatsh.* **59,** 161 (1932); Späth, Simon, *ibid.* **67,** 344 (1936); Fujita, Furuya, *J. Pharm. Soc. Japan* **74,** 795 (1954); **76,** 535 (1956); Svendsen, Ottestad, *Pharm. Acta Helv.* **32,** 457 (1957); Svendsen, *C.A.* **52,** 2173g (1958); Svendsen *et al., Planta Med.* **7,** 113 (1959); *Pharm. Acta Helv.* **34,** 33 (1959); Jastrzebski, *C.A.* **54,** 1205b (1960). Synthesis: M. W. Reed, H. W. Moore, *J. Org. Chem.* **53,** 4166 (1988).

Off-white needles from methylene chloride/hexane, mp 119°. Practically insol in water. Sol in alcohol. Absorption spectrum: Wessely, Koltan, *Monatsh.* **86,** 430 (1955).

7592. Pinacidil. *(±)-N-Cyano-N'-4-pyridinyl-N''-(1,2,2-trimethylpropyl)guanidine monohydrate;* P-1134; Pindac. $C_{13}H_{19}N_5.H_2O$; mol wt 263.34. C 59.29%, H 8.04%, N 26.59%, O 6.08%. Potassium channel opening vasodilator. Prepn: H. J. Petersen, **Ger. pat. 2,557,438;** *idem,* **U.S. pat. 4,057,636** (1976, 1977 both to Leo Pharm.); H. J. Petersen *et al., J. Med. Chem.* **21,** 773 (1978). Mechanism of action: E. Arrigoni-Martelli *et al., Experientia* **36,** 445 (1980); K. M. Bray *et al., Brit. J. Pharmacol.* **91,** 421 (1987). Metabolism: E. Eilertsen *et al., Xenobiotica* **12,** 187 (1982). Bioavailability: *eidem, ibid,* 177. Determn in plasma: M. Hamilton *et al., J. Chromatog.* **375,** 359 (1986). Pharmacokinetics and hypotensive effects: J. W. Ward *et al., Eur. J. Clin. Pharmacol.* **26,** 603 (1984). Clinical comparison with hydralazine, *q.v.:* R. L. Byyny *et al., Clin. Pharmacol. Ther.* **42,** 50 (1987). Review of pharmacology and mechanism of action: M. L. Cohen: *Drug Develop. Res.* **9,** 249-258 (1986).

Crystals, mp 164-165°. LD_{50} in mice, rats (mg/kg): 600, 570 orally (Petersen, 1978).

THERAP CAT: Antihypertensive.

7593. Pinacol. *2,3-Dimethyl-2,3-butanediol;* pinacone; tetramethylethylene glycol. $C_6H_{14}O_2$; mol wt 118.18. C 60.98%, H 11.94%, O 27.08%. Prepd by the reduction of acetone: Holleman, *Rec. Trav. Chim.* **25,** 206 (1906); R. Adams, E. W. Adams, *Org. Syn. coll. vol. I,* 459 (2nd ed., 1941); **Ger. pat. 233,894** in *Frdl.* **10,** 1000. Convenient lab procedure: L. F. Fieser, *Experiments in Organic Chemistry* (Boston, 3rd ed., 1955) p 101. Crystal structure: G. A. Jeffrey, A. Robbins, *Acta Crystallogr.* **B34,** 3817 (1978).

Hexahydrate, $C_6H_{14}O_2.6H_2O$, four-sided plates from water. mp 45.4°; d^{15} 0.967 (supercooled liquid). The anhydr compd crystallizes in needles from alc or ether, mp 41.1°; bp_{760} 174.4°. Freely sol in hot water, in alc, in ether; slightly sol in cold water, carbon disulfide.

Dimethyl ether, $C_8H_{18}O_2$, liquid. Agreeable odor. bp 144°.

7594. Pinacolone. *3,3-Dimethyl-2-butanone; tert-butyl methyl ketone; pinacolin.* $C_6H_{12}O$; mol wt 100.16. C 71.95%, H 12.08%, O 15.97%. Prepd by distillation of pinacol hydrate with dil H_2SO_4: Fittig, *Ann.* **114,** 56 (1860); Hill, Flosdorf, *Org. Syn. coll. vol.* **I** (2nd ed., 1941) p 462; L. F. Fieser, K. L. Williamson, *Organic Experiments* (D. C. Heath and Co., Lexington, Mass., 5th ed., 1983) p 332. Almost quantitative yields are obtained when 56.5 g pinacol hydrate is boiled for 3 hrs with 0.5 liter 25% H_2SO_4 and the product is steam distilled: Boeseken, van Tonningen, *Rec. Trav. Chim.* **39,** 189 (1920).

Liquid. Peppermint or camphor-like odor. d_{25}^{25} 0.7250. mp —52.5°. bp_{760} 106.2°. n_D^{25} 1.3939. Volatile in steam. Moderately sol in water (2.44% at 15°). Soluble in alcohol, ether, acetone.

Oxime, $C_6H_{13}NO$, needles from aq alcohol. mp 78°. bp_{748} 171.6°. Sol in alc, ether, petr ether, benzene, chloroform.

7595. Pinaverium Bromide. *4-[(2-Bromo-4,5-dimethoxyphenyl)methyl]-4-[2-[2-(6,6-dimethylbicyclo[3.1.1]hept-2-yl)ethoxy]ethyl]morpholinium bromide;* 4-(6-bromoveratryl)-4-[2-[2-(6,6-dimethyl-2-norpinyl)ethoxy]ethyl]morpholinium bromide; LAT-1717; Dicetel. $C_{26}H_{41}Br_2NO_4$; mol wt 591.42. C 52.80%, H 6.99%, Br 27.02%, N 2.37%, O 10.82%. Spasmolytic agent with low incidence of anticholinergic effects. Prepn: **Belg. pat. 769,469;** R. Baronnet, U.S. pat. **3,845,048** (1971, 1974 both to Societe Berri-Balzac). Synthesis and pharmacology: R. Baronnet *et al., Eur. J. Med. Chem.* **9,** 182 (1974). Pharmacodynamics: J. Bretaudeau *et al., Therapie* **30,** 919 (1975). Synthesis of ^{14}C pinaverium bromide and pharmacokinetics: C. Jacquot *et al., Eur. J. Med. Chem.* **13,** 61 (1978). Mechanism of action: J. Bretaudeau, O. Foussard-Blanpin, *J. Pharmacol.* **11,** 233 (1980). Inhibition of gastrointestinal contractile activity in dogs: Z. Itoh, T. Takahashi, *Arzneimittel-Forsch.* **31,** 1450 (1981). Effects in humans on gastric emptying and transit time: J. Bertrand *et al., Therapie* **36,** 555 (1981).

Crystals from methyl ethyl ketone, mp 181°; also reported as mp ~170° (Baronnet, 1974). LD_{50} in mice (mg/kg): 1400 orally; 66 i.v. (Baronnet, 1974); also reported as LD_{50} i.v. in mice: 37 ±2.4 mg/kg (Baronnet *et al.*).

THERAP CAT: Antispasmodic.

7596. Pinazepam. *7-Chloro-1,3-dihydro-5-phenyl-1-(2-propynyl)-2H-1,4-benzodiazepin-2-one;* 7-chloro-1-propargyl-5-phenyl-3H-1,4-benzodiazepin-2(1H)-one; Z-905; Domar; Duna. $C_{18}H_{13}ClN_2O$; mol wt 308.77. C 70.02%, H 4.24%, Cl 11.48%, N 9.07%, O 5.18%. Prepn: F. Tenconi *et al.,* **Ger.** pat. **2,339,790** (1974 to Zambeletti), *C.A.* **80,** 133492k (1974); C. Podesva, K. Vagi, U.S. pat. **3,842,094** (1974 to Delmar). Metabolism study: A. Trebbi *et al., J. Chromatog.* **110,** 309 (1975). Pharmacological and toxicological studies: F. Scrollini *et al., Arzneimittel-Forsch.* **25,** 934 (1975). Pharmacokinetics: P. M. Boselli, F. Scrollini, *Boll. Chim. Farm.* **116,** 363 (1977). Psychopharmacological properties: F. Scrollini *et al., Arzneimittel-Forsch.* **28,** 423 (1978). Physico-chemical profile: G. Filipi, A Trebbi, *Boll.*

Chim. Farm. **118,** 105 (1979). Clinical study: V. Bertoncelli *et al., Clin. Ter.* **94,** 641 (1980).

Cryst from methanol/water, mp 140-142°. LD_{50} in mice, rats (mg/kg): 1355, 5819 orally; 266, 622 i.p. (Scrollini, 1975). Also reported as LD_{50} in mice (mg/kg): 670 orally (Podesva, Vagi).

Note: This is a controlled substance (depressant) listed in the U.S. Code of Federal Regulations, Title 21 Part 1308.14 (1995).

THERAP CAT: Anxiolytic.

7597. Pindolol. *1-(1H-Indol-4-yloxy)-3-[(1-methylethyl)amino]-2-propanol;* 4-[2-hydroxy-3-(isopropylamino)propoxy]indole; prinodolol; LB-46; Betapindol; Blocklin L; Calvisken; Decreten; Durapindol; Glauco-Visken; Pectobloc; Pinbetol; Pynastin; Visken. $C_{14}H_{20}N_2O_2$; mol wt 248.33. C 67.72%, H 8.12%, N 11.28%, O 12.89%. β-Adrenergic blocker with partial agonist activity. Prepn: Troxler, **Swiss** pats. **469,002** and **472,404** (both 1969 to Sandoz). HPLC determn in plasma: H. Smith, *J. Chromatog.* **415,** 95 (1987). Symposium on clinical studies: *Am. Heart J.* **104,** Suppl. 2, pt. 2, 333-520 (1982); *Brit. J. Clin. Pharmacol.* **13,** Suppl. 2, 143S-450S (1982). Brief review: W. H. Frishman, *N. Engl. J. Med.* **308,** 940-944 (1983).

Crystals from ethanol, mp 171-173°.

THERAP CAT: Antihypertensive, antianginal, antiarrhythmic, antiglaucoma agent.

7598. Pindone. *2-(2,2-Dimethyl-1-oxopropyl)-1H-indene-1,3(2H)-dione; 2-pivaloyl-1,3-indandione;* 2-pivalyl-1,3-indandione; 2-trimethylacetyl-1,3-indandione; pivalyl indandione; pivaldione; Pival; Pivalyl Valone; Tri-Ban. $C_{14}H_{14}O_3$; mol wt 230.26. C 73.03%, H 6.13%, O 20.84%. Prepn: Kilgore *et al., Ind. Eng. Chem.* **34,** 494 (1942); Zelmens, Vanags, *C.A.* **53,** 21830d (1959). Toxicity study: T. B. Gaines, *Toxicol. Appl. Pharmacol.* **14,** 515 (1969).

Bright yellow crystals from ethanol, mp 108-110°. LD_{50} orally in male rats: 280 mg/kg (Gaines).

Sodium salt, *Pivalyn.* Bright yellow crystals, mp 205-210°. Schwarz, U.S pat. **2,880,132** (1959 to Morton Chem.). Sol in water.

Caution: Potential symptoms of overexposure are epistaxis, excess bleeding from minor cuts and bruises; smoky urine, black tarry stools; abdominal and back pain. *See*

NIOSH Pocket Guide to Chemical Hazards (DHHS/NIOSH 90-117, 1990) p 184.

USE: Rodenticide; insecticide.

7599. α-Pinene. *2,6,6-Trimethylbicyclo[3.1.1]hept-2-ene; 2-pinene;* pinene. $C_{10}H_{16}$; mol wt 136.24. C 88.16%, H 11.84%. Obtained from oil of turpentine which contains 58-65% α-pinene along with 30% β-pinene, *q.v.:* E. Gildemeister, F. Hoffmann, *Die ätherischen Oele* **Band IV** (Akademie-Verlag, Berlin, 4th ed., 1956) p 39. α-Pinene in North American oils is dextrorotatory, in most European oils it is levorotatory. Constituent of many volatile oils. Isoln of *d*-α-pinene from Port Oxford cedar wood oil *(Chamaecyparis lawsoniana* Parl., *Pinaceae):* Thurber, Roll, *Ind. Eng. Chem.* **19**, 739 (1927). Isoln of *l*-α-pinene from mandarin peel oil *(Citrus reticulata* Blando, *Rutaceae):* Kugler, Kováts, *Helv. Chim. Acta* **46**, 1480 (1963). Total synthesis of α- and β-forms: Komppa, *Ann. Acad. Sci. Fennicae* **A59**, 3 (1943), *C.A.* **41**, 425 (1947); Thomas, Fallis, *Tetrahedron Letters* **1973**, 4687; *eidem, J. Am. Chem. Soc.* **98**, 1227 (1976). Review: Palmer, *Ind. Eng. Chem.* **34**, 1028 (1942); J. L. Simonsen, *The Terpenes* vol. **II** (Cambridge Univ. Press, 2nd ed., 1949) pp 105-191; D. V. Banthorpe, D. Whittaker, *Chem. Rev.* **66**, 643-654 (1966); *Food Cosmet. Toxicol.* **16**, Suppl. I, 853-857 (1978).

dl-Form, liquid, characteristic odor of turpentine. bp_{760} 155-156°; bp_{20} 52.5°. d_4^{20} 0.8592. n_D^{20} 1.4664. Practically insol in water. Sol in alc, chloroform, ether, glacial acetic acid.

Hydrochloride, $C_{10}H_{17}Cl$, mp 132°.

d-Form, bp_{760} 155-156°. d_4^{20} 0.8591. n_D^{20} 1.4663. $[\alpha]_D^{20}$ +51.14°.

Hydrochloride, mp 132°. $[\alpha]_D^{20}$ +33.52° (alcohol).

l-Form, bp_{760} 155-156°. d_4^{20} 0.8590. n_D^{20} 1.4662. $[\alpha]_D^{20}$ −51.28°.

Hydrochloride, mp 132°. $[\alpha]_D^{20}$ −33.24° (alcohol).

Caution: Toxic effects similar to turpentine, *q.v.*

USE: Manufacture of camphor, insecticides, solvents, plasticizers, perfume bases, synthetic pine oil.

7600. β-Pinene. *6,6-Dimethyl-2-methylenebicyclo-[3.1.1]heptane;* nopinene. $C_{10}H_{16}$; mol wt 136.24. C 88.16%, H 11.84%. Found in most essential oils which contain α-pinene, but in far smaller proportions; the *l*-form occurs most commonly. Initial identification: A. von Baeyer, *Ber.* **29**, 25 (1896). Isoln of the *d*-form from *Ferula galbaniflua* Boiss. et Buhse, *Umbelliferae:* B. N. Rutovski, I. V. Vinogradova, *J. Prakt. Chem.* **120**, 41 (1928); from *Cynomarathrum nuttallii* A. Gray, *Umbelliferae:* E. K. Nelson, *J. Am. Chem. Soc.* **55**, 3400 (1933). Irreversible isomerization of β-pinene to α-pinene occurs on shaking with platinum black satd with hydrogen: F. Richter, W. Wolff, *Ber.* **59**, 1733 (1926). Synthesis: G. Bonnet, *Bull. Inst. Pin* **1938**, 217; **1939**, 1, *C.A.* **33**, 4223⁴ (1939); K. J. Crowley, *Proc. Chem. Soc. (London)* **1962**, 245; *Tetrahedron* **21**, 1001 (1965); L. M. Harwood, M. Julia, *Synthesis* **1980**, 456. For general refs *see* α-pinene.

dl-Form, bp_{760} 165-166°.

d-Form, bp_{760} 164-166°. d_{20}^{20} 0.8654. n_D^{20} 1.4739. $[\alpha]_D$ +28.59° (Nelson). Also reported as bp_{760} 162-163°. d_{20}^{20} 0.8662. n_D^{20} 1.4745. $[\alpha]_D$ +20.75° (Rutovski, Vinogradova).

l-Form, bp_{760} 162-163°. d^{15} 0.874. n_D^{15} 1.4872, $[\alpha]_D$ −22.4°.

7601. Pine Oil. Yarmor. An oil from *Pinus palustris* Mill. and certain other species of pines, *Pinaceae.* It is obtained from pitch-soaked pine wood by steam distillation or solvent extraction followed by steam distillation and also by destructive distillation. It consists mainly of isomeric tertiary and secondary, cyclic terpene alcohols.

Colorless to pale yellow liquid, turpentine-like odor. d about 0.9; bp 200-220°. Insol in water. Sol in the usual organic solvents.

Caution: Irritating to skin, mucous membranes. Large doses may cause CNS depression.

USE: Pharmaceutic aid (flavor and perfume). Manuf terpin hydrate and other terpin products; as a solvent, disinfectant and deodorant; in textile scouring; for flotation of lead and zinc ores.

7602. Pine Tar. A product obtained by destructive distillation of wood of *Pinus palustris* Mill., or other species of pine, *Pinaceae.*

Blackish-brown, viscous liquid; heavier than water; empyreumatic odor and sharp taste. Slightly sol in water; sol in alc, chloroform, ether, acetone, glacial acetic acid, fixed and volatile oils, and in solns of caustic alkalies. Principal constituents: turpentine, resin, guaiacol, creosol, methylcreosol, phenol, phlorol, toluene, xylene, and other hydrocarbons.

THERAP CAT: Topical antieczematic; rubefacient.

THERAP CAT (VET): Mild irritant, antiseptic in chronic skin conditions. Expectorant.

7603. Pinguinain. A proteolytic enzyme obtained from the juice of the fruit of *Bromelia pinguin* Plum. ex L., *Bromeliaceae* (pineapple family): Asenjo, *Science* **95**, 48 (1942); Bloch, Messing, U.S. pat. **2,977,287** (1961 to Ethicon). Immunochemical studies: E. Toro-Goyco, I. Rodriguez-Costas, *Arch. Biochem. Biophys.* **175**, 359 (1976). Structure studies: E. Toro-Goyco *et al., Biochim. Biophys. Acta* **622**, 151 (1980).

Aq solns are capable of digesting necrotic tissue, but do not attack viable tissue. Optimum proteolytic activity is at pH 5.2 to 5.5 and also at pH 7.3. Inactivated at temps above 80°.

7604. Pinosylvin. *(E)-5-(2-Phenylethenyl)-1,3-benzenediol; E-3,5-stilbenediol;* 5-styrylresorcinol; *trans*-3,5-dihydroxystilbene. $C_{14}H_{12}O_2$; mol wt 212.25. C 79.23%, H 5.70%, O 15.08%. Occurs together with its monomethyl and dimethyl ethers in the heartwood of pine and other woody plants. Naturally occurring pinosylvins have the *trans* configuration. Isoln from *Pinus sylvestris* L., *Pinaceae:* H. Erdtman, *Ann.* **539**, 116 (1939); from other *Pinus* species: G. Lindstedt, *Acta Chem. Scand.* **3**, 755-772 (1949); J. C. Alvarez-Novoa *et al., ibid.* **4**, 444 (1950); from *Alnus sieboldiana, Betulaceae:* Y. Asakawa, *Bull. Chem. Soc. Japan* **44**, 2761 (1971); from *Polygonum nodosum, Polygonaceae:* M. Kuroyanagi *et al., Chem. Pharm. Bull.* **30**, 1602 (1982). Synthesis of pinosylvin: E. Späth, F. Liebherr, *Ber.* **74**, 869 (1941); of monomethyl ether: E. Späth, K. Kromp, *ibid.* 1424; of dimethyl ether: G. Aulin-Erdtman, H. Erdtman, *ibid.* 50; of pinosylvin and derivatives: A. A. Loman, L. R. Snowdon, *Can. J. Chem.* **48**, 1554 (1970). Biosynthesis: Birch, *Fortschr. Chem. Org. Naturst.* **14**, 186 (1957). Toxicological study: K. O. Frykholm, *Nature* **155**, 454 (1945). Use as antimicrobial agent: E. H. Sheers, Ger. pat. **1,952,-451;** *idem,* U.S. pat. **3,577,230** (1970, 1971 both to Arizona Chem. Co.). Deterrent to feeding behavior of snowshoe hare: J. P. Bryant *et al., Science* **222**, 1023 (1983).

Fine needles from glacial acetic acid, mp 155.5-156°. uv max (ethanol): 305 nm (log ε 4.49). Practically insol in water. Sol in benzene, acetone, chloroform, glacial acetic acid.

Monomethyl ether, $C_{15}H_{14}O_2$, crystals, mp 122-123°. uv max (ethanol): 303 nm (log ϵ 4.26). More sol in benzene than pinosylvin. Also sol in methanol, glacial acetic acid.

Dimethyl ether, $C_{16}H_{16}O_2$, crystals from methanol-water, mp 55-56°. uv max (ethanol): 305 nm (log ϵ 4.39).

7605. Pioglitazone. (±)-5-[[4-[2-(5-Ethyl-2-pyridinyl)-ethoxy]phenyl]methyl]-2,4-thiazolidinedione; (±)-5-[p-[2-(ethyl-2-pyridyl)ethoxy]benzyl]-2,4-thiazolidinedione; AD-4833. $C_{19}H_{20}N_2O_3S$; mol wt 356.45. C 64.02%, H 5.66%, N 7.86%, O 13.47%, S 9.00%. Insulin sensitizer. Prepn: K. Meguro, T. Fujita, **Eur. pat. Appl.** 193,256; *eidem*, **U.S. pat.** 4,687,777 (1986, 1987 both to Takeda); T. Sohda *et al.*, *Arzneimittel-Forsch.* **40**, 37 (1990). Pharmacology: H. Ikeda *et al.*, *ibid.* 156. HPLC determn in serum: W. Z. Zhong, D.B. Lakings, *J. Chromatog.* **490**, 377 (1989). Mechanism of action: C. Hofmann *et al.*, *Endocrinology* **129**, 1915 (1991); M. Kobayashi *et al.*, *Diabetes* **41**, 476 (1992). Effect on adipocyte differentiation: T. Sandouk *et al.*, *Am J. Physiol.* **264**, C1600 (1993).

Colorless needles from DMF + water, mp 183-184°.
Hydrochloride, $C_{19}H_{20}N_2O_3S.HCl$, *U-72107A*. Colorless prisms from ethanol, mp 193-194°.
THERAP CAT: Antidiabetic.

7606. Pipacycline. [4S-(4α,4aα,5aα,6β,12aα)]-4-(Dimethylamino)-1,4,4a,5,5a,6,11,12a-octahydro-3,6,10,12,12a-pentahydroxy-N-[[4-(2-hydroxyethyl)-1-piperazinyl]methyl]-6-methyl-1,11-dioxo-2-naphthacenecarboxamide; N-[[4-(2-hydroxyethyl)-1-piperazinyl]methyl]tetracycline; N-[4-(β-hydroxyethyl)diethylenediamino-1-methyl]tetracycline; mepicycline; mepiciclina; Ambra-Vena; Sieromicin; Valtomicina. $C_{29}H_{38}N_4O_9$; mol wt 586.64. C 59.38%, H 6.53%, N 9.55%, O 24.55%. Semi-synthetic broad spectrum antibiotic related to tetracycline. Prepn: Pedrazzoli *et al.*, *Boll. Chim. Farm.* **98**, 516 (1959), *C.A.* **54**, 3856a (1960); Gradnik *et al.*, **Brit. pat.** 888,968 corresp to **U.S. pat.** 3,149,114 (1962 and 1964 to E.R.A.S.M.E.). Properties: *eidem*, *Pharm. Acta Helv.* **35**, 529 (1960). Pharmacokinetic studies: A. Scalvini, A. Delmonte, *Gazz. Med. Ital.* **131**, 1 (1972).

Yellow cryst powder, dec 162-163°. $[\alpha]_D^{20}$ −195° (c = 0.5). $[\alpha]_D^{20}$ −175° (c = 0.5 in methanol). uv max (10 γ/ml 0.1N HCl): 286, 355 nm. pH of 2% aq soln, 7.2-7.4. Freely sol in water, methanol, formamide; slightly sol in ethanol, isopropanol. Practically insol in ether, benzene, chloroform. Sensitive to light, heat, and air. LD_{50} i.v. in white mice: 188 mg/kg (Scalvini, Delmonte).
THERAP CAT: Antibacterial.

7607. Pipamazine. 1-[3-(2-Chloro-10H-phenothiazin-10-yl)propyl]-4-piperidinecarboxamide; 1-[3-(2-chlorophenothiazin-10-yl)propyl]isonipecotamide; 10-[3-(4-carbamoyl-piperidin-1-yl)propyl]-2-chlorophenothiazine; 2-chloro-10-[3-(4-carbamoylpiperidinyl)propyl]phenothiazine; 10-[3-(4-carbamoylpiperidino)propyl]-2-chlorophenothiazine; SC-9387; Nausidol; Mornidine. $C_{21}H_{24}ClN_3OS$; mol wt 401.96. C 62.75%, H 6.02%, Cl 8.82%, N 10.45%, O 3.98%, S 7.98%. Prepn: Cusic *et al.*, **U.S. pat.** 2,957,870 (1960 to Searle).

Crystals from 2-propanol + petr ether, mp about 139°.
Hydrochloride, crystals, mp about 196-197° with formation of bubbles.
THERAP CAT: Antiemetic.

7608. Pipamperone. 1'-[4-(4-Fluorophenyl)-4-oxobutyl]-[1,4'-bipiperidine]-4'-carboxamide; 1'-[3-(p-fluorobenzoyl)propyl]-[1,4'-bipiperidine]-4'-carboxamide; 1-(p-fluorophenyl)-4-(4-piperidino-4-carbamoylpiperidino)-1-butanone; 1-[γ-(4-fluorobenzoyl)propyl]-4-piperidinopiperidine-4-carboxamide; 4'-fluoro-4-[N-[4-(N-piperidino)-4-carbamido]-piperidino]butyrophenone; floropipamide; R-3345. $C_{21}H_{30}FN_3O_2$; mol wt 375.49. C 67.17%, H 8.05%, F 5.06%, N 11.19%, O 8.52%. Prepn of the dihydrochloride by reaction of γ-chloro-4-fluorobutyrophenone and 4-piperidinopiperidine-4-carboxamide: Janssen, **Belg. pat.** 610,830 (1962 to Janssen), *C.A.* **57**, 13740b (1962).

Dihydrochloride, $C_{21}H_{30}FN_3O_2.2HCl$, *Dipiperon, Piperonil, Propitan*. Crystals, mp 124.5-126.0°.
THERAP CAT: Antipsychotic.

7609. Pipazethate. 10H-Pyrido[3,2-b][1,4]benzothiadiazine-10-carboxylic acid 2-(2-piperidinoethoxy)ethyl ester; 2-(2-piperidinoethoxy)ethyl 10H-pyrido[3,2-b][1,4]benzothiadiazine-10-carboxylate; thiophenylpyridylamino-10-carboxylic acid piperidinoethoxyethyl ester; 2-(2-piperidinoethoxy)ethyl 10-thia-1,9-diazaanthracene-10-carboxylate; 1-azaphenothiazine-10-carboxylic acid 2-(2-piperidinoethoxy)ethyl ester; D-254. $C_{21}H_{25}N_3O_3S$; mol wt 399.51. C 63.13%, H 6.31%, N 10.52%, O 12.01%, S 8.03%. Prepn: Schuler *et al.*, *Ann.* **673**, 102 (1964); Schuler, **U.S. pat.** 2,989,529 (1961 to Degussa).

Hydrochloride, $C_{21}H_{25}N_3O_3S.HCl$, *Lenopect, Selvigon, Selvjgon, Theratuss*. Pale yellow crystals from isopropanol, mp 160-161°. Soluble in water, methanol. Practically insol in acetone, petr ether. LD_{50} orally in rats: 560 mg/kg (Schuler).
THERAP CAT: Antitussive.

7610. Pipebuzone. 4-Butyl-4-[(4-methyl-1-piperazinyl)-methyl]-1,2-diphenyl-3,5-pyrazolidinedione; 1,2-diphenyl-3,5-dioxo-4-[(4-methyl-1-piperazinyl)methyl]-4-butylpyrazolidine; LD-4644; Elarzone. $C_{25}H_{32}N_4O_2$; mol wt 420.55. C 71.40%, H 7.67%, N 13.32%, O 7.61%. The 4-[(4-methyl-1-piperazinyl)methyl]- deriv of phenylbutazone, *q.v.* Prepn:

Ger. pat. **1,958,722** and **Fr.** pat. **M7914** (both 1970 to Dausse), *C.A.* **73**, 25460v (1970); **76**, 140797j (1972).

Crystals from isopropanol, mp 129°.
THERAP CAT: Anti-inflammatory; antipyretic; analgesic.

7611. Pipecolic Acid. *2-Piperidinecarboxylic acid;* pipecolinic acid; hexahydropicolinic acid; homoproline; dihydrobaikiaine. $C_6H_{11}NO_2$; mol wt 129.16. C 55.80%, H 8.58%, N 10.84%, O 24.77%. The *l*-form occurs in plants: Phillips, *Chem. & Ind. (London)* **1953**, 127. Prepn: A. Ladenburg, *Ber.* **24**, 640 (1891); Stevens, Ellman, *J. Biol. Chem.* **182**, 75 (1950); V. Asher *et al., Tetrahedron Letters* **22**, 141 (1981). Synthesis of L-pipecolic acid from L-lysine: Fujii, Miyoshi, *Bull. Chem. Soc. Japan* **48**, 1341 (1975). Synthesis of racemate: R. T. Shuman *et al., J. Org. Chem.* **55**, 738 (1990).

l-Form, needles by sublimation, mp 270°. $[\alpha]_D^{25}$ −34.9°. Sol in water, dil alcohol. Sparingly sol in abs alcohol, acetone, chloroform. Insol in ether.
d-Form, platelets from alcohol, mp 270°. $[\alpha]_D^{25}$ +35.7°. Sol in water. Somewhat sol in alcohol.
dl-Form, leaflets from water, mp 264°. Sol in water, boiling alcohol.
dl-Hydrochloride, $C_6H_{11}NO_2$·HCl, crystals (warts) from alcohol + benzene, mp 258-262°.

7612. Pipecurium Bromide. *4,4'-[(2β,3α,5α,16β,17β)-3,17-Bis(acetyloxy)androstane-2,16-diyl]bis(1,1-dimethylpiperazinium)dibromide;* 4,4'-(3α,17β-dihydroxy-5α-androstan-2β,16β-ylene)bis(1,1-dimethylpiperazinium)dibromide diacetate; pipecuronium bromide; RGH-1106; Arduan. $C_{35}H_{62}Br_2N_4O_4$; mol wt 762.71. C 55.12%, H 8.19%, Br 20.95%, N 7.35%, O 8.39%. Non-depolarizing curare-like agent; deriv of androstane, *q.v.*, related structurally to pancuronium bromide, *q.v.* Prepn: Z. Tuba *et al.*, **Ger.** pat. **2,337,882** (1974 to Gedeon Richter), *C.A.* **80**, 121210d (1974); Z. Tuba, *Arzneimittel-Forsch.* **30**, 342 (1980). Toxicity study: E. Kárpáti, K. Biro, *ibid.* 346. Series of articles on pharmacology, properties, disposition, pharmacokinetics, safety tests, clinical studies: *ibid.* 346-394. Determn in human serum: G. Szabo, E. Tassonyi, *ibid.* **31**, 1013 (1981).

Crystals from methylene dichloride/acetone, mp 262-264° (dec). $[\alpha]_D^{25}$ +8.1° (c = 1 in water). LD$_{50}$ in mice, rats

(mg/kg): 29.7, 172.6 i.v.; 70.6, 449.6 i.p.; 60.5, 455.8 s.c. (Kárpáti, Biro).
THERAP CAT: Muscle relaxant (skeletal).

7613. Pipemidic Acid. *8-Ethyl-5,8-dihydro-5-oxo-2-(1-piperazinyl)pyrido[2,3-d]pyrimidine-6-carboxylic acid;* piperamic acid; 1489-RB; Filtrax; Memento 400; Pi-Coli; Pipeacid; Pipedac; Pipemid; Pipurin; Tractur; Uropimid; Urosten; Uroval. $C_{14}H_{17}N_5O_3$; mol wt 303.32. C 55.44%, H 5.65%, N 23.09%, O 15.82%. Quinolone antibacterial. Prepn: S. Minami *et al.*, **Ger.** pat. **2,341,146**; *eidem*, **U.S.** pat. **3,887,557** (1974, 1975 both to Dainippon Pharm.); Pesson *et al., C.R. Acad. Sci., Ser. C* **278**, 1169 (1974); De Lajudie *et al., ibid. Ser. D* **279**, 1931 (1974); Matsumoto, Minami, *J. Med. Chem.* **18**, 74 (1975). Characterization: Ficicchia, *Farmaco Ed. Prat.* **30**, 207 (1974). Antibacterial activity: Ficicchia, De Lajudie, *ibid.* 252. Pharmacology: Shimizu *et al., Antimicrob. Ag. Chemother.* **7**, 441 (1975).

Yellowish-white, odorless, bitter-tasting crystals, mp 253-255°. Hygroscopic. Yellows slowly in light. Sol in acid soln and alkaline. Very slightly sol in water, alcohol; slightly sol in chloroform (0.5%), methanol (0.4%). Practically insol in ether, benzene. LD$_{50}$ in mice (mg/kg): 4000 orally; 1000 i.p., 50 i.v. (Ficicchia).
Trihydrate, *Deblaston, Dolcol, Pipram, Solupemid.* Nearly colorless needles, mp 253-255°.
THERAP CAT: Antibacterial.
THERAP CAT (VET): Antibacterial.

7614. Pipenzolate Bromide. *1-Ethyl-3-[(hydroxydiphenylacetyl)oxy]-1-methylpiperidinium bromide; 1-ethyl-3-hydroxy-1-methylpiperidinium bromide benzilate;* benzilic acid 1-ethyl-3-piperidyl ester methyl bromide; *N*-ethyl-3-piperidyl benzilate methobromide; pipenzolate methylbromide; pipenzolone bromide; JB-323; Piptal. $C_{22}H_{28}BrNO_3$; mol wt 434.37. C 60.83%, H 6.50%, Br 18.40%, N 3.22%, O 11.05%. Prepn from *N*-ethyl-3-chloropiperidine and benzilic acid: Biel *et al., J. Am. Chem. Soc.* **74**, 1485 (1952); **U.S.** pat. **2,918,406** (1959 to Lakeside).

Crystals from methyl ethyl ketone, mp 179-180°. Soluble in water.
Note: *N*-Ethyl-3-piperidyl benzilate is a controlled substance (hallucinogen) listed in the U.S. Code of Federal Regulations, Title 21, Part 1308.11 (1995).
THERAP CAT: Anticholinergic.

7615. Piperacetazine. *1-[10-[3-[4-(2-Hydroxyethyl)-1-piperidinyl]propyl]-10H-phenothiazin-2-yl]ethanone; 10-[3-[4-(2-hydroxyethyl)piperidino]propyl]phenothiazin-2-yl ketone;* 2-acetyl-10-[3-[4-(β-hydroxyethyl)piperidino]propyl]phenothiazine; 2-acetyl-10-[3-[γ-(2-hydroxyethyl)piperidino]propyl]phenothiazine; PC-1421; Psymod; Quide. $C_{24}H_{30}N_2O_2S$; mol wt 410.58. C 70.21%, H 7.36%, N 6.82%, O 7.79%, S 7.81%. Prepn: **Brit.** pat. **861,807** (1961 to Searle).

Hydrochloride, mp 100-110°.
THERAP CAT: Antipsychotic.
THERAP CAT (VET): Antipsychotic.

7616. Piperacillin. *[2S-[2α,5α,6β(S*)]]-6-[[[[(4-Ethyl-2,3-dioxo-1-piperazinyl)carbonyl]amino]phenylacetyl]amino]-3,3-dimethyl-7-oxo-4-thia-1-azabicyclo[3.2.0]heptane-2-car-boxylic acid;* (2S,5R,6R)-6-[(R)-2-(4-ethyl-2,3-dioxo-1-piper-azinecarboxamido)-2-phenylacetamido]-3,3-dimethyl-7-oxo-4-thia-1-azabicyclo[3.2.0]heptane-2-carboxylic acid; 6-D(-)-α-(4-ethyl-2,3-dioxo-1-piperazinylcarbonylamino)-α-phenyl-acetamidopenicillanic acid; 4-ethyl-2,3-dioxopiperazine car-bonyl ampicillin. $C_{23}H_{27}N_5O_7S$; mol wt 517.56. C 53.38%, H 5.26%, N 13.53%, O 21.64%, S 6.20%. Broad spectrum semi-synthetic antibiotic related to pencillin. Prepn: I. Saikawa *et al.*, Ger. pat. **2,519,400**; *eidem*, U.S. pat. **4,087,-424** (1976, 1978 both to Toyama). *In vitro* studies: G. P. Bodey, B. LeBlanc, *Antimicrob. Ag. Chemother.* **14**, 78 (1978); R. Wise *et al., ibid.* 549. *In vitro* and *in vivo:* N. A. Kuck, G. S. Redin, *J. Antibiot.* **31**, 1175 (1978). Pharmaco-kinetics: M. A. Evans *et al., J. Antimicrob. Chemother.* **4**, 255 (1978). Metabolism: K. Iida *et al., Antimicrob. Ag. Chemother.* **14**, 257 (1978). Clinical study: T. Saito, Y. Yamada, *Japan. J. Antibiot.* **30**, 835 (1977). Toxicity study: A. Takai *et al., Chemother. (Tokyo)* **25**, 816 (1977). Review of antibacterial activity, pharmacokinetics: G. L. Mandell, *Clin. Ther.* **7**, Suppl. B, 28-35 (1985). Review of clinical experience in urinary tract infections: S. J. Childs, *ibid.* 36-45. Series of articles on antibacterial activity and clinical use: *Chemotherapy (Tokyo)* **36**, Suppl. 7, 1-85 (1988).

Sodium salt, $C_{23}H_{26}N_5NaO_7S$, *CL-227193, T-1220, Isipen, Pentcillin, Pipracil, Pipril.* mp 183-185° (dec). LD$_{50}$ in mice, rats, dogs, monkeys (g/kg): 5, 2.7, >6, >4 i.v. (Takai).
THERAP CAT: Antibacterial.

7617. Piperazine. Hexahydropyrazine; piperazidine; diethylenediamine. $C_4H_{10}N_2$; mol wt 86.14. C 55.78%, H 11.70%, N 32.52%. Prepn: Cloez, *Jahresber.* **1853**, 468; Wolff, *Ber.* **26**, 724 (1893); Kyrides, U.S. pat. **2,267,686** (1941); Martin, Martell, *J. Am. Chem. Soc.* **70**, 1817 (1948); MacKenzie, Turbin, U.S. pat. **2,901,482** (1959 to Dow); Moss, Godfrey, U.S. pat. **3,037,023** (1962 to Jefferson Chem.). Prepn of salts: Hefferren *et al., J. Am. Pharm. Assoc.* **44**, 678 (1955). Determn in animal feeds: J. D. McLean, O. L. Daniels, *J. Assoc. Off. Anal. Chem.* **54**, 555 (1971). Veterinary trial in combination with levamisole: J. H. Drudge *et al., Am. J. Vet. Res.* **35**, 67 (1974). Clinical efficacy in roundworm infection: A. S. Nanivadekar *et al., J. Postgrad. Med.* **30**, 144 (1984).

Leaflets from alc, mp 106°. Salty taste. bp$_{760}$ 146°. Strong base: pKa: 4.19, absorbs water and CO_2 from air. Absorption spectrum: Purvis, *J. Chem. Soc.* **103**, 2286, 2293 (1913). Freely sol in water, glycerol, glycols; one gram dissolves in 2 ml of 95% alcohol. Insol in ether. pH of a 10% aq soln 10.8-11.8. Forms a sol compd with theophyl-line. *Keep tightly closed and protect from light.*

Hexahydrate, $C_4H_{10}N_2.6H_2O$, *Dietelmin, Eraverm, Helmi-fren, Parid, Paravermin, Tasnon (elixir), Upixon, Uvilon.* Crystals from water (contg 44.34% anhydr piperazine), mp 44°. bp 125-130°. The piperazine of commerce is usually this hydrate. Freely sol in water; sol in alc (∼1:2). Practi-cally insol in ether. pH of a 10% aq soln 10.8-11.8.

Citrate, $3C_4H_{10}N_2.2C_6H_8O_7$, *tripiperazine dicitrate, Helme-zine, Oxucide, Parazine, Pinozan, Pipizan Citrate, Pipracid (syrup), Rhomex, Ta-Verm, Worm Away.* Crystals, dec 182-187°. Freely sol in water. Practically insol in alcohol, ether, chloroform. pH of a 10% aq soln 5.0-6.0.

Edetate calcium, $C_{14}H_{24}CaN_4O_8$, *piperazine calcium eda-thamil, Perin.* Prepn: Schlesinger *et al.,* U.S. pat. **2,834,782** (1958 to Endo). Occurs as the dihydrate. Crystals; slightly salty taste. Freely sol in water. Very slightly sol in alcohol, chloroform. Practically insol in ether. pH of 20% aq soln 4.3-5.4.

Phosphate, $C_4H_{10}N_2.H_3PO_4$, *Antepar (tabl.), Pincets, Pin-rou, Pinsirup, Piperverm, Piperazate, Tasnon (tabl.).* Minute crystals. Very slightly sol in water. pH of saturated aq soln, 6.5.

Tartrate, $C_4H_{10}N_2.C_4H_6O_6$, *Noxiurotan, Piperate.* Crys-tals, dec 258-263°. Soly in g/100 ml at 25°: water 26; alcohol 0.01; chloroform 0.01. pH of 1% soln: 4.8.

N^1,N^4-Dibenzoylpiperazine, $C_{18}H_{18}N_2O_2$, prepd from pip-erazine and benzoyl chloride in dil NaOH. Crystals from alc, mp 191°.
THERAP CAT: Anthelmintic (Nematodes).
THERAP CAT (VET): Anthelmintic (Nematodes).

7618. Piperazine Adipate. *Hexanedioic acid compd with piperazine (1:1);* Entacyl; Oxyzin (tabl.); Vermicom-pren (tabl.); Nometan; Oxypaat; Pipadox; Oxurasin. $C_{10}H_{20}N_2O_4$; mol wt 232.28. C 51.71%, H 8.68%, N 12.06%, O 27.55%. $C_4H_{10}N_2.C_6H_{10}O_4$. The neutral salt of piperazine and adipic acid: Davies *et al., J. Pharm. Pharmacol.* **6**, 707 (1954). Prepn in methanolic medium: Forrest, Petrow, U.S. pat. **2,799,617** (1957 to British Drug Houses); **Brit.** pat. **767,826**. Pharmacology: B. G. Cross *et al., J. Pharm. Pharmacol.* **6**, 711 (1954).

Prisms, mp 256-257°. Stable to heat and air. Pleasant, slightly acid taste. Dissolves slowly. Soly in 100 ml water at 20°: 5.53 g, at 30°: 6.61 g, at 56.3°: 10.14 g; in 100 g methanol at 25°: 0.02 g. Practically insol in abs ethanol, isopropanol, dioxane. Aq solns of 0.2-0.01M have pH 5.45 which is only slightly affected by increases in ionic strength caused by the addition of simple neutral salts. LD$_{50}$ in mice, rats (g/kg): 11.4, 7.9 orally (Cross).
THERAP CAT: Anthelmintic (Nematodes).
THERAP CAT (VET): Anthelmintic (Nematodes).

7619. 2,5-Piperazinedione. Glycine anhydride; cyclo-glycylglycine; α,γ-diacipiperazine; glycylglycine lactam; di-glycolyldiamide; 2,5-diketopiperazine; 2,5-dioxopiperazine. $C_4H_6N_2O_2$; mol wt 114.10. C 42.11%, H 5.30%, N 24.55%, O 28.04%. Can exist in 5 enol forms: Richardson *et al., J. Am. Chem. Soc.* **51**, 3074 (1929). Prepn from glycine ethyl ester hydrochloride: Fischer, *Ber.* **39**, 2930 (1906).

Platelets from water, needles by sublimation. Dec 311-312° (begins to sublime at 260° and sinters at 305°). Sparingly sol in water; sol in HCl (d 1.19) from which it can be precipitated by the addition of alcohol. Acts as a weak base. Hydrolyzed to glycylglycine by acids and alkalies.

Hydrochloride, $C_4H_6N_2O_2 \cdot HCl$, crystals, mp 129-130°.

7620. Piperic Acid. *5-(1,3-Benzodioxol-5-yl)-2,4-penta-dienoic acid; 5-(3,4-methylenedioxyphenyl)-2,4-pentadienoic acid.* $C_{12}H_{10}O_4$; mol wt 218.21. C 66.05%, H 4.62%, O 29.33%. Early literature treats the title compound without specifying stereochemistry; however, four isomers exist: piperinic acid *(E,E)*; isochavicinic acid *(E,Z)*; isopiperinic acid *(Z,E)*; chavicinic acid *(Z,Z)*. Physical constants given in early references for piperic acid agree with those of piperinic acid. *See Beilstein* **19**, 281; suppl. II, 300. Synthesis of *(E,E)*-form: L. von Babo, E. Keller, *J. Prakt. Chem.* **72**, 53 (1857); A. Ladenburg, M. Sholtz, *Ber.* **27**, 2958 (1894); of *(E,Z)*-form: H. Lohaus, H. Gall, *Ann.* **517**, 278 (1935); of *(Z,E)*-form: E. Ott, F. Eichler, *Ber.* **55**, 2653 (1922); of *(Z,Z)*-form: *idem, loc. cit.;* H. Lohaus, H. Gall, *loc. cit.;* of all isomers: R. Grewe *et al., Ber.* **103**, 3752 (1970); R. De Cleyn, A. Verzele, *Bull. Soc. Chim. Belg.* **81**, 529 (1972).

(E,E)-form

(E,E)-Form, *piperinic acid, piperinsäure (German).* Needles from alc. Colorless when freshly prepd, rapidly turns yellow on exposure to light. mp 216-217°. Sublimes as yellow needles with partial decompn. uv max (methanol): 340 nm (ε 28800). Sol in 50 parts boiling alcohol, 275 parts abs alcohol at 25°. Practically insol in water, ether, benzene, carbon disulfide.

(E,Z)-Form, *isochavicinic acid.* Yellow crystals from methanol/water, sublimes as needles. mp 134-136° (Lohaus, Gall); 143° (Grewe). uv max (methanol): 335 nm (ε 14500). Sol in methanol, benzene.

(Z,E)-Form, *isopiperinic acid.* Needles from benzene. mp 145° (Ott, Eichler); 153° (Grewe); 138° (DeCleyn, Verzele). uv max (methanol): 328 nm (ε 22000). Sol in benzene.

(Z,Z)-Form, *chavicinic acid.* Amorphous yellow granules from benzene. Wide disparities in mp have been reported: 200-202° (Ott, Eichler); 130° (Grewe); 120° (De Cleyn, Verzele). uv max (methanol): 335 nm (ε 17500). 0.55 g sol in 16 g boiling 95% alcohol, and in 65 g boiling benzene.

7621. Piperidine. Hexahydropyridine. $C_5H_{11}N$; mol wt 85.15. C 70.53%, H 13.02%, N 16.45%. Found in small quantities in *Piper nigrum* L., *Piperaceae* (black pepper). May be obtained from piperine by heating with alcoholic KOH, or from 1,5-diaminopentane hydrochloride by cyclization. Usually prepd by electrolytic reduction of pyridine. Forms complexes with salts of heavy metals. Because of its reactivity, piperidine is useful in the prepn of cryst derivatives of aromatic nitro compds contg nuclear halogen atoms: Seikel, *J. Am. Chem. Soc.* **62**, 750 (1940). Toxicity study: H. F. Smyth *et al., Am. Ind. Hyg. Assoc. J.* **23**, 95 (1962). Review of physical constants of piperidine and N-alkyl piperidines: Magnusson, Schierz, *Univ. Wyoming Pub.* **7**, 1 (1940).

Liquid. Characteristic odor. Soapy feel. Solidifies −13° to −17°; mp −7°; bp$_{760}$ 106°; bp$_{20}$ 18°; d_4^{20} 0.8622; n_D^{20} 1.4534. Infrared absorption spectrum: Freymann, *Compt. Rend.* **205**, 852 (1937); *Ann. Chim.* **11**, 11 (1939). Ultraviolet and Raman spectra: Lecomte, *Compt. Rend.* **207**, 395 (1938). Strong base: pK (25°): 2.80. Misc with water. Sol in alcohol, benzene, chloroform. LD$_{50}$ orally in rats: 0.52 ml/kg (Smyth).

Hydrochloride, $C_5H_{11}N \cdot HCl$, orthorhombic prisms from alcohol, mp 247°. Freely sol in water, alcohol.

Nitrate, $C_5H_{11}N \cdot HNO_3$, hygroscopic plates, sublimes 75° (10 mm), mp 110°. Freely sol in water, alcohol, ether. Absorption spectrum: Harper, Macbech, *J. Chem. Soc.* **107**, 91 (1915).

Bitartrate, $C_5H_{11}N \cdot C_4H_6O_6$, crystals, freely sol in water.

Aurichloride, $C_5H_{11}N \cdot HAuCl_4$, yellow crystals, mp 219°.

Platinichloride, $2C_5H_{11}N \cdot H_2PtCl_6$, yellow monoclinic prisms from water, mp 202°. Freely sol in water, slightly in alcohol.

N-Benzoylpiperidine, long needles, mp 44-48°. *Ref: Org. Syn.* **coll. vol. I** (2nd ed., 1941) p 99.

Phosphate, prepn from piperidine + phosphoric acid: Abood, U.S. pat. **3,035,977** (1962). Crystals, mp 204-206°.

7622. Piperidione. *3,3-Diethyl-2,4-piperidinedione; 3,3-diethyl-2,4-dioxopiperidine;* dihyprylone; Sedulon; Tusseval. $C_9H_{15}NO_2$; mol wt 169.22. C 63.88%, H 8.93%, N 8.28%, O 18.91%. Prepn: Tsukita, *J. Pharm. Soc. Japan* **69**, 194 (1949).

Crystals, mp 102-107°. Bitter taste. Freely sol in water, alcohol, chloroform.

THERAP CAT: Sedative, antitussive.

7623. Piperidolate. *α-Phenylbenzeneacetic acid 1-ethyl-3-piperidinyl ester; diphenylacetic acid 1-ethyl-3-piperidyl ester;* N-ethyl-3-piperidyl diphenylacetate; JB-305. $C_{21}H_{25}$-NO_2; mol wt 323.44. C 77.99%, H 7.79%, N 4.33%, O 9.89%. Prepd from diphenylacetyl chloride and N-ethyl-3-hydroxypiperidine: Biel *et al., J. Am. Chem. Soc.* **74**, 1485 (1952); U.S. pat. **2,918,407** (1959 to Lakeside Labs.).

Liquid. bp$_{0.18}$ 191-192°.

Hydrochloride, $C_{21}H_{25}NO_2 \cdot HCl$, *Crapinon, Dactil.* Crystals, mp 195-196°. Soluble in water.

THERAP CAT: Anticholinergic.

7624. Piperilate. *α-Hydroxy-α-phenylbenzeneacetic acid 2-(1-piperidinyl)ethyl ester;* β-piperidylethyl benzilate; pipethanate; benzilic acid 1-piperidineethanol ester; 1-piperidineethanol benzilate; 2-(1-piperidino)ethyl benzilate. C_{21}-$H_{25}NO_3$; mol wt 339.43. C 74.31%, H 7.42%, N 4.13%, O 14.14%. Prepn: Ford-Moore, Ing, *J. Chem. Soc.* **1947**, 55.

Hydrochloride, $C_{21}H_{25}NO_3 \cdot HCl$, *Daipisate, Norticon, Pensanate, Pipenale.* Crystals from acetone or ethanol, mp 170-171°.

Consult the Name Index before using this section.

Ethyl bromide, $C_{23}H_{30}BrNO_3$, *Panpurol.*

THERAP CAT: Anticholinergic; antispasmodic.

7625. Piperine. *(E,E)-1-[5-(1,3-Benzodioxol-5-yl)-1-oxo-2,4-pentadienyl]piperidine; (E,E)-1-piperoylpiperidine.* $C_{17}H_{19}NO_3$; mol wt 285.34. C 71.56%, H 6.71%, N 4.91%, O 16.82%. Isolated from black pepper *(Piper nigrum* L.); also in *P. longum* L., *P. retrofractum* Vahl. *(P. officinarum* C.D.C.), and *P. clusii* C.D.C.; in root bark of *Piper geniculatum* Sw., *Piperaceae.* Extraction procedure: Cazeneuve, Caillot, *Bull. Soc. Chim.* [2] **27**, 291 (1877). Synthesis: Rugheimer, *Ber.* **15**, 1390 (1882); Newman, *Chem. Products* **16**, 379 (1953); Normant, Feugeas, *Compt. Rend.* **258**, 2846 (1964). Spectroscopic structural elucidation and preparative separation of piperine and its stereoisomers isopiperine, isochavicine and chavicine, *q.v.:* R. De Cleyn, M. Verzele, *Bull. Soc. Chim. Belg.* **84**, 435 (1975). Synthesis of isomers: R. Grewe *et al., Ber.* **103**, 3752 (1970); of piperine and isochavicine: S. Tsuboi *et al., Tetrahedron Letters* **1979**, 1043. Stereoselective synthesis of piperine: R. A. Olsen, G. O. Spessard, *J. Agr. Food Chem.* **29**, 942 (1981). More toxic to houseflies than pyrethrum: Harvill *et al., Contrib. Boyce Thompson Inst.* **13**, 87 (1943).

Monoclinic prisms from alcohol, mp 130°. Tasteless at first, but burning aftertaste. Neutral to litmus. pK (18°): 12.22. Almost insol in water (40 mg/liter at 18°), in petr ether. One gram dissolves in 15 ml alcohol, 1.7 ml chloroform, 36 ml ether. Sol in benzene, acetic acid.

(E,Z)-Form, *isochavicine.* Crystals from chloroform + hexane, mp 89° (Grewe), 103° (De Cleyn). uv max (methanol): 333 nm (ε 16300).

(Z,E)-Form, *isopiperine.* Crystals from chloroform + hexane, mp 110° (Grewe), 86° (De Cleyn). uv max (methanol): 332 nm (ε 21800).

USE: To impart pungent taste to brandy. As insecticide.

7626. Piperitone. *3-Methyl-6-(1-methylethyl)-2-cyclohexen-1-one; p-menth-1-en-3-one;* 4-isopropyl-1-methyl-1-cyclohexen-3-one. $C_{10}H_{16}O$; mol wt 152.24. C 78.90%, H 10.59%, O 10.51%. Isoln of *d*-form from oil of *Cymbopogon sennaarensis* Chiov., *Gramineae:* Roberts, *J. Chem. Soc.* **107**, 1465 (1915); from oil of *Andropogon iwarancusa* Jones: Simonsen, *ibid.* **119**, 1644 (1921); from oil of *Mentha* spp., *Labiatae:* Reitsma, *J. Am. Pharm. Assoc.* **47**, 265, 267 (1958). Isoln of *l*-form from Sitka spruce oil: von Rudloff, *Can. J. Chem.* **42**, 1057 (1964). Isoln of *dl*-form from oil of *Eucalyptus dives* Schau., *Myrtaceae:* Read, Smith, *J. Chem. Soc.* **119**, 779 (1921); from peppermint oil: Katsuragi, *Koryo* No. **24**, 16 (1953). Synthesis of *dl*-form: Misrock, Church, *Ind. Eng. Chem.* **49**, 822 (1957); Bain *et al.,* U.S. pat. **2,972,-632** (1961 to Glidden); Wiemann, Dubois, *Bull. Soc. Chim. France* **1962**, 1813; Stepanov, Myrsina, *Zh. Obshch. Khim.* **34**, 3092 (1964).

d-Form, liquid. Peppermint odor. bp 232-235°; bp$_{20}$ 116-118.5°. d$_4^{20}$ 0.9344. [α]$_D^{20}$ +49.13°. n$_D^{20}$ 1.4848.

l-Form, liquid. bp$_{15}$ 109.5-110.5°. d$_4^{20}$ 0.9324. [α]$_D^{20}$ −15.9°. n$_D^{20}$ 1.4823. Practically insol in water. Sol in alcohol, oils.

dl-Form, liquid. bp$_{769}$ 232-233°; bp$_{16}$ 116-118°. d$_4^{20}$ 0.9331. n$_D^{24}$ 1.4823. uv max (ethanol): 232.5 nm (ε 13350).

USE: In masking odors in dentifrices.

7627. Piperocaine. *2-Methyl-1-piperidinepropanol benzoate;* γ-(2-methylpiperidyl)propyl benzoate; (2-methylpiperidino)propyl benzoate; 3-benzoxy-1-(2-methylpiperidino)-propane; benzoyl-γ-(2-methylpiperidino)propanol; Neothesin; Metycaine. $C_{16}H_{23}NO_2$; mol wt 261.36. C 73.53%, H 8.87%, N 5.36%, O 12.24%. Prepn of the hydrochloride from 2-methylpiperidine and γ-chloropropyl benzoate: U.S. pat. **1,784,903** (1930).

Hydrochloride, $C_{16}H_{23}NO_2$·HCl, crystals, mp 172-175°. Slightly bitter taste followed by numbness of the mouth. Stable in air. One gram dissolves in 1 ml water. Sol in alc, chloroform. Insol in ether, olive oil. A 1:10 aq soln is faintly acid to litmus. Alkali carbonates and hydroxides precipitate piperocaine base as an oil from aq solns of the hydrochloride. LD$_{50}$ in rats: 1.3 g/kg s.c.; 20 mg/kg i.v.

THERAP CAT: Anesthetic (local).

THERAP CAT (VET): Anesthetic (local).

7628. Piperonal. *1,3-Benzodioxole-5-carboxaldehyde;* 3,4-(methylenedioxy)benzaldehyde; heliotropin; piperonylaldehyde; dioxymethyleneprotocatechuic aldehyde. $C_8H_6O_3$; mol wt 150.13. C 64.00%, H 4.03%, O 31.97%. Prepn: Blair, U.S. pat. **2,916,499** (1959 to Welsbach Corp.); Holum, *J. Org. Chem.* **26**, 4814 (1961); Feugeas, *Bull. Soc. Chim. France* **1964**, 1892. Toxicity: Hagan *et al., Toxicol. Appl. Pharmacol.* **7**, 18 (1965).

Colorless, lustrous crystals, mp 37°. Heliotrope odor. bp ~263°; bp$_{0.5}$ 88°. Sol in 500 parts water; freely sol in alcohol, ether. *Keep in cool place protected from light.* LD$_{50}$ orally in rats: 2700 mg/kg (Hagan).

USE: In perfumery, in cherry and vanilla flavors, in organic syntheses.

THERAP CAT: Has been used as pediculicide.

7629. Piperonyl Butoxide. *5-[[2-(2-Butoxyethoxy)ethoxy]methyl]-6-propyl-1,3-benzodioxole;* α-[2-(2-butoxyethoxy)ethoxy]-4,5-methylenedioxy-2-propyltoluene; [3,4-(methylenedioxy)-6-propylbenzyl] butyl diethyleneglycol ether; 6-propylpiperonyl butyl diethylene glycol ether; butylcarbityl (6-propylpiperonyl) ether; ENT-14250; Butacide. $C_{19}H_{30}O_5$; mol wt 338.44. C 67.43%, H 8.93%, O 23.64%. Prepn: H. Wachs, U.S. pats. **2,485,681; 2,550,737** (1949, 1951 both to U.S. Industrial Chemicals); *idem, Science* **105**, 530 (1947). Toxicity: T. B. Gaines, *Toxicol. Appl. Pharmacol.* **14**, 515 (1969). Review of carcinogenic risk: *IARC Monographs* **30**, 183-195 (1983).

Liquid. d 1.04-1.07. bp$_{1.0}$ 180°. n$_D^{20}$ 1.50. Flash pt 340°F. Miscible with methanol, ethanol, benzene, Freons, Geons, other organic solvents, oils. LD$_{50}$ orally in female, male rats: 6150, 7500 mg/kg (Gaines).

Available in various mixtures with synthetic pyrethroids, *e.g., Derringer, Duracide, Grovex, Prentox, Scourge.*

In combination with rotenone, *q.v.,* **PB-NOX, Chem-Fish, Rotacide.**

Caution: Large doses have caused vomiting, diarrhea: Sarles *et al., Am. J. Trop. Med.* **29**, 151 (1949).

USE: Insecticide synergist, especially for pyrethroids and rotenone.

7630. Piperonylic Acid. *1,3-Benzodioxole-5-carboxylic acid;* 3,4-methylenedioxybenzoic acid; protocatechuic acid methylene ether. $C_8H_6O_4$; mol wt 166.13. C 57.84%, H 3.64%, O 38.52%. Occurs in Paracoto bark. Prepd by permanganate oxidation of piperonal: Shriner, Kleiderer, *Org. Syn.* **10**, 82 (1930).

Prisms (by sublimation), needles from alc, feathery crystals from water. mp 229°. Sublimes around 210°. Slightly sol in water, chloroform, cold alcohol, ether. Absorption spectrum: Dobbie, Lauder, *J. Chem. Soc.* **83**, 621 (1903). Methyl ester, $C_9H_8O_4$, mp 53°. Sublimes easily. Volatile in steam. Freely sol in alcohol, ether.

7631. Piperoxan. *1-[(2,3-Dihydro-1,4-benzodioxin-2-yl)methyl]piperidine;* 2-piperidinomethyl-1,4-benzodioxan; 2-(1-piperidylmethyl)-1,4-benzodioxan; benzodioxane; Benodaine. $C_{14}H_{19}NO_2$; mol wt 233.31. C 72.07%, H 8.21%, N 6.00%, O 13.72%. α-Adrenergic blocker. Prepn: Fourneau, U.S. pat. 2,056,046 (1936 to Rhône-Poulenc).

bp_{17} 193°.
dl-Form hydrochloride, $C_{14}H_{19}NO_2 \cdot HCl$, *compd 933F, Fourneau 933.* Crystals, mp 232-234° (darkens at 220°). uv max: 275 nm ($E_{1cm}^{1\%}$ 82); min: 240 nm ($E_{1cm}^{1\%}$ < 5). Freely sol in water, pH of 1% soln ~5. Alkalies liberate the water-insol base. Soluble in acid solns. Soly in isopropanol: ~10.8 mg/g at 25°. Crystals are not hygroscopic and are stable to light, air, and normal storage temps. An aq soln at its own pH (5) is stable to autoclaving and to many months of storage at room temp.
THERAP CAT: Antihypertensive. Diagnostic aid (pheochromocytoma).

7632. Piperylone. *4-Ethyl-1,2-dihydro-2-(1-methyl-4-piperidinyl)-5-phenyl-3H-pyrazol-3-one;* 4-ethyl-1-(1-methyl-4-piperidyl)-3-phenyl-3-pyrazolin-5-one; 1-(N-methylpiperid-4-yl)-3-phenyl-4-ethylpyrazol-5-one; PR-66. $C_{17}H_{23}N_3O$; mol wt 285.39. C 71.55%, H 8.12%, N 14.72%, O 5.61%. Prepn: Ebnother *et al., Helv. Chim. Acta* **42**, 1201 (1959); Jucker *et al.,* U.S. pat. 2,903,460 (1959 to Sandoz).

Crystals from methanol + ether, dec 162-163°. pK_1 5.90; pK_2 9.15. Very sparingly sol in water. Sol in ethanol, isopropanol, acetone, ethyl acetate.
Ingredient of *Palerol (Pelerol).*
Hydrobromide, $C_{17}H_{23}N_3O \cdot HBr$, crystals from methanol + ether, dec 194-197°. Sol in water.
Tartrate, crystals from methanol, mp 145-147°.
THERAP CAT: Analgesic.

7633. PIPES. *1,4-Piperazinediethanesulfonic acid;* piperazine-N,N'-bis(2-ethanesulfonic acid); 1,4-piperazinebis(ethanesulfonic acid). $C_8H_{18}N_2O_6S_2$; mol wt 302.37. C 31.78%, H 6.00%, N 9.26%, O 31.75%, S 21.21%. One of the zwitterionic N-substituted aminosulfonic acids known as "Good" buffers; active in the pH range 6-8.5. Prepn: N. Good *et al., Biochemistry* **5**, 467 (1966). Interference with Lowry protein determn: H. M. Himmel, W. Heller, *J. Clin.*

Chem. Clin. Biochem. **25**, 909 (1987). Use as biological buffer: R. Salema, I. Brandao, *J. Submicros. Cytol.* **5**, 79 (1973); S. Haviernick *et al., J. Micros.* **135**, 83 (1984).

Monosodium salt, $C_8H_{17}N_2NaO_6S_2$, crystals from water and alcohol, mp > 300 (dec). pKa_1 ~3, pKa_2 (20°): 6.82 (0.1M); 6.82 (0.2M); 6.96 (0.01M). $\Delta pKa/°C$ −0.0085. Disodium salt, $C_8H_{16}N_2Na_2O_6S_2$, *sodium pipesate.*
USE: Biological buffer.

7634. Pipobroman. *1,4-Bis-(3-bromo-1-oxopropyl)piperazine;* 1,4-bis-(3-bromopropionyl)piperazine; A-8103; NSC-25154; Amedel; Vercyte. $C_{10}H_{16}Br_2N_2O_2$; mol wt 356.06. C 33.73%, H 4.53%, Br 44.88%, N 7.87%, O 8.99%. Prepn: Horrom, Carbon, **Ger.** pat. 1,138,781 (1962 to Abbott), *C.A.* **58**, 7955c (1963); Groszkowski, *Roczniki Chem.* **38**, 229 (1964), *C.A.* **60**, 14506a (1964). Structure-cytostatic effect studies: Groszkowski *et al., J. Med. Chem.* **11**, 621 (1968); Oteleanu, Retezeanu, *Farmacia (Bucharest)* **16**, 279 (1968), *C.A.* **69**, 33591w (1968).

Crystals, mp 106-107°.
THERAP CAT: Antineoplastic.

7635. Piposulfan. *1,4-Bis[3-[(methylsulfonyl)oxy]-1-oxopropyl]piperazine;* 1,4-dihydracryloylpiperazine dimethanesulfonate; 1,4-bis(3-hydroxypropionyl)piperazine dimethanesulfonate; N,N'-bis(3-methanesulfonyloxypropionyl)piperazine; N,N'-bis(3-methylsulfonyloxypropionyl)piperazine; A-20968; NSC-47774; Ancyte. $C_{12}H_{22}N_2O_8S_2$; mol wt 386.45. C 37.30%, H 5.74%, N 7.25%, O 33.12%, S 16.60%. Prepn: Horrom, Carbon, **Ger.** pat. 1,177,162 (1964 to Abbott), *C.A.* **61**, 13329a (1964).

Crystals from water, mp 175-177°.
THERAP CAT: Antineoplastic.

7636. Pipotiazine. *10-[3-[4-(2-Hydroxyethyl)-1-piperidinyl]propyl]-N,N-dimethyl-10H-phenothiazine-2-sulfonamide;* 2-[1-[3-[2-[(dimethylamino)sulfonyl]-10H-phenothiazin-10-yl]propyl]-4-piperidinyl]ethanol; pipothiazine; RP-19366; Piportil. $C_{24}H_{33}N_3O_3S_2$; mol wt 475.68. C 60.60%, H 6.99%, N 8.83%, O 10.09%, S 13.48%. Prepn: **Fr.** pat. M7835 (1970 to Rhône-Poulenc), *C.A.* **78**, 43499x (1973). Pharmacokinetics: P. J. De Schepper *et al., Arzneimittel-Forsch.* **29**, 1056 (1979); and HPLC determn in plasma: D. A. Ogden *et al., J. Pharm. Biomed. Anal.* **7**, 1273 (1989). Series of articles on pharmacology and clinical use: *Acta Psychiat. Scand.,* Suppl. 241, 9-138 (1973). Toxicity: L. Julou *et al., ibid.* 9.

LD$_{50}$ in mice (mg/kg): 108 i.p.; 360 s.c.; 440 orally (Julou).

Palmitic ester, C$_{40}$H$_{63}$N$_3$O$_4$S$_2$, *pipotiazine palmitate, RP-19552, Piportil L4.*

Undecylenic ester, C$_{35}$H$_{51}$N$_3$O$_4$S$_2$, *pipotiazine undecylenate, RP-19551, Piportil M2.*

THERAP CAT: Antipsychotic.

7637. Pipoxolan Hydrochloride. *5,5-Diphenyl-2-[2-(1-piperidinyl)ethyl]-1,3-dioxolan-4-one hydrochloride;* 2-(β-N-piperidylethyl)-4,4-diphenyl-1,3-dioxolan-5-one hydrochloride; BR-18; Rowapraxin. C$_{22}$H$_{26}$ClNO$_3$; mol wt 387.91. C 68.12%, H 6.76%, Cl 9.14%, N 3.61%, O 12.37%. Prepn: M. Pailer *et al., Monatsh.* **99**, 891 (1968); **Brit.** pat. **1,109,-959** and **Belg.** pat. **719,230** (1968, 1969 to Rowa-Wagner). Pharmacology and acute toxicity: K. Morsdorf, H. Wengenroth, *Pharmacology* **3**, 193 (1970).

Crystals from isopropyl alc, mp 207-209°. Soluble in water; stable to mild attack by acids or bases. LD$_{50}$ in rats, mice (mg/kg): 1500, 700 orally; 60, 35 i.v.; 130, 130 i.p.; in rats (mg/kg): > 300 s.c. (Morsdorf).

THERAP CAT: Antispasmodic.

7638. Pipradrol. *α,α-Diphenyl-2-piperidinemethanol;* α-(2-piperidyl)benzhydrol; alpha-pipradrol; pipradol; MRD-108; C$_{18}$H$_{21}$NO; mol wt 267.37. C 80.86%, H 7.92%, N 5.24%, O 5.98%. Prepd by hydrogenation of α,α-diphenyl-2-pyridinemethanol: Tilford *et al., J. Am. Chem. Soc.* **70**, 4001 (1948); Werner, Tilford, **U.S.** pat. **2,624,739** (1953 to Merrell).

Hydrochloride, C$_{18}$H$_{21}$NO.HCl, *Detaril, Meratran, Meratonic, Stimolag Fortis.* Crystals from butanone, dec 308-309°. Slightly bitter taste. One gram dissolves in 60 ml of hot water.

Note: This is a controlled substance (stimulant) listed in the U.S. Code of Federal Regulations, Title 21 Part 1308.14 (1995).

THERAP CAT: CNS stimulant.

7639. Piprozolin. *[3-Ethyl-4-oxo-5-(1-piperidinyl)-2-thiazolidinylidene]acetic acid ethyl ester;* 3-ethyl-4-oxo-5-piperidino-Δ^{2,α}-thiazolidineacetic acid ethyl ester; Gö-919; W-3699; Coleflux; Epsyl; Probilin; Secrebil. C$_{14}$H$_{22}$N$_2$O$_3$S; mol wt 298.41. C 56.35%, H 7.43%, N 9.39%, O 16.08%, S 10.75%. Prepn: G. Satzinger, *Ann.* **665**, 150 (1963); G. Satzinger *et al.*, **Ger.** pat. **2,414,345** (1975 to Goedecke),

corresp to **U.S.** pat. **3,971,794** (1976 to Warner-Lambert). Metabolism and pharmacokinetics: K. O. Vollmer, F. W. Koss, *Arch. Int. Pharmacodyn. Ther.* **198**, 312 (1972). Mechanism of action: F. W. Koss *et al., ibid.* 333. Series of articles on synthesis, pharmacology, toxicology and clinical studies: *Arzneimittel-Forsch.* **27**, 463-526 (1977). Toxicity data: M. Herrmann *et al., ibid.* 467.

Colorless cryst, mp 86-87°. Practically insol in water. Sol in dil aq acids and most organic solvents. uv max (methanol): 245, 285 nm (ε 8200, 20000) (Vollmer). LD$_{50}$ in mice, rats (mg/kg): 1070, 3256 orally (Herrmann).

THERAP CAT: Choleretic.

7640. Pipsyl Chloride. *4-Iodobenzenesulfonyl chloride;* p-iodophenyl sulfonyl chloride. Usually prepd from iodide ion and p-diazobenzenesulfonic acid, followed by treatment with phosphorus pentachloride. A 5-10-fold excess reacts quantitatively with amino acids as indicated by the disappearance of amino nitrogen. Radioactive pipsyl chloride is used in the analysis of proteins: Keston *et al., J. Am. Chem. Soc.* **68**, 1390 (1946).

7641. Piracetam. *2-Oxo-1-pyrrolidineacetamide;* 2-pyrrolidoneacetamide; 2-pyrrolidinoneacetamide; 2-ketopyrrolidine-1-ylacetamide; 1-acetamido-2-pyrrolidinone; UCB-6215; Avigilen; Axonyl; Cerebroforte; Encetrop; Gabacet; Genogris; Geram; Nootron; Nootrop; Nootropil; Nootropyl; Norzetam; Normabraïn; Pirroxil. C$_6$H$_{10}$N$_2$O$_2$; mol wt 142.16. C 50.69%, H 7.09%, N 19.71%, O 22.51%. Prepn: H. Morren, **Neth.** pat. **Appl. 6,509,994**; *eidem,* **U.S.** pat. **3,459,738** (1966, 1969 both to U.C.B.). Pharmacology: Giurgea *et al., Arch. Int. Pharmacodyn. Ther.* **166**, 238 (1967); Giurgea, Moyersoons, *ibid.* **188**, 401 (1970); Giurgea *et al., Psychopharmacologia* **20**, 160 (1971). Metabolism and biochemical studies: Gobert, *J. Pharm. Belg.* **27**, 281 (1972). Clinical studies: W. J. Oosterveld, *Arzneimittel-Forsch.* **30**, 1947 (1980); G. Chouinard *et al., Psychopharmacol. Bull.* **17**, 129 (1981); in dyslexia: M. Di Ianni *et al., J. Clin. Psychopharmacol.* **5**, 272 (1985).

Crystals from isopropanol, mp 151.5-152.5°.

THERAP CAT: Nootropic.

7642. Pirarubicin. *[8S-[8α,10α(S*)]]-10-[[3-Amino-2,-3,6-trideoxy-4-O-(tetrahydro-2H-pyran-2-yl)-α-L-lyxo-hexopyranosyl]oxy]-7,8,9,10-tetrahydro-6,8,11-trihydroxy-8-(hydroxyacetyl)-1-methoxy-5,12-naphthacenedione;* 4'-O-tetrahydropyranyl doxorubicin; (2''R)-4'-O-tetrahydropyranyladriamycin; (8S,10S)-10-[[3-amino-2,3,6-trideoxy-4-O-(tetrahydro-2H-pyran-2-yl)-α-L-lyxo-hexopyranosyl]oxy]-8-glycoloyl-7,8,9,10-tetrahydro-6,8,11-trihydroxy-1-methoxy-5,-12-naphthacenedione; tepirubicin; THP; THP-ADM; THP-adriamycin; 1609-RB; Pinorubicin; Theprubicin; Therarubicin. C$_{32}$H$_{37}$NO$_{12}$; mol wt 627.65. C 61.24%, H 5.94%, N 2.23%, O 30.59%. Structural analog of doxorubicin, *q.v.* Prepn of (2''R) and (2''S)-diastereomers: H. Umezawa *et al., J. Antibiot.* **32**, 1082 (1979); H. Umezawa *et al.,* **Eur.** pat. **Appl. 14,853**; *eidem,* **U.S.** pat. **4,303,785** (1980, 1981 both to Microbiochem. Res. Found. Japan). Absolute configuration: H. Umezawa *et al., J. Antibiot.* **37**, 1094 (1984). HPLC determn in biological fluids: Y. Matsushita *et al., J. Antibiot.* **36**, 880 (1983). Preliminary toxicology: D. Dantchev *et al., ibid.* **32**, 1085 (1979). *In vitro* antitumor activity: T. Tsuruo *et al., Cancer Res.* **42**, 1462 (1982); *in vivo* activity: T. Hisamatsu *et al., Japan. J. Cancer Res.* **76**, 1008

(1985). Cellular uptake and inhibition of DNA synthesis: S. Kunimoto *et al., J. Antibiot.* **36**, 312 (1983); and cytostatic activity: S. Kunimoto *et al., ibid.* **37**, 1697 (1984). Pharmacokinetics and tissue distribution: H. Iguchi *et al., Cancer Chemother. Pharmacol.* **15**, 132 (1985). Clinical pharmacology: A. A. Miller, C. G. Schmidt, *Cancer Res.* **47**, 1461 (1987). Clinical studies in breast cancer: G. Mathe *et al., Biomed. Pharmacother.* **40**, 376 (1986); in leukemia and malignant lymphoma: R. Ohno *et al., Cancer Chemother. Pharmacol.* **20**, 230 (1987). Brief review of early clinical evaluations: H. Majima, K. Ohta, *Biomed. Pharmacother.* **41**, 237-243 (1987).

Red solid, mp 188-192° (dec). $[\alpha]_D^{25}$ 175 ±25° (c = 0.2 in CHCl$_3$). uv and visible max (methanol): 234, 252, 290, 498, 531.5, 580 nm (E$_{1cm}^{1\%}$ 480, 350, 110, 140, 100, 45). Sol in ethyl acetate, chloroform, and ethanol; slightly sol in water, *n*-hexane, petr ether. Ethanolic and acidic solutions are red in color; give a positive ninhydrin reaction and do not reduce Fehling's solution. LD$_{50}$ i.v. in mice: 27.8 mg/kg (Umezawa, 1979).

THERAP CAT: Antineoplastic.

7643. Pirazolac. *4-(4-Chlorophenyl)-1-(4-fluorophenyl)-1H-pyrazole-3-acetic acid;* ZK-76604. C$_{17}$H$_{12}$ClFN$_2$O$_2$; mol wt 330.75. C 61.74%, H 3.66%, Cl 10.72%, F 5.74%, N 8.47%, O 9.67%. Prepn: **Belg.** pat. **844,972;** H. Ahrens *et al.,* **U.S.** pat. **4,042,706** (both 1977 to Schering AG); H. Biere *et al., Eur. J. Med. Chem.-Chim. Ther.* **17**, 27 (1982). Pharmacokinetics: U. Täuber *et al., Eur. J. Drug Metabol. Pharmacokinet.* **10**, 41 (1985). Effect on arachidonic acid metabolism in human synovia: M. Kurowski *et al., Curr. Med. Res. Opin.* **10**, 494 (1987). Preliminary clinical trial in rheumatoid arthritis: D. Symmons *et al., ibid.* **9**, 542 (1985).

Crystals from ethanol + water, mp 148-149°. Also reported as second crystalline modification from toluene, mp 135-136°.

THERAP CAT: Anti-inflammatory.

7644. Pirbuterol. α^6-*[[(1,1-Dimethylethyl)amino]methyl]-3-hydroxy-2,6-pyridinedimethanol;* 2-hydroxymethyl-3-hydroxy-6-(1-hydroxy-2-*tert*-butylaminoethyl)pyridine. C$_{12}$H$_{20}$N$_2$O$_3$; mol wt 240.30. C 59.98%, H 8.39%, N 11.66%, O 19.97%. Analog of albuterol, *q.v.,* with β_2-adrenergic stimulating activity. Prepn: W. E. Barth, **Ger.** pat. **2,204,195;** *idem,* **U.S.** pat. **3,700,681** (both 1972 to Pfizer). Stability study: P. C. Bansal, D. C. Monkhouse, *J. Pharm. Sci.* **66**, 819 (1977). Biotransformation: H. M. McIlhenny, M. S. D. Ghaly, *Fed. Proc.* **38**, 1130 (1979). Pharmacokinetics

and cardiopulmonary effects in dogs: J. W. Constantine *et al., J. Pharmacol. Exp. Ther.* **208**, 371 (1979). Comparative study in respiratory disease: A. J. Dyson, A. D. Mackay, *Brit. J. Dis. Chest* **74**, 70 (1980). Use in treatment of cardiac failure: N. A. Awan *et al., Clin. Res.* **28**, 17A (1980); W. S. Colucci *et al., N. Engl. J. Med.* **305**, 185 (1981); G. I. Nelson *et al., Eur. Heart J.* **3**, 238 (1982); K. T. Weber *et al., Circulation* **66**, 1262 (1982). Review of pharmacology and efficacy in bronchospastic disease: D. M. Richards, R. N. Brogden, *Drugs* **30**, 6-21 (1985).

Dihydrochloride, C$_{12}$H$_{22}$Cl$_2$N$_2$O$_3$, *CP-24314-1, Broncocor, Exirel.* Crystals from ethanol/isopropyl ether, mp 182° (dec).

Monoacetate, C$_{14}$H$_{24}$N$_2$O$_5$, *Maxair, Spirolair.*

THERAP CAT: Bronchodilator.

7645. Pirenoxine. *1-Hydroxy-5-oxo-5H-pyrido[3,2-a]-phenoxazine-3-carboxylic acid;* 1-hydroxy-5H-pyrido[3,2-a]-phenoxazin-5-one-3-carboxylic acid; 1-hydroxy-3-carboxy-5H-pyrido[3,2-a]phenoxazin-5-one; pirfenoxone; Catalin. C$_{16}$H$_8$N$_2$O$_5$; mol wt 308.25. C 62.34%, H 2.62%, N 9.09%, O 25.95%. Prepn: S. Ogino, **Japan.** pat. **59 2227,** *C.A.* **54**, 11058i (1960); S. Ishii, **Japan.** pat. **61 1782,** *C.A.* **55**, 21494e (1961) (1959, 1961 both to Chizu Drug). Prepn of the sodium salt: S. Ishii, K. Ogata, **Japan.** pat. **73 02,672** (1973 to Senju), *C.A.* **80**, 6959t (1974). Pharmacological studies: F. Ikemoto *et al., Oyo Yakuri* **8**, 937 (1974), *C.A.* **83**, 71663t (1975). Toxicological studies: *eidem, ibid.* **911**, 923, *C.A.* **82**, 51617g, 68262k (1975). Influence on carbohydrate metabolism in the lens: I. Korte *et al., Ophthalmic Res.* **7**, 282 (1975). Effect on NADH, NADPH: *eidem, ibid.* **440**. Effect on senile cataracts: T. Murata, *Folia Ophthalmol. Japan* **31**, 1217 (1980). Clinical trial in treatment of cataracts: S. K. Angra *et al., Indian J. Ophthalmol.* **31**, 5 (1983).

Orange-yellow powder, mp 247-248° (dec).
Sodium salt, C$_{16}$H$_7$N$_2$NaO$_5$, *Clarvisan.* Very sol in water. LD$_{50}$ in mice (mg/kg): > 10000 orally; > 5000 s.c.; 2120-2250 i.p. LD$_{50}$ in rats (mg/kg) i.p.: 2400 (males); 1460 (females) (Ikemoto).

THERAP CAT: Treatment of cataracts.

7646. Pirenzepine. *5,11-Dihydro-11-[(4-methyl-1-piperazinyl)acetyl]-6H-pyrido[2,3-b][1,4]benzodiazepin-6-one;* LS-519. C$_{19}$H$_{21}$N$_5$O$_2$; mol wt 351.41. C 64.94%, H 6.02%, N 19.93%, O 9.11%. Tricyclic gastric-acid inhibitor. Prepn: **Fr.** pat. **1,505,795** (1967 to Thomae), *C.A.* **70**, 4154w (1969). Pharmacology: W. Eberlein *et al., Arzneimittel-Forsch.* **27**, 356 (1977). Pharmacokinetics: R. Hammes *et al., ibid.* **928.** Mechanism of action: G. Heller *et al., Verh. Deut. Ges. Inn. Med.* **84**, 991 (1978), *C.A.* **90**, 132984s (1979). Human pharmacology: H. Brunner *et al., Arzneimittel-Forsch.* **27**, 684 (1977). Multicenter controlled clinical trial: *Scand. J. Gastroenterol.* **17**, Suppl. 81, 1-42 (1982). Radioimmunoassay determn in human plasma and urine: C. A. Homon *et al., Ther. Drug Monit.* **9**, 236 (1987). Symposium: *ibid.* Suppl. 72, 1-273. Review of pharmacology and therapeutic efficacy: A. A. Carmine, R. N. Brogden, *Drugs* **30**, 85-126 (1985). Comprehensive description: H. A. El-Obeid *et al.,* in *Analytical Profiles of Drug Substances* vol. 16, K. Florey, Ed. (Academic Press, New York, 1987) pp 445-506.

Dihydrochloride, $C_{19}H_{21}N_5O_2 \cdot 2HCl$, *LS-519-Cl2, Duogastral, Durapirenz, Gasteril, Gastrozepin, Leblon, Maghen, Renzepin, Tabe, Ulcuforton, Ulcosan.* Sol in water, slightly sol in methanol. Practically insol in ether.
THERAP CAT: Antiulcerative.

7647. Piretanide. *3-(Aminosulfonyl)-4-phenoxy-5-(1-pyrrolidinyl)benzoic acid;* 4-phenoxy-3-(1-pyrrolidinyl)-5-sulfamoylbenzoic acid; HOE-118; S-73-4118; Arelix; Diumax; Eurelix; Tauliz. $C_{17}H_{18}N_2O_5S$; mol wt 362.41. C 56.34%, H 5.01%, N 7.73%, O 22.07%, S 8.85%. High-ceiling loop diuretic, structurally related to bumetanide, *q.v.* Prepn: D. Bormann *et al.,* Ger. pat. 2,419,970; *eidem,* U.S. pat. 4,010,273 (1975, 1977 both to Hoechst). Chemistry and pharmacology: W. Merkel *et al., Eur. J. Med. Chem.* **11,** 399 (1976). Diuretic activity in man: N. Pozet *et al., Brit. J. Clin. Pharmacol.* **9,** 577 (1980); T. Saruta, E. Kato, *Arzneimittel-Forsch.* **30,** 1807 (1980). Vascular effects of piretanide: E. Klaus *et al., ibid.* **33,** 1273 (1983); renal effects: M. Omosu *et al., ibid.* 1277. Review of pharmacology and therapeutic efficacy: S. P. Clissold, R. N. Brogden, *Drugs* **29,** 489-530 (1985).

Pale yellow platelets from methanol/water, mp 225-227°. Exhibits intense light blue fluorescence at 366 nm. LD_{50} in rats, mice (mg/kg): 5601, 3672 orally (Merkel).
THERAP CAT: Diuretic.

7648. Piribedil. *2-[4-(1,3-Benzodioxol-5-ylmethyl)-1-piperazinyl]pyrimidine;* 2-(4-piperonyl-1-piperazinyl)pyrimidine; 2-[4-(3,4-methylenedioxybenzyl)piperazino]pyrimidine; 1-(2-pyrimidyl)-4-piperonylpiperazine; 1-(2''-pyrimidyl)-4-(methylene-3',4'-dioxybenzyl)piperazine; ET-495; EU-4200; Trivastal. $C_{16}H_{18}N_4O_2$; mol wt 298.34. C 64.41%, H 6.08%, N 18.78%, O 10.73%. Central dopaminergic agonist. Prepn: Neth. pat. Appl. 6,413,349 corresp to G. Regnier *et al.,* U.S. pat. 3,299,067; *eidem,* Brit. pat. 1,101,425 (1965, 1967, 1968, all three to Sci. Union et Cie-Soc. Franc. Rech. Med.); G. Regnier *et al., J. Med. Chem.* **11,** 1151 (1968). Activity in man: R. Royer, *Proc. Int. Pharmacol. Meet., 3rd, 1966,* **3,** R. K. Richards, Ed. (Pergamon Press, New York, 1968) pp 45-55. Pharmacology: M. Laubie *et al., Eur. J. Pharmacol.* **6,** 75 (1969). Metabolism: D. B. Campbell *et al., Advan. Neurol.* **3,** 199 (1973).

Crystals from anhydr ethanol, mp 98°. LD_{50} i.p. in mice: 690.3 mg/kg.
THERAP CAT: Vasodilator (peripheral).

7649. Piridocaine. *2-Piperidineethanol 2-aminobenzoate (ester);* anthranilic acid 2-(2-piperidyl)ethyl ester; 2-piperidineethanol anthranilate; β-(2-piperidyl)ethyl o-aminobenzoate; PT-14; Lucaine. $C_{14}H_{20}N_2O_2$; mol wt 248.33. C 67.72%, H 8.12%, N 11.28%, O 12.89%. Prepn: Walter, Fosbinder, *J. Am. Chem. Soc.* **61,** 1713 (1939); U.S. pat. 2,229,533 (1941 to Maltbie Chem.).

Hydrochloride, $C_{14}H_{21}ClN_2O_2$, crystals from abs alcohol, mp 209-211°.
THERAP CAT: Anesthetic (local).

7650. Pirifibrate. *2-(4-Chlorophenoxy)-2-methylpropanoic acid [6-(hydroxymethyl)-2-pyridinyl]methyl ester;* 2,6-pyridinedimethanol mono-p-chlorophenoxyisobutyrate; EL-466; Bratenol. $C_{17}H_{18}ClNO_4$; mol wt 335.79. C 60.81%, H 5.40%, Cl 10.56%, N 4.17%, O 19.06%. Hypolipemic agent related structurally to clofibrate, *q.v.* Prepn: D. Humbert, R. Ratouis, Ger. pat. 2,432,322; *eidem,* U.S. pat. 3,971,798 (1975, 1976 both to Roussel). Multicenter study in hyperlipoproteinemias: A. J. Domingo *et al., Clin. Ther.* **3,** 219 (1980).

Crystals from isopropyl ether, mp 46°. LD_{50} i.p. in mice: 915-1098 mg/kg (Humbert, Ratouis, 1975).
Hydrochloride, $C_{17}H_{18}ClNO_4 \cdot HCl$, crystals, mp 110°. LD_{50} i.p. in mice: 1098-1281 mg/kg (expressed as free base) (Humbert, Ratouis 1975).
THERAP CAT: Antihyperlipoproteinemic.

7651. Pirimicarb. *Dimethylcarbamic acid 2-(dimethylamino)-5,6-dimethyl-4-pyrimidinyl ester;* 2-(dimethylamino)-5,6-dimethyl-4-pyrimidinyl dimethylcarbamate; 5,6-dimethyl-2-dimethylamino-4-pyrimidinyl dimethylcarbamate; 5,6-dimethyl-2-dimethylamino-4-dimethylcarbamoyloxypyrimidine; PP-062; ENT-27766; Aphox; Fernos; Pirimor. $C_{11}H_{18}N_4O_2$; mol wt 238.29. C 55.45%, H 7.61%, N 23.51%, O 13.43%. Selective aphicide. Prepn: F. L. C. Baranyovits *et al.,* S. Afr. pat. 67 01,588; *eidem,* U.S. pat. 3,493,574 (1968, 1970 to ICI). Activity: F. L. C. Baranyovits, R. Ghosh, *Chem. & Ind. (London)* **1969,** 1018. Degradn in soil: I. R. Hill, *ACS Symp. Ser.* **29,** 358 (1976). GLC analysis: P. D. Bland, *J. Assoc. Off. Anal. Chem.* **64,** 1315 (1981).

Crystalline solid, mp 90.5°. Vapor pressure at 30°: 3×10^{-5} mm Hg. Soly in water at 25°: 2.7 g/l. Sol in most organic solvents. Dec by prolonged boiling with acids or alkali. Aq solns are unstable to light. LD_{50} orally in female rats: 147 (mg/kg) (Baranyovits, Ghosh).

USE: Insecticide.

7652. Pirimiphos-ethyl. *O-[2-(Diethylamino)-6-methyl-4-pyrimidinyl]phosphorothioic acid O,O-diethyl ester; O,O-diethyl O-[2-(diethylamino)-6-methyl-4-pyrimidinyl]phosphorothioate; PP-211; Fernex; Primicid.* $C_{13}H_{24}N_3O_3PS$; mol wt 333.39. C 46.83%, H 7.26%, N 12.60%, O 14.40%, P 9.29%, S 9.62%. Broad spectrum contact and fumigant insecticide. Prepn: B. K. Snell, S. P. Sharpe, **Brit.** pat. **1,205,000** (1970 to ICI), *C.A.* **74**, 52482f (1971). Persistence and degradn in soil: D. L. Suett, *Pestic. Sci.* **6**, 385 (1975). Analysis: D. J. W. Bullock, *Anal. Methods Pestic. Plant Growth Regul.* **8**, 171 (1976).

Straw-colored liquid, dec >130°. d^{20} 1.14; n_D^{25} 1.520. Vapor pressure at 25°: 2.9×10^{-4} mm Hg. Soly in water at 30°: <1 mg/l. Miscible with most organic solvents. LD_{50} orally in rats: 140 mg/kg, *RTECS* **vol. II**, R. J. Lewis, R. L. Tatken, Eds. (1980) p 338.

O,O-Dimethyl analog, $C_{11}H_{20}N_3O_3PS$, *pirimiphos-methyl, PP-511, Actellic, Actellifog, Blex, Silosan.* Straw-colored liquid, d^{30} 1.157; n_D^{25} 1.527. Vapor press at 30°: $\sim 1 \times 10^{-4}$ mm Hg. Soly in water at 30°: 5 mg/l.

USE: Insecticide.

7653. Piritramide. *1'-(3-Cyano-3,3-diphenylpropyl)-[1,4'-bipiperidine]-4'-carboxamide;* 1-(3,3-diphenyl-3-cyanopropyl)-4-piperidino-4-piperidinecarboxamide; 2,2-diphenyl-4-(4-piperidino-4-carbamoylpiperidino)butyronitrile; pirinitramide; A-65; R-3365; Dipidolor; Piridolan.* $C_{27}H_{34}N_4O$; mol wt 430.59. C 75.31%, H 7.96%, N 13.01%, O 3.72%. Prepn: **Belg.** pat. **606,850;** P. A. J. Janssen, **U.S.** pat. **3,080,366** (1961, 1963 both to Janssen). Crystal structure: C. Humblet *et al., Acta Crystallogr.* **B33**, 1615 (1977).

Crystals from acetone, mp 149-150°.

Note: This is a controlled substance (opiate) listed in the U.S. Code of Federal Regulations, Title 21 Part 1308.11 (1995).

THERAP CAT: Analgesic (narcotic).

7654. Piritrexim. *6-[(2,5-Dimethoxyphenyl)methyl]-5-methylpyrido[2,3-d]pyrimidine-2,4-diamine;* 2,4-diamino-6-(2,5-dimethoxybenzyl)-5-methylpyrido[2,3-d]pyrimidine; BW-301U; NSC-351521.* $C_{17}H_{19}N_5O_2$; mol wt 325.37. C 62.76%, H 5.89%, N 21.52%, O 9.83%. Lipid soluble dihydrofolate reductase inhibitor. Prepn: E. M. Grivsky *et al., J. Med. Chem.* **23**, 327 (1980); E. M. Grivsky *et al., Eur.* pat. **Appl. 21,292** (1981 to Wellcome); E. M. Grivsky, **U.S.** pat. **4,959,474** (1990 to Burroughs Wellcome). Biochemistry and antitumor activity: D. S. Duch *et al., Cancer Res.* **42**, 3987 (1982). Cytotoxicity: I. W. Taylor *et al., ibid* **45**, 978 (1985). Pharmacology and toxicology: C. W. Sigel *et al., NCI Monogr.* **5**, 111 (1987). Antipneumocystis and antitoxoplasma activities: J. A. Kovacs *et al., Antimicrob. Ag. Chemother.* **32**, 430 (1988). HPLC determn in plasma: R. G. Foss, C. W. Sigel, *J. Pharm. Sci.* **71**, 1176 (1982).

Determn by protein binding assay in plasma: J. L. Woolley, Jr., *et al., ibid* **78**, 749 (1989). Crystal structure and conformational analysis: P. A. Sutton, V. Cody, *J. Am. Chem. Soc.* **110**, 6219 (1988). Clinical pharmacology in children: P. C. Adamson *et al., Cancer Res.* **50**, 4464 (1990). Clinical evaluation in lung cancer: M. G. Kris *et al., Cancer Treat. Rep.* **71**, 763 (1987); in advanced squamous head and neck cancer: W.-C. Uen *et al., Cancer,* **69**, 1008 (1992).

Yellow powder from ethanol/water, mp 252-254°. log P (octanol/water) = 1.74.

Isethionate, $C_{17}H_{19}N_5O_2 \cdot C_2H_6O_4S$. LD_{10} orally in rats: 764 mg/kg; LD_{50}: 1168; LD_{90}: 1572 (Sigel).

THERAP CAT: Antineoplastic.

7655. Pirlimycin. *(2S-cis)Methyl 7-chloro-6,7,8-trideoxy-6-[[(4-ethyl-2-piperidinyl)carbonyl]amino]-1-thio-L-threo-α-D-galactooctopyranoside.* $C_{17}H_{31}ClN_2O_5S$; mol wt 410.96. C 49.69%, H 7.60%, Cl 8.63%, N 6.82%, O 19.47%, S 7.80%. Semi-synthetic lincosaminide antibiotic; structural analog of clindamycin, *q.v.* Prepn: R. D. Birkenmeyer, **Ger.** pat. **3,043,502;** *idem,* **U.S.** pat. **4,278,789** (both 1981 to Upjohn); R. D. Birkenmeyer *et al., J. Med. Chem.* **27**, 216 (1984). *In vitro* activity vs aerobic bacteria: V. I. Ahonkhai *et al., Antimicrob. Ag. Chemother.* **21**, 902 (1982); vs anaerobic bacteria: S. M. H. Qadri *et al., J. Antibiot.* **36**, 42 (1983). HPLC determn in biological fluids: J. A. Shah, D. J. Weber, *J. Chromatog.* **309**, 95 (1984). Ion-pairing LC determn in pharmaceutic prepns: D. L. Theis *et al., ibid.* **402**, 335 (1987). Metabolism and residue studies in dairy cows: R. E. Hornish *et al., ACS Symposium Series* **503**, 132-147 (1992). Evaluation in bovine mastitis: W. E. Owens *et al., Agri-Pract.* **15**, 19 (1994).

Hydrochloride, *U-57903E, Pirsue.* Crystals from water, mp 222-224°. $[\alpha]_D^{25}$ +176° (Birkenmeyer, **U.S.** pat.). Also reported as mp 210-212°. $[\alpha]_D^{25}$ +181°. LD_{50} i.p. in mice: 600 mg/kg (Birkenmeyer, 1984).

THERAP CAT (VET): Antibacterial.

7656. Pirmenol. *cis-(±)-α-[3-(2,6-Dimethyl-1-piperidinyl)propyl]-α-phenyl-2-pyridinemethanol;* (±)-cis-2,6-dimethyl-α-phenyl-α-2-pyridyl-1-piperidinebutanol; (±)-1-phenyl-1-(2-pyridyl)-4-(cis-2,6-dimethyl-1-piperidyl)butanol.* $C_{22}H_{30}N_2O$; mol wt 338.49. C 78.06%, H 8.93%, N 8.28%, O 4.73%. Prepn: R. W. Fleming, **Ger.** pat. **2,806,654;** *idem,* **U.S.** pat. **4,112,103** (both 1978 to Parke, Davis). Antiarrhythmic profile in dogs: T. E. Mertz, T. J. Steffe, *J. Cardiovasc. Pharmacol.* **2**, 527 (1980). Toxicology study: J. L. Schardein *et al., Toxicol. Appl. Pharmacol.* **56**, 294 (1980). HPLC determn in biological fluids: E. L. Johnson, L. A. Pachla, *J. Pharm. Sci.* **73**, 754 (1984). Pharmacokinetics in humans: S. C. Hammill *et al., Circulation* **65**, 369 (1982); S. W. Sanders *et al., J. Clin. Pharmacol.* **23**, 113 (1983). Hemodynamic effects in cardiac patients: M. S. Nieminen *et al., Eur. Heart J.* **7**, 150 (1986). Clinical evaluations in ventricular arrhythmias: L. K. Toivonen *et al., J. Cardiovasc. Pharmacol.* **8**, 156 (1986); E. M. Hampton *et al., Eur. J. Clin. Pharmacol.* **31**, 15 (1986). Chronic toxicity study: J. R. Watkins *et al., Drug Invest.* **3**, 141 (1991). Evaluation of carcinogenicity: G. E. Macallum *et al., ibid.* **278**. Symposium on pharmacology and clinical efficacy: *Am. J. Cardiol.* **59**, Suppl., 1H-57H (1987).

Crystals from petroleum ether, mp 70-71°.
Monohydrochloride, $C_{22}H_{30}N_2O.HCl$, *CI-845.* mp 171-172°. LD_{50} in mice, rats, dogs (mg/kg): 20.8, 23.6, > 7.0 i.v.; 215.5, 359.9, > 40.0 orally (Schardein).

THERAP CAT: Antiarrhythmic (class IA).

7657. Piroctone. *1-Hydroxy-4-methyl-6-(2,4,4-trimethylpentyl)-2(1H)-pyridinone.* $C_{14}H_{23}NO_2$; mol wt 237.34. C 70.85%, H 9.77%, N 5.90%, O 13.48%. Pyridone deriv related structurally to ciclopirox, *q.v.* Prepn: G. Lohaus, W. Dittmar, **Ger.** pat. **2, 214,608** corresp to U.S. pat. **3,972,888** (1973, 1976 both to Hoechst); *eidem, Arzneimittel-Forsch.* **31**, 1311 (1981). Evaluation of efficacy as anti-dandruff agent: E. Futterer, *J. Soc. Cosmet. Chem.* **32**, 327 (1981).

Crystals, mp 108°.
Ethanolamine salt (1:1), $C_{16}H_{30}N_2O_3$, *Octopirox.*

THERAP CAT: Antiseborrheic.

7658. Piroheptine. *3-(10,11-Dihydro-5H-dibenzo[a,d]cyclohepten-5-ylidene)-1-ethyl-2-methylpyrrolidine.* $C_{22}H_{25}N$; mol wt 303.45. C 87.08%, H 8.30%, N 4.62%. Prepn: **Neth.** pat. **Appl. 6,609,280**; Y. Deguchi *et al.,* U.S. pat. **3,454,595** (1967, 1969 both to Fujisawa). Crystal structure: Y. Tokuma *et al., Bull. Chem. Soc. Japan* **44**, 2665 (1971). Pharmacological studies: M. Hitomi *et al., Arzneimittel-Forsch.* **22**, 953, 961 (1972). *In vitro* study: T. Ohashi *et al., ibid.* 966. Metabolism: Y. Tokuma *et al., Bull. Chem. Soc. Japan* **48**, 294 (1975). Clinical pharmacology: A. Barbeau, *Ann. Rev. Pharmacol.* **14**, 91 (1974). Toxicity study: M. Hitomi *et al., Arzneimittel-Forsch.* **22**, 961 (1972).

Liquid, bp_4 167°. uv max (95% ethanol): 240 nm (ε 12100).
Hydrochloride, $C_{22}H_{25}N.HCl$, *Trimol.* Crystals, mp 250-253°. LD_{50} in male mice, rats (mg/kg): 153, 600 orally; 19, 17 i.v.; 95, 110 i.p.; 109, 330 s.c. (Hitomi).

THERAP CAT: Antiparkinsonian.

7659. Piromen. Pyromen; desacchromin dispersion. Pyrogenic *pseudomonas* polysaccharide-nucleic acid complex. Prepd by proteolyzing the bacterial organism and separating the complex from inactive material by dialyzing: N. M. Nesset, L. G. Ginger, **Brit.** pat. **699,663** (1953 to Baxter Labs.), *C.A.* **48**, 6083 (1954). Contains deoxyribonucleic acid, ribonucleic acid, and hexosamine, the pyrogenic reducing sugar: N. M. Nesset *et al., J. Am. Pharm. Assoc.* **39**, 456 (1950). Use in detn of bone marrow granulocyte reserves: B. C. Korbitz *et al., Curr. Ther. Res., Clin. Exp.*

11, 491 (1969). Effects on spinal cord regeneration: M. A. Matthews *et al., Neuropathol. Appl. Neurobiol.* **5**, 161 (1979); *eidem, Acta Neurobiol. Exp.* **40**, 489 (1980).

7660. Piromidic Acid. *8-Ethyl-5,8-dihydro-5-oxo-2-(1-pyrrolidinyl)pyrido[2,3-d]pyrimidine-6-carboxylic acid;* 5,8-dihydro-8-ethyl-5-oxo-2-pyrrolidinopyrido[2,3-d]pyrimidine-6-carboxylic acid; PD-93; Bactramyl; Enterol; Gastrurol; Panacid; Pirodal; Purim; Reelon; Septural; Uropir. $C_{14}H_{16}N_4O_3$; mol wt 288.31. C 58.32%, H 5.59%, N 19.43%, O 16.65%. Quinolone antibacterial. Prepn: S. Minami *et al.,* **Japan.** pat. **67 25,912**; *eidem,* **Brit.** pat. **1,129,358** (1967, 1968 both to Dainippon Pharm.); *eidem, Chem. Pharm. Bull.* **19**, 1426 (1971). Activity studies: M. Shimizu *et al., Antimicrob. Ag. Chemother.* **1970**, 117. Metabolism: *eidem, ibid.* 123.

Crystals from ethanol-chloroform, mp 314-316°. LD_{50} in male, female mice, male, female rats (mg/kg): 287, 268, 177, 158 i.v.; all > 4000 orally, s.c. and i.p. (Shimizu).

THERAP CAT: Antibacterial.

7661. Piroxicam. *4-Hydroxy-2-methyl-N-2-pyridinyl-2H-1,2-benzothiazine-3-carboxamide 1,1-dioxide;* 3,4-dihydro-2-methyl-4-oxo-N-2-pyridyl-2H-1,2-benzothiazine-3-carboxamide 1,1-dioxide; CP-16171; Artroxicam; Baxo; Bruxicam; Caliment; Erazon; Feldene; Flogobene; Geldene; Improntal; Larapam; Pirkam; Piroflex; Reudene; Riacen; Roxicam; Roxiden; Sasulen; Solocalm; Zunden. $C_{15}H_{13}N_3O_4S$; mol wt 331.35. C 54.37%, H 3.95%, N 12.68%, O 19.31%, S 9.68%. Non-steroidal anti-inflammatory with long half-life. Prepn (keto form): J. Lombardino, **Ger.** pat. **1,943,265**; *idem,* U.S. pat. **3,591,584** (1970, 1971 to Pfizer). Synthesis and biological properties: J. Lombardino, E. Wiseman, *J. Med. Chem.* **15**, 848 (1972); J. Lombardino *et al., ibid.* **16**, 493 (1973). Pharmacology: E. Wiseman *et al., Arzneimittel-Forsch.* **26**, 1300 (1976). Evaluation of ulcerogenic effects in comparison with droxicam, *q.v.:* G. Palacios *et al., Method Find. Exp. Clin. Pharmacol.* **9**, 353 (1987). Clinical pharmacology: L. Martinez *et al., ibid.* **10**, 729 (1988). *Review: eidem,* in *Pharmacological and Biochemical Properties of Drug Substances* vol. 3, M. E. Goldberg, Ed. (Am. Pharm. Assoc., Washington, DC, 1981) pp 324-346. Review of pharmacology and therapeutic efficacy: R. N. Brogden *et al., Drugs* **22**, 165-187 (1981); *eidem, ibid.* **28**, 292-323 (1984). Symposium on clinical efficacy and safety: *Am. J. Med.* **81**, Suppl. 5B, 1-55 (1986). Comprehensive description: M. Mihalic *et al.* in *Analytical Profiles of Drug Substances* vol. 15, K. Florey, Ed. (Academic Press, New York, 1986) pp 509-531.

Crystals from methanol, mp 198-200°. pKa 6.3 (2:1 dioxane-water). LD_{50} orally in mice: 360 mg/kg (Wiseman).
Cinnamic acid ester, $C_{24}H_{19}N_3O_5S$, *piroxicam cinnamate, cinnoxicam, SPA-S-510, Sinartrol, Zelis, Zen.*
Compd with β-cyclodextrin, $C_{57}H_{83}N_3O_{39}S$, *Brexin, Cicladol, Cycladol.*

THERAP CAT: Anti-inflammatory.

7662. Pirozadil. *3,4,5-Trimethoxybenzoic acid 2,6-pyridinediylbis(methylene)ester;* 2,6-pyridinedimethanol bis(3,4,-5-trimethoxybenzoate); 722 D; Pemix. $C_{27}H_{29}NO_{10}$; mol wt

527.53. C 61.47%, H 5.54%, N 2.66%, O 30.33%. Hypolipidemic agent that inhibits platelet aggregation. Prepn: J. P. Cochs, **Ger. pat. 2,411,902** (1974 to Inst. Int. Ter.), *C.A.* **82**, 4136q (1975). Effect on cerebral metabolic blood flow in rabbits: J. Balasch, L. Palacios, *Arch. Farmacol. Toxicol.* **3**, 137 (1977). Efficacy in exptl atherosclerosis: M. R. Parwaresch *et al., Atherosclerosis* **31**, 395 (1978). Toxicological and histopathological study: J. Roca *et al., Arch. Farmacol. Toxicol.* **6**, 41 (1980). Clinical trials in hyperlipoproteinemia: M. Shinomiya *et al., Arzneimittel-Forsch.* **37**, 1069 (1987); R. Tapounet, I. Marti Ragué, *Drugs Exp. Clin. Res.* **13**, 447 (1987).

White cryst powder, mp 119-126°. Very sol in chloroform; sol in dioxane, acetonitrile. Practically insol in ether, water.

THERAP CAT: Antihyperlipoproteinemic.

7663. Pirprofen. *3-Chloro-4-(2,5-dihydro-1H-pyrrol-1-yl)-α-methylbenzeneacetic acid; 3-chloro-4-(3-pyrrolin-1-yl)-hydratropic acid;* Su-21524; Rangasil 400; Rengasil; Seflenyl. $C_{13}H_{14}ClNO_2$; mol wt 251.71. C 62.03%, H 5.61%, Cl 14.08%, N 5.56%, O 12.71%. Non-steroidal anti-inflammatory. Prepn: R. W. Carney, G. DeStevens, **U.S. pat. 3,641,040** (1972 to Ciba); R. W. Carney *et al., Experientia* **29**, 938 (1973). Metabolism. R. C. Luders *et al., Clin. Pharmacol. Ther.* **21**, 721 (1977). Pharmacology: G. Wilhelmi, *Pharmacology* **16**, 268 (1978). Pharmacokinetics: M. Beth *et al., J. Clin. Pharmacol.* **15**, 563 (1975). Comparative efficacy: R. J. Saykaly *et al., ibid.* **19**, 59 (1979). Review of pharmacology and therapeutic efficacy: P. A. Todd, R. Beresford, *Drugs* **32**, 509-537 (1986).

Crystals from benzene-hexane, mp 98-100°.
THERAP CAT: Anti-inflammatory.

7664. Piscidic Acid. *2,3-Dihydroxy-2-[(4-hydroxyphenyl)methyl]butanedioic acid; (p-hydroxybenzyl)tartaric acid.* $C_{11}H_{12}O_7$; mol wt 256.21. C 51.57%, H 4.72%, O 43.71%. Isoln from root bark of *Piscidia piscipula* (L.) Sarg. (*P. erythrina* L.), *Leguminosae* (Jamaica dogwood): Freer, Clover, *Am. Chem. J.* **25**, 390 (1901); from *Narcissus poeticus* L., *Amaryllidaceae:* Smeby *et al., J. Am. Chem. Soc.* **76**, 6127 (1954). Structure: Bridge *et al., J. Chem. Soc.* **1948**, 257. Synthesis of *p-O*-methyl deriv: Buckle *et al., ibid.* **1954**, 3981.

Elongated prisms. mp 186-187°. $[\alpha]_D^{20}$ +41.02° (c = 2.65). Sol in water, ethyl acetate; practically insol in CHCl$_3$.
p-O-Methyl derivative, $C_{12}H_{14}O_7$, *(p-methoxybenzyl)tartaric acid.* Crystals, mp 169-170°. $[\alpha]_D^{23}$ +44.01° ±5.0 (c = 1.262).

7665. Pithecolobine. *19-Heptyl-10-hydroxy-1,5,10,14-tetraazacyclononadecan-15-one.* $C_{22}H_{46}N_4O_2$; mol wt 398.63. C 66.29%, H 11.63%, N 14.05%, O 8.03%. Occurs in the bark of *Samanea saman* Merr. (formerly *Pithecolobium saman* Benth.), bark and seed of *P. bigeminum* Mart. and *P. lobatum* Benth., *Leguminosae.* Isoln: Greshoff, *Ber.* **23**,

3541 (1890); K. Wiesner *et al., Can. J. Chem.* **30**, 761 (1952). Structure: *eidem, J. Am. Chem. Soc.* **75**, 6348 (1953); D. E. Orr, K. Wiesner, *Chem. & Ind. (London)* **1959**, 672; K. Wiesner, D. E. Orr, *Tetrahedron Letters* no. 16, 11 (1960); K. Wiesner *et al., Can. J. Chem.* **46**, 1886, 3617 (1968).

Crystals, mp 67-69° or oily liquid, bp$_{0.007}$ 230°. Sublimes at 135° at 0.007 mm pressure. Sol in water, alcohol, chloroform, ether, petr ether.

7666. Pituitary, Posterior. Pituamin; Di-Sipidin; Pituitrin. Desiccated hypophysis. The cleaned, dried, and powdered posterior lobe of pituitary body of domesticated animals used for food by man. Contains both oxytocin and vasopressin, *q.q.v.*
Yellowish or grayish, amorphous powder; characteristic odor. Partially sol in water. *Keep well closed and in a cool place.*

THERAP CAT: Oxytocic; antidiuretic.
THERAP CAT (VET): Oxytocic; antidiuretic.

7667. Pivalic Acid. *2,2-Dimethylpropanoic acid;* α,α-dimethylpropionic acid; trimethylacetic acid. $C_5H_{10}O_2$; mol wt 102.13. C 58.80%, H 9.87%, O 31.33%. $HOOCC(CH_3)_3$. Prepd by the reaction of *tert*-butylmagnesium chloride and carbon dioxide: Bouveault, *Compt. Rend.* **138**, 1108 (1904); Puntambeker, Zoellner, *Org. Syn.* **8**, 104 (1928); other methods: *ibid.* 108. Forms higher esters (e.g., isobutyl ester) only with difficulty.
Needles, mp 35.5°. bp$_{760}$ 163.8°; d^{50} 0.905; $n_D^{36.5}$ 1.3931. pKa (25°): 5.01. One gram dissolves in 40 ml water. Freely sol in alcohol, ether.
Ethyl ester, $C_7H_{14}O_2$, liquid. d$_4^{18}$ 0.8580; bp 118.2°; n_D^{18} 1.3922.

7668. Pivalylbenzhydrazine. *2,2-Dimethylpropanoic acid 2-(phenylmethyl)hydrazide; pivalic acid 2-benzylhydrazide;* N^1-pivaloyl-N^2-benzylhydrazine; 1-benzyl-2-trimethylacetylhydrazine; 1-benzyl-2-pivalylhydrazine; 1-(trimethylacetyl)-2-benzylhydrazine; pivhydrazine; Ro-4-1634; Tersavid. $C_{12}H_{18}N_2O$; mol wt 206.29. C 69.87%, H 8.79%, N 13.58%, O 7.76%. $C_6H_5CH_2NHNHOCC(CH_3)_3$. Prepn: **Brit. pat. 883,379** (1961 to Hoffmann-La Roche).
Crystals, mp 68-69°.
THERAP CAT: Monoamine oxidase inhibitor.

7669. Pivampicillin. *[2S-[2α,5α,6β(S*)]]-6-[(Aminophenylacetyl)amino]-3,3-dimethyl-7-oxo-4-thia-1-azabicyclo-[3.2.0]heptane-2-carboxylic acid (2,2-dimethyl-1-oxopropoxy)methyl ester; hydroxymethyl 6-(2-amino-2-phenylacetamido)-3,3-dimethyl-7-oxo-4-thia-1-azabicyclo[3.2.0]heptane-2-carboxylate pivalate (ester);* 6-[D-α-aminophenylacetamido]penicillanic acid pivaloyloxymethyl ester; pivaloyloxymethyl D-α-aminobenzylpenicillinate; ampicillin pivaloyloxymethyl ester; pivaloyloxymethyl ampicillinate; MK-191. $C_{22}H_{29}N_3$-O_6S; mol wt 463.55. C 57.00%, H 6.31%, N 9.06%, O 20.71%, S 6.92%. Semi-synthetic antibiotic related to penicillin. Prepn: E. K. Frederiksen, W. O. Godtfredsen, **S. Afr. pat. 68 05,952;** *eidem,* **U.S. pat. 3,660,575** (1969, 1972 to Lövens Kemiske Fabrik); von Daehne *et al., J. Med. Chem.* **13**, 607 (1970). Pharmacology: *eidem, loc. cit.;* Jordan *et al., Antimicrob. Ag. Chemother.* **1970**, 438; Foltz *et al., ibid.* 442. Pharmacokinetics in man: M. Ehrnebo *et al., J. Pharmacokinet. Biopharm.* **7**, 429 (1979). Toxicity: von Daehne *et al., Antimicrob. Ag. Chemother.* **1970**, 430.

Hydrochloride, $C_{22}H_{29}N_3O_6S.HCl$, *Alphacilina, Alphacillin, Berocillin, Centurina, Diancina, Inacilin, Maxifen, Pivatil, Pondocil, Pondocillin, Pondocillina, Sanguicillin.* Microcrystalline powder, mp 155-156° (decomp). $[\alpha]_D^{20}$ +196° (c = 1 in water). Weak uv max (water): 268, 262, 256 nm ($E_{1cm}^{1\%}$ ~3.9, 5.7, 6.3). pKa ~7.0. Relatively stable in acid soln; ester hydrolyzes slowly in neutral soln. pH of 0.5 g/100 ml water: ~4.5. Very sol in water and chloroform; freely sol in ethanol; sparingly sol in *n*-propanol, *tert*-butanol, and ethyl ether. LD_{50} in mice, rats (g/kg): 3.34, 5.00 orally; 3.60, 4.50 s.c. (von Daehne).

THERAP CAT: Antibacterial.

7670. Pivcefalexin. *[6R-[6α,7β(R*)]]-7-[(Aminophenylacetyl)amino]-3-methyl-8-oxo-5-thia-1-azabicyclo[4.2.0]oct-2-ene-2-carboxylic acid (2,2-dimethyl-1-oxopropoxy)methyl ester;* 7-(D-2-amino-2-phenylacetamido)desacetoxycephalosporanic acid pivaloyloxymethyl ester; pivcephalexin. $C_{22}H_{27}N_3O_6S$; mol wt 461.54. C 57.25%, H 5.90%, N 9.10%, O 20.80%, S 6.95%. Orally active semi-synthetic cephalosporin antibiotic; deriv of cephalexin, *q.v.* Prepn: W. O. Godtfredsen, E. T. Binderup, **Ger.** pat. **1,951,012** (1970 to Lövens Kemiske Fabrik), *C.A.* **72**, 132701v (1970). Pharmacology: P. Foresta *et al.*, *Arzneimittel-Forsch.* **27**, 819 (1977). Absorption and excretion: E. Trabucchi *et al.*, *Clin. Ther.* **81**, 299 (1977). Clinical trial: C. Vittorini *et al.*, *Arzneimittel-Forsch.* **31**, 1163 (1981).

Hydrochloride, $C_{22}H_{27}N_3O_6S.HCl$, *ST-21, Ausocef, Bencef, Cefalen, Cefalex, Pivacef, Sigmacef.*

THERAP CAT: Antibacterial.

7671. Pizotyline. *4-(9,10-Dihydro-4H-benzo[4,5]cyclohepta[1,2-b]thien-4-ylidene)-1-methylpiperidine;* 4-(1-methyl-4-piperidylidene)-9,10-dihydro-4*H*-benzo[4,5]cyclohepta-[1,2-*b*]thiophene; pizotifen; pizotifan; BC-105; Litec. $C_{19}H_{21}NS$; mol wt 295.45. C 77.24%, H 7.16%, N 4.74%, S 10.85%. Serotonin antagonist structurally related to cyproheptadine, *q.v.* Prepn: Jucker *et al.*, **Belg.** pat. **636,717** and **U.S.** pat. **3,272,826** (1964, 1966 to Sandoz); Bastian *et al.*, *Helv. Chim. Acta* **49**, 214 (1966). Toxicity data: Bagdon, Dorado, *Pharmacologist* **12**, No. 2, 297 (1970). Mechanism of action study: E. Müller-Schweinitzer, *J. Cardiovasc. Pharmacol.* **8**, 805 (1986). Clinical trial in migraine: D. Crowder, W. P. Maclay, *Curr. Med. Res. Opin.* **9**, 280

(1984). Review of pharmacology and therapeutic efficacy: Speight, Avery, *Drugs* **3**, 159-203 (1972).

Hydrochloride, crystals from isopropanol-ether, mp 261-263° (dec).

Malate, $C_{19}H_{21}NS.C_4H_6O_5$, *Sandomigran, Sanmigran, Sanomigran, Mosegor.* Crystals from methanol, mp 185-186° (dec).

THERAP CAT: Antimigraine; appetite stimulant.

7672. Placenta. Lioplacentyl; Tissulina; Inplacen; Placentafil; Tissural. Human placenta prepd for injection or topical use. Has been used to stimulate regeneration of damaged tissue.

7673. Plafibride. *2-(4-Chlorophenoxy)-2-methyl-N-[[(4-morpholinylmethyl)amino]carbonyl]propanamide;* N-2-(*p*-chlorophenoxy)isobutyryl-*N'*-morpholinomethylurea; ITA-104; Idonor; Perifunal. $C_{16}H_{22}ClN_3O_4$; mol wt 355.82. C 54.01%, H 6.23%, Cl 9.96%, N 11.81%, O 17.99%. Analog of clofibrate, *q.v.* Prepn: J. Iniesta-Pons, **Ger.** pat. **2,716,-374** (1977 to Investigacion Tecnica y Aplicada), *C.A.* **88**, 37810g (1978). Toxicity data: J. Zapatero, L. Brugeghini, *Arch. Farmacol. Toxicol.* **4**, 137 (1978). Series of articles on antiplatelet aggregation activity, hypolipemic activity, pharmacology, and toxicology: *ibid.*, 132-142.

Crystals, mp 100-102°. Sol in acetone. Slightly sol in alcohol. Practically insol in water, petr ether. LD_{50} in mice, rats, guinea pigs (mg/kg): 3569, >4000, 2168 orally (Zapatero, Brugeghini).

THERAP CAT: Antithrombotic.

7674. Plantago Seed. Psyllium seed; plantain seed; flea seed. Seed from *Plantago ovata* Forsk., *Plantaginaceae,* known as blond or Indian plantago seed, *P. psyllium* L. known as Spanish psyllium seed, or *P. indica* L. (*P. arenaria* Waldst. & Kit.) known as French psyllium seed. *Habit.* Mediterranean countries, India. Cultivated in India, Pakistan, France. *Psyllium husk* is the epidermis from seeds of any of the above plantago species; *Ispaghula husk* refers specifically to the epidermis of *P. ovata* seeds. The husk contains a water-soluble mucilage consisting of a highly branched, acidic arabinoxylan: J. F. Kennedy *et al.*, *Carbohyd. Res.* **75**, 265 (1979). Use in chronic constipation: M. Borgia *et al.*, *J. Int. Med Res.* **11**, 124 (1983). Review: J. N. BeMiller in *Industrial Gums,* R. L. Whistler, Ed. (Academic Press, New York, 2nd ed., 1973) pp 345-354. Review of therapeutic potential in reducing blood cholesterol: H. Lipsky *et al.*, *J. Clin. Pharmacol.* **30**, 699-703 (1990).

Small, dark reddish-brown, odorless, almost tasteless seeds. Mixed with an equal bulk of water, forms a mucilaginous mass.

Ispaghula husk, *Fibrolax, Fybrogel, Isogel, Regulan.*

Psyllium hydrophilic mucilloid, psyllium hydrocolloid, psyllium seed gum, Effer-Syllium, Fiberall, Metamucil, Perdiem Fiber, Serutan. Obtained from epidermis of *P. ovata* seeds. White to cream-colored, slightly granular powder. Slightly acid taste; little or no odor.

THERAP CAT: Laxative.

7675. Plantisul. Banabins. Orally effective antidiabetic principle contained in aq extracts from the leaves and fruits of the banaba tree, *Lagerstroemia speciosa (L.) Pers., Lythraceae.* Extraction procedure: Faustino Garcia, *Philippine J. Sci.* **76**, no. 3, 3-21 (1944); *J. Philippine Med. Assoc.* **31**, 216-224 and 276-282 (1955). *Constit.* of banaba extracts: Carew, Chin, *Nature* **190**, 1108 (1961).

7676. Plasma (in Physics). In physics and in nuclear chemistry plasma means highly ionized gases, i.e., a mixture of ions, electrons and at times neutral particles at high energy levels. Books: J. G. Linhart, *Plasma Physics* (North Holland Publ., Amsterdam, 1960); *Topics in Current Chemistry,* S. Veprek, M. Venugopalan, Eds., **Vols. 89, 90** and **94,**

entitled *Plasma Chemistry I, II* and *III* (Springer-Verlag, New York, 1980) 143 pp, 121 pp and 125 pp, resp.

7677. Plasmalogens. Aldehydogenic lipids characteristic of the animal kingdom; references to their existence in plants and bacteria are rare. Plasmalogens contain an aldehydogenic chain linked to glycerol as an α,β-unsaturated ether. Although most are aldehydogenic phosphatides, nonphosphatide or neutral plasmalogens have been detected in animal tissues. "Native plasmalogens", when deacylated, yield lysoderivatives, *lysoplasmalogens*, contg the aldehydogenic chain linked to glycerol as an α,β-unsaturated ether. The nomenclature *"phosphatidal ethanolamine"*, *"phosphatidal choline"*, etc. has been proposed for native plasmalogen phosphatides, and *"lysophosphatidal ethanolamine"*, *"lysophosphatidal choline"*, etc. for corresponding deacylated derivatives. This nomenclature minimizes the confusion arising from less precise terms, such as "plasmalogen" or *"ethanolamine plasmalogen"*, *"choline plasmalogen"*, etc. to designate either native compds, lysoderivatives, or other structures which may occur (molecules with two α,β-unsaturated ether chains or with one saturated ether and one α,β-unsaturated ether chain, and true cyclic glyceryl acetal derivatives). First isoln of a pure native plasmalogen (phosphatidal choline): Gottfried, Rapport, *Fed. Proc.* **20**, 278 (1961); *eidem, J. Biol. Chem.* **237**, 329 (1962). *Ref:* Rapport, Norton, *Ann. Rev. Biochem.* **31**, 103 (1962). Synthesis: Piantadosi *et al., J. Org. Chem.* **28**, 2425 (1963); Chacko *et al., ibid.* **32**, 3698 (1967); Slotboom *et al., Chem. Phys. Lipids* **1**, 192 (1967); Gigg, Gigg, *J. Chem. Soc. (C)* **1968**, 16, 2030; Vtorov *et al., Tetrahedron Letters* **1971**, 4605. *Reviews:* E. Klenk, H. Debuch, "Plasmalogens" in Holman *et al., Progr. Chem. Fats Lipids* **vol. 6**, (Macmillan, New York, 1963) pp 1-29; Piantadosi, Snyder, *J. Pharm. Sci.* **59**, 283-297 (1970).

7678. Plasmin. Fibrinolysin; serum tryptase; Actase; Thrombolysin. Mol wt about 90,000. Trypsin-like proteolytic enzyme which cleaves fibrin, fibrinogen, *q.q.v.*, and other plasma proteins. Component of the mammalian fibrinolytic system specifically responsible for the dissolution of fibrin clots. Exists in plasma as an inactive precursor, plasminogen, *q.v.* Converted to the active enzyme at the clot site by tissue plasminogen activator, *q.v.* Also activated by streptokinase and urokinase, *q.q.v.* Rapidly inactivated in plasma by α₂-antiplasmin, a glyceroprotein with high specific binding affinity for plasmin. Prepn: Christensen, MacLeod, *J. Gen. Physiol.* **28**, 599 (1945); E. C. Loomis, U.S. pat. **2,624,691** (1953 to Parke, Davis); J. H. Hink, J. K. McDonald, U.S. pat. **3,234,106** (1966 to Cutter Labs); K. C. Robbins, L. Summaria, *J. Biol. Chem.* **238**, 952 (1963). Converted from plasminogen by proteolysis of a single arg-val bond: K. C. Robbins *et al., ibid.* **242**, 2333, 4279 (1967). Structure consists of two polypeptide chains connected by two disulfide bonds. The heavy (A) chain (MW ~65,000) originates from the amino-terminal portion of plasminogen and contains the binding sites. The light (B) chain (MW ~25,000) originates from the carboxy-terminus and contains the active site. Isolation and characterization of heavy and light chains: L. Summaria *et al., J. Biol. Chem.* **242**, 5046 (1967); *eidem, ibid.* **246**, 2143 (1971). Amino acid sequence studies: W. R. Groskopf *et al., ibid.* **244**, 3590 (1969); Hartley, *Phil. Trans. Roy. Soc. London Ser. B* **257**, 77 (1970); S. Nagasawa, T. Suzuki, *Biochem. Biophys. Res. Commun.* **41**, 562 (1970). Amino acid sequence of active site: K. C. Robbins *et al., J. Biol. Chem.* **248**, 1631 (1973). Review of structural studies: F. J. Castellino, *Semin. Throm. Hemostasis.* **10**, 18-23 (1984). Enzyme specificity for lysine and arginine peptide bonds: W. Troll *et al., J. Biol. Chem.* **208**, 85 (1954). Review of biochemistry: D. Ogston, *J. Clin. Pathol.* **33**, Suppl. 14, 5-9 (1980). *Reviews:* D. Collen, *Thromb. Haemostasis* **43**, 77-89 (1980); K. C. Robbins *et al., Methods Enzymol.* **80**, 379-387 (1981); K. K. Kane, *Ann. Clin. Lab. Sci.* **14**, 443-449 (1984).
THERAP CAT: Thrombolytic enzyme.

7679. Plasminogen. Profibrinolysin; plasma trypsinogen. Mol wt ~90,000. The circulating plasma precursor (zymogen) or inactive form of plasmin, *q.v.* Prepn from blood plasma: Loomis *et al., Arch. Biochem.* **12**, 1 (1947); Oncley *et al., J. Am. Chem. Soc.* **71**, 541 (1949). Purifica-

tion: Christensen, Smith, *Proc. Soc. Exp. Biol. Med.* **74**, 840 (1950); Kline, *J. Biol. Chem.* **204**, 949 (1953); Hagan *et al., ibid.* **235**, 1005 (1960); Hagan *et al.,* U.S. pat. **3,066,079** (1962 to Am. Cyanamid); Derechin *et al., Biochem. J.* **84**, 336 (1962); Mertz, Chan, *Can. J. Biochem. Physiol.* **41**, 1811 (1963); D. G. Deutsch, E. T. Mertz, *Science* **170**, 1095 (1970). Physical properties: Davies, Englert, *J. Biol. Chem.* **235**, 1011 (1960). Converted to plasmin by natural activators such as streptokinase, urokinase, or tissue plasminogen activator, *q.q.v.*, through the cleavage of a single arg-val bond. Mechanism of activation: K. C. Robbins *et al., ibid.* **242**, 2333, 4279 (1967). Native plasminogen is a single chain glycopeptide. Two major molecular forms exist, differing in carbohydrate content and separable by affinity chromatography. Mol wt studies: Barlow *et al., ibid.* **244**, 1138 (1969). Sequence studies: K. C. Robbins *et al., ibid.* **247**, 6757 (1972); F. J. Castellino *et al., Biochem. Biophys. Res. Commun.* **53**, 845 (1973). Carbohydrate comp: M. L. Hayes, F. J. Castellino, *J. Biol. Chem.* **254**, 8768, 8772, 8777 (1979). Amino acid sequence and biochemistry: F. J. Castellino, *Semin. Thromb. Hemostasis* **10**, (1984). Biosynthesis and secretion by cultured hepatocytes: J. F. Bohmfalk, G. M. Fuller, *Science* **209**, 408 (1980). Therapeutic use with streptokinase: I. D. Walker *et al., Thromb. Haemostasis* **51**, 204 (1984). Review of fibrinolytic system: D. Collen, *ibid.* **43**, 77-89 (1980). *Reviews: Fibrinolysis,* D. L. Kline, K. N. N. Reddy, Eds. (CRC, Boca Raton, 1980) 256 pp; F. J. Castellino, *Chem. Rev.* **81**, 431-446 (1981).
Soluble below pH 5 and above pH 9 and only sparingly sol at intermediate pH values. Resistant to heat below pH 5. Displays maximum stability at pH 2-3.

7680. Plasmocid. *N,N-Diethyl-N'-(6-methoxy-8-quinolinyl)-1,3-propanediamine;* 8-(3-diethylaminopropylamino)-6-methoxyquinoline; 6-methoxy-8-(3-diethylaminopropylamino)quinoline; 710-F; SN-3115; Fourneau 710; Antimalarine; Rhodoquine. $C_{17}H_{25}N_3O$; mol wt 287.41. C 71.05%, H 8.77%, N 14.62%, O 5.57%. Prepn: Fourneau *et al., Ann. Inst. Pasteur* **44**, 503 (1930); **46**, 514 (1931); **50**, 731 (1933); Magidson, Strukov, *Arch. Pharm.* **271**, 359 (1933); Strukov, Russian pat. **39,105** (1934), *C.A.* **30**, 3445 (1936); Magidson *et al., J. Appl. Chem. USSR* **9**, 304 (1936); Giral, *Ciencia (Mexico)* **6**, 253 (1945), *C.A.* **40**, 3563 (1946); Yanko *et al., J. Am. Chem. Soc.* **67**, 664 (1945).

Oily liquid, d_4^{24} 1.0569. $bp_{1.0}$ 182°. n_D^{24} 1.5855.
Dihydrochloride, $C_{17}H_{27}Cl_2N_3O$, yellow crystals, mp 218-220°. Slightly sol in water, alcohol.
Diphosphate, yellow crystals, mp 169-171°. Slightly sol in water, alcohol.
Note: The name Rhodoquine was first applied to 8-(3-dimethylaminopropylamino)-6-methoxyquinoline.
THERAP CAT: Antimalarial.

7681. Plastoquinones. PQs. Family of lipid-soluble benzoquinone derivatives involved in photosynthetic electron transport. Structurally and functionally analogous to the ubiquinones, *q.v.* Structures are characterized by a side chain of repeating isoprenoid units with 9 units being the most common. Originally discovered by Kofler in 1946. Isoln of PQ-9 from alfalfa: F. L. Crane, *Plant Physiol.* **34**, 546 (1959). Localization in chloroplasts and role in photosynthesis: N. I. Bishop, *Proc. Nat. Acad. Sci.* **45**, 1696 (1959). Structure: M. Kofler *et al., Helv. Chim. Acta* **42**, 2252 (1959); N. R. Trenner *et al., J. Am. Chem. Soc.* **81**, 2026 (1959). Synthesis: C. H. Shunk *et al., ibid.* 5000. Identification of plastoquinone analogs B and C: L. P. Kegel *et al., Biochem. Biophys. Res. Commun.* **8**, 294 (1962). Review of role in photosynthetic electron flow: U. Siggel, *Bioelectrochem. Bioenerg.* **3**, 302-318 (1976); A. Trebst in *Coenzyme Q,* G. Lenaz, Ed. (John Wiley & Sons, Chichester, 1985) pp 257-284.

Plastoquinone 9

Plastoquinone 9, $C_{53}H_{80}O_2$, *(all-E)-2,3-dimethyl-5-(3,7,11,-15,19,23,27,31,35-nonamethyl-2,6,10,14,18,22,26,30,34-hexatriacontanonaenyl)-2,5-cyclohexadiene-1,4-dione, 2,3-dimethyl-5-solanesyl-1,4-benzoquinone, Kofler's quinone, plastoquinone A, plastoquinone 45, Q-254, PQ-9.* Yellow crystals, mp 48-49°. uv max (isooctane): 254, 262 nm (E $^{1\%}_{1cm}$ 247, 226).

7682. Platelet Activating Factor.

PAF; 1-*O*-alkyl-2-acetyl-*sn*-glycero-3-phosphorylcholine; acetyl glyceryl ether phosphorylcholine; AGEPC; PAF-acether; antihypertensive polar renomedullary lipid; APRL. Phospholipid mediator of platelet aggregation, inflammation, and anaphylaxis. Produced in response to specific stimuli by a variety of cell types, including neutrophils, basophils, platelets, and endothelial cells. Several molecular species of PAF have been identified which vary in the length of the *O*-alkyl side chain. *In vitro* identification of a soluble platelet activating factor produced by antigen stimulated leukocytes: P. M. Henson, *J. Exp. Med.* **131**, 287 (1970). Isoln from rabbit basophils, characterization of PAF response and potential role in immune complex disease: J. Benveniste *et al., ibid.* **136**, 1356 (1972). Isoln from human leukocytes: J. Benveniste, *Nature* **249**, 581 (1974). Release of PAF *in vivo* during anaphylaxis: R. N. Pinckard *et al., J. Immunol.* **123**, 1847 (1979). Structural study: J. Benveniste *et al., Nature* **269**, 170 (1977); and synthetic approaches: J. Benveniste *et al., Compt. Rend. Ser. D* **289**, 1037 (1979); C. A. Demopoulos *et al., J. Biol. Chem.* **254**, 9355 (1979). Identity with APRL and antihypertensive activity: M. L. Blank *et al., Biochem. Biophys. Res. Commun.* **90**, 1194 (1979). Structures of 1-*O*-octadecyl- and 1-*O*-hexadecyl-PAF, two of the predominant molecular forms: D. J. Hanahan *et al., ibid.* **255**, 5514 (1980). Total synthesis of 1-*O*-octadecyl-PAF: J. Godfroid *et al., FEBS Letters* **116**, 161 (1980). Enantiomeric synthesis of C_{16}- and C_{18}-PAF: M. Ohno *et al., Chem. Pharm. Bull.* **33**, 572 (1985). Molecular heterogeneity of naturally occurring PAF: R. N. Pinckard *et al., Biochem. Biophys. Res. Commun.* **122**, 325 (1984); H. W. Mueller *et al., J. Biol. Chem.* **259**, 14554 (1984). Ulcerogenic effects: A.-C. Rosam *et al., Nature* **319**, 54 (1986). Effects on human pulmonary and cardiovascular function: F. M. Cuss *et al., Lancet* **2**, 189 (1986). Review of isoln and analytical methods: D. J. Hanahan, S. T. Weintraub, *Meth. Biochem. Anal.* **31**, 195-219 (1985); of chemical and biochemical characteristics: D. J. Hanahan, R. Kumar, *Prog. Lipid Res.* **26**, 1-28 (1987). Reviews of biosynthesis, biological activities and PAF antagonists: P. Braquet *et al., Pharmacol. Rev.* **39**, 97-145 (1987); P. J. Barnes *et al., J. Allergy Clin. Immunol.* **81**, 919-934 (1988). Books: *Platelet-Activating Factor and Related Lipid Mediators,* F. Snyder, Ed. (Plenum Press, New York, 1987) 471 pp; *Prog. Biochem. Pharmacol.* **22** entitled "Biologically Active Ether Lipids" by P. Braquet *et al.,* Eds. (Karger, Basel, 1988).

n usually represents 15 or 17

1-*O*-Hexadecyl PAF, $C_{26}H_{54}NO_7P$, *(R)-7-(acetyloxy)-4-hydroxy-N,N,N-trimethyl-3,5,9-trioxa-4-phosphapentacosan-1-aminium hydroxide inner salt 4-oxide,* C_{16}-PAF. White,

amorphous solid, mp 247° (dec). $[\alpha]^{21}_D$ −3.66° (c = 0.71 in chloroform).

1-*O*-Octadecyl-PAF, $C_{28}H_{58}NO_7P$, *(R)-7-(acetyloxy)-4-hydroxy-N,N,N-trimethyl-3,5,9-trioxa-4-phosphaheptacosan-1-aminium hydroxide inner salt 4-oxide,* C_{18}-PAF. mp 212-215° (dec). $[\alpha]^{20}_D$ −4.00° (c = 0.71 in chloroform).

7683. Platelet-Derived Growth Factor.

PDGF. Cationic glycoprotein mitogen for fibroblasts, smooth muscle cells and glial cells. Mol wt ~30,000 daltons. Principal factor in serum required for the growth and proliferation of mesenchymal derived cells in tissue culture. Originally isolated from blood platelets, PDGF-like peptides have subsequently been found to be synthesized and secreted by a number of normal and neoplastic cell types. Identification of platelet-derived mitogen for smooth muscle cells: R. Ross *et al., Proc. Nat. Acad. Sci. USA* **71**, 1207 (1974); for fibroblasts: N. Kohler, A. Lipton, *Exp. Cell Res.* **87**, 297 (1974); for glial cells: B. Westermark, A. Wasteson, *Adv. Metabol. Dis.* **8**, 85 (1975). Isoln of *human serum growth factor*: H. N. Antoniades *et al., Proc. Nat. Acad. Sci. USA* **72**, 2635 (1975); and immunological identity with PDGF: H. N. Antoniades, C. D. Scher, *ibid.* **74**, 1973 (1977). Purification: H. N. Antoniades *et al., ibid.* **76**, 1809 (1979); C.-H. Heldin *et al., ibid.* 3722. PDGF is stored in platelet α-granules and released at sites of tissue injury upon platelet activation. Intracellular localization and potential role in wound repair: D. R. Kaplan *et al., Blood* **53**, 1043 (1979). Chemotactic effect on smooth muscle cells: G. R. Grotendorst *et al., Proc. Nat. Acad. Sci. USA* **78**, 3669 (1981); on monocytes and neutrophils: T. F. Deuel *et al., J. Clin. Invest.* **69**, 1046 (1982). PDGF is a disulfide-linked dimer of two polypeptide chains, designated A and B, which may be assembled as a heterodimer and/or as AA or BB homodimers. Proposed structure as AB heterodimer: C.-H. Heldin *et al., Biochem. J.* **193**, 907 (1981); A. Johnsson *et al., Biochem. Biophys. Res. Commun.* **104**, 66 (1982). PDGF-B is structurally homologous to *p28sis*, the transforming protein of the simian sarcoma virus: M. D. Waterfield *et al., Nature* **304**, 35 (1983); R. F. Doolittle *et al., Science* **221**, 275 (1983). *Osteosarcoma-derived growth factor* or *ODGF*, a growth factor isolated from human osteosarcoma cells, is structurally similar to a PDGF-A homodimer: C.-H. Heldin *et al., Nature* **319**, 511 (1986). Comparison of biological properties and transforming potential of A and B chains: M. P. Beckmann *et al., Science* **241**, 1346 (1988). Identification of PDGF receptor: C.-H. Heldin *et al., Proc. Nat. Acad. Sci. USA* **78**, 3664 (1981). Correlation of receptor binding with mitogenic activity: D. F. Bowen-Pope, R. Ross, *J. Biol. Chem.* **257**, 5161 (1982). Molecular mechanisms of action: S. R. Coughlin *et al., Prog. Clin. Biol. Res.* **266**, 39 (1988); L. T. Williams, *Science*, **243**, 1564 (1989). Review of purification, properties and biological activities: T. F. Deuel, J. S. Huang, *Prog. Hematol.* **13**, 201-221 (1983). Review of relationship to oncogenes and malignant transformation: H. N. Antoniades, *Biochem. Pharmacol.* **33**, 2823-2828 (1984); C.-H. Heldin *et al., J. Cell. Sci.* **Suppl. 3**, 65-76 (1985); of role in atherosclerosis: R. Ross, *N. Engl. J. Med.* **314**, 488-500 (1986); in arteriosclerosis: H. R. Baumgartner, M. Hosang, *Experientia* **44**, 109-112 (1988). Reviews: C.-H. Heldin *et al., Mol. Cell. Endocrinol.* **39**, 169-187 (1985); R. Ross *et al., Cell* **46**, 155-169 (1986); T. F. Deuel *et al., J. Cell Physiol.* **Suppl. 5**, 95-99 (1987); R. Ross, *Ann. Rev. Med.* **38**, 71-79 (1987).

7684. Platinic Chloride.

Hydrogen hexachloroplatinate(IV); acid platinic chloride; hexachloroplatinic(IV) acid; chloroplatinic acid. Cl_6H_2Pt; mol wt 409.81. Cl 51.91%, H 0.49%, Pt 47.60%. H_2PtCl_6. Exists as hexahydrate.

Hexahydrate, brownish-yellow, very deliquesc, cryst mass. d 2.431; mp 60°. Easily sol in water, alcohol. *Keep tightly closed, protected from light.*

USE: In platinum plating, photography, platinum mirrors, platinum luster on glass and porcelain, platinized carbon for acetic acid manuf; platinizing pumice stone or asbestos, as catalyst in manuf of SO_3; indelible ink; relief etching of zinc for artistic and commercial purposes; fixing microscopic preps, etc.

Caution: May cause asthma or dermatitis, E. Browning,

Toxicity of Industrial Metals (Appleton-Century-Crofts, New York, 2nd ed., 1969) pp 270-275.

7685. Platinic Iodide. I₄Pt; mol wt 702.70. I 72.24%, Pt 27.76%. PtI₄. Prepn: Argue, Banewicz, *J. Inorg. Nucl. Chem.* **25**, 923 (1963).
Brownish-black to black powder. Sol in water.

7686. Platinic Oxide. Platinum dioxide; Adams' catalyst. O₂Pt; mol wt 227.08. O 14.09%, Pt 85.91%. PtO₂. Prepd by the interaction of platinic chloride and excess NaOH: Wöhler, *Z. Anorg. Chem.* **40**, 434 (1904); Bellucci *ibid.* **44**, 171 (1905); by the reduction of chloroplatinic acid with formaldehyde: Willstätter, Waldschmidt-Leitz, *Ber.* **54**, 113 (1921); Feulgen, *ibid.* **360**; by fusing chloroplatinic acid with sodium nitrate: Adams *et al., Org. Syn.* **8**, 92 (1928). Determination of orthorhombic (CaCl₂) crystal structure of β-PtO₂: Siegel *et al., J. Inorg. Nucl. Chem.* **31**, 3803 (1969).
Several hydrates have been reported; the compd generally used for catalysis is the monohydrate. Black powder. When freshly pptd it is sol in concd acids, also in dil H₃PO₄, esp when warmed. Easily sol in dilute solns of potassium hydroxide.
USE: As catalyst in hydrogenations. The actual catalyst is platinum black which is formed *in situ* by reduction of the PtO₂ by the hydrogen used for the hydrogenation. Especially useful for reduction at room temp and hydrogen pressures up to 4 atmospheres. Suitable for the reduction of double and triple bonds, aromatic nuclei, carbonyl groups, nitro groups, and nitriles.

7687. Platinous Chloride. Cl₂Pt; mol wt 265.99. Cl 26.66%, Pt 73.34%. PtCl₂. Prepn from hexachloroplatinic acid, H₂PtCl₆.6H₂O: Cohen, *Inorg. Syn.* **6**, 209 (1960).
Grayish-green to brown powder. d 5.87. Insol in water, alcohol, ether. Sol in hydrochloric acid. Combines with PCl₃ to form a compd sol in benzene or chloroform.

7688. Platinous Iodide. Platinum diiodide. I₂Pt; mol wt 448.89. I 56.54%, Pt 43.46%. PtI₂. Prepn: Argue, Banewicz, *J. Inorg. Nucl. Chem.* **25**, 923 (1963).
Heavy, black powder, dec 325°. Insol in water or alkali iodides.

7689. Platinum. Pt; at. wt 195.08; at. no. 78; valences 2,4; seldom 1, 5, 6. Six naturally occurring isotopes: 190 (0.01%); 192 (0.8%); 194 (32.9%); 195 (33.8%); 196 (25.2%); 198 (7.2%); 190 is radioactive: T₁/₂ 6.9 × 10¹¹ years. Artificial, radioactive isotopes: 173-189; 191; 193; 197; 199-201. Abundance in earth's crust about 0.01 ppm. Believed to be mentioned by Pliny under the name *"alutiae"*. Has been known and used in South America as "platina del Pinto". Reported by Ulloa in 1735; brought to Europe by Wood, and described by Watson in 1741. Occurs native alloyed with one or more members of its group (iridium, osmium, palladium, rhodium, and ruthenium) in gravels and sands. Prepn: Wichers *et al., Trans. Amer. Inst. Min. Met. Eng.* **76**, 602 (1928). Reviews of prepn, properties and chemistry of platinum and other platinum metals: Gilchrist, *Chem. Rev.* **32**, 277-372 (1943); Beamish *et al.,* in *Rare Metals Handbook,* C. A. Hampel, Ed. (Reinhold, New York, 1956) pp 291-328; Livingstone in *Comprehensive Inorganic Chemistry,* Vol. 3, J. C. Bailar, Jr. *et al.,* Eds. (Pergamon Press, Oxford, 1973) pp 1163-1189, 1330-1370; F. R. Hartley, *The Chemistry of Platinum and Palladium with Particular Reference to Complexes of the Elements* (Halsted Press, New York, 1973).
Silver-gray, lustrous, malleable and ductile metal; face-centered cubic structure. Also prepd in the form of a black powder (platinum black) and as spongy masses (platinum sponge). mp 1773.5 ± 1°; Roeser *et al., Nat. Bur. Stand. J. Res.* **6**, 1119 (1931); bp about 3827°. d 21.447 (calcd). Brinell hardness: 55. Sp heat 0.0314 cal/g at 0°. Electrical resistivity (20°) 10.6 μohm-cm. Does not tarnish on exposure to air. Absorbs hydrogen at a red heat and retains it tenaciously at ord. temp, gives off the gas at a red heat *in vacuo.* Occludes carbon monoxide, carbon dioxide, nitrogen. Volatilizes considerably when heated in air above 1000°. The heated metal absorbs oxygen; gives it off on cooling. Not affected by water or by single mineral acids. Reacts with boiling aqua regia with formation of chloroplatinic

acid, also with molten alkali cyanides. Attacked by halogens, by fusion with caustic alkalies, alkali nitrates, alkali peroxides; by arsenates and phosphates in presence of reducing agents.
Caution: Inhalation of the dust of sol platinum salts is irritating and may cause rhinorrhea, sneezing, tightness of chest, shortness of breath, cyanosis, wheezing, cough. May cause dermatitis: E. Browning, *Toxicity of Industrial Metals* (Appleton-Century-Crofts, New York, 2nd ed., 1969) pp 270-275.
USE: Manuf apparatus for laboratory and industrial use, thermocouples, platinum resistance thermometers, acid-proof containers, electrodes, etc. In dentistry; jewelry; electroplating. As oxidation catalyst in manuf acetic acid, nitric acid from ammonia, manuf sulfuric acid; control of automotive emissions.

7690. Platonin. 2,2'-[3-[(3-Heptyl-4-methyl-2(3H)-thiazolylidene)ethylidene]-1-propene-1,3-diyl]bis(3-heptyl-4-methylthiazolium) diiodide; 4,4',4''-trimethyl-3,3',3''-triheptyl-8-(2''-thiazolyl)-2,2'-pentamethinethiazolocyanine 3,3''-diiodide; 3,3',3''-triheptyl-4,4',4''-trimethyl-7-(2''-methylthiazolyl-2,2'-trimethine)thiazolcyanine 3,3''-diiodide; platonin J; NK-19; Kankohso 101; Photosensitizer 101. C₃₈H₆₁I₂N₃S₃; mol wt 909.93. C 50.16%, H 6.76%, I 27.89%, N 4.62%, S 10.57%. Photosensitizing cyanine dye with antimicrobial and immunomodulating activities. General prepn and use in rheumatoid arthritis: I. Yamamoto, **Belg. pat. 894,635;** *idem,* U.S. pat. **4,464,383** (1983, 1984). Antifungal properties: K. Ito, P. K. Kuroda, *Bull. Pharm. Res. Inst. Osaka Med. Coll.* **3**, 20 (1952); C. A. **47**, 11335f (1953). Effect on experimentally induced leukopenia: H. Iwata *et al., Folia Pharmacol. Japon.* **50**, 169 (1954), *C.A.* **49**, 10518b (1955). Immunomodulating effect in animals: H. Ichihashi, T. Kondo, *Gann* **58**, 529 (1967); Y. Oyanagui, *Arch. Int. Pharmacodyn. Ther.* **266**, 162 (1983). Enzyme immunoassay in plasma: I. Yamamoto, K. Morishita, *J. Immunoassay* **7**, 17 (1986). Toxicity study: T. Kimoto, K. Nishitani, *Kanko Shikiso* **85**, 43 (1977), *C.A.* **86**, 84374m (1977).

Brilliant green crystalline powder, mp 204°. Sol in water, alcohol. LD₅₀ in male, female mice (mg/kg): 46.9, 50.5 i.p.; in male, female rats (mg/kg): 1539, 1571 orally (Kimoto, Nishitani).
THERAP CAT: Immunomodulator.

7691. Platyphylline. (1α)-1,2-Dihydro-12-hydroxysenecionan-11,16-dione; platifillin. C₁₈H₂₇NO₅; mol wt 337.42. C 64.07%, H 8.07%, N 4.15%, O 23.71%. Toxic pyrrolizidine alkaloid from *Senecio platyphyllus* DC. and other *Senecio* spp., Compositae: Orékhov, Tiedebel, *Ber.* **68**, 650 (1935); Orékhov *et al., ibid.* **1886**; Konovalova, Orékhov, *ibid.* **69**, 1908 (1936); *Bull. Soc. Chim. France* [5] **4**, 2037 (1937); Allen, *Am. J. Pharm.* **117**, 110 (1945). Structure: Dry *et al., J. Chem. Soc.* **1955**, 63. Comprehensive reviews of platyphylline and other pyrrolizidine alkaloids: L. Bull *et* *al., The Pyrrolizidine Alkaloids* (North-Holland, Amsterdam, 1968) 293 pp; F. L. Warren in *The Alkaloids,* vol. **12,** R. H. F. Manske, Ed. (Academic Press, New York, 1970) pp 245-331.

Crystals, mp 129°. $[\alpha]_D^{20}$ $-56.4°$ (c = 0.7 in chloroform). Practically insol in water. Sol in alcohol, chloroform, ether, dil acids.

Bitartrate, $C_{18}H_{27}NO_5.C_4H_6O_6$, crystals, mp 199°. $[\alpha]_D^{25}$ $-40°$ (c = 2). Sol in 10 parts of water, in 5 parts of hot water. Sparingly sol in alc, in 42 parts of boiling alc. Nearly insol in chloroform, ether.

7692. Plaunotol. *(Z,E,E)-2-(4,8-Dimethyl-3,7-nonadienyl)-6-methyl-2,6-octadiene-1,8-diol; (E,Z,E)-7-hydroxymethyl-3,11,15-trimethyl-2,6,10,14-hexadecatetraen-1-ol;* CS-684; Kelnac. $C_{20}H_{34}O_2$; mol wt 306.49. C 78.38%, H 11.18%, O 10.44%. Acyclic diterpene alcohol isolated from a Thai medicinal plant identified as *Croton sublyratus* Kurz, *Euphorbiaceae.* Isoln, synthesis and anti-ulcer activity: H. Mishima *et al.,* **Japan. Kokai 77 62,213;** *eidem,* **U.S. pat. 4,059,641** (both 1977 to Sankyo); A. Ogiso *et al., Chem. Pharm. Bull.* **26,** 3117 (1978). Pharmacology: S. Kobayashi *et al., Oyo Yakuri* **24,** 599 (1982).

Light yellow oil. Aromatic odor, bitter taste. Sol in benzene, acetone, alcohols, ethers. Insol in water. LD_{50} in male, female mice, rats (μl/kg): 8800, 8100, 10900, 11200 orally (Kobayashi).

THERAP CAT: Antiulcerative.

7693. Plegatil. *N,N-Diethyl-N-methyl-2-[2-(trimethylammonio)ethoxy]-1-propanaminium diiodide; N,N-diethyl-N,N',N',N'-tetramethyl-N,N'-(2-methyl-3-oxapentamethylene)bis[ammonium iodide]; 2-(2-dimethylaminoethoxy)-N,-N-diethylpropylamine dimethiodide; β-dimethylaminoethyl β'-diethylamino-α'-methylethyl ether dimethiodide; 2-dimethylaminoethyl 2'-diethylaminoisopropyl ether bismethiodide;* HL-8731. $C_{13}H_{32}I_2N_2O$; mol wt 486.22. C 32.11%, H 6.63%, I 52.20%, N 5.76%, O 3.29%. Ganglionic blocking agent. Prepd but not claimed: G. Pakleppa, D. Lenke, **East Ger. pat. 11,363** (1956), *C.A.* **53,** 2094i (1959). Pharmacology: D. Lenke, *Arzneimittel-Forsch.* **7,** 616 (1957); *ibid.* **11,** 571 (1961).

Crystals, mp 178-180°. LD_{50} in mice (mg/kg): 39.1 i.v., 1125 orally (Lenke, 1957).

7694. Pleuromutilin. *[3aS-(3aα,4β,5α,6α,8β,9α,9aβ,-10S*)]-Hydroxyacetic acid 6-ethenyldecahydro-5-hydroxy-4,6,9,10-tetramethyl-1-oxo-3a,9-propano-3aH-cyclopentacyclo-octen-8-yl ester; glycolic acid 8-ester with octahydro-5,8-di-hydroxy-4,6,9,10-tetramethyl-6-vinyl-3a,9-propano-3aH-cyclo-pentacycloocten-1(4H)-one;* drosophilin B; BC-757. $C_{22}H_{34}O_5$; mol wt 378.51. C 69.81%, H 9.05%, O 21.13%. Antibiotic substance produced by the basidiomycetes *Pleurotus mutilus* (Fr.) Sacc. and *P. passeckerianus* Pilat.: Kavanagh *et al., Proc. Nat. Acad. Sci. USA* **37,** 570 (1951); **38,** 555 (1952). Isoln procedure: Anchel, *J. Biol. Chem.* **199,** 133 (1952). Structure: D. Arigoni, *Gazz. Chim. Ital.* **92,** 884 (1962). A. J. Birch *et al., Chem. & Ind. (London)* **1963,** 374; *Tetrahedron* **Suppl. 8,** 359 (1966). Stereochemistry: D. Arigoni, *Pure Appl. Chem.* **17,** 331 (1968). Fermentation and biosynthetic study: F. Knauseder, E. Brandl, *J. Antibiot.* **29,** 125 (1976). Mechanism of action: G. Hoegenauer, *Antibiotics* vol. **5,** pt. 1, F. E. Hahn, Ed. (Springer-Verlag, New York, 1979) p 344. Synthetic studies: E. G. Gibbons, *J. Org. Chem.* **45,** 1540 (1980); M. Kahn, *Tetrahedron Letters* **21,** 4547 (1980). Total synthesis: E. G. Gibbons, *J. Am. Chem. Soc.* **104,** 1767 (1982).

Crystals from ethyl acetate + Skellysolve B, mp 170-171°. $[\alpha]_D^{24}$ +20° (c = 3 in abs ethanol). uv max (5 mg/ml in 95% ethanol): 290 nm. Soluble in methanol, ethanol, ethyl acetate, chloroform. LD_{50} in mice: > 60 mg/kg (Kavanagh, 1952).

Diacetate, $C_{26}H_{38}O_7$, crystals from abs ethanol, mp 145.5°.

7695. Pleurotin(e). *(2aα,4aβ,5β,6β,8aα,12bβ,12cβ,-12dβ)-(−)-2a,3,4,4a,5,6,7,8a,12b,12c-Decahydro-6-methyl-2H-5,12d-ethanofuro[4',3',2':4,10]anthra[9,1-bc]oxepin-2,9,-12-trione.* $C_{21}H_{22}O_5$; mol wt 354.40. C 71.17%, H 6.26%, O 22.57%. Antibiotic substance produced by the fungus *Pleurotus griseus:* Robbins *et al., Bull. Torrey Bot. Club* **72,** 165 (1945). Isoln: Robbins *et al., Proc. Nat. Acad. Sci. USA* **33,** 171 (1947). Partial structure: Huls, *Publ. Univ. Congo Elisabethville* **4,** 109 (1962), *C.A.* 3894f (1965). Structure and stereochemistry: J. Grandjean, R. Huls, *Tetrahedron Letters* **1974,** 1893. Synthesis of racemate: D. J. Hart, H.-C. Huang, *J. Am. Chem. Soc.* **110,** 1634 (1988).

Amber crystals from ether + chloroform, mp 200-215°. $[\alpha]_D^{23}$ $-20°$ (chloroform). Neutral reaction. Sparingly sol in water which inactivates it. More stable at acid pH, but still loses ∼50% of its activity when boiled for 10 min at pH 3. Aq solns also are inactivated on exposure to light. Moderately sol in alcohol, ether. Freely sol in chloroform.

7696. Plicamycin. *Mithramycin;* aureolic acid; mitramycin; aurelic acid; antibiotic LA-7017; Mithracin. $C_{52}H_{76}O_{24}$; mol wt 1085.16. C 57.56%, H 7.06%, O 35.39%. Oligosaccharide antibiotic produced by *Streptomyces argillaceus* n. sp. and *S. tanashiensis.* Isoln: Grundy *et al., Antibiot. & Chemother.* **3,** 1215 (1953); Philip, Schenck, *ibid.* 1218; Rao *et al., ibid.* **12,** 182 (1962). Identity of aureolic acid with mithramycin: Berlin *et al., Nature* **218,** 193 (1968). Prepn: Gado *et al.,* **Hung. pat. 155,679** (1969 to Gyogyszerkutato Intezet), *C.A.* **70,** 118093f (1969). Similar to the chromomycins and olivomycins, q.q.v. Structural elucidation of the aglycone, **chromomycinone,** and the carbohydrate moieties: Bakhaeva *et al., Tetrahedron Letters* **1968,** 3595; Berlin *et al., Khim. Prir. Soedin.* **1972,** 537, 542, *C.A.* **78,** 16436t, 16443t (1973). Revised structure: J. Thiem, B. Meyer, *Tetrahedron* **37,** 551 (1981). NMR and fluorometric characterization: N. R. Krishna *et al., J. Antibiot.* **43,** 1543 (1990). Metabolic studies: Kennedy *et al., Cancer Res.* **27,** 1534 (1967). Mode of action studies: Kushch *et al., Antibiotiki* **17,** 504 (1972). Toxicity study: M. Slavik, S. K. Carter, *Adv. Pharmacol. Chemother.* **12,** 1 (1975). *Review:* J. D. Skarbek, M. K. Speedie in *Antitumor Compounds of Natural Origin* vol. **1,** A. Aszalos, Ed. (CRC Press, Boca Raton, 1981) pp 191-235.

Yellow solid, mp 180-183°. $[\alpha]_D^{20}$ −51° (c = 0.4 in ethanol). Sol in lower alcohols, acetone, ethyl acetate, water. Moderately sol in chloroform. Slightly sol in ether, benzene. LD$_{50}$ in mice, rats (mg/kg): 2.14, 1.74 i.v. (Slavik, Carter).

THERAP CAT: Antineoplastic.

7697. Plumbagin. *5-Hydroxy-2-methyl-1,4-naphthalenedione; 5-hydroxy-2-methyl-1,4-naphthoquinone.* $C_{11}H_8O_3$; mol wt 188.18. C 70.21%, H 4.28%, O 25.51%. Found in the roots of *Plumbago europaea* L., *P. zeylanica*, *P. rosea* L., *Plumbaginaceae*. Isoln: Dulong d'Astafort, *J. Pharm. Chim.* **14**, 441 (1828); Wefers-Bettink, *Rec. Trav. Chim.* **8**, 319 (1889); Ray, Dutt, *J. Indian Chem. Soc.* **5**, 419 (1928). Structure and synthesis: Fieser, Dunn, *J. Am. Chem. Soc.* **58**, 572 (1936). Biosynthesis: Durand, Zenk, *Tetrahedron Letters* **1971**, 3009. Chemotherapeutic properties: Vichkanova *et al.*, *C.A.* **78**, 66906s (1973). Efficient syntheses: A. Ichihara *et al.*, *Agr. Biol. Chem.* **44**, 211 (1980); G. Wurm *et al.*, *Arch. Pharm.* **314**, 861, 1055 (1981); H. Möhrle, H. Foltmann, *ibid.* **321**, 259 (1988). Toxicity data: M. Debray *et al.*, *Plant Med. Phytother.* **7**, 77 (1973), *C.A.* **79**, 14255e (1973).

Yellow needles from dil alcohol, mp 78-79°. Sublimes. Volatile with steam. Slightly sol in hot water; sol in alc, acetone, chloroform, benzene, acetic acid. LD$_{50}$ i.p. in mice: ∼0.015 g/kg (Debray).

7698. Plumericin. *[3aS-(3E,3aα,4aβ,7aβ,9aR*,9bβ)]-3-Ethylidene-3,3a,7a,9b-tetrahydro-2-oxo-2H,4aH-1,4,5-trioxadicyclopent[a,hi]indene-7-carboxylic acid methyl ester.* $C_{15}H_{14}O_6$; mol wt 290.27. C 62.07%, H 4.86%, O 33.07%. A sesquiterpenoid, exhibiting *in vitro* activity against fungi, some bacteria including *Mycobacterium tuberculosis* 607. Isoln from roots of *Plumeria multiflora* Muell.-Arg., *Apocynaceae*, also from roots of *P. rubra* var. *alba*: J. E. Little, D. B. Johnstone, *Arch. Biochem.* **30**, 445 (1951); from *Allamanda cathartica* Linn., *Apocynaceae*: B. R. Pai *et al.*, *Indian J. Chem.* **8**, 851 (1970). Structure, physical properties and IR, ¹HNMR spectra: G. Albers-Schönberg, H. Schmid, *Helv. Chim. Acta* **44**, 1447 (1961). Synthetic approach: J. K. Whitesell *et al.*, *Syn. Commun.* **7**, 355 (1977). Biomimetic total synthesis of (±)-form: B. M. Trost *et al.*, *J. Am. Chem. Soc.* **105**, 6755 (1983); *eidem, ibid.* **108**, 4974 (1986). Alternate synthesis of (±)-form: *eidem, ibid.* 4965; K. E. B. Parkes, G. Pattenden, *J. Chem. Soc. Perkin Trans. I* **1988**, 1119.

Narrow, rectangular plates from alc, toluene or methylene chloride + ether, dec 211.5-212.5°. Sublimes in high vacuum at 160-180°. $[\alpha]_D^{25}$ +197.5° ±2 (c = 0.982 in chloroform). uv max (ethanol): 214-215 nm (log ε 4.24). Sol in chloroform; slightly sol in methanol, alc, ether, acetone, benzene. Practically insol in petr ether, water.

7699. Plumieride. *[1S-[1α,4aα,7α(R*),7aα]]-1-(β-D-Glucopyranosyloxy)-4a,7a-dihydro-4'-(1-hydroxyethyl)-5'-oxospiro[cyclopenta[c]pyran-7(1H),2'(5'H)-furan]-4-carboxylic acid methyl ester;* agoniadin. $C_{21}H_{26}O_{12}$; mol wt 470.43. C 53.62%, H 5.57%, O 40.81%. Found in bark of *Plumeria lancifolia* Muell.-Arg., *Apocynaceae*, also in *P. acutifolia* and *P. rubra* var. *alba*: Peckolt, *Arch. Pharm.* [2] **142**, 40 (1870); *Merck's Jahresber.* **1895**, 11; Boorsma, *Mededeel. Lands Plant.* **13**, 27 (1894); **31**, 132 (1899); Franchimont, *Rec. Trav. Chim.* **18**, 334, 477 (1899); **19**, 350 (1900); Halpern, Schmid, *Helv. Chim. Acta* **41**, 1109 (1958). Structure: Albers-Schönberg, Schmid, *ibid.* **44**, 1447 (1961). Biosynthesis: D. A. Yeowell, H. Schmid, *Experientia* **20**, 250 (1964); K. Inoue *et al.*, *Chem. Pharm. Bull.* **27**, 3115 (1979).

Monohydrate, bitter crystals, mp 156-158°. When anhydrous, mp 224-225°. $[\alpha]_D^{20}$ −114°; $[\alpha]_D^{20}$ − 80° (methanol). Broad uv absorption: 217-238 nm. Sol in water, alcohol, ethyl acetate.

7700. Plutonium. Pu; at. no. 94; valences 3, 4, 5, 6, 7. No stable nuclides, known isotopes (mass numbers): 232-246. Longest-lived known isotopes: ²⁴²Pu (T$_{1/2}$ 3.76 × 10⁵ years, α-emitter, rel. at. mass 242.0587), 244 (T$_{1/2}$ 8.26 × 10⁷ years, α-emitter, rel. at. mass 244.0642). Commercially useful isotopes: ²³⁸Pu (T$_{1/2}$ 87.74 years; α-emitter); ²³⁹Pu (T$_{1/2}$ 2.41 × 10⁴ years; α-emitter). Occurrence in the earth's crust: 10⁻²²%. Discovery of isotope ²³⁸Pu: G. T. Seaborg *et al.*, *Phys. Rev.* **69**, 366, 367 (1946); of isotope ²³⁹Pu: J. W. Kennedy *et al.*, *ibid.* **70**, 555 (1946). Isoln of ²³⁹Pu from pitchblende: G. T. Seaborg, M. L. Perlman, *J. Am. Chem. Soc.* **70**, 1571 (1948). Prepn of metal: B. B. Cunningham, L. B. Werner, *ibid.* **71**, 1521 (1949). Chemical properties: Seaborg, Wahl, *ibid.* 1128; Harvey *et al.*, *J. Chem. Soc.* **1947**, 1010. Reviews: J. M. Cleveland, *The Chemistry of Plutonium* (Gordon & Breach, New York, 1970) 653 pp; C. Keller, *The Chemistry of the Transuranium Elements* (Verlag Chemie, Weinheim, English Ed., 1971) pp 333-484; *Comprehensive Inorganic Chemistry* vol. 5, J. C. Bailar, Jr. *et al.*, Eds. (Pergamon Press, Oxford, 1973) *passim;* Handb. Exp. Pharmakol. **36**, 307-688 (1973); F. Weigel in Kirk-Othmer *Encyclopedia of Chemical Technology* vol. 18 (Wiley-Interscience, New York, 3rd ed., 1982) pp 278-301; *Plutonium Chemistry,* W. T. Carnall, G. R. Choppin, Eds. (Am. Chem. Soc., Washington, D.C., 1983) 484 pp; F. Weigel *et al.* in *The Chemistry of the Actinide Elements* vol. 1, J. J. Katz *et al.*,

Eds. (Chapman and Hall, New York, 1986) pp 499-886. Review of toxicology: W. J. Bair, R. C. Thompson, *Science* **183**, 715-722 (1974); and health effects: *Toxicological Profile for Plutonium* (PB91-180406, 1990) 206 pp.

Silvery-white metal. Highly reactive. Oxidizes readily in dry air and oxygen, rate increases in presence of moisture. Six allotropic forms: simple monoclinic α-form, d^{21} 19.86, transforms to β-form at 122 ±4°; body-centered monoclinic β-form, d^{190} 17.70, transforms to γ-form at 207 ±5°; face-centered orthorhombic γ-form, d^{235} 17.14, transforms to δ-form at 315 ±3°; face-centered cubic δ-form, d^{320} 15.92, transforms to δ'-form at 457 ±2°; body-centered tetragonal δ'-form, d^{405} 16.00, transforms to ϵ-form at 479 ±4°; body-centered cubic ϵ-form, d^{490} 16.51, transforms to liquid at mp 640 ±2°.

Trivalent plutonium, weak reducing agent. Stable in soln in absence of air. Slowly oxidized to the tetravalent plutonium by atmospheric oxygen, by permanganate in acid soln in the cold; oxidized to the hexavalent form by permanganate at 60°. Trivalent salts are blue; form complexes very readily; form a series of double sulfates. Crystal structure of the complex and double salts: Zachariasen, *J. Am. Chem. Soc.* **70**, 2147 (1948).

Tetravalent plutonium, reduced in aq soln to the trivalent form by sulfur dioxide, hydroxylamine hydrochloride, hydrazine hydrochloride, the uranous ion, the iodide ion; by shaking with mercury in chloride soln; electrolytically at a platinum cathode. Tetravalent salts are pink or greenish; form complexes very readily.

Hexavalent plutonium, obtained by the action of strong oxidizing agents (ceric salts, dichromates, permanganates, or hot bromate soln contg nitric acid) on the tri- or tetravalent form. Reduced to tri- or tetravalent plutonium by sulfur dioxide or ferrocyanide.

Caution: Radiation hazard; handling requires special equipment and shielding facilities. Animal studies have indicated that inhaled ^{239}Pu as particulate matter may remain in the lungs, or move to the bones, liver or other body organs, and over a period of time may give rise to neoplasms due to its α-emitting radioactive decay. *See* Katz *et al., loc. cit.,* vol. 2, p. 1128, 1182-1188. Max permissible concn of ^{238}Pu in air: 7×10^{-13} μCurie/cc; of ^{239}Pu in air: 6×10^{-13} μCurie/cc: *National Bureau of Standards Handbook* **69**, 87 (1959). *See also Handb. Exp. Pharmakol., loc. cit.*

USE: ^{238}Pu as heat source; as radioisotope thermoelectric generator; in radionuclide batteries for pacemakers. ^{239}Pu as fuel in atomic weapons and nuclear power reactors.

7701. Podocarpic Acid. *[1S-(1α,4aα,10aβ)]-1,2,3,4,4a,-9,10,10a-Octahydro-6-hydroxy-1,4a-dimethyl-1-phenanthrenecarboxylic acid; 12-hydroxypodocarpa-8,11,13-trien-16-oic acid.* $C_{17}H_{22}O_3$; mol wt 274.36. C 74.42%, H 8.08%, O 17.49%. Chief acidic constituent of the resin of the Javanese *Podocarpus cupressina* (L'Hérit.) Pers., *Coniferae,* esp. in *var. imbricata.* Also in New Zealand kahikatea resin (from *Podocarpus dacrydioides)* and in rimu resin *(Dacrydium cupressinum).* A related component, nimbiol, *q.v.,* occurs in the nim tree, *Azadirachta indica* Juss. (*Melia azadirachta* L.), *Meliaceae.* Isoln from *Podocarpus cupressina:* A. C. Oudemans, *Ber.* **6**, 1122 (1873); from *Dacrydium cupressinum:* I. R. Sherwood, W. F. Short, *J. Chem. Soc.* **1938**, 1008. Structure: Campbell, Todd, *J. Am. Chem. Soc.* **64**, 928 (1942). Conversion to nimbiol as proof of structure: Bible, *Tetrahedron* **11**, 22 (1960). Synthesis from desoxypodocarpic acid: Wenkert, Jackson, *J. Am. Chem. Soc.* **80**, 217 (1958). Total synthesis of *dl*-form: Meyer, Maheshwari, *Tetrahedron Letters* **1964**, 2175; P. R. Kanjilal *et al., Syn. Commun.* **11**, 795 (1981). Synthesis: Pelletier *et al., Tetrahedron Letters* **1971**, 4179.

Platelets from dil alcohol, mp 193.5°. $[\alpha]_{546}^{20}$ +165° (c = 4 in abs ethanol). Sol in methanol, ethanol, ether, acetic acid. Practically insol in water, chloroform, benzene, carbon disulfide.

Methyl ester, $C_{18}H_{24}O_3$, crystals from alcohol, mp 208°.

7702. Pododacric Acid. *[1S-(1α,4aα,10aβ)]-1,2,3,4,4a,-9,10,10a-Octahydro-6-hydroxy-7-[2-hydroxy-1-(hydroxymethyl)ethyl]-1,4a-dimethyl-1-phenanthrenecarboxylic acid; 12-hydroxy-13-[2-hydroxy-1-(hydroxymethyl)ethyl]podocarpa-8,-11,13-trien-16-oic acid.* $C_{20}H_{28}O_5$; mol wt 348.44. C 68.94%, H 8.10%, O 22.96%. From heartwood of *Podocarpus dacrydioides; P. totara:* Briggs *et al., Tetrahedron* **7**, 270 (1959); Cambie, Mander, *ibid.* **18**, 465 (1962). Structure: Cambie, Mathai, *Chem. Commun.* **1971**, 154.

Colorless needles from 30% methanol, mp 213-214°. $[\alpha]_D^{25}$ +118° (c = 0.9 in 5:1 chloroform-methanol). uv max: 225, 284 nm (ϵ 5500, 3100). pK 8.44 (in methyl cellosolve system).

7703. Podophyllic Acids. $C_{22}H_{24}O_9$; mol wt 432.43. C 61.11%, H 5.59%, O 33.30%. Since the prepn of a "podophyllic acid", mp 145-150° (crude), 163-165° (pure), by Borsche, Niemann, *Ann.* **494**, 126 (1932), studies have been made of two isomers: the 2,3-*trans*-hydroxy acid and the 2,3-*cis*-hydroxy acid. Prepn of the *trans*-acid: Kuhn, Wartburg, *Experientia* **19**, 391 (1963); **Neth. pat.** **6,405,480** (1964 to Sandoz), *C.A.* **62**, 9083c (1965); and *cis*-acid: Renz *et al., Ann.* **681**, 207 (1965). Prepn of a DL-stereoisomer: Gensler *et al., J. Am. Chem. Soc.* **76**, 315 (1954). Configuration and nomenclature: Rutschmann, Renz, *Helv. Chim. Acta* **42**, 890 (1959).

Podophyllinic Acid

2,3-*trans*-Hydroxy acid, *podophyllinic acid.* Crystals from acetone + ether, mp 164-168°. $[\alpha]_D^{20}$ −199.9° (c = 0.463 in ethanol), −292.6° (c = 0.671 in pyridine).

2,3-*cis*-Hydroxy acid, *picropodophyllic acid.* Needles from methanol + ether, double mp 150-155° and 200-232°. $[\alpha]_D^{??}$ −100.4° (c = 0.615 in ethanol), −185° (c = 0.746 in pyridine).

2,3-*trans*-Hydroxy acid hydrazide, *podophyllinic acid hydrazide.* Prisms from methanol, mp 198-199°. $[\alpha]_D$ −202° (c = 0.4 in ethanol). Prepn and properties: Rutschmann, U.S. pat. **2,977,359** (1961 to Sandoz).

trans-2-Ethylhydrazide, *podophyllinic acid 2-ethylhydrazide, SP-I, Proresid.* Prepn: Rutschmann, U.S. pat. **3,-054,802** (1962 to Sandoz). *Note:* Patent also claims prepn of *picropodophyllinic acid 2-ethylhydrazide* or *1-ethyl-2-picropodophyllinic acid hydrazide.* Amorphous powder, pptd from chloroform + petr ether. $[\alpha]_D$ −154° (c = 0.5 in chloroform).

2,3-*cis*-Hydroxy acid hydrazide, *picropodophyllinic acid hydrazide, picropodophyllic acid hydrazide.* Rhombic scales, mp 152-154°. $[\alpha]_D$ −106° (c = 0.4 in ethanol).

THERAP CAT: Podophyllinic acid 2-ethylhydrazide as antineoplastic.

Consult the Name Index before using this section.

7704. Podophyllotoxin. *[5R-(5α,5aβ,8aα,9α)]-5,8,8a,9-Tetrahydro-9-hydroxy-5-(3,4,5-trimethoxyphenyl)furo[3',4':6,-7]naphtho[2,3-d]-1,3-dioxol-6(5aH)-one;* 1-hydroxy-2-hydroxymethyl-6,7-methylenedioxy-4-(3',4',5'-trimethoxyphenyl)-1,2,3,4-tetrahydronaphthalene-3-carboxylic acid lactone; podophyllinic acid lactone; podofilox; Condyline; Condylox; Wartec; Warticon. $C_{22}H_{22}O_8$; mol wt 414.41. C 63.76%, H 5.35%, O 30.89%. Precursor of antineoplastics etoposide, teniposide, *q.q.v.* Found in the rhizomes of North American *Podophyllum peltatum* L., *Podophyllaceae:* V. Podwyssotski, *Arch. Exp. Pathol. Pharmakol.* **13,** 29 (1880); also in *P. emodi* Wall.: von Wartburg *et al., Helv. Chim. Acta* **40,** 1331 (1957); in *Juniperus virginiana* L., *Cupressaceae:* Kupchan *et al., J. Pharm. Sci.* **54,** 659 (1965). Structure and absolute configuration: Schrecker, Hartwell, *J. Org. Chem.* **21,** 381 (1956). Crystal structure: T. J. Petcher *et al., J. Chem. Soc., Perkin Trans. II* **1973,** 288. Synthesis: Gensler, Gatsonis, *J. Am. Chem. Soc.* **84,** 1748 (1962); T. Kaneko, H. Wong, *Tetrahedron Letters* **28,** 517 (1987). Asymmetric total synthesis: R. C. Andrews *et al., J. Am. Chem. Soc.* **110,** 7854 (1988); E. J. Bush, D. W. Jones, *Chem. Commun.* **1993,** 1200. Toxicity data: F. S. Phillips *et al., Fed. Proc.* **7,** 249 (1948). ELISA determn in plants: K.-J. Yoo, J. R. Porter, *J. Nat. Prod.* **56,** 715 (1993). Clinical evaluation for treatment of genital warts: A. Lassus, *Lancet* **2,** 513 (1987); K. R. Beutner *et al., ibid.* **1,** 831 (1989). Review of early literature: Hartwell, Schrecker in *Fortschr. Chem. Org. Naturst.* **15,** 98-121 (1958). Review of chemistry and antineoplastic activity: I. Jardine, *Anticancer Agents Based on Natural Product Models,* J. M. Cassady, J. D. Douros, Eds. (Academic Press, New York, 1980) pp 319-351. Review of syntheses: R. S. Ward, *Synthesis* **1992,** 719-730. Review: D. L. Sackett, *Pharmacol. Ther.* **59,** 163-228 (1993).

Solvated crystals. mp 114-118° (effervescence). Several polymorphic modifications. After drying: mp 183.3-184.0°. $[α]_D^{20}$ −132.7° (chloroform). Soly in water at 23°: 120 mg/l. Sol in alcohol, chloroform, acetone, warm benzene, glacial acetic acid. LD_{50} in rats (mg/kg): 8.7 i.v.; 15 i.p. (Phillips).

Podophyllotoxin-β-D-glucoside, $C_{28}H_{32}O_{13}$, hygroscopic amorphous white flakes, mp 152-154°. $[α]_D^{20}$ −76.4° (c = 0.576 in methanol); $[α]_D^{20}$ −117.0° (c = 0.668 in pyridine).

THERAP CAT: Antiviral (topical).

7705. Podophyllum. May apple; mandrake root; Indian apple; vegetable calomel. Dried rhizome and roots of *Podophyllum peltatum* L., *Berberidaceae.* **Habit.** North America. **Constit.** 3-6% resin, 0.2-1% podophyllotoxin, picropodophyllin, quercetin, peltatins. Sixteen well-characterized compds have been isolated. They fall into two chemical classes: lignans and flavonol pigments. History, isoln procedures, structures: Hartwell, Schrecker, *Fortschr. Chem. Org. Naturst.* **15,** 83-166 (1958), a review with 227 refs.

THERAP CAT: Cathartic.

THERAP CAT (VET): Has been used as a purgative.

7706. Podophyllum, Indian. Rhizome and roots of *Podophyllum emodi* Wall., *Berberidaceae.* **Habit.** Himalaya Mountains, Kashmir. **Constit.** 6-12% resin, 1-4% podophyllotoxin; berberine.

THERAP CAT: Cathartic.

7707. Podophyllum Resin. Podophyllin. A dry, alcoholic extract of the rhizomes and roots of *Podophyllum peltatum* L., *Berberidaceae:* "Extract powdered podophyl-

lum (1 kilo) by slow percolation until it is exhausted of its resin, using alcohol as the menstruum. Concentrate the percolate by evaporation until the residue has the consistency of a thin syrup, and pour this with constant stirring, into one liter of water contg 10 ml concd HCl and previously cooled to a temp < 10°. Allow the precipitate to settle, decant the clear liq, and wash the precipitate with two 1000 ml portions of cold water. Dry the resin and powder it." **Constit.** Podophyllotoxin (up to 20%), α- and β-peltatin, desoxypodophyllotoxin, dehydropodophyllotoxin. These constits are originally present in the plant as β-D-glucosides. The picropodophyllin often reported in the literature is not present in the fresh plant, but forms during isoln of the resin by epimerization of podophyllotoxin and may be regarded as the more stable *cis*-form of the natural *trans*-podophyllotoxin: von Wartburg *et al., Helv. Chim. Acta* **40,** 1331 (1957). *Review:* Kelly, Hartwell, *J. Natl. Cancer Inst.* **14,** 967 (1954).

Light brown to greenish-yellow powder, or small, yellowish, bulky, fragile lumps becoming darker on exposure to light or heat (> 25°); faint odor; acrid, bitter taste. Sol in alcohol, usually with a slight opalescence, also sol in dil caustic. Not less than 75% is sol in ether and not less than 65% in chloroform.

Caution: Extremely caustic to the skin and mucous membranes; dust inflames the eyes.

THERAP CAT: Caustic.

THERAP CAT (VET): Externally as a caustic for warts. Has been used internally as a purgative.

7708. Poi. The root of the taro plant, *Colocasia esculenta* (L.) Schott, *Araceae.* Used as a cereal substitute, particularly in allergy cases and potentially allergic infants.

THERAP CAT: Nutrient.

7709. Poison Ivy. Poison vine; markweed. *Toxicodendron radicans* (L.) Kuntze, *Anacardiaceae.* Erroneously called poison oak. Vigorous woody vine, shrub, or subshrub with trifoliate, alternate leaves. **Habit.** All states of the United States east of the Cascade Mountains, the Great Basin, and the Mojave Desert (absent in Nevada); all states of Mexico except in the Yucatan Peninsula and northern Baja California. Its southern limit is Huehuetenango Department of Guatemala and its northern limit is the 52nd parallel of latitude. It is found in Bermuda and several islands in the Bahamas, in Japan, and in the middle elevations in the mountains of Taiwan and of central and western China. *Review:* Gillis in J. M. Kingsbury, *Poisonous Plants of the United States and Canada* (Prentice-Hall, Englewood Cliffs, N. J., 1964) pp 209-214. **Constit.** Toxic constituent: urushiol *q.v.:* Dawson, *Trans. N.Y. Acad. Sci.* [2] **18,** 427 (1956).

Caution: Can cause severe allergic dermatitis.

THERAP CAT: Extract as antiallergic (hyposensitization therapy).

7710. Poison Oak. Western poison oak. *Toxicodendron diversilobum* (T. & G.) Greene, *Anacardiaceae.* Similar to poison ivy in that it has three leaflets, grows as a shrub or vine and produces dermatitis in man. Interbreeds with poison ivy. **Habit.** Western North America from southern British Columbia to northern Baja California. **Constit.** Poisons presumably closely related if not identical to those in poison ivy.

Eastern poison oak. *Toxicodendron quercifolium* (Michx.) Greene, *Anacardiaceae.* Differs significantly from poison ivy. Never climbs or produces aerial roots and rarely hybridizes with poison ivy. **Habit.** Southern New Jersey to Florida, west to eastern Texas and Kansas. **Constit.** Poisons may be the same as in poison ivy or Western poison oak.

Extract of *T. quercifolium, Anergex.*

Caution: Direct contact can cause severe allergic dermatitis. See M.J. Ellenhorn, D.G. Barceloux, *Medical Toxicology: Diagnosis and Treatment of Human Poisoning* (Elsevier, New York, 1988) pp 1301-1306.

THERAP CAT: Extract as antiallergic (hyposensitization therapy).

7711. Poison Sumac. Poison elder. *Toxicodendron vernix* (L.) Kuntze, *Anacardiaceae.* Tall, rangy shrub with leaves compounded with seven to eleven leaflets. Rachis is bright red and leaflets have no teeth on their margins, thus differing from nonpoisonous sumacs. *Habit.* Southern Quebec to central Florida predominantly east of Mississippi River. Found only in bogs, swamps and wet bottom lands.

Caution: Can cause severe allergic dermatitis.

THERAP CAT: Extract as antiallergic (hyposensitization therapy).

7712. Polaprezinc. *[N-β-Alanyl-L-histidinato(2—)-N,-NN,Oα]zinc; catena-(S)-[μ-[Nα-(3-aminopropionyl)-L-histidinato(2—)-N1,N2,O:NT]zinc];* zinc L-carnosine; β-alanyl-L-histidinato zinc; N-(3-aminopropionyl)-L-histidinato zinc; Z-103; Promac. Polymeric zinc(II) complex with L-carnosine, corresponding to the formula, $(C_9H_{12}N_4O_3Zn)_n$. Exhibits antioxidant and gastroprotective properties. Prepn of crystalline complex: T. Matsukura *et al.,* **Eur. pat. Appl. 303,380;** *eidem,* U.S. pat. **4,981,846** (1989, 1991 both to Hamari). Prepn and structure elucidation: T. Matsukura *et al., Chem. Pharm. Bull.* **38,** 3140 (1990). Pharmacology: T. Arakawa *et al., Dig. Dis. Sci.* **35,** 559 (1990). Mechanism of action study: T. Yoshikawa *et al., Biochim. Biophys. Acta* **1115,** 15 (1991). Series of articles on metabolism: *Arzneimittel-Forsch.* **41,** 965-995 (1991); on toxicology: *ibid.* 1033-1057. Acute toxicity: K. Matsuda *et al., ibid.* 1033.

White or pale yellow crystalline powder. Insol in water and common organic solvents. LD$_{50}$ in male, female mice, rats (mg/kg): 220, 165, 405, 422 i.p.; 758, 874, > 5000, > 5000 s.c.; 1269, 1331, 8441, 7375 orally (Matsuda).

THERAP CAT: Antiulcerative.

7713. Polar® Yellow. *5-Chloro-2-[4,5-dihydro-3-methyl-4-[[4-[[(4-methylphenyl)sulfonyl]oxy]phenyl]azo]-5-oxo-1H-pyrazol-1-yl]benzenesulfonic acid monosodium salt;* 4-p-hydroxybenzeneazo-1-p-chloro-o-sulfophenyl-3-methyl-5-hydroxypyrazole toluene-p-sulfonyl ester sodium salt; C.I. Acid Yellow 40; C.I. 18950. $C_{23}H_{18}ClN_4NaO_7S_2$; mol wt 584.99. C 47.22%, H 3.10%, Cl 6.06%, N 9.58%, Na 3.93%, O 19.14%, S 10.96%. Prepn from diazotized *p*-aminophenol and 1-(*o*-sulfo-*p*-chlorophenyl)-3-methyl-5-pyrazolone: B. Richard, U.S. pat. **1,067,881** (Geigy). *See also: Colour Index* vol. **4** (3rd ed., 1971) p 4128.

Yellowish-brown powder. Sol in water giving a pure yellow soln; in alcohol (yellow color); in alkalies; in concd sulfuric acid. Slightly sol in ethanol, acetone. Insol in other org solvents. Precipitated by dil acids.

USE: To dye wool. In the isoln of polymyxin.

7714. Poldine Methylsulfate. *2-[[(Hydroxydiphenylacetyl)oxy]methyl]-1,1-dimethylpyrrolidinium methyl sulfate (salt);* 2-hydroxymethyl-1,1-dimethylpyrrolidinium methyl sulfate benzilate; 2-benziloyloxymethyl-1,1-dimethylpyrrolidinium methyl sulfate; (1-methyl-2-pyrrolidinyl)methyl benzilate methylmethosulfate; IS-499; McN-R-726-47; Nacton; Nactate. $C_{22}H_{29}NO_7S$; mol wt 451.54. C 58.52%, H 6.47%, N 3.10%, O 24.80%, S 7.10%. Prepn: Doyle *et al., J. Chem. Soc.* **1958,** 4458. Prepn of the base: Blicke, Lu, *J. Am. Chem. Soc.* **79,** 29 (1955).

Needles from methyl ethyl ketone + ethanol + ether, mp 154-155°. Soluble in water.

THERAP CAT: Anticholinergic.

7715. Pole Reagent Paper. Unsized paper impregnated with a soln of 1-2 g phenolphthalein in 10 ml alcohol and 110 ml water and then passed, while still wet, through a soln of 20 g Na$_2$SO$_4$ (cryst) in 100 ml water; the paper is then dried.

USE: For recognizing the negative poles of batteries (the reaction at the negative pole coloring the paper red). The ends of the wires are placed about ½ inch apart on the moistened paper.

7716. Polidexide. *Sephadex 2-(diethylamino)ethyl ether;* dextran 2-(diethylamino)ethyl 2-[[2-(diethylamino)-ethyl]diethylammonio]ethyl ether chloride hydrochloride epichlorohydrin crosslinked; poly[2-(diethylamino)ethyl] polyglycerylene dextran hydrochloride; PDX chloride; DEAE-Sephadex; Secholex. An anion exchange resin containing quaternary ammonium groups, which reduces serum cholesterol by binding the bile acids in the intestine, thus facilitating their increased faecal excretion. This depletion of the bile acids is countered by synthesis of more bile acids from cholesterol. Prepn: **Brit. pat. 1,013,585** (1965 to Pharmacia AB). Activity studies: Parkinson, *J. Lipid Res.* **8,** 24 (1967); Howard, Hyams, *Brit. Med. J.* **3,** 25 (1971); Evans *et al., Angiology* **24,** 22 (1973); Ritland *et al., Scand. J. Gastroenterol.* **10,** 791 (1975).

Tasteless powder; forms gel in water.

USE: Ion exchange resin.

THERAP CAT: Anticholesteremic.

7717. Polidocanol. *α-Dodecyl-ω-hydroxypoly(oxy-1,2-ethanediyl);* polyethylene glycol (9) monododecyl ether; dodecyl alcohol polyoxyethylene ether; hydroxypolyethoxy-dodecane; laureth 9; polyoxyethylene lauryl ether; Aethoxysklerol; Aetoxisclerol; Atlas G-4829; Hetoxol L-9; Lipal 9LA; Thesit. $C_{12}H_{25}(OCH_2CH_2)_nOH$. Contains an average of nine ethylene oxide units and has an average mol wt ~600. Prepd by reaction of ethylene oxide and dodecyl alcohol: Pertsemlides, Soehring, *Arzneimittel-Forsch.* **10,** 990 (1960). Toxicology: H. S. Zipf *et al., ibid.* **7,** 162 (1957). Review of clinical experience: P. M. Goldman, *J. Dermatol. Surg. Oncol.* **15,** 204-209 (1989).

Sol in water, ethanol, toluene. Miscible with hot mineral, natural and synthetic oils; with fats and fatty alcohols. LD$_{50}$ in mice (mg/kg): 1170 orally, 125 i.v. (Zipf).

USE: Solvent; non-ionic emulsifier; pharmaceutic aid (surfactant); spermaticide.

THERAP CAT: Anesthetic (topical); antipruritic; sclerosing agent.

7718. Polonium. Po; at. wt of naturally occurring isotope 210; at. no. 84; valence 4, occasionally 2, rarely 6. Isotopes range in mass number from 193-218; all are radioactive. The first radioactive substance discovered by Mme. Curie in 1898. A product of disintegration of radium; one gram is contained in ~25,000 tons of pitchblende or in 7.5 kg radium that is more than 30 years old. Separated in form of a deposit on a bismuth plate immersed in a soln of the chloride: Marckwald, *Ber.* **35**, 2285 (1902); using a silver, gold or nickel plate: Curie, Joliot, *J. Chim. Phys.* **28**, 201 (1931); Haissinsky, *ibid.* **33**, 97 (1936); Rollier, *Gazz. Chim. Ital.* **66**, 797 (1936); Ziv, *C. R. Acad. Sci. URSS* **25**, 743 (1939). Obtained in the metallic form by volatilization from nickel on a collodion film: Rollier *et al., J. Chim. Phys.* **4**, 648 (1936). The only readily accessible isotope is the penultimate member of the radium decay series, $^{210}_{84}$Po, also called **Radium F** (Ra-F). Decays by α-emission ($T_{\frac{1}{2}}$ 138.4 days) to $^{206}_{82}$Pb. Comprehensive reviews: K. W. Bagnall, *Chemistry of the Rare Radioelements* (Butterworths, London, 1957); *idem, Endeavour* **22**(86), 61 (May 1963); *idem,* "Selenium, Tellurium and Polonium" in *Comprehensive Inorganic Chemistry* vol. **2**, J. C. Bailar, Jr., et al., Eds. (Pergamon Press, Oxford, 1973) pp 935-1008.

Two allotropic forms; coexist between 18° and 54°; d (α-form) 9.196; d (β-form) 9.398. mp 254°; bp 962°. Latent heat of vaporization: 24.597 kcal/mole. Resistivity: α-Po = 42 μohm-cm at 0°; β-Po = 44 μohm-cm at 0°. Physical properties: Maxwell, *J. Chem. Phys.* **17**, 1288 (1949). Chemically resembles tellurium and bismuth. Forms a volatile, unstable hydride, PoH$_2$; forms a polonide, Na$_2$Po; a carbonyl PoCO; a hydroxide Po(OH)$_4$.

Caution: Radiation hazard; alpha emitter. Sol compds hazard to kidneys, spleen; insol, airborne compds hazard to lungs. Max permissible concn of insol $^{210}_{84}$Po in air: 7 × 10^{-11} μ-Curie/cc, *National Bureau of Standards Handbook* **69**, 79 (1959).

7719. Polonium Dioxide. O$_2$Po; mol wt 240.98. O 13.28%, Po 86.72%. PoO$_2$. Formed from the elements at 250°: Martin, *J. Phys. Chem.* **58**, 911 (1954).

Two crystal modifications: yellow, low temperature form, face-centered cubic symmetry; red, high-temperature form, tetragonal symmetry. Darkens in color on heating, chocolate brown at sublimation temp of 885°. Dec into the elements at 500° under vacuum; slowly reduced to the metal in hydrogen at 200°. Heat of formation ~30 kcal/mole. Sol in aq solns of ammonium carbonate, phosphoric acid.

7720. Polonium Tetrachloride. Cl$_4$Po; mol wt 350.79. Cl 40.43%, Po 59.57%. PoCl$_4$. May be prepd by dissolving metallic polonium in hydrochloric acid; by heating polonium dioxide in carbon tetrachloride vapor at 200°; by heating the metal in dry chlorine at 200°; also by heating the dioxide in dry hydrogen chloride, thionyl chloride vapor, or with phosphorus pentachloride: *Review:* K. W. Bagnall, *Chemistry of the Rare Radioelements* (Butterworths, London, 1957) p 64.

Hygroscopic, bright yellow, cryst solid. Monoclinic or triclinic. Hydrolyzed by moist air, forming a white solid of indefinite composition. mp ~300° (in chlorine), turns scarlet red at 350°. bp 390°. Vapors are purple-brown, turning blue-green >500°. Sol in water (fairly slow hydrolysis). Freely sol in hydochloric acid, thionyl chloride. Moderately sol in ethanol, acetone. Dec by dil nitric acid. Forms a complex with two moles of tributyl phosphate.

7721. Poloxalene. *Methyloxirane polymer with oxirane.* Poly(oxyethylene)-poly(oxypropylene)-poly(oxyethylene) polymer; bis[hydroxyethylpoly(ethyleneoxy)ethyl]polypropyleneglycol; dipolyoxyethylated polypropyleneglycol ether; oxyethylene oxypropylene polymer; SKF-18667; Bloat Guard; Therabloat. A block polymer of ethylene oxide and propylene oxide, with a mol wt ~3000. Example of prepn: Lundsted, U.S. pat. **2,674,619** (1954 to Wyandotte Chem.). Field trial in wheat pasture bloat of cattle: E. E. Bartley *et al., J. Animal Sci.* **41**, 752 (1975). Absorption and excretion in rats: J. B. Rogers *et al., Drug Metab. Dispos.* **12**, 631 (1984). Hypocholesterolemic effect in rabbits: *idem, J. Clin. Invest.* **71**, 1490 (1983); *idem, Atherosclerosis (Shannon, Ire.)* **64**, 37 (1987). *Review:* Stanton,

Am. Perfumer Aromat. **72**, No. 4, 54, 56, 58 (1958). *See also* Poloxamers.

$$HO(CH_2CH_2O)_a[CH(CH_3)CH_2O]_b(CH_2CH_2O)_cH$$

average values for a, b, c are:

a = 12; b = 34; c = 12

A liquid nonionic surfactant polymer.
USE: Pharmaceutic aid (surfactant).
THERAP CAT (VET): Prevention of bloat in cattle.

7722. Poloxamers. *Methyl oxirane polymers, polymer with oxirane; polyethylenepolypropylene glycols, polymers;* α-hydro-ω-hydroxypoly(oxyethylene)poly(oxypropylene)poly(oxyethylene) block copolymers; Pluronics. Series of nonionic surfactants with the structure HO(CH$_2$CH$_2$O)$_a$(CH(CH$_3$)CH$_2$OH)$_b$(CH$_2$CH$_2$O)$_c$H where b is at least 15 and (CH$_2$CH$_2$O)$_a$ + c is varied from 20 to 90% by weight. Mol wt ranges from 1000 to >16,000. The poly(oxypropylene) segment is hydrophobic; the poly(oxyethylene) segment hydrophilic. Comprehensive reviews: I. R. Schmolka, *Am. Perfumer Cosmet.* **82**(7), 25-30 (1967); *idem* "Polyalkylene Oxide Block Copolymers" in *Nonionic Surfactants*, M. Schick, Ed. (Dekker, New York, 1967) pp 300-371. *See also* Poloxalene.

Mobile liquids, pastes or flakeable solids. Relatively non-hygroscopic. Vary from water-insol to very water-sol compds; more sol in cold than hot water. In general, sol in aromatic solvents (benzene, toluene, xylene), chlorinated solvents, acetone, alc, propylene or hexylene glycol, butyl cellosolve, butyl carbitol, methyl ethyl ketone, cyclohexanone. Insol in ethylene glycol, kerosene, mineral oil. Low-foaming properties. Stable to acids, alkalies, metallic ions.

Poloxamer 182LF, Pluronic L62LF. a = 7, b = 30, c = 7; av. mol wt 2450. Low-foaming liquid. d$^{25}_{25}$ 1.035. Brookfield viscosity (25°) 375 cp; cloud pt (10% aq soln): 24°.

Poloxamer 188, poloxalkol (obsolete), Exocorpol, Pluronic F68. a = 75, b = 30, c = 75; av. mol wt 8350. Flakeable solid. mp 50° (minimum); cloud pt (10% aq soln): >100°.

Poloxamer 331, Pluronic L101. a = 7, b = 54, c = 7; av. mol wt 3800. Liquid. d$^{25}_{25}$ 1.018; Brookfield viscosity (25°): 756 cp; cloud pt (10% aq soln): 11°.

USE: Food additives; defoamers; antistatic agents; demulsifiers; detergents; wetting agents; gelling agents; emulsifiers; foam controllers; dispersants; dye levelers.

THERAP CAT: Poloxamer 182LF as pharmaceutic aid; 188 as cathartic.

7723. Polybenzarsol. *(4-Hydroxyphenyl)arsonic acid polymer with formaldehyde;* Benzodol. A polymeric mixture obtained by adding formaldehyde to *p*-hydroxybenzenearsonic acid. Prepn: H. E. Faith, *J. Am. Chem. Soc.* **72**, 837 (1950). Toxicity studies: W. R. Jones *et al., Antibiot. & Chemother.* **8**, 400 (1958).

White powder. Somewhat sol in water; sol in alcoholic NaOH. LD$_{50}$ i.p. in male mice: 235 mg/kg (Jones).

THERAP CAT: Antiamebic.

7724. Polybrominated Biphenyls. PBBs; brominated biphenyls; polybromobiphenyls. Mixtures of isomers and congeners with structures similar to polychlorinated biphenyls, *q.v.*, where each X = H or Br. Once widely used commercially. Prepn: H. Hahn *et al.,* Ger. pat. **1,161,547** (1964 to Chem. Fabrik Kalb); G. A. Burk, U.S. pat. **3,733,-366** (1973 to Dow); L. C. Mitchell, D. R. Breckenridge, U.S. pats. **3,763,248** and **3,833,674** (1973, 1974 both to Ethyl Corp.). Persistence in soils: L. W. Jacobs *et al., J. Agr. Food Chem.* **24**, 1198 (1976). Photodegradation: L. O. Ruzo *et al., ibid.* 1062. Reviews of environmental hazards: K. Kay, *Environ. Res.* **13**, 74-93 (1977); F. J. DiCarlo *et al., Environ. Health Perspect.* **23**, 351-365 (1978); of toxicology and mechanism of action: S. Safe, *Crit. Rev. Toxicol.* **13**, 319-395 (1984); of carcinogenicity: E. M. Silberhorn *et al., ibid.* **20**, 439-496 (1990).

Firemaster BP-6, major component is *2,2',4,4',5,5'-hexabromobiphenyl.* Softens at 72°, dec >300°. Low vapor pressure; degraded by uv light. Very sol in benzene, toluene. Insol in water.

Caution: The 1973 "Michigan Incident" in which BP-6 was accidentally added to animal feed, and resulted in the contamination of the food chain and wide-spread destruction of contaminated farm animals, led to the removal of BP-6 from the market: L. J. Carter, *Science* **192,** 240 (1976). Reported symptoms of overexposure include impaired immune system; hypothyroidism; neurological effects, headache, joint stiffness, memory loss; chloracne-like lesions (Safe). *See also Patty's Industrial Hygiene and Toxicology* vol. **2D,** G. D. Clayton, F. E. Clayton, Eds. (John Wiley & Sons, New York, 4th ed., 1994) p 2433-2504. These substances may reasonably be anticipated to be carcinogens: *Seventh Annual Report on Carcinogens* (PB95-109781, 1994) p 322.
USE: Formerly as flame retardant.

7725. Polychlorinated Biphenyls. PCBs; chlorinated biphenyls; chlorobiphenyls; Aroclor; Clophen; Fenclor; Kanechlor; Pyralene. Once widely used industrial chemicals whose high stability contributed to both their commercial usefulness and their long-term deleterious environmental and health effects. Synthesis: H. Schmidt, G. Schulz, *Ann.* **207,** 338 (1881). Commercially available since 1930: C. Penning, *Ind. Eng. Chem.* **22,** 1180 (1930). Commercial PCBs are mixtures of various isomers and congeners. The Aroclors are characterized by four digit numbers. The first two digits indicate that the mixture contains biphenyls (12), triphenyls (54) or both (25, 44); the last two digits give the weight percent of chlorine in the mixture (e.g. Aroclor 1242 contains biphenyls with ~42% chlorine). Accumulation of airborne PCBs in foliage: E. H. Buckley, *Science* **216,** 520 (1982). *Reviews:* H. L. Hubbard in *Kirk-Othmer Encyclopedia of Chemical Technology* vol. **5** (Interscience, New York, 2nd ed., 1964) pp 289-297; O. Hutzinger *et al., The Chemistry of PCBs* (CRC Press, Cleveland, Ohio, 1974) 269 pp; J. W. Lloyd *et al., J. Occup. Med.* **18,** 109-113 (1976). Reviews of environmental impact and toxicity: L. Fishbein, *Ann. Rev. Pharmacol.* **14,** 139-156 (1974); *National Conference on Polychlorinated Biphenyls,* Nov. 19-21, 1975 (EPA-560/6-75-004, 1976) 487 pp; R. D. Kimbrough, *Crit. Rev. Toxicol.* **2,** 445-498 (1974); S. H. Safe, *ibid.* **24,** 87-149 (1994). Reviews of carcinogenicity: *IARC Monographs* **18,** 43-103 (1978); E. M. Silberhorn *et al., Crit. Rev. Toxicol.* **20,** 439-496 (1990); of toxicology and mechanism of action: S. Safe, *ibid.* **13,** 319-395 (1984); of toxicology and human exposure: *Toxicological Profile for Polychlorinated Biphenyls* (PB93-182517, 1993) 23 p.

X = H or Cl

Aroclor 1242, clear, mobile liquid; av. number Cl/molecule: 3.10. d_4^{25} 1.381, $d_4^{15.5}$ 1.392. Distillation range 325-366°. Flash point (open cup) 348-356°F. n_D^{20} 1.627-1.629. Dielectric constant (1000 cycles) 5.6 (25°), 4.9 (100°).
Aroclor 1254, light yellow, viscous liquid; av. number Cl/molecule: 4.96. d_4^{65} 1.495; $d_4^{15.5}$ 1.505. Distillation range 365-390°. No open cup flash point to boiling. n_D^{20} 1.629-1.641. Dielectric constant (1000 cycles) 5.0 (25°), 4.3 (100°). LD_{50} orally in weanling rats: 1295 mg/kg (Kimbrough).
Aroclor 1260, light yellow, soft, sticky resin; av. number Cl/molecule: 6.30. d_4^{90} 1.555; $d_4^{15.5}$ 1.566. Distillation range 385-420°. No open cup flash point to boiling. n_D^{20} 1.647-1.649. Dielectric constant (1000 cycles) 4.3 (25°); 3.7 (100°). LD_{50} orally in weanling rats: 1315 mg/kg (Kimbrough).
Caution: In Japan, 1968, oral intoxication to humans due to accidental contamination of rice bran oil with Kanechlor 400 led to an outbreak of what became known as "Yusho disease". Symptoms of oral intoxication in humans included nausea, lethargy, chloracne, brown pigmentation of skin and nails, subcutaneous edema of the face, distinctive hair follicles, excessive eye discharge, swelling of eyelids, visual

disturbances, GI disturbances and jaundice. *See* M. Kuratsune *et al.,* (EPA-560/6-75-004, 1976) p 14. Potential symptoms of occupational overexposure are chloracne, dermal lesions; hepatic injury; decreased pulmonary function; decreased birth weight in offspring of exposed mothers; eye irritation (Safe, 1994). *See also Patty's Industrial Hygiene and Toxicology* vol. **2D,** G. D. Clayton, F. E. Clayton, Eds. (John Wiley & Sons, Inc., New York, 4th ed., 1994) 2433-2504. These substances may reasonably be anticipated to be carcinogens: *Seventh Annual Report on Carcinogens* (PB95-109781, 1994) p 324.
USE: In electrical capacitors, electrical transformers, gas-transmission turbines, vacuum pumps. Formerly used in U.S. as hydraulic fluids, plasticizers, adhesives, fire retardants, wax extenders, dedusting agents, pesticide extenders, inks, lubricants, cutting oils, in heat transfer systems, carbonless reproducing paper.

7726. Polydextrose. Litesse. Randomly bonded condensation polymer of dextrose with small amounts of bound sorbitol and citric acid. Functions to replace the bulk and mouthfeel of sugar and/or fat in reduced calorie foods. Prepn: H. H. Rennhard, U.S. pat. **3,766,165** (1973 to Pfizer). Improved process: D. B. Guzek *et al.,* **Eur.** pat. **Appl. 473,333** (1992 to Pfizer); *C.A.* **116,** 237725s (1992). Reviews of physical properties and applications in foods: A. Torres, R. D. Thomas, *Food Technol.* **35,** 44-49 (1981); F. K. Moppett in *Food Sci. Technol.* **48,** entitled "Alternative Sweeteners", L. O. Nabors, R. C. Gelardi, Eds. (1991) pp 401-421.
White to light tan, amorphous powder, mp >130°. Bland, non-sweet taste. Hygroscopic. pH of 10% w/w aq soln: 2.5-3.5. Viscosity of 50% aq soln: 35 cps. Very sol in water (to ~80%). Partially sol in glycerin, propylene glycol. Insol in ethanol. Caloric utilization value in humans: 1 kcal/g.
USE: Bulking agent for reduced calorie foods.

7727. Polyestradiol Phosphate. *Estradiol phosphate polymer;* PEP; Estradurin. Polymeric ester of phosphoric acid and estradiol. Mol wt ~26,000. Prepn: Diczfalusy, *Endocrinology* **54,** 471 (1954); Fernö *et al., Acta Chem. Scand.* **12,** 1675 (1958); Diczfalusy *et al.,* U.S. pat. **2,928,-849** (1960 to AB Leo). Clinical pharmacology: P. O. Gunnarsson, B. J. Norlén, *Prostate* **13,** 299 (1988). Clinical trial in prostatic carcinoma: J. Aro, *ibid.* **18,** 131 (1991).

n is approx. 80

Solid, mp 195-202°. Very sol in aq pyridine; sol in aq alkali; very slightly sol in ethanol, ethanol + water (1:1), water, dioxane, acetone, chloroform. Intrinsic viscosity [η] in 0.25*M* NaCl soln at pH 7.5 = 0.04.
THERAP CAT: Antineoplastic (hormonal).

7728. Polyethylene. *Ethene homopolymer;* Agilene; Alathon; Alkathene; Courlene; Lupolen; Platilon; Polythene; Pylen; Reevon. Mol wt about 1500-100,000. C 85.7%, H 14.3%. Prepd by polymerization of liq ethylene at high temps and high or low pressure. *Reviews:* Aggarwal, Sweeting, *Chem. Rev.* **57,** 665-742 (1957); Raff, Allison, *Polyethylene,* vol. XI of High Polymers series (Interscience, New York, 1956); Faith *et al., Industrial Chemicals* (Wiley, New York, 3rd ed., 1965) pp 624-630.

Plastic solid of milky transparency. d_4^{20} 0.92. Tough and flexible at room temps, mp 85-110°. Breaks with cryst fracture at −50°. Good electrical insulator. Surface resistivity: 10^{14} ohms. Will burn, but hardly supports combustion. Stable to water, non-oxidizing acids and alkalies, alcohols, ethers, ketones, esters at ordinary temps. Attacked by oxidizing acids such as nitric acid and perchloric acid, free halogens, benzene, petr ether, gasoline and lubricating oils, aromatic and chlorinated hydrocarbons.

USE: Laboratory tubing; in making prostheses; electrical insulation; packaging materials; kitchenware; tank and pipe linings; paper coatings; textile stiffeners.

7729. Polyethylene Glycol. α-Hydro-ω-hydroxypoly-(oxy-1,2-ethanediyl); macrogol; PEG; Carbowax; Pluracol E; Poly-G; Polyglycol E. Liquid and solid polymers of the general formula $H(OCH_2CH_2)_nOH$, where n is greater than or equal to 4. In general, each PEG is followed by a number which corresponds to its average mol wt. Synthesis: Fordyce, Hibbert, *J. Am. Chem. Soc.* **61**, 1905, 1910 (1939). *Reviews: Glycols*, G. O. Curme, Jr., F. Johnston, Eds., A.C.S. Monograph Series no. **114** (Reinhold, New York, 1952) pp 176-202; Kastens in *High Polymers*, H. Mark *et al.*, Eds., **vol. 13** entitled *Polyethers*, part **1** (Interscience, New York, 1963) pp 169-189, 274-291; G. M. Powell, III in *Handbook of Water-Soluble Gums & Resins*, R. L. Davidson, Ed. (McGraw-Hill, New York, 1980) pp 18/1-18/31.

Clear, viscous liquids or white solids which dissolve in water forming transparent solns. Sol in many organic solvents. Readily sol in aromatic hydrocarbons. Only slightly sol in aliphatic hydrocarbons. Do not hydrolyze or deteriorate on storage, will not support mold growth. Solvent action on some plastics. Polyethylene glycols are compds of low toxicity: Smyth *et al.*, *J. Am. Pharm. Assoc.*, *Sci. Ed.* **39**, 349 (1950). Toxicity data (PEG 400): W. Bartsch *et al.*, *Arzneimittel-Forsch.* **26**, 1581 (1976).

Polyethylene glycol 200, average value of n is 4, mol wt range 190-210. Viscous, hygroscopic liq; slight characteristic odor; d_{25}^{25} 1.127. Viscosity (210°F): 4.3 centistokes. Supercools upon freezing.

Polyethylene glycol 400, average value of n between 8.2 and 9.1, mol wt range 380-420. Viscous, slightly hygroscopic liq; slight characteristic odor; d_{25}^{25} 1.128. mp 4-8°. Viscosity (210°F): 7.3 centistokes. LD_{50} orally in rats: 30 ml/kg (Bartsch).

Polyethylene glycol 600, average value of n between 12.5 and 13.9, mol wt range 570-630. Viscous, slightly hygroscopic liq; characteristic odor; d_{25}^{25} 1.128. mp 20-25°. Viscosity (210°F): 10.5 centistokes.

Polyethylene glycol 1500, average value of n between 29 and 36, mol wt range 1300-1600. White, free-flowing powder; d_{25}^{25} 1.210. mp 44-48°. Viscosity (210°F): 25-32 centistokes.

Polyethylene glycol 4000, average value of n between 68 and 84, mol wt range 3000-3700. White, free-flowing powder or creamy-white flakes; d_{25}^{25} 1.212. mp 54-58°. Viscosity (210°F): 76-110 centistokes. LD_{50} orally in rats (divided doses): 59 g/kg (Smyth).

Polyethylene glycol 6000, average value of n between 158 and 204, mol wt range 7000-9000. Powder or creamy-white flakes; d_{25}^{25} 1.21. mp 56-63°. Viscosity (210°F): 470-900 centistokes. LD_{50} orally in rats: > 50 g/kg (Smyth).

USE: As water-soluble lubricants for rubber molds, textile fibers, and metal-forming operations. In food and food packaging. In hair prepns, in cosmetics in general. Pharmaceutic aid (ointment and suppository base). As a stationary phase in gas chromatography. Also in water paints, paper coatings, polishes and in the ceramics industry.

THERAP CAT (VET): Ointment base.

7730. Polyethylene Terephthalates. PET. Fiber forming polyesters prepd from terephthalic acid, *q.v.* or its esters and ethylene glycol: Whinfield, Dickson, U.S. pat. **2,465,-319** (1949 to du Pont). Review of structures, definition of trade names: R. W. Moncrieff, *Man-Made Fibres* (John Wiley & Sons, New York, 4th ed., 1963) pp 361-389, 707-723.

Dacron   R = OH
Terylene   R = OCH₃

R = OH, *Dacron, Amilar, Fiber V.* Solid, dec at approx 250°. Sp gr 1.38. Sol in hot *m*-cresol, trifluoroacetic acid, *o*-chlorophenol, a mixture of 7 parts of trichlorophenol and 10 parts (by wt) of phenol, a mixture of 2 parts of tetrachloroethane and 3 parts (by wt) of phenol. Fiber has good resistance to weak acids even at boiling temp, to strong acids in the cold, to weak alkalies, to bleaches, to most alcohols, ketones, soaps, detergents, and dry cleaning agents. Fabric has good resistance to creasing, abrasion, heat aging, and sunlight when behind glass. When "heat-set", fabric will not shrink in either boiling water or boiling drycleaning solvent. Fabric burns, but local melting generally prevents spread of fire. Insects cannot thrive on the fiber, but some can cut through it. Molds, mildew, and fungi may grow on applied finishes, but do not attack fiber.

R = OCH₃, *Terylene.* For physical properties, *see* Dacron above. Other similar products: *Diolen, Enkalene, Fortrel, Tergal, Terital, Tertenka, Trevira, Mylar.*

USE: In fabric manufacture; as films; as base for magnetic coatings. Surgical aid (arterial grafts).

7731. Polyferose. Jefron. An iron carbohydrate chelate contg approx 45% iron in which the metallic (Fe) ion is sequestered within a polymerized carbohydrate derived from sucrose.

THERAP CAT: Hematinic.

7732. Polygodial. [1R-(1α,4aβ,8aα)]-1,4,4a,5,6,7,8,8a-Octahydro-5,5,8a-trimethyl-1,2-naphthalenedicarboxaldehyde; tadeonal. $C_{15}H_{22}O_2$; mol wt 234.34. C 76.88%, H 9.46%, O 13.65%. Widely distributed drimane sesquiterpene with insect antifeedant properties; naturally occurring as the (−)-form. Isoln from *Polygonum hydropiper* L., *Polygonaceae* (Australia) and structure: C. S. Barnes, J. W. Loder, *Aust. J. Chem.* **15**, 322 (1962); from the bark of *Warburgia stuhlmanni* Engl. or *W. ugandensis, Canellaceae* (E. Africa): I. Kubo *et al.*, *Chem. Commun.* **1976**, 1013; from nudibranch *Dendrodons limbata* (Mediterranean): G. Cimino *et al.*, *Science* **219**, 1237 (1983); from nudibranchs *D. nigra, D. tuberculosa* (Hawaii) and *D. krebsii* (Mexico): R. K. Okuda *et al.*, *J. Org. Chem.* **48**, 1866 (1983). Relationship between structure and antifeedant-activity: K. Nakanishi, I. Kubo, *Isr. J. Chem.* **16**, 28 (1977); M. D'Ischia *et al.*, *Tetrahedron Letters* **1982**, 3295. Synthesis of racemate: T. Kato *et al.*, *Tetrahedron Letters* **1971**, 1961; S. C. Howell *et al.*, *Chem. Commun.* **1981**, 507. Synthesis of (−)-form: I. Razmilic *et al.*, *Synth. Comm.* **17**, 95 (1987).

Colorless needles from petroleum (40-60°), mp 57° (Barnes, Loder). $[\alpha]_D^{24}$ −131° (c = 0.96 in ethanol). uv max (ethanol): 231, 295 nm (ε 11800, 76).

(±)-Form, mp 93-94° (Tanis, Nakanishi).

7733. Polylysine. L-*Lysine, homopolymer.* A lysine polypeptide or homopolymer, the chain length of which varies with the method of prepn. Prepn: Katchalski *et al.*, *J. Am. Chem. Soc.* **69**, 2564 (1947); **70**, 2094 (1948); Fasman *et al.*, *ibid.* **83**, 709 (1961); Sela *et al.*, *Biopolymers* **1**, 517

(1963); Strojny, White, U.S. pat. **3,215,684** (1965 to Dow). For structure *see* Lysine.

L-Form hydriodide, average dp (or n) = 32. Transparent, solid, film-like polymer. Readily sol in water; practically insol in the usual organic solvents. Transition of high-mol-wt poly-L-lysine (dp 1500) in aq soln from a helical to a randomly coiled conformation under the influence of decreasing pH or increasing temp: Applequist, Doty, *C.A.* **58**, 6925b (1963).

7734. Polymyxin. Antibiotic complex produced by *Bacillus polymyxa:* Brownlee, Jones, *Biochem. J.* **43**, XXV (1948). Prepn: Ainsworth, Pope, U.S. pat. **2,565,057** (1951 to Burroughs Wellcome); Petty, U.S. pat. **2,595,605** (1952 to Am. Cyanamid); Benedict, Stodola, U.S. pat. **2,771,397** (1956 to USDA). Purification: Hastings *et al., Brit.* pat. **782,926** (1957 to Distillers Co.). Polymyxins A, B, C, D, E, F, K, M, P, S and T have been identified. Isoln of polymyxins A, B, C and E: Few, Schulman, *Biochem. J.* **54**, 171 (1953); of *polymyxin D:* Stansley *et al., Bull. Johns Hopkins Hosp.* **81**, 43 (1947); of *polymyxin F:* W. L. Parker *et al, J. Antibiot.* **30**, 767 (1977); of *polymyxin K:* Kimura, **Japan.** pat. **16,152('71),** *C.A.* **75**, 62180r (1971); of *polymyxin M:* Khokhlov *et al., C.A.* **55**, 5653h (1961); of *polymyxin P:* Kimura *et al., J. Antibiot.* **22**, 449 (1969); of *polymyxin S_1* and *polymyxin T_1:* J. Shoji *et al., ibid.* **30**, 1029 (1977). Resolution of polymyxin B into B_1 and B_2: Hausmann, Craig, *J. Am. Chem. Soc.* **76**, 4892 (1954). Structure and synthesis of polymyxin B_1: Wilkinson, Lowe, *Nature* **202**, 1211 (1964); Vogler *et al., Helv. Chim. Acta* **48**, 1161 (1965). Structure of polymyxin B_2: Wilkinson, Lowe, *Nature* **204**, 993 (1964). Separation of polymyxin D into D_1 and D_2 and structures: Hayashi *et al., Experientia* **22**, 354 (1966). Structure of polymyxin S_1: J. Shoji *et al., J. Antibiot.* **30**, 1035 (1977); of polymyxin T_1: *eidem, ibid.* 1042. *Review:* Vogler, Studer, *Experientia* **22**, 345-354 (1966); Paulus, "Polymyxins" in *Antibiotics* **II**, D. Gottlieb, P. Shaw, Eds. (Springer-Verlag, New York, 1967) pp 254-267.

DAB = L-α,γ-diaminobutyric acid

Polymyxin B_1	R = (+)-6-methyloctanoyl	X = Phe	Y = Leu	Z = DAB
B_2	R = 6-methylheptanoyl	X = Phe	Y = Leu	Z = DAB
D_1	R = (+)-6-methyloctanoyl	X = Leu	Y = Thr	Z = D-Ser
D_2	R = 6-methylheptanoyl	X = Leu	Y = Thr	Z = D-Ser

Obtained as the hydrochloride. Nearly colorless powder, dec 228-230°. $[\alpha]_D^{23}$ −40° (c = 1.05). The hydrochloride is very sol (more than 40%) in water and methanol. The soly decreases in the higher alcs. Practically insol in the usual ethers, esters, ketones, hydrocarbons, and the chlorinated solvents. Forms water insol salts with a number of ppts such as picric acid, helianthic acid, Reinecke salt. The free base is slightly sol in water; almost insol in alc.

Polymyxin B. Mixture of polymyxins B_1 and B_2. $[\alpha]_{5461}$ −106.3° (1*N* HCl).

Polymyxin B sulfate, *Aerosporin, Mastimyxin.* Solubilities: Weiss *et al., Antibiot. & Chemother.* **7**, 374 (1957).

Polymyxin B sulfate mixture with trimethoprim, *Polytrim.*

Polymyxin B_1. $C_{56}H_{98}N_{16}O_{13}$.

Polymyxin B_1 pentahydrochloride, $C_{56}H_{98}N_{16}O_{13} \cdot 5HCl$, white powder. $[\alpha]_D^{25}$ −85.11° (c = 2.33 in 75% ethanol).

Polymyxin B_2. $C_{55}H_{96}N_{16}O_{13}$. $[\alpha]_{5461}^{22}$ −112.4° (2% acetic acid).

Polymyxin D_1. $C_{50}H_{93}N_{15}O_{15}$.
Polymyxin D_2. $C_{49}H_{91}N_{15}O_{15}$.
Polymyxin E, see Colistin.

THERAP CAT: Antibacterial.
THERAP CAT (VET): Antibacterial.

7735. Polynoxylin. *Urea polymer with formaldehyde;* poly[methylenedi(hydroxymethyl)urea]; oxymethyleneurea; polynoxyline; polyoxymethyleneurea; Anaflex; Larex; Ponoxylan. A condensation product of formaldehyde and urea, substantially linear chains, possibly lightly cross-linked. Prepn: Haler, Aebi, *Nature* **190**, 734 (1961); **Brit.** pat. **905,-195** (1962 to Ed. Geistlich Soehne).

Amorphous powder, dec without melting. Soly in water: 0.28-0.31%.

THERAP CAT: Topical antibacterial.

7736. Polyoxins. Agricultural antifungal antibiotic complex produced by *Streptomyces cacaoi* var *asoensis* and *S. piomogenus.* Polyoxins A through O are known, all except C and I having specific activity against phytopathogenic fungi by inhibiting cell wall chitin synthesis. Isoln and characterization of polyoxin A: S. Suzuki *et al., J. Antibiot.* **18A**, 131 (1965); of A and B: K. Isono *et al., Agr. Biol. Chem.* **29**, 848 (1965); S. Suzuki *et al.,* **Japan.** pat. **66 15,-520**; of C through K: K. Isono *et al., Agr. Biol. Chem.* **30**, 817 (1966); **31**, 190 (1967); **32**, 792 (1968); of M: K. Isono, S. Suzuki, *Tetrahedron Letters* **1970**, 425. Isoln of polyoxins N and O: S. Suzuki *et al.,* **Japan. Kokai** 72 **23,596** (to Inst. Phys. Chem. Res., and Hokko Chem. Ind.), *C.A.* **78**, 41566t (1973); production of N and O: *eidem,* **Japan.** pat. 77 20,-555 (1977 to Inst. Phys. Chem. Res.), *C.A.* **87**, 150183x (1977). Isoln and structure of N: M. Uramoto *et al., Nucleic Acids Res., Spec. Publ.* **5**, 327 (1978). Structural elucidation of polyoxins A through L: K. Isono *et al., J. Am. Chem. Soc.* **91**, 7490 (1969). Revised structure of A: S. Hanessian *et al., Tetrahedron Letters* **34**, 4153 (1993). Total synthesis of J: H. Kuzuhara *et al., ibid.* **1973**, 5055. Biosynthetic studies: K. Isono *et al., Biochemistry* **14**, 2992 (1975); S. Funayama, K. Isono, *ibid.* 5568; K. Isono, R. J. Suhadolnick, *Arch. Biochem. Biophys.* **173**, 141 (1976); S. Funayama, K. Isono, *Biochemistry* **16**, 3121 (1977); K. Isono *et al., J. Am. Chem. Soc.* **100**, 3937 (1978). Mode of action: N. Ohta *et al., Agr. Biol. Chem.* **34**, 1224 (1970); M. Hori *et al., ibid.* **35**, 1280 (1971); **38**, 691, 699 (1974). *Review:* R. J. Suhadolnick, *Nucleoside Antibiotics* (Wiley-Interscience, New York, 1970) pp 218-234; K. Isono, S. Suzuki, *Heterocycles* **13**, 333-351 (1979).

Polyoxin A

Polyoxin A, $C_{23}H_{32}N_6O_{14}$, *[S-(Z)]-1-[5-[[2-amino-5-O-(aminocarbonyl)-2-deoxy-L-xylonoyl]amino]-1,5-dideoxy-1-[3,4-dihydro-5-(hydroxymethyl)2,4-dioxo-1(2H)-pyrimidinyl]-β-D-allofuranuronoyl]-3-ethylidene-2-azetidinecarboxylic acid.* Colorless needles from aq ethanol, dec > 180°. $[\alpha]_D^{20}$ −30°. uv max (0.05*N* HCl): 262 nm (log ε 3.94); (0.05*N* NaOH): 264 nm (log ε 3.80).

Polyoxin B, $C_{17}H_{25}N_5O_{13}$, *5-[[2-amino-5-O-(aminocarbonyl)-2-deoxy-L-xylonoyl]amino]-1,5-dideoxy-1-[3,4-dihydro-5-(hydroxymethyl)-2,4-dioxo-1(2H)-pyrimidinyl]-β-D-allofuranuronic acid, Polyoxin AL.* Amorphous powder from aq ethanol. $[\alpha]_D^{20}$ +34°. uv max (0.05*N* HCl): 262 nm (log ε 3.94); (0.05*N* NaOH): 264 nm (log ε 3.82). LD_{50} in mice:

800 mg/kg orally, *RTECS* Vol. II, R. J. Lewis, R. L. Tatken, Eds. (1980) p 409. *See also* Suzuki, *loc. cit.* (1965).

USE: Fungicide, esp for *Alternaria* leaf spot of many agricultural products.

7737. Polyoxyethylene Alcohols. Polyethylene glycol fatty alcohol ethers; ethoxylated fatty alcohols; macrogol fatty alcohol ethers; alcohol ethoxylates; POE alcohol ethers. $R(OCH_2CH_2)_nOH$ where R is a long chain alkyl group or mixture of alkyl groups. Nonionic surfactants prepared by ethoxylation of fatty alcohols with ethylene oxide. Prepn and reaction mechanism: J. D. Malkemus, *J. Am. Oil Chem. Soc.* **33**, 571 (1956); W. Satkowski, C. G. Hsu, *Ind. Eng. Chem.* **49**, 1875 (1957); R. D. Fine, *J. Am. Oil Chem. Soc.* **35**, 542 (1958). Compounds with a broad range of properties can be prepared by varying the fatty alcohol (lipophile) used and the degree of polymerization of the polyethylene glycol (hydrophile) segment. CFTA-assigned names based on fatty alcohol segment include *ceteth* (cetyl alc), *laureth* (lauryl alc), *myreth* (myristyl alc), *oleth* (oleyl alc), *steareth* (stearyl alc), *trideth* (tridecyl alc). The average number of ethylene oxide units in the polyethylene glycol segment is indicated by an appended number (e.g. ceteth-20). Additional products prepared from fatty alcohol mixtures include *ceteareth* (cetyl/stearyl alcs), *laneth* (lanolin alcs). Trademarks for some commercially available series of compounds: *Alfonic, Bio Soft EA, Brij, Dehydol, Emulphogene, Ethosperse, Eumulgin* (B, EP, O, WM), *Ethoxyal, Lipal* (CSA, LA, MA, OA, TD), *Lipocol, Macol* (LA, OA, SA, TD), *Polychol, Siponic* (C, E, L, TD, Y), *Tergitol* (L, S), *Trycol, Volpo.* Comprehensive description: W. B. Satkowski *et al.,* "Polyoxyethylene Alcohols" in *Nonionic Surfactants* M. J. Schick, Ed. (Dekker, New York, 1967) pp 86-141. Effects of structure on properties: T. Kuwamura, *ACS Symposium Ser.* **253**, 27-47 (1984). Review of properties and uses in household cleaning products: K. W. Dillan *et al., Household Pers. Prod. Ind.* **23**, 32 (1986). Liquids to waxy solids. Compds with one to five moles ethylene oxide are sol in oil and many hydrocarbons. Water soly increases with increasing ethylene oxide content. Properties of Brij surfactants: G. King, *Drug Cosmet. Ind.* **90**, 24 (1962).

Laureth 9, see Polidocanol.
Ceteth 20, polyoxyethylene (20) cetyl ether, Brij 58, Lipocol C-20. $C_{16}H_{33}(OCH_2CH_2)_nOH$ where the average value of n is 20. Waxy solid. Pour point ~39°; cloud point ~45°. Sol in water; 2% ethanol.
USE: Used as emulsifiers, wetting agents, antistats, solubilizers, defoamers, detergents, lubricants in pharmaceutical, cosmetic and other industrial applications.

7738. Polyoxyethylene Fatty Acid Esters. Polyethylene glycol esters of fatty acids; PEG fatty acid esters; POE fatty acid esters; ethoxylated fatty acid esters; macrogol fatty acid esters. $RCOO(CH_2CH_2O)_nH$ or $RCOO(CH_2CH_2O)_nOCR$ where R is a long chain alkyl group or mixture of alkyl groups. Nonionic surfactants prepared commercially by esterification of fatty acids with ethylene oxide or with polyethylene glycol. Wide range of properties achieved by changing the hydrophobic fatty acid segment and/or varying the degree of polymerization of the hydrophilic polyoxyethylene segment. Trademarks for some commercially available series of compounds include *Emerest* (2600, 2700 series), *Emulphor* (M, VN), *Emulsynt, Ethofat, Lipal* (DL, DS, S), *Lipopeg, Mapeg, Myrj, Nonisol, Nopalcol, Pegosperse, Renex 20, Varonic* (DL, MS). Review of prepn, properties and uses: W. B. Satkowski *et al.,* "Polyoxyethylene Esters of Fatty Acids" in *Nonionic Surfactants*, M. J. Schick, Ed. (Dekker, New York, 1967) pp 142-174. Liquids, soft solids, solids or flakes. Solubility properties depend on length of polyoxyethylene (POE) segment added to a specific fatty acid. Sol in oil and hydrocarbon solvents when < 8 ethylene oxide units are added. Soly in water begins when 12-15 ethylene oxide units are added. Specific gravity and viscosity increase with increasing ethylene oxide content.
Polyoxyl 8 stearate, polyethylene glycol 400 monostearate, PEG-400 monostearate, POE (8) stearate, macrogol ester 400, Emerest 2711, Lipal 400S, Lipopeg 4-S, Myrj 45, Mapeg 400 MS, Pegosperse 400-MS. $C_{17}H_{35}COO(CH_2CH_2)_nH$ where

the average value of n is 8. Prepn and properties: R. L. Birkmeier, J. D. Brandner, *J. Agr. Food. Chem.* **6**, 471 (1958).
Polyoxyl 40 stearate, polyoxyethylene (40) stearate, POE (40) monostearate, Myrj 52, Myrj 52S, Pegosperse 1750-MS. The average number of oxyethylene units is 40. Waxy, white to light tan solid; odorless or faint fat-like odor. Sol in water, alc, ether, acetone. Insol in mineral oil, vegetable oils.
USE: As antistats, emulsifiers, defoamers, wetting agents, solubilizers, conditioning agents, lubricants, detergents. Have wide range of cosmetic, pharmaceutical and other industrial applications. Polyoxyl 8 and 40 stearates as pharmaceutic aid (surfactant).

7739. Polyphosphazenes. High polymers containing an inorganic backbone of alternating nitrogen and phosphorus atoms. They exist as glasses, flexible solids, or rubbery solids with a low tendency for crystallization; nonflammable and more elastic than carbon-backbone polymers. Prepn of the first polyphosphazene, hexachlorocyclotriphosphazene: J. Liebig, *Ann.* **11**, 139 (1834). Improved synthesis and basis of modern mfg methods: R. Schenck, G. Römer, *Ber.* **57B**, 1343 (1924). Review of syntheses, properties, chemistry and applications: H. R. Allcock, *Science*, **193**, 1214-1219 (1976); *idem, Angew. Chem. Int. Ed.* **16**, 147-156 (1977); E. N. Peters, "Inorganic High Polymers", in Kirk-Othmer *Encyclopedia of Chemical Technology* vol. **13** (Wiley-Interscience, New York, 3rd ed., 1981) pp 398-413.

$$\left[\begin{array}{c} X \\ | \\ P=N \\ | \\ X \end{array} \right]_n$$

USE: In waterproofing; as flame retardants; in gaskets, o-rings, hydrocarbon fuel hoses.

7740. Polyphosphoric Acid. Pholpholeum; tetraphosphoric acid. May be prepd by heating H_3PO_4 with sufficient phosphoric anhydride to give the resulting product an 82-85% P_2O_5 content: Bell, *Ind. Eng. Chem.* **40**, 1464 (1949); Van Wazer, Holst, *J. Am. Chem. Soc.* **72**, 639 (1950); Kennard, *Org. Chem. Bull.* **29**, no. 1 (1957). Consists of about 55% tripolyphosphoric acid, the remainder being H_3PO_4 and other polyphosphoric acids. Typical analysis: 83.0% P_2O_5; ortho equivalent 115.0%.
Viscous liquid at room temps. Conveniently fluid at 60°. Solidifies to a glass at low temps. Sol in water with evolution of heat and hydrolysis to H_3PO_4.
Caution: In strong concns moderately irritating to skin, mucous membranes.
USE: In organic synthesis for cyclizations and acylations.

7741. Polypropylene. *1-Propene homopolymer;* propylene polymer. Class of compounds also known as olefins. Three forms are possible. *Isotactic* (fiber-forming): methyl groups are all on same side of plane of zig-zag carbon atom chain. *Syndiotactic:* methyl groups are on alternate sides of plane of carbon atom chain. *Atactic* (not fiber-forming, amorphous): methyl groups are in a random arrangement with respect to plane of carbon atom chain. Early synthesis of isotactic form with Ziegler catalyst and comparison with atactic form: Natta *et al., J. Chem. Soc.* **77**, 1708 (1955); Natta, *J. Polymer Sci.* **16**, 143 (1955). *Reviews:* N. G. Gaylord, H. F. Mark, *Linear and Stereoregular Addition Polymers* (Interscience, New York, 1959) pp 54-65; R. W. Moncrieff, *Man-Made Fibres* (Wiley, New York, 4th ed., 1963) pp 500-510; J. G. Cook, *Handbook of Textile Fibres* (Merrow Publishing Co., England, 3rd ed., 1964) pp 369-379; G. Crespi, L. Luciani, "Olefin Polymers (Polypropylene)" in Kirk-Othmer *Encyclopedia of Chemical Technology* Vol. **16** (Wiley-Interscience, New York, 3rd ed., 1981) pp 453-469. Review of thermal decomposition and toxicity: V. Purohit, R. A. Orzel, *J. Am. Coll. Toxicol.* **7**, 221-242 (1988).

Isotactic form, *Amco, Amerfil, Beamette, Courlene PY, DLP, Gerfil, Herculon, Lambeth, Meraklon, Moplen, Olane, Prolene, Tuff-Lite, Ulstron*. Solid material, softens at ~155°, mp ~165°. Low flammability comparable to that of wool. Keeps strength down to −100°. d 0.90-0.92. Practically insol in cold org solvents. Sol in hot decalin, hot tetralin, boiling tetrachloroethane. Shrinks in boiling trichloroethylene. Resistant to acids, alkalies; attacked by strong oxidizing agents, *e.g.*, hydrogen peroxide. Good resistance to abrasion ("pilling"). Tendency to develop static charges. Unstabilized material has poor resistance to sunlight. Difficult to dye, lacks dye-attracting polar groups in structure.

USE: Isotactic form: for fishing gear, ropes, filter cloths, laundry bags, protective clothing, blankets, fabrics, carpets, yarns, etc.

7742. Polysorbates. Polyoxyethylene sorbitan esters; POE sorbitan esters. Nonionic surfactants derived from sorbitan esters, *q.v.* Comprehensive description: P. Becher, "Polyol Surfactants" in *Nonionic Surfactants*, M. J. Schick, Ed. (Dekker, New York, 1967) pp 247-299. Description of prepn and uses: L. R. Chislett, J. Walford, *Int Flavours Food Addit.* **7,** 61 (1976). Pharmacology of polysorbate 80: R. K. Varma *et al., Arzneimittel-Forsch.* **35,** 804 (1985). Determn in foods: H. Kato *et al., J. Assoc. Off. Anal. Chem.* **72,** 27 (1989).

Polysorbate 80
(Sum of w, x, y, and z is 20)

Polysorbate 80, polyoxyethylene (20) sorbitan monooleate, POE (20) sorbitan monooleate, Emsorb 6900, Liposorb O-20, Monitan, Sorlate, T-Maz 80, Tween 80. Lemon- to amber-colored, oily liquid. d 1.06-1.09. Viscosity (25°): 300-500 centistokes. Very sol in water; sol in alcohol, cottonseed oil, corn oil, ethyl acetate, methanol, toluene. Insol in mineral oil. pH of 5% aq soln between 6 and 8. LD$_{50}$ in mice, rats (ml/kg): 7.5, 6.3 i.p. (Varma).

USE: As emulsifiers and dispersing agents in medicinal products; as defoamers and emulsifiers in foods. Pharmaceutic aid (surfactant).

7743. Polytetrafluoroethylene. *Tetrafluoroethene homopolymer; tetrafluoroethylene polymer;* polytetrafluoroethylene resin; polytef; PTFE; Fluon; Teflon; Tetran. A highly stable thermoplastic tetrafluoroethylene homopolymer. Composed of at least 20,000 C$_2$F$_4$ monomer units linked into very long unbranched chains. Prepd by polymerization of tetrafluoroethylene: Plunkett, U.S. pat. 2,230,654 (1941 to Kinetic Chem.); Brubaker, U.S. pat. 2,393,967; Joyce, U.S. pat. **2,394,243** (both 1946 to du Pont); Hanford, Joyce, *J. Am. Chem. Soc.* **68,** 2082 (1946); Renfrew, Lewis, *Ind. Eng. Chem.* **38,** 870 (1946); Renfrew, U.S. pat. 2,534,058 (1950 to du Pont); C. E. Schildknecht, *Vinyl and Related Polymers* (Wiley, New York, 1952) pp 483-494. Account of discovery by Roy J. Plunkett: A. B. Garrett, *J. Chem. Ed.* **39,** 288 (1962). *Reviews:* R. W. Moncrieff, *Man-Made Fibres* (John Wiley, New York, 4th ed., 1963) pp 512-517; McCane in *Encyclopedia of Polymer Science and Technology* vol. **13,** N. M. Bikales, Ed. (Interscience, New York, 1970) pp 623-654; S. V. Gangal in Kirk-Othmer *Encyclopedia of Chemical Technology* vol. **11** (Wiley-Interscience, New York, 3rd ed., 1980) pp 1-24.

n ≥ 20,000

Nonflammable, high polymer. White translucent to opaque solid (depending on thickness). Very inert chemically. Useful temp range from cryogenic to +260°. Melts to an extremely viscous gel at 327° and reverts to the gaseous monomer at temperatures above 400°. d 2.2. Shore hardness 55-56. Tensile strength 3500-4500 psi. Flexural modulus ~80,000-90,000 psi at room temp. Brittle point below −80°. Dielectric constant (at 60 to 3 × 10^9 cycles) 2.0-2.05. Not affected by water, aqua regia, chlorosulfonic acid, acetyl chloride, boron fluoride, hot nitric acid, boiling solns of sodium hydroxide, and organic solvents. Not wetted by water. No substance has been found which will dissolve the polymer at moderate temperatures, but prolonged contact with fluorine, hot plasticizers and polymeric waxes is not recommended. Is subject to cold flow at high pressure. Because of its high melt viscosity molding and sintering techniques similar to those used in powder metallurgy are normally used for fabrication.

Caution: The finished compound is inert under ordinary conditions. There have been reports of polymer fume fever, the symptoms of which resemble those of an attack of influenza, in humans exposed to the heated polymer under conditions of inadequate ventilation. Contamination of smoking tobacco with polytetrafluoroethylene, even in small amounts, is to be avoided.

USE: As tubing and sheets for chemical laboratory and process work; for lining reaction vessels; for gaskets and pump packings, sometimes mixed with graphite or glass fibers; as electrical insulator esp in high frequency applications; filtration fabrics; protective clothing. Prosthetic aid.

7744. Polythiazide. *6-Chloro-3,4-dihydro-2-methyl-3-[[(2,2,2-trifluoroethyl)thio]methyl]-2H-1,2,4-benzothiadiazine-7-sulfonamide 1,1-dioxide;* 2-methyl-3-(β,β,β-trifluoroethylthiomethyl)-6-chloro-7-sulfamyl-3,4-dihydro-1,2,4-benzothiadiazine 1,1-dioxide; 6-chloro-3,4-dihydro-2-methyl-7-sulphamoyl-3-(2,2,2-trifluoroethylthiomethyl)-2H-benzo-1,2,4-thiadiazine 1,1-dioxide; Drenusil; Nephril; Renese. C$_{11}$H$_{13}$ClF$_3$N$_3$O$_4$S$_3$; mol wt 439.89. C 30.04%, H 2.98%, Cl 8.06%, F 12.96%, N 9.55%, O 14.55%, S 21.87%. Prepn: J. M. McManus, U.S. pat. **3,009,911** (1961 to Pfizer). Comprehensive description: T. Negendra Varo Prasad in *Analytical Profiles of Drug Substances* vol. **20,** K. Florey Ed. (Academic Press, New York, 1991) pp 665-692.

Crystals from isopropanol, mp 202.5°. Sol in methanol and acetone. Practically insol in water and chloroform. Sol in aq solns made alkaline with carbonates or hydroxides of the alkali metals. Rate of decompn increases with increase in pH.

THERAP CAT: Diuretic, antihypertensive.

7745. Polyvinyl Alcohol. *Ethenol homopolymer;* PVA; Akwa Tears; Elvanol; Gelvatol; Liquifilm; Mowiol; Polyviol; Sno Tears; Vinarol; Vinol. A polymer prepd from polyvinyl acetates by replacement of the acetate groups with hydroxyl groups. The alcoholysis proceeds most rapidly in a methanol + methyl acetate mixture in the presence of catalytic amounts of alkali or mineral acids: Hermann, Haehnel, *Ber.* **60,** 1658 (1927). *Monograph:* C. E. Schildknecht, *Vinyl and Related Polymers* (Wiley, New York, 1952). The head-to-tail or 1,3-glycol structure is favored: Staudinger *et al., Ber.* **60,** 1782 (1927); *J. Prakt. Chem.* **155,**

261 (1940); Marvel, Denoon, *J. Am. Chem. Soc.* **60**, 1045 (1938); McDowell, Kenyon, *ibid.* **62**, 415 (1940); Marvel, Inskeep, *ibid.* **65**, 1710 (1943). *Reviews:* M. Leeds in Kirk-Othmer *Encyclopedia of Chemical Technology* **vol. 21** (Wiley-Interscience, New York, 2nd ed., 1970) pp 353-368; *Polyvinyl Alcohol,* A. C. Finch, Ed. (Wiley, New York, 1973) 640 pp; A. S. Dunn, *Chem. & Ind. (London)* **1980**, 801-806. Vinyl alcohol monomer has not been isolated. *Review:* R. B. Seymour, G. B. Kauffman, *J. Chem. Educ.* **71**, 582 (1994).

$$\left[\!\!\begin{array}{c} CH_2-CH \\ | \\ OH \end{array}\!\!\right]_n$$

Dry, unplasticized polyvinyl alcohol powders are white to cream colored, soften at about 200° with decompn. Commercial polyvinyl alcohols have different contents of residual acetyl groups and therefore different viscosity characteristics. The first code number following the trade name indicates the degree of hydrolysis, while the second set of numbers indicates the approx viscosity in centipoises (4% aq soln at 20°). Polyvinyl alcohols are essentially sol in hot and cold water, but those coded 20-105 require alcohol-water mixtures. Aq solns are colloidal and compatible with lower alcohols. Pure aq solns are neutral or faintly acid and subject to mold growth. Insol in petroleum solvents.

USE: In the plastics industry in molding compds, surface coatings, films resistant to gasoline, textile sizes and finishing compositions; can be compounded to yield elastomers to be used in manuf artificial sponges, fuel hoses, etc., also in printing inks for plastics and glass, in pharmaceutical finishing, cosmetics, water-sol film and sheeting. Pharmaceutic aid (viscosity increasing agent); ophthalmic lubricant.

7746. Polyvinyl Chloride. *Chloroethene homopolymer; chloroethylene polymer;* PVC; Geon; Breon; Welvic; Movyl; Tevilon; Koroseal; Marvinol. Polyvinyl chloride fibers are marketed under the names: *Rhovyl, Fibravyl, Thermovyl, Isovyl, Retractyl, Crinovyl, Envilon, Nip.* Average mol wt ~60,000 to 150,000. Prepn: Baumann, *Ann.* **163**, 308 (1872); Schoenfeld, U.S. pat. **2,168,808** (1937 to B. F. Goodrich). Structure: Natta, Rigamonti, *Atti accad. Lincei Classe sci. fis., mat. nat.* **24**, 381 (1936); *C.A.* **31**, 4563⁹ (1937); Marvel *et al., J. Am. Chem. Soc.* **61**, 3241 (1939). *Reviews:* C. E. Schildknecht, *Vinyl and Related Polymers* (Wiley, New York, 1952) pp 392-442. Technology: W. S. Penn, *PVC Technology* (Maclaren, London, 1962); J. A. Davidson, K. L. Gardner, "Vinyl Polymers (PVC)" in Kirk-Othmer *Encyclopedia of Chemical Technology* **vol. 23** (Wiley-Interscience, New York, 3rd ed., 1983) pp 886-936. Book: Sarvetnick, *Polyvinyl Chloride* (Van Nostrand, Reinhold, New York, 1969).

$$\left[\!\!\begin{array}{c} CH_2-CH \\ | \\ Cl \end{array}\!\!\right]_n$$

Plastic solid. d 1.406; n 1.54. Stabilizers are necessary to prevent discoloration from exposure to light or heat. Solvents for unmodified polyvinyl chloride of high mol wt: cyclohexanone, methyl cyclohexanone, dimethyl formamide, nitrobenzene, tetrahydrofuran, isophorone, mesityl oxide. Solvents for lower polymers: dipropyl ketone, methyl amyl ketone, methyl isobutyl ketone, acetonylacetone, methyl ethyl ketone, dioxane, methylene chloride.

USE: Rubber substitutes, electric wire and cable-coverings, pliable thin sheeting, film finishes for textiles, nonflammable upholstery, raincoats, tubing, belting, gaskets, shoe soles.

7747. Pomegranate. Granatum. Dried bark of stem or root of *Punica granatum* L., *Punicaceae. Habit.* Mediterranean region; Eastern, Western, and Southern Asia; cultivated in subtropical countries. *Constit.* 0.5-1% alkaloids consisting of pelletierine, methylpelletierine, pseudopelletier-

ine (granatonine), and isopelletierine; mannite, about 20% tannin. Rind of fruit contains about 30% tannin.

THERAP CAT: Formerly as teniafuge.

7748. Ponasterone A. *(2β,3β,5β,22R)-2,3,14,20,22-Pentahydroxycholest-7-en-6-one;* 25-deoxyecdysterone; 25-deoxy-20-hydroxyecdysone. $C_{27}H_{44}O_6$; mol wt 464.64. C 69.79%, H 9.54%, O 20.66%. Polyhydroxylated steroid with strong moulting hormone activity; first phytoecdysteroid to be isolated. Isoln with ponasterones B, C, D, from plant source, *Podocarpus nakaii* Hay., *Podocarpaceae* and structure determn: K. Nakanishi *et al., Chem. Commun.* **1966**, 915; isoln from various Japanese ferns: T. Takemoto *et al., Chem. Pharm. Bull.* **21**, 2336 (1973). Isoln from crustaceans: J. F. McCarthy, *Steroids* **34**, 799 (1979). Stereochemical elucidation: H. Moriyama, K. Nakanishi, *Tetrahedron Letters* **1968**, 1111; using ¹³C-NMR: H. Hikino *et al., Chem. Pharm. Bull.* **23**, 125 (1975). Chromatographic sepn: M. Hori, *Steroids* **14**, 33 (1969). HPLC analysis: I. D. Wilson *et al., J. Chromatog.* **238**, 97 (1982); R. E. Isaac *et al., ibid.* **246**, 317 (1982). Synthesis: G. Hüppi, J. B. Siddall, *Tetrahedron Letters* **1968**, 1113. Moulting hormone activity on houseflies, silkworms: M. Kobayashi *et al., Steroids* **9**, 529 (1967); on various insects: W. E. Robbins *et al., ibid.* **16**, 105 (1970). Hormonal regulation to increase yield from silkworms: T. Okauchi *et al.,* U.S. pat. **3,941,879** (1976 to Takeda). Used to characterize ecdysteroid receptors in *Drosophilia* cells: P. Maroy *et al., Proc. Natl. Acad. Sci. USA* **75**, 6035 (1978); B. A. Sage *et al., J. Biol. Chem.* **257**, 6373 (1982). *Review of ponasterones:* K. Nakanishi, *Bull. Soc. Chim. France* **1969**, 3475. *See also* Ecdysteroids.

Crystals from ethanol, mp 259-260° (dcc). $[\alpha]_D^{15}$ +90° (methanol). uv max (methanol): 244, 326 nm (ε 12400, 130).

7749. Ponceau 3R. *3-Hydroxy-4-[(2,4,5-trimethylphenyl)azo]-2,7-naphthalenedisulfonic acid disodium salt;* sodium cumeneazo-β-naphthol disulfonate; C.I. Food Red 6; C.I. 16155; FD & C Red no. 1; Ext. D & C Red no. 15. $C_{19}H_{16}$-$N_2Na_2O_7S_2$; mol wt 494.46. C 46.15%, H 3.26%, N 5.67%, Na 9.30%, O 22.65%, S 12.97%. Prepn: *Colour Index* **vol. 4** (3rd ed., 1971) p 4092. Toxicity studies: Hansen *et al., Toxicol. Appl. Pharmacol.* **5**, 105 (1963).

Dark red powder. Soluble in water with cherry-red color, slightly in alcohol. Addition of HCl to its aq soln does not change the color, but NaOH produces a yellow ppt. It dissolves in concd H_2SO_4 to a cherry-red soln which does not change color on dilution.

Note: Delisted by the FDA for use in foods, drugs, and cosmetics.

USE: Dyeing wool.

7750. Ponceau SX. *3-[(2,4-Dimethyl-5-sulfophenyl)-azo]-4-hydroxy-1-naphthalenesulfonic acid disodium salt;*

FD & C Red No. 4; C.I. Food Red 1; C.I. 14700. $C_{18}H_{14}$-$N_2Na_2O_7S_2$; mol wt 480.43. C 45.00%, H 2.94%, N 5.83%, Na 9.57%, O 23.31%, S 13.35%. Prepn: E. Nölting, O. Kohn, *Ber.* **19**, 137 (1886). Metabolism: J. L. Radomski, T. J. Mellinger, *J. Pharmacol. Exp. Ther.* **136**, 259 (1962). Toxicity studies: F. C. Lu, A. Lavalle, *Can. Pharm. J.* **97**, 30 (1964); K. J. Davis *et al., Toxicol. Appl. Pharmacol.* **8**, 306 (1966). *Review: IARC Monographs* **8**, 207-214 (1975). *See also: Colour Index* vol. **4** (3rd ed., 1971) p 4068.

Red crystals. Absorption max (0.02 N CH_3COONH_4): 500 nm. Sol in water; slightly sol in ethanol. Practically insol in vegetable oils. Deep red soln in conc H_2SO_4, changing to red ppt on dilution. Orange soln in conc HNO_3, turning yellow. LD_{50} orally in rats: > 2 g/kg (Lu, Lavalle).

USE: In externally applied drugs and cosmetics; formerly in maraschino cherries. Delisted by FDA for use in food: *Fed. Reg.* **41**, 41852 (1976).

7751. Poppy Capsules. Poppy heads. Fully grown, dried capsules of *Papaver somniferum* L., *Papaveraceae. Habit.* Europe, Asia. *Constit.* Capsules: 0.15-0.5% morphine and small amounts of other opium alkaloids. Seed: Fixed oil (poppy oil), albuminoids.

USE: For preparing emulsions—only the white seeds should be used. Poppy seeds are chiefly used for making the oil, and in baking; the bluish-black variety is generally used in the U.S.A. for baking.

7752. Poppy Oil. Poppy-seed oil. Expressed from poppy seeds. Contains no morphine or other opium alkaloids. *See also* Ethiodized Oil; Iodized Oil.

Pale yellow, drying oil; pleasant odor and taste. d 0.924-0.927. Solidif about $-18°$. n_D^{20} 1.4766-1.4774. Sapon. no. 189-197. Iodine no. 133-158.

USE: Manuf paints, varnishes, and soaps. Edible grades are marketed in Europe and Asia.

7753. Populin. *2-(Hydroxymethyl)phenyl-β-D-glucopyranoside 6-benzoate;* populoside; salicin benzoate. $C_{20}H_{22}O_8$; mol wt 390.39. C 61.53%, H 5.68%, O 32.79%. In bark and leaves of *Populus tremula* L., *P. nigra* L., *P. nigra* L. var. *italica* Duroi, *P. canadensis* Moench., *P. grandidentate* Michx., and *P. tremuloides* Michx., *Salicaceae*, perhaps also in *Salix helix, Salix purpureae* L. var *helix* (L.) Koch. Isoln: Pearl *et al., J. Org. Chem.* **27**, 2685 (1962). May be made from salicin by melting with benzoic anhydride, or from salicin and benzoyl chloride in presence of potassium hydroxide: Richtmyer, Yeakel, *J. Am. Chem. Soc.* **56**, 2495 (1934). Alkaline cleavage produces benzoic acid and salicin. Enzymatic hydrolysis using taka-diastase *(q.v.)* gives salicyl alcohol and benzoyl glucose: Kitasato, *Biochem. Z.* **190**, 109 (1927).

Dihydrate, needles from water, sweet taste, like licorice. Becomes anhydr at 100°. mp 179°. $[\alpha]_D^{20}$ $-2.0°$ (c = 5 in pyridine); $[\alpha]_D^{25}$ $-29.7°$ (c = 5 in acetone). One gram dissolves in about 2 liters water, in 42 ml boiling water, in about 100 ml alcohol. Practically insol in ether.

7754. Poractant Alfa. PLS; Curosurf. Porcine lung extract composed of 99% polar lipids and 1% apoproteins. The phospholipid component consists primarily of phos-phatidylcholine and its derivatives. Differs from other pulmonary surfactants in its 9.9% sphingomyelin content. Isolation: T. Curstedt *et al., Eur. J. Biochem.* **168**, 255 (1987). Characterization of phospholipid content: E. Redenti *et al., Farmaco* **49**, 285 (1994). Evaluation as surfactant replacement in rabbit: B. Robertson *et al., Eur. J. Pediatr.* **147**, 168 (1988); in comparison with synthetic surfactants: S. Bongrani *et al., Biol. Neonate* **65**, 406 (1994). Multicenter clinical trial in RDS in infants: Collab. Eur. Multicent. Study Group, *Pediatrics* **82**, 683 (1988); H. L. Halliday *et al., Arch. Dis. Child.* **69**, 276 (1993).

THERAP CAT: Pulmonary surfactant; in treatment of respiratory distress syndrome.

7755. Porfimer Sodium. *Photofrin II;* CL-184116. Light sensitive polyporphyrin oligomer linked via ethers and esters; purification product of hematoporphyrin derivative. Forms aggregates with combined mol wt ~10000. Prepn: T. J. Dougherty *et al.,* "Photoradiation Therapy - Clinical and Drug Advances" in *Porphyrin Photosensitization,* D. Kessel, T. J. Dougherty, Eds. (Plenum Press, New York, 1983) pp 3-13; T. J. Dougherty *et al.,* U.S. pat. 4,-649,151 (1987 to Health Res.). Structural studies: T. J. Dougherty, *Photochem. Photobiol.* **46**, 569 (1987); R. K. Pandey *et al., Biomed. Environ. Mass Spectrom.* **19**, 405 (1990). Photobleaching: J. Moan, D. Kessel, *J. Photochem. Photobiol.* **1**, 429 (1988); J. D. Spikes, *Photochem. Photobiol.* **55**, 797 (1992). Pharmacokinetics in mice: D. A. Bellnier *et al., ibid.* **50**, 221 (1989). Clinical evaluation in treatment of esophageal cancer: T. Okunaka *et al., Surg. Endosc.* **4**, 150 (1990). Imaging study for premalignant lesions: C. Liebow *et al., Proc. Nat. Acad. Sci. USA* **90**, 1897 (1993). Clinical use as imaging agent for lung cancer: S. Lam *et al., Chest* **97**, 333 (1990).

Orange red color.

USE: Fluorescent imaging of malignancies.

THERAP CAT: Antineoplastic (photosensitizer).

7756. Porfiromycin. *[1aS-(1aα,8β,8aα,8bα)]-6-Amino-8-[[(aminocarbonyl)oxy]methyl]-1,1a,2,8,8a,8b-hexahydro-8a-methoxy-1,5-dimethylazirino[2',3':3,4]pyrrolo[1,2-a]indole-4,7-dione; N*-methylmitomycin C; U-14743. $C_{16}H_{20}N_4$-O_5; mol wt 348.36. C 55.17%, H 5.79%, N 16.08%, O 22.96%. Antibiotic substance isolated from a *Streptomyces ardus* fermentation broth: Herr *et al., Antimicrob. Ag. Ann.* **1960**, 23. Isoln from *S. verticillatus* and structure: Webb *et al., J. Am. Chem. Soc.* **84**, 3185, 3187 (1962). Production process: Bohonos *et al.,* U.S. pat. **3,219,530** (1965 to Am. Cyanamid). Synthesis of (±)-form: F. Nakatsubo *et al., J. Am. Chem. Soc.* **99**, 8115 (1977). For stereochemistry and other synthetic studies, *see* Mitomycins.

Dark purple triclinic crystals, dec 201-201.5°. $[\alpha]_D^{25}$ +275° ±55° (c = 0.1% in methanol); $[\alpha]_D^{25}$ +242° ±100° (c = 0.045% in methanol). uv max (methanol): 217, 360, 555 nm (ε 24600, 23000, 209). Slightly sol in water, moderately sol in polar organic solvents. Practically insol in hydrocarbon solvents.

THERAP CAT: Antibacterial; antineoplastic.

7757. Porofor® BSH. *Benzenesulfonic acid hydrazide;* benzenesulfohydrazide; phenylsulfohydrazide. $C_6H_8N_2O_2S$; mol wt 172.21. C 41.85%, H 4.68%, N 16.27%, O 18.58%, S 18.62%. $C_6H_5SO_2NHNH_2$. *Ref:* Lober, *Angew. Chem.* **64**, 65 (1952).

Crystals, dec 103-104° with the evolution of nitrogen. May be stored indefinitely at temps up to 80°. Sensitive to moist oxidizing agents.

USE: Gas generating agent for use in making foam rubber and foam plastics.

7758. Porphine. *21H,23H-Porphine;* porphin. $C_{20}H_{14}$-N_4; mol wt 310.36. C 77.40%, H 4.55%, N 18.05%. Parent substance of the *porphyrins,* a group of compounds found in all living matter which are the basis of respiratory pigments in animals and plants. In porphyrins, side chains are substituted for the hydrogens in the porphine pyrrole rings. *Chlorins* are dihydroporphyrins. *See also:* chlorophyll, hemoglobin, vitamin B_{12}, hematin. Prepn of porphine: Fischer, Gleim, *Ann.* **521,** 157 (1935); Rothemund, *J. Am. Chem. Soc.* **58,** 625 (1936); Krol, *J. Org. Chem.* **24,** 2065 (1959). Study of porphyrin analogs: C. L. Honeybourne *et al., Tetrahedron* **36,** 1833 (1980). Review of biosyntheses of porphyrins and chlorins: A. R. Battersby, E. McDonald in *Porphyrins Metalloporphyrins,* K. M. Smith, Ed. (Elsevier, New York, 1975) pp 61-122. Review of porphyrin syntheses: R. P. Evstigneeva, *Pure Appl. Chem.* **53,** 1129-1140 (1981). Comprehensive seven volume treatise: *The Porphyrins,* D. Dolphin, Ed. (Academic Press, New York, 1978).

Dark red, shiny leaflets from chloroform-methanol. Darkens at 360° but does not melt. The absorption bands are those characteristic of the substituted porphyrins, details in Fischer-Orth, *Die Chemie des Pyrrols* vol. II, 1, 175 (1937). Sol in pyridine, dioxane, and phenol; slightly sol in chloroform, bromoform, glacial acetic acid. Almost insol in acetone, methanol, ether. HCl number: 1.7 (Fischer); 3.3 (Rothemund).
Iron salt, $C_{20}H_{12}N_4$·FeCl, brown cubes from ether.
Magnesium salt, $C_{20}H_{12}MgN_4$, red needles.
Copper salt, $C_{20}H_{12}N_4Cu$, brown needles.

7759. Porphobilinogen. *5-(Aminomethyl)-4-(carboxymethyl)-1H-pyrrole-3-propanoic acid;* 2-aminomethylpyrrol-3-acetic acid 4-propionic acid. $C_{10}H_{14}N_2O_4$; mol wt 226.23. C 53.09%, H 6.24%, N 12.38%, O 28.29%. An intermediate in the biosynthesis of heme, *q.v.,* found in the urine of patients with acute porphyria: Westall, *Nature* **170,** 614 (1952). Isoln and structure: Cookson, Rimington, *Biochem. J.* **57,** 476 (1954). Synthesis: Jackson, MacDonald, *Can. J. Chem.* **35,** 715 (1957). Chemistry of conversion into porphyrins: Mathewson, Corwin, *J. Am. Chem. Soc.* **83,** 135 (1961).

Monohydrate, minute pink crystals from dil ammonium acetate soln at pH 4. Dec 172-175°. pK' = 3.70; 4.95; 10.1. Slightly sol in water. Converted to uroporphyrins by hot dil HCl or by blood hemolyzates.
Hydrochloride monohydrate, fine triclinic needles from dil HCl, dec 165-170°. Soluble in water.

7760. Porphyrillic Acid. *1,3-Dihydro-4,10-dihydroxy-8-methyl-3-oxoisobenzofuro[5,4-b]benzofuran-7-carboxylic acid;* 1,7-dihydroxy-9-hydroxymethyl-3-methyl-4,8-dibenzofurandicarboxylic acid (8,9)-γ-lactone. $C_{16}H_{10}O_7$; mol wt 314.25. C 61.15%, H 3.21%, O 35.64%. Constituent of lichens. Isoln from *Haematomma coccineum* Dicks., *Lecanoreae,* and structure studies: Wachtmeister, *Acta Chem. Scand.* **8,** 1433 (1954); **10,** 1404 (1956).

Needles from ethanol or dioxane, mp 280-283° (dec) after darkening from 270°. Soluble in concd sulfuric acid, sodium carbonate, sodium hydroxide; sparingly sol in sodium bicarbonate.

7761. Porphyropsin. Photoreceptor protein found in the retinal rod cells of fresh water and migrating fish, lampreys, and certain amphibians. Absorption maximum approx 520-530 nm. Composed of the chromophore, 11-*cis*-3-dehydroretinal, *q.v.,* bound to scotopsin, the specific protein component of rod pigments (see Opsins). Biological activity is similar to that of rhodopsin, *q.v.* Isoln from retinas: G. Wald, *Nature* **139,** 1017 (1937); *idem, J. Gen. Physiol.* **22,** 775 (1939). Prepn from 3-dehydroretinal and opsin: *idem, Ann. Rev. Biochem.* **22,** 497 (1953). Methods of purification, prepn and assay: R. Hubbard *et al.,* "Methodology of Vitamin A and Visual Pigments", in *Methods in Enzymology* vol. **18,** D. B. McCormick, L. D. Wright, Eds. (Academic Press, New York, 1971) pp 615-653. Exposure to light initiates the conversion of porphyropsin through a series of distinct intermediates to yield opsin and *trans*-3-dehydroretinal. Bleaching kinetics and photochemistry: T. Yoshizawa in *Handbook of Sensory Physiology* Vol. VII(2), H. J. A. Dartnall, Ed. (Springer-Verlag, New York, 1972) pp 146-179. Porphyropsin may co-exist with rhodopsin in the retinas of certain amphibians and fish: T. E. Reuter *et al., J. Gen. Physiol.* **58,** 351 (1971). Environmental effect on visual pigment formation and interconversion: P. Witkovsky *et al., ibid.* **72,** 821 (1978); A. T. C. Tsin, D. D. Beatty, *Exp. Eye Res.* **30,** 143 (1980). *Review:* G. Wald, *Science* **162,** 230-239 (1968).

7762. Potash Sulfurated. Liver of sulfur; sulfurated potassa; hepar sulfuris. Mixture of potassium polysulfides, chiefly trisulfide, and potassium thiosulfate, containing not less than 12.8% sulfur as sulfide. *U.S.D.* (5th ed., 1955) p 1092. Has been incorrectly called "potassium sulfide".
Yellowish-brown lumps; fresh fracture is liver-brown, slight H_2S odor. Dec on exposure to air, forming free sulfur and potassium carbonate and becomes yellow to gray. Freely sol in water, usually with slight residue; partially sol in alcohol. *Keep tightly closed.*
Incompat. Acids, alcohol, carbonated waters, acid salts. *Caution:* Avoid metal bathtubs, metal spoons, and water containing much CO_2.
USE: Producing color effects on brass, bronze, nickel, gunmetal.
THERAP CAT: Sulfide source (skin diseases).
THERAP CAT (VET): Externally for chronic and parasitic skin diseases.

7763. Potassium. Kalium. K; at. wt 39.0983; at. no. 19; valence 1. Alkali metal. Occurrence in earth's crust: 2.59% by wt. Natural isotopes: 39 (93.22%); 40 (0.012%); 41 (6.77%); ^{40}K is radioactive: $T_{1/2}$ 1.26 × 19^9 years; isotopes range in mass number from 37 to 47. Found mainly as the chloride (sylvite); also in the aluminosilicates *orthoclase,* and *microcline* ($KAlSi_3O_8$), and as *carnallite* (KCl·$MgCl_2$·$6H_2O$). First prepd in free form by Davy in 1807 by electrolysis of fused potassium hydroxide. Prepns: Hackspill, *Helv. Chim. Acta* **11,** 1003 (1928); Jackson, Werner, U.S. pat. **2,480,655** (1949 to Mine Safety Appliances Co.). NMR spectrum of potassium anion (K^-): P. P. Edwards *et al., Nature* **317,** 242 (1985). *Review:* Whaley, "Sodium, Potassium, Rubidium, Cesium, and Francium" in *Comprehensive Inorganic Chemistry* vol. **1,** J. C. Bailar, Jr. *et al.,* Eds. (Pergamon Press, Oxford, 1973) pp 369-529; J. S. Greer *et al.,* in Kirk-Othmer *Encyclopedia of Chemical Technology* vol. **18** (Wiley-Interscience, New York, 3rd ed., 1982) pp 912-920.

Consult the Name Index before using this section.

Soft, silvery-white metal; body-centered cubic structure; tarnishes on exposure to air; becomes brittle at low temps; mp 63.2°. bp 765.5°. d^{20} 0.856. Specific heat (0°): 0.176 cal/g deg. Thermal conductivity (cal/sec °C cm): 0.23 (21°); 0.956 (400°). Sol in liq ammonia, ethylenediamine, aniline; sol in several metals, forming alloys. One of the most active metals; E^0 (aq) K/K^+ 2.922 V. Reacts vigorously with oxygen; with water even at $-100°$; with acids; with the halogens, igniting with bromine and iodine. Molten metal reacts with sulfur; with hydrogen sulfide. Reacts with hydrogen slowly at 200°, rapidly at 350-400°. Reacts slowly with anhyd hydrogen halides at room temp; molten metal ignites in the reaction. Reduces silicates, sulfates, nitrates, carbonates, phosphates, oxides and hydroxides of the heavy metals, often with the separation of the metal. Reacts with organic compds containing active groups. Inert to saturated aliphatic and to aromatic hydrocarbons. *Keep under liquid containing no oxygen*, e.g., liquid petrolatum, petroleum, etc.

USE: In synthesis of inorganic potassium compds; in organic syntheses involving condensation, dehalogenation, reduction, and polymerization reactions. As heat transfer medium together with sodium: *Chem. & Eng. News* **33**, 648 (1955).

7764. Potassium Acetate. $C_2H_3KO_2$; mol wt 98.14. C 24.48%, H 3.08%, K 39.84%, O 32.60%. CH_3COOK.
Colorless, lustrous, rapidly deliquesc crystals or white cryst powder or flakes. d^{25} 1.57. mp 292°. One gram dissolves in 0.5 ml water, 0.2 ml boiling water, 2.9 ml alcohol. The aq soln is alkaline to litmus. pH of 0.1 molar aq soln 9.7. *Keep tightly closed.* LD_{50} orally in rats: 3.25 g/kg, H. F. Smyth *et al.*, *Am. Ind. Hyg. Assoc. J.* **30**, 470 (1969).
THERAP CAT: Alkalizer.
THERAP CAT (VET): Has been used in cardiac arrhythmias, as an expectorant. Diuretic.

7765. Potassium Aluminate. *Aluminum potassium oxide.* $Al_2K_2O_4$; mol wt 196.16. Al 27.51%, K 39.86%, O 32.63%.
Trihydrate, hard, lustrous crystals. Freely sol in water. Insol in alcohol. The aq soln is strongly alkaline.

7766. Potassium p-Aminobenzoate. *p-Aminobenzoic acid potassium salt;* potassium *para*-aminobenzoate; KPABA; Potaba. $C_7H_6KNO_2$; mol wt 175.23. C 47.98%, H 3.45%, K 22.31%, N 7.99%, O 18.26%. Prepn: E. A. Meyers *et al.*, *J. Am. Chem. Soc.* **89**, 3565 (1967). Crystal structure: G. S. Ciminago, *Compt. Rend.* Ser. C **267**, 1402 (1968). Antifibrotic efficacy in idiopathic pulmonary fibrosis: U. H. Cegla *et al.*, *Pneumonologie* **152**, 75 (1975); in Peyronie's disease: A. J. Riley, *Brit. J. Sex. Med.* **6**, 29 (1970); G. Williams, N. A. Green, *Brit. J. Urol.* **52**, 392 (1980).

Crystals from alcohol. Saline taste. Slightly alkaline to litmus. pH of 1% soln about 7. Very freely sol in water, less sol in alc. Practically insol in ether. Reported to cause less gastric irritation than the free acid or the sodium salt.
USE: Catalyst in the manuf of condensation polymers of polyglycol ethers.
THERAP CAT: Antifibrotic.

7767. Potassium Arsenate. Potassium acid arsenate; potassium dihydrogen arsenate; Macquer's salt. AsH_2KO_4; mol wt 180.03. As 41.62%, H 1.12%, K 21.72%, O 35.55%. KH_2AsO_4.
Colorless crystals or white, cryst mass or powder. *Poisonous!* d 2.8. Sol in 5.5 parts cold, more sol in hot water, slowly in 1.6 parts glycerol. Insol in alcohol.
USE: In the textile, tanning, and paper industries. In insecticidal formulations (especially fly paper).

7768. Potassium Arsenite. Somewhat variable composition; the article of commerce is of a composition corresponding to approx. $KAsO_2.HAsO_2$.
White, hygroscopic powder; gradually dec on exposure to air (by the CO_2). *Very poisonous!* Sol in water. *Keep well closed.* LD_{50} orally in rats: 14 mg/kg, *Handbook of Toxicology* vol. **1**, W. S. Spector, Ed. (Saunders, Philadelphia, 1956) pp 240-241.
USE: In manuf of mirrors to reduce the silver salt to metallic silver.

7769. Potassium Arsenite Solution. Fowler's soln; arsenical soln. Made from arsenic trioxide 10 g; potassium bicarbonate 7.6 g; alcohol 30 ml; distilled water to one liter.
Colorless and odorless, clear liquid. It may mold. *Very poisonous! Incompat.* Alkaloidal salts, hypophosphites and sulfites in acid soln; salts of Fe and of most other heavy metals; tannic acid.
THERAP CAT: Antineoplastic; dermatologic.
THERAP CAT (VET): Has been used as an alterative or tonic, in pulmonary emphysema, chronic coughs, anemias, chronic skin diseases.

7770. Potassium Bicarbonate. Potassium acid carbonate; Kafylox; K-Lyte. $CHKO_3$; mol wt 100.12. C 12.00%, H 1.01%, K 39.05%, O 47.94%. $KHCO_3$. Contains not less than 99% $KHCO_3$.
Colorless, transparent crystals, white granules or powder. Sol in 2.8 parts water, 2 parts water at 50°. Almost insol in alcohol. pH: 8.2 (in 0.1 molar concn).
USE: In baking powders, effervescent salts.
THERAP CAT: Potassium supplement.

7771. Potassium Bifluoride. Potassium acid fluoride; potassium hydrogen fluoride. F_2HK; mol wt 78.10. F 48.65%, H 1.29%, K 50.06%. KF.HF. Prepd according to the eq $KOH + 2HF = KHF_2 + H_2O$: Lange, Eichler, *Z. Physik. Chem.* **129**, 285 (1927); Kwasnik in *Handbook of Preparative Inorganic Chemistry* vol. **1**, G. Brauer, Ed. (Academic Press, New York, 2nd ed., 1963) p 237. Made commercially from potassium carbonate and hydrofluoric acid.
Tetragonal crystals. *Poisonous!* d 2.37. mp 238.7°. Transformation pt 195°. Soly in water (g/100 ml): 30.1 (10°); 39.2 (20°); 114.0 (80°). Sol in dil alc. Insol in abs alc.
Caution: Corrosive and irritating to skin, mucous membranes.
USE: In the prepn of pure potassium fluoride; as an electrolyte in the manuf of fluorine; frosting glass; treating coal to prevent slag formation; flux for silver solders; catalyst in the alkylation of benzene with olefins.

7772. Potassium Binoxalate. Potassium acid oxalate; salt of sorrel; sal acetosella. C_2HKO_4; mol wt 128.13. C 18.75%, H 0.79%, K 30.52%, O 49.95%. KOOCCOOH. Incorrectly *"salt of lemon".* The same synonyms apply to potassium tetraoxalate.
Monohydrate, white, odorless crystals. *Poisonous!* d 2.0. Sol in 40 parts cold, 6 parts boiling water, slightly in alcohol. pH of 0.1 molar aq soln: 2.7.
USE: Removing ink stains, scouring metals, cleaning wood; photography; as mordant in dyeing; bleaching stearic acids.

7773. Potassium Biphthalate. *Phthalic acid potassium acid salt;* potassium acid phthalate; potassium hydrogen phthalate; acid potassium phthalate. $C_8H_5KO_4$; mol wt 204.22. C 47.05%, H 2.47%, K 19.14%, O 31.34%. HOOC-C_6H_4COOK. Prepd by half-neutralization of a phthalic anhydride soln: F. J. Welcher, *Organic Analytical Reagents* vol. **2** (Van Nostrand, New York, 1947) pp 75-79.
Orthorhombic crystals, stable in air. d^{25} 1.636. Acid reaction; pH of 0.05M aq soln at 25° = 4.005 (glass electrode). Sol in about 12 parts cold water, 3 parts boiling water; slightly sol in alcohol.
USE: As primary standard for preparing volumetric alkali solns, also as a buffer in pH determinations.

7774. Potassium Bisulfate. Potassium acid sulfate; potassium hydrogen sulfate; sal enixum. HKO_4S; mol wt 136.17. H 0.74%, K 28.71%, O 47.00%, S 23.55%. $KHSO_4$.

White, deliquesc crystals, pieces, or granules. d 2.24. mp 197°; at higher temp loses water and is converted into pyrosulfate. Sol in 1.8 parts water, 0.85 part boiling water. *Keep well closed.*

USE: As flux in analysis of ores and siliceous compds.

THERAP CAT: Cathartic.

7775. Potassium Bisulfide. Potassium hydrosulfide; potassium hydrogen sulfide; potassium sulfhydrate. HKS; mol wt 72.17. H 1.40%, K 54.17%, S 44.43%. KHS. Prepd industrially from $Ca(HS)_2$ and K_2SO_4: Hene, **Ger.** pat. **380,385** (1922); from H_2S and K_2S: Bassett, **U.S.** pat. **1,662,735** (1925); Strosacker, Jones, **U.S.** pat. **1,771,384** (1926 to Dow). Prepn of pure material by the action of dry H_2S upon potassium metal dissolved in abs ethanol: Rule, *J. Chem. Soc.* **99,** 558, 564 (1911); West, *Z. Kryst.* **88,** 102 (1934).

Colorless, deliquescent crystals or white, strongly hygroscopic, cryst mass. Usually present as the hemihydrate. Trigonal system. d 1.70. Rapidly becomes yellow on exposure to air with formation of polysulfides and H_2S. Becomes anhydr at 175-200°. mp 450-510° forming a dark red liquid. Heat of formation: +62.5 kcal. Heat of soln at 17°: +0.77 kcal, for hemihydrate at 16°: +0.62 kcal. Freely sol in water, alcohol.

7776. Potassium Bitartrate. Potassium acid tartrate; acid potassium tartrate; potassium hydrogen tartrate; cream of tartar; cremor tartari; faecula; faecla. $C_4H_5KO_6$; mol wt 188.18. C 25.53%, H 2.68%, K 20.78%, O 51.01%. $KH-C_4H_4O_6$. Obtained from the sediments in the manuf of wine, known as argols or wine lees. The salt is at least 99.5% pure. *See also* Argol and Tartaric Acid.

Colorless crystals or white, cryst powder; pleasant acidulous taste. One gram dissolves in 162 ml water, in 16 ml boiling water, 8820 ml alcohol; readily sol in dil mineral acids, in solns of alkalies or borax. Soly in water also given as about 0.4% at 10° to about 6% at 100°.

USE: Largely in baking powders; coloring metals, galvanic tinning of metals; reducer of CrO_3 in mordants for wool.

THERAP CAT: Cathartic.

THERAP CAT (VET): Laxative, diuretic.

7777. Potassium Borohydride. *Potassium tetrahydroborate.* BH_4K; mol wt 53.94. B 20.04%, H 7.47%, K 72.48%. KBH_4. Prepd from potassium tetramethoxyborohydride and diborane at low temps: Schlesinger *et al., J. Am. Chem. Soc.* **75,** 199 (1953). Commercial process from sodium borohydride and NaOH in water: Banus *et al., ibid.* **76,** 3848 (1954). Review of potassium and other metal tetrahydroborates: James, Wallbridge, *Prog. Inorg. Chem.* **11,** 99-231 (1970).

Non-hygroscopic cryst (difference from $NaBH_4$). Stable to air. d 1.11. Dec at about 500° without melting. Supports combustion. Thermally more stable and less reactive to oxidation than $NaBH_4$. Negative heat of soln in H_2O = 6.3 kcal/mol. Soly (w/w) in water at 25°: 19%; liq ammonia at 25°: 20%; ethylenediamine at 75°: 3.9%; methanol at 20°: 0.7%; DMF at 20°: 15.0%. Soly in a 4:1 water-methanol mixture: 13 g/100 g. Alkaline aq solutions are stable.

USE: As of sodium borohydride.

7778. Potassium Borotartrate. Potassium tartratoborate; soluble cream of tartar; borated cream of tartar; potassium sodium borotartrate. Made by evaporating a soln of 2 parts borax and 7 parts potassium bitartrate.

White, odorless powder. Freely sol in water.

USE: Has been used in photography as a retarder for alkaline developers.

7779. Potassium Bromate. $BrKO_3$; mol wt 167.00. Br 47.85%, K 23.41%, O 28.74%. $KBrO_3$.

White crystals or granules. d 3.27. mp about 350°, decomposing at about 370° with evolution of oxygen. Sol in 12.5 parts water, 2 parts boiling water. Almost insol in alc.

Caution: Ingestion may cause vomiting, diarrhea, methemoglobinemia, renal injury.

USE: Bread- and flour-improving agent; in analytical chemistry.

7780. Potassium Bromide. BrK; mol wt 119.00. Br 67.14%, K 32.86%. KBr. Continuous electrolytic process of prepn: Maylott, Elkins, **U.S.** pat. **2,989,450** (1961 to Dow).

Colorless crystals or white granules or powder. d 2.75. mp 730°. One gram dissolves in 1.5 ml water, 1 ml boiling water, 250 ml alc, 4.6 ml glycerol. The aq soln is neutral.

Caution: Large doses cause CNS depression. Prolonged intake may cause mental deterioration, acneform skin eruptions.

USE: Manuf photographic papers and plates; process engraving.

THERAP CAT: Sedative, anticonvulsant.

THERAP CAT (VET): Sedative.

7781. Potassium Carbonate. Salt of tartar; pearl ash. CK_2O_3; mol wt 138.21. C 8.69%, K 56.58%, O 34.73%. K_2CO_3.

Hygroscopic, odorless granules or granular powder. d 2.29; mp 891°. Sol in 1 part cold, 0.7 part boiling water. Practically insol in alcohol. Its aq soln is strongly alkaline. pH 11.6. *Keep tightly closed.* LD_{50} orally in rats: 1.87 g/kg, H. F. Smyth *et al., Am. Ind. Hyg. Assoc. J.* **30,** 470 (1969).

Sesquihydrate, small granular crystals. When it contains the full amount of water (16.36%) it is not hygroscopic. Sol in less than 1 part water. Practically insol in alcohol. The aq soln is strongly alkaline.

Caution: Irritant, caustic.

USE: Manuf soap, glass, pottery, smalts and many potassium salts; in process engraving and lithography; tanning and finishing leather; liq shampoos; for removal of water from organic liqs; in anal. chemistry.

THERAP CAT: Alkalizer, diuretic.

7782. Potassium Chlorate. Potcrate. $ClKO_3$; mol wt 122.55. Cl 28.93%, K 31.90%, O 39.17%. $KClO_3$. Contains at least 99% $KClO_3$.

Colorless, lustrous crystals, or white granules or powder. d 2.32. mp 368°; above this temp it dec into perchlorate and oxygen. One gram dissolves slowly in 16.5 ml water, 1.8 ml boiling water, about 50 ml glycerol. Almost insol in alcohol. *Keep out of contact with organic matter or other oxidizable substances. Caution:* Explodes with sulfuric acid; inflames with explosion if triturated with any organic substances, sulfur, phosphorus, sulfite, hypophosphite, and other oxidizable substances. *Incompat.* Iodides, tartaric acid.

Caution: Irritating to G.I. tract, kidney; can cause hemolysis of red blood cells and methemoglobinemia: *Clinical Toxicology of Commercial Products,* R. E. Gosselin *et al.,* Eds. (Williams & Wilkins, Baltimore, 5th ed., 1984) Section II, p 112; Section III, pp 74-77.

USE: Explosives; fireworks; matches; printing and dyeing cotton and wool black; manuf aniline black and other dyes; source of oxygen; in chemical analyses.

THERAP CAT: Formerly as topical antiseptic.

THERAP CAT (VET): In dilute soln as antiseptic mouthwash.

7783. Potassium Chloride. Chloropotassuril; Diffu-K; Enseal; Kaleorid; Kalitabs; Kalium-Duriles; Kaon-Cl; Kaskay; Kayback; Kay-Cee-L; K-Contin; Klor-Con; K-Norm; K-Tab; Lento-Kalium; Micro K; Nu-K; Peter-Kal; PfiKlor; Rekawan; Repone K; Slow-K; Span-K. ClK; mol wt 74.55. Cl 47.55%, K 52.45%. KCl. Occurs in nature as the mineral *sylvine* or *sylvite.* Industrial prepns: Faith, Keyes & Clark's *Industrial Chemicals,* F. A. Lowenheim, M. K. Moran, Eds. (Wiley-Interscience, New York, 4th ed., 1975) pp 666-673.

White crystals or crystalline powder. d 1.98. mp 773°. One gram dissolves in 2.8 ml water, 1.8 ml boiling water, 14 ml glycerol, about 250 ml alcohol. Insol in ether, acetone. Hydrochloric acid, sodium or magnesium chlorides diminish its soly in water. d of saturated aq soln at 15° is 1.172. pH: about 7.

Caution: Large doses by mouth can cause G.I. irritation, purging, weakness and circulatory disturbances.

USE: In photography. In buffer solns, electrode cells.

THERAP CAT: Electrolyte replenisher.
THERAP CAT (VET): Potassium supplement.

7784. Potassium Chromate(VI). Neutral potassium chromate. CrK_2O_4; mol wt 194.19. Cr 26.78%, K 40.27%, O 32.96%. K_2CrO_4.

Lemon-yellow crystals; d 2.73; mp 975°. Sol in 1.6 parts cold, 1.2 parts boiling water. Insol in alcohol. The aq soln is alkaline to litmus or phenolphthalein.

USE: Has a limited application in enamels, finishing leather, rustproofing of metals, being replaced by the sodium salt; as reagent in analytical chemistry.

7785. Potassium Citrate. Urocit-K. $C_6H_5K_3O_7$; mol wt 306.40. C 23.52%, H 1.64%, K 38.28%, O 36.55%. $K_3C_6H_5O_7$. It is at least 99% pure.

Monohydrate, white crystals, granules or powder. Loses its water at 180°. One gram dissolves in 0.65 ml water; very slowly in 2.5 ml glycerol. Practically insol in alcohol. The aq soln is alkaline to litmus; pH about 8.5.

THERAP CAT: Antiurolithic. Antacid.
THERAP CAT (VET): Diuretic.

7786. Potassium Citrate, Monobasic. Monopotassium citrate. $C_6H_7KO_7$; mol wt 230.22. C 31.30%, H 3.06%, K 16.98%, O 48.65%. $KH_2C_6H_5O_7$.

White, cryst powder. Sol in water; the soln is subject to molding.

USE: A 0.05 molal solution as standard for pH scale (pH at 25° 3.776): Staples, Bates, *J. Res. Nat. Bur. Stand.* **73A**, 37 (1969).

7787. Potassium Cobaltous Selenate. $CoK_2O_8Se_2$; mol wt 423.05. Co 13.93%, K 18.48%, O 30.26%, Se 37.33%. $K_2Co(SeO_4)_2$. Prepd by evaporating a soln of the component salts: von Hauer, *Sitzungsber. Akad. Wien* **39**, 839 (1860).

Hexahydrate, garnet-red monoclinic crystals. d 2.514. Stable in air.

7788. Potassium Cyanate. CKNO; mol wt 81.12. C 14.81%, K 48.20%, N 17.27%, O 19.72%. Inhibitor of the sickling of erythrocytes *in vitro*: Cerami, Manning, *Proc. Nat. Acad. Sci. USA* **68**, 1180 (1971). *See also* Sodium Cyanate. Pharmacology: Cerami *et al.*, *J. Pharmacol. Exp. Ther.* **185**, 653 (1973).

White, cryst powder. d 2.05. Sol in water, very slightly in alcohol. LD_{50} i.p. in mice: 320 mg/kg.

7789. Potassium Cyanide. CKN; mol wt 65.12. C 18.45%, K 60.04%, N 21.51%. KCN. The article of commerce contains about 95% KCN. Toxicity study: Hayes, *Toxicol. Appl. Pharmacol.* **11**, 327 (1967).

White, deliquesc, granular powder or fused pieces; odor of HCN. *Violent poison!* On exposure to air it is gradually dec by CO_2 and moisture. d 1.52; mp 634°. Sol in 2 parts cold, 1 part boiling water, 2 parts glycerol, 100 parts alcohol, 25 parts methanol. The aq soln is strongly alkaline and rapidly dec. pH of 0.1N aq soln: 11.0. *Keep tightly closed and protected from light. Incompat.* Acids and acid syrups; alkaloids, chloral hydrate, iodine, metallic salts, permanganates, chlorates, peroxides. LD_{50} orally in rats: 10 mg/kg (Hayes).

Caution: Potential symptoms of overexposure are weakness, headache and confusion; nausea, vomiting; increased rate of respiration; slow gasping respiration; irritation of eyes and skin; asphyxia and death can occur. *See NIOSH Pocket Guide to Chemical Hazards* (DHHS/NIOSH 90-117, 1990) p 76.

USE: Similar to sodium cyanide.

7790. Potassium Dichromate(VI). Potassium bichromate. $Cr_2K_2O_7$; mol wt 294.18. Cr 35.35%, K 26.58%, O 38.07%. $K_2Cr_2O_7$. In the U.S.A. it is usually prepared by the reaction of potassium chloride on sodium dichromate: Vetter in Kirk-Othmer *Encyclopedia of Chemical Technology* **vol. 3** (Interscience, New York, 1949) p 951; Hartford, Copson, *ibid.* **vol. 5** (2nd ed., 1964) pp 484-488. In Germany it is obtained from potassium chromate produced by roasting the chrome ore with KOH. *Ref:* Müller, Glissmann in *Ullmann's Encyklopädie der Technischen Chemie*, **vol. 5** (Munich, 3rd ed., 1954) p 580.

Bright orange-red crystals. Not hygroscopic or deliquescent (difference from sodium dichromate). Crystal habit: prismatic. Crystal system: triclinic pinacoidal, transition to monoclinic at 241.6°. d_4^{25} 2.676. Bulk density: 100 lbs/cu ft. mp 398°. Dec at about 500°. Heat of fusion 29.8 cal/g. Heat of soln -62.5 cal/g. Specific heat 0.186 at 16° $-98°$. Soluble in water. A satd aq soln contains at 0°: 4.3%, at 20°: 11.7%, at 40°: 20.9%, at 60°: 31.3%, at 80°: 42.0%, at 100°: 50.2%. Acid reaction: A 1% aq soln has a pH of 4.04 and a 10% soln has a pH of 3.57.

Caution: Intern. a corrosive poison. Industrial contact may result in ulceration of hands, destruction of mucous membranes and perforation of nasal septum. *See* E. Browning, *Toxicity of Industrial Metals* (Appleton-Century Crofts, New York, 2nd ed., 1969) pp 119-131. *See also* Chromium.

USE: In tanning leather, dyeing, painting, decorating porcelain, printing, photolithography, pigment-prints, staining wood, pyrotechnics, safety matches; for bleaching palm oil, wax, and sponges; waterproofing fabrics; as oxidizer in the manuf of organic chemicals; in electric batteries; as depolarizer for dry cells. As corrosion inhibitor in preference to sodium dichromate where lower soly is advantageous. Pharmaceutic aid (oxidizing agent).

THERAP CAT (VET): Caustic.

7791. Potassium Dicyanoaurate(I). Gold potassium cyanide; potassium aurocyanide. C_2AuKN_2; mol wt 288.10. C 8.34%, Au 68.37%, K 13.57%, N 9.72%. $KAu(CN)_2$. Prepd by electrolysis of Au in KCN: Glassford, Napier, *Phil. Mag.* **25**, 61 (1844).

Dihydrate, cryst powder. One gram dissolves in 7 ml water, 0.5 ml boiling water; slightly sol in alcohol. Practically insol in ether.

USE: For electroplating.

7792. Potassium Ferricyanide. *Tripotassium hexakis-(cyano-C)ferrate(3—)*; *potassium hexacyanoferrate(III)*; red prussiate of potash. $C_6FeK_3N_6$; mol wt 329.25. C 21.89%, Fe 16.96%, K 35.63%, N 25.53%. $K_3Fe(CN)_6$.

Ruby-red crystals. d 1.89. Slowly sol in 2.5 parts cold water, in 1.3 parts boiling water; slightly sol in alc; dec by acids. The aq soln dec slowly on standing. *Protect from light.*

USE: Chiefly for blueprints; in photography; also for staining wood, dyeing wool, calico printing, as etching liquid (Mercer's liquor), tempering iron and steel; in electroplating; as a mild oxidizing agent in organic synthesis; in analytical chemistry.

7793. Potassium Ferrocyanide. *Tetrapotassium hexakis(cyano-C)ferrate(4—)*; *potassium hexacyanoferrate(II)*; yellow prussiate of potash. $C_6FeK_4N_6$; mol wt 368.34. C 19.56%, Fe 15.16%, K 42.46%, N 22.82%. $K_4Fe(CN)_6$. Review of properties, chemistry and syntheses: *The Chemistry of Ferrocyanides*, American Cyanamid Co. (Beacon Press, New York, 1953) 112 p.

Trihydrate, soft, slightly efflorescent crystals; begins to lose water at 60°, becomes anhyd at 100°. d 1.85.

7794. Potassium Fluoride. FK; mol wt 58.10. F 32.70%, K 67.30%. KF. Prepd by thermal decompn of KHF_2 or by neutralizing HF with K_2CO_3: Lange, Eichler, *Z. Physik. Chem.* **129**, 285, 286 (1927); Kwasnik in *Handbook of Preparative Inorganic Chemistry* **vol. 1**, G. Brauer, Ed. (Academic Press, New York, 2nd ed., 1963) p 236.

Cubic crystals (NaCl lattice). Usually obtained as white, deliquesc powder or solid. *Poisonous!* d 2.481. mp 859.9°. bp 1505°. Soly in water (g/100 ml): 92.3 (18°); 96.4 (21°). Very freely sol in boiling water. Also sol in aq HF, liq NH_3. Insol in alcohol unless water is present. May be stored in aluminum containers. *Caution:* Attracts moisture from the air. Aq solns corrode glass and porcelain. MLD orally in guinea pigs: 250 mg/kg.

Dihydrate, monoclinic crystals, mp 41°. Soly in water (18°): 349.3 g/100 ml.

Tetrahydrate, crystals, mp 19.3°.

Caution: Irritating to skin, eyes, mucous membranes.

USE: In the fluorination of organic compds; in flux for hard solder; to prevent unwanted fermentations; in insecticide formulations; for frosting glass.

7795. Potassium Formate. $CHKO_2$; mol wt 84.12. C 14.28%, H 1.20%, K 46.48%, O 38.04%. HCOOK.

Colorless, deliquesc granules. d 1.91. mp 167°, dec at higher temp with evolution of H_2. Sol in 0.4 part water. The aq soln is practically neutral. *Keep tightly closed.*

7796. Potassium Gluconate. *Gluconic acid potassium salt;* Gluconsan K; Kalimozan; Kaon; Katorin; Potasoral; Potassuril; K-IAO; Tumil-K. $C_6H_{11}KO_7$; mol wt 234.25. C 30.76%, H 4.73%, K 16.69%, O 47.81%. The normal potassium salt of gluconic acid.

$$HOH_2C-\overset{\overset{\displaystyle H}{|}}{C}-\overset{\overset{\displaystyle H}{|}}{C}-\overset{\overset{\displaystyle OH}{|}}{C}-\overset{\overset{\displaystyle H}{|}}{C}-COO^- \quad K^+$$
$$\underset{\displaystyle OH}{|} \quad \underset{\displaystyle OH}{|} \quad \underset{\displaystyle H}{|} \quad \underset{\displaystyle OH}{|}$$

Yellowish-white crystals. Stable in air. Mild, slightly saline taste. Dec 180°. Freely sol in water. Practically insol in abs alcohol, ether, benzene, chloroform. Aq solns are alkaline to litmus and have a pH of 7.5-8.5.

THERAP CAT: Replenisher (electrolyte).

THERAP CAT (VET): Potassium supplement.

7797. Potassium Glycerophosphate. $C_3H_7K_2O_6P$; mol wt 248.26. C 14.51%, H 2.84%, K 31.50%, O 38.67%, P 12.48%. $K_2C_3H_7PO_6$.

Trihydrate, colorless to slightly yellow, viscous mass. Very sol in water. Alkaline to phenolphthalein. Usually marketed as a 50-75% aq soln of more or less syrupy consistency and colorless to pale yellow.

THERAP CAT: Tonic.

7798. Potassium Guaiacolsulfonate. *3-Hydroxy-4-methoxybenzenesulfonic acid monopotassium salt mixt with monopotassium 4-hydroxy-3-methoxybenzenesulfonate;* sulfogaiacol; Thiocol; Orthocoll. $C_7H_7KO_5S$; mol wt 242.29. C 34.70%, H 2.91%, K 16.14%, O 33.02%, S 13.23%. A mixture of the potassium salts of 4- and 5-guaiacolsulfonic acid.

White, odorless crystals or cryst powder; faint bitter taste. Gradually turns pink on exposure to air and light. Soluble in 7.5 parts water. Almost insol in alcohol. Insol in ether. The aq soln is neutral to litmus. *Keep well closed and protected from light.*

THERAP CAT: Expectorant.

7799. Potassium Hexachloroosmate(IV). Osmium potassium chloride. Cl_6K_2Os; mol wt 481.14. Cl 44.21%, K 16.25%, Os 39.54%. $K_2[OsCl_6]$. Preparation: Turner et al., *Analyt. Chem.* **30**, 1708 (1958).

Dark red to almost black, cubic crystals. Freely sol in water; sparingly sol in alcohol.

7800. Potassium Hexachloroplatinate(IV). Platinic potassium chloride; potassium platinic chloride; potassium chloroplatinate. Cl_6K_2Pt; mol wt 485.99. Cl 43.77%, K 16.09%, Pt 40.14%. $K_2[PtCl_6]$.

Orange-yellow crystals or yellow powder. d 3.50. Slightly sol in cold water; sol in hot water. Almost insol in alcohol.

USE: In photography.

7801. Potassium Hexacyanocobaltate(III). *Tripotassium hexakis(cyano-C)cobaltate(3—);* potassium cyanocobaltate(III); potassium cobalticyanide; potassium cobaltihexacyanide; cobalt potassium cyanide. $C_6CoK_3N_6$; mol wt 332.33. C 21.68%, Co 17.73%, K 35.29%, N 25.29%. $K_3[Co(CN)_6]$. Prepd by reacting cobalt acetate or cyanide with potassium cyanide, followed by air-oxidation of the cobaltocyanide formed: Biltz, Biltz, *Z. Anorg. Allgem. Chem.* **50**, 108 (1906); Grube, *Z. Elektrochem.* **32**, 561 (1926); Benedetti-Pichler, *Z. Anal. Chem.* **70**, 258 (1927); Bigelow, *Inorg. Syn.* **2**, 225 (1946).

Faintly yellow, monoclinic crystals from water. d 1.906. Melts with decompn forming an olive-green mass. Freely sol in water, acetic acid solns. Insol in alcohol. Very slightly sol in liq ammonia. Dec by strong mineral acids. Aq solns are slightly yellow and are regarded as non-poisonous when freshly prepd: Zwenger, *Ann.* **62**, 178 (1847). An odor of HCN develops after about one week's standing in the dark. Exposure to light causes rapid development of a yellow color, but hydrolysis is slight.

7802. Potassium Hexafluoromanganate(IV). Potassium manganese hexafluoride; potassium hexafluomanganite. F_6K_2Mn; mol wt 247.13. F 46.13%, K 31.64%, Mn 22.23%. $K_2[MnF_6]$. Prepn: Jenssen, Bandte, *Angew. Chem.* **65**, 304 (1953).

Golden-yellow, hexagonal platelets; become brown on heating, regain the original color on cooling. Hydrolyzed by water with precipitation of MnO_2.

7803. Potassium Hexafluorosilicate. Potassium fluosilicate; potassium silicofluoride. F_6K_2Si; mol wt 220.27. F 51.75%, K 35.50%, Si 12.75%. $K_2[SiF_6]$.

White, fine powder or crystals. d 2.27. Slightly sol in cold water. Insol in alcohol. In hot water it hydrolyzes to KF, HF, and silicic acid. pH of 1% aq soln 3.4.

Caution: Strong irritant. Ingestion can cause vomiting, diarrhea.

USE: In the manuf of opalescent glass, in porcelain enamels, in insecticides. Also used in aluminum metallurgy.

7804. Potassium Hexafluorozirconate(IV). Potassium zirconium fluoride; zirconium potassium fluoride. F_6K_2Zr; mol wt 283.41. F 40.22%, K 27.59%, Zr 32.19%. $K_2[ZrF_6]$.

Crystals. Slightly sol in cold water, sol in hot water.

USE: For manuf of metallic zirconium.

7805. Potassium Hexathiocyanatoplatinate(IV). *Dipotassium hexakis(thiocyanato-S)platinate(2—);* platinic potassium thiocyanate; potassium platinic thiocyanate. $C_6K_2N_6PtS_6$; mol wt 621.78. C 11.59%, K 12.58%, N 13.52%, Pt 31.37%, S 30.94%. $K_2[Pt(SCN)_6]$.

Carmine-red crystals. Sol in water.

7806. Potassium Hydroxide. Potassium hydrate; caustic potash; potassa. HKO; mol wt 56.11. H 1.80%, K 69.69%, O 28.52%. KOH. Prepd industrially by electrolysis of potassium chloride: Faith, Keyes & Clark's *Industrial Chemicals*, F. A. Lowenheim, M. K. Moran, Eds. (Wiley-Interscience, New York, 4th ed., 1975) pp 674-678. Toxicity: H. F. Smyth et al., *Am. Ind. Hyg. Assoc. J.* **30**, 470 (1969).

White or slightly yellow lumps, rods, pellets. *Very caustic to tissue.* Rapidly absorbs moisture and CO_2 from the air and deliquesces. mp about 360°; mp 380° when anhyd. Sol in 0.9 part water, about 0.6 part boiling water, 3 parts alcohol, 2.5 parts glycerol. When dissolved in water or alcohol or when the soln is treated with an acid, much heat is generated. *Keep tightly closed and do not handle with bare hands.* A $0.1M$ aq soln has a pH of 13.5. LD_{50} orally in rats: 1.23 g/kg (Smyth).

Caution: Extremely corrosive. Ingestion may produce violent pain in throat and epigastrium, hematemesis, collapse. If not immediately fatal, stricture of esophagus may develop, cf. *Clinical Toxicology of Commercial Products*, R. E. Gosselin et al., Eds. (Williams & Wilkins, Baltimore, 4th ed., 1976) Section III, pp 206-212.

USE: Manuf liq soap; mordant for wood; absorbing CO_2; mercerizing cotton; paint and varnish removers; electroplating, photoengraving and lithography; printing inks; in analytical chemistry and in organic syntheses. Pharmaceutic aid (alkalizer).

THERAP CAT (VET): Caustic. In disbudding calves' horns. In aq solution to dissolve scales and hair in skin scrapings.

7807. Potassium Hypophosphite. H_2KO_2P; mol wt 104.09. H 1.94%, K 37.56%, O 30.74%, P 29.76%. KH_2PO_2.

White crystals or granular, deliquesc powder. Odorless; pungent, saline taste. When strongly heated it dec with evolution of phosphine which ignites spontaneously in air. *It explodes when triturated with chlorates or other oxidizing agents.* One gram dissolves in 0.6 ml water, 9 ml alc, 5 ml boiling alc. The aq soln is neutral or slightly alkaline. *Keep well closed.*

7808. Potassium Iodate. IKO_3; mol wt 214.00. I 59.30%, K 18.27%, O 22.43%. KIO_3.

White, odorless crystals or cryst powder. d 3.89. mp 560° with partial decompn. Slowly sol in 12 parts water, in 3.1 parts boiling water. Insol in alcohol.

USE: Oxidizing agent in volumetric chemical analysis; as maturing agent and dough conditioner.

THERAP CAT: Topical antiseptic (mucous membrane).
THERAP CAT (VET): In feeds as a source of iodine.

7809. Potassium Iodide. Jodid; Thyroblock; Thyrojod. IK; mol wt 166.00. I 76.45%, K 23.55%. KI. Potassium iodide of commerce contains about 99.5% KI. Prepd from HI and $KHCO_3$. Purification by melting in dry hydrogen: Lingane, Kolthoff, *Inorg. Syn.* **1**, 163 (1939). Continuous electrolytic process for large scale industrial prepn: Morylott, Elkins, U.S. pat. 2,989,450 (1961 to Dow). Toxicity data: Hildebrandt, *Arch. Exp. Pathol. Pharmakol.* **96**, 292 (1923). Use in the treatment of radiation poisoning resulting from a nuclear accident: W. K. Waterfall, *Brit. Med. J.* **281**, 988 (1980); *Bull. N.Y. Acad. Med.* **57**, 395 (1981).
Colorless or white, cubical crystals, white granules, or powder. Slightly deliquescent in moist air; on long exposure to air becomes yellow due to liberation of iodine, and small quantities of iodate may be formed; light and moisture accelerate the decompn. Aq solns also become yellow in time due to oxidation, but a small amount of alkali prevents it. d 3.12. mp 680° (volatilizes at higher temp). One gram dissolves in 0.7 ml water, 0.5 ml boiling water, 22 ml alcohol, 8 ml boiling alcohol, 51 ml abs alcohol, 8 ml methanol, 75 ml acetone, 2 ml glycerol, about 2.5 ml glycol. Potassium iodide solns readily dissolve elemental iodine. The aq soln is neutral or, usually, slightly alkaline. pH: 7-9. 30 g KI with 21 ml water gives 30 ml of a saturated soln at 25°. Approx LD i.v. in rats: 285 mg/kg (Hildebrandt).
Incompat. Alkaloidal salts, chloral hydrate, tartaric and other acids, calomel, potassium chlorate, metallic salts.
USE: Manuf photographic emulsions; in animal and poultry feeds to the extent of 10-30 parts per million; in table salt as a source of iodine and in some drinking water; also in anal. chemistry.
THERAP CAT: Antifungal; expectorant; iodine supplement.
THERAP CAT (VET): In actinobacillosis, actinomycosis. For simple goiter. As expectorant. In iodine deficiency and in chronic poisoning with lead or mercury. Orally only, not by injection. Externally for treatment of bursal enlargements.

7810. Potassium Manganate(VI). K_2MnO_4; mol wt 197.13. K 39.67%, Mn 27.87%, O 32.46%. Prepn: Scholder, Waterstradt, *Z. Anorg. Allgem. Chem.* **277**, 172 (1954).
Dark green crystals; dec at 190°. Sol in water. Sol and stable in KOH solns. It is an oxidizing agent. With HCl it gives free chlorine.

7811. Potassium Metabisulfite. *Potassium pyrosulfite.* $K_2O_5S_2$; mol wt 222.33. K 35.17%, O 35.98%, S 28.85%. $K_2S_2O_5$. The article of commerce contains ~95% $K_2S_2O_5$.
White crystals or cryst powder; sulfur dioxide odor; acid reaction; liberates SO_2 with acids; oxidizes in air to sulfate, more readily in presence of moisture. It may catch fire if much heat develops in powdering it. Freely sol in water; insol in alcohol. *Keep dry and well closed.*
USE: As antifermentative in breweries and wineries; bleaching straw; preservative for fruits and vegetables.

7812. Potassium Metaphosphate. Potassium Kurrol's salt; potassium polymetaphosphate; potassium polyphosphate. $(KPO_3)_x$. High mol wt polymer; degree of polymerization dependent upon preparative conditions. Prepd by dehydration of KH_2PO_4: Pfansteil, Iler, *J. Am. Chem. Soc.* **74**, 6059 (1952). Structural studies: Jost, *Acta Cryst.* **16**, 623 (1963); Jost, Schulze, *ibid.* **25B**, 1110 (1969); *eidem, ibid.* **27B**, 1345 (1971). Reviews of metaphosphates: J. R. Van Wazer, *Phosphorus and its Compounds* vol 1 (Interscience, New York, 1958) pp 601-678; Thilo, *Advan. Inorg. Chem. Radiochem.* **4**, 1-75 (1962).
White, monoclinic crystals. d^{20} 2.45. Insol in pure water. Sol in aq solns of alkali metal (except potassium) salts.

7813. Potassium Methyl Sulfate. CH_3KO_4S; mol wt 150.20. C 8.00%, H 2.01%, K 26.03%, O 42.61%, S 21.35%. KCH_3SO_4.
Hemihydrate, white crystals. Sol in water, alcohol.
USE: In organic syntheses.

7814. Potassium Molybdate(VI). K_2MoO_4; mol wt 238.13. K 32.84%, Mo 40.29%, O 26.87%.

Pentahydrate, white, deliquesc, cryst powder. d 2.3; mp 919°. Sol in 0.6 part water; insol in alc. *Keep well closed.*

7815. Potassium Nitrate. Saltpeter; niter. KNO_3; mol wt 101.10. K 38.67%, N 13.85%, O 47.47%. Contains ~99.5% KNO_3.
Colorless transparent prisms, white granular or cryst powder; cooling, saline, pungent taste. d 2.11; mp 333°; dec at 400° with evolution of O_2. One gram dissolves in 2.8 ml water, 0.5 ml boiling water, 620 ml alc. Sol in glycerol; insol in abs alc. Dissolves in water with a lowering of the temp. pH ~7. LD_{50} orally in rabbits: 1.166 g anion/kg, Dollahite, Rowe, *Southwest Vet.* **27**, 246 (1974).
Caution: Ingestion of large quantities may cause violent gastroenteritis. Prolonged exposure to small amts may produce anemia, methemoglobinemia, nephritis.
USE: In fireworks, fluxes, pickling meats; manuf glass, matches, gunpowder, blasting powders; freezing mixtures; impregnating candle wicks; treating tobacco to make it burn evenly; tempering steel.
THERAP CAT: Diuretic.

7816. Potassium Nitrite. KNO_2; mol wt 85.10. K 45.94%, N 16.46%, O 37.60%. The nitrite of commerce usually contains ~85% KNO_2, the remainder consisting chiefly of nitrate.
White or slightly yellow, deliquesc granules or rods. Dec even by weak acids with evolution of brown fumes of nitrous anhydride. d 1.915; mp 441° (decompn starts at 350°). Sol in 0.35 part water, slightly in alc. The aq soln is alkaline. *Keep well closed.* LD_{50} orally in rabbits: 108 mg anion/kg, Dollahite, Rowe, *Southwest Vet.* **27**, 246 (1974).
USE: In analytical chemistry.
THERAP CAT: Vasodilator; antidote (cyanide poisoning).

7817. Potassium Nitroprusside. *Dipotassium pentakis-(cyano-C)nitrosylferrate(2—); potassium nitrosylpentacyanoferrate(II).* $C_5FeK_2N_6O$; mol wt 294.14. C 20.42%, Fe 18.99%, K 26.59%, N 28.57%, O 5.44%. $K_2Fe(CN)_5(NO)$.
Dihydrate, garnet-red, hygroscopic crystals. Sol in 1 part water, in alcohol. *Keep well closed.*

7818. Potassium Oleate. *Oleic acid potassium salt.* Approx $C_{18}H_{33}KO_2$.
Yellowish or brownish, soft mass. Sol in water, alc. The aq soln is alkaline to phenolphthalein.
USE: Detergent.

7819. Potassium Osmate(VI). K_2O_4Os; mol wt 332.42. K 23.52%, O 19.25%, Os 57.23%. K_2OsO_4.
Dihydrate, violet, hygroscopic crystals. *Poisonous!* Sol in water; insol in alcohol, ether. Slowly dec in aq solns with formation of the tetroxide. *Keep well closed.*

7820. Potassium Oxalate. $C_2K_2O_4$; mol wt 166.22. C 14.45%, K 47.05%, O 38.50%. $K_2C_2O_4$.
Monohydrate, colorless, odorless crystals; efflorescent in warm dry air. *Poisonous!* d 2.13. Loses its water at ~160°; when ignited is converted into carbonate without appreciable charring. Sol in 3 parts water.
USE: Cleaning and bleaching straw, removing stains in photography; *in vitro* blood anticoagulant; also in analytical chemistry.

7821. Potassium Percarbonate. $C_2K_2O_6$; mol wt 198.22. C 12.12%, K 39.45%, O 48.43%. $K_2C_2O_6$. Prepn of practically anhydr compd: Partington, Fathallah, *J. Chem. Soc.* **1950**, 1934.
Monohydrate, white, granular mass. Sol in water with evolution of oxygen. One part potassium percarbonate is sol in 15 parts of cold water; dec in boiling water; 100 parts water dissolve 6.5 parts potassium percarbonate at ordinary temp. *Keep dry and protected from light.*
Caution: Strong irritant. Causes vomiting if swallowed. Large quantities can be fatal.
USE: Has been used in microscopy for detecting tubercle bacilli stained with fuchsin in smears; in photography under the name *Anti-hypo,* to remove last traces of sodium thiosulfate; also as oxidizing agent in chem analyses, but is no longer favored.

7822. Potassium Perchlorate. Peroidin; Perchloracap. $ClKO_4$; mol wt 138.55. Cl 25.59%, K 28.22%, O 46.19%. $KClO_4$.

Colorless crystals or white, cryst powder. Dec at 400°; also dec by organic matter, oxidizable substances and on concussion, but is less reactive than the chlorate. d 2.52. Sol in 65 parts cold water, 15 parts boiling water; practically insol in alcohol.

USE: In explosives, pyrotechnics and photography, in analytical chemistry.

7823. Potassium Periodate. Potassium metaperiodate. IKO_4; mol wt 230.00. I 55.18%, K 17.00%, O 27.82%. KIO_4. Prepd by oxidizing potassium iodate with chlorine in alkaline soln: Hill, *J. Am. Chem. Soc.* **50**, 2678 (1928); *Inorg. Syn.* **1**, 171 (1939).

Colorless tetragonal crystals, d_4^{15} 3.618. mp 582°. Soly in water (g/100 g H_2O): 0.168 at 0°; 0.42 at 20°; 0.93 at 40°; 2.16 at 60°; 4.44 at 80°; 7.87 at 100°; also given as 0.66 at 13°. Sparingly sol in aq KOH.

Caution: Highly irritating to skin, eyes, mucous membranes.

USE: Powerful oxidizer in acid soln, oxidizing manganese compds to permanganate; used for this purpose in analytical chemistry (colorimetric estimation of Mn), also for the oxidation of some organic compds.

7824. Potassium Permanganate. *Permanganic acid potassium salt;* chameleon mineral. $KMnO_4$; mol wt 158.03. K 24.74%, Mn 34.76%, O 40.50%. Prepn from manganese ore by electrolytic oxidation: Faith, Keyes & Clark's *Industrial Chemicals,* F. A. Lowenheim, M. K. Moran, Eds. (Wiley-Interscience, New York, 4th ed., 1975) pp 679-683. Toxicity study: H. F. Smyth *et al., Am. Ind. Hyg. Assoc. J.* **30**, 470 (1969).

Dark purple or bronze-like, odorless crystals. Almost opaque by transmitted light and of a blue metallic luster by reflected light. Sweet with astringent aftertaste; stable in air. Dec ~240° with evolution of oxygen. d 2.7. Soluble in 14.2 parts cold, 3.5 parts boiling water. Dec by alc and many other organic solvents, also by concd acids with liberation of oxygen; with HCl, chlorine is liberated. Readily dec by many reducing substances, such as ferrous salts, iodides, oxalates, etc., especially in the presence of an acid. *Caution:* Take great care in handling as explosions may occur if it is brought into contact with organic or other readily oxidizable substances, either in soln or in the dry state. *Incompat.* Alcohol, arsenites, bromides, iodides, hydrochloric acid, charcoal; organic substances generally; ferrous or mercurous salts, hypophosphites, hyposulfites, sulfites, peroxides, oxalates. LD_{50} orally in rats: 1.09 g/kg (Smyth).

Caution: Dilute solns are mildly irritating and high concns are caustic.

USE: Bleaching resins, waxes, fats, oils, straw, cotton, silk and other fibers and chamois skins; dyeing wood brown; printing fabrics; washing CO_2 in manuf mineral waters; exterminating *Oidium tuckeri;* photography; tanning leathers; purifying water; with formaldehyde soln to expel formaldehyde gas for disinfecting; as an important reagent in analytical and synthetic organic chemistry.

THERAP CAT: Anti-infective (topical).

THERAP CAT (VET): Antiseptic (topical), astringent, deodorant.

7825. Potassium Persulfate. $K_2O_8S_2$; mol wt 270.32. K 28.93%, O 47.35%, S 23.72%. $K_2S_2O_8$. The article of commerce contains 93-97% $K_2S_2O_8$.

Colorless or white, odorless crystals. Gradually dec, losing available oxygen; dec more quickly at higher temps; completely dec ~100°. A powerful oxidizing agent. Sol in ~50 parts water, 25 parts water at 40°; insol in alc; the aq soln dec at ordinary temp and more rapidly on warming. The aq soln is acid. *Keep well closed, in a cool place.*

USE: Bleaching fabrics, soaps; in photography under the name *Anthion* to remove last traces of thiosulfate from plates and paper; in analytical chemistry.

7826. Potassium Phenolsulfonate. *Hydroxybenzenesulfonic acid potassium salt;* potassium sulfocarbolate. C_6-

H_5KO_4S; mol wt 212.27. C 33.95%, H 2.37%, K 18.42%, O 30.15%, S 15.11%. $C_6H_4(OH)SO_3K$.

Monohydrate, white, odorless crystals. d 1.87. Sol in water, alcohol. The aq soln is practically neutral to litmus.

7827. Potassium Phenoxide. *Phenol potassium salt;* potassium phenate; potassium phenylate; potassium carbolate. C_6H_5KO; mol wt 132.20. C 54.51%, H 3.81%, K 29.57%, O 12.10%. C_6H_5OK. Prepd from phenol and KOH in dil methanol: Kornblum, Lurie, *J. Am. Chem. Soc.* **81**, 2710 (1959).

White to reddish, hygroscopic, cryst lumps. Very sol in water; sol in alcohol. The aq soln is strongly alkaline. *Keep tightly closed.*

7828. Potassium Phosphate, Dibasic. Dipotassium phosphate; dikalium phosphate; DKP; dipotassium hydrogen phosphate. HK_2O_4P; mol wt 174.18. H 0.58%, K 44.90%, O 36.74%, P 17.78%. K_2HPO_4.

White, somewhat hygroscopic granules. Very sol in water, slightly in alcohol. 100 g will dissolve rapidly and completely in 67 g of cold water. Converted into pyrophosphate by ignition. The aq soln is slightly alkaline to phenolphthalein. *Keep well closed.*

USE: Buffering agent in antifreeze solns; nutrient in the culturing of antibiotics; ingredient of instant fertilizers; as sequestrant in the prepn of non-dairy powdered coffee creams.

THERAP CAT: Cathartic.

7829. Potassium Phosphate, Monobasic. Potassium biphosphate; potassium acid phosphate; potassium dihydrogen phosphate; monopotassium phosphate; Sörensen's potassium phosphate. H_2KO_4P; mol wt 136.09. H 1.48%, K 28.73%, O 47.03%, P 22.76%. KH_2PO_4.

Colorless crystals or white, granular powder; permanent in air; at 400° loses H_2O, forming metaphosphate. d 2.34. Sol in ~4.5 parts water. Insol in alcohol. pH 4.4-4.7.

USE: In buffers for determination of pH. Pharmaceutic aid (buffering agent).

7830. Potassium Phosphate, Tribasic. Tripotassium phosphate. K_3O_4P; mol wt 212.27. K 55.26%, O 30.15%, P 14.59%. K_3PO_4. Purification: Jänecke, *Z. Physik. Chem.* **127**, 75 (1927); Simon, Schulze, *Z. Anorg. Allgem. Chem.* **242**, 331 (1939).

Deliquescent, orthorhombic crystals. d_4^{17} 2.564. mp 1340°. Soly in water: 43.7% at 0°; 50.8% at 25°; 59.7% at 45.1°. Insol in alcohol. Aq solns are strongly alkaline. Octahydrate, flat, rectangular platelets, mp 45.1°.

7831. Potassium Phosphite. HK_2O_3P; mol wt 158.18. H 0.64%, K 49.44%, O 30.34%, P 19.58%. K_2HPO_3.

White, deliquesc powder. Slowly oxidizes in the air to phosphate; dec by heat. Very sol in water; insol in alcohol. *Keep well closed.*

7832. Potassium Pyroantimonate, Acid. *Potassium antimonate(V).* Approx $K_2H_2Sb_2O_7.4H_2O$.

White granules or cryst powder. Moderately sol in cold water; insol in alcohol.

USE: Formerly used for detection of sodium.

7833. Potassium Pyrophosphate. *Diphosphoric acid tetrapotassium salt.* $K_4O_7P_2$; mol wt 330.34. K 47.34%, O 33.90%, P 18.75%. $K_4P_2O_7$. Manuf: Faith, Keyes & Clark's *Industrial Chemicals,* F. A. Lowenheim, M. K. Moran, Eds. (Wiley-Interscience, New York, 4th ed., 1975) pp 684-687.

Trihydrate, colorless, deliquesc granules or cryst mass; freely sol in water; insol in alcohol. The aq soln is alkaline.

USE: In detergents and surfactants; in water treatment; in drilling muds as a clay thinner.

7834. Potassium Pyrosulfate. *Disulfuric acid dipotassium salt;* "anhydrous" potassium acid sulfate. $K_2O_7S_2$; mol wt 254.32. K 30.75%, O 44.04%, S 25.22%. $K_2S_2O_7$.

Colorless, fused pieces. d 2.28. mp ~325°. Sol in water. The aq soln is strongly acid.

7835. Potassium Salicylate. $C_7H_5KO_3$; mol wt 176.21. C 47.71%, H 2.86%, K 22.19%, O 27.24%. HOC_6H_4COOK.

White, odorless powder. Becomes pink on exposure to light. Very sol in water, alcohol. One gram dissolves in

0.85 ml H_2O. A satd aq soln contains 55.82% w/w at 28.5°. *See* Sidgwick, Ewbank, *J. Chem. Soc.* **121**, 1847, 1850 (1922). The aq soln is neutral or slightly acid to litmus. *Keep well closed and protected from light.*

7836. Potassium Selenate. K_2O_4Se; mol wt 221.15. K 35.36%, O 28.94%, Se 35.70%. K_2SeO_4.
Colorless crystals or white powder. d 3.07. Sol in about one part water.
USE: Reagent.

7837. Potassium Selenide. K_2Se; mol wt 157.16. K 49.76%, Se 50.24%. Prepd by heating selenium with excess potassium: Fonzes-Diacon, *Contribution a l'Etudes des Séléniures Métalliques,* Montpellier (1901); by adding selenium to a soln of potassium in liq ammonia: Hugot, *Recherches sur l'Action du Sodammonium et du Potassammonium sur Quelque Métalloides,* Paris (1900); Feher in *Handbook of Preparative Inorganic Chemistry* vol. **1**, G. Brauer, Ed. (Academic Press, New York, 2nd ed., 1963) p 421; by igniting potassium selenite or selenate in an atm of hydrogen: Berzelius, cited in *Mellor's* vol. **10**, 767 (1930).
Crystalline mass. d 2.29. Reddens on exposure to air. Turns brownish-black when heated. Deliquescent. Forms potassium selenite with selenious acid. Forms potassium hydroselenide with hydrogen selenide. Sol in water; insol in ammonia.
Nonahydrate, needle-like cryst. Freely sol in water.

7838. Potassium Silicate. Soluble potash glass; soluble potash water glass. Variable composition: $K_2Si_2O_5$ to $K_2Si_3O_7$; may also contain water.
Colorless or yellowish, translucent to transparent, hygroscopic, glass-like pieces; strong alkaline reaction. Usually very slowly sol in cold water, or depending on the composition, almost insol. More readily sol in water when heated with it under pressure. Insol in alcohol; dec by acids with precipitation of silica. *Keep well closed.*
USE: As binder (*e.g.*, in carbon electrodes, lead pencils, protective coatings, insol pigments); detergent, in glass and ceramics manuf.

7839. Potassium Silver Cyanide. *Potassium bis(cyano-C)argentate(1—); potassium dicyanoargentate(I);* silver potassium cyanide. C_2AgKN_2; mol wt 199.00. C 12.07%, Ag 54.20%, K 19.65%, N 14.08%. $KAg(CN)_2$.
White crystals; sensitive to light. *Poisonous!* Sol in water; acids ppt silver cyanide from the soln. *Protect from light.*
USE: In silver plating; as bactericide.

7840. Potassium Sodium Tartrate. Rochelle salt; Seignette salt. $C_4H_4KNaO_6$; mol wt 210.16. C 22.86%, H 1.92%, K 18.60%, Na 10.94%, O 45.68%. $KNaC_4H_4O_6$.
Tetrahydrate, translucent crystals or white, cryst powder; cooling saline taste. Slightly effloresces in warm air. d 1.79; mp 70-80°; at 100° loses $3H_2O$; becomes anhydr at 130-140°; at 220° begins to dec. Sol in 0.9 part water. Almost insol in alcohol. The aq soln is slightly alkaline to litmus. pH 7-8. *Incompat.* Acids, calcium or lead salts, magnesium sulfate, silver nitrate.
USE: Manuf of mirrors; as a constituent of Fehling's soln. For the control of radio frequencies, and wherever piezoelectric crystals are used.
THERAP CAT: Cathartic.

7841. Potassium Sorbate. *2,4-Hexadienoic acid potassium salt; sorbic acid potassium salt;* potassium 2,4-hexadienoate; BB powder. $C_6H_7KO_2$; mol wt 150.22. C 47.97%, H 4.70%, K 26.03%, O 21.30%. $CH_3CH=CHCH=CH-COOK$. Prepd from sorbic acid and KOH: Probst, Oehme, *U.S. pat.* **3,173,948** (1965 to Hoechst).
Crystals, d_{20}^{25} 1.363. Dec above 270°. Soly in water at 20°: 58.2%; in alc: 6.5%.
USE: As mold and yeast inhibitor, like sorbic acid, esp where greater soly in water is desirable.

7842. Potassium Stannate(IV). K_2O_3Sn; mol wt 244.90. K 31.93%, O 19.60%, Sn 48.47%.

Trihydrate, colorless crystals. d 3.197. Sol in one part water. Insol in alcohol. The aq soln is alkaline.
USE: In textile dyeing and printing, in tin plating.

7843. Potassium Stannosulfate. Marignac's salt. $K_2O_8S_2Sn$; mol wt 389.03. K 20.10%, O 32.90%, S 16.48%, Sn 30.51%. $K_2Sn(SO_4)_2$.
White crystals. Partly dec by water; sol in dil fixed alkali hydroxide solns.

7844. Potassium Stearate. *Stearic acid potassium salt.* Approx $KC_{18}H_{35}O_2$. The article of commerce contains a considerable proportion of palmitate.
White powder; usually with slight odor of fat. Slowly sol in cold, readily in hot water or alcohol. The aq soln is strongly alkaline to litmus or phenolphthalein, but the alcoholic soln is only slightly alkaline to phenolphthalein.
USE: In the manuf of textile softeners.

7845. Potassium Sulfate. Sal polychrestum; arcanum duplicatum; tartarus vitriolatus. K_2O_4S; mol wt 174.26. K 44.87%, O 36.73%, S 18.40%. K_2SO_4. Contains at least 99% K_2SO_4.
Colorless or white, odorless, hard, bitter crystals, or white granules or powder; permanent in air. d 2.66; mp 1067°. One gram dissolves in 8.3 ml water, 4 ml boiling water, 75 ml glycerol. Insol in alcohol. Its soly in water is decreased by KCl or $(NH_4)_2SO_4$ and is practically insol in a saturated soln of the latter. The aq soln is neutral. pH about 7.
Caution: Swallowing large doses causes severe G.I. irritation.
USE: Technical grades are used in fertilizers for manuf of potassium alum, potassium carbonate and glass; the reagent grade is used in the Kjeldahl determination of nitrogen.
THERAP CAT: Cathartic.

7846. Potassium Sulfide. Potassium monosulfide. K_2S; mol wt 110.26. K 70.92%, S 29.08%. Best prepd from the elements in liq ammonia: Klemm *et al., Z. Anorg. Allgem. Chem.* **241**, 281 (1939); Feher in *Handbook of Preparative Inorganic Chemistry* vol. **1**, G. Brauer, Ed. (Academic Press, New York, 2nd ed., 1963) pp 360-361.
White, cubic crystals or fused plates. Discolors in air. Very hygroscopic. Unstable. May explode on percussion or rapid heating. d 1.74. mp 912°. Freely sol in water. Aq solns are strongly alkaline.
Pentahydrate, colorless rhombs. Odor of hydrogen sulfide. Discolors upon exposure to light and air (yellow to yellowish red). mp 60°. Freely sol in water, alc, glycerol. Insol in ether. Densities of aq solutions at 25° (calcd as % $K_2S.5H_2O$): 1.82%: 1.009; 10.90%: 1.049; 21.80%: 1.100; 29.07%: 1.136; 39.97%: 1.192; 50.88%: 1.250; 81.77%: 1.432. Aq solns are very caustic and prone to decompn. *See* the hydrolysis reactions *under* Sodium Sulfide.

7847. Potassium Sulfite. K_2O_3S; mol wt 158.26. K 49.41%, O 30.33%, S 20.26%. K_2SO_3. The sulfite of commerce is 90-95% pure.
Dihydrate, white crystals or cryst powder; gradually oxidizes in the air to sulfate. Soluble in about 3.5 parts water, slightly in alcohol; dec by dil acids with evolution of sulfur dioxide. pH about 8. *Keep well closed in a cool place.*
USE: In photographic developers.
THERAP CAT: Cathartic, diuretic.

7848. Potassium Sulfobenzoate. *2-Sulfobenzoic acid dipotassium salt.* $C_7H_4K_2O_5S$; mol wt 278.37. C 30.20%, H 1.45%, K 28.09%, O 28.74%, S 11.52%. $KSO_3C_6H_4COOK$.
White powder. Freely sol in water or alcohol.

7849. Potassium Tartrate. Soluble tartar. $C_4H_4K_2O_6$; mol wt 226.27. C 21.23%, H 1.78%, K 34.56%, O 42.43%.
Hemihydrate, white crystals or granular powder; loses its water at about 150°. d 1.98. Sol in about 0.5 part water. Almost insol in alc. The aq soln is slightly alkaline to litmus; pH 7-8.
THERAP CAT: Cathartic.

7850. Potassium Tellurate(IV). Potassium tellurite. K_2O_3Te; mol wt 253.79. K 30.81%, O 18.91%, Te 50.28%. K_2TeO_3.
White, granular powder. Sol in water to an alkaline soln.

USE: In 1:50,000 soln for detecting presence of living pathogenic bacteria in sera and vaccines, such organisms forming characteristic black reduction compds.

7851. Potassium Tellurate(VI). K_2O_4Te; mol wt 269.79. K 28.98%, O 23.72%, Te 47.30%. K_2TeO_4. Crystallizes with 3 or $5H_2O$.
Trihydrate, white, cryst powder. Sol in 4 parts water.
USE: Anhidrotic.

7852. Potassium Tetraborate. Potassium biborate; potassium borate. $B_4K_2O_7$; mol wt 233.44. B 18.52%, K 33.50%, O 47.98%. $K_2B_4O_7$.
Pentahydrate, white, cryst powder. Sol in 4 parts water, slightly in alc.

7853. Potassium Tetrabromoaurate(III). Gold potassium bromide; potassium auribromide. $AuBr_4K$; mol wt 555.68. Au 35.45%, Br 57.52%, K 7.04%. Prepn: Block, *Inorg. Syn.* **4**, 14 (1953).
Dihydrate, violet crystals. Sol in water, alcohol. *Protect from light.*

7854. Potassium Tetrachloroaurate(III). Gold potassium chloride; potassium aurichloride. $AuCl_4K$; mol wt 377.88. Au 52.12%, Cl 37.53%, K 10.35%.
Dihydrate, yellow, monoclinic crystals. Sol in water. *Protect from light.*
USE: Photography, painting on porcelain and glass; prepn of several other gold compds.

7855. Potassium Tetrachloroplatinate(II). Platinous potassium chloride; potassium platinochloride; potassium chloroplatinate. Cl_4K_2Pt; mol wt 415.09. Cl 34.16%, K 18.84%, Pt 47.00%. $K_2[PtCl_4]$.
Ruby-red crystals. Sol in water.
USE: In photography, in acid toning baths.

7856. Potassium Tetracyanomercurate(II). *Dipotassium tetrakis(cyano-C)mercurate(2—)*; mercuric potassium cyanide. $C_4HgK_2N_4$; mol wt 382.86. C 12.55%, Hg 52.39%, K 20.42%, N 14.63%. $K_2[Hg(CN)_4]$.
Colorless or white crystals. Sol in water. *Violent poison!*
USE: In manuf of mirrors to prevent the silver coating from becoming yellow; as reagent in testing for free acids.

7857. Potassium Tetracyanonickelate(II). *Dipotassium tetrakis(cyano-C)nickelate(2—)*; nickel potassium cyanide. $C_4K_2N_4Ni$; mol wt 240.96. C 19.94%, K 23.25%, Ni 24.36%. $K_2[Ni(CN)_4]$.
Monohydrate, orange-yellow, cryst powder. *Poison!* Loses H_2O at about 100°. Sol in water.

7858. Potassium Tetracyanoplatinate(II). *Dipotassium tetrakis(cyano-C)platinate(2—)*; platinous potassium cyanide; potassium platinocyanide. $C_4K_2N_4Pt$; mol wt 377.35. C 12.73%, K 20.72%, N 14.85%, Pt 51.70%. $K_2[Pt(CN)_4]$.
Crystallizes as a tri- or a dodecahydrate.
Trihydrate, almost colorless, rhombic prisms; blue in direction of principal axis. Sol in hot water.

7859. Potassium Tetrafluoroborate. Potassium borofluoride; potassium fluoborate; avogadrite. BF_4K; mol wt 125.90. B 8.59%, F 60.36%, K 31.05%. KBF_4. Prepared according to the eq $H_3BO_3 + 4HF + KOH = KBF_4 + 4H_2O$: Vorländer *et al.*, *Ber.* **65**, 535 (1932); Kwasnik in *Handbook of Preparative Inorganic Chemistry* vol. 1, G. Brauer, Ed. (Academic Press, New York, 2nd ed., 1963) p 223. For other methods of prepn see H. S. Booth, D. R. Martin, *Boron Trifluoride and its Derivatives* (New York, 1949) pp 99-106.
Orthorhombic bipyramidal or cubic crystals. d_4^{20} 2.505. mp 530°. Soly in water (g/100 g): 0.3 (3°); 0.448 (20°); 0.55 (25°); 1.4 (40°); 6.27 (100°). Index of refraction of solns is lower than that of water. A satd soln (0.6%) on heating gives color effects, if excess crystals are added. With a 10% aq soln a transparent blue color appears at 100°, turning green at 90°, and yellow at 60°. In a concd fluoboric acid soln these phenomena occur with variations in room temp. Aq solns of KBF_4 are at first neutral to litmus, but upon standing, diluting, or heating become acidic without etching their glass containers. Slightly sol in boiling alc.

USE: Has been proposed as a flux for soldering and brazing; filler in resin-bonded grinding wheels.

7860. Potassium Tetraiodoaurate(III). Gold potassium iodide; potassium auric iodide. AuI_4K; mol wt 743.68. Au 26.49%, I 68.26%, K 5.26%.
Black, lustrous crystals. Sol in water with decompn and liberation of iodine. *Protect from light.*

7861. Potassium Tetraiodocadmate. *Dipotassium tetraiodocadmate(2—)*; cadmium potassium iodide; potassium cadmium iodide; potassium iodocadmate; Marme's reagent. CdI_4K_2; mol wt 698.23. Cd 16.10%, I 72.70%, K 11.20%. $K_2[CdI_4]$. Dihydrate prepd from aq solns of stoichiometric amounts of CdI_2 and KI; must be recrystallized from water at low temps: Croft, *Phil. Mag.* **21**, 355 (1842); *J. Prakt. Chem.* **58**, 399 (1856); Hittorf, *Pogg. Ann.* **106**, 525 (1859); Eder, *Photogr. Mitt.* **13**, 67 (1867); Rimbach, *Ber.* **38**, 1562 (1905).
Dihydrate, large, water-clear, somewhat distorted octahedra. Deliquesc upon standing in air. Turns yellow with age. d_2^{21} 3.359. One part (w/w) dissolves at 15° in 0.73 part water, in 1.4 part alcohol, in 24.5 parts ether, in 4.5 parts of a 1:1 alcohol-ether mixture; also sol in ethyl acetate.
USE: In testing for alkaloids, amines, imines, heterocyclic nitrogens, peroxides in ether. In chelatometric quantitative organic analysis: *Chim. Anal. (Paris)* **43**, 449 (1961).

7862. Potassium Tetraiodomercurate(II). Mercuric potassium iodide. HgI_4K_2; mol wt 786.40. Hg 25.51%, I 64.55%, K 9.94%. Also crystallizes with 1-2 mols H_2O.
Sulfur-yellow crystals; deliquesc in moist air. *Poison!* Very sol in water; sol in alcohol, ether, acetone. *Keep well closed.*
THERAP CAT: Anti-infective (topical); disinfectant.
THERAP CAT (VET): Antiseptic (topical); disinfectant.

7863. Potassium Tetroxalate. Potassium quadroxalate; sal acetosella; salt of sorrel. $C_4H_3KO_8$; mol wt 218.16. C 22.02%, H 1.39%, K 17.92%, O 58.67%. $KHC_2O_4 \cdot H_2C_2O_4$. Incorrectly called *"essential salt of lemons".*
Dihydrate, colorless or white crystals; permanent in air. Sol in 60 parts cold, 12 parts boiling water; slightly sol in alcohol.
USE: Removing rust and ink spots; in metal polishes; in analytical chemistry.

7864. Potassium Thioantimonate(V). Potassium sulfantimonate. K_3S_4Sb; mol wt 367.32. K 31.93%, S 34.92%, Sb 33.15%. K_3SbS_4.
Heminonahydrate, colorless to yellowish crystals. Sol in water. Insol in alcohol. The aq soln is strongly alkaline. *Keep well closed in a cool place.*

7865. Potassium Thiocarbonate. Potassium sulfocarbonate; potassium trithiocarbonate. CK_2S_3; mol wt 186.41. C 6.44%, K 41.95%, S 51.61%. K_2CS_3. Monograph: K. N. Johri, *Chemical Analysis without H_2S using Potassium Trithiocarbonate* (Asia Pub., New York, 1963) 107 pp.
Yellowish-red, deliquesc granules or crystals. Very sol in water. The aq soln is strongly alkaline. *Keep tightly closed. Caution:* Irritant, caustic.
USE: In analytical chemistry.

7866. Potassium Thiocyanate. Potassium sulfocyanate; potassium rhodanide; Rhocya. CKNS; mol wt 97.18. C 12.36%, K 40.23%, N 14.41%, S 33.00%. KSCN.
Colorless, deliquesc crystals. d 1.89; mp $\sim$173°, the fused salt turning brown, then green, blue, and white again on cooling. One gram dissolves in 0.5 ml acetone, 12 ml alcohol, 8 ml boiling alcohol. When dissolved in its own wt of water, the temp drops about 30°. The aq soln is neutral. *Keep well closed.* LD_{50} orally in mice, rats: 594, 854 mg/kg, Andersen, Chen, *J. Am. Pharm. Assoc.* **29**, 152 (1940).
Caution: May cause skin eruptions, psychosis.
USE: Manuf artificial mustard oil; printing and dyeing textiles; in photography as intensifier; in analytical chemis-

---

try. The sodium salt now is replacing it for most of these uses.

THERAP CAT: Hypotensive.

7867. Potassium Thiosulfate. Potassium hyposulfite. $K_2O_3S_2$; mol wt 190.33. K 41.09%, O 25.22%, S 33.70%. $K_2S_2O_3$.

Colorless, hygroscopic crystals. Sol in water. Insol in alcohol. May crystallize with 0.33 to 1.5 mol H_2O. *Keep well closed.*

7868. Potassium Titanyl Oxalate. *Dipotassium bis[ethanedioato(2−)-O,O']oxotitanate(2−);* titanium potassium oxalate; titanyl potassium oxalate. $C_4K_2O_9Ti$; mol wt 318.10. C 15.10%, K 24.58%, O 45.27%, Ti 15.05%. $K_2TiO(C_2O_4)_2$.

Crystals or cryst powder. Very sol in water.
Dihydrate, *potassium oxodioxalatodiaquotitanate(IV).*
USE: As mordant in dyeing cotton and leather.

7869. Potassium Triiodide. I_3K; mol wt 419.81. I 90.69%, K 9.31%. Monohydrate prepd by adding the stoichiometric amount of iodine to a hot soln of KI and cooling to 0°: H. L. Wells, H. L. Wheeler, *Z. Anorg. Allgem. Chem.* **1**, 453 (1892); H. W. Foote, W. C. Chalker, *J. Am. Chem. Soc.* **39**, 565 (1908). Prepn of anhydrous: N. S. Grace, *J. Chem. Soc.* **1931**, 594.

Reasonably stable only as the monohydrate, dark brown, hygroscopic, monoclinic prisms. d_4^{15} 3.498. mp 38° (closed tube). Gives off iodine at 225°, leaving KI. "Incongruently" sol in water, sol (with partial decompn) in alcohol, ether.

7870. Potassium Triiodomercurate(II) Solution. Mercuric potassium iodide soln; potassium mercuriiodide soln; soln potassium iodohydrargyrate; Channing's soln; Thoulet's soln. Prepd by dissolving 1 g HgI_2 and 0.8 g KI in water to make 100 ml. *Poison!*

USE: As reagent for alkaloids.
THERAP CAT: Antiseptic (topical), disinfectant.

7871. Potassium Triiodozincate. Potassium zinc iodide; zinc potassium iodide. I_3KZn; mol wt 485.20. I 78.46%, K 8.06%, Zn 13.48%. $KZnI_3$.

Very hygroscopic crystals. Very sol in water. *Keep well closed.*

7872. Potassium Tungstate(VI). K_2O_4W; mol wt 326.03. K 23.98%, O 19.63%, W 56.39%. K_2WO_4. Crystallizes also with $2H_2O$.

Heavy, deliquesc, cryst powder. d 3.12. mp 921°. Sol in about 2 parts cold, about 0.7 part boiling water. Insol in alcohol. *Keep well closed.*

7873. Potassium Uranate(VI). Potassium diuranate; uranium oxide orange. $K_2O_7U_2$; mol wt 666.25. K 11.74%, O 16.81%, U 71.45%. $K_2U_2O_7$.

Orange powder. Insol in water; sol in acids.
USE: Painting on porcelain.

7874. Potassium Uranyl Nitrate. Uranyl potassium nitrate. $KN_3O_{11}U$; mol wt 495.14. K 7.90%, N 8.49%, O 35.54%, U 48.07%. $K(UO_2)(NO_3)_3$.

Greenish-yellow, cryst powder. Sol in about 1 part water.

7875. Potassium Uranyl Sulfate. Uranyl potassium sulfate. $K_2O_{10}S_2U$; mol wt 540.35. K 14.47%, O 29.61%, S 11.87%, U 44.05%. $K_2(UO_2)(SO_4)_2$.

Dihydrate, greenish-yellow, cryst powder. Freely sol in water.

7876. Potassium Xanthogenate. *Carbonodithioic acid O-ethyl ester potassium salt; ethylxanthic acid potassium salt;* potassium ethyldithiocarbonate; potassium ethylxanthogenate; potassium ethylxanthate. $C_3H_5KOS_2$; mol wt 160.30. C 22.48%, H 3.14%, K 24.39%, O 9.98%, S 40.01%. C_2H_5O-CS_2K. Made by treating an alcoholic soln of CS_2 with alcoholic KOH. Usually contains 8-10% H_2O.

White to pale yellow crystals or cryst powder. Very sol in water; sol in alcohol. The aq soln is strongly alkaline. *Keep well closed and protected from light.*

USE: As reagent in analytical chemistry.

7877. Potassium Zinc Sulfate. Zinc potassium sulfate. $K_2O_8S_2Zn$; mol wt 335.71. K 23.29%, O 38.13%, S 19.10%, Zn 19.48%. $K_2Zn(SO_4)_2$.

Hexahydrate, crystals. Sol in water.

7878. Potassium Zirconium Sulfate. Zirconium potassium sulfate. $K_4O_{16}S_4Zr$; mol wt 631.87. K 24.75%, O 40.51%, S 20.30%, Zr 14.44%. $K_4Zr(SO_4)_4$.

Trihydrate, $K_4Zr(SO_4)_4.3H_2O$, cryst powder. Sparingly sol in water; the aq soln is acid to litmus.

7879. Povidone. *1-Ethenyl-2-pyrrolidinone polymers; 1-vinyl-2-pyrrolidinone polymers;* poly[1-(2-oxo-1-pyrrolidinyl)ethylene]; polyvinylpyrrolidone; polyvidone; P.V.P.; RP-143; Kollidon; Periston; Plasdone; Plasmosan; Protagent; Subtosan; Vinisil. Solns are known as *Haemodyn.* Produced commercially as a series of products having mean mol wts ranging from ∼10,000 to 700,000. Prepared by Reppe's process: 1,4-butanediol obtained in the Reppe butadiene synthesis is dehydrogenated over copper at 200° forming γ-butyrolactone; reaction with ammonia yields pyrrolidone. Subsequent treatment with acetylene gives the vinyl pyrrolidone monomer. Polymerization is carried out by heating in the presence of H_2O_2 and NH_3. *Cf.* P. B. reports **163; 1288;** also DeBell *et al., German Plastics Practice* (Springfield, 1946); Hecht, Weese, *Münch. Med. Wochenschr.* **1943**, 11; Weese, *Naturforschung & Medizin* **62**, 224 (Wiesbaden, 1948), and the corresp vol. of *FIAT Review of German Science.* Monographs: General Aniline and Film Corp., *PVP* (New York, 1951); W. Reppe, *Polyvinylpyrrolidon* (Monographie zu *"Angewandte Chemie"* no. 66, Weinheim/Bergstr., 1954).

Faintly yellow solid resembling albumin, but does not give the reactions of albumin. Sol in water giving a colloidal soln. Also sol in alcohol, chloroform. Practically insol in ether. A 3.5% soln develops an osmotic pressure of about 400 mm water.

USE: Pharmaceutic aid (dispersing and suspending agent). Proposed as clarifying agent in wines.

7880. Povidone-Iodine. *1-Ethenyl-2-pyrrolidinone homopolymer compd with iodine; 1-vinyl-2-pyrrolidinone polymers, iodine complex;* iodine-polyvinylpyrrolidone complex; polyvinylpyrrolidone-iodine complex; PVP-I; Betadine; Betaisodona; Braunol; Braunosan H; Disadine D.P.; Disphex; Efodine; Inadine; Isodine; Proviodine; Traumasept; Videne. An iodophor, *q.v.,* prepd by Beller, Hosmer, U.S. pat. **2,706,701;** Hosmer, U.S. pat. **2,826,532;** Siggia, U.S. pat. **2,900,305** (1955, 1958, and 1959, all to GAF). Prepn, history and use: Shelanski, Shelanski, *J. Int. Coll. Surgeons* **25**, 727 (1956).

Yellowish-brown, amorphous powder with slight characteristic odor. Aq solns have a pH near 2 and may be made more neutral (but less stable) by the addition of sodium bicarbonate. Sol in alc, water. Practically insol in chloroform, carbon tetrachloride, ether, solvent hexane, acetone. Solns do not give the familiar starch test when freshly prepared.

THERAP CAT: Anti-infective (topical).
THERAP CAT (VET): Anti-infective (topical).

7881. PPACK. *D-Phenylalanyl-N-[4-[(aminoiminomethyl)amino]-1-(chloroacetyl)butyl]-L-prolinamide;* D-phenylalanyl-prolyl-arginine chloromethyl ketone; FPRMeCl. $C_{21}H_{31}ClN_6O_3$; mol wt 450.97. C 55.93%, H 6.93%, Cl 7.86%, N 18.64%, O 10.64%. Selective irreversible inhibitor of thrombin: C. A. Kettner, E. N. Shaw, *Thromb. Res.* **14**, 969 (1979). Synthesis: E. N. Shaw, C. A. Kettner, U.S. pat. **4,318,904** (1982 to Research Corp.). Anticoagulant and antithrombotic activity: J. Hauptman, F. Markwardt, *Thromb. Res.* **20**, 347 (1980). *In vivo* studies: D. Collen *et al., J. Lab. Clin. Med.* **99**, 76 (1982).

D-Phe–Pro–ArgCH₂Cl

Note: Commercially available as the dihydrochloride.
USE: As a research tool; as a clinical reagent to measure the level of thrombin in blood.

7882. Practolol. *N-[4-[2-Hydroxy-3-[(1-methylethyl)-amino]propoxy]phenyl]acetamide; 4'-[2-hydroxy-3-(isopropylamino)propoxy]acetanilide;* 1-(4-acetamidophenoxy)-3-isopropylamino-2-propanol; AY-21011; ICI-50172; Dalzic; Eraldin. C₁₄H₂₂N₂O₃; mol wt 266.34. C 63.14%, H 8.33%, N 10.52%, O 18.02%. β-Adrenergic blocker. Prepn: Howe, Smith, **Neth. pat. Appl. 6,512,676** corresp to **U.S. pat. 3,408,387** (1966, 1968, to I.C.I.). Synthesis of *R(+)*-form: Danilewicz, Kemp, *J. Med. Chem.* **16**, 168 (1973). Pharmacological studies: Dunlop, Shanks, *Brit. J. Pharmacol. Chemother.* **32**, 201 (1968); Barrett, *Postgrad. Med. J., Suppl.* **47**, 7 (1971). Metabolic studies: Scales, Cosgrove, *J. Pharmacol. Exp. Ther.* **175**, 338 (1970).

mp 134-136° (BuOAc). Sol in warm isopropanol.
Hydrochloride monohydrate, mp 140-142°,
R(+)-Form, crystals from dioxane, mp 130-131.5°. $[\alpha]_{365}^{25}$ +4.3°, $[\alpha]_{578}^{25}$ +3.5° (ethanol).
R(+)-Hydrochloride, $[\alpha]_{436}^{25}$+26.0°, $[\alpha]_{578}^{25}$ +14.0°.
THERAP CAT: Antiarrhythmic.

7883. Prajmaline. *(17R,21α)-17,21-Dihydroxy-4-propylajmalanium; N⁴-propylajmalinium;* prajmalium; N-propylajmaline. [C₂₃H₃₃N₂O₂]⁺. Prepn: J. Keck, *Z. Naturforsch.* **18b**, 177 (1963); A. Bonati, E. Bombardelli, *Farmaco Ed. Sci.* **18**, 851 (1963); J. Keck, **Ger. pat. 1,154,120** (1963 to Thomae); *idem,* **U.S. pat. 3,414,577** (1968 to Boehringer, Ing.). Pharmacology: R. Koch, *Arzneimittel-Forsch.* **22**, 2079 (1972); H. M. Mertens *et al., ibid.* **23**, 642 (1973); K.-W. Diederich, D. Boyksen, *ibid.* 1302. Toxicology: G. Von Philipsborn, B. Stalder, *ibid.* **22**, 2085 (1972). Pharmacokinetics: A. T. Trompler *et al., ibid.* **33**, 436 (1983). Mechanism of action: H. Langenfeld *et al., J. Cardiovasc. Pharmacol.* **15**, 338 (1990). GC/MS determn in urine: H. H. Maurer, *Arch. Toxicol.* **64**, 218 (1990). Clinical comparison with tocainide, *q.v.:* B. Schwartzkopff *et al., Arzneimittel-Forsch.* **33**, 153 (1983).

Hydrogen tartrate, C₂₃H₃₃N₂O₂.C₄H₅O₆, *prajmaline bitartrate, NPAB, GT-1012, Neo-Gilurytmal.* White crystals from ethanol + ether, mp 149-152°. LD₅₀ in mice (mg/kg): 43 orally; 1.7 i.v. (Von Philipsborn, Stalder).
THERAP CAT: Antiarrhythmic.

7884. Pralidoxime Chloride. *2-[(Hydroxyimino)methyl]-1-methylpyridinium chloride; 2-formyl-1-methylpyridinium chloride oxime;* 1-methyl-2-formylpyridinium chloride oxime; N-methylpyridinium-2-aldoxime chloride; 2-pyridine aldoxime methyl chloride; 2-PAM chloride; Protopam chloride. C₇H₉ClN₂O; mol wt 172.61. C 48.71%, H 5.26%, Cl 20.54%, N 16.23%, O 9.27%. Prepn of salts: I. B. Wilson *et al.,* **U.S. pat. 2,816,113** (1957 to U.S. Sec'y. of Army); A. A. Kondritzer *et al., J. Pharm. Sci.* **50**, 109 (1961). Manufacturing processes: R. I. Ellin *et al.,* **U.S.**

pat. **3,140,289** (1964 to U.S. Dept. of the Army); McDowell, **U.S. pat. 3,155,674** (1964 to Olin Mathieson). Commercial prepn: R. I. Ellin, *Ind. Eng. Chem. Prod. Res. Develop.* **3** (1), 20 (1964). Toxicology: Christensen, Richter, *Arch. Environ. Health* **15**, 599 (1967). Toxicity data: Fleisher *et al., Toxicol. Appl. Pharmacol.* **16**, 40 (1970); R. I. Ellin, J. H. Wills, *J. Pharm. Sci.* **53**, 1143 (1964). Comprehensive description: U. V. Banakar, U. N. Patel in *Analytical Profiles of Drug Substances* vol. **17**, K. Florey, Ed. (Academic Press, New York, 1988) pp 533-569.

Crystals from alcohol + ether, mp 235-238° (dec). Soly (g/100 ml) 25°: acetone 0, isopropanol 0.09, ethanol 0.89, methanol 8.5, water 65.5. LD₅₀ in rats (mg/kg): 96 i.v. (Fleisher). LD₅₀ in rabbits (mg/kg): 95 i.v.; LD₅₀ in mice (mg/kg): 115 i.v., 205 i.p., 4100 orally (Ellin, Wills).
Pralidoxime iodide, C₇H₉IN₂O, *2-pyridine aldoxime methiodide, 2-PAM.* Yellow crystals from alcohol, mp 225-226°. Soly at 25°: 48 mg/ml. Very sol in water, fairly sol in hot alcohol, poorly sol in cold alcohol. Insol in ether, acetone. LD₅₀ in mice (mg/kg): 140-178 i.v., 136-260 i.p., 290-340 s.c., 1500-4000 orally (Ellin, Wills).
Pralidoxime mesylate, C₈H₁₂N₂O₄S, *Contrathion.* Very hygroscopic crystals from ethanol, mp 155°. pKa 8.0. Soly in water: 1 g in 2 ml. LD₅₀ in mice (mg/kg): 118-122 i.v., 216 i.p., 3700 orally; in rats (mg/kg): 109 i.v., 262 i.p. (Ellis, Wills).
THERAP CAT: Cholinesterase reactivator. Antidote (nerve gases and cholinesterase-inhibitor type insecticides).

7885. Pramipexole. *(S)-4,5,6,7-Tetrahydro-N⁶-propyl-2,6-benzothiazolediamine;* (S)-2-amino-4,5,6,7-tetrahydro-6-(propylamino)benzothiazole. C₁₀H₁₇N₃S; mol wt 211.33. C 56.83%, H 8.11%, N 19.88%, S 15.17%. Dopamine D₂-receptor agonist. Prepn of racemate: G. Griss *et al.,* **Eur. pat. Appl. 186,087**; *eidem,* **U.S. pat. 4,886,812** (1986, 1989 both to Thomae); of enantiomers: C. S. Schneider, J. Mierau, *J. Med. Chem.* **30**, 494 (1987). Pharmacology: J. Mierau, G. Schingnitz, *Eur. J. Pharmacol.* **215**, 161 (1992). Clinical safety and pharmacodynamics: J. C. Schilling *et al., Clin. Pharmacol. Ther.* **51**, 541 (1992).

Dihydrochloride, C₁₀H₁₇N₃S.2HCl, *SND-919.* Crystals from methanol, mp 296-298°. $[\alpha]_D^{20}$ −67.2° (c = 1 in methanol).
THERAP CAT: Antiparkinsonian.

7886. Pramiracetam. *N-[2-[Bis(1-methylethyl)amino]-ethyl]-2-oxo-1-pyrrolidineacetamide;* N-[2-(diisopropylamino)ethyl]-2-oxo-1-pyrrolidineacetamide; amacetam. C₁₄H₂₇-N₃O₂; mol wt 269.39. C 62.42%, H 10.10%, N 15.60%, O 11.88%. Cognition enhancer structurally related to piracetam, *q.v.* Prepn: Y. J. L'Italien, I. C. Nordin, **Ger. pat. 2,808,067**; *eidem,* **U.S. pat. 4,145,347** (1978, 1979 both to Parke, Davis); D. E. Butler *et al., J. Med. Chem.* **27**, 684 (1984). Pharmacology: B. P. H. Poschel *et al., Drugs Exptl. Clin. Res.* **9**, 853 (1983). GC determn in plasma: T. Chang, R. M. Young, *J. Chromatog.* **274**, 346 (1983). Clinical pharmacokinetics: T. Chang *et al., J. Clin. Pharmacol.* **25**, 291 (1985). Clinical evaluation in Alzheimer's disease: R. J. Branconnier *et al., Psychopharmacol. Bull.* **19**, 726 (1983); in various cognitive disorders: R. Dejong, *Curr. Ther. Res.* **41**, 254 (1987).

bp$_{0.15}$ 162-164°. Monohydrate, mp 47-48°.

Sulfate, C$_{14}$H$_{27}$N$_3$O$_2$.H$_2$SO$_4$, *CI-879*. LD$_{50}$ in male, female mice (mg/kg): 5434, 4355 orally (Poschel).

THERAP CAT: Nootropic.

7887. Pramiverin. *N-(1-Methylethyl)-4,4-diphenylcyclo-hexanamine; N-isopropyl-4,4-diphenylcyclohexylamine;* pramiverine; primaverine; propaminodiphen. C$_{21}$H$_{27}$N; mol wt 293.45. C 85.95%, H 9.27%, N 4.77%. Prepn: **Neth.** pat. **Appl. 6,515,046**; R. Unger *et al.*, U.S. pat. **3,376,312** (1966, 1968 both to E. Merck). Pharmacology: H. J. Enenkel *et al.*, *Arzneimittel-Forsch.* **26**, 690 (1976). Toxicological study: M. Von Eberstein *et al.*, *ibid.*, 703. Series of articles on pharmacology, pharmacokinetics, GC determn, clinical studies: *ibid.*, 686-752. Use in management of labor: G. LoDico *et al.*, *Minerva Ginecol.* **31**, 683 (1979).

Liq, bp$_{0.05}$ 164-165°; also reported as a solid, mp 70° (Enenkel).

Hydrochloride, C$_{21}$H$_{27}$N.HCl, *EMD-9806, HSP-2986, Monoverin, Sistalgin*. Off-white crystalline powder, mp 230°; also reported as 234-237° (Enenkel). Soly at 25° (g/100 ml): in water 0.3; in ethanol 4; in chloroform 5. Practically insol in ether. LD$_{50}$ (after 14 days) in mice, rats (mg/kg): 346, 623 orally; 25, 26 i.v. (Von Eberstein).

THERAP CAT: Antispasmodic.

7888. Pramoxine. *4-[3-(4-Butoxyphenoxy)propyl]morpholine; p-butoxyphenyl γ-morpholinopropyl ether; γ-morpholinopropyl 4-n-butoxyphenyl ether; pramocaine; proxazocain.* C$_{17}$H$_{27}$NO$_3$; mol wt 293.41. C 69.59%, H 9.28%, N 4.77%, O 16.36%. Prepn: Wright, Moore, *J. Am. Chem. Soc.* **73**, 2281 (1951); **76**, 4396 (1954); U.S. pat. **2,870,151** (1959 to Abbott).

bp$_6$ 196°; bp$_{2.8}$ 183-184°.

Hydrochloride, C$_{17}$H$_{27}$NO$_3$.HCl, *Proctofoam, Tronolane, Tronothane*. Crystals, mp 181-183°. Sol in water. LD$_{50}$ i.v. in mice: 79.5 mg/kg.

THERAP CAT: Anesthetic (local).

7889. Pranlukast. *N-[4-Oxo-2-(1H-tetrazol-5-yl)-4H-1-benzopyran-8-yl]-4-(4-phenylbutoxy)benzamide; 8-[4-(4-phenylbutoxy)benzamido]-2-(tetrazol-5-yl)-4H-1-benzopyran-4-one; 8-[p-(4-phenylbutoxy)benzoyl]amino-2-(5-tetrazolyl)-4-oxo-4H-1-benzopyran;* ONO-1078; ONO-RS-411; Onon. C$_{27}$H$_{23}$N$_5$O$_4$; mol wt 481.51. C 67.35%, H 4.81%, N 14.54%, O 13.29%. Leukotriene antagonist. Prepn: M. Toda *et al.*, **Eur.** pat. **Appl. 173,516**; *eidem*, U.S. pat. **4,780,469** (1986, 1988 both to Ono); H. Nakai *et al.*, *J. Med. Chem.* **31**, 84 (1988). Pharmacology: T. Obata *et al.*, *Adv. Prostaglandin Thromboxane Leukot. Res.* **15**, 229 (1985); *idem et al.*, *ibid.* **17**, 540 (1987). Clinical evaluations in asthma: Y. Tani-

guchi *et al.*, *J. Allergy Clin. Immunol.* **92**, 507 (1993); H. Yamamoto *et al. Am. J. Respir. Crit. Care Med.* **150**, 254 (1994).

Occurs as the hemihydrate. Crystals, mp 244-245°. LD$_{50}$ in male mice (mg/kg): > 1000 i.v. (Toda).

THERAP CAT: Antiasthmatic.

7890. Pranoprofen. *α-Methyl-5H-[1]benzopyrano[2,3-b]pyridine-7-acetic acid;* 2-(5H-[1]benzopyrano[2,3-b]pyridin-7-yl)propionic acid; Y-8004; Niflan;. C$_{15}$H$_{13}$NO$_3$; mol wt 255.27. C 70.58%, H 5.13%, N 5.49%, O 18.80%. Prepn: M. Nakanishi *et al.*, **Ger.** pat. **2,337,052**; *eidem*, U.S. pat. **3,931,205** (1974, 1976 both to Yoshitomi); *eidem, Yakugaku Zasshi* **96**, 99 (1976), *C.A.* **84**, 135515b (1976). Physicochemical properties: M. Nobutoki, Y. Ota, *Iyakuhin Kenkyu* **7**, 200 (1976), *C.A.* **88**, 110468h (1978). Absorption, metabolism, distribution, excretion: Y. Kato *et al.*, *Yakuri To Chirgo* **4**, 1473 (1976), *C.A.* **88**, 130710j (1978). Pharmacological study: Y. Maruyama *et al., Nippon Yakurigaku Zasshi* **73**, 113 (1977), *C.A.* **88**, 276m (1978). Toxicity study: M. Edanaga *et al., Iyakuhin Kenkyu* **7**, 211 (1976), *C.A.* **88**, 131015y (1978).

Cryst from aq dioxane, mp 182-183°. LD$_{50}$ in male mice, rats (mg/kg): 447.3, 87.3 orally (Edanaga).

THERAP CAT: Anti-inflammatory.

7891. Praseodymium. Pr; at. wt 140.90765; at. no. 59; valences 3, 4. A lanthanide; belongs to the cerium group of rare earth metals. Naturally occurring isotope (mass number): 141; known artificial radioactive isotopes: 126, 129-140; 142-152. Prepn of radioactive isotopes by bombardment: Seaborg, *Chem. Rev.* **27**, 199 (1940); Huber *et al., Helv. Phys. Acta* **18**, 221 (1945). Abundance in earth's crust 5.5-9.1 ppm; found in rare earth minerals. Reported in 1843 by Mosander as *didymium* which was a mixture of Pr and Nd; separated in 1885 by von Welsbach. Separated from other rare earth elements by fractional crystn. Also by ion exchange: Spedding *et al., J. Am. Chem. Soc.* **69**, 2786, 2812 (1947). Prepn of metal by electrolysis: Muthmann, Weiss, *Ann.* **331**, 1 (1904); Canneri, Rossi, *Gazz. Chim. Ital.* **62**, 1160 (1932); Mazza, *Atti X Congr. Int. Chim.* **3**, 604 (1939); Trombe, Mahn, *Compt. Rend.* **220**, 778 (1945). Toxicity study: Haley *et al., Toxicol. Appl. Pharmacol.* **6**, 614 (1964). Reviews of prepn, properties and compds: *The Rare Earths*, F. H. Spedding, A. H. Daane, Eds. (Krieger, Huntington, N.Y., 1971, reprint of 1961 ed.) 641 pp; Hulet, Bode, "Separation Chemistry of the Lanthanides and Transplutonium Actinides" in *MTP Int. Rev. Sci.: Inorg. Chem., Ser. One*, Vol. 7, K. W. Bagnall, Ed. (Univ. Park Press, Baltimore, 1972) pp 1-45; Moeller, "The Lanthanides", *Comprehensive Inorganic Chemistry*, Vol. 4, J. C. Bailar, Jr. *et al.*, Eds. (Pergamon Press, Oxford, 1973) pp 1-101; F. H. Spedding in Kirk Othmer *Encyclopedia of Chemical Technology* vol. 19 (John Wiley & Sons, New York, 3rd ed., 1982) pp 833-854; *Chemistry of the Elements*, N. N. Greenwood, A. Earnshaw, Eds. (Pergamon Press, New York, 1984) pp 1423-1449. Brief review of properties: G. T. Seaborg, *Radiochim. Acta* **61**, 115-122 (1993).

Yellowish metal, forms oxide film on exposure to moist air. Two crystalline forms: hexagonal α-form, d 6.475, transforms to β-form at 798°; body-centered cubic β-form, d 6.64, exists at > 798°. mp 935°. bp 3020°; also reported as bp 3520° (Spedding). E^0 (aq) Pr^{3+}/Pr −2.47 V (calc). Experimental reduction potentials (referred to a normal calomel electrode): −1.875, −1.990 V: Noddack, Brukl, *Angew. Chem.* **50**, 362 (1937). Heat of sublimation (25°): 355.6 kJ/mol. Heat of fusion: 6.912 kJ/mol (Spedding, 1982). Also reported as heat of fusion: 11.3 ±2.1 kJ/mol (Greenwood, Earnshaw).

Oxide, Pr_2O_3, d 7.07, prepd by reducing with hydrogen the oxides PrO_2 or Pr_6O_{11}. Oxidizes to Pr_6O_{11} on heating in air; forms the dioxide, PrO_2, on fusing with potassium chlorate.

Hydroxide, $Pr(OH)_3$, a gelatinous pale green precipitate, obtained by the action of alkali hydroxide on a soln of a praseodymium salt; a purple powder when obtained by the action of water on praseodymium carbide.

Chloride, $PrCl_3$, heptahydrate, green crystals; mp in its water of crystn at 111°; on partial dehydration yields the hexa-, tri-, or monohydrate; on heating to 180-200° in a stream of hydrogen chloride yields the anhyd chloride, green needles, mp 769-782°, sol in water, alcohol. LD_{50} in mice: 600 mg/kg i.p.; 4.5 g/kg orally (Haley).

Sulfate, $Pr_2(SO_4)_3$, light green crystals, prepd by evaporating a soln of the oxide in H_2SO_4 and cooling the resulting product over phosphorus pentoxide. Several hydrates and double sulfates are known.

7892. Prasterone. *(3β)-3-Hydroxyandrost-5-en-17-one;* dehydroepiandrosterone; dehydroisoandrosterone; *trans*-dehydroandrosterone; $Δ^5$-androsten-3β-ol-17-one; DHEA; Astenile; Deandros; Diandrone; Psicosterone. $C_{19}H_{28}O_2$; mol wt 288.43. C 79.12%, H 9.78%, O 11.09%. Major secretory steroidal product of the adrenal gland; occurs in serum unconjugated or as the sulfate. May have estrogen- or androgen-like effects depending on the hormonal milieu. Isoln from male urine: Butenandt, Tscherning, *Z. Physiol. Chem.* **229**, 167 (1934); Butenandt, Dannenbaum, *ibid.* 192. Prepn from cholesterol: Butenandt *et al., ibid.* **237**, 57 (1935); Ruzicka, Wettstein, *Helv. Chim. Acta* **18**, 986 (1935); Wallis, Fernholz, *J. Am. Chem. Soc.* **57**, 1379, 1504 (1935); Schoeller *et al., Naturwiss.* **23**, 337 (1935); from sitosterol: Oppenauer, *Nature* **135**, 1039 (1935). High yield prepn: H. Hosoda *et al., J. Org. Chem.* **38**, 4209 (1973). Metabolism study: P. Knapstein *et al. Acta Endocrinol.* **58**, 261 (1968). Study of the sulfate as an indicator of adrenal androgen function: R. A. Lobo *et al., Obstet. Gynecol.* **57**, 69 (1981). Toxicity study of the sulfate: M. Yahara *et al., J. Toxicol. Sci.* **2**, 161 (1977). Review of physiological importance: P. Ebeling, V. A. Koivisto, *Lancet* **343**, 1479-1481 (1994).

Dimorphous, needles, mp 140-141°; leaflets, mp 152-153°. $[α]_D^{18}$ +10.9° (c = 0.4 in alc). Pptd by digitonin. Sol in benzene, alcohol, ether; sparingly sol in chloroform, petr ether.

Sodium sulfate, $C_{19}H_{27}NaO_5S$, *sodium dehydroepiandrosterone sulfate, DHA-S, Mylis.* White cryst powder, mp 154° (dec). Sol in methanol, slightly sol in water, abs ethanol. Practically insol in acetone, chloroform, benzene. LD_{50} in male mice, rats (mg/kg): > 10,000 orally; 899, 1005 s.c.; 460, 559 i.p.; 274, 468 i.v., *Japan. Med. Gaz.* **18**(5), 10 (1981).

THERAP CAT: Androgen.

7893. Pratensein. *5,7-Dihydroxy-3-(3-hydroxy-4-methoxyphenyl)-4H-1-benzopyran-4-one; 3′,5,7-trihydroxy-4′-methoxyisoflavone.* $C_{16}H_{12}O_6$; mol wt 300.27. C 64.00%, H 4.03%, O 31.97%. Isoln from red clover *(Trifolium pratense L., Leguminosae),* structure and synthesis: E. Wong, *J. Org. Chem.* **28**, 2336 (1963). Alternate syntheses: A. C. Jain, P. K. Bambab, *Indian J. Chem.* **26B**, 488 (1987).

Needles from ethanol, mp 272-273°. uv max (ethanol): 263 nm (log ε 4.53).

Triacetylpratensein, $C_{22}H_{18}O_9$, white needles from chloroform, mp 175-177°.

7894. Pravastatin Sodium. *[1S-[1α(βS*,δS*),2α,6α,8β-(R*),8aα]]-1,2,6,7,8,8a-Hexahydro-β,δ,6-trihydroxy-2-methyl-8-(2-methyl-1-oxobutoxy)-1-naphthaleneheptanoic acid monosodium salt;* sodium (+)-(3R,5R)-3,5-dihydroxy-7-[(1S,2S,6S,8S,8aR)-6-hydroxy-2-methyl-8-[(S)-2-methylbutyryloxy]-1,2,6,7,8,8a-hexahydro-1-naphthyl]heptanoate; eptastatin sodium; 3β-hydroxycompactin sodium salt; CS-514; SQ-31000; Elisor; Lipostat; Mevalotin; Oliprevin; Pravachol; Pravaselect; Selectin; Selipran; Vasten. $C_{23}H_{35}NaO_7$; mol wt 446.52. C 61.87%, H 7.90%, Na 5.15%, O 25.08%. HMG-CoA reductase inhibitor; bioactive metabolite of mevastatin, *q.v.* Prepn by microbial hydroxylation: A. Terahara, M. Tanaka, Ger. pat. **3,122,499**; *eidem*, U.S. pat. **4,346,227** (1981, 1982 both to Sankyo); N. Serizawa *et al., J. Antibiot.* **36**, 604 (1983). Structure elucidation: H. Haruyama *et al., Chem. Pharm. Bull.* **34**, 1459 (1986). Effect on serum lipid concentration: N. Nakaya *et al., Atherosclerosis* **61**, 125 (1986); on hepatic metabolism of cholesterol: E. Reihner *et al., N. Engl. J. Med.* **323**, 224 (1990). Preliminary clinical trial in primary hypercholesterolemia: T. Kazumi *et al., Horm. Metabol. Res.* **18**, 654 (1986). Clinical comparison with probucol, *q.v.:* G. Yoshino *et al., Lancet* **2**, 740 (1986).

Amorphous powder. uv max (methanol): 230, 237, 245 nm.

Lactone, $C_{23}H_{34}O_6$, colorless plate crystals, mp 138-142°. $[α]_D^{22}$ +194.0° (c = 0.51 in methanol). uv max (methanol): 230, 237, 245 nm.

THERAP CAT: Antihyperlipoproteinemic.

7895. Prazepam. *7-Chloro-1-(cyclopropylmethyl)-1,3-dihydro-5-phenyl-2H-1,4-benzodiazepin-2-one;* 1-(cyclopropylmethyl)-5-phenyl-7-chloro-1H-1,4-benzodiazepin-2(3H)-one; W-4020; Centrax; Demetrin; Lysanxia; Prazene; Sedapran; Settima; Trepidan; Verstran. $C_{19}H_{17}ClN_2O$; mol wt 324.81. C 70.26%, H 5.28%, Cl 10.91%, N 8.62%, O 4.93%. Prepn: McMillan, Pattison, Fr. pat. **1,394,287**, corresp to U.S. pat. **3,192,199** (both 1965 to Warner-Lambert); Wuest, U.S. pat. **3,192,200** (1965); Inaba *et al., Chem. Pharm. Bull.* **17**, 1263 (1969). Pharmacology: Robichaud *et al., Arch.*

Int. Pharmacodyn. Ther. **185**, 213 (1970). Metabolism: DiCarlo *et al., J. Pharm. Sci.* **58**, 960 (1969).

Crystals from methanol, mp 145-146°.

Note: This is a controlled substance (depressant) listed in the U.S. Code of Federal Regulations, Title 21 Part 1308.14 (1995).

THERAP CAT: Anxiolytic.

7896. Praziquantel. *2-(Cyclohexylcarbonyl)-1,2,3,6,7,-11b-hexahydro-4H-pyrazino[2,1-a]isoquinolin-4-one;* EMBAY 8440; Biltricide; Cesol; Droncit. $C_{19}H_{24}N_2O_2$; mol wt 312.41. C 73.05%, H 7.74%, N 8.97%, O 10.24%. Prepn: J. Seubert *et al., Ger.* pat. **2,362,539** corresp to U.S. pat. **4,001,411** (1975, 1977, both to E. Merck, W. Ger.); D. Frehel *et al., Heterocycles* **20**, 1731 (1983). Properties: J. Seubert *et al., Experientia* **33**, 1036 (1977). *In vitro* study: C. J. Chavasse *et al., Z. Parasitenk.* **58**, 169 (1979). Effect on ultrastructure of trematodes: H. Mehlhorn *et al., Arznei-mittel-Forsch.* **33**, 91 (1983). Pharmacodynamics: G. C. Coles, *J. Helminthol.* **53**, 31 (1979). Ovicidal activity in *Echinococcus:* A. S. Thakur *et al., Exp. Parasitol.* **47**, 131 (1979). Clinical pharmacology: G. Leopold *et al., Eur. J. Clin. Pharmacol.* **14**, 281 (1978). Efficacy in dogs: F. L. Anderson *et al., Am. J. Vet. Res.* **40**, 700 (1979). Mutagenicity study: H. Bartsch *et al., Mutat. Res.* **58**, 133 (1978). Tolerance study: P. Muermann *et al., Vet. Med. Rev.* **1976** (2), 142. Symposium on African schistosomiasis: *Arznei-mittel-Forsch.* **31**, 535-618 (1981). Clinical studies (human schistosomiasis): T. E. Nash *et al., Am. J. Trop. Med. Hyg.* **31**, 977 (1982); R. N. H. Pugh, C. H. Teesdale, *Brit. Med. J.* **286**, 429 (1983); C. H. Schutte *et al., S. Afr. Med. J.* **64**, 7 (1983). Clinical trial in human parenchymal brain cysticercosis: J. Sotelo *et al., N. Engl. J. Med.* **310**, 1001 (1984).

Crystals, mp 136-138°. Soly (g/100 ml): ethanol 9.7; chloroform 56.7; water 0.04. LD_{50} in mice, rats (mg/kg): 2000-3000 orally; > 3000 s.c. (Muermann).

THERAP CAT: Anthelmintic (Schistosoma).

THERAP CAT (VET): Anthelmintic.

7897. Prazosin. *1-(4-Amino-6,7-dimethoxy-2-quinazo-linyl)-4-(2-furanylcarbonyl)piperazine;* 2-[4-(2-furoyl)piper-azin-1-yl]-4-amino-6,7-dimethoxyquinazoline; furazosin. $C_{19}H_{21}N_5O_4$; mol wt 383.41. C 59.52%, H 5.52%, N 18.27%, O 16.69%. α_1-Adrenergic blocker. Prepn: *Brit.* pat. **1,156,-973**; Hess, U.S. pat. **3,511,836** (1969, 1970 both to Pfizer); Neth. pat. Appl. **7,206,067** (1972 to Brocades-Stheeman), *C.A.* **78**, 72180s (1973); E. Honkanen *et al., J. Heterocycl. Chem.* **17**, 797 (1980). Pharmacology and clinical data: Scriabine *et al., Experientia* **24**, 1150 (1968); Cohen, *J. Clin. Pharmacol. J. New Drugs* **10**, 408 (1970). Pharmacokinetics: P. Jaillon, *Clin. Pharmacokinet.* **5**, 365 (1980). HPLC determn in biological fluids: R. K. Bhamra *et al., J. Chromatog.* **380**, 216 (1986). Clinical evaluation in Raynaud's phenomenon: I. J. Russell, J. A Lessard, *J. Rheumatol.* **12**, 94 (1985); in dysuria with benign prostatic hypertrophy: A. LeDuc *et al., Urol. Int.* **45**, Suppl. 1, 56 (1990). Book: *Prazosin: Pharmacology, Hypertension and Congestive Heart Failure,* M. D. Rawlins, Ed. (Grune & Stratton, New York, 1981) 143 pp. Review of pharmacology and therapeutic use: W. F. Stanaszek *et al., Drugs* **25**, 339-384 (1983); J. L. Reid, J. Vincent, *Cardiology* **73**, 164-174 (1986). Comprehensive description: L. J. Kostek in *Analytical Profiles of Drug Substances* vol. 18, K. Florey, Ed. (Academic Press, New York, 1989) pp 351-378.

Crystals, mp 278-280°.

Hydrochloride, $C_{19}H_{21}N_5O_4 \cdot HCl$, *CP-12299-1, Alpress LP, Duramipress, Eurex, Hypovase, Minipress, Peripress, Sinetens.* uv max (methanolic 0.01N HCl): 246, 329 nm (a_M 137 ±3, 27.6 ±0.3). Soly at ambient temp (mg/ml): acetone 0.0072, methanol 6.4, ethanol 0.84, dimethylformamide 1.3, dimethylacetamide 1.2, water (pH~3.5) 1.4, chloroform 0.041.

THERAP CAT: Antihypertensive.

7898. Precocenes. Anti-JH. Anti-juvenile hormones found in plants that induce reversible precocious metamorphosis and sterilization in insects by suppressing the function of the corpora allata gland: W. S. Bowers *et al., Science* **193**, 542 (1976); W. S. Bowers, R. Martinez-Pardo, *ibid.* **197**, 1369 (1977). Isoln from *Ageratum mexicanum* Sw., *Compositae* and structure: A. R. Alertsen, *Acta Chem. Scand.* **9**, 1725 (1955). Synthesis of precocene I: R. Livingstone, R. B. Watson, *J. Chem. Soc.* **1957**, 1509; of precocene II: R. Huls, *Bull. Soc. Chim. Belg.* **66**, 409 (1957); **67**, 22 (1958); of I and II: J. R. Hlubucek *et al., Aust. J. Chem.* **24**, 2347 (1971); A. Banerji, N. C. Goomer, *Indian J. Chem.* **20B**, 144 (1981); V. K. Ahluwalia *et al., Chem. & Ind. (London)* **1982**, 369. Biosynthesis: A. V. Vyas, N. B. Mulchandani, *Phytochemistry* **19**, 2597 (1980). Metabolism studies: D. M. Soderlund *et al., J. Agr. Food Chem.* **1980**, 724; B. J. Bergot *et al., Pestic Biochem. Physiol.* **13**, 95 (1980). Mechanism of action: W. S. Bowers, *Am. Zool.* **21**, 737 (1981).

Precocene I   R = H
Precocene II   R = OCH₃

Precocene I, $C_{12}H_{14}O_2$, *7-methoxy-2,2-dimethyl-2H-1-benzo-pyran, 7-methoxy-2,2-dimethylchromene, 6-demethoxyagera-tochromene.* Oil, bp_6 120°. n_D^{21} 1.5548. uv max (ethanol): 280, 304 nm (ε 6644, 5844).

Precocene II, $C_{13}H_{16}O_3$, *6,7-dimethoxy-2,2-dimethyl-2H-1-benzopyran, 6,7-dimethoxy-2,2-dimethylchromene, agerato-chromene.* Crystals, mp 47.5°, bp_6 136°. uv max (ethanol): 280, 323 nm (ε 5500, 9200).

7899. Prednicarbate. *(11β)-17-[(Ethoxycarbonyl)oxy]-11-hydroxy-21-(1-oxopropoxy)pregna-1,4-diene-3,20-dione;* 11β,17,21-trihydroxypregna-1,4-diene-3,20-dione 17-(ethyl carbonate) 21-propionate; prednisolone 17-ethylcarbonate 21-propionate; HOE-777; S-77-0777; EsCort; Dermatop; Prednitop; Regenit. $C_{27}H_{36}O_8$; mol wt 488.58. C 66.38%, H 7.43%, O 26.20%. Topical anti-inflammatory agent. Prepn: U. Stache *et al., Ger.* pat. **2,735,110;** *eidem,* U.S. pat. **4,242,-334** (1979, 1980 both to Hoechst AG); U. Stache *et al., Arzneimittel-Forsch.* **35**, 1753 (1985). Topical and systemic activity in rats: H. G. Alpermann *et al., ibid.* **32**, 633 (1982). Series of articles on pharmacology, pharmacokinetics and clinical efficacy: *Z. Hautkr.* **61**, Suppl. 1, 1-96 (1986).

Crystals from ethanol + diethylether, mp 110-112°; second crystalline form, mp 183°. uv max (ethanol): 241 nm (ε 15000). $[\alpha]_D^{20}$ +63° (c = 0.1 in ethanol).

THERAP CAT: Glucocorticoid.

7900. Prednimustine. *(11β)-21-[4-[4-[Bis(2-chloroethyl)amino]phenyl]-1-oxobutoxy]-11,17-dihydroxypregna-1,4-diene-3,20-dione; 11β,17,21-trihydroxypregna-1,4-diene-3,20-dione 21-[4-[p-[bis(2-chloroethyl)amino]phenyl]butyrate]; prednisolone 21-[4-[p-[bis(2-chloroethyl)amino]phenyl]butyrate]; Leo-1031; Sterecyt; Stereocyt.* $C_{35}H_{45}Cl_2NO_6$; mol wt 646.65. C 65.01%, H 7.01%, Cl 10.97%, N 2.17%, O 14.85%. Prednisolone ester of chlorambucil, *q.v.* Prepn: H. F. Fex *et al.,* Ger. pat. **2,001,305** corresp to U.S. pat. **3,732,-260** (1970, 1973 both to AB Leo). *In vitro* study: A. H. Evenaar *et al., Eur. J. Cancer* **9,** 773 (1973). Metabolism: R. Y. Kirdani *et al., Oncology* **35,** 47 (1978). Clinical studies: L. Hakarsson *et al., ibid.* 103; J. E. Johnsson *et al., Cancer Treat. Rep.* **63,** 421 (1970). Toxicity and antitumor activity study: K. R. Harrap *et al., Eur. J. Cancer* **13,** 873 (1977). Series of articles on pharmacology, toxicology and clinical efficacy: *Semin. Oncol.* **13,** Suppl. 1, 1-44 (1986).

Crystals from methanol-water, mp 163-164°. $[\alpha]_D^{24}$ +92.9° (c = 1.06 in chloroform).

THERAP CAT: Antineoplastic.

7901. Prednisolone. *(11β)-11,17,21-Trihydroxypregna-1,4-diene-3,20-dione; 1,4-pregnadiene-3,20-dione-11β,17α,-21-triol; 1,4-pregnadiene-11β,17α,21-triol-3,20-dione; 3,20-dioxo-11β,17α,21-trihydroxy-1,4-pregnadiene; metacortandralone; delta F; Δ¹-dehydrocortisol; Δ¹-hydrocortisone; Δ¹-dehydrohydrocortisone; hydroretrocortine; Codelcortone; Cortalone; Decaprednil; Decortin H; Delta-Cortef; Deltacortril Enteric; Deltastab; Deltasolone; Flamasone; Hydeltra; Hydrodeltalone; Klismacort; Meticortelone; Paracortol; Precortancyl; Precortilon; Precortisyl; Prednelan; Prednicen; Predniretard; Predonine; Solone; Sterolone.* $C_{21}H_{28}O_5$; mol wt 360.45. C 69.98%, H 7.83%, O 22.19%. Synthetic corticosteroid; metabolically interconvertible with prednisone, *q.v.* Prepn: A. Nobile *et al., J. Am. Chem. Soc.* **77,** 4184 (1955); A. Nobile, U.S. pats. **2,837,464; 3,134,718** (1958, 1964, both to Schering); Herzog *et al., Tetrahedron* **18,** 581 (1962). HPLC determn in urine: B. M. Frey, F. J. Frey, *J. Chromatog.* **229,** 283 (1982). Clinical use in advanced cancer: R. G. Twycross, D. Guppy, *Practitioner* **229,** 57 (1985); in inflammatory bowel disease: C. J. Hawkey, *Neth. J. Med.* **35** Suppl 1, S21-S26 (1989). Review of clinical pharmacology: W. J. Jusko, J. Q. Rose, *Ther. Drug Monit.* **2,** 169-176 (1980); of pharmacokinetics and metabolism: B. M. Frey, F. J. Frey, *Clin. Pharmacokinet.* **19,** 126-146 (1990). Comprehensive description: S. L. Ali, *Anal. Profiles Drug Subs. Excip.* **21,** 415-500 (1992).

Crystals, dec 240-241°. $[\alpha]_D^{25}$ +102° (dioxane). uv max (methanol): 242 nm (ε 15000; $A_{1cm}^{1\%}$ 414). Very slightly sol in water. One gram dissolves in about 30 ml of alcohol, in about 180 ml of chloroform, in about 50 ml of acetone. Sol in methanol, dioxane.

21-Acetate, $C_{23}H_{30}O_6$, *Ak-Tate, Hostacortin H, Inflanefran, Pred Forte, Pred Mild, Scherisolon, Sterane.* Crystals, decomp 237-239°. $[\alpha]_D^{25}$ +116° (dioxane).

21-*tert*-Butylacetate, $C_{27}H_{38}O_6$, *prednisolone tebutate, Codelcortone-T.B.A., Hydeltra-T.B.A., Predalone T.B.A.* Prepn: Sarett, U.S. pat. **2,736,734** (1956 to Merck & Co.). Crystals from ethanol, mp 266-273°.

21-Hydrogen succinate, $C_{25}H_{32}O_8$, *prednisolone hemisuccinate, Fiasone (amp).*

21-Succinate sodium salt, $C_{25}H_{31}NaO_8$, *prednisolone sodium succinate, Di-Adreson-F, Meticortelone Soluble, Solu-Decortin-H.* Prepn: Shull, Kita, Ger. pat. **1,045,400** (1958 to Pfizer), *C.A.* **55,** 2746f (1961).

21-Stearoylglycolate, $C_{41}H_{64}O_8$, *prednisolone steaglate, Sintisone.* Prepn: Giraldi, Nannini, U.S. pat. **3,171,846** (1965 to Carlo Erba); Giraldi *et al., Arzneimittel-Forsch.* **16,** 162 (1966). Crystals from dilute alcohol or butyl ether, mp 105-107°. uv max (methanol): 242 nm ($E_{1cm}^{1\%}$ 212 ±10). $[\alpha]_D^{20}$ +57-63°.

21-*m*-Sulfobenzoate sodium salt, $C_{28}H_{31}NaO_9S$, *(11β)-11,-17-dihydroxy-21-[(3-sulfobenzoyl)oxy]pregna-1,4-diene-3,20-dione monosodium salt, prednisolone sodium metasulfobenzoate, Cortico-Sol, Predenema, Predfoam, Solupred.* Prepn: Allais, Girault, U.S. pat. **3,032,568;** Joly, Warnant, U.S. pat. **3,037,034** (both 1962 to Roussel-UCLAF). Crystals from water, dec 293-295°. $[\alpha]_D^{20}$ +170° (water).

21-Trimethylacetate, $C_{26}H_{36}O_6$, *prednisolone 21-pivalate, Ultracortenol.* Prepn: Vischer *et al., Helv. Chim. Acta* **38,** 1502 (1955); Joly, Warnant, U.S. pat. **3,037,034** (1962 to Roussel-UCLAF). Crystals from acetone, mp 233-236°. $[\alpha]_D^{26}$ +103° (c = 1.208 in chloroform); $[\alpha]_D^{20}$ +97.5° (c = 1 in chloroform). uv max: 244 nm (ε 14700).

THERAP CAT: Glucocorticoid.

THERAP CAT (VET): Glucocorticoid.

7902. Prednisolone 21-Diethylaminoacetate. *N,N-Diethylglycine (11β)-11,17-dihydroxy-3,20-dioxopregna-1,4-dien-21-yl ester; 11β,17,21-trihydroxypregna-1,4-diene-3,20-dione 21-(N,N-diethylglycine) ester; prednisolone 21-N,N-diethylglycinate; prednisolamate.* $C_{27}H_{39}NO_6$; mol wt 473.61. C 68.47%, H 8.30%, N 2.96%, O 20.27%. Prepd from prednisolone and *N,N*-diethylaminoacetyl chloride hydrochloride: Pancrazio, Sbarigia, *Farmaco Ed. Prat.* **16,** 190 (1961); from prednisolone and chloroacetyl chloride followed by reaction with diethylamine: **Brit.** pat. **862,370** (1961 to Pfizer).

Crystals from acetone + hexane, mp 175-177°.

Hydrochloride, $C_{27}H_{39}NO_6 \cdot HCl$, *Deltacortril DA.* Crystals from ethanol, mp 237.4-239.8°. $[\alpha]_D$ +120.7° (water).

THERAP CAT: Glucocorticoid.

7903. Prednisolone Sodium Phosphate. *11,17-Dihydroxy-21-(phosphonooxy)pregna-1,4-diene-3,20-dione disodium salt; 11β,17,21-trihydroxypregna-1,4-diene-3,20-dione 21-(dihydrogen phosphate) disodium salt; 21-prednisolonephosphoric acid disodium salt; prednisolone phosphate disodium; disodium prednisolone 21-phosphate; Ak-Pred; Codelsol; Hefasolon; Hydeltrasol; Inflamase; Metreton; Prednesol; Predsol; Solucort; Solu-Predalone.* $C_{21}H_{27}Na_2O_8P$; mol wt 484.39. C 52.07%, H 5.62%, Na 9.49%, O 26.42%, P 6.39%. $RPO(ONa)_2$ where R is the 21-prednisolone radical. Prepn: Sarett, U.S. pat. **2,789,117** (1957 to Merck & Co.). Alternate synthesis: Poos *et al., Chem. & Ind. (London)*

1958, 1260; Elks, Phillips, U.S. pat. **2,936,313** (1960 to Glaxo).

White powder. Slightly hygroscopic. Stable at room temp. $[\alpha]_D^{25}$ +102.5°. uv max (methanol): 243 nm ($A_{1cm}^{1\%}$ 308). Sol in water, methanol, ethanol. pH of 1% aq soln 7.5 to 8.5.

THERAP CAT: Glucocorticoid.

THERAP CAT (VET): Glucocorticoid.

7904. Prednisone. *17,21-Dihydroxypregna-1,4-diene-3,11,20-trione;* 1,4-pregnadiene-17α,21-diol-3,11,20-trione; Δ¹-dehydrocortisone; Δ¹-cortisone; deltacortisone; delta E; metacortandracin; retrocortine; NSC-10023; Ancortone; Colisone; Cortancyl; Dacortin; Decortancyl; Decortin; Delcortin; Deltacortone; Deltasone; Deltison; Di-Adreson; Encorton; Meticorten; Nurison; Orasone; Paracort; Prednilonga; Pronison; Rectodelt; Sone; Ultracorten. $C_{21}H_{26}O_5$; mol wt 358.43. C 70.37%, H 7.31%, O 22.32%. Prepn: Oliveto, Gould, U.S. pat. **2,897,216** (1959 to Schering). Microbiological prepn: Nobile *et al., J. Am. Chem. Soc.* **77,** 4184 (1955); Nobile, U.S. pat. **2,837,464** and **3,134,718** (1958, 1964 both to Schering); Herzog *et al., Tetrahedron* **18,** 581 (1962). Structure: Herzog *et al., Science* **121,** 176 (1955); *cf.* Djerassi *et al.,* U.S. pat. **2,579,479** (1951 to Syntex).

Crystals, dec 233-235°. $[\alpha]_D^{25}$ +172° (dioxane). uv max (methanol): 238 nm (ε 15500). Very slightly sol in water. One gram dissolves in about 150 ml alcohol, in about 200 ml chloroform. Slightly sol in methanol, dioxane.

21-Acetate, $C_{23}H_{28}O_6$, *Delta-Cortelan, Hostacortin.* Crystals, dec 226-232°. $[\alpha]_D^{25}$ +186° (dioxane). uv max (ethanol): 238 nm (ε 16100).

THERAP CAT: Glucocorticoid.

THERAP CAT (VET): Adrenocortical steroid. Glucocorticoid, anti-inflammatory.

7905. Prednival. *(11β)-11,21-Dihydroxy-17-[(1-oxopentyl)oxy]pregna-1,4-diene-3,20-dione; 11β,17,21-trihydroxypregna-1,4-diene-3,20-dione 17-valerate;* prednisolone 17-valerate; W-4869. $C_{26}H_{36}O_6$; mol wt 444.57. C 70.24%, H 8.16%, O 21.59%. Prepn: Vitali, Ercoli, *Tetrahedron Letters* **1961,** 448; Ercoli, Gardi, U.S. pat. **3,152,154** (1964 to Vismara).

Crystals from aq methanol, mp 210-213°. $[\alpha]_D$ +3.5° (dioxane).

21-Acetate, $C_{28}H_{38}O_7$, *Acepreval.*

THERAP CAT: Glucocorticoid.

7906. Prednylidene. *(11β)-11,17,21-Trihydroxy-16-methylenepregna-1,4-diene-3,20-dione;* Δ¹·⁴-pregnadiene-16-methylene-11β,17α,21-triol-3,20-dione; 16-methylene-11β,-17α,21-trihydroxypregna-1,4-diene-3,20-dione; 16-methyleneprednisolone; Dacortilen; Decortilen; Sterocort. $C_{22}H_{28}-O_5$; mol wt 372.46. C 70.94%, H 7.58%, O 21.48%. Prepn: Mannhardt *et al., Tetrahedron Letters* no. 16, 21 (1960). Prepn of the 21-diethylaminoacetate: *idem et al.,* Ger. pat. **1,134,074** (1962 to E. Merck), *C.A.* **58,** 568a (1963).

Solid, mp 233-235°. $[\alpha]_D^{23}$ +31° (dioxane). uv max: 243 nm (ε 15900).

21-Diethylaminoacetate hydrochloride, $C_{28}H_{39}NO_6 \cdot HCl$, *Decortilen soluble.* Solid, mp 245-246°. $[\alpha]_D$ +45° (water). uv max (water): 246-247 nm (E$_{1cm}^{1\%}$ 300).

THERAP CAT: Glucocorticoid.

7907. Pregnane. *5β-Pregnane;* 17β-ethyletiocholane. $C_{21}H_{36}$; mol wt 288.52. C 87.42%, H 12.58%. Prepd by reduction of etiocholyl methyl ketone or of pregnanedione: Butenandt, *Ber.* **64,** 2529 (1931); Marker *et al., J. Am. Chem. Soc.* **60,** 1067 (1938); Steiger, Reichstein, *Helv. Chim. Acta* **21,** 161 (1938).

Monoclinic scales, plates from methanol; mp 83.5°. d$_4^{15}$ 1.032. $[\alpha]_D^{19}$ +20° (c = 2 in chloroform).

7908. Pregnanediol. *(3α,5β,20S)-Pregnane-3,20-diol.* $C_{21}H_{36}O_2$; mol wt 320.52. C 78.70%, H 11.32%, O 9.98%. A metabolite of progesterone, present in large amounts in pregnancy urine. Isoln from pregnancy urine of women: Marrian, *Biochem. J.* **23,** 1090 (1929), Butenandt, *Ber.* **63,** 659 (1930); of cows, mares, and chimpanzees: Fish *et al., J. Biol. Chem.* **143,** 716 (1942). Prepn by reduction of pregn-16-ene-3,20-dione: Marker *et al.,* U.S. pat. **2,352,852** (1944 to Parke, Davis). Conversion to progesterone: Butenandt, Schmidt, *Ber.* **67,** 1893, 1901 (1934). Conversion to 3α-hydroxypregnan-20-one: Marker, U.S. pat. **2,223,377** (1940 to Parke, Davis). Prepn of the 3-acetate: Hirschmann, *J. Biol. Chem.* **140,** 797 (1941); Ralls *et al., ibid.* **210,** 709 (1954). Prepn of the 20-acetate: Hirschmann, *loc. cit.* Prepn of the diacetate: Johnson *et al., J. Chem. Soc.* **1954,** 1302. Crystal structure: Haner, Norton, *Acta Cryst.* **16,** 707 (1963). Review of metabolism, bioactivity and assay during pregnancy: P. J. Keller, *Contrib Gynecol. Obstet.* **2,** 75-91 (1976).

Crystals from acetone or ethanol, mp 239°. $[\alpha]_D^{20}$ +27.4° (c = 0.7 in alc). Sparingly sol in organic solvents. Not precipitated by digitonin.

3-Acetate, $C_{23}H_{38}O_3$, *3α-acetoxypregnan-20α-ol.* Crystals, mp 132°. $[\alpha]_D^{25}$ +45° (CHCl₃).

20-Acetate, $C_{23}H_{38}O_3$, *20α-acetoxypregnan-3α-ol.* Crystals, mp 174°.

Diacetate, $C_{25}H_{40}O_4$, *3α,20α-diacetoxypregnane.* Crystals from light petroleum, mp 180°, also reported as mp 182-183°. $[\alpha]_D^{15}$ +35° (c = 1.1 in $CHCl_3$).

USE: In manuf of progesterone.

7909. 3,20-Pregnanedione. *(5β)-Pregnane-3,20-dione.* $C_{21}H_{32}O_2$; mol wt 316.48. C 79.70%, H 10.19%, O 10.11%. From pregnancy urine of mares. Prepn from other steroids: Butenandt, *Ber.* **63**, 659 (1930); Butenandt, Fleischer, *Ber.* **68**, 2094 (1935); Marker *et al., J. Am. Chem. Soc.* **59**, 1595 (1937); Shoppee, Reichstein, *Helv. Chim. Acta* **24**, 356 (1941); U.S. pats. **2,160,719; 2,352,852; 2,323,276; 2,229,-818.**

Needles from dil alc, mp 123°. Insol in water. Freely sol in the usual organic solvents.

Dioxime, $C_{21}H_{34}N_2O_2$, dec above 250°.

7910. Pregnan-3α-ol-20-one. *(3α,5β)-3-Hydroxypregnan-20-one;* 3α-hydroxy-5β-pregnan-20-one; pregnanolone; eltanolone; KABI 2213. $C_{21}H_{34}O_2$; mol wt 318.50. C 79.19%, H 10.76%, O 10.05%. Naturally occurring metabolite of progesterone, *q.v.* Isoln from urine of pregnant women: R. E. Marker, O. Kamm, *J. Am. Chem. Soc.* **59**, 1373 (1937); from bile of pregnant cows: W. H. Pearlman, E. Cerceo, *J. Biol. Chem.* **176**, 847 (1948). Prepn: R. E. Marker *et al., J. Am. Chem. Soc.* **59**, 1841 (1937); L. Gyermek *et al., J. Med. Chem.* **11**, 117 (1968); T. L. G. Lemos, J. D. McChesney, *J. Nat. Prod.* **53**, 152 (1990). Pharmacokinetics and pharmacodynamics: P. Carl *et al., Acta Anaesthesiol. Scand.* **38**, 734 (1994). Comparative clinical evaluation: J. Van Hemelrijck *et al., Anesthesiology* **80**, 36 (1994); H. Eriksson *et al., Acta Anaesthesiol. Scand.* **39**, 479 (1995). Hemodynamic effects in humans: J. W. Sear *et al., J. Clin. Anesthesia* **7**, 126 (1995).

Needles from aq methanol, mp 148-148.5° (Pearlman, Cerceo); also reported as cryst from hexane, mp 131-132° (Lemos, McChesney). $[\alpha]_D^{23}$ +59.6° (c = 0.3 in chloroform). $[\alpha]_D^{26}$ +108.5 ±1° (c = 9.23 mg/1.23 ml abs ethanol). LD_{50} in mice, rats (mg/kg): 66 ± 10, 27.5 ± 2.4 i.v. (Gyermek).

THERAP CAT: Anesthetic (local).

7911. 4-Pregnene-20,21-diol-3,11-dione. *20,21-Dihydroxypregn-4-ene-3,11-dione;* Reichstein's substance T. $C_{21}H_{30}O_4$; mol wt 346.47. C 72.80%, H 8.73%, O 18.47%. Isoln from adrenal glands: Reichstein, von Euw, *Helv. Chim. Acta* **22**, 1222 (1939).

Crystals, mp ~210°. $[\alpha]_D^{20}$ +176° (acetone).
Diacetate, $C_{25}H_{34}O_6$, crystals, mp 212-213°.

7912. 4-Pregnene-11β,17α,20β,21-tetrol-3-one. *11β,-17,20β,21-Tetrahydroxypregn-4-en-3-one;* 17-(1,2-dihydroxyethyl)androsten-3-one-11,17-diol; Reichstein's substance E. $C_{21}H_{32}O_5$; mol wt 364.48. C 69.20%, H 8.85%, O 21.95%. Occurs in adrenal cortex. Isoln: Reichstein, *Helv. Chim. Acta* **19**, 29 (1936); **20**, 953 (1937); Reichstein, von Euw, *ibid.* **24**, 247E (1941).

Hydrated crystals from dil acetone, dec 125°. $[\alpha]_D^{20}$ +87° (alc). $[M]_D$ +317°. uv max: 240 nm.
20,21-Diacetate, crystals, dec 229-230°. $[\alpha]_D^{20}$ +162.7°. $[M]_D$ +730° (acetone).

7913. 4-Pregnene-17α,20β,21-triol-3,11-dione. *(20R)-17,20,21-Trihydroxypregn-4-ene-3,11-dione;* Reichstein's substance U. $C_{21}H_{30}O_5$; mol wt 362.47. C 69.59%, H 8.34%, O 22.07%. Isoln from adrenal glands: Reichstein, von Euw, *Helv. Chim. Acta* **24**, 247E (1941).

Clusters of needles from acetone + ether, mp 208-209°.
20,21-Diacetate, $C_{25}H_{34}O_7$, pointed needles from acetone + ether or chloroform + ether, mp 253°. $[\alpha]_D^{21}$ +178.5° (c = 0.924 in acetone). uv max (alc): 239 nm (log ε 4.1). Less sol than the 20,21-diacetate of 4-pregnene-11β,17α,20,21-tetrol-3-one.

7914. 4-Pregnene-17α,20β,21-triol-3-one. *(20R)-17,-20,21-Trihydroxypregn-4-en-3-one;* 17-(1,2-dihydroxyethyl)-Δ^4-androsten-3-on-17α-ol; 17α-pregnenetriolone. $C_{21}H_{32}O_4$; mol wt 348.48. C 72.38%, H 9.26%, O 18.36%. Prepd from 17-ethynyltestosterone by hydrogenation with palladium in pyridine, allylic rearrangement, and hydroxylation with osmium tetroxide: Ruzicka, Müller, *Helv. Chim. Acta* **22**, 755 (1939); Logemann, *Naturwiss.* **27**, 196 (1939); from a 3-enol ester of a 17,21-diacyloxyprogesterone by reduction followed by saponification: Swiss pat. **207,496** (1940), *C.A.* **36**, 3636 (1942).

Crystals from methanol. mp 190°. Sol in dioxane, chloroform, methanol. $[\alpha]_D$ +63° (c = 1 in dioxane). uv max: 240 nm (log ϵ 4.1).

20,21-Diacetate, $C_{25}H_{36}O_6$, crystals from acetone + ether. Polymorphic; mp 170° and 194°; $[\alpha]_D^{20}$ +125° (dioxane). Reaction with zinc in toluene yields 17-isodesoxycorticosterone acetate.

7915. Pregnenolone. *(3β)-3-Hydroxypregn-5-en-20-one;* Δ^5-*pregnen-3β-ol-20-one;* 17β-*(1-ketoethyl)-*Δ^5-*androsten-3β-ol.* $C_{21}H_{32}O_2$; mol wt 316.48. C 79.70%, H 10.19%, O 10.11%. Prepn from stigmasterol: Butenandt *et al., Ber.* **67**, 1611 (1934); Butenandt, Fleischer, *Ber.* **70**, 96 (1937); from androstenolone: H. Butenandt, J. Schmidt-thome, *Ber.* **72**, 182 (1939); S. Danishefsky *et al., J. Org. Chem.* **40**, 1989 (1975); from Δ^5-3-acetoxyetiocholenic acid chloride: Wettstein, *Helv. Chim. Acta* **23**, 1373 (1940); from diosgenin: Marker, Krueger, *J. Am. Chem. Soc.* **62**, 3349 (1940); Marker *et al., ibid.* **69**, 2173 (1947); from nologenin: *eidem, ibid.* 2395; by treating a 21-halo-Δ^5-pregnen-3-ol-20-one with a reducing agent: **Swiss** pat. **215,139** (1941), *C.A.* **42**, 3144 (1948). Crystal structure: J. Bordner *et al., Cryst. Struct. Commun.* **7**, 513 (1978).

Needles from dil alc, mp 193°. $[\alpha]_D^{20}$ +28° (alc). Very sparingly sol in water. Soly (g/100 ml of soln): carbon tetrachloride 0.5; petr ether 0.1; ethyl acetate 1.1; acetone 0.6; chloroform 17.0; ethanol 1.9; benzene 0.9; isopropanol 1.5. Soly (g/100 ml of solvent): propylene glycol 0.1; dioxane 3.1; benzyl alcohol 8.1. On refluxing with methyl alcohol yields the 17-isopregnenolone, mp 172-173°, $[\alpha]_D^{20}$ −140.5° (alcohol).

Acetate, $C_{23}H_{34}O_3$, needles from alcohol, mp 149-151°. $[\alpha]_D^{20}$ +22° (alcohol). Soly (g/100 ml of soln): carbon tetrachloride 5.0; petr ether 1.0; ethyl acetate 7.9; acetone 2.7; chloroform 55.0; ethanol 2.5; benzene 26.0; isopropanol 2.0. Soly (g/100 ml of solvent): propylene glycol 0.1; dioxane 20.2; benzyl alcohol 11.1; benzyl benzoate 9.1.

Methyl ether, $C_{22}H_{34}O_2$, crystals from abs or dil methanol, mp 123.5°. $[\alpha]_D^{18}$ +18° (c = 1.085 in chloroform).

7916. Prelog-Djerassi Lactone. *[2S-[2α(S*),3β,5β]]-Tetrahydro-α,3,5-trimethyl-6-oxo-2H-pyran-2-acetic acid;* (+)-Prelog-Djerassi lactonic acid; 6-(1-carboxyethyl)-3,4,5,6-tetrahydro-3,5-dimethyl-2-pyranone. $C_{10}H_{16}O_4$; mol wt 200.23. C 59.98%, H 8.05%, O 31.96%. Intermediate in the synthesis of macrolide antibiotics. Isoln as degradation product of narbomycin, *q.v.:* R. Anliker *et al., Helv. Chim. Acta* **39**, 1785 (1956); of methymycin, *q.v.:* C. Djerassi, J. A. Zderic, *J. Am. Chem. Soc.* **78**, 6390 (1956). Abs config: R. W. Rickards, R. M. Smith, *Tetrahedron Letters* **1970**, 1025. Synthesis: P. A. Grieco *et al., J. Am. Chem. Soc.* **101**, 4749 (1979); D. A. Evans, J. Bartroli, *Tetrahedron Letters* **23**, 807 (1982). Synthesis of (±)-form: S. Masamune *et al., J. Am. Chem. Soc.* **97**, 3512 (1975); C. Santelli-Rouvier *et al., Tetrahedron Letters* **35**, 6101 (1994). Review of stereo-

selective syntheses: S. F. Martin, D. E. Guinn, *Synthesis* **1991**, 245.

Crystals, mp 124-125°. $[\alpha]_D$ +33° (c = 0.797 in CHCl$_3$).

7917. Prenalterol. *(S)-4-[2-Hydroxy-3-[(1-methylethyl)-amino]propoxy]phenol;* (−)-*(S)*-1-(p-hydroxyphenoxy)-3-(isopropylamino)-2-propanol. $C_{12}H_{19}NO_3$; mol wt 225.29. C 63.98%, H 8.50%, N 6.22%, O 21.31%. A β_1-adrenergic agonist. Prepn: K. A. Jaeggi *et al.,* **Ger.** pat. **2,503,968** corresp to **U.S.** pats. **3,978,041** and **4,049,797** (1974, 1976, 1977, all to Ciba-Geigy). Pharmacologic study: E. Carlsson *et al., Arch. Pharmacol.* **300**, 101 (1977). Metabolism, hemodynamic effects, pharmacokinetics in man: O. Rönn *et al., Eur. J. Clin. Pharmacol.* **17**, 81 (1980). Cardiovascular effects: D. H. Scott *et al., Brit. J. Clin. Pharmacol.* **7**, 365 (1979). Clinical study in coronary heart disease: I. Hutton *et al., Brit. Heart J.* **43**, 134 (1980). *See also* T. P. Kenakin, D. Beek, *J. Pharmacol. Exp. Ther.* **213**, 406 (1980) for a discussion of the selectivity of action. Prepns of the racemic mixture: **Neth.** pat. **Appl. 6,409,883** corresp to H. Köppe *et al.,* **U.S.** pat. **3,637,852** (1965, 1972 both to Boehringer, Ing.); **Neth.** pat. **Appl. 301,580** corresp to A. F. Crowther, L. H. Smith, **U.S.** pat. **3,501,769** (1965, 1970 both to ICI); A. F. Crowther *et al., J. Med. Chem.* **12**, 638 (1969). Symposium: *Acta. Med. Scand.,* suppl. 659, 1-325 (1982).

Crystals from ethyl acetate, mp 127-128°. $[\alpha]_D^{20}$ −1° ±1°; $[\alpha]_{Hg}^{20}$ +2° ±1° (c = 0.940 in methanol).

Hydrochloride, $C_{12}H_{19}NO_3$.HCl, *H-133/22, CGP-7760B,* (−)*H-80/62, Hyprenan, Varbian.*

THERAP CAT: Cardiotonic.

7918. Prenoxdiazine Hydrochloride. *1-[2-[3-(2,2-Diphenylethyl)-1,2,4-oxadiazol-5-yl]ethyl]piperidine monohydrochloride;* 3-(β,β-diphenylethyl)-5-(β-piperidinoethyl)-1,2,4-oxadiazole hydrochloride; HK-256; Libexin; Lomapect; Tibexin. $C_{23}H_{28}ClN_3O$; mol wt 397.95. C 69.42%, H 7.09%, Cl 8.91%, N 10.56%, O 4.02%. Prepn: K. Harsanyi *et al.,* **Hung.** pat. **151,748** (1964 to Chinoin), *C.A.* **62**, 11821f (1965). Pharmacology: L. Tardos, I. Erdély, *Arzneimittel-Forsch.* **16**, 617 (1966). Stability study: E. Pandula *et al., Acta Pharm. Hung.* **38**, 68 (1968). Review of pharmacology and clinical studies: K. Harsanyi *et al., Boll. Chim. Farm.* **112**, 691 (1973).

Crystals from ethanol, mp 192-193°. LD$_{50}$ in mice, rats (mg/kg): 920, > 2000 orally; 34, 32 i.v. (Tardos, Erdély).

THERAP CAT: Antitussive.

7919. Prenylamine. *N-(1-Methyl-2-phenylethyl)-γ-phenylbenzenepropanamine;* N-(3,3-diphenylpropyl)-α-methylphenethylamine; N-(3'-phenyl-2'-propyl)-1,1-diphenyl-3-propylamine; 1-phenyl-2-[1',1'-diphenylpropyl-3'-amino]-propane; B-436; Elecor. $C_{24}H_{27}N$; mol wt 329.49. C 87.49%, H 8.26%, N 4.25%. Prepn: G. Ehrhart *et al.,* **Ger.** pat. **1,100,031**, *C.A.* **56**, 3413h (1962) and **Ger.** pat. **1,111,-**

642 corresp to **U.S.** pat. **3,152,173** (1961, 1961, 1964 all to Hoechst); G. Erhart, *Arch. Pharm.* **295**, 196 (1962). Series of articles on pharmacology and chemistry: *Arzneimittel-Forsch.* **10**, 569-588 (1960). Metabolism: M. Volz, *ibid.* **21**, 1320 (1971). *Review:* J. E. Murphy, *J. Int. Med. Res.* **1**, 204-209 (1973).

mp 36.5-37.5°.

Lactate, $C_{27}H_{33}NO_3$, *Angormin, Bismetin, Carditin-Same, Coredamin, Corontin, Crepasin, Daxauten, Hostaginan, Incoran, Irrorin, Lactamin, Plactamin, Reocorin, Roinin, Seccidin, Sedolatan, Segontin, Synadrin.* mp 140-142°. Sparingly sol in water (~0.5%); sol in organic solvents. uv max (chloroform): 260 nm ($E_{1cm}^{1\%}$ 170).

THERAP CAT: Vasodilator (coronary).

7920. Prephenic Acid. *1-Carboxy-4-hydroxy-α-oxo-2,5-cyclohexadiene-1-propanoic acid; 1-carboxy-4-hydroxy-2,5-cyclohexadiene-1-pyruvic acid.* $C_{10}H_{10}O_6$; mol wt 226.19. C 53.10%, H 4.46%, O 42.44%. Non-aromatic biosynthetic intermediate that represents a secondary branch-point in the pathway from chorismic acid to phenylalanine and tyrosine, *q.q.v.*, in many organisms. Isoln from cultures of mutant *Escherichia coli:* B. D. Davis, *Science* **118**, 251 (1953). Characterized as cryst Ba salt: U. Weiss *et al., ibid.* **119**, 774 (1954); R. L. Metzenberg, H. K. Mitchell, *Arch. Biochem. Biophys.* **64**, 51 (1956). Structure: H. Plieninger, G. Keilich, *Z. Naturforsch.* **16b**, 81 (1961). Synthetic approaches: H. Plieninger, *Angew. Chem. Int. Ed.* **1**, 367 (1962). Total synthesis of disodium salt: S. Danishevsky, M. Hirama, *J. Am. Chem. Soc.* **99**, 7740 (1977); S. Danishevsky *et al., ibid.* **101**, 7013 (1979); W. Gramlich, H. Plieninger, *Ber.* **112**, 1571 (1979). *Review:* U. Weiss, J. M. Edwards, *The Biosynthesis of Aromatic Compounds* (Wiley, New York, 1980) pp 144-184. *See also* shikimic acid.

Barium salt monohydrate, $C_{10}H_8BaO_6 \cdot H_2O$, crystals from water + methanol or water + pyridine.

7921. Pretilachlor. *2-Chloro-N-(2,6-diethylphenyl)-N-(2-propoxyethyl)acetamide; N-propoxyethyl-N-chloroacetyl-2,6-diethylaniline; 2,6-diethyl-N-(2′-n-propoxyethyl)chloroacetanilide; CGA-26423; CG-113; Rifit; Solnet.* $C_{17}H_{26}$-$ClNO_2$; mol wt 311.85. C 65.48%, H 8.40%, Cl 11.37%, N 4.49%, O 10.26%. Selective herbicide for control of perennial weeds in transplanted rice. Prepn, herbicidal properties: **Neth.** pat. **Appl. 7,307,584**; C. Vogel, R. Aebi, **U.S.** pat. **4,168,965** (1973, 1979 both to Ciba-Geigy). Residue analysis, stability: H. Egli, *J. Agr. Food Chem.* **30**, 861 (1982). Effect on leaf elongation in rice, use in combination with a safening agent (CGA 123,407): R. A. Christ, *Weed Res.* **25**, 193 (1985). Brief account: J. Rufener, M. Quadranti, *Proc. 10th Conf. Int. Congr. Plant Prot.* **1**, 332 (1983).

Colorless liquid, $bp_{0.001}$ 135°. Vapor pressure at 20°: 1×10^{-6} mm Hg. Almost insol in water (50 mg/l at 20°). Sol in most organic solvents. n_D^{20} 1.5204. LD_{50} in rats (mg/kg): 6099 orally; >3100 dermally (Rufener, Quadranti). LC_{50} in rainbow trout, carp, catfish: 3.0, 3.0, 2.6 ppm (Vogel, Aebi).

USE: Herbicide for use in rice paddys.

7922. Pridinol. *α,α-Diphenyl-1-piperidinepropanol; 1,-1-diphenyl-3-piperidino-1-propanol; 3-piperidino-1,1-diphenyl-1-propanol; 1,1-diphenyl-3-(1-piperidyl)-1-propanol; 3-(N-piperidyl)-1,1-diphenyl-1-propanol; ridinol; C-238.* $C_{20}H_{25}NO$; mol wt 295.42. C 81.31%, H 8.53%, N 4.74%, O 5.42%. May be prepd from ethyl 1-piperidinepropionate and phenylmagnesium bromide: Adamson, **Brit.** pat. **624,-118** (1949 to Wellcome Found.). Toxicity study: R. W. Cunningham *et al., J. Pharmacol. Exp. Ther.* **96**, 151 (1949).

Crystals, mp 120-121°. Soluble in acetone.

Hydrochloride, $C_{20}H_{25}NO \cdot HCl$, *Parks 12.* Crystals, dec 238°. Sol in alc. LD_{50} in mice, rats (mg/kg): 35, 33 i.v.; 131, 91 i.p. (Cunningham).

Methanesulfonate, $C_{20}H_{25}NO \cdot CH_3SO_3H$, *pridinol mesylate, Konlax, Loxeen, Lyseen, Mitanoline.* Crystals, mp 152.5-155.0°. Sparingly sol in water.

THERAP CAT: Antiparkinsonian; anticholinergic.

7923. Prifinium Bromide. *3-(Diphenylmethylene)-1,1-diethyl-2-methylpyrrolidinium bromide; 3-(diphenylmethylene)-1-ethyl-2-methylpyrrolidine ethyl bromide; pyrodifenium bromide; Padrin; Riabal.* $C_{22}H_{28}BrN$; mol wt 386.38. C 68.39%, H 7.30%, Br 20.68%, N 3.63%. Prepn: Sadao Oki, **Japan.** pat. **22,462('65)**, *C.A.* **64**, 3489a (1966); of the free base: *idem,* **Japan.** pat. **17,015('65)**, *C.A.* **63**, 18034h (1965). Ganglion blocking agent. Metabolism: Nakai *et al., Arzneimittel-Forsch.* **20**, 1112 (1970). Toxicity studies: Kumada *et al., ibid.* **22**, 706 (1972).

Crystals, mp 216-218°. (Free base, $bp_{0.15}$ 183-185°). LD_{50} in male mice (mg/kg): 11 i.v.; 43 i.p.; 30 s.c.; 330 orally (Kumada).

THERAP CAT: Antispasmodic.

7924. Prilocaine. *N-(2-Methylphenyl)-2-(propylamino)-propanamide; 2-(propylamino)-o-propionotoluidide; N-(α-propylaminopropionyl)-o-toluidine; α-propylamino-2-methylpropionanilide; propitocaine.* $C_{13}H_{20}N_2O$; mol wt 220.31. C 70.87%, H 9.15%, N 12.72%, O 7.26%. Prepn: N. Löfgren, C. Tegner, *Acta Chem. Scand.* **14**, 486 (1960); **Brit.** pat. **839,943**; N. Löfgren, C. Tegner, **U.S.** pat. **3,160,662** (1960, 1964 both to Astra).

Needles, mp 37-38°. $bp_{0.1}$ 159-162°. n_D^{20} 1.5298.

Hydrochloride, $C_{13}H_{20}N_2O.HCl$, *L-67, Citanest, Xylonest.* Crystals from ethanol + isopropyl ether, mp 167-168°. Readily sol in water.

THERAP CAT: Anesthetic (local).

7925. Primaquine. N^4-*(6-Methoxy-8-quinolinyl)-1,4-pentanediamine; 8-(4-amino-1-methylbutylamino)-6-methoxyquinoline;* SN-13272. $C_{15}H_{21}N_3O$; mol wt 259.35. C 69.47%, H 8.16%, N 16.20%, O 6.17%. Prepn following the synthesis of pamaquine: Elderfield *et al., J. Am. Chem. Soc.* **68,** 1525 (1946); improved procedure: Elderfield *et al., ibid.* **77,** 4816 (1955). *Review:* Olenick in *Antibiotics* **vol. 3,** J. W. Corcoran, F. E. Hahn, Eds. (Springer-Verlag, New York, 1975) pp 516-520.

Viscous liquid, $bp_{0.2}$ 175-179°. Soluble in ether.
Diphosphate, $C_{15}H_{21}N_3O.2H_3PO_4$, yellow crystals from 90% ethanol, mp 197-198°. Moderately sol in water.
Oxalate, $C_{15}H_{21}N_3O.C_2H_2O_4$, yellow crystals from 80% ethanol, mp 182.5-185°.

THERAP CAT: Antimalarial.

7926. Primeverose. *6-O-β-D-Xylopyranosyl-D-glucose; 6-(β-D-xylosido)-D-glucose.* $C_{11}H_{20}O_{10}$; mol wt 312.27. C 42.31%, H 6.46%, O 51.24%. By enzymatic hydrolysis of gaultherin (monotropitoside), primeverin, rhamnicoside and other glycosides: Goris, Vischniac, *Compt. Rend.* **169,** 871 (1919); Bridel, *ibid.* **179,** 991 (1924); Bridel, Charaux, *ibid.* **180,** 857 (1925); Richter, *J. Chem. Soc.* **1936,** 1701. Synthesis: Helferich, Rauch, *Ber.* **59,** 2655 (1926); *Ann.* **455,** 168 (1927); McCloskey, Coleman, *J. Am. Chem. Soc.* **65,** 1778 (1943).

Crystals from methanol or 80% alcohol. Darkens at 190°, mp 209-210° (copper block). Shows mutarotation. $[\alpha]_D^{20}$ +23° → −3.2° (c = 5 in water). Sol in water, methanol, 80% alcohol. Is hydrolyzed to 1 mol D-glucose and 1 mol D-xylose by boiling 2% H_2SO_4 for 5 hrs. Reduces Fehling's soln slowly in the cold, instantly when hot.

7927. Primidone. *5-Ethyldihydro-5-phenyl-4,6(1H,5H)-pyrimidinedione;* 5-ethyl-5-phenylhexahydropyrimidine-4,6-dione; 5-phenyl-5-ethylhexahydropyrimidine-4,6-dione; 2-desoxyphenobarbital; Liskantin; Mylepsin; Mysoline; Resimatil; Sertan. $C_{12}H_{14}N_2O_2$; mol wt 218.26. C 66.04%, H 6.47%, N 12.84%, O 14.66%. Prepd by electrolytic reduction of phenobarbital or by catalytic desulfuration of the corresp 2-thiobarbituric acid: Boon *et al.,* **Brit.** pat. **666,027** (1952 to I.C.I.); Bogue, Carrington, *Brit. J. Pharmacol.* **8,** 230 (1953). Comprehensive description: R. D. Daley in *Anal. Profiles Drug Subs.* **2,** 409-437 (1973); A. A. Al-Badr, H. A. El-Obeid, *ibid.* **17,** 749-795 (1988).

Crystals, mp 281-282°. Practically tasteless. Has no acidic properties. Sparingly sol in water (0.6 g/l at 37°); in most organic solvents.

THERAP CAT: Anticonvulsant.
THERAP CAT (VET): Anticonvulsant. Chiefly to control epileptiform seizures.

7928. Primisulfuron-methyl. *2-[[[[[4,6-Bis(difluoromethoxy)-2-pyrimidinyl]amino]carbonyl]amino]sulfonyl]-benzoic acid methyl ester;* 2-[3-(4,6-bis(difluoromethoxy)pyrimidin-2-yl)ureidosulfonyl]benzoic acid methyl ester; *N*-(2-methoxycarbonylphenylsulfonyl)-*N'*-[4,6-bis(difluoromethoxy)pyrimidin-2-yl]urea; CGA-136872; Beacon. $C_{15}H_{12}F_4$-N_4O_7S; mol wt 468.34. C 38.47%, H 2.58%, F 16.23%, N 11.96%, O 23.91%, S 6.85%. Post-emergence sulfonylurea herbicide. Prepn: W. Meyer *et al.,* **Eur.** pat. **Appl. 84,020;** *eidem,* **U.S.** pat. **4,478,635** (1983, 1984 both to Ciba-Geigy). Comprehensive description: W. Maurer *et al., Proc. Brit. Crop Prot. Conf. - Weeds* **1987,** 41-48. Field trial in corn: C. L. Foy, H. L. Witt, *Weed Technol.* **4,** 615 (1990).

Colorless crystals, mp 203.1° (Maurer); also reported as mp 186-188° (Meyer). Vapor pressure (20°): $< 7.5 \times 10^{-12}$ mm Hg. Soly in water (pH 7): 0.07 g/l at 20°. LD_{50} in rats (mg/kg): >4000 orally; >2000 dermally (Maurer).

USE: Herbicide.

7929. Primocarcin. *5-(Acetylamino)-4-oxo-5-hexenamide;* 4-acetamido-4-penten-3-one-1-carboxamide. C_8H_{12}-N_2O_3; mol wt 184.19. C 52.17%, H 6.57%, N 15.21%, O 26.06%. Antineoplastic antibiotic produced by the actinomycete, *Nocardia fukaya.* Isoln: Y. Sumiki *et al., J. Antibiot.* **13A,** 416 (1960); Isono, Suzuki, *ibid.* **15A,** 77 (1962). Isoln from *Streptomyces diastatochromogens* var. *luteus:* H. Abe *et al., ibid.* **20A,** 167 (1967). Structure: Isono, *ibid.* **15A,** 80 (1962). Synthesis: Bowman *et al., J. Chem. Soc.* **1965,** 470.

Needles from hot methanol, mp 130-131°. Sol in water, lower alcohols, acetone, pyridine; slightly sol in most other organic solvents. Aq soln stable from pH 5-8. uv max (water): 253 nm (ε 3420); in 0.1N NaOH: 215, 347 nm (ε 3230, 304). LD_{50} in mice (mg/kg): 50-60 i.p.; 300 i.v. (Isono, Suzuki).

Dihydroprimocarcin, $C_8H_{14}N_2O_3$, plates from acetone, mp 137-141°.

Tetrahydroprimocarcin, $C_8H_{16}N_2O_3$, needles from methanol + acetone, mp 183°.

7930. Primulaverin. $C_{20}H_{28}O_{13}$; mol wt 476.43. C 50.42%, H 5.92%, O 43.66%. Isoln from species of *Primulaceae:* Thieme, Winkler, *Pharmazie* **26,** 434 (1971). Synthesis: Chaudhury *et al., J. Chem. Soc.* **1948,** 2220.

Dihydrate, star-shaped clusters from ethyl acetate, needles from alc, mp 163°. $[\alpha]_D^{20}$ −67°. Sol in water, alc, acetone.

7931. Primycin. Macrolide antibiotic complex of more than 20 components produced by actinomycetes found in the intestinal tract of the wax moth *(Galeria melonella).* Nine primary components, in 3 major groups designated A, B, and C, represent 90% of the total material. Originally thought to be a single entity which was subseqently identified as the major component, primycin A_1. Isoln from cultures of *Streptomyces primicini:* Vályi-Nagy *et al., Nature* **174**, 1105 (1954); **Hung.** pats. **146,332** and **151,197** (1962, 1964 both to Hung. Acad. Sci.). Production by *Thermopolyspora galeriensis* (also designated *Micromonospora galeriensis):* T. Vályi-Nagy *et al.,* **Hung.** pat. **153,593;** *eidem,* U.S. pat. **3,498,884** (1967, 1970 both to Chinoin). Structure of A_1: J. Aberhart *et al., J. Am. Chem. Soc.* **92**, 5816 (1970); *eidem, J. Chem. Soc. Perkin Trans. I* **1974**, 816, 836. TLC sepn of components: I. Szilagyi *et al., J. Chromatog.* **295**, 141 (1984). Structures of the nine primary components: J. Frank *et al., Tetrahedron Letters* **28**, 2759 (1987). Antibacterial activity and toxicity: T. Vályi-Nagy *et al., Pharmazie* **11**, 304 (1956). Antifungal activity *in vitro:* J. V. Uri, P. Actor, *J. Antibiot.* **32**, 1207 (1979). Mode of action: I. Horvath *et al., Arch. Microbiol.* **121**, 135 (1979). Clinical studies in dermatological infections: C. Mészáros, K. Vezekényi, *Ther. Hung.* **35**, 77 (1987); J. Biro, V. Várkonyi, *ibid.* 136; G. Bálint, *ibid.* 140. *Review:* J. V. Uri, *Acta Microbiol. Hung.* **33**, 141 (1986).

Primycin A_1

Sulfate, mp 192-195° (dec) (Aberhart, 1974); also reported as 202-206° (dec) (Uri, 1986). Infrared and ultraviolet spectra: Szilágyi *et al., Nature* **193**, 243 (1962). Fairly soluble in methanol (2.5%), less sol in the higher alcohols, sparingly sol in pyridine, glacial acetic acid, water. LD_{50} in mice, guinea pigs, rats, rabbits (mg/kg): 2.5, 5.0, 10.0, 10.0 i.p. (Vályi-Nagy, 1956).

THERAP CAT: Antibacterial (topical).

7932. Pristane. *2,6,10,14-Tetramethylpentadecane;* norphytane; Robuoy. $C_{19}H_{40}$; mol wt 268.53. C 84.99%, H 15.01%. Isoprenoid alkane obtained from the unsaponifiable fraction of shark liver oil where it occurs to an extent of 14%: Tsujimoto, *J. Soc. Chem. Ind.* **51**, 317T (1932); Sörensen, Mehllum, *Acta Chem. Scand.* **2**, 140 (1948). Identity with norphytane: Pliva, Sörensen, *ibid.* **4**, 846 (1950). Isoln from petroleum crude oils: Bendoraitis *et al., Anal. Chem.* **34**, 49 (1962); from wool wax: Mold *et al., Nature* **199**, 283 (1963). Synthesis from phytol: Sörensen, Sörensen, *Acta Chem. Scand.* **3**, 939 (1949). Metabolism: McKenna, Kallio, *Proc. Nat. Acad. Sci. USA* **68**, 1552 (1971). Use in inducing murine plasmacytomas: P. N. Anderson, M. Potter, *Nature* **222**, 994 (1969). Effect on ascites tumor formation and monoclonal antibody production: N. J. Hoogenraad, C. J. Wraight, *Methods Enzymol.* **121**, 375 (1986).

Mobile, transparent, stable liq. d_4^{20} 0.78267. Congealing point −100°. bp_{760} 296°; bp_{10} 158°; $bp_{0.001}$ 68° (bath temp). n_D^{20} 1.43848. Acid no. 0-5. Iodine no. 0-7.5. Sapon no. 0-5. Viscosity at 25°: 5 cps. Soluble in ether, petr ether, benzene, chloroform, carbon tetrachloride.

USE: Lubricant; transformer oil. Anti-corrosion agent. Biological marker. In experimental systems to induce plasmacytomas; in production of monoclonal antibodies.

7933. Pristinamycin. RP-7293; Pyostacine. An antibiotic mixture produced by *Streptomyces pristinaespiralis* (NRRL 2958). Isoln: Mancy *et al.,* **Fr.** pat. **1,301,857** corresp to U.S. pat. **3,154,475** (1962, 1964 both to Rhône-Poulenc); Preud'homme *et al., Compt. Rend.* **260**, 1309

(1965). Consists of pristinamycins I_A, I_B, I_C, which are identical to vernamycins B_α, B_β, B_γ *(q.v.)* respectively, and pristinamycins II_A, II_B, which are identical to virginiamycin M_1, *q.v.,* and 26,27-dihydrovirginiamycin M_1. Separation of components and structure: Preud'homme *et al., Bull. Soc. Chim. France* **1968**, 585. Nomenclature: Crooy, De Neys, *J. Antibiot.* **25**, 371 (1972).

THERAP CAT: Antibacterial.

7934. Probenecid. *4-[(Dipropylamino)sulfonyl]benzoic acid; p-(dipropylsulfamoyl)benzoic acid;* p-(dipropylsulfamyl)benzoic acid; Benemid; Probecid; Proben. $C_{13}H_{19}NO_4S$; mol wt 285.36. C 54.72%, H 6.71%, N 4.91%, O 22.43%, S 11.24%. Prepn from p-carboxybenzenesulfonyl chloride and dipropylamine: C. S. Miller, U.S. pat. **2,608,507** (1952 to Sharp & Dohme). Study of metabolites: Z. H. Israili *et al., J. Med. Chem.* **15**, 709 (1972). Pharmacokinetics in man: R. F. Cunningham *et al., Clin. Pharmacokinet.* **6**, 135 (1981). Comprehensive description: A. A. Al-Badr, H. A. El-Obeid in *Analytical Profiles of Drug Substances* vol. 10, K. Florey, Ed. (Academic Press, New York, 1981) pp 639-663.

Crystals from dilute alcohol, mp 194-196°. uv max (0.1*N* NaOH): 242.5 nm. pKa 5.8. Slightly bitter taste, pleasant aftertaste. Sol in chloroform, in dil solns of NaOH buffered to pH 7.4. Nearly insol in water. LD_{50} orally in rats: 1.6 g/kg.

THERAP CAT: Uricosuric.

7935. Probucol. *4,4′-[(1-Methylethylidene)bis(thio)]bis-[2,6-bis(1,1-dimethylethyl)phenol]; 4,4′-(isopropylidenedithio)bis[2,6-di-tert-butylphenol];* acetone bis(3,5-di-tert-butyl-4-hydroxyphenyl)mercaptole; DH-581; Lorelco; Lurselle; Sinlestal. $C_{31}H_{48}O_2S_2$; mol wt 516.85. C 72.04%, H 9.36%, O 6.19%, S 12.41%. Prepn: M. B. Neuworth, **Fr.** pat. **1,561,853;** *eidem,* U.S. pat. **3,576,883** (1969, 1971 both to Consolidation Coal Co.); and use as a cholesterol-lowering agent: J. W. Barnhart, P. J. Shea, U.S. pat. **3,862,332** (1975 to Dow). Prepn and activity studies: M. B. Neuworth *et al., J. Med. Chem.* **13**, 722 (1970). Pharmacological studies: J. W. Barnhart *et al., Am. J. Clin. Nutr.* **23**, 1229 (1970); Drake *et al., Circulation* **40**, Suppl. 3, 73 (1969). Clinical studies: *eidem, Metab. Clin. Exp.* **18**, 916 (1969); Kalams *et al., Curr. Ther. Res.* **13**, 692 (1971). Review of pharmacology and therapeutic use: R. C. Heel *et al., Drugs* **15**, 409-428 (1978). Symposium on mechanism of action, clinical efficacy and safety: *Am. J. Cardiol.* **57**, 1H-54H (1986).

White crystalline solid from ethanol, mp 124.5-126°; fine, yellow crystals from isopropanol, mp 125-126.5°.

THERAP CAT: Antihyperlipoproteinemic.

7936. Procainamide Hydrochloride. *4-Amino-N-[2-(diethylamino)ethyl]benzamide monohydrochloride;* procaine amide hydrochloride; Amisalin; Novocamid; Procamide; Procan-SR; Procapan; Pronestyl. $C_{13}H_{22}ClN_3O$; mol wt 271.79. C 57.45%, H 8.16%, Cl 13.04%, N 15.46%, O 5.89%. Prepn: M. Yamazaki *et al., J. Pharm. Soc. Japan* **73**, 294 (1953); Y. Tashika, M. Kuranari, *ibid.* 1069. Comprehensive description: R. B. Poet, H. Kadin in *Analytical Profiles of Drug Substances* vol. 4, K. Florey, Ed. (Academic Press, New York, 1975) pp 333-383.

Crystals, mp 165-169°. uv max: 278 nm. Freely sol in water. Sol in alc; slightly sol in chloroform; very sparingly sol in benzene, ether. The pH of a 10% aq soln is 5.5. Commercially available aq solns are preserved with 0.9% benzyl alcohol and 0.09% sodium bisulfite.

THERAP CAT: Antiarrhythmic (class IA).

7937. Procaine. *4-Aminobenzoic acid 2-(diethylamino)ethyl ester;* *p*-aminobenzoyldiethylaminoethanol; 2-diethylaminoethyl *p*-aminobenzoate. $C_{13}H_{20}N_2O_2$; mol wt 236.31. C 66.07%, H 8.53%, N 11.85%, O 13.54%. Benzoic acid derivative with anesthetic activity. Prepn: A. Einhorn, U.S. pat. **812,554** (1906); *idem, Ann.* **371**, 125 (1909); A. Einhorn, E. Uhlfelder, *ibid.* 131. CNS effects: C. G. Peterson, *Anesthesiology* **16**, 976 (1955). Intravenous pharmacokinetics in humans: A. B. Seifen *et al., Anesth. Analg. (Cleveland)* **58**, 382 (1979). Clinical evaluation as antiarrhythmic and cough suppressant during anesthesia: D. S. Thompson *et al., Am. J. Surg.* **138**, 798 (1979). Stabilization of vascular smooth muscle *in vitro:* K. Kitamura *et al., Drugs Exp. Clin. Res.* **12**, 773 (1986). Toxicity data: W. C. North, K. F. Urbach, *J. Am. Pharm. Assoc. Sci. Ed.* **45**, 382 (1956); E. I. Goldenthal, *Toxicol. Appl. Pharmacol.* **18**, 185 (1971).

Hygroscopic, anhydr plates, tablets from ligroin or ether, mp 61°. When freshly precipitated, one gram dissolves in 200 ml water. Sol in alc, ether, benzene, chloroform. LD_{50} in mice (mg/kg): 195 i.p.; 45 i.v. (North, Urbach).

Dihydrate, needles from aq alc, mp 51°. Slightly bitter taste; applied to the tongue causes transitory numbing sensation.

Nitrate, $C_{13}H_{20}N_2O_2 \cdot HNO_3$, crystals, mp 100-102°. Sol in water, alcohol. The aq soln is neutral. Particularly useful with silver nitrate because no precipitate forms.

Butyrate, $C_{13}H_{20}N_2O_2 \cdot C_4H_8O_2$, *Probutylin.* Hygroscopic crystals. Soluble in water, alcohol, vegetable oils.

Hydrochloride, $C_{13}H_{20}N_2O_2 \cdot HCl$, *Anestil, Enpro, Gero, Jenacaine, Medaject, Naucaine, Neocaine, Novocain, Omnicain, Planocaine, Rocain, Syntocain.* Six-sided plates, monoclinic or triclinic. mp 153-156°. Numbing taste. Stable in air. One gram dissolves in 1 ml water and in 30 ml alcohol. Slightly sol in chloroform. Almost insol in ether. The pH of a 0.1*M* aq soln is 6.0. Aq solns may be sterilized by boiling. LD_{50} in mice (mg/kg): 660 ± 60 s.c. (Goldenthal).

THERAP CAT: Anesthetic (local).

THERAP CAT (VET): Anesthetic (local).

7938. Procarbazine. *N-(1-Methylethyl)-4-[(2-methylhydrazino)methyl]benzamide; N-isopropyl-α-(2-methylhydrazino)-p-toluamide; N-4-isopropylcarbamoylbenzyl-N'-methylhydrazine; p-(N¹-methylhydrazinomethyl)-N-isopropylbenzamide; ibenzmethyzin; MIH; Ro-4-6467.* $C_{12}H_{19}N_3O$; mol wt 221.30. C 65.13%, H 8.65%, N 18.99%, O 7.23%. Prepn: **Belg.** pat. **618,638** corresp to W. Bollag *et al.,* U.S. pat. **3,520,926** (1962, 1970 to Hoffmann-La Roche). Toxicity data: E. I. Goldenthal, *Toxicol. Appl. Pharmacol.* **18**, 185 (1971). Comprehensive description: R. J. Rucki, *Anal. Profiles Drug Subs.* **5**, 403-427 (1976).

Hydrochloride, $C_{12}H_{19}N_3O \cdot HCl$, *Matulane, Natulan.* Crystals from methanol, mp 223-226°. LD_{50} orally in rats: 785 ±34 mg/kg (Goldenthal).

Hydrobromide, $C_{12}H_{19}N_3O \cdot HBr$, crystals from methanol + ether, dec 216-217°.

Note: Procarbazine hydrochloride may reasonably be anticipated to be a carcinogen: *Seventh Annual Report on Carcinogens* (PB95-109781, 1994) p 335.

THERAP CAT: Antineoplastic.

7939. Procaterol. *(R*,S*)-(±)-8-Hydroxy-5-[1-hydroxy-2-[(1-methylethyl)amino]butyl]-2(1H)-quinolinone; (±)-erythro-8-hydroxy-5-[1-hydroxy-2-(isopropylamino)butyl]-carbostyril.* $C_{16}H_{22}N_2O_3$; mol wt 290.36. C 66.18%, H 7.64%, N 9.65%, O 16.53%. Sympathomimetic amine with selective β_2-adrenergic agonist activity. Prepn: K. Nakagawa *et al.,* **Belg.** pat. **823,841;** *eidem,* U.S. pat. **4,026,897** (1975, 1977 both to Otsuka); S. Yoshizaki *et al., J. Med. Chem.* **19**, 1138 (1976). Prepn of optical isomers: *eidem, ibid.* **20**, 1103 (1977). Assessment of selective action: H. Himori, N. Taira, *Brit. J. Pharmacol.* **61**, 9 (1977). Metabolism: Y. Yasuda *et al., Arzneimittel-Forsch.* **29**, 261 (1979). Pharmacokinetics: M. Ishigami *et al., ibid.* 266. Series of reproduction studies: *Iyakuhin Kenkyu* **10**, 68-111 (1979), *C.A.* **90**, 197728f-30a (1979). Antigenicity test: N. Nakagiri, S. Tei, *Oyo Yakuri* **17**, 363 (1979), *C.A.* **91**, 186577a (1979).

Hydrochloride hemihydrate, $C_{16}H_{22}N_2O_3 \cdot HCl \cdot \frac{1}{2}H_2O$, *OPC-2009, Lontermin, Masacin, Meptin, Onsukil, Pro-Air, Procadil, Promaxol, Propulm.* Off-white crystalline powder, mp 193-197° (dec). Sol in methanol; slightly sol in ethanol. Practically insol in acetone, ether, ethyl acetate, chloroform, benzene. Colors on exposure to light. LD_{50} of the hydrochloride in male rats (mg/kg): 2600 orally, 80 i.v. (Nakagawa).

THERAP CAT: Bronchodilator.

7940. Procerin. *2-Hydroxy-5-(3-methyl-2-butenyl)-4-(1-methylethenyl)-2,4,6-cycloheptatrien-1-one.* $C_{15}H_{18}O_2$; mol wt 230.31. C 78.23%, H 7.88%, O 13.89%. A tropolone isolated from *Juniperus procera* Hochst., *Cupressaceae,* obtained in Kenya: Petterson, Runeberg, *Acta Chem. Scand.* **15**, 713 (1961). Structure: Runeberg, *ibid.* 645. Synthesis: Kitahara, Kato, *Bull. Chem. Soc. Japan* **37**, 895 (1964).

Crystals from dil alc, mp 71-72°. Sol in alc, petr ether.

7941. Prochloraz. *N-Propyl-N-[2-(2,4,6-trichlorophenoxy)ethyl]-1H-imidazole-1-carboxamide;* 1-[N-propyl-N-[2-

(2,4,6-trichlorophenoxy)ethyl]carbamoyl]imidazole; BTS-40542; Sportak. $C_{15}H_{16}Cl_3N_3O_2$; mol wt 376.67. C 47.83%, H 4.28%, Cl 28.24%, N 11.16%, O 8.50%. Broad spectrum fungicide for use on cereal crops, fruit and vegetables. Prepn from imidazole: R. F. Brookes *et al.*, **Ger.** pat. **2,429,523**; *eidem*, **U.S.** pat. **3,991,071** (1975, 1976 both to Boots). Fungicidal properties: R. J. Birchmore *et al.*, *Proc. Brit. Crop. Prot. Conf.-Pests Dis.* **1977**, 593; R. G. Harris *et al.*, *ibid.* **1979**, 53. Efficacy against cereal powdery mildew: D. M. Weighton *et al.*, *ibid.* **1977**, 25. Protective effect against blotch in winter wheat and barley: R. G. Harris, G. Barnes, *ibid.* **1981**, 267. Effect on cytochrome P-450 and sterol biosynthesis in Japanese quail: J.-L. Riviere *et al.*, *Pestic. Sci.* **15**, 317 (1984). *Review:* A. de Saint-Blanquat, J. My, *Def. Veg.* **37**, 121 (1983).

White crystalline solid, mp 38.5-41°. Technical product is pale yellow viscous oil. Vapor pressure at 20°: 0.57×10^{-9} torr. Almost insol in water (0.0055 g/l). Soly in chloroform, diethyl ether, toluene, xylene: 2500 g/l; in acetone: 3500 g/l. LD_{50} in rats (mg/kg): 1600 orally; > 5000 s.c.; 400-800 i.p.; LC_{50} (96 hour) in rainbow trout, bluegill (mg/l): 1, 2.2 (de Saint-Blanquat, My). Relatively nontoxic to bees.

USE: Fungicide.

7942. Prochlorperazine. *2-Chloro-10-[3-(4-methyl-1-piperazinyl)propyl]-10H-phenothiazine; 3-chloro-10-[3-(4-methyl-1-piperazinyl)propyl]phenothiazine;* 2-chloro-10-[3-(1-methyl-4-piperazinyl)propyl]phenothiazine; *N*-[γ-(4'-methylpiperazinyl-1')propyl]-3-chlorophenothiazine; chlormeprazine; prochlorpemazine; proclorperazine; Bayer A 173; RP-6140; SKF-4657. $C_{20}H_{24}ClN_3S$; mol wt 373.95. C 64.24%, H 6.47%, Cl 9.48%, N 11.24%, S 8.57%. Prepn: R. J. Horclois, **Brit.** pat. **780,193**; **Fr.** pat. **1,167,627**; **U.S.** pat. **2,902,484** (1957, 1958, 1959 all to Rhône-Poulenc). Pharmacology and toxicity: S. Courvoisier *et al.*, *C.R. Soc. Biol.* **152**, 1371 (1958). Pharmacokinetics: W. B. Taylor, D. N. Bateman, *Brit. J. Clin. Pharmacol.* **23**, 137 (1987). Clinical efficacy as antiemetic and antipsychotic: J. Lapierre *et al.*, *Can. Psychiat. Assoc. J.* **14**, 267 (1969). Clinical trial in treatment of nausea, dizziness and vertigo: A. E. Ward, *Brit. J. Clin. Pract.* **42**, 228 (1988).

Dimaleate, $C_{20}H_{24}ClN_3S \cdot 2C_4H_4O_4$, *Buccastem, Compazine, Meterazine, Stemetil, Vertigon.* Minute crystals, mp 228°. Very slightly sol in water (< 0.1% at 20°). Slightly sol in methanol, ethanol. Practically insol in ether, benzene, chloroform. LD_{50} in mice (mg/kg): 400 s.c.; 120 i.p.; 90 i.v., 400 orally (Courvoisier).

Dimethanesulfonate, $C_{20}H_{24}ClN_3S \cdot 2CH_4O_3S$, *prochlorperazine mesylate, Novamin, Tementil.*

THERAP CAT: Antiemetic; antipsychotic. In treatment of vertigo.

THERAP CAT (VET): Antiemetic.

7943. Procodazole. *1H-Benzimidazole-2-propanoic acid;* β-(2-benzimidazolyl)propionic acid; 2-(2-carboxyethyl)benzimidazole; propazol; AL-1241; Estimulocel. $C_{10}H_{10}N_2O_2$; mol wt 190.20. C 63.15%, H 5.30%, N 14.73%, O

16.82%. Non-specific, active immunoprotective agent against viral and bacterial infections. Prepn: J. Maier, *Ann.* **327**, 17 (1903); R. Meyers, H. Lüders, *ibid.* **415**, 29 (1918); B. Chatterjee, *J. Chem. Soc.* **1929**, 2966; **Span.** pat. **407,882** (1972 to Lafarquim). Pharmacological studies: C. Fernández *et al.*, *Rev. Clin. Esp.* **135**, 539 (1974); *eidem, ibid.* **141**, 51 (1976). Clinical studies: M. Pérez Tascon, J. M. Monturio, *Med. Klin.* **181**, 78 (1976).

Silky white needles from water, mp 228° (dec). Sol in alc, warm water. Practically insol in ether, benzene. The water soln is extremely sweet-tasting.

THERAP CAT: Immunopotentiator (non-specific).

7944. Procyclidine. α-*Cyclohexyl-α-phenyl-1-pyrrolidinepropanol;* 1-cyclohexyl-1-phenyl-3-(1-pyrrolidinyl)-1-propanol; 1-cyclohexyl-1-phenyl-3-pyrrolidino-1-propanol. $C_{19}H_{29}NO$; mol wt 287.45. C 79.39%, H 10.17%, N 4.87%, O 5.57%. Prepn of free base and hydrochloride: Adamson *et al.*, *J. Chem. Soc.* **1951**, 52; Adamson, **U.S.** pat. **2,891,890** (1959 to Burroughs Wellcome). Prepn of methochloride: Bottorff, **U.S.** pat. **2,826,590** (1958 to Lilly); Harfenist, Magnien, **U.S.** pat. **2,842,555** (1958 to Burroughs Wellcome).

Crystals from petr ether, mp 85.5-86.5°. uv max (0.17% in ethanol): 258.5 nm (ε 233).

Hydrochloride, $C_{19}H_{29}NO \cdot HCl$, *Arpicolin, Kemadrin, Osnervan.* Crystals from ethanol + ethyl acetate, dec 226-227°. Moderately sol in water (about 3.0 g/100 ml); more sol in alcohol, chloroform; very slightly sol in ether.

Methochloride, $C_{19}H_{29}NO \cdot CH_3Cl$, *1-(3-cyclohexyl-3-hydroxy-3-phenylpropyl)-1-methylpyrrolidinium chloride, tricyclamol chloride, Lergine, Tricoloid.* Crystals from nitroethane, mp 159-164°. Moderately sol in water, alcohol. Practically insol in ether.

Methosulfate, $C_{21}H_{35}NO_5S$, crystals, mp ~100°. Soluble in water (about 2% at 25°), alcohol. Practically insol in ether.

THERAP CAT: Anticholinergic.

7945. Procymate. α-*Ethylcyclohexanemethanol carbamate; carbamic acid 1-cyclohexylpropyl ester;* 1-cyclohexylpropyl carbamate; T-3033; Equipax. $C_{10}H_{19}NO_2$; mol wt 185.27. C 64.83%, H 10.34%, N 7.56%, O 17.27%. Prepn: Swierkot, **Brit.** pat. **979,236** (1960 to Comp. Franc. Matières Colorantes).

Crystals, mp 128-129°.

THERAP CAT: Tranquilizer.

7946. Procymidone. *3-(3,5-Dichlorophenyl)-1,5-dimethyl-3-azabicyclo[3.1.0]hexane-2,4-dione; N-(3,5-dichlorophenyl)-1,2-dimethyl-1,2-cyclopropanedicarboximide;* dicyclidine; S-7131; Sumisclex; Sumilex. $C_{13}H_{11}Cl_2NO_2$; mol wt 284.14. C 54.95%, H 3.90%, Cl 24.95%, N 4.93%, O 11.26%. Dicarboximide fungicide systemically active against *Botrytis* and *Sclerotinia* spp. on fruits and vegetables. Prepn: **Neth.** pat. **Appl. 7,003,836**; A. Fujinami *et al.*, **U.S.**

pat. **3,903,090** (1970, 1975 both to Sumitomo). Antimicrobial spectrum and systemic activity: Y. Hisada *et al.*, *J. Pestic. Sci.* **1**, 145 (1976). Mechanism of action: A. C. Pappas, D. J. Fisher, *Pestic. Sci.* **10**, 239 (1979). Field evaluation: I. F. Jackson, B. N. Smith, *Proc. 32nd N.Z. Weed Pest Control Conf.* 278 (1979). HPLC determn in wine musts and extracts: P. Cabras *et al.*, *J. Chromatog.* **256**, 176 (1983). *Review:* N. Mikami, J. Miyamoto, *Rev. Plant Protect. Res.* **14**, 85-95 (1981).

Crystalline solid, mp 165-167°. d^{25} 1.42-1.46. Vapor pressure: 1.32×10^{-4} mm Hg. uv max: 207.5, 275 nm (ϵ 4.2×10^4, 4.1×10^2). Soly in water (25°): 4.5 ppm. Highly sol in acetonitrile, acetone, ether, chloroform. Moderately sol in benzene, toluene. Stable in solvents. Unstable in alkaline media. LD_{50} in male rats (mg/kg): 6800 orally, > 10000 dermally (Jackson). Also reported as LD_{50} in male, female rats (g/kg): 7.8, 9.1 orally (Mikami).

USE: Systemic agricultural fungicide.

7947. Prodiamine. *2,4-Dinitro-N^3,N^3-dipropyl-6-(trifluoromethyl)-1,3-benzenediamine; α,α,α-trifluoro-3,5-dinitro-N^4,N^4-dipropyltoluene-2,4-diamine;* 5-dipropylamino-α,α,α-trifluoro-4,6-dinitro-*o*-toluidine; 2,6-dinitro-N^1,N^1-dipropyl-4-trifluoromethyl-*m*-phenylenediamine; N^3,N^3-di-*n*-propyl-2,4-dinitro-6-trifluoromethyl-1,3-phenylenediamine; USB-3153; CN-11-2936; Marathon; Blockade; Endurance; Rydex. $C_{13}H_{17}F_3N_4O_4$; mol wt 350.30. C 44.57%, H 4.89%, F 16.27%, N 15.99%, O 18.27%. Broad-spectrum, pre-emergence dinitroaniline herbicide. Prepn: D. L. Hunter *et al.*, **Ger.** pat. **2,013,510**; *eidem*, U.S. pat. **3,764,623** (1970, 1973 both to U.S. Borax and Chemical Corp.). Weed control in container grown plants: T. A. Fretz, W. J. Sheppard, *HortScience* **15**, 489 (1980); S. A. Duray, F. T. Davies, *J. Environ. Hort.* **5**, 82 (1987). Field studies: M. G. Sybouts, *Proc. West. Soc. Weed Sci.* **40**, 169 (1987). Evaluation of herbicidal activity: W. Bond, *Crop Prot.* **7**, 75 (1988). Review of herbicidal activity: S. J. Bowe, *Proc. West. Soc. Weed Sci.* **39**, 216-218 (1986).

Orange needles from 95% ethanol, mp 124-125°.
USE: Herbicide.

7948. Prodigiosin. *4-Methoxy-5-[(5-methyl-4-pentyl-2H-pyrrol-2-ylidene)methyl]-2,2'-bi-1H-pyrrole;* 2,2'-[3-methoxy-4'-amyl-5'-methyl-5-(2''-pyrryl)]dipyrrylmethene; prodigiosine. $C_{20}H_{25}N_3O$; mol wt 323.44. C 74.27%, H 7.79%, N 12.99%, O 4.95%. Antibiotic pigment produced by *Chromobacterium prodigiosum (Serratia marcescens)*. Exhibits antimicrobial and cytotoxic properties. Isoln: Wrede, Hettche, *Ber.* **62**, 2678 (1929); Lasseur, Georges, *Trav. Lab. Microbiol. Nancy* **9**, 47 (1936); Lasseur, Melcion, *ibid.* **13**, 192 (1944). Purification: Morgan, Tanner, *J. Chem. Soc.* **1955**, 3305. Structure: Wrede, Rothhaas, *Z. Physiol. Chem.* **226**, 95 (1934). Revised structure and synthesis: H. H. Wasserman *et al.*, *J. Am. Chem. Soc.* **82**, 506 (1960); H. Rapoport, K. G. Holden, *ibid.* 5510; *ibid.* **84**, 635 (1962); A. J. Castro *et al.*, *J. Org. Chem.* **28**, 857 (1963). Total synthesis and *in vitro* cytotoxic activity: D. L. Boger, M. Patel, *Tetrahedron Letters* **28**, 2499 (1987); *eidem*, *J. Org. Chem.* **53**, 1405 (1988). NMR studies: R. J. Cushley *et al.*, *Can. J. Chem.* **53**, 148 (1975). Antimalarial activity: A. J. Castro, *Nature* **213**, 903 (1967). *Review:* R. P. Williams, W. R. Hearn in *Antibiotics* **vol. 2**, D. Gottlieb, P. D. Shaw, Eds. (Springer-Verlag, New York, 1967) pp 410-432, 449-451. Review of

synthesis: A. H. Jackson, K. M. Smith, in *The Total Synthesis of Natural Products*, **vol. 1**, J. ApSimon, Ed. (Wiley-Interscience, New York, 1973) pp 227-232; of biological activity and production: V. Alonzo, *Ig. Mod.* **81**, 557-564 (1984), *C.A.* **101**, 126398h (1984).

Lustrous square pyramids (dark red with green reflex) from petr ether, mp 151-152°. Almost insol in water. Moderately sol in alcohol, ether; freely sol in chloroform, bromoform, benzene. Alkaline or neutral solns are orange-yellow, acid solns are red. Absorption max (isopropanol): 466 nm (ϵ 43000); 336, 280 nm.

Hydrochloride, $C_{20}H_{26}ClN_3O$, magenta crystals from benzene + petr ether, dec 148.5-150°. Absorption max (isopropanol): 540, 294 nm (ϵ 70700, 10800).

7949. Prodipine. *1-(1-Methylethyl)-4,4-diphenylpiperidine; 1-isopropyl-4,4-diphenylpiperidine.* $C_{20}H_{25}N$; mol wt 279.43. C 85.97%, H 9.02%, N 5.01%. CNS stimulating agent with antidepressant and antiparkinson properties. Prepn: H. G. Menge, J. Klosa, **Ger.** pat. **1,936,452**; *eidem*, U.S. pat. **4,016,280** (1971, 1977 both to Byk Gulden). Evaluation of MAO-inhibitory properties *in vivo* and *in vitro:* G. Planz *et al.*, *Arzneimittel-Forsch.* **23**, 281 (1973). Effects on 5-hydroxytryptamine uptake in platelets: M. Eltze *et al.*, *ibid.* **30**, 1129 (1980). Comparative study of effects in Parkinson's disease: E. Gründig, F. Gerstenbrand, *Wien Klin. Wochenschr.* **92**, 868 (1980).

Oil, $bp_{0.01}$ 117-125°.
Hydrochloride, $C_{20}H_{25}N \cdot HCl$, *Anthen.* Crystals from isopropanol, mp 267°.
THERAP CAT: Antiparkinsonian.

7950. Prodlure. *(E,Z)-9,11-Tetradecadien-1-ol acetate; cis-9,trans-11-tetradecadienyl acetate; cis-9,trans-11-TDDA;* litlure A. $C_{16}H_{28}O_2$; mol wt 252.40. C 76.14%, H 11.18%, O 12.68%. Major component of sex pheromone of female *spodoptera litura* (F.) and Egyptian cotton leafworm, *S. littoralis* (Boisd.). Isoln and prepn: Y. Tamaki *et al.*, *Appl. Entomol. Zool.* **8**, 200 (1973); B. F. Nesbitt *et al.*, *Nature New Biol.* **244**, 208 (1973). Prepn: T. Yushimo *et al.*, **Ger.** pat. **2,406,259** (1974 to Natl. Inst. Agr. Sci.), *C.A.* **82**, 97681b (1975). Stereoselective synthesis: D. R. Hall *et al.*, *Chem. & Ind. (London)* **1975**, 216; G. Goto *et al.*, *Chem. Letters* **1975**, 103; G. Decodts *et al.*, *Synthesis* **7**, 510 (1979). Activity studies in presence of minor component, *(Z,E)-9,-12-tetradecadien-1-ol acetate:* S. Neumark *et al.*, *Environ. Letters* **6**, 219 (1974); Y. Tamaki, T. Yushima *et al.*, *J. Insect. Physiol.* **20**, 1005 (1974); M. Kehat *et al.*, *Appl. Entomol. Zool.* **11**, 45 (1976).

Colorless liquid, $bp_{0.2}$ 147-148°, $bp_{0.003}$ 85-86°. uv max (hexane): 232 nm (ϵ 27300).
USE: Insect sex attractant.

7951. Pro-Drone®. *1-(8-Methoxy-4,8-dimethylnonyl)-4-(1-methylethyl)benzene;* 2-methoxy-9-(p-isopropylphenyl)-

2,6-dimethylnonane; MV-678; AI-3-36206. $C_{21}H_{36}O$; mol wt 304.52. C 82.83%, H 11.92%, O 5.25%. Synthetic analog of juvenile hormones, *q.v.* Prepn: M. Schwarz *et al.*, **Ger. pat. 2,536,298** corresp to **U.S. pat. 4,002,769** (1976, 1977 to Stauffer). Insect maturation inhibiting activity: *eidem, J. Econ. Entomol.* **67,** 598 (1974); J. E. Wright *et al., ibid.* **69,** 79 (1976); C. H. Schaefer *et al., ibid.* 119.

USE: Insecticide.

7952. Producer Gas. Blow gas. Obtained by blowing air and steam through incandescent coke. *Composition:* 14% CO_2; 10% CO; 75% N_2; 1% Ar.
Caution: Asphyxiant. *See* Carbon Monoxide.
USE: In the manuf of ammonia as source of nitrogen. *cf.* Water gas.

7953. Proflavine. *3,6-Acridinediamine;* 3,6-diamino-acridine; 2,8-diaminoacridine. $C_{13}H_{11}N_3$; mol wt 209.25. C 74.62%, H 5.30%, N 20.08%. Prepn: M. Schöpff, *Ber.* **27,** 2320 (1894); **Ger. pat. 230,412** (1910 to Cassella), *C.A.* **5,** 2734 (1911); W. P. Thompson, **Brit. pat. 137,214** (1919 to Poulenc Frères), *C.A.* **14,** 1445 (1920); A. Albert, *J. Chem. Soc.* **1941,** 121, 484; **1947,** 244. Toxicity study: S. D. Rubbo, *Brit. J. Exp. Pathol.* **28,** 1 (1947). *Review: The Acridines,* A. Albert, Ed. (St. Martin's Press, New York, 2nd ed., 1966) pp 300-302; *Acridines,* R. M. Acheson, Ed. (Interscience, New York, 1956) pp 341-344.

Yellow needles from alc, mp 281° (Schöpff), mp 288° (Albert). Sol in water, ethanol. Practically insol in benzene, ether. pKa 9.7. Solns are brownish and when diluted are fluorescent. *Solns should be discarded when they become turbid.* LD_{50} s.c. in mice: 0.14 g/kg (Rubbo).
Dihydrochloride, $C_{13}H_{11}N_3.2HCl$, orange-yellow needles. pH of 0.1% soln 2.5-3.0.
Sulfate, $C_{13}H_{11}N_3.H_2SO_4$, *proflavine hemisulfate, neutral proflavine sulfate.* Occurs as the hydrate. Red needles, hygroscopic. Sol in 300 parts cold water, in 1 part boiling water; slightly sol in alc. Practically insol in ether, $CHCl_3$. pH of satd soln is 6-8. pH of 0.1% soln is 2.5.
Mixture with 3,6-diamino-10-methylacridinium chloride: *See* Acriflavine.
THERAP CAT: Topical antiseptic.
THERAP CAT (VET): Topical antiseptic.

7954. Profluralin. *N-(Cyclopropylmethyl)-2,6-dinitro-N-propyl-4-(trifluoromethyl)benzenamine; N-(cyclopropylmethyl)-α,α,α-trifluoro-2,6-dinitro-N-propyl-p-toluidine; N-*cyclopropylmethyl-*N-n*-propyl-4-trifluoromethyl-2,6-dinitroaniline; CGA-10832; ER-5461; B-4576; Tolban; Pregard. $C_{14}H_{16}F_3N_3O_4$; mol wt 347.29. C 48.42%, H 4.64%, F 16.41%, N 12.10%, O 18.43%. Selective pre-planting herbicide. Prepn: **Neth. pat. Appl. 6,911,565** corresp to L. L. Maravetz, **U.S. pat. 3,546,295** (1969, 1970 to Esso). Persistence in soil: J. M. Kennedy *et al., Weed Sci.* **25,** 373 (1977).

Yellow-orange crystals or liquid, mp 33-36°. Soly in water at 20°: 0.1 mg/l. Readily sol in most organic solvents. LD_{50} orally in rats: 10000 mg/kg, *RTECS* Vol. II, R. J. Lewis, R. L. Tatken, Eds. (1979) p 628.
USE: Herbicide.

7955. Progabide. *4-[[(4-Chlorophenyl)(5-fluoro-2-hydroxyphenyl)methylene]amino]butanamide;* 4-[[α-(*p*-chlorophenyl)-5-fluorosalicylidene]amino]butyramide; 4-[[α-(*p*-chlorophenyl)-5-fluoro-2-hydroxybenzylidene]amino]butyramide; halogabide; SL-76.002; Gabren(e). $C_{17}H_{16}ClFN_2O_2$; mol wt 334.78. C 60.99%, H 4.82%, Cl 10.59%, F 5.67%, N 8.37%, O 9.56%. Gamma-aminobutyric acid (GABA) antagonist with anti-epileptic activity. Prepn: J.-P. Kaplan *et al.,* **Ger. pat. 2,634,288;** *eidem,* **U.S. pat. 4,094,992** (1977, 1978 to Synthelabo); *eidem, J. Med. Chem.* **23,** 702 (1980). Use as analgesic: J.-P. Kaplan, **U.S. pat. 4,361,583** (1982 to Synthelabo). Pharmacokinetics: I. Johno *et al., J. Pharm. Sci.* **71,** 633 (1982). HPLC determn in biological fluids: P. Padovani *et al., J. Chromatog.* **308,** 229 (1984). Double-blind clinical trial in therapy resistant epilepsy: P. Loiseau *et al., Epilepsia* **24,** 703 (1983); in spasticity: K. Mondrup, E. Pedersen, *Acta Neurol. Scand.* **69,** 200 (1984).

Crystals from cyclohexane and toluene, mp 133-135°; possible second crystalline form, mp 142.5° (Kaplan, *J. Med. Chem.*). uv max in methanol: 332, 250, 210 (ε 4200, 10800, 24000). LD_{50} i.p. in mice: 900 mg/kg (Kaplan).
THERAP CAT: Anticonvulsant.

7956. Progesterone. *Pregn-4-ene-3,20-dione;* $Δ^4$-pregnene-3,20-dione; corpus luteum hormone; luteohormone; Corlutina; Corluvite; Cyclogest; Gestiron; Gestone; Lipo-Lutin; Lutocyclin M; Lutogyl; Lutromone; Progestasert; Progestin; Progestogel; Progestol; Progeston; Prolidon; Proluton; Syngesterone; Utrogestan. $C_{21}H_{30}O_2$; mol wt 314.47. C 80.21%, H 9.62%, O 10.18%. Active principle of the corpus luteum, secreted during the latter half of the menstrual cycle. If pregnancy ensues, secretion continues. Exerts an antiovulatory effect when administered during days 5 to 25 of the normal menstrual cycle. Isoln from corpus luteum of pregnant sows: Butenandt, Westphal, *Ber.* **67,** 1440 (1934); Wintersteiner, Allen, *J. Biol. Chem.* **107,** 321 (1934). Structure: Butenandt *et al., Ber.* **67,** 1611 (1934). Synthesis of DL-form: Johnson *et al., J. Am. Chem. Soc.* **93,** 4332 (1971). Prepn from other steroids in review by W. H. Strain in Gilman's *Organic Chemistry* vol. II (Wiley, New York, 2nd ed., 1943) pp 1487-1489. Numerous patents, e.g., **U.S. pats. 2,379,832; 2,232,438; 2,314,185.** Anesthetic effect and toxicity: H. Selye, *Proc. Soc. Exp. Biol. Med.* **46,** 116 (1941). Review of physiology: Csapo, *Sci. Am.* **198,** 40-46 (April, 1958); Rothchild, *Vitam. Horm. (New York)* **23,** 209-327 (1965). Book: *Progesterone and Progestins,* C. W. Bardin *et al.,* Eds. (Raven Press, New York, 1982) 462 pp.

Exists in two cryst forms of equal physiologic activity and which are readily interconverted. The α-form is orthorhombic (prisms from dil alc) with a:b:c = 0.750:1:0.905. Crystals show (011), (110), (010). Poor (011) cleavage. d^{23} 1.166. mp 127-131°. The β-form is orthorhombic (needles) with a:b:c = 0.563:1:0.275. Crystals acicular with parallel extinction and negative elongation. Cleavage on (001) and (110). d^{20} 1.171. mp 121°. $[\alpha]_D^{20}$ +172° to +182° (c = 2 in dioxane). uv max: 240 nm. Insol in water. Sol in alcohol, acetone, dioxane, concd H_2SO_4. Sparingly sol in vegetable oils. 1 mg = 1 I.U.

Note: This substance may reasonably be anticipated to be a carcinogen: *Seventh Annual Report on Carcinogens* (PB95-109781, 1994) p 337.

THERAP CAT: Progestogen.

THERAP CAT (VET): Progestogen. Has been used to control habitual abortion, to suppress or synchronize estrus.

7957. Proglumetacin. (±)-*1-(4-Chlorobenzoyl)-5-methoxy-2-methyl-1H-indole-3-acetic acid 2-[4-[3-[[4-(benzoylamino)-5-(dipropylamino)-1,5-dioxopentyl]oxy]propyl]-1-piperazinyl]ethyl ester;* (±)-*N-[2-[1-(p-chlorobenzoyl)-5-methoxy-2-methyl-3-indolylacetoxy]ethyl]-N'-[3-(N-benzoyl-N',N'-di-n-propyl-DL-isoglutaminoyl)oxypropyl]piperazine.* $C_{46}H_{58}ClN_5O_8$; mol wt 844.45. C 65.43%, H 6.92%, Cl 4.20%, N 8.29%, O 15.16%. Deriv of indomethacin, *q.v.* Prepn: F. Makovec *et al.,* **Ger. pat. 2,535,799** corresp to **U.S. pat. 3,985,878** (both 1976 to Rotta). Series of articles on pharmacology, mechanism of action, safety: *Arzneimittel-Forsch.* **29,** 1116-1129 (1979). Bioavailability study: A. A. Bignamini, P. L. Casula, *Curr. Med. Res. Opin.* **6,** 299 (1979). Clinical evaluation: J. Münzenberg, S. Tachibana, *Pharmatherapeutica* **2,** 279 (1980); P. Loizzi *et al., ibid.* 285. Mutagenicity study: R. Vidal y Plana *et al., Farmaco Ed. Prat.* **33,** 543 (1978). Toxicity data for dimaleate: A. L. Rovati *et al., Arzneimittel-Forsch.* **29,** 1116 (1979).

Dimaleate, $C_{46}H_{58}ClN_5O_8 \cdot 2C_4H_4O_4$, *protacine, CR-604, Afloxan, Miridacin, Protaxon, Proxil.* Crystals from ethanol, mp 146-148°. LD_{50} in male mice, rats (mg/kg): 262, 450 (Rovati).

THERAP CAT: Anti-inflammatory.

7958. Proglumide. (±)-*4-(Benzoylamino)-5-(dipropylamino)-5-oxopentanoic acid;* DL-*4-benzamido-N,N-dipropylglutaramic acid; N-benzoyl-N',N'-di-n-propyl-DL-isoglutamine;* xylamide; CR-242; W-5219; Milid; Milide; Promid. $C_{18}H_{26}N_2O_4$; mol wt 334.42. C 64.65%, H 7.84%, N 8.38%, O 19.14%. Prepn: **Neth. pat. Appl. 6,510,006; S. Afr. pat. 65/4065** (both 1966 to Rotta), *C.A.* **65,** 3793b (1966). Pharmacological activity: A. L. Rovati *et al., Minerva Med.* **58,** 3653 (1967); T. Umetzu *et al., Eur. J. Pharmacol.* **64,** 69 (1980). Pharmacokinetic study: A. A. Bignamini *et al., Arzneimittel-Forsch.* **29,** 639 (1979). Clinical study in duodenal ulcer: W. Bergemann *et al., Med. Klin.* **76,** 226 (1981). Cholecystokinin receptor antagonist activity: W. F. Hahne *et al., Proc. Nat. Acad. Sci. USA* **78,** 6304 (1981). Selective blockade of cholecystokinin CNS effects: L. A. Chiodo, B. S. Bunney, *Science* **219,** 1449 (1983).

Crystals, mp 142-145°. LD_{50} in mice (mg/kg): 2211-2649 i.v.; 7350-8861 orally (Rovati).

THERAP CAT: Anticholinergic.

7959. Proheptazine. *Hexahydro-1,3-dimethyl-4-phenyl-1H-azepin-4-ol propanoate (ester);* DL-α-*1,3-dimethyl-4-phenyl-4-propionoxyazacycloheptane;* 1,3-dimethyl-4-phenyl-4-propionoxyhexamethylenimine; 4-phenyl-4-propionoxy-1,3-dimethylazacycloheptane; 4-propionoxy-1,3-dimethyl-4-

phenylhexamethylenimine; dimepheprimine; Wy-757. $C_{17}H_{25}NO_2$; mol wt 275.39. C 74.14%, H 9.15%, N 5.09%, O 11.62%. Prepn: Diamond, Bruce, **U.S. pat. 2,775,589** (1956 to American Home Prod.). Structure and synthesis: Diamond *et al., J. Med. Chem.* **7,** 57 (1964).

Liquid, $bp_{0.3}$ 126°. n_D^{28} 1.5182. n_D^{21} 1.5215.

Hydrobromide, $C_{17}H_{26}BrNO_2$, crystals from acetone + methanol, dec 207-207.5°. Soly in water at 25°: 1-2%.

Hydrochloride, $C_{17}H_{26}ClNO_2$, crystals, dec 207°.

Note: This is a controlled substance (opiate) listed in the U.S. Code of Federal Regulations, Title 21 Part 1308.11 (1995).

THERAP CAT: Analgesic (narcotic).

7960. Proinsulin. Single chain insulin precursor consisting of the insulin A and B chains and a connecting polypeptide (C-peptide), which contains 30-35 amino acids; the number and sequence of these amino acids are species dependent. Its presence was discovered in a human islet cell adenoma: D. F. Steiner, P. E. Oyer, *Proc. Nat. Acad. Sci. USA* **57,** 473 (1967). Conversion of proinsulin to insulin has a half-time of about 1 hour in rat islets *in vitro;* it is postulated that proteolytic enzymes cleave proinsulin at the sites where two amino acids connect the C-peptide to the insulin chain.This cleavage results in the production of insulin and the C-peptide, both of which are retained in the secretory granules of the beta cells and discharged in equimolar amounts during exocytosis of the granules. *See* W. Kemmler, D. F. Steiner, *Biochem. Biophys. Res. Commun.* **41,** 1223 (1970). *In vitro* conversion of proinsulin to insulin with trypsin and carboxypeptidase B: W. Kemmler *et al., J. Biol. Chem.* **246,** 6786(1971), **248,** 4544 (1973). Isoln of bovine proinsulin: D. F. Steiner *et al., Diabetes* **17,** 725 (1968). Structural studies on mammalian proinsulin: R. E. Chance *et al., Science* **161,** 165 (1968) (porcine); C. Nolan *et al., J. Biol. Chem.* **246,** 2780 (1971) (bovine); P. E. Oyer *et al., ibid.* 1365 (human, monkey); J. D. Petersen *et al., ibid.* **247,** 4866 (1972) (monkey, sheep, dog); H. S. Tager, D. F. Steiner, *ibid.* 7936 (rat, horse); D . E. Massey, D. G. Smyth, *ibid.* **250,** 628 (1975) (guinea pig). Proposed three-dimensional structure: C. R. Snell, D. G. Smyth, *ibid.* 6291. Syntheses of C-peptides and human proinsulin: N. Yanaihara *et al., Diabetes* **27**(Suppl. 1), 149 (1978). Synthesis of rat proinsulin in bacteria: L. Villa-Komaroff *et al.,* in *Polypeptide Hormones,* R. F. Beers, E. G. Bassett, Eds. (Raven Press, New York, 1980) pp 49-65. *Reviews:* D. F. Steiner *et al.* in *Diabetes, 8th Proc. Congr. Int. Diabetes Fed.,* W. J. Malaisse, J. Pirart, Eds. (Excerpta Med., Amsterdam, 1974) pp 119-133; A. H. Rubenstein *et al.,* in *Rec. Progr. Horm. Res.* **33,** R. O. Greep, Ed. (Academic Press, New York, 1977) pp 435-475; A. E. Kitabshi, *Metabolism* **26,** 547-587 (1977). Book: *Proinsulin, Insulin, C-Peptides,* S. Baba *et al.,* Eds. (Excerpta Medica, Amsterdam, 1979) 468 pp.

7961. Prolactin. *Pituitary lactogenic hormone;* adenohypophysial luteotropin; anterior pituitary luteotropin; lactogen; galactin; mammotropin; luteotropic hormone; luteotropin; LTH; Ferolactan. Polypeptide hormone of mol wt about 23,000; active principle of adenohypophysial gland essential in the induction of lactation in mammals at parturition. Its synergistic action with estrogen, *q.v.* promotes mammary gland proliferation. Also brings about the release of progesterone, *q.v.* from lutein cells which renders the uterine mucosa suited for the imbedding of the ovum, should fertilization occur. Isoln procedures from adenohypophyseal tissue or whole pituitary glands of ox, sheep, and swine: Lyons, *Cold Spring Harbor Symposia Quant. Biol.* **5,** 198 (1937); Li *et al., J. Biol. Chem.* **146,** 627 (1942); White et al., ibid. **143**, 447 (1942). Isoln from sheep pituitaries: Reisfeld et al., J. Am. Chem. Soc. **83**, 3719 (1961); from other mammalian pituitaries: Nelson; Eppstein, U.S. pats. **3,265,580** and **3,317,392** (1966, 1967, both to Upjohn); from human pituitaries: Lewis et al., Biochem. Biophys. Res. Commun. **44**, 1169 (1971). Amino acid sequence of ovine prolactin: Li et al., Nature **224**, 695 (1969); Li, Dixon, Arch. Biochem. Biophys. **146**, 233 (1971). Review of structural studies of human prolactin and relationship to somatotropin, q.v.: H. D. Niall et al., Recent Progr. Horm. Res. **29**, 387 (1973). Amino acid sequence of human prolactin: B. Shome, A. F. Parlow, J. Clin. Endocrinol. Metab. **45**, 1112 (1977). Effects of prolactin on the murine immune system: E. Nagy et al., Acta Endocrinol. **102**, 351 (1983). Symposium on clinical endocrinology: Horm. Res. **22**, 129-252 (1985). General reviews: Li, Evans, Hormones **1**, 631 (1948); White, Vitam. Horm. (New York) **7**, 253 (1949); Voss, Arzneimittel-Forsch. **4**, 467 (1954); Apostalakis, Vitam. Horm. (New York) **26**, 197 (1968). Books: Prolactin vols. 1-8, D. F. Horrobin, Ed. (Eden Press, Quebec, 1973-1981); Prolactin: Physiology, Pharmacology and Clinical Findings, O. Hutzinger et al., Eds. (Springer-Verlag, New York, 1982) 224 pp; Prolactin and Prolactinomas, G. Tolis et al., Eds. (Raven Press, New York, 1983) 478 pp.

Crystals. Isoelectric point 5.73. $[\alpha]_D^{25}$ −40.5° (c = 1 in phosphate buffer of pH 7). Practically insol in water (0.102 g/l) unless electrolytes are present. Forms a water-soluble hydrochloride. Sol in abs methanol or ethanol, if a small amount of acid is present. These data apply to prolactin obtained from ox glands. Prolactin from sheep glands is slightly different: In 0.357M NaCl at pH 2.25 the sheep hormone has a soly of 0.506 g/l, while the soly of the ox hormone is only 0.316 g. In citrate buffer (1M, pH 6.36) and in alcohol the ox protein is more sol than the sheep protein. In the absence of salt, prolactin shows little loss of potency after boiling for one hour at pH 8.0 or at 60° for 5 hours; in the presence of salts complete destruction may occur. An 0.04% soln was stable in a boiling water bath for 15 minutes at pH 1 to 9, but lost activity rapidly at pH 11 and 13. As a rule the hormone is more stable in acid than in alkaline solns. 1 mg = 30 I.U.

THERAP CAT: Lactation stimulating hormone.

7962. Prolamins. A group of simple proteins. Yield only amino acids upon cleavage by enzymes or acids. Examples: gliadin, hordein, zein from grains.

Insoluble in water or neutral salt solvents. Sol in dil acids and alkalies and in 70-90% alcohol.

7963. Proline. L-Proline; Pro; P; (S)-2-pyrrolidinecarboxylic acid. $C_5H_9NO_2$; mol wt 115.13. C 52.16%, H 7.88%, N 12.17%, O 27.79%. Non-essential amino acid for human development. Only imino acid of the 20 amino acids commonly found in proteins. Shows cis-trans isomerism. First synthesized: R. Willstätter, Ber. **33**, 1160 (1900); prior to identification in casein: E. Fischer, Ber **34**, 454 (1901). Named in 1904 by Fischer. Early chemistry and biochemistry: Amino Acids and Proteins, D. M. Greenberg, Ed. (Charles C. Thomas, Springfield, IL, 1951) 950 pp., passim; J. P. Greenstein, M. Winitz, Chemistry of the Amino Acids vols 1-3 (John Wiley and Sons, Inc., New York, 1961) pp. 2178-2201, passim. Stereospecific synthesis: S. L. Titouani et al., Tetrahedron **36**, 2961 (1980). HPLC determn: A. Carisano, J. Chromatog. **318**, 132 (1985). HPLC separation of cis-trans isomers of peptides: S. Friebe et al., ibid A. **661**, 7 (1994). Review of biosynthesis and degradation: E. Adams, L. Frank, Ann. Rev. Biochem. **49**, 1005-1061 (1980). Review of effects on protein structure and biological function: A. Yaron, F. Naider, Crit. Rev. Biochem. Mol. Biol. **28**, 31-81 (1993); G. Vanhoof et al., FASEB J. **9**, 736-744 (1995).

Flat needles from alcohol + ether, prisms from water, dec 220-222°. $[\alpha]_D^{23.4}$ −85.0°; $[\alpha]_D^{20}$ −52.6° (c = 0.57 in 0.50N

HCl); $[\alpha]_D^{20}$ −93.0° (c = 2.42 in 0.6N KOH). pI 6.30. pK_1 1.99; pK_2 10.60. Soly in 100 ml water: 127.4 g at 0°; 162.3 g at 25°; 206.7 g at 50°; 239 g at 65°. Sol in alcohol 1.55% at 35°. Insol in ether, butanol, isopropanol.

DL-Form, monohydrate, crystals, mp 190° (when anhydr, dec 205°). Sol in water, alc; sparingly sol in acetone, chloroform, benzene; insol in ether.

7964. Prolintane. 1-[1-(Phenylmethyl)butyl]pyrrolidine; 1-(α-propylphenethyl)pyrrolidine; 1-phenyl-2-pyrrolidylpentane; phenylpyrrolidinopentane; SP-732. $C_{15}H_{23}N$; mol wt 217.35. C 82.89%, H 10.67%, N 6.44%. Prepn: Brit. pat. **807,835**; Seeger, Kottler, Ger. pat. **1,093,799** (1959, 1960 both to Thomae), C.A. **55**, 19950c (1961). Pharmacological studies: R. Kadatz, E. Pötzsch, Arzneimittel Forsch. **7**, 344 (1957); K. Takagi et al., Oyo Yakuri **5**, 5 (1971), C.A. **76**, 94575k (1972).

$bp_{0.5}$ 105°; bp_{16} 153°.

Hydrochloride, $C_{15}H_{23}N·HCl$, Katovit, Promotil. Crystals from methyl ether, mp 133-134°. LD_{50} orally in mice: 257 mg/kg (Kadatz, Pötzsch).

THERAP CAT: CNS stimulant; antidepressant.

7965. Prolonium Iodide. 2-Hydroxy-N,N,N,N',N',N'-hexamethyl-1,3-propanediaminium diiodide; (2-hydroxytri methylene)bis[trimethylammonium] iodide; 1,3-bis(trimethyl ylamino)-2-propanol diiodide; hexamethyldiaminoisopropanol diiodide; di(iodohexamethyl)diaminoisopropanol; iodisan; Endojodin. $C_9H_{24}I_2N_2O$; mol wt 430.11. C 25.13%, H 5.62%, I 59.01%, N 6.51%, O 3.72%. Prepn: Callsen, U.S. pat. **1,526,627** (1925).

White, crystalline powder. mp ∼275° with decompn, but becomes brown at 240°. Freely sol in water, slightly in alcohol. Practically insol in ether, acetone.

THERAP CAT: Iodine source.

7966. Promazine. N,N-Dimethyl-10H-phenothiazine-10-propanamine; 10-(3-dimethylaminopropyl)phenothiazine; RP-3276; Wy-1094; A-145. $C_{17}H_{20}N_2S$; mol wt 284.43. C 71.79%, H 7.09%, N 9.85%, S 11.27%. Prepd by heating a xylene soln of phenothiazine and 3-dimethylamino-1-chloropropane in the presence of sodamide: Charpentier, U.S. pat. **2,519,886** (1950 to Rhône-Poulenc).

Oily liq. Amine odor. Alkaline reaction. $bp_{0.3}$ 203-210°. Hydrochloride, $C_{17}H_{20}N_2S·HCl$, Liranol, Promwill, Prazine, Protactyl, Sparine, Talofen. White to slightly yellow crystals, dec 181° (microblock). Oxidizes upon prolonged exposure to air and acquires a blue or pink color. Hygroscopic. One gram dissolves in about 3 ml water. Sol in methanol, ethanol, chloroform. Practically insol in ether, benzene. Aq solns are slightly acid to litmus. Incompatible with alkalies, oxidizing agents, heavy metals.

THERAP CAT: Antipsychotic.

THERAP CAT (VET): Tranquilizer.

7967. Promecarb. 3-Methyl-5-(1-methylethyl)phenol methylcarbamate; methylcarbamic acid m-cym-5-yl ester; m-cym-5-yl methylcarbamate; 3-methyl-5-isopropyl *N*-methyl-carbamate; SN-34615; Carbamult. $C_{12}H_{17}NO_2$; mol wt 207.27. C 69.54%, H 8.27%, N 6.76%, O 15.44%. Prepn: **Brit.** pat. **913,707**, *C.A.* **58**, 7316c (1963); **U.S.** pat. **3,167,-472** (1962, 1965 both to Schering). Toxicity study: N. Miyao *et al.*, *Kagoshima Daigaku Nogakubu Gakujsitsu Hoboku* **22**, 131 (1972), *C.A.* **78**, 119931e (1973). *Review:* Jaeger, *Z. Angew. Entomol.* **58**, 188 (1966).

Colorless crystals, mp 87.0-87.5°. $bp_{0.1}$ 117°. Sol in polar organic solvents. Practically insol in water. Hydrolyzed in alkalies. LD_{50} in mice, rats (mg/kg): 39.5, 60 orally (Miyao).
Caution: Cholinesterase inhibitor.
USE: Insecticide.

7968. Promedol. *1,2,5-Trimethyl-4-phenyl-4-piperidin-ol propanoate (ester);* 1,2,5-trimethyl-4-phenyl-4-propionyl-oxypiperidine; 1,2,5-trimethyl-4-phenyl-4-piperidyl propion-ate; dimethylmeperidine. $C_{17}H_{25}NO_2$; mol wt 275.39. C 74.14%, H 9.15%, N 5.09%, O 11.62%. Prepn: Nazarov *et al.*, *J. Gen. Chem. USSR* **26**, 3117 (1956); Nazarov, Shves-tov, *ibid.* 3533. Conformation studies: Prostakov, Mik-heeva, *ibid.* **31**, 108 (1961); **33**, 2931 (1963); *eidem, Russ. Chem. Rev.* **31**, 556 (1962). Of the four possible isomers α and γ are shown.

α-isomer

γ-isomer

α-Isomer hydrochloride, $C_{17}H_{25}NO_2$·HCl, *α-promedol.* Crystals from benzene, mp 153-154°. Has been also report-ed as mp 106-107° or 126-131° (Prostakov, Mikheeva, 1961).
β-Isomer hydrochloride, *isopromedol.* Crystals, mp 183-184°, Nazarov, Shvestov, *Bull. Acad. Sci. USSR Phys. Ser.* **1959**, 2059.
γ-Isomer hydrochloride, *trimeperidine,* γ-*promedol.* Crys-tals from acetone, mp 222-223°.
Note: This is a controlled substance (opiate) listed in the U.S. Code of Federal Regulations, Title 21 Part 1308.11 (1995).

THERAP CAT: Analgesic (narcotic).

7969. Promegestone. *(17β)-17-Methyl-17-(1-oxoprop-yl)estra-4,9-dien-3-one; 17α-methyl-17-propionylestra-4,9-dien-3-one;* 17α-methyl-17β-propionyl-19-nor-4,9-androsta-dien-3-one; 17α,21-dimethyl-19-norpregna-4,9-diene-3,20-dione; R-5020; RU-5020; Surgestone. $C_{22}H_{30}O_2$; mol wt 326.48. C 80.94%, H 9.26%, O 9.80%. Synthetic progestin with no androgenic activity and with high affinity for the progesterone receptor. Prepn: J. Warnant, A. Farcilli, **Belg.** pat. **763,099**; *eidem,* **U.S.** pats. **3,679,714, 3,761,591** (1971, 1972, 1973 all to Roussel-UCLAF). Inhibition of gonadotropin secretion and lack of androgenic activity: F. Labrie *et al.*, *Fert. Steril.* **28**, 1104 (1977). Binding studies in mouse uterus: D. Philibert, J.-P. Raynaud, *Steroids* **22**, 89 (1973); *eidem, Endocrinology* **94**, 627 (1974). Binding to human endometrium: *eidem, Contraception* **10**, 457 (1974); M. Haukkamaa, T. Luukkainen, *J. Steroid Biochem.* **5**, 447 (1974). Review and possible use in treatment of hormone-dependent breast cancer: J.-P. Raynaud, T. Ojasoo, *J. Gynecol. Obstet. Biol. Reprod.* **12**, 697 (1983).

Colorless crystals from isopropyl ether, mp 152°. Sol in acetone, benzene. Insol in water. $[\alpha]_D^{20}$ $-262°$ (c = 0.5 in ethanol). uv max in ethanol: 215, 305 nm ($E_{1cm}^{1\%}$ 202, 648).
USE: As radioligand for the progestin receptor.
THERAP CAT: Progestogen.

7970. Promethazine. *N,N,α-Trimethyl-10H-phenothia-zine-10-ethanamine; 10-(2-dimethylaminopropyl)phenothia-zine;* 10-(2-dimethylamino-2-methylethyl)phenothiazine; *N-(2'-dimethylamino-2'-methyl)ethylphenothiazine; proaza-mine;* RP-3277. $C_{17}H_{20}N_2S$; mol wt 284.43. C 71.79%, H 7.09%, N 9.85%, S 11.27%. Prepn from 10-phenothiazine-propyl chloride and dimethylamine in presence of Cu: Charpentier, *Compt. Rend.* **225**, 306 (1947); **U.S.** pat. **2,530,451** (1950 to Rhône-Poulenc); from Grignard com-plexes of dimethylaminopropyl halide and phenothiazine: Berg, Ashley, **U.S.** pat. **2,607,773** (1952 to Rhône-Poulenc). Structure and isomerism: Edge, Wragg, *J. Pharm. Pharma-col.* **5**, 279 (1953). Metabolic studies: Huang *et al.*, *J. Pharm. Sci.* **59**, 772 (1970). Toxicity: Rajsner, *Coll. Czech. Chem. Commun.* **34**, 1019 (1969). Comprehensive descrip-tion: C. M. Shearer, S. M. Miller in *Analytical Profiles of Drug Substances* vol. **5**, K. Florey, Ed. (Academic Press, New York, 1976) pp 429-465.

Crystals, mp 60°. bp_3 190-192°.
Hydrochloride, $C_{17}H_{20}N_2S$·HCl, *RP-3389, Atosil, Dorme, Duplamin, Fellozine, Fenazil, Genphen, Hiberna, Lergigan, Phencen, Phenergan, Prorex, Prothazine, Provigan, Remsed.* Crystals from ethylene dichloride, mp 230-232° (some dec). Turns blue on prolonged exposure to air and moisture. Lower mp reported in the literature are caused by admix-ture with isopromethazine, *q.v.* uv max (water): 249, 297 nm (ε 28770, 3400). pH of 10% aq soln 5.3. Freely sol in water. Sol in alcohol, chloroform. Practically insol in acetone, ether, ethyl acetate. LD_{50} i.v. in mice: 55.0 mg/kg (Rajsner).
Compd with 8-chlorotheophylline, $C_{17}H_{20}N_2S$·C_7H_7Cl-N_4O_2, *promethazine teoclate, Avomine.*

THERAP CAT: Antihistaminic.
THERAP CAT (VET): Antihistaminic, antiemetic, CNS de-pressant.

7971. Promethium. Pm; at. no. 61; valence 3. No stable nuclides, known radioactive isotopes (mass numbers): 132-155. Longest-lived known isotope: 145 (rel. at. mass 144.9127, $T_{1/2}$ 17.7 years, α-emitter). Most available isotope: 147 (rel. at. mass 146.9151, $T_{1/2}$ 2.6234 years, β-emitter). Abundance in the earth's crust reported as 4.5×10^{-20} ppm; present at low levels in uranium ores. Independent discovery in rare earth concentrate claimed by: J. A. Har-ris, B. S. Hopkins, *J. Am. Chem. Soc.* **48**, 1585 (1926); J. A. Harris *et al., ibid.* 1594; and by: L. Rolla, L. Fernandez, *Gazz. Chim. Ital.* **56**, 435 (1926). Synthetic prepn: H. B. Law *et al., Phys. Rev.* **59**, 936 (1941). Positive identification from fission products of ^{235}U by ion-exchange chromato-graphy: J. A. Marinsky *et al., J. Am. Chem. Soc.* **69**, 2781 (1947). Metal prepd by reduction of halides: Weigel, *Angew. Chem. Int. Ed.* **2**, 326 (1963); E. J. Wheelwright, *J.*

Phys. Chem. **73**, 2867 (1969). Review of prepn, properties and compds: Boyd, J. Chem. Ed. **36**, 3-14 (1959); Weigel, Fortschr. Chem. Forsch. **12**, 539-621 (1969); F. H. Spedding in Kirk Othmer Encyclopedia of Chemical Technology **vol. 19** (John Wiley & Sons, New York, 3rd ed., 1982) pp 833-854; Chemistry of the Elements, N. N. Greenwood, A. Earnshaw, Eds. (Pergamon Press, New York, 1984) pp 1423-1449. Review of chemistry, toxicology and industrial uses: Promethium Technology, E. J. Wheelwright, Ed. (American Nuclear Society, Hindsdale, IL, 1973) 395 pp. Brief review of properties: G. T. Seaborg, Radiochim. Acta **61**, 115-122 (1993).

[147]Pm, metallic solid; d 7.22. mp 1080° (Weigel); also reported as 1169° (Wheelwright). A number of salts have been prepared including the trihalides (PmX$_3$), the sesquioxide (Pm$_2$O$_3$), the hydroxide [Pm(OH)$_3$] and the nitrate [Pm(NO$_3$)$_3$.xH$_2$O].

USE: [147]Pm as energy source for nuclear powered batteries, β-particle source for thickness gauges; in the prepn of self-luminous compds; as portable x-ray sources.

7972. Prometon. 6-Methoxy-N,N'-bis(1-methylethyl)-1,-3,5-triazine-2,4-diamine; 2-methoxy-4,6-bis(isopropylamino)-s-triazine; methoxypropazine; prometone; G-31435; Gesafram; Primatol; Pramitol. C$_{10}$H$_{19}$N$_5$O; mol wt 225.29. C 53.31%, H 8.50%, N 31.09%, O 7.10%. Nonselective pre- and post-emergent herbicide for use on noncrop land. General preparative information and use as herbicide: H. Gysin, F. Knüsli, U.S. pat. **2,909,420** (1959 to Geigy); see also: M. A. Priola, U.S. pat. **3,713,806** (1973 to Ciba-Geigy). Alternate process: H. V. Lemaster, U.S. pat. **3,663,542** (1972 to Geigy). Comparison with other triazine herbicides: H. Gysin, E. Knüsli, Adv. Pest Contr. Res. **3**, 289 (1960). Metabolism in rats: J. E. Bakke et al., J. Agr. Food Chem. **15**, 628 (1967). GC determn in human urine: ibid. **27**, 740 (1979). Toxicology study in sheep and cattle: A. E. Johnson et al., Am. J. Vet. Res. **33**, 1433 (1972). Review of properties and analytical methods: B. G. Tweedy, R. A. Kahrs, Anal. Methods Pestic. Plant Growth Regul. **10**, 493 (1978).

Crystalline solid, mp 91-92°. Soly in water (20°): 750 ppm. Very sol in organic solvents. Vapor pressure at 20°: 2.3×10^{-6} Torr. LD$_{50}$ in mice, rats (mg/kg): 2160, 2980 orally (Gysin, Knüsli, 1960). LC$_{50}$ (96 hour) in bluegill sunfish, rainbow trout (ppm): >32, 20 (Tweedy, Kahrs).

USE: Nonselective herbicide.

7973. Prometryn. N,N'-Bis(1-methylethyl)-6-methyl-thio-1,3,5-triazine-2,4-diamine; 2,4-bis(isopropylamino)-6-(methylthio)-s-triazine; 2-methylthio-4,6-bis(isopropylamino)-s-triazine; G-34161; Gesagard; Caparol. C$_{10}$H$_{19}$N$_5$S; mol wt 241.36. C 49.76%, H 7.93%, N 29.02%, S 13.29%. Selective pre- and post-emergence herbicide. General preparative information and use as herbicide: H. Gysin, E. Knüsli, Swiss pat. **337,019**; eidem, U.S. pat. **2,909,420** (both 1959 to Geigy). See also: M. A. Priola, U.S. pat. **3,713,806** (1973 to Ciba-Geigy). Alternate process: Fr. pat. **1,372,-089**; E. Knüsli, W. Stammbach, U.S. pat. **3,207,756** (1964, 1965 both to Geigy). Activity: H. Gysin, Chem. & Ind. (London) **1962**, 1393. Metabolism in rats and rabbits: C. Böhme, F. Bär, Food Cosmet. Toxicol. **5**, 23 (1967). Toxicity: Geigy, Toxicology Data on Prometryne, July, 1964. Review of properties and analytical methods: B. G. Tweedy, R. A. Kahrs, Anal. Methods Pestic. Plant Growth Regul. **10**, 493 (1978).

Crystals, mp 118-120°. Vapor pressure at 20°: 1×10^{-6} mm Hg. Soly in water at 20°: 48 ppm. Readily sol in org solvents. Stable in neutral or slightly acid or alkaline media. Hydrolyzed under stronger acidic or basic conditions. LD$_{50}$ orally in rats: 3.75 g/kg (Geigy). LC$_{50}$ (96 hour) in bluegill sunfish, rainbow trout (ppm): 10.0, 2.5 (Tweedy, Kahrs).

USE: Herbicide.

7974. Pronethalol. α-[[(1-Methylethyl)amino]methyl]-2-naphthalenemethanol; α-[(isopropylamino)methyl]-2-naphthalenemethanol; [2-hydroxy-2-(2-naphthyl)ethyl]isopropyl-amine; 2-isopropylamino-1-(2-naphthyl)ethanol; 1-(2'-naph-thyl)-2-isopropylaminoethanol; nethalide. C$_{15}$H$_{19}$NO; mol wt 229.32. C 78.56%, H 8.35%, N 6.11%, O 6.98%. β-Adrenergic blocker. Prepd by reduction of 2-naphthacyl bromide with NaBH$_4$ followed by reaction with isopropyl-amine: Stephenson, **Brit.** pat. **909,357** (1962 to ICI). Pharmacology: Black et al., Brit. J. Pharmacol. Chemother. **25**, 577 (1965). Carcinogenicity in mice: R. Howe, Nature **207**, 594 (1965). Metabolism: W. G. Stillwell, M. G. Horning, Res. Commun. Chem. Pathol. Pharmacol. **9**, 601 (1974).

Crystals, mp 108°.
Hydrochloride, C$_{15}$H$_{19}$NO.HCl, *ICI-38174, Alderlin.* Crystals, mp 184°. LD$_{50}$ in mice (mg/kg): 512 orally, 46 i.v. (Black).

THERAP CAT: Antianginal; antiarrhythmic; antihypertensive.

7975. Pro-opiomelanocortin. Proopiocortin; ACTH-β-lipotropin common precursor; precursor to ACTH-LPH-β-endorphin; 31K-precursor; POMC. A precursor protein of mol wt ~30,000, synthesized in the hypothalamus, pituitary gland, brain, and several peripheral tissues that incorporates the amino acid sequences of the pituitary hormones ACTH and β-lipotropin. These two hormones, in turn, contain biologically active component peptides: α-MSH, *corticotropin-like intermediate lobe peptide (CLIP)*, α-LPH, β-MSH, endorphins, and met-enkephalin. First description of the presence of ACTH and LPH in the same molecule: P. J. Lowry et al., Int. Congr. Ser.-Excerpta Med. **402**, 71 (1976). Confirmation of the common precursor by radioimmunoassay: R. E. Mains et al., Proc. Nat. Acad. Sci. USA **74**, 3014 (1977); J. L. Roberts, E. Herbert, ibid. 4826, 5300. Isoln of rat pro-opiomelanocortin: M. Rubenstein et al., ibid. **75**, 669 (1978). Partial sequence analysis using cDNA from mouse: J. L. Roberts et al., ibid. **76**, 2153 (1979). Complete sequence of bovine POMC using cloned cDNA: S. Nakanishi et al., Nature **278**, 423 (1979). The precursor has also been found in human ACTH ectopic tumor: X. Bertagna et al., Proc. Nat. Acad. Sci. USA **75**, 5160 (1978). Primary structure of the NH$_2$-terminal glycopeptide of human pituitary POMC: N. G. Seidah et al., J. Biol. Chem. **256**, 7977 (1981). Reviews: M. Chrétien, N. G. Seidah, Mol. Cell. Biochem. **34**, 101-127 (1981); D. De Wied, J. Jolles, Physiol. Rev. **62**, 976-1059 (1982); Y. P. Loh et al., Peptides **3**, 397-404 (1982).

7976. Propacetamol. N,N-Diethylglycine 4-(acetyla-mino)phenyl ester; N,N-diethylglycine ester with 4'-hydroxy-acetanilide; 4-acetamidophenyl (diethylamino)acetate. C$_{14}$-H$_{20}$N$_2$O$_3$; mol wt 264.32. C 63.62%, H 7.63%, N 10.60%, O

18.16%. Injectable prodrug of acetaminophen, *q.v.* Prepn and pharmacology: J. C. Cognacq, **Belg.** pat. **854,376;** *idem,* **U.S.** pat. **4,127,671** (1977, 1978 both to Hexachimie). Clinical comparison with injectable aspirin: R. De Marneffe, L. Mokassa, *Compt. Rend. Ther. Pharm. Clin.* **3,** 23 (1985). Clinical studies: P. Delacroix *et al., Sem. Hop.* **61,** 2739 (1985); J. Modai, *ibid.* **62,** 587 (1986).

Thick oil.

Hydrochloride, $C_{14}H_{20}N_2O_3$.HCl, *UP-34101, Pro-Dafalgan.* mp 228°. Sol in water.

THERAP CAT: Analgesic; antipyretic.

7977. Propachlor. *2-Chloro-N-(1-methylethyl)-N-phenylacetamide; 2-chloro-N-isopropylacetanilide; N-isopropyl-α-chloroacetanilide;* CP-31393; Bexton; Ramrod. $C_{11}H_{14}$-ClNO; mol wt 211.69. C 62.41%, H 6.67%, Cl 16.75%, N 6.62%, O 7.56%. Selective pre-emergence herbicide. Prepn: P. C. Hamm, A. J. Speziale, **U.S.** pat. **2,863,752** (1958 to Monsanto). Activity studies: Duke, *Diss. Abstr.* B **28,** 1315 (1967); Jaworski, *J. Agr. Food Chem.* **17,** 165 (1969); Lamoureaux *et al., ibid.* **19,** 346 (1971); Dhillon, Anderson, *Weed Res.* **12,** 182 (1972). Metabolism: J. E. Bakke *et al., Science* **210,** 433 (1980). Toxicity data: E. E. Kenaga, *Down Earth* **35,** 25 (1979).

Light tan solid, mp 67-76°. Vapor pressure at 110°: 0.03 mm Hg. Soly in water at 20°: 700 mg/l. Sol in common organic solvents except aliphatic hydrocarbons. LD_{50} orally in rats: 710 mg/kg (Kenaga).

USE: Herbicide.

7978. Propafenone. *1-[2-[2-Hydroxy-3-(propylamino)-propoxy]phenyl]-3-phenyl-1-propanone; 2'-[2-hydroxy-3-(propylamino)propoxy]-3-phenylpropiophenone;* SA-79. $C_{21}H_{27}$-NO_3; mol wt 341.45. C 73.87%, H 7.97%, N 4.10%, O 14.06%. Prepn: R. Sachse, **Ger.** pat. **2,001,431** (1971 to Helopharm), *C.A.* **75,** 151538f (1971). Pharmacology: H.-J. Hapke, E. Prigge, *Arzneimittel-Forsch.* **26,** 1849 (1976). Use in arrhythmia: O. A. Beck, H. Hochrein, *Deut. Med. Wochenschr.* **103,** 1261 (1978). Pharmacokinetics: H. Blanke *et al., ibid.* **104,** 587 (1979); M. Hollmann *et al., Arzneimittel-Forsch.* **33,** 763 (1983). Brief review: M. Fischer, *Med. Klin.* **75,** 39 (1980).

Hydrochloride, $C_{21}H_{27}NO_3$.HCl, *Arythmol, Pronon, Rythmol, Rytmonorm.* Fine white crystals. Slightly bitter taste. Sol in basic lower aliphatic alcohols, CCl_4, hot water; slightly sol in cold water. Insol in ether. LD_{50} in rats (mg/kg): 18.8 i.v.; 700 orally (Hapke, Prigge).

THERAP CAT: Antiarrhythmic (class IC).

7979. Propagermanium. *3,3'-(1,3-Dioxo-1,3-digermoxanediyl)bispropanioic acid;* 2-(carboxyethyl)germanium sesquioxide; 3-oxygermylpropionic acid polymer; poly-*trans*-[(2-carboxyethyl)germasesquioxane]; proxigermanium; re-

pagermanium; GE-132; SK-818; Serocion. ($GeCH_2CH_2CO-OH)_2O_3$. Biologic response modifier. Polymeric organogermanium compound. Orginally synthesized at Asai Germanium Institute in 1967. Prepn and crystal structure: M. Tsutsui *et al., J. Am. Chem. Soc.* **98,** 8287 (1976). Toxicology and clinical pharmacokinetics: K. Miyao *et al., Curr. Chemother. Infect. Dis.* **2,** 1527 (1979). TLC quantitation: X. Peishan *et al., J. Planar Chromatog.* **3,** 141 (1990). Electrochemical behavior and determn: K. Hasebe *et al., Electroanalysis* **6,** 779 (1994). Antiviral activity: Y. Ishiwata *et al., Arzneimittel-Forsch.* **44,** 357 (1994). Review of biological activities and mechanism of antitumor action: R. R. Brutkiewicz, F. Suzuki, *In Vivo* **1,** 189-204 (1987). *Review:* K. Miyao, N. Tanaka, *Drugs Future* **13,** 441-453 (1988). Review of clinical use as an immunostimulant: H. Fukazawa *et al., Head-Neck* **16,** 30-38 (1994).

Colorless, monoclinic crystals or crystalline powder having no odour but a slightly acidic taste. mp 270° (dec). pKa 3.6. Sol in water at 1.09% at 20°. Insol or very slightly sol in almost all organic solvents. Very sol in water under alkaline conditions; >10% at pH 7.4. Aq solns exhibit acid reactions. Stable at pH 2-12, especially at pH 7.4 especially where decomposition does not occur at 110° for 5 min. LD_{50} i.p. in mice: 2.8 g/kg (Brutkiewicz).

THERAP CAT: Antineoplastic; immunostimulant.

7980. Propallylonal. *5-(2-Bromo-2-propenyl)-5-(1-methylethyl)-2,4,6(1H,3H,5H)pyrimidinetrione; 5-(2-bromo-allyl)-5-isopropylbarbituric acid;* Noctal. $C_{10}H_{13}BrN_2O_3$; mol wt 289.13. C 41.54%, H 4.53%, Br 27.64%, N 9.69%, O 16.60%. Prepn: **U.S.** pat. **1,622,129** (1927); *cf.* Herzog, *Arch. Pharm.* **263,** 216 (1925). Acute toxicity: Maloney, *J. Pharmacol. Exp. Ther.* **42,** 267 (1931).

Crystals. mp 177-179°. Slightly bitter taste. Slightly sol in water; freely sol in alcohol, glacial acetic acid, acetone, alkalies. Sparingly sol in ether, chloroform, benzene. MLD orally in rabbits: 300-350 mg/kg (Maloney).

Caution: May be habit forming. This is a controlled substance (depressant) listed in the U.S. Code of Federal Regulations, Title 21 Parts 329.1 and 1308.13 (1995).

THERAP CAT: Sedative, hypnotic.

7981. Propamidine. *4,4'-[1,3-Propanediylbis(oxy)]bis-benzenecarboximidamide; 4,4'-(trimethylenedioxy)dibenzamidine;* 4,4'-diamidino-α,ω-diphenoxypropane. $C_{17}H_{20}N_4O_2$; mol wt 312.37. C 65.37%, H 6.45%, N 17.94%, O 10.24%. Prepn: A. J. Ewins *et al.,* **Brit.** pat. **507,565** (1939 to May & Baker); J. N. Ashley *et al., J. Chem. Soc.* **1942,** 103; of isethionate: G. Newbery, A. P. T. Easson, **U.S.** pat. **2,394,-003** (1946 to May & Baker). Trypanocidal activity: E. M. Lourie, W. Yorke, *Ann. Trop. Med. Parasitol.* **33,** 289 (1939). Preliminary pharmacology: R. Wien, *ibid.* **37,** 1 (1943). Determn in biological fluids: D. P. Jackson *et al., J. Biol. Chem.* **167,** 377 (1947). Mode of action: M. J. Pine, *Biochem. Pharmacol.* **17,** 75 (1968). Activity in fibrinolytic systems: J. D. Geratz, *Thromb. Diath. Haemorrh.* **29,** 154 (1973). Clinical use in treatment of *Acanthamoeba* keratitis: D. L. Easty, *Brit. Med. J.* **296,** 228 (1988); J. J. Wiens, W. B. Jackson, *Can. J. Ophthalmol.* **23,** 107 (1988). Early review of pharmacology, mode of action and clinical applications: E. B. Schoenbach, E. M. Greenspan, *Medicine* **27,** 327-377 (1948).

Isethionate, $C_{21}H_{32}N_4O_{10}S_2$, *M & B 782, Brolene Drops*. Hygroscopic, very bitter crystals or granular powder, mp ~235°. Soluble in water (~1 in 5), glycerol, 95% alcohol (~1 in 33). Practically insol in ether, chloroform, fixed oils, liq petrolatum. pH of a 5% w/v soln in water = 4.5 to 6.5.

THERAP CAT: Antiprotozoal (Trypanosoma); antiamebic.

THERAP CAT (VET): Anti-infective (topical). Formerly used as antiprotozoal (Trypanosoma, Babesia).

7982. Propane. Dimethylmethane; propyl hydride. C_3H_8; mol wt 44.10. C 81.71%, H 18.29%. $CH_3CH_2CH_3$. Constituent of natural gas and of crude petroleum. Obtained by the so-called "stabilization process" using fractional distillation under pressure: Francis, Robbins, *J. Am. Chem. Soc.* **55**, 4339 (1933). Many syntheses, e.g., by using butyronitrile and sodium: Timmermans, *J. Chim. Phys.* **18**, 133 (1920).

Gas. Odorless when pure. Burns with a luminous, smoky flame. Explosive limits, % by vol in air: 2.37-9.5. Heavier than air. One liter weighs 2.0200 g at 0° and 760 mm; 1.8324 g at 25° and 760 mm. Liquefies at −42°; solid at −187.7°. bp (1 atm) −42.1°; bp (2 atm) −25.6°; bp (5 atm) +1.4°; bp (10 atm) 26.9°; bp (20 atm) 58.1°; bp (30 atm) 78.7°; bp (40 atm) 94.8°. Crit temp 96.81°; crit press. 42.01 atm. Heat of combustion (const vol) 528.4 cal, (const pressure) 553.5 cal. 100 vols water dissolve 6.5 vols at 17.8° and 753 mm pressure; 100 vols abs alc dissolve 790 vols at 16.6° and 754 mm pressure; 100 vols ether dissolve 926 vols at 16.6° and 757 mm pressure; 100 vols chloroform dissolve 1299 vols at 21.6° and 757 mm pressure; 100 vols benzene dissolve 1452 vols at 21.5° and 757 mm pressure; 100 vols turpentine dissolve 1587 vols at 17.7° and 757 mm pressure.

Caution: Potential symptoms of overexposure are dizziness, disorientation, excitation and frostbite. *See NIOSH Pocket Guide to Chemical Hazards* (DHHS/NIOSH 90-117, 1990) p 186.

USE: As fuel gas, sometimes mixed with butane. In organic syntheses. As refrigerant.

7983. 1-Propanearsonic Acid. *n-Propylarsonic acid*. $C_3H_9AsO_3$; mol wt 168.02. C 21.45%, H 5.40%, As 44.59%, O 28.57%. $C_3H_7AsO(OH)_2$. White needles, mp 125°. Freely sol in water; sol in alcohol. Insol in ether.

USE: For the determination of zirconium.

7984. 1,3-Propanedithiol. 1,3-Dimercaptopropane; trimethylenedithioglycol; dithiotrimethyleneglycol; trimethylenedimercaptan. $C_3H_8S_2$; mol wt 108.23. C 33.29%, H 7.45%, S 59.26%. $HSCH_2CH_2CH_2SH$. Prepd by alkaline hydrolysis of propylene-1,3-diisothiuronium dihydrochloride: Grogan *et al., J. Org. Chem.* **20**, 50 (1955).

Oil. Disagreeable odor. d_4^{20} 1.0772. bp_{760} 169-170°; bp_{759} 170-171°; bp_{56} 92-98°. n_D^{20} 1.5392. Volatile with steam. Slightly sol in water. Miscible with alcohol, ether, chloroform and benzene.

7985. Propanethial S-Oxide. *Thiopropionaldehyde S-oxide;* thiopropanal *S*-oxide. C_3H_6OS; mol wt 90.15. C 39.97%, H 6.71%, O 17.75%, S 35.57%. $CH_3CH_2CH=SO$. Lachrymatory factor of the onion, *Allium cepa L.*, found as a 95% *cis*- and 5% *trans*- mixture. Early structure studies: W. D. Niegisch, W. H. Stahl, *Food Res.* **21**, 657 (1956); C. G. Spare, A. I. Virtanen, *Acta Chem. Scand.* **17**, 641 (1963). Structure: M. H. Brodnitz, J. V. Pascale, *J. Agr. Chem.* **19**, 269 (1971). Stereochemistry: E. Block *et al., J. Am. Chem. Soc.* **101**, 2200 (1979); *eidem, Tetrahedron Letters* **21**, 1277 (1980). Chemistry: *eidem, J. Am. Chem. Soc.* **102**, 2490 (1980).

7986. Propanidid. *4-[2-(Diethylamino)-2-oxoethoxy]-3-methoxybenzeneacetic acid propyl ester; [4-[(diethylcarbamoyl)methoxy]-3-methoxyphenyl]acetic acid propyl ester;* [3-methoxy-4-[(N,N-diethylcarbamido)methoxy]phenyl]acetic acid *n*-propyl ester; propyl 4-[(diethylcarbamoyl)methoxy]-3-methoxyphenyl]acetate; Bayer 1420; FBA-1420; Epontol; Sombrevin. $C_{18}H_{27}NO_5$; mol wt 337.42. C 64.07%, H 8.07%, N 4.15%, O 23.71%. Prepn: R. Hiltmann *et al.,* **Ger.** pat. **1,134,981;** U.S. pat. **3,086,978** (1962, 1963 both to Bayer). Toxicity study: E. I. Goldenthal, *Toxicol. Appl. Pharmacol.* **18**, 185 (1971).

Pale yellow oil, $bp_{0.7}$ 210-212°. Practically insol in water; sol in alcohol, chloroform. LD_{50} orally in rats: > 10,000 mg/kg (Goldenthal).

THERAP CAT: Anesthetic (intravenous).

7987. Propanil. *N-(3,4-Dichlorophenyl)propanamide; 3′,4′-dichloropropionanilide; N-(3,4-dichlorophenyl)propion-amide;* DPA; FW-734; Stam; Stampede; Rogue; Chem Rice; Surcopur. $C_9H_9Cl_2NO$; mol wt 218.08. C 49.57%, H 4.16%, Cl 32.51%, N 6.42%, O 7.34%. Selective contact herbicide. Prepn: W. Schäfer *et al.,* **Ger.** pat. **1,039,779** (1958 to Bayer), *C.A.* **54**, 20060i (1960); Huffman, Allen, *J. Agr. Food Chem.* **8**, 298 (1960). Use in nematocide formulations: Fielding, Stoddard, U.S. pat. **3,108,038** (1963 to du Pont). Toxicity study: G. W. Bailey, J. L. White, *Residue Rev.* **10**, 97 (1965).

White crystalline solid, mp 91-93°. Soly in water at room temp: 225 ppm. LD_{50} orally in rats: 1384 mg/kg (Bailey, White).

USE: Herbicide, nematocide.

7988. Propanocaine. *α-[2-(Diethylamino)ethyl]benzenemethanol benzoate (ester); α-[2-(diethylamino)ethyl]benzyl alcohol benzoate;* 3-diethylamino-1-phenylpropyl benzoate; α-(2-diethylaminoethyl)benzyl benzoate; 1-benzoyloxy-3-diethylamino-1-phenylpropane. $C_{20}H_{25}NO_2$; mol wt 311.42. C 77.14%, H 8.09%, N 4.50%, O 10.27%. Prepd from acetophenone and trioxymethylene in the presence of diethylamine hydrochloride followed by catalytic hydrogenation and reaction with benzoyl chloride: Detrie, **Fr.** pat. **1,234,-701** (1960 to Clin-Byla), *C.A.* **55**, 24682g (1961).

Yellow oil.

THERAP CAT: Anesthetic (local).

7989. Propantheline Bromide. *N-Methyl-N-(1-methylethyl)-N-[2-[(9H-xanthen-9-ylcarbonyl)oxy]ethyl]-2-propanaminium bromide; (2-hydroxyethyl)diisopropylmethylammonium bromide xanthene-9-carboxylate;* β-diisopropylaminoethyl 9-xanthenecarboxylate methobromide; Corrigast; Ercotina; Pro-Banthine; Neo-Metantyl; Pantheline. $C_{23}H_{30}BrNO_3$; mol wt 448.40. C 61.61%, H 6.74%, Br 17.82%, N 3.12%, O 10.70%. Prepn: Cusic, Robinson, *J. Org. Chem.* **16**, 1921 (1951); U.S. pat. **2,659,732** (1953 to Searle). Metabolic studies: Beermann *et al., Clin. Pharmacol. Ther.* **13**, 212 (1972).

Crystals from isopropanol + ether, mp 159-161°. Very soluble in water, alcohol, chloroform. Practically insol in ether, benzene.

THERAP CAT: Anticholinergic.

7990. Propaquizafop. *(R)-2-[4-[(6-Chloro-2-quinoxalinyl)oxy]phenoxy]propanoic acid 2-[[(1-methylethylidene)-amino]oxy]ethyl ester;* 2-isopropylideneaminooxyethyl *(R)*-2-[4-(6-chloroquinoxalin-2-yloxy)phenoxy]propionate; Ro-17-3664; Agil; Falcon; Shogun. $C_{22}H_{22}ClN_2O_5$; mol wt 443.89. C 59.53%, H 5.00%, Cl 7.99%, N 9.47%, O 18.02%. Post-emergence graminicide for use in broadleaf crops. Prepn: G. Frater *et al.,* U.S. pat. **4,687,849** (1987 to Hoffmann-La Roche); R. Klaus *et al., ACS Symposium Series* **443,** 226-235 (1991). Comprehensive description: P. F. Bocion *et al., Proc. Brit. Crop Prot. Conf. - Weeds* 1987, 55-62. Review of comparative field trials: J. D. A. Wevers, *Meded. Fac. Landbouwwet. Rijksuniv. Gent.* **56,** 611-615 (1991).

Crystals from ether/hexane, mp 62-64°. $[\alpha]_D^{20}$ +29.7° (c = 0.93% in chloroform). d^{20} 1.29. Vapor pressure (20°): 1.3 × 10^{-8} Pa. Partition coefficient (*n*-octanol/water at pH 7): 3.5. Soly in water (25°): 2 ppm (pH 7). LD_{50} in rats (mg/kg): > 5000 orally; > 2000 dermally (Bocion).

USE: Herbicide.

7991. Proparacaine. *3-Amino-4-propoxybenzoic acid 2-(diethylamino)ethyl ester;* 2-(diethylamino)ethyl 3-amino-4-propoxybenzoate; proxymetacaine. $C_{16}H_{26}N_2O_3$; mol wt 294.39. C 65.28%, H 8.90%, N 9.52%, O 16.30%. Prepn of the hydrochloride: Clinton *et al., J. Am. Chem. Soc.* **74,** 592 (1952). Pharmacology: McIntyre, Sievers, *J. Pharmacol. Exp. Ther.* **63,** 369 (1938). Comprehensive description: D. B. Whigan in *Analytical Profiles of Drug Substances* vol. 6, K. Florey, Ed. (Academic Press, New York, 1977) pp 423-456.

Hydrochloride, $C_{16}H_{26}N_2O_3 \cdot HCl$, *Ak-Taine, Alcaine, Ophthaine, Ophthetic.* Prisms from abs alcohol + ethyl acetate, mp 182.0-183.3°. uv max (methanol): 225, 270, 300 nm. Sol in water, warm alcohol, methanol. Insol in ether, benzene. Solns are neutral to litmus. pKa 3.2.

THERAP CAT: Topical anesthetic (ophthalmic).

7992. Propargite. *Sulfurous acid 2-[4-(1,1-dimethylethyl)phenoxy]cyclohexyl 2-propynyl ester; sulfurous acid 2-(p-tert-butylphenoxy)cyclohexyl 2-propynyl ester;* 2-(p-tert-butylphenoxy)cyclohexyl propargyl sulfite; cyclosulfyne; propargil; BPPS; ENT-27226; DO-14; Omite; Comite. $C_{19}H_{26}O_4S$; mol wt 350.48. C 65.11%, H 7.48%, O 18.26%, S 9.15%. Acaricide for control of mites on crops. Prepn and insecticidal properties: R. A. Covey *et al.,* U.S. pat. **3,272,854** (1966 to U.S. Rubber). Field tests, comparison with other

acaricides in control of Pacific spider mite: E. M. Stafford, *J. Econ. Entomol.* **61,** 1641 (1968). Efficacy against citrus red mite: L. R. Jeppson *et al., ibid.* **62,** 531 (1969). Residue determn: J. M. Devine, H. R. Sisken, *J. Agr. Food Chem.* **20,** 59 (1972). Brief account of properties, analysis: G. M. Stone, *Anal. Methods Pestic. Plant Growth Regul.* **7,** 355 (1973). Toxicity studies: T. B. Gaines, *Toxicol. Appl. Pharmacol.* **14,** 515 (1969).

Viscous liquid, distils with decomp, stabilized by adding 0.5-1.0% propylene oxide. Soluble in most organic solvents. Practically insoluble in water (10.5 ppm). LD_{50} in male, female rats (mg/kg): 1480, 1480 orally; 250, 680 dermally (Gaines). LC_{50} in rainbow trout, bluegill sunfish (ppb): > 100, 31 (Stone).

USE: Acaricide.

7993. Propargyl Alcohol. *2-Propyn-1-ol.* C_3H_4O; mol wt 56.06. C 64.27%, H 7.19%, O 28.54%. $HC \equiv CCH_2OH$. Prepd by heating β-bromoallyl alcohol with conc KOH: Henry, *Ber.* **5,** 456, 569 (1872); **6,** 729 (1873); **14,** 404 (1881); from formaldehyde and sodium acetylide: Henne, Greenlee, *J. Am. Chem. Soc.* **67,** 484 (1945); from epichlorohydrin + sodium: Eglinton *et al., J. Chem. Soc.* **1952,** 2873; from acetylene and formaldehyde: Reppe, *Ann.* **596,** 1 (1955).

Moderately volatile liquid. Mild geranium odor. d_4^{20} 0.9715. mp —48° to —52°. bp_{760} 114-115°; $bp_{490.3}$ 100°; $bp_{147.6}$ 70°; $bp_{35.4}$ 40°; $bp_{20.6}$ 30°; $bp_{11.6}$ 20°. n_D^{20} 1.43064. Viscosity at 20° = 1.68 cp. Flash pt 33°C. Spec heat: 0.616 cal/g. Misc with water, benzene, chloroform, 1,2-dichloroethane, ethanol, ether, acetone, dioxane, tetrahydrofuran, pyridine. Appreciable heat is evolved on mixing with pyridine. Moderately sol in carbon tetrachloride. Immiscible with aliphatic hydrocarbons. With water, propargyl alcohol forms an azeotrope, bp 97.5°. This mixture has a composition of 39.5 parts by weight of propargyl alcohol and 60.5 parts of water. Polymerized by heat or caustic. Acidified aq solns are resistant to polymerization. LD_{50} orally in rats: 0.07 g/kg, company information, G.A.F. Corp.

Caution: Irritating to skin, mucous membranes.

7994. Propargyl Chloride. *3-Chloro-1-propyne.* C_3H_3Cl; mol wt 74.51. C 48.36%, H 4.06%, Cl 47.58%. Prepd by treating propargyl alcohol with phosphorus trichloride: Henry, *Ber.* **7,** 761 (1874); **8,** 398 (1875); Kirsmann, *Bull. Soc. Chim. France* [4] **39,** 698 (1926).

Liquid. d_4^{25} 1.0306. mp —78°. bp_{760} 57°. Flash pt < 60°. Practically insol in water, glycerol. Miscible with benzene, carbon tetrachloride, ethanol, ethylene glycol, ether, ethyl acetate. Reacts with hydroxy compds to form ethers; with sulfides, ammonia, amines or metal hypoiodites to give the corresp propargyl compds; with aldehydes and ketones to give β-acetylenic alcohols. Undergoes isomerization.

USE: Intermediate in organic synthesis.

7995. Propatyl Nitrate. *2-Ethyl-2-[(nitrooxy)methyl]-1,3-propanediol dinitrate (ester);* 2-ethyl-2-(hydroxymethyl)-1,3-propanediol trinitrate; 1,1,1-tris(nitratomethyl)propane; ethyltrimethylolmethane trinitrate; trimethylolethylmethane trinitrate; 2,2-bis(hydroxymethyl)-1-butanol trinitrate; 1-hydroxy-2,2-bis(hydroxymethyl)butane trinitrate; ettriol trinitrate; ETTN; Win-9317; Atrilon 5; Etrynit; Gina; Ginapect; Vasangor. $C_6H_{11}N_3O_9$; mol wt 269.17. C 26.77%, H 4.12%, N 15.61%, O 53.50%. $(O_2NOCH_2)_3CCH_2CH_3$. Prepn: Médard, *Mém. poudres* **35,** 113 (1953); Bourjol, *ibid.* **36,** 79 (1954); Hensinger, *Fr.* pat. **1,103,113** (1955).

White powder, mp 51-52°. d 1.49. Readily sol in acetone, alcohol. Practically insol in water. Lowest explosive temp:

220°. Heat of combustion 829.2 kcal/mol. Explosive but only slightly sensitive to shock.

THERAP CAT: Vasodilator (coronary).

7996. Propazine. *6-Chloro-N,N'-bis(1-methylethyl)-1,3,5-triazine-2,4-diamine; 2-chloro-4,6-bis(isopropylamino)-s-triazine;* 2,4-bis(isopropylamino)-6-chloro-s-triazine; G-30028; Gesamil; Milogard; Prozinex. $C_9H_{16}ClN_5$; mol wt 229.71. C 47.06%, H 7.02%, Cl 15.43%, N 30.49%. Selective pre-emergence herbicide. Prepn: Gysin, Knüsli, **Swiss** pats. **342,784-5** (both 1960 to Geigy). Toxicity study: G. W. Bailey, J. L. White, *Residue Rev.* **10**, 97 (1965).

Crystals, mp 213°. Solubility in water at 20° = 8.6 ppm; difficultly sol in organic solvents. LD_{50} orally in rats: > 5000 mg/kg (Bailey, White).

USE: Herbicide.

7997. Propentofylline. *3,7-Dihydro-3-methyl-1-(5-oxohexyl)-7-propyl-1H-purine-2,6-dione;* 3-methyl-1-(5-oxohexyl)-7-propylxanthine; 1-(5'-oxohexyl)-3-methyl-7-propylxanthine; HWA-285; Albert-285; HOE-285; Hextol; Karsivan. $C_{15}H_{22}N_4O_3$; mol wt 306.36. C 58.81%, H 7.24%, N 18.29%, O 15.67%. Peripheral vasodilator which inhibits cyclic AMP phosphodiesterase. Prepn: W. Mohler *et al.*, **Ger.** pat. **2,330,742;** *eidem,* **U.S. pat. 4,289,776** (1975, 1981 both to Hoechst). Cardiovascular effects in animals: O. Hudlicka *et al., Brit. J. Pharmacol.* **72**, 723 (1981). Effect on cAMP phosphodiesterase: K. Nagata *et al., Arzneimittel-Forsch.* **35**, 1034 (1985). Inhibition of adenosine uptake: B. B. Fredholm, K. Lindström, *Acta Pharmacol. Toxicol.* **58**, 187 (1986). Cerebrovascular effects in animals: J. J. Grome *et al., Drug Dev. Res.* **5**, 111 (1985). Effect on postischemic brain edema: B. B. Mrsulja *et al., ibid.* **6**, 339 (1985). Effect on cognitive function in humans: I. Hindmarch, Z. Subhan, *ibid.* **5**, 379 (1985). Acute toxicity: M. Inazu *et al., Oyo Yakuri* **31**, 357 (1986), *C.A.* **104**, 218960a (1986).

Crystals from diisopropyl ether, mp 69-70°. Soly in water at 25°: 3.2%; in ethanol: > 10%; in DMSO: > 10%. LD_{50} in male, female mice, male, female rats (mg/kg): 900, 780, 1150, 940 orally; 168, 170, 180, 195 i.v.; 375, 346, 199, 196 i.p.; 450, 508, 400, 338 s.c. (Inazu).

THERAP CAT: Nootropic.

THERAP CAT (VET): Vasodilator (peripheral and cerebral).

7998. Propenzolate. *α-Cyclohexyl-α-hydroxybenzeneacetic acid 1-methyl-3-piperidinyl ester; α-phenylcyclohexaneglycolic acid 1-methyl-3-piperidyl ester;* 1-methyl-3-piperidyl α-phenylcyclohexaneglycolate; N-methyl-3-piperidyl phenylcyclohexylglycolate. $C_{20}H_{29}NO_3$; mol wt 331.46. C 72.47%, H 8.82%, N 4.23%, O 14.48%. Prepn: Biel *et al., J. Org. Chem.* **26**, 4096 (1961); Biel, **U.S. pat. 2,995,492** (1961 to Lakeside).

Hydrochloride, $C_{20}H_{29}NO_3 \cdot HCl$, *NDR-263, Delinal.* Crystals, mp 222°.

Note: The (+)-1-methyl-3-piperidinyl ester is designated as *oxyclipine* or *oxiclipine.*

THERAP CAT: Anticholinergic.

7999. Properdin. Mol wt 223,000. A highly basic serum protein believed to be a factor in natural immunity against germ and virus diseases, perhaps also against cancer: L. Pillemer *et al., Science* **120**, 279 (1954). Combines with zymosan, *q.v.,* the insol cell wall residue from yeast: L. Pillemer, O. A. Ross, *ibid.* **121**, 732 (1955). Participates also in a nonspecific manner in a variety of immunological reactions of normal serum. Isoln from human serum: L. Pillemer *et al., J. Exp. Med.* **103**, 1 (1956). Purification: Spicer *et al.*, **U.S. pat. 3,038,838** (1962 to Merck & Co.). Characterization of highly purified human properdin: J. Pensky *et al., J. Immunol.* **100**, 142 (1968). Activity studies: O. Götze, H. J. Müller-Eberhard, *J. Exp. Med.* **139**, 44 (1974). The human properdin system has been studied extensively; it was the subject of long years of scientific controversy. Components of the human properdin system that have been identified and characterized are: properdin, factor B, factor D and C3. Comprehensive review of L. Pillemer's work and history of the properdin system discovery: I. H. Lepow, *J. Immunol.* **125**, 471 (1980). The rabbit properdin system: G. B. Naff, *ibid.* **124**, 2625 (1980).

8000. Propetamphos. *(E)-3-[[(Ethylamino)methoxyphosphinothioyl]oxy]-2-butenoic acid 1-methylethyl ester; (E)-3-hydroxycrotonic acid isopropyl ester O-ester with O-methyl ethylphosphoramidothioate; (E)-1-methylethyl 3-[[(ethylamino)methoxyphosphinothioyl]oxy]-2-butenoate; (E)-O-2-isopropoxycarbonyl-1-methylvinyl O-methyl ethylphosphoramidothioate;* SAN-322I; Blotic; Safrotin. $C_{10}H_{20}NO_4PS$; mol wt 281.31. C 42.70%, H 7.17%, N 4.98%, O 22.75%, P 11.01%, S 11.40%. Prepn: J. P. Leber, K. Lutz, **Ger.** pat. **2,035,103** corresp to U.S. pat. **3,758,645** (1971, 1973 to Sandoz). Activity: W. Berg, R. Gothe, *Proc. Brit. Crop Prot. Conf.-Pests Dis.* **1979**, 517.

Yellowish liquid, $bp_{0.005}$ 87-89°. n_D^{20} 1.495. d_4^{20} 1.1294. Soly in water at 24°: 110 mg/l. Sol in most organic solvents. LD_{50} orally in male rats: 82 mg/kg (Berg, Gothe).

USE: Insecticide.

THERAP CAT (VET): Ectoparasiticide.

8001. Propham. *Phenylcarbamic acid 1-methylethyl ester; carbanilic acid isopropyl ester;* N-phenyl isopropyl carbamate; isopropyl carbanilate; O-isopropyl N-phenyl carbamate; INPC; IPC; isoPPC. $C_{10}H_{13}NO_2$; mol wt 179.22. C 67.02%, H 7.31%, N 7.82%, O 17.85%. Prepn: Allen, **U.S. pat. 2,615,916** (1952 to Columbia-Southern Chem. Corp.); Kovalenko, *Zh. Obshch. Khim.* **24**, 1041 (1954); J. L. Hermanson, R. Olson, *Trans. Kansas Acad. Sci.* **64**, 231 (1961). Acute toxicity: T. B. Gaines, R. E. Linder, *Fundam. Appl. Toxicol.* **7**, 299 (1986).

Crystals, mp 90°. Practically insol in water. Sol in most organic solvents. LD_{50} in male, female rats (mg/kg): 3724, 4315 orally (Gaines, Linder).

USE: Herbicide, applied as a spray to the soil.

8002. Propicillin. *[2S-(2α,5α,6β)]-3,3-Dimethyl-7-oxo-6-[(1-oxo-2-phenoxybutyl)amino]-4-thia-1-azabicyclo[3.2.0]-heptane-2-carboxylic acid;* 6-(α-phenoxybutyramido)penicillanic acid; α-phenoxypropylpenicillin; levopropylcillin. $C_{18}H_{22}N_2O_5S$; mol wt 378.45. C 57.13%, H 5.86%, N 7.40%, O

21.14%, S 8.47%. Semi-synthetic antibiotic related to penicillin. Prepn from 6-aminopenicillanic acid: **Brit.** pat. **877,-120** (1961 to Beecham); of salts: Perron *et al., J. Am. Chem. Soc.* **82,** 3934 (1960); Glombitza, *Ann.* **673,** 166 (1964). Metabolism in humans: M. Cole *et al., Antimicrob. Ag. Chemother.* **3,** 463 (1973). TLC determn: S. Hendrickx *et al., J. Chromatog.* **291,** 211 (1984).

Potassium salt, $C_{18}H_{21}KN_2O_5S$, *BRL-284, PA-248, Baycillin, Brocillin, Cetacillin, Oricillin, Trescillin, Ultrapen.* Crystals, dec 195-197°. Soluble at 20° in 1.2 parts water, 23 parts 95% (w/v) alcohol. pH of 1% w/v soln: 5-7.5.

THERAP CAT: Antibacterial.

8003. Propiconazole. *1-[[2-(2,4-Dichlorophenyl)-4-propyl-1,3-dioxolan-2-yl]methyl]-1H-1,2,4-triazole;* proconazole; CGA-64250; Banner; Desmel; Orbit; Radar; Tilt. $C_{15}H_{17}Cl_2N_3O_2$; mol wt 342.22. C 52.65%, H 5.01%, Cl 20.72%, N 12.28%, O 9.35%. Systemic foliar fungicide. Prepn: G. Van Reet *et al.,* **Ger.** pat. **2,551,560;** *eidem,* **U.S.** pat. **4,079,062** (1976, 1978 both to Janssen). Physicochemical properties, toxicity and antifungal activity: P. A. Urech *et al., Proc. Brit. Crop Prot. Conf. - Pests Dis.* **1979,** 508. Efficacy vs cereal diseases: G. Eyries, *Phytiatr. Phytopharm.* **30,** 37 (1981). GC determn in soil, water, plant material: B. Büttler, *J. Agr. Food Chem.* **31,** 762 (1983).

Yellowish, viscous liquid. $bp_{0.1mm}$ 180°. Vapor pressure at 20°: $< 3 \times 10^{-6}$ mm Hg. Soly in water at 20°: 110 mg/l. Sol in most organic solvents. LD_{50} orally in rats: 1517 mg/kg (Urech).

USE: Agricultural fungicide.

8004. Propineb. *[[(1-Methyl-1,2-ethanediyl)bis[carbamodithioato]](2—)]zinc; [propylenebis(dithiocarbamato)]-zinc;* zinc 1,2-propylene bisdithiocarbamate; mezineb; methylzineb; Bayer 46131; Antracol. $C_5H_8N_2S_4Zn$; mol wt 289.79. C 20.72%, H 2.78%, N 9.67%, S 44.26%, Zn 22.56%. Prepn and use as fungicide: **Ital.** pat. **611,046** (1960 to Montecatini); Lehmann *et al.,* **Belg.** pats. **611,960** and **628,114** (1962 and 1963 to Bayer). Chemistry, biological activity and toxicology: Grewe, *Pflanzenschutz-Nachr.* **20,** 581 (1967). Fungicidal activity: A. G. Channon, *Ann. Appl. Biol.* **65,** 481 (1965); E. J. S. Reddy *et al., Pesticides* **19,** 67 (1985). Distribution and degradation in soil: W. Mittelstaedt, F. Fuhr, *Landwirtsch. Forsch.* **30,** 221 (1977); in fruit: K. Vogeler *et al., Pflanzenschutz-Nachr.* **30,** 72 (1977).

White to yellowish powder, practically odorless. Decomposes and turns brown above 160°. Decomposes in strongly acid or alkaline media. Practically insol in all conventional

solvents. LD_{50} in male rats, rabbits, cats (mg/kg): 8500, > 2500, > 2500 orally (Grewe).

USE: Agricultural fungicide.

8005. β-Propiolactone. *2-Oxetanone; hydracrylic acid β-lactone;* β-propionolactone; propanolide; NSC-21626; Betaprone. $C_3H_4O_2$; mol wt 72.06. C 50.00%, H 5.59%, O 44.40%. Prepd by the condensation of ketene with formaldehyde: Küng, **U.S.** pat. **2,356,459** (1941). Purification: Gresham, Jansen, **U.S.** pat. **2,602,802** (1952 to B. F. Goodrich). Use in Diels-Alder diene synthesis: Gresham *et al., J. Am. Chem. Soc.* **76,** 609 (1954). Review of carcinogenic risk: *IARC Monographs* **4,** 259-269 (1974).

Liquid. d_4^{20} 1.1460; d_4^{25} 1.1420; d_{20}^{20} 1.1490. mp −33.4°. bp_{760} 162° (dec); bp_{750} 150° (dec); bp_{20} 61°; bp_{10} 51°. Flash pt 70° (158°F). n_D^{20} 1.4131; n_D^{25} 1.4110. Dipole moment 3.8. Slowly hydrolyzed to hydracrylic acid. Stable when stored at 5° in glass containers. Soly in water: 37% v/v. Misc with alcohol, acetone, ether, chloroform.

Caution: Potential symptoms of overexposure are skin irritation, blistering and burns; corneal opacity; frequent urination; dysuria; hematuria. See *NIOSH Pocket Guide to Chemical Hazards* (DHHS/NIOSH 90-117, 1990) p 186. This substance may reasonably be anticipated to be a carcinogen: *Seventh Annual Report on Carcinogens* (PB95-109781, 1994) p 342.

USE: Versatile intermediate in organic synthesis.

THERAP CAT: Disinfectant.

8006. Propiolic Acid. *2-Propynoic acid;* acetylenecarboxylic acid; propargylic acid. $C_3H_2O_2$; mol wt 70.05. C 51.44%, H 2.88%, O 45.68%. $CH{\equiv}CCOOH$. Prepn: Wilson, Wenzke, *J. Am. Chem. Soc.* **57,** 1265 (1935); Owen, Sultanbawa, *J. Chem. Soc.* **1949,** 3109; Wolf, *Ber.* **86,** 735 (1953); **87,** 668 (1954). Manuf: Wolf, **U.S.** pat. **2,786,022** (1957); Pachter, **U.S.** pat. **2,799,703** (1957 to Ethyl Corp.). Review on its occurrence and biological activity: Reisch, *Pharmazie* **20,** 194 (1965).

Liquid at room temp (crystals from carbon disulfide, mp 9°). bp 144° (dec); $bp_{10.5}$ 54-55°, bp_{50} 70-75°. d_4^{15} 1.1435, d_4^{20} 1.1380, d_4^{25} 1.1325. $n_D^{20.4}$ 1.4302. Dipole moment at 25°, 2.08D (in dioxane): Wilson, Wenzke, *loc. cit.*

8007. Propiomazine. *1-[10-[2-(Dimethylamino)propyl]-10H-phenothiazin-2-yl]-1-propanone;* 10-dimethylaminoisopropyl-2-propionylphenothiazine; propionylpromethazine; Wy-1359. $C_{20}H_{24}N_2OS$; mol wt 340.49. C 70.55%, H 7.10%, N 8.23%, O 4.70%, S 9.42%. Prepn: Schmitt *et al., Bull. Soc. Chim. France* **1957,** 1474; Schmitt, **Fr.** pat. **addn. 71,342** (addn. to **Fr.** pat. **1,176,919** to Clin-Byla). Comprehensive description of the hydrochloride: K. B. Crombie, L. F. Cullen, in *Analytical Profiles of Drug Substances* vol. **2,** K. Florey, Ed. (Academic Press, New York, 1973) pp 439-466. GC determn in plasma and urine: U. Ahs, G. Wickström, *J. Chromatog.* **183,** 229 (1980). Clinical use as sedative: W. F. Powell *et al., Anesth. Analg.* **49,** 132 (1970); as hypnotic: M. Viukari, P. Miettinen, *Neuropsychobiology* **12,** 134 (1984).

$bp_{0.5}$ 235-245°.

Maleate, $C_{20}H_{24}N_2OS.C_4H_4O_4$, *CB-1678, Indorm, Propavan.* Crystals from isopropanol, mp 160-161°.

Hydrochloride, $C_{20}H_{24}N_2OS.HCl$, *Largon.* Yellow, practically odorless powder. Very soluble in water; freely sol in alcohol. Insol in benzene.

THERAP CAT: Sedative, hypnotic.
THERAP CAT (VET): Tranquilizer.

8008. Propionaldehyde. *Propanal;* methylacetaldehyde; propylaldehyde. C_3H_6O; mol wt 58.08. C 62.04%, H 10.41%, O 27.55%. CH_3CH_2CHO. Prepd by treating propyl alcohol with a bichromate oxidizing mixture: C. D. Hurd, R. N. Meinert, *Org. Syn. coll. vol.* II, 541 (1943); by passing propyl alcohol vapor over copper at high temp: Sabatier, Senderens, *Compt. Rend.* **136**, 923 (1903); Willimott, *Analyst* **50**, 13; *Chem. Zentr.* **1925**, I, 2097. Toxicity study: H. F. Smyth *et al., Arch. Ind. Hyg. Occup. Med.* **4**, 119 (1951).

Liq. Suffocating odor. d_4^0 0.8432; $d_4^{9.7}$ 0.8192; d_4^{20} 0.8071; d_4^{33} 0.7898. mp −81°. bp_{760} 49°; bp_{740} 47°; bp_{687} 45°. $n_D^{16.6}$ 1.3695; n_D^{19} 1.36460. Flash pt, open cup: <20°F (<−6°C). Absorption spectrum: Kwiecinski, Marchlewski, *Bull. Soc. Chim.* [4] **45**, 608 (1929). Sol in 5 vols water at 20°. Misc with alc and ether. LD_{50} orally in rats: 1.4 g/kg; LC for rats in air: 8000 ppm (Smyth).

8009. Propionamide. *Propanamide;* propionic acid amide. C_3H_7NO; mol wt 73.09. C 49.30%, H 9.65%, N 19.16%, O 21.89%. Prepd by heating ammonium propionate under pressure: Hofmann, *Ber.* **15**, 981 (1882); by dropping propionyl chloride into cooled ammonia water: Aschan, *Ber.* **31**, 2347 (1898).

Orthorhombic platelets from benzene. d_4^{20} 1.0335; d_4^{80} 0.9597; mp 79°, bp 222.2°. Volatile with steam. n_D^{110} 1.4160. Freely sol in water, alcohol, ether, chloroform. Lethal concn for rats in air: 8000 ppm.

8010. Propionic Acid. *Propanoic acid;* methylacetic acid; ethylformic acid. $C_3H_6O_2$; mol wt 74.08. C 48.64%, H 8.16%, O 43.20%. CH_3CH_2COOH. Occurs in dairy products in small amounts. Can be obtained from wood pulp waste liquor by a fermentation process using bacteria of the genus *Propionibacterium:* Wayman *et al.,* U.S. pat. 3,067,-107 (1962 to Columbia Cellulose). Prepn from ethylene, carbon monoxide and steam: Reppe, *Angew. Chem.* **1956**, 46; Larson, U.S. pat. 2,448,375 (1948 to du Pont); from ethanol and carbon monoxide using a boron trifluoride catalyst: Loder, U.S. pats. 2,135,448; 2,135,451; 2,135,453 (all 1939 to du Pont); by oxidation of propionaldehyde: Hasche, U.S. pat. 2,294,984 (1942 to Kodak); from natural gas by the Fischer-Tropsch process; as a byproduct in the pyrolysis of wood; by the action of microorganisms on a variety of materials in small yields. Very pure propionic acid can be obtained from propionitrile. Toxicity study: H. F. Smyth *et al., Am. Ind. Hyg. Assoc. J.* **23**, 95 (1962).

Oily liquid. Slightly pungent, disagreeable, rancid odor. d_4^{20} 0.99336. mp −21.5°. bp_{760} 141.1°; bp_{400} 122.0°; bp_{100} 85.8°; $bp_{1.0}$ 4.6°. n_D^{25} 1.3848. Flash pt, open cup: 136°F (58°C). Viscosity (cp) at 15°: 1.175; at 25°: 1.020; at 30°: 0.956; at 60°: 0.668; at 90°: 0.495. Surface tension in dynes/cm at 15°: 27.21. Ka at 25°: 1.34×10^{-5}. Misc with water. Can be salted out of water solns by the addn of $CaCl_2$ or other salts. Sol in alcohol, ether, chloroform. Azeotrope with water, bp 99.98°, contains 17.7% acid; with toluene, bp 110.45°, contains 3% acid; with *o*-xylene, bp 135.4°, contains 43% acid; with ethylbenzene, bp 131.1°, contains 28% acid. LD_{50} orally in rats: 4.29 g/kg (Smyth).

Barium salt monohydrate, $C_6H_{10}BaO_4.H_2O$, *barium propionate.* Powder, usually with slight odor. *Poisonous!* Freely sol in water; slightly sol in alcohol.

USE: Esterifying agent; in the production of cellulose propionate (thermoplastic) and other propionates, e.g., calcium propionate, used as mold inhibitors and preservatives; in the manuf of ester solvents, fruit flavors, and perfume bases.
THERAP CAT: Antifungal.

8011. Propionic Anhydride. *Propanoic acid anhydride;* propanoic anhydride; methylacetic anhydride. $C_6H_{10}O_3$; mol wt 130.14. C 55.37%, H 7.74%, O 36.88%. $(CH_3CH_2CO)_2O$. Obtained by dehydration of the acid or by carbonylation of its esters: Reppe, Friederich, U.S. pat 2,730,546 (1956 to Badische Anilin- & Soda-Fabrik); from propionaldehyde by air oxidation in the presence of cobalt and copper acetate catalysts: McFarlane, U.S. pat. 2,491,572 (1949 to Celanese). Other syntheses, e.g., from ethanol and carbon

monoxide. Toxicity study: H. F. Smyth *et al., Arch. Ind. Hyg. Occup. Med.* **10**, 61 (1954).

Liquid. Odor more pungent than that of the acid. d_4^0 1.0336; d_4^{15} 1.0169; d_4^{20} 1.0125; d_4^{40} 0.98974; d_4^{50} 0.97913. 8.4 lbs/gal at 20°. mp −45°. bp_{760} 167.0°; bp_{400} 146.0°; bp_{200} 127.8°; bp_{100} 107.2°; bp_{60} 94.5°; bp_{40} 85.6°; bp_{20} 70.4°; bp_{10} 57.7°; bp_5 45.3°; $bp_{1.0}$ 20.6°. Flash pt, open cup: 165°F (74°C). n_D^{17} 1.4041; n_D^{20} 1.4038. Viscosity (cp): 1.144 at 20°; 0.978 at 30°; 0.853 at 40°; 0.7511 at 50°. Dec by water. Sol in methanol, ethanol, ether, chloroform. LD_{50} orally in rats: 2.36 g/kg (Smyth).

USE: Esterifying agent for certain perfume oils, fats, oils, and especially cellulose. In the production of alkyd resins, dyestuffs and drugs. Has been used as a dehydrating agent in some sulfonations and nitrations.

8012. Propionitrile. *Propanenitrile;* ethyl cyanide. C_3H_5N; mol wt 55.08. C 65.42%, H 9.15%, N 25.43%. CH_3CH_2CN. May be prepd by dehydration of propionamide (or propionic acid + NH_3) or by distilling ethyl sulfate and concd aq KCN, also by reduction of acrylonitrile. Has been shown to be a duodenal ulcerogen in rats: S. Szabo *et al., Res. Commun. Chem. Pathol. Pharmacol.* **16**, 311 (1977); L. M. Lichtenberger *et al., Gastroenterology* **73**, 1305 (1977). Toxicity study: H. F. Smyth *et al., Arch. Ind. Hyg. Occup. Med.* **4**, 119 (1951).

Liquid. *Poisonous when heated to decompn or on contact with acids.* Pleasant, ethereal, sweetish odor. d_4^0 0.8020; d_4^{20} 0.7818; d_4^{30} 0.7716; d_4^{56} 0.7515; $d_4^{70.2}$ 0.7291. mp −91.8°. bp_{760} 97.2°; bp_{400} 77.7°; bp_{200} 58.2°; bp_{100} 41.4°; bp_{60} 30.1°; bp_{40} 22.0°; bp_{20} +8.8°; bp_{10} −3.0°; bp_5 −13.6°; $bp_{1.0}$ −35.0°. n_D^{15} 1.36812; n_D^{20} 1.36585; n_D^{30} 1.36132. Soly in water at 40° = 11.9 g/100 g H_2O; at 100° = 29 g/100 g H_2O. Misc with alcohol, ether, DMF. LD_{50} orally in rats: 39 mg/kg (Smyth).

8013. Propionyl Chloride. *Propanoyl chloride.* C_3H_5-ClO; mol wt 92.52. C 38.94%, H 5.45%, Cl 38.32%, O 17.29%.

Liquid, pungent odor. d_4^{20} 1.065; mp −94°; bp 80°; n_D^{20} 1.4051. Vigorously dec and dissolved by water or alcohol.

8014. Propionylpromazine. *1-[10-[3-(Dimethylamino)-propyl]-10H-phenothiazin-2-yl]-1-propanone;* 3-propionyl-10-(γ-dimethylaminopropyl)phenothiazine; 10-(3-dimethylaminopropyl)-2-propionylphenothiazine; propiopromazine; 1497-CB. $C_{20}H_{24}N_2OS$; mol wt 340.49. C 70.55%, H 7.10%, N 8.23%, O 4.70%, S 9.42%. Prepn: Schmitt *et al., Bull. Soc. Chim. France* **1957**, 938, 1474; Schmitt, Fr. pat. addn. 71,342 (1956 to Clin-Byla), *C.A.* **56**, 3486h (1962).

Crystals, mp 69-70°.
Hydrochloride, $C_{20}H_{24}N_2OS.HCl$, *Tranvet.*
Maleate, $C_{20}H_{24}N_2OS.C_4H_4O_4$, crystals from acetone, mp 135°.
Methiodide, $C_{20}H_{24}N_2OS.CH_3I$, yellow crystals from isopropanol, mp 79-80°.
THERAP CAT (VET): Tranquilizer.

8015. Propiophenone. *1-Phenyl-1-propanone;* ethyl phenyl ketone; propionylbenzene; phenyl ethyl ketone. C_9-$H_{10}O$; mol wt 134.18. C 80.56%, H 7.51%, O 11.92%. Prepared from propionyl chloride and benzene in the presence of anhydr aluminum chloride: Read, *J. Am. Chem. Soc.* **44**, 1751 (1922).

Leaflets or tabular crystals. Strong, persistent, agreeable, flowery odor. Usually supplied as a liquid. d_4^0 (solid) 1.157; d_4^{20} (liq) 1.0105; d_{20}^{20} 1.0118; d_{25}^{25} 1.0087; $d_4^{41.8}$ 0.9934; $d_4^{61.2}$ 0.9776; $d_4^{85.5}$ 0.9572. mp 21°. Typical freezing point: 18.6°. bp$_{760}$ 218.0°; bp$_{400}$ 194.2°; bp$_{200}$ 170.2°; bp$_{100}$ 149.3°; bp$_{60}$ 135.0°; bp$_{40}$ 124.3°; bp$_{20}$ 107.6°; bp$_{10}$ 92.2°; bp$_5$ 77.9°. $n_D^{15.9}$ 1.5290; n_D^{20} 1.5269. uv max (hexane): 250, 280, 323 nm. Dipole moment: 2.7. Parachor 328.7. Flash pt 99° (210°F). Misc with methanol, anhydr ethanol, ether, benzene, toluene. Insol in water, glycerol, ethylene glycol, propylene glycol.

USE: In perfumery; in the synthesis of ephedrine and related compds.

8016. Propipocaine. *3-(1-Piperidinyl)-1-(4-propoxyphenyl)-1-propanone; 3-piperidino-4'-propoxypropiophenone; β-piperidinoethyl 4-propoxyphenyl ketone; 4-n-propoxy-β-(1-piperidyl)propiophenone; 4-propoxyphenyl piperidine ethyl ketone.* $C_{17}H_{25}NO_2$; mol wt 275.39. C 74.14%, H 9.15%, N 5.09%, O 11.62%. Prepd from *p*-propoxyacetophenone by condensation with formaldehyde and piperidine HCl: Profft, *Chem. Tech. (Leipzig)* **4**, 241 (1952), *C.A.* **47**, 10531 (1953). *See also:* Dyclonine.

Hydrochloride, $C_{17}H_{25}NO_2$.HCl, *falicain, Urocomb.* Crystals, mp 166°. Sol in water, alcohol, acetone. Phenol coefficient 3.6. Aq solns should not be sterilized by autoclaving.

THERAP CAT: Anesthetic (local).

8017. Propiram. *N-[1-Methyl-2-(1-piperidinyl)ethyl]-N-2-pyridinylpropanamide; N-(1-methyl-2-piperidinoethyl)-N-2-pyridylpropionamide; N-propionyl-2-(1-piperidinoisopropyl)aminopyridine; N-propionyl-N-(2-pyridyl)-1-piperidino-2-aminopropane.* $C_{16}H_{25}N_3O$; mol wt 275.39. C 69.78%, H 9.15%, N 15.26%, O 5.81%. Prepn: Hiltmann *et al.*, **Brit. pat. 939,947** corresp to **U.S. pat. 3,163,654;** **Fr. pat. M 2283** (1963, 1964, 1964 all to Bayer). Resolution of isomers: **Neth. pat. Appl. 6,610,362;** Wollweber *et al.*, **U.S. pat. 3,594,477** (1967, 1971 both to Bayer). Pharmacology: Jasinski *et al.*, *Clin. Pharmacol. Ther.* **12**, 613 (1971). Series of articles on chemistry, pharmacology, toxicology, pharmacokinetics, clinical trials: *Arzneimittel-Forsch.* **24**, 581-722 (1974). Toxicity data: D. Tettenborn, *ibid.* 624. Absolute configuration and analgesic activity of enantiomers: W. Wollweber, *Eur. J. Med. Chem.-Chim. Ther.* **17**, 125 (1982).

bp 162-163°.
Fumarate, $C_{20}H_{29}N_3O_5$, *FBA-4503, Bay 4503, Algeril.* LD$_{50}$ in mice, rats (mg/kg): 874, 1657 orally (Tettenborn). *Note:* This is a controlled substance (opiate) listed in the U.S. Code of Federal Regulations, Title 21 Part 1308.11 (1995).

THERAP CAT: Analgesic (narcotic).

8018. Propiverine. *α-Phenyl-α-propoxybenzeneacetic acid 1-methyl-4-piperidinyl ester; α,α-diphenyl-α-propoxyacetic acid 1-methyl-4-piperidyl ester; 1-methyl-4-piperidyl diphenylpropoxyacetate.* $C_{23}H_{29}NO_3$; mol wt 367.49. C

75.17%, H 7.95%, N 3.81%, O 13.06%. Prepn: C. Starke *et al.*, **E. Ger. pat. 106,643** (1974), *C.A.* **82**, 155841s (1975). GC-MS determn in plasma and urine: T. Marunaka *et al.*, *J. Chromatog.* **420**, 43 (1987). Pharmacology: A. Haruno, *Arzneimittel-Forsch.* **42**, 815, 1459 (1992). Acute toxicity: K. Yamashita *et al.*, *J. Toxicol. Sci.* **14**, Suppl. 2, 1 (1989), *C.A.* **112**, 172060 (1990). Clinical pharmacokinetics: C. Müller *et al.*, *Eur. J. Drug Metab. Pharmacokinet.* **18**, 265 (1993). Clinical trial in urinary incontinence: D. Mazur *et al.*, *Urologe A*, **33**, 447 (1994).

Hydrochloride, $C_{23}H_{29}NO_3$.HCl, *P-4, Mictonorm, Mictonetten.* mp 216-218°. Sparingly sol in water and ethanol. LD$_{50}$ in male, female mice, male, female rats (mg/kg): 36, 36, 22, 25 i.v.; 223, 283, 1632, 1411 s.c.; 410, 323, 1000, 1092 orally (Yamashita).

THERAP CAT: Anticholinergic; in treatment of urinary incontinence.

8019. Propizepine. *6-[2-(Dimethylamino)propyl]-1,6-dihydro-5H-pyrido[2,3-b][1,5]benzodiazepin-5-one; 6,11-dihydro-6-[2-(dimethylamino)-2-methylethyl]-5H-pyrido[2,3-b]-[1,5]benzodiazepin-5-one.* $C_{17}H_{20}N_4O$; mol wt 296.37. C 68.90%, H 6.80%, N 18.90%, O 5.40%. Prepn: **Neth. pat. Appl. 6,600,065** (1966 to Labs. U.P.S.A.); Hoffmann, Faure, *Bull. Soc. Chim. France* **1966**, 2316. Psychopharmacology: Lwoff *et al.*, *Thérapie* **26**, 451 (1971).

mp 122°.
Hydrochloride, $C_{17}H_{20}N_4O$.HCl, *UP-106, Vagran.* mp 235°.

THERAP CAT: Antidepressant.

8020. Propofol. *2,6-Bis(1-methylethyl)phenol; 2,6-diisopropylphenol;* disoprofol; ICI-35868; Diprivan; Disoprivan; Rapinovet. $C_{12}H_{18}O$; mol wt 178.27. C 80.85%, H 10.18%, O 8.97%. Prepn: A. J. Kolka *et al.*, *J. Org. Chem.* **21**, 712 (1956); **22**, 642 (1957); G. G. Ecke, A. J. Kolka, **U.S. pat. 2,831,898** (1958 to Ethyl Corp.); T. J. Kealy, D. D. Coffman, *J. Org. Chem.* **26**, 987 (1961); B. E. Firth, T. J. Rosen, **U.S. pat. 4,447,657** (1984 to Universal Oil Products). Chromatographic study: J. K. Carlton, W. C. Bradbury, *J. Am. Chem. Soc.* **78**, 1069 (1956). Animal studies: J. B. Glen, *Brit. J. Anaesth.* **52**, 731 (1980). Pharmacokinetics: H. K. Adam *et al.*, *ibid.* 743; *idem, ibid.* **55**, 97 (1983). Determn in blood: *eidem, J. Chromatog.* **223**, 232 (1981). Comparative studies vs other injectable anesthetics: B. Kay, D. K. Stephenson, *Anaesthesia* **35**, 1182 (1980); D. V. Rutter *et al.*, *ibid.* 1188. Use in i.v. anesthesia: E. Major *et al.*, *ibid.* **37**, 541 (1982). Cardiovascular effects: D. Al-Khudhairi *et al.*, *ibid.* 1007. Pharmacology of emulsion formulation: J. B. Glen, S. C. Hunter, *Brit. J. Anaesth.* **56**, 617 (1984). Series of articles on pharmacology and clinical experience: *Postgrad. Med. J.* **61**, Suppl. 3, 1-169 (1985).

bp$_{30}$ 136°. bp$_{17}$ 126°. mp 19°. n_D^{20} 1.5134. n_D^{25} 1.5111. d$_{20}$ 0.955.

THERAP CAT: Anesthetic (intravenous).

THERAP CAT (VET): Intravenous anesthetic (dogs and cats).

8021. Propolis. Bee bread; hive dross. A resinous substance found in beehives. Collected by bees from buds. Isoln of caffeic acid from propolis: Cizmárik, Matel, *Experientia* **26**, 713 (1970). Antimicrobial constituents of propolis: J. Metzner *et al.*, *Pharmazie* **30**, 799 (1975); E. M. Schneidewind *et al.*, *ibid.* 803. Review on the origin, chemical constituents and therapeutic activity: M. H. Haydak, *State of Iowa*, *Repts. State Apiarist 1953*, p 74-87; M. Vanhaelen, R.Vanhaelen-Fastre, *J. Pharm. Belge* **34**, 253 (1979). Greenish-brown, sticky mass. Aromatic odor. d 1.2. mp 64°. Becomes brittle when cooled below 15°. Extraction with alcohol gives *propolis wax*. The residue from the alcohol extraction is called *propolis resin*, yielding *propolis balsam* on extraction with hot petr ether. Propolis balsam has a hyacinth odor and is said to contain 10% cinnamyl alcohol.

8022. Propoxur. 2-(1-Methylethoxy)phenol methylcarbamate; o-isopropoxyphenyl N-methylcarbamate; aprocarb; Bay 39007; Bay 9010; Baygon; Difen; Blattanex; Propyon; Suncide; Unden. C$_{11}$H$_{15}$NO$_3$; mol wt 209.25. C 63.14%, H 7.23%, N 6.69%, O 22.94%. Prepn: U.S. pat. 3,111,539 (1963 to Bayer; Chemagro Corp.). Properties: *Pflanzenschutz Nachr. Bayer* **18**, 53 (1965). Toxicity data: T. B. Gaines, *Toxicol. Appl. Pharmacol.* **14**, 515 (1969). Efficacy vs fleas and ticks in domestic animal collars: R. G. Hughes, *Vet. Med. Rev.* **1**, 80 (1985). Teratogenicity study: K. D. Courtney *et al.*, *J. Environ. Sci. Health* **B20**, 373 (1985).

Minute crystals, mp 91.5°. Dec at high temp forming methyl isocyanate. Sol in methanol, acetone and many organic solvents, but only slightly sol in cold hydrocarbons. Water soly about 0.2% at 20°. Unstable in highly alkaline media. LD$_{50}$ orally in male, female rats: 83, 86 mg/kg (Gaines).

USE: Insecticide.

8023. Propoxycaine Hydrochloride. 4-Amino-2-propoxybenzoic acid 2-diethylaminoethyl ester hydrochloride; 2-diethylaminoethyl 4-amino-2-propoxybenzoate hydrochloride; 2-diethylaminoethyl 2-propoxy-4-aminobenzoate hydrochloride; Ravocaine Hydrochloride; Blockain Hydrochloride. C$_{16}$H$_{27}$ClN$_2$O$_3$; mol wt 330.85. C 58.08%, H 8.23%, Cl 10.72%, N 8.47%, O 14.51%. Prepn: Clinton, Laskowski, U.S. pat. 2,689,248 (1954 to Sterling Drug).

White, odorless crystals, mp 148-150°. Discolors upon prolonged exposure to light and to air. Freely sol in water;

sol in ethanol, chloroform. Sparingly sol in ether. Practically insol in acetone, chloroform. pH of a 2% aq soln 5.4.

THERAP CAT: Anesthetic (local).

8024. Propoxyphene. [S-(R*,S*)]-α-[2-(Dimethylamino)-1-methylethyl]-α-phenylbenzeneethanol propanoate (ester); α-d-4-dimethylamino-3-methyl-1,2-diphenyl-2-butanol propionate; (+)-1,2-diphenyl-2-propionoxy-3-methyl-4-dimethylaminobutane; (+)-4-dimethylamino-1,2-diphenyl-3-methyl-2-propionyloxybutane; α-d-propoxyphene; dextropropoxyphene. C$_{22}$H$_{29}$NO$_2$; mol wt 339.48. C 77.84%, H 8.61%, N 4.13%, O 9.43%. Prepn of racemate: Pohland, Sullivan, *J. Am. Chem. Soc.* **75**, 4458 (1953); Pohland, U.S. pat. 2,728,779 (1955 to Lilly). Prepn of (+)-form: Pohland, Sullivan, *J. Am. Chem. Soc.* **77**, 3400 (1955). Stereochemistry: Sullivan *et al.*, *J. Org. Chem.* **28**, 2381 (1963); Casy, Myers, *J. Pharm. Pharmacol.* **16**, 455 (1964). Stereospecific synthesis: Pohland *et al.*, *J. Org. Chem.* **28**, 2483 (1963). Metabolism: S. L. Due *et al.*, *Biomed. Mass Spectrom.* **3**, 217 (1976). The α-dl- and d-diastereoisomers possess marked analgesic activity in contrast to the β-diastereoisomers which are substantially inactive. Toxicity: E. I. Goldenthal, *Toxicol. Appl. Pharmacol.* **18**, 185 (1971); J. L. Emerson *et al.*, *ibid.* **19**, 445 (1971). Comprehensive description: B. McEwan, *Anal. Profiles Drug Subs.* **1**, 301-318 (1972). Symposium on pharmacology, toxicology, and clinical efficacy of propoxyphene alone and in combination with acetaminophen: *Human Toxicol.* **3**, Suppl., 1S-238S (1984).

Crystals from petr ether, mp 75-76°. [α]$_D^{25}$ +67.3° (c = 0.6 in chloroform).

Hydrochloride, C$_{22}$H$_{29}$NO$_2$.HCl, *Antalvic, Darvon, Deprancol, Develin, Dolene, Dolocap, Erantin, Femadol, Harmar, Propoxychel, Proxagesic.* Bitter crystals from methanol + ethyl acetate, mp 163-168.5°. [α]$_D^{25}$ +59.8° (c = 0.6 in water). Sol in water, alc, chloroform, acetone. Practically insol in benzene, ether. LD$_{50}$ in mice, rats (mg/kg): 28, 15 i.v.; 111, 58 i.p.; 211, 134 s.c.; 282, 230 orally (Emerson).

Napsylate monohydrate, C$_{22}$H$_{29}$NO$_2$.C$_{10}$H$_8$O$_3$S.H$_2$O, *Darvon-N, Doloxene.* LD$_{50}$ orally in female rats: 990 mg/kg (Goldenthal).

α-l-Form, *see* Levopropoxyphene.

α-dl-Form, *racemic propoxyphene, diméprotane.*

α-dl-Form hydrochloride, crystals from methanol + ethyl acetate, mp 170-171°. Soluble in water, alcohol, chloroform. Practically insol in benzene, ether.

β-dl-Form, crystals from acetone + ether. mp 187-188°. More soluble than the α-form.

Note: Bulk dextropropoxyphene (non-dosage forms) is a controlled substance (opiate) listed in the U.S. Code of Federal Regulations, Title 21 Part 1308.12 (1995); dextropropoxyphene is a controlled substance (narcotic), *ibid.*, Title 21 Part 1308.14.

THERAP CAT: Analgesic (narcotic).

8025. Propranolol. 1-[(1-Methylethyl)amino]-3-(1-naphthalenyloxy)-2-propanol; 1-(isopropylamino)-3-(1-naphthyloxy)-2-propanol; Avlocardyl; Euprovasin. C$_{16}$H$_{21}$NO$_2$; mol wt 259.35. C 74.10%, H 8.16%, N 5.40%, O 12.34%. β-Adrenergic blocker. Prepn: Belg. pats. 640,312 and 640,-313; Crowther, Smith, U.S. pats. 3,337,628 and 3,520,919 (1964, 1964, 1967, 1970 all to I.C.I.). Description of optical isomers: Howe, Shanks, *Nature* **210**, 1336 (1966). Biological studies: Barrett, Cullum, *Brit. J. Pharmacol.* **34**, 43 (1968). Metabolism: Bond, *Nature* **213**, 721 (1967). Multicenter clinical trial in myocardial infarction: V. Hansteen *et al.*, *Brit. Med. J.* **284**, 155 (1982). Use as migraine prophylactic: S. Diamond, E. Millstein, *J. Clin. Pharmacol.* **28**, 193

(1988); S. Diamond *et al.*, *Headache* **27**, 70 (1987). Toxicity data: M. Martin, P. Linee, *Eur. J. Med. Chem.* **9**, 563 (1974). Review of pharmacokinetics: P. A. Routledge, D. G. Shand, *Appl. Pharmacokinet.* **1980**, 464-485; of pharmacology: J. D. Fitzgerald in *Pharmacology of Antihypertensive Drugs,* A. Scriabine, Ed. (Raven Press, New York, 1980) pp 195-208. Series of articles on use in hypertension: *Drugs* **37**, Suppl. 2, 42-76 (1989).

Crystals from cyclohexane, mp 96°.

Hydrochloride, $C_{16}H_{21}NO_2.HCl$, *AY-64043, ICI-45520, NSC-91523, Angilol, Apsolol, Bedranol, Beprane, Berkolol, Beta-Neg, Beta-Tablinen, Beta-Timelets, Cardinol, Caridolol, Deralin, Dociton, Duranol, Efektolol, Elbrol, Frekven, Inderal, Indobloc, Intermigran, Kemi S, Oposim, Prano-Puren, Prophylux, Propranur, Pylapron, Rapynogen, Sagittol, Servanolol, Sloprolol, Sumial, Tesnol.* Crystals from *n*-propanol, mp 163-164°. Sol in water, alc. Practically insol in ether, benzene, ethyl acetate. LD_{50} in mice (mg/kg): 565 orally; 22 i.v.; 107 i.p. (Martin, Linee).

THERAP CAT: Antihypertensive; antianginal; antiarrhythmic (class II).

8026. Propyl Acetate. *Acetic acid n-propyl ester.* $C_5H_{10}O_2$; mol wt 102.13. C 58.80%, H 9.87%, O 31.33%. $CH_3COOCH_2CH_2CH_3$. Prepn from acetic acid and *n*-propyl alcohol: Wagner, *J. Chem. Ed.* **27**, 245 (1950). Manuf from acetic acid and mixture of propene + propane in the presence of $ZnCl_2$ catalyst: Biller, **Brit.** pat. **872,876** (1961). Toxicity study: P. M. Jenner *et al.*, *Food Cosmet. Toxicol.* **2**, 327 (1964).

Liquid, odor of pears, bp 101.6°. mp −92°. d_4^{20} 0.836, d_{20}^{20} 0.887. n_D^{20} 1.3844. Flash pt, closed cup: 58°F (14°C). Soly in water at 16°: 1.6:100. Misc with alcohol, ether. LD_{50} in rats, mice (mg/kg): 9370, 8300 orally (Jenner).

Caution: Potential symptoms of overexposure are irritation of eyes, nose and throat; narcosis; dermatitis. *See NIOSH Pocket Guide to Chemical Hazards* (DHHS/NIOSH 90-117, 1990) p 188.

USE: Mfg flavors, perfumes. Solvent for resins, cellulose derivatives, plastics.

8027. n-Propyl Alcohol. *1-Propanol;* propylic alcohol; Optal. C_3H_8O; mol wt 60.10. C 59.96%, H 13.42%, O 26.62%. $CH_3CH_2CH_2OH$. Discovered by Chancel in 1853 in crude fusel oil and obtained therefrom by fractionation; available as a byproduct of the reaction between carbon monoxide and hydrogen. Toxicity study: Smyth *et al.*, *Arch. Ind. Hyg. Occup. Med.* **10**, 61 (1954).

Liquid; alcoholic and slightly stupefying odor. mp −127°. bp 97.2°. d_4^{20} 0.8053; d_4^{25} 0.8016: Mumford, Phillips, *J. Chem. Soc.* **1950**, 75. Flash pt 22°. n_D^{20} 1.3862. Miscible with water, alcohol, ether. LD_{50} orally in rats: 1.87 g/kg (Smyth).

Caution: Potential symptoms of overexposure are mild irritation of eyes, nose and throat; dry cracking skin; drowsiness, headache; ataxia, gastrointestinal pain; abdominable cramps; nausea, vomiting and diarrhea. *See NIOSH Pocket Guide to Chemical Hazards* (DHHS/ NIOSH 90-117, 1990) p 188.

USE: As a solvent for resins and cellulose esters, etc.

8028. Propylamine. *1-Propanamine;* 1-aminopropane. C_3H_9N; mol wt 59.11. C 60.96%, H 15.35%, N 23.70%. $CH_3CH_2CH_2NH_2$. Prepn from propionaldehyde + ammonia with a Raney nickel catalyst: Olin, Schwoegler, U.S. pat. **2,373,705** (1945 to Sharples); by low pressure catalytic hydrogenation of nitropropane: Iffland, Cassis, *J. Am. Chem. Soc.* **74**, 6284 (1952); by catalytic hydrogenation of propionitrile: Rylander, Kaplan, U.S. pat. **3,117,162** (1964 to Englehard). Toxicity study: H. F. Smyth *et al.*, *Am. Ind. Hyg. Assoc. J.* **23**, 95 (1962).

Colorless, alkaline liq; strong ammonia odor. d_{20}^{20} 0.719; mp −83°; bp 48-49°; n_D^{20} 1.389. Flash pt, closed cup: 10°F (−12°C). Misc with water, alcohol, ether. *Keep tightly closed.* LD_{50} orally in rats: 0.57 g/kg (Smyth).

Hydrochloride, $C_3H_9N.HCl$, deliquesc crystals, mp 157-158°. Soluble in 0.4 part water, 1.5 parts chloroform. *Keep well closed.*

Caution: Strong irritant, possible skin sensitizer.

8029. n-Propylbenzene. 1-Phenylpropane. C_9H_{12}; mol wt 120.19. C 89.94%, H 10.06%. $C_6H_5CH_2CH_2CH_3$. Prepd by the action of diethyl sulfate on benzylmagnesium chloride: H. Gilman, W. E. Catlin, *Org. Syn.* coll. vol. I, 471 (2nd ed., 1941). Toxicity study: P. M. Jenner *et al.*, *Food Cosmet. Toxicol.* **2**, 327 (1964).

Liquid. d_4^{20} 0.8621. mp −99.2°. bp_{760} 159.2°; bp_{400} 135.7°; bp_{200} 113.5°; bp_{100} 94°; bp_{40} 71.6°; bp_{20} 56.8°; bp_{10} 43.4°; bp_5 31.3°; $bp_{1.0}$ 6.3°. n_D^{20} 1.4919. Very slightly sol in water (0.06 g/l); sol in alcohol, ether. LD_{50} orally in rats: 6040 mg/kg (Jenner).

USE: In textile dyeing and printing; as solvent for cellulose acetate.

8030. Propyl Bromide. *1-Bromopropane.* C_3H_7Br; mol wt 122.99. C 29.30%, H 5.74%, Br 64.97%. $CH_3CH_2CH_2Br$.

Colorless liquid. d_{20}^{20} 1.353; mp −110°; bp 71°; n_D^{20} 1.4341. Sol in 400 parts water; miscible with alcohol, etc.

8031. Propyl Butyrate. *Butanoic acid propyl ester.* $C_7H_{14}O_2$; mol wt 130.19. C 64.58%, H 10.84%, O 24.58%. $CH_3CH_2CH_2COOC_3H_7$.

Colorless liquid. d_4^{15} 0.879; mp −95°; bp 143°; n_D^{20} 1.4005. Slightly sol in water; misc with alcohol, ether. LD_{50} orally in rats: 15,000 mg/kg, P. M. Jenner *et al.*, *Food Cosmet. Toxicol.* **2**, 327 (1964).

8032. Propyl Chloride. *1-Chloropropane.* C_3H_7Cl; mol wt 78.54. C 45.88%, H 8.98%, Cl 45.14%. $CH_3CH_2CH_2Cl$. Prepd from propyl alc and PCl_3 in the presence of $ZnCl_2$.

Colorless liquid. d_{20}^{20} 0.890; mp −122° to −123°; bp 46-47°; n_D^{20} 1.3886. Sol in ~300 parts water; miscible with alcohol, ether.

8033. Propyl Chlorocarbonate. *Carbonochloridic acid propyl ester;* propyl chloroformate. $C_4H_7ClO_2$; mol wt 122.55. C 39.20%, H 5.76%, Cl 28.93%, O 26.11%. C_3H_7O-$COCl$.

Colorless liquid. d^{20} 1.090; bp 114-116°. Gradually dec by water or alc; miscible with benzene, chloroform, ether.

Caution: Vapors strongly irritating to eyes and mucous membranes.

8034. Propylene. *1-Propene;* methylethylene; methylethene. C_3H_6; mol wt 42.08. C 85.63%, H 14.37%. $H_2C=CHCH_3$. Obtained from petr oils during the refining of gasoline. Catalytic or thermal cracking of hydrocarbons always yields propylene. Can be obtained by catalytic dehydrogenation of propane. Chronic toxicity studies: J. A. Quest *et al.*, *Toxicol. Appl. Pharmacol.* **76**, 288 (1984). Reviews: R. F. Goldstein, *The Petroleum Chemicals Industry* (New York-London, 1949) p 114 sqq.; Sherwood, *Ind. Chemist* **1960**, 542-546; *Chim. et Ind.* **1961**, 576-587; Haney, "Ethylene, Propylene and 1-Butene" in *Vinyl and Diene Monomers,* E. C. Leonard, Ed. (Interscience, New York, 1971) pp 577-689; M. R. Schoenberg *et al.*, in Kirk-Othmer *Encyclopedia of Chemical Technology* vol. **19** (Wiley-Interscience, New York, 3rd ed., 1982) pp 228-246. Review of carcinogenic risk: *IARC Monographs* **60**, 161-180 (1994).

Flammable gas. Burns with yellow sooty flame. d 1.49 (air = 1.0). mp (triple pt) −185°. bp_{760} −48°. Critical temp 91.8°. Critical pressure 45.6 atm. Heat of fusion 717.6 cal/mol. Liquefies at 7-8 atm. d_4^{20} (liq) 0.5139. Flammable limits in air: 2.4-10.3% (by volume). Latent heat of vaporization at bp: 104.62 cal/g. Dipole moment 0.35. n_D^{-40} 1.3567. Surface tension at 90°: 16.70 dynes/cm. Shipped as a liquefied gas in low pressure steel cylinders under its own vapor pressure of ~136 pounds per square inch. Contaminants are propane, ethane, carbon dioxide.

Caution: Gas may act as a simple asphyxiant and mild anesthetic. Direct contact with liquid may cause skin burns. *See Patty's Industrial Hygiene and Toxicology* Vol.

2B, G. D. Clayton, F. E. Clayton, Eds. (John Wiley & Sons, Inc., New York, 4th ed., 1994) p 1244-1245.

USE: In polymerized form as polypropylene for plastics and carpet fibers. Chemical intermediate in the manuf of acetone, isopropylbenzene, isopropanol, isopropyl halides, propylene oxide, acrylonitrile, cumene.

8035. Propylene Chlorohydrin. *2-Chloro-1-propanol;* 2-chloropropyl alcohol. C_3H_7ClO; mol wt 94.54. C 38.11%, H 7.46%, Cl 37.50%, O 16.92%. $CH_3CHClCH_2OH$. Colorless liquid; pleasant odor. d^{20} 1.103; bp 133-134°; n_D^{20} 1.4362. Sol in water, alcohol, etc. LD_{50} orally in rats: 0.22 ml/kg; by skin penetration in rabbits: 0.48 ml/kg, Smyth *et al., Am. Ind. Hyg. Assoc. J.* **30,** 470 (1969).

USE: In prepn of propylene oxide *(q.v.).*

8036. sec-Propylene Chlorohydrin. *1-Chloro-2-propanol;* 1-chloroisopropyl alcohol. C_3H_7ClO; mol wt 94.54. C 38.11%, H 7.46%, Cl 37.50%, O 16.92%. $ClCH_2CH(OH)CH_3$. Colorless liquid. d^{20} 1.115; bp 126-127°; n_D^{20} 1.4392. Sol in water, alcohol, etc.

8037. Propylenediamine. *1,2-Propanediamine.* $C_3H_{10}N_2$; mol wt 74.13. C 48.61%, H 13.60%, N 37.79%. $CH_3CH(NH_2)CH_2NH_2$. Prepd from propylene dibromide and alcoholic ammonia at 100°.
Extremely hygroscopic, strongly alkaline liq. Rapidly absorbs moisture to form a hemihydrate. d^{15} 0.878 in anhydr form. bp 119-120°. Very sol in water. *Keep tightly closed.*

USE: In conjunction with cupric sulfate it is a very sensitive reagent for mercury.

8038. Propylene Dibromide. *1,2-Dibromopropane.* $C_3H_6Br_2$; mol wt 201.89. C 17.85%, H 3.00%, Br 79.16%. $CH_3CHBrCH_2Br$. Prepd from propyl bromide and Br in the presence of $AlCl_3$ or $AlBr_3$.
Colorless liquid. mp −55°; bp 140-142°; n_D^{20} 1.5203; d^{20} 1.933. Slightly sol in water; miscible with organic solvents.

8039. Propylene Dichloride. *1,2-Dichloropropane.* $C_3H_6Cl_2$; mol wt 112.99. C 31.89%, H 5.35%, Cl 62.76%. $CH_3CHClCH_2Cl$. Prepd from propyl chloride and Sb_2Cl_5. Toxicity data: H. F. Smyth *et al., Am. Ind. Hyg. Assoc. J.* **30,** 470 (1969).
Flammable, mobile liq. Odor of chloroform. d_{25}^{25} 1.159; bp 95-96°. Solidifies below −70°. n_D^{20} 1.4388. Flash point (ASTM open cup) 21° (70°F). Despite the low flash pt it does not catch fire readily in industrial applications. Fire pt 38°. Slightly sol in water; miscible with organic solvents. LD_{50} orally in rats: 1.19 ml/kg (Smyth).
Caution: Potential symptoms of overexposure are eye and skin irritation; drowsiness, lightheadedness. Potential occupational carcinogen. *See NIOSH Pocket Guide to Chemical Hazards* (DHHS/NIOSH 90-117, 1990) p 188.

USE: Oil and fat solvent; in dry cleaning fluids; in degreasing. In insecticidal fumigant mixtures.

8040. Propylene Glycol. *1,2-Propanediol;* methyl glycol; 1,2-dihydroxypropane. $C_3H_8O_2$; mol wt 76.10. C 47.35%, H 10.60%, O 42.05%. $CH_3CHOHCH_2OH$. Prepn from glycerol: Raschig, Prahl, *Ber.* **61,** 185 (1928). Prepn of levorotatory propylene glycol from hydroxyacetone by yeast reduction: Levene, Walti, *Org. Syn. coll. vol.* II, 545 (1943). Synthesis of S-(+)-form: C. Melchiorre, *Chem. Ind. (London)* **1976,** 218. Manuf from propylene oxide by hydration: Faith, Keyes & Clark's *Industrial Chemicals,* F. A. Lowenheim, M. K. Moran, Eds. (Wiley-Interscience, New York, 4th ed., 1975) pp 688-691. Taken internally, propylene glycol is harmless, probably because its oxidation yields pyruvic and acetic acids, *cf.* Whitmore, *Organic Chemistry* (New York, 1951). Toxicity data: W. Bartsch *et al., Arzneimittel-Forsch.* **26,** 1581 (1976). Review on toxicity, metabolism and biochemistry: Ruddick, *Toxicol. Appl. Pharmacol.* **21,** 102 (1972).
dl-Form, hygroscopic, viscous liquid. Slightly acrid taste. d_4^{25} 1.036. mp −59°. bp$_{760}$ 188.2°; bp$_{400}$ 168.1°; bp$_{200}$ 149.7°; bp$_{100}$ 132.0°; bp$_{60}$ 119.9°; bp$_{40}$ 111.2°; bp$_{20}$ 96.4°; bp$_{10}$ 83.2°; bp$_5$ 70.8°; bp$_{1.0}$ 45.5°. Flash pt, open cup: 210°F (99°C). Miscible with water, acetone, chloroform. Sol in ether. Will dissolve many essential oils, but is immiscible with fixed oils. It is a good solvent for rosin. Under ordinary conditions propylene glycol is stable, but at high temps it tends to oxidize giving rise to products such as propionaldehyde, lactic acid, pyruvic acid and acetic acid. LD_{50} orally in rats: 25 ml/kg (Bartsch).
l-Form, bp$_{12}$ 88-90°, bp$_{760}$ 187-189°. [α]$_D^{20}$ −15.0°.
d-Form, bp$_{14}$ 94-96°. [α]$_D^{20}$ +15.84% (neat). d^{25} 1.04.

USE: As nontoxic antifreeze in breweries and dairy establishments. Substitute for ethylene glycol and glycerol. In the manuf of synthetic resins. As inhibitor of fermentation and mold growth. As mist to disinfect air. As emulsifier in foods. Pharmaceutic aid (humectant; solvent).

THERAP CAT (VET): Glucogenic (orally) in ruminants.

8041. Propylene Oxide. *Methyloxirane;* propene oxide. C_3H_6O; mol wt 58.08. C 62.04%, H 10.41%, O 27.55%. Results from the action of KOH (aq) on propylene chlorohydrin. Toxicity: H. F. Smyth *et al., J. Ind. Hyg. Toxicol.* **23,** 259 (1941). Toxicological comparison with ethylene oxide, *q.v.:* E. Agurell *et al., Mutat. Res.* **250,** 229 (1991). *Reviews:* Holden in *Glycols,* G. O. Curme, F. Johnston, Eds., A.C.S. Monograph Series no. **114** (Reinhold, New York, 1952) pp 250-261; Faith, Keyes & Clark's *Industrial Chemicals,* F. A. Lowenheim, M. K. Moran, Eds. (Wiley-Interscience, New York, 4th ed., 1975) pp 692-697; R. O. Kirk, T. J. Dempsey in Kirk-Othmer *Encyclopedia of Chemical Technology* vol. **19** (Wiley-Interscience, New York, 3rd ed., 1982) pp 246-274. Review of manufacturing processes: *Chem. & Eng. News* **70,** 9-12 (Mar. 2, 1992); of carcinogenic risk: *IARC Monographs* **60,** 181-213 (1994).

Colorless ethereal liquid. *Extremely flammable.* d_4^0 0.859. mp −112.13°. bp 34.23°. Flash pt, closed cup: −31°F (−35°C). Soly in water (20°): 40.5% by wt; soly of water in propylene oxide: 12.8% by wt; miscible with alcohol, ether. LD_{50} orally in rats: 1.14 g/kg (Smyth).
Caution: Potential symptoms of overexposure are irritation of eyes, upper respiratory system and lungs; skin irritation, blistering and burns. *See NIOSH Pocket Guide to Chemical Hazards* (DHHS/ NIOSH 90-117, 1990) p 190. This substance may reasonably be anticipated to be a carcinogen: *Seventh Annual Report on Carcinogens* (PB95-109781, 1994) p 344.

USE: Chemical intermediate in prepn of polyethers to form polyurethanes; in prepn of urethane polyols and propylene and dipropylene glycols; in prepn of lubricants, surfactants, oil demulsifiers. Also as solvent; fumigant; soil sterilant.

8042. Propyl Ether. *1,1'-Oxybispropane;* dipropyl ether. $C_6H_{14}O$; mol wt 102.18. C 70.53%, H 13.81%, O 15.66%. $C_3H_7OC_3H_7$. Obtained by heating propyl alcohol with benzenesulfonic acid.
Mobile liquid. *Extremely flammable.* d^{20} 0.7360; mp −122°; bp 89-91°; n_D^{20} 1.3807. Flash pt, open cup: −5°F (−20°C). Slightly sol in water; sol in alcohol, ether. Highly volatile. Tends to form explosive peroxides, esp when anhyd. Do not allow to evaporate to near dryness.

8043. Propyl Formate. *Formic acid propyl ester.* $C_4H_8O_2$; mol wt 88.11. C 54.53%, H 9.15%, O 36.32%. HCOO-C_3H_7.
Colorless liquid; pleasant odor. d^{20} 0.901; mp −93°; bp 81-82°. Flash pt, closed cup: 27°F (−3°C). n_D^{20} 1.3771. Sol in 45 parts water; misc with alcohol, ether. LD_{50} orally in rats: 3980 mg/kg, P. M. Jenner *et al., Food Cosmet. Toxicol.* **2,** 327 (1964).

8044. Propyl Gallate. *3,4,5-Trihydroxybenzoic acid propyl ester; n-propyl gallate;* gallic acid propyl ester; PG; Progallin P; Tenox PG. $C_{10}H_{12}O_5$; mol wt 212.20. C 56.60%, H 5.70%, O 37.70%. Spectrophotometric determn: C. S. Sastry *et al., Talanta* **29,** 917 (1982). Effects on survival of *Saccharomyces cerevisiae:* V. L. Eubanks, L. R. Beuchat, *J. Food Prot.* **46,** 29 (1983). Antioxidant effectiveness: M. A. Augustin, S. K. Berry, *J. Am. Oil Chem. Soc.* **60,** 105 (1983). Comprehensive review of biological effects and toxicology: *J. Am. Coll. Toxicol.* **4,** 23-64 (1985).

Crystals, mp 150°. Soly at 25° in water = 0.35 g/100 ml; in alcohol = 103 g/100 g; in ether = 83 g/100 g. Soly in cottonseed oil at 30° = 1.23 g/100 g; in lard at 45° = 1.14 g/100 g. Darkens in the presence of iron and iron salts. Synergic with acids, BHA, BHT. pKa 8.11. Partition coefficient (oleyl alcohol:water) 17. Partition coefficient (octanol:water) 32. pH (0.05% aq soln) 6.3; (0.1% aq soln) 5.9; (0.2% aq soln) 5.7. LD_{50} in mice, rats, hamsters, rabbits (g/kg): 1.70-3.50, 2.1-7, 2.48, 2.75 orally; LD_{50} i.p. in rats: 0.38 g/kg (J. Am. Coll. Toxicol.).

USE: Antioxidant for cosmetics, foods, fats, oils, ethers, emulsions, waxes, transformer oils.

8045. Propylhexedrine. *N,α-Dimethylcyclohexaneethanamine;* 1-cyclohexyl-2-methylaminopropane; hexahydrodesoxyephedrine; Benzedrex. $C_{10}H_{21}N$; mol wt 155.28. C 77.35%, H 13.63%, N 9.02%. Prepd by catalytic hydrogenation of the phenyl analog using Adams platinum catalyst and glacial acetic acid as the solvent: Zenitz *et al., J. Am. Chem. Soc.* **69**, 1117 (1947); Ullyot, U.S. pat. **2,454,746** (1948 to SK & F).

dl-Form, oily liq, d_4^{25} 0.8501, amine odor, bp_{760} 205°; bp_{20} 92-93°; volatilizes slowly at room temp. n_D^{20} 1.4600. Very slightly sol in water. Absorbs carbon dioxide from air and its aq solns are alkaline to litmus. Miscible with alcohol, chloroform, ether.

dl-Form hydrochloride, $C_{10}H_{21}N.HCl$, crystals, dec 127-128°. Sol in water.

d-Form, oily liq, bp_{10} 82-83°; n_D^{20} 1.4588.

d-Form hydrochloride, crystals, dec 138-139°; $[\alpha]_D^{26}$ +14.73°. Sol in water.

l-Form, oily liq, bp_9 80-81°. n_D^{20} 1.4590.

l-Form ethylphenylbarbiturate, $C_{22}H_{33}N_3O_3$, *l-1-cyclohexyl-2-(methylamino)propane ethylphenylbarbiturate, phenobarbital compd with 1-propylhexedrine, barbexaclone, Maliasin.*

l-Form hydrochloride, crystals, dec 138-139°; $[\alpha]_D^{26}$ −14.74°. Sol in water.

THERAP CAT: Adrenergic (vasoconstrictor); decongestant (nasal).

8046. Propylidene Chloride. *1,1-Dichloropropane.* $C_3H_6Cl_2$; mol wt 112.99. C 31.89%, H 5.35%, Cl 62.76%. $CH_3CH_2CHCl_2$. Obtained by the action of PCl_5 on propionaldehyde. Toxicity data: Smyth *et al., Arch. Ind. Hyg. Occup. Med.* **10**, 61 (1954).

Liquid. d^{10} 1.143; bp 87°. Very slightly sol in water; sol in many organic solvents. LD_{50} orally in rats: 6.5 g/kg (Smyth *et al.*).

8047. Propyl Iodide. *1-Iodopropane.* C_3H_7I; mol wt 169.99. C 21.20%, H 4.15%, I 74.65%. $CH_3CH_2CH_2I$. Prepd by heating propyl alcohol with iodine and red phosphorus.

Colorless or slightly yellow liquid. d_4^{20} 1.747; mp ~ −98°; bp 102-103°; n_D^{20} 1.5051. Sol in 575 parts water; miscible with alcohol, ether.

8048. Propyliodone. *3,5-Diiodo-4-oxo-1(4H)-pyridineacetic acid propyl ester;* 3,5-diiodo-4-pyridone-*N*-acetic acid propyl ester; propyl 3,5-diiodo-4-pyridone-*N*-acetate; propiodone; Dionosil. $C_{10}H_{11}I_2NO_3$; mol wt 447.01. C 26.87%, H 2.48%, I 56.78%, N 3.13%, O 10.74%. Prepn: Branscombe, Brit. pat. **517,382** (1940 to I.C.I.). Description: Tomich *et al., Brit. J. Pharmacol.* **8**, 166 (1953).

Crystals, dec 186-187°. Slightly sol in saline, serum. Solubility in water at 15°: 0.014 g/100 ml; at 35°: 0.020 g/100 ml; at 95°: 0.020 g/100 ml. LD_{50} i.v. in mice: 300 mg/kg (Tomich).

THERAP CAT: Diagnostic aid (radiopaque medium).

8049. *n*-Propyl Nitrate. *Nitric acid propyl ester.* $C_3H_7NO_3$; mol wt 105.09. C 34.29%, H 6.71%, N 13.33%, O 45.67%. $CH_3CH_2CH_2ONO_2$. Prepd by nitration of propanol with nitric acid, usually in the presence of urea and ammonium nitrate or sulfuric acid: Vogel, *J. Chem. Soc.* **1948**, 1847.

Pale yellow liquid. Sweet, sickly odor. d_4^{20} 1.0538. bp_{762} 110°. *Heating may cause it to explode.* n_D^{20} 1.3979. Dipole moment: 2.98. Azeotrope with water contg 75% $C_3H_7NO_3$, bp_{760} 84.8°. Very slightly sol in water. Sol in alcohol, ether. More physical data (especially concerning combustion and propulsion): Penner, Ducarme, *The Chemistry of Propellants* (Pergamon Press, 1960).

USE: Fuel ignition promoter, in rocket fuel formulations, as organic intermediate.

8050. *n*-Propyl Nitrite. *Nitrous acid n-propyl ester.* $C_3H_7NO_2$; mol wt 89.09. C 40.44%, H 7.92%, N 15.72%, O 35.92%. $CH_3CH_2CH_2ONO$. Prepd from silver nitrite and *n*-propyl bromide: Reynolds, Adkins, *J. Am. Chem. Soc.* **51**, 279 (1929); from *n*-propanol, sodium nitrite, and dil sulfuric acid: Cowley, Partington, *J. Chem. Soc.* **1933**, 1252.

Liquid. d_4^{20} 0.8864. bp_{760} 46-48°. n_D^{20} 1.3592; also reported as n_D^{20} 1.3613. Soluble in alcohol, ether.

Caution: Inhalation causes vasodilation, smooth muscle relaxation, hypotension.

USE: Jet propellant.

8051. Propylparaben. *4-Hydroxybenzoic acid propyl ester;* propyl *p*-hydroxybenzoate; Nipasol M; Solbrol P; Propyl Parasept. $C_{10}H_{12}O_3$; mol wt 180.20. C 66.65%, H 6.71%, O 26.64%. Prepn: Stohmann, *J. Prakt. Chem.* **36**, 368 (1887); L. Nobli, *Giorn. Farm. Chim.* **84**, 168 (1935), *C.A.* **30**, 4223[9] (1936).

White crystals, mp 96-97°. Sol in 2000 parts water; freely sol in alcohol, ether; slightly sol in boiling water.

USE: Pharmaceutic aid (antifungal). Preservative in foods.

8052. Propyl Propionate. *Propanoic acid propyl ester.* $C_6H_{12}O_2$; mol wt 116.16. C 62.04%, H 10.41%, O 27.55%. $CH_3CH_2COOC_3H_7$.

Liquid. d^{20} 0.883; mp −76°; bp 122-124°; n_D^{20} 1.3935. Sol in 200 parts water; miscible with alcohol, ether.

8053. Propyl Sulfide. *1,1'-Thiobispropane;* dipropyl sulfide. $C_6H_{14}S$; mol wt 118.24. C 60.95%, H 11.93%, S 27.12%. $C_3H_7SC_3H_7$. Prepd from propyl bromide and an alcoholic soln of Na_2S.

Colorless liquid. d^{17} 0.814; mp ~ −102°; bp 142°. Insol in water; sol in alcohol, ether.

8054. Propylthiouracil. *2,3-Dihydro-6-propyl-2-thioxo-4(1H)pyrimidinone;* 6-propyl-2-thiouracil; 2-thio-4-oxo-6-propyl-1,3-pyrimidine; 2-thio-6-propyl-1,3-pyrimidin-4-one; Propacil; Propycil; Propyl-Thyracil; Thyreostat II. $C_7H_{10}N_2OS$; mol wt 170.24. C 49.39%, H 5.92%, N 16.46%, O 9.40%, S 18.84%. Prepd by the condensation of ethyl β-oxocaproate with thiourea: Anderson *et al., J. Am. Chem. Soc.* **67**, 2197 (1945). Comprehensive description: H. Y. Aboul-Enein, *Anal. Profiles Drug Subs.* **6**, 457-486 (1977).

Review of pharmacology and clinical experience: D. S. Cooper *et al., N. Engl. J. Med.* **311**, 1353-1362 (1984).

White, bitter cryst powder of starch-like appearance to the eye and to the touch, mp 219-221°. uv max (methanol): 275, 214 nm (ε 15800, 15600); (methanolic KOH): 315.5, 260, 207.5 nm (ε 10900, 10700, 15400). One part dissolves in ~900 parts water at 20°, in 100 parts boiling water, in 60 parts ethyl alcohol, in 60 parts acetone. Practically insol in ether, chloroform, benzene. Freely sol in aq solns of ammonia and alkali hydroxides. A satd aq soln is neutral or slightly acid to litmus.

Note: This substance may reasonably be anticipated to be a carcinogen: *Seventh Annual Report on Carcinogens* (PB95-109781, 1994) p 348.

THERAP CAT: Antihyperthyroid.

THERAP CAT (VET): Antihyperthyroid. Has been used to promote fattening.

8055. Propylure. *(E)-10-Propyl-5,9-tridecadien-1-ol acetate;* trans-1-acetoxy-10-(*n*-propyl)trideca-5,9-diene; 10-propyl-*trans*-5,9-tridecadienyl acetate. $C_{18}H_{32}O_2$; mol wt 280.45. C 77.09%, H 11.50%, O 11.41%. Once thought to be the sex attractant of the pink bollworm moth, *Pectinophora gossypiella* (Saunders), a destructive cotton pest, H. E. Hummel *et al., Science* **181**, 873 (1973), *cf.* Gossyplure. Reported as the first known natural product possessing di *n*-propyl branching. Isoln, structure, and synthesis: Jones *et al., Science* **152**, 1516 (1966). Improved total syntheses: Pattenden, *J. Chem. Soc. (C)* **1968**, 2385; Meyers, Collington, *Tetrahedron* **27**, 5979 (1971); Vig *et al., J. Indian Chem. Soc.* **50**, 39 (1973); K. Utimoto *et al., Tetrahedron Letters* **1975**, 4233. Photochemical synthesis: Kossanyi *et al., ibid.* **1973**, 3459. Stereoselective synthesis: A. Alexakis *et al., ibid.* **1978**, 2027. *trans*-Propylure is rendered inactive by the presence of >15% of the *cis*-isomer: Jacobson, *Science* **163**, 190 (1969).

Colorless liquid having no detectable odor. $bp_{0.1}$ 135°. n_D^{25} 1.4635. Strong IR band at 970 cm^{-1}.

USE: Insect sex attractant.

8056. Propyphenazone. *1,2-Dihydro-1,5-dimethyl-4-(1-methylethyl)-2-phenyl-3H-pyrazol-3-one; 4-isopropylantipyrine;* 4-isopropyl-2,3-dimethyl-1-phenyl-3-pyrazolin-5-one; 2,3-dimethyl-1-phenyl-4-isopropylpyrazolone; isopropylphenazone; Budirol; Causyth; Cibalgina; Eufibron; Isoprochin P. $C_{14}H_{18}N_2O$; mol wt 230.31. C 73.01%, H 7.88%, N 12.16%, O 6.95%. Prepn: Stenz, U.S. pat. **1,972,036** (1934 to Hoffmann-La Roche); Sawa, *J. Pharm. Soc. Japan* **57**, 953 (1937), *C.A.* **32**, 2533 (1938).

Slightly bitter crystals, mp 103°. Readily sol in alcohol, ether. Soly in water: 0.24 g/100 ml at 16.5°.

THERAP CAT: Analgesic; antipyretic; anti-inflammatory.

8057. Propyromazine. *1-Methyl-1-[1-methyl-2-oxo-2-(10H-phenothiazin-10-yl)ethyl]pyrrolidinium bromide; 1-methyl-1-[1-(phenothiazin-10-ylcarbonyl)ethyl]pyrrolidinium bromide;* 10-[2-(1-methyl-1-pyrrolidinyl)propionyl]phenothiazine bromide; 10-[2-(1-pyrrolidinyl)propionyl]phenothiazine methobromide; 10-(2-methyl-1-oxo-2-pyrrolidinylethyl)phenothiazine methobromide; SD-104-19; Diaspasmyl. $C_{20}H_{23}BrN_2OS$; mol wt 419.39. C 57.28%, H 5.53%, Br 19.05%, N 6.68%, O 3.81%, S 7.65%. Prepd by *N*-acylation of phenothiazine: Boissier *et al., Thérapie* **13**, 989 (1958). Prepn of the free base from 10-(α-bromopropionyl)-phenothiazine and pyrrolidine: Dahlbom, Ekstrand, *Acta Chem. Scand.* **5**, 102 (1951); *eidem,* U.S. pat. **2,615,886** (1952 to Astra).

Crystals, mp 228-229°.

Free base, $C_{19}H_{20}N_2OS$, crystals from petr ether, mp 94.5-95.5°.

THERAP CAT: Anticholinergic, antispasmodic.

8058. Propyzamide. *3,5-Dichloro-N-(1,1-dimethyl-2-propynyl)benzamide;* pronamid; RH-315; Kerb. $C_{12}H_{11}Cl_2$-NO; mol wt 256.13. C 56.27%, H 4.33%, Cl 27.68%, N 5.47%, O 6.25%. Selective pre-emergence herbicide. Prepn: B. W. Horrom *et al., S. Afr. pat.* **68 00,090**; *eidem,* U.S. pats. **3,534,098**; and **3,640,699** (1969, 1970, 1972 all to Rohm & Haas). Activity: K. L. Viste *et al., Science* **167**, 280 (1970); C. Swithenbank *et al., J. Agr. Food Chem.* **19**, 417 (1971). Metabolism: R. Y. Yih *et al., Weed Sci.* **18**, 604 (1970); R. Y. Yih, C. Swithenbank, *J. Agr. Food Chem.* **19**, 314, 320 (1971); J. D. Fisher, *ibid.* **22**, 606 (1974). Carcinogenicity study: M. D. Reuber, *Environ. Res.* **23**, 1 (1980).

White solid, mp 155-156°. Soly in water at 25°: 15 ppm. Sol in aliphatic, aromatic solvents. Vapor pressure at 25°: 8.5×10^{-5} mm Hg. LD_{50} in male, female rats (mg/kg): 8350, 5620 orally (Viste).

USE: Herbicide.

8059. Proquazone. *7-Methyl-1-(1-methylethyl)-4-phenyl-2(1H)-quinazolinone; 1-isopropyl-7-methyl-4-phenyl-2(1H)-quinazolinone;* RU-43715; Sandoz 43-715; Biarison. $C_{18}H_{18}N_2O$; mol wt 278.35. C 77.67%, H 6.52%, N 10.06%, O 5.75%. Prepn: H. Ott, M. Denzer, Ger. pat. **1,805,501** corresp to U.S. pats. **3,845,128** and **3,925,548** (1969, 1974 and 1975, all to Sandoz); R. V. Coombs *et al., J. Med. Chem.* **16**, 1237 (1973). Metabolism: M. B. Zucker, *Proc. Soc. Exp. Biol. Med.* **156**, 209 (1977); H. Ott, J. Meier, *Scand. J. Rheumatol. (Suppl.)* **21**, 12 (1978). Pharmacology: H. U. Gubler, M. Baggiolini, *ibid.* 8; H. Ott, *ibid.* 5. Clinical studies: P. Sfikakis, P. Tsachalos, *Ther. Umsch.* **34**, 730 (1977); series of articles in *Scand. J. Rheumatol. (Suppl.)* **21**, 15-39 (1978). Review of pharmacodynamics, pharmacokinetics and therapeutic efficacy: S. P. Clissold, R. Beresford, *Drugs* **33**, 478-502 (1987).

Yellow crystals from ethyl acetate, mp 137-138°. Sol in chloroform. Insol in water.

THERAP CAT: Anti-inflammatory.

8060. Proscillaridin. *(3β)-3-[(6-Deoxy-α-L-mannopyranosyl)oxy]-14-hydroxybufa-4,20,22-trienolide;* 14-hydroxy-3β-(rhamnosyloxy)bufa-4,20,22-trienolide; 3β-rhamnosido-14β-hydroxy-Δ4,20,22-bufatrienolide; proscillaridin A; desglucotransvaaline; scillarenin 3β-rhamnoside; Caradrin; Cardion; Carmazon; Proscillan; Prostosin; Proszine; Protasin; Purosin-TC; Sandoscill; Scillacrist; Simeon; Solestril; Stellarid; Talucard; Talusin; Urgilan; Wirnesin. $C_{30}H_{42}O_8$; mol wt 530.66. C 67.90%, H 7.98%, O 24.12%. Prepn by acid cleavage of scillaren A: Stoll *et al., Helv. Chim. Acta* **16**, 703 (1933); by enzymic decompn of glucoscillaren A with strophanthobiase: Stoll *et al., ibid.* **35**, 2495 (1952); from *Urginea burkei* Baker, *Liliaceae:* Zoller, Tamm, *ibid.* **36**, 1744 (1953); from *U. (Scilla) maritima* (L.) Baker, *Liliaceae:* Görlich, *Arzneimittel-Forsch.* **10**, 770 (1960). Structure: Stoll *et al., Helv. Chim. Acta* **35**, 1934 (1952). Pharmacology: Lenke, Brock, *Arzneimittel-Forsch.* **20**, 1 (1970). Metabolic studies: Davis *et al., Arch. Int. Pharmacodyn.* **177**, 231 (1969); Nakano *et al., ibid.* **183**, 199 (1970). Clinical studies: Several authors, *Minerva Med.* **58**, 4243-4322 (1967). Toxicity study: E. I. Goldenthal, *Toxicol. Appl. Pharmacol.* **18**, 185 (1971). Series of articles on prepn, pharmacology, toxicology, pharmcokinetics, metabolism of 4'-methyl ether: *Arzneimittel-Forsch.* **28**, 493-573 (1978).

Prisms from methanol, mp 219-222°. $[\alpha]_D^{20}$ — 91.5° (CH_3OH). LD_{50} in male, female rats (mg/kg): 56, 76 orally (Goldenthal).

Proscillaridin-4-methyl ether, $C_{31}H_{44}O_8$, *meproscillarin, Clift.* mp 213-217°. $[\alpha]_D^{20}$ — 94° (CH_3OH). uv max (CH_3OH): 297 nm (log ε 3.79), (1N KOH/CH_3OH): 355 nm (log ε 4.65). Sol in methanol, ethanol, THF, dioxane; slightly sol in $CHCl_3$, CH_2Cl_2, acetone. Practically insol in water, nonpolar organics.

THERAP CAT: Cardiotonic.

8061. Prostacyclin. *(5Z,9α,11α,13E,15S)-6,9-Epoxy-11,15-dihydroxyprosta-5,13-dien-1-oic acid;* (5Z)-9-deoxy-6,9α-epoxy-Δ5-$PGF_{1α}$; epoprostenol; prostaglandin I_2; prostaglandin X; PGI_2; PGX; U-53217. $C_{20}H_{32}O_5$; mol wt 352.47. C 68.15%, H 9.15%, O 22.70%. A prostaglandin produced by enzymatic transformation of prostaglandin en-

doperoxides (PGG_2, PGH_2), which dilates blood vessels and is approximately 30 times more potent than prostaglandin E_1, *q.v.,* in inhibiting platelet aggregation. Evidence for its occurrence during biosynthetic conversion of arachidonic acid by rat stomach homogenates: C. Pace-Asciak, L. S. Wolfe, *Biochemistry* **10**, 3657 (1971). Isoln from microsomes of pig and rabbit aorta by J. R. Vane and co-workers: S. Moncada *et al., Nature* **263**, 663 (1976). Prepn: S. Moncada, N. Whittaker, **Ger.** pat. 2,720,999; S. Moncada, **U.S.** pat. 4,539,333 (1977, 1985 both to Wellcome Found.). PGI_2 is also synthesized in bovine coronary arteries as well as human arteries and veins: *eidem, Lancet* **1**, 18 (1977); G. J. Dusting *et al., Prostaglandins* **13**, 3 (1977); by cultured human and bovine endothelial cells: B. B. Weksler *et al., Proc. Nat. Acad. Sci. USA* **74**, 3922 (1977); by pig aortic endothelial cells: D. E. MacIntyre *et al., Nature* **271**, 549 (1978). It has been suggested that endoperoxides released by platelets can be converted to PGI_2 by vascular tissue and that a balance between formation of PGI_2 and release of thromboxane A_2, *q.v.,* which induces platelet aggregation, controls the formation of thrombi in blood vessels. It has also been postulated that PGI_2 acts to stimulate platelet adenylate cyclase and to prevent the action of thrombi on phospholipid breakdown as well as platelet aggregation. Structure: R. A. Johnson *et al., Prostaglandins* **12**, 915 (1976). Synthesis: E. J. Corey *et al., J. Am. Chem. Soc.* **99**, 3006 (1977); of sodium salt and stereochemistry: R. A. Johnson *et al., ibid.* 4182. Additional syntheses: I. Tomoskozi *et al., Tetrahedron Letters* **1977**, 2627; N. Whittaker, *ibid.* 2805; K. Nicolaou, *Chem. Commun.* **1977**, 630. Synthesis of the 5E-isomer: E. J. Corey *et al., Tetrahedron Letters* **1977**, 3529. Chemical stability in aq solns: M. J. Cho, M. A. Allen, *Prostaglandins* **15**, 943 (1978). Biosynthetic study: V. Tomasi *et al., Nature* **273**, 670 (1978). Biological properties: R. J. Gryglewski *et al., Prostaglandins* **12**, 685 (1976). Preliminary clinical study: A. E. S. Gimson *et al., Lancet* **1**, 173 (1980). Antimetastatic effects: K. V. Honn *et al., Science* **212**, 1270 (1981); *eidem, ibid.* **217** 542 (1982). Preliminary study of effect of PGX infusion in patients with acute myocardial infarction: O. Edhag *et al., N. Engl. J. Med.* **308**, 1032 (1983). Review of biological properties: S. Moncada, J. R. Vane, *Clin. Sci.* **61**, 369-372 (1981); of therapeutic potential: *eidem, Advan. Pharmacol. Ther.* **4**, 215-233 (1982); of physiological role: J. R. Vane *et al., Int. Rev. Exp. Pathol.* **23**, 161-207 (1982). General reviews: S. Moncada, J. R. Vane, *Fed. Proc.* **38**, 66-71 (1979); J. C. McGiff, *Ann. Rev. Pharmacol. Toxicol.* **21**, 479-509 (1981); S. Moncada *et al., Advan. Pharmacol. Ther.* **6**, 39-47 (1982). Books: *Prostacyclin,* J. R. Vane, S. Bergstrom, Eds. (Raven Press, New York, 1979) 453 pp; *Prostaglandins in Cardiovascular and Renal Function,* A. Scriabine *et al.,* Eds. (Spectrum Publications, New York, 1980) 498 pp.

Chemically unstable in aq soln. Hydrolyzes to 6-oxo-$PGF_{1α}$. Half-life at 4° is approx 14.5 min when total phosphate is 0.165 *M*. Anti-aggregating activity disappears within 0.25 min on boiling or within 10 min at 37°.

Sodium salt, $C_{20}H_{31}NaO_5$, U-53217A, *Cyclo-Prostin, Flolan.* Hygroscopic, free-flowing white powder. Stable for 2 months if kept dry at −30°.

THERAP CAT: Platelet aggregation inhibitor.

8062. Prostaglandin(s). A family of biologically potent lipid acids first discovered in seminal fluid and extracts of accessory genital glands of man and sheep: von Euler, *Arch. Exp. Pathol. Pharmakol.* **175**, 78 (1934); *Klin. Wochenschr.* **14**, 1182 (1935). Isoln: Bergstrom, Sjovall, **U.S.** pats. 3,-069,322 and 3,598,858 (1962, 1971); Samuelsson, *J. Biol. Chem.* **238**, 3229 (1963). Also found in lower concns in other organs: *idem, Biochim. Biophys. Acta* **84**, 707 (1964). The single non-mammalian source of prostaglandin intermediates, or *syntons*, is the gorgonian sea whip or sea fan, *Plexaura homomalla:* Weinheimer, Spraggins, *Tetrahedron Letters* **1969**, 5185; Schneider *et al., J. Am. Chem. Soc.* **94**, 2122 (1972). Prostaglandins are named as derivatives of *prostanoic acid.* Prostaglandins are divided into the types E, F, A, B, C, and D based on functions in the cyclopentane ring. Numerical subscripts refer to the number of unsaturations in the side chains; α or β subscripts refer to the configuration of substituents in the ring. Six naturally occurring prostaglandins, E₁, E₂, E₃, F₁α, F₂α, F₃α, are considered primary in that no one is derived from another in the living organism. First structural and stereochemical elucidations: Bergstrom *et al., Acta Chem. Scand.* **16**, 501 (1962); *eidem, J. Biol. Chem* **238**, 3555 (1963). Absolute config: Nugteren *et al., Nature* **212**, 38 (1966). First total synthesis of racemic PGE₁ and PGF₁α: Corey *et al., J. Am. Chem. Soc.* **90**, 3245 (1968). Review of synthetic studies: Pike, *Fortschr. Chem. Org. Naturst.* **28**, 313 (1970); Axen *et al.,* in *The Total Synthesis of Natural Products,* vol. 1, J. ApSimon, Ed. (Wiley-Interscience, New York, 1973) pp 81-143; Clarkson in *Progress in Organic Chemistry,* vol. 8, W. Carruthers, J. K. Sutherland, Eds. (Wiley, New York, 1973) pp 1-28. Book: J. S. Bindra, R. Bindra, *Prostaglandin Synthesis* (Academic Press, New York, 1977). Biosynthesis occurs by enzymatic conversion of unsaturated twenty-carbon fatty acids. Review of biosynthetic studies: Samuelsson, *Progr. Biochem. Pharmacol.* **5**, 109 (1969). Review of metabolism: Samuelsson *et al., Ann. N.Y. Acad. Sci.* **180**, 138 (1971). Biological activities include stimulation of smooth muscle, dilation of small arteries, bronchial dilation, lowering of blood pressure, inhibition of gastric secretion, of lipolysis, and of platelet aggregation, induction of labor, abortion, and menstruation, and increase in ocular pressure. Implicated also in dysmenorrhea, inflammatory reactions, nasal vasoconstriction, kidney function, and in autonomic neurotransmission. Reviews of pharmacological and biochemical aspects: Horton, *Experientia* **21**, 113 (1965); Weeks, *Ann. Rev. Pharmacol.* **12**, 317 (1972); Hinman, *Ann. Rev. Biochem.* **41**, 161 (1972). Review of biological activities of synthetic prostaglandins: Ramwell *et al., Nature* **221**, 1251 (1969). Review of analytical and preparative techniques: *Methods Enzymol.* **86**, entitled "Prostaglandins and Arachidonate Metabolites", W. E. M. Landis, W. L. Smith, Eds. (Academic Press, New York, 1982) 705 pp. General reviews: Bergstrom, *Science* **157**, 382 (1967); Bergstrom *et al., Pharmacol. Rev.* **20**, 1 (1968); Ramwell, Shaw, *Recent Progr. Horm. Res.* **26**, 139 (1970); Pike, *Sci. Amer.* **225**, 84 (Nov., 1971); Bindra, Bindra, *Progress in Drug Research,* vol. 17 (Birkhäuser Verlag, Basel, 1973) pp 410-487. Books: *The Prostaglandins,* vols. 1, 2, P. Ramwell, Ed. (Plenum Press, New York, 1973, 1974); *Prostaglandins in Cardiovascular and Renal Function,* A. Scriabine *et al.,* Eds. (Spectrum Publications, New York, 1980) 498 pp; *Cardiovascular Pharmacology of the Prostaglandins,* A. Herman, Ed. (Raven Press, New York, 1982) 472 pp.

Prostanoic acid

8063. Prostaglandin E₁. *(11α,13E,15S)-11,15-Dihydroxy-9-oxoprost-13-en-1-oic acid; 3-hydroxy-2-(3-hydroxy-1-octenyl)-5-oxocyclopentaneheptanoic acid;* alprostadil; PGE₁; U-10136; Caverject; Liple; Minprog; Palux; Prostandin; Prostine VR; Prostivas. C₂₀H₃₄O₅; mol wt 354.49. C 67.77%, H 9.67%, O 22.57%. A primary prostaglandin; easily crystallized from purified biological extracts. Isoln from sheep seminal vesicle tissue, and structure: Bergstrom *et al., Acta Chem. Scand.* **16**, 501 (1962); *eidem, J. Biol. Chem.* **238**, 3555 (1963). Enzymic conversion from 8,11,14-eicosatrienoic acid: Nugteren *et al., Rec. Trav. Chim.* **85**, 405 (1966). Synthesis of the *dl*-form: Corey *et al., J. Am. Chem. Soc.* **90**, 3245, 3247 (1968); Schneider *et al., ibid.* 5895; **91**, 5372 (1969); Axen *et al., Chem. Commun.* **1969**,

303; Taub *et al., ibid.* **1970**, 1258; Slates *et al., ibid.* **1972**, 304; Kuo *et al., Tetrahedron Letters* **1972**, 5317; Taub *et al., Tetrahedron* **29**, 1447 (1973); Miyano, Stealey, *Chem. Commun.* **1973**, 180; Finch *et al., J. Org. Chem.* **38**, 4412 (1973). Synthesis of natural form: Corey *et al., J. Am. Chem. Soc.* **91**, 535 (1969); **92**, 2586 (1970); Sih *et al., ibid.* **94**, 3643 (1972); **95**, 1676 (1973); Schaaf, Corey, *J. Org. Chem.* **37**, 2921 (1974); Slates *et al., Tetrahedron* **30**, 819 (1974). Metabolism in guinea pigs: Anggard, Samuelsson, *J. Biol. Chem.* **239**, 4097 (1964). Metabolism in humans: Hamberg, Samuelsson, *ibid.* **246**, 6713 (1971). Review of biological activities: Berti *et al., Progr. Biochem. Pharmacol.* **3**, 110 (1967). Comparative pharmacology with respect to other prostaglandins: Weeks, *Ann. Rev. Pharmacol.* **12**, 317 (1972). Clinical use to maintain patency of ductus arteriosus in neonatal cardiac problems: P. M. Olley *et al., Adv. Prostaglandin Thromboxane Res.* **7**, 913 (1980); J. S. Donahoo *et al., J. Thoracic Cardiovasc. Surg.* **81**, 227 (1981); E. D. Silove *et al., Circulation* **63**, 682 (1981). Use in non-atherosclerotic vasculopathy: D. L. Wooster *et al., J. Am. Med. Assoc.* **245**, 1846 (1981). For general refs see Prostaglandins.

Crystals from ethyl acetate + heptane, mp 115-116°. [α]₅₇₈ −61.6° (c = 0.56 in tetrahydrofuran). Easily dehydrated in soln at pHs <4 or >8.

THERAP CAT: Vasodilator (peripheral).

8064. Prostaglandin E₂. *(5Z,11α,13E,15S)-11,15-Dihydroxy-9-oxoprosta-5,13-dien-1-oic acid; 7-[3-hydroxy-2-(3-hydroxy-1-octenyl)-5-oxocyclopentyl]-5-heptenoic acid;* dinoprostone; PGE₂; U-12062; Cerviprost; Minprostin E₂; Prepidil; Propess; Prostarmon-E; Prostin E₂. C₂₀H₃₂O₅; mol wt 352.47. C 68.15%, H 9.15%, O 22.70%. The most common and most biologically potent of mammalian prostaglandins. Isoln from sheep prostate: S. Bergström, J. Sjövall, **Brit.** pat. 851,827; *eidem,* U.S. pat. 3,598,858 (1960, 1971); from sheep seminal vesicle tissue: S. Bergström *et al., Acta Chem. Scand.* **16**, 501 (1962). Total synthesis of the *dl*-form: W. P. Schneider, *Chem. Commun.* **1969**, 304; E. J. Corey *et al., J. Am. Chem. Soc.* **91**, 5675 (1969); E. J. Corey *et al., Tetrahedron Letters* **1970**, 307; W. P. Schneider, **Ger.** pat. 2,011,-969 (1970 to Upjohn), *C.A.* **74**, 87486n (1971); J. Fried *et al., J. Am. Chem. Soc.* **94**, 4342 (1972). Synthesis of naturally occurring form: E. J. Corey *et al., ibid.* **92**, 397, 2586 (1970); J. B. Heather *et al., Tetrahedron Letters* **1973**, 2313; from *Plexaura homomalla* prostaglandin intermediates: G. L. Bundy *et al., J. Am. Chem. Soc.* **94**, 2123 (1972); W. P. Schneider *et al., Chem. Commun.* **1973**, 254. Biosynthesis: D. A. Van Dorp *et al., Biochim. Biophys. Acta* **90**, 204 (1964); S. Bergström *et al., ibid.* 207; **Neth.** pat. **Appl. 6,-**505,799 (1965 to Unilever), *C.A.* **65**, 7584h (1966). Metabolism: E. Anggard, B. Samuelsson, *Mem. Soc. Endocrinol.,* no. 14, 107 (1966); M. Hamberg, B. Samuelsson, *J. Biol. Chem.* **246**, 6713 (1971). Several reviews in *Prostaglandin Symp. Worcester Found. Exp. Biol.,* P. Ramwell, Ed. (Interscience, New York, 1968). For general refs see Prostaglandins.

Natural form, colorless crystals. mp 66-68°. $[\alpha]_D^{26}$ −61° (c = 1 in tetrahydrofuran). Easily dehydrated in soln at pHs <4 or >8.

THERAP CAT: Oxytocic; abortifacient.

8065. Prostaglandin $F_{2\alpha}$. *(5Z,9α,11α,13E,15S)-9,11,15-Trihydroxyprosta-5,13-dien-1-oic acid; 7-[3,5-dihydroxy-2-(3-hydroxy-1-octenyl)cyclopentyl]-5-heptenoic acid;* dinoprost; $PGF_{2\alpha}$; U-14583; Enzaprost F; Glandin; Prostarmon F. $C_{20}H_{34}O_5$; mol wt 354.49. C 67.77%, H 9.67%, O 22.57%. One of the most biologically studied of the primary prostaglandins. Closely related to prostaglandin E_2 (PGE_2) in that both prostaglandins are biosynthesized from the same precursors and that $PGF_{2\alpha}$ is the synthetic reduction product of PGE_2. For refs to synthesis of *dl* and natural forms *see* Prostaglandin E_2. Prepn of the tromethamine salt: W. Morozowich, **Ger. pat. 2,126,127**; idem, **U.S. pat. 3,657,327** (1971, 1972 both to Upjohn). Alternate synthesis of natural $PGF_{2\alpha}$: Schneider, Murray, *J. Org. Chem.* **38**, 397 (1973); R. B. Woodward *et al., J. Am. Chem. Soc.* **95**, 6853 (1973); G. Stork *et al., ibid.* **100**, 8272 (1978); K. Kondo *et al., Tetrahedron Letters* **1978**, 3927; N. R. A. Beeley *et al., Tetrahedron* **37**, Suppl. 9, 411 (1981); R. J. Cave *et al., J. Chem. Soc. Perkin Trans. I* **1981**, 646. Causes vasocontraction and exhibits luteolytic activity; is most commonly associated with its role in pregnancy: Karim *et al., J. Obstet. Gynaecol. Brit. Commonw.* **78**, 172 (1971). Metabolism in female subjects: Granstrom, Samuelsson, *J. Biol. Chem.* **246**, 5254 (1971). Toxicity data: T. Fujita *et al., Iyakuhin Kenkyo* **9**, 261 (1978), *C.A.* **89**, 71399k (1978). For general refs *see* Prostaglandins.

Natural form, crystals, mp 25-35°. $[\alpha]_D^{25}$ +23.5° (c = 1 in tetrahydrofuran). Freely sol in methanol, abs ethanol, ethyl acetate, chloroform; slightly sol in water. Stable for two years in light resistant containers at 5-15°. Degrades in one week when exposed to sunlight or in three months at 40°. LD_{50} in rabbits (mg/kg): 2.5-5.0 i.v.; 2.5-5.0 i.m. (Fujita).

Tromethamine salt, $C_{20}H_{34}O_5 \cdot C_4H_{11}NO_3$, Lutalyse, Prostin F_2 Alpha. White or off-white cystalline powder, mp 100-101°. Readily sol in water to at least 200 mg/ml.

THERAP CAT: Oxytocic; abortifacient.
THERAP CAT (VET): Oxytocic.

8066. Prostalene. *(9α,11α,13E,15R)-(±)-9,11,15-Trihydroxy-15-methylprosta-4,5,13-trien-1-oic acid methyl ester;* (±)-methyl 7-[3,5-dihydroxy-2-[(E)-3-hydroxy-3-methyl-1-octenyl]cyclopentyl]-4,5-heptadienoate; RS-9390; Synchrocept. $C_{22}H_{36}O_5$; mol wt 380.52. C 69.44%, H 9.54%, O 21.02%. Synthetic analog of prostaglandin $F_{2\alpha}$, *q.v.* Prepn: P. Crabbé, J. H. Fried, **Ger. pat. 2,258,668** corresp to **U.S. pat. 3,879,438** (1973, 1975 both to Syntex). Use in induction of parturition in sows: W. Holtz *et al., J. Anim. Sci.* **49**, 367 (1979). Pharmacodynamics in mares: R. G. Loy *et al., J. Reprod. Fertil. (Suppl)* **27**, 229 (1979).

THERAP CAT (VET): Luteolytic.

8067. Prosulfuron. *N-[[(4-Methoxy-6-methyl-1,3,5-triazin-2-yl)amino]carbonyl]-2-(3,3,3-trifluoropropyl)benzenesulfonamide;* 1-(4-methoxy-6-methyl-1,3,5-triazin-2-yl)-3-[2-(3,3,3-trifluoropropyl)phenylsulfonyl]urea; 1-[2-(3,3,3-trifluoropropyl)phenylsulfonyl]-3-(4-methoxy-6-methyl-1,3,5-triazin-2-yl)urea; CGA-152005; Exceed; Peak. $C_{15}H_{16}F_3N_5$-

O_4S; mol wt 419.38. C 42.96%, H 3.85%, F 13.59%, N 16.70%, O 15.26%, S 7.65%. Post-emergence sulfonylurea herbicide for use in maize; acetolactate synthase inhibitor. Prepn: W. Meyer, K. Oertle, **Eur. pat. Appl. 120,814;** *eidem,* **U.S. pat. 4,671,819** (1984, 1987 both to Ciba-Geigy); J. G. Dingwall, *Pestic. Sci.* **41**, 259 (1994). Comprehensive description: M. Schulte *et al., Proc. Brighton Crop Prot. Conf. - Weeds* **1993**, 53-59.

Colorless, odorless crystals, mp 155° (dec). Vapor pressure (25°): <3.5 × 10⁻⁶ Pa. Soly in water (25°): 4000 mg/l (pH 6.8). log *P* (*n*-octanol/water): −0.21 (pH 6.9). LD_{50} orally in rats: 986 mg/kg; dermally in rabbits: >2000 mg/kg. LC_{50} (4 hr) by inhalation in rats: >5000 mg/m³ (Schulte).

USE: Herbicide.

8068. Prosultiamine. *N-[(4-Amino-2-methyl-5-pyrimidinyl)methyl]-N-[4-hydroxy-1-methyl-2-(propyldithio)-1-butenyl]formamide;* 2-(2-methyl-4-aminopyrimidin-5-yl)methylformamido-5-hydroxy-2-penten-3-yl propyl disulfide; vitamin B_1 propyl disulfide; thiamine propyl disulfide; dithiopropylthiamine; DTPT; TPD; Alinamin; Binova. $C_{15}H_{24}N_4$-O_2S_2; mol wt 356.51. C 50.54%, H 6.79%, N 15.72%, O 8.98%, S 17.99%. Synthesis: Matsukawa, Kawasuki, *J. Pharm. Soc. Japan* **73**, 216 (1953), *C.A.* **48**, 2071 (1954); Matsukawa *et al., J. Vitaminol.* **1**, 13 (1954); Fujiwara *et al.,* **U.S. pat. 2,833,768** (1958 to Takeda), *cf.* **Fr. pat. 1,068,459** (1954 to Takeda). Structural studies: Nishikawa *et al., Chem. Pharm. Bull.* **17**, 932 (1969). Metabolism: Suzuoki-Ziro *et al., J. Biochem.* **58**, 279 (1965); Nishikawa *et al., J. Pharmacol. Exp. Ther.* **157**, 589 (1967).

E-form

Prisms from benzene, dec 128-129°. Sparingly soluble in water. Sol in organic solvents and lipids. Better absorbed upon oral ingestion by man, than thiamine hydrochloride.

Hydrochloride, $C_{15}H_{24}N_4O_2S_2 \cdot HCl$, crystals, dec 160-161°.

THERAP CAT: Vitamin (enzyme co-factor).

8069. Protactinium. Protoactinium. Pa; at. no. 91; at. wt 231.03588 (characteristic naturally occurring isotopic mixture); valences 3, 4, 5. Known isotopes (mass numbers): 216, 222-238; naturally occurring isotopes: 231, 234m, 234. First isotope discovered, ²³⁴ᵐPa ($T_{1/2}$ 1.17 minutes); called *brevium, uranium X_2* or UX_2, natural decay product of ²³⁸U. Modes of decay: γ by isomeric transition to yield ²³⁴Pa; β⁻ to yield ²³⁴U. Longest-lived isotope, ²³¹Pa ($T_{1/2}$ 3.276 × 10⁴ years, rel. at. mass 231.0359); natural decay product of ²³⁵U. Decays by α emission; parent of ²²⁷Ac. ²³⁴Pa ($T_{1/2}$ 6.75 hrs); called *Uranium Z* or *UZ*; natural decay product of ²³⁸U. Decays by β⁻ emission. Discovery of ᵐ²³⁴Pa: K. Fajans, O. H. Göhring, *Naturwiss.* **1**, 339 (1913), *C.A.* **7**, 3916 (1913); *eidem, Phys. Z.* **14**, 877 (1913), *C.A.* **7**, 297 (1913); of ²³¹Pa: O. Hahn, L. Meitner, *ibid.* **19**, 208 (1918); and independently by F. Soddy, J. A. Cranston, *Proc. Roy. Soc.* **94A**, 384 (1918); of ²³⁴Pa: O. Hahn, *Ber.* **54B**, 1131 (1921). Prepn of metal: A. V. Grosse, M. Agrass, *J. Am. Chem. Soc.* **56**, 2200 (1934). Review of discovery: K. Fajans, D. F. C. Morris, *Nature* **244**, 137-138 (1973); of use in nuclear fuel cycle: O. L. Keller, *Radiochim. Acta* **25**, 211-223 (1978). *Reviews: Comprehensive Inorganic Chemistry* vol. 5, J. C. Bailar, Jr. *et al.,* Eds. (Pergamon Press, Oxford, 1973) pas-

sim; H. W. Kirby in *The Chemistry of the Actinide Elements* **vol. 1**, J. J. Katz *et al.,* Eds. (Chapman and Hall, New York, 1986) pp 102-168.

Shiny, silvery, malleable, ductile metal. Body-centered tetragonal crystal structure. Easily tarnished in air to an undetermined oxide. mp 1560°; also reported as 1575° (Bailar). bp 4227°. d^{25} 15.37. Reacts with H_2 at 250-300° to form PaH_3. In dil solns of HF, is deposited (10-96%) on Be, Al, Mn, Zn, and Pl; gives small deposits on Cr, Ta, Fe, Cd, Ni, Cu, Hg, W: Camarcat *et al., J. Chim. Phys.* **46**, 153-157 (1949).

Caution: Radiation hazard; handling requires special equipment and shielding facilities (Katz *et al., loc cit.,* **vol 2**, p 1128). Inhalation hazard in insol form; general hazard if absorbed systemically. Max permissible concn of insoluble form in air: 4×10^{-11} μCi/cc; of sol form in air: 4×10^{-13} μCi/cc; *Natl. Bur. Stand. Handbook* **69**, 83 (1959).

USE: ^{233}Pa as intermediate in production of fissile ^{233}U in thorium breeder reactors.

8070. Protein C. *Blood-coagulation factor XIV.* Vitamin K-dependent plasma protein involved in the regulation of hemostasis. Two-chain, multi-domain glycoprotein, mol wt ~62,000. Circulates as a zymogen; activated by thrombin complexed with thrombomodulin, *q.v.* Activated protein C is a serine protease that controls coagulation by the selective inactivation of factors Va and VIIIa. This activity is potentiated by factor V and by *Protein S*, a vitamin K-dependent protein unrelated to the serine proteases. Identification of *autoprothrombin II-A*, an anticoagulant obtained from purified prothrombin complex: E. F. Mammen *et al., Thromb. Diath. Haemorrh.* **5**, 218 (1960). Purification: W. H. Seegers *et al., Thromb. Res.* **1**, 443 (1972). Characterization: J. Stenflo, *J. Biol. Chem.* **251**, 355 (1976). Identity with autoprothrombin II-A: W. H. Seegers *et al., Thromb. Res.* **8**, 543 (1976). Isoln from human plasma: W. Kisiel, *J. Clin. Invest.* **64**, 761 (1979). Amino acid sequence of bovine protein C: P. Fernlund, J. Stenflo, *J. Biol. Chem.* **257**, 12170, 12180 (1982). Review of role in hemostasis: L. H. Clouse, P. C. Comp, *N. Engl. J. Med.* **314**, 1298-1304 (1986); C. T. Esmon, *J. Biol. Chem.* **264**, 4743-4746 (1989); of preparative methods and mechanism of activation: C. T. Esmon *et al., Methods Enzymol.* **222**, 359-385 (1993). Review of protein C system and role in thromboembolic disease: B. Dahlbäck, *Thromb. Res.* **77**, 1-43 (1995).

8071. Protein Hydrolysates. Amigen; Aminokrovin; Aminosol; Bioplex; Dekamin; Parenamine; Parentamin; Protigényl; Travamin. Sterile solution of amino acids and short-chain peptides which represent the approx nutritive equivalent of the casein, lactalbumin, plasma, fibrin or other suitable protein from which it is derived by acid, enzymatic or other method of hydrolysis. It may be modified by partial removal and restoration or addn of one or more amino acids. It may contain alcohol, dextrose, or other carbohydrate suitable for i.v. infusion. Not <50% of the total nitrogen present is in the form of α-amino nitrogen.

THERAP CAT: Replenisher (fluid and nutrient).

THERAP CAT (VET): Parenteral nutrient.

8072. Protheobromine. *3,7-Dihydro-1-(2-hydroxypropyl)-3,7-dimethyl-1H-purine-2,6-dione; 1-(2-hydroxypropyl)-theobromine;* Tebe; Bonicor. $C_{10}H_{14}N_4O_3$; mol wt 238.25. C 50.41%, H 5.92%, N 23.52%, O 20.15%. Prepn: Rojahn, Fegler, *Arch. Pharm.* **268**, 570 (1930); **Brit. pat. 816,299** (1959 to Chemiewerk Homburg), *cf.* **Ger. pat. 1,067,025**. Improved procedure: Carstens, Donat, **East Ger. pat. 26757** (1964). Pharmacology: Giertz *et al., Arzneimittel-Forsch.* **6**, 457 (1956). Toxicity study: R. Taugner *et al., Arzneimittel-Forsch.* **6**, 601 (1956).

Crystals from isopropanol, mp 140-142°. Freely sol in water at neutral or slightly acid pH. Sol in chloroform, hot ethanol, warm glycerol. Practically insol in ether. LD_{50} s.c. in mice: 580 mg/kg (Taugner).

THERAP CAT: Diuretic.

8073. Prothipendyl. *N,N-Dimethyl-10H-pyrido[3,2-b]-[1,4]benzothiazine-10-propanamine; 10-(3-dimethylaminopropyl)-10H-pyrido[3,2-b][1,4]benzothiazine;* 10-(γ-dimethylaminopropyl)-1-azaphenothiazine; *N*-(3-dimethylaminopropyl)thiophenylpyridylamine; 2,3-pyridino-(5',6')-5,6-benzo-4-(3''-dimethylaminopropyl)-1,4-thiazine; $C_{16}H_{19}N_3S$; mol wt 285.41. C 67.33%, H 6.71%, N 14.72%, S 11.23%. Prepn: Yale, Bernstein, **U.S. pat. 2,943,086** (1960 to Olin Mathieson); von Schlichtegroll, *Proc. Int. Congr. Neuro-Pharm., 1st, Rome,* **1958**, 408 (1959), *C.A.* **54**, 13400g (1960); **Fr. pat. 1,173,134** (1959 to Rhône-Poulenc).

Liquid, $bp_{0.7}$ 217-219°; $bp_{0.5}$ 195-198°.

Hydrochloride, $C_{16}H_{19}N_3S.HCl$, *D-206, Dominal, Tolnate.* mp 177-178° with sintering ~176°. Monohydrate as crystals, mp 108-112°. Freely sol in water, methanol. Practically insol in ether, petr ether.

Dihydrochloride, $C_{16}H_{19}N_3S.2HCl$, crystals from acetonitrile, mp 205-207°.

THERAP CAT: Antipsychotic.

8074. Prothrombin. *Blood-coagulation factor II;* Factor II; prothrombase; serozyme; thrombogen. Mol wt 69,000-74,000. Coagulation proenzyme present in highest concentration in blood. Prothrombin is one of the vitamin-K dependent blood coagulation factors. It is converted to thrombin by the action of factor X_a, factor V and phospholipid in the presence of Ca^{2+} ions. Accounts for < 0.2% of total plasma protein. Prepn of human and bovine prothrombin: Goldstein *et al., J. Biol. Chem.* **234**, 2857 (1959); Lanchantin *et al., ibid.* **238**, 238 (1963). Chemistry of activation: Magnussen, *Biochem. J.* **115**, 2P (1969). Synthesized normally by liver parenchymal cells in a cyclic asynchronous manner. Dicoumarol derivatives halt the synthesis, while vitamin K_1 stimulates synchronized activity of all the liver parenchymal cells: Barnhart, Anderson, *Biochem. Pharmacol.* **9**, 23 (1962). The glycoprotein structure probably consists of a single polypeptide chain containing between 8 and 10% carbohydrate: Magnussen, *Thromb. Diath. Haemorrh.* **suppl. 54**, 31 (1973). Reviews: W. H. Seegers, *Prothrombin* (Harvard Univ. Press, 1962) 728 pp; Magnussen "Prothrombin and Thrombin," in *The Enzymes,* **Vol. III**, P. D. Boyer, Ed. (Academic Press, New York, 3rd ed., 1971) pp 277-321; K. G. Mann, *Methods Enzymol.* **45B**, 123-156 (1976). Review on prothrombin activation: several authors in *Ann. N.Y. Acad. Sci.* **370**, 336-528 (1981).

Most stable within the range pH 4-9.5. Isoelec pt pH 4.2. Very sol in water but pptd at pH 4.2-4.5. Solns tend to activate spontaneously. There is little loss of activity when drying from the frozen state but the dry material alters in a few months. Drying with organic solvents destroys activity.

8075. Protiofate. *3,4-Dihydroxy-2,5-thiophenedicarboxylic acid dipropyl ester;* dipropyl 3,4-dihydroxy-2,5-thiophenedicarboxylate; Atrimycon. $C_{12}H_{16}O_6S$; mol wt 288.32. C 49.99%, H 5.59%, O 33.29%, S 11.12%. Thiophene derivative with antimycotic activity: L. Riviera *et al., Chemioterapia* **3**, 116 (1984). Prepn: **Fr. pat. 2,255,063** (1975 to Bioindustria), *C.A.* **84**, 89986f (1976).

Crystals from ethanol-water, mp 96-97°.
THERAP CAT: Topical fungicide.

8076. Protionamide. *2-Propyl-4-pyridinecarbothio-amide; 2-propylthioisonicotinamide;* prothionamide; 2-propyl-4-thiocarbamoylpyridine; Ektebin; Peteha; Trevintix. $C_9H_{12}N_2S$; mol wt 180.27. C 59.96%, H 6.71%, N 15.54%, S 17.79%. Prepn: Libermann *et al., Compt. Rend.* **242,** 2409 (1956); **Brit.** pat. **800,250** (1958 to Chimie et Atomistique); L. Bernardi, G. Palamidessi, **Fr.** pat. **2,091,338** (1972 to Societa Farm. Ital); L. N. Yakhontov *et al., Khim.-Farm. Zh.* **10,** 96 (1976), *C.A.* **85,** 32787h (1976). Pharmacology: A. M. Il'in, *Farmakol. Toksikol.* **38,** 471 (1975).

Crystals, mp 142°. Sol in ethanol, methanol; slightly sol in ether, chloroform. Practically insol in water. LD_{50} in mice, rats, cats: 1.0, 1.32 mg/kg, >1 g/kg i.p.
THERAP CAT: Antibacterial (tuberculostatic).

8077. Protizinic Acid. (±)-*7-Methoxy-α,10-dimethyl-10H-phenothiazine-2-acetic acid;* 7-methoxy-α,10-dimethyl-phenothiazine-3-acetic acid; 2-(7-methoxy-10-methyl-2-phenothiazinyl)propionic acid; Pirocrid. $C_{17}H_{17}NO_3S$; mol wt 315.39. C 64.74%, H 5.43%, N 4.44%, O 15.22%, S 10.17%. Prepn: **Brit.** pat. **1,048,680** corresp to Farge *et al.,* **U.S.** pat. **3,450,698;** *eidem,* **S. Afr.** pat. **68 00,435,** *C.A.* **70,** 68394g (1969), (1966, 1969, 1968, all to Rhône-Poulenc).

Crystals from diisopropyl ether or acetonitrile, mp 124-125°.
THERAP CAT: Anti-inflammatory.

8078. Protoanemonin. *5-Methylene-2(5H)-furanone; 4-hydroxy-2,4-pentadienoic acid γ-lactone;* 5-methylene-2-oxo-dihydrofuran; Isomycin. $C_5H_4O_2$; mol wt 96.09. C 62.50%, H 4.20%, O 33.30%. An antibacterial principle of *Anemone pulsatilla* L., *Ranunculaceae:* Seegal, Holden, *Science* **101,** 413 (1945). Exists as a glucoside, *ranunculin,* in the intact plant, and is released by an enzymatic process on maceration of the plant tissue: R. Hill, R. van Heyningen, *Biochem. J.* **49,** 332 (1951). Isoln: Asahina, Fujita, *Acta Phytochim. Japan* **1,** 1 (1922); Baer *et al., J. Biol. Chem.* **162,** 65 (1946). Structure and synthesis: Asahina, Fujita, *loc. cit.;* Muskat *et al., J. Am. Chem. Soc.* **52,** 326 (1930); Fox, Jr., *Proc. Soc. Exp. Biol.* **51,** 102 (1942); Shaw, *J. Am. Chem. Soc.* **68,** 2510 (1946). Industrial prepns: **Brit.** pat. **759,999** (1956 to Olin Mathieson), *C.A.* **51,** 9678f (1957); Reicheneder *et al.,* **Ger.** pat. **1,088,047** (BASF), *C.A.* **56,** 14086i (1962); Sakuma, Hirano, **U.S.** pat. **3,203,863** (1965 to Lion Dentifrice).

Pale yellow oil; volatile with steam. $bp_{1.5}$ 45°. Sol in ethylene dichloride, chloroform. Solubility in water ~1%.

Stable in water. When pure the compd turns to a hard polymer which, when ground and extracted with boiling ethyl acetate, yields anemonin, *q. v.*
THERAP CAT: Antibacterial.

8079. Protocatechualdehyde. *3,4-Dihydroxybenzalde-hyde;* 3,4-dihydroxybenzenecarbonal; protocatechuic aldehyde; rancinamycin IV. $C_7H_6O_3$; mol wt 138.12. C 60.87%, H 4.38%, O 34.75%. Prepn from catechol: Reimer, Tiemann, *Ber.* **9,** 1268 (1876); Tiemann, Koppe, *ibid.* **14,** 2015 (1881); from vanillin: Tiemann, Haarmann, *ibid.* **7,** 620 (1874); from veratric aldehyde: Dreyfus, **Ger.** pat. **193,958;** *Frdl.* **9,** 161 (1908-10); from piperonal: Hoering, Baum, *Ber.* **41,** 1914 (1908); Barger, *J. Chem. Soc.* **93,** 563 (1908); Buck, Zimmerman, *Org. Syn.* coll. vol. II, 549 (1943).

Platelets from water or toluene. Dimorphic. Dec 153-154°. K at 25° = 2.8×10^{-8}. Soly in water (g/100 ml): 5 (20°); 33 (99°); in ethanol: 79 (78°). Freely sol in ether.

8080. Protocatechuic Acid. *3,4-Dihydroxybenzoic acid.* $C_7H_6O_4$; mol wt 154.12. C 54.55%, H 3.92%, O 41.52%. Minute amounts are found in wheat grains, in wheat seedlings, and in many other plants: L. Hörhammer, A. Scherm, *Arch. Pharm.* **288,** 441 (1955). Prepd by the alkaline fusion of vanillin: Tiemann, Haarmann, *Ber.* **7,** 617 (1874); Pearl, *J. Am. Chem. Soc.* **68,** 2180 (1946); *Org. Syn.* coll. vol. III, 745 (1955).

White to brownish, cryst powder; discolors in air. d 1.54. mp ~200° with decompn. Sol in 50 parts water; sol in alcohol, ether. *Keep well closed.*

8081. Protokosin. From dried flowers of *Hagenia abyssinica* J. J. Gmel. (*Brayera anthelmintica* Kunth.), *Rosaceae.* Isoln: Leichsenring, *Arch. Pharm.* **232,** 58 (1894); Lobeck, *ibid.* **239,** 682 (1901); Hems, Todd, *J. Chem. Soc.* **1937,** 562. Structural studies: Birch, Todd, *ibid.* **1952,** 3102; Orth, Riedl, *Ann.* **663,** 83 (1963). Exists in several tautomeric forms and is a mixt of isobutyryl, isovaleryl, and 2-methylbutyryl side chain homologues. Structural elucidation: Lounasmaa *et al., Acta Chem. Scand.* **28B,** 1200 (1974).

$$R = CH(CH_3)_2, \quad CH_2CH(CH_3)_2, \quad CH(CH_3)CH_2CH_3$$

Colorless needles from acetone, mp 181-183°. $[\alpha]_D^{25}$ +13.9° (c = 0.610 in chloroform). uv max (cyclohexane): 224, 285 nm (ε 28,200, 36,400). Practically insol in water; slightly sol in alcohol, light petroleum; freely sol in ether, acetone, ethyl acetate, chloroform.

8082. Protokylol. *4-[2-[[2-(1,3-Benzodioxol-5-yl)-1-methylethyl]amino]-1-hydroxyethyl]-1,2-benzenediol; α-[(α-methyl-3,4-methylenedioxyphenethylamino)methyl]protocatechuyl alcohol; α-(3,4-dihydroxyphenyl)-β-[2-(3,4-methylenedioxyphenyl)isopropylamino]ethanol; 1-(3,4-dihydroxyphenyl)-2-(α-methyl-3,4-methylenedioxyphenethylamino)ethanol; N-[β-(3,4-methylenedioxyphenyl)isopropyl]-β-(3,4-dihydroxyphenyl)-β-hydroxyethylamine; N-[2-(3,4-methylenedioxyphenylisopropyl)]norepinephrine.* $C_{18}H_{21}NO_5$; mol wt 331.37. C 65.24%, H 6.39%, N 4.23%, O 24.14%. β-Adrenergic agonist. Prepn: Biel *et al., J. Am. Chem. Soc.* **76**, 3149 (1954); Biel, U.S. pat. **2,900,415** (1959 to Lakeside Labs.). Toxicity: E. I. Goldenthal, *Toxicol. Appl. Pharmacol.* **18**, 185 (1971).

Hydrochloride, $C_{18}H_{21}NO_5 \cdot HCl$, *JB-251, Caytine, Ventaire.* Crystals from isopropanol, mp 126-127°. Sol in water. LD_{50} orally in rats: 938 ±96 mg/kg (Goldenthal).
THERAP CAT: Bronchodilator.
THERAP CAT (VET): Bronchodilator.

8083. Protopine. *4,6,7,14-Tetrahydro-5-methyl-bis[1,3]benzodioxolo[4,5-c:5',6'-g]azecin-13(5H)-one; 7-methyl-2,3:9,10-bis(methylenedioxy)-7,13a-secoberbin-13a-one;* fumarine; macleyine. $C_{20}H_{19}NO_5$; mol wt 353.37. C 67.98%, H 5.42%, N 3.96%, O 22.64%. From opium: Hesse, *Ber.* **4**, 693 (1871); also from the herb *Fumaria officinalis* L., *Chelidonium majus* L., and many other *Papaveraceae* and *Fumariaceae;* Manske in *The Alkaloids* vol. **IV**, R. H. F. Manske, H. L. Holmes, Eds. (Academic Press, New York, 1954) pp 157-159. Structure: Perkin, Jr., *J. Chem. Soc.* **109**, 815 (1916); Gadamer, Bruchhausen, *Arch. Pharm.* **260**, 97 (1922); Mottus *et al., Can. J. Chem.* **31**, 1144 (1953); Anet, Marion, *ibid.* **32**, 452 (1954). Synthesis: Haworth, Perkin, *J. Chem. Soc.* **1926**, 1769. Crystal structure: Hall, Ahmed, *Acta Cryst.* **24B**, 337 (1968).

Monoclinic prisms from alcohol + chloroform, mp 208°. d 1.399 (calc). uv max (95% ethanol): 293 nm (log ε 3.93). Sol in 15 parts chloroform, 900 parts alc, 1000 parts ether. Slightly sol in ethyl acetate, carbon disulfide, benzene, petr ether. Practically insol in water.
Hydrochloride, $C_{20}H_{19}NO_5 \cdot HCl$, prisms from alcohol, sol in 143 parts water at 13°, sol in alcohol. Also a hexahydrate, needles from water.
Methiodide, $C_{20}H_{19}NO_5 \cdot CH_3I$, twinned crystals from methanol, dec 217°.

8084. Protoporphyrin IX. *7,12-Diethenyl-3,8,13,17-tetramethyl-21H,23H-porphine-2,18-dipropanoic acid; 3,7,12,17-tetramethyl-8,13-divinyl-2,18-porphinedipropionic acid;* 1,3,5,8-tetramethyl-2,4-divinylporphine-6,7-dipropionic acid; ooporphyrin; Kammerer's porphyrin. $C_{34}H_{34}N_4O_4$; mol wt 562.67. C 72.58%, H 6.09%, N 9.96%, O 11.37%. Biological precursor of blood and plant pigments. Prepd from hemin: Fischer-Orth, *Die Chemie des Pyrrols* **II**, 1, 396 (Leipzig, 1937); Ramsey, *Biochem. Prepn.* **3**, 39 (1953). Structure: Sparatore, Mauzerall, *J. Org. Chem.* **25**, 1073 (1960). Synthesis: Carr *et al., J. Chem. Soc. (C)* **1971**, 487. Crystal and molecular structure: W. S. Caughey, J. A. Ibers, *J. Am. Chem. Soc.* **99**, 6639 (1977). Chelates with metals, esp iron, in the ferrous state to form heme *(q.v.),* in the ferric state to form hematin *(q.v.). Review:* Rimington,

Kennedy, in M. Florkin, H. S. Mason, *Comparative Biochemistry* (Academic Press, New York, 1962) pp 557-614.

Monoclinic, brownish-yellow prisms from ether. Absorption max (25% HCl): 602.4, 582.2, 557.2 nm. Freely sol in chloroform, glacial acetic acid, alcohol contg HCl, ether contg some glacial acetic acid, hydrochloric acid. Somewhat sol in dil alkalies, aniline, pyridine. Forms sparingly sol disodium and dipotassium salts.
Disodium salt, $C_{34}H_{32}N_4Na_2O_4$, *Depocolin-S.*
Dimethyl ester, $C_{36}H_{38}N_4O_4$, crystals from chloroform + methanol, mp 228-230°. Absorption max (25% HCl): 601, 556, 409 nm. Soluble in chloroform, slightly sol in methanol. Insol in sodium carbonate solns.
THERAP CAT: In liver disease.

8085. Protostephanine. *6,7,8,9-Tetrahydro-2,3,10,12-tetramethoxy-7-methyl-5H-dibenz[d,f]azonine.* $C_{21}H_{27}NO_4$; mol wt 357.45. C 70.56%, H 7.61%, N 3.92%, O 17.90%. A member of the hasubanan alkaloids. First alkaloid known to possess the unique dibenz[d,f]azonine structure. Isoln from *Stephania japonica,* Miers, *Menispermaceae:* H. Kondo, T. Sanada, *J. Pharm. Soc. Japan* **541**, 177 (1927), *C.A.* **21**, 2700[4] (1927); H. Kondo, T. Watanabe, *ibid.* **58**, 268 (1938), *C.A.* **32**, 5403[5] (1938). Structure: K. Takeda, *Bull. Agr. Chem. Soc. Japan* **20**, 165 (1956). Synthesis: *idem, C.A.* **60**, 5570f (1964); B. Pecherer, A. Brossi, *J. Org. Chem.* **32**, 1053 (1967); A. R. Battersby *et al., J. Chem. Soc. Perkin Trans. I* **1981**, 2002. Biosynthesis: *eidem, ibid.* 2010, 2016, 2030.

White crystalline solid from benzene, mp 84-86° (Pecherer). Also reported as mp 75° (Kondo).

8086. Protoveratrines. From the rhizome of *Veratrum album* L., Liliaceae: Salzberger, *Arch Pharm.* **228**, 462 (1890); Poethke, *ibid.* **275**, 357 (1937); Craig, Jacobs, *J. Biol. Chem.* **143**, 427 (1942); **149**, 271 (1943). Mixture of protoveratrines A and B: Glen *et al., Nature* **170**, 932 (1952); Klohs *et al., J. Am. Chem. Soc.* **74**, 5107 (1952); Nash, Brooker, *ibid.* **75**, 1942 (1953); Stoll, Seebeck, *Helv. Chim. Acta* **36**, 718 (1953); Nash, Brooker, U.S. pat. **2,929,812** (1960 to Allied Labs). Structure of protoveratrines A and B: Kupchan, Ayres, *J. Am. Chem. Soc.* **82**, 2252 (1960). Comparative toxicity: O. Krayer *et al., J. Pharmacol. Exp. Ther.* **82**, 167 (1944); K. Tanaka, *ibid.* **113**, 89 (1955).

Protoveratrine A   R = H
Protoveratrine B   R = OH

Slightly bitter, sternutative crystals from alc, dec 266-267°. $[\alpha]_D^{25}$ −38.6° (pyridine). $[\alpha]_D^{25}$ −8.5° (c = 1.99 in chloroform). Soluble in chloroform; dil aq acidic solns; slightly sol in ether. Practically insol in water, petr ether. LD$_{50}$ i.v. in mice: 0.048 mg/kg (Krayer).

Protoveratrine A, C$_{41}$H$_{63}$NO$_{14}$, *[3β(S),4α,6α,7α,15α(R),-16β]-4,9-epoxycevane-3,4,6,7,14,15,16,20-octol 6,7-diacetate 3-(2-hydroxy-2-methylbutanoate) 15-(2-methylbutanoate), Protalba.* Crystals from acetone, dec 267-269°. $[\alpha]_D^{25}$ −40.5° (pyridine); $[\alpha]_D^{25}$ −10.5° (chloroform). Soluble in chloroform, pyridine, hot alcohol. LD$_{50}$ s.c. in male mice: 0.29 mg/kg (Tanaka).

Protoveratrine B, C$_{41}$H$_{63}$NO$_{15}$, *[3β(2R,3R),4α,6α,7α,15α-(R),16β]-4,9-epoxycevane-3,4,6,7,14,15,16,20-octol 6,7-diacetate 3-(2,3-dihydroxy-2-methylbutanoate) 15-(2-methylbutanoate), veratetrine, neoprotoveratrine.* Crystals from acetone, dec 268-270°. $[\alpha]_D^{25}$ −37° (pyridine); $[\alpha]_D^{25}$ −3.5° (chloroform). Soluble in chloroform, pyridine, hot alcohol. LD$_{50}$ s.c. in male mice: 0.21 mg/kg (Tanaka).

Mixture of protoveratrine A and B, *Provell, Tensatrin, Veralba.*

THERAP CAT: Antihypertensive.

8087. Protoverine. *(3β,4α,6α,7α,15α,16β)-4,9-Epoxycevane-3,4,6,7,14,15,16,20-octol.* C$_{27}$H$_{43}$NO$_9$; mol wt 525.64. C 61.70%, H 8.25%, N 2.66%, O 27.39%. Obtained by alkaline hydrolysis of protoveratrine A: Poethke, *Arch. Pharm.* **275**, 571 (1937); Stoll, Seebeck, *Helv. Chim. Acta* **36**, 718 (1953). Structure: Kupchan *et al., Chem. & Ind. (London)* **1958**, 1626; *J. Am. Chem. Soc.* **81**, 1009 (1959); **82**, 2242 (1960). Comparative toxicity: O. Krayer *et al., J. Pharmacol. Exp. Ther.* **82**, 167 (1944); K. Tanaka, *ibid.* **113**, 89 (1955).

Fine needles from methanol, dec 220-222°. $[\alpha]_D^{20}$ −15.7° (c = 1.1 in pyridine). Sol in CHCl$_3$, methanol, pyridine. LD$_{50}$ i.v. in mice: 194.0 mg/kg (Krayer). LD$_{50}$ s.c. in male mice: 520 mg/kg (Tanaka).

8088. Protriptyline. *N-Methyl-5H-dibenzo[a,d]cyclo-heptene-5-propanamine;* 5-(3-methylaminopropyl)-5H-di-

benzo[a,d]cycloheptene; 7-(3-methylaminopropyl)-1,2:5,6-dibenzocycloheptatriene; amimetilina. C$_{19}$H$_{21}$N; mol wt 263.38. C 86.65%, H 8.04%, N 5.32%. Prepn: Engelhardt, Christy, **Belg. pat.** **617,967** (1962 to Merck & Co.), *C.A.* **59**, 517f (1963); Tishler *et al.,* U.S. pats. **3,244,748** and **3,271,-451** (both 1966 to Merck & Co.); Engelhardt *et al., J. Med. Chem.* **11**, 325 (1968). Metabolism in man, pig and dog: Sisenwine *et al., J. Pharmacol. Exp. Ther.* **175**, 51 (1970). Use in treatment of sleep apnea: R. W. Clark *et al., Neurology* **29**, 1287 (1979); L. G. Brownell *et al., N. Engl. J. Med.* **307**, 1037 (1982).

Hydrochloride, C$_{19}$H$_{21}$N.HCl, *MK-240, Concordin, Maximed, Triptil, Vivactil.* Crystals from isopropanol-ethyl ether, mp 169-171°. uv max: 290 nm (ε 13311). Freely sol in water. pKa 8.2.

THERAP CAT: Antidepressant.

8089. Pro-Urokinase. *Prourokinase (enzyme-activating);* single-chain urokinase-type plasminogen activator; single-chain pro-urokinase; scu-PA; pro-UK; pro u-PA; PUK; Sandolase; Thombolyse; Tomieze. Single-chain proenzyme form of urokinase, *q.v.* Produced by the kidney; present in urine and blood. Consists of 411 amino acid residues, mol wt ~54,000 daltons. Converted by plasmin or kallikrein to active two-chain urokinase by proteolytic cleavage of the Lys 158 - Ile 159 peptide bond. Identification of proenzyme activity from cultured human kidney cells: M. B. Bernik, *J. Clin. Invest.* **52**, 823 (1973). Isoln from culture media: C. Nolan *et al., Biochim. Biophys. Acta* **496**, 384 (1977). Purification from human plasma: T.-C. Wun *et al., J. Biol. Chem.* **257**, 3276 (1982); from human urine: S. S. Husain *et al., Arch. Biochem. Biophys.* **220**, 31 (1983). Amino acid sequence and cleavage site: S. Kasai *et al., J. Biol. Chem.* **260**, 12382 (1985). Clot-specific activity: V. Gurewich *et al., J. Clin. Invest.* **73**, 1731 (1984). Clinical pharmacology: G. Trübestein *et al., Haemostasis* **17**, 238 (1987). Clinical trials in acute myocardial infarction: C. Bode *et al., Am. J. Cardiol.* **61**, 971 (1988); in combination with t-PA: V. Gurewich, *J. Am. Coll. Cardiol.* **10**, 16B (1987). Review of mechanism of action studies: V. Gurewich, R. Pannell, *Sem. Thromb. Hemostas.* **13**, 146-151 (1987); and physicochemical properties: V. Gurewich, *ibid.* **14**, 110-115 (1988).

Saruplase, prourokinase (enzyme-activating) (human clone pUK4/pUK18 protein moiety reduced), CG-4509. Nonglycosylated human prourokinase produced in *E. coli* by recombinant DNA technology. Prepn: H. L. Heyneker *et al.,* **Eur. pat. Appl.** **92182;** eidem, U.S. pat. **5,112,755** (1983, 1992 both to Genentech); W. E. Holmes *et al., Biotechnology* **3**, 923 (1985). Series of articles on thrombolytic activity: *Thromb. Haemostas.* **52**, 19-33 (1984). Pharmacokinetics: A. de Boer *et al., ibid.* **70**, 320 (1993). Clinical trial: PRIMI Trial Study Group, *Lancet* **1**, 863 (1989); J. Schofer *et al., Eur. Heart J.* **14**, 958 (1993).

THERAP CAT: Thrombolytic.

8090. Proxazole. *N,N-Diethyl-3-(1-phenylpropyl)-1,2,4-oxadiazole-5-ethanamine;* 5-[2-(diethylamino)ethyl]-3-(α-ethylbenzyl)-1,2,4-oxadiazole; propaxoline. C$_{17}$H$_{25}$N$_3$O; mol wt 287.41. C 71.05%, H 8.77%, N 14.62%, O 5.57%. Smooth muscle relaxant. Prepn: **Brit. pat.** **924,608;** Palazzo, Silvestrini, U.S. pat. **3,141,019** (1963, 1964 both to Angelini Francesco). Pharmacology: Silvestrini, Pozzatti, *Arzneimittel-Forsch.* **13**, 798 (1963). Separation and pharmacology of the enantiomers: De Feo *et al., Farmaco Ed. Sci.* **26**, 370 (1971). Clinical trial in cerebrovascular insufficiency: G. Esposito, M. De Gregorio, *Arzneimittel-Forsch.* **24**, 1692 (1974).

bp$_{0.2}$ 132°.

Citrate, $C_{17}H_{25}N_3O.C_6H_8O_7$, *AF-634, Flou, Pirecin, Toness.* LD$_{50}$ in rats (mg/kg): 39 i.p., 60 orally (Silvestrini, Pozzatti).

Nitrate, crystals, mp 127-128°.

THERAP CAT: Antispasmodic.

THERAP CAT (VET): Antispasmodic.

8091. Proxibarbal. *5-(2-Hydroxypropyl)-5-(2-propenyl)-2,4,6(1H,3H,5H)-pyrimidinetrione; 5-allyl-5-(2-hydroxypropyl)barbituric acid;* 5-allyl-5-(β-hydroxypropyl)barbituric acid; 5-allyl-5-(β-hydroxypropyl)malonylurea; proxibarbital; HH-184; Axeen; Centralgol; Centralgyl; Ipronal. $C_{10}H_{14}$-N_2O_4; mol wt 226.23. C 53.09%, H 6.24%, N 12.38%, O 28.29%. Prepn: Bobranski *et al., Roczniki Chem.* **30,** 175 (1956), *C.A.* **51,** 438f (1957); **Brit.** pat. 953,387 (1964 to Hommel A.G.), *C.A.* **61,** 3123a (1964); Smissman *et al., J. Med. Chem.* **14,** 853 (1971). Metabolism: Bobranski *et al., Arch. Immunol. Ther. Exp.* **10,** 895 (1962); B. Lambrey *et al., Eur. J. Med. Chem.-Chim. Ther.* **15,** 463 (1980). Pharmacokinetics: *eidem, ibid.* **12,** 565 (1977).

Crystals from benzene + ethanol, mp 157-158°; also reported as mp 166.5-168.5° from acetone + chloroform (Smissman). Moderately sol in water.

Note: This is a controlled substance (depressant) listed in the U.S. Code of Federal Regulations, Title 21 Part 1308.13 (1995).

THERAP CAT: Sedative, hypnotic.

8092. Proxyphylline. *3,7-Dihydro-7-(2-hydroxypropyl)-1,3-dimethyl-1H-purine-2,6-dione; 7-(2-hydroxypropyl)theophylline;* Brontyl; Proxy-Retardoral; Purophyllin; Spantin; Spasmolysin; Thean; Theon. $C_{10}H_{14}N_4O_3$; mol wt 238.25. C 50.41%, H 5.92%, N 23.52%, O 20.15%. Smooth muscle relaxant. Prepd by refluxing 1-chloro-2-propanol with theophylline in an alkaline medium: Rice, **U.S.** pat. 2,715,-125 (1955 to Gane's Chem. Works).

Crystals from abs ethanol, mp 135-136°. One gram dissolves in about 1 ml water, in 14 ml abs ethanol. More sol in boiling ethanol. pH of a 5% aq soln 5.5 to 7.0. Solns may be sterilized by heating.

THERAP CAT: Bronchodilator; vasodilator.

8093. Prozapine. *1-(3,3-Diphenylpropyl)hexahydro-1H-azepine; 1-(3,3-diphenylpropyl)hexamethylenimine;* 1,1-diphenyl-3-hexamethyleneiminopropane; hexadiphane. C_{21}-$H_{27}N$; mol wt 293.45. C 85.95%, H 9.27%, N 4.77%.

Prepn: Janssen, de Jongh, **U.S.** pat. **2,881,165** (1959 to N. V. Nederland. Comb. Chem. Ind.).

bp$_1$ 170-174°.

Mixture of hydrochloride with sorbitol, *Norbiline.*

THERAP CAT: Antispasmodic.

8094. Prunetin. *5-Hydroxy-3-(4-hydroxyphenyl)-7-methoxy-4H-1-benzopyran-4-one; 4',5-dihydroxy-7-methoxyisoflavone;* prunusetin. $C_{16}H_{12}O_5$; mol wt 284.27. C 67.60%, H 4.25%, O 28.14%. Isoln from *Prunus* spp., *Rosaceae:* Hasegawa, Shirato, *J. Am. Chem. Soc.* **79,** 450 (1957); Hasegawa, *ibid.* 1738; Goel, Seshadri, *Tetrahedron* **5,** 91 (1959); Plouvier, *Compt. Rend.* **250,** 594 (1960). Identity with prunusetin: King, Jurd, *J. Chem. Soc.* **1952,** 3211. Structure: Shrimer, Hull, *J. Org. Chem.* **10,** 288 (1945). Synthesis: Bradbury, White, *J. Chem. Soc.* **1953,** 871.

Needles from ethanol, mp 240°.

Diacetate, $C_{20}H_{16}O_7$, rods from methanol, mp 222.5°.

4'-Glucoside, $C_{22}H_{22}O_{10}$, prunitrin. Isoln: Finnemore, *Pharm. J.* **31,** 604 (1910). Structure and synthesis: Zemplén, Farkas, *Ber.* **90,** 836 (1957). Needles, mp 235-236°. Sol in hot water, ethyl acetate.

8095. Pseudoaconitine. *(1α,3α,6α,14α,16β)-20-Ethyl-1,6,16-trimethoxy-4-(methoxymethyl)aconitane-3,8,13,14-tetrol 8-acetate 14-(3,4-dimethoxybenzoate);* acraconitine; feraconitine; nepaline; veratroylaconine; "English" aconitine; "Nepal" aconitine. $C_{36}H_{51}NO_{12}$; mol wt 689.80. C 62.68%, H 7.45%, N 2.03%, O 27.83%. From tubers of *Aconitum ferox* Wall., *Ranunculaceae. Ref:* Wright, Luff, *J. Chem. Soc.* **33,** 151 (1878); Dunstan, Carr, *ibid.* **71,** 350 (1897); Henry, Sharp, *ibid.* **1928,** 1105, 3094; K. Paech, M. V. Tracey, *Modern Methods of Plant Analysis* Vol. IV (Springer-Verlag, Berlin, 1955) p 375. Structure: Gilman, Marion, *Tetrahedron Letters* **1962,** 923. Stereochemistry: Tsuda, Marion, *Can. J. Chem.* **41,** 1485 (1963).

White crystals, or amorphous, syrupy mass. *Very poisonous!* mp 214°. [α]$_D^{20}$ +17° (alc); the salts are levorotatory. Practically insol in water. Sol in alcohol, ether.

Hydrobromide trihydrate, $C_{36}H_{52}BrNO_{12}.3H_2O$, triangular prisms, mp 199°. [α]$_D^{20}$ −18.5° (water).

8096. Pseudobaptigenin. *3-(1,3-Benzodioxol-5-yl)-7-hydroxy-4H-1-benzopyran-4-one; 7-hydroxy-3',4'-(methylenedioxy)isoflavone.* $C_{16}H_{10}O_5$; mol wt 282.25. C 68.09%, H 3.57%, O 28.34%. The aglycon of pseudobaptisin. Prepn by hydrolysis of natural pseudobaptisin: Gorter, *Arch. Pharm.*

Pseudococaine

235, 494 (1897). Structure: Späth, Schmidt, *Monatsh.* **53**, 454 (1929). Syntheses: Späth, Lederer, *Ber.* **63**, 743 (1930); Mahal *et al.*, *J. Chem. Soc.* **1934**, 1771; Baker *et al.*, *ibid.* **1937**, 805; **1953**, 1852; Farkas, *Ber.* **91**, 2858 (1958); Dhoubhadel, Joshi, *J. Indian Chem. Soc.* **52**, 440 (1975).

Long, felted needles from alc, dec 296-298°. Sublimes in high vacuum at ~220°. Sparingly sol in the usual solvents. 7-Rhamnoglucoside trihydrate, $C_{28}H_{30}O_{14}.3H_2O$, *pseudobaptisin*. Crystals, becomes anhyd at 120°. mp 148-150° (evac tube). If heating is continued, it resolidif at 180-210° and melts again at 249-251°. $[\alpha]_D^{14}$ —98.1° (c = 1.1 in methanol). Freely sol in methanol, hot water, hot acetone.

8097. Pseudococaine. *[1R-(2-endo,3-exo)]-3-(Benzoyloxy)-8-methyl-8-azabicyclo[3.2.1]octane-2-carboxylic acid methyl ester; 3β-hydroxy-1αH,5αH-tropane-2α-carboxylic acid methyl ester benzoate; 2α-carbomethoxy-3β-benzoxytropane; depsococaine; dextrocaine; isocaine; Delcaine.* $C_{17}H_{21}$-NO_4; mol wt 303.36. C 67.31%, H 6.98%, N 4.62%, O 21.10%. Cocaine diastereomer with greater local anesthetic activity than the natural substance. Synthesis: Einhorn, Marquardt, *Ber.* **23**, 468, 981 (1890); Willstätter, Bode, *Ann.* **326**, 42 (1903); Willstätter, Bommer, *ibid.* **422**, 34 (1921); Willstätter *et al.*, *ibid.* **434**, 138 (1923); **Brit. pat. 210,050** (1923 to E. Merck). Configuration: S. P. Findlay, *J. Am. Chem. Soc.* **76**, 2855 (1954). Conformational analysis and ¹H, ¹³C NMR studies: F. I. Carroll *et al.*, *J. Org. Chem.* **1982**, 13. Pharmacokinetics: A. L. Misra *et al.*, *Experientia* **32**, 895 (1976). Interaction with sodium channels: J C. Matthews, A. Collins, *Biochem. Pharm.* **32**, 455 (1983). Pharmacology of cocaine and pseudocaine: G. Schmidt *et al.*, *Arch. Exp. Pathol. Pharmakol.* **240**, 523 (1961).

Prisms, mp 47°. $[\alpha]_D^{20}$ +42° (c = 5 in chloroform). Freely sol in ether, chloroform, benzene, petr ether. Slightly sol in water.
Hydrochloride, $C_{17}H_{21}NO_4.HCl$, crystals from alcohol, mp 210°. $[\alpha]_D^{20}$ +41° (c = 5), less sol in water than cocaine hydrochloride.
Tartrate, $C_{17}H_{21}NO_4.C_4H_6O_6$, *Psicaine*. Crystals, mp 139°. $[\alpha]_D^{20}$ +43° (c = 5 in water). Sol in 4 parts water, in alcohol. The aq soln is stable and may be sterilized by boiling, without decompn. pH about 3.7 (2% soln).
Sodium tartrate, *Psicaine N*. More sol in water than the tartrate; esp useful where a neutral soln is desired.
n-Propyl ester analog, $C_{19}H_{25}NO_4$, *Neopsicaine*.
THERAP CAT: Formerly as anesthetic (local).

8098. Pseudocodeine. *(5α,8β)-6,7-Didehydro-4,5-epoxy-3-methoxy-17-methylmorphinan-8-ol; ψ-codeine; neoisocodeine.* $C_{18}H_{21}NO_3$; mol wt 299.37. C 72.22%, H 7.07%, N 4.68%, O 16.03%. An isomer of codeine. Review and bibliography: K. W. Bentley, *The Chemistry of the Morphine Alkaloids* (Oxford, 1954).

White needles, mp 181-182°. $[\alpha]_D$ —96.6° (alc). Slightly sol in water; sol in alcohol.

8099. Pseudoconhydrine. *(3S-trans)-6-Propyl-3-piperidinol;* 5-hydroxy-2-propylpiperidine; 5-hydroxyconiine. C_8-$H_{17}NO$; mol wt 143.23. C 67.09%, H 11.96%, N 9.78%, O 11.17%. In *Conium maculatum* L., *Umbelliferae* (hemlock). Extraction procedure: Ladenburg *et al.*, *Ber.* **24**, 1671 (1891); Braun, *Ber.* **38**, 3108 (1905); Löffler, *Ber.* **42**, 116 (1909). Structure: E. Späth *et al.*, *Ber.* **66**, 591 (1933). Stereochemistry: R. K. Hill, *J. Am. Chem. Soc.* **80**, 1611 (1958); K. Tadano *et al.*, *J. Carbohydr. Chem.* **4**, 129 (1985).

Hygroscopic needles from abs ether, mp 106° (monohydrate, scales, mp 60° from moist ether). bp 236°. $[\alpha]_D^{20}$ +11° (c = 10 in alc). pK (18°): 3.70. Soluble in water and most organic solvents.
Hydrochloride, $C_8H_{17}NO.HCl$, crystals from alcohol, mp 213°. Freely sol in water, sparingly sol in alcohol, acetone.

8100. Pseudocumene. *1,2,4-Trimethylbenzene;* pseudocumol; asymmetrical trimethylbenzene. C_9H_{12}; mol wt 120.19. C 89.94%, H 10.06%. Occurs in coal tar and in many petroleums. Physical properties: Hirschler, Falconer, *J. Am. Chem. Soc.* **68**, 210 (1946). Toxicity: *Handbook of Toxicology* **vol. 1**, W. S. Spector, Ed. (Saunders, Philadelphia, 1956) pp 306-307.

Liquid. d_4^{20} 0.8761. bp 169-171°. mp —43.78° (Hirschler, Falconer). n_D^{21} 1.5044. Practically insol in water; sol in alc, benzene, ether. LD i.p. in rats: ~2.0 ml/kg (Spector).
Caution: CNS depressant; respiratory irritant.
USE: Sterilizing catgut by heating one hour at 160°; manuf dyes, perfumes and resins. Oxidation yields trimellitic anhydride.

8101. Pseudoionone. *6,10-Dimethyl-3,5,9-undecatriene-2-one;* citrylideneacetone; 2,6-dimethylhendeca-2,6,8-trien-10-one. $C_{13}H_{20}O$; mol wt 192.30. C 81.20%, H 10.48%, O 8.32%. Intermediate in the synthesis of α- and β-ionone. Prepn from citral and acetone: Tiemann, Krüger, *Ber.* **26**, 2692 (1893); Stiehl, *J. Prakt. Chem.* [2] **58**, 84 (1898); Tiemann, *Ber.* **32**, 115 (1899); Hibbert, Cannon, *J. Am. Chem. Soc.* **46**, 119 (1924); A. Russel, R. L. Kenyon, *Org. Syn.* **coll. vol. III**, 747 (1955). A less pure product is obtained by the treatment of oil of lemon grass and acetone with bleaching powder, cobalt nitrate, and alcohol: Ziegler, *J. Prakt. Chem.* [2] **57**, 493 (1898); Tiemann, *Ber.* **31**, 2313 (1898); Haarmann, Reimer & Co., **Ger. pat. 73,089**; *Frdl.* **3**, 889. Synthesis: T. Onishi *et al.*, *Synthesis* **1980**, 651.

Consult the Name Index before using this section.

Pale yellow oil, bp$_2$ 114-116°; bp$_4$ 124-126°; bp$_{12}$ 143-145°; d^{20} 0.8984; n$_D^{20}$ 1.53346.

8102. Pseudojervine. *(3β,23β)-17,23-Epoxy-3-(β-D-glucopyranosyloxy)veratraman-11-one.* C$_{33}$H$_{49}$NO$_8$; mol wt 587.75. C 67.44%, H 8.40%, N 2.38%, O 21.78%. Glucoside of jervine. Isoln from *Veratrum album* L., *Liliaceae:* Wright, Luff, *J. Chem Soc.* **35**, 405 (1879); Poethke, *Arch. Pharm.* **276**, 170 (1938); from *V. viride* Ait., *Liliaceae:* Jacobs, Craig, *J. Biol. Chem.* **155**, 565 (1944); from *V. eschscholtzii* Gray, *Liliaceae:* Klohs *et al., J. Am. Chem. Soc.* **75**, 2133 (1953).

Lustrous leaflets, mp 300-301° (dec). [α]$_D^{25}$ −133° (c = 0.48 in 1:3 ethanol-chloroform). Sol in chloroform, benzene; slightly sol in alcohol. Almost insol in ether.

8103. Pseudomonic Acids. A group of antibacterial antibiotics produced by *Pseudomonas fluorescens* NCIB 10586 that have unusual structural features. Four members of the group are known: *pseudomonic acid A*, the major component; *pseudomonic acid B*, the 3,4,5-trihydroxy analog of A (also referred to as *pseudomonic acid I*); *pseudomonic acid D*, the 4-nonenoic acid analog of A; and *pseudomonic acid C*, in which the epoxide oxygen is replaced by a double bond. Isoln and characterization of A and B: A. T. Fuller *et al., Nature* **234**, 416 (1971); K. D. Barrow, G. Mellows, **Ger.** pat. **2,227,739;** *eidem,* **U.S.** pat. **3,977,943** (1973, 1976 both to Beecham). Structure of B: E. B. Chain, G. Mellows, *J. Chem. Soc. Perkin Trans. I* **1977**, 318. Prepn of C: N. H. Rogers, P. J. O'Hanlon, **Eur.** pat. **Appl. 3069;** *eidem,* **U.S.** pat **4,205,002** (1979, 1980 both to Beecham). Isoln, structure, configuration of C: J. P. Clayton *et al., Tetrahedron Letters* **21**, 881 (1980). Total syntheses of naturally occurring (+)-form of C: A. P. Kozikowski *et al., J. Am. Chem. Soc.* **102**, 6577 (1980); G. E. Keck *et al., J. Org. Chem.* **49**, 1462 (1984); J. C. Barrish *et al., ibid.* **52**, 1372 (1987); of (±)-A and (±)-C: B. B. Snider, G. B. Phillips, *J. Am. Chem. Soc.* **104**, 1113 (1982); B. B. Snider *et al., J. Org. Chem.* **48**, 3003 (1983). Prepn of D: P. J. O'Hanlon, **Eur.** pat. **Appl. 68,680** (1983 to Beecham), *C.A.* **98**, 159135t (1983); P. J. O'Hanlon *et al., J. Chem. Soc. Perkin Trans. I* **1983**, 2655. Antimycoplasmal activity in vitro: R. M. Banks *et al., J. Antibiot.* **41**, 609 (1988).

Pseudomonic Acid C

Isolated as a mixture of sodium salts. Can be stored at 0° for several months with no activity loss. Stable within pH 4-9 at 37° for 24 hrs. Hemolytic; inactivated by serum at conc > 50%.

Pseudomonic Acid A, *See* Mupirocin.

Pseudomonic Acid C, C$_{26}$H$_{44}$O$_8$, *[2S-[2α(E),3β,4β,5α(2E,-4S*,5R*)]]-9-[[3-methyl-1-oxo-4-[tetrahydro-3,4-dihydroxy-5-(5-hydroxy-4-methyl-2-hexenyl)-2H-pyran-2-yl]-2-butenyl]-oxy]nonanoic acid.* [α]$_D^{25}$ +7.64° (c = 0.78 in chloroform). uv max (ethanol): 222 nm (ε 14100).

Pseudomonic Acid D, C$_{26}$H$_{42}$O$_9$, *[2S-[2α[E(E)],3β,4β,5α-[2R*,3R*(1R*,2R*)]]]-9-[[3-Methyl-1-oxo-4-[tetrahydro-3,4-dihydroxy-5-[[3-(2-hydroxy-1-methylpropyl)oxiranyl]methyl]-2H-pyran-2-yl]-2-butenyl]oxy]-4-nonenoic acid.* Oil. uv max (ethanol): 220 nm (ε 15499).

8104. Pseudomorphine. *(5'α,6'α)-7,7',8,8'-Tetrahydro-4,5:4',5'-diepoxy-17,17'-dimethyl-[2,2'-bimorphinan]-3,-3',6,6'-tetrol;* 2,2'-bimorphine; pseudomorphine (C34 alkaloid). C$_{34}$H$_{36}$N$_2$O$_6$; mol wt 568.67. C 71.81%, H 6.38%, N 4.93%, O 16.88%. A dimolecular base formed by the gentle oxidation of morphine in alkaline soln. *Refs:* Broockmann, Polstorff, *Ber.* **13**, 88 (1880); Bentley, Dyke, *J. Chem. Soc.* **1959**, 2574. Review and bibliography: K. W. Bentley, *The Chemistry of the Morphine Alkaloids* (Oxford, 1954) p 54.

Trihydrate, crystalline powder; levorotatory in acid soln. Becomes anhydr at 150° and dec at about 327°. Insol in cold water, alcohol, ether, dil H$_2$SO$_4$. Slightly sol in cold, more sol in hot ammonia; sol in fixed alkali hydroxide solns, in pyridine, aniline, benzyl alcohol, guaiacol.

Note: Name pseudomorphine is also used for the C17 alkaloid base.

8105. Pseudopelletierine. *9-Methyl-9-azabicyclo[3.3.1]-nonan-3-one; 9-methyl-3-granatanone;* pseudopunicine. C$_9$-H$_{15}$NO; mol wt 153.22. C 70.55%, H 9.87%, N 9.14%, O 10.44%. In root bark of *Punica granatum* L., *Punicaceae.* Extraction procedure: Hess, Eichel, *Ber.* **50**, 380, 1391, 1395 (1917); *ibid.* **52**, 1012 (1919). Synthesis under physiological conditions from calcium acetonedicarboxylate, glutardialdehyde, and methylamine: Menzies, Robinson, *J. Chem. Soc.* **1924**, 2163; *cf.* Schöpf, *Angew. Chem.* **50**, 779, 797 (1937). Alternate syntheses: Cope *et al., Org. Syn.* **37**, 73 (1957); Robinson, Hunt, *J. Pharm. Pharmacol.* **22**, 29S (1970); Bottini, Gal, *J. Org. Chem.* **36**, 1718 (1971). Conformation: Chen, LeFevre, *J. Chem. Soc. (B)* **1966**, 539.

Orthorhombic prisms from petr ether, mp 54°; bp 246°. Volatile. Strong base. One gram dissolves in about 2.5 ml water, 10 ml ether. Freely sol in alcohol and chloroform; sparingly sol in petr ether.

Dihydrate, plates from water.

Hydrochloride, C$_9$H$_{15}$NO.HCl, rhombohedra. One gram dissolves in about 1 ml water.

Sulfate tetrahydrate, 2C$_9$H$_{15}$NO.H$_2$SO$_4$.4H$_2$O, crystals, sol in water.

8106. Pseudotropine. *exo-8-Methyl-8-azabicyclo[3.2.1]-octan-3-ol; 1αH,5αH-tropan-3β-ol; 3β-tropanol; ψ-tropine;* 3-pseudotropanol. C$_8$H$_{15}$NO; mol wt 141.21. C 68.04%, H 10.71%, N 9.92%, O 11.33%. Stereoisomeric with tropine. Prepn, structure, separation from tropine, and stereochemical configuration: *See* Tropine. Stereochemical synthesis: J. J. Tufariello *et al., J. Am. Chem. Soc.* **101**, 2435 (1979).

Orthorhombic, bipyramidal crystals from petr ether + benzene. mp 109°; bp 241°; pK (15°): 3.80. pH of 0.05 M soln 11.5. Freely sol in water, alcohol, benzene.

Hydrochloride, $C_8H_{15}NO.HCl$, needles from alcohol, dec 282°. Soluble in water, hot alcohol.

Tropate, $C_8H_{15}NO.C_9H_8O_2$, cryst from benzene, mp 132°.

8107. Pseudoyohimbine. *(3β,16α,17α)-17-Hydroxy-yohimban-16-carboxylic acid methyl ester.* $C_{21}H_{26}N_2O_3$; mol wt 354.45. C 71.16%, H 7.39%, N 7.90%, O 13.54%. Present in the bark of *Corynanthe johimbe* K. Schum., *Rubiaceae.* Obtained from the residues of commercial isoln procedures for the manuf of yohimbine: Karrer, Salomon, *Helv. Chim. Acta* **9**, 1059 (1926). Structure and stereochemistry: Janot *et al., Bull. Soc. Chim. France* **1952**, 1085; **1961**, 637. Synthesis: van Tamelen *et al., J. Am. Chem. Soc.* **80**, 5006 (1958); **91**, 7315 (1969); Stork, Guthikonda, *ibid.* **94**, 5109 (1972).

Rhombic platelets, mp 293° (corr, Maquenne block), mp 268° (open capillary), browns at 250°. Also reported as crystals from methanol, mp 252-256° (van Tamelen). $[\alpha]_D^{19}$ +27° (pyridine). uv max (methanol): 225, 281, 290 nm (log ϵ 4.54, 3.86, 3.80).

Hydrochloride, $C_{21}H_{26}N_2O_3.HCl$, needles from alcohol + ether, dec 258°. $[\alpha]_D^{20} - 10°$ (c = 1 in water).

8108. Psicofuranine. *9-β-D-Psicofuranosyl-9H-purin-6-amine; 9β-D-psicofuranosyladenine;* 6-amino-9-D-psicofuranosylpurine; angustmycin C; U-9586. $C_{11}H_{15}N_5O_5$; mol wt 297.27. C 44.44%, H 5.09%, N 23.56%, O 26.91%. Nucleoside antibiotic produced by *Streptomyces hygroscopicus* var. *decoyicus:* Eble *et al., Antibiot. & Chemother.* **9**, 419 (1959); Yüntsen, *J. Antibiot.* **11A**, 244 (1958); Eble, Lewis, **U.S.** pat. **3,020,274** (1962 to Upjohn). Antibacterial and antitumor activity: N. Tanaka *et al., J. Antibiot.* **14A**, 98 (1961). Structure: Schroeder, Hoeksema, *J. Am. Chem. Soc.* **81**, 1767 (1959); Garrett, *ibid.* **82**, 827 (1960). Synthesis: Farkas, Sorm, *Coll. Czech. Chem. Commun.* **28**, 882 (1963); L. A. Aleksandrova, F. W. Lichtenthaler, *Nucleic Acids Symp. Ser.* **9**, 263 (1981). Biosynthesis: Sugimori, Suhadolnik, *J. Am. Chem. Soc.* **87**, 1136 (1965). Inhibits the conversion of xanthosine-5′-phosphate to guanosine-5′-phosphate: Slechta, *Biochem. Pharmacol.* **5**, 96 (1960).

Crystals, dec 212-214°. $[\alpha]_D^{25} - 68°$ (dimethylformamide). uv max (0.01N acid): 259 nm ($E_{1cm}^{1\%}$ 508); in 0.01N base: 261 nm ($E_{1cm}^{1\%}$ 527). Soly at 25° (mg/ml): water 8; methanol 8; ethanol 6; butanol 2; ethyl acetate 0.23.

8109. D-Psicose. D-*ribo*-2-Ketohexose; D-ribohexulose; D-allulose; D-erythrohexulose; pseudofructose. $C_6H_{12}O_6$; mol wt 180.16. C 40.00%, H 6.71%, O 53.28%. Occurrence and identification as a non-fermentable substance in cane molasses: Zerban, Sattler, *Ind. Eng. Chem.* **34**, 1180 (1942). Structure: Ohle, Just, *Ber.* **68**, 601 (1935). Prepn starting with D-ribono-γ-lactone: Wolfrom *et al., J. Am. Chem. Soc.*

67, 1793 (1945); from D-allose: Steiger, Reichstein, *Helv. Chim. Acta* **19**, 184 (1937); from D-glucose: Hough *et al., J. Chem. Soc.* **1953**, 2005.

Sweet syrupy liquid. $[\alpha]_D^{25}$ +4.7° (c = 4.3 in water; no detectable mutarotation). Soluble in water, methanol, ethanol. Practically insol in acetone.

Phenylosazone, $C_{18}H_{22}N_4O_4$, yellow crystals from water, mp about 162-163° (dec); also reported mp 178°.

8110. Psilocin. *3-[2-(Dimethylamino)ethyl]-1H-indol-4-ol;* 4-hydroxy-*N,N*-dimethyltryptamine; psilocyn. $C_{12}H_{16}N_2O$; mol wt 204.27. C 70.56%, H 7.89%, N 13.71%, O 7.83%. The minor hallucinogenic component of Teonanácatl, the sacred mushroom of Mexico. Isolated in trace amounts from the fruiting bodies of *Psilocybe mexicana* Heim, *Agaricaceae:* Hofmann *et al., Experientia* **14**, 107 (1958); Heim *et al., Helv. Chim. Acta* **42**, 1557 (1959). Prepn: Heim *et al., Ger.* pat. **1,087,321** (1960 to Sandoz). Synthetic precursor of psilocybin: Hofmann, Troxler, **U.S.** pat. **3,075,992** (1963 to Sandoz). Psilocin, the 4-hydroxy analog of psilocybin, is formed by metabolic dephosphorylation of psilocybin and is the active species in the central nervous system: Horita, Weber, *Toxicol Appl. Pharmacol.* **4**, 730 (1962). Crystal structure: T. J. Petcher, H. P. Weber, *J. Chem. Soc. Perkin Trans. II* **1974**, 946. Review: Hofmann, *Bull. Narcotics* **23**, 3 (1971).

Plates from methanol, mp 173-176°. Amphoteric substance. Unstable in soln, esp. akaline soln. Very slightly sol in water. uv max: 222, 260, 267, 283, 293 nm (log ϵ 4.6, 3.7, 3.8, 3.7, 3.6).

Note: This is a controlled substance (hallucinogen) listed in the U.S. Code of Federal Regulations, Title 21 Part 1308.11 (1995).

8111. Psilocybin. *3-[2-(Dimethylamino)ethyl]-1H-indol-4-ol dihydrogen phosphate ester; O*-phosphoryl-4-hydroxy-*N,N*-dimethyltryptamine; Indocybin. $C_{12}H_{17}N_2O_4P$; mol wt 284.25. C 50.71%, H 6.03%, N 9.86%, O 22.51%, P 10.90%. The major of two hallucinogenic components of Teonanácatl, the sacred mushroom of Mexico, the other component being psilocin, *q.v.* from the fruiting bodies of *Psilocybe mexicana* Heim, *Agaricaceae:* Hofmann *et al., Experientia* **14**, 107 (1958); Heim *et al., Helv. Chim. Acta* **42**, 1557 (1959); Heim *et al., Ger.* pat. **1,087,321** (1960 to Sandoz). Structure: Hofmann *et al., Experientia* **14**, 397 (1958). Synthesis: Hofmann, Troxler, **U.S.** pat. **3,075,992** (1963 to Sandoz). Crystal structure: H. P. Weber, T. J. Petcher, *J. Chem. Soc., Perkin Trans. II* **1974**, 942. Converted to psilocin *in vivo.* Toxicity data: E. Usdin, D. H. Efron, *Psychotropic Drugs and Related Compounds* (National Institute of Mental Health, Rockville, Md., 2nd ed., 1972) p 138. *Reviews:* Hofmann, *Proc. 1st Int. Congr. Neuro-Pharm., Rome* **1958**, 446; Cerletti, *Deut. Med. Wochenschr.* **84**, 2317 (1959); Hofmann, *Bull. Narcotics* **23**, 3 (1971).

Crystals from boiling water, mp 220-228°; from boiling methanol, mp 185-195°. uv max (methanol): 220, 267, 290 nm (log ε 4.6, 3.8, 3.6). pH 5.2 in 50% aq ethanol. Sol in 20 parts boiling water, 120 parts boiling methanol; difficultly sol in ethanol. Practically insol in chloroform, benzene. LD_{50} in mice, rats, rabbits (mg/kg): 285, 280, 12.5 i.v. (Usdin, Efron).

Note: This is a controlled substance (hallucinogen) listed in the U.S. Code of Federal Regulations, Title 21 Part 1308.11 (1995).

THERAP CAT: Psychomimetic.

8112. PSK®. KRESTIN. Protein-bound polysaccharide having immunostimulant and anticancer properties. Isolated from a basidiomycete, *Coriolus versicolor* (Fr.) Quel.: S. Otsuka *et al., Japan.* pat. 73 08,489 (1973 to Kureha), *C.A.* **80**, 41025g (1974). Antitumor activity: S. Tsukagoshi *et al., Prog. Chemother. (Antibacterial, Antiviral, Antineoplast.), Proc. Int. Congr. Chemother., 8th, Athens 1973,* vol. **3**, G. K. Daikos, Ed. (Hellen. Soc. Chemother., Athens, 1974) pp 799-803. Efficacy vs mouse sarcoma: S. Tsukagoshi, F. Ohashi, *Gann* **65**, 557 (1974), *C.A.* **82**, 119037 (1975). Immunochemotherapy study: J. Akiyama *et al., Cancer Res.* **37**, 3042 (1977). Comparative *in vitro* study: B. T. Nguyen, S. Stadtsbaeder, *Adv. Exp. Med. Biol.* **121A**, 255 (1979). Effects on immunological parameters and expt'l infections: P. Mayer, J. Drews, *Infection* **8**, 13 (1980). Acute toxicity: *Japan. Med. Gaz.* **13**(9), 11 (1976). *Review:* S. Tsukagoshi *et al., Cancer Treat. Rev.* **11**, 131-155 (1984).

Brownish powder. Tasteless, but has a slight odor. Sol in water. Practically insol in methanol, pyridine, chloroform, benzene, hexane. pH of 1% soln is 6.6-7.2. LD_{50} in mice, rats (mg/kg): >20,000 orally; >5,000 i.p.; >5000 s.c.; >1300, >600 i.v. (*Japan. Med. Gaz.*).

THERAP CAT: Antineoplastic.

8113. Psoralen. *7H-Furo[3,2-g][1]benzopyran-7-one;* 6-hydroxy-5-benzofuranacrylic acid δ-lactone; furo[3,2-g]-coumarin; ficusin. $C_{11}H_6O_3$; mol wt 186.17. C 70.97%, H 3.25%, O 25.78%. One of a group of furocoumarins occurring naturally in more than two dozen plant sources, including *Rutaceae* (*e.g.* bergamot, limes, cloves), *Umbelliferae* (*e.g.* celery, parsnips), *Leguminosae* (*e.g. Psoralen coryfolia*), and *Moraceae* (*e.g.* figs). Isoln: H. S. Jois *et al., J. Indian Chem. Soc.* **10**, 41 (1933); A. Stoll *et al., Helv. Chim. Acta* **33**, 1637 (1950); F. E. King *et al., J. Chem. Soc.* **1954**, 1392. Synthesis: E. Späth *et al., Ber.* **69**, 1087 (1936); R. C. Esse, B. E. Christensen, *J. Org. Chem.* **25**, 1565 (1960); O. Dann, D. Volz, *Arch. Pharm.* **308**, 121 (1975); V. K. Ahluwalia *et al., Monatsch.* **111**, 877 (1980). Psoralens are *phytoalexins;* they are used by plants in a defensive response to attacks by fungi and insects: M. Berenbaum, P. Feeny, *Science* **212**, 927 (1981). They have also shown photosensitizing and phototoxic effects in animals and humans and have been used in photochemotherapy for management of vitiligo, psoriasis, and mycosis fungoides, *cf.* T. F. Anderson, J. J. Voorhees, *Ann. Rev. Pharmacol. Toxicol.* **20**, 235 (1980); A. Kornhauser *et al., Science* **217**, 733 (1982). Review of psoralen photochemistry: B. J. Parsons, *Photochem. Photobiol.* **32**, 813-821 (1980). Review of genetic toxicity of psoralen and uv radiation in human cells: *Acta Derm. Venereol.* **Suppl. 104**, 4-40 (1982). *See* Methoxsalen, Trioxsalen, Bergapten for additional refs.

Crystals from ether, mp 163-164°; 169-179° (Späth). Absorption spectra: Wessely, Kaltan, *Monatsh.* **86**, 430 (1955). USE: As photochemical probe in biological systems. P.-S. Song, C.-N. Ou, *Ann. N.Y. Acad. Sci.* **346**, 355 (1980).

8114. Psoralidin. *3,9-Dihydroxy-2-(3-methyl-2-buten-yl)-6H-benzofuro[3,2-c][1]benzopyran-6-one;* 6-(3-methylbut-2-enyl)coumestrol. $C_{20}H_{16}O_5$; mol wt 336.34. C 71.42%, H 4.79%, O 23.78%. Extracted from the seed kernel of *Psoralea corylifolia* Linn., *Leguminosae.* Isoln: Chakravarti, *J. Sci. Ind. Res.* **7B**, 24 (1948). Structure: Duttagupta *et al., Chem. & Ind. (London)* **1960**, 937. Synthesis: Khastgir *et al., Tetrahedron* **14**, 275 (1961).

Crystals from acetone, mp 290-292°. Shows violet fluorescence in dil ethanol soln. Absorption spectra: Khastgir *et al., loc. cit.*

Diacetate, $C_{24}H_{20}O_7$, crystals from ethyl acetate, mp 221-223°.

8115. Psychotrine. *1',15-Didehydro-7',10,11-trimeth-oxyemetan-6'-ol.* $C_{28}H_{36}N_2O_4$; mol wt 464.60. C 72.39%, H 7.81%, N 6.03%, O 13.77%. Minor alkaloid found in ipecac, the ground roots of *Uragoga ipecacuanha* (Brot.) Baill. [*Cephaelis ipecacuanha* (Brot.) A. Rich.], *Rubiaceae:* F. H. Carr, F. L. Pyman, *J. Chem. Soc.* **105**, 1591 (1914); O. Hesse, *Ann.* **405**, 1 (1914). Structure and stereochemistry: A. R. Battersby *et al., J. Chem. Soc.* **1959**, 2704, 3512. Mass spectrum: H. Budzikiewicz *et al., Tetrahedron* **20**, 399 (1964); $^1H, ^{13}C$-nmr of *O*-methylpsychotrine: T. Fujii *et al., Chem. Pharm. Bull.* **30**, 598 (1982); ^{13}C-nmr of psychotrine: eidem, *ibid.* **31**, 2583 (1983). Proposed alternate structure with exocyclic double bond: E. E. van Tamelen *et al., J. Am. Chem. Soc.* **79**, 4817 (1957); C. Schuij *et al., J. Chem. Soc. Perkin Trans. I* **1979**, 970. *Review:* M. Janot in Manske, Holmes, *The Alkaloids* vol. **3** (Academic Press, New York, 1953) pp 363-394.

Tetrahydrate, yellow prisms with blue fluorescence from dil acetone or alcohol. *Very bitter taste, produces nausea instantly.* The anhyd material sinters at 120°, becomes transparent at 120-126° and melts completely at 128°. $[\alpha]_D^{15}$ +69.3° (c = 2 in alcohol, calcd as the tetrahydrate). Sparingly sol in water, benzene, petr ether, ether. More sol in alcohol, acetone, chloroform. uv max (0.1N HCl): 240, 288, 306, 356 nm (ε 13900, 5700, 6250, 6800).

Sulfate trihydrate, $C_{28}H_{36}N_2O_4 \cdot H_2SO_4 \cdot 3H_2O$, pale yellow scales from water, dec 214-217° when anhydr. $[\alpha]_D^{20}$ +39.2°.

O-Methylpsychotrine has a methoxy group in place of the hydroxyl group. mp 123-124°. $[\alpha]_D^{20}$ +43.2° (alc). uv max (water) pH 1: 241.5, 288.5, 305, 354 (log ε 4.26, 3.86, 3.92, 3.91); pH 13: 226, 278.5, 307 (log ε 4.43, 3.96, 3.77).

8116. Pteridine. Pyrazino[2,3-*d*]pyrimidine; pyrimido-[4,5-*b*]pyrazine; pyrimidine-4',5':2,3-pyrazine; 1,3,5,8-tetraazanaphthalene; azinepurine; benzotetrazine. $C_6H_4N_4$; mol wt 132.12. C 54.54%, H 3.05%, N 42.40%. Prepn from

4,5-diaminopyrimidine and polyglyoxal or glyoxal sodium bisulfite: Albert *et al., J. Chem. Soc.* **1951**, 474.

Yellow crystals from benzene, mp 138-138.5°. uv max (water): 298, 309 nm (log ε 3.87, 3.83). Sol in 7.2 parts water at 20-25°.

Caution: Related to folic acid antagonists.

8117. Pterocarpin. *(6aR-cis)-6a,12a-Dihydro-3-methoxy-6H-[1,3]dioxolo[5,6]benzofuro[3,2-c][1]benzopyran.* $C_{17}H_{14}O_5$; mol wt 298.30. C 68.45%, H 4.73%, O 26.82%. From red sandalwood (*Pterocarpus santalinus* L.f., *Leguminosae):* Cazeneuve, Hugouneng, *Compt. Rend.* **104,** 1722 (1887); **107,** 737 (1888); Leonhardt, Fay, *Arch. Pharm.* **273,** 53 (1935); McGookin *et al., J. Chem. Soc.* **1940,** 787; from *Pterocarpus* spp: King *et al., ibid.* **1953,** 3693. Structure: Bredenberg, Shoolery, *Tetrahedron Letters* **1961,** 285. Synthesis of the *dl*-form: Fukui, Nakayama, *ibid.* **1966,** 1805; *eidem, Bull. Chem. Soc. Japan* **42,** 1408 (1969); Uchiyama, Matsui, *Agr. Biol. Chem.* **31,** 1490 (1967). Stereochemistry: Ito *et al., Chem. Commun.* **1965,** 595; Pachler, Underwood, *Tetrahedron* **23,** 1817 (1967).

Crystalline plates from petr ether or from ethanol, mp 164.5°. $[\alpha]_D$ −214.5° and $[\alpha]_{5461}^{20.5}$ −207.5° (c = 0.53 in chloroform). Practically insol in water, cold alcohol or ether. Sol in hot alcohol, chloroform.

8118. Pteroic Acid. *4-[[(2-Amino-1,4-dihydro-4-oxo-6-pteridinyl)methyl]amino]benzoic acid; p-[(2-amino-4-hydroxy-6-pteridylmethyl)amino]benzoic acid.* $C_{14}H_{12}N_6O_3$; mol wt 312.29. C 53.85%, H 3.87%, N 26.91%, O 15.37%. Intermediate in the synthesis of folic acids. Prepn: Waller *et al., J. Am. Chem. Soc.* **70,** 19 (1948); *cf.* Hultquist *et al., ibid.* 23; Boothe, U.S. pat. **2,472,462** (1949 to Am. Cyanamid); C. Temple *et al., J. Org. Chem.* **46,** 3666 (1981). Synthesis: M. G. Nair *et al., ibid.* 3152.

Crystals from dil HCl. Sol in aq NaOH solns.

8119. Pteropterin. *N-[N-[N-[4-[[(2-Amino-1,4-dihydro-4-oxo-6-pteridinyl)methyl]amino]benzoyl]-L-γ-glutamyl]-L-glutamyl]-L-glutamic acid; N-[N-(N-pteroyl-γ-glutamyl)-γ-glutamyl]glutamic acid;* fermentation L. casei Factor; pteroyl-γ-glutamyl-γ-glutamylglutamic acid; pteroyldi-γ-glutamylglutamic acid; pteroyltriglutamic acid; PTGA; Teropterin. $C_{29}H_{33}N_9O_{12}$; mol wt 699.63. C 49.79%, H 4.75%, N 18.02%, O 27.44%. Isoln from aerobic culture of *Corynebacterium:* Hutchings *et al., J. Am. Chem. Soc.* **70,** 1 (1948). Identification, structure and synthesis: Boothe *et al., ibid.* 1099; *ibid.* **71,** 2304 (1949). Prepn: Cosulich, U.S. pat. **2,563,707** (1951 to Am. Cyanamid).

Forms a crystalline barium salt.

Monohydrate, crystals from water adjusted to pH 2.8 with HCl and contg some NaCl. Has the general properties of a polypeptide. Absorption spectrum, Hutchings, *loc. cit.* Soly in water at 5°: 0.10 mg/ml; at 80°: 3.00 mg/ml. Sol in NaOH solns.

Methyl ester, crystals from methanol contg NaCl. Soly in water at 5°: 0.12 mg/ml; at 80°: 1.00 mg/ml; in methanol at −5°: 0.30 mg/ml; at 60°: 5.00 mg/ml.

THERAP CAT: Formerly as antineoplastic.

8120. Pteroylhexaglutamylglutamic Acid. Pteroylheptaglutamic acid; vitamin Bc conjugate; Bc conjugate; PHGA. $C_{49}H_{61}N_{13}O_{24}$; mol wt 1216.10. C 48.40%, H 5.06%, N 14.97%, O 31.58%. One of the naturally occurring polyglutamates of folic acid. Isoln from yeast: Pfiffner *et al., Science* **102,** 228 (1945); *J. Am. Chem. Soc.* **68,** 1392 (1946). Solid phase synthesis: C. L. Krumdieck, C. M. Baugh, *Biochemistry* **8,** 1568 (1969); *eidem, Methods in Enzymology* **66,** 523 (1980). Chain length characterization of pteroylpolyglutamates: D. G. Priest *et al., Anal. Biochem.* **115,** 163 (1981).

Rosettes of minute needles from 5% NaCl soln. Has no definite mp. When heated on the hot stage it begins to darken from about 200° and partially melts at 230-260°, and remains partially melted up to 360°. The ultraviolet absorption curves are practically identical with those of folic acid. Sparingly sol in water contg NaCl. Sol in dil NaOH soln.

8121. Puberulic Acid. *3,4,6-Trihydroxy-5-oxo-1,3,6-cycloheptatriene-1-carboxylic acid.* $C_8H_6O_6$; mol wt 198.13. C 48.50%, H 3.05%, O 48.45%. Antibiotic substance produced by *Penicillium puberulum, P. aurantiovirens, P. johannioli, P. cyclopium viridicatum.* Isoln: Birkinshaw, Raistrick, *Biochem. J.* **26,** 441 (1932); Barger, Dorrer, *ibid.* **28,** 11 (1934). Structure: McGowan, *Chem. & Ind. (London)* **1947,** 205; Corbett *et al., ibid.* **1949,** 626; *J. Chem. Soc.* **1950,** 6. Synthesis: Johns *et al., ibid.* **1954,** 198. Biosynthesis: Ferretti, Richards, *Proc. Nat. Acad. Sci. USA* **46,** 1438 (1960). *Review:* Nozoe, *Fortschr. Chem. Org. Naturst.* **13,** 232-301 (1956).

Cream-colored powder from water, mp 316-318°. Sublimes in high vacuum. Dibasic acid. uv max (water): 270, 350 nm (log ε 4.55, 3.86). Sol in hot water. Forms a sodium salt which is freely sol in water.

8122. Puberulonic Acid. *5,6,7-Trihydroxy-1H-cyclohepta[c]furan-1,3,4-trione; 4,5,6-trihydroxy-7-oxo-1,3,5-cycloheptatriene-1,2-dicarboxylic anhydride.* $C_9H_4O_7$; mol wt 224.13. C 48.23%, H 1.80%, O 49.97%. Antibiotic substance produced by *Penicillium puberulum, P. aurantiovirens, P. johannioli, P. cyclopium viridicatum.* Isoln: Birkinshaw, Raistrick, *Biochem. J.* **26,** 441 (1932); Barger, Dorrer, *ibid.* **28,** 11 (1934). Structure: McGowan, *Chem. & Ind. (London)* **1947,** 205; Aulin-Erdtman, *Acta Chem. Scand.* **4,** 1325 (1950). Johnson *et al., J. Chem. Soc.* **1951,** 1139. Synthesis: Nozoe *et al., Bull. Chem. Soc. Japan* **33,** 1071 (1960). Biosynthesis: Ferritti, Richards, *Proc. Nat. Acad. Sci. USA* **46,** 1438 (1960); A. I. Scott, E. Lee, *Chem. Commun.* **1972,** 655. *Review:* Nozoe, *Fortschr. Chem. Org. Naturst.* **13,** 232-301 (1956).

Yellow plates, dec 296°. Sol in water. The yellow color of an aq soln changes to pink on neutralization and disappears altogether when the soln is made alkaline.
Sodium salt, crystals from ethanol, freely sol in water.

8123. Pukateine. *(R)-6,7,7a,8-Tetrahydro-7-methyl-5H-benzo[g]-1,3-benzodioxolo[6,5,4-de]quinolin-12-ol; 1,2-(methylenedioxy)aporphin-11-ol;* 1,2-methylenedioxy-11-hydroxyaporphine. $C_{18}H_{17}NO_3$; mol wt 295.34. C 73.20%, H 5.80%, N 4.74%, O 16.25%. From bark of *Laurelia novaezelandiae* A. Cunn, *Lauraceae.* Isoln: Aston, *J. Chem. Soc.* **97**, 1381 (1910); Bernauer, *Helv. Chim. Acta* **50**, 1583 (1967). Structure: Barger, Giradet, *ibid.* **14**, 481, 504 (1931); Barger, Schlittler, *ibid.* **15**, 381 (1932). Total synthesis: Zymalkowski, Happel, *Tetrahedron Letters* **1969**, 219; *Ber.* **102**, 2959 (1969); Kametani *et al., J. Chem. Soc. Perkin Trans. I* **1972**, 1435.

dl-Form, crystals from absolute alcohol, mp 232-233°.
l-Form, crystals from alcohol, mp 208-212°. $[\alpha]_D^{15}$ −240° (c = 0.097 in alcohol). Practically insol in water. Sol in alc, ether, chloroform, pyridine; slightly sol in petr ether.

8124. Pulegone. *(R)-5-Methyl-2-(1-methylethylidene)-cyclohexanone; R-(+)-p-menth-4(8)-en-3-one;* 1-methyl-4-isopropylidene-3-cyclohexanone. $C_{10}H_{16}O$; mol wt 152.24. C 78.90%, H 10.59%, O 10.51%. Found in oils derived from plants of the *Labiatae* family as (+)-form. Readily isolated in quantity from the pennyroyal oils from *Mentha pulegium* L., *M. longifolia* (L.) Huds., and *Hedeoma pulegioides* (L.) Pers., *Labiatae:* Gildemeister, Hoffmann, *Die ätherischen Öle* Vol. I, p 560 (1928); Simonsen, *The Terpenes* Vol. I (2nd ed, 1947) p 370. Synthesis: Kuhn, Schinz, *Helv. Chim. Acta* **36**, 161 (1953). Synthesis of the (±)-form: Black *et al., J. Chem. Soc.* **1956**, 2971; Wolinsky *et al., J. Org. Chem.* **30**, 3207 (1965); of the S(−)-form: E. J. Corey *et al., ibid.* **41**, 380 (1976). Improved synthesis of the R(+)-form: T. Sato *et al., Tetrahedron Letters* **1980**, 3377. Conversion of the R(+) to the S(−)-form: H. E. Ensley, R. V. C. Carr, *ibid.* **1977**, 513. Biosynthetic study: A. Akhila, D. V. Banthorpe, *Z. Pflanzenphysiol.* **99**, 277 (1980), *C.A.* **94**, 2073r (1981).

Oil. Pleasant odor, midway between peppermint and camphor. d_4^{15} 0.9346. bp$_{760}$ 224°; bp$_{100}$ 151-153°; bp$_{17}$ 103°; bp$_6$ 84°. $[\alpha]_D^{20}$ +21°; $[\alpha]_{546}^{20}$ +28.2°. n_D^{20} 1.4894. Practically insol in water. Miscible with alcohol, ether, chloroform.

(±)-Form. Liquid. bp$_7$ 78-80°. n_D^{16} 1.4856. uv max (alc): 253.3 nm (log ε 3.86).
(−)-Form. Liquid. bp$_{20}$ 104-108°. $[\alpha]_D^{23}$ −22.5° (neat).

8125. Pulsatilla. Pasque flower; wind flower; meadow anemone; Easter flower. Dried herb of *Anemone pulsatilla* L. *(Pulsatilla vulgaris* Mill.), *A. pratensis* L. *(P. pratensis* (L.) Mill.), or *A. patens* L. *(P. patens* (L.) Mill.), *Ranunculaceae. Habit.* Europe, Asia. *Constit.* Anemone camphor, volatile oil, tannin.

8126. Pumactant. ALEC. Artificial lung expanding compound composed of *dipalmitoylphosphatidylcholine* (DPPC) and egg *phosphatidylglycerol* in a 7:3 (w/w) ratio. Description of prepn and clinical evaluation in premature babies: C. J. Morley *et al., Lancet* **1**, 64 (1981). Clinical trials in premature babies: Ten Centre Study Group, *Brit. Med. J.* **294**, 991 (1987). Review of physical properties and design of mixture: A. D. Bangham *et al.,* Easy Breathing and the Design of Artificial Lung Expanding Compounds (ALEC) in *Topics in Lipid Research,* R. Klein, B. Schmitz, Eds (Royal Society of Chemistry, London, Great Britain, 1986) 268-275; of various surfactant treatments: C. J. Morley, in *Surfactant Replacement Therapy,* D. L. Shapiro, R. H. Notter, Eds (Alan R. Liss, Inc., NY, NY, 1989) 219-234.
Fine crystalline powder.
THERAP CAT: Pulmonary surfactant in treatment of respiratory distress syndrome in infants.

8127. Pumice. Light, hard, rough, porous, gray masses or a gritty, gray-colored powder of volcanic origin. Found chiefly in the Lipari islands and in the Greek archipelagos. Consists mainly of complex silicates of Al, K, and Na. It is insol in water and is not attacked by acids.
USE: Abrasive in metal polishes; in fireproofing and insulating compds; also as a carrier for metal catalysts. In pharmacy as filtering medium and dispersant. In cosmetics for removing rough skin.

8128. Pumpkin Seed. Pepo. Dried ripe seeds of cultivated varities of *Cucurbita pepo* L., *Cucurbitaceae. Habit.* Southern Asia, Europe, America. *Constit.* Fixed oil, acrid resin, myosin, vitellin, sugar.
THERAP CAT: Anthelmintic.

8129. 1H-Purine. *7H-Imidazo[4,5-d]pyrimidine.* C_5-H_4N_4; mol wt 120.11. C 50.00%, H 3.36%, N 46.64%. Prepd by heating 4,5-diaminopyrimidine with anhydr formic acid in a current of carbon dioxide: Isay, *Ber.* **39**, 251 (1906); by boiling 2,6-diiodopurine with zinc dust and water under carbon dioxide: Fischer, *ibid.* **31**, 2551, 2564 (1898).

Needles from toluene or alcohol, mp 216-217° (partial sublimation on rapid heating). Freely sol in water, hot alcohol. Slightly sol in hot ethyl acetate, acetone. Practically insol in ether, chloroform. Aq solns are neutral to litmus. Forms salts with acids and bases.

8130. Puromycin. *(S)-3'-[[2-Amino-3-(4-methoxyphenyl)-1-oxopropyl]amino]-3'-deoxy-N,N-dimethyladenosine; 3'-(α-amino-p-methoxyhydrocinnamamido)-3'-deoxy-N,N-dimethyladenosine;* 6-dimethylamino-9-[3-deoxy-3-(p-methoxy-L-phenylalanylamino)-β,D-ribofuranosyl]-β-purine; 6-dimethylamino-9-[3'-(p-methoxy-L-phenylalanylamino)-3'-deoxy-β,D-ribofuranosyl]purine; CL-13900; P-638; 3123-L; Stylomycin. $C_{22}H_{29}N_7O_5$; mol wt 471.52. C 56.04%, H 6.20%, N 20.79%, O 16.97%. Antibiotic substance produced by the soil actinomycete *Streptomyces alboniger.* Isoln: Porter *et al., Antibiot. & Chemother.* **2**, 409 (1952); *eidem,* U.S. pat. **2,763,642;** Szumski, Goodman, U.S. pat. **2,797,187** (1956, 1957 both to Am. Cyanamid). Structure: Waller *et al., J. Am. Chem. Soc.* **75**, 2025 (1953); Fryth, Waller, *ibid.* **76**, 4044 (1954); **77**, 12 (1955). Synthesis: Baker *et al., ibid.* **76**, 4044 (1954); **77**, 12 (1955). Conformation: Jardetzky, *J. Am. Chem. Soc.* **85**, 1823 (1963). Puromycin is known to interfere with protein formation by interfering with the function of RNA in the cells involved. *Review:* Nathans in *Antibiotics* vol. I, D. Gottlieb, P. D. Shaw, Eds. (Springer-Verlag, New York, 1967) pp 259-277.

Crystals, mp 175.5-177°. $[\alpha]_D^{25}$ −11° (ethanol). uv max (0.13N NaOH): 275 nm (ϵ 20300); in 0.1N HCl: 267.5 nm (ϵ 19500). It is a diacidic base and readily forms a water-sol dihydrochloride or monosulfate. LD$_{50}$ in mice (mg/kg): 350 i.v.; 525 i.p.; 675 orally (Porter, 1952).

Note: Puromycin was once referred to by the trademark "Achromycin" which has since 1953 been reassigned to tetracycline hydrochloride.

THERAP CAT: Antineoplastic; antiprotozoal (Trypanosoma).

8131. Purothionin. A low mol wt wheat protein composed of α_1-, α_2- and β-purothionins. Contains about 20% cystine, 10% arginine and 10% lysine. Isoln from unbleached patent flour: A. K. Balls, W. S. Hale, *Cereal Chem.* **17**, 243 (1940); A. K. Balls *et al., ibid.* **19**, 279 (1942). Inhibits the growth of bacteria and yeasts: L. S. Stuart, T. H. Harris, *ibid.* 288. Toxic effects in animals: E. J. Coulson *et al., ibid.* 301. Sepn and characterization of α- and β-purothionin: D. G. Redman, N. Fisher, *J. Sci. Food Agr.* **19**, 651 (1968); C. C. Nimmo *et al., ibid.* **25**, 607 (1974). Amino acid sequences of α_1- and β-purothionin: S. Ohtani *et al., Agr. Biol. Chem.* **39**, 2269 (1975); *eidem, J. Biochem.* **82**, 753 (1977); of β-purothionin: A. S. Mak, B. L. Jones, *Can. J. Chem.* **54**, 835 (1976); of α_1 and α_2-purothionin: B. L. Jones, A. S. Mak, *Cereal Chem.* **54**, 511 (1977). *Review:* D. D. Kasarda *et al.,* "Wheat Proteins" in *Advances in Cereal Science and Technology* vol. 1 (Am. Cereal Chem., St. Paul, 1976) pp 208-210.

Forms a cryst hydrochloride. Freely sol in water. Easily digested by proteolytic enzymes including cryst chymotrypsin, chymopapain, papain.

α_1-*Purothionin*, $C_{198}H_{340}N_{68}O_{56}S_8$, *5-L-arginine-27-glycine-33-L-isoleucine-34-L-serine-42-glycine-purothionin A I (reduced), purothionin A II.*

α_2-*Purothionin*, $C_{202}H_{348}N_{68}O_{59}S_8$, *5-L-arginine-6-L-threonine-18-L-serine-26-L-serine-27-L-threonine-42-glycine-purothionin A I (reduced).*

β-*Purothionin*, $C_{203}H_{339}N_{67}O_{59}S_8$, *purothionin A I.*

8132. Purpurin. *1,2,4-Trihydroxy-9,10-anthracenedione; 1,2,4-trihydroxyanthraquinone;* C.I. Natural Red 8; C.I. Natural Red 16; C.I. 58205; C.I. 75410. $C_{14}H_8O_5$; mol wt 256.21. C 65.63%, H 3.15%, O 31.22%. Occurs as glycoside in the madder root (*Rubia tinctorum* L., *Rubiaceae*) of commerce. Is formed during storage; no appreciable amount in the fresh root: Hill, Richter, *J. Chem. Soc.* **1936**, 1714. Although a dye itself, it is usually considered as an undesirable contaminant of alizarin extracted from madder. May be prepd from alizarin by oxidation with ammonium persulfate: Wacker, *J. Prakt. Chem.* [2] **54**, 90 (1896); also by Friedel-Crafts condensation of hydroxyhydroquinone with phthalic anhydride: Dimroth, Fick, *Ann.* **411**, 321 (1916).

Long orange needles with 1 H_2O from dil alcohol, anhydr at 100°. Anhydr red needles from abs alcohol or by sublimation around 150° in high vacuum (less than 2 mm Hg). mp 257°. Absorption spectrum: Meek, *J. Chem. Soc.* **111**, 969 (1917); Ezaby, *ibid. (B)* **1970**, 1293. More sol in boiling water than alizarin (yellow color with yellowish hue). Freely sol in alcohol (red), in ether (intensely yellow with fluorescence). Soluble in benzene, toluene, xylene (dark yellow), in boiling alum soln (red).

2-Methyl ether, $C_{15}H_{10}O_5$, orange crystals from benzene, mp 240°.

2,4-Dimethyl ether, $C_{16}H_{12}O_5$, orange needles, mp 186-189°.

USE: Forms colored "lakes" with various metal salts and is a fast dye for cotton printing. Now used mostly in the manuf of acid and chrome dyes. Reagent for the detection of boron; for detection of insol calcium salts in the cell contents of histological material and as a nuclear stain.

8133. Purpurogallin. *2,3,4,6-Tetrahydroxy-5H-benzo-cyclohepten-5-one.* $C_{11}H_8O_5$; mol wt 220.18. C 60.01%, H 3.66%, O 36.33%. The aglycone of several glycosides from various nutgalls. Prepn by oxidation of pyrogallol: Perkin, Steven, *J. Chem. Soc.* **83**, 192 (1903); Perkin, *ibid.* **101**, 803 (1912); Nierenstein, Spiers, *Ber.* **46**, 3151 (1913); Evans, Dehn, *J. Am. Chem. Soc.* **52**, 3647 (1930). Structure: Willstätter, Heiss, *Ann.* **433**, 17 (1923); Barltrop, Nicholson, *J. Chem. Soc.* **1948**, 116. Synthesis: Caunt *et al., ibid.* **1950**, 1631; **1951**, 1313.

Deep red needles from glacial acetic acid, dec 274-275°. Sparingly sol in most solvents.

Diglucoside, $C_{23}H_{28}O_{15}$, *dryophantin.* Coloring matter of "red pea" gall produced by *Dryophanta divisa* Ald. on *Quercus pedunculata* Ehrh., *Fagaceae:* Nierenstein, *J. Chem. Soc.* **115**, 1328 (1919); Nierenstein, Swanton, *Biochem. J.* **38**, 373 (1944). Dark red needles with bronze luster, mp 220-221°. Slightly sol in water, cold alcohol; sol in boiling alcohol; in methanol, acetone.

USE: As an additive to edible or inedible fats or oils, hydrocarbon fuels or lubricants, retards oxidation or metal contamination: Thompson, U.S. pat. **2,770,545** (1956 to Universal Oil Prod.).

8134. Putrescine. *1,4-Butanediamine;* tetramethylenediamine. $C_4H_{12}N_2$; mol wt 88.15. C 54.50%, H 13.72%, N 31.78%. $NH_2(CH_2)_4NH_2$. Biogenic polyamine and precursor of spermidine, *q.v.,* initially detected in decaying animal tissues, but now known to be present in all cells and certain bacterial cultures. It is essential for both normal and neoplastic tissue growth. Formed via decarboxylation of ornithine or by decarboxylation of arginine, followed by hydrolysis. Prepn: A. Ladenburg, *Ber.* **19**, 780 (1886); G. Ciamician, C. U. Zanetti, *ibid.* **22**, 1970 (1889); R. Willstätter, W. Heubner, *ibid.* **40**, 3871 (1907); of the dihydrochloride: *Org. Syn.* coll. vol. **IV** (1963) p 819. Role in cell growth processes: C. W. Tabor, H. Tabor, *Ann. Rev. Biochem.* **45**, 285 (1976); J. Janne *et al., Biochim. Biophys. Acta* **473**, 241 (1978). Formation and interconversion of putrescine and spermidine in mammalian cells: A. E. Pegg *et al., Advan. Enzyme Regul.* **19**, 427 (1980). Biosynthetic study in fungi: L. Stevens, *Med. Biol.* **59**, 308 (1981). Regulation of tRNA methyl transferase activity: M. Mach *et al., Biochem. J.*

202, 153 (1982). Alteration of DNA conformation in rat brain tumor cells by depletion of intracellular putrescine: D. T. Hung *et al., Science* **221**, 368 (1983). Use of labeled putrescine as a positron-emission tomographic tracer in brain tumors: N. Volkow *et al., ibid.* 673. Reviews of early literature: M. Guggenheim, *Die biogenen Amine* (S. Karger, Basel, 1951, 4th ed.) 619 pp; H. Tabor *et al., Ann. Rev. Biochem.* **30**, 579-604 (1961). Review of formation of GABA, the major inhibitory neurotransmitter in vertebrate brains, from putrescine: N. Seiler, *Physiol. Chem. Phys.* **12**, 411-429 (1980). Review of metabolism: T. L. Sourkes, K. Missala, *Agents Actions* **11**, 20-27 (1981). Book: *Polyamines in Biology and Medicine*, D. R. Morris, L. J. Marton, Eds. (Dekker, New York, 1981) 512 pp.

Colorless oil, bp 158-160°. Cryst on cooling, mp 23-24°. Strong piperidine-like odor. Very sol in water.

Dihydrochloride, $C_4H_{12}N_2.2HCl$, cryst from 85% alc, mp > 275°.

USE: As a tool in biochemical research.

8135. Pygeum Africanum Extract. Tadenan. Extract from the bark, wood, flowers and fruit of the tree *Pygeum africanum* Hook, also known as *Prunus africana* (Hook) F. Kalkm (Rosaceae-Amygdaleae), which is native to tropical Asia and Africa. Isoln: J. Debat, **Fr.** pat. **1,578,711;** *idem,* **Fr.** pat. **M6703;** *idem,* **U.S.** pat. **3,856,946** (1969, 1969, 1974 all to Labs. du Debat). Pharmacology: L. Thieblot *et al., Thérapie* **26**, 575 (1971); of an active principle, 1-docosanol: *idem et al., ibid.* **32**, 99 (1977). Characterization of components: R. Longo, S. Tira, *Farmaco Ed. Prat.* **38**, 287 (1983); E. M. Martinelli *et al., J. High Resolut. Chromatog. Chromatog. Commun.* **9**, 106 (1986). Clinical study in BPH: A. Barlet *et al., Wein Klin. Wochenschrift* **102**, 667 (1990).

1-Docosanol, $C_{22}H_{46}O$, *behenic alcohol, behenyl alcohol, docoxyl alcohol, IK-2, V-1326.* Mechanism of action study: J. Müntzing *et al., Invest. Urol.* **17**, 176 (1979). Identification and GC-MS determn in extract: N. Pierini *et al., Boll. Chim. Farm.* **121**, 27 (1982).

THERAP CAT: In treatment of micturitional disorders due to benign prostatic hypertrophy.

8136. Pymetrozine. *(E)-4,5-Dihydro-6-methyl-4-[(3-pyridinylmethylene)amino]-1,2,4-triazin-3(2H)-one;* CGA-215944; Chess; Sterling; Fulfill. $C_{10}H_{11}N_5O$; mol wt 217.23. C 55.29%, H 5.10%, N 32.24%, O 7.37%. Pyridine azomethine antifeedant. Prepn: H. Kristinsson, **Eur.** pat. **Appl.** **314,615;** *idem,* **U.S.** pat. **4,931,439** (1989, 1990 both to Ciba-Geigy). Comprehensive description: C. R. Flückiger *et al., Brighton Crop Prot. Conf.-Pests Dis.* **1992**, 43-50. Field trials against aphids and white flics: *idem et al., ibid.* 1187. Historical account of discovery, synthesis, and development: H. Kristinsson, *Spec. Publ.-Royal Soc. Chem.* **147**, 85-102 (1994).

Crystalline, mp 234.4°. Vapor pressure at 20°: ≤ 9.7 × 10^{-8} Pa. Soly at 20° (g/l): water 0.270; ethanol 2.25; hexane < 0.001. Partition coefficient (*n*-octanol/water): 0.2. LD_{50} orally in rats: 5820 mg/kg (Flückiger).

USE: Insecticide.

8137. Pyocyanine. *5-Methyl-1(5H)-phenazinone.* $C_{13}H_{10}N_2O$; mol wt 210.24. C 74.27%, H 4.79%, N 13.32%, O 7.61%. Produced by *Pseudomonas aeruginosa (Ps. pyocyanea, Bacillus pyocyaneus).* Isoln from broth cultures: Wrede, Strack, *Z. Physiol. Chem.* **140**, 1 (1924). Isoln from agar cultures: Schoental, *Brit. J. Exp. Pathol.* **22**, 137 (1941). Synthesis by dimethyl sulfate methylation of α-hydroxyphenazine and alkaline treatment of the resulting methosulfate: Wrede, Strack, *Z. Physiol. Chem.* **181**, 58 (1929); *Ber.* **62**, 2051 (1929); Surrey, *Org. Syn.* **26**, 86 (1946); **coll. vol. III,** 753 (1955).

Dark blue needles from water (usually with 1 H_2O which is lost at 50° over P_2O_5 *in vacuo*). mp 133°. Upon further heating it sublimes with decompn. Absorption spectrum: Nitzsche, *Ber.* **77**, 337 (1944). Freely sol in chloroform. Sol in nitrobenzene, pyridine, phenol, acetic acid, hot water, hot alcohol. Slightly sol in cold water and benzene. The blue water soln, made alkaline with Na_2CO_3, is easily rendered colorless by reduction with glucose or sodium hydrosulfite. Acidic $KMnO_4$ solns are decolorized by pyocyanine. An alkaline water soln of pyocyanine acquires a maroon color upon heating.

Aurichloride, dec 185°.

8138. Pyracarbolid. *3,4-Dihydro-6-methyl-N-phenyl-2H-pyran-5-carboxamide; 3,4-dihydro-6-methyl-2H-pyran-5-carboxanilide; 5,6-dihydro-2-methyl-3-(phenylcarbamoyl)-4H-pyran;* HOE-13764; Sicarol. $C_{13}H_{15}NO_2$; mol wt 217.27. C 71.87%, H 6.96%, N 6.45%, O 14.73%. Prepn: **Fr.** pat. **2,000,552;** O. Scherer *et al.,* **U.S.** pat. **3,632,821** (1969, 1972 both to Hoechst). Activity: B. Jank, F. Grossman, *Pestic. Sci.* **2**, 43 (1971).

Colorless solid, mp 110-111°. Vapor press: 7.66×10^{-3} mm Hg. Soly in water at 40°: 0.6 g/l. Sol in most organic solvents; slightly sol in benzene, xylene, cyclohexane.

USE: Agricultural fungicide.

8139. Pyrantel. *(E)-1,4,5,6-Tetrahydro-1-methyl-2-[2-(2-thienyl)ethenyl]pyrimidine.* $C_{11}H_{14}N_2S$; mol wt 206.31. C 64.04%, H 6.84%, N 13.58%, S 15.54%. Prepn: **Belg.** pat. **658,987** (1965 to Pfizer); *C.A.* **64**, 8192c (1966); Austin *et al., Nature* **212**, 1273 (1966); Kasubrick, McFarland, **S. Afr.** pat. **68 00,516;** *eidem,* **U.S.** pat. **3,502,661** (1968, 1970 both to Pfizer). Structure-activity studies: McFarland *et al., J. Med. Chem.* **12**, 1066 (1969). Pharmacology and animal trials: Cornwell, *Vet. Rec.* **79**, 590 (1966); Howes, Lynch, *J. Parasitol.* **53**, 1085 (1967); Eyre, *J. Pharm. Pharmacol.* **22**, 26 (1970).

Crystals from methanol, mp 178-179°. Tartrate, $C_{11}H_{14}N_2S.C_4H_6O_6$, *CP-10423-18, Banminth, Strongid.* White crystals from hot methanol, mp 148-150°. uv max (water): 312 nm (log ε 4.27). Pamoate, $C_{34}H_{30}N_2O_6S$, *pyrantel embonate, CP-10423-16, Antiminth, Cobantril, Combantrin, Early Bird, Helmex, Helmintox, Piranver.* Tasteless, yellow crystalline powder. Insol in water.

THERAP CAT: Anthelmintic (Nematodes).

THERAP CAT (VET): Anthelmintic (Nematodes).

8140. Pyrazinamide. *Pyrazinecarboxamide;* pyrazinoic acid amide; pyrazine carboxylamide; D-50; Pezetamid; Pyrafat; Pirilène; Piraldina; Tebrazid; Unipyranamide; Zinamide. $C_5H_5N_3O$; mol wt 123.11. C 48.78%, H 4.09%, N 34.13%, O 13.00%. Prepd by ammonolysis of methyl pyrazinoate (from quinoxaline): Dalmer, Walter, **Ger.** pat. **632,-257** (1936 to E. Merck); **Brit.** pat. **451,304; U.S.** pat. **2,149,-**

279 (1939 to Merck & Co.); Hall, Spoerri, *J. Am. Chem. Soc.* **62,** 664 (1940); Kushner *et al., ibid.* **74,** 3617 (1952); Williams, Kushner, U.S. pat. **2,677,641** (1954 to Am. Cyanamid); by thermal decarboxylation of pyrazine-2-carboxamide-3-carboxylic acid: Webb, Arlt, U.S. pat. **2,780,624** (1957 to Am. Cyanamid). Pharmacology: I. M. Weiner, J. P. Tinker, *J. Pharmacol. Exp. Ther.* **180,** 411 (1972). Comprehensive description: E. Feldner, D. Pitre, *Anal. Profiles Drug Subs.* **12,** 433-462 (1983).

Crystals from water or alcohol, mp 189-191°. Begins to sublime at 60°. pKa 0.5. uv max: 269 nm ($E_{1cm}^{1\%}$ 660). Soly (mg/ml): water 15 ; methanol 13.8; abs ethanol 5.7; isopropanol 3.8; ether 1.0; isooctane 0.01; chloroform 7.4. Aq solns are neutral.
THERAP CAT: Antibacterial (tuberculostatic).

8141. Pyrazine. 1,4-Diazine; paradiazine. $C_4H_4N_2$; mol wt 80.09. C 59.99%, H 5.03%, N 34.98%. Prepn: S. Gabriel, G. Pinkus, *Ber.* **26,** 2197 (1893); P. Brandes, C. Stoehr, *J. Prakt. Chem.* [2] **54,** 481 (1896). *See also: Beilstein* **XXIII,** 91 (1936). Crystal structure: G. De With *et al., Acta Crystallogr.* **B32,** 3178 (1976).

Crystals or wax-like solid. Strong pyridine-like odor. d_4^{61} 1.031; mp 53°; bp 115-118°; n_D^{61} 1.4953. Freely sol in water, alcohol, ether. Volatile with steam.

8142. 2,3-Pyrazinedicarboxylic Acid. $C_6H_4N_2O_4$; mol wt 168.11. C 42.87%, H 2.40%, N 16.66%, O 38.07%. Prepd by permanganate oxidation of quinoxaline: Gabriel, Sonn, *Ber.* **40,** 4850 (1907); Sausville, Spoerri, *J. Am. Chem. Soc.* **63,** 3153 (1941); Jones, McLaughlin, *Org. Syn.* **30,** 86 (1950).

Dihydrate, prisms from water. Becomes anhydr at 100°, then melts at 183-185° (with evolution of CO_2). The dry dust provokes sneezing. Freely sol in water. Soluble in methanol, acetone, ethyl acetate. Slightly sol in ethanol, ether, chloroform, benzene, petr ether.
Dimethyl ester, $C_8H_8N_2O_4$, crystals, mp 50°. Sol in water, acetone. Insol in benzene, petr ether.
Diamide, $C_6H_6N_4O_2$, crystals, dec 240°. Sol in boiling water.

8143. Pyrazinoic Acid. *Pyrazinecarboxylic acid;* pyrazinemonocarboxylic acid. $C_5H_4N_2O_2$; mol wt 124.10. C 48.39%, H 3.25%, N 22.57%, O 25.78%. Prepn by thermal decompn of 2,3-pyrazinedicarboxylic acid: Hall, Spoerri, *J. Am. Chem. Soc.* **62,** 664 (1940); by oxidation of alkylpyrazines: Gainer, *J. Org. Chem.* **24,** 691 (1959).

Fine white needles from water, dec 225-229°. Sublimes. K at 25° = 1.2×10^{-3}. Slightly sol in cold water, more sol in hot water. One gram dissolves in 120 g abs ethanol at 25°. Practically insol in ether, chloroform, benzene.

Methyl ester, crystals from ether, mp 59°.
Amide, *see* Pyrazinamide.

8144. Pyrazole. 1,2-Diazole. $C_3H_4N_2$; mol wt 68.08. C 52.93%, H 5.92%, N 41.15%. Prepn from acetylene and diazomethane: H. v. Pechmann, *Ber.* **31,** 2950 (1898). Two tautomeric forms. ^{1}H-NMR: J. F. K. Wilshire, *Aust. J. Chem.* **19,** 1935 (1966); ^{13}C-NMR: W. M. Litchman, *J. Am. Chem. Soc.* **101,** 545 (1979); ^{15}N-NMR: I. I. Schuster *et al., J. Org. Chem.* **44,** 1765 (1979). Spectral studies: D. Dumanovic *et al., Talanta* **22,** 819 (1975). Inhibition of alcohol dehydrogenase: W. K. Lelbach, *Experientia* **25,** 816 (1969). Induction of thyroid necrosis: S. Szabo *et al., Science* **199,** 1209 (1979). Toxicity: G. Magnussen *et al., Experientia* **28,** 1198 (1972).

Needles or prisms from petr ether; pyridine-like odor, bitter taste. mp 69.5-70°. $bp_{757.9}$ 186-188°. K at 25° = 3.0×10^{-12}. uv max (12M H_2SO_4): 215 nm (log ϵ 3.76); at pH 6.1: 310 nm (log ϵ 3.61). Sol in water, alcohol, ether, benzene. LD_{50} rats, mice (mmol/kg) 24 hr mortality: 19, 21 i.v.; 21, 22 orally (Magnussen).
USE: Chelating agent; in organic synthesis.

8145. 2-Pyrazoline. *4,5-Dihydro-1H-pyrazole.* $C_3H_6N_2$; mol wt 70.09. C 51.41%, H 8.63%, N 39.97%. Prepn from ethylene and diazomethane: Azzarello, *Atti Acad. Lincei* [5] **14,** II, 285 (1905).

Liquid. Faint amine odor. An odor resembling chocolate has been detected in some prepns. d_4^{17} 1.0200. bp_{760} 144°. Volatile with steam. $n_{587}^{17.2}$ 1.4796. Miscible with water, alc.

8146. Pyrazophos. *2-[(Diethoxyphosphinothioyl)oxy]-5-methylpyrazolo[1,5-a]pyrimidine-6-carboxylic acid ethyl ester; 2-hydroxy-5-methylpyrazolo[1,5-a]pyrimidine-6-carboxylic acid ethyl ester, O-ester with O,O-diethyl phosphorothioate;* HOE-2873; Afugan; Curamil. $C_{14}H_{20}N_3O_5PS$; mol wt 373.37. C 45.04%, H 5.40%, N 11.25%, O 21.43%, P 8.30%, S 8.59%. Systemic fungicide. Prepn: Neth. pat. Appl. **6,602,131;** O. Scherer, H. Mildenberger, U.S. pat. **3,632,757** (1966, 1972 both to Hoechst). Account of fungicidal properties and toxicity: F. M. Smit, *Meded. Rijksfac. Landbouwwetensch. Gent* **34,** 763 (1969), *C.A.* **73,** 76038w (1970). In control of rusty spot on peaches: R. H. Daines, *Plant Dis. Res.* **58,** 254 (1974). Effectiveness against powdery mildew: D. J. Butt *et al., Ann. Appl. Biol.* **75,** 217 (1973); F. O. Riordain, *Plant Dis. Rep.* **58,** 12 (1974); R. T. Burchill, M. E. Cook, *Plant Pathol.* **24,** 194 (1975). Enzymatic detection by TLC: G. F. Ernst, F. Schuring, *J. Chromatog.* **49,** 325 (1970). Absorption and metabolism by plants: S. Gorbach *et al., Environ. Qual. Saf. Suppl.* **3,** 840 (1975).

Yellow oil, mp 38-40°C. LD_{50} in rats, marmots, rabbits (mg/kg): 140, 184, 435 orally (Smit).
USE: Fungicide.

8147. Pyrene. Benzo[*def*]phenanthrene. $C_{16}H_{10}$; mol wt 202.26. C 95.02%, H 4.98%. Occurs in coal tar. Isoln: Kruber, *Ber.* **64,** 84 (1931). Also obtained by the destructive hydrogenation of hard coal: I. G. Farben., **Ger.** pats. **639,240; 640,580; 654,201.** Purification by chromatography: Winterstein *et al., Z. Physiol. Chem.* **230,** 162 (1934). Synthesis from *o,o'*-ditolyl: Weitzenböck, *Monatsh.* **34,** 193 (1913). From *peri*-trimethylenenaphthalene and malonyl chloride with $AlCl_3$: Fleischer, Retze, *Ber.* **55**, 3280 (1922). From α-tetralone: Braun, Rath, *Ber.* **61**, 956 (1928). From 4-keto-1,2,3,4-tetrahydrophenanthrene by a Reformatsky reaction: Cook, Hewett, *J. Chem. Soc.* **1934**, 366.

Monoclinic prismatic tablets from alcohol or by sublimation. d^{23} 1.271. Pure pyrene is colorless, the usual contaminant which gives it a yellow color is tetracene. Solid and solns have slight blue fluorescence. mp 156°. bp 404°. Absorption spectrum: Clar, *Ber.* **69**, 1677 (1936); Seshan, *Proc. Indian Acad. Sci.* **A3**, 148 (1936). Fluorescence maxima: Sannié, Poremski, *Bull. Soc. Chim.* [5] **3**, 1139 (1936). Insol in water. Fairly sol in organic solvents.

8148. Pyrethrins. Active insecticidal constituents of pyrethrum flowers. Isoln: Staudinger, Ruzicka, *Helv. Chim. Acta* **7**, 177 (1924). Prepn by reconstitution from pyrethrolone and chrysanthemic acid: Elliott, Janes, *Chem. & Ind. (London)* **1969**, 270. Structure: Crombie *et al., J. Chem. Soc.* **1956**, 3963; Godin *et al., ibid.* (C) **1966**, 332. Stereochemistry: Begley *et al., Chem. Commun.* **1972**, 1276. Biosynthesis: Crowley *et al., Biochim. Biophys. Acta* **60**, 312 (1962). *Review:* Crombie, Elliott, *Fortschr. Chem. Org. Naturst.* **19**, 120 (1961).

Pyrothrin I   R = CH_3
Pyrethrin II   R = $COOCH_3$

Pyrethrin I, $C_{21}H_{28}O_3$, *[1R-[1α[S*(Z)],3β]]-2,2-dimethyl-3-(2-methyl-1-propenyl)cyclopropanecarboxylic acid 2-methyl-4-oxo-3-(2,4-pentadienyl)-2-cyclopenten-1-yl ester, chrysanthemummonocarboxylic acid pyrethrolone ester.* Viscous liquid. $bp_{0.0005}$ 146-150°. n_D^{20} 1.5242. $[α]_D^{20}$ −14° (isooctane). uv max (95% ethanol): 225 nm (ε 36400). Oxidizes readily and becomes inactive in air. Should be refrigerated and stored in darkness. Stability data of pyrethrins I and II: Godin, *Pyrethrum Post* **9**, 17 (1968), *C.A.* **71**, 12031q (1969). Practically insol in water. Sol in alcohol, petr ether, kerosene, carbon tetrachloride, ethylene dichloride, nitromethane.

Pyrethrin II, $C_{22}H_{28}O_5$, *[1R-[1α[S*(Z)],3β(E)]]-3-(3-methoxy-2-methyl-3-oxo-1-propenyl)-2,2-dimethylcyclopropanecarboxylic acid 2-methyl-4-oxo-3-(2,4-pentadienyl)-2-cyclopenten-1-yl ester, chrysanthemumdicarboxylic acid monomethyl ester pyrethrolone ester.* Viscous liquid. Oxidizes rapidly and becomes inactive in air. $bp_{0.007}$ 192-193°. n_D^{20} 1.5355. $[α]_D^{19}$ +14.7° (isooctane-ether). uv max (95% ethanol): 229 nm (ε 45850). Practically insol in water. Sol in alc, petr ether (less sol than pyrethrin I), kerosene, carbon tetrachloride, ethylene dichloride, nitromethane.

Caution: Direct skin contact or inhalation may cause severe allergic attacks in sensitive people. Potential symptoms of overexposure are dermatitis, asthma, vasomotor rhinitis, anaphylactic reactions; numbness of lips and tongue, sneezing, vomiting, diarrhea; tinnitus, headache, restlessness, incoordination, clonic convulsions, stupor, prostration; death due to respiratory paralysis. *See Clinical Toxicology of Commercial Products,* R. E. Gosselin *et al.,* Eds. (Williams & Wilkins, Baltimore, 5th ed., 1984) Section III, pp 352-355.

USE: Insecticide.

8149. Pyrethrosin. *[1aR-(1aR*,4E,6R*,6aR*,9aS*,-10aR*)]-6-(Acetyloxy)-1a,3,6,6a,7,9a,10,10a-octahydro-4,10a-dimethyl-7-methyleneoxireno[8,9]cyclodeca[1,2-b]furan-*

8(2H)-one; 1,10-epoxy-6,8-dihydroxygermacra-4,11(13)-dien-12-oic acid 12,8-lactone acetate.* $C_{17}H_{22}O_5$; mol wt 306.36. C 66.65%, H 7.24%, O 26.11%. From flowers of *Chrysanthemum cinerariaefolium* Vis., *Compositae.* Isoln: Thoms, *Pharm. Ztg.* **1891**, 503. Structure: Barton, de Mayo, *J. Chem. Soc.* **1957**, 150; Barton *et al., ibid.* **1960**, 2263. Revised structure: S. Iriuchijima, S. Tamura, *Agr. Biol. Chem.* **34**, 204 (1970); E. J. Gabe *et al., Chem. Commun.* **1971**, 559.

Crystals from ethanol, ethyl acetate or benzene + light petroleum, mp 198-200°. $[α]_D$ −31° (c = 1.73 in chloroform). uv max: 204, 210, 220, 230 nm (ε 14,600; 12,200; 6000; 1700). Practically insol in water. Sol in hot alcohol, chloroform; slightly sol in ether or petr ether.

8150. Pyrethrum Flowers. Dalmatian insect powder; Persian insect powder. Flowers of *Chrysanthemum (Pyrethrum) cinerariaefolium* Vis., and of *C. coccineum* Willd. (*C. roseum* Adam), *Compositae.* Habit. Dalmatia, Montenegro, Western Asia. Constit. About 1-1.5% volatile oil; pyrethrin, pyretol, pyrethrotoxic acid, pyrethrosin, chrysanthemine, chrysathemumic acid.

USE: In insecticide prepns.

THERAP CAT: Scabicide.

THERAP CAT (VET): Ectoparaciticide.

8151. Pyridate. *O-(6-Chloro-3-phenyl-4-pyridazinyl)-carbonothioic acid S-octyl ester;* fenpyrate; CL-11344; Lentagran. $C_{19}H_{23}ClN_2O_2S$; mol wt 378.92. C 60.23%, H 6.12%, Cl 9.36%, N 7.39%, O 8.44%, S 8.46%. Contact herbicide for control of weeds in cereals. Prepn: **Belg. pat.** 816,574; R. Schonbeck *et al.,* **U.S. pat.** 3,953,445 (1974, 1976 to Chemie Linz). Description: A. Diskus *et al., Proc. Brit. Crop. Prot. Conf. - Weeds* **1976**, 717; M. Chéroux, J.-M. Moncorge, *Def. Veg.* **38**, 310 (1984). Residue analysis by HPLC: W. Lindner, H. Ruckendorfer, *Int. J. Environ. Anal. Chem.* **16**, 205 (1983).

Oily brown liquid, $bp_{0.1}$ 220°. mp 27°. n_D^{20} 1.568. d^{20} 1.555. Flash pt (open cup): 200°C. Almost insol in water (90 ppm). Sol in organic solvents. LD_{50} in male rats, female rats, mice (mg/kg): 1970, 2400, > 10,000 orally; in rabbits (mg/kg): > 3450 dermally. LC_{50} (96 hr) in carp, trout: > 100, 81 mg/l (Chéroux, Moncorge).

USE: Herbicide.

8152. Pyridazine. *1,2-Diazine; orthodiazine; oizine.* $C_4H_4N_2$; mol wt 80.09. C 59.99%, H 5.03%, N 34.98%. Prepd from maleic hydrazide: Mizzoni, Spoerri, *J. Am. Chem. Soc.* **73**, 1873 (1951). Resonance energy and position of double bonds: Maccoll, *J. Chem. Soc.* **1946**, 671.

Liquid. d_4^{18} 1.107; $d_4^{23.5}$ 1.1035. mp −8°. bp_{760} 208°; bp_{14} 87°; $bp_{1.0}$ 48°. $n_D^{23.5}$ 1.52311. Dipole moment in dioxane at 35° = 3.94. uv max: 338 nm. Resonance energy: 22 kcal.

Miscible with water, benzene, DMF. Freely sol in methanol, ethanol, ether. Practically insol in petr ether.
Hydrochloride, $C_4H_4N_2 \cdot HCl$, yellow solid, mp 161-163°.

8153. Pyridine. C_5H_5N; mol wt 79.10. C 75.92%, H 6.37%, N 17.71%. Discovered in coal tar: T. Anderson, *Ann.* **60**, 86 (1846). General review: G. L. Goe in Kirk-Othmer *Encyclopedia of Chemical Technology* vol. 19 (Wiley-Interscience, New York, 3rd ed., 1982) pp 454-483. Books: H. Maier-Bode, *Das Pyridin und Seine Derivate in Wissenschaft und Technik* (Edwards Bros., Ann Arbor, Mich., 1943); Hill in *Chemistry of Coal Utilization* vol 2, H. H. Lowry, Ed. (Wiley, New York, 1945), Chapter 27; *The Chemistry of Heterocyclic Compounds* vol. **14**, a series of books entitled "Pyridine and its Derivatives": Parts 1-4, E. Klingsberg, Ed. (1960-1964); Suppl. Parts 1-4, R. A. Abramovitch, Ed. (1974-1975); Part 5, G. R. Newkome, Ed. (1984) (Wiley-Interscience, New York). Physical data: B. A. Middleton, J. R. Partington, *Nature* **141**, 516 (1938). Acute toxicity: H. F. Smyth Jr. *et al.*, *Arch. Ind. Hyg. Occup. Med.* **4**, 119 (1951). Human toxicity: E. Browning, *Toxicity and Metabolism of Industrial Solvents* (Elsevier, New York, 1965) pp 304-309. Brief review of commercial synthesis and uses of pyridine and derivatives: D. J. Berry, *Spec. Chem.* **3**, 13 (1983). Review of occurrence of pyridine and derivatives in foods, tobacco and essential oils: G. Vernin, *Perfum. Flavor.* **7**, 23-26 (1982); of physical, chemical, toxicological properties and environmental fate: A. Jori *et al.*, *Ecotoxicol. Environ. Saf.* **7**, 251 (1983); of use in the study of surface properties of transition metal oxides: M. C. Kung, H. H. Kung, *Catal. Rev. Sci. Eng.* **27**, 425-460 (1985).

Flammable, colorless liq; characteristic disagreeable odor; sharp taste. d_4^{20} 0.98272. Flash pt, closed cup: 68°F (20°C). mp −41.6°. bp 115.2-115.3°. n_D^{20} 1.50920. Dipole moment in benzene: 2.26. Forms an azeotropic mixture with 3 mols water, boiling at 92-93°. Volatile with steam. Miscible with water, alcohol, ether, petr ether, oils and many other organic liquids. Good solvent for many organic and inorganic compds. Weak base; forms salts with strong acids. pKa 5.19. pH of 0.2 molar soln in H_2O: 8.5. LD_{50} orally in rats: 1.58 g/kg (Smyth).
Caution: Potential symptoms of overexposure are headache, nervousness, dizziness and insomnia; nausea, anorexia; frequent urination; eye irritation; dermatitis; liver and kidney damage. *See NIOSH Pocket Guide to Chemical Hazards* (DHHS/NIOSH 90-117, 1990) p 190.
USE: As solvent for anhydr mineral salts. Synthetic intermediate in laboratory and industry.

8154. 3-Pyridineacetic Acid. Lioxone; 3PAA: Minedil. $C_7H_7NO_2$; mol wt 137.14. C 61.31%, H 5.14%, N 10.21%, O 23.33%. Metabolite of nicotine, *q.v.* Prepn: K. Miescher, H. Kägi, *Helv. Chim. Acta* **24**, 1471 (1941); S. Carboni, *Gazz. Chim. Ital.* **85**, 1194 (1955). Manuf. process: M. Hartmann *et al.*, U.S. pat. **2,408,020** (1946 to Ciba). Pharmacology: E. Ginoulhiac, L. T. Tenconi, *Minerva Med.* **51**, 1166 (1960). Clinical evaluation as hypocholesteremic: F. Pupita, *ibid.* 1140. Thermal and photodecarboxylation: F. R. Stermitz, W. H. Huang, *J. Am. Chem. Soc.* **93**, 3427 (1971). Determn in biological fluids by HPLC: G. A. Kyerematen *et al.*, *J. Chromatog.* **419**, 191 (1987); by thermospray LC-MS: K. T. McManus *et al.*, *J. Chromatog. Sci.* **28**, 510 (1990).

Crystals from ethyl acetate or alcohol, mp 144°.

Note: Often confused with homonicotinic acid, (4-methyl-3-pyridinecarboxylic acid), in early literature.
THERAP CAT: Antihyperlipoproteinemic.

8155. Pyridine *N*-Oxide. Pyridine *N*-oxide. C_5H_5NO; mol wt 95.10. C 63.15%, H 5.30%, N 14.73%, O 16.82%. Prepd industrially by heating pyridine in glacial acetic acid at 80° with 30% hydrogen peroxide. The acid is neutralized and the oxide recovered by distillation: Katritzky *et al.*, *J. Chem. Soc.* **1957**, 1769. Laboratory procedure using peracetic acid: H. S. Mosher *et al.*, *Org. Syn.* **coll. vol. IV**, 828 (1963).

Deliquescent crystals, mp 66°; $bp_{1.0}$ 100-105°.
Hydrochloride, $C_5H_5NO \cdot HCl$, crystals from isopropanol, mp 179.5-181°.
USE: Synthetic intermediate.

8156. Pyridinium Bromide Perbromide. *Hydrogen tribromide compd with pyridine (1:1).* $C_5H_6Br_3N$; mol wt 319.82. C 18.78%, H 1.89%, Br 74.95%, N 4.38%. C_5H_5-$N^+HBr_3^-$. Prepd from pyridine, hydrobromic acid, and bromine: L. F. Fieser, M. Fieser, *Reagents for Organic Chemistry* vol. **1** (John Wiley, New York, 1967) p 967.
Orange-needles from acetic acid. Odorless. Nonvolatile. Sparingly sol in acetic acid. Dissociates in the presence of a bromine acceptor, such as an alkene, to liberate one mole of bromine.
USE: In small-scale brominations, where it is much more convenient and agreeable to measure and use than elemental bromine.

8157. Pyridinium Chlorochromate. *Chlorotrioxochromate(1−), hydrogen, compd with pyridine (1:1);* PCC. C_5-$H_6ClCrNO_3$; mol wt 215.56. C 27.86%, H 2.81%, Cl 16.45%, Cr 24.12%, N 6.50%, O 22.27%. Oxidizing agent for the efficient conversion of primary or secondary alcohols to carbonyl compounds. Prepn from chromyl chloride and pyridine: R. J. Meyer, R. Best, *Z. Anorg. Allgem. Chem.* **22**, 192 (1899); J. Bernard, M. Camelot, *C.R. Acad. Sci.* **258**, 5881 (1964). Prepn from chromium trioxide in HCl and pyridine, use as an oxidant: E. J. Corey, J. W. Suggs, *Tetrahedron Letters* **1975**, 2647. Oxidation, ring enlargement of furans to pyranones: G. Piancatelli *et al.*, *ibid.* **1977**, 2199; oxidation of tertiary allylic alcohols: W. G. Dauben, D. M. Michno, *J. Org. Chem.* **42**, 682 (1977). Role in oxidative cationic cyclization: E. J. Corey, D. L. Boger, *Tetrahedron Letters* **1978**, 2461. *Review:* G. Piancatelli *et al.*, *Synthesis* **1982**, 245-258.

Orange-yellow solid, mp 205° (dec). Insol in dichloromethane.
USE: Oxidizing agent in organic synthesis.

8158. Pyridinol Carbamate. *2,6-Pyridinedimethanol bis(methylcarbamate) (ester);* 2,6-pyridinylenebis[methyl-*N*-methylcarbamate]; pyricarbate; H-3749; Anginin; Angioxine; Aterosan; Atover; Cicloven; Colesterinex; Duvaline; Movecil; Prodectin; Ravenil; Sospitan; Vasagin; Vasapril; Vasocil; Vasoverin. $C_{11}H_{15}N_3O_4$; mol wt 253.26. C 52.17%, H 5.97%, N 16.59%, O 25.27%. Prepd by the reaction of methylamine with 2,6-pyridinedimethanol bis[phenylcarbonate] in methanol: Matsumoto, **Japan.** pat. **66 22,185** (1966 to Banyu), *C.A.* **66**, 75907x (1967). Pharmacology and clinical evaluation: T. Shimamoto *et al.*, *Am. Heart J.* **71**, 297 (1966). Metabolism: Mallein *et al.*, *Therapie* **28**, 115 (1973). Effects on lipid metabolism and antiatherogenic effect in animals: V. Orbetzova *et al.*, *Artery* **8**, 560 (1980).

HPLC determn in biological fluids: A. Suenaga *et al.*, *Acta Pharm. Suec.* **23**, 245 (1986).

Needles from methanol or acetone, mp 136-137°. uv max (methanol): 264-265 nm. Sparingly sol in cold water; freely sol in hot water. LD$_{50}$ in mice, rats, rabbits, dogs (mg/kg): 4500, 3400, 5200, 1000 orally (Shimamoto).

THERAP CAT: Antiarteriosclerotic.

8159. Pyridofylline. *3,7-Dihydro-1,3-dimethyl-7-[2-(sulfooxy)ethyl]-1H-purine-2,6-dione compd with 5-hydroxy-6-methyl-3,4-pyridinedimethanol (1:1); 7-(2-hydroxyethyl)-theophylline hydrogen sulfate compound with pyridoxol; β-[1,3-dimethyl-2,6(1H,3H)-purinedion-7-yl]ethyl sulfate of 5-hydroxy-6-methyl-3,4-pyridinedimethanol; pyridoxol compd with 7-(2-hydroxyethyl)theophylline hydrogen sulfate; theophyllineethyl sulfate of pyridoxine; Atherophylline.* C$_{17}$H$_{23}$N$_5$O$_9$S; mol wt 473.46. C 43.13%, H 4.90%, N 14.79%, O 30.41%, S 6.77%. Prepn and therapeutic properties: A. Debarge, **Fr. pat.** **M828** (1961), *C.A.* **58**, 9098g (1963).

Crystals from abs alcohol, mp 144-146°. LD$_{50}$ in mice (g/kg): 1 i.v.; 1.6 orally (Debarge).

THERAP CAT: Respiratory stimulant; vasodilator (coronary).

8160. Pyridomycin. *[5R-(2Z,5R*,6S*,9S*,10S*,11R*)]-3-Hydroxy-N-[10-hydroxy-5,11-dimethyl-2-(1-methylpropylidene)-3,7,12-trioxo-9-(3-pyridinylmethyl)-1,4-dioxa-8-azacyclo-dodec-6-yl]-2-pyridinecarboxamide.* C$_{27}$H$_{32}$N$_4$O$_8$; mol wt 540.57. C 59.99%, H 5.97%, N 10.36%, O 23.68% Antimycobacterial antibiotic substance produced by *Streptomyces albidofuscus* Okami et Umezawa, nov sp., renamed *S. pyridomyceticus.* Isoln: Maeda *et al.*, *J. Antibiot.* **6A**, 140 (1953); K. Yagishita, *ibid.* **7A**, 143 (1954). Degradation studies: Maeda, *ibid.* **10A**, 94 (1957). Structure: Koyama *et al.*, *Tetrahedron Letters* **1967**, 3587; H. Ogawara *et al.*, *Chem. Pharm. Bull.* **16**, 679 (1968). Synthetic study: M. Kinoshita, M. Awamura, *Bull. Chem. Soc. Japan* **51**, 869 (1978); M. Kinoshita *et al.*, *ibid.* 3595. Production: **Japan.** pats. **54 7048; 55 1349; 56 9566** (1954, 1955, 1956 all to Nippon Antibiotic Subst. Sci. Assoc.).

Crystals from ethanol, mp 222°. uv max (ethanol): 303 nm (E$_{1cm}^{1\%}$ 209). Soluble in lower alcohols, ethyl or butyl acetate, benzene, acetone, dioxane, tetrahydrofuran. Practically insol in water.

Hydrochloride, crystals, mp 194-196°. Freely sol in water. [α]$_D^{16}$ −53.2° (water). pH of aq solns about 2.0. LD$_{50}$ i.p. in mice: >2100 mg/kg (Yagishita).

8161. Pyridostigmine Bromide. *3-[[(Dimethylamino)-carbonyl]oxy]-1-methylpyridinium bromide; 3-hydroxy-1-methylpyridinium bromide dimethylcarbamate; 1-methyl-3-hydroxypyridinium bromide dimethylcarbamate; 3-(dimethylcarbamyloxy)-1-methylpyridinium bromide; Ro-1-5130; Kalymin; Mestinon; Regonol.* C$_9$H$_{13}$BrN$_2$O$_2$; mol wt 261.12. C 41.40%, H 5.02%, Br 30.60%, N 10.73%, O 12.25%. Reversible inhibitor of acetylcholinesterase. Prepn: Urban, **U.S. pat. 2,572,579** (1951 to Hoffmann-La Roche). Mechanism of protective effect in soman poisoning: X. Deyi *et al.*, *Fundam. Appl. Toxicol.* **1**, 217 (1981). Evaluation of effect on neuromuscular function: M. Glikson *et al.*, *ibid.* **16**, 288 (1991). Evaluation of side effects profile under desert conditions: J. E. Cook *et al.*, *Military Med.* **157**, 250 (1992). Review of prophylactic effect in nerve agent poisoning: R. M. Dawson, *J. Appl. Toxicol.* **14**, 317 (1994).

Shiny, hygroscopic crystals from abs ethanol, mp 152-154°. Freely sol in water, alcohol. Practically insol in ether, acetone, benzene. Aq solns may be sterilized by autoclaving with steam.

THERAP CAT: Cholinergic; in treatment of myasthenia gravis. Pre-exposure antidote to chemical warfare agents.

8162. Pyridoxal. *3-Hydroxy-5-(hydroxymethyl)-2-methyl-4-pyridinecarboxaldehyde; 3-hydroxy-5-(hydroxymethyl)-methylisonicotinaldehyde;* 2-methyl-3-hydroxy-4-formyl-5-hydroxymethylpyridine. C$_8$H$_9$NO$_3$; mol wt 167.16. C 57.48%, H 5.43%, N 8.38%, O 28.71%. Synthesis and structure: Harris *et al.*, *J. Biol. Chem.* **154**, 315 (1944); *J. Am. Chem. Soc.* **66**, 2088 (1944).

Hydrochloride, C$_8$H$_9$NO$_3$.HCl, rhombic crystals, mp approx 165° with decompn. Soluble in water (1 g/2 ml); sol in 95% ethanol (1.7 g/100 ml); pH of 1% water soln = 2.65. The water solns are sensitive to heat. uv max: 292.5 nm (E mol 7600). Can be reduced to pyridoxine hydrochloride (mp 206-208°).

Monoethylacetal hydrochloride, C$_{10}$H$_{14}$ClNO$_3$, *1,3-dihydro-1-ethoxy-6-methylfuro[3,4-c]pyridin-7-ol hydrochloride.* Crystals from alcohol and ether, mp 142-143°.

8163. Pyridoxal 5-Phosphate. *3-Hydroxy-2-methyl-5-[(phosphonooxy)methyl]-4-pyridinecarboxaldehyde;* pyridoxal 5-monophosphoric acid ester; 3-hydroxy-5-(hydroxymethyl)-2-methylisonicotinaldehyde 5-phosphate; codecarboxylase; Pyromijin; Sechvitan; Vitazechs. C$_8$H$_{10}$NO$_6$P; mol wt 247.14. C 38.88%, H 4.08%, N 5.67%, O 38.84%, P 12.53%. Prepn by the action of adenosine triphosphate on pyridoxal: Gunsalus *et al.*, *J. Biol. Chem.* **155**, 685 (1944); by the action of phosphorus oxychloride on pyridoxal in aq soln: Gunsalus *et al.*, *ibid.* **161**, 743 (1945); Umbreit *et al.*, *Arch. Biochem.* **7**, 189 (1945); by phosphorylation of pyridoxamine with 100% H$_3$PO$_4$ followed by oxidation: Wilson, Harris, *J. Am. Chem. Soc.* **73**, 4693 (1951). Alternate route: Schorre, **U.S. pat. 3,124,587** (1964 to E. Merck). Isoln in pure form as the oxime and as the *O*-methyl oxime and structure: Heyl *et al.*, *J. Am. Chem. Soc.* **73**, 3430 (1951). Activity as a cotransaminase in the synthesis of amino acids: Lichstein *et al.*, *J. Biol. Chem.* **161**, 311 (1945). Activity of natural and synthetic material: Umbreit *et al.*, *loc. cit.*

Colorless in acid soln, bright-yellow in alkaline soln. uv max (alkaline soln): 390 nm (E_m 3.7); in acid soln: 295 nm (E_m 5.1). Gives a negative 2,6-dichloroquinone chlorimide test. On oxidation with H_2O_2 in alkaline solution yields [(2-methyl-3,4-dihydroxy-5-pyridyl)methyl]phosphoric acid.

Oxime, $C_8H_{11}N_2O_6P$, crystals, dec 229-230°. Practically insol in water, alcohol, ether.

O-Methyloxime, $C_9H_{13}N_2O_6P$, crystals, dec 212-213°. Practically insol in water, alcohol, ether.

Calcium salt, *Aderoxal*. Yellow precipitate. Practically insol in water, alcohol or ether.

THERAP CAT: Vitamin (enzyme co-factor).

8164. Pyridoxamine Dihydrochloride. *4-(Aminomethyl)-5-hydroxy-6-methyl-3-pyridinemethanol dihydrochloride;* 2-methyl-3-hydroxy-4-aminomethyl-5-hydroxymethylpyridine dihydrochloride. $C_8H_{14}Cl_2N_2O_2$; mol wt 241.12. C 39.85%, H 5.85%, Cl 29.41%, N 11.62%, O 13.27%. $C_8H_{12}N_2O_2$.2HCl. Synthesis and structure: Harris *et al.*, *J. Biol. Chem.* **154**, 315 (1944); *J. Am. Chem. Soc.* **66**, 2088 (1944).

Platelets, mp 226-227° with dec. Soluble in water (approx 1 g/2 ml); sol in 95% alc (0.65 g/100 ml). Reasonably stable at room temp; shows no decompn in a few days at 60°. Liquefies on exposure to 80% relative humidity. pH of a 1% water soln 2.4. uv max (pH 1.94): 287.5 nm (E mol 91,000).

Free base, crystals from alc; mp 193-193.5°; sol in alc.

Pyridoxamine phosphate, $C_8H_{13}N_2O_5P$, *2-methyl-3-hydroxy-4-aminomethyl-5-pyridylmethylphosphoric acid.* Prepn in soln by autoclaving codecarboxylase with glutamic acid: McNutt, Snell, *J. Biol. Chem.* **182**, 557 (1950); prepn by direct phosphorylation of pyridoxamine in aq soln with phosphorus oxychloride: Heyl *et al.*, *J. Am. Chem. Soc.* **73**, 3430 (1951).

8165. 4-Pyridoxic Acid. *3-Hydroxy-5-(hydroxymethyl)-2-methyl-4-pyridinecarboxylic acid; 3-hydroxy-5-(hydroxymethyl)-2-methylisonicotinic acid;* 2-methyl-3-hydroxy-4-carboxy-5-hydroxymethylpyridine. $C_8H_9NO_4$; mol wt 183.16. C 52.46%, H 4.95%, N 7.65%, O 34.94%. Occurs in urine. It is the chief metabolic product of pyridoxine pyridoxal, and pyridoxamine. Isoln from human urine: Singal, Sydenstricker, *Science* **78**, 545 (1941). Isoln and synthesis: Huff, Perlzweig, *J. Biol. Chem.* **155**, 345 (1944).

Wedge-shaped crystals, mp 247-248°. Slightly sol in water, alcohol, pyridine. Insoluble in ether and in aq acid soln, but completely sol in aq alkaline soln. Possesses two acidic groups, one a phenolic and the other a carboxyl having pK values of 9.75 and 5.50, respectively. Characteristic blue fluorescence, max at pH 3 to 4. The fluorescence disappears on reduction with hydrosulfite and is restored to

the original intensity with H_2O_2. Is adsorbed on zeolite from aq solns at pH 4 to 5 and can be eluted with 25% KCl; butanol extracts of neutral eluates also show characteristic blue fluorescence which is increased by a trace of acetic acid. Stable to boiling with dil alkali (1*N*); but upon being heated with 0.5*N* acid for a few minutes it is converted to the lactone.

Lactone, $C_8H_7NO_3$, *β-pyracine*. mp 263-265°, exhibits much stronger fluorescence. Is more easily followed in the course of biochemical investigations.

8166. Pyridoxine Hydrochloride. *5-Hydroxy-6-methyl-3,4-pyridinedimethanol hydrochloride; pyridoxol hydrochloride;* vitamin B_6 hydrochloride; pyridoxinium chloride; adermine hydrochloride; 2-methyl-3-hydroxy-4,5-bis(hydroxymethyl)pyridine hydrochloride; 5-hydroxy-6-methyl-3,4-pyridinedicarbinol hydrochloride; 3-hydroxy-4,5-dimethylol-α-picoline hydrochloride; Bonasanit; Hexabione hydrochloride; Hexabetalin; Hexavibex; Pyridipca; Pyridox; Béclan; Benadon; Hexermin; Campoviton 6; Hexobion. C_8H_{12}-ClNO$_3$; mol wt 205.64. C 46.73%, H 5.88%, Cl 17.24%, N 6.81%, O 23.34%. One of the vitamins of the B_6 complex; see also Codecarboxylase; Pyridoxal; Pyridoxamine Dihydrochloride. Present in many foodstuffs; esp good sources are yeast, liver and cereals. Isoln: Keresztesy, Stevens, *Proc. Soc. Exp. Biol. Med.* **38**, 64 (1938); György, *J. Am. Chem. Soc.* **60**, 983 (1938); Kuhn, Wendt, *Ber.* **71**, 780, 1118 (1938). Structure: E. T. Stiller *et al.*, *J. Am. Chem. Soc.* **61**, 1237 (1939); R. Kuhn *et al.*, *Ber.* **72**, 305 sqq. (1939). uv absorption spectrum: E. A. Peterson, H. A. Sober, *J. Am. Chem. Soc.* **76**, 169 (1954). Synthesis: Harris, Folkers, *J. Am. Chem. Soc.* **61**, 1242, 1245, 3307 (1939); P. G. Stevens, U.S. pats. **2,680,743** and **2,734,063** (1954, 1956 both to Gen. Aniline); P. I. Pollak, U.S. pats. **2,904,551, 3,024,244, 3,024,245** (1959, 1962, 1962 all to Merck & Co.); E. E. Harris *et al.*, *J. Org. Chem.* **27**, 2705 (1962); W. Böll, H. König, *Ann.* **1979**, 1657; T. Shono *et al.*, *Chem. Letters* **1981**, 1121. Biosynthesis: Hiel, Spenser, *Science* **169**, 773 (1970). *Review: The Vitamins* vol. 2, W. H. Sebrell, R. S. Harris, Eds. (Academic, New York, 2nd ed., 1968) pp 1-117.

Platelets or thick, birefringent rods from alcohol + acetone. Reasonably stable to light and air. Dec 205-212°. Sublimes. (Free base, mp 160°.) uv max (0.1*N* HCl): 290 nm (ε 8400); (phosphate buffer, pH 7): 253, 325 nm (ε 3700, 7100). One gram dissolves in about 4.5 ml water, 90 ml alcohol. Sol in propylene glycol. Sparingly sol in acetone. Insol in ether, chloroform. pH of a 10% w/v soln in water: 3.2. Acidic aq solns are stable and may be heated for 30 min at 120° without decompn.

THERAP CAT: Vitamin (enzyme cofactor).

THERAP CAT (VET): Nutritional factor.

8167. Pyrifenox. *1-(2,4-Dichlorophenyl)-2-(3-pyridinyl)ethanone O-methyloxime;* 2′,4′-dichloro-2-(3-pyridyl)acetophenone O-methyloxime; ACR-3453A; Ro-15-1297; Dorado. $C_{14}H_{12}Cl_2N_2O$; mol wt 295.17. C 56.97%, H 4.10%, Cl 24.02%, N 9.49%, O 5.42%. Ergosterol biosynthesis inhibitor for use on crops. Prepn: F. Dorn, Ger. pat. **3,310,148;** idem, U.S. pat. **4,605,656** (1983, 1986 both to Hoffmann-La Roche). Review of physical properties, toxicity and bioactivity: P. Zobrist *et al.*, *Proc. Brit. Crop Prot. Conf. - Pests*

Dis. **1986**, 47-53; of field trials: G. Neumann *et al., Meded. Fac. Landbouwwet. Rijksuniv. Gent* **54**, 635-642 (1989).

Slightly viscous, tan liquid with mild aromatic odor. $bp_{0.1}$ > 150°. vapor pressure at 25°: 1.4×10^{-5} torr. Soly (at 20°) in water at pH 7: 115 mg/l; in hexane: < 1 g/l; in acetone, ethyl acetate, chloroform, diethyl ether, DMF, isopropyl alcohol, toluene: > 250 g/l. LD_{50} in rats (mg/kg): 950 i.p.; 2900 orally; > 5000 dermally; LC_{50} by inhalation in rats: > 2.05 mg/l (Zobrist).

USE: Agricultural fungicide.

8168. Pyrilamine. *N-[(4-Methoxyphenyl)methyl]-N',N'-dimethyl-N-2-pyridinyl-1,2-ethanediamine; 2-[(2-dimethyl-aminoethyl)(p-methoxybenzyl)amino]pyridine; N-p-methoxy-benzyl-N',N'-dimethyl-N-α-pyridylethylenediamine; mepyramine; pyranisamine; RP-2786.* $C_{17}H_{23}N_3O$; mol wt 285.39. C 71.55%, H 8.12%, N 14.72%, O 5.61%. Prepn: Bovet *et al., C.R. Soc. Biol.* **138**, 99 (1944); Huttrer *et al., J. Am. Chem. Soc.* **68**, 1999 (1946); Viaud, *Prod. Pharm.* **2**, 53 (1947); Horclois, U.S. pat. **2,502,151** (1950 to Rhône-Poulenc). Toxicity: A. von Schlichtergroll, *Arzneimittel-Forsch.* **7**, 237 (1957); of the maleate: F. Hunziker *et al., ibid.* **13**, 324 (1963).

Oily liquid. bp_5 201°; $bp_{0.06}$ 168-172°. n_D^{25} 1.5760-1.5765. LD_{50} orally in mice: 312 mg/kg (Schlichtergroll).

Hydrochloride, $C_{17}H_{23}N_3O\cdot HCl$, crystals, mp 143-143.5°. Very sol in water.

Maleate, $C_{17}H_{23}N_3O\cdot C_4H_4O_4$, *Antamine, Anthisan, Dorantamin, Enrumay, Histalon, Histan, Histapyran, Histatex, Neo-Antergan, Paraminyl, Parmal, Pyramal, Stamine, Stangen, Thylogen.* Crystals. Bitter saline taste. Stable in air. mp 100-101°. uv max: 244 nm ($E_{1cm}^{1\%}$ 420). One gram dissolves in about 0.4 ml water, in about 15 ml abs alc. pH of 10% soln ~5.1. On raising the pH to 7.5 or 8.0 pptn of the oily, free base begins. Slightly sol in benzene, ether. LD_{50} orally in mice: 338 mg/kg (Hunziker).

THERAP CAT: Antihistaminic.

THERAP CAT (VET): Antihistaminic.

8169. Pyrimethamine. *5-(4-Chlorophenyl)-6-ethyl-2,4-pyrimidinediamine; 2,4-diamino-5-(p-chlorophenyl)-6-ethyl-pyrimidine; RP-4753; Chloridin; Daraprim; Malocide; Tindurin.* $C_{12}H_{13}ClN_4$; mol wt 248.71. C 57.95%, H 5.27%, Cl 14.25%, N 22.53%. Prepn: Russel, Hitchings, *J. Am. Chem. Soc.* **73**, 3763 (1951); Logemann *et al., Ber.* **87**, 435 (1954). Commercial process: Chase, Walker, *J. Chem. Soc.* **1953**, 3518; Hitchings *et al.,* U.S. pats. **2,576,939; 2,579,259,** and **2,602,794** (1951 and 1952 to Burroughs Wellcome); Jacob, U.S. pat. **2,680,740** (1954 to Rhône-Poulenc). Review of antimicrobial activity and mechanism of action: Burchall in *Antibiotics* vol. 3, J. W. Corcoran, F. E. Hahn, Eds. (Springer-Verlag, New York, 1975) pp 312-320. Comprehensive description: M. A. Loutfy, H. Y. Aboul-Enein in *Analytical Profiles of Drug Substances* vol. 12, K. Florey, Ed. (Academic Press, New York, 1983) pp 463-482. Clinical evaluations in toxoplasmosis in AIDS patients: C. Leport *et al., Am. J. Med.* **84**, 94 (1988); B. Dannemann *et al., Ann. Int. Med.* **116**, 33 (1992).

Crystals, mp 233-234° (capillary); mp 240-242° (copper block). Practically insol in water. Slightly sol in ethanol,

(about 9 g/l), in dil HCl (about 5 g/l); sol in boiling ethanol (about 25 g/l). Very sparingly sol in propylene glycol and dimethylacetamide at 70°.

Combination with sulfadoxine, *Fansidar.*

THERAP CAT: Antiprotozoal (Toxoplasma); antimalarial.

THERAP CAT (VET): Antiprotozoal (Toxoplasma).

8170. Pyrimidine. 1,3-Diazine; metadiazine; miazine. $C_4H_4N_2$; mol wt 80.09. C 59.99%, H 5.03%, N 34.98%. Prepd by reducing 2,4,6-trichloropyrimidine with zinc dust: Gabriel, *Ber.* **33**, 3666 (1900); by reducing tetrachloropyrimidine: Emery, *ibid.* **34**, 4180 (1901); by reducing 2,4-dichloropyrimidine with magnesium oxide and palladium-charcoal: Wittaker, *J. Chem. Soc.* **1953**, 1646.

Liquid or cryst mass; penetrating odor. mp 20-22°. bp_{762} 123-124°. uv max (water): 240 nm (ε 2400). Soluble in water, alcohol, ether.

Methiodide, $C_4H_4N_2\cdot CH_3I$, yellow plates from ethanol, mp 136-137°.

8171. Pyriminil. *N-(4-Nitrophenyl)-N'-(3-pyridinyl-methyl)urea; N-3-pyridylmethyl-N'-p-nitrophenylurea; pyrinuron; RH-787; DLP-787; Vacor.* $C_{13}H_{12}N_4O_3$; mol wt 272.26. C 57.35%, H 4.44%, N 20.58%, O 17.63%. Prepn: J. E. Ware *et al.,* Ger. pat. **2,409,686** (1974 to Rohm & Haas). Evaluation as rodenticide: D. L. Peardon, *Pest Contr.* **42**, 14, 16, 18, 27 (1974); J. E. Brooks, P. T. Htun, *J. Hyg.* **80**, 401 (1978). Mode of action: D. L. Peardon, J. E. Ware, *Pest Contr.* **45**, 49 (1977).

mp 223-225° (dec). LD_{50} orally in male, female rats: 6.2, 7.2 mg/kg (Brooks, Htun).

USE: Rodenticide.

8172. Pyrimithate. *Phosphorothioic acid O-[2-(dimethylamino)-6-methyl-4-pyrimidinyl] O,O-diethyl ester; O,O-diethyl O-[2-(dimethylamino)-4-methyl-6-pyrimidinyl] phosphorothioate; ICI-29661; Diothyl.* $C_{11}H_{20}N_3O_3PS$; mol wt 305.34. C 43.27%, H 6.60%, N 13.76%, O 15.72%, P 10.14%, S 10.50%. Prepn: McHattie, Brit. pat. **1,019,227** corresp to U.S. pat. **3,287,453** (both in 1966 to ICI).

Liquid, $bp_{0.04}$ 128-132°. Sp gr 1.165. Practically insol in water. Sol in alcohols, acetone, benzene, solvent naphtha.

USE: Insecticide, acaricide.

8173. Pyrinoline. *α-[3-(Di-2-pyridinylmethylene)-1,4-cyclopentadien-1-yl]-α-2-pyridinyl-2-pyridinemethanol; 3-(di-2-pyridylmethylene)-α,α-di-2-pyridyl-1,4-cyclopentadiene-1-methanol; di-2-pyridyl-(6,6-di-2-pyridylfulven-2-yl)meth-anol; McN-1210; Surexin.* $C_{27}H_{20}N_4O$; mol wt 416.48. C 77.87%, H 4.84%, N 13.45%, O 3.84%. Prepn: Brit. pat. **1,009,012** (1965 to McNeil).

Crystals, mp 146.5-147.5°.

THERAP CAT: Antiarrhythmic (cardiac depressant).

8174. Pyriproxyfen. *2-[1-Methyl-2-(4-phenoxyphen-oxy)ethoxy]pyridine;* 4-phenoxyphenyl *(RS)*-2-(2-pyridyl-oxy)propyl ether; S-9318; S-31183; Sumilarv. $C_{20}H_{19}NO_3$; mol wt 321.38. C 74.75%, H 5.96%, N 4.36%, O 14.94%. Juvenile hormone mimic. Prepn: S. Nishida *et al.,* **Brit.** pat. **Appl. 2,140,010;** *eidem,* **U.S.** pat. **4,751,225** (1984, 1988 both to Sumitomo). Ovicidal activity: K. R. S. Ascher, M. Eliyahu, *Phytoparasitica* **16**, 15 (1988). Environmental persistence study: C. H. Schaefer *et al., Ecotoxicol. Environ. Safety* **21**, 207 (1991). Field trial to control tsetse flies: J. W. Hargrove, P. A. Langley, *Bull. Entomol. Res.* **83**, 361 (1993).

Pale yellow liquid. $n_D^{20.5}$ 1.5823. Solidifies upon standing to give crystals, mp 49.7°.

USE: Insecticide.

8175. Pyrisuccideanol. *Butanedioic acid 2-(dimethyl-amino)ethyl [5-hydroxy-4-(hydroxymethyl)-6-methyl-3-pyri-dinyl]methyl ester; pyridoxol 3-[2-(dimethylamino)ethyl suc-cinate];* 2-(dimethylamino)ethyl [5-hydroxy-4-(hydroxy-methyl)-6-methyl-3-pyridyl]succinic acid methyl ester; pirisudanol. $C_{16}H_{24}N_2O_6$; mol wt 340.38. C 56.46%, H 7.11%, N 8.23%, O 28.20%. Prepn: Esanu, **Ger.** pat. **2,102,831** (1971 to Soc. d'Etudes de Prods. Chim.) corresp to **U.S.** pat. **3,717,636** (1973). Activity studies: Hugelin *et al., C.R. Soc. Biol.* **166**, 1435 (1972).

Dimaleate, $C_{16}H_{24}N_2O_6 \cdot 2C_4H_4O_4$, *Mentium, Nadex, Stiv-ane,* mp 134°.

THERAP CAT: Antidepressant.

8176. Pyrithiamine. *1-[(4-Amino-2-methyl-5-pyrimi-dinyl)methyl]-3-(2-hydroxyethyl)-2-methylpyridinium bro-mide monohydrobromide;* 1-(2-methyl-4-amino-5-pyrimidyl)-methyl-2-methyl-3-hydroxyethylpyridinium bromide hydro-bromide; neopyrithiamine. $C_{14}H_{20}Br_2N_4O$; mol wt 420.15. C 40.02%, H 4.80%, Br 38.04%, N 13.34%, O 3.81%. Thia-mine antagonist prepd by condensation of 2-methyl-3-(β-hydroxyethyl)pyridine with the pyrimidine moiety of vita-min B_1: Tracy, Elderfield, *J. Org. Chem.* **6**, 54 (1941); im-proved procedure: Wilson, Harris, *J. Am. Chem. Soc.* **71**, 2231 (1949); **U.S.** pat. **2,587,262** (1952 to Merck & Co.). *Cf.* Woolley, *J. Am. Chem. Soc.* **72**, 5763 (1950).

Crystals from acetone, dec 218-220°. uv max (water): 238, 271 nm. Sol in water.

Chloride hydrochloride, $C_{14}H_{20}Cl_2N_4O$, dec 234-236°. uv max (water): 235, 273 nm. Sol in water.

8177. Pyrithiobac. *2-Chloro-6-[(4,6-dimethoxy-2-pyri-midinyl)thio]benzoic acid;* KIH-8921. $C_{13}H_{11}ClN_2O_4S$; mol wt 326.76. C 47.79%, H 3.39%, Cl 10.85%, N 8.57%, O 19.59%, S 9.81%. Pyrimidinylsalicylic acid herbicide for use in cotton; acetolactate synthase inhibitor. Prepn: Y. Saito *et al.,* **Eur.** pat. **Appl. 315,889;** *eidem,* **U.S.** pat. **4,923,501** (1989, 1990 both to Kumiai Chem. Ind.; Ihara Chem. Ind.). Field trials in cotton: J. W. Keeling *et al. Weed Technol.* **7**, 930 (1993). Mechanism of action: T. Shimizu *et al., J. Pestic. Sci.* **19**, 59 (1994). Comprehensive description: S. Takahashi *et al., Brighton Crop Prot. Conf. - Weeds* **1991**, 57-62.

Ochre powder, mp 148-151°.

Sodium salt, $C_{13}H_{10}ClN_2NaO_4S$; *KIH-2031, DPX-PE350, Staple.* White solid, mp 247.7° (dec). Soly in water (10°): 760 g/l. LD_{50} in male, female rats (mg/kg): 1000-3000, 3000-5000 orally. LD_{50} in rats (mg/kg): > 2000 dermally (Takahashi).

USE: Herbicide.

8178. Pyrithione. *1-Hydroxy-2(1H)-pyridinethione; 2-pyridinethiol 1-oxide;* 2-mercaptopyridine 1-oxide; PTO; Omadine. C_5H_5NOS; mol wt 127.17. C 47.23%, H 3.96%, N 11.01%, O 12.58%, S 25.22%. Prepn: Shaw *et al., J. Am. Chem. Soc.* **72**, 4362 (1950); Semenoff, Dolliver, **U.S.** pat. **2,745,826** (1956 to Olin Mathieson). Prepn of zinc deriv: **Brit.** pat. **761,171** (1956 to Olin Mathieson). Activity of zinc deriv: Karsten *et al.,* and Judge *et al.,* **U.S.** pats. **3,-236,733** and **3,281,366** (both 1966 to Procter and Gamble).

Sodium salt, C_5H_4NNaOS, *Fonderma.*

Zinc derivative, $C_{10}H_8N_2O_2S_2Zn$, *zinc pyrithione, zinc pyr-idinethione, bis(2-pyridylthio)zinc 1,1'-dioxide, bis-(1-hydr-oxy-2(1H)-pyridinethionato-O,S)zinc, Desquaman.* Ingredi-ent in *Head & Shoulders.*

Dimer, $C_{10}H_8N_2O_2S_2$, *dipyrithione, 2,2'-dithiobispyridine 1,1'-dioxide, OMDS, Omadine Disulfide.*

USE: Fungicide, bactericide.

THERAP CAT: Antibacterial; antifungal. Zinc deriv also as antiseborrheic.

8179. Pyrithyldione. *3,3-Diethyl-2,4-(1H,3H)pyridine-dione;* 3,3-diethyl-2,4-dioxotetrahydropyridine; 2,4-dioxo-3,3-diethyltetrahydropyridine; 2,4-dioxo-3,3-diethyl-1,2,3,4-tetrahydropyridine; Presidon; Persedon; Tetridin; Bene-dorm. $C_9H_{13}NO_2$; mol wt 167.21. C 64.65%, H 7.84%, N 8.38%, O 19.14%. Prepn: Schnider, *Festschrift Emil Barell* **1936**, 195; Preiswerk, Schnider, **U.S.** pat. **2,090,068** (1937 to Hoffmann-La Roche); Strukov *et al., Med. Prom. SSSR* **13**, no. 9, 9 (1959). Alternate procedure: Rechen *et al.*, U.S. pat. 3,019,230 (1962 to Hoffmann-La Roche).

Exists in three cryst modifications: I, mp 92-93°; II, mp 97-98°; III, mp 81-86° [Scheibl, Wachter, *Arch. Toxikol.* **18**, 253 (1960)]. bp_{14} 187-189°. Moderately sol in water; freely sol in the usual organic solvents except petr solvents. Aq solns are just acid to litmus.

THERAP CAT: Sedative, hypnotic.

8180. Pyritinol. *3,3'-[Dithiobis(methylene)]bis[5-hydroxy-6-methyl-4-pyridinemethanol];* bis(4-hydroxymethyl-5-hydroxy-6-methyl-3-pyridylmethyl) disulfide; bis[(3-hydroxy-4-hydroxymethyl-2-methyl-5-pyridyl)methyl] disulfide; dipyridoxolyldisulfide; pyridoxine-5-disulfide; pyrithioxin; Bonifen; Epocan. $C_{16}H_{20}N_2O_4S_2$; mol wt 368.48. C 52.15%, H 5.47%, N 7.60%, O 17.37%, S 17.40%. Prepn: Zima, Schorre, U.S. pat. 3,010,966 (1961 to E. Merck); Iwanami *et al.*, *Bitamin* **36**, 122 (1967); *J. Vitaminol. (Kyoto)* **14**, 321, 326 (1968). HPLC determn in urine: K. Kitao *et al.*, *Chem. Pharm. Bull.* **25**, 1335 (1977). Pharmacokinetics and metabolism: Darge *et al.*, *Arzneimittel-Forsch.* **19**, 5, 9, (1969); Nowak, Schorre, *ibid.* 11. Clinical trial in dementia: S. Hoyer *et al.*, *ibid.* **27**, 671 (1977); A. J. Cooper, R. V. Magnus, *Pharmacotherapeutica* **2**, 317 (1980); in cerebrovascular disorders: Y. Tazaki *et al.*, *J. Int. Med. Res.* **8**, 118 (1980).

Crystals, mp 218-220°.
Dihydrochloride monohydrate, $C_{16}H_{20}N_2O_4S_2 \cdot 2HCl \cdot H_2O$, *Biocefalin, Enbol, Encephabol, Enerbol, Life.* mp 184°.
Note: Has no vitamin B_6 activity.

THERAP CAT: Nootropic.

8181. Pyrocalciferol. *(3β,10α,22E)-Ergosta-5,7,22-trien-3-ol;* 9α-lumisterol; 9α-lumista-5,7,22-trien-3β-ol. $C_{28}H_{44}O$; mol wt 396.66. C 84.79%, H 11.18%, O 4.03%. Differs from ergosterol in the steric configuration at C-9 and C-10. Formation via isopyrocalciferol: Askew *et al.*, *Proc. Roy. Soc. London* **B109**, 488 (1932); Windaus *et al.*, *Nachr. Ges. Wiss. Göttingen, math.-phys. Klasse* **1932**, 150. Dehydrogenation with selenium gives Diel's hydrocarbon (3'-methyl-1,2-cyclopentenophenanthrene). Fieser, Fieser, *Steroids* (New York, 1959) pp 136-143, argue the configuration at C-9 and C-10. See also Castells *et al.*, *J. Chem. Soc.* **1959**, 1159. Critical examination and comparison with other unnatural steroids: Castells *et al.*, *ibid.* **1962**, 2907.

Needles from methanol, mp 93-95°. $[\alpha]_D^{20}$ +512°, $[\alpha]_{546}^{20}$ +624° (c = 0.15 in alcohol). uv max: 274, 294 nm (Askew).
Acetate, $C_{30}H_{46}O_2$, mp 81-82°, $[\alpha]_D$ +403° (chloroform).
3,5-Dinitrobenzoate, $C_{35}H_{46}N_2O_6$, mp 168-170°. $[\alpha]_D^{21}$ +195°, $[\alpha]_{546}^{20}$ +249° (both c = 2 in chloroform).

8182. Pyrocarbonic Acid Diethyl Ester. *Dicarbonic acid diethyl ester; oxydiformic acid diethyl ester;* diethyl pyrocarbonate; diethyl oxydiformate; DEPC; Ue-5908; Baycovin. $C_6H_{10}O_5$; mol wt 162.14. C 44.45%, H 6.22%, O 49.34%. $C_2H_5OOCOCOOC_2H_5$. Prepd according to the equation $R_1OOCCl + NaOCOOR_2 \rightarrow R_1OOCOCOOR_2 +$ NaCl: Kovalenko, *Zh. Obshch. Khim.* **22**, 1546 (1952); Thoma, Rinke, *Ann.* **624**, 30 (1959); alternate method: Degering *et al.*, *J. Am. Pharm. Assoc. Sci. Ed.* **39**, 626 (1950). *Review:* *Handbook of Food Additives*, T. A. Furia, Ed. (Chemical Rubber Co., Cleveland, 1968) pp 181-185.
Viscous liquid. Fruity odor. d_4^{20} 1.12. Viscosity at 20° = 1.97 cp. Soluble in alcohols, esters, ketones, hydrocarbons. A 50% w/w soln in ethanol may be prepd. Slowly sol in water with hydrolytic decompn yielding ethanol and CO_2.
Caution: Concd DEPC is irritating to eyes, mucous membranes and skin.
USE: Gentle esterifying agent. Preservative for wines, soft drinks, fruit juices.

8183. Pyrocatechol. *1,2-Benzenediol;* pyrocatechin; catechol; 1,2-dihydroxybenzene. $C_6H_6O_2$; mol wt 110.11. C 65.45%, H 5.49%, O 29.06%. Prepd by treating salicylaldehyde with hydrogen peroxide, or from its monomethyl ether (guaiacol) by treatment with hydrobromic acid: Dakin, *Org. Syn.* **coll. vol. I**, 149 (2nd ed., 1941). Toxicity data: A. J. Lehman *et al.*, *Adv. Food Res.* **3**, 197 (1951). Carcinogenicity study: M. Hirose *et al.*, *Carcinogenesis* **14**, 525 (1993). *Review:* J. Varagnat, "Hydroquinone, Resorcinol, and Catechol", in Kirk-Othmer *Encyclopedia of Chemical Technology* **vol. 13** (Wiley-Interscience, New York, 3rd ed., 1981) pp 39-69.

Monoclinic tablets, prisms from toluene. Discolors in air and light. d 1.344; mp 105°; bp_{760} 245.5°; bp_{400} 221.5°; bp_{200} 197.7°; bp_{100} 176°; bp_{60} 161.7°; bp_{40} 150.6°; bp_{20} 134°; bp_{10} 118.3°; bp_5 104°. Sublimes. Volatile with steam. K at 18° = 3.3 × 10⁻¹⁰. Sol in 2.3 parts water, in alcohol, benzene, chloroform, ether; very sol in pyridine, aq alkalies. Its aq solns soon turn brown. LD_{50} in mice (mg/kg): 260 orally; 190 i.p. (Lehman).
Note: Catechol also refers to catechin, *q.v.*
Caution: Direct skin contact can cause eczematous dermatitis. Systemic effects similar to phenol, *q.v. See Patty's Industrial Hygiene and Toxicology* vol. 2A, G. D. Clayton, F. E. Clayton, Eds. (Wiley-Interscience, New York, 3rd ed., 1981) pp 2584-2586.
USE: In photography; dyeing fur; as reagent.
THERAP CAT: Antiseptic.

8184. Pyrogallol. *1,2,3-Benzenetriol;* 1,2,3-trihydroxybenzene; pyrogallic acid. $C_6H_6O_3$; mol wt 126.11. C 57.14%, H 4.80%, O 38.06%. Observed by Scheele in 1786; prepd by Braconot in 1818. Prepn from gallic acid: Marsh, **Brit.** pat. 144,897 (1919); Rinderknecht, Niemann, *J. Am. Chem. Soc.* **70**, 2605 (1948); from *p-tert*-butylphenol: Stevens, U.S. pat. 2,603,662 (1952 to Gulf). Synthesis from aliphatic sources: Shipchandler *et al.*, *J. Chem. Soc. Perkin Trans. I* **1975**, 1400. Isoln from *Penicillium patulum*: Tanenbaum, Bassett, *Biochim. Biophys. Acta* **28**, 21 (1958). Acute toxicity: J. W. Dollahite *et al.*, *Am. J. Vet. Res.* **23**, 1264 (1962).

White, odorless crystals, mp 131-133°. Becomes grayish on exposure to air and light. *Poisonous!* d 1.45; bp 309°. Sublimes when slowly heated. One gram dissolves in 1.7 ml water, 1.3 ml alc, 1.6 ml ether; slightly sol in benzene, chloroform, carbon disulfide. The aq soln darkens on exposure to air, quite rapidly when alkaline. *Keep well closed and protected from light. Incompat.* Alkalies, ammonium hydroxide, antipyrine, camphor, phenol, menthol. LD_{50} orally in rabbits: 1.6 g/kg (Dollahite).

Monoacetate, $C_8H_8O_4$, *Eugallol*. White or brownish liquid, bp_{23} ~185°. Sol in water, alcohol, chloroform, ether, acetone and castor oil. Marketed as a 67% soln in acetone.

Triacetate, $C_{12}H_{12}O_6$, *acetpyrogall, Lenigallol*. White, crystalline powder, mp 165°. Slightly sol in water; sol in alcohol. Dec by alkali hydroxide solns.

Caution: Ingestion may cause G.I. irritation, renal and hepatic damage, hemolysis, methemoglobinemia, convulsions, circulatory collapse, death. Poisoning and death have occurred from percutaneous absorption. *See: Clinical Toxicology of Commercial Products,* R. E. Gosselin *et al.,* Eds. (Williams & Wilkins, Baltimore, 5th ed., 1981) section II, p 190.

USE: Developer in photography; making colloidal solns of metals; as mordant for wool, staining leather, process engraving; manuf various dyes; dyeing furs, hair, etc.; in analytical chemistry as a reagent for antimony and bismuth; as an active reducer for gold, silver and mercury salts; especially for absorption of oxygen in gas analysis.

THERAP CAT: Has been used as antipsoriatic.

8185. L-Pyroglutamic Acid. *5-Oxo-L-proline; 5-oxo-2-pyrrolidinecarboxylic acid;* 2-pyrrolidone-5-carboxylic acid; glutimic acid; glutiminic acid; α-aminoglutaric acid lactam; glutamic acid lactam; pyroGlu. $C_5H_7NO_3$; mol wt 129.12. C 46.51%, H 5.46%, N 10.85%, O 37.17%. Cyclized internal amide of L-glutamic acid found in vegetables, fruits, grasses, and molasses. Easily prepd from L-glutamic acid by autoclaving with an equal wt of water at 135-140°: Dearborn, Stekol, U.S. pat. **2,528,267**; purification: Blish, U.S. pat. **2,738,353** (1950, 1956, both to International Minerals and Chem.). *Review:* Orlowski, Meister, in *The Enzymes,* vol. 4, P. D. Boyer, Ed. (Academic Press, New York, 3rd ed., 1971) pp 123-151; C. Moret, M. Briley, *Trends Pharmacol. Sci.* **9**, 278-279 (1988).

Orthorhombic bisphenoidal crystals from alcohol + petr ether. mp 162-163°. $[\alpha]_D^{20}$ −11.9° (c = 2); $[\alpha]_D^{25}$ −23.6° (c = 5 at pH 7). Sol in water, alcohol, acetone.

USE: In the resolution of racemic amines.

8186. Pyrolan®. *Dimethylcarbamic acid 3-methyl-1-phenyl-1H-pyrazol-5-yl ester;* 1-phenyl-3-methyl-5-pyrazolyl dimethylcarbamate; 3-methyl-1-phenyl-5-pyrazolyl dimethylcarbamate; ENT-17588; G-22008. $C_{13}H_{15}N_3O_2$; mol wt 245.28. C 63.66%, H 6.16%, N 17.13%, O 13.05%. Prepn from 1-phenyl-3-methyl-5-pyrazolone and dimethylcarbamoyl chloride: **Swiss pat. 279,553** (1952 to Geigy); **Brit. pat. 681,376.** Anticholinesterase activity: Pulver, Domenjoz, *Experientia* **7**, 306 (1951); Ferguson, Alexander, *J. Agr. Food Chem.* **1**, 888 (1953); Müller, Spindler, *Experientia* **10**, 91 (1954). Toxicity study: E. W. Schafer, *Toxicol. Appl. Pharmacol.* **21**, 315 (1972).

Crystals, mp 50°. $bp_{0.2}$ 160-162°. Possesses water and lipoid solubility. LD_{50} orally in rats: 62 mg/kg (Schafer).

Caution: Cholinesterase inhibitor.

USE: Insecticide.

8187. Pyroligneous Acid. Wood vinegar; pyroligneous vinegar. Obtained by destructive distillation of wood.

Yellowish, acid liquid; empyreumatic odor; contains about 6% acetic acid. Miscible with water, alcohol.

USE: Largely for smoking meats.

8188. Pyromellitic Acid. *1,2,4,5-Benzenetetracarboxylic acid.* $C_{10}H_6O_8$; mol wt 254.15. C 47.26%, H 2.38%, O 50.36%. Prepd by heating mellitic (benzenehexacarboxylic) acid with $KHSO_4$ and H_2SO_4: Silberrad, *J. Chem. Soc.* **89**, 1795 (1906) (actually the dianhydride, mp 286°, is obtained by this method). Laboratory prepn from pine or spruce charcoal: Philippi, Thelen, *Org. Syn.* **coll. vol. II**, 551 (1943); by oxidation of benzene derivatives contg substituents in the 1,2,4,5 positions: Jacobsen, *Ber.* **17**, 2516 (1884).

Dihydrate, triclinic plates from water. When anhydr, mp 276°. Distills with anhydride formation. 1.5 g (anhydr) dissolves in 100 ml water. Freely sol in alc.

Tetramethyl ester, $C_{14}H_{14}O_8$, leaflets from alc, mp 141.5°.

8189. Pyronine B. *3,6-Bis(diethylamino)xanthylium chloride; [6-(diethylamino)-3H-xanthen-3-ylidene]diethylammonium chloride;* C.I. 45010. $C_{21}H_{27}ClN_2O$; mol wt 358.91. C 70.28%, H 7.58%, Cl 9.88%, N 7.81%, O 4.46%. Prepn of metallic complex: **Ger. pat. 54,190** (1889 to Bayer); **Ger. pat. 59,003** (1891 to A. Leonhardt); Biehringer, *Ber.* **27**, 3299 (1894); *J. Prakt. Chem.* **54**, 217 (1896); Albert, *J. Chem. Soc.* **1947**, 244. Structure: Chamberlin *et al., J. Org. Chem.* **27**, 2263 (1962). The commercial product is a complex with iron as depicted below.

Ferric chloride complex, green metallic needles, mp 176-178°. Absorption max (50% ethanol): 555 nm ($E_{1cm}^{1\%}$ 2324).

USE: Stain for bacteria, molds, ribonucleic acids.

8190. Pyronine Y. *3,6-Bis(dimethylamino)xanthylium chloride; N-[6-(dimethylamino)-3H-xanthene-3-ylidene]-N-methylmethanaminium chloride; [6-(dimethylamino)-3H-xanthen-3-ylidene]dimethylammonium chloride;* tetramethyldiaminoxanthylium chloride; C.I. 45005; pyronine G. $C_{17}H_{19}ClN_2O$; mol wt 302.80. C 67.43%, H 6.32%, Cl 11.71%, N 9.25%, O 5.28%. Prepn from dimethyl-*m*-aminophenol: Mohlau, Koch, *Ber.* **27**, 2887 (1894); **Ger. pats. 58,955; 59,003; 63,081** (1892 to Leonhardt); *Frdl.* **3**, 92-94. *See also: Colour Index* vol. 4 (3rd ed., 1971) p 4417.

Ferric chloride complex, lustrous green crystals from alc. Soly in water at 26°: 8.96% giving a red soln. Soly in alc at 26°: 0.60%. Solns show a yellowish fluorescence. Absorption max about 552 nm; curves: Stotz *et al., Stain Technol.* **25**, 57 (1950).

USE: Bacterial and biological stain.

8191. Pyrophosphoric Acid. *Diphosphoric acid.* $H_4O_7P_2$; mol wt 177.98. H 2.27%, O 62.93%, P 34.81%. Prepd according to the equation $5H_3PO_4 + POCl_3 \rightarrow 3H_4P_2O_7 + 3HCl$: Geuther, *J. Prakt. Chem.* [2] **8**, 359 (1874); Partington, Wallsom, *Chem. News* **136**, 97 (1928); by heating H_3PO_4: Bell, *Ind. Eng. Chem.* **40**, 1464 (1948).

Consult the Name Index before using this section.

Hygroscopic glass, seldom acicular crystals, mp 61°. K_1 at 18° = 0.14; K_2 = 0.011; K_3 = 2.1 × 10^{-7}; K_4 = 4.1 × 10^{-10}. Forms normal salts, such as $Na_4P_2O_7$, and dihydrogen salts, such as $Na_2H_2P_2O_7$. Solubility in water (23°): 709 g/100 ml. Quickly converted to phosphoric acid when dissolved in hot water. Sol in alcohol and ether.

8192. Pyrosulfuric Acid. *Disulfuric acid.* $H_2O_7S_2$; mol wt 178.14. H 1.13%, O 62.87%, S 36.00%. $H_2S_2O_7$.
Colorless to slightly yellow, very hygroscopic crystals, fuming strongly in air. d 1.89; mp 35°. Very sol in water with violent hissing and liberation of much heat. *Keep tightly closed and handle with caution.*

8193. Pyrosulfuryl Chloride. *Disulfuryl chloride;* disulfur pentoxydichloride; chlorosulfonic anhydride. $Cl_2O_5S_2$; mol wt 215.03. Cl 32.97%, O 37.20%, S 29.82%. $S_2O_5Cl_2$.
Colorless, mobile, very refractive, fuming liquid. d 1.837; mp −37°; bp 151°. Violently dec by water into H_2SO_4, and HCl. *Keep tightly closed and handle with caution.*

8194. Pyrovalerone. *1-(4-Methylphenyl)-2-(1-pyrrolidinyl)-1-pentanone;* 4'-methyl-2-(1-pyrrolidinyl)valerophenone; α-pyrrolidino-p-methylvalerophenone; 1-(1-pyrrolidinyl)butyl p-tolyl ketone; 1-(p-tolyl)-1-oxo-2-pyrrolidino-n-pentane; 1-(p-tolyl)-2-pyrrolidino-1-pentanone. $C_{16}H_{23}NO$, mol wt 245.36. C 78.32%, H 9.45%, N 5.71%, O 6.52%. Prepn: Brit. pat. 927,475 (1963 to Wander); Brit. pat. 933,507; E. Seeger, U.S. pat. 3,314,970 (1963, 1967 both to Thomae); Heffe, *Helv. Chim. Acta* **47**, 1289 (1964). Pharmacology: Stille *et al.*, *Arzneimittel-Forsch.* **13**, 871 (1963). Metabolism: Michaelis *et al.*, *J. Med. Chem.* **13**, 497 (1970).

$bp_{0.08}$ 104°.
Hydrochloride, $C_{16}H_{23}NO$.HCl, *F-1983, Centroton, Thymergix.* Crystals from 2-butanone or from methanol + acetone + diethyl ether, mp 178°. LD_{50} orally in mice: 350 mg/kg (Stille).
Note: This is a controlled substance (stimulant) listed in the U.S. Code of Federal Regulations, Title 21 Part 1308.15 (1995).
THERAP CAT: CNS stimulant.

8195. Pyroxylin. *Cellulose nitrate; nitrocellulose;* collodion cotton; soluble gun cotton; collodion wool; colloxylin; xyloidin; celloidin; Parlodion. Variable mixture which consists chiefly of cellulose tetranitrate. *Review:* R. T. Bogan *et al.* in Kirk-Othmer *Encyclopedia of Chemical Technology* vol. 5 (Wiley-Interscience, New York, 3rd ed., 1979) pp 129-143.
Yellowish-white, matted mass of filaments, having the appearance of raw cotton. *Highly flammable;* pyroxlin with higher nitrogen content *may explode!* Flash pt 40°F (closed cup); ignites at 160-170°. When kept in well-closed containers and exposed to light it dec. Sol in 25 parts of a mixture of 1 vol alcohol and 3 vols ether; also sol in methanol, acetone, glacial acetic acid, amyl acetate. *Keep loosely packed in cartons and protected from light and moisture.* Can be shipped with safety only when wet with 25-30% water or alcohol.
USE: In manuf of collodions; in lacquer coatings, inks, adhesives. Cellulose hexanitrate is used in explosives and

propellants. Celloidin is used for embedding sections in microscopy; in electrotechnics, photography, galvanoplasty.
THERAP CAT: Topical protectant.

8196. Pyrrobutamine. *1-[4-(4-Chlorophenyl)-3-phenyl-2-butenyl]pyrrolidine; 1-[γ-p-(chlorobenzyl)cinnamyl]pyrrolidine;* 1-p-chlorophenyl-2-phenyl-4-pyrrolidyl-2-butene; Pyronil. $C_{20}H_{22}ClN$; mol wt 311.85. C 77.03%, H 7.11%, Cl 11.37%, N 4.49%. Prepn: J. Mills, U.S. pat. 2,655,509 (1953 to Eli Lilly). Pharmacology and toxicity: H. M. Lee *et al.*, *Proc. Soc. Exp. Biol. Med.* **80**, 458 (1952). Pharmacology and physical properties: N. G. Lordi, J. E. Christian, *J. Am. Pharm. Assoc.* **45**, 300 (1956). Structure-activity and stereochemical study: A. F. Casy, R. R. Ison, *J. Pharm. Pharmacol.* **22**, 270 (1970). GC-MS determn in urine: H. Maurer, K. Pfleger, *J. Chromatog.* **430**, 31 (1988).

Oily liquid, $bp_{0.3}$ 190-195°. On standing gives crystals, mp 48-49°. d^{25} 1.1052. uv max (95% ethanol): 360, 243 nm (α 112, 9500), (c = 0.538, 0.00538 mg/ml).
Diphosphate, $C_{20}H_{28}ClNO_8P_2$, crystals from alcohol + ether, mp 129.5-130°. pKa_1 8.77, pKa_2 5.23. Sol in warm water to the extent of 10%. Soly in alcohol at 25° about 5%. Soly at pH 7.4 (37.5°): 0.00087 M/l. Practically insol in chloroform, ether. LD_{50} in mice, guinea pigs (mg/kg): 1270 ±136, 1241 ±165 s.c.; 836.8 +95, 625.6 ±41.9 i.m.; 1116 ±73, 992.6 ±107 orally; in mice (mg/kg): 53.53 ±1.61 i.v. (Lee).
Hydrochloride, $C_{20}H_{22}ClN$.HCl, crystals from alcohol + ether, mp 227-228°.
Hydrobromide, $C_{20}H_{22}ClN$.HBr, crystals from alcohol + ether, mp 228-229°.
THERAP CAT: Antihistaminic.

8197. Pyrrocaine. *N-(2,6-Dimethylphenyl)-1-pyrrolidineacetamide; 1-pyrrolidineaceto-2',6'-xylidide;* 2-(1-pyrrolidinyl)-2',6'-acetoxylidide; 1-pyrrolidinoaceto-2,6-dimethylanilide; EN-1010; NSC-52644; Endocaine; Dynacaine. $C_{14}H_{20}N_2O$; mol wt 232.33. C 72.38%, H 8.68%, N 12.06%, O 6.89%. Prepn: Schlesinger, Gordon, U.S. pat. 2,813,861 (1957 to Endo); Löfgren *et al.*, *Acta Chem. Scand.* **11**, 1724 (1957).

Crysts from hexane or petr ether + dibutyl ether, mp 83°.
Hydrochloride, $C_{14}H_{20}N_2O$.HCl, crystals from isopropanol, mp 205°. Soluble in water, alcohol, isopropyl alcohol. Practically insol in chloroform, ether.
THERAP CAT: Anesthetic (local).

8198. 1H-Pyrrole. *Azole; imidole; divinylenimine.* C_4H_5N; mol wt 67.09. C 71.61%, H 7.51%, N 20.88%. A constituent of coal tar and bone oil: Runge, *Ann. Phys.* **31**, 67 (1834). Prepd industrially by fractional distillation of bone oil, or by the thermal decompn of ammonium mucate with glycerol or mineral oil: McElvain, Bolliger, *Org. Syn. coll. vol. I* (2nd ed., 1941) p 473; Blicke, Powers, *Ind. Eng. Chem.* **19**, 1334 (1927). Also formed on heating of albumin; on heating sheep's wool with aq barium hydroxide soln; by pyrolysis of gelatin. Alternate prepns from acetaldehyde and ammonia: Tschitschibabin, *Chem. Zentr.* **1916**, I, 920; from succindialdehyde with ammonia and acetic acid: Harries, *Ber.* **34**, 1496 (1901); **35**, 1183 (1902); distilling succinimide with zinc or sodium: Bell, Bernthsen, *Ber.* **13**, 877, 1049 (1880). Purification and physical properties: R. V.

Helm *et al., J. Phys. Chem.* **62**, 858 (1958). *Review:* Fischer-Orth, *Die Chemie des Pyrrols* (Leipzig, 1934-1940); E. Vittort, L. R. Anderson in Kirk-Othmer *Encyclopedia of Chemical Technology* vol. **19** (Wiley-Interscience, New York, 3rd ed., 1982) pp 499-520.

Liquid. Agreeable empyreumatic odor resembling that of chloroform. Colorless when freshly distilled, darkens unless every trace of oxygen is removed. d_4^{20} 0.9691. bp_{760} 129.8°. Best distilled *in vacuo*. n_D^{20} 1.5085. Flash pt, closed cup: 102° F (390° C). Absorpion spectrum: Menczel, *Phys. Chem.* **125**, 161; *Chem. Zentr.* **1927**, I, 2510. Sparingly sol in water; freely sol in alcohol, benzene, ether. Insol in aq alkalies. Sol in dil acids with decompn. Solns in dil HCl yield pyrrole red, an amorphous, orange-colored substance; also polymerization takes place under the influence of acids and glycols.

8199. Pyrrolidine. Tetrahydropyrrole. C_4H_9N; mol wt 71.12. C 67.55%, H 12.75%, N 19.69%. Found in tobacco and carrot leaves. Probable biosynthesis from ornithine and putrescine. Usually prepd by reduction of pyrrole.

Almost colorless liquid; unpleasant ammonia-like odor. Fumes in air. bp 88.5-89°. $d_D^{22.5}$ 0.8520. n_D^{28} 1.4402. Strong base. K at 25° = 1.3 × 10^{-3}. Miscible with water. Soluble in alcohol, ether, chloroform.

8200. 2-Pyrrolidone. *2-Pyrrolidinone;* 2-oxopyrrolidine; α-pyrrolidone; 2-ketopyrrolidine. C_4H_7NO; mol wt 85.11. C 56.45%, H 8.29%, N 16.46%, O 18.80%. Prepd on a large scale from butyrolactone by a Reppe process: **Ger. pat. 1,085,525** (to BASF). Other prepns: Metzger, Seelert, *Angew. Chem.* **75**, 919 (1963); Copenhaver, Ney, **U.S. pat 3,095,423** (1963 to Minnesota Mining & Manuf); Lidov, **U.S. pat. 3,109,005** (1963 to Halcon International).

Liquid above 25°. bp_{760} 245°; $bp_{9.2}$ 113-114°; $bp_{0.2}$ 76°. d_4^{25} 1.116. Flash pt, open cup: 265°F (129°C). Viscosity at 25° = 13.3 cps. Dipole moment: 2.3. In the presence of the stoichiometric amount of water a cryst monohydrate, mp 30°, can be formed. Non-corrosive. Good chemical stability. Miscible with water, ethanol, ether, chloroform, benzene, ethyl acetate, carbon disulfide.
USE: Intermediate in the manuf of polyvinylpyrrolidone and polypyrrolidone (a polymer, formed in the presence of alkaline catalysts). Also used as high-boiling solvent in petroleum processing, acrylonitrile manuf. Industrial solvent for polymers, chlordane, DDT, sorbitol, glycerol, iodine, sugars. In specialty printers inks. As plasticizer and coalescing agent for acrylic-styrene emulsion-type floor polishes.

8201. 3-Pyrroline. *2,5-Dihydro-1H-pyrrole.* C_4H_7N; mol wt 69.11. C 69.52%, H 10.21%, N 20.27%. Prepd by reduction of pyrrole with zinc and glacial acetic or hydrochloric acid.

Almost colorless liquid. Unpleasant, ammonia-like odor. Fumes in air. Very hygroscopic, also absorbs CO_2. bp_{748} 90-91°; d_D^{20} 0.9097; n_D^{20} 1.4664. Strong base. Miscible with water. Sol in alcohol, ether, chloroform.

8202. Pyrrolnitrin. *3-Chloro-4-(3-chloro-2-nitrophenyl)pyrrole;* 3-chloro-4-(2'-nitro-3'-chlorophenyl)pyrrole; PN; Pyroace. $C_{10}H_6Cl_2N_2O_2$; mol wt 257.08. C 46.72%, H 2.35%, Cl 27.58%, N 10.90%, O 12.45%. Antifungal antibiotic isolated from *Pseudomonas pyrrocinia* n. sp.: K. Arima *et al., Agr. Biol. Chem.* **28**, 575 (1964); *eidem, J. Antibiot.* **18**, 201 (1965). Structure: Imanaka *et al., ibid.* 207. Synthesis: S. Umio *et al.,* **Belg. pat. 670,427**; *eidem,* **U.S. pat. 3,428,648** (1964, 1969 both to Fujisawa); Nakano *et al., Tetrahedron Letters* **1966**, 737; S. Umio *et al.* (10 publications) *Chem. Pharm. Bull.* **17**, 559-628 (1969); Gosteli, *Helv. Chim. Acta* **55**, 451 (1972). Pharmacological studies: M. Nishida *et al., J. Antibiot.* **18**, 211 (1965).

Pale yellow crystals from hot cyclohexane, mp 124.5°. Gradually changes to a red or brown color on exposure to sunlight and loses antibiotic activity. uv max: 252 nm (ε 7500). Slightly sol in water, petr ether, cyclohexane; sol in methanol, ethanol, butanol, acetone, ethyl acetate, benzene, ether, chloroform, carbon tetrachloride, pyridine, acetic acid. LD_{50} in rats, rabbits (mg/kg): 68, 105 i.p. (Nishida).
THERAP CAT: Antifungal.

8203. Pyruvaldehyde. *2-Oxopropanal;* methylglyoxal; 2-ketopropionaldehyde; acetylformaldehyde. $C_3H_4O_2$; mol wt 72.06. C 50.00%, H 5.59%, O 44.40%. CH_3COCHO. Obtained by warming isonitrosoacetone with dilute sulfuric acid: v. Pechmann, *Ber.* **20**, 3213 (1887); by distilling a dilute solution of dihydroxyacetone from calcium carbonate: Bernhauer, Görlich, *Biochem. Z.* **212**, 462 (1929). There are several biochemical methods of prepn.
Yellow, mobile liquid, pungent odor. Polymerizes very readily. d^{24} 1.0455. $n_D^{17.5}$ 1.4002. Hygroscopic. bp_{760} 72° forming a yellowish-green vapor. May be kept in a sealed tube for several days. The polymer, a viscous solid, dissolves in water with the evolution of heat, giving monomeric solns. Faintly acid to litmus. Sol in alc, giving a colorless soln. Sol in ether, benzene, giving yellow solns.
Oxime, *see* Isonitrosoacetone.

8204. Pyruvate Decarboxylase. *Pyruvic decarboxylase;* pyruvic carboxylase; α-carboxylase; 2-oxo-acid carboxy-lyase. An enzyme found in yeast, in plants, in wheat germ and in bacteria: Neuberg, Rosenthal, *Biochem. Z.* **51**, 142 (1913); Holzer *et al., ibid.* **327**, 331 (1956); Singer, Pensky, *J. Biol. Chem.* **196**, 375 (1952); King, Chelelin, *ibid.* **208**, 821 (1954); Howells, Lindstrom, *J. Bacteriol.* **75**, 305 (1958). Cofactors: Diphosphothiamine, divalent metal ions. Catalyzes decarboxylation of α-ketoacids to aldehydes + CO_2: Juni, *J. Biol. Chem.* **236**, 2302 (1961). Structure and mechanism of action of the yeast enzyme: Schellenberger, *Angew. Chem. Int. Ed.* **6**, 1024 (1967). *Review:* Utter, *The Enzymes* vol. **5**, P. D. Boyer *et al.,* Eds. (Academic Press, New York, 2nd ed., 1961) pp 320-323.

8205. Pyruvic Acid. *2-Oxopropanoic acid;* α-ketopropionic acid; acetylformic acid; pyroracemic acid; brenztraubensäure (German). $C_3H_4O_3$; mol wt 88.06. C 40.92%, H 4.58%, O 54.50%. $CH_3COCOOH$. Intermediate in sugar metabolism and in enzymatic carbohydrate degradation (alcoholic fermentation) where it is converted to acetaldehyde and CO_2 by carboxylase, *see* Nord, *Chem. Rev.* **26**, 423 (1940). In muscle, pyruvic acid (derived from glycogen) is reduced to lactic acid during exertion, which is reoxidized and partially retransformed to glycogen during rest. The liver can convert pyruvic acid to alanine by amination. Pyruvic acid has been isolated from cane sugar fermentation broth by fixation with β-naphthylamine giving α-methyl-β-naphthocinchonic acid: Grab, *Biochem. Z.* **123**, 84 (1921). The most practical method of preparing pyruvic acid is by distillation of tartaric acid in presence of potassium acid sulfate as dehydrating agent. The distillate must be fractionated under reduced pressure: Erlenmeyer, *Ber.* **14**, 321 (1881); Döbner, *Ann.* **242**, 269 (1887); Howard, Fraser, *Org. Syn.* **coll. vol. I** (2nd ed., 1941) p 475.

Liquid. Odor resembling that of acetic acid. d_4^{15} 1.267; mp 11.8°; bp_{760} 165° (dec); bp_{100} 106.5°; bp_{40} 85.3°; bp_{20} 70.8°; bp_{10} 57.9°; bp_5 45.8°; $bp_{1.0}$ 21.4°; n_D^{20} 1.4138. Absorption spectrum: Henri, Fromageot, *Bull. Soc. Chim.* [4] **37**, 846 (1925). K at 25°: 3.2 × 10⁻³. Miscible with water, alcohol, ether. Polymerizes and dec on standing unless pure and kept in container with airtight closure.

Methyl ester, $C_4H_6O_3$, liquid, bp_{760} 134-137°.

Ethyl ester, $C_5H_8O_3$, liquid, bp_{760} 155°; bp_{750} 147.5°; bp_{42} 69-71°; bp_{20} 66° also 56°. Prepn from ethyl lactate by oxidation with $KMnO_4$: Cornforth, *Org. Syn.* **coll. vol. IV**, 467 (1963).

8206. Pyrvinium Pamoate. *6-(Dimethylamino)-2-[2-(2,5-dimethyl-1-phenyl-1H-pyrrol-3-yl)ethenyl]-1-methylquinolinium salt with 4,4'-methylenebis[3-hydroxy-2-naphthalenecarboxylic acid] (2:1); bis[6-(dimethylamino)-2-[2-(2,5-dimethyl-1-phenylpyrrol-3-yl)vinyl]-1-methylquinolinium] 4,4'-methylenebis(3-hydroxy-2-naphthoate);* 6-(dimethylamino)-2-[2-(2,5-dimethyl-1-phenyl-3-pyrryl)vinyl]-1-methylquinolinium salt of 2,2'-dihydroxy-1,1'-dinaphthylmethane-3,3'-dicarboxylic acid; pyrvinium embonate; viprynium embonate; Molevac; Neo-Oxypaat; Pamovin; Poquil; Povan; Povanyl; Pyrcon; Tru; Vanquin; Vermitiber. $C_{75}H_{70}N_6O_6$; mol wt 1151.42. C 78.24%, H 6.13%, N 7.30%, O 8.34%.

Cyanine dye. Prepn: E. F. Elslager, D. F. Worth, U.S. pat. 2,925,417 (1960 to Parke, Davis). Determn in pharmaceutical prepn: H. D. Beckstead, S. J. Smith, *J. Pharm. Sci.* **56**, 390 (1967); by HPLC: U. G. G. Hennig *et al.*, *Arch. Toxicol.* **60**, 278 (1987). Clinical trial: R. A. Buchanan *et al.*, *Clin. Pharmacol. Ther.* **16**, 716 (1974). Clinical pharmacology: T. C. Smith *et al.*, *ibid* **19**, 802 (1976). Review of use in *Enterobius vermicularis* infections: G. C. Cook, *Gut* **35**, 1159-1162 (1994).

Bright orange or orange-red to almost black precipitate, mp 210-215° (softens at 190°). Absorption max: 236, 356, 503 nm. Stable to heat, light and air. Slightly sol in chloroform, methoxyethanol; very slightly sol in alcohol. Practically insol in water, ether.

Pyrvinium chloride, $C_{26}H_{28}ClN_3$, *viprynium chloride, SN-4395*. Prepn: E. Van Lare, L. G. S. Brooker, U.S. pat. 2,515,912 (1950 to Eastman Kodak). May occur as dihydrate, deep-red powder, dec 249-251°. Sparingly sol in water.

THERAP CAT: Anthelmintic (Nematodes).

Q

8207. Q-Enzyme®. *α-Glucan branching glycosyltrans-ferase;* potato branching enzyme; enzyme Q; branching factor; glucosan transglycosylase. A widely distributed enzyme which transfers part of a 1,4-glucosan chain from a 4- to a 6-position in polysaccharides. Isoln from potatoes: Barker *et al., J. Chem. Soc.* **1949**, 1705. Crystallization: Gilbert, Patrick, *Biochem. J.* **51**, 181 (1952). Improved isoln procedure: Griffen, Wu, *Biochemistry* **7**, 3063 (1968). Although Q-enzyme is unstable in aq soln, it can be stored for long periods without appreciable loss in activity when it is freeze-dried. *Review:* Barker, Bourne, *Quart. Rev. Chem. Soc.* **7**, 65-68 (1953); Dixon, Webb, *Enzymes* (Academic Press, New York, 1958) *passim.*

8208. Quassia. Bitter wood; bitter ash. The wood of *Picrasma excelsa* (Sw.) Planch, or of *Quassia amara* L., *Simaroubaceae.* The first is known in commerce as Jamaica quassia, the second as Surinam quassia. *Habit. Picrasma escelsa* inhabits Jamaica and the Caribbean Islands; *Quassia amara* is a native of Brazil and Guiana and is cultivated in Colombia, Panama, and the West Indies. Quassin and neoquassin are the bitter principles of Surinam quassia; picrasmin, that of Jamaica quassia. These bitter pinciples are obtained in yields of 0.1-0.2% and appear commercially under the name of quassin.

Unground quassia occurs usually in chips, raspings, or shavings, occasionally in billets; yellowish-white to bright yellow with a few light gray pieces; coarsely grained, fibrous. Slight odor; very bitter taste. The powdered form is pale yellow in color.

USE: The extract is used for fly poison on flypaper; to imitate hops.

THERAP CAT: Anthelmintic.

THERAP CAT (VET): Has been used as a bitter, as an anthelmintic.

8209. Quassin. *2,12-Dimethoxypicrasa-2,12-diene-1,-11,16-trione;* 3aβ,6aβ,7,7aα,8,11a,11bα,11c-octahydro-2,10-dimethoxy-3,8α,11aβ,11cβ-tetramethylphenanthro[10,1-bc]pyran-1,5,11(4H)-trione; nigakilactone D. $C_{22}H_{28}O_6$; mol wt 388.46. C 68.02%, H 7.27%, O 24.71%. One of the bitter constituents of the wood of *Quassia amara* L., *Simaroubaceae* known in commerce as Surinam quassia. Obtained by the resolution of the mixture of bitter constituents of quassia wood: E. P. Clark, *J. Am. Chem. Soc.* **59**, 927 (1937); London *et al., J. Chem. Soc.* **1950**, 3431. Structure: Valenta *et al., Tetrahedron Letters* no. **20**, 25 (1960); Carman, Ward, *ibid.* **1961**, 317; Valenta *et al., Tetrahedron* **15**, 100 (1961). Stereochemistry: Valenta *et al., ibid.* **18**, 1433 (1962). Identity with nigakilactone D: Murae *et al., ibid.* **27**, 1545 (1971). Synthetic approach: Stojanac *et al., Can. J. Chem.* **53**, 619 (1975); P. A. Grieco *et al., Tetrahedron Letters* **21**, 1619 (1980). Total synthesis of *dl*-form: *eidem, J. Am. Chem. Soc.* **102**, 7586 (1980).

Very bitter rectangular plates from dilute methanol, mp 222°. $[\alpha]_D^{20}$ +34.5° (c = 5.09 in CHCl₃). uv max: ∼255 nm (ε ∼11,650). Sol in benzene, alc, acetone, chloroform, pyridine, acetic acid, hot ethyl acetate. Sparingly sol in ether, petr ether. Bitterness threshold 1:60,000.

8210. Quatrimycin. *[4R-(4α,4aβ,5aβ,6α,12aβ)]-4-(Dimethylamino)-1,4,4a,5,5a,6,11,12a-octahydro-3,6,10,12,12a-pentahydroxy-6-methyl-1,11-dioxo-2-naphthacenecarboxamide;* epitetracycline. $C_{22}H_{24}N_2O_8$; mol wt 444.44. C

59.45%, H 5.44%, N 6.30%, O 28.80%. Prepn by epimerization of tetracycline: McCormick *et al., J. Am. Chem. Soc.* **79**, 2849 (1957); Kaplan *et al., Antibiot. Chemother. (Basel)* **7**, 569 (1957); Remmers *et al., J. Pharm. Sci.* **52**, 752 (1963). Acute toxicity: L. Kung, H. Sun, *Yao Hsueh Hsueh Pao* **13**, 244 (1966), *C.A.* **65**, 7869e (1966).

Monohydrate, crystals, mp 178° dec. $[\alpha]_D^{25}$ −335° (c = 0.5 in 0.03N HCl). uv max (0.01N H₂SO₄): 216, 255, 270, 355 (ε 13,900, 16,400, 15,200, 14,700). LD₅₀ i.v. in mice: 85.8 mg/kg (Kung, Sun).

Ammonium salt monohydrate, $C_{22}H_{27}N_3O_8 \cdot H_2O$, yellow crystals, mp 170° dec. $[\alpha]_D^{25}$ −321° (c = 0.5 in 0.03N HCl). Soly in water > 50 mg/kg.

Hydrochloride, $C_{22}H_{25}ClN_2O_8$, amorphous yellow solid. $[\alpha]_D^{25}$ −325° (c = 0.58 in 0.2N HCl). Soly in water > 100 mg/ml.

Methiodide, $C_{23}H_{27}IN_2O_8$, crystals, mp 161-162°. $[\alpha]_D^{25}$ −265° (c = 0.5 in 0.03N HCl).

8211. Quazepam. *7-Chloro-5-(2-fluorophenyl)-1,3-di-hydro-1-(2,2,2-trifluoroethyl)-2H-1,4-benzodiazepine-2-thio-ne;* Sch-16134; Doral; Dormalin; Oniria; Prosedar; Quazium; Selepam. $C_{17}H_{11}ClF_4N_2S$; mol wt 386.80. C 52.79%, H 2.87%, Cl 9.17%, F 19.65%, N 7.24%, S 8.29%. Benzodiazepine hypnotic. Prepn: M. Steinman, Ger. pat. **2,138,773;** *idem,* U.S. pat. **3,845,039** (1972, 1974 both to Schering Corp.); M. Steinman *et al., J. Med. Chem.* **16**, 1354 (1973). Pharmacology: A. Barnett *et al., Arzneimittel-Forsch.* **32**, 1452 (1982); E. Ongini *et al., ibid.* 1456. GC determn in plasma: J. M. Hilbert *et al., J. Pharm. Sci.* **73**, 516 (1984). Pharmacokinetics: M. Chung *et al., Clin. Pharm. Ther.* **35**, 520 (1984). Metabolism: N. Zampaglione *et al., Drug Metabol. Dispos.* **13**, 25 (1985). Clinical trials in insomnia: J. W. Goethe, G. Kader, *Curr. Ther. Res.* **32**, 150 (1982); M. Mamelak *et al., J. Clin. Pharmacol.* **24**, 65 (1984). Toxicology: H. E. Black *et al., Arzneimittel-Forsch.* **37**, 906 (1987).

Crystals from methylene chloride-hexane, mp 137.5-139°. LD₅₀ in mice (mg/kg): > 1370 i.v., > 5000 orally (Orgini). LD₅₀ in male, female mice, rats (mg/kg): 845, 921, 3072, 2749 i.p. (Black).

Note: This is a controlled substance (depressant) listed in the U.S. Code of Federal Regulations, Title 21 Part 1308.14 (1995).

THERAP CAT: Sedative, hypnotic.

8212. Quebrachamine. *7-Ethyl-1,4,5,6,7,8,9,10-octahy-dro-2H-3,7-methanoazacycloundecino[5,4-b]indole;* kamassin. $C_{19}H_{26}N_2$; mol wt 282.43. C 80.80%, H 9.28%, N 9.92%. From the bark of *Aspidosperma quebracho blanco* Schlecht., *Apocynaceae:* Hesse, *Ann.* **211**, 249 (1882); Field, *J. Chem. Soc.* **125**, 1444 (1924). Identity with kamassin: Gellert, Witkop, *Helv. Chim. Acta* **35**, 114 (1952). Structure: Witkop, *J. Am. Chem. Soc.* **79**, 3193 (1957); Kny, Witkop, *J. Org. Chem.* **25**, 635 (1960); Biemann, Spiteller, *J. Am. Chem. Soc.* **84**, 4578 (1962). Crystal structure: C. Puglisi *et al., Acta Crystallogr.* **B32**, 1900 (1976). Total synthesis of *dl*-form: Stork, Dolfini, *J. Am. Chem. Soc.* **85**,

2872 (1963); Ziegler *et al.*, *ibid.* **91**, 2342 (1969); Kutney *et al.*, *ibid.* **92**, 1727 (1970); V. S. Giri *et al.*, *J. Heterocycl. Chem.* **17**, 1133 (1980); S. Takano *et al.*, *Heterocycles* **16**, 247 (1981). Enantioselective synthesis: *eidem, Chem. Commun.* **1980**, 616; **1981**, 1153.

Bitter leaflets, mp 145-147°. $[\alpha]_D^{20}$ −109 to −110° (acetone). uv max (methanol): 230, 287, 293 nm (log ε 4.55, 3.85, 3.84). Sol in acetone, alc, chloroform, ether, dil acids.

8213. Quebracho Colorado. Red quebracho. Wood of *Loxopterygium lorentzii* Griseb., *Anacardiaceae. Habit.* Argentine Republic. *Constit.* Tannin, coloring matter, loxopterygine.

USE: In dyeing and tanning.

8214. Queen Substance. *(E)-9-Oxo-2-decenoic acid.* $C_{10}H_{16}O_3$; mol wt 184.24. C 65.19%, H 8.75%, O 26.05%. Secreted in the mandibular gland of queen honey bees *(Apis mellifera, A. florea, A. cerana, A. dorsata)*; inhibits the development of ovaries in worker bees, prevents queen cell formation and attracts male bees (drones) to virgin queens for the purpose of mating: Butler, *Experientia* **13**, 256 (1957); Sannasi, Rajulu, *Life Sci.* **10**, part 2, 195 (1971). Similarity with the ovary inhibiting hormone of prawns *(Leander serratus):* Carlisle, Butler, *Nature* **177**, 276 (1956). Extraction and purification: Carlisle, Butler, *loc. cit.;* Butler *et al.*, *Nature* **184**, 1871 (1959). Synthesis: Barbier *et al.*, *Compt. Rend.* **251**, 1133 (1960); Jaeger, Robinson, *Tetrahedron* **14**, 320 (1961); Y. Naoshima *et al.*, *Agr. Biol. Chem.* **48**, 2151 (1984); U. P. Dhokte, A. S. Rao, *Synth. Comm.* **17**, 355 (1987).

Transparent elongated plates from ether + petr ether or aq methanol, mp 54.5-55.5°. Stable to heat, acids, less stable to alkalies. Sol in acetone, alcohol. IR spectrum: Butler *et al.*, *loc. cit.*

8215. Quercetagetin. *2-(3,4-Dihydroxyphenyl)-3,5,6,7-tetrahydroxy-4H-1-benzopyran-4-one; 3,3′,4′,5,6,7-hexahydroxyflavone;* 6-hydroxycyanidenolon 1555. $C_{15}H_{10}O_8$; mol wt 318.24. C 56.61%, H 3.17%, O 40.22%. From flowers of French marigold, *Tagetes patula* Linn., *Compositae:* Perkin, *J. Chem. Soc.* **103**, 209 (1913). Synthesis: Baker *et al.*, *ibid.* **1929**, 74; Rao, Seshadri, *Proc. Indian Acad. Sci.* **23A**, 23 (1946), *C.A.* **40**, 5052² (1946).

Dihydrate, pale yellow needles from dil alcohol, mp 318°. uv max (alc): 259, 361 nm (log ε 4.23, 4.34). Sol in hot alcohol; sparingly sol in boiling water.

Hexaacetate, $C_{27}H_{22}O_{14}$, needles from alcohol + acetic acid, mp 209-211°. Sparingly sol in alc.

7-Glucoside, $C_{21}H_{20}O_{13}$, *quercetagitrin.* From flowers of the African marigold, *Tagetes erecta* L., *Compositae:* Rao, Seshadri, *Proc. Indian Acad. Sci.* **14A**, 289 (1941), *C.A.* **36**, 2555³ (1942); from *Chrysanthemum coronarium* L., *Compositae:* Anyas, Steelink, *Arch. Biochem. Biophys.* **90**, 63 (1960). Structure: Rajagopalan, Seshadri, *Proc. Indian*

Acad. Sci. **28A**, 31 (1948), *C.A.* **43**, 4265b (1949). Crystals from aqueous pyridine, dec 236-238°. uv max (95% ethanol): 260, 272, 362 nm.

8216. Quercetin. *2-(3,4-Dihydroxyphenyl)-3,5,7-trihydroxy-4H-1-benzopyran-4-one; 3,3′,4′,5,7-pentahydroxyflavone;* meletin; sophoretin; cyanidenolon 1522. $C_{15}H_{10}O_7$; mol wt 302.24. C 59.61%, H 3.33%, O 37.06%. The aglucon of quercitrin, of rutin, and of other glycosides. Widely distributed in the plant kingdom, esp in rinds and barks, in clover blossoms and in ragweed pollen. Isoln from *Rhododendron cinnabarinum* Hook, *Ericaceae:* Rangaswami *et al.*, *Proc. Indian Acad. Sci.* **56A**, 239 (1962), *C.A.* **58**, 9414a (1963). Structure: Underhill *et al.*, *Can. J. Biochem. Physiol.* **35**, 219 (1957). Biosynthesis: Watkin *et al.*, *ibid.* 229; Grisebach, *Biochem. J.* **85**, 3p (1962); Patschke *et al.*, *Z. Naturforsch.* **21b**, 201 (1966). Synthesis: Shakhova *et al.*, *Zh. Obshch. Khim.* **32**, 390 (1962), *C.A.* **58**, 1426f (1963). Metabolism: Nakagawa *et al.*, *Biochim. Biophys. Acta* **97**, 233 (1965). Toxicity data: M. Sullivan *et al.*, *Proc. Soc. Exp. Biol. Med.* **77**, 269 (1951). *See also* Bioflavonoids.

Dihydrate, yellow needles from dil alcohol. Becomes anhydr at 95-97°. When anhydr dec 314°. uv max (alc): 258, 375 nm (log ε 2.75, 2.75). One gram dissolves in 290 ml abs alc, in 23 ml boiling alc. Soluble in glacial acetic acid; in aq alkaline solns with yellow color. Practically insol in water. Alcoholic solns taste very bitter. LD_{50} orally in mice: 160 mg/kg (Sullivan).

Pentabenzyl ether, $C_{50}H_{40}O_7$, *3,3′,4′,5,7-pentakis(benzyloxy)flavone, penta-O-benzylquercetin, Parletrope.* Prepn: Chopin, Chadenson, *Compt. Rend. Ser. C* **263**, 729 (1966); Binovic, **Ger.** pat. **2,122,514** (1972 to Biosedra), *C.A.* **76**, 113072n (1972). Crystals, mp 123-125°. uv max (chloroform): 249, 343 nm (log ε 4.43, 4.14).

3-β-D-Galactoside hemipentahydrate, $C_{21}H_{20}O_{12}$.2½H_2O, *hyperin, hyperoside.* From *Acacia melanoxylon* R. Br., *Leguminosae:* Falco, de Vries, *Naturwiss.* **51**, 462 (1964). Yellow needles from ethanol, dec 227-230°. $[\alpha]_D^{20}$ −83° (c = 0.2 in pyridine). uv max: 259, 364 nm (log ε 4.31, 4.39).

THERAP CAT: Capillary protectant.

8217. Quercimeritrin. *2-(3,4-Dihydroxyphenyl)-7-(β-D-glucopyranosyloxy)-3,5-dihydroxy-4H-1-benzopyran-4-one;* quercetin-7-D-glucoside; *3,3′,4′,5,7-pentahydroxyflavone-7-D-glucoside.* $C_{21}H_{20}O_{12}$; mol wt 464.38. C 54.32%, H 4.34%, O 41.34%. Found in flowers of *Gossypium herbaceum* L., *Malvaceae:* Perkin, *J. Chem. Soc.* **95**, 2181 (1909); from leaves of *Chrysanthemum ségetum* L. and *C. coronarium* L., *Compositae:* Geissman, Steelink, *J. Org. Chem.* **22**, 946 (1957); Anyas, Steelink, *Arch. Biochem. Biophys.* **90**, 63 (1960). In the mother liquor from quercimeritrin the glucosides gossypitrin and isoquercitrin *q.v.*, are also found. Structure: Attree, Perkin, *J. Chem. Soc.* **1927**, 234; Rao, Seshadri, *Proc. Indian Acad. Sci.* **9A**, 365 (1939), *C.A.* **34**, 107¹ (1940); Pacheco, Grouiller, *Compt. Rend.* **253**, 1178 (1961).

Trihydrate, yellow plates from aq pyridine. The water of crystn is given up at 100°, the anhydr material is hygroscopic, mp 247-249°. uv max (ethanol): 372, 257 nm (log ε 4.33, 4.38). Practically insol in cold water, more sol in hot water; sol in methanol. Sol in aq alkaline solns with deep yellow color. Is hydrolyzed by 7% H_2SO_4 yielding 1 mol quercetin and 1 mol D-glucose.

Gossypitrin, $C_{21}H_{20}O_{13}$, orange-yellow needles melting at 200-202°; slightly sol in alcohol and acetic acid.

8218. *d*-Quercitol. *2-Deoxy-D-chiro-inositol;* D-1-deoxy-muco-inositol; "acorn sugar"; (+)-protoquercitol; 1,2,3,4,5-cyclohexanepentol. $C_6H_{12}O_5$; mol wt 164.16. C 43.90%, H 7.37%, O 48.73%. Found in the acorns of various spp of

Quercus, Fagaceae: Prunier, *Ann. Chim. Phys.* **15**, 1 (1878); from leaves of the European palm (*Chaemerops humilis):* Muller, *J. Chem. Soc.* **91**, 1766 (1907). Configuration: Posternak, *Helv. Chim. Acta* **19**, 1007 (1936). Synthesis: G. E. McCasland *et al., J. Org. Chem.* **33**, 4220 (1968).

Sweet crystals. mp 234-235°. $[\alpha]_D^{20}$ +24 to +26°. Sol in water; slightly sol in hot, almost insol in cold alcohol; practically insol in ether.

8219. Quercitrin. *3-[(6-Deoxy-α-L-mannopyranosyl)-oxy]-2-(3,4-dihydroxyphenyl)-5,7-dihydroxy-4H-1-benzopyran-4-one;* quercitroside; quercimelin; quercetin-3-L-rhamnoside; thujin. $C_{21}H_{20}O_{11}$; mol wt 448.38. C 56.25%, H 4.50%, O 39.25%. From *Aesculus hippocastanum* L., Hippocastanaceae: Hörhammer *et al., Arch. Pharm.* **292**, 113 (1959). Structure: Zemplén *et al., Ber.* **61**, 2486 (1928); Marchlewski, Skarzynski, *Biochem. Z.* **297**, 56 (1938); Wolfrom, Thompson in R. L. Whistler, M. L. Wolfrom, *Methods in Carbohydrate Chemistry* vol. 1 (Academic Press, New York, 1962) p 202.

Yellow crystals from dil methanol or ethanol, mp 176-179°; from water, mp 167°. uv max (ethanol): 350, 258 nm (log ε 4.18, 4.30). Practically insol in cold water, ether; sol in alc; moderately sol in hot water; sol in aq alkaline solns with intense yellow color which is oxidized by air to brown.

USE: Has been used as textile dye. *Flavine yellow shade* is prepared by extracting quercitron bark with high pressure steam and consists mainly of quercitrin: Tisdale, *Can. Textile J.* **57**, 44 (1941).

8220. Quercus. White oak. Dried inner bark of trunk and branches of *Quercus alba* L., Fagaceae. *Habit.* Eastern U.S. and Canada. *Constit.* Tannic acid, oak-red, resin, pectin, levulin, quercitol.

THERAP CAT: Astringent.

8221. Quillaic Acid. *(3β,4α,16α)-3,16-Dihydroxy-23-oxoolean-12-en-28-oic acid;* quillaja sapogenin. $C_{30}H_{46}O_5$; mol wt 486.69. C 74.04%, H 9.53%, O 16.44%. Prepn from quillaja saponin: Windaus *et al., Z. Physiol. Chem.* **160**, 301 (1926); Elliott, Kon, *J. Chem. Soc.* **1939**, 1130. Structure: Windaus, *loc. cit.;* Ruzicka, van Veen, *Z. Physiol. Chem.* **184**, 69 (1929); Ruzicka *et al., Coll. Czech. Chem. Commun.* **15**, 893 (1950).

Crystals from ethyl acetate, mp 292-293°; also reported as 258-265° (dec): Kubota *et al., Tetrahedron Letters* **1969**, 771. $[\alpha]_D^{20}$ +56.1° (c = 2.9 in pyridine). Soluble in alcohol, ether, acetone, ethyl acetate, glacial acetic acid.

Diacetate, $C_{34}H_{50}O_7$, needles from glacial acetic acid, mp 180-187°.

Methyl ester, $C_{31}H_{48}O_5$, needles from dil methanol, mp 223-225°; also reported as 200-205° (dec), Kubota *et al., loc. cit.* $[\alpha]_D^{20}$ +40.5° (c = 4 in pyridine).

8222. Quillaja. Soap bark; quillay bark; Panama bark; China bark; Murillo bark. Inner dried bark of *Quillaja saponaria* Molina, Rosaceae. *Habit.* South America (Peru, Chile); cultivated in Northern Hindustan. *Constit.* Quillaic acid, quillajasaponin, sucrose, tannin.

USE: Manuf of saponin; in mineral water industry, shampoo liquids, etc.; as foam producer.

8223. Quillaja Saponin. The saponin of quillay bark. Isolation: Kobert, *Arch. Exp. Pathol. Pharmakol.* **23**, 233 (1887); Pachorukow, *Arbeiten des Pharmakologischen Instituts zu Dorpat* **I**, 5 (1888); Cofman-Nicoresti, *Pharm. J.* **111**, 103 (1923), found the total saponin content of quillay bark to be ∼9 or 10%. The saponin is built from quillaic acid, a sugar and possibly other substances.

Amorphous, deliquescent powder which causes sneezing when dispersed in the air. Freely sol in dil alc. Foams easily when shaken with water, the foam being relatively stable.

8224. Quinacillin. *[2S-(2α,5α,6β)]-6-[[(3-Carboxy-2-quinoxalinyl)carbonyl]amino]-3,3-dimethyl-7-oxo-4-thia-1-azabicyclo[3.2.0]heptane-2-carboxylic acid;* 3-carboxy-2-quinoxalinylpenicillin. $C_{18}H_{16}N_4O_6S$; mol wt 416.41. C 51.92%, H 3.87%, N 13.45%, O 23.05%, S 7.70%. Semisynthetic antibiotic related to penicillin. Prepd by condensation of quinoxaline-2,3-dicarboxylic anhydride with 6-aminopenicillanic acid: Richards *et al., Nature* **199**, 354 (1963).

Disodium salt, $C_{18}H_{14}N_4Na_2O_6S$, crystals, dec 261-262°. $[\alpha]_D^{23}$ +183.5° (water). Very hygroscopic. uv max (containing 9.2% water): 242, 326 nm (ε 32,100; 7280). Acquires a bright yellow color on exposure to strong sunlight but is stable at 100° for at least 3 months. Freely sol in water; a 25% aq soln is stable for 2 months at 0°. Antimicrobial activity is highest against *Staphylococcus aureus.*

Bistriethylammonium salt monohydrate, $C_{30}H_{46}N_6O_6S.H_2O$, crystals from acetone, dec 135-137°. $[\alpha]_D^{20}$ +142° (c = 0.376 in water).

THERAP CAT: Antibacterial.

8225. Quinacrine. N^4-*(6-Chloro-2-methoxy-9-acridinyl)-N¹,N¹-diethyl-1,4-pentanediamine; 6-chloro-9-[[4-(diethylamino)-1-methylbutyl]amino]-2-methoxyacridine;* 3-chloro-7-methoxy-9-(1-methyl-4-diethyl aminobutylamino)acridine; 2-chloro-5-(ω-diethylamino-α-methylbutylamino)-7-methoxyacridine; mepacrine. $C_{23}H_{30}ClN_3O$; mol wt 399.96. C 69.07%, H 7.56%, Cl 8.86%, N 10.51%, O 4.00%. Prepd by condensing 1-diethylamino-4-aminopentane with 3,9-dichloro-7-methoxyacridine: F. Mietzsch, H. Mauss, **Ger.** pats. **553,072; 571,499** (1934); *eidem,* **U.S. pat. 2,113,357** (1938 to Winthrop Chemical); *cf.* Drosdov, Cherutzov, *J. Gen. Chem. USSR* **5**, 1576, 1736 (1935); **8**, 1192 (1938); Magidson, Grigorovskii, *Khim. Farm. Prom.* **1933**, 187; Jensch, Eisleb, **U.S. pat. 1,782,727** (1930); Schulemann *et al.,* **U.S. pat. 1,889,704** (1932). Review of quinacrine and other acridines: A. O. Wolfe in *Antibiotics* vol. 3, J. W. Corcoran, F. E. Hahn, Eds. (Springer-Verlag, New York, 1975) pp 203-233.

Dihydrochloride dihydrate, $C_{23}H_{30}ClN_3O.2HCl.2H_2O$, *RP-866, SN-390, Atabrine hydrochloride.* Bitter, bright yellow crystals. Dec 248-250° (mp poorly discernible). One gram dissolves in ∼35 ml water. Much more sol in hot water. Slightly sol in ethanol, somewhat more sol in methanol. Insol in ether, benzene, acetone. pH of a 1% aq soln

~4.5. Under uv light the yellow aq solns exhibit a vivid fluorescence which is detectable in a dilution of 1:5,000,000.

Methanesulfonate monohydrate, $C_{25}H_{38}ClN_3O_7S_2.H_2O$, bitter, bright yellow crystals. One part dissolves in 3 parts water at 15.5°, in 36 parts 95% alcohol at 15.5°. pH of 2.0% w/v soln in water 3.0-5.0.

THERAP CAT: Anthelmintic (Cestodes); antimalarial.

THERAP CAT (VET): Antiprotozoal, teniacide.

8226. Quinagolide. *(3α,4aα,10aβ)-(±)-N,N-Diethyl-N'-(1,2,3,4,4a,5,10,10a-octahydro-6-hydroxy-1-propylbenzo[g]-quinolin-3-yl)sulfamide;* (±)-1-*n*-propyl-3α-diethylsulfa-moylamino-6-hydroxy-1,2,3,4,4aα,5,10,10aβ-octahydroben-zo[g]quinoline. $C_{20}H_{33}N_3O_3S$; mol wt 395.57. C 60.73%, H 8.41%, N 10.62%, O 12.13%, S 8.11%. Nonergot dopamine D_2-receptor agonist. Prepn: R. Nordmann, T. J. Petcher, **Eur. pat. Appl. 77,754;** *eidem,* **U.S. pat. 4,565,818** (1983, 1986 both to Sandoz); *eidem, J. Med. Chem.* **28,** 367 (1985). Receptor binding study *in vivo:* A. Closse *et al., Brain Res.* **440,** 123 (1988). Clinical pharmacology: R. C. Gaillard, J. Brownell, *Life Sci.* **43,** 1355 (1988). Clinical evaluation in hyperprolactinemia: C. Rasmussen *et al., Acta Endocrinol.* **125,** 170 (1991).

Beige powder, mp 122.5-124°.

Hydrochloride, $C_{20}H_{33}N_3O_3S.HCl$, *CV-205-502, SDZ-205-502, Norprolac.* Crystals from methylene chloride + methanol, mp 234-236°.

THERAP CAT: Prolactin inhibitor.

8227. Quinaldic Acid. *2-Quinolinecarboxylic acid;* quinaldinic acid. $C_{10}H_7NO_2$; mol wt 173.17. C 69.36%, H 4.07%, N 8.09%, O 18.48%.

Dihydrate, crystals. mp 155-157° (anhydr). Moderately sol in water; sol in alcohol and in alkali solns.

USE: For the determination of copper, zinc and uranium with which it forms insol salts.

8228. Quinaldine. *2-Methylquinoline.* $C_{10}H_9N$; mol wt 143.19. C 83.88%, H 6.34%, N 9.78%. Occurs in coal tar; made from aniline, acetaldehyde and HCl. Lab procedure: A. I. Vogel, *Practical Organic Chemistry* (Longmans, London, 3rd ed., 1959) p 831; Gattermann-Wieland, *Praxis des Organischen Chemikers* (de Gruyter, Berlin, 40th ed., 1961) p 318. Pharmacology: Brown *et al., Comp. Biochem. Physiol. A* **42,** 223 (1972). Toxicity data: Smyth *et al., Arch. Ind. Hyg. Occup. Med.* **4,** 119 (1951); L. L. Marking, *Invest. Fish Contr.* **23,** 3 (1969).

Colorless, oily liquid; quinoline odor; becomes reddish-brown on exposure to air. d ~1.06. bp 246-247°. Practically insol in water. Sol in chloroform, ether. *Keep tightly closed and protected from light.* LD_{50} orally in rats: 1.23 g/kg (Smyth).

Sulfate, $C_{10}H_9N.H_2SO_4$. Prepn: S. Hoogewerff, W. A. Van Dorp, *Rec. Trav. Chim.* **3,** 345 (1884); J. L. Allen, J. B. Sills, *Invest. Fish Contr.* **47,** 3 (1973). Efficacy as anesthetic in fish: P. A. Gilderhus *et al., ibid.* **49,** 3 (1973); G. C. Blasiola Jr., *J. Fish Biol.* **10,** 113 (1977). Toxicity: L. L. Marking, V. K. Dawson, *Invest. Fish Contr.* **48,** 3 (1973).

Light yellow crystalline powder, mp 211-214°. uv max (0.1*N* H_2SO_4): 236, 317 nm. Soly (g/100 ml) in water 104.05, in methanol 7.44, in ethanol 2.27, in acetone 0.08. Practically insol in ether, benzene, hexane. LC_{50} (96 hour) in largemouth bass: 6.80 mg/l; in carp: 72.5 mg/l (Marking, Dawson).

USE: As anesthetic in transport and handling of fish.

8229. Quinaldine Blue. *1-Ethyl-2-[3-(1-ethyl-2(1H)-quinolinylidene)-1-propenyl]quinolinium chloride;* bis[1-ethylquinoline-(2)]trimethinecyanine chloride; 1,1'-diethyl-2,2'-trimethinequinocyanine chloride; pinacyanol chloride; Vernitest Reagent. $C_{25}H_{25}ClN_2$; mol wt 388.94. C 77.20%, H 6.48%, Cl 9.12%, N 7.20%. Prepn: Fischer, *J. Prakt. Chem.* **98,** 204 (1918). Properties and use as histological stain: F. Proescher, *Proc. Soc. Exp. Biol. Med.* **31,** 79 (1933); H. P. Klinger *et al., Stain Technol.* **46,** 43 (1971); R. K. J. Narayan, *ibid.* **55,** 9 (1980).

Bright blue-green alcohol-containing prisms or needles from alcohol. Drying at 110° expels alcohol of crystn. Dec at ~263°. Moderately sol in water with a violet-red color, in alcohol with a blue color. Solns are dichroic.

USE: Histological stain (chromosomes).

8230. Quinaldine Red. *2-[2-[4-(Dimethylamino)phen-yl]ethenyl]-1-ethylquinolinium iodide; 2-[p-(dimethylamino)-styryl]-1-ethylquinolinium iodide;* 2-(*p*-dimethylaminostyr-yl)quinoline ethiodide; α-(*p*-dimethylaminophenylethylene)-quinoline ethiodide; Eastman no. 1361. $C_{21}H_{23}IN_2$; mol wt 430.33. C 58.61%, H 5.39%, I 29.49%, N 6.51%. Prepd by condensing quinaldine ethyliodide and *p*-dimethylamino-benzaldehyde in alc in the presence of piperidine: König, *J. Prakt. Chem.* [2] **86,** 172 (1912); **Ger. pat. 294,744;** *Frdl.* **14,** 735.

Very dark red powder. Sparingly sol in water. Freely sol in alcohol, giving a dark red soln. Solns are slowly dec by light. pKb 11.25. Colorless at pH 1.4; red at pH 3.2. In the presence of chlorides the transition interval is from pH 1.2 to pH 3.0.

USE: As an indicator.

8231. Quinalizarin. *1,2,5,8-Tetrahydroxy-9,10-anthra-cenedione; 1,2,5,8-tetrahydroxyanthraquinone;* alizarin bordeaux B; C.I. Mordant Violet 26; C.I. 58500. $C_{14}H_8O_6$; mol wt 272.21. C 61.77%, H 2.96%, O 35.27%. From hemipic (hemipinic) acid and hydroquinone with H_2SO_4: Liebermann, Kostanecki, *Ann.* **240,** 245 (1887); Liebermann, Wense, *Ber.* **20,** 862 (1887). From alizarin or quin-izarin by treatment with 80% oleum, then boiling with caustic: Gattermann, *J. Prakt. Chem.* [2] **43,** 246 (1891); Schmidt, *ibid.* **237,** 242 (1891); *Bull. Soc. Ind. Mulhouse* **84,** 409 (1914). See also: *Colour Index* vol. 4 (3rd ed., 1971) p 4519.

Red needles with green metallic luster from acetic acid or from nitrobenzene or by sublimation *in vacuo.* mp >275°. Absorption spectrum: Meek, Watson, *J. Chem. Soc.* **109,** 544 (1916); Meek, *ibid.* **111,** 969 (1917). Insol in water. Very slightly sol in most other solvents. Dissolves in aq solns of alkalies with reddish-violet, in acetic acid with yellow, in sulfuric acid with blue-violet color.

USE: Dyes cotton mordanted with Al salts dark red, cotton mordanted with Cr salts bluish-violet. No longer used as a dye, except occasionally in printing cotton.

8232. Quinamine. *[1S-[1α,2α(3aS*,8aR*),4α,5β]]-8a-(5-Ethenyl-1-azabicyclo[2.2.2]oct-2-yl)-2,3,8,8a-tetrahydro-3aH-furo[2,3-b]indol-3a-ol.* $C_{19}H_{24}N_2O_2$; mol wt 312.41. C 73.05%, H 7.74%, N 8.97%, O 10.24%. In Cinchona barks, especially those of *Cinchona pubescens* Vahl. (*C. succirubra* Pav.) and *C. officinalis* L. (*C. ledgeriana* Moens), *Rubiaceae*: Hesse, *Ber.* **5,** 265 (1872); **10,** 2157 (1877); *Ann.* **207,** 288 (1881); Oudemans, *Ann.* **197,** 49 (1879). Isoln from quinine sulfate mother liquors: De Vrij, *Pharm. J.* **4,** 609 (1874); Howard, *ibid.* **5,** 1 (1875); Henry *et al., J. Chem. Soc.* **1945,** 524. Reduction with lithium aluminum hydride yields cinchonamine: Goutarel *et al., Helv. Chim. Acta* **33,** 150 (1950). Structure: Witkop, *J. Am. Chem. Soc.* **72,** 2311 (1950).

Needles from boiling benzene, mp 185-186°. $[\alpha]_D$ +116° or +104° (c = 0.5 in alcohol). uv max (methanol): 242, 301 nm (log ε 3.93, 3.39). One gram dissolves in 1500 ml water, 50 ml ether, about 120 ml 80% alcohol. Sol in abs alc, hot ether, hot benzene, hot petr ether.

Hydrochloride monohydrate, $C_{19}H_{25}ClN_2O_2 \cdot H_2O$, crystals, mp 166-167°. $[\alpha]_D$ +102.8°.

Hydriodide, $C_{19}H_{25}IN_2O_2$, crystals, mp 224°. $[\alpha]_D$ +84.82° (c = 0.5 in alcohol).

Nitrate, $C_{19}H_{25}N_3O_5$, crystals, mp 186-188°. $[\alpha]_D$ +94.9°.

8233. Quinapril. *[3S-[2[R*(R*)],3R*]]-2-[2-[[1-(Ethoxycarbonyl)-3-phenylpropyl]amino]-1-oxopropyl]-1,2,3,4-tetrahydro-3-isoquinolinecarboxylic acid; (S)-2-[(S)-N-[(S)-1-carboxy-3-phenylpropyl]alanyl]-1,2,3,4-tetrahydro-3-isoquinolinecarboxylic acid 1-ethyl ester.* $C_{25}H_{30}N_2O_5$; mol wt 438.52. C 68.47%, H 6.90%, N 6.39%, O 18.24%. Angiotensin converting enzyme (ACE) inhibitor. Prepn: M. L. Hoefle, S. Klutchko, *Eur. pat.* **Appl. 49,605;** *eidem,* U.S. pat. **4,344,949** (both 1982 to Warner-Lambert); of quinapril and its active diacid metabolite, quinaprilat, and activity: S. Klutchko *et al., J. Med. Chem.* **29,** 1953 (1986). Synthesis of novel crystalline and purified form: O. P. Goel, U. Krolls, U.S. pat. **4,761,479** (1988 to Warner-Lambert). Pharmacology: H. R. Kaplan *et al., Fed. Proc.* **43,** 1326 (1984). Antihypertensive activity in animals: M. J. Ryan *et al., ibid.* 1330. Hemodynamic effects in congestive heart failure: P. Holt *et al., Eur. J. Clin. Pharmacol.* **31,** 9 (1986). Clinical evaluation in essential hypertension: P. Säynävälammi *et al., J. Cardiovasc. Pharmacol.* **12,** 88 (1988). HPLC determn of quinapril and quinaprilat in human plasma and urine: H. Hengy, M. Most, *J. Liq. Chromatog.* **11,** 517 (1988). Toxicity data: H. R. Kaplan *et al., Angiology* **40,** 335 (1989). Series of articles on pharmacology, pharmacokinetics, clinical evaluation and safety: *ibid.* 329-415.

Hydrochloride, $C_{25}H_{30}N_2O_5 \cdot HCl$, *CI-906, PD-109452-2, Accupril, Accuprin, Accupro, Acequin, Acuitel, Korec, Quinazil.* Crystals from ethyl acetate-toluene, mp 120-130°. $[\alpha]_D^{23}$ +14.5° (c = 1.2 in ethanol) (Klutchko). Also reported as white crystalline solid from acetonitrile, mp 119-121.5°. $[\alpha]_D^{25}$ +15.4° (c = 2 in methanol) (Goel, Krolls). LD_{50} in male, female mice, rats (mg/kg): 1739, 1840, 4280, 3541 orally; 504, 523, 158, 107 i.v. (Kaplan, 1989).

Diacid, $C_{23}H_{26}N_2O_5$, *quinaprilat, CI-928.* Hydrate, crystals from methanol-ethyl ether, mp 166-168°. $[\alpha]_D^{23}$ +20.9° (c = 1 in methanol).

Mixture with hydrochlorothiazide, *Accuretic, Acequide, Koretic.*

THERAP CAT: Antihypertensive.

8234. Quinapyramine. *4-Amino-6-[(2-amino-1,6-dimethyl-4(1H)-pyrimidinylidene)amino]-1,2-dimethylquinolinium conjugate monoacid;* 4-amino-6-[(2-amino-6-methyl-4-pyrimidinyl)amino]-1-methylquinaldinium methosalts; 4-amino-6-[(2-amino-1,6-dimethyl-4-pyrimidinyl)amino]-1,2-dimethylquinoline salts; 4-amino-6-(2-amino-6-methyl-4-pyrimidylamino)quinaldine-1,1'-dimethosalts; M-7555; Antrycide. Prepn: F. H. S. Curd, D. G. Davey, *Brit. J. Pharmacol.* **5,** 25 (1950); Curd, U.S. pat. **2,585,917** (1952 to I.C.I.); Ainley *et al., J. Chem. Soc.* **1953,** 59.

Dimethosulfate, $C_{19}H_{28}N_6O_8S_2$, creamy white crystals from aq methanol, mp 265-266°. Freely sol in water. LD_{50} i.v. in mice: 10-15 mg/kg (Curd, Davey).

Diiodide, pale cream needles, mp 312-313° (dec). Sparingly sol in water.

Dichloride, crystals from water, mp 316-317° (dec). Sparingly sol in water. LD_{50} i.v. in mice: 10-15 mg/kg (Curd, Davey).

THERAP CAT: Antiprotozoal (Trypanosoma).

8235. Quinazoline. *1,3-Benzodiazine;* benzo[a]pyrimidine; 5,6-benzopyrimidine; phenmiazine. $C_8H_6N_2$; mol wt 130.15. C 73.83%, H 4.65%, N 21.52%. Prepn from 2-nitrobenzylidenebis(formamide): Riedel, **Ger. pat. 174,941;** *Chem. Zentr.* **1906,** II, 1372; *Frdl.* **8,** 1238; Bogert, McColm, *J. Am. Chem. Soc.* **49,** 2650 (1927).

Leaflets from petr ether; odor of quinoline. Slightly bitter taste. mp 48.0-48.5°. $bp_{772.5}$ 243°; bp_{764} 241.5°. Freely sol in water. Neutral reaction. Sol in many organic solvents.

8236. Quinbolone. *(17β)-17-(1-Cyclopenten-1-yloxy)androsta-1,4-dien-3-one;* 1-dehydrotestosterone 17-cyclopent-1'-enyl ether; Anabolicum Vister. $C_{24}H_{32}O_2$; mol wt 352.52. C 81.77%, H 9.15%, O 9.08%. Prepn from 17β-hydroxyandrosta-1,4-dien-3-one: Ercoli *et al., Chem. & Ind. (London)* **1962,** 1284.

Crystals from ethanol, mp 133-135°. $[\alpha]_D^{20}$ +61° (dioxane). uv max: 244-245 nm ($E_{1cm}^{1\%}$ 430-450). Practically insol in water. Sol in many organic solvents; sparingly sol in ethanol, hexane. Soly in sesame oil: 40-45 mg/ml.

THERAP CAT: Anabolic.

8237. Quince Seed. Gum quince seed; semen cydonia; golden apple seed; cydonia seed. Seed of *Cydonia oblonga* Mill. (*C. vulgaris* Pers.), *Rosaceae. Habit.* Southern Asia, Europe; widely cultivated. *Constit.* Amygdalin, emulsin, about 15% fatty oil, about 20% of a mucilage named cydonin. Brief review of seed and gum uses: BeMiller in *Industrial Gums*, R. L. Whistler, Ed. (Academic Press, New York, 2nd ed., 1973) pp 339-345.

USE: Gum from the seeds as suspending agent, stabilizer; in hair and cosmetic prepns.

8238. Quinestradiol. *(16α,17β)-3-(Cyclopentyloxy)estra-1,3,5(10)-trien-16,17-diol;* quinestradol; estriol 3-cyclopentyl ether; Colpovis; Pentovis. $C_{23}H_{32}O_3$; mol wt 356.51. C 77.49%, H 9.05%, O 13.46%. Prepn from estriol and cyclopentylbromide: Ercoli, **Brit. pat.** 909,662 (1962 to Vismara).

Crystals, mp 98-100°.
THERAP CAT: Estrogen.

8239. Quinestrol. *(17α)-3-(Cyclopentyloxy)-19-norpregna-1,3,5(10)-trien-20-yn-17-ol;* 17α-ethinylestradiol 3-cyclopentyl ether; W-3566; Estrovis. $C_{25}H_{32}O_2$; mol wt 364.53. C 82.37%, H 8.85%, O 8.78%. Prepn: Ercoli, Gardi, *Chem. & Ind. (London)* **1961**, 1037; Ercoli, **U.S. pat.** 3,159,543; Ercoli *et al.*, **U.S. pat.** 3,231,567 (1964, 1966, both to Vismara).

Crystals, mp 107-108°. $[\alpha]_D^{25}$ +5° (c = 0.5 in dioxane).
THERAP CAT: Estrogen.

8240. Quinethazone. *7-Chloro-2-ethyl-1,2,3,4-tetrahydro-4-oxo-6-quinazolinesulfonamide;* 7-chloro-2-ethyl-6-sulfamoyl-1,2,3,4-tetrahydro-4-quinazolinone; 7-chloro-2-ethyl-1,2,3,4-tetrahydro-4-oxo-6-sulfamoylquinazoline; CL-36010; Hydromox; Aquamox. $C_{10}H_{12}ClN_3O_3S$; mol wt 289.74. C 41.45%, H 4.17%, Cl 12.24%, N 14.50%, O 16.57%, S 11.07%. Prepn: Cohen *et al., J. Am. Chem. Soc.* **82**, 2731 (1960); Cohen, Vaughan, Jr., **U.S. pat.** 2,976,289 (1961 to Am. Cyanamid).

Fibrous crystals from 50% acetone, mp 250-252°. Sol in acetone, alcohol.

THERAP CAT: Diuretic, antihypertensive.

8241. Quinfamide. *2-Furancarboxylic acid 1-(dichloroacetyl)-1,2,3,4-tetrahydro-6-quinolinyl ester;* 2-furoic acid ester with 1-(dichloroacetyl)-1,2,3,4-tetrahydro-6-quinolinol; 1-(dichloroacetyl)-6-(2-furoyloxy)-1,2,3,4-tetrahydroquinoline; Win-40014; Amenox; Amenide. $C_{16}H_{13}Cl_2NO_4$; mol wt 354.19. C 54.26%, H 3.70%, Cl 20.02%, N 3.95%, O 18.07%. Prepn: D. M. Bailey, **U.S. pat.** 3,997,542 (1976 to Sterling); D. M. Bailey *et al., J. Med. Chem.* **22**, 600 (1979). Amebicidal activity and toxicological evaluation: R. G. Slighter *et al., Parasitology* **81**, 157 (1980). Distribution and metabolism in rats: J. F. Baker, *Arch. Int. Pharmacodyn.* **258**, 29 (1982). Clinical evaluation in adults with chronic amebiasis: L. Guevara *et al., Clin. Therap.* **6**, 43 (1983); in children: F. A. Rojas *et al., ibid.* 47.

Crystals from ethyl acetate, mp 150.5-151°.
THERAP CAT: Antiamebic.

8242. Quinhydrone. *2,5-Cyclohexadiene-1,4-dione compd with 1,4-benzenediol (1:1);* green hydroquinone. $C_{12}H_{10}O_4$; mol wt 218.21. C 66.05%, H 4.62%, O 29.33%. An addn compd of one mol hydroquinone and one mol quinone. Prepn from hydroquinone and quinone: Gattermann-Wieland, *Praxis des Organischen Chemikers* (de Gruyter, Berlin, 40th ed., 1961) p 270. Alternate prepn by the action of ferric ammonium sulfate on hydroquinone: A. I. Vogel, *Practical Organic Chemistry*, (Longmans, London, 3rd ed., 1959) p 747. Toxicity study: Woodward *et al., Fed. Proc.* **8**, 348 (1949).

Green crystals with metallic luster; reddish-brown by transmitted light. d 1.40. mp 171°; sublimes with partial decompn. Slightly sol in cold water; sol in hot water, ammonia, alc, ether. Insol in petr ether. LD_{50} orally in rats: 225 mg/kg (Woodward).

USE: In pH determinations (quinhydrone electrode).

8243. Quinic Acid. *[1R-(1α,3α,4α,5β)]-1,3,4,5-Tetrahydroxycyclohexanecarboxylic acid;* chinic acid; kinic acid; hexahydro-1,3,4,5-tetrahydroxybenzoic acid. $C_7H_{12}O_6$; mol wt 192.17. C 43.75%, H 6.29%, O 49.95%. Found in cinchona bark, particularly in South American barks; also in many other plants, such as tobacco leaves, carrot leaves, apples, peaches, pears, plums, etc. Structure and configuration: Fischer, Dangschat, *Ber.* **65**, 1009 (1932). Total synthesis: Grewe *et al., ibid.* **87**, 793 (1954); Smissman, Oxman, *J. Am. Chem. Soc.* **85**, 2184 (1963). Stereospecific synthesis: Wolinsky *et al., J. Org. Chem.* **29**, 3596 (1964). *Review:* Bohm, *Chem. Rev.* **65**, 435 (1965).

White crystals; strong acid taste. d 1.64. mp 162-163°; at higher temps forms a lactone. $[\alpha]_D^{20}$ −42° to −44° in water. Sol in 2.5 parts water, in alcohol, glacial acetic acid.

8244. Quinidine. *(9S)-6'-Methoxycinchonan-9-ol;* α-(6-methoxy-4-quinolyl)-5-vinyl-2-quinuclidinemethanol; con-quinine; pitayine; β-quinine. $C_{20}H_{24}N_2O_2$; mol wt 324.42. C 74.05%, H 7.46%, N 8.63%, O 9.86%. A dextrorotatory stereoisomer of quinine, q.v. Present in cinchona barks to the extent of 0.25-3.0%. Found in quinine sulfate mother liquors. Review of structural elucidation and early synthetic studies: R. B. Turner, R. B. Woodward, in *The Alkaloids*, **vol. 3**, 1-63 (1953). Configuration: Prelog, Zalán, *Helv. Chim. Acta* **27**, 535 (1944); Prelog, Häfliger, ibid. **33**, 2021 (1950); Roth, *Pharmazie* **16**, 257 (1961). Crystal and molecular structure: R. Doherty et al., *J. Pharm. Sci.* **67**, 1698 (1978). Rotatory dispersion studies: Lyle, Gaffield, *Tetrahedron Letters* **1963**, 1371. Prepn by isomerization of quinine: W. E. Doering et al., *J. Am. Chem. Soc.* **69**, 1700 (1949). Total synthesis: J. Gutzwiller, M. Uskokovic, ibid. **92**, 204 (1970); eidem, *Helv. Chim. Acta* **56**, 1494 (1973); eidem, *J. Am. Chem. Soc.* **100**, 576 (1973). Toxicity data: C. Turba et al., *Arzneimittel-Forsch.* **18**, 1127 (1968); K. Dietmann et al., ibid. **27**, 589 (1977). Comprehensive description of the sulfate: M. A. Loutfy et al., *Anal. Profiles Drug Subs.* **12**, 483-546 (1983). Clinical evaluation in severe malaria: R. E. Phillips et al., *N. Engl. J. Med.* **312**, 1273 (1985). Review of pharmacology and clinical efficacy in cardiac arrhythmias: J. W. Mason, L. M. Hondeghem, *Ann. N.Y. Acad. Sci.* **432**, 162-176 (1984); A. R. Leon, J. D. Merlino, *Heart Dis. Stroke*, **2**, 407-413 (1993).

Triboluminescent. mp 174-175° after drying of solvated crystals. $[\alpha]_D^{15}$ +230° (c = 1.8 in chloroform), $[\alpha]_D^{17}$ +258° (alc), $[\alpha]_D^{17}$ +322° (c = 1.6 in 2M HCl). pK$_1$ (20°) 5.4; pK$_2$ 10.0. Blue fluorescence in dil H_2SO_4. The uv absorption spectrum is identical with that of quinine. One gram dissolves in about 2000 ml cold, 800 ml boiling water, 36 ml alcohol, 56 ml ether, 1.6 ml chloroform; very sol in methanol. Practically insol in petr ether. LD$_{50}$ in rats (mg/kg): 30 i.v., 263 orally (Dietmann).

Hemipentahydrate, prisms from dil alcohol, loses ½ H_2O in air, mp ~168°.

Hydrogen sulfate tetrahydrate, $C_{20}H_{24}N_2O_2 \cdot H_2SO_4 \cdot 4H_2O$, *quinidine bisulfate, Chinidin-Duriles, Kiditard, Kinichron, Kinidin Durules, Quiniduran.* Rods, sol in 8 parts water with blue fluorescence.

Sulfate dihydrate, $(C_{20}H_{24}N_2O_2)_2 \cdot H_2SO_4 \cdot 2H_2O$, *Cin-Quin, Quinidex Extentabs, Quinicardine, Quinora.* White, very bitter, odorless, fine crystals, frequently cohering in masses. Darkens on exposure to light. Does not lose all of its water below 120°. $[\alpha]_D^{25}$ ~+212° (95% alcohol); ~+260° (dil HCl). pKa 4.2, 8.8. pH (1% aq soln): 6.0-6.8. One gram dissolves in about 90 ml water, 15 ml boiling water, 10 ml alcohol, 3 ml methanol, 12 ml chloroform. Insol in ether, benzene. *Protect from light.* LD$_{50}$ in mice, rats (mg/kg): 700, 455.8 orally; 83, 56 i.v. (Turba).

Gluconate, $C_{26}H_{36}N_2O_9$, *gluconic acid quinidine salt, Duraquin, Quinaglute.* Crystals, mp 175-176.5°. Sol in 9 parts water, 60 parts alcohol.

Polygalacturonate, *Cardioquin, Galactoquin, Naticardina.* $(C_{20}H_{24}N_2O_2 \cdot C_6H_{10}O_7 \cdot H_2O)_x$. Prepn: A. Halpern et al., *Am. J. Pharm.* **130**, 190 (1958). Pharmacology: A. Halpern et al., *Antibiot. Chemother.* **9**, 97 (1959). Amorphous powder, mp 180° (dec). Anhydr product is insol in methanol, ethanol, chloroform, ether, acetone, dioxane. Soly in hot 40% methanol or ethanol: 12%; in water at 25°: ~2%. LD$_{50}$ in rats, mice (mg/kg): 3200 ± 350, 2680 ± 210 orally (Halpern, 1959).

THERAP CAT: Antiarrhythmic (class IA); antimalarial.
THERAP CAT (VET): Antiarrhythmic.

8245. Quinine. *(8α,9R)-6'-Methoxycinchonan-9-ol.* $C_{20}H_{24}N_2O_2$; mol wt 324.42. C 74.05%, H 7.46%, N 8.63%, O 9.86%. Primary alkaloid of various species of *Cinchona* (Rubiaceae), see Cinchona. Representative samples of dried bark contain ~0.8 to 4% quinine. Optical isomer of quinidine, q.v. Isoln: Pelletier, Caventau, *Ann. Chem. Phys.* [2], **15**, 291 (1820). Extraction procedure: Jucker, Stoll, in *Ullmann's Enzyklopädie der technischen Chemie* **3**, 213-218 (1953). Configuration: Prelog, Zalán, *Helv. Chim. Acta* **27**, 535 (1944); Prelog, Häfliger, ibid. **33**, 2021 (1950); Roth, *Pharmazie* **16**, 257 (1961). Synthesis: Woodward, Doering, *J. Am. Chem. Soc.* **66**, 849 (1944); **67**, 860 (1945); Taylor, Martin, ibid. **94**, 6218 (1972); Gutzwiller, Uskokovic, ibid. **100**, 576 (1978); G. Grethe et al., ibid. **589**; T. Imanishi et al., *Chem. Pharm. Bull.* **30**, 1925 (1982). Review of structural elucidation and early synthetic studies: R. B. Turner, R. B. Woodward in *The Alkaloids*, **vol. 3**, 1-63 (1953); of bioactivity: F. E. Hahn, Ed. in *Antibiotics* vol. 5(pt. 2) (Springer-Verlag, New York, 1979) pp 353-362. Comprehensive description of the hydrochloride: F. J. Muhtadi et al., *Anal. Profiles Drug Subs.* **12**, 547-621 (1983). LC determn in soft drinks: L. P. Valenti, *J. Assoc. Off. Anal. Chem.* **68**, 782 (1985). HPLC determn in blood: V. K. Dua et al. *J. Chromatog.* **614**, 87 (1993). Clinical evaluation to relieve nocturnal leg cramps: P. S. Connolly et al., *Arch. Intern Med.* **152**, 1877 (1992). Clinical efficacy in malaria: P. G. Kremsner et al., *J. Infect. Dis.* **169**, 467-470 (1994).

Triboluminescent, orthorhombic needles from abs alcohol, mp 177° (some decompn). Sublimes in high vacuum at 170-180°. $[\alpha]_D^{15}$ −169° (c = 2 in 97% alcohol), $[\alpha]_D^{17}$ −117° (c = 1.5 in chloroform), $[\alpha]_D^{15}$ −285° (c = 0.4M in 0.1N H_2SO_4). pK$_1$ (18°) 5.07; pK$_2$ 9.7. pH of satd aq soln 8.8. Absorption spectra: Dobbie, Lauder, *J. Chem. Soc.* **99**, 1260 (1911); Dobbie, Fox, ibid. **101**, 78 (1912). Fluorescence: Rabe, Marschall, *Ann.* **382**, 362 (1911). The blue fluorescence is especially strong in dil H_2SO_4. One gram dissolves in 1900 ml water, 760 ml boiling water, 0.8 ml alcohol, 80 ml benzene (18 ml benzene at 50°), in 1.2 ml chloroform; 250 ml dry ether, 20 ml glycerol, 1900 ml of 10% ammonia water. Almost insol in petr ether.

Trihydrate, microcrystalline powder, mp 57°, efflorescent, loses one H_2O in air, two H_2O over H_2SO_4, anhydr at 125°.

Bisulfate heptahydrate, $C_{20}H_{24}N_2O_2 \cdot H_2SO_4 \cdot 7H_2O$, *Biquinate, Dentojel, Quinbisan.* Very bitter crystals or cryst powder; efflorescent on exposure to air and darkens on exposure to light. One gram dissolves in 9 ml water, 0.7 ml boiling water, 23 ml alcohol, 0.7 ml alcohol at 60°, 625 ml chloroform, 2500 ml ether, 15 ml glycerol. mp: 3.5.

Dihydrochloride, $C_{20}H_{24}N_2O_2 \cdot 2HCl$, *quinine dichloride, quinine bimuriate, acid quinine hydrochloride.* Very bitter powder or crystals. One gram dissolves in about 0.6 ml water, in about 12 ml alcohol. Slightly sol in chloroform, very slightly sol in ether. Aq solns are strongly acid to litmus paper (pH about 2.6).

Hydrochloride dihydrate, $C_{20}H_{24}N_2O_2 \cdot HCl \cdot 2H_2O$. Bitter, silky needles. Effloresces on exposure to warm air. Does not lose all its water below 120°. One gram dissolves in 16 ml water, in 0.5 ml boiling water, in 1.0 ml alcohol, in about 7.0 ml glycerol, in about 1 ml chloroform, in about 350 ml ether. pH (1% aq soln): 6.0-7.0. Bitterness threshold 1:30000. *Protect from light.*

Sulfate dihydrate, $(C_{20}H_{24}N_2O_2)_2 \cdot H_2SO_4 \cdot 2H_2O$, *Quinamm, Quine, Quinate, Quinsan.* Dull needles or rods, making a light and readily compressible mass. Becomes brownish on exposure to light. Loses its water of crystn at about 100°. $[\alpha]_D^{15}$ −220° (5% soln in about 0.5N HCl). One gram dissolves in 810 ml water, 32 ml boiling water, 120 ml alcohol, 10 ml alcohol at 78°. Slightly sol in chloroform, ether, but freely sol in a mixture of 2 vols chloroform and 1 vol abs alcohol. Aq solns are neutral to litmus, pH of satd soln 6.2.

USE: Flavor in carbonated beverages.

THERAP CAT: Antimalarial; muscle relaxant (skeletal).

THERAP CAT (VET): Antiprotozoal for fish.

8246. Quinine Iodosulfate. Herapathite; iodoquinine sulfate. $C_{80}H_{104}I_6N_8O_{20}S_3$; mol wt 2355.37. C 40.80%, H 4.45%, I 32.33%, N 4.76%, O 13.59%, S 4.08%. $4C_{20}H_{24}N_2O_2 \cdot 3H_2SO_4 \cdot 2HI \cdot I_4$. Named after its discoverer, Herapath, an English physician.

Hexahydrate, plate-like crystals of pale olive-green color by transmitted light and of a brilliant green to reddish green by reflected light. The crystals polarize light 5 times as much as tourmaline. Loses its water at 100° and becomes red. Almost insol in water. Sol in about 1000 parts boiling water; in 800 parts cold, 50 parts boiling alcohol, in 60 parts hot glacial acetic acid.

USE: In the manuf of polarizing glasses and plastics.

8247. Quinine Oleate. Usually with an excess of oleic acid. Contains about 21% anhydr quinine.

Brown, thick liquid. Insol in water. Sol in alcohol, ether, oils, petr ether.

USE: Has been credited with opacity to ultraviolet rays and suggested as prophylactic against sunburn, applied as a 5-10% ointment. A soln of the oleate without excess of oleic acid in petr naphtha has been recommended for mothproofing of fabrics.

8248. Quininic Acid. *6-Methoxy-4-quinolinecarboxylic acid;* 6-methoxycinchoninic acid. $C_{11}H_9NO_3$; mol wt 203.20. C 65.02%, H 4.46%, N 6.89%, O 23.62%.

Pale yellow crystals. mp about 280° with dec. Slightly sol in water, cold alcohol or ether; sol in about 80 parts boiling abs alcohol; sol in aq alkalies.

8249. Quininone. *(8α)-6'-Methoxycinchonan-9-one.* $C_{20}H_{22}N_2O_2$; mol wt 322.41. C 74.51%, H 6.88%, N 8.69%, O 9.92%. By careful oxidation of quinine or quinidine: Rabe, Kuliga, *Ann.* **364,** 346, 349 (1909); Woodward *et al., J. Am. Chem. Soc.* **67,** 1425 (1945). Alternate procedure: Rabe, Kindler, *Ber.* **51,** 466 (1918). Prepn of amorphous epimeric mixture of quininone and quinidinone from quinotoxine: Gutzwiller, Uskokovic, *Helv. Chim. Acta* **56,** 1494 (1973).

Crystals from ether, mp 108° (rapid heating). Shows mutarotation, final $[\alpha]_D^{20}$ +76° (c = 2 in alcohol). Alkaline reaction to litmus. Freely sol in alcohol, ether, chloroform, benzene. Almost insol in water and petr ether.

Hydrochloride, $C_{20}H_{22}N_2O_2 \cdot HCl$, hygroscopic crystals, mp 212°. Final $[\alpha]_D^{18}$ +59° (c = 2 in abs alcohol).

8250. Quinizarin. *1,4-Dihydroxy-9,10-anthracenedione; 1,4-dihydroxyanthraquinone;* C.I. 58050. $C_{14}H_8O_4$; mol wt 240.22. C 70.00%, H 3.36%, O 26.64%. Prepn from p-chlorophenol and phthalic anhydride: Reynolds, Bigelow, *J. Am. Chem. Soc.* **48,** 420 (1926); U.S. pat. **1,845,632,** *C.A.* **26,** 2203; *Org. Syn.* coll. vol. I, 476 (New York, 1941). Prepn from hydroquinone: Gattermann-Wieland, *Praxis des Organischen Chemikers* (de Gruyter, Berlin, 40th ed., 1961) p 299. Also prepd from diazotized p-chloroaniline and phthalic anhydride: **Brit. pat. 373,999,** *C.A.* **27,** 3946 (1933); by treating anthraquinone with ammonium persulfate in sulfuric acid: Wacker, *J. Prakt. Chem.* [2] **54,** 90 (1896). Purification from contaminating purpurin: *Org. Syn. (loc. cit.).* See also: *Colour Index* vol. 4 (3rd ed., 1971) p 4515.

Orange crystals from acetic acid, mp 200-203° (*Org. Syn.*). Orange plates from ether. Deep red needles from alcohol, benzene, toluene, xylene. mp 196°. Sublimes in high vacuum. Absorption spectrum: Meek, Watson, *J. Chem. Soc.* **109,** 544 (1916); Meek, *ibid.* **111,** 969 (1917). K at 18° = 3.1×10^{-10}. Moderately sol in alcohol with red color. Sol in ether with brown color and yellow fluorescence. Sol with violet color in aq alkalies and in ammonia. Black precipitate with CO_2. One gram dissolves in about 13 g of boiling glacial acetic acid.

Dimethyl ether, $C_{16}H_{12}O_4$, mp 177°.

8251. Quinizarin Green SS. *1,4-Bis[(4-methylphenyl)-amino]-9,10-anthracenedione; 1,4-di-p-toluidinoanthraquinone;* D & C Green No. 6; C.I. 61565. $C_{28}H_{22}N_2O_2$; mol wt 418.49. C 80.36%, H 5.30%, N 6.69%, O 7.65%. Discovered by R. E. Schmidt in 1894: *Colour Index* vol. 4 (3rd ed., 1971) p 4541.

Dark violet needles, mp 218°.   Blue soln in conc H_2SO_4 giving a blue-green ppt on dilution.

USE: For sutures: *Fed. Regist.* **40**, 18167 (1975). Permitted for use in externally applied drugs and cosmetics: *ibid.* **47**, 14138 (1982).

8252.   Quinocide.   *N¹-(6-Methoxy-8-quinolinyl)-1,4-pentanediamine;* 8-(4-aminopentylamino)-6-methoxyquinoline; 8-[(4-amino-4-methylbutyl)amino]-6-methoxyquinoline; 6-methoxy-8-(4-aminopentylamino)quinoline; chinocide; khinocyde. $C_{15}H_{21}N_3O$; mol wt 259.35. C 69.47%, H 8.16%, N 16.20%, O 6.17%.   Prepn: Braude, Stavrovskaya, *J. Gen. Chem. USSR* **26**, 999 (1956); modified method: *idem, Med. Prom.* **11**(7), 19 (1957), *C.A.* **52**, 11043 (1958); B. Balkrishen *et al., Chem. & Ind. (London)* **1983**, 899.

bp$_1$ 183-186°.   mp 46°.
Hydrochloride, $C_{15}H_{21}N_3O.HCl$, crystals from abs alcohol, mp 224-224.5°.
Dihydrochloride, $C_{15}H_{21}N_3O.2HCl$, mp 227-227.5°.
Diphosphate, $C_{15}H_{27}N_3O_9P_2$, mp 174-176°.
THERAP CAT: Antimalarial.

8253.   Quinoline.   Leucoline; chinoleine; 1-benzazine; benzo[b]pyridine. C_9H_7N; mol wt 129.16. C 83.69%, H 5.46%, N 10.84%.   Occurs in small amounts in coal tar. Prepd by the Skraup synthesis by heating aniline with glycerol and nitrobenzene in presence of sulfuric acid: Clarke, Davis, *Org. Syn.* **coll. vol. I**, 478 (2nd ed., 1941) 478; alternate syntheses: Manske, *Chem. Rev.* **30**, 113 (1942); Bergstrom, *ibid.* **35**, 150 (1944).   Process based on the interaction of aniline with acetaldehyde and a formaldehyde hemiacetal: Cislak, Wheeler, U.S. pat. **3,020,281** (1962 to Reilly Tar & Chem.). Toxicity data: Smyth *et al., Arch. Ind. Hyg. Occup. Med.* **4**, 119 (1951).

Hygroscopic liquid.   Absorbs as much as 22% water. Darkens on storage in ordinary, stoppered bottle. Penetrating odor, not as offensive as pyridine. Volatile with steam. d_4^{25} 1.0900; mp −15°; bp$_{760}$ 237.7°; bp$_{100}$ 163.2°; bp$_{40}$ 136.7°; bp$_{20}$ 119.8°; bp$_{10}$ 103.8°; bp$_5$ 89.6°; bp$_{1.0}$ 59.7°. n_D^{20} 1.62683. Absorption spectrum: Hantzsch, *Ber.* **44**, 1824 (1911). Weak base (neutral to phenolphthalein), forms water-soluble salts with strong acids: pK 9.5. Difficultly sol in cold water, more easily in hot water.   Miscible with alcohol, ether, carbon disulfide. Dissolves sulfur, phosphorus, arsenic trioxide. *Protect from light and moisture.* LD$_{50}$ orally in rats: 460 mg/kg (Smyth).
Bisulfate, $C_9H_7N O_4 S$, white to grayish-white, crystalline powder. mp 163-165°. Freely sol in water; one gram dissolves in 50 ml cold, in 9 ml boiling abs alcohol. *Protect from light.*
Hydrochloride, $C_9H_7N.HCl$, white deliquesc crystals. mp 93-94°. Freely sol in water, alcohol, hot benzene, chloroform, sparingly in cold ether. *Keep well closed and protected from light.*
Salicylate, $C_{16}H_{13}NO_3$, reddish-gray crystalline powder. Sol in 80 parts water; freely sol in alcohol, benzene, ether, glycerol, oils. *Protect from light.*
Tartrate, $C_{43}H_{45}N_3O_{24}$, white crystals, pungent odor, sharp taste. mp 125° (dec). Sol in 80 parts water, 150 parts alcohol. Insol in ether.

USE: Manuf dyes; prepn hydroxyquinoline sulfate, *q.v.*, niacin. As preservative for anatomical specimens. Solvent for resins, terpenes.
THERAP CAT: Antimalarial.

8254.   8-Quinolineboronic Acid. $C_9H_8BNO_2$; mol wt 172.98. C 62.49%, H 4.66%, B 6.25%, N 8.10%, O 18.50%. Prepd by reacting 8-bromoquinoline with butyllithium and tributyl borate: Letsinger, Dandegaonker, *J. Am. Chem. Soc.* **81**, 498 (1959).

Crystals from alc, mp > 300°.
Butyl diester, $C_{17}H_{24}BNO_2$, bp$_4$ 180°. n_D^{25} 1.4840.
Chloroethyl diester, $C_{13}H_{14}BCl_2NO_2$, crystals from toluene, mp 193-194°.

8255.   8-Quinolinecarboxylic Acid. $C_{10}H_7NO_2$; mol wt 173.17. C 69.36%, H 4.07%, N 8.09%, O 18.48%. Prepd by heating 2-aminobenzoic acid with 2-nitrobenzoic acid, glycerol and sulfuric acid: Schlosser, Skraup, *Monatsh.* **2**, 530 (1881); by oxidizing 8-quinolylaldehyde with chromic acid: Howitz, *Ber.* **35**, 1275 (1902).

Needles from water.   mp 186-187.5°.   Sublimes above mp. Slightly sol in cold water, appreciably sol in hot water and in alcohol.   Freely sol in acids and alkalies.
USE: Detection of cadmium, copper, iron, lead, mercury, silver, thallium; quantitative determination of copper.

8256.   Quinoline Yellow.   *C.I. Acid Yellow 3;* D & C Yellow No. 10; Acid Yellow 3; Food Yellow 13; C.I. 47005. Synthetic dye.   A mixture of the sodium salts of the mono- and disulfonic acids of quinoline yellow spirit soluble, *q.v.* Principal constituents: sodium salts of *2-(2,3-dihydro-1,3-dioxo-1H-indene-2-yl)-6-quinolinesulfonic acid* and *2-(2,3-dihydro-1,3-dioxo-1H-indene-2-yl)-8-quinolinesulfonic acid.* Prepn of sulfonic acids: M. C. Traub, *Ber.* **16**, 297 (1883); *Colour Index* **vol. 4** (3rd ed., 1971) p 4435. HPLC determn: J. Chudy *et al., J. Chromatog.* **154**, 306 (1978). Purification: N. A. Ambrosiano *et al.,* U.S. pat. **4,398,916** (1983 to Sterling Drug). Stability: H. Delonca *et al., Pharm. Acta Helv.* **58**, 332 (1983).   ¹³C-NMR spectrum: M. Gelbcke *et al., Bull. Soc. Chim. Belg.* **91**, 237 (1982).   TLC identification: J. A. Steele, *J. Assoc. Off. Anal. Chem.* **67**, 540 (1984). Composition and skin sensitivity: Y. Sata *et al., Contact Dermatitis* **10**, 30 (1984).   The unsulfonated quinoline yellow spirit soluble may remain as an impurity and is limited by the FDA: *Code of Federal Regulations,* Title 21, Parts 1-99, Sec. 74.1710(b), p 307 (1985). LC determn of unsulfonated impurity: A. L. Goldberg, *J. Assoc. Off. Anal. Chem.* **68**, 477 (1985).   Toxicity: F. C. Lu, A. Lavalle, *Can. Pharm. J.* **97**, 30 (1964).
Bright greenish yellow.   Orange in conc H_2SO_4. Sol in water, slightly sol in ethanol.   Practically insol in vegetable oils. LD$_{50}$ in rats (g/kg): > 2 orally (Lu, Lavalle).
USE: Textile dye for wool, nylon, silk.   Paper dye.   Barium salt in printing inks. Color for food, drugs, cosmetics.   Approved by the FDA for use in drugs, cosmetics, except for use in the area of the eye: *Fed. Reg.* **48**, 39217 (1983); *ibid.* **49**, 27744 (1984).

8257.   Quinoline Yellow Spirit Soluble.   *C.I. Solvent Yellow 33;* quinoline yellow base; quinoline yellow A; D & C Yellow No. 11; C.I. 47000.   Synthetic dye consisting principally of *quinophthalone (2-(2-quinolinyl)-1H-indene-1,3-(2H)-dione).* Prepn: M. C. Traub, *Ber.* **16**, 297 (1883); E. Jacobsen, C. L. Reimer, *ibid.* 1082; *Colour Index* **vol. 4**

(3rd ed., 1971) p 4435. ^{13}C-NMR spectrum: M. Gelbcke *et al., Bull. Soc. Chim. Belg.* **91**, 237 (1982). Composition and skin sensitivity: Y. Sato *et al., Contact Dermatitis* **10**, 30 (1984); S. Kita *et al., ibid.* **11**, 210 (1984). Inhalation toxicity in animals: T. C. Marrs *et al., Human Toxicol.* **3**, 289 (1984).

Quinophthalone

Bright greenish yellow. Insol in water. Slightly sol in ethanol (yellow), linseed oil, mineral oil, oleic acid, paraffin wax, stearic acid, turpentine. Sol in acetone, chloroform, benzene, toluene. Yellow brown in conc H_2SO_4, yellow flocculent ppt in dilution.

USE: In spirit lacquers, polystyrenes, polycarbonates, polyamides, and acrylic resins. In colored smokes. Occasionally in hydrocarbon solvents. Approved by the FDA for use in externally applied drugs and cosmetics: *Code of Federal Regulations*, Title 21, Parts 1-99, Secs. 74.1711, 74.3106, pp 307, 315 (1985).

8258. Quinolinic Acid. *2,3-Pyridinedicarboxylic acid.* $C_7H_5NO_4$; mol wt 167.12. C 50.31%, H 3.02%, N 8.38%, O 38.29%. Metabolite of tryptophan, *q.v.* Prepn: S. Hoogewerff, W. A. van Dorp, *Ber.* **12**, 747 (1879); W. Koenigs, *ibid.* 983; O. Fischer, E. Renouf, *ibid.* **17**, 755 (1884); E. Sucharda, *ibid.* **58B**, 1727 (1925); V. Yu. Stiks, S. A. Bulgach, *ibid.* **65B**, 11 (1932); A. F. Lindenstruth, C. A. VanderWeif, *J. Am. Chem. Soc.* **71**, 3020 (1949). Metabolism studies: L. M. Henderson *et al., J. Biol. Chem.* **181**, 667, 677, 687, 731 (1949); H. P. Sarett, *ibid.* **193**, 627 (1951). Neuroexcitatory activity and possible role in neurodegenerative disorders: R. Schwarcz *et al., Science* **219**, 316 (1983).

Odorless crystals. mp 190° when rapidly heated, with decompn into CO_2 and nicotinic acid. Sol in 180 parts water, in alkalies, slightly in alcohol. Almost insol in ether or benzene.

8259. Quinone. *2,5-Cyclohexadiene-1,4-dione;* p-quinone; 1,4-benzoquinone; 1,4-cyclohexadienedione. $C_6H_4O_2$; mol wt 108.10. C 66.67%, H 3.73%, O 29.60%. Made by oxidation of aniline with sodium dichromate in presence of sulfuric acid. Laboratory prepn from hydroquinone: Vliet, *Org. Syn.* **2**, 85 (1922); also *Org. Syn.* **coll. vol. I** (2nd ed., 1941). *Cf.* Underwood, Walsh, *ibid.* **16**, 73 (1936). Toxicity study: Woodard *et al., Fed. Proc.* **8**, 348 (1949).

Yellow monoclinic prisms from water or petr ether. Penetrating odor resembling that of chlorine. Irritating vapors. d_4^{20} 1.318. mp 115.7°. Sublimes [sublimation velocities *in vacuo*: Kempf, *J. Prakt. Chem.* [2] **78**, 236 (1908)]. Volatile with steam. Absorption spectrum: Hantzsch, *Ber.* **49**, 522 (1916). Dipole moment: 0.67. Polemic over correct values: Paoloni, *J. Am. Chem. Soc.* **80**, 3879 (1958). Slightly sol in water; sol in alcohol, ether, hot petr ether, alkalies. LD_{50} orally in rats: 130 mg/kg (Woodard).

Caution: Potential symptoms of overexposure are eye irritation, conjunctivitis; keratitis; skin irritation. *See NIOSH*

Pocket Guide to Chemical Hazards (DHHS/NIOSH 90-117, 1990) p 192.

USE: Oxidizing agent; in photography; manuf dyes; manuf hydroquinone; tanning hides; making gelatin insol; strengthening animal fibers; as reagent.

8260. Quinovic Acid. *(3β)-3-Hydroxyurs-12-ene-27,28-dioic acid;* quinovaic acid; chinovic acid; chinova acid. $C_{30}H_{46}O_5$; mol wt 486.69. C 74.04%, H 9.53%, O 16.44%. Isoln from cinchona bark as the glycoside quinovin: Hlasiwetz, *Ann.* **79**, 129 (1851); **111**, 182 (1859); Votocek, Rác, *Coll. Czech. Chem. Commun.* **1**, 234 (1929); Tschesche *et al., Ann.* **667**, 151 (1963); from *Zygophyllum coccineum* L., *Zygophyllaceae:* Soliman, *J. Chem. Soc.* **1939**, 1760; from bark of *Mitragyna inermis* Kuntze and leaves of *M. ciliata* Aubrev & Pellegr. and *M. rubrostipulacea* Havil., *Rubiaceae:* Badger *et al., ibid.* **1950**, 867. Structure: Brossi *et al., Helv. Chim. Acta* **34**, 244 (1951); Barton, de Mayo, *J. Chem. Soc.* **1953**, 3111. *Review:* J. Simonsen, W. C. J. Ross, *The Terpenes* vol. 5 (University Press, Cambridge, 1957) pp 75-113.

Prisms from dil pyridine, decomp 297°. Very bitter taste. $[\alpha]_D^{20}$ +99° (c = 2.48 in pyridine). Practically insol in water and other solvents except pyridine.

O-Acetylquinovic acid, $C_{32}H_{48}O_6$, needles from dilute acetone, mp 282-284°.

Dimethyl ester, $C_{32}H_{50}O_5$, needles from dil acetone, mp 175°. $[\alpha]_D^{19}$ +117.4° (c = 2.2 in $CHCl_3$). Sol in organic solvents.

8261. Quinovin. Quinova-bitter; chinovin. Isoln from cinchona bark *(cortex chinae):* Hlasiwetz, *Ann.* **111**, 182 (1859); Liebermann, Giesel, *Ber.* **16**, 926 (1883). A mixture of three glycosides, 60% A, 5% B and 30% C: Tschesche *et al., Ann.* **667**, 151 (1963).

Glycoside A, $C_{36}H_{56}O_9$, *quinovic acid β-D-quinovoside.* Crystals from methanol + water, mp 237-238°. $[\alpha]_D^{21}$ +57° ± 2° (ethanol).

Glycoside B, $C_{36}H_{56}O_9$, *cincholic acid β-D-quinovoside.* Crystals from methanol + water, mp 193-195°. $[\alpha]_D^{19}$ +78° ± 3° (c = 0.78 in ethanol).

Glycoside C, $C_{36}H_{56}O_{10}$, *quinovic acid β-D-glucoside.* Crystals from methanol + water, mp 247-250°. $[\alpha]_D^{20}$ +62° ±2° (ethanol).

8262. Quinovose. *6-Deoxy-D-glucose;* D-glucomethylose; D-isorhamnose; D-epirhamnose; isorhodeose; epifucose; chinovose. $C_6H_{12}O_5$; mol wt 164.16. C 43.90%, H 7.37%, O 48.73%. Isoln from cinchona bark: Freudenerg, *Ber.* **62**, 373 (1929). Identity with D-glucomethylose: Votocek, Rác, *Coll. Czech. Chem. Commun.* **1**, 239 (1929). Structure: Karrer, Boettcher, *Helv. Chim. Acta* **36**, 570 (1953). Synthesis from 6-*O*-p-tolylsulfonyl-D-glucose: Schmidt, "6-Deoxy-α-D-glucose" in *Methods in Carbohydrate Chemistry* vol. I, R. L. Whistler, M. L. Wolfrom, Eds. (Academic Press, New York, 1962) pp 198-201. Alternate facile synthesis: V. K. Srivastava, L. M. Lerner, *Carbohydr. Res.* **64**, 263 (1978).

Crystals from ethyl acetate, mp 146°. $[\alpha]_D^{20}$ +73° (5 min) → +30° (3 hr, final) (c = 8.3 in water). Sol in water, ethanol. Practically insol in ether, acetone.

8263. Quinoxaline. 1,4-Benzodiazine; benzo[a]pyrazine; benzoparadiazine; phenpiazine. $C_8H_6N_2$; mol wt 130.15. C 73.83%, H 4.65%, N 21.52%. Prepd from o-phenylenediamine and glyoxal or glyoxal sodium bisulfite: Hinsberg, *Ber.* **17**, 320 (1884); *Ann.* **237**, 334 (1887); Jones, McLaughlin, *Org. Syn.* **30**, 86 (1950).

Crystals, mp 29-30°. Odor of quinoline when cool, odor of piperidine when hot. d_4^{48} (liq) 1.1334. $bp_{760.3}$ 229.5°. bp_{12} 108-111°. n_D^{48} 1.6231. Very freely sol in water, alc, ether, benzene.
Monohydrate, crystals, mp 37°.
Sulfate, $C_8H_6N_2.H_2SO_4$, leaflets, mp 186-187°. Freely sol in water. Less sol in alcohol.

8264. Quintozene. *Pentachloronitrobenzene;* PCNB; terrachlor; PKhNB; Botrilex; Brassicol; Folosan; Pentagen; Terraclor; Tilcarex. $C_6Cl_5NO_2$; mol wt 295.34. C 24.40%, Cl 60.02%, N 4.74%, O 10.83%. Prepd by treating pentachlorobenzene with fuming nitric acid: Jungfleisch, *Ann. Chim.* [4] **15**, 286 (1868); A. Roedig, K. Kiepert, *Ann.* **593**, 71 (1955). Persistence of residues in foods: P. B. Baker, B. Flaherty, *Analyst* **97**, 378 (1972); A. R. P. Paxton, D. Purser, *Pestic. Sci.* **13**, 401 (1982); in soil: J. Beck, K. E. Hansen, *ibid.* **5**, 41 (1974). Metabolism by sheep and goats: P. W. Aschbacher, V. J. Feil, *J. Agr. Food Chem.* **31**, 1150 (1983). Toxicity data: J. K. Finnegan *et al.*, *Arch. Int. Pharmacodyn. Ther.* **114**, 38 (1958). Review: *WHO Environmental Health Criteria* **41**, 1-38 (1984). Comparative review of the effects on soil organisms: E. R. Ingham, *Crop. Prot.* **4**, 3-32 (1985).

Fine needles from alcohol, platelets from carbon disulfide. d_4^{25} 1.718. mp 144°. bp_{760} 328° (some dec). Practically insol in water, cold alcohol. Freely sol in carbon disulfide, benzene, chloroform. LD_{50} in male, female rats (g/kg): 1.71 ±0.20, 1.65 ±0.17 by gavage (Finnegan).
USE: Fungicide for seed and soil treatment.

8265. Quinuclidine. *1-Azabicyclo[2.2.2]octane;* 1,4-ethylenepiperidine. $C_7H_{13}N$; mol wt 111.19. C 75.62%, H 11.78%, N 12.60%. Prepn: Löffler, Stietzel, *Ber.* **42**, 124 (1909); Meisenheimer *et al.*, *Ann.* **420**, 191 (1920); Clemo, Metcalfe, *J. Chem. Soc.* **1937**, 1989; Prelog, U.S. pat. **2,192,-840** (1940); Wawzonek *et al.*, *J. Am. Chem. Soc.* **73**, 2806 (1951); Leonard, Elkin, *J. Org. Chem.* **27**, 4635 (1962).

Prisms from petr ether. Sublimes. mp 156° (sealed tube). Very sol in water and the usual organic solvents.
Hydrochloride, $C_7H_{13}N.HCl$, crystals from abs ethanol, mp 364-365°.

8266. 3-Quinuclidinol. *1-Azabicyclo[2.2.2]octan-3-ol;* 3-hydroxyquinuclidine. $C_7H_{13}NO$; mol wt 127.19. C 66.11%, H 10.30%, N 11.01%, O 12.58%. Prepn and resolution of dl-forms: Sternbach, Kaiser, *J. Am. Chem. Soc.* **74**, 2215 (1952); Sternbach, U.S. pat. **2,648,667** (1953 to Hoffmann-La Roche); of free alcohol, acetate and/or benzoate esters: Grob *et al.*, *Helv. Chim. Acta* **40**, 2170 (1957);

Mikhalina, Rubtsov, *Zh. Obshch. Khim.* **30**, 163 (1960), *C.A.* **54**, 22632h (1960). Structure activity of esters: M. D. Mashkovsky in *Proc. 1st Int. Pharmacol. Mtg. Stockholm, 1961* vol. 7 (Macmillan, NewYork, 1963) pp 359-366.

dl-Form, crystals from benzene or acetone, mp 221-223° also reported as 225-227°. Sublimes at 120° and 20 mm Hg. Very sol in water.
dl-Form hydrochloride, prisms from methanol + acetone, mp >300°.
dl-Form acetate (ester), $C_9H_{15}NO_2$, *aceclidine, 3-acetoxyquinuclidine, 3-quinuclidinyl acetate.* Oily liq, $bp_{0.4}$ 73-74°, bp_{11} 113-115°. n_D^{25} 1.4675.
dl-Form acetate hydrochloride, $C_9H_{15}NO_2.HCl$, *Glaucostat, Glaucotat, Glaudin.* Crystals, mp 166°.
dl-Form benzilate (ester), $C_{21}H_{23}NO_3$, *3-quinuclidinyl benzilate, BZ, QNB, Ro-2-3308.* Prepn: Sternbach, Kaiser, *loc. cit.*, 2219. Crystals from acetone-ether, mp 164-165°.
dl-Form benzoate (ester), $C_{14}H_{17}NO_2$, *3-benzoyloxyquinuclidine, 3-quinuclidinyl benzoate.* Oily liquid, $bp_{0.3}$ 148-150°.
dl-Form benzoate hydrochloride, $C_{14}H_{17}NO_2.HCl$, *oksilidin, oxylidine.* Crystals, mp 238-240°.
l-Form, prisms from acetone, mp 220-222°. $[\alpha]_D^{25}$ −2.0° (c = 6.5), −43.8° (c = 3 in 1N HCl).
USE: Benzilate (ester) has been used as an incapacitating agent in chemical warfare: *Health Aspects of Chemical and Biological Weapons* (WHO, Geneva, 1970) pp 49-51.
THERAP CAT: Hypotensive; acetate (ester) as cholinergic.

8267. Quinupramine. *5-(1-Azabicyclo[2.2.2]oct-3-yl)-10,11-dihydro-5H-dibenz[b,f]azepine;* 10,11-dihydro-5-(3-quinuclidinyl)-5H-dibenz[b,f]azepine; LM-208; Kinupril; Kevopril. $C_{21}H_{24}N_2$; mol wt 304.44. C 82.85%, H 7.95%, N 9.20%. Analog of imipramine, q.v. Prepn: C. Gueremy, P. C. Wirth, Ger. pat. **2,030,492** (1971 to Sogeras), *C.A.* **74**, 141581e (1971). Animal studies: W. Van Dorsser, A. Dresse, *Arch. Int. Pharmacodyn. Ther.* **208**, 373 (1974); eidem, *ibid.* **220**, 164 (1976). Clinical study: R. Volmat *et al.*, *Clin. Neurol. Psychiat.* **239**, 445 (1978).

Crystals, mp 150°.
THERAP CAT: Antidepressant.

8268. Quisqualic Acid. *(S)-α-Amino-3,5-dioxo-1,2,4-oxadiazolidine-2-propanoic acid;* L-quisqualic acid; β-(3,5-dioxo-1,2,4-oxadiazolidin-2-yl)-L-alanine. $C_5H_7N_3O_5$; mol wt 189.13. C 31.75%, H 3.73%, N 22.22%, O 42.30%. Excitatory amino acid (EAA) used to identify a specific subset of EAA receptors; consequently, the receptors are known as quisqualate receptors. See also NMDA, kainic acid. Isoln from the seeds of *Quisqualis chinesis* and anthelmintic activity: Y.-C. Tuan *et al.*, *Yao Hsueh Hsueh Pao* **5**, 87 (1957), *C.A.* **56**, 14896b (1958); from *Q. indica*: S.-T. Fang, J.-H. Chu, *Hua Hsueh Hsueh Pao* **30**, 226 (1964), *C.A.* **61**, 7359f (1962); from *Q. fructus*: T. Takemoto *et al.*, *Yakugaku Zasshi* **95**, 176 (1975), *C.A.* **82**, 152211a (1975). Enzymic synthesis: I. Murakoshi *et al.*, *Chem. Pharm. Bull.* **22**, 473 (1974). Total synthesis: J. E. Baldwin *et al.*, *Chem. Commun.* **1985**, 256. Crystal structure: J. L. Flippen, R. D. Gilardi, *Acta Crystallogr.* **B32**, 951 (1976). Identification as a neuroexcitant: H. Shinozaki, I. Shibuya, *Neuropharmacology* **13**, 665 (1974). Receptor binding studies: K. Koshiya, *Life Sci.* **37**, 1373 (1985); J. T. Greenamyre, *J. Pharmacol. Exp. Ther.* **233**, 254 (1985). Review of isolation of quisqualic acid and other EAAs: T. Takemoto, in *Kainic Acid as a Tool in Neurobiology*, R. G. McGeer *et al.*, Eds. (Raven Press, New York, 1978) pp 1-15.

Crystals from water-ethanol, mp 190-191°. $[\alpha]_D^{20}$ +17.0° (c = 2.0 in 6*M* HCl).

8269. Quizalofop-Ethyl. *2-[4-[(6-Chloro-2-quinoxalinyl)oxy]phenoxy]propanoic acid ethyl ester;* ethyl 2-[4-(6-chloro-2-quinoxalinyloxy)phenoxy]propionate; quinofopethyl; DPX-Y6202; NCI-96683; NC-302; Assure; Targa; Pilot. $C_{19}H_{17}ClN_2O_4$; mol wt 372.81. C 61.21%, H 4.60%, Cl 9.51%, N 7.51%, O 17.17%. Post-emergence herbicide for control of grassy weeds in broad-leaved crops. Prepn and herbicidal activity: Y. Ura *et al.*, **Ger. pat. 3,004,770** (1980 to Nissan), *C.A.* **94**, 103421h (1981). Prepn and structure-activity relationships: G. Sakata *et al.*, *J. Pestic. Sci.* **10**, 61 (1985). Pharmacological effects in laboratory animals: K. Inokuchi *et al.*, *Oyo Yakuri* **30**, 509 (1985), *C.A.* **104**, 30101s (1986). Site of action studies: T. Ikai *et al.*, *Proc. Brit. Crop Prot. Conf.-Weeds* **1985**, 163. Account of properties, toxicity, crop tolerance: G. Sakata *et al.*, *Proc. 10th Conf. Int. Congr. Plant Prot.* **1**, 315 (1983).

White crystals, mp 92-93°. $bp_{0.2}$ 220°. Practically insol in water (0.3 × 10⁻⁶ g/ml at 20°). Soly in acetone, ethanol, benzene, xylene at 20° (g/ml): 0.11, 0.009, 0.29, 0.12. LD_{50} in male, female rats, male, female mice (mg/kg): 1670, 1480, 2350, 2360 orally; all 10,000 dermally. LC_{50} (96 hr) in rainbow trout: 10.7 mg/l (Sakata).

USE: Herbicide.

8270. Quorn®. Myco-protein. High protein, meat substitute derived from the biomass of the filamentous fungus, *Fusarium graminearum* (Schwabe). Description of production: J. Edelman *et al.*, *Nutr. Abstr. Rev.* **53**, 471 (1983). *See also:* R. A. Marsh, **U.S. pat. 4,555,485** (1985 to Ranks Hovis McDougall). Nutritional profile: M. Sadler, *Nutr. Food Sci.* **112**, 9 (1988). Metabolic study in comparison with meat: W. H. Turnbull *et al.*, *Am. J. Clin. Nutr.* **52**, 646 (1990). Effect on blood lipids: *eidem, ibid.* **55**, 415 (1992). *Review:* A. P. J. Trinci, *Mycol. Res.* **96**, 1-13 (1992).

Very mild, wheaty taste; slight aroma of mushrooms. Fibrous texture resembling that of meat.

USE: Meat substitute.

R

8271. R-11. *1,5a,6,9,9a,9b-Hexahydro-4a(4H)-dibenzofurancarboxaldehyde;* 2,3,4,5-bis(2-butenylene)tetrahydrofurfural; MGK-11; MGK Repellent 11. $C_{13}H_{16}O_2$; mol wt 204.27. C 76.44%, H 7.90%, O 15.67%. Prepd by heating furfuraldehyde with butadiene and water under pressure: Hillyer, Nicewander, U.S. pat. **2,683,151** (1954 to Phillips Petroleum). Clinical efficacy vs. sand flies: F. P. Fossati, M. Maroli, *Trans. Roy. Soc. Trop. Med. Hyg.* **80**, 771 (1986).

Liquid, bp 307°. d_4^{20} 1.10. mp −80°. n_D^{20} 1.5254. Practically insol in water.
Oxime, mp 97.2°.
Dinitrophenylhydrazone, mp 152.8°.
USE: Insect repellent.

8272. Rabeprazole. *2-[[[4-(3-Methoxypropoxy)-3-methyl-2-pyridinyl]methyl]sulfinyl]-1H-benzimidazole;* pariprazole. $C_{18}H_{21}N_3O_3S$; mol wt 359.45. C 60.15%, H 5.89%, N 11.69%, O 13.35%, S 8.92%. Partially reversible gastric proton pump inhibitor. Prepn: S. Souda *et al.*, **Eur. pat. Appl. 268,956;** *eidem*, U.S. pat. **5,045,552** (1988, 1991 both to Eisai). Pharmacology: H. Goto *et al.*, *Arzneimittel-Forsch.* **41**, 635 (1991). Mode of action: M. Morii, N. Takeguchi, *J. Biol. Chem.* **268**, 21553 (1993). HPLC determn in plasma: H. Nakai *et al.*, *J. Chromatog. B* **660**, 211 (1994). Clinical pharmacokinetics: S. Yasuda *et al.*, *Int. J. Clin. Pharmacol. Ther.* **32**, 466 (1994).

White crystals from CH_2Cl_2/ether, mp 99-100° (dec).
Sodium salt, $C_{18}H_{20}N_3NaO_3S$, *E-3810*. White crystals from ether, mp 140-141° (dec).
THERAP CAT: Antiulcerative.

8273. Racefemine. *(±)-α-Methyl-N-(1-methyl-2-phenoxyethyl)benzeneethanamine;* dl-threo-α-methyl-N-(1-methyl-2-phenoxyethyl)phenethylamine; *dl-threo-α-methyl-N-(1-phenoxy-2-propyl)phenethylamine;* dl-threo-N-(1-methyl-2-phenylethyl)-N-(2-phenoxy-1-methylethyl)amine; CB-3697. $C_{18}H_{23}NO$; mol wt 269.39. C 80.26%, H 8.61%, N 5.20%, O 5.94%. Smooth muscle relaxant. Prepn from amphetamine and phenoxyacetone with isomer separation: **Neth. pat. Appl. 6,407,309** (1964 to Clin-Byla), *C.A.* **63**, 353c (1965).

Isomers: From *dl*-amphetamine, a yellow liq, $bp_{0.05}$ 132-135°, was obtained which was separated into isomers I and II. *I-Fumarate:* mp 162°. *I-Hydrochloride:* mp 156-157°. *II-Hydrochloride:* mp 167-168°. From *d*-amphetamine, isomers III, $[\alpha]_D^{21}$ +41° (c = 0.01 in ethanol), and IV, $[\alpha]_D^{22}$ +22° (c = 0.005 in ethanol), were obtained. *III-Fumarate:* mp 164-165°, $[\alpha]_D^{20}$ +19° (c = 0.01 in ethanol). *III-Hydrochloride:* mp 186-187°, $[\alpha]_D^{21}$ +22° (c = 0.01 in ethanol). *IV-Hydrochloride:* mp 160-163°, $[\alpha]_D^{20}$ +22° (c = 0.01 in ethanol). From *l*-amphetamine, isomers V, $[\alpha]_D^{24}$ −24° (c = 0.01 in ethanol), and VI, $[\alpha]_D^{24}$ −41° (c = 0.01 in ethanol), were obtained. *V-Hydrochloride:* mp 189-190°, $[\alpha]_D^{21}$ −22° (c =

0.01 in ethanol). *VI-Fumarate:* mp 164-164.5°, $[\alpha]_D^{22}$ −15° (c = 0.01 in ethanol). *VI-Hydrochloride:* mp 186-187°, $[\alpha]_D^{21}$ −21.5° (c = 0.01 in ethanol).
Fumarate, $C_{22}H_{27}NO_5$, *Dysmalgine.*
THERAP CAT: Antispasmodic.

8274. Racemethorphan. *(±)-3-Methoxy-17-methylmorphinan;* dl-cis-1,3,4,9,10,10a-hexahydro-6-methoxy-11-methyl-2H-10,4a-iminoethanophenanthrene; *dl-cis-*1,2,3,9,-10,10a-hexahydro-6-methoxy-11-methyl-4H-10,4a-iminoethanophenanthrene; deoxydihydrothebacodine; methorphan. $C_{18}H_{25}NO$; mol wt 271.40. C 79.66%, H 9.28%, N 5.16%, O 5.90%. Prepn: Schnider, Grüssner, U.S. pat. **2,676,177** (1954 to Hoffmann-La Roche). Prepn of *d*-form: *eidem*, *ibid.*; Häfliger *et al.*, *Helv. Chim. Acta* **39**, 2053 (1956). Prepn of *l*-form hydrobromide: Corrodi *et al.*, *ibid.* **42**, 215 (1959). Crystal structure and absolute configuration of *d*-form hydrobromide: L. Gylbert, D. Carlstrom, *Acta Crystallogr. B,* 2833 (1977).

Hydrobromide, $C_{18}H_{25}NO.HBr$, *Ro-1-5470.* Crystals, mp 124-126°.
d-Form hydrobromide, **dextromethorphan hydrobromide, demorphan hydrobromide, Ro-1-5470/5, Benylin DM, Canfodion, Cosylan, Hihustan M.** Occurs as the monohydrate, crystals, mp 122-124°. $[\alpha]_D^{20}$ +27.6° (c = 1.5 in water). Approx soly in water: 1.5% at 25°, 5% at 50°, 25% at 85°. Soly (w/w): 25% in 95% ethanol at room temp, 10% in glycerol. Sol in propylene glycol, chloroform. Practically insol in ether. pH of a 1% aq soln: 5.2-6.5. Long range stability of aq solns obtained by adjusting pH within 4-5.6. Reacts with alkalies forming the free base which is practically insol in water.
l-Form hydrobromide, **levomethorphan hydrobromide, Ro-1-5470/6, Ro-1-7788.** Occurs as the dihydrate, crystals, mp 124-126°. $[\alpha]_D^{20}$ −26.3°.
Note: Racemethorphan and levomethorphan are controlled substances (opiates) listed in the U.S. Code of Federal Regulations, Title 21 Part 1308.12 (1995).
THERAP CAT: Antitussive.

8275. Ractopamine. *4-Hydroxy-α-[[[3-(4-hydroxyphenyl)-1-methylpropyl]amino]methyl]benzenemethanol;* 1-(4-hydroxyphenyl)-2-[1-methyl-3-(4-hydroxyphenyl)-propylamino]ethanol; *N-[2-(4-hydroxyphenyl)-2-hydroxyethyl]-1-methyl-3-(4-hydroxyphenyl)propylamine.* $C_{18}H_{23}NO_3$; mol wt 301.39. C 71.73%, H 7.69%, N 4.65%, O 15.93%. β-Adrenergic agonist; repartitioning agent. Prepn (stereochemistry unspecified): J. van Dijk, H. D. Moed, *Rec. Trav. Chim.* **92**, 1281 (1973). Prepn of *R,R*-isomer: J. Mills *et al.*, **Eur. pat. Appl. 7,205** (1980 to Lilly). Prepd (not claimed): D. B. Anderson *et al.*, U.S. pat. **4,690,951** (1987 to Lilly). LC determn in animal feeds: M. P. Turberg *et al.*, *J. AOAC Int.* **77**, 840 (1994). Metabolism and tissue residue studies: J. E. Dalidowicz *et al.*, in *ACS Symposium Series* **503**, entitled "Xenobiotics and Food Producing Animals," D. H. Hutson *et al.*, Eds. (ACS, Washington, DC, 1992) pp 234-243. Effect on growth, carcass characteristics and meat quality in swine: B. E. Uttaro *et al.*, *J. Anim. Sci.* **71**, 2439 (1993). Effect on β-receptor affinity and density in pigs: M. E. Spurlock *et al.*, *ibid.* **72**, 75 (1994).

Consult the Name Index before using this section.

Hydrochloride, $C_{18}H_{23}NO_3$, *EL-737, LY-031537, Paylean.* Mixture of 4 stereoisomers in approx equal proportions. Product containing 51% *RR,SS-* and 49% *RS,SR-*diastereomers, mp 124-129°.

*R,R-*Form Hydrochloride, *butopamine hydrochloride.* Crystals from ethanol + diethyl ether, mp 176-176.5° (dec). $[\alpha]_D$ —22.7°; $[\alpha]_{365}$ —71.2° (c = 3.7 mg/ml in methanol).

THERAP CAT (VET): Animal growth promotant.

8276. Radicinin. *[2S-[2α,3β,7(E)]]-2,3-Dihydro-3-hydroxy-2-methyl-7-(1-propenyl)-4H,5H-pyrano[4,3-b]pyran-4,5-dione;* stemphylone. $C_{12}H_{12}O_5$; mol wt 236.22. C 61.01%, H 5.12%, O 33.86%. A mold metabolite from the plant pathogen *Stemphylium radicinum:* Clark, Nord, *Arch. Biochem. Biophys.* **45**, 469 (1953); **59**, 269 (1955). Structure and identity with stemphylone: Grove, *J. Chem. Soc.* **1964**, 3234. Absolute configuration: M. Nukina, S. Marumo, *Tetrahedron Letters* **1977**, 3271. Synthesis of *dl*-form: Kato *et al., Chem. Commun.* **1969**, 95.

Needles from ethanol, dec 238-240°. $[\alpha]_D^{27}$ —217.4° (c = 2.37 in pyridine); $[\alpha]_D^{27}$ —2° (c = 0.2 in ethanol); $[\alpha]_D^{27}$ —208° (c = 1.25 in chloroform). uv max: 343, 280, 270 nm (log ϵ 4.27, 3.62, 3.79). Soluble in alkali. Practically insol in dil acids, sodium bicarbonate, sodium carbonate.

Monoacetate, $C_{14}H_{14}O_6$, crystals from methanol, mp 197°. $[\alpha]_D^{27}$ —267° (c = 0.69 in pyridine).

8277. Radium. Ra; at. wt 226 (mass number of most stable isotope); at. no. 88; valence 2. A radioactive alkaline earth metal. Occurrence in earth's crust: approx 10^{-11}% by wt. Natural isotopes: 223, *actinium X;* 224, *thorium X;* 226; 228, *mesothorium I.* ^{226}Ra is a product of disintegration of uranium and is present in all ores contg uranium. Separated in the form of a salt by P. and M. S. Curie from the pitchblende of Joachimsthal, Bohemia: Curie *et al., Compt. Rend.* **127**, 1215 (1898). Isoln of the element by electrolysis of an aq soln of radium chloride: Curie, Debierne, *ibid.* **151**, 523 (1910). ^{228}Ra ($T_{1/2}$ 6.7 years); produced by disintegration of thorium (^{232}Th); discovered in 1907 by O. Hahn in monazite residues from isolating thorium. Zaire (Congo) is the main producer of radium, Canada next. Clinical evaluation in brachytherapy of tongue: J. Horiuchi *et al., Int. J. Rad. Oncol. Biol. Phys.* **8**, 82 (1982); M. Hoshina *et al., Brit. J. Radiol.* **62**, 59 (1989); in intracavitary radiation of uterus: D. A. Jones, R. Stout, *Clin. Radiol.* **37**, 169 (1986). Review of radiotherapy in cervical carcinoma: P. R. Reddi *et al., Obstet. Gynecol.* **43**, 238-247 (1974); in rectal carcinoma: *Dis. Colon Rectum* **29**, 600-614 (1986), reprint of C. Gordon-Watson, *Brit. J. Surg.* **17**, 649-669 (1930). Comprehensive reviews: K. W. Bagnall, *Chemistry of the Rare Radioelements* (New York, Academic Press, 1957); Goodenough, Stenger, "Magnesium, Calcium, Strontium, Barium, and Radium" in *Comprehensive Inorganic Chemistry* vol. 1, J. C. Bailar, Jr. *et al.,* Eds. (Pergamon Press, Oxford, 1973) pp 591-664.

Brilliant, white metal; body-centered cubic structure; blackens on exposure to air. mp 700°; bp 1737°; d 5.5. $T_{1/2}$ 1600 yrs (^{226}Ra). One gram of radium evolves about 1000 kcal per year. Undergoes spontaneous disintegration with formation of radon. One gram of radium produces about 0.0001 ml of radon per day at normal temp and pressure. Radium is produced and used in the form of its salts: the chloride, bromide, carbonate, sulfate. Its compds closely resemble those of barium; the element itself is more volatile than barium. Radium salts impart a carmine-red color to a flame.

Bromide, Br_2Ra. White or slightly brownish crystals. d 5.79. mp 728°. Sublimes at 900°. Sol in water. The salt of commerce is usually a mixture with barium bromide.

Chloride, Cl_2Ra. White or slightly brownish crystals. d 4.91. mp 1000°. Sol in water. The salt of commerce is usually a mixture with barium chloride.

Caution: Inhalation, ingestion or body exposure may result in lung cancer, osteogenic sarcoma, osteitis, blood dyscrasias, skin injury.

USE: In physical research. In radiography of metals because the penetration of gamma rays is more pronounced than that of x-rays. As source of radon. No longer used to make luminous paints.

THERAP CAT: Antineoplastic (radiation source).

8278. Radon. Rn; at. no. 86. Obsolete synonyms: *niton,* symbol Nt, and *radium emanation* or *emanation,* symbol Em. Group VIIIA(18), also known as Group 0. A noble gas characterized by an electronic structure in which the outer *p* subshell is entirely filled. No stable nuclides; naturally occurring isotope (mass number): 222 (longest-lived known isotope, $T_{1/2}$ 3.825 days, rel. at. mass 222.0176, α-emitter, member of decay chain of ^{238}U). Known artificial radioactive isotopes: 200-221, 223-226. Isotope ^{220}Rn ($T_{1/2}$ 55.6 sec, α-emitter, formed in decay chain of ^{232}Th), unofficial name *Thoron,* symbol Tn, was discovered by Owens and Rutherford in 1900. ^{222}Rn was discovered by Dorn in 1900 and isolated by Rutherford and Sody in 1902. Isotope ^{219}Rn ($T_{1/2}$ 3.96 sec, α-emitter, formed in decay chain of ^{235}U), unofficial name *actinon,* symbol An, was discovered by Debierne and Giesel in 1902. Concentration in air: 6×10^{-14} ppm by vol; in igneous rock of the earth's crust: 1.7×10^{-10} ppm by wt. Prepn of fluoride: Fields *et al., J. Am. Chem. Soc.* **84**, 4164 (1962); Stein, *Science* **168**, 362 (1970). Prepn of gas from radium salt: Jennings, Russ, *Radon: Its Technique and Use* (Murray, London, 1948) pp 79-95. Reviews of chemistry and compds: *Argon, Helium and the Rare Gases,* G. A. Cook, Ed. (Interscience, New York, 1961); Haissinsky, Adloff, *Radiochemical Survey of the Elements* (Elsevier, New York, 1965) pp 126-128; Cockett, Smith, "The Monatomic Gases" in *Comprehensive Inorganic Chemistry* vol. 1, J. C. Bailar, Jr. *et al.,* Eds. (Pergamon Press, Oxford, 1973) pp 139-211; Bartlett, Sladky, *ibid.,* pp 213-330; S.-C. Hwang, W. R. Weltmer, Jr. in Kirk Othmer *Encyclopedia of Chemical Technology* vol. 13 (John Wiley & Sons, 4th ed., 1995) pp 1-38; G. J. Schrobilgen, J. M. Whalen, *ibid.* pp 38-53; *Chemistry of the Elements,* N. N. Greenwood, A. Earnshaw, Eds. (Pergamon Press, New York, 1984) pp 1042-1059. Reviews of sources and health effects of indoor radon exposure: F. P. Radford, *Environ. Health Perspect.* **62**, 281-287 (1985); D. L. Henshaw, *Contemp. Phys.* **34**, 31-48 (1993).

Colorless, odorless, tasteless, inert, monatomic gas; will form compds with higly electronegative elemnts such as O, F, Cl. Soly of gas in water (20°) 230 cm³/l water. Soluble in organic solvents. Strongly adsorbed on various surfaces. Heat capacity at 25° and 1 atm: 4.9860 cal/g-atom/°K. Triple point temp 202°K, press 70 kPa. Critical temp 378°K, critical press 6280 kPa. Gas: d^0 (101.3 kPa) 9.73 kg/m³. Liquid: normal bp 211°K, d (normal bp) 4400 kg/m³, heat of vaporization (normal bp) 18,100 J/mol. Heat of fusion (triple point) 3247 J/mol. Solid form exists as face-centered cubic crystals at normal pressure.

Caution: Toxicity due to ionizing radiation (α-emitter). Increased incidences of lung cancer have been reported due to occupational exposure to high doses of radon, especially in hard rock miners. Max permissible concn of ^{222}Rn in air: 10^{-8} μ-Curie/cc: *National Bureau of Standards Handbook* **69**, 79 (1959). The U.S. EPA has established an action level of 4 pCi/l (0.02 working levels) for home indoor radon and radon decay product measurements. *See: Interim Protocols for Screening and Follow-up Radon and Radon Decay Product Measurements* (PB89-224265, EPA-520/1-86-014-1, 1987) 22 p. Radon and its isotopic forms, ^{222}Rn and ^{220}Rn, have been listed as known carcinogens: *Seventh Annual Report on Carcinogens* (PB95-109781, 1994) p 65.

USE: To initiate and influence chemical reactions, as a surface label in the study of surface reactions; in the determination of radium or thorium; in the study of the behavior of filters; in combination with Be or other light materials as a source of neutrons.

THERAP CAT: Antineoplastic (radiation source).

8279. Raffinose. *β-D-Fructofuranosyl-O-α-D-galactopyranosyl-(1→6)-α-D-glucopyranoside;* gossypose; melitose; melitriose. $C_{18}H_{32}O_{16}$; mol wt 504.44. C 42.86%, H 6.39%,

O 50.75%. A trisaccharide built from 1 mol each of D-galactose, D-glucose, and D-fructose which are obtained from it by acid hydrolysis. Invertase splits it into melibiose and saccharose. Occurs in Australian manna (from *Eucalyptus* spp, *Myrtaceae);* in cottonseed meal. Prepn: C. S. Hudson, T. S. Harding, *J. Am. Chem. Soc.* **36**, 2110 (1914); E. P. Clark, *ibid.* **44**, 210 (1922); Harding, *Sugar* **25**, 82 (1923); Hungerford, Nees, *Ind. Eng. Chem.* **26**, 462 (1934). Configuration: Haworth *et al., J. Chem. Soc.* **1923**, 3125; Charlton *et al.;* Haworth *et al., ibid.* **1927**, 1527, 3146. Structure: Hassid, Ballou in W. Pigman, *The Carbohydrates* (Academic Press, New York, 1957) p 517. Synthesis: Suami *et al., Carbohyd. Res.* **26**, 234 (1973). Review: E. B. Rathbone, *Dev. Food Carbohyd.* **2**, 145-185 (1980).

Pentahydrate, crystals in clusters from dil alc. Indifferent taste. d 1.465. mp 80°. Loses water of crystn upon slow heating to 100°. The anhydrous form dec 118-119°. $[\alpha]_D^{20}$ +105.2° (c = 4). One gram dissolves in 7 ml water (soly table; Hungerford, Nees), in 10 ml methanol. Sol in pyridine, slightly sol in alc. Does not form an osazone and does not reduce Fehling's soln.

8280. Rafoxanide. *N-[3-Chloro-4-(4-chlorophenoxy)-phenyl]-2-hydroxy-3,5-diiodobenzamide; 3'-chloro-4'-(p-chlorophenoxy)-3,5-diiodosalicylanilide;* MK-990; Bovanide; Duofas; Flukanide; Ranide. $C_{19}H_{11}Cl_2I_2NO_3$; mol wt 626.02. C 36.45%, H 1.77%, Cl 11.33%, I 40.54%, N 2.24%, O 7.67%. Prepn: **Neth. pat. Appl. 6,815,783; Belg. pat. 724,668** (both 1969 to Merck & Co.); Mrozik *et al., Experientia* **25**, 883 (1969). Trials in sheep and cattle: Snijders, Horak, Louw, *J. S. Afr. Vet. Med. Assoc.* **43**, 397 (1972); **44**, 251 (1973); **46**, 265 (1975); Horak, Snijders, *Vet. Rec.* **94**, 12 (1974).

Crystals, mp 168-170°. Moderately sol in acetone and acetonitrile. Practically insol in water.

THERAP CAT (VET): Fasciolicide; anthelmintic.

8281. Raloxifene. *[6-Hydroxy-2-(4-hydroxyphenyl)-benzo[b]thien-3-yl][4-[2-(1-piperidinyl)ethoxy]phenyl]methanone;* keoxifene; LY-139481. $C_{28}H_{27}NO_4S$; mol wt 473.59. C 71.01%, H 5.75%, N 2.96%, O 13.51%, S 6.77%. Nonsteroidal estrogen receptor mixed agonist/antagonist. Prepn: C. D. Jones, **Eur. pat. Appl. 62,503** (1982 to Lilly); C. D. Jones *et al., J. Med. Chem.* **27**, 1057 (1984). Pharmacology: B. L. Neubauer *et al., Prostate* **23**, 245 (1993); R. Fuchs-Young *et al., Ann. N.Y. Acad. Sci.* **761**, 355 (1995). Effects on bones of ovariectomized rats: M. Sato *et al., J. Bone Min. Res.* **9**, 715 (1994). Disposition and metabolism in animals: T. D. Lindstrom *et al., Xenobiotica* **14**, 841 (1984). Clinical pharmacology: M. W. Draper *et al., Pharmacology* **50**, 209 (1995).

Crystals from acetone, mp 143-147°. uv max (ethanol): 290 nm (ϵ 34000).

Hydrochloride, $C_{28}H_{27}NO_4S.HCl$, *LY-156758.* Crystals from methanol/water, mp 258°. uv max (ethanol): 286 nm (ϵ 32800).

THERAP CAT: Antiosteoporotic.

8282. Ramifenazone. *1,2-Dihydro-1,5-dimethyl-4-[(1-methylethyl)amino]-2-phenyl-3H-pyrazol-3-one;* 4-isopropyl-amino-2,3-dimethyl-1-phenyl-3-pyrazolin-5-one; 4-isopropylaminoantipyrine; isopropylaminophenazone; isopyrin. $C_{14}H_{19}N_3O$; mol wt 245.32. C 68.54%, H 7.81%, N 17.13%, O 6.52%. Prepn: E. Skita, W. Stühmer, **Ger. pats. 930,328; 932,677** (both 1955), *C.A.* **52**, 16372i, 20200h (1958). Acute toxicity: E. Tubaro *et al., Arzneimittel-Forsch.* **20**, 1024 (1970). Clinical evaluation of combination with phenylbutazone, *q.v.,* in migraine: W. Sacks, *S. Afr. Med. J.* **58**, 444 (1980).

Crystals from acetone + glacial acetic acid, mp 80°. LD_{50} in mice (mg/kg): 843 i.p.; 1070 orally (Tubaro).

Hydrochloride monohydrate, $C_{14}H_{19}N_3O.HCl.H_2O$.

Mixture with phenylbutazone, *Tomanol.*

THERAP CAT: Analgesic, antipyretic, anti-inflammatory.

8283. Ramipril. *[2S-[1[R*(R*)],2α,3aβ,6aβ]]-1-[2-[[1-(Ethoxycarbonyl)-3-phenylpropyl]amino]-1-oxopropyl]octahy-drocyclopenta[b]pyrrole-2-carboxylic acid;* N-(1S-carboeth-oxy-3-phenylpropyl)-S-alanyl-cis,endo-2-azabicyclo[3.3.0]-octane-3S-carboxylic acid; (2S,3aS,6aS)-1-[(S)-N-[(S)-1-carboxy-3-phenylpropyl]alanyl]octahydrocyclopenta[b]pyr-role-2-carboxylic acid 1-ethyl ester; HOE-498; Altace; Cardace; Delix; Pramace; Quark; Ramace; Triatec; Tritace; Unipril; Vesdil. $C_{23}H_{32}N_2O_5$; mol wt 416.52. C 66.32%, H 7.74%, N 6.73%, O 19.21%. Angiotensin converting enzyme (ACE) inhibitor. Prepn: V. Teetz *et al,* **Eur. pat. Appl. 79,022** (1983 to Hoechst A.G.), *C.A.* **100**, 52012h (1984); E. H. Gold *et al., U.S. pat. 4,587,258* (1986 to Schering Corp.); V. Teetz *et al., Arzneimittel-Forsch.* **34**, 1399 (1984). Converted to active, diacid metabolite, *ramiprilat.* General pharmacology: M. Omosu *et al., ibid.* **38**, 1309 (1988). Series of articles on pharmacology, pharmacokinetics and preliminary clinical studies: *ibid.* **34**, 1402-1454 (1984). Radioimmunoassay determn in human serum and plasma: H. G. Eckert *et al., ibid.* **35**, 1251 (1985). Toxicology: H. H. Donaubauer, D. Mayer, *ibid.* **38**, 14 (1988). Symposium on pharmacology and clinical efficacy: *Am. J. Cardiol.* **59**, 1D-177D (1987).

Felty needles from ether, mp 109°. $[\alpha]_D^{24}$ +33.2° (c = 1 in 0.1N ethanolic HCl). LD_{50} (14 day) in male, female mice, male, female rats (mg/kg): 1194, 1158, 687, 608 i.v.; 10933, 10048, > 10000, > 10000 orally (Donaubauer, Mayer).

THERAP CAT: Antihypertensive.

8284. Raney Nickel®. Raney nickel catalyst. Prepd by fusing 50 parts nickel with 50 parts aluminum: Raney, U.S. pats. **1,628,190** (1927); **1,915,473** (1933); pulverizing the alloy and dissolving out most of the aluminum with NaOH soln: Covert, Adkins, J. Am. Chem. Soc. **54**, 4116 (1932); Ruggli, Preiswerk, Helv. Chim. Acta **22**, 494 (1939); Mozingo, Org. Syn. **21**, 15 (1941). Absorption studies: Kokes, Emmett, J. Am. Chem. Soc. **83**, 29 (1961). The residual aluminum, which amounts to several percent, appears to be necessary for proper catalytic activity. Review: J. S. Pizey, Synthetic Reagents vol. **2** (John Wiley, New York, 1974) pp 175-311.

Grayish-black powder or cubic crystals. Ignites on contact with air. Contains hydrogen and has been attributed the formula Ni_2H. Generally stored under alcohol, although ether, water, methylcyclohexane, and dioxane may be used. Raney nickel loses its hydrogen slowly on storage and becomes inactive. Properly prepd and stored it should remain active for 6 months.

USE: Catalyst for the hydrogenation of organic compds with gaseous hydrogen. Usually from 1 to 10% of the substance to be reduced is employed. Is active at room temp and 1 atm pressure, but is also used at high temps and high pressures, see H. B. Adkins, Reactions of Hydrogen (Madison, 1937).

8285. Ranimustine. Methyl 6-[[[(2-chloroethyl)nitroso-amino]carbonyl]amino]-6-deoxy-α-D-glucopyranoside; methyl N-carbamyl-N'-(2-chloroethyl)-N'-nitroso-6-amino-6-deoxy-α-D-glucopyranoside; 6-[3-(2-chloroethyl)-3-nitrosoureido]-6-deoxy-α-D-glucopyranoside; 3-(methyl-α-D-glucopyranos-6-yl)-1-(2-chloroethyl)-1-nitrosourea; ranomustine; MCNU; NSC-0270516; Cymerin; Thymerin. $C_{10}H_{18}ClN_3O_7$; mol wt 327.72. C 36.65%, H 5.54%, Cl 10.82%, N 12.82%, O 34.17%. Chloroethylnitrosourea derivative with antitumor activity. Similar to carmustine, chlorozotocin, lomustine, nimustine, q.q.v. Prepn: Neth. pat. **7,507,973**; G. Kimura, J. Sekine, U.S. pat. **4,057,684**; Neth. pat. Appl. **7,800,920**; G. Kimura, U.S. pat. **4,156,777** (1976, 1977, 1978, 1979, all to Tokyo Tanabe). Antitumor activity in rodents, comparison with other nitrosoureas: S. Sekido et al., Cancer Treat. Rep. **63**, 961 (1979); S. Fujimoto, M. Ogawa, Cancer Chemother. Pharmacol. **9**, 134 (1982); S. Fujimoto et al., Gann **75**, 937 (1984). Mechanism of action: R. Kanamaru et al., Curr. Chemother. Immunother. **2**, 1377 (1982). Effect of sugar alcohols on toxicity in mice: T. Tashiro et al., Cancer Chemother. Pharmacol. **8**, 183 (1982). Series of articles on metabolic fate in rats: Y. Esumi et al., Iyakuhin Kenkyu **16**, 381-428 (1985), C.A. **103**, 115620q, 115621r, 134369f (1985). Clinical study in hematological malignant diseases: T. Masaoka et al., Chemotherapy (Tokyo) **33**, 271 (1985).

Pale yellowish needles from anhydrous ethanol-ethyl ether (1:1), mp 111-112°. $[\alpha]_D^{20}$ +93.2° (c = 0.5 in methanol) (Kimura). Also reported as 101-103° (dec) from isopropanol. $[\alpha]_D^{25}$ +73.2° (c = 0.3 in methanol) (Kimura, Sekine). Solubility in water: 900 mg/ml at 25°. LD_{50} in male rats (mg/kg): 42 i.p., 42 i.v., 50 orally (Kimura).

THERAP CAT: Antineoplastic.

8286. Ranitidine. N-[2-[[[-5-[(Dimethylamino)methyl]-2-furanyl]methyl]thio]ethyl]-N'-methyl-2-nitro-1,1-ethenedi-amine. $C_{13}H_{22}N_4O_3S$; mol wt 314.41. C 49.66%, H 7.05%, N 17.82%, O 15.27%, S 10.20%. Histamine H_2-receptor

antagonist which inhibits gastric acid secretion. Prepn: B. J. Price et al., Fr. pat. **2,384,765**; eidem, U.S. pat. **4,128,658** (both 1978 to Allen & Hanburys). HPLC determn in plasma: P. F. Carey, L. F. Martin, J. Liq. Chromatog. **1979**, 1291. Pharmacological studies: J. Bradshaw et al., Brit. J. Pharmacol. **66**, 464 (1979); M. J. Daly et al., Gut **21**, 408 (1980). Efficacy in treatment of duodenal ulcers: A. Berstad et al., Scand. J. Gastroenterol. **15**, 637 (1980); R. P. Walt et al., Gut **22**, 49 (1981). Review of pharmacology and therapeutic use: R. N. Brogden et al., Drugs **24**, 267-303 (1982). Comprehensive description: M. Hohnjec et al., Anal. Profiles Drug Subs. **15**, 533-561 (1986).

Solid, mp 69-70°.

Hydrochloride, $C_{13}H_{22}N_4O_3S.HCl$, AH-19065, Azantac, Melfax, Noctone, Raniben, Ranidil, Raniplex, Sostril, Taural, Terposen, Trigger, Ulcex, Ultidine, Zantac, Zantic. Off-white solid, mp 133-134°. Freely sol in acetic acid and water, sol in methanol, sparingly sol in ethanol. Practically insol in chloroform.

Bismuth citrate, $C_{13}H_{22}N_4O_3S.C_6H_5BiO_7$, ranitidine bismutrex, GR-122311X, Pylorid. Pharmacology and activity vs Helicobacter sp: R. Stables et al., Aliment. Pharmacol. Ther. **7**, 237 (1993).

THERAP CAT: Antiulcerative.

8287. Ranolazine. (±)-N-(2,6-Dimethylphenyl)-4-[2-hydroxy-3-(2-methoxyphenoxy)propyl]-1-piperazineacetamide; (±)-4-[2-hydroxy-3-(o-methoxyphenoxy)propyl]-1-piper-azineaceto-2',6'-xylidide; (±)-1-[3-(2-methoxyphenoxy)-2-hydroxypropyl]-4-[N-(2,6-dimethylphenyl)carbamoylmeth-yl]piperazine. $C_{24}H_{33}N_3O_4$; mol wt 427.54. C 67.42%, H 7.78%, N 9.83%, O 14.97%. Anti-ischemic agent which modulates myocardial metabolism. Prepn: A. F. Kluge et al., Eur. pat. Appl. **126,449**; eidem, U.S. pat. **4,567,264** (1984, 1986 both to Syntex). HPLC resolution of enantiomers: E. Delée et al., Chromatographia **24**, 357 (1987). Protective effect in myocardial ischemia: M. C. Allely, B. J. Alps, Brit. J. Pharmacol. **99**, 5 (1990). Clinical pharmacology: D. Jain et al., Eur. J. Clin. Pharmacol. **38**, 111 (1990). Clinical evaluation in angina: G. Cocco et al., J. Cardiovasc. Pharmacol. **20**, 131 (1992).

Dihydrochloride, $C_{24}H_{33}N_3O_4.2HCl$, RS-43285. White crystalline powder from methanol/ether, mp 164-166°. Readily sol in water.

THERAP CAT: Antianginal.

8288. Rapamycin. RAPA; RPM; sirolimus; AY-22989; NSC-226080; Rapamune. $C_{51}H_{79}NO_{13}$; mol wt 914.19. C 67.01%, H 8.71%, N 1.53%, O 22.75%. Triene macrolide antibiotic isolated from Streptomyces hygroscopicus. Name derived from the native word for Easter Island, Rapa Nui. Isoln: S. N. Sehgal et al., Ger. pat. **2,347,682**; eidem, U.S. pat. **3,929,992** (1974, 1975 both to Ayerst McKenna Harrison); purification and characterization: C. Vézina et al., J. Antibiot. **28**, 721 (1975); S. N. Sehgal et al., ibid. 727. Inhibition of immune response: R. R. Martel et al., Can. J. Physiol. Pharmacol. **55**, 48 (1977); of graft rejection in mice: C. P. Eng et al., Transplant. Proc. **23**, 868 (1991). HPLC determn: K. L. Napoli, B. D. Kahan, Clin. Chem. **37**, 294 (1991). Pharmacokinetics in rabbit: N. Honcharik et al., Ther. Drug Monit. **14**, 475 (1992). Potential utility as antiarthritic agent: R. P. Carlson et al., J. Pharmacol. Exp. Therap. **266**, 1125 (1993). Total synthesis:

K. C. Nicolaou *et al.*, *J. Am. Chem. Soc.* **115**, 4419 (1993); D. Romo *et al.*, *ibid.* 7906. Brief review of pharmacology: J. Y. Chang, S. N. Sehgal, *Brit. J. Rheumatol.* **30**, Suppl. 2, 62-65 (1991). Review of mechanism of action: N. H Sigal, F. J. Dumont, *Ann. Rev. Immunol.* **10**, 519-560 (1992).

Colorless crystalline solid from ether, mp 183-185°. uv max (95% ethanol): 267, 277, 288 nm ($E_{1cm}^{1\%}$ 417, 541, 416). $[\alpha]_D^{25}$ −58.2° (methanol). Sol in ether, chloroform, acetone, methanol and DMF; very sparingly sol in hexane and petr ether. Substantially insol in water. LD_{50} in mice (mg/kg): ∼600 i.p.; > 2,500 orally (Vézina).

USE: Tool for immunochemistry.

THERAP CAT: Immunosuppressant.

8289. Rapeseed Oil. Colza oil. Oil expressed from seeds of *Brassica campestris* L., *Cruciferae.*

Pale yellow, rather viscid liq. d 0.913-0.917; n_D^{20} 1.4720-1.4752. Solidif −2° to −10°. Sapon no. 170-177. Iodine no. 97-105. Sol in chloroform, ether, CS_2.

USE: Lubricant; manuf rubber substitutes, margarine, soft soaps, blown oils; oiling woolens.

8290. Raspberry. Fresh ripe fruit of varieties of *Rubus idaeus* L., or *R. stigosus* Michx., *Rosaceae. Habit.* Europe, Asia, cultivated in Canada and U.S. *Constit.* Sugar, malic and citric acids.

USE: Flavoring. Pharmaceutic aid (flavor).

8291. Raubasine. *(19α)-16,17-Didehydro-19-methyl-oxayohimban-16-carboxylic acid methyl ester;* δ-yohimbine; pytetrahydroserpentine; tetrahydroserpentine; ajmalicine; Circolene; Hydrosarpan; Isoarteril; Lamuran. $C_{21}H_{24}N_2O_3$; mol wt 352.43. C 71.57%, H 6.86%, N 7.95%, O 13.62%. α_1-Adrenergic blocker isolated from bark of *Corynanthe johimbe* K. Schum., *Rubiaceae:* H. Heinemann, *Ber.* **67**, 15 (1934); from roots of *Rauwolfia serpentina* (L.) Benth., *Apocynaceae:* S. Siddiqui, R. H. Siddiqui, *J. Indian Chem. Soc.* **8**, 667 (1931); A. H. Popelak *et al., Naturwiss.* **40**, 625 (1953); A. Hofmann, *Helv. Chim. Acta* **37**, 849 (1954); M. W. Klohs *et al., J. Am. Chem. Soc.* **76**, 1332 (1954). Review of early literature: R. E. Woodson *et al., Rauwolfia: Botany, Pharmacognosy, Chemistry and Pharmacology* (Little, Brown and Co., Boston, 1957) 147 pp. Structure: Goutarel, Le Hir, *Bull. Soc. Chim. France* **18**, 909 (1951). Stereochemistry: Wenkert *et al., J. Am. Chem. Soc.* **83**, 5037 (1961); Shamma, Richey, *ibid.* **85**, 2507 (1963). Total synthesis of *dl*-form: van Tamelen, Placeway, *ibid.* **83**, 2594 (1961); van Tamelen *et al., ibid.* **91**, 7359 (1969); J. Gutzwiller *et al., Helv. Chim. Acta* **64**, 1663 (1981); T. Kametani *et al., J. Chem. Soc. Perkin Trans. I* **1981**, 3168. Biosynthesis: N. Nagakura *et al., ibid.* **1979**, 2308; M. Rueffer *et al., Chem. Commun.* **1979**, 1016. Pharmacokinetics: A. Marzo *et al., Farmaco Ed. Prat.* **36**, 173 (1981). Clinical evaluation of platelet anti-aggregant activity: J. Neuman *et al., Arzneimittel-Forsch.* **36**, 1394 (1986). Evaluation of combination

with almitrine, *q.v.*, in cerebral ischemia in rats: M. G. Borzeix, J. Cahn, *ibid.* **37**, 491 (1987).

Prisms from methanol, dec 257°. $[\alpha]_D^{20}$ −60° (c = 0.5 in chloroform); $[\alpha]_D^{20}$ −45° (c = 0.5 in pyridine); $[\alpha]_D^{20}$ −39° (c = 0.25 in methanol). uv max (methanol): 227, 292 nm (log ε 4.61, 3.79).

Hydrochloride, $C_{21}H_{24}N_2O_3$·HCl, leaflets from ethanol, mp 290° (dec). $[\alpha]_D^{20}$ −17° (c = 0.5 in methanol). Sparingly sol in water or dil HCl.

Hydrobromide, $C_{21}H_{24}N_2O_3$·HBr, diamond-shaped platelets from methanol, mp 295-296°.

THERAP CAT: Antihypertensive, anti-ischemic (cerebral and peripheral).

8292. Raunescine. *(3β,16β,17α,18β,20α)-17-Hydroxy-18-[(3,4,5-trimethoxybenzoyl)oxy]-yohimban-16-carboxylic acid methyl ester;* 3,4,5-trimethoxybenzoyl methyl raunescate. $C_{31}H_{36}N_2O_8$; mol wt 564.64. C 65.94%, H 6.43%, N 4.96%, O 22.67%. Isoln from *Rauwolfia canescens* L., *R. tetraphylla* and other *Rauwolfia* spp, *Apocynaceae.* Isoln and structure: Hosansky, Smith, *J. Am. Pharm. Assoc. Sci. Ed.* **44**, 639 (1955); Huebner, Schlittler, *J. Am. Chem. Soc.* **79**, 250 (1957); van Tamelen, Taylor, *ibid.* **79**, 5256 (1957). Extraction: **Brit.** pat. **833,149** (1960 to Ciba).

Monohydrate, hexagonal prisms from 90% methanol, mp 160-170°. $[\alpha]_D^{25}$ −74° (chloroform). uv max (U.S.P. alc): 218, 271 nm (log ε 4.77, 4.26).

Nitrate, $C_{31}H_{36}N_2O_8$·HNO$_3$, mp 223-225°; (mp 208-210°). $[\alpha]_D^{25}$ −80° (5N acetic acid).

Monoacetate monohydrate, $C_{33}H_{38}N_2O_9$·H$_2$O, mp 159-162°. $[\alpha]_D^{20}$ −157° (chloroform).

8293. Rauwolfia serpentina. Hiwolfia; Koglucoid; Raudixin; Rauserpol; Rauverid; Rauwoldin; Rauwolfemms; Serfolia; Serpina; Wolfina. Powdered, dried root of *Rauwolfia serpentina* (L.) Benth. *Apocynaceae*, a small shrub native to the Orient from India to Sumatra. Contains many indole alkaloids of which reserpine, rescinnamine and deserpidine are the primary constituents. Others include: reserpiline, yohimbine, ajmaline, serpentine, serpetinine. Review and bibliography: Monachino, *Econ. Bot.* **8**, 349-365 (1954). Clinical trial in hypertension: M. C. Srivastava *et al., Indian Pract.* **20**, 279 (1967); in combination with chlorothiazide: W. M. Smith, *Circ. Res.* **40**, Suppl I, I98 (1977). LC determn of reserpine-rescinnamine content: U. R. Cieri, *J. Assoc. Off. Anal. Chem.* **70**, 540 (1987).

Alseroxylon, Austrawolf, Egalin, Gendon, Rautensin, Rauwiloid, Ra-Valeas, Rivadescin. Alkaloidal extract of the

dried root. Red, amorphous powder with characteristic odor. Contains 0.15-0.2% reserpine-rescinnamine alkaloids.
THERAP CAT: Antihypertensive.

8294. Rayon. Regenerated cellulose. The Code of Federal Regulations defines rayon as a manufactured fiber composed of regenerated cellulose, as well as manufactured fibers composed of regenerated cellulose in which substituents have replaced not more than 15 percent of the hydrogens of the hydroxyl groups. Now produced almost exclusively by the viscose process. *Review:* J. Lundberg, A. Turbak in Kirk-Othmer *Encyclopedia of Chemical Technology* **Vol. 19** (Wiley-Interscience, New York, 3rd ed., 1982) pp 855-880.
Purified rayon, Avisco Rayon. A medicinal grade of rayon.
USE: Purified rayon as surgical aid.

8295. Razoxane. *4,4'-(1-Methyl-1,2-ethanediyl)bis-2,6-piperazinedione;* (±)*-4,4'-propylenedi-2,6-piperazinedione;* (±)-(3,5,3',5'-tetraoxo)-1,2-dipiperazinopropane; (±)-1,2-bis(3,5-dioxopiperazinyl)propane; ICI-59118; ICRF-159; NSC-129943; Razoxin. $C_{11}H_{16}N_4O_4$; mol wt 268.27. C 49.25%, H 6.01%, N 20.88%, O 23.86%. Cyclized analog of EDTA, *q.v.;* exhibits intracellular iron chelating activity. Plays role in preventing the formation of doxorubicin-iron complexes which generate radical oxygen species that are associated with cardiotoxicity. Prepn: A. M. Creighton *et al., Nature* **222,** 384 (1969); A. M. Creighton, Brit. pat. **1,234,935** eidem, U.S. pat. **3,941,790** (1971, 1976 both to Natl. Res. Dev. Corp.). Mode of action: H. B. A. Sharpe *et al., Nature* **226,** 524 (1970). Metabolism: R. E. Bellet *et al., Eur. J. Cancer* **13,** 2293 (1977). Pharmacology: A. Atherton, *ibid.* **11,** 383 (1975). Clinical studies: M. T. Bakowski *et al., Int. J. Radiat. Oncol. Biol. Phys.* **4,** 115 (1978); H. W. Bruckner *et al., Cancer Treat. Rep.* **66,** 1713 (1982). Toxicity study: E. Hassenstein, K. Renner, *Strahlentherapie* **154,** 122 (1978). Review of antineoplastic activity and radiosensitization: M. T. Bakowski, *Cancer Treat. Rev.* **3,** 95-107 (1976); K. Hellmann, W. Rhomberg, *ibid.* **18,** 225-240 (1991).

Pale cream microcrystalline solid, mp 237-239°.
(+)-Form, *dexrazoxane, ADR-529, ICRF-187, NSC-169780, Zinecard.* Clinical pharmacokinetics: R. H. Earhart *et al., Cancer Res.* **42,** 5255 (1982). HPLC determn in plasma and urine: R. C. Lewis *et al., Pharm. Res.* **9,** 101 (1992). Clinical study of protection vs. doxorubicin-associated cardiomyopathy: J. L. Speyer *et al., J. Clin. Oncol.* **10,** 117 (1992). Review of pharmacology and clinical studies: J. Koning *et al., Cancer Treat. Rep.* **18,** 1-19 (1991). Crystals from aq methanol/ether, mp 193°. $[\alpha]_D$ +11.35° (c = 5 in DMF). Soly (mg/ml): 10-12 water, 35-43 0.1N HCl, 25-34 0.1N NaOH, 6.7-10 10% ethanol, 1 methanol, 7.1-10 H_2O/DMA (1:1), 9.7-14.5 0.1M citrate buffer (pH 4), 8.7-13 0.1M borate buffer (pH 9).
(+)-Form hydrochloride, *Cardioxane, Eucardion.*
THERAP CAT: Racemate as antineoplastic. (+)-Form as cardioprotectant.

8296. Rebamipide. α-[(4-Chlorobenzoyl)amino]-1,2-dihydro-2-oxo-4-quinolinepropanoic acid; (±)-α-(p-chlorobenzamido)-1,2-dihydro-2-oxo-4-quinolinepropionic acid; 2-(4-chlorobenzoylamino)-3-[2(1H)-quinolinon-4-yl]propionic acid; proamipide; OPC-12759; Mucosta. $C_{19}H_{15}ClN_2O_4$; mol wt 370.79. C 61.55%, H 4.08%, Cl 9.56%, N 7.56%, O 17.26%. Gastric cytoprotectant. Prepn: M. Uchida *et al.,* Ger. pat. **3,324,034;** eidem, U.S. pat. **4,578,381;** (1984, 1986 both to Otsuka). Synthesis and pharmacology: M. Uchida *et al., Chem. Pharm. Bull.* **33,** 3775 (1985); of enantiomers:

eidem, ibid. **35,** 853 (1987). Antiulcer activity in rats: K. Yamasaki *et al., Eur. J. Pharmacol.* **142,** 23 (1987); K. Yamasaki *et al., Japan. J. Pharmacol.* **49,** 441 (1989). HPLC determn in plasma and urine: Y. Shioya, T. Shimizu, *J. Chromatog.* **434,** 283 (1988).

White powder from DMF-water, mp 288-290° (dec) as hemihydrate.
(−)-Form, colorless needles from DMF, mp 305-306° (dec). $[\alpha]_D^{20}$ −116.7° (c = 1.0 in DMF).
(+)-Form, colorless needles from DMF, mp 305-306° (dec). $[\alpha]_D^{20}$ +116.9° (c = 1.0 in DMF).
THERAP CAT: Antiulcerative.

8297. Reductic Acid. *2,3-Dihydroxy-2-cyclopenten-1-one;* 2-cyclopenten-2,3-diol-1-one. $C_5H_6O_3$; mol wt 114.10. C 52.63%, H 5.30%, O 42.07%. Antioxidant obtained from pectic substances: Reichstein, Oppenauer, *Helv. Chim. Acta* **16,** 988 (1933); **17,** 390 (1934); Goldstein, U.S. pat. **2,854,-484** (1958). Alternate syntheses: Hesse *et al., Ann.* **563,** 31 (1949); **592,** 137 (1955); **736,** 134 (1970). Crystal structure: D. Semmingsen, *Acta Chem. Scand.* B **31,** 81 (1977). Similar to vitamin C in structure but not activity.

Pale yellow crystals (has been obtained colorless) from ethyl acetate, dec 213-213.5° (also reported as 207.5° and 211°). Soluble in water, methanol, ethanol. Sparingly sol in ether, ethyl acetate, acetone. Insol in benzene.
5-Methylreductic acid, mp 71°. Stronger reducing agent than reductic acid.
USE: As antioxidant, like isoascorbic or ascorbic acid.

8298. Reinecke Salt. *Ammonium diamminetetrakis-(thiocyanato-N)chromate(1−); ammonium reineckate;* ammonium tetrathiocyanodiammonochromate. $C_4H_{10}CrN_7S_4$; mol wt 336.43. C 14.28%, H 3.00%, Cr 15.46%, N 29.14%, S 38.12%. $NH_4[Cr(NH_3)_2(SCN)_4]$. Exists as monohydrate; made by fusing ammonium thiocyanate with ammonium dichromate. Prepn: Dakin, *Org. Syn.* **coll. vol. II,** 555 (1943).
Monohydrate, dark red crystals or red cryst powder. Sparingly sol in cold water; sol in hot water, alcohol. Dec in aq soln with formation of a blue color and free HCN; dec occurs in about 2 weeks at room temp, rapidly above 65°. A similar decompn takes place in boiling alcohol.
Note: The K-salt is also known as Reinecke salt.
USE: Precipitant for primary and secondary amines, proline, hydroxyproline, and certain amino acids; also a reagent for mercury with which it gives a red color or precipitate.

8299. Relaxin. Cervilaxin; Releasin (formerly). A polypeptide hormone secreted by the corpora lutea of many mammalian species during pregnancy; it is also produced in several non-mammalians, including the shark. Relaxin facilitates the birth process by causing a softening and lengthening of the pubic symphysis and cervix; it also inhibits contraction of the uterus and may play a role in timing of parturition. First discovered in estrogen-primed guinea pigs: F. L. Hisaw, *Proc. Soc. Exp. Biol. Med.* **23,** 661 (1926). Extraction from corpora lutea: Fevold *et al., J. Am. Chem. Soc.* **52,** 3340 (1930); Albert *et al., Endocrinology* **40,** 370 (1947); from pregnant rabbit serum: Abramowitz *et al., Anat. Rec.* **84,** 456 (1942); *cf.* Hall, Newton, *J. Physiol.* **106,**

18 (1947). Can be obtained commercially from pregnant sows' ovaries. Isoln procedure and purification by resin chromatography: Lehrman et al., J. Am. Pharm. Assoc. **44**, 206 (1955); Kroc, Phillips; Phillips, U.S. pats. **2,852,431-2** (both 1958 to Warner-Lambert). Isoln from ovarian tissue using glacial acetic acid: Cohen, U.S. pat. **2,930,737** (1960 to Princeton Laboratories). Alternate method: Keck, U.S. pat. **3,008,878** (1961 to Thomae). Prepn from hog ovaries: Doczi, U.S. pat. **3,096,246** (1963 to Warner-Lambert). Porcine relaxin has a mol wt of approx 6000 and consists of two peptide chains, A and B, of 22 and 31 residues respectively, linked covalently by one intra- and two inter-chain disulfide bonds. Purification, characterization of porcine relaxin: O. D. Sherwood, E. O'Byrne, Arch. Biochem. Biophys. **160**, 185 (1974). Structure of the A chain: C. Schwabe et al., Biochem. Biophys. Res. Commun. **70**, 397 (1976); of the B chain: eidem, ibid. **75**, 503 (1977). Relaxin is structurally homologous to insulin and related growth factors: R. A. Bradshaw, Rev. Biochem. **47**, 191 (1978); S. Bedarkar et al., Nature **270**, 449 (1977); N. Isaacs et al., ibid. **271**, 278 (1978). Complete amino acid sequence of porcine relaxin: R. James et al., ibid. **267**, 544 (1977); of rat: M. J. John et al., Endocrinology **108**, 726 (1981). Isoln, characterization of relaxin from the sand tiger shark (Odontaspis taurus): J. W. Reinig et al., ibid. **109**, 537 (1981); structure: L. K. Gowan et al., FEBS Letters **129**, 80 (1981). Demonstration of synthesis of rat relaxin as a preprorelaxin molecule with a connecting peptide of 105 amino acid residues, using molecular cloning: P. Hudson et al., Nature **291**, 127 (1981). Structure of a genomic clone from which the amino acid sequence of biologically active human relaxin was predicted: eidem, ibid. **301**, 628 (1983). Reviews: C. Schwabe et al., Recent. Progr. Horm. Res. **34**, 123-211 (1978); Ann. N.Y. Acad. Sci. **380**, entitled "Relaxin: Structure, Function, and Evolution", B. G. Steinetz et al., Eds. (1982) pp 1-244.

Amorphous powder. Diffusion constant, $cm^2/sec = 12.6 \times 10^{-7}$. Sedimentation constant, $sec = 1.6 \times 10^{-13}$. Isoelec pt around pH 7.0. uv max (water pH 7.0): 277.5 nm ($E_{1cm}^{1\%}$ 7.20). Slightly sol in water and 95% alcohol. Sol in acid or alkaline solns. Insol in abs alcohol, ether, acetone, petr ether, benzene. Dec above pH 9. Neutral or acid solns which are free from oxidizing agents, are stable. Ampuled solns at pH 7.0 and 3.2 have been kept in refrigerator for one year without loss of activity.

THERAP CAT: Hormone (ovarian).

8300. Remifentanil. 4-(Methoxycarbonyl)-4-[(1-oxopropyl)phenylamino]-1-piperidinepropanoic acid methyl ester; 4-carboxy-4-(N-phenylpropionamido)-1-piperidinepropionic acid dimethyl ester; remifentanyl; GI-87084. $C_{20}H_{28}N_2O_5$; mol wt 376.45. C 63.81%, H 7.50%, N 7.44%, O 21.25%. Synthetic mu-opioid agonist. Prepn: P. L. Feldman et al., Eur. pat. Appl. **383,579**; eidem, U.S. pat. **5,019,583** (1990, 1991 both to Glaxo); eidem et al., J. Med. Chem. **34**, 2202 (1991). Opioid receptor activity: M. K. James et al., J. Pharmacol. Exp. Ther. **259**, 712 (1991). Pharmacology: W. E. Hoffman et al., Anesthesiology **79**, 107 (1993). Clinical pharmacokinetics: T. D. Egan et al., ibid. 881; C. L. Westmoreland et al., ibid. 893. HPLC determn in blood: K. Selinger et al., J. Pharm. Biomed. Anal. **12**, 243 (1994).

Hydrochloride, $C_{20}H_{28}N_2O_5$·HCl, GI-87084B.
THERAP CAT: Analgesic (narcotic).

8301. Remoxipride. (S)-3-Bromo-N-[(1-ethyl-2-pyrrolidinyl)methyl]-2,6-dimethoxybenzamide; (−)-N-ethyl-2-(3-bromo-2,6-dimethoxybenzamidomethyl)pyrrolidine. $C_{16}H_{23}BrN_2O_3$; mol wt 371.27. C 51.76%, H 6.24%, Br 21.52%, N 7.55%, O 12.93%. Specific dopamine D_2-receptor antagonist. Prepn: G. L. Florvall, S. O. Ogren, Eur. pat. Appl. **4,831**; eidem, U.S. pat. **4,232,037** (1979, 1980 both to Astra); eidem, J. Med. Chem. **25**, 1280 (1982). Antidopa-

minergic activity: S. O. Ogren et al., Eur. J. Pharmacol. **102**, 459 (1984). Radiometric determn in urine: A. C. Veltkamp et al., J. Chromatog. **384**, 357 (1987). Clinical pharmacology and pharmacokinetics: L. Farde et al., Psychopharmacology **95**, 157 (1988). Clinical evaluation in tardive dyskinesia: U. Andersson et al., ibid. **94**, 167 (1988); in schizophrenia: R. G. McCreadie et al., Acta Psychiatr. Scand. **78**, 49 (1988).

$[\alpha]_D^{20}$ −64° (c = 2 in ethanol).
Hydrochloride monohydrate, $C_{16}H_{23}BrN_2O_3$·HCl·H$_2$O, A-33547, FLA-731, Roxiam. White to off-white crystalline solid. Loss of water at 105°; mp 173°. $[\alpha]_D^{20}$ −11° (c = 2 in water). pKa 8.9. uv max in water, ethanol, 0.1N HCl: 286, 287, 286 nm (ϵ 2280, 2260, 2280). Soly (mg/ml): water 300; ethanol 400; dichloromethane 80; acetone 20. LD_{50} i.p. in rats: 794 μmol/kg (Florvall, Ogren, 1982).
THERAP CAT: Antipsychotic.

8302. Renin. Highly specific aspartyl proteinase of mol wt about 40000 Da, produced and secreted by the kidney. Found also in amniotic fluid. Conc. of renin in the human kidney is about 20 times less than in hog kidney. Purification from hog kidneys: Haas et al., Arch. Biochem. Biophys. **42**, 368 (1953); from kidneys of various species: Haas et al., ibid. **48**, 256 (1954). Prepn of human renin: Haas et al., ibid. **110**, 534 (1965). Exists in α, β and γ configurations: Haas et al., ibid. **44**, 79 (1953). Purification on DEAE cellulose columns: Passananti, Biochim. Biophys. Acta **34**, 246 (1959); Maier, Morgan, ibid. **128**, 193 (1966). Has been observed to exist in high molecular weight inactive forms which upon acid treatment decrease in molecular size and increase in enzymic activity. See review of purification studies: E. Haber, E. E. Slater, Circ. Res. **40**, Suppl. I, I36 (1977). Renin is secreted by juxtaglomerular cells and acts on the plasma substrate, angiotensinogen, to split off the inactive decapeptide angiotensin I which is converted to the active pressor agent angiotensin II, q.v. Renin itself has no activity. Secretion of renin is stimulated by constriction of a renal artery, blood loss, low sodium levels, adrenal insufficiency. Increased levels of renin shown also in pregnancy. Cloning and sequence analysis of human renin cDNA: T. Imai et al., Proc. Nat. Acad. Sci. USA **80**, 7405 (1983). Crystal structure of human recombinant renin: A. R. Sielecki et al., Science **243**, 1346 (1989). Review: W. S. Peart, "The Renin-Angiotensin System" in Pharmacol. Rev. **17**, 143 (1965); idem, Proc. Roy. Soc. Ser. B **173**, 317 (1969); Smeby, Bumpus, "Renin" in Renal Hypertension, I. Page, J. McCubbin, Eds. (Year Book Medical Publishers, Chicago, 1968) pp 14-61. Books: M. R. Lee, Renin and Hypertension (Williams & Wilkins, Baltimore, 1969); Renin vol. 5, S. Oparil et al., Eds. (Eden Press, Quebec, 1980) 368 pp. Review of the renin-angiotensin system: J. C. Romero, F. G. Knox, Hypertension **11**, 724-738 (1988).

8303. Rennin. Chymosin; rennase; lab; abomasal enzyme; Lab Ferment (German). Mol wt about 31,000. The predominant milk-clotting enzyme from the true stomach or abomasum of the suckling calf: Tauber, Kleiner, J. Biol. Chem. **96**, 745 (1932). Secreted as an inactive precursor called prorennin and converted in the acid environment of the stomach to the active enzyme. Obtained by extracting dried strips of calf stomach with 5 to 10% NaCl soln contg boric acid for 5 days; improved process: Keil, U.S. pat. **2,339,931** (1942). Structure consists of a single polypeptide chain with internal disulfide bridges. Amino acid composition: Schwander et al., Helv. Chim. Acta **35**, 553 (1952). Determn of N-terminal amino acid sequence: Foltmann, Phil. Trans. Roy. Soc. London **257B**, 147 (1970). Homologies in amino acid sequences of rennin and pepsin: Tang, Hartley, Biochem. J. **118**, 611 (1970). Reviews: Berridge,

Advan. Enzymol. **15**, 423 (1954); Foltmann, *Milk Proteins* **2**, 217 (1971).

Yellowish powder, grains or scales. Peculiar, not unpleasant odor. Characteristic, slightly salty taste. Slightly hygroscopic. Can be purified enough to crystallize. The dry substance (sometimes diluted with NaCl or lactose) is stable, aq solns are not. Partially sol in water, dil alcohol. Insol at pH values near its isoelectric point of pH 4.5. pH of aq soln 5.8 (acceptable range 5.3-6.3). Strongly affected by ultraviolet (sunlight). Causes milk to coagulate, optimum temp 37-43°, optimum pH of milk 5.8. Not active below 15° nor above 55°. When a commercial prepn is labeled 1:10,000 it means that one gram will curdle 10 liters of milk at 35° within 40 minutes. Other strengths are 1:100,000 to 1:7,600,000.

USE: *Rennet* (a dried extract contg rennin) is used in the manuf of cheese and rennet casein, also sold retail for the making of junket or rennet custards.

THERAP CAT: Digestive enzyme.

8304. Repaglinide. *(S)-2-Ethoxy-4-[2-[[3-methyl-1-[2-(1-piperidinyl)phenyl]butyl]amino]-2-oxoethyl]benzoic acid;* (+)-2-ethoxy-α-[[(S)-α-isobutyl-o-piperidinobenzyl]carbamoyl]-p-toluic acid; *(S)-(+)-2-ethoxy-4-[N-[1-(2-piperidinophenyl)-3-methyl-1-butyl]aminocarbonylmethyl]benzoic acid;* AG-EE-623 ZW; Novonorm. $C_{27}H_{36}N_2O_4$; mol wt 452.59. C 71.65%, H 8.02%, N 6.19%, O 14.14%. Nonsulfonylurea oral hypoglycemic agent. Prepn: W. Grell *et al.*, PCT Int. pat. Appl. WO 93 00,337 (1993 to Thomae). HPLC determn of racemate in human plasma: A. Greischel *et al.*, *J. Chromatog.* **568**, 246 (1991). Clinical pharmacokinetics: F. J. Ampudia-Blasco *et al.*, *Diabetologia* **37**, 703 (1994). Comparative clinical evaluation in NIDDM: B. H. R. Wolffenbuttel *et al.*, *Eur. J. Clin. Pharmacol.* **45**, 113 (1993).

White, odorless, crystalline powder. Crystals from ethanol/water (2:1), mp 126-128°. Crystals from neutralized water, mp 130-131°. $[\alpha]_D^{20}$ +6.97° (c = 0.975 in methanol); $[\alpha]_D^{20}$ +7.45° (c = 1.06 in methanol). LD_{50} orally in rats: > 1 g/kg (Grell).

THERAP CAT: Antidiabetic.

8305. Repirinast. *5,6-Dihydro-7,8-dimethyl-4,5-dioxo-4H-pyrano[3,2-c]quinoline-2-carboxylic acid 3-methylbutyl ester;* isopentyl 5,6-dihydro-7,8-dimethyl-4,5-dioxo-4H-pyrano[3,2-c]quinoline-2-carboxylate; MY-5116; Romet. $C_{20}H_{21}NO_5$; mol wt 355.39. C 67.59%, H 5.96%, N 3.94%, O 22.51%. Prepn: **Belg.** pat. **876,751;** Y. Morinaka, K. Takahashi, **U.S.** pat. **4,298,610** (1979, 1981 both to Mitsubishi Yuka); Y. Morinaka *et al.*, *Eur. J. Med. Chem. - Chim. Ther.* **16**, 251 (1981). General pharmacology in animals: K. Takahashi *et al.*, *Oyo Yakuri* **32**, 233 (1986), C.A. **105**, 218691j (1986). Anti-allergic effects: S. Kono, K. Ohata, *Arerugi* **35**, 1105 (1986), C.A. **106**, 165870z (1987). Toxic-

ity: M. Nagase *et al.*, *Iyakuhin Kenkyu* **17**, 545 (1986), C.A. **106**, 108121r (1987).

Crystals from chloroform + *n*-hexane, mp 236-241°. LD_{50} in rats, mice (mg/kg): > 5000 orally, > 5000 s.c. both species (Nagase).

THERAP CAT: Antiallergic.

8306. Reposal. *5-Bicyclo[3.2.1]oct-2-en-3-yl-5-ethyl-2,-4,6(1H,3H,5H)-pyrimidinetrione; 5-bicyclo[3.2.1]oct-2-en-3-yl-5-ethylbarbituric acid;* WT-161; Reposamal. $C_{14}H_{18}N_2O_3$; mol wt 262.31. C 64.11%, H 6.92%, N 10.68%, O 18.30%. Prepn: Taub, **U.S.** pat. **2,808,408** (1956 to Calanda Stiftung). Metabolism: Nielsen, Tarding, *Acta Pharmacol. Toxicol.* **26**, 521 (1968). Toxicity: Frey, *Arzneimittel-Forsch.* **12**, 389 (1962).

Needles from isopropanol, mp 213°. pKa 7.4-7.5°. LD_{50} i.p. in mice: 175 mg/kg (Frey).

Note: This is a controlled substance (depressant) listed in the U.S. Code of Federal Regulations, Title 21 Part 1308.13 (1995).

THERAP CAT: Sedative; hypnotic.

8307. Reproterol. *7-[3-[[2-(3,5-Dihydroxyphenyl)-2-hydroxyethyl]amino]propyl]-3,7-dihydro-1,3-dimethyl-1H-purine-2,6-dione; 7-[3-[(β,3,5-trihydroxyphenethyl)amino]propyl]theophylline;* D-1959. $C_{18}H_{23}N_5O_5$; mol wt 389.41. C 55.52%, H 5.95%, N 17.98%, O 20.54%. Theophylline deriv which has selective activity on β_2 receptors. Prepn: **Fr.** pat. **M5969** (1968 to Degussa), C.A. **71**, 70644c (1969). *See also* Fr. pat. **Addn. 0308** (1970 to Degussa), C.A. **78**, 72216h (1973). Synthesis: K. H. Klingler, *Arzneimittel-Forsch.* **27**, 4 (1977). Toxicology: S. Habersang, *ibid.* 45. Series of articles on metabolism, pharmacology and clinical trials: *ibid.* 15-76. Other clinical studies: D. Nolte *et al.*, *Deut. Med. Wochenschr.* **102**, 619 (1977); H. Budmiger *et al.*, *Schweiz. Med. Wochenschr.* **108**, 1190 (1978).

Hydrochloride, $C_{18}H_{23}N_5O_5$·HCl, *W-2946M, Asmaterolo, Bronchodil, Bronchospasmin.* Crystals, mp 249-250°. LD_{50} in mice (mg/kg): 148 i.v., > 10,000 orally (Habersang).

THERAP CAT: Bronchodilator.

8308. Resacetophenone. *1-(2,4-Dihydroxyphenyl)ethanone; 2′,4′-dihydroxyacetophenone.* $C_8H_8O_3$; mol wt 152.15. C 63.15%, H 5.30%, O 31.55%. Made by heating resorcinol, glacial acetic acid, and anhydr zinc chloride at 145-150°: Cooper, *Org. Syn.* **21**, 103 (1941); from resorcinol, acetic anhydride and boron trifluoride: Killelea, Lindwall, *J. Am. Chem. Soc.* **70**, 428 (1948).

Needles or leaflets, mp 145-147°. Gradually dec by water; sol in pyridine, in warm alcohol, in glacial acetic acid; practically insol in benzene, ether, chloroform.

USE: Reagent for iron as a 10% alcoholic soln. A red color is obtained with ferric ions in slightly acid soln: Cooper, *Ind. Eng. Chem. Anal. Ed.* **9**, 334 (1937).

8309. Resazurin. *7-Hydroxy-3H-phenoxazin-3-one 10-oxide;* "diazoresorcinol"; resazoin. $C_{12}H_7NO_4$; mol wt 229.19. C 62.89%, H 3.08%, N 6.11%, O 27.92%. Prepn from resorcinol: Rast *et al., Ger.* pat. **961,829** (1957 to Bayer), *C.A.* **53**, 15098 (1959).

Dark red, small crystals with greenish luster. Insol in water, ether; sparingly sol in alcohol, glacial acetic acid; sol in dil alkali hydroxides.

USE: As indicator, using a soln of 0.1 g in 20 ml 0.1N NaOH, and water to make 500 ml. pH: 3.8 orange, 6.5 dark violet. In the detection of hyposulfite (sulfoxylate). In food research (reductase test).

8310. Rescimetol. *[3β,16β,17α,18β(E),20α]-18-[[3-(4-Hydroxy-3-methoxyphenyl)-1-oxo-2-propenyl]oxy]-11,17-dimethoxyyohimban-16-carboxylic acid methyl ester;* methyl O-(4-hydroxy-3-methoxycinnamoyl)reserpate; CD-3400; WHO-4939; Toscarna. $C_{33}H_{38}N_2O_8$; mol wt 590.67. C 67.10%, H 6.48%, N 4.74%, O 21.67%. Analog of rescinnamine, *q.v.* Prepn: **Belg.** pat. **782,661** corresp to T. Kametani, **U.S.** pat. **3,898,215** (1972, 1975 both to Nippon Chemiphar); T. Kametani *et al., J. Med. Chem.* **15**, 686 (1972). Mass spectrum: *eidem, J. Chem. Soc. Perkin Trans. I* **1975**, 932. Antihypertensive activity: H. Kato *et al., Japan. J. Pharmacol.* **27**, 87 (1977). Effects on respiration, cardiovascular system, renal function: T. Kamishiro *et al., Oyo Yakuri* **20**, 441 (1980), *C.A.* **96**, 28430w (1982). Toxicity study: T. Fukuda *et al., ibid.* **18**, 71 (1979), *C.A.* **92**, 104392v (1980).

Pale yellow needles from methanol, mp 259-260°.
THERAP CAT: Antihypertensive.

8311. Rescinnamine. *(3β,16β,17α,18β,20α)-11,17-Dimethoxy-18-[[1-oxo-3-(3,4,5-trimethoxyphenyl)-2-propenyl]oxy]-3,20-yohimban-16-carboxylic acid methyl ester;* 3,4,5-trimethoxycinnamic acid ester of methyl reserpate; methyl 3,4,5-trimethoxycinnamoyl reserpate; reserpinine; Anaprel; Apoterin S; Cartric; Cinnaloid; Moderil. $C_{35}H_{42}N_2O_9$; mol wt 634.73. C 66.23%, H 6.67%, N 4.41%, O 22.69%. Found in *Rauwolfia serpentina* Benth., *Apocynaceae:* Haack *et al., Naturwiss.* **41**, 214 (1954); Ordway, Guercio, **U.S.** pat. **2,876,228** (1959 to Pfizer); Klohs *et al.,* **U.S.** pat. **2,974,144** (1961 to Riker); from *Rauwolfia* spp: Banes *et al., J. Am. Pharm. Assoc.* **47**, 625 (1958). Identity with reserpinine: Haack *et al., Naturwiss.* **42**, 47 (1955). Structure: Klohs *et al., J. Am. Chem. Soc.* **77**, 2241 (1955).

Fine needles from benzene, mp 238-239° (vac). $[\alpha]_D^{24}$ −97° (c = 1 in chloroform). uv max (methanol): 228, 302 nm (log ε 4.79, 4.48). Practically insol in water. Moderately sol in methanol, benzene, chloroform, other organic solvents.
THERAP CAT: Antihypertensive.

8312. Reserpic Acid. *(3β,16β,17α,18β,20α)-18-Hydroxy-11,17-dimethoxyyohimban-16-carboxylic acid;* reserpinolic acid. $C_{22}H_{28}N_2O_5$; mol wt 400.47. C 65.98%, H 7.05%, N 7.00%, O 19.98%. Obtained by controlled alkaline hydrolysis of reserpine: Dorfman *et al., Helv. Chim. Acta* **37**, 59 (1954); Schlittler, **U.S.** pat. **2,824,874** (1958).

Crystals from methanol, mp 241-243°.
Hydrochloride hemihydrate, $C_{22}H_{28}N_2O_5 \cdot HCl \cdot \frac{1}{2}H_2O$, crystals, mp 257-259°. $[\alpha]_D^{23}$ −81°.
Methyl ester, $C_{23}H_{30}N_2O_5$, needles from methanol, mp 235-240°. $[\alpha]_D^{25}$ −106°.
Methyl ester hydrochloride, mp ∼228°.

8313. Reserpiline. *(3β,19α,20α)-16,17-Didehydro-10,-11-dimethoxy-19-methyloxayohimban-16-carboxylic acid methyl ester.* $C_{23}H_{28}N_2O_5$; mol wt 412.49. C 66.97%, H 6.84%, N 6.79%, O 19.39%. Isoln from *Rauwolfia serpentina* Benth., *Apocynaceae:* Klohs *et al., Chem. & Ind. (London)* **1954**, 1264; from *R. canescens* L.: Stoll *et al., Helv. Chim. Acta* **38**, 270 (1955). Stereochemistry: Shamma, Richey, *J. Am. Chem. Soc.* **85**, 2507 (1963). Partial synthesis: S.-I. Sakai *et al., Heterocycles* **17**, 99 (1982).

Amorphous powder. $[\alpha]_D^{20}$ −38° (ethanol); −14° (c = 1.5 in pyridine); −12° (c = 1.7 in chloroform). uv max (ethanol): 229, 300-304 nm (log ε 4.57, 4.03). Freely sol in ethanol, acetone, chloroform, benzene.
Hydrochloride, $C_{23}H_{28}N_2O_5 \cdot HCl$, crystals, mp 205-207°. $[\alpha]_D^{24}$ −40° (c = 0.44 in ethanol).
Reserpilic acid dimethylaminoethyl ester dihydrochloride, $C_{26}H_{35}N_3O_5 \cdot 2HCl$, *Paratensiol, Resporisan.*
THERAP CAT: Antihypertensive.

8314. Reserpine. *(3β,16β,17α,18β,20α)-11,17-Dimethoxy-18-[(3,4,5-trimethoxybenzoyl)oxy]yohimban-16-carboxylic acid methyl ester;* 3,4,5-trimethoxybenzoyl methyl reserpate; Crystoserpine; Eskaserp; Rau-sed; Reserpoid; Rivasin; Sandril; Sedaraupin; Serfin; Serpasil; Serpasol; Serpine; Serpiloid. $C_{33}H_{40}N_2O_9$; mol wt 608.69. C 65.12%, H 6.62%, N 4.60%, O 23.66%. Found in *Rauwolfia* spp. Isoln from the roots of *Rauwolfia serpentina* L. Benth., *Apocynaceae:* J. M. Müller *et al.*, *Experientia* **8**, 338 (1952); Dorfman *et al.*, *Helv. Chim. Acta* **37**, 59 (1954); Schwyzer, Mueller, U.S. pat. **2,833,771** (1958 to Ciba). Structure: Dorfman *et al.*, *loc. cit.*; Neuss *et al.*, *J. Am. Chem. Soc.* **76**, 2463 (1954). Stereochemistry: Diassi *et al.*, *ibid.* **77**, 2028, 4687 (1955); Huebner, Wenkert, *ibid.* 4180; van Tamelen, Hance, *ibid.* 4693; Aldrich *et al.*, *ibid.* **81**, 2481 (1959); Ban, Yonemitsu, *Tetrahedron* **20**, 2877 (1964). Total synthesis: R. B. Woodward *et al.*, *J. Am. Chem. Soc.* **78**, 2023 (1956); *eidem, Tetrahedron* **2**, 1-57 (1958); B. A. Pearlman, *J. Am. Chem. Soc.* **101**, 6404 (1979); P. A. Wender *et al.*, *ibid.* **102**, 6157 (1980). Stereoselective synthesis: G. Stork, *Pure Appl. Chem.* **61**, 439 (1989). Comprehensive description: R. E. Schirmer, *Anal. Profiles Drug Subs.* **4**, 384-430 (1975). *Review:* A. Scriabine, Ed. in *Pharmacology of Antihypertensive Drugs* (Raven, New York, 1980) pp 119-125. Review of carcinogenicity studies: *IARC Monographs* **10**, 217-229 (1976); R. M. Diener *et al.*, *Toxicol. Pathol.* **8**, 1-21 (1980).

Long prisms from dil acetone, dec 264-265° (dec 277-277.5° in evac tube). $[\alpha]_D^{23}$ −118° (CHCl$_3$); $[\alpha]_D^{26}$ −164° (c = 0.96 in pyridine); $[\alpha]_D^{26}$ −168° (c = 0.624 in DMF). uv max (CHCl$_3$): 216, 267, 295 nm (ε 61700, 17000, 10200). Weak basc. pK 6.6. Very sparingly sol in water. Freely sol in chloroform (∼1 g/6 ml), methylene chloride, glacial acetic acid. Sol in benzene, ethyl acetate. Slightly sol in acetone, methanol, alcohol (1 g/1800 ml), ether, and in aq solns of acetic and citric acids. Upon standing most solns acquire a yellow color and pronounced fluorescence; esp after the addition of acid or upon exposure to light.

Hydrochloride hydrate, $C_{33}H_{40}N_2O_9 \cdot HCl \cdot H_2O$, crystals, dec 224°.

Note: Reserpine may reasonably be anticipated to be a carcinogen: *Seventh Annual Report on Carcinogens* (PB95-109781, 1994) p 349.

THERAP CAT: Antihypertensive.

THERAP CAT (VET): Hypotensive, tranquilizer. Has been used to prevent aortic rupture in turkeys.

8315. Resibufogenin. (3β,5β,15β)-14,15-Epoxy-3-hydroxy-5-bufa-20,22-dienolide; Respigon. $C_{24}H_{32}O_4$; mol wt 384.52. C 74.97%, H 8.39%, O 16.64%. Cytotoxic constituent of toad venom. Isoln: Meyer, *Helv. Chim. Acta* **35**, 2444 (1952); Linde, Meyer, *Pharm. Acta Helv.* **33**, 327 (1958). Structure: Thiessen, *Chem. & Ind. (London)* **1958**,

440; Linde, Meyer, *Experientia* **14**, 238 (1958); *eidem, Helv. Chim. Acta* **42**, 807 (1959). Synthesis: Pettit *et al.*, *J. Org. Chem.* **36**, 3736 (1971); Haede *et al.*, *Ann.* **1973**, 5. Stereochemical study of cytotoxic properties: Y. Kamono *et al.*, *J. Chem. Res. (S)* **1977**, 78. Pharmacology: Leigh, Caldwell, *J. Pharm. Pharmacol.* **21**, 708 (1969).

Crystals from acetone + water, mp 113-140°/155-168°. $[\alpha]_D^{22}$ −7.1° (c = 1.259 in chloroform). Also obtained as an amorphous solid, $[\alpha]_D^{16}$ −5.4° (c = 2.030 in chloroform).

Hydrochloride, $C_{24}H_{33}ClO_4$, crystals from acetone, dec 230-232°. $[\alpha]_D^{15}$ +15.1° (c = 0.5302 in chloroform). uv max (alc): 298 nm (log ε 3.74).

THERAP CAT: Cardiotonic.

8316. Resilin. Highly amorphous rubberlike protein found in insect cuticle. Isoln and characterization: Weis-Fogh, *J. Exp. Biol.* **37**, 889 (1960). The structure is a cross-linked random network of flexible protein chains, incorporating 15 of the more common amino acids and involving covalent bonds. Cross-links have been identified as di- and trityrosine. Biosynthesis: Coles, *J. Insect Physiol.* **12**, 679 (1966). Similar to elastin *(q.v.)* but does not form fibers; occurs as masses or typically as 2-5 μ layers. Elasticity approaches that of the ideal cross-linked rubber. Responsible for various springlike actions of winged insects. In flea: Rothschild, *Nature* **239**, 45 (1972). *Review:* Andersen, *Acta Physiol. Scand.* **66**, Suppl. 263 (1966); *idem, "Resilin"* in *Comprehensive Biochemistry* vol. 26C, M. Florkin, E. H. Stotz, Eds. (Elsevier, New York, 1971) pp 633-657.

The colorless material is insol in water <140° and in all other solvents not rupturing peptide bonds. Swells in water and protein solvents; shrinks and becomes hard in absolute methanol, ethanol, dioxane, acetone. Shows no tendency to crystallize at all. Fluorescent in uv light.

8317. Resin Ipomea. Resin of Mexican scammony. Made by extracting the dried roots of *Ipomoea orizabensis* Ledenois, *Convolvulaceae* with alcohol and pptg with water.

Brown, translucent, brittle fragments or pale brown powder. Insol in water; sol in alc, chloroform, partially in ether.

THERAP CAT: Cathartic.

8318. Resin Jalap. Prepd by extracting the powdered root of *Exogonium purga* (Hayne) Lindl. (*E. jalapa* Baill., *Ipomoea purga* Hayne), *Convolvulaceae* with alcohol and pptg with water.

Yellow to brown, amorphous mass or powder. Insol in water, benzene, carbon disulfide, oils; freely sol in alc.

THERAP CAT: Cathartic.

8319. Resin Scammony. Made by extracting the tubers of *Convolvulus scammonia* L., *Convolvulaceae* with alcohol and pptg with water.

Brown, amorphous mass. Freely sol in alc; not less than 95% sol in ether and in caustic alkalies (with gentle heat).

THERAP CAT: Cathartic.

8320. Resistomycin. 3,5,7,10-Tetrahydroxy-1,1,9-trimethyl-2H-benzo[cd]pyrene-2,6(1H)-dione; X-340. $C_{22}H_{16}O_6$; mol wt 376.37. C 70.21%, H 4.28%, O 25.51%. Unusually stable antibiotic substance produced by *Streptomyces resistomycificus:* Brockmann, Schmidt-Kastner, *Naturwiss.* **38**, 479 (1951); *Ber.* **87**, 1460 (1954); Brockmann, Ger. pat. **888,918** (1953). Structure: Brockmann, *Angew. Chem.* **76**, 863 (1964). Revised structure: Bailey *et al.*, *Chem. Commun.* **1968**, 374; Brockmann *et al.*, *Ber.* **102**, 1224 (1969). Synthetic studies: J. F. Kingston, L. Weiler, *Can. J. Chem.* **55**, 785 (1977); K. James, R. A. Raphael, *Tetrahedron Letters* **1979**, 3895. Total synthesis: B. A. Keay, R. Rodrigo, *J. Am. Chem. Soc.* **104**, 4725 (1982). Isoln from *Streptomyces griseoflavus* B71, ^{13}C-NMR and biosynthesis: G. Höfle, H. Wolf, *Ann.* **1983**, 835.

Yellow needles from dioxane, dec 315°. Sublimes at 200-205° (10^{-4} mm) without activity loss. Stable to hot conc H_2SO_4 or hot N KOH. Weakly acid. Absorption max: 268, 290, 320, 337, 366, 457 nm (ϵ 24000, 23000, 14400, 13900, 11000, 15400). Slight soly in water; fair in ether, benzene, alcohol, acetone, acetic acid. Orange-red in sodium hydroxide and piperidine; red in pyridine. Fluoresces in alcohol, benzene, and acetone solns, also in sulfuric acid.
Note: The name Resistomycin is also used as a trademark for Kanamycin A sulfate.

8321. Resodec. A polycarboxylic cation exchange resin. Almost inert, nonabsorbable powder, intended to aid the dietary restriction of salt by removing sodium from the contents of the intestinal tract.
THERAP CAT: Ion exchange resin (sodium).

8322. Resorantel. *N-(4-Bromophenyl)-2,6-dihydroxybenzamide; 4'-bromo-γ-resorcylanilide;* 4'-bromo-2,6-dihydroxybenzanilide; 2-6-dihydroxybenzoic acid 4'-bromoanilide; resorcylam; HOE-296V; Terenol. $C_{13}H_{10}BrNO_3$; mol wt 308.13. C 50.67%, H 3.27%, Br 25.93%, N 4.55%, O 15.58%. Prepn: LeMaire *et al., J. Pharm. Sci.* **50,** 831 (1961); **Brit. pat.** **1,124,613;** H. Ruschig *et al.,* U.S. pat. **3,449,420** (1968, 1969 to Hoechst). Animal studies: Düwel; Behrens, Matschullat; Pfeiffer, *Deut. Tierärztl. Wochenschr.* **77,** 97-107 (1970); Christ *et al., Berlin. München. Tierärztl. Wochenschr.* **83,** 61 (1970).

Colorless powder, mp 229-230°. Also reported as mp 183-186° (LeMaire). Stable at room temp for at least two years. Very resistant to acids and alkalies. Highly sensitive to iron compds which stain it deep red or violet. Very soluble in DMF; slightly sol in lower alcohols. Insol in water, hydrocarbons, vegetable oils.
THERAP CAT (VET): Anthelmintic (sheep and other ruminants).

8323. Resorcinol. *1,3-Benzenediol; m-dihydroxybenzene;* resorcin. $C_6H_6O_2$; mol wt 110.11. C 65.45%, H 5.49%, O 29.06%. Is at least 99.5% pure. Made by fusing *m*-benzenedisulfonic acid with excess NaOH. *Review:* J. Varagnat in Kirk-Othmer *Encyclopedia of Chemical Technology* vol. 13 (Wiley-Interscience, New York, 3rd ed., 1981) pp 39-69.

White, needle-like crystals; sweetish taste. Becomes pink on exposure to light and air, or by contact with iron. d 1.272. mp 109-111°. bp 280°, but volatilizes at lower temp and is slightly volatile with steam. One gram dissolves in 0.9 ml water, 0.2 ml water at 80°, 0.9 ml alcohol; freely sol in ether, glycerol; slightly sol in chloroform. pH: 5.2. *Protect from light. Incompat.* Acetanilide, albumin, alkalies,

antipyrine, camphor, ferric salts, menthol, spirit nitrous ether, urethan.
Monoacetate, $C_8H_8O_3$, *acetylresorcinol, Euresol.* Golden-yellow, thick, syrupy, oily liquid. bp ~283° with decompn. Insol in water. Miscible with alc, benzene, chloroform, acetone. Sol in solns of alkali hydroxides.
Caution: Irritating to skin, mucous membranes. Absorption can cause methemoglobinemia, cyanosis, convulsions, death, *cf.* Patty's *Industrial Hygiene and Toxicology* vol. 2A, G. D. Clayton, F. E. Clayton, Eds. (Wiley-Interscience, New York, 3rd ed.,1981) pp 2586-2589.
USE: In tanning; manuf resins, resin adhesives, hexylresorcinol, *p*-aminosalicylic acid, explosives, dyes; in cosmetics; dyeing and printing textiles; as a reagent for zinc.
THERAP CAT: Keratolytic; antiseborrheic.
THERAP CAT (VET): Topical antipruritic and antiseptic. Has been used as an intestinal antiseptic.

8324. β-Resorcylaldehyde. *2,4-Dihydroxybenzaldehyde;* 2,4-dihydroxybenzenecarbonal. $C_7H_6O_3$; mol wt 138.12. C 60.87%, H 4.38%, O 34.75%. Prepd by saturating with HCl a soln of resorcinol and anhyd hydrocyanic acid in abs ether and boiling the precipitated resorcylaldimide-HCl with water: Gattermann, Köbner, *Ber.* **32,** 278 (1899); **Ger. pat.** **106,508;** *Chem. Zentr.* **1900,** I, 742. By adding $POCl_3$ to a soln of resorcinol and formanilide in ether, then boiling with dil NaOH, acidifying with H_2SO_4 and extracting with ether: Dimrath, Zöppritz, *Ber.* **35,** 995 (1902); by passing HCl into a mixture of resorcinol, ether, anhyd zinc chloride and cyanogen bromide, then boiling the precipitate with water: Karrer, *Helv. Chim. Acta* **2,** 92 (1919).

Needles from water or ether + ligroin. mp 135-136°. bp_{22} 226°. Freely sol in water, alcohol, ether, chloroform, glacial acetic acid. Difficultly sol in cold benzene. β-Resorcylaldehyde is easily dec by acids and alkalies. Prolonged exposure to moist air converts it to a brown, amorphous powder which is insol in ether.

8325. β-Resorcylic Acid. *2,4-Dihydroxybenzoic acid;* 2,4-dihydroxybenzenecarboxylic acid; BRA. $C_7H_6O_4$; mol wt 154.12. C 54.55%, H 3.92%, O 41.52%. Prepared from resorcinol and $KHCO_3$ in glycerol or water by heating and passing CO_2 through the mixture: Brunner, *Ann.* **351,** 320 (1907). *Cf.* Bistrycki, Kostanecki, *Ber.* **18,** 1984 (1885); Nierenstein, Clibbens, *Org. Syn.* coll. vol. II, 557 (1943).

Hydrated crystals from water. Becomes anhydr at 100°, mp 213° (rapid heating). K at 25° = 5.00 × 10^{-4}. Sol in hot water, alcohol, ether, olive oil. Boiling with water, acids or salt solns results in loss of CO_2.
USE: Intermediate for dyestuffs and drugs; spot test reagent for iron.

8326. Resveratrol. *(E)-5-[2-(4-hydroxyphenyl)ethenyl]-1,3-benzenediol; trans-*resveratrol; *(E)*-5-(*p*-hydroxystyryl)resorcinol; 3,4',5-stilbenetriol; 3,5,4'-trihydroxystilbene. $C_{14}H_{12}O_3$; mol wt 228.25. C 73.67%, H 5.30%, O 21.03%. Phytoalexin found in a variety of plants; active ingredient of Asian folk medicine "Kojo-Kon" which is the powdered root of the Japanese knotweed. Isoln from white hellebore and structure: M. Takaoka, *J. Faculty Sci., Hokkaido Imp.*

Univ., (Ser. III), **3**, 1 (1940); *C.A.* **34**, 7887 (1940); from fescue grass: R. G. Powell *et al., Phytochemistry* **35**, 335 (1994). Isoln of monomers, oligomers, isomers and glucosides from wine: F. Mattivi *et al., J. Agric. Food. Chem.* **43**, 1820 (1995). GC-MS determn: D. M. Goldberg *et al., J. Am. Enol. Viticult.* **46**, 159 (1995). Produced as stress metabolite in response to fungal infection or injury: G. H. Dai *et al., Physiolog. Mol. Plant Pathol.* **46**, 177 (1995); P. Jeandet *et al., J. Phytopathol.* **143**, 135 (1995). Antiplatelet aggregating activity: A. A. E. Bertelli *et al., Int. J. Tissue React.* **17**, 1 (1995); and implications in coronary heart disease: C. R. Pace-Asciak *et al., Clin. Chim. Acta* **235**, 207 (1995).

Off-white powder from MeOH, mp 253°-255°.

3-β-mono-D-glucoside, $C_{20}H_{22}O_8$, *polydatin, piceid.* Isoln from fresh root of *Polygonum cuspidatum* Sieb. & Zucc., *Polygonaceae,* and structure: Nonomura *et al., Yakugaku Zasshi* **83**, 988 (1963).

(*Z*)-Form, *cis-resveratrol.* GC-MS determn in wine: D. M. Goldberg *et al., J. Agric. Food Chem.* **43**, 1245 (1995).

8327. Retamine. *Dodecahydro-7,14-methano-2H,6H-dipyrido[1,2-a:1',2'-e][1,5]diazocin-1-ol.* $C_{15}H_{26}N_2O$; mol wt 250.38. C 71.96%, H 10.47%, N 11.19%, O 6.39%. From bark and young branches of *Genista sphaerocarpa* Lam. (*Retama sphaerocarpa* (Lam.) Boiss.), *Leguminosae.* Isoln: Battandier, Malosse, *Compt. Rend.* **125**, 360, 450 (1897); Ribas *et al., Anales Fis. y Quim.* **42**, 516 (1946), *C.A.* **41**, 4894c (1947). Structure: Bohlmann *et al., Ber.* **98**, 659 (1965). Synthesis: Bohlmann *et al., ibid.* 653. Stereochemistry: Ribas *et al., Tetrahedron Letters* **1965**, 3181.

Bitter crystals from alc, mp 165-166°. $[\alpha]_D^{20}$ +46.2° (c = 1.017 in abs ethanol). Practically insol in water; slightly sol in ether; sol in chloroform, alcohol, methanol; very sol in benzene.

Hydrochloride, mp 272-273°.

8328. Retene. *1-Methyl-7-(1-methylethyl)phenanthrene.* $C_{18}H_{18}$; mol wt 234.34. C 92.26%, H 7.74%. Occurs in pine tar, in fossilized pine, in high-boiling tar oils. Formed by Se or Pd dehydrogenation of abietic acid: Ruzicka, Waldmann, *Helv. Chim. Acta* **16**, 842 (1933). Synthesis starting with methyl β-(6-isopropyl-2-naphthoyl)propionate: Bardhan, Sengupta, *J. Chem. Soc.* **1932**, 2520, 2798; *cf.* Bogert, *Science* **77**, 289 (1933). Review and bibliography: Fieser, Fieser, *Natural Products Related to Phenanthrene* (New York, 3rd ed., 1949).

Plates, scales from alc, mp 99°. bp$_{10}$ 208°; bp$_{760}$ 390-394°. Absorption spectrum: Askew, *J. Chem. Soc.* **1935**, 509. Insol in water. Sol in hot alcohol, benzene, hot ether, carbon disulfide. Upon treatment with CrO_3 in acetic acid forms retenequinone, $C_{18}H_{16}O_2$, mp 197.5°.

8329. Reticulin. One of the connective tissue proteins occurring wherever connective tissue forms a boundary: Jacobson, *Nature and Structure of Collagen,* J. T. Randall,

Ed. (Butterworths, London, 1953) pp 6-13, *C.A.* **48**, 10814e (1954). Found together with collagen and elastin. Particularly abundant in membranes of the glomeruli and tubules of the kidney and in supportive structure of the vasculature and lung alveoli. Isoln from kidney tissue slices: Kramer, Little, *ibid.* pp 33-43, *C.A.* **48**, 10815a (1954). Contains about 85% protein, the amino acid content of which is very similar to that of collagen. Also contains about 4.2% carbohydrate and approx 11% bound lipid which yields, on hydrolysis, 95% myristic acid and 5% palmitic acid: Windrum *et al., Brit. J. Exp. Pathol.* **36**, 49 (1955). Lacks hydroxyproline; glycine content is that of normal proteins: Pras, Glynn, *ibid.* **54**, 449 (1973).

Stable in boiling water and in 1*N* HCl. Can be dispersed in distilled water but precipitated by physiological saline (0.15 mol/l).

8330. Reticuline. *1,2,3,4-Tetrahydro-1-[(3-hydroxy-4-methoxyphenyl)methyl]-6-methoxy-2-methyl-7-isoquinolinol; 1,2,3,4-tetrahydro-6-methoxy-2-methyl-1-vanillyl-7-isoquinolinol; 2-methyl-7-hydroxy-6-methoxy-1-(3-hydroxy-4-methoxybenzyl)-1,2,3,4-tetrahydroisoquinoline; coclanoline.* $C_{19}H_{23}NO_4$; mol wt 329.40. C 69.28%, H 7.04%, N 4.25%, O 19.43%. Precursor of many aporphine and morphine-type alkaloids. Isoln of *d*-form from *Anona reticulata* Linn., *Anonaceae:* Gopinath *et al., Ber.* **92**, 776 (1959); *dl*-form from opium: Brochmann-Hanssen, Furuya, *J. Pharm. Sci.* **53**, 575 (1964); *Planta Med.* **12**, 328 (1964). Structure: Brochmann-Hanssen, Nielson, *Tetrahedron Letters* **1965**, 1271. Configuration: Battersby, Evans, *ibid.* 1275. Synthesis of amorphous *dl*-form: Gopinath *et al., Ber.* **92**, 1657 (1959); P. Kerekes *et al., Acta Chim. Acad. Sci. Hung.* **98**, 491 (1978). Synthesis of crystalline *dl*-form: Baxter *et al., J. Chem. Soc. (C)* **1965**, 3645; Chan, Maitland, *ibid.* **1966**, 753; K. C. Rice, A. Brossi, *J. Org. Chem.* **45**, 592 (1980); G. Dornyei *et al., Tetrahedron Letters* **1982**, 2913. Asymmetric synthesis of (*S*)-reticuline: Konda *et al., Chem. Pharm. Bull.* **23**, 1063 (1975). Biosynthesis: D. S. Dewan *et al., J. Chem. Soc. Perkin Trans. I* **1977**, 1662; A. Barrett *et al., ibid.* **1979**, 652.

dl-Form, pink crystals, mp 146°. uv max: 284 nm (log ε 3.85). Sol in aqueous buffer of pH < 7.5 or > 11. Practically insol in water at pH 8-10.

(*S*)-Form perchlorate, $C_{19}H_{23}NO_4 \cdot HClO_4$, colorless prisms from ethanol, mp 203-204°. $[\alpha]_D^{18}$ +88.3° (c = 0.21 in alc).

8331. Retinal. Retinaldehyde; retinene$_1$; vitamin A aldehyde; axerophthal. $C_{20}H_{28}O$; mol wt 284.44. C 84.45%, H 9.92%, O 5.62%. Carotenoid component of the visual pigments. Of the 16 possible stereoisomers, 13 have been synthesized. The 11-*cis* isomer is the chromophore of the majority of naturally occurring opsins, *q.v.* Isoln from retinas: G. Wald, *J. Gen. Physiol.* **19**, 351 (1935); recognition as vitamin A aldehyde: R. A. Morton, *Nature* **153**, 69 (1944). Prepd by the oxidation of vitamin A: Ball *et al., Biochem. J.* **42**, 516 (1948); by the oxidation of β-carotene: Wendler *et al., J. Am. Chem. Soc.* **72**, 234 (1950); from β-ionone: Eiter, Truscheit, U.S. pat. **3,060,229** (1962 to Bayer); Jacobs *et al., Rec. Trav. Chim.* **84**, 1113 (1965). Synthesis *in vitro* from vitamin A$_1$: G. Wald, R. Hubbard, *Proc. Nat. Acad. Sci. USA* **36**, 92 (1950). Synthesis of 11-*cis* isomer: J. M. Dieterle, C. D. Robeson, *Science* **120**, 219 (1954). X-ray crystallography of retinals: R. Gilardi *et al., Nature* **232**, 187 (1971); H. Matsumoto *et al., J. Am. Chem. Soc.* **102**, 4259 (1980). Exposure of opsin-bound retinal to light initiates the *cis-trans* isomerization and resultant dissociation of retinal from the apoprotein. This catalyzes an enzyme cascade which leads to visual excitation: R. Hubbard, A. Kropf,

Proc. Natl. Acad. Sci. USA **44**, 130 (1958); B. Honig et al., ibid. **76**, 2503 (1979). Bioregulation of absorption maxima: B. Honig et al., J. Am. Chem. Soc. **101**, 7084 (1979). HPLC determn in biological material: T. Suzuki, M. Makino-Tasaka, Anal. Biochem. **129**, 111 (1983). Review of syntheses: V. Balogh-Nair, K. Nakanishi, Methods in Enzymology vol. **88**, L. Packer, Ed. (Academic Press, New York, 1982) pp 496-506; R. Liu, A. Asata, ibid. pp 506-516. Role in vision: D. F. O'Brien, Science **218**, 961-966 (1982); P. S. Zurer, Chem. & Eng. News **61**, 24-35 (Nov. 28, 1983). Book: The Retinoids, Vol. **1-2**, M. B. Sporn et al., Eds. (Academic Press, New York, 1984).

11-cis Retinal

all-trans Retinal

trans-Isomer, orange colored crystals from petr ether, mp 61-64°. uv max (hexane): 368 nm (ε 48000). Sol in ethanol, chloroform, cyclohexane, petr ether, oils. Practically insol in water.

11-cis-Isomer, orange prisms from petr ether, mp 63.5-64.4°. uv max (hexane): 365 nm (ε 26360).

8332. Retine. Inhibitor of malignant growth, extracted along with promine, a cancer growth-promoting and sterilizing substance, from calves' thymus: Szent-Györgyi et al., Proc. Nat. Acad. Sci. USA **48**, 1439 (1962). Sterilizing action of promine caused by another substance, infertine: Hegyeli et al., ibid. **49**, 230 (1963). Prepn of retine from human thymus: Hegyeli et al., Science **142**, 1571 (1963); from calves' aortas: Szent-Györgyi et al., ibid. **140**, 1391 (1963); from both sources: eidem, U.S. pat. 3,297,533 (1967 to U.S. Dept. of HEW). Identification studies: Szent-Györgyi, Proc. Nat. Acad. Sci. USA **57**, 1642 (1967). Isoln of promine and retine from calf thymus and liver and their identification as ε-N-methylated lysines and guanidino-methylated arginines, resp.: E. Tyihak, A. Patthy, Acta Agron. (Budapest) **22**, 445 (1973). Polemic on the findings on Tyihak and Patthy: T. Nakajima, Acta Agron. Acad. Sci. Hung. **23**, 236 (1974); C. Marmasse, ibid. 216.

8333. Retinoic Acid. (all-E)-3,7-Dimethyl-9-(2,6,6-trimethyl-1-cyclohexen-1-yl)-2,4,6,8-nonatetraenoic acid; vitamin A acid; tretinoin; Aberel; Airol; Aknoten; Cordes Vas; Dermairol; Epi-Aberel; Eudyna; Retin-A; Vesnaroid. C$_{20}$H$_{28}$O$_2$; mol wt 300.44. C 79.96%, H 9.39%, O 10.65%. Physiological metabolite of retinol, q.v., occurring primarily as the all-trans form. Prepn: C. D. Robeson et al., J. Am. Chem. Soc. **77**, 4111 (1955); Pommer, Sarnecki, U.S. pat. 3,006,939 (1961 to BASF). Prepn by oxidation of retinal: Lakshmanan et al., Biochem. J. **90**, 569 (1964). Single-step process for all-trans and 13-cis isomers: R. Marbet, Ger. pat. 2,061,507; U.S. pat. 3,746,730 (1971, 1973 both to Hoffmann-La Roche). Structure: Stam, MacGillavry, Acta Cryst. **16**, 62 (1963). Toxicology: J. J. Kamm, J. Am. Acad. Dermatol. **6**, 652 (1982). Clinical evaluation in treatment of photoaged skin: J. S. Weiss et al., J. Am. Med. Assoc. **259**, 527 (1988); in promyelocytic leukemia: R. P. Warrell, Jr. et al., Leukemia **8**, 929 (1994). Book: The Retinoids, Vols. **1-2**, M. B. Sporn et al., Eds. (Academic Press, New York, 1984). Reviews: D. B. Ott, P. A. Lachance, Am. J. Clin. Nutr. **32**, 2522-2531 (1979); D. S. Goodman, N. Engl. J. Med. **310**, 1023-1031 (1984); G. D. Goss et al., Crit. Rev. Clin. Lab. Sci. **29**, 185-215 (1992).

all-trans Retinoic Acid

Crystals from ethanol, mp 180-182°. uv max (methanol): 351 nm (ε 45000). LD$_{50}$ (10 day) in mice, rats (mg/kg): 790, 790 i.p.; 2200, 2000 orally (Kamm).

13-cis-Form, isotretinoin, Ro-4-3780, Accutane, Isotrex, Roaccutane. Stereoselective process: R. Lucci, Eur. pat. Appl. **111,325**; U.S. pat. 4,556,518 (1984, 1985 both to Hoffmann-La Roche). Use in severe cystic acne: G. L. Peck et al., N. Engl. J. Med. **300**, 329 (1979). Teratogenicity study: E. J. Lammer et al., ibid. **313**, 837 (1985). Reddish-orange plates from isopropyl alcohol, mp 174-175°. uv max: 354 nm (ε 39800). LD$_{50}$ (20 day) in mice, rats (mg/kg): 904, 901 i.p.; 3389, > 4000 orally (Kamm).

α-Tocopheryl ester, see Tocoretinate.

THERAP CAT: Keratolytic; all-trans-form also as antineoplastic.

8334. Retronecine. (1R-trans)-2,3,5,7a-Tetrahydro-1-hydroxy-1H-pyrrolizine-7-methanol. C$_8$H$_{13}$NO$_2$; mol wt 155.20. C 61.91%, H 8.44%, N 9.03%, O 20.62%. The most common base portion of pyrrolizidine alkaloids. "Necine" bases are 1-methylpyrrolizidines of different stereochemical configurations and degrees of hydroxylation that occur in the form of esters in alkaloids of Senecio, Crotalaria and a number of genera of the Boraginaceae. Retronecine occurs in nature in the (+)-form. It is the necine base of monocrotaline, senecionine, q.q.v., senecyphylline, retrorsine and numerous other hepatotoxic pyrrolizidine alkaloids. Structure: R. Adams, E. F. Rogers, J. Am. Chem. Soc. **63**, 228 (1941). Total synthesis: T. A. Geissman, A. C. Waiss, J. Org. Chem. **27**, 139 (1962). Stereospecific synthesis: E. Vedejs, G. R. Martinez, J. Am. Chem. Soc. **102**, 7993 (1980); H. Niwa et al., Tetrahedron Letters **27**, 4605 (1986). Total synthesis of (−)-form: J. Cooper et al., Chem. Commun. **1988**, 509. Toxicity data: P. N. Harris et al., J. Pharmacol. Exp. Ther. **75**, 78 (1942). Biosynthesis: D. J. Robins, J. R. Sweeney, Chem. Commun. **1979**, 120; G. Grue Sorensen, I. D. Spenser, J. Am. Chem. Soc. **103**, 3208 (1981). Comprehensive reviews on retronecine and other necine bases: F. L. Warren, Fortschr. Chem. Org. Naturst. **12**, 198-269 (1955); **24**, 329-406 (1966); D. J. Robins, ibid. **41**, 115-203 (1982); F. L. Warren in Alkaloids Vol. **XII**, R. H. F. Manske, Ed. (Academic Press, New York, 1970) pp 245-331.

Crystals from acetone, mp 119-120°. [α]$_D^{20}$ +4.95° (c = 0.58 in alc). LD$_{50}$ i.v. in mice: 634.0 ±26.0 mg/kg (Harris). (±)-Form, crystals from acetone, mp 130-131°.

8335. Retrorsine. 12,18-Dihydroxysenecionan-11,16-dione; β-longilobine. C$_{18}$H$_{25}$NO$_6$; mol wt 351.40. C 61.52%, H 7.17%, N 3.99%, O 27.32%. Hepatotoxic pyrrolizidine alkaloid; common constituent of Senecio species. Isoln from Senecio retrorsus DC, Compositae: R. H. F. Manske, Can. J. Chem. **5**, 651 (1931); G. Barger et al., J. Chem. Soc. **1935**, 11; from Crotalaria usaramoensis E. G. Baker, Leguminosae: C. C. J. Culvenor, L. W. Smith, Aust. J. Chem. **19**, 2127 (1966). Structure: S. M. H. Christie et al., J. Chem. Soc. **1949**, 1700; E. C. Leisegang, F. L. Warren, ibid. **1950**, 702. Identity with β-longilobine: F. L. Warren et al., J. Am. Chem. Soc. **72**, 1421 (1950). Review and evaluation of toxicity and carcinogenicity studies: IARC Monographs **10**, 303-312, 333-342 (1976). Comprehensive reviews of pyrrolizidine alkaloids: L. B. Bull et al., The Pyrrolizidine Alkaloids (North Holland, Amsterdam, 1968) 293 pp; F. L.

Warren in *The Alkaloids* **vol. 12**, R. H. F. Manske, Ed. (Academic Press, New York, 1970) pp 245-331.

Crystals from ethyl acetate, mp 212° (Barger *et al.*); 216-216.5° (Bull *et al.*). $[\alpha]_D^{18}$ −17.6° (c = 1.99 in ethanol). uv max (water): 217 nm (log ε 3.85). Readily sol in alcohol, chloroform; slightly sol in water, acetone, ethyl acetate; practically insol in ether.

N-Oxide, $C_{18}H_{25}NO_7$, *isatidine*, *12,18-dihydroxysenecionan-11,16-dione-4-oxide*, *retrorsine N-oxide*. Isolated from *Senecio* species. Review and evaluation of carcinogenicity and toxicity studies: *IARC Monographs* **10**, 269-273 (1976). Crystals from ethanol, mp 140.5-141.5° (Christie *et al.*).

8336. Reviparin Sodium. LU-47311; Clivarin(e). Low molecular weight heparin, prepared by controlled nitrous acid degradation of a standard heparin extracted from porcine intestinal mucosa. Mean molecular weight 3900 Da (range: 3500-4500 Da). Biochemical and pharmacological profile: W. Jeske *et al.*, *Sem. Thromb. Hemostas.* **19**, Suppl 1, 229 (1993). Clinical pharmacokinetics: K. Andrassy *et al.*, *Thromb. Res.* **73**, 95 (1994). Symposium on pharmacology and clinical efficacy in the prevention of thromboembolism and restenosis: *Blood Coagulat. Fibrinol.* **4**, S3-S60 (1993).

THERAP CAT: Antithrombotic.

8337. Rhamnetin. *2-(3,4-Dihydroxyphenyl)-3,5-dihydroxy-7-methoxy-4H-1-benzopyran-4-one; 3,3',4',5-tetrahydroxy-7-methoxyflavone;* 7-methylquercetin; β-rhamnocitrin; cyanidenolon-7-methyl ether 1537. $C_{16}H_{12}O_7$; mol wt 316.27. C 60.76%, H 3.82%, O 35.41%. The aglucon of xanthorhamnin. From fruit of *Rhamnus cathartica* L., *Rhamnaceae:* Krassowski, *J. Russ. Phys. Chem. Soc.* **40**, 1510 (1909); from commercial xanthorhamnin: Nystrom *et al.*, *J. Org. Chem.* **22**, 1272 (1957). Structure: Oesch, Perkin, *J. Chem. Soc.* **105**, 2354 (1914); Jurd, Horowitz, *J. Org. Chem.* **22**, 1618 (1957). Synthesis: Kuhn, Low, *Ber.* **77**, 211 (1944); Jurd, *J. Am. Chem. Soc.* **80**, 5531 (1958); U.S. pat. **2,892,845** (1959 to USDA); Anand *et al.*, *J. Sci. Ind. Res.* **21B**, 322 (1962); *C.A.* **57**, 13712f (1962); Kawano *et al.*, *Chem. Pharm. Bull.* **15**, 711 (1967).

Yellow needles from acetone + methanol, mp 292-294°. uv max (ethanol): 371, 256 nm (log ε 4.41, 4.40). Sol in hot phenol. Slightly sol in hot water, hot alcohol, hot glacial acetic acid, hot acetone. Freely sol in dil alkalies with intense yellow color.

Tetraacetate, $C_{24}H_{20}O_{11}$, needles from acetone + methanol, mp 189-190°.

USE: Has been used for dyeing wool and cotton.

8338. Rhamnose. *6-Deoxy-L-mannose;* L-rhamnose; L-mannomethylose; isodulcit. $C_6H_{12}O_5$; mol wt 164.16. C 43.90%, H 7.37%, O 48.73%. Occurs free in poison sumac *(Rhus toxicodendron* L., *Anacardiaceae),* combined in the form of glycosides of many plants. Preparation: Clark, *J. Biol. Chem.* **38**, 255 (1919). Structure and configuration: Fischer, Morrell, *Ber.* **27**, 384 (1894); Fischer, Zach, *ibid.*

45, 3762 (1912); Hirst, Macbeth, *J. Chem. Soc.* **1926**, 23; Avery, Hirst, *ibid.* **1929**, 2466. Isoln from walls of gram-negative bacteria: Salton, *Biochim. Biophys. Acta* **45**, 364 (1960); from *Afraegle paniculata* Engl., *Rutaceae:* Torto, *J. Chem. Soc.* **1961**, 5234; from rabbit skin: Malawista, Davidson, *Nature* **192**, 871 (1961); from leaves of *Solanum chacoense* Bitter, *Solanaceae:* Kuhn, Löw, *Ber.* **94**, 1088 (1961).

α-L-Rhamnose

α-Form, always obtained by crystn from H_2O or EtOH. Monohydrate, holohedric rods from water, hemihedric monoclinic columns from alcohol. Loses water of crystn on heating and partially changes to the β-modification. Very sweet taste. mp 82-92°: Ghosh, *Proc. Roy. Soc. Edinburgh* **36**, 216 (1915/16). Sublimes at 105° and 2 mm Hg. d_4^{20} 1.4708. Shows mutarotation. $[\alpha]_D^{20}$ −7.7° → +8.9°: Hudson, Yanovsky, *J. Am. Chem. Soc.* **39**, 1032 (1917).

β-Form, prepd by heating α-rhamnose monohydrate on a steam bath; crystallized from anhydr acetone + alcohol. Needles, mp 122-126° (rapid heating). $[\alpha]_D^{20}$ +31.5° (1 min, p = 10). After a short time the rotation adjusts to the same final value as α-rhamnose. The β-form is hygroscopic and changes into crystals of the α-modification upon exposure to moist air.

8339. Rhamnus cathartica. Buckthorn bark. Dried, ripe fruit of *Rhamnus cathartica* L., *Rhamnaceae.* Habit. Europe, Northern Africa to Middle Asia. *Constit.* Rhamnocathartin, rhamnotannic acid, rhamnin, rhamnetin.

THERAP CAT: Cathartic.

8340. Rhapontin. *(E)-3-Hydroxy-5-[2-(3-hydroxy-4-methoxyphenyl)ethenyl]phenyl β-D-glucopyranoside; 4'-methoxy-3,3',5-stilbenetriol 3-glucoside;* rhaponticin; ponticin. $C_{21}H_{24}O_9$; mol wt 420.42. C 60.00%, H 5.75%, O 34.25%. From root of *Rheum rhaponticum* L., *Polygonaceae:* Hesse, *J. Prakt. Chem.* **77**, 321 (1908); Schürhoff, Plettner, *Arch. Pharm.* **275**, 281 (1937). On hydrolysis yields rhapontigenin and glucose: Kawamura, *J. Pharm. Soc. Japan* **58**, 405 (1938), *C.A.* **32**, 6655[4] (1938). Constitution of rhapontigenin: Takaoka, *Proc. Imp. Acad. (Tokyo)* **16**, 408 (1940), *C.A.* **35**, 1399[2] (1941).

β-D-glucose—O

Crystals, dec 236-237°. $[\alpha]_D^{32}$ −59.5° (acetone). Exhibits bright blue fluorescence. Sol in dil alcohol, hot acetone, hot water; slightly sol in ether, alc, acetone, cold water; practically insol in benzene, petr ether, chloroform.

8341. Rheadine. *(8β)-8-methoxy-16-methyl-2,3:10,11-bis[methylenebis(oxy)]rheadan;* rhoeadine. $C_{21}H_{21}NO_6$; mol wt 383.40. C 65.79%, H 5.52%, N 3.65%, O 25.04%. Isoln from the seed capsules of corn poppy *(Papaver rhoeas* L., *Papaveraceae):* Hesse, *Ann.* **140**, 145 (1866); *Arch. Pharm.* **228**, 7 (1890); Awe, *ibid.* **274**, 439 (1936); Awe, Winkler, *ibid.* **290**, 367 (1957); Nemecková, Santavy, *Coll. Czech. Chem. Commun.* **27**, 1210 (1962). Structure and stereochemistry: Santavy *et al.*, *ibid.* **30**, 3479 (1965); **32**, 4452 (1967). Total synthesis: Klotzer *et al.*, *Helv. Chim. Acta* **54**, 2057 (1971); **55**, 2228 (1972); Irie *et al.*, *J. Chem. Soc. Perkin Trans. I* **1972**, 2986.

Crystals from ethanol, mp 252-254°. $[\alpha]_D^{23}$ +235° (c = 1.01 in chloroform); $[\alpha]_D^{22}$ +174° (c = 0.69 in pyridine). uv max (ethanol): 239, 290 nm (ε 9150, 9180). Slightly sol in ethyl acetate, methylene chloride. Practically insol in water, alcohol, chloroform, ether, benzene.

8342. Rhein. *9,10-Dihydro-4,5-dihydroxy-9,10-dioxo-2-anthracenecarboxylic acid; 1,8-dihydroxyanthraquinone-3-carboxylic acid; 4,5-dihydroxyanthraquinone-2-carboxylic acid; chrysazin-3-carboxylic acid; monorhein; rheic acid; cassic acid; parietic acid; rhubarb yellow.* $C_{15}H_8O_6$; mol wt 284.22. C 63.39%, H 2.84%, O 33.77%. Found in the free state and as glucoside in *Rheum* spp, *Polygonaceae* (rhubarb) and in Senna leaves; also in several spp of *Cassia (Leguminosae).* Diacetate used as antirheumatic. Isoln from Chinese rhubarb: J. Schlossberger, O. Doepping, *Ann.* **50**, 196 (1844); O. Hesse, *ibid.* **309**, 32 (1899); F. Tutin, H. W. B. Clewer, *J. Chem. Soc.* **99**, 946 (1911); from aloeemodin, q.v.: O. A. Oesterle, *Arch. Pharm.* **241**, 604 (1903); from *Cassia alata* L.: H. Hauptmann, L. L. Nazario, *J. Am. Chem. Soc.* **72**, 1492 (1950); from *C. fistula:* V. K. Murty *et al., Tetrahedron* **23**, 515 (1967). Prepn from chrysophanic diacetate: Fischer *et al., J. Prakt. Chem.* [2] **83**, 208; **84**, 369 (1911). Structure: H. Nawa *et al., J. Org. Chem.* **26**, 979 (1961). Use in arthritis: C. A. Friedmann, *Ger. pat.* **2,711,-493;** *idem,* U.S. pat. **4,244,968** (1977, 1981 both to Proter). Spectrophotometric study: A. A. Habib, N. A. El-Sebakhy, *J. Nat. Prod.* **43**, 452 (1980). Determn by gas chromatography, mass spectroscopy: G. W. Van Eijk, H. J. Roeijmans, *J. Chromatog.* **295**, 497 (1984); HPLC: A. J. J. Van den Berg, R. P. Labadie, *ibid.* **329**, 311 (1985). Pharmacological effects on colonic mucosa: R. Wanitschke, *Pharmacology* **20**, Suppl. 1, 21 (1980); K. Ewe, *ibid.* **27**. Metabolism study: J. Lemli, L. Lemmons, *ibid.* **50**. In electron transport: P. Egerer *et al., Z. Physiol. Chem.* **363**, 627 (1982). Inhibition of proteases: L. Raimondi *et al., Pharmacol. Res. Commun.* **14**, 103 (1982). Effect on prostaglandin biosynthesis: S. Franchi-Micheli *et al., J. Pharm. Pharmacol.* **35**, 262 (1983).

Yellow needles by sublimation, mp 321-322°, dec 330°. Absorption max (methanol): 229, 258, 435 nm (ε 36800, 20100, 11100). Practically insol in water. Sol in alkalies, pyridine; slightly sol in alc, benzene, chloroform, ether, petr ether. Forms a red potassium salt and a pink sodium salt. Red precips are also obtained with $Ca(OH)_2$ and $Ba(OH)_2$. Diacetate, *See* Diacerein.

8343. Rhenium. Re; at. wt 186.207; at. no. 75; valences 1-7; the heptavalent state is the most stable. Two natural isotopes: 185 (37.07%); 187 (62.93%); the latter is radioactive, $T_{1/2}$ ~10^{11} years. Artificial, radioactive isotopes: 177-184; 186; 188-192. Occurs in gadolinite, molybdenite, columbite, rare earth minerals, and some sulfide ores. Average concn in earth's crust 1 × 10^{-9} (0.001 ppm). Discovered by Noddack *et al., Naturwiss.* **13**, 567, 571 (1925). Prepn of metallic rhenium by reduction of potassium perrhenate or ammonium perrhenate: Hurd, Brim, *Inorg. Syn.* **1**, 175 (1939). Prepn of high purity rhenium: Rosenbaum *et al., J. Electrochem. Soc.* **103**, 518 (1956). *Reviews:* Melaven in *Rare Metals Handbook,* C. A. Hampel, Ed. (Rein-

hold, New York, 1954) pp 347-364; Peacock in *Comprehensive Inorganic Chemistry* vol. 3, J. C. Bailar, Jr. *et al.,* Eds. (Pergamon Press, Oxford, 1973) pp 905-978; P. M. Treichel in Kirk-Othmer *Encyclopedia of Chemical Technology* vol. **20** (Wiley-Interscience, New York, 3rd ed., 1982) pp 249-258.

Hexagonal close-packed crystals, black to silver-gray. d 21.02. mp 3180°. bp 5900° (estimated). Specific heat 0-20° 0.03263 cal/g/°C. Specific electrical resistance: 0.21 × 10^{-4} ohm/cm at 20°. Brinell hardness: 250. Latent heat of vaporization 152 kcal/mol. Reacts with oxidizing acids, nitric and concd sulfuric; not with HCl.

Caution: No toxic manifestations have been reported in exptl animals or in man. *See* E. Browning, *Toxicity of Industrial Metals* (Appleton-Century-Crofts, New York, 1969) pp 276-277.

USE: Electron tube and semiconductor applications, in alloys for electrical contacts, as catalyst; possibly in high temp thermocouples and to improve the workability of tungsten and molybdenum alloys; plating jewelry, medical instruments, high vac equipment, mirror backings.

8344. Rhenium Heptoxide. Dirhenium heptoxide. O_7-Re_2; mol wt 484.41. O 23.12%, Re 76.88%. Re_2O_7. Prepd from rhenium metal or its lower oxides in a stream of air or oxygen at 400-425°: Noddack, Noddack, *Naturwiss.* **17**, 93 (1929); *Z. Anorg. Allgem. Chem.* **181**, 1 (1929); **215**, 129 (1933); Briscoe *et al., Nature* **129**, 618 (1932); Melaven *et al., Inorg. Syn.* **3**, 188 (1950). Crystal structure: Krebs *et al., Chem. Commun.* **1968**, 263.

Canary-yellow, very deliquescent crystals. Begins to sublime at 250°. mp 300.3°; bp 360.3°: Smith *et al., J. Am. Chem. Soc.* **74**, 4964 (1952). Freely sol in water, alcohol, ether, ethyl acetate, dioxane, pyridine. Readily absorbs water forming perrhenic acid, $HReO_4$.

8345. Rhenium Hexafluoride. F_6Re; mol wt 300.20. F 37.97%, Re 62.03%. ReF_6. Prepd by direct fluorination of rhenium metal: Ruff, Kwasnik, *Z. Anorg. Allgem. Chem.* **209**, 113 (1932); **219**, 65 (1934); Malm, Selig, *J. Inorg. Nucl. Chem.* **20**, 189 (1961). The product obtained by Ruff was probably a mixture of ReF_6 and ReF_7: eidem, *J. Am. Chem. Soc.* **82**, 1510 (1960).

Yellow, cubic crystals or liquid; colorless gas; extremely hygroscopic. Forms bluish vapors on contact with air, discolors to a dark purple liquid. mp 18.5°. bp 33.7°. Sol in nitric acid. Soln in anhydr HF: 1.75 moles/1000 g HF, Frlec, Hyman, *Inorg. Chem.* **6**, 1596 (1967). Does not attack dried Pyrex glass or silica. May be handled in copper up to 150°.

8346. Rhenium Oxychloride. *Rhenium chloride oxide; perrhenyl chloride; rhenium trioxide chloride.* ClO_3Re; mol wt 269.66. Cl 13.15%, O 17.80%, Re 69.05%. ReO_3Cl. Prepd by direct chlorination of rhenium(VI) oxide: Wolf *et al., J. Am. Chem. Soc.* **79**, 4257 (1957).

Clear, colorless liquid. mp +4.5°. bp_{760} 128°; also reported to be 131°: Bruckl, Ziegler, *Ber.* **65**, 916 (1932). Distills *in vacuo* without decompn. Can be stored in sealed glass ampuls. Any purple color developing is due to residual ReO_3. Sol in CCl_4. Hydrolyzed by water to $HReO_4$ and HCl. Extremely reactive toward mercury, silver and a large number of organic substances.

8347. Rhenium Trioxide. *Rhenic anhydride.* O_3Re; mol wt 234.21. O 20.49%, Re 79.51%. ReO_3. Prepd by reduction of Re_2O_7 with dioxane: Nechamkin, Hiskey, *Inorg. Syn.* **3**, 186 (1950); with CO: Melaven *et al., ibid.* 187.

Red cubic crystals; green luster by transmitted light; the color of the powder may appear blue (*rhenium blue*). At 400° disproportionates *in vacuo* to Re_2O_7 and ReO_2. bp 750°. d 6.9-7.4. Practically insol in water, alkalies, nonoxidizing acids; oxidized by HNO_3 to $HReO_4$.

8348. Rhizopterin. *4-[[(2-Amino-1,4-dihydro-4-oxo-6-pteridinyl)methyl]formylamino]benzoic acid; p-[N-[(2-amino-4-hydroxy-6-pteridinyl)methyl]formamido]benzoic acid; p-[N-(2-amino-4-hydroxypyrimido[4,5-b]pyrazin-6-ylmethyl)-formamido]benzoic acid; S.L.R. factor; formylpteroic acid.* $C_{15}H_{12}N_6O_4$; mol wt 340.30. C 52.94%, H 3.55%, N 24.70%, O 18.81%. A folic acid factor that possesses high growth-promoting activity for *Streptococcus lactis* R *(S. faecalis* R*),* but almost none for *Lactobacillus casei*, chick and rat: Keresztesy *et al., Science* **97**, 465 (1943). Isoln from charcoal adsorbate obtained from the purification of *Rhizopus nigricans* fumaric acid fermentation liquors: Rickes *et al., J. Am. Chem. Soc.* **69**, 2749 (1947); Keresztesy, Rickes, U.S. pat. 2,478,404 (1949 to Merck & Co.). Structure: Wolf *et al., J. Am. Chem. Soc.* **69**, 2753 (1947).

Light yellow platelets from aq ammonium acetate soln. Does not melt below 300°. Absorption spectrum: Rickes *et al., J. Am. Chem. Soc.* **69**, 2749, 2751 (1947). Practically insol in water and the usual organic solvents. Sol in ammonia water, in aq solns of alkali hydroxides, and in pyridine.

8349. Rhodamine B. *9-(2-carboxyphenyl)-3,6-bis(diethylamino)xanthylium chloride; N-[9-(2-Carboxyphenyl)-6-(diethylamino)-3H-xanthen-3-ylidene]-N-ethylethanaminium chloride;* tetraethylrhodamine; D & C Red No. 19; C.I. Basic Violet 10; C.I. 45170. $C_{28}H_{31}ClN_2O_3$; mol wt 479.02. C 70.21%, H 6.52%, Cl 7.40%, N 5.85%, O 10.02%. Prepn: *Colour Index* vol. **4** (3rd ed., 1971) p 4420. Toxicity study: J. M. Webb *et al., Toxicol. Appl. Pharmacol.* **3**, 696 (1961).

Green crystals or reddish-violet powder. Very sol in water with a bluish-red color, dil solns being strongly fluorescent; very sol in alcohol; slightly sol in HCl, NaOH. LD_{50} i.v. in rats: 89.5 mg/kg (Webb).
USE: As a dye, especially for paper; as a reagent for antimony, bismuth, cobalt, niobium, gold, manganese, mercury, molybdenum, tantalum, thallium, tungsten; as biological stain. Provisionally listed for use in drugs and cosmetics.

8350. Rhodanilic Acid. *Hydrogen bis(benzenamine-N)-tetrakis(thiocyanato-N)chromate(1—).* $C_{16}H_{15}CrN_6S_4$; mol wt 471.60. C 40.75%, H 3.21%, Cr 11.03%, N 17.82%, S 27.20%. [Cr(SCN)$_4$(C$_6$H$_5$NH$_2$)$_2$]H. Prepared from chrome alum, potassium thiocyanate and aniline: Bergmann, *J. Biol. Chem.* **110**, 471 (1935).
Ammonium salt sesquihydrate, $C_{16}H_{18}CrN_7S_4$, *ammonium rhodanilate.* Rose-colored platelets. Sol in methanol.
USE: In the separation of amino acids, *cf.* Reinecke Salt.

8351. Rhodanine. *2-Thioxo-4-thiazolidinone;* 2-thio-4-ketothiazolidine; 4-oxo-2-thionothiazolidine; rhodanic acid. $C_3H_3NOS_2$; mol wt 133.19. C 27.05%, H 2.27%, N 10.52%, O 12.01%, S 48.15%. Prepd by the action of sodium chloroacetate on ammonium dithiocarbamate: Julian, Sturgis, *J. Am. Chem. Soc.* **57**, 1126 (1935); Redemann *et al., Org. Syn.* coll. vol. **III**, 763 (1955).

Pale yellow crystals (long blades) from glacial acetic acid or water. d 0.868. mp 168.5° (capillary); mp 170.5-171° (hot stage). May explode on rapid heating. K at 25° = 3

× 10^{-6}. Freely sol in boiling water, in methanol, ethanol, DMF, ether, alkalies, ammonia, hot acetic acid.
USE: In the synthesis of phenylalanine.

8352. Rhodinol. *S-(—)-3,7-Dimethyl-7-octen-1-ol;* α-citronellol. $C_{10}H_{20}O$; mol wt 156.27. C 76.86%, H 12.90%, O 10.24%. Name that has been used to describe a group of related chemicals as well as a single chemical. The term, at times, was applied to what was considered the most important alcohol in geranium oil, *1-citronellol. See:* Barbier, Bouveault, *Ber.* **29**, 289 (1896). Prepn in pure form: Sutherland, *J. Am. Chem. Soc.* **73**, 2385 (1951). Synthesis from *cis*-pinane: Rienäcker, *Chimia* **27**, 97 (1973).

Oily liquid, odor of rose. d_4^{20} 0.8549. bp_{12} 114-115°. n_D^{20} 1.4556. $[\alpha]_D^{20}$ —2.88°. uv max: 186-189 nm (ε 9000). Very slightly sol in water, miscible with alcohol, ether.
USE: In perfumery. *See also* β-Citronellol.

8353. Rhodium. Rh; at. wt 102.9055; at. no. 45; valences 1-6; most common states 1, 3. One naturally occurring isotope: 103; artificial radioactive isotopes: 97-102; 104-110. Belongs to the platinum group of metals. One of the rarest elements: constitutes about 1 × 10^{-7}% of the earth's crust; found in small quantities associated with all native platinum; in the minerals rhodite, sperrylite, iridosmine; in some nickel-copper ores. Discovered in 1803 by Wollaston, *Phil. Trans.* **94**, 419 (1804). Prepn: Vauquelin, etc., cited by Mellor, *A Comprehensive Treatise on Inorganic and Theoretical Chemistry* **15**, 546 (1936). Practical methods of separation: Wichers, Gilchrist, *Trans. Am. Inst. Mining Metallurg. Eng.* **76**, 619 (1928). Reviews of prepn, properties and chemistry of rhodium and other platinum metals: Gilchrist, *Chem. Rev.* **32**, 277-372 (1943); Beamish *et al.,* in *Rare Metals Handbook,* C. A. Hampel, Ed. (Reinhold, New York, 1956) pp 291-328; W. P. Griffith, *The Chemistry of the Rarer Platinum Metals* (John Wiley, New York, 1967) pp 1-41, 313-430; Livingstone in *Comprehensive Inorganic Chemistry,* Vol. **3**, J. C. Bailar Jr. *et al.,* Eds. (Pergamon Press, Oxford, 1973) pp 1163-1189, 1233-1253.

Silvery-white, soft, ductile, malleable metal; face-centered cubic structure. mp 1966°: Roeser, Wensel, *Nat. Bur. Stand. J. Res.* **12**, 519 (1934); d^{20} 12.41. Electr resistivity (0°) 4.51 µ-ohms-cm. Brinell hardness: 100. Not attacked by acids even aqua regia when in compact form; the finely divided metal reacts with aqua regia. Absorbs oxygen when melted; at a red heat is slowly oxidized to the sesquioxide. Converted to the trihalide by chlorine or bromine at a red heat; not attacked by fluorine.
Caution: No toxic effects have been reported in exptl animals or in man, E. Browning, *Toxicity of Industrial Metals* (Appleton-Century-Crofts, New York, 1969) pp 278-279.
USE: As an alloy with platinum; as a corrosion-resistant electroplate for protecting silverware from tarnishing; for making high-reflectivity mirrors for cinema projectors, searchlights. Spongy or black rhodium is used as a catalyst in various organic hydrogenation and oxidation reactions.

8354. Rhodium Carbonyl Chloride. *Tetracarbonyldi-µ-chlorodirhodium;* dichlorotetracarbonyldirhodium; chlorodicarbonylrhodium(I) dimer. $C_4Cl_2O_4Rh_2$; mol wt 388.76. C 12.36%, Cl 18.24%, O 16.46%, Rh 52.94%. [Rh(CO)$_2$Cl]$_2$. Prepn: Hieber, Lagally, *Z. Anorg. Allgem. Chem.* **251**, 96 (1943); McCleverty, Wilkinson, *Inorg. Syn.* **8**, 211 (1966); Colton *et al., Aust. J. Chem.* **23**, 1351 (1970); Cramer, *Inorg. Syn.* **15**, 14 (1974). Structure: Dahl *et al., J. Am. Chem. Soc.* **83**, 1761 (1961).
Orange-red crystalline solid or red crystalline sublimate. mp 124-125°. Sol in most organic solvents (except aliphatic

hydrocarbons). Stable to dry air. Solutions in organic solvents decompose when exposed to air.

USE: Catalyst.

8355. Rhodium Chloride. Rhodium trichloride. Cl_3-Rh; mol wt 209.26. Cl 50.82%, Rh 49.18%. $RhCl_3$. Similar to the chlorides of other platinum group metals, rhodium chloride readily forms double salts with alkali chlorides. The chloride is obtainable also as a water-sol hydrate with variable amounts of water. Prepn of trihydrate: Anderson, Basolo, *Inorg. Syn.* **7,** 214 (1963). Toxicity data: Landolt *et al., Toxicol. Appl. Pharmacol.* **21,** 589 (1972).

Red powder. Insol in water. Sol in alkali hydroxide or cyanide solns. LD_{50} in rats (mg/kg): 198 i.v. (Landolt).

8356. Rhododendrin. *3-(4-Hydroxyphenyl)-1-methylpropyl β-D-glucopyranoside;* betuloside. $C_{16}H_{24}O_7$; mol wt 328.36. C 58.53%, H 7.37%, O 34.11%. From leaves of *Rhododendron* spp., *Ericaceae:* Archangelski, *Arch. Exp. Pathol. Pharmakol.* **46,** 313 (1901); Kawaguchi *et al., J. Pharm. Soc. Japan* **62,** 4 (1942), *C.A.* **44,** 9634a (1950). Structure: Kim, *ibid.* 455, *C.A.* **45,** 4222h (1951). Identity with betuloside: Kim, *ibid.* **63,** 103 (1943), *C.A.* **45,** 4222i (1951). Chemotaxonomic significance: Thieme, Winkler, *Pharmazie* **24,** 703 (1969).

Bitter crystals, mp 187°. Sol in hot water, alcohol; slightly sol in chloroform, ether.

Pentaacetate, needles, mp 96-97°.

Methyl ether, prisms from warm water, mp 102-103°. $[\alpha]_D^{21}$ −17.1° (96% alcohol).

USE: In systematic classification of *Rhododendron* species.

8357. Rhodomycins. Antibiotic substance produced by *Streptomyces purpurascens* nov. sp.: Brockmann, Bauer, *Naturwiss.* **37,** 492 (1950); Brockmann *et al., Ber.* **84,** 700 (1951); **Brit.** pat. **708,749** (1954 to Bayer). Separation of rhodomycin A and rhodomycin B: Brockmann, Patt, *Ber.* **88,** 1455 (1955). Isoln and structural elucidation of additional rhodomycins produced by *Streptomyces stammes:* Biedermann, Brauniger, *Pharmazie* **27,** 782 (1972); Brockmann *et al., Tetrahedron Letters* **1973,** 3699. Total synthesis of racemic rhodomycinones (the aglycones of the rhodomycins): A. S. Kende, Y.-G. Tsay, *Chem. Commun.* **1977,** 140.

Rhodomycin A

Rhodomycin A, $C_{36}H_{48}N_2O_{12}$, *[7R-(7α,8β,10β)]-8-ethyl-7,-8,9,10-tetrahydro-1,6,8,11-tetrahydroxy-7,10-bis[[2,3,6-trideoxy-3-(dimethylamino)-α-L-lyxo-hexopyranosyl]oxy]-5,12-naphthacenedione, β-rhodomycin II.* Dec on mild acidic hydrolysis into 1 mole β-rhodomycinone and 2 moles rhodosamine: Brockmann, Spohler, *Naturwiss.* **48,** 716 (1961). Structure of β-rhodomycinone: Brockmann, Niemeyer, *ibid.* 570. Structure of rhodosamine: Brockmann, Spohler, *ibid.*

42, 154 (1955). Total structure: Brockmann *et al., Tetrahedron Letters* **1969,** 415.

Dihydrochloride, $C_{36}H_{48}N_2O_{12}.2HCl$, red prisms from ethanol+ isopropanol, mp 205°. Freely sol in water, lower alcohols; sparingly sol in benzene, chloroform; practically insol in ether, petr ether.

Perchlorate, $C_{36}H_{48}N_2O_{12}.2HClO_4$, minute red needles, mp 188°. Freely sol in methanol, acetone; sparingly sol in water, butanol.

Rhodomycin B, $C_{28}H_{31}NO_9$. *[7R-(7α,8β,10β)]-8-ethyl-7,8,-9,10-tetrahydro-1,6,7,8,11-pentahydroxy-10-[[2,3,6-trideoxy-3-(dimethylamino)-α-L-lyxo-hexopyranosyl]oxy]-5,12-naphthacenedione.* Dec on mild acidic hydrolysis into 1 mole β-rhodomycinone and 1 mole rhodosamine: Brockmann, Spohler, *ibid.* **48,** 716 (1961).

Hydrochloride, red prisms, mp 180°. Freely sol in lower alcohols; sparingly sol in water, acetone. Practically insol in chloroform, benzene.

8358. Rhodopin. *1,2-Dihydro-1-hydroxy-ψ,ψ-carotene; 1,2-dihydro-1-hydroxylycopene.* $C_{40}H_{58}O$; mol wt 554.90. C 86.58%, H 10.54%, O 2.88%. Pigment from *Rhodovibrio* and *Thiocystis* bacteria: Karrer, Solmssen, *Helv. Chim. Acta* **18,** 1306 (1935); **19,** 1019 (1936); Karrer *et al., ibid.* **21,** 454 (1938); from *Rhodomicrobium vannielii:* Volk, Pennington, *J. Bacteriol.* **59,** 169 (1950); Ryvarden, Jensen, *Acta Chem. Scand.* **18,** 643 (1964). Structure: Jensen, *ibid.* **13,** 842, 2142 (1959). Synthesis: Bonnett *et al., ibid.* **18,** 1739 (1964); Surmatis *et al., J. Org. Chem.* **31,** 186 (1966). Synthesis of deuterated rhodopin: Johansen, Liaaen-Jensen, *Acta Chem. Scand.* **28B,** 301 (1974).

Aggregates of dark red crystals from carbon disulfide + petr ether, mp 171° (preliminary sintering). Absorption max (chloroform): 521, 486, 453 nm.

8359. Rhodopsin. Visual purple. Mol wt ~40,000. A photoreceptor protein found in the rods of the retina of the eye. Composed of 11-*cis* retinal, *q.v.,* bound as a protonated Schiff base to a lysine moiety in the apoprotein opsin, *q.v.* Prepn from retinal and opsin: G. Wald, P. K. Brown, *Proc. Natl. Acad. Sci. USA* **36,** 84 (1950); R. Hubbard, G. Wald, *ibid.* **37,** 69 (1951). Purification and amino acid compn of bovine rhodopsin: Shields *et al., Biochim. Biophys. Acta* **147,** 238 (1967). Synthesis of labelled rhodopsin and studies on the active site: Hirtenstein, Akhtar, *Biochem. J.* **119,** 359 (1970). Characterization in synthetic membranes: W. L. Hubbell, *Accounts Chem. Res.* **8,** 85 (1975). Amino acid sequence of bovine rhodopsin: P. A. Hargrave *et al., Biophys. Struct. Mech.* **9,** 235 (1983). Studies on transmembrane molecular structure: E. A. Dratz, P. A. Hargrave, *Trends Biochem. Sci.* **8,** 128 (1983). Isoln and sequence of gene encoding bovine rhodopsin: J. Nathans, D. S. Hogness, *Cell* **34,** 807 (1983); of gene encoding human rhodopsin: *eidem, Proc. Natl. Acad. Sci. USA* **81,** 4851 (1984). Exposure to light initiates the conversion of rhodopsin through a series of distinct intermediates to yield opsin + *trans*-retinal: R. Hubbard, A. Kropf, *Proc. Natl. Acad. Sci. USA* **44,** 130 (1958); R. Hubbard *et al., Nature* **183,** 142 (1959); B. Honig *et al., Proc. Natl. Acad. Sci. USA* **76,** 2503 (1979). Review of visual transduction and bioregulation of absorption maxima: K. Nakanishi, *Pure Appl. Chem.* **57,** 769-776 (1985). *Reviews:* T. G. Ebrey, B. Honig, *Quart. Rev. Biophys.* **8,** 129-184 (1975); D. F. O'Brien, *Science* **218,** 961-966 (1982); A. Maeda, T. Yoshizawa, *Photochem. Photobiol.* **35,** 891-898 (1982); P. S. Zurer, *Chem. & Eng. News* **61,** 24-35 (Nov. 28, 1983). *See also: Methods in Enzymology* **81** and **88,** Lester Packer, Ed. (Academic Press, New York, 1982).

8360. Rhodoquinone. *2-Amino-5-(3,7,11,15,19,23,27,-31,35,39-decamethyl-2,6,10,14,18,22,26,30,34,38-tetracontadecaenyl)-3-methoxy-6-methyl-2,5-cyclohexadiene-1,4-dione;* rhodoquinone-10. $C_{58}H_{89}NO_3$; mol wt 848.35. C 82.12%,

H 10.57%, N 1.65%, O 5.66%. Isoln from *Rhodospirillum rubrum:* Gloves, Threlfall, *Biochem. J.* **85,** 14P (1962). Structure and synthesis: Moore, Folkers, *J. Am. Chem. Soc.* **87,** 1409 (1965); **88,** 567 (1966); Daves *et al., ibid.* **90,** 5587 (1968).

Violet needles, mp 70-71°. uv max (cyclohexane): 280, 500 nm ($E_{1cm}^{1\%}$ 154, 14).

8361. Rhodoviolascin. *3,3',4,4'-Tetradehydro-1,1',2,2'-tetrahydro-1,1'-dimethoxy-ψ,ψ-carotene; 3,3',4,4'-tetradehydro-1,1',2,2'-tetrahydro-1,1'-dimethoxylycopene;* spirilloxanthin. $C_{42}H_{60}O_2$; mol wt 596.94. C 84.51%, H 10.13%, O 5.36%. Carotenoid pigment isolated from *Rhodovibrio* and *Thioceptis* bacteria: Karrer, Solmssen, *Helv. Chim. Acta* **18,** 1306 (1935). Identity with spirilloxanthin: Zechmeister *et al., Arch. Biochim.* **5,** 243 (1944). Structure: Karrer, Koenig, *Helv. Chim. Acta* **23,** 460 (1940); Barber *et al., Proc. Chem. Soc.* **1959,** 96. Synthesis: Surmatis, Ofner, *J. Org. Chem.* **28,** 2735 (1963); Surmatis, U.S. pat. **3,160,658** (1964 to Hoffman-La Roche).

Dark red, spindle-shaped crystals from benzene, mp 218°. Absorption max (chloroform): 544, 507, 476 nm. Almost insol in petr ether, ligroin, methanol, somewhat more sol in hot benzene.

8362. Rhodoxanthin. *4',5'-Didehydro-4,5-retro-β,β-carotene-3,3'-dione.* $C_{40}H_{50}O_2$; mol wt 562.84. C 85.36%, H 8.95%, O 5.69%. Carotenoid pigment found widely distributed in nature, but in very small amounts only. One of very few carotenoids with retro configuration. Best isolated from the arils of *Taxus baccata* L., *Taxaceae:* Kuhn, Brockmann, *Ber.* **66,** 828 (1933). Occurs also in the feathers of birds such as *Phoenicircus nigricollis, Megaloprepia magnifica.* Isoln: Volker, *Z. Physiol. Chem.* **292,** 75 (1953). Structure: Kuhn, Brockmann, *loc. cit.;* Karrer, Solmssen, *Helv. Chim. Acta* **18,** 477 (1935). Synthesis: Mayer *et al., ibid.* **50,** 1606 (1967); Surmatis, Walser, U.S. pats. **3,466,331; 3,624,105** (1969, 1971 to Hoffmann-La Roche); P. R. Ellis *et al., Helv. Chim. Acta* **64,** 1092 (1981); E. Widmer *et al., ibid.* **65,** 944 (1982). *Review:* Liaaen-Jensen in *Aspects Terpenoid Chem. Biochem., Proc. Phytochem. Soc. Symp.,* 2nd, 1970, T. W. Goodwin, Ed. (Academic Press, London, 1971) p 223-254.

Rosettes of deep purple, lancet-shaped crystals from benzene:methanol (1:4). mp 219° (evacuated tube). Absorption max (chloroform): 546, 510, 482 nm. Freely sol in pyridine; sol in benzene, chloroform. Sparingly sol in ethanol and methanol. Practically insol in hexane, petr ether.
USE: Coloring material for food, beverages, pharmaceuticals and cosmetics.

8363. Rhubarb. Rhizome and roots of *Rheum officinale* Baill., *R. palmatum* L., or other *Rheum* spp, *Polygonaceae. Habit.* Central Asia; cultivated in Europe, Southern Siberia, North America. *Constit.* Chrysophanic acid, emodin, rhein, rheotannic acid, erythroretin, methylchrysophan-

ic acid, rhabarberon, cinnamic and gallic acids, calcium oxalate. Comprehensive studies: Workman, Hiner, *J. Am. Pharm. Assoc.* **49,** 118 (1960). *Incompat.* Mineral acids, iron salts, lead acetate, zinc salts, lime water; infusions of catechu, cinchona, and nutgall; tartar emetic, tannic acid.
THERAP CAT: Cathartic.
THERAP CAT (VET): Has been used as a laxative.

8364. Rhynchophylline. *(7β,16E,20α)-16,17-Didehydro-17-methoxy-2-oxocorynoxan-16-carboxylic acid methyl ester;* rhyncophylline; mitrinermine. $C_{22}H_{28}N_2O_4$; mol wt 384.48. C 68.73%, H 7.34%, N 7.29%, O 16.65%. From stems and roots of *Uncaria rynchophylla* Miq. [*Ourouparia rynchophylla* (Miq.)] Matsum, *Rubiaceae:* Kondo *et al., J. Pharm. Soc. Japan* **48,** 54 (1928); from *Mitragyna rotundifolia* (Roxb.), *Rubiaceae:* Barger *et al., J. Org. Chem.* **4,** 418 (1939). Structure: Seaton, Marion, *Can. J. Chem.* **35,** 1102 (1957); **38,** 1035 (1960). Stereochemistry: Nozoye, *Chem. Pharm. Bull.* **6,** 309 (1958); Ban, Oishi, *Tetrahedron Letters* **1961,** 791; eidem, *Chem. Pharm. Bull.* **11,** 441, 446, 451 (1963). Partial synthesis: Finch, Taylor, *J. Am. Chem. Soc.* **84,** 1318, 3871 (1962). Total synthesis: Ban *et al., Tetrahedron Letters* **1972,** 2113.

Crystals from methanol, mp 216°; *dl*-form reported as colorless pillars from ethyl acetate-ether, mp 197-199° (Ban *et al., loc. cit.*). $[\alpha]_D^{13}$ – 14.7° (c = 2.5 in chloroform). pKa 6.4. uv max: 245, 280 nm (log ε 4.24, 3.15). Sol in chloroform; moderately sol in acetone, alcohol, benzene; sparingly sol in ether, ethyl acetate. Practically insol in petr ether.

8365. Ribavirin. *1-β-D-Ribofuranosyl-1H-1,2,4-triazole-3-carboxamide;* RTCA; ICN-1229; Viramid; Virazid; Virazole. $C_8H_{12}N_4O_5$; mol wt 244.21. C 39.35%, H 4.95%, N 22.94%, O 32.76%. The first synthetic, non-interferon-inducing, broad-spectrum antiviral nucleoside. Preliminary information: *Chem. & Eng. News* **50,** 26 (April 17, 1972). Synthesis: J. T. Witkowski *et al.,* 163rd Am. Chem. Soc. Meeting (Boston, April 1972), *Abstracts of Papers* MEDI 19; eidem, *J. Med. Chem.* **15,** 1150 (1972); eidem, *J. Carbohyd. Nucl. Nucl.* **5,** 363 (1978). Regioselective synthesis: Y. Ito et al., *Tetrahedron Letters* **1979,** 2521; R. R. Schmidt, D. Heermann, *Ber.* **114,** 2825 (1981). Structure and conformation studies: Kreishman *et al., J. Am. Chem. Soc.* **94,** 5894 (1972); Prusiner, Sundaralingam, *Nature New Biol.* **244,** 116 (1973). Activity studies: Sidwell *et al., Science* **177,** 705 (1972); Huffman *et al., Antimicrob. Ag. Chemother.* **3,** 235 (1973); Sidwell *et al., ibid.* **242;** Khare *et al., ibid.* **517.** *In vitro* inhibition of HIV-1 (HTLV-III/LAV) virus replication: J. B. McCormick *et al., Lancet* **2,** 1367 (1984). Toxicity data: J. T. Witkowski *et al., J. Med. Chem.* **15,** 1150 (1972). Teratogenicity studies: V. H. Ferm *et al., Teratology* **17,** 93 (1978); D. M. Kochhar *et al., Toxicol. Appl. Pharmacol.* **52,** 99 (1980). Controlled clinical trial in infants with respiratory syncytial viral infection: C. B. Hall *et al., N. Engl. J. Med.* **308,** 1443 (1983); in Lassa fever: J. B. McCormick *et al., ibid.* **314,** 20 (1986). *Review:* R. Sidewell *et al., Pharmacol. Ther.* **6,** 123-146 (1979); F. E. Hahn, Ed.

in *Antibiotics* vol. **5**, pt. 2 (Springer-Verlag, New York, 1979) pp 439-458.

Colorless, water-soluble, stable material. Exists in two polymorphic forms: mp 166-168° (aq ethanol); mp 174-176° (ethanol). $[\alpha]_D^{25}$ −36.5°. LD_{50} i.p. in mice: 1.3 g/kg; orally in rats: 5.3 g/kg (Witkowski).

THERAP CAT: Antiviral.

8366. α-Ribazole. *5,6-Dimethyl-1-α-D-ribofuranosyl-1H-benzimidazole.* $C_{14}H_{18}N_2O_4$; mol wt 278.31. C 60.42%, H 6.52%, N 10.07%, O 23.00%. Nucleoside moiety of vitamin B_{12}, *q.v.* Isoln by acid hydrolysis of vitamin B_{12}: N. G. Brink *et al., J. Am. Chem. Soc.* **72**, 1866 (1950); N. G. Brink, K. Folkers, *ibid.* **74**, 2856 (1952). Syntheses of α- and β-anomers: F. W. Holly *et al., ibid.* 4521; R. S. Wright *et al., ibid.* **80**, 2004 (1958); J. D. Stevens *et al., J. Org. Chem.* **33**, 1806 (1968). Biosynthetic study: J. Hörig, P. Renz, *FEBS Letters* **80**, 337 (1977). Spectroscopic characterization: K. L. Brown *et al., Inorg. Chem.* **23**, 1463 (1984).

Crystals from water, mp 198-199°. $[\alpha]_D^{23}$ +14° (c = 0.9 in pyridine).

8367. Riboflavin. Vitamin B_2; lactoflavine; vitamin G; 7,8-dimethyl-10-(D-*ribo*-2,3,4,5-tetrahydroxypentyl)isoalloxazine; 7,8-dimethyl-10-ribitylisoalloxazine; Beflavin; Flavaxin. $C_{17}H_{20}N_4O_6$; mol wt 376.37. C 54.25%, H 5.36%, N 14.89%, O 25.51%. Nutritional factor found in milk, eggs, malted barley, liver, kidney, heart, leafy vegetables. Richest natural source is yeast. Minute amounts present in all plant and animal cells. Occurs in the free form only in the retina of the eye, in whey, and in urine. Bioactive forms occurring in tissues and cells are riboflavin monophosphate and flavine-adenine dinucleotide, *q.q.v.* Syntheses: Karrer *et al., Helv. Chim. Acta* **18**, 426, 522 (1935); Kuhn *et al., Naturwiss.* **23**, 260 (1935); Howe, U.S. pat. **2,807,611** (1957 to Merck & Co.); F. Yoneda *et al., J. Chem. Soc. Perkin Trans. I*, **1978**, 348. May also be produced by fermentation: Malzahn *et al.*, U.S. pat. **2,876,169** (1959 to Grain Processing Corp.). Toxicity studies: Kuhn, Boulanger, *Z. Physiol. Chem.* **241**, 233 (1936); K. Unna, J. G. Greslin, *J. Pharmacol. Exp. Ther.* **76**, 75 (1942). Early reviews: Rosenberg in *The Vitamins* (New York, 1945); Sebrell, Harris, *The Vitamins* vol. **III** (Academic Press, New York, 1954) pp 299-402; Vogel-Knobloch, *Chemie und Technik der Vitamine* vol. **II**, part 1 (Enke, Stuttgart, 1955). Review of metabolism: Rivlin, *N. Engl. J. Med.* **283**, 463 (1970). Review of extraction and assay methods: Pearson, *The Vitamins* vol. **VII**, P. György, W. N. Pearson, Eds. (Academic Press, New York, 2nd ed., 1967) pp 99-136. Comprehensive reviews: *ibid.* vol. **V**, pp 1-96 (1972); F. J. Al-Shammary *et al.* in *Anal. Profiles Drug Subs.* **19**, 429-476 (1990).

Fine orange-yellow needles from 2N acetic acid, alcohol, water, or pyridine. Dec at 278-282° (darkens at about 240°). Three different crystal forms having different solubilities in water: Dale, U.S. pats. **2,603,633** and **2,797,215**. $[\alpha]_D^{25}$ −112° to −122° (50 mg in 2 ml 0.1N alcoholic NaOH dil to 10 ml with water). Absorption max: 220-225, 266, 371, 444, 475 nm. One gram dissolves in from 3000 to about 15,000 ml of water, the variations in the soly being due to differences in the crystal structure. Soly in alcohol (27.5°): 0.0045 g/100 ml of abs ethanol. Slightly sol in cyclohexanol, amyl acetate, benzyl alcohol, phenol; very sol in dil alkalies (with decompn). Insol in ether, chloroform, acetone, benzene. pKa 10.2; pKb 1.7. pH of satd aq soln: ∼6. Isoelectric pt: pH 6. Aq solns are yellow showing a green fluoresence with max at 565 nm. Sensitive to alkalies; stable to mineral acids in the dark. Visible or ultraviolet irradiation of alkaline solns causes the formation of lumiflavine, *q.v.*, whereas irradiation of acid or neutral solns gives rise to the production of lumichrome, *q.v.*, together with varying amounts of lumiflavine. LD_{50} in rats (g/kg): > 10 orally, 5.0 s.c., 0.56 i.p. (Unna, Greslin).

2′,3′,4′,5′-Tetrabutyrate, $C_{33}H_{44}N_4O_{10}$, *Bituvitan, Eyekas, Hibon, Lacflavin, Ribolact, Wakaflavin L, Viras.*

THERAP CAT: Vitamin (enzyme cofactor).

THERAP CAT (VET): Vitamin (enzyme cofactor).

8368. Riboflavin Monophosphate. *Riboflavin 5′-(dihydrogen phosphate);* flavin mononucleotide; FMN; vitamin B_2 phosphate. $C_{17}H_{21}N_4O_9P$; mol wt 456.35. C 44.74%, H 4.64%, N 12.28%, O 31.55%, P 6.79%. One of the bioactive forms of riboflavin, *q.v.* Prepd by phosphorylation of riboflavin with chlorophosphoric acid: L. A. Flexser, W. G. Farkas, U.S. pat. **2,610,179** (1952 to Hoffmann-La Roche); with pyrophosphoric acid: Breivogel, Ridge, U.S. pat. **2,535,385** (1950 to White Labs.); with metaphosphoric acid: Viscontini *et al., Helv. Chim. Acta* **35**, 457 (1952); U.S. pat. **2,740,775** (1956 to Hoffmann-La Roche); with pyrocatechol cyclic phosphate: Ukita *et al.*, U.S. pat. **3,118,876** (1964 to Takeda).

Monosodium salt, $C_{17}H_{20}N_4NaO_9P$, *Hyryl, Ribo.* Occurs as the dihydrate. Yellow crystals. Soly in water at pH 6.9 = 112 mg/ml; at pH 5.6 = 68 mg/ml; at pH 3.8 = 43 mg/ml. *Protect from light.*

THERAP CAT: Vitamin (enzyme cofactor).

THERAP CAT (VET): Vitamin (enzyme cofactor).

8369. Ribonuclease. RNase. An enzyme that digests RNA. First isolated from beef pancreas in 1920. Crystallization from acid extracts of pancreas: Kunitz, *J. Gen. Physiol.* **24**, 15 (1940). Isoln: Kunitz, McDonald, *Biochem. Prepn.* **3**, 9 (1953). Bovine pancreatic RNase A is a single-chain peptide of 124 amino acid residues. Primary structure: C. H. W. Hirs *et al., J. Biol. Chem.* **235**, 633 (1960). Amino acid sequence: D. H. Spackman *et al., ibid.* 648. Pictorial representation of entire structure: Stein, Moore, *Sci. Am.* **204**, no. 2, 81-92 (Feb. 1961). Ribonuclease from plant leaves has slightly different characteristics: Markham, Strominger, *Biochem. J.* **64**, 46P (1956). Can be obtained as a by-product of microbial erythromycin production: **Japan.** pat. **26,938**('63) (to Shionogi). Chemical synthesis of materials possessing partial RNase enzyme activity: Gutte, Merrifield, *J. Am. Chem. Soc.* **91**, 501 (1969); Denkewalter, Hirschmann *et al., ibid.* 502. Series of articles describing the total synthesis of a protein having the full enzymic activity of bovine pancreatic RNase: N. Fujii, H. Yajima, *J. Chem. Soc. Perkin Trans. I* **1981**, 789-841. Specifically catalyzes the cleavage of the phosphodiester bond between the 3′ and 5′ positions of the ribose moieties in RNA with the formation of oligonucleotides terminating in 2′,3′-cyclic phosphate derivs: Roberts *et al., Proc. Nat. Acad. Sci. USA* **62**, 1151 (1969). *Reviews:* Anfinsen, White, *The Enzymes*, vol. **5**, P. D. Boyer *et al.*, Eds. (Academic Press, New York, 2nd ed., 1961) pp 95-122; Richards, Wyckoff, *ibid.* vol. **4** (3rd ed., 1971) pp 647-806.

Crystals. $[\alpha]_D^{25}$ (per mg N) −0.47° (c = 5). uv max (0.1M KCl soln): 277.5 nm (ε 9700). Isoelec pt about pH 8.0. Freely sol in water. Aq solns of cryst ribonuclease are quite stable at temps below 25°. The region of maximum stability is between pH 2 and 4.5. Stable for years stored as refrigerated dry powder or in frozen soln. Aggregates on lyophilization and storage. Shows affinity for glass surfaces. The optimum temp for digestion of yeast ribonucleic acid is 65°. The optimum pH is 7.7. Inhibited by heavy metal ions; by magnesium ions at concns as low as 0.0005M. Competitively inhibited by DNA, the effect of denatured DNA being greater than that of the native nucleic acid. Ribonuclease is precipitated by trichloroacetic acid, and does not diffuse through collodion or cellophane.

8370. Ribonucleic Acid. RNA; yeast nucleic acid. Polynucleotide directly involved in protein synthesis; found in both the nucleus and the cytoplasm of cells. Description of components of RNA: *see* Nucleic Acids. The four primary nucleosides are adenosine, guanosine, cytidine and uridine; minor nucleosides are also found. The nucleosides are linked by phosphate diester bonds from the 3'-hydroxyl of one D-ribose to the 5'-hydroxyl of the next. The secondary structure of RNA is that of an incompletely organized single-stranded polynucleotide consisting of some areas with helical structure alternating with nonhelical lengths. *Compare* Deoxyribonucleic Acid (DNA). Structure: Brown, Todd in *The Nucleic Acids* vol. 1, E. Chargaff, J. N. Davidson, Eds. (Academic Press, NewYork, 1955) pp 409-439; Spirin, *Progr. Nucleic Acid Res.* **1**, 301-345 (1963); J. N. Davidson, *The Biochemistry of Nucleic Acids* (Academic Press, New York, 7th ed., 1972) pp 106-128. Review of NMR studies: B. R. Reid, *Ann. Rev. Biochem.* **50**, 969-996 (1981). Book: *The Ribonucleic Acids,* P. R. Stewart, D. S. Letham, Eds. (Springer-Verlag, New York, 2nd ed., 1977). Comprehensive reviews and monographs: *see* Nucleic Acids.

Ribosomal RNA, rRNA. Metabolically stable form; about 80% of the RNA found in cells; an important component of ribosomes *(q v)* which play a central role in protein synthesis. Two rRNA species of high molecular weight have been isolated from ribosomes: mol wts approx 0.6×10^6 and 1.1×10^6 in bacterial cells; higher in eukaryotic cells. In addition, at least one rRNA species of low mol wt, 5S RNA, has been identified; approx 120 nucleotides. *Reviews:* Attardi, Amaldi, *Ann. Rev. Biochem.* **39**, 183-226; Möller, Garrett, "The Ribonucleic Acids of the Bacterial Ribosome" in *Protein Synthesis* vol. 1, E. H. McConkey, Ed. (Dekker, New York, 1971) pp 229-272; Craig, "Ribosomal RNA Synthesis in Eukaryotes and its Regulation" in *MTP Int. Rev. Sci.: Biochem., Ser. One* vol. 6, K. Burton, Ed. (University Park Press, Baltimore, 1974) pp 255-288.

Messenger RNA, mRNA, informational RNA. A short-lived form of high molecular weight; acts as template for protein synthesis in the cell; complementary to one strand of DNA. A linear relationship exists between the sequence of amino acids in a polypeptide and the sequence of nucleotides in the corresponding mRNA and DNA. The genetic code has been proposed to explain the necessary translation process. The four purine and pyrimidine bases are treated as letters which can be combined to form 3-letter words or *codons;* 4^3 or 64 codons can be formed. 61 have been found to code for specific amino acids; the remaining three are probably chain termination codons. Reviews of mRNA and the genetic code: Lipmann, "Messenger Ribonucleic Acid" in *Progr. Nucleic Acid Res.* **1**, 135-161 (1963); Crick, "The Recent Excitement in the Coding Problem" *ibid.,* pp 163-217; Woese, "The Present Status of the Genetic Code" in *Progr. Nucleic Acid Res. Mol. Biol.* **7**, 107-172 (1967); Hadjiolov, "Ribonucleic Acids and Information Transfer in Animal Cells" *ibid.* 195-242; Jukes, Gatlin, "Recent Studies Concerning the Coding Mechanism" *ibid.* **11**, 303-350 (1971); J. N. Davidson, *The Biochemistry of the Nucleic Acids* (Academic Press, New York, 7th ed., 1972) pp 290-383; Mathews, "Mammalian Messenger RNA" in *Essays in Biochemistry* vol. 9, P. N. Campbell, F. Dickens, Eds. (Academic Press, New York, 1973) pp 59-102; Brawerman, *Ann. Rev. Biochem.* **43**, 621-642 (1974); Watts, Watts, "The Genetic Code" in *MTP Int. Rev. Sci.: Biochem., Ser. One* vol. 7, H. R. V. Arnstein, Ed. (University Park Press, Baltimore, 1975) pp 255-294.

Transfer RNA, tRNA, soluble RNA, sRNA. Low mol wt: 23,000-27,000; approx 75-85 nucleotides. Each tRNA is specific for and binds with a particular amino acid; more than one may exist for each amino acid. Performs three functions during protein synthesis: binds with its specific amino acid; recognizes the corresponding codon on mRNA and places the amino acid in the correct position for attachment to the polypeptide chain being formed; binds the polypeptide to the ribosome. First determination of total structure of a transfer RNA (yeast alanine tRNA): Holley *et al., Science* **147**, 1462 (1965). Reviews of structure and function: Miura, "Specificity in the Structure of Transfer RNA" in *Progr. Nucleic Acid Res. Mol. Biol.* **6**, 39-82 (1967); Cramer, "Three-Dimensional Structure of tRNA", *ibid.* **11**, 391-421 (1971); *Nucleic Acid Sequence Analysis,* S. Mandeles (Columbia University Press, New York, 1972) pp 256-280; Nishimura, "Transfer RNA: Structure and Biosynthesis" in *MTP Int. Rev. Sci.: Biochem., Ser. One* vol. 6, K. Burton, Ed. (University Park Press, Baltimore, 1974) pp 289-322; A. Rich, V. L. Raj Bhandary, *Ann. Rev. Biochem.* **45**, 805-860 (1976); P. F. Agris, *The Modified Nucleosides of Transfer RNA, II* (A. R. Liss, New York, 1983) 220 pp.

8371. D-Ribose. $C_5H_{10}O_5$; mol wt 150.13. C 40.00%, H 6.71%, O 53.28%. Prepd by hydrolysis of yeast-nucleic acid: Levene, Jacobs, *Ber.* **42**, 1201, 3247 (1909); Levene, Clark, *J. Biol. Chem.* **46**, 19 (1921); Bredereck, *Ber.* **71**, 408 (1938); Bredereck *et al., ibid.* **73**, 956 (1940); Phelps, U.S. pat. **2,152,662** (1939 to U.S. Gov't); by ion-exchange resin chromatography: Cohn, *Science* **109**, 377 (1949); *J. Am. Chem. Soc.* **71**, 2275 (1949); **72**, 1471 (1950). From glucose: Karrer, *Helv. Chim. Acta* **18**, 1435 (1935); Austin, Humoller, *J. Am. Chem. Soc.* **56**, 1152 (1934); Kuhn *et al., Ber.* **68**, 1765 (1935); from nucleosides: Laufer, Charney, U.S. pats. **2,379,913-4** (1945 to Schwarz Labs.); from D-erythrose: Sowden, *J. Am. Chem. Soc.* **72**, 808 (1950); from L-glutamic acid: Koga *et al., Tetrahedron Letters* **1971**, 263. Reduction of D-ribonic acid: van Ekenstein, Blanksma, *Chem. Zentr.* **1913, II**, 1562; Steiger, *Helv. Chim. Acta* **19**, 189 (1936). *Review:* Overend, Stacey, in *Nucleic Acids* vol. I, E. Chargaff, J. N. Davidson, Eds. (Academic Press, New York, 1955) pp 1-80.

α-D-Ribofuranose

Plates from abs alcohol, mp 87°. Shows complex mutarotation: Phelps *et al., J. Am. Chem. Soc.* **56**, 748 (1934). Final $[\alpha]_D^{24}$ −25° (water). Sol in water, slightly sol in alc.

Phenylosazone, $C_{17}H_{20}N_4O_3$, yellow needles from pyridine + water, mp 163-164°.

Methyl-D-riboside, crystals from ethyl acetate, mp 83-84°. $[\alpha]_D^{20}$ −113.6° (p = 3), Minsaas, *Ann.* **512**, 286 (1934).

8372. D-Ribose-5-phosphoric Acid. *5-(Dihydrogen phosphate)-D-ribose;* D-ribofuranose-5-phosphoric acid; D-ribose-5-phosphate. $C_5H_{11}O_8P$; mol wt 230.11. C 26.10%, H 4.82%, O 55.62%, P 13.46%. From inosinic acid: Levene, Jacobs, *Ber.* **41**, 2703 (1908); **44**, 746 (1911); Levene, Mori, *J. Biol. Chem.* **81**, 215 (1929); Levene, Stiller, *ibid.* **104**, 299 (1934); LePage, Umbreit, *ibid.* **148**, 255 (1943); Marmur *et al., Arch. Biochem. Biophys.* **34**, 209 (1951). Purification: Groth *et al., J. Biol. Chem.* **199**, 389 (1952). Differentiation between ribose-5-phosphate and ribose-3-phosphate by means of the orcinol-pentose reaction: Albaum, Umbreit, *ibid.* **167**, 369 (1947).

Ba salt hemihendecahydrate, $C_5H_9BaO_8P.5\frac{1}{2}H_2O$, hexagonal plates from water. Sparingly soluble in cold water. Reduces Fehling's soln.

8373. Ribostamycin. *O-2,6-Diamino-2,6-dideoxy-α-D-glucopyranosyl-(1→4)-O-[β-D-ribofuranosyl-(1→5)]-2-deoxy-D-streptamine;* SF-733 antibiotic. $C_{17}H_{34}N_4O_{10}$; mol wt

454.48. C 44.93%, H 7.54%, N 12.33%, O 35.20%. Amino-cyclitol antibiotic belonging to the neomycin, *q.v.*, group of antibiotics. Produced by *Streptomyces ribosidificus* (former-ly called *S. thermoflavus*). Taxonomy, isoln and toxicity data: T. Shomura *et al., J. Antibiot.* **23**, 155 (1970). Prepn: *eidem,* **Ger.** pat. **1,814,735** corresp to U.S. pats. **3,661,892, 3,799,842** (1969, 1972, 1974 all to Meiji). Structure: E. Akita *et al., J. Antibiot.* **23**, 173 (1970). Synthesis from neamine, *q.v.*: Ito *et al., Antimicrob. Ag. Chemother.* **1970**, 33; *eidem, Agr. Biol. Chem.* **34**, 980 (1970); V. Kumar, W. A. Remers, *J. Org. Chem.* **43**, 3327 (1978); **46**, 4298 (1981). Total synthesis: H. Fukami *et al., Tetrahedron Letters* **1976**, 545; *eidem, Agr. Biol. Chem.* **41**, 1689 (1977). *In vitro* anti-bacterial activity: E. Yourassowsky, M. P. Vander Linden, *Arzneimittel-Forsch.* **26**, 184 (1976).

Colorless needles from methanol, mp 192-195°. [Also reported as the monohydrate from water-methanol-ethanol, mp 175-180° (dec)]. $[\alpha]_D^{23}$ +42°. pKa' 7.70. Stable in neu-tral and alkaline solns; slightly unstable in acidic media. Sol in water; slightly sol in methanol. Practically insol in acetone, *n*-butanol, ethyl acetate, benzene, hexane, ether.

Sulfate, **Ibistacin, Landamycin, Ribostamin, Ribomycine, Vistamycin.** White or yellowish powder. $[\alpha]_D^{20}$ +39° (c = 1). LD_{50} i.v. in mice: 225 mg/kg (Shomura).

THERAP CAT: Antibacterial.

8374. D-Ribulose. *D-erythro-2-Pentulose;* D-adonose; D-*erythro*-2-ketopentose. $C_5H_{10}O_5$; mol wt 150.13. C 40.00%, H 6.71%, O 53.28%. Prepn from D-arabinose: Glatthaar, Reichstein, *Helv. Chim. Acta* **18**, 80 (1935); from 2-keto-D-gluconic acid: Hall *et al., Biochem. J.* **60**, 271 (1955); from D-fructose: P. M. Collins *et al., J. Chem. Soc. Perkin Trans. I* **1980**, 277. Stereoselective synthesis: M. Yamaguchi, T. Mukaiyama, *Chem. Letters* **1981**, 1005; K. Suzuki *et al., ibid.* 1529.

Sweet syrup. $[\alpha]_D^{24}$ −15° (c = 0.5 in water).

Diphosphate, $C_5H_{12}O_{11}P_2$. Prepn from D-ribose-5-phos-phate: Horecker *et al., Biochem. Prep.* **6**, 83 (1958).

8375. Rice Bran Oil. Oil of rice bran; rice oil. Ex-tracted from bran obtained in making white or polished rice (which is actually the endosperm of the seed of *Oryza sativa* L., *Gramineae*). Rice bran separated by milling of hulled rice consists of bran, aleurone layer and germ, and contains about 15% oil. The oil is usually obtained by solvent extrac-tion, *see* J. R. Loeb, N. J. Morris: *Abstract Bibliography of the Chemistry, Processing, and Utilization of Rice Bran and Rice Bran Oil,* U.S. Dept. of Agriculture, Ms. no **AIC-328** (1952 and suppl.). Rice bran contains an unusually active lipase which raises the free fatty acid content on storage. A variety of methods to inhibit this rise can be found in the literature: Williams *et al., J. Am. Oil Chem. Soc.* **42**, 151 (1965). The free fatty acid content of the oil should not exceed 4 to 5%, but may run as high as 70%. Acceptable rice bran oil contains 15-20% of satd and 80-85% unsatd fatty acids as glycerides: Myristic 0.4-1%, palmitic 12-18%, stearic 1-3%, C_{20}-C_{22} satd 1% oleic 40-50%, linoleic 29-42%,

linolenic trace to 1%, palmitoleic 0.2-0.4%. *See also* Sreeni-vasan, *J. Am. Oil Chem. Soc.* **45**, 263 (1968).

Golden yellow oil, difficult to bleach. Not affected by temporary heating to 160°. d_{25}^{25} 0.916-0.921. n_D^{25} 1.470-1.473; n_D^{40} 1.465-1.468. Titer (solidification pt) 24-28. Sa-ponification value 181-189. Acid value 4-120 (see above comments on the lipase content of the bran). Iodine value 99-108. Thiocyanogen value 69-76. Hydroxyl value 5-14. Reichert-Meissl value <0.5. Polenské value <0.5. Hehner value 95.3. Unsaponifiable matter 3-5%. The high refrac-tive index is ascribed to the presence of squalene. Also contains other antioxidants such as tocopherols. Miscible with hexane and other fat solvents.

USE: Soaps, hydrogenated shortenings. Edible oil, also suitable for cosmetics and pharmaceuticals when low in free fatty acids and all the lipase has been deactivated.

8376. Ricin. A toxic lectin and hemagglutinin isolated from castor bean, *Ricinus communis* L., *Euphorbiaceae:* Stillmark, *Arb. Pharmak. Inst. Dorpat* **1889**, 59; T. Osborne *et al., Am. J. Physiol.* **14**, 259 (1905); H. L. Craig *et al.,* **U.S.** pat. **3,060,165** (1962 to U.S. Dept. Army). Purification of toxic ricin protein: M. Ishiguro *et al., J. Biochem.* **55**, 587 (1964). The hemagglutinating activity of ricin was initially believed to be the cause of its high toxicity, but later studies have shown that separate proteins are responsible for the toxicity and hemagglutination: S. Olsnes, A. Pihl, *Biochem-istry* **12**, 3121 (1973). "Ricin", the original term for the mixed extract, is now used in various ways. Two ricin agglutinins and two toxins have been identified: **RCL I, RCL II** (agglutinins), **RCL III** or **Ricin D** and **RCL IV** (tox-ins), *cf.* T. T. Lin, S. S. Li, *Eur. J. Biochem.* **105**, 453 (1980). All four lectins consist of two different polypeptide chains joined by a disulfide bond; the toxins are dimers of an A-chain (30,000 Da) and a B-chain (33,000 Da) and the agglu-tinins occur as a tetramer composed of two 30,000 and two 33,000 mol wt subunits. Ricin has been shown to have anti-tumor properties: J. Y. Lin *et al., Nature* **227**, 92 (1970); P. E. Thorp *et al., ibid.* **297**, 594 (1982). Synergistic effect with daunorubicin, cisplatin, and vincristine, *q.q.v.,* in systemic L 1210 leukemia: O. Fodstad, A. Pihl, *Cancer Res.* **42**, 2152 (1982). Conjugates of ricin and cell-binding antigens or antibodies, called immunotoxins, have been used in cancer therapy: E. S. Vitetta *et al., Science* **219**, 644 (1983). Exper-imental study with ricin-A chain conjugate in AIDS: M. A. Till *et al., Science* **242**, 1166 (1988). Radioimmunoassay in blood: A. Godal *et al., J. Toxicol. Environ. Health* **8**, 409 (1981). *Reviews:* M. Funatsu, "The Structure and Toxic Function of Ricin" in *Proteins, Structure and Function* vol. **2,** M. Funatsu *et al.,* Eds. (Kodansha, Tokyo, Wiley, New York, 1972) pp 103-139; G. A. Balint, *Toxicology* **2**, 77-122 (1974). *See also* Abrin, Lectins.

Isoelectric pt 7.1. uv max: 280 nm (ε 85000). MLD i.p. in mice at 48 hrs: 0.001 µg ricin nitrogen/g body wt (Ishi-guro).

Caution: Ricin is among the most toxic compounds known. Seeds of *R. communis,* if thoroughly masticated, can produce serious poisoning and death: J. M. Kingsbury, *Poisonous Plants of the United States and Canada* (Prentice-Hall, New Jersey, 1964) pp 194-197.

USE: As a tool in studies of cell-surface properties; exptly in cancer research.

8377. Ricinine. *1,2-Dihydro-4-methoxy-1-methyl-2-oxo-3-pyridinecarbonitrile; 1,2-dihydro-4-methoxy-1-methyl-2-oxonicotinonitrile;* ricidine. $C_8H_8N_2O_2$; mol wt 164.16. C 58.53%, H 4.91%, N 17.06%, O 19.49%. From seeds and leaves of the castor plant, *Ricinus communis* L., *Euphorbia-ceae.* Extraction procedure: Böttcher, *Ber.* **51**, 673 (1918). Synthesis starting with polymerization of cyanoacetyl chlo-ride: Schroeter *et al., Ber.* **65**, 432 (1932). Biosynthesis:

Waller, Henderson, *J. Biol. Chem.* **236**, 1186 (1961); Essery *et al., Can. J. Chem.* **41**, 1142 (1963).

Prisms, needles from alc, mp 201.5°. Sublimes at 170-180° under 20 mm pressure. Sparingly sol in water, alc, chloroform, ether. Neutral to litmus. Does not form salts with acids.

Caution: Ingestion may cause nausea, vomiting, hemorrhagic gastroenteritis, hepatic and renal damage, convulsions, coma, hypotension, respiratory depression, death.

8378. Ricinoleic Acid. *[R-(Z)]-12-Hydroxy-9-octadecenoic acid; d-12-hydroxyoleic acid.* $C_{18}H_{34}O_3$; mol wt 298.47. C 72.44%, H 11.48%, O 16.08%. $CH_3(CH_2)_5CH-(OH)CH_2CH=CH(CH_2)_7COOH$. Found primarily in oils from the seeds of *Ricinus* spp, *Euphorbiaceae.* Accounts for about 90% of the triglyceride fatty acids of castor oil, and up to about 40% of the glyceride fatty acids of ergot oil. Bibliography on its isoln: Ralston, *Fatty Acids* (New York, 1948) p 189. Also isolated from *Linum mucronatum* (flax), *Linaceae:* Kleiman, Spencer, *Lipids* **6**, 962 (1971). Structure: Goldsobel, *Ber.* **27**, 3121 (1894). Mechanism of biosynthesis: Morris, *Biochem. Biophys. Res. Commun.* **29**, 311 (1967).

Liquid. $d_4^{27.4}$ 0.940; mp +5.5°; bp_{10} 245°. $[\alpha]_D^{22}$ +6.67°; $[\alpha]_D^{26}$ +7.15° (c = 5 in acetone). n_D^{20} 1.4716. Neutralization value 187.98; iodine value 85.05. Sol in alcohol, acetone, ether, chloroform (*cf.* the solubilities of castor oil).

Acid sulfate, $C_{18}H_{34}O_6S$, *ricinolsulfuric acid.* Obtained by the action of chlorosulfonic acid. Viscous brown liquid with weak blue fluorescence. Sol in water (about 10%), alcohol, ether, chloroform.

Sodium salt, *Soricin, Colidosan.* Sodium salts of the fatty acids from castor oil. White or slightly yellow, odorless or almost odorless powder. Sol in water or alcohol. The aq soln is alkaline.

USE: In textile finishing; sometimes added to Turkey red oil, dry-cleaning soaps.

THERAP CAT: Has been used in contraceptive jellies. The sodium salt has been used as sclerosing agent.

8379. Ridogrel. *(E)-5-[[[3-Pyridinyl[3-(trifluoromethyl)phenyl]methylene]amino]oxy]pentanoic acid;* R-68070. $C_{18}H_{17}F_3N_2O_3$; mol wt 366.34. C 59.02%, H 4.68%, F 15.56%, N 7.65%, O 13.10%. Combined thromboxane A_2 synthase inhibitor and thromboxane A_2/prostaglandin endoperoxide receptor antagonist. Prepn: E. J. E. Freyne *et al., Eur.* pat. **Appl. 221,601**; *eidem,* **U.S. pat. 4,963,573** (1987, 1990 both to Janssen). Clinical pharmacology: B. Hoet *et al., Thromb. Haemostasis* **64**, 87 (1990); C. Weber *et al., ibid.* **68**, 214 (1992). Clinical study in peripheral arterial obstructive disease: J. De Cree *et al., Int. Angiol.* **12**, 59 (1993); as adjunct to thrombolysis in acute myocardial infarction: RAPT Investigators, *Circulation* **89**, 588 (1994).

Crystals from diisopropyl ether/hexane (2:1), mp 70.3°.
THERAP CAT: Antithrombotic.

8380. Rifabutin. *1',4-Didehydro-1-deoxy-1,4-dihydro-5'-(2-methylpropyl)-1-oxorifamycin XIV;* (9S,12E,14S,15R,-16S,17R,18R,19R,20S,21S,22E,24Z)-6,16,18,20-tetrahydroxy-1'-isobutyl-14-methoxy-7,9,15,17,19,21,25-heptamethyl-spiro[9,4-(epoxypentadeca[1,11,13]trienimino)-2H-furo-[2',3':7,8]naphth[1,2-d]imidazole-2,4'-piperidine]-5,10,26-(3H,9H)-trione-16-acetate; 4-deoxo-3,4-[2-spiro-(N-isobutyl-4-piperidyl)]-(1H)-imidazo-(2,5-dihydro)rifamycin S; 4-N-isobutylspiropiperidylrifamycin S; LM-427; Ansatipine; Mycobutin.* $C_{46}H_{62}N_4O_{11}$; mol wt 847.02. C 65.23%, H 7.38%, N 6.61%, O 20.78%. Semisynthetic derivative of rifamycin S that inhibits nucleic acid synthesis. Prepn: L. Marsili *et al., Ger.* pat. **2,825,445** (1979 to Farmitalia); *eidem,* **U.S. pat. 4,219,478** (1980 to Archifar Labs). *In vitro* and *in vivo* antibacterial activity: A. Sanfilippo *et al., J. Antibiot.* **33**, 1193 (1980); C. Della Bruna *et al., ibid.* **36**,

1502 (1983). Mechanism of action: D. Ungheri *et al., Drugs Exp. Clin. Res.* **10**, 681 (1984). Comparative *in vitro* antimycobacterial spectrum: J. M. Dickinson, D. A. Mitchison, *Tubercle* **68**, 177 (1987). *In vitro* inhibition of HIV-1 replication: R. Anand *et al., Antimicrob. Ag. Chemother.* **32**, 684 (1988). Clinical pharmacokinetics: M. H. Skinner *et al., ibid.* **33**, 1237 (1989). Pharmacology and clinical efficacy in mycobacterial infections: R. J. O'Brien *et al., Rev. Infect. Dis.* **9**, 519 (1987).

Violet-red crystalline powder. Highly sol in chloroform, sol in methanol, slightly sol in ethanol, minimally sol in water. uv max (methanol): 493, 315, 274, 238 nm.
THERAP CAT: Antibacterial (tuberculostatic).

8381. Rifamide. *4-O-[2-(Diethylamino)-2-oxoethyl]-rifamycin; 2-[(1,2-dihydro-5,6,17,19,21-pentahydroxy-23-methoxy-2,4,12,16,18,20,22-heptamethyl-1,11-dioxo-2,7-(epoxypentadeca[1,11,13]trienimino)naphtho[2,1-b]furan-9-yl)oxy]-N,N-diethylacetamide 21-acetate; N,N-diethylrifomycin B amide; rifamycin B diethylamide; M-14; Rifocin M.* $C_{43}H_{58}-N_2O_{13}$; mol wt 810.94. C 63.69%, H 7.21%, N 3.45%, O 25.65%. Prepn: **Belg.** pat. **632,770** corresp to Sensi, Maggi, **U.S. pat. 3,313,804** (1963 and 1967 to Lepetit); Sensi *et al., J. Med. Chem.* **7**, 596 (1964). Physical and chemical properties: Maggi *et al., Farmaco Ed. Prat.* **20**, 147 (1965). Activity data: Pallanza *et al., Arzneimittel-Forsch.* **15**, 800 (1965); Monnier, Bourse, *Pathol. Biol.* **16**, 901 (1968). Metabolic studies: Fürész *et al., Arzneimittel-Forsch.* **15**, 802 (1965); Maffii, Shiatti, *Toxicol. Appl. Pharmacol.* **8**, 138 (1966). Toxicology: Dezulian *et al., ibid.* 126.

Yellow-orange ppt, crystallized from benzene + hexane. No definite mp, begins to soften at 140°, melts completely at 170° (dec). $[\alpha]_D^{20}$ −48.7° (c = 0.4 in methanol). Absorption spectrum in phosphate buffer (pH 7.38): 222, 302, 421 nm (ε 42820, 20770, 16200). LD_{50} in mice, rats (mg/kg): 2450, >4000 orally; 640, 2500 s.c.; 320, 535 i.p.; 315, 380 i.v. (Dezulian).
THERAP CAT: Antibacterial.

8382. Rifampin. *3-[[(4-Methyl-1-piperazinyl)imino]-methyl]rifamycin; 5,6,9,17,19,21-hexahydroxy-23-methoxy-2,-4,12,16,18,20,22-heptamethyl-8-[N-(4-methyl-1-piperazinyl)-*

formimidoyl]-2,7-(epoxypentadeca[1,11,13]trienimino)naphtho[2,1-b]furan-1,11(2H)-dione 21-acetate; rifampicin; rifaldazine; rifamycin AMP; R/AMP; Abrifam; Eremfat; Rifa; Rifadin(e); Rifaldin; Rifaprodin; Rifoldin; Rimactan(e). $C_{43}H_{58}N_4O_{12}$; mol wt 822.95. C 62.76%, H 7.10%, N 6.81%, O 23.33%. Semisynthetic antibiotic obtained by reacting 3-formylrifamycin SV with 1-amino-4-methylpiperazine in tetrahydrofuran. Prepn and structure: Maggi *et al., Chemotherapia* **11**, 285 (1966); **Neth. pat. Appl. 6,509,961**; Maggi, Sensi, U.S. pat. **3,342,810** (1966, 1967 both to Lepetit). Chemical and biological properties: Fürész, *Antibiot. & Chemother. (Basel)* **16**, 316 (1970). Activity studies and clinical survey: Arioli *et al., Arzneimittel-Forsch.* **17**, 523 (1967); Pallanza *et al., ibid.* 529; Bergamini, *ibid.* **20**, 1546 (1970); Dans *et al., Am. J. Med. Sci.* **259**, 120 (1970). Metabolism: Meyer-Brunot *et al., Int. Congr. Chemother. Proc., 5th, Vienna 1967* **1**(2), 763; Fürész *et al., Arzneimittel-Forsch.* **17**, 534 (1967); Maggi *et al., ibid.* **19**, 651 (1969). Inhibition of protein synthesis in mammalian cells: W. C. Buss *et al., Science* **200**, 432 (1978). Comprehensive reviews: Binda *et al., Arzneimittel-Forsch.* **21**, 1907-1978 (1971); Lester, *Ann. Rev. Microbiol.* **26**, 88-102 (1972). Comprehensive description: G. G. Gallo, P. Radaelli, in *Analytical Profiles of Drug Substances* vol. 5, K. Florey, Ed. (Academic Press, New York, 1976) pp 467-513. Symposium on the use of rifampin in the treatment of nontuberculous infections: *Rev. Infect. Dis.* **5**, Suppl. 3, S399-S632 (1983).

Red to orange platelets from acetone, dec 183-188°. Absorption max (pH 7.38): 237, 255, 334, 475 nm (ϵ 33200, 32100, 27000, 15400). Rifampin is a "zwitterion" with pKa 1.7 related to the 4-hydroxy and pKa 7.9 related to the 3-piperazine nitrogen. Very stable in DMSO; rather stable in water. Freely sol in CH_3Cl, DMSO; sol in ethyl acetate, methanol, tetrahydrofuran; slightly sol in water (pH <6), acetone, carbon tetrachloride. LD_{50} in mice, rats (mg/kg): 885, 1720 orally; 260, 330 i.v.; 640, 550 i.p. (Fürész).

THERAP CAT: Antibacterial (tuberculostatic).

8383. Rifamycins. Rifomycins. Group of antibiotics characterized by a natural ansa structure (chromophoric naphthohydroquinone group spanned by a long aliphatic bridge) not previously found in other known antibiotics. Isoln from the fermentation broths of *Streptomyces mediterranei:* P. Sensi *et al., Antibiot. Ann.* **1959-1960**, 262. Among rifamycins, rifamycin B, O, S, SV and X are the more studied members. Prepn of rifamycin B derivs: P. Sensi *et al.,* U.S. pat. **3,313,804** (1967 to Lepetit). Structure: Prelog, *Pure Appl. Chem.* **7**, 551 (1963); *Chemotherapia* **7**, 133 (1963); Oppolzer *et al., Experientia* **20**, 336 (1964); Oppolzer, Prelog, *Helv. Chim. Acta* **56**, 2287 (1973). Total synthesis of rifamycin S: H. Nagaoka *et al., J. Am. Chem. Soc.* **102**, 7962 (1980); H. Iio *et al., ibid.* 7965; H. Nagaoka, Y. Kishi, *Tetrahedron* **37**, 3873 (1981); S. Hanessian *et al., J. Am. Chem. Soc.* **104**, 6164 (1982). Review of chemistry of ansamycin antibiotics: K. L. Rinehart *et al., Fortschr. Chem. Org. Naturst.* **33**, 231-307 (1976). General reviews: P. Sensi, *Research Progress in Organic-Biological & Medicinal Chemistry* vol. **1**, U. Gallo, L. Santamaria, Eds. (Società Editoriale Farmaceutica, Milan, Italy, 1964) pp 337-421; Wehrli, Staehelin, in *Antibiotics* vol. 3, J. W. Corcoran, F. E. Hahn, Eds. (Springer-Verlag, New York, 1975) pp 252-268; C. Gurgo *et al.,* "Rifamycins" in *Handbook of Experimental Pharmacology* vol. **61**, G. V. R. Born *et al.,* Eds., entitled

"Chemotherapy of Viral Infections", P. E. Came, L. A. Caliguiri, Eds. (Springer-Verlag, New York, 1982) pp 519-555.

	R	R'
Rifamycin B	—OCH_2COOH	—OH
Rifamycin O	—(1,3-dioxolan-4-on)-2-yl	==O
Rifamycin S	==O	==O
Rifamycin X	==N⁺=N⁻	==O

Rifamycin AMP, see Rifampin.

Rifamycin B, $C_{39}H_{49}NO_{14}$, *4-O-(carboxymethyl)rifamycin, nancimycin.* Yellow prismatic needles from benzene, mp 300° (dec 160-164°). $[\alpha]_D^{20}$ −11° (methanol). Absorption max (phosphate buffer soln pH 7.3): 223, 304, 425 nm ($E_{1cm}^{1\%}$ 555, 275, 220). Dibasic acid. Very stable. Solubilities: water 0.027% (w/w), methanol 2.62%, ethanol 0.44%. LD_{50} in mice (mg/kg): 2040 i.v., >3000 i.p., s.c., and orally (Sensi, review).

Rifamycin O, $C_{39}H_{47}NO_{14}$, *4-O-(carboxymethyl)-1-deoxy-1,4-dihydro-4-hydroxy-1-oxorifamycin γ-lactone.* Prepn: P. Sensi *et al., Farmaco Ed. Sci.* **15**, 228 (1960); Umezawa, **Japan. pat. 15,518**('66), *C.A.* **66**, 1583v (1967). Pale yellow crystals from methanol, mp 300° (dec 160°). Also reported as mp 180-185° (Umezawa). $[\alpha]_D^{20}$ +71.5° (c = 1 in dioxane). uv max (methanol contg 5% acetate buffer soln pH 4.62): 226, 273, 370 nm ($E_{1cm}^{1\%}$ 365, 440, 60). Weak acid. Slowly sol in alkaline soln with red-violet color. Sol in acetone, tetrahydrofuran; slightly sol in methanol, ethanol, ethyl acetate. Practically insol in ether, petr ether, water, dilute acids.

Rifamycin S, $C_{37}H_{45}NO_{12}$, *1,4-dideoxy-1,4-dihydro-1,4-dioxorifamycin.* Activation product found in solns of rifamycin B and rifamycin O: P. Sensi *et al., Experientia* **16**, 412 (1960). Yellow-orange crystals from methanol, dec 179-181°. $[\alpha]_D^{20}$ +476° (c = 0.1 in methanol). Absorption max (phosphate buffer soln pH 7.3): 317, 525 nm ($E_{1cm}^{1\%}$ 426, 62). LD_{50} in mice (mg/kg): 122 i.v.; 258 i.p.; 3000 orally (Sensi, review).

Rifamycin SV, see separate entry.

Rifamycin X, $C_{37}H_{45}N_3O_{11}$, Prepn: Greco *et al., Farmaco Ed. Sci.* **16**, 766 (1961). Yellow crystals, no definite mp, dec 135-140°. Unstable to light. $[\alpha]_D^{20}$ +491.8° (c = 0.981 in dioxane). Absorption max (acetate buffer soln): 286, 317, 402 nm ($E_{1cm}^{1\%}$ 400, 362, 195). Practically insol in water. Sol in methanol, ethanol, ethyl acetate, benzene.

Rifamycin AG, a condensation product of rifamycin O and aminoguanidine: P. Sensi *et al., Antibiot. & Chemother.* **12**, 448 (1962).

8384. Rifamycin SV. *5,6,9,17,19,21-Hexahydroxy-23-methoxy-2,4,12,16,18,20,22-heptamethyl-2,7-(epoxypentadeca-[1,11,13]trienimino)naphtho[2,1-b]furan-1,11(2H)-dione 21-acetate;* rifomycin SV; rifamicine SV. $C_{37}H_{47}NO_{12}$; mol wt 697.78. C 63.69%, H 6.79%, N 2.01%, O 27.51%. Semisynthetic antibiotic derived from Rifamycin S. Prepn: Sensi *et al., Experientia* **16**, 412 (1960); *Farmaco, Ed. Sci.* **16**, 165 (1961). Comprehensive review: Bergamini, Fowst, *Arzneimittel-Forsch.* **15**, 951-1002 (1965). For general refs *see* Rifamycins.

Yellow-orange crystals, mp 300° (dec 140°). $[\alpha]_D^{20}$ −4° (methanol). uv max (phosphate buffer soln pH 7.3): 223, 314, 445 nm ($E_{1cm}^{1\%}$ 586, 322, 204). Acid reaction. Slightly sol in water, petr ether; sol in ether, bicarbonate soln; very sol in methanol, ethanol, acetone, ethyl acetate. A reducing substitute, such as ascorbic acid, should be added to aq solns of rifamycin SV to prevent its transformation to rifamycin S. LD_{50} in mice (mg/kg): 550 i.v.; 625 i.p.; 2120 orally (Bergamini, Fowst).

Sodium salt, *Rifamastene, Rifocin.* Orange-red crystals. Soly in water pH 7.2: ~5 g/100 ml.

THERAP CAT: Antibacterial.

THERAP CAT (VET): Antibacterial.

8385. Rifapentine. *3-[[(4-Cyclopentyl-1-piperazinyl)-imino]methyl]rifamycin;* MDL 473; DL-473; R-773. $C_{47}H_{64}N_4O_{12}$; mol wt 877.04. C 64.37%, H 7.36%, N 6.39%, O 21.89%. Semi-synthetic rifamycin. Prepn: R. Cricchio, V. Arioli, **Ger. pat. 2,608,218**; *eidem,* **U.S. pat. 4,002,752** (1976, 1977 both to Lepetit). Antibacterial activity: V. Arioli *et al., J. Antibiot.* **34,** 1026 (1981); H. C. Neu, *Antimicrob. Ag. Chemother.* **24,** 457 (1983). Pharmacokinetics: A. Assandri *et al., J. Antibiot.* **37,** 1066 (1984). HPLC determn in plasma: E. Riva *et al., J. Chromatog.* **553,** 35 (1991). Clinical trial in tuberculosis: Natl. Co-op Group Clin. Study Rifapentine, *Chin. J. Antibiot.* **17,** 313 (1992), *B.A.* **95,** 70840 (1993).

Crystals from ethyl acetate, mp 179-180°. uv max: 475, 334 nm (ε 15200, 26700). LD_{50} in mice (mg/kg): > 2000 orally; 750 i.p. (Cricchio). Also reported as LD_{50} in mice (mg/kg): 3300 orally; 710 i.p. (Aroli).

THERAP CAT: Antibacterial.

8386. Rifaximin. *[2S-(2R*,16Z,18E,20R*,21R*,22S*,-23S*,24S*,25R*,26S*,27R*,28E)]-25-(Acetyloxy)-5,6,21,23-tetrahydroxy-27-methoxy-2,4,11,16,20,22,24,26-octamethyl-2,-7-[epoxypentadeca(1,11,13)trienimino]benzofuro[4,5-e]pyrido-[1,2-a]benzimidazole-1,15(2H)-dione;* 5,6,21,23,25-penta-hydroxy-27-methoxy-2,4,11,16,20,22,24,26-octamethyl-2,7-(epoxypentadeca[1,11,13]trienimino)benzofuro[4,5-e]pyrido-[1,2-a]benzimidazole-1,15(2H)-dione 25-acetate; 4-deoxy-4'-methylpyrido[1',2'-1,2]imidazo[5,4-c]rifamycin SV; rifaxidin; L-105; L-105SV; α-0817185; Fatroximin; Normix; Rifacol. $C_{43}H_{51}N_3O_{11}$; mol wt 785.89. C 65.72%, H 6.54%, N 5.35%, O 22.39%. Nonabsorbable intestinal disinfectant

related to rifamycin SV, *q.v.* Prepn: **Belg.** pat. **888,895**; E. Marchi, L. Montecchi, **U.S. pat. 4,341,785** (1981, 1982 both to Alfa); E. Marchi *et al., J. Med. Chem.* **28,** 960 (1985); prepn and NMR study: M. Brufani *et al., J. Antibiot.* **37,** 1611 (1984). Structural study: M. Brufani *et al., ibid.* 1623. Antibacterial spectrum: C. Eftimiadi *et al., Drugs Exp. Clin. Res.* **10,** 691 (1984); *in vitro* and *in vivo* antibacterial activity: A. P. Venturini, E. Marchi, *Chemioterapia* **5,** 257 (1986). Pharmacodynamics in rats and dogs: A. P. Venturini, *Chemotherapy (Basel)* **29,** 1 (1983). Absorption profile: A. P. Venturini *et al., Drugs Exp. Clin. Res.* **13,** 233 (1987). Toxicological study: G. Borelli, D. Bertoli, *Chemioterapia* **4,** 263 (1986). Clinical evaluation in infectious diarrhea: M. Vinci *et al., Curr. Ther. Res.* **36,** 92 (1984); in hepatic encephalopathy: F. De Marco *et al., ibid.* 668; in intestinal infections: V. Alvisi *et al., J. Int. Med. Res.* **15,** 49 (1987).

Red orange powder, mp 200-205° (dec). uv max: 232, 260, 292, 320, 370, 450 nm ($E_{1cm}^{1\%}$ 489, 339, 295, 216, 119, 159). Sol in alcohols, ethyl acetate, chloroform, toluene. Insol in water. LD_{50} orally in rats: > 2000 mg/kg (Borelli, Bertoli).

THERAP CAT: Antibacterial.

THERAP CAT (VET): Antibacterial.

8387. Rilmazafone. *5-[[(Aminoacetyl)amino]methyl]-1-[4-chloro-2-(2-chlorobenzoyl)phenyl]-N,N-dimethyl-1H-1,2,4-triazole-3-carboxamide;* 1-(2-o-chlorobenzoyl-4-chlorophenyl)-5-glycylaminomethyl-3-dimethylcarbamoyl-1H-1,2,4-triazole; 2',5-dichloro-2-(3-dimethylcarbamoyl-5-glycyl-aminomethyl-1H-1,2,4-triazol-1-yl)benzophenone; $C_{21}H_{20}Cl_2N_6O_3$; mol wt 475.33. C 53.06%, H 4.24%, Cl 14.92%, N 17.68%, O 10.10%. Ring-opened triazolobenzodiazepine derivative. Prepn: K. Hirai, H. Sugimoto, **Ger. pat. 2,725,-164**; *eidem,* **U.S. pat. 4,159,374** (1977, 1979 both to Shionogi); and CNS activity: K. Hirai *et al., J. Heterocycl. Chem.* **19,** 1363 (1982). Receptor binding study of rilmazafone and its closed ring active metabolites: M. Fujimoto *et al., Biochem. Pharmacol.* **33,** 1645 (1984). Combined HPLC determn and enzyme immunoassay in plasma: G. Kominami *et al., J. Chromatog.* **417,** 216 (1987). Series of articles on teratogenicity studies: *Oyo Yakuri* **30,** 749-833 (1985), *C.A.* **104,** 81885h-81887k; 102328w (1986).

Monohydrochloride dihydrate, $C_{21}H_{20}Cl_2N_6O_3 \cdot HCl \cdot 2H_2O$, *450191-S, Rhythmy.* Solid from 95% ethanol, mp 107°. LD_{50} orally in mice: > 1500 mg/kg (Hirai, 1979).
THERAP CAT: Sedative, hypnotic.

8388. Rilmenidine. *N-(Dicyclopropylmethyl)-4,5-dihydro-2-oxazolamine;* 2-[N-(dicyclopropylmethyl)amino]-oxazoline; oxaminozoline; S-3341. $C_{10}H_{16}N_2O$; mol wt 180.25. C 66.64%, H 8.95%, N 15.54%, O 8.88%. α_2-Adrenoceptor agonist. Prepn: C. Malen *et al., Ger.* pat. **2,362,754;** *eidem,* U.S. pat. **4,102,890** (1974, 1978 both to Sci. Union et Cie. - Soc. Franc. Recher. Med.). Adrenoceptor binding study: P. Guicheney *et al., J. Pharmacol.* **12,** 255 (1981). Pharmacokinetics in hypertensive subjects: J. Velly *et al., ibid.* **13,** 413 (1982). Animal pharmacology: M. Laubie *et al., ibid.* **16,** 259 (1985); P. A. van Zwieten *et al., Arch. Int. Pharmacodyn.* **279,** 130 (1986). Hypotensive effect in humans: K. Weerasuriya *et al., Eur. J. Clin. Pharmacol.* **27,** 281 (1984). Mechanism of action study: R. E. Gomez *et al., Eur. J. Pharmacol.* **195,** 181 (1991). *Review:* T. J. Verbeuren *et al., Cardiovasc. Drug Rev.* **8,** 56-70 (1990).

Crystals from hexane, mp 106-107°.
Phosphate, $C_{10}H_{16}N_2O \cdot H_3PO_4$, *S-3341-3, Hyperium.* pKa 9.3. Soly in H_2O: ~19% w/v; in methanol: ~7% w/v; in chloroform and ethanol: ~0.7% w/v. LD_{50} orally in mice, rats: 375, 295 mg/kg (Verbeuren).
Fumarate, $C_{14}H_{20}N_2O_5$, crystals from ethanol, mp 170°.
THERAP CAT: Antihypertensive.

8389. Riluzole. *6-(Trifluoromethoxy)-2-benzothiazolamine;* 2-amino-6-(trifluoromethoxy)benzothiazole; PK-26124; RP-54274. $C_8H_5F_3N_2OS$; mol wt 234.20. C 41.03%, H 2.15%, F 24.34%, N 11.96%, O 6.83%, S 13.69%. Modulates glutamatergic transmission. Prepn: L. M. Yagupol'-skii, L. Z. Gandel'sman, *Zh. Obshch. Khim.* **33,** 2301 (1963), *C.A.* **60,** 692a (1964). Prepd but not claimed: J. Mizoule, **Eur.** pat. **Appl. 50,551;** *idem,* U.S. pat. **4,370,338** (1982, 1983 both to Pharmindustrie). Pharmacology: J. Mizoule *et al., Neuropharmacology* **24,** 767 (1985); F. Wahl *et al., Eur. J. Pharmacol.* **230,** 209 (1993). Mechanism of action studies: M.-W. Debono *et al., ibid.* **235,** 283; D. Martin *et al., ibid.* **250,** 473. Use in treatment of motor nerve diseases: E. Louvel, **Eur.** pat. **Appl. 558,861** (1993 to Rhône-Poulenc Rorer). Clinical study in amyotrophic lateral sclerosis: G. Bensimon *et al., N. Engl. J. Med.* **330,** 585 (1994).

Crystals from ethanol/water (1:1), mp 119°. LD_{50} in mice (mg/kg): 46 i.p.; 67 orally (Mizoule).
THERAP CAT: Neuroprotective.

8390. Rimantadine. *α-Methyltricyclo[3.3.1.1^{3,7}]decane-1-methanamine;* α-methyl-1-adamantanemethylamine; remantadin(e). $C_{12}H_{21}N$; mol wt 179.31. C 80.38%, H 11.80%, N 7.81%. Deriv of adamantane, *q.v.* Prepn: **Neth.** pat. **Appl. 6,408,505;** W. W. Prichard, U.S. pat. **3,352,912** (1965, 1967 both to du Pont). Antiviral activity: A. Tsunoda *et al., Antimicrob. Ag. Chemother.* **1965,** 553. Effects on influenza in mice: J. W. McGahen *et al., Ann. N.Y. Acad. Sci.* **173,** 557 (1970). Mechanism of action study: A. Bukrinskaya *et al., Arch. Virol.* **66,** 275 (1980); *eidem, Gen. Virol.* **60,** 49 (1982). Pharmacokinetics in humans: R. J. Wills *et al., Antimicrob. Ag. Chemother.* **31,** 826 (1987). Clinical trial in prophylaxis of influenza A infection: R. Dolin *et al., N. Engl. J. Med.* **307,** 580 (1982). Comparative toxicity of rimantadine and amantadine in healthy adults: F. G. Hayden *et al., Antimicrob. Ag. Chemother.* **19,** 226 (1981). Controlled study of CNS effects: V. M. Millet *et al., ibid.* **21,** 1 (1982). Review of studies in the USSR on

exptl and clinical pharmacology: D. M. Zlydnikov *et al., Rev. Infect. Dis.* **3,** 408-421 (1981).

Hydrochloride, $C_{12}H_{21}N \cdot HCl$, *EXP-126, Flumadine, Meradan(e), Roflual.* White crystals, mp 373-375° (sealed tube).
THERAP CAT: Antiviral.

8391. Rimazolium Metilsulfate. *3-(Ethoxycarbonyl)-6,7,8,9-tetrahydro-1,6-dimethyl-4-oxo-4H-pyrido[1,2-a]pyrimidinium methyl sulfate;* 3-carboxy-6,7,8,9-tetrahydro-1,6-dimethyl-4-oxo-4H-pyrido[1,2-a]pyrimidinium methyl sulfate, ethyl ester; 1,6-dimethyl-3-carbethoxy-4-oxo-6,7,8,9-tetrahydrohomopyrimidazole methyl sulfate; MZ-144; MZ-0780; Ro-11-780; Dolcuran; Probon; Probonal; Rimagina. $C_{14}H_{22}N_2O_7S$; mol wt 362.40. C 46.40%, H 6.12%, N 7.73%, O 30.90%, S 8.85%. Prepn: Z. Mészáros *et al., Hung.* pat. **Teljes 519** (1970 to Chinoin), *C.A.* **74,** 42374h (1971); *eidem, Arzneimittel-Forsch.* **22,** 815 (1972). Anti-inflammatory and analgesic effects in animals: K. Gyires *et al., Drugs Exptl. Clin. Res.* **11,** 493 (1985). Clinical analgesic effect: H. Graber, *Int. J. Clin. Pharmacol. Ther. Toxicol.* **6,** 354 (1972); M. Haataja *et al., Curr. Ther. Res.* **22,** 784 (1977). Series of articles on pharmacology, pharmacodynamics and toxicology: *Arzneimittel-Forsch.* **21,** 717-738 (1971). Toxicity data: J. Knoll *et al., ibid.* 719.

White crystals, mp 165-166°. uv max: 336, 258 nm (ϵ 3630, 27500). Readily sol in water. LD_{50} in rats (mg/kg): 220 i.v.; 720 i.p.; 790 s.c.; 1600 orally (Knoll).
THERAP CAT: Analgesic.

8392. Rimexolone. *(11β,16α,17β)-11-Hydroxy-16,17-dimethyl-17-(1-oxopropyl)androsta-1,4-dien-3-one;* 11β-hydroxy-16α,17α-dimethyl-17-propionylandrosta-1,4-dien-3-one; 11β-hydroxy-16α,17α,21-trimethylpregna-1,4-diene-3,20-dione; trimexolone; Org-6216; Rimexel; Vexol. $C_{24}H_{34}O_3$; mol wt 370.53. C 77.80%, H 9.25%, O 12.95%. Prepn: **Neth.** pat. **Appl. 7,300,313;** G. F. Woods *et al.,* U.S. pat. **3,947,478** (1973, 1976 both to Akzo); J. Cairns *et al., J. Chem. Soc. Perkin Trans. I* **1981,** 2306. Pharmacology: P. K. Fox *et al., Arzneimittel-Forsch.* **30,** 55 (1980). Clinical trial in rheumatoid arthritis: E. van Vliet-Daskalopoulou *et al., Brit. J. Rheumatol.* **26,** 450 (1987). Clinical pharmacokinetics: G. Gevers *et al., Clin. Rheumatol.* **13,** 103 (1994).

Crystals, mp 258-268°. $[\alpha]_D$ +100° (c = 0.92 in pyridine). uv max: 244 nm (ϵ 14600).
THERAP CAT: Anti-inflammatory (local).

8393. Rimiterol. *4-(Hydroxy-2-piperidinylmethyl)-1,2-benzenediol; erythro-α-(3,4-dihydroxyphenyl)-2-piperidinemethanol; erythro-3,4-dihydroxyphenyl-2-piperidinylcarbinol.* $C_{12}H_{17}NO_3$; mol wt 223.27. C 64.55%, H 7.67%, N

6.27%, O 21.50%. Prepn: Sankey, Whiting, **Ger.** pat. **2,024,049** corresp to **U.S.** pat. **3,910,934** (1970, 1975 to 3M); Kaiser, Ross, **Ger.** pat. **2,047,937** corresp to **U.S.** pat. **3,705,169** (1971, 1972 to SK & F); Sankey, Whiting, *J. Heterocycl. Chem.* **9**, 1049 (1972). Pharmacology: Carney *et al., Arch. Int. Pharmacodyn. Ther.* **194**, 334 (1971); Griffin, Turner, *J. Clin. Pharmacol.* **11**, 280 (1971); Bowman, Rodger, *Brit. J. Pharmacol.* **45**, 574 (1972).

Crystals from ethyl acetate, mp 203-204°.

Hydrobromide, $C_{12}H_{17}NO_3 \cdot HBr$, *R-798, WG-253, Asmaten, Pulmadil*. White powder, mp 220° (dec).

THERAP CAT: Bronchodilator.

8394. Rimocidin. $C_{39}H_{61}NO_{14}$; mol wt 767.91. C 61.00%, H 8.01%, N 1.82%, O 29.17%. Polyene antifungal antibiotic produced by *Streptomyces rimosus* along with Terramycin (oxytetracycline, *q.v.*). Recovered from fermentation broth by extracting the mycelium with butanol. Isoln and antifungal activity: Davisson *et al., Antibiot. & Chemother.* **1**, 289 (1951); Seneca *et al.,* ibid. **2**, 435 (1952); Davisson *et al.,* **U.S.** pat. **2,963,401** (1960 to Pfizer). Partial structure: Cope *et al., J. Am. Chem. Soc.* **87**, 5452 (1965). Structure: L. Falkowski *et al., J. Antibiot.* **29**, 197 (1976); R. Pandey, K. L. Rinehart, ibid. **30**, 146 (1977).

Dec above 110°. $[\alpha]_D^{25}$ +116° (pyridine). uv max (80% methanol): 279, 291, 304, 318 nm. Slightly sol in water, acetone, lower alcohols. A cryst sodium salt was prepd by reaction with NaOH in methanol.

Sulfate heptahydrate, large plates from dil methanol, dec 151°. $[\alpha]_D^{25}$ +75.2° (methanol). Sol in water. LD_{50} i.v. in mice: 20 mg/kg (Seneca).

8395. Rioprostil. *(11α,13E)-1,11,16-Trihydroxy-16-methylprost-13-en-9-one; (2R,3R,4R)-4-hydroxy-2-(7-hydroxyheptyl)-3-[(E)-(4RS)-(4-hydroxy-4-methyl-1-octenyl)]-cyclopentanone; 16-methyl-1,11α,16RS-trihydroxyprost-13E-en-9-one; Bay o 6893; ORF-15927; TR-4698; Rostil.* $C_{21}H_{38}O_4$; mol wt 354.53. C 71.15%, H 10.80%, O 18.05%. Prostaglandin E_1 analog with cytoprotective and gastric antisecretory activity. Prepn: H. C. Kluender *et al.,* **U.S.** pat. **4,132,738** (1979 to Miles). Pharmacology: D. A. Shriver *et al., Arzneimittel-Forsch.* **35**, 839 (1985). Effect on human gastric secretion: P. Demol *et al.,* ibid. 861; B. Vaona *et al., Adv. Prostaglandin Thromboxane Leukotriene Res.* **17**, 328 (1987). Clinical evaluation in duodenal ulcer: H. G. Dammann *et al.,* ibid. 303.

$[\alpha]_D$ −58.6° (c = 1 in chloroform).

THERAP CAT: Antiulcerative.

8396. Risedronic Acid. *[1-Hydroxy-2-(3-pyridinyl)ethylidene]bisphosphonic acid.* $C_7H_{11}NO_7P_2$; mol wt 283.11. C 29.70%, H 3.92%, N 4.95%, O 39.56%, P 21.88%. Bisphosphonate inhibitor of bone resorption. Prepn: J. J. Benedict, C. M. Perkins, **Eur.** pat. **Appl.** **186,405** (1986 to Proctor & Gamble); *see also* R. R. Dauer *et al.,* **PCT Int.** pat. **Appl.** **95 06,052** (1995 to Merck & Co.). Pharmacology: R. W. Boyce *et al., J. Bone Min. Res.* **10**, 211 (1995); W. S. S. Jee *et al.,* ibid. 963. Clinical evaluation in multiple myeloma: C. Roux *et al., Bone* **15**, 41 (1994). Preliminary report of clinical efficacy in Paget's disease: J. P. Brown *et al., Semin. Arthritis Rheum.* **23**, 272 (1994).

Monosodium salt, $C_7H_{10}NNaO_7P_2$, *risedronate sodium, NE-58095.*

THERAP CAT: Bone resorption inhibitor.

8397. Risperidone. *3-[2-[4-(6-Fluoro-1,2-benzisoxazol-3-yl)-1-piperidinyl]ethyl]-6,7,8,9-tetrahydro-2-methyl-4H-pyrido[1,2-a]pyrimidin-4-one;* R-64766; Risperdal. $C_{23}H_{27}FN_4O_2$; mol wt 410.49. C 67.30%, H 6.63%, F 4.63%, N 13.65%, O 7.80%. Combined serotonin ($5\text{-}HT_2$) and dopamine (D_2) receptor antagonist. Prepn: L. E. J. Kennis, J. Vandenberk, **Eur.** pat. **Appl.** **196,132**; *eidem*, **U.S.** pat. **4,804,663** (1986, 1989 both to Janssen). Pharmacology: P. A. J. Janssen *et al., J. Pharmacol. Exp. Ther.* **244**, 685 (1988). Receptor binding studies: J. E. Leysen *et al.,* ibid. **247**, 661 (1988). Clinical study in psychoses: Y. G. Gelders *et al., Pharmacopsychiatry.* **23**, 206 (1990). Brief review: M. G. Livingston, *Lancet* **343**, 457-460 (1994). Review of pharmacology and therapeutic potential: S. Grant, A. Fitton, *Drugs* **48**, 253-273 (1994).

Crystals from DMF + 2-propanol, mp 170.0°. LD_{50} in male, female mice, rats, dogs (mg/kg): 29.7, 26.9, 34.3, 35.4, 14.1, 18.3 i.v.; 82.1, 63.1, 113, 56.6, 18.3, 18.3 orally (Janssen, 1988).

THERAP CAT: Antipsychotic.

8398. Ristocetin. Ristomycin; Spontin; Riston. Glycopeptide antibiotic complex produced by the actinomycete *Nocardia lurida*. Ristocetin A and the more active ristocetin B differ in the number of glucose, mannose, rhamnose and D-arabinose groups in their side chains. Isoln, crystallizn and chemical properties: J. E. Philip *et al., Antibiot. Ann.* **1956-57**, p 699; *eidem*, **U.S.** pat. **2,990,329** (1961 to Abbott). Structural studies of ristocetin A: Fehlner *et al., Proc. Nat. Acad. Sci. USA* **69**, 2420 (1972); C. M. Harris *et al., J. Am. Chem. Soc.* **101**, 437 (1979); F. Sztaricskai *et al.,* ibid. **102**, 7093 (1980). Total structure determn of ristocetin A: D. H. Williams *et al., Chem. Commun.* **1979**, 906; J. R. Kalman, D. H. Williams, *J. Am. Chem. Soc.* **102**, 897 (1980). Identity of ristocetin A with ristomycin A: D. H. Williams *et al., J. Chem. Soc. Perkin Trans. I* **1979**, 787. ^{13}C-NMR studies: F. Sztaricskai *et al., Tetrahedron Letters* **21**, 2983 (1980); M. P. Williamson, D. H. Williams, *J. Chem. Soc. Perkin Trans. I* **1981**, 1483. Revised configuration: C. M. Harris, T. M. Harris, *J. Am. Chem. Soc.* **104**, 363 (1982). Biosynthesis: S. J. Hammond *et al., Chem. Commun.* **1983**, 116. *Review:* D. C. Jordan, *Antibiotics* vol.

1, D. Gottlieb, P. Shaw, Eds. (Springer-Verlag, New York, 1967) pp 84-89.

Ristocetin A

Crystalline sulfates. Sol in acidic aq solns; much less sol in the neutral pH range. Generally insol in organic solvents. Both components show good stability in aq acidic solns, but are readily inactivated above pH 7.0. Commercial prepns are mixtures of both with >90% ristocetin A.

Ristocetin A, $C_{95}H_{110}N_8O_{44}$, *ristomycin A*. Cryst sulfate, $[\alpha]_D$ − 120° to − 133° (water).

Ristocetin B, *ristomycin B*. Cryst sulfate, $[\alpha]_D$ − 144° to − 149° (water).

USE: Tool for investigation of platelet aggregation: Howard, Firkin, *Thromb. Diath. Haemorrh.* **26**, 362 (1971).

THERAP CAT: Antibacterial.

8399.   Ritanserin. *6-[2-[4-[4-[Bis(4-fluorophenyl)methylene]-1-piperidinyl]ethyl]-7-methyl-5H-thiazolo[3,2-a]pyrimidin-5-one;* R-55667; Tiserton. $C_{27}H_{25}F_2N_3OS$; mol wt 477.58. C 67.90%, H 5.28%, F 7.96%, N 8.80%, O 3.35%, S 6.71%. Selective serotonin (5-HT$_2$) receptor antagonist. Prepn: L. E. J. Kennis *et al., Eur. pat.* **Appl.** 110,435; *eidem, U.S. pat.* **4,533,665** (1984, 1985 both to Janssen). Pharmacological profile: F. Awouters *et al., Drug Dev. Res.* **15**, 61 (1988). GC/MS determn in plasma and pharmacokinetics: P. Timmerman *et al, Biomed. Environ. Mass Spectrom.* **18**, 498 (1989). Clinical studies: G. Nappi *et al., Headache* **30**, 439 (1990); G. Bersani *et al., Acta Psychiat. Scand.* **83**, 244 (1991); J. M. Monti *et al., Sleep* **16**, 647 (1993).

Crystals from acetonitrile, mp 145.5°. LD$_{50}$ in male, female mice, rats, dogs (mg/kg): 28.2, 28.2, 20.0, 22.2, 24.1, 33.2 i.v.; 626, 993, 956, 515, ~1280, 640-1280 orally (Awouters).

L-Tartrate, $C_{27}H_{25}F_2N_3OS \cdot C_4H_6O_6$, solid from 2-propanol, mp 198.7°.

THERAP CAT: Anxiolytic; antidepressant.

8400.   Ritipenem. *[5R-[5α,6α(R*)]]-3-[[(Aminocarbonyl)oxy]methyl]-6-(1-hydroxyethyl)-7-oxo-4-thia-1-azabicyclo[3.2.0]hept-2-ene-2-carboxylic acid;* (5R,6S,8R)-6-α-hydroxyethyl-2-carbamoyloxymethyl-2-penem-3-carboxylic acid; (5R,6S)-6-[(1R)-1-hydroxyethyl]-3-(hydroxymethyl)-7-oxo-4-thia-1-azabicyclo[3.2.0]hept-2-ene-2-carboxylic acid, 3-carbamate. $C_{10}H_{12}N_2O_6S$; mol wt 288.28. C 41.66%, H 4.20%, N 9.72%, O 33.30%, S 11.12%. Prepn: M. Alpegiani *et al.,* **Ger. pat.** 3,245,270; M. Foglio *et al., U.S. pat.* **4,482,-565** (1983, 1984 both to Carlo Erba). Synthesis: G. Franceschi *et al., J. Antibiot.* **36**, 938 (1983). Total synthesis: W. Cabri *et al., Tetrahedron Letters* **34**, 3491 (1993). Toxicity study: M. Brughera *et al., J. Antimicrob. Chemother.* **23**, Suppl. C, 129 (1989). Series of articles on synthesis, *in vitro* activity, metabolism: *ibid.,* 1-204 (1989). Clinical pharmacokinetics of acid and ester forms: S. R. Norrby *et al., ibid.* **25**, 371 (1990); A. M. Lovering *et al., ibid.* **29**, 179 (1992). HPLC determn in serum and urine: R. Mendez *et al., J. Chromatog.* **579**, 115 (1992).

Sodium salt, $C_{10}H_{11}N_2NaO_6S$, *FCE-22101*. $[\alpha]_D^{20}$ +140°. uv max (H$_2$O): 258, 306 nm (ε 4150, 6030). LD$_{50}$ in male, female mice, male, female rats (mg/kg): 3872, 4393, 2000, 2201 i.v. (Brughera).

Acetoxymethyl ester, $C_{13}H_{16}N_2O_8S$, *ritipenem acoxil*, *FCE-22891*. LD$_{50}$ in male, female mice, male, female rats (mg/kg): 4363, 6167, > 5000, > 5000 orally (Brughera).

THERAP CAT: Antibacterial.

8401.   Ritodrine. *(R*,S*)-4-Hydroxy-α-[1-[[2-(4-hydroxyphenyl)ethyl]amino]ethyl]benzenemethanol; erythro-p-hydroxy-α-[1-[(p-hydroxyphenethyl)amino]ethyl]benzyl alcohol;* N-[2-(p-hydroxyphenyl)ethyl]-N-[2-(p-hydroxyphenyl)-2-hydroxy-1-methylethyl]amine; 1-(4-hydroxyphenyl)-2-[2-(4-hydroxyphenyl)ethylamino]propanol; N-(p-hydroxyphenylethyl)-4-hydroxynorephedrine. $C_{17}H_{21}NO_3$; mol wt 287.36. C 71.06%, H 7.37%, N 4.87%, O 16.70%. β$_2$-Adrenergic agonist. Prepn: **Belg. pat.** 660,244 (1965 to N.V. Philips); Claassen *et al., U.S. pat.* **3,410,944** (1968 to No. Am. Philips). Clinical investigations: Coutinho *et al., Am. J. Obstet. Gynecol.* **104**, 1053 (1969); Landesman *et al., ibid.* **110**, 111 (1971); Wesselius-De Casparis *et al., Brit. Med. J.* **3**, 144 (1971). Clinical efficacy in treatment of preterm labor: J. F. Larsen *et al., Obstet. Gynecol.* **67**, 607 (1986).

Base, resinous mass, mp 88-90°.

Hydrochloride, $C_{17}H_{21}NO_3 \cdot HCl$, *DU-21220, Miolene, Prempar, Pre-Par, Utemerin, Utopar, Yutopar.* mp 193-195° (dec) from ethanol-ether. uv max: 267.5 nm (ε 3310).

THERAP CAT: Tocolytic.

8402.   Ritonavir. *[5S-(5R*,8R*,10R*,11R*)]-10-Hydroxy-2-methyl-5-(1-methylethyl)-1-[2-(1-methylethyl)-4-thiazolyl]-3,6-dioxo-8,11-bis(phenylmethyl)-2,4,7,12-tetraazatridecan-13-oic acid 5-thiazolylmethyl ester;* (2S,3S,5S)-5-[N-[N-methyl-N-[(2-isopropyl-4-thiazolyl)methyl]amino]carbonyl]valinyl]amino]-2-[N-[(5-thiazolyl)methoxycarbonyl]amino]-1,6-diphenyl-3-hydroxyhexane; A-84538; Abbott 84538; ABT-538. $C_{37}H_{48}N_6O_5S_2$; mol wt 720.96. C 61.64%, H 6.71%, N 11.66%, O 11.10%, S 8.90%. Peptidomimetic HIV-1 protease inhibitor. Prepn: D. J. Kempf *et al.,* **PCT**

Int. pat. Appl. **94 14,436** (1994 to Abbott). Antiretroviral spectrum, pharmacokinetics: *idem et al., Proc. Nat. Acad. Sci. USA* **92**, 2484 (1995). Structural model for drug resistance: M. Markowitz *et al., J. Virol.* **69**, 701 (1995). Evaluation in HIV-infected patients: D. D. Ho *et al., Nature* **373**, 123 (1995).

THERAP CAT: Antiviral.

8403. Robenidine. *Bis[(4-chlorophenyl)methylene]carbonimidic dihydrazide; 1,3-bis[(p-chlorobenzylidene)amino]guanidine.* $C_{15}H_{13}Cl_2N_5$; mol wt 334.21. C 53.91%, H 3.92%, Cl 21.22%, N 20.96%. Prepn: Tomcufcik, **Ger. pat. 1,933,112** (1970 to Am. Cyanamid), *C.A.* **72**, 90113c (1970). Activity studies: Kantor *et al., Science* **168**, 373 (1970); Wong *et al., Biochem. Biophys. Res. Commun.* **46**, 621 (1972). Metabolism: Zulalian *et al.,* 163rd Am. Chem. Soc. Meeting (Boston, April 1972) *Abstracts of Papers,* PEST 12, 13. Animal studies: Millard, *Res. Vet. Sci.* **11**, 394 (1970); Joyner, Norton, *ibid.* **13**, 279 (1972).

Hydrochloride, $C_{15}H_{13}Cl_2N_5 \cdot HCl$, *robenzidene, Cycostat, Robenz.* Crystals from ethanol, mp 289-290°.

THERAP CAT (VET): Coccidiostat.

8404. Robinin. *3-[[6-O-(6-Deoxy-α-L-mannopyranosyl)-β-D-galactopyranosyl]oxy]-7-[(6-deoxy-α-L-mannopyranosyl)oxy]-5-hydroxy-2-(4-hydroxyphenyl)-4H-1-benzopyran-4-one;* kaempferol 3-robinoside 7-rhamnoside. $C_{33}H_{40}O_{19}$; mol wt 740.67. C 53.51%, H 5.44%, O 41.04%. Dimorphic flavanoid isolated from the leaves and flowers of *Robinia pseudoacacia* L., *Leguminosae:* C. Zwenger, F. Dronke, *Ann. suppl.* **1**, 257 (1861); C. Sando, *J. Biol. Chem.* **94**, 675 (1932). Structure: Zemplén, Bognár, *Ber.* **74B**, 1783 (1941). Total synthesis and structure: L. Farkas *et al., Phytochemistry* **15**, 215 (1976).

β-Form. Yellow crystals, mp 250-254° (Farkas); also reported as straw-yellow needles from alc, mp 249-250° (Sando). uv max (ethanol): 352, 368 nm (log ε 4.14, 4.18), Jurd, Horowitz, *J. Org. Chem.* **22**, 1619 (1957). Sol in hot water, hot alc; practically insol in ether. On hydrolysis yields kaempferol, *q.v.*
α-Form. Obtained by crystallization from water and dehydrating, mp 195-197° (Sando). Also reported as hydrate, yellow needles from aq methanol, mp 196-199° (Farkas).

8405. Roccellic Acid. *[S-(R*,S*)]-2-Dodecyl-3-methylbutanedioic acid; (2R,3S)-2-dodecyl-3-methylsuccinic acid; d-α-dodecyl-β-methylsuccinic acid; d-α-methyl-α'-dodecylsuccinic acid.* $C_{17}H_{32}O_4$; mol wt 300.44. C 67.96%, H 10.74%, O 21.30%. Occurs in lichens. Isoln from *Lecanora sordida*

(Pers.) Th. Fries, *Parmeliaceae:* Hesse, *J. Prakt. Chem.* **58**, 497 (1898); Kennedy *et al., Sci. Proc. Roy. Dublin Soc.* **21**, 557 (1937); from *Roccella montagnei, Graphidaceae:* Subbaraya, Seshadri, *Proc. Indian Acad. Sci.* **12A**, 640 (1940); from *Crocynia membranacae* (Dicks.) Zahlbr., *Chrysotrichaceae:* Akermark *et al., Acta Chem. Scand.* **13**, 1855 (1959). Structure: Kennedy *et al., loc. cit.* Absolute configuration: Akermark, *Acta Chem. Scand.* **16**, 599 (1962).

Rectangular rods from acetone, mp 132-133°. $[\alpha]_D^{20}$ +18° (c = 1.84 in ethanol). Practically insol in water. Freely sol in alcohol, ether; sol in aq sodium bicarbonate solns. Forms a water-sol sodium salt.

8406. Rociverine. *1-Hydroxy[1,1'-bicyclohexyl]-2-carboxylic acid 2-(diethylamino)-1-methylethyl ester;* 2-(diethylamino)-1-methylethyl *cis*-1-hydroxy[bicyclohexyl]-2-carboxylate; LG-30158; Rilaten. $C_{20}H_{37}NO_3$; mol wt 339.52. C 70.75%, H 10.98%, N 4.13%, O 14.14%. Spasmolytic agent with balanced neurotropic and myotropic properties. Prepn: L. Turbanti, **S. Afr. pat. 67 05649,** *C.A.* **70**, 47117d (1969) and **U.S. pat. 3,700,675** (1968, 1972 both to Guidotti). Antispasmodic activity *in vitro* and *in vivo:* G. Toson *et al., Arzneimittel-Forsch.* **28**, 1130 (1978). Effect in cystitis or bladder spasm: A. Manganelli, *Farmaco Ed. Prat.* **34**, 384 (1979). Clinical studies: M. Petrillo *et al., Curr. Med. Res. Opin.* **7**, 73 (1980); R. Assisi, S. deStefano, *Acta Ther.* **6**, 353 (1980); F. Marsala, *Minerva Med.* **73**, 2179 (1982).

Oil, $bp_{0.1}$ 148-150°. n_D^{20} 1.4820. Sol in alc, ether, chloroform, benzene, dil mineral acids. Insol in water.

THERAP CAT: Antispasmodic.

8407. Rocuronium. *1-[(2β,3α,5α,16β,17β)-17-(Acetyloxy)-3-hydroxy-2-(4-morpholinyl)androstan-16-yl]-1-(2-propenyl)pyrrolidinium;* 1-allyl-1-(3α,17β-dihydroxy-2β-morpholino-5α-androstan-16β-yl)pyrrolidinium 17-acetate. $[C_{32}H_{53}N_2O_4]^+$. Non-depolarizing neuromuscular blocking agent. Prepn: D. S. Savage *et al.,* **Eur. pat. Appl. 287,150;** T. Sleigh *et al.,* **U.S. pat. 4,894,369** (1988, 1990 both to Akzo). Pharmacology: A. W. Muir *et al., Brit. J. Anaesth.* **63**, 400 (1989); K. Khuenl-Brady *et al., Anesthesiology* **72**, 669 (1990). Clinical pharmacodynamics: T. J. Quill *et al., Anesth. Analg.* **72**, 203 (1991); and pharmacokinetics in the elderly: R. S. Matteo *et al., ibid.* **77**, 1193 (1993). Comparative clinical trial: T. Magorian *et al., Anesthesiology* **79**, 913 (1993). HPLC determn: U. W. Kleef *et al., J. Chromatog.* **621**, 65 (1993). Review: T. C. Wicks, *J. Am. Assoc. Nurse Anesthet.* **62**, 33-38 (1994).

Bromide, $C_{32}H_{53}BrN_2O_4$, *Org-9426, Esmeron, Zemuron.*
Crystals, mp 161-169°. $[\alpha]_D^{20}$ +18.7° (c = 1.03 in $CHCl_3$).
THERAP CAT: Muscle relaxant (skeletal).

8408. Rokitamycin. *Leucomycin V 4B-butanoate 3B-propanoate;* 3″-propionylleucomycin A5; 5-[O-(2,6-dideoxy-hexopyranosyl)-(1→4)-3,6-dideoxy-3-dimethylamino-β-D-glucopyranosyloxy]-6-formylmethyl-3,9-dihydroxy-4-methoxy-8-methyl-10,12-hexadecadien-15-olide; rikamycin; M-19-Q; TMS-19Q; Ricamycin; Rokital. $C_{42}H_{69}NO_{15}$; mol wt 828.01. C 60.92%, H 8.40%, N 1.69%, O 28.98%. Macrolide antibiotic active against Mycoplasma and macrolide resistant strains of *Staphylococcus aureus* and *Streptococcus pyogenes.* Prepn: H. Sakakibara *et al.,* Ger. pat. **2,918,954** corresp to U.S. pat. **4,242,504** (1979, 1980 both to Toyo Jozo); *eidem, J. Antibiot.* **34,** 1001 (1981). Series of articles on antibacterial activity, pharmacology, metabolism, toxicology: *Chemotherapy (Japan)* **32,** Suppl. 6, 1-627 (1984). Toxicity data: K. Matsumoto *et al., ibid.* **138.**

mp 116°. $[\alpha]_D^{20}$ −71° (c = 1.0 in chloroform). uv max (ethanol): 232 nm (ε 28000).
THERAP CAT: Antibacterial.

8409. Rolicyprine. *5-Oxo-N-(2-phenylcyclopropyl)-2-pyrrolidinecarboxamide;* D-N-(*trans*-2-phenylcyclopropyl)-L-5-pyrrolidone-2-carboxamide; rolicypram; EX-4883; RMI-83027; Cypromin. $C_{14}H_{16}N_2O_2$; mol wt 244.29. C 68.83%, H 6.60%, N 11.47%, O 13.10%. Prepn: (isomer obtained not specified) **Brit.** pat. **961,313** (1964 to Lakeside Labs.); **Brit.** pat. **1,000,895** (1965 to Colgate-Palmolive); Biel, U.S. pat. **3,192,229** (1965 to Colgate-Palmolive). Prepn of other stereoisomers: Biel, *loc. cit.* Toxicity study: E. I. Goldenthal, *Toxicol. Appl. Pharmacol.* **18,** 185 (1971).

Crystals from water, mp 144-147°. $[\alpha]_D^{25}$ +104.28° (DMF). LD_{50} orally in rats: 96±8 mg/kg (Goldenthal).
THERAP CAT: Antidepressant.

8410. Rolipram. *4-[3-(Cyclopentyloxy)-4-methoxyphenyl]-2-pyrrolidinone;* ZK-62711. $C_{16}H_{21}NO_3$; mol wt 275.35. C 69.79%, H 7.69%, N 5.09%, O 17.43%. Pyrrolidone antidepressant; selective type IV inhibitor of cyclic AMP phosphodiesterase. Prepn: **Belg.** pat. **826,923;** R. Schmiechen *et al.,* U.S. pat. **4,193,926** (1975, 1980 both to Schering AG). Selective enzyme inhibition and enhancement of cAMP accumulation in rat brain tissue: U. Schwabe *et al., Mol. Pharmacol.* **12,** 900 (1976). Mechanism of action studies: H. Wachtel, *Neuropharmacol.* **22,** 267 (1983); H. Wachtel, H. H. Schneider, *ibid.* **25,** 1119 (1986). Pharmacokinetics: W. Krause, G. Kühne, *Xenobiotica* **18,** 561 (1988). Clinical evaluation in depression: E. Zeller *et al., Pharmacopsychiatry* **17,** 188 (1984). Comparative clinical trial with amitriptyline, *q.v.,* in severe depression: F. Eckmann *et al., Curr. Ther. Res.* **43,** 291 (1988). Pharmacology and efficacy in exptl autoimmune encephalomyelitis: N. Sommer *et al., Nature Medicine* **1,** 244 (1995).

Crystals from ethyl acetate, mp 132°.
THERAP CAT: Antidepressant.

8411. Rolitetracycline. *[4S-(4α,4aα,5aα,6β,12aα)]-4-(Dimethylamino)-1,4,4a,5,5a,6,11,12a-octahydro-3,6,10,12,-12a-pentahydroxy-6-methyl-1,11-dioxo-N-(1-pyrrolidinylmethyl)-2-naphthacenecarboxamide;* N-(pyrrolidinomethyl)-tetracycline; N-(1-pyrrolidinylmethyl)tetracycline; Reverin; Syntetrin; Tetraverin; Transcycline. $C_{27}H_{33}N_3O_8$; mol wt 527.57. C 61.47%, H 6.30%, N 7.96%, O 24.26%. Semisynthetic antibiotic prepd from tetracycline: Siedel *et al., München. Med. Wochenschr.* **100,** 661 (1958); Lindner *et al.,* **S. Afr.** pat. **3169/57.** Alternate procedure and structure: Gottstein *et al., J. Am. Chem. Soc.* **81,** 1198 (1959); Cheney *et al.,* U.S. pat. **3,104,240** (1963 to Bristol-Myers). Toxicity: E. I. Goldenthal, *Toxicol. Appl. Pharmacol.* **18,** 185 (1971).

Fine, pale yellow needles, dec 162-165°. Amphoteric. More sol than tetracycline and tetracycline hydrochloride. Soly in water at 25°: 1.25 g/ml. Freely sol in alc. Sol in dil acids and alkalies.
Nitrate sesquihydrate, $C_{27}H_{33}N_3O_8·HNO_3·1½H_2O$, *Bristacin, Pyrrocycline-N, Syntetrex, Tetrim, Tetriv.* LD_{50} i.v. in rats: 91 mg/kg (Goldenthal).
Compd with chloramphenicol succinate, $C_{27}H_{33}N_3O_8·C_{15}H_{16}Cl_2N_2O_8$, *cafrolicycline, gradocycline, levocycline, senociclin, Cloruciclin, Crovicina, Metilcaf, Proterciclina, Reicaf, Tecaf, Tetrafenicol.* Prepn: **Belg.** pat. **636,234;** Scevola, U.S. pat. **3,218,335** (1963, 1965 both to Lab. Pro-Ter). Bright yellow powder, dec 140-144°. Bitter taste. Very sol in water. Practically insol in ether, ligroin, hexane.
THERAP CAT: Antibacterial.

8412. Romurtide. *N²-[N²-[N-(N-Acetylmuramoyl)-L-alanyl]-D-α-glutaminyl]-N⁶-(1-oxooctadecyl)-L-lysine;* N²-(N-acetylmuramoyl-L-alanyl-D-isoglutaminyl)-N⁶-stearoyl-L-lysine; MDP-Lys (L18); murocatasin; DJ-7041; Nopia. $C_{43}H_{78}N_6O_{13}$; mol wt 887.12. C 58.22%, H 8.86%, N 9.47%, O 23.45%. Derivative of muramyl dipeptide, *q.v.,* with immunostimulating activity. Prepn: T. Shiba *et al., Eur. pat. Appl.* **21,367;** *eidem,* U.S. pat. **4,317,771** (1981, 1982 both to Daiichi Seiyaku). Immunopotentiating effect *in vitro:* Y. Osada *et al., Infect. Immun.* **38,** 848 (1982); *in vivo:* K. Matsumoto *et al., ibid.* **39,** 1029 (1983); M. Akasaki *et al., Agents Actions* **22,** 144 (1987). Enzyme immunoassay in plasma: H. Masayasu *et al., Chem. Pharm. Bull.* **33,** 5522 (1985). Supplement on pharmacology, toxicology and clinical studies: *Arzneimittel-Forsch.* **38,** Suppl. 7A, 951-1074 (1988). Acute toxicity: Y. Ono *et al., ibid.* 1022.

Odorless, tasteless white powder, mp 176-178°. $[\alpha]_D^{20}$ +24.9° (c = 1 in methanol). pKa 5.70. Soly at 20°: DMF 1 g/72 ml; methanol 1 g/190 ml; ethanol 1 g/330 ml. Practically insol in acetone, acetonitrile, chloroform, water. LD_{50} in male, female mice, male, female rats, dogs (mg/kg): 436, 625, 761, 801, >200 s.c. (Ono).

THERAP CAT: Immunostimulant.

8413. Ronidazole. *1-Methyl-5-nitroimidazole-2-methanol carbamate (ester);* carbamic acid (1-methyl-5-nitroimidazol-2-yl) ester; 1-methyl-2-[(carbamoyloxy)methyl]-5-nitroimidazole; (1-methyl-5-nitroimidazole-2-yl)methyl carbamate; MCMN; Dugro; Ridzol. $C_6H_8N_4O_4$; mol wt 200.15. C 36.01%, H 4.03%, N 27.99%, O 31.97%. Preparation and antiprotozoal activity: **Neth. pats. 6,609,552; 6,609,553** (both 1967 to Merck & Co.), *C.A.* **67,** 54123u, 11487y (1967); Verdi; Kollonitsch, **U.S. pats. 3,450,710; 3,459,764** (both 1969 to Merck & Co.).

Pale yellow crystals, mp 167-169°. Soly in water (at pH 6.5) ~2.9 mg/ml at room temp. More sol in acid solns. Unstable in alkaline solns. pKa 1.2. Freely sol in acetone. Sol in methanol, ethanol, chloroform and ethyl acetate.

THERAP CAT (VET): Antimicrobial.

8414. Ronifibrate. *3-Pyridinecarboxylic acid 3-[2-(4-chlorophenoxy)-2-methyl-1-oxopropoxy]propyl ester;* 3-(nicotinoyloxy)propyl *p*-chlorophenoxyisobutyrate; 3-[α-(*p*-chlorophenoxy)isobutyryloxy]propyl nicotinate; I-612; Cloprane. $C_{19}H_{20}ClNO_5$; mol wt 377.82. C 60.40%, H 5.34%, Cl 9.38%, N 3.71%, O 21.17%. Diester of clofibric and nicotinic acids, *q.q.v.* Prepn: Y. Hirata *et al.,* **Japan. Kokai 73 40,777** (1973 to Yamanouchi), *C.A.* **79,** 66180w (1973); H. Shindo *et al.,* **Japan. Kokai 74 30,377** (1974 to Kowa Pharm.), *C.A.* **81,** 135984s (1974). Clinical trial in hyperdyslipidemia: A. Bucalossi *et al., Clin. Ter.* **89,** 127 (1979). Efficacy and tolerability: G. Buzzelli *et al., ibid.* 251.

LD_{50} orally in mice: 4.08 g/kg (Bucalossi).

THERAP CAT: Antihyperlipoproteinemic.

8415. Ronnel. *Phosphorothioic acid, O,O-dimethyl O-(2,4,5-trichlorophenyl)ester;* fenchlorphos; dimethyl trichlorophenyl thiophosphate; Trolene; Etrolene; Nankor; Korlan; Viozene; Ectoral. $C_8H_8Cl_3O_3PS$; mol wt 321.55. C 29.88%, H 2.51%, Cl 33.08%, O 14.93%, P 9.63%, S 9.97%. Cholinesterase inhibitor. Prepn: Schrader, **Ger. pat. 814,-152** (1948 to Bayer); Moyle, **U.S. pat. 2,599,516** (1952 to Dow). Description of mfg process: *Can. Dept. Agr. Bull.,* (3rd ed., Oct. 1957) p 136. Toxicity study: T. B. Gaines, *Toxicol. Appl. Pharmacol.* **14,** 515 (1969).

White powder, mp 41°. Vapor pressure at 25°: 8×10^{-4} mm Hg. Practically insol in water (0.004 g/100 ml at 25°). Freely sol in acetone, carbon tetrachloride, ether, methylene chloride, toluene, kerosene. LD_{50} in male, female rats (mg/kg): 1250, 2630 orally at 25°).

Caution: Potential symptoms of overexposure in animals are cholinesterase inhibition; eye irritation; liver and kidney

damage. *See NIOSH Pocket Guide to Chemical Hazards* (DHHS/NIOSH 90-117, 1990) p 192.

USE: Insecticide.

THERAP CAT (VET): Insecticide (systemic).

8416. Ropinirole. *4-[2-(Dipropylamino)ethyl]-1,3-dihydro-2H-indol-2-one;* 4-[2-(di-*n*-propylamino)ethyl]-2(3H)-indolone; SKF-101468. $C_{16}H_{24}N_2O$; mol wt 260.38. C 73.81%, H 9.29%, N 10.76%, O 6.14%. Selective dopamine D_2-receptor agonist. Prepn: G. Gallagher, Jr., U.S. pat. **4,452,808** (1984 to Smithkline Beckman); *idem et al., J. Med. Chem.* **28,** 1533 (1985). HPLC determn in plasma: J. E. Swagzdis, B. A. Mico, *J. Pharm. Sci.* **75,** 90 (1986). Pharmacology and receptor binding study: R. J. Eden *et al., Pharmacol. Biochem. Behav.* **38,** 147 (1991). Clinical pharmacology, pharmacokinetics: C. de Mey *et al., Brit. J. Clin. Pharmacol.* **32,** 483 (1991). Preliminary evaluation in Parkinson's disease: M. J. Vidailhet *et al., Lancet* **336,** 316 (1990).

Hydrochloride, $C_{16}H_{24}N_2O \cdot HCl$, *SKF-101468A.* Crystals from acetonitrile, mp 241-243°.

THERAP CAT: Antiparkinsonian.

8417. Ropivacaine. *(S)-N-(2,6-dimethylphenyl)-1-propyl-2-piperidinecarboxamide;* (S)-(−)-1-propyl-2',6'-pipecoloxylidide; *l-N-n*-propylpipecolic acid-2,6-xylidide; LEA-103. $C_{17}H_{26}N_2O$; mol wt 274.41. C 74.41%, H 9.55%, N 10.21%, O 5.83%. Prepn: A. F. Thuresson, C. Bovin, **PCT Int. pat. Appl. 85 00,599** (1985 to Apothekernes); H.-J. Federsel *et al., Acta Chem. Scand.* **B41,** 757 (1987). Physicochemical properties: G. R. Strichartz *et al., Anesth. Analg.* **71,** 158 (1990). HPLC determn in human plasma: Z. Yu *et al., J. Chromatog. B* **654,** 221 (1994). *In vitro* metabolism: Y. Oda *et al., Anesthesiology* **82,** 214 (1995). Clinical pharmacokinetics: D. J. Kopacz *et al., ibid.* **81,** 1139 (1994). Toxicity study in sheep: A. C. Santos *et al., ibid.* **82,** 734 (1995). Clinical evaluation in relief of surgical pain: I. Cederholm *et al., Regional Anesth.* **19,** 18 (1994); B. Johansson *et al., Anesth. Analg.* **78,** 210 (1994); labor pain: R. Stienstra *et al., ibid.* **80,** 285 (1995).

Crystals from toluene, mp 144-146°. $[\alpha]_D^{25}$ −82.0° (c = 2 in methanol). pKa 8.16. Distribution coefficient (1-octanol/aq buffer, pH 7.4): 115.0.

Monohydrochloride, $C_{17}H_{26}N_2O \cdot HCl$, crystals from isopropyl alcohol, mp 260-262°. $[\alpha]_D^{25}$ −6.6° (c = 2 in water).

Monohydrochloride monohydrate, crystals from acetone + water, mp 269.5-270.6°. $[\alpha]_D^{20}$ −7.28° (c = 2 in water).

THERAP CAT: Anesthetic (local).

8418. Roquinimex. *1,2-Dihydro-4-hydroxy-N,1-dimethyl-2-oxo-N-phenyl-3-quinolinecarboxamide;* N-methyl-N-phenyl-1,2-dihydro-4-hydroxy-1-methyl-2-oxoquinoline-3-carboxamide; 1,2-dihydro-4-hydroxy-N,1-dimethyl-2-oxo-3-quinolinecarboxanilide; LS-2616; Linomide. $C_{18}H_{16}N_2O_3$; mol wt 308.34. C 70.12%, H 5.23%, N 9.09%, O 15.57%. Prepn: E. Eriksoo *et al., Eur. pat. Appl.* **59,698;** *eidem,* U.S. pat. **4,738,971** (1982, 1988 both to AB Leo). Immunomodulating effects in autoimmune mice: A. Tarkowski *et al., Arthritis Rheum.* **29,** 1405 (1986); *eidem, Immunology* **59,** 589 (1986); in murine metastatic melanoma: R. Harning, J. Szalay, *Invest. Ophthalmol. Vis. Sci.* **29,** 1505 (1988). Mechanism of action study: E.-L. Larsson *et al., Int J. Immunopharmacol.* **9,** 425 (1987).

Crystals from pyridine, mp 200-204°.
THERAP CAT: Antineoplastic.

8419. Rosaprostol. *2-Hexyl-5-hydroxycyclopentaneheptanoic acid;* 9-hydroxy-19,20-bisnorprostanoic acid; C-83; IBI-C83; Rosal. $C_{18}H_{34}O_3$; mol wt 298.47. C 72.44%, H 11.48%, O 16.08%. Prostaglandin analog. Prepn, hypolipemic, platelet aggregation inhibitory activity: U. Valcavi, Ger. pat. **2,535,343,** eidem, U.S. pat. **4,073,938** (1976, 1978 both to Ist. Biochim. Ital.). Alternate process: V. Marotta, G. Zabban, **Eur.** pat. **Appl. 155,392** (1985 to Ist. Biochim. Ital.). Gastric antisecretory, cytoprotective activity: U. Valcavi et al., *Arzneimittel-Forsch.* **32,** 657 (1982). Effect on mucus and gastrin secretion in duodenal ulcer: D. Foschi et al., *Prostaglandins Leukotrienes Med.* **15,** 147 (1984). Comparison with cimetidine, q.v.: eidem, *Drugs Exp. Clin. Res.* **10,** 427 (1984). Clinical evaluation in treatment of ulcers: G. P. Tincani et al., *Minerva Med.* **78,** 847 (1987).

Oil.
Sodium salt, $C_{18}H_{33}NaO_3$, white solid. LD_{50} in mice, rats: ~3000 mg/kg orally (Valcavi, 1978), > 5 g/kg orally (Valcavi, 1982).
THERAP CAT: Antiulcerative.

8420. Rosaramicin. *4'-Deoxycirramycin A_1;* 3-ethyl-7-hydroxy-2,8,12,16-tetramethyl-5,13-dioxo-9-[[3,4,6-trideoxy-3-(dimethylamino)-β-D-xylo-hexopyranosyl]oxy]-4,17-dioxabicyclo[14.1.0]heptadec-14-ene-10-acetaldehyde; antibiotic 67-694; juvenimicin A_3; rosamicin; M-4365A2; Sch-14947. $C_{31}H_{51}NO_9$; mol wt 581.75. C 64.00%, H 8.84%, N 2.41%, O 24.75%. Macrolide antibiotic isolated from fermentations of *Micromonospora rosaria* NRRL-3718. Isoln and properties: M. J. Weinstein et al., **S. Afr.** pat. **71 00,-402;** Fr. pat. **2,081,448** (1971, 1972 both to Sherico), *C.A.* **76,** 139065n (1972); **78,** 109310n (1973); G. H. Wagman et al., *J. Antibiot.* **25,** 641 (1972). Biological studies: J. A. Waitz et al., ibid. 647. Structure: H. Reimann, R. S. Jaret, *Chem. Commun.* **1972,** 1270. Crystal structure: A. K. Ganguly et al., *Tetrahedron Letters* **21,** 4699 (1980). Biosynthesis: A. K. Ganguly et al., *J. Antibiot.* **29,** 976 (1976). Identity with juvenimicin A_3: T. Kishi et al., ibid. 1171. Isoln from *M. capillata* MCRL 0940 and identity with antibiotic M-4365A2: A. Kinumaki et al., ibid. **30,** 450 (1977). *In vitro* activity: S. Feltham et al., *J. Antimicrob. Chemother.* **5,** 731 (1979). Use in experimental pneumococcal meningitis: C. M. Nolan et al., *Antimicrob. Ag. Chemother.* **16,** 776 (1979).

Crystals from chloroform, mp 119-122°. $[\alpha]_D^{26}$ −35° (ethanol). uv max (methanol): 240 nm (ε 14600). Very sol in methanol, acetone, chloroform, benzene. Sparingly sol in ether. Slightly sol in water. LD_{50} in mice (mg/kg): 625 s.c.; 350 i.p.; 155 i.v. (Wagman).
THERAP CAT: Antibacterial.

8421. Rose Bengal. *4,5,6,7-Tetrachloro-3',6'-dihydroxy-2',4',5',7'-tetraiodospiro[isobenzofuran-1(3H),9'-[9H]xanthen]-3-one dipotassium salt; 4,5,6,7-tetrachloro-2',4',5',7'-tetraiodofluorescein dipotassium salt;* C.I. Acid Red 94; Rose Bengale B; C.I. 45440. $C_{20}H_2Cl_4I_4K_2O_5$; mol wt 1049.86. C 22.88%, H 0.19%, Cl 13.51%, I 48.35%, K 7.45%, O 7.62%. Discovered by Gnehm 1882; tetraiodinate of 4,5,6,7-tetrachlorofluorescein (from resorcinol and tetrachlorophthalic anhydride) converted to potassium salt: *Colour Index* vol. 4 (3rd ed., 1971) 4428. Labeling with ^{131}I: Liebster, Andrysek, *Nature* **184,** 913 (1959). See also: *H. J. Conn's Biological Stains,* R. D. Lillie, Ed., (Williams & Wilkins, Baltimore, 9th ed., 1977) p 350. Diagnostic use in hepatic function: K. S. Nijran, D. C. Barber, *Phys. Med. Biol.* **31,** 563 (1986).

Bright bluish pink. Soluble in water (bluish red without fluorescence); brown soln in concd H_2SO_4, on dilution gives flesh pink ppt.
Disodium salt, $C_{20}H_2Cl_4I_4Na_2O_5$, *Rose Bengal Extra.*
^{131}I-Labeled disodium salt, $C_{20}H_2Cl_4I_4Na_2O_5$, *sodium rose bengal I 131, rose bengal sodium I 131, Robenogatope I-131.*
Note: C.I. Acid Red 94 and C.I. 45440 also used for disodium salt.
USE: As a dye, biological stain; in coloring straw, wood chips, and inks; for coloring edible products and cosmetics.
THERAP CAT: Diagnostic aid (corneal trauma indicator). Sodium Rose Bengal I 131 as diagnostic aid (hepatic function).

8422. Rose Hips. Hipberries. The fruits or berries from wild rose bushes, notably *Rosa canina* L., *R. gallica* L., *R. condita* Scop. and *R. rugosa* Thunb., Rosaceae. Rich in ascorbic acid: M. Szczyglowa et al., *Roczniki Panstwowego Zakladu Hig.* **1,** 523-532 (1950), *C.A.* **46,** 1177d. Ascorbic acid content of hips of *R. canina* reported as 19-27 mg/g dry weight: Å. Gustafsson, J. Schröderheim, *Nature* **153,** 196 (1944); as 33.1 mg/g dry weight: I. Roubani et al., *J. Hort. Sci.* **51,** 375 (1976). Comparison of methods for determn of ascorbic acid in hips: K. Gliniecki et al., *Pharm. Z.* **127,** 823 (1982).

8423. Rosemary. Garden rosemary. Flowers and leaves of *Rosmarinus officinalis* L., Labiatae. Habit. Mediterranean Basin; cultivated in gardens. *Constit.* Flowers: About 1% volatile oil, resin, bitter principle. Leaves: Volatile oil, tannin.
USE: In perfumery; manuf oil rosemary.
THERAP CAT: Emmenagogue.

8424. Rosin. Colophony; yellow resin; abietic anhydride. Residue left after distilling off the volatile oil from the oleoresin obtained from *Pinus palustris* and other species of *Pinus, Pinaceae.* Offered as wood rosin (from Southern pine stumps), gum rosin (the exudate from incisions in the living tree, *P. palustris* and *P. caribaea*), and tall oil rosin, see Tall Oil. Rosin is chiefly produced in the U.S.A. *Constit.* About 90% resin acids and 10% neutral matter. Of the resin acids about 90% are isomeric with abietic acid ($C_{20}H_{30}O_2$); the other 10% is a mixture of dihydroabietic acid ($C_{20}H_{32}O_2$) and dehydroabietic acid ($C_{20}H_{28}O_2$).
Pale yellow to amber, translucent fragments; brittle fracture at ordinary temp; slight turpentine odor and taste. Readily fusible when heated. d 1.07-1.09. Acid no. not less

than 150. Insol in water. Freely sol in alc, benzene, ether, glacial acetic acid, oils, carbon disulfide; also sol in dil solns or fixed alkali hydroxides.

USE: Manuf varnishes, varnish and paint driers, printing inks, cements, soap, sealing wax, wood polishes, floor coverings, paper, plastics, fireworks, tree wax, sizes, rosin oil; for water-proofing cardboard, walls, etc. Pharmaceutic aid (stiffening agent).

8425. Rosin Oil. Rosinol. Obtained by dry distillation of rosin.

Yellow, viscid, fluorescent, oily liquid. bp > 280°. Insol in water; sol in ether, oil turpentine and other oils. It dissolves phosphorus, sulfur, camphor, phenols, and many other organic compds.

USE: Manuf of carbon black for lithography and printing inks; in varnishes, retinol colors, lacquers, brewers' pitch, axle greases.

8426. Rosoxacin. *1-Ethyl-1,4-dihydro-4-oxo-7-(4-pyridinyl)-3-quinolinecarboxylic acid;* acrosoxacin; Win-35213; Eracine; Eradacil; Eradacin; Winuron. $C_{17}H_{14}N_2O_3$; mol wt 294.31. C 69.38%, H 4.79%, N 9.52%, O 16.31%. Quinolone antibacterial. Prepn: G. Y. Lesher, P. M. Carabateas, **Ger.** pat. **2,224,090;** *eidem,* **U.S.** pat. **3,753,993** (1972, 1973 both to Sterling). HPLC determn in plasma and urine: M. P. Kullberg *et al., J. Chromatog.* **173,** 155 (1979). Pharmacological studies: S. Maigaard *et al., Urol. Res.* **8,** 113 (1980); *eidem, Invest. Urol.* **17,** 149 (1979). Clinical study in gonorrhea: B. M. Limson *et al., Curr. Ther. Res.* **26,** 842 (1979).

Yellow crystals from DMF, mp 290°. Stable in dry heat at 70°. Light sensitive. Solns and suspensions stable at pH 2.0-9.5.

THERAP CAT: Antibacterial.

8427. Rotenone. *[2R-(2α,6aα,12aα)]-1,2,12,12a-Tetrahydro-8,9-dimethoxy-2-(1-methylethenyl)-[1]benzopyrano[3,4-b]furo[2,3-h][1]benzopyran-6(6aH)-one;* Canex. $C_{23}H_{22}O_6$; mol wt 394.42. C 70.04%, H 5.62%, O 24.34%. Principal insecticidal constituent of derris root, cubé, etc. Powerful inhibitor of mitochondrial electron transport. Isoln from *Lonchocarpus nicou* (Aubl.) DC., *Leguminosae:* Geoffrey, *Ann. Inst. Colon. Marseille* **2,** 1 (1895). Review of structure: La Forge *et al., Chem. Rev.* **12,** 181 (1933); Butenandt, McCartney, *Ann.* **494,** 17 (1932); King, *Ann. Rep. Progr. Chem. (Chem. Soc. London)* **29,** 186 (1932). Absolute configuration: Büchi *et al., J. Chem. Soc.* **1961,** 2843; Nakazaki, Arakawa, *Bull. Chem. Soc. Japan* **34,** 1246 (1961); Begley *et al., Chem. Commun.* **1975,** 850. NMR spectrum: Crombie, Lown, *ibid.* **1962,** 775. Total synthesis: Miyano *et al., Agr. Biol. Chem.* **25,** 673 (1961); Miyano, *J. Am. Chem. Soc.* **87,** 3958 (1965). Alternate synthesis: Crombie *et al., J. Chem. Soc. Perkin Trans I* **1973,** 1277, *eidem, Chem. Commun.* **1979,** 1142; I. Sasaki, K. Yamashita, *Agr. Biol. Chem.* **43,** 137 (1979). Synthesis of stereoisomers: Unai, Yamamoto, *ibid.* **37,** 897 (1973). Toxicology: Santi, Toth, *Farmaco Ed. Sci.* **20,** 270 (1965). Toxicity data: J.-I. Fukami *et al., Science* **155,** 713 (1967). *Review:* H. Fukami, M. Nakajima in *Naturally Occurring Insecticides,* M. Jacobson, D. G. Crosby, Eds. (Dekker, New York, 1971) pp 71-97; S. B. Soloway, *Environ. Health Persp.* **14,** 109-117 (1976); T. J. Haley, *J. Environ. Pathol. Toxicol.* **1.** 315-337 (1978).

Orthorhombic, six-sided plates from trichloroethylene, mp 165-166° (dimorphic form, mp 185-186°). $[\alpha]_D^{20}$ − 228° (c = 2.22 in benzene). uv spectra: Büchi *et al., loc. cit.* Practically insol in water. Sol in alcohol, acetone, carbon tetrachloride, chloroform, ether, and many other organic solvents. Dec upon exposure to light and air. Colorless solns in organic solvents oxidize upon exposure and become yellow, orange and then deep red and may deposit crystals of dehydrorotenone and rotenonone which are toxic to insects. LD_{50} i.p. in mice: 2.8 mg/kg (Fukami); in rats (mg/kg): 132 orally; 6 i.v. (Soloway).

Caution: Inhalation or ingestion of large doses may cause numbness of oral mucous membrane, nausea and vomiting, muscle tremors, tachypnea. With lethal doses respiratory paralysis occurs. Death may be preceded by convulsions. Chronic poisoning may produce fatty changes in liver, kidney. Direct contact occasionally causes mild irritation of skin or conjunctiva. More toxic when inhaled than when ingested. *See Clinical Toxicology of Commercial Products,* R. E. Gosselin *et al.,* Eds. (Williams & Wilkins, Baltimore, 4th ed., 1976) Section III, pp 293-295.

USE: Pesticide.

THERAP CAT (VET): Acaricide, ectoparasiticide.

8428. Rotraxate. *trans-4-[[4-(Aminomethyl)cyclohexyl]carbonyl]benzenepropanoic acid;* 3-[p-(trans-4-aminomethylcyclohexylcarbonyl)phenyl]propionic acid; *p*-[[trans-4-(aminomethyl)cyclohexyl]carbonyl]hydrocinnamic acid; traxaprone. $C_{17}H_{23}NO_3$; mol wt 289.37. C 70.56%, H 8.01%, N 4.84%, O 16.59%. Gastric cytoprotectant structurally similar to cetraxate, *q.v.* Prepn: T. Takeshita *et al.,* **Eur.** pat. **Appl. 44,541;** *eidem,* **U.S.** pat. **4,402,975** (1982, 1983 both to Teijin); T. Takeshita *et al., Chem. Pharm. Bull.* **33,** 5059 (1985). Crystal and molecular structure: K. Koyano *et al., ibid.* **34,** 3900 (1986). Antiulcerative effects in animals: K. Hoshina *et al., Arzneimittel-Forsch.* **35,** 493 (1985). Effects on gastric mucosa: *eidem, ibid.* **37,** 1154, 1284, 1289 (1987). Acute toxicity: Y. Izawa *et al., Yakuri to Chiryo* **16,** 357 (1988), *C.A.* **109,** 16846k (1988). Series of articles on toxicology: *eidem, ibid.* 371-599, *C.A.* **109,** 558g-565g; 16847n (1988).

Hydrochloride, $C_{17}H_{23}NO_3$.HCl, *TEI-5103, TG-51, Cumelon.* Crystals from water, mp 245° (dec) (Takeshita, 1985); also reported as mp 221-227° (Takeshita, 1983). LD_{50} in male, female rats (mg/kg): 9800, 9800 orally; 862, 835 i.p.; 5000, 5000 s.c. (Izawa).

THERAP CAT: Antiulcerative.

8429. Rottlerin. *(E)-1-[6-[(3-Acetyl-2,4,6-trihydroxy-5-methylphenyl)methyl]-5,7-dihydroxy-2,2-dimethyl-2H-1-benzopyran-8-yl]-3-phenyl-2-propen-1-one;* 5,7-dihydroxy-2,2-dimethyl-6-(2,4,6-trihydroxy-3-methyl-5-acetylbenzyl)-8-cinnamoyl-1,2-chromene; mallotoxin. $C_{30}H_{28}O_8$; mol wt 516.55. C 69.76%, H 5.46%, O 24.78%. Principal phenolic component of kamala, *q.v.,* an anthelmintic dye obtained from *Mallotus philippinensis* (Lam.) Muell. Arg. (also known as *Rottlera tinctoria* Roxb.), *Euphorbiaceae.* Isoln: H. Telle,

Arch. Pharm. **244**, 446 (1906); S. Dutt, *J. Chem. Soc.* **127**, 2044 (1925); A. McGookin *et al.*, *ibid.* **1937**, 748. Structure: Brockmann, Maier, *Ann.* **535**, 149 (1938); A. McGookin *et al.*, *J. Chem. Soc.* **1939**, 1579. Synthesis of tetrahydrorottlerin: Backhouse *et al.*, *ibid.* **1948**, 113. Review of isoln and chemistry: M. Lounasmaa *et al.*, *Planta Med.* **28**, 16 (1975). Activity as protein kinase inhibitor: M. Gschwendt *et al.*, *Biochem. Biophys. Res. Commun.* **199**, 93 (1994). HPTLC determn in serum and pharmacokinetics: V. D. Mody *et al.*, *J. Pharm. Technol.* **10**, 71 (1994).

Light reddish-brown plates or needles with golden lustre from ethyl acetate, mp 212° (McGookin). Also reported as brownish-yellow plates from toluene, mp 206-207° (Dutt). Sol in ether, chloroform, benzene, ethyl acetate; sparingly sol in cold alc, acetic acid. Practically insol in water.

5,7-Dimethyl ether, $C_{32}H_{32}O_8$, yellow crystals from ethyl acetate + acetone or chloroform + methanol, dec 245-246°. Sol in chloroform, pyridine, hot glacial acetic acid; slightly sol in cold methanol, ethyl acetate, acetone, benzene, ether.

Pentamethyl ether, $C_{35}H_{38}O_8$, crystals from petr ether, methanol or 90% alcohol, mp 144°. Very sol in acetone, glacial acetic acid, benzene, ethyl acetate.

Pentaacetate, $C_{40}H_{38}O_{13}$, leaflets from benzene + alcohol or acetone + alcohol, prisms from ethyl acetate, mp 212°.

Tetrahydrorottlerin, $C_{30}H_{32}O_8$, yellow prisms from alcohol, mp 214°.

8430. Roxarsone. *4-Hydroxy-3-nitrophenylarsonic acid; 4-hydroxy-3-nitrobenzenearsonic acid; 3-nitro-4-hydroxyphenylarsonic acid; 2-nitro-1-hydroxybenzene-4-arsonic acid; nitrophenolarsonic acid; NSC-2101; Ren-o-sal.* C_6H_6-$AsNO_6$; mol wt 263.04. C 27.40%, H 2.30%, As 28.48%, N 5.32%, O 36.50%. Prepared by treating sodium *p*-hydroxyphenylarsonate with a mixture of nitric and sulfuric acids at 0°: Benda, Bertheim, *Ber.* **44**, 3446 (1911); **Ger. pat. 224,-953**, *C.A.* **5**, 156 (1911). Toxicity study: Kerr *et al.*, *Toxicol. Appl. Pharmacol.* **5**, 507 (1963).

Tufts of pale-yellow needles or rhombohedral plates from water. Puffs up and deflagrates on heating. Slightly sol in cold water; sol in ~30 parts boiling water; freely sol in methanol, ethanol, acetic acid, acetone, alkalies; sparingly sol in dil mineral acids. Insol in ether, ethyl acetate. Forms mono-, di-, and trisodium salts. LD_{50} in rats, chickens (mg/kg): 155, 110-123 orally; 66, 34 i.p. (Kerr).

USE: Formerly in manuf arsphenamines; as reagent for zirconium.

THERAP CAT: Antibacterial.

THERAP CAT (VET): Control of enteric infections. To improve growth and feed efficiency.

8431. Roxatidine Acetate. *2-(Acetyloxy)-N-[3-[3-(1-piperidinylmethyl)phenoxy]propyl]acetamide;* aceroxatidine. $C_{19}H_{28}N_2O_4$; mol wt 348.44. C 65.49%, H 8.10%, N 8.04%, O 18.37%. Histamine H_2-receptor antagonist. Prepn and anti-ulcerative activity: K. Shibata *et al.*, *Eur. pat.* **24,510**; *eidem*, U.S. pat. **4,293,557** (both 1981 to Teikoku Hormone). Pharmacology in animals: M. Tarutani *et al.*, *Arzneimittel-Forsch.* **35**, 703 (1985). Comparison with cimetidine of effect on gastric acid secretion and ulcer formation in animals: *eidem, ibid.* 844. Metabolism in humans:

S. Honma *et al.*, *Oyo Yakuri* **30**, 555 (1985); *C.A.* **104**, 61453n (1985).

mp 59-60°. LD_{50} orally in male mice: 1000 mg/kg (Shibata).

Hydrochloride, $C_{19}H_{28}N_2O_4$.HCl, *pifatidine, HOE-760, TZU-0460, Altat, Gastralgin, Neo H2, Roxit.* mp 145-146°.

THERAP CAT: Antiulcerative.

8432. Roxindole. *3-[4-(3,6-Dihydro-4-phenyl-1(2H)-pyridinyl)butyl]-1H-indol-5-ol; 5-hydroxy-3-[4-(1,2,3,6-tetrahydro-4-phenyl-1-pyridyl)butyl]indole.* $C_{23}H_{26}N_2O$; mol wt 346.47. C 79.73%, H 7.56%, N 8.09%, O 4.62%. Dopamine D_2-receptor agonist. Synthesis: H.-H. Hausberg *et al.*, *Acta Pharm. Suec.* **1983**, Suppl 2, 213; H. Böttcher *et al.*, *J. Med. Chem.* **35**, 4020 (1992). Pharmacology: C. A. Seyfried *et al.*, *Eur. J. Pharmacol.* **160**, 31 (1989). Clinical pharmacology: K. Wiedemann, M. Kellner, *Exp. Clin. Endocrinol.* **102**, 284 (1994). Clinical evaluation in schizophrenia: H. Wetzel *et al.*, *Am. J. Psychiat.* **151**, 1499 (1994); in depression: G. Gründer *et al.*, *Psychopharmacology* **111**, 123 (1993). Neuroendocrinology in schizophrenics: G. Gründer *et al.*, *ibid.* **117**, 472 (1995).

Hydrochloride, $C_{23}H_{26}N_2O$.HCl, *EMD-38362.* Monoclinic crystals, mp 274°.

Methanesulfonate, *roxindole mesylate, EMD-49980.*

THERAP CAT: Antidepressant.

8433. Roxithromycin. *Erythromycin 9-[O-[(2-methoxyethoxy)methyl]oxime]; oxacyclotetradecane erythromycin deriv; 9-(2',5'-dioxahexyloxyimino)erythromycin;* RU-28965; RU-965; Assoral; Claramid; Forilin; Overal; Rossitrol; Rotramin; Rulid; Surlid. $C_{41}H_{76}N_2O_{15}$; mol wt 837.06. C 58.83%, H 9.15%, N 3.35%, O 28.67%. Semisynthetic erythromycin derivative. Prepn: S. Gouin d'Ambrieres *et al.*, *Fr. pat.* **2,473,525**; *eidem*, U.S. pat. **4,349,545** (1981, 1982 both to Roussel-UCLAF). *In vitro* antibacterial spectrum: R. N. Jones *et al.*, *Antimicrob. Ag. Chemother.* **24**, 209 (1983); Y. J. Drabu *et al.*, *Drugs Exptl. Clin. Res.* **13**, 201 (1987). Antiprotozoal activity in mice: B. J. Luft, *Eur. J. Clin. Microbiol.* **6**, 479 (1987); H. R. Chang, J.-C. R. Pechere, *Antimicrob. Ag. Chemother.* **31**, 1147 (1987). Pharmacokinetics: R. Wise *et al.*, *ibid.* 1051. Bioassay in biological fluids: A. L. Barry, R. R. Packer, *Eur. J. Clin. Microbiol.* **5**, 536 (1986). Clinical evaluations in respiratory tract infections: J. M. Lachat *et al.*, *Schweiz. Med. Wochenschr.* **116**, 1739 (1986); C. Grassi *et al.*, *Chemioterapia* **6**, 41 (1987). Series of articles on antimicrobial activity, pharmacokinetics and clinical efficacy: *J. Antimicrob. Chemother.* **20**, Suppl. B, 1-185 (1987).

$[\alpha]_D^{25}$ −77.5 ±2° (c = 0.45 in chloroform).
THERAP CAT: Antibacterial.

8434. Royal Jelly. Queen bee jelly; apilak; Weiselfuttersaft (German); Gelée royale (French). Secretion from the salivary glands of the worker honey bee which is essential for the development of queen bees. Production from bee hives: Ritschel, *Oesterr. Drogisten-Ztg.* **12**, 4-7 (1958). Synthetic mixture fed to bee larvae maintains life, but does not produce queens. The presence of hormones affecting mammals has not been demonstrated: Hinglais, Gautherie, *Compt. Rend.* **242**, 2483 (1956). No practical utility in human nutrition because of the very large amounts required for any definite effect: Moreaux, *Bull. Soc. Sci. Nancy* **14**, 49-53 (1955), *C.A.* **50**, 13214f (1956). Review of composition and biological activity: A. D. Dayan, *J. Pharm. Pharmacol.* **12**, 377-383 (1960). Monographs: B. deBelvefer, *Royal Jelly* (Paris, Librairie Maloine, 1958) 270 pp; H. Rembold, *Biologically Active Substances in Royal Jelly* in *Vitamins and Hormones* **23**, 359-382 (1965).
Milky white, highly viscous secretion. Analysis of fresh specimen (% wt at pH 5): moisture 65-70, protein 15-20, carbohydrate 10-15, lipid 1.7-6, ash 0.7-2.0; elemental analysis: P up to 0.5, S up to 0.6; trace elements present: Na, K, Fe, Cu, Mg, Mn, Ca. Vitamins ($\mu g/g$): thiamine 2, riboflavine 10, pyridoxine 2, nicotinic acid 75, biotin 2, folic acid 0.3, inositol 100, pantothenic acid 250, ascorbic acid 3-5, vitamin D trace, vitamin E trace. When stored at room temp, changes to a lightly yellow gum, and after some weeks, to a brittle amber solid.
Royal jelly acid, $C_{10}H_{18}O_3$, *trans*-10-hydroxy-Δ^2-decenoic *acid.* $HOCH(CH_2)_7CH=CHCOOH$. Constitutes ~10% of the dried royal jelly. Isoln: Townsend, Lucas, *Biochem. J.* **34**, 1155 (1940); Butenandt, Rembold, *Z. Physiol. Chem.* **308**, 284 (1957). Synthesis: Fray *et al., Tetrahedron Letters* **4**, 15 (1960); Smissman *et al., J. Org. Chem.* **29**, 3517 (1964); Bestmann *et al., Ann.* **699**, 33 (1966); J. Tsuji *et al., Bull. Chem. Soc. Japan* **50**, 2507 (1977); T. Fujisawa *et al., Chem. Letters* **1982**, 219; R. Chiron, *J. Chem. Ecol.* **8**, 709 (1982). Leukemia prevention in mice: Townsend *et al., Nature* **183**, 1270 (1959). Prisms from ether + petr ether or methanol + water, mp 64-65°. uv max: 211 nm (ϵ 12000).
See also Queen Substance.

8435. Rubber. Caoutchouc; India rubber. $(C_5H_8)_n$. Primarily obtained by coagulating the milk juice (latex) of several tropical trees, chiefly *Hevea brasiliensis* Muell.-Arg., *Euphorbiaceae. Habit.* Brazil, East Indies, Java, etc. Rubber also occurs in a number of plants, among which *guayule* (*Parthenium argentatum* Gray), a shrub which grows primarily in the northern Mexican desert, represents a potentially useful renewable source. Guayule produces rubber which is chemically identical to *Hevea* rubber in amounts ranging from 10-20% (dry basis) distributed in the roots and stems. *See* E. Campos-Lopez, *J. Polymer Sci., Polym. Lett. Ed.* **14**, 649 (1976); *J. Polymer Sci., Polym. Chem. Ed.* **14**, 1561 (1976); J. D. Johnson, C. W. Hinman, *Science* **208**, 460 (1980). Possible biosynthesis from acetate through mevalonic acid and isopentenyl pyrophosphate: Archer *et al., Nature* **184**, 268 (1959). Comprehensive review of chemistry: Ellis, *Chem. Ind. (London)* **1962**, 1447. Book: L. G. Polhamus, *Rubber* (Wiley, New York, 1962). *Review:* Jir-

genson's, *Natural Organic Macromolecules* (Pergamon Press, New York, 1962) pp 115-124; D. R. St. Cyr in Kirk-Othmer *Encyclopedia of Chemical Technology* **vol. 20** (Wiley-Interscience, New York, 3rd ed., 1982) pp 468-491. Natural rubber is defined as a *cis*-1,4-polyisoprene with a molecular weight varying from 100,000 to one million.

The best grades of raw rubber (pale crepe or smoked sheet) contain ~95% rubber hydrocarbon. The rest consists of proteins (2-3%), acetone-sol resins and fatty acids (2%), small amounts of sugar and a little mineral matter. Vulcanization, which consists of heating rubber with 1-3% of sulfur, introduces cross links between chains to produce a three-dimensional lattice of improved elasticity, strength, and temp sensitivity. Accelerators such as zinc dimethyldithiocarbamate greatly decrease the time or lower the temp required for vulcanization.
Pure rubber is nearly colorless and transparent in thin layers; odorless and tasteless. It is very elastic and lighter than water. Dec at about 120°. Burns with smoky flame. Emits characteristic offensive odor while burning. Practically insol in water, alcohol, dil acids, or alkali; sol in abs ether, chloroform, most fixed and volatile oils, petr ether, carbon disulfide, oil of turpentine.

8436. Rubeanic Acid. Ethanedithioamide; dithiooxamide. $C_2H_4N_2S_2$; mol wt 120.20. C 19.99%, H 3.35%, N 23.31%, S 53.35%. $NH_2CSCSNH_2$. Review of rubeanic acid and derivs as chelating ligands and analytical reagents: Ray, Xavier, *J. Indian Chem. Soc.* **38**, 535 (1961).
Red crystals or cryst powder; dec ~200°. Slightly sol in water; sol in alcohols. Insol in ether.
USE: As a reagent for copper, cobalt, and nickel. As a stabilizer of ascorbic acid solns: Smoczkiewicz, Grochmalicka, *Nature* **192**, 16 (1961).

8437. Ruberythric Acid. 1-Hydroxy-2-[(6-O-β-D-xylopyranosyl-β-D-glucopyranosyl)oxy]-9,10-anthracenedione; 1-hydroxy-2-anthraquinonyl 6-O-β-D-xylopyranosyl-β-D-glucopyranoside; β-2-alizarin primeveroside; ruberythrinic acid; rubian; rubianic acid. $C_{25}H_{26}O_{13}$; mol wt 534.47. C 56.18%, H 4.90%, O 38.92%. Isoln from root of *Rubia tinctorium* L., *Rubiaceae:* Hill, Richter, *Proc. Roy. Soc. (London) Ser. B* **121**, 547 (1937). Structure: Richter, *J. Chem. Soc.* **1936**, 1701. Synthesis: Zemplén, Bognár, *Ber.* **72B**, 913 (1939). Metabolism: Maehner, Dulce, *Z. Klin. Chem. Klin. Biochem.* **6**, 99 (1968).

Golden-yellow, silky, lustrous prisms or long needles. mp 259-261° (from water). Sol in hot water, alkalies; slightly sol in alc, ether. Practically insol in benzene.

8438. Rubiadin. 1,3-Dihydroxy-2-methyl-9,10-anthracenedione; 1,3-dihydroxy-2-methylanthraquinone. $C_{15}H_{10}O_4$; mol wt 254.24. C 70.86%, H 3.96%, O 25.17%. From *Rubia tinctorum* L., *Coprosma* var., *Morinda citrifolia* Linn, *Rubiaceae:* Schunck, *Ann.* **87**, 344 (1853); Briggs, Nicholls, *J. Chem. Soc.* **1949**, 1241; Briggs *et al., ibid.* **1952**, 1718; Borvie, Cooke, *Aust. J. Chem.* **15**, 332 (1962). Structure: Marchlewski, *J. Chem. Soc.* **63**, 1137 (1893); Schunck, Marchlewski, *ibid.* **65**, 182 (1894); Jones, Robertson, *ibid.* **1930**, 1699. Synthesis: Joshi *et al., J. Sci. Ind. Res.* **14B**, 87 (1955); Hirase, *Chem. Pharm. Bull.* **8**, 417 (1960).

Yellow plates from glacial acetic acid, mp 302°; yellow slender plates from alc, mp 290°. Absorption max (ethanol): 246, 280, 415 nm (log ε 4.39, 4.52, 3.87). Sol in alc, ether. Practically insol in water, alkalies.

Diacetate, yellow rods from acetic anhydride, mp 228°.

Dimethyl ether, yellow needles from alc, mp 160.5°.

8439. Rubidium. Rb; at. wt 85.4678; at. no. 37; valence 1. Alkali metal. Widely distributed in very small quantities in earth's crust: 0.0034% by wt. Natural isotopes: 85 (72.15%); 87 (27.85%); [87]Rb is radioactive: $T_{\frac{1}{2}}$ 4.8×10^{10} years; isotopes range in mass number from 79 to 95. Found with other alkali metals in rhodizite (borate), lepidolite (aluminosilicate), rubidium carnallite (chloride); in sea water; in mineral springs and salt lakes. Discovered by Bunsen and Kirchhoff in 1861. Prepn: Hackspill, *Helv. Chim. Acta* **11**, 1003 (1928). *Review:* Whaley, "Sodium, Potassium, Rubidium, Cesium and Francium" in *Comprehensive Inorganic Chemistry* vol. I, J. C. Bailar Jr. *et al.,* Eds. (Pergamon Press, Oxford, 1973) pp 369-529; F. B. White, W. G. Lidman in Kirk-Othmer *Encyclopedia of Chemical Technology* vol. 20 (Wiley-Interscience, New York, 3rd ed., 1982) pp 492-499.

Lustrous, silvery-white, soft metal; body-centered cubic structure; rapidly tarnishes on exposure to air; mp 39°; bp 688°; d^{20} 1.532. Sp heat 0.0802 cal/g deg. One of the most active metals. E° (aq) Rb/Rb+ 2.924 V. In its chemical properties closely resembles potassium. Dec water rapidly, even at −108°. Ignites spontaneously in oxygen; when molten readily takes fire in the air. Reacts vigorously with the halogens. Forms a series of solid solns with potassium, cesium, sodium. Combines vigorously with mercury. *Keep under benzene, petroleum, or other liq not containing oxygen.*

USE: In making rubidium salts; as a reagent in making zeolite catalysts; in photoelectric cells.

8440. Rubidium Bromide. BrRb; mol wt 165.37. Br 48.32%, Rb 51.68%. RbBr.

White, cryst powder; d 3.35; mp 682°; bp 1340°. Sol in about 1 part cold water, 0.5 part boiling water. The aq soln is neutral.

8441. Rubidium Chloride. Rubinorm. ClRb; mol wt 120.92. Cl 29.32%, Rb 70.68%. RbCl. Causes shock-induced aggression in rats: B. S. Eichelman, *Psychopharmacol. Bull.* **9**, 21 (1973). Biochemical, behavioral and metabolic studies in humans: R. R. Fieve *et al., Am. J. Psychiatry* **130**, 55 (1973). Clinical evaluations in depression: C. Paschalis *et al., J. Roy. Soc. Med.* **71**, 343 (1978); G. Placidi *et al., J. Clin. Psychopharmacol.* **8**, 184 (1988).

White, crystalline powder. d 2.76. mp 715°. bp 1390°. One gram dissolves in 1 ml cold, 0.7 ml boiling water, 90 ml methanol, 1650 ml alc. The aq soln is neutral.

USE: As catalyst; as gasoline octane number improver.

THERAP CAT: Antidepressant.

8442. Rubidium Hydroxide. HORb; mol wt 102.48. H 0.98%, O 15.61%, Rb 83.40%. RbOH. Book: L. S. Itkina, *Lithium, Rubidium and Cesium Hydroxide* (Nauka, Moscow, USSR, 1973) 136 pp. Toxicological study: G. T. Johnson *et al., Toxicol. Appl. Pharmacol.* **32**, 239 (1975).

Grayish-white, deliquesc mass. d 3.203. Melts at red heat. Sol in 0.6 part water; sol in alcohol. It is a stronger base than potassium hydroxide. *Keep well closed.*

USE: Catalyst in oxidative chlorination.

8443. Rubidium Iodide. IRb; mol wt 212.37. I 59.76%, Rb 40.24%. RbI.

White crystals or cryst powder; discolors on exposure to air and light; d 3.55; mp 642°; bp 1300°. Sol in 0.66 part water; sol in alcohol. The aq soln is neutral or slightly alkaline. *Keep well closed and protected from light.*

THERAP CAT: Iodine source.

8444. Rubijervine. *(3β,12α)-Solanid-5-ene-3,12-diol;* Δ^5-3β,12α-dihydroxysolanidene; rubigervine. $C_{27}H_{43}NO_2$; mol wt 413.64. C 78.40%, H 10.48%, N 3.39%, O 7.74%. From various species of *Veratrum, Liliaceae.* Isoln: Wright, Luff, *J. Chem. Soc.* **35**, 405 (1879); Salzberger, *Arch. Pharm.* **228**, 462 (1890); Jacobs, Craig, *J. Biol. Chem.* **148**, 41 (1943); Sato, Jacobs, *ibid.* **179**, 623 (1949); Pelletier, Locke, *J. Am. Chem. Soc.* **79**, 4531 (1957). Stereochemistry: Höhne *et al., Tetrahedron* **22**, 673 (1966). Comparative toxicity: O. Krayer *et al., J. Pharmacol. Exp. Ther.* **82**, 167 (1944).

Solvated needles from alc, mp 240-246°. $[\alpha]_D^{25}$ +19.0° (ethanol). Very sparingly sol in water. Sol in ethanol, methanol, benzene, chloroform. Slightly sol in ether, petr ether. Red color in concd H_2SO_4. Precipitated by digitonin. LD_{50} i.v. in mice: 70 mg/kg (Krayer).

Hydrobromide, $C_{27}H_{43}NO_2 \cdot HBr$, needles from methanol + acetone, dec 265-270°.

THERAP CAT: Antifungal.

8445. Rubixanthin. *(3R)-β,ψ-Caroten-3-ol.* $C_{40}H_{56}O$; mol wt 552.88. C 86.90%, H 10.21%, O 2.89%. Carotenoid isomeric with cryptoxanthin. Isolated from hipberries (rose hips, *Rosa rubiginosa* L., *Rosaceae*): Kuhn, Grundmann, *Ber.* **67**, 339, 1133 (1934); from the flowers of *Tagetes patula* L., *Compositae.* Found also in other *Rosa* varieties. Structural studies: Brown, Weedon, *Chem. Commun.* **1968**, 382. Abs config: L. Bartlett *et al., J. Chem. Soc. (C)* **1969**, 2527. Biosynthesis studies: McDermott *et al., Biochem. J.* **134**, 1115 (1973). Has no vitamin A activity.

Deep red needles with metallic luster from benzene + methanol. Orange crystals from benzene + petr ether, mp 160°. Absorption max (chloroform): 509, 474, 439 nm. Sol in benzene, chloroform; slightly sol in petr ether, alcohol.

8446. Rubus. Blackberry bark. Dried bark of rhizome and roots of the section *Eubatus* Focke of the genus *Rubus* L. or *R. nigrobaccus* Bailey, *Rosaceae.* Habit. Eastern U.S. Constit. Tannin, gallic acid, villosin (a saponin).

THERAP CAT: Antidiarrheal.

8447. Rufigallol. *1,2,3,5,6,7-Hexahydroxy-9,10-anthracenedione; 1,2,3,5,6,7-hexahydroxyanthraquinone;* rufigallic acid; C.I. 58600. $C_{14}H_8O_8$; mol wt 304.21. C 55.28%, H 2.65%, O 42.07%. Prepd by the action of H_2SO_4 on gallic acid: Robiquet, *Ann.* **19**, 204 (1836); by dry distn of gallic acid: Kunz-Krause, Manicke, *Ber.* **53**, 190 (1920). *See also: Colour Index* vol. 4 (3rd ed., 1971) p 4520.

Consult the Name Index before using this section.

Red needles. Does not melt, but sublimes with partial decompn upon heating. Absorption spectrum: Treibs, Steinmetz, *Ann.* **506,** 171, 191 (1933). Practically insol in water. Freely sol in acetone; slightly sol in alcohol, ether, with yellow color. Sol in alkali hydroxide solns with violet color, but is soon dec by oxidation.

Hexamethyl ether, $C_{20}H_{20}O_8$, yellow needles, mp 240°.

USE: For color reactions with Zr and Hf.

8448. Rufloxacin. *9-Fluoro-2,3-dihydro-10-(4-methyl-1-piperazinyl)-7-oxo-7H-pyrido[1,2,3-de]-1,4-benzothiazine-6-carboxylic acid;* MF-934. $C_{17}H_{18}FN_3O_3S$; mol wt 363.41. C 56.19%, H 4.99%, F 5.23%, N 11.56%, O 13.21%, S 8.82%. Fluorinated quinolone; structurally similar to ofloxacin, *q.v.* Prepn: G. Mascellani *et al.,* **Austral. pat.** *Appl.* **85 39,142;** *eidem),* U.S. pat **4,684,647** (1985, 1987 both to Mediolanum); V. Cecchetti *et al., J. Med. Chem.* **30,** 465 (1987). *In vitro* activity: R. Wise *et al., Antimicrob. Ag. Chemother.* **29,** 649 (1992). Mechanism of action and resistance: L. J. V. Piddock *et al., J. Antimicrob. Chemother.* **31,** 855 (1993). Pharmacokinetics and safety: J. C. Kisicki *et al., ibid.* **36,** 1296 (1992). HPLC determn in biological fluids: F. Lombardi *et al., J. Chromatog.* **576,** 129 (1992). Clinical study in cystitis and urinary tract infections: R. Mattina *et al., Infection* **21,** 106 (1993); in bronchitis: W. Klietmann *et al., Antimicrob. Ag. Chemother.* **37,** 2298 (1993).

Hydrochloride, $C_{17}H_{18}FN_3O_3S \cdot HCl$, *ISF-09334, Qari, Monos, Tebraxin.* Crystals from ethanol/water, mp 322-324° (Cecchetti). LD_{50} in rats, mice (mg/kg): 285, 224 i.v.; LD_{50} in rabbits, male, female rats (mg/kg): 660, 631, 501 orally (Cecchetti).

THERAP CAT: Antibacterial.

8449. Rugulovasines. $C_{16}H_{16}N_2O_2$; mol wt 268.32. C 71.62%, H 6.01%, N 10.44%, O 11.93%. Mixture of the interconvertible diastereomers, rugulovasine A and rugulovasine B. Isolated as the hydrochlorides from surface cultures of *Penicillium concavo-rugulosum* Abe and other *penicillium* strains: Abe *et al., Agr. Biol. Chem.* **33,** 469 (1969); R. J. Cole *et al., Can. J. Microbiol.* **22,** 741 (1976). Prepn and separation of isomers: Abe, **Ger.** pat. **1,919,255** (1969 to Takeda), *C.A.* **72,** 30311f (1970). Structure: Yamatodani *et al., Agr. Biol. Chem.* **34,** 485 (1970). Abs config and mechanism of interconversion of the isomers: R. J. Cole *et al., Tetrahedron Letters* **1976,** 3849. Total synthesis: J. Rebek, Y. K. Shue, *J. Am. Chem. Soc.* **102,** 5426 (1980). Hypotensive and cardiovascular activity studies of the hydrochloride isomers: Nagaoka *et al., Arzneimittel-Forsch.* **22,** 137, 143 (1972).

Rugulovasine A

Rugulovasine A, cis-(±)-3,4-dihydro-4'-methyl-4-(methyl-amino)spiro[benz[cd]indole-5(1H),2'(5'H)-furan]-5'-one. Colorless prisms from benzene, chloroform, or acetonitrile, mp 138° (dec). $[\alpha]_{436}^{22}$ −3.0° (pyridine). uv max (ethanol): 224, 277, 288, 295 nm (log ε 4.37, 3.70, 3.78, 3.77). Sparingly sol in water, petr ether. Moderately sol in ether, chloroform, benzene, acetonitrile. Readily sol in ethyl acetate, acetone, methanol, ethanol, pyridine, and dil acids.

Hydrochloride, $C_{16}H_{16}N_2O_2 \cdot HCl$, colorless prisms from water, mp 225° (dec).

Rugulovasine B, trans analog of Rugulovasine A. Difficult to crystallize from common solvents. Separated from benzene as colorless resinous oil. $[\alpha]_{436}^{22}$ +1.4° (pyridine). uv max (ethanol): 227, 278, 288, 295 nm (log ε 4.16, 3.68, 3.73, 3.72). Same solubilities as rugulovasine A.

Hydrochloride, $C_{16}H_{16}N_2O_2 \cdot HCl$, white prisms from water, mp 187° (dec).

8450. Rumex. Yellow dock; curled dock. Dried root of *Rumex crispus* L., or of *R. obtusifolius* L., *Polygonaceae.* *Habit.* Europe, North America. *Constit.* Chrysophanic acid, emodin, tannin, calcium oxalate, lapathin.

THERAP CAT: Cathartic, astringent.

8451. Rutecarpine. *8,13-Dihydroindolo[2',3':3,4]pyrido[2,1-b]quinazolin-5(7H)-one;* rutaecarpine. $C_{18}H_{13}N_3O$; mol wt 287.32. C 75.25%, H 4.56%, N 14.62%, O 5.57%. From fruit of *Evodia rutaecarpa* Hook. & Thoms. and *Hortia arborea* Engl., *Rutaceae:* Asahina, Kashiwaki, *J. Pharm. Soc. Japan* **1915,** 1293; Pachter *et al., J. Am. Chem. Soc.* **82,** 5187 (1960). Structure: Y. Asahina, *J. Pharm. Soc. Japan* **1924,** 1. Synthesis: Y. Asahina *et al., J. Pharm. Soc. Japan* **1927,** 1708; T. Kametani *et al., J. Am. Chem. Soc.* **99,** 2306 (1977); *eidem, Chem. Pharm. Bull.* **26,** 1922 (1978); H. Möhrle *et al., Arch. Pharm.* **313,** 990 (1980); J. Bergman, S. Bergman, *Heterocycles* **16,** 347 (1981). Simple synthesis: J. Kökösi *et al., Tetrahedron Letters* **22,** 4861 (1981). Synthesis under physiological conditions: Schöpf, *Angew. Chem.* **50,** 779, 797 (1937). Biosynthesis: M. Yamazaki *et al., Tetrahedron Letters* **1966,** 3221; **1967,** 3317. Mass spec.: J. Tamas *et al., Acta Chim. Acad. Sci. Hung.* **89,** 85 (1976).

Needles from ethyl acetate, mp 259.5-260°. uv max (ethanol): 278, 290, 332, 345, 364 nm (log ε 3.83, 3.88, 4.49, 4.54, 4.44). Sol in alc, benzene, chloroform, ether. Practically insol in water.

8452. Ruthenium. Ru; at. wt 101.07; at. no. 44; valences 1-8; most common states 2, 3, 4. Naturally occurring isotopes: 96 (5.46%); 98 (1.87%); 99 (12.63%); 100 (12.53%); 101 (17.02%); 102 (31.6%); 104 (18.87%); artificial radioactive isotopes: 93-95; 97; 103; 105-108. Belongs to the platinum group of metals. Found in the minerals osmiridium, laurite, in platinum ores, in some copper-nickel ores. Discovered in 1828 by Osann; prepd in the pure state in 1845 by Klaus. Constitutes about 0.0004 ppm of the crust of the earth. Prepn from osmiridium: Fremy *et al.,* cited by Mellor, *A Comprehensive Treatise on Inorganic and Theoretical Chemistry* **15,** 499 (1936). Reviews of prepn, properties and chemistry of ruthenium and other platinum metals: Gilchrist, *Chem. Rev.* **32,** 277-372 (1943); W. P. Griffith, *The Chemistry of The Rarer Platinum Metals* (John Wiley, New York, 1967) pp 1-41, 126-226; Livingstone in *Comprehensive Inorganic Chemistry* vol. 3, J. C. Bailar, Jr. *et al.,* Eds. (Pergamon Press, Oxford, 1973) pp 1163-1209.

Lustrous, hard metal; hexagonal, close-packed structure. d_4^{20} 12.45. mp about 2450°; bp about 4150°. Sp heat (0°): 0.057 cal/g/°C. Does not react with acids, even aqua regia. Not oxidized by air in the cold; on heating combines readily with oxygen; the powdered metal forms the dioxide on igniting in air. Superficially attacked by concd alkaline hypochlorites. The powdered metal is attacked by chlorine above 200°; by bromine between 300-700°. Oxidized by fused alkali hydroxides. Forms alloys with platinum, palladium, cobalt, nickel, tungsten; forms definite compds with zinc and with tin.

USE: As substitute for platinum in jewelry; for pen nibs; as hardener in electrical contact alloys, electrical filaments; in ceramic colors; catalyst in synthesis of long chain hydrocarbons.

8453. Ruthenium Red. Ruthenium oxychloride ammoniated; C.I. 77800. $Cl_6H_{42}N_{14}O_2Ru_3$; mol wt 786.35. Cl 27.05%, H 5.38%, N 24.94%, O 4.07%, Ru 38.56%. $[(NH_3)_5Ru—O—Ru(NH_3)_4—O—Ru(NH_3)_5]Cl_6$. Exists as tetrahydrate. Identification: A. Joly, *C.R. Acad. Sci.* **115**, 1299 (1893). Prepn and structure: Fletcher *et al., J. Chem. Soc.* **1961**, 2000. Mössbauer study: Clausen *et al., Inorg. Nucl. Chem. Letters* **7**, 485 (1971). Alternate structure and composition: Sterling, *Am. J. Bot.* **57**, 172 (1970). Use as microscopic stain: L. Mangin, *C.R. Acad. Sci.* **116**, 653 (1893); in electron microscopy: R. Dierichs, *Histochemistry* **64**, 171 (1979). Differential staining: P. R. Blanquet, *ibid.* **47**, 175 (1976); M. G. Gutierrez-Gonzalvez *et al., J. Microscopy* **145**, 333 (1987). Series of articles on chemistry, mechanism of action, specific uses: *Anat. Rec.* **171**, 347-442 (1971). Conversion *in situ* to *ruthenium violet*: J. R. Luft, *ibid.* 347; M. G. Guiterrez-Gonzalvez *et al., loc. cit.*
Brownish-red powder. Sol in water, ammonia.
USE: In microscopy as a general stain.

8454. Ruthenium Tetroxide. Ruthenium(VIII) oxide. O_4Ru; mol wt 165.07. O 38.77%, Ru 61.23%. RuO_4. Prepd by fusing Ru with $KMnO_4$ and KOH in a proportion of 1:2:20. Lab procedures: Ruff, Vidic, *Z. Anorg. Allgem. Chem.* **136**, 49 (1924); Grube in *Handbook of Preparative Inorganic Chemistry* vol. **2**, G. Brauer, Ed. (Academic Press, New York, 2nd ed., 1965) p 1599. Crystal structure: Tréhoux *et al., Compt. Rend. Ser. C* **268**, 246 (1969). Review of use as oxidizing agent: P. N. Rylander, *Organic Syntheses with Noble Metal Catalysts* (Academic Press, New York, 1973) pp 134-144.
Golden-yellow monoclinic prisms. Very volatile, sublimes at room temp. *Handle in hood only. Vapors irritating to eyes and respiratory tract.* mp 25.4°. bp 40°. Sparingly sol in water, about 2.0% (w/v) at 20°. Freely sol in carbon tetrachloride, other chlorinated solvents. Also sol in bromine, liquid SO_2. Reacts explosively with alcohol and filter paper.
USE: Oxidizing agent similar to osmium tetroxide, but more difficult to handle. The solvents normally employed in OsO_4 oxidations (ether, benzene, pyridine) cannot be used because of their violent reaction with RuO_4. Only CCl_4 is recommended: Djerassi, Engle, *J. Am. Chem. Soc.* **75**, 3838 (1953).

8455. Ruthenium Trichloride. Cl_3Ru; mol wt 207.43. Cl 51.27%, Ru 48.73%. $RuCl_3$. Two forms, α and β, are known. Prepn: Fletcher *et al., Nature* **199**, 1089 (1963).
α-$RuCl_3$, black lustrous crystals. Insol in alcohol, water.
β-$RuCl_3$, dark-brown, fluffy, hexagonal crystals. Sol in alcohol.

8456. Rutin. *3-[[6-O-(6-Deoxy-α-L-mannopyranosyl)-β-D-glucopyranosyl]oxy]-2-(3,4-dihydroxyphenyl)-5,7-dihydroxy-4H-1-benzopyran-4-one;* rutoside; quercetin-3-rutinoside; 3,3′,4′,5,7-pentahydroxyflavone-3-rutinoside; melin; phytomelin; eldrin; ilixathin; sophorin; globularicitin; paliuroside; osyritrin; osyritin; myrticolorin; violaquercitrin; Birutan. $C_{27}H_{30}O_{16}$; mol wt 610.53. C 53.12%, H 4.95%, O 41.93%. Identity with ilixanthin: Schindler, Herb, *Arch. Pharm.* **288**, 372 (1955). Found in many plants, especially the buckwheat plant *(Fagopyrum esculentum* Moench., *Polygonaceae)* which contains about 3% (dry basis): Couch *et al., Science* **103**, 197 (1946). From tobacco *(Nicotiana tabacum* L., *Solanaceae)* Couch, Krewson, *U.S. Dept. Agr., Eastern Regional Res. Lab.,* AIC-52 (1944). In forsythia [*Forsythia suspensa* (Thunb.) Vahl. var. *fortunei* (Lindl.) Rehd., *Oleaceae*], in hydrangea *(Hydrangea paniculata* Sieb., *Saxifragaceae),* in pansies *(Viola* sp., *Violaceae).* General extraction procedure: *Beilstein* XXXI, 376. From leaves of *Eucalyptus macroryncha* F. v. Muell., *Myrtaceae:* Attree, Perkin, *J. Chem. Soc.* **1927**, 234. Industrial production from *Eucalyptus* spp.: Humphreys, *Econ. Bot.* **18**, 195 (1964). Structure: Zemplén, Gerecs, *Ber.* **68B**, 1318 (1935). Synthesis: Shakhova *et al., Zh. Obshch. Khim.* **32**, 390 (1962), *C.A.* **58**, 1426e (1963). Rutin is hydrolyzed by rhamnodiastase from the seed of *Rhamnus utilis* Decne, *Rhamnaceae* (Chinese buckthorn); emulsin is not effective: Bridel, Charaux, *Compt. Rend.* **181**, 925 (1925). Toxicity data: Harrison *et al., J. Am. Pharm. Assoc.* **39**, 557 (1950). Book: J. Q. Griffith, Jr., *Rutin and Related Flavonoids* (Mack, Eas-

ton, Pa., 1955). Comprehensive description: T. I. Khalifa *et al.* in *Analytical Profiles of Drug Substances* vol. **12**, K. Florey, Ed. (Academic Press, New York, 1983) pp 623-681.

Pale yellow needles from water, gradual darkening on exposure to light. The crystals contain 3 H_2O and become anhydr after 12 hrs at 110° and 10 mm Hg. Anhydr rutin browns at 125°, becomes plastic at 195-197°, and dec 214-215° (with effervescence). $[\alpha]_D^{23}$ +13.82° (ethanol); $[\alpha]_D^{23}$ −39.43° (pyridine). Anhydr rutin is hygroscopic. One gram dissolves in about 8 liters water, about 200 ml boiling water, 7 ml boiling methanol. Sol in pyridine, formamide and alkaline solns; slightly sol in alcohol, acetone, ethyl acetate. Practically insol in chloroform, carbon bisulfide, ether, benzene, petr solvents. Dil solns give green color with ferric chloride. Rutin is colored brown by tobacco enzyme under experimental conditions: Neuberg, Kobel, *Naturwiss.* **23**, 800 (1935). LD_{50} i.v. in mice: 950 mg/kg (propylene glycol soln) (Harrison).
THERAP CAT: Capillary protectant.

8457. Rutinose. *6-O-(6-Deoxy-α-L-mannopyranosyl)-D-glucose;* 6-O-α-L-rhamnosyl-D-glucose. $C_{12}H_{22}O_{10}$; mol wt 326.30. C 44.17%, H 6.80%, O 49.03%. A disaccharide present in glycosides. Prepn from rutin by enzymatic hydrolysis using rhamnodiastase: Charaux, *Compt. Rend.* **178**, 1312 (1924); *Bull. Soc. Chim. Biol.* **6**, 634 (1924); Zemplén, Gerecs, *Ber.* **68B**, 1320 (1935). Synthesis: *eidem, Ber.* **67B**, 2049 (1934); **68B**, 1318 (1935). Structure: Gorin, Perlin, *Can. J. Chem.* **37**, 1930 (1959).

Very hygroscopic powder from alcohol and ether turning into a syrup on exposure to air. The freshly pptd material loses about 6% of its wt at 100°. The dried rutinose softens around 140°, dec 189-192°. $[\alpha]_D^{20}$ +3.2° → 0.8° (c = 4 in water). $[\alpha]_D^{20}$ −10° (95% alc). Very sol in water. Sol in alc. Hydrolysis with dil HCl yields 1 mol L-rhamnose and 1 mol D-glucose.
Heptaacetylrutinose, mp 169-170°. $[\alpha]_D^{20}$ −27.7° (chloroform).

8458. Ryania. A genus of tropical American shrubs and trees belonging to the *Flacourtiaceae* family. The wood of various species is insecticidal. Ground stem wood of *Ryania speciosa* Vahl., *Flacourtiaceae* is employed in the commercial insecticide formulations **Ryanex, Ryanicide** (formerly). *See* Folkers *et al.,* U.S. pat. **2,400,295** (1946 to Merck & Co.); Pepper, Carruth, *J. Econ. Entomol.* **38**, 59 (1945); Heal, *Agr. Chem.* **4**, 37 (May, 1949). Insecticidal components such as ryanodine, *q.v.,* are extractable by water, chloroform, or methanol. Toxicity data: Kuna, Heal, *J. Pharmacol. Exp. Ther.* **93**, 407 (1948).
LD_{50} orally in rats, mice, rabbits, guinea pigs (mg/kg): 1200, 650, 650, 2500 (Kuna, Heal).
Caution: In exptl animals causes weakness, deep and slow respiration, vomiting, diarrhea, tremors, convulsions, coma; death may occur.
USE: Insecticide.

8459. Ryanodine. *Ryanodol 3-(1H-pyrrole-2-carboxylate).* $C_{25}H_{35}NO_9$; mol wt 493.55. C 60.84%, H 7.15%, N 2.84%, O 29.18%. Insecticidal principle isolated from *Ryania speciosa* Vahl., *Flacourtiaceae:* E. F. Rogers *et al., J. Am. Chem. Soc.* **70**, 3086 (1948); R. B. Kelly *et al., Can. J. Chem.* **29**, 905 (1951). Structure: D. R. Babin *et al., Ex-*

perientia **21**, 425 (1965); J. Santroch *et al., ibid.* **21**, 730

(1965); K. Wiesner *et al., Tetrahedron Letters* **1967**, 221; K. Wiesner, *Advan. Org. Chem.* **8**, 295 (1972). Crystal structure: S. N. Srivastava, M. Przybylska, *Can. J. Chem.* **46**, 795 (1968). ¹H-NMR: A. L. Waterhouse *et al., Chem. Commun.* **1984**, 1265. Synthetic studies: P. Deslongchamps, *Pure Appl. Chem.* **49**, 1329 (1977). Effect on calcium ion uptake by cardiac sarcoplasmic reticulum vesicles: L. R. Jones *et al., J. Pharmacol. Exp. Ther.* **209**, 48 (1979); L. R. Jones, S. E. Cala, *J. Biol. Chem.* **256**, 11809 (1981).

Crystals, dec 219-220°. $[\alpha]_D^{25}$ +26° (methanol). uv max (alc): 268.5 nm (log ϵ 4.18). Sol in water, alcohol, acetone, ether, chloroform. Practically insol in benzene, petr ether.

USE: Insecticide.

S

8460. Sabadilla. Cevadilla; caustic barley. The dried ripe seeds of *Schoenocaulon officinale* (Schlecht. & Cham.) A. Gray (*Sabadilla officinarum* Brandt, *Asagraea officinalis* Lindl.), *Liliaceae*. Habit. The Andes of Mexico, Guatemala, Venezuela. *Constit.* About 3 to 6% total alkaloids of which cevadine (cryst veratrine) and veratridine (amorphous veratrine) are the most important; other alkaloids are sabadinine, sabadilline, and sabadine; also contains sabadillic and veratric acids; fatty oil, resin. Toxicity data: E. D. Swiss, R. O. Bauer, *Proc. Soc. Exp. Biol. Med.* **76**, 847 (1951).

LD$_{50}$ i.p. in mice: 7.5 mg/kg (Swiss, Bauer).

THERAP CAT: Pediculicide.

THERAP CAT (VET): Has been used as an insecticide.

8461. Sabadine. (3β,4α,16β)-*Cevane-3,4,12,14,16,17,-20-heptol 3-acetate;* sabatine. C$_{29}$H$_{47}$NO$_8$; mol wt 537.69. C 64.78%, H 8.81%, N 2.60%, O 23.80%. Isolated from Sabadilla seed: Merck, *Arch. Pharm.* **229**, 164 (1891); Hennig *et al., J. Am. Pharm. Assoc.* **40**, 168 (1951). Structure: Auterhoff, Schwartz, *Arch. Pharm.* **288**, 549 (1955); Möhrle, Auterhoff, *ibid.* **292**, 337 (1959); Kupchan *et al., J. Med. Pharm. Chem.* **5**, 690 (1962). NMR data: Ito *et al., Tetrahedron* **20**, 913 (1964).

Prisms, mp 256-258° (dec). [α]$_D^{25}$ −11° (c = 1.93 in alc).

8462. Sabeluzole. (±)-*4-(2-Benzothiazolylmethylamino)-α-[(4-fluorophenoxy)methyl]-1-piperidineethanol; N-methyl-N-[1-[3-(4-fluorophenoxy)-2-hydroxypropyl]piperidin-4-yl]benzothiazole-2-amine;* R-58735; Reminyl. C$_{22}$H$_{26}$-FN$_3$O$_2$S; mol wt 415.53. C 63.59%, H 6.31%, F 4.57%, N 10.11%, O 7.70%, S 7.72%. Prepn: R. A. Stokbroekx *et al.,* **Eur. pat. Appl. 184,257** (1986 to Janssen). Pharmacology: A. Wauquier *et al., Drug Dev. Res.* **8**, 373 (1986). Comparative activity of racemate and enantiomers: L. Werbrouck *et al., ibid.* **24**, 41 (1991). Clinical pharmacokinetics: P. P. DeDeyn *et al., Eur. J. Clin. Pharmacol.* **43**, 661 (1992). Clinical evaluation of effects on memory functions: G. H. C. Clincke *et al., Psychopharmacology* **94**, 52 (1988). Clinical studies in age-associated memory impairment: L. Tritsmans *et al., Drug Dev. Res.* **20**, 473 (1990); G. Clincke *et al., ibid.* **23**, 301 (1991); in memory functions of epileptics: A. P. Aldenkamp *et al., Neuropsychobiology* **32**, 37 (1995).

mp 101.7°.

THERAP CAT: Nootropic.

8463. Saccharin. *1,2-Benzisothiazol-3(2H)-one 1,1-dioxide; 2,3-dihydro-3-oxobenzisosulfonazole;* 1,2-dihydro-2-ketobenzisosulfonazole; saccharin insoluble; benzosulfimide;

o-sulfobenzimide; benzoic sulfimide; o-sulfobenzoic acid imide; Gluside; Glucid; Garantose; Saccharinol; Saccharinose; Saccharol; Saxin; Sykose; Hermesetas. C$_7$H$_5$NO$_3$S; mol wt 183.19. C 45.90%, H 2.75%, N 7.65%, O 26.20%, S 17.50%. Prepn from o-sulfamoylbenzoic acid: Remsen, Fahlberg, *Ber.* **12**, 470 (1879). History and comparison with other sweetening agents: R. W. Moncrieff, *The Chemical Senses* (Wiley, New York, 1946). Mfg processes: O. Beyer, *Handbuch der Saccharinfabrikation* (Rascher, Zürich, 1923); *FIAT Rep.,* **PB 901** (1945); *Chem. Eng. (New York)* **61**, 128, 150 (July 1954). Maumee Chemical's process: *Chem. & Eng. News* **41**, 76-78 (Dec. 9, 1963). Stability studies: DeGarmo *et al., J. Am. Pharm. Assoc. (Sci. Ed.)* **41**, 17 (1952). Toxicity data: J. D. Taylor *et al., Food Cosmet. Toxicol.* **6**, 313 (1968). Series of articles on toxicology and metabolism: *Food Chem. Toxicol.* **23**, 419-546 (1985). Review of toxicology and carcinogenicity studies of saccharin and its salts: *IARC Monographs* **22**, 111-170 (1980); D. L. Arnold, *Fundam. Appl. Toxicol.* **4**, 674-685 (1984). *Review:* R. Mazur, "Sweeteners" in Kirk-Othmer *Encyclopedia of Chemical Technology* vol. **22** (John Wiley & Sons, New York, 3rd ed., 1983) pp 448-464.

Monoclinic crystals, mp 228.8-229.7°. Twinning on (001). Perfect 100 cleavage. Acicular crystals by vacuum sublimation. In dil aq soln it is 500 times as sweet as sugar; the sweet taste is still detectable in 1:100,000 dilution. Bitter, metallic aftertaste. d 0.828. Heat of combustion at constant volume: 4753.1 cal/g. uv max (0.1N NaOH): 267.3 nm (ε 1570). One gram dissolves in 290 ml water, 25 ml boiling water, 31 ml alcohol, 12 ml acetone, about 50 ml glycerol; freely sol in solns of alkali carbonates; slightly sol in chloroform, ether. pH of 0.35% aq soln 2.0. Hydrolysis products of saccharin are o-sulfamoylbenzoic acid (alkaline hydrolysis) and ammonium o-sulfobenzoic acid (acid hydrolysis).

Ammonium salt, C$_7$H$_8$N$_2$O$_3$S, *saccharin ammonium, Daramin, Sucline.*

Sodium salt dihydrate, C$_7$H$_4$NNaO$_3$S.2H$_2$O, *soluble saccharin, saccharin sodium, Kristallose, Crystallose, Dagutan, Sucaryl, Sucromat.* Crystalline powder; effloresces in dry air. In dil aq soln it is 300-500 times as sweet as sugar. One gram dissolves in 1.2 ml water, in about 50 ml alcohol. Aq solns are neutral or alkaline to litmus, but not alkaline to phenolphthalein. LD$_{50}$ in mice, rats (g/kg): 6.3, 7.1 i.p.; 17.5, 17.0 orally (Taylor). Anhydrous form, *Sucrédulcor.*

Note: Saccharin may reasonably be anticipated to be a carcinogen: *Seventh Annual Report on Carcinogens* (PB95-109781, 1994) p 352.

USE: Non-caloric sweetener; pharmaceutic aid (flavor). In formulations for electroplating-bath brighteners.

8464. L-Saccharopine. *N-(5-Amino-5-carboxypentyl)-L-glutamic acid;* ε-*N-(L-glutar-2-yl)-L-lysine.* C$_{11}$H$_{20}$N$_2$O$_6$; mol wt 276.29. C 47.82%, H 7.30%, N 10.14%, O 34.74%. A lysine precursor in the aminoadipic acid-lysine pathway in yeast: Darling, Larsen, *Acta Chem. Scand.* **15**, 743 (1961); Kjaer, Larsen, *ibid.* 750; Kuo *et al., Biochem. Biophys. Res. Commun.* **8**, 227 (1962); Trupin, Broquist, *J. Biol. Chem.* **240**, 2524 (1965); Jones, Broquist, *ibid.* **241**, 3430 sqq. (1966). Isoln from mycelium of the yeast *Candida utilis:* Marimoto, Yamano, *Biochem. Z.* **340**, 155 (1964).

Hydrated crystals; loses water of crystn over P_2O_5 at 100°. Dec 240-248°, when anhydr. $[\alpha]_D^{23}$ +33.6° (c = 1 in 0.5N HCl); $[M]_D$ +93°. pK_2 2.6; pK_3 4.1; pK_4 9.2; pK_5 10.3. Sparingly sol in water, ethanol. Readily sol in alkaline solns and in strong acids.

8465. Safflower Oil. The oil from the seed of *Carthamus tinctorius* L., *Compositae*. Milling and extraction procedures: Winter, *J. Am. Oil Chem. Soc.* **27**, 82 (1950). Purification and stabilization: Freedman, Shapiro, U.S. pat. **2,978,381** (1961). Monograph: R. E. Woodward, G. M. Severson, *Industrial Survey of Safflower*, Chemurgy Dept. Report no. S-3 (Agricult. Expt. Sta., Lincoln, Nebraska, 1951). Fatty acids present as glycerides: palmitic 6.4%, stearic 3.1%, arachidic 0.2%, oleic 13.4%, linoleic 76.6-79.0%, linolenic 0.04-0.13%.

Edible drying oil, intermediate between soybean and linseed oil. d_{25}^{25} 0.9211-0.9215. Titer: 15-18°. n_D^{25} 1.472-1.475. n_D^{40} 1.4690-1.4692. Acid value 1.0-9.7. Saponif value 188-194. Iodine value 140-150. Thiocyanogen value 82.5-86.0. Reichert-Meissl value below 0.5. Hydroxyl value 2.9-6.0. Unsaponifiable below 1.5%. Thickens and becomes rancid on prolonged exposure to air. Sol in the usual oil and fat solvents.

USE: As linseed oil in paints. For salad oil blends, in hydrogenated state as shortening.

THERAP CAT: Dietary supplement in hypercholesteremia (and possible prophylaxis and treatment of atherosclerosis).

8466. Saffron. Crocus; Spanish saffron; French saffron. Stigmas of *Crocus sativus* L., *Iridaceae*. *Habit*. Western Asia, Southern Europe. *Constit*. About 1% volatile oil, picrocrocin—a bitter glucoside, crocin (polychroit), fixed oil, wax.

Flattish-tubular, almost thread-like stigmas ~1.25 inches (3 cm) long; orange-brown color; strong, peculiar, aromatic odor; bitterish, aromatic taste.

USE: Almost entirely used for coloring and flavoring.

8467. Safranal. *2,6,6-Trimethyl-1,3-cyclohexadiene-1-carboxaldehyde;* 2,2,6-trimethyl-4,6-cyclohexadien-1-aldehyde. $C_{10}H_{14}O$; mol wt 150.22. C 79.96%, H 9.39%, O 10.65%. Prepn from picrocrocin: Kuhn, Winterstein, *Ber.* **67**, 354 (1934). Synthesis: Kuhn, Wendt, *Ber.* **69**, 1549 (1936); Könst *et al.*, *Tetrahedron Letters* **1974**, 3175. Biogenetic synthesis: T. Kametani *et al.*, *Chem. Pharm. Bull.* **29**,105 (1981).

Liquid. d_4^{19} 0.9734. $bp_{1.0}$ 70° (bath temp). n_D^{19} 1.5281. Sol in methanol, ethanol, petr ether, glacial acetic acid.

8468. Safrole. *5-(2-Propenyl)-1,3-benzodioxole; 4-allyl-1,2-methylenedioxybenzene;* allylcatechol methylene ether; allyldioxybenzene methylene ether; *m*-allylpyrocatechin methylene ether. $C_{10}H_{10}O_2$; mol wt 162.19. C 74.06%, H 6.21%, O 19.73%. Constituent of several essential oils, notably of sassafras in which it is present to the extent of about 75%. Has been used as topical antiseptic; pediculicide. Metabolism studies: Oswald *et al.*, *Biochim. Biophys. Acta* **230**, 237 (1971). Toxicity data: Hagan *et al.*, *Toxicol. Appl. Pharmacol.* **7**, 18 (1965). Review of carcinogenic risk: *IARC Monographs* **10**, 231-241 (1976).

Colorless or slightly yellow liq; sassafras odor. d^{20} 1.096. mp ~11°. bp 232-234°. n_D^{20} 1.5383. Insol in water. Very sol in alcohol. Miscible with chloroform, ether. LD_{50} in rats, mice (mg/kg): 1950, 2350 orally (Hagan).

Note: This substance may reasonably be anticipated to be a carcinogen: *Seventh Annual Report on Carcinogens* (PB95-109781, 1994) p 355.

USE: In perfumery; denaturing fats in soap manuf; also in manuf of heliotropin.

8469. SAICAR. *N-[[5-Amino-1-(5-O-phosphono-β-D-ribofuranosyl)-1H-imidazol-4-yl]carbonyl]-L-aspartic acid; N-[(5-amino-1-ribofuranosylimidazol-4-yl)carbonyl]aspartic acid 5'-phosphate;* 5-amino-N-(1,2-dicarboxyethyl)-1-ribofuranosylimidazole-4-carboxamide 5'-phosphate; 5-amino-4-imidazole-N-succinocarboxamide ribonucleotide; 5-amino-4-imidazole-N-succinocarboxamide ribotide; N-[5-amino-1-(5'-phosphoribofuranosyl)-4-imidazolecarbonyl]aspartic acid; succino-AICAR. $C_{13}H_{19}N_4O_{12}P$; mol wt 454.29. C 34.37%, H 4.22%, N 12.33%, O 42.26%, P 6.82%. Important purine precursor. Synthesis by enzymic reaction according to the equation AIR + ATP + aspartic acid + CO_2 → SAICAR + ADP + orthophosphate: Lukens, Buchanan, *J. Biol. Chem.* **234**, 1791 (1959). Production by adenine-requiring mutants of *Bacillus subtilis*: Ohmura *et al.*, U.S. pat. **3,280,-007** (1966 to Takeda). Synthesis of SAICAR and derivatives: Burrows, Shaw, *J. Chem. Soc. (C)* **1967**, 1088; Burrows *et al.*, *ibid.* **1968**, 40.

Dibarium salt, $C_{13}H_{15}Ba_2N_4O_{12}P$. uv max (pH 1): 269, 244 nm (ε 11850, 9580). If the Bratton-Marshall procedure [*J. Biol. Chem.* **128**, 537 (1939)] is modified by keeping the sample in ice during, and for 10 minutes after the addition of the reagents, SAICAR produces a purple chromophore having max absorption at 550 nm. The diazonium salt of SAICAR is very unstable and dec before it can couple under the usual test conditions.

8470. Sakuranetin. *2,3-Dihydro-5-hydroxy-2-(4-hydroxyphenyl)-7-methoxy-4H-1-benzopyran-4-one; 4',5-dihydroxy-7-methoxyflavanone;* naringen 7-methyl ether. $C_{16}H_{14}O_5$; mol wt 286.28. C 67.13%, H 4.93%, O 27.94%. Isoln by hydrolysis of 5-glucoside obtained from bark of *Prunus yedoensis* Matsum., *Rosaceae*: Asahina, *Arch. Pharm.* **246**, 259 (1908); from wood of various *Prunus* species: Hasegawa, Shirato, *J. Am. Chem. Soc.* **76**, 5559 (1954); **77**, 3557 (1955); Hasegawa, *ibid.* **79**, 1738 (1957). Structure: Asahina *et al.*, *J. Pharm. Soc. Japan* No. **550**, 1007 (1927); Shinoda, Sato, *ibid.* **48**, 220 (1928). Synthesis: Zemplén *et al.*, *Ber.* **75B**, 1432 (1942). Absolute configuration of (+)-isomer: Arakawa, Nakazaki, *Ann.* **636**, 111 (1960).

Needles, mp 152°. Also reported as dihydrate, mp 131-132° (Arakawa, Nakazaki, *loc. cit.*). (+)-Isomer, $[\alpha]_D^{16}$ +8.0° (c = 7.92 in acetone); $[\alpha]_D^{12}$+8.4° (c = 6.28 in methanol). Sol in alcohol, ether, chloroform, benzene, ethyl acetate, pyridine; slightly sol in boiling water. Practically insol in cold water.

5-Glucoside, $C_{22}H_{24}O_{10}$, *4',5-dihydroxy-7-methoxyflavanone 5-glucoside, sakuranin.* Tetrahydrate, bitter needles from dil alcohol. Begins to lose water at 100°. mp 212-214°. $[\alpha]_D^{28}$ −106.6° (tetrahydrate in acetone); $[\alpha]_D^{28}$ −123.2° (anhydr compd in acetone); $[\alpha]_D^{16}$ −97.4° (c = 1.58 for anhydr compd in 75% acetone). Very sol in dil alcohol, pyridine; slightly sol in abs alcohol, hot water, ethyl acetate. Practically insol in ether, cold water.

8471. Salacetamide. *N-Acetyl-2-hydroxybenzamide; N-acetylsalicylamide;* acetsalicylamide; Actylamide. $C_9H_9NO_3$; mol wt 179.18. C 60.33%, H 5.06%, N 7.82%, O 26.79%. Prepn: Hicks, *J. Chem. Soc.* **97**, 1033 (1910); Anshutz, Mitarbeiter, *Ann.* **442**, 25 (1925); **Belg. pat.** **496,438** (1950 to Soc. Belge de l'Azote Prod. Chim. Marly); Prost, Charlier, *Experientia* **18**, 319 (1962). Metabolism: Rayet *et al.*, *Arch. Int. Pharmacodyn. Ther.* **88**, 159 (1951).

mp 148°.
THERAP CAT: Analgesic, antipyretic, anti-inflammatory.

8472. Salatrim. Family of low-calorie fats composed of mixtures of long-chain saturated fatty acids (predominantly stearic) and short-chain saturated fatty acids (acetic, propionic, and/or butyric) esterified to the glycerol backbone. Name is an acronym derived from *s*hort *a*nd *l*ong acyl*tri*glyceride molecule. Prepd but not claimed: T. A. Pelloso *et al.*, **PCT Int. pat. Appl. 94 18,290**; *eidem*, **U.S. pat. 5,434,278** (1994, 1995 both to Nabisco). Series of articles on prepn, characterization, HPLC determn, pharmacology, and clinical assessment: *J. Agric. Food Chem.* **42**, 432-604 (1994). GC/MS determn in foods: A. S. Huang *et al.*, *ibid.* **43**, 1834 (1995).
Average caloric availability in humans: 4.9 kcal/g.
USE: Alternative fat used in foods.

8473. Salazosulfadimidine. *5-[[4-[[(4,6-Dimethyl-2-pyrimidinyl)amino]sulfonyl]phenyl]azo]-2-hydroxybenzoic acid; 5-[p-[(4,6-dimethyl-2-pyrimidinyl)sulfamoyl]phenylazo]salicylic acid;* 4'-(4,6-dimethylpyrimidin-2-ylsulfamoyl)-4-hydroxyazobenzene-3-carboxylic acid; 5-[p-[(4,6-dimethyl-2-pyrimidinyl)aminosulfonyl]phenylazo]salicylic acid; salicylazosulfadimidine; salicylazosulfamethazine; Azudimidine. $C_{19}H_{17}N_5O_5S$; mol wt 427.44. C 53.39%, H 4.01%, N 16.38%, O 18.72%, S 7.50%. Prepn: Korkuczanski, *Przem. Chem.* **37**, 162 (1958), *C.A.* **52**, 13727 (1958); Malatesta, Lotti, *Ann. Chim. (Rome)* **50**, 114 (1960), *C.A.* **56**, 8613g (1962).

Brown crystals, mp 207°.
THERAP CAT: Antibacterial.

8474. Salen. *(E,E)-2,2'-[1,2-Ethanediylbis(nitrilomethylidyne)]bisphenol; N,N'-bis[(2-hydroxyphenyl)methylene]-1,2-ethanediamine; N,N'-ethylenebis(salicylideneimine);* H_2sal_2en. $C_{16}H_{16}N_2O_2$; mol wt 268.32. C 71.62%, H 6.01%, N 10.44%, O 11.93%. Tetradentate Schiff base chelating ligand. Prepn: A. T. Mason, *Ber.* **20**, 267 (1887); and spectral analysis of Cu(II) complex: G. V. Panova *et al.*, *J. Gen. Chem. USSR* **53**, 1452 (1983). Prepn of transition metal complexes: P. Pfeiffer *et al.*, *Ann.* **505**, 84 (1933); and their physicochemical properties: A. Z. El-Sonbati *et al.*, *Transition Met. Chem.* **17**, 66 (1992); M. M. Bhadbhade, D. Srinivas, *Inorg. Chem.* **32**, 6122 (1993). Mass spectra of transition metal complexes: S. M. Schildcrout *et al.*, *ibid.* **34**, 4117 (1995). Thermodynamic properties of salen and its

metal complexes: F. Lloret *et al.*, *Inorg. Chim. Acta* **189**, 195 (1991). Use of Mn(III) complex as catalyst in epoxidation of olefins: K. Srinivasan *et al.*, *J. Am. Chem. Soc.* **108**, 2309 (1986); as specific DNA cleaving agent: D. J. Gravert, J. H. Griffin, *J. Org. Chem.* **58**, 820 (1993).

Crystals from alcohol, mp 125-126°. Insol in water.
USE: As stabilizing ligand for a variety of oxidation states of metal ions.

8475. Salep. Dried tubers of several species of *Orchis, Orchidaceae.* *Habit.* Europe, Asia Minor. *Constit.* Mucilage, starch.
USE: Nutrient, demulcent; as a vehicle for acrid remedies.

8476. Salicin. *2-(Hydroxymethyl)phenyl-β-D-glucopyranoside;* salicoside; salicyl alcohol glucoside; saligenin-β-D-glucopyranoside. $C_{13}H_{18}O_7$; mol wt 286.28. C 54.54%, H 6.34%, O 39.12%. Usually obtained by making hot water extracts from the ground bark of poplar *(Populus)* and willow *(Salix);* also found in the leaves and female flowers of the willow. Isoln from root bark of *Viburnum prunifolium* L., *Caprifoliaceae:* Evans *et al.*, *J. Am. Pharm. Assoc.* **34**, 207 (1945). Structure and synthesis: Irvine, Rose, *J. Chem. Soc.* **89**, 814 (1906); Kunz, *J. Am. Chem. Soc.* **48**, 262 (1926). Hydrolysis by acids: Moelwyn-Hughes, *Trans. Faraday Soc.* **25**, 503 (1929). Enzymatic hydrolysis: Pigman, *J. Res. Nat. Bur. Stand.* **27**, 6 (1941). Enzymatic hydrolysis in heavy water: Stacie, *Z. Phys. Chem.* **28B**, 236 (1935).

Orthorhombic crystals from water, mp 199-202°. $[\alpha]_D^{25}$ −62° to −67° (c = 3). $[\alpha]_D^{20}$ −45.6° (c = 0.6 in abs alc), Brauns, *J. Am. Chem. Soc.* **47**, 1292 (1925). One gram dissolves in 23 ml water, 3 ml boiling water, 90 ml alcohol, 30 ml alcohol at 60°. Soluble in alkalies, pyridine, glacial acetic acid. Practically insol in ether, chloroform. Aq solns are neutral to litmus and have a bitter taste.
USE: Standard substrate in evaluating enzyme prepns contg β-glucosidase: Weidenhagen, *Z. Ver. Zucker-Ind.* **79**, 591 (1929).
THERAP CAT: Analgesic.
THERAP CAT (VET): Has been used as a bitter stomachic, as an antirheumatic and as an analgesic.

8477. Salicyl Alcohol. *2-Hydroxybenzenemethanol;* saligenin; saligenol; *o*-hydroxybenzyl alcohol. $C_7H_8O_2$; mol wt 124.14. C 67.73%, H 6.50%, O 25.78%. Prepd by the action of emulsin on salicin; by heating phenol with methylene chloride and aq NaOH.

Plates or cryst powder. d 1.16. mp 86-87°; sublimes at 100°. Soluble in 15 parts water; very sol in alcohol, chloroform, ether, sol in benzene. Gives a red color with H_2SO_4.
THERAP CAT: Anesthetic (local).

8478. Salicylaldehyde. *2-Hydroxybenzaldehyde;* salicylic aldehyde. $C_7H_6O_2$; mol wt 122.12. C 68.85%, H

4.95%, O 26.20%. Made by heating sodium phenolate and chloroform with NaOH.

HC=O, OH (structure)

Clear, colorless, oily liq; bitter almond-like odor; burning taste. d_4^{20} 1.167; mp −7°; bp 196-197°; n_D^{20} 1.5735. Slightly sol in water; sol in alc, ether. Gives an orange color with H_2SO_4. MLD s.c. in rats: 1.0 g/kg, *Handbook of Toxicology*, W. S. Spector, Ed. (Saunders, Philadelphia, 1956) p 262.

USE: In perfumery.

8479. Salicylaldoxime. *2-Hydroxybenzaldehyde oxime;* Saldox. $C_7H_7NO_2$; mol wt 137.14. C 61.31%, H 5.14%, N 10.21%, O 23.33%. $HOC_6H_4CH=NOH$.
Prisms, mp 57°. Slightly sol in cold water; more sol in hot water; freely sol in alcohol, ether, benzene, dil HCl. Insol in petr ether. On heating it dec into salicylaldehyde and hydroxylamine.

USE: As a reagent for copper and nickel.

8480. Salicylamide. *2-Hydroxybenzamide;* Salamid; Samid; Cidal; Salizell; Salymid; Urtosal. $C_7H_7NO_2$; mol wt 137.14. C 61.31%, H 5.14%, N 10.21%, O 23.33%. Prepn: A. Cahours, *Ann.* **48**, 60 (1843); D. S. Hoffenberg, C. R. Hauser, *J. Org. Chem.* **20**, 1496 (1955); J. B. Chattopadhyaya, A. V. Rama Rao, *Tetrahedron* **30**, 2899 (1974). Toxicity data: Hart, *J. Pharmacol. Exp. Ther.* **89**, 205 (1947). Clinical efficacy in headache: W. J. Murray, *J. Clin. Pharmacol. J. New Drugs* **7**, 150 (1967). Pharmacokinetics. A. G. de Boer *et al.*, *Biopharm. Drug Dispos.* **4**, 321 (1983).

O, NH₂, OH (structure)

White or slightly pink, crystalline powder, somewhat bitter taste. Sensation of warmth on tongue. mp 140°. Soly in water at 30° = 0.2%, at 47° = 0.8%; in glycerol at 5° = 2.0%, at 39° = 5.0%, at 60° = 10.0%; in propylene glycol at 5° = 10.0%. Sol in hot water, alcohol, chloroform, ether. pH of satd aq soln at 28° about 5. Forms a water-soluble sodium salt at pH 9. LD_{50} orally in mice: 1.4 g/kg (Hart).

THERAP CAT: Analgesic.

8481. Salicylamide O-Acetic Acid. *[2-(Aminocarbonyl)phenoxy]acetic acid;* o-(carbamylphenoxy)acetic acid; α-(2-carbamoylphenoxy)acetic acid. $C_9H_9NO_4$; mol wt 195.17. C 55.39%, H 4.65%, N 7.18%, O 32.79%. Prepn: Merriman, *J. Chem. Soc.* **103**, 1838 (1913); Tzofin, *J. Gen. Chem. USSR* **3**, 17 (1933); Klosa, *Arch. Pharm.* **288**, 389 (1955); Izumi *et al.*, **Japan.** pat. **56 874** (1956 to Yoshitomi). Pharmacokinetics and efficacy: D. Loew *et al.*, *Int. J. Clin. Pharmacol. Ther. Toxicol.* **30**, 509 (1992).

HOOC, O, O, NH₂ (structure)

Crystals, mp 221°. Sol in aq alkali.
Sodium salt, $C_9H_8NNaO_4$, *Salizell (amp.).* Crystals, mp 212-215°.
Diethylamine salt, $C_{13}H_{22}N_2O_4$, *Akistin.*
THERAP CAT: Analgesic; anti-inflammatory; antipyretic.

8482. Salicylanilide. *2-Hydroxy-N-phenylbenzamide;* N-phenylsalicylamide; Salinidol. $C_{13}H_{11}NO_2$; mol wt 213.24. C 73.23%, H 5.20%, N 6.57%, O 15.01%. Usually

made by the reaction of salicylic acid with aniline in the presence of PCl_3 at an elevated temp. Better yields are obtained by using an inert organic solvent such as toluene or carbon tetrachloride as a reaction diluent. Novel process using ion-exchange resins: Majewski, Skelly, U.S. pat. **3,221,051** (1965). Alternate process: Majewski *et al.*, U.S. pat. **3,231,611** (1966 to Dow).

N, H, O, OH (structure)

Odorless leaflets. mp 135.8-136.2°. Slightly sol in water; freely sol in alcohol, ether, chloroform, benzene.
Caution: In concd form may cause irritation of skin, mucous membranes. *See also* Salicylic Acid.

USE: Anti-mildew, fungicide.
THERAP CAT: Antifungal (topical).
THERAP CAT (VET): Antifungal (topical).

8483. Salicylhydroxamic Acid. *N,2-Dihydroxybenzamide;* 2-hydroxybenzhydroxamic acid. $C_7H_7NO_3$; mol wt 153.14. C 54.90%, H 4.61%, N 9.15%, O 31.34%. $HOC_6H_4CONHOH$ or $HOC_6H_4C(OH)=NOH$. Prepn: Jeanrenaud, *Ber.* **22**, 1272 (1889); Urbanski, *Nature* **166**, 267 (1950). Trypanocidal activity: F. Opperdoes *et al.*, *Exp. Parasitol.* **40**, 198 (1976); A. B. Clarkson, F. H. Brohn, *Science* **194**, 204 (1976); A. J. Barnicoat *et al.*, *Experientia* **37**, 1291 (1981).
Needles from acetic acid, mp 168° (slow heating), mp 176-178° (quick heating). Sublimes. K at 25° = 6.4 × 10⁻⁶.
Sodium salt, $C_7H_6NNaO_3$, plates. Freely sol in water. pH of 0.1N soln 7.7.

USE: Complexing agent.

8484. Salicylic Acid. *2-Hydroxybenzoic acid;* Keralyt; Occlusal; Verrugon. $C_7H_6O_3$; mol wt 138.12. C 60.87%, H 4.38%, O 34.75%. Occurs in the form of esters in several plants, notably in wintergreen leaves and the bark of sweet birch. Made synthetically by heating sodium phenolate with carbon dioxide under pressure. Large scale process details: Faith *et al.*, *Industrial Chemicals* (Wiley, New York, 3rd ed., 1965) pp 652-655. Novel method by microbial oxidation of naphthalene: Zajic, Dunlap, U.S. pat. **3,274,074** (1966 to Kerr-McGee). Toxicity data: K. Sota *et al.*, *J. Pharm. Soc. Japan* **89**, 1392 (1969).

COOH, OH (structure)

Acicular crystals, or crystalline powder, mp 157-159°. Gradually discolors in sunlight. d 1.44. bp_{20} ∼211°. Sublimes at 76°; when rapidly heated at atm pressure it dec into phenol and CO_2. One gram dissolves in 460 ml water, 15 ml boiling water, 2.7 ml alcohol, 3 ml acetone, 42 ml chloroform, 3 ml ether, 135 ml benzene, 52 ml oil turpentine, about 60 ml glycerol, about 80 ml fats or oils. Soly in water increased by sodium phosphate, borax, alkali acetates or citrates. pH of satd aq soln 2.4. *Keep protected from light.* LD_{50} i.v. in mice: 500 mg/kg (Sota).
Salicylic acid or its salts are colored reddish even by merest traces of ferric salts. *Incompat:* Iron salts, spirit nitrous ether, lead acetate, iodine.
Lithium salt, $C_7H_5LiO_3$, *lithium salicylate.* White or grayish-white, odorless, sweetish powder. Deliquesc in moist air. Very sol in water, alcohol; the aq soln is slightly acid to litmus. *Keep well closed and protected from light.*
Silver salt, $C_7H_5AgO_3$, *silver salicylate.* White to reddish-white crystals. Slightly sol in water, alc. *Protect from light.*
Caution: Ingestion of large amounts can cause vomiting, abdominal pain, increased respiration, acidosis, mental disturbances. May cause skin rashes in sensitive individuals. *See: Clinical Toxicology of Commercial Products,* R. E.

Gosselin et al., Eds. (Williams & Wilkins, Baltimore, 5th ed., 1984) pp 368-375.

USE: As preservative of food products, but its use for this purpose is forbidden in some countries; manuf methyl salicylate, acetylsalicylic acid and other salicylates, dyes; as reagent in analytical chemistry.

THERAP CAT: Keratolytic (topical).

THERAP CAT (VET): Externally as an antiseptic and antifungal agent, and for various skin conditions. Keratolytic. Has been used internally in equine gastric tympany.

8485. 4-Salicyloylmorpholine. *4-(2-Hydroxybenzoyl)-morpholine;* salicyl morpholide; L-1102; Tardisal. $C_{11}H_{13}NO_3$; mol wt 207.23. C 63.76%, H 6.32%, N 6.76%, O 23.16%. Prepn: Carron, **Fr.** pat. **1,096,209** (1955 to Robert & Carrière); Schweitzer, U.S. pat. **2,906,728** (1959 to Dow).

Crystals, mp 175°. Soly (g/100 ml): water 0.41; alcohol 3.3; ether 0.22. pH of satd aq soln 6.2.

THERAP CAT: Choleretic.

8486. Salicylsulfuric Acid. *2-(Sulfooxy)benzoic acid;* salicylic acid, acid sulfate; salicylic acid sulfuric acid ester. $C_7H_6O_6S$; mol wt 218.19. C 38.53%, H 2.77%, O 44.00%, S 14.70%. Prepd by treating salicylic acid with chlorosulfonic acid in pyridine: Loeper et al., Compt. Rend. Soc. Biol. **135**, 917 (1941); Chim. Ind. (Paris) **49**, 99 (1943).

Monosodium salt, *sodium salicylsulfate, Salcyl, Salcylix.* Fine needles. Sol in water. Insol in organic solvents.

THERAP CAT: Analgesic, anti-inflammatory.

8487. Salinazid. *[(2-Hydroxyphenyl)methylene]hydrazide 4-pyridinecarboxylic acid;* 1-isonicotinoyl-2-salicylidenehydrazine; isonicotinic acid salicylidenehydrazide; saliniazid; N'-o-hydroxybenzylidenepyridine-4-carbohydrazide; o-hydroxybenzal isonicotinylhydrazone; Nupa-Sal; Acozid. $C_{13}H_{11}N_3O_2$; mol wt 241.25. C 64.72%, H 4.60%, N 17.42%, O 13.26%. Prepd by the interaction of isoniazid and salicylaldehyde in water: Hart et al., Antibiot. & Chemother. **4**, 803 (1954); cf. Yale et al., J. Am. Chem. Soc. **75**, 1933 (1953). Tuberculostatic activity and toxicity: Bavin et al., J. Pharm. Pharmacol. **7**, 1032 (1955).

Crystals from ethanol, mp 232-233° (also given as 251°). Soly at 25° in 100 ml water: 0.005 g; in 100 ml abs ethanol: 0.18 g; in 100 ml propylene glycol: 0.212 g. Sol in dil aq acids and alkalies forming a yellow soln.

THERAP CAT: Antibacterial (tuberculostatic).

8488. Salinomycin. $C_{42}H_{70}O_{11}$; mol wt 751.01. C 67.17%, H 9.39%, O 23.43%. Polyether ionophoric antibiotic having a unique tricyclic spiroketal ring system and an unsaturated 6-membered ring in the molecule. Produced by a strain of *Streptomyces albus* (FERM-P No. 419 and ATCC 21838). Production: Y. Miyazaki et al., **Japan. Kokai 72 25392** (1972 to Kaken Chem.), C.A. **78**, 41561 (1973). Structure: H. Kinashi et al., Tetrahedron Letters **1973**, 4955. Taxonomy, production, isolation and physicochemical and biological properties: Y. Miyazaki et al., J. Antibiot. **27**, 814 (1974). Use as a coccidiostat: Y. Tanaka

et al., **Ger.** pat. **2,253,031**; eidem, U.S. pat. **3,857,948** (1973, 1974 to Kaken Chem.). Anticoccidial efficacy: T. T. Migaki et al., Poultry Sci. **58**, 1192 (1979). Total synthesis: Y. Kishi et al., Front. Chem., Plenary Keynote Lect. IUPAC Congr., 28th **1981**, K. J. Laidler, Ed. (Pergamon, Oxford, 1982) pp 287-304; R. C. D. Brown, P. J. Kocienski, Synlett **1994**, 415, 417. HPLC determn in animal feeds: M. R. LaPointe, H. Cohen, J. Assoc. Offic. Anal. Chem. **71**, 480 (1988).

mp 112.5-113.5°. pKa' 6.4 (DMF). $[\alpha]_D^{25}$ −63° (c = 1 in ethanol). uv max (ethanol-water, 2:1): 284 nm (ε 126). LD_{50} in mice (mg/kg): 18 i.p.; 50 orally (Miyazaki).

Sodium salt, $C_{42}H_{69}NaO_{11}$, *Ovicox, Sacox, Salgain.* mp 140-142°. $[\alpha]_D^{25}$ −37° (c = 1 in ethanol).

THERAP CAT (VET): Anticoccidial agent.

8489. Salmeterol. *(±)-4-Hydroxy-α¹-[[[6-(4-phenylbutoxy)hexyl]amino]methyl]-1,3-benzenedimethanol;* (±)-4-hydroxy-α¹-[[[6-(4-phenylbutoxy)hexyl]amino]methyl]-m-xylene-α,α¹-diol; GR-33343X. $C_{25}H_{37}NO_4$; mol wt 415.57. C 72.26%, H 8.97%, N 3.37%, O 15.40%. β₂-Adrenergic agonist; structural analog of albuterol, q.v. Prepn: I. F. Skidmore et al., **Ger.** pat. **3,414,752**; eidem, U.S. pat. **4,992,474** (1984, 1991 both to Glaxo). Pharmacology: M. Johnson, Lung **168**, Suppl., 115 (1990). Preliminary evaluations in asthma: A. Ullman et al., Am. Rev. Resp. Dis. **142**, 571 (1990); O. P. Twentyman et al., Lancet **336**, 1338 (1990). Review of pharmacology and therapeutic use: R. N. Brogden, D. Faulds, Drugs **42**, 895-912 (1991).

mp 75.5-76.5°.

1-Hydroxy-2-naphthoate, $C_{25}H_{37}NO_4 \cdot C_{11}H_8O_3$, *salmeterol xinafoate, GR-33343G, Arial, Salmetedur, Serevent.* mp 137-138°. Freely sol in methanol; slightly sol in ethanol, chloroform, isopropanol; sparingly sol in water.

THERAP CAT: Bronchodilator.

8490. Salmine Sulfate. A protamine found in the sperm of salmon. Contains arginine, proline, serine, glycine, valine, leucine, alanine, threonine, isoleucine, lysine, histidine, aspartic and glutamic acids. Mol wt 6000 to 7000. N about 25%. Prepn and properties: Fisher, Scott, J. Pharmacol. Exp. Ther. **58**, 78 (1936); Felix, Am. Sci. **43**, 431 (1955); Callanan et al., J. Biol. Chem. **229**, 279 (1957); Carroll et al., ibid. **234**, 2314 (1959). Separation of components and amino acid sequence of major component: Ando, Watanabe, Int. J. Protein Res. **1**, 221 (1969). Toxicity study: Starbuck et al., Arch. Int. Pharmacodyn. Ther. **165**, 374 (1967).

Salmine base is a powder sol in disodium phosphate buffer; the sulfate separates from water in the cold as a clear, immiscible liquid. LD_{50} i.v. in rats: 75 mg/kg (Starbuck). *See also* Protamines.

8491. Salsalate. *2-Hydroxybenzoic acid 2-carboxyphenyl ester;* disalicylic acid; salicylic acid bimolecular ester; salicyloxysalicylic acid; salicylsalicylic acid; NSC-49171; Disalcid; Disalgesic; Mono-Gesic; Salflex. $C_{14}H_{10}O_5$; mol wt 258.23. C 65.12%, H 3.90%, O 30.98%. Nonacetylated aspirin analog. Prepn: **Ger.** pats. **211,403** and **214,044** (1909, both to Boehringer, Mann.), Frdl. **9**, 928 and C.A. **4**, 368 (1910); W. Baker et al., J. Chem. Soc. **1951**, 201. Metabolism: S. M. Dromgoole et al., J. Pharm. Sci. **73**, 1657 (1984). Clinical evaluation in arthritis: T. C. McPherson, Clin. Ther. **6**, 388 (1984). Mechanism of action studies: C. A. Divincenzo, F. R. Venezio, Curr. Ther. Res. **42**, 720 (1987). HPLC determn in plasma and urine: L. I. Harrison *et al., J. Pharm. Sci.* **69**, 1268 (1980). Review of chemistry, pharmacokinetics, safety and clinical efficacy: P. T. Singleton, *Clin. Ther.* **3**, 80-102 (1980).

Crystals from benzene. mp 148-149°. Practically insol in water but gradually hydrolyzed by it into 2 mols of salicylic acid. Sol in alc, ether; sparingly sol in benzene.

THERAP CAT: Analgesic, anti-inflammatory.

8492. Salsoline. *(R)-1,2,3,4-Tetrahydro-7-methoxy-1-methyl-6-isoquinolinol;* 6-hydroxy-7-methoxy-1-methyl-1,2,-3,4-tetrahydroisoquinoline; (+)-salsoline. $C_{11}H_{15}NO_2$; mol wt 193.25. C 68.37%, H 7.82%, N 7.25%, O 16.56%. In *Salsola richteri* Karel., *Chenopodiaceae.* Extraction procedure: A. Orechoff, N. Proskurnina, *Ber.* **66**, 841 (1933); N. Proskurnina, A. Orekhoff, *Bull. Soc. Chim. France* **1937**, 1265. Structure and synthesis: E. Späth *et al., Ber.* **67**, 1214 (1934); O. Kovács, G. Fodor, *Ber.* **84**, 795 (1951). Abs config: A. R. Battersby, T. P. Edwards, *J. Chem. Soc.* **1960**, 1214. Biosynthetic studies: I. McFarlane, M. Slaytor, *Phytochemistry* **11**, 235 (1972). Synthesis of enantiomers: S. Teitel *et al., J. Med. Chem.* **17**, 134 (1974). HPLC study: J. Strömbom, J. G. Bruhn, *J. Chromatog.* **147**, 513 (1978).

Crystals from alc, mp 221°. $[\alpha]_D^{20}$ +34.5° (c = 1 in 0.1*N* HCl). Sol in chloroform, hot alc, dil NaOH. Slightly sol in water, benzene. Almost insol in ether, petr ether.
Hydrochloride, mp 174-175°. $[\alpha]_D$ +31.0° (methanol). uv max (isopropanol): 204, 227, 284, 286 nm (ε 39400, 5900, 3540, 3530). Sol in water, hot alcohol. Very sparingly sol in acetone, chloroform. LD_{50} in mice (mg/kg): 140 i.v.; > 1000 orally (Teitel).
(−)-Form, *(S)-1,2,3,4-tetrahydro-7-methoxy-1-methyl-6-isoquinolinol.* mp 214-215°.

8493. Salutaridine. *5,6,8,14-Tetradehydro-4-hydroxy-3,6-dimethoxy-17-methylmorphinan-7-one;* floripavine. $C_{19}H_{21}NO_4$; mol wt 327.38. C 69.71%, H 6.47%, N 4.28%, O 19.55%. Intermediate in morphine synthesis; naturally occurring in (+)-form. Isolated from various species of *Croton* and *Papaver:* Barton *et al., J. Chem. Soc.* **1965**, 2423; Pfeifer, Kühn, *Pharmazie* **23**, 267 (1968). Structure and identity with floripavine: Mndzhoyan *et al., C.A.* **68**, 114787w (1968). Synthesis: T. Kametani *et al., J. Chem. Soc. (C)* **1969**, 2030; G. Horvath, S. Makleit, *Acta Chim. Acad. Sci. Hung.* **106**, 37 (1981); of racemate: T. Kametani *et al., J. Chem. Soc. Perkin Trans. I* **1972**, 1435; C. Szántay *et al., J. Org. Chem.* **47**, 594 (1982); of racemate and (−)-form: W. Ludwig, H. J. Schafer, *Angew. Chem. Int. Ed.* **25**, 1025 (1986). *Review:* Stuart, *Chem. Revs.* **1971**, 47.

Rods from ethyl acetate, mp 197-198°. $[\alpha]_D^{12}$ +114° (c = 0.5 in methanol). uv max (methanol): 236, 279 nm (log ε 4.23, 3.76).

(−)-Form, *Sinoaculine.* $[\alpha]_D^{20}$ −98° (c = 0.55 g/100 ml CH_3OH). uv max (methanol): 276, 240 nm (log ε 3.76, 4.24).

8494. Salverine. *2-[2-(Diethylamino)ethoxy]-N-phenylbenzamide;* 2-[2-(diethylamino)ethoxy]benzanilide. $C_{19}H_{24}$-N_2O_2; mol wt 312.41. C 73.05%, H 7.74%, N 8.97%, O 10.24%. Prepn: **Austrian** pat. **239,779** (1965 to Montavit), *C.A.* **63**, 4211h (1965).

Hydrochloride, $C_{19}H_{24}N_2O_2$·HCl, crystals, mp 173°.
Ingredient in *Montamed.*
THERAP CAT: Analgesic.

8495. Salvia. Sage. Dried leaves of *Salvia officinalis* L., *Labiatae.* Habit. Southern Europe; cultivated in U.S., England, France. *Constit.* 1-2.5% volatile oil; resin, tannin, bitter principles.
USE: Flavoring, condiment.
THERAP CAT: Antisecretory agent.

8496. Samaderins. Antitumor agents isolated from the bark and seed of *Samadera indica* Gaertn, *Simaroubaceae:* van der Marck, *Arch. Pharm.* **239**, 96 (1901). Separable into three components, samaderins A, B and C: Polonsky *et al., Bull. Soc. Chim. France* **1962**, 1715. According to Zylber *et al., ibid.* **1963**, 1322, samaderin C is identical with *samaderoside A* isolated by Mitra, Gregg, *Naturwiss.* **14**, 327 (1962). Structure of B and C: Zylber, Polonsky, *Bull. Soc. Chim. France* **1964**, 2016. Structure of A: M. C. Wani *et al., Chem. Commun.* **1977**, 295. Crystal structure of A: K. D. Onan, A. T. McPhail, *J. Chem. Res.(S)* **1978**, 14.

| Samaderin A | Samaderin B | =O |
| | Samaderin C | —OH |

Samaderin A, $C_{18}H_{18}O_6$, mp 253-255°. $[\alpha]_D^{25}$ −31.3° (c = 0.259 in pyridine). uv max (methanol): 288 nm (ε 13400).
Samaderin B, $C_{19}H_{22}O_7$, prisms from ethyl acetate, mp 235-240°.
Samaderin C, $C_{19}H_{24}O_7$, shiny plates from ethyl acetate+ alcohol, mp 265-268°. Sol in pyridine. Practically insol in ether. Very sparingly sol in other org solvents.

8497. Samandarine. *1α,4α-Epoxy-3-aza-A-homoandrostan-16β-ol.* $C_{19}H_{31}NO_2$; mol wt 305.46. C 74.71%, H 10.23%, N 4.59%, O 10.48%. Constitutes together with samandarone approx 75% of the alkaloid content in the skin glands of the European fire and Alpine salamanders: *Salamandra maculosa* resp. *atra* Laur., *Salamandridae.* Isoln: Schöpf, Braun, *Ann.* **514**, 69 (1934). Structure and configuration: Wölfel *et al., Ber.* **94**, 2361 (1961).

Needles from abs methanol or 50% acetone, mp 187-188°. Solvated crystals from dil methanol. $[\alpha]_D^{17}$ +43.3° (acetone). Freely sol in most organic solvents. Practically insol in water and NaOH solns.

Hydrochloride, $C_{19}H_{31}NO_2 \cdot HCl$, crystals from methanol, mp 312-313°; from water, mp 321-322° (monohydrate).

Samandarone, $C_{19}H_{29}NO_2$, ketone corresponding to the secondary alcohol in samandarine. Synthesis: Hara, Oka, *J. Am. Chem. Soc.* **89**, 1041 (1967). Needles from methyl ethyl ketone, mp 191-192°; from 35% acetone, mp 189-191°. $[\alpha]_D^{21}$ −115.7° (acetone).

8498. Samarium. Sm; at. wt 150.36; at. no. 62; valences 2, 3. A lanthanide belonging to the cerium group of rare earth metals; named for the mineral "samarskite" from which it was isolated. Naturally occurring isotopes (mass numbers): 144 (3.1%); 147 (15.0%), radioactive, $T_{1/2}$ 1.06 × 10^{11} years, α-emitter; 148 (11.3%), radioactive, $T_{1/2}$ 7 × 10^{15} years, α-emitter; 149 (13.8%); 150 (7.4%); 152 (26.7%); 154 (22.7%). Known artificial radioactive isotopes: 133-143; 145; 146; 151; 153; 155-158. Abundance in earth's crust: 6.47-7.0 ppm. Commercially important sources are the rare earth minerals monazite and bastnaesite; also occurs in samarskite, cerite, orthite, ytterbite, and fluorspar. Isoln: L. de Boisbaudran, *Compt. Rend.* **88**, 322 (1879); **89**, 212 (1880). Sepn by crystn of the nitrates: Demarcay, *ibid.* **122**, 728 (1896); Feit, Przibylla, *Z. Anorg. Chem.* **43**, 203 (1905). Sepn of metal: Schumacher, Harris, *J. Am. Chem. Soc.* **48**, 3108 (1926); by reduction of salts: Marsh, *J. Chem. Soc.* **1942**, 398, 523; **1943**, 8; Daane *et al.*, *J. Am. Chem. Soc.* **75**, 2272 (1953); Onstott, *ibid.* **75**, 5128 (1953); 77, 812 (1955). Toxicity study: Haley, *J. Pharm. Sci.* **54**, 663 (1965). Reviews of prepn, properties and compds: Prandtl, *Z. Anorg. Allgem. Chem.* **238**, 321-334 (1938); *The Rare Earths*, F. H. Spedding, A. H. Daane (Krieger, Huntington, N.Y., 1971, reprint of 1961 ed) 641 pp; Hulet, Bode, "Separation Chemistry of the Lanthanides and Transplutonium Actinides" in *MTP Int. Rev. Sci: Inorg. Chem., Ser. One*, Vol. 7, K. W. Bagnall, Ed. (University Park Press, Baltimore, 1972) pp 1-45; Moeller, "The Lanthanides" in *Comprehensive Inorganic Chemistry*, Vol. 4, J. C. Bailar Jr. *et al.*, Eds. (Pergamon Press, Oxford, 1973) pp 1-101; F. H. Spedding in Kirk-Othmer *Encyclopedia of Chemical Technology* vol. 19, (John Wiley & Sons, New York, 3rd ed., 1982) pp 833-854; *Chemistry of the Elements*, N. N. Greenwood, A. Earnshaw, Eds. (Pergamon Press, New York, 1984) pp 1423-1449. Brief review of properties: G. T. Seaborg, *Radiochim Acta* **61**, 115-122 (1993).

Yellow metal; tarnishes on exposure to air. Hardest metal of the cerium group. Crystalline forms: rhombohedral α-form, d 7.536, transforms to β-form at 917°; body-centered cubic β-form exists at >917°. mp 1074°. bp 1794°. Heat of fusion: 8.623 kJ/mol. Heat of sublimation (25°): 206.7 kJ/mol. E°(aq) Sm^{3+}/Sm −2.41 V (calc).

Oxide, Sm_2O_3, yellowish-white powder. d 8.347.

Hydroxide, $Sm(OH)_3$, gelatinous precipitate.

Trichloride, $SmCl_3$, white-yellowish powder. d 4.465, mp 686°. Forms addition compds with ammonia. Forms a hexahydrate, $SmCl_3 \cdot 6H_2O$, d 2.382, yellow crystalline plates. By reducing the anhydr trichloride at high temps with hydrogen, ammonia or aluminum powder, samarium dichloride is obtained. LD_{50} in mice (mg/kg): 585 i.p.; >2000 orally (Haley).

Dichloride, $SmCl_2$, dark brown crystalline mass. d^{22} 3.687. Practically insol in alcohol. Dec by water.

Sulfate, $Sm_2(SO_4)_3$, octahydrate, light yellow crystals, d^{18} 2.930. Sparingly sol in water.

USE: Oxide in control rods of some commercial nuclear power reactors. Alloys with cobalt to produce extremely stong permanent magnets.

8499. Sambucus. Elder; American elder; sweet elder. Flowers of *Sambucus canadensis* L., or of *S. nigra* L., Caprifoliaceae. *Habit.* U.S., east of Rocky Mts. *Constit.* Volatile oil, resin, fat, mucilage, tannin.

THERAP CAT: Diaphoretic, diuretic, cathartic.

8500. Sancycline. *[4S-(4α,4aα,5aα,12aα)]-4-(Dimethylamino)-1,4,4a,5,5a,6,11,12a-octahydro-3,10,12,12a-tetrahydroxy-1,11-dioxo-2-naphthacenecarboxamide;* 6-demethyl-6-deoxytetracycline; norcycline; Bonomycin. $C_{21}H_{22}N_2O_7$;

mol wt 414.41. C 60.86%, H 5.35%, N 6.76%, O 27.03%. Semi-synthetic antibiotic related to tetracycline, *q.v.* Prepn: Beereboom *et al.*, *J. Am. Chem. Soc.* **82**, 1003 (1960); McCormick *et al.*, *ibid.* 3381; McCormick, Jensen, U.S. pat. **3,019,260** (1962 to Am. Cyanamid); **Brit. pat. 901,209** (1962 to Pfizer). Total synthesis: Conover *et al.*, *J. Am. Chem. Soc.* **84**, 3222 (1962).

Hydrochloride hemihydrate, $C_{21}H_{22}N_2O_7 \cdot HCl \cdot \frac{1}{2}H_2O$, crystals from ethanol, dec 215-220°. uv max (MeOH/0.01N HCl): 267, 347 nm (ε 19300, 15500); in 0.1N H_2SO_4: 217, 268, 343 nm (ε 13400, 19000, 14600).

THERAP CAT: Antibacterial.

8501. Sandalwood, White. White saunders; yellow saunders. Heartwood of *Santalum album* L., Santalaceae. *Habit.* East Indies. *Constit.* Volatile oil, resin, tannin. A source of Oil of Santal, *q.v.*

USE: As incense and fumigant.

8502. Sandarac. Resin from *Callitris quadrivalvis* Vent., Pinaceae. *Habit.* Morocco. *Constit.* About 80% pimaric acid, about 10% callitrolic acid; sandaricinic acid.

Light yellow, brittle, elongated tears; translucent with vitreous fracture; crumbles to powder when masticated. Insol in water, benzene, petr ether. Sol in alcohol, ether, acetone, hot caustic alkalies; partially sol in chloroform, volatile oils, oil turpentine, carbon disulfide.

USE: Tooth cements, lacquers, varnishes; as an incense. Pharmaceutic aid (in ointments and plasters).

8503. Sanguinaria. Bloodroot; red puccoon; red root; puccoon root; tetterwort. Dried rhizome and roots of *Sanguinaria canadensis* L., Papaveraceae. *Habit.* North America. *Constit.* Sanguinarine, chelerythrine, protopine, homochelidonine, resin. An extract of *Sanguinaria canadensis* has been used in cough syrup; used experimentally as toothpaste base, in gingivitis and in periodontal disease: P. A. Ladanyi, U.S. pat. **4,145,412** (1979 to Vipont Chem. Co.).

8504. Sanguinarine. *13-Methyl[1,3]benzodioxolo[5,6-c]-1,3-dioxolo[4,5-i]phenanthridinium;* pseudochelerythrine; ψ-chelerythrine. $[C_{20}H_{14}NO_4]^+$. Principal alkaloid in sanguinaria extract. Isolated from the root of *Sanguinaria canadensis* L., and other *Papaveraceae:* Schmidt *et al.*, *Arch. Pharm.* **231**, 145 (1893). Widely distributed in poppy-fumaria species; constituent of argemone oil: S. A. E. Hakim *et al.*, *Nature* **189**, 198 (1961). Identity with pseudochelerythrine: Gadamer, Stichel, *ibid.* **262**, 488 (1924). Structure: Späth, Kuffner, *Ber.* **64**, 370, 2034 (1931); Beke, *Acta Chim. Acad. Sci. Hung.* **17**, 463 (1958). Biosynthesis: Leete, *J. Am. Chem. Soc.* **85**, 473 (1963). Synthesis: Dyke *et al.*, *Tetrahedron Letters* **1968**, 3933; Sainsbury *et al.*, *J. Chem. Soc. (C)* **1970**, 1797; Onda *et al.*, *Chem. Pharm. Bull.* **17**, 404 (1969); **19**, 31 (1971); M. Hanaoka *et al.*, *Chem. Letters* **22**, 739 (1986). Purification procedure: Stipanovic *et al.*, *J. Heterocycl. Chem.* **9**, 1453 (1972). Studies of effect of sanguinarine and argemone oil on intraocular pressure: S. A. E. Hakim, *Brit. J. Ophthalmol.* **38**, 193 (1954); G. C. Dobbie, M. E. Langham, *ibid.* **45**, 81 (1961). Anti-inflammatory activity in rats: J. Lenfeld *et al.*, *Planta Med.* **43**, 161 (1981). Clinical study of sanguinarine oral rinse as antiplaque agent: G. L. Southard *et al.*, *J. Am. Dental Assn.* **108**, 338 (1984). Acute toxicity study: P. J. Becci *et al.*, *J. Toxicol. Environ. Health* **20**, 199 (1987). *Review:* M. Shamma, *The Isoquinoline Alkaloids* (Academic Press, New York, 1972) pp 315-343. Symposium on chemistry, antimicrobial effects, clinical studies and toxicology: *J. Can. Dental Assoc.* **56**, Suppl 7, 5-47 (1990).

Monohydrate, $C_{20}H_{13}NO_4 \cdot H_2O$, crystals from water, mp 278-280° (Stipanovic). Soluble in alcohol, chloroform, acetone, ethyl acetate.

Chloride, $C_{20}H_{14}ClNO_4$, *Viadent.* LD_{50} in male mice (mg/kg): 15.9 i.v.; in female mice (mg/kg): 102.0 s.c. (Lenfeld). LD_{50} in rats (mg/kg): 29 i.v.; 1658 orally (Becci).

Chloride dihydrate, $C_{20}H_{14}ClNO_4 \cdot 2H_2O$, fine orange needles, mp 273-274° (dec). uv max (methanol): 234, 283, 325 nm (log ε 4.50, 4.52, 4.18).

USE: In mouthwash and toothpaste.

8505. α-Santalol. *5-(2,3-Dimethyltricyclo[2.2.1.0^{2,6}]-hept-3-yl)-2-methyl-2-penten-1-ol.* $C_{15}H_{24}O$; mol wt 220.35. C 81.76%, H 10.98%, O 7.26%. A sesquiterpene alcohol which together with β-santalol comprises about 90% of commercial sandalwood oil: F. W. Semmler, K. Bode, *Ber.* **40**, 1124 (1907); E. Gruenther, *The Essential Oils* vol. 2 (Van Nostrand, New York, 1949) pp 266-269. Structure: F. W. Semmler, *Ber.* **43**, 1893 (1910). Separation of α- and β-santalols of high purity: A. E. Bradfield *et al., J. Chem. Soc.* **1935**, 309. Synthesis of α-santalol and conversion to β-santalol: S. C. Bhattacharyya, *Sci. Cult.* **13**, 207 (1947), *C.A.* **43**, 5484gh (1949); *see also* V. M. Sathe *et al., Indian J. Chem.* **4**, 393 (1966); S. Y. Kamat *et al., Tetrahedron* **23**, 4487 (1967). Stereochemistry studies: G. Brieger, *Tetrahedron Letters* **1963**, 2123. Natural α-santalol has a cis-(+)-configuration. Total synthesis: R. G. Lewis *et al., ibid.* **1967**, 401. Stereospecific synthesis: E. J. Corey *et al., J. Am. Chem. Soc.* **92**, 226, 6314 (1970); M. Julia, P. Ward, *Bull. Soc. Chim. France* **1973**, 3065; M. Tamura, G. Suzukamo, *Tetrahedron Letters* **22**, 577 (1981). Synthesis and separation of *cis-* and *trans-*isomers: K. Sato *et al., Bull. Chem. Soc. Japan* **49**, 3351 (1976). Review: J. Simonsen, D. H. R. Barton, *The Terpenes* vol. 3 (Univ. Press, Cambridge, 1961) pp 178-188; A. Bhati, *Flavour Ind.* **1**, 235-251 (1970).

Liquid. bp_{14} 166-167°. d_{25}^{25} 0.9770. n_D^{25} 1.5017. α_{5461} +10.3°. $[\alpha]_D^{20}$ +17.20° (c = 0.8 in chloroform). Sol in alcohol; slightly sol in propylene glycol, glycerine. Practically insol in water.

USE: In perfumes, soaps and detergents.

8506. β-Santalol. *[1S-[1α,2α(Z),4α]]-2-Methyl-5-(2-methyl-3-methylenebicyclo[2.2.1]hept-2-yl)-2-penten-1-ol; 2-methyl-5-(2-methyl-3-methylene-2-norbornyl)-2-penten-1-ol.* $C_{15}H_{24}O$; mol wt 220.35. C 81.76%, H 10.98%, O 7.26%. A sesquiterpene alcohol from sandalwood oil. Isoln and purification and additional refs: *see* α-santalol. Structure: L. Ruzicka, G. Thomann, *Helv. Chim. Acta* **18**, 355 (1935). Synthesis: S. C. Bhattacharyya, *Sci. Cult.* **13**, 209 (1957), *C.A.* **43**, 5385a (1949); H. C. Kretschmar, W. F. Erman, U.S. pats. **3,662,008, 3,679,756** (1972 to Procter & Gamble). Natural β-santalol has a cis-(−)-configuration. Total synthesis: *eidem, Tetrahedron Letters* **1970**, 41. Synthesis of dl-form: P. A. Christenson, B. J. Willis, *J. Org. Chem.* **44**, 2012 (1979); M. Baumann, W. H. Hoffman, *Ann.* **1979**, 743; K. Sato *et al., Chem. Letters* **1981**, 1183.

Liquid. bp_{17} 177-178°. d_{25}^{25} 0.9717. n_D^{25} 1.5100. α_{5461} −87.1°. Sol in alcohol; practically insol in water.

dl-Form, $bp_{0.1}$ 101-103°.

USE: In perfumes, soaps and detergents.

8507. Santonica. Levant wormseed. Dried, unexpanded flower heads of *Artemisia maritima* L., sens. lat., *Compositae.* Habit. Iran, Turkestan. *Constit.* 2-4% santonin, 2-3% volatile oil; artemisin, resin.

THERAP CAT: Anthelmintic.

8508. Santonic Acid. *[1R-[1α,1(S*),3aβ,4α,5α,7aβ]]-Octahydro-α,3a,5-trimethyl-6,8-dioxo-1,4-methano-1H-indene-1-acetic acid; hexahydro-α,3a,5-trimethyl-6,8-dioxo-1,4-methanoindan-1-acetic acid.* $C_{15}H_{20}O_4$; mol wt 264.32. C 68.16%, H 7.63%, O 24.21%. Prepd from santonin by the action of hot concd aq bases, preferably potassium hydroxide: v. Oettingen, *Dissertation*, Göttingen, 1913; Woodward *et al., J. Am. Chem. Soc.* **70**, 4216 (1948). Structure: Wedekind, Engel, *J. Prakt. Chem.* **139**, 115 (1934); Woodward *et al., loc. cit.*; Woodward, Yates, *Chem. & Ind. (London)* **1954**, 1391. Chemistry: Hortmann, Daniel, *J. Org. Chem.* **37**, 4446 (1972).

Crystals from water or alcohol, mp 170-172°; bp_{15} 285°; $[\alpha]_D^{20}$ −74.0° (chloroform). Soluble in 190 parts of water at 17°. Freely sol in alc, chloroform, ether, glacial acetic acid.

8509. α-Santonin. *[3S-(3α,3aα,5aβ,9bβ)]-3a,5,5a,9b-Tetrahydro-3,5a,9-trimethylnaphtho[1,2-b]furan-2,8(3H,4H)-dione; 1,2,3,4,4a,7-hexahydro-1-hydroxy-α,4a,8-trimethyl-7-oxo-2-naphthaleneacetic acid γ-lactone; l-santonin.* $C_{15}H_{18}O_3$; mol wt 246.31. C 73.15%, H 7.37%, O 19.49%. Anthelmintic isolated from the dried unexpanded flower heads of *Artemisia maritima* L., sens. lat., *Compositae* [Levant wormseed] and other species of *Artemisia* found principally in Russian and Chinese Turkestan and the Southern Ural region. Structure: Clemo *et al., J. Chem. Soc.* **1930**, 1110; Ruzicka, Steiner, *Helv. Chim. Acta* **17**, 614 (1934). Stereochemistry: Corey, *J. Am. Chem. Soc.* **77**, 1044 (1955); Cocker, McMurry, *Tetrahedron* **8**, 181 (1960); Asher, Sim, *Proc. Chem. Soc.* **1962**, 111, 335; Nakazaki, Arakawa, *Bull. Chem. Soc. Japan* **37**, 464 (1964); Pregosin *et al., J. Chem. Soc. Perkin I* **1972**, 299. Total synthesis: Abe *et al., Proc. Japan Acad.* **30**, 116 (1954); *J. Am. Chem. Soc.* **78**, 1422 (1956); U.S. pat. **2,836,604** (1958 to Takeda); J. A. Marshall, P. G. M. Wuts, *J. Org. Chem.* **43**, 1086 (1978). Mass spectral studies of the santonins and derivs: Woseda *et al., Tetrahedron* **23**, 4623 (1967). Review: C. H. Heathcock in *The Total Synthesis of Natural Products* vol. 2, J. W. ApSimon, Ed. (Wiley, New York, 1973) pp 315-324.

(−)-Form, tabular crystals, orthorhombic sphenoidal, mp 170-173°. Almost tasteless with bitter aftertaste. $[\alpha]_D^{25}$

−170° to −175° (c = 2 in alc). Becomes yellow on exposure to light. Irritating to mucous membranes. d 1.187. One part dissolves in 5000 parts of cold water, in 250 parts of boiling water, in 280 parts of 50% alcohol at 17°, in 10 parts of boiling 50% alcohol, in 44 parts of cold 90% alcohol, in 3 parts of boiling 90% alcohol, in 125 parts of cold ether, in 72 parts of boiling ether, in 4.3 parts of cold chloroform.

(±)-Form, colorless plates from methanol, mp 181°. uv max (ethanol): 241 nm (log ε 4.10).

(+)-Form, colorless plates from methanol, mp 172°. $[\alpha]_D^{20}$ +165.9° (c = 1.92 in ethanol).

4-Hydroxysantonin, see Artemisin.

THERAP CAT: Anthelmintic (Nematodes).

THERAP CAT (VET): Has been used as an anthelmintic.

8510. Saperconazole. *4-[4-[4-[4-[[2-(2,4-Difluorophenyl)-2-(1H-1,2,4-triazol-1-ylmethyl)-1,3-dioxolan-4-yl]methoxy]phenyl]-1-piperazinyl]phenyl]-2,4-dihydro-2-(1-methylpropyl)-3H-1,2,4-triazol-3-one;* (±)-1-*sec-*butyl-4-[p-[4-[p-[[(2R*,4S*)-2-(2,4-difluorophenyl)-2-(1H-1,2,4-triazol-1-ylmethyl)-1,3-dioxolan-4-yl]methoxy]phenyl]-1-piperazinyl]-phenyl]-Δ²-1,2,4-triazolin-5-one; R-66905. $C_{35}H_{38}F_2N_8O_4$; mol wt 672.74. C 62.49%, H 5.69%, F 5.65%, N 16.66%, O 9.51%. Orally active, fluorinated triazole antifungal. Prepn: J. Heeres *et al.,* **Eur. pat. Appl.** 283,992; *eidem,* **U.S. pat.** 4,916,134 (1988, 1990 both to Janssen). *In vitro* antifungal activity: F. C. Odds, *J. Antimicrob. Chemother.* **24,** 533 (1989); D. W. Denning *et al., Eur. J. Clin. Microbiol. Infect. Dis.* **9,** 693 (1990). *In vivo* efficacy vs *Aspergillus* species: J. Van Cutsem *et al., Antimicrob. Ag. Chemother.* **33,** 2063 (1989).

Crystals from acetonitrile, mp 189.5°. Poorly sol in water.

THERAP CAT: Antifungal.

8511. Saponaria. Soapwort; soaproot; fuller's herb; bruisewort; bouncing bet. Herb and root of *Saponaria officinalis* L., *Caryophyllaceae. Habit.* Europe to Middle Asia; naturalized in U.S. *Constit.* Saponin, sapotoxin, saponarin.

8512. Saponarin. *6-β-D-Glucopyranosyl-7-(β-D-glucopyranosyloxy)-5-hydroxy-2-(4-hydroxyphenyl)-4H-1-benzopyran-4-one.* $C_{27}H_{30}O_{15}$; mol wt 594.53. C 54.55%, H 5.09%, O 40.37%. From leaves of *Saponaria officinalis* L., *Caryophyllaceae:* Barger, *J. Chem. Soc.* **89,** 1210 (1906); from *Hibiscus syriacus* L., *Malvaceae:* Nakoaki, *J. Pharm. Soc. Japan* **64,** 304 (1944), *C.A.* **46,** 108d (1952); from *Lemna (Spirodela) oligorrhiza* Kurz., *Lemnaceae:* Jurd *et al., Arch. Biochem. Biophys.* **67,** 284 (1957). Structure: Seikel, Geissman, *ibid.* **71,** 17 (1957). Revised structure: Hörhammer *et al., Tetrahedron Letters* **1965,** 1707. On acidic hydrolysis yields the aglycone, *saponaretin.*

Monohydrate, pale yellow granules, mp 228°. $[\alpha]_D$ −7.9° (aq pyridine). uv max (ethanol): 335, 272 nm. Practically insol in cold water. Sparingly sol in hot water, alcohol; sol in alkalies with yellow color and in concd H_2SO_4 with blue fluorescence; sol in pyridine.

8513. Saponins. Sapogenin glycosides. A type of glycoside widely distributed in plants. Each saponin consists of a sapogenin which constitutes the aglucon moiety of the molecule, and a sugar. The sapogenin may be a steroid or a triterpene and the sugar moiety may be glucose, galactose, a pentose, or a methylpentose. Poisonous towards the lower forms of life and used for killing fish by the aborigines of South America. *Review and bibliography:* R. J. McIlroy, *The Plant Glycosides* (Edward Arnold & Co., London, 1951) Chapter IX; Y. Birk, I. Peri in *Toxic Const. Plant Foodst.,* I. E. Liener, Ed. (Academic Press, New York, 2nd ed., 1980) pp 161-182.

Bitter taste. All saponins foam strongly when shaken with water. They form oil-in-water emulsions and act as protective colloids.

Caution: Although practically non-toxic to man upon oral ingestion, they act as powerful hemolytics when injected into the blood stream, dissolving the red corpuscles even at extreme dilutions.

8514. Sapphire. The corundum modification of aluminum oxide, Al_2O_3. Mined in Montana, Siam, Burma, Ceylon. The highest priced variety of the native gem has the blue color of cornflowers. Easily prepd synthetically. Hydrothermal synthesis: Laudise, Ballman, *J. Am. Chem. Soc.* **80,** 2655 (1958).

USE: In jewelry, as phonograph needles, industrial abrasive, watch and instrument bearings.

8515. Sapropterin. *[6R-[6R*(1R*,2S*)]]-2-amino-6-(1,2-dihydroxypropyl)-5,6,7,8-tetrahydro-4(1H)-pteridinone;* (6R)-L-*erythro*-tetrahydrobiopterin; dapropterin; *R-*THBP; $6R$-BH_4. $C_9H_{15}N_5O_3$; mol wt 241.25. C 44.81%, H 6.27%, N 29.03%, O 19.90%. Natural cofactor of the aromatic amino acid hydroxylases required for catecholamine and serotonin biosynthesis. Identification of cofactor activity: S. Kaufman, *Proc. Nat. Acad. Sci. USA* **50,** 1085 (1963). Prepn of (6R,S)-BH_4: B. Schircks *et al., Helv. Chim. Acta* **61,** 2731 (1978). Chromatographic separation of diastereoisomers: S. W. Bailey, J. E. Ayling, *J. Biol. Chem.* **253,** 1598 (1978). Absolute configuration of natural isomer: W. L. F. Armarego *et al., Aust. J. Chem.* **35,** 785 (1982). Stereospecific synthesis: S. Matsuura *et al., Heterocycles* **23,** 3115 (1985); H. Sakai, T. Kanai, **Eur. pat. Appl.** 191,335; *eidem,* **U.S. pat.** 4,713,454 (1986, 1987 both to Shiratori; Suntory). Bioavailability: G. Kapatos, S. Kaufman, *Science* **212,** 955 (1981). Effect on neurotransmitter monoamine biosynthesis: S. Miwa *et al., Arch. Biochem. Biophys.* **239,** 234 (1985). LC determn in biological samples: Y. Tani, T. Ishihara, *Life Sci.* **46,** 373 (1990). Therapeutic potential in hyperphenylalaninemia: S. Kaufman, *J. Nutr. Sci. Vitaminol., Suppl.,* 601 (1992).

pK' 5.05. uv max (0.1 N HCl): 265 nm (ε 14000). Dihydrochloride, $C_9H_{15}N_5O_3$.2HCl, SUN-0588, Biopten. Crystals from HCl, mp 245-246° (dec). $[\alpha]_D^{25}$ −6.81° (c = 0.665 in 0.1 M HCl). uv max (2 M HCl): 264 nm (ε 16770).

THERAP CAT: In treatment of hyperphenylalaninemia.

8516. Saquinavir. *[3S-[2[1R*(R*),2S*],3α,4aβ,8aβ]]-N¹[3-[3-[[(1,1-Dimethylethyl)amino]carbonyl]octahydro-2(1H)-isoquinolinyl]-2-hydroxy-1-(phenylmethyl)propyl]-2-[(2-quinolinylcarbonyl)amino]butanediamide;* (S)-N-[(αS)-α-[(1R)-2-[(3S,4aS,8aS)-3-(tert-butylcarbamoyl)octahydro-2(1H)-isoquinolyl]-1-hydroxyethyl]phenethyl]-2-quinaldamido succinamide; *N-tert-*butyldecahydro-2-[2(R)-hydroxy-4-phenyl-3(S)-[[N-(2-quinolinylcarbonyl)-L-asparaginyl]amino]-butyl](4aS,8aS)-isoquinoline-3(S)-carboxamide; Ro-31-8959. $C_{38}H_{50}N_6O_5$; mol wt 670.85. C 68.04%, H 7.51%, N 12.53%, O 11.92%. Selective HIV proteinase inhibitor. Prepn: J. A. Martin, S. Redshaw, **Eur. pat. Appl.** 432,695; *eidem,* **U.S. pat.** 5,196,438 (1991, 1993 both to Hoffmann-LaRoche); K. E. B. Parkes *et al., J. Org. Chem.* **59,** 3656 (1994). *In vitro*

HIV proteinase inhibition: N. A. Roberts *et al.*, *Science* **248**, 358 (1990). Antiviral properties: J. C. Craig *et al.*, *Antiviral Res.* **16**, 295 (1991); S. Galpin *et al.*, *Antiviral Chem. Chemother.* **5**, 43-45 (1994). Clinical evaluation of tolerability and activity: V. S. Kitchen *et al.*, *Lancet* **345**, 952 (1995). Brief review of clinical evaluations: N. A. Roberts, *Int. Antiviral News* **3** (1), 2-3 (1995).

White crystalline solid. $[\alpha]_D^{20}$ $-55.9°$ (c = 0.5 in methanol). Soly (21°): 0.22 g/100 ml water.

Monomethanesulfonate salt, $C_{38}H_{50}N_6O_5 \cdot CH_4O_3S$, *saquinavir mesylate, Ro-31-8959/003, Invirase.*

THERAP CAT: Antiviral.

8517. Sarafloxacin. *6-Fluoro-1-(4-fluorophenyl)-1,4-dihydro-4-oxo-7-(1-piperazinyl)-3-quinolinecarboxylic acid.* $C_{20}H_{17}F_2N_3O_3$; mol wt 385.37. C 62.33%, H 4.45%, F 9.86%, N 10.90%, O 12.46%. Fluorinated quinolone antibacterial. Prepn: D. T. W. Chu, **Eur. pat. Appl. 131,839;** *idem*, **U.S. pat. 4,730,000** (1985, 1988 both to Abbott); D. T. W. Chu *et al.*, *J. Med. Chem.* **28**, 1558 (1985); H. Narita *et al.*, **Japan. Kokai 85 237,069** (1985 to Toyama). Comparative *in vitro* antibacterial spectrum: A. Digranes, W. L. Dibb, *Chemotherapy* **34**, 298 (1988). *In vivo* antibacterial activity: P. B. Fernandes *et al.*, *Antimicrob. Ag. Chemother.* **29**, 201 (1986). HPLC determn: J. F. Bauer *et al.*, *Pharm. Res.* **7**, 1177 (1990).

Monohydrochloride, $C_{20}H_{17}F_2N_3O_3 \cdot HCl$, *Abbott 56620, A-56620.* mp as monohydrate > 275°.

THERAP CAT (VET): Antibacterial.

8518. Saralasin. *1-(N-Methylglycine)-5-L-valine-8-L-alanineangiotensin II;* N-[1-[N-[N-[N-[N²-(N-methylglycyl)-L-arginyl]-L-valyl]-L-tyrosyl]-L-valyl]-L-histidyl]-L-prolyl]-L-alanine; 1-sar-8-ala-angiotensin II. $C_{42}H_{65}N_{13}O_{10}$; mol wt 912.06. C 55.31%, H 7.18%, N 19.96%, O 17.54%. A specific antagonist of angiotensin II. Prepn: Sipos *et al.*, **Ger. pat. 2,127,393** corresp to **U.S. pat. 3,751,404** (1972, 1973 to Norwich Pharmacal). Activity studies: Pals *et al.*, *Circ. Res.* **29**, 673 (1971); Pals, Fulton, *Arch. Int. Pharmacodyn. Ther.* **204**, 20 (1973); Solomon, Buckley, *J. Pharm. Sci.* **63**, 1109 (1974); Streeten *et al.*, *N. Engl. J. Med.* **292**, 657 (1975). Comprehensive review of the acetate: C. T. Huang *et al.*, in *Pharmacological and Biochemical Properties of Drug Substances* vol. 3, M. E. Goldberg, Ed. (Am. Pharm. Assoc., Washington, DC, 1981) pp 176-225.

Sar–Arg–Val–Tyr–Val–His–Pro–Ala

Hydrated acetate, *saralasin acetate, P-113, Sarenin.* Described as $C_{42}H_{65}N_{13}O_{10} \cdot xC_2H_4O_2 \cdot xH_2O$. Fluffy white powder, mp 256°. Sol in water, 5% aq dextrose, 90-95% aq alcohol. LD_{50} i.v. in male mice: 1171 mg/kg (Huang).

THERAP CAT: Antihypertensive, diagnostic aid (renin-dependent hypertension).

8519. Sarcosine. *N-Methylglycine;* N-methylaminoacetic acid; Sar; N-methylglycocoll; methylaminoethanoic acid. $C_3H_7NO_2$; mol wt 89.09. C 40.44%, H 7.92%, N 15.72%, O 35.92%. CH_3NHCH_2COOH. Has been found in starfish and sea urchins: A. Kossel, S. Edlbacher, *Z. Physiol. Chem.* **94**, 264 (1915); in rock lobsters: L. Novellie, H. M. Schwartz, *Nature* **173**, 450 (1954). Formation from caffeine by dec with barium hydroxide: Paulmann, *Arch. Pharm.* **232**, 601 (1894). Prepd on a large scale from formaldehyde, sodium cyanide, and methylamine: Eschweiler, *Ann.* **279**, 39 (1894); Baumann, *J. Biol. Chem.* **21**, 563 (1915); Caverly, **U.S. pat. 2,720,540** (1955 to du Pont); Leake, Brakebill, **U.S. pat. 3,009,954** (1961 to Allied Chem.). Synthesis: M. Ebata *et al.*, *Bull. Chem. Soc. Japan* **39**, 2535 (1966). Review of syntheses: T. Shirai, *Synthetic Production and Utilization of Amino Acids*, T. Kaneko *et al.*, Eds. (Wiley, New York, 1974) pp 184-186.

Orthorhombic, deliquescent crystals from dil methanol. Dec 212°. Sweetish taste. pK_1' 2.23; pK_2' 10.01. Sol in water: 5 ml of a satd aq soln contain 2.1412 g sarcosine. Slightly sol in alc.

Sodium salt, usually sold as aq soln, pH 12.0.

Hydrochloride, $C_3H_7NO_2 \cdot HCl$, needles from alcohol, dec 171°. Freely sol in water; slightly sol in alcohol, ether.

USE: Intermediate in the synthesis of antienzyme agents for toothpaste.

8520. Sarin. *Methylphosphonofluoridic acid 1-methylethyl ester;* isopropoxymethylphosphoryl fluoride; isopropyl methylphosphonofluoridate; GB. $C_4H_{10}FO_2P$; mol wt 140.09. C 34.29%, H 7.19%, F 13.56%, O 22.84%, P 22.11%. Nerve gas: Holmstedt, *Acta Physiol. Scand.* **25**, suppl. 90, p 106 (1951); *Protar* **14**, 113 (1948); **16**, 131 (1950). Additional syntheses: Schrader, *B.I.O.S.* **714**, 41 (1947); B. C. Saunders, *Some Aspects of the Chemistry and Toxic Action of Organic Compounds Containing Phosphorus and Fluorine* (Cambridge, 1957) p 92 sqq; Bryant *et al.*, *J. Chem. Soc.* **1960**, 1553. Absolute configuration: Benschop *et al.*, *Rec. Trav. Chim.* **87**, 387 (1968). Toxicity study: B. Holmstedt, *Pharmacol. Rev.* **11**, 567 (1959). Brief review: Schrader, *Die Entwicklung neuer insektizider Phosphorsäure-Ester* (Verlag Chemie, Weinheim, 1963) p 4. *See also* Tabun.

Liquid, d_4^{20} 1.10. mp $-57°$. bp_{760} 147°; bp_{16} 56°. Miscible with and hydrolyzed by water. Rapidly hydrolyzed by dil aq sodium hydroxide or sodium carbonate forming relatively non-toxic products. Water alone removes the fluorine atom producing the non-toxic acid $CH_3PO[OCH(CH_3)_2]$-OH. LD_{50} i.p. in mice: 0.42 mg/kg (Holmstedt, 1959). Lethal dose for man may be as low as 0.01 mg/kg, *Chem. & Eng. News* **31**, 4676 (1953).

Caution: Extremely active cholinesterase inhibitor. Toxic effects similar to, but more severe than, parathion, *q.v.*

USE: Chemical warfare agent.

8521. Sarkomycin. Sarcomycin. Antibiotic and antitumor substance produced by *Streptomyces erythrochromogenes*, strain W-115-C, from soil found at Kamakura, Japan: Umezawa *et al.*, *J. Antibiot.* **6A**, 101, 147, 153 (1953); *idem*, *Antibiot. & Chemother.* **4**, 514 (1954). Structure of sarkomycin A, the active principle: Hooper *et al.*, *ibid.* **5**, 585 (1955). Characterization of sarkomycins A, A' and B: Maeda, Kondo, *J. Antibiot.* **11A**, 37 (1958). Sarkomycin reacts with H_2S to form sarkomycins S_1, S_2 and S_3: Tatsuoka *et al.*, *ibid.* **9B**, 107, 110 (1956). Structure of S_1 and S_2: *idem*, *ibid.* **11B**, 275 (1958). Synthesis of *d*- and *l*-sarkomycin A: Toki, *ibid.* **10A**, 35, 226 (1957). Synthesis of *dl*-sarkomycin A: Toki, *Bull. Chem. Soc. Japan* **30**, 450 (1957); R. K. Boeckman, Jr. *et al.*, *J. Org. Chem.* **45**, 752 (1980); J. N. Marx, G. Minaskanian, *ibid.* **47**, 3306 (1982); B. A. Wexler *et al.*, *ibid.* 3333; A. P. Kozikowski, P. D.

Stein, *J. Am. Chem. Soc.* **104**, 4023 (1982). Abs config of A: Sato *et al.*, *Chem. Pharm. Bull.* **11**, 829 (1963). Revised config of A: Hill *et al.*, *J. Org. Chem.* **32**, 2330 (1967). *Review:* S. A. Waksman, H. A. Lechevalier, *The Actinomycetes* **vol. III** (Williams & Wilkins, Baltimore, 1962) pp 362-364; Sung in *Antibiotics*, **vol. I**, D. Gottlieb, P. Shaw, Eds. (Springer-Verlag, New York, 1967) pp 156-165.

Sarkomycin A

Oily liquid. $[\alpha]_D^{15}$ −32.5° (c = 1 in methanol). uv max (water): 230 nm. Sol in water, methanol, ethanol, butanol, ethyl acetate. Sparingly sol in petr ether.

Sarkomycin A, $C_7H_8O_3$, *(R)-2-methylene-3-oxocyclopentanecarboxylic acid.*

8522. Sarmentogenin. *(3β,5β,11α)-3,11,14-Trihydroxy-card-20(22)-enolide.* $C_{23}H_{34}O_5$; mol wt 390.52. C 70.74%, H 8.78%, O 20.48%. An aglucon from sarmentocymarin isolated from the seeds of *Strophanthus sarmentosus* DC. var. *senegambiae* (A. DC.) Monachino, *Apocynaceae:* Jacobs, Heidelberger, *J. Biol. Chem.* **81**, 765 (1929); Katz, *Helv. Chim. Acta* **31**, 993 (1948). Structure: Callow, Taylor, *J. Chem. Soc.* **1952**, 2299; Repke, Klesczewski, *Arch. Exp. Pathol. Pharmakol.* **239**, 131 (1960).

Prisms from methanol or methanol + acetone, mp 278-282°. $[\alpha]_D^{19}$ +21.1° ± 4° (c = 0.521 in methanol). Absorption max (98% H_2SO_4): 230, 415 nm ($E_{1cm}^{1\%}$ 460, 430). Sol in alcohol, methanol, pyridine; sparingly sol in acetone, chloroform; practically insol in benzene, ether. Dissolves in concd H_2SO_4 with a bright golden color which slowly turns green and then indigo. In dil pyridine soln gives a deep red color with alkaline nitroprusside soln. When recrystallized from pyridine, it forms long microplatelets contg ~1 mol of the solvent and melting at 258° with effervescence. Is not precipitated with digitonin. Yields sarmentogenone on oxidation.

Diacetate, $C_{27}H_{38}O_7$, needles from abs ether, mp 135-155°. Not affected by CrO_3 in acetic acid.

Dibenzoate, $C_{37}H_{42}O_7$, flat hexagonal prisms from acetone, mp 281°. $[\alpha]_D^{20}$ +14° (acetone). Sparingly sol in alc; practically insol in ether. Not affected by CrO_3 in acetic acid.

8523. Sarmentose. *2,6-Dideoxy-3-O-methyl-xylo-hexose.* $C_7H_{14}O_4$; mol wt 162.19. C 51.84%, H 8.70%, O 39.46%. A methyl ether of a 2-desoxyhexomethylose, isomeric with cymarose. Obtained on hydrolysis of sarmentocymarin, a glycoside isolated from seeds of *Strophanthus sarmentosus* DC., *Apocynaceae* by the enzymic method: Jacobs, Heidelberger, *J. Biol. Chem.* **81**, 765 (1929); Jacobs, Bigelow, *ibid.* **96**, 355 (1932). Synthesis: H. Hauenstein, T. Reichstein, *Helv. Chim. Acta* **33**, 446 (1950).

Prisms, plates from ether + petr ether, mp 78-79°. Shows mutarotation. $[\alpha]_D^{20}$ +12° (20 min) → +15.8° (24 hrs c = 1.08): Elderfield, *Advan. Carbohyd. Chem.* **1**, 172 (1945).

8524. Sarpagine. *Sarpagan-10,17-diol;* raupine. $C_{19}H_{22}N_2O_2$; mol wt 310.40. C 73.52%, H 7.14%, N 9.03%, O 10.31%. Isoln from *Rauwolfia serpentina* (L.) Benth., *Apocynaceae:* Stoll, Hofmann, *Helv. Chim. Acta* **36**, 1143 (1953). Structure: Stauffacher *et al.*, *ibid.* **40**, 508 (1957); Poisson *et al.*, *Bull. Soc. Chim. France* **1957**, 610; Biemann, *J. Am. Chem. Soc.* **83**, 4801 (1961). Stereochemistry: Bartlett *et al.*, *ibid.* **84**, 622 (1962); Ohashi *et al.*, *Tetrahedron* **19**, 2241 (1963).

Needles from methanol or ethanol, long plates from acetone, dec > 320°. $[\alpha]_D^{20}$ +54° (c = 0.5 in pyridine). uv max (ethanol): 230, 278 nm (log ε 4.30, 3.92). One gram dissolves in 60 ml boiling ethanol, 250 ml boiling methanol, 400 ml boiling acetone. Practically insol in chloroform.

Hydrochloride, $C_{19}H_{22}N_2O_2$·HCl, needles from alcohol, dec > 220°.

8525. Sarsaparilla. Dried root of *Smilax aristolochiaefolia* Mill. (*S. medica* Cham. & Schlecht.) (Mexican Sarsap.), or *S. regelii* Killip & Morton (*S. ornata* Hook f.) (Jamaica or Central American Sarsap.), *Liliaceae*. *Habit.* Honduras, Jamaica, Mexico (Vera Cruz), Brazil, Guatemala. *Constit.* Sarsaponin, smilacin (parillin), paroaparic acid, resin, volatile oil.

USE: As flavor in beverages. Pharmaceutic aid (flavor).

8526. Sarsasapogenin. *(3β,5β,25S)-Spirostan-3-ol;* parigenin. $C_{27}H_{44}O_3$; mol wt 416.64. C 77.84%, H 10.64%, O 11.52%. Steroid sapogenin from *Smilax ornata* Hooker, fil., *Liliaceae* (Sarsaparilla): Power, Salway, *J. Chem. Soc.* **105**, 201 (1914); Jacobs, Simpson, *J. Biol. Chem.* **105**, 501 (1934); **109**, 573 (1935). Partial structure: Hirschmann *et al.*, *J. Org. Chem.* **20**, 572 (1955). Stereochemistry: Taylor, *Chem. & Ind. (London)* **1954**, 1066; Wall, Serota, *J. Am. Chem. Soc.* **76**, 2850 (1954); Callow, Massey-Beresford, *J. Chem. Soc.* **1957**, 4482; Rosen *et al.*, *J. Am. Chem. Soc.* **81**, 1687 (1959). *Reviews:* Turner, *The Steroidal Sapogenins*, *Univ. Microfilms* (Ann Arbor, Mich.), Pub. no. **361**, 249 pp (1941); Shabica, *Sarsasapogenin and Related Compounds*, *Univ. Microfilms* (Ann Arbor, Mich.), Pub. no. **549**, 89 pp (1943); L. F. Fieser, M. Fieser, *Steroids* (Reinhold, New York, 1959) p 810. *See also* Smilagenin.

Large prismatic needles from acetone, mp 199-199.5°. $[\alpha]_D^{25}$ −75°, $[\alpha]_{546}^{25}$ −89° (c = 0.5 in $CHCl_3$). uv spectrum: Smith, Eddy, *Anal. Chem.* **31**, 1539 (1959). Sol in alcohol, acetone, benzene, chloroform. Is precipitated by digitonin.

Acetate, $C_{29}H_{46}O_4$, flat needles from methanol, mp 144-145°. $[\alpha]_D^{25}$ −70.2°; $[\alpha]_{546}^{25}$ −83.1° (c = 1.18 in $CHCl_3$).

USE: In the manuf of compds of the pregnane series.

8527. Sarverogenin. *(3β,5β,7β,11α)-7,8-Epoxy-3,11,14-trihydroxy-12-oxo-card-20(22)-enolide.* $C_{23}H_{30}O_7$; mol wt 418.49. C 66.01%, H 7.23%, O 26.76%. The aglycone of sarveroside isolated from seeds of *Strophanthus* spp, *Apocy-*

naceae: von Euw *et al., Festschrift Prof. Paul Casparis* (Zürich, 1949) p 178; Buzas *et al., Helv. Chim. Acta* **33**, 465 (1950); Rothrock *et al., J. Am. Chem. Soc.* **72**, 3827 (1950); von Euw, Reichstein, *Helv. Chim. Acta* **33**, 1006 (1950); Taylor, *J. Chem. Soc.* **1952**, 4832. Structural studies: Taylor, *Chem. & Ind. (London)* **1953**, 62; Schindler, *Helv. Chim. Acta* **39**, 375 (1956); Taylor, *J. Chem. Soc. (C)* **1966**, 790. Revised structure: Fuhrer *et al., Helv. Chim. Acta* **52**, 616 (1969).

Crystallizes in various forms with different mps, but [according to Taylor, *J. Chem. Soc.* **1952**, 4832] on recrystn from methanol all of these forms yield the one melting at 232-234°. $[\alpha]_D^{15}$ +44.5° ± 2° (c = 1.1683 in methanol); $[\alpha]_D^{23}$ +49.5° (methanol). uv max: 218, 277.5 nm (log ε 4.2, 1.83). Sol in water, methanol, chloroform.

Diacetate, crystals, mp 254-256° (corr). $[\alpha]_D^{22}$ +24.2° (chloroform).

Dibenzoate, cryst powder from methanol, mp 189° (Taylor, *loc. cit.*); mp 187-191° (micro block, Rothrock *et al., loc. cit.*); $[\alpha]_D^{23}$ +30.7° ± 1° (c = 1.301 in acetone). Sol in acetone; slightly sol in methanol.

8528. Sassafras. Saxifrax; ague tree; cinnamon wood; saloop. Bark of root of *Sassafras albidum* (Nutt.) Nees [*S. variifolium* (Salisb.) Kuntze, *S. officinale* Nees & Eberm.], *Lauraceae. Habit.* North America. *Constit.* Root: ~2% volatile oil. Bark of root: 6-9% volatile oil; sassafrid, wax.

USE: Aromatic, flavoring agent.

THERAP CAT: Sudorific.

8529. Sassy Bark. Saucy bark; Mancona bark; ordeal bark; red-water tree bark; casca bark; Saxon bark, doom bark; teli, bondou. Bark of *Erythrophleum guineense* G. Don, *Leguminosae. Habit.* Central and Western Africa. *Constit.* Erythrophleine, tannin, resin.

8530. Satumomab. B72.3; CYT-099. Murine monoclonal antibody (MAb) of the IgG₁ subclass that targets a high molecular weight tumor-associated glycoprotein, *TAG-72,* expressed on the surface of certain human colon, breast and ovarian carcinoma cell lines. Prepn and characterization of MAb: D. Colcher *et al., Proc. Nat. Acad. Sci. USA* **78**, 3199 (1981); J. Schlom *et al.,* U.S. pat. **4,612,282** (1986 to USA Secy. Dept. Health Human Services). Pharmacokinetics of ^{111}In labeled MAb: B. A. Brown *et al., Cancer Res.* **47**, 1149 (1987). Clinical studies of immunoscintigraphy in colorectal cancer: B. D. Collier *et al., Radiology* **185**, 179 (1992); in ovarian cancer: E. A. Surwit *et al., Gynecol. Oncol.* **48**, 285 (1993). Review of prepn and diagnostic use: R. T. Maguire *et al., Cancer* **72**, 3453-3462 (1993). Book: *Diagnosis of Colorectal and Ovarian Carcinoma,* R. T. Maguire, D. Van Nostrand, Eds. (Marcel Dekker, Inc., New York, 1992) 237 pp.

Indium ^{111}In satumomab pendetide, ^{111}In-CYT-103, monoclonal antibody B72.3-GYK-DTPA-In-111, B72.3-glycyl-tyrosyl-N-ε-diethylenetriaminepentaacetic acid-In-111, OncoScint.

THERAP CAT: ^{111}In-labeled form as diagnostic aid (radioactive tumor-targeting agent).

8531. Saunders, Red. Red sandalwood; ruby wood. Heartwood of *Pterocarpus santalinus* L. f., *Leguminosae. Habit.* East Indies. *Constit.* Santalin (santalinic acid), santal pterocarpin, tannin.

USE: For coloring tinctures and similar medicinal prepns; formerly used as a dye.

8532. Savin. Young shoots of *Juniperus sabina* L., *Cupressaceae. Habit.* Europe, Northern Asia, North America south to New York, and Montana. *Constit.* 2-4% volatile oil, tannin, savinin, podophyllotoxin. Isoln and structure of *savinin:* Hartwell *et al., J. Am. Chem. Soc.* **75**, 235 (1953); Schrecker, Hartwell, *ibid.* **76**, 4896 (1954).

THERAP CAT: Emmenagogue. Anthelmintic. Antirheumatic.

8533. Saxitoxin. Mussel poison; clam poison; paralytic shellfish poison; gonyaulax toxin; STX. $[C_{10}H_{17}N_7O_4]^{2+}$. Powerful neurotoxin produced by the dinoflagellates *Gonyaulax catenella,* or *G. tamarensis,* the consumption of which causes the California sea mussel *Mytilus californianus,* the Alaskan butterclam *Saxidomus giganteus* and the scallop to become poisonous: Sommer *et al., Arch. Pathol.* **24**, 537, 560 (1937); Schantz *et al., Can. J. Chem.* **39**, 2117 (1961); Ghazarossian *et al., Biochem. Biophys. Res. Commun.* **59**, 1219 (1974). These poisonous shellfish have been connected to instances of toxic "red-tides" where the high concentrations of algae discoloring the water were of the *Gonyaulax* genus. Isoln and partial characterization: Schantz *et al., J. Am. Chem. Soc.* **79**, 5230 (1957); Mold *et al., ibid.* 5235. Purifn and biochemical data: Schantz *et al., Biochemistry* **5**, 1191 (1966); Schantz, *Ann. N.Y. Acad. Sci.* **90**, 843 (1960). pKa values: R. S. Rogers, H. Rapoport, *J. Am. Chem. Soc.* **102**, 7335 (1980). Toxicity: G. S. Wiberg, N. R. Stephenson, *Toxicol. Appl. Pharmacol.* **2**, 607 (1960). Structural studies: Schuett, Rapoport, *J. Am. Chem. Soc.* **84**, 2266 (1962); Wong *et al., ibid.* **93**, 4633, 7344 (1971). Revised structure: Schantz *et al., ibid.* **97**, 1238 (1975); Bordner *et al., ibid.* **97**, 6008 (1975). Pharmacology including study of sodium transport inhibition: Cheymol *et al., Arch. Int. Pharmacodyn. Ther.* **174**, 393 (1968); Evans, *Brit. Med. Bull.* **25**, 263 (1969). Stereospecific total synthesis of *dl*-saxitoxin: H. Tanino *et al., J. Am. Chem. Soc.* **99**, 2818 (1977); Y. Kishi, *Heterocycles* **14**, 1477 (1980). *Reviews:* Kao, *Pharmacol. Rev.* **18**, 997 (1966); Kao, *Fed. Proc., Fed. Am. Soc. Exp. Biol.* **31**, 1117 (1972); Narahashi, *ibid.* 1124, Evans, *Int. Rev. Neurobiol.* **15**, 83 (1972); Y. Shimizu in *Progress in the Chemistry of Natural Products* vol. **45**, W. Herz *et al.,* Eds. (Springer-Verlag, New York, 1984) pp 239-246. *See also* Brevetoxins.

Dihydrochloride, white, hygroscopic solid. pKa in water: 8.24, 11.60. $[\alpha]_D^{25}$ +130°. Very sol in water, methanol; sparingly sol in ethanol, glacial acetic acid. Practically insol in lipid solvents. Stable in acid solns; decomposes rapidly in alkaline media. Boiling 3-4 hrs at pH 3 causes loss of activity. LD₅₀ in mice (μg/kg): 10 i.p.; 263 orally; 3.4 i.v. (Wiberg, Stephenson).

USE: As a tool in neurochemical research.

8534. SBR Rubber. Styrene-butadiene rubber; GR-S; Government Rubber Styrene. General-purpose synthetic rubbers that were originally produced in government owned plants as GR-S. They are copolymers obtained by the polymerization of butadiene and styrene in a ratio of approx 3:1. The chains contain a random sequence of the two monomers. Very similar in composition to *Buna S, Kraton, Solprene, Stereon. Review:* P. Schneider *et al.,* in Ullmann, *Encyklopädie der technischen Chemie,* 3 Aufl., Bd. IX (München-Berlin, 1957) p 331; M. Morton, *Introduction to Rubber Technology* (Reinhold, 1959) pp 56 and 256-284; R. G. Bauer in Kirk-Othmer *Encyclopedia of Chemical Technology* vol. **8** (Wiley-Interscience, New York, 3rd ed., 1979) pp 608-625.

USE: In tires, adhesives, chewing gum.

8535. Scabiolide. *2,3-Dihydroxy-2-methylpropanoic acid 6-[(acetyloxy)methyl]-2,3,3a,4,7,8,11,11a-octahydro-10-*

methyl-3-methylene-2-oxocyclodeca[b]furan-4-yl ester. C_{21}-$H_{28}O_8$; mol wt 408.45. C 61.75%, H 6.91%, O 31.34%. A sesquiterpenic lactone, the bitter principle of *Centaurea scabiosa* (L.) Presl., *Compositae.* Isolation: Suchy *et al., Coll. Czech. Chem. Commun.* **27**, 1905 (1962). Structure: *eidem, ibid.* **27**, 2398 (1962). Revised structure: *eidem, ibid.* **30**, 3473 (1965). Further structural revisions: *eidem, ibid.* **33**, 2238 (1968).

Crystals, mp 120°. $[\alpha]_D^{20}$ +101.0° (c = 3.67 in chloroform). Slightly sol in ether, diisopropyl ether, acetone; practically insol in petr ether.

8536. Scammony Root. Root of *Convolvulus scammonia* L., *Convolvulaceae,* yielding not less than 8% resin. *Habit.* Eastern Mediterranean region, especially Syria. *Constit.* 8-13% resin; dihydroxycinnamic acid, β-methylesculetin, ipuranol, a reducing sugar, starch.
THERAP CAT: Cathartic.

8537. Scandium. Sc; at. wt 44.955910; at. no. 21; valence 3. Group IIIb(3). A rare earth metal. Naturally occurring isotope (mass number): 45; known artificial radioactive isotopes: 40-44, 46-51. Abundance in earth's crust: 5-25 ppm. Widely dispersed in nature. Occurs in the minerals thortveitite [(Sc,Y)$_2$Si$_2$O$_7$] and in other rare earth minerals such as davidite, ytterbite, orthite and cerrite; frequently associated with tin or zirconium. Predicted and called "ekaboron" by Mendeleev. Discovered by Nilson: *Ber.* **12**, 551, 554 (1879); **13**, 1430, 1439 (1880). Sepn from wolframite: Lukens, *J. Am. Chem. Soc.* **35**, 1470 (1913); on basis of solubility: Fischer, Bock, *Z. Anorg. Chem.* **249**, 146 (1942). Toxicity of the chloride: Haley *et al., J. Pharm. Sci.* **51**, 1043 (1962). Review of isolns including ion-exchange techniques: F. H. Spedding *et al., J. Electrochem. Soc.* **105**, 683-686 (1958). Review of prepn, properties and compds: R. C. Vickery, *The Chemistry of Yttrium and Scandium* (Pergamon Press, New York, 1960) 123 pp; *idem,* "Scandium, Yttrium and Lanthanum" in *Comprehensive Inorganic Chemistry* vol. **3**, J. C. Bailar, Jr. *et al.,* Eds. (Pergamon Press, Oxford, 1973) pp 329-353; T. Moeller, "The Lanthanides", *ibid.* vol. **4**, pp 1-101; F. H. Spedding in Kirk Othmer *Encyclopedia of Chemical Technology* vol. **19** (John Wiley & Sons, New York, 3rd ed., 1982) pp 833-854; *Chemistry of the Elements,* N. N. Greenwood, A. Earnshaw, Eds. (Pergamon Press, New York, 1984) pp 1102-1110, 1423-1449.

Metal. Reported to be dimorphic. Crystalline forms: hexagonal close-packed α-form, d 2.9890; evidence of existence of face-centered cubic β-form, d 3.19, is inconclusive. mp 1541°. bp 2836°. Heat of fusion: 14.10 kJ/mol. Heat of sublimation (25°): 377.8 kJ/mol. E°(aq) Sc³⁺/Sc −2.08 V (calc). Salts are hydrolyzed in aq soln.

Oxide, O$_3$Sc$_2$, *scandia,* fine white powder, d 3.864. Obtained by igniting the metal or its compds. Readily sol in hot or concd acids.

Hydroxide, Sc(OH)$_3$, white gelatinous precipitate forming a hard, horny mass on exposure to air; obtained by the action of alkalies on solns of the salts; dissolves readily in dil acids.

Chloride, ScCl$_3$, white deliquescent solid. Prepd by the action of a mixture of sulfur chloride and chlorine on the heated oxide; mp 960°; crystallizes with 6 mols of water. Sol in water. Practically insol in alc. LD$_{50}$ in mice: 755 mg/kg i.p.; 4 g/kg orally (Haley).

Sulfate, Sc$_2$(SO$_4$)$_3$, pentahydrate, d 2.519. Most sol of the sulfates of the rare earths (54.6 g/100 ml at 25°). Converted to the dihydrate on heating above 100°.

Nitrate, Sc(NO$_3$)$_3$, crystallizes as the tetrahydrate, prismatic deliquescent crystals, readily dissolves in water or alc.

8538. Scarlet Red. *1-[[2-Methyl-4-[(2-methylphenyl)-azo]phenyl]azo]-2-naphthalenol;* C.I. Solvent Red 24; *o*-tolylazo-*o*-tolylazo-β-naphthol; 1-(4-*o*-tolylazo-*o*-tolylazo)-2-naphthol; C.I. 26105; Biebrich Scarlet Red; Sudan IV; Fat Ponceau R. C$_{24}$H$_{20}$N$_4$O; mol wt 380.45. C 75.77%, H 5.30%, N 14.73%, O 4.21%. Prepd by diazotizing *o*-amino-azotoluene and coupling with β-naphthol: *Colour Index* vol. **4** (3rd ed., 1971) p 4227. Review of carcinogenicity studies: *IARC Monograph* **8**, 217 (1975).

Dark brown powder. Softens at 175°; mp 181-188°; dec completely at 260°. Practically insol in water. One gram dissolves in 15 ml chloroform. Sol in oils, fats, warm petrolatum, paraffin, phenol. Slightly sol in acetone, alcohol, benzene.
USE: Fat stain.
THERAP CAT: Has been used to promote wound healing.
THERAP CAT (VET): Has been used to stimulate healing of wounds.

8539. Schiff Bases. R-CH=N-C$_6$H$_5$. Condensation products of aldehydes and ketones with primary amines. The compounds are stable if there is at least one aryl group on the nitrogen or carbon. H. Schiff, *Ann.* Suppl. **3**, 343 (1864-1865). *Reviews:* N. V. Sidgewick, *Organic Chemistry of Nitrogen* (Clarendon Press, Oxford, 3rd ed., 1966) p 164-166; R. L. Reeves, *Chemistry of the Carbonyl Group,* S. Patai, Ed. (Wiley, New York, 1966) pp 608-614; J. Fastrez, *Ind. Chim. Belg.* **34**, 835 (1969); A. Bruylants, E. Feymantis-de Medicis, *Chemistry of the Carbon-Nitrogen Double Bond,* S. Patai, Ed. (Wiley, New York, 1970) pp 465-502. Metal complexes: S. Yamada, *Coord. Chem. Res.* **1**, 415 (1966); E. Sinn, C. M. Harris, *ibid.* **4**, 391 (1969); L. F. Lindsy, *Quart. Rev.* **25**, 379 (1971).

8540. Schradan. *Octamethyldiphosphoramide; octamethyl pyrophosphoramide;* bis[bisdimethylaminophosphonous] anhydride; bis-*N,N,N',N'*-tetramethylphosphorodiamidic anhydride; OMPA; Pestox III; Sytam. C$_8$H$_{24}$N$_4$-O$_3$P$_2$; mol wt 286.25. C 33.57%, H 8.45%, N 19.57%, O 16.77%, P 21.64%. Synthesis: Schrader, *B.I.O.S. Final Report no.* **1808** (1948); Gardiner, Kilby, *J. Chem. Soc.* **1950**, 1769; *Biochem. J.* **51**, 79 (1952); Hartley *et al., J. Sci. Food Agr.* **2**, 303, 310 (1951); **3**, 60 (1952); Toy, Costello, U.S. pat. **2,717,249** (1955 to Victor Chem.); Toy, Walsh, *Inorg. Syn.* **7**, 73 (1963). Has been found to inhibit peripheral cholinesterase without pronounced effects on the central nervous system: DuBois *et al., J. Pharmacol. Exp. Ther.* **99**, 376 (1950). Toxicity study: T. B. Gaines, *Toxicol. Appl. Pharmacol.* **14**, 515 (1969).

Viscous liquid, d$_4^{25}$ 1.09. bp$_{0.5}$ 120-125°. bp$_{2.0}$ 154°. mp 14-20°. n_D^{25} 1.462. Vapor pressure at 25°: 1 × 10⁻³ mm Hg. Misc with water. Sol in most organic solvents, including ketones, nitriles, esters, aromatic hydrocarbons and alcohols. Practically insol in higher aliphatic hydrocarbons. Hydrolyzed by acids. LD$_{50}$ in male, female rats (mg/kg): 9.1, 42 orally; 15, 44 dermally (Gaines).
USE: Insecticide.

8541. Schweizer's Reagent. *Tetraamminecopper dihydroxide.* Approximately [Cu(NH$_3$)$_4$](OH)$_2$. Usually applied in the form of an ammoniacal soln which is easily obtained by dissolving $Cu(OH)_2$ in 20% ammonia water. The soln is an intense azure color. Evaporation yields long, blue needles: Schweizer, *J. Prakt. Chem.* **72,** 109, 344 (1857).

USE: Dissolves cellulose; used in rayon production.

8542. Scillabiose. *6-Deoxy-4-O-β-D-glucopyranosyl-L-mannose;* 4-O-D-glucopyranosyl-L-rhamnose; glucosido-rhamnose. $C_{12}H_{22}O_{10}$; mol wt 326.30. C 44.17%, H 6.80%, O 49.03%. Obtained by acid hydrolysis of scillaren A: Stoll *et al., Helv. Chim. Acta* **16,** 703 (1933); Zemplén, *C.A.* **33,** 4202 (1939); from glucoscilliphäosid: Stoll *et al., Helv. Chim. Acta* **35,** 2495 (1952). Structure: Stoll, Renz, *Enzymologia* **7,** 362 (1939), *C.A.* **34,** 6298³ (1940).

Syrup. $[\alpha]_D^{20}$ −24.8° (0.83 g in 25 ml). Soluble in water, pyridine.
Phenylosazone, $C_{24}H_{32}N_4O_8$, crysts from dil alc, mp 165°.
Hexacetylscillabiose, $C_{24}H_{34}O_{16}$, needles from alc, mp 97°.

8543. Scillaren. A mixture of glycosides, scillaren A and B in the proportions in which they occur in fresh squill, *Urginea (Scilla) maritima* (L.) Baker, *Liliaceae,* about 2 parts of A to 1 part of B. Isolation of scillaren and separation of A and B: Stoll *et al., Helv. Chim. Acta* **16,** 703 (1933).

scillabiose—O

Scillaren A

Granular, very bitter powder. On drying in high vacuum at 78° for 15 hrs it loses not more than 6% of its wt. $[\alpha]_D^{20}$ −25° to −35° (0.5 g in 25 ml of 75% w/w alcohol). One gram dissolves in 3000 ml water, in 5 ml abs alcohol, in 5 ml methanol; practically insol in ether and chloroform. Aq solns are neutral to litmus. LD i.v. in cats: 0.18-0.62 mg/kg; LD orally in rabbits: 0.9 mg/kg; LD s.c. in rats: 10 mg/kg, *Handbook of Toxicology,* W. S. Spector, Ed. (Saunders, Philadelphia, 1956) p 266.

Scillaren A, $C_{36}H_{52}O_{13}$, *glucoproscillaridin A, transvaalin.* mol wt 692.80. C 62.41%, H 7.57%, O 30.02%. Characterization and structure: Stoll *et al., Z. Physiol. Chem.* **222,** 24 (1933); *Helv. Chim. Acta* **17,** 641, 1334 (1934); **18,** 82, 120, 401, 644, 1247 (1935); **24,** 1380 (1941); Zoller, Tamm, *ibid.* **36,** 1744 (1953). Very bitter taste. Two crystal modifications from methanol: prisms, mp 184-186°; leaflets, mp 208-211°. $[\alpha]_D^{23}$ −71.9° (c = 1.011 in methanol) (Zoller, Tamm, *loc. cit.*). Tendency to form solvated crystals. Six-sided plates, tablets contg 1 mol CH_3OH + 1 mol H_2O from dil methanol, mp 230-240°. Anhydr from 85% alcohol, solvent-free: mp 270°; $[\alpha]_D^{20}$ −72° to −78° (c = 1 to 3 in 75% alcohol), $[\alpha]_D^{20}$ −73.8° (Stoll, *primo loc. cit.*). Soluble in 350 parts alcohol, in 80 parts methanol, in 40 parts dil alcohol (4 vols ethanol + 1 vol water). Practically insol in chloroform, ether. Sparingly sol in water. LD i.v. in cats: 0.143 mg/kg.

Scillaren B. A water-sol mixture of glucosides remaining after extraction of scillaren A. Amorphous, granular powder. Very bitter taste. Scillaren B dried in a high vacuum at 78°C for 15 hrs loses not more than 5% of its weight. Completely dried scillaren B contains approx 99.5% active glyco-

sidal substance. $[\alpha]_D^{20}$ +35° to 41° (0.5 g in 25 ml of 75% w/w ethanol). Freely sol in water. Soluble in alcohol, methanol (about 1 in 5); very slightly sol in chloroform (about 1 in 10,000). Practically insol in ether. Aq solns are neutral to litmus.

THERAP CAT: Cardiotonic.

8544. Scillarenin. *(3β)-3,14-Dihydroxybufa-4,20,22-trienolide.* $C_{24}H_{32}O_4$; mol wt 384.52. C 74.97%, H 8.39%, O 16.64%. Prepn by adaptive enzymic decompn of proscillaridin A: Stoll *et al., Helv. Chim. Acta* **34,** 2301 (1951); from *Urginea burkei* Baker, *Liliaceae:* Zoller, Tamm, *ibid.* **36,** 1744 (1953). Structure: Stoll *et al., ibid.* **35,** 1934 (1952). Toxicity study: Chen, Henderson, *J. Pharmacol. Exp. Ther.* **111,** 365 (1954).

Prisms from methanol, mp 232-238°. $[\alpha]_D^{20}$ −16.8° (c = 0.357 in methanol); $[\alpha]_D^{20}$ +17.9° (c = 0.39 in chloroform). uv max: 300 nm (log ε 3.72). Mean LD i.v. in cats: 0.1567 mg/kg (Chen, Henderson).
3-Acetate, $C_{26}H_{34}O_5$, crystals from methanol, dec 240-243°. $[\alpha]_D^{20}$ −23.4° (c = 1.365 in chloroform).

THERAP CAT: Cardiotonic.

8545. Scilliroside. *(3β,6β)-6-(Acetyloxy)-3-(β-D-glucopyranosyloxy)-8,14-dihydroxybufa-4,20,22-trienolide.* $C_{32}H_{44}O_{12}$; mol wt 620.69. C 61.92%, H 7.15%, O 30.93%. Glycoside from red squill, the red variety of *Urginea maritima* (L.) Baker, *Liliaceae:* Stoll, Renz, *Helv. Chim. Acta* **25,** 43 (1942); Wichtt, Fuchs, *Arch. Pharm.* **295,** 361 (1962). Structure: Stoll *et al., Helv. Chim. Acta* **26,** 648 (1943); von Wartburg, Renz, *ibid.* **42,** 1620 (1959). Toxicity data: F. Dybing *et al., Acta Pharmacol. Toxicol.* **8,** 391 (1952).

β-D-glucose—O

Long prisms from dilute methanol. The crystals are solvated and lose about 8% of their wt in high vacuum, but retain ½ mol H_2O. For the hemihydrate, indistinct mp 168-170°, dec 200°. $[\alpha]_D^{20}$ −59° to −60° (methanol). Absorption max (96% ethanol): 300 nm (log ε 3.73); (98% H_2SO_4): 295, 505 nm ($E_{1cm}^{1\%}$ 260, 315). Freely sol in the lower alcohols, in ethylene glycol, dioxane, glacial acetic acid. Slightly sol in water, acetone, chloroform, ethyl acetate. Practically insol in ether, petr ether. LD_{50} in mice (mg/kg): 0.471 s.c.; 0.440 orally (Dybing).

Tetraacetyl scilliroside, $C_{40}H_{52}O_{16}$, rosettes of long needles from methanol, mp 199°. Sol in chloroform. The four additional acetyl groups are on the sugar.

8546. Scoparin. *8-β-D-Glucopyranosyl-5,7-dihydroxy-2-(4-hydroxy-3-methoxyphenyl)-4H-1-benzopyran-4-one;* 8-glycosyl-4',5,7-trihydroxy-3'-methoxyflavone; scoparoside. $C_{22}H_{22}O_{11}$; mol wt 462.41. C 57.14%, H 4.80%, O 38.06%. From leaves and branches of *Spartium scoparium* L., *Leguminosae:* Stenhouse, *Ann.* **78**, 15 (1851); Mascré, Paris, *Compt. Rend.* **204**, 1270 (1937). Structural study: Hörhammer *et al.*, *Naturwiss.* **49**, 393 (1962). Revised structure: Prox, *Tetrahedron* **24**, 3697 (1968).

Needles from 80% methanol, mp 253°. uv max (methanol): 345, 270 nm (log ε 4.27, 4.18). Practically insol in cold water, ether, chloroform, benzene. Sol in hot water, ethanol, methanol, acetic acid, ethyl acetate, acetone, pyridine.
Heptaacetate, $C_{36}H_{36}O_{18}$, crystals, mp 240-241°.

8547. Scoparius. Broom; green broom; Scotch broom; Irish broom; hogweed; bannal. Dried tops of *Cytisus scoparius* (L.) Link, *Leguminosae.* The seeds and flowers are also used, though not officially. *Habit.* Western Asia, Southern and Western Europe; cultivated in U.S.A., particularly in Oklahoma. *Constit.* Sparteine, scoparin, genisteine, sarothamnine, volatile oil, tannin, fat, wax, sugar.
THERAP CAT: Diuretic, cathartic.

8548. Scoparone. *6,7-Dimethoxy-2H-1-benzopyran-2-one; 6,7-dimethoxycoumarin;* esculetin dimethyl ether. $C_{11}H_{10}O_4$; mol wt 206.20. C 64.07%, H 4.89%, O 31.04%. Isolation from *Zanthoxylum* and *Artemisia* spp: Araki, Miyashita, *J. Pharm. Soc. Japan* **48**, 437 (1928); Sera, Shibuye, *C.A.* **24**, 5742 (1930); Parihar, Dutt, *Proc. Indian Acad. Sci.* **25A**, 153 (1947); King *et al.*, *J. Chem. Soc.* **1954**, 1392; Singh *et al.*, *J. Sci. Ind. Res. (India)* **15B**, 190 (1956); Ban'kovskaya, Ban'kovskaya, *C.A.* **55**, 17776c (1961). Structure and synthesis: Singh *et al.*, *Chem. & Ind. (London)* **1954**, 1294. Improved synthesis: R. S. Mali *et al.*, *Indian J. Chem.* **21B**, 759 (1982).

Crystals, mp 145-146°. Absorption spectra: Cingolani, *Gazz. Chim. Ital.* **89**, 985, 999 (1959).

8549. Scopola. Dried rhizome of *Scopola carniolica* Jacq., *Solanaceae. Habit.* Germany (Bavaria), Hungary, Russia. *Constit.* Scopolamine (hyoscine), atropine, hyoscyamine; total about 0.7% alkaloids.
THERAP CAT: Anticholinergic.

8550. Scopolamine. *[7(S)-(1α,2β,4β,5α,7β)]-α-(Hydroxymethyl)benzeneacetic acid 9-methyl-3-oxa-9-azatricyclo[3.3.1.0²,⁴]non-7-yl ester; 6β,7β-epoxy-1αH,5αH-tropan-3α-ol (−)-tropate; 6β,7β-epoxy-3α-tropanyl S-(−)-tropate;* 6,7-epoxytropine tropate; scopine tropate; tropic acid ester with scopine; hyoscine; *l*-scopolamine; Scop; Scopoderm TTS; Transcop; Transderm Scop. $C_{17}H_{21}NO_4$; mol wt 303.36. C 67.31%, H 6.98%, N 4.62%, O 21.10%. Tropane alkaloid isolated from *Datura metel* L., *Scopola carniolica* Jacq. and other *Solanaceae.* Constituent of impure duboisine from *Duboisia myoporoides* R. Br., pure duboisine is *l*-hyoscyamine, *q.v.* Isoln: A. Ladenburg, *Ann.* **206**, 274 (1881); E. Schmidt, *Arch. Pharm.* **230**, 207 (1892). Identity with hyoscine: O. Hesse, *Ann.* **271**, 100 (1892); *idem, J. Prakt. Chem.*

66, 194 (1902). Absorption spectra: J. J. Dobbie, J. J. Fox, *J. Chem. Soc.* **103**, 1194 (1913). Resolution of isomers and review of early literature: H. King, *ibid.* **1919**, 476. Extraction procedure: F. Chemnitius, *J. Prakt. Chem.* **120**, 221 (1928). Structural studies: J. Gadamer, F. Hammer, *Arch. Pharm.* **259**, 122 (1921); K. Hess, O. Wahl, *Ber.* **55**, 1979 (1922); R. Willstätter, E. Berner, *ibid.* **56**, 1079 (1923); W. Steffens, *Arch. Pharm.* **262**, 205 (1924). Configuration: G. Fodor, *Nature* **170**, 278 (1952); J. Meinwald, *J. Chem. Soc.* **1953**, 712. Review of stereochemistry: G. Fodor, *Tetrahedron* **1**, 86 (1957). Absolute configuration of tropic acid moiety: G. Fodor, G. Csepreghy, *Tetrahedron Letters* **1959**, 16; *eidem, J. Chem. Soc.* **1961**, 3222. Synthesis of DL-form: G. Fodor *et al.*, *Chem. & Ind.* **1956**, 764; P. Dobo *et al.*, *J. Chem. Soc.* **1959**, 3461. Clinical evaluation in motion sickness: J. J. Brand, P. Whittingham, *Lancet* **2**, 232 (1970); in peripheral vertigo: T. Rahko, P. Karma, *J. Laryngol. Otol.* **99**, 653 (1985). Acute toxicity of the hydrobromide: Stockhaus, Wick, *Arch. Int. Pharmacodyn. Ther.* **180**, 155 (1969). Review of CNS effects in humans: D. J. Safer, R. P. Allen, *Biol. Psychiat.* **3**, 347-355 (1971); of use in anesthesia: L. E. Shutt, J. B. Bowles, *Anaesthesia* **34**, 476-490 (1979); of pharmacology and clinical efficacy: S. P. Clissold, R. C. Heel, *Drugs* **29**, 189-207 (1985). Comprehensive description: F. J. Muhtadi, M. M. A. Hassan in *Analytical Profiles of Drug Substances* vol. 19, K. Florey, Ed. (Academic Press, New York, 1990) pp 477-551.

Viscous liquid. $[α]_D^{20}$ −28° (c = 2.7). Sol in 9.5 parts water at 15°. Forms a cryst monohydrate, mp 59°. Freely sol in hot water, in alcohol, ether, chloroform, acetone. Sparingly sol in benzene, petr ether. Easily hydrolyzed by acids or alkalies. Dec on standing.
Hydrobromide trihydrate, $C_{17}H_{21}NO_4 \cdot HBr \cdot 3H_2O$, *scopolammonium bromide, Scopos.* Orthorhombic sphenoidal crystals from water, slightly efflorescent in dry air. mp 195° (after drying at 105° for 3 hours). $[α]_D^{25}$ −24° to −26° (c = 5, calculated on anhydrous basis). uv max (methanol) 246, 252, 258, 264 nm ($A_{1cm}^{1\%}$ 3.5, 4.0, 4.5, 3.0). pH of 0.05M soln 5.85. One gram dissolves in 1.5 ml water, 20 ml alcohol. Slightly sol in chloroform. Practically insol in ether. LD_{50} in rats (mg/kg): 3800 s.c. (Stockhaus, Wick).
Hydrochloride, $C_{17}H_{21}NO_4 \cdot HCl$, crystals from acetone, mp 200°. Dihydrate, prisms from water, melts in water of crystn at 80°. Very sol in water and alcohol. pH of 0.05M soln 5.85.
Methyl bromide, *See* Methscopolamine Bromide.
Methyl nitrate, $C_{17}H_{21}NO_4 \cdot CH_3NO_3$, *methscopolamine nitrate, Skopyl.* Crystals. Freely sol in water, dil alcohol; slightly sol in abs alcohol.
DL-Form, *atroscine.* Dihydrate, chisel-shaped prisms from ethanol + water, mp 38-40°. Monohydrate, efflorescent crystals, mp 55-57°. Anhydrous, long prisms, mp 82-83°. Very slightly sol in water; sol in alc, chloroform, ether, oils.
THERAP CAT: Anticholinergic. Treatment of motion sickness.
THERAP CAT (VET): Has been used as a sedative, a preanesthetic and to control motion sickness.

8551. Scopolamine N-Oxide. *[7(S)-(1α,2β,4β,5α,7β)]-α-(Hydroxymethyl)benzeneacetic acid 9-methyl-3-oxa-9-azatricyclo[3.3.1.0²,⁴]non-7-yl ester N-oxide; 6β,7β-epoxy-1αH,5αH-tropan-3α-ol (−)-tropate 8-oxide;* scopolamine aminoxide. $C_{17}H_{21}NO_5$; mol wt 319.36. C 63.94%, H 6.63%, N 4.39%, O 25.05%. Prepn: Polonovski, Polonovski, *Compt. Rend.* **180**, 1755 (1925). Absolute configuration: Huber *et al.*, *Can. J. Chem.* **49**, 3258 (1971).

White powder, mp ~80°. $[\alpha]_D^{20}$ −14° in water. Hydrobromide, $C_{17}H_{22}BrNO_5$, **Genoscopolamine.** Prisms from water, mp 135-138°. $[\alpha]_D^{25}$ −25° (c = 2). Soly in water about 10 g/100 ml. Slightly sol in alcohol, acetone. pH of 3% aq soln about 3.2.

THERAP CAT: Anticholinergic.

8552. Scopoletin. *7-Hydroxy-6-methoxy-2H-1-benzopyran-2-one; 7-hydroxy-6-methoxycoumarin;* 6-methoxyumbelliferone; *β*-methylesculetin; chrysatropic acid; gelseminic acid. $C_{10}H_8O_4$; mol wt 192.17. C 62.50%, H 4.20%, O 33.30%. The aglucone of scopolin. Occurs in root of *Scopolia japonica* Maxim., *Scopolia carniolica* Jacq., *Atropa belladonna* L., *Solanaceae*, *Convolvulus scammonia* L., *Convolvulaceae.* Isoln: Eykman, *Ber.* **17** III, 442 (1884). Synthesis: Crosby, *J. Org. Chem.* **26**, 1215 (1961); Desai, Desai, *J. Indian Chem. Soc.* **40**, 456 (1963).

Needles or prisms from chloroform or acetic acid, mp 204°. uv max: 230, 254, 260, 298, 346 nm (log ε 4.11, 3.68, 3.63, 3.68, 4.07), Ballantyne *et al.*, *Tetrahedron* **27**, 871 (1971). Slightly sol in water or cold alcohol; sol in hot alcohol or hot glacial acetic acid; moderately sol in chloroform. Practically insol in benzene. Its alcoholic soln has a blue fluorescence. Reduces Fehling's soln.

8553. Scopolin. *7-(β-D-Glucopyranosyloxy)-6-methoxy-2H-1-benzopyran-2-one.* $C_{16}H_{18}O_9$; mol wt 354.31. C 54.24%, H 5.12%, O 40.64%. Mono-β-glucopyranoside of scopoletin. From root of *Scopolia japonica* Maxim., *S. carniolica* Jacq., *Solanaceae* and *Nerium odorum* Sol., *Apocynaceae.* Isoln: Eykman, *Ber.* **17** III, 442 (1884); Ritter *et al.*, *Helv. Chim. Acta* **36**, 434 (1953). Synthesis: Merz, *Arch. Pharm.* **270**, 476 (1932). Possible identity with *murrayin*: Bose, Mookerjee, *J. Indian Chem. Soc.* **14**, 489 (1937). Chromatography and spectrum of murrayin: Chakraborty, Bose, *ibid.* **33**, 905 (1956); Chakraborty, Chakraborty, *Trans. Bose Res. Inst. (Calcutta)* **24** (1), 15 (1961), *C.A.* **55**, 24228c (1961).

Needles. mp 218°. Soluble in water, alcohol. Practically insol in chloroform, ether.

Tetraacetate, $C_{24}H_{26}O_{13}$, polymorphous crystals, prisms, mp 166° or flat plates, mp 184-185°.

8554. Scopoline. *Hexahydro-4-methyl-2,5-methano-2H-furo[3,2-b]pyrrol-6-ol; 3β,7β-epoxy-1βH,5βH-tropan-6α-ol;* 3α,6α-epoxy-7β-hydroxytropane; Oscine. $C_8H_{13}NO_2$; mol wt 155.20. C 61.91%, H 8.44%, N 9.03%, O 20.62%. Decomposition product of scopolamine: Fodor, Kovács, *J. Chem. Soc.* **1953**, 2341. Prepn from teloidine: Zeile, Heusner, *Ber.* **90**, 2800, 2809 (1957); **Ger.** pats. **1,056,617; 1,062,-703** (1959 to Boehringer, Ing.). Structure: Fodor *et al.*, *J. Chem. Soc.* **1955**, 3504. Stereochemistry: Fodor, *Tetrahedron* **1**, 86 (1957).

Crystals from petr ether, mp 107-109°. bp 248°. Sol in water, alcohol, acetone, ether.

Hydrochloride, $C_8H_{13}NO_2 \cdot HCl$, crystals from ethanol, mp 255-257°.

Acetate, $C_{10}H_{15}NO_3$, crystals from petr ether, mp 56-57°.

8555. Scotophobin. L-Seryl-L-aspartyl-L-asparaginyl-L-asparaginyl-L-glutaminyl-L-glutaminylglycyl-L-lysyl-L-seryl-L-alanyl-L-glutaminyl-L-glutaminylglycylglycyl-L-tyrosinamide. $C_{62}H_{97}N_{23}O_{26}$; mol wt 1580.59. C 47.11%, H 6.19%, N 20.38%, O 26.32%. Pentadecapeptide, isolated from the brains of rats with acquired fear of the dark, which induces dark avoidance in untrained mice. (From the Greek *skotos*, dark and *phobos*, fear.) Thought to be the first isolation, identification and synthesis of one of the "chemical code words" that control memory and learning. *See* Ungar *et al.*, *Nature* **238**, 198 (1972). Original training, testing and extraction procedures: Ungar *et al.*, *ibid.* **217**, 1259 (1968); *eidem*, *Proc. West. Pharmacol. Soc.* **13**, 149 (1970). Structural elucidation: Desiderio *et al.*, *Chem. Commun.* **1971**, 432. Synthesis of preliminary structures: Ali *et al.*, *Experientia* **27**, 1138 (1971); *eidem*, *Int. J. Peptide Protein Res.* **4**, 395 (1972). Solid phase synthesis of revised structure: Parr, Holzer, *Z. Physiol. Chem.* **352**, 1043 (1971); *eidem*, *Experientia* **28**, 884 (1972). Passive transfer of dark avoidance to mice: Malin, Guttman, *Science* **178**, 1219 (1972); to goldfish: Guttman *et al.*, *Nature New Biol.* **235**, 126 (1972); Bryant *et al.*, *Science* **177**, 635 (1972). Chemistry and critical analysis of published data on scotophobin: Stewart, *Nature* **238**, 202-210 (1972).

Ser–Asp–Asn–Asn–Gln–Gln–Gly–Lys–Ser–Ala–Gln–Gln–Gly–Gly–TyrNH₂

8556. Scutellarein. *5,6,7-Trihydroxy-2-(4-hydroxyphenyl)-4H-1-benzopyran-4-one; 4′,5,6,7-tetrahydroxyflavone;* 6-hydroxypelargidenon 1465. $C_{15}H_{10}O_6$; mol wt 286.24. C 62.94%, H 3.52%, O 33.54%. By hydrolysis of scutellarin: Molisck, Goldschmidt, *Monatsh.* **22**, 679 (1901); Marsh, *Biochem. J.* **59**, 58 (1955). From *Digitalis lanata* Ehrh., *Scrophulariaceae*: Rangaswami, Rao, *Proc. Indian Acad. Sci.* **54A**, 51 (1961), *C.A.* **56**, 10076f (1962). Structure: Wessely, Moser, *Monatsh.* **56**, 97 (1930). Synthesis: Sastri, Seshadri, *Proc. Indian Acad. Sci.* **23A**, 262 (1946), *C.A.* **41**, 449a (1947); Zemplén *et al.*, *Acta Chim. Acad. Sci. Hung.* **16**, 445 (1958); Jouanne, Mentzer, *Compt. Rend. Ser. C* **263**, 1022 (1966).

Yellow leaflets from methanol, does not melt below 300°. uv max (ethanol): 286, 339 nm (ε 16600; 18300).

Tetraacetate, $C_{23}H_{18}O_{10}$, prisms from ethyl acetate, mp 235-237°.

Glucuronide, $C_{21}H_{18}O_{12}$, *scutellarin.* From leaves of *Scutellaria altissima* Linn., *Labiatae*: Goldschmiedt, Zerner, *Monatsh.* **31**, 439 (1910); from *Centaurea scabiosa* L., *Compositae*: Charaux, Rabaté, *J. Pharm. Chim.* **9**, 155 (1940); from *Scutellaria* spp.: Marsh, *Biochem. J.* **59**, 58 (1955); *Nature* **183**, 1824 (1959). Synthesis: Farkas *et al.*, *Ber.* **107**, 3878 (1974). Needles from alc, darkens above 230°, does not melt below 300°. $[\alpha]_D^{18}$ −14° (water); $[\alpha]_D^{20}$ −139° (pyridine). uv max (ethanol): 285, 335 nm (ε 20000; 26100).

Practically insol in water. Sol in alkali hydroxides, glacial acetic acid; slightly sol in organic solvents.

8557. Scutellaria. Skullcap; helmet flower. Dried overground portion of *Scutellaria lateriflora* L., *Labiatae. Habit.* North America. *Constit.* Scutellarin, volatile oil, tannin.

8558. Sebacic Acid. *Decanedioic acid;* 1,8-octanedicarboxylic acid. $C_{10}H_{18}O_4$; mol wt 202.25. C 59.39%, H 8.97%, O 31.64%. $HOOC(CH_2)_8COOH$. Obtained on a large scale by heating castor oil or ricinoleic acid with sodium hydroxide. Convenient lab procedures: Topchiev, Pavlov, *C.A.* **47**, 8002h (1953); Dominguez *et al., J. Chem. Ed.* **29**, 446 (1952). Also prepd by ozonization of undecylenic acid: Noorduyn, *Rec. Trav. Chim.* **38**, 326 (1919); Verkade *et al., ibid.* 385. Toxicity study: P. M. Jenner *et al., Food Cosmet. Toxicol.* **2**, 327 (1964). Review and bibliography: Jones, *Chimia* **5**, 169-173 (1951).

Monoclinic prismatic tablets, leaflets from acetone + petr ether (commercial product is a white, free-flowing powder having a mild fatty acid odor). d_4^{20} 1.207. mp 134.5°. bp_{100} 294.5°; bp_{50} 273°; bp_{15} 243.5°; bp_{10} 232°. Sp heat (solid) 0.48 cal/g; sp heat (liq) 0.53 cal/g. Sublimes slowly at 750 mm when heated to mp. n_D^{34} 1.422. pK_1 = 4.59; pK_2 = 5.59. Aq solns are neutral to methyl orange. Soly in water: 0.004% at 0°; 0.10% at 20°; 0.42% at 65°; 2.0% at 100°. Freely sol in alcohols, esters, ketones. Soly in ether: 0.1% at 17°. Sparingly sol in hydrocarbons and chlorinated hydrocarbons.

Ethyl ester, $C_{14}H_{26}O_4$, *ethyl sebacate, diethyl sebacate.* Colorless to yellowish fluid. d_4^{20} 0.965. bp ~307° with some decompn. n_D^{20} 1.4369. Sol in about 700 parts cold, 50 parts boiling water; misc with alc, ether and other organic solvents. LD_{50} orally in rats: 14470 mg/kg (Jenner).

USE: Raw material in the manuf of synthetic resins of the alkyd or polyester type, non-migrating plasticizers, polyester rubbers, synthetic fibers of the polyamide type.

8559. Sebacil. *1,2-Cyclodecanedione.* $C_{10}H_{16}O_2$; mol wt 168.24. C 71.39%, H 9.59%, O 19.02%. Prepd by oxidation of sebacoin with chromium trioxide in acetic acid: Blomquist *et al., J. Am. Chem. Soc.* **74**, 3640 (1952); *cf.* Prelog *et al., Helv. Chim. Acta* **35**, 1610 (1952); Blomquist, Goldstein, *Org. Syn.* coll. **vol. IV**, 838 (1963).

Crystals, mp 40-41°. bp_{18} 120-125°; bp_{10} 104-105°.

8560. Sebacoin. *2-Hydroxycyclodecanone;* 1-cyclodecanol-2-one. $C_{10}H_{18}O_2$; mol wt 170.25. C 70.55%, H 10.66%, O 18.79%. Prepd by cyclization of methyl or ethyl sebacate with sodium metal: Hansley, U.S. pat. **2,228,268** (1941 to du Pont); Prelog *et al., Helv. Chim. Acta* **35**, 1598 (1952); Allinger, *Org. Syn.* coll. **vol. IV**, 840 (1963).

Prisms from ether + petr ether, mp 42°. bp_6 113-114°; $bp_{0.1}$ 77°.

8561. Secalonic Acids. $C_{32}H_{30}O_{14}$; mol wt 638.58. C 60.19%, H 4.74%, O 35.08%. Toxic mold metabolites which are diastereoisomeric dimeric hydroxanthone derivatives manifest as yellow pigments. First isolation studies: Stoll *et al., Helv. Chim. Acta* **35**, 2022 (1952). Seven isomers have been isolated: *secalonic acid A, secalonic acid B,* and *secalonic acid C* from ergot: Aberhart *et al., Tetrahedron* **21**, 1417 (1965); Franck *et al., Ber.* **99**, 3842, 3875 (1966); *secalonic acid D* from *Penicillium oxalicum:* Steyn, *Tetrahedron* **26**, 51 (1970); *secalonic acid E* from *Phoma terrestris:* Han-

sen, Howard *et al., Chem. Commun.* **1973**, 464; *secalonic acid F* from *Aspergillus aculeatus:* R. Andersen *et al., J. Org. Chem.* **42**, 352 (1977); *secalonic acid G* from *Pyrenochaeta terrestris:* I. Kurobane, L. C. Vining, *Tetrahedron Letters* **1978**, 4633. Secalonic acid A is enantiomeric with D; secalonic acid B is enantiomeric with E. Chromatographic separation of the ergot pigments: Franck, *Angew. Chem. Int. Ed.* **8**, 251 (1969); structures: ApSimon *et al., J. Chem. Soc.* **1965**, 4144. Revised structures of secalonic acids A, B, C, D: Hooper *et al., Chem. Commun.* **1971**, 111; *eidem, J. Chem. Soc. (C)* **1971**, 3580. Crystal and molecular structure of secalonic acid A: C. C. Howard *et al., J. Chem. Soc. Perkin Trans. I* **1976**, 1820. Identity of secalonic acid A with *entothein*: Yoshioka *et al., Chem. Pharm. Bull.* **16**, 2090 (1968). Biosynthesis of secalonic acid A: I. Kurobane *et al., Tetrahedron Letters* **1978**, 1379. Review of secalonic acids and other ergochromes: Franck, Flasch in *Fortschr. Chem. Org. Naturst.* **30**, 151-206 (1973).

Secalonic acid A

Secalonic acid A, *[3R-[3α,4β,4aβ,7(3'R*,4'S*,4'aS*)]]-2,- 2'3,3',4,4',9,9'-octahydro-1,1',4,4',8,8'-hexahydroxy-3,3'- dimethyl-9,9'-dioxo-[7,7'-bi-4aH-xanthene]-4a,4'a-dicarboxylic acid dimethyl ester; 6,6',7,7'-tetrahydro-1,1',5β,5'β,8,8'- hexahydroxy-6α,6'α-dimethyl-9,9'-dioxo-[2,2'-bixanthene]- 10aβ,10'aβ(5H,5'H)-dicarboxylic acid dimethyl ester; ergochrome AA(2,2').* Fine lemon-yellow needles from dioxane + petr ether + chloroform, mp 246-248° (dec); also reported as mp 260° (Yoshioka *et al., loc. cit.*). $[α]_D$ −73° (CHCl₃). uv max (methanol): 247, 340 nm (ε 18140, 31200). pK_{DMF} 8.7. Readily soluble in pyridine, dimethylformamide, dioxane; moderately sol in chloroform, methylene chloride, acetone. Gives a wine-red color reaction with 5% $FeCl_3$ in methanol.

8562. Secnidazole. *α,2-Dimethyl-5-nitro-1H-imidazole-1-ethanol;* 1-(2-hydroxypropyl)-2-methyl-5-nitroimidazole; 1-(2-methyl-5-nitroimidazol-1-yl)-2-propanol; PM-185184; RP-14539; Flagentyl. $C_7H_{11}N_3O_3$; mol wt 185.18. C 45.40%, H 5.99%, N 22.69%, O 25.92%. Analog of metronidazole, *q.v.* Prepn: Fr. pat. **M3270** (1965 to Rhône-Poulenc), *C.A.* **63**, 11571d (1965); C. Cosar *et al., Arzneimittel-Forsch.* **16**, 23 (1966). Anti-amebic and trichomonacidal activities: F. Benazet, L. Guillaume, *Bull. Soc. Pathol. Exot. Ses Fil.* **69**, 309 (1976), *C.A.* **90**, 145922v (1979). Serum half-life: J. Symonds, *J. Antimicrob. Chemother.* **5**, 484 (1979). Therapeutic use: D. Videau *et al., Brit. J. Vener. Dis.* **54**, 77 (1978).

Cryst from toluene, mp 76° (Cosar).
THERAP CAT: Antiamebic. Antiprotozoal (Trichomonas).

8563. Secobarbital Sodium. *5-(1-Methylbutyl)-5-(2-propenyl)-2,4,6(1H,3H,5H)-pyrimidinetrione monosodium salt; sodium 5-allyl-5-(1-methylbutyl)barbiturate;* 5-allyl-5-(1-methylbutyl)barbituric acid sodium salt; 5-allyl-5-(1-methylbutyl)malonylurea sodium salt; meballymal sodium; quinalbarbitone sodium; Barbosec; Immenoctal; Pramil; Quinalspan; Sedutain; Seconal Sodium; Seotalnatrium. C_{12}-

$H_{17}N_2NaO_3$; mol wt 260.27. C 55.38%, H 6.58%, N 10.76%, Na 8.83%, O 18.44%. Prepn: U.S. pat. 1,954,429 (1934). Comprehensive description: I. Comer, *Anal. Profiles Drug Subs.* 1, 343-365 (1972). Toxicity data: E. W. Schafer, *Toxicol. Appl. Pharmacol.* 21, 315 (1972).

White, hygroscopic powder. Bitter taste. Very sol in water. Sol in alc. Practically insol in ether. Aq solns are alkaline to litmus. The soly in elixirs is increased by the presence of methenamine. LD_{50} orally in rats: 125 mg/kg (Schafer).

Free acid, $C_{12}H_{18}N_2O_3$, mp 100°.

Caution: May be habit forming. This is a controlled substance (depressant) listed in the U.S. Code of Federal Regulations, Title 21 Parts 329.1, 1308.12 and 1308.13 (1995).

THERAP CAT: Sedative; hypnotic.

THERAP CAT (VET): Short-acting hypnotic; sedative. Preanesthetic.

8564. Secretin. Secretine; Sekretolin. A strongly basic polypeptide gastrointestinal hormone, produced in the duodenum and jejunum of many animal species and man. Its primary biological activity is stimulation of exocrine pancreatic secretion. Discovered by W. M. Bayliss and E. H. Starling, *Ergeb. Physiol.* 5, 664 (1905). Isoln of porcine secretin: J. E. Jorpes, V. Mutt, *Acta Chem. Scand.* 15, 1790 (1961). Structure: J. E. Jorpes *et al.*, *Biochem. Biophys. Res. Commun.* 9, 275 (1962); V. Mutt *et al.*, *Biochemistry* 4, 2358 (1965). Synthesis: M. Bodanszky *et al.*, *J. Am. Chem. Soc.* 89, 685, 6753 (1967); M. A. Ondetti *et al.*, *ibid.* 90, 4711 (1968); M. Bodanszky *et al.*, *U.S. pat.* 3,767,639 (1973 to Squibb); E. Wuensch *et al.*, *Ber.* 104, 2430 (1971); 105, 2508 (1972); E. Wuensch, *Naturwiss.* 59, 239 (1972); B. Hemmasi, E. Bayer, *Int. J. Peptide Protein Res.* 9, 63 (1977); D. H. Coy *et al.*, *Peptides* 3, 137 (1982). Isoln of bovine secretin and amino acid sequence (identical to porcine): M. Carlquist *et al.*, *FEBS Letters* 127, 71 (1981). Stability study: H. Beyerman *et al.*, *Life Sci.* 29, 885 (1981). Comparative effects of synthetic and natural secretin on pancreatic secretion and on secretin, insulin and glucagon levels in man: C. Beglinger *et al.*, *Dig. Dis. Sci.* 27, 231 (1982). Review of radioimmunoassay methods: T. M. Chang, W. Y. Chey, *ibid.* 25, 529-552 (1980). General reviews: W. H. Häcki, *Clin. Gastroenterol.* 9, 609-632 (1980); K. G. Wormsley, *Scand. J. Gastroenterol.* 15, 513-517 (1980); V. Mutt, *Vitam. Horm.* 39, 231-426 (1982).

```
1
His-Ser-Asp-Gly-Thr-Phe-Thr-Ser-Glu-Leu-Ser-Arg-Leu-Arg
                                                       |
27
Val-Leu-Gly-Gln-Leu-Leu-Arg-Gln-Leu-Arg-Ala-Ser-Asp

              Porcine Secretin
```

Porcine secretin, $C_{130}H_{220}N_{44}O_{41}$, secretin [pig], porcine hormone secretin, secretin [ox].

THERAP CAT: Hormone (pancreatic).

8565. Securinine. Securinan-11-one. $C_{13}H_{15}NO_2$; mol wt 217.27. C 71.87%, H 6.96%, N 6.45%, O 14.73%. From leaves and roots of *Securinega suffruticosa* Rehder, *Euphorbiaceae* found in the Ussuri region: Murav'eva, Ban'kovskii, *C.A.* 50, 17335c (1956); *eidem, Doklady Akad. Nauk. SSSR* 110, 998 (1956), *C.A.* 51, 8121a (1957). Improved extraction process: Kogan *et al.*, **Brit.** pat. 1,169,471 (1959 to All-Union Sci. Res. Inst. of Medicinal Plants), *C.A.* 72, 47349x (1970) corresp to U.S. pat. 3,538,103 (1970). Structure: Satoda *et al.*, *Tetrahedron Letters* 1962, 1199; Mukherjee *et al.*, *Naturwiss.* 50, 155 (1963); Saito *et al.*, *Tetrahedron* 19, 2085 (1963). Stereochemistry: Nakano *et al.*, *Chem. & Ind.*

(London) 1963, 1034; Parello *et al.*, *Bull. Soc. Chim. France* 1963, 898; Nakano *et al.*, *J. Org. Chem.* 28, 2619 (1963); Horii *et al.*, *Tetrahedron* 19, 2101 (1963); Imado *et al.*, *Chem. & Ind. (London)* 1964, 1691. Synthesis of the racemate: Saito *et al.*, *Chem. Pharm. Bull.* 14, 313, 1059 (1966); Horii *et al.*, *Tetrahedron* 23, 1165 (1967). Toxicology study: S. L. Friess *et al.*, *Toxicol. Appl. Pharmacol.* 3, 347 (1961).

Yellow crystals from alcohol, mp 142-143°. $[\alpha]_D^{20} -1042°$ (c = 1 in alc). uv max (alc): 256, 330 nm (log ϵ 4.27, 3.30). Hydrochloride, mp 230°. $[\alpha]_D^{20} -259.2°$ (alcohol). Nitrate, mp 205°. $[\alpha]_D^{20} -312.12°$ (alcohol). LD_{50} i.v. in mice: 3.5 ± 0.9 mg/kg (Friess).

8566. Sedecamycin. [1S-(1R*,2S*,3E,5E,7R*,9E,11E,-13R*,15S*,19S*)]-N-[13-(Acetyloxy)-7-hydroxy-1,4,10,19-tetramethyl-17,18-dioxo-16-oxabicyclo[13.2.2]nonadeca-3,5,-9,11-tetraen-2-yl]-2-oxopropanamide; N-(7,13-dihydroxy-1,4,-10,19-tetramethyl-17,18-dioxo-16-oxabicyclo[13.2.2]nonadeca-3,5,9,11-tetraen-2-yl)pyruvamide 13-acetate; Antibiotic T-2636A; bundlin B; lankacidin A; lankacidin C 14-acetate; T-2636A; Takelan. $C_{27}H_{35}NO_8$; mol wt 501.58. C 64.66%, H 7.03%, N 2.79%, O 25.52%. One of 7 components of antibiotic T-2636. First isolated as bundlin B from *Streptomyces griseofuscus:* J. M. J. Sakamoto *et al.*, *J. Antibiot.* A15, 98 (1962). Isoln from *S. rochei* var *volubilis:* Neth. pat. Appl. 6,807,119; E. Higashide *et al.*, U.S. pat. 3,626,055 (1968, 1971 both to Takeda); *eidem, J. Antibiot.* 24, 1 (1971); S. Harada *et al.*, *ibid.* 13. Similar to lankacidin which is isolated from *S. violaceoniger* with lankamycin, *q.v.* Structure, absolute configuration and identity with bundlin B: H. Harada *et al.*, *Tetrahedron Letters* 1969, 2239; K. Kamiya *et al.*, *ibid.* 2245; M. Uramoto *et al.*, *ibid.* 2249. Enzymatic conversion to other lankacidins: T. Fugono *et al.*, *Experientia* 26, 26 (1970). Antibacterial activity: K. Tsuchiya *et al.*, *J. Antibiot.* 24, 29 (1971). Mass spectra: M. Uramoto *et al.*, *Agr. Biol. Chem.* 38, 855 (1974). Biosynthetic studies: *eidem, J. Am. Chem. Soc.* 100, 3616 (1978); K. Kakinima *et al.*, *Tetrahedron Letters* 23, 5303 (1982). Synthetic study: M. J. Fray, E. J. Thomas, *Tetrahedron* 40, 673 (1984). Activity against *Treponema hyodysenteriae* (swine dysentery): **Belg.** pat. 891,275; N. Narukawa *et al.*, U.S. pat. 4,425,356 (1982, 1984 both to Takeda).

Neutral lipophilic colorless plates from ether-ethyl acetate, mp 207-210° (dec). $[\alpha]_D^{25} -235°$ (c = 1.0 in ethanol). uv max (ethanol): 228 mu (ϵ 46900). Slightly sol in *n*-hexane, petroleum ether. Sol in ethyl acetate, acetone, chloroform, methanol. LD_{50} in mice (g/kg): up to 10 orally, 8-10 i.p. (Harada*).

THERAP CAT (VET): Antibacterial. In swine dysentery.

8567. Seidlitz Mixture. A mixture of 3 parts Rochelle salt and 1 part sodium bicarbonate. Ten grams of the mixture are employed with 2.17 g tartaric acid for one Seidlitz powder.

THERAP CAT: Cathartic.

8568. Selectins. Family of calcium-dependent cell adhesion molecules (CAMs) that initiate leukocyte adhesion to activated platelets or vascular endothelium in reponse to tissue injury or infection. Involved in lymphocyte homing to lymph nodes and in the extravasation of neutrophils into inflamed tissues. Binds to specific carbohydrate ligands on the surface of target cells. Three types have been identified and are designated according to the cell type from which they were originally discovered: L-selectin, from lymphocytes; E-selectin, from endothelial cells; and P-selectin, from platelets. Structures consist of an amino terminal lectin domain, an epidermal growth factor type repeat, and a discreet number of short consensus repeats similar to those found in complement-regulatory proteins (CR domains). The term "selectin" was proposed to highlight the presence of the lectin domain and to emphasize their selective expression and function. Reviews: D. Vestweber, *Semin. Cell Biol.* **3**, 211-220 (1992); M. P. Bevilacqua, R. M. Nelson, *J. Clin. Invest.* **91**, 379-387 (1993); R. P. McEver, *Curr. Opin. Immunol.* **6**, 75-84 (1994). Review of role in inflammation: M. A. Jutila, *Adv. Pharmacol.* **25**, 235-262 (1994); as potential therapeutic targets: J. K. Welply *et al.*, *Biochim. Biophys. Acta* **1197**, 215-226 (1994). Review of selectin ligands: A. Varki, *Proc. Nat. Acad. Sci. USA* **91**, 7390-7397 (1994).

L-Selectin, lymphocyte adhesion molecule-1, LAM-1, leukocyte-endothelial cell adhesion molecule-1, LECAM-1, leu-8. Constitutively expressed on the surface of almost all leukocytes; rapidly shed upon cell activation. Binds to activated endothelial cells and to the high endothelial venule of lymph nodes. Human form contains 2 CR domains. Originally described as a lymphocyte homing receptor: W. M. Gallatin *et al.*, *Nature* **304**, 30 (1983).

E-Selectin, endothelial-leukocyte adhesion molecule-1, ELAM-1. Expressed by endothelial cells following stimulation by endotoxin or inflammatory cytokines such as IL-1 or TNF. Binds to neutrophils, monocytes, eosinophils, certain lymphocytes and some tumor cells. Human form contains 6 CR domains. Identification: M. P. Bevilacqua *et al.*, *Proc. Nat. Acad. Sci. USA* **84**, 9238 (1987). Crystal structure of lectin domain: W. I. Weis, *Structure* **2**, 147 (1994).

P-Selectin, α-granule membrane protein 140, GMP-140, platelet activation-dependent granule-external membrane protein, PADGEM. Synthesized constitutively by endothelial cells and platelets; stored in platelet α-granules or in Weibel-Palade bodies of endothelial cells. Rapidly mobilized to the cell surface following activation by histamine or thrombin. Binds to neutrophils, monocytes, and certain lymphocyte subsets. Human form contains 9 CR domains. Identification on activated platelets: S.-C. Hsu-Lin *et al.*, *J. Biol. Chem.* **259**, 9121 (1984).

8569. Selegiline. (*R*)-*N*,α-*Dimethyl-N-2-propynylbenzeneethanamine; L-(−)-N,α-dimethyl-N-2-propynylphenethylamine;* (−)-deprenil; L-deprenyl. $C_{13}H_{17}N$; mol wt 187.28. C 83.37%, H 9.15%, N 7.48%. Monoamine oxidase-B inhibitor related structurally to pargyline, *q.v.* Prepn of racemate: *Fr. pat.* **M 2635** (1964 to Chinoin), *C.A.* **63**, 2927e (1965); of (−)-form: *Neth. pat. Appl.* **6,605,956** (1966 to Chinoin), *C.A.* **67**, 21611y (1967); J. S. Fowler, *J. Org. Chem.* **42**, 2637 (1977). *See also:* Z. Ecsery *et al.*, *U.S. pat.* **4,564,706** (1986 to Chinoin). Effects of racemate on CNS: J. Knoll *et al.*, *Arch. Int. Pharmacodyn.* **155**, 154 (1965). Comparative pharmacology of optical isomers: K. Magyar *et al.*, *Acta Physiol. Acad. Sci. Hung.* **32**, 377 (1967). Preliminary study in Alzheimer's disease: L. S. Schneider *et al.*, *Am. J. Psychiatr.* **150**, 321 (1993). Review of pharmacology: E. H. Heinonen, R. Lammintausta, *Acta Neurol. Scand.* **84**, Suppl 136, 44-59 (1991); of efficacy in Parkinson's disease: D. W. Robin, *Am. J. Med. Sci.* **302**, 392-395 (1991). Symposium: *Mov. Disord.* **8**, Suppl. 1, S1-S44 (1993).

Oil, $bp_{0.8}$ 92-93°. n_D^{20} 1.5180. $[\alpha]_D^{20}$ −11.2°.

Hydrochloride, $C_{13}H_{17}N.HCl$, *Déprényl, Eldéprine, Eldepryl, Jumex, Movergan, Plurimen.* Crystals, mp 141-142°. $[\alpha]_D^{25}$ −10.8° (c = 6.48 in water). LD_{50} in rats (mg/kg): 81 i.v., 280 s.c. (Magyar).

(±)-Form, *deprenyl, phenylisopropyl-N-methylpropynylamine.* Oil, bp_5 103-110°. n_D^{20} 1.5224.

(±)-Form hydrochloride, *E-250.* LD_{50} in rats (mg/kg): 63 i.v., 126 s.c., 385 orally (Knoll).

THERAP CAT: Antiparkinsonian.

8570. Selenic Acid. H_2O_4Se; mol wt 144.97. H 1.39%, O 44.14%, Se 54.47%. H_2SeO_4. Prepd by treating lead selenate with hydrogen sulfide and concentrating the filtered soln by evaporation: Mitscherlich, *Pogg. Ann.* **9**, 623 (1827); **11**, 327 (1827); *Ann. Chim. Phys.* [2] **36**, 100 (1827); by treating calcium selenate with cadmium oxalate and hydrogen sulfide: von Hauer, *J. Prakt. Chem.* [1] **80**, 214, 317 (1860); by treating a soln of silver selenite with bromine and evaporating the filtered soln: Thomsen, *Ber.* **2**, 598 (1869); by oxidizing selenium oxide with 30% hydrogen peroxide and removing water: Gilbertson, King, *J. Am. Chem. Soc.* **58**, 180 (1936); *Inorg. Syn.* **3**, 137 (1950).

Hexagonal prisms; mp 58°; d_4^{15} 2.9508; bp 260°. Very sol in water; sol in sulfuric acid; insol in ammonia; dec in alcohol. Very deliquescent. Reduced by hydrobromic acid, hydriodic acid, hydrogen sulfide, hydroxylamine hydrochloride, phenylhydrazine, formic acid, oxalic acid, malonic acid, pyruvic acid, acetyl chloride, and several metals.

8571. Selenious Acid. Monohydrated selenium dioxide; selenous acid. H_2O_3Se; mol wt 128.97. H 1.56%, O 37.22%, Se 61.22%. H_2SeO_3. Prepd by dissolving selenium dioxide in hot water and cooling: Berzelius, *Acad. Handl. Stockholm* **39**, 13 (1818).

Deliquescent hexagonal prisms; d_4^{15} 3.004. Vapor press. (mm Hg): 2 at 15°; 4.5 at 35°; 7 at 40.3°. K_1 = 0.0024; K_2 = 4.8 × 10^{-9}. Gives off water upon heating and selenium oxide sublimes. Oxidized to selenic acid by strong oxidizing agents such as ozone, hydrogen dioxide, chlorine. Reduced to selenium by most reducing agents including hydriodic acid, sulfurous acid, sodium hyposulfite, hydroxylamine salts, hydrazine salts, hypophosphorous acid, phosphorous acid. 90 parts dissolve in 100 parts water at 0°, 400 parts at 90°; very sol in alcohol; insol in ammonia.

USE: As a reagent for alkaloids; as oxidizing agent.

8572. Selenium. Se; at. wt 78.96; at. no. 34; valences 2, 4, 6. Six stable isotopes: 74 (0.87%); 76 (9.02%); 77 (7.58%); 78 (23.52%); 80 (49.82%); 82 (9.19%); artificial, radioactive isotopes: 70-73; 75; 79; 81; 83-85; 87. Discovered in 1817 by Berzelius. Constitutes about 0.09 ppm of the earth's crust. Occurs in nature usually in the sulfide ores of the heavy metals; found in small quantities in pyrite; in the minerals *clausthalite* (PbSe), *naumannite* [(Ag,Pb)Se], *tiemannite* (HgSe); in selenosulfur. Prepn: Waitkins *et al.*, *Ind. Eng. Chem.* **34**, 899 (1942). Purification: Nielsen, *Heritage, J. Electrochem. Soc.* **106**, 39 (1959). Purification by distn from a mixture contg a small amount of Mg: Oberbacher, Schlier, *U.S. pat.* **2,930,678** (1960 to Norddeutsche Raffinerie). Essential trace element in animal diets at levels of approx 0.1 ppm. Toxic element at higher concentrations. Reviews of nutritional and toxic properties of selenium: I. Rosenfeld, O. A. Beath, *Selenium* (Academic Press, New York, 1964) 411 pp; E. Browning, *Toxicity of Industrial Metals* (Appleton-Century-Crofts, New York, 2nd ed., 1969) pp 286-295; E. J. Underwood, *Trace Elements in Human and Animal Nutrition* (Academic Press, New York, 1971) pp 323-368; Frost, *Crit. Rev. Toxicol.* **1**, 467-514 (1972); Allaway, *Cornell Vet.* **63**, 151-170 (1973); Frost, Lish, *Ann. Rev. Pharmacol.* **15**, 259-284 (1975). Review of selenium biochemistry: T. C. Stadtman, *Ann. Rev. Biochem.* **49**, 93-110 (1980); idem, *FASEB J.* **1**, 375-379 (1987). Symposium on organic selenium and tellurium compds: Y. Okamoto, W. H. H. Gunther, Eds., *Ann. N.Y. Acad. Sci.* **192**, 1-225 (1972). Reviews of selenium and its compds: K. W. Bagnall, *The Chemistry of Selenium, Tellurium and Polonium* (Elsevier, New York, 1966) 200 pp; Bagnall in *Comprehensive Inorganic Chemistry* vol. 2, J. C. Bailar, Jr. *et al.*, Eds. (Pergamon Press, Oxford, 1973) pp 935-1008; *Selenium,* R. A. Zingaro, W. C. Cooper, Eds.

(Van Nostrand, Reinhold, New York, 1974) 835 pp; E. M. Elkin in Kirk-Othmer *Encyclopedia of Chemical Technology* vol. 20 (Wiley-Interscience, New York, 3rd ed., 1982) pp 575-601.

Exists in several allotropic forms: amorphous, cryst or red, and gray or metallic. Liq is a brownish red, boils at 685° forming dark red vapors. Sol in dil aq caustic alkali solns; in aq potassium cyanide soln, in potassium sulfite soln. Burns in air with a bright blue flame forming the dioxide and emitting a characteristic odor resembling rotten horseradish. Combines directly with hydrogen, with the halogens (excluding iodine). Oxidized to selenious acid by nitric acid, to selenic acid by sulfuric acid. Reduces hot aqueous solutions of silver and gold salts with formation of silver selenide and metallic gold, respectively. Reacts with many metals.

Amorphous forms: Vitreous, black selenium; dark red-brown to bluish-black solid; formed when molten Se is cooled rapidly. d 4.28. Softens at 50-60° and becomes elastic at 70°. Red, amorphous form; formed by reduction of selenious acid in water; by condensation of Se vapor on a cold surface. d 4.26. Review of structural studies: Richter, Breitling, *Z. Naturforsch.* **26B**, 1699-1708, 2074-2075 (1971). When freshly precipitated, reacts with water at 50° forming selenious acid and hydrogen. Sol in carbon disulfide, methylene iodide, benzene, or quinoline.

Crystalline or red: Two monoclinic forms; dark red, transparent crystals. α-Form prepd by slow evaporation of CS_2 soln of Se; β-form by rapid evaporation of soln. mp below 200°. d (α-form) 4.46. Structure of α-form: Cherin, Unger, *Acta Cryst.* **28B**, 513 (1972); *see also* Bagnall, *loc. cit.* Both forms are metastable and change into gray form on heating.

Gray or metallic: Lustrous gray to black hexagonal crystals. The most stable form. d_4^{20} 4.81. mp 217°. Mohs' hardness 2.0. Latent heat of fusion 16.4 cal/g. Latent heat of vaporization 20.6 kcal/mol. Linear coefficient of thermal expansion per degree C = 37 $\times$ 10^{-6}. Specific heat (28°): 0.084 cal/g/°C. Surface tension (217°): 92.5 dynes/cm. Thermal conductivity (25°): 0.0007-0.00183 cal/(cm)(°C)(sec). Insol in water, alcohol; very slightly sol in carbon disulfide (2 mg/100 ml at ord temp). Sol in ether. Conducts electricity and rectifies alternating current; the conductivity increases up to a thousand times on exposure to light.

Caution: Potential symptoms of overexposure are irritation of eyes, nose and throat; visual disturbance; headache; chills, fever; dyspnea, bronchitis; metallic taste, garlic breath, gastrointestinal disturbance; dermatitis; skin and eye burns. *See NIOSH Pocket Guide to Chemical Hazards* (DHHS/NIOSH 90-117, 1990) p 194.

USE: As ingredient of toning baths in photography; as pigment in manuf ruby-, pink-, orange-, or red-colored glass; as metallic base in making electrodes for arc lights, electrical instruments and apparatus, as rectifier in radio and television sets; in selenium photocells, in semiconductor fusion mixtures, selenium cells, telephotographic apparatus; as vulcanizing agent in processing of rubber; as catalyst in determination of nitrogen by Kjeldahl method; for dehydrogenation of organic compds.

THERAP CAT (VET): Nutritional factor (interrelationship with vitamin E). Chiefly to prevent muscle degenerative diseases.

8573. Selenium Bromide. Selenium monobromide. Br_2Se_2; mol wt 317.73. Br 50.30%, Se 49.70%. Se_2Br_2. Prepd by mixing bromine and selenium in the correct proportions: Schneider, *Pogg. Ann.* **128**, 327 (1866); **129**, 450, 634 (1866); by treating selenium dioxide with hydrobromic acid: Lenher, Kao, *J. Am. Chem. Soc.* **47**, 772 (1925).

Dark red liquid. Unpleasant odor. d_4^{15} 3.604. Dec when heated. Dec in water and moist air. Sol in chloroform, ethyl bromide, carbon disulfide.

8574. Selenium Chloride. Selenium monochloride. Cl_2Se_2; mol wt 228.83. Cl 30.99%, Se 69.01%. Se_2Cl_2. Prepd from the elements: Sacc *et al.,* cited in *Mellor's* vol. X, 894 (1930); by the action of PCl_5 on selenium, on the selenide of phosphorus or antimony, on SeO_2 or on $SeOCl_2$: Baudrimont, *Ann. Chim. Phys.* [4] **2**, 5 (1864); Michaelis, *Z.*

Chem. [2] **6**, 460 (1870); by the action of HCl on a soln of selenium in fuming sulfuric acid: Divers, Shimose, *J. Chem. Soc.* **45**, 194, 198, 201 (1884); from Se, SeO_2 and HCl: Lenher, Kao, *J. Am. Chem. Soc.* **47**, 772 (1925); **48**, 1550 (1926).

Deep red, oily liquid. bp_{733} 127° (dec). d_4^{25} 2.7741. mp −85°. Sol in chloroform, benzene, carbon tetrachloride, carbon disulfide, fuming sulfuric acid; dec in water.

8575. Selenium Hexafluoride. F_6Se; mol wt 192.95. F 59.08%, Se 40.92%. SeF_6. Prepd by passing gaseous fluorine over finely divided selenium in a copper vessel: Klemm, Henkel, *Z. Anorg. Allgem. Chem.* **207**, 74 (1932); Yost, Claussen, *J. Am. Chem. Soc.* **55**, 885 (1933); Yost, *Inorg. Syn.* **1**, 121 (1939). *Review:* Kemmitt, Sharp, *Advan. Fluorine Chem.* **4**, 235 (1965).

Gas. mp −50.8°; sublimes −63.8°. Vapor press. (mm Hg): 651.2 (−48.7°); 213.1 (−64.8°); 30.4 (−87.5°). Insol in water. Reacts with ammonia gas at 200° to give selenium, nitrogen, and hydrogen fluoride. Covalently satd, does not attack glass.

USE: Gaseous electric insulator.

8576. Selenium Oxide. Selenium dioxide; selenious anhydride. O_2Se; mol wt 110.96. O 28.84%, Se 71.16%. SeO_2. Prepd by burning selenium in oxygen: Berzelius, *Acad. Handl. Stockholm* **39**, 13 (1818); *Ann. Chim. Phys.* [2] **9**, 160, 225, 337 (1818); by burning selenium in oxygen and nitrogen dioxide: Lenher, *J. Am. Chem. Soc.* **20**, 559 (1898); by oxidizing selenium with nitric acid: A. P. Julien, *ibid.* **47**, 1799 (1925). Toxicology: E. A. Cerwenka, W. C. Cooper, *Arch. Environ. Health* **3**, 189 (1961). *Review:* Waitkins, Clark, *Chem. Rev.* **36**, 235-289 (1945).

Lustrous, tetragonal needles. Acidic taste; leaves a burning sensation. Its yellowish green vapor has a pungent, sour smell. mp 340°; d_{15}^{15} 3.954. Vapor pressure (mm Hg): 12.5 at 70°; 20.2 at 94°; 39.0 at 181°; 760.0 at 315°; 848.0 at 320°. n_D^{20} < 1.76. Soly (parts/100 parts solvent) in water: 38.4 (14°); in methanol: 10.16 (11.8°); in 93% ethanol: 6.67 (14°); in acetone: 4.35 (15.3°); in acetic acid: 1.11 (13.9°). Sol in concd H_2SO_4. Stable to light and heat. Rapidly absorbs dry hydrogen fluoride, hydrogen chloride, hydrogen bromide, and hydrogen iodide to form the corresponding selenium oxyhalide. Reacts with ammonia to form nitrogen and selenium; in alcohol soln forms ammonium ethyl selenite, $HN_4(C_2H_5)SeO_3$. Yields nitrogen and black amorphous selenium with hydrazine; nitrogen and reddish-brown amorphous selenium with hydroxylamine hydrochloride. Forms selenic acid with nitric acid. Reduced by carbon and other organic substances.

Caution: Direct contact may cause intense local irritation of skin and eyes (Cerwenka, Cooper).

USE: In the manuf of other selenium compounds; as a reagent for alkaloids; as oxidizing agent: L. F. Fieser, M. Fieser, *Reagents for Organic Chemistry* vol. 1 (New York, Wiley, 1967) p 992.

8577. Selenium Oxybromide. *Seleninyl bromide;* selenyl bromide. Br_2OSe; mol wt 254.77. Br 62.73%, O 6.28%, Se 30.99%. $SeOBr_2$. Prepd by the reaction of selenium dioxide, fused selenium, and fused calcium bromide: Lenher, *J. Am. Chem. Soc.* **44**, 1668 (1922).

Red-yellow solid. mp 41.5-41.7°; d^{50} 3.38; bp_{740} 217° (dec). Dec in water. Chlorine displaces the bromine. Dec by sulfur and hydrogen sulfide. Sol in sulfuric acid, carbon disulfide, chloroform, benzene, toluene, xylene, carbon tetrachloride. Reacts explosively with sodium and potassium.

8578. Selenium Oxychloride. *Seleninyl chloride.* Cl_2OSe; mol wt 165.86. Cl 42.75%, O 9.65%, Se 47.61%. $SeOCl_2$. Prepd by passing chlorine into SeO_2 suspended in carbon tetrachloride: Lenher, *J. Am. Chem. Soc.* **42**, 2498 (1920); Hönigschmid, Görnhardt, *Z. Naturforsch.* **1**, 661 (1946); by dehydration of dichloroselenious acid: Smith, Jackson, *Inorg. Syn.* **3**, 130 (1950); from $SOCl_2$ and SeO_2: Paetzold, Aurich, *Z. Anorg. Allgem. Chem.* **315**, 72 (1962); from $AlCl_3$ and SeO_2: Drago, Whitten, *Inorg. Chem.* **5**, 677 (1966). Acute toxicity: C. G. Wilber, *Clin. Toxicol.* **17**, 171 (1980).

Nearly colorless or yellowish, corrosive liquid; fumes in the air. d_4^{16} 2.44. Solidif $\sim 5°$. bp 180°. n_D^{20} 1.651. Dec by water into HCl and selenious acid. Miscible with carbon tetrachloride, chloroform, carbon disulfide, benzene, toluene. It is an excellent solvent for many substances including metals. LD_{50} s.c. in rabbits: 7 mg/kg (Wilber).

Caution: Strong irritant, vesicant: *Clinical Toxicology of Commercial Products,* R. E. Gosselin *et al.,* Eds. (Williams & Wilkins, Baltimore, 5th ed., 1984) Section II, p 129.

8579. Selenium Oxyfluoride. *Seleninyl fluoride;* selenyl fluoride. F_2OSe; mol wt 132.96. F 28.58%, O 12.03%, Se 59.39%. F_2SeO. Prepd from selenium dioxide and fluorine: Aynsley *et al., J. Chem. Soc.* **1952,** 1231; from selenium oxychloride and sodium fluoride: Tullock, Coffman, *J. Org. Chem.* **25,** 2016 (1960); from selenium tetrafluoride and tellurium oxide: J. Carre *et al., J. Fluorine Chem.* **14,** 139 (1979).

Fuming liquid. Ozone-like odor. mp 15.0°; bp 126°. $d^{21.5}$ 2.8. Rapidly attacks glass.

Caution: Strong irritant, vesicant. Can cause fatal pulmonary edema.

8580. Selenium Sulfides. Prepn from selenious acid and H_2S; product may be a mixture of the elements: *Gmelin's Selenium* (8th ed.) **10** (part B), 159-176 (1949). Prepn by fusion of the elements; solid solns of sulfur (S_8), selenium (Se_8) and selenium sulfides (S_nSe_{8-n}; $n = 1-7$) formed: Fergusson *et al., J. Inorg. Nucl. Chem.* **24,** 157 (1962); Hawes, *Nature* **198,** 1267 (1963); Cooper, Culka, *J. Inorg. Nucl. Chem.* **27,** 755 (1965); **29,** 1217 (1967).

Se_4S_4, red tabular crystals from benzene, d 3.20. mp 113° (dec). Sol in CS_2. Solubility in benzene (20°): 0.4 g/l. Se_2S_6, light-orange needles from benzene, d 2.44. mp 121.5°. Sol in CS_2. Solubility in benzene (20°): 12 g/l. Selenium disulfide, SeS_2, *Exsel, Selsun.* Prepn of shampoo: M. M. Baldwin, A. P. Young, Jr., U.S. pat. **2,694,669** (1954 to Abbott). Toxicity studies: L. M. Cummins, E. T. Kimura, *Toxicol. Appl. Pharmacol.* **20,** 89 (1971). Review of use as antidandruff agent: Matson, *J. Soc. Cosmet. Chem.* **7,** 459-466 (1956). Insol in water, in 0.01N HCl. LD_{50} orally in rats: 138 mg/kg (Cummins, Kimura).

Note: Selenium sulfide (as SeS) may reasonably be anticipated to be a carcinogen: *Seventh Annual Report on Carcinogens* (PB95-109781, 1994) p 357. *See also Bioassay of Selenium Sulfide (Gavage) for Possible Carcinogenicity* (PB82-164955, 1980) 113p.

THERAP CAT: Disulfide as topical antiseborrheic.

THERAP CAT (VET): Topically in eczemas and dermatomycoses.

8581. Selenium Tetrabromide. Br_4Se; mol wt 398.58. Br 80.19%, Se 19.81%. $SeBr_4$. Prepd by adding excessive bromine to selenium: Schneider, *Pogg. Ann.* **128,** 327 (1866); **129,** 450, 634 (1866); by dissolving selenium dioxide in hydrobromic acid: Muthmann, Schäfer, *Ber.* **26,** 1008 (1893).

Red-brown, cryst powder. Unpleasant odor. Dec 70-80°. Dec in moist air and water. Sol in carbon disulfide, chloroform, ethyl bromide.

8582. Selenium Tetrachloride. Cl_4Se; mol wt 220.77. Cl 64.23%, Se 35.77%. $SeCl_4$. Prepd by the action of an excess of chlorine on selenium: Berzelius, cited in *Mellor's* vol X, 898 (1930); by the action of thionyl chloride or phosphorus trichloride on selenium oxychloride and by the action of phosphorus pentachloride on selenium oxide: Michaelis, *Z. Chem.* [2] **6,** 460 (1870); by the action of anhydr selenic acid on acetyl chloride at 0°: Lamb, *Am. Chem. J.* **30,** 209 (1903). Crystal structure: Shoemaker, Abrahams, *Acta Cryst.* **18,** 296 (1965).

White to pale yellow crystals. d 2.6. Sublimes when heated. Dec in water and moist air. Insol in liquid bromine. Dec by dry ammonia.

8583. Selenium Tetrafluoride. F_4Se; mol wt 154.95. F 49.04%, Se 50.96%. SeF_4. Fluorinating agent. Prepn from $SeCl_4$ and AgF: Prideaux, Cox, *J. Chem. Soc.* **1927,** 928; **1928,** 1603; from Se and F_2: Aynsley *et al., ibid.* **1952,** 1231; J. Carre *et al., J. Fluorine Chem.* **14,** 139 (1979); from Se and AgF: Glemser *et al., Naturwiss.* **52,** 130 (1965); from Se_2Cl_2 and F_2: Goggin, *J. Inorg. Nucl. Chem.* **28,** 661 (1966); from

Se and ClF: Pitts, Jache, U.S. pat. **3,373,000** (1968 to Olin Mathieson); from Se and ClF_3: Olah *et al., J. Am. Chem. Soc.* **96,** 925 (1974).

Colorless liquid; fumes in air. mp $-9.5°$. bp 106°. d^{18} 2.75. Forms a white hygroscopic solid. Hydrolyzed violently by water. Attacks glass slowly when perfectly dry. Completely miscible with ether, ethanol, iodine pentafluoride, sulfuric acid; sol in chloroform, carbon tetrachloride.

USE: Fluorination of alcohols, carboxylic acids and carbonyl compds: Olah *et al., loc. cit.*

8584. Selenocysteine. *3-Selenyl-L-alanine;* selenium cysteine. $C_3H_7NO_2Se$; mol wt 168.05. C 21.44%, H 4.20%, N 8.33%, O 19.04%, Se 46.98%. Naturally occurring amino acid found in the active site of certain prokaryotic and eukaryotic enzymes. Highly oxidizable making synthesis difficult. Easily prepared from its oxidized form *selenocystine.* Prepn of *dl*-selenocystine: A. Fredga, *Svensk. Kem. Tidskrift.* **48,** 160 (1936); E. P. Painter, *J. Chem. Soc.* **69,** 229 (1947); of *l*-form: H. Tanaka, K. Soda, *Methods in Enzymology* **143,** 240 (1987). Reduction of selenocystine and separation of L-selenocysteine: J. N. Burnell *et al., J. Inorg. Biochem.* **12,** 343 (1980). Chemical properties: R. E. Huber, R. S. Criddle, *Arch. Biochem. Biophys.* **122,** 164 (1967). Determn by HPLC/MS: H. E. Ganther *et al., Methods in Enzymology* **107,** 582 (1984); colorimetric determn: N. Esaki, K. Soda, *ibid.* **143,** 148 (1987). Identification of Se as a component in protein: S.-H. Oh *et al., Biochemistry* **13,** 1825 (1974). Characterization of selenocysteine as the Se containing amino acid moiety in various enzymes: J. E. Cone *et al., Proc. Nat. Acad. Sci. USA* **73,** 2659 (1976); J. B. Jones *et al., Arch. Biochem. Biophys.* **195,** 255 (1979); presence in the catalytic site of mammalian glutathione peroxidase: J. W. Forstrom *et al., Biochemistry* **17,** 2639 (1978); J. J. Zabrowski *et al., Biochem. Biophys. Res. Commun.* **84,** 248 (1978). Metabolism in animals: H. Tanaka *et al., Curr. Topics Cell Reg.* **27,** 487 (1985). Identification of selenocysteine-specific tRNA: W. C. Hawkes *et al., Biochim. Biophys. Acta* **699,** 183 (1982); W. Leinfelder *et al., Nature* **331,** 323 (1988). *Review:* T. C. Stadtman, *FASEB J.* **1,** 375-379 (1987).

pK 5.2. Unstable to acid hydrolysis.

8585. Selenomethionine. *2-Amino-4-(methylseleno)butanoic acid;* α-amino-γ-(methylseleno)butyric acid. $C_5H_{11}NO_2Se$; mol wt 196.11. C 30.62%, H 5.65%, N 7.14%, O 16.32%, Se 40.26%. $CH_3SeCH_2CH_2CH(NH_2)COOH$. Selenium analog of methionine. Prepn of DL-form: E. P. Painter, *J. Am. Chem. Soc.* **69,** 232 (1949); H. J. Klosterman, E. P. Painter, *ibid.* 2009; G. Zdansky, *Ark. Kemi* **19,** 559 (1962), *C.A.* **58,** 6919h (1963). Prepn of L-form: *idem, ibid.,* **29,** 437 (1968); C. S. Pande *et al., J. Org. Chem.* **35,** 1440 (1970). Biosynthesis of radioactive (^{75}Se) form: M. Blau, *Biochim. Biophys. Acta* **49,** 389 (1961). Metabolism: M. Blau, J. F. Holland in *Radioactive Pharmaceuticals,* G. A. Andrews *et al.,* Eds. (USAEC Symposium Series #6, 1976) pp 409-421. Review of use in pancreatic scanning: J. E. Agnew *et al., Brit. J. Radiol.* **49,** 979-995 (1976).

DL-Form, transparent, hexagonal sheets or plates from methanol and water, metallic luster; mp 265° (dec). IR spectrum: Zdansky, *loc. cit.* (1962).

L-Form, crystals from aqueous acetone; mp 266-268° (dec). $[\alpha]_D^{22}$ +21.6° (c = 0.5 in 2N HCl).

^{75}Se-Injection, *Sethotope.*

THERAP CAT: ^{75}Se as diagnostic aid (radioactive imaging agent).

8586. Sematilide. *N-[2-(Diethylamino)ethyl]-4-[(methylsulfonyl)amino]benzamide;* CK-1752. $C_{14}H_{23}N_3O_3S$; mol wt 313.42. C 53.65%, H 7.40%, N 13.41%, O 15.31%, S 10.23%. Prepn: D. D. Davey *et al.,* U.S. pat. **4,544,654** (1985 to Schering AG); W. C. Lumma, Jr. *et al., J. Med. Chem.* **30,** 755 (1987). Pharmacology: S. S. Greenberg *et al., Drug Dev. Res.* **25,** 283 (1992). Pharmacokinetics and

Consult the Name Index before using this section.

HPLC determn in plasma: S. Dancik *et al., J. Pharm. Sci.* **80**, 157 (1991). Clinical pharmacology: W. Wong *et al., Am. J. Cardiol.* **69**, 206 (1992). Electrophysiologic profile: P. T. Sager *et al., Circulation* **88**, 1072 (1993). *Review:* T. A. Argentieri, *Cardiovasc. Drug Rev.* **10**, 182-198 (1992).

Monohydrochloride, $C_{14}H_{23}N_3O_3S.HCl$, *CK-1752A*. Crystals from acetone/methanol. mp 141-142° (Lumma). Also reported as having two polymorphic crystalline forms, mp 137° and mp 142° (Argentieri). Apparent pKa_1 7.60; pKa_2 9.51. Distribution coefficient (octanol/pH 7.4 buffer): 0.104 at 10 mM. uv max (0.1*N* NaOH): 289 nm (ϵ 1.91×10^4). uv max (0.1*N* HCl): 254 nm (ϵ 1.78×10^4). LD_{50} i.p. in mice: 250-300 mg/kg (Lumma). LD_{50} in mice, rats, dogs (mg/kg): 96, 92, 143-175 i.v.; 1,800, 3,200, 500-1,000 orally (Argentieri).

THERAP CAT: Antiarrhythmic (class III).

8587. Semduramicin. *[3R,4S,5S,6R,7S,22S(2S,5S,6R)]-23,27-Didemethoxy-2,6,22-tridemethyl-5,11-di-O-demethyl-6-methoxy-22-[(tetrahydro-5-methoxy-6-methyl-2H-pyran-2-yl)-oxy]lonomycin A;* UK-61689. $C_{45}H_{76}O_{16}$; mol wt 873.09. C 61.91%, H 8.77%, O 29.32%. Semi-synthetic polyether ionophore antibiotic. Prepn: A. C. Goudie, N. D. A. Walshe, **Eur. pat. Appl.** 255,335; *eidem,* U.S. pat. **4,804,680** (1988, 1989 both to Pfizer). LC determn in poultry liver: J. F. Ericson *et al., J. AOAC Int.* **77**, 577 (1994). Metabolism and tissue residue studies in broiler chickens: M. J. Lynch *et al., ACS Symposium Series* **503**, 49-69 (1992). Anticoccidial effects in chickens: M. E. McKenzie *et al., Poultry Sci.* **72**, 2052 (1993); N. B. Logan *et al., ibid.,* 2058.

Sodium salt, $C_{45}H_{75}NaO_{16}$; *UK-61689-2, Aviax.* Crystals from isopropyl ether. mp 175-176°. $[\alpha]_D^{25}$ +19.3° (c = 0.5 in methanol).

THERAP CAT (VET): Coccidiostat.

8588. Semicarbazide Hydrochloride. *Hydrazinecarboxamide monohydrochloride;* aminourea hydrochloride; carbamylhydrazine hydrochloride. CH_6ClN_3O; mol wt 111.53. C 10.77%, H 5.42%, Cl 31.79%, N 37.68%, O 14.35%. $NH_2NHCONH_2.HCl$. Prepd by electrolytic reduction of nitrourea with cathodes of copper, nickel, lead, and mercury in hydrochloric acid solution: Ingersoll *et al., Org. Syn.* **5**, 93 (1925). Commercial prepn from hydrazine hydrate, iron carbonyl, and carbon monoxide: Sampson, U.S. pat. **2,589,289** (1952 to du Pont).

Prisms from dil alc, dec 175-185°. Freely sol in water with acid reaction. Dissociation consts of free base: Kb at 20° = 29 × 10^{-11}; Ka = 1.6 × 10^{-11}. Very slightly sol in hot alcohol, insol in anhydr ether. Ammonolysis with liq NH_3 gives a 95% yield of the free base; mp 96°: Audrieth, *J. Am. Chem. Soc.* **52**, 1250 (1930).

Acetone semicarbazone, $C_4H_9N_3O$, needles from water, mp 192° (heated stage); mp 188° (open capillary tube).

USE: As a reagent for ketones and aldehydes with which it affords cryst compds having characteristic melting points.

8589. Semioxamazide. *Aminooxoacetic acid hydrazide;* aminooxamide; oxamic acid hydrazide. $C_2H_5N_3O_2$; mol wt 103.08. C 23.30%, H 4.89%, N 40.76%, O 31.04%. $H_2N-COCONHNH_2$. Made from hydrazine sulfate, ethyl oxamate and KOH. *Ref:* Wilson, Pickering, *J. Chem. Soc.* **123**, 394 (1923); Leonard, Boyer, *J. Org. Chem.* **15**, 42 (1950).

Lustrous leaflets, mp ~220° with decompn. Soluble in 400 parts water; more sol in hot water; easily sol in acids and alkalies with formation of salts; insol in alcohol, ether.

USE: As a reagent for aldehydes and ketones.

8590. Semotiadil. *(R)-2-[2-[3-[[2-(1,3-Benzodioxol-5-yloxy)ethyl]methylamino]propoxy]-5-methoxyphenyl]-4-methyl-2H-1,4-benzothiazin-3(4H)-one; (+)-(R)-3,4-dihydro-2-[5-methoxy-2-[3-[N-methyl-[N-2-[3,4-(methylenedioxy)phenoxy]ethyl]amino]propoxy]phenyl]-4-methyl-3-oxo-2H-1,4-benzothiazine;* sesamodil; DS-4823; $C_{29}H_{32}N_2O_6S$; mol wt 536.65. C 64.91%, H 6.01%, N 5.22%, O 17.89%, S 5.98%. Benzothiazine calcium antagonist. Prepn: J. Iwao *et al.,* **PCT Int.** pat. **Appl.** 87 838; *eidem,* U.S. pat. **4,786,635** (1987, 1988 both to Santen). Synthesis: M. Fujita *et al., J. Med. Chem.* **33**, 1898 (1990). LC determn in plasma: K. Morishima *et al., J. Chromatog.* **527**, 381 (1990). Molecular conformation and crystal structure: A. Ota *et al., Chem. Pharm. Bull.* **41**, 1681 (1993). Electrophysiology: N. Miyawaki *et al., J. Cardiovasc. Pharmacol.* **16**, 769 (1990); and antiarrhythmic effects: *eidem, Drug Devel. Res.* **22**, 293 (1991). Antihypertensive effects in rats: A. Kanda, H. Hashimoto, *Japan. J. Pharmacol.* **63**, 121 (1993). Mechanism of action study: K. Murakami *et al., J. Cardiovasc. Pharmacol.* **25**, 262 (1995).

Fumarate, $C_{29}H_{32}N_2O_6S.C_4H_4O_4$, *SD-3211.* Crystals from EtOH, mp 134-135°. $[\alpha]_D^{25}$ +195° (in DMSO).

THERAP CAT: Antihypertensive; antianginal.

8591. Sempervirine. *2,3,4,13-Tetrahydro-1H-benz[g]indolo[2,3-a]quinolizin-6-ium; 3,4,5,6,14,15,20,21-octadehydro-yohimbanium;* sempervirene. $C_{19}H_{16}N_2$; mol wt 272.35. C 83.79%, H 5.92%, N 10.29%. Found in rhizome and roots of Carolina jessamine (yellow jessamine) *Gelsemium sempervirens* (L.) Ait. f., *Loganiaceae:* Stevenson, Sayre, *J. Am. Pharm. Assoc.* **4**, 60 (1915); **8**, 708 (1919); Hasenfratz, *Bull. Soc. Chim. France* [4] **53**, 1084 (1933); Forsyth *et al., J. Chem. Soc.* **1945**, 579; Prelog, *Helv. Chim. Acta* **31**, 588 (1948). Believed to exist primarily as inner salt depicted below: Woodward, Witkop, *J. Am. Chem. Soc.* **71**, 379 (1949). Synthesis: Woodward, McLamore, *ibid.* 379; Swan, *J. Chem. Soc.* **1958**, 2039; Ban, Seo, *Tetrahedron* **16**, 11 (1961); Potts, Mattingly, *J. Org. Chem.* **33**, 3985 (1968).

Monohydrate, yellow needles from chloroform, mp 228°. Brownish-yellow leaflets from dil ethanol, mp 258°. uv max (ethanol): 243, 249, 297, 345, 387 nm (log ϵ 4.58, 4.57, 4.20, 4.26, 4.24). Dipole moment: 8.5 (dioxane); 7.5 (benzene). Soluble in alcohol, chloroform, pyridine. Slightly sol in acetone. Practically insol in ether, benzene.

Methochloride, $C_{20}H_{19}ClN_2$, minute yellow needles from ethanol, mp 330-332°. Soluble in water giving yellow solns with purple fluorescence. uv max: 241, 292, 330, 395 nm (log ε 4.56, 4.20, 4.28, 4.22).

8592. Senecialdehyde. *3-Methyl-2-butenal; 3,3-dimethylacrolein;* 3-methylcrotonaldehyde; senecioaldehyde; *β,β-dimethylacrolein.* C_5H_8O; mol wt 84.12. C 71.39%, H 9.59%, O 19.02%. $(CH_3)_2C=CHCHO$. Prepn from isoamyl alcohol: Fischer *et al., Ber.* **64**, 30 (1931); Fischer, Löwenberg, *Ann.* **494**, 272 (1932); *cf.* Burkhardt *et al.*, **Brit.** pat. **512,465** (1939); Montavon, Saucy, **U.S.** pat. **2,902,515** (1959 to Hoffmann-La Roche).
Liquid. Odor more pungent than that of isovaleraldehyde. Autooxidizes easily, but the pure substance can be stored under vacuum. d_4^{20} 0.8722; bp_{730} 133°; n_D^{20} 1.4526. Absorption spectrum: Forbes, Skilton, *J. Org. Chem.* **24**, 436 (1959).
Caution: Irritant to mucous membranes. Narcotic in high concentrations.

8593. Senecic Acid. *[R-[R*,R*-(Z)]]-5-Ethylidene-2-hydroxy-2,3-dimethylhexanedioic acid.* $C_{10}H_{16}O_5$; mol wt 216.23. C 55.55%, H 7.46%, O 37.00%. One of the most widely studied *necic acids* which are constituent acids of the hepatotoxic pyrrolizidine alkaloids and are isolated from these alkaloids by hydrolysis or by hydrogenolysis. Necic acids are C_5- to C_{10}-acids; many are γ- or δ-hydroxy acids and have structures that may be dissected into isoprene units. Synthesis of senecic acid: C. C. J. Culvenor, T. A. Geissman, *J. Am. Chem. Soc.* **83**, 1647 (1961). uv spectra of senecic acid and its isomers: V. Simanek *et al., Collect. Czech. Chem. Commun.* **34**, 1832 (1969). Biosynthesis: D. H. G. Crout *et al., J. Chem. Soc. Perkin Trans. I* **1972**, 671. Comprehensive reviews on necic acids: F. L. Warren, *Fortschr. Chem. Org. Naturst.* **12**, 198-269 (1955); **24**, 329-406 (1966); D. J. Robins *ibid.* **41**, 115-203 (1982); F. L. Warren in *Alkaloids*, **Vol. XII**, R. H. F. Manske, Ed. (Academic Press, New York, 1970) pp 245-331.

HO   CH₃

HOOC

CH₃   COOH

CH₃

Needles from benzene or ethyl acetate, mp 145-146°. Sublimes readily at ~125°.

8594. Senecio. Golden ragwort; squaw weed; life root. Dried plant of *Senecio aureus* L., *Compositae.* **Habit.** Canada and Eastern U.S. *Constit.* The alkaloids, senecifoline, senecine, etc., resins.
THERAP CAT: Emmenagogue.

8595. Senecionine. *12-Hydroxysenecionan-11,16-dione;* aureine. $C_{18}H_{25}NO_5$; mol wt 335.40. C 64.46%, H 7.51%, N 4.18%, O 23.85%. Hepatotoxic pyrrolizidine alkaloid from whole plant of *Senecio vulgaris* L., *Compositae* and several other *Senecio* spp. Isoln: Barger, Blackie, *J. Chem. Soc.* **1936**, 743. Structure: Adams, Govindachari, *J. Am. Chem. Soc.* **71**, 1953 (1949); Koekmoer, Warren, *J. Chem. Soc.* **1955**, 63. Conformation: Culvenor, *Tetrahedron Letters* **1966**, 1091. Toxicity study: P. N. Harris *et al., J. Pharmacol. Exp. Ther.* **75**, 69 (1942). Review of toxicology of senecionine and other pyrrolizidine alkaloids: E. K. McLean, *Pharmacol. Rev.* **22**, 429-483 (1970). Comprehensive reviews of pyrrolizidine alkaloids: L. Bull *et al., The Pyrrolizidine Alkaloids* (North-Holland, Amsterdam, 1968) 293 pp; F. L. Warren in *The Alkaloids* **vol. 12**, R. H. F. Manske, Ed.

HO   CH₃

H₃C

CH₃   O

O   O

H

N

(Academic Press, New York, 1970) pp 245-331; D. J. Robins, *Fortschr. Chem. Org. Naturst.* **41**, 115-203 (1982).
Bitter plates, mp 236°. $[\alpha]_D^{25}$ −55.1° (c = 0.034 in chloroform). Practically insol in water. Freely sol in chloroform; slightly sol in alcohol, ether. LD_{50} i.v. in mice: 64.12 ±2.24 mg/kg (Harris)
C_{15}-trans Isomer, *integerrimine, squalidine, [15(20)E] 12-hydroxysenecionan-11,16,dione.* Isolated from several *Senecio* and *Crotalaria* spp.: L. Bull *et al., loc. cit.;* F. L. Warren, *loc. cit.* Identity of squalidine and integerrimine: A. G. Gonzalez, A. Calero, *Chem. & Ind. (London)* **1958**, 126. mp 172-172.5°; $[\alpha]_D^{28}$ +4.3° (c = 0.8 in methanol).

8596. Seneciphylline. *13,19-Didehydro-12-hydroxysenecionan-11,16-dione;* jacodine; α-longilobine. $C_{18}H_{23}NO_5$; mol wt 333.38. C 64.85%, H 6.95%, N 4.20%, O 24.00%. Hepatotoxic pyrrolizidine alkaloid, common constituent of *Senecio* species. Isoln from *Senecio platyphllus* DC. *Compositae:* A. Orechoff, W. Tiedebel, *Ber.* **68**, 650 (1935); from *S. jacobaea* L.: G. Barger, J. J. Blackie, *J. Chem. Soc.* **1937**, 584; from *Crotalaria juncea* L. *Leguminosae:* R. Adams, M. Gianturco, *J. Am. Chem. Soc.* **78**, 1919 (1956). Identity with α-longilobine: R. Adams, J. H. Looker, *ibid.* **73**, 134 (1951). Identity with jacodine: R. B. Bradbury, C. C. J. Culvenor, *Chem. & Ind. (London)* **1954**, 1021. Structural study: R. Adams *et al., J. Am. Chem. Soc.* **74**, 700 (1952). Revised structure: S. Masume, *Chem. & Ind. (London)* **1959**, 21. Review and evaluation of toxicity and carcinogenicity studies: *IARC Monographs* **10**, 319-325, 333-342 (1976). Comprehensive reviews of seneciphylline and other pyrrolizidine alkaloids: L. Bull *et al., The Pyrrolizidine Alkaloids* (North-Holland, Amsterdam, 1968) 293 pp; F. L.Warren in *The Alkaloids* **vol. 12**, R. H. F. Manske, Ed. (Academic Press, New York, 1970) pp 245-331.

HO   CH₃

H₃C

O   O

CH₂   O

O   O

H

N

Small rhombic platelets from hot alcohol or acetone, mp 217-218°. $[\alpha]_D$ −128° (chloroform). Easily sol in chloroform, ethylene chloride; less sol in alc, acetone. Difficultly sol in ether, ligroin.

8597. Senega. Senega snakeroot; Seneca snakeroot; rattlesnake root. Dried root of *Polygala senega* L., *Polygalaceae.* **Habit.** North America. *Constit.* Polygalic acid, senegin (polygalin)—a saponin, senegenin *q.v.*, fixed and volatile oils. Quantitative analysis of senega saponins: Kita *et al., Yakugaku Zasshi* **89**, 1111 (1969).
THERAP CAT: Expectorant, emetic.
THERAP CAT (VET): Has been used as expectorant.

8598. Senegenin. *(2β,3β,4α,12α)-12-(Chloromethyl)-2,3-dihydroxy-27-norolean-13-ene-23,28-dioic acid.* $C_{30}H_{45}ClO_6$; mol wt 537.14. C 67.08%, H 8.44%, Cl 6.60%, O 17.87%. From root of *Polygala senega* Linn., *Polygalaceae:* Jacobs, Isler, *J. Biol. Chem.* **119**, 155 (1937). Structure studies: Shamma, Reiff, *Chem. & Ind. (London)* **1960**, 1272; Dugan *et al., Can. J. Chem.* **42**, 491 (1964). Structure: Dugan *et al., Proc. Chem. Soc.* **1964**, 264. Formation from presenegenin: Pelletier *et al., Chem. Commun.* **1966**, 727.

Needles from dil alc, mp 281-283°. $[\alpha]_D$ +19° (c = 0.70 in ethanol). uv max: 205 nm (ϵ 6950). Sol in alcohol, acetone, acetic acid; practically insol in chloroform, benzene.

Monoacetate, $C_{32}H_{47}ClO_7$, crystals from methanol + chloroform, mp 209-211°. $[\alpha]_D$ +23° (c = 1.32). uv max (methanol): 205 nm (ϵ 6400).

Diacetate, $C_{34}H_{49}ClO_8$, dec 257-263°. $[\alpha]_D$ +29° (c = 1.09 in methanol).

8599. Senna. Dried leaflets of *Cassia senna* L. (*C. acutifolia* Delile), (Alexandria senna), or of *C. angustifolia* Vahl (India or Tinnevelly senna), *Leguminosae*. *Habit.* Egypt and neighboring districts, Southern India (Tinnevelly). *Constit.* Sennosides A and B; glucosides of rhein and chrysophanic acid: Becker, *Planta Med.* 7, 390 (1959); Fairbairne, *ibid.* 12, 260 (1964). Prepn of senna powder: Friedmann, Ryan, **Brit. pat.** 832,017 (1960 to Westminster Labs.); Corran, **Brit. pat.** 852,343 (1960 to C. E. Fulford); Baetz, **Ger. pat.** 1,158,211 (1963 to Ludwig Heumann).

Incompat: Mineral acids, carbonates, infusion cinchona, lime water, salts of heavy metals, tartar emetic.

THERAP CAT: Cathartic.

THERAP CAT (VET): Has been used as a purgative.

8600. Sennosides. Colonorm; Glysennid; Nytilax; Pursennid. $C_{42}H_{38}O_{20}$; mol wt 862.75. C 58.47%, H 4.44%, O 37.09%. Anthraquinone glucosides found in senna in equal amounts and in the rhubarbs where sennoside A predominates. Isoln from Tinnevelly senna (*Cassia angustifolia* Vahl, *Leguminosae*): Stoll *et al.*, *Helv. Chim. Acta* 32, 1892 (1949); **Brit. pat.** 804,232 (1958 to Byk-Gulden); Khorana, Sanghavi, *J. Pharm. Sci.* 53, 110 (1964); Menssen *et al.*, **U.S. pat.** 3,517,269 (1970 to Nattermann & Cie). Isoln from the rhizome of *Rheum palmatum* L.: Zwaving, *Planta Med.* 13, 474 (1965). Isoln of sennoside A from rhubarb: Miyamoto *et al.*, *Yakugaku Zasshi* 87, 1040 (1967). Sennosides B, C, D, E and F have also been isolated from rhubarb, *see* Oshio *et al.*, *Chem. Pharm. Bull.* 22, 823 (1974). Structure: Stoll *et al.*, *Helv. Chim. Acta* 33, 313 (1950). Sennoside A is built up from the dextrorotatory aglucon sennidin A and D-glucose. Sennoside B is built up from the intramolecularly compensated *meso*-sennidin B and D-glucose. *Review:* Stoll, Becker, *Fortschr. Chem. Org. Naturst.* 7, 248 (1950). Transport and mechanism of action studies: Dobbs *et al.*, *Farmaco Ed. Sci.* 30, 147 (1975).

Sennoside A, (R*,R*)-5,5'-Bis(β-D-glucopyranosyloxy)-9,9',-10,10'-tetrahydro-4,4'-dihydroxy-10,10'-dioxo[9,9'-bianthracene]-2,2'-dicarboxylic acid. Rectangular yellow plates from dil acetone, dec 200-240°. $[\alpha]_D^{20}$ −164° (c = 0.1 in 60% acetone); $[\alpha]_D^{20}$ −147° (c = 0.1 in 70% acetone); $[\alpha]_D^{20}$ −24° (c = 0.2 in 70% dioxane). Insol in water, benzene, ether,

chloroform. Sparingly sol in methanol, Carbitol, acetone, dioxane. The soly increases tremendously in water-miscible organic solvents contg an optimum of 30% w/w of water. Sol in aq solns of sodium bicarbonate. Sennoside A can be slowly isomerized to B in $NaHCO_3$ soln at 80°.

Sennoside B, (R*,S*)-5,5'-Bis(β-D-glucopyranosyloxy)-9,9',-10,10'-tetrahydro-4,4'-dihydroxy-10,10'-dioxo[9,9'-bianthracene]-2,2'-dicarboxylic acid. Light yellow prisms from dil acetone; fine needles from water, dec 180-186°. $[\alpha]_D^{20}$ −100° (c = 0.2 in 70% acetone); $[\alpha]_D^{20}$ −67° (c = 0.4 in 70% dioxane). uv max (0.5% $NaHCO_3$): 270, 308, 355 nm. Has the same soly characteristics as sennoside A, but is more sol. Can be recrystallized from large amounts of hot water. *Keep protected from air and light.*

THERAP CAT: Cathartic.

8601. Sepia. Cuttle-fish bone. Calcareous substance found under the skin of the back of *Sepia officinalis* L., *Cephalopoda*. *Habit.* Mediterranean Sea, Atlantic and Pacific Oceans. *Constit.* Calcium carbonate and phosphate, gluten.

USE: As a polishing agent and in tooth powders.

8602. Sepiomelanin. The eumelanin isolated from ink sacs of the cuttlefish, *Sepia officinalis* L., *Cephalopoda*. Present in the raw ink as a suspension of small, dark granules in a colorless plasma. Structure is a macromolecule or probably a mixture of polyacid polymers in which the predominant chemical unit is of the indole type. Structure studies: Piattelli, Nicolaus, *Tetrahedron* 15, 66 (1961); Piattelli *et al.*, *ibid.* 19, 2061 (1963). *Review:* R. A. Nicolaus, *Melanins* (Hermann, Paris, 1968) pp 68-91.

In the dry form, an amorphous, hygroscopic, black powder. Insol in all solvents.

USE: Pigment.

8603. Seratrodast. (±)-ζ-(2,4,5-Trimethyl-3,6-dioxo-1,-4-cyclohexadien-1-yl)benzeneheptanoic acid; (±)-7-(3,5,6-trimethyl-1,4-benzoquinon-2-yl)-7-phenylheptanoic acid; AA-2414; Abbott 73001. $C_{22}H_{26}O_4$; mol wt 354.45. C 74.55%, H 7.39%, O 18.06%. Thromboxane A_2-receptor antagonist. Prepn: S. Terao, Y. Maki, **Eur. pat. Appl.** 171,251; eidem, **U.S. pat.** 5,180,742 (1986, 1993 both to Takeda); M. Shiraishi *et al.*, *J. Med. Chem.* 32, 2214 (1989). Effect on thromboxane A_2/prostaglandin endoperoxidase-receptor: Y. Imura *et al.*, *Japan. J. Pharmacol.* 52, 35 (1990). Clinical effect in asthma: M. Fujimura *et al.*, *J. Allergy Clin. Immunol.* 87, 23 (1991). Clinical pharmacokinetics: Z. Hussein *et al.*, *Clin. Pharmacol. Ther.* 55, 441 (1994).

Crystals from ethanol, mp 128-129°.
THERAP CAT: Antiasthmatic.

8604. Serine. L-*Serine*; Ser; S; 2-amino-3-hydroxypropionic acid; β-hydroxyalanine; (S)-2-amino-3-hydroxypropanoic acid; α-amino-β-hydroxypropionic acid. C_3H_7NO$_3$; mol wt 105.09. C 34.29%, H 6.71%, N 13.33%, O 45.67%. A non-essential amino acid for human development. A major intracellular source of one carbon units for *de novo* purine synthesis. Found in the active site of serine proteases such as trypsin. Isoln from the silk protein, sericine: E. Cramer, *Prakt. Chem.* 96, 76 (1865). Structure and synthesis as DL-form: E. Fischer, H. Leuchs, *Ber.* 35, 3787 (1902). Early chemistry and biochemistry: *Amino Acids and Proteins*, D. M. Greenberg, Ed. (Charles C. Thomas, Springfield, IL, 1951) 950 pp., *passim*; J. P. Greenstein, M. Winitz, *Chemistry of the Amino Acids* vols 1-3 (John Wiley and Sons, Inc., New York, 1961) pp. 2202-2237, *passim*. Enzymatic determn in biological samples: R. D. Hurst *et* al., Anal. Biochem. **117**, 339 (1981). Synthetic reviews: K. Toi in *Synthetic Production and Utilization of Amino Acids*, T. Kaneko *et al.*, Eds. (Halsted, New York, 1974) pp 187-195; of DL-forms: L. Bassignani *et al., Ber.* **112**, 148-160 (1979). Review of role in enzyme active sites: B. S. Hartley, *Annals N. Y. Acad. Sci.* **227**, 438-445 (1974); J. Lamotte-Brasseur *et al., Biotech. Genet. Eng. Rev.* **12**, 189-230 (1994). Review of metabolism: K. Snell, D. A. Fell, *Adv. Enzyme Regul.* **30**, 13-32 (1990).

HO— [structure with NH₂ and COOH] —OH, NH₂

Hexagonal plates or prisms. Sweetish taste, insipid aftertaste. Dec 228°. Sublimes at 150° in high vac (10⁻⁴ mm Hg). $[\alpha]_D^{20}$ −6.83° (c = 10.41); $[\alpha]_D^{25}$ +14.95° (c = 9.34 in 1N HCl). Absorption spectrum: *Z. Physiol. Chem.* **176**, 257 (1928). Sol in water. Insol in common neutral solvents.

DL-Form, monoclinic prismatic leaflets from water. d 1.537. Dec 246° (closed capillary, bath preheated to 225°). Sublimes at 150° in high vac (10⁻⁴ mm Hg). pK_1' 2.21; pK_2' 9.15. Soly in water (g/l) at 0° = 22.04; at 25° = 50.23; at 50° = 103; at 75° = 192; at 100° = 322. Insol in common neutral solvents.

D-Form. Presence in rat brain: A. Hashimoto *et al., FEBS Letters* **296**, 33 (1992). As synaptic modulator: M. J Schell *et al., Proc. Nat. Acad. Sci. USA* **92**, 3948 (1995).

8605. Sermorelin. Human growth hormone-releasing factor(1-29)amide; human pancreatic somatoliberin(1-29)amide; GRF(1-29)NH₂; hpGRF(1-29)NH₂; SM-8144; Geref; Groliberin. $C_{149}H_{246}N_{44}O_{42}S$; mol wt 3357.93. C 53.30%, H 7.38%, N 18.35%, O 20.01%, S 0.95%. Amidated fragment of human somatoliberin, q.v. Characterization of *in vitro* activity: J. Rivier *et al., Nature* **300**, 276 (1982). Prepn: N. Ling *et al., Biochem. Biophys. Res. Commun.* **123**, 854 (1984); J. E. F. Rivier *et al.,* U.S. pat. **4,703,035** (1987 to Salk Inst. Biol. Stud.). Clinical pharmacology: M. Losa *et al., Klin. Wochenschr.* **62**, 1140 (1984). Clinical evaluation in growth hormone deficiency: M. B. Ranke *et al., Eur. J. Pediatr.* **145**, 485 (1986); R. Hümmelink *et al., Acta Paediatr. Scand. Suppl.* **331**, 48 (1987). Field trial in pigs and dairy heifers: D. Petitclerc *et al., J. Anim. Sci.* **65**, 996 (1987).

1
Tyr–Ala–Asp–Ala–Ile–Phe–Thr–Asn–Ser–Tyr–Arg–Lys–Val–Leu–Gly

29
NH₂Arg–Ser–Met–Ile–Asp–Gln–Leu–Leu–Lys–Arg–Ala–Ser–Leu–Gln

$[\alpha]_D^{20}$ −63.1° (c = 1 in 30% acetic acid).
THERAP CAT: Growth hormone releasing factor.

8606. Seroquel®. *2-[2-(4-Dibenzo[b,f][1,4]thiazepin-11-yl-1-piperazinyl)ethoxy]ethanol (E)-2-butenedioate (2:1) salt;* 11-[4-[2-(2-hydroxyethoxy)ethyl]-1-piperazinyl]dibenzo[b,f][1,4]thiazepine, hemifumarate; ICI-204636. $C_{46}H_{54}N_6O_8S_2$; mol wt 883.10. C 62.56%, H 6.16%, N 9.52%, O 14.49%, S 7.26%. $2C_{21}H_{25}N_3O_2S.C_4H_4O_4$. Benzothiazepine with mixed serotonin (5HT₂) and dopamine (D₂) receptor antagonistic properties. Prepn: E. J. Warawa, B. M. Migler, **Eur.** pat. **Appl. 240,228**; *eidem,* U.S. pat. **4,879,288** (1987, 1989 both to ICI). HPLC and GC/MS determn in

[structure of Seroquel with · C₄H₄O₄ and subscript 2]

plasma: R. H. Pullen *et al., J. Chromatog.* **573**, 49 (1992). Series of articles on pharmacology: *Psychopharmacology* **112**, 285-307 (1993). Clinical evaluation in schizophrenia: H. Wetzel *et al., ibid.* **119**, 231 (1995).
Crystals from ethanol, mp 172-173°.
THERAP CAT: Antipsychotic.

8607. Serotonin. *3-(2-Aminoethyl)-1H-indol-5-ol;* 5-hydroxytryptamine; 3-(β-aminoethyl)-5-hydroxyindole; 5-hydroxy-3-(β-aminoethyl)indole; enteramine; thrombocytin; thrombotonin; 5-HT. $C_{10}H_{12}N_2O$; mol wt 176.22. C 68.16%, H 6.86%, N 15.90%, O 9.08%. Vasoactive amine found in tissues and fluids of vertebrates and invertebrates. Extraction of enteramine from rabbit tissue, pharmacology: V. Erspamer, *Arch. Exp. Pathol. Pharmakol.* **196**, 343 (1940). Isoln of 5-HT from beef serum: M. M. Rapport *et al., Science* **108**, 329 (1948); *eidem, J. Biol. Chem.* **176**, 1237, 1243 (1948); M. M. Rapport, *ibid.* **180**, 961 (1949). Identity with enteramine, the enterochromaffin cell hormone: V. Erspamer, B. Asero, *Nature* **169**, 800 (1952). Ultrastructural localization in enterochromaffin cells: R. D. Dey, J. Hoffpauir, *J. Histochem. Cytochem.* **32**, 661 (1984). Synthesis starting with 5-benzyloxyindole: M. E. Speeter *et al., J. Am. Chem. Soc.* **73**, 5514 (1951); K. E. Hamlin, U.S. pat. **2,715,129** (1955 to Abbott). Alternate routes: R. Justoni, R. Pessina, U.S. pat. **2,947,757** (1960 to Vismara). Pharmacology: G. Reid, M. Rand, *Nature* **169**, 801 (1952). Existence of multiple serotonin receptors: S. J. Peroutka, S. H. Snyder, *Molec. Pharmacol.* **16**, 687 (1979); G. Engel *et al., Arch. Pharmacol.* **324**, 116 (1983); W. Feniuk *et al., Eur. J. Pharmacol.* **96**, 71 (1983). Extraneural synthesis in CNS: C. Maruki *et al., J. Neurochem.* **43**, 316 (1984). Effect on smooth muscle: W. Feniuk *et al., loc. cit.* Activity as neurotransmitter: S. G. Griffith *et al., Brain Res.* **247**, 388 (1982). Role in hemostasis: F. De Clerck *et al., Agents Actions* **15**, 627 (1984). Series of articles on role in hypertension, Raynaud's phenomenon and platelet activation: *J. Cardiovasc. Pharmacol.* **7**, Suppl. 7, 1-182 (1985). Review of role in migraine: E. T. MacKenzie *et al., Cephalalgia* **5**, 69-78 (1985). *Review:* V. Erspamer in E. Jucker, *Progress in Drug Research* **3**, 151-367 (1961). Books: *Serotonin in Health and Disease* vols. I-V, W. B. Essman, Ed. (Spectrum, New York, 1978); *Adv. Exp. Med.* **133**, entitled "Serotonin: Current Aspects of Neurochemistry and Function," B. Haber, Ed. (1981) 824 pp; *Serotonin and the Cardiovascular System*, P. M. Vanhoutte, Ed. (Raven Press, New York, 1985) 280 pp. *See also* Bufotenine and Ketanserin.

[structure of serotonin with H, N, HO, NH₂]

Hydrochloride, $C_{10}H_{12}N_2O.HCl$, hygroscopic crystals, sensitive to light, mp 167-168°. Water sol. Aq solns are stable at pH 2-6.4.
Complex with creatinine sulfate, monohydrate, $C_{14}H_{21}N_5O_6S.H_2O$, *Antemovis.* Plates, 215° dec. uv max (water at pH 3.5): 275 nm (ε 15000). pK_1' = 4.9; pK_2' = 9.8. pH of 0.01 molar aq soln: 3.6. Sol in glacial acetic acid. Very sparingly sol in methanol, 95% ethanol. Insol in abs ethanol, acetone, pyridine, chloroform, ethyl acetate, ether, benzene.

8608. Serpentaria. Snakeroot; snakeweed; sangrel; birthwort. Dried rhizome and roots of *Aristolochia serpentaria* L. (Virginia snakeroot), and of *Aristolochia reticulata* Nutt. (Texas snakeroot), *Aristolochiaceae.* Habit. U.S. *Constit.* Aristolochic acid, volatile oil (∼1%), resin, tannin.
THERAP CAT: Bitter tonic.

8609. Serpentine (Alkaloid). *3,4,5,6,16,17-Hexadehydro-16-(methoxycarbonyl)-19α-methyloxayohimbanium.* $C_{21}H_{20}N_2O_3$; mol wt 348.40. C 72.40%, H 5.79%, N 8.04%, O 13.78%. From roots of *Rauwolfia serpentina* (L.) Benth., *Apocynaceae:* Schlittler, Schwarz, *Helv. Chim. Acta* **33**, 1463 (1950). Structure: Schlittler *et al., ibid.* **37**, 1912 (1954);

Klohs *et al., J. Am. Chem. Soc.* **76**, 1332 (1954); Wenkert, Roychaudhuri, *ibid.* **80**, 1613 (1958). Stereoisomer of alstonine *(q.v.)*. Stereochemistry: Fritz, *Ann.* **655**, 148 (1962); Shamma, Richey, *J. Am. Chem. Soc.* **85**, 2507 (1963).

Solvated, yellow rods or leaflets from abs ethanol, mp 158° (air-dried), mp 175° (after drying at 120° in a high vacuum, but still not entirely solvent-free). Dec and turns red on drying at 150°. $[\alpha]_D^{25}$ +292° (c = 0.27 in methanol); $[\alpha]_D^{25}$ +267° (c = 0.21 in ethanol). uv max (ethanol): 252, 308, 370 nm (log ϵ 4.49, 4.30, 3.61). Freely sol in methanol. Sol in 10% acetic acid, in other organic solvents.

Nitrate, $C_{21}H_{20}N_2O_3 \cdot HNO_3$, yellow crystals, mp 170-172°. Only slightly sol in water.

Hydrochloride monohydrate, $C_{21}H_{20}N_2O_3 \cdot HCl \cdot H_2O$, yellow crystals, mp 246-248°. $[\alpha]_D$ +178°. Sol in water.

Perchlorate, $C_{21}H_{20}N_2O_3 \cdot HClO_4$, yellow crystals, mp 255-256°. $[\alpha]_D^{22}$ +185° (c = 0.5 in acetone). uv max (methanol): 252, 306-308, 365-370 nm (log ϵ 4.54, 4.35, 3.68).

8610. Sertaconazole. *1-[2-[(7-Chlorobenzo[b]thien-3-yl)methoxy]-2-(2,4-dichlorophenyl)ethyl]-1H-imidazole;* (±)-1-[2,4-dichloro β-[(7-chlorobenzo[b]thien-3-yl)methoxy]phenethyl]imidazole; 7-chloro-3-[1-(2,4-dichlorophenyl)-2-(1H-imidazol-1-yl)ethoxy-methyl]benzo[b]thiophene; FI-7045. $C_{20}H_{15}Cl_3N_2OS$; mol wt 437.78. C 54.87%, H 3.45%, Cl 24.30%, N 6.40%, O 3.65%, S 7.32%. Prepn: R. Foguet *et al., Eur. pat.* **Appl. 151,477;** *eidem,* U.S. pat. **5,135,943** (1985, 1992 both to Ferrer). Synthesis, antifungal activity, pharmacology, toxicology and clinical evaluations: Arzneimittel-Forsch. **42**, 689-774 (1992). Physical props: C. Albet *et al., ibid.* 695. HPLC determn: *eidem, J. Pharm. Biomed. Anal.* **10**, 205 (1992).

mp 146-147°.

Nitrate, $C_{20}H_{15}Cl_3N_2OS \cdot HNO_3$, *FI-7056, Dermofix, Zalain.* Odorless, white crystalline powder, mp 158-160°. Fairly sol in ethanol (1.7%); chloroform (1.5%); slightly sol in acetone (0.95%); very slightly sol in *n*-octanol (0.069%). Practically insol in water (< 0.01%). pKb: 7.26. uv max (methanol): 302.3 ($A_{1cm}^{1\%}$ 79.8), 292.9 260.3 nm.

THERAP CAT: Antifungal.

8611. Sertindole. *1-[2-[4-[5-Chloro-1-(4-fluorophenyl)-1H-indol-3-yl]-1-piperidinyl]ethyl]-2-imidazolidinone;* 5-chloro-1-(4-fluorophenyl)-3-[1-[2-(2-imidazolidinon-1-yl)ethyl]-4-piperidyl]-1H-indole; LU-23-174. $C_{24}H_{26}ClFN_4O$; mol wt 440.95. C 65.37%, H 5.94%, Cl 8.04%, F 4.31%, N 12.71%, O 3.63%. Limbic-selective dopamine D_2-receptor antagonist and serotonin $5HT_2$-receptor antagonist. Prepn: J. K. Perregaard, **Eur.** pat. **Appl. 200,322;** *idem,* U.S. pat. **4,710,500** (1986, 1987 both to Lundbeck); *idem et al., J. Med. Chem.* **35**, 1092 (1992). Neuropharmacology and receptor binding study: C. Sanchez *et al., Drug Dev. Res.* **22**, 239 (1991).

Crystals from acetone, mp 154-155°. Second crystal modification from 2-propanol + ethyl acetate, mp 166°.

THERAP CAT: Antipsychotic.

8612. Sertraline. *(1S-cis)-4-(3,4-Dichlorophenyl)-1,2,3,-4-tetrahydro-N-methyl-1-naphthalenamine; (1S,4S)-4-(3,4-dichlorophenyl)-1,2,3,4-tetrahydro-N-methyl-1-naphthyl-amine.* $C_{17}H_{17}Cl_2N$; mol wt 306.23. C 66.68%, H 5.60%, Cl 23.15%, N 4.57%. Selective serotonin uptake inhibitor. Prepn: W. M. Welch *et al., Eur.* pat. **Appl. 30,081;** *eidem,* U.S. pat. **4,536,518** (1981, 1985 both to Pfizer). Prepn and stereospecific activity: W. M. Welch *et al., J. Med. Chem.* **27**, 1508 (1984). Pharmacology in animals: B. K. Koe *et al., J. Pharmacol. Exp. Ther.* **226**, 686 (1983). Psychopharmacology and pharmacodynamics in humans: B. Saletu *et al., J. Neural Transm.* **67**, 241 (1986). GC determn in human plasma: H. G. Fouda *et al., J. Chromatog.* **417**, 197 (1987). Clinical evaluation in depression: F. W. Reimherr *et al., Psychopharmacol. Bull.* **24**, 200 (1988).

Hydrochloride, $C_{17}H_{17}Cl_2N \cdot HCl$, *CP-51974-1, Lustral, Zoloft.* Crystals, mp 243-245°. $[\alpha]_D^{23}$ +37.9° (c = 2 in methanol).

THERAP CAT: Antidepressant.

8613. Serum Albumin. Most abundant of the plasma proteins (comprising ~60%). Structure characterized as a single, carbohydrate-free polypeptide chain consisting of 580-585 amino acid residues; stabilized by seventeen S-S bridges. Main biological functions include maintenance of osmotic pressure and the transport of sparingly soluble physiological and pharmacological substances, esp long-chain fatty acids and bilirubin. Trace amounts present in normal urine, excreted as the major protein constit in the condition of proteinuria. Review of determn in serum and plasma: P. G. Hill, *Ann. Clin. Biochem.* **22**, 565-578 (1985); of biosynthesis and biological function: M. A. Rothschild *et al., Hepatology* **8**, 385-401 (1988); of structural studies: D. C. Carter, J. X. Ho, *Adv. Protein Chem.* **45**, 153-203 (1994).

Human serum albumin, HSA, Albuminate, Albuminar, Albumisol, Albuspan, Buminate, Pro-Bumin, Proserum. Mol wt 65,000. Net charge per molecule (pH 7.4) −19. High solubility at neutrality: 30% solns prepd in water, 40% solns prepd in methanol or ethanol. Below pH 3, sol in 80-100% methanol, ethanol or acetone. Low viscosity. Stable, even at high temperatures. Review of biochemistry and clinical experience: *Proceedings of the Workshop on Albumin,* J. T. Sgouris, A. René, Eds. (DHEW Publication No. (NIH) 76-925, 1975) 355 pp; of therapeutic use: G. E. Hastings, P. G. Wolf, *Arch. Fam. Med.* **1**, 281-287 (1992).

¹³¹I-labelled form, *iodinated (¹³¹I) human serum albumin, ¹³¹I HSA, Albumotope-¹³¹I, Macroscan-131, Risa-131-H.* Normal human serum albumin mildly iodinated with radioactive iodine (¹³¹I) which has a half-life of 8 days, and emits beta and gamma rays. Contains not more than one atom of iodine per molecule of albumin. Physical and chemical properties essentially the same as those of albumin. Stable at 10° for at least 3 weeks. Supplied for injection in aq isotonic sodium chloride soln.

Sonicated human serum albumin, Albunex. Suspension of air filled albumin microspheres in a 5% (w/v) human albumin soln. Prepn: M. W. Keller *et al., J. Am. Coll. Cardiol.* **12**, 1039 (1988); D. Cerny *et al.,* **Eur.** pat. **Appl.** **359,246**; eidem, **U.S.** pat. **4,957,656** (both 1990 to Molecular Biosystems). Pharmacology: M. W. Keller *et al., J. Am. Soc. Echo.* **2**, 48 (1989). Diagnostic use in echocardiography: L. J. Crouse *et al., J. Am. Coll. Cardiol.* **22**, 1494 (1993). Physical and biochemical properties: C. Christiansen *et al., Biotechnol. Appl. Biochem.* **19**, 307 (1994). *Review:* N. Sponheim *et al., IEEE Conf. Publ., Acoustic Sensing and Imaging* **369**, 103-108 (1993). pH 6.79-6.86. Osmolality 253-264 mmol/kg. Viscosity 1.44.

THERAP CAT: Plasma volume expander; treatment of hypoproteinemia. ¹³¹I-labelled form as diagnostic aid (blood volume determination). Sonicated form as diagnostic aid (ultrasound contrast agent).

8614. Sesame Oil. Benne oil; teel oil; gingilli oil. From the seeds of cultivated varieties of *Sesamum indicum* L., *Pedaliaceae. Constit.* Olein, stearin, palmitin, myristin, linolein, sesamin, sesamolin. *Review:* Budowski, Markley, *Chem. Rev.* **48**, 125 (1951).

Pale yellow oil; almost odorless; bland taste. d 0.916-0.920. Solidif about −5°. [α]$_D^{25}$ +1° to +9°. n_D^{40} 1.4650-1.4665. Sapon no. 188-193. Iodine no. 103-122. Soluble in chloroform, ether, petr ether, carbon disulfide; slightly sol in alcohol. Insol in water.

USE: Manuf oleomargarine, cosmetics, iodized oil, etc. Pharmaceutic aid (solvent).

THERAP CAT (VET): Formerly as emollient, pediculicide.

8615. Sesamex. *5-[1-[2-(2-Ethoxyethoxy)ethoxy]ethoxy]-1,3-benzodioxole; acetaldehyde 2-(2-ethoxyethoxy)ethyl 3,4-(methylenedioxy)phenyl acetal;* 2-(3,4-methylenedioxyphenoxy)-3,6,9-trioxaundecane; Sesoxane. C$_{15}$H$_{22}$O$_6$; mol wt 298.34. C 60.39%, H 7.43%, O 32.18%. Prepn: M. Beroza, **U.S.** pat. **2,832,792** (1958 to U.S. Secy. Agr.). Use as a synergist: S. A. El-Aziz *et al., J. Econ. Entomol.* **62**, 318 (1969); H. T. Gordon, L. T. Jao, *ibid.* **64**, 546 (1971).

Straw-colored liquid, bp$_{0.08}$ 137-141°. Sol in kerosene, Freon 11, Freon 12; other organic solvents. n_D^{25} 1.4938.

USE: Insecticide, especially as synergist for pyrethrins, allethrin, methoxychlor.

8616. Sesamin. *5,5′-(Tetrahydro-1H,3H-furo[3,4-c]furan-1,4-diyl)bis-1,3-benzodioxole; tetrahydro-1,4-bis[3,4-(methylenedioxy)phenyl]-1H,3H-furo[3,4-c]furan;* 2,6-bis(3,4-methylenedioxyphenyl)-3,7-dioxabicyclo[3.3.0]octane. C$_{20}$H$_{18}$O$_6$; mol wt 354.36. C 67.79%, H 5.12%, O 27.09%. Isolated from the bark of various *Fagara* species; from sesame oil: Bertram *et al., Biochem.* **197**, 1 (1928); from fruit of *Piper lowong* Blume, *Piperaceae:* Peinemann, *Arch. Pharm.* **234**, 238 (1896). Structure: Cohen, *Rec. Trav. Chim.* **57**, 653 (1938). Synthesis of *dl*-form: Beroza, Schechter, *J. Am. Chem. Soc.* **78**, 1242 (1956); Freudenberg, *Naturwiss.* **43**, 16 (1956). Identity of *dl*-form with *fagarol:* Carnmalm, *Acta Chem. Scand.* **9**, 1111 (1955); identity of *d*-form with *pseudocubebin: idem, ibid.* **10**, 134 (1956). Stereochemistry of the *d*-isomer: Freudenberg, Sidhu, *Ber.* **94**, 851 (1961); Becker, Beroza, *Tetrahedron Letters* **1962**, 157. *Review:* Budowski, *J. Am. Oil Chem. Soc.* **41**, 280 (1964).

(+)-Sesamin

d-Form, needles from ethanol, mp 122-123°. [α]$_D^{20}$ +64.5° (c = 1.75 in chloroform).

dl-Form, crystals from ethanol, mp 125-126°. Freely sol in chloroform, benzene, acetic acid, acetone. Practically insol in water, alkaline solns and hydrochloric acid.

USE: Insecticide synergist.

8617. Sesamolin. *5-[4-(1,3-Benzodioxolol-5-yloxy)tetrahydro-1H,3H-furo[3,4-c]furan-1-yl]-1,3-benzodioxole; tetrahydro-1-[3,4-(methylenedioxy)phenoxy]-4-[3,4-(methylenedioxy)phenyl]-1H,3H-furo[3,4-c]furan.* C$_{20}$H$_{18}$O$_7$; mol wt 370.36. C 64.86%, H 4.90%, O 30.24%. Constituent of sesame seeds, the seeds of *Sesamum indicum* L., *Pedaliaceae.* Isoln: Canzoneri, Perciabosco, *Gazz. Chim. Ital.* **33**, II, 253 (1907). Structure and synthesis: J. Boseken, W. D. Cohen, *Rec. Trav. Chim.* **55**, 815 (1936); E. Haslam, R. D. Haworth, *J. Chem. Soc.* **1955**, 827. Stereochemistry: E. Haslam, *ibid.* **1970**, 2332.

White plates from ethanol, mp 93-94°. [α]$_D^{20}$ +212° (chloroform).

USE: Synergist for pyrethrum insecticides.

8618. Sesin. *Benzoic acid 2,4-dichlorophenoxyethyl ester; 2-(2,4-dichlorophenoxy)ethanol benzoate;* 2,4-dichlorophenoxyethyl benzoate. C$_{15}$H$_{12}$Cl$_2$O$_3$; mol wt 311.16. C 57.90%, H 3.89%, Cl 22.79%, O 15.43%. Prepn: Lambrech, **U.S.** pat. **2,765,224** (1956 to Union Carbide).

Crystals, mp 66°. bp$_{1.5}$ 185°.

USE: In weed control.

8619. Setastine. *1-[2-[1-(4-Chlorophenyl)-1-phenylethoxy]ethyl]hexahydro-1H-azepine;* 1-[2-[(p-chloro-α-methyl-α-phenylbenzyl)oxy]ethyl]hexahydro-1H-azepine; 1-[2-(4-chloro-α-methyl-α-phenylbenzyloxy)ethyl]azepam; *N-[2-(α-methyl-p-chlorobenzhydryloxy)ethyl]hexamethyleneimine; N-[2-[(4-chloro-α-methyl-α-phenylbenzyl)oxy]ethyl]hexamethylenimine.* C$_{22}$H$_{28}$ClNO; mol wt 357.92. C 73.83%, H 7.89%, Cl 9.91%, N 3.91%, O 4.47%. Nonsedating-type histamine H$_1$-receptor antagonist; structurally similar to clemastine, q.v. Prepn: I. Beck *et al.,* **Ger.** pat. **2,528,194** (1976 to EGYT), *C.A.* **85**, 46439h (1976). *In vitro* antihistaminic activity: A. Németh, Z. Huszti, *Agents Actions* **23**, 194 (1988). Clinical pharmacology: Z. Vezekényi *et al., Acta Med. Hung.* **44**, 55 (1987).

Hydrochloride, $C_{22}H_{28}ClNO \cdot HCl$, *EGIS-2062, EGYT-2062, Loderix.*
THERAP CAT: Antihistaminic.

8620. Sethoxydim. *2-[1-(Ethoxyimino)butyl]-5-[2-(ethylthio)propyl]-3-hydroxy-2-cyclohexen-1-one;* BAS-9052; NP-55; Poast. $C_{17}H_{29}NO_3S$; mol wt 327.49. C 62.35%, H 8.93%, N 4.28%, O 14.66%, S 9.79%. Selective post-emergence herbicide. Prepn: **Japan. Kokai 81 75,408** (1981 to Nippon Soda), *C.A.* **95**, 110175e (1981). Activity: C. M. Knott, *Proc. Brit. Crop. Prot. Conf.—Weeds* 1980, 39; G. H. Ingram *et al., ibid.* 91.

USE: Herbicide.

8621. Sevoflurane. *1,1,1,3,3,3-Hexafluoro-2-(fluoromethoxy))propane; fluoromethyl 2,2,2-trifluoro-1-(trifluoromethyl)ethyl ether;* fluoromethyl 1,1,1,3,3,3-hexafluoro-2-propyl ether; Sevofrane. $C_4H_3F_7O$; mol wt 200.06. C 24.02%, H 1.51%, F 66.48%, O 8.00%. $(CF_3)_2CHOCH_2F$. Prepn: B. M. Regan, J. C. Longstreet, **Ger. pat. 1,954,268;** *eidem*, **U.S. pat. 3,689,571** (1970, 1972 both to Baxter Labs.). Pharmacology and physicochemical properties: R. F. Wallin *et al., Anesth. Analg.* **54**, 758 (1975). Clinical pharmacology and metabolism: D. A. Holaday, F. R. Smith, *Anesthesiology* **54**, 100 (1981). Toxicity study: D. P. Strum *et al., Anesth. Analg.* **66**, 769 (1987). Minimum alveolar concentration (MAC) in man: T. Katoh, K. Ikeda, *Anesthesiology* **66**, 301 (1987). Preliminary clinical evaluation in pheochromocytoma: M. Doi, K. Ikeda, *ibid*, **70**, 360 (1989).
Volatile liquid, bp_{751} 58.1-58.2°; bp_{760} 58.5°. d_4^{23} 1.505. Vapor pressure (25°): 200 torr; (20°): 160 torr. Nonflammable in air at ambient temps; flammability limit in O_2: 11 vol. percent; in N_2O: 10 vol. percent. Partition coefficient (37°): blood:gas 0.60±0.07; water:gas 0.36±0.01; corn oil:gas 41.6±2.1.
THERAP CAT: Anesthetic (inhalation).

8622. Shark Liver Oil. Robecote. Oil expressed from the livers of sharks and other *Elasmobranchii* species. *Constit:* Squalene, pristane, vitamins A and D, esters of fatty acids (esp. palmitic and oleic), glycerol ethers, triglycerides, cholesterol and fatty alcohols. Isoln: A. Ehrenreich, **Brit. pat. 284,657** (1927); C. E. Brown, S. A. Reed, **Brit. pat. 1,021,542** (1966 to Marfleet Refining Co. Ltd.). Composition: N. A. Sörensen, J. Mehlum, *Acta Chem. Scand.* **2**, 140 (1948); E. Gelpi, J. Oro, *J. Am. Oil Chem. Soc.* **45**, 144 (1968); R. Lombardi *et al., Ann. Pharm. Franc.* **29**, 429 (1971). Comparison with cod liver oil: A. O. Banjo, *J. Food Technol.* **14**, 107 (1979).
Clear, yellow oil. Phys props of liver oil from basking shark: n_D^{20} 1.4784; d^{20} 0.922; sapon no. 140-146; iodine no. 140-145 (Lombardi).
THERAP CAT: Topical protectant.

8623. Shellac. Lacca; lac. A resinous excretion of the insect *Laccifer (Tachardia) lacca* Kerr, order *Homoptera,* family *Coccidae.* Different resiniferous trees of India serve as host trees as the season progresses. The insects suck the juice of the tree and excrete "stick-lac" almost continuously. Whitest shellac is produced while the kusum tree *(Schleichera trijuga)* is the host. Most shellac is produced in the

Central and in the United provinces of India and in the states of Bihar and Orissa. The major component of lac is a resin which upon mild hydrolysis gives a complex mixture of aliphatic and alicyclic hydroxy acids and their polyesters. The composition of the hydrolysate depends on the lac source and the time of collection. The major component of the aliphatic fraction is aleuritic acid, *q.v.;* the major component of the alicyclic fraction is shellolic acid, *q.v.:* Yates, Field, *Tetrahedron* **26**, 3135 (1970). Physical and chemical properties of shellac: Cockeran, Levine, *J. Soc. Cosmet. Chem.* **12**, 316 (1961). Trends in shellac chemistry: *eidem, Am. Ink Maker* **39** (7), 26 (1961); E. Hicks, *Shellac, Its Origin and Applications* (Chemical Publ. Co., New York, 1961). Review: J. Martin in Kirk-Othmer *Encyclopedia of Chemical Technology* vol. **20** (Wiley-Interscience, New York, 3rd ed., 1982) pp 737-747.
Brittle, yellowish, transparent sheets or crushed pieces or powder. d 1.035-1.140. mp 115-120°. Sapon no. 185-210; iodine no. 10-18. Solubility in alcohol 85-95% w/w (very slowly sol); in ether 13-15%; in benzene 10-20%; in petr ether 2-6%. Sparingly sol in oil of turpentine. Insol in water; sol in aq solns of ethanolamines, of alkalies or borax with slightly purple color.
USE: Chiefly in lacquers and varnishes; also in manuf buttons, grinding wheels, sealing wax, cements, inks, phonograph records, paper; for stiffening hats; in electrical machines; coating confections and medicinal tablets; finishing leather.

8624. Shellolic Acid. *[3R-(3α,3aβ,4α,7β,8α,8aα)]-2,3,4,7,8,8a-Hexahydro-4-hydroxy-8-(hydroxymethyl)-8-methyl-1H-3a,7-methanoazulene-3,6-dicarboxylic acid; 10β,13-dihydroxycedr-8-ene-12,15-dioic acid.* $C_{15}H_{20}O_6$; mol wt 296.32. C 60.80%, H 6.80%, O 32.40%. Major component of the alicyclic fraction of shellac hydrolysate of which it usually constitutes 5-8%. Isoln: Harries, Nagel, *Ber.* **55**, 3833 (1922); Nagel, Mertens, *ibid.* **70**, 2173 (1937); Kirk *et al., J. Am. Chem. Soc.* **63**, 1243 (1941). Structure studies: Carruthers *et al., J. Chem. Soc.* 1961, 5251; Cookson *et al., Tetrahedron* **18**, 547, 1321 (1962); Yates, Field, *ibid.* **26**, 3135 (1970). Stereochemistry: Yates *et al., ibid.* **26**, 3159 (1970).

Crystals, mp 204-207°.
Dimethyl ester, crysts from ethyl acetate, mp 153-154.5°. uv max (ethanol): 231 nm (ε 6200).

8625. Shikimic Acid. *[3R-(3α,4α,5β)]-3,4,5-Trihydroxy-1-cyclohexene-1-carboxylic acid.* $C_7H_{10}O_5$; mol wt 174.15. C 48.28%, H 5.79%, O 45.93%. Naturally occurring (−)-form is a major biosynthetic precursor of phenylalanine, tyrosine, and tryptophan and hence of the majority of plant alkaloids. It is also involved in the biosynthesis of lignin, *q.v.,* flavonoids and other important aromatic compounds. Isoln from the fruit of the oriental plant *Illicium religiosum* Sieb. et Zucc., *Magnoliaceae* (called in Japanese shikimi-no-ki): J. F. Eykman, *Rec. Trav. Chim.* **4**, 32 (1885); *idem, Ber.* **24**, 1278 (1891). Structural study: H. O. L. Fischer, G. Dangshat, *Helv. Chim. Acta* **17**, 1200 (1934). Configuration: *eidem, ibid.* **18**, 1206 (1935); **20**, 705 (1937). Conformation in soln: L. D. Hall, *J. Org. Chem.* **29**, 297 (1964). Enzymatic synthesis: P. R. Srinivasan *et al., J. Am. Chem. Soc.* **77**, 4943 (1955). Stereochemistry: R. McCrindle *et al., J. Chem. Soc.* 1960, 1560; E. E. Smissman *et al., J. Am. Chem. Soc.* **84**, 1040 (1962). Improved synthesis: J. L. Pawlak, G. A. Berchtold, *J. Org. Chem.* **52**, 1765 (1987). Synthesis of the (±)-form: R. Grewe, I. Hinrichs, *Ber.* **97**, 443 (1964); R. Grewe, S. Kersten, *Angew. Chem.* **77**, 859 (1965). Carcinogenicity study: I. A. Evans, M. A. Osman, *Nature* **250**, 348 (1974). Reviews: B. A. Bohm, *Chem. Rev.* **65**, 435-466 (1965); B. Ganem, *Tetrahedron* **34**, 3353-3383 (1978); J. B. Harborne, *Biosynthesis* **6**, 40-75 (1980). Books: E. Haslam, *The Shikimate Pathway* (Halsted Press, New York, 1974) 316 pp; *idem*, "Shikimic Acid Metabolites" in *Comprehensive Organic Chemistry* **vol. 5** (Pergamon Press, Oxford, 1979) pp 1167-1205; U. Weiss, J. M. Edwards, *The Biosynthesis of Organic Compounds* (Wiley, New York, 1980) 728 pp.

Needles from methanol/ethyl acetate, mp 183-184.5°. Sublimes. $[\alpha]_D -161°$ (c = 0.57 in methanol). $[\alpha]_D^{18}$ $-183.8°$ (c = 4.03 in water). uv max (alcohol): 213 nm (ϵ 8900). K at 14.1° = 7.1 × 10^{-5}. Soly in water about 18 g/100 ml. Soly at 23° (g/100 m): 2.25 in abs alcohol: 0.015 in anhydr ether. Practically insol in chloroform, benzene, petr ether. LD_5 i.p. in mice: 1000 mg/kg (Evans, Osman).

(±)-Form, needles from methanol/ethyl acetate, mp 191-192°. uv max (ethanol): 212 nm (ϵ 8200).

8626. Shionone. *D:A-Friedo-18,19-secolup-19-en-3-one; 3β,5α,8,17aβ-tetramethyl-3-(4-methyl-3-pentenyl)-D-homoandrostan-17-one.* $C_{30}H_{50}O$; mol wt 426.73. C 84.44%, H 11.81%, O 3.75%. From *Aster tartaricus* L., *Compositae*: Nakaoki, *J. Pharm. Soc. Japan* **52**, 499 (1932); *C.A.* **26**, 4821[4] (1932); Takahashi *et al.*, *ibid.* **79**, 1281 (1959); *C.A.* **54**, 4667g (1960). Structure: Y. Tanahashi *et al.*, *Bull. Soc. Chim. France* **1966**, 1670. Conformation: Y. Moriyama *et al.*, *ibid.* **1968**, 2890.

Crystals from alc, mp 161-162°. $[\alpha]_D^{26.5} -56.1°$ (c = 1.07 in chloroform). uv max (chloroform): 290 nm (log ϵ 1.47).

8627. Showdomycin. *3-β-D-Ribofuranosyl-1H-pyrrole-2,5-dione; 2-β-D-ribofuranosylmaleimide.* $C_9H_{11}NO_6$; mol wt 229.19. C 47.17%, H 4.84%, N 6.11%, O 41.89%. Nucleoside antibiotic isolated from *Streptomyces showdoensis*: Nishimura *et al.*, *J. Antibiot.* **17A**, 148 (1964); Nishimura, Fr. pat. M2751 corresp to U.S. pat. 3,316,149 (1964, 1967, both to Shionogi). Structure: Nakagawa *et al.*, *Tetrahedron Letters* **1967**, 4105; Darnall *et al.*, *Proc. Nat. Acad. Sci. USA* **57**, 548 (1967). Synthesis: Kalvoda *et al.*, *Tetrahedron Letters* **1970**, 2297; Trummlitz, Moffatt, *J. Org. Chem.* **38**, 1841 (1973); T. Sato *et al.*, *Tetrahedron Letters* **1978**, 1829; J. G. Buchanan *et al.*, *J. Chem. Soc. Perkin Trans. I* **1979**, 225; T. Inoue, I. Kuwajima, *Chem. Commun.* **1980**, 251; A. P. Kozikowski, A. Ames, *J. Am. Chem. Soc.* **103**, 3923 (1981); Y. Araki *et al.*, *Tetrahedron Letters* **29**, 351 (1988). Exhibits antitumor activity: Shinzo *et al.*, *J. Antibiot.* **17A**, 234 (1964). Review: D. W. Visser, S. Roy-Burman in *Antibiotics* **vol. 5**(pt. 2), F. E. Hahn, Ed. (Springer-Verlag, New York, 1979) pp 363-371.

Leaflets from acetone + benzene, mp 153-154°. $[\alpha]_D^{22.5}$ +49.9° (H_2O). uv max: 220-221 nm ($E_{1cm}^{1\%}$ 422). Sol in water, alcohol, acetone, dioxane; insol in ethyl ether, benzene, petr ether. More stable in acid media than in neutral or alkaline. LD_{50} in mice (mg/kg): 25 i.p.; 18 s.c.; 110 i.v. (Nishimura).

8628. Sialic Acids. Nonulosaminic acids. Family of amino sugars contg 9 or more carbon atoms, N- and O-substituted derivatives of neuraminic acid, *q.v.* Widely distributed throughout the animal kingdom; appear to be regular components of all types of mucoproteins, mucopolysaccharides and certain mucolipids. Bovine and ovine submaxillary gland sialic acids are diacetyl derivs. Pig submaxillary mucin, and pig, horse, and ox erythrocytes contain sialic acids having an N-glycolyl group. Reviews: A. Gottschalk, *The Chemistry and Biology of Sialic Acids and Related Substances* (Cambridge U. Press, 1960); *idem, Rev. Pure Appl. Chem.* **12**, 46 (1962). Books: *Biological Roles of Sialic Acid*, A. Rosenberg, C.-L. Schengrund, Eds. (Plenum Press, New York, 1976) 393 pp; *Sialic Acids: Chemistry, Metabolism & Function*, R. Schauer, Ed. (Springer-Verlag, New York, 1983) 346 pp.

N-Acetylneuraminic acid

N-Acetylneuraminic acid, $C_{11}H_{19}NO_9$, *5-(acetylamino)-3,5-dideoxy-D-glycero-D-galacto-2-nonulosonic acid, NAN, NANA, lactaminic acid, O-sialic acid.* mol wt 309.27. The most important of the 5 sialic acids. Isoln: Klenk, Faillard, *Z. Physiol. Chem.* **298**, 230 (1954). Prepn from sheep submaxillary mucins: Blix, *Methods Carbohyd. Chem.* **1**, 246 (1962). Synthesis: Cornforth *et al.*, *Biochem. J.* **68**, 57 (1958); R. Csuk *et al.*, *Helv. Chim. Acta* **71**, 609 (1988). First synthesis from non-carbohydrate sources: M. P. DeNinno, S. J. Danishevsky, *J. Org. Chem.* **51** 2615 (1986). Configuration: Kuhn, Brossmer, *Angew. Chem. Int. Ed.* **1**, 218 (1962). Crystals from water/acetic acid (0.8:12) at 5°, mp 180-182° (dec). $[\alpha]_D^{20} -33.1°$ (c = 0.9 in water).

8629. Sibutramine. *1-(4-Chlorophenyl)-N,N-dimethyl-α-(2-methylpropyl)cyclobutanemethanamine; N-1-[1-[1-(4-chlorophenyl)cyclobutyl]-3-methylbutyl-N,N-dimethylamine.* $C_{17}H_{26}ClN$; mol wt 279.85. C 72.96%, H 9.36%, Cl 12.67%, N 5.01%. Monoamine reuptake inhibitor. Prepn: J. E. Jeffery *et al.*, **Ger. pat. 3,212,682** (1982 to Boots); J. E. Jeffery, D. Whybrow, **Brit. pat. 2,184,122;** J. E. Jeffery, U.S. pat. 4,929,629 (1987, 1990 both to Boots). Pharmacology of sibutramine and its metabolites: G. P. Luscombe *et al.*, *Neuropharmacology* **28**, 129 (1989). Clinical pharmacology: D. J. King, N. Devaney, *Brit. J. Clin. Pharmacol.* **26**, 607 (1988). Clinical efficacy in weight control: M. Weintraub *et al.*, *Clin. Pharmacol. Ther.* **50**, 330 (1991).

Hydrochloride monohydrate, $C_{17}H_{26}ClN \cdot HCl \cdot H_2O$, *BTS-54524, Reductil, Meridia.* Crystals from water, mp 193-195.5°.

THERAP CAT: Anorexic.

8630. Siccanin. *[4aS-(4aα,6aα,11bα,13aR*,13bα)]-1,2,-3,4,4a,5,6,6a,11b,13b-Decahydro-4,4,6a,9-tetramethyl-13H-benzo[a]furo[2,3,4-mn]xanthen-11-ol;* Tackle. $C_{22}H_{30}O_3$;

mol wt 342.48. C 77.16%, H 8.83%, O 14.01%. Antibiotic obtained from the cultured broth of *Helminthosporium siccans* Drechsler, a parasitic organism of rye grass (*Lolium multiflorum* Lam). Isoln: K. Ishibashi, **Japan.** pat. **62 3548** (1962 to Sankyo); *idem, J. Antibiot.* **15A**, 161 (1962). The *cis* formation of the decahydronaphthalene system may be the first example of the naturally occurring drimane skeleton. Structure: K. Hirai *et al., Tetrahedron Letters* **1967**, 2177; K. Hirai *et al., Tetrahedron* **27**, 6057 (1971). Approach to synthesis: H. Oida *et al., Chem. Pharm. Bull.* **20**, 2634 (1972); T. Yoshida *et al., ibid.* 2642. Total synthesis of racemate: M. Kato *et al., J. Am. Chem. Soc.* **103**, 2434 (1981); *eidem, Tetrahedron* **43**, 711 (1987). CD spectral study: S. Nozoe *et al., Tetrahedron* **30**, 2773 (1974). *In vitro* activity studies: M. Arai *et al., Antimicrob. Ag. Chemother.* **1969**, 247; M. G. Bellotti, L. Riviera, *Chemioterapia* **4**, 431 (1985); *in vivo* studies: S. Sugarwara, *Antimicrob. Ag. Chemother.* **1969**, 253. Toxicology: Y. Suzuki *et al., Sankyo Kenkyusho Nempo* **21**, 120 (1969), *C.A.* **72**, 130939k (1970). Mode of action: K. Nose, A. Endo, *J. Bacteriol.* **105**, 176 (1971). Biosynthesis: K. Suzuki *et al., Bioorg. Chem.* **3**, 72 (1974). Clinical evaluation in surface mycosis: D. Crippa *et al., G. Ital. Dermatol. Venereol.* **120**, 57 (1985).

White to pale yellow odorless crystals, mp 139-140°. $[\alpha]_D^{20}$ −136° (c = 2 in chloroform). uv max (ethanol): 210, 285 nm (ε 45690, 1717). pK 10.9. Very sol in chloroform, dimethylformamide, benzene; freely sol in acetone, ether, ethyl acetate; sol in alcohol. Practically insol in water. LD$_{50}$ in mice, rats (mg/kg): > 6000, > 1000 orally; > 3000, > 600 s.c. or i.p. (Suzuki).

THERAP CAT: Antifungal.

8631. Siduron. *N-(2-Methylcyclohexyl)-N'-phenylurea;* Tupersan. $C_{14}H_{20}N_2O$; mol wt 232.33. C 72.38%, H 8.68%, N 12.06%, O 6.89%. Prepn: Luckenbaugh, **Belg.** pat. **631,-289** (1963 to du Pont), *C.A.* **60**, 15775h (1964), corresp to **Brit.** pat. **1,028,818.** Prepn of *trans*-form: Hayman *et al., J. Pharm. Pharmacol.* **16**, 538 (1964).

Solid, mp 133-138°. Sol in water (18 ppm); sol to the extent of 10% or more in ethanol, dimethylacetamide, dimethylformamide, methylene chloride.

trans-Form, crystals, mp 157-159°.

USE: Herbicide.

8632. Sikkimotoxin. *3a,4,9,9a-Tetrahydro-4-hydroxy-6,7-dimethoxy-9-(3,4,5-trimethoxyphenyl)naphtho[2,3-c]-furan-1(3H)-one;* 1-hydroxy-2-hydroxymethyl-6,7-dimethoxy-4-(3',4',5'-trimethoxyphenyl)-1,2,3,4-tetrahydronaphthalene-3-carboxylic acid lactone. $C_{23}H_{26}O_8$; mol wt 430.45. C 64.18%, H 6.09%, O 29.73%. Isoln from roots and rhizomes of *Podophyllum sikkimensis* R. Chatterjee et Mukerjee, *Berberidaceae:* Chatterjee, Datta, *Indian J. Physiol. Allied Sci.* **4**, 61 (1950). Structure: Chatterjee, Chakravarti, *J. Am. Pharm. Assoc. Sci. Ed.* **41**, 415 (1952); Schreier, *Helv. Chim. Acta* **47**, 1529 (1964).

Crystals from ethanol + benzene, mp 120°. $[\alpha]_D^{25}$ −91.8°. Freely sol in ethanol, methanol, chloroform. Sparingly sol in benzene. Practically insol in water and petr ether.

8633. Silane. Silicon tetrahydride; silicane; monosilane. H_4Si; mol wt 32.12. H 12.55%, Si 87.45%. SiH_4. Prepd from aluminum silicide and hydrochloric acid or by electrolysis of an aq soln of sodium, ammonium, manganese, or ferrous chloride using an aluminum-silicon alloy as the positive pole: Buff, Wöhler, *Ann.* **102**, 128 (1857); from magnesium silicide and HCl: Wöhler, *ibid.* **106**, 56 (1858); Stock, Somiesky, *Ber.* **49**, 111 (1916); from zinc silicide and acid: Deville, Caron, *Compt. Rend.* **45**, 163 (1857); from lithium silicide and acid: Moissan, *ibid.* **135**, 1284 (1902); from magnesium silicide and ammonium bromide or chloride in liquid ammonia: Johnson, Isenberg, *J. Am. Chem. Soc.* **57**, 1349 (1935); Clasen, **Ger.** pat. **926,069** (1952); *Angew. Chem.* **70**, 179 (1958); from lithium aluminum hydride in ether and silicon tetrachloride: Finholt *et al., J. Am. Chem. Soc.* **69**, 2692 (1947); from silica gel, aluminum oxide and hydrogen: Jackson, U.S. pat. **3,068,069** (1962 to du Pont); from silicon tetrachloride and sodium aluminum hydride: Mason, Kelly, U.S. pat. **3,043,664** (1962 to Allied Chem.). Purification: Jacob, Trenner, U.S. pat. **3,019,087** (1962 to Merck & Co.); Shoemaker, Sumrell, U.S. pat. **3,041,141** (1962 to Baker Chem.).

Gas, repulsive odor. d^{-185} (liq) 0.68. Solidifies at approx −200°. mp −185°; bp −112°. Stable at ordinary temps; completely dec ~400° into silicon and hydrogen; dec by an electric discharge; ignites in an atm of chlorine; ignites in air by raising the temp. Slowly dec in water; practically insol in alc, ether, benzene, chloroform, silicochloroform and silicon tetrachloride. Dec in potassium hydroxide solns.

Caution: Irritating to respiratory tract.

USE: Source of hyperpure silicon for semiconductors.

8634. Silicic Acid. Precipitated silica. Approx H_2SiO_3. Occurs in nature as *opal.* Monograph: E. A. Hauser, *Silicic Science* (Van Nostrand, Princeton, 1958). Short review of chemistry: S. A. Greenberg, *J. Chem. Ed.* **36**, 218-219 (1959).

White, amorphous powder. Insol in water or acids except hydrofluoric (*see under* Silicon Dioxide); sol in hot fixed alk hydroxide solns.

Silica gel, Dri-Die. Pptd silicic acid in form of lustrous granules, esp prepd and adapted for absorption of various vapors.

USE: Silica gel as insecticide.

8635. Silicon. Si; at. wt 28.0855; at. no. 14; valence 4, also 2. Three naturally occurring isotopes: 28 (92.18%); 29 (4.71%); 30 (3.12%); artificial isotopes: 25-27; 31; 32. Does not occur free in nature; found as silica (quartz, sand, sandstone) or as silicate (feldspar or orthoclase, kaolinite, anorthite, etc.). Constitutes about 27.6% of the earth's crust; second most abundant element on earth, oxygen being first. Prepd industrially by carbon reduction of silica in an electric arc furnace. Purification by zone refining (*see ref. under* Germanium). Very pure silicon is obtained by decompn of silicon tetraiodide: Litton, Anderson, *J. Electrochem. Soc.* **101**, 287 (1954); *Chem. & Eng. News* **34**, 5007 (1956). From silicon tetrachloride: Lyon *et al., Trans. Electrochem. Soc.* **96**, 359 (1949); Klyuchnikov, *J. Appl. Chem. USSR* **29**, 139 (1956). By thermal decompn of a chlorosilane: Schering, U.S. pat. **3,041,144** (1962 to Siemens-Schuckertwerke). Review of silicon and its compounds: Rochow, "Silicon" in

Comprehensive Inorganic Chemistry vol. **1**, J. C. Bailar, Jr. *et al.*, Eds. (Pergamon Press, Oxford, 1973) pp 1323-1467; W. Runyan in Kirk-Othmer *Encyclopedia of Chemical Technology* vol. **20** (Wiley-Interscience, New York, 3rd ed., 1982) pp 826-845. The uses of silicon compounds in organic chemistry: E. W. Colvin, *Chem. Soc. Rev.* **7**, 15 (1978); I. Fleming, *ibid.* **10**, 83 (1981); L. A. Paquette, *Science* **217**, 793 (1982).

Black to gray, lustrous, needle-like crystals or octahedral platelets (cubic system). The amorphous form is a dark brown powder. Poor conductor of electricity. d_4^{25} 2.33. mp 1410°. Average heat capacity (16-100°): 0.1774 cal/g/°C. Lattice constant (25°): 5.41987 × 10^{-8} cm. Compressibility (V/V$_0$) at 25 × 10^3 kg/cm^2: 0.978; at 100 × 10^3 kg/cm^2: 0.940, *Gmelin's Silicon* (8th ed.) **15B** (1959) p 57. Dielectric const: 13. Covalent bond ionization energy at 0°K = 1.2 ev. Band gap: 1.106 ev. Impurity atom ionization energy: ~0.04 ev. Intrinsic resistivity at 300°K = 0.23 megohm. Electron mobility at 300°K: 1500 cm^2/volt/sec. Hole mobility at 300°K: 500 cm^2/volt/sec. Intrinsic charge density at 300°K: 1.5 × 10^{10}. Electron diffusion constant at 300°K: 38. Hole diffusion constant at 300°K: 13. Practically insol in water. Attacked by hydrofluoric or a mixture of hydrofluoric and nitric acids (depending upon cryst modifications). Soluble in molten alkali oxides. Burns in fluorine, chlorine.

USE: In making silanes and silicones, the Si—C bond being about as strong as a C—C bond. In the manuf of transistors, silicon diodes and similar semiconductors. For making alloys such as ferrosilicon, silicon bronze, silicon copper. As a reducing agent like aluminum in high temp reactions.

8636. Silicon Carbide. CSi; mol wt 40.10. C 29.96%, Si 70.04%. SiC. Made by heating coke and sand (and salt as flux) in electric furnace.

Exceedingly hard, green to bluish-black, iridescent, sharp crystals; hardness = 9.5. d 3.23. Dielectric constant: 7.0. Electron mobility: > 100 cm^2/volt-sec. Hole mobility: > 20 cm^2/volt-sec. Band gap energy: 2.8 ev.

USE: Polishing glass, granite; smoothing bisque ware; in sharpening-stones; abrasive; manuf porcelain, "emery" paper and wheels, shoe soles, refractory brick, furnace linings; antiskid pavements; in semiconductor technology.

8637. Silicon Dioxide. *Silica;* silicic anhydride. O$_2$Si; mol wt 60.08. O 53.26%, Si 46.74%. SiO$_2$. Occurs in nature as *agate, amethyst, chalcedony, cristobalite, flint, quartz, sand, tridymite.* *Reviews:* Several authors in Kirk-Othmer *Encyclopedia of Chemical Technology* vol. **18** (Interscience, New York, 2nd ed., 1969) pp 46-111; Rochow in *Comprehensive Inorganic Chemistry* vol. **1**, J. C. Bailar, Jr. *et al.*, Eds. (Pergamon Press, Oxford, 1973) pp 1388-1402. Toxicology: L. T. Fairhall, *Industrial Toxicology* (Hafner, New York, 1969) pp 105-107.

Transparent, tasteless crystals, or amorphous powder. d (amorphous) 2.2. d^0 (quartz) 2.65. Melts to a glass. Silica has the lowest coefficient of expansion by heat of any known substance. It is practically insol in water or acids, except hydrofluoric acid in which it readily dissolves forming the gas silicon tetrafluoride; it is also slowly attacked by heating with concd phosphoric acid. The crystallized forms of silica are scarcely attacked by alkalies, while the amorphous is sol, especially when finely divided. *See also* Infusorial Earth.

Caution: Potential symptom of overexposure by inhalation of dust is pneumoconiosis (silicosis). *See NIOSH Pocket Guide to Chemical Hazards* (DHHS/NIOSH 90-117, 1990) p 194; Fairhall, *loc. cit.*

USE: Manuf glass, water glass, refractories, abrasives, ceramics, enamels; decolorizing and purifying oils, petroleum products, etc.; in scouring- and grinding-compounds, ferrosilicon, molds for castings; as anticaking and defoaming agent.

8638. Silicon Disulfide. S$_2$Si; mol wt 92.22. S 69.54%, Si 30.46%. SiS$_2$. Prepd from SiO$_2$ and Al$_2$S$_3$: Tiede, *Ber.* **59**, 1703 (1926); Zintl, Loosen, *Z. Phys. Chem.* **A174**, 301 (1935).

White, fibrous mass or needles. Yields H$_2$S on contact with moist air. d 2.02. mp 1090° (sublimes). Burns when held in flame. Sol with decompn in water, alcohol, alkaline solns. Insol in benzene.

8639. Silicones. Broad family of synthetic polymers containing a repeating silicon-oxygen backbone with organic side groups attached via carbon-silicon bonds. Classified into three major types: fluids, resins and elastomers. The name was coined by F. S. Kipping under the supposition that the compounds were analogous to ketones; now used to designate a variety of complex polymeric siloxanes which may be linear, branched or cross-linked. Structural unit is represented by the formula, $[R_nSiO_{(4-n/2)}]_m$, where n = 1-3 and m > 1. R groups are usually methyl, longer alkyl, fluoroalkyl, phenyl, vinyl, alkoxy or alkylamino. Silicones are characterized by thermal and oxidative stability, chemical inertness, resistance to weathering, good dielectric strength and low surface tension. Prepd commercially by the direct reaction of silicon with an organic halide, hydrolysis of the resulting organohalosilanes and condensation of the unstable diols. Review of chemistry, properties and uses: B. Hardman, A. Torkelson in *Encyclopedia of Polymer Science and Engineering* vol. **15** (Wiley-Interscience, New York, 1989) pp 204-308. Review of silicone surfactants: B. Grüning, G. Koerner, *Tenside Surfact. Deterg.* **26**, 312-317 (1989); and their uses in cosmetics and toiletries: C. Gould, *Spec. Chem.* **11**, 354-359 (1991). Clinical use of silicone elastomers in reconstructive surgery: M. B. Habal in *Advances in Biomedical Polymers*, C. G. Gebelein, Ed. (Plenum Press, New York, 1987) pp 17-29. Review of properties and commercial uses of silicone fluids: D. H. Demby *et al.* in *Synthetic Lubricants and High-Performance Functional Liquids*, R. L. Shubkin, Ed. (Marcel Dekker, New York, 1993) pp 183-203.

Polydimethylsiloxane, *see* Dimethicone.

Polyether copolymers, *silicone polyethers, dimethicone copolyols, silicone glycol copolymers, silicone glycols.* Branched siloxanes modified at the chain ends with a polyether, *e.g.* polyoxyethylene and/or polyoxypropylene. Trade marks for such products include: *Abil, Stilwet L.*

USE: As lubricants; wetting agents; defoamers; surfactants; hydraulic oils; dielectric oils; protective coatings; adhesives and caulking compounds; release agents; in paints, enamels, and varnishes; water repellents for textiles, paper, masonry, and concrete; damping fluids; electrical insulation; dental molds; medical implants; contact lenses; and in cosmetic and pharmaceutical formulations. A specialty product known as "bouncing putty" or "silly putty" has unusual rheological properties which allow it to be stretched, shattered, molded, or to rebound like rubber when dropped on hard surface.

8640. Silicon Monoxide. OSi; mol wt 44.08. O 36.29%, Si 63.71%. SiO. Prepd by high vacuum sublimation of a mixture of silicon and quartz: Schenk in *Handbook of Preparative Inorganic Chemistry* vol. **1**, G. Brauer, Ed. (Academic Press, New York, 2nd ed., 1963) pp 696-697. Existence below 1200° in doubt; may disproportionate to a mixture of Si and SiO$_2$: Benyon, *Vacuum* **20**, 293 (1970). *Review:* Rochow in *Comprehensive Inorganic Chemistry* vol. **1**, J. C. Bailar, Jr. *et al.*, Eds. (Pergamon Press, Oxford, 1973) pp 1353-1355.

Brownish-black scales by sublimation. Has been crystallized. Cubic system. About as hard as silicon. Nonconductor of electricity even at red heat. d 2.18.

8641. Silicon Nitride. Trisilicon tetranitride. N$_4$Si$_3$; mol wt 140.28. N 39.94%, Si 60.06%. Si$_3$N$_4$. Non-oxide ceramic material of high thermal stability. Manuf by direct nitridation of pure silicon. Prepn by pyrolysis of silicon diimide (Si(NH$_2$)$_2$): O. Glemser, P. Naumann, *Z. Anorg. Allgem. Chem.* **298**, 134 (1954); and characterization: K. S. Mazdiyasni, C. M. Cooke, *J. Am. Ceram. Soc.* **56**, 628 (1973); *eidem,* U.S. pat. **3,959,446** (1976 to USA); by pyrolysis of silazanes: D. Seyferth *et al., J. Am. Ceram. Soc.* **66**, C-13 (1983); *eidem,* U.S. pat. **4,397,828** (1983 to M.I.T.). Brief review: D. L. Segal, *Chem. & Ind. (London)* **1985**, 544.

Exists in two crystalline forms, α-Si$_3$N$_4$ and β-Si$_3$N$_4$. α-Form occurs when powdered Si$_3$N$_4$ is heated to 1200° > 4 hr; increasing the temperature to > 1450°C for 2 hr produces the β-form. White when pure but often brown or black powder. mp 2173°K (sublimes). d 3.2 kg/dm³. Coefficient of thermal expansion K⁻¹: 2.5×10^{-6}.

USE: As high temp engineering material for use in gas turbines, diesel engines.

8642. Silicon Tetraacetate. *Acetic acid tetraanhydride with silicic acid.* $C_8H_{12}O_8Si$; mol wt 264.26. C 36.36%, H 4.58%, O 48.43%, Si 10.63%. $Si(CH_3COO)_4$. Prepd from $SiCl_4 + (CH_3CO)_2O$: Balthis, *Inorg. Syn.* **4**, 45 (1953).

Extremely hygroscopic cryst mass. Hydrolyzed rapidly by moist air. mp 110°. Dec at 160-170° with evolution of acetic anhydride. $bp_{6.0}$ 148°. Violent reaction on contact with water. Forms SiO_2 and ethyl acetate on contact with alcohol. Moderately sol in acetone, benzene.

8643. Silicon Tetrabromide. Silicon bromide. Br_4Si; mol wt 347.70. Br 91.92%, Si 8.08%. $SiBr_4$. Prepd from the elements: Gattermann, *Ber.* **22**, 189 (1889).

Colorless, fuming liq; disagreeable odor; becomes yellow on exposure to air. d 2.8. Solidif −12°; mp 5°; bp 153°. Dec by water into silicic acid and HBr with great evolution of heat. Reacts violently with metallic potassium.

8644. Silicon Tetrachloride. Silicon chloride. Cl_4Si; mol wt 169.90. Cl 83.47%, Si 16.53%. $SiCl_4$. Prepd from the elements: Schenk in *Handbook of Preparative Inorganic Chemistry* vol. 1, G. Brauer, Ed. (Academic Press, New York, 2nd ed., 1963) pp 682-683.

Colorless, clear, mobile, fuming liq; suffocating odor. d_4^0 1.52. mp −70°; bp 59°. Dec by water with much heat into silicic acid and HCl. Miscible with benzene, ether, chloroform, petr ether.

Caution: May be irritating to eyes, respiratory tract.

USE: Producing smoke screens in warfare. In the prepn of pure silicon.

8645. Silicon Tetrafluoride. Silicon fluoride. F_4Si; mol wt 104.08. F 73.02%, Si 26.98%. SiF_4. Commercial prepn from waste gases of phosphate fertilizer production: Molstad, U.S. pat. **2,833,628** (1958 to W. R. Grace). Lab prepns: Booth, Swinehart, *J. Am. Chem. Soc.* **57**, 1337 (1935); Hoffman, Gutowski, *Inorg. Syn.* **4**, 145 (1953); Belf, *Chem. & Ind. (London)* **1955**, 1296.

Colorless gas; very pungent odor similar to that of hydrogen chloride. Forms heavy clouds with moist air. Dec by water into silicic acid and HF. Sublimes −95.7°; mp −90.2° (under pressure): d (liq; −80°) 1.590. Critical temp −1.5°; critical pressure 50 atm.

8646. Silicotungstic Acid. *Tungstosilicic acid.* $H_4O_{40}SiW_{12}$; mol wt 2878.17. H 0.14%, O 22.24%, Si 0.98%, W 76.65%. $H_4[SiO_4.(W_3O_9)_4]$. Prepn of hydrated compd: Grüttner, Jander in *Handbook of Preparative Inorganic Chemistry* vol. 2, G. Brauer, Ed. (Academic Press, New York, 2nd ed., 1965) pp 1717-1718.

Hydrate, white to slightly yellow, deliquesc crystals. Very sol in water, alcohol. *Keep tightly closed.*

USE: Preparing heavy solns for separating minerals; as a mordant for basic aniline dyes; as a reagent for pptn and determination of alkaloids.

8647. Silver. Ag; at. wt 107.868; at. no. 47; valence 1, 2. Occurrence in the earth's crust: 0.1 ppm; also present in seawater: 0.01 ppm. Natural isotopes: 107 (51.35%); 109 (48.65%); artificial isotopes (mass numbers): 100-106, 108, 110-117. One of the earliest known metals. Found native or associated with copper, gold and lead. Principle ores are argentite *q.v., cerargyrite* or *horn silver* (mixture of halides), *proustite* $(3Ag_2S.As_2S_3)$, *pyrargyrite* $(Ag_2S.Sb_2S_3)$. Extraction from ores: Percy *et al.*, cited by Mellor, *A Comprehensive Treatise on Inorganic and Theoretical Chemistry* **3**, 301 (1928); *Silver, Its Economics, Extraction, Use,* A. Butts, C. D. Coxe, Eds. (Van Nostrand, Princeton, 1967) 480 pp. Reviews of silver and silver compounds: Thompson, "Silver" in *Comprehensive Inorganic Chemistry,* vol. 3, J. C. Bailar, Jr. *et al.,* Eds. (Pergamon Press, Oxford, 1973) pp 79-128; G. H. Sistare and H. B. Lockhart in Kirk-Othmer *Encyclopedia of Chemical Technology* vol. 21 (Wiley-Interscience, New York, 3rd ed., 1983) pp 1-32.

White metal, face-centered cubic structure. More malleable and ductile than any other metal except gold; excellent conductor of heat and electricity. mp 960.5°. bp ~2000°.

d^{15} 10.49. Not attacked by water or atmospheric oxygen; blackened by ozone, by hydrogen sulfide, by sulfur. Inert to most acids; readily reacts with dil nitric acid, hot concd sulfuric acid; superficially attacked by hydrochloric acid. Sol in fused alkali hydroxides in presence of air, in fused alkali peroxides, in alkali cyanides in presence of air or oxygen. Most silver salts are light-sensitive.

Caution: Potential symptoms of overexposure are blue-gray eyes, nasal septum, throat and skin; irritation of skin, ulceration; gastrointestinal disturbance. *See NIOSH Pocket Guide to Chemical Hazards* (DHHS/NIOSH 90-117, 1990) p 196. Blue-gray discoloration, known as argyria or argyrosis, results from chronic exposure to silver or silver salts. *See Patty's Industrial Hygiene and Toxicology* vol. 2A, G. D. Clayton, F. E. Clayton, Eds. (Wiley-Interscience, New York, 3rd ed., 1981) pp 1881-1894.

USE: For coinage, most frequently alloyed with copper or gold; for manuf tableware, mirrors, jewelry, ornaments; for electroplating; for making vessels and apparatus used in manuf medicinal chemicals, in processing foods and beverages, in handling organic acids; as catalyst in hydrogenation and oxidation processes; as ingredient of dental alloys. Has been used for purification of drinking water because of toxicity to bacteria and lower forms of life. Some salts used in photography.

8648. Silver Acetate. $C_2H_3AgO_2$; mol wt 166.91. C 14.39%, H 1.81%, Ag 64.63%, O 19.17%. CH_3COOAg.

White to slightly grayish, lustrous needles or cryst powder. d 3.26. Sol in 100 parts of cold water, 35 parts of boiling water; freely sol in dil nitric acid. *Protect from light.*

USE: Oxidizing agent for use in liquid ammonia: Kline, Kershner, *Inorg. Chem.* **5**, 932 (1966).

8649. Silver Bromide. AgBr; mol wt 187.77. Ag 57.45%, Br 42.55%. Yellowish, odorless powder; darkens on exposure to light. d 6.47; mp 432°. Soly in water (25°): 0.135 mg/l. Insol in alcohol, or most acids. Slightly sol in dil ammonia, but moderately sol in concd ammonia. One liter of 10% ammonia dissolves 3.3 g at 12°; sol in solns of alkali cyanides, sparingly sol in solns of thiocyanates or thiosulfates; sol in 220 parts saturated NaCl, in 35 parts saturated KBr solns; slightly sol in ammonium carbonate soln. *Protect from light.*

USE: In photography.

THERAP CAT: Topical anti-infective; astringent.

8650. Silver Carbonate. CAg_2O_3; mol wt 275.75. C 4.36%, Ag 78.24%, O 17.41%. Ag_2CO_3. Prepn from silver nitrate and sol carbonate soln: Poyer *et al., Inorg. Syn.* **5**, 19 (1957).

Light yellow powder when freshly pptd, but becomes darker on drying and on exposure to light. Dec at about 220° into Ag_2O and CO_2 and at higher temp into metal Ag. d 6.08. Sol in 30,000 parts cold water, 2000 parts boiling water; readily sol in dil nitric acid, ammonia, or alkali cyanides. *Protect from light.*

8651. Silver Chlorate. Argentous chlorate. $AgClO_3$; mol wt 191.32. Ag 56.38%, Cl 18.53%, O 25.09%. Prepd from $AgNO_3$ and $NaClO_3$: Nicholson, Holley, Jr., *Inorg. Syn.* **2**, 4 (1946).

White, tetragonal crystals. d_4^{20} 4.430. mp 230°, dec 270° forming silver chloride and oxygen. Solubility in water (g/100 ml): 10 (15°); 20 (27°); 50 (80°). Slightly sol in alc. Darkens upon exposure to light due to slow decompn. *Keep away from light, organic vapors and oxidizable substances.*

USE: In organic synthesis as an effective oxidizing agent for certain organic compds.

8652. Silver Chloride. AgCl; mol wt 143.32. Ag 75.26%, Cl 24.74%. White powder; darkens on exposure to light. d 5.56; mp 455°; bp 1550°. Soly in water (25°): 1.93 mg/l; hydrochloric acid increases its soly. Sol in 250 parts of concd HCl, in 13 parts of 10% ammonia; more sol in stronger ammonia and also at higher temps. Sol in solns of alkali cyanides, thiosulfates, ammonium carbonates; appreciably sol in concd aq solns of ammonium chloride; mercuric nitrate, and silver nitrate. The freshly pptd chloride

dissolves more readily than the dried precipitate. Insol in alcohol, or dil acids. *Protect from light.*
USE: In silver plating, in making antiseptic silver prepns.

8653. Silver Chromate(VI). Ag$_2$CrO$_4$; mol wt 331.73. Ag 65.03%, Cr 15.67%, O 19.29%. Dark brownish-red, cryst powder. d^{25} 5.625. Soly in water (0°): 0.0014%. Sol in nitric acid and ammonia.
USE: Catalyst for formation of aldol from alcohol; formed at end point of Mohr titration of halides.

8654. Silver Citrate. *2-Hydroxy-1,2,3-propanetricarboxylic acid silver salt;* Itrol. C$_6$H$_5$Ag$_3$O$_7$; mol wt 512.71. C 14.06%, H 0.98%, Ag 63.12%, O 21.84%.
White, odorless, heavy, crystalline powder; darkens in light. Sol in 3500 parts water, more sol in boiling water; readily sol in dil HNO$_3$, ammonia. *Protect from light.*
USE: Anti-infective dusting powder.

8655. Silver Cyanide. CAgN; mol wt 133.89. C 8.97%, Ag 80.57%, N 10.46%. AgCN.
White or grayish, odorless powder; stable in dry air; darkens on exposure to light; dec at 320°. *Poisonous!* d 3.95. Insol in water, alcohol, or dil acids; sol in alkali cyanides and in boiling concd nitric acid; converted by dil HCl into hydrocyanic acid and silver chloride; sparingly sol in dil, more in concd ammonia. *Protect from light.*
USE: For silver plating; formerly used for extemporaneous prepn of dil hydrocyanic acid by treatment with HCl.

8656. Silver Difluoride. Argentic fluoride. AgF$_2$; mol wt 145.87. Ag 73.95%, F 26.05%. Prepd by the action of fluorine on silver: Ebert *et al., J. Am. Chem. Soc.* **55**, 3056 (1933); by the interaction of fluorine and silver halides: Ruff, Giese, *Z. Anorg. Allgem. Chem.* **219**, 143 (1934); v. Wartenberg, *ibid.* **242**, 406 (1939); Struve *et al., Ind. Eng. Chem.* **39**, 353 (1947). Laboratory procedure according to the equation 2AgCl + 2F$_2$ → 2AgF$_2$ + Cl$_2$: Priest, Swinehart, *Inorg. Syn.* **3**, 176 (1950); Kwasnik in *Handbook of Preparative Inorganic Chemistry,* vol. **1**, G. Brauer, Ed. (Academic Press, New York, 2nd ed., 1963) pp 241-242. Prepd from silver chloride and chlorine trifluoride: Rochow, Kukin, *J. Am. Chem. Soc.* **74**, 1615 (1952).
White when pure. Usually obtained as a gray-black or brownish, amorphous solid showing a yellow bloom. Sensitive to light. d 4.7. mp 690°. Very hygroscopic, is converted to a greasy black mass on exposure to atmospheric moisture. Violent reaction with water (instant hydrolytic cleavage). Powerful oxidizing agent giving off some ozone when treated with dil acids. Liberates iodine from iodides. May be stored in quartz or iron ampuls.
Caution: Symptoms due to fluoride. Highly toxic.
USE: In the fluorination of hydrocarbons.

8657. Silver Fluoride. Argentous fluoride; silver monofluoride. AgF; mol wt 126.87. Ag 85.02%, F 14.98%. Prepared according to the equation Ag$_2$CO$_3$ + 2HF → 2AgF + CO$_2$ + H$_2$O; also prepd by reduction of AgF$_2$ with H$_2$: Ruff, *Die Chemie des Fluors* (Berlin, 1920) p 37; Emeléus in *J. H. Simons' Fluorine Chemistry,* vol. I (Academic Press, New York, 1950) p 32; Kwasnik in *Handbook of Preparative Inorganic Chemistry,* vol. **1**, G. Brauer, Ed. (Academic Press, New York, 2nd ed., 1963) pp 240-241.
ξ Flexible leaflets (cubic, NaCl lattice). Very hygroscopic. Darkens on exposure to light. d 5.852. mp 435°. bp ~1150°. Soly in water when freshly prepd: 182 g/100 ml at 15.5°. In moist air it gradually becomes insol because of basic fluoride formation. Aq solns are neutral and are solvents for silver oxide which renders them alkaline. Also sol in HF, NH$_3$, CH$_3$CN. Forms several hydrates. The dihydrate is stable to 39.5° and the tetrahydrate is stable from −14° to +18.7°. In addition, an acid fluoride, AgF.3HF can be prepd by cooling a soln of AgF in HF. This loses HF at 0°, forming AgF.HF.
Caution: Prolonged absorption may cause mottling of teeth, skeletal changes.
USE: To convert organic Br and Cl compds to their fluoro analogs; as antiseptic.
THERAP CAT: Anti-infective.

8658. Silver Iodate. AgIO$_3$; mol wt 282.77. Ag

38.15%, I 44.88%, O 16.97%. White, cryst powder. d 5.53. mp >200°. Sol in 1875 parts water at 25°, in about 1000 parts of 35% nitric acid, in 2.5 parts of 10% ammonia. *Protect from light.*
USE: As reagent for determining small quantities of chloride, e.g., in blood: Sendroy, *J. Biol. Chem.* **120**, 335-445 (1937).

8659. Silver Iodide. AgI; mol wt 234.77. Ag 45.95%, I 54.05%. Prepd according to the equation AgNO$_3$ + KI → AgI + KNO$_3$: Kolkmeijer, van Hengel, *Z. Kristallogr.* **A88**, 317 (1934).
Light yellow, odorless powder; slowly darkened by light. Crystals are hexagonal or cubic. d 5.67; mp 552°. Practically insol in water (0.03 mg/l); in acid (except concd HI in which it dissolves readily on heating); in ammonium carbonate. Freely sol in solns of alkali cyanides or iodides; 35 mg dissolve in a liter of 10% ammonia; appreciably sol in concd solns of alkali bromides, chlorides, thiocyanates, thiosulfates, mercuric and silver nitrates. It is slowly attacked by boiling concd acids, but not affected by hot solns of alkali hydroxides.
USE: In cloud precipitation (rain-making).
THERAP CAT: Local anti-infective.
THERAP CAT (VET): In colloidal suspensions as local antiseptic for mucous membranes.

8660. Silver Lactate. *2-Hydroxypropanoic acid silver salt.* C$_3$H$_5$AgO$_3$; mol wt 196.94. C 18.30%, H 2.56%, Ag 54.77%, O 24.37%. CH$_3$CH(OH)COOAg.
Monohydrate, white or slightly gray, cryst powder readily affected by light. Sol in about 15 parts water; slightly sol in alcohol. *Protect from light.*
THERAP CAT: Topical anti-infective; astringent.

8661. Silver Nitrate. AgNO$_3$; mol wt 169.87. Ag 63.50%, N 8.25%, O 28.26%. Silver nitrate of commerce is, with the exception of 0.1-0.2% water, practically 100% pure.
Colorless, odorless, transparent, large crystals or white small crystals. *Poisonous!* Not photosensitive when pure; presence of trace amounts of organic material promotes photoreduction. Reduced by H$_2$S in the dark. d 4.35. mp 212°, forming a yellowish liq solidifying to a white, cryst mass on cooling. Dec at 440° into metallic silver, nitrogen, oxygen and nitrogen oxides. One gram dissolves in 0.4 ml water, 0.1 ml boiling water, 30 ml alc, 6.5 ml boiling alc, 253 ml acetone; readily sol in ammonia water, slightly in ether. The aq and alc solns are neutral to litmus. pH ~6.
Incompat: Alkalies, antimony salts, arsenites, bromides, carbonates, chlorides, iodides, thiocyanates, ferrous salts, hypophosphites, morphine salts, oils, creosote, phosphates, tannic acid, tartrates, vegetable decoctions, and extracts.
Caution: Caustic and irritating to skin, mucous membranes. Swallowing can cause severe gastroenteritis that may be fatal. *See also* Silver.
USE: In photography; manuf of mirrors; other silver salts; silver plating; in sympathetic and indelible inks; dyeing hair; coloring porcelain; etching ivory; as a very important and extensively used reagent in analytical chemistry.
THERAP CAT: Topical anti-infective.
THERAP CAT (VET): Astringent, antiseptic, caustic, styptic. In dilute solution as eye lotion. Intramammary to destroy secretory tissue.

8662. Silver Nitrate, Toughened. Lunar caustic; fused silver nitrate; molded silver nitrate. Consists of about 97-98% silver nitrate, remainder is silver chloride to toughen it.
White or grayish, hard rods or thin small cones. Darkens on exposure to light. *Protect from light.*
THERAP CAT: Caustic.
THERAP CAT (VET): Locally as escharotic.

8663. Silver Nitrite. AgNO$_2$; mol wt 153.87. Ag 70.10%, N 9.10%, O 20.80%.
Pale yellow, odorless needles; non-hygroscopic; becomes gray in light. d 4.45. Dec at 140°. Sol in 300 parts water; more sol in boiling water; less sol in aqueous solns of silver nitrate; partly dec on prolonged boiling with water; insol in alcohol; dec by dil acids. *Protect from light.*
USE: For prepn of standard NaNO$_2$ soln for water analysis; prepn of aliphatic nitro-compds; as reagent for primary, secondary, and tertiary alcohols.

8664. Silver Oxalate. *Ethanedioic acid silver salt.* C₂-Ag₂O₄; mol wt 303.76. C 7.91%, Ag 71.02%, O 21.07%. AgOOCCOOAg.

White, cryst powder; at 140° dec violently. d 5.03. Sol in 24,000 parts water; sol in moderately concd nitric acid, in ammonia.

USE: Exptl photographic emulsion (low sensitivity).

8665. Silver Oxide. Argentous oxide. Ag₂O; mol wt 231.74. Ag 93.10%, O 6.90%. Prepd by making sol silver salt solns alkaline: Madsen, *Z. Anorg. Allgem. Chem.* **79**, 197 (1913). Toxicity study: H. F. Smyth *et al.*, *Am. Ind. Hyg. Assoc. J.* **30**, 470 (1969).

Brownish-black, heavy, odorless powder. d₄²⁵ 7.22. Begins to dec at about 200° and at 250-300° decompn is rapid; it also breaks up into its constituents in sunlight. Reduced by hydrogen, carbon monoxide and most metals. Moist silver oxide absorbs carbon dioxide. Sol in 40,000 parts water; freely sol in dil nitric acid, in ammonia; somewhat sol in NaOH solns. Practically insol in alcohol. *Protect from light.* LD₅₀ orally in rats: 2.82 g/kg (Smyth).

USE: As catalyst; in the purification of drinking water; in the glass industry (polishing, coloring glass yellow).

THERAP CAT (VET): Has been used as germicide and parasiticide.

8666. Silver(II) Oxide. Argentic oxide; silver peroxide; silver suboxide; Divasil. AgO; mol wt 123.87. Ag 87.08%, O 12.92%. Is actually a silver(I)-silver(III) oxide: McMillan, *Chem. Rev.* **62**, 65 (1962). Prepd by the oxidation of silver nitrate with potassium peroxydisulfate in an alk medium: Hammer, Kleinberg, *Inorg. Syn.* **4**, 12 (1953). Prepn of single crystals: *Chem. & Eng. News* **47**, 32 (Aug 11, 1969).

Charcoal-gray powder. *Avoid contact with skin, organic matter, strong ammonia and alkalies.* Malleable. Cubic or orthorhombic system when cryst. d₄²⁵ 7.483. The dry solid is stable at 100° for 18 hrs, dec above 100° into silver and oxygen. Possesses semiconductor properties and is diamagnetic. *Strong oxidizing agent.* Ka 7.9 × 10⁻¹³. Practically insol in water. Sol in alkalies and NH₄OH (with decompn and evolution of N₂). In dil acids oxygen is evolved immediately, in concd acid intensely colored solns are formed (brown in nitric acid and olive green in sulfuric acid).

Caution: Highly irritating to skin, eyes, mucous membranes, respiratory tract.

USE: In the manuf of silver oxide-zinc alkali batteries.

8667. Silver Perchlorate. AgClO₄; mol wt 207.32. Ag 52.03%, Cl 17.10%, O 30.87%. Prepn from NOClO₄ + AgBr: Markowitz *et al., J. Inorg. Nucl. Chem.* **16**, 159 (1960).

Deliquescent crystals, dec 486°. d²⁵ 2.806. Freely sol in water: 557 g/100 ml; a satd aq soln contains 84.8% w/w at 25°: Smith, Ring, *J. Am. Chem. Soc.* **59**, 1889 (1937). Much less sol in 60% perchloric acid (5.63% w/w). Forms a monohydrate stable to 43°. Sol in many organic solvents, e.g., aniline, pyridine, benzene, toluene, nitromethane, glycerol, nitrobenzene, and chlorobenzene. Solvated crystals have been obtained contg 6 mols aniline, 4 mols pyridine, 1 mol benzene, 1 mol toluene. These compds explode readily when struck: Brinkley, *J. Am. Chem. Soc.* **62**, 3524 (1940).

Caution: Irritating to skin, mucous membranes.

USE: In the explosives industry.

8668. Silver Permanganate. Silver manganate(VII). AgMnO₄; mol wt 226.80. Ag 47.56%, Mn 24.22%, O 28.22%. Prepn: Büssem, Herrmann, *Z. Kristallogr.* **A74**, 459 (1930).

Violet, cryst powder; dec in light. d 4.49. Soly in water at room temps about 9 g/l. More sol in hot water; dec by alcohol. *Protect from light.*

USE: In gas masks.

8669. Silver Phosphate. Silver orthophosphate. Ag₃-O₄P; mol wt 418.58. Ag 77.31%, O 15.29%, P 7.40%. Ag₃PO₄.

Yellow, odorless powder; darkens in light. d 6.37. mp 849°. Reduced by hydrogen. Soluble in 15,500 parts water, the soly in water is decreased by presence of silver nitrate;

slightly sol in dil acetic acid; freely sol in dil nitric acid, in ammonia, ammonium carbonate, also in alkali cyanides and thiosulfates. *Protect from light.*

USE: In photography for collodion emulsions.

8670. Silver Picrate. *2,4,6-Trinitrophenol silver salt; picric acid silver derivative;* silver trinitrophenolate; Picragol; Picrotol. C₆H₂AgN₃O₇; mol wt 335.97. C 21.45%, H 0.60%, Ag 32.11%, N 12.51%, O 33.34%. (O₂N)₃C₆H₂OAg.

Monohydrate, yellow crystals. Sol in about 50 parts water; sparingly sol in alc; slightly sol in acetone or glycerol. Insol in chloroform, ether.

THERAP CAT: Antiprotozoal (Trichomonas).

THERAP CAT (VET): Has been used in bovine granular vaginitis.

8671. Silver Protein. Argentoproteinum; protargin; silver proteinate; silver nucleate; silver nucleinate. Group of compounds characterized as colloidal combinations of silver and protein. Variably prepd from silver oxide, silver nitrate or other silver salts with gelatin, serum albumin, casein or peptone. Examples of prepns: **Ger.** pat. **105,866** (1897 to Bayer); **Brit.** pat. **18,478** (1897); H. P. Kaufmann, *Arzneimittel-Synthese* (Springer, 1953) p 588. Generally classified as mild or strong based on their germicidal and irritant properties. Differentiated by the amount of ionizable silver present: Van Deripe, Konnerth, *Am. J. Pharm.* **111**, 65 (1939). Use to stain nerve fibers: D. Bodian, *Anat. Rec.* **65**, 89 (1936). Neuronal selectivity and use in light and electron microscopy: I. S. Zagon, J. H. Haring, *Acta Anat.* **114**, 193 (1982); in Alzheimer's disease: C. Duyckaerts *et al., Acta Neuropathol.* **73**, 167 (1987).

Mild silver protein, argentum vitellinum, Argyrol, Silvol. Brown, dark-brown, or almost black, odorless, lustrous scales or granules. Contains ∼20% silver. Somewhat hygroscopic. Freely sol in water. Almost insol in alcohol, chloroform, ether. *Keep well closed and protected from light.*

Strong silver protein, albumose silver, Protargol. Pale yellowish-orange to brownish-black, odorless powder. Contains ∼8% silver, most of which is present in ionic form. Somewhat hygroscopic. Freely sol in water. Almost insol in alcohol, ether, chloroform. *Keep well closed and protected from light.*

USE: Neuronal-selective histological stain; general contrast medium for electron microscopy.

THERAP CAT: Antiseptic.

THERAP CAT (VET): Antiseptic.

8672. Silver Selenate. Ag₂O₄Se; mol wt 358.69. Ag 60.14%, O 17.84%, Se 22.01%. Ag₂SeO₄. Prepd by treating silver carbonate with selenic acid: Mitscherlich, Pogg. *Ann.* **9**, 623 (1827); *Ann. Chim. Phys.* [2] **36**, 100 (1827).

Orthorhombic crystals. d 5.72. Solubility in water (20°): 1.182 g/l.

8673. Silver Selenide. Ag₂Se; mol wt 294.70. Ag 73.21%, Se 26.79%. Occurs in nature as the mineral naumannite. Prepd by melting silver and selenium together and by passing hydrogen selenide into a soln of silver nitrate: Berzelius, cited by Mellor, *A Comprehensive Treatise on Inorganic and Theoretical Chemistry* **10**, 771 (1930); by combination of ions or elements in solns or melts: Kulifay, U.S. pat. **3,026,175** (1962 to Monsanto).

Gray, hexagonal, microscopic needles. mp 880°. d¹⁵ 8.216. Two forms exist; transition temp at 133°. Forms metallic silver and selenium oxide when heated in oxygen; transformed by chlorine into silver and selenium chlorides; oxidized to silver selenite by fuming nitric acid. Sol in molten silver or bismuth without chemical change. Practically insol in water.

8674. Silver Selenite. Ag₂O₃Se; mol wt 342.69. Ag 62.95%, O 14.01%, Se 23.04%. Ag₂SeO₃. Prepd by dissolving silver selenide in boiling nitric acid and cooling: Berzelius, cited by Mellor, *A Comprehensive Treatise on Inorganic and Theoretical Chemistry* **10**, 824 (1930); from a soln of silver carbonate in selenious acid: Thomsen, *Ber.* **2**, 598 (1869).

Needles. d 5.9297. Stable to mp 530°; dec above this temp to silver, selenium oxide and oxygen. Sparingly sol in cold water; freely in hot water; sol in nitric acid.

8675. Silver Subfluoride. Disilver fluoride. Ag_2F; mol wt 234.73. Ag 91.91%, F 8.09%. Prepd by electrolytic reduction of argentous fluoride, AgF: Hettich, *Z. Anorg. Allgem. Chem.* **167**, 67 (1927); Scholder, Traulsen, *ibid.* **197**, 57 (1931). By the reaction of elemental silver with argentous fluoride at 50-90°: Güntz, *Compt. Rend.* **110**, 1337 (1890); **112**, 861 (1891); Poyer *et al., Inorg. Syn.* **5**, 18 (1957).

Hexagonal, bronze-colored crystals with green luster, d 8.57. Turns grayish-black on prolonged exposure to air. Thermal decompn to Ag and AgF begins at 100° and is complete at 200°. Good electrical conductor. On contact with water almost instant hydrolysis with precipitation of Ag powder. Apparently stable to abs ethanol; insensitive to light. When dry may be kept in a stoppered container for weeks without evidence of decompn.

8676. Silver Sulfate. Ag_2O_4S; mol wt 311.80. Ag 69.19%, O 20.53%, S 10.28%. Ag_2SO_4. Conveniently prepd from silver nitrate and sulfuric acid: Richards, Jones, *Z. Anorg. Allgem. Chem.* **55**, 72 (1907).

Crystals or cryst powder; slowly darkens on exposure to light. mp 657°; dec 1085°. d 5.45. Slowly sol in 125 parts water, 71 parts boiling water; sol in nitric acid, ammonia, concd sulfuric acid. Soly in $AgNO_3$ solns: Lietzke, Stoughton, *J. Inorg. Nucl. Chem.* **28**, 1877 (1966). *Protect from light.*

8677. Silver Sulfide. Argentous sulfide. Ag_2S; mol wt 247.80. Ag 87.06%, S 12.94%. Occurs in nature as the mineral *argentite.* Usually prepd from the elements: Hönigschmid, Sachtleben, *Z. Anorg. Allgem. Chem.* **195**, 207 (1931).

Grayish-black, heavy powder. The cryst structure is orthorhombic, changing to cubic when heated above 179°. d_4^{20} 7.234. mp 845°. Insol in water. Sol in nitric acid, in solns of alkali cyanides.

USE: In ceramics.

8678. Silver Tetraiodomercurate(II). Mercuric silver iodide. Ag_2HgI_4; mol wt 923.94. Ag 23.35%, Hg 21.71%, I 54.94%.

Deep yellow thermochromic powder. Insol in water or dil acids; sol in solns of alkali iodides or cyanides.

USE: Thermal sensor to detect overheating in journal bearings, etc., becomes blood-red at 40-50° and yellow again on cooling.

8679. Silvex. *2-(2,4,5-Trichlorophenoxy)propionic acid;* fenoprop; 2,4,5-TC. $C_9H_7Cl_3O_3$; mol wt 269.51. C 40.11%, H 2.62%, Cl 39.46%, O 17.81%. Selective pre- and post-emergence herbicide. Prepn: Toothill *et al., Ann. Appl. Biol.* **44**, 547 (1956). Toxicity: G. W. Bailey, J. W. White, *Residue Rev.* **10**, 97 (1965).

Crystals, mp 181.6°. Soly at 25° in water 0.014%; acetone 15.2%; benzene 0.16%; carbon tetrachloride 0.024%; ether 7.13%; heptane 0.017%; methanol 10.5%. LD_{50} orally in rats: 650 mg/kg (Bailey, White).

Triethanolamine salt, *silvexamine.* Sol in water.

Esters with butyl ethers of mono-, di-, or polypropylene glycols, **Kuron.** Prepn: Williams, U.S. pat. **2,749,360** (1956 to Dow).

Caution: Irritating to eyes, skin, mucous membranes.

Note: In March 1985, the E.P.A. terminated all registrations for the use of this herbicide on rice fields, orchards, sugarcane, rangeland and other noncrop sites. This follows the 1979 emergency action of the agency in banning almost all uses of the pesticide: *Chem. & Eng. News* **63**, 6 (Mar. 25, 1985).

USE: Formerly in the control of woody plants on uncropped land.

8680. Silymarin. Apihepar; Laragon; Legalon; Pluropon; Silarine; Silepar; Silirex; Silliver; Silmar. Antihepatotoxic principle isolated from the seeds of the milk thistle,

Silybum marianum (L.) Gaertn. (*Carduus marianus* L.). Comprised mainly of three isomers: *silidianin, silicristin* and the major component, *silybin* (formerly called silymarin). Characterized as a new class of substances, flavonolignans, almost certainly produced in the plant by a radical coupling of a flavonoid and coniferyl alcohol: Hänsel *et al., Deut. Apotheker Ztg.* **108**, 1988 (1968). Isoln and chemistry: Wagner *et al., Arzneimittel-Forsch.* **18**, 688 (1968); **24**, 466 (1974). Structural work: Janiak, Hänsel, *Planta Med.* **8**, 71 (1960); Hänsel, Schöpflin, *Tetrahedron Letters* **1967**, 3645; Pelter, Hänsel, *ibid.* **1968**, 2911. Pharmacology and toxicology: Hahn *et al., Arzneimittel-Forsch.* **18**, 698 (1968); H. M. Rauen, H. Schriewer, *ibid.* **23**, 148-170 (1973); Halbach, Trost, *ibid.* **24**, 866 (1974).

Silybin

Silybin, $C_{25}H_{22}O_{10}$, *[2R-[2α,3β,6(2R*,3R*)]]-2-[2,3-dihydro-3-(4-hydroxy-3-methoxyphenyl)-2-(hydroxymethyl)-1,4-benzodioxin-6-yl]-2,3-dihydro-3,5,7-trihydroxy-4H-1-benzopyran-4-one, silibinin, silybum substance E_6, silymarin I.* Revised structure and synthetic studies: R. Hänsel *et al., Chem. Commun.* **1972**, 195; *eidem, Ber.* **110**, 3664 (1977). Total synthesis: L. Merlini *et al., Chem. Commun.* **1979**, 695; *eidem, J. Chem. Soc. Perkin Trans. I* **1980**, 775. Anhydrous subst, mp 158°, dec 180°. $[α]_D^{20}$ +11° (c = 0.25 in acetone + alcohol). uv max (methanol): 288 nm (log ε 4.33). Sol in acetone, ethyl acetate, methanol, ethanol; sparingly sol in $CHCl_3$. Practically insol in water. Also occurs as the monohydrate, crystals from acetone + petr ether, mp 167°, dec 180°; from methanol + water, mp 180°.

THERAP CAT: Hepatoprotectant.

8681. Simazine. *6-Chloro-N,N'-diethyl-1,3,5-triazine-2,4-diamine; 2-chloro-4,6-bis(ethylamino)-s-triazine;* 2,4-bis-(ethylamino)-6-chloro-s-triazine; G-27692; Aquazine; Gesatop; Primatol S; Princep; Simanex. $C_7H_{12}ClN_5$; mol wt 201.66. C 41.69%, H 6.00%, Cl 17.58%, N 34.73%. Selective pre-emergence herbicide. Prepn: Hofmann, *Ber.* **18**, 2776 (1885); Pearlman, Banks, *J. Am. Chem. Soc.* **70**, 3726 (1948); Thurston *et al., ibid.* **73**, 2981 (1951). As herbicide: Gysin, Knüsli, U.S. pat. **2,891,855** (1959 to Geigy). Toxicity studies: G. W. Bailey, J. L. White, *Residue Rev.* **10**, 97 (1965).

Crystals from ethanol or methyl Cellosolve, mp 226-227°. Practically insol in water. Slightly sol in dioxane, ethyl Cellosolve. LD_{50} orally in rats: 5000 mg/kg (Bailey, White).

USE: Herbicide.

8682. Simetride. *1,4-Bis[(2-methoxy-4-propylphenoxy)-acetyl]piperazine;* AP2; Kyorin AP2. $C_{28}H_{38}N_2O_6$; mol wt 498.62. C 67.45%, H 7.68%, N 5.62%, O 19.25%. Prepn: Fr. pat. **M2936** (1964 to Kyorin Pharmaceutical Co., Ltd.); T. Irikura *et al.,* U.S. pat. **3,184,463** (1965). Pharmacological studies: T. Irikura *et al., Yakugaku Zasshi* **83**, 785 (1963), *C.A.* **60**, 1008b (1964). Analgesic activity: K. Nishino *et al., Yakugaku Zasshi* **85**, 715 (1965), *C.A.* **63**, 16957b (1965).

Crystals, mp 128-130° (**Fr.** pat.); also reported as mp 138.5-139° (Irikura, 1963). Practically insoluble in water, ethanol. Sol in $CHCl_3$. LD_{50} i.p. in mice: 15000 mg/kg (Irikura).

THERAP CAT: Analgesic.

8683. Simetryn. *N,N'-Diethyl-6-(methylthio)-1,3,5-tri-azine-2,4-diamine; 2,4-bis(ethylamino)-6-(methylthio)-s-triazine;* 2,4-bis(ethylamino)-6-(methylthio)-1,3,5-triazine; 2-methylthio-4,6-bis(monoethylamino)-s-triazine; G-32911; Gy-bon. $C_8H_{15}N_5S$; mol wt 213.31. C 45.05%, H 7.09%, N 32.83%, S 15.03%. Prepn: Gysin, Knüsli, **Swiss** pat. 337,-019 (1959 to Geigy), *C.A.* **57**, 14226c (1962). Crystal structure: A. J. Graham *et al.*, *Cryst. Struct. Commun.* **7**, 227 (1978). Toxicity study: E. F. Edson *et al.*, *World Rev. Pest Contr.* **1**, 36 (1965).

Crystals, mp 82-83°. LD_{50} orally in rats: 1830 mg/kg (Edson).

USE: Herbicide.

8684. Simfibrate. *2-(4-Chlorophenoxy)-2-methylpropanoic acid 1,3-propanediyl ester; 2-(p-chlorophenoxy)-2-methylpropionic acid trimethylene ester;* 1,3-propanediol bis[α-(p-chlorophenoxy)isobutyrate]; 1,3-propanediol bis[2-(4-chlorophenoxy)-2-methylpropionate]; sinfibrate; CLY-503; Cholesolvin; Liposolvin. $C_{23}H_{26}Cl_2O_6$; mol wt 469.36. C 58.86%, H 5.58%, Cl 15.11%, O 20.45%. Prepn: **Neth.** Appl. **6,600,044** corresp to Nakanishi *et al.*, **U.S.** pat. **3,494,957** (1966, 1970 to Yoshitomi). Activity and toxicology studies: Nakanishi *et al.*, *J. Pharm. Soc. Japan* **90**, 267, 921, 926 (1970). Activity on postheparin lipoprotein lipase: L. Cavallini *et al.*, *Clin. Ter.* **92**, 25 (1980).

Crystals, mp 51-53°. $bp_{0.03}$ 197-200°, $bp_{0.15}$ 220-230°. LD_{50} in mice, rats (g/kg): 3.3-3.5; 7.3-8.0 orally (Nakanishi, *J. Pharm. Soc. Japan).*

THERAP CAT: Anticholesteremic.

8685. Simplesse®. Protein-based fat substitute. Prepn from microparticulated, denatured dairy whey protein: N. S. Singer *et al.*, **Eur.** pat. **Appl. 250,623;** *eidem*, **U.S.** pat. **4,734,287** (both 1988 to John LaBatt Ltd.). Brief description: D. D. Duxbury, *Food Process.* **49**, 21 (1988).

USE: Fat substitute for use in foods.

8686. Simvastatin. *[1S-[1α,3α,7β,8β(2S*,4S*),8aβ]]-2,-2-Dimethylbutanoic acid 1,2,3,7,8,8a-hexahydro-3,7-dimethyl-8-[2-(tetrahydro-4-hydroxy-6-oxo-2H-pyran-2-yl)ethyl]-1-naphthalenyl ester;* 2,2-dimethylbutyric acid 8-ester with (4R,6R)-6-[2-[(1S,2S,6R,8S,8aR)-1,2,6,7,8,8a-hexahydro-8-hydroxy-2,6-dimethyl-1-naphthyl]ethyl]tetrahydro-4-hydroxy-2H-pyran-2-one; synvinolin; MK-733; Denan; Liponorm; Lodalès; Simovil; Sinvacor; Sivastin; Zocor; Zocord. $C_{25}H_{38}O_5$; mol wt 418.57. C 71.74%, H 9.15%, O 19.11%. Competitive inhibitor of HMG-CoA reductase; synthetic analog of lovastatin, *q.v.* Prepn: A. K. Willard *et al.*, **Eur.** pat. **Appl. 33,538;** W. F. Hoffman *et al.*, **U.S.** pat. **4,444,784** (1981, 1984 both to Merck & Co.); W. F. Hoffman *et al.*, *J.*

Med. Chem. **29**, 849 (1986). Clinical evaluation in hypercholesterolemia: M. J. Mol *et al.*, *Lancet* **2**, 936 (1986); L. A. Simons *et al.*, *Med. J. Aust.* **147**, 65 (1987). Effect on survival in CHD: Scandinavian Simvastatin Survival Study Group, *Lancet* **344**, 1383 (1994).

Crystals from *n*-butyl chloride + hexane, mp 135-138°.

THERAP CAT: Antihyperlipidemic.

8687. Sinalbin. Sinapine glucosinalbate. $C_{30}H_{42}N_2O_{15}S_2$; mol wt 734.80. C 49.04%, H 5.76%, N 3.81%, O 32.66%, S 8.73%. Mustard oil glucoside from white or yellow mustard *(Sinapis alba* L., *Cruciferae):* Gadamer, *Arch. Pharm.* **235**, 38, 83, 570 (1897); Viehoever, Nelson, *J. Assoc. Offic. Agr. Chem.* **21**, 488 (1938); *Am. J. Pharm.* **110**, 411 (1938). Structure: Schulz, Gmelin, *Z. Naturforsch.* **7b**, 500 (1952); Ettlinger, Lundeen, *J. Am. Chem. Soc.* **78**, 4172 (1956).

Needles from 95% ethanol, mp 100-102°. (Also reported as 83-84° for the hydrated, 138-140° for the anhydr.) $[\alpha]_D^{20}$ -8.76° (c = 0.29). Soluble in hot alcohol, cold water.

8688. Sinapine. *2-[[3-(4-Hydroxy-3,5-dimethoxyphenyl)-1-oxo-2-propenyl]oxy]-N,N,N-trimethylethanaminium;* sinapic acid choline ester. $[C_{16}H_{24}NO_5]^+$; mol wt 310.38. From black mustard seeds, the seeds of *Brassica nigra* Koch, *Cruciferae*. Extraction procedure: Remsen, Coale, *Am. Chem. J.* **6**, 52 (1884). Synthesis: Späth, *Monatsh.* **41**, 271 (1920).

Chloride, $(C_{16}H_{24}NO_5)Cl$, does not crystallize, very sol in water, alcohol.

Bromide trihydrate, $(C_{16}H_{24}NO_5)Br.3H_2O$, needles, mp 92°; mp 107-115° when anhydr, very sol in water.

Iodide trihydrate, $(C_{16}H_{24}NO_5)I.3H_2O$, crystals, becomes anhydr over H_2SO_4, mp 179° (anhydr). Sparingly sol in water.

Acid sulfate trihydrate, $(C_{16}H_{24}NO_5)HSO_4.3H_2O$, tablets, dec 127°; dec 187° when anhydr. Freely sol in water, hot alcohol; very sparingly sol in ether.

Neutral sulfate pentahydrate, $(C_{16}H_{24}NO_5)_2SO_4.5H_2O$, tablets from hot alcohol, mp 193° (anhydr). Not very stable; very sol in water, sparingly in alcohol.

8689. Sincalide. *1-De(5-oxo-L-proline)-2-de-L-glutamine-5-L-methioninecaerulein;* cholecystokinin *C*-terminal octapeptide; CCK *C*-terminal octapeptide; pancreozymin *C*-terminal octapeptide; SQ-19844; Kinevac. $C_{49}H_{62}N_{10}O_{16}S_3$; mol wt 1143.29. C 51.48%, H 5.47%, N 12.25%, O 22.39%, S 8.41%. The *C*-terminal octapeptide of cholecystokinin-pancreozymin (*see* cholecystokinin). Structurally similar to caerulein, *q.v.,* it evokes a variety of biological responses similar to CCK-PZ, including smooth-muscle contraction of the gall bladder and small intestine, relaxation of the chole-dochoduodenal junction, protein secretion by the pancreas and acid secretion by the stomach. It is more active than cholecystokinin on a weight or molar basis. Initial isoln: V. Mutt, J. E. Jorpes, *Eur. J. Biochem.* **6**, 156 (1968). Prepn: M. A. Ondetti *et al.*, **Ger.** pat. **1,922,185** corresp to U.S. pat. **3,723,406** (1969, 1970, both to Squibb); *eidem, J. Am. Chem. Soc.* **92**, 195 (1970). Mechanical and metabolic effects: K. E. Andersson *et al., Acta Physiol. Scand.* **96**, 495 (1976). *In vitro* study: M. Deschodt-Lanckman *et al., Gastroenterology* **68**, 819 (1975). Use in clinical radiology: W. M. Thompson, J. R. Amberg, *Gastrointest. Radiol.* **3**, 191 (1978). Effect on food intake: M. A. Della-Fera, C. A. Baile, *Science* **206**, 471 (1979). Review of synthesis: L. Balaspiri, *Acta Physiol. Acad. Sci. Hung.* **47**, 299 (1976). Review of pharmacology: S. L. Engel *et al.*, in *Pharmacological and Biochemical Properties of Drug Substances* vol. 2, M. E. Goldberg, Ed. (Am. Pharm. Assoc., Washington, DC, 1979) pp 516-526.

$$SO_3H$$
$$|$$
Asp—Tyr—Met—Gly—Trp—Met—Asp—PheNH$_2$

Solid. $[\alpha]_D^{23}$ −18.4° (c = 0.7 in 1N NH$_3$). uv max (0.1N NaOH): 280, 288 nm (ε 4850, 4230).

THERAP CAT: Choleretic.

8690. Sinefungin. *6,9-Diamino-1-(6-amino-9H-purin-9-yl)-1,5,6,7,8,9-hexadeoxy-D-glycero-α-L-talo-decofuranuronic acid;* Compd 57926; Antibiotic 32232RP; A-9145. C$_{15}$-H$_{23}$N$_7$O$_5$; mol wt 381.39. C 47.24%, H 6.08%, N 25.71%, O 20.98%. Adenine-containing antibiotic produced by *Streptomyces griseoleus* NRRL 3739. Isoln: R. L. Hamill, M. M. Hoehn, U.S. pat. **3,758,681** (1973 to Lilly); *eidem, J. Antibiot.* **26**, 463 (1973). Antifungal activity: R. S. Gordee, T. F. Butler, *ibid.* 466. Antiprotozoal activity: D. K. Dube *et al., Am. J. Trop. Med. Hyg.* **32**, 31 (1983); A. Ferrante *et al., Trans. Roy. Soc. Trop. Med. Hyg.* **78**, 837 (1984); P. Paolantonacci *et al., Antimicrob. Ag. Chemother.* **28**, 528 (1985); E. Zweygarth *et al., Trop. Med. Parasit.* **37**, 255 (1986); *in vivo* antileishmanial activity: J. L. Avila *et al., Am. J. Trop. Med. Hyg.* **43**, 139 (1990). Synthesis: M. Geze *et al., J. Am. Chem. Soc.* **105**, 7638 (1983); stereoselective synthesis: M. P. Maguire *et al., J. Org. Chem.* **55**, 948 (1990). Inhibition of methyltransferase: R. T. Borchardt *et al., Biochem. Biophys. Res. Commun.* **89**, 919 (1979).

White powder. Sol in water, very slightly sol in methanol and ethanol. Insol in organic solvents. Stable at room temp in aq soln from pH 1 to 11. pKa (66% DMF): 2.9, 3.9, 8.9 10.2. uv max in neutral solns: 206 nm, 258 nm (E$_{1cm}^{1\%}$ 520, 325); in basic solns: 256 nm (E$_{1cm}^{1\%}$ 325). $[\alpha]_D^{25}$ −2.61° (c = 5 in water). Also reported as $[\alpha]_D^{23}$ +12° ±2° (c = 0.227 in water). LD$_{50}$ s.c. in mice: 185 mg/kg (Hamill, Hoehn, *J. Antibiot.*)

8691. Sinigrin. *1-Thio-β-D-glucopyranose 1-[N-(sulfooxy)-3-butenimidate] monopotassium salt;* sinigroside; myronate potassium; potassium myronate; allyl glucosinolate. C$_{10}$H$_{16}$KNO$_9$S$_2$; mol wt 397.47. C 30.22%, H 4.06%, K 9.84%, N 3.52%, O 36.23%, S 16.14%. A β-glucopyranoside isolated from black mustard seeds and from the horseradish root *Alliaria officinalis* Andrz., *Cruciferae*: A. Bussy, *J. Pharm. Chim.* **26**, 39 (1840); Gadamer, *Arch. Pharm.* **235**, 44 (1897); Stoll, Seebeck, *Helv. Chim. Acta* **31**, 1432 (1948). Structure: Ettlinger, Lundeen, *J. Am. Chem. Soc.* **78**, 4173 (1956); **79**, 1764 (1957). Hydrolysis: Ettlinger *et al., Proc.*

Nat. Acad. Sci. USA **47**, 1875 (1961). Crystal structure: Waser, Watson, *Nature* **198**, 1297 (1963). Synthesis: Benn, Ettlinger, *Chem. Commun.* **1965**, 445; A. Kjaer, *Acta Chem. Scand.* **22**, 3324 (1968).

Monohydrate, crystals, mp 127-129°. When anhydr mp 179°. $[\alpha]_D^{18}$ −16.4°. Freely sol in water, hot alcohol; insol in benzene, chloroform, ether.

Tetraacetate, C$_{18}$H$_{24}$KNO$_{13}$S$_2$, crystals, mp 193-195°. $[\alpha]_D^{26}$ −16° (c = 0.14).

8692. Sinomenine. *(9α,13α,14α)-7,8-Didehydro-4-hydroxy-3,7-dimethoxy-17-methylmorphinan-6-one;* cuculine; coculine; kukoline. C$_{19}$H$_{23}$NO$_4$; mol wt 329.40. C 69.28%, H 7.04%, N 4.25%, O 19.43%. An optical isomer of methoxythebainone. Configuration at the asymmetric centers, C$_9$, C$_{13}$, and C$_{14}$, is the mirror image of those in morphine. From root of *Sinomenium acutum* (Thunb.) Rehd. & Wils. (*Cocculus diversifolius* DC.), *Menispermaceae*: Goto, *J. Chem. Soc. Japan* **44**, 795 (1923), *C.A.* **18**, 2710 (1924); Ohta, *Kitasato Arch. Exp. Med. (Tokyo)* **6**, 259 (1925), *C.A.* **21**, 3233^2 (1927). Structure: Kondo, Ochiai, *Ann.* **470**, 224 (1929); Sasaki, *Yakugaku Zasshi* **80**, 270 (1960), *C.A.* **54**, 11702g (1960). Biogenesis: Cohen, *Chem. & Ind. (London)* **1956**, 1391. Pharmacology: S. Fu *et al., Yao Hsueh Hsueh Pao* **10**, 673 (1963), *C.A.* **60**, 8525h (1964). Reviews: Holmes in R. H. F. Manske, H. L. Holmes, *The Alkaloids* **vol. II** (Academic Press, New York, 1952) pp 219-260; K. W. Bentley, *The Chemistry of the Morphine Alkaloids* (Oxford, 1954).

Clusters of needles from benzene, mp 161°. After melting once, the mp is raised to 182°. $[\alpha]_D^{26}$ −71° (c = 2.1 in alc). Sol in alc, acetone, chloroform, dil alkali; slightly sol in water, ether, benzene. LD$_{50}$ orally in mice: 580 mg/kg (Fu).

Hydrochloride dihydrate, C$_{19}$H$_{24}$ClNO$_4$.2H$_2$O, long prisms from water, dec 231° when anhydr. $[\alpha]_D^{17}$ −82° (c = 4.4). Sol in about 1.5 parts water.

8693. Sintropium Bromide. *(endo,syn)-8-Methyl-8-(1-methylethyl)-3-[(1-oxo-2-propylpentyl)oxy]-8-azoniabicyclo-[3.2.1]octane bromide;* (8r)-3α-hydroxy-8-isopropyl-1αH,-5αH-tropanium bromide 2-propylvalerate; *endo-8-methyl-8-syn-(1-methyl)ethyl-8-azoniabicyclo[3.2.1]octane-3-(2-propyl)-pentanoate bromide;* VAL-480; VAL-4000. C$_{19}$H$_{36}$-BrNO$_2$; mol wt 390.40. C 58.45%, H 9.29%, Br 20.47%, N 3.59%, O 8.20%. Prepn: V. Bernareggi *et al.,* **Eur.** pat. **Appl. 42,365;** *eidem,* **U.S.** pat. **4,411,902** (1981, 1983 both to Valeas). Clinical trial: M. Galeone, *Minerva Med.* **76**, 1605 (1985); V. Colosi *et al., Minerva Urol. Nefrol.* **38**, 271 (1986).

mp > 280° (dec). LD$_{50}$ i.p. in male mice: 77 mg/kg (Bernareggi).

THERAP CAT: Antispasmodic.

8694. Sisal. *Agave sisalana* Perrine, *Amaryllidaceae;* henequem; yaxci; potosina. One of the plants native to the American desert. Process of treating juice and extracting steroidal material and waxes: Spray, U.S. pat. **3,019,193** (1962). *Habit.* Principally the Mexican tableland. Abundant in the Mexican states Yucatan and Chiapas. *Constit.* Hemp, sapogenins such as hecogenin and tigogenin. *Review: Chem. Week* (April 20, 1957) pp 41-48.

Caution: The dust is irritating to respiratory tract.

USE: Manuf sisal ropes, steroid intermediates.

8695. Sisomicin. *O-3-Deoxy-4-C-methyl-3-(methylamino)-β-L-arabinopyranosyl-(1→6)-O-[2,6-diamino-2,3,4,6-tetradeoxy-α-D-glycero-hex-4-enopyranosyl-(1→4)]-2-deoxy-D-streptamine; (2S-cis)-4-O-[3-amino-6-(aminomethyl)-3,4-dihydro-2H-pyran-2-yl]-2-deoxy-6-O-[3-deoxy-4-C-methyl-3-(methylamino)-β-L-arabinopyranosyl]-D-streptamine;* rickamicin; antibiotic 6640; Sch-13475. C$_{19}$H$_{37}$N$_5$O$_7$; mol wt 447.53. C 50.99%, H 8.33%, N 15.65%, O 25.03%. Gentamicin-like aminoglycoside antibiotic produced by *Micromonospora inyoesis* (NRRL 3292): M. J. Weinstein *et al.* **Ger. pat. 1,932,309;** *eidem,* **U.S. pat. 3,832,286** (1970, 1974 both to Schering); *eidem, J. Antibiot.* **23,** 551 (1970). Isoln: G. H. Wagman *et al., ibid.* 555. Chemical and physical studies: Cooper *et al., Chem. Commun.* **1971,** 285, 924; Kugelman *et al., J. Antibiot.* **26,** 394 (1973). Structure consists of 2-deoxystreptamine, *q.v.,* linked to two saccharide units, *garosamine* and *sisosamine:* Keiman *et al., J. Org. Chem.* **39,** 1451 (1974). Synthesis: D. H. Davies *et al., J. Med. Chem.* **21,** 189 (1978). Synthesis of sisosamine derivs: Cleophax *et al., Chem. Commun.* **1975,** 11. Has broad spectrum antibiotic activity: Waitz *et al., J. Antibiot.* **23,** 559 (1970). Biotransformation of sisomicin to gentamicin: B. K. Lee *et al., Antimicrob. Ag. Chemother.* **12,** 335 (1977). Clinical studies: R. Lorber *et al., Clin. Ther.* **4,** 263 (1981); J. P. Sculier *et al., J. Antimicrob. Chemother.* **9,** 63 (1982). *Review:* P. Noone, *Drugs* **27,** 548-578 (1984).

sisosamine

2-deoxystreptamine

garosamine

Monohydrate, needles from ethanol, mp 198-201°. [α]$_D^{26}$ +189° (c = 0.3).

Penta-*N*-acetyl deriv, C$_{29}$H$_{47}$N$_5$O$_{12}$, amorphous white powder from methanol-ethyl ether, mp 188-198° (dec). [α]$_D^{26}$ +200° (c = 0.3).

Sulfate, (C$_{19}$H$_{37}$N$_5$O$_7$)$_2$·5H$_2$SO$_4$, *Baymicin, Extramycin, Mensiso, Siseptin, Sisobiotic, Sisolline, Sisomin.* LD$_{50}$ in mice (mg/kg): 34 i.v., 221 i.p., 288 s.c. (Weinstein).

THERAP CAT: Antibacterial.

8696. α$_1$-Sitosterol. *(3β,4α,5α,24Z)-4-Methylstigmasta-7,24(28)-dien-3-ol;* citrostadienol. C$_{30}$H$_{50}$O; mol wt 426.73. C 84.44%, H 11.81%, O 3.75%. Occurs in plants. Isoln from wheat germ oil: Anderson *et al., J. Am. Chem. Soc.* **48,** 2987 (1926); Wallis, Fernholz, *ibid.* **58,** 2446 (1936). Structure: Bernstein, Wallis, *ibid.* **61,** 2308 (1939); Mazur, Sondheimer, *ibid.* **80,** 6293, 6296 (1958). Identity with citrostadienol: Schreiber, Osske, *Experientia* **19,** 69 (1963); *eidem, Tetrahedron* **20,** 2575 (1964). Stereochemistry: C. Brooks *et al., Steroids* **20,** 487 (1972).

Needles from alcohol, mp 164-166°. [α]$_D^{28}$ −1.7° (c = 2 in chloroform). Is pptd by digitonin.

8697. β-Sitosterol. *(3β)-Stigmast-5-en-3-ol;* 22:23-dihydrostigmasterol; α-dihydrofucosterol; Δ^5-stigmasten-3β-ol; 24β-ethyl-Δ^5-cholesten-3β-ol; α-phytosterol; cinchol; cupreol; rhamnol; quebrachol; sitosterin; Harzol; Prostasal; Sito-Lande. C$_{29}$H$_{50}$O; mol wt 414.72. C 83.99%, H 12.15%, O 3.86%. Common sterol in plants. Isoln from wheat germ oil, corn oil: Anderson *et al., J. Am. Chem. Soc.* **48,** 2987 (1926); from rye germ oil: Gloyer, Schuette, *ibid.* **61,** 1901 (1939); from cottonseed oil: Wallis, Chakravorty, *J. Org. Chem.* **2,** 335 (1937); from tall oil: Sandqvist, Bengtsson, *Ber.* **64,** 2167 (1931). Also occurs in soy and calabar beans, in rice embryos: Tanaka, *J. Biochem.* **17,** 483 (1933); in cascara, cinchona bark and cinchona wax: Dirscherl, *Z. Physiol. Chem.* **235,** 1 (1935). Prepn from tall oil: G. I. Fujimoto, A. E. Jacobson, *J. Org. Chem.* **29,** 3377 (1964). Identity with 22:23-dihydrostigmasterol: W. Dirscherl, N. Nahm, *Ann.* **555,** 57 (1944). Identity with cinchol: *eidem, Ann.* **558,** 231 (1947). Structure: Bernstein, Wallis, *J. Org. Chem.* **2,** 341 (1937); Bergmann, Low, *ibid.* **12,** 67 (1947); Shoppee, *J. Chem. Soc.* **1948,** 1032. Stereospecific synthesis: W. Sucrow, M. Slopianka, *Ber.* **108,** 3721 (1975). Clinical efficacy in treatment of type II hyperlipoproteinemia: R. S. Lees, A. M. Lees, in *Lipoprotein Metabolism,* H. Greten, Ed. (Springer-Verlag, New York, 1976) pp 119-124; P. Oster *et al., ibid.* pp 125-130. Studies on inhibition of cholesterol absorption: S. M. Grundy, H. Y. I. Mok, *ibid.* pp 112-118; I. Ikeda, M. Sugano, *Biochim. Biophys. Acta* **732,** 651 (1983). Inhibition of induced carcinogenesis: N. D. Nigro *et al., J. Natl. Cancer Inst.* **69,** 103 (1982). Clinical trial in treatment of prostatic adenoma: H.-P. Szutrely, *Med. Klin.* **77,** 520 (1982). Book: *Monographs on Atherosclerosis* Vol. **10,** T. B. Clarkson *et al.,* Eds. entitled "Sitosterol" by O. J. Pollak, D. Kritchevsky (Karger, Basel, 1981) 219 pp.

Plates from alcohol, mp 140°. $[\alpha]_D^{25}$ $-37°$ (c = 2 in chloroform).

Acetate, $C_{31}H_{52}O_2$, mp 127-128°. $[\alpha]_D^{25}$ $-41°$ (c = 2 in chloroform).

THERAP CAT: Anticholesteremic. Treatment of prostatic adenoma.

8698. γ-Sitosterol. *(3β,24S)-Stigmast-5-en-3-ol;* clionasterol. $C_{29}H_{50}O$; mol wt 414.72. C 83.99%, H 12.15%, O 3.86%. Principal sterol of soybean oil: Bonstedt, *Z. Physiol. Chem.* **176,** 269 (1928). Isoln from corn oil: Anderson, Shriner, *J. Am. Chem. Soc.* **48,** 2976 (1926); from venom of Chinese toad, *Bufo gargarizans:* Rukstuhl, Meyer, *Helv. Chim. Acta* **40,** 1270 (1957). Differs from β-sitosterol in spatial configuration at C-24: Bergmann, Low, *J. Org. Chem.* **12,** 67 (1947). Structure: Pavanaram, *Helv. Chim. Acta* **46,** 1377 (1963). Stereospecific synthesis: W. Sucrow, M. Slopianka, *Ber.* **108,** 3721 (1975).

Monohydrate, plates from alcohol. When dry, mp 147-148°. $[\alpha]_D^{20}$ $-43°$ (c = 1.9 in chloroform). Table of mp's and $[\alpha]_D$'s: Dirscherl, *Z. Physiol. Chem.* **257,** 242 (1939).

Acetate, $C_{31}H_{52}O_2$, mp 143-144°. $[\alpha]_D^{20}$ $-45.3°$ (c = 1.9 in chloroform).

8699. Sizofiran. Schizophyllan; SPG; Sonifilan. Immunostimulant polysaccharide produced by the fungus *Schizophyllum commune* Fries. Polymer of repeating tetrasaccharide units composed of three β-(1→3)-linked D-glucopyranose residues, to one of which is attached a single β-(1→6)-linked D-glucopyranosyl side chain. Forms triple-stranded helix of mol wt ∼450,000 in neutral aqueous solution. Prepn by fermentation: S. Kikumoto *et al., Japan.* pat. **67 12,000** (1967 to Taito), *C.A.* **67,** 107386r (1967). Synthesis of the 3 possible tetrasaccharide units: K. Takeo, S. Tei, *Carbohydrate Res.* **145,** 293 (1986). Host-mediated antitumor activity: N. Komatsu *et al., Gann* **60,** 137 (1969). Pharmacology, toxicology and antitumor activity in mice: T. Matsuo *et al., Arzneimittel-Forsch.* **32,** 647 (1982). Macrophage stimulant activity *in vitro:* I. Sugawara *et al., Cancer Immunol. Immunother.* **16,** 137 (1984). Clinical trial in cervical cancer: K. Okamura *et al., Cancer* **58,** 865 (1986).

White, amorphous powder. Sol in water, forming highly viscous soln.

THERAP CAT: Antineoplastic.

8700. Skatole. *3-Methyl-1H-indole.* C_9H_9N; mol wt 131.18. C 82.41%, H 6.92%, N 10.68%. Constituent of feces, beetroot, nectandra wood, and coal tar. Obtained by fusing egg albumin with KOH.

White to brownish scales; fecal odor; mp 95°; bp 265-266°. Sol in hot water, alc, benzene, chloroform, ether. With potassium ferrocyanide and sulfuric acid it gives a violet color. MLD s.c. in frogs: 1.0 g/kg, *Handbook of Toxicology,* W. S. Spector, Ed. (Saunders, Philadelphia, 1956) p 268.

8701. Skimmianine. *4,7,8-Trimethoxyfuro[2,3-b]quinoline;* β-fagarine; 7,8-dimethoxydictamnine. $C_{14}H_{13}NO_4$; mol wt 259.26. C 64.86%, H 5.05%, N 5.40%, O 24.68%. In *Skimmia japonica* Thunb., *Fagara* spp., *Glycosmis pentaphyl-*

la Corr., *Ruta graveolens* L., *Rutaceae:* Honda, *Arch. Exp. Pathol. Pharmakol.* **52,** 83 (1904); Paris, Moyse-Mignon, *Ann. Pharm. Franc.* **5,** 410 (1947), *C.A.* **42,** 3909h (1948); Chatterjee, Majumdar, *J. Am. Chem. Soc.* **76,** 2459 (1954); Schneider, *Arzneimittel-Forsch.* **14,** 435 (1964). Identity with β-fagarine: Deulofeu *et al., J. Am. Chem. Soc.* **64,** 2326 (1942). Structure: Asahina, Inubuse, *Ber.* **63,** 2052 (1930).

Pyramids, octahedral rods from alcohol. mp 178°. uv max (ethanol): 212, 251, 321, 331 nm (log ε 4.10, 4.91, 3.89, 3.91). Neutral to litmus. Sol in alcohol, chloroform; slightly in ether, amyl alcohol and carbon disulfide; practically insol in water and petr ether.

8702. Skimmin. *7-(β-D-Glucopyranosyloxy)-2H-1-benzopyran-2-one; 7-(glucosyloxy)coumarin;* 7-hydroxycoumarin-7-glucoside; umbelliferone glucoside. $C_{15}H_{16}O_8$; mol wt 324.29. C 55.56%, H 4.97%, O 39.47%. In bark and wood of *Skimmia japonica* Thunb., *Rutaceae.* Isoln, structure and synthesis: Späth, Neufeld, *Rec. Trav. Chim.* **57,** 535 (1938).

Crystals with one H_2O from water. After drying at 100° and 10 mm: mp 219-221° (evac tube). $[\alpha]_D^{18}$ $-80°$ (c = 10 in pyridine). uv max (ethanol): 320 nm. Moderately sol in hot water, alcohol; practically insol in ether, chloroform.

8703. Skunk Oil. A fatty oil from the winter fat of the skunk [*Mephitis mephitis* Schreb. (North America) or *Putorius putorius* L. (Europe and Asia), Fam. *Mustelidae*]: Weil, *Am. Druggist* **59,** 8 (1912), *C.A.* **6,** 1655 (1912); Lobachev, *Sovet. Med.* **7,** no. 10, 21 (1943), *C.A.* **39,** 4193 (1945); Stevens, *J. Am. Chem. Soc.* **67,** 407 (1945).

Neutral or faintly acid oil, of only slightly disagreeable odor. Solidifies around 21°. On standing, a deposit of stearic acid may separate which redissolves on heating. At 75° a pungent odor is evolved. d_{15}^{15} 0.931; iodine no. 83.48 (Hanus). Sapon no. 89.32 (Weil); 201 (Lobachev). Insol fatty acids mp 30°; their Hehner no. 93.45. The oil may be used directly or as an emulsion; the emulsion should be protected from the bacterial deterioration by an added antiseptic.

8704. Slendid®. Natural food ingredient made from pectin extracted from citrus peels. Analysis of interaction with flavor compounds: J. P. Schirle-Keller *et al., J. Food Sci.* **59,** 813 (1994). Brief review: D. D. Duxbury, *Food Process.* **52,** 54-55 (1991).

Light cream to light tan, granular free-flowing powder. Tasteless and odorless. Completely sol in 25 parts of water at 60° with adequate agitation.

USE: Fat replacer in foods.

8705. Smilagenin. *(25R)-Spirostan-3β-ol;* isosarsasapogenin. $C_{27}H_{44}O_3$; mol wt 416.64. C 77.84%, H 10.64%, O 11.52%. Steroid sapogenin from *Smilax ornata* Hooker, *Liliaceae* (Sarsaparilla): Ashew *et al., J. Chem. Soc.* **1936,** 1399; from *Agave lecheguilla* Tou., *Amaryllidaceae:* Nord, U.S. pat. **2,897,192** (1959). C_{25} isomer of sarsasapogenin: Scheer *et al., J. Am. Chem. Soc.* **77,** 641 (1955). Obtained by acid isomerization of sarsasapogenin: Marker, Rohrmann, *ibid.* **61,** 846 (1939); Wall *et al., ibid.* **77,** 3086 (1955). For structure and stereochemistry see Sarsasapogenin.

Silky needles from acetone, mp 185°. $[\alpha]_D^{25}$ $-69°$ (c = 0.5 in chloroform). Absorption spectrum: Smith, Eddy, *Anal. Chem.* **31,** 1539 (1959).

Acetate, $C_{29}H_{46}O_4$, needles from methanol, mp 152°.
USE: In making compds of the pregnane series.

8706. Soap. Any salt of a fatty acid, usually made by saponification of a vegetable oil with caustic soda. Hard soap consists largely of sodium oleate (low titer) or of sodium palmitate or stearate (high titer); soft soap, of potassium and sodium salts of coconut and/or palm oil fatty acids. Rosin acid salts have detergent properties and are often incorporated in laundry soaps. Shaving soaps contain glycerol and gum to prevent rapid drying. Medicated soaps contain antiseptics such as phenols or mercury salts. Soaps also may be prepd from the fatty acid and an amine, e.g., triethanolamine. Nonalkali-metal salts of the fatty acids are water insol and as such have certain uses.

Hard soap (olive-oil castile) is a white or yellowish white powder or bar, sol in water, alc. The aq soln is alkaline due to hydrolysis; the alc soln is only slightly alkaline. Transparent hard soap is obtained by adding additional glycerol during saponification. Floating soaps are made by adding air after saponification during the so-called "crutching" process. Soft soap (green soap) is a yellowish-white to brownish-yellow (or green) soft mass, more sol than hard soap.

USE: Hard soap—detergent; in soln as a vehicle for liniments; as ingredient of pills contg resinous drugs like aloe, etc. Soft soap—detergent; extern. as vehicle for active medicaments applied in ointment form or liniment. Insoluble soaps—fungicides (Cu), disinfectants (Pb, Cu, Hg), face powders and ointments (Zn), waxes and polishes (Al), greases (Al, Ca), and surface coatings (Ca).

THERAP CAT: Topical anti-infective. Antidote (mineral acid and heavy metal poisoning).

THERAP CAT (VET): Topically as detergent (cleansing agent) and mild antiseptic. In liniments as counterirritant. Has been used internally as laxative, antacid and as an antidote for mineral acids and heavy metals.

8707. Sobrerol. *5-Hydroxy-α,α,4-trimethyl-3-cyclohexene-1-methanol; p-menth-6-ene-2,8-diol*; 6,8-carvomenthene-diol; 1-p-menthene-6,8-diol; pinol hydrate. $C_{10}H_{18}O_2$; mol wt 170.25. C 70.55%, H 10.66%, O 18.79%. An oxidation product of oil of turpentine formed by the autooxidation of α-pinene, q.v., in the presence of water and especially in sunlight. First obtained by Margueron in 1797. Confused with terpin hydrate, q.v., until fully characterized by Sobrero in 1851. The naturally occurring form is the *trans*-form. For a review of early structure and prepn work, see J. L. Simonsen, *The Terpenes* vol. II (University Press, Cambridge, 2nd ed., 1949) pp 137-138. Summary of general methods of prepn: Lombard, Heywang, *Bull. Soc. Chim. France* **1954**, 1210. Prepn of *dl-trans*-form: Ward, *J. Am. Chem. Soc.* **60**, 325 (1938); Schenck *et al.*, *Ann.* **584**, 177 (1953); Moore *et al.*, *J. Am. Chem. Soc.* **78**, 1173 (1956); Brook, Wright, *J. Org. Chem.* **22**, 1314 (1957); of *l-trans*-form: Klein, U.S. pat. **2,815,378** (1958 to Glidden). Prepn of *cis*- and *trans*-forms and analysis of stereochemistry: Blumann, Wood, *J. Chem. Soc.* **1952**, 4420; H. Schmidt, *Ber.* **86**, 1437 (1953). Pharmacology and chemistry of *dl-trans*-form: **Brit.** pat. **1,176,817**; C. Corvi-Mora, U.S. pat. **3,592,908** (1970, 1971 to Camillo Corvi); V. Dalla Valle, *Boll. Chim. Farm.* **109**, 761 (1970). Clinical evaluation: C. F. Marchioni, G. Monzali, *Clin. Ter.* **60**, 135 (1972). Activity on mucociliary transport: P. Braga, *Clin. Trials J.* **18**, 30 (1981). Pharmacokinetics: P. C. Braga *et al.*, *Eur. J. Clin. Pharmacol.* **24**, 209 (1983). Metabolism: P. Ventura *et al.*, *Xenobiot.* **13**, 139 (1983); *eidem, ibid.* **15**, 317 (1985). GLC determn in human plasma: U. A. Shulka, *J. Chromatog.* **308**, 189 (1984).

CH₃

HO.

H₃C   CH₃
     OH

(−)-trans form

dl-trans-Form, **Lysmucol, Sobrepin.** Crystals, mp 130-131.5°. bp 270-271°. Soly in water: 3.3 g/100 ml (15°). LD₅₀ in rats, mice (mg/kg): both 580 i.v. (Dalla Valle). *d-* or *l-trans*-Form, mp 150° from ethanol. $[\alpha]_D^{20}$ (±)150° resp. (c = 10 in ethanol) (Schmidt).

THERAP CAT: Mucolytic.

8708. Sobuzoxane. *1,2-Ethanediylbis[(2,6-dioxo-4,1-piperazinediyl)methylene]carbonic acid bis(2-methylpropyl) ester*; 4,4'-ethylenebis[1-(hydroxymethyl)-2,6-piperazinedione] bis(isobutyl carbonate) (ester); 4,4'-(1,2-ethanediyl)bis-(1-isobutoxycarbonyloxymethyl)-2,6-piperazinedione; 1,2-bis(4-isobutoxycarbonyloxymethyl-3,5-dioxopiperazin-1-yl)-ethane; MST-16; Perazolin. $C_{22}H_{34}N_4O_{10}$; mol wt 514.53. C 51.36%, H 6.66%, N 10.89%, O 31.09%. Noncleavable complex-stabilizing DNA topoisomerase II inhibitor. Prepn: J.-C. Cai, M. Takase, **Eur.** pat. **140,327**; *eidem,* **U.S.** pat. **4,650,799** (1985, 1987 both to Zenyaku Kogyo); *idem et al., Chem. Pharm. Bull.* **37**, 2976 (1989). Antitumor activity: T. Narita *et al., Cancer Chemother. Pharmacol.* **28**, 235 (1991). Clinical evaluation in non-Hodgkin's lymphoma: R. Ohno *et al., J. Nat. Cancer Inst.* **84**, 435 (1992); in adult T-cell leukemia-lymphoma: *idem et al., Cancer* **71**, 2217 (1993). Acute toxicity study: T. Tokiwa *et al., Iyakuhin Kenkyu* **22**, 435 (1991). *C.A.* **115**, 222916k (1991).

Crystals from ethylene glycol monomethylether, mp 128-130°. Also reported as crystals from ethanol, mp 132-133° (Cai, 1989). Practically insol in water. LD₅₀ in male, female mice, rats (mg/kg): 807, 960, 877, 567 i.p.; 400, 673, 3025, 2821 s.c.; > 5 g/kg orally in all species (Tokiwa).

THERAP CAT: Antineoplastic.

8709. Soda Lime. Sodium hydroxide with lime. A mixture of calcium oxide with 5-20% sodium hydroxide and contg 6-18% water. It absorbs 25-35% its wt of CO_2.

White or gray-white granules. Rapidly deteriorates on exposure to air through absorption of carbon dioxide. *Keep tightly closed.*

Caution: Strongly corrosive and irritating to skin, mucous membranes, eyes. Ingestion can cause severe damage to G.I. tract, death.

USE: To absorb carbon dioxide in basal metabolism tests, in rebreathing anesthesia systems, in submarines, and in carbon determinations.

8710. Sodium. Natrium. Na; at. wt 22.98977; at. no. 11; valence 1. Alkali metal. Occurrence in earth's crust: 2.83% by wt; principal cation in hydrosphere. Natural isotopes: 23 (100%); radioactive isotopes (mass number): 20-22; 24-26. Prepd by Davy in 1807 by electrolysis of fused sodium hydroxide. Found in form of its compds, halides, silicates, carbonates; does not occur free. Industrial prepns: primarily in Downs cells; also in Castner cells: Batsford, *Chem. Met. Eng.* **26**, 888, 932 (1932); Regelsberger, *Chemische Technologie der Leichtmetalle*, Leipzig, 1926; Hardie, *Ind. Chemist* **30**, 161 (1954). Reviews: M. Sittig, *Sodium* A.C.S. Monograph Series, no. 133 (Reinhold, New York, 1956); Whaley, "Sodium, Potassium, Rubidium, Cesium and Francium" in *Comprehensive Inorganic Chemistry*, **Vol. 1**, J. C. Bailar, Jr. *et al.*, Eds. (Pergamon Press, Oxford, 1973) pp 369-529.

Light, silvery-white metal; body-centered cubic structure; lustrous when freshly cut; tarnishes on exposure to air, becoming dull and gray. Soft at ordinary temp, fairly hard at −20°. mp 97.82°; bp 881.4°; d²⁰ 0.968. Heat capacity of solid: 0.292 cal/g deg; heat capacity of liquid at mp: 0.331 cal/g deg. Heat of fusion: 27.05 cal/g; thermal conductivity (cal/sec °C cm): 0.205 (97.82°) 0.170 (400°). E⁰ (aqueous) Na/Na⁺ 2.714 V. Violently decomposes water, forming sodium hydroxide and hydrogen which may ignite spontaneously. Decomposes alc. Reacts vigorously with oxy-

gen, burning with a yellow flame. Combines directly with the halogens, with phosphorus. Reduces most oxides to the elemental state, reduces metallic chlorides. Dissolves in liq ammonia to give a blue soln; when heated in ammonia gas yields sodamide. Dissolves in mercury, forming sodium amalgam. *Keep under liquids contg no oxygen, such as kerosene, naphtha.*
Caution: Extremely caustic to all tissue.
USE: Manuf of sodium compds, such as the cyanide, azide, peroxide, etc.; in manuf of tetraethyllead; in org syntheses; for photoelectric cells; in sodium lamps. Alloyed with potassium in heat transfer media.

8711. Sodium Acetate. $C_2H_3NaO_2$; mol wt 82.03. C 29.28%, H 3.69%, Na 28.02%, O 39.01%. CH_3COONa. Crystal structure of the trihydrate: K.-T. Wei, D. L. Ward, *Acta Crystallogr.* **B33**, 522 (1977).
Hygroscopic powder. Very sol in water; moderately sol in alcohol.
Trihydrate, transparent crystals or granules. Efflorescent in warm air. d 1.45. mp 58°; becomes anhyd at 120°; dec at higher temp. One gram dissolves in 0.8 ml water, 0.6 ml boiling water, 19 ml alcohol. pH of 0.1 molar aq soln at 25°: 8.9.
USE: Anhydrous form as auxiliary in acetylations. Trihydrate in photography; as reagent in anal. chemistry to eliminate effect of strong acids (buffer); for foot warmers and milk-bottle warmers. Technical grades are used as a mordant in dyeing. As acidulant in food. Pharmaceutic aid (in dialysis solutions).
THERAP CAT (VET): Has been used in bovine ketosis.

8712. Sodiumacetic Acid Sodium Salt. *(Carboxymethyl)sodium sodium salt;* sodium acetate sodium derivative; sodium α-sodioacetate; α-sodioacetic acid sodium salt. $C_2H_2Na_2O_2$; mol wt 104.02. C 23.09%, H 1.94%, Na 44.20%, O 30.76%. $NaCH_2COONa$. Prepd by fusing anhydr sodium acetate with sodamide in an atm of nitrogen at 200°: DePree, Closson, *J. Am. Chem. Soc.* **80**, 2311 (1958).
Light gray solid. Stable in dry air. Dec 280°. Insol in hydrocarbons and ethers. Dec by water into sodium acetate and sodium hydroxide.

8713. Sodium Acid Pyrophosphate. Disodium dihydrogen pyrophosphate. $H_2Na_2O_7P_2$; mol wt 221.94. H 0.91%, Na 20.72%, O 50.46%, P 27.91%. $Na_2H_2P_2O_7$. Prepn: Bell, *Inorg. Syn.* **3**, 98 (1950).
White, fused masses or powder. Dec at 220°. d (hexahydrate) 1.86. Sol in water, the soln having an acid reaction.
USE: Chiefly in baking powders.

8714. Sodium Alizarinesulfonate. *9,10-Dihydro-3,4-dihydroxy-9,10-dioxo-2-anthracenesulfonic acid monosodium salt;* alizarine S; alizarine carmine; C.I. Mordant Red 3; C.I. 58005. $C_{14}H_7NaO_7S$; mol wt 342.26. C 49.13%, H 2.06%, Na 6.72%, O 32.72%, S 9.37%. Prepn: *Colour Index* vol. 4 (3rd ed., 1971) p 4514.

Monohydrate, orange-yellow powder. Freely sol in water with yellow color; sol in alcohol.
USE: As acid-base indicator and in the determination of fluorine, usually used as a 1% aq soln. pH: 3.7 yellow, 5.2 purple; as a reagent for aluminum and stain in microscopy.

8715. Sodium Aluminate. *Aluminum sodium oxide.* $AlNaO_2$; mol wt 81.97. Al 32.92%, Na 28.05%, O 39.04%.
White granular mass. mp 1650°. Very sol in water. Insol in alcohol. The aq soln is strongly alkaline.
USE: Printing on fabrics; manuf lake colors; sizing paper; manuf milk-glass, soap; hardening building stones; water softener.

8716. Sodium Amalgam. An alloy of sodium and mercury. Readily prepd by adding sodium metal in small portions at a time to mercury. The combination takes place with evolution of much heat. The sodium contents of the amalgam can be varied, as desired, up to about 20% sodium, corresponding to approx Na_2Hg. The consistency and mp vary with the sodium content; the amalgam corresponding to $NaHg_2$, 5.4% Na, has the highest mp, +360°.
Sodium amalgams are dec by water similarly to sodium metal, but more slowly. *Keep tightly closed.*

8717. Sodium Amide. Sodamide. H_2NNa; mol wt 39.01. H 5.17%, N 35.90%, Na 58.93%. $NaNH_2$. Prepd from sodium metal and gaseous ammonia (or liq ammonia with ferric nitrate catalyst): Dennis, Browne, *Inorg. Syn.* **1**, 74 (1939); Greenlee, Henne, *ibid.* **2**, 128 (1946); Schenk in *Handbook of Preparative Inorganic Chemistry* vol. **1**, G. Brauer, Ed. (Academic Press, New York, 2nd ed., 1963) pp 465-468. Alternate method: Bergstrom, *Org. Syn.* coll. vol. **III**, 778 (1955).
Crystals, mp 210°. The commercial product may be a white to olive-green solid with sea-shell fracture. Begins to volatilize at 400° and dec into its elements between 500 and 600°. Heat of formation of solid (18°; 1 atm): −32.26 kcal/mole. Heat of soln (21°): −31.06 kcal/mole. Soly in liq ammonia at 20° = about 0.1%. Reacts violently with water forming NaOH and NH_3. The reaction with alcohol is considerably slower. Should be stored in sealed containers which prevent all contact with air during storage: Bergstrom, Fernelius, *Chem. Rev.* **12**, 63, 75, 78 (1933). When exposed to the atmosphere, sodium amide rapidly absorbs H_2O and CO_2. When only limited absorption takes place, as in poorly sealed containers, products are fomed which render the resulting mixture highly explosive. The formation of oxidation products is accompanied by the development of a yellow or brownish color. If such a change is noticed, the substance should be destroyed at once. This is conveniently accomplished by covering with much benzene, toluene, or kerosene and slowly adding dil ethanol with stirring.
Caution: Intensely irritating to skin, eyes, mucous membranes.
USE: Dehydrating agent. In the production of indigo and hydrazine. Intermediate in the prepn of sodium cyanide. In ammonolysis reactions, in Claisen condensations, alkylation of nitriles and ketones, synthesis of ethynyl compds, acetylenic carbinols. Fused $NaNH_2$ dissolves metallic Mg, Zn, Mo, W, quartz, glass, silicates and other substances.

8718. Sodium Amylosulfate. Sulfated potato amylopectin sodium salt; sodium amylopectin sulfate; SN-263; Depepsen. This product has substantially 1.6 sulfate groups per glucose unit and a mol wt of $2.0-8.0 \times 10^7$ and contains 15-16.5% of sulfur. Prepn: P. S. Camarata, S. Eich, **Belg.** pat. **645,762** corresp to U.S. pat. **3,271,388** (1964, 1966 to Searle).
THERAP CAT: Enzyme inhibitor.

8719. Sodium Arsanilate. *(4-Aminophenyl)arsonic acid sodium salt; arsanilic acid sodium salt;* sodium aminarsonate; sodium *p*-aminophenylarsonate; sodium anilarsonate; Arsamin; Atoxyl; Nuarsol; Protoxyl; Soamin; Sonate; Piglet Pro-Gen V; Trypoxyl. $C_6H_7AsNNaO_3$; mol wt 239.04. C 30.15%, H 2.95%, As 31.34%, N 5.86%, Na 9.62%, O 20.08%.

Tetrahydrate, white, odorless, cryst powder. *Poisonous!* Soluble in about 6 parts water, about 100 parts alcohol. The aq soln is moderately acid to litmus.
THERAP CAT: Formerly as antisyphilitic.
THERAP CAT (VET): Has been used as an anthelmintic in swine dysentery and other enteric conditions of pigs and poultry.

8720. Sodium Arsenate, Dibasic. Disodium arsenate. $AsHNa_2O_4$; mol wt 185.91. As 40.30%, H 0.54%, Na 24.73%, O 34.42%. Na_2HAsO_4. The salt of commerce is about 99% pure. Toxicity: Franke, Moxon, *J. Pharmacol. Exp. Ther.* **58**, 454 (1936).

Powder. Very sol in water; slightly sol in alcohol.

Heptahydrate, odorless crystals; effloresces in warm air. *Poisonous,* but less so than the arsenite. Loses 5 H_2O (29%) at about 50°; becomes anhydr at 100°. At 150° or higher it is converted into pyroarsenate. d 1.87. mp 57° when rapidly heated. Sol in 1.3 parts water; sol in glycerol, slightly in alcohol. The aq soln is alkaline to litmus. LD_{75} i.p. in rats: 14-18 mg As/kg (Franke, Moxon).

USE: The technical grade, about 98% pure, is used in dyeing with Turkey-red oil and in printing fabrics; manuf of other arsenates.

THERAP CAT: Formerly as antimalarial; dermatologic.

THERAP CAT (VET): Has been used in parasitism (internally and externally), also for nonparasitic skin and blood diseases, in rheumatism, asthma and heaves, and as an alterative.

8721. Sodium Arsenite. Sodium meta-arsenite. Approx $NaAsO_2$. The product of commerce is 95-98% pure. Toxicity study: H. F. Smyth *et al., Am. Ind. Hyg. Assoc. J.* **30**, 470 (1969).

White or grayish-white powder; somewhat hygroscopic. Absorbs CO_2 from air. *Very poisonous!* Freely sol in water, slightly in alcohol. *Keep well closed.* LD_{50} orally in rats: 0.041 g/kg (Smyth).

USE: The technical grade, 90-95% pure, is used in manuf of arsenical soap for use on skins, for treating vines against certain scale diseases; as insecticide especially for termites.

THERAP CAT (VET): Topical acaricide.

8722. Sodium Arsphenamine. *4,4'-Arsenobis[2-aminophenol] disodium salt;* arsphenamine sodium. $C_{12}H_{10}As_2N_2Na_2O_2$; mol wt 410.05. C 35.15%, H 2.46%, As 36.54%, N 6.83%, Na 11.21%, O 7.80%. Usually contains some water and inert salts (NaCl) and hence only about 20% arsenic (As).

Bright yellow powder. *Poisonous!* Very unstable in air. Freely sol in water with alkaline reaction. *Keep as described for Neoarsphenamine.* Solution for injection must be prepared immediately after opening the container and must be administered promptly.

THERAP CAT: Formerly as antisyphilitic.

8723. Sodium Ascorbate. *Ascorbic acid sodium derivative;* vitamin C sodium; Ascorbin; Sodascorbate; Natrascorb; Cenolate; Ascorbicin; Cebitate. $C_6H_7NaO_6$; mol wt 198.11. C 36.38%, H 3.56%, Na 11.60%, O 48.46%. One mg of the sodium salt is equivalent to 0.8890 mg of ascorbic acid, or one mg of the acid is equivalent to 1.1248 mg of sodium ascorbate. Preparation: Holland, U.S. pat. **2,442,-005** (1948); ascorbic acid is dissolved in water and an equiv amount of sodium bicarbonate is added. After cessation of effervescence the sodium ascobate is precipitated by the addn of isopropanol.

CH_2OH
HC—OH

Na^+ O OH

Minute crystals. Dec 218°. $[\alpha]_D^{20}$ +104.4°. Freely sol in water at 25°: 62 g/100 ml H_2O. Even more sol in warm water (78 g/100 ml H_2O at 75°). pH of aq soln 5.6-7.0 or even higher. A 10% soln, made from a commercial grade, may have a pH of 7.4-7.7. Aq solns are unstable and subject to quick oxidation by air at pH > 6.0. Solns can be buffered with ascorbic acid solns which have a pH of 2.3-2.5.

USE: In vitamin C prepns; antioxidant in chopped meat and other food, also in curing meat.

THERAP CAT: Vitamin C source.

8724. Sodium Azide. Smite. N_3Na; mol wt 65.01. N 64.64%, Na 35.36%. NaN_3. Cytochrome oxidase inhibitor. Prepd from $NaNH_2 + N_2O$: Dennis, Browne, *Z. Anorg. Allgem. Chem.* **40**, 95 (1904); Schenk in *Handbook of Preparative Inorganic Chemistry* Vol. 1, G. Brauer, Ed. (Academic Press, New York, 2nd ed., 1963) pp 474-475. Alternate procedures: *Inorg. Syn.* **1**, 79 (1939); **2**, 139 (1946). Large scale manuf processes: B. T. Fedoroff *et al., Encyclopedia of Explosives and Related Items,* Vol. 1 (Picatinny Arsenal, Dover, N.J., 1960) pp A601-A619. *Review:* L. E. Audrieth, *Chem. Rev.* **15**, 169 (1934). Potent vasodilator; has been used therapeutically to control blood pressure. Review of toxicity, mutagenicity and carcinogenicity: K. A. Frederick, J. G. Babish, *Regul. Toxicol. Pharmacol.* **2**, 308-322 (1982).

Colorless hexagonal crystals, d 1.846. On heating dec into sodium and nitrogen. Highly sol in water. Rapidly converted into hydrazoic acid, *q.v.* Soly in water: 40.16% at 10°, 41.7% at 17°. pK = 4.8, aq solns contains HN_3 which escapes readily at 37°. Slightly sol in alcohol. Insol in ether. Sol in liquid ammonia. LD_{50} in rats (mg/kg): 45 orally (Frederick, Babish).

Caution: Highly toxic. May cause hypotension, tachycardia, tachypnea, hypothermia, convulsions and severe headache: *Clinical Toxicology of Commercial Products,* R. E. Gosselin *et al.,* Eds. (Williams & Wilkins, Baltimore, 5th ed., 1984), Section II, p 114; *Prudent Practices for Handling Hazardous Chemicals in Laboratories* (National Academy Press, Washington, D.C., 1981) pp 145-147.

USE: In organic syntheses; in the preparation of hydrazoic acid, lead azide, pure sodium. In the differential selection of bacteria; in automatic blood counters; as preservative for laboratory reagents. Propellant for inflating automotive safety bags. Agricultural nematocide; herbicide; in fruit rot control.

8725. Sodium Benzoate. $C_7H_5NaO_2$; mol wt 144.11. C 58.34%, H 3.50%, Na 15.95%, O 22.21%. Toxicity: Smyth, Carpenter, *J. Ind. Hyg. Toxicol.* **30**, 63 (1948).

COO^- Na^+

White, odorless granules or crystalline powder; sweetish, astringent taste. One gram dissolves in 1.8 ml water, 1.4 ml boiling water, about 75 ml alcohol, in 50 ml of a mixture of 47.5 ml alcohol and 3.7 ml water. The aq soln is slightly alkaline to litmus. pH about 8. *Incompat:* Acids, ferric salts. LD_{50} orally in rats: 4.07 g/kg (Smyth, Carpenter).

USE: Antimicrobial agent, flavoring agent and adjuvant in food; not to exceed a maximum level of 0.1% in food as listed in the U.S. Code of Federal Regulations, Title 21 Parts 184.1733 and 582.3733 (1992). Antifungal and bacteriostatic preservative in pharmaceuticals at concentrations of ~0.1%. Clinical reagent (bilirubin assay).

THERAP CAT: Diagnostic aid (hepatic function).

8726. Sodium Bicarbonate. Sodium hydrogen carbonate; sodium acid carbonate; baking soda. $CHNaO_3$; mol wt 84.01. C 14.30%, H 1.20%, Na 27.37%, O 57.14%. $NaHCO_3$. The bicarbonate of commerce is about 99.8% pure. Prepd from sodium carbonate, water and carbon dioxide. Manuf: Faith, Keyes & Clark's *Industrial Chemicals,* F. A. Lowenheim, M. K. Moran, Eds. (Wiley-Interscience, New York, 4th ed., 1975) pp 702-705.

White cryst powder or granules. Begins to lose CO_2 at about 50° and at 100° it is converted into Na_2CO_3. Readily dec by weak acids. In aq soln it begins to break up into carbon dioxide and sodium carbonate at about 20° and completely on boiling. Sol in 10 parts water at 25°, in 12 parts water at about 18°. Insol in alcohol. Its aq soln prepd with cold water and without agitation is only slightly alkaline to litmus or phenolphthalein; on standing or rise in temp the alkalinity increases. pH of freshly prepd 0.1 molar aq soln at 25°: 8.3.

USE: Manuf many sodium salts; source of CO_2; ingredient of baking powder, effervescent salts and beverages; in fire extinguishers, cleaning compds.

THERAP CAT: Antacid, urinary and systemic alkalizer.

THERAP CAT (VET): Antacid, systemic and urinary alkalizer. Locally in burns, erythema, to dissolve mucus, exudates, scabs.

8727. Sodium Bifluoride. F_2HNa; mol wt 61.99. F 61.29%, H 1.63%, Na 37.08%. NaF.HF.

White, cryst powder. Sol in water. The aq soln corrodes glass.

USE: As a "sour" in laundering.

8728. Sodium Bismuthate(V). $BiNaO_3$; mol wt 279.97. Bi 74.64%, Na 8.21%, O 17.14%. $NaBiO_3$. The bismuthate of commerce contains about 85% $NaBiO_3$; the balance is chiefly water and Bi_2O_3.

Yellow to yellowish-brown, somewhat hygroscopic. Slowly dec on keeping; decompn accelerated by moisture and higher temp. Insol in cold, dec by hot water forming Bi_2O_3, NaOH, and liberating oxygen; dec by acids; with HCl chlorine is formed; with oxy-acids oxygen is liberated. LD_{100} orally in rats: 720 mg/kg, Hanzlik *et al., J. Pharmacol. Exp. Ther.* **62**, 372 (1938).

USE: For the determination of manganese in iron and steel, etc., the manganese being oxidized by it in hot HNO_3 or H_2SO_4 soln to permanganate.

8729. Sodium Bisulfate. Sodium acid sulfate; sodium hydrogen sulfate; sodium pyrosulfate. $HNaO_4S$; mol wt 120.06. H 0.84%, Na 19.15%, O 53.30%, S 26.71%. $NaHSO_4$.

Fused $NaHSO_4$, hygroscopic pieces. d 2.435. mp ~315°. Sol in 2 parts water, 1 part boiling water; dec by alcohol into sodium sulfate and free H_2SO_4. *Keep well closed.*

Monohydrate, odorless crystals. When strongly heated it changes into pyrosulfate. Sol in about 0.8 part water; dec by alcohol into sodium sulfate and free H_2SO_4. The aq soln is strongly acid. pH of 0.1 molar soln: 1.4.

USE: Fusion of minerals to make them sol for analysis; for liberating CO_2 in carbonic acid baths. Technical grades are used for pickling metals, carbonizing wool, bleaching and swelling leather, manuf magnesia cements, etc.

8730. Sodium Bisulfide. Sodium sulfhydrate; sodium hydrosulfide; sodium hydrogen sulfide. NaHS; mol wt 56.06. H 1.80%, Na 41.01%, S 57.20%. NaSH. Prepd from sodium ethylate and hydrogen sulfide: Rule, *J. Chem. Soc.* **99**, 558 (1911); Teichert, Klemm, *Z. Anorg. Allgem. Chem.* **243**, 86 (1939); Eibeck, *Inorg. Syn.* **7**, 128 (1963). The technical grade may be obtained by reacting sodium bisulfate with calcium sulfide in the cold or by saturating NaOH solns with H_2S.

Rhombohedric-cubic crystals. White to colorless. Odor of hydrogen sulfide. Very hygroscopic. Readily hydrolyzed in moist air to NaOH and Na_2S. d 1.79. Turns yellow upon heating in dry air, changing to orange at higher temps. mp 350° forming a black liquid. Sol in water, alcohol, ether. Gives a blue-green soln in dimethylformamide.

Dihydrate, needles or flakes, mp 55°. Completely and rapidly sol in water, alcohol, ether. *Note:* The commercial product is usually the dihydrate. Can be shipped in lacquer-lined steel drums.

Trihydrate, shiny rhombs, mp 22°.

USE: Dehairing hides; desulfurizing viscose rayon; in the manuf of sulfur-contg dyes and other thio compds such as thioamides, thiourea, thioglycolic acid, thio- and dithiobenzoic acids, sodium thiosulfate.

8731. Sodium Bisulfite. Sodium acid sulfite. $HNaO_3S$; mol wt 104.06. H 0.97%, Na 22.09%, O 46.12%, S 30.81%. $NaHSO_3$. The bisulfite of commerce consists chiefly of sodium metabisulfite, $Na_2S_2O_5$, and for all practical purposes possesses the same properties as the true bisulfite. Toxicity: Hoppe, Goble, *J. Pharmacol. Exp. Ther.* **101**, 101 (1951).

White, crystalline powder; SO_2 odor; disagreeable taste; on exposure to air it loses some SO_2 and is gradually oxidized to sulfate. d 1.48. Sol in 3.5 parts cold water, 2 parts

boiling water, in about 70 parts alcohol. Its aq soln is acid. *Keep well closed and in a cool place. Incompat:* Acids, oxidizers. LD_{50} i.v. in rats: 115 mg/kg (Hoppe, Goble).

Caution: Concd solns are irritating to skin, mucous membranes.

USE: As disinfectant and bleach, particularly for wool; in dyeing for preparing hot and cold indigo vats; in papermaking in place of sodium hyposulfite to remove Cl from bleached fibers; as stripper (reducer) in laundering; to remove permanganate stains from skin and clothing; to render certain dyes sol; manuf sodium hydrosulfite; coagulating rubber latex; as preservative for deteriorative liqs or solns used for technical purposes; as antiseptic in fermentation industries; as preservative and bleach in food. Pharmaceutic aid (antioxidant).

8732. Sodium Bitartrate. Sodium acid tartrate. C_4H_5-NaO_6; mol wt 172.07. C 27.92%, H 2.93%, Na 13.36%, O 55.79%. $NaHC_4H_4O_6$.

Monohydrate, white crystals. Sol in about 9 parts water, 2 parts boiling water. Almost insol in alcohol. The aq soln is acid.

USE: For detecting potassium; in nutrient media.

8733. Sodium Borate. Sodium biborate; sodium pyroborate; sodium tetraborate. $B_4Na_2O_7$; mol wt 201.22. B 21.49%, Na 22.85%, O 55.66%. $Na_2B_4O_7$. Toxicity: H. F. Smyth *et al., Am. Ind. Hyg. Assoc. J.* **30**, 470 (1969).

Anhydrous, *fused sodium borate, borax glass, fused borax.* Powder or glass-like plates becoming opaque on exposure to air. Slowly sol in water.

Decahydrate, *borax, Jaikin.* Hard odorless crystals, granules or cryst powder; efflorescent in dry air, the crystal often being coated with white powder. d 1.73. when rapidly heated at 75°; at 100° loses $5H_2O$; at 150° loses $9H_2O$; becomes anhydr at 320°. One gram dissolves in 16 ml water, 0.6 ml boiling water, about 1 ml glycerol. Insol in alcohol. The aq soln is alkaline to litmus and phenolphthalein. pH about 9.5. Borax dissolves many metallic oxides when fused with them. *Incompat:* Acids, alkaloidal and metallic salts. LD_{50} orally in rats: 5.66 g/kg (Smyth).

Caution: Ingestion of 5 to 10 g by young children can cause severe vomiting, diarrhea, shock, death, *see Clinical Toxicology of Commercial Products,* R. E. Gosselin *et al.,* Eds. (Williams & Wilkins, Baltimore, 4th ed., 1976) Section III, pp 63-66. *See also* Boric Acid.

USE: Soldering metals; manuf glazes and enamels; tanning; in cleaning compds; artificially aging wood; as preservative, either alone or with other antiseptics against wood fungus; fireproofing fabrics and wood; curing and preserving skins; in cockroach control. Pharmaceutic aid (alkalizer).

THERAP CAT (VET): Has been used as antiseptic, detergent, astringent for mucous membranes.

8734. Sodium Borate Solution Compound. Dobell's soln. Made from 1.5 g sodium borate, 1.5 g sodium bicarbonate, 0.3 ml liquefied phenol, 3.5 ml glycerol and water to make 100 ml.

Yellowish, clear liquid.

THERAP CAT: Wash for mucous membranes.

THERAP CAT (VET): Has been used as a nonirritant wash for mucous membranes.

8735. Sodium Borohydride. *Sodium tetrahydroborate.* BH_4Na; mol wt 37.83. B 28.58%, H 10.66%, Na 60.77%. $NaBH_4$. Prepared from methyl borate and sodium hydride at elevated temps: Schlesinger *et al., J. Am. Chem. Soc.* **75**, 205 (1953). Review of sodium and other metal tetrahydroborates: James, Walbridge, *Prog. Inorg. Chem.* **11**, 99-231 (1970).

Hygroscopic, cubic crystals forming a dihydrate, mp 36-37°. The anhydr material (d 1.074) is stable in dry air to 300°; dec slowly at 400° and rapidly at 500°. Supports combustion. Soly (w/w) in water at 25°: 55%; at 60°: 88.5%; liq ammonia at 25°: 104%; ethylenediamine at 75°: 22%; morpholine at 25°: 1.4%; pyridine at 25°: 3.1%; methanol at 20°: 16.4% (reacts); ethanol at 20°: 4.0% (reacts slowly); tetrahydrofurfuryl alcohol at 20°: 14.0% (reacts slowly); tetrahydrofuran at 20°: 0.1%; diglyme at 25°: 5.5%; dimethylformamide at 20°: 18.0%. Aq solns are most stable in the presence of small amounts of NaOH (0.2% for a nearly satd soln contg 44% NaBH$_4$) and can be kept for several days. Solns are rapidly dec by boiling.

USE: Reducing agent for aldehydes, ketones and Schiff bases in nonaqueous solvents. Also reduces acids, esters, acid chlorides, disulfides, nitriles, inorganic anions. Further used to generate diborane, as foaming agent, as scavenger for traces of aldehyde, ketone and peroxides in organic chemicals.

8736. Sodium Bromate. BrNaO$_3$; mol wt 150.89. Br 52.95%, Na 15.24%, O 31.81%. NaBrO$_3$. The article of commerce contains about 99% NaBrO$_3$.

Colorless, odorless crystals, white granules or cryst powder. d 3.34. mp 381° with dec and liberation of oxygen. Sol in 2.5 parts water, 1.1 parts boiling water. The aq soln is neutral. *Keep from contact with organic matter.*

Caution: See Potassium Bromate.

USE: As a mixture with sodium bromide for dissolving gold from its ores.

8737. Sodium Bromide. Sedoneural. BrNa; mol wt 102.89. Br 77.66%, Na 22.34%. NaBr. Prepd commercially by adding some excess bromine to a sodium hydroxide soln forming a mixture of bromide and bromate. The reaction products are evaporated to dryness and treated with carbon to reduce the bromate to bromide. Prepn: J. H. van der Meulen, U.S. pat. **1,775,598** (1930). Toxicity study: Smith, Hambourger, *J. Pharmacol. Exp. Ther.* **55,** 200 (1935). Brief review of physical and chemical properties and pharmacology: T. O. Soine, C. O. Wilson, *Roger's Inorganic Pharmaceutical Chemistry* (Lea & Febiger, Philadelphia, 8th ed., 1967) pp 213-216.

White crystals, granules or powder; saline, feebly bitter taste. Absorbs moisture from air but is not deliquescent. d 3.21. mp 755°; volatilizes at somewhat higher temp. One gram dissolves in 1.1 ml water, about 16 ml alcohol, 6 ml methanol. The aq soln is practically neutral. pH 6.5-8.0. *Keep well closed.* From water of room temp, sodium bromide crystallizes with 2H$_2$O in the form of colorless crystals. *Incompat.* Acids, alkaloidal and heavy metal salts. LD$_{50}$ orally in rats: 3.5 g/kg (Smith, Hambourger).

USE: In photography.

THERAP CAT: Sedative; hypnotic; anticonvulsant.

THERAP CAT (VET): Sedative. Has been used to control convulsions, chorea, hysteria.

8738. Sodium Cacodylate. *[(Dimethylarsino)oxy]sodium As-oxide;* sodium dimethylarsonate; Arsecodile; Arsicodile; Arsycodile; Rad-e-cate; Silvisar. C$_2$H$_6$AsNaO$_2$; mol wt 159.98. C 15.02%, H 3.78%, As 46.83%, Na 14.37%, O 20.00%. An organic compd of arsenic yielding inorganic, trivalent arsenic in the body (is excreted partly unchanged, also yields dimethylarsine oxide. Prepd by the distillation of a mixture of arsenic trioxide and potassium acetate which yields Cadet's liquid, contg mostly dimethylarsine oxide. This is oxidized with mercuric oxide yielding crystals of cacodylic acid which is neutralized with Na$_2$CO$_3$ or NaOH: Cadet de Gassicourt, *Mem. savants étrangers* **3,** 633 (1760); Valeur, Gaillot, *Compt. Rend.* **185,** 956 (1927). Use in treatment of psoriasis: S. I. Dawes, H. C. Jackson, *J. Am. Med. Assoc.* **48,** 2020 (1903). Metabolism and excretion: A. Heffter, *Arch. Exp. Pathol. Pharmakol.* **46,** 230 (1901). Use as herbicide: M. A. Sprague, U.S. pat. **3,056,663** (1962 to Ansul). Possible use in radiation dosimetry: H. S. Levinson, E. B. Garber, *Nature* **207,** 751 (1965).

$$H_3C-\underset{\underset{CH_3}{|}}{\overset{\overset{O}{\parallel}}{As}}-O^-\ Na^+$$

Trihydrate, crystals, granules. Slight odor. Liquefies in its water of hydration at about 60°. Becomes anhydr at 120°. Burns with a bluish flame, emitting a garlic-like odor. One gram dissolves in 0.5 ml water, 2.5 ml alcohol. pH about 8-9. *Keep well closed.*

Caution: More toxic by mouth than by injection due to rapid release of inorganic arsenic by gastric acid. Large doses may cause nephritis, albuminuria, hematuria: *Martin-*

dale The Extra Pharmacopoeia, N. W. Blacow, Ed. (Hazell, Watson & Viney, Aylesbury, Bucks, 1972) p 219.

USE: Herbicide.

THERAP CAT (VET): Has been used in chronic eczema, anemia, as a tonic.

8739. Sodium Carbonate. CNa$_2$O$_3$; mol wt 105.99. C 11.33%, Na 43.38%, O 45.29%. Na$_2$CO$_3$. Occurs in nature as the hydrate, *thermonatrite,* and the decahydrate, *natron* or *natrite.* Produced by the ammonia-soda or Solvay processes: Faith, Keyes & Clark's *Industrial Chemicals,* F. A. Lowenheim, M. K. Moran, Eds. (Wiley-Interscience, New York, 4th ed., 1975) pp 706-715. Toxicity: C. Norde *et al., Compt. Rend.* **257,** 791 (1963). *Reviews:* Bailey in *Mellor's* vol II, suppl II, *The Alkali Metals* (part 1), 1058-1205 (1961).

Anhydrous, *Solvay soda.* The technical grade (about 99% pure) is known as *soda ash.* Odorless, hygroscopic powder; alkaline taste; d 2.53. mp 851° but begins to lose CO$_2$ even at 400°. On exposure to air it will gradually absorb one mol water—about 15%. Sol in glycerol; in 3.5 parts water at room temp, 2.2 parts water at 35°. Insol in alcohol. Dec by acids with effervescence. Combines with water with evolution of heat. Its aq soln is strongly alkaline. pH 11.6. *Keep well closed.* LD$_{50}$ (30 day) i.p. in mice: 116.6 mg/kg (Norde).

Monohydrate, odorless, small crystals or cryst powder; alkaline taste. Stable at ordinary temps and atmospheric conditions; dries out somewhat in warm, dry air or above 50°; becomes anhydr at 100°. d 2.25; also reported as 1.55. Sol in 3 parts water, 1.8 parts boiling water, 7 parts glycerol. Insol in alcohol.

Decahydrate, *Nevite, Soda.* The technical product is known as *sal soda* or *washing soda.* Transparent crystals; readily effloresces on exposure to air. d 1.46. mp 34°. Sol in 2 parts cold, 0.25 part boiling water, in glycerol. Insol in alcohol. The aq soln is strongly alkaline to litmus. *Keep well closed and in a cool place.*

Caution: Potential symptoms of overexposure to dusts or vapors of sodium carbonate are irritation of mucous membranes with subsequent coughing and shortness of breath. Chronic overexposure may lead to perforation of the nasal septum. Direct contact may cause skin irritation and redness, with concentrated solutions causing erythema. Chronic skin exposures may cause dermatitis and ulceration. See *Patty's Industrial Hygiene and Toxicology* vol. 2A, G. D. Clayton, F. E. Clayton, Eds. (Wiley-Interscience, New York, 4th ed., 1993) pp 769-771.

USE: In manuf of Na salts, glass, soap; for washing wool; textiles, etc.; in bleaching linen, cotton; general cleanser; in water-softening; in photography; as reagent in analytical chemistry. Pharmaceutic aid (alkalizer).

THERAP CAT (VET): Has been used as an emetic. In solution to cleanse skin, in eczema, to soften scabs of ringworm.

8740. Sodium Cellulose Phosphate. SCP; Calcibind; Calcisorb. Non-absorbable ion exchange resin with high affinity for calcium ions. Prepn: G. P. Touey, U.S. pat. **2,759,924** (1956 to Eastman Kodak). Use in treatment of hypercalcemia: C. Y. C. Pak *et al., J. Clin. Endocrinol. Metab.* **28,** 1828 (1968). Mechanism of action: C. Y. C. Pak, *J. Clin. Pharmacol.* **13,** 15 (1973). Clinical pharmacology: *idem, ibid.* **19,** 451 (1979). Use in treatment of calcium urolithiasis: U. Backman *et al., J. Urol.* **123,** 9 (1980); C. Y. C. Pak, *Invest. Urol.* **19,** 187 (1981).

Powder. Insol in water.

USE: As ion exchange resin.

THERAP CAT: Treatment of calcium urolithiasis.

8741. Sodium Chlorate. Atlacide; Defol; Dervan. ClNaO$_3$; mol wt 106.44. Cl 33.31%, Na 21.60%, O 45.09%. NaClO$_3$. The chlorate of commerce is about 99% pure. Produced from sodium chloride by electrolysis: Faith, Keyes & Clark's *Industrial Chemicals,* F. A. Lowenheim, M. K. Moran, Eds. (Wiley-Interscience, New York, 4th ed., 1975) pp 716-721. Toxicity: Ulrich, *J. Pharmacol. Exp. Ther.* **35,** 1 (1929).

Colorless, odorless crystals or white granules. d 2.5. mp 248°; at about 300° liberates oxygen; entirely dec at higher temp. Sol in about 1 ml cold, 0.5 ml boiling water, about

130 ml alcohol, about 50 ml boiling alcohol, 4 ml glycerol. Sodium chloride diminishes its soly in water. The aq soln is neutral. *Keep out of contact with organic matter or other oxidizable substances.* LD orally in rats: 12000 mg/kg (Ulrich).

USE: An oxidizer, like potassium chlorate, in manuf of dyes; explosives and matches; dyeing and printing fabrics; tanning and finishing leather. Herbicide. Pharmaceutic aid (oxidizing agent).

8742. Sodium Chloride. Salt; common salt. ClNa; mol wt 58.44. Cl 60.66%, Na 39.34%. NaCl. The article of commerce is also known as *table salt, rock salt* or *sea salt.* Occurs in nature as the mineral *halite.* Produced by mining (rock salt), by evaporation of brine from underground salt deposits and from sea water by solar evaporation: Faith, Keyes & Clark's *Industrial Chemicals,* F. A. Lowenheim, M. K. Moran, Eds. (Wiley-Interscience, New York, 4th ed., 1975) pp 722-730. Toxicity studies: E. M. Boyd, M. N. Shanas, *Arch. Int. Pharmacodyn.* **144,** 86 (1963). Comprehensive monograph: D. W. Kaufmann, *Sodium Chloride,* ACS Monograph Series no. **145** (Reinhold, New York, 1960) 743 pp.

Cubic, white crystals, granules, or powder; colorless and transparent or translucent when in large crystals. d 2.17. The salt of commerce usually contains some calcium and magnesium chlorides which absorb moisture and make it cake. mp 804° and begins to volatilize at a little above this temp. One gram dissolves in 2.8 ml water at 25°, in 2.6 ml boiling water, in 10 ml glycerol; very slightly sol in alcohol. Its soly in water is decreased by HCl. Almost insol in concd HCl. Its aq soln is neutral. pH: 6.7-7.3. d of satd aq soln at 25° is 1.202. A 23% aq soln of sodium chloride freezes at −20.5°C (5°F). LD_{50} orally in rats: 3.75 ±0.43 g/kg (Boyd, Shanas).

Note: **Blusalt,** a brand of sodium chloride contg trace amounts of cobalt, iodine, iron, copper, manganese, zinc is used in farm animals.

Caution: Not generally considered poisonous. Accidental substitution of NaCl for lactose in baby formulas has caused fatal poisoning.

USE: Natural salt is the source of chlorine and of sodium as well as of all, or practically all, their compds, e.g., hydrochloric acid, chlorates, sodium carbonate, hydroxide, etc.; for preserving foods; manuf soap, dyes—to salt them out; in freezing mixtures; for dyeing and printing fabrics, glazing pottery, curing hides; metallurgy of tin and other metals.

THERAP CAT: Electrolyte replenisher; emetic; topical anti-inflammatory.

THERAP CAT (VET): Essential nutrient factor. May be given orally as emetic, stomachic, laxative or to stimulate thirst (prevention of calculi). Intravenously as isotonic solution to raise blood volume, to combat dehydration. Locally as wound irrigant, rectal douche.

8743. Sodium Chlorite. ClNaO$_2$; mol wt 90.44. Cl 39.20%, Na 25.42%, O 35.38%. NaClO$_2$. Prepd on a commercial scale by passing chlorine dioxide into a soln of sodium hydroxide contg carbonaceous matter and lime: Vincent, U.S. pats. **2,092,944/5** (1937 to Mathieson). Review of manuf and properties: Kesting, *Pulp Paper Mag. Can.* **53,** no. 8, 99-104 (1952).

Slightly hygroscopic crystals or flakes, does not cake. Dec 180-200°. Powerful oxidizer, but will not explode on percussion unless in contact with oxidizable material. Soly in water (g/100 g soln): at 5°: 34; at 17°: 39; at 30°: 46; at 45°: 53; at 60°: 55.

Trihydrate, triclinic leaflets, becomes anhydr at 38° or in desiccator over KOH at room temp.

USE: In the preparation of chlorine dioxide for immediate use; in water purification; as bleaching agent for textiles, paper pulp.

8744. Sodium 6-Chloro-5-nitrotoluene-3-sulfonate. *4-Chloro-5-nitro-m-toluenesulfonic acid sodium salt.* C$_7$H$_5$Cl-NNaO$_5$S; mol wt 273.63. C 30.73%, H 1.84%, Cl 12.96%, N 5.12%, Na 8.40%, O 29.24%, S 11.72%.

White or slightly yellow, cryst powder. Moderately sol in water, slightly in alcohol.

USE: As reagent for determination of potassium.

8745. Sodium Chromate(VI). Neutral sodium chromate. CrNa$_2$O$_4$; mol wt 161.97. Cr 32.10%, Na 28.39%, O 39.51%. Na$_2$CrO$_4$. Crystallizes as a tetra- or decahydrate. Manuf: Faith, Keyes & Clark's *Industrial Chemicals,* F. A. Lowenheim, M. K. Moran, Eds. (Wiley-Interscience, New York, 4th ed., 1975) pp 731-736. Diagnostic use of labeled compound in GI bleeding: A. Lussier *et al., Sem. Arthritis Rheum.* **17,** Suppl. 2, 40-45 (1988). Review of use as radiolabel for blood elements: *Diagnostic and Investigational Uses of Radiolabeled Blood Elements,* R. J. Davey, M. E. Wallace, Eds., (Am. Assoc. Blood Banks, Arlington, 1987) 118 pp.

Tetrahydrate, yellow, somewhat deliquesc crystals. Sol in about 1 part water, slightly in alcohol. The aq soln is alkaline. *Keep well closed.* The decahydrate is unstable with respect to water content. mp ∼20°.

^{51}Cr-labeled compound, ^{51}CrNa$_2$O$_4$, *sodium chromate-^{51}Cr, sodium radio-chromate(^{51}Cr), Chromitope Sodium, Rachromate-51.* Prepd from radioactive chromium (^{51}Cr) which has a half-life of 27.7 days. The emission of gamma rays is applicable to biological tagging and tracing. Other properties identical to those of unlabeled sodium chromate.

USE: Protection of iron against corrosion and rusting.

THERAP CAT: ^{51}Cr-labeled compound as diagnostic aid (blood volume determination; blood cell survival).

8746. Sodium Citrate. Trisodium citrate; Citrosodine; Citnatin; Urisal. C$_6$H$_5$Na$_3$O$_7$; mol wt 258.07. C 27.92%, H 1.95%, Na 26.72%, O 43.40%. Toxicity data: Gruber, Halbeisen, *J. Pharmacol. Exp. Ther.* **94,** 65 (1948).

Dihydrate, white, odorless crystals, granules or powder; cool, saline taste. Stable in air; becomes anhydrous at 150°. Sol in 1.3 parts water, 0.6 part boiling water. Insol in alcohol. The aq soln is slightly alkaline to litmus. pH about 8. LD_{50} i.p. in rats: 6.0 mmoles/kg (Gruber, Halbeisen).

Pentahydrate, relatively large, colorless crystals or white granules. Not as stable as the dihydrate, drying out on exposure to air and also caking. *Keep well closed.*

USE: In photography; as sequestering agent to remove trace metals; *in vitro* anticoagulant; as emulsifier, acidulant and sequestrant in foods.

THERAP CAT: Systemic alkalizer; diuretic; expectorant; sudorific.

THERAP CAT (VET): Anticoagulant for collection of blood.

8747. Sodium Citrate, Acid. Disodium citrate; disodium hydrogen citrate; Alkacitron. C$_6$H$_6$Na$_2$O$_7$; mol wt 236.09. C 30.52%, H 2.56%, Na 19.48%, O 47.44%.

Sesquihydrate, white powder, saline taste. One gram dissolves in slightly less than 2 ml water; pH of a 3% w/v soln in water: 4.9 to 5.2.

USE: Anticoagulant, generally in solution with glucose, to prevent the clotting of blood intended for transfusion. Preferable to sodium citrate, since it prevents carmelization of glucose on sterilization because of its acidity. A suitable soln contains 1.7 to 2%, and 2.5% dextrose; 120 ml of this soln prevents the clotting of 420 ml blood.

8748. Sodium Cobaltinitrite. *Trisodium hexakis(nitrato-N)cobaltate(3−); sodium hexanitrocobaltate(III).* CoN$_6$-Na$_3$O$_{12}$; mol wt 403.94. Co 14.59%, N 20.81%, Na 17.07%, O 47.53%. Na$_3$Co(NO$_2$)$_6$.

Yellow to brownish-yellow, cryst powder. Very sol in water, slightly in alc. Dec by mineral acids, but unaffected by dil acetic or similar organic acids. The aq soln dec gradually but if a few drops of acetic acid are added it may be kept for about 3 months.

USE: For the detection of potassium with which it forms a slightly sol compd.

8749. Sodium Cyanate. *Cyanic acid sodium salt.* CN-NaO; mol wt 65.01. C 18.48%, N 21.55%, Na 35.37%, O 24.61%. NaOCN. Prepn and properties: *Gmelin's, Sodium* (8th ed.) **21**, 799-801 (1928) and supplement, part 4, 1382-1386 (1967). Used experimentally in treatment of sickle cell anemia. Effect of cyanate on sickling: May *et al., Lancet* **1**, 658 (1972); P. N. Gillette *et al., N. Engl. J. Med.* **290**, 654 (1974). Pharmacology and toxicology: Cerami *et al. J. Pharmacol. Exp. Ther.* **185**, 653 (1973). Clinical studies: Peterson *et al., ibid.* **189**, 577 (1974).

Colorless needles from alcohol. d_4^{20} 1.893. mp 550°. Sol in water; decomposes to form Na_2CO_3 and urea. Soly in alc (0°): 0.22 g/100 g solvent. Insol in ether. LD_{50} i.p. in mice: 260 mg/kg (Cerami).

8750. Sodium Cyanide. Cyanogran. CNNa; mol wt 49.01. C 24.51%, N 28.58%, Na 46.91%. NaCN. This cyanide of commerce is 95-98% pure. Mixtures of sodium cyanide with sodium chloride or carbonate for special uses are also marketed. Toxicity study: Smyth *et al., Am. Ind. Hyg. Assoc. J.* **30**, 470 (1969).

White granules or fused pieces. *Violent poison!* Odorless when perfectly dry; somewhat deliquesc in damp air and emits slight odor of HCN. mp 563°. Freely sol in water, slightly in alcohol. The aq soln is strongly alkaline and rapidly decomposes; the soln readily dissolves gold and silver in presence of air. *Keep well closed.* LD_{50} orally in rats: 15 mg/kg (Smyth).

Caution: Potential symptoms of overexposure are weakness, headache and confusion; nausea, vomiting; increased rate of respiration; slow gasping respiration; irritation of eyes and skin; asphyxia and death can occur. *See NIOSH Pocket Guide to Chemical Hazards* (DHHS/NIOSH 90-117, 1990) p 76.

USE: Extracting gold and silver from ores; electroplating baths; fumigating citrus and other fruit trees, ships, railway cars, warehouses, etc.; manuf hydrocyanic acid and many other cyanides; case hardening of steel.

8751. Sodium Cyanoborohydride. *Sodium (cyano-C)-trihydroborate(1—);* sodium borocyanohydride; sodium cyanohydridoborate. CH_3BNNa; mol wt 62.84. C 19.11%, H 4.81%, B 17.20%, N 22.29%, Na 36.58%. $NaBH_3CN$. Reducing agent prepared from $NaBH_4$ and HCN: R. C. Wade *et al., Inorg. Chem.* **9**, 2146 (1970); R. C.Wade, Ger. pat. **2,028,695** corresp to U.S. pat. **3,667,923** (1971, 1972 both to Ventron). Review of prepn, properties and use: C. F. Lane, *Synthesis* **1975**, 135-146.

White, hygroscopic powder, mp 240-242° (dec). d^{28} 1.199. Soly (g/100 g solvent) in water (29°): 212; in THF (28°): 37.2; in diglyme (25°): 17.6. Very sol in methanol; slightly sol in ethanol, isopropylamine; insol in ethyl ether, benzene, hexane. Stable in acid to pH 3; undergoes rapid hydrolysis in 12N HCl. Rate of hydrolysis 10^{-8} that of $NaBH_4$.

USE: Selective reducing agent for aldehydes, ketones, oximes, enamines; does not reduce amides, ethers, lactones, nitriles, nitro compds and epoxides. Also used for reductive amination of ketones and aldehydes, reductive alkylation of amines and hydrazines, reductive displacement of halides and tosylates, deoxygenation of aldehydes and ketones. *See* Lane, *loc. cit.*

8752. Sodium Diacetate. Sodium acid acetate; Dykon. $CH_3COONa.CH_3COOH$. Described as a "bound" compd of sodium acetate and acetic acid.

White powder, dec above 150°. Sol in water, liberating 42.25% available acetic acid.

USE: Acetic acid in solid form; as an inhibitor of molds and rope-forming bacteria in bread: Glabe, *Food Inds.* **14**, no. 2, 46 (1942); as sequestrant.

8753. Sodium Dibunate. *2,6-Bis(1,1-dimethylethyl)-1-naphthalenesulfonic acid sodium salt;* L-1633; 1633 Labaz; Becantal; Becantex; Dibunafon; Keuten; Linctussal. $C_{18}H_{23}NaO_3S$; mol wt 342.43. C 63.14%, H 6.77%, Na 6.71%, O 14.02%, S 9.36%. Although most sources refer to this compound as sodium 2,6-di-*tert*-butylnaphthalene sulfonate, it is a mixture of at least two isomers. Prepn and separation of isomers: Menard *et al., Can. J. Chem.* **39**, 729 (1961).

Slightly hygroscopic crystals, dec > 300°. Slightly sol in cold water (0.5-1.0%); freely sol in hot water; sol in methanol, less sol in ethanol. Aq solns are stable to boiling.

THERAP CAT: Antitussive.

8754. Sodium Dichromate(VI). Sodium bichromate; bichromate of soda. $Cr_2Na_2O_7$; mol wt 261.97. Cr 39.70%, Na 17.55%, O 42.75%. $Na_2Cr_2O_7$. Usually prepd from Na_2CrO_4 and H_2SO_4. Description of industrial processes: Müller, Glissmann in *Ullmann's Encyklopädie der Technischen Chemie* vol. **5** (Munich, 3rd ed., 1954) p 575; Faith, Keyes & Clark's *Industrial Chemicals*, F. A. Lowenheim, M. K. Moran, Eds. (Wiley-Interscience, New York, 4th ed., 1975) pp 731-736.

Dihydrate, reddish to bright orange, somewhat deliquescent crystals. Crystal system: monoclinic sphenoidal. Crystal habit: elongated prismatic. d_4^{25} 2.348. Bulk density: 96 lbs/cu ft. Becomes anhydr on prolonged heating at ~100°. The anhydr salt mp 356.7° and starts to dec at ~400°. Heat of soln −28.2 cal/g. Very sol in water. A satd aq soln contains at 0°: 70.6% $Na_2Cr_2O_7.2H_2O$; at 20°: 73.18%; at 40°: 77.09%; at 60°: 82.04%; at 80°: 88.39%; at 100°: 91.43%. A 20% soln freezes at −3.5°, a 30% soln at −6°, a 60% soln at −26°, a 69% soln at −48°. Specific heat of 20% soln at 25°: 0.85 cal/g/°C. Solns are acidic: pH of 1% soln: 4.0; pH of 10% soln: 3.5.

Caution: Irritant and caustic to skin, mucous membranes.

USE: Oxidizing agent in manuf of dyes, many other synthetic organic chemicals, inks, etc.; in chrome-tanning of hides; in electric batteries; bleaching fats, oils, sponges, resins; refining petroleum; manuf chromic acid, other chromates and chrome pigments; in corrosion-inhibitors, corrosion-inhibiting paints; in many metal treatments; electroengraving of copper; mordant in dyeing; for hardening gelatin; for the defoliation of cotton plants and other plants and shrubs, La Lande, U.S. pat. **2,760,854** (1956 to Pennsylvania Salt).

THERAP CAT: Anti-infective (topical).

8755. Sodium Dicyanoaurate(I). Gold sodium cyanide; sodium aurocyanide. C_2AuN_2Na; mol wt 271.99. C 8.83%, Au 72.42%, N 10.30%, Na 8.45%. $NaAu(CN)_2$. White, cryst powder. Sol in water. *Poison!*

USE: Goldplating.

8756. Sodium Dithionate. *Dithionous acid disodium salt.* $Na_2O_6S_2$; mol wt 206.11. Na 22.31%, O 46.58%, S 31.12%. Prepd according to the equations $MnO_2 + 2SO_2 \rightarrow MnS_2O_6$ and $MnS_2O_6 + Na_2CO_3 \rightarrow MnCO_3 + Na_2S_2O_6$: de Baat, *Rec. Trav. Chim.* **45**, 237 (1926); Pfanstiel, *Inorg. Syn.* **2**, 170 (1946).

Dihydrate, colorless, water-clear, orthorhombic crystals. Very stable in air. d 2.189. Loses all of its water of crystn at 110°. When heated to 267° it is dissociated into Na_2SO_4 and SO_2. Soly in water at 0°: 6.05% (w/w); at 20°: 13.39%; at 30°: 17.32%. Insol in alc.

8757. Sodium Dodecylbenzenesulfonate. *Dodecylbenzenesulfonic acid sodium salt;* dodecylbenzene sodium sulfonate; Santomerse #1; Conoco C-50; Conoco SD 40; Conoco C-60. $C_{18}H_{29}NaO_3S$; mol wt 348.48. C 62.04%, H 8.39%, Na 6.60%, O 13.77%, S 9.20%. $C_{12}H_{25}C_6H_4SO_3Na$. Manuf: *Chem. Eng.* **61**, no. 6, 372 (1954); Huber *et al., J. Am. Oil Chem. Soc.* **33**, 57 (1956); **Brit.** pat. **761,095;** Seaton, U.S. pat. **2,782,230** (1956, 1957 both to Monsanto); **Brit.** pat. **773,423;** Gerhart, Karwacki, U.S. pat. **2,820,056**

(1957, 1958 both to Continental Oil). Toxicity study: Hopper *et al.*, *J. Am. Pharm. Assoc. Sci. Ed.* **38**, 428 (1949).

LD_{50} in mice: 2 g/kg orally; 105 mg/kg i.v. (Hopper).

Caution: May cause skin irritation. If swallowed will cause vomiting.

USE: Anionic detergent.

8758. Sodium Ethoxide. Sodium ethylate; caustic alcohol. C_2H_5NaO; mol wt 68.05. C 35.30%, H 7.41%, Na 33.78%, O 23.51%.

White or yellowish, hygroscopic powder. Dec on exposure to air and becomes darker on keeping. Dec by water into NaOH and alcohol. Sol without decompn in abs alc. *Keep tightly closed, protected from light and in a cool place.*

8759. Sodium Ethyl Sulfate. Sodium sulfovinate. $C_2H_5NaO_4S$; mol wt 148.12. C 16.22%, H 3.40%, Na 15.52%, O 43.21%, S 21.65%. $NaC_2H_5SO_4$.

Monohydrate, white, very hygroscopic crystals. Sol in 0.7 part water, in alcohol. *Keep well closed.*

USE: In organic syntheses.

8760. Sodium Ferrocyanide. *Tetrasodium hexakis(cyano-C)ferrate(4—); sodium hexacyanoferrate(II);* yellow prussiate of soda; sodium prussiate yellow. $C_6FeN_6Na_4$; mol wt 303.91. C 23.71%, Fe 18.38%, N 27.65%, Na 30.26%. $Na_4Fe(CN)_6$. Review of properties, chemistry and syntheses: *The Chemistry of Ferrocyanides*, American Cyanamid Co. (Beacon Press, New York, 1953) 112 pp.

Decahydrate, pale yellow, monoclinic, slightly efflorescent crystals. Steady dehydration occurs > 50°. Becomes anhydr at 81.5°. Dec 435°, forming sodium cyanide, iron, carbon, and nitrogen. Soly in water at 1°: 10.2% (calcd as the anhydr salt); at 17°: 14.7%; at 25°: 17.6%; at 53°: 28.1%; at 85°: 39%; at 96.6°: 39.7%. Practically insol in most organic solvents.

Caution: Do not mix with hot or concd acids and do not expose solns to sunlight for any length of time to avoid generation of hydrogen cyanide. Waste ferrocyanides in streams and lakes should not exceed 2 ppm because irradiated solns become toxic to fish, G. E. Burdick, M. Lipscheutz, *C.A.* **44**, 10939f (1950). *Toxicity:* Because of strong chemical bondage between the cyanide groups and the iron, ferrocyanides have a low order of toxicity.

USE: Addition of sodium ferrocyanide solns to slightly acidic solns of iron salts causes precipitation of insol Prussian blue (ferric ferrocyanide), $Fe_4[Fe(CN)_6]_3$. Alkaline solns yield sol Prussian blue, $NaFe[Fe(CN)_6]$. Sodium ferrocyanide forms gels with heavy metals in general. Used in ore flotation. In photography for bleaching, toning, and fixing. To prevent caking of rock salt and foods. Additive to pickling baths. Peptizing agent in rubber. Arc stabilizer in welding rod coatings. Emulsion polymerization catalyst.

8761. Sodium Fluoborate. *Sodium tetrafluoroborate;* sodium borofluoride. BF_4Na; mol wt 109.79. B 9.85%, F 69.21%, Na 20.94%. $NaBF_4$. Prepd according to the equation $2H_3BO_3 + 8HF + Na_2CO_3 \rightarrow 2NaBF_4 + 7H_2O + CO_2$: Balz, Wilke-Dörfurt, *Z. Anorg. Allgem. Chem.* **159**, 197 (1927); Kwasnik in *Handbook of Preparative Inorganic Chemistry*, vol. **1**, G. Brauer, Ed. (Academic Press, New York, 2nd ed., 1963) p 222.

Orthorhombic, stout rectangular prisms d^{20} 2.47. mp 384° (slight decompn). Does not etch glass when absolutely dry. Soly in water (g/100 ml): 108 (26°); 210 (100°). Sparingly sol in alcohol. Aq solns have a bitter taste and are acid to litmus.

USE: Fluorinating agent, *see* Lawton, Levy, *J. Am. Chem. Soc.* **77**, 6083 (1955).

8762. Sodium Fluoride. Chemifluor; Dentalfluoro; Duraphat; Fluoros; Luride-SF; Villiaumite; Florocid; Flura-Drops; Karidium; Lemoflur; Ossalin; Ossin; Osteo-F; Osteofluor; Slow-Fluoride; Zymafluor. FNa; mol wt 41.99. F 45.25%, Na 54.75%. NaF. Prepd by fusing cryolite with NaOH; by adding equiv amounts of NaOH or Na_2CO_3 to 40% HF (precipitation is instantaneous and crystal size depends on pH, but too much HF yields sodium bifluoride, $NaHF_2$): Müller, *Chem.-Ztg.* **52**, 5 (1928); Kwasnik in *Handbook of Preparative Inorganic Chemistry*, vol. **1**, G. Brauer, Ed. (Academic Press, New York, 2nd ed., 1963) pp 235-236. Technical grades are 90% and 95% NaF, light (37

cu in/lb) and dense (23 cu in/lb), and 98%. The impurities are mainly sodium and aluminum fluosilicates. Pharmacology: Caruso *et al.*, *Handb. Exp. Pharmakol.* **XX** (Part 2), F. Smith, Ed. (Springer, Berlin, 1970) pp 144-165. Toxicity: H. F. Smyth *et al.*, *Am. Ind. Hyg. Assoc. J.* **30**, 470 (1969). Carcinogenicity studies: J. R. Bucher *et al.*, *Int. J. Cancer* **48**, 733 (1991). Review of toxicology: D. W. Banting, *J. Am. Dent. Assoc.* **122**, 86-91 (1991). Review of clinical efficacy in prevention of dental caries: L. G. Petersson, *Caries Res.* **27**, Suppl 1, 35-42 (1993); of clinical experience in osteoporosis: M. Kleerekoper, D. B. Mendlovic, *Endocr. Rev.* **14**, 312-323 (1993).

Cubic or tetragonal crystals (NaCl lattice). d 2.78. mp 993°. bp 1704°. *Poisonous!* Soly in water (g/100 ml): 4.0 (15°); 4.3 (25°); 5.0 (100°). Insol in alc. Aq solns have an alkaline reaction caused by partial hydrolysis. pH of freshly prepd satd soln 7.4. Aq solns etch glass, but the dry crystals or powder may be kept in glass bottles. Sodium fluoride sold as household insecticide must be tinted Nile Blue. LD_{50} orally in rats: 0.18 g/kg (Smyth).

Caution: Severe symptoms have occured from ingestion of less than one gram; death from 5 to 10 g. Acute ingestion of large doses may lead to abrupt abdominal pain, intense vomiting and diarrhea, hematemesis and melena, dehydration and thirst. This is followed by symptoms of systemic toxicity including muscular weakness, convulsions, tetany; lethargy, coma, respiratory arrest; arrhythmia, hypotension, cardiovascular collapse; hypocalcemia, hypomagnesemia, hypokalemia; metabolic and/or respiratory acidosis; albuminuria, anuria. Chronic ingestion may cause mottling of tooth enamel and osteosclerosis. *See Clinical Toxicology of Commercial Products,* R. E. Gosselin *et al.*, Eds. (Williams & Wilkins, Baltimore, 5th ed., 1984) Section III, pp 185-193.

USE: As insecticide, particularly for roaches and ants; in other pesticide formulations; constituent of vitreous enamel and glass mixes; as a steel degassing agent; in electroplating; in fluxes; in heat-treating salt compositions; in the fluoridation of drinking water; for disinfecting fermentation apparatus in breweries and distilleries; preserving wood, pastes and mucilage; manuf of coated paper; frosting glass; in removal of HF from exhaust gases to reduce air pollution. Dental caries prophylactic.

THERAP CAT: In treatment of osteoporosis.

THERAP CAT (VET): Anthelmintic, pediculicide, acaricide.

8763. Sodium Folate. *Folic acid sodium salt;* sodium pteroylglutamate; sodium Folvite. $C_{19}H_{18}N_7NaO_6$; mol wt 463.39. C 49.25%, H 3.92%, N 21.16%, Na 4.96%, O 20.72%.

Sold only as sterile soln in ampuls. Clear, mobile liquid. Yellow to orange-yellow color. pH between 8.5 and 11.0. For spectrophotometric data *see* Folic Acid.

THERAP CAT: Water-soluble hematopoietic vitamin.

8764. Sodium Formaldehydesulfoxylate. *Hydroxymethanesulfinic acid sodium salt;* formaldehyde sodium sulfoxylate; formaldehydesulfoxylic acid sodium salt; sodium hydroxymethanesulfinate; sodium methanalsulfoxylate; Aldanil; Rongalite; Rongalite C. CH_3NaO_3S; mol wt 118.09. C 10.17%, H 2.56%, Na 19.47%, O 40.65%, S 27.15%. $Na[HOCH_2SO_2]$. Prepn: Heyl, Greer, *Am. J. Pharm.* **94**, 80 (1922); Binns, U.S. pat. **2,013,125** (1935 to Virginia Smelting); Postnikov, Kunin, *J. Appl. Chem. USSR* **13**, 185 (1940). Structure of dihydrate: Truter, *J. Chem. Soc.* **1955**, 3064; **1962**, 3400. Of limited value in treatment of mercuric chloride poisoning: Modell *et al.*, *J. Pharmacol. Exp. Ther.* **61**, 66 (1937).

Dihydrate, crystals, mp 63-64°, dec at higher temp. Odorless when freshly prepd but quickly develops a characteristic (garlic) odor. Freely sol in water; practically insol in abs alcohol, ether, benzene. Readily dec by dil acids. Aq soln is practically neutral. *Keep well closed in a cool place.* LD s.c. in mice, 4.0 g/kg: Rosenthal, *Public Health Rept. (U.S.)* **49**, 908 (1934).

USE: In vat color printing pastes: Borstelmann, Fordemwalt, U.S. pat. **2,597,281** (1952 to Am. Cyanamid). In polymerization of ethylenic compds: **Brit.** pats. **816,252**; **852,593** (1959 to Hercules Powder; 1960 to Air Reduction).

In manuf of arsphenamines: Krumwiede, *J. Am. Pharm. Assoc.* **8**, 795 (1919); Heyl, Miller, *ibid.* **11**, 432 (1922).
THERAP CAT: Treatment of mercury poisoning.

8765. Sodium Formate. CHNaO$_2$; mol wt 68.01. C 17.66%, H 1.48%, Na 33.80%, O 47.05%. HCOONa.
White, deliquesc granules or cryst powder; slight odor of formic acid. d 1.92. mp 253°; at higher temp dec into sodium oxalate and hydrogen, then into sodium carbonate. Sol in ~1.3 parts water; sol in glycerol, slightly in alcohol. The aq soln is neutral. pH ~7. Has buffering action. *Keep well closed.*
USE: In dyeing and printing fabrics; also in anal. chemistry as a precipitant for the "noble" metals. Solubilizes trivalent metal ions in soln by forming complex ions. Buffering action adjusts the pH of strong mineral acids to higher values.
THERAP CAT: Caustic, astringent.

8766. Sodium Gluconate. *Gluconic acid sodium salt.* C$_6$H$_{11}$NaO$_7$; mol wt 218.14. C 33.04%, H 5.08%, Na 10.54%, O 51.34%. The normal sodium salt of gluconic acid.

$$
\begin{array}{c}
COO^- \; Na^+ \\
| \\
HC{-}OH \\
| \\
HO{-}CH \\
| \\
HC{-}OH \\
| \\
HC{-}OH \\
| \\
CH_2OH
\end{array}
$$

Crystals. The technical grade may have a pleasant odor. Soly in water at 25°: 59 g/100 ml. Sparingly sol in alcohol. Insol in ether. Aq solns are stable to short boiling periods.
USE: As sequestering agent forming water-sol complexes with calcium in alkaline media and with iron in near neutral solns. Used in metal plating, mineral tanning of hides, mordanting fabrics, and in water-paste paints. Has been suggested as a photographic processing aid.

8767. Sodium Glycerophosphate. *1,2,3-Propanetriol mono(dihydrogen phosphate) disodium salt.* C$_3$H$_7$Na$_2$O$_6$P; mol wt 216.04. C 16.68%, H 3.27%, Na 21.28%, O 44.43%, P 14.34%. Three isomers exist: The β-glycerophosphoric acid disodium salt ((HOCH$_2$)$_2$CHOPO$_3$Na$_2$) and D(+)- and L(−)-α-glycerophosphoric acid disodium salt (HOCH$_2$CH-(OH)CH$_2$OPO$_3$Na$_2$). Prepn: H. King, F. L. Pyman, *Pharm. J.* **92**, 511 (1914). Exptl use in diagnosis of prostatic carcinoma: M. K. Schwartz *et al.*, *Ann. N.Y. Acad. Sci.* **166**, 775 (1969). Chronic toxicity study of β-form: K. L. Raheja *et al.*, *Toxicology* **8**, 115 (1977). Efficacy of β-form as cariostatic agent: T. H. Grenby, J. M. Bull. *Arch. Oral Biol.* **20**, 717 (1975). GC determn: Y. Handa *et al.*, *J. Chromatog.* **206**, 387 (1981).
β-Form hemiundecahydrate, white, odorless, scale-like crystals; dec > 130°. Sol in ~1.5 parts water; more sol in hot water. (pH of aq soln: ~9.5). Insol in alcohol.
THERAP CAT: Tonic.
THERAP CAT (VET): Has been used as a tonic.

8768. Sodium Hexachloroplatinate(IV). *Disodium hexachloroplatinate(2−);* sodium platinichloride; sodium chloroplatinate. Cl$_6$Na$_2$Pt; mol wt 453.78. Cl 46.88%, Na 10.13%, Pt 42.99%. Na$_2$[PtCl$_6$]. Prepn: Grube in *Handbook of Preparative Inorganic Chemistry* vol. 2, G. Brauer, Ed. (Academic Press, New York, 2nd ed., 1965) pp 1571-1572; Cox, Peters, *Inorg. Syn.* **13**, 173 (1971).
Yellow, hygroscopic crystals. Easily forms hexahydrate at 25° and relative humidity > 50% (reconverted to anhydr salt by heat at 110° for one hour). uv max (1 formal HCl): 262 nm (ε 24500). Sol in water, alcohol.
USE: Catalyst.

8769. Sodium Hexafluorosilicate. Sodium fluosilicate; sodium silicofluoride; Salufer. F$_6$Na$_2$Si; mol wt 188.06. F 60.62%, Na 24.45%, Si 14.93%. Na$_2$SiF$_6$. The average grade of commerce is ~98% pure.

White, granular powder. d 2.68. Melts at red heat with decompn. Sol in 150 parts cold, 40 parts boiling water; insol in alc. The soln in cold water is neutral. LD$_{50}$ orally in rats: 125 mg/kg, *Handbook of Toxicology* vol. 1, W. S. Spector, Ed. (Saunders, Philadelphia, 1956) pp 278-279.
USE: In enamels for china and porcelain; manuf opal glass; as insect exterminator and poison for rodents; mothproofing of woolens.
THERAP CAT (VET): Pediculicide.

8770. Sodium Hydride. HNa; mol wt 24.00. H 4.20%, Na 95.80%. NaH. Prepd by passing hydrogen into molten sodium dispersed in oil or mixed with a catalyst such as anthracene above 250°: Hansley, Carlisle, *Chem. & Eng. News* **23**, 1332 (1945). Laboratory procedure by hydrogenating sodium dispersions: Mattson, Whaley, *Inorg. Syn.* **5**, 10 (1957). Book: J. Plesek, S. Hermanek, *Sodium Hydride, Its Use in the Laboratory and in Technology* (Iliffe Books, London, 1968) 185 pp.
Silvery needles; the commercial product is a gray-white powder. d 1.396. Dec 425°. Reacts explosively with water, violently with lower alcohols, ignites spontaneously on standing in moist air. Sol in molten sodium hydroxide, insol in liq ammonia but forms sodamide at moderate temps.
USE: At low temps where reducing properties of sodium are undesirable as in the condensation of ketones and aldehydes with acid esters; in soln with molten sodium hydroxide for the reduction of oxide scale on metals; at high temps as a reducing agent and reduction catalyst.

8771. Sodium Hydrosulfite. Sodium sulfoxylate; sodium dithionite. Na$_2$O$_4$S$_2$; mol wt 174.11. Na 26.41%, O 36.76%, S 36.83%. Na$_2$S$_2$O$_4$. The hydrosulfite of commerce contains 85-90% Na$_2$S$_2$O$_4$.
White or grayish-white, cryst powder; slight characteristic odor. Oxidizes in air (more readily so in presence of moisture or when in soln) to bisulfite and bisulfate and acquires an acid reaction. Very sol in water, slightly in alcohol.
USE: As reducing agent, particularly in dyeing with indigo and vat dyes; bleaching soaps, straw; removing dyes from dyed fabrics.
Note: The name sodium hydrosulfite is applied also to NaHSO$_2$, mol wt 88.06, sol in water, alcohol. Still more confusion results when "sodium hyposulfite" is applied to this compd (Na$_2$S$_2$O$_4$) *see* 1957 Subject Index to *Chem. Abstracts*, p 2218s *under* sodium dithionite.

8772. Sodium Hydroxide. Caustic soda; soda lye; sodium hydrate. HNaO; mol wt 40.00. H 2.52%, Na 57.48%, O 40.00%. NaOH. By reacting calcium hydroxide with sodium carbonate; from sodium chloride by electrolysis; from sodium metal and water vapor at low temp. Description of industrial processes: Faith, Keyes & Clark's *Industrial Chemicals*, F. A. Lowenheim, M. K. Moran, Eds. (Wiley-Interscience, New York, 4th ed., 1975) pp 737-745. Toxicity: Fazekas, *Arch. Exp. Pathol. Pharmakol.* **184**, 587 (1937).
Fused solid with crystalline fracture. Rapidly absorbs carbon dioxide and water from the air. Very corrosive (caustic) to animal and vegetable tissue and to aluminum metal in the presence of moisture. Sold as lumps, sticks, pellets, chips, etc. When kept in tight containers, the usual grades contain 97-98% NaOH. mp 318°. d^{25} 2.13. One gram dissolves in 0.9 ml water, 0.3 ml boiling water, 7.2 ml abs alcohol, 4.2 ml methanol, also sol in glycerol. Generates considerable heat while dissolving, or when the soln is mixed with an acid. Volumetric NaOH solns used in the laboratory must be protected from air to avoid formation of carbonate. Concentrated NaOH solns contain practically no sodium carbonate. The pH of a 0.05% w/w soln ~12, of a 0.5% soln ~13, of a 5% soln ~14. Density, boiling and freezing pt data for (w/w) water solns. d$_4^{15}$: 5% 1.056, 10% 1.111, 20% 1.222, 30% 1.333, 40% 1.434, 50% 1.530. bp: 5% 102°, 10% 105°, 20% 110°, 30% 115°, 40% 125°, 50% 140°. fp: 5% −4°, 10% −10°, 20% −26°, 30% 1°, 40% 15°, 50% 12°. LD orally in rabbits: 500 mg/kg (10% soln) (Fazekas).
Caution: Potential symptoms of overexposure are irritation of nose; pneumonitis; burns eyes and skin; temporary loss of hair. *See* NIOSH *Pocket Guide to Chemical Hazards* (DHHS/NIOSH 90-117, 1990) p 198.

USE: NaOH solutions are used to neutralize acids and make sodium salts, *e.g.*, in petroleum refining to remove sulfuric and organic acids; to treat cellulose in making viscose rayon and cellophane; in reclaiming rubber to dissolve out the fabric; in making plastics to dissolve casein. NaOH solns hydrolyze fats and form soaps; they precipitate alkaloids (bases) and most metals (as hydroxides) from water solns of their salts. Pharmaceutic aid (alkalizer).

THERAP CAT (VET): Caustic; dehorning of calves.

8773. Sodium Hypochlorite. ClNaO; mol wt 74.44. Cl 47.62%, Na 30.88%, O 21.49%. NaClO. Pentahydrate prepd from NaOH and Cl_2 in the presence of water: Sanfourche, Gardent, *Bull. Soc. Chim.* [4] **35**, 1089 (1924); Schmeisser in *Handbook of Preparative Inorganic Chemistry* vol. 1, G. Brauer, Ed. (Academic Press, New York, 2nd ed., 1963) pp 309-310.

Pentahydrate, highly unstable. Crystals, mp 18°. Dec by CO_2 from air. Anhyd NaClO may be obtained by freeze-drying in a vacuum (over concd H_2SO_4). Anhyd NaClO is very explosive. Soly of the pentahydrate at 0°: 29.3 g/100 ml H_2O. The hypochlorite ion in aq soln is remarkably stable. Names for aq solns are *Eau de Labarraque; Clorox; Dazzle.*

Caution: Ingestion may cause corrosion of mucous membranes, esophageal or gastric perforation, laryngeal edema. Inhalation may produce severe bronchial irritation, pulmonary edema. Prolonged skin contact may result in irritation, *cf. Clinical Toxicology of Commercial Products,* R. E. Gosselin *et al.,* Eds. (Williams & Wilkins, Baltimore, 4th ed., 1976) Section III, pp 174-176.

USE: Aq soln as bleach, disinfectant.

THERAP CAT: Disinfectant.

8774. Sodium Hypochlorite Solution, Diluted. Modified Dakin's soln; surgical chlorinated soda soln; Carrel-Dakin soln. A soln of sodium hypochlorite contg 0.45-0.50 g of the salt in 100 ml. Prepd by diluting with distilled water a soln of sodium hypochlorite, adding a 5% soln of sodium bicarbonate, and adjusting to the proper strength and concn according to the procedure described in N.F. May be prepd also from 14.4 g chlorinated lime (30% available chlorine), 7.7 g anhyd sodium carbonate, and 6.4 g sodium bicarbonate per liter; this soln is then adjusted as in the previous method.

Colorless or slightly yellow liquid; faint chlorine-like odor. d of the soln should not exeed 1.025. Does not produce red color with phenolphthalein.

Caution: See Sodium Hypochlorite.

THERAP CAT: Anti-infective (topical).

THERAP CAT (VET): Antiseptic for wound irrigation.

8775. Sodium Hypophosphite. *Phosphinic acid, sodium salt.* H_2NaO_2P; mol wt 87.98. H 2.29%, Na 26.13%, O 36.37%, P 35.21%. NaH_2PO_2. Solubility data: Palit, *J. Am. Chem. Soc.* **69**, 3120 (1947).

Monohydrate, white, odorless, deliquesc granules; saline taste. When strongly heated, it dec with evolution of phosphine which ignites spontaneously in the air. *It explodes when triturated with chlorates or other oxidizing agents.* Sol in 1 part water, 0.15 part boiling water; freely sol in glycerol and in boiling alcohol; sol in cold alcohol, slightly in abs alcohol. Insol in ether. Soly of anhyd NaH_2PO_2 at 25° in ethylene glycol: 33.0 g/100 g; in propylene glycol: 9.7 g/100 g. The aq soln is neutral. *Keep well closed.*

USE: As reagent for arsenic and iodates; prepn of hypophosphites syrup.

8776. Sodium Iodate. $INaO_3$; mol wt 197.89. I 64.13%, Na 11.62%, O 24.25%. $NaIO_3$. The article of commerce contains about 99% $NaIO_3$.

White, cryst powder. d 4.28. Sol in ~11 parts water, 3 parts boiling water; insol in alc. The aq soln is neutral. LD i.v. in dogs: 200 mg/kg, *Handbook of Toxicology* vol. 1, W. S. Spector, Ed. (Saunders, Philadelphia, 1956) pp 274-275.

THERAP CAT: Antiseptic (mucous membranes).

8777. Sodium Iodide. Ioduril; Anayodin. INa; mol wt 149.89. I 84.66%, Na 15.34%. NaI. U.S.P. NaI is at least 99% pure.

White, odorless, deliquesc crystals or granules. Gradually absorbs up to about 5% (½ mol) moisture on exposure to air. Slowly becomes brown in the air due to liberation of iodine. Its aq soln is similarly affected. d 3.67. mp 651°. One gram dissolves in 0.5 ml water, ~2 ml alc, 1 ml glycerol; sol in acetone. It is made slightly alkaline to render it more stable. pH: 8-9.5. *Keep well closed and protected from light.* At ordinary room temp crystallizes from water with $2H_2O$ in the form of colorless, prismatic crystals. *Incompat.* As of potassium iodide. MLD i.v. in rats: 1.3 g/kg, Loeser, Konwiser, *J. Lab. Clin. Med.* **15**, 35 (1929).

THERAP CAT: Iodine supplement; expectorant.

THERAP CAT (VET): Actinobacillosis, actinomycosis. Expectorant. Has been used for ringworm, hyperplastic fibrous lesions, paraplegia from pachymeningitis of dogs.

8778. Sodium Iodide, Radioactive. Sodium radioiodide (^{131}I); sodium iodide—^{131}I; Na^{131}I; Iodotope; Oriodide; Radiocaps-131; Theriodide-131. Prepd from radioactive iodine (^{131}I) which has a half-life of 8 days and emits beta and gamma rays. Other properties identical with those of ordinary sodium iodide. Dispensed as carrier-free sodium radioiodide in capsules for oral use or in aq soln for oral or parenteral administration.

THERAP CAT: Diagnostic aid (thyroid function); antineoplastic.

8779. Sodium Iodomethamate. *1,4-Dihydro-3,5-diiodo-1-methyl-4-oxo-2,6-pyridinedicarboxylic acid disodium salt;* 3,5-diiodo-1-methylchelidamic acid sodium salt; disodium N-methyl-3,5-diiodo-4-pyridone-2,6-dicarboxylate; iodoxyl; D-40; Neo-Iopax; Pyelectan; Uropac. $C_8H_3I_2NNa_2O_5$; mol wt 492.90. C 19.49%, H 0.61%, I 51.49%, N 2.84%, Na 9.33%, O 16.23%. Prepn: Chelidonic acid is converted to chelidamic acid by the action of NH_3; chelidamic acid is iodinated with iodine in a boiling aq alkaline soln, the iodinated product is methylated at the nitrogen with dimethyl sulfate in hot alkaline soln, *cf.* Dohrn, Diedrich, *Ann.* **494**, 284 (1932); **Ger. pats.** 545,916; 545,266; 556,142; **U.S. pat. 1,919,417** (1933).

Crystals, dec around 200° with effervescence. Freely sol in water (1:1). Practically insol in chloroform, ether, acetone, petr ether.

THERAP CAT: Diagnostic aid (radiopaque medium).

8780. Sodium Isopropyl Xanthate. *Isopropylxanthic acid sodium salt;* Good-Rite Nix. $C_4H_7NaOS_2$; mol wt 158.22. C 30.37%, H 4.46%, Na 14.53%, O 10.11%, S 40.53%. $(CH_3)_2CHOCSSNa$.

Deliquescent, white to yellowish powder or lumps, dec 150°. Slightly unpleasant odor. Soly in water: 30% at 4°; 46% at 24°; 54% at 35°.

Caution: Irritating to skin, eyes, mucous membranes, respiratory tract.

USE: Control of annual weeds in bean and pea fields.

8781. Sodium Lactate. Lacolin. $C_3H_5NaO_3$; mol wt 112.06. C 32.15%, H 4.50%, Na 20.52%, O 42.83%. Commercially available as a mixture with water containing 70-80% sodium lactate. *Ref:* Shaw, U.S. pat. 2,856,326 (1958 to Nat. Dairy Prod. Corp.).

Colorless or almost colorless, thick, odorless liquid. Miscible with water, alcohol. The soln is neutral.

USE: Instead of glycerol in calico printing; as a plasticizer for casein; as a corrosion inhibitor in alc antifreeze mixture.

THERAP CAT: Electrolyte replenisher; systemic and urinary alkalizer.

THERAP CAT (VET): Has been used in bovine ketosis.

8782. Sodium Lauryl Sulfate. *Sulfuric acid monododecyl ester sodium salt;* sodium dodecyl sulfate; SDS; Irium. $C_{12}H_{25}NaO_4S$; mol wt 288.38. C 49.98%, H 8.74%, Na 7.97%, O 22.19%, S 11.12%. $CH_3(CH_2)_{10}CH_2OSO_3Na$. Anionic detergent prepd by sulfation of lauryl alcohol, followed by neutralization with sodium carbonate: A. Lotter-moser, F. Stoll, *Kolloid-Z.* **63**, 50 (1933). Surfactant proper-ties: J. Powney, C. C. Addison, *Trans. Faraday Soc.* **33**, 1244 (1937); E. E. Dreger *et al., Ind. Eng. Chem.* **36**, 610 (1944). Use in electrophoretic sepn and mol wt estimation of proteins: A. L. Shapiro *et al., Biochem. Biophys. Res. Commun.* **28**, 815 (1967); K. Weber, M. Osborn, *J. Biol. Chem.* **244**, 4406 (1969); of glycopolypeptides: B. S. Leach *et al., Biochemistry* **19**, 5734 (1980). Toxicity study: A. I. T. Walker *et al., Food Cosmet. Toxicol.* **5**, 763 (1967). Re-view of toxicology: Ch. Gloxhuber, *Arch. Toxicol.* **32**, 245-270 (1974).

White or cream-colored crystals, flakes, or powder. Faint odor of fatty substances. Smooth feel. Neutral reaction. One gram dissolves in 10 ml water, giving an opalescent soln. Lowers the surface tension of aq solns. Emulsifies fats. LD_{50} orally in rats: 1288 mg/kg (Walker).

USE: Wetting agent, detergent, esp in the textile industry. Electrophoretic separation of proteins and lipids. Ingredi-ent of toothpastes.

8783. Sodium-Lead Alloy. Lead-sodium alloy; sodi-um-lead; Drynap; Dri-Na. Usually contains a minimum of 9.5% active sodium. For tetraethyllead manufacture the sodium-lead alloy is produced in large quantities by making a melt of 90 parts of lead with 10.5 parts of sodium (w/w). The reaction is strongly exothermic and starts at 225°. Prepn of the alloy on a laboratory scale: Soroos, *Ind. Eng. Chem., Anal. Ed.* **11**, 657 (1939).

Obtained in brittle lumps which can be stored in an air-tight container. It reacts only slowly with air or water, yet dries ether as completely as sodium wire. Residues of ether or other liquids still containing some active alloy can be destroyed safely by the addition of water, as the reaction never reaches the violence observed with pure sodium metal. The alloy may be ground to a very fine powder under the surface of a non-polar solvent such as kerosine or ether.

Caution: Finely ground powder, if not protected by a suitable liquid, may react with excess moisture from the air sufficiently to catch fire.

USE: In manuf of tetraethyllead by reaction of alloy with ethyl chloride; for drying ether and for reductions: Tabei *et al., Bull. Chem. Soc. Japan* **40**, 1538 (1967); as a sodium substitute in many chemical reactions where the presence of lead is not objectionable.

8784. Sodium Metabisulfite. *Sodium pyrosulfite.* $Na_2O_5S_2$; mol wt 190.11. Na 24.19%, O 42.08%, S 33.73%. $Na_2S_2O_5$. Crystallizes from cold water with $7H_2O$.

White crystals or powder; odor of SO_2. Freely sol in water, glycerol; slightly sol in alcohol. The aq soln is acid.

USE: Pharmaceutic aid (antioxidant).

8785. Sodium Metaborate. $BNaO_2$; mol wt 65.80. B 16.43%, Na 34.94%, O 48.63%. $NaBO_2$. Obtained by fusing equivalent mol wts of borax and sodium carbonate.

White pieces or powder. mp 966°. Sol in water, the soln being strongly alkaline.

8786. Sodium Metaperiodate. *Sodium periodate.* $INaO_4$; mol wt 213.89. I 59.33%, Na 10.75%, O 29.92%. $NaIO_4$. Synthesis starting with sodium iodide: *Inorg. Syn.* **2**, 212 (1946) and **1**, 170 (1939).

White, tetragonal crystals, d_4^{16} 3.865. Dec ~300°. Sol in cold water, sulfuric, nitric, acetic acids.

Trihydrate, white, efflorescent, trigonal crystals, d_4^{18} 3.219. Dec 175°. One gram dissolves in 8 ml water at 20°.

8787. Sodium Metaphosphate. $(NaPO_3)_n$ where *"n"* may be 2 or more. Prepd by dehydration of sodium phos-phates (e.g. $Na_2H_2P_2O_7$, NaH_2PO_4, Na_2HPO_4). Several different forms are known. *Reviews:* J. R. Van Waser, *Phosphorus and Its Compounds* **vol. 1** (Interscience, New York, 1958) pp 601-800; Pantony in *Mellor's* **vol. II** supple-ment II, *The Alkali Metals* (Part 1), 1316-1349 (1961); Thilo in *Advan. Inorg. Chem. Radiochem.* **4**, 1-75 (1962). *See also* Sodium Hexametaphosphate; Sodium Trimetaphosphate; Maddrell's Salt.

8788. Sodium Metasilicate. Na_2O_3Si; mol wt 122.06. Na 37.67%, O 39.32%, Si 23.01%. Na_2SiO_3. Usually pre-pared from sand (SiO_2) and soda ash (Na_2CO_3) by fusion:

Schwarz, *Z. Anorg. Allgem. Chem.* **126**, 62 (1923); Faith, Keyes & Clark's *Industrial Chemicals*, F. A. Lowenheim, M. K. Moran, Eds. (Wiley-Interscience, New York, 4th ed., 1975) pp 755-761. *Review: Gmelin's, Sodium* (8th ed. sup-plement) **21**, pp 1474-1478 (1967).

Usually obtained as a glass; also orthorhombic crystals. d 2.614. mp 1089°. n_D^{25} (glass) 1.520. Specific heat (20°): 0.217. Heat of formation: −371.2 kcal/mol. Heat of soln (cryst) −7.45 kcal/mol. Heat of fusion: 10.3 kcal/mol. Sol in cold water, hydrolyzed by hot water. Insol in alco-hol, acids, salt solns.

Nonahydrate, efflorescent, orthorhombic bipyramidal platelets, mp 48° in water of crystn. Heat of hydration: −24.15 kcal/mol.

8789. Sodium Methoxide. Sodium methylate. CH_3NaO; mol wt 54.02. C 22.23%, H 5.60%, Na 42.55%, O 29.62%. CH_3ONa. Prepn: Burness, *Org. Syn.* **39**, 51 (1959).

White, free-flowing powder. Sensitive to air and mois-ture. Dec by water. Sol in methanol, ethanol. Also exists in solvated form, $CH_3ONa.2CH_3OH$, white powder. Ap-parent density of solvent-free material ~4.6 lb/gal.

USE: In organic syntheses.

8790. Sodium Molybdate(VI). $MoNa_2O_4$; mol wt 205.92. Mo 46.59%, Na 22.33%, O 31.08%. Na_2MoO_4. Dihydrate, $MoNa_2O_4.2H_2O$, *Molyhibit 100.* Cryst pow-der. Loses its water of crystn at 100°. Sol in 1.7 parts cold water, ~0.9 part boiling water. pH of 5% aq soln at 25° = 9.0-10.0. Has been reported to be less toxic than the corresponding compds of group 6B in the periodic table: Fairhall *et al., Public Health Bull.* no. 293 (1945).

USE: Manuf of inorganic and organic pigments, corrosion inhibitor, bath additive for metals finishing, reagent for alkaloids, micronutrient for plants and animals.

8791. Sodium β-Naphthoquinone-4-sulfonate. *3,4-Di-hydro-3,4-dioxo-1-naphthalenesulfonic acid sodium salt;* sodium 1,2-naphthoquinone-4-sulfonate; 1,2-naphthoqui-none-4-sulfonic acid sodium salt; β-naphthoquinone-4-sul-fonic acid sodium salt. $C_{10}H_5NaO_5S$; mol wt 260.20. C 46.16%, H 1.94%, Na 8.84%, O 30.74%, S 12.32%. Prepn: Folin, *J. Biol. Chem.* **51**, 377 (1922).

Yellow crystals from dil alc. Readily sol in water; slight-ly sol in 95% alcohol; moderately sol in acetone; practically insol in ether, chloroform, carbon disulfide, benzene, petr ether. Aq solns fade slowly on exposure to light but may be stabilized by HCl.

USE: Colorimetric determn of amino acids and amines.

8792. Sodium Nitrate. Chile saltpeter; cubic niter; soda niter. $NNaO_3$; mol wt 84.99. N 16.48%, Na 27.05%, O 56.47%. $NaNO_3$. The purified grade contains at least 99% $NaNO_3$. Occurs as a mineral in Chile. Toxicity data: Dollahite, Rowe, *Southwest Vet.* **27**, 246 (1974).

Colorless, transparent crystals, white granules or powder; deliquesc in moist air. d 2.26. mp 308°. One gram dis-solves in 1.1 ml water, 0.6 ml boiling water, 125 ml alcohol, 52 ml boiling alcohol, 3470 ml abs alcohol, 300 ml abs methanol. When dissolved in water the temp of the soln is lowered. The aq soln is neutral. *Keep well closed.* LD_{50} orally in rabbits: 1.955 g anion/kg (Dollahite, Rowe).

USE: Manuf of nitric acid and as catalyst in the manuf of sulfuric acid. Manuf sodium nitrite, glass, enamels for pot-tery; in matches; for improving burning properties of tobac-co; pickling meats; as color fixative in meats. Clinical re-agent (parasites). The technical grade is used as fertilizer.

8793. Sodium Nitrite. *Nitrous acid sodium salt;* erini-trit. $NNaO_2$; mol wt 69.00. N 20.30%, Na 33.32%, O 46.38%. Contains 96-98% $NaNO_2$. Thermodynamic properties: Plekhotkun, *Zh. Prikl. Khim.* **40**, 1843 (1967), *C.A.* **68**, 33797x (1968). Toxicity studies: Faccini *et al., Ind. Aliment.* **8**, 77 (1969), *C.A.* **71**, 100583b (1969); H. F. Smyth *et al., Am. Ind. Hyg. Assoc. J.* **30**, 470 (1969). Review of chemistry of $NaNO_2$ in a biological system as related to meat curing: Bard, Townsend, *Science of Meat & Meat Products,* J. F. Price, B. S. Schweigert, Eds. (W. H. Freeman, 2nd ed. 1971) pp 452-470.

White or slightly yellow, hygroscopic granules, rods, or powder. Very slowly oxidizes to nitrate in air. d 2.17. mp 271°; dec above 320°. Sol in 1.5 parts cold water, 0.6 part boiling water, slightly in alc. Dec even by weak acids with evolution of brown fumes of N_2O_3. The aq soln is alkaline. pH ~9. *Keep well closed.* LD_{50} orally in rats: 180 mg/kg (Smyth).

Incompat: Acetanilide, antipyrine, chlorates, hypophosphites, iodides, mercury salts, permanganate, sulfites, tannic acid, vegetable astringent decoctions, infusions or tinctures.

Caution: Consult latest Government regulation to determine max amounts allowable in food.

USE: Manuf diazo dyes, nitroso compds, and in many other processes of manuf of organic chemicals; dyeing and printing textile fabrics; bleaching flax, silk, and linen; photography. In meat curing, coloring and preserving; in processing smoked chub. Also as reagent in anal. chemistry.

THERAP CAT: Vasodilator; antidote (cyanide poisoning).

THERAP CAT (VET): In cyanide poisoning. Has been used as a vasodilator, as a circulatory (blood pressure) depressant and to relieve smooth muscle spasm.

8794. Sodium Nitroprusside. *Pentakis(cyano-C)nitrosylferrate(2—) disodium; sodium nitrosylpentacyanoferrate-(III);* sodium nitroferricyanide; sodium nitroprussiate; Nipruss. $C_5FeN_6Na_2O$; mol wt 261.92. C 22.93%, Fe 21.32%, N 32.09%, Na 17.55%, O 6.11%. $Na_2[Fe(CN)_5NO]$. Prepn: L. Playfair, *Proc. Roy. Soc. London* **5**, 846 (1849). Pharmacology: Fernandez *et al., Arch. Inst. Farmacol. Exp. Madrid* **23**, 1-51 (1971). *Review:* I. H. Tuzel, *J. Clin. Pharmacol.* **14**, 494-503 (1974). Review of pharmacology, toxicology and therapeutic uses: J. H. Tinker, J. D. Michenfelder, *Anesthesiol.* **45**, 340 (1976). Comprehensive description: R. Rucki, *Anal. Profiles Drug Subs.* **6**, 487-513 (1977); A. Bult *et al., ibid.* **15**, 781-792 (1986).

Dihydrate, *Nipride, Nitropress.* Ruby-red, practically odorless, transparent crystals. Sol in ~2.3 parts water, slightly in alcohol. Slowly dec in aq soln.

USE: Reagent for the detection of many organic compds, e.g., acetone, aldehydes, also of alkali sulfides, zinc, SO_2.

THERAP CAT: Antihypertensive.

8795. Sodium Oxalate. Ethanedioic acid disodium salt. $C_2Na_2O_4$; mol wt 134.00. C 17.93%, Na 34.31%, O 47.76%. $Na_2C_2O_4$.

White, odorless, cryst powder. Sol in 27 parts water, 16 parts boiling water; insol in alcohol. The aq soln is practically neutral.

Caution: Ingestion of concd solns may cause severe G.I. irritation, hematemesis, CNS and cardiac depression, death. Dilute solns produce little G.I. distress but may cause weakness, muscular twitchings, rarely convulsions, coma, death. Chronic ingestion of small amounts may cause hypocalcemic tetany, urinary calculi, *cf. Clinical Toxicology of Commercial Products,* R. E. Gosselin *et al.,* Eds. (Williams & Wilkins, Baltimore, 4th ed., 1976) Section III, pp 260-263.

USE: Finishing textiles, tanning and finishing leather; for standardizing potassium permanganate soln.

8796. Sodium Oxide. Sodium monoxide. Na_2O; mol wt 61.98. Na 74.19%, O 25.81%.

White, amorphous pieces or powder. d 2.27. Melts at a dull red heat and begins to dec > 400° into sodium peroxide and metal. It is very reactive and combines violently with water, forming sodium hydroxide. *Handle with tongs and not with bare hands, and keep tightly closed.*

USE: As a dehydrating agent; in certain chemical reactions as a polymerizing or condensing agent.

8797. Sodium Perborate. Dexol. $BNaO_3$; mol wt 81.80. B 13.22%, Na 28.11%, O 58.68%. $NaBO_3$. Contains, when reasonably fresh, ~95% of the perborate corresp to 9.9% available oxygen. Prepn from sodium metaborate and

hydrogen peroxide: Leblon, Lambert, U.S. pat. **3,109,706** (1963 to Solvay & Cie.).

Tetrahydrate, white, odorless, cryst powder; saline taste. Stable when kept cool and dry, but is dec with liberation of oxygen in warm or moist air. Dec > 60°. Sol in ~40 parts water, the soln being alkaline and dec with the liberation of H_2O_2, and then of oxygen. In the presence of acids, H_2O_2 is formed. *Keep well closed and in a cool place.*

Caution: Prevent swallowing of soln.

USE: Bleaching straw and other fibers, ivory, sponges, bristles, waxes, textiles; in laundering, dentifrices, soaps.

THERAP CAT: Antiseptic (topical).

THERAP CAT (VET): Mouthwash.

8798. Sodium Perchlorate. Irenat. $ClNaO_4$; mol wt 122.44. Cl 28.96%, Na 18.78%, O 52.27%. $NaClO_4$.

Monohydrate, white, deliquesc crystals. Dec ~130°. d 2.02. Very sol in water. *Keep well closed.*

USE: In the explosives industry.

THERAP CAT: Thyroid inhibitor.

8799. Sodium Permanganate. $MnNaO_4$; mol wt 141.93. Mn 38.71%, Na 16.20%, O 45.09%. $NaMnO_4$.

Trihydrate, reddish-black, very hygroscopic granules. Very sol in water; dec by alcohol.

8800. Sodium Peroxide. Sodium dioxide; sodium superoxide; Solozone. Na_2O_2; mol wt 77.98. Na 58.96%, O 41.04%. The product of commerce contains 90-95% Na_2O_2. Prepd by heating sodium metal to 300° in aluminum vessels with a current of air from which carbon dioxide has been removed. Prepn of the octahydrate: Penneman, *Inorg. Syn.* **3**, 1 (1950).

Yellowish-white, granular powder. Absorbs water and CO_2 from the air. Freely sol in water, forming sodium hydroxide and hydrogen peroxide, the latter quickly dec into oxygen and water. With dil acids H_2O_2 is formed which remains stable. In contact with organic matter or readily oxidizable substances ignition and explosion may take place. *Keep tightly closed and protected from contact with organic or oxidizable substances.*

Caution: Irritant and corrosive. See Sodium Hydroxide.

USE: Bleaching animal and vegetable fibers, feathers, bones, ivory, wood, wax, sponges, coral; rendering air charged with CO_2 respirable as in torpedo boats, submarines, diving bells, etc.; purifying air in sick rooms; dyeing and printing textiles; chemical analysis. General oxidizing agent.

8801. Sodium Persulfate. *Sodium peroxydisulfate.* $Na_2O_8S_2$; mol wt 238.11. Na 19.31%, O 53.76%, S 26.93%. $Na_2S_2O_8$.

White, cryst powder. Gradually dec; decompn is promoted by moisture and higher temp. Initial soly in water at 20°: 549 g/l; dec by alcohol and silver ions. MLD i.v. in rabbits: 178 mg/kg, *Handbook of Toxicology* vol. **1**, W. S. Spector, Ed. (Saunders, Philadelphia, 1956) pp 278-279.

Caution: Highly irritating to skin, mucous membranes.

USE: Bleaching and oxidizing agent; promoter for emulsion polymerization reactions.

8802. Sodium Pertechnetate ^{99m}Tc. Pertscan; Ultra-Technekow. $NaO_4{}^{99m}$Tc. $Na^{99m}TcO_4$. Prepn: Keller, Kanellakopulos, *Radiochim. Acta* **1**, No. 2, 107 (1963), *C.A.* **59**, 1256a (1963); Kanellakopulos, *AEC Accession* No. **31424**, Rept. No. **KFK-197**, 73 pp (1964), *C.A.* **62**, 7350d (1965). Clinical application for labelling red blood cells: D. Ducassou *et al., Brit. J. Radiol.* **49**, 344 (1976). Diagnostic use in Meckel's diverticulum: D. R. Cooney *et al., J. Pediatr. Surg.* **17**, 611 (1982); in thyroid neoplasm: M. Vorne, K. Jarve, *Eur. J. Nucl. Med.* **13**, 362 (1987). Review of diagnostic use in brain scanning: J. G. McAfee *et al., J. Nucl. Med.* **5**, 811-827 (1964); in thyroid function: M. S. Sucupira *et al., Int. J. Nucl. Med. Biol.* **10**, 29-33 (1983).

THERAP CAT: Diagnostic aid (radioactive imaging agent).

8803. Sodium Phenolsulfonate. *Hydroxybenzenesulfonic acid sodium salt;* sodium sulfocarbolate. $C_6H_5NaO_4S$; mol wt 196.16. C 36.74%, H 2.57%, Na 11.72%, O 32.63%, S 16.35%. $HOC_6H_4SO_3Na$.

Dihydrate, white, odorless crystals; slightly bitter taste; somewhat efflorescent in dry air. One gram dissolves in 4.2

ml water, 0.8 ml boiling water, 140 ml alcohol, 13.5 ml boiling alcohol, 5 ml glycerol. The aq soln is neutral.

THERAP CAT: Intestinal antiseptic.

THERAP CAT (VET): Has been used as an intestinal antiseptic, in dusting powders for ulcers, slowly granulating wounds and in dilute solution in the eye.

8804. Sodium Phenoxide. Sodium phenate; sodium carbolate; sodium phenolate; phenol sodium. C_6H_5NaO; mol wt 116.09. C 62.08%, H 4.34%, Na 19.80%, O 13.78%. C_6H_5ONa. Prepn from phenol and NaOH in dil methanol: Kornblum, Lurie, *J. Am. Chem. Soc.* **81**, 2710 (1959).

White to reddish, deliquescent rods or granules. Decomposed by the CO_2 of the air. Very sol in water; sol in alcohol. The aq soln is caustic.

8805. Sodium Phosphate, Dibasic. Dibasic sodium phosphate; disodium hydrogen phosphate; disodium orthophosphate; disodium phosphate; DSP; phosphate of soda; secondary sodium phosphate. HNa_2O_4P; mol wt 141.96. H 0.71%, Na 32.39%, O 45.08%, P 21.82%. Na_2HPO_4. Industrial production: *Faith, Keyes, & Clark's Industrial Chemicals* (John Wiley, New York, 4th ed., 1975) pp 746-754. Toxicity of heptahydrate: H. F. Smyth *et al.*, *Am. Ind. Hyg. Assoc. J.* **30**, 470 (1969).

Anhydr, *exsiccated sodium phosphate.* Hygroscopic powder. On exposure to air will absorb from 2 to 7 mols H_2O, depending on the humidity and temp. Sol in ∼8 parts water, much more sol in hot water. Soly per 100 gal water increases from ∼14 lbs at slightly >0° to over 900 lbs at 95°. Insol in alc. pH of 1% aq soln at 25°: 9.1. *Keep well closed.*

Dihydrate, *Sorensen's phosphate, Sorensen's sodium phosphate.*

Heptahydrate, crystals or granular powder. Stable in the air. d ∼1.7. Sol in 4 parts water, more sol in boiling water; practically insol in alcohol. The aq soln is alkaline, pH ∼9.5. LD_{50} orally in rats: 12.93 g/kg (Smyth).

Dodecahydrate, translucent crystals or granules; readily loses 5 mols of water on exposure to air at ordinary temp. mp 34-35° (when it contains the full 12 mols of H_2O). d ∼1.5. Sol in 3 parts water; practically insol in alcohol. Aq soln is alkaline, pH ∼9.5. *Keep well closed and in a cool place.*

Incompat: Alkaloids, antipyrine, chloral hydrate, lead acetate, pyrogallol, resorcinol.

Caution: Anhydr form may cause mild irritation to skin, mucous membranes; intern. causes purging.

USE: As sequestrant, emulsifier and buffer in foods. As mordant in dyeing; for weighting silk; in tanning; in manuf of enamels, ceramics, detergents, boiler compds; as fireproofing agent; in soldering and brazing instead of borax; as reagent and buffer in analytical chemistry.

THERAP CAT: Cathartic.

THERAP CAT (VET): Laxative.

8806. Sodium Phosphate, Monobasic. Sodium biphosphate; sodium dihydrogen phosphate; acid sodium phosphate; monosodium orthophosphate; primary sodium phosphate. H_2NaO_4P; mol wt 119.98. H 1.68%, Na 19.16%, O 53.34%, P 25.82%. NaH_2PO_4. It is about 99% pure.

Monohydrate, white, odorless, slightly deliquesc crystals or granules. At 100° loses all its water; when ignited it converts into metaphosphate. Freely sol in water; practically insol in alcohol. The aq soln is acid. pH of 0.1 molar aq soln at 25°: 4.5.

Dihydrate, orthorhombic bisphenoidal colorless crystals, mp 60°. d 1.915. At room temp crystallizes with $2H_2O$. Directions for max yield: Beans, Kiehl, *J. Am. Chem. Soc.* **49**, 1878 (1927).

USE: In baking powders; in boiler water treatment; as dry acidulant and sequestrant for foods: Tidridge, Pals, U.S. pat. **3,030,213** (1962 to FMC).

THERAP CAT: Urinary acidifier.

THERAP CAT (VET): Urinary acidifier.

8807. Sodium Phosphate, Radioactive. Sodium phosphate—^{32}P; radioactive sodium phosphate; sodium radiophosphate (^{32}P); Phosphotope.

Aq soln of mixed radioactive phosphates with a pH range of 5.0-6.0. Contains radioactive monobasic sodium phos-

phate ($NaH_2{}^{32}PO_4$) and radioactive dibasic sodium phosphate ($Na_2H{}^{32}PO_4$). ^{32}P is a pure beta emitter with a half-life of 14.3 days.

THERAP CAT: Antineoplastic; antipolycythemic; diagnostic aid (neoplasm).

8808. Sodium Phosphate, Tribasic. Trisodium orthophosphate; trisodium phosphate; TSP; Oakite. Na_3O_4P; mol wt 163.94. Na 42.07%, O 39.04%, P 18.89%. Na_3PO_4. Crystallizes with 8 and 12 mols of H_2O.

Dodecahydrate, colorless or white crystals. When rapidly heated melts at ∼75°. Does not lose the last mol of water even on moderate ignition. d 1.6. Sol in 3.5 parts water, 1 part boiling water; insol in alcohol. The aq soln is strongly alkaline. pH of 0.1% soln: 11.5; of 0.5% soln: 11.7; of 1.0% soln: 11.9. Technical crystals are sometimes made with excess alkali to prevent caking and give more alkaline solutions. LD_{50} orally in rats: 7.40 g/kg, H. F. Smyth *et al.*, *Am. Ind. Hyg. Assoc. J.* **30**, 470 (1969).

USE: In photographic developers; clarifying sugar; removing boiler scale, softening water; manuf paper; laundering; tanning leather; in detergent mixture.

8809. Sodium Phosphite. HNa_2O_3P; mol wt 125.96. H 0.80%, Na 36.50%, O 38.11%, P 24.59%. Na_2HPO_3.

Pentahydrate, white, hygroscopic cryst powder. Heat of formation (25°): −684.2 kcal/mole. Freely sol in water. *Keep well closed.*

8810. Sodium Phosphomolybdate. *Sodium molybdophosphate.* $Mo_{12}Na_3O_{40}P$; mol wt 1891.20. Mo 60.88%, Na 3.65%, O 33.84%, P 1.64%. $Na_3PO_4 \cdot 12MoO_3$.

White crystals. Freely sol in water.

USE: As reagent in chemical analysis.

8811. Sodium Phosphotungstate. *Sodium tungstophosphate.* Approx $2Na_2O \cdot P_2O_5 \cdot 12WO_3 \cdot 18H_2O$.

White, granular powder. Sol in water.

USE: As reagent for alkaloids, uric acid, potassium.

8812. Sodium Plumbate(IV). Na_2O_3Pb; mol wt 301.18. Na 15.27%, O 15.94%, Pb 68.80%. Na_2PbO_3.

Trihydrate, light yellow, fused, hygroscopic lumps. Sol in water, but gradually dec with separation of PbO_2. *Keep well closed.*

8813. Sodium Polyanetholesulfonate. Polyanetholesulfonic acid sodium salt; anetholesulfonic acid sodium salt polymer. A polymer of anetholesulfonic acid. Originally developed as an anticoagulant, it was soon found that it possesses anticomplement action and lowers the bactericidal action of blood. *Ref:* Demole, Reinert, *Arch. Exp. Pathol. Pharmakol.* **158**, 211 (1930); Friedmann, *Klin. Wochenschr.* **14**, 215 (1935); Stuart, *J. Clin. Path.* **1**, 311 (1948); Hoffmann-La Roche *Biochemicals Catalog.*

Light brown powder. Insol in alcohol. Swells in water and slowly goes in soln with neutral reaction. Aq solns are stable to heat, dil alkalies and dil acids.

USE: Under the trademark *Liquoid* to inhibit blood coagulation *in vitro*, and as diagnostic reagent to encourage the growth of pathogens in blood. Also to stabilize colloidal solns such as milk and gelatin. (Not to be confused with Liquoid, registered trademark of Johnson & Johnson for castor and olive oil emulsion.)

8814. Sodium Polymetaphosphate. Graham's salt; "sodium hexametaphosphate"; glassy sodium metaphosphate; Hy-Phos. $(NaPO_3)_x$. A mixture of polymeric metaphosphates; not a hexamer. Prepd by rapidly chilling molten sodium metaphosphate: Bell, *Inorg. Syn.* **3**, 103 (1950). *Reviews: see* Sodium Metaphosphate.

Clear, hygroscopic glass. mp 628°. Sol in water, but dissolves slowly. Depolymerizes in aqueous soln to form sodium trimetaphosphate and sodium orthophosphates.

Sodium hexametaphosphate detergents, Calgon, Giltex, Quadrafos, Hagan phosphate, Micromet. Mixtures contg

Graham's salt as the principal agent. Supplied in the form of a powder, flakes, and as small, broken, glass-like particles. Sol in water (pH adjusted to 8-8.6). Insol in organic solvents. Possess dispersing and deflocculating properties, coagulate albumins, and inhibit the crystn of slightly sol compds such as calcium carbonate and calcium sulfate.
USE: Water softeners and detergents. For leather tanning, dyeing, laundry work, textile processing; for the "threshold treatment" of softening industrial water supplies.

8815. Sodium Polystyrene Sulfonate. Resonium A; Kayexalate. A cation exchange resin charged with sodium. Marketed as a powder, insol in water; also as an emulsion with methyl cellulose.
THERAP CAT: Ion-exchange resin (potassium).

8816. Sodium Propionate. *Propionic acid sodium salt;* Impedex. $C_3H_5NaO_2$; mol wt 96.06. C 37.51%, H 5.25%, Na 23.93%, O 33.31%. CH_3CH_2COONa.
Transparent crystals, granules. Deliquescent in moist air. Neutral or slightly alkaline reaction to litmus. One gram dissolves in ~1 ml water, in ~0.65 ml boiling water, in ~24 ml alcohol at 25°. Most active at acid pH: Wolford, Andersen, *Food Ind.* **17**, 622 (1945); Olsen, Macy, *J. Dairy Sci.* **29**, 173 (1946).
USE: Fungicide, mold preventative.
THERAP CAT: Antifungal (topical).
THERAP CAT (VET): In ketoses of ruminants (glucose precursor). Antifungal agent. Has been used in dermatoses, wound infections, conjunctivitis.

8817. Sodium Pyroantimonate. Sodium antimonate. Approx $Na_2H_2Sb_2O_7$.
Monohydrate, white, granular powder. Slightly sol in water.
USE: Manuf opaque glass and opaque glazes.

8818. Sodium Rhodizonate. *5,6-Dihydroxy-5-cyclohexene-1,2,3,4-tetrone disodium salt;* [(3,4,5,6-tetraoxo-1-cyclohexen-1,2-ylene)dioxy]disodium. $C_6Na_2O_6$; mol wt 214.04. C 33.67%, Na 21.48%, O 44.85%.

Violet crystals. Sol in water with an orange-yellow color; slightly sol in soda soln; insol in alc. Solns are unstable even in the refrigerator, and must be prepd fresh every other day.
USE: As a reagent for barium and strontium.

8819. Sodium Salicylate. *2-Hydroxybenzoic acid monosodium salt;* Alysine; Idocyl; Enterosalicyl; Enterosalil. $C_7H_5NaO_3$; mol wt 160.10. C 52.51%, H 3.15%, Na 14.36%, O 29.98%. It is at least 99.5% pure.

White, odorless crystals, scales or powder; becomes pinkish on long exposure to light. One gram dissolves in 0.9 ml water, 9.2 ml alcohol, 4 ml glycerol. The aq soln is slightly acid. pH 5-6. A satd aq soln contains 20.06% w/w at −1.5°. *See* Sidgwick, Ewbank, *J. Chem. Soc.* **121**, 1847, 1850 (1922). *Protect from light.* LD_{50} i.p. in rats: 780 mg/kg, E. I. Goldenthal, *Toxicol. Appl. Pharmacol.* **18**, 185 (1971).
Incompat: Ferric salts, lime water, spirit nitrous ether, mineral acids, iodine, lead acetate, silver nitrate, sodium phosphate in powder.
USE: As preservative for paste, mucilage, glue, hides.

THERAP CAT: Analgesic, antipyretic.
THERAP CAT (VET): Analgesic, antipyretic.

8820. Sodium Selenate. Na_2O_4Se; mol wt 188.94. Na 24.34%, O 33.87%, Se 41.79%. Na_2SeO_4.
Decahydrate, white crystals; very sol in water. LD_{100} (anhydrous) orally in rabbits: 4 mg/kg, *Handbook of Toxicology* vol. **1**, W. S. Spector, Ed. (Saunders, Philadelphia, 1956) pp 278-279.
USE: Insecticide in some horticultural applications.

8821. Sodium Selenide. Na_2Se; mol wt 124.94. Na 36.80%, Se 63.20%. Prepd by adding selenium to a soln of sodium in liquid ammonia: Hugot, *Compt. Rend.* **129**, 299 (1899); *Ann. Chim. Phys.* [7] **21**, 34 (1900); Feher in *Handbook of Preparative Inorganic Chemistry* vol. **1**, G. Brauer, Ed. (Academic Press, New York, 2nd ed., 1963) p 421.
Amorphous crystals. d^{10} 2.625. mp > 875°. Turns red on exposure to air and deliquesces. Dec in water. Insol in ammonia.
Hemienneahydrate, fine needles. Turns red on exposure to air and deliquesces.
Decahydrate, needles. Turns red and then brown on exposure to air.
Hexadecahydrate, prisms. mp 40°. Dec in air to sodium carbonate, selenium and a small amount of sodium selenide.

8822. Sodium Selenite. *Selenious acid disodium salt.* Na_2O_3Se; mol wt 172.94. Na 26.59%, O 27.75%, Se 45.66%. Na_2SeO_3. Prepd by evaporating an aqueous solution of sodium hydroxide and selenious acid between 60° and 100°: Krak, *J. Am. Ceram. Soc.* **12**, 530 (1929); by heating a mixture of sodium chloride and selenium oxide: Cameron, Macallan, *Proc. Roy. Soc.* **46**, 13 (1890). Metabolism: M. Sandholm, *Acta Pharmacol. Toxicol.* **33**, 6 (1973); H. W. Symonds *et al.*, *Brit. J. Nutr.* **45**, 117 (1981). Mutagenicity study: M. Nodo *et al.*, *Mutat. Res.* **66**, 175 (1979). Toxicity study: Cummins, Kimura, *Toxicol. Appl. Pharmacol.* **20**, 89 (1971).
Tetragonal prisms. Stable in air. Freely sol in water. Insol in alcohol. LD_{50} orally in rats: 7 mg/kg (Cummins, Kimura).
Pentahydrate, acicular crystals. Loses water of crystn in dry air.
USE: Removing green color from glass during its manuf; alkaloidal reagent.

8823. Sodium Sesquicarbonate. Urao; trona. $C_2H-Na_3O_6$; mol wt 190.00. C 12.64%, H 0.53%, Na 36.30%, O 50.53%. $Na_2CO_3.NaHCO_3$. Found in nature as the dihydrate, e.g., Owens Lake, Searles Lake (U.S.A.); Lake Magadi (Kenya). Produced on a large scale from sodium carbonate and a slight excess of sodium bicarbonate: Schenk in *Winnacker-Weingaertner, Chemische Technologie* vol. **I** (München, 1950) p 427.
Dihydrate, monoclinic needles, d 2.112. Crystals are stable in air. Soly in water (g/100 ml) at 0°: 13; at 100°: 42. Aq solns are mildly alkaline. pH of 0.1M soln = 10.1.
Caution: Irritating to skin, mucous membranes.
USE: Chiefly in laundering in conjunction with soap.

8824. Sodium Silicate. Water glass; soluble glass. Manuf: Faith, Keyes & Clark's *Industrial Chemicals,* F. A. Lowenheim, M. K. Moran, Eds. (Wiley-Interscience, New York, 4th ed., 1975) pp 755-761.
The compositions of the commonly available sodium silicates in dry form are: Na_2SiO_3, $Na_6Si_2O_7$, and $Na_2Si_3O_7$, with variable amounts of water, the first-named containing approx 5H_2O. They are in the form of colorless to white or grayish-white, crystal-like pieces or lumps. These silicates are very slightly sol or almost insol in cold water. They are best brought into soln by heating with water under pressure. They are less readily sol in large amounts of water than in small amounts of water, and the anhydr dissolve with more difficulty than the hydrated silicates; also, the silicates containing more sodium dissolve more readily. The aq solns are strongly alkaline. The dry sodium silicates are used relatively little.
See also Sodium Metasilicate.
Caution: Irritating and caustic to skin, mucous membranes. If swallowed causes vomiting and diarrhea.

USE: Lining Bessemer converters, acid concentrators. Manuf grindstones, abrasive wheels (as binder only).

8825. Sodium Silicate Solution. Sodium silicate solns of varying composition with respect to the ratio between sodium and silica, as well as of various densities, are available. One of the most commonly used sodium silicate solns, also known as "egg preserver", contains ~40% $Na_2Si_3O_7$. Solns of sodium silicate are strongly alkaline and are readily dec by acids with separation of silicic acid. The greater the ratio of Na_2O to SiO_2 (the greater the alkalinity) the more tacky is the soln.

USE: For preserving eggs; fireproofing fabrics; as a detergent in soaps; as adhesive; waterproofing walls; in cements; in cold-water paints; manuf of abrasive wheels; weighting silk, etc.

8826. Sodium Stannate(IV). *Sodium tin oxide.* Na_2O_3-Sn; mol wt 212.69. Na 21.62%, O 22.57%, Sn 55.81%. Na_2SnO_3.

Trihydrate, white or colorless crystals; gradually dec in the air; dec by weak acids. Sol in ~1.7 parts water; insol in alcohol. The aq soln is alkaline. *Keep well closed.*

USE: As mordant in dyeing and printing calico; for fireproofing of curtains, etc.

8827. Sodium Stearate. *Stearic acid sodium salt.* Approx $C_{18}H_{35}NaO_2$. Usually contains sodium palmitate.

White powder; soapy feel; slight, tallow-like odor. Slowly soluble in cold water or cold alcohol; freely sol in the hot solvents. The aq soln is strongly alkaline, due to hydrolysis; the alcohol soln is practically neutral.

USE: Pharmaceutic aid (emulsifying and stiffening agent). In glycerol suppositories; also in toothpaste; as waterproofing agent.

8828. Sodium Succinate. Succinic acid sodium salt; disodium succinate; Soduxin. $C_4H_4Na_2O_4$; mol wt 162.05. C 29.65%, H 2.49%, Na 28.37%, O 39.49%.

Hexahydrate, granules or crystalline powder; stable in air. Loses all its water at 120°. Sol in ~5 parts water; insol in alcohol. The aq soln is neutral or slightly alkaline. LD_{50} i.v. in mice: 4.5 g/kg, *Handbook of Toxicology* vol. 1, W. S. Spector, Ed. (Saunders, Philadelphia, 1956) pp 278-279.

THERAP CAT: Respiratory stimulant, analeptic; urinary alkalizer, diuretic; cathartic.

8829. Sodium Sulfate. Na_2O_4S; mol wt 142.04. Na 32.37%, O 45.06%, S 22.57%. Na_2SO_4. Occurs in nature as the minerals *mirabilite, thenardite.* Industrial production: Faith, Keyes & Clark's *Industrial Chemicals,* F. A. Lowenheim, M. K. Moran, Eds. (Wiley-Interscience, New York, 4th ed., 1975) p 762-768.

Anhydrous form, *salt cake* (technical grade). Powder or orthorhombic bipyramidal crystals, mp ~800°. d 2.7. Sol in ~3.6 parts water. Max soly at 33°: 1 in 2. Above this temp the soly gradually decreases and at 100° requires 2.4 parts water. Insol in alcohol.

Decahydrate, $Na_2O_4S.10H_2O$, *Glauber's salt.* Odorless, efflorescent crystals or granules, mp 32.4°. d 1.46. Loses all its water at 100°. Sol in 1.5 parts water at 25°, in 3.3 parts water at 15°. Soly in water decreased by NaCl. Sol in glycerol; insol in alcohol. The aq soln is neutral. pH 6-7.5. *Keep well closed in a cool place.*

USE: For standardizing dyes; in freezing mixtures; in dyeing and printing textiles. The *anhydrous* form for drying organic liquids; in Kjeldahl nitrogen determination; in manuf of glass, ultramarine, paper pulp.

THERAP CAT: Cathartic.

THERAP CAT (VET): Purgative.

8830. Sodium Sulfide. Sodium monosulfide; sodium sulfuret. Na_2S; mol wt 78.05. Na 58.91%, S 41.09%. Best prepd from the elements in liq ammonia, also obtained by dehydration of the nonahydrate: Courtois, *Compt. Rend.* **207**, 1220 (1938); Klemm *et al., Z. Anorg. Allgem. Chem.* **241**, 281 (1939); Feher in *Handbook of Preparative Inorganic Chemistry,* vol 1, G. Brauer, Ed. (Academic Press, New York, 2nd ed., 1963) pp 358-360. *Review:* C. Drum in Kirk-Othmer *Encyclopedia of Chemical Techology* vol. **21** (Wiley-Interscience, New York, 3rd ed., 1983) pp 256-262.

Cubic crystals or granules. Extremely hygroscopic. Discolors upon exposure to air. d_4^{14} 1.856. mp 1180° *(in vacuo)*; also reported as mp 920°. Soly in water (g/100 g H_2O): 8.1 (−9.0°); 12.4 (0°); 18.6 (20°); 29.0 (40°); 35.7 (48°); 39.0 (50°). Slightly sol in alcohol. Insol in ether. Aq solns are strongly alkaline.

Pentahydrate, flat, shiny, four-sided, prismatic crystals. Loses 3 mols water at 100°. mp 120° (with loss of all water of crystn). Freely sol in water. Also sol in alcohol. Aq solns are strongly alkaline. Insol in ether. Dec by acids with evolution of H_2S.

Nonahydrate, tetragonal, deliquescent crystals. Odor of hydrogen sulfide. Discolors upon exposure to light and air (first yellow, then brownish-black). d_4^{16} 1.427. mp ~50°. One gram dissolves in 0.5 ml water, also reported as 18 g dissolve in 100 ml water at 25°. Slightly sol in alcohol, insol in ether. Dec by acids, even by carbonic acid. Aqueous solns are very alkaline and upon standing in contact with air are slowly converted according to the equation: $2Na_2S + 2O_2 + H_2O \rightarrow Na_2S_2O_3 + 2NaOH$. Exposure of the crystals to air produces hydrogen sulfide according to the equation $Na_2S + H_2O + CO_2 \rightarrow Na_2CO_3 + H_2S$. *Keep well closed and in a cool place. Do not handle with bare hands.*

USE: In dehairing hides and wool pulling; desulfurizing viscose rayon; in the manuf of rubber, sulfur dyes; in ore flotation, metal refining, engraving, cotton printing, as chemical intermediate, as laboratory reagent; in paper-pulping process.

8831. Sodium Sulfite. Na_2O_3S; mol wt 126.04. Na 36.48%, O 38.08%, S 25.44%. Na_2SO_3.

Small crystals or powder. It is fairly stable and does not oxidize as readily as the hydrated sulfite. Sol in 3.2 parts water; sol in glycerol; almost insol in alcohol. pH about 9. *Keep well closed.* LD_{50} i.v. in mice: 175 mg/kg, Hoppe, Goble, *J. Pharmacol. Exp. Ther.* **101**, 101 (1951).

Heptahydrate, efflorescent crystals. Unstable, oxidizing in the air to sulfate. Sol in 1.6 parts water, about 30 parts glycerol; sparingly sol in alc. The aq soln is alkaline and dissolves sulfur. pH ~9. *Keep well closed in a cool place. Note:* The commercial heptahydrate (about 90% pure, the remainder consisting chiefly of the sulfate) has been largely replaced by the commercial anhydr salt (about 96-99% pure) which is more stable.

USE: Chiefly in photographic developers and instead of "hypo" for fixing prints; bleaching wool, straw, silk; generating SO_2; as reducer in manuf dyes; silvering glass; removing traces of Cl in bleached textiles and paper; preserving meat, egg yolks, etc.

8832. Sodium β-Sulfopropionitrile. *2-Cyanoethanesulfonic acid sodium salt;* sodium 2-cyanoethanesulfonate. $C_3H_4NNaO_3S$; mol wt 157.13. C 22.93%, H 2.57%, N 8.91%, Na 14.63%, O 30.55%, S 20.41%. $NaO_3SCH_2CH_2CN$.

Crystals, mp 243-244°. Sol in water, hot methanol, glacial acetic acid. Insol in most other organic solvents.

USE: In organic synthesis for introducing the sulfonic acid group into a variety of molecules, especially surface active agents.

8833. Sodium Tartrate. $C_4H_4Na_2O_6$; mol wt 194.05. C 24.76%, H 2.08%, Na 23.69%, O 49.47%. $Na_2C_4H_4O_6$.

Dihydrate, white crystals or granules. Loses its water at about 120°. d 1.82. Sol in about 3 parts cold water, 1.5 parts boiling water; insol in alcohol. The aq soln is slightly alkaline to litmus, pH 7 to 9.

USE: Dihydrate as standard for standardizing Karl Fischer reagent (determination of water).

THERAP CAT: Cathartic.

8834. Sodium Tellurate(IV). Sodium tellurite. Na_2O_3-Te; mol wt 221.58. Na 20.75%, O 21.66%, Te 57.59%. Na_2TeO_3.

White powder. Soluble in water. LD_{50} orally in mice: 20 mg/kg, *Toxic Substances List,* H. E. Christensen, Ed. (1973) p 906.

8835. Sodium Tellurate(VI). Na_2O_4Te; mol wt 237.58. Na 19.35%, O 26.94%, Te 53.71%. Na_2TeO_4.

White powder. Sol in 130 parts cold water, 50 parts boiling water. MLD i.p. in rats: 37.2-55.8 mg/kg, *Handbook of*

Toxicology vol. **1,** W. S. Spector, Ed. (Saunders, Philadelphia, 1956) pp 280-281.

8836. Sodium Tetrachloroaluminate. Aluminum sodium chloride; sodium chloroaluminate. $AlCl_4Na$; mol wt 191.78. Al 14.07%, Cl 73.94%, Na 11.99%. $AlCl_3.NaCl$. Prepn: *Gmelin's, Aluminum* (8th ed.) **35B,** p 376 (1934); Good, Batha, U.S. pat. **2,867,499** (1959 to Olin Mathieson). Yellowish powder. Sol in water.
USE: Catalyst for organic reactions.

8837. Sodium Tetrachloroaurate(III). Gold sodium chloride; sodium chloroaurate(III). $AuCl_4Na$; mol wt 361.77. Au 54.45%, Cl 39.20%, Na 6.35%. Prepd by evapn of a soln of $HAuCl_4$ and NaCl.
Dihydrate, orange-yellow, rhombic, bipyramidal crystals. Stable to 100°. Dec before all H_2O of hydration lost. Freely sol in water; sol in alcohol, ether.
USE: For paper-toning in photography, in electroplating baths.

8838. Sodium Tetradecyl Sulfate. *7-Ethyl-2-methyl-4-undecanol hydrogen sulfate sodium salt;* 7-ethyl-2-methyl-4-hendecanol sulfate sodium salt; sodium 2-methyl-7-ethyl-4-undecyl sulfate; sodium 7-ethyl-2-methylundecyl-4-sulfate; Sotradecol; Tergitol 4; Trombavar; Trombovar. $C_{14}H_{29}$-$NaSO_4$; mol wt 316.44. C 53.14%, H 9.24%, Na 7.27%, S 10.13%, O 20.22%.

White, waxy solid. Sol in water, alcohol, ether. The pH of a 5% soln is from 6.5 to 9.0. Surface tension (dynes/cm) of aq soln at 25°: 56.5 dynes/cm (0.05% w/w); 52 (0.10%); 47 (0.20%); 40 (0.50%); 35 (1.0%). LD_{50} orally in rats: 4.95 g/kg, H. F. Smyth, C. P. Carpenter, *J. Ind. Hyg. Toxicol.* **30,** 63 (1948).
USE: Wetting agent.
THERAP CAT: Sclerosing agent.

8839. Sodium Tetraphenylborate. Tetraphenylboron sodium; Kalignost. $C_{24}H_{20}BNa$; mol wt 342.22. C 84.23%, H 5.89%, B 3.16%, Na 6.72%. $Na[B(C_6H_5)_4]$. Prepn: Wittig, Raff, *Ann.* **573,** 195 (1950); U.S. pat. **2,853,525** (1958); Heyl, **Brit.** pat. **705,719** (1954); Kozlova, Pal'm, *Zh. Obshch. Khim.* **31,** 2922 (1961); Holtzapfel, Richter, *ibid.* **32,** 1358 (1962).
Snow-white crystals from chloroform. Freely sol in water, acetone. Less sol in ether, chloroform. Practically insol in petr ether. Aq solns should be adjusted to a pH of about 5 and can be stored at room temp or lower. Such solns have been stored at 45° for 5 days without deterioration. The soly in polar solvents increases as the temp decreases.
USE: As reagent for determination of potassium, ammonium, rubidium and cesium ions: Barnard, Buechl, *Chemist Analyst* **48,** 44, 49 (1959); Montequi, Serrano, *Anales Real Acad. Farm.* **26,** 107 (1960).

8840. Sodium Thioantimonate(V). Sodium sulfantimonate; Schlippe's salt. Na_3S_4Sb; mol wt 318.99. Na 21.62%, S 40.21%, Sb 38.17%. Na_3SbS_4.
Nonahydrate, colorless or light yellow, large crystals; strong alk reaction. On exposure to air becomes covered with a reddish-brown coat of antimony sulfide. Sol in 3 parts cold water, 1 part boiling water; insol in alc; dec by weak acids.

8841. Sodium Thiocarbonate. Sodium sulfocarbonate. CNa_2S_3; mol wt 154.19. C 7.79%, Na 29.82%, S 62.39%. Na_2CS_3.
USE: Usually supplied as a 40°Bé soln for destroying insects injurious to plants, especially vines.

8842. Sodium Thioglycolate. *Mercaptoacetic acid sodium salt;* sodium mercaptoacetate. $C_2H_3NaO_2S$; mol wt

114.10. C 21.05%, H 2.65%, Na 20.15%, O 28.04%, S 28.10%. $HSCH_2COONa$.
Hygroscopic crystals. Slight characteristic odor. Discolors on exposure to air or iron. Freely sol in water; slightly sol in alc. LD_{50} i.p. in rats: 148 mg/kg, Freeman, Rosenthal, *Fed. Proc.* **11,** 347 (1952).
USE: In cold-waving of hair; as depilatory; in bacteriology for the prepn of thioglycolate media; as analytical reagent, *see* Thioglycolic Acid.

8843. Sodium Thiophosphate. Sodium phosphorothioate; sodium monothiophosphate. Na_3O_3PS; mol wt 180.01. Na 38.31%, O 26.66%, P 17.21%, S 17.81%. Na_3PO_3S.
Conveniently prepd from $NaPO_3 + Na_3S$: Zintl, Bertram, *Z. Anorg. Allgem. Chem.* **245,** 16 (1940); from $PSCl_3 + NaOH$: Tridot, Tudo, *Bull. Soc. Chim. France* **1960,** 1231. Several alternate routes, e.g. from P_2S_5 and NaOH, *see* Klement in *Handbook of Preparative Inorganic Chemistry* vol. **1,** G. Brauer, Ed. (Academic Press, New York, 2nd ed., 1963) pp 569-570.
Dodecahydrate, thin, six-sided leaflets from water. Effloresces in dry air. mp 60°. Freely sol in warm water. Aq solns are strongly alkaline.

8844. Sodium Thiosulfate. Sodium hyposulfite; "hypo"; antichlor; Sodothiol; Sulfothiorine; Ametox. $Na_2O_3S_2$; mol wt 158.11. Na 29.08%, O 30.36%, S 40.56%. $Na_2S_2O_3$.
Review of mfg processes: Faith, Keyes & Clark's *Industrial Chemicals,* F. A. Lowenheim, M. K. Moran, Eds. (Wiley-Interscience, New York, 4th ed., 1975) pp 769-773. Crystal structure of the pentahydrate: A. A. Uraz, N. Armagan, *Acta Crystallogr.* **B33,** 1396 (1977). Toxicity of the pentahydrate: C. Voegtlin *et al., J. Pharmacol. Exp. Ther.* **25,** 297 (1925).
Powder. Sol in water; practically insol in alcohol.
Pentahydrate, odorless crystals or granules, mp 48° when rapidly heated. Effloresces in warm dry air; slightly deliquesces in moist air. d 1.69. Loses all its water at 100°; dec at higher temp. Sol in water; insol in alcohol. Slowly dec in aq soln at ordinary temp, more rapidly when heated. The aq soln is practically neutral. pH 6.5-8.0. It dissolves silver halides and many other salts of silver. *Incompat:* Iodine, acids; lead, mercury, and silver salts. LD i.v. in rats: > 2.5 g/kg (Voegtlin).
USE: To remove chlorine from solns according to the eq:
$Na_2S_2O_3 + 4Cl_2 + 5H_2O \rightarrow 2NaHSO_4 + 8HCl$ and $Na_2S_2O_3 + 2HCl \rightarrow 2NaCl + H_2O + S + SO_2$. As "antichlor" in bleaching of paper pulp; as fixer in photography; for extraction of silver from ores; as mordant in dyeing and printing textiles; reducer in chrome dyeing; manuf leather; bleaching bone, straw, ivory; also as reagent in analytical chemistry.
THERAP CAT: Antidote (cyanide).
THERAP CAT (VET): Antidote (cyanide). Has been used as a "general detoxifier", in bloat, and externally in ringworm, mange.

8845. Sodium Trimetaphosphate. *Metaphosphoric acid trisodium salt.* $Na_3O_9P_3$; mol wt 305.89. Na 22.55%, O 47.07%, P 30.38%. $(NaPO_3)_3$. Prepd by tempering sodium hexametaphosphate at about 500° for 8-12 hrs; by heating Na_2HPO_4 and NH_4NO_3 to 320° or by heating NaH_2PO_4 at 530°: Fleitmann, Henneberg, *Ann.* **65,** 304 (1848); v. Knoore, *Z. Anorg. Allgem. Chem.* **24,** 381 (1900); Karbe, Jander, *Kolloid-Beihefte* **54,** 35 (1942); Bell, *Inorg. Syn.* **3,** 103 (1950); Lee, Bond, *J. Appl. Chem.* **18,** 345 (1968). Crystal structure of anhydrate: Caglioti *et al., Atti Acad. Italia Rend.* **3** [7], 761 (1942); *C.A.* **40,** 6927[b] (1946); of anhydrate: Ondik, *Acta Cryst.* **18,** 226 (1965). Use in detergent processing: Shen, *J. Am. Oil Chem. Soc.* **45,** 510 (1968). Toxicity study: Behrens, Seelkopf, *Arch. Exp. Pathol. Pharmakol.* **169,** 238 (1932). *Reviews: see* Sodium Metaphosphate.
Hexahydrate, efflorescent, triclinic-rhombohedral prisms. d 1.786. mp 53°. Loses water on storage at 20°, becomes definitely anhydr at 100°. d (anhydr) 2.49. One gram dissolves in 4.5 ml water. Insol in alc. LD orally in mice: > 100 mg/kg (Behrens, Seelkopf).
USE: In detergent processing.

8846. Sodium Tripolyphosphate. *Triphosphoric acid pentasodium salt;* sodium triphosphate; tripolyphosphate; STPP; poly; pentasodium triphosphate. $Na_5O_{10}P_3$; mol wt

367.86. Na 31.25%, O 43.49%, P 25.26%. $Na_5P_3O_{10}$. Prepd by molecular dehydration of mono- and disodium phosphates: Audrieth, Bell, *Inorg. Syn.* **3**, 85 (1950). Toxicity study: H. F. Smyth *et al.*, *Am. Ind. Hyg. Assoc. J.* **30**, 470 (1969).

Slightly hygroscopic granules. Soly in water (g/100 ml) at 25°: 20; at 100°: 86.5. pH of 1% soln at 25° = 9.7-9.8. Its properties are between those of tetrasodium pyrophosphate and sodium metaphosphate. Stability: With prolonged heating of STPP solns, it tends to revert to the orthophosphate. More stable than the higher (i.e., meta-) phosphates, but less stable than tetrasodium pyrophosphate. LD_{50} orally in rats: 6.50 g/kg (Smyth).

Caution: Moderately irritating to skin, mucous membranes. Ingestion can cause violent purging.

USE: In water softening (calcium and magnesium hardness is sequestered from soln without precipitation); peptizing agent, emulsifier and dispersing agent, ingredient of cleansers in drilling fluids to control mud viscosity in oil fields; as preservative, sequestrant and texturizer in foods.

8847. Sodium Tungstate(VI). Na_2O_4W; mol wt 293.82. Na 15.65%, O 21.78%, W 62.57%. Na_2WO_4.

Dihydrate, colorless crystals or white, cryst powder; effloresces in dry air. Loses its water at 100°. Sol in about 1.1 parts water. Insol in alcohol. The aq soln is slightly alkaline, pH 8-9. LD orally in guinea pigs: 990 mg/kg, *Handbook of Toxicology* vol. **1**, W. S. Spector, Ed. (Saunders, Philadelphia, 1956) pp 280-281.

USE: Fireproofing and waterproofing fabrics; preparing complex compds, such as phosphotungstate, silicotungstate, etc.; as a reagent for biological products; precipitant for alkaloids, etc.

8848. Sodium Uranate(VI). Sodium diuranate; uranium oxide yellow; uranium yellow. $Na_2O_7U_2$; mol wt 634.03. Na 7.25%, O 17.66%, U 75.08%. $Na_2U_2O_7$.

Monohydrate, yellow powder. Insol in water; sol in acids.

USE: Manuf yellowish-green fluorescent glass; painting on porcelain and enameling.

8849. Sodium Vanadate(V). Sodium metavanadate. NaO_3V; mol wt 121.93. Na 18.85%, O 39.37%, V 41.78%. $NaVO_3$.

Tetrahydrate, yellowish-white, cryst powder. Sol in hot water. LD_{75} i.p. in rats: 4-5 mg V/kg, Franke, Moxon, *J. Pharmacol. Exp. Ther.* **58**, 454 (1936).

USE: In photography and manuf of inks.

8850. Sofalcone. *[5-[(3-Methyl-2-butenyl)oxy]-2-[3-[4-[(3-methyl-2-butenyl)oxy]phenyl]-1-oxo-2-propenyl]phenoxy]-acetic acid;* 2'-carboxymethoxy-4,4'-bis(3-methyl-2-butenyloxy)chalcone; Su-88; Solon. $C_{27}H_{30}O_6$; mol wt 450.53. C 71.98%, H 6.71%, O 21.31%. Substituted chalcone with anti-ulcer activity. Prepn: K. Kyogoku *et al.*, *Ger. pat.* **2,705,603**; *eidem*, U.S. pat. **4,085,135** (1977, 1978 both to Taisho); *eidem*, *Chem. Pharm. Bull.* **27**, 2943 (1979). Effects on exptl ulcer models: R. Saziki *et al.*, *Oyo Yakuri* **18**, 579 (1979), *C.A.* **92**, 174433h (1980). Absorption, distribution, excretion: T. Nozu *et al.*, *ibid.* 815, *C.A.* **92**, 191017t (1980). Reproduction studies: T. Yamada *et al.*, *ibid.* **19**, 543 (1980), *C.A.* **93**, 230785j (1980). Toxicity study: Y. Tarumoto *et al.*, *Yakuri To Chirgo* **10**, 61 (1982), *C.A.* **96**, 193205j (1982). Effect on gastric blood flow: Y. Matsuo *et al.*, *Arzneimittel-Forsch.* **33**, 242 (1983). Gastric cytoprotective effect in rats: T. Suwa *et al.*, *Japan. J. Pharmacol.* **35**, 47 (1984); S. Kishimoto *et al.*, *Arzneimittel-Forsch.* **31**, 944 (1987).

Light yellow needles from ethanol, mp 143-144°. LD_{50} in mice, rats: >10 g/kg orally (Tarumoto).

THERAP CAT: Antiulcerative.

8851. Solan. *N-(3-Chloro-4-methylphenyl)-2-methylpentanamide;* 3'-chloro-2-methyl-p-valerotoluidide; Niagara 4512. $C_{13}H_{18}ClNO$; mol wt 239.74. C 65.13%, H 7.57%, Cl 14.79%, N 5.84%, O 6.67%. Prepn: Dorschner *et al.*, **Brit. pat.** 869,169; U.S. pat. 3,020,142 (1961, 1962 both to FMC). Toxicity study: G. W. Bailey, J. L. White, *Residue Rev.* **10**, 97 (1965).

Crystals, mp 79-80°. Sol in pine oil, diisobutylketone, isophorone, xylene. Practically insol in water. LD_{50} orally in rats: 10000 mg/kg (Bailey, White).

USE: Herbicide.

8852. Solanidine. *(3β)-Solanid-5-en-3-ol;* solatubine. $C_{27}H_{43}NO$; mol wt 397.64. C 81.55%, H 10.90%, N 3.52%, O 4.02%. From *Solanum* spp., Solanaceae. Potato sprouts contain about 0.008%. Obtained by hydrolysis of solanine. Extraction procedure: Soltys, Wallenfels, *Ber.* **69**, 811 (1936). Review on structure: Reichstein, Reich, *Ann. Rev. Biochem.* **15**, 155 (1946). Stereochemistry: Rosen, Rosen, *Chem. & Ind. (London)* **1954**, 1581. Revised stereochemistry: Höhne *et al.*, *Tetrahedron* **22**, 673 (1966). Synthesis: Schreiber, Roensch, *Ber.* **97**, 2362 (1964); *Tetrahedron* **21**, 645 (1965); Kessar *et al.*, *ibid.* **27**, 2153 (1971).

Long needles from chloroform-methanol, mp 218-219°. Sublimes near mp with slight decompn. $[\alpha]_D^{21}$ −29° (c = 0.5 in chloroform). Freely sol in benzene, chloroform; slightly in alcohol, methanol; almost insol in ether, water.

Hydrochloride, $C_{27}H_{43}NO.HCl$, prisms from 80% alcohol, dec 345°.

Methyliodide, $C_{27}H_{43}NO.CH_3I$, crystals from 50% alcohol, dec 286°.

Acetylsolanidine, $C_{29}H_{45}NO_2$, crystals from alc, mp 208°.

8853. Solanine. Solatunine. $C_{45}H_{73}NO_{15}$; mol wt 868.07. C 62.26%, H 8.48%, N 1.61%, O 27.65%. From *Solanum* species, especially from *S. tuberosum* L. (potato), *S. nigrum* L. (woody nightshade), and *Lycopersicon esculentum* Mill., Solanaceae (tomato). Fresh potato sprouts contain about 0.04%. Extraction procedure: Soltys, Wallenfels, *Ber.* **69**, 811 (1936). Separated by chromatography into six components: α-, β-, γ-solanine and α-, β-, γ-chaconine. Main constituent α-solanine: Kuhn, Löw, *Angew. Chem.* **66**, 639 (1954); Kuhn *et al.*, *Ber.* **88**, 1492 (1955); Paseshnichenko, Guseva, *Biokhimiya* **21**, 585 (1956); *eidem*, *ibid.* **22**, 843 (1957). Pharmacology: Nishie *et al.*, *Toxicol. Appl. Pharmacol.* **19**, 81 (1971).

Slender needles from 85% alc, browns and sinters about 190°, dec about 285°. $[\alpha]_D^{20}$ −60° (pyridine). pK at 15° = 6.66; K = 2.2 × 10⁻⁷. Readily sol in hot alc; practically insol in water (25 mg/l at pH 6.0), ether, chloroform.

Hydrochloride, $C_{45}H_{73}NO_{15}$·HCl, usually amorphous and gummy, but has been crystallized, dec about 212°. Sol in water. Has been used as an agricultural insecticide. LD₅₀ i.p. in mice: 42 mg/kg.

8854. Solanocapsine. *(3β,5α,16α,22α,23β,25β)-3-Amino-16,23-epoxy-16,28-secosolanidan-23-ol; 3β-amino-22,26-imino-16β,23-oxido-5α,22α,23β,25α-cholestan-23-ol; 3β-amino-22,26-imino-16β,23-oxido-5α,22R,-23S,25R-cholestan-23-ol.* $C_{27}H_{46}N_2O_2$; mol wt 430.67. C 75.30%, H 10.77%, N 6.50%, O 7.43%. From *Solanum pseudocapsicum* L., *S. capsicastrum* Link, *S. hendersonii* hort., *Solanaceae*. Isolation: Barger, Fraenkel-Conrat, *J. Chem. Soc.* **1936**, 1537; K. Schreiber, H. Ripperger, *Z. Naturforsch.* **17B**, 217 (1962). Structure: *eidem, Experientia* **16**, 536 (1960). Revised structure and synthesis: *eidem, Ann.* **723**, 159 (1969). Stereochemistry: *eidem, Tetrahedron Letters* no. 27, 9 (1960); *Ann.* **655**, 114 (1962). Synthesis: Ripperger *et al., Tetrahedron Letters* **1970**, 5251; *eidem, Tetrahedron* **28**, 1629 (1972).

Monohydrate, needles from methanol, mp 213-215°. $[\alpha]_D^{20}$ +26.3° (c = 1.80 in methanol); $[\alpha]_D^{20}$ +24.9° (c = 2.13 in methanol).

N'-Acetylsolanocapsine, $C_{29}H_{48}N_2O_3$, needles from ethanol, mp 233-234°. $[\alpha]_D^{23}$ −33.7° (c = 1.59 in pyridine).

N,N'-Diacetylsolanocapsine, $C_{31}H_{50}N_2O_4$, crystals from chloroform + acetone, mp 281-283°. $[\alpha]_D^{22}$ −38.6° (c = 1.31 in pyridine).

N,N,N'-Trimethylsolanocapsine, $C_{30}H_{52}N_2O_2$, needles from abs methanol, mp 222-224°. $[\alpha]_D^{22}$ +53.2° (c = 1.09 in pyridine).

N-Benzylidenesolanocapsine, $C_{34}H_{50}N_2O_2$, needles from pyridine + abs methanol, mp 249-251°. $[\alpha]_D^{18}$ +31.4° (c = 0.78 in pyridine).

8855. Solanone. *[S-(E)]-8-Methyl-5-(1-methylethyl)-6,8-nonadien-2-one;* L-(+)-2-methyl-5-isopropyl-1,3-nonadien-8-one. $C_{13}H_{22}O$; mol wt 194.32. C 80.35%, H 11.41%, O 8.23%. Stereochemically unique terpene. Isoln from tobacco, structure and synthesis: Johnson, Nicholson *J. Org. Chem.* **30**, 2918 (1965). Stereoselective synthesis: A. Kohda, T. Sato, *Chem. Commun.* **1981**, 951.

Liquid. bp₁ 60°. n_D^{20} 1.4755. d_4^{20} 0.870. $[\alpha]_D^{23}$ +13.6° (neat). uv max (ethanal and hexane): 230 nm (log ε 4.07).

8856. Solanum. Bull nettle; radical weed; sand-brier; horse nettle. Air-dried ripe fruit of *Solanum carolinense* L., *Solanaceae*. *Habit.* South America, Florida and other sections of U.S.A. *Constit.* Solanine, solanidine.

THERAP CAT: Sedative, anticonvulsant.

8857. Solasodine. *(3β,22α,25R)-Spirosol-5-en-3-ol; solasod-5-en-3β-ol;* Δ⁵-20β_F,22α_F,25α_F,27-azaspirosten-3β-ol; solancarpidine; solanidine-S; purapuridine. $C_{27}H_{43}NO_2$; mol wt 413.64. C 78.40%, H 10.48%, N 3.39%, O 7.74%. Steroidal alkaloid isolated from various *Solanum* species. By hydrolysis of solasonine: Rochelmeyer, *Arch. Pharm.* **277**, 329 (1939). *See also* ref *under* Solasonine and Solanidine. Structure: Briggs *et al., J. Chem. Soc.* **1950**, 3013. Synthesis: Uhle, *J. Org. Chem.* **27**, 656 (1962); Schreiber, Rönsch, *Tetrahedron* **20**, 1939 (1964); Kessar *et al., ibid.* **27**, 2869 (1971).

Hexagonal plates from methanol or by sublimation in high vacuum, mp 200-202°. $[\alpha]_D^{25}$ −98° (c = 0.14 in methanol); $[\alpha]_D$ −113° (CHCl₃). Alkaline reaction to litmus in alcoholic soln. pKb 6.30. Absorption spectrum: Rochelmeyer, *loc. cit.* Freely sol in benzene, pyridine, and chloroform; moderately sol in alcohol, methanol, and acetone; slightly sol in water; practically insol in ether.

USE: Starting material for steroidal drugs.

8858. Solasonine. *(3β,22α,25R)-Spirosol-5-en-3-yl O-6-deoxy-α-L-mannopyranosyl-(1→2)-O-[β-D-glucopyranosyl-(1→3)]-β-D-galactopyranoside;* solanine-S; purapurine. $C_{45}H_{73}NO_{16}$; mol wt 884.07. C 61.14%, H 8.32%, N 1.58%, O 28.96%. From *Solanum aviculare* Forst F., *S. sodomeum* L., *S. xanthocarpum* Schrad & Wendl., *Solanaceae*: Bell, Briggs, *J. Chem. Soc.* **1942**, 1; Briggs *et al., ibid.* **1942**, 3; Kuhn, Löw, *Ber.* **88**, 289 (1955). Isoln from other *S.* spp: Briggs, Cambie, *J. Chem. Soc.* **1958**, 1422; Briggs *et al., ibid.* **1961**, 4645. Paper chromatography: Szendey, *Arch. Pharm.* **290**, 563 (1957). Structure: Briggs *et al., J. Chem. Soc.* **1963**, 2848. Enzymic decompn: Guseva, Pasechnichenko, *Biokhimiya* **24**, 563 (1959).

Consult the Name Index before using this section.

Needles from methanol, sinters 296°, mp 301-303°. $[\alpha]_D^{23}$ −88° (c = 1.01 in pyridine); $[\alpha]_D^{22}$ −74.5° (c = 0.51 in methanol). Sol in hot alc, hot dioxane; slightly sol in hot water, dil acetic acid. Practically insol in chloroform, ether.

8859. Solasulfone. *1,1'-[Sulfonylbis(4,1-phenyleneimino)]bis[3-phenyl-1,3-propanedisulfonic acid] tetrasodium salt;* di(γ-phenylpropylamino)-4,4'-diphenylsulfone α,α',γ,γ'-sodium tetrasulfonate; tetrasodium 4,4'-bis(γ-phenylpropylamino)diphenylsulfone α,γ,α',γ'-tetrasulfonate; phenopryldiasulfone sodium; solapsone; RP-3668; Cimedone; Sulphetrone. $C_{30}H_{28}N_2Na_4O_{14}S_5$; mol wt 892.85. C 40.36%, H 3.16%, N 3.14%, Na 10.30%, O 25.09%, S 17.96%. Prepared by reacting 4,4'-diaminodiphenyl sulfone with more than 2 mols cinnamic aldehyde in the cold and treating the resulting 4,4'-dicinnamylideneaminodiphenyl sulfone with $NaHSO_3$ soln: Gray, Henry, **Brit.** pat. 491,265 (1938 to Wellcome Foundation); Buttle *et al., Biochem. J.* **32**, 1101 (1938). Improved process: Henry, Gray, **Brit.** pat. 562,216 (1944 to Wellcome Foundation). Properties and estimation: Brownlee *et al., Brit. J. Pharmacol.* **3**, 15 (1948).

Crystals contg 5-7% H_2O from alc. Becomes anhydr at 110° *in vacuo.* Exceedingly sol in cold water, giving a neutral soln. 20% and 40% (w/v) aq solns are stable when neutral or slightly alkaline and may be autoclaved. Practically insol in cold alc. Solns are incompatible with acids. Absorption spectrum of dye obtained by diazotization and coupling to *N*-(1-naphthyl)ethylenediamine hydrochloride: Brownlee, *loc. cit.*

THERAP CAT: Antibacterial (leprostatic).

8860. Soman. *Methylphosphonofluoridic acid 1,2,2-trimethylpropyl ester;* pinacoloxymethylphosphoryl fluoride; pinacolyl methylphosphonofluoridate; GD. $C_7H_{16}FO_2P$; mol wt 182.18. C 46.15%, H 8.85%, F 10.43%, O 17.56%, P 17.00%. Nerve gas. Prepn: Holmstedt, *Acta Physiol. Scand.* **25**, suppl. 90, p 106 (1951); Protar **16**, 131 (1950); G. Schrader, *Die Entwicklung neuer Insektizide auf Grundlage organischer Fluor- und Phosphor-Verbindungen* [Monographie Nr. 62 zu Angewandte Chemie und Chemie-Ingenieur-Technik (Verlag Chemie 1951)]; B. C. Saunders, *Some Aspects of the Chemistry and Toxic Action of Organic Compounds Containing Phosphorus and Fluorine* (Cambridge, 1957) p 94. Toxicity study: Loomis, Salafsky, *Toxicol. Appl. Pharmacol.* **5**, 685 (1963).

LD_{50} in mice (mg/100 g): 0.062 i.p.; 0.78 dermally (Loomis, Salafsky). Lethal dose for man may be as low as 0.01 mg/kg, *Chem. & Eng. News* **31**, 4676 (1953).

Caution: A potent cholinesterase inhibitor. Toxic effects similar to, but more intense than, parathion, *q.v. See also* Sarin and Tabun.

USE: Chemical warfare agent.

8861. Somatoliberin. *Growth hormone-releasing factor;* GH-RF; GH-RH; GRF; growth hormone-releasing hormone; hGRF; hpGRF; somatocrinin. Stimulatory growth-hormone releasing factor of the hypothalamus that mediates, together with somatostatin, *q.v.,* the neuroregulation of somatotropin secretion. The concept of hypothalamic regulation of growth hormone release was postulated on the basis of physiological and biochemical evidence, *cf.* A. V. Schally *et al., Recent. Progr. Horm. Res.* **24**, 497 (1968); J. B. Martin in *Frontiers in Neuroendocrinology,* L.

Martini, W. F. Ganong, Eds. (Raven Press, New York, 1976) pp 129-168. Demonstration of the existence of a GH-RH: R. Deuben, J. Meites, *Endocrinology* **74**, 408 (1964); A. V. Schally *et al., ibid.* **82**, 271 (1968); E. Dickermann *et al., Neuroendocrinology* **4**, 75 (1969). Isoln of a decapeptide originally believed to be GRF: A. V. Schally *et al., Endocrinology* **84**, 1493 (1969). Review of early literature: *eidem, Science* **179**, 341 (1973). Isoln and characterization of hpGRF, a fully bioactive, 44 amino acid peptide from human pancreatic tumor: R. Guillemin *et al., Science* **218**, 585 (1982); J. Rivier *et al., Nature* **300**, 276 (1982). Cloning and sequence analysis of cDNA for hpGRF precursor: U. Gubler *et al., Proc. Natl. Acad. Sci. USA* **80**, 4311 (1983). Expression-cloning of cDNA for hpGRF: K. E. Mayo *et al., Nature* **306**, 86 (1983). Isoln, primary structure and synthesis of native human hypothalamic GRF (hGRF), identity with hpGRF and comparison with GRF from other species: N. Ling *et al., Proc. Natl. Acad. Sci. USA* **81**, 4302 (1984). Immunohistochemical detection of GRF in brain: B. Bloch *et al., ibid.* **301**, 607 (1983). Potent interaction of glucocorticoids with GRF *in vivo:* W. B. Wehrenberg *et al., Science* **221**, 556 (1983). Series of articles on clinical pharmacology: *Hormone Res.* **22**, 32-57 (1985). Clinical evaluation in treatment of hypopituitary dwarfism: M. O. Thorner *et al., N. Engl. J. Med.* **312**, 4 (1985). *Reviews:* R. Guillemin *et al., Recent Progr. Horm. Res.* **40**, 233-299 (1984); W. B. Wehrenberg *et al., Ann. Rev. Pharmacol. Toxicol.* **25**, 463-483 (1985).

Porcine GRF is destroyed by trypsin, chymotrypsin, pepsin. Biological activity remains after incubation at pH 8.5 for 4 hours at 37°, but is destroyed after similar incubation at pH 2.

8862. Somatomedins. A family of peptide hormones with mitogenic properties and insulin-like effects that mediates the action of growth hormone (somatotropin, *q.v.*) on skeletal tissue. They are synthesized in liver, pituitary, brain and in various fetal organs. Discovered and originally termed *"sulphation factors"* because of their ability to stimulate in cartilage the incorporation of sulfate into chondroitin sulfate, *cf.* W. D. Salmon, W. H. Daughaday, *J. Lab. Clin. Med.* **49**, 825 (1957). The designation *somatomedin* was later proposed to connote the hormonal and intermediary relationship to somatotropin: W. H. Daughaday *et al., Nature* **235**, 107 (1972). Within the group, several somatomedins have been described, all closely related chemically and biologically and having mol wts of about 7,500: *somatomedin A (SM-A),* *somatomedin C (SM-C),* *insulin-like growth factor I (IGF-I),* *insulin-like growth factor II (IGF-II),* and *multiplication stimulating activity (MSA).* The two IGF peptides have also been referred to as *NSILA-S,* or nonsuppressible insulin-like acting substance. Biological activity and purification of human SM-A: K. Hall, *Acta Endocrinol.* **Suppl. 163,** 1 (1972); K. Uthne, *ibid.* **Suppl. 175,** 1 (1973); L. Fryklund *et al., Biochem. Biophys. Res. Commun.* **61**, 957 (1974). Purification of MSA from rat liver culture medium: N. C. Dulak, H. M. Temin, *J. Cell. Physiol.* **81**, 153 (1973). Purification of SM-C from human plasma: J. J. Van Wyk *et al., Recent Progr. Horm. Res.* **30**, 259 (1974); M. E. Svoboda *et al., Biochemistry* **19**, 790 (1980). Isoln, chemical characterization, biological properties of IGF-I and IGF-II: E. Rinderknecht, R. E. Humbel, *Proc. Nat. Acad. Sci. USA* **73**, 2365, 4379 (1976). Amino acid sequence of IGF-I: *eidem, J. Biol. Chem.* **253**, 2769 (1978); of IGF-II: *eidem, FEBS Letters* **89**, 283 (1978). Homology of rat MSA with human IGF-II: H. Marquardt *et al., J. Biol. Chem.* **256**, 6859 (1981). Identity of SM-C with IGF-I: D.G. Klapper *et al., Endocrinology* **112**, 2215 (1983). Total synthesis of human IGF-I: C. H. Li *et al., Proc. Natl. Acad. Sci. USA* **80**, 2216 (1983); of human IGF-II: *eidem, Biochem. Biophys. Res. Commun.* **127**, 420 (1985). All the somatomedins circulate in plasma in high molecular weight forms because of their binding to specific carrier proteins, *cf.* A. C. Moses *et al., Endocrinology* **104**, 536 (1979); D. J. Knauer *et al., J. Supramol. Struct.* **15**, 177 (1981). Although the serum concentrations of somatomedins are growth hormone-dependent, there is evidence that somatotropin levels may be influenced, in turn, by somatomedins through a "negative feedback loop" by stimulating release of somatostatin, *q.v.,* a GH-release-inhibiting polypeptide: M. Berelowitz *et* al., Science 212, 1279 (1981); see also G. S. Tannenbaum et al., ibid. 220, 77 (1983). Developmental patterns of IGF-I and IGF-II synthesis and regulation in rat fibroblasts: S. O. Adams et al., Nature 302, 150 (1983). Reviews: M. Binoux, Ann. Endocrinol. 41, 157-192 (1980); D. R. Clemmons, J. J. Van Wyk, Handbook Exp. Pharmacol. 57, 161-208 (1981); C. D. Scher et al., Progr. Cancer Res. Ther. 23, 57-71 (1982); K. A. Thomas, Ann. Rep. Med. Chem. 17, 219-228 (1982); E. R. Froesch et al., Ann. Rev. Physiol. 47, 443-467 (1985); W. E. Russell, Semin. Liver Dis. 5, 46-58 (1985).

8863. Somatostatin. Growth hormone-release inhibiting factor; GH-RIF; somatotropin release inhibiting factor; SRIF; SRIF-14. $C_{76}H_{104}N_{18}O_{19}S_2$; mol wt 1637.90. C 55.73%, H 6.40%, N 15.39%, O 18.56%, S 3.92%. Widely occurring cyclic tetradecapeptide that mediates, together with somatoliberin, q.v., the neuroregulation of somatotropin secretion. Inhibits release of growth hormone, insulin and glucagon. Also a potent inhibitor in a number of systems, including central and peripheral neural, gastrointestinal, and vascular smooth muscle. Isoln from ovine hypothalamic extracts: P. Brazeau et al., Science 179, 77 (1973). Structure: Burgus et al., Proc. Nat. Acad. Sci. USA 70, 684 (1973); Ling et al., Biochem. Biophys. Res. Commun. 50, 127 (1973). Synthesis: Rivier et al., Compt. Rend. Ser. D 276, 2737 (1973); Sarantakis, McKinley, Biochem. Biophys. Res. Commun. 54, 234 (1973); Yamashiro, Li, ibid. 882; Coy et al., ibid. 1267; A. M. Felix et al., Int. J. Peptide Protein Res. 15, 342 (1980); B. Hartrodt et al., Pharmazie 37, 403 (1982). Synthesis and bacterial expression of somatostatin gene: K. Itakura et al., Science 198, 1056 (1977); K. Itakura, A. D. Riggs, Belg. pat. 871,782; K. Itakura, U.S. pat. 4,356,270; A. D. Riggs, U.S. pat. 4,366,246 (1979, 1982, 1982 all to Genentech). Isoln of a 28-amino acid somatostatin, termed SRIF-28, from porcine gastrointestinal tract: L. Pradayrol et al., Biochem. Biophys. Res. Commun. 85, 701 (1978). Primary structure: eidem, FEBS Letters 109, 55 (1980). Synthesis: E. Wunsch, Z. Naturforsch. B 35, 911 (1980). Isoln of a 25-amino acid somatostatin, SRIF-25, and structure: P. Brazeau et al., Compt. Rend. Ser. D 290, 1369 (1980). Both of these peptides consist of an N-terminal extension of SRIF-14. They are more potent than SRIF-14 in inhibition of insulin release but less potent in inhibition of glucagon release: L. Mandarino et al., Nature 291, 76 (1981). Synthetic linear SRIF-14 exhibits the same bioactivity as the natural cyclic form: P. Brazeau et al., Endocrinology 94, 184 (1974). Activity studies in humans: T. M. Siler et al., J. Clin. Endocrinol. Metab. 37, 632 (1973). Inhibitory effects on the secretion of thyrotropin: Vale et al., Fed. Proc. 32, 211 (1973); of insulin: K. G. Alberti et al., Lancet 2, 1299 (1973); Koeker et al., Science 184, 482 (1974). Role in diabetes: Maugh, ibid. 188, 920 (1975). Clinical trial in controlling acute variceal hemorrhage: S. A. Jenkins et al., Brit. Med. J. 290, 275 (1985). Review of distribution, secretion, physiology of gastrointestinal somatostatin: C. H. S. McIntosh, Life Sci. 37, 2043-2058 (1985). Symposium of biosynthesis, bioactivity, and clinical applications: Adv. Exp. Med. Biol. 188, 1-524 (1985). Reviews: Vale et al., Recent Progr. Horm. Res. 31, 365-397 (1975); Guillemin, Gerich, Ann. Rev. Med. 27, 379-388 (1976); R. L. Moss, Ann. Rev. Physiol. 41, 617 (1979); A. Arimura, J. B. Fishback, Neuroendocrinology 33, 246-256 (1981); several authors in Gut Hormones, S. R. Bloom, J. M. Polack, Eds. (Churchill, New York, 2nd ed., 1981) 605 pp; S. M. McCann, Ann. Rev. Pharmacol. Toxicol. 22, 491-515 (1982); N. Bethge et al., J. Clin. Chem. Clin. Biochem. 20, 603-613 (1982). Book: Somatostatin Vol. 2, M. T. McQuillan, Ed. (Eden Press, Quebec, 1980) 238 pp.

Ala–Gly–Cys–Lys–Asn–Phe–Phe–Trp
| |
Cys–Ser–Thr–Phe–Thr–Lys

Acetate, SRIF-A, Aminopan, Modustatina, Somatofalk, Stilamin.

THERAP CAT: Treatment of severe, acute hemorrhage of gastro-duodenal ulcers. Treatment of erosive or hemorrhagic gastritis. Experimental antidiabetic. Growth hormone inhibitor.

8864. Somatotropin. Adenohypophyseal growth hormone; GH; hypophyseal growth hormone; anterior pituitary growth hormone; growth hormone; phyone; pituitary growth hormone; somatotropic hormone; STH. Species-specific anabolic protein that promotes somatic growth, stimulates protein synthesis, and regulates carbohydrate and lipid metabolism. Increases serum levels of somatomedins, q.v. Secreted by the anterior pituitary under the regulation of the hypothalamic hormones, somatoliberin and somatostatin, q.q.v. Growth hormones from various species differ in amino acid sequence, antigenicity, isoelectric point, and in the range of animals in which they can produce biological responses. Isoln from bovine anterior pituitary substance: C. H. Li et al., J. Biol. Chem. 159, 353 (1945); A. E. Wilhelmi et al., ibid. 176, 735 (1948); C. H. Li, U.S. pat. 3,118,815 (1964 to Upjohn). Isoln from human pituitaries: Lewis, Brink, U.S. pat. 2,974,088 (1961 to Merck & Co., Inc.); Reisfeld et al., Endocrinology 71, 559 (1962). Amino acid sequence of human growth hormone (HGH): C. H. Li et al., J. Am. Chem. Soc. 88, 2050 (1966); eidem, Arch. Biochem. Biophys. 133, 70 (1969). Total synthesis: C. H. Li, D. Yamashiro, J. Am. Chem. Soc. 92, 7608 (1970). Revised sequence of HGH: H. D. Niall, Nature New Biol. 230, 90 (1971). Review of structural studies and comparison with the lactogenic hormones: H. D. Niall et al., Recent Progr. Horm. Res. 29, 387-416 (1973). Review of protein chemistry: C. H. Li, Mol. Cell Biochem. 46, 31-41 (1982). Cloning and expression of cDNA for HGH in E. coli: D. V. Goeddel et al., Nature 281, 544 (1979); D. V. Goeddel, H. L. Heyneker, Belg. pat. 884,012; eidem, U.S. pat. 4,342,832 (1980, 1982 both to Genentech); in mammalian cells: G. N. Pavakis et al., Proc. Nat. Acad. Sci. USA 78, 7398 (1981). Purification and bioactivity of recombinant HGH: K. C. Olson et al., Nature 293, 408 (1981). Clinical comparison of natural and recombinant HGH: R. L. Hintz et al., Lancet 1, 1276 (1982). Use of recombinant bovine GH to increase milk production in cows: D. E. Bauman et al., J. Dairy Sci. 68, 1352 (1985). Discussion of use of recombinant GH in applied animal agriculture: J. L. Burton, B. W. McBride, J. Agric. Ethics 2, 129-159 (1989). Review of GH gene studies: D. D. Moore et al., Recent Progr. Horm. Res. 38, 197-225 (1982). Review of bioregulation of HGH secretion: D. G. Johnston et al., J. Roy. Soc. Med. 78, 319-327 (1985); of mechanism of action on target cells: O. G. P. Isaksson et al., Ann. Rev. Physiol. 47, 483-499 (1985). Symposium on pharmacology and clinical efficacy: Hormone Research 33, Suppl. 4, 1-107 (1990). Books: Growth and Growth Hormone, A. Pecile, E. Muller, Eds. (Excerpta Medica, Amsterdam, 1972); Human Growth Hormone and Gonadotrophins in Health and Disease, P. Franchimont, H. Burger, Eds. (Elsevier, New York, 1975) 494 pp; Growth Hormone and Other Biologically Active Peptides, A. Pecile, E. Muller, Eds. (Elsevier, New York, 1980); Evaluation of Growth Hormone Secretion, Z. Laron, O. Butenandt, Eds. (S. Karger, New York, 1983); Use of Somatotropin in Livestock Production, K. Sejrsen et al., Eds. (Elsevier, New York, 1989).

Human growth hormone, $C_{990}H_{1529}N_{263}O_{299}S_7$; HGH, somatropin, CB-311, Asellacrin, Crescormon, Genotropin, Grorm, Humatrope, Nanormon, Norditropin, Nutropin, Saizen, Umatrope. A single polypeptide chain of 191 amino acids having a molecular weight of 22,124. Isoelectric point 4.9. $[\alpha]^{25}$ −38.7° (0.1M acetic acid).

Methionyl human growth hormone, $C_{995}H_{1537}N_{263}O_{301}S_8$, somatrem, met-HGH, Protropin, Somatonorm. Produced in bacteria from recombinant DNA. Contains the complete amino acid sequence of the natural hormone plus an additional N-terminal methionine.

Bovine somatotropin. One of the four naturally occurring molecular variants is known as somavubove, $C_{976}H_{1533}N_{265}O_{286}S_8$. Several variants have been produced by recombinant DNA technology: somagrebove, $C_{987}H_{1550}N_{268}O_{291}S_9$, CL-291894, Quest; sometribove, $C_{978}H_{1537}N_{265}O_{286}S_9$; somidobove, $C_{1020}H_{1596}N_{274}O_{302}S_9$, EL-349, LY-177837, Optiflex.

Porcine somatotropin, Somacton. Several variants have been produced by recombinant DNA technology: somalapor, $C_{977}H_{1527}N_{265}O_{287}S_7$; somenopor, $C_{938}H_{1469}N_{255}O_{275}S_7$; sometripor, $C_{979}H_{1527}N_{265}O_{287}S_8$; somfasepor, $C_{938}H_{1465}N_{257}O_{278}S_6$, Grolean, Leanstar.

THERAP CAT: Growth stimulant.

THERAP CAT (VET): Growth stimulant. Bovine somatotropin as galactopoietic.

8865. Songorine. *(1α,15β)-21-Ethyl-1,15-dihydroxy-4-methyl-16-methylene-7,20-cycloveatchan-12-one;* napellonine; zongorine. $C_{22}H_{31}NO_3$;mol wt 357.49. C 73.92%, H 8.74%, N 3.92%, O 13.43%. From *Aconitum songoricum* Popov, *Ranunculaceae:* Yunusov, *J. Gen. Chem. USSR* **18**, 515 (1948); Kuzovkov, *ibid.* **23**, 504 (1953); **25**, 2006 (1955). Identity with napellonine: Kuzovkov, *Zh. Obshch. Khim.* **28**, 2283 (1958); **29**, 1728 (1959). Structure: Sugasawa, *Chem. Pharm. Bull.* **9**, 889, 897 (1961). Absolute configuration: Okamoto *et al., ibid.* **13**, 1270 (1965). Synthesis of the aromatic intermediate: Wiesner *et al., Can. J. Chem.* **51**, 3978 (1973). Pharmacology: Sadritdinov, *Farmakol. Alk.* No. 312 (1965), *C.A.* **66**, 93772d (1967). Mass spectra data: Yunusov *et al., Khim. Prir. Soedin.* **6**, 101 (1970), *C.A.* **73**, 131178u (1970).

Crystals, mp 201-202°. $[\alpha]_D^{20}$ −135.4°. uv max: 290 nm (log ε 2.6). LD_{50} in mice: 142.5 mg/kg.

Hydrochloride, crystals, mp 257-258°. $[\alpha]_D^{20}$ −114° (c = 2 in water).

8866. Sophorabioside. *3-[4-[[2-O-(6-Deoxy-α-L-mannopyranosyl)-β-D-glucopyranosyl]oxy]phenyl]-5,7-dihydroxy-4H-1-benzopyran-4-one;* genistein-4'-glucosidorhamnoside. $C_{27}H_{30}O_{14}$; mol wt 578.53. C 56.06%, H 5.23%, O 38.72%. From fruits of *Sophora japonica* L., *Leguminosae.* Isoln and structure: Zemplén, Bognár, *Ber.* **75B**, 482 (1942). The biose is not identical with rutinose.

O-β-D-glucose-L-rhamnose

The anhydr substance mp 248° (slight decompn). $[\alpha]_D^{19}$ −73° (0.27 g in 10 ml pyridine). Freely sol in pyridine; sol in hot alcohol, hot acetone; slightly sol in boiling water. The alcoholic soln gives a purple color with ferric chloride.

Trihydrate, needles from dil alcohol, mp 156-160°. The water of crystn can be removed by drying at 100° over P_2O_5 *in vacuo* for 12 hours.

8867. Sophoricoside. *3-[4-(β-D-Glucopyranosyloxy)-phenyl]-5,7-dihydroxy-4H-1-benzopyran-4-one;* 4',5,7-trihydroxyisoflavone-4'-D-glucoside; genistein-4'-glucoside. $C_{21}H_{20}O_{10}$; mol wt 432.38. C 58.33%, H 4.66%, O 37.00%. From the green pods of *Sophora japonica* L., *Leguminosae:* Charaux, Rabate, *Bull. Soc. Chim. Biol.* **20**, 454 (1938). Structure: Zemplén *et al., Ber.* **76**, 267 (1943); Bognár, *C.A.* **46**, 8104h (1952). Synthesis: Bognár, Szabo, *Acta Chim. Acad. Sci. Hung.* **4**, 383 (1954); *Chem. & Ind. (London)* **1954**, 518. Hydrolyzed by emulsin or by dil acids.

O-β-D-glucose

Crystals from alcohol, mp 298°. $[\alpha]_D^{20}$ −47° (pyridine). $[\alpha]_D^{20}$ −32° (10% aq pyridine). uv max (abs ethanol): 262

nm. Sparingly sol in water, alc, acetic acid; more sol in hot alc, hot acetic acid; sol in pyridine, dil alkalies; practically insol in ethyl acetate, acetone. Ferric chloride gives the color of red wine to alcoholic solns which turn orange after addition of a few drops of Na-carbonate soln.

8868. Sophorose. *2-O-β-D-Glucopyranosyl-α-D-glucose.* $C_{12}H_{22}O_{11}$; mol wt 342.30. C 42.11%, H 6.48%, O 51.41%. From pods of *Sophora japonica* L., *Leguminosae:* Rebaté, *Bull. Soc. Chim. France* **7**, 565 (1940); Clancy, *J. Chem. Soc.* **1960**, 4213; Clancy in *Methods in Carbohydrate Chemistry* vol. I, R. L. Whistler, M. L. Wolfrom, Eds. (Academic Press, New York, 1962) pp 345-349. Structure and synthesis: Coxon, Fletcher, *J. Org. Chem.* **26**, 2892 (1961); Koeppen, *Carbohyd. Res.* **7**, 410 (1968). Crystal structure: J. Ohanessian *et al., Acta Crystallogr.* **B34**, 3666 (1978).

Monohydrate, needles from 80% aq methanol, mp 196-198°. $[\alpha]_D^{18}$ +19° (c = 1.2 in water).

Octa-O-acetyl-β-sophorose, $C_{28}H_{38}O_{19}$, needles from ethanol, mp 193-194°. $[\alpha]_D^{18}$ −3.2° (c = 2.5 in chloroform).

8869. Sorbic Acid. *(E,E)-2,4-Hexadienoic acid;* 2-propenylacrylic acid. $C_6H_8O_2$; mol wt 112.13. C 64.27%, H 7.19%, O 28.54%. $CH_3CH=CHCH=CHCOOH$. May be obtained from berries of the mountain ash, *Sorbus aucuparia* L., *Rosaceae* where it occurs as the lactone, called parasorbic acid: Hofmann, *Ann.* **110**, 129 (1859). Synthesis by condensing crotonaldehyde and malonic acid in pyridine soln: Doebner, *Ber.* **33**, 2140 (1900); Allen, Van Allan, *Org. Syn. coll. vol.* **III**, 783 (1955); by condensing crotonaldehyde and ketene in the presence of boron trifluoride: Hagemeyer, Jr., *Ind. Eng. Chem.* **41**, 768 (1949). Prepn from 1,1,3,5-tetraalkoxyhexane: Parker, MacLean, U.S. pat. **2,921,090** (1960 to Celanese). Additional syntheses: Fernholz, Mundlos, **Ger.** pat. **1,049,852;** U.S. pat. **3,021,365** (1959, 1962 both to Hoechst). Prepn of sodium salt: Horn, Fernholz, **Ger.** pat. **1,045,390** (1958 to Hoechst); of calcium salt: C. M. Gooding, U.S. pat. **3,139,378** (1964 to Corn Products Co.). Purification: Fernholz, **Ger.** pat. **1,044,803** (1958 to Hoechst). The *trans,trans*-isomer is usually obtained as the commercial product. Toxicity study: Smyth, Carpenter, *J. Ind. Hyg. Toxicol.* **30**, 63 (1948).

Needles from water, mp 134.5°. Should be stored at temps below 40°. bp 228° (dec). Vapor pressure at 20° <0.01 mm, at 143° 50 mm. Flash pt 260°F (127°C). pK (25°) = 4.76. Soly in water at 30° 0.25%, at 100° 3.8%, in propylene glycol at 20° 5.5%, in abs ethanol or methanol 12.90%, in 20% ethanol 0.29%, in glacial acetic acid 11.5%, in acetone 9.2%, in benzene 2.3%, in carbon tetrachloride 1.3%, in cyclohexane 0.28%, in dioxane 11.0%, in glycerol 0.31%, in isopropanol 8.4%, in isopropyl ether 2.7%, in methyl acetate 6.1%, in toluene 1.9%. LD_{50} orally in rats: 7.36 g/kg (Smyth, Carpenter).

USE: Mold and yeast inhibitor. Fungistatic agent for foods, especially cheeses. To improve the characteristics of drying oils. In alkyd type coatings to improve gloss. To improve milling characteristics of cold rubber. *See also* Potassium Sorbate.

8870. Sorbic Alcohol. *2,4-Hexadien-1-ol;* 1-hydroxy-2,4-hexadiene; hexadenol; Hexene-Ol; Hexakose. $C_6H_{10}O$; mol wt 98.14. C 73.43%, H 10.27%, O 16.30%. $CH_3-CH=CHCH=CHCH_2OH$. Prepd from sorbic aldehyde by means of aluminum isopropylate in isopropanol: Reichstein *et al., Helv. Chim. Acta* **15**, 264 (1932).

Long needles. Agreeable odor of new mown grass. mp 30.5-31.5°. bp$_{12}$ ~80°. Volatile with steam. Insol in water. Sol in alcohol, ether, oils. Not stable to air; must be sealed in evacuated ampuls for storage.

Diphenylurethan, C$_{19}$H$_{19}$NO$_2$, crystals from petr ether, mp 78-79°.

3,5-Dinitrobenzoate, yellowish needles from petr ether, mp 85°.

8871. Sorbinil. (S)-6-Fluoro-2,3-dihydrospiro[4H-1-benzopyran-4,4'-imidazolidine]-2',5'-dione; (+)-(4S)-6-fluoro-spiro[chroman-4,4'-imidazolidine]-2',5'-dione; CP-45634. C$_{11}$H$_9$FN$_2$O$_3$; mol wt 236.20. C 55.94%, H 3.84%, F 8.04%, N 11.86%, O 20.32%. Spirohydantoin aldose reductase inhibitor. Prepn of racemate and resolution of isomers: R. Sarges, Ger. pat. **2,821,966;** idem, U.S. pat. **4,130,714** (both 1978 to Pfizer). Synthesis: R. Sarges et al., J. Org. Chem. **47,** 4081 (1982); N. L. Dirlam et al., J. Org. Chem. **52,** 3587 (1987). Absolute configuration: R. Sarges et al., J. Med. Chem. **28,** 1716 (1985). Effect on polyol accumulation in diabetic rats: M. J. Peterson et al., Metabolism **28,** Suppl. 1, 456 (1979). Pharmacokinetics in humans: G. Foulds et al., Clin. Pharmacol. Ther. **30,** 693 (1981). HPLC determn in human lens and plasma: P. Lloyd, M. J. C. Crabbe, J. Chromatog. **343,** 402 (1985). Preliminary clinical trials in diabetic neuropathy: R. J. Young et al., Diabetes **32,** 938 (1983); J. Jaspan et al., Lancet **2,** 758 (1983). Symposium on pharmacology and clinical efficacy: Metabolism **35,** Suppl 1, 1-121 (1986).

Crystals from ethanol, mp 241-243°. [α]$_D^{25}$ +54.0° (c = 1 in methanol).

8872. Sorbitan Esters. Sorbitan fatty acid esters; SFAE. Nonionic surface active agents; partial esters of the common fatty acids (lauric, palmitic, stearic and oleic) and hexitol anhydrides derived from sorbitol. Commercial products may be mixtures of fatty acid esters of 1,4- and 1,5 anhydrosorbitol and 1,4,3,6-dianhydrosorbitol, but generally conform to the structure depicted below. Prepn: K. R. Brown, U.S. pat. **2,322,820** (1943 to Atlas Powder Co.). Improved process: G. J. Stockburger, U.S. pat. **4,297,290** (1981 to ICI). Comprehensive description: P. Becher, "Polyol Surfactants" in Nonionic Surfactants, M. J. Schick, Ed. (Dekker, New York, 1967) pp 247-299. Description of prepn and uses: L. R. Chislett, J. Walford, Int. Flavours Food Addit. **7,** 61 (1976). GLC determn of sorbitan monolaurate in plasma: S. H. Giovanetto, Anal. Letters **16,** 867 (1983). Analysis by HPLC: N. Garti et al., J. Am. Oil Chem. Soc. **60,** 1151 (1983). Series of toxicity studies: Food Cosmet. Toxicol. **16,** 519-542 (1978). Review: Cosmetic, Toiletry and Fragrance Assoc., J. Am. Coll. Toxicol. **4,** 65-121 (1985).

Sorbitan Laurate	R = OOC(C$_{11}$H$_{23}$)
Sorbitan Stearate	R = OOC(C$_{17}$H$_{35}$)
Sorbitan Oleate	R = OOC(C$_{17}$H$_{33}$)

Polyoxyethylene derivatives, see Polysorbates.

Sorbitan Laurate, C$_{18}$H$_{34}$O$_6$, sorbitan monolaurate, Alka-muls SML, Arlacel 20, Emsorb 2515, Glycomul L, Span 20. Yellow liquid. Acid value: <8. Saponification value: 150-165. Hydroxyl value: 330-360. Sol in mineral oil, cotton-

seed oil, methanol, ethanol, isopropyl alc, ethylene glycol. Insol in water, propylene glycol.

Sorbitan Stearate, C$_{24}$H$_{46}$O$_6$, sorbitan monostearate, Alka-muls SMS, Arlacel 60, Glycomul S, Span 60. White to tan waxy solid, mp 49-65°. Acid value: 5-11. Saponification value: 140-157. Hydroxyl value: 230-260. Sol in alcohols, carbon tetrachloride, toluene. Insol in water, acetone, mineral spirits.

Sorbitan Oleate, C$_{24}$H$_{44}$O$_6$, sorbitan monooleate, Alkamuls SMO, Arlacel 80, Capmul O, Emsorb 2500, Glycomul O, Span 80. Yellow to amber colored oily liquid. Acid number: 5-8. Saponification value: 140-160. Hydroxyl value: 193-215. Sol in ethanol, isopropyl alc, mineral oil, vegetable oil. Insol in water, propylene glycol.

use: As emulsifiers, stabilizers, and thickeners in foods, cosmetics, and medicinal products. In the textile industry as fiber lubricants and softeners. Pharmaceutic aid (surfactant).

8873. Sorbitol. D-Glucitol; D-sorbitol; L-gulitol; sorbit; Cystosol; Resulax; Sorbilax; Sorbitur; Sorbo; Sorbostyl; Sorbilande. C$_6$H$_{14}$O$_6$; mol wt 182.17. C 39.56%, H 7.75%, O 52.70%. First found in the ripe berries of the mountain ash Pyrus aucuparia Ehrh. (L.) (Sorbus aucuparia L.), Rosaceae. Occurs also in many other berries (except grapes) and in cherries, plums, pears, apples, seaweed and algae. Has been detected in blackstrap molasses. Isoln from berries: Embden, Griesbach, Z. Physiol. Chem. **91,** 268 (1914). Prepd industrially from glucose by high pressure hydrogenation or by electrolytic reduction: Boye, Chem.-Ztg. **82,** 657 (1958); Fedor et al., Ind. Eng. Chem. **52,** 282 (1960); from dextrose by catalytic hydrogenation: Faith, Keyes & Clark's Industrial Chemicals, F. A. Lowenheim, M. K. Moran, Eds. (Wiley-Interscience, New York, 4th ed., 1975) pp 774-778. Review of uses: Kempf, Die Stärke **6,** 269-274 and 303-306 (1954); **9,** 234-237 (1955).

Needles with ½ or 1H$_2$O. Sweet taste, ~60% as sweet as sugar (w/w). In the healthy human organism 1.0 g of sorbitol yields 3.994 calories which is comparable to 3.940 calories from 1.0 g of cane sugar. Seventy percent of orally ingested sorbitol is converted to CO$_2$ without appearing as glucose in the blood: Adcock, Gray, Biochem. J. **65,** 554 (1957). The hydrated crystals melt somewhat below 100°. When completely anhyd mp 110-112°. [α]$_D^{20}$ −2.0° (H$_2$O). In the presence of molybdate the rotation is reversed and increased to +56°. Freely sol in water (up to 83%). High % sorbitol solns are much more viscous than corresp glycerol solns. Quite sol in hot alcohol, sparingly sol in cold alcohol. Also sol in methanol, isopropanol, butanol, cyclohexanol, phenol, acetone, acetic acid, DMF, pyridine, acetamide solns. Practically insol in most other organic solvents. Not attacked in the cold when mixed with dil acids, alkalies or mild oxidizing substances. Ka at 17.5° = 2.5 × 10^{-14}. pH about 7.0. A commercial 70% aq soln may have the following characteristics: d$_{20}^{20}$ 1.2879; n$_D^{25}$ 1.45831; [α]$_D^{20}$ −2.10°; bp$_{760}$ 105°; pH between 6 and 7; viscosity (25°): 110 cp. d$_4^{20}$ for various % solns: 5% 1.014; 10% 1.038; 25% 1.099; 50% 1.198; 60% 1.249; 70% 1.299; 83% 1.391. Viscosity in cp at 20°: 5% soln 1.230; 10% 1.429; 25% 2.689; 50% 11.09; 60% 35.73; 70% 185; 83% >10,000.

USE: In manuf of sorbose, ascorbic acid, propylene glycol, synthetic plasticizers and resins; as humectant (moisture conditioner) on printing rolls, in leather, tobacco. In writing inks to insure a smooth flow and to prevent crusting on the point of the pen. In antifreeze mixtures with glycerol or glycols. In candy manuf to increase shelf life by retarding the solidification of sugar; as humectant and softener in shredded coconut and peanut butter; as texturizer in foods;

as sequestrant in soft drinks and wines. Used to reduce the undesirable aftertaste of saccharin in foodstuffs; as sugar substitute for diabetics. Pharmaceutic aid (flavor; tablet excipient); to increase absorption of vitamins and other nutrients in pharmaceutical preparations: *Chem. & Eng. News* **36**, 59 (Feb. 24, 1958).

THERAP CAT: Laxative.

THERAP CAT (VET): In ruminant ketosis, osmotic diuretic, laxative.

8874. Sorbose. *L-Sorbose*; sorbin; sorbinose. $C_6H_{12}O_6$; mol wt 180.16. C 40.00%, H 6.71%, O 53.28%. From sorbitol by fermentation with *Acetobacter suboxydans*: Perlman, *J. Chem. Ed.* **36**, 60 (1959); Lockwood, *Methods in Carbohydrate Chemistry* **1**, 151 (Academic Press, New York, 1962).

$$CH_2OH$$
$$|$$
$$C=O$$
$$|$$
$$HO-CH$$
$$|$$
$$HC-OH$$
$$|$$
$$HO-CH$$
$$|$$
$$CH_2OH$$

Orthorhombic, bisphenoidal crystals. About as sweet as sucrose. d^{15} 1.65. mp 165°. $[\alpha]_D^{30}$ −42.7° (c = 5). Ka at 17.5° = 2.8 × 10^{-12}. Freely sol in water; almost insol in alcohol. Reduces Fehling's soln.

USE: In the manuf of vitamin C (accounts for nearly 1000 tons of ascorbic acid produced every year). For conversion of L-sorbose to 2-keto-L-gulonic acid *see* Reichstein, Grüssner, *Helv. Chim. Acta* **17**, 311 (1934).

8875. Sorivudine. *(E)-1-β-D-Arabinofuranosyl-5-(2-bromoethenyl)-2,4(1H,3H)-pyrimidinedione*; 1-β-D-arabinofuranosyl-*(E)*-5-(2-bromovinyl)uracil; 5-bromovinyl-araU; brovavir; BV-araU; BVAU; YN-72; SQ-32756; Usevir. $C_{11}H_{13}$-BrN_2O_6; mol wt 349.14. C 37.84%, H 3.75%, Br 22.89%, N 8.02%, O 27.50%. Orally active antiviral agent; inhibits viral DNA synthesis. Prepn: H. Machida, S. Sakata, *Eur. pat. Appl.* **31,128**; *eidem, U.S. pat.* **4,386,076** (1981, 1983 both to Yamassa Shoyu); S. Sakata *et al., Nucleic Acids Symp. Ser.* **8**, s39 (1980); R. Busson *et al., ibid.* **9**, 49 (1981). Antiviral activity and toxicity: H. Machida, S. Sakata, *Antiviral Res.* **4**, 135 (1984). Mechanism of action studies: T. Yokota *et al., Mol. Pharmacol.* **36**, 312 (1989); H. Machida, *Adv. Exp. Med. Biol.* **278**, 255 (1990). Clinical trial in herpes zoster infection: M. Niimura, *ibid.* 267.

mp 182° (Sakata, 1980); also reported as white crystals from ethanol, mp 195-200° (dec) (Machida, Sakata, 1983). $[\alpha]_D^{25}$ +0.5° (1N NaOH). LD_{50} in mice (mg/kg): ~3300 i.p.; >5000 s.c.; >10000 orally (Machida, Sakata, 1984).

THERAP CAT: Antiviral.

8876. Sotalol. *N-[4-[1-Hydroxy-2-[(1-methylethyl)amino]ethyl]phenyl]methanesulfonamide; 4'-[1-hydroxy-2-(isopropylamino)ethyl]methanesulfonanilide*. $C_{12}H_{20}N_2O_3S$; mol wt 272.37. C 52.92%, H 7.40%, N 10.29%, O 17.62%, S 11.77%. β-Adrenergic blocker. Prepn: Uloth *et al., J. Med. Chem.* **9**, 88 (1966). Pharmacology and toxicology: Larsen, Lish, *Nature* **203**, 1283 (1964); Lish *et al., J. Pharmacol. Exp. Ther.* **149**, 161 (1965); Stanton *et al., ibid.* 175; Kvam *et al., ibid.* 183. Activity of the *d*- and *l*-forms: Somani, Watson, *ibid.* **164**, 317 (1968); Somani, Bachand, *Eur. J.*

Pharmacol. **7**, 239 (1969). HPLC determn in body fluids: W. P. Gluth *et al., Arzneimittel-Forsch.* **38**, 408 (1988). Review of pharmacology and therapeutic use: B. N. Singh *et al., Drugs* **34**, 311-349 (1987); S. H. Hohnloser, R. L. Woosley, *N. Engl. J. Med.* **331**, 31-38 (1994). Comprehensive description: R. T. Foster, R. A. Carr in *Analytical Profiles of Drug Substances and Excipients* vol. 21, H. G. Brittain, Ed. (Academic Press, San Diego, 1992) pp 501-533.

Crystals. uv max (chloroform): 242.2, 275.2 nm. pK_1 8.2, pK_2 9.8.

Hydrochloride, $C_{12}H_{20}N_2O_3S$.HCl, *MJ-1999, Beta-Cardone, Betapace, Darob, Sotacor, Sotalex*. White crystalline solid, mp 206.5-207° (dec), also reported as mp 218-219° (Singh). Freely sol in water; slightly sol in chloroform. Partition coefficient (water/*n*-octanol): 0.24; (octan-1-ol/pH 7.4 phosphate buffer) at 37°: 0.09. LD_{50} in male mice, rats (mg/kg): 2600, 3450 orally; 670, 680 i.p.; LD_{50} orally in rabbits: 1000 mg/kg; LD_{50} i.p. in dogs: 330 mg/kg (Lish).

THERAP CAT: Antianginal; antiarrhythmic (class II + III); antihypertensive.

8877. Soterenol. *N-[2-Hydroxy-5-[1-hydroxy-2-[(1-methylethyl)amino]ethyl]phenyl]methanesulfonamide; 2'-hydroxy-5'-[1-hydroxy-2-(isopropylamino)ethyl]methanesulfonanilide*; MJ-1992. $C_{12}H_{20}N_2O_4S$; mol wt 288.37. C 49.98%, H 6.99%, N 9.71%, O 22.19%, S 11.12%. β-Adrenergic agonist. Prepn: Larsen *et al., J. Med. Chem.* **10**, 462 (1967). Pharmacological study: K. W. Dungan *et al., J. Pharmacol. Exp. Ther.* **164**, 290 (1968).

Monohydrochloride, $C_{12}H_{20}N_2O_4S$.HCl, crystals from methanol-isopropyl ether, mp 195.5-196.5° (dec). LD_{50} (7 day) in mice (mg/kg): 41 i.v.; 315 i.p.; 660 orally (Dungan).

THERAP CAT: Bronchodilator.

8878. Soybean. Soya bean; soja bean; Lincoln bean; Manchurian bean; Chinese pea. Seed of *Glycine max* (L.) Merrill [*G. soja* Sieb. & Zucc., *G. hispida* (Moench) Maxim., *Soja hispida* Moench], Leguminosae. Habit. Eastern Asia, especially Manchuria. Cultivated in the midwestern U.S.A., Brazil, some European countries, such as Italy and Yugoslavia. Constit. Proteins 40%, carbohydrates 17%, oil 18%, ash 4.6%. The ash content equals K 1.67%, Na 0.34%, Ca 0.28%, Mg 0.22%, Cl 0.024%, I 0.000054%, Fe 0.0097%, Cu 0.0012%, Mn 0.0028%, Zn 0.0022%, P 0.66%, S 0.41%, Al 0.0007%. The chief proteins are glycinin, a globulin, phaseolin, another globulin, and 2 albumins: legumelin and soy legumelin. The carbohydrates are sucrose, raffinose, stachyose, and pentosans. Phosphorus compds (about 3%) are phospholipids, nucleic acid phosphates, phytin, and inorganic phosphates. The phospholipids contain lecithin, cephalin, and inositol. The vitamin content is moderate: Vitamin A 110 i.u./100g; thiamine 1.14 mg/100 g; riboflavin 0.31 mg/100 g; niacin 2.1 mg/100 g; ascorbic acid: trace. Food energy: 350 cal/100 g. Amino acid analysis of the protein fraction: Glutamic acid 18.4%, leucine 8.1%, arginine 7.5%, lysine 6.7%, valine 5.4%, isoleucine 5.3%, phenylalanine 5.2%, threonine 3.9%, histidine 2.3%, tryptophan 1.6%, methionine 1.4%. Ten kg of soybeans yield 15 g of the isoflavone glycoside genistin, the mother liquor contains daidzin. Soybeans contain a small amount of enzymes such as lipoxidase, urease, uricase, protease, antienzymes such as antitrypsin, and a growth-inhibiting factor. Bitter principles and saponins are also found. *Reviews:* K. S. Markley, *Soybeans and Soybean Products,* 2 vols. (Interscience, New York, 1951); Cowan in Kirk-Othmer *Encyclopedia of Chemical Technology* vol. 18 (Interscience, New York, 2nd ed., 1969) pp 599-614.

Whitish or yellowish-green to brownish-black, ovoid beans, about 8 mm long, 7 mm wide, and 6 mm thick. Most varieties have a wt of 20 g/100 seeds. USE: In the production of soybean oil. As food and feedstuff. Debittered soybean flour contains practically no starch and is widely used in dietetic foods. Soybean meal obtained after expressing the oil is a preferred source of protein for feedstuffs. Other products are soybean lecithin, genistin, monosodium glutamate, soybean milk, soy sauce; *tofu,* a soybean curd; *miso,* a fermented mixture with barley or rice; *natto,* which is soybean cheese. Soybean proteins are used also in the adhesive and plastics industries.

THERAP CAT: Nutrient.

8879. Soybean Oil. Obtained from soybeans by solvent extraction using petroleum hydrocarbons or, to a lesser extent, by expression using continuous screw press operations. *Constit.* Triglycerides of oleic acid 26%, of linoleic acid 49%, of linolenic acid 11%, of saturated acids 14%. Free fatty acids are usually less than 1%. Phospholipids (lecithin) 1.5-4%. Another 0.8% consists of stigmasterol, sitosterols, and tocopherols. The phospholipids and sterols are removed by refining with alkali. Reviews and bibliographies: K. S. Markley, *Soybeans and Soybean Products,* 2 vols. (Interscience, New York, 1951); E. W. Eckey, *Vegetable Fats and Oils* (Reinhold, New York, 1954); W. J. Wolf in Kirk-Othmer, *Encyclopedia of Chemical Technology* vol. 21 (Wiley-Interscience, New York, 3rd ed., 1983) pp 417-442.

Pale yellow to brownish-yellow oil. Slight characteristic odor and taste. The flavor of crude soybean oil has been described as paint-like and grass-like, and is somewhat distasteful to the palate of occidentals. d_{25}^{25} 0.916-0.922. Flash pt 540°F (282°C). Ignition temp 833°F (445°C). n_D^{25} 1.471-1.475. Titer 22-27. Solidifies at -10 to $-16°$. Viscosity in centipoises: 172.9 at 0°; 99.7 at 10°; 50.09 at 25°; 28.86 at 40°. Acid value 0.3-3.0. Saponification value 189-195. Iodine value 127-138. Thiocyanogen value 77-85. Diene no. 0.7. Hydroxyl value 4-8. Reichert-Meissl value 0.2-0.7. Polenske value 0.2-1.0. Miscible with abs alcohol, ether, petr ether, chloroform, carbon disulfide.

USE: Directly as food in the Orient, also in blends with olive oil. In the manuf of margarine, shortenings, candy, soap. In drying oil industries (paints, varnishes, linoleum, printing ink). The foots or residue from refining has been used to obtain soybean fatty acids which can be used in the manuf of alkyd resins.

8880. Soy Sauce. Soy sauce is a hydrolysis product of soybeans. A combination of mold fermentaion and acid hydrolysis is used. The molds employed are *Aspergillus flavus, A. niger,* and *A. oryzae.* Soy sauce consists of a mixture of amino acids, peptides, polypeptides, peptones, simple proteins, purines, carbohydrates, and lesser organic compds suspended in an 18% sodium chloride soln. Adenine, arginine, choline, lysine, betaine, and glutamic acid have been isolated. In the manuf of soy sauce, during the fermentation process, 60-70% of the vitamins present in soybeans are destroyed. Review and bibliography: K. S. Markley, *Soybeans and Soybean Products,* 2 vols. (Interscience, N.Y., 1951).

8881. Sozoiodolic Acid. *4-Hydroxy-3,5-diiodo-benzenesulfonic acid;* 2,6-diiodophenol-4-sulfonic acid; Optojod. $C_6H_4I_2O_4S$; mol wt 425.97. C 16.92%, H 0.95%, I 59.58%, O 15.02%, S 7.53%. Made by iodizing potassium *p*-phenol-

sulfonic acid by means of ICl, or by KI and KIO_3. Prepn of salts: Rupp, Herman, *Arch. Pharm.* **254,** 488 (1916).

Trihydrate, white, almost odorless needles, becoming anhydr at 100°. mp 120° when anhydr; dec at 190°. Freely sol in water, alcohol, ether, glycerol.

Mercury deriv, $C_6H_2HgI_2O_4S$, *Sozoiodole-mercury.* Fine, orange-yellow powder. *Poisonous!* Practically insol in water, alcohol, ether, glycerol. Sol in NaCl or KI soln.

Sodium salt dihydrate, $C_6H_3I_2NaO_4S.2H_2O$, *Sozoiodole-sodium, Dijozol.* Crystals. Sol in water, alcohol, glycerol. Practically insol in ether.

Zinc salt hexahydrate, $C_{12}H_6I_4O_8S_2Zn.6H_2O$, *Sozoiodole-zinc.* Needles, sol in water, alc, glycerol. Practically insol in ether.

USE: In a test for albumin.

THERAP CAT: Diagnostic aid (radiopaque medium in retrograde pyelography). Hg deriv formerly in syphilis.

8882. Spandex. Lycra; Vyrene. An elastic, segmented polyurethane fiber obtained by the interaction of a diisocyanate with a glycol. The term "segmented" indicates that it is made from a block copolymer in which reasonably long flexible chains are joined to shorter stiff chains through the urethane linkages. *Ref: Modern Textiles Mag.* **40,** 38 (Dec. 1959); Hicks Jr., *Am. Dyestuff Reptr.* **52,** no. 1, 33 (1963). Book: Moncrieff, *Man-made Fibres* (Heywood, London, 4th ed., 1963) pp 396-403. Moncrieff cites **U.S.** pats. **2,692,873** (1954 to du Pont) and **2,751,363** (1956 to U.S. Rubber) in describing the historical development of spandex fibers. Compare also Smith; Frazer, Shivers, **U.S.** pats. **3,061,574** and **3,071,557** (1962, 1963 to du Pont). Historical review of industrial production with extensive patents list: L. Rose, *Repts. Progr. Appl. Chem. (London)* **51,** 609-612 (1966). Monograph: M. McDonald, *Spandex Manufacture* (Noyes, Park Ridge, N.J., 1970) 190 pp.

All linkages are hydrolytically stable to acids, alkalies. *Caution:* Can cause allergic contact dermatitis if any diisocyanate is present as impurity.

USE: Snap-back fiber, stronger and lighter than rubber. Used mainly for elastic garments such as belts, girdles, corsets, brassieres, garters, surgical stockings and sock tops. Can be used bare, giving lighter fabrics and eliminating the costly covering process needed for rubber.

8883. Sparassol. *2-Hydroxy-4-methoxy-6-methylbenzoic acid methyl ester; 4-methoxy-2,6-cresotic acid methyl ester;* everninic acid methyl ester; orsellinic acid methyl ester 4-methyl ether. $C_{10}H_{12}O_4$; mol wt 196.20. C 61.22%, H 6.16%, O 32.62%. Antibiotic substance produced by the fungus *Sparassis ramosa:* Falck, *Ber.* **56,** 2555 (1923). Also obtained in methanol extracts of the lichen *Evernia prunasti:* Stenhouse, *Ann.* **68,** 55 (1848); Spåth, Jeschki, *Ber.* **57,** 471 (1924). Structure: Fischer, Hoesch, *Ann.* **391,** 347 (1912); Wedekind, Fleischer, *Ber.* **56,** 2556 (1923). Synthesis: G. Nicollier *et al., Helv. Chim. Acta* **61,** 2899 (1978).

Prisms from water, mp 67-68°. Slightly sol in hot water; freely sol in acetone, ether, chloroform; moderately sol in methanol, ethanol, petr ether. Alkaline hydrolysis yields free everninic acid, mp 170°.

8884. Sparfloxacin. *(cis)-5-Amino-1-cyclopropyl-7-(3,5-dimethyl-1-piperazinyl)-6,8-difluoro-1,4-dihydro-4-oxo-3-quinolinecarboxylic acid;* AT-4140; CI-978; PD-131501; Spara; Zagam. $C_{19}H_{22}F_2N_4O_3$; mol wt 392.41. C 58.16%, H 5.65%, F 9.68%, N 14.28%, O 12.23%. Fluorinated quinolone antibacterial. Prepn: J. Matsumoto *et al., Eur.* pat. **Appl. 221,463;** *eidem,* **U.S.** pat. **4,795,751** (1987, 1989 both to Dainippon). Synthesis and structure-activity relationship: T. Miyamoto *et al., J. Med. Chem.* **33,** 1645 (1990). Antibacterial activity: S. Nakamura *et al., Antimicrob. Ag. Chemother.* **33,** 1167 (1989). Comparative *in vitro* antimicrobial spectrum: T. Kojima *et al., ibid.* 1980. Pharmacokinetics: S. Nakamura *et al., ibid.* **34,** 89 (1990). Subacute toxicity studies: M. Yasuba *et al., Chemotherapy* **39,** Suppl. 4, 180 (1991), *C.A.* **115,** 247561d (1991); M. Iida *et al., ibid.* 195, *C.A.* **115,** 247562e (1991).

Crystals from chloroform + ethanol, mp 266-269° (dec). pKa₁ 6.25, pKa₂ 9.30.

THERAP CAT: Antibacterial.

8885. Sparsiflorine. *5,6,6a,7-Tetrahydro-2-methoxy-4H-dibenzo[de,g]quinoline-1,10-diol; 2-methoxy-6aβ-noraporphine-1,10-diol.* $C_{17}H_{17}NO_3$; mol wt 283.33. C 72.07%, H 6.05%, N 4.94%, O 16.94%. From the leaves of *Croton sparsiflorus* Morung, *Euphorbiaceae:* Saha, *Sci. Cult.* **24,** 572 (1959), *C.A.* **54,** 1580a (1960). Structure: Chatterjee *et al., Tetrahedron Letters* **1965,** 1539.

Needles from alc, dec 228°. Sol in methanol, ethanol, ethyl acetate, acetone; slightly sol in ether; practically insol in benzene, petr ether, hexane.

Hydrochloride, $C_{17}H_{17}NO_3$.HCl, crystals from water, dec 283-284°. $[\alpha]_D^{30}$ +43° (water). uv max (ethanol): 226, 266, 275, 310 nm (log ε 4.54, 4.09, 4.23, 3.95).

N-Methyl methiodide, $C_{18}H_{19}NO_3$.CH₃I, crystals, dec 236-238°. uv max (ethanol): 225, 267, 277, 310 nm (log ε 4.57, 4.08, 4.0, 3.85).

8886. Sparsomycin. *[R-[R*,S*-(E)]]-N-[1-(Hydroxymethyl)-2-[[(methylthio)methyl]sulfinyl]ethyl]-3-(1,2,3,4-tetrahydro-6-methyl-2,4-dioxo-5-pyrimidinyl)-2-propenamide; (E)-(1S)-1,2,3,4-tetrahydro-N-[1-(hydroxymethyl)-2-[[(methylthio)methyl]sulfinyl]ethyl]-6-methyl-2,4-dioxo-5-pyrimidineacrylamide;* (+)-sparsomycin; NSC-59729; U-19183. $C_{13}H_{19}N_3O_5S_2$; mol wt 361.44. C 43.20%, H 5.30%, N 11.63%, O 22.13%, S 17.74%. Protein synthesis inhibitor with antibiotic and antitumor activity. Isoln from fermentation broth of *Streptomyces sparsogenes* var *sparsogenes:* S. P. Owen *et al., Antimicrob. Ag. Chemother.* **1962,** 772; **Brit.** pat. **974,541** (1964 to Upjohn); from *S. cuspidosporus:* E. Higashide *et al., Takeda Kenkyusho Nempo* **25,** 1 (1966), *C.A.* **66,** 54238q (1967). Characterization and purification: A. D. Argoudelis, R. R. Herr, *Antimicrob. Ag. Chemother.* **1962,** 780. Structural elucidation: P. F. Wiley, F. A. MacKellar, *J. Org. Chem.* **41,** 1858 (1976). Total synthesis: H. C. J. Ottenheijm *et al., ibid,* **46,** 3273 (1981); R. M. J. Liskamp *et al., ibid,* 5408. HPLC determn in plasma and urine: B. Winograd *et al., J. Chromatog.* **275,** 145 (1983). Pharmacokinetics and toxicology: Z. Zylicz *et al., Cancer Chemother. Pharmacol.* **20,** 115 (1987). Ribosomal binding studies: E. Lazaro *et al., Antimicrob. Ag. Chemother.* **35,** 10 (1991). Biosynthetic studies: R. J. Parry *et al., J. Am. Chem. Soc.* **114,** 5946 (1992). Review of chemistry and biological activity: H. C. J. Ottenheijm *et al., Prog. Med. Chem.* **23,** 220-268 (1986).

Crystals, mp 208-209° (dec). $[\alpha]_D^{25}$ +69° (c = 0.5 in water). pKa' 8.67 in water, 9.05 in 40% ethanol. uv max: 302 nm (water or 0.1*N* aq sulfuric acid), 328 nm (0.1*N* aq potassium hydroxide). Slightly sol in water and lower alcohols. Insol in less polar organic solvents. LD_{50} in dogs, rats, mice (mg/kg): 0.5-1.0, 2.25, 4.32 i.v. (Ottenheijm, 1986). LD_{50} i.p. in mice: 2.4 mg/kg (Owen).

USE: Research tool for studying protein biosynthesis.

8887. Sparteine. *[7S-(7α,7aα,14α,14aβ)]-Dodecahydro-7,14-methano-2H,6H-dipyrido[1,2-a:1',2'-e][1,5]diazocine; l-*sparteine; lupinidine. $C_{15}H_{26}N_2$; mol wt 234.38. C 76.87%, H 11.18%, N 11.95%. In yellow and black lupin beans, *Lupinus luteus* L. and *L. niger* Hort.; also in *Cytisus scoparius* (L.) Link. and *Anagyris foetida* L., *Leguminosae.* Extraction procedure: Karrer *et al., Helv. Chim. Acta* **11,** 1062 (1928). Structure: Clemo *et al., J. Chem. Soc.* **1931,** 429; Ing, *ibid.* **1933,** 504; Schirm, Besendorf, *Arch. Pharm.* **280,** 64 (1942); Galinovsky, Stern, *Ber.* **77B,** 132 (1944); Clemo *et al., J. Chem. Soc.* **1949,** 663. Biosynthesis: Anet *et al., Nature* **165,** 35 (1950); Schöpf *et al., Angew. Chem.* **65,** 161 (1953); **69,** 69 (1957); van Tamelen, Foltz, *J. Am. Chem. Soc.* **82,** 2400 (1960). Absolute configuration: Okuda *et al., Chem. & Ind. (London)* **1961,** 1116. Conformation: Bohlmann *et al., Tetrahedron Letters* **1965,** 2705; Wiewiorowski *et al., Can. J. Chem.* **45,** 1447 (1967). Synthesis of racemate: Bohlmann *et al., Ber.* **106,** 3026 (1973); N. Takatsu *et al., Chem. Pharm. Bull.* **35** 4990 (1987). Chemistry: Binnig, *Arzneimittel-Forsch.* **24,** 752 (1974). Studies of cardiovascular effects: Raschak, *ibid.* 753.

Viscous, oily liquid. bp₈ 173°. Volatile with steam. $[\alpha]_D^{21}$ —16.4° (c = 10 in abs alc). n_D^{20} 1.5312. d_4^{20} 1.020. pK_1 at 20°: 2.24; pK_2: 9.46; pH of 0.01 molar soln 11.6. One gram dissolves in 325 ml water. Freely sol in alcohol, chloroform or ether.

Sulfate pentahydrate, $C_{15}H_{26}N_2$.H_2SO_4.5H_2O, *Depasan, Tocosamine.* Columnar crystals, loses water of crystn at 100° turning brown, dec 136°. pH of 0.05 molar soln 3.3. One gram dissolves in 1.1 ml water, 3 ml alcohol. Practically insol in chloroform and ether.

Adenylate, *Spartopan.*

THERAP CAT: Oxytocic.

8888. Spasmolytol. *2-[(3,5-Dibromo-2-methoxyphenyl)methoxy]-N,N-diethylethanamine; 2-(3,5-dibromo-2-methoxybenzyloxy)triethylamine; β-diethylaminoethyl ether of 3,5-dibromo-2-methoxybenzyl alcohol; 3,5-dibromo-2-methoxybenzyl alcohol β-diethylaminoethyl ether; A-124.* $C_{14}H_{21}Br_2NO_2$; mol wt 395.13. C 42.56%, H 5.36%, Br 40.44%, N 3.54%, O 8.10%. Preparation: Felton, U.S. pat. **2,766,238** (1956 to Hynson, Westcott & Dunning). Pharmacology and toxicology: Bryant *et al., J. Pharmacol. Exp. Ther.* **121**, 210 (1957).

Liquid, bp$_3$ 185-195°.

Hydrochloride, $C_{14}H_{21}Br_2NO_2 \cdot HCl$, crystals from methanol + ether, mp 124-127°.

THERAP CAT: Antispasmodic.

8889. Spearmint. Mint. Dried leaves and tops of *Mentha spicata* L. (*M. viridis* L.), *Labiatae.* Habit. Europe, Asia; widely cultivated in U.S. Principal active constit is a volatile oil which contains at least 50% carvone.

USE: As flavor.

8890. Spectinomycin. *[2R-(2α,4aβ,5aβ,6β,7β,8β,9α,-9aα,10aβ)]-Decahydro-4a,7,9-trihydroxy-2-methyl-6,8-bis-(methylamino)-4H-pyrano[2,3-b][1,4]benzodioxin-4-one;* actinospectacin; espectinomicina; CHX-3101; M-141. $C_{14}H_{24}$-N_2O_7; mol wt 332.35. C 50.59%, H 7.28%, N 8.43%, O 33.70%. Antibiotic isolated from fermentation broth of *Streptomyces spectabilis:* D. J. Mason *et al., Antibiot. & Chemother.* **11**, 118 (1961); M. E. Bergy *et al., ibid.* 661; M. E. Bergy, C. De Boer, U.S. pat. 3,234,092 (1966 to Upjohn). Purification and crystallization: H. K. Jahnke, U.S. pat. 3,206,360; V. J. Peters, U.S. pat. 3,272,706 (1965, 1966 both to Upjohn). Isoln and characterization: A. C. Sinclair, A. F. Winfield, *Antimicrob. Ag. Chemother.* **1961**, 503. Structure: H. Hoeksema *et al., J. Am. Chem. Soc.* **84**, 3212 (1962); P. F. Wiley *et al., ibid.* **85**, 2652 (1963). Solubility data: J. R. Marsh, P. J. Weiss, *J. Assoc. Offic. Anal. Chem.* **50**, 457 (1967). Stereochemistry and abs config: T. G. Cochran *et al., Chem. Commun.* **1972**, 494. Biosynthesis: L. A. Mitscher *et al., ibid.* **1971**, 1541; H. Otsuka *et al., J. Am. Chem. Soc.* **102**, 6817 (1980). Enantioselective synthesis: D. R. White *et al., Tetrahedron Letters* **1979**, 2737; S. Hanessian, R. Roy, *Can. J. Chem.* **63**, 163 (1985). Mechanism of action study: M. F. Brink *et al., Nucleic Acids Res.* **22**, 325 (1994). HPLC determn in animal plasma: N. Haagsma *et al., J. Chromatog.* **615**, 289 (1993). Clinical pharmacokinetics: J. G. Wagner *et al., Int. Z. klin. Pharmakol. Ther. Toxikol.* **1**, 261 (1968). Clinical evaluation in gonorrhea: Y. H. Kouri *et al., Genitourin. Med.* **65**, 342 (1989); in chancroid: M. Guzmán *et al., Sex. Transm. Dis.* **19**, 291 (1992). Co-treatment with lincomycin, *q.v.,* for footrot in sheep: C. M. Venning *et al., Aust. Vet. J.* **67**, 258 (1990). *Reviews:* W. M. McCormack, M. Finland, *Ann. Intern. Med.* **84**, 712-716 (1976); W. J. Holloway, *Med. Clin. North Am.* **66**, 169-173 (1982).

Amorphous powder. $[α]_D^{25}$ −20° (water). pKa'$_1$ 6.95, pKa'$_2$ 8.70. Sol in water, methanol, ethanol. Insol in acetone, hydrocarbon solvents.

Dihydrochloride pentahydrate, $C_{14}H_{24}N_2O_7 \cdot 2HCl \cdot 5H_2O$, *Spectam, Spectogard, Stanilo, Togamycin, Trantan, Trobicin.* Exists as a ketone hydrate and not in the carbonyl form (Cochran *et al.*). Colorless needles from aq acetone, mp 205-207° (dec). $[α]_D$ +14.8° (c = 0.42 in water). Soly in water, methanol, propylene glycol, formamide, DMSO, 0.1*N* NaOH, 0.1*N* HCl: >20 mg/ml. Soly (mg/ml): chloroform 0.042; acetone 0.015; diethyl ether 0.010.

Sulfate tetrahydrate, $C_{14}H_{24}N_2O_7 \cdot H_2SO_4 \cdot 4H_2O$, white crystals from aq acetone, mp ~185° (dec). $[α]_D^{25}$ +17.0° (c = 1 in water). pKa'$_1$ 7.00, pKa'$_2$ 8.75. Soly in water, DMSO: 225, 5-10 (mg/ml). Practically insol in pyridine, chloroform, ethanol, ethyl acetate, cyclohexane, benzene, acetone, DMF, dioxane, acetonitrile.

THERAP CAT: Antibacterial.

THERAP CAT (VET): Antibacterial.

8891. Spectrin. Tektin A. Major protein component of the membrane cytoskeleton of mammalian and avian erythroid cells. Responsible for maintaining the normal shape, strength and stability of the erythrocyte. A ubiquitous family of proteins, spectrin analogs have also been isolated from a variety of non-erythroid cells. Given the name spectrin because of its original isolation from hemoglobin-free red cell membranes known as erythrocyte "ghosts": V. T. Marchesi, E. Steers, *Science* **159**, 203 (1968). Comparison with actin, *q.v.:* E. Steers, V. Marchesi, *J. Gen. Physiol.* **54**, 65 S (1969). Isoln from human red cells and species comparison: T. W. Tillack *et al., Biochim. Biophys. Acta* **200**, 125 (1970). Spectrin is a heterodimer composed of 2 high mol wt polypeptide subunits referred to as band 1 or α-spectrin (mol wt ~240 kDa) and band 2 or β-spectrin (mol wt ~220 kDa). Structural studies: M. Clarke, *Biochem. Biophys. Res. Commun.* **45**, 1063 (1971); T. L. Steck, *J. Cell Biol.* **62**, 1 (1974); J. M. Anderson, *J. Biol. Chem.* **254**, 939 (1979); D. W. Speicher *et al., ibid.* **257**, 9093 (1982). The β-subunit of spectrin is bound to the red cell membrane by proteins known as *ankyrin:* V. Bennett, P. J. Stenbuck, *ibid.* **254**, 2533 (1979); or *syndeins:* J. Yu, S. R. Goodman, *Proc. Nat. Acad. Sci. USA* **76**, 2340 (1979). The spectrin heterodimers aggregate on the membrane surface and, together with actin and other proteins, form a filamentous network that covers the surface of the cytoplasmic membrane: S. E. Lux, *Nature* **281**, 426 (1979); *idem, Sem. Hematol.* **16**, 21 (1979). Identification of spectrin analogs from nonerythroid cells: S. R. Goodman *et al., Proc. Nat. Acad. Sci. USA* **78**, 7580 (1981). Isoln of *fodrin*, also known as brain or neural cell spectrin: J. Levine, M. Willard, *J. Cell Biol.* **90**, 631 (1981); V. Bennett *et al., Nature* **299**, 126 (1982). Isoln of *TW 260/240*, a spectrin analog from chicken intestinal epithelial cells: J. R. Glenney *et al., Cell* **28**, 843 (1982). Comparison of erythroid and non-erythroid spectrins: J. R. Glenney *et al., Proc. Nat. Acad. Sci. USA* **79**, 4002 (1982); J. R. Glenney, P. Glenney, *Eur. J. Biochem.* **144**, 529 (1984). Review of erythrocyte spectrin: S. R. Goodman, K. Shiffer, *Am. J. Physiol.* **244**, C121-C141 (1983); of neural cell spectrin: S. R. Goodman, I. S. Zagon, *ibid.* **250**, C347-C360 (1986). Comprehensive review: S. R. Goodman *et al., Crit. Rev. Biochem.* **23**, 171-234 (1988).

8892. Spermaceti. Cetaceum; Spermwax. A waxy substance from the head of the sperm whale. *Constit.* Chiefly cetyl palmitate; free cetyl alcohol present in appreciable amounts; esters of lauric, stearic, and myristic acids; esters of higher alcohols also present.

White, somewhat translucent, slightly unctuous masses with crystalline fracture and pearly luster; almost odorless and tasteless but becomes yellow and rancid on long exposure to air. d 0.938-0.944; mp 42-50°; n_D^{80} ~1.4330. Sapon no. 120-136. Iodine no. 3-4.4. Insol in water or cold alcohol. Sol in chloroform, ether, carbon disulfide, oils, boiling alcohol; slightly sol in petr ether.

USE: As a base for ointments, cerates, etc., and as emulsion with egg yolk or expressed almond oil. In manuf of candles, soaps, cosmetics, laundry wax; finishing and lustering linens. Emollient.

8893. Spermidine. *N-(3-Aminopropyl)-1,4-butanediamine; N-(γ-aminopropyl)tetramethylenediamine.* $C_7H_{19}N_3$; mol wt 145.25. C 57.89%, H 13.18%, N 28.93%. NH_2-$(CH_2)_4NH(CH_2)_3NH_2$. Biogenic polyamine formed from putrescine; a precursor of spermine, *q.q.v.* First detected in human sperm, but occurs widely in nature. Essential for both normal and neoplastic tissue growth. Synthesis: H. W. Dudley *et al., Biochem. J.* **20**, 1082 (1926); **21**, 97 (1927); J. V. Braun, W. Pinkernelle, *Ber.* **70**, 1234 (1937); M. Danzig, H. P. Schultz, *J. Am. Chem. Soc.* **74**, 1836 (1952). For reviews of the early literature, *see* M. Guggenheim, *Die biogenen Amine* (S. Karger, Basel, 4th ed., 1951) 619 pp; H. Tabor *et al., Ann. Rev. Biochem.* **30**, 579-604 (1961). Role in cell growth processes: C. W. Tabor, H. Tabor, *ibid.* **45**, 285 (1976); J. Janne *et al., Biochim. Biophys. Acta* **473**, 241 (1978); C. W. Porter, R. J. Bergeron, *Science* **219**, 1083 (1983). Formation and interconversion of spermidine and putrescine in mammalian cells: A. E. Pegg *et al., Advan. Enzyme Regul.* **19**, 427 (1980). Effect on polypeptide chain elongation *in vitro:* A. K. Abraham, A. Pihl, *Eur. J. Biochem.* **106**, 257 (1980). Regulation of tRNA methyltransferase activity: M. Mach *et al., Biochem. J.* **202**, 153 (1982). HPLC study: C. E. Prussak, D. H. Russell, *J. Chromatog.*

229, 47 (1982). Interaction with actin, *q.v.:* C. Oriol-Audit, *Biochem. Biophys. Res. Commun.* **105**, 1096 (1982). Studies on use as a biochemical tool in cancer research: A. Thyss *et al., Eur. J. Cancer Clin. Oncol.* **18**, 611 (1982); V. Quemener *et al., J. Nat. Prod.* **45**, 608 (1982); C. W. Porter *et al., Cancer Res.* **42**, 4072 (1982). Toxicity study: P. R. Langford *et al., J. Antibiot.* **35**, 1387 (1982). *Reviews:* T. C. Theoharides, *Life Sci.* **27**, 703-713 (1980); O. Heby, *Differentiation* **19**, 1-20 (1981); L. Stevens, *Med. Biol.* **59**, 308-313 (1981). Book: *Polyamines in Biology and Medicine,* D. R. Morris, L. J. Marton, Eds. (Dekker, New York, 1981) 512 pp.

Liq, bp_{14} 128-130°. Sol in water, ethanol, ether.

Trihydrochloride, $C_7H_{22}Cl_3N_3$, cryst from ethanolic HCl (65:2). mp 256-258°.

USE: As a tool in biochemical research.

8894. Spermine. *N,N'-Bis(3-aminopropyl)-1,4-butanediamine; N,N'*-bis(3-aminopropyl)tetramethylenediamine; gerontine; musculamine; neuridine. $C_{10}H_{26}N_4$; mol wt 202.34. C 59.36%, H 12.95%, N 27.69%. $H_2N(CH_2)_3NH$-$(CH_2)_4NH(CH_2)_3NH_2$. Biogenic polyamine formed from spermidine, *q.v.,* and occurring in almost all tissues. Essential for both normal and neoplastic tissue growth. First observed in human semen and described as the cryst phosphate salt by A. von Leeuwenhoek, *Phil. Trans. Roy. Soc. London* **12**, 1040 (1678). For a description of the history, occurrence, formation, and early prepns of spermine, *see Beilstein* **vol. IV**, Suppl. 1, 704; M. Guggenheim, *Die biogenen Amine* (S. Karger, Basel, 4th ed., 1951) 619 pp; H. Tabor *et al., Ann. Rev. Biochem.* **30**, 579-604 (1961). Prepn of spermine and its tetrahydrochloride: Israel *et al., J. Med. Chem.* **7**, 710 (1964). Role in cell growth processes: C. W. Tabor, H. Tabor, *Ann. Rev. Biochem.* **45**, 285 (1976); J. Janne *et al., Biochim. Biophys. Acta* **473**, 241 (1978). Modulation of calcium-dependent immune processes: T. C. Theoharides, *Life Sci.* **27**, 703 (1980). Biosynthesis in fungi: L. Stevens, *Med. Biol.* **59**, 308 (1981). HPLC study: C. E. Prussak, D. H. Russell, *J. Chromatog.* **229**, 47 (1982). Use as a biochemical marker for malignant tumors: Y. Horn *et al., Cancer Res.* **42**, 3248 (1982). Metabolic study: A. E. Pegg *et al., Biochemistry* **21**, 5082 (1982). Review of role in cell proliferation and differentiation: O. Heby, *Differentiation* **19**, 1-20 (1981). Book: *Polyamines in Biology and Medicine,* D. R. Morris, L. J. Marton, Eds. (Dekker, New York, 1981) 512 pp.

Needles, mp 55-60°. Liq, $bp_{0.5}$ 141-142°. Strong base, absorbs carbon dioxide from air. *Keep well closed.* Sol in water, lower alcohols, chloroform. Practically insol in ether, benzene, petr ether.

Diphosphate hexahydrate, $C_{10}H_{32}N_4O_8P_2.6H_2O$, *spermine phosphate.* Cryst from water, mp 230-234° (dec). Soly in water: 0.037% at 20°; 1% at 100°. Insol in alc, ether and other organic solvents. Sol in dilute acids and alkali. Known as *Charcot-Neumann crystals,* also found in spleen, blood, bone marrow in leukemia and secretions in asthma.

Tetrahydrochloride, $C_{10}H_{30}Cl_4N_4$, cryst from ethanol, mp 312-314.5°.

USE: As a tool in biochemical research.

8895. Sperm Oil. From sperm whale. *Constit.* Esters of fatty acids; small quantities of spermaceti.

Yellow, thin liquid; slight fishy odor if not of good quality. d 0.875-0.884. Sapon no. 123-147. Iodine no. 80-84. Insoluble in water, cold alcohol, or petr ether; sol in chloroform, ether.

USE: As lubricant, in lamps, hardening steel, manuf soap.

8896. Spherophysine. *N-(4-Aminobutyl)-N-(3-methyl-2-butenyl)guanidine;* sphaerophysine. $C_{10}H_{22}N_4$; mol wt 198.31. C 60.57%, H 11.18%, N 28.25%. Ganglion blocking agent. Isolated from the following Central Asiatic *Leguminosae: Swainsona salsula* (Pall.) Taub. (*Sphaerophysa salsula* (Pall.) DC.); *Smirnowia turkestana* Bunge; *Eremosparton flaccidum* Litw.: Rubinshtein, Menshikov, *J. Gen. Chem. USSR* **14**, 161 (1944); Rjabinin, *ibid.* **17**, 2265 (1947); Merlis, *ibid.* **22**, 347 (1952). Structure: Heesing, Eckard, *Ber.* **103**, 534 (1970).

Strong, diacidic base.

Dibenzoate, $C_{24}H_{34}N_4O_4$, crystals, mp 149-150°.

8897. Sphingofungins. Family of six antifungal agents, sphingofungins A-F, isolated from *Aspergillus fumigatus* of which sphingofungin C is the most abundant. Specific inhibitors of serine palmitoyltransferase. Isoln and charaterization of sphingofungins A-D: F. VanMiddlesworth *et al., J. Antibiot.* **45**, 861 (1992); of sphingofungins E-F: W. S. Horn *et al., ibid.* 1692. Stereochemistry of A-D: F. VanMiddlesworth *et al., Tetrahedron Letters* **33**, 297 (1992). Mechanism of action: M. M. Zweerink *et al., J. Biol. Chem.* **267**, 25032 (1992). Synthesis of D: K. Mori, K. Otaka, *Tetrahedron Letters* **35**, 9207 (1994).

Sphingofungin C

Sphingofungin C, $C_{22}H_{41}NO_7$, *5-(acetyloxy)-2-amino-3,4,-14-trihydroxy-6-eicosenoic acid.* Cream colored solid.

8898. Sphingomyelins. Rowamyelin. Sphingosine phosphatides occurring in the myelin sheaths of nerves. Prepn of beef heart sphingomyelin: Rapport, Lerner, *J. Biol. Chem.* **232**, 63 (1958). Isoln from human and rat brain: Hausheer *et al., Helv. Chim. Acta* **46**, 601 (1963). Structure: Fujino, *J. Biochem. (Japan)* **39**, 45 (1952); Rouser *et al., J. Am. Chem. Soc.* **75**, 310 (1953); Marinetti *et al., ibid.* 313. Structure of honey bee *(Apis mellifera)* sphingomyelin: Karlander *et al., Biochim. Biophys. Acta* **270**, 117 (1972). Synthesis of dihydrosphingomyelins: Shapiro *et al., J. Am. Chem. Soc.* **80**, 2339 (1958). Synthesis of sphingomyelins: Shapiro *et al., ibid.* **81**, 3743, 4360 (1959). Configuration: Shapiro, Flowers, *ibid.* **84**, 1047 (1962). Monograph: S. R. Korey, *The Biology of Myelin* (Hoeber-Harper, New York, 1959).

R = different fatty acids such as stearic, palmitic, lignoceric, nervonic acids.

Slightly hygroscopic powder. Occurs as a hydrate. Sol in hot abs alcohol, acetic acid, chloroform; slightly sol in pyridine. Practically insol in ether, acetone, water.

8899. Sphingosine. *[R-[R*,S*-(E)]]-2-Amino-4-octadecene-1,3-diol; (E)-D-erythro-4-octadecene-1,3-diol;* 1,3-dihydroxy-2-amino-4-octadecene; 4-sphingenine. $C_{18}H_{37}NO_2$; mol wt 299.50. C 72.19%, H 12.45%, N 4.68%, O 10.68%. Important membrane component; long-chain amino-dialcohol; moiety of certain phosphatides, such as sphingomyelins, cerebrosides, and gangliosides. Natural sphingosine is D(+)-*erythro*-1,3-dihydroxy-2-amino-4-*trans*-octadecene. Does not exist in the free state in animals, plants or microorganisms. First obtained by hydrolysis of cerebrosides from brains: J. L. W. Thudichum, *Die Konstitution des Gehirns des Menschen und der Tiere* (Tübingen, 1901). Separation procedures: Carter *et al., J. Biol. Chem.* **170**, 269

(1947); Wittenberg, *ibid.* **216,** 379 (1955); Tipton, *Biochem. Prepns.* **9,** 127 (1962). Stereospecific synthesis of DL-*erythro-trans*-form and isomers: C. A. Grob, F. Gadient, *Helv. Chim. Acta* **40,** 1145 (1957). Stereoselective synthesis: H. Newman, *J. Am. Chem. Soc.* **95,** 4098 (1973); B. Bernet, A. Vasella, *Tetrahedron Letters* **24,** 5491 (1983). Comprehensive review: C. A. Grob, *Record Chem. Progr. (Kresge-Hooker Sci. Lib.)* **18,** 55-66 (1957); D. Shapiro, *Chemistry of Sphingolipids* (Hermann, Paris, 1969) 111 pp. Bibliography: *Rodd's Chemistry of Carbon Compounds,* **Vol. I,** part E, S. Coffey, Ed. (Elsevier, New York, 2nd ed., 1976) pp 394-397.

DL-Sphingosine, waxy crystals from ether + pentane, mp 67°.

DL-Triacetylsphingosine, $C_{24}H_{43}NO_5$, crystals from pentane + ether, mp 91-92°.

Triacetyl derivative of natural sphingosine, $C_{24}H_{43}NO_5$, crystals, mp 101-102°. $[\alpha]_D^{25}$ $-11.7°$ (chloroform).

8900. Spigelia. Pinkroot; Indian pink; Carolina pink; Maryland pink; wormgrass. Dried rhizome and roots of *Spigelia marilandica* L., *Loganiaceae*. *Habit.* North America (New Jersey to Florida and west to Wisconsin). *Constit.* Spigeline, resin, tannin, bitter principle, volatile oil.

THERAP CAT: Anthelmintic.

8901. α-Spinasterol. *(3β,5α,22E)-Stigmasta-7,22-dien-3-ol;* α-spinasterin; bessisterol; hitodesterol. $C_{29}H_{48}O$; mol wt 412.70. C 84.40%, H 11.72%, O 3.88%. Stereoisomeric with chondrillasterol, *q.v.* Extracted from spinach leaves: Hart, Heyl, *J. Biol. Chem.* **95,** 311 (1932); from *Citrullus colocynthis* Schrad., *Cucurbitaceae*: Hamilton, Kermack, *J. Chem. Soc.* **1952,** 5051. Structure: Fieser *et al., J. Am. Chem. Soc.* **71,** 2226 (1949). Prepn: Kircher, Rosenstein, *J. Org. Chem.* **38,** 2259 (1973); M. Anastasia *et al., J. Chem. Soc. Perkin Trans. I* **1981,** 2561.

Crystals from alcohol + light petr, mp 168-169°. $[\alpha]_D^{25}$ $-3.6°$ (c = 2.8 in chloroform).

Benzoate, $C_{36}H_{52}O_2$, plates from ethyl acetate, mp 200°. $[\alpha]_D^{25}$ $+1.8°$ (c = 1.7 in chloroform).

8902. Spinulosin. *2,5-Dihydroxy-3-methoxy-6-methyl-2,5-cyclohexadiene-1,4-dione; 2,5-dihydroxy-3-methoxy-6-methyl-p-benzoquinone;* 3,6-dihydroxy-5-methoxy-*p*-toluquinone; 3,6-dihydroxy-4-methoxy-2,5-toluquinone; hydroxyfumigatin. $C_8H_8O_5$; mol wt 184.15. C 52.18%, H 4.38%, O 43.44%. Metabolic product of the fungi *Penicillium spinulosum* Thom. and *Aspergillus fumigatus* Fres.: Birckinshaw, Raistrick, *Philos. Trans.* **B220,** 245 (1931); Anslow, Raistrick, *Biochem. J.* **32,** 2288 (1938); Pettersson, *Acta Chem. Scand.* **17,** 1771 (1963). Structure: Aulin, Erdtman, *Svensk Kem. Tidskr.* **49,** 208 (1937). Synthesis: Anslow, Raistrick, *Biochem. J.* **32,** 803 (1938). Biosynthesis: Pettersson, *Acta Chem. Scand.* **19,** 1016 (1965).

Purple-black crystals, mp 202°. Slightly sol in cold water; more sol in hot water; sol in dil NaOH soln giving a purple soln with a bluish hue.

8903. Spiperone. *8-[4-(4-Fluorophenyl)-4-oxobutyl]-1-phenyl-1,3,8-triazaspiro[4.5]decan-4-one; 8-[3-(p-fluorobenzoyl)propyl]-1-phenyl-1,3,8-triazaspiro[4.5]decan-4-one;* 4-phenyl-8-[3-(4-fluorobenzoyl)propyl]-1-oxo-2,4,8-triazaspiro-[4.5]decane; Spiropitan. $C_{23}H_{26}FN_3O_2$; mol wt 395.48. C 69.85%, H 6.63%, F 4.80%, N 10.63%, O 8.09%. Prepn: Janssen, U.S. pats. **3,155,669; 3,155,670** and **3,161,644** (all 1964 to Janssen).

Crystals, mp 190-193.6°.

THERAP CAT: Antipsychotic.

8904. Spiramycin. RP-5337; Selectomycin; Rovamicina; Rovamycin. Antibiotic substance classified in the erythromycin-carbomycin group and produced by *Streptomyces ambofaciens* from soil of northern France: Cosar *et al., Compt. Rend. Soc. Biol.* **234,** 1498 (1952); Pinnert-Sindico *et al., Antibiot. Ann.* **1954-1955,** 724; Ninet, Verrier, U.S. pat. **2,943,023** (1960 to Rhône-Poulenc), *see also* U.S. pat. **3,000,785** (1961 to Rhône-Poulenc). Antibacterial activity and toxicity: H. Sous *et al., Arzneimittel-Forsch.* **8,** 386 (1958). Separation into 3 components named spiramycin I, II and III: Preud'homme, Charpentier, U.S. pats. **2,978,380** and **3,011,947** (1961 to Rhône-Poulenc). Structure: Kuehne, Benson, *J. Am. Chem. Soc.* **87,** 4660 (1965). Revised structure: Omura *et al., ibid.* **91,** 3401 (1969); Mitscher *et al., J. Antibiot.* **26,** 55 (1973). Revised configuration at C-9: Freiberg *et al., J. Org. Chem.* **39,** 2474 (1974). Symposium on pharmacology, antibacterial spectrum, and clinical efficacy: *J. Antimicrob. Chemother.* **22,** Suppl. B, 1-213 (1988).

Spiramycin I	R = H
Spiramycin II	R = COCH_3
Spiramycin III	R = COCH_2CH_3

Amorphous base, slightly sol in water. $[\alpha]_D^{20}$ $-80°$ (methanol). uv max (ethanol): 231 nm. Sol in most organic solvents. Active on gram-positive bacteria and rickettsiae. Cross resistance between microorganisms resistant to erythromycin and carbomycin. LD$_{50}$ in rats (mg/kg): 9400 orally; 1000 s.c.; 170 i.v. (Sous).

Embonate, *Spira 200.*

Hexanedioate, *spiramycin adipate, Stomamycin, Suanovil.*

Spiramycin I, C$_{43}$H$_{74}$N$_2$O$_{14}$, *Foromacidin A.* Crystals, mp 134-137°. [α]$_D^{20}$ —96°.

Spiramycin I triacetate, crystals, mp 140-142°. [α]$_D^{20}$ —92.5°.

Spiramycin II, C$_{45}$H$_{76}$N$_2$O$_{15}$, *Foromacidin B.* Crystals, mp 130-133°. [α]$_D^{20}$ —86°.

Spiramycin II diacetate, crystals from cyclohexane, mp 156-160°. [α]$_D^{20}$ —98.4°.

Spiramycin III, C$_{46}$H$_{78}$N$_2$O$_{15}$, *Foromacidin C.* Crystals, mp 128-131°. [α]$_D^{20}$ —83°.

Spiramycin III diacetate, crystals from cyclohexane, mp 140-142°. [α]$_D^{20}$ —90.4°.

THERAP CAT: Antibacterial.

THERAP CAT (VET): Antibacterial; growth promotant.

8905. Spirapril. *[8S-[7[R*(R*)],8R*]]-7-[2-[[1-(ethoxy-carbonyl)-3-phenylpropyl]amino]-1-oxopropyl]-1,4-dithia-7-azaspiro[4.4]nonane-8-carboxylic acid;* (8S)-7[(S)-N-[(S)-1-carboxy-3-phenylpropyl]alanyl]-1,4-dithia-7-azaspiro[4.4]-nonane-8-carboxylic acid 1-ethyl ester. C$_{22}$H$_{30}$N$_2$O$_5$S$_2$; mol wt 466.62. C 56.63%, H 6.48%, N 6.00%, O 17.14%, S 13.74%. Angiotensin-converting enzyme (ACE) inhibitor. Prepn: E. H. Gold *et al.,* U.S. pat. **4,470,972** (1984 to Schering); E. M. Smith *et al., J. Med. Chem.* **32,** 1600 (1989). Pharmacology: E. J. Sybertz *et al., Arch. Int. Pharmacodyn. Ther.* **286,** 216 (1987); T. Baum *et al., ibid.* 230. Radioimmunoassay in plasma: M. Hossein-Nia *et al., Ther. Drug Monit.* **14,** 234 (1992). Pharmacokinetics and evaluation in CHF: S. A. J. van den Broek *et al., J. Cardiovasc. Pharmacol.* **18,** 614 (1991). Clinical evaluation in essential hypertension: G. P. Reams *et al., J. Clin. Pharmacol.* **33,** 348 (1993). *Review:* P. Jerie, H.-J. Kremer, *Cor Vasa* **34,** 82-87 (1992).

Hemihydrate, white foam. [α]$_D^{26}$ —29.5° (c = 0.2 in ethanol).

Hydrochloride, C$_{22}$H$_{30}$N$_2$O$_5$S$_2$.HCl, *Sch-33844, TI-211-950, Renormax, Renpress, Sandopril.* White solid, mp 192-194° (dec). [α]$_D^{26}$ —11.2° (c = 0.4 in ethanol).

Diacid, C$_{20}$H$_{26}$N$_2$O$_5$S$_2$, *spiraprilat, spiraprilic acid, Sch-33861.* Hemihydrate, mp 163-165°. [α]$_D^{26}$ +4.1° (c = 0.4 in ethanol).

THERAP CAT: Antihypertensive.

8906. Spirit of Ammonia, Aromatic. Spirit of hartshorn, aromatic. A soln contg 34 g ammonium carbonate, 90 ml 10% ammonia water, 10 ml lemon oil, 1 ml oil of lavender, 1 ml oil of myristica, 700 ml alcohol and a sufficient quantity of water to make 1 liter. Absolute alc by vol, ~66%.

Almost colorless to slightly yellow liquid with an aromatic, pungent odor and the taste of ammonia. d 0.90. Forms an opalescent mixture with water; miscible with alcohol.

THERAP CAT: Reflex respiratory stimulant.

THERAP CAT (VET): By inhalation: respiratory and circulatory stimulant. Internally: expectorant, diaphoretic, antacid, carminative. Externally: counterirritant, and in diluted form to relieve the irritation of insect stings and bites.

8907. Spirit of Camphor. A soln of camphor in alcohol contg 10 g camphor per 100 ml soln.

Colorless liquid; camphor odor.

THERAP CAT: Carminative; analgesic (topical); antipruritic.

THERAP CAT (VET): Has been used as rubefacient, stimulant, expectorant, carminative.

8908. Spirit of Chloroform. An alcoholic soln of chloroform contg 6% by vol of chloroform, corresponding to 10.5% by wt, and ~89% abs alcohol by vol.

Colorless, clear liquid; chloroform odor. d ~0.85.

THERAP CAT: Carminative.

8909. Spirit of Ether. Hoffmann's drops. A mixture of alcohol and ether. Made by mixing 325 ml ether U.S.P. with sufficient U.S.P. alcohol (~690 ml) to make 1000 ml. Absolute alc by vol, ~66%.

Colorless, volatile liquid. d 0.785.

THERAP CAT: Carminative.

8910. Spirit of Ether Compound. Hoffmann's anodyne. A mixture of 325 ml ether U.S.P., 25 ml ethereal oil and sufficient alcohol (~665 ml) to make 1000 ml. Absolute alc by vol, ~64%.

Colorless, clear liquid. d ~0.785.

THERAP CAT: Carminative.

THERAP CAT (VET): Has been used as a stomachic, carminative, in colic.

8911. Spirit of Ethyl Nitrite. Spirit of nitrous ether; sweet spirit of niter. An alcoholic soln of ethyl nitrite, contg 3.5-4.5% ethyl nitrite. *Incompat:* Antipyrine, tannin, acetanilide, acetophenetidin, iodides, tincture guaiac, morphine salts, carbonates, acacia.

Pale yellow or faintly greenish-yellow, clear, mobile, volatile, flammable liquid; ethereal, pungent odor; sharp, burning taste. The ethyl nitrite evaporates rapidly; dec on exposure to air; dec in light. d not above 0.823 at 25°. Miscible with water and alcohol. *Keep tightly closed, protected from light, in a cool place, remote from fire.*

THERAP CAT: Diaphoretic, diuretic.

8912. Spirit of Formic Acid. Spirit of ants. Composed of 40 ml of 25% formic acid and 225 ml water in sufficient alcohol to make one liter; corresponds to ~1% formic acid and ~70% abs alcohol by vol.

Colorless liquid. Miscible with water, alcohol.

8913. Spirit of Glyceryl Trinitrate. Spirit of nitroglycerin; spirit of trinitroglycerin; spirit of glonoin; soln of glyceryl trinitrate. An alcoholic soln contg 1.0-1.1% glyceryl trinitrate. *Incompat:* Alkalies, carbonates, HCl, HI.

Colorless, clear liquid. d 0.814-0.820. Miscible with alcohol, chloroform, ether; 1 ml dissolves in 6 ml almond oil; very slightly sol in water; miscible with chloroform, ether.

Caution: Likely to produce violent headache when tasted or applied to the skin.

Note: When spilled, immediately pour NaOH soln over it or dry residue may explode by friction or shock.

THERAP CAT: Vasodilator.

8914. Spirit of Peppermint. An alcoholic soln contg per liter 100 ml oil of peppermint and the alcohol-soluble principles from 10 g of powdered peppermint previously macerated with water.

THERAP CAT: Carminative.

THERAP CAT (VET): Has been used as a carminative.

8915. Spirit of Spearmint. An alcoholic soln contg per liter 100 ml oil of spearmint and the alcohol-soluble principles from 10 g coarsely powdered spearmint leaves previously macerated with water.

THERAP CAT: Carminative.

THERAP CAT (VET): Has been used as a carminative.

8916. Spirogermanium. *8,8-Diethyl-N,N-dimethyl-2-aza-8-germaspiro[4.5]decane-2-propanamine;* 2-[3-(dimethyl-amino)propyl]-8,8-diethyl-2-aza-8-germaspiro[4.5]decane. C$_{17}$H$_{36}$GeN$_2$; mol wt 341.10. C 59.86%, H 10.64%, Ge 21.29%, N 8.21%. Cytostatic germanium deriv. Prepn: L. M. Rice, Ger. pat. **2,243,550;** *eidem,* U.S. pat. **3,825,546** (1973, 1974 both to Geschickter Fund for Med. Res.); L. M. Rice *et al., J. Heterocycl. Chem.* **11,** 1041 (1974). Preclinical toxicological evaluation: M. C. Henry, C. D. Port, *U.S. NTIS Rep.* **PB-264117** (1977) 213 pp, *C.A.* **87,** 177620 (1977). Early clinical studies: P. S. Schein *et al., Cancer Treat. Rep.* **64,** 1051 (1980); C. Tropie *et al., ibid.* **65,** 119 (1981). Toxicity study: M. C. Henry *et al., ibid.* **64,** 1207 (1980).

Oil, $bp_{0.03}$ 106-109°.

Dihydrochloride, $C_{17}H_{36}GeN_2 \cdot 2HCl$, NSC-192965, Spiro-32. Crystals, mp 287-288°. LD_{50} in mice (mg/m^2): 128.4 i.v.; 401.1 i.m. (Henry, Port). Also reported as LD_{50} in mice (mg/kg): 324 orally; 150 i.v. (Rice et al.).

THERAP CAT: Antineoplastic.

8917. Spironolactone. *(7α,17α)-7-(Acetylthio)-17-hydroxy-3-oxopregn-4-ene-21-carboxylic acid γ-lactone; 17-hydroxy-7α-mercapto-3-oxo-17α-pregn-4-ene-21-carboxylic acid γ-lactone, acetate;* 3-(3-oxo-7α-acetylthio-17β-hydroxy-4-androsten-17α-yl)propionic acid γ-lactone; SC-9420; Aldace; Aldactone; Aldopur; Almatol; Altex; Aquareduct; Deverol; Diatensec; Dira; Duraspiron; Euteberol; Lacalmin; Lacdene; Laractone; Nefurofan; Osiren; Osyrol; Sagisal; Sincomen; Spiretic; Spiroctan; Spiroderm; Spirolone; Spiro-Tablinen; Supra-Puren; Suracton; Urusonin; Verospiron; Xenalon. $C_{24}H_{32}O_4S$; mol wt 416.58. C 69.20%, H 7.74%, O 15.36%, S 7.70%. Aldosterone antagonist. Prepn: Cella, Tweit, *J. Org. Chem.* **24**, 1109 (1959); U.S. pat. **3,013,012** (1961 to Searle); Tweit et al., *J. Org. Chem.* **27**, 3325 (1962); Kiprianov et al., *Khim. Farm. Zh.* **3**, 10 (1969), *C.A.* **71**, 70779g (1969). Activity and metabolic studies: Gerhards, Engelhardt, *Arzneimittel-Forsch.* **13**, 972 (1963); Rosenthale et al., *Proc. Soc. Exp. Biol. Med.* **118**, 806 (1965); Karim, Brown, *Steroids* **20**, 41 (1972). Crystal and molecular structure: Dideberg, Dupont, *Acta Crystallogr. Sect. B* **28**, 3014 (1972). Comprehensive description: J. L. Sutter, E. P. K. Lau, in *Analytical Profiles of Drug Substances* vol. 4, K. Florey, Ed. (Academic Press, New York, 1975) pp 431-451.

Crystals from methanol, mp 134-135° (resolidifies and dec 201-202°). $[α]_D^{20}$ −33.5° (chloroform). uv max: 238 nm (ε 20200). Practically insol in water. Sol in most organic solvents.

THERAP CAT: Diuretic.

THERAP CAT (VET): Has been used as diuretic.

8918. Spizofurone. *5-Acetylspiro[benzofuran-2(3H),1'-cyclopropan]-3-one;* AG-629; Maon. $C_{12}H_{10}O_3$; mol wt 202.21. C 71.28%, H 4.98%, O 23.74%. Prepn: H. Sugihara et al., *Eur. pat. Appl.* **3,084**; eidem, U.S. pat. **4,284,644** (1979, 1981 both to Takeda); M. Kawada et al., *Chem. Pharm. Bull.* **32**, 3532 (1984). One step synthesis: M. Watanabe et al., *ibid.* 3373. Anti-ulcerative effect in animals: N. Inatomi et al., *Arzneimittel-Forsch.* **35**, 1553 (1985). Cytoprotective effects on rat gastric mucosa: eidem, *Eur. J. Pharmacol.* **112**, 81 (1985). Stimulation of endogenous prostaglandin synthesis in vitro: I. Inada et al., *ibid.* **124**, 149 (1986).

Crystals from ethanol, mp 102-104° (Kawada); also reported as mp 106-107° (Watanabe).

THERAP CAT: Antiulcerative.

8919. Splenin. Thymopoietin III; SP. Polypeptide hormone of the spleen which induces differentiation and maturation of both T- and B-cell precursors. Originally detected by its cross reactivity with thymopoietin, *q.v.*, in radioimmunoassay. Human splenin (hSP) contains 48 amino acids; bovine splenin (bSP) contains 49. *Splenopentin*, a pentapeptide corresponding to residues 32-36, exhibits full bioactivity. The amino acid sequence of splenin is highly homologous with that of thymopoietin; the differences in bioactivity are attributed to a single amino acid substitution at position 34. Unlike the thymic hormone, splenin exhibits no neuromuscular effects. Isoln from bovine spleen and complete amino acid sequence: T. Audhya et al., *Biochemistry* **20**, 6195 (1981). Synthesis: T. Abiko et al., *Chem. Pharm. Bull.* **34**, 2133 (1986). Isoln and sequence of human splenin: T. Audhya et al., *Proc. Nat. Acad. Sci. USA* **84**, 3545 (1987). Biological activities of bSP and comparison with thymopoietin: T. Audhya et al., *ibid.* **81**, 2847 (1984). Effect on antibody production in vivo: W. Diezel et al., *Biomed. Biochim. Acta* **45**, 1349 (1986). Effect of hSP on B-cell number in vitro: T. Abiko, H. Sekino, *Chem. Pharm. Bull.* **37**, 391 (1989).

Arg–Lys–Ala–Val–Tyr

Human Splenopentin

Note: The term, splenin, formerly referred to a crude mixture of bioactive substances extracted from mammalian spleen: G. Ungar, *Endocrinology* **37**, 329 (1945).

8920. Sporidesmins. Toxic fungal mixture from *Pithomyces chartarum* Ellis (*Sporidesmium bakeri* Sydow), composed of sporidesmins A-H and J, as causative agent of "facial eczema" in sheep. First isoln of A (major metabolite): R. L. M. Synge, E. P. White, *Chem. & Ind. (London)* **1959**, 1546. Isoln of A and B: J. W. Ronaldson et al., *J. Chem. Soc. (C)* **1963**, 3172. Structure of A: J. Fridrichsons, A. M. Mathieson, *Tetrahedron Letters* **1962**, 1265. Abs config of A: A. F. Beecham et al., *ibid.* **1966**, 3131. Isoln and structure of D and F: W. D. Jamieson et al., *J. Chem. Soc. (C)* **1969**, 1564; of E: R. Rahman et al., *ibid.* 1665; of G: E. Francis et al., *J. Chem. Soc. Perkin Trans. I* **1972**, 470; of H and J: R. Rahman et al., *ibid.* **1978**, 1476. Total synthesis of A: Y. Kishi et al., *J. Am. Chem. Soc.* **95**, 6493 (1973).

Sporidesmin A

Sporidesmin A, $C_{18}H_{20}ClN_3O_6S_2$, *(3α,5aα,10bα,11β,11aα)-9-chloro-2,3,5a,6,10b,11-hexahydro-10b,11-dihydroxy-7,8-dimethoxy-2,3,6-trimethyl-3,11a-epidithio-11aH-pyrazino[1',2':1,5]pyrrolo[2,3-b]indole-1,4-dione, sporidesmin.* Colorless needles with a faint green sheen, from aq methanol, mp 179° (dependent on rate of heating). $[α]_D^{20}$ −45° (c = 0.98 in methanol). uv max: 219, 253, 305 nm ($E_{1cm}^{1\%}$ 700, 220, 40). Very slightly sol in water, light petroleum, CCl_4. Readily sol in most organic solvents.

8921. Sporidesmolides. Cyclodepsipeptides containing both D and L amino acids, isolated from the pasture fungus *Sporidesmium bakeri* Syd. (*Pithomyces chartarum* (Berk. & Curt.) Ellis): D. W. Russell, *Biochim. Biophys. Acta* **45**, 411 (1960); W. S. Bertaud et al., *J. Gen. Microbiol.* **32**, 385 (1963); E. Bishop, D. W. Russell, *Biochem. J.* **92**, 19P (1964).

Consult the Name Index before using this section.

cyclo–Hiv–Val–MeLeu–Hiv–D-Val–D-Leu

Sporidesmolide I

Hiv = α-hydroxyisovaline

Sporidesmolide I. $C_{33}H_{58}N_4O_8$. Structure: D. W. Russell, *J. Chem. Soc.* **1962**, 753. Synthesis: M. M. Shemyakin *et al., Tetrahedron* **19**, 995 (1963). Biosynthetic study: D. W. Russell, *Biochim. Biophys. Acta* **261**, 469 (1972). MS determn: B. C. Das *et al., J. Antibiot.* **32**, 569 (1979). Needles from 70% acetic acid , mp 261-263°. $[\alpha]_D^{17}$ −217° (c = 1.5 in chloroform). Practically insol in water; very sol in chloroform; sparingly sol in other common organic solvents.
Sporidesmolide II. $C_{34}H_{60}N_4O_8$. *2-D-Alloisoleucine sporidesmolide I.* Structure and synthesis: M. M. Shemyakin *et al., Tetrahedron Lett.* **1963**, 1927. Crystals, mp 228-230°. $[\alpha]_D^{20}$ −195° (c = 0.6 in chloroform).
Sporidesmolide III. $C_{32}H_{56}N_4O_8$. *6-L-Leucine sporidesmolide I.* Synthesis: Y. A. Ovchinnikov *et al., ibid.* **1965**, 1111. Crystals, mp 277-278°. $[\alpha]_D$ −80° (c = 1.6 in acetic acid).
Sporidesmolide IV. $C_{34}H_{60}N_4O_8$. *4-(4-Methyl-L-2-hydroxypentanoic acid) sporidesmolide I.* Structure and synthesis: A. A. Kiryushkin *et al., ibid.* **1965**, 143; E. Bishop, D. W. Russell, *J. Chem. Soc. (C)* **1967**, 634. Crystals, mp 227-228°. $[\alpha]_D$ −195° (c = 1 in chloroform). Practically insol in water, sol in chloroform, moderately sol in common organic solvents.

8922. Squalamine. *(3β,5α,7α)-3-[[3-((4-Aminobutyl)-amino)propyl]amino]-cholestane-7,24-diol 24-hydrogen sulfate;* 3β-(N-[3-aminopropyl]-1,4-butanediamine)-7α,24ζ-dihydroxy-5α-cholestane 24-sulfate; 3β-N-1-[N-[3-(4-aminobutyl)]-1,3-diaminopropane]-7α,24ζ-dihydroxy-5α-cholestane 24-sulfate. $C_{34}H_{67}N_3O_5S$; mol wt 629.99. C 64.82%, H 10.72%, N 6.67%, O 12.70%, S 5.09%. Broad-spectrum aminosterol antibiotic present in shark tissues; novel indication for steroid as vertebrate host-defense agent. Isoln from the stomach of the spiny dogfish shark, *Squalus acanthias:* M. Zasloff *et al.,* U.S. pat. **5,192,756** (1993 to Children's Hospital of Penn.). Structural determn and antimicrobial activity: K. S. Moore *et al., Proc. Nat. Acad. Sci. USA* **90**, 1354 (1993). NMR spectral studies: S. L. Wehrli *et al., Steroids* **58**, 370 (1993). Synthesis of trihydrochloride: R. M. Moriarty *et al.,* PCT Int. pat. **Appl. 19,366** (1994 to Magainin Pharm.).

Sol in water.

8923. Squalane. *2,6,10,15,19,23-Hexamethyltetracosane;* perhydrosqualene; dodecahydrosqualene; spinacane; Cosbiol; Robane. $C_{30}H_{62}$; mol wt 422.82. C 85.22%, H 14.78%. $[(CH_3)_2CHCH_2CH_2CH_2CH(CH_3)CH_2CH_2CH_2CH-(CH_3)CH_2CH_2-]_2$. Prepn by complete hydrogenation of squalene, *q.v.:* Chapman, *J. Chem. Soc.* **123**, 770 (1923); Heilbron *et al., ibid.* **126**, 3135. Commercial grades are obtained by direct hydrogenation of shark liver oil and may contain some batyl alcohol: Tsujimoto, *Chem. Umschau Fette* **34**, 256 (1927), *C.A.* **21**, 4081 (1927). Synthesis: J. W. Scott, D. Valentine, *Org. Prep. Proced. Int.* **12**, 7 (1980); T. Mandai *et al., Tetrahedron Letters* **22**, 763 (1981).
Oil. Stable to air and oxygen. d_4^{15} 0.8115. mp ∼−38°. bp_{760} ∼350°; bp_{10} 263°; bp_5 248°. Flash pt 425°F (218°C). n_D^{15} 1.4530. Specific heat at 20° ∼0.62. Viscosity (Engler) at 20° ∼6.08. Readily sol in ether, gasoline, petr ether, benzene, chloroform, oils. Slightly sol in methanol, ethanol, acetone, glacial acetic acid. Concd H_2SO_4 at 70° is discolored, but the squalane remains unchanged.

USE: Lubricant, transformer oil. Ingredient of watch and chronometer oils. Perfume fixative. In pharmacy and cosmetics as skin lubricant, ingredient of suppositories, carrier of lipid-soluble drugs.

8924. Squalene. *(all-E)-2,6,10,15,19,23-Hexamethyl-2,-6,10,14,18,22-tetracosahexaene;* Spinacene; Supraene. $C_{30}H_{50}$; mol wt 410.73. C 87.73%, H 12.27%. All trans isoprenoid contg six isoprene units. Found in large quantities in shark liver oil. Occurs in smaller amounts (0.1 to 0.7%) in olive oil, wheat germ oil, rice bran oil, and yeast. Intermediate in biosynthesis of cholesterol. Isoln: Heilbron *et al., J. Chem. Soc.* **1926**, 1630. Structure: *eidem, ibid.* **1929**, 873; Heilbron, Thompson, *ibid.* 883; Karrer *et al., Helv. Chim. Acta* **13**, 1084 (1930); Karrer, Helfenstein, *ibid.* **14**, 78 (1931); Karrer *et al., ibid.* 435. Crystal structure: J. Ernst *et al., Angew. Chem. Int. Ed.* **15**, 778 (1976). Synthesis: Trippett, *Chem. & Ind. (London)* **1956**, 80; Dicker, Whiting, *J. Chem. Soc.* **1958**, 1994; Cornforth *et al., ibid.* **1959**, 2539; Johnson *et al., J. Am. Chem. Soc.* **92**, 741 (1970); Hirai *et al., Tetrahedron Letters* **1971**, 4359; P. A. Grieco, Y. Masaki, *J. Org. Chem.* **39**, 2135 (1974). Synthesis of squalene and trans-cis-trans-trans isomer: Biellmann, Ducep, *Tetrahedron* **27**, 5861 (1971).

Oil. Faint, agreeable odor. Absorbs oxygen and becomes viscous like linseed oil. d^{20} 0.8584; bp_{25} 285°; bp_2 240°; $bp_{0.15}$ 203°. mp ∼−75°. n_D^{20} 1.4965. Viscosity at 25°: 12 cps. Iodine no. 360-380. Flash pt ∼200°. Practically insol in water. Freely sol in ether, petr ether, CCl_4, acetone, other fat solvents; sparingly sol in alc, glacial acetic acid.
USE: Bactericide; intermediate in manuf of pharmaceuticals, organic coloring materials, rubber chemicals, aromatics and surface active agents.

8925. Squill. Sea onion; Bulbus Scillae; Meerzwiebel. The fleshy inner bulb scales of the white variety of *Urginea maritima* (L.) Baker (*Scilla maritima* L.), Liliaceae. Habit. Lands of the Mediterranean seacoast. Constit. Glucoscillaren A (scillarenin + rhamnose + glucose + glucose); scillaren A (scillarenin + rhamnose + glucose); proscillaridin A (scillarenin + rhamnose); scillaridin A; scilliglaucoside; scillipheoside; glucoscillipheoside; scillicyanoside; scillicoeloside; scilliazuroside; scillicryptoside. *Review:* G. Baumgarten, W. Förster, *Die Herzwirksamen Glykoside* (Thieme, Leipzig, 1963) pp 70-75 *et passim.*
USE: Rodenticide.
THERAP CAT: Diuretic, emetic, expectorant, cardiotonic.
THERAP CAT (VET): Has been used as expectorant, emetic.

8926. Stachydrine. *(S)-2-Carboxy-1,1-dimethylpyrrolidinium inner salt;* methyl hygrate betaine; hygric acid methylbetaine. $C_7H_{13}NO_2$; mol wt 143.19. C 58.72%, H 9.15%, N 9.78%, O 22.35%. Occurs widely in nature, especially in alfalfa, chrysanthemum, citrus and stachys species. May be prepd by methylation of proline. Isoln: Planta, Schulze, *Ber.* **26**, 939 (1893); Jahns, *ibid.* **29**, 2065 (1896); Yoshimura, *Z. Physiol. Chem.* **88**, 334 (1913); Vickery, *J. Biol. Chem.* **61**, 117 (1924). Structure: Schulze, Trier, *Z. Physiol. Chem.* **67**, 59 (1910). Synthesis: Karrer, Widmer, *Helv. Chim. Acta* **8**, 364 (1925). Biosynthesis: Robertson, Marion, *Can. J. Chem.* **38**, 396 (1960). *Review:* Marion in *The Alkaloids* vol. 1, R. H. F. Manske, H. L. Holmes, Eds. (Academic Press, New York, 1950) pp 101-103.

Monohydrate, deliquescent crystals, sweetish taste, mp 235° when anhydr. Isomerizes at the mp to methyl hygrate. Sol in water, alcohol, dil acids. Practically insol in ether, chloroform.

Hydrochloride, $C_7H_{13}NO_2 \cdot HCl$, large prisms from abs alcohol, dec 235°, very sol in water, sol in 13 parts alcohol.

Acid oxalate, $C_7H_{13}NO_2 \cdot C_2H_2O_4$, needles, mp 106°. Practically insol in abs alcohol.

Aurichloride, $C_7H_{13}NO_2 \cdot HAuCl_4$, yellow leaflets, mp 225° (rapid heating). Practically insol in cold water. Quite sol in hot water.

Platinichloride tetrahydrate, $(C_7H_{13}NO_2)_2 \cdot H_2PtCl_6 \cdot 4H_2O$, orange crystals, dec 210-220° (rapid heating), very sol in water and dil alc. Also obtained with 2 mols H_2O of crystn.

8927. Stallimycin. *N-[5-[[(3-Amino-3-iminopropyl)-amino]carbonyl]-1-methyl-1H-pyrrol-3-yl]-4-[[[4-(formyl-amino)-1-methyl-1H-pyrrol-2-yl]carbonyl]amino]-1-methyl-1H-pyrrole-2-carboxamide; β-[1-methyl-4-[1-methyl-4-(1-methyl-4-formylaminopyrrole-2-carboxamido)pyrrole-2-carboxamido]pyrrole-2-carboxamido]propionamidine; distamy-cin A; F.I. 6426.* $C_{22}H_{27}N_9O_4$; mol wt 481.51. C 54.88%, H 5.65%, N 26.18%, O 13.29%. Antibiotic substance with antiviral and oncolytic properties, obtained from *Streptomyces distallicus:* Arcamone *et al.,* Ger. pat. 1,039,198 (1958 to Farmaceutici Italia), *C.A.* 55, 2012f (1961). Structure and synthesis: Arcamone *et al., Nature* 203, 1064 (1964); *eidem, Gazz. Chim. Ital.* 97, 1097, 1110 (1967). Total synthesis: M. Bialer *et al., Tetrahedron* 1978, 2389; L. Grehn, U. Ragnarsson, *J. Org. Chem.* 46, 3492 (1981). Toxicity and antitumor activity: A. DiMarco *et al., Cancer Chemother. Rep.* 18, 15 (1962). *Review:* Hahn in *Antibiotics* vol. 3, J. W. Corcoran, F. E. Hahn, Eds. (Springer-Verlag, New York, 1975) pp 79-100.

mp 154-156°.

Hydrochloride, $C_{22}H_{27}N_9O_4 \cdot HCl$, *Herperal.* Crystals from dil HCl, mp 184-187° (Arcamone, 1967). Also reported as 189-193° from ethanol-ethyl acetate (Bialer). uv max (96% ethanol): 237, 303 nm (ε 30000, 37000). LD_{50} in mice (mg/kg): 75 i.v.; 500 i.p. (DiMarco).

THERAP CAT: Antiviral.

8928. Stannic Bromide. Tin tetrabromide. Br_4Sn; mol wt 438.33. Br 72.92%, Sn 27.08%. $SnBr_4$.

White, cryst mass; fumes strongly on exposure to air. d 3.34; mp 31°; bp 202°. Very sol in water with evolution of heat; also sol in alcohol. *Keep tightly closed.*

USE: In metallurgical separation of minerals.

8929. Stannic Chloride. Tin tetrachloride; fuming spirit of Libavius. Cl_4Sn; mol wt 260.52. Cl 54.43%, Sn 45.57%. $SnCl_4$. Improperly called *"tin bichloride".*

Fuming, caustic liquid. d 2.26; mp −33°; bp 114°. Sol in water and evolution of much heat; sol in alcohol, carbon tetrachloride, benzene, toluene, acetone, kerosene, gasoline. *Keep tightly closed.*

Pentahydrate, white or slightly yellow crystals or fused small lumps; slight HCl odor. Very sol in H_2O; sol in alc.

Caution: May be highly irritating to eyes, mucous membranes.

USE: As mordant; reviving colors; stabilizer for colors and perfumes in soap; in dyeing of fabrics, weighting silk, tinning vessels; dehydrating agent in organic syntheses; in ceramics to produce abrasion-resistant or light-reflecting coatings.

8930. Stannic Chromate(VI). Cr_2O_8Sn; mol wt 350.70. Cr 29.65%, O 36.50%, Sn 33.85%. $Sn(CrO_4)_2$.

Brownish-yellow, cryst powder; dec when heated. Sol in water.

USE: Decorating porcelain and china in rose and violet colors.

8931. Stannic Fluoride. Tin tetrafluoride. F_4Sn; mol wt 194.70. F 39.03%, Sn 60.97%. SnF_4. Lewis acid. Prepd from $SnCl_4$ and HF: Ruff, Plato, *Ber.* 37, 673 (1904); or from stannous fluoride and chlorine or bromine: Forbes,

Anderson, *J. Am. Chem. Soc.* 67, 1911 (1945); from stannous oxide or sulfide and fluorine: Haendler *et al., J. Am. Chem. Soc.* 76, 2179 (1954). *Review:* Kemmitt, Sharp, *Advan. Fluorine Chem.* 4, 185-186 (1965).

Snow-white, tetragonal crystals. Very hygroscopic. d_4^{19} 4.78. Sublimes at 705°. Hydrolyzes readily, but is more resistant to water than stannic chloride. Forms complexes with donor molecules.

USE: Friedel-Crafts catalyst.

8932. Stannic Iodide. Tin tetraiodide. I_4Sn; mol wt 626.33. I 81.05%, Sn 18.95%. SnI_4.

Yellow to reddish crystals. d 4.46. mp ~143°; sublimes at about 180°. bp 340°. Dec by water; sol in alcohol, benzene, chloroform, ether, carbon disulfide. MLD i.v. in rats: 200 mg/kg, *Handbook of Toxicology* vol. 1, W. S. Spector, Ed. (Saunders, Philadelphia, 1956) pp 282-283.

8933. Stannic Oxide. White tin oxide; tin dioxide; stannic anhydride; flowers of tin. O_2Sn; mol wt 150.71. O 21.23%, Sn 78.77%. SnO_2. Occurs in nature as the mineral *cassiterite.* The commercial grade is also known as *polishing powder, putty powder,* or *tin ash.*

White or slightly gray powder. d 6.95. Insol in water, alcohol, or cold acids. Slowly sol in hot concd potassium or sodium hydroxide soln.

USE: Polishing glass and metals; manuf milk-colored, ruby and alabaster glass, enamels, pottery, putty; mordant in printing and dyeing fabrics; in fingernail polishes.

8934. Stannic Selenide. Tin diselenide. Se_2Sn; mol wt 276.63. Se 57.09%, Sn 42.91%. $SnSe_2$. Prepd by passing the vapor of selenium over heated tin: Little, *On Selenium and Some of the Metallic Selenides,* Göttingen (Thesis, 1859); by treating a soln of stannic chloride with hydrogen selenide: Berzelius, cited in *Mellor's* vol. X, 785 (1930); by treating a soln of an alkali selenostannate or sulfoseleno-stannate with hydrochloric acid: Ditte, *Compt. Rend.* 95, 641 (1882).

Red-brown crystals. d 5.133 (Little); d 4.85 [Schneider, *Pogg. Ann.* 127, 624 (1866)]. mp 650°. Soluble in alkali, concd sulfuric acid, aqua regia, aq ammonia. Insol in water, dilute acids. Dec in nitric acid. Forms potassium selenostannate with potassium selenide; sodium selenostan-nate with sodium selenide.

8935. Stannic Selenite. O_6Se_2Sn; mol wt 372.63. O 25.76%, Se 42.38%, Sn 31.86%. $Sn(SeO_3)_2$. Prepd by treating stannic oxide with selenious acid: Berzelius, cited in *Mellor's* vol. X, 832 (1930).

Cryst powder. Insol in water. Sol in excess warm HCl.

8936. Stannic Sulfide. Tin disulfide; mosaic gold; tin bronze. S_2Sn; mol wt 182.84. S 35.08%, Sn 64.92%. SnS_2.

Golden leaflets with metallic luster; fatty feel to the touch. d 4.5. Insol in water or dil acids. Sol in aqua regia, in solns of alkali hydroxides or sulfides.

Note: The term "mosaic gold" is also used to designate an alloy consisting of 65.3% copper and 34.7% zinc.

USE: Gilding and bronzing metals, gypsum, wood and paper, usually by suspending in lacquer or varnish.

8937. Stannous Acetate. $C_4H_6O_4Sn$; mol wt 236.80. C 20.29%, H 2.55%, O 27.03%, Sn 50.13%. $Sn(C_2H_3O_2)_2$. Prepd by refluxing granulated tin with 98% glacial acetic acid: Colonna, *Gazz. Chim. Ital.* 35 II, 224 (1905); by refluxing SnO with 50% (v/v) acetic acid under nitrogen: Donaldson *et al., J. Chem. Soc.* 1964, 5942.

White, orthorhombic crystals; dec by water. mp 182.5-183°. d 2.31. Sol in dil HCl. *Keep well closed.*

USE: Reducing agent.

8938. Stannous Bromide. Tin dibromide. Br_2Sn; mol wt 278.52. Br 57.38%, Sn 42.62%. $SnBr_2$.

Yellowish powder; oxidizes in air. d 5.12; mp 215°; bp 623°. Sol in little water, gradually dec by much water; sol in alcohol, ether, acetone. *Keep tightly closed and protected from light.*

8939. Stannous Chloride. Tin dichloride; tin protochloride; Stannochlor. Cl_2Sn; mol wt 189.62. Cl 37.39%, Sn 62.61%. $SnCl_2$. Prepn: Stephen, *J. Chem. Soc.* 1930, 2786; Williams, *Org. Syn. coll.* vol. III, 627 (1955).

Orthorhombic cryst mass or flakes; fatty appearance. bp 247°; d 3.95. Sol in water, ethanol, acetone, ether, methyl acetate, methyl ethyl ketone, isobutyl alcohol. Practically insol in mineral spirits, petr naphtha, xylene. LD_{50} i.p. in mice: 66 mg/kg, *Toxic Substances List*, H. E. Christensen *et al.*, Eds. (1974) p 765.

Dihydrate, crystals; absorbs oxygen from air and forms insol oxychloride. d 2.71. mp 37-38° when rapidly heated; dec on strong heating. Sol in less than its own wt of water; with much water it forms an insol basic salt; very sol in dil or in concd hydrochloric acid; also sol in alcohol, ethyl acetate, glacial acetic acid, sodium hydroxide soln. *Keep tightly closed, in a cool place.*

USE: Powerful reducing agent, particularly in manuf of dyes; in tinning by galvanic methods; in liquor finishing of wire; in sensitizing of glass and plastics before metallizing; as soldering flux; as mordant in dyeing with cochineal; in manufacture of tin chemicals, color pigments, pharmaceuticals, sensitized paper, lubricating oil additives; as tanning agent; in removing ink stains; in yeast revivers; as reagent in analytical chemistry; as catalyst in organic reactions.

8940. Stannous Fluoride. Tin difluoride; Fluoristan. F_2Sn; mol wt 156.71. F 24.25%, Sn 75.75%. SnF_2. Prepd by evaporating a soln of stannous oxide in hydrofluoric acid in the absence of oxygen: Gay-Lussac, Thénard, *Mém. Phys. Chim.* **2**, 317 (1809); Nebergall *et al.*, *J. Am. Chem. Soc.* **74**, 1604 (1952); from tin and hydrogen fluoride: Muetterties, *Inorg. Chem.* **1**, 342 (1962). *Review:* Kemmitt, Sharp, *Advan. Fluorine Chem.* **4**, 186 (1965).

Monoclinic, lamellar plates. mp 213°. d^{25} 4.57. Sol in water (about 30%). Forms an oxyfluoride, $SnOF_2$, on exposure to air.

USE: Dental caries prophylactic.

8941. Stannous Hexafluorozirconate(IV). Stannous fluozirconate(IV). F_6SnZr; mol wt 323.92. F 35.19%, Sn 36.65%, Zr 28.16%. $SnZrF_6$. Prepd from ZrF_4 and SnF_2: Muhler, U.S. pat. 3,266,996 (1966 to Indiana University Foundation).

Crystals. d 4.21. Sol in water.

USE: In anticaries preparations.

8942. Stannous Iodide. Tin diiodide. I_2Sn; mol wt 372.52. I 68.13%, Sn 31.87%. SnI_2. Preparation and crystal structure: Moser, Trevena, *Chem. Commun.* **1969**, 25.

Red, cryst powder or needles. d 5.28; mp 320°; bp 720° with decompn. Slightly sol in and dec by water; sol in solns of alkali chlorides or iodides, in benzene, chloroform, carbon disulfide. *Keep tightly closed.*

8943. Stannous Oxalate. C_2O_4Sn; mol wt 206.73. C 11.62%, O 30.96%, Sn 57.42%. SnC_2O_4.

White, heavy powder. d 3.56. Insol in water. Sol in dil hydrochloric acid.

USE: Dyeing and printing textiles.

8944. Stannous Oxide. Tin monoxide; tin protoxide. OSn; mol wt 134.71. O 11.88%, Sn 88.12%. SnO. Prepn of high purity SnO: Kwestroo, Vromans, *J. Inorg. Nucl. Chem.* **29**, 2187 (1967). Prepn of metastable, red, orthorhombic form: Donaldson *et al.*, *J. Chem. Soc.* **1961**, 839.

Brownish-black powder; burns to SnO_2 on heating in air. d 6.45. Insol in water or alcohol. Sol in acids, in concd sodium or potassium hydroxide solns.

USE: Reducing agent; prepn of stannous salts.

8945. Stannous Pyrophosphate. Diphosphoric acid tin-(2+) salt (1:2); ditin diphosphate; ditin pyrophosphate. $O_7P_2Sn_2$; mol wt 411.36. O 27.23%, P 15.06%, Sn 57.72%. $Sn_2P_2O_7$. Prepd by thermal dehydration of stannous hydrogen phosphate: Jablczynski, Wieckowski, *Z. Anorg. Allgem. Chem.* **152**, 207 (1926); from $Na_2H_2P_2O_7$ and $SnCl_2$: Klement, Haselbeck, *Ber.* **96**, 1022 (1958); from pyrophosphoric acid and stannous chloride: Nelson, U.S. pat. 3,401,012 (1968 to Monsanto). Clinical application for labelling red blood cells: D. Ducasson *et al.*, *Brit. J. Radiol.* **49**, 344 (1976). Chemical characterization of technetium complexes: J. Kroesbergen *et al.*, *Nucl. Med. Biol.* **15**, 209 (1988). Diagnostic application in bone scanning: N. S. Anderton *et al.*, *Am. J. Roentgenol. Rad. Ther. Nucl. Med.* **124**, 625 (1975);

in myocardial infarction: D. E. Jansen *et al.*, *Circulation* **75**, 611 (1987).

Amorphous powder. $d^{16.4}$ 4.009. Insol in water. Sol in concd acid, excess alkali.

^{99m}Technetium complex, ^{99m}Tc-PPi, ^{99m}Tc-PYP, *TechneScan PYP.*

USE: Ingredient in caries-preventing toothpaste: Norris, Schweizer, U.S. pat. 2,946,725 (1960 to Proctor & Gamble).

THERAP CAT: ^{99m}Tc complex as diagnostic aid (radioactive imaging agent).

8946. Stannous Selenide. Tin monoselenide. SeSn; mol wt 197.67. Se 39.95%, Sn 60.05%. SnSe. Prepd by direct fusing of the elements: Ditte, *Compt. Rend.* **95**, 641 (1882); **96**, 1792 (1883); **97**, 44 (1883); Berzelius cited in *Mellor's* vol. X, 784 (1930); by adding powdered selenium to molten, anhyd stannous chloride: Schneider, *Pogg. Ann.* **127**, 624 (1866).

Steel-gray prisms. d^0 6.18; mp 861°. Insol in water. Sol in aqua regia and alkali sulfides and selenides.

8947. Stannous Sulfate. O_4SSn; mol wt 214.77. O 29.80%, S 14.93%, Sn 55.27%. $SnSO_4$. Prepn: Donaldson, Moser, *J. Chem. Soc.* **1960**, 4000.

Snow-white, orthorhombic crystals; dec at 378° to SnO_2 and SO_2. Sol in water, the soln soon decomposing with pptn of a basic sulfate; sol in dil H_2SO_4. *Keep well closed.*

USE: In tin plating; prepn of stannous salts.

8948. Stannous Sulfide. Tin monosulfide; tin protosulfide. SSn; mol wt 150.78. S 21.27%, Sn 78.73%. SnS. Dark-gray crystals or black, amorphous powder. d 5.08. Insol in water or alkali hydroxide or alkali sulfide soln. Sol in concd HCl, hot concd H_2SO_4.

USE: Polymerization catalyst.

8949. Stannous Tartrate. $C_4H_4O_6Sn$; mol wt 266.78. C 18.01%, H 1.51%, O 35.98%, Sn 44.50%. $SnC_4H_4O_6$.

White powder. Sol in water, dil HCl. *Keep well closed.*

USE: Dyeing and printing textiles.

8950. Stanolone. (5α,17β)-17-Hydroxyandrostan-3-one; 17β-hydroxy-3-androstanone; 3-oxo-17β-hydroxyandrostane; androstan-17β-ol-3-one; 4-dihydrotestosterone; androstanolone; Anabolex; Andractim; Androlone. $C_{19}H_{30}O_2$; mol wt 290.45. C 78.57%, H 10.41%, O 11.02%. Prepd by hydrogenation of testosterone: Butenandt *et al.*, *Ber.* **68**, 2097 (1935); from dehydroepiandrosterone: Ruzicka, Kagi, *Helv. Chim. Acta* **20**, 1557 (1937); Ruzicka *et al.*, *ibid.* **24**, 1151 (1941); from 3,17-androstandione: Oliveto, Hershburg, U.S. pat. 2,927,921 (1960 to Schering).

Crystals from ethyl acetate + hexane. Sublimes$_{0.01}$ 135°. mp 181°. $[\alpha]_D^{20}$ +32.4° (alcohol). Infrared absorption data: *J. Am. Chem. Soc.* **75**, 903 (1953). Sol in acetone, ether, alcohol, ethyl acetate. Practically insol in water.

Note: This is a controlled substance (anabolic steroid) in the U.S. Code of Federal Regulations, Title 21 Part 1308.13, as defined in Part 1308.02 (1995).

THERAP CAT: Androgen.

8951. Stanozolol. (5α,17β)-17-Methyl-2'H-androst-2-eno[3,2-c]pyrazol-17-ol; 1,2,3,3a,3b,4,5,5a,6,8,10,10a,10b,-11,12,12a-hexadecahydro-1,10a,12a-trimethylcyclopenta-[7,8]phenanthro[2,3-c]pyrazol-1-ol; 17β-hydroxy-17α-methylandrostano[3,2-c]pyrazole; androstanazole; stanazol; NSC-43193; Win-14833; Stromba; Strombaject; Winstrol. $C_{21}H_{32}N_2O$; mol wt 328.50. C 76.78%, H 9.82%, N 8.53%, O 4.87%. Prepn: Clinton *et al.*, *J. Am. Chem. Soc.* **81**, 1513 (1959); **83**, 1478 (1961); Manson, U.S. pat. 3,030,358 (1962 to Sterling Drug).

Crystals from alc, mp 229.8-242.0°. $[\alpha]_D$ +35.7° (chloroform); $[\alpha]_D$ +48.6° (methanol). uv max: 223 nm (ϵ 4740).

Note: This is a controlled substance (anabolic steroid) in the U.S. Code of Federal Regulations, Title 21 Part 1308.13, as defined in Part 1308.02 (1995).

THERAP CAT: Androgen.

THERAP CAT (VET): Anabolic steroid.

8952. Staphisagria. Stavesacre. Ripe seed of *Delphinium staphisagria* L., *Ranunculaceae. Habit.* Mediterranean basin; cultivated in France, Italy. *Constit.* Delphinine, delphinoidine, delphisine, staphisagrine, staphisagroine, malic acid, fixed oil.

THERAP CAT: Parasiticide (external).

8953. Star Anise. Chinese anise. Fruit of *Illicium verum* Hook. f., *Magnoliaceae. Habit.* Southeastern Asia and subtropical countries; commercial supply chiefly from China. *Constit.* About 5% volatile and fixed oils; anisic acid, tannin, resin, pectin.

Note: Japanese star anise is *Illicium anisatum* L. (*I. religiosum* Sieb. & Zucc.; *I. japonicum* Sieb.) and contains a toxic lactone called anisatin. Chinese star anise does not contain this toxic principle. Shikimic acid has been found in both.

USE: Manufacture of liqueurs and the volatile oil. The fruit as source of oil of anise.

THERAP CAT: Hemostatic.

8954. Starch. Amylum. $(C_6H_{10}O_5)_n$. Stored by plants; analogous to storage of fats by animals. Occurs as discrete granules in the mature grain of corn, *Zea mays* Linné, *Gramineae* or of wheat, *Triticum aestivum* Linné, *Gramineae* or tubers of potato, *Solanum tuberosum* Linné, *Solanaceae* or rice, *Oryza sativa* Linné, *Gramineae.* Starches are mixtures of two polymers: *amylose,* a linear $(1 \to 4)$-α-D-glucan and *amylopectin,* a branched D-glucan with mostly α-D-$(1 \to 4)$ and approx 4% α-D-$(1 \to 6)$ linkages. The starch in corn contains approx 27% amylose and 73% amylopectin, with these two polymers so associated in the crystal lattice that they are practically insol in cold water or alcohol. *Refs:* J. N. BeMiller, "Starch Amylose" in *Industrial Gums,* R. L. Whistler, Ed. (Academic Press, New York, 2nd ed., 1973) pp 545-566; E. L. Powell, "Starch Amylopectin", *ibid.,* pp 567-576.

Although hydrolysis will not take place in cold water, and starch is comparatively resistant to naturally occurring enzymes, the reaction may be brought about by the use of acids or enzymes (α-amylase, β-amylase, amyloglucosidase). The hydrolysis reaction follows a different path depending on whether acids or enzymes are used. While acid hydrolysis produces a mixture of saccharides, the enzymes give more specific products. β-Amylase, for example, breaks off mostly maltose units, and amyloglucosidase yields mainly D-glucose. Chemistry and technology: R. L. Whistler, E. F. Paschall, Eds., *Starch Chemistry and Technology,* 2 vols. (Academic Press, New York, 1965); J. A. Radley, Ed., *Starch and Its Derivatives* (Chapman & Hall, London, 4th ed., 1968).

USE: Starching and sizing fabrics, etc.; paste; as indicator in iodometric analyses. In the food industry. Pharmaceutic aid (tablet disintegrant, filler, binder); dusting powder. Dietetic grades of corn starch are marketed as *Maizena; Mondamin.*

THERAP CAT: Antidote (iodine poisoning).

THERAP CAT (VET): Internally: demulcent, mild astringent, in diarrhea, as an antidote for iodine poisoning. Externally: absorbent, emollient, in dusting powders and in ointments.

8955. Starch, Soluble. Amylodextrin; amylogen. Prepd by treating potato or corn starch with dilute hydrochloric acid.

White, odorless, tasteless powder. Readily soluble in hot water; forms transparent mobile liquid.

USE: For determination of diastatic power of malt, etc.; as indicator in iodometric analyses.

8956. Statine. *[S-(R*,R*)]-4-Amino-3-hydroxy-6-methylheptanoic acid;* AHMHA. $C_8H_{17}NO_3$; mol wt 175.23. C 54.84%, H 9.78%, N 7.99%, O 27.39%. Amino acid present in pepstatin, *q.v.* Synthesis: H. Morishima *et al., J. Antibiot.* **26**, 115 (1973). Abs config and stereospecific synthesis of all four isomers: M. Kinoshita *et al., ibid.* 249. Crystal structure: H. Nakamura *et al., ibid.* 255. Biosynthesis: H. Morishima *et al., ibid.* **27**, 267 (1974). Alternate syntheses: M. Kinoshita *et al., Bull. Chem. Soc. Japan* **48**, 570 (1975); W.-S. Liu, G. I. Glover, *J. Org. Chem.* **43**, 754 (1978); D. H. Rich *et al., ibid.* 3624; K. E. Rittle *et al., ibid.* **47**, 3016 (1982). Distribution in rats: D. A. Grant *et al., Biochem. Pharmacol.* **31**, 2302 (1982).

mp 201-203° (dec). $[\alpha]_D^{15}$ $-20°$ (c = 0.64 in water).

8957. Statolon. An antiviral substance which appears to be a macromolecular polyanionic polysaccharide composed of galacturonic acid, galactose, galactosamine, glucose, arabinose, xylose, and rhamnose. Prophylactically active against a wide range of viruses, including those causing canine distemper, lymphomatosis in fowl, shipping fever in cattle, transmissible gastroenteritis in swine, and coryza and other upper respiratory illnesses, as well as against ECHO viruses, enteroviruses in monkeys, MM neurotropic virus, Semliki Forest virus, and NEF 1 poliomyelitis virus. Antitumor activity demonstrated in experimental leukemia and sarcoma. Produced by submerged culture fermentation, using *Penicillium stoloniferum* var. ATCC 14586: Stark *et al.,* U.S. pat. **3,108,047** (1963 to Lilly).

THERAP CAT: Antiviral.

8958. Stavudine. *2′,3′-Didehydro-3′-deoxythymidine; 1-(2,3-dideoxy-β-glycero-pent-2-enofuranosyl)thymine;* 3′-deoxy-2′-thymidinene; D4T; BMY-27857; Zerit. $C_{10}H_{12}N_2O_4$; mol wt 224.22. C 53.57%, H 5.39%, N 12.49%, O 28.54%. Analog of thymidine, *q.v.;* reverse transcriptase inhibitor. Prepn: J. P. Horwitz *et al., J. Org. Chem.* **31**, 205 (1966); J. W. Beach *et al., ibid.* **57**, 3887 (1992). Large scale production: J. E. Starrett *et al.,* Eur. pat. Appl. **334,368**; *eidem,* U.S. pat. **5,130,421** (1989, 1992 both to Bristol-Myers). *In vitro* activity against HIV and toxicology: M. M. Mansuri *et al., Antimicrob. Ag. Chemother.* **34**, 637 (1990). Mechanism of action study: H.-T. Ho, M. J. M. Hitchcock, *ibid.* **33**, 844 (1989). Disposition and metabolism: E. M. Cretton *et al., ibid.* **37**, 1816 (1993). Clinical pharmacokinetics: M. N. Dudley *et al., J. Infec. Dis.* **166**, 480 (1992). HPLC determn in plasma and urine: J. S. Janiszewski *et al., J. Chromatog.* **577**, 151 (1992). Clinical evaluation in AIDS and ARC: M. J. Browne *et al., J. Infec. Dis.* **167**, 21 (1993).

Colorless, granular solid from ethanol/benzene, mp 165-166° (Horwitz). Also reported as crystals from ethanol-ether, mp 174° (Beach). uv max (water): 266 nm (ϵ 10149).

$[\alpha]_D^{25}$ −39.4° (c = 0.701 in water); $[\alpha]_D^{20}$ −46.1° (c = 0.7 in water).

THERAP CAT: Antiviral.

8959. Stearic Acid. *Octadecanoic acid;* Emersol 132. $C_{18}H_{36}O_2$; mol wt 284.48. C 76.00%, H 12.76%, O 11.25%. $CH_3(CH_2)_{16}COOH$. Occurs as a glyceride in tallow and other animal fats and oils, as well as in some vegetable oils; also prepd synthetically by hydrogenation of cottonseed and other vegetable oils.

White leaflets. d^{70} 0.847; mp 69-70°; bp 383°; n_D^{80} 1.4299. Slowly volatilizes at 90-100°. Very slightly sol in water. One gram dissolves in 21 ml alcohol, 5 ml benzene, 2 ml chloroform, 26 ml acetone, 6 ml carbon tetrachloride, 3.4 ml carbon disulfide; also sol in amyl acetate, toluene. LD_{50} i.v. in mice, rats: 23±0.7, 21.5±1.8 mg/kg, L. Orö, A. Wretlind, *Acta Pharmacol. Toxicol.* **18**, 141 (1961).

U.S.P. stearic acid consists chiefly of a mixture of stearic and palmitic acids. It is in the form of white or slightly yellow, crystal masses, or a white to slightly yellow powder; slight tallow-like odor. Does not congeal below 54°.

Ethyl ester, $C_{20}H_{40}O_2$, *ethyl stearate.* White, cryst solid; odorless or practically so. mp 33-35°. bp 224°. Ethyl stearate of commerce solidifies at 20-24°; bp_4 180°. Insol in water; sol in alcohol or ether.

Methyl ester, $C_{19}H_{38}O_2$, *methyl stearate.* White crystals. mp 38-39°. bp_{15} 215°. Insol in water; sol in alcohol, ether.

USE: For suppositories, coating enteric pills, ointments, and for coating bitter remedies. Manuf stearates of aluminum, zinc, and other metals, stearin soap for opodeldoc, candles, phonograph records, insulators, modeling compds; impregnating plaster of Paris; in vanishing creams and other cosmetics.

8960. Stearyl Alcohol. *1-Octadecanol;* stenol. $C_{18}H_{38}$O; mol wt 270.50. C 79.93%, H 14.16%, O 5.91%. CH_3-$(CH_2)_{16}CH_2OH$. The official substance is a mixture of solid alcohols consisting chiefly of stearyl alcohol. Preparation from ethyl stearate: Brown, Rao, *J. Am. Chem. Soc.* **78**, 2582 (1956); Hesse, Schrödel, *Ann.* **607**, 24 (1954). Prepn of technical grade from sperm whale oil: Maiorov *et al., Zh. Prikl. Khim.* **37**, 1344 (1964).

Unctuous white flakes or granules, mp 56-60° (the pure substance, mp 59.4-59.8°, bp_{15} 210°). Sol in alcohol, ether, benzene, acetone.

USE: Substitute for cetyl alc in pharmaceutical dispensing, in cosmetic creams, for emulsions, textile oils and finishes, as antifoam agent, lubricant, and chemical raw material.

8961. Stellar®. Starch based fat replacer produced by controlled acid hydrolysis of corn starch. Shears to create a firm, deformable creme consisting of loosely aggregated submicron-sized starch particles. Consists of branched amylodextrins; average mol wt < 20000. Prepn: D. W. Harris, K. D. Stanly, *Eur. pat. Appl.* **529,891** (1992 to Staley Manuf.). Functional properties in relationship to structure: D. W. Harris, G. A. Day, *Starch/Stärk* **45**, 221 (1993). Brief description: *Prepared Foods* **161**, 77 (1992).

Bland taste, slick fat-like mouthfeel. Hydrophilic. pH of 10% slurry, 3.0-5.0. Caloric value: 16 kJ/g. Exhibits thixotropic rheological behavior similar to hydrogenated vegetable shortening.

USE: Fat substitute for reduced calorie foods.

8962. Stenbolone. *(5α,17β)-17-Hydroxy-2-methylandrost-1-en-3-one;* 2-methyl-5α-androst-1-en-17β-ol-3-one; 2-methyl-17β-hydroxy-5α-androst-1-en-3-one; stenobolone (rescinded USAN). $C_{20}H_{30}O_2$; mol wt 302.46. C 79.42%, H 10.00%, O 10.58%. Preparation of free alcohol and acetate: Mauli, *J. Am. Chem. Soc.* **82**, 5494 (1960); Kaspar *et al.,*

Ger. pat. **1,096,356** (1961 to Schering AG), *C.A.* **55**, 27440b (1961); Counsell *et al., J. Org. Chem.* **27**, 248 (1962); Brit. pat. **925,849** (1963 to Syntex).

Crystals from acetone + hexane, mp 155-158°. $[\alpha]_D$ +52° (chloroform), $[\alpha]_D^{26}$ +47° (chloroform). uv max (95% ethanol): 241 nm (log ε 3.99).

Acetate, $C_{22}H_{32}O_3$, *Anatrofin.* Crystals from acetone + hexane, mp 146-149°. $[\alpha]_D$ +32° (chloroform), $[\alpha]_D^{26}$ +60° (chloroform): see Mauli *et al.,* and Counsell *et al.* uv max (95% ethanol): 241 nm (log ε 4.03).

THERAP CAT: Acetate as anabolic.

8963. Stepronin. *N-[1-Oxo-2-[(2-thienylcarbonyl)thio]-propyl]glycine;* 2-(α-thenoylthio)propionylglycine; *N*-(2-mercaptopropionyl)glycine 2-thiophenecarboxylate (ester); TTPG; prostenoglycine; Tiase. $C_{10}H_{11}NO_4S_2$; mol wt 273.33. C 43.94%, H 4.06%, N 5.12%, O 23.41%, S 23.46%. Prepn, properties and toxicology: F. Bolasco, Belg. pat. **875,186**; *eidem,* U.S. pat. **4,242,354** (1979, 1980 to Mediolanum Farmaceutici). Clinical trials in bronchitis: M. Lingetti *et al., Int. J. Clin. Pharm. Res.* **1**, 273 (1981); F. Dotta *et al., Clin. Ter.* **105**, 307 (1983); G. C. Morandini *et al., Curr. Ther. Res.* **35**, 783 (1984). Effect on human lung mucus: M. C. Morale *et al., Int. J. Tiss. Reac.* **5**, 231 (1983).

Colorless crystals from acetonitrile, mp 168-170°. uv max: 292 nm. LD_{50} in mice (mg/kg): > 2500 orally, > 1250 i.v.; in rats (mg/kg): > 2500 orally, 1801 i.m. (Bolasco).

Lysine salt, *Masor, Mucodil.*

Sodium salt, $C_{10}H_{10}NNaO_4S_2$, *Broncoplus, Tioten.*

THERAP CAT: Mucolytic.

8964. Stevioside. *(4α)-13-[(2-O-β-D-Glucopyranosyl-β-D-glucopyranosyl)oxy]kaur-16-en-18-oic acid β-D-glucopyranosyl ester;* steviosin. $C_{38}H_{60}O_{18}$; mol wt 804.88. C 56.71%, H 7.51%, O 35.78%. Isolated from leaves of *Stevia rebaudiana* (Bert.) Hemsl. (*Eupatorium rebaudianum* Bert.), *Compositae,* also called *yerba dulce. Habit.* Paraguay. Isoln and structure: Wood *et al., J. Org. Chem.* **20**, 875 (1955). Review of botany: Samaniego, *Rev. farm. (Buenos Aires)* **88**, 199 (1946). Review of chemistry and use as sweetening agent: Bell, *Chem. & Ind. (London)* **1954**, 897; Fletcher, Jr., *Chemurgic Digest* **14**, 7, 18 (July-Aug. 1955); *Chem. & Eng. News* **34**, 124 (1956). Total synthesis: *eidem, Tetrahedron* **36**, 2641 (1980). Synthesis and sensory evaluation of analogs: G. E. Dubois *et al., J. Med. Chem.* **24**, 1269 (1981).

Hygroscopic crystals, mp 198°. $[\alpha]_D^{25}$ —39.3° (c = 5.7 in H_2O). 300 times as sweet as cane sugar. One gram dissolves in 800 ml water. Sol in dioxane. Slightly sol in alc.

Steviol, (4α)-13-Hydroxykaur-16-en-18-oic acid, hydroxydehydrostevic acid. Aglucon of stevioside, the sweet glucoside abundant in leaves and stems of *Stevia rebaudiana* Bert. Absolute configuration: Mosettig *et al., ibid.* **85**, 2305 (1963). Biosynthesis: Ruddat *et al., Arch. Biochem. Biophys.* **110**, 496 (1965). Total synthesis of *dl*-form: Mori *et al., Tetrahedron Letters* **1970**, 2411; Cook, Knox, *ibid.* 4091; *eidem, Tetrahedron* **28**, 3217 (1972). Needles from methanol or ethanol + water, mp 215°. $[\alpha]_D$ —94.7°.

USE: Has been proposed as non-nutritive sweetening agent.

8965. Stibine. Antimony hydride. H_3Sb; mol wt 124.78. H 2.42%, Sb 97.58%. SbH_3. Conveniently prepd by dissolving zinc-antimony or magnesium-antimony alloy in dil HCl: Hurd, *Chemistry of the Hydrides* (Wiley, New York, 1952) p 132. Detailed directions (including prepn of the alloy from powdered Sb and Mg): Schenk in *Handbook of Preparative Inorganic Chemistry* vol. 1, G. Brauer, Ed. (Academic Press, New York, 2nd ed., 1963) pp 606-608. Review of preparative methods: Jolly, Norman, "Hydrides of Groups IV and V" in *Preparative Inorganic Reactions* vol. 4, W. L. Jolly, Ed. (Interscience, New York, 1968) pp 1-58. Colorless gas. *Intensely poisonous.* Disagreeable odor. mp —88°. bp —18.4°. d 2.204 g/ml at bp. Heat of formation +34.68 kcal/mole: Gunn, Green, *J. Phys. Chem.* **65**, 779 (1961). Thermally less stable than arsine. Dec slowly on standing at room temp. Quickly destroyed at 200°. The decomposition products are hydrogen and metallic antimony, generally deposited in the form of a mirror. The gas is slightly sol in water. Freely sol in alcohol, carbon disulfide, other organic solvents. Lethal concn in air for mice: about 100 ppm.

Caution: Potential symptoms of overexposure are headache, weakness; nausea, abdominal pain; lumbar pain, hemoglobinuria, hematuria and hemolytic anemia; jaundice; irritation of lung. *See NIOSH Pocket Guide to Chemical Hazards* (DHHS/NIOSH 90-117, 1990) p 198.

USE: Has been used as fumigating agent.

8966. Stibocaptate. *2,2'-[(1,2-Dicarboxy-1,2-ethanediyl)bis(thio)]bis-1,3,2-dithiastibolane-4,5-dicarboxylic acid hexasodium salt; 2,3-dimercaptosuccinic acid cyclic ester with antimonic acid, diester with 2,3-dimercaptosuccinic acid hexasodium salt;* 2,3-dimercaptosuccinic acid cyclic thioantimonate (III) S,S-diester with 2,3-dimercaptosuccinate hexasodium salt; sodium antimony-2,3-mercaptosuccinate; "antimony dimercaptosuccinate"; TWSb; Ro-4-1544/6; SB-58; Astiban. $C_{12}H_6Na_6O_{12}S_6Sb_2$; mol wt 916.03. C 15.73%, H 0.66%, Na 15.06%, O 20.96%, S 21.00%, Sb 26.58%. Prepn: Friedheim, U.S. pat. **2,880,222** (1959). Prepn and antischistosomal activity: J.-K. Hsu, M.-K. Ch'en, *K'o Hsueh Tung Pao* **1966**, 978, *C.A.* **65**, 17580e (1966). Toxicity data: Ercoli, *Proc. Soc. Exp. Biol. Med.* **129**, 284 (1968).

White or slightly yellowish-green powder. Hygroscopic, unstable when moistened. Sol in water. LD_{50} s.c. in mice: 500 mg Sb/kg (Ercoli).

Note: Stibocaptate described as the tripotassium salt: Friedheim *et al., Am. J. Tropic. Med. Hyg.* **3**, 714 (1954).

THERAP CAT: Anthelmintic (Schistosoma).

8967. Stibophen. *(T-4)-Bis[4,5-dihydroxy-1,3-benzenedisulfonato(4—)-O⁴,O⁵]-antimonate(5—) pentasodium heptahydrate;* sodium antimony bis(pyrocatechol-2,4-disulfonate) heptahydrate; antimony pyrocatechol sodium disulfonate heptahydrate; Sdt-91; Fuadin; Fouadin; Fantorin; Neoantimosan; Repodral. $C_{12}H_4Na_5O_{16}S_4Sb.7H_2O$; mol wt 895.23. C 16.10%, H 2.03%, Na 12.84%, O 41.11%, S 14.33%, Sb 13.60%. Prepn of potassium salt: **Brit. pat. 213,285** (1923

to Heyden). Prepn from sodium pyrocatechol-3,5-disulfonate and antimony trioxide in alkaline soln: H. Schmidt, **Ger. pat. 448,800** (1924 to I. G. Farbenind.), *Friedl.* **16**, 2546 (1931); *idem,* U.S. pat. **1,549,154** (1925); *idem, Z. Angew. Chem.* **43**, 963 (1930); *idem,* U.S. pat. **1,873,668** (1932 to Winthrop). Toxicity: H. Eagle *et al., J. Pharmacol. Exp. Therap.* **89**, 196 (1947). Clinical trial in schistosomiasis: J. Azar *et al., Am. J. Trop. Med.* **29**, 595 (1949); in leishmaniasis: H. S. Girgla *et al., Brit. J. Dermatol.* **97**, 307 (1977). UV spectrophotometric analysis: A. Besada *et al., Anal. Letters* **21**, 447 (1988).

Fine crystals. Almost insol in abs alcohol, ether, chloroform, acetone, petr ether. Readily sol in cold water. Unless the colorless neutral water soln is acidified it acquires a yellowish tint and finally reaches lemon-yellow color. LD_{50} i.v. in rabbits: ~90 mg/kg (Eagle).

Potassium salt, *Antimosan, Heyden 611.*

THERAP CAT: Anthelmintic (Schistosoma).

THERAP CAT (VET): Has been used in schistosomiasis.

8968. Stigmastanol. *(3β,5α)-Stigmastan-3-ol;* dihydro-β-sitosterol; β-sitostanol; 24α-ethylcholestanol; fucostanol. $C_{29}H_{52}O$; mol wt 416.73. C 83.58%, H 12.58%, O 3.84%. Occurs along with β-sitosterol and stigmasterol. Formation from β-sitosterol: Bernstein, Wallis, *J. Org. Chem.* **2**, 341 (1937); from β-sitostenone: Marker, Wittle, *J. Am. Chem. Soc.* **59**, 2704 (1937). Antihypercholesteremic activity in rabbits: I. Ikeda *et al., J. Nutr. Sci. Vitaminol.* **27**, 243 (1981).

Monohydrate, crystals, mp 138-139°. When dry mp 144-145°. $[\alpha]_D^{20}$ +25° (c = 1.1 in chloroform).

Acetate, $C_{31}H_{54}O_2$, mp 137-138°. $[\alpha]_D^{20}$ +14° (c = 1.8 in chloroform).

8969. Stigmasterol. *(3β,22E)-Stigmasta-5,22-dien-3-ol;* 3β-hydroxy-24-ethyl-Δ⁵,²²-cholestadiene. $C_{29}H_{48}O$; mol wt 412.70. C 84.40%, H 11.72%, O 3.88%. Usually isolated from the phytosterol mixture from soy or calabar beans, *cf.* Thornton *et al., J. Am. Chem. Soc.* **62**, 2006 (1940); Byerrum, Ball, *Biochem. Prepns.* **7**, 86 (1959). Structure: Fernholz, *Ann.* **507**, 128 (1933); **508**, 215 (1933); Fernholz, Chakravorty, *Ber.* **67**, 2021 (1934). Identity with *guinea-pig-anti-stiffness factor:* H. Rosenkrantz *et al., Proc. Soc. Exp. Biol. Med.* **76**, 408 (1951). Synthesis: W. Sucrow, M. Slopianka, *Ber.* **108**, 3721 (1975). HPLC separation from plant oils: H. Colin *et al., Anal. Chem.* **51**, 1661 (1979).

Monohydrate, crystals from alcohol. When dry, mp 170°. $[\alpha]_D^{22}$ −51° (c = 2 in chloroform). Insol in water. Sol in the usual organic solvents.

Acetate, $C_{31}H_{50}O_2$, mp 144°. $[\alpha]_D^{20}$ −55.6° (c = 2 in chloroform).

p-Nitrobenzoate, $C_{36}H_{51}NO_4$, mp 203°. $[\alpha]_D^{17}$ −13° (c = 2 in alcohol).

8970. Stilbamidine. *4,4'-(1,2-Ethenediyl)bisbenzenecarboximidamide; 4,4'-stilbenedicarboxamidine;* 4,4'-diamidinostilbene. $C_{16}H_{16}N_4$; mol wt 264.33. C 72.70%, H 6.10%, N 21.20%. Prepn: A. J. Ewins, J. N. Ashley, **Brit. pat.** 510,097 (1939 to May & Baker); J. N. Ashley *et al., J. Chem. Soc.* **1942,** 103; of isethionate: G. Newbery, A. P. T. Easson, **U.S. pat. 2,394,003** (1946 to May & Baker). Trypanocidal activity: E. M. Lourie, W. Yorke, *Ann. Trop. Med. Parasitol.* **33,** 289 (1939). Preliminary pharmacology: R. Wien, *ibid.* **37,** 1 (1943). Mechanism of action studies: M. J. Pine, *Biochem. Pharmacol.* **17,** 75 (1968); G. Weissmann *et al., ibid.* **19,** 1251 (1970). Crystal structure: C. Courseille *et al., C.R. Acad. Sci. Ser. C* **272,** 1115 (1971). DNA binding study: N. Gresh, B. Pullman, *Mol. Pharmacol.* **25,** 452 (1984). Early review of pharmacology, mode of action and clinical applications: E. B. Schoenbach, E. M. Greenspan, *Medicine* **27,** 327-377 (1948).

Dihydrochloride, $C_{16}H_{18}Cl_2N_4$, needles from water. LD_{50} in mice (mg/g): 0.031 i.v.; 0.18 s.c. (Wien).

Isethionate, $C_{20}H_{28}N_4O_8S_2$, *M & B 744.* Crystals, discolored by light, dec 290°. uv max: 330 nm ($E_{1cm}^{1\%}$ 750). Soly in water: approx 1 g in 2.5 to 3.0 ml H_2O; in methanol: 1.5 g/100 ml. pH of 0.5% aq soln 5.5-6.5.

THERAP CAT: Antiprotozoal (Leishmania, Trypanosoma).

8971. Stilbazium Iodide. *1-Ethyl-2,6-bis[2-[4-(1-pyrrolidinyl)phenyl]ethenyl]pyridinium iodide; 1-ethyl-2,6-bis[p-(1-pyrrolidinyl)styryl]pyridinium iodide;* 2,6-bis(p-1-pyrrolidinylstyryl)-1-ethylpyridinium iodide; BW-61-32; Monopar. $C_{31}H_{36}IN_3$; mol wt 577.55. C 64.47%, H 6.28%, I 21.97%, N 7.28%. Prepn: Phillips, Burrows, **U.S. pat. 3,085,935** (1963 to Burroughs Wellcome). Toxicity data: R. B. Burrows, *Prog. Drug Res.* **17,** 110 (1973).

mp 282-283°. Nearly insol in hot methanol. LD_{50} in mice (mg/kg): 1360 orally; 7 i.p. (Burrows).

THERAP CAT: Anthelmintic (Nematodes).

8972. Stilbene. *1,1'-(1,2-Ethenediyl)bis[benzene];* α,β-diphenylethylene; bibenzal; bibenzylidene. $C_{14}H_{12}$; mol wt 180.25. C 93.29%, H 6.71%. *trans*-Form prepd by Clemmensen reduction of benzoin: Shriner, Berger, *Org. Syn.* **coll. vol. III,** 786 (1955); *cis*-form by copper-chromite decarboxylation of α-phenylcinnamic acid: Buckles, Wheeler,

ibid. **coll. vol. IV,** 857 (1963). Synthesis of *cis*- and *trans*-forms by the Wittig reaction and by decarboxylation of phenylcinnamic acids: Wheeler, Batlle de Pabon, *J. Org. Chem.* **30,** 1473 (1965).

trans-Form   cis-Form

trans-Form, crystals from 95% ethanol, mp 124°. bp_{760} 306-307°. uv max (95% ethanol): 296, 305 nm (ε 28100; 26700). Volatile with steam. Practically insol in water. Sol in 90 parts cold alc, 13 parts boiling alc, freely in benzene, ether.

cis-Form, liquid, solidifies at −5°. bp_{10} 135°; $bp_{1.0}$ 96°. n_D^{25} 1.6188. uv max (95% ethanol): 278 nm (ε 10200). Completely sol in cold abs alcohol.

8973. Stillingia. Queen's root; yaw root; silver leaf. Dried roots of *Stillingia sylvatica* L., *Euphorbiaceae. Habit.* Southeastern U.S. *Constit.* Acrid resin (sylvacrol), fixed and volatile oils, a glucoside.

THERAP CAT: Emetic, cathartic.

8974. Storax. Styrax; sweet oriental gum. Balsam obtained from the trunk of *Liquidambar orientalis* Mill., known as Levant Storax, or of *L. styraciflua* L., known as American Storax (family *Hamamelidaceae). Habit.* Levant storax is native to Asia Minor. American storax is produced chiefly along the Atlantic coast from Connecticut to Central America. *Constit.* 33-50% α- and β-storesin and its cinnamic ester; 5-10% styracin; 10% phenylpropyl cinnamate; small amounts of ethyl cinnamate; benzyl cinnamate; 5-15% free cinnamic acid; styrene; 0.4% levorotatory oil; $C_{10}H_{16}O$, and traces of vanillin.

Semiliquid, grayish, sticky, opaque mass (Levant storax) or a semisolid, sometimes solid mass softened by warming (American storax). Storax is transparent in thin layers, has characteristic taste and odor, and is denser than water. Insol in water. Almost completely sol in 1 part warm alcohol, ether, acetone, carbon disulfide.

Ingredient of Compound Benzoin Tincture, *see* Benzoin.

USE: In fumigating pastilles and powders; in perfumery; as imbedding material in microscopy.

THERAP CAT: Topical protectant. Expectorant.

THERAP CAT (VET): As a component of Compound Benzoin Tincture. Has been used as a parasiticide.

8975. Stramonium. Thorn apple; Jamestown weed; Jimson weed; Jimson weed; stinkweed; devil's apple; apple of Peru. Dried leaves and flowering tops of *Datura stramonium* L., *Solanaceae. Habit.* Europe, Asia, America. *Constit.* Leaves: 0.25-0.45% alkaloids consisting of atropine, hyoscyamine, and scopolamine; proteins, albumin. Seed: the same alkaloids but in lesser amounts; fixed oil, malic acid proteins. Comprehensive monograph: A. F. Blakeslee, *The Genus Datura* (Ronald Press, New York, 1959) 289 pp. *See also* Datura.

THERAP CAT: Anticholinergic.

THERAP CAT (VET): Same action as belladonna. Has been used to relieve signs of heaves in horses.

8976. Strepogenin. A name for the biologically active principle capable of stimulating growth of certain microorganisms, and originally found present in products of natural origin, such as liver extract, flour, yeast, and tomato juice. Its peptide nature was indicated by its presence in hydrolyzates of purified proteins. To compare the activities of various protein hydrolyzates, a strepogenin unit was introduced and defined as: 1 mg of a standard liver extract (Wilson's liver fraction L) has the strepogenin activity of one unit (using *Lactobacillus casei* as test organism). The different structures of strepogenin-active peptides subsequently obtained from natural sources or by synthesis, indicate that strepogenin is not a distinct, chemically defined, biologically active peptide, but rather represents a biological phenomenon of a multitude of peptides. From the large number of strepogenin-active peptides known, no conclusion can be drawn with regard to structure-function relationships; these peptides have been suspected to be nothing more than easily accessible amino acids, particularly since they (peptides) are not essential for the growth of the microorganisms. Early isolation from liver extract and from tryptic digests of pure proteins: Sprince, Woolley, *J. Exp. Med.* **80**, 213 (1944); *eidem, J. Am. Chem. Soc.* **67**, 1734 (1945). *Review:* E. Schröder, K. Lübke, *The Peptides* vol. II (Academic Press, New York, 1966) pp 267-270.

8977. Streptidine. *N,N'-Bis(aminoiminomethyl)streptamine; N,N'-diamidinostreptamine; N,N'''-(2,4,5,6-tetrahydroxy-1,3-cyclohexanediyl)bisguanidine; 1,3-diguanido-2,4,5,-6-cyclohexanetetrol.* $C_8H_{18}N_6O_4$; mol wt 262.27. C 36.64%, H 6.92%, N 32.04%, O 24.40%. The aglycone component of the streptomycin molecule joined with streptobiosamine through a glycosidic linkage. Obtained from streptomycin by acid hydrolysis: Peck *et al., J. Am. Chem. Soc.* **68**, 29 (1946); Fried *et al., J. Biol. Chem.* **162**, 391 (1946). Structure: Carter *et al., Science* **103**, 53, 540 (1946). Synthesis: Wolfrom *et al., J. Am. Chem. Soc.* **72**, 1724 (1950). Absolute configuration: Dyer, Todd, *ibid.* **85**, 3896 (1963); Tatsuoka, Horii, *J. Antibiot.* **17A**, 88 (1964). Biosynthesis studies: Bruton *et al., J. Biol. Chem.* **242**, 813 (1967).

Optically inactive, diacidic base.

Dipicrate dihydrate, $C_{20}H_{24}N_{12}O_{18} \cdot 2H_2O$, yellow needles from water, dec 283-284° (becomes anhydrous *in vacuo* at 56°, dec 284-285°).

Sulfate, $C_8H_{18}N_6O_4 \cdot H_2SO_4$, crystals from dilute sulfuric acid + acetone, not melted at 300°. Forms solvated crystals from methanol.

Dihydrochloride, $C_8H_{18}N_6O_4 \cdot 2HCl$, hygroscopic, amorphous powder, dec 170-210°. Forms solvated crystals from methanol.

8978. Streptobiosamine. *5-Deoxy-2-O-[2-deoxy-2-(methylamino)-α-L-glucopyranosyl]-3-C-formyl-L-lyxose.* $C_{13}H_{23}NO_9$; mol wt 337.33. C 46.29%, H 6.87%, N 4.15%, O 42.69%. A disaccharide representing two-thirds of the streptomycin molecule. It is built up from streptose and N-methyl-L-glucosamine. Structure and formation of derivatives by acid cleavage of streptomycin: Brink *et al., J. Am. Chem. Soc.* **68**, 2096, 2405, 2557, 2679 (1946); Lemieux *et al., ibid.* **69**, 1838 (1947); Kuehl *et al., ibid.* **70**, 2613 (1948); **71**, 1445 (1948).

R = CH_2OH
R' = $NHCH_3$

8979. Streptodornase. *Streptococcal deoxyribonuclease.* DNAase produced by various strains of β-hemolytic streptococci. Identification in culture medium: W. S. Tillett *et al., Proc. Soc. Exp. Biol. Med.* **68**, 184 (1948); M. McCarty, *J. Exp. Med.* **88**, 181 (1948). Purification: J. H. Mowat *et al.,* U.S. pat. 2,753,291 (1956 to Am. Cyanamid). Clinical use with streptokinase in wound debridement: J. Poulsen *et al., Acta Chir. Scand.* **149**, 245 (1983).

Mixture with streptokinase, *Ernodasa, Varidase.*
THERAP CAT: Debriding agent.

8980. Streptogramin. Antibiotic complex produced by *Streptomyces graminofaciens* from Texas soil: Charney *et al., Antibiot. & Chemother.* **3**, 1283 (1953); Brit. pat. 776,035 (1957 to Merck & Co.). Consists of an antibiotic factor, streptogramin A, capable of inhibiting protein synthesis; and a synergizing factor, streptogramin B, which is a cyclic polypeptide: Chabbert, Acar, *Ann. Inst. Pasteur* **107**, 777 (1964). The term streptogramin more broadly refers to a family of closely related antibiotics of which streptogramin itself is a member. For other members see Mikamycin, Pristinamycin, Vernamycin, Virginiamycin. Nomenclature: P. Crooy, R. De Neys, *J. Antibiot.* **25**, 371 (1972). Toxicity study: Charney *et al., Antibiot. Annual* **1953**, 171. *Review:* Vazquez, *Antibiotics* vol. 1, D. Gottlieb, P. D. Shaw, Eds. (Springer-Verlag, New York, 1967) pp 387-403; vol. 3, J. W. Corcoran, F. E. Hahn, Eds. (1975) pp 521-534.

Solid spherules, mp ~155°. $[\alpha]_D$ −134°. Sol in methanol, ethanol, acetone, ethyl acetate; sparingly sol in water, dil acid. Practically insol in petr ether. LD_{50} i.p. in mice: 450 mg/kg (Charney).

Streptogramin A, see Virginiamycin M_1.
Streptogramin B, see Mikamycin B.

8981. Streptokinase. Streptococcal fibrinolysin; plasminokinase; Kabikinase; Streptase. Single-chain protein produced by β-hemolytic streptococci; mol wt ~47 kDa. Activates plasminogen to produce plasmin which dissolves fibrin. Discovery: W. S. Tillet, R. L. Garner, *J. Exp. Med.* **58**, 485 (1933). Isoln from streptococcal culture filtrates: L. R. Christensen, *J. Gen. Physiol.* **28**, 363 (1945). Purification: M. Siegel *et al.,* U.S. pat. 3,042,586 (1962 to Merck & Co.); P. K. Siiteri, R. D. Mills, U.S. pat. 3,226,304 (1965 to Am. Cyanamid); and characterization: E. C. De Renzo *et al., J. Biol. Chem.* **242**, 533 (1967). Review of isoln, assay and properties: F. B. Taylor, R. H. Tomar, *Methods Enzymol.* **19**, 807-821 (1970); F. Castellino, Jr. *et al., ibid.* **45B**, 244-257 (1976); of mechanism of action: K. N. N. Reddy in *Fibrinolysis,* D. L. Kline, K. N. N. Reddy, Eds. (CRC, Boca Raton, 1980) pp 71-94. Clinical trials in acute myocardial infarction: J. L. Anderson *et al., N. Engl. J. Med.* **308**, 1312 (1983); ISIS-2 Collaborative Group, *Lancet* **2**, 349 (1988). Clinical comparison with alteplase (tPA): GISSI-2 and Int. Study Group, *Eur. Heart J.* **13**, 1692 (1992). Pharmacokinetics: D. S. Grierson, T. D. Bjornsson, *Clin. Pharmacol. Ther.* **41**, 304 (1987). Review of clinical efficacy in deep venous thrombosis: L. Q. Rogers, C. L. Lutcher, *Am. J. Med.* **88**, 389-395 (1990); in myocardial infarction: K. L. Goa *et al., Drugs* **39**, 693-719 (1990).

THERAP CAT: Thrombolytic.

8982. Streptolydigin. Portamycin. $C_{32}H_{44}N_2O_9$; mol wt 600.71. C 63.98%, H 7.38%, N 4.66%, O 23.97%. Antibiotic isolated from culture filtrates of *Streptomyces lydicus* n. sp.: Eble *et al., Antibiot. Annual* **1955-1956**, 893; Brit. pat. 779,570 (1957 to Upjohn). Structure: Rinehart *et al., J. Am. Chem. Soc.* **85**, 4083 (1963). Stereochemistry: Duchamp *et al., ibid.* **95**, 4077 (1973). RNA-polymerase inhibitor: Cassani *et al., Nature New Biol.* **230**, 197 (1971). NMR studies: V. J. Lee, K. L. Rinehart, *J. Antibiot.* **34**,

408 (1980). Toxicity study: DeBoer *et al.*, *Antibiot. Annual* **1955-1956**, 886.

Crystals from acetone + water, mp 147-148°. $[\alpha]_D^{25}$ −93° (c = 1.6 in chloroform). uv max (0.01*N* ethanolic KOH): 261, 291, 336 nm ($E_{1cm}^{1\%}$ 223.6, 270.7, 331.0); uv max (0.01*N* ethanolic H$_2$SO$_4$): 234, 357.5, 370 nm ($E_{1cm}^{1\%}$ 139.1, 590.5, 560.3). Sol in ethanol, ethyl acetate, ether, chloroform. Practically insol in water, hydrocarbons. LD$_{50}$ i.p. in mice: 533 mg/kg (DeBoer).

Sodium salt, C$_{32}$H$_{43}$N$_2$NaO$_9$, mp 225°. $[\alpha]_D^{25}$ +153° (c = 1.35 in chloroform). Sol in the usual organic solvents and in water. Practically insol in hydrocarbons.

THERAP CAT: Antibacterial.

8983. Streptomycin. *O-2-Deoxy-2-(methylamino)-α-L-glucopyranosyl-(1→2)-O-5-deoxy-3-C-formyl-α-L-lyxofuranosyl-(1→4)-N,N'-bis(aminoiminomethyl)-D-streptamine;* streptomycin A. C$_{21}$H$_{39}$N$_7$O$_{12}$; mol wt 581.58. C 43.37%, H 6.76%, N 16.86%, O 33.01%. Antibiotic substance produced by the soil Actinomycete *Streptomyces griseus* (Krainsky) Waksman et Henrici (Fam. *Actinomycetaceae).* Isolation: Schatz *et al., Proc. Soc. Exp. Biol. Med.* **55**, 66 (1944). Production by aerobic fermentation and purification: Tishler in *Streptomycin,* Selman A. Waksman, Ed. (Williams & Wilkins, Baltimore, 1949) pp 32-54. Isoln and purification by ion exchange: Bartels *et al., Chem. Eng. Progr.* **54** (8), 49-51 (Aug. 1958); Bartels *et al.,* U.S. pat. **2,868,779** (1959 to Olin Mathieson). Structure: Brink, Folkers, *J. Am. Chem. Soc.* **69**, 1234 (1947); Wolfrom *et al., ibid.* **76**, 3675 (1950). Total synthesis: Umezawa *et al., J. Antibiot.* **27**, 997 (1974). Mechanism of action: B. J. Wallace *et al.,* in *Antibiotics* vol. 5(pt. 1), F. E. Hahn, Ed. (Springer-Verlag, New York, 1979) pp 272-303. Review: Lemieux, Wolfrom, "The Chemistry of Streptomycin" in W. W. Pigman, M. L. Wolfrom, *Advan. Carbohyd. Chem.* **3**, 337-384 (1948). Comprehensive description: J. Mossa *et al.,* in *Analytical Profiles of Drug Substances* vol. 16, K. Florey, Ed. (Academic Press, New York, 1986) pp 507-609.

R = CH$_2$OH
R' = NHCH$_3$

Streptomycin is usually available as the trihydrochloride, trihydrochloride-calcium chloride double salt, phosphate, or sesquisulfate, which occur as granules or powder. Odorless or nearly so, with a slightly bitter taste. Most salts are hygroscopic and deliquesce on exposure to air, but are not affected by air or light. The salts are very sol in water; but almost insol in alc, chloroform, ether. Solns are levorotatory.

Trihydrochloride, C$_{21}$H$_{39}$N$_7$O$_{12}$·3HCl, *streptomycin hydrochloride.* $[\alpha]_D^{25}$ −84°. Solubilities as determined by Weiss *et al., Antibiot. & Chemother.* **7**, 374 (1957) in mg/ml at about 28°: water > 20; methanol > 20; ethanol 0.90; isopropanol 0.12; isoamyl alcohol 0.117; petr ether 0.02; carbon tetrachloride 0.042; ether 0.01.

Trihydrochloride-calcium chloride double salt, (C$_{21}$H$_{39}$N$_7$O$_{12}$·3HCl)$_2$CaCl$_2$, *streptomycin hydrochloride-calcium chloride complex.* Prepn from the trihydrochloride: Peck, U.S. pat. **2,446,102** (1948 to Merck & Co.). Very hygroscopic, dec about 200°. $[\alpha]_D^{25}$ −76°.

Pantothenate, *Streptothenat.* Prepn: **Brit.** pat. **771,338** (1957 to Grünenthal). The commercial prepn may contain the sulfate.

Sesquisulfate, (C$_{21}$H$_{39}$N$_7$O$_{12}$)$_2$·3H$_2$SO$_4$, *streptomycin sulfate, AgriStrep, Streptobrettin, Vetstrep.* White to light gray or pale buff powder with faint amine-like odor. Solubilities as determined by Weiss *et al., loc. cit.,* in mg/ml at about 28°: water > 20; methanol 0.85; ethanol 0.30; isopropanol 0.01; petr ether 0.015; carbon tetrachloride 0.035; ether 0.035.

THERAP CAT: Antibacterial (tuberculostatic).
THERAP CAT (VET): Antibacterial.

8984. Streptomycin B. Mannosidostreptomycin; mannosylstreptomycin. C$_{27}$H$_{49}$N$_7$O$_{17}$; mol wt 743.72. C 43.60%, H 6.64%, N 13.18%, O 36.57%. Isoln from streptomycin concentrates: Fried, Titus, *J. Biol. Chem.* **168**, 391 (1947); **174**, 57 (1948); Fried, Stavely, *ibid.* **70**, 3615 (1948); O'Keeffe *et al., ibid.* **71**, 2452 (1949); *see also* Langlykke, Perlman, U.S pat. **2,493,489** (1950 to Squibb). Structure: Fried, Stavely, 113th Meeting *Am. Chem. Soc.,* Chicago, Ill. (April 20, 1948) *Abstracts of Papers* p 25C; Peck *et al., J. Am. Chem. Soc.* **70**, 3968 (1948); Stavely, Fried, *ibid.* **71**, 135 (1949); **74**, 5461 (1952).

R = CH$_2$OH
R' = NHCH$_3$

On catalytic hydrogenation absorbs approx one mole of hydrogen and forms dihydrostreptomycin B. Yields maltol on alkaline hydrolysis. When treated with dil acids liberates streptidine (or an isomer of it) isolated as the sulfate monohydrate, C$_8$H$_{18}$N$_6$O$_4$·H$_2$SO$_4$·H$_2$O.

Trihydrochloride monohydrate, C$_{27}$H$_{49}$N$_7$O$_{17}$·3HCl·H$_2$O, white, amorphous powder; mp 190-200° (corr, decompn) after drying *in vacuo* at 100°. $[\alpha]_D^{25}$ −47° (c = 1.35 in water). Appears to have about one-fourth of the antibiotic activity of streptomycin trihydrochloride.

8985. Streptonicozid. *4-Pyridinecarboxylic acid hydrazide, hydrazone with O-2-deoxy-2-(methylamino)-α-L-glucopyranosyl-(1→2)-O-5-deoxy-3-C-formyl-α-L-lyxofuranosyl-(1→4)-N,N'-bis(aminoiminomethyl)-D-streptamine sulfate (2:3) (salt); isonicotinic acid hydrazide, hydrazone with streptomycin, sulfate (salt) (2:3);* streptomyclidene isonicotinyl hydrazine sulfate; streptoniazide; Strazide; Streptohydrazid. C$_{54}$H$_{94}$N$_{20}$O$_{36}$S$_3$; mol wt 1695.65. C 38.25%, H 5.59%, N 16.52%, O 33.97%, S 5.67%. Semi-synthetic antibiotic prepd by the condensation of streptomycin salts and isonicotinic acid hydrazide in methanol: Pennington *et al., J. Am. Chem. Soc.* **75**, 2261 (1953).

• 3 H₂SO₄

$$\cdot\ 3\ H_2SO_4$$

```
R  = CH₂OH
R' = NHCH₃
R" = HC=NNHC
              ‖
              O
```

Crystals, dec ~230°. Freely sol in water with dissocn in dil solns. Solubilities determined by Weiss *et al., Antibiot. & Chemother.* **7**, 374 (1957) in mg/ml at about 28°: water > 20; ethanol 0.115; carbon tetrachloride 0.025; ether 0.31.
THERAP CAT: Antibacterial (tuberculostatic).

8986. Streptonigrin. *5-Amino-6-(7-amino-5,8-dihydro-6-methoxy-5,8-dioxo-2-quinolyl)-4-(2-hydroxy-3,4-dimethoxyphenyl)-3-methyl-2-pyridinecarboxylic acid; 5-amino-6-(7-amino-5,8-dihydro-6-methoxy-5,8-dioxo-2-quinolyl)-4-(2-hydroxy-3,4-dimethoxyphenyl)-3-methylpicolinic acid;* bruneomycin; NSC-45383; Nigrin. $C_{25}H_{22}N_4O_8$; mol wt 506.47. C 59.29%, H 4.38%, N 11.06%, O 25.27%. Antitumor antibiotic produced by *Streptomyces flocculus:* Rao, Cullen, *Antibiot. Annual* **1959-1960**, 950. Structure: Rao *et al., J. Am. Chem. Soc.* **85**, 2532 (1963). Crystal and molecular structure: Y. Y. H. Chiu, W. N. Lipscomb, *ibid.* **97**, 2525 (1975). Identity with bruneomycin: Brazhnikova *et al., Antibiotiki* **13**, 99 (1968); *C.A.* **68**, 89890q (1968). Synthetic studies: Kametani *et al., C.A.* **76**, 14272w (1972); **79**, 31817g (1973). Total synthesis: F. Z. Basha *et al., J. Am. Chem. Soc.* **102**, 3962 (1980); A. S. Kende *et al., ibid.* **103**, 1271 (1981); S. M. Weinreb *et al., ibid.* **104**, 536 (1982). Biosynthesis: S. J. Gould, D. E. Cane, *ibid.* 343. *Reviews:* N. S. Mizuno in *Antibiotics* vol. 5, pt. 2, F. E. Hahn, Ed. (Springer-Verlag, New York, 1979) pp 372-384; S. J. Gould, S. M. Weinreb, *Fortschr. Chem. Org. Naturst.* **41**, 77-111 (1982).

Coffee-brown to almost black rectangular plates from acetone or dioxane, dec 275°; brown needles, mp 262-263° (Brazhnikova). uv max (methanol): 248, 375-380 nm (ε 38400, 17400). Weak acid. pKa 6.2-6.4 (1:1 aq dioxane). Sol in dioxane, pyridine, dimethylformamide, aq sodium bicarbonate with decompn; slightly sol in water, lower alcohols, ethyl acetate, chloroform.

Caution: Causes severe and prolonged bone marrow depression. Toxicity studies: Wilson *et al., Antibiot. Chemother.* **11**, 147 (1961) *sqq.*
THERAP CAT: Antineoplastic.

8987. L-Streptose. *5-Deoxy-3-C-formyl-L-lyxose;* 3-C-formyl-5-deoxy-L-lyxofuranose. $C_6H_{10}O_5$; mol wt 162.14. C 44.45%, H 6.22%, O 49.34%. A sugar which is part of the streptomycin molecule: Kuehl *et al., J. Am. Chem. Soc.* **68**, 2096 (1946). Structure: *eidem, ibid.* **68**, 2679 (1946). Configuration: Wolfrom, DeWalt, *ibid.* **70**, 3148 (1948); Kuehl *et al., ibid.* **71**, 1445 (1949). Biosynthesis: Hough, Jones, *Nature* **167**, 180 (1951). Synthesis of α- and β-L-streptose: Dyer *et al., J. Am. Chem. Soc.* **87**, 654 (1965); Paulsen *et al., Ber.* **105**, 1978 (1972).

Colorless, strongly hygroscopic glass. $[\alpha]_D^{20}$ −18° (c = 0.65).

8988. Streptothricins. Racemomycins; yazumycins. A mixture of antibiotics, originally thought to be a single substance, differing only in the number of repeating residues in the peptide side chain. Streptothricins F, E, D, C, B, A, and X are known and have 1 to 7 β-lysine residues resp.: Khokhlov, Shutova, *J. Antibiot.* **25**, 501 (1972). Separation of isomers: Khokhlov, Reshetov, *J. Chromatog.* **14**, 495 (1964). Identity with racemomycins: H. Taniyama *et al., Chem. Pharm. Bull.* **19**, 1627 (1971). Identity with yazumycins: *eidem, J. Antibiot.* **24**, 390 (1971). Activity studies: Germanova, Goncharskaya, *Antibiotiki (Moscow)* **14**, 48, 137 (1969); *C.A.* **70**, 66547, 85906t (1969). Biosynthetic studies: Voronina *et al., ibid.* 1063, *C.A.* **72**, 77387a (1970); Y. Sawada *et al., Chem. Pharm. Bull.* **26**, 885 (1978). Partial syntheses: M. Kinoshita, Y. Suzuki, *Bull. Chem. Soc. Japan* **50**, 2375 (1977); S. Kusumoto *et al., Chem. Letters* **1981**, 1317. Prepn of semi-synthetic racemomycins: Y. Sawada, H. Taniyama, *Chem. Pharm. Bull.* **25**, 1302 (1977). *Streptolin* was formerly used for the streptothricin components where n equals 2 or 3.

Streptothricin F, $C_{19}H_{34}N_8O_8$, *racemomycin A, streptothricin VI, yazumycin A.* n = 1. Originally known as streptothricin. Produced by variants of *Actinomyces lavendulae.* Isoln: Waksman, Woodruff, *Proc. Soc. Exp. Biol. Med.* **49**, 207 (1942). Purification: Peck *et al., J. Am. Chem. Soc.* **68**, 772 (1946); Kocholaty, Junowicz-Kocholaty, *Arch. Biochem.* **15**, 55 (1947). Identity with streptothricin VI: Hutchinson *et al., Arch. Biochem.* **22**, 16 (1949); Swart, *J. Am. Chem. Soc.* **71**, 2942 (1949). Prepn: Carter *et al., J. Am. Chem. Soc.* **76**, 566 (1954). Structure: van Tamelen *et al., ibid.* **83**, 4295 (1961); Johnson, Westley, *J. Chem. Soc.* **1962**, 1642. Revised structure: S. Kusumoto *et al., J. Antibiot.* **35**, 925 (1982). Biosynthesis: S. J. Gould *et al., J. Am. Chem. Soc.* **103**, 2871 (1981). Total synthesis: S. Kusumoto *et al., Tetrahedron Letters* **23**, 2961 (1982). Amorphous, white powder. Thermostable. Water sol; slightly sol in most organic solvents. Stable between pH 1 and 8.5. LD_{50} i.v. in mice: 300 mg/kg (Taniyama).

Hydrochloride, white powder. $[\alpha]_D^{25}$ −51.3° (c = 1.4). Sol in water and dil mineral acids. Practically insol in ether, petr ether, chloroform.

8989. Streptovaricin. Streptovarycin. A complex of ansamycin antibiotics consisting of streptovaricins A, B, C, D, E, F, G, J and K of which streptovaricin C is the major component. Isolation from *Streptomyces spectabilis:* Siminoff *et al., Am. Rev. Tuberc. Pulm. Dis.* **75**, 576 (1957); Whitfield *et al., ibid.* 584; and resolution of components: Dietz *et al.,* U.S. pat. 3,116,202 (1963 to Upjohn). Characterization of complex: Rinehart *et al., Chem. Commun.* **1974**, 861. Preliminary structural studies: Rinehart *et al., J. Am. Chem. Soc.* **88**, 3149, 3150 (1966); **90**, 6241 (1968); Rinehart *et al., ibid.* **93**, 6273 (1971). Revised structure of streptovaricin C: Sasaki *et al., J. Antibiot.* **25** 68 (1972); Rinehart, Antosz, *ibid.* 71. Activity studies: Quintrell, McAuslan, *J. Virol.* **6**, 485 (1970); Tan, McAuslan, *Biochem. Biophys. Res. Commun.* **42**, 230 (1971); Carter *et al., Nature New Biol.* **232**, 212, 214 (1971). Biosynthetic study: P. V. Deshmukh *et al., J. Am. Chem. Soc.* **98**, 870 (1976). 13C-NMR studies: K. B. Kakinuma *et al., J. Org. Chem.* **41**, 1358 (1976). Synthesis of a key intermediate for the total synthesis of streptovaricin A: P. A. McCarthy, *Tetrahedron Letters* **23**, 4199 (1982). Synthesis of the naphthalene core of streptovaricin D: B. M. Trost, W. H. Pearson, *ibid.* **24**, 269 (1983). Review of the chemistry of streptovaricins: K. L. Rinehart *et al., Fortschr. Chem. Org. Naturst.* **33**, 231-307 (1976).

Streptovaricin C

The complex yields yellow crystals from ethyl acetate hexane. Neutral or weakly acidic substance which hydrolyzes on treatment with dil alkali. Stable at room temp for 3-4 days at pH 2.0-6.0. Gradually loses its antibiotic activity while standing at room temp for 2-4 days at pH 7.8.
Streptovaricin A, $C_{42}H_{53}NO_{16}$, mol wt 827.88. mp 194-196°. $[\alpha]_D$ in chloroform +61°. uv max (95% ethanol): 245, 260, 320, 430 nm ($E_{1cm}^{1\%}$ 418.9, 352.1, 139.8, 136.2).
Streptovaricin B, $C_{42}H_{53}NO_{15}$, mol wt 811.88. mp 185-187°. $[\alpha]_D$ in chloroform +576° (Dietz, +454°). uv max (95% ethanol): 245, 266, 320, 432 nm ($E_{1cm}^{1\%}$ 408.3, 338.5, 140.1, 109.7).
Streptovaricin C, $C_{40}H_{51}NO_{14}$, mol wt 769.84. mp 189-191° (Dietz, 168-171°). $[\alpha]_D$ in chloroform +602° (Dietz, +317°). uv max (95% ethanol): 245.5, 260, 320, 430 nm ($E_{1cm}^{1\%}$ 443.5, 389.7, 153.1, 121.9). Solubility data: March, Weiss, *J. Assoc. Offic. Anal. Chem.* **50**, 457 (1967).
Streptovaricin D, $C_{40}H_{51}NO_{13}$, mol wt 753.84. mp 167-170° (Dietz, 115-118°). $[\alpha]_D$ in chloroform +436° (Dietz, +102°). uv max (95% ethanol): 246, 264, 320, 433 nm ($E_{1cm}^{1\%}$ 431.4, 370.3, 147.7, 105.3).
Streptovaricin E, $C_{40}H_{49}NO_{14}$, mol wt 767.83. mp 102-105°. $[\alpha]_D^{24}$ in chloroform +164°, also reported as +613°. uv max (95% ethanol): 245, 273, 320, 437 nm ($E_{1cm}^{1\%}$ 382.2, 346.9, 190.4, 91.9).
Streptovaricin F, $C_{39}H_{47}NO_{14}$, mol wt 753.80.
Streptovaricin G, $C_{40}H_{51}NO_{15}$, mol wt 785.84. mp 190-192°. $[\alpha]_D$ in chloroform +473°.
Streptovaricin J, $C_{42}H_{53}NO_{15}$, mol wt 811.88. mp 177-180°. $[\alpha]_D$ +326° (c = 0.25 in ethanol); +436° (c = 0.094 in chloroform).

8990. Streptovirudin. Nucleoside antibiotic complex produced by *Streptomyces griseoflavus* subsp. *thuringiensis,* strain JA 10124, that has inhibitory activity vs gram-positive bacteria, mycobacteria, and several RNA and DNA viruses. Isoln: K. Eckardt *et al., Z. Allg. Mikrobiol.* **13**, 625

(1973); *eidem, J. Antibiot.* **28**, 274 (1975). Taxonomy, fermentation, production: H. Thrum *et al., ibid.* 514. Streptovirudin is related chemically to tunicamycin, *q.v.,* and contains at least 10 components, of which 8 have been isolated in pure form. The complex is divided into 2 series of compounds: series I (streptovirudins A_1 through D_1) contains dihydrouracil and series II (streptovirudins A_2 through D_2) contains uracil. Differentiation of streptovirudin and tunicamycin by gel chromatography, HPLC, and hydrolysis: K. Eckardt *et al., ibid.* **33**, 908 (1980). Structures of streptovirudins and identity of components B_{2a} and C_2 with tunicamycin components: *eidem, ibid.* **34**, 1631 (1981); W. Ihn *et al., Tetrahedron* **38**, 1781 (1982). Biological activities of isolated streptovirudin and tunicamycin components, identity of streptovirudin C_2 with tunicamycin A: R. W. Keenan *et al., Biochemistry* **20**, 2968 (1981). Chemical and biological properties of streptovirudins: A. D. Elbein *et al., ibid.* 4210. Viral inhibition study: M. S. Kang *et al., Biochem. Biophys. Res. Commun.* **99**, 422 (1981).

Streptovirudin A_1

White amorphous solid. uv max (methanol): 212 nm. Sol in methanol, pyridine. Slightly sol in water, ethanol. Insol in chloroform, acetone, benzene, ethyl acetate. Stable up to 5 hours at 90°. No loss of activity was observed after standing at 37° for 4 weeks. LD_{50} in mice (mg/kg) of the antibiotic complex from culture filtrate: 250 orally; 15 s.c.; 15-17.5 i.p.; 17.5 i.v.; of the complex from mycelium: 150 orally; 5-10 s.c.; 3.5-7.5 i.p. (Thrum).

8991. Streptozocin. 2-Deoxy-2-[[(methylnitrosoamino)-carbonyl]amino]-D-glucopyranose; 2-deoxy-2-(3-methyl-3-nitrosoureido)-D-glucopyranose; streptozotocin; NSC-85998; U-9889; Zanosar. $C_8H_{15}N_3O_7$; mol wt 265.22. C 36.23%, H 5.70%, N 15.84%, O 42.23%. Isoln from *Streptomyces achromogenes* fermentation broth: Herr *et al., Antibiot. Ann.* 1959-1960, 236; Bergy *et al.,* Fr. pat. 1,434,920 (1966 to Upjohn), *C.A.* **65**, 17661h (1966). Structure and synthesis: Herr *et al., J. Am. Chem. Soc.* **89**, 4808 (1967); Hardegger *et al., Helv. Chim. Acta* **52**, 2555 (1969); Hessler, Jahnke, *J. Org. Chem.* **35**, 245 (1970); P. F. Wiley *et al., ibid.* **44**, 9 (1979). Diabetogenic effect: Rakieten *et al., Cancer Chemother. Rep.* No. 29, 91 (May 1963). Comparison of streptozotocin- and alloxan-induced diabetes: C. Rerup, F. Tarding, *Eur. J. Clin. Pharmacol.* **7**, 89 (1969). Antileukemic activity: Bhuyan *et al., Cancer Chemother. Rep. Part 1* **56**, 709 (1972). Biosynthetic study: S. Singaram *et al., J. Antibiot.* **32**, 379 (1979). Review and evaluation of studies of carcinogenicity in laboratory animals: *IARC Monographs* **4**, 221-227 (1974). Toxicity data: M. Iwasaki *et al., J. Med. Chem.* **19**, 918 (1976). *Review:* B. Rudas, *Arzneimittel-Forsch.* **22**, 830-861 (1972); P. F. Wiley, *Anticancer Agents Based on Natural Product Models,* J. M. Cassidy, J. D. Douros, Eds. (Academic Press, New York, 1980) pp 127-200. Book: *Streptozotocin: Fundamentals and Therapy,* M. K. Agarwal, Ed. (Elsevier/North Holland Biomedical Press, New York, 1981) 309 pp.

Pointed platelets or prisms, from 95% ethanol, mp 115° (dec). Sol in H_2O, lower alcohols and ketones. uv max (ethanol): 228 nm (ϵ 6360). A mixture of α and β anomers; aq solns rapidly undergo mutarotation to an equilibrium value of $[\alpha]_D^{25}$ +39°. LD_{50} in female mice (mg/kg): 360 i.p.; 275 i.v.; in male dogs (mg/kg): 50 i.v. (Iwasaki).

Tetraacetate, $C_8H_{11}N_3O_7(COCH_3)_4$, crystals, mp 111-114° (dec). $[\alpha]_D^{25}$ +41° (c = 0.78 in 95% ethanol).

Note: Streptozocin may reasonably be anticipated to be a carcinogen: *Seventh Annual Report on Carcinogens* (PB95-109781, 1994) p 366.

USE: Production of experimental diabetes in laboratory animals.

THERAP CAT: Antineoplastic.

8992. Strigol. *[3aR-[3E(R*),3aα,5β,8bα]]-3-[[(2,5-Di-hydro-4-methyl-5-oxo-2-furanyl)oxy]methylene]-3,3a,4,5,6,7,-8,8b-octahydro-5-hydroxy-8,8-dimethyl-2H-indeno[1,2-b]-furan-2-one.* $C_{19}H_{22}O_6$; mol wt 346.38. C 65.88%, H 6.40%, O 27.71%. Potent seed germination stimulant for the root parasite, witchweed, *Striga lutea* Lour. isolated from root exudates of cotton, *Gossypium hirsutum* L.: C. E. Cook *et al., Science* **154**, 1189 (1966). Structure: *eidem, J. Am. Chem. Soc.* **94**, 6198 (1972). Crystal structure: P. Coggan *et al., J. Chem. Soc. Perkin Trans. 2* **1973**, 465. Synthesis: J. B. Heather *et al., J. Am. Chem. Soc.* **98**, 3661 (1976); of (±)-form: G. A. MacAlpine *et al., Chem. Commun.* **1974**, 834; *eidem, J. Chem. Soc. Perkin Trans. 1* **1976**, 410. Synthesis of analogs: A. W. Johnson *et al., ibid.* **1981**, 1734. Witchweed seed germination stimulating activity: A. L. Hsiao *et al., Weed Sci.* **29**, 101 (1981); A. B. Pepperman *et al., ibid.* **30**, 561 (1982).

White needles from benzene-hexane, mp 200-202° (dec). $[\alpha]_D^{25}$ +293° (c = 0.15 in $CHCl_3$). uv max: 234 nm (ϵ 17700). Sol in acetone, methylene chloride. Moderately sol in benzene; insol in hexane.

8993. Strobane®. Terpene polychlorinates; Dichloricide Aerosol; Dichloricide Mothproofer; Insecticide 3960-X14. A mixture of chlorinated terpene isomers. Preparation by chlorinating a mixture of camphene and pinenes contg dipentene: Schultz, Bloor, U.S. pat. **2,767,115** (1956 to B. F. Goodrich).

Viscous, amber liquid. d_{25} 1.6267. Misc with aromatic and aliphatic hydrocarbons. Slightly sol in alcohol.

Caution: May be mildly irritating to skin.

USE: Insecticide. *Compare* Toxaphene.

8994. Strontium. Sr; at. wt 87.62; at. no. 38; valence 2. An alkaline earth metal. Isotopes: 88 (82.56%); 86 (9.86%); 87 (7.02%); 84 (0.56%). Occurs as the sulfate, celestine; or the carbonate, strontianite; found in small quantities associated with calcium or barium minerals. First prepared in 1807 by Davy. Prepn: Glascock, *J. Am. Chem. Soc.* **32**, 1222 (1910); Matignon, *Compt. Rend.* **177**, 1116 (1923); *J. Chem. Soc.* **126**[ii], 44 (1924); Guntz *et al.,* cited in *Gmelin's,*

Strontium (8th ed.) **29**, 35 (1931). Review of strontium and its compounds: Goodenough, Stenger, "Magnesium, Calcium, Strontium, Barium and Radium" in *Comprehensive Inorganic Chemistry,* vol. **1**, J. C. Bailar, Jr. *et al.,* Eds. (Pergamon Press, Oxford, 1973) pp 591-774.

Silvery-white metal; face-centered cubic structure; rapidly becomes yellow on exposure to air and assumes an oxide film. The finely divided metal ignites spontaneously in air. d 2.6; mp 757° ± 1°; bp 1366°. E^0 (aq) Sr^{2+}/Sr −2.89 V. For a description of reactions which are characteristic of alkaline earth metals *see* Calcium. *Keep under liquid containing no oxygen.* The heated metal combines with hydrogen to form *strontium hydride* and with nitrogen to form *strontium nitride.* Strontium salts impart brilliant red color to a flame.

Note: The problems of internal radiation hazards from radiostrontium are discussed by A. Engström *et al., Bone and Radiostrontium* (Wiley, New York, 1958).

USE: In fireworks, in red signal flares; on tracer bullets. The artificial isotope ^{90}Sr, with a half life of 28 years, is being considered as a source of electric power. Heat liberated by decay would be changed to electrical energy.

8995. Strontium Acetate. $C_4H_6O_4Sr$; mol wt 205.71. C 23.36%, H 2.94%, O 31.11%, Sr 42.59%. $Sr(C_2H_3O_2)_2$. Toxicity: V. V. Cole *et al., J. Pharmacol. Exp. Ther.* **71**, 1 (1941).

Hemihydrate, white crystalline powder. Loses its water at 150°; on ignition is converted into $SrCO_3$. Sol in 2.5 parts water; slightly sol in alcohol. The aq soln is practically neutral to litmus. LD i.v. in rats: 1.16 mmol/kg (Cole).

8996. Strontium Bromate. Br_2O_6Sr; mol wt 343.42. Br 46.53%, O 27.95%, Sr 25.51%. $Sr(BrO_3)_2$.

Monohydrate, white to slightly yellow, hygroscopic, cryst powder. Sol in 3 parts water. *Keep well closed.*

8997. Strontium Bromide. Br_2Sr; mol wt 247.43. Br 64.59%, Sr 35.41%. $SrBr_2$. Acute toxicity: K. W. Cochran *et al., Arch. Ind. Hyg. Occup. Med.* **1**, 637 (1950).

Hexahydrate, colorless, deliquesc crystals or white granules; bitter saline taste. mp 88° when rapidly heated; mp 643° when anhydr. Loses all its water at 180°. Sol in 0.35 part water; sol in alcohol. Insol in ether. The aq soln is neutral. *Incompat:* Soluble sulfates. LD_{50} i.p. in rats: 1000 mg/kg (Cochran).

THERAP CAT: Has been used as anticonvulsant.

THERAP CAT (VET): Has been used as sedative, anticonvulsant.

8998. Strontium Carbonate. CO_3Sr; mol wt 147.63. C 8.14%, O 32.51%, Sr 59.35%. $SrCO_3$. Occurs in nature as the mineral *strontianite.*

White, odorless, tasteless powder. d 3.5. Dec at 1100° into SrO and CO_2. Sol in 100,000 parts water, in about 1000 parts of water saturated with CO_2; sol in dil acids.

USE: In pyrotechnics; manuf iridescent glass; refining sugar.

8999. Strontium Chlorate. Cl_2O_6Sr; mol wt 254.52. Cl 27.86%, O 37.72%, Sr 34.43%. $Sr(ClO_3)_2$.

Colorless or white crystals. d 3.15. mp 120° with decomposition and evolution of O_2. Sol in 0.6 part water, slightly in alcohol. *Handle with caution like potassium chlorate.*

USE: In pyrotechnics to produce red fire.

9000. Strontium Chloride. Cl_2Sr; mol wt 158.53. Cl 44.73%, Sr 55.27%. $SrCl_2$. Acute toxicity: I. B. Syed, F. Hosain, *Toxicol. Appl. Pharmacol.* **22**, 150 (1972).

Hexahydrate, colorless, odorless crystals or white granules. Effloresces in air; deliquesces in moist air. At 100° loses $5H_2O$; at 150° all its H_2O; d 1.96; mp 61° when rapidly heated; the anhydr salt melts at 868°. Soluble in 0.8 part water, 0.5 part boiling water; sol in alcohol. The aq soln is neutral. *Keep well closed.* LD_{50} i.v. in mice: 147.6 mg/kg (Syed, Hosain).

USE: Manuf other strontium salts; in pyrotechnics; as dental desensitizer under the name *Elecol.*

9001. Strontium Chromate(VI). CrO_4Sr; mol wt 203.61. Cr 25.54%, O 31.43%, Sr 43.03%. $SrCrO_4$.

Yellow powder. d 3.89. Soluble in 840 parts cold water, about 5 parts boiling water; freely sol in dil hydrochloric, nitric or acetic acids.

USE: Corrosion inhibitor in pigments; in electrochemical processes to control sulfate concn of solns.

9002. Strontium Fluoride. F_2Sr; mol wt 125.62. F 30.25%, Sr 69.75%. SrF_2. Prepd by dissolving $SrCO_3$ in an excess of 40% HF and evaporating to dryness: Berzelius, *Pogg. Ann.* **1**, 20 (1824); final drying should be done at 150° *in vacuo.*

White powder or cubic crystals. d 4.24. mp ~1400°. bp 2460°. Stable in air up to 1000°. Above this temp it is oxidized to strontium oxide. Heat of evaporation 78.2. Trouton const 28.6. Soly in water (18°) 11.7 mg/100 ml. More sol in dil acids. Dec by strong acids. May be stored in glass bottles.

9003. Strontium Formate. $C_2H_2O_4Sr$; mol wt 177.66. C 13.52%, H 1.13%, O 36.02%, Sr 49.32%. $Sr(HCOO)_2$.

Dihydrate, white crystals or granules. d 2.69. Sol in water.

9004. Strontium Hydroxide. Strontium hydrate. H_2O_2Sr; mol wt 121.63. H 1.66%, O 26.31%, Sr 72.04%. $Sr(OH)_2$.

Octahydrate, colorless, deliquesc crystals or white powder; absorbs CO_2 from air forming carbonate. Loses part of its water at about 100°. Sol in 50 parts water, 2.1 parts boiling water. The soln is very alkaline. pH about 13.5. *Keep tightly closed.*

USE: Refining beet sugar; separating crystallizable sugar from molasses.

9005. Strontium Iodide. I_2Sr; mol wt 341.43. I 74.34%, Sr 25.66%. SrI_2. The article of commerce is hydrated. Toxicity data: K. W. Cochran *et al., Arch. Ind. Hyg. Occup. Med.* **1**, 637 (1950).

Hexahydrate, colorless to yellowish, deliquesc, fused masses or granules with bitterish saline taste; becomes yellow on exposure to air and light due to liberation of iodine. Melts when rapidly heated, at about 120°. mp 402° (anhydrous). d 4.42. Sol in 0.2 part water; sol in alcohol. The aq soln is practically neutral. *Keep well closed and protected from light.* LD_{50} i.p. in rats: 800 mg/kg (Cochran).

THERAP CAT: Iodine source.

THERAP CAT (VET): Has been used in iodide therapy.

9006. Strontium Lactate. *2-Hydroxypropanoic acid strontium salt.* $C_6H_{10}O_6Sr$; mol wt 265.76. C 27.12%, H 3.79%, O 36.12%, Sr 32.97%. $(CH_3CHOHCOO)_2Sr$. Acute toxicity: K. W. Cochran *et al., Arch. Ind. Hyg. Occup. Med.* **1**, 637 (1950).

Trihydrate, *Strontolac.* White, odorless, granular powder. Loses its water of crystn at 120°. Sol in 3 parts water, 0.5 part boiling water; slightly sol in alcohol. The aq soln is practically neutral. LD_{50} i.p. in rats: 900 mg/kg (Cochran).

9007. Strontium Nitrate. N_2O_6Sr; mol wt 211.63. N 13.24%, O 45.36%, Sr 41.40%. $Sr(NO_3)_2$.

White granules or powder. d 2.99. mp 570°. Sol in 1.5 parts water; slightly sol in alcohol or acetone. The aq soln is neutral. At low temps strontium nitrate crystallizes with $4H_2O$ (25.4%). LD_{50} i.p. in rats: 540 mg/kg, Cochran *et al., Arch. Ind. Hyg.* **1**, 637 (1950).

USE: In pyrotechnics (red fire), signal lights, marine signals, railroad flares, matches.

9008. Strontium Oxalate. *Ethanedioic acid strontium salt.* C_2O_4Sr; mol wt 175.64. C 13.68%, O 36.44%, Sr 49.89%. SrC_2O_4.

Monohydrate, white, odorless, cryst powder. Soluble in 20,000 parts water; in 1900 parts of 3.5% acetic acid, in 1115 parts of the 23% acid, but less sol in the 35% acid; readily sol in dil HCl or HNO_3.

9009. Strontium Oxide. Strontia; strontium monoxide. OSr; mol wt 103.62. O 15.44%, Sr 84.56%. SrO.

White to grayish-white, porous, caustic mass. d 4.7. mp 2430°. When treated with water, forms the hydroxide with evolution of much heat. *Keep tightly closed.*

USE: Manuf of strontium salts.

9010. Strontium Peroxide. O_2Sr; mol wt 119.62. O 26.75%, Sr 73.25%. SrO_2. The peroxide of commerce contains about 85% SrO_2.

White, odorless, tasteless powder; gradually dec on exposure to air. Almost insol in water, but is gradually dec by it with evolution of oxygen; forms hydrogen peroxide with dil acids. *Keep well closed.*

THERAP CAT: Antiseptic.

9011. Strontium Phosphate, Tribasic. $O_8P_2Sr_3$; mol wt 452.80. O 28.27%, P 13.68%, Sr 58.05%. $Sr_3(PO_4)_2$.

White, odorless powder. Insol in water; sol in acids.

9012. Strontium Selenate. O_4SeSr; mol wt 230.58. O 27.76%, Se 34.24%, Sr 38.00%. $SrSeO_4$. Prepared by roasting strontium carbonate with selenium or selenium oxide: Lenher, Wechter, *J. Am. Chem. Soc.* **47**, 1522 (1925).

Orthorhombic crystals. d 4.25. Sol in hot hydrochloric acid; insol in water.

9013. Strontium Sulfate. O_4SSr; mol wt 183.68. O 34.84%, S 17.46%, Sr 47.70%. $SrSO_4$. Occurs in nature as the mineral *celestine* or *celestite*.

White, odorless, crystalline powder. d 3.96. One gram dissolves in ~8800 ml water, 800 ml 2% HCl, 700 ml 3% HNO_3; appreciably sol in alkali chloride solns.

USE: In ceramics, pyrotechnics.

9014. Strontium Sulfide. SSr; mol wt 119.69. S 26.79%, Sr 73.21%. SrS. The article of commerce contains 65-75% SrS. Strontium sulfide made by reduction of the sulfate is not luminous; the luminous sulfide is prepared by heating strontium hydroxide with sulfur.

Gray powder; has the odor of H_2S in moist air. d 3.70. Slightly sol in water; sol in acids with decompn. *Keep dry.*

USE: In luminous paints; as a depilatory.

9015. Strophanthidin. *($3\beta,5\beta$)-3,5,14-Trihydroxy-19-oxocard-20(22)-enolide;* apocynamarin; convallatoxigenin; corchorin; cymarigenin; cynotoxin; corchorgenin; corchsularin. $C_{23}H_{32}O_6$. mol wt 404.50. C 68.29%, H 7.97%, O 23.73%. By acid or enzymic hydrolysis of glycosides present in several species of *Strophanthus, Apocynaceae.* Isoln: Jacobs, Heidelberger, *J. Biol. Chem.* **54**, 253 (1922). Structure: Kon, *Chem. & Ind. (London)* **53**, 593, 956 (1934); Jacobs, Elderfield, *Science* **80**, 533 (1934). Identity with corchorin, corchorgenin and corchsularin: Sen *et al., Helv. Chim. Acta* **40**, 588 (1957). Synthesis from pregnenolone acetate, *q.v.*: E. Yoshii *et al., J. Org. Chem.* **43**, 3946 (1978). Toxicity study: Graebner, Geisel, *Arzneimittel-Forsch.* **22**, 1854 (1972). *Reviews:* Elderfield, *Chem. Rev.* **17**, 187 (1935); Tschesche, *Ergeb. Physiol.* **38**, 31 (1936); Stoll, *The Cardiac Glycosides* (Pharmaceutical Press, London, 1937); Strain in *Organic Chemistry,* **vol. II**, Gilman, Ed. (New York, 2nd ed., 1943); Fieser, Fieser, *Steroids* (Reinhold, New York, 1959) pp 729-730, 736-750.

Orthorhombic tablets from 5 parts methanol and 10 parts water. *Very poisonous!* Contains ½ mol H_2O which is given up at 110° over P_2O_5 at 20 mm Hg. mp 171-175° with effervescence. Occasionally a few isolated crystals are found which melt at ~230°. $[\alpha]_D^{25}$ +43.1° (c = 2.8 in methanol). Sol in alcohol, acetone, chloroform, benzene, and glacial

acetic acid. Practically insol in water, ether, petr ether. Color reaction with H_2SO_4 and Ac_2O like cholesterol. LD_{100} i.v. in cats: 0.337 mg/kg (Graebner, Geisel).
3-Benzoylstrophanthidin, $C_{30}H_{36}O_7$, mp 230°. $[\alpha]_D^{26}$ +47.8° (c = 1.07 in acetone).
3-(p-Bromobenzoyl)strophanthidin, $C_{30}H_{35}BrO_7$, mp 222-224°. $[\alpha]_D^{20}$ +42.0° (c = 1.094 in acetone).
Dihydrostrophanthidin, $C_{23}H_{34}O_6$, mp 100-103°, resolidifies and again melts at 190-195°. $[\alpha]_D^{23}$ +34.9° (methanol).
16-Hydroxystrophanthidin, $C_{23}H_{32}O_7$, *strophadogenin*. Long prisms from ethanol, dec 238-241°. $[\alpha]_D^{23}$ +52.9° (c = 1.24 in 96% alc). uv max: 219 nm (log ε 4.3).

9016. Strophanthin. K-Strophanthin; K-strophanthoside; Kombetin. A glycoside or a mixture of glycosides obtained from *Strophanthus kombé* Oliv., *Apocynaceae.* Properly speaking K-strophanthoside contains α-glucose 19%, β-glucose 19%, cymarose 15%, strophanthidin 47%. K-Strophanthin-β (Strophosid) contains β-glucose, cymarose, and strophanthidin: Geissberger, *Schweiz. Med. Wochenschr.* **91**, 241 (1961). Pharmacology and acute toxicity: H. Mehnert, *Arch. Pharmacol.* **184**, 181 (1937).
White or yellowish powder contg as much as 10% water, which it does not lose entirely without decompn. *Very poisonous!* Stable in air, but affected by light. Soluble in water and in dil alcohol; less sol in abs alcohol. Nearly insol in chloroform, ether, benzene. Aq solns are neutral to litmus. When bioassayed, strophanthin shall possess a potency per mg equivalent to 0.5 mg of U.S.P. Ouabain Reference Standard. A deviation of 20% is permitted. MLD in cats, rats (mg/kg): 0.11, 9.4 i.v. (Mehnert).
G-strophanthin, *see* Ouabain.
THERAP CAT: Cardiotonic.
THERAP CAT (VET): Has been used as a cardiotonic.

9017. Strophanthobiose. *2,6-Dideoxy-4-O-β-D-glucopyranosyl-3-O-methyl-D-ribo-hexose;* periplobiose. $C_{13}H_{24}O_9$; mol wt 324.33. C 48.14%, H 7.46%, O 44.40%. Obtained by hydrolysis of K-strophanthin-β, isolated from seeds of *Strophanthus kombé* Oliv., *Apocynaceae:* Jacobs, Hoffmann, *J. Biol. Chem.* **69**, 153 (1926); Stoll *et al., Helv. Chim. Acta* **20**, 1484 (1937). Identity with periplobiose: Barbier, Schindler, *ibid.* **42**, 1065 (1959). Structure: Lichti, von Wartburg, *ibid.* **44**, 238 (1961).

Needles from methanol, mp 144-146°. $[\alpha]_D^{20}$ +33.8° (c = 2.068 in water).
Pentaacetate, $C_{23}H_{34}O_{14}$, needles, mp 174-175°. $[\alpha]_D^{20}$ +11.9° (c = 1.05 in chloroform).

9018. Strophanthus. Dried ripe seeds of *Strophanthus kombé* Oliv., or of *S. hispidus* DC., *Apocynaceae,* deprived of the awns. *Habit.* East and Central Africa. *Constit.* 2-5% Strophanthin, kombic acid, choline, trigonelline, ~30% oil. *Poisonous!*
USE: As arrow poison by African natives.
THERAP CAT: Cardiotonic.
THERAP CAT (VET): Has been used as a cardiotonic.

9019. Struvite. Magnesium ammonium phosphate hexahydrate; guanite. $H_6MgNO_5P.6H_2O$; mol wt 263.42. H 6.89%, Mg 9.23%, N 5.32%, O 66.81%, P 11.76%. $(NH_4)MgPO_4.6H_2O$. Phosphate mineral named in honor of H. C. G. von Stürve, a Russian naturalist, in 1845 by Ulex, a Swedish geologist. A component of urinary calculi and bat guano also found as a contaminant in canned foods, in waste management systems. Production by urease-splitting bacteria: H. Robinson, *Proc. Cambridge Phil. Soc.* **6**, 360 (1889); J. Beavon, N. G. Heatley, *J. Gen. Microbiol.* **31**, 167 (1962); M. T. Gonzalez-Munoz *et al., Chemosphere* **26**, 1881 (1993). Identification in tinned seafood: M. T. Gilles,

Seafood Processing in *Food Processing Review* **22**, (Noyes Data Corp., Park Ridge, NJ, 1971) pp 31-58; in home-canned beef: J. G. Sebranek *et al., J. Muscle Foods* **4**, 81 (1993). Precipitation in sludge digesters: C. Maqueda *et al., Water Res.* **28**, 411 (1994). Crystal growth: R. J. C. McLean *et al., Urol. Res.* **18**, 39 (1990). Thermodynamics of formation: J. R. Buchanan *et al., Trans. Am. Soc. Agric. Engin.* **37**, 617 (1994). Chemical control in waste systems: J. R. Buchanan *et al., ibid.* 1301. Review of struvite urolithiasis: formation, detection and dissolution: C.A. Osborne *et al., Adv. Vet. Sci. Compar. Med.* **29**, 1-101 (1985). Review of struvite calculi: M. J. Gleeson, D. P. Griffith, *Brit. J. Urol.* **71**, 503-511 (1993).
Glass like crystals.

9020. Strychnine. *Strychnidin-10-one.* $C_{21}H_{22}N_2O_2$; mol wt 334.42. C 75.42%, H 6.63%, N 8.38%, O 9.57%. Occurs most abundantly in seeds of *Strychnos nux-vomica* L., *Loganiaceae* and beans of *S. ignatti,* Berg. One of the first alkaloids isolated in pure form. Isoln: Pelletier, Caventou, *Ann. Chim. Phys.* **8**, 323 (1818); *ibid.* **10**, 142 (1819). Extraction procedure: C. Srinivasulu *et al., Res. Ind.* **23**, 224 (1978). Structure elucidation: L. H. Briggs *et al., J. Chem. Soc.* **1946**, 903; H. T. Openshaw, R. Robinson, *Nature* **157**, 438 (1946); R. B. Woodward *et al., J. Am. Chem. Soc.* **69**, 2250 (1947). Total synthesis: R. B. Woodward *et al., ibid.* **76**, 4749 (1954); *eidem, Tetrahedron* **19**, 247 (1963). Abs config: A. F. Peerdeman, *Acta Cryst* **9**, 824 (1956); K. Nagarajan *et al., Helv. Chim. Acta* **46**, 1212 (1963). Stereoselective total synthesis: S. D. Knight *et al., J. Am. Chem. Soc.* **115**, 9293 (1993). ^{1}H- and ^{13}C-NMR analysis: E. Wenkert *et al., J. Org. Chem.* **43**, 1099 (1978). Pharmacology and pharmacokinetics: S. Weiss, R. A. Hatcher, *J. Pharmacol. Exp. Ther.* **19**, 419 (1921). Acute toxicity: I. Setnikar, M. J. Magistretti, *Arzneimittel-Forsch.* **14**, 996 (1964). HPLC determn in urine and tissues: T. Egloff *et al., J. Clin. Chem. Clin. Biochem.* **20**, 203 (1982). *Review:* J. B. Hendrickson in *The Alkaloids* vol. **VI**, R. H. F. Manske, Ed. (Academic Press, New York, 1960) pp 179-195; G. F. Smith, *ibid.* **vol. VIII** (1965) pp 591-671. Comprehensive description: F. J. Muhtadi, M. S. Hifnawy, *Anal. Profiles Drug Subs.* **15**, 563-646 (1986). Brief review of synthesis: U. Beifuss, *Angew. Chem. Int. Ed.* **33**, 1144-1149 (1994).

Brilliant, colorless cubes from chloroform-ether, mp 275-285°. d^{18} 1.359. $[\alpha]_D^{18}$ -104.3° (c = 0.254 in alc); $[\alpha]_D^{25}$ -139° (c = 0.4 in chloroform). pKa (25°) 8.26: A. J. Everett *et al., J. Chem. Soc.* **1957**, 1120. uv max (95% ethanol): 2550, 2800, 2900 Å ($E_{1 cm}^{1\%}$ 377, 130, 101): A. I. Biggs, *J. Pharm. Pharmacol.* **13**, 547 (1952). One gram dissolves in 182 ml ethanol, 6.5 ml chloroform, 150 ml benzene, 250 ml methanol, 83 ml pyridine; very slightly soluble in ether, water. Bitter taste in soln contg 1 part in 700,000 of water. LD_{50} i.v. (slow infusion) in rats: 0.96 mg/kg (Setnikar, Magistretti).
Acetate, $C_{23}H_{26}N_2O_4$, crystals, sol in 20 parts water, sol in alcohol.
Gluconate pentahydrate, $C_{27}H_{34}N_2O_9.5H_2O$, crystals, darken above 80°, sol in 2 parts water, ~40 parts alcohol; the aq soln is neutral.
Glycerophosphate hexahydrate, $C_{45}H_{53}N_4O_{10}P.6H_2O$, one gram dissolves in ~350 ml water, ~310 ml alcohol. Slightly sol in chloroform; very slightly in ether.
Hydrochloride dihydrate, $C_{21}H_{23}ClN_2O_2.2H_2O$, efflorescent, trimetric prisms. One gram dissolves in ~40 ml water, ~80 ml alc. Insol in ether. pH of 0.01*M* soln 5.4.
Nitrate, *see* separate entry.
Phosphate, *see* separate entry.

Salicylate, $C_{28}H_{28}N_2O_5$, leaflets, sol in ~30 parts water; sparingly sol in alcohol.

Sulfate, *see separate entry*.

Caution: Extremely poisonous, *see:* G. D. Osweiler, *Curr. Vet. Ther.* **6**, 115 (1977). Potential symptoms of overexposure are stiff neck and facial muscles; restlessness, apprehension and increased acuity of perception; increased reflex excitability; cyanosis; tetanic convulsions with opisthotonos. *See NIOSH Pocket Guide to Chemical Hazards* (DHHS/NIOSH 90-117, 1990) p 198.

USE: Chiefly in poison baits for rodents.

THERAP CAT (VET): Has been used as a tonic and central stimulant.

9021. Strychnine Nitrate. $C_{21}H_{23}N_3O_5$; mol wt 397.43. C 63.47%, H 5.83%, N 10.57%, O 20.13%.

Colorless, odorless needles or white, cryst powder. *Violent poison!* One gram dissolves in 42 ml water, 10 ml boiling water, 150 ml alcohol, 80 ml alcohol at 60°, 105 ml chloroform, 50 ml glycerol; insol in ether. pH ~5.7. *Protect from light.*

THERAP CAT (VET): Has been used as tonic, stimulant.

9022. Strychnine N^6-Oxide. Genostrychnine; Movellan (tabl.). $C_{21}H_{22}N_2O_3$; mol wt 350.42. C 71.98%, H 6.33%, N 7.99%, O 13.70%. Prepn: Pictet, Mattison, *Ber.* **38**, 2782 (1905); Bailey, Robinson, *J. Chem. Soc.* **1948**, 703; Zellner, U.S. pat. **2,758,113** (1956 to Donau-Pharmazie).

Monoclinic prisms from water, dec 207°. pK 5.17. Freely sol in alc, glacial acetic acid, chloroform; fairly sol in water; sparingly sol in benzene. Practically insol in ether, petr ether.

Hydriodide, light yellow plates from hot water, dec 253°.

Nitrate, small transparent prisms from water, dec 250°.

Strychnic acid sodium salt, *Movellan (amps.).*

9023. Strychnine Phosphate. $C_{21}H_{25}N_2O_6P$; mol wt 432.41. C 58.33%, H 5.83%, N 6.48%, O 22.20%, P 7.16%. Usually as $C_{21}H_{22}N_2O_2 \cdot H_3PO_4 \cdot 2H_2O$.

Dihydrate, colorless or white crystals or white powder. *Violent poison!* One gram dissolves slowly in ~30 ml water; more sol in hot water; slightly sol in alcohol. The aq soln is acid.

THERAP CAT (VET): Has been used as tonic, stimulant.

9024. Strychnine Sulfate. $C_{42}H_{46}N_4O_8S$; mol wt 766.92. C 65.78%, H 6.05%, N 7.31%, O 16.69%, S 4.18%. Usually crystallizes as $2C_{21}H_{22}N_2O_2 \cdot H_2SO_4 \cdot 5H_2O$. Also crystallizes with $6H_2O$ (12.36% water).

Incompatibility of all strychnine salts: alkalies, alkali carbonates and bicarbonates, benzoates, dichromates, bromides, iodides, tannic and picric acids, salicylates, borax, gold chloride and other alkaloid precipitants, piperazine, potassium-mercuric iodide (not if acacia is present).

Pentahydrate, colorless, odorless, very bitter crystals, or white, cryst powder. *Violent poison!* Effloresces in dry air; loses all its water of crystn at 100°. mp when anhydr ~200° with decompn. One gram dissolves in 35 ml water, 7 ml boiling water, 81 ml alcohol, 26 ml alcohol at 60°, 220 ml chloroform, 6 ml glycerol; insol in ether. pH 5.5 in 1:100 soln. *Protect from light.* LD_{50} orally in rats: 5 mg/kg, E. W. Schafer, *Toxicol. Appl. Pharmacol.* **21**, 315 (1972).

THERAP CAT (VET): Has been used as tonic, stimulant.

9025. Stylopine. *6,7,12b,13-Tetrahydro-4H-bis[1,3]benzodioxolo[5,6-a:4′,5′-g]quinolizine; 2,3:9,10-bis(methylenedioxy)-13a-berbine;* tetrahydrocoptisine. $C_{19}H_{17}NO_4$; mol wt 323.35. C 70.58%, H 5.30%, N 4.33%, O 19.79%. From root of *Stylophorum diphyllum* (Michx.) Nutt, *Corydalis cava* Schwg. and *Chelidonium majus* L., *Papaveraceae:* Schlotterbeck, Watkins, *Ber.* **35**, 7 (1902); Späth, Julian, *Ber.* **64**, 1131 (1931); Manske, *Can. J. Res.* **B20**, 53 (1942); Slavík, *Coll. Czech. Chem. Commun.* **20**, 198 (1955); Bandelin, Malesh, *J. Am. Pharm. Assoc.* **45**, 702 (1956); Slavík, *Coll. Czech. Chem. Commun.* **26**, 2933 (1961).

dl-Form, crystals from chloroform + methanol, mp 222-224°.

Hydrochloride, $C_{19}H_{18}ClNO_4$, needles from methanol + HCl, mp 266-269°.

l-Form, needles from chloroform + ethanol, mp 203°. $[\alpha]_D^{25}$ −300° (c = 0.53 in chloroform). Practically insol in water; sol in alc, glacial acetic acid; slightly sol in dil acids.

d-Form, crystals from alcohol, mp 203-204°. $[\alpha]_D^{25}$ +310° (c = 0.88 in chloroform). Sol in chloroform; slightly sol in ether, ethanol, methanol.

9026. Styphnic Acid. *2,4,6-Trinitro-1,3-benzenediol; 2,-4,6-trinitroresorcinol;* 2,4-dihydroxy-1,3,5-trinitrobenzene. $C_6H_3N_3O_8$; mol wt 245.11. C 29.40%, H 1.23%, N 17.14%, O 52.22%. May be prepd from natural sources, such as Pernambuco wood extract or quebracho extract by the action of nitric acid: Einbeck, Jablonski, *Ber.* **54**, 1086 (1921). Prepd industrially by first sulfonating, then nitrating resorcinol: Merz, Zetter, *Ber.* **12**, 681, 2037 (1879); Datta, Varma, *J. Am. Chem. Soc.* **41**, 2043 (1919). Prepd by nitrating 2,4-dinitroresorcinol: Kametani, Ogasawara, *Chem. Pharm. Bull.* **15**, 893 (1967).

Hexagonal yellow crystals from dil alc. Astringent, but not bitter taste. Solvated crystals from acetic acid. Sublimes in high vacuum. Becomes almost colorless upon vacuum sublimation, but turns deep yellow on contact with air. When dry, mp 175.5°; also reported as mp 179-180°. One gram dissolves in 156 ml water at 14°, in 88 ml water at 62°. Freely sol in alcohol, ether. Acid to litmus. Dibasic acid. Deflagrates on rapid heating.

USE: Has been used in the manuf of explosives.

9027. Styramate. *1-Phenyl-1,2-ethanediol 2-carbamate;* carbamic acid β-hydroxyphenethyl ester; 2-phenyl-2-hydroxyethyl carbamate; 2-hydroxy-2-phenylethyl carbamate; Sinaxar; Linaxar. $C_9H_{11}NO_3$; mol wt 181.19. C 59.66%, H 6.12%, N 7.73%, O 26.49%. Prepn: Brit. pat. **841,626**; C. D. Bossinger, K. G. Taylor, U.S. pat. **3,265,728** (1960, 1966 both to Armour). Pharmacology: De Salva *et al., J. Pharmacol. Exp. Ther.* **126**, 318 (1959).

Crystals from chloroform, mp 111-112°. LD_{50} orally in mice: 1240 mg/kg (De Salva).

THERAP CAT: Muscle relaxant (skeletal).

9028. Styrene. *Ethenylbenzene;* styrol; styrolene; cinnamene; cinnamol; phenylethylene; vinylbenzene. C_8H_8; mol wt 104.15. C 92.26%, H 7.74%. $C_6H_5CH{=}CH_2$. Isolated from storax by Bonastre in 1831. Manuf from benzene and ethylene: Faith, Keyes & Clark's *Industrial Chemicals,* F. A. Lowenheim, M. K. Moran, Eds. (Wiley-Interscience, New York, 4th ed., 1975) pp 779-785. Synthesis starting with 1-phenylethanol and leading to polystyrene: Wilen *et al., J. Chem. Ed.* **38**, 304 (1961). When heated to 200° it is converted into the polymer, polystyrene, which is a clear plastic having excellent insulating properties even at ultra-high radio frequencies. Toxicity data: H. J. Meyer, R. Kretzschmar, *Arzneimittel-Forsch.* **19**, 617 (1969). Monograph: W. C. Teach, G. C. Kiessling, *Polystyrene* (Reinhold, New York, 1960). Reviews of styrene monomer and polymers: Boyer *et al.,* "Styrene Polymers" in *Encyclopedia of Polymer Science and Technology* vol. **13** (Interscience, New York, 1970) pp 128-447; Coulter *et al.,* "Styrene and Related Monomers" in *Vinyl and Diene Monomers* (part 2), E. C. Leonard, Ed. (Wiley-Interscience, New York, 1971) pp 479-576. Review of neurotoxicity: R. Pahwa, J. Kalra, *Vet. Hum. Toxicol.* **35**, 516-520 (1993); of toxicology and carcinogenic risk: *IARC Scientific Publ.* **127**, 1-412 (1993); of toxicology and human exposure: *Toxicological Profile for Styrene* (PB93-110849, 1992) 164 p.

Colorless to yellowish, very refractive, oily liq; penetrating odor. On exposure to light and air it slowly undergoes polymerization and oxidation with formation of peroxides, etc. d^{20} 0.9059. mp $-30.6°$. bp 145-146°. n_D^{20} 1.5463. Flash pt, closed cup: 87° F (31° C). Sparingly sol in water; sol in alcohol, ether, methanol, acetone, carbon disulfide. LD_{50} in mice (mg/kg): 660 ± 44.3 i.p.; 90 ± 5.2 i.v. (Meyer, Kretzschmar).

Polystyrene, Dylene, Trycite. Physical properties of unmodified polystyrene: d_4^{20} 1.04-1.065; n_D^{25} 1.60; water-clear solid plastic, begins to soften at ~85°. Dielectric constant at 100 megacycles: 2.4-2.65.

Caution: Potential symptoms of overexposure are irritation of eyes and nose; drowsiness, weakness and unsteady gait; narcosis; defatting dermatitis. *See NIOSH Pocket Guide to Chemical Hazards* (DHHS/NIOSH 90-117, 1990) p 200.

USE: Manuf plastics; synthetic rubber; resins; insulator.

9029. Styrene Glycol. *1-Phenyl-1,2-ethanediol;* phenylglycol; phenylethyleneglycol; α,β-dihydroxyethylbenzene. $C_8H_{10}O_2$; mol wt 138.17. C 69.55%, H 7.30%, O 23.16%. Prepd by reduction of phenylglyoxylic acid with $LiAlH_4$: Nystrom, Brown, *J. Am. Chem. Soc.* **69**, 2548 (1947); by reduction of styrene peroxide: Russel, *J. Am. Chem. Soc.* **75**, 5011 (1953). Prepn of optically active styrene glycol: Wilhelm, Bright, *Helv. Chim. Acta* **37**, 221 (1954).

Needles from ligroin, mp 67-68°. bp_{755} 272-274°. Freely sol in water, alcohol, benzene, ether, chloroform, acetic acid; slightly sol in ligroin.

USE: Esters as plasticizers.

9030. Subathizone. *2-[[4-(Ethylsulfonyl)phenyl]methylene]hydrazinecarbothioamide; p-ethylsulfonylbenzaldehyde thiosemicarbazone;* 4-ethylsulfonylbenzaldehyde 3-thiosemicarbazone; Ethizone. $C_{10}H_{13}N_3O_2S_2$; mol wt 271.36. C 44.26%, H 4.83%, N 15.48%, O 11.79%, S 23.63%. Prepn from 4-ethylsulfonyltoluene: Bernstein *et al., J. Am. Chem. Soc.* **73**, 906 (1951); other prepns are described in **Japan.** pats. **1119('52), 1768('52), 3977('52), 1872('53), 6032('54), 6427('54);** in U.S. pat. **2,621,484** (1952 to Schenley).

Crystals from aq *n*-propanol, mp 234-235° (dec).

THERAP CAT: Antibacterial (tuberculostatic).

9031. Suberic Acid. *Octanedioic acid;* 1,6-hexanedicarboxylic acid. $C_8H_{14}O_4$; mol wt 174.20. C 55.16%, H 8.10%, O 36.74%. $COOH(CH_2)_6COOH$. Made by heating castor oil or ricinoleic acid with nitric acid: Dominguez, Lopez, *Ciencia* **20**, 73 (1959), *C.A.* **54**, 18990f (1960). Prepn from diethyl malonate + ω-bromocapronitrile: Cason *et al., J. Org. Chem.* **14**, 37 (1949); by oxidation of oleic acid with nitric acid: Cavanaugh, Weir, U.S. pat. **2,560,156** (1951 to du Pont).

Crystals, mp 140-144°. bp_{100} 279°. Sublimes at 300° without decompn. One gram dissolves in 625 ml water, 172 ml ether; sol in alcohol; almost insol in chloroform.

USE: In the plastics industry.

9032. Substance P. SP. $C_{63}H_{98}N_{18}O_{13}S$; mol wt 1347.65. C 56.15%, H 7.33%, N 18.71%, O 15.43%, S 2.38%. An undecapeptide belonging to a group of proteins named tachykinins characterized by contractile action on extravascular smooth muscle. Discovered by von Euler and Gaddum, *J. Physiol.* **72**, 74 (1931). Present in the brain of all vertebrates including man, in spinal ganglia, and in the intestines, esp the duodenum and jejunum. Isoln from cattle brain: Zuber, Jaques, *Angew. Chem. Int. Ed.* **1**, 160 (1962); from horse intestine: Studer *et al., Helv. Chim. Acta* **56**, 860 (1973). Purification from bovine hypothalami and amino acid composition: N. N. Chang, S. E. Leeman, *J. Biol. Chem.* **245**, 4784 (1970). Amino acid sequence and solid-phase synthesis: Chang *et al., Nature New Biol.* **232**, 86 (1971); Tregear *et al., ibid.* 87; G. H. Fisher *et al., J. Med. Chem.* **17**, 843 (1974). Additional syntheses: Yajima, Kitigawa, *Chem. Pharm. Bull.* **21**, 682 (1973); Bayer, Mutter, *Ber.* **107**, 1344 (1974); K. Neubert *et al., Pharmazie* **36**, 10 (1981); A. Fournier *et al., J. Med. Chem.* **25**, 64 (1981). Substance P acts as a vasodilator, a depressant, stimulates salivation and produces increased capillary permeability. It is also capable of producing both analgesia and hyperalgesia in animals, depending on dose and pain responsiveness of the animal, see R. C. A. Frederickson *et al., Science* **199**, 1359 (1978); P. Oehme *et al., ibid.* **208**, 305 (1980); role in sensory transmission and pain perception: T. M. Jessell, *Advan. Biochem. Psychopharmacol.* **28**, 189 (1981). Mechanism of action studies: Stern *et al., Arch. Pharmacol.* **281**, 233 (1974); B. G. Livett *et al., Nature* **278**, 256 (1979). *Reviews:* Haefeli, Huelimann, *Experientia* **18**, 297 (1962); Lembeck, Zetler, in *International Encyclopedia of Pharmacology and Therapeutics,* sect. **72**, vol. 1, J. M. Walker, Ed. (Pergamon Press, New York, 1971) pp 29-71; J. L. Barker, *Physiol. Rev.* **56**, 435 (1976); D. R. Brown, R. J. Miller, *Ann. Rep. Med. Chem.* **17**, 271-280 (1982). Books: U.S. Von Euler, B. Pernow, Eds., *Substance P* (Raven Press, New York, 1977) 360 pp; *Substance P* vol. 2, P. Skrabanek, D. Powell, Eds. (Eden Press, Quebec, 1980) 175 pp; *Substance P in The Nervous System: Ciba Foundation Symposium* **91**, R. Porter, M. O'Connor, Eds. (Pitman, London, 1982) 350 pp.

Arg–Pro–Lys–Pro–Gln–Gln–Phe–Phe–Gly–Leu–MetNH₂

Behaves during isoln as a basic polypeptide (R_f value = 0.5 in the thin layer chromatogram: silica gel—*n*-butanol/pyridine/AcOH/H_2O = 30:20:6:24). Upon high voltage electrophoresis (pH 1.9; acetic acid/formic acid/H_2O = 15:3:100) it migrates approx as far as glutamic acid or serine, and at pH 9.5 (0.25*M* triethylammonium carbonate buffer) it behaves in a manner similar to arginine. Biological activity is destroyed by pepsin and chymotrypsin. Aqueous solns of highly purified material lose biological activity in a few minutes. This loss is prevented by storage at low pH, under nitrogen, or with various antioxidants. Addition of Tween 80, gelatin, human plasma albumin, or bovine γ-globulin increases stability of aq SP solns. Crude SP solns are stable though rapidly destroyed above pH 8.

9033. Subtilin. Polypeptide antibiotic structurally similar to nisin, *q.v.,* but contains no methionine. Produced by *Bacillus subtilis* (NRRL no. B-543): E. F. Jansen, D. J. Hirshmann, *Arch. Biochem.* **4**, 297 (1944); Lewis *et al., ibid.* **14**, 415 (1947); Dimick *et al., ibid.* **15**, 1 (1947). Isoln by partition chromatography on silica gel: Alderton, Snell, *J. Am. Chem. Soc.* **81**, 701 (1959). Early structural studies: Stracher, Craig, *ibid.* 696. Structure is a heterodetic pentacyclic peptide consisting of 32 amino acid residues and containing unusual amino acids such as lanthionine, β-methyllanthionine, D-alanine, dehydroalanine and dehydrobutyrine: E. Gross *et al., Z. Physiol. Chem.* **354**, 810 (1973). Biosynthetic studies: C. Nishio *et al., Biochem. Biophys. Res. Commun.* **116**, 751 (1983); S. Banerjee, J. N. Hansen, *J. Biol. Chem.* **263**, 9508 (1988).

Amorphous powder. $[\alpha]_D^{23}$ −29° to −35° (c = 1 in 1% acetic acid). Diffuses quickly through cellophane. Readily sol in dilute acids; sparingly sol (less than 5 mg/ml) in water at pH 6-9. Sol in methanol, 80% alc. Soly in butanol satd with water: about 5 mg/ml. Easily salted out of soln by NaCl.

USE: Food preservative.

9034. Succimer. *(R*,S*)-2,3-Dimercaptobutanedioic acid; meso-2,3-dimercaptosuccinic acid;* DMS; DMSA; Ro-1-7977; Chemet. $C_4H_6O_4S_2$; mol wt 182.22. C 26.37%, H 3.32%, O 35.12%, S 35.19%. HOOCCH(SH)CH(SH)-COOH. A water soluble chelating agent. Prepn: L. N. Owen, M. U. S. Sultanbawa, *J. Chem. Soc.* **1949**, 3109. Pharmacology, toxicity and activity in heavy metal poisoning in animals: E. Friedheim, C. Corvi, *J. Pharm. Pharmacol.* **27**, 624 (1975); J. H. Graziano *et al., J. Pharmacol. Exp. Ther.* **207**, 1051 (1978). GC determn in urine: J. J. Knudsen, E. L. McGown, *J. Chromatog.* **424**, 231 (1988). Clinical evaluations in heavy metal poisoning: E. Friedheim *et al., Lancet* **2**, 1234 (1978); L. Fournier *et al., Med. Toxicol.* **3**, 499 (1988). Diagnostic use of complex with technetium in urinary tract imaging: I. G. Verber *et al., Arch. Dis. Child.* **63**, 1320 (1988); in tumor imaging: H. Ohta *et al., Nucl. Med. Commun.* **9**, 105 (1988). Review: H. V. Aposhian, "Biological Chelation: 2,3-Dimercaptopropanesulfonic Acid and Meso-Dimercaptosuccinic Acid" in *Advan. Enzyme Reg.* **20**, G. Weber, Ed. (Pergamon Press, Oxford, 1982) pp 301-319. Review of pharmacology and clinical use in heavy metal poisoning: idem, *Ann. Rev. Pharmacol. Toxicol.* **23**, 193-215 (1983); J. H. Graziano, *Med. Toxicol.* **1**, 155-162 (1986).

White crystals from aqueous methanol, mp 192-194°. LD_{50} i.p. in mice: > 3000 mg/kg (Friedheim, Corvi).

USE: Chelating agent.

THERAP CAT: Antidote (heavy metal poisoning). ^{99m}Tc complex as diagnostic aid (radioactive imaging agent).

9035. Succinamide. *Butanediamide.* $C_4H_8N_2O_2$; mol wt 116.12. C 41.37%, H 6.94%, N 24.12%, O 27.56%. $H_2NOCCH_2CH_2CONH_2$. Made by the action of ammonia water on the dimethyl or diethyl ester of succinic acid.

Needles. mp 260° with decompn; also stated as 242°. Sol in 220 parts cold, 9 parts boiling water; insol in alc, ether.

9036. Succinanil. *1-Phenyl-2,5-pyrrolidinedione; N-phenylsuccinimide;* N-phenylbutanimide. $C_{10}H_9NO_2$; mol wt 175.19. C 68.56%, H 5.18%, N 8.00%, O 18.27%. Prepd by treating succinanilic acid with acetyl chloride: Ruggli, *Ann.* **412**, 4 (1916); Warren, Briggs, *Ber.* **64**, 29 (1931); L. F. Fieser, *Experiments in Organic Chemistry* (Boston, 1955) p 106.

Monoclinic prisms from water or alcohol, mp 154.5°. d_4^{20} 1.356. bp$_{760}$ ~400°. Sol in ether, boiling alcohol. Slightly sol in boiling water.

9037. Succinanilic Acid. *4-Oxo-4-(phenylamino)butanoic acid;* succinic acid monoanilide; N-phenylsuccinamic acid; N-phenylbutanedioic acid monoamide. $C_{10}H_{11}NO_3$; mol wt 193.20. C 62.17%, H 5.74%, N 7.25%, O 24.84%. Prepd by boiling succinic anhydride with aniline in benzene: Menschutkin, *Ann.* **162**, 176 (1872); Warren, Briggs, *Ber.* **64**, 29 (1931); L. F. Fieser, *Experiments in Organic Chemistry* (Boston, 3rd ed., 1955) p 105.

Needles from benzene, mp 150°. K at 25° = 2.03 × 10⁻⁵. Soluble in alcohol, ether, boiling water (some dec).

9038. Succinic Acid. *Butanedioic acid;* amber acid; ethylenesuccinic acid; Bernsteinsäure (German); Asuccin. $C_4H_6O_4$; mol wt 118.09. C 40.68%, H 5.12%, O 54.19%. HOOCCH₂CH₂COOH. Observed by Agricola in 1546, in the distillate from amber. Occurs in fossils, fungi, lichens, etc. Prepn from acetic acid: Coffman *et al., J. Am. Chem. Soc.* **80**, 2864 (1958). Manuf: Crosby, Braunwarth, U.S. pat. **2,862,028** (1958 to Pure Oil); Chafetz, Patterson, U.S. pat. **2,978,473** (1961 to Texaco). Toxicity study: H. F. Smyth *et al., Arch. Ind. Hyg. Occup. Med.* **4**, 119 (1951).

Odorless, monoclinic prisms; very acid taste; d 1.56; mp 185-187°; bp 235° with partial conversion into the anhydride. One gram dissolves in 13 ml cold water, 1 ml boiling water, 18.5 ml alcohol, 6.3 ml methanol, 36 ml acetone, 20 ml glycerol, 113 ml ether. Practically insol in benzene, carbon disulfide, carbon tetrachloride, petr ether. pH of 0.1 molar aq soln 2.7.

Potassium salt trihydrate, $C_4H_4K_2O_4 \cdot 3H_2O$, *potassium succinate.* Hygroscopic crystalline powder. d 1.564. Freely sol in water; the aq soln is practically neutral. *Keep tightly closed.*

Diethyl ester, $C_8H_{14}O_4$, *ethyl succinate.* Colorless, clear liq; faint, pleasant odor. d_4^{20} 1.040. bp 217-218°. mp −21°. n_D^{20} 1.4201. Insol in water. Miscible with alcohol, ether. LD_{50} orally in rats: 8.53 g/kg (Smyth).

Dimethyl ester, $C_6H_{10}O_4$, *methyl succinate.* Colorless liquid. d_4^{18} 1.1202. mp 19.5°. bp$_{760}$ 195.3°; bp$_{25}$ 103.5°. Sol in 120 parts water, 35 parts alcohol.

USE: Manuf lacquers, dyes, esters for perfumes, succinates; in photography.

9039. Succinic Anhydride. *Dihydro-2,5-furandione;* succinic acid anhydride; succinyl oxide; 2,5-diketotetrahydrofuran; butanedioic anhydride; Bernsteinsäureanhydrid (German). $C_4H_4O_3$; mol wt 100.07. C 48.01%, H 4.03%, O 47.96%. Usually prepd by warming succinic acid with acetic anhydride, acetyl chloride, or phosphorus oxychloride: Shriner, Struck, *Org. Syn.* **12**, 66 (1932); L. F. Fieser, *Experiments in Organic Chemistry* (Boston, 3rd ed., 1955) p 105.

Orthorhombic prisms from abs alc. d 1.503 (also given as 1.104 and 1.234). Rate of sublimation: Kempf, *J. Prakt. Chem.* [2] **78**, 257 (1908). mp 119.6°; bp$_{760}$ 261°; bp$_{400}$ 237°; bp$_{200}$ 212°; bp$_{100}$ 189°; bp$_{60}$ 174°; bp$_{40}$ 163°; bp$_{20}$ 145°; bp$_{10}$ 128°. Sublimes at 115° and 5 mm pressure, at 92° and 1.0 mm pressure. Soluble in chloroform, carbon tetrachloride, alcohol; very slightly sol in ether and water.

9040. Succinimide. *2,5-Pyrrolidinedione;* butanimide; 2,5-diketopyrrolidine; 3,4-dihydropyrrole-2,5-dione; dihydro-3-pyrroline-2,5-dione; 2,5-dioxopyrrolidine; Orotric. $C_4H_5NO_2$; mol wt 99.09. C 48.49%, H 5.09%, N 14.14%, O 32.29%. Prepd by rapid distillation of ammonium succinate: Clarke, Behr, *Org. Syn.* coll. vol. II (1943) p 562; Fehling, *Ann.* **49**, 198 (1844); Bunge, *Ann. Suppl.* **7**, 118 (1870); Menschutkin, *Ann.* **162**, 166 (1872). Prepn by heating succinic acid with ammonia: Fehling, loc. cit.; Franchimont, Friedmann, *Rec. Trav. Chim.* **25**, 79 (1906); with urea: Ma, Sah, *C.A.* **28**, 6108 (1934); see also Tafel, Stern, *Ber.* **33**, 2232 (1900); Koller, *Ber.* **37**, 1598 (1904). Prevention of oxalic lithiasis in rats: J. M. Melon *et al., Therapie* **26**, 991 (1971).

Orthorhombic bipyramidal crystals from acetone or alc. mp 125-127°. d 1.41. bp 287-289° (slight decompn). pKa 9.5. One gram dissolves in 3 ml water, 0.7 ml boiling water, 24 ml alc, 5 ml alc at 60°. Insol in ether, chloroform. LD_{50} orally in rats: 14 g/kg (Melon).

THERAP CAT: Antiurolithic.

9041. Succinonitrile. *Butanedinitrile;* ethylene dicyanide; succinic acid dinitrile; *sym*-dicyanoethane; ethylene cyanide; Dinile; Suxil. $C_4H_4N_2$; mol wt 80.09. C 59.99%, H 5.03%, N 34.98%. $NCCH_2CH_2CN$. Prepd from ethylene bromide and KCN in alcohol: Fauconnier, *Bull. Soc. Chim. France* **50**, 214 (1888).

Waxy isometric crystals. d_4^{45} 1.023. mp 57.15°. d_4^{60} (liq) 0.9868. n_D^{60} (liq) 1.41734. May be heated to 200° for 72 hrs without decompn. bp_{760} 265-267°; bp_{60} 185°; bp_{20} 158-160°. Sol in acetone, chloroform, dioxane. Slightly sol in water, ethanol, benzene, ether, carbon disulfide.

9042. Succinyl Chloride. *Butanedioyl chloride;* succinyl dichloride. $C_4H_4Cl_2O_2$; mol wt 154.98. C 31.00%, H 2.60%, Cl 45.75%, O 20.65%. $ClOCCH_2CH_2COCl$. Fuming, highly refractive liq; d_4^{15} 1.395; n_D^{15} 1.473; mp 17°; also stated as 20°; bp 192-193°. Solidif at 0° to cryst leaflets.

9043. Succinylcholine Bromide. *2,2'-[(1,4-Dioxo-1,4-butanediyl)bis(oxy)]bis[N,N,N-trimethylethanaminium] dibromide;* bis[2-dimethylaminoethyl]succinate bis(methyl iodide]; 3,8-dioxadecane-4,7-dione-1,10-bis[trimethylammonium bromide]; suxamethonium bromide; IS-370; compd 48/268; LT-1; M & B 2207; Brevidil M. $C_{14}H_{30}Br_2N_2O_4$; mol wt 450.21. C 37.35%, H 6.72%, Br 35.50%, N 6.22%, O 14.22%. Prepd by reacting β-bromoethyl succinate with trimethylamine: Glick, *J. Biol. Chem.* **137**, 357 (1941); Fusco *et al., Gazz. Chim. Ital.* **79**, 129, 837 (1949); Walker, *J. Chem. Soc.* **1950**, 193. Alternate prepns: Phillips, *J. Am. Chem. Soc.* **71**, 3264 (1949); Tammelin, *Acta Chem. Scand.* **7**, 185 (1953).

Slightly hygroscopic crystals, mp 225°. Freely sol in water or normal saline, giving solns which are very slightly acidic. Aq solns undergo progressive hydrolysis with corresponding loss of activity and increase in acidity. *Prepare solns just before using.* The rate of this decompn increases with temp and autoclaving or prolonged exposure to warmth should be avoided.

THERAP CAT: Muscle relaxant (skeletal).

9044. Succinylcholine Chloride. *2,2'-[(1,4-Dioxo-1,4-butanediyl)bis(oxy)]bis[N,N,N-trimethylethanaminium] dichloride;* bis[2-dimethylaminoethyl]succinate bis[methochloride]; 2-dimethylaminoethyl succinate dimethochloride; diacetylcholine dichloride; suxamethonium chloride; choline succinate dichloride; succinic acid bis[β-dimethylaminoethyl] ester dimethochloride; choline chloride succinate (2:1); listenon; Anectine; Lysthenon; Midarine; Quelicin; Scoline; Sucostrin. $C_{14}H_{30}Cl_2N_2O_4$; mol wt 361.31. C 46.54%, H 8.37%, Cl 19.62%, N 7.75%, O 17.71%. Prepn: R. Fusco *et al., Gazz. Chim. Ital.* **79**, 129 (1949); Tammelin, *Acta Chem. Scand.* **7**, 185 (1953); O. Schmid, **Austrian** pat. **171,411** (1952 to OSSW), *C.A.* **47**, 4902f (1953); C. H. Wang *et al., Org. Prep. Proc. Int.* **11**, 93 (1979). Crystal structure: B. Jensen, *Acta Chem. Scand. B* **30**, 1002 (1976). Toxicity data: P. Anttila, P. Ertama, *Med. Biol.* **56**, 152 (1978). Comprehensive description: P. R. B. Foss, S. A. Benezra, in *Analytical Profiles of Drug Substances* vol. **10**, K. Florey, Ed. (Academic Press, New York, 1981) pp 691-704.

Exists as a dihydrate at room temp, crystals, mp 156-163°. Anhydr form mp ~190°. Slightly bitter taste. Freely sol in water (about 1 g/1 ml water). Soly in 95% ethanol: 0.42 g/100 ml. Sparingly sol in benzene, chloroform. Practically insol in ether. The pH of a 2-5% aq soln may vary from 4.5 to 3.0. Solns without stabilizers (e.g., 0.1% methyl *p*-hydroxybenzoate) must be kept refrigerated, they are incompatible with alkaline agents, such as thiopental sodium. LD_{50} i.v. in mice: 0.45 mg/kg (Anttila, Ertama).

THERAP CAT: Muscle relaxant (skeletal).
THERAP CAT (VET): Muscle relaxant (skeletal).

9045. Succinylcholine Iodide. *O,O*-Succinyldicholine iodide; diacetylcholine iodide; diacetylcholine diiodide; suxamethonium iodide; bis(β-dimethylaminoethyl)succinate bis(methyl iodide); Celocurine. $C_{14}H_{30}I_2N_2O_4$; mol wt 544.21. C 30.90%, H 5.56%, I 46.64%, N 5.15%, O 11.76%. Prepd by reacting β-dimethylaminoethyl succinate with trimethylamine: Glick, *J. Biol. Chem.* **137**, 357 (1941); Fusco *et al., Gazz. Chim. Ital.* **79**, 129, 837 (1949); Walker, *J. Chem. Soc.* **1950**, 193. Succinic acid chloride may be coupled with choline chloride directly or with dimethylaminoethanol, followed by quaternization with methyl iodide: Tammelin, *Acta Chem. Scand.* **7**, 185 (1953). A third method is to start with the diethyl ester of succinic acid, then bring about the exchange reaction with dimethylaminoethanol, and quaternize with methyl iodide: Phillips, *J. Am. Chem. Soc.* **71**, 3264 (1949).

Slightly hygroscopic crystals, mp 243-245°. Freely sol in water or normal saline, giving solns which are very slightly acidic. Most stable at pH 4-5. Aq solns undergo progressive hydrolysis with corresp loss of activity and increase in acidity. *Prepare solns just before using.* Incompatible with solns of alkaline salts.

THERAP CAT: Muscle relaxant (skeletal).
THERAP CAT (VET): Muscle relaxant (skeletal).

9046. Succinylsulfathiazole. *4-Oxo-4-[[4-[(2-thiazolylamino)sulfonyl]phenyl]amino]butanoic acid; 4'-(2-thiazolylsulfamoyl)succinanilic acid; p-2-thiazolylsulfamylsuccinanilic acid; 2-(N⁴-succinylsulfanilamido)thiazole; Sulfasuxidine.* $C_{13}H_{13}N_3O_5S_2$; mol wt 355.40. C 43.94%, H 3.69%, N 11.82%, O 22.51%, S 18.05%. Prepd by refluxing sulfathiazole with a slight excess of succinic anhydride in alcohol: Moore, Miller, *J. Am. Chem. Soc.* **64**, 1572 (1942); U.S. pats. **2,324,013; 2,324,014** (both 1943); by refluxing *p*-succinimidobenzenesulfonyl chloride with 2-aminothiazole in benzene: **Brit.** pat. **578,004** (1946). Toxicology: A. D. Welch *et al., J. Pharmacol. Exp. Ther.* **75**, 231 (1942).

Monohydrate, crystals with unclear mp. Has been reported as mp 184-186°; also reported as mp 192-195°. Emits pungent fumes on heating to mp. One gram dissolves in about 4800 ml water. Soluble in solns of alkali hydroxides and in solns of sodium bicarbonate with the evolution of carbon dioxide; sparingly sol in alcohol and in acetone. In-

Consult the Name Index before using this section.

sol in chloroform and ether. LD_{50} i.p. in mice: ~ 5.7 g/kg (Welch).

THERAP CAT: Antibacterial.

THERAP CAT (VET): Antibacterial, esp in enteric infections.

9047. Succisulfone. *4-[[4-[(4-Aminophenyl)sulfonyl]-phenyl]amino]-4-oxobutanoic acid; 4'-sulfanilylsuccinanilic acid;* 4-amino-4'-(β-carboxypropionylamino)phenylsulfonyl-benzene; 4-amino-4'-β-carboxypropionylaminodiphenylsul-fone; 4-(β-carboxypropionylamino)-4'-aminodiphenyl sul-fone; 4-succinylamido-4'-aminodiphenylsulfone; F-1500; Fourneau 1500; Exosulfonyl. $C_{16}H_{16}N_2O_5S$; mol wt 348.38. C 55.16%, H 4.63%, N 8.04%, O 22.96%, S 9.20%. Prepn: Fourneau, Tréfouel, **Fr.** pat. **866,619** (1941 to Rhône-Poulenc); Kharasch, Reinmuth, U.S. pat. **2,268,754** (1942 to Lilly); Bauer, *J. Am. Chem. Soc.* **70**, 2254 (1948); Rohls, Behnish, **Ger.** pat. **895,600** (1953 to Bayer); Liberman, **Fr.** pat. **1,020,713** (1953 to Chimie et Atomistique).

Crystals, mp 157°. Soluble in ammonia.

2,2'-Iminodiethanol salt, $C_{20}H_{27}N_3O_7S$. Prepn: **Fr.** pat. **992,112** (1951 to Theraplix).

THERAP CAT: Antibacterial (leprostatic).

9048. Suclofenide. *3-Chloro-4-(2,5-dioxo-3-phenyl-1-pyrrolidinyl)benzenesulfonamide; 3-chloro-4-(phenylsuccin-imido)benzenesulfonamide;* 1-(α-phenylsuccinimido)-2-chlo-ro-4-sulfamoylbenzene; 1-(2-chloro-4-sulfamoylphenyl)-3-phenylsuccinimide; CGP-8426; GS-385; Sulfalepsin. $C_{16}H_{13}ClN_2O_4S$; mol wt 364.81. C 52.68%, H 3.59%, Cl 9.72%, N 7.68%, O 17.54%, S 8.79%. Anti-epileptic agent. Prepn: R. W. Pfirrmann, **Belg.** pat. **752,109;** *idem*, U.S. pat. **3,789,-056** (1970, 1974 both to Geistlich). Pharmacology: *Arznei-mittel-Forsch.* **27**, 1942 (1977). HPLC determn in biological materials: F. Mikes *et al., ibid.* **29**, 1583 (1979).

Crystals from methanol/ether-petroleum ether, mp 205-207°. LD_{50} orally in mice: > 5000 mg/kg (Mikes).

THERAP CAT: Anticonvulsant.

9049. Sucralfate. *Hexadeca-μ-hydroxytetracosahydroxy-[μ_8-[1,3,4,6-tetra-O-sulfo-β-D-fructofuranosyl-α-D-glucopyran-oside tetrakis(hydrogen sulfato)(8—)]]hexadecaaluminum;* β-D-fructofuranosyl-α-D-glucopyranoside octakis(hydrogen sulfate) aluminum complex; sucrose octakis(hydrogen sul-fate) aluminum complex; Antepsin; Carafate; Citogel; Hexa-gastron; Keal; Succosa; Sucralfin; Sucrate; Sugast; Sulcrate; Ulcar; Ulcerlmin; Ulcogant. $C_{12}H_{54}Al_{16}O_{75}S_8$; mol wt 2086.75. C 6.91%, H 2.61%, Al 20.69%, O 57.50%, S 12.29%. A basic aluminum sucrose sulfate complex which inhibits peptic hydrolysis and stomach acidity. Prepn: M. Nametaka *et al., Yakugaku Zasshi* **87**, 889 (1967), *C.A.* **68**, 20831d (1968); **Fr.** pat. **1,500,571** corresp to N. Yoshihiro *et al.*, U.S. pat. **3,432,489** (1967, 1969 both to Chugai). Struc-ture and properties: R. Nagashima, N. Yoshida, *Arznei-mittel-Forsch.* **29**, 1668 (1979). *In vitro* and *in vivo* study of antipeptic and antiulcer activity: L. E. Borella, *et al., ibid.* 793. Series of articles on mode of action: *ibid.* **30**, 73-88 (1980). Pharmacokinetics, metabolism and selective bind-ing studies: K. Steiner *et al., ibid.* **32**, 512 (1982). Clinical studies: J. F. Mayberry *et al., Brit. J. Clin. Pract.* **32**, 291 (1978). Symposium on clinical studies: *J. Clin. Gastroen-terol.* **3**, Suppl. 2, 103-184 (1981); *Scand. J. Gastroenterol* **18**, Suppl. 83, 1-82 (1983). Review of pharmacology and clini-cal studies: R. N. Brogden *et al., Drugs* **27**, 194 (1984).

Series of articles on clinical efficacy in peptic ulcer disease and gastritis: *Am. J. Med.* **79**, Suppl. 2C, 1-64 (1985).

$$R = SO_3[Al_2(OH)_5]$$

White amorphous powder. Sol in dil HCL and NaOH solns. Practically insol in water, ethanol, $CHCl_3$. pKa = 0.43 to 1.19. Dissolution of aluminum occurs at pH < 3; of sucrose sulfate at pH > 4.

THERAP CAT: Antiulcerative.

9050. Sucralose. *1,6-Dichloro-1,6-dideoxy-β-D-fructo-furanosyl-4-chloro-4-deoxy-α-D-galactopyranoside; 4,1',6'-tri-chloro-4,1',6'-trideoxy-galacto-sucrose; 1',4,6'-trichloro-galactosucrose;* TGS. $C_{12}H_{19}Cl_3O_8$; mol wt 397.64. C 36.25%, H 4.82%, Cl 26.75%, O 32.19%. Chlorinated su-crose derivative with enhanced sweetness. Prepn: P. H. Fairclough *et al., Carbohyd. Res.* **40**, 285 (1975). Prepn of crystalline anhydrous and pentahydrate: M. R. Jenner, D. Waite, **Eur.** pat. **Appl. 30,804** (1981 to Tate & Lyle; Talres Dev.); *eidem*, U.S. pat. **4,343,934** (1982 to Talres Dev.). Use as non-nutritive sweetener: **Belg.** pat. **850,180;** L. Hough *et al.*, U.S. pat. **4,435,440** (1977, 1984 both to Tate & Lyle). *In vitro* activity vs cariogenic bacteria: J. Verran, D. B. Drucker, *Arch. Oral Biol.* **27**, 693 (1982). Structure-sweetness relationship: M. Mathlouthi *et al., Carbohyd. Res.* **152**, 47 (1986). *Review:* L. Hough, R. Khan, *Trends Biochem. Sci.* **3**, 61-63 (1978).

Syrup, $[\alpha]_D$ +68.2° (c = 1.1 in ethanol). Anhydrous crys-talline form: orthorhombic, needlelike crystals, mp 130°. Intensely sweet taste. LD_{50} in mice: > 16 g/kg (Hough, Khan).

Pentahydrate, mp 36.5°.

9051. Sucrose. *β-D-Fructofuranosyl-α-D-glucopyrano-side; α-D-glucopyranosyl-β-D-fructofuranoside;* sugar; sac-charose; cane sugar; beet sugar. $C_{12}H_{22}O_{11}$; mol wt 342.30. C 42.11%, H 6.48%, O 51.41%. Obtained from sugar cane *(Saccharum officinarum* L., *Gramineae)* and sugar beet *(Beta valgaris* L., *Chenopodiaceae).* Sugar cane contains from 15-20% and sugar beet from 10-17% sucrose. Struc-ture: Avery *et al., J. Chem. Soc.* **1927**, 2308; Beevers, Coch-rane, *Proc. Roy. Soc.* **190A**, 257 (1947). Synthesis: Pictet, Vogel, *Helv. Chim. Acta* **11**, 436 (1928); Lemieux, Huber, *J. Am. Chem. Soc.* **78**, 4117 (1956). Ref. with extensive biblio-graphy: Bates, *Polarimetry, Saccharimetry, and the Sugars,* National Bureau of Standards Circular C440, Washington, 1942; W. Pigman, *The Carbohydrates* (Academic Press, New York, 1957) pp 501-506. *Reviews:* M. R. Jenner, *Dev. Food Carbohyd.* **2**, 91-143 (1980); R. Khan, *Pure Appl. Chem.* **56**, 833 (1984).

Monoclinic sphenoidal crystals, cryst masses, blocks, or powder. Sweet taste. Stable in air. Finely divided sugar is hygroscopic and absorbs up to 1% moisture which is given up on heating to 90°. d_4^{25} 1.587. Dec 160-186°. Chars and emits characteristic odor of caramel. $[\alpha]_D^{20}$ not less than +65.9° (c = 26); usual value $[\alpha]_D^{25}$ +66.47° to +66.49°. One gram dissolves in 0.5 ml water; in slightly more than 0.2 ml boiling water, in 170 ml alcohol; in about 100 ml methanol. Moderately sol in glycerol, pyridine. pKa 12.62. d_4^{20} of water solns (g/100 g): 2% 1.0060; 6% 1.0219; 10% 1.0381; 20% 1.0810; 30% 1.1270; 40% 1.1764; 50% 1.2296; 60% 1.2865; 70% 1.3471; 76% 1.3854. n_D^{20} of 10% soln 1.34783. Sucrose does not reduce Fehling's soln, form an osazone, or show mutarotation. It is hydrolyzed to glucose and fructose by dil acids and by invertase, a yeast enzyme. Upon hydrolysis the optical rotation falls and is negative when the hydrolysis is complete. The mixture of glucose and fructose is known as "invert sugar." Sucrose is fermentable, but resists bacterial decompn when in high concentrations.

USE: Sweetening agent and food. Starting material in the fermentative production of ethanol, butanol, glycerol, citric and levulinic acids. Used in pharmaceuticals as a flavor, as a preservative, as an antioxidant (in the form of invert sugar), as a demulcent, as substitute for glycerol, as granulation agent and excipient for tablets, as coating for tablets. In the plastics and cellulose industry, in rigid polyurethane foams, manuf of ink and of transparent soaps.

9052. Sucrose Octaacetate. $C_{28}H_{38}O_{19}$; mol wt 678.60. C 49.56%, H 5.64%, O 44.80%. Prepn from sucrose: Linstead *et al., J. Am. Chem. Soc.* **62**, 3260 (1940). Synthesis: Lemieux, Huber, *ibid.* **78**, 4117 (1956).

Hygroscopic, intensely bitter needles from alcohol, mp 89°; dec above 285°; bp_1 260°. $[\alpha]_D^{25.4}$ +58.5° (c = 2.56 in abs alcohol). n_D 1.4660. Sol in 1100 parts water, 1.1 parts acetal, 0.7 part glacial acetic acid, 0.3 part acetone, 11 parts alcohol, 0.6 part benzene, 22 parts carbon tetrachloride, about 0.5 part methyl acetate, 7 parts paraldehyde, about 0.5 part toluene.

USE: Adhesive; impregnating and insulating papers; in lacquers and plastics; as a denaturant for alcohol.

9053. Sucrose Polyester. Olestra; SPE. Non-absorbable lipid; substitute for fat in foods to reduce cholesterol level. Mixture of hexa-, hepta- and octa-fatty acid esters of sucrose. Analysis by gel permeation chromatography: C. G. Birch, F. E. Crowe, *J. Am. Oil Chem. Soc.* **53**, 581 (1976). Effect on absorption of dietary cholesterol in rats: F. H. Mattson *et al., J. Nutr.* **106**, 747 (1976); L. Aust *et al., Ann. Nutr. Metab.* **25**, 255 (1981). Effects as a dietary agent for lowering plasma cholesterol in man: R. W. Fallat *et al., Am. J. Clin. Nutr.* **29**, 1024 (1976); C. J. Glueck *et al., ibid.* **32**, 1636 (1979). Effects on cholesterol metabolism in man: J. R. Crouse, S. M. Grundy, *Metab., Clin. Exp.* **28**, 994 (1979); R. J. Jandacek *et al., Am. J. Clin. Nutr.* **33**, 251 (1980). Caloric dilution study in obese patients: C. J. Glueck *et al., ibid.* **35**, 1352 (1982).

USE: As supplement in dietary foods.

9054. Sudan III. *1-[[4-(Phenylazo)phenyl]azo]-2-naphthalenol;* 1-(p-phenylazophenylazo)-2-naphthol; tetrazobenzene-β-naphthol; D & C Red No. 17; C.I. Solvent Red 23; C.I. 26100; Sudan Red BK; Tony Red. $C_{22}H_{16}N_4O$; mol wt 352.40. C 74.98%, H 4.58%, N 15.90%, O 4.54%. Prepn from diazotized aminoazobenzene and 2-naphthol: *Colour Index* vol. 4 (3rd ed., 1971) p 4227. *Review:* H. J. Conn's *Biological Stains*, R. D. Lillie, Ed. (Williams & Wilkins, Baltimore, 9th ed., 1977) pp 168-169, 576.

Reddish-brown powder. Insol in water. Sol in chloroform, glacial acetic acid; moderately sol in alcohol, ether, acetone, petr ether, fixed and volatile oils, hot glycerol.

USE: Coloring oils, spirit lacquers, etc.; also as stain for zoological, pathological, and vegetable objects like wax, cutin, resin, contents of lactiferous ducts, etc., which are colored red, while cellulose membranes remain uncolored. Approved by FDA for external use only.

9055. Suet, Prepared. Mutton suet. The internal fat of abdomen of sheep purified by melting and straining.

White, solid fat with slight odor and taste; bland if fresh, but rancid after long exposure to air. mp 45-50°; n_D^{60} 1.449-1.451. Sapon no. 192-195. Iodine no. 33-46. Insol in water or cold alcohol; one gram dissolves in 45 ml boiling alcohol, about 60 ml ether.

USE: Preparing ointments.

9056. Sufentanil. *N-[4-(Methoxymethyl)-1-[2-(2-thienyl)ethyl]-4-piperidinyl]-N-phenylpropanamide; N-[4-(methoxymethyl)-1-[2-(2-thienyl)ethyl]-4-piperidyl]propionanilide; sufentanyl; R-30730. $C_{22}H_{30}N_2O_2S$; mol wt 386.56. C 68.36%, H 7.82%, N 7.25%, O 8.28%, S 8.30%. Potent deriv of fentanyl, *q.v.* Prepn: P. A. J. Janssen, G. H. P. Van Daele, Ger. pat. 2,610,228; *eidem,* U.S. pat. 3,998,834 (both 1976 to Janssen); G. H. P. Van Daele *et al., Arzneimittel-Forsch.* **26**, 1521 (1976). *In vitro* binding properties to mu-opiate receptor: J. E. Leysen, W. Gommeren, *Arch. Int. Pharmacodyn. Ther.* **260**, 287 (1982); J. E. Leyson *et al., Eur. J. Pharmacol.* **87**, 209 (1983). Analgesic activity and safety assessment: W. F. M. Van Bever *et al., Arzneimittel-Forsch.* **26**, 1548 (1976); C. J. E. Niemegeers *et al., ibid.* 1551. GC determn in biological fluids: R. Woestenborghs *et al., J. Chromatog.* **224**, 122 (1981); T. J. Gillespie *et al., J. Anal. Toxicol.* **5**, 133 (1981). Radioimmunoassay: M. Michaels *et al., J. Pharm. Pharmacol.* **35**, 86 (1983). Comparison to fentanyl for coronary artery surgery: S. de Lange *et al., Anesthesiology* **56**, 112 (1982). EEG effects in man: J. G. Bovill *et al., Brit. J. Anaesth.* **54**, 45 (1982). Review of pharmacokinetics: L. E. Mather, *Clin. Pharmacokinet.* **8**, 422-426 (1983).

Crystals from petr ether, mp 96.6°. LD_{50} i.v. in mice: 18.7 mg/kg (Van Bever).

Citrate, $C_{22}H_{30}N_2O_2S \cdot C_6H_8O_7$, R-33800, Sufenta.

Note: This is a controlled substance (opiate) listed in the U.S. Code of Federal Regulations, Title 21 Part 1308.12 (1995).

THERAP CAT: Analgesic (narcotic).

9057. Suint. Suint is that portion of the sheep's fleece which is sol in cold water after the wax has been removed. It is a complex mixture of metallic ions, organic acids, peptides, weak bases, neutral substances, and inorganic cations. The following acids have been identified: acetic, propionic, butyric, valeric, oxalic, succinic, and glutaric. Paper chromatography indicates the presence of adipic and pimelic acids: Deane, Truter, *Biochim. Biophys. Acta* **18**, 435 (1955); *J. Chem. Soc.* **1959**, 2746. Determination of ion content: Mohsin, Shah, *Pak. J. Sci. Ind. Res.* **12**, 286 (1970), *C.A.* **73**, 4778q (1970).

9058. Sulbactam. (2S-cis)-3,3-Dimethyl-7-oxo-4-thia-1-azabicyclo[3.2.0]heptane-2-carboxylic acid 4,4-dioxide; penicillanic acid sulfone; penicillanic acid 1,1-dioxide; CP-45899. $C_8H_{11}NO_5S$; mol wt 233.25. C 41.20%, H 4.75%, N 6.01%, O 34.30%, S 13.75%. Semi-synthetic β-lactamase inhibitor. Prepn and use with β-lactam antibiotics: **Belg.** pat. 867,859; W. E. Barth, U.S. pat. 4,234,579 (1978, 1980 both to Pfizer); R. A. Volkmann *et al., J. Org. Chem.* **47**, 3344 (1982). β-Lactamase activity and antibacterial spectrum *in vitro:* A. R. English *et al., Antimicrob. Ag. Chemother.* **14**, 414 (1978); R. N. Jones *et al., Diagn. Microbiol. Infect. Dis.* **3**, 489 (1985). HPLC determn in human plasma and urine: J. Haginaka *et al., J. Chromatog.* **341**, 115

(1985). Pharmacokinetics in humans: G. Foulds *et al.*, *Antimicrob. Ag. Chemother.* **23**, 692 (1983). Clinical study of synergistic effect with ampicillin: S. Mehtar *et al.*, *J. Antimicrob. Chemother.* **17**, 389 (1986); B. V. Stromberg *et al.*, *Surg. Gynecol. Obstet.* **162**, 575 (1986). Review of activity and therapeutic use of sulbactam with ampicillin: D. M. Campoli-Richards, R. N. Brogden, *Drugs* **33**, 577-609 (1987).

White, crystalline solid, mp 148-151° (Barth); also reported as mp 154-155.5° (dec) (Barth); also reported as mp 170° (dec) (Volkmann). $[\alpha]_D^{20}$ +251° (c = 0.01 in pH 5.0 buffer). Sol in water.

Sodium salt, $C_8H_{10}NNaO_5S$, *CP-45899-2*, *Betamaze*.
Mixture of sodium salt with ampicillin sodium, *Bethacil (inj.)*, *Loricin*, *Unacid*, *Unacim (inj.)*.
Mixture of sodium salt with cefoperazone sodium, *Sulperazone*.
Compd with ampicillin, *see* Sultamicillin.
THERAP CAT: In combination with β-lactam antibiotics as antibacterial.

9059. Sulbenicillin. *[2S-[2α,5α,6β(S*)]]-3,3-Dimethyl-7-oxo-6-[(phenylsulfoacetyl)amino]-4-thia-1-azabicyclo-[3.2.0]heptane-2-carboxylic acid;* α-sulfobenzylpenicillin, sulfocillin. $C_{16}H_{18}N_2O_7S_2$; mol wt 414.46. C 46.37%, H 4.38%, N 6.76%, O 27.02%, S 15.47%. Semi-synthetic antibiotic related to penicillin. Prepn: S. Morimoto *et al.*, *Ger. pat.* **1,948,943** corresp to U.S. pat. **3,660,379** (1970, 1972 both to Takeda); *eidem*, *J. Med. Chem.* **15**, 1105, 1108 (1972). *In vitro* activity studies: Tsuchiya *et al.*, *J. Antibiot.* **24**, 607 (1971); *in vivo:* Yamazaki, Tsuchiya, *ibid.* 620. Metabolism: Tsuchiya *et al.*, *ibid.* **25**, 336 (1972). Toxicity: Y. Murata *et al.*, *Takeda Kenkyusho Ho* **30**, 262 (1971), *C.A.* **76**, 147x (1972). Pharmacokinetics: A. Montanari *et al.*, *Clin. Ter.* **89**, 163 (1979); *eidem*, *Int. J. Clin. Pharmacol. Ther. Toxicol.* **18**, 225 (1980).

Disodium salt, $C_{16}H_{16}N_2Na_2O_7S_2$, *Kedacillina*, *Sulpelin*, *Lilacillin*. Prepd as a 3:1 mixture of D(−) and L(+) isomers. Yellowish-white powder, mp 195-198° (dec). $[\alpha]_D^{22}$ +169-173°. uv max: 257, 262, 268 nm. Very sol in water; sol in methanol. Almost insol in *n*-propanol, acetone, chloroform, benzene, ethyl acetate. LD_{50} in male, female mice, male, female rats (mg/kg): 7900, 8000, 6000, 6200 i.v.; 9600, 10000, 7200, 7500 i.p.; 11500, 13500, 11000, 11800 s.c.; 11000, 10500, 8300, 8600 i.m.; all > 15000 orally (Murata).
THERAP CAT: Antibacterial.

9060. Sulbenox. *(4,5,6,7-Tetrahydro-7-oxobenzo[b]thien-4-yl)urea;* CL-206576; Vigazoo. $C_9H_{10}N_2O_2S$; mol wt 210.26. C 51.41%, H 4.79%, N 13.32%, O 15.22%, S 15.25%. Prepn: Neth. pat. Appl. **7,610,270** (1977 to Am. Cyanamid), *C.A.* **88**, 22607x (1978). Activity: G. Asato, R. D. Wilbur, *Experientia* **35**, 1458 (1979).

Crystals, mp 245-246°. LD_{50} orally in rats: > 5000 mg/kg (Asato, Wilbur).
THERAP CAT (VET): Growth stimulant.

9061. Sulbentine. *Tetrahydro-3,5-bis(phenylmethyl)-2H-1,3,5-thiadiazine-2-thione; 3,5-dibenzyltetrahydro-2H-1,-3,5-thiadiazine-2-thione;* 2-thioxo-3,5-dibenzyltetrahydro-1,-3,5-thiadiazine; dibenzthione. Fungiplex; Refungine. $C_{17}H_{18}N_2S_2$; mol wt 314.48. C 64.93%, H 5.77%, N 8.91%, S 20.39%. Prepn: Rieche *et al.*, *Arch. Pharm.* **293**, 957 (1960); East Ger. pat. **20,634** (1961), *C.A.* **56**, 487e (1962).

Crystals from acetone or methanol, mp 101-102°.
THERAP CAT: Antifungal.

9062. Sulconazole. *1-[2-[[(4-Chlorophenyl)methyl]-thio]-2-(2,4-dichlorophenyl)ethyl]-1H-imidazole;* (±)-1-[2,4-dichloro-β-[(p-chlorobenzyl)thio]phenethyl]imidazole. $C_{18}H_{15}Cl_3N_2S$; mol wt 397.75. C 54.35%, H 3.80%, Cl 26.74%, N 7.04%, S 8.06%. Prepn: K. A. M. Walker, M. Marx, *Ger. pat.* **2,541,833**; *eidem*, U.S. pat. **4,038,409** (1976, 1977 both to Syntex). HPLC determn in plasma: M. Fass *et al.*, *J. Pharm. Sci.* **70**, 1338 (1981). Mechanism of action study: W. H. Beggs, *Biochem. Arch.* **10**, 117 (1994). Clinical trial in tinea pedis: W. A. Akers *et al.*, *J. Am. Acad. Dermatol.* **21**, 686 (1989). Review of pharmacology and clinical efficacy: P. Benfield, S. P. Clissold, *Drugs* **35**, 143-153 (1988).

Mononitrate, $C_{18}H_{15}Cl_3N_2S.HNO_3$, *RS-44872*, *Exelderm*, *Myk*, *Sulcosyn*. Colorless crystals from acetone, mp 130.5-132°.
THERAP CAT: Antifungal.

9063. Sulcotrione. *2-[2-Chloro-4-(methylsulfonyl)ben-zoyl]-1,3-cyclohexanedione;* ICIA-0051; SC-0051; Mikado. $C_{14}H_{13}ClO_5S$; mol wt 328.77. C 51.15%, H 3.99%, Cl 10.78%, O 24.33%, S 9.75%. Triketone bleaching herbicide for use in maize. Inhibits 4-hydroxyphenylpyruvate dioxygenase, an enzyme involved in the biosynthesis of plastoquinones. Prepn: W. J. Michaely, G. W. Kraatz, *Eur. pat.* **Appl. 137,963**; *eidem*, U.S. pat. **4,780,127** (1983, 1988 both to Stauffer Chem.). Field trials in corn: J. S. Wilson, C. L. Foy, *Weed Technol.* **4**, 731 (1990). Soil adsorption: *eidem*, *ibid.* **6**, 583 (1992). Mode of action study: J. Secor, *Plant Physiol.* **106**, 1429 (1994). Comprehensive description: J. M. Beraud *et al.*, *Brighton Crop Prot. Conf. - Weeds* **1991**, 51-56.

Light tan solid, mp 139°. Vapor pressure (25°): 4 × 10⁻⁸ mm Hg. Soly in water (25°): 165 mg/l. Sol in acetone, chlorobenzene.

USE: Herbicide.

9064. Sulfabenz. *4-Amino-N-phenylbenzenesulfon-amide; sulfanilanilide; p-aminobenzenesulfonanilide; p-ami-nophenylsulfonamidobenzene; N¹-phenylsulfanilamide.* $C_{12}H_{12}N_2O_2S$; mol wt 248.31. C 58.05%, H 4.87%, N 11.28%, O 12.89%, S 12.91%. Prepn: Gelmo, *J. Prakt. Chem.* **77**, 369 (1908); Mastryukova *et al., Tetrahedron* **19**, 357 (1963).

Needles from dil alc, mp 200°. Practically insol in water; readily sol in warm alcohol, acetone. pKa: 10.94 (in 50% aq alc); 11.59 (in 80% aq alc).

THERAP CAT (VET): Antibacterial; coccidiostat (for poultry).

9065. Sulfabenzamide. *N-[(4-Aminophenyl)sulfonyl]-benzamide; N-sulfanilylbenzamide; N¹-benzoylsulfanilamide; N-(p-aminobenzenesulfonyl)benzamide; Sulfabenzide.* $C_{13}H_{12}N_2O_3S$; mol wt 276.32. C 56.51%, H 4.38%, N 10.14%, O 17.37%, S 11.60%. Prepn: Crossley *et al., J. Am. Chem. Soc.* **61**, 2950 (1939); Dvornikoff, U.S. pat. **2,240,496** (1941 to Monsanto); **Brit.** pat. **541,958** (1942 to Schering AG); Siebenmann, Schnitzer, *J. Am. Chem. Soc.* **65**, 2126 (1943).

Long, hexagonal prisms from 60% alcohol, mp 181.2-182.3°. pKa 4.57; Ka at 25°: 2.7 × 10⁻⁵. Substantially insol in water. One gram dissolves in 3225 ml of water at 30°, in 33 ml of 95% ethanol, in 9 ml of acetone. Soly in 1N NaOH or KOH: ~16 g/100 ml; pH just above 7.0. Aq solns of the water-soluble sodium salt have the same pH.

THERAP CAT: Antibacterial.
THERAP CAT (VET): Antimicrobial.

9066. Sulfabromomethazine. *4-Amino-N-(5-bromo-4,6-dimethyl-2-pyrimidinyl)benzenesulfonamide; N¹-(5-bromo-4,-6-dimethyl-2-pyrimidinyl)sulfanilamide; 5-bromo-4,6-di-methyl-2-sulfanilamidopyrimidine; 2-sulfanilamide-5-bro-mo-4,6-dimethyl pyrimidine; 5-bromosulfamethazine; SN-3517.* $C_{12}H_{13}BrN_4O_2S$; mol wt 357.23. C 40.35%, H 3.67%, Br 22.37%, N 15.68%, O 8.96%, S 8.98%. Prepn: English *et al., J. Am. Chem. Soc.* **68**, 453 (1946). Pharmacology: C. M. Stowe *et al., Am. J. Vet. Res.* **29**, 345 (1958).

Crystals, dec 250-252°. uv max (methanol): 238, 272 nm (A 428, 635); min: 250 nm (A 290). Sol in alkaline solns. Sodium salt monohydrate, $C_{12}H_{12}BrN_4O_2S.H_2O$, *Sulfabrom.* Cream-colored powder. Sol in water.

THERAP CAT (VET): Antibacterial.

9067. Sulfacetamide. *N-[(4-Aminophenyl)sulfonyl]-acetamide; N-sulfanilylacetamide; N¹-acetylsulfanilamide; p-aminobenzenesulfonoacetamide.* $C_8H_{10}N_2O_3S$; mol wt 214.25. C 44.85%, H 4.70%, N 13.08%, O 22.40%, S 14.97%. Incorrectly called *sulfacetimide.* Not to be confused with N^4-acetylsulfanilamide, the substance formed in the animal body by acetylation of sulfanilamide. Prepn of base and sodium salt from sulfanilamide and acetic anhydride: Crossley *et al., J. Am. Chem. Soc.* **61**, 2950 (1939). *See also* Dohrn, Diedrich, *Münch. Med. Wochenschr.* **85**, 2017 (1938); U.S. pat. **2,411,495** (1946 to Schering). Toxicity: R. S. Fisher, H. B. Haag, *J. Urol.* **47**, 183 (1942). Stability study of ophthalmic solns: T. Ahmad, I. Ahmad, *Pharmazie* **36**, 619 (1981). Antimicrobial activity: R. D. Houlsby *et al., J. Pharm. Sci.* **72**, 1401 (1983).

Prisms, mp 182-184°. Soluble in 150 parts water at 20°, in 15 parts alcohol, in 7 parts acetone. Insol in ether. An aq soln is acid to litmus. LD₅₀ orally in dogs: 8000 mg/kg (Fisher).

Sodium salt monohydrate, $C_8H_9N_2NaO_3S.H_2O$, *soluble sulfacetamide, Ak-Sulf, Albucid, Antébor, Beocid-Puroptal, Bleph-10, Cetamide, Isopto Cetamide, Locula, Op-Sulfa, Prontamid, Sebizon, Sodium Sulamyd, Sulf-10, Sulten-10.* Minute prisms from dil alcohol, mp 257°. Slightly bitter taste. Sol in 1.5 parts water. Sparingly sol in alc, acetone. pH of 10% (w/v) aq soln: 9.0.

THERAP CAT: Antibacterial.
THERAP CAT (VET): Antibacterial.

9068. Sulfachlorpyridazine. *4-Amino-N-(6-chloro-3-pyridazinyl)benzenesulfonamide; N¹-(6-chloro-3-pyridazinyl)-sulfanilamide; 3-chloro-6-sulfanilamidopyridazine; 3-sulfa-nilamido-6-chloropyridazine; 3-(p-aminophenylsulfonami-do)-6-chloropyridazine; Ciba 10370; Ba-10370; Cosulid; Cosumix; Nefrosul; Sonilyn.* $C_{10}H_9ClN_4O_2S$; mol wt 284.73. C 42.18%, H 3.19%, Cl 12.45%, N 19.68%, O 11.24%, S 11.26%. Prepn: Lester, English, U.S. pat. **2,790,-798** (1957 to Am. Cyanamid).

Sodium salt, $C_{10}H_8ClN_4NaO_2S$, *Prinzone, Vetisulid.*
THERAP CAT: Antibacterial.
THERAP CAT (VET): Antibacterial agent. In enteric infections.

9069. Sulfachrysoidine. *3,5-Diamino-2-[[4-(aminosul-fonyl)phenyl]azo]benzoic acid; 3,5-diamino-2-[p-(sulfamoyl-phenyl)azo]benzoic acid; carboxysulfamidochrysoidine; Azo Compd No. 4; Rubiazol; Collubiazol.* $C_{13}H_{13}N_5O_4S$; mol wt 335.34. C 46.56%, H 3.91%, N 20.88%, O 19.08%, S 9.56%. Prepn: Gley, Girard, *Compt. Rend. Soc. Biol.* **125**, 1027 (1937).

Deep red crystals. mp above 300°.
THERAP CAT: Antibacterial.

9070. Sulfacytine. *4-Amino-N-(1-ethyl-1,2-dihydro-2-oxo-4-pyrimidinyl)benzenesulfonamide; N'-(1-ethyl-1,2-dihydro-2-oxo-4-pyrimidinyl)sulfanilamide; 1-ethyl-N-sulfanilylcytosine; N-sulfanilyl-1-ethylcytosine; CL-636; Renoquid.* $C_{12}H_{14}N_4O_3S$; mol wt 294.33. C 48.97%, H 4.79%, N 19.04%, O 16.31%, S 10.89%. Prepn: **Neth. pat. Appl.** **6,610,815** corresp to Doub, Krolls, **U.S. pat. 3,375,247** (1967, 1968 to Parke, Davis). Prepn and activity: Doub *et al., J. Med. Chem.* **13**, 242 (1970).

Crystals from butyl alcohol, methanol, mp 166.5-168°, (monohydrate, mp 104°). uv max (methanol): 263, 297 nm ($E_{1cm}^{1\%}$ 584, 762). Equilibrium soly in pH 5 buffer at 37°: approx 175 mg/100 ml. Soly increases with increasing pH. pK' 6.9.
THERAP CAT: Antibacterial.

9071. Sulfadiazine. *4-Amino-N-2-pyrimidinylbenzenesulfonamide; 2-sulfanilamidopyrimidine; N'-2-pyrimidinylsulfanilamide; N'-2-pyrimidylsulfanilamide; 2-sulfanilylaminopyrimidine; sulfapyrimidine; Adiazine; Diazyl; Sulfolex.* $C_{10}H_{10}N_4O_2S$; mol wt 250.28. C 47.99%, H 4.03%, N 22.39%, O 12.79%, S 12.81%. Prepd by condensing 2-aminopyrimidine with acetylsulfanilyl chloride followed by hydrolysis of the acetyl group with NaOH: Roblin *et al., J. Am. Chem. Soc.* **62**, 2002 (1940); Barber, **Brit. pat. 557,055** (1943); Sprague, **U.S. pat. 2,407,966** (1946 to Sharp & Dohme); Winnek, Roblin, **U.S. pat. 2,410,793** (1946 to Am. Cyanamid). Toxicological, chemotherapeutic and pharmacokinetic studies: Böhni *et al., Chemotherapy* **14**, 195-226 (1969). *In vitro* activity of silver salt: Carr *et al., Antimicrob. Ag. Chemother.* **4**, 585 (1973). Use of silver salt in burns: C. L. Fox, *Arch. Surg.* **96**, 184 (1968). Comprehensive description: H. Stober, W. DeWitte, *Anal. Profiles Drug Subs.* **11**, 523-551 (1982).

White or slightly yellow powder. mp 252-256°. Sparingly sol in water at 37°: 13 mg/100 ml at pH 5.5; 200 mg/100 ml at pH 7.5. Sparingly sol in alcohol, acetone. One gram dissolves in about 620 ml of human serum at 37°. Freely sol in dil mineral acids and in solns of potassium and sodium hydroxides and in ammonia water.
Mixture with trimethoprim (usually 5:1), *co-trimazine, Geatrim-Boli, Norodine, Scorprin, Synutrim, Tribrissen, Triglobe, Vesuprim.*
Sodium salt, $C_{10}H_9N_4NaO_2S$, *sulfadiazine sodium, soluble sulfadiazine.* White powder. On prolonged exposure to humid air, it absorbs carbon dioxide with the liberation of sulfadiazine and becomes incompletely sol in water. One gram dissolves in about 2 ml of water. Slightly sol in alc. Solns are alkaline to phenolphthalein (pH 9-11).
Silver salt, *Flamazine, Flammazine, Silvadene.*

THERAP CAT: Antibacterial.
THERAP CAT (VET): Antibacterial.

9072. Sulfadicramide. *N-[(4-Aminophenyl)sulfonyl]-3-methyl-2-butenamide; 3-methyl-N-sulfanilylcrotonamide; N'-senecioylsulfanilamide; N-sulfanilylseneciamide; N'-dimethylacroylsulfanilamide; N-sulfanilyl-β,β-dimethylacrylamide; Irgamide.* $C_{11}H_{14}N_2O_3S$; mol wt 254.31. C 51.95%, H 5.55%, N 11.02%, O 18.87%, S 12.61%. Prepd from β,β-dimethylacrylamide by treatment with sodamide and p-nitrobenzenesulfonyl chloride; the resulting 4-nitro-N-(β,β-dimethylacrylyl)benzenesulfonamide is reduced with Fe and dil AcOH: Martin, Häfliger, **U.S. pat. 2,417,005** (1947), *cf.* **U.S. pat. 2,383,874** (1945).

Crystals from alc. mp 184-185°. Slightly sol in water, ether. Freely sol in alcohol, acetone. For the prepn of solns the water-soluble sodium salt is marketed. A 20% aq soln of the sodium salt has a pH of 8.3.
THERAP CAT: Antibacterial.

9073. Sulfadimethoxine. *4-Amino-N-(2,6-dimethoxy-4-pyrimidinyl)benzenesulfonamide; N'-(2,6-dimethoxy-4-pyrimidinyl)sulfanilamide; 2,6-dimethoxy-4-sulfanilamidopyrimidine; 2,4-dimethoxy-6-sulfanilamido-1,3-diazine; 6-sulfanilamido-2,4-dimethoxypyrimidine; 2,6-dimethoxy-4-(p-aminobenzenesulfonamido)pyrimidine; Agribon; Albon; Arnosulfan; Bactrovet; Diasulfa; Madribon; Maxulvet; Neostreptal; Sudine; Suldixine; Sulfabon; Sulxin; Symbio; Ultrasulfon.* $C_{12}H_{14}N_4O_4S$; mol wt 310.33. C 46.44%, H 4.55%, N 18.05%, O 20.62%, S 10.33%. Prepn: Bretschneider, Klötzer, **U.S. pat. 2,703,800** (1955 to Oesterr. Stickstoffwerke); *eidem, Monatsh.* **87**, 136 (1956); Langley, **Brit. pat. 866,843** (1961 to I.C.I.); Bretschneider *et al., Monatsh.* **92**, 128 (1961); **U.S. pat. 3,127,398** (1964 to Hoffmann-La Roche); Shepherd *et al., J. Org. Chem.* **26**, 2764 (1961). Toxicity data: Seki *et al., Arzneimittel-Forsch.* **15**, 1441 (1965). Toxicological, chemotherapeutic and pharmacokinetic studies: Böhni *et al., Chemotherapy* **14**, 195-226 (1969); *see also* Aviado *et al., ibid.* 37.

Crystals from dil alc, mp 201-203°. Sol in dil HCl and in aq solns of sodium carbonate. Soly in water at 37° (mg/100 ml): 4.6 at pH 4.10; 29.5 at pH 6.7; 58.0 at pH 7.06; 5170 at pH 8.71. LD_{50} orally in mice: > 10 g/kg (Seki).
Mixture with trimethoprim, *Maxutrim, Prazil, Trivalbon.*
Sodium salt, $C_{12}H_{13}N_4NaO_4S$, crystals from methanol + ether. Freely sol in water. pH of 5% solution: 8.1; of 10% soln: 8.6.
Sodium salt mixture with trimethoprim, *Sulfaprim, Vetiprim.*
THERAP CAT: Antibacterial.
THERAP CAT (VET): Antibacterial.

9074. Sulfadoxine. *4-Amino-N-(5,6-dimethoxy-4-pyrimidinyl)benzenesulfonamide; N'-(5,6-dimethoxy-4-pyrimidinyl)sulfanilamide; 6-(4-aminobenzenesulfonamido)-4,5-dimethoxypyrimidine; 4-sulfanilamido-5,6-dimethoxypyrimidine; sulforthomidine; sulphormethoxine; Ro-4-4393; Fanasil; Fanzil.* $C_{12}H_{14}N_4O_4S$; mol wt 310.33. C 46.44%, H 4.55%, N 18.05%, O 20.62%, S 10.33%. Prepn: **Belg. pat. 618,639;** Bretschneider *et al.*, **U.S. pat. 3,132,139** (1962, 1964 both to Hoffmann-La Roche). Toxicological, chemotherapeutic and pharmacokinetic studies: E. Böhne *et al.*,

Chemotherapy **14**, 195-226 (1969); *see also* Aviado *et al., ibid.* 37. Comprehensive description: V. K. Kapoor in *Analytical Profiles of Drug Substances* **vol. 17**, K. Florey, Ed. (Academic Press, New York, 1988) pp 571-605.

Crystals from 50% aq alc, mp 190-194°. Practically insol in ether. Very slightly sol in water, slightly sol in alcohol, methanol. Sol in dilute mineral acids, solutions of alkali hydroxides and carbonates. LD_{50} in mice (microcrystals, mg/kg): 5200 orally, 2900 s.c, 2900 i.p. (Böhne).
Mixture with trimethoprim, *Animar, Borgal.*
THERAP CAT: Antibacterial.
THERAP CAT (VET): Antibacterial.

9075. Sulfaethidole. *4-Amino-N-(5-ethyl-1,3,4-thiadiazol-2-yl)benzenesulfonamide; N^1-(5-ethyl-1,3,4-thiadiazol-2-yl)sulfanilamide;* 5-ethyl-2-sulfanilamido-1,3,4-thiadiazole; 2-(p-aminobenzenesulfonamido)-5-ethylthiadiazole; sulfaethylthiadiazole; VK-55; Sul-Spantab. $C_{10}H_{12}N_4O_2S_2$; mol wt 284.36. C 42.24%, H 4.25%, N 19.70%, O 11.25%, S 22.55%. The commercial prepn (Schering AG) is described in P.B. reports, no. 245, p 5, and no. 1361, p 73. Prepn: Wojahn, Wuckel, *Arch. Pharm.* **284**, 53 (1951).

Crystals, mp 185.5-186.0°. One gram dissolves in 4000 ml water, in 40 g methanol, in 30 g ethanol, in 10 g acetone, in 1350 g ether, in 2800 g chloroform, in 20,000 g benzene. Faintly acid to litmus.
Sodium salt. pH of 20% soln: 7.5.
THERAP CAT: Antibacterial.
THERAP CAT (VET): Antimicrobial.

9076. Sulfaguanidine. *4-Amino-N-(aminoiminomethyl)benzenesulfonamide; 4-amino-N-(diaminomethylene)benzenesulfonamide; N^1-amidinosulfanilamide; N^1-guanylsulfanilamide;* p-aminobenzenesulfonylguanidine; sulfanilylguanidine; RP-2275; Diacta; Ganidan; Guanicil; Resulfon; Shigatox. $C_7H_{10}N_4O_2S$; mol wt 214.25. C 39.24%, H 4.70%, N 26.15%, O 14.94%, S 14.97%. The official product is the monohydrate. Prepd by condensing acetylsulfanilyl chloride with guanidine nitrate in the presence of much NaOH in water-acetone: Winnek, U.S. pats. **2,218,490** (1940); **2,229,784** (1941); **2,233,569** (1941); or by fusing N^4-acetylsulfanilamide and dicyanodiamide: Haworth, Rose, Brit. pat. **551,524** (1943). Also made from p-nitrobenzenesulfonyl chloride: Winnek, *loc. cit.* Several other syntheses, *cf.* E. H. Northey, *Sulfonamides,* A.C.S. monograph No. **106** (New York, 1948). Revised structure: G. R. Sullivan, J. D. Roberts, *J. Org. Chem.* **42**, 1095 (1977).

Monohydrate, needles. When anhydr, mp 190-193°. One gram dissolves in about 1000 ml water at 25° and in about 10 ml at 100°. Sparingly sol in alcohol or acetone. Freely sol in dil mineral acids. Insol in NaOH solns at room temp. LD_{100} i.p. in mice: 1.0 g/kg.

THERAP CAT: Antibacterial.
THERAP CAT (VET): Antimicrobial. In enteric infections.

9077. Sulfaguanole. *4-Amino-N-[[(4,5-dimethyl-2-oxazolyl)amino]iminomethyl]benzenesulfonamide; N^1-[(4,5-dimethyl-2-oxazolyl)amidino]sulfanilamide;* 1-(4,5-dimethyloxazol-2-yl)-3-sulfanilylguanidine; 1-(p-aminophenylsulfonyl)-3-(4,5-dimethyl-2-oxazolyl)guanidine; sulfadimethyloxazolylguanidine; Enterocura. $C_{12}H_{15}N_5O_3S$; mol wt 309.35. C 46.59%, H 4.89%, N 22.64%, O 15.52%, S 10.37%. Prepn: Loop *et al.,* **Brit.** pat. **1,185,139,** *C.A.* **73**, 3909w (1970); *eidem,* **U.S.** pat. **3,562,258** (1970, 1971 both to Nordmark-Werke); Loop, Kohlmann, *Arzneimittel-Forsch.* **23**, 171 (1973). Toxicology: J. Kuhne *et al., ibid.* 178. Series of articles on pharmacology: *ibid.* 172-192.

Almost colorless crystals from 9:1 acetone-water, mp 233-236°; from water-methanol, mp 228-230°. pKa 7.76. Practically insol in water. Sol in dil NaOH solns. No deaths in mice or rats for oral doses of 5 g/kg (Kuhne).
THERAP CAT: Antibacterial.

9078. Sulfalene. *4-Amino-N-(3-methoxypyrazinyl)benzenesulfonamide; N^1-(3-methoxy-2-pyrazinyl)sulfanilamide;* 2-(p-aminobenzenesulfonamido)-3-methoxypyrazine; 3-methoxy-2-sulfanilamidopyrazine; sulfamethopyrazine; 2-sulfanilamido-3-methoxypyrazine; sulfamethoxypyrazine; sulfapyrazinemethoxyine; Farmitalia 204/122; Dalysep; Kelfizina; Kelfizine W; Longum; Policydal; Vetkelfizina. $C_{11}H_{12}N_4O_3S$; mol wt 280.31. C 47.13%, H 4.32%, N 19.99%, O 17.12%, S 11.44%. Prepn: B. Camerino, G. Palamidessi, *Gazz. Chim. Ital.* **90**, 1815 (1960); **Brit.** pat. **928,151;** U.S. pat. **3,098,069** (both 1963 to Farmitalia). *Review:* Gasparini, *Veterinaria (Milan)* **20**, 302 (1971). Also used in combination with trimethoprim, *q.v.* Toxicology: V. De Pascale *et al., Boll. Chim. Farm.* **116**, 155 (1977). Pharmacodynamics: D. S. Reeves *et al., J. Antimicrob. Chemother.* **6**, 647 (1980). Clinical assessment: H. Harazim *et al., J. Int. Med. Res.* **11**, 197 (1983); M. L. Colombo *et al., Pharmatherapeutica* **3**, 556 (1984). Review of clinical efficacy: F. Celotti *et al., J. Int. Med. Res.* **14**, 236-241 (1986).

Crystals from alcohol, mp 176°. LD_{50} in mice (g/kg): 2.164 orally; 1.41 i.v. (U.S. pat.).
Mixture with trimethoprim, *Kelfiprim.* LD_{50} in mice, rats (mg/kg): 3500; 3550 orally (De Pascale).
THERAP CAT: Antibacterial.
THERAP CAT (VET): Antibacterial.

9079. Sulfallate. *Diethylcarbamodithioic acid 2-chloro-2-propenyl ester; diethyldithiocarbamic acid 2-chloroallyl ester;* 2-chloroallyl diethyldithiocarbamate; CDEC; CP-4742; Vegadex. $C_8H_{14}ClNS_2$; mol wt 223.79. C 42.94%, H 6.31%, Cl 15.84%, N 6.26%, S 28.66%. Selective pre-planting or pre-emergence herbicide. Prepn: Harman, D'Amico, *J. Am. Chem. Soc.* **75**, 4081 (1953); **Brit.** pat. **769,222** (1957 to Monsanto). Toxicity data: G. W. Bailey, J. L. White, *Residue Rev.* **10**, 97 (1965). Review of carcinogenic risk: *IARC Monographs* **30**, 283-291 (1983).

Amber liquid, bp_1 128-130°. n_D^{25} 1.5822. d^{25} 1.088. Soly in water at 25°: 100 ppm. Sol in most organic solvents. LD_{50} orally in rats: 850 mg/kg (Bailey, White).

Caution: Prolonged contact with skin and eyes may cause moderate irritation. *See Clinical Toxicology of Commercial Products,* R. E. Gosselin *et al.,* Eds. (Williams & Wilkins, Baltimore, 5th ed., 1984) Section II, p 313. This substance may reasonably be anticipated to be a carcinogen: *Seventh Annual Report on Carcinogens* (PB95-109781, 1994) p 368.

USE: Herbicide.

9080. Sulfaloxic Acid. *2-[[[4-[[[[(Hydroxymethyl)amino]carbonyl]amino]sulfonyl]phenyl]amino]carbonyl]benzoic acid;* 4'-(carbamoylsulfamoyl)phthalanilic acid hydroxymethyl derivative; 4'-(carbamoylsulfamoyl)-N-(hydroxymethyl)phthalanilic acid; formophthaloylsulfanilyl urea; 2-[4-(hydroxymethylureidosulfonyl)phenylcarbamoyl]benzoic acid; sulphaloxic acid. $C_{16}H_{15}N_3O_7S$; mol wt 393.38. C 48.85%, H 3.84%, N 10.68%, O 28.47%, S 8.15%. Prepn: Wiedemann, Strassburger, **Ger.** pat. **960,190** (1957 to von Heyden), *C.A.* **53,** 6547i (1959).

Crystals, mp 160-165°. Soluble in dil bases. Calcium salt, $C_{32}H_{28}CaN_6O_{14}S_2$, *Enteromide, Intestin-Euvernil.*

THERAP CAT: Antibacterial.

9081. Sulfamerazine. *4-Amino-N-(4-methyl-2-pyrimidinyl)benzenesulfonamide;* N^1-(4-methyl-2-pyrimidyl)sulfanilamide; N^1-(4-methyl-2-pyrimidinyl)sulfanilamide; 2-sulfanilamido-4-methylpyrimidine; sulfamethyldiazine; RP-2632; Mesulfa; Percoccide. $C_{11}H_{12}N_4O_2S$; mol wt 264.31. C 49.99%, H 4.58%, N 21.20%, O 12.11%, S 12.13%. Prepd by condensing 2-amino-4-methylpyrimidine with acetylsulfanilyl chloride followed by hydrolysis of the acetyl group: Roblin *et al., J. Am. Chem. Soc.* **62,** 2002 (1940); Sprague *et al., ibid.* **63,** 3028 (1941); Sprague, U.S. pat. **2,407,966** (1946 to Sharp & Dohme). For prepn of 2-amino-4-methylpyrimidine *see* Benary, *Ber.* **63,** 2601 (1930); Backer, Grevenstuk, *Rec. Trav. Chim.* **61,** 291 (1942); *cf.* E. H. Northey, *Sulfonamides* (Reinhold, New York, 1948). Antimicrobial activity: Gill *et al., Indian J. Vet. Sci.* **32,** 240 (1962); Vaichulis, Vedros, *Chemotherapia* **11,** 315 (1966). Toxicity studies: Simunek *et al., Vet. Med. (Prague)* **13,** 619 (1968). Kinetics of sulfamerazine decompn: Zajac, *Diss. Pharm. Pharmacol.* **22,** 455 (1970). Comprehensive description: R. D. G. Woolfenden in *Analytical Profiles of Drug Substances* vol. 6, K. Florey, Ed. (Academic Press, New York, 1977) pp 515-577.

Crystals, mp 234-238°. uv max (water): 243, 257 nm ($E_{1cm}^{1\%}$ 875, 822); (0.1M HCl): 243, 307 nm ($E_{1cm}^{1\%}$ 625, 200); (ethanol): 271 nm ($E_{1cm}^{1\%}$ 835). Slowly darkens on exposure to light. Soly in water at 37°: 35 mg/100 ml at pH 5.5; 170 mg/100 ml at pH 7.5. Readily sol in dil mineral acids and in solns of potassium, ammonium and sodium hydroxides. Sparingly sol in acetone, slightly sol in alcohol, very slightly sol in ether, chloroform.

Monosodium salt, $C_{11}H_{11}N_4NaO_2S$, *soluble sulfamerazine, Solumédine.* Crystals. Bitter, caustic taste. Hygroscopic. On prolonged exposure to humid air, it absorbs CO_2 with the liberation of sulfamerazine and becomes incompletely sol in water. Its solns are alkaline to phenolphthalein (pH 10 or more). One gram dissolves in 3.6 ml water. Slightly sol in alc. Insol in ether, chloroform.

THERAP CAT: Antibacterial.

THERAP CAT (VET): Antibacterial.

9082. Sulfameter. *4-Amino-N-(5-methoxy-2-pyrimidinyl)benzenesulfonamide;* N^1-(5-methoxy-2-pyrimidinyl)sulfanilamide; sulfa-5-methoxypyrimidine; 2-sulfanilamido-5-methoxypyrimidine; 2-(p-aminobenzenesulfonamido)-5-methoxypyrimidine; 5-methoxysulfadiazine; sulfamethoxydiazine; sulfametin (rescinded USAN); sulfametorine; methoxypyrimal; AHR-857; I-2586; Bayrena; Durenat; Kinecid; Kiron; Kirocid; Sulla; Supramid; Ultrax. $C_{11}H_{12}N_4O_3S$; mol wt 280.31. C 47.13%, H 4.32%, N 19.99%, O 17.12%, S 11.44%. Prepn: **Ger.** pat. **1,101,428**; P. Diedrich, U.S. pat. **3,214,335** (1961, 1965 both to Schering AG); Horstmann *et al., Arzneimittel-Forsch.* **11,** 682 (1961); Budesinky *et al., Experientia* **17,** 129 (1961); **Czech.** pat. **98,818**; **Belg.** pat. **633,513** (1963 to SPOFA); *Coll. Czech. Chem. Commun.* **29,** 2980 (1964). Toxicity study: G. Hecht *et al., Arzneimittel-Forsch.* **11,** 695 (1961). Toxicology, chemotherapeutics and pharmacokinetics: Böhne *et al., Chemotherapy* **14,** 195-226 (1969).

Minute, bitter crystals, mp 214-216°. uv max: 230, 271 nm ($E_{1cm}^{1\%}$ 562, 726). Very sparingly sol in water, alcohol, ether. Sol in dil acids, alkalies.

Sodium salt, water soluble. pK about 6.8. pH of 1% aq soln: about 8.9. LD_{50} in mice, rats (g/kg): 1.1 ±0.2, 1.2 i.v.; 1.5, 1.1 i.p.; 3.0, 1.0 orally (Hecht).

THERAP CAT: Antibacterial.

9083. Sulfamethazine. *4-Amino-N-(4,6-dimethyl-2-pyrimidinyl)benzenesulfonamide;* N^1-(4,6-dimethyl-2-pyrimidinyl)sulfanilamide; 4,6-dimethyl-2-sulfanilamidopyrimidine; sulfamezathine; sulfadimerazine; sulfadimidine; sulfamidine; sulfadimethylpyrimidine; Diazil; Sulfadine; S-Dimidine; Dimidin-R; Neazina; Sulmet. $C_{12}H_{14}N_4O_2S$; mol wt 278.33. C 51.78%, H 5.07%, N 20.13%, O 11.50%, S 11.52%. Prepn: Caldwell *et al., J. Am. Chem. Soc.* **63,** 2188 (1941); Roblin *et al., ibid.* **64,** 567 (1942); Weisner, Katscher, **Brit.** pat. **546,-158** (1942 to Ward, Blenkinsop); Sprague, U.S. pat. **2,407,-966** (1946 to Sharp & Dohme). Alternate prepn: Haworth, Rose, **Brit.** pat. **552,887** (1943 to I.C.I.); Garzia, U.S. pat. **3,119,818** (1964 to Ist. Chemioterap. Ital.). Anticoccidiosis activity: Zarin, Feodorova, *C.A.* **66,** 114, 435p-436q (1967). Metabolic studies: Turco *et al., Clin. Pharmacol. Ther.* **7,** 603 (1966); Parker *et al., Hum. Hered.* **19,** 402 (1969). Solubility studies: Shkadova *et al., Farm. Zh. (Kiev)* **24,** 39 (1969). Acute toxicity: B. Bobranski *et al., Arch. Immunol. Ther. Exp.* **16,** 804 (1968). Comprehensive description: C. Papostephanou, M. Frantz in *Analytical Profiles of Drug*

Substances vol. 7, K. Florey, Ed. (Academic Press, New York, 1978) pp 401-422.

Crystals from dioxane-water, mp 176° (also reported as mp 178-179°; mp 198-199°; mp 205-207°). uv max (water, pH 6.6): 241 nm ($E_{1cm}^{1\%}$ 670); (0.01N NaOH): 243, 257 nm ($E_{1cm}^{1\%}$ 765, 776); (0.01N HCl): 241, 297 nm ($E_{1cm}^{1\%}$ 561, 266). Soly in water at 29°: 150 mg/100 ml; at 37°: 192 mg/100 ml at pH 7.0. pK_1 7.4±0.2, pK_2 2.65±0.2. The soly increases rapidly with an increase in pH. LD_{50} i.p. in mice: 1.06 g/kg (Bobranski).

Sodium salt, $C_{12}H_{13}N_4NaO_2S$, **Intradine, Sulfoxine 33, Vesadin.**

THERAP CAT: Antibacterial.
THERAP CAT (VET): Antibacterial.

9084. Sulfamethizole. *4-Amino-N-(5-methyl-1,3,4-thiadiazol-2-yl)benzenesulfonamide; N^1-(5-methyl-1,3,4-thiadiazol-2-yl)sulfanilamide;* 2-sulfanilamido-5-methyl-1,3,4-thiadiazole; 5-methyl-2-sulfanilamido-1,3,4-thiadiazole; 2-(p-aminobenzenesulfonamido)-5-methylthiadiazole; sulfamethylthiadiazole; Famet; Lucosil; Methazol; Renasul; Rufol; Salimol; Sulfapyelon; Thidicur; Thiosulfil; Urolucosil. $C_9H_{10}N_4O_2S_2$; mol wt 270.34. C 39.99%, H 3.73%, N 20.72%, O 11.84%, S 23.72%. Prepn: Hübner, U.S. pat. **2,447,702** (1948 to Lundbeck).

Crystals from water, mp 208°. pKa 5.45. One gram dissolves in 4000 ml water at pH 6.5, in 5 ml water at pH 7.5, in 40 g methanol, in 30 g ethanol, in 10 g acetone, in 1370 g ether, in 2800 g chloroform. Practically insol in benzene.

THERAP CAT: Antibacterial.
THERAP CAT (VET): Antibacterial.

9085. Sulfamethomidine. *4-Amino-N-(6-methoxy-2-methyl-4-pyrimidinyl)benzenesulfonamide; N^1-(6-methoxy-2-methyl-4-pyrimidinyl)sulfanilamide;* 2-methyl-4-methoxy-6-sulfanilamidopyrimidine; 4-sulfanilamido-2-methyl-6-methoxypyrimidine; 2-methyl-4-methoxy-6-sulfanilamido-1,3-diazine; 4-(p-aminobenzenesulfonyl)amino-2-methyl-6-methoxypyrimidine; Duroprocin; Methofadin; Télémid. $C_{12}H_{14}N_4O_3S$; mol wt 294.33. C 48.97%, H 4.79%, N 19.04%, O 16.31%, S 10.89%. Prepn: Loop, Lührs, *Ann.* **580**, 225 (1953); Loop, **Ger. pat. 926,131** (1955 to Nordmark).

Crystals, mp 146°. Also obtained as the monohydrate.
THERAP CAT: Antibacterial.

9086. Sulfamethoxazole. *4-Amino-N-(5-methyl-3-isoxazolyl)benzenesulfonamide; N^1-(5-methyl-3-isoxazolyl)sulfanilamide;* 5-methyl-3-sulfanilamidoisoxazole; 3-sulfanilamido-5-methylisoxazole; 3-(p-aminophenylsulfonyl)-5-methylisoxazole; sulfisomezole; sulfamethylisoxazole; sulfamethoxizole; Gantanol; Sinomin. $C_{10}H_{11}N_3O_3S$; mol wt 253.28. C 47.42%, H 4.38%, N 16.59%, O 18.95%, S 12.66%. Prepn starting with ethyl 5-methylisoxazole-3-carbamate: Kano et al., U.S. pat. **2,888,455** (1959 to Shionogi); *Ann. Rep. Shionogi Res. Lab.* **7**, 1 (1957), C.A. **51**, 17889 (1957). Toxicity data: Yamamoto et al., *Chemotherapy (Tokyo)* **21**, 187 (1973), C.A. **79**, 73738n (1973). Clinical trial of mixture with trimethoprim in *Pneumocystis carinii* pneumonia: J. M. Wharton et al., *Ann. Int. Med.* **105**, 37

(1986). Comprehensive description: B. C. Rudy, B. Z. Senkowski, *Anal. Profiles Drug Subs.* **2**, 467-486 (1973). Review of antibacterial activity and clinical efficacy of mixture with trimethoprim: G. P. Wormser et al., *Drugs* **24**, 459-518 (1982). Symposium on clinical intravenous therapy: *Rev. Infect. Dis.* **9**, Suppl. 2, S152-S229 (1987).

Bitter crystals from dil ethanol, mp 167°. LD_{50} orally in mice: 3662 mg/kg (Yamamoto).

Mixture with trimethoprim, *co-trimoxazole, Abacin, Apo-Sulfatrim, Bactramin, Bactrim, Baktar, Chemotrim, Comox, Drylin, Eusaprim, Fectrim, Gantaprim, Gantrim, Imexim, Kepinol, Laratrim, Linaris, Microtrim, Nopil, Oraprim, Septra, Septrin, Sigaprim, Sulfotrim, Sulprim, Sumetrolim, Supracombin, Suprim, Teleprim, Thiocuran, Trigonyl, Trimesulf, Uroplus, Uro-Septra.* LD_{50} orally in mice: 5513 mg/kg (Yamamoto).

N^4-Acetylsulfamethoxazole, crystals from alcohol, mp 209-210°.

THERAP CAT: Antibacterial; antipneumocystis.
THERAP CAT (VET): Antibacterial.

9087. Sulfamethoxypyridazine. *4-Amino-N-(6-methoxy-3-pyridazinyl)benzenesulfonamide; N^1-(6-methoxy-3-pyridazinyl)sulfanilamide;* 6-methoxy-3-sulfanilamidopyridazine; 3-(p-aminobenzenesulfamido)-6-methoxypyridazine; CL-13494; RP-7522; Kynex; Lederkyn; Midicel; Midikel; Sulfalex; Sulfdurazin; Sultirene. $C_{11}H_{12}N_4O_3S$; mol wt 280.31. C 47.13%, H 4.32%, N 19.99%, O 17.12%, S 11.44%. Prepn: Clark, U.S. pat. **2,712,012** (1955 to Am. Cyanamid). Toxicity: Seki et al., *Arzneimittel-Forsch.* **15**, 1441 (1965). Toxicological, chemotherapeutic and pharmacokinetic studies: Böhni et al., *Chemotherapy* **14**, 195-226 (1969).

Bitter crystals from water, mp 182-183°. pKa 6.7. Soly in water at 37° (mg/100 ml): 110 at pH 5; 120 at pH 6; 147 at pH 6.5. Slightly sol in methanol, ethanol (about 1:200); more sol in acetone (1:50), in dimethylformamide (1 g/1 ml). Freely sol in aq solns of alkali hydroxides. LD_{50} orally in mice: 1750 mg/kg, (Seki).

Sodium salt, $C_{11}H_{11}N_4NaO_3S$, **Davosin, Sulfoxine LA.**
THERAP CAT: Antibacterial.
THERAP CAT (VET): Antibacterial.

9088. Sulfamethylthiazole. *4-Amino-N-(4-methyl-2-thiazole)benzenesulfonamide; 4-methyl-2-sulfanilamidothiazole; N^1-(4-methyl-2-thiazolyl)sulfanilamide;* 2-(p-aminobenzenesulfonamido)-4-methylthiazole. $C_{10}H_{11}N_3O_2S_2$; mol wt 269.35. C 44.59%, H 4.12%, N 15.60%, O 11.88%, S 23.81%. Prepn: Fosbinder, Walter, *J. Am. Chem. Soc.* **61**, 2032 (1939). Alternate routes of synthesis: E. H. Northey, *The Sulfonamides and Allied Compounds* (Reinhold, New York, 1948) p 89.

Crystals from alcohol, mp 238-240°. Slightly sol in water (0.2 g dissolves in 100 ml H_2O). Insol in ether. Readily sol in solns of alkali hydroxides and carbonates and in dil mineral acids.

Sodium salt, *Ultraseptyl.* Sol in water.

THERAP CAT: Antibacterial.

9089. Sulfametrole. *4-Amino-N-(4-methoxy-1,2,5-thiadiazol-3-yl)benzenesulfonamide;* N^1-(4-methoxy-1,2,5-thiadiazol-3-yl)sulfanilamide; 3-methoxy-4-(4'-aminobenzenesulfonamido)-1,2,5-thiadiazole. $C_9H_{10}N_4O_3S_2$; mol wt 286.34. C 37.75%, H 3.52%, N 19.57%, O 16.76%, S 22.40%. Anti-infective sulfanilamide, related structurally to sulfadiazine, *q.v.* Prepn: **Belg. pat. 629,551** corresp to K. Menzl, U.S. pat. **3,247,193** (1963, 1966 both to OSSW). Improved prepn: *idem,* U.S. pat. **4,151,164** (1979 to Chemie Linz). Bacteriological study (of mixture with trimethoprim): G. Nabert-Bock, H. Grims, *Arzneimittel-Forsch.* **27,** 1109 (1977). Tolerance, therapeutic effect, pharmacokinetics: S. Breyer *et al., Wien Med. Wochenschr.* **130,** 448 (1980). Renal excretion mechanism, pharmacokinetics: T. B. Vree *et al., Eur. J. Clin. Pharmacol.* **20,** 283 (1981). Pharmacokinetics of sulfametrole and its metabolite in man: Y. A. Hester *et al., J. Antimicrob. Chemother.* **8,** 133 (1981). Crystal and molecular structures: C. H. Koo *et al., Bull. Korean Chem. Soc.* **3,** 9 (1982), *C.A.* **96,** 208771 (1982). Single-dose therapy of chancroid with trimethoprim-sulfametrole: F. A. Plummer *et al., N. Engl. J. Med.* **309,** 67 (1983).

Cryst from dilute acetic acid, mp 149-150°.

Mixture with trimethoprim, *Lidaprim, Maderan.*

THERAP CAT: Antibacterial.

9090. Sulfamic Acid. Amidosulfonic acid. H_3NO_3S; mol wt 97.09. H 3.11%, N 14.43%, O 49.43%, S 33.03%. H_2NSO_3H. Obtained from chlorosulfonic acid and ammonia, or by heating urea with H_2SO_4. Purification: Sisler *et al., Inorg. Syn.* **2,** 178 (1946). Toxicity data: Ambrose, *J. Ind. Hyg. Toxicol.* **25,** 26 (1943). *Reviews:* Audrieth *et al., Chem. Rev.* **26,** 49 (1940); Burton, Nickless, "Amido- and Imido-Sulfonic Acids" in *Inorganic Sulphur Chemistry,* G. Nickless, Ed. (Elsevier, New York, 1968) pp 607-627, 661-667; E. B. Bell in Kirk-Othmer *Encyclopedia of Chemical Technology* vol. 21 (Wiley-Interscience, New York, 3rd ed., 1983) pp 949-960.

Orthorhombic crystals. d 2.15. mp ∼205° (dec). Stable when dry but in soln it slowly hydrolyzes forming ammonium bisulfate. Sol in 6.5 parts water at 0°, in about 2 parts water at 80°. Sulfuric acid decreases its soly in water. It is sparingly sol in alcohol, methanol; slightly sol in acetone. Insol in ether. Freely sol in nitrogenous bases, e.g., liq ammonia, also in nitrogen contg organic solvents, e.g., pyridine, formamide, dimethylformamide. A strong acid; pH of a 1% soln at 25° 1.18. Can be titrated with bases by means of indicators showing color change between pH 4.5 to 9. MLD orally in rats: 1.6 g/kg (Ambrose).

Caution: Moderately irritating to skin, mucous membranes.

USE: As standard in alkalimetry; in acid cleaning; in nitrite removal; in chlorine stabilization for use in swimming pools, cooling towers, paper mills. The acid or its ammonium salt has been recommended for flameproofing fabrics and wood. Metal salts are used in electroplating. Ammonium sulfamate, *q.v.,* is also widely used as a weed killer.

9091. Sulfamide. *Sulfuryl amide.* $H_4N_2O_2S$; mol wt 96.11. H 4.19%, N 29.15%, O 33.29%, S 33.36%. H_2N-SO_2NH_2. Prepd from sulfuryl chloride and gaseous ammonia: Traube, *Ber.* **25,** 2427 (1892); **26,** 610 (1893); Hantzsch, Holl, *Ber.* **34,** 3430 (1902); Schenk in *Handbook of Preparative Inorganic Chemistry* vol. 1, G. Brauer, Ed. (Academic Press, New York, 2nd ed., 1963) pp 482-483.

Orthorhombic, tasteless plates from abs alcohol, mp 93°; dec 250°. Dipole moment: 3.9. Freely sol in water, hot

alcohol, acetone. Sparingly sol in cold alc. Solid and solns are stable.

Note: Not to be confused with ammonium sulfamate or with sulfanilamide.

USE: Similar to urea in many of its reactions, but is more acidic, and can act as a dibasic acid.

9092. Sulfamidochrysoidine. *4-[(2,4-Diaminophenyl)-azo]benzenesulfonamide;* 2,4-diaminoazobenzene-4'-sulfonamide; 4'-sulfamyl-2,4-diaminoazobenzene. $C_{12}H_{13}N_5O_2S$; mol wt 291.33. C 49.47%, H 4.50%, N 24.04%, O 10.98%, S 11.01%. Prototype compd leading to the development of sulfonamide antibacterials. Prepn: F. Mietzsch, J. Klarer, **Ger. pat. 607,537** (1935 to I. G. Farbenind.); *eidem,* U.S. pat. **2,085,037** (1937 to Winthrop). Antibacterial activity *in vivo:* G. Domagk, *Deut. Med. Wochenschr.* **61,** 573 (1935); (for English translation *see Rev. Inf. Dis.* **8,** 163 (1986)). Identification of sulfanilamide as active metabolite: J. Tréfouel *et al., Compt. Rend. Soc. Biol.* **120,** 756 (1935); L. Colebrook *et al., Lancet* **2,** 1323 (1936). Polymorphic forms in commercial prepns: L. Kofler, A. Kofler, *Monatsh.* **81,** 321 (1950). Use to visualize carbonic anhydrase during isoelectric focusing in polyacrylamide gels: W. Siffert *et al., J. Biochem. Biophys. Methods* **8,** 331 (1983). Historical review: M. H. Bickel, *Gesnerus* **45,** 67-86 (1988).

Hydrochloride, $C_{12}H_{13}N_5O_2S.HCl$, *Prontosil, Prontosil flavum, Prontosil rubrum, Rubiazol I, Septosan, Streptocide, Streptozon.* Orange-red crystals, mp 248-250°. One gram dissolves in 400 ml water; much more sol in hot water. Sol in alcohol, acetone, fats, oils.

USE: Specific stain for carbonic anhydrase in polyacrylamide gels.

THERAP CAT: Antibacterial.

9093. Sulfamoxole. *4-Amino-N-(4,5-dimethyl-2-oxazolyl)benzenesulfonamide;* N^1-(4,5-dimethyl-2-oxazolyl)sulfanilamide; 2-(p-aminobenzenesulfonamido)-4,5-dimethyloxazole; 4,5-dimethyl-2-sulfanilamidooxazole; p-aminobenzenesulfonyl-2-amino-4,5-dimethyloxazole; sulfadimethyloxazole; Justamil; Sulfmidil; Sulfuno; Tardamide. $C_{11}H_{13}N_3O_3S$; mol wt 267.31. C 49.43%, H 4.90%, N 15.72%, O 17.96%, S 12.00%. Prepn: Loop *et al.,* U.S. pat. **2,809,966** (1957 to Nordmark-Werke). Series of articles on antimicrobial activity, pharmacology and clinical efficacy of mixture with trimethoprim: *Arzneimittel-Forsch.* **26,** 595-684 (1976). Toxicity data: F. Lagler *et al., ibid.* 634.

Crystals, mp 193-194°. Soly in mg/100 ml at 20°: Water 85; 0.01N HCl 163; 0.01N NaOH 196; methanol 2315; chloroform 240. uv max (4.91 μg/ml methanol): 210, 250, 270 nm (ε 740, 546, 857). LD₅₀ in mice, rats (g/kg): > 10.0, > 12.5 orally; 1.80, ∼2.50 i.p. (Lagler).

Mixture with trimethoprim, *co-trifamole, CN-3123, Co-Fram, Dibactil, Nevin, Supristol.* LD₅₀ in mice, rats (g/kg): > 12.0, 14.0 orally; 1.87, 2.00 i.p. (Lagler).

THERAP CAT: Antibacterial.

9094. Sulfanilamide. *4-Aminobenzenesulfonamide;* p-anilinesulfonamide; p-sulfamidoaniline; 1162-F; Albexan; Deseptyl; Prontalbin; Prontosil album; Prontylin; Septoplix; Streptocid album; Streptocide. $C_6H_8N_2O_2S$; mol wt 172.21. C 41.85%, H 4.68%, N 16.27%, O 18.58%, S 18.62%. Prepd by the action of ammonia on acetylsulfanilyl chloride; the resulting N^4-acetylsulfanilamide is then hydrolyzed: Gelmo,

J. Prakt. Chem. **77**, 369 (1908). Lab procedures: L. F. Fieser, *Experiments in Organic Chemistry* (Boston, 3rd ed., 1955) p 147; Hurdis, Yang, *J. Chem. Ed.* **46**, 697 (1969). Large-scale production and purification: Mietzsch, Klarer, U.S. pats. **2,132,178** (1938); **2,276,664** (1942). Alternate prepn: Simons, U.S. pat. **2,237,372** (1941).

Crystals, mp 164.5-166.5°. pKa 10.43; pKb 11.63. Neutral to litmus. pH (0.5% aq soln): 5.8-6.1. uv max: 257, 313 nm. Soly in water (g/liter): 2.6 (10°); 7.5 (25°); 17.0 (40°); 40 (60°); 477 (100°). One gram dissolves in about 37 ml alcohol; in about 5 ml acetone. Also sol in glycerol, propylene glycol, HCl, solns of K and Na hydroxides. Practically insol in chloroform, ether, benzene, petr ether. LD$_{50}$ orally in dogs: 2.0 g/kg.

THERAP CAT: Antibacterial.
THERAP CAT (VET): Antimicrobial.

9095. 4-Sulfanilamidosalicylic Acid. *p*-Sulfanilamidosalicylic acid; 4-(*p*-aminobenzenesulfonamido)-2-hydroxybenzoic acid; Metrasil. C$_{13}$H$_{12}$N$_2$O$_5$S; mol wt 308.31. C 50.64%, H 3.92%, N 9.09%, O 25.95%, S 10.40%. Prepn: Benedikt, Eibl, *Monatsh.* **81**, 419 (1950); Goldberg, Salame, **Brit.** pat. **648,467** (1951 to Ward, Blenkinsop).

Crystals from dil alcohol, dec 220°. The water soly is described as high between pH 5 and 8.

THERAP CAT: Antibacterial.

9096. Sulfanilic Acid. *4-Aminobenzenesulfonic acid; p*-anilinesulfonic acid. C$_6$H$_7$NO$_3$S; mol wt 173.19. C 41.61%, H 4.07%, N 8.09%, O 27.71%, S 18.51%. Readily prepd from aniline and sulfuric acid: I. G. Farben, *FIAT* Final Rept. **1313** (I), 255 (1948); Fierz-David, Blangey, *Fundamental Processes of Dye Chemistry* (Interscience, New York, 1949) pp 126-128; A. I. Vogel, *Practical Organic Chemistry* (Longmans, London, 3rd ed., 1959) p 586; R. Q. Brewster et al., *Unitized Experiments in Organic Chemistry* (Van Nostrand, Princeton, 2nd ed., 1964) pp 162-163. Improved procedure: Jacobs et al., *Ind. Eng. Chem.* **35**, 321-323 (1943).

Monohydrate, orthorhombic plates from water (very slow crystn can yield the dihydrate), becomes anhydr at around 100°. Dec without melting at ~288° (also reported as > 360°). pK$_2$ (25°): 3.23. Slowly sol in water: ~1% at 20°, 1.45% at 30°, 1.94% at 40° (w/w anhydr). Almost insol in ethanol, benzene, ether. Slightly sol in hot methanol.

Sodium salt dihydrate, C$_6$H$_6$NNaO$_3$S.2H$_2$O, orthorhombic bipyramidal plates. Freely sol in water, sol in hot methanol.

Zinc salt tetrahydrate, C$_{12}$H$_{12}$N$_2$O$_6$S$_2$Zn.4H$_2$O, *zinc sulfanilate tetrahydrate, Nizin.*

Amide see Sulfanilamide.

USE: Manuf various dyes and organic chemicals; as reagent in anal. chemistry (Ehrlich's reagent, detn of nitrites).

THERAP CAT: Antibacterial.

9097. 2-*p*-Sulfanilylanilinoethanol. *2-[[4-[(4-Aminophenyl)sulfonyl]phenyl]amino]ethanol;* 4-amino-4'-(β-hydroxyethylamino)diphenyl sulfone; *p*-aminophenyl *p*-(2-hydroxyethylamino)phenyl sulfone; Smith's sulfone. C$_{14}$H$_{16}$-N$_2$O$_3$S; mol wt 292.36. C 57.52%, H 5.52%, N 9.58%, O 16.42%, S 10.97%. Preparation: Jackson, *J. Am. Chem. Soc.* **70**, 680 (1948); Weijlard, Swanezy, *ibid.* **71**, 4134 (1949); Jackson, *ibid.* **72**, 395 (1950); Anand et al., *J. Sci. Ind. Res. (India)* **13B**, 260 (1954); Onishi, Fukagawa, **Japan.** pat. **2,975('56)** (to Yoshitomi Drug).

Exists in diamorphic forms, crystals, mp 130.5-131.5° and crystals, mp 144°. The lower melting crystals' soly in water: 74 mg/100 ml at 37°. The higher melting form is readily sol in acetone; sol in ethanol, dioxane, cold 1% HCl; fairly sol in hot water; slightly sol in cold water.

THERAP CAT: Antibacterial (Mycobacterium).

9098. *p*-Sulfanilylbenzylamine. *4-[(4-Aminophenyl)-sulfonyl]benzenemethanamine; p*-sulfanilidobenzylamine; 4-aminomethyl-4'-aminodiphenyl sulfone; 4-aminophenyl-*p*-aminomethylphenyl sulfone; Alphamide. C$_{13}$H$_{14}$N$_2$O$_2$S; mol wt 262.33. C 59.52%, H 5.38%, N 10.68%, O 12.20%, S 12.22%. Prepn: Dewing, *J. Chem. Soc.* **1946**, 466; Wenner, *J. Org. Chem.* **22**, 1508 (1957).

Needles from water, mp 159°.
Monohydrochloride monohydrate, C$_{13}$H$_{14}$N$_2$O$_2$S.HCl.-H$_2$O, crystals, mp 195°. Very sol in water.
Dihydrochloride, C$_{13}$H$_{14}$N$_2$O$_2$S.2HCl, mp 285°.
THERAP CAT: Antibacterial.

9099. Sulfanilyl Fluoride. *p-Aminobenzenesulfonyl fluoride.* C$_6$H$_6$FNO$_2$S; mol wt 175.18. C 41.14%, H 3.45%, F 10.84%, N 8.00%, O 18.27%, S 18.30%. Prepd by heating potassium fluoride with 4-acetamidobenzenesulfonyl chloride: Parker, Hofmann, U.S. pat. **2,576,037** (1951 to Am. Cyanamid).

Crystals, mp 68-69°.
USE: In the prepn of dyes which pick up light readily.

9100. N⁴-Sulfanilylsulfanilamide. *4'-Sulfamoylsulfanilanilide;* 4-aminobenzenesulfono-*p*-sulfamoylanilide; 4-(4'-aminobenzenesulfonamido)benzenesulfonamide; DB-32; Diseptal C; Uliron C; Disulon; Neosanamid II; Albasil C; Disulfan. C$_{12}$H$_{13}$N$_3$O$_4$S$_2$; mol wt 327.39. C 44.03%, H 4.00%, N 12.84%, O 19.55%, S 19.59%. Prepn: **Fr.** pat. **817,034** (to I. G. Farben.), *C.A.* **32**, 1714i (1938); Sakai, Yamamoto, *J. Pharm. Soc. Japan* **58**, 683 (1938); Golovchinskaya, *J. Appl. Chem. USSR* **18**, 647 (1945); Sasa, *J. Soc. Org. Syn. Chem. Japan* **12**, 211 (1954), *C.A.* **51**, 2780h (1957); Grigorovskii, Dykhanov, *J. Appl. Chem. USSR* **30**, 1284 (1957).

Needles from water, mp 133-134°. Absorption max: Maschka *et al.*, *Monatsh.* **84,** 1071 (1953). Slightly sol in cold water; considerably more sol in hot water. Sol in methanol, ethanol, ether, dil NH$_3$, dil HCl; practically insol in petr ether, chloroform.

Note: The name Disulon is also used for disulfanilamide, H$_2$NC$_6$H$_4$SO$_2$NHSO$_2$C$_6$H$_4$NH$_2$, esp. in the French literature.

THERAP CAT: Antibacterial (topical).

9101. Sulfanilylurea. *4-Amino-N-(aminocarbonyl)-benzenesulfonamide; N-sulfanilylcarbamide; sulfacarbamide; sulfaurea; Euvernil; Uractyl; Uramid; Urenil; Urosulfan.* C$_7$H$_9$N$_3$O$_3$S; mol wt 215.23. C 39.06%, H 4.21%, N 19.52%, O 22.30%, S 14.90%. Prepd by treating N^4-acetylsulfanilamide with potassium cyanate or with carbamyl chloride or with urea (or with nitrourea and sodium carbonate) in 80% alcohol. The *p*-AcNHC$_6$H$_4$SO$_2$NHCONH$_2$ is saponified by slight warming with dil KOH and then acidified: Martin *et al.*, U.S. pat. **2,411,661** (1946 to Geigy). By boiling sulfanilamide with urea and sodium carbonate in 75% alcohol: Haack, *Alien Prop. Custodian, Serial* **369,** 118 (1943). By warming calcium acetylsulfanilylcyanamide with dil HCl: Winnek *et al.*, *J. Am. Chem. Soc.* **64,** 1684 (1942); improved procedure: Leitch *et al.*, *Can. J. Res.* **23B,** 139 (1945).

Crystals from water. mp 146-148° (slight dec). Solubility in water at 37°: 811 mg/100 ml. Soluble in alkalies. Forms a very soluble sodium salt.

Monohydrate, mp 125-127°.

THERAP CAT: Antibacterial.

9102. N-Sulfanilyl-3,4-xylamide. *N-[(-4-Aminophenyl)sulfonyl]-3,4-dimethylbenzamide; N¹-(3,4-dimethylbenzoyl)sulfanilamide; Geigy 867; Irgafen.* C$_{15}$H$_{16}$N$_2$O$_3$S; mol wt 304.37. C 59.19%, H 5.30%, N 9.20%, O 15.77%, S 10.54%. Prepd from *p*-nitrobenzenesulfonamide and 3,4-dimethylbenzoyl chloride; the resulting *p*-nitro-N-(3,4-dimethylbenzoyl)benzenesulfonamide is reduced with Fe and dil AcOH: Martin *et al.*, U.S. pat. **2,383,874** (1945).

Needles from alcohol, mp 222-223°. Sparingly sol in water. Marketed as the water-soluble sodium salt for making solns; a 5% aq soln having a pH of 8.2.

THERAP CAT: Antibacterial.

9103. Sulfanitran. *N-[4-[[(4-Nitrophenyl)amino]sulfonyl]phenyl]acetamide; 4'-[(p-nitrophenyl)sulfamoyl]acetanilide; N⁴-acetyl-N¹-(p-nitrophenyl)sulfanilamide; 4-acetaminobenzenesulfon-4'-nitroanilide; N-(p-acetylaminobenzene sulfonyl)-p-nitroaniline; APNPS.* C$_{14}$H$_{13}$N$_3$O$_5$S; mol wt 335.34. C 50.14%, H 3.91%, N 12.53%, O 23.86%, S 9.56%. Prepn: Webster, Powers, *J. Am. Chem. Soc.* **60,** 1553 (1938); Kaufmann, Bückmann, *Arch. Pharm.* **279,** 194 (1941); Shepherd, *J. Org. Chem.* **12,** 275 (1947).

Crystals from dil alcohol, mp 239-240° (Kaufmann); mp 264° (Shepherd). Freely sol in acetone; sol in hot ethanol, methanol; sparingly sol in water, ether.

Note: Ingredient of *Unistat, Novastat W.*

THERAP CAT (VET): Antibacterial; coccidiostat (poultry).

9104. Sulfaperine. *4-Amino-N-(5-methyl-2-pyrimidinyl)benzenesulfonamide; N¹-(5-methyl-2-pyrimidinyl)sulfanilamide; 5-methyl-2-sulfanilamidopyrimidine; 2-sulfanilamido-5-methylpyrimidine; isosulfamerazine; 5-methylsulfadiazine; Pallidin; Retardon; Rexulfa; Sintosulfa; Sulfatreis.* C$_{11}$H$_{12}$N$_4$O$_2$S; mol wt 264.31. C 49.99%, H 4.58%, N 21.20%, O 12.11%, S 12.13%. Prepd by condensing 2-amino-5-methylpyrimidine with acetylsulfanilyl chloride followed by hydrolysis of the acetyl group with NaOH: Sprague, U.S. pat. **2,407,966** (1946 to Sharp & Dohme).

Minute, cream-colored crystals, mp 262-263°. Very sparingly sol in water, ethanol; ~40 mg/100 ml H$_2$O at pH 5.5. Sol in aq solns of acids and alkalies. Forms a water-sol sodium salt.

THERAP CAT: Antibacterial.

9105. Sulfaphenazole. *4-Amino-N-(1-phenyl-1H-pyrazol-5-yl)benzenesulfonamide; N'-(1-phenylpyrazol-5-yl)sulfanilamide; 1-phenyl-5-sulfanilamidopyrazole; 3-(p-aminobenzenesulfonamido)-2-phenylpyrazole; 5-sulfanilamido-1-phenylpyrazole; Isarol V; Orisul; Orisulf.* C$_{15}$H$_{14}$N$_4$O$_2$S; mol wt 314.37. C 57.31%, H 4.49%, N 17.82%, O 10.18%, S 10.20%. Synthesis: Schmidt, Druey, *Helv. Chim. Acta* **41,** 309 (1958); Druey, Schmidt, U.S. pat. **2,858,309** (1958 to Ciba). Toxicity study: Seki *et al.*, *Arzneimittel-Forsch.* **15,** 1441 (1965).

Crystals from alcohol, mp 179-183°. Very sparingly sol in water: 0.15 g/100 ml H$_2$O at pH 7.0 and 25°. More sol in methanol, ethanol, glacial acetic acid. LD$_{50}$ orally in mice: 5800 mg/kg (Seki).

Sodium salt monohydrate, C$_{15}$H$_{13}$N$_4$NaO$_2$S.H$_2$O, white powder. Sol in water. A 10% soln (w/w) has a pH of ~9.

THERAP CAT: Antibacterial.

THERAP CAT (VET): Antibacterial.

9106. Sulfaproxyline. *p-Isopropoxy-N-sulfanilylbenzamide; N¹-(p-isopropoxybenzoyl)sulfanilamide; N¹-(4-isopropoxybenzoyl)-p-aminobenzenesulfonamide; sulphaproxyline.* C$_{16}$H$_{18}$N$_2$O$_4$S; mol wt 334.40. C 57.47%, H 5.43%, N 8.38%, O 19.14%, S 9.59%. Prepn: Gysin, U.S. pat. **2,503,820** (1950 to Geigy).

Crystals, mp 172-173°.
Ingredient of *Dosulfin* which also contains sulfamerazine in equal parts.
THERAP CAT: Antibacterial.

9107. Sulfapyrazine. *4-Amino-N-(pyrazinyl)benzenesulfonamide; N¹-2-pyrazinylsulfanilamide;* 2-sulfanilamidopyrazine. $C_{10}H_{10}N_4O_2S$; mol wt 250.28. C 47.99%, H 4.03%, N 22.39%, O 12.79%, S 12.81%. Prepd by reacting acetylsulfanilyl chloride with 2-aminopyrazine in pyridine, followed by hydrolysis and neutralization: Ellingson, U.S. pat. 2,420,703 (1947). Biological properties: Ghione et al., *Chemotherapia* **6**, 344 (1962).

Crystals, dec 250-254°. Very slightly sol in alc; slightly sol in acetone. Sol in aq solns of sodium, potassium, and barium hydroxides, in ammonia water, and in dil and concd mineral acid solns. Practically insol in water (5 mg/100 ml at 25°, and 5.2 mg/100 ml at 37°).
Sodium salt monohydrate, $C_{10}H_9N_4NaO_2S.H_2O$, bitter powder. Freely soluble in water (1 g/3.33 ml at 25°). Alkaline to litmus. pH of 10% aq soln 9.1. Very sol in acetone, slightly sol in alc. Insol in ether, chloroform. Aq solns of sulfapyrazine sodium may absorb carbon dioxide which causes precipitation of sulfapyrazine.
THERAP CAT: Antibacterial.
THERAP CAT (VET): Antibacterial.

9108. Sulfapyridine. *4-Amino-N-2-pyridinylbenzenesulfonamide; N¹-2-pyridylsulfanilamide;* 2-sulfanilamidopyridine; 2-(p-aminobenzenesulfonamido)pyridine; M & B 693; Dagenan; Eubasin; Pyriamid; Coccoclase; Septipulmon. $C_{11}H_{11}N_3O_2S$; mol wt 249.29. C 53.00%, H 4.45%, N 16.86%, O 12.84%, S 12.86%. Prepn: A. J. Ewins, M. A. Phillips, Brit. pat. 512,145 (1939); *eidem*, U.S. pat. 2,275,354 (1942); R. Winterbottom, *J. Am. Chem. Soc.* **62**, 160 (1940). Metabolism: R. Wien, J. Hampton, *J. Pharmacol. Exp. Ther.* **84**, 211 (1945). Toxicology: R. Wien et al., *ibid.* 203. GC-MS determn in edible animal tissues: A. E. Mooser, H. Koch, *J. AOAC Int.* **76**, 976 (1993). Efficacy and proposed mechanism of action in dermatological disease: O. J. Stone, *Med. Hypotheses* **31**, 99 (1990).

Crystals from alc, mp 190-191°. One gram dissolves in ~3500 ml water, 440 ml alcohol, 65 ml acetone. Freely sol in dil mineral acids and in aq solns of KOH and NaOH. More sol in warm sugar solns than in water alone. The aq soln is neutral. LD_{50} orally in mice: 7.5 mg/g (Wien).
Sodium salt monohydrate, $C_{11}H_{10}N_3NaO_2S.H_2O$, *soluble sulfapyridine, Izopiridina, Soludagenan.* On prolonged exposure to humid air it absorbs CO_2, liberates sulfapyridine, becomes incompletely sol in water. One gram dissolves in ~1.5 ml water, 10 ml alcohol. pH of 5% aq soln: 11.4. LD_{50} orally in mice: 2.7 mg/g (Wien).

THERAP CAT: In treatment of dermatitis herpetiformis; antibacterial.
THERAP CAT (VET): Antibacterial.

9109. Sulfaquinoxaline. *4-Amino-N-2-quinoxalinylbenzenesulfonamide; N¹-(2-quinoxalinyl)sulfanilamide;* 2-sulfanilamidoquinoxaline; N¹-(2-quinoxalyl)sulfanilamide; sulfabenzpyrazine; Compd 3-120; S.Q.; Sulquin. $C_{14}H_{12}N_4O_2S$; mol wt 300.34. C 55.99%, H 4.03%, N 18.65%, O 10.65%, S 10.68%. Prepd by treating 2-aminoquinoxaline with acetylsulfanilyl chloride in the presence of pyridine and hydrolyzing the resulting acetyl deriv: Weijlard et al., *J. Am. Chem. Soc.* **66**, 1957 (1944); Weijlard, Tishler, U.S. pat. 2,404,199 (1946 to Merck & Co.). Anticoccidial spectrum: G. F. Mathis et al., *Poultry Sci.* **63**, 1149 (1984). Evaluation of antibacterial and anticoccidial efficacy of mixture with trimethoprim: G. White, R. B. Williams, *Vet. Rec.* **113**, 608 (1983); D. W. T. Piercy et al., *ibid.* **114**, 60 (1984). Drug residue study in poultry muscle and liver: M. Patthy, J. *Chromatog.* **275**, 115 (1983). HPLC determn in rabbit plasma and urine: J. G. Eppel, J. J. Thiessen, *J. Pharm. Sci.* **73**, 1635 (1984).

Minute crystals, mp 247-248°. uv max (pH 6.6 in H_2O): 252, 360 nm ($E_{1cm}^{1\%}$ 1110, 275). Solubility in water at pH 7: 0.75 mg/100 ml; in 95% alcohol: 73 mg/100 ml; in acetone: 430 mg/100 ml. Sol in aq Na_2CO_3 and NaOH solns.
Sodium salt, $C_{14}H_{11}N_4NaO_2S$, *Aviochina*. Very sol in water; pH of 1% soln ~10. The amorphous salt is deliquescent and absorbs CO_2 which liberates the practically insol sulfaquinoxaline.
THERAP CAT (VET): Coccidiostat.

9110. Sulfarside. *[2-Amino-4-(aminosulfonyl)phenyl]arsinous acid; 4-sulfamoyl-o-arsanilic acid;* 2-amino-4-aminosulfonylphenylarsinic acid; RP-4482; Bemarside. C_6H_9-AsN_2O_5S; mol wt 296.14. C 24.34%, H 3.06%, As 25.30%, N 9.46%, O 27.01%, S 10.83%. Prepn: Trefouel, U.S. pat. 2,616,913 (1953 to Rhône-Poulenc).

THERAP CAT: Sodium salt as antiamebic.

9111. Sulfarsphenamine. *[1,2-Diarsenediylbis[(6-hydroxy-3,1-phenylene)imino]]bismethanesulfonic acid disodium salt;* disodium 3,3'-diamino-4,4'-dihydroxyarsenobenzene N-dimethylenesulfonate; sulfarsenobenzene; Karsulphan; Myosalvarsan; Myarsenol; Metarsenobillon; Thiosarmine. $C_{14}H_{14}As_2N_2Na_2O_8S_2$; mol wt 598.23. C 28.11%, H 2.36%, As 25.05%, N 4.68%, Na 7.69%, O 21.40%, S 10.72%. Prepn: C. Voegtlin, J. M. Johnson, *J. Am. Chem. Soc.* **44**, 2573 (1922); W. J. C. Dyke, H. King, *J. Chem. Soc.* **1935**, 1745.

Yellow, odorless, or almost odorless powder. Very sol in water, slightly in alcohol. The aq soln is usually slightly acid. *Keep in evacuated or in inert gas-filled ampuls.*

THERAP CAT: Formerly as antisyphilitic.

THERAP CAT (VET): Has been used as antibacterial.

9112. Sulfasalazine. *2-Hydroxy-5-[[4-[(2-pyridinylamino)sulfonyl]phenyl]azo]benzoic acid; 5-[p-(2-pyridylsulfamoyl)phenylazo]salicylic acid;* salazosulfapyridine; 5-[4-(2-pyridylsulfamoyl)phenylazo]-2-hydroxybenzoic acid; 4-(pyridyl-2-amidosulfonyl)-3'-carboxy-4'-hydroxyazobenzene; salicylazosulfapyridine; sulphasalazine; Azulfidine; Colo-Pleon; Salazopyrin. $C_{18}H_{14}N_4O_5S$; mol wt 398.40. C 54.27%, H 3.54%, N 14.06%, O 20.08%, S 8.05%. Conjugate of 5-aminosalicylic acid and sulfapyridine, *q.v.* Prepn and bactericidal effect: E. E. A. Askelof *et al.*, U.S. pat. 2,396,145 (1946 to Pharmacia); Doraswamy, Guha, *J. Indian Chem. Soc.* **23**, 278 (1946). Physicochemical properties: B. Nygard *et al.*, *Acta Pharm. Suecica* **3**, 313-342 (1966). Metabolism studies: H. Schroder, D. E. S. Campbell, *Clin. Pharmacol. Ther.* **13**, 539 (1972); M. A. Peppercorn, P. Goldman, *J. Pharmacol. Exp. Ther.* **181**, 555 (1972). HPLC determn in biological fluids: R. A. Hogezand *et al.*, *J. Chromatog.* **305**, 470 (1984). Clinical trial in rheumatoid arthritis: D. E. Bax, R. S. Amos, *Ann. Rheum. Dis.* **44**, 194 (1985). Review of clinical pharmacokinetics: U. Klotz, *Clin. Pharmacokinet.* **10**, 285-302 (1985); of pharmacology and clinical use in inflammatory bowel disease: M. A. Peppercorn, *Ann. Intern. Med.* **3**, 377 (1984); G. Watkinson, *Drugs* **32**, Suppl. 1, 1-11 (1986). Comprehensive description: J. P. McDonnell, *Anal. Profiles Drug Subs.* **5**, 515-532 (1976).

Minute, brownish-yellow crystals, dec 240-245°. uv max: 237 ($E_{1cm}^{1\%}$ ~658) and 359 nm. Slightly sol in alc. Practically insol in water, benzene, chloroform, ether.

THERAP CAT: Treatment of ulcerative colitis and Crohn's disease.

THERAP CAT (VET): Has been used in granulomatous colitis.

9113. Sulfasomizole. *4-Amino-N-(3-methyl-5-isothiazolyl)benzenesulfonamide; 3-methyl-5-sulfanilamidoisothiazole;* 5-p-aminobenzenesulfonamido-3-methylisothiazole; 5-sulfanilamido-3-methylisothiazole; *N*^{1}-(3-methylisothiazol-5-yl)sulfanilamide; sulphasomizole; Bidizole; Amidozol. $C_{10}H_{11}N_3O_2S_2$; mol wt 269.35. C 44.59%, H 4.12%, N 15.60%, O 11.88%, S 23.81%. Preparation from *N*-acetylsulfanilyl chloride and 5-amino-3-methylisothiazole in pyridine followed by deacetylation in 2N NaOH: Adams, Slack, *J. Chem. Soc.* **1959**, 3070; Adams *et al.*, *Nature* **186**, 221 (1960); **Brit.** pat. 835,753 (1960 to May & Baker).

Cream-colored needles from H_2O, mp 192-192.5°. Soly in H_2O at pH 6.0: 248 mg/100 ml; at pH 7.0: 2.26 g/100 ml. Sodium salt monohydrate, $C_{10}H_{10}N_3NaO_2S_2.H_2O$, crystals, when anhydr dec 345°. Freely sol in water. Aq solns of almost neutral pH contg more than 50% w/v of the sodium salt can be made.

THERAP CAT: Antibacterial.

9114. Sulfasymazine. *4-Amino-N-(4,6-diethyl-1,3,5-triazin-2-yl)benzenesulfonamide; N*^{1}-(4,6-diethyl-s-triazin-2-yl)sulfanilamide; 2-sulfanilamido-4,6-diethyl-1,3,5-triazine; 4,6-diethyl-2-sulfanilamido-1,3,5-triazine; sulphsymazine; Symasul; Prosymasul. $C_{13}H_{17}N_5O_2S$; mol wt 307.38. C 50.80%, H 5.57%, N 22.78%, O 10.41%, S 10.43%. Prepn by nucleophilic displacement of methoxy groups from 2-methoxy-4,6-disubstituted *s*-triazines with sulfanilamide anion: Taft *et al.*, *J. Med. Chem.* **8**, 784 (1965); Bader, U.S. pat. 3,344,137 (1967 to Am. Cyanamid). Pharmacological stud-

ies: Frisk, Hultman, *Antimicrob. Ag. Chemother.* **1965**, 672; Kruger-Thiemer *et al.*, *Chemotherapia* **10**, 325 (1966).

White precipitate from ethanol, mp 186.5-187.5° (Taft); 190-190.5° (Bader). Tasteless. pKa (30% acetone) = 5.4. Sol in water, acetate buffer (1 mg/ml at pH 5.9); also sol in human urine at 37° (2 mg/ml at pH 6; >10 mg/ml at pH 7).

THERAP CAT: Antibacterial.

9115. Sulfathiazole. *4-Amino-N-2-thiazolylbenzenesulfonamide; N*^{1}-2-thiazolylsulfanilamide;* 2-sulfanilamidothiazole; 2-(sulfanilylamino)thiazole; 2-(p-aminobenzenesulfonamido)thiazole; norsulfazole; RP-2090; M & B 760; Thiazamide; Cibazol; Enterobiocine; Duatok; Sulfamul; Sulfavitina; Sulzol. $C_9H_9N_3O_2S_2$; mol wt 255.32. C 42.34%, H 3.55%, N 16.46%, O 12.53%, S 25.12%. Prepn: R. J. Fosbinder, L. A. Walter, *J. Am. Chem. Soc.* **61**, 2032 (1939); W. A. Lott, F. H. Bergeim, *ibid.* 3593; G. Newbery, P. Viaud, **Brit.** pat. **517,272**; G. Newbery, U.S. pat. **2,362,087** (1940, 1944 both to May & Baker). Antibacterial activity *in vivo*: O. W. Barlow, E. Homburger, *Proc. Soc. Exp. Biol. Med.* **42**, 792 (1939). Toxicology: H. B. van Dyke *et al.*, *ibid.* 410; P. H. Long *et al.*, *ibid.* **43**, 328 (1940). Review of early literature: E. H. Northey, *ACS Monograph Series* **106**, entitled "Sulfonamides and Allied Compounds" (Reinhold, New York, 1948) 660 pp. HPLC determn in milk: V. K. Agarwal, *J. Liq. Chromatog.* **16**, 3793 (1993). Oral bioavailability in ruminants: K. Weijkamp *et al.*, *Vet. Quart.* **16**, 33 (1994).

mp 202-202.5°. pKa 7.2. Soly at 26° (mg/100 ml): water 60 (pH 6.03); alcohol 525. Sol in acetone, dil mineral acids, KOH and NaOH solns, ammonia water.

Sodium salt sesquihydrate, $C_9H_8N_3NaO_2S_2.1\frac{1}{2}H_2O$, *soluble sulfathiazole*. Crystals or white powder or granules. Also occurs as the monohydrate and pentahydrate. One gram dissolves in ~2.5 ml water, in ~15 ml alc. pH of 1% aq soln 9.35; of 10% soln 10.2. LD_{50} s.c. in mice: 1.45 g/kg (van Dyke); also reported as 1.95 g/kg (Long).

Polymer with formaldehyde, *formosulfathiazole, formaldehyde-sulfathiazole*, Forbina, Formo-Cibazol, Socatil. ($C_9H_9N_3O_2S_2.CH_2O)_x$. Contains ~11% formaldehyde. Prepn: J. Druey, A. Becker, *Helv. Chim. Acta* **31**, 2184 (1948). Intrauterine use in cows: D. P. Dobson, D. E. Noakes, *Vet. Rec.* **127**, 128 (1990). Amorphous powder. Practically insol in water.

THERAP CAT: Antibacterial.

THERAP CAT (VET): Antibacterial.

9116. Sulfathiourea. *4-Amino-N-(aminothioxomethyl)-benzenesulfonamide; 1-sulfanilyl-2-thiourea;* sulfathiocarbamide; RP-2255; Badional; Fontamide. $C_7H_9N_3O_2S_2$; mol wt 231.30. C 36.35%, H 3.92%, N 18.17%, O 13.83%, S 27.73%. Prepd by reacting acetylsulfanilylcyanamide with hydrogen sulfide. Has also been prepd directly from sulfanilyl cyanamide and ammonium sulfide: Leitch *et al.*, *Can. J. Res.* **23B**, 139 (1945); alternate method: Földi *et al.*, U.S. pat. 2,332,906 (1943).

Crystals, dec 171.5-172°. Soly in water at 37°: 1.1 g/100 ml.

Sodium salt, $C_7H_8N_3NaO_2S$, dec 245-245.5°. Very sol in water. pH of 1% aq soln: 7.

Water sol form, *Solufontamide*, marketed as a slightly viscous soln contg 33% of 1-sulfanilyl-2-thiourea. d_4^{15} 1.16. pH 6.8-7.2.

THERAP CAT: Antibacterial.

9117. Sulfatolamide. *4-Amino-N-(aminothioxomethyl)-benzenesulfonamide compd with 4-(aminomethyl)benzenesulfonamide (1:1); 1-sulfanilyl-2-thiourea compd with α-amino-p-toluenesulfonamide;* α-amino-*p*-toluenesulfonamide salt with 1-sulfanilyl-2-thiourea; 4-homosulfanilamide salt with 1-sulfanilyl-2-thiourea; 1-sulfanilylthiocarbamide salt with 4-aminomethylbenzenesulfonamide; Marbadal. $C_{14}H_{19}N_5O_4S_3$; mol wt 417.53. C 40.27%, H 4.59%, N 16.77%, O 15.33%, S 23.04%. Prepn: Behnisch, Klarer, U.S. pat. **2,696,454** (1954 to Schenley Ind.).

Crystals, mp 179-181°. Slightly sol in water (to 0.78%) with neutral reaction. Freely sol in dil hydrochloric acid; sol in dil sodium hydroxide and ammonia; slightly sol in alcohol and acetone. Practically insol in ether.

THERAP CAT: Antibacterial.

9118. Sulfazamet. *4-Amino-N-(3-methyl-1-phenyl-1H-pyrazol-5-yl)benzenesulfonamide; N^1-(3-methyl-1-phenylpyrazol-5-yl)sulfanilamide;* 5-(*p*-aminobenzenesulfonamido)-3-methyl-1-phenylpyrazole; 3-(*p*-aminobenzenesulfonamido)-2-phenyl-5-methylpyrazole; 3-methyl-1-phenyl-5-(sulfanilamido)pyrazole; sulfamethylphenazole; sulfapyrazole; Vesulong. $C_{16}H_{16}N_4O_2S$; mol wt 328.39. C 58.52%, H 4.91%, N 17.06%, O 9.74%, S 9.76%. Prepn: Crippa, Guarneri, *Gazz. Chim. Ital.* **85**, 199 (1955); **Brit. pat. 848,627** (1960 to Ciba); Seydel, Krueger-Theimer, *Arzneimittel-Forsch.* **14**, 1294 (1964). Amino-imino tautomerism: *eidem, ibid.*

Crystals from alc, mp 195° (Seydel, Krueger-Theimer, *loc. cit.*); mp 181-182° (**Brit. pat.** *loc. cit.*). pK'a 5.69.

THERAP CAT (VET): Antibacterial.

9119. Sulfazecin. *(R)-D-γ-Glutamyl-N-(3-methoxy-2-oxo-1-sulfo-3-azetidinyl)-D-alaninamide;* 3-γ-D-glutamyl-D-alanylamino-3-methoxyazetidin-2-one-1-sulfonic acid; (3R)-3-(γ-D-glutamyl-D-alanylamino)-3-methoxy-2-oxoazetidine-1-sulfonic acid; antibiotic G-6302. $C_{12}H_{20}N_4O_9S$; mol wt 396.38. C 36.36%, H 5.09%, N 14.13%, O 36.33%, S 8.09%. Novel monocyclic β-lactam (monobactam) antibiotic, produced by *Pseudomonas acidophila* and *P. mesoacidophila*. Isoln, description of physico-chemical and bacteriostatic properties: A. Imada *et al.*, **Ger. pat. 2,855,949** corresp to U.S. pat **4,229,436** (1979, 1980 both to Takeda); *eidem, Nature* **289**, 590 (1981); M. Asai *et al., J. Antibiot.* **34**, 621 (1981). X-ray crystallographic structure determn: K.

Kamiya *et al., Acta Crystallogr.* **B37**, 1626 (1981). Prepn of fluorinated sulfazecin analogs: K.Yoshioka *et al., J. Org. Chem.* **49**, 1427 (1984); P. F. Bevilacqua *et al., ibid.* 1430.

Crystallizes from methanol/water as the alcoholate-hemihydrate; colorless needles, mp 168-170°. $[\alpha]_D^{25}$ +82° (c = 1.0 in water). pKa: 3.4 (COO−); 9.2 (NH_3^+). Sol in water, DMF, DMSO. Slightly sol in methanol, THF. Practically insol in ethanol, acetone, ethyl acetate, chloroform and other organic solvents. Relatively stable in neutral and weakly acidic solns; unstable in alkaline and strongly acidic solns.

Sodium salt monohydrate, $C_{12}H_{19}N_4NaO_9S.H_2O$, colorless powder, browns at 170°, no sharp mp. $[\alpha]_D^{20}$ +85° (c = 0.37 in water).

9120. Sulfinalol. *4-Hydroxy-α-[[[3-(4-methoxyphenyl)-1-methylpropyl]amino]methyl]-3-(methylsulfinyl)benzenemethanol;* 4-hydroxy-α-[[[3-(*p*-methoxyphenyl)-1-methylpropyl]amino]methyl]-3-(methylsulfinyl)benzyl alcohol. $C_{20}H_{27}NO_4S$; mol wt 377.50. C 63.63%, H 7.21%, N 3.71%, O 16.95%, S 8.49%. β-Adrenergic blocker. Prepn: R. E. Philion, **Ger. pat. 2,728,641** (1978 to Sterling), *C.A.* **90**, 137642 (1979). Pharmacological activity profile: P. H. Hernández, *Fed. Proc.* **38**, 738 (1979). Chromatographic determn in human plasma and urine: G. B. Park *et al., Anal. Chem.* **53**, 604 (1981). Metabolism, disposition in animals: D. P. Benziger *et al., Drug. Metab. Dispos.* **9**, 493 (1981).

Hydrochloride, $C_{20}H_{27}NO_4S.HCl$, *Win-40808-7, Perifadil*. Crystals, mp 172-175°.

THERAP CAT: Antihypertensive.

9121. Sulfinpyrazone. *1,2-Diphenyl-4-[2-(phenylsulfinyl)ethyl]-3,5-pyrazolidinedione;* 1,2-diphenyl-3,5-dioxo-4-(2-phenylsulfinylethyl)pyrazolidine; 4-(phenylsulfoxyethyl)-1,2-diphenyl-3,5-pyrazolidinedione; 4-(2-benzenesulfinylethyl)-1,2-diphenylpyrazolidine-3,5-dione; sulfoxyphenylpyrazolidine; G-28315; Anturan; Anturane; Anturano; Enturen. $C_{23}H_{20}N_2O_3S$; mol wt 404.49. C 68.30%, H 4.98%, N 6.93%, O 11.87%, S 7.93%. Prepn: F. Häfliger, U.S. pat. **2,700,671** (1955 to Geigy AG); R. Pfister, F. Häfliger, *Helv. Chim. Acta* **44**, 232 (1961). Uricosuric effect: J. J. Burns *et al., J. Pharmacol. Exp. Ther.* **119**, 418 (1957). Clinical trial in prevention of sudden cardiac death: Anturane Reinfarction Trial Research Group, *N. Engl. J. Med.* **298**, 289 (1978); **302**, 250 (1980); as antithrombotic: R. D. Sautter *et al., J. Am. Med. Assoc.* **250**, 2649 (1983); D. T. Domoto *et al., Thromb. Res.* **62**, 737 (1991). *Review:* E. H. Margulies, A. M. White, in *Pharmacological and Biochemical Properties of Drug Substances* vol. 2, M. E. Goldberg, Ed. (Am. Pharm. Assoc., Washington, DC, 1979) pp 255-278; E. H. Margulies *et al., Drugs* **20**, 179-197 (1980).

Crystals from chloroform + heptane, mp 136-137°. Stable to light and air. uv max (1.0N NaOH): 255 nm. Sol in ethyl acetate, chloroform. Slightly sol in water, alcohol, ether, mineral oils, fats.

d-Form, crystals from ethanol, mp 130-133°. $[\alpha]_D^{22}$ +67.1° (c = 2.04 in ethanol); $[\alpha]_D^{25}$ +109.3° (c = 0.5 in CHCl$_3$).

l-Form, crystals, mp 130-133°. $[\alpha]_D^{23}$ −64.2° (c = 2.14 in ethanol); $[\alpha]_D^{26}$ −104.5° (c = 0.5 in CHCl$_3$).

THERAP CAT: Uricosuric; antithrombotic.

9122. 4,4′-Sulfinyldianiline. *4,4′-Sulfinylbisbenzeneamine;* di(*p*-aminophenyl) sulfoxide; 4,4′-diaminodiphenyl sulfoxide; bis(*p*-aminophenyl) sulfoxide; Medapsol. C$_{12}$H$_{12}$-N$_2$OS; mol wt 232.31. C 62.04%, H 5.21%, N 12.06%, O 6.89%, S 13.80%. Analog of dapsone, *q.v.* Prepn: Gazdar, Smiles, *J. Chem. Soc.* **93,** 1835 (1908); **Brit. pat.** 509,415 (1939 to I. G. Farbenind.), *C.A.* **34,** 3764 (1940); Berti *et al., Rev. Brasil. Leprol.* **20,** 104 (1952), *C.A.* **50,** 11520 (1956). *In vivo* activity vs *Mycobacterium leprae:* L. Levy, *Antimicrob. Ag. Chemother.* **14,** 791 (1978).

Prisms from alc or water, mp 175°. Slightly sol in water.

THERAP CAT: Antibacterial (leprostatic).

9123. Sulfiram. *Tetraethylthiodicarbonic diamide; bis-(diethylthiocarbamoyl)sulfide;* tetraethylthiuram monosulfide; monosulfiram; TTMS; Kutkasin; Tetmosol. C$_{10}$H$_{20}$-N$_2$S$_3$; mol wt 264.48. C 45.41%, H 7.62%, N 10.59%, S 36.37%. Prepn: Davies, Sexton, *Biochem. J.* **40,** 331 (1946); Ritter, U.S. pat. **2,524,081** (1950 to Sharples Chem.); Sahasrabudhey, Radhakrishnan, *J. Indian Chem. Soc.* **31,** 853 (1954); Zbirovsky, Ettel, *Chem. Listy* **51,** 2094 (1957); **52,** 95 (1958).

Crystals from benzene + petr ether, mp 30-32°.

USE: As vulcanization agent.

THERAP CAT: Ectoparasiticide.

THERAP CAT (VET): Ectoparasiticide.

9124. Sulfisomidine. *4-Amino-N-(2,6-dimethyl-4-pyrimidinyl)benzenesulfonamide; N^1-(2,6-dimethyl-4-pyrimidinyl)sulfanilamide;* 2,6-dimethyl-4-sulfanilamidopyrimidine; 6-sulfanilamido-2,4-dimethylpyrimidine; 6-(sulfanilamido)-2,4-dimethyl-1,3-diazine; 6-(*p*-aminophenylsulfonamido)-2,4-dimethylpyrimidine; 6-(*p*-aminobenzenesulfonyl)amino-2,4-dimethylpyrimidine; sulfadimetine; sulfaisodimidine; sulphasomidine; Elkosin; Elcosine; Elkosil; Domian; Aristamid. C$_{12}$H$_{14}$N$_4$O$_2$S; mol wt 278.33. C 51.78%, H 5.07%, N 20.13%, O 11.50%, S 11.52%. Prepn: **Fr. pat. 886,009** (1943 to Nordmark); Gysin, U.S. pat. **2,351,333** (1944 to Geigy); Hartmann *et al.,* U.S. pats. **2,386,852; 2,429,184** (1945, 1947, both to Ciba); Matsukawa *et al., J. Pharm. Soc. Japan* **70,** 283 (1950); Loop, Lührs, *Ann.* **580,** 225 (1953).

Needles from ethanol, mp 243°. Soly in water at 15°: 0.12 g/100 ml, at 30°: 0.30 g/100 ml. More sol in hot water (about 1:60). Aq solns are neutral to litmus. Soly in urine at 37°: 360 mg/100 ml at pH 5.5 to 1100 mg/100 ml at pH 7.5. Slightly sol in alcohol, acetone. Practically insol in benzene, ether, chloroform. Freely sol in dil HCl and NaOH. Compared with sulfanilamide it is but slightly acetylated in the body. *Do not confuse with sulfamethazine.*

THERAP CAT: Antibacterial.

THERAP CAT (VET): Antimicrobial.

9125. Sulfisoxazole. *4-Amino-N-(3,4-dimethyl-5-isoxazolyl)benzenesulfonamide; N^1-(3,4-dimethyl-5-isoxazolyl)-sulfanilamide;* 3,4-dimethyl-5-sulfanilamidoisoxazole; 5-(4-aminophenylsulfonamido)-3,4-dimethylisoxazole; 5-(*p*-aminobenzenesulfonamido)-3,4-dimethylisooxazole; sulfafurazole; sulphafurazole; Gantrisin; Sosol; Soxisol; Soxomide; Sulfalar; Sulfazin; Sulfoxol; Sulsoxin. C$_{11}$H$_{13}$N$_3$O$_3$S; mol wt 267.31. C 49.43%, H 4.90%, N 15.72%, O 17.96%, S 12.00%. Prepn: H. M. Wuest, M. Hoffer, U.S. pat. **2,430,-094** (1947 to Hoffmann-La Roche). Toxicity study: Seki *et al., Arzneimittel-Forsch.* **15,** 1441 (1965). Comprehensive description: B. C. Rudy, B. Z. Senkowski, *Anal. Profiles Drug Subs.* **2,** 487-506 (1973). HPLC determn in biological fluids: D. Jung, S. Oie, *Clin. Chem.* **26,** 51 (1980). Clinical trial in otitis media: P. A. M. Bernard *et al., Pediatrics* **88,** 215 (1991).

White to slightly yellowish crystalline powder. Bitter taste. mp 194°. pKa 5. Soly in water (25°): 0.13 mg/ml. Sol in alcohol. LD$_{50}$ orally in mice: 6800 mg/kg (Seki).

Diethanolamine salt, C$_{11}$H$_{13}$N$_3$O$_3$S.C$_4$H$_{11}$NO$_2$, *sulfisoxazole diolamine, Suladrin, Sulfium.* White to off-white, odorless, crystalline powder. Sol in alcohol; freely sol in water. A 4% soln is about isotonic with tears.

Acetyl sulfisoxazole, C$_{13}$H$_{15}$N$_3$O$_4$S, *N-[(4-Aminophenyl)-sulfonyl]-N-(3,4-dimethyl-5-isoxazolyl)acetamide, N^1-monoacetyl sulfisoxazole.* Prepn: M. Hoffer, U.S. pat. **2,721,200** (1955 to Hoffmann-La Roche). Should not be confused with N^4-acetyl sulfisoxazole which is a metabolite. Tasteless crystals, mp 193-194°. Soly (mg/ml): 0.07 in water; 4.93 in methanol; 5.7 in 95% ethanol; 0.94 in ether; 29.0 in chloroform.

THERAP CAT: Antibacterial.

THERAP CAT (VET): Antibacterial.

9126. Sulfoacetic Acid. Sulfoethanoic acid. C$_2$H$_4$O$_5$S; mol wt 140.12. C 17.14%, H 2.88%, O 57.09%, S 22.89%. HO$_3$SCH$_2$COOH.

Monohydrate, hygroscopic crystals. mp 84-86° when anhydr. bp 245° with decompn. Sol in water, alcohol; insol in ether, chloroform. *Keep well closed.* LD$_{50}$ orally in rats: 3.16 g/kg, Smyth *et al., J. Ind. Hyg. Toxicol.* **31,** 60 (1949).

9127. Sulfobromophthalein Sodium. *3,3′-(4,5,6,7-Tetrabromo-3-oxo-1(3H)-isobenzofuranylidene)bis[6-hydroxybenzenesulfonic acid] disodium salt;* disodium phenoltetrabromophthalein sulfonate; bromosulfophthalein; BSP; Bromsulphalein; Bromthalein; Hepartest. C$_{20}$H$_8$Br$_4$Na$_2$O$_{10}$-S$_2$; mol wt 838.01. C 28.67%, H 0.96%, Br 38.14%, Na 5.49%, O 19.09%, S 7.65%. Prepd by the condensation of phenol and tetrabromophthalic acid or its anhydride, followed by sulfonation and conversion to the disodium salt.

Pharmacology and use as liver function test: S. M. Rosenthal, E. C. White, *J. Pharmacol.* **24**, 265 (1924); W. Häcki *et al., J. Lab. Clin. Med.* **88**, 1019 (1976).

Hygroscopic crystals. Bitter taste. Sol in water. Insol in alcohol, acetone. Alkaline solns have an intense bluish-purple color.

THERAP CAT: Diagnostic aid (hepatic function).

THERAP CAT (VET): Diagnostic aid (hepatic function).

9128. Sulfolane. *Tetrahydrothiophene 1,1-dioxide;* tetrahydrothiophene 1-dioxide; thiophan sulfone. $C_4H_8O_2S$; mol wt 120.17. C 39.98%, H 6.71%, O 26.63%, S 26.68%. Prepn by catalytic hydrogenation of sulfolene oxides: Mahan, Fauske, **U.S.** pat. **2,578,565** (1951 to Phillips Petroleum); from butadiene and sulfur dioxide: Staaterman *et al., Chem. Eng. Progr.* **43**, 148 (1947); Evans, Morris, **U.S.** pat. **2,360,859** (1944 to Shell). Toxicity study: H. F. Smyth *et al., Am. Ind. Hyg. Assoc. J.* **30**, 470 (1962).

Liquid, d_4^{30} 1.2606. mp 27.4-27.8°. bp$_{760}$ 285°. n_D^{30} 1.481. Viscosity at 30° = 10.34 cp. Flash pt, open cup: 350°F (176°C). At 30°, miscible with water, acetone, toluene; partially misc with octanes, olefins and naphthenes. LD$_{50}$ orally in rats: 1.54 ml/kg (Smyth).

USE: Selective solvent for liquid-vapor extractions.

9129. 3-Sulfolene. *2,5-Dihydrothiophene 1,1-dioxide;* 1-thia-3-cyclopentene 1,1-dioxide; butadiene sulfone. $C_4H_6O_2$-S; mol wt 118.16. C 40.66%, H 5.12%, O 27.08%, S 27.14%. Preparation: Staudinger, Rizenthaler, *Ber.* **68**, 455 (1935); de Roy van Zuydewijn, *Rec. Trav. Chim.* **56**, 1047 (1937); Hooker *et al.,* **U.S.** pat. **2,395,050** (1946 to Dow); Morris, Finch, **U.S.** pat. **2,420,834** (1947 to Shell).

Crystals, mp 64-65.5°. Sol in water, organic solvents.

9130. Sulfometuron Methyl. *2-[[[[(4,6-Dimethyl-2-pyrimidinyl)amino]carbonyl]amino]sulfonyl]benzoic acid methyl ester;* N-[(4,6-dimethylpyrimidin-2-yl)aminocarbonyl]-2-methoxycarbonylbenzenesulfonamide; 2-[3-(4,6-dimethylpyrimidin-2-yl)ureidosulfonyl]benzoic acid methyl ester; Aa-5648; DPX-5648; Oust. $C_{15}H_{16}N_4O_5S$; mol wt 364.38. C 49.44%, H 4.43%, N 15.38%, O 21.95%, S 8.80%. Nonselective pre- and post-emergence sulfonylurea herbicide for noncropland use; inhibits branched chain amino acid biosynthesis. Prepn: G. Levitt, **Eur.** pat. **Appl. 7,687;** *idem,* **U.S.** pat. **4,394,506** (1980, 1983 both to Du Pont); *idem, Pestic. Chem.: Hum. Welfare Environ., Proc. 5th Int. Congr. Pestic.* **1**, 243 (1983). Herbicidal activity and toxicity: R. H. Harding *et al., Proc. West. Soc. Weed Sci.* **34**, 120 (1981). Mechanism of action: T. B. Ray, *Proc. Brit. Crop Prot. Conf. - Weeds* **1985**, 131. HPLC determn in soil and water: E. W. Zahnow, *J. Agr. Food Chem.* **33**, 479 (1985). Degradation and bioaccumulation studies: J. Harvey *et al., ibid.* 590; J. J. Anderson, J. J. Dulka, *ibid.* 596. Field trials: R. L. Atkins *et al., Proc. South. Weed Sci. Soc.* **36**, 300

(1983); in afforested areas: A. Nir, Z. Arenstein, *Proc. Brit. Crop Prot. Conf. - Weeds* **1987**, 787.

White solid, mp 198-202°. Soly in water at 25° (ppm): 10 at pH 5, 300 at pH 7. pKa 5.7. LD$_{50}$ in male, female rats (mg/kg): > 5000, > 5000 orally (Harding).

USE: Herbicide.

9131. Sulfonethylmethane. *2,2-Bis(ethylsulfonyl)butane;* diethylsulfonmethylethylmethane; methylsulfonal; Trional. $C_8H_{18}O_4S_2$; mol wt 242.36. C 39.65%, H 7.49%, O 26.41%, S 26.46%. Prepd by condensing ethylmercaptan with methyl ethyl ketone and oxidizing the resulting mercaptol with KMnO$_4$: Fromm, *Ann.* **253**, 135 (1889); *cf.* Baumann, *Ber.* **18**, 883 (1885); *Arch. Pharm.* **226**, 511 (1888); Bayer, *Pharm. Z.* **34**, 98 (1889); **Ger.** pat. **46,333;** Moragas, Uthoff, *Chim. Ind. (Paris)* **17**, 284 (1927); Ramberg, Samén, *C.A.* **28**, 6103 (1934); Suter, *Organic Chemistry of Sulfur* (New York, 1944) pp 735, 742.

Lustrous leaflets, scales from water. Bitter taste. mp 74-76°. Upon further heating dec with evolution of SO$_2$. One gram dissolves in 200 ml cold, about 30 ml boiling water, in 8 ml alc. Sol in ether. A satd aq soln is neutral to litmus.
Caution: May be habit forming. This is a controlled substance (depressant) listed in the U.S. Code of Federal Regulations, Title 21 Parts 329.1 and 1308.13 (1995).

THERAP CAT: Sedative, hypnotic.

THERAP CAT (VET): Has been used as a hypnotic.

9132. Sulfoniazide. *4-Pyridinecarboxylic acid [(3-sulfophenyl)methylene]hydrazide; isonicotinic acid m-sulfobenzylidene hydrazide;* isonicotinyl hydrazonotoluene-*m*-sulfonic acid; G-605. $C_{13}H_{11}N_3O_4S$; mol wt 305.31. C 51.14%, H 3.63%, N 13.76%, O 20.96%, S 10.50%. Prepn: Girard, **U.S.** pat. **2,727,041** (1955 to Lab. Français Chimiother.); Sugimoto, **Japan.** pat. **57 1117** (1957 to Tanake), *C.A.* **52**, 4696e (1958).

Needles, dec 250-253°. Slightly sol in water.
Sodium salt trihydrate, $C_{13}H_{10}N_3NaO_4S.3H_2O$, *Sulfon-Niazone.* Crysts from water (at 18°): 1 g/33 cc.

THERAP CAT: Antibacterial (tuberculostatic).

9133. Sulfonmethane. *2,2-Bis(ethylsulfonyl)propane;* diethylsulfondimethylmethane; propane-diethyl sulfone; sulfonal. $C_7H_{16}O_4S_2$; mol wt 228.33. C 36.82%, H 7.06%, O 28.03%, S 28.09%. Prepd by condensing ethyl mercaptan with acetone, then oxidizing with KMnO$_4$: Fromm, *Ann.* **253**, 140 (1889); *cf.* Baumann, *Ber.* **18**, 883 (1885); *Arch. Pharm.* **226**, 511 (1888); Bayer, *Pharm. Z.* **34**, 98 (1889); **Ger.** pat. **46,333;** Moragas, Uthoff, *Chim. Ind. (Paris)* **17**, 284 (1927); Ramberg, Samén, *C.A.* **28**, 6103 (1934); Suter, *Organic Chemistry of Sulfur* (New York, 1944) pp 735, 742. Soly data: Falck, *J. Pharm. Chim.* [7] **21**, 279 (1920).

Crystals, almost tasteless, mp 124-126°. bp 300°. One gram dissolves in 365 ml water, 16 ml boiling water, 60 ml alcohol, 3 ml boiling alcohol, 64 ml ether, 11 ml chloroform. Soluble in benzene. Insol in glycerol.

Note: This is a controlled substance (depressant) listed in the U.S. Code of Federal Regulations, Title 21 Part 1308.13 (1995).

THERAP CAT: Hypnotic.

THERAP CAT (VET): Has been used as a hypnotic.

9134. Sulfonyldiacetic Acid. Sulfonediacetic acid; dimethylenesulfone-α,α'-dicarboxylic acid. $C_4H_6O_6S$; mol wt 182.15. C 26.38%, H 3.32%, O 52.70%, S 17.60%. HOOC-$CH_2SO_2CH_2COOH$. Prepd by oxidation of thiodiglycolic acid: Lovén, *Ber.* **17**, 2818 (1884).

Tabular crystals, mp 182°. Freely sol in water. Sol in alc. Slightly sol in ether.

USE: Detection of barium, lead, mercury and silver.

9135. Sulforaphane. *(R)-1-Isothiocyanato-4-(methylsulfinyl)butane;* 4-methylsulfinylbutyl isothiocyanate. C_6H_{11}-NOS_2; mol wt 177.29. C 40.65%, H 6.25%, N 7.90%, O 9.02%, S 36.17%. Potent, selective inducer of phase II detoxification enzymes with anticarcinogenic properties. Occurs naturally in broccoli. Prepn: H. Schmid, P. Karrer, *Helv. Chim. Acta* **31**, 1497 (1948). Absolute configuration: K. Mislow *et al., J. Am. Chem. Soc.* **87**, 665 (1965). Isoln from SAGA broccoli and enzyme inducing activities: Y. Zhang *et al., Proc. Nat. Acad. Sci. USA* **89**, 2399 (1992). Asymmetric synthesis: J. K. Whitesell, M.-S. Wong, *J. Org. Chem.* **59**, 597 (1994). *In vivo* antitumor activity: Y. Zhang *et al., Proc. Nat. Acad. Sci. USA* **91**, 3147 (1994).

$[\alpha]_D^{22}$ −79.3 ±1° (c = 1.223 in chloroform). $[\alpha]_D^{25}$ −78.6° (c = 1.19 in chloroform). uv max (water): 238 nm (ϵ 910 $M^{-1}cm^{-1}$); upon addition of 0.1M NaOH uv max: 226 nm (ϵ 15300 $M^{-1}cm^{-1}$).

9136. Sulforaphen. *4-Isothiocyanato-1-(methylsulfinyl)-1-butene;* isothiocyanic acid 4-(methylsulfinyl)-3-butenyl ester; raphanin. $C_6H_9NOS_2$; mol wt 175.28. C 41.12%, H 5.18%, N 7.99%, O 9.13%, S 36.59%. From seeds of radish, *Raphanus sativus* L., *Cruciferae:* Ivanovics, Horvath, *Naturе* **160**, 297 (1947); *Proc. Soc. Exp. Biol. Med.* **66**, 625 (1947). Identity with raphanin: Koczka, Ivanovics, *C.A.* **44**, 5538c (1950). Structure: Schmid, Karrer, *Helv. Chim. Acta* **31**, 1017 (1948). Synthesis of (±)-form: Balenovic *et al., Tetrahedron* **22**, 2139 (1966).

Liquid. bp$_{0.015}$ 125-130°. $[\alpha]_D^{19}$ −107° (c = 1.37 in chloroform); $[\alpha]_D^{14}$ −136° (c = 1.38 in alcohol). Sol in water, methanol, chloroform; moderately sol in ether; practically insol in petr ether.

9137. Sulforidazine. *10-[2-(1-Methyl-2-piperidinyl)ethyl]-2-(methylsulfonyl)-10H-phenothiazine;* 3-(methylsulfonyl)-10-[2-(1-methyl-2-piperidyl)ethyl]phenothiazine; thioridazine-2-sulfone; TPN-12; Imagotan; Inofal. $C_{21}H_{26}N_2O_2S_2$; mol wt 402.58. C 62.65%, H 6.51%, N 6.96%, O 7.95%, S 15.93%. Dopamine receptor blocker; metabolite of thioridazine and mesoridazine, *q.q.v.* Prepn: Renz *et al., Fr. pats.* **1,363,683** and **1,459,476** (1964, 1966, both to Sandoz). Pharmacology: Maruyama *et al., C.A.* **68**, 20799z (1968). Effects on dopaminergic function in comparison with mesoridazine and thioridazine: D. M. Niedzwiecki *et al., J. Phar-*

macol. Exp. Ther. **228**, 636 (1984); C. D. Kilts *et al., ibid.* **231**, 334 (1984). Clinical evaluation as antipsychotic: R. Axelsson, *Curr. Ther. Res.* **21**, 587 (1977). HPLC determn in plasma: D. A. Ganes, K. K. Midha, *J. Chromatog.* **423**, 227 (1987).

Crystals from acetone, mp 121-123°.

THERAP CAT: Antipsychotic.

9138. Sulfosalicylic Acid. *2-hydroxy-5-sulfobenzoic acid; 3-carboxy-4-hydroxybenzenesulfonic acid;* 5-sulfosalicylic acid; 2-hydroxybenzoic-5-sulfonic acid; salicylsulfonic acid. $C_7H_6O_6S$; mol wt 218.19. C 38.53%, H 2.77%, O 44.00%, S 14.70%. Prepn: Pusca *et al., Rev. Chim. (Bucharest)* **13**, no. 1, 49-50 (1962), *C.A.* **57**, 14994b (1962).

Dihydrate, white crystals or cryst powder. Colored pink by traces of iron. mp ~120° when anhydr; at higher temp dec into phenol and salicylic acid. Very sol in water or alcohol; sol in ether. Generally sol in polar solvents. *Keep well closed and protected from light.*

Caution: Irritating to skin, mucous membranes.

USE: Clinical reagent for albumin in urine; colorimetric reagent for ferric ion with which it gives a violet color. It is a trifunctional aromatic compound, undergoing reactions typical of phenols, carboxylic and sulfonic acid groups. Suggested industrial uses: metal chelating agent, intermediate in the manuf of surface-active compounds, organic catalysts and grease additives.

9139. Sulfotep. *Thiodiphosphoric acid tetraethyl ester; thiopyrophosphoric acid tetraethyl ester;* sulfotepp; thiotepp; dithio; dithione; dithiophos; TEDP; ASP-47; Bayer E 393; ENT-16273; Bladafum. $C_8H_{20}O_5P_2S_2$; mol wt 322.32. C 29.81%, H 6.25%, O 24.82%, P 19.22%, S 19.90%. Prepn: A. D. F. Toy, *J. Am. Chem. Soc.* **73**, 4670 (1951); G. Schrader, R. Mühlmann, *Ger. pat.* **848,812** (1952 to Bayer), *C.A.* **47**, 5426a (1953); R. Appel, H. Einig, *Z. Anorg. Allgem. Chem.* **414**, 236 (1975). Activity: F. F. Smith *et al., J. Econ. Entomol.* **43**, 627 (1950). Toxicity data: B. Holmstedt, *Pharmacol. Rev.* **11**, 567 (1959).

Pale yellow liquid, bp$_2$ 136-139°. *Caution: Poison!* Vapor press. at 20°: 1.7×10^{-4} mm Hg. d$_4^{25}$ 1.196. n$_D^{25}$ 1.4753. Soly in water: 25 mg/l. Miscible with most organic solvents. Corrosive to iron. LD$_{50}$ s.c. in mice: 8 mg/kg (Holmstedt).

Caution: Potential symptoms of overexposure are eye pain, blurred vision, lacrimation and rhinorrhea; headache; cyanosis; anorexia, nausea, vomiting and diarrhea; local sweating, weakness, twitching, paralysis, Cheyne-Stokes respiration, convulsions, low blood pressure and cardiac irreg-

ularities; skin and eye irritation. *See NIOSH Pocket Guide to Chemical Hazards* (DHHS/NIOSH 90-117, 1990) p 204.
USE: Insecticide; miticide.

9140. Sulfoxide. *5-[2-(Octylsulfinyl)propyl]-1,3-benzodioxole; 1,2-(methylenedioxy)-4-[2-(octylsulfinyl)propyl]benzene; n-octylsulfoxide of isosafrole; 1-methyl-2-(3,4-methylenedioxyphenyl)ethyl octyl sulfoxide; Sulfox-Cide.* $C_{18}H_{28}O_3S$; mol wt 324.48. C 66.63%, H 8.70%, O 14.79%, S 9.88%. Prepn: M. E. Synerholm *et al., Contribs. Boyce Thompson Inst.* **15**, 35 (1947); M. E. Synerholm, U.S. pat. **2,486,579** (1949 to Boyce Thompson Inst. for Plant Res.). Synergistic use with pyrethrum, allethrin, rotenone, ryania, etc.: *idem*, U.S. pat. **2,486,445** (1949 to Boyce Thompson Inst. for Plant Res.). Acute toxicity: T. B. Gaines, R. E. Linder, *Fundam. Appl. Toxicol.* **7**, 299 (1986).

Pale yellow, sweet-smelling, viscous oil. Practically insol in water. Sol in most organic solvents except petr ether. LD_{50} in adult male, female rats (mg/kg): 3957, 3477 orally (Gaines, Linder).
USE: Insecticide.

9141. Sulfoxone Sodium. *Disodium[sulfonylbis(p-phenylenimino)]dimethanesulfinate;* 4,4'-diaminodiphenylsulfone disodium formaldehyde sulfoxylate; disodium formaldehydesulfoxylate-diaminodiphenylsulfone; aldesulfone sodium; Diazon; Novotrone; Diasone. $C_{14}H_{14}N_2Na_2O_6S_3$; mol wt 448.45. C 37.50%, H 3.15%, N 6.25%, Na 10.25%, O 21.41%, S 21.45%. Prepn: H. Bauer, *J. Am. Chem. Soc.* **61**, 617 (1939); S. M. Rosenthal, H. Bauer, U.S. pat. **2,234,981** (1941); G. W. Raiziss *et al., J. Am. Pharm. Assoc.* **33**, 43 (1944); U.S. pat. **2,256,575** (1941). Therapeutic use in dermatitis herpetiformis: T. Cornbleet, *Arch. Dermatol. Syphilol.* **64**, 684 (1951); S. I. Katz, *Arch. Dermatol.* **118**, 809 (1982); in leprosy: J. H. Peters *et al., Leprosy Rev.* **46**, 171 (1975). Comprehensive description: V. K. Kapoor, *Anal. Profiles Drug Subs.* **19**, 553-573 (1990).

Dihydrate, needles or amorphous powder. After drying at 100-110° it dec 263-265°. Easily sol in water. Slightly sol in alcohol. Practically insol in the usual organic solvents. Must be stored in vacuum-sealed ampuls. The addition of 10% sodium bicarbonate helps to prevent oxidative changes.
THERAP CAT: Antibacterial (leprostatic).

9142. Sulfur. Brimstone; sulphur. S; at. wt 32.064 ± 0.003; at. no. 16; valences 2, 4, 6. Four naturally occurring isotopes: 32 (95.0%); 33 (0.76%); 34 (4.22%); 36 (0.014%); artificial, radioactive isotopes: 29-31; 35; 37; 38. Has been known from very early times. Occurs both in the free state and in combination, mainly as sulfides and sulfates; constitutes 0.05% of the crust of the earth. Industrial prepn by Frasch process: Faith, Keyes & Clark's *Industrial Chemicals,* F. A. Lowenheim, M. K. Moran, Eds. (Wiley-Interscience, New York, 4th ed., 1975) pp 786-794. Exists in several allotropic modifications; at S.T.P. only the orthorhombic, cyclooctasulfur (S_α) is thermodynamically stable: Meyer, *Chem. Rev.* **64**, 429 (1964). Prepn of cyclohexasulfur: Bartlett *et al., J. Am. Chem. Soc.* **83**, 10 (1961); of cyclohepta- and cyclooctasulfur: Schmidt *et al., Angew. Chem. Int. Ed.* **7**, 632 (1968); of cyclononasulfur: Schmidt, Wilhelm, *Chem. Commun.* **1970**, 1111; of cyclododecasulfur: *eidem, Angew. Chem. Int. Ed.* **5**, 964 (1966). Review of structures of elemental sulfur: Meyer in *Advan. Inorg. Chem. Radiochem.* **18**, 287-317 (1976). Reviews of sulfur

and its compds: W. N. Tuller, *The Sulphur Data Book* (McGraw-Hill, New York, 1954); B. Meyer, *Elemental Sulfur* (Interscience-Wiley, 1965); *Inorganic Sulphur Chemistry,* G. Nickless, Ed. (Elsevier, New York, 1968); Schmidt, Siebert, in *Comprehensive Inorganic Chemistry* vol. 2, J. C. Bailar, Jr. *et al.,* Eds. (Pergamon, Oxford, 1973) pp 795-933. *See also* Sulfur, Pharmaceutical.

Sulfur is insol in water. Sparingly sol in alcohol, in ether; sol in carbon disulfide (one gram/2 ml); sol in benzene (about 2.4% at 30°, much more at higher temps.), in toluene. Liquid ammonia (anhydr) dissolves 38.5% S at −78°; acetone dissolves 2.65% at 25°; methylene iodide dissolves 9.1% at 10°; chloroform dissolves about 1.5% at 18°. For other solvents *see* Tuller, *op. cit.* Ignites in air above 261°, in oxygen below 260°, burning to the dioxide; combines readily with hydrogen; combines in the cold with fluorine, chlorine, and bromine; combines with carbon at high temperatures; reacts with silicon, phosphorus, arsenic, antimony and bismuth at their melting points; combines with nearly all metals; with lithium, sodium, potassium, copper, mercury and silver in the cold on contact with the solid; with magnesium, zinc and cadmium very slightly in the cold, more readily on heating; with other metals at high temperatures. Does not react with iodine, nitrogen, tellurium, gold, platinum and iridium.

α-Sulfur, orthorhombic cyclooctasulfur. Amber-colored crystals; the stable form at ordinary temperature. d 2.06; when heated to 94.5° becomes opaque owing to the formation of monoclinic sulfur.

β-Sulfur, monoclinic cyclooctasulfur. Opaque, light-yellow, brittle, needle-like crystals; stable between 94.5-120°. Passes slowly into the rhombic form on standing. d 1.96; mp 115.21°: West, *J. Am. Chem. Soc.* **81**, 29 (1959). bp ~444.6°.

γ-Sulfur, *mother-of-pearl sulfur.* A second monoclinic form, mp 106.8°.

Polymeric sulfur, *Crystex.* Amorphous form; mol wt approx 200,000. Metastable, gradually reverts to the α-form. Insol in solvents used for orthorhombic form. *Reviews:* Meyer, *loc. cit.* (1964); A. V. Tobolsky, W. J. MacKnight, *Polymeric Sulfur and Related Polymers* (Interscience, New York, 1965) pp 87-97.

Caution: May cause irritation of skin, mucous membranes.
USE: In mfg sulfuric acid, carbon disulfide, sulfites, insecticides, plastics, enamels, metal-glass cements; in vulcanizing rubber; in syntheses of dyes; in making gunpowder, matches; for bleaching wood pulp, straw, wool, silk, felt, linen.

9143. Sulfur Chloride. Disulfur dichloride; sulfur monochloride; sulfur subchloride. Cl_2S_2; mol wt 135.04. Cl 52.51%, S 47.49%. S_2Cl_2. Prepared by passing chlorine into molten sulfur, can be purified by fractional distillation: Fehér in *Handbook of Preparative Inorganic Chemistry* vol. 1, G. Brauer, Ed. (Academic Press, New York, 2nd ed., 1963) pp 371-372. Hydrolysis: H. L. Olin, *J. Am. Chem. Soc.* **48**, 167 (1926).

Non-flammable, light amber to yellowish red, fuming, oily liquid. Penetrating odor. $d_{15.5}^{15.5}$ 1.6885. n_D^{20} 1.670. mp −77°. bp_{760} 138°; bp_{100} 72.0°; bp_{10} 19.1°. Dielectric constant 4.9 at 22°. Dipole moment 1.60. Sol in alcohol, benzene, ether, carbon disulfide, carbon tetrachloride, oils. Readily dissolves sulfur (up to 67% at room temp). Dec by water yielding sulfur, hydrogen chloride, sulfur dioxide, hydrogen sulfide, sulfite, thiosulfate. In acid solns pentathionic and other polythionic acids are formed. *Keep tightly closed and out of contact with water.*

Caution: Potential symptoms of overexposure are lacrimation; coughing; eye and skin burns; pulmonary edema. *See NIOSH Pocket Guide to Chemical Hazards* (DHHS/NIOSH 90-117, 1990) p 200.
USE: Intermediate and chlorinating agent in the manuf of organic chemicals, sulfur dyes, insecticides, synthetic rubbers; in cold vulcanization of rubber; as polymerization catalyst for vegetable oils; for hardening soft woods.

9144. Sulfur Dioxide. Sulfurous anhydride; sulfurous oxide. O_2S; mol wt 64.06. O 49.95%, S 50.05%. SO_2.

Colorless, nonflammable gas; strong suffocating odor. Condenses at $-10°$ and ordinary pressure to a colorless liquid. d (liq) 1.5. mp $-72°$. bp $-10°$. Mixed with oxygen and passed over red-hot Pt, it is converted into SO_3. With water forms sulfurous acid (H_2SO_3). Bleaches vegetable colors. Soly: in water 17.7% at 0°, 11.9% at 15°, 8.5% at 25°, 6.4% at 35°; in alcohol 25%; in methanol 32%; also sol in chloroform or ether. It is supplied compressed in cylinders.

Caution: Potential symptoms of overexposure are irritation of eyes, nose and throat, rhinorrhea; choking, coughing; reflex bronchoconstriction; eye and skin burns. *See NIOSH Pocket Guide to Chemical Hazards* (DHHS/NIOSH 90-117, 1990) p 200.

USE: Preserving fruits, vegetables, etc.; disinfectant in breweries and food factories; bleaching textile fibers, straw, wicker ware, gelatin, glue, beet sugars. Liquid SO_2 used as solvent: Waddington, *Non-aqueous Solvent Systems*, T. C. Waddington, Ed. (Academic Press, New York, 1965) pp 253-284.

9145. Sulfuretin. (Z)-2-[(3,4-Dihydroxyphenyl)methylene]-6-hydroxy-3(2H)-benzofuranone; 2-(3,4-dihydroxybenzylidene)-6-hydroxy-3(2H)-benzofuranone; 3',4',6-trihydroxyaurone; 3',4',6-trihydroxybenzalcoumaranone. $C_{15}H_{10}O_5$; mol wt 270.24. C 66.67%, H 3.73%, O 29.60%. Flavonoid pigment responsible for the yellow color of certain species of *Compositae*. Isoln from *Cosmos sulphureus* Cav., *Compositae*: Shimokoriyama, Hattori, *J. Am. Chem. Soc.* **75**, 1900 (1953); from *Dahlia variabilis* Desf., *Compositae*: Nordström, Swain, *Arch. Biochem. Biophys.* **60**, 329 (1956). Prepn by condensation of 6-hydroxy-3-coumaranone and protocatechuic alcohol: Nordström, Swain, *ibid.* Prepn and structure studies: Geissman, Jurd, *J. Am. Chem. Soc.* **76**, 4475 (1954); Farkas *et al., Ber.* **92**, 2847 (1959). Structure: Shimokoriyama, Geissman, *J. Org. Chem.* **25**, 1956 (1960).

Orange crystals from dil alc, mp 280-285° (Shimokoriyama), mp 315° (dec) (Nordström), mp 302-304° (Farkas). uv max: 398 nm.

Triacetate, $C_{21}H_{16}O_8$, pale yellow needles from methanol, mp 191-194° (Shimokoriyama), mp 167-168° (Nordström).

9146. Sulfur Hexafluoride. F_6S; mol wt 146.06. F 78.05%, S 21.95%. SF_6. Prepd by direct fluorination of sulfur or sulfur dioxide: Moissan, Lebeau, *Compt. Rend.* **130**, 865, 984 (1900); *eidem, Ann. Chim. Phys.* [7] **26**, 147 (1902); Schumb, *Inorg. Syn.* **3**, 119 (1950); Kwasnik in *Handbook of Preparative Inorganic Chemistry*, Vol. **1**, G. Brauer, Ed. (Academic Press, New York, 2nd ed., 1963) pp 169-170. *Reviews:* Cady, "Fluorine-Containing Compounds of Sulfur" in *Advan. Inorg. Chem. Radiochem.* **2**, 105-157 (1960); Kemmitt, Sharp, *Advan. Fluorine Chem.* **4**, 218-219 (1965).

Colorless, odorless gas. One of the heaviest known gases; density approx 5 times that of air. mp $-50.8°$. Sublimes at $-63.8°$. Crit temp 45.6°. d (liq, $-50.8°$) 1.88. Sparingly sol in water, somewhat more in alcohol. At 25° and 1 atm 0.297 ml SF_6 dissolves in 1.0 ml of transformer oil. Thermodynamically unstable but kinetically stable gas. This stability explained by symmetrical, octahedral structure of the molecule. Inert to nucleophilic attack. Does not attack glass. No fluorine exchange with HF. Stable to silent electrical discharge. Unchanged at 500°.

USE: In electrical circuit interrupters. In electronic ultra-high frequency piping.

9147. Sulfuric Acid. Oil of vitriol. H_2O_4S; mol wt 98.08. H 2.06%, O 65.25%, S 32.69%. H_2SO_4. Prepd by the Contact Process according to the reactions $2SO_2 + O_2 \rightarrow 2SO_3$, and $SO_3 + H_2O \rightarrow H_2SO_4$; by the Chamber Process according to the reactions $2NO + O_2 \rightarrow 2NO_2$, and $NO_2 +$

$SO_2 + H_2O \rightarrow H_2SO_4 + NO$. Sulfuric acid of commerce contains 93-98% H_2SO_4; the remainder is water. Monograph: W. W. Duecker, J. R. West, *The Manufacture of Sulfuric Acid* (Reinhold, New York, 1959) 515 pp. Review of manuf: Pearce, "Sulphuric Acid: Physico-Chemical Aspects of Manufacture" in *Inorganic Sulphur Chemistry*, G. Nickless, Ed. (Elsevier, New York, 1968) pp 535-561; Faith, Keyes & Clark's *Industrial Chemicals*, F. A. Lowenheim, M. K. Moran, Eds. (Wiley-Interscience, New York, 4th ed., 1975) pp 795-806. Toxicity data: H. F. Smyth *et al., Am. Ind. Hyg. Assoc. J.* **30**, 470 (1969).

Clear, colorless, odorless, oily liquid. *Very corrosive!* Has a very great affinity for water, abstracting it from the air and also from many organic substances; hence it chars sugar, wood, etc. d $\sim$1.84. bp $\sim$290°; dec 340° into sulfur trioxide and water. mp 10° (anhydrous acid). 98% H_2SO_4 freezes at $+3°$; 93% at $-32°$; 78% at $-38°$; 74% at $-44°$; 65% at $-64°$. Misc with water and alcohol with the generation of much heat and with contraction in vol. *When diluting, the acid should be added to the diluent. Keep tightly closed. Handle with caution. Avoid contact with skin.* LD_{50} orally in rats: 2.14 g/kg (Smyth).

Sulfuric acid, fuming. H_2SO_4 with free SO_3, designated in commerce as *oleum*. Available grades contain up to about 80% free SO_3. Colorless or slightly colored, viscous liquid, emitting choking fumes of sulfur trioxide. *Extremely corrosive. Handle with great care. Avoid contact with skin. Keep tightly closed in glass-stoppered bottles.*

Caution: Potential symptoms of overexposure are eye, nose and throat irritation; pulmonary edema, bronchitis; emphysema; conjunctivitis; stomatis; dental erosion; tracheobronchitis; skin and eye burns; dermatitis. *See NIOSH Pocket Guide to Chemical Hazards* (DHHS/NIOSH 90-117, 1990) p 200. *See also Clinical Toxicology of Commercial Products,* R. E. Gosselin *et al.,* Eds. (Williams & Wilkins, Baltimore, 5th ed., 1984) section III, pp 8-12.

USE: In manuf of fertilizers, explosives, dyestuffs, other acids, parchment paper, glue, purification of petroleum, pickling of metal.

THERAP CAT: Dil acid formerly in treatment of gastric hypoacidity. Concd acid formerly as a topical caustic.

9148. Sulfur Iodide. Iodosulfane. Approx SI. Obtained by fusing together 4 parts iodine with 1 part sulfur. It probably contains free iodine as well as free sulfur.

Grayish-black masses of metallic luster and iodine odor. Insol in water; sol in carbon disulfide, alcohol, ether, in about 50 parts glycerol. KI soln dissolves a portion of the iodine leaving free sulfur. *Keep well closed.*

THERAP CAT: Dermatologic.

THERAP CAT (VET): Has been used in chronic eczema, ringworm, mange.

9149. Sulfurous Acid. Sulfur dioxide soln. A soln of $\sim$6% sulfur dioxide in water.

Colorless, clear, acid liquid; suffocating odor of sulfur dioxide. d $\sim$1.03. Gradually oxidizes in the air to sulfuric acid. *Keep in nearly full, tightly closed containers, in a cool place.*

USE: Dental bleach.

THERAP CAT: Antiseptic.

9150. Sulfur, Pharmaceutical. Three forms of sulfur of 99.5% purity or better are recognized in pharmacy. *Precipitated sulfur,* also known as milk of sulfur, made by boiling sulfur with lime and pptg the filtered soln with hydrochloric acid. *Sublimed sulfur,* also known as flowers of sulfur. *Washed sulfur,* made by treating sublimed sulfur with ammonia to dissolve impurities, particularly arsenic (the U.S.A. sulfur contains practically no arsenic) and to remove traces of acid.

Sublimed and washed sulfur are in the form of a fine, yellow crystalline powder, with only a faint odor and taste. Both are slowly and usually incompletely sol in carbon disulfide. Precipitated sulfur is in the form of a very fine, pale yellow, amorphous or microcryst powder, odorless and tasteless. It dissolves much more quickly and usually more completely in carbon disulfide than the other forms. Soly in anhydr lanolin at 45°: 0.38%; in olive oil at 15°: 2.2%, at 30°: 4.1%, at 100°: 20.0%.

THERAP CAT: Scabicide.

THERAP CAT (VET): Externally: antiseptic and parasiticidal; in lotions, ointments and dips, often with lime. Internally: has been used as a laxative, coccidiostat.

9151. Sulfur Tetrafluoride. F_4S; mol wt 108.06. F 70.33%, S 29.67%. SF_4. First prepd from cobalt trifluoride and elemental sulfur: Fischer, Jaenckner, *Angew. Chem.* **42**, 810 (1929); Kwasnik in *Handbook of Preparative Inorganic Chemistry* vol. 1, G. Brauer, Ed. (Academic Press, New York, 2nd ed., 1963) pp 168-169. Best prepared from SCl_2 and NaF suspended in acetonitrile; also from sulfur, chlorine and sodium fluoride: Tullock *et al., J. Am. Chem. Soc.* **82**, 539 (1960); Arth, Fried, U.S. pat. **3,046,094** (1962 to Merck & Co.); Fawcett, Tullock, *Inorg. Syn.* **7**, 119 (1963). Reviews of prepn and chemistry: Smith, *Angew. Chem. Int. Ed.* **1**, 467-475 (1962); Kemmitt, Sharp, *Advan. Fluorine Chem.* **4**, 220-222 (1965); Martin, *Ann. N. Y. Acad. Sci.* **145**, 161-168 (1967).

Colorless gas. Thermostable to 600°. mp −121.0°. bp −38°. d (liq; −78°) 1.95. d (solid; −183°) 2.349. Freely sol in benzene. Violent reaction with water. Dec by concd H_2SO_4. Attacks glass but not quartz or mercury.

Caution: About as toxic as phosgene. Strong irritant, corrosive.

USE: Selective fluorinating agent.

9152. Sulfur Trioxide. Sulfuric anhydride. O_3S; mol wt 80.06. O 59.95%, S 40.05%. SO_3. Exists in 3 modifications. Prepd by the contact process, *i.e.*, by the action of oxygen on sulfur dioxide in the presence of catalysts such as platinized asbestos, platinized magnesium sulfate, ferric oxide, or vanadium compds. Extensive review and bibliography: Fairlie, *Sulfuric Acid Manufacture*, A.C.S. Monograph Series no. 69 (New York, 1936). May be prepd in the laboratory by heating fuming sulfuric acid and collecting the sublimate in a cooled receiver. If the vapor is condensed above 27°, the γ-form is obtained as a liquid. If the vapor is condensed below 27° and in the presence of a trace of moisture, a mixture of all 3 forms is obtained. The 3 forms can be separated by fractional distillation. The α-form is the stable modification, the β- and γ-forms are metastable.

Melted SO_3 exists in the γ-form and on solidifying tends to the α-form. The difference in vapor pressures explains the so-called "alpha explosion" of sulfur trioxide. Heating of high-melting SO_3 in glass vessels should be avoided to prevent possible shattering of the container. Absolutely dry SO_3 is not corrosive to metals and shows no acid reaction. On exposure to air, it absorbs moisture rapidly, emitting dense white fumes. Combines with water with explosive violence (heat of dilution 504 cals/g) forming sulfuric acid. Due to this avidity for water, SO_3 chars many organic substances. On contact with wood shavings the heat produced by dehydration is sufficient to cause fire.

α-Form, asbestos-like needles, mp 62.3°. Vapor pressure at 25° = 73 mm.

β-Form, asbestos-like needles, mp 32.5°. Vapor pressure at 25° = 344 mm.

γ-Form, ice-like mass, mp 16.8° or liquid, d 1.9224. bp_{760} 44.8°. Vapor pressure at 25° = 433 mm.

Caution: Irritant and corrosive to mucous membranes. May cause coughing, choking and severe discomfort in a concn of 1 ppm.

USE: Intermediate in sulfuric acid manuf; in sulfonations for formation of addition compds with amines; in the manuf of explosives.

9153. Sulfuryl Chloride. Cl_2O_2S; mol wt 134.97. Cl 52.53%, O 23.71%, S 23.76%. SO_2Cl_2. Prepd by passing a mixture of dry sulfur dioxide and chlorine through activated charcoal (other catalysts, such as camphor, can be used): Danneel, *Angew. Chem.* **39**, 1553 (1926); Durrans, *J. Soc. Chem. Ind.* **45**, 347 (1926); Meyer, *Angew. Chem.* **44**, 41 (1931); Danneel, Hesse, *Z. Anorg. Allgem. Chem.* **212**, 214 (1933); Allen, Maxson, *Inorg. Syn.* **1**, 114 (1939).

Colorless, mobile liquid, very pungent odor. *Even the vapors are corrosive to human skin and mucous membranes.* Turns yellow upon prolonged standing because of slight dissociation into SO_2 and Cl_2. d_4^{20} 1.6674. d_4^0 1.7045. mp −54.1° (also given as −46°). bp 69.3°. n_D^{20} 1.4437. Dipole moment 1.86. Trouton constant 20.7. Slowly dec by water, forming H_2SO_4 and HCl. With ice-cold water it forms a

hydrate, $SO_2Cl_2.15H_2O$, resembling camphor in appearance. Violent reaction on contact with alkalies. Miscible with benzene, toluene, ether, glacial acetic acid, other organic solvents.

USE: Chlorinating and sulfonating or chlorosulfonating agent in organic syntheses, e.g., in the manufacture of chlorophenol and chlorothymol. Has been used in war gas formulations.

9154. Sulfuryl Fluoride. Vikane. F_2O_2S; mol wt 102.06. F 37.23%, O 31.35%, S 31.42%. SO_2F_2. Prepd by heating $Ba(SO_3F)_2$: Trautz, Ehrmann, *J. Prakt. Chem.* [2] **142**, 91 (1935); Kwasnik in *Handbook of Preparative Inorganic Chemistry*, vol. 1, G. Brauer, Ed. (Academic Press, New York, 2nd ed., 1963) pp 173-174. Alternate prepn from AgF_2 and SO_2: Muetterties, *Inorg. Syn.* **6**, 158 (1960). *Review:* Kemmitt, Sharp, *Advan. Fluorine Chem.* **4**, 225-227 (1965).

Odorless, colorless gas. Not very reactive. Stable to 400°. mp −135.82°. bp −55.38°. Density and thermodynamic data: Bockhoff *et al., J. Chem. Phys.* **32**, 799 (1960). Soly (ml gaseous F_2SO_2/100 ml solvent): 4-5 (water); 24-27 (alcohol); 210-220 (toluene); 136-138 (CCl_4). Not hydrolyzed by water. Hydrolyzed by NaOH solns. May be stored in compressed form in steel cylinders or in a gasometer over H_2SO_4.

Caution: Potential symptoms of overexposure are conjunctivitis, rhinorrhea and pharyngitis; paresthesia. *See NIOSH Pocket Guide to Chemical Hazards* (DHHS/NIOSH 90-117, 1990) p 202.

USE: Fumigant.

9155. Sulindac. *(Z)-5-Fluoro-2-methyl-1-[[4-(methylsulfinyl)phenyl]methylene]-1H-indene-3-acetic acid; cis-5-fluoro-2-methyl-1-[p-(methylsulfinyl)benzylidene]indene-3-acetic acid;* MK-231; Aflodac; Algocetil; Arthrocine; Artribid; Citireuma; Clinoril; Clisundac; Imbaral; Reumofil; Reumyl; Sudac; Sulinol; Sulreuma. $C_{20}H_{17}FO_3S$; mol wt 356.42. C 67.40%, H 4.81%, F 5.33%, O 13.47%, S 9.00%. Non-steroidal anti-inflammatory. Prepn: T.-Y. Shen *et al.,* Ger. pat. **2,039,426**; *eidem,* U.S. pat. **3,654,349**; D. F. Hinkley, J. B. Conn, U.S. pat. **3,647,858** (1971, 1972, 1972, all to Merck & Co.). Stereospecific synthesis: R. F. Shuman *et al., J. Org. Chem.* **42**, 1914 (1977). ^{13}C-NMR study: A. W. Douglas, *Can. J. Chem.* **56**, 2129 (1978). Pharmacological properties and metabolism: Van Arman *et al.,* and H. B. Hucker *et al, Fed. Proc.* **31**, 577 (1972); H. B. Hucker *et al., Drug Metab. Dispos.* **1**, 721 (1973). Renal metabolism: M. J. Miller *et al., Adv. Prostaglandin Thromboxane Leukotriene Res.* **11**, 487 (1983); D. C. Brater *et al., Kidney Int.* **27**, 66 (1985). HPLC determn in biological fluids: D. G. Musson *et al., J. Pharm. Sci.* **73**, 1270 (1984). Review of pharmacology and efficacy in rheumatic disease: R. N. Brogden *et al., Drugs* **16**, 97-114 (1978). Books: *Clinoril in the Treatment of Rheumatic Disorders,* E. C. Huskisson, P. Franchimont, Eds. (Raven Press, New York, 1976) 198 pp; *Current Concepts on Anti-inflammatory Drugs,* K. Miehlke, Ed. (Biomedical Information Corp., New York, 1980) 240 pp.

Yellow odorless crystals from ethyl acetate, mp 182-185° (dec). uv max (methanolic 0.1N HCl): 327, 285, 256, 226 nm ($E_{1cm}^{1\%}$ 375, 420, 410, 540). pKa (25°) 4.7. Sparingly sol in methanol, U.S.P. alcohol; slightly sol in ethyl acetate. Practically insol in water at pH <4.5. Soly increases with

rising pH to ~3.0 mg/ml at pH 7. Stable in aq acid and base. Solid stable for at least three days in air at 100°.
THERAP CAT: Anti-inflammatory.

9156. Sulisatin. *1,3-Dihydro-7-methyl-3,3-bis[4-(sulfooxy)phenyl]-2H-indol-2-one;* 3,3-bis(p-hydroxyphenyl)-7-methyl-2-indolinone bis(hydrogen sulfate) (ester). $C_{21}H_{17}$-NO_9S_2; mol wt 491.50. C 51.32%, H 3.49%, N 2.85%, O 29.30%, S 13.05%. Deriv of isatin, *q.v.*, which stimulates motility of the large intestine. Prepn: A. Hosta Pujol, S. Brosa Rabassa, **Ger.** pat. 2,521,966 corresp to **U.S.** pat. 4,053,483 (1975, 1977 both to Andreu); F. Garrido *et al.*, *Eur. J. Med. Chem.* **10,** 143 (1975). Mechanism of action: J. Queralt *et al.*, *Arch. Farmacol. Toxicol.* **1,** 137 (1975). Laxative properties: M. Moreto *et al.*, *Eur. J. Pharmacol.* **36,** 221 (1976). Metabolism: *eidem, Arzneimittel-Forsch.* **29,** 1561 (1979).

Disodium salt, $C_{21}H_{15}NNa_2O_9S_2$, *DAN-603, Laxitex.* White crystals, mp > 360°. Slightly hygroscopic.
THERAP CAT: Laxative.

9157. Sulisobenzone. *5-Benzoyl-4-hydroxy-2-methoxybenzenesulfonic acid;* 3-benzoyl-4-hydroxy-6-methoxybenzenesulfonic acid; 2-benzoyl-5-methoxy-1-phenol-4-sulfonic acid; 2-hydroxy-4-methoxybenzophenone-5-sulfonic acid; benzophenone-4; NSC-60584; Spectra-Sorb UV 284; Sungard; Uval; Uvinul MS-40. $C_{14}H_{12}O_6S$; mol wt 308.31. C 54.54%, H 3.92%, O 31.14%, S 10.40%. Prepn: A. J. Cofrancesco, **Brit.** pat. 1,136,525 (1968 to GAF). Activity: J. M. Knox *et al.*, *J. Invest. Dermatol.* **34,** 51 (1960).

Light-tan powder, mp 145°. 1 g dissolves in 2 ml methanol, 3.3 ml alc, 4 ml water, 100 ml ethyl acetate.
USE: Ultraviolet stabilizer in wool, cosmetics, pesticides, and lithographic plate coatings.
THERAP CAT: Ultraviolet screen.

9158. Sulmarin. *4-Methyl-6,7-bis(sulfooxy)-2H-1-benzopyran-2-one; 6,7-dihydroxy-4-methylcoumarin disulfate;* 4-methylesculetin bis(hydrogen sulfate); 4-methyl-6,7-dihydroxycoumarindisulfate; 4-methylesculetindisulfonic acid; 4-methylesculetin-6,7-disulfuric ester; MG-143; Idro P₂. C_{10}-$H_8O_{10}S_2$; mol wt 352.30. C 34.09%, H 2.29%, O 45.41%, S 18.20%. Prepn: Cavallini, Mazzucchi, *Farm. Sci. e tec. (Pavia)* **3,** 297 (1948), *C.A.* **42,** 8900d (1948). Description: Maggiorelli, *Giorn. Med. Militare* **101,** 365 (1951), *C.A.* **47,** 6412d (1951); Banchetti, *Farmaco Ed. Sci.* **10,** 970 (1955). Polarographic method for determn: A. M. Contri, *Farmaco Ed. Prat.* **25,** 231 (1970).

Disodium salt trihydrate, $C_{10}H_6Na_2O_{10}S_2.3H_2O$, crystals from 60% alc, dec 252-253°. uv max (pH 11.85): 304 nm.
THERAP CAT: Hemostatic.

9159. Sulmazole. *2-[2-Methoxy-4-(methylsulfinyl)phenyl]-1H-imidazo[4,5-b]pyridine;* AR-L 115BS; Vardax. C_{14}-$H_{13}N_3O_2S$; mol wt 287.34. C 58.52%, H 4.56%, N 14.62%, O 11.14%, S 11.16%. Orally active non-glycoside, non-adrenergic inotropic agent. Prepn: **Neth.** pat. **Appl. 7,401,-254;** E. Kutler *et al.*, **U.S.** 3,985,891 (1974, 1976 both to Thomae). Toxicity study: W. Diederen, R. Kadatz, *Arzneimittel-Forsch.* **31,** 141 (1981). Series of articles on pharmacology, pharmacokinetics, metabolism, clinical studies: *ibid.* 129-278. Brief review of pharmacology: H. Koch, *Pharm. Int.* **3,** 5 (1982).

Solid, mp 203-205°. LD_{50} in albino mice (mg/kg): 560 orally; 163 i.v. (Diederen, Kadatz).
THERAP CAT: Cardiotonic.

9160. Suloctidil. *4-[(1-Methylethyl)thio]-α-[1-(octylamino)ethyl]benzenemethanol; erythro-p-*(isopropylthio)-α-[1-(octylamino)ethyl]benzyl alcohol; 1-(4-isopropylthiophenyl)-2-octylaminopropanol; CP-556S; MJF-12637; Bemperil; Cerebro; Circleton; Dulasi; Duloctil; Euvasal; Fluversin; Fluvisco; Hemoantin; Iangene; Loctidon; Locton; Octamet; Polivasal; Sudil; Sulocton; Sulodene. $C_{20}H_{35}NOS$; mol wt 337.57. C 71.16%, H 10.45%, N 4.15%, O 4.74%, S 9.50%. Prepn: G. Lambelin *et al.*, **Ger.** pat. 2,334,404 (1974 to Continental Pharma), *C.A.* **82,** 97820w (1975). Pharmacology: R. Roncucci *et al.*, *Naturwiss.* **62,** 141 (1975); J. Roba *et al.*, *Eur. J. Pharmacol.* **37,** 265 (1976); *eidem. Arch. Int. Pharmacodyn.* **221,** 54 (1976); T. Godfraind, *ibid.* 342.

Crystals from *n*-pentane or methanol-water, mp 62-63°. LD_{50} orally in mice: 3700 mg/kg (Lambelin).
THERAP CAT: Vasodilator (peripheral).

9161. Sulphan Blue. *N-[4-[[4-(Diethylamino)phenyl]-(2,4-disulfophenyl)methylene]-2,5-cyclohexadien-1-ylidene]-N-ethylethanaminium inner salt, sodium salt;* C.I. Acid Blue 1; [4-[α-[p-(diethylamino)phenyl]-2,4-disulfobenzylidene]-2,5-cyclohexadien-1-ylidene]diethylammonium hydroxide inner salt sodium salt; anhydro-4,4'-bis(diethylamino)triphenylmethanol-2'',4''-disulfonic acid monosodium salt; C.I. Food Blue 3; C.I. 42045; Disulphine Blue. $C_{27}H_{31}N_2$-NaO_6S_2; mol wt 566.67. C 57.23%, H 5.51%, N 4.94%, Na 4.06%, O 16.94%, S 11.32%. Prepn from 4-formylbenzene-1,3-disulfonic acid and diethylaniline: *Colour Index* vol. 4 (3rd ed., 1971) p 4382.

Violet powder. One gram dissolves in 20 ml water at 20°. Dilute aq solns are blue and turn yellow upon the addition of concd hydrochloric acid. In the absence of strong acid or caustic the blue color is stable over a wide pH range. Partly sol in alcohol.

2,5-Disulfophenyl isomer, *isosulfan blue, Lymphazurin.*

USE: Coloring medicinal products; in dyeing and printing wool, silk.

THERAP CAT: Isosulfan blue as diagnostic aid (lymphangiography).

9162. Sulphenone®. *1-Chloro-4-(phenylsulfonyl)benzene; p-chlorophenyl phenyl sulfone;* 4-chlorodiphenyl sulfone; sulfenone; R-242. $C_{12}H_9ClO_2S$; mol wt 252.72. C 57.03%, H 3.59%, Cl 14.03%, O 12.66%, S 12.69%. The commercial product contains 80% sulfenone and 20% related diaryl sulfones. Prepn: Bender, Pitt, U.S. pat. **2,593,001** (1952 to Stauffer Chem.). Toxicology: L. W. Hazleton *et al., J. Agr. Food Chem.* **3,** 836 (1955).

Dimorphic crystals, mp 94°. Slight aromatic odor. Tasteless. Soly at 20° (g/100 ml): acetone 74.4; dioxane 65.6; isopropanol 21; hexane 0.4; benzene 44.4; toluene 29.4; xylene 18.2; carbon tetrachloride 4.9. Practically insol in water. Slightly sol in petr oils. Stable to oxidizing and reducing agents, acids and alkalies found in spray formulations. LD_{50} orally in mice: 2.7 g/kg (Hazleton).

USE: Acaricide.

9163. Sulpiride. *5-(Aminosulfonyl)-N-[(1-ethyl-2-pyrrolidinyl)methyl]-2-methoxybenzamide; N-[(1-ethyl-2-pyrrolidinyl)methyl]-5-sulfamoyl-o-anisamide; N-[(1-ethyl-2-pyrrolidinyl)methyl]-2-methoxy-5-sulfamoylbenzamide;* Abilit; Aiglonyl; Coolspan; Dobren; Dogmatil; Dogmatyl; Dolmatil; Guastil; Meresa; Miradol; Mirbanil; Misulvan; Neogama; Omperan; Pyrikappl; Sernevin; Splotin; Sulpitil; Sursumid; Synédil; Trilan. $C_{15}H_{23}N_3O_4S$; mol wt 341.43. C 52.77%, H 6.79%, N 12.31%, O 18.74%, S 9.39%. Dopamine D_2 and D_3-receptor antagonist. Prepn: C. S. Miller *et al.,* U.S. pat. **3,342,826** (1964 to Soc. Etudes Sci. Ind. l'Ile-de-France); of *l*-form: F. Mauri, Ger. pat. **2,903,891** (1979 to Ravizza), *C.A.* **91,** 211259h (1979). Crystal structure: L. Y. Y. Ma *et al., Acta Crystallogr.* **B38,** 2861 (1982). Mechanism of action: P. Jenner *et al., J. Pharm. Pharmacol.* **32,** 39 (1980). Structure-activity study: P. Dostert *et al., Eur. J. Med. Chem. - Chim. Ther.* **17,** 437 (1982). Physical properties: D. Pitrè, E. Valoti, *Arch. Pharmacol.* **320,** 859 (1987). Comprehensive description: D. Pitrè *et al., Anal. Profiles Drug Subs.* **17,** 607-641 (1988). Review of clinical efficacy in psychiatry: A. J. Wagstaff *et al., CNS Drugs* **2,** 313-333 (1994); of levosulpiride in gastroenterology: M. Guslandi, *Curr. Ther. Res.* **53,** 484-501 (1993).

White, odorless crystalline powder, mp 178-180°. Sparingly sol in methanol. Practically insol in water, ether, chloroform, benzene. pKa_1 9.00, pKa_2 10.19. $[\alpha]_D^{25}$ −66.8° (c = 0.5 in DMF). LD_{50} in mice (mg/kg): 170 i.p.; 2250 orally (Dostert).

l-Form, *levosulpiride, S(−)-sulpiride, Levobren, Levopraid.* Monomorphic crystalline solid, mp 185-187°.

THERAP CAT: Antipsychotic; antidepressant; antiemetic.

9164. Sulprofos. *O-Ethyl O-[4-(methylthio)phenyl]-phosphorodithioic acid S-propyl ester; O-ethyl O-[4-(methylmercapto)phenyl]-S-n-propylphosphorothionothiolate; O-ethyl O-[4-(methylthio)phenyl] S-propylphosphorodithio-*

ate; BAY NTN 9306; Bolstar; Helothion. $C_{12}H_{19}O_2PS_3$; mol wt 322.45. C 44.70%, H 5.94%, O 9.92%, P 9.61%, S 29.83%. Prepn: S. Kishino *et al.,* Ger. pat. **2,111,414;** *eidem,* U.S. pat. **3,947,529** (1971, 1976 both to Bayer). Metabolism: D. L. Bull, G. W. Ivie, *J. Agr. Food Chem.* **24,** 143 (1976); G. W. Ivie *et al., ibid.* 147; D. L. Bull *et al., ibid.* 601. Photodegradation: G. W. Ivie, D. L. Bull, *ibid.* 1053.

Tan liquid, phosphorus odor, $bp_{0.1}$ 155-158°. d_{20}^{20} 1.20, n_D^{20} 1.5859. Low soly in water, sol in organic solvents. LD_{50} orally in rats: 227 mg/kg (Bull, Ivie).

Caution: Toxic to fish and wildlife.

USE: Insecticide.

9165. Sulprostone. *[1R-[1α(Z),2β(1E,3R*),3α]]-7-[3-Hydroxy-2-(3-hydroxy-4-phenoxy-1-butenyl)-5-oxocyclopentyl]-N-(methylsulfonyl)-5-heptenamide;* CP-34089; SHB-286; ZK-57671; Nalador. $C_{23}H_{31}NO_7S$; mol wt 465.57. C 59.34%, H 6.71%, N 3.01%, O 24.06%, S 6.89%. Analog of prostaglandin E_2, *q.v.,* with uterine stimulant activity. Prepn: J. Bindra, M. R. Johnson, Ger. pat. **2,355,540** (1974 to Pfizer), *C.A.* **81,** 49330u (1974); *eidem,* U.S. pat. **4,024,-179** (1977 to Pfizer). Antifertility effects: H. Hess *et al., Experientia* **33,** 1076 (1977). Influence on platelet function: R. C. Briel, T. H. Lippert, *Adv. Prostaglandin Thromboxane Res.* **6,** 351 (1980). Biological action and half-life: *eidem, Prostaglandins Med.* **6** , 1 (1981). Induction of abortion: K. Schmidt-Gollwitzer, *Int. J. Fertil.* **26,** 86 (1981).

Colorless oil.

THERAP CAT: Abortifacient.

9166. Sultamicillin. *[2S-[2α(2R*,5S*),5α,6β(S*)]]-6-[(Aminophenylacetyl)amino]-3,3-dimethyl-7-oxo-4-thia-1-azabicyclo[3.2.0]heptane-2-carboxylic acid [[(3,3-dimethyl-7-oxo-4-thia-1-azabicyclo[3.2.0]hept-2-yl)carbonyl]oxy]methyl ester, S,S-dioxide;* 1,1-dioxopenicillanoyloxymethyl 6-(D-α-amino-α-phenylacetamido)penicillanate; 6'-(2-amino-2-phenylacetamido)penicillanoyloxymethyl penicillanate 1,1-dioxide; CP-49952; VD-1827. $C_{25}H_{30}N_4O_9S_2$; mol wt 594.67. C 50.49%, H 5.08%, N 9.42%, O 24.21%, S 10.78%. Orally absorbed double ester of sulbactam and ampicillin, *q.q.v.* Prepn: E. C. Bigham, Ger. pat. **3,018,590;** *eidem,* U.S. pat. **4,244,951** (1980, 1981 both to Pfizer); and pharmacology: B. Baltzer *et al., J. Antibiot.* **33,** 1183 (1980). HPLC determn and pharmacokinetics: H. J. Rogers *et al., J. Antimicrob. Chemother.* **11,** 435 (1983). Clinical trial in pediatric otitis media: K. H. Chan *et al., Pediatr. Infect. Dis. J.* **12,** 24 (1993). Review of antibacterial activity, pharmacokinetics and therapeutic efficacy: H. A. Friedel *et al., Drugs* **37,** 491-522 (1989).

mp 190°.

Tosylate, $C_{25}H_{30}N_4O_9S_2 \cdot C_7H_8O_3S$, *Bacimex, Bethacil orale, Unacid PD oral, Unacim orale, Unasyn.*
THERAP CAT: Antibacterial.

9167. Sulthiame. *4-(Tetrahydro-2H-1,2-thiazin-2-yl)-benzenesulfonamide S,S-dioxide; N-(4'-sulfamylphenyl)-1,4-butanesultam; tetrahydro-2-(p-sulfamoylphenyl)-1,2-thiazine 1,1-dioxide; 2-(p-aminosulfonylphenyl)tetrahydro-1,2-thiazine dioxide; 1-(p-amidosulfonylphenyl)-2-thiapiperidine 2,2-dioxide; N-(p-aminosulfonylphenyl)-1,4-butanesultam;* Elisal; Ospolot; Trolone. $C_{10}H_{14}N_2O_4S_2$; mol wt 290.36. C 41.37%, H 4.86%, N 9.65%, O 22.04%, S 22.09%. Prepn of this type of compd: Helferich, Behnisch, U.S. pat. **2,916,-489** (1959 to Schenley).

Crystalline powder, mp 180-182°. Practically insol in cold water. Partly sol in boiling water; slightly sol in alcohol, acids; readily sol in alkalies.
THERAP CAT: Anticonvulsant.

9168. Sultopride. *N-[(1-Ethyl-2-pyrrolidinyl)methyl]-5-(ethylsulfonyl)-2-methoxybenzamide; N-[(1-ethyl-2-pyrrolidinyl)methyl]-5-(ethylsulfonyl)-o-anisamide.* $C_{17}H_{26}N_2O_4S$; mol wt 354.47. C 57.60%, H 7.39%, N 7.90%, O 18.05%, S 9.05%. Dopamine D_2-receptor antagonist. Prepn: C. S. Miller *et al.*, **Fr. pat. M 5,916**; *see also:* G. Bulteau *et al.*, U.S. pat. **3,975,434** (1966, 1976 both to Soc. Etudes Sci. Ind. l'Ile-de-France). Pharmacology: C. Lavelle, J. Margarit, *J. Pharmacol.* **5**, Suppl. 2, 58 (1974); B. Bruguerolle *et al., ibid.* **12**, 27 (1981). Clinical trial in psychosis: D. Morel, *Sem. Hop.* **59**, 2337 (1983). GC determn in serum: A. Kamizono *et al., J. Chromatog.* **567**, 113 (1991). Pharmacokinetics of enantiomers: *idem et al., Biol. Pharm. Bull.* **16**, 1121 (1993).

Hydrochloride, $C_{17}H_{26}N_2O_4S \cdot HCl$, *LIN-1418, Barnetil, Barnotil.* Crystals from methyl ethyl ketone, mp 181-182° (Miller); also reported as crystals from ethanol, mp 190° (Bulteau).
THERAP CAT: Antipsychotic.

9169. Sultosilic Acid. *2-Hydroxy-5-[[(4-methylphenyl)-sulfonyl]oxy]benzenesulfonic acid;* 2-hydroxy-5-tosyloxybenzenesulfonic acid; 2,5-dihydroxybenzenesulfonic acid 5-p-toluenesulfonate; $C_{13}H_{12}O_7S_2$; mol wt 344.37. C 45.34%, H 3.51%, O 32.52%, S 18.62%. Prepn: **Neth. pat. Appl. 7,-305,916** (1973 to Lab. del Esteve); J. Esteve *et al., Eur. J. Med. Chem.* **11**, 43 (1976); of piperazine salt: A. Esteve-Subirana, U.S. pat. **3,954,767** (1976). Toxicology: L. Rodriguez *et al., Arch. Farmacol. Toxicol.* **5**, 281 (1979). Metabolism: S. G. Wood *et al., Xenobiotica* **12**, 165 (1982). Clinical evaluation: H. Vinazzer, J. C. Farine, *Atherosclerosis* **49**, 109 (1983). HPLC determn and pharmacokinetics: R. Roser *et al., Eur. J. Drug Metab. Pharmacokinet.* **11**, 1 (1986).

Piperazine salt, $C_{13}H_{12}O_7S_2 \cdot C_4H_{10}N_2$, *diethylenediamine sultosylate, piperazine sultosylate, A-585, Mimedran.* Crystals from ethanol, mp 174°. LD_{50} i.p. in beagles, male rats, female rats: 605, 833.6, 1272 mg/kg; orally in rats: > 11 g/kg (Rodriguez).
THERAP CAT: Antihyperlipoproteinemic.

9170. Sultroponium. *endo-(±)-3-(3-Hydroxy-1-oxo-2-phenylpropoxy)-8-methyl-8-(3-sulfopropyl)-8-azoniabicyclo-[3.2.1]octane inner salt; 3α-hydroxy-8-(3-sulfopropyl)-1αH,5αH-tropanium hydroxide, (±)-tropate, inner salt;* sulfobetaine of atropine; tropyl tropate sulfo-betaine; A-118; Sultropan. $C_{20}H_{29}NO_6S$; mol wt 411.52. C 58.37%, H 7.10%, N 3.40%, O 23.33%, S 7.79%. Prepn: Savini *et al., Therapie* **22**, 843 (1967); Wahl, *Prod. Probl. Pharm.* **25**, 260 (1970); Raudnitz *et al.*, Belg. pat. **672,926** (1966) and Brit. pat. **1,082,445** (1967).

Crystals from alc, mp 220° (dec). Sol in water and warm ethanol; insol in ether, acetone, benzene, chloroform. uv max (water): 251.5, 257.5, 263.5 nm.
THERAP CAT: Anticholinergic, antispasmodic.

9171. Sumach. Name used for several plants of the *Rhus* species. Leaves of *Rhus glabra* L. and *Rhus typhina* L., *Anacardiaceae* are known commercially as **North American Sumach.** They contain fisetin, dihydrofisetin, 27% tannin and gallic acid esters. Constituents of the *Rhus typhina* fruit: J. Tischer, *Pharmazie* **15**, 83 (1960).
USE: North American sumach as black dye; as scenting agent for tobacco.

9172. Sumatriptan. *3-[2-(Dimethylamino)ethyl]-N-methyl-1H-indole-5-methanesulfonamide;* GR-43175. $C_{14}H_{21}N_3O_2S$; mol wt 295.41. C 56.92%, H 7.17%, N 14.22%, O 10.83%, S 10.85%. Serotonin $5HT_1$-receptor agonist. Prepn: M. D. Dowle, I. H. Coates, Ger. pat. **3,320,521**; *eidem*, U.S. pat. **4,816,470**; A. W. Oxford, Brit. pat. Appl. **2,162,522** (1983, 1989, 1986 all to Glaxo). Receptor binding studies: P. P. A. Humphrey *et al., Brit. J. Pharmacol.* **94**, 1123 (1988); P. Schoeffter, D. Hoyer, *Arch. Pharmacol.* **340**, 135 (1989). LC-MS determn in plasma: J. Oxford, M. S. Lant, *J. Chromatog.* **496**, 137 (1989). Clinical evaluations in migraine: A. Doenicke *et al., Lancet* **1**, 1309 (1988); Subcutaneous Sumatriptan International Study Group, *N. Engl. J. Med.* **325**, 316 (1991); in acute cluster headache: Sumatriptan Cluster Headache Study Group, *ibid.* 322. Review of pharmacology and clinical experience: S. J. Peroutka, *Headache* **30** (Suppl. 2), 554-560 (1990).

mp 169-171°.
Succinate, $C_{14}H_{21}N_3O_2S \cdot C_4H_6O_4$, *GR-43175C, Imigran, Imitrex.* mp 165-166°.
THERAP CAT: Antimigraine.

9173. Sumatrol. *[2R-(2α,6aα,12aα)]-1,2,12,12a-Tetrahydro-5-hydroxy-8,9-dimethoxy-2-(1-methylethenyl)-[1]benzopyrano[3,4-b]furo[2,3-h][1]benzopyran-6(6aH)-one.* $C_{23}H_{22}O_7$; mol wt 410.42. C 67.31%, H 5.40%, O 27.29%. Present in resin from *Derris* sp., *Leguminosae:* Robertson, Rusby, *J. Chem. Soc.* **1937**, 497. Isolated also from *Tephrosia toxi-*

caria Pers., *Papilionaceae:* Harper, *ibid.* **1940,** 1178. Structure and stereochemistry: Harper, *loc. cit.;* Crombie, Peace, *ibid.* **1961,** 5445.

Needles from alcohol, mp 195-196°. Needles from acetone, mp 183°; melts again at 194° after keeping for several days. $[\alpha]_D^{20} -182°$ (c = 2.84 in benzene). Absorption max: 670 nm (ε 4660). Practically insol in water; sparingly sol in methanol, cold acetic acid, moderately sol in benzene, ethyl acetate; sol in chloroform.

9174. Sumbul. Musk root. Dried rhizome and roots of *Ferula sumbul* (Kauffm.) Hook. f. *(Euryangium sumbul* Kauffm.), or of other closely related species of *Ferula* having a musk-like odor *(Umbelliferae). Habit.* Central Asia, East Indies. *Constit.* Angelic (sumbulic) acid, valeric acid, methylcrotonic acid, resin, 0.2-0.4% volatile oil.

THERAP CAT: Sedative.

9175. Sunflower Seed Oil. The oil obtained by milling the seeds of *Helianthus annuus* L., *Compositae.* Classified as a semidrying oil. Alternately classified as an oleic-linoleic acid oil. Produced on a large scale in U.S.S.R., the Baltic region, India, Egypt, Canada, Argentina. Composition: palmitic acid 6.4%, stearic acid 1.3%, arachidic acid 4.0%, behenic acid 0.8%, oleic acid 21.3%, linoleic acid 66.2% (the glycerides in sunflower oil consist mainly of mixed triglycerides, each contg one or two linoleic acid radicals), linolenic acid <0.1%. Vitamin E (mixed tocopherols) 75 mg/100 g. This approaches the tocopherol content of wheat germ oil which is 103 mg/100 g. *Reviews:* E. W. Eckey, *Vegetable Fats and Oils* (Reinhold, New York, 1954) pp 772-777; L. H. Bailey, *Standard Cyclopedia of Horticulture* vol. II (Macmillan, New York, 1953) pp 1445-1449.

Pale yellow oil. Bland, agreeable taste. d_{15}^{15} 0.922-0.926. d_{25}^{25} 0.915-0.919. mp -18°. Titer: 16-20°. Acid value: 0.6. Saponif value: 188-194. Iodine value: 125-136. Thiocyanogen value: 78. Hydroxyl value: 14-16. Reichert-Meissl value below 0.5. Polenske value below 0.5. Unsaponifiable below 1.5%. n_D^{25} 1.472-1.474. n_D^{40} 1.466-1.468. Slightly sol in alc. Miscible with benzene, chloroform, carbon tetrachloride. Forms a "skin" after exposure to air for 2-3 weeks.

USE: Food and salad oil, in candy manuf, in oleomargarine. Like wheat germ oil in dietary supplements. Industrially in oil-modified alkyd resins and soap manuf.

9176. Sunset Yellow FCF. *6-Hydroxy-5-[(4-sulfophenyl)azo]-2-naphthalenesulfonic acid disodium salt;* 1-*p*-sulfophenylazo-2-naphthol-6-sulfonic acid disodium salt; FD & C Yellow No. 6; C.I. Food Yellow 3; C.I. 15985. $C_{16}H_{10}N_2$-$Na_2O_7S_2$; mol wt 452.38. C 42.48%, H 2.23%, N 6.19%, Na 10.16%, O 24.76%, S 14.18%. Prepn: P. Griess, *Ber.* **11,** 2191 (1878). Metabolism: J. L. Radomski, T. J. Mellinger, *J. Pharmacol. Exp. Ther.* **136,** 259 (1962). Toxicology: I. F. Gaunt *et al., Food Cosmet. Toxicol.* **5,** 747 (1967); **7,** 9 (1969); **12,** 1 (1974). *Review: IARC Monographs* **8,** 257-266 (1967). *See also: Colour Index* vol. **4** (3rd ed., 1971) p 4087.

Orange-red crystals. Absorption max (0.02 *N* CH₃COO-NH₄): 480 nm. Sol in water; slightly sol in ethanol. Reddish-orange soln in conc H_2SO_4, changing to yellow on dilution. LD_{50} in rats, mice (g/kg): >10, >6 orally (Gaunt, 1967).

USE: Provisionally listed for use in food, drugs and cosmetics.

9177. Superoxide Dismutase. SOD. Family of naturally occurring metalloenzymes that act as free oxygen radical scavengers and protect against the deleterious effects of biologically generated superoxide oxygen radicals (O_2^-) by dismutation to hydrogen peroxide (H_2O_2) + molecular oxygen (O_2). Three metallotypes have been described; distinguished by the metal found at the active site: cupro-zinc containing (Cu/Zn SOD), manganese containing (Mn SOD), and iron containing (Fe SOD). Widely distributed, present in almost all aerobic organisms and some anaerobes. Isoln and purification of Cu/Zn SOD from bovine erythrocytes and identity with *erythrocuprein (hemocuprein):* J. M. McCord, I. Fridovich, *J. Biol. Chem.* **244,** 6049 (1969); of Mn SOD from *E. coli:* B. B. Keele *et al., ibid.* **245,** 6176 (1970); of Fe SOD from *E. coli:* F. J. Yost, Jr., I. Fridovich, *ibid.* **248,** 4905 (1973). Enzyme immunoassay for determn of human Cu/Zn SOD: N. Kurobe *et al., Clin. Chim. Acta* **187,** 11 (1990); of human Mn SOD: *idem et al., ibid.* **192,** 171 (1990). Review of biological activity and therapeutic use: A. Petkau, *Cancer Treat. Rev.* **13,** 17-44 (1986); of protective effect vs. radiation injury: *eidem, Br. J. Cancer* **55,** Suppl. 8, 87-95 (1987); of structure and function: J. V. Bannister *et al., CRC Crit. Rev. Biochem.* **22,** 111-180 (1987); of chemical and biological characteristics: B. A. Omar *et al., Adv. Pharmacol.* **23,** 109-161 (1992). Books: *Superoxide and Superoxide Dismutases,* A. M. Michelson *et al.,* Eds. (Academic Press, New York, 1977) 568 p.; *Superoxide Dismutase* vols. I, II, III, L. W. Oberley, Ed. (CRC Press, Inc., Boca Raton, 1982, 1982, 1985).

Orgotein, Artrolasi, Ormetein (rescinded), Ontosein, Oxinorm, Palosein, Peroxinorm. Water-soluble Cu/Zn SOD produced from beef liver. Mol wt ~33,000 with a compact conformation maintained by about 4 gram atoms of chelated divalent metal. Isoln: W. Huber, S. Afr. pat. **69 02,983;** *idem,* U.S. pat. **3,579,495** (1969, 1971 both to Diagnostic Data). Purification: *idem,* Ger. pat. **2,101,866;** *idem,* U.S. pat. **3,624,251** (both 1971 to Diagnostic Data). Toxicology: S. Carson *et al., Toxicol. Appl. Pharmacol.* **26,** 184 (1973). Clinical pharmacokinetics: G. Jadot *et al., Clin. Pharmacokinet.* **28,** 17 (1995). Clinical studies in osteoarthritis: K. Lund-Olesen, K. B. Menander-Huber, *Arzneimittel-Forsch.* **33,** 1199 (1983); in Peyronie's disease: G. Primus, *Int. Urol. Nephrol.* **25,** 169 (1993). Clinical use of liposomal form vs. radiation-induced fibrosis: S. Delanian *et al., Radiother. Oncol.* **32,** 12 (1994). Therapeutic use in animals: D. E. Breshears *et al., Mod. Vet. Pract.* **55,** 85 (1974); S. Ahlengard *et al., Equine Vet. J.* **10,** 122 (1978); J. R. Coffman *et al., J. Am. Vet. Med. Assoc.* **174** 261 (1979). Review of clinical trials: L. Flohé, *Molec. Cell. Biochem.* **84,** 123-131 (1988). LD_{50} in mice (mg/kg): >5800 s.c., >60 i.p., >4000 i.v.; in rats (mg/kg): >400 s.c., >284 i.p. (Jadot).

Pegorgotein, PEG-SOD, Win-22118, Dismutec. Orgotein polyethylene glycol conjugate; mol wt 71-105 kDa. Prepn and anti-inflammatory activity: P. S. Pyatak *et al., Res. Commun. Chem. Pathol. Pharmacol.* **29,** 113 (1980). Pharmacokinetics: E. Boccu *et al., Pharmacol. Res. Commun.* **14,** 113 (1982). Toxicology: A. T. Viau *et al., J. Free Rad. Biol. Med.* **2,** 283 (1986). Clinical trials in head injury: J. P. Muizelaar, *Adv. Exp. Med. Biol.* **366,** 389 (1994).

Recombinant human form, *rhSOD.* Pharmacokinetics: C. Tsao *et al., Clin. Pharmacol. Ther.* **50,** 713 (1991). Clinical study in renal transplantation: W. Land *et al., Transplantation* **57,** 211 (1994). Review of cloning and expression: D. Tuoati, *Free Rad. Biol. Med.* **5,** 393-402 (1988).

USE: Research probe for O_2^- in biological and chemical reactions.

THERAP CAT: Anti-inflammatory; radioprotective agent.

THERAP CAT (VET): Anti-inflammatory; antirheumatic.

9178. Suplatast Tosylate. *[3-[[4-(3-Ethoxy-2-hydroxypropoxy)phenyl]amino]-3-oxopropyl]dimethylsulfonium salt with 4-methylbenzenesulfonic acid;* (±)-[2-[[p-(3-ethoxy-2-hydroxypropoxy)phenyl]carbamoyl]ethyl]dimethylsulfonium *p*-toluenesulfonate; IPD-1151T. $C_{16}H_{26}NO_4S.C_7H_7O_3$-S; mol wt 499.65. C 55.29%, H 6.66%, N 2.80%, O 22.41%, S 12.84%. Selective inhibitor of the IgE antibody response; inhibits interleukin-4 gene expression and mast cell differentiation. Prepn: A. Koda *et al.,* Ger. pat. 3,408,708; *eidem,* U.S. pat. 4,556,737 (1984, 1985 both to Taiho). Suppression of IL-4 production and the IgE response: Y. Yanagihara *et al., Japan. J. Pharmacol.* **61,** 23, 31 (1993). Effect on mast cell differentiation: S. Konno *et al., Eur. J. Pharmacol.* **259,** 15 (1994). Series of articles on toxicology: *J. Toxicol. Sci.* **17,** Suppl. 2, 1-205, (1992); *C.A.* **117,** 103839-103846; 103995 (1992). Acute toxicity: K. Yamishita *et al., ibid.* 1.

Crystals, mp 70-73°. LD_{50} in male, female mice, male, female rats (mg/kg): 81, 96, 96, 93 i.v.; in mice, rats (g/kg): > 12.5, > 10 orally (Yamashita).

THERAP CAT: Antiallergic; antiasthmatic.

9179. Suprasterol II. *(2α,7α,8R,19α,22E)-7,19:8,19-Dicyclo-9,10-secoergosta-5(10),22-dien-2-ol.* $C_{28}H_{44}O$; mol wt 396.66. C 84.79%, H 11.18%, O 4.03%. Ultraviolet irradiation product of vitamin D_2. Isoln: Windaus *et al., Ann.* **483,** 17 (1930). Structure and stereochemistry: Dauben, Baumann, *Tetrahedron Letters* **1961,** 565. Crystal structure: C. P. Saunderson *et al., ibid.* 573. Improved isoln and spectral data: T. Kobayashi *et al., J. Nutr. Sci. Vitaminol.* **23,** 291 (1977).

Rough prisms from acetone or methanol, mp 110°. $bp_{0.005}$ 190°. $[\alpha]_D^{19}$ +62.9° (methanol); $[\alpha]_D^{22}$ +47.8° (methanol). Sol in all organic solvents.

9180. Suprofen. *α-Methyl-4-(2-thienylcarbonyl)benzeneacetic acid;* *p*-(2-thenoyl)hydratropic acid; sutoprofen; R-25061; Masterfen; Srendam; Sulprotin; Supranol; Suprocil; Suprol; Topalgic. $C_{14}H_{12}O_3S$; mol wt 260.31. C 64.60%, H 4.65%, O 18.44%, S 12.32%. Prostaglandin biosynthesis inhibitor. Prepn: P. A. Janssen *et al.,* Ger. pat. 2,353,357; *eidem,* U.S. pat. 4,035,376 (1974, 1977 both to Janssen); P. G. H. Van Daele *et al., Arzneimittel-Forsch.* **25,** 1495 (1975). Pharmacology: C. J. E. Niemegeers *et al., ibid.,* 1537. HPLC determn in plasma and urine: H. Muller *et al., ibid.* **32,** 257 (1982). Series of articles on pharmacology, toxicology and clinical trials: *ibid.* **25,** 1501-1542 (1975); *ibid.* **33,** 1322-1338 (1983); on pharmacology, efficacy and safety: *Pharmacology* **27,** Suppl. 1, 1-96 (1983); on pharmacoki-

netics and clinical efficacy: *Arzneimittel-Forsch.* **35,** 738-759 (1985); *ibid.* **36,** 941-971 (1986).

White to slightly yellow microcryst powder, mp 124.3°. pKa 3.91. uv max (0.01N HCl-90% 2-propanol): 266, 292 nm (ε 15700, 15600). Freely sol in methanol, ethanol, chloroform, acetone, polyethylene glycol, 1.0N NaOH. Sol in ether. Slightly sol in 0.1N NaOH. Practically insol in *n*-hexane. Very slightly sol in water. LD_{50} (7 days postdrug) in mice, rats, guinea pigs, dogs (mg/kg): 590, 353, 280, 160 orally (Niemegeers).

THERAP CAT: Anti-inflammatory; analgesic.

9181. Suramin Sodium. *8,8'-[Carbonylbis[imino-3,1-phenylenecarbonylimino(4-methyl-3,1-phenylene)carbonylimino]]bis-1,3,5-naphthalenetrisulfonic acid hexasodium salt;* hexasodium *sym*-bis(*m*-aminobenzoyl-*m*-amino-*p*-methylbenzoyl-1-naphthylamino-4,6,8-trisulfonate) carbamide; Bayer 205; Fourneau 309; Antrypol; Germanin; Moranyl; Naganol; Naphuride. $C_{51}H_{34}N_6Na_6O_{23}S_6$; mol wt 1429.19. C 42.86%, H 2.40%, N 5.88%, Na 9.65%, O 25.75%, S 13.46%. Discovered in 1917 by O. Dressel and R. Kothe: J. Dressel, *J. Chem. Ed.* **38,** 620 (1961). Prepn: E. Fourneau *et al., Compt. Rend.* **178,** 675 (1924); J. Trefouel, E. Fourneau, **Brit.** pat. 224,849 (1923); B. Heymann, *Angew. Chem.* **37,** 585 (1924). Pharmacology, toxicology and clinical antiparasitic activity: F. Hawking, *Adv. Pharmacol. Chemother.* **15,** 289-322 (1978). Inhibition of reverse transcriptase *in vitro:* E. De Clercq, *Cancer Letters* **8,** 9 (1979); vs HIV: H. Mitsuya *et al., Science* **226,** 172 (1984). HPLC determn in plasma: R. W. Klecker, J. M. Collins, *J. Liq. Chromatog.* **8,** 1685 (1985). Pharmacokinetics: J. M. Collins *et al., J. Clin. Pharmacol.* **26,** 22 (1986). Pharmacology and virustatic effect in AIDS: S. Broder *et al., Lancet* **2,** 627 (1985); A. M. Levine *et al., Ann. Int. Med.* **105,** 32 (1986). Clinical trial in onchocerciasis: H. Schultz-Key *et al., Trop. Med Parasit.* **36,** 244 (1985); in prostate cancer: C. Myers *et al., J. Clin. Oncol.* **10,** 881 (1992). *Review:* Olenick in *Antibiotics* **vol. 3,** J. W. Corcoran, F. E. Hahn, Eds. (Springer-Verlag, New York, 1975) pp 699-703; R. La Rocca *et al., Cancer Cells* **2,** 106-115 (1990).

White or slightly pink or cream-colored powder. Slightly bitter taste. Hygroscopic. Freely sol in water, in physiological saline; sparingly sol in 95% alcohol. Insol in benzene, ether, petr ether, chloroform. Aq solns are neutral to litmus. LD_{50} in mice (mg/kg): ~620 i.v. (Hawking).

THERAP CAT: Anthelmintic (Nematodes); antiprotozoal (Trypanosoma).

THERAP CAT (VET): Antiprotozoal (Trypanosoma).

9182. Suriclone. *4-Methyl-1-piperazinecarboxylic acid 6-(7-chloro-1,8-naphthyridin-2-yl)-2,3,6,7-tetrahydro-7-oxo-5H-1,4-dithiino[2,3-c]pyrrol-5-yl ester;* 6-(7-chloro-1,8-naphthyridin-2-yl)-5-(4-methylpiperazin-1-yl)carbonyloxy-7-oxo-2,3,6,7-tetrahydro-1,4-dithiino[2,3-c]pyrrole; RP-31264; Celexane; Clexane; Suril. $C_{20}H_{20}ClN_5O_3S_2$; mol wt 478.00. C 50.26%, H 4.22%, Cl 7.42%, N 14.65%, O 10.04%, S 13.42%. Cyclopyrrolone derivative; structural analog of zopiclone, *q.v.* Prepn: C. Jeanmart *et al.,* **Ger.** pat. 2,360,-

362; *eidem,* U.S. pat. **3,948,917** (1974, 1976 both to Rhone-Poulenc). Benzodiazepine receptor binding studies: J.-C. Blanchard, L. Julou, *J. Neurochem.* **40,** 601 (1983); R. R. Trifiletti, S. H. Snyder, *Mol. Pharmacol.* **26,** 458 (1984); J. L. Zundel *et al., Life Sci.* **36,** 2247 (1985). Pharmacology: L. Julou *et al., Pharmacol. Biochem. Behav.* **23,** 653 (1985). Clinical evaluation as an anxiolytic: Y. D. Lapierre, K. L. Oyewumi, *Prog. Neuro-Psychopharmacol. Biol. Psychiat.* **7,** 805 (1983); W. E. Falk *et al., Psychopharmacol. Bull.* **23,** 134 (1987).

Crystals from acetonitrile, mp 280°.
THERAP CAT: Anxiolytic.

9183. Surinamine. *N-Methyl-L-tyrosine;* andirine; angeline; geoffroyine; ratanhine. $C_{10}H_{13}NO_3$; mol wt 195.22. C 61.53%, H 6.71%, N 7.17%, O 24.59%. Isoln from bark of *Andira retusa* Kunth. (*Geoffroya retusa* Lam.), *Leguminosae:* Hiller-Bombein, *Arch. Pharm.* **230,** 513 (1892). Prepn of inactive compd: Kanevskaya, *J. Prakt. Chem.* **124,** 48 (1929). Prepn of L-form: Kanao, *J. Pharm. Soc. Japan* **66,** 4 (1946); Huguenin, Boissonnas, *Helv. Chim. Acta* **44,** 213 (1961). Prepn of D-form: Izumiya, Nagamatsu, *Bull. Chem. Soc. Japan* **25,** 265 (1952).

Crystals, mp 292-295°. $[\alpha]_D^{21}$ +19.6° (c = 3.8 in 10% HCl). Almost insol in water. Slightly sol in alcohol; sol in dil acids.
D-Form, crystals, mp 273-274°. $[\alpha]_D^{15}$ −18.9° (c = 1.74 in 3*N* HCl).
DL-Form, crystals from water.

9184. Suxethonium Bromide. *2,2'-[(1,4-Dioxo-1,4-butanediyl)bis(oxy)]bis[N-ethyl-N,N-dimethylethanaminium] dibromide;* bis[2-dimethylaminoethyl]succinate bis[ethobromide]; 3,8-dioxadecane-4,7-dione-1,10-bis[ethyldimethylammonium bromide]; IS-362; M-115; M & B 2210; Brevidil E. $C_{16}H_{34}Br_2N_2O_4$; mol wt 478.27. C 40.18%, H 7.17%, Br 33.41%, N 5.86%, O 13.38%. Neuromuscular blocker. Clinical use during endotracheal intubation: S. Day, *Brit. J. Anaesth.* **48,** 1011 (1976).

Slightly hygroscopic crystals, mp 158°. Freely sol in water or normal saline, giving solns which are slightly acidic. Hydrolyzed by alkali.
THERAP CAT: Muscle relaxant (skeletal).

9185. Suxibuzone. *Butanedioic acid mono[(4-butyl-3,5-dioxo-1,2-diphenyl-4-pyrazolidinyl)methyl] ester; succinic acid monoester with 4-butyl-4-(hydroxymethyl)-1,2-diphenyl-3,5-pyrazolidinedione;* 4-butyl-4-(hydroxymethyl)-1,2-diphenyl-3,5-pyrazolidinedione hydrogen succinate (ester); 1,2-diphenyl-4-*n*-butyl-4-hydroxymethyl-3,5-dioxopyrazoli-

dine hemisuccinate; 4-hydroxymethylbutazolidine hemisuccinate; AE-17; Calibène; Danilon; Flogos; Solurol. $C_{24}H_{26}N_2O_6$; mol wt 438.48. C 65.74%, H 5.98%, N 6.39%, O 21.89%. Prodrug of phenylbutazone, *q.v.* Prepd (not claimed) and use as anti-inflammatory: A. Esteve, Ger. pat. **1,936,747;** *idem,* U.S. pat. **3,752,894** (1970, 1973 both to Lab. Esteve). Pharmacology: A. Esteve *et al., Quim. Ind. (Madrid)* **17,** 107 (1971). HPLC determn in plasma and urine: T. Marunaka *et al., J. Pharm. Sci.* **69,** 1258 (1980). Metabolism in humans: Y. Yasuda *et al., ibid.* **71,** 565 (1982). Comparative clinical study with phenylbutazone in rheumatoid arthritis: Y. Mizushima *et al., Int. J. Tissue React.* **5,** 35 (1983).

White, bitter crystalline powder from alcohol, mp 126-127°. Sol in most organic solvents. Insol in water. LD_{50} orally in mice: 5.683 micromol/kg (Esteve, 1973).
THERAP CAT: Anti-inflammatory.
THERAP CAT (VET): Analgesic, antipyretic, anti-inflammatory.

9186. Swertiamarin. *[4aR-(4aα,5β,6α)]-5-Ethenyl-6-(β-D-glucopyranosyloxy)-4,4a,5,6-tetrahydro-4a-hydroxy-1H,3H-pyrano[3,4-c]pyran-1-one;* 4,4a,5,6-tetrahydro-4aα-hydroxy-1-oxo-5β-vinyl-1*H,3H*-pyrano[3,4-*c*]pyran-6-yl β-D-glucopyranoside. $C_{16}H_{22}O_{10}$; mol wt 374.34. C 51.34%, H 5.92%, O 42.74%. Bitter principle of *Swertia japonica* (Maxim.) Makino, *Gentianaceae.* Yields erythrocentaurin on hydrolysis with emulsin. Isoln: Kubota, Tomita, *Chem. & Ind. (London)* **1958,** 229; Inouye *et al., Chem. Pharm. Bull.* **18,** 1856 (1970). Occurrence in gentianaceous plants: Inouye, Nakamura, *J. Pharm. Sci. Japan* **91,** 755 (1971), *C.A.* **75,** 95431b (1971). Structure: Kubota, Tomita, *Tetrahedron Letters* **1961,** 176; *Bull. Chem. Soc. Japan* **34,** 1345 (1961); Koch *et al., Bull. Soc. Chim. France* **1964,** 405. Stereochemistry: Inouye *et al., Tetrahedron Letters* **1968,** 4429. Biosynthesis studies: Inouye *et al., Chem. Pharm. Bull.* **18,** 2043 (1970).

Plates from ethanol + chloroform + ether, mp 113-114°. $[\alpha]_D^{20}$ −127° (c = 1 in 96% ethanol). uv max (methanol): 238 nm (log ε 3.93).
Tetraacetate, $C_{24}H_{30}O_{14}$, prisms, mp 190-191°. $[\alpha]_D$ −100.3° (CHCl$_3$). uv max: 206, 234 nm (log ε 3.20, 4.00).

9187. Sydnones. A term originally applied to the compd obtained by action of acetic anhydride on *N*-nitrosophenylglycine: Eade, Earl, *J. Chem. Soc.* **1946,** 591; Earl, *Chem. & Ind. (London)* **1953,** 746. Sydnones are 5-membered heterocyclic mesoionic compounds that are best represented by polar resonance structures: W. Baker, W. D. Ollis, *Quart. Rev. Chem. Soc.* **11,** 15 (1957); Y. Noel, *Bull. Soc. Chim. France* **1964,** 173. Photochromic properties: S. Nespurek, M. Sorm, *Coll. Czech. Chem. Commun.* **42,** 811 (1977). ^{14}N-NMR study: L. Stefaniak, *Tetrahedron* **33,** 2571 (1977). ^{15}N-NMR study: L. Stefaniak *et al., Org. Magn. Res.* **13,** 274 (1980). Regioselectivity study: A. Padwa *et al., J. Org. Chem.* **47,** 786 (1982). Review of chemistry: F. H. C. Stewart, *Chem. Rev.* **64,** 129-147 (1964).

or

9188. Symclosene. *1,3,5-Trichloro-1,3,5-triazine-2,4,6-(1H,3H,5H)-trione;* trichloroiminocyanuric acid; trichloroisocyanuric acid; ACL-85; Chloreal. $C_3Cl_3N_3O_3$; mol wt 232.41. C 15.50%, Cl 45.76%, N 18.08%, O 20.65%. Prepd by chlorinating cyanuric acid in NaOH soln: Hands, Whitt, *J. Soc. Chem. Ind.* **67**, 66 (1948); Hardy, U.S. pat. **2,607,738** (1952 to Monsanto); Christian, U.S. pat. **2,956,056** (1960 to W. R. Grace). Purification by dissolving in concd H_2SO_4 and diluting with ice water: Lorenz, U.S. pat. **2,828,308** (1958 to Purex). Structure studies and prepn: Petterson *et al., J. Org. Chem.* **25**, 1595 (1960).

Needles from ethylene chloride, mp 246-247° (dec). May be stored in the dry state for at least a year. Releases hypochlorous acid on contact with water. Available chlorine about 90%. pH of aq solns about 4.4. Soly in water at 25° about 0.2%. Soluble in chlorinated and highly polar solvents.
Caution: Moderately irritating to eyes, skin, mucous membranes.
USE: Chlorinating agent, disinfectant, industrial deodorant. In household cleansers, such as *Bab-O*.
THERAP CAT: Anti-infective (topical).

9189. Synephrine. *4-Hydroxy-α-[(methylamino)methyl]benzenemethanol; p-hydroxy-α-[(methylamino)methyl]benzyl alcohol;* 1-(4-hydroxyphenyl)-2-methylaminoethanol; *p*-methylaminoethanolphenol; β-methylamino-α-(4-hydroxyphenyl)ethyl alcohol; methylaminomethyl 4-hydroxyphenyl carbinol; Analeptin; Ethaphene; Oxedrine; Parasympatol; Simpalon; Synephrin; Synthenate. $C_9H_{13}NO_2$; mol wt 167.21. C 64.65%, H 7.84%, N 8.38%, O 19.14%. Prepd by hydrogenating ω-methylamino-4-hydroxyacetophenone in water in the presence of Pt or Pd: Ger. pat. **566,578** (1931 to Boehringer, Ing.), *Frdl.* **18**, 3025.

Crystals, mp 184-185°. Stable to air and light.
Hydrochloride, $C_9H_{13}NO_2 \cdot HCl$, crystals, mp 151-152°. Freely sol in water.
Tartrate, $2C_9H_{13}NO_2 \cdot C_4H_6O_6$, *Corvasymton, Simpadren, Sympathol.* Crystals, mp 188-190° (some dec). Freely sol in water; sol in alcohol.
Tartaric acid monoester, $C_{13}H_{17}NO_7$, *p-methylaminoethanolphenol tartrate, Neupentedrin, Pentedrin.*
THERAP CAT: Adrenergic; vasopressor.

9190. Synhexyl. *3-Hexyl-7,8,9,10-tetrahydro-6,6,9-trimethyl-6H-dibenzo[b,d]pyran-1-ol;* 1-hydroxy-3-hexyl-6,6,9-trimethyl-7,8,9,10-tetrahydro-6H-dibenzo[b,d]pyran; parahexyl; pyrahexyl. $C_{22}H_{32}O_2$; mol wt 328.49. C 80.44%, H 9.82%, O 9.74%. Psychotomimetic synthetic analog of the tetrahydrocannabinols, *q.v.* Prepn: Adams *et al., J. Am. Chem. Soc.* **63**, 1971 (1941); Adams, U.S. pat. **2,419,935**

(1947); Hughes *et al.,* U.S. pat. **3,576,887** (1971 to Am. Home Products). Comparison of physiological effects with those of tetrahydrocannabinol: Hollister *et al., Clin. Pharmacol. Ther.* **9**, 783 (1968). Physical constants: Farmilo *et al., Bull. Narcotics* **6**, 7 (1954). Toxicity data: Loewe, *J. Pharmacol. Exp. Ther.* **88**, 154 (1946).

Liquid, $bp_{1.0}$ 190-192°; n_D^{20} 1.5504. uv and ir spectra: *Bull. Narcotics* **6**, 27 (1954); **7**, 42 (1955). LD_{50} in mice, dogs, rabbits (mg/kg): 170, 223, 143 i.v. (Loewe).
Note: This is a controlled substance (hallucinogen) listed in the U.S. Code of Federal Regulations, Title 21 Part 1308.11 (1995).

9191. Syringaldehyde. *4-Hydroxy-3,5-dimethoxybenzaldehyde;* syringic aldehyde; 3,5-dimethoxy-4-hydroxybenzene carbonal; gallaldehyde 3,5-dimethyl ether. $C_9H_{10}O_4$; mol wt 182.18. C 59.34%, H 5.53%, O 35.13%. Widely distributed in plants. Hydrolysis product of the naturally occurring glycosyringic aldehyde: Körner, *Gazz. Chim. Ital.* **18**, 215 (1888). Prepn from heat-treated beechwood and lignin: Kratzl, Silbernagel, *C.A.* **50**, 6040 (1956). Synthesis from pyrogallol 1,3-dimethyl ether: Pearl, *J. Am. Chem. Soc.* **70**, 1746 (1948); *cf.* Graebe, Martz, *Ber.* **36**, 1032 (1903); Allen, Leubner, *Org. Syn.* **31**, 92 (1951); **coll. vol. IV**, 866 (1963).

Very pale yellow needles from petr ether, mp 113°. bp_{14} 192-193°. uv max (dioxane): 305 nm. Very sparingly sol in water, petr ether. Sol in alcohol, ether, chloroform, hot benzene, glacial acetic acid. Forms yellow sodium and potassium salts.

9192. Syringin. *(E)-4-(3-Hydroxy-1-propenyl)-2,6-dimethoxyphenyl-β-D-glucopyranoside;* 4-(3-hydroxypropenyl)-2,6-dimethoxyphenyl-D-glucoside; syringoside; ligustrin; lilacin; methoxyconiferine. $C_{17}H_{24}O_9$; mol wt 372.37. C 54.83%, H 6.50%, O 38.67%. First isolated in 1841 from bark of *Syringa vulgaris* L. (lilac): Meillet in 1841 *Ann.* **40**, 319 (1841). Prepn: Pauly, Strassberger, *Ber.* **62**, 2277 (1929); Freudenberg, Schraube, *Ber.* **88**, 16 (1955). Isoln from lilac bark: Freudenberg *et al., Ber.* **84**, 472 (1951); from various plants: Plouvier, *Compt. Rend.* **254**, 4196 (1962); from cambial sap of spruce: Freudenberg, Harkin, *Phytochemistry* **2**, 189 (1963).

Monohydrate, crystals from water, mp 192°. $[\alpha]_D^{20}$ −8,2° (c = 2.43 in chloroform), −17.25° (water). Slightly sol in cold water; sol in hot water, alc. Practically insol in ether.

9193. Syrosingopine. *18-[[4-[(Ethoxycarbonyl)oxy]-3,5-dimethoxybenzoyl]oxy]-11,17-dimethoxyyohimban-16-carboxylic acid methyl ester;* methyl carbethoxysyringoyl reserpate; carbethoxysyringoyl methyl reserpate; methyl *O-(O′-carbethoxysyringoyl)reserpate;* syringopine; Su-3118; Isotense; Londomin; Raunova; Seniramin; Singoserp; Siringina. $C_{35}H_{42}N_2O_{11}$; mol wt 666.73. C 63.05%, H 6.35%, N 4.20%, O 26.40%. Prepn: Lucas *et al., J. Am. Chem. Soc.* **81,** 1928 (1959); **U.S.** pat. **2,813,871** (1957 to Ciba).

Crystals from acetone, mp 175-179°.

THERAP CAT: Antihypertensive.

T

9194. **2,4,5-T.** *(2,4,5-Trichlorophenoxy)acetic acid;* Esteron 245; Trioxone; Weedone. $C_8H_5Cl_3O_3$; mol wt 255.48. C 37.61%, H 1.97%, Cl 41.63%, O 18.79%. Post-emergence herbicide. Prepd from 2,4,5-trichlorophenol: Pokorny, *J. Am. Chem. Soc.* **63**, 1768 (1941); from benzenehexachloride: Galat, *ibid.* **74**, 3890 (1952). Activity: C. L. Hamner, T. B. Tukey, *Science* **100**, 154 (1944). Contains trace levels of TCDD, *q.v.*, as a contaminant: J. Smith, *Science* **203**, 1090 (1979); *Chem. & Eng. News* **59**, 6 (Jan. 5, 1981). Toxicity: V. A. Rowe, T. A. Hymas, *Am. J. Vet. Res.* **15**, 622 (1954). *See also* 2,4-D.

Crystals from benzene, mp 153°. d_{20}^{20} 1.80. Soly in water at 30°: 238 mg/kg. Sol in alcohol. Forms water-soluble sodium and alkanolamine salts. Commercial products are usually in the form of amines or esters, often in mixture with 2,4-D. LD_{50} orally in mice, rats: 389, 500 mg/kg (Rowe, Hymas).
Caution: Potential symptoms of overexposure in animals are ataxia; skin irritation, acne-like rash. *See NIOSH Pocket Guide to Chemical Hazards* (DHHS/NIOSH 90-117, 1990) p 202.
Note: In March 1985 the E.P.A. terminated all registrations for the use of this herbicide on rice fields, orchards, sugarcane, rangeland and other noncrop sites. This follows the 1970 action of the Department of Agriculture halting the use of the pesticide on all food crops except rice: *Chem. & Eng. News* **63**, 6 (Mar. 25, 1985).
USE: Formerly as herbicide.

9195. **Tabernanthine.** *13-Methoxyibogamine.* $C_{20}H_{26}N_2O$; mol wt 310.44. C 77.38%, H 8.44%, N 9.02%, O 5.15%. Indole alkaloid isolated from root of *Tabernanthe iboga* Baill., *Apocynaceae:* Delourme-Houdé, *Ann. Pharm. Franc.* **4**, 30 (1946); Dickel *et al., J. Am. Chem. Soc.* **80**, 123 (1958). Also in *Tabernaemontana* and *Stemmadenia* spp.; usually found in ibogaine mother liquors: Walls *et al., Tetrahedron* **2**, 173 (1958). Isoln from genus *Conopharingia, Apocynaceae:* Renner, Prins, U.S. pat. **3,008,954** (1961 to Geigy). Structure: Bartlett *et al., J. Am. Chem. Soc.* **80**, 126 (1958). Mass spectrum: Biemann, Friedmann-Spiteller, *ibid.* **83**, 4805 (1961). Derivs: Taylor, U.S. pat. **2,877,229** (1959 to Ciba). Interaction with benzodiazepine receptors: J.-H. Trouvin *et al., Eur. J. Pharmacol.* **140**, 303 (1987).

Needles or shiny leaflets from ethanol, mp 213.5-215°. Sublimes at 160° (0.005 mm pressure). $[\alpha]_D^{20}$ —40° (acetone). pKa 6.04 in 80% methylcellosolve. uv max (ethanol): 228, 271, 299 nm (log ϵ 4.53, 3.64, 3.77). Sol in alcohol, benzene, ether, chloroform. Practically insol in water.
Hydrochloride, $C_{20}H_{26}N_2O.HCl$, crystals from water, dec 275-277°. $[\alpha]_D^{25}$ —66° (methanol, Dickel, *loc. cit.*); mp 210°, $[\alpha]_D^{20}$ —76.5° (methanol, Delourme-Houdé). Sol in water. More sol in chloroform than ibogaine hydrochloride.

9196. **Tabun.** *Dimethylphosphoramidocyanidic acid, ethyl ester;* ethyl *N*-dimethylphosphoramidocyanidate; dimethylamidoethoxyphosphoryl cyanide; GA. $C_5H_{11}N_2O_2P$; mol wt 162.13. C 37.04%, H 6.84%, N 17.28%, O 19.74%, P 19.10%. Military nerve gas; prepd from dimethylamidophosphoryl dichloride and sodium cyanide in the presence of ethanol: Holmstedt, *Acta Physiol. Scand.* **25**, Suppl. 90, 26 (1951). The synthesis of dimethylamidophosphoryl di-

chloride is also described by Michaelis, *Ann.* **326**, 129 (1903). Alternate synthetic route: B. C. Saunders, *Some Aspects of the Chemistry and Toxic Action of Organic Compounds Containing Phosphorus and Fluorine* (Cambridge, 1957) p 91. Toxicity study: B. Holmstedt, *Pharmacol. Rev.* **11**, 567 (1959). Brief review: Schrader, *Die Entwicklung neuer insektizider Phosphorsaüre-Ester* (Verlag Chemie, Weinheim, 1963) p 3.

Liquid. Fruity odor reminiscent of bitter almonds. d 1.077. mp —50°. bp$_{760}$ 240°; bp$_{10}$ 120°; bp$_9$ 100-108°. n_D^{20} 1.4250. IR absorption: *Acta Chem. Scand.* **5**, 1179 (1951). Readily sol in organic solvents. Miscible with water, but quickly hydrolyzed. Bleaching powder (chlorinated lime) destroys Tabun, but gives rise to cyanogen chloride. *Extremely poisonous!* LD_{50} i.p. in mice: 0.6 mg/kg (Holmstedt). The lethal dose for man may be as low as 0.01 mg/kg, *Chem. & Eng. News* **31**, 4676 (1953).
Caution: Potent cholinesterase inhibitor. Toxic not only by inhalation but by absorption through skin and eyes. Inhalation produces constriction of pupils of the eye, difficulty in breathing followed by bronchial constriction, convulsions, death.
USE: Chemical warfare agent.

9197. **Tacalcitol.** *(1α,3β-5Z,7E,24R)-9,10-Secocholesta-5,7,10(19)-triene-1,3,24-triol;* 1α,24(R)-dihydroxycholecalciferol; 1α,24R-dihydroxyvitamin D_3; TV-02; Bonalfa. $C_{27}H_{44}O_3$; mol wt 416.64. C 77.84%, H 10.64%, O 11.52%. Bioactive, synthetic vitamin D_3 analog; exhibits antiproliferative effect on keratinocytes. Prepn: T. Takeshita *et al.,* Ger. pat. **2,526,981**; *eidem,* U.S. pat. **4,022,891** (1976, 1977 both to Teijin); M. Morisaki *et al., J. Chem. Soc, Perkin Trans. I* **1975**, 1421; K. Ochi *et al., ibid.* **1979**, 165. Pharmacology: T. Matsunage *et al., J. Dermatol.* **17**, 135 (1990). Clinical evaluation in psoriasis: M. J. P. Gerritsen *et al., Brit. J. Dermatol.* **131**, 57 (1994). *Review:* M. Nishimura *et al., Eur. J. Dermatol.* **3**, 255-261 (1993).

White solid. uv max (ethanol): 265 nm.
THERAP CAT: Antipsoriatic.

9198. **Tachysterol.** *(3β,6E,22E)-9,10-Secoergosta-5(10),6,8,22-tetraen-3-ol.* $C_{28}H_{44}O$; mol wt 396.66. C 84.79%, H 11.18%, O 4.03%. From ergosterol or lumisterol by ultraviolet irradiation: Windaus *et al., Ann.* **492**, 226 (1932); *Ann.* **499**, 188 (1932); Dimroth, *Ber.* **70**, 1631 (1937). From calciferol by adsorption on acid clay: Thibaudet, *Compt. Rend.* **220**, 751 (1945). From precalciferol: Velluz, Goffinet, U.S. pat. **2,847,426** (1958 to UCLAF). Structure: Grundmann, *Z. Physiol. Chem.* **252**, 151 (1938); Thibaudet, *loc. cit.* Stereochemistry of the tachysterol system: Inhoffen, *Ber.* **88**, 1424 (1955); Verloop, *Rec. Trav. Chim.* **76**, 689 (1957); Delaroff *et al., Bull. Soc. Chim. France* **1963**, 1739.

Oil. $[\alpha]_D^{18}$ −70° (24.6 mg in 2 ml petr ether); $[\alpha]_{546}^{18}$ −86.3° (petr ether). uv max: 280 nm. Not pptd by digitonin. Insol in water. Sol in organic solvents, but not in methanol. Very easily oxidized by air.

4-Methyl-3,5-dinitrobenzoate, $C_{36}H_{48}N_2O_6$, pale yellow crystals, mp 155°.

Dihydro derivative, see dihydrotachysterol.

9199. Tacrine. *1,2,3,4-Tetrahydro-9-acridinamine; 9-amino-1,2,3,4-tetrahydroacridine; 5-amino-1,2,3,4-tetrahydroacridine; 1,2,3,4-tetrahydro-5-aminoacridine.* $C_{13}H_{14}N_2$; mol wt 198.27. C 78.75%, H 7.12%, N 14.13%. Centrally active anticholinesterase; cognition enhancer. Prepn: Braun *et al., Ber.* **64,** 227 (1931); Albert, Gledhill, *J. Soc. Chem. Ind.* **64,** 169T (1945); Petrow, *J. Chem. Soc.* **1947,** 634; Moore, Kornreich, *Tetrahedron Letters* **1963,** 1277. Pharmacology and toxicity: S. Gershon, F. H. Shaw, *J. Pharm. Pharmacol.* **10,** 638 (1958). Enzyme inhibiting activity: P. N. Kaul, *ibid.* **14,** 243 (1962). Neuropharmacology: B. Drukarch *et al., Life Sci.* **42,** 1011 (1988). Clinical trial in severe Alzheimer's disease: W. K. Summers *et al., N. Engl. J. Med.* **315,** 1241 (1986). *See also: ibid.* **316,** 1603-1606. Review of pharmacology and clinical uses: W. K. Summers *et al., Clin. Toxicol.* **16,** 269-281 (1980).

Octahedra from very dil alc, mp 183-184°.

Hydrochloride, $C_{13}H_{14}N_2 \cdot HCl$, *Cognex, THA.* Yellow needles from concd hydrochloric acid, mp 283-284°. Bitter taste. Sol in water. pH of 1.5% soln: 4.5-6.

THERAP CAT: Nootropic. Antidote to curare. Respiratory stimulant.

9200. Tacrolimus. *[3S-[3R*[E(1S*,3S*,4S*)],4S*,5R*,-8S*,9E,12R*,14R*,15S*,16R*,18S*,19S*,26aR*]]-5,6,8,11,-12,13,14,15,16,17,18,19,24,25,26,26a-Hexadecahydro-5,19-dihydroxy-3-[2-(4-hydroxy-3-methoxycyclohexyl)-1-methylethenyl]-14,16-dimethoxy-4,10,12,18-tetramethyl-8-(2-propenyl)-15,19-epoxy-3H-pyrido[2,1-c][1,4]oxaazacyclotricosine-1,7,-20,21(4H,23H)-tetrone;* 17-allyl-1,14-dihydroxy-12-[2-(4-hydroxy-3-methoxycyclohexyl)-1-methylvinyl]-23,25-dimethoxy-13,19,21,27-tetramethyl-11,28-dioxa-4-azatricyclo-[22.3.1.0^{4,9}]octacos-18-ene-2,3,10,16-tetraone; FK-506; FR-900506; Prograf. $C_{44}H_{69}NO_{12}$; mol wt 804.03. C 65.73%, H 8.65%, N 1.74%, O 23.88%. Macrolide isolated from *Streptomyces tsukubaensis* no. 9993: M. Okuhara *et al.,* **Eur.** pat. **Appl. 184,162** (1986 to Fujisawa); and characterization: T. Kino *et al., J. Antibiot.* **40,** 1249 (1987). Structure determn: H. Tanaka *et al., J. Am. Chem. Soc.* **109,** 5031 (1987). Total synthesis of (−)-form: T. K. Jones *et al., J. Am. Chem. Soc.* **111,** 1157 (1989). *In vitro* immunosuppressant activity in comparison with cyclosporin, *q.v.*: T. Kino *et al., J. Antibiot.* **40,** 1256 (1987); N. Yoshimura *et al., Transplantation* **47,** 351 (1989). Toxicity studies: K. Ohara *et al., Transplantation Proc.* **22,** 83 (1990). Symposium on pharmacology, pharmacokinetics, toxicology, determn in biological fluids, experimental and clinical studies in transplant therapy, clinical use in autoimmune diseases: *Transplantation Proc.* **23,** 2709-3376 (1991). Review of mechanism of action: G. Wiederrecht *et al., Ann. N.Y. Acad. Sci.* **696,** 9-19 (1993). *Reviews:* W. H. Parsons *et al., ibid.* **685,** 22-36; D. H. Peters *et al., Drugs* **46,** 746-794 (1993).

Monohydrate, $C_{44}H_{69}NO_{12} \cdot H_2O$, colorless prisms from acetonitrile, mp 127-129°. $[\alpha]_D^{23}$ −84.4° (c = 1.02 in chloroform). Sol in methanol, ethanol, acetone, ethyl acetate, chloroform, diethyl ether; sparingly sol in hexane, petroleum ether. Insol in water. LD_{50} i.p. in mice: > 200 mg/kg (Kino). LD_{50} in male, female rats (mg/kg): 57.0, 23.6 i.v.; 134, 194 orally (Ohara).

THERAP CAT: Immunosuppressant.

9201. Tacryl®. Acrylic fiber of a specific multichain type. The structure could be described as a spider molecule with up to six long straight linear legs which can orient independently and build up a fibrous structure. The molecular structure is built up by a controlled cross-linking process, so that Tacryl is like wool as it has a specific chain molecular interlinking which contributes to its mechanical and elastic properties. Synthesis of multichain polymers: Schaefgen, Flory, *J. Am. Chem. Soc.* **70,** 2709 (1948). Tacryl has a higher shear modulus and a higher strength than linear acrylics with the same degrees of orientation. It undergoes very slow hydrolysis under hot acid aq conditions, and is very resistant to dry and wet heat (neutral conditions): Sundén, *Tappi* **41,** 173A (1958).

9202. D-Tagatose. D-*lyxo*-Hexulose. $C_6H_{12}O_6$; mol wt 180.16. C 40.00%, H 6.71%, O 53.28%. Occurs in *Sterculia setigera* gum, a partially acetylated acidic polysaccharide: Hirst *et al., J. Chem. Soc.* **1949,** 3145. Prepn from D-galactose: Reichstein, Bosshard, *Helv. Chim. Acta* **17,** 753 (1934); Jones, Wall, *Can. J. Chem.* **38,** 2290 (1960); by biochemical oxidation of D-talitol: Totton, Lardy, *J. Am. Chem. Soc.* **71,** 3076 (1949); from D-galacturonic acid: Gorin *et al., Can. J. Chem.* **33,** 1116 (1955); from lactose in heated milk: Adachi, *Nature* **181,** 840 (1958). Synthesis: Wolfrom, Bennett, *J. Org. Chem.* **30,** 1284 (1965); Al-Gobore *et al., Carbohyd. Res.* **16,** 466 (1971). Structure: Takagi, Rosenstein, *ibid.* **11,** 156 (1969). Stereoselective synthesis of L-*tagatose*, antipode of the naturally occurring D-tagatose: T. Mukaiyama *et al., Chem. Letters* **1982,** 1169.

Crystals from dil alc, mp 134-135°. $[\alpha]_D^{20}$ −2.3° (c = 2.19 in water); $[\alpha]_D^{25}$ −5° (c = 1 in water).

Phenylosazone, $C_{18}H_{22}N_4O_4$, crystals from dil ethanol, dec 187-188°. $[\alpha]_D^{23}$ +47° (c = 0.82 in 2-methoxyethanol).

9203. Taka-Diastase. α-*Amylase (Aspergillus oryzae);* Koji; Aspergillus diastase; Sanzyme. A purified multienzyme produced by the microorganism *Aspergillus oryzae* (Ahl.) Cohn, *Aspergillaceae,* grown on sterilized wheat bran or on rice hulls. Represents more than 30 different enzymatic functions. It is not only amylolytic but digests proteins and fats also.

Whitish-yellow, very hygroscopic powder. Converts 450 times its wt of starch into maltose. *Keep tightly closed.*

USE: Preparing the Japanese national drink "sake"; in converting maizes into sugar in manuf of whiskey.

THERAP CAT: Amylolytic.

9204. Talampicillin. *[2S-[2α,5α,6β(S*)]]-6-[(Aminophenylacetyl)amino]-3,3-dimethyl-7-oxo-4-thia-1-azabicyclo-[3.2.0]heptane-2-carboxylic acid 1,3-dihydro-3-oxo-1-isobenzofuranyl ester;* (2S,5R,6R)-6-[(R)-2-amino-2-phenylacetamido]-3,3-dimethyl-7-oxo-4-thia-1-azabicyclo[3.2.0]-heptane-2-carboxylic acid ester with 3-hydroxyphthalide; 6-[D(−)-α-aminophenylacetamido]penicillanic acid phthalide ester; ampicillin 1-oxo-1,3-dihydroisobenzofuran-3-yl ester; phthalidyl D-α-aminobenzylpenicillanate. $C_{24}H_{23}N_3$-O_6S; mol wt 481.53. C 59.86%, H 4.81%, N 8.73%, O 19.94%, S 6.66%. Semi-synthetic antibiotic related to penicillin. Prepn: H. Ferres, M. P. Clayton, Ger. pat. **2,228,012;** *eidem,* U.S. pat. **3,860,579** (1972, 1975 both to Beecham). *See also:* M. Murakami *et al.,* U.S. pat. **3,951,-954** (1976 to Yamanouchi). Pharmacology: Clayton *et al.,* Antimicrob. Ag. Chemother. **5,** 670 (1974). HPLC determn in urine and comparison of pharmacokinetics with other aminopenicillins: T. Uno *et al., Chem. Pharm. Bull.* **29,** 1957 (1981). Clinical comparison with ampicillin, *q.v.:* H. N. Williams *et al., Brit. J. Clin. Pract.* **35,** 147 (1981).

Hydrochloride, $C_{24}H_{23}N_3O_6S.HCl$, *BRL-8988, Talat, Talpen, Yamacillin.* White powder, mp 154-157° (dec).

THERAP CAT: Antibacterial.

9205. Talastine. *2-[2-(Dimethylamino)ethyl]-4-(phenylmethyl)-1(2H)-phthalazinone; 4-benzyl-2-[2-(dimethylamino)ethyl]-1(2H)-phthalazinone.* $C_{19}H_{21}N_3O$; mol wt 307.40. C 74.24%, H 6.89%, N 13.67%, O 5.20%. Prepn: Engelbrecht *et al.,* Ger. pat. **1,046,625;** *eidem,* U.S. pat. **3,017,411** (1958, 1962, both to VEB Deutsches Hydrierwerk Rodleben). Pharmacology: Lenke, *Arzneimittel-Forsch.* **7,** 678 (1957).

Liquid, $bp_{0.3}$ 215-222°.

Hydrochloride, $C_{19}H_{21}N_3O.HCl$, *HL-2186, Ahanon.* Crystals, mp 178°. LD_{50} i.p. in mice: 116 mg/kg.

THERAP CAT: Antihistaminic.

9206. Talbutal. *5-(1-Methylpropyl)-5-(2-propenyl)-2,4,-6(1H,3H,5H)-pyrimidinetrione; 5-allyl-5-sec-butylbarbituric acid;* 5-allyl-5-(1-methylpropyl)barbituric acid; Lotusate. $C_{11}H_{16}N_2O_3$; mol wt 224.26. C 58.91%, H 7.19%, N 12.49%, O 21.40%. Prepn: Volwiler, *J. Am. Chem. Soc.* **47,** 2236 (1925). Acute toxicity: E. W. Schafer, *Toxicol. Appl. Pharmacol.* **21,** 315 (1972).

Crystals from water or dil alcohol, mp 108-110°. Slightly bitter taste. Practically insol in water and petr ether. Sol in alcohol, chloroform, ether, acetone, glacial acetic acid, also in solns of fixed alkali hydroxides. A satd aq soln is acid to litmus. LD_{50} orally in rats: 57.5 mg/kg (Schafer).

Caution: May be habit forming. This is a controlled substance (depressant) listed in the U.S. Code of Federal Regulations, Title 21 Parts 329.1 and 1308.13 (1995).

THERAP CAT: Sedative, hypnotic.

9207. Talc. Talcum; French chalk. The lumps are also known as *soapstone* or *steatite.* Finely powdered native hydrous magnesium silicate.

White to grayish-white, very fine odorless, crystalline powder; unctuous, and adheres readily to the skin. Insol in water, cold acids or in alkalies.

Caution: Potential symptom of overexposure is fibriotic pneumoconiosis. *See NIOSH Pocket Guide to Chemical Hazards* (DHHS/NIOSH 90-117, 1990) p 202.

USE: Dusting powder, either alone or with starch or boric acid, for medicinal and toilet prepns; excipient and filler for pills, tablets and for dusting tablet molds; clarifying liquids by filtration. As pigment in paints, varnishes, rubber; filler for paper, rubber, soap; in fireproof and cold-water paints for wood, metal and stone; lubricating molds and machinery; glove and shoe powder; electric and heat insulator.

9208. Talinolol. *(±)-N-Cyclohexyl-N′-[4-[3-[(1,1-dimethylethyl)amino]-2-hydroxypropoxy]phenyl]urea;* (±)-1-[p-[3-(tert-butylamino)-2-hydroxypropoxy]phenyl]-3-cyclohexylurea; 02-115; Cordanum. $C_{20}H_{33}N_3O_3$; mol wt 363.50. C 66.09%, H 9.15%, N 11.56%, O 13.20%. β-Adrenergic blocker structurally related to practolol, *q.v.* Prepn: R. Eckardt *et al.,* Ger. pat. **2,153,024** (1972 to Arzneimittelwerke VEB), *C.A.* **77,** 61639b (1972); M. Wilhelm, Swiss pat. **566,322** corresp to U.S. pats. **3,935,259** and **4,038,313** (1975, 1976, 1977, all to Ciba-Geigy); R. Eckardt *et al., Pharmazie* **30,** 633 (1975). Series of articles on pharmacology, toxicology: *ibid.* 638-683. Acute toxicity: K. Femmer *et al., ibid.* 642. Clinical pharmacology: K. O. Haustein *et al., Int. J. Clin. Pharmacol. Biopharm.* **17,** 465 (1979). Cardiovascular, β-blocking effects: S. Heer, K. Femmer, *Pharmazie* **34,** 317 (1980).

Cryst from isopropanol, mp 142-144°. LD_{50} in rats, mice (mg/kg): 1180, 593 orally; 54.3, 74.7 i.p.; 29.7, 25.0 i.v. (Femmer).

THERAP CAT: Antihypertensive; antiarrhythmic.

9209. Talipexole. *5,6,7,8-Tetrahydro-6-(2-propenyl)-4H-thiazolo[4,5-d]azepin-2-amine;* 6-allyl-2-amino-5,6,7,8-tetrahydro-4H-thiazolo[4,5-d]azepine. $C_{10}H_{15}N_3S$; mol wt 209.32. C 57.38%, H 7.22%, N 20.08%, S 15.32%. α₂-Adrenoceptor and dopamine D_2-receptor agonist. Prepn: G. Griss *et al.,* Ger. pat. **2,040,510** (1972 to Thomae), U.S. pat. **3,804,849** (1974 to Boehringer, Ing.). Dopaminergic effects: P. A. Johansen *et al., Life Sci.* **43,** 515 (1988); S. K. Kulkarni, K. Chopra, *Meth. Find. Exp. Clin. Pharmacol.* **12,** 99 (1990). Clinical evaluation in Parkinson's disease: Y. Mizuno *et al., Drug Invest.* **5,** 186 (1993).

Dihydrochloride, $C_{10}H_{15}N_3S.2HCl$, **B-HT-920, Domin.** mp 245° (dec).

THERAP CAT: Antiparkinsonian.

9210. Tall Oil. Liquid rosin; Acintol C; tallol; talleol. ["Tall" is Swedish for "pine".] A by-product of the wood pulp industry. Usually recovered from pine wood "black liquor" of the sulfate or kraft paper process. Contains rosin acids, oleic and linoleic acids. Long chain alcohols and small amounts of sterols, especially phytosterol, have also been found. Comprehensive collection of 1660 abstracts: J. Weiner, *Tall Oil* (The Institute of Paper Chemistry, Appleton, Wisconsin, 3rd ed., 1959) 450 pp; J. Weiner, J. Byrne, 1st supplement (1965).

Dark brown liquid. Acrid odor similar to that of burnt rosin. d 0.95 to 1.0. $n_D^{20} \sim 1.5$. Acid no. 170-180°. Sapon no. 172-185. Iodine no. 120-188. Fatty acids 50-60%. Rosin acids 34-40%. Unsaponifiable matter 5-10%.

USE: Mfg soap pastes, flotation agents, greases, paint, alkyd resins, linoleum, soaps, fungicides, asphalt emulsions, rubber formulations, cutting oils, sulfonated oils. Review of possible uses: Cannon, *Chem. Eng.* **61,** 142 (June 1954).

9211. Tallow. In North America designates the fat from the fatty tissue of bovine cattle and sheep only. It may be offered separately as beef tallow and as sheep or mutton tallow. The term horse tallow is generally no longer admitted. *Oleo stock* is the highest grade of beef tallow. Contains (as glycerides): Oleic acid (37-43%), palmitic (24-32%), stearic (20-25%), myristic (3-6%), linoleic (2-3%). Minor constituents are cholesterol, arachidonic, elaidic, and vaccenic acids. Perhaps the most observed characteristic of tallow is its titer (solidif pt) which ranges from 40° to 46°.

9212. Tallow Alcohol. A name for commercial mixtures of *n*-octadecanol and *n*-hexadecanol.

Fatty crystalline mass, mp 46-47°.

USE: Defoaming agent, emollient, intermediate for surface active agents.

9213. Tallysomycin. Talisomycin; BU-2231. Antitumor antibiotic complex and third generation analog of bleomycins, *q.v.*, produced by *Streptoalloteichus hindustanus* E 465-94. Prodn, isoln, properties of the major components, tallysomycins A and B: H. Kawaguchi *et al.*, **Belg.** pat. **845,513;** *eidem*, **U.S.** pat. **4,051,237** (both 1977 to Bristol-Myers); *eidem*, *J. Antibiot.* **30,** 779 (1977). Structure of A and B, based on originally proposed bleomycin structure:

Tallysomycin A   R =

Tallysomycin B   R =

M. Konishi *et al.*, *ibid.* 789; *cf.* bleomycin for revised structure. Antitumor activity: H. Imanishi *et al.*, *ibid.* **31,** 667 (1978). Radioimmunoassay: A. Broughton *et al.*, *Cancer Treat. Rep.* **63,** 1829 (1979). Pharmacokinetics: J. E. Strong *et al.*, *ibid.* 1821. ^{13}C-NMR spectra of tallysomycin and its zinc complex: F. T. Greenaway *et al.*, *Org. Magn. Reson.* **13,** 270 (1980). Biosynthetic derivs: T. Miyaki *et al.*, *J. Antibiot.* **34,** 658, 665 (1981). Relative pulmonary toxicity: A. Broughton *et al.*, *Cancer Treat. Rep.* **64,** 659 (1980). *Review:* S. T. Crooke *et al.*, *Recent Results Cancer Res.* **76,** 83-90 (1981).

Tallysomycin A, $C_{68}H_{110}N_{22}O_{27}S_2$, N^1-*[4-amino-6-[[3-[(4-aminobutyl)amino]propyl]amino]-6-oxohexyl]-13-[(4-amino-4,6-dideoxy-α-L-talopyranosyl)oxy]-19-demethyl-12-hydroxybleomycinamide; BU-2231A.* The term "talisomycin" has also been used to refer to tallysomycin A. White amorphous solid. No definite mp; gradually dec > 210°. Sol in water, methanol, DMF; practically insol in other organic solvents. $[\alpha]_D^{23}$ $-21°$ (c = 0.5 in water). uv max (water): 290 nm ($E_{1cm}^{1\%}$ 67). LD_{50} s.c. in mice: 28 mg/kg (Kawaguchi, U.S. pat. **4,051,237**).

Tallysomycin B, $C_{62}H_{98}N_{20}O_{26}S_2$, N^1-*[3-[(4-aminobutyl)-amino]propyl]-13-[(4-amino-4,6-dideoxy-α-L-talopyranosyl)-oxy]-19-demethyl-12-hydroxybleomycinamide; talisomycin B; BU-2231B.* Mp, soly similar to tallysomycin A. $[\alpha]_D^{23}$ $-19°$ (c = 0.5 in water). uv max (water): 289.5 ($E_{1cm}^{1\%}$ 77). LD_{50} s.c. in mice: > 50 mg/kg (Kawaguchi, U.S. pat. **4,051,237**).

9214. Talniflumate. *2-[[3-(Trifluoromethyl)phenyl]-amino]-3-pyridinecarboxylic acid 1,3-dihydro-3-oxo-1-isobenzofuranyl ester;* phthalidyl 2-(3-trifluoromethylanilino)-nicotinate; phthalidyl 2-(α,α,α-trifluoro-*m*-toluidino)nicotinate; BA-7602-06; Somalgen. $C_{21}H_{13}F_3N_2O_4$; mol wt 414.34. C 60.88%, H 3.16%, F 13.76%, N 6.76%, O 15.45%. Deriv of niflumic acid, *q.v.* Prepn: **Belg.** pat. **858,864** (1978 to Bago), *C.A.* **89,** 109104 (1978); S. Bago, U.S. pat. **4,168,313** (1979). Synthesis and pharmacologic study: M. Los *et al.*, *Farmaco Ed. Sci.* **36,** 372 (1981).

White or pale yellow crystalline powder, mp 165-166°. uv max (chloroform): 287, 357 nm (ε 25600, 7800). LD_{50} orally in rats: 12000 mg/kg (Los).

THERAP CAT: Anti-inflammatory; analgesic.

9215. Tamarind. Partially dried ripe fruit of *Tamarindus indica* L., *Leguminosae,* preserved in sugar or syrup. *Habit.* East Indies, India, Africa; naturalized in West Indies. *Constit.* The pulp contains about 10% tartaric acid, also some citric and malic acids; 25-40% invert sugar, pectin. *Review:* Rao, Srivastava, in *Industrial Gums,* R. L. Whistler, Ed. (Academic Press, New York, 2nd ed., 1973) pp 369-411.

USE: The pulp as souring agent in Indian curries. The seed kernel powder with water as sizing agent.

9216. Tamoxifen. *(Z)-2-[4-(1,2-Diphenyl-1-butenyl)-phenoxy]-N,N-dimethylethanamine;* 1-*p*-β-dimethylaminoethoxyphenyl-*trans*-1,2-diphenylbut-1-ene. $C_{26}H_{29}NO$; mol wt 371.52. C 84.06%, H 7.87%, N 3.77%, O 4.31%. Nonsteroidal estrogen antagonist. Prepn: **Belg.** pat. **637,389** (1964 to ICI). Identification and separation of isomers: G. R. Bedford, D. N. Richardson, *Nature* **212,** 733 (1966); **Belg.** pat. **678,807;** M. J. K. Harper *et al.*, **Brit.** pat. **1,064,-629;** *eidem,* **U.S.** pat. **4,536,516** (1966, 1967, 1985 all to ICI). Synthesis of isomers: D. W. Robertson, J. A. Katzenellenbogen, *J. Org. Chem.* **47,** 2387 (1982). Stereospecific synthesis of isomers: R. B. Miller, M. S. Al-Hassan, *ibid.* **50,** 2121 (1985). Synthesis of *cis* and *trans* forms: M. I. Al-Hassan, *Synth. Commun.* **17,** 1247 (1987). Comparative activity of isomers: M. J. K. Harper, A. L. Walpole, *Nature* **212,** 87 (1966). Pharmacology: *eidem, J. Endocrinol.* **37,** 83

(1967). Review of biochemistry: R. I. Nicholson *et al.,* *Adv. Sex Horm. Res.* **4,** 119-152 (1980); of pharmacology and use in breast cancer: V. C. Jordan *et al., Cancer Treat. Rep.* **64,** 745-759 (1980); of pharmacokinetics and metabolism: V. C. Jordan, *Breast Cancer Res. Treat.* **2,** 123-138 (1982); of chemistry, pharmacology and clinical uses: B. J. A. Furr, V. C. Jordan, *Pharmacol. Ther.* **25,** 127-205 (1984). Brief review of use in mastalgia: I. S. Fentiman, *Drugs* **32,** 477-480 (1986). Review of toxicities and drug resistance: V. C. Jordan, *Ann. Rev. Pharmacol. Toxicol.* **35,** 195-211 (1995).

Crystals from petr ether, mp 96-98°.
Citrate, $C_{26}H_{29}NO.C_6H_8O_7$, *ICI-46474, Kessar, Noltam, Nolvadex, Nourytam, Tamofen, Tamoxasta, Zemide.* Fine, white, odorless crystalline powder, mp 140-142°. Slightly sol in water; sol in ethanol, methanol, acetone. Hygroscopic at high relative humidities. Sensitive to uv light. LD_{50} in mice, rats (mg/kg): 200, 600 i.p.; 62.5, 62.5 i.v.; 3000-6000, 1200-2500 orally (Furr, Jordan).
cis-Form base, mp 72-74° from methanol.
cis-Form citrate, $C_{26}H_{29}NO.C_6H_8O_7$, *ICI-47699.* mp 126-128°.
THERAP CAT: Antiestrogen. Palliative treatment of breast cancer.

9217. Tamsulosin. *(R)-5-[2-[[2-(2-Ethoxyphenoxy)ethyl]amino]propyl]-2-methoxybenzenesulfonamide;* amsulosin. $C_{20}H_{28}N_2O_5S$; mol wt 408.52. C 58.80%, H 6.91%, N 6.86%, O 19.58%, S 7.85%. Specific α_1-adrenoceptor antagonist. Prepn: K. Imai *et al.,* **Eur.** pat. **Appl. 34,432;** *eidem,* **U.S.** pat. **4,703,063** (1981, 1987 both to Yamanouchi). Pharmacology: K. Honda *et al., Arch. Pharmacol.* **328,** 264 (1985); of racemate and enantiomers: K. Honda *et al., ibid.* **336,** 295 (1987). HPLC determn in plasma: Y. Soeishi *et al., J. Chromatog.* **533,** 291 (1990). Clinical trials in micturition difficulty: K. Kawabe, T. Niijima, *Urol. Int.* **42,** 280 (1987); in benign prostatic hypertrophy: K. Kawabe *et al., J. Urol.* **144,** 908 (1990).

(±)-Form hydrochloride, $C_{20}H_{28}N_2O_5S.HCl$, *YM-12617.* mp 254-256°.
(*R*)-Form hydrochloride, *LY-253351, YM-12617-1, YM-617, Harnal.* mp 228-230°. $[\alpha]_D^{24}$ −4.0° (c = 0.35 in methanol).
(*S*)-Form hydrochloride, *YM-12617-2.* mp 228-230°. $[\alpha]_D^{24}$ +4.2° (c = 0.36 in methanol).
THERAP CAT: In treatment of benign prostatic hypertrophy.

9218. Tanacetin. *[3aS-(3aα,5aβ,6β,9aα,9bβ)]-Decahydro-6,9a-dihydroxy-5a-methyl-3,9-bis(methylene)naphtho-[1,2-b]furan-2(3H)-one;* 1β,5α-dihydroxy-6β,7αH-selina-4(15),11(13)-dien-6,12-olide. $C_{15}H_{20}O_4$; mol wt 264.32. C 68.16%, H 7.63%, O 24.21%. Isoln from seed, herb, and flowers of *Tanacetum vulgare* L., *Compositae:* Homolle, *J. Pharm. Chim.* **7,** 57 (1845); Jaretzky, Kühne, *Arch. Pharm.* **271,** 353 (1933); Suchy, *Coll. Czech. Chem. Commun.* **27,**

1058 (1962). Structure and absolute config: Samek *et al., ibid.* **38,** 1971 (1973).

Crystals, mp 205°. $[\alpha]_D^{22}$ +179.5° (c = 2.3 in ethanol).

9219. Tandospirone. *(3aα,4β,7β,7aα)-Hexahydro-2-[4-[4-(2-pyrimidinyl)-1-piperazinyl]butyl]-4,7-methano-1H-isoindole-1,3(2H)-dione; (1R*,2S*,3R*,4S*)-N-[4-[4-(2-pyrimidinyl)-1-piperazinyl]butyl]-2,3-bicyclo[2.2.1]heptanedicarboximide.* $C_{21}H_{29}N_5O_2$; mol wt 383.49. C 65.77%, H 7.62%, N 18.26%, O 8.34%. Serotonin (5-HT_{1A}) receptor agonist. Prepn: K. Ishizumi *et al.,* **Eur.** pat. **Appl. 82,402;** *eidem,* **U.S.** pat. **4,507,303** (1983, 1985 both to Sumitomo); *idem et al., Chem. Pharm. Bull.* **39,** 2288 (1991). Behavioral pharmacology: C. A. Sannerud *et al., Drug Alc. Depend.* **32,** 195 (1993). Clinical efficacy in treatment of bulimia: H. Tamai *et al., Int. J. Obesity* **14,** 289 (1990). Clinical evaluation of potential adverse effects: M. Suzuki *et al., Japan. J. Psychopharmacol.* **13,** 213 (1993); of abuse liability: S. M. Evans *et al., J. Pharmacol. Exp. Ther.* **271,** 683 (1994). Review of pharmacology: P. A. Seymour *et al., Prog. Clin. Biol. Res.* **361,** 453-460 (1990).

Crystals from toluene/*n*-hexane, mp 112-113.5°.
Citrate, $C_{21}H_{29}N_5O_2.C_6H_8O_7$, *SM-3997,* mp 169.5-170°.
Hydrochloride, $C_{21}H_{29}N_5O_2.HCl$, crystals from isopropanol, mp 227-229°.
THERAP CAT: Anxiolytic; antidepressant.

9220. Tanghinigenin. *(3β,5β,7β)-7,8-Epoxy-3,14-dihydroxycard-20(22)-enolide.* $C_{23}H_{32}O_5$; mol wt 388.50. C 71.11%, H 8.30%, O 20.59%. Isoln from glucosides: Sigg *et al., Helv. Chim. Acta* **38,** 166 (1955). Structure: Flury, Reichstein, *Ann. Chim. (Rome)* **53,** 23 (1963); Flury *et al., Helv. Chim. Acta* **48,** 1113 (1965). Toxicity study: Chen, Henderson, *J. Pharmacol. Exp. Ther.* **111,** 365 (1954).

Prisms from acetone + petr ether, mp 187-188°. $[\alpha]_D^{25}$ +14.1° (c = 1.138 in chloroform). uv max: 217 nm (log ε 4.22). LD_{50} i.v. in cats: 1 mg/kg (Chen, Henderson).
Acetate, $C_{25}H_{34}O_6$, *acetyltanghinigenin.* Prisms from acetone + petr ether, mp 241-243°. $[\alpha]_D^{24}$ +14.9° (c = 1.075 in chloroform).
3-[(2-*O*-Acetyl-6-deoxy-3-*O*-methyl-α-L-glucopyranosyl)-oxy], $C_{32}H_{46}O_{10}$, *tanghinin.* From the seed of *Tanghinia*

madagascariensis Pet., *Apocynaceae* and *Tanghinia venenifera* Poir., *Apocynaceae*. Isoln: Arnaud, *Compt. Rend.* **108**, 1255 (1889); Frèrejacque *et al.*, *ibid.* **226**, 268 (1948). Leaflets from methanol + ether, mp 128-131°. $[\alpha]_D^{21}$ −81.5° (c = 1.092 in methanol). LD_{50} i.v. in cats: 0.4 mg/kg (Chen, Henderson).

9221. Tannic Acid. Tannin; gallotannin; gallotannic acid. Incorrectly *"digallic acid"*. Tannic acid of commerce usually contains about 10% H_2O. Occurs in the bark and fruit of many plants, notably in the bark of the oak species, in sumac and myrobalan. It is produced from Turkish or Chinese nutgall, the former contg 50-60%, the latter about 70%. The chemistry of the tannins is most complex and non-uniform. Tannins may be divided into 2 groups: *(a)* derivatives of flavanols, so-called condensed tannins and *(b)* hydrolyzable tannins (the more important group) which are esters of a sugar, usually glucose, with one or more trihydroxybenzenecarboxylic acids. The structure given here is that of a tannin named *corilagin*: Schmidt *et al.*, *Ann.* **587**, 67 (1954). The empirical formula of corilagin is $C_{27}H_{22}O_{18}$. For the commercial tannic acid, whose specifications follow, the empirical formula is usually given as $C_{76}H_{52}O_{46}$. Toxicity study: Robinson, Graessle, *J. Pharmacol. Exp. Ther.* **77**, 63 (1943). Comprehensive reviews: M. Nierenstein, *The Natural Organic Tannins* (London, 1934); O. Th. Schmidt, "Gallotannine" in *Fortschr. Chem. Org. Naturst.* **13**, 70-136 (1956); *Symposium on the Chemistry of Vegetable Tannins* (Soc. Leather Trades Chemists, Croydon 1956).

Corilagin

Yellowish-white to light brown, amorphous, bulky powder or flakes, or spongy masses; faint characteristic odor; astringent taste. Gradually darkens on exposure to air and light; at 210-215° dec mostly into pyrogallol and CO_2. Gives insol ppts with albumin, starch, gelatin, most alkaloidal and metallic salts; produces a bluish-black color or precip with ferric salts. One gram dissolves in 0.35 ml water, 1 ml warm glycerol; very sol in alc, acetone. Practically insol in benzene, chloroform, ether, petr ether, carbon disulfide, carbon tetrachloride. *Keep well closed and protected from light.* LD_{100} orally in mice: 6.0 g/kg (Robinson, Graessle).

USE: Mordant in dyeing; manuf ink; sizing paper and silk; printing fabrics; with gelatin and albumin for manuf of imitation horn and tortoise shell; tanning; clarifying beer or wine; in photography; as coagulant in rubber manuf; manuf gallic acid and pyrogallol; as reagent in analytical chemistry.

THERAP CAT: Astringent.

THERAP CAT (VET): Astringent, hemostatic, in solutions for burns. Has been used internally as an astringent and as a heavy metal antidote.

9222. Tannoform. *Methyleneditannin;* tannin-formadehyde; Helgotan. Prepd by condensing one mole formaldehyde with two moles tannin: Chemnitius, *Pharm. Zentralh.* **68**, 273 (1927); Schwyzer, *Pharm. Ztg.* **74**, 1334 (1929).

Reddish, odorless, tasteless, bulky powder, mp ~230° with decompn. Practically insol in water; sol in alcohol, alkaline fluids.

THERAP CAT: Astringent.

THERAP CAT (VET): Externally as astringent, antiseptic (skin lesions and otorrhea). Has been used internally for diarrhea.

9223. Tantalum. Ta; at. wt 180.9479; at. no. 73; valence 5, also 4, 3, 2. Two naturally occurring isotopes: 181 (99.9877%); 180 (0.0123%; $T_{1/2}$ > 10^{12} years); artificial radioactive isotopes: 172-179; 182-186. Occurs almost invariably with niobium; less abundant than niobium. Found in the minerals columbite, *q.v.*, *tantalite* [(Fe,Mn)(Ta,Nb)$_2O_6$] and *microlite* [(Na,Ca)$_2$Ta$_2O_6$(O,OH,F)]. Discovered by Ekeberg in 1802; first obtained pure by Bolton: *Z. Elektrochem.* **11**, 45 (1905). Prepn: Schoeller, Powell, *J. Chem. Soc.* **119**, 1927 (1921). Reviews of tantalum and its compounds: G. L. Miller, *Tantalum and Niobium* (Academic Press, New York, 1959) 767 pp; Brown, "The Chemistry of Niobium and Tantalum" in *Comprehensive Inorganic Chemistry* vol. 3, J. C. Bailar, Jr. *et al.*, Eds. (Pergamon Press, Oxford, 1973) pp 553-622.

Gray, very hard, malleable, ductile metal; can readily be drawn in fine wires. mp 2996°. bp 5429°. d 16.69. Spec heat (0°): 0.036 cal/g/°C. Electrical resistivity (18°): 12.4 μohm-cm. Insol in water. Very resistant to chemical attack; not attacked by acids other than hydrofluoric; not attacked by aq alkalies; slowly attacked by fused alkalies. Reacts with fluorine, chlorine, and oxygen only on heating. At high temps absorbs several hundred times its volume of hydrogen; combines with nitrogen, with carbon.

USE: In pen points; analytical weights; apparatus and instruments for chemical, surgical, and dental use instead of platinum, in tantalum capacitors (a type of electrolytic condenser, trademarked "Tantalytic").

9224. Tantalum Pentachloride. Cl_5Ta; mol wt 358.21. Cl 49.49%, Ta 50.51%. $TaCl_5$. Prepn: Rolsten, *J. Am. Chem. Soc.* **80**, 2952 (1958). Review of tantalum halides: Fairbrother in *Halogen Chemistry* vol. 3, V. Gutmann, Ed. (Academic Press, New York, 1967) pp 123-178. Acute toxicity: K. W. Cochran *et al.*, *Arch. Ind. Hyg. Occup. Med.* **1**, 637 (1950).

White or light yellow, cryst powder; monoclinic; dec in moist air. d 3.68; mp 216.5-220°. Begins to volatilize at 144°, bp 239.3°. Dec by water; sol in abs alcohol. LD_{50} in rats (mg/kg): 75 i.p.; 1900 orally (Cochran).

9225. Tantalum Pentafluoride. F_5Ta; mol wt 275.94. F 34.42%, Ta 65.58%. TaF_5. Prepd from tantalum pentachloride by the halide exchange method according to the equation $TaCl_5 + 5HF \rightarrow TaF_5 + 5HCl$: Ruff, Zedner, *Ber.* **42**, 492 (1909); Ruff, Schiller, *Z. Anorg. Allgem. Chem.* **72**, 329 (1911); Kwasnik in *Handbook of Preparative Inorganic Chemistry* vol. 1, G. Brauer, Ed. (Academic Press, New York, 2nd ed, 1963) pp 255-256. Prepn from the elements: Fairbrother, Frith, *J. Chem. Soc.* **1951**, 3051. Review of transition metal pentafluorides: Peacock, *Advan. Fluoride Chem.* **7**, 113-145 (1973).

Deliquescent, strongly refractive prisms. d^{20} 4.74. mp 96.8°. Also reported as 95.1°: Fairbrother, Frith, *loc. cit.* bp 229.5°. Sol in water and ether with formation of oxyfluoro complexes. Also sol in concd nitric acid, more sol in fuming nitric acid. Sparingly sol in hot carbon disulfide and hot carbon tetrachloride. Etches glass slowly.

USE: Friedel-Crafts catalyst.

9226. Tantalum Pentoxide. Tantalic acid anhydride. O_5Ta_2; mol wt 441.89. O 18.10%, Ta 81.90%. Ta_2O_5. Acute toxicity: K. W. Cochran *et al.*, *Arch. Ind. Hyg. Occup. Med.* **1**, 637 (1950).

White, microcrystalline, infusible powder. Insol in water, alcohol, mineral acids. Sol in HF. Dec by fusing with $KHSO_4$ or KOH, forming potassium tantalate with the latter. LD_{50} orally in rats: 8000 mg/kg (Cochran).

9227. Taprostene. *[3aR-[2Z,3aα,4α(1E,3S*)-5β,6aα]]-3-[[4-(3-Cyclohexyl-3-hydroxy-1-propenyl)hexahydro-5-hydroxy-2H-cyclopenta[b]furan-2-ylidene]methyl]benzoic acid;* α-[(2Z,3aR,4R,5R,6aS)-4-[(1E,3S)-3-cyclohexyl-3-hydroxypropenyl]hexahydro-5-hydroxy-2H-cyclopenta[b]furan-2-ylidene]-*m*-toluic acid; [(5Z,13E,9α,11α,15S)-2,3,4-trinor-1,-5-inter-*m*-phenylene-6,9-epoxy-11,15-dihydroxy-15-cyclohexyl-16,17,18,19,20-pentanor]prosta-5,13-dienoic acid. $C_{24}H_{30}O_5$; mol wt 398.50. C 72.34%, H 7.59%, O 20.07%. Prostacyclin analog; platelet aggregation inhibitor. Prepn: U. Seipp *et al.*, *Eur. pat. Appl.* **45,842**; *eidem*, U.S. pat. **4,372,971** (1982, 1983 both to Grünenthal); L. Flohé *et al.*,

Arzneimittel-Forsch. **33**, 1240 (1983). As adjuvant to thrombolytic therapy in acute myocardial infarction: F. W. Bär *et al., Eur. Heart J.* **14**, 1118 (1993). Review of pharmacology and clinical experience: J. Schneider *et al., Cardiovasc. Drug Rev.* **11**, 479-500 (1993).

Sodium salt, $C_{24}H_{29}NaO_5$, *CG-4203, Rheocyclan.* $[\alpha]_D^{22}$ +249° (c = 0.68 in methanol). LD_{50} in mice, rats (mg/kg): 164, 20 i.v. (Schneider).

THERAP CAT: Antithrombotic.

9228. Taraxacum. Dandelion; lion's tooth. Dried rhizome and roots of *Taraxacum palustre* (Lyons) Lam. & DC. (*T. officinale* Weber, *Leontodon taraxacum* L.), *Compositae. Habit.* Europe; naturalized in North America. *Constit.* Taraxerol, choline, levulin, inulin, pectin.

9229. Taraxasterol. *(3β,18α,19α)-Urs-20(30)-en-3-ol;* taraxast-20(30)-en-3β-ol; anthesterin; α-lactucerol; taraxasterin. $C_{30}H_{50}O$; mol wt 426.73. C 84.44%, H 11.81%, O 3.75%. A monohydroxy triterpene. Isoln from *Taraxacum officinale*, Wiggers, *Compositae:* Power, Browning, *J. Chem. Soc.* **101**, 2411 (1912). Structure and configuration: Ames *et al., ibid.* **1954**, 1905. Identity with anthesterin: Power, Browning *ibid.* **105**, 1829 (1914); with α-lactucerol: Zellner, *Monatsh.* **47**, 681 (1926).

Needles from alcohol, mp 221-222°. $[\alpha]_D$ +96.3° (CHCl₃). Very sol in alcohol, ether, petr ether; slightly sol in chloroform, benzene, carbon disulfide, acetone.

Acetate, $C_{32}H_{52}O_2$, *lactucerin, lactucon.* Hexagonal plates. mp 251-252° (from ethyl acetate + alcohol). $[\alpha]_D$ +100.5°.

9230. Taraxein. A protein complex isolated from the blood serum of schizophrenics by chromatography on diethylaminoethyl cellulose. The taraxein fraction precedes the ceruloplasmin fraction. Contains copper bound to protein. Method of isolation: Heath *et al., Am. J. Psychiat.* **114**, 14 (1957); *(Lippincott's) Medical Science* **6**, 401 (1959). Processing and identification: Heath *et al., Proc. 3rd World Congr. Psychiat., Montreal, 1961* **1**, 619 (1962). Studies of taraxein in schizophrenics: Heath *et al., Arch. Gen. Psychiat.* **16**, 1, 10, 24 (1967).

9231. Taraxerol. *(3β)-D-Friedoolean-14-en-3-ol;* isoolean-14-en-3β-ol; skimmiol; alnulin; tiliadin. $C_{30}H_{50}O$; mol wt 426.73. C 84.44%, H 11.81%, O 3.75%. Found in *Tilia cordata* Mill., *Tiliaceae:* Bräutigam, *Arch. Pharm.* **238**, 555 (1900); in *Alnus glutinosa* (L.) Gaertn., *Betulaceae:* Zellner, Weiss, *Sitzber. Akad. Wiss Wien* **132**, 258 (1923); in *Taraxacum officinale* Weber, *Compositae:* Burrows, Simpson, *J. Chem. Soc.* **1938**, 2042; in *Litsea dealbata* Nees, *Lauraceae:* Dunstan *et al., Aust. J. Chem.* **6**, 321 (1953); from *Befaria*

racemosa (Vent.), *Ericaceae:* Euda *et al., J. Org. Chem.* **26**, 271 (1961). Structure: Beaton *et al., J. Chem. Soc.* **1955**, 2131. Partial synthesis from β-amyrin: *eidem, Chem. & Ind. (London)* **1955**, 35.

Plates from chloroform + methanol, needles from benzene, mp 282-285°. uv max (ethanol): 210, 215, 220, 223 nm (ε 3900, 2400, 700, 250). Sol in benzene, chloroform, ether, ethyl acetate, acetic anhydride, acetic acid, phenol, pyridine, xylene; less sol in alcohol.

Acetate, $C_{32}H_{52}O_2$, plates from chloroform + methanol, mp 303-305°. $[\alpha]_D$ +10.5° (c = 1.8 in chloroform).

Benzoate, $C_{37}H_{54}O_2$, needles from benzene or chloroform + alc, mp 292-293°. $[\alpha]_D^{23}$ +35.7° (c = 0.7 in chloroform).

9232. Tar Oil. Volatile oil distilled from wood tar. *Principal constit.* Phenolic substances and hydrocarbons.

Almost colorless liquid when fresh, but soon becomes dark brownish-red. d 0.860-0.900. Insol in water; sol in alcohol, ether.

9233. Tar Oil, Rectified. Pine tar oil. The volatile oil from pine tar rectified by steam distillation. Chief active constituents are phenolic substances.

Dark reddish-brown, thin liquid; strong empyreumatic odor and taste. d_{25}^{25} 0.960-0.990. Insol in water; miscible with alcohol.

THERAP CAT: Antiseptic (topical); dermatologic.

THERAP CAT (VET): Antiseptic, antipruritic. For chronic skin conditions and in hoof dressings. Has been used internally as an expectorant.

9234. Tarragon. Estragon. The dried leaves and flowering tops of the perennial herb, *Artemisia dracunculus* L., *Compositae. Habit.* Siberia, Caspian Sea region; cultivated in Western Europe. Yields up to 0.8% oil of tarragon. *Constit.* p-Allylanisole (estragole; methyl chavicol); ocimene; myrcene; phellandrene(?); p-methoxycinnamaldehyde.

USE: The herb for culinary purposes, the oil as flavoring agent in liqueurs, soups, sauces and salad dressings. In perfumery to improve the note of chypre type perfumes.

9235. D-Tartaric Acid. *[S-(R*,R*)]-2,3-Dihydroxybutanedioic acid;* unusual tartaric acid; unnatural tartaric acid; l-tartaric acid; (−)-tartaric acid; levotartaric acid; D-threo-2,3-dihydroxysuccinic acid. $C_4H_6O_6$; mol wt 150.09. C 32.01%, H 4.03%, O 63.96%. Levorotatory tartaric acid having a dextro configuration. Although termed "unnatural," its occurrence in nature has been demonstrated. Obtained in small amounts from racemic tartaric acid through biochemical cleavage using *Penicillium notatum, Aspergillus griseus, A. niger* or other microorganisms: Pasteur, *Compt. Rend.* **51**, 298 (1860). Alternate route using salt formation with *d*-methylamphetamine: Walton, *J. Soc. Chem. Ind.* **64**, 219 (1945). Monograph: K. Freudenberg, *Stereochemie* **I**, (1933), reprinted by J. W. Edwards (Ann Arbor, 1945). Crystallographic data: A. N. Winchell, *The Optical Properties of Organic Compounds* (Academic Press, New York, 2nd ed., 1954) p 47.

Monoclinic sphenoidal prisms. d_4^{20} 1.7598. mp 168-170°. $[\alpha]_D^{20}$ −12.0° (c = 20 in H_2O). pKa_1 2.93; pKa_2 4.23. One gram dissolves in 0.75 ml water at room temp, in 0.5 ml boiling water, 1.7 ml methanol, 3 ml ethanol, 0.5 ml propanol, 250 ml ether. Also sol in glycerol. Insol in chloroform. Maximum soly in water at 20°: 139 g/100 ml.

9236. DL-Tartaric Acid. *2,3-Dihydroxybutanedioic acid;* racemic tartaric acid; racemic acid; *dl*-tartaric acid; resolvable tartaric acid; uvic acid; paratartaric acid; *dl*-Weinsäure (German); Vogesensäure (German); Traubensäure (German). $C_4H_6O_6$; mol wt 150.09. C 32.01%, H 4.03%, O 63.96%. Probably never a natural product, although sometimes found in small amounts during winemaking. Prepn from L-tartaric acid by boiling with aq NaOH (*meso*-tartaric acid is obtained as a byproduct): Holleman, *Org. Syn.* **6**, 82 (1926); **coll. vol. I** (2nd ed., 1941) p 497. Synthesis by oxidation of fumaric acid: Milas, Terry, *J. Am. Chem. Soc.* **47**, 1412 (1925); Milas, Sussman, *ibid.* **58**, 1302 (1936); U.S. pat. **2,000,213** (1935 to Standard Brands). From maleic acid: Church, Blumberg, *Ind. Eng. Chem.* **43**, 1780 (1951).

COOH
|
HC—OH
|
HC—OH
|
COOH

Anhyd acid, triclinic pinacoidal crystals from abs alc, from water above 73°, or by drying the monohydrate at 100°. mp 206°. pKa_1 2.96; pKa_2 4.24. Less soluble in water than L-tartaric acid. pH of 0.1M aq soln: 2.0. Soly in alcohol (g/100 g): 2.006 at 0°; 3.153 at 15°; 5.01 at 25°; 6.299 at 40°. Soly in ether about 1%.
Monohydrate, triclinic pinacoidal crystals from water. d_4^{20} 1.697. One hundred parts (w/w) of water dissolve 14.00 parts at 10°; 20.60 at 20°; 29.10 at 30°; 43.32 at 40°; 99.88 at 70°; 184.91 at 100°.

9237. L-Tartaric Acid. *[R-(R*,R*)]-2,3-Dihydroxybutanedioic acid;* ordinary tartaric acid; natural tartaric acid; *d*-tartaric acid; (+)-tartaric acid; dextrotartaric acid; L-2,3-dihydroxybutanedioic acid; *d*-α,β-dihydroxysuccinic acid; Weinsäure (German); Weinsteinsäure (German). $C_4H_6O_6$; mol wt 150.09. C 32.01%, H 4.03%, O 63.96%. Dextrorotatory tartaric acid having a levo configuration. Widely distributed in nature, classified as a fruit acid. Occurs in many fruits, free and combined with potassium, calcium or magnesium. Observed in antiquity as the acid potassium salt found deposited as a fine crystalline crust during fermentation of grape juice or tamarind juice and termed *faecula* (little yeast) by the Romans. The derivation from *Tartarus* is of medieval, alchemical origin. In modern processes the acid potassium tartrate obtained during wine-making is first converted to calcium tartrate which is then hydrolyzed to tartaric acid and calcium sulfate: Metzner, *Chem. Eng. Progress* **43**, 160 (1947); several modifications, e.g., **Ital. pat. 490,221** (1954 to Procedimenti Chimici), *C.A.* **50**, 11607c (1956). Extraction from tamarind pulp in about 10% yield: **Indian pat. 52,167** (1955), *C.A.* **50**, 5249g (1956). Synthesis by hydroxylation of maleic acid: Church, Blumberg, *Ind. Eng. Chem.* **43**, 1780 (1951). Practically all of the L-tartaric acid sold today is a byproduct of the wine industry. Monograph: U. Roux, *La Grande Industrie des Acides Organiques* (Dounod, Paris, 1939). Example of a modern process: Dabul, **U.S. pat. 3,114,770** (1963 to Orandi & Massera).

COOH
|
HC—OH
|
HO—CH
|
COOH

Monoclinic sphenoidal prisms, mp 168-170°. Stable to air and light. Strong acid taste. Refreshing when in dil aq soln. d_4^{20} 1.7598. Odor of burnt sugar when heated to mp. $[\alpha]_D^{20}$ +12.0° (c = 20 in H_2O). Strong organic acid. At 25° pKa_1 2.98; pKa_2 4.34. pH of 0.1N soln: 2.2. Heat of com-

bustion: −275.1 kcal/mol. Specific heat: 0.288 cal/g/°C at 21 to 51°; 0.296 at 0 to 99.6°. Dielectric constant 36.0 for 1200 cm waves. Freely sol in water. d_4^{15} of aq solns (w/w at 15°): 1% 1.0045; 10% 1.0469; 20% 1.0969; 30% 1.1505; 40% 1.2078; 50% 1.2696. Max soly in water in g/100 ml at various temps: 0° = 115; 10° = 126; 20° = 139; 30° = 156; 40° = 176; 50° = 195; 60° = 217; 70° = 244; 80° = 273; 90° = 307; 100° = 343. One gram dissolves in 0.75 ml water at room temp, in 0.5 ml boiling water, 1.7 ml methanol, 3 ml ethanol, 10.5 ml propanol, 250 ml ether. Also sol in glycerol. Insol in chloroform.
Caution: Can cause local irritation. *See Patty's Industrial Hygiene and Toxicology* **vol. 2C**, G. D. Clayton, F. E. Clayton, Eds. (Wiley-Interscience, New York, 3rd ed., 1982) p 4937, 4943-4945.
USE: In the soft drink industry, confectionery products, bakery products, gelatin desserts, as an acidulant. In photography, tanning, ceramics, manuf tartrates. The common commercial esters are the diethyl and dibutyl derivs used for lacquers and in textile printing. Pharmaceutic aid (buffering agent).

9238. *meso*-Tartaric Acid. Mesotartaric acid; internally compensated tartaric acid; unresolvable tartaric acid; Antiweinsäure (German). $C_4H_6O_6$; mol wt 150.09. C 32.01%, H 4.03%, O 63.96%. Prepd by boiling L-tartaric acid with alkali; as byproduct of racemization: Winther, *Z. Physik. Chem.* **56**, 507 (1906); Holleman, *Org. Syn.* **coll. vol. I** (2nd ed., 1941) p 497; Milas, **U.S. pat. 2,414,385** (1947). Microbial prepn: Martin, Foster, *J. Bacteriol.* **70**, 405 (1955); Foster, **U.S. pat. 2,947,665** (1960). Explanation of optical inactivity: Noller, *Science* **102**, 508 (1945); C. R. Noller, *Chemistry of Organic Compounds* (Philadelphia, 2nd ed., 1957) p 339.

COOH
|
HO—CH
|
HO—CH
|
COOH

Monohydrate, rectangular plates, d_4^{20} 1.666 (also reported as 1.737). mp 140° (also reported as 159-160°). pKa_1 3.11; pKa_2 4.80. Maximum soly in water at 20°: 125 g/100 ml.

9239. Tartrazine. *4,5-Dihydro-5-oxo-1-(4-sulfophenyl)-4-[(4-sulfophenyl)azo]-1H-pyrazole-3-carboxylic acid trisodium salt; C.I. acid yellow 23;* 3-carboxy-5-hydroxy-1-*p*-sulfophenyl-4-*p*-sulfophenylazopyrazole trisodium salt; 5-hydroxy-1-(*p*-sulfophenyl)-4-[(*p*-sulfophenyl)azo]pyrazole-3-carboxylic acid trisodium salt; hydrazine yellow; C.I. 19140; FD & C Yellow No. 5; C.I. Food Yellow 4. $C_{16}H_9N_4Na_3O_9S_2$; mol wt 534.37. C 35.96%, H 1.70%, N 10.48%, Na 12.91%, O 26.95%, S 12.00%. Prepn: **U.S. pat. 2,457,823** (1949 to Ilford); Freeman *et al., J. Assoc. Offic. Agr. Chem.* **33**, 937 (1950). *See also: Colour Index* **vol. 4** (3rd ed., 1971) p 4132.

Bright orange-yellow powder. Freely sol in water. The aq soln is not changed by HCl but becomes redder with sodium hydroxide.
USE: As a dye for wool and silk. In biochemistry as an adsorption-elution indicator for chloride estimations. Approved by FDA for use in food, ingested drugs, externally applied drugs and cosmetics. FDA requires labelling of FD & C No. 5 in food and drugs: *Fed. Regist.* **44**, 37212 (1979); **50**, 35774 (1985).

9240. Tartronic Acid. *Hydroxypropanedioic acid;* hydroxymalonic acid. $C_3H_4O_5$; mol wt 120.06. C 30.01%, H

3.36%, O 66.63%. HOCH(COOH)₂. Prepd by ozonization of malonic acid in aq soln: Dobinson, *Chem. & Ind. (London)* **1959**, 853.

Colorless, odorless crystals. Crystallizes from water with ½ and 1H₂O. Becomes anhydr at 60°. Sublimes 110-120°. mp 158-160° with decompn (CO₂ evolved). pK₁ 2.42; pK₂ 4.54, Grandjean, *Bull. Soc. Roy. Sci. Liege* **38**, 288 (1969), *C.A.* **72**, 42684t (1970). Very sol in water, alcohol; the anhydrous acid is sol in ether.

9241. Taurine. *2-Aminoethanesulfonic acid.* C₂H₇NO₃S; mol wt 125.15. C 19.19%, H 5.64%, N 11.19%, O 38.35%, S 25.62%. NH₂CH₂CH₂SO₃H. Conditionally essential nutrient, important during mammalian development, especially for cells of the cerebellum and retina. Conjugates bile acids, present in most milk (but only minimally in that of dairy cows); occurs also in lungs and flesh extract of oxen, in shark blood, in mussels, in oysters. Isoln from ox bile: Hammarsten, *Z. Physiol. Chem.* **32**, 456 (1901); from the large muscle of abalone (*Haliotis*): Schmidt, Watson, *J. Biol. Chem.* **33**, 499 (1918). Synthesis starting with 2-bromoethanesulfonate: C. S. Marvel, C. F. Bailey, *Org. Syn. coll. vol. II*, 563 (1943). Synthesis by sodium sulfite sulfonation of ethylene chloride followed by ammonolysis with anhyr NH₃ or with aq NH₃ and ammonium carbonate: Schick, Degering, *Ind. Eng. Chem.* **39**, 906 (1947). Importance to retinal function in cats: A. R. Rabin *et al.*, *Invest. Ophthalmol.* **12**, 694 (1973); K. C. Ha *et al.*, *Science* **188**, 949 (1975); E. L. Berson *et al.*, *Invest. Ophthalmol.* **15**, 52 (1976); K. C. Hayes, *Can. Vet. J.* **23**, 2 (1982). Comparison of taurine depletion in humans and other mammals: K. C. Hayes, J. A. Sturman, *Adv. Exp. Med. Biol.* **139**, 79 (1982). Presence in human milk: G. E. Gaull, *J. Pediatr. Gastroenterol. Nutr.* **2**, Suppl. 1, 266 (1983); *idem, Acta Paediatr. Scand., Suppl.* **296**, 38 (1982). Concn in plasma and urine of breast-fed and formula-fed infants: A.-L. Järvenpää *et al.*, *Pediatrics* **70**, 221 (1982). Association of retinal dysfunction with low plasma taurine levels in adults: H. S. Geggel *et al.*, *N. Engl. J. Med.* **312**, 142 (1985). Books: *Taurine*, R. Huxtable, A. Barbeau, Eds. (Raven Press, New York, 1975) 480 pp; *Taurine: Nutritional Value and Mechanisms of Action*, J. B. Lombardini *et al.*, Eds. (Plenum Press, New York, 1992) 456 pp.

Large monoclinic prismatic rods, dec about 300°. pK₁' 1.5; pK₂' 8.74. Sol in 15.5 parts of water at 12°. 100 parts of 95% alc dissolve 0.004 parts at 17°. Insol in abs alc.

Note: Some infant formulas are being reformulated to increase the concn of taurine to that found in human milk, *Chem. & Eng. News* **62**, 22 (July 30, 1984).

9242. Taurocholic Acid. *2-[[(3α,5β,7α,12α)-3,7,12-Trihydroxy-24-oxocholan-24-yl]amino]ethanesulfonic acid; N-choloyltaurine;* cholaic acid; cholyltaurine. C₂₆H₄₅NO₇S; mol wt 515.71. C 60.55%, H 8.80%, N 2.72%, O 21.72%, S 6.22%. The product of conjugation of cholic acid with taurine. Its sodium salt is the chief ingredient of the bile of carnivorous animals. Prepn from dog bile: Hammarsten in Abderhalden, *Handbuch der Biol. Arbeitsmethoden*, Abt. I, Teil 6, p 219 (1925). Separation from bile acids: Ahrens, Craig, *J. Biol. Chem.* **195**, 763 (1952). Enzymic synthesis: Siperstein, Murray, *Science* **123**, 377 (1956). Prepn of the sodium salt: Cortese, *J. Am. Chem. Soc.* **59**, 2532 (1937); Norman, *Arkiv. Kemi* **8**, 331 (1955); of the barium salt: Kazuno, Yanazaki, *Z. Physiol. Chem.* **224** 160 (1934). Binding of taurocholic acid to serum proteins: Burke *et al.*, *Clin.*

Chim. Acta **32**, 207 (1971). Toxicity data: Klaassen, *Toxicol. Appl. Pharmacol.* **24**, 37 (1973).

Clusters of slender, four-sided prisms from alcohol + ether, stable to air. (Commercial prepns are usually amorphous, very hygroscopic, and of yellow color.) Dec about 125°. [α]D¹⁸ +38.8° (c = 2 in alcohol). pK 1.4. Freely sol in water, sol in alcohol. Almost insol in ether, ethyl acetate. Is hydrolyzed to cholic acid and taurine by acids and alkalies. LD₅₀ in newborn rats: 380 mg/kg (Klaassen).

Sodium salt, C₂₆H₄₄NNaO₇S, *sodium taurocholate*. Crystals with 1.5 and 2 mols H₂O. Sweet taste with bitter aftertaste. Dec about 230°. [α]D²⁰ +24° (c = 3). Very freely sol in water or alcohol. Solvent action on cholesterol: Rosin, *Z. Physiol. Chem.* **124**, 282 (1923).

Barium salt, C₅₂H₈₈BaN₂O₁₄S₂, *barium taurocholate*. Crystals with 5 mols H₂O, dec 225-227°. [α]D²⁰ +25.6°. Converted to taurocholic acid by treatment with sulfuric acid.

USE: Sodium salt is a lipase accelerator. *See also* Ox Bile Extract.

THERAP CAT: Choleretic.

9243. Taurolidine. *4,4'-Methylenebis(tetrahydro-1,2,4-thiadiazine) 1,1,1',1'-tetraoxide;* 4,4'-methylenebis(perhydro-1,2,4-thiadiazine 1,1-dioxide); bis(1,1-dioxoperhydro-1,2,4-thiadiazin-4-yl)methane; Drainasept; Taurolin; Tauroflex. C₇H₁₆N₄O₄S₂; mol wt 284.36. C 29.57%, H 5.67%, N 19.70%, O 22.51%, S 22.55%. Broad spectrum, synthetic formaldehyde carrier formed by the condensation of two molecules of taurine and three molecules of formaldehyde. Prepn: Fr. pat. **1,458,701**; R. W. Pfirrmann, U.S. pat. **3,423,408** (1966, 1969 both to Ed. Geistlich Söhne). Antibacterial activity in mice: M. K. Browne *et al.*, *J. Appl. Bacteriol.* **41**, 363 (1976). Anti-endotoxin activity in lab animals: R. W. Pfirrmann, G. B. Leslie, *ibid.* **46**, 97 (1979). Mechanism of action: E. Myers *et al.*, *ibid.* **48**, 89 (1980). HPLC determn of metabolites in plasma: A. D. Woolfson *et al.*, *Int. J. Pharm.* **49**, 135 (1989). Pharmacokinetics: C. Steinbach-Lebbin *et al.*, *Arzneimittel-Forsch.* **32**, 1542 (1982). Metabolism in humans: B. I. Knight *et al.*, *Brit. J. Clin. Pharmacol.* **12**, 695 (1981). Clinical trials in peritonitis: M. K. Browne *et al.*, *Surg. Gynecol. Obstet.* **146**, 721 (1978); G. Wesch *et al.*, *Fortschr. Med.* **101**, 545 (1983); in wound sepsis: A. K. Halsall *et al.*, *Pharmatherapeutica* **2**, 673 (1981); in pleural infection: A. A. Conlan *et al.*, *S. Afr. Med. J.* **64**, 653 (1983).

White crystals, mp 154-158°. Sol in water. LD₅₀ in mice, rats (g/kg): 1.5 i.p.; 2.0 i.v.; 2.5 orally (Browne, 1978).

THERAP CAT: Antibacterial.

9244. Taxicatin. 3,5-Dimethoxyphenol glucoside. C₁₄H₂₀O₈; mol wt 316.31. C 53.16%, H 6.37%, O 40.47%. Isoln from leaves of *Taxus baccata* L., *Taxaceae*: Lefebvre, *Arch. Pharm.* **245**, 486 (1907); Merz, Preuss, *ibid.* **279**, 134 (1941). Structure and synthesis from 3,5-dimethoxyphenol and α-acetobromoglucose: Merz, Preuss, *loc. cit.*

Crystals from ethyl acetate, mp 170-170.5°. [α]D²⁰ −72° (0.12 g/10 ml of aq soln). Sol in water, alcohol, ethyl acetate, methanol, acetic acid, acetone, dioxane, pyridine.

9245. Taxicins. Acyl-free polyols from which the taxines are prepared by partial esterification: Baxter *et al.*, *J. Chem. Soc.* **1962**, 2964. Structure of taxicin-I and -II: Eyre *et al.*, *Proc. Chem. Soc.* **1963**, 271. Structure and stereochemical studies of taxicin-I and -II: Dukes *et al.*, *Tetrahedron Letters* **1965**, 4765; *J. Chem. Soc. (C)* **1967**, 448; Eyre *et al., ibid.* 452. Structure of *O*-cinnamoyltaxicin-II triacetate (taxinine): Kurono *et al., Tetrahedron Letters* **1963**, 2153; Nakanishi, Kurono, *ibid.* **1963**, 2161; Ueda *et al., ibid.* **1963**, 2167; Uyeo *et al., J. Pharm Soc. Japan* **84**, 762 (1964). Stereochemistry of taxinine: Kurono *et al., Tetrahedron Letters* **1965**, 1917; Shiro *et al., Chem. Commun.* **1966**, 97.

	R	R'	R"
Taxicin-I	OH	H	H
O-cinnamoyltaxicin-I	OH	H	COCH=CHC₆H₅
O-cinnamoyltaxicin-I triacetate	OH	COCH₃	COCH=CHC₆H₅
Taxicin-II	H	H	H
O-cinnamoyltaxicin-II triacetate	H	COCH₃	COCH=CHC₆H₅

Taxicin-I. $C_{20}H_{30}O_6$, *[3S-(3α,4aα,5α,6α,11β,12α,12aβ)]-1,3,4,4a,5,6,7,11,12,12a-decahydro-3,5,6,11,12-pentahydroxy-9,12a,13,13-tetramethyl-4-methylene-6,10-methanobenzocyclodecen-8(2H)-one.*
O-Cinnamoyltaxicin-I, $C_{29}H_{36}O_7$. mp 233-234°. $[α]_D^{21}$ +285° (chloroform). uv max (alc): 282 nm (ε 28200).
O-Cinnamoyltaxicin-I triacetate, $C_{35}H_{42}O_{10}$. Prisms from alc. mp 237-239°. $[α]_D^{18}$ +218° (chloroform).
Taxicin-II. $C_{20}H_{30}O_5$.
O-Cinnamoyltaxicin-II triacetate, $C_{35}H_{42}O_9$, *taxinine.* Prisms from alc, mp 265-267°. $[α]_D^{18}$ +137° (chloroform). uv max: 217, 223, 278 nm (ε 20500; 16000; 28600).

9246. Taxine. A mixture of alkaloids from needles and berries of the yew tree, *Taxus baccata* L., *Taxaceae*: Lucas, *Arch. Pharm.* **85**, 145 (1856); Winterstein, Iatrides, *Z. Physiol. Chem.* **117**, 240 (1921); Winterstein, Guyer, *ibid.* **128**, 175 (1923); Callow *et al., J. Chem. Soc.* **1931**, 2138. Structural studies of *taxine-I*: Baxter *et al., ibid.* **1962**, 2964e; of *taxine-II*: Dukes *et al., ibid.* **1967**, 448; Eyre *et al., ibid.* 452. Sepn of cryst *taxine A* and amorph *taxine B* from the amorph mixture: E. Graf, H. Bertholdt, *Pharm. Zentral.* **96**, 385 (1957). Structure of taxine A: E. Graf *et al., Ann.* **1982**, 376. *Review:* B. Lythgoe, *The Alkaloids* vol. X, R. H. F. Manske, Ed. (Academic Press, New York, 1968) pp 597-626. Toxicity data: Y. Tekol, *Vet. Hum. Toxicol.* **33**, 337 (1991).

Taxine A

Granular amorphous powder, mp 121-124°. $[α]_D^{17}$ +95.7° (c = 4.59 in ethanol). Sol in ether, chloroform, alcohol. Practically insol in water, petr ether. LD₅₀ (as the sulfate salt) in mice (mg/kg): 19.72 orally, 21.88 i.p.; in rats (mg/kg): 20.18 s.c. (Tekol).
Taxine A, $C_{35}H_{47}NO_{10}$, *[1S-[1R*,2S*,3E,5R*(αS*,βR*),-7R*,8R*,10S*,13R*]]-β-(dimethylamino)-α-hydroxy benzenepropanoic acid 2,13-bis(acetyloxy)-7,10-dihydroxy-8,12,15,15-*

tetramethyl-9-oxotricyclo[9.3.1.1⁴,⁸]hexadeca-3,11-dien-5-yl ester. mp 204-206°. $[α]_D$ −140° (CHCl₃).
Caution: Taxine is undoubtedly responsible for the poisonous properties of the yew. Fatalities among domestic animals due to yew poisoning are not uncommon. Human fatal symptoms are those of gastrointestinal irritation, cardiac and respiratory failure.

9247. Taxodione. *(4bS-trans)-4b,5,6,7,8,8a-Hexahydro-4-hydroxy-4b,8,8-trimethyl-2-(1-methylethyl)-3,9-phenanthrenedione; 11-hydroxy-13-isopropylpodocarpa-7,9(11),13-triene-6,12-dione.* $C_{20}H_{26}O_3$; mol wt 314.42. C 76.40%, H 8.33%, O 15.27%. Diterpenoid quinone methide with tumor-inhibitory properties. Isoln of naturally occurring (+)-form from *Taxodium distichum* Rich, *Taxodiaceae*: S. M. Kupchan *et al., J. Am. Chem. Soc.* **90**, 5923 (1968). Structure: *eidem, J. Org. Chem.* **34**, 3912 (1969). Total synthesis of the racemate: K. Mori, M. Matsui, *Tetrahedron* **26**, 3467 (1970); T. Matsumoto *et al., Bull. Chem. Soc. Japan* **44**, 2766 (1971); **50**, 1575 (1977); D. L. Snitman *et al., Tetrahedron Letters* **1979**, 2477; R. V. Stevens, G. S. Bisacchi, *J. Org. Chem.* **47**, 2396 (1982). Total synthesis of the (+)-form: T. Matsumoto *et al., Bull. Chem. Soc. Japan* **50**, 266 (1977); R. H. Burnell *et al., Can. J. Chem.* **65**, 775 (1987). Antitumor activity studies: Hanson *et al., Science* **168**, 378 (1970).

Golden plates from methanol, mp 115-116°. $[α]_D^{28}$ +56° (c = 1 in CHCl₃). uv max (methanol): 320, 332, 400 nm (ε 25000, 26000, 2000).

9248. Tazanolast. *Oxo[[3-(1H-tetrazol-5-yl)phenyl]amino]acetic acid butyl ester;* butyl 3'-(1H-tetrazol-5-yl)oxanilate; TO-188; WP-833; Tazalest; Tazanol. $C_{13}H_{15}N_5O_3$; mol wt 289.29. C 53.97%, H 5.23%, N 24.21%, O 16.59%. Mast-cell degranulation inhibitor. Prepn: **Japan. Kokai 82 11,975** (1982 to Wakamoto), *C.A.* **96**, 217856a (1982). Immunopharmacology: M. Agata *et al., Japan. J. Pharmacol.* **32**, 689 (1982). Series of articles on pharmacology and mechanism of action: *Arzneimittel-Forsch.* **38**, 70-92 (1988).

THERAP CAT: Antiallergic.

9249. Tazarotene. *6-[(3,4-Dihydro-4,4-dimethyl-2H-1-benzothiopyran-6-yl)ethynyl]-3-pyridinecarboxylic acid ethyl ester;* ethyl 6-[2-(4,4-dimethylthiochroman-6-yl)ethynyl] nicotinate; AGN-190168. $C_{21}H_{21}NO_2S$; mol wt 351.47. C 71.76%, H 6.02%, N 3.99%, O 9.10%, S 9.12%. Synthetic acetylenic retinoid. Prepn: R. A. S. Chandraratna, **Eur. pat. Appl. 284,288;** *idem,* **U.S. pat. 5,089,509** (1988, 1992 both to Allergan). Pharmacokinetics in rats: P.-H. Hsyu *et al., Biopharm. Drug Dispos.* **15**, 347 (1994). Clinical evaluation as a topical in psoriasis: T. Esgleyes-Ribot *et al., J. Am. Acad. Dermatol.* **30**, 581 (1994).

White solid.
THERAP CAT: Antiacne; antipsoriatic.

9250. Tazettine. Sekisanine; sekisanoline; ungernine. $C_{18}H_{21}NO_5$; mol wt 331.37. C 65.24%, H 6.39%, N 4.23%, O 24.14%. From *Narcissus tazetta* L., *Lycoris radiata* Herb., *Ungernia sewerzowi* (Rgl.) Fedtsch., and other *Amaryllidaceae*: Späth, Kahovec, *Ber.* **67**, 1501 (1934). Structure and stereochemistry: Ikeda *et al., J. Chem. Soc.* **1956**, 4749. Abs config: Highet, Highet, *Tetrahedron Letters* **1966**, 4099. Synthesis: Hendrickson *et al., J. Am. Chem. Soc.* **92**, 5538 (1970); Tsuda *et al., Tetrahedron Letters* **1972**, 3153. Biosynthesis: Fales, Wildman, *J. Am. Chem. Soc.* **86**, 294 (1964). Identity with sekisanine and sekisanoline: Ikeda *et al., loc. cit.* Stereospecific total synthesis: Hendrickson *et al., J. Am. Chem. Soc.* **96**, 7781 (1974); S. Danishefsky *et al., ibid.* **102**, 2838 (1980); **104**, 7591 (1982).

Crystals, mp 210-211° (evac tube); racemate reported as mp 237-238° (Tsuda) and mp 175-176° (Danishefsky). $[\alpha]_D^{25}$ +150.3° (82 mg in 2 ml chloroform). Sol in methanol, ethanol, choroform. Sparingly sol in ether.
Hydrochloride, crystals, mp 206°, water soluble.
Methiodide, crystals, dec 220° (evacuated tube).

9251. Tazobactam. *[2S-(2α,3β,5α)]-3-Methyl-7-oxo-3-(1H-1,2,3-triazol-1-ylmethyl)-4-thia-1-azabicyclo[3.2.0]heptane-2-carboxylic acid 4,4-dioxide; 2β-[(1,2,3-triazol-1-yl)-methyl]-2α-methylpenam-3α-carboxylic acid 1,1-dioxide;* YTR-830H; CL-298741. $C_{10}H_{12}N_4O_5S$; mol wt 300.30. C 40.00%, H 4.03%, N 18.66%, O 26.64%, S 10.68%. β-Lactamase inhibitor. Prepn: R. G. Micetich *et al.,* Eur. pat. Appl. 97,446; *eidem.* U.S. pat. 4,562,073 (1984, 1985 both to Taiho); R. G. Micetich *et al., J. Med. Chem.* **30**, 1469 (1987). Degradation in solution: T. Marunaka *et al., Chem. Pharm. Bull.* **36**, 4478 (1988); in solid state: E. Matsushima *et al., ibid.* 4593. β-Lactamase inhibiting activity in comparison with clavulanic acid and sulbactam, *q.q.v.,* vs aerobes: M. R. Jacobs *et al., Antimicrob. Ag. Chemother.* **29**, 980 (1986); vs anaerobes: P. C. Appelbaum *et al., ibid.* **30**, 789. HPLC determn in biological materials: T. Marunaka *et al., J. Chromatog.* **431**, 87 (1988). Clinical trial in combination with piperacillin, *q.v.*: I. M. Gould *et al., Drugs Exp. Clin. Res.* **17**, 187 (1991).

Sodium salt, $C_{10}H_{11}N_4NaO_5S$, *YTR-830, CL-307579.* Amorphous solid, mp > 170° (dec).
Combination of sodium salt with piperacillin sodium, *Tazocilline, Tazocin, Zosyn.*

THERAP CAT: In combination with β-lactam antibiotics as antibacterial.

9252. TCDD. *2,3,7,8-Tetrachlorodibenzo[b,e][1,4]dioxin; 2,3,7,8-tetrachlorodibenzo-p-dioxin; 2,3,6,7-tetrachlorodibenzodioxin; dioxin;* TCDBD. $C_{12}H_4Cl_4O_2$; mol wt 321.97. C 44.77%, H 1.25%, Cl 44.04%, O 9.94%. Highly toxic contaminant; produced as a by-product during the manuf of chlorinated phenols (2,4,5-trichlorophenol, *q.v.*) and phenoxyherbicides (2,4-D and 2,4,5-T, *q.q.v.*), chlorine bleaching of paper pulp and combustion of chlorine-containing waste. Prepn: W. Sandermann, *Ber.* **90**, 690 (1957); M. Tomita *et al., Yakugaku Zasshi* **79**, 186 (1959), *C.A.* **53**, 13152d (1959). Crystal structure: F. P. Boer *et al., Acta Crystallogr.* **28B**, 1023 (1972). Toxicity and metabolism: B. A. Schwetz *et al.,* in *Chlorodioxins-Origin and Fate,* E. H. Blair, Ed., *Advances in Chemistry Series* 120 (A.C.S., Washington, D.C., 1973) pp 55-69; A. Poland, A. Kende, *Fed. Proc.* **35**, 2404 (1976). Environmental degradation: D. G. Crosby, A. S. Wong, *Science* **195**, 1337 (1976). Comprehensive review of formation, chemistry, and toxic and environmental effects: *Chlorodioxins-Origin and Fate, loc. cit.* 141 pp; *Dioxin-Toxicological and Chemical Aspects,* F. Cattabeni *et al.,* Eds. (Wiley, New York, 1978) 222 pp; special issue, *Chem. & Eng. News* **61** (June 6, 1983). Review of toxicology and human exposure: *Toxicological Profile for 2,3,7,8-Tetradichlorodibenzo-p-dioxin* (PB89-214522, 1989) 135 pp; of receptor binding and mechanism of toxicity: J. P. Whitlock, Jr., *Ann. Rev. Pharmacol. Toxicol.* **30**, 251-277 (1990); of epidemiological data: L. Tollefson, *Regul. Toxicol. Pharmacol.* **13**, 150-169 (1991); of carcinogenicity: J. Huff *et al., Ann. Rev. Pharmacol. Toxicol.* **34**, 343-372 (1994).

Needles, mp 295° (Tomita); crystals from anisole, mp 320-325° (Sandermann). LD_{50} in male, female rats (mg/kg): 0.022, 0.045 orally (Schwetz).
Note: An industrial accident during the manufacture of 2,4,5-trichlorophenol in Seveso, Italy on July 10, 1976 caused the release of an estimated two to ten pounds of TCDD into the environment. Concentrations as high as 51.3 ppm TCDD were found in some samples: R. Rawls, D. A. O'Sullivan, *Chem. & Eng. News* **54**, 27 (Aug. 23, 1976); A. Hay, *Nature* **262**, 636 (1976).
TCDD, as a contaminant created in the manufacture of *Agent Orange,* a widely used defoliant in Vietnam during the 1960's, has also been implicated as the causative agent of various symptoms described by veterans exposed to the defoliant, *see* C. Holden, *Science* **205**, 770 (1979).
Caution: Toxic effects in animals include the wasting syndrome, gastric ulcers, immunotoxicity, hepatotoxicity, hepatoporphyria, vascular lesions, chloracne, teratogenicity, fetotoxicity, impaired reproductive performance, endometriosis and delayed death. Industrial workers exposed to TCDD have developed chloracne, porphyrinuria and porphyria cutanea tarda. *See* Poland, Kende, *loc. cit.;* C. D. Carter *et al., Science* **188**, 738 (1975). This substance may reasonably be anticipated to be a carcinogen: *Seventh Annual Report on Carcinogens* (PB95-109781, 1994) p 369.

9253. Tebuconazole. *(±)-α-[2-(4-Chlorophenyl)ethyl]-α-(1,1-dimethylethyl)-1H-1,2,4-triazole-1-ethanol; (RS)-1-(4-chlorophenyl)-4,4-dimethyl-3-(1H-1,2,4-triazol-1-ylmethyl)-pentan-3-ol; ethyltrianol;* fenetrazole; terbuconazole; terbutrazole; BAY HWG 1608; HWG-1608; Corail; Elite; Folicur; Horizon; Lynx; Raxil; Silvacur. $C_{16}H_{22}ClN_3O$; mol wt 307.82. C 62.43%, H 7.20%, Cl 11.52%, N 13.65%, O 5.20%. Ergosterol biosynthesis inhibitor. Prepn: G. Holmwood *et al.,* Eur. pat. Appl. **40,345**; *eidem,* U.S. pat. **4,723,984** (1981, 1988 both to Bayer). Synthesis of enantiomers: J. Kaulen, *Agnew. Chem. Int. Ed. Engl.* **28**, 462 (1989). Photodegradation: H. Wamhoff *et al., Z. Naturforsch.* **49b**, 280 (1994). GC determn in plant material, soil and water: W. Maasfeld, *Pflanzenschutz-Nachr. Bayer (Eng. Ed.)* **40**, 29 (1987). Review of chemistry and biochemistry: D. Berg *et* *al., ibid.* 111-132. Series of articles on field trials: *ibid.* **42**, 91-222 (1989); *BCPC Monograph 57*, 97-108 (1994).

Colorless crystals, mp 104.7°. Vapor pressure (20°) $< 10^{-5}$ mbar. Soly (g/l at 20°): water 0.032, dichloromethane > 200, *n*-hexane 2-5, 2-propanol, 100-200, toluene 50-100.

USE: Fungicide.

9254. Tebufenozide. *3,5-Dimethylbenzoic acid 1-(1,1-dimethylethyl)-2-(4-ethylbenzoyl)hydrazide; N-tert-butyl-N'-(4-ethylbenzoyl)-3,5-dimethylbenzoylhydrazide; RH-5992;* Confirm; Mimic. $C_{22}H_{28}N_2O_2$; mol wt 352.48. C 74.97%, H 8.01%, N 7.95%, O 9.08%. Synthetic nonsteroidal ecdysone agonist causing premature molting; novel insect growth regulator specific to lepidopteran species. Prepn: A. C. Hsu, H. E. Aller, **Austral.** pat. **64,289;** *eidem,* **U.S.** pat. **4,985,461** (1986, 1991 both to Rohm & Haas). Properties and field evaluation: J. J. Heller *et al., Brighton Crop Prot. Conf. - Pests Dis.* **1992,** 59. Determn by LC in environmental samples: K. M. S. Sundaram *et al., J. Assoc. Off. Anal. Chem. Int.* **76,** 668 (1993); by GC and LC in formulations: *idem, et al., J. Chromatog. A* **687,** 323 (1994). Soil adsorption: *idem, J. Environ. Sci. Health* **B29,** 415 (1994). Insecticidal activity: G. Smagghe, D. Degheele, *Pest. Sci.* **42,** 85 (1994). Mechanism of action: A. Retnakaran *et al., Insect Biochem. Molec. Biol.* **25,** 109 (1995). Aquatic toxicology: D. P. Kreutzweiser *et al., Ecotoxicol. Environ. Safety* **28,** 14 (1994). Review of biological activity and use: H. Oberlander *et al., Arch. Insect Biochem. Physiol.* **28,** 209-223 (1995).

mp 191°. Also reported as mp 186-188° (Sundaram, 1081). Vapor pressure (25°): 3 × 10^{-8} mm Hg. Also reported as vapor pressure (25°): 1.7 × 10^{-6} mm Hg (Sundaram, 415). Soly (25°): 0.83 μg/ml water. Partition coefficient (octanol/water): 3.2 × 10^4. LD$_{50}$ orally in rats, mice: > 5000 mg/kg; dermally in rats: > 5000 mg/kg; LD$_{50}$ in honey bees (96 hr, contact): > 234 μg/bee; LC$_{50}$ in mallard duck (8-day dietary): > 5000 mg/kg; LC$_{50}$ in rainbow trout (96 hr): 5.7 mg/l (Heller).

USE: Insecticide.

9255. Tebuthiuron. *N-[5-(1,1-Dimethylethyl)-1,3,4-thiadiazol-2-yl]-N,N'-dimethylurea; 1-(5-tert-butyl-1,3,4-thiadiazol-2-yl)-1,3-dimethylurea; EL-103; Graslan; Spike; Perflan.* $C_9H_{16}N_4OS$; mol wt 228.32. C 47.35%, H 7.06%, N 24.54%, O 7.01%, S 14.04%. Broad spectrum pre- and post-emergent herbicide. Prepn: **Brit.** pat. **1,266,172** (1972 to Air Prod. Chem.); E. V. P. Tao, **Belg.** pat. **799,575;** *idem,* **U.S.** pat. **3,803,164** (1973, 1974 both to Lilly); *idem, Synth. Commun.* **4,** 249 (1974). Herbicidal activity: A. T. Perkins *et al., Proc. South Weed Sci. Soc.* **30,** 326 (1977). Toxicity study: G. C. Todd *et al., Food Cosmet. Toxicol.* **12,** 461 (1974). Metabolism in animals: D. M. Morton, D. G. Hoffman, *J. Toxicol. Environ. Health* **1,** 757 (1976). Determn by HPLC: J. H. Kennedy, *J. Chromatog. Sci.* **15,** 79 (1977); by GC: A. Loh *et al., J. Agr. Food Chem.* **26,** 410 (1978). Field trial against woody shrubs: C. H. Herbel *et al., J. Range Management* **38,** 391 (1985).

mp 160-163°. Moderately sol in water (2500 ppm); sol in methanol (170,000 ppm) (company literature). LD$_{50}$ in mice, rats, rabbits (mg/kg): 579, 644, 286 orally (Todd).

USE: Herbicide.

9256. Technetium. Tc; at. wt (longest-lived isotope) 98; at. no. 43. Usual valences 4 and 7. Trivalent Tc less common. Radioactive element. Discovery claimed by Noddack, Tacke, and Berg who called it "masurium"; the existence of masurium has never been confirmed by isoln of the element. Element no. 43 is the first artificially produced element. Named from the Greek word for "artificial"; separated from a molybdenum plate that had been bombarded for a few months with a strong beam of deuterons in the Berkeley cyclotron: Perrier, Segré, *Nature* **140,** 193 (1937); *eidem, J. Chem. Phys.* **5,** 712 (1937); Cacciapuoti, Segré, *Phys. Rev.* **52,** 1252 (1937). The most commonly available isotope, ^{99}Tc, has an at. wt of 98.9062; T$_{1/2}$ 2.12 × 10^5 years. Other long-lived isotopes are ^{97}Tc, T$_{1/2}$ 2.6 × 10^6 years; ^{98}Tc, T$_{1/2}$ 1.5 × 10^6 years. Isotopes have been prepared with mass numbers 92-107. Prepn of metal: Cobble *et al., J. Am. Chem. Soc.* **74,** 1852 (1952). Comprehensive reviews: Boyd, *J. Chem. Ed.* **36,** 3-14 (1959); Schwochau, *Angew. Chem.* **76,** 9-19 (1964); Kotegov *et al.,* in *Advan. Inorg. Chem. Radiochem.* **11,** 1-90 (1968); Peacock in *Comprehensive Inorganic Chemistry,* vol. 3, J. C. Bailar Jr. *et al.,* Eds. (Pergamon Press, Oxford, 1973) pp 877-903. Pharmacokinetics and organ distribution of radiopharmaceuticals: O. P. D. Noronha, K. S. Venkateswarlu, *Eur. J. Nucl. Med.* **6,** 121 (1980).

Close-packed hexagonal structure; isomorphous with rhodium, ruthenium, and osmium. The element obtained by hydrogen reduction of ammonium pertechnate is a silver-gray spongy mass which tarnishes slowly in moist air. mp 2250 ±50°. In its chemical behavior resembles rhenium. The element is precipitated by hydrogen sulfide from hydrochloric acid (diluted up to 5N); for higher acid concns the precipitation is not complete. Is oxidized by hydrogen peroxide in alkaline soln into sol anions: Perrier, Segré, *Atti R. accad. Lincei* [6] **27,** 579 (1938); Segré, *Nature* **143,** 460 (1939). Reacts with dil or concd nitric acid, in aqua regia, in concd sulfuric acid; not with HCl. Burns in fluorine to form penta- and hexafluorides. Combines with sulfur at high temp to form disulfide; with carbon to form TcC.

USE: Minute quantities of TcO$_4^-$ ion exert remarkable inhibition of the corrosion of soft iron in neutral aq soln: Cartledge, *J. Am. Chem. Soc.* **77,** 2658 (1955).

THERAP CAT: ^{99m}Tc as diagnostic aid (radioactive imaging agent).

9257. Technetium ^{99m}Tc Bicisate. *(SP-5-35)-[[Diethyl-N,N'-1,2-ethanediylbis[L-cysteinato]](3—)-N,N',S,S']oxotechnetium-^{99m}Tc; [N,N'-ethylenedi-L-cysteinato(3—)]oxo[^{99m}Tc]-technetium(V) diethyl ester; technetium ^{99m}Tc ethyl cysteinate dimer; ^{99m}Tc-L,L-ECD; RP-217; Neurolite.* $C_{12}H_{21}N_2$-O_5S_2^{99m}Tc. Neutral, lipophilic complex which rapidly crosses blood-brain barrier. Prepn: P. L. Bergstein *et al.,* **Eur.** pat. **Appl.** 279,417 (1988 to DuPont); *eidem,* **U.S.** pat. **5,279,811** (1994 to DuPont Merck). Pharmacology: R. C. Walovitch *et al., J. Nucl. Med.* **30,** 1892 (1989). Clinical pharmacokinetics: S. Vallabhajosula *et al., ibid.* 599. Metabolism in humans: R. C. Walovitch *et al., Neuropharmacology* **30,** 283 (1991). TLC determn of radiochemical purity: J. M. Green *et al., J. Nucl. Med. Technol.* **22,** 21 (1994). Series of articles on clinical pharmacology and diagnostic use in SPECT brain imaging: *J. Cerebral Blood Flow Metab.* **14,** Suppl. 1, S1-S120 (1994).

THERAP CAT: Diagnostic aid (radioactive imaging agent).

9258. Technetium ⁹⁹ᵐTc Mertiatide. *(SP-5-25)-[N-[N-[N-(Mercaptoacetyl)glycyl]glycyl]glycinato(5—)-N,N',N'',S]-oxotechnetate(2—)-⁹⁹ᵐTc disodium;* ⁹⁹ᵐTc-MAG₃; ⁹⁹ᵐTc mercaptoacetyltriglycine; TechneScan MAG3. $C_8H_8N_3Na_2O_6S^{99m}Tc$. Nonisomeric mercaptide compound complexed with technetium; radioactive renal function imaging agent. Prepn and biodistribution: A. R. Fritzberg *et al., J. Nucl. Med.* **27,** 111 (1986); *idem et al., Eur.* pat. **Appl.** 284,071 (1988 to Neorx). Stability and HPLC determn of radiochemical purity: A. M. Millar *et al., Nucl. Med. Commun.* **11,** 405 (1990). Clinical pharmacokinetics: B. Bubeck *et al., J. Nucl. Med.* **31,** 1285 (1990). Diagnostic use in captopril renography: I. E. Datseris *et al., ibid.* **35,** 251 (1994); in renal transplant evaluation: J. P. O'Malley *et al., Clin. Nucl Med.* **18,** 22 (1993). Review of synthesis and diagnostic use: D. Eshima, A. Taylor, Jr., *Sem. Nucl. Med.* **22,** 61-73 (1992).

THERAP CAT: Diagnostic aid (radioactive imaging agent).

9259. Technetium ⁹⁹ᵐTc Sestamibi. *(OC-6-11)-Hexakis(1-isocyano-2-methoxy-2-methylpropane)technetium(1+)-⁹⁹ᵐTc;* hexakis(2-methoxy-2-methylpropyl isocyanide)[⁹⁹ᵐTc] technetium(1); ⁹⁹ᵐTc hexakis(2-methoxyisobutylisonitrile); ⁹⁹ᵐTc-HEXAMIBI; ⁹⁹ᵐTc-MIBI; ⁹⁹ᵐTc-RP-30A; Cardiolite. $C_{36}H_{66}N_6O_6{}^{99m}Tc$. Lipophilic, cationic isonitrile compound complexed with technetium; radioactive myocardial perfusion imaging agent. Prepn: P. L. Bergstein, N. Subramanyam, **Eur.** pat. **Appl.** 233,368 (1987 to Du Pont); P. Angelberger, E. Zbiral, *Oesterr. Forschungszent. Seibersdorf* 4411, I (1987). Preparative TLC purification: A. Proulx *et al., Appl. Radiat. Isot.* **40,** 95 (1989). Biodistribution and comparative clinical evaluation as myocardial perfusion imaging agent: F. J. Th. Wackers *et al., J. Nucl. Med.* **30,** 301 (1989). Clinical evaluation as parathyroid imaging agent: A. J. Coakley *et al., Nucl. Med. Commun.* **10,** 791 (1989). Symposium on pharmacology and clinical studies: *Am. J. Cardiol.* **66,** 1E-96E (1990). Review of cardiac imaging: J. A. Leppo *et al., J. Nucl. Med.* **32,** 2012-2022 (1991).

THERAP CAT: Diagnostic aid (radioactive imaging agent).

9260. Technetium ⁹⁹ᵐTc Teboroxime. *(TPS-7-1-232'4'-54)-[Bis[(1,2-cyclohexanedione dioximato)(1—)-O] [(1,2-cyclohexanedione dioximato)(2—)-O]methylborato(2—)-N,N',-N'',N''',N'''',N''''']chlorotechnetium-⁹⁹ᵐTc;* CDO-MeB; SQ-30217; Cardiotec. $C_{19}H_{29}BClN_6O_6{}^{99m}Tc$. Neutral, lipophilic technetium complex classified as a boronic acid adduct of technetium dioxime (BATO); radioactive myocardial perfusion imaging agent. Prepn: A. D. Nunn *et al., Eur.* pat. **Appl.** 199,260; *eidem,* U.S. pat. **4,705,849** (1986, 1987 both to Squibb); and structural characterization: E. N. Treher *et al., Inorg. Chem.* **28,** 3411 (1989). Pharmacokinetics: R. E. Stewart *et al., J. Nucl. Med.* **31,** 1183 (1990). Reviews of physiologic properties and pharmacology: D. J. Meerdink, J. A. Leppo, *Am. J. Cardiol.* **66,** 9E-15E (1990); of clinical pharmacokinetics and comparative studies of myocardial imaging: L. L. Johnson, D. W. Seldin, *ibid.* 63E-67E; of diagnostic use in coronary artery disease: J. A. Leppo *et al., J. Nucl. Med.* **32,** 2012-2022 (1991).

THERAP CAT: Diagnostic aid (radioactive imaging agent).

9261. Teclothiazide. *6-Chloro-3,4-dihydro-3-trichloromethyl-2H-1,2,4-benzothiadiazine-7-sulfonamide 1,1-dioxide;* 6-chloro-3,4-dihydro-7-sulfamoyl-3-trichloromethyl-2H-1,2,4-benzothiadiazine 1,1-dioxide; 3-trichloromethylhydrochlorothiazide; tetrachlormethiazide. $C_8H_7Cl_4N_3O_4S_2$; mol wt 415.10. C 23.15%, H 1.70%, Cl 34.16%, N 10.12%, O 15.42%, S 15.45%. Prepn: Close *et al., J. Am. Chem. Soc.* **82,** 1132 (1960); Novello *et al., J. Org. Chem.* **25,** 970 (1960). Pharmacology and toxicology: M. Auclair *et al., Therapie* **18,** 131, 137 (1963).

Crystals, mp 300-303° (Close); mp 287° (Novello).
Potassium salt, $C_8H_6Cl_4KN_3O_4S_2$, *PS-207, K-33, Depleil.* LD_{50} i.p. in mice: 4.75 g/kg (Auclair).
THERAP CAT: Diuretic.

9262. Teclozan. *N,N'-[1,4-Phenylenebis(methylene)]-bis[2,2-dichloro-N-(2-ethoxyethyl)acetamide];* N,N'-bis(ethoxyethyl)-N,N'-bis(dichloroacetyl)-1,4-xylylenediamine; N,-N'-bis(dichloroacetyl)-N,N'-bis(2-ethoxyethyl)-1,4-bis-(aminomethyl)benzene; teclosan; teclosine; teclozine; NSC-107433; Win-13146; Win-AM-13146; Falmonox. $C_{20}H_{28}Cl_4N_2O_4$; mol wt 502.26. C 47.83%, H 5.62%, Cl 28.23%, N 5.58%, O 12.74%. Prepn: Surrey, Mayer, *J. Med. Pharm. Chem.* **3,** 409 (1961). Amebicidal activity and toxicity: Berberian *et al., Am. J. Trop. Med. Hyg.* **10,** 503 (1961).

Crystals, mp 137.6-143.9°. LD_{50} orally in mice: > 8000 mg/kg (Berberian).
THERAP CAT: Antiamebic.

9263. Tecomanine. *[4R-(4α,7β,7aβ)]-1,2,3,4,7,7a-Hexahydro-2,4,7-trimethyl-6H-2-pyrindin-6-one;* tecomine. $C_{11}H_{17}NO$; mol wt 179.26. C 73.70%, H 9.56%, N 7.81%, O 8.93%. Principal alkaloid isolated from *Tecoma stans* (L.) H.B.K. *(Bignonia stans* L.), Bignoniaceae: Y. Hammouda,

M. M. Motawi, *Egypt. Pharm. Bull.* **41**, 73 (1959), *C.A.* **54**, 21646c (1960). Structure: G. Jones *et al., Tetrahedron Letters* **1963**, 397. Stereochemistry, abs config and crystal structure: G. Jones *et al., Chem. Commun.* **1971**, 994; G. Ferguson, W. C. Marsh, *J. Chem. Soc. Perkin Trans. II* **1975**, 1124. Hypoglycemic activity of salts: Y. Hammouda *et al., J. Pharm. Pharmacol.* **16**, 833 (1964); Y. Hammouda, M. S. Amer, *J. Pharm. Sci.* **55**, 1452 (1966). Stereoselective total synthesis of ±-form: T. Imanishi *et al., Tetrahedron Letters* **22**, 667 (1981). Facile synthesis of (+)-form: T. Kametani *et al., Heterocycles* **26**, 1491 (1987).

Liquid. $bp_{0.1}$ 125°. $[\alpha]_D^{24}$ −175° (c = 1.17 in CHCl$_3$). uv max (alc): 226 nm (log ϵ 4.10). Aver. lethal dose in mice: 300 mg/kg (Hammouda).

Methiodide, $C_{12}H_{20}INO$, crystals, dec 240-242°.

Methoperchlorate, $C_{12}H_{20}ClNO_5$, crystals from methanol, mp 242°.

9264. Tectorigenin. *5,7-Dihydroxy-3-(4-hydroxyphenyl)-6-methoxy-4H-1-benzopyran-4-one; 4',5,7-trihydroxy-6-methoxyisoflavone.* $C_{16}H_{12}O_6$; mol wt 300.27. C 64.00%, H 4.03%, O 31.97%. The aglucone of tectoridin: Shibata, *J. Pharm. Soc. Japan* No. **543**, 380 (1927), *C.A.* **21**, 3050[8] (1927). Structure: Asahina *et al., J. Pharm. Soc. Japan* **48**, 1087 (1928), *C.A.* **23**, 2718[1] (1929); Shriner *et al., J. Am. Chem. Soc.* **61**, 2322 (1939). Synthesis: Farkas, Várady, *Ber.* **93**, 1269 (1960); Baker *et al., Tetrahedron Letters* no. 5, 6 (1960); Várady, *ibid.* **1965**, 4273.

Yellow needles from 30% alc, dec 230°.

Triacetate, $C_{22}H_{18}O_9$, prisms from acetone, mp 190°.

7-Glucoside, $C_{22}H_{22}O_{11}$, *tectoridin, shekanin.* From rhizomes of *Iris tectorum* Maxim., *Iridaceae:* Shibata, *loc. cit.* Identity with shekanin and structure: Mannich *et al., Arch. Pharm.* **275**, 317 (1937). Synthesis: Várady, *Acta Chim. Acad. Sci. Hung.* **48**, 181 (1966). Needles from alc, dec 257-258°. $[\alpha]_D^{20}$ −29.4° (pyridine). Sol in pyridine, alcohol, water. Practically insol in organic solvents.

9265. Teflurane. *2-Bromo-1,1,1,2-tetrafluoroethane; 1,1,1,2-tetrafluoro-2-bromoethane; Abbott 16900; DA-708; Terflurane.* C_2HBrF_4; mol wt 180.93. C 13.28%, H 0.56%, Br 44.16%, F 42.00%. FBrCHCF$_3$. Prepn: Larsen, U.S. pat. **2,971,990** (1961 to Dow). Pharmacology: Black *et al., Brit. J. Anaesth.* **41**, 288 (1969).

Gas, bp 8°. Nonexplosive; nonflammable.

THERAP CAT: Anesthetic (inhalation).

9266. Tefluthrin. *[1α,3α(Z)]-(±)-3-(2-Chloro-3,3,3-trifluoro-1-propenyl)-2,2-dimethylcyclopropanecarboxylic acid (2,3,5,6-tetrafluoro-4-methylphenyl)methyl ester; (±)-cis-4-methyltetrafluorobenzyl 3-(2-chloro-3,3,3-trifluoro-prop-1-en-1-yl)-2,2-dimethylcyclopropane carboxylate; 2,3,5,6-tetrafluoro-4-methylbenzyl-(Z)-(1RS,3RS)-3-(2-chloro-3,3,3-trifluoro-1-propenyl)-2,2-dimethylcyclopropanecarboxylate; JF-6064; PP-993; R-151993; Force; Forza; Komet RP.* $C_{17}H_{14}ClF_7O_2$; mol wt 418.74. C 48.76%, H 3.37%, Cl 8.47%, F 31.76%, O 7.64%. Prepn: N. Punja, Eur. pat. Appl. **31,199;** *idem,* U.S. pat. **4,405,640** (1981, 1983 both to ICI). Evaluation as seed treatment: G. J. Marrs, R. F. S. Gordon, *Brit. Crop Prot. Council Monogr.* **39**, 17 (1988); for sugar beet: D. Cooke *et al., Brit. Sugar Beet Rev.* **59**, 30

(1991). Effects on non-target organisms: J. M. Coulson *et al., Brighton Crop Prot. Conf. — Pests Dis.* **1990**, 975; A. M. Dewar *et al., ibid.* 987. Review of properties and insecticidal activity: M. Konradt, C. Hemmen, *Gesunde Pflanz.* **44**, 64-68 (1992).

Z-(1S)-form

Solid, mp 44°. d 1.48 g/cm^3. $bp_{1.33hPa}$ 156°. Vapor pressure (20°): 8×10^{-6} mPa. Soly in water: 2×10^{-2} mg/l; in acetone, dichloromethane, hexane and toluene: > 500 g/l. LD_{50} technical grade in rats (mg/kg): 35 orally; 200-1000 dermally (Marrs, Gordon).

USE: Insecticide.

9267. Tegafur. *5-Fluoro-1-(tetrahydro-2-furanyl)-2,4-(1H,3H)-pyrimidinedione; 5-fluoro-1-(tetrahydro-2-furyl)-uracil; N_1-(2'-furanidyl)-5-fluorouracil; FT-207; MJF-12264; NSC-148958; Citofur; Coparogin; Exonal; Fental; Franrose; Ftorafur; Fulaid; Fulfeel; Furafluor; Furofutran; Futraful; Lamar; Lifril; Neberk; Nitobanil; Riol; Sinoflurol; Sunfural; Tefsiel C.* $C_8H_9FN_2O_3$; mol wt 200.17. C 48.00%, H 4.53%, F 9.49%, N 13.99%, O 23.98%. Prepn: S. A. Hillers *et al., Doklady Akad. Nauk SSSR* **176**, 332 (1967), *C.A.* **68**, 29664j (1968); *eidem,* Brit. pat. **1,168,391** (1969), *C.A.* **72**, 43715r (1970) and Fr. pat. **1,574,684** (1969), *C.A.* **73**, 77281g (1970). Alternate synthesis: T. Kametani *et al., J. Heterocycl. Chem.* **14**, 473 (1977). Synthesis, antitumor activity, toxicity study: M. Yasumoto *et al., J. Med. Chem.* **21**, 738 (1978). Crystal structure: Y. Nakai, *Chem. Pharm. Bull.* **30**, 2629 (1982). Pharmacokinetics, metabolism in man: J. L. Au *et al., Cancer Treat. Rep.* **63**, 343 (1979). Evaluation of efficacy and toxicity: C. R. Smart *et al., Cancer* **36**, 103 (1975). Evaluation in colorectal cancer: T. Buroker *et al., Cancer* **44**, 48 (1979). *In vivo* and *in vitro* studies on Walker carcinoma in rats: J. Mattern *et al., Arzneimittel-Forsch.* **30**, 981 (1980).

Crystals from ethanol, mp 164-165°. uv max: 270 nm [ϵ 8460 (pH 2); ϵ 8050 (pH 7); ϵ 6700 (pH 12)]. Easily sol in hot water, alcohol, DMF. Practically insol in ether. LD_{50} in mice (mg/kg): 900 orally (3 days) (Yasumoto); 750 i.p. (Hillers, Fr. pat.), also reported as 1150 i.p. (Smart).

THERAP CAT: Antineoplastic.

9268. Teichoic Acids. Major components of walls and membranes of a number of bacteria, accounting for 20 to 60% of the dry weight of cell walls. Teichoic acids vary considerably in structure but are all rich in phosphodiester linkages. Depending on their location, they can be divided into two classes: membrane and cell-wall teichoic acids. *Reviews:* Baddiley, *Endeavour* **23**, 33 (1964); *idem, Proc. Roy. Soc. London* **170B**, 331 (1968); *idem, Accounts Chem. Res.* **3**, 98 (1970); M. Duckworth in *Surface Carbohydrates of the Procaryotic Cell,* I. W. Sutherland, Ed. (Academic Press, London, 1977) pp 177-208.

Membrane teichoic acids contain polyglycerol phosphate chains linking positions 1 and 3 on adjacent glycerol units through the phosphodiesters, with glycosyl substituents and alanine residues on some or all of the 2 positions. Structural studies: Kelemen, Baddiley, *Biochem. J.* **80**, 246 (1961).

Biosynthetic studies: Burger, Glaser, *J. Biol. Chem.* **239**, 3168, 3187 (1964); **241**, 494 (1966). Synthesis of a membrane teichoic acid fragment of *Staphylococcus aureus:* J. Oltvoort *et al., Rec. Trav. Chim.* **101**, 87 (1982).

Wall teichoic acids have greater structural diversity and include also polyribitol phosphate chains linking positions 1 and 5 on adjacent ribitol residues. Polymers may contain 6 to 20 repeating units. Structural studies: Armstrong *et al., Biochem. J.* **76**, 610 (1960); Baddiley, *ibid.* **85**, 49 (1962). Biosynthetic studies: Glaser, *J. Biol. Chem.* **239**, 3178 (1964). Molecular arrangement in cell wall of *Staphylococcus lactis:* Archibald *et al., Nature New Biol.* **241**, 29 (1973).

9269. Teicoplanin. *Teicoplanin A₂;* teichomycin A₂; MDL-507; Targocid; Targosid. Glycopeptide antibiotic complex produced by *Actinoplanes teichomyceticus* nov. sp.; structurally related to vancomycin, *q.v.* Comprised of 5 major components differentiated by a specific fatty acid moiety. Inhibits peptidoglycan synthesis in the cell wall of gram-positive bacteria. Isoln: **Belg.** pat. **839,259;** C. Coronelli *et al.,* **U.S.** pat. **4,239,751** (1976, 1980 both to Lepetit); F. Parenti *et al., J. Antibiot.* **31**, 276 (1978); M. R. Bardone *et al., ibid.* 170. Separation of components: A. Borghi *et al.,* **Ger.** pat. **3,320,342;** *eidem,* **U.S.** pat. **4,542,018** (1983, 1985 both to Lepetit); *eidem, J. Antibiot.* **37**, 615 (1984). Structural studies: C. Coronelli *et al., ibid.* 621; J. C. J. Barna *et al., J. Am. Chem. Soc.* **106**, 4895 (1984). Comparative *in vitro* activity: M. H. Cynamon, P. A. Granato, *Antimicrob. Ag. Chemother.* **21**, 504 (1982); H. C. Neu, P. Labthavikul, *idem.* **24**, 425 (1983). Human pharmacokinetics: L. Verbist *et al., ibid.* **26**, 881 (1984); P. L. Carver *et al., ibid.* **33**, 82 (1989). Mechanism of action: S. Somma *et al., ibid.* **26**, 917 (1984). Synergism with rifampin, *q.v.:* C. Watanakunakorn, *J. Antimicrob. Chemother.* **19**, 439 (1987). HPLC determn in human serum: J. Levy *et al., ibid.* 533. Clinical trial in gram-positive bacterial infections: Y. Glupczynski *et al., Antimicrob. Ag. Chemother.* **29**, 52 (1986). Brief review: A. H. Williams, R. N. Grüneberg, *J. Antimicrob. Chemother.* **14**, 441 (1984). Symposium on chemistry, pharmacology, activity, and clinical trials: *J. Hosp. Infect.* 7, Suppl. A, 47-112 (1986).

Teicoplanin A₂-1   (Z)-4-decenoic acid
         A₂-2   8-methylnonanoic acid
         A₂-3   n-decanoic acid
         A₂-4   8-methyldecanoic acid
         A₂-5   9-methyldecanoic acid

Amorphous powder, mp 260° (dec). uv max in 0.1N HCl: 278 ($E_{1cm}^{1\%}$ 53); in 0.1N NaOH: 297 ($E_{1cm}^{1\%}$ 74). Sol in aq soln at pH 7.0; partially sol in methanol, ethanol. Insol in dil mineral acids, in non-polar organic solvents.

Teicoplanin A₂-1, $C_{88}H_{95}Cl_2N_9O_{33}$, white amorphous powder, darkens at 220°, dec 255°.

Teicoplanin A₂-2, $C_{88}H_{97}Cl_2N_9O_{33}$, white amorphous powder, darkens at 210°, dec 250°.

Teicoplanin A₂-3, $C_{88}H_{97}Cl_2N_9O_{33}$, white amorphous powder, darkens at 210°, dec 250°.

Teicoplanin A₂-4, $C_{89}H_{99}Cl_2N_9O_{33}$, white amorphous powder, darkens at 210°, dec 250°.

Teicoplanin A₂-5, $C_{89}H_{99}Cl_2N_9O_{33}$, white amorphous powder, darkens at 210°, dec 250°.

THERAP CAT: Antibacterial.

9270. Telenzepine. *4,9-Dihydro-3-methyl-4-[(4-methyl-1-piperazinylacetyl)]-10H-thieno[3,4-b][1,5]benzodiazepin-10-one.* $C_{19}H_{22}N_4O_2S$; mol wt 370.48. C 61.60%, H 5.99%, N 15.12%, O 8.64%, S 8.66%. Gastric secretion inhibitor; selective muscarinic M₁-receptor antagonist. Prepn: V. Figala *et al.,* **Eur.** pat. **Appl. 35,519;** G. Rainer, **U.S.** pat. **4,381,301** (1981, 1983 both to Byk-Gulden). Pharmacology: M. Eltze *et al., Eur. J. Pharmacol.* **112**, 211 (1985). GC-MS determn in serum: E. Sturm, A. Junker, *J. Chromatog.* **430**, 43 (1988). Clinical pharmacology: W. Londong *et al., Gut* **28**, 888 (1987); R.W. Stockbrügger *et al., Z. Gastroenterol.* **26**, 297 (1988). Comparative clinical trial with ranitidine, *q.v.:* B. Simon *et al., ibid.* **28**, 90 (1990).

Crystals from ethanol, mp 263-264°.
THERAP CAT: Antiulcerative.

9271. Teleocidins. Indole alkaloid tumor promoters isolated from the mycelia of *Streptomyces.* Consists of *teleocidin A* (two isomers) and, the major component, *teleocidin B* (four isomers). The mixture is known as telocidin; the name is derived from its toxicity to fish (Teleostei). Isoln: M. Takashima, H. Sakai, *Bull. Agric. Chem. Soc. Japan* **24**, 647 (1960). Resolution into A and B and their isomers: H. Fujiki, T. Sugimura, *Cancer Surveys* **2**, 539 (1983). Structure: S. Sakai *et al., Chem. Pharm. Bull.* **32**, 354 (1984). Abs config: Y. Endo *et al., ibid* 358. Structure determn of B isomers: Y. Hitotsuyanagi *et al., ibid* 4233. Structure-activity study: T. Kawai *et al., J. Med. Chem.* **35**, 2248 (1992). Tumor promotion: H. Fujiki *et al., Biochem. Biophys. Res. Commun.* **90**, 976 (1979); *eidem, Carcinogenesis* **3**, 895 (1982). *Review:* H. Fujiki, T. Sugimura, *Adv. Cancer Res.* **49**, 233-264 (1987).

Teleocidin B₄

White-crystal like powder, dec > 61°. Easily sol in methanol, ethanol, ether, acetone, ethyl-acetate, benzene, chloroform, carbon tetrachloride; slightly sol in petroleum-ether, petroleum-benzine, ligroin. Almost insol in water, %5 HCl and 5% NaOH solution. LD₅₀ orally in mice: 2 mg/kg (Takashima).

Teleocidin B₄, $C_{28}H_{41}N_3O_2$, *[4S-(4R*,7R*,10S*,13S*)]-13-ethenyl-1,3,4,5,7,8,10,11,12,13-decahydro-4-(hydroxymethyl)-8,10,13-trimethyl-7,10-bis(1-methylethyl)-6H-benzo[g][1,4]-diazonino[7,6,5-cd]indol-6-one, olivoretin D.*

Dihydroteleocidin B. Structure: H. Harada *et al., Bull. Chem. Soc. Japan* **39**, 1773 (1966). Tumor promoting activity: H. Fujiki *et al., Proc. Nat. Acad. Sci. USA* **78**, 3872 (1981).

use: Tumor promoting agent.

9272. Telluric(VI) Acid. Orthotelluric acid. H_6O_6Te; mol wt 229.64. H 2.63%, O 41.80%, Te 55.56%. $Te(OH)_6$. Most stable form. Prepd by oxidizing tellurium or its dioxide with chromic acid or potassium permanganate in nitric acid: Staudenmaier, *Z. Anorg. Chem.* **10**, 189 (1895); Mathers *et al., Inorg. Syn.* **3**, 145 (1950); from tellurium by oxidation with chloric acid: Meyer, Franke, *Z. Anorg. Allgem. Chem.* **193**, 191 (1930); Meyer, Holowatyi, *Ber.* **81**, 119 (1948); Fehér in *Handbook of Preparative Inorganic Chemistry* vol. 1, G. Brauer, Ed. (Academic Press, New York, 2nd ed., 1963) pp 451-453. Forms *polymetatelluric acid* $(H_2TeO_4)_n$ when heated in air at 100-200°; *n* approx 11. Concd soln of polymerized form called *allotelluric acid*; formed when orthotelluric acid is heated in a sealed tube and dissolves in its water of constitution. *Review:* Datton, Cooper, *Chem. Rev.* **66**, 657-675 (1966).

White solid; dimorphic: monoclinic, d 3.068; cubic, d 3.163: Avinens, Petit, *Compt. Rend. Ser. C* **266**, 981 (1968). Crystallizes as tetrahydrate at temp below 10°. Very weak acid: $K_1 = 2 \times 10^{-8}$; $K_2 = 1 \times 10^{-11}$. Soly in water about 33% at 30°: Mylius, *Ber.* **34**, 2208 (1901). Strong tendency to polymerize (like stannic acid) as the mol wt increases the soly becomes less and less and aq solns become truly colloidal. Sparingly sol in concd nitric acid. Soly in dil nitric acid: *Inorg. Syn. (loc. cit.).*

9273. Tellurium. Aurum paradoxum; metallum problematum. Te; at. wt 127.60; at. no. 52; valence 2, 4, 6. Diatomic (Te_2) in the vapor state. Eight stable isotopes: 120 (0.089%); 122 (2.46%); 123 (0.87%); 124 (4.61%); 125 (6.99%); 126 (18.71%); 128 (31.79%); 130 (34.48%); artificial radioactive isotopes: 114-119; 121; 127; 129; 131-134. Present in the earth's crust to the extent of 0.002 ppm. Discovered by von Reichenstein in 1782; named by Klaproth in 1798. Occurs as tellurides in combination with metals in the minerals tetradymite, altaite, coloradolite; found as the dioxide, tellurite; found also native, associated with silver and gold. Prepn: Kracek, *J. Am. Chem. Soc.* **63**, 1989 (1941); Fehér in *Handbook of Preparative Inorganic Chemistry* vol. 1, G. Brauer, Ed. (Academic Press, New York, 2nd ed., 1963) pp 437-438. Prepn of spectrally pure Te for semiconductor devices: Weidel, *Z. Naturforsch.* **9a**, 697 (1954). Symposium on organic selenium and tellurium compds: Y. Okamoto, W. H. H. Gunther, Eds., *Ann. N.Y. Acad. Sci.* **192**, 1-225 (1972). Reviews: Stone, Caron in *Rare Metals Handbook*, C. A. Hampel, Ed. (Reinhold, New York, 1954) pp 405-415; Bagnall in *Comprehensive Inorganic Chemistry* vol. 2, J. C. Bailar, Jr. *et al.*, Eds. (Pergamon Press, Oxford, 1973) pp 935-1008; E. M. Elkin in Kirk-Othmer *Encyclopedia of Chemical Technology* vol. 22 (Wiley-Interscience, New York, 3rd ed., 1983) pp 658-679.

Grayish-white, lustrous, brittle, crystalline solid, hexagonal, rhombohedral structure, or dark-gray to brown, amorphous powder with metal characteristics. d (cryst) 6.11-6.27. mp 449.8°. bp 989.9°. Electrical resistivity (19.6°): 200,000 μ-ohms-cm. Latent heat of fusion: 4.27 kcal/mole. Linear coefficient of thermal expansion: 16.8×10^{-6}/°C. Magnetic susceptibility (18°): -0.31×10^{-6} cgs. Hardness (Mohs): 2.3. Modulus of elasticity: 6,000,000 psi. Specific heat (solid): 0.047 cal/g/°C. Thermal conductivity: 0.014 at 20°. Burns in air with a greenish-blue flame, forming the dioxide. Insol in water, in benzene, in carbon disulfide. Not attacked by hydrochloric acid; reacts with nitric acid; with concd or fuming sulfuric acids, forming a red soln; in presence of air dissolves in potassium hydroxide with formation of a deep-red soln. Combines with the halogens; does not react with sulfur or selenium.

Caution: Potential symptoms of overexposure are garlic odor on breath and sweat; dry mouth, metal taste; somnolence; anorexia, nausea and no sweating; dermatitis. *See NIOSH Pocket Guide to Chemical Hazards* (DHHS/NIOSH 90-117, 1990) p 204.

USE: As coloring agent in chinaware, porcelains, enamels, glass; reagent in producing black finish on silverware; in

manuf special alloys of marked electrical resistance; in semiconductor research.

9274. Tellurium Dibromide. Tellurous bromide. Br_2-Te; mol wt 287.41. Br 55.60%, Te 44.40%. $TeBr_2$. Prepd from $CBrF_3$ and Te: Aynsley, Watson, *J. Chem. Soc.* **1955**, 2603.

Chocolate-brown. Sol in ether; less sol in chloroform. Decomposed by water.

9275. Tellurium Dichloride. Tellurous chloride. Cl_2-Te; mol wt 198.51. Cl 35.72%, Te 64.28%. $TeCl_2$. Prepn from CCl_2F_2 and Te: Aynsley, *J. Chem. Soc.* **1953**, 3016.

Black, amorphous solid; black liquid; purple vapor. mp 208°. bp 328°. Disproportionates in ether, dioxane, dibutyl ether. Insol in CCl_4.

9276. Tellurium Dioxide. O_2Te; mol wt 159.60. O 20.05%, Te 79.95%. TeO_2. Prepd by oxidation of Te by HNO_3: Norris, *J. Am. Chem. Soc.* **28**, 1675 (1906); Marshall, *Inorg. Syn.* **3**, 143 (1950); Fehér in *Handbook of Preparative Inorganic Chemistry* vol. 1, G. Brauer, Ed. (Academic Press, New York, 2nd ed., 1963) pp 447-449. The orthorhombic form occurs in nature as the mineral *tellurite*. *Review:* Dutton, Cooper, *Chem. Rev.* **66**, 657-675 (1966).

White crystals, dimorphic: tetragonal, d 5.75; orthorhombic, d 6.04. Turns yellow on heating, mp 733°, forming a deep yellow liquid. Soly in water about 1:150,000. Sol in sodium hydroxide solns and in hydrochloric acid. Practically insol in ammonia water.

9277. Tellurium Hexafluoride. F_6Te; mol wt 241.59. F 47.18%, Te 52.82%. TeF_6. Prepd by direct fluorination of tellurium metal: Prideaux, *J. Chem. Soc.* **89**, 322 (1906); Klemm, Henkel, *Z. Anorg. Allgem. Chem.* **207**, 74 (1932); Yost, Claussen, *J. Am. Chem. Soc.* **55**, 885 (1933); Yost, *Inorg. Syn.* **1**, 121 (1939).

Colorless gas. Repulsive odor. mp $-37.6°$. Sublimes at $-38.9°$. Critical temperature 83°. d (solid; $-191°$) 4.006; d (liq; $-10°$) 2.499. Not as inert chemically as SeF_6 and SF_6 because the covalence maximum of tellurium is higher than 6. Slowly absorbed by water with hydrolysis to telluric acid, H_6TeO_6; more quickly hydrolyzed by aq KOH. Does not attack glass when pure. Corrodes mercury.

Caution: Potential symptoms of overexposure are headache; dyspnea; garlic odor on breath. *See NIOSH Pocket Guide to Chemical Hazards* (DHHS/NIOSH 90-117, 1990) p 204.

9278. Tellurium Tetrabromide. Telluric bromide. Br_4-Te; mol wt 447.22. Br 71.47%, Te 28.53%. $TeBr_4$. Prepd from the elements: Fehér in *Handbook of Preparative Inorganic Chemistry* vol. 1, G. Brauer, Ed. (Academic Press, New York, 2nd ed., 1963) pp 445-446.

Orange crystals when cold, red when hot. d 4.3. mp ~380°. bp 414-420°, dec into dibromide and bromine; can be sublimed without dec at 300° in vacuum. Soluble in a little water, but hydrolyzed by much water. Sol in HBr, ether, glacial acetic acid. *Keep well closed.*

9279. Tellurium Tetrachloride. Telluric chloride. Cl_4-Te; mol wt 269.41. Cl 52.64%, Te 47.36%. $TeCl_4$. Prepd from the elements: Suttle, Smith, *Inorg. Syn.* **3**, 140 (1950).

White, very hygroscopic, cryst solid. d 3.01. mp 225°. Melts into a yellow liquid, becoming dark red at higher temp. bp 380° without decompn. Dec by water into TeO_2 and HCl. Sol in abs alcohol and toluene. *Keep tightly closed.*

9280. Tellurium Tetraiodide. I_4Te; mol wt 635.22. I 79.91%, Te 20.09%. TeI_4. Prepd from $Te(OH)_6$ and HI, Gutbier, Flury, *Z. Anorg. Allgem. Chem.* **32**, 108 (1902); Damiens, *Ann. Chimie* [9] **19**, 44 (1923); Fehér in *Handbook of Preparative Inorganic Chemistry* vol. 1, G. Brauer, Ed. (Academic Press, New York, 2nd ed., 1963) p 447.

Gunmetal-gray crystals. Stable in moist air. d_4^{15} 5.05. mp 280°. Gives off I_2 when heated. Hydrolyzed slowly by cold, quickly by hot water, forming TeO_2 and HI. Sol in HI; somewhat sol in acetone.

9281. Tellurous Acid. *Tellurium(IV) acid.* H_2O_3Te; mol wt 177.61. H 1.13%, O 27.02%, Te 71.84%. H_2TeO_3. White crystals or cryst powder. d 3.0. Slightly sol in water; sol in dil acids or alkalies. Its potassium salt is reduced by many microorganisms, producing dark-colored solns.

9282. Telomerase. Telomere terminal transferase. RNA-containing enzyme found in unicellular organisms and germline cells. Also produced by tumor cells and by immortalized cell lines; not found in normal somatic cells. Specialized reverse transcriptase that adds short, tandemly repeated segments of DNA onto the ends of eukaryotic chromosomes (telomeres) using its RNA component as a template. Replenishes the small amount of telomeric DNA that is normally lost during replication. In the absence of telomerase, this incremental loss of DNA continues with each cell division until the chromosome can no longer replicate and ultimately results in cell death. Activation of telomerase is associated with the unlimited proliferation of tumor cells. Discovery in the ciliate, *Tetrahymena thermophila:* C. W. Greider, E. H. Blackburn, *Cell* **43,** 405 (1985). Overview and potential physiological significance: C. W. Greider, *BioEssays* **12,** 363-369 (1990). Proposed role in cell immortalization: C. M. Counter *et al., EMBO J.* **11,** 1921 (1992). Detection in human ovarian carcinoma cells: *idem et al., Proc. Nat. Acad. Sci. USA* **91,** 2900 (1994). Specific association with malignant cells: N. W. Kim *et al., Science* **266,** 2011 (1994). Characterization in yeast: M. Cohn, E. H. Blackburn, *Science* **269,** 396 (1995). Structural studies: A. Bhattacharyya, E. H. Blackburn, *EMBO J.* **13,** 5721 (1994); M. McCormick-Graham, D. P. Romero, *Nucl. Acids Res.* **23,** 1091 (1995). Effect of reverse transcriptase inhibitors on telomerase activity in *T. thermophila:* C. Strahl, E. H. Blackburn, *ibid.* **22,** 893 (1994). Models for telomerase activation in leukemogenesis: C. M. Counter *et al., Blood* **85,** 2315 (1995).

9283. Telomycin. $C_{59}H_{77}N_{13}O_{19}$; mol wt 1272.34. C 55.70%, H 6.10%, N 14.31%, O 23.89%. Polypeptide antibiotic produced by *Streptomyces* spp. from Florida soil: Misiek *et al., Antibiot. Annual* **1957-1958,** 852. Structure: Sheehan *et al., J. Am. Chem. Soc.* **85,** 2867 (1963); **90,** 462 (1968). NMR spectrum and conformation: Kumar, Urry, *Biochemistry* **12,** 3811, 4392 (1973). Antibacterial spectrum: A. Gourevitch *et al., Antibiot. Ann.* **1957-1958,** 856. Pharmacology: D. E. Tisch *et al., ibid.* 863.

Asp–Ser–Thr–a–Thr–Ala–Gly–t-3-Hyp–e-3-Hyl
c-3-Hyp–Δ-Trp–β-MeTrp

a-Thr = allo-threonine
t-3-Hyp = trans-3-hydroxyproline
c-3-Hyp = cis-3-hydroxyproline
Δ-Trp = α,β-didehydrotryptophan
e-3-Hyl = erythro-3-hydroxyleucine

Amorphous, gray solid. $[\alpha]_D^{28}$ −133° (c = 1 in 2:1 methanol:water). uv max (ethanol:water, 2:1): 222.5, 277, 290, 339 nm (ε 63732; 13746; 11890; 22058). Minimum soly in water at pH 3.0 to 3.3 = 4 mg/ml. Maximum soly is above pH 8.5 = >150 mg/ml. Soly in 10% sodium chloride soln less than 1 mg/ml. Moderately sol in methanol, ethanol. Very slightly sol in acetone, ethyl acetate. Insol in ether, chloroform, hydrocarbons. Aq solns are stable to heat. LD_{50} in mice (mg/kg): >1000 orally, i.v., i.p., i.m. (Tisch).

9284. Temafloxacin. *1-(2,4-Difluorophenyl)-6-fluoro-1,4-dihydro-7-(3-methyl-1-piperazinyl)-4-oxo-3-quinoline-carboxylic acid;* T-1258; A-63004; Teflox; Temac; Omniflox. $C_{21}H_{18}F_3N_3O_3$; mol wt 417.39. C 60.43%, H 4.35%, F 13.66%, N 10.07%, O 11.50%. Trifluorinated quinolone antibacterial. Prepn: D. T. W. Chu, **Eur. pat. Appl. 131,839;** *idem,* **U.S. pat. 4,730,000** (1985, 1988 to Abbott). Antibacterial activities: D. J. Hardy *et al., Antimicrob. Ag. Chemother.* **31,** 1768 (1987); K. V. I. Rolston *et al., Eur. J. Clin. Microbiol. Infect. Dis.* **7,** 684 (1988); in comparison with other fluoroquinolones: A. L. Barry, R. N. Jones, *J. Antimicrob. Chemother.* **23,** 527 (1989).

Hydrochloride, $C_{21}H_{18}F_3N_3O_3$·HCl, *A-62254.*

9285. Temazepam. *7-Chloro-1,3-dihydro-3-hydroxy-1-methyl-5-phenyl-2H-1,4-benzodiazepin-2-one;* 3-hydroxydiazepam; *N*-methyloxazepam; oxydiazepam; ER-115; K-3917; Ro-5-5345; Wy-3917; Euhypnos; Euipnos; Gelthix; Levanxene; Levanxol; Normison; Perdorm; Planum; Remestan; Restoril. $C_{16}H_{13}ClN_2O_2$; mol wt 300.74. C 63.90%, H 4.36%, Cl 11.79%, N 9.31%, O 10.64%. Pharmacologically active metabolite of diazepam, *q.v.* Prepn: S. C. Bell, S. J. Childress, *J. Org. Chem.* **27,** 1691 (1962); S. C. Bell, U.S. pat. **3,197,467** (1965 to Am. Home. Prod.). *See also:* E. Reeder *et al.,* U.S. pats. **3,340,253** and **3,374,225** (1967, 1968, both to Hoffmann-La Roche). Metabolism: H. J. Schwandt *et al., Xenobiotica* **4,** 733 (1974); S. H. Curry *et al., Brit. J. Pharmacol.* **57,** 427P (1976). Pharmacology: L. O. Randall *et al., Arch. Int. Pharmacodyn.* **185,** 135 (1970); S. Garattini *et al.,* "Metabolic Studies on Benzodiazepines in Various Animal Species" in *Benzodiazepines,* S. Garattini, Ed. (Raven Press, New York, 1973) pp 73-97. Pharmacology and toxicity study: L. O. Randall *et al., Curr. Ther. Res.* **7,** 590 (1965). Clinical study: P. Sarteschi *et al., Arzneimittel-Forsch.* **22,** 93 (1972). Review of pharmacology and therapeutic efficacy: R. C. Heel *et al., Drugs* **21,** 321-340 (1981).

Crystals from cyclohexane, mp 119-121°.
Note: This is a controlled substance (depressant) listed in the U.S. Code of Federal Regulations, Title 21 Part 1308.14 (1995).

THERAP CAT: Sedative, hypnotic.

9286. Temephos. *O,O'-(Thiodi-4,1-phenylene)phosphorothioic acid O,O,O',O'-tetramethyl ester; O,O'-(thiodi-4,1-phenylene)bis(O,O'-dimethylphosphorothioate); O,O,O',O'-tetramethyl O,O'-thiodi-p-phenylene phosphorothioate;* phosphorothioic acid *O,O-dimethyl ester O,O-diester with* 4,4'-thiodiphenol; ENT-27165; AC-52160; Abate; Biothion. $C_{16}H_{20}O_6P_2S_3$; mol wt 466.48. C 41.20%, H 4.32%, O 20.58%, P 13.28%, S 20.62%. Prepn: J. B. Lovell, R. W. Baer, Belg. pat. 648,531, *C.A.* **63,** 11433e (1965); *eidem,* U.S. pat. **3,317,636** (1964, 1967 both to Am. Cyanamid). Metabolism: R. C. Blinn, *J. Agr. Food Chem.* **17,** 118 (1969). Toxicity study: T. B. Gaines, *Toxicol. Appl. Pharmacol.* **14,** 515 (1969).

Crystalline solid, mp 30.0-30.5°. Optimum stability at pH 5-7. Sol in acetonitrile, carbon tetrachloride, ether, dichloroethane, toluene. Almost insol in water, hexane. LD_{50} in male, female rats (mg/kg): 8600, 13000 orally (Gaines).

USE: Insecticide.

9287. Temocapril. *[2S-[2α,6β(R*)]]-6-[[1-(Ethoxycarbonyl)-3-phenylpropyl]amino]tetrahydro-5-oxo-2-(2-thienyl)-1,4-thiazepine-4(5H)-acetic acid;* (+)-(2S,6R)-6-[[(1S)-1-carboxy-3-phenylpropyl]amino]tetrahydro-5-oxo-2-(2-thienyl)-1,4-thiazepine-4(5H)-acetic acid, 6-ethyl ester. $C_{23}H_{28}N_2O_5S_2$; mol wt 476.62. C 57.96%, H 5.92%, N 5.88%, O 16.78%, S 13.46%. Angiotensin-converting enzyme (ACE) inhibitor. Hydrolyzed *in vivo* to the active diacid metabolite. Prepn: H. Yanagisawa *et al.*, Eur. pat. Appl. **161,801**; *eidem*, U.S. pat. **4,699,905** (1985, 1987 both to Sankyo); *idem et al.*, *J. Med. Chem.* **30**, 1984 (1987). Pharmacology: K. Oizumi *et al.*, *Japan. J. Pharmacol.* **48**, 349 (1988). GC/MS determn in human plasma and urine: H. Shioya *et al.*, *J. Chromatog.* **496**, 129 (1989). Clinical evaluation: M. Arita *et al.*, *Clin. Exp. Pharmacol. Physiol.* **21**, 195 (1994); and pharmacokinetics: S. Tokunaga *et al.*, *Drug Invest.* **7**, 161 (1994). Review: H. Koike *et al.*, *Ann. Rep. Sankyo Res. Lab.* **44**, 1-81 (1992).

Cryst powder, mp 168°. $[\alpha]^{23}$ +40° (c = 1.1 in DMF). Hydrochloride, *CS-622*. Solid from ethanol/ethyl acetate, mp 187° (dec). $[\alpha]_D^{25}$ +47.7° (c = 1 in DMF). LD_{50} in mice, rats, dogs (mg/kg): > 5000, > 5000, > 800 orally (Koike). Diacid, $C_{21}H_{24}N_2O_5S_2$, *temocaprilat, RS-5139*. Solid from ethanol, mp 246° (dec). $[\alpha]_D^{25}$ +63.4° (c = 1 in DMF).

THERAP CAT: Antihypertensive.

9288. Temocillin. *[2S-(2α,5α,6α)]-6-[(Carboxy-3-thienylacetyl)amino]-6-methoxy-3,3-dimethyl-7-oxo-4-thia-1-azabicyclo[3.2.0]heptane-2-carboxylic acid;* (6S)-6-[2-carboxy-2-(3-thienyl)acetamido]-6-methoxypenicillanic acid. $C_{16}H_{18}N_2O_7S_2$; mol wt 414.46. C 46.37%, H 4.38%, N 6.76%, O 27.02%, S 15.47%. Semi-synthetic injectable penicillin deriv with high activity vs a large number of gram-negative bacteria, but with little activity vs gram-positive organisms. Prepn: J. P. Clayton, P. H. Bentley, Ger. pat. **2,600,866** (1976 to Beecham), *C.A.* **86**, 55420t (1977); P. H. Bentley *et al.*, *J. Chem. Soc. Perkin Trans. I* **1979**, 2455. *In vitro* antibacterial activity and β-lactamase susceptibility: I. Phillips *et al.*, *J. Antimicrob. Chemother.* **10**, 271 (1982). Comparison to other penicillins vs *H. influenzae* and intestinal gram-negative rods: H. Y. Chen, J. D. Williams, *ibid.* 279. Pharmacokinetics and tissue penetration in healthy volunteers: R. M. Brown *et al.*, *ibid.* 295. *In vivo* and *in vitro* comparison to ampicillin: R. Yogev *et al.*, *Antimicrob. Ag. Chemother.* **23**, 182 (1983). Series of articles on microbiology, pharmacology and clinical studies: *Drugs* **29**, Suppl. 5, 1-243 (1985).

Disodium salt, $C_{16}H_{16}N_2Na_2O_7S_2$, *BRL-17421, Temopen*. Amorphous solid.

THERAP CAT: Antibacterial.

9289. Temozolomide. *3,4-Dihydro-3-methyl-4-oxoimidazo[5,1-d]-1,2,3,5-tetrazine-8-carboxamide;* 8-carbamoyl-3-methylimidazo[5,1-d]-1,2,3,5-tetrazin-4(3H)-one; methazolastone; M & B 39831; CCRG-81045; NSC-362856. C_6H_6-N_6O_2; mol wt 194.15. C 37.12%, H 3.11%, N 43.29%, O 16.48%. Antitumor imidazotetrazine. Prepn: E. Lunt *et al.*, Ger. pat. **3,231,255** (1983 to May & Baker), *C.A.* **98**, 198285m (1983); M. F. G. Stevens *et al.*, *J. Med. Chem.* **27**, 196 (1984). Crystal structure and structure-activity study: P. R. Lowe *et al.*, *ibid.* **35**, 3377 (1992). Mechanism of action study: C. Dive *et al.*, *Cancer Chemother. Pharmacol.* **25**, 149 (1989). Clinical evaluation and pharmacokinetics: E. S. Newlands *et al.*, *Brit. J. Cancer* **65**, 287 (1992).

Crystals from dichloromethane, mp 212° (dec). uv max (95% ethanol): 327 nm.

THERAP CAT: Antineoplastic.

9290. Tenidap. *(Z)-5-Chloro-2,3-dihydro-3-(hydroxy-2-thienylmethylene)-2-oxo-1H-indole-1-carboxamide; 5-chloro-2,3-dihydro-2-oxo-3-(2-thienylcarbonyl)-1H-indole-1-carboxamide;* 5-chloro-3-(2-thenoyl)-2-oxindole-1-carboxamide; CP-66248. $C_{14}H_9ClN_2O_3S$; mol wt 320.76. C 52.42%, H 2.83%, Cl 11.05%, N 8.73%, O 14.96%, S 10.00%. Inhibitor of 5-lipoxygenase and interleukin-1 (IL-1) activity. Prepn: S. B. Kadin, Eur. pat. Appl. **156,603**; *idem*, U.S. pat. **4,556,-672** (both 1985 to Pfizer). Effect on 5-lipoxygenase activity *in vitro*: K. Fogh *et al.*, *Arch. Dermatol. Res.* **280**, 430 (1988). Effect on IL-1 activity in patients with rheumatoid arthritis: B. McDonald *et al.*, *Arthritis Rheum.* **31**, Suppl., S52 (1988). Clinical evaluation: P. Katz *et al.*, *ibid.* S52.

Fluffy, yellow crystals from acetic acid, mp 230° (dec). Sodium salt, $C_{14}H_8ClN_2NaO_3S$, *CP-66248-2*. Crystals from methanol-isopropanol, mp 237-238°.

THERAP CAT: Anti-inflammatory.

9291. Teniposide. *[5R-[5α,5aβ,8aα,9β(R*)]]-5,8,8a,9-Tetrahydro-5-(4-hydroxy-3,5-dimethoxyphenyl)-9-[[4,6-O-(2-thienylmethylene)-β-D-glucopyranosyl]oxy]furo[3',4':6,7]-naphtho[2,3-d]-1,3-dioxol-6(5aH)-one; 4'-demethylepipodophyllotoxin 9-(4,6-O-2-thenylidene-β-D-glucopyranoside);* 4'-demethylepipodophyllotoxin-β-D-thenylidine glucoside; ETP; NSC-122819; VM-26; Vehem-Sandoz; Vumon. $C_{32}H_{32}O_{13}S$; mol wt 656.66. C 58.53%, H 4.91%, O 31.67%, S 4.88%. Semi-synthetic derivative of podophyllotoxin, *q.v.* Prepn: A. Von Wartburg, S. Afr. pat. **66 07,585**; C. Keeler-Juslen *et al.*, U.S. pat. **3,524,844** (1968, 1970 both to Sandoz). Mechanism of action: H. Stählen, *Eur. J. Cancer* **6**, 303 (1970). Pharmacology: M. Hacker, D. Roberts, *Cancer Res.* **37**, 3287 (1977); S. M. Sieber *et al.*, *Teratology* **18**, 31 (1978); T. J. Vietti *et al.*, *Cancer Treat. Rep.* **62**, 1313 (1978). Metabolism: L. Allen, *Drug Metab. Rev.* **8**, 119 (1978); *Cancer Res.* **38**, 2549 (1978). Clinical studies: N. M. Gadel-Mawla *et al.*, *Cancer Treat. Rep.* **62**, 993 (1978); R. E. Bellet *et al.*, *ibid.* 445. Studies on delayed toxicity in mice after i.p. injections: M. Hacker, D. Roberts, *Cancer Res.* **35**, 1756 (1975); H. Stählin, *Eur. J. Cancer* **12**, 925 (1976). Review of pharmacology, pharmacokinetics and assay methods: P. I. Clark, M. L. Slevin, *Clin. Pharmacokinet.* **12**, 223-252 (1987). Comprehensive description: J. J. Kettenes-van den Bosch *et al.* in *Analytical Profiles of Drug Substances* vol. 19, K. Florey, Ed. (Academic Press, New York, 1990) pp 575-600.

Crystals from abs ethanol, mp 242-246°. $[\alpha]_D^{20}$ —107° (9:1 chloroform/methanol). uv max (methanol): 283 nm ($E_{1cm}^{1\%}$ 64.1). pKa 10.13.

THERAP CAT: Antineoplastic.

9292. Tenonitrozole. *N-(5-Nitro-2-thiazolyl)-2-thiophenecarboxamide;* 2-(α-thenoylamino)-5-nitrothiazole; thenitrazole; TC-109; Atrican; Moniflagon. $C_8H_5N_3O_3S_2$; mol wt 255.28. C 37.64%, H 1.97%, N 16.46%, O 18.80%, S 25.12%. Prepd from 2-thenoyl chloride and 2-amino-5-nitrothiazole: **Fr. pat. M715** (1961 to Chantereau), *C.A.* **59**, 7533g (1963).

Crystals from dioxane or DMF, mp 255-256°.

THERAP CAT: Antiprotozoal (Trichomonas); antifungal.

9293. Tenoxicam. *4-Hydroxy-2-methyl-N-2-pyridinyl-2H-thieno[2,3-e]-1,2-thiazine-3-carboxamide 1,1-dioxide;* Ro-12-0068; Alganex; Dolmen; Liman; Mobiflex; Rexalgan; Tilatil; Tilcotil. $C_{13}H_{11}N_3O_4S_2$; mol wt 337.38. C 46.28%, H 3.29%, N 12.45%, O 18.97%, S 19.01%. Nonsteroidal anti-inflammatory agent. Prepn: O. Hromatka *et al.*, **Ger. pat. 2,537,070** (1976 to Hoffmann-La Roche), *C.A.* **85**, 63077 (1976). Pharmacology: Y. Tanaka *et al.*, *Nippon Yakurigaku Zasshi* **77**, 531 (1981), *C.A.* **95**, 35473 (1981). HPLC determn in plasma: M. E. Pickup *et al.*, *J. Chromatog.* **225**, 493 (1981). Comparative study vs aspirin in normal volunteers: H. A. Bird *et al.*, *Curr. Med. Res. Opin.* **8**, 9 (1982). Preliminary review of pharmacology, therapeutic efficacy and mechanism of action: J. P. Gonzalez, P. A. Todd, *Drugs* **34**, 289-310 (1987).

Crystals from xylene, mp 209-213° (dec).

THERAP CAT: Anti-inflammatory; analgesic.

9294. Tenuazonic Acid. *3-Acetyl-1,5-dihydro-4-hydroxy-5-(1-methylpropyl)-2H-pyrrol-2-one;* 3-acetyl-5-sec-butyl-4-hydroxy-3-pyrrolin-2-one; 3-acetyl-5-sec-butyltetramic acid. $C_{10}H_{15}NO_3$; mol wt 197.23. C 60.90%, H 7.67%, N 7.10%, O 24.34%. A metabolite found in a strain of the fungus *Alternaria tenuis* Auct.: Rosett *et al.*, *Biochem. J.* **67**, 390 (1957); Kaczka *et al.*, *Biochem. Biophys. Res. Commun.* **14**, 54 (1964); Gitterman *et al.*, *Cancer Res.* **24**, 440 (1964). Structure: Stickings, *Biochem. J.* **72**, 332 (1959); Biosynthesis: Stickings, Townsend, *ibid.* **78**, 412 (1961). Chemical synthesis from diketene and L-isoleucine: Lacey, *J. Chem. Soc.* **1954**, 850; Harris *et al.*, *J. Med. Chem.* **8**, 478 (1965).

Pale brown, viscous, gummy substance, $[\alpha]_{5461}^{20}$ —136 ±5°; —132 ± 2° (c = 0.2, 0.5 in $CHCl_3$). $bp_{0.035}$ 117°. Readily sol in organic solvents incl petr ether; sparingly sol in water. On long standing, changes into the crystalline *iso*-form.

Copper salt monohydrate, $C_{20}H_{28}CuN_2O_6 \cdot H_2O$, green needles, mp ~175°. $[\alpha]_{5461}^{19}$ —124 ± 5° (c = 0.2 in methanol). Sol in chloroform, methanol, ethanol, acetone; sparingly sol in water; practically insol in benzene.

Sodium salt, $C_{10}H_{14}NNaO_3$, *sodium L-tenuazonate.* $[\alpha]_{546}$ —96.7° (c = 2 in methanol). uv max (pH 7): 280, 241 nm ($E_{1cm}^{1\%}$ 573, 400).

THERAP CAT: Antineoplastic.

9295. Tephrosin. *(7aR-cis)-13,13a-Dihydro-7a-hydroxy-9,10-dimethoxy-3,3-dimethyl-3H-bis[1]benzopyrano-[3,4-b:6',5'-e]pyran-7(7aH)-one;* hydroxydeguelin; toxicarol. $C_{23}H_{22}O_7$; mol wt 410.42. C 67.31%, H 5.40%, O 27.29%. Occurs in leaves of *Tephrosia vogelii* Hook. f., *Leguminosae,* in derris root, cubé root: Hanriot, *Compt. Rend.* **144**, 150 (1907); Clark, *J. Am. Chem. Soc.* **53**, 729 (1931). Structure: Butenandt, Hilgetag, *Ann.* **495**, 172 (1932).

Prisms. mp 198° (218-220°). Practically insol in water. Sol in chloroform, ether, acetone; sparingly sol in methanol.

9296. Teprenone. *6,10,14,18-Tetramethyl-5,9,13,17-nonadecatetraen-2-one;* geranylgeranylacetone; GGA; E-0671; E36U31; Selbex. $C_{23}H_{38}O$; mol wt 330.55. C 83.57%, H 11.59%, O 4.84%. Acyclic polyisoprenoid 2:3 mixture of *(5Z,9E,13E)* and all *trans (5E,9E,13E)* isomers. Prepn from geranylgeranyl bromide: L. Ruzicka, L. Castro, *Helv. Chim. Acta* **28**, 590 (1945); from substituted acetylenes: K. Sato *et al.*, *J. Org. Chem.* **35**, 565 (1970). Anti-ulcer and toxicity studies: M. Murakami *et al.*, *Arzneimittel-Forsch.* **31**, 799 (1981). *In vivo* effects on aspirin induced gastric ulcer: *eidem, Japan. J. Pharmacol.* **32**, 299 (1982); on stress induced gastric ulcer: *eidem, ibid.* **33**, 549 (1983). Mechanism of action studies: Y. Nishizawa *et al.*, *Biochem. Biophys. Res. Commun.* **103**, 706 (1981); K. Oketani *et al., Japan. J. Pharmacol.* **33**, 593 (1983). Metabolism in rats: Y. Nishizawa *et al., Xenobiotica* **17**, 575 (1987). Determn in serum by GC-MS: M. Tanaka *et al., J. Chromatog.* **231**, 301 (1982); by HPLC: T. Seki *et al., ibid.* **424**, 410 (1988).

all trans-form

Yellow oil, $bp_{0.01}$ 155-160°. $d_4^{20.5}$ 0.9081. n_D^{20} 1.4947.

THERAP CAT: Antiulcerative.

9297. Terazosin. *1-(4-Amino-6,7-dimethoxy-2-quinazolinyl)-4-[(tetrahydro-2-furanyl)carbonyl]piperazine;* 2-[(4-tetrahydro-2-furoyl)-1-piperazinyl]-4-amino-6,7-dimethoxyquinazoline. $C_{19}H_{25}N_5O_4$; mol wt 387.44. C 58.90%, H 6.50%, N 18.08%, O 16.52%. α_1-Adrenergic blocker related to prazosin, *q.v.* Prepn: M. Winn *et al.*, **Ger.** pat. **2,646,-186**; *eidem*, **U.S.** pat. **4,026,894** (both 1977 to Abbott); of the hydrochloride dihydrate: R. Roteman, **Ger.** pat. **2,831,-112**; *idem*, **U.S.** pat. **4,251,532** (1979, 1981 both to Abbott). HPLC determn in biological fluids: S. E. Patterson, *J. Chromatog.* **311**, 206 (1984). Toxicity in rats: F. L. Fort *et al.*, *Drug Chem. Toxicol.* **7**, 435 (1984). Clinical study in essential hypertension: P. A. Abraham *et al.*, *Pharmacotherapy* **5**, 285 (1985); in treatment of benign prostatic hyperplasia: H. Lepor *et al.*, *J. Urol.* **148**, 1467 (1992). Symposium on pharmacology, pharmacokinetics, and clinical efficacy in hypertension: *Am. J. Med.* **80**, Suppl. 5B, 1-105 (1986). Comprehensive description: Z. L. Chang, J. F. Bauer in *Analytical Profiles of Drug Substances* vol. **20**, K. Florey Ed. (Academic Press, New York, 1991) pp 693-727. Review of pharmacology and therapeutic efficacy: R. Achari, A. Laddu, *J. Clin. Pharmacol.* **32**, 520-523 (1992).

Fine, white, odorless crystals, mp 272.6-274. Soly at 25°C (mg/ml): methanol 33.7; water 29.7; 95% ethanol 4.1; 0.1*N* HCl 3.8; chloroform 1.2; acetone 0.01. Practically insol in hexane. uv max (0.005% in water): 212, 245, 330 nm (a 65.7, 127.5, 24.0). pKa (0.1*N* NaOH): 7.1.

Hydrochloride, $C_{19}H_{25}N_5O_4 \cdot HCl$, hygroscopic crystals from isopropyl alc, mp 278-279°. Soly in water: 761.2 mg/ml. LD_{50} in mice (mg/kg): 259.3 i.v. (Winn).

Hydrochloride dihydrate, $C_{19}H_{25}N_5O_4 \cdot HCl.2H_2O$, *Abbott 45975, Heitrin, Hytracin, Hytrin, Hytrinex, Itrin, Urodie, Vasocard, Vasomet, Vicard.* mp 271-274°. Soly in water: 24.2 mg/ml. LD_{50} in male, female rats (mg/kg): 277, 293 i.v. (Fort).

THERAP CAT: Antihypertensive. In treatment of benign prostatic hyperplasia.

9298. Terbacil. *5-Chloro-3-(1,1-dimethylethyl)-6-methyl-2,4(1H,3H)-pyrimidinedione; 3-tert-butyl-5-chloro-6-methyluracil;* Du Pont Herbicide 732; Sinbar. $C_9H_{13}ClN_2O_2$; mol wt 216.67. C 49.89%, H 6.05%, Cl 16.36%, N 12.93%, O 14.77%. Prepn: Loux, **U.S.** pat. **3,235,357** (1966 to du Pont). Herbicidal activity and physical properties: G. D. Hill, Jr. *et al.*, *Symp. N. Herbic., Eur. Weed Res. Counc., 2nd, Paris* **1965**, 313, *C.A.* **66**, 18234b (1967). Metabolism and mode of action: Herholdt, *Diss. Abstr. Int. B* **30**, 1978 (1969).

Crystals, mp 175-177°. Soly in water at 25°: 710 ppm. Sol in DMF, dimethylacetamide, cyclohexanone; moderately sol in methyl isobutyl ketone, butyl acetate, xylene. Approx. lethal dose orally in rats: > 5000 mg/kg (Hill).

USE: Herbicide.

9299. Terbinafine. *(E)-N-(6,6-Dimethyl-2-hepten-4-yn-yl)-N-methyl-1-naphthalene methanamine; trans-N-methyl-N-(1-naphthylmethyl)-6,6-dimethylhept-2-en-4-ynyl-1-amine;* SF-86-327; Lamisil. $C_{21}H_{25}N$; mol wt 291.44. C 86.55%, H 8.65%, N 4.81%. Antimycotic allylamine related to naftifine, *q.v.* Prepn: A. Stütz, **Eur.** pat. **Appl. 24,587** (1981 to Sandoz); A. Stütz, G. Petranyi, *J. Med. Chem.* **27**, 1539 (1984). Specific inhibitor of squalene epoxidase, a key enzyme in fungal ergosterol biosynthesis: G. Petranyi *et al.*, *Science* **224**, 1239 (1984). N. S. Ryder, *Antimicrob. Ag. Chemother.* **27**, 252 (1985). *In vitro* antifungal activity: S.

Shadomy *et al.*, *Sabouraudia* **23**, 125 (1985). Series of articles on activity, pharmacokinetics and metabolism: *Proc. 13th Int. Congr. Chemother., Vienna 1983*, **6**, 116/1-116/59. Toxicology: U. Ganzinger *et al.*, *ibid.* 116/52. Symposium on pharmacology and clinical trials: *J. Am. Acad. Dermatol.* **23**, Suppl., 775-812 (1990).

Hydrochloride, $C_{21}H_{25}N \cdot HCl$, crystals from 2-propanol + diethyl ether, mp 195-198° (change in crystal structure begins ~150°). LD_{50} in mice, rats (mg/kg): 4000, 4000 orally; 393, 213 i.v. (Ganzinger).

THERAP CAT: Antifungal (topical).

9300. Terbium. Tb; at. wt 158.92534; at. no. 65; valences 3, 4. A lanthanide; belongs to the yttrium group of rare earth metals. Naturally occurring isotope (mass number): 159; known artificial radioactive isotopes: 144-158; 160-164. Abundance in earth's crust: 0.91-1.2 ppm; occurs in small quantities in monazite, cerite, gadolinite and other rare earth minerals. Discovered by Mosander, *Skand. Naturför. Förh.* **3**, 387 (1842); *Phil. Mag.* [3] **23**, 241 (1843). Sepn by fractional crystn and precipitation: Urbain, *Compt. Rend.* **139**, 736 (1904); **141**, 521 (1905); James, Bissel, *J. Am. Chem. Soc.* **36**, 2060 (1914); by ion exchange: Spedding *et al.*, *ibid.* **76**, 2557 (1954). Prepn of metal by electrodeposition: *eidem*, *J. Electrochem. Soc.* **100**, 442 (1953). Absorption spectrum: Urbain, *loc. cit.* Toxicity study: Haley, *J. Pharm. Sci.* **54**, 663 (1965). Reviews of prepn, properties and compds: *The Rare Earths*, F. H. Spedding, A. H. Daane, Eds. (Krieger, Huntington, N.Y., 1971, reprint of 1961 ed.) 641 pp; Hulet, Bode, "Separation Chemistry of the Lanthanides and Transplutonium Actinides" in *MTP Int. Rev. Sci.: Inorg. Chem., Ser. One*, vol. 7, K. W. Bagnall, Ed. (University Park Press, Baltimore, 1972) pp 1-45; Moeller, "The Lanthanides" in *Comprehensive Inorganic Chemistry*, vol. **4**, J. C. Bailar, Jr. *et al.*, Eds. (Pergamon Press, Oxford, 1973) pp 1-101; F. H. Spedding in Kirk Othmer *Encyclopedia of Chemical Technology* vol. **19** (John Wiley & Sons, New York, 3rd ed., 1982) pp 833-854; *Chemistry of the Elements*, N. N. Greenwood, A. Earnshaw, Eds. (Pergamon Press, New York, 1984) pp 1423-1449. Brief review of properties: G. T. Seaborg, *Radiochim. Acta* **61**, 115-122 (1993).

Silver-gray metal, easily oxidized in air. Hexagonal close-packed crystals at room temp; d 8.27 (Spedding, Daane, *loc. cit.* p. 183). mp 1356°. bp 3230°; also reported as bp 2480° (Moeller).

Oxide, O_3Tb_2, *terbia*, a white solid.

Oxide, non-stoichiometric, approx composition, Tb_4O_7, for this formula Tb^{3+} and Tb^{4+} present in equal amounts. Dark brown or black solid; obtained by igniting the oxalate or the sulfate. Dissolves in hot concd acids with formation of salts; loses oxygen on heating; forms the oxide when heated in hydrogen.

Nitrate, $Tb(NO_3)_3$. Occurs as the hexahydrate, monoclinic crystals, mp 89.3°. *See* Urbain, cited by Mellor, *A Comprehensive Treatise on Inorganic and Theoretical Chemistry* **5**, 695 (1929). LD_{50} in rats (mg/kg): 260 i.p.; > 5000 orally (Haley).

Chloride hexahydrate, $TbCl_3.6H_2O$, prismatic deliquesc crystals. Very sol in water; forms supersaturated solns. Dehydrated on heating in hydrogen chloride at 180-200°. The anhyd chloride, crystals, d^0 4.35; mp 588°; dissolves in water without hydrolysis. LD_{50} in mice (mg/kg): 550 i.p.; 5100 orally (Haley).

9301. Terbufos. *Phosphorodithioic acid S-[[(1,1-dimethylethyl)thio]methyl] O,O-diethyl ester; phosphorodithioic acid S-[(tert-butylthio)methyl] O,O-diethyl ester;* AC-92100; Counter. $C_9H_{21}O_2PS_3$; mol wt 288.44. C 37.48%, H 7.34%, O 11.09%, P 10.74%, S 33.35%. Organophosphate insecticide. Prepn: Ikeda *et al.;* Takahashi *et al.*, **Japan.** pats. 66

16,799; 67 1743; 67 4455 (all to Hokko), *C.A.* 66, 18510p, 85467p, 67, 11205e (all 1967). As insecticidal soil treatment: F. M. Gordon, **Ger.** pat. **2,258,528** (1973 to Am. Cyanamid), *C.A.* 79, 62601m (1973). Toxicity to birds: E. F. Hill, M. B. Camardese, *Echotoxicol. Environ. Saf.* 8, 551 (1984).

Technical product (85 to 88% purity): clear, colorless to pale yellow liq, d^{24} 1.105. $bp_{0.01}$ 69°. mp −29.2°. Flash pt 88° (tag open cup). Sol in acetone, alcs, aromatic and chlorinated hydrocarbons. Soly in water: 10-15 ppm. LD_{50} orally in quail: 15 mg/kg (Hill, Camardese).
Caution: Cholinesterase inhibitor.
USE: Soil insecticide.

9302. Terbutaline. *5-[2-[(1,1-Dimethylethyl)amino]-1-hydroxyethyl]-1,3-benzenediol; α-[(tert-butylamino)methyl]-3,5-dihydroxybenzyl alcohol;* 1-(3,5-dihydroxyphenyl)-2-(*tert*-butylamino)ethanol. $C_{12}H_{19}NO_3$; mol wt 225.29. C 63.98%, H 8.50%, N 6.22%, O 21.31%. β-Adrenergic agonist. Prepn: K. Wetterlin, L. A. Svensson, **Belg.** pat. **704,-932;** *eidem,* **U.S.** pat. **3,937,838** (1968, 1976 both to Draco). Pharmacology: Bergman *et al., Experientia* 25, 899 (1969). Resolution of isomers and activity studies: K. Wetterlin, *J. Med. Chem.* 15, 1182 (1972). Clinical study in treatment of preterm labor: S. N. Caritis *et al., Am. J. Obstet. Gynecol.* 150, 7 (1984). Review: J. J. McPhillips in *Pharmacological and Biochemical Properties of Drug Substances* vol. 1, M. E. Goldberg, ed. (Am. Pharm. Assoc., Washington, DC, 1977) pp 311-328. Review of clinical toxicology: J. D. Truwit, *Crit. Care Clinics* 7, 639-657 (1991). Comprehensive description: S. Ahuja, J. Ashman in *Analytical Profiles of Drug Substances* vol. 19, K. Florey, Ed. (Academic Press, New York, 1990) pp 601-625.

Crystals from abs ether, mp 119-122°.
Sulfate, $(C_{12}H_{19}NO_3)_2 \cdot H_2SO_4$, *Brethaire, Brethine, Bricanyl, Butaliret, Monovent, Terbasmin, Terbul.* mp 246-248°. uv max (0.1N HCl): 276 nm ($A_{1cm}^{1\%}$ 67.6). pKa_1 8.8, pKa_2 10.1, pKa_3 11.2. Soly at 25° (mg/ml): water > 20; 0.1N HCl > 20; 0.1N NaOH > 20; ethanol 1.2; 10% ethanol > 20; methanol 2.7.
THERAP CAT: Bronchodilator; tocolytic.

9303. Terconazole. *cis-1-[4-[[2-(2,4-Dichlorophenyl)-2-(1H-1,2,4-triazol-1-ylmethyl)-1,3-dioxolan-4-yl]methoxy]-phenyl]-4-(1-methylethyl)piperazine;* triaconazole; R-42470; Fungistat; Gyno-Terazol; Terazol; Tercospor. $C_{26}H_{31}Cl_2N_5$-O_3; mol wt 532.47. C 58.65%, H 5.87%, Cl 13.32%, N 13.15%, O 9.01%. Topical triazole antifungal. Prepn: J. Heeres *et al.,* **Ger.** pat. **2,804,096;** *eidem,* **U.S.** pats. **4,144,-346; 4,223,036** (1978, 1979, 1980 all to Janssen); *eidem, J. Med. Chem.* 26, 611 (1983). Pharmacology: J. Van Cutsem *et al., Chemotherapy* 29, 322 (1983). Clinical comparison with clotrimazole, *q.v.,* in vaginal candidiasis: A. Kjaeldgaard, *Pharmacotherapeutica* 4, 525 (1986).

Crystals from isopropyl ether, mp 126.3°.
THERAP CAT: Antifungal.

9304. Terebene. A mixture of dipentene and other hydrocarbons obtained by shaking oil of turpentine with successive quantities of sulfuric acid: Howard, *Pharm. J.* 103, 76 (1919).
Colorless liquid; thyme-like odor, bp 160-172°. Resinifies on exposure to air and light. d_{25}^{25} 0.860-0.865. Practically optically inactive. Almost insol in water. Miscible with chloroform, ether, abs alcohol; 1 ml dissolves in 3 ml 95% alcohol. *Keep well closed and protected from light.*
USE: Treatment of cellulosic matter with terebene to render it water and oil resistant.
THERAP CAT: Expectorant; antiseptic.
THERAP CAT (VET): Orally or by inhalation: antiseptic and expectorant.

9305. Terebic Acid. *Tetrahydro-2,2-dimethyl-5-oxo-3-furancarboxylic acid; tetrahydro-2,2-dimethyl-5-oxo-3-furoic acid;* terebinic acid; (1-hydroxy-1-methylethyl)succinic acid γ-lactone. $C_7H_{10}O_4$; mol wt 158.15. C 53.16%, H 6.37%, O 40.47%. Prepared from fumaric or maleic acid: Schenck, Steinmetz, *Tetrahedron Letters* no. 21, 1 (1960); Lipp *et al., Ann.* 644, 37 (1961). Prepn of optical isomers: Fredga, *C.A.* 42, 123g (1948); Delépine, Badoche, *Compt. Rend.* 235, 1069 (1952).

Crystals, mp 174-175°, but begins to volatilize at 100°. d 0.815. Slightly sol in cold water, freely in boiling water or warm alcohol.
(+)-Form, $[\alpha]_D^{25}$ +13.2° (c = 0.03 in acetone).
(−)-Form, mp 201-205° (dec). $[\alpha]_D^{25}$ −13.2° (c = 0.03 in acetone).

9306. Terephthalic Acid. *1,4-benzenedicarboxylic acid; p*-phthalic acid; Tephthol. $C_8H_6O_4$; mol wt 166.13. C 57.84%, H 3.64%, O 38.52%. Prepd by oxidation of *p*-methylacetophenone: Koelsch, *Org. Syn.* **coll. vol. III,** 791 (1955). Manuf processes: **U.S.** pat. **3,014,961** (1959 to VEB Chemie Werke Buna); Sherwood, *Chem. & Ind. (London)* **1960,** 1096. Review: Faith, Keyes & Clark's *Industrial Chemicals,* F. A. Lowenheim, M. K. Moran, Eds. (Wiley-Interscience, New York, 4th ed., 1975) pp 807-813; A. G. Bemis *et al.,* "Phthalic Acids" in Kirk-Othmer *Encyclopedia of Chemical Technology* vol. 17 (Wiley-Interscience, New York, 3rd ed., 1982) pp 732-777.

Crystals. Sublimes at 402°. Practically insol in water, chloroform, ether, acetic acid; slightly sol in cold alcohol, more in hot alcohol; sol in alkalies.
Dimethyl ester, $C_{10}H_{10}O_4$, *dimethyl terephthalate, DMT.* White crystals, mp 140.6°. bp 288°.
Caution: Mild irritant.

USE: Forms polyesters with glycols which are made into plastic films and sheets; in analytical chemistry.

9307. Terfenadine. α-[4-(1,1-Dimethylethyl)phenyl]-4-(hydroxydiphenylmethyl)-1-piperidinebutanol; 1-(p-tert-butylphenyl)-4-[4'-(α-hydroxydiphenylmethyl)-1'-piperidyl]-butanol; α-(p-tert-butylphenyl)-4-(α-hydroxy-α-phenylbenzyl)-1-piperidinebutanol; MDL-9918; Allerplus; Cyater; Nebralin; Seldane; Teldane; Teldanex; Terdin; Terfex; Ternadin; Triludan. $C_{32}H_{41}NO_2$; mol wt 471.68. C 81.49%, H 8.76%, N 2.97%, O 6.78%. Nonsedating-type histamine H_1-receptor antagonist. Prepn: A. A. Carr, C. R. Kinsolving, Ger. pat. **2,303,306;** eidem, U.S. pat. **3,878,217** (1973, 1975 both to Richardson-Merrell); A. A. Carr, D. R. Meyer, Arzneimittel-Forsch. **32,** 1157 (1982). Antihistamine activity: A. A. Carr et al., Pharmacologist **15,** 221 (1973); C. R. Kinsolving et al., ibid. 221. Metabolism in rats: G. A. Leeson, Fed. Proc. **34,** 2911 (1975). Bioavailability: C. R. Kinsolving, N. L. Munro, Drug Metab. Rev. **4,** 285 (1975). Clinical study: M. L. Brandon, M. Weiner, Ann. Allergy **44,** 71 (1980). Acute toxicity: C. R. Kinsolving et al., Pharmacologist **15,** 221 (1973). Series of articles on chemistry, pharmacology, clinical studies, toxicity studies: Arzneimittel-Forsch. **32,** 1153-1218 (1982). Review of pharmacology, clinical efficacy: J. T. Connell, Pharmacotherapy **5,** 201-208 (1985). Comprehensive description: A. A. Badwan et al. in Analytical Profiles of Drug Substances vol. **19,** K. Florey, Ed. (Academic Press, New York, 1990) pp 627-662.

Crystals from acetone, mp 146.5-148.5°; may exist in three polymorphic forms, mp 149-152°, 146-148°, 142-144°. Soly at 30° (g/100ml): water 0.001; ethanol 3.780; methanol 3.750; hexane 0.034; 0.1M HCl 0.012; 0.1M citric acid 0.110; 0.1M tartaric acid 0.045. uv max (methanol): 260 nm (A 660.4); (ethanol): 260 nm (A 671.4); (dichloromethane): 260 nm (A 762.2). LD_{50} in rats, mice, guinea pigs (mg/kg): > 2000 orally (Kinsolving).

THERAP CAT: Antihistaminic.

9308. Terguride. N,N-Diethyl-N'-[(8α)-6-methylergolin-8-yl]urea; N-(D-6-methyl-8-isoergolin-1-yl)-N',N'-diethylurea; 6-methyl-8α-(diethylcarbamoylamino)ergoline; 9,10α-dihydrolisuride; 9,10-transdihydrolisuride; TDHL; $C_{20}H_{28}N_4O$; mol wt 340.47. C 70.56%, H 8.29%, N 16.46%, O 4.70%. Ergot derivative; dihydrogenated analog of lisuride, q.v. Exhibits dopamine agonist and antagonist activity. Prepn: V. Zikán et al., Coll. Czech. Chem. Commun. **37,** 2600 (1972); eidem, Ger. pat. **2,238,540;** idem, U.S. pat. **3,953,454** (1973, 1976 both to Spofa). Physical properties: A. Cerny et al., Coll. Czech. Chem. Commun. **52,** 1331 (1987). Pharmacology in animals and humans: H. Wachtel, R. Dorow, Life Sci. **32,** 421 (1983). Receptor binding studies in rat brain: W. Kehr et al., Acta Pharm. Suec. **1983,** Suppl. 2, 98; M. W. Valchár et al., Eur. J. Pharmacol. **136,** 97 (1987). Evaluation in animal models of Parkinson's disease: W. C. Koller, G. Herbster, Neurology **37,** 723 (1987); T. Brücke et al., Eur. J. Pharmacol. **148,** 445 (1988). Radioreceptor assay in biological fluids: R. Lapka et al., J. Pharmacol. Methods **11,** 263 (1984). Pharmacokinetics in humans: W. Krause et al., Eur. J. Clin. Pharmacol. **27,** 335 (1984). Clinical evaluation in Huntington's disease: S. Bassi et al., Neurology **36,** 984 (1986); in Parkinson's disease: T. Brücke et al., Advan. Neurol. **45,** 573 (1986); I. Suchy et al., ibid. 577; in hyperprolactinemia and acromegaly: D. Dallabonzana et al., J. Clin. Endocrinol. Metab. **63,** 1002 (1986).

Crystals from ethanol, mp 203-204° (dec). $[\alpha]_D^{20}$ +30° (c = 1 in pyridine). Also reported as crystals from ethanol, mp 205-207° (dec) (Cerny). $[\alpha]_D^{20}$ +29.0 (c = 0.2 in pyridine). uv max: 292, 281, 224 nm (log ε 3.72, 3.81, 4.42). Practically insol in water.

Hydrogen maleate, $C_{24}H_{32}N_4O_5$, SH-406, VUFB-6638, ZK-31224, Dironyl, Mysalfon. Crystals from ethanol, mp 190-191°.

Hydrogen maleate hydrate, crystals from ethanol, ml 150-153°. $[\alpha]_D^{20}$ −15.0° (c = 0.1 in H_2O). Soly in water: 1.26 mg/ml.

THERAP CAT: Antiparkinsonian; antihyperprolactinemic.

9309. Teriparatide Acetate. hPTH 1-34 acetate; MN-10T; Parathar. $C_{181}H_{292}N_{55}O_{51}S_2 \cdot yC_2H_4O_2 \cdot xH_2O$. Synthetic, biologically active polypeptide consisting of the 1-34 amino-terminal fragment of human parathyroid hormone (hPTH), q.v. Amino acid sequence: H. B. Brewer et al., Proc. Nat. Acad. Sci. USA **69,** 3585 (1972). Prepn: R. H. Andreatta et al., Helv. Chim. Acta **56,** 470 (1973). Revised structure: H. D. Niall et al., Proc. Nat. Acad. Sci. USA **71,** 384 (1974). Prepn: R. L. Colescott, Ger. pat. **2,649,727;** idem, Ger. pat. **2,649,848;** G. W. Tregear, U.S. pat. **4,086,196** (1977, 1977, 1978 all to Armour Pharm.). Solid phase synthesis and biological activity: G. W. Tregear et al., Z. Physiol. Chem. **355,** 415 (1974). Solution synthesis and biological activity: M. Takai et al., Peptide Chem. **17,** 187 (1980). Clinical pharmacology: R. M. Neer et al., J. Clin. Endocrinol. Metab. **38,** 420 (1977). RIA determn in plasma: J. M. Zanelli et al., J. Immunoassay **1,** 289 (1980). Pharmacokinetics in humans: G. N. Kent et al., Clin. Sci. **68,** 171 (1985). Clinical evaluation in osteoporosis: J. Reeve et al., Brit. Med. J. **280,** 1340 (1980); D. M. Slovik et al., J. Clin. Invest. **68,** 1261 (1981). Diagnostic use: T. Igarashi et al., Pharmatherapeutica **3,** 79 (1982); L. E. Mallette et al., J. Clin. Endocrinol. Metab. **67,** 964 (1988). Review of diagnostic use in modified Ellsworth-Howard test: L. E. Mallette, Ann. Int. Med. **109,** 800-804 (1988).

THERAP CAT: Treatment of osteoporosis. Diagnostic aid (hypocalcemia).

9310. Terlipressin. N-[N-(N-Glycylglycyl)glycyl]-8-L-lysinevasopressin; N^α-glycylglycylglycylvasopressin; 1-triglycyl-8-lysinevasopressin; Glypressin. $C_{52}H_{74}N_{16}O_{15}S_2$; mol wt 1227.39. C 50.89%, H 6.08%, N 18.26%, O 19.55%, S 5.23%. Analog of lypressin, q.v. Prepn: E. Kasafirek et al., Coll. Czech. Chem. Commun. **31,** 4581 (1966); Z. Procházka et al., ibid. **43,** 1285 (1978). Pharmacological studies: J. Kyncl et al., Eur. J. Pharmacol. **28,** 294 (1974); P. O. B. Sjöquist et al., Acta Pharmacol. Toxicol. **40,** 369 (1977). Use in treatment of uterine bleeding: V. Pavlin et al., Brit. J. Obstet. Gynaecol. **85,** 801 (1978). Hemostatic effects in cirrhosis patients: C. V. Prowse et al., Eur. J. Clin. Invest. **10,** 49 (1980). Conversion to lysine vasopressin in man: W. I. Forsling et al., J. Endocrinol. **85,** 237 (1980). Uterine effects in first trimester of pregnancy: T. Laudanski, M. Aakerlund, Contraception **22,** 199 (1980). Single-dose effects on renal potassium excretion in humans: H. Nadvornikova, O. Schueck, Int. J. Clin. Pharmacol. Ther. Toxicol. **20,** 155 (1982).

Gly–Gly–Gly–Cys–Tyr–Phe–Gln–Asn–Cys–Pro–Lys–GlyNH₂

Diacetate pentahydrate, $C_{56}H_{82}N_{16}O_{19}S_2 \cdot 5H_2O$, Glycylpressin. $[\alpha]_D^{25}$ −82° (c = 0.2 in 1M acetic acid).

THERAP CAT: Vasopressor (treatment of uterine and esophogeal bleeding).

9311. Terodiline. *N-(1,1-Dimethylethyl)-α-methyl-γ-phenylbenzenepropanamine; N-tert-butyl-1-methyl-3,3-diphenylpropylamine; N-(1-methyl-3,3-diphenylpropyl)-tert-*butylamine; 4,4-diphenyl-2-(*tert-*butylamino)butane. $C_{20}H_{27}N$; mol wt 281.44. C 85.35%, H 9.67%, N 4.98%. Calcium antagonist with anticholinergic and vasodilatory activity. Prepn and use in treatment of angina: **Brit. pat. 923,-942**; S. Carlsson, **U.S. pat. 3,371,014** (1963, 1968 both to AB Recip). Pharmacokinetics: B. Karlen *et al., Eur. J. Clin. Pharmacol.* **23**, 267 (1982). Series of articles on pharmacology and clinical efficacy in urinary frequency and motor urge incontinence: *Scand. J. Urol. Nephrol.* **Suppl. 87**, 1-61 (1984). Clinical trial in urinary incontinence: U. Ulmsten *et al., Am. J. Obstet. Gynecol.* **153**, 619 (1985).

Yellow liquid, $bp_{1.0}$ 130-132°.
Hydrochloride, $C_{20}H_{27}N.HCl$, *Bicor, Mictrol, Micturin, Micturol.* Crystals from water, mp 178-180°. Sol in ethanol. Slightly sol in ether.
THERAP CAT: Antianginal. In treatment of urinary incontinence.

9312. Terofenamate. *2-[(2,6-Dichloro-3-methylphenyl)-amino]benzoic acid ethoxymethyl ester; N-(2,6-dichloro-m-tolyl)anthranilic acid ethoxymethyl ester;* etoclofene; Etofen. $C_{17}H_{17}Cl_2NO_3$; mol wt 354.23. C 57.64%, H 4.84%, Cl 20.02%, N 3.95%, O 13.55%. Anti-inflammatory deriv of anthranilic acid, *q.v.* Prepn: **Brit. pat. 1,199,386**; E. Manghisi, **U.S. pat. 3,642,864** (1970, 1972 both to Lusofarmaco); A. Salembeni *et al., Farmaco Ed. Sci.* **30**, 276 (1975). Biological properties: G. B. Fregnan *et al., ibid.* 353. Absorption, distribution, excretion: *eidem, Boll. Chim. Farm.* **114**, 85 (1975). Pharmacological and toxicity studies: T. Chieli *et al., ibid.* **115**, 41 (1976).

Crystals from ethanol, mp 73-74°. LD_{50} in mice, rats (mg/kg): 918, 307 orally; 300, 274 i.p. (Chieli).
THERAP CAT: Anti-inflammatory; analgesic.

9313. Terpenylic Acid. *Tetrahydro-2,2-dimethyl-5-oxo-3-furanacetic acid; 3-(1-hydroxy-1-methylethyl)glutaric acid γ-lactone;* tetrahydro-5-keto-2,2-dimethyl-3-furanacetic acid; terpenolic acid. $C_8H_{12}O_4$; mol wt 172.18. C 55.81%, H 7.02%, O 37.17%. Prepd from α-terpineol, terebic acid, or terpin hydrate: Suga, Sakoda, *J. Sci. Hiroshima Univ. Ser. A* **22**, 69 (1958); *C.A.* **53**, 10273f; Lipp *et al., Ann.* **644**, 37 (1961). Resolution of optical isomers: Fredga, Sandberg, *C.A.* **52**, 11747g (1958).

Monohydrate, prisms, mp 57°. When anhydrous, melts at 90°. Sublimes 130-140°. Moderately sol in cold water; very sol in hot water.

(+)-Form, mp 92-94°, $[\alpha]_D^{25}$ +56.3°.
(−)-Form, mp 92-94°, $[\alpha]_D^{25}$ −56.5°.

9314. Terpin. *4-Hydroxy-α,α,4-trimethylcyclohexane-methanol; p-menthane-1,8-diol;* dipenteneglycol. $C_{10}H_{20}O_2$; mol wt 172.27. C 69.72%, H 11.70%, O 18.58%. Both *cis-* and *trans-*modifications are known. The *cis-*compd is obtained most readily in the hydrated form, *cis-*terpin hydrate. Prepn of *cis-*form from oil of turpentine: Hempel, *Ann.* **180**, 71 (1876); Wallach, *Ann.* **230**, 225 (1885); Schmitt, *Mfg. Chemist* **26**, 350 (1955). From *d-*limonene: Sword, *J. Chem. Soc.* **127**, 1632 (1925). Prepn of *trans-*form from 1,8-cineole, α-terpineol or *cis-*terpin hydrate: Matsuura *et al., Bull. Chem. Soc. Japan* **31**, 990 (1958); Lombard, Ambroise, *Bull. Soc. Chim. France* **1961**, 230. Structure of *cis-* and *trans-*forms: Baeyer, *Ber.* **26**, 2861 (1893); Wagner, *ibid.* **27**, 1636 (1894).

cis-form

*cis-*Form hydrate, *terpin hydrate, terpinol.* Rhombic pyramids from water, mp 116-117°; sublimes at ~100° when heated slowly; slight characteristic odor and slightly bitter taste; efflorescent in dry air. Anhydr *cis-*form: mp 104-105°; bp 258°; rapidly re-forms hydrate on exposure to air. One gram dissolves in 34 ml boiling water, 13 ml alcohol, 3 ml boiling alcohol, 135 ml chloroform, 140 ml ether, ~1 ml boiling glacial acetic acid. At 20°, one gram dissolves in 13 ml methanol, 13 ml ethyl acetate, 250 ml water, 77 ml benzene, 290 ml carbon tetrachloride, 250 ml carbon disulfide. Practically insol in petr ether.
*trans-*Form, monoclinic prisms, mp 158-159°. One gram dissolves at 20° in 11 ml methanol, 20 ml ethyl acetate, 100 ml water, 250 ml benzene, 250 ml carbon tetrachloride, 500 ml carbon disulfide.
THERAP CAT: *cis-*Form hydrate as expectorant.
THERAP CAT (VET): Expectorant.

9315. Terpinene. $C_{10}H_{16}$; mol wt 136.24. C 88.16%, H 11.84%. Mixture of three isomeric hydrocarbons: α-terpinene and γ-terpinene which occur naturally and β-terpinene which has been prepd synthetically. Isoln of α-terpinene from cardamom and marjoram oils: Weber, *Ann.* **238**, 101 (1887); Beltz, *Ber.* **32**, 996 (1899). Isoln of α-terpinene from oils of *Mosla japonica, M. grosserrata* and *Cupressus macrocarpa:* Richter, Wolff, *Ber.* **60**, 477 (1927); Briggs, Sutherland, *J. Org. Chem.* **7**, 397 (1942). Prepn of β-terpinene from sabinine: Wallach, *Ann.* **357**, 64 (1907); **362**, 285 (1908). Structure: Wallach, *ibid.* **362**, 293 (1908). *Review:* J. L. Simonsen, *The Terpenes* vol. **I** (University Press, Cambridge, 2nd ed., 1947) pp 172-193.

α-Terpinene, oil, pleasant odor of lemons. bp 173.5-174.8°; $bp_{13.5}$ 65.4-66°; $d_4^{19.6}$ 0.8375. n_D^{20} 1.4784. Practically insol in water. Miscible with alcohol, ether.
β-Terpinene, oil, bp 173-174°. d^{22} 0.838. n_D^{22} 1.4754.

γ-Terpinene, oil, bp 183°. d_4^{15} 0.853. $n_D^{15.6}$ 1.4754. Dihydrochloride, $C_{10}H_{18}Cl_2$, crystals, mp 51-52°.

9316. α-Terpineol. *α,α,4-Trimethyl-3-cyclohexene-1-methanol; p-menth-1-en-8-ol.* $C_{10}H_{18}O$; mol wt 154.25. C 77.87%, H 11.76%, O 10.37%. Terpineol exists as three isomers, α-, β-, and γ-terpineol: J. L. Simonsen, *The Terpenes* vol. I (University Press, Cambridge, 2nd ed., 1947) pp 256-274. Isoln of d-α-terpineol from petitgrain oil: Walbaum, Hüthig, *J. Prakt. Chem.* **67**, 322 (1903). Isoln from l-α-terpineol from long leaf pine oil: Teeple, *J. Am. Chem. Soc.* **30**, 412 (1908). Isoln of dl-α-terpineol from cajeput oil: Voiry, *Compt. Rend.* **106**, 1540 (1888). Synthesis of d-α-terpineol: Cologne, Crabalona, *Bull. Soc. Chim. France* **1960**, 102. Synthesis of l-α-terpineol: *eidem, ibid.* **1959**, 1505. Stereochemistry: Henbest, McElkinney, *J. Chem. Soc.* **1959**, 1834. *Review:* Wagner, *Mfg. Chemist* **22**, 98, 153 (1951).

d-Form, liquid. bp$_{4.5}$ 81-82°; bp$_{731}$ 206-207°. d_4^{20} 0.9338. n_D^{20} 1.4818. $[α]_D^{20}$ +92.45°. Solidifies at 31°.

Phenylurethan, $C_{17}H_{23}NO_2$, crystals from petr ether, mp 111°. $[α]_D^{20}$ +30.50° (benzene).

l-Form, liquid. bp$_5$ 80-81.5°. d_4^{20} 0.935. n_D^{20} 1.4820. $[α]_D^{20}$ −100° (c = 20 in alc). Solidifies at 36.4°.

Dinitrobenzoate, $C_{17}H_{20}N_2O_6$, crystals from alcohol, mp 101.5°. $[α]_D^{20}$ −31° (c = 9.5 in carbon tetrachloride).

dl-Form, liquid. bp$_3$ 85°; bp$_{752}$ 218.8-219.4°, d^{15} 0.9386. n_D^{20} 1.4831.

Phenylurethan, $C_{17}H_{23}NO_2$, crystals, mp 113°.

USE: Perfumes; denaturing fats for soap manufacture.

THERAP CAT: Antiseptic.

9317. Terreic Acid. *(1R)-3-Hydroxy-4-methyl-7-oxabicyclo[4.1.0]hept-3-ene-2,5-dione;* 2-hydroxy-3-methyl-1,4-benzoquinone 5,6-epoxide; 5,6-epoxy-3-hydroxy-p-toluquinone. $C_7H_6O_4$; mol wt 154.12. C 54.55%, H 3.92%, O 41.52%. Antibiotic metabolite produced by the mold *Aspergillus terreus*; naturally occurring as the (−)-form: Wilkins, Harris, *Brit. J. Exp. Pathol.* **23**, 166 (1942); Abraham, Florey, in H. W. Florey *et al.*, *Antibiotics* vol. I (Oxford Univ. Press, New York, 1949) p 337; Kaplan *et al.*, *Antibiot. & Chemother.* **4**, 746 (1954). Structure: Sheehan *et al.*, *J. Am. Chem. Soc.* **80**, 5536 (1958). Synthesis of the racemate: Rashid, Read, *J. Chem. Soc. (C)* **1967**, 1323. Alternate synthesis and resolution of isomers: Sheehan, Lo, *J. Med. Chem.* **17**, 371 (1974). Improved synthesis: A. Enhsen *et al.*, *J. Org. Chem.* **55**, 1177 (1990).

Pale yellow plates from benzene or hexane. Easily sublimed *in vacuo.* mp 127-127.5°. Rotation varies considerably with the solvent: $[α]_D^{20}$ −16.6° (chloroform); $[α]_D^{22}$ −28.6° (methanol-benzene 1:1); $[α]_D^{22}$ +74.3° (pH 7 phosphate buffer). uv max (ethanol): 214, 316 nm (log ε 4.03, 3.88). Enol-type acid, pKa 4.5. Slightly sol in water. Soluble in ether, lower alcohols, acetone, hot cyclohexane. Moderately stable to mineral acid, but dec rapidly in alkaline soln.

9318. Tertatolol. *(±)-1-[[(3,4-Dihydro-2H-1-benzothiopyran-8-yl)oxy]-3-[(1,1-dimethylethyl)amino]-2-propanol;* (±)-1-tert-butylamino-3-(1-thiachroman-8-yloxy)-2-pro-

panol; dl-8-[2-hydroxy-3-[(tert-butylamino)propyl]oxy]thiochromane. $C_{16}H_{25}NO_2S$; mol wt 295.45. C 65.05%, H 8.53%, N 4.74%, O 10.83%, S 10.85%. Nonselective β-adrenergic blocker. Prepn: C. Malen, M. Laubie, Ger. pat. **2,115,201**; *eidem*, U.S. pat. **3,960,891** (1971, 1976 both to Sci. Union et Cie). Pharmacology in animals: M. Laubie *et al.*, *Arch. Int. Pharmacodyn. Ther.* **201**, 323, 334 (1973); B. R. Walker *et al.*, *J. Cardiovasc. Pharmacol.* **7**, 1193 (1985). Synergistic effect with indapamide, *q.v.:* E. Marmo *et al.*, *Drugs Exptl. Clin. Res.* **11**, 709 (1985). GC-MS determn in plasma and urine: S. Staveris *et al.*, *J. Chromatog.* **339**, 97 (1985). Pharmacology in humans and mechanism of action study: A. De Blasi *et al.*, *Clin. Pharmacol. Ther.* **39**, 245 (1986). Preliminary evaluation in hypertension: J. P. Degaute *et al.*, *Am. J. Hypertens.* **1**, 263S (1988).

Crystals from hexane, mp 70-72°.

Hydrochloride, $C_{16}H_{25}NO_2S$.HCl, S-2395, SE-2395, Artex, Artexal, Prenalex. Crystals from acetonitrile, mp 180-183°. pKa 9.8. LD$_{50}$ in rats, mice (mg/kg): 40, 37 i.v.; 90, 120 i.p. (Laubie).

THERAP CAT: Antihypertensive.

9319. α-Terthienyl. *2,2':5'2''-Terthiophene;* 5-(2-thienyl)-2,2'-bithiophene. $C_{12}H_8S_3$; mol wt 248.39. C 58.03%, H 3.25%, S 38.73%. Biocidal constituent of various species of marigolds. Isoln from *Tagetes erecta* L., *Compositae:* L. Zechmeister, J. W. Sease, *J. Am. Chem. Soc.* **69**, 273 (1947). Isoln and distribution in *Tageteae:* K. R. Downum, G. H. N. Towers, *J. Nat. Prod.* **46**, 98 (1983). Synthesis: W. Steinkopf *et al.*, *Ann.* **546**, 180 (1941); H. J. Kooreman, H. Wynberg, *Rec. Trav. Chim.* **86**, 37 (1967); J.-P. Beny *et al.*, *J. Org. Chem.* **47**, 2201 (1982). Biological activity as nematocide: J. H. Uhlenbroek, J. D. Bijloo, *Rec. Trav. Chim.* **77**, 1004 (1958); J. D. Bijloo *et al.*, Ger. pat. **1,075,891**; *eidem*, U.S. pat. **3,050,442** (1960, 1962, both to North American Philips); J. Bakker *et al.*, *J. Biol. Chem.* **254**, 1841 (1979); as herbicide: J. Harvey, Jr., U.S. pat. **3,086,854** (1963 to Du Pont); G. Campbell *et al.*, *J. Chem. Ecol.* **8**, 961 (1982); as antimicrobial: J. R. Kagan *et al.*, *Photochem. Photobiol.* **31**, 465 (1980); T. Arnason *et al.*, *ibid.* **33**, 821 (1981); F. DiCosmo *et al.*, *Pestic. Sci.* **13**, 589 (1982).

Yellow-orange plates from methanol, mp 93-94°. uv max (methanol): 254, 350 nm (ε 7100, 21300). Sol in carbon bisulfide, ether, benzene, acetone, petr ether; slightly sol in methanol, ethanol. Insol in water.

9320. Tertiomycins. Antibiotic substances produced by *Streptomyces eurocidicus*. Isoln of **Tertiomycin A** and **Tertiomycin B**: Osato *et al.*, *J. Antibiot.* **8A**, 105, 161 (1955). Tertiomycin A also isolated from *S. albireticuli*: Miyake *et al.*, *ibid.* **12A**, 59 (1959). Manuf by culture of *S. eurocidicus*: Umezawa *et al.*, Japan. pat. **5200('67)** (to Sanraku Ocean Co.), *C.A.* **67**, 10386c (1967). Belongs to the erythromycin-carbomycin group of antibiotics.

9321. Testolactone. *D-Homo-17a-oxaandrosta-1,4-diene-3,17-dione; 13-hydroxy-3-oxo-13,17-secoandrosta-1,4-dien-17-oic acid δ-lactone;* 1,2,3,4,4a,4b,7,9,10,10a-decahydro-2-hydroxy-2,4b-dimethyl-7-oxo-1-phenanthrenepropionic acid δ-lactone; delta-1-testololactone; 1-dehydrotestololactone; 17α-oxo-D-homo-1,4-androstadiene-3,17-dione; Δ^1-testolactone; NSC-23759; SQ-9538; Fludestrin; Teslac. $C_{19}H_{24}O_3$; mol wt 300.40. C 75.97%, H 8.05%, O 15.98%. Obtained by microbial transformation of progesterone, Reichstein's substance S, or testosterone: Fried *et al.*, *J. Am. Chem. Soc.* **75**, 5764 (1953); *eidem*, U.S. pat. **2,744,120** (1956 to Olin Mathieson); Brannon *et al.*, *J. Org. Chem.* **30**,

760 (1965). Comprehensive description: K. Florey, *Anal. Profiles Drug Subs.* **5**, 533-553 (1976).

Crystals from acetone, mp 218-219°. $[\alpha]_D^{23}$ −45.6° (c = 1.24 in chloroform). uv max (ethanol): 242 nm (ε 15800).

Note: This is a controlled substance (anabolic steroid) in the U.S. Code of Federal Regulations, Title 21 Part 1308.13, as defined in Part 1308.02 (1995).

THERAP CAT: Antineoplastic.

9322. Testosterone. *(17β)-17-Hydroxyandrost-4-en-3-one;* Δ⁴-*androsten-17β-ol-3-one; trans-testosterone;* Andro; Mertestate; Oreton; Testoderm; Testolin; Testro AQ; Virosterone. $C_{19}H_{28}O_2$; mol wt 288.43. C 79.12%, H 9.78%, O 11.09%. Principal hormone of the testes, produced by the interstitial cells. Major circulating androgen; converted by 5α-reductase in androgen-dependent target tissues to *5α-dihydrotestosterone* which is required for normal male sexual differentiation. Also converted by aromatization to estradiol, *q.v.* Isoln from bull testes: David *et al., Z. Physiol. Chem.* **233**, 281 (1935). Prepn from cholesterol and confirmation of structure: A. Butenandt, G. Hanisch, *Ber.* **68**, 1859 (1935); *eidem, Z. Physiol. Chem.* **237**, 89 (1935); from dehydroandrosterone: L. Ruzicka, A. Wettstein, *Helv. Chim. Acta* **18**, 1264 (1935); from mixed esters: L. Ruzicka *et al., ibid.* 1478. Crystal structure: P. J. Roberts *et al., J. Chem. Soc., Perkin Trans. II* **1973**, 1978. Historical review: J. M. Hoberman, C. E. Yesalis, *Sci. Am.* **272**, 76-81 (Feb. 1995). Review of role in aging males: F. E. Kaiser, J. E. Morley, *Neurobiol. Aging* **15**, 559-563 (1994); of clinical relevance in females: R. S. Rittmaster, *Am. J. Med.* **98**, Suppl. 1A, 17S-21S (1995).

Needles from dil acetone, mp 155°. $[\alpha]_D^{24}$ +109° (c = 4 in alc). uv max: 238 nm. Insol in water. Sol in alcohol, ether, and other organic solvents.

Acetate, $C_{21}H_{30}O_3$, mp 140-141°.

17β-Cyclopentanepropionate, $C_{27}H_{40}O_3$, *testosterone cypionate, depAndro, Depotest, Depo-Testosterone, Depovirin, Pertestis, Virilon.* Pharmacology: A. C. Ott *et al., J. Clin. Endocrinol. Metabol.* **12**, 15 (1952). Crystals, mp 101-102°. $[\alpha]_D^{25}$ +87° (CHCl₃). Sol in oils.

Enanthate, $C_{26}H_{40}O_3$, *Andro LA, Androtardyl, Delatestryl, Everone, Primoteston, Testinon, Testo-Enant.* Prepn: Junkmann *et al.,* U.S. pat. **2,840,508** (1958 to Schering AG). Comprehensive description: K. Florey, *Anal. Profiles Drug Subs.* **4**, 452-465 (1975). Crystals, mp 36-37.5°.

Isobutyrate, $C_{23}H_{34}O_3$, *Perandren M.*

Propionate, $C_{22}H_{32}O_3$, *Anertan, Enarmon, Neo-Hombreol (amp.), Orchisterone, Perandren (amp.), Synandrol, Synerone, Testex, Testoviron, Virormone.* Prepn: H. Koester *et al.,* U.S. pat. **2,236,574** (1941 to Schering). Stout prisms from alcohol + water, mp 118-122°. $[\alpha]_D^{25}$ +83° to +90° (100 mg in 10 ml dioxane). Insol in water. Freely sol in alcohol, ether, pyridine, in other organic solvents. Sol in vegetable oils.

Undecanoate, $C_{29}H_{48}O_3$, *testosterone 17β-undecylate, Andriol, Pantestone, Restandol.*

Note: This is a controlled substance (anabolic steroid) in the U.S. Code of Federal Regulations, Title 21 Part 1308.13, as defined in Part 1308.02 (1995).

THERAP CAT: Androgen.
THERAP CAT (VET): Androgen.

9323. Tetraamminecopper Sulfate. Cuprammonium sulfate; ammonium cupric sulfate; cupric sulfate, ammoniated; Eau Celeste. $CuH_{12}N_4O_4S$; mol wt 227.73. Cu 27.90%, H 5.31%, N 24.60%, O 28.10%, S 14.08%. $[Cu(NH_3)_4]SO_4$. Prepd by dissolving copper sulfate in ammonia water and pptg with alcohol: Mazzi, *Acta Cryst.* **8**, 137 (1955). Crystal structure of monohydrate: Morosin, *ibid* **25B**, 19 (1969).

Monohydrate, large, dark blue crystals. d_4^{20} 1.81. Ammonia odor, dec in air. Loses H_2O and $2NH_3$ on heating to 120°, remaining $2NH_3$ at 160°. Soly in water at 21.5°: 18.5 g/100 ml. Practically insol in the lower alcohols.

USE: In textile printing, especially in calico finishing. As fungicide.

9324. Tetrabarbital. *5-Ethyl-5-(1-ethylbutyl)-2,4,6(1H,-3H,5H)-pyrimidinetrione; 5-ethyl-5-(1-ethylbutyl)barbituric acid;* 5-ethyl-5-(1-ethylbutyl)malonylurea; JL-991; Butysal; Butysedal. $C_{12}H_{20}N_2O_3$; mol wt 240.30. C 59.98%, H 8.39%, N 11.66%, O 19.97%. Prepd from ethyl 2-ethyl-2-(1-ethylbutyl)malonate + urea: Kopp, Tchoubar, *Bull. Soc. Chim. France* **1951**, 30; Tchoubar, Fr. pat. **1,020,357** (1953), *C.A.* **53**, 1389a.

Crystals, mp 122°.

Note: This is a controlled substance (depressant) listed in the U.S. Code of Federal Regulations, Title 21 Part 1308.13 (1995).

THERAP CAT: Sedative; hypnotic.

9325. Tetrabenazine. *1,3,4,6,7,11b-Hexahydro-9,10-dimethoxy-3-(2-methylpropyl)-2H-benzo[a]quinolizin-2-one;* 2-oxo-3-isobutyl-9,10-dimethoxy-1,2,3,4,6,7-hexahydro-11bH-benzo[a]quinolizine; Ro-1-9569. $C_{19}H_{27}NO_3$; mol wt 317.43. C 71.89%, H 8.57%, N 4.41%, O 15.12%. Dopamine depleting agent. Prepn: Brossi *et al., Helv. Chim. Acta* **41**, 119 (1958); U.S. pat. **2,830,993** (1958 to Hoffmann-La Roche); Osbond, *J. Chem. Soc.* **1961**, 4711. Pharmacology: Leusen *et al., Arch. Int. Pharmacodyn.* **119**, 225 (1959). Metabolism: Schwartz *et al., Biochem. Pharmacol.* **15**, 645 (1966). Clinical trials in hyperkinetic movement disorders: J. Janovic, *Ann. Neurol.* **11**, 41 (1982); J. Janovic, J. Orman, *Neurology* **38**, 391 (1988).

Prisms from methanol, mp 125-126°.

Oxime, crystals from dil alcohol, mp 158°.

Hydrochloride, $C_{19}H_{27}NO_3$.HCl, crystals, mp 208-210°. Soluble in hot water. Practically insol in acetone. uv max (alcohol): 230, 284 nm (ε 7780, 3820).

Methanesulfonate, $C_{19}H_{27}NO_3$.CH₃SO₃H, *Nitoman.* Bitter crystals, sensitive to light, mp 126-130°. Sparingly sol in water, sol in alcohol. Practically insol in ether.

THERAP CAT: Antidyskinetic. Antipsychotic.

9326. Tetraborane(10). Tetraboron decahydride; borobutane. B_4H_{10}; mol wt 53.32. B 81.10%, H 18.90%. Prepd by the reaction of magnesium boride with hydrochloric or phosphoric acid: Stock, Kuss, *Ber.* **56B**, 789 (1923); from dihydropentaborane: Burg, Schlesinger, *J. Am. Chem. Soc.* **55**, 4009 (1933); from diborane: Stock, Mathing, *Ber.* **69B**,

1456 (1936); Klein *et al., J. Am. Chem. Soc.* **80**, 4149 (1958). *Review:* Greenwood in *Comprehensive Inorganic Chemistry,* vol. **1**, J. C. Bailar, Jr. *et al.,* Eds. (Pergamon Press, Oxford, 1973) pp 785-791.

Gas; disagreeable odor; mp $-120°$; bp $18°$. Vapor pressure 580 mm Hg at $6°$; 388 mm at $0°$. Dec at room temp in a few hours. Dec rapidly at $100°$. Ignites spontaneously in air or oxygen. Hydrolyzes in water to boric acid and hydrogen. Reacts with ammonia to form a tetraammoniate.

9327. 3,4,5,6-Tetrabromo-*o*-cresol. $C_7H_4Br_4O$; mol wt 423.72. C 19.84%, H 0.95%, Br 75.43%, O 3.78%. Prepd by bromination of *o*-cresol or its methyl or ethyl ether: Zincke, Hedenström, *Ann.* **350**, 269 (1906); Bonneaud, *Bull. Soc. Chim. France* **7**, 776 (1910); Treacy, U.S. pat. **2,319,960** (1943 to Merck & Co.).

White to buff, cryst powder, mp 205-208° with dec. Practically insol in water. Sol in alcohol, ether, alkali hydroxides.

Acetate, $C_9H_6Br_4O_2$, needles from dil alcohol, mp 154°. Slightly sol in alcohol and glacial acetic acid.

Caution: Irritating to skin, mucous membranes.

USE: Fungicide.

9328. *sym*-Tetrabromoethane. *1,1,2,2-Tetrabromoethane;* acetylene tetrabromide; Muthmann's liquid. $C_2H_2Br_4$; mol wt 345.65. C 6.95%, H 0.58%, Br 92.47%. Br_2CH-$CHBr_2$. Manuf by bromination of acetylene: **Brit. pat. 889,649** (1962 to Associated Ethyl Co.). Toxicity studies: Gray, *Arch. Ind. Hyg. Occup. Med.* **2**, 407 (1950); D. L. Wolff, *Acta Biol. Med. Ger.* **41**, 945 (1982).

Yellowish, heavy, very refractive liquid; odor of camphor and iodoform. d 2.964; bp₅₄ 151°; mp 0°; n_D^{20} 1.638. Insol in water. Miscible with alc, chloroform, ether, aniline, glacial acetic acid. LD₅₀ i.p. in mice: 443.3 mg/kg (Wolff).

Caution: Potential symptoms of overexposure are irritation of eyes and nose; anorexia, nausea; severe headaches; abdominal pain; jaundice; monocytosis. *See NIOSH Pocket Guide to Chemical Hazards* (DHHS/NIOSH 90-117, 1990) p 32.

USE: In microscopy, as solvent, separating minerals by density.

9329. 3′,3″,5′,5″-Tetrabromophenolphthalein. *3,3-Bis-(3,5-dibromo-4-hydroxyphenyl)-1(3H)-isobenzofuranone.* $C_{20}H_{10}Br_4O_4$; mol wt 633.91. C 37.89%, H 1.59%, Br 50.42%, O 10.10%. Prepd by direct bromination of phenolphthalein: Blicke *et al., J. Am. Chem. Soc.* **54**, 1465 (1932). Diagnostic use: E. A. Graham, W. H. Cole, *J. Am. Med. Assoc.* **250**, 2975 (1984), reprint of *J. Am. Med. Assoc.* **82**, 613 (1924).

Crystals from alcohol, glacial acetic acid, or ether, mp 295-297° (dec). Practically insol in water. Slightly sol in alc, glacial acetic acid; sol in ether, in alkali with a violet color.

Disodium salt, $C_{20}H_8Br_4Na_2O_4$, *tetrabromophenolphthalein sodium, Brom-Tetragnost, Cholegnostyl, Tetrabrom.* Blue crystals, dec on exposure to air becoming partially

insol. Freely sol in water; slightly sol in alcohol. Insol in ether. pH of aq soln is 9-10. *Keep well closed and protected from light.*

THERAP CAT: Disodium salt as diagnostic aid (radiopaque medium).

9330. Tetracaine Hydrochloride. *4-(Butylamino)benzoic acid 2-(dimethylamino)ethyl ester monohydrochloride; p-*butylaminobenzoyl-2-dimethylaminoethanol hydrochloride; 2-dimethylaminoethyl 4-*n*-butylaminobenzoate hydrochloride; dicain; amethocaine hydrochloride; Anethaine; Decicain; Pantocaine; Pontocaine Hydrochloride; Tonexol. $C_{15}H_{25}ClN_2O_2$; mol wt 300.83. C 59.89%, H 8.38%, Cl 11.79%, N 9.31%, O 10.64%. Prepn: U.S. pat. **1,889,645** (1932); **Brit. pat. 815,144** (1959 to Abbott). Prepn of pharmaceutical dosage forms: Shupe, U.S. pat. **3,272,700** (1966 to Sterling Drug). Mechanism of action studies: Y.-W. Leung *et al., J. Infect. Dis.* **136**, 679 (1977). Toxicity study: Dawes, *Brit. J. Pharmacol. Chemother.* **1**, 90 (1946). Acute toxicity: B. A. Bopp *et al., J. Pharm. Sci.* **67**, 882 (1978). Comprehensive description: M. Riaz in *Analytical Profiles of Drug Substances* vol. **18**, K. Florey, Ed. (Academic Press, New York, 1989) pp 379-411.

Faintly bitter crystals producing transient numbness of the tongue. mp 147-150°. pKa 8.39. uv max (water): 225, 310 nm (ϵ 14108, 26352); (0.1*N* H_2SO_4): 229, 281, 312 nm (E₁cm^1% 509, 55, 76); (methanol): 226, 310 nm (ϵ 7586, 29512); (chloroform): 308 nm (ϵ 27542). Sol at 20° in 7.5 parts water, 40 parts alcohol; 30 parts chloroform. Practically insol in ether, benzene, acetone. The aq soln is neutral to litmus. Aq solns are stable and may be sterilized by brief boiling. LD₅₀ i.p. in mice: 70 mg/kg (Dawes); also reported as LD₅₀ in female mice (mg/kg): 13 i.v.; 35 s.c. (Bopp).

THERAP CAT: Anesthetic (local).

THERAP CAT (VET): Anesthetic (topical).

9331. Tetrachloroethane. *1,1,2,2-Tetrachloroethane; sym*-tetrachloroethane; acetylene tetrachloride; Cellon; Bonoform. $C_2H_2Cl_4$; mol wt 167.85. C 14.31%, H 1.20%, Cl 84.49%. $Cl_2CHCHCl_2$. Manuf by catalytic addition of chlorine to acetylene: Peters, Neumann, *Angew. Chem.* **45**, 261 (1932); by chlorination of ethylene: Pye, U.S. pat. **2,752,402** (1956 to Dow); by catalytic chlorination of ethane: Joseph, U.S. pat. **2,752,401** (1956 to Dow); by chlorination of 1,2-dichloroethane: Conrad, U.S. pat. **2,725,412** (1955 to Ethyl Corp.); Fox, U.S. pat. **2,846,484** (1958 to Monsanto). Toxicity study: H. F. Smyth *et al., Am. Ind. Hyg. Assoc. J.* **30**, 470 (1969).

Nonflammable, heavy, mobile liquid. Sweetish, suffocating, chloroform-like odor. d₄²⁵ 1.58658. mp $-44°$. bp₇₆₀ 146.5°. n_D^{20} 1.49419. Very sparingly sol in water. At 25° one gram dissolves in 350 ml H_2O. Misc with methanol, ethanol, benzene, ether, petr ether, carbon tetrachloride, chloroform, carbon disulfide, dimethylformamide, oils. Has the highest solvent power of the chlorinated hydrocarbons. LD₅₀ orally in rats: 0.20 ml/kg (Smyth).

Caution: Potential symptoms of overexposure are nausea, vomiting and abdominal pain; finger tremors; jaundice, enlarged tender liver; dermatitis; monocytosis; kidney damage. *See NIOSH Pocket Guide to Chemical Hazards* (DHHS/NIOSH 90-117, 1990) p 208.

USE: Nonflammable solvent for fats, oils, waxes, resins, cellulose acetate, rubber, copal, phosphorus, sulfur. As solvent in certain types of Friedel-Crafts reactions or phthalic anhydride condensations. In the manuf of paint, varnish, and rust removers. In soil sterilization and weed killer and insecticide formulations. In the determination of theobromine in cacao. As immersion fluid in crystallography. In the biological laboratory to produce pathological changes in gastrointestinal tract, liver, and kidneys. Intermediate in the manuf of trichloroethylene and other chlorinated hydrocarbons having two carbon atoms.

9332. Tetrachloroethylene. *Tetrachloroethene;* perchloroethylene; ethylene tetrachloride; tetrachlorethylene; Nema; Tetracap; Tetropil; Perclene; Ankilostin; Didakene. C_2Cl_4; mol wt 165.83. C 14.49%, Cl 85.51%. $Cl_2C=CCl_2$. Prepd by Faraday in 1821. Manuf by catalytic oxidation of 1,1,2,2-tetrachloroethane: Ellsworth, Vancamp, U.S. pat. **2,951,103** (1960 to Columbia-Southern Chem.); Feathers, Rogerson, U.S. pat. **3,040,109** (1962 to Pittsburgh Plate Glass); by catalytic chlorination of acetylene: Thermet, Parvi, U.S. pat. **2,938,931** (1960 to Société d'électrochimie, d'électrométallurgie et des aciéries électriques d'Ugine). Review of mfg processes: Faith, Keyes & Clark's *Industrial Chemicals,* F. A. Lowenheim, M. K. Moran, Eds. (Wiley-Interscience, New York, 4th ed., 1975) pp 604-611. Physical properties: Mumford, Phillips, *J. Chem. Soc.* **1950,** 75. Toxicity data: Dybing, *Acta Pharmacol. Toxicol.* **2,** 223 (1946); Lazarew, *Arch. Exp. Pathol. Pharmacol.* **141,** 19 (1929).

Colorless, nonflammable liq; ethereal odor; d_4^{15} 1.6311; d_4^{20} 1.6230. bp 121°. mp $\sim -22°$. n_D^{20} 1.5055. Sol in about 10,000 vol water; misc with alcohol, ether, chloroform, benzene. LD_{50} orally in mice: 8.85 g/kg (Dybing). LC for mice in air: 5925 ppm (Lazarew).

Caution: Potential symptoms of overexposure are irritation of eyes, nose and throat; nausea; flushing of face and neck; vertigo, dizziness, incoordination; headache, somnolence; skin erythema; liver damage. *See NIOSH Pocket Guide to Chemical Hazards* (DHHS/NIOSH 90-117, 1990) p 208. This substance may reasonably be anticipated to be a carcinogen: *Seventh Annual Report on Carcinogens* (PB95-109781, 1994) p 372.

USE: Dry cleaning; textile processing; degreasing metals; solvent; chemical intermediate in production of fluorocarbons. Insulating fluid and cooling gas in electrical transformers.

THERAP CAT: Anthelmintic (Nematodes, Trematodes).

THERAP CAT (VET): Anthelmintic.

9333. 3,3',4',5-Tetrachlorosalicylanilide. *3,5-Dichloro-N-(3,4-dichlorophenyl)-2-hydroxybenzamide;* Irgasan BS200. $C_{13}H_7Cl_4NO_2$; mol wt 351.01. C 44.48%, H 2.01%, Cl 40.40%, N 3.99%, O 9.12%. Prepn of polyhalosalicylanilides: Bindler, Model, U.S. pat. **2,703,332** (1955 to Geigy). Bacteriostat in the manuf of thermoplastic articles: Teller, U.S. pat. **3,005,720** (1961 to Weco Products).

Crystals, mp 161°. Fluoresces under ultraviolet light. Practically insol in water. Sol in alkaline aq solns and in solns of wetting agents. Sol in many organic solvents.

USE: Bacteriostat in formulations of surgical soaps, laundry soaps, rinses, polishes, shampoos, deodorants. Also as preservative in textile finishes, certain petroleum products, cellulose esters, cutting oils, coolants. Banned by FDA from use in cosmetics.

9334. Tetrachlorvinphos. *Phosphoric acid (Z)-2-chloro-1-(2,4,5-trichlorophenyl)ethenyl dimethyl ester;* 2,4,5-trichloro-α-(chloromethylene)benzyl phosphate; 2-chloro-1-(2,4,5-trichlorophenyl)vinyl dimethyl phosphate; stirofos; ENT-25841; SD-8447; Gardona; Rabon; Rabond. $C_{10}H_9Cl_4O_4P$; mol wt 365.96. C 32.82%, H 2.48%, Cl 38.75%, O 17.49%, P 8.46%. Organophosphate insecticide; cholinesterase inhibitor. Prepn: D. D. Phillips, L. F. Ward, Jr., U.S. pat. **3,102,842;** D. E. Ramey, U.S. pat. **3,553,297** (1963, 1971 both to Shell). Properties: Whetstone *et al., J. Agr. Food Chem.* **14,** 352 (1966). Soil persistence: N. P. Agnihotri *et al., Indian J. Agric. Chem.* **14,** 27 (1981). Metabolism: M. H. Akhtar, T. S. Foster, *J. Agric. Food Chem.* **28,** 698 (1980); *eidem, ibid.* **29,** 766 (1981). Determn of residues in foods: V. Leoni *et al., J. A. O. A. C. Int.* **75,** 511 (1992). Toxicology studies: T. B. Gaines, *Toxicol. Appl. Pharmacol.*

14, 515 (1969); A. I. T. Walker *et al., Pesticide Sci.* **3,** 517 (1972). Review of agricultural use: A. G. Jennings, *Int. Pest Contr.* **12,** 28-33 (1970); of carcinogenic risk: *IARC Monographs* **30,** 197-206 (1983).

mp 97-98°. Vapor pressure at 20° $= 4.2 \times 10^{-8}$. Soly in water: 11 ppm; in xylene: $< 15\%$; 40-50% in chloroform at room temp. LD_{50} in male, female rats (mg/kg): 1100, 1125 orally (Gaines).

USE: Insecticide.

THERAP CAT (VET): Insecticide.

9335. Tetracosamethylhendecasiloxane. *Tetracosamethylundecasiloxane.* $C_{24}H_{72}O_{10}Si_{11}$; mol wt 829.77. C 34.74%, H 8.75%, O 19.28%, Si 37.23%. Prepd by reaction of hexamethyldisiloxane with octamethylcyclotetrasiloxane and sulfuric acid: Patnode, Wilcock, *J. Am. Chem. Soc.* **68,** 362 (1946); by cohydrolysis of ethoxytrimethylsilane and diethoxydimethylsilane: Hyde, U.S. pat. **2,457,677** (1948 to Corning Glass).

Liquid; $bp_{4.7}$ 201°; $bp_{0.5}$ 152°; d_4^{25} 0.9247; n_D^{20} 1.3994. Flash pt 188.33°. Stable. Inert to most chemical reagents and rubber. Maintains about the same viscosity over a wide temperature range. Sol in benzene and the lighter hydrocarbons; slightly sol in alcohol and the heavy hydrocarbons.

USE: As a basis for silicone oils or fluids designed to withstand extremes of temperature; as a foam suppressant in petroleum lubricating oil.

9336. Tetracyanoethylene. *Ethenetetracarbonitrile;* 1,-1,2,2-tetracyanoethene; $\Delta^{2,2'}$-bimalononitrile; TCNE. C_6N_4; mol wt 128.09. C 56.26%, N 43.74%. $(NC)_2C=C(CN)_2$. Prepd by debromination of dibromomalononitrile with copper powder in boiling benzene: Cairns *et al., J. Am. Chem. Soc.* **80,** 2775 (1958).

Crystals, mp 200°. Begins to sublime at 120°.

USE: In the synthesis of spiro compds, in modified Diels-Alder reactions, as aromatizing agent: Longone, Smith, *Tetrahedron Letters* **1962,** 205.

9337. Tetracycline. *[4S-(4α,4aα,5aα,6β,12aα)]-4-(Dimethylamino)-1,4,4a,5,5a,6,11,12a-octahydro-3,6,10,12,12a-pentahydroxy-6-methyl-1,11-dioxo-2-naphthacenecarboxamide;* deschlorobiomycin; tsiklomitsin; Abricycline; Ambramycin; Bio-Tetra; Cyclomycin; Dumocyclin; Tetradecin. $C_{22}H_{24}N_2O_8$; mol wt 444.44. C 59.45%, H 5.44%, N 6.30%, O 28.80%. Antibiotic substance produced by *Streptomyces* spp. Prepn: J. H. Boothe *et al., J. Am. Chem. Soc.* **75,** 4621 (1953); L. H. Conover *et al., ibid.* **4622;** Conover, U.S. pat. **2,699,054** (1955). Production by *Streptomyces viridifaciens:* Gourevitch, Lein, Heinemann *et al.,* U.S. pats. **2,712,517; 2,886,595** (1955, 1959 both to Bristol Labs.); by *S. aureofaciens:* Miller, Arishima, Sekizwa, U.S. pats. **3,005,023; 3,019,173** (1961, 1962 both to Am. Cyanamid). Purification: Kaplan, Granatek, U.S. pat. **3,301,899** (1967 to Bristol-Myers). Prepn of phosphate complex: Seiger, Weidenheimer, U.S. pat. **3,053,892** (1962 to Am. Cyanamid). Total synthesis of tetracyclines: J. H. Boothe *et al., J. Am. Chem. Soc.* **81,** 1006 (1959); L. H. Conover *et al., ibid.* **84,** 3222 (1962). Graphic outline of Woodward synthesis: *Chem. & Eng. News* **40,** 36 (Oct. 8, 1962). Abs config: V. N. Dobrynin *et al., Tetrahedron Letters* **1962,** 901. Solubility studies: Weiss *et al., Antibiot. & Chemother.* **7,** 374 (1957). Toxicity: E. I. Goldenthal, *Toxicol. Appl. Pharmacol.* **18,** 185 (1971). *Review:* "Tetracycline" in *The Technology of the Tetracyclines,* **vol. I,** R. C. Evans, Ed. (Quadrangle Press, New

York, 1968) pp 209-426. Mechanism of action: A. Kaji, M. Ryoji in *Antibiotics* vol. **5** (pt. 1), F. E. Hahn, Ed. (Springer-Verlag, New York, 1979) pp 304-328. Review of biosynthesis: C. R. Hutchinson in *Antibiotics* vol. **4**, J. W. Corcoran, Ed. (Springer-Verlag, New York, 1981) pp 1-12.

Trihydrate, crystals. Swells at 165°. Dec 170-175°. Becomes anhydr by drying *in vacuo* at 60° for 8 hrs. $[\alpha]_D^{25}$ −257.9° (0.1N HCl); $[\alpha]_D^{25}$ −239° (methanol). uv max (0.1N HCl): 220, 268, 355 nm (ε 13000, 18040, 13320). pKa (50% aq DMF): 8.3, 10.2. Stable in neutral and in alkaline soln. Soly at about 28°: 1.7 mg/ml water; > 20 mg/ml methanol. LD_{50} in rats, mice (mg/kg): 807, 808 orally (Goldenthal).

Hydrochloride, $C_{22}H_{24}N_2O_8 \cdot HCl$, *Achro, Achromycin, Achromycin V, Ala Tet, Ambracyn, Ambramicina, Bristacilina, Cefracycline, Criseociclina, Cyclopar, Diocyclin, Helvecyclin, Hostacyclin, Imex, Mediletten, Mephacyclin, Panmycin, Partrex, Polycycline, Purocyclina, Quadracyclin, Remicyclin, Riocyclin, Ro-Cycline, Sanclomycine, Steclin, Sumycin, Supramycin, Sustamycin, Tefilin, Tetrabakat, Tetrabid, Tetrablet, Tetrabon, Tetrachel, Tetracompren, Tetracyn, Tetrakap, Tetralution, Tetramavan, Tetramycin, Tetrosol, Topicycline, Totomycin, Triphacyclin, Unicin, Vetquamycin-324.* Crystals from butanol +HCl, dec 214°. $[\alpha]_D^{25}$ −257.9° (c = 0.5 in 0.1N HCl). Freely sol in water, sol in methanol, ethanol. Insol in ether, hydrocarbons. pH (2% aq soln): 2.1-2.3. LD_{50} orally in rats: 6443 mg/kg (Goldenthal).

Phosphate complex, *Panmycin P, Telotrex, Tetradecin Novum, Tetrex, Upcyclin.* Yellow, odorless powder. Sparingly sol in water; slightly sol in ethanol.

Lauryl sulfate, *Lauracycline.*

THERAP CAT: Antiamebic; antibacterial; antirickettsial.

THERAP CAT (VET): Antibacterial.

9338. Tetradecamethylhexasiloxane. $C_{14}H_{42}O_5Si_6$; mol wt 459.00. C 36.64%, H 9.22%, O 17.43%, Si 36.71%. Prepd by hydrolysis of dimethyldichlorosilane and trimethylchlorosilane: Patnode, Wilcock, *J. Am. Chem. Soc.* **68**, 358 (1946); by reaction of hexamethyldisiloxane with octamethylcyclotetrasiloxane and sulfuric acid: *eidem, ibid.* 362; Patnode, U.S. pat. **2,469,888** (1949 to General Electric); by cohydrolysis of ethoxytrimethylsilane and diethoxydimethylsilane: Hunter *et al., J. Am. Chem. Soc.* **68**, 2284 (1946); Hyde, U.S. pat. **2,457,677** (1948 to Corning Glass).

Liquid; bp_{20} 142°; d_4^{20} 0.8910; n_D^{20} 1.3948; flash pt 118.33°. Freezes below −100°. Stable. Inert to most chemical reagents and rubber. Maintains about the same viscosity over a wide temp range. Sol in benzene and the lighter hydrocarbons; slightly sol in alcohol and the heavy hydrocarbons.

USE: As a basis for silicone oils or fluids designed to withstand extremes of temp; as a foam suppressant in petroleum lubricating oil.

9339. Tetradifon. *1,2,4-Trichloro-5-[(4-chlorophenyl)-sulfonyl]benzene; p-chlorophenyl 2,4,5-trichlorophenyl sulfone;* 2,4,5,4'-tetrachlorodiphenyl sulfone; tedion; Tedion-V18. $C_{12}H_6Cl_4O_2S$; mol wt 356.06. C 40.48%, H 1.70%, Cl 39.83%, O 8.99%, S 9.01%. Prepn: Meltzer, Huisman, U.S. pat. **2,812,281** (1957 to Phillips); and soly data: Huisman *et al., Rec. Trav. Chim.* **77**, 103 (1958). Toxicity study: Bendyke *et al., World Rev. Pest Contr.* **9**, 119 (1970).

Crystals from benzene; mp 146.5-147.5°. Stable to concd and dil alkalies, mineral acids, high temp, and u.v. light. LD_{50} orally in rats: 556 mg/kg (Ben-dyke).

USE: Acaricide. Ovicide on deciduous fruits, citrus, cotton and other crops.

9340. Tetraethylammonium Bromide. *N,N,N-Triethylethanaminium bromide;* TEAB; TMD-10; Etylon; Etambro; Sympatektoman; Tetranium. $C_8H_{20}BrN$; mol wt 210.16. C 45.72%, H 9.59%, Br 38.02%, N 6.66%. $(C_2H_5)_4NBr$. Ganglion blocking agent. Prepd from triethylamine and ethyl bromide: Hofmann, *Ann.* **78**, 263 (1851). Review of the pharmacology of the tetraethylammonium ion: Moe, Freyburger, *Pharmacol. Rev.* **2**, 61-95 (1950).

Deliquesc crystals. Freely sol in water, alc, chloroform, acetone. Slightly sol in benzene. pH of a 10% aq soln 6.5. The pH is not changed by heating for 28 hrs at 95°.

9341. Tetraethylammonium Chloride. Etamon chloride; T.E.A. chloride; $C_8H_{20}ClN$; mol wt 165.71. C 57.99%, H 12.17%, Cl 21.39%, N 8.45%. $(C_2H_5)_4NCl$. Ganglion blocking agent. *See* Tetraethylammonium Bromide.

Deliquescent crystals. d_4^{21} 1.0801. Freely sol in water, alcohol, chloroform, acetone; slightly sol in benzene. pH of 10% aq soln 6.48. The pH is not changed by heating for 28 hrs at 95°.

Tetrahydrate, monoclinic prismatic crystals. mp 37.5°. d 1.084.

9342. Tetraethylammonium Hydroxide. $C_8H_{21}NO$; mol wt 147.26. C 65.25%, H 14.37%, N 9.51%, O 10.86%. $(C_2H_5)_4NOH$. Made from the corresp halide by treating with silver oxide or with a soln of potassium hydroxide in methanol.

Marketed as an aq soln. A 10% soln has a d_4^{25} ∼1.01. The free base is known only in soln or as hydrates. Dec on boiling. It is a very strong base readily absorbing CO_2 from the air. The aq soln is colorless, odorless, bitter, caustic, strongly alkaline, and imparts a soapy feel to the skin. *Keep well closed.*

Tetrahydrate, mp 49-50°.

Hexahydrate, mp 55°.

9343. Tetraethyllead. *Tetraethylplumbane;* lead tetraethyl; TEL. $C_8H_{20}Pb$; mol wt 323.45. C 29.71%, H 6.23%, Pb 64.06%. $Pb(C_2H_5)_4$. Prepd by the action of PbCl₂ on zinc ethyl or on a Grignard reagent; by heating C_2H_5Cl and sodium-lead alloy in an autoclave. The production from lead, ethylene, and hydrogen using triethylaluminum as intermediate was first described by K. Ziegler at the 14th International Congress of Pure and Applied Chemistry (July 1955): *Chem. & Eng. News* **33**, 3486 (1955). Alternate synthesis using nonhalide compds: Pearson *et al., Advances in Chemistry Series* **23**, 299-305 (1959). Toxicity study: Schroeder *et al., Experientia* **28**, 923 (1972).

Colorless liq; burns with an orange-colored flame with green margin. *Extremely poisonous!* d^{20} 1.653. bp ∼200° also stated as 227.7° with decompn. n_D^{20} 1.5198. Practically insol in water. Sol in benzene, petr ether, gasoline, slightly in alcohol. LD_{50} orally in rats: 12.3 mg/kg (Schroeder).

Caution: Potential symptoms of overexposure are insomnia, lassitude and anxiety; tremor, hyper-reflexia and spasticity; bradycardia, hypotension, hypothermia, pallor, nausea, anorexia and weight loss; disorientation, hallucinations, psychosis, mania, convulsions and coma; eye irritation. *See NIOSH Pocket Guide to Chemical Hazards* (DHHS/NIOSH 90-117, 1990) p 208. *See also* Patty's *Industrial Hygiene and Toxicology* vol. **2A**, G. D. Clayton, R. E. Clayton, Eds. (Wiley-Interscience, New York, 3rd ed., 1981) pp 1687-1728; *Clinical Toxicology of Commercial Compounds*, R. E. Gosselin *et al.,* Eds. (Williams & Wilkins, Baltimore, 5th ed., 1984) Section II, p 139, Section III, pp 226-239.

USE: As a gasoline additive to prevent "knocking" in motors. *See also* Milde, Beatty, "Chemical Reactions of

Tetraethyllead" in *Advances in Chemistry Series* **23**, 306-318 (1959).

9344. N,N,N',N'-Tetraethylphthalamide. *N,N,N',N'-Tetraethyl-1,2-benzenedicarboxamide;* orthophthalic acid didiethylamide; *o*-phthalic acid bis[diethylamide]; tetraethylbis-(phthalamide); Analetil; Neo-Cardiamine; Neospiran; Unispiran. $C_{16}H_{24}N_2O_2$; mol wt 276.38. C 69.53%, H 8.75%, N 10.14%, O 11.58%. Prepd by treating phthalyl chloride with diethylamine: **Fr. pat.** 785,428; **Brit. pat.** 443,396; **U.S. pat.** 2,057,145 (1936 to Chem. Fabrik Grünau); by heating sodium phthalate with diethylamine phosphate: **Fr. pat.** 866,229 (1941 to Corbière).

Crystals, mp 39°. bp 175-180°. Soluble in water, physiol saline.

THERAP CAT: Analeptic.

9345. Tetraethyl Pyrophosphate. *Diphosphoric acid tetraethyl ester; pyrophosphoric acid tetraethyl ester;* bis-*O,O*-diethylphosphoric anhydride; TEPP; Bladan; Nifos T; Kilmite 40; Vapotone; Tetron; Killax; Mortopal. $C_8H_{20}O_7P_2$; mol wt 290.19. C 33.11%, H 6.95%, O 38.59%, P 21.35%. Cholinesterase inhibitor. Prepd commercially by controlled hydrolysis of *O,O*-diethylphosphoric acid chloride: Kosolapoff, **U.S. pat.** 2,479,939 (1947 to Monsanto); Toy, *J. Am. Chem. Soc.* **70**, 3882 (1948). Chemical history and comparison of various syntheses: G. Schrader, *Die Entwicklung Neuer Insektizider Phosphorsäure-Ester* (Verlag Chemie, Weinheim, 3rd ed., 1963) pp 68-79. Toxicity study: T. B. Gaines, *Toxicol. Appl. Pharmacol.* **14**, 515 (1969).

Mobile liquid. Agreeable odor. Hygroscopic. d_4^{20} 1.185. Thermal decompn range 170-213° with copious formn of ethylene. $bp_{0.05}$ 82°; $bp_{1.0}$ 124°; $bp_{2.3}$ 138°. Vapor pressure at 30° = 4.7 × 10^{-4} mm Hg. n_D^{20} 1.4196. Misc with water, but quickly hydrolyzed by it (half life at 25° about 7 hrs in a 50 v/v mixt.). Also misc with acetone, methanol, ethanol, benzene, chloroform, carbon tetrachloride, glycerol, ethylene glycol, propylene glycol, toluene, xylene. Not misc with petr ether, kerosene, other petr oils. LD$_{50}$ orally in male rats: 1.1 mg/kg (Gaines).

Caution: Potential symptoms of overexposure are eye pain, blurred vision, lacrimation, rhinorrhea, headache, tight chest and cyanosis; anorexia, nausea, vomiting and diarrhea; weakness, twitching, paralysis, Cheyne-Stokes respiration and convulsions; low blood pressure, cardiac irregularities; local sweating. *See NIOSH Pocket Guide to Chemical Hazards* (DHHS/ NIOSH 90-117, 1990) p 206.

USE: Insecticide.

9346. 2,2,3,3-Tetrafluoro-1-propanol. C_3-Fluoroalcohol. $C_3H_4F_4O$; mol wt 132.06. C 27.29%, H 3.05%, F 57.55%, O 12.12%. HCF$_2$CF$_2$CH$_2$OH. Prepn: Bestian, Rehn, **Ger. pat.** 1,007,771 (1957 to Hoechst).

Liquid. d_4^{20} 1.4853. mp −15°. bp_{760} 109-110°. n_D^{20} 1.3197. Surface tension at 20° = 27.6 dyn/cm.

p-Nitrobenzoate, mp 47°.

USE: To introduce fluoroalkyl groups into an organic molecule. Proposed intermediate for plastics, surface active agents, lubricants, elastomers.

9347. Tetraglycine Hydroperiodide. Globaline. $C_{16}H_{42}I_7N_8O_{16}$; mol wt 1490.89. C 12.89%, H 2.84%, I 59.58%, N 7.52%, O 17.17%. 2[(NH$_2$CH$_2$COOH)$_4$HI]·2½I$_2$. Prepn: Frost, Eddy, *J. Am. Chem. Soc.* **74**, 1346 (1952); Morris *et al., Ind. Eng. Chem.* **45**, 1013 (1953).

Flat needles with brassy-bronze metallic luster in reflected light, dec 162-167°. Soly in water at 25° = 380 g/l.

USE: Decontamination of drinking water in emergencies. Used in amounts sufficient to yield 8 ppm of active iodine. A tablet contg 20 mg plus 96 mg Na$_2$HPO$_7$ plus 4 mg talc will decontaminate one quart of water. Such tablets after 7 days' storage at 60° retained 60% of their original active iodine. Less stable than aluminum hexaurea sulfate triiodide.

9348. Tetraglyme. *2,5,8,11,14-Pentaoxapentadecane;* tetraethylene glycol dimethyl ether; dimethoxytetraethylene glycol. $C_{10}H_{22}O_5$; mol wt 222.28. C 54.04%, H 9.98%, O 35.99%. CH$_3$O(CH$_2$CH$_2$O)$_4$CH$_3$. Prepd from ethylene glycol methyl ether and 2,2'-dichlorodiethyl ether: Zellhoefer, **U.S. pat.** 2,111,234 (1935); *Ind. Eng. Chem.* **29**, 550 (1937). Purification: Vogel, *J. Chem. Soc.* **1948**, 618. Toxicity study: H. F. Smyth *et al., J. Ind. Hyg. Toxicol.* **23**, 259 (1941).

Liquid. d_4^{20} 1.0087; d_4^{86} 0.9514. mp −27°. bp_{760} 275.3°; bp_2 118°. n_D^{20} 1.4325. Sol in water. Misc with hydrocarbon solvents. LD$_{50}$ orally in rats: 5.14 g/kg (Smyth).

USE: Solvent.

9349. Tetrahydrocannabinols. *Tetrahydro-6,6,9-trimethyl-3-pentyl-6H-dibenzo[b,d]pyran-1-ol.* $C_{21}H_{30}O_2$; mol wt 314.47. C 80.21%, H 9.62%, O 10.18%. Active constituents of marihuana (hashish). The Δ¹-3,4-*trans* isomer, also referred to as Δ⁹-THC, is the only major active constituent in hashish; the Δ⁶-3,4-*trans* isomer, although physiologically active, represents no more than 1%: R. Mechoulam *et al., Science* **169**, 611 (1970). Isoln of Δ¹-3,4-*trans*-form from marihuana: Gaoni, Mechoulam, *J. Am. Chem. Soc.* **86**, 1646 (1964). Isoln of Δ⁶-3,4-*trans*-form: Hively *et al., ibid.* **88**, 1832 (1966). Synthesis of *dl*-Δ¹-3,4-*trans*-form: Fahrenholtz *et al., ibid.* **89**, 5934 (1967); Razden *et al., ibid.* **96**, 5860 (1974); *eidem, Experientia* **31**, 16 (1975); of *dl*-Δ⁶-3,4-*trans*-form: Taylor *et al., J. Am. Chem. Soc.* **88**, 367 (1966). Stereospecific synthesis of (−)-Δ¹-3,4-*trans*-form and (−)-Δ⁶-3,4-*trans*-form: Mechoulam *et al., ibid.* **89**, 4552 (1967). Abs config of naturally occurring (−)-Δ¹-3,4-*trans* form: Mechoulam, Gaoni, *Tetrahedron Letters* **1967**, 1109. IR, NMR, mass spec data for Δ⁹-THC: Petrzilka, Sikemeier, *Helv. Chim. Acta* **50**, 2111 (1967); for Δ⁶-THC: *eidem, ibid.* 1416. Metabolism in mice: H. D. Christensen *et al., Science* **172**, 165 (1971); in man: M. M. Halldin *et al., Arzneimittel-Forsch.* **32**, 764 (1982). Clinical studies of Δ⁹-THC as an anti-emetic in cancer patients: S. E. Sallan *et al., N. Engl. J. Med.* **302**, 135 (1980); A. E. Chang *et al., Cancer* **47**, 1746 (1981); D. S. Poster *et al., J. Am. Med. Assoc.* **245**, 2047 (1981). Topical use in hypertensive glaucomas: J. C. Merrit *et al., J. Pharm. Pharmacol.* **33**, 40 (1981). Effects of long-term THC treatment on the menstrual cycle of rhesus monkeys: C. G. Smith *et al., Science* **219**, 1453 (1983). Toxicity studies: R. N. Phillips *et al., Proc. Soc. Exp. Biol. Med.* **136**, 260 (1971); H. Rosenkranz *et al., Toxicol. Appl. Pharmacol.* **28**, 18 (1974); H. Yoshimura *et al., J. Med. Chem.* **21**, 1079 (1978). Review of analytical methods: L. Vollner *et al., Reg. Toxicol. Pharmacol.* **6**, 348-358 (1986). *See also* Cannabis and Hashish.

Δ⁹-THC

(−)-Δ¹-3,4-*trans*-Form, Δ¹-*THC*, Δ⁹-*THC*, *dronabinol*, *QCD-84924, Marinol.* $bp_{0.02}$ 200°. $[\alpha]_D^{20}$ −150.5° (c = 0.53 in CHCl$_3$). uv max (ethanol): 283, 276 nm (log ε 3.21, 3.20). LD$_{50}$ in Fischer rats (mg/kg): 1270 (males), 730 (females) orally, sesame oil vehicle; 800 (males) orally, sesame oil, 1% polysorbate 80, saline emulsion; 40 (males, females) i.v.; 105.7 (males, females) inhalation, corrected for particulate losses and pulmonary absorption to 42 mg/kg (Rosenkranz).

(−)-Δ⁶-3,4-*trans*-Form, Δ⁶-*THC*, Δ⁸-*THC*. bp$_{0.001}$ 200°. [α]$_D^{18}$ −264° (c = 0.11 in ethanol). uv max (ethanol): 282, 275 nm (log ε 3.22, 3.22); shoulder at 230 nm (log ε 4.07). LD$_{50}$ i.v. in mice: 27.5 mg/kg (Yoshimura).

Note: This is a controlled substance (hallucinogen) listed in the U.S. Code of Federal Regulations, Title 21 Part 1308.11 (1995).

THERAP CAT: Antiemetic.

9350. Tetrahydrocortisone. *(3α,5β)-3,17,21-Trihydroxypregnane-11,20-dione;* 3α,17α,21-pregnanetriol-11,20-dione; 11,20-pregnanedione-3α,17α,21-triol; THE. C$_{21}$H$_{32}$O$_5$; mol wt 364.48. C 69.20%, H 8.85%, O 21.95%. A normal mammalian metabolite of cortisone: Schneider, *J. Biol. Chem.* **183**, 365 (1950). Prepn by microbial reduction using a *Streptomyces* sp.: Barkemeyer *et al.*, *Appl. Microbiol.* **8**, 237 (1960). Prepn of the triacetate: **Brit.** pat. 737,291 (1955 to Merck & Co.). Prepn of the 21-acetate: Julian *et al.*, U.S. pat. 2,752,339 (1956 to Glidden).

Crystals from ethyl acetate, mp 190°. [α]$_D^{25}$ +85.5° (abs ethanol).

21-Acetate, C$_{23}$H$_{34}$O$_6$, crystals from acetone, mp 227°.

Triacetate, C$_{27}$H$_{38}$O$_8$, crystals from methanol, mp 150-152°; solvated crystals from ethyl acetate, mp 112-118°.

9351. Tetrahydrofuran. Diethylene oxide; tetramethylene oxide. C$_4$H$_8$O; mol wt 72.11. C 66.63%, H 11.18%, O 22.19%. Prepn from 1,4-butanediol: Schmoyer, Case, *Nature* **187**, 592 (1960). Manuf by catalytic hydrogenation of maleic anhydride: Gilbert, Howk, U.S. pat. 2,772,293 (1956 to du Pont); of furan: Banford, Manes, U.S. pat. 2,846,449 (1958 to du Pont); Manly, U.S. pat. 3,021,342 (1962 to Quaker Oats). Stabilization to prevent excessive peroxide formation on storage with 0.05-1.0% *p*-cresol, 0.05-0.1% hydroquinone, or less than 0.01-0.1% 4,4'-thiobis(6-*tert*-butyl-*m*-cresol): Bordner, Hinegardner, and Campbell, U.S. pats. 2,489,260, 2,525,410, and 3,029,257 (1949, 1950, 1962 all to du Pont). Review of toxicology and biological effects: D. E. Moody, *Drug Chem. Toxicol.* **14**, 319-342 (1991).

Liquid. Ether-like odor. mp −108.5°. d$_4^{20}$ 0.8892. bp$_{760}$ 66°; bp$_{176}$ 25°. Flash pt 1°F. n$_D^{20}$ 1.4070. Dipole moment: 1.70. uv cut-off for spectro grade: 220 nm. Miscible with water, alcohols, ketones, esters, ethers, and hydrocarbons. *Caution:* Distil only in presence of a reducing agent, such as ferrous sulfate; peroxide explosions have occurred: *Angew. Chem.* **68**, 182 (1956).

Caution: Potential symptoms of overexposure are irritation of eyes and upper respiratory system; nausea, dizziness and headache; may cause skin irritation. *See* NIOSH *Pocket Guide to Chemical Hazards* (DHHS/NIOSH 90-117, 1990) p 210.

USE: Solvent for high polymers, esp polyvinyl chloride. As reaction medium for Grignard and metal hydride reactions. In the synthesis of butyrolactone, succinic acid, 1,4-butanediol diacetate. Solvent in histological techniques. May be used under Federal Food, Drug & Cosmetic Act for fabrication of articles for packaging, transporting, or storing of foods if residual amount does not exceed 1.5% of the film: *Fed. Reg.* **27**, 3919 (Apr. 25, 1962).

9352. 2,5-Tetrahydrofurandimethanol. *THF glycol; 2,5-bis(hydroxymethyl)tetrahydrofuran.* C$_6$H$_{12}$O$_3$; mol wt 132.16. C 54.53%, H 9.15%, O 36.32%. Prepd from diallyl by oxidation with perbenzoic acid in chloroform, boiling with dil sulfuric acid and hydrolyzing the reaction product with KOH soln: Böeseken, *Rec. Trav. Chim.* **45**, 838 (1926); by Raney nickel reduction of 5-hydroxymethylfurfural or of dimethyl furan-2,5-dicarboxylate: Cope, Baxter, *J. Am. Chem. Soc.* **77**, 393 (1955). The usual form obtained is the *cis* form, described here.

Hygroscopic liquid. Faint odor. *Avoid contact with eyes.* d$_4^0$ 1.1719; d$_4^{25}$ 1.1542; d$_4^{50}$ 1.1359. mp below −50°. bp$_{760}$ 265°; bp$_{96}$ 200°; bp$_{11}$ 155°; bp$_{0.25}$ 105°. n$_D^{25}$ 1.4766. Viscosity: 1926 cp at 0°; 225 at 25°; 51.9 at 50°. Coefficient of expansion: 0.00063 per °C at 25°. Specific heat 0.5 cal/g/°C. Heat of vaporization 115 cal/g. Miscible with water, methanol, ethanol, acetone, benzene, methyl acetate, methyl ethyl ketone, chloroform. Moderately sol in ether, toluene. Almost insol in heptane, methylcyclohexane.

Caution: Highly irritating to eyes, skin, mucous membranes.

USE: Solvent, softener, humectant. In the synthesis of plasticizers, resins, surfactants, agricultural chemicals.

9353. Tetrahydrofurfuryl Alcohol. *Tetrahydro-2-furanmethanol;* tetrahydro-2-furancarbinol; tetrahydro-2-furylmethanol; THFA. C$_5$H$_{10}$O$_2$; mol wt 102.13. C 58.80%, H 9.87%, O 31.33%. Prepn by catalytic hydrogenation of furfuryl alcohol: Lukes, Nelson, *J. Org. Chem.* **21**, 1096 (1956). Manuf by catalytic hydrogenation of furfural or furfuryl alcohol: Dunlop, Schegulla, U.S. pat. 2,838,523 (1958 to Quaker Oats). Occurs in two isomeric forms: D-isomer (levorotatory), L-isomer (dextrorotatory). Abs config: Gagnaire, Butt, *Bull. Soc. Chim. France* **1961**, 312; Hartman, Barker, *J. Org. Chem.* **29**, 873 (1964). *Review:* A. P. Dunlop, F. N. Peters, *The Furans* (Reinhold, New York, 1953).

Liquid. Hygroscopic. d$_{20}^{20}$ 1.0543; d$_{24}^{24}$ 1.0511; d$_{31}^{31}$ 1.0450. Melts below −80°. bp$_{760}$ 178°. n$_D^{20}$ 1.4520; n$_D^{25}$ 1.4499. Flash pt, open cup: 183°F (84°C). Flammability in air: Upper limit 9.7% by vol, lower limit 1.5% by vol. Heat capacity at 30-37°: 0.432 cal/g/°C. Heat of combustion at constant vol: 708.6 cal/g mole. Viscosity at 20°: 6.24 cp. Surface tension at 25°: 37 dyn/cm. Octane no. 82.5. Evaporation rate: 7 (*n*-butyl acetate = 100). Kauributanol value 71.5. Dilution ratio (lacquer ingredients): 4.5. Dielectric constant at 23°: 13.6. Miscible with water, alcohol, ether, acetone, chloroform, benzene.

L-Isomer, prepn *see* Hartman, Barker, *loc. cit.* [α]$_D^{24}$ +14.9° (c = 5.0 in nitromethane).

Caution: Moderately irritating to skin, mucous membranes.

USE: Solvent for fats, waxes, resins. In organic synthesis: Undergoes the reactions of a primary alcohol, while the ring exhibits characteristics of a saturated cyclic ether.

9354. Tetrahydropalmatine. *5,8,13,13a-Tetrahydro-2,3,9,10-tetramethoxy-6H-dibenzo[a,g]quinolizine; 2,3,9,10-tetramethoxyberbine;* 2,3,9,10-tetramethoxydibenzo[a,g]quinolizidine; hyndarin. C$_{21}$H$_{25}$NO$_4$; mol wt 355.43. C 70.96%, H 7.09%, N 3.94%, O 18.01%. Synthesis of *dl*-form: Haworth *et al.*, *J. Chem. Soc.* **1927**, 548; Bradsher, Dutta, *J. Org. Chem.* **26**, 2231 (1961); T. Kametani, M. Ihara, *J. Chem. Soc. (C)* **1967**, 530; G. D. Pandey, K. P. Tiwari, *Indian J. Chem.* **18B**, 545 (1979); Z. Kiparissides *et al.*, *Can. J. Chem.* **58**, 2770 (1980); N. S. Narasimhan *et al.*, *Tetrahedron Letters* **22**, 2797 (1981). Biosynthesis: D. S. Bhakuni *et al.*, *Tetrahedron* **36**, 2491 (1980). *See also* Palmatine. Both optically active forms are found in plants.

l-Form, *caseanine, gindarine, rotundine.* Crystals from dil methanol, mp 147°. $[\alpha]_D^{20}$ −291° (c = 0.8 in 95% alcohol). Also a hydrate, mp 115° (effervescence).

Hydrochloride, $C_{21}H_{25}NO_4 \cdot HCl$, crystals.

d-Form, crystals from methanol, mp 141-142° (evac tube). $[\alpha]_D^{14}$ +292° (c = 0.8 in 95% alcohol).

Hydrochloride, crystals, mp 266°.

dl-Form, crystals from methanol upon addition of water. mp 148-149°.

Hydrochloride, needles from methanol, mp 215-216°.

9355. Tetrahydropapaveroline. *1-[(3,4-Dihydroxyphenyl)methyl]-1,2,3,4-tetrahydro-6,7-isoquinolinediol; 1-(3,4-dihydroxybenzyl)-1,2,3,4-tetrahydro-6,7-isoquinolinediol;* 6,7-dihydroxy-1-(3,4-dihydroxybenzyl)-1,2,3,4-tetrahydroisoquinoline; norlaudanosoline; THP. $C_{16}H_{17}NO_4$; mol wt 287.32. C 66.89%, H 5.96%, N 4.88%, O 22.27%. Alkaloid deriv of dopamine and a biosynthetic precursor of morphine, *q.v.* Prepn of the racemic hydrochloride: F. L. Pyman, *J. Chem. Soc.* **95**, 1610 (1909); of the (+)- and (−)-form hydrochlorides: S. Teitel *et al., J. Med. Chem.* **15**, 845 (1972). Chromatographic study: K. D. McMurtrey *et al., J. Liq. Chromatog.* **3**, 663 (1980). Possible role of THP in the biochemical mediation of alcohol addiction: V. E. Davis, M. J. Walsh, *Science* **167**, 1005 (1970); G. Cohen, M. Collins, *ibid.* 1749; R. D. Meyers, C. L. Melchior, *ibid.* **196**, 554 (1977); M. Sandler *et al., Prog. Clin. Biol. Res.* **90**, 215 (1982). Biosynthetic study: D. K. Choudhary, B. L. Kaul, *Indian Drugs* **19**, 229 (1982). *In vivo* and *in vitro* effects on rat pituitary function: D. R. Britton *et al., Biochem. Pharmacol.* **31**, 1205 (1982). Effect on dopaminergic neurons: I. S. Hoffman, L. X. Cubeddu, *J. Pharmacol. Exp. Ther.* **220**, 16 (1982). Comparison to opioid effects in brain regions: G. R. Siggins *et al., Prog. Clin. Biol. Res.* **90**, 275 (1982).

(±)-Form hydrochloride, $C_{16}H_{18}ClNO_4$, colorless microscopic prisms, mp 291-293° (dec). Practically insol in water, alc.

(−)-Form hydrochloride, cryst from 6*N* HCl, mp 285-286°. $[\alpha]_D^{25}$ −32.4° (c = 1 in water). uv max (ethanol): 230, 286 nm (ε 11100, 6700).

(+)-Form hydrochloride, cryst, mp 285-286°. $[\alpha]_D^{25}$ +32.1° (c = 1 in water).

USE: As a research tool in neurological biochemistry.

9356. Tetrahydropyran. Pentamethylene oxide. $C_5H_{10}O$; mol wt 86.13. C 69.72%, H 11.70%, O 18.58%. Prepd by hydrogenation of dihydropyran (from furfural): Paul, *Bull. Soc. Chim. France* [4] **53**, 1489 (1933); Andrus, Johnson, *Org. Syn.* **23**, 90 (1943); Cass, *Ind. Eng. Chem.* **40**, 219 (1948).

Mobile, flammable liquid. Pungent, sweetish odor. d_4^{20} 0.8814. mp −49.2°. bp 88°. n_D^{20} 1.4211. Dipole moment

1.87. Flash pt −4.0°F. Azeotrope with water, bp 71°, contains 8.5% H_2O. Sol in water [relative solubilities: Bennett, Philip, *J. Chem. Soc.* **1928**, (1939)]. Miscible with alcohol, ether, many other organic solvents. Forms peroxides on exposure to air. All technical tetrahydropyran is stabilized against peroxide formation.

9357. Tetrahydrothiophene. Tetramethylene sulfide; thiacyclopentane; THT; THTP. C_4H_8S; mol wt 88.17. C 54.49%, H 9.15%, S 36.37%. Prepn: J. v. Braun, A. Trümpler, *Ber.* **43**, 545 (1910); J. K. Lawson *et al., Org. Syn. Coll. Vol.* **IV**, 892 (1963). Method for odorizing liquified natural gas: D. K. Mulliner, *Proc., Opec. Sect., Am. Gas Assoc.* **D**, 27 (1974). GC/MS determn in water: G. Carlucci *et al., J. Chromatog.* **287**, 425 (1984). On-site GC determn in natural gas: R. C. M. de Nijs *et al., J. High Resolut. Chromatog.* **16**, 379 (1993).

Colorless liquid, bp 119-121°. n_D^{25} 1.5000-1.5014. LD_{50} by inhalation in mice: 26.7 mg/l (Carlucci).

USE: Natural gas odorant.

9358. Tetrahydrozoline. *4,5-Dihydro-2-(1,2,3,4-tetrahydro-1-naphthalenyl)-1H-imidazole; 2-(1,2,3,4-tetrahydro-1-naphthyl)-2-imidazoline;* tetryzoline. $C_{13}H_{16}N_2$; mol wt 200.28. C 77.96%, H 8.05%, N 13.99%. Prepn from ethylenediamine, ethylenediamine hydrochloride, and methyl 1,2,3,4-tetrahydro-1-naphthoate: Synerholm *et al.,* U.S. pat. **2,731,471** (1956 to Sahyun Labs.). Use as potentiator for veterinary depressants: Gardocki *et al.,* U.S. pat. **2,842,478** (1958 to Pfizer).

Hydrochloride, $C_{13}H_{16}N_2 \cdot HCl$, *Rhinopront, Tinarhinin, Tyzanol, Tyzine, Visine, Yxin.* Crystals from alcohol, dec 256-257°. uv max: 264.5, 271.5 nm ($A_{1cm}^{1\%}$ 17.5, 15.5). Freely sol in water, alcohol. Very slightly sol in chloroform. Practically insol in ether. pH of a 1% aq soln 5.0 to 6.5.

THERAP CAT: Adrenergic (vasoconstrictor); nasal decongestant.

9359. Tetraiodoethylene. *Tetraiodoethene;* diiodoform; ethylene periodide; ethylene tetraiodide. C_2I_4; mol wt 531.64. C 4.52%, I 95.48%. $I_2C = CI_2$. Prepd by the action of iodine on diiodoacetylene obtained from calcium carbide and iodine. Crystal and molecular structure: B. C. Haywood, R. Shirley, *Acta Crystallogr.* **B33**, 1765 (1977).

Light yellow, heavy, small, practically odorless crystals; characteristic odor. On exposure to light turns brown. d 2.98. mp 187°. Insol in water; sol in benzene, chloroform, toluene, CS_2, slightly in ether. *Protect from light.*

9360. Tetralin®. *1,2,3,4-Tetrahydronaphthalene;* Tetranap. $C_{10}H_{12}$; mol wt 132.21. C 90.85%, H 9.15%. Prepd by catalytic hydrogenation of purified naphthalene. Toxicity study: Smyth *et al., Arch. Ind. Hyg. Occup. Med.* **4**, 119 (1951). See ref *under* Decalin.

Liquid. Odor resembling that of a mixture of benzene and menthol. d_4^{20} 0.9702; d_4^{25} 0.9662. Volatile with steam; mp −31.0°; bp_{760} 207.2°; bp_{400} 181.8°; bp_{200} 157.2°; bp_{100} 135.3°; bp_{60} 121.3°; bp_{40} 110.4°; bp_{20} 93.8°; bp_{10} 79.0°; bp_5 65.3°; $bp_{1.0}$ 38.0°. n_D^{20} 1.54135; n_D^{25} 1.53919. Flash pt, open cup 171°F (77°C), closed cup 180°F (82°C). Insol in water; miscible with ethanol, butanol, acetone, benzene, ether, chloroform, petr ether, Decalin; soluble in methanol: 50.6%

w/w. Prolonged, intimate contact with air may cause the formn of tetralin peroxide which may cause explosion of tetralin distn residues. Peroxide formn is prevented by the addn of an antioxidant, such as hydroquinone. LD_{50} orally in rats: 2.86 g/kg (Smyth).

Caution: Irritating to skin, eyes, mucous membranes, and, in high concns, narcotic. In exptl animals has produced cataracts: E. Browning, *Toxicity and Metabolism of Industrial Solvents* (Elsevier, New York, 1965) pp 119-124.

USE: Degreasing agent. Solvent for naphthalene, fats, resins, oils, waxes, used instead of turpentine in lacquers, shoe polishes, floor waxes.

9361. Tetralol.
1,2,3,4-Tetrahydro-2-naphthalenol; ac-tetrahydro-β-naphthol; *ac*-β-tetralol. $C_{10}H_{12}O$; mol wt 148.20. C 81.04%, H 8.16%, O 10.80%. Prepd by hydrogenation of 2-naphthol in the presence of a palladium catalyst: Foreman, Stork, U.S. pat. 2,526,859 (1950 to Lakeside Labs.); by reduction of 2-naphthol with sodium: Hueckel *et al., Ann.* 645, 162 (1961). Toxicity study: Draize *et al., J. Pharmacol. Exp. Ther.* 93, 26 (1948).

Liquid, bp_{12} 140°; mp 15.5°. Crystallizes on prolonged storage at −50° and then at −15°. LD_{50} orally in rats: 1.0 ml/kg (Draize).

p-Tosylate, $C_{10}H_{11}O_3S.C_6H_4CH_3$, crystals from petr ether + benzene, mp 86°.

Methyl ether, $C_{10}H_{11}OCH_3$, bp_{11} 114.5°, d_4^{20} 1.0239, n_D^{20} 1.5326.

9362. Tetramethrin.
2,2-Dimethyl-3-(2-methyl-1-propenyl)cyclopropanecarboxylic acid (1,3,4,5,6,7-hexahydro-1,3-dioxo-2H-isoindol-2-yl)methyl ester; 2,2-dimethyl-3-(2-methylpropenyl)cyclopropanecarboxylic acid ester with N-(hydroxymethyl)-1-cyclohexene-1,2-dicarboximide; N-(3,4,5,6-tetrahydrophthalimide)methyl-cis,trans-chrysanthemate; N-(chrysanthemoxymethyl)-1-cyclohexene-1,2-dicarboximide; phthalthrin; FMC-9260; SP-1103; Neo-Pynamin. $C_{19}H_{25}$-NO_4; mol wt 331.41. C 68.86%, H 7.60%, N 4.23%, O 19.31%. Potent synthetic pyrethroid insecticide. Prepn of racemic mixture: T. Kato *et al., Japan.* pat. 65 8535; *eidem,* U.S. pat. 3,268,398 (1965, 1966 both to Sumitomo). Activity: *eidem, Agr. Biol. Chem.* 28, 914 (1965). Comparative activity of isomers: Y. Okuno *et al., Ger.* pat. 2,348,930; *eidem,* U.S. pat. 3,934,023 (1973, 1976 both to Sumitomo). Metabolism: J. Miyamoto *et al., Agr. Biol. Chem.* 32, 628 (1968). Photodecompn: Y.-L. Chen, J. E. Casida, *J. Agr. Food Chem.* 17, 208 (1969). Toxicity study: T. Kato *et al., Agr. Biol. Chem.* 28, 914 (1965).

(1R-trans)-form

The commercial product is a mixture of isomers. White crystalline solid, mp 65-80°. d_{20}^{20} 1.108; $n_D^{21.5}$ 1.5175. LD_{50} orally in mice: 1000 mg/kg (Kato).

USE: Insecticide.

9363. Tetramethylammonium Hydroxide.
N,N,N-Trimethylmethanaminium hydroxide. $C_4H_{13}NO$; mol wt 91.15. C 52.71%, H 14.37%, N 15.37%, O 17.55%. $(CH_3)_4NOH$.

Usually marketed in 10% aq soln. Strong ammonia-like odor. d_4^{25} ~1.00. The free base is known only in soln or as a solid pentahydrate, forming colorless, deliquesc needles, mp 63°. On distillation, it dec to trimethylamine and CH_3OH. It is a very strong base, readily absorbing CO_2 from the air. *Keep well closed.*

9364. Tetramethylammonium Iodide.
$C_4H_{12}IN$; mol wt 201.05. C 23.90%, H 6.02%, I 63.12%, N 6.97%. $(CH_3)_4$-NI.

Pale yellow crystals. Begins to dec at about 230°. d 1.84. Sparingly sol in water, freely in abs alcohol; insol in chloroform, ether.

USE: Emergency disinfection of drinking water. Required dosage: 8 ppm of iodine.

9365. Tetramethyldiaminobutane.
N,N,N',N'-Tetramethyl-1,4-butanediamine; N,N,N',N'-tetramethylputrescine. $C_8H_{20}N_2$; mol wt 144.26. C 66.61%, H 13.97%, N 19.42%. $(CH_3)_2N(CH_2)_4N(CH_3)_2$. From root and herb of *Hyoscyamus reticulatus* L. and *H. muticus* L., *Solanaceae:* Willstätter, Heuber, *Ber.* 40, 3869 (1907); Konowalowa, Magidson, *Arch. Pharm.* 266, 449 (1928). Synthesis: Lunsford *et al., J. Org. Chem.* 22, 1225 (1957); Solov'ev, Skoldinov, *Zh. Obshch. Khim.* 33, 1821 (1963), *C.A.* 59, 7360f (1963).

White crystals; penetrating odor and sharp, scratching taste. bp 169°; bp_{28} 78-80°; bp_7 43°. n_D^{20} 1.4280. d^{20} 0.7861. Volatile with steam. Sol in water, alc, ether.

Dihydrochloride, $C_8H_{20}N_2.2HCl$, crystals from alc, mp 273°.

Dipicrate, $C_8H_{20}N_2.2C_6H_3N_3O_7$, needles from hot water, mp 203-205°.

9366. Tetramethylenedisulfotetramine.
2,6-Dithia-1,-3,5,7-tetraazatricyclo[3.3.1.1^{3,7}]decane 2,2,6,6-tetraoxide; 2,6-dithia-1,3,5,7-tetraazaadamantane 2,2,6,6-tetraoxide. C_4H_8-$N_4O_4S_2$; mol wt 240.26. C 20.00%, H 3.36%, N 23.32%, O 26.64%, S 26.69%. Prepd from sulfamide, $H_2NSO_2NH_2$ and formaldehyde in 60% H_2SO_4: Hecht, Henecka, *Angew. Chem.* 61, 365 (1949). Toxicity: Hagen, *Deut. Med. Wochenschr.* 75, 183 (1950).

Cubic crystals from acetone, dec 255-260°. *Violent convulsive poison!* Stable to acids and alkalies in dilutions up to 0.1N. Dec upon prolonged boiling of aq solns. Soly in water about 0.25 mg/ml. Slightly sol in acetone. Insol in methanol, ethanol. LD in mice (mg/kg): 0.20 orally or s.c. (Hagen).

9367. Tetramethyl-*p*-phenylenediamine.
N,N,N',N'-Tetramethyl-1,4-benzenediamine; Wurster's reagent; Wurster's blue. $C_{10}H_{16}N_2$; mol wt 164.25. C 73.13%, H 9.82%, N 17.06%. Prepn: Meyer, *Ber.* 36, 2979 (1903); Cox, Smith, *J. Org. Chem.* 29, 488 (1964).

Crystals from petr ether, mp 51-52°. bp 260°. Slightly sol in cold water; more sol in hot water; freely sol in alc, chloroform, ether, petr ether.

USE: In the form of the hydrochloride as a reagent in analytical chemistry.

9368. Tetramethylurea.
TMU; Temur. $C_5H_{12}N_2O$; mol wt 116.16. C 51.70%, H 10.41%, N 24.12%, O 13.77%. $[(CH_3)_2N]_2CO$. Prepn: Lawson, Croom, *J. Org. Chem.* 28, 232 (1963). Review of prepn, manuf, properties and use as solvent and as reagent: Lüttringhaus, Dirksen, *Angew. Chem. Int. Ed.* 3, 260 (1964).

Liquid with faint, pleasant odor, bp 176.5°, bp_{740} 174.5°, bp_{12} 63-64°. mp −1.2°. Flash pt ~75°. d_4^{20} 0.9687. n_D^{25} 1.4493. Dipole moment: 3.47 D in benzene. pKb 2. uv max: 217.5 nm (ε 1940). Miscible with water and with all common organic solvents including petr ether. LD_{50} i.v. in rats: 1.1 g/kg (Lüttringhaus, Dirksen).

USE: As solvent and reagent.

9369. Tetrandrine. *(1β)-6,6',7,12-Tetramethoxy-2,2'-dimethylberbaman.* $C_{38}H_{42}N_2O_6$; mol wt 622.76. C 73.29%, H 6.80%, N 4.50%, O 15.41%. From root of *Stephania tetrandra* S. Moore, *Menispermaceae.* Isoln: Kondo, Yano, *Ann.* **497**, 90 (1932). Structure: Fujita, Murai, *J. Pharm. Soc. Japan* **71**, 1039 (1951). Synthesis: Kataoka, *C.A.* **51**, 16501i (1957). Total synthesis: Inubushi *et al., Tetrahedron Letters* **1968**, 3399; eidem, *J. Chem. Soc. (C)* **1969**, 1547. Present in the Chinese drug han-fang-chi.

Needles. mp 217-218°. $[\alpha]_D^{26}$ +252.4° (chloroform). Practically insol in water, petr ether; sol in ether and some other organic solvents.

l-Form, **Phaenthine.** mp 210°. $[\alpha]_D^{20}$ −278° (chloroform).

THERAP CAT: Analgesic; antipyretic.

9370. Tetranectin. Tetrameric protein isolated from human plasma. Enhances plasminogen activation catalyzed by tissue plasminogen activator, *q.v.* Composed of four identical, non-covalently bound, polypeptide chains each containing 181 amino acids, with mol wt 20,100 daltons. Isoln from human serum: I. Duhl Clemmensen, C. Kluft, **Eur. pat. Appl. 206,400** (1986 to Ned. Cent. Org. Toegepast-Natuurwetenschappelijk Onderzoek). Purification, characterization and plasminogen binding activity: I. Clemmensen *et al., Eur. J. Biochem.* **156**, 327 (1986). Primary structure: J. Fuhlendorff *et al., Biochemistry* **26**, 6757 (1987). Enzyme-linked immunoassay in human plasma: B. A. Jensen *et al., J. Lab. Clin. Med.* **110**, 612 (1987). Distribution in human endocrine tissue: L. Christensen *et al., Histochemistry* **87**, 195 (1987). Possible role in cancer metastasis: B. A. Jensen, I. Clemmensen, *Cancer* **62**, 869 (1988). Isoelectric point 5.8. $E_{280nm}^{1\%}$ 12.5.

9371. Tetranitromethane. CN_4O_8; mol wt 196.03. C 6.13%, N 28.58%, O 65.29%. $C(NO_2)_4$. Prepd by nitration of acetic anhydride with anhydrous nitric acid: Liang, *Org. Syn. coll. vol. III*, p 803 (1955).

Pale yellow liquid. d_4^{25} 1.6229; d_4^{25} 1.638 (tech). mp +13.8°. bp_{760} 126°; $bp_{25.8}$ 40°; $bp_{14.9}$ 30°; $bp_{8.4}$ 20°; $bp_{5.7}$ 13.8°; $bp_{1.9}$ 0°. n_D^{20} 1.4384; n_D^{25} 1.4358. Viscosity at 20° = 1.76 cp. Energy of decompn: Tschinkel, *Ind. Eng. Chem.* **48**, 732 (1956). Insol in water. Freely sol in alcohol, ether, alcoholic KOH. Attacks iron, copper, brass, zinc, rubber. Highly explosive in the presence of impurities.

Caution: Potential symptoms of overexposure are irritation of eyes, nose and throat; dizziness, headache; chest pain, dyspnea; methemoglobinuria, cyanosis; skin burns. *See NIOSH Pocket Guide to Chemical Hazards* (DHHS/NIOSH 90-117, 1990) p 286. See also Patty's Industrial Hygiene and Toxicology vol. 2A, G. D. Clayton, F. E. Clayton, Eds. (John Wiley & Sons, New York, 4th ed., 1994) 614-617. This substance may reasonably be anticipated to be a carcinogen: *Seventh Annual Report on Carcinogens* (PB95-109781, 1994) p 376.

USE: Oxidizer in rocket propellants. As explosive in admixture with toluene. To increase cetane number of diesel fuels. Reagent for detecting the presence of double bonds in organic compds. Has been proposed as irritant war gas.

9372. Tetrantoin. *3',4'-Dihydrospiro[imidazolidine-4,-2'(1'H)-naphthalene]-2,5-dione; 7,8-benzo-1,3-diazaspiro-[4.5]decane-2,4-dione; S-2-676; Spirodon; Spirodon.* $C_{12}H_{12}N_2O_2$; mol wt 216.24. C 66.65%, H 5.59%, N 12.95%, O 14.80%. Prepn: Novelli, *Anales Farm. Bioquim.* **21**, 81 (1954), *C.A.* **50**, 4922 (1956); Faust *et al., J. Am. Pharm. Assoc.* **46**, 118

(1957); Jules *et al., U.S. pat.* **2,716,648** (1955 to Cutter Labs.).

Crystals from ethanol or glacial acetic acid, mp 267-268°.

THERAP CAT: Anticonvulsant.

9373. Tetraphenylarsonium Bromide. $C_{24}H_{20}AsBr$; mol wt 463.25. C 62.23%, H 4.35%, As 16.17%, Br 17.25%. $(C_6H_5)_4AsBr$.

Dihydrate, crystals. *Poisonous!* Loses all its water over sulfuric acid or at 100°. mp 281-284°. Sol in ~60 parts water; sol in alcohol or methanol, sparingly in acetone.

9374. Tetraphenylarsonium Chloride. $C_{24}H_{20}AsCl$; mol wt 418.80. C 68.83%, H 4.81%, As 17.89%, Cl 8.47%. $(C_6H_5)_4AsCl$. Willard, Smith, *Ind. Eng. Chem. Anal. Ed.* **11**, 186 (1939).

Dihydrate, crystals. *Poisonous!* Loses all its water at 100°. mp 258-260°. Freely sol in water; sol in alcohol or methanol, sparingly in acetone.

USE: As a reagent for Cd, Hg, Zn, perchlorate, periodate, and other ions.

9375. Tetraphosphorus Trisulfide. Phosphorus sesquisulfide; trisulfurated phosphorus. P_4S_3; mol wt 220.09. P 56.29%, S 43.71%. Conveniently prepd by fusing red phosphorus with sulfur: Stock, *Ber.* **43**, 150 (1910); also from white phosphorus and sulfur in a high-boiling solvent such as α-chloronaphthalene: Frary, *Ger. pat.* **309,618** (1918); *Chem. Zentr.* **1919**, II, 55.

Yellowish-green, long, rhombic needles from benzene. Stable to air. d_4^{20} 2.03. mp 172.5°. bp 407.5°. Insol in cold water. Dec by hot water, yielding H_2S. Sol in carbon disulfide (20°): about 60% (w/w). Sol in benzene, similar hydrocarbons, phosphorus trichloride. LD orally in rabbits: 100 mg/kg, *Handbook of Toxicology*, vol. 1, W. S. Spector, Ed. (Saunders, Philadelphia, 1956) pp 236-237.

USE: In match tips.

9376. Tetrasilane. Tetrasilicon decahydride; silicobutane; tetrasilicane; tetrasilicobutane. $H_{10}Si_4$; mol wt 122.42. H 8.23%, Si 91.77%. Si_4H_{10}. Prepn by the action of hydrochloric acid on magnesium silicide: Stock, Somiesky, *Ber.* **49**, 111 (1916); **54B**, 524 (1921); **56B**, 247 (1923); Stock *et al., Ber.* **56**, 1695 (1923); Emeleus, Maddock, *J. Chem. Soc.* **1946**, 1131.

Liquid. mp ~−90°; bp 109°; vapor pressure 7.8 mm Hg at 0° (Stock *et al., loc. cit.*). mp −84.3°; bp (calc) 107.4°; vapor pressure 9.1 mm Hg at 0° (Emeleus, Maddock, *loc. cit.*). d^0 0.825. Dec at room temp; explodes in air. Reacts vigorously with CCl_4 and $CHCl_3$. Dec in water.

9377. Tetrasodium Pyrophosphate. TSPP; pyro; sodium pyrophosphate. $Na_4O_7P_2$; mol wt 265.90. Na 34.58%, O 42.12%, P 23.30%. $Na_4P_2O_7$. Available alkalinity as Na_2O 4.4%, total alkalinity 22.7%. Produced by molecular dehydration of dibasic sodium phosphate at 500°: Bell, *Inorg. Syn.* **3**, 98 (1950).

Crystals, d 2.534. mp 988°. Soly in water (g/100 ml) at 0°: 2.61; at 25°: 6.70; at 100°: 42.2. pH of a 1% soln = 10.2. Hydrolyzes to orthophosphate in aq soln, but the rate of hydrolysis is much slower than for the more acid pyrophosphate. No noticeable hydrolysis within 60 hrs at 70°: Bell, *Ind. Eng. Chem.* **39**, 136 (1947).

Decahydrate, crystals, d 1.82. mp 79.5°. Slight efflorescence in dry air. Soly in water (g/100 ml) at 0°: 3.16; at 20°: 6.23; at 25°: 8.14; at 60°: 21.83; at 80°: 30.04. pH of 1% soln at 25° = 10.2. Insol in alc.

USE: In cleansing compds, oil-well drilling, water treatment, cheese emulsification, as general sequestering agent, to remove rust stains, as ingredient of one-fluid ink eradicators, in electrodeposition of metals.

9378. Tetrasulfur Tetranitride. Schwefelstickstoff. N_4S_4; mol wt 184.29. N 30.40%, S 69.60%. S_4N_4. Prepd by the interaction of disulfur dichloride and ammonia: Becke-Goehring, *Inorg. Syn.* **6**, 124 (1960).

Orange-red, monoclinic needles from benzene, mp 178°. Additional purification by sublimation in high vacuum, (bath temp 100°), mp 180°. bp$_{760}$ ~185°. Further heating results in deflagration and explosion. Practically insol in cold water, hydrolyzed by boiling water. Slightly sol in benzene, abs ethanol, carbon disulfide. *Handle with caution:* May dec explosively on striking or at temps much above 100°.

9379. Tetrazepam. *7-Chloro-5-(1-cyclohexen-1-yl)-1,3-dihydro-1-methyl-2H-1,4-benzodiazepin-2-one;* 7-chloro-5-(1-cyclohexenyl)-1-methyl-2-oxo-2,3-dihydro-1*H*-[1,4]benzo[*f*]-diazepine; CB-4261; Musaril; Myolastan. C$_{16}$H$_{17}$ClN$_2$O; mol wt 288.78. C 66.55%, H 5.93%, Cl 12.28%, N 9.70%, O 5.54%. Prepn: Schmitt, **Neth.** pat. **Appl. 6,600,095** and **U.S.** pats. **3,426,014; 3,551,412** (1966, 1969, 1970 to Clin-Byla); Schmitt *et al., Chim. Ther.* **2**, 254 (1967). Spectroscopic and chromatographic studies: Lafargue *et al., Ann. Pharm. Franc.* **28**, 343, 477 (1970). GC-MS determn in urine: H. Maurer, K. Pfleger, *J. Chromatog.* **422**, 85 (1987). Pharmacokinetics: H. Bun *et al., Arzneimittel-Forsch.* **37**, 199 (1987). Pharmacology and benzodiazepine receptor study: P. E. Keane *et al., J. Pharmacol Exp. Ther.* **245**, 692, 699 (1988).

Yellow-brown crystals from ethyl acetate, mp 144°. uv max (ethanol): 227 nm (ε 28500). LD$_{50}$ in mice (mg/kg): 415 i.p.; 2000 orally (Schmitt).

Note: This is a controlled substance (depressant) listed in the U.S. Code of Federal Regulations, Title 21 Part 1308.14 (1995).

THERAP CAT: Muscle relaxant (skeletal).

9380. Tetrazolium Blue. *3,3'-(3,3'-Dimethoxy[1,1'-biphenyl]-4,4'-diyl)bis[2,5-diphenyl-2H-tetrazolium] dichloride;* 3,3'-dianisolebis[4,4'-(3,5-diphenyl)tetrazolium chloride]; blue tetrazolium; dimethoxy neotetrazolium; ditetrazolium chloride; BT. C$_{40}$H$_{32}$Cl$_2$N$_8$O$_2$; mol wt 727.65. C 66.03%, H 4.43%, Cl 9.74%, N 15.40%, O 4.40%. Prepn: A. M. Rutenburg *et al., Cancer Res.* **10**, 113 (1950); L. J. Pannone, J. B. Rust, U.S. pat. **2,713,581** (1955 to Montclair Res. Corp. and Ellis-Foster Co.).

Lemon-yellow crystals, dec 242-245°. Freely sol in methanol, ethanol, chloroform. Slightly sol in water. Insol in ethyl acetate, acetone, ether. Reduction potential about −0.08 volt. Yields a dark blue diformazan pigment in the presence of a reducing agent.

USE: For research in seed germination, as stain for bacteria and molds, in histochemical studies, to demonstrate oxidation-reduction enzymes in normal and cancerous tissues. *See also* Triphenyltetrazolium Chloride.

9381. Tetrin. Polyene antifungal antibiotic produced by Streptomyces *Illinois* #155-2: Pote, *Diss. Abstr.* **19**, 2778 (1959); Gottlieb, Pote, *Phytopathology* **50**, 817 (1960). Isoln

of the two tetraenes, tetrin A and B: Rinehart *et al., Ann.* **668**, 77 (1963); German, *Diss. Abstr.* **25**, 97 (1964). Structure of tetrin A: Pandey *et al., J. Am. Chem. Soc.* **93**, 3738 (1971); of tetrin B: Rinehart *et al., ibid.* 3747. Revised structure: R. C. Pandey, K. L. Rinehart, *J. Antibiot.* **29**, 1035 (1976). Mode of action of tetrin A: van Etten, Gottlieb, *J. Gen. Microbiol.* **46**, 377 (1967).

Tetrin A   R = H
Tetrin B   R = OH

Tetrin A. C$_{34}$H$_{51}$NO$_{13}$. Fine, colorless needles from methanol or aqueous *n*-butanol, mp > 350° (dec). [α]$_D^{21}$ +8.3° (c = 0.72 in pyridine). [α]$_D^{28}$ +27.5° (c = 1.0 in pyridine). uv max: 214, 278, 290, 303, 318 nm (E$_{1cm}^{0.1\%}$ 19.4, 44.2, 81.2, 115.0, 110.9). Monobasic, pKa' 8.30 in 60% ethanol. Sol in pyridine, dil alkalies, dil mineral acids; moderately sol in lower alcohols; practically insol in acetone, ether, water.

Tetrin B. C$_{34}$H$_{51}$NO$_{14}$. Brown, amorphous powder, mp > 360° (darkens at 160-165°, blackens at 250-295°). [α]$_D^{24}$ +43.5° (c = 0.14 in methanol); [α]$_D^{28}$ +45° (c = 0.3 in pyridine). uv max: 214, 278, 290, 303, 318 nm (E$_{1cm}^{0.1\%}$ 18.6, 51.4, 80.1, 112.8, 108.9). Readily sol in ethanol + water, dioxane + water; fairly sol in water, lower alcohols, dioxane, pyridine, dimethyl sulfoxide; slightly sol in acetone. Practically insol in ethyl acetate, chloroform, ether, ethylene dichloride.

9382. Tetrodotoxin. *Octahydro-12-(hydroxymethyl)-2-imino-5,9:7,10a-dimethano-10aH-[1,3]dioxocino[6,5-d]pyrimidine-4,7,10,11,12-pentol;* maculotoxin; spheroidine; tarichatoxin; tetrodontoxin; tetrodotoxin; fugu poison; TTX. C$_{11}$H$_{17}$N$_3$O$_8$; mol wt 319.27. C 41.38%, H 5.37%, N 13.16%, O 40.09%. Toxin from the ovaries and liver of many species of *Tetraodontidae,* esp the globe fish *(Spheroides rubripes):* Yokoo, *J. Chem. Soc. Japan* **71**, 590 (1950); *C.A.* **45**, 6759c (1951). Identity with tarichatoxin: Buchwald *et al., Science* **143**, 474 (1963); with maculotoxin: D. D. Sheumack *et al., ibid.* **199**, 188 (1978). Structure studies: Goto *et al., Tetrahedron Letters* **1963**, 2105, 2115; **1964**, 779, 1831. Structure: Woodward, *Pure Appl. Chem.* **9**, 49 (1964); Tsuda *et al., Chem. Pharm. Bull.* **12**, 1357 (1964); Goto *et al., Tetrahedron* **21**, 2059 (1965). Synthetic studies: Kishi *et al., Tetrahedron Letters* **1970**, 5127, 5129. Total synthesis: Kishi *et al., J. Am. Chem. Soc.* **94**, 9219 (1972). Pharmacology: Evans, *Brit. Med. Bull.* **25**, 263 (1969); Kao, *Fed. Proc.* **31**, 1117 (1972). Mechanism of action: Narahashi, *ibid.* 1124. Toxicity study: C. Y. Kao, F. A. Fuhrman, *J. Pharmacol. Exp. Ther.* **140**, 31 (1963). *Review:* Scheuer, *Fortschr. Chem. Org. Naturst.* **22**, 265 (1964); Mosher *et al., Science* **144**, 1100 (1964); Kao, *Pharmacol. Rev.* **18**, 997 (1966); Evans, *Int. Rev. Neurobiol.* **15**, 83 (1972).

Darkens above 220° without dec. [α]$_D^{25}$ −8.64° (c = 8.55 in dil acetic acid). pKa: 8.76 (water); 9.4 (50% alc). Sol in dil acetic acid; slightly sol in water, dry alc, ether. Practically insol in other organic solvents. Toxin destroyed in strong acids and in alkaline solns. LD$_{50}$ i.p. in mice: 10 μg/kg (Kao, Fuhrman).

9383. Tetrofosmin. *6,9-Bis(2-ethoxyethyl)-3,12-dioxa-6,9-diphosphatetradecane;* ethylenebis[bis(2-ethoxyethyl)-phosphine]; P53. $C_{18}H_{40}O_4P_2$; mol wt 382.46. C 56.53%, H 10.54%, O 16.73%, P 16.20%. Prepn: J. D. Kelly *et al.,* **Eur. pat. Appl.** 337,654; *eidem,* **U.S. pat.** 5,045,302 (1989, 1991 both to Amersham). Pharmacology and determn of radiochemical purity: *idem et al., J. Nucl. Med.* **34,** 222 (1993). Clinical biodistribution: B. Higley *et al., ibid.* 30. Clinical trial as a myocardial perfusion imaging agent: B. L. Zaret *et al., Circulation* **91,** 313 (1995).

^{99m}Tc-Complex, $C^{36}H_{80}O_{10}P_4^{99m}$Tc, ^{99m}Tc tetrofosmin, [^{99m}Tc(tetrofosmin)$_2$O$_2$]$^+$, PPN1011, Myoview.

THERAP CAT: ^{99m}Tc-Complex as diagnostic aid (radioactive imaging agent).

9384. Tetronasin. *(6S,7S,8R,12S,15R,16R,19S,22S,-23R,24S,26R,27S)-4-Demethylene-22,24-dimethyltetronomy-cin;* antibiotic M139603; ICI-139603; M-139603. $C_{35}H_{54}O_8$; mol wt 602.81. C 69.74%, H 9.03%, O 21.23%. Polyether antibiotic produced by *Streptomyces longisporoflavus* NCIB 11426. Possesses a biosynthetically rare acid grouping in the form of an acyl tetronic acid moiety. Isoln and use in ruminants: D. H. Davies, G. L. F. Norris, **Brit. pat. Appl.** 2,027,013; *eidem,* **U.S. pat.** 4,279,894 (1980, 1981 both to ICI). Physical data and crystal structure: D. H. Davies *et al., Chem. Commun.* **1981,** 1073. Solution conformation and cation-binding properties: J. Grandjean, P. Laszlo, *Tetrahedron Letters* **24,** 3319 (1983). Synthetic studies: A. M. Doherty, S. V. Ley, *ibid.* **27,** 105 (1986). Biosynthetic studies: J. M. Bulsing *et al., Chem. Commun.* **1984,** 1301; D. M. Doddrell *et al., ibid.* 1302; A. K. Demetriadou *et al., ibid.* **1985,** 408. Antimicrobial activity: C. J. Newbold *et al., Appl. Environ. Microbiol.* **54,** 544 (1988). Effect on gain efficiency in cattle: S. J. Bartle *et al., J. Anim. Sci.* **66,** 1502 (1988).

Sodium salt, $C_{35}H_{53}NaO_8$. mp 176-178°. $[\alpha]_D^{23}$ −82° (c = 0.2 in methanol). uv max (ethanol): 234, 270 nm (ε 13000, 11000). pKa 1.8 ±0.3 (methanol/H$_2$O 1:9). Sol in most organic solvents. Insol in water.

THERAP CAT (VET): Ruminant performance enhancer.

9385. Tetroquinone. *2,3,5,6-Tetrahydroxy-2,5-cyclohex-adiene-1,4-dione; tetrahydroxy-p-benzoquinone;* tetrahydroxy-quinone; THQ; HPEK-1; NSC-112931; Kelox. $C_6H_4O_6$; mol wt 172.09. C 41.88%, H 2.34%, O 55.78%. Prepn from glyoxal: A. J. Fatiadi, W. F. Sager, *Org. Syn.* **coll. vol.** V, 1011 (1973).

Blue-black crystals appearing yellow in transmitted light. Slightly sol in cold water; freely sol in hot water or in alcohol; slightly sol in ether. It acts like a strong dibasic acid.

Disodium salt, $Na_2C_6H_2O_6$, almost black crystals with a green metallic luster, sparingly sol in water.

USE: Disodium salt is used as an indicator in the volumetric determination of sulfate by means of barium chloride solution.

THERAP CAT: Keratolytic (systemic).

9386. Tetroxoprim. *5-[[3,5-Dimethoxy-4-(2-methoxy-ethoxy)phenyl]methyl]-2,4-pyrimidinediamine;* 2,4-diamino-5-[3,5-dimethoxy-4-(2-methoxyethoxy)benzyl]pyrimidine; HE-781. $C_{16}H_{22}N_4O_4$; mol wt 334.38. C 57.47%, H 6.63%, N 16.76%, O 19.14%. Analog of trimethoprim, *q.v.* Prepn: W. Liebenow, J. Prikryl, **Fr. pat.** 2,221,147; *eidem,* **U.S. pat.** 3,992,379 (1974, 1976 both to Heumann). Series of articles on tetroxoprim and other antibacterial folate inhibitors: *J. Antimicrob. Chemother.* **5,** Suppl. B, 1-239 (1979); on synthesis of radioactive tetroxoprim, tissue distribution, kinetics, HPLC determn: *Arzneimittel-Forsch.* **30,** 307-319 (1980). Effect on bacterial growth kinetics: J. K. Seydel, E. Wempe, *Chemotherapy* **26,** 361 (1980). *In vitro* activity of the tetroxoprim-sulfadiazine combination: H. Hahn, A. Kirov, *Arzneimittel-Forsch.* **30,** 1047 (1980).

Crystals from water, mp 153-156° (Liebenow, Prikryl); also reported as mp 160.1° (*eidem, J. Antimicrob. Chemother.* **5,** Suppl B, 15 (1979)). Soly at 30° (mg/ml): water 2.65; chloroform 69; *n*-octanol 1.61. pKb 8.25. LD$_{50}$ orally in rats: 1357 mg/kg (Liebenow, Prikryl).

Mixture with sulfadiazine, *co-tetroxazine, Biroxin, Steri-nor, Tibirox, Troximin.*

THERAP CAT: Antibacterial.

9387. Tevenel®. *[R-(R*,R*)]-N-[2-[4-(Aminosulfonyl)-phenyl]-2-hydroxy-1-(hydroxymethyl)ethyl]-2,2-dichloroacet-amide; D-threo-(−)-2,2-dichloro-N-[β-hydroxy-α-(hydroxy-methyl)-p-sulfamoylphenethyl]acetamide; D-threo-1-(p-sul-famoylphenyl)-2-(2′,2′-dichloroacetamido)-1,3-propanediol;* AMP-3; D-AMP-3. $C_{11}H_{14}Cl_2N_2O_5S$; mol wt 357.21. C 36.99%, H 3.95%, Cl 19.85%, N 7.84%, O 22.39%, S 8.98%. Sulfamoyl analog of chloramphenicol, *q.v.* Prepn: Gregory, **U.S. pat.** 2,680,135 (1954 to Du Pont). Toxicology: R. M. Jiji *et al., Arch. Int. Med.* **111,** 70 (1963).

Crystals from acetonitrile, mp 155-158°. Sol in ethanol, propylene glycol. Slightly sol in water, dil aq solns of sodium bicarbonate, acetic acid. Aq solns are neutral.

THERAP CAT: Anti-infective.

9388. TFM. *4-Nitro-3-(trifluoromethyl)phenol; α,α,α-trifluoro-4-nitro-m-cresol;* 3-trifluoromethyl-4-nitrophenol; HOE-2770; Lamprecid. $C_7H_4F_3NO_3$; mol wt 207.11. C 40.60%, H 1.95%, F 27.52%, N 6.76%, O 23.18%. A substance toxic to the parasitic sea lamprey, *Petromyzon marinus,* which preys upon commercial fish species of the Great Lakes: Scherer *et al.,* **Ger. pat.** 1,068,505, *C.A.* **55,** 9774d (1961) corresp to **U.S. pat.** 3,157,571 (1959, 1964, both to Hoechst). Toxicological studies in fish, and bibliography: Kawatski, McDonald, *Comp. Gen. Pharmacol.* **5,** 67 (1974). Physical properties: Smith, *J. Chem. Eng. Data* **6,** 607 (1961). *Review:* Schnick, *Investigations in Fish Control,* no. 44 (Sport Fisheries and Wildlife Bureau) 31 pp.

mp 76°. pK 6.07. uv max (acid): 280 nm (ϵ 1930); in 1% NaOH: 300, 395 nm (ϵ 4650, 13,130); in 95% ethanol: 290 nm (ϵ 14,700).

USE: Lamprey killer.

9389. Thalicarpine. *[S-(R*,R*)]-9-[4,5-Dimethoxy-2-[(1,2,3,4-tetrahydro-6,7-dimethoxy-2-methyl-1-isoquinolinyl)-methyl]phenoxy]-5,6,6a,7-tetrahydro-1,2,10-trimethoxy-6-methyl-4H-dibenzo[de,g]quinoline.* $C_{41}H_{48}N_2O_8$; mol wt 696.84. C 70.67%, H 6.94%, N 4.02%, O 18.37%. Tumor-inhibitory alkaloid from *Thalictrum dasycarpum* Fisch. & Lall., *Ranunculaceae:* Kupchan *et al., J. Pharm. Sci.* **52**, 985 (1963). Structure: Kupchan, Yokoyama, *J. Am. Chem. Soc.* **86**, 2177 (1964); Tomita *et al., Tetrahedron Letters* **1965**, 4309. Total synthesis: Kupchan, Liepa, *Chem. Commun.* **1971**, 599; Kupchan *et al., J. Am. Chem. Soc.* **95**, 2995 (1973). Pharmacology: Herman, Chadwick, *Toxicol. Appl. Pharmacol.* **26**, 137 (1973); S. M. Sieber *et al., Cancer Treat. Rep.* **60**, 1127 (1976). Antimicrobial and hypotensive activity: W. N. Wu *et al., Lloydia* **40**, 508 (1977). Biosynthesis: D. S. Bhakuni, S. Jain, *Tetrahedron* **38**, 729 (1982).

Needles from ethyl acetate, mp 160-161°. $[\alpha]_D^{25}$ +133° (c = 0.83 in methanol); $[\alpha]_D^{25}$ +89° (c = 0.88 in chloroform). uv max (methanol): 282, 302 nm (ϵ 17,000, 13,000).

9390. Thalidomide. *2-(2,6-Dioxo-3-piperidinyl)-1H-iso-indole-1,3(2H)-dione; N-(2,6-dioxo-3-piperidyl)phthalimide;* α-phthalimidoglutarimide; 3-phthalimidoglutarimide; 2,6-dioxo-3-phthalimidopiperidine; *N*-phthalylglutamic acid imide; *N*-phthaloylglutamimide; K-17; Distaval; Softenon; Sedalis; Talimol; Pantosediv; Neurosedyn; Contergan. $C_{13}H_{10}N_2O_4$; mol wt 258.23. C 60.47%, H 3.90%, N 10.85%, O 24.78%. Selective inhibitor of tumor necrosis factor α. Prepn: Brit. pat. 768,821 (1957 to Chemie Grünenthal). Teratogenicity studies: I. D. Fratta *et al., Toxicol. Appl. Pharmacol.* **7**, 268 (1965); H. Schumacher *et al., J. Pharmacol. Exp. Ther.* **160**, 189 (1968). Mechanism of teratogenic activity: G. B. Gordon *et al., Proc. Nat. Acad. Sci. USA* **78**, 2545 (1981); of immunomodulatory activity: S. M. McHugh *et al., Clin. Exp. Immunol.* **99**, 160 (1995). *In vitro* activity vs HIV-1: S. Makonkawkeyoon *et al., Proc. Nat. Acad. Sci. USA* **90**, 5974 (1993). HPLC determn in plasma: A. Delon *et al., J. Liq. Chromatog.* **18**, 297 (1995). Clinical use vs aphthous ulceration related to HIV infection: D. L. Paterson *et al., Clin. Infect. Dis.* **20**, 250 (1995). Review of chemistry, pharmacokinetics and clinical safety: V. Günzler, *Drug Safety* **7**, 116-134 (1992); of clinical use in immunological disorders: U. Schuler, G. Ehninger, *ibid.* **12**, 364-369 (1995); in leprosy: C. L. Crawford, *Adverse Drug React. Toxicol. Rev.* **13**, 177-192 (1994).

Needles, mp 269-271°. uv max (neutral soln): 220, 300 nm. Soly in water: $\sim 2 \times 10^{-4}$ mol/L; 45-60 mg/L. Sparingly sol in water, methanol, ethanol, acetone, ethyl acetate, butyl acetate, glacial acetic acid. Very sol in dioxane, DMF, pyridine. Practically insol in ether, chloroform, benzene.

THERAP CAT: Immunomodulator. Formerly as sedative, hypnotic.

9391. Thallium. Tl; at. wt 204.383; at. no. 81; valence 1, 3. Natural isotopes: 203 (29.50%), 205 (70.50%); artificial, radioactive isotopes: 191-202; 204; 206-210. Occurs in crookesite, $(Cu,Tl,Ag)_2Se$, found in Sweden; in lorandite, $TlAgS_2$, found in Greece; in hutchinsonite, $(Tl,Cu,Ag)_2S$-$PbS.2As_2S_3$, found in Switzerland. Occurrence in the earth's crust: 0.7 ppm. Discovered by Crookes in 1861. Prepn: Sanderson, *Can. Mining J.* **65**, 624 (1944). Use in organic syntheses: McKillop *et al., Tetrahedron Letters* **1970**, 5281; Taylor *et al., ibid.* 5285. *Review:* Wade, Banister in *Comprehensive Inorganic Chemistry* vol. 1, J. C. Bailar, Jr. *et al.,* Eds. (Pergamon Press, Oxford, 1973) pp 997-1000, 1119-1172. Review of toxicity: E. Browning, *Toxicity of Industrial Metals* (Appleton-Century-Crofts, New York, 2nd ed., 1969) pp317-322.

Bluish-white, very soft, inelastic, easily fusible, heavy metal; leaves a streak on paper. Oxidizes superficially in air forming a coating of Tl_2O. Forms alloys with other metals and readily amalgamates with mercury. d 11.85. Begins to volatilize at 174°. mp 303.5°. bp 1457°. Specific heat at 20° 0.031 cal/g/°C. Latent heat of fusion 5.04 cal/g. Brinell hardness: 2. May be distilled in a stream of hydrogen. Insol in water; reacts with nitric or sulfuric acid; difficultly with hydrochloric acid.

Caution: Potential symptoms of overexposure are nausea, diarrhea, abdominal pain and vomiting; ptosis, strabismus; peripheral neuritis, tremor; retrosternal tightness, chest pain and pulmonary edema; seizure, chorea and psychosis; liver and kidney damage; alopecia; paresthesia of legs. *See NIOSH Pocket Guide to Chemical Hazards* (DHHS/NIOSH 90-117, 1990) p 212. *See also Patty's Industrial Hygiene and Toxicology* vol. 2A, G. D. Clayton, F. E. Clayton, Eds. (Wiley-Interscience, New York, 3rd ed., 1981) 1914-1931.

USE: In admixture with 97-98% of inert substances the salts are used as poison for rats and other rodents. In semiconductor research. Alloyed with mercury for switches and closures which operate at subzero temps.

9392. Thallium Acetate. Thallous acetate. $C_2H_3O_2Tl$; mol wt 263.43. C 9.12%, H 1.15%, O 12.15%, Tl 77.59%. White, deliquesc crystals. *Poisonous!* Sol in water, alcohol. *Keep well closed.* MLD orally in dogs: 18.5 mg/kg, *Handbook of Toxicology* vol. 1, W. S. Spector, Ed. (Saunders, Philadelphia, 1956) pp 294-295.

9393. Thallium Bromide. Thallous bromide. BrTl; mol wt 284.29. Br 28.11%, Tl 71.89%. TlBr. Pale yellow, cryst powder. *Poisonous!* d 7.5. mp $\sim$460°. Sol in 2360 parts water.

9394. Thallium Carbonate. Thallous carbonate. CO_3-Tl_2; mol wt 468.78. C 2.56%, O 10.24%, Tl 87.20%. Tl_2CO_3. White crystals. *Poisonous!* d 7.1. mp 272°. Soluble in 24 parts water, 3.7 parts boiling water; insol in alc. USE: Manuf imitation diamonds.

9395. Thallium Chloride. Thallous chloride. ClTl; mol wt 239.84. Cl 14.78%, Tl 85.22%. TlCl. White, cryst powder. *Poisonous!* d 7.0. mp 430°. Sol in 260 parts cold water, 70 parts boiling water. Insol in alcohol. HCl decreases its soly in water. USE: As catalyst in chlorinations.

9396. Thallium Cyanide. Thallous cyanide. CNTl; mol wt 230.40. C 5.21%, N 6.08%, Tl 88.71%. TlCN. Prepn: *Gmelin's, Thallium (8th ed.)* **38**, 390 (1940); E. C. Taylor *et al., J. Org. Chem.* **43**, 2280 (1978).

White, hexagonal platelets. d 6.523. Soly in water: 16.8 g/100 ml. Sol in alcohol, acid. Aqueous soln is alkaline.
USE: In organic synthesis.

9397. Thallium Fluoride. Thallous fluoride. FTl; mol wt 223.38. F 8.50%, Tl 91.50%. TlF. Prepared from Tl_2CO_3 and HF: Ketelaar, *Z. Kristallogr.* **92**, 30 (1935); Hayek, *Z. Anorg. Allgem. Chem.* **225**, 47 (1935); Barrow *et al., Trans. Faraday Soc.* **51**, 1650 (1955); from Tl and HF: Keneshea, Cubicciotti, *J. Phys. Chem.* **69**, 3910 (1965); Tranquard, *Bull. Soc. Chim. France* **1967**, 2578.

Hard, shiny, crystals which deliquesce when breathed upon, but which resolidify immediately in dry air. Not hygroscopic in the usual sense. Orthorhombic (deformed NaCl lattice). d_4^{25} 8.36. mp 322°. Begins to sublime at 300°. Very freely sol in water. Concd solns show strong alkalinity.
USE: In the prepn of fluoro esters.

9398. Thallium Hydroxide. Thallous hydroxide. HOTl; mol wt 221.39. H 0.46%, O 7.23%, Tl 92.32%. TlOH.

Yellow needles. *Poisonous!* Very sol in water; the soln is strongly alkaline and turns turmeric paper brown.

9399. Thallium Iodide. Thallous iodide. ITl; mol wt 331.29. I 38.31%, Tl 61.69%. TlI.

Yellow, cryst powder. *Poisonous!* d 7.1; mp 440°; bp 824°. Almost insol in water. Insol in alcohol. Sol in KI soln.

9400. Thallium Nitrate. Thallous nitrate. NO_3Tl; mol wt 266.39. N 5.26%, O 18.02%, Tl 76.72%. $TlNO_3$.

White crystals. *Poisonous!* d 5.55. mp 206°; dec at 450°. Sol in 10 parts cold water, 0.3 part boiling water. Insol in alcohol. LD orally in dogs: 45 mg/kg, *Handbook of Toxicology* vol. **1**, W. S. Spector, Ed. (Saunders, Philadelphia, 1956) pp 294-295.
USE: As a reagent in analytical chemistry, esp for the determination of iodine in presence of Br and Cl; also with $KClO_3$, HgCl and resin for green fire for signalling at sea.

9401. Thallium Oxide. Thallous oxide. OTl_2; mol wt 424.77. O 3.77%, Tl 96.23%. Tl_2O.

Black powder. *Poisonous!* mp about 300°. Sol in water, forming the hydroxide; also sol in alcohol. On exposure to air it gradually oxidizes to thallic oxide and becomes insol.
USE: In manuf of glass of a high coefficient of refraction for optical purposes (thallium flint glass) and for artificial gems.

9402. Thallium Selenate. Thallous selenate. O_4SeTl_2; mol wt 551.72. O 11.60%, Se 14.31%, Tl 74.09%. Tl_2SeO_4. Prepd from a soln of thallous carbonate in selenic acid: Kuhlmann, *Bull. Soc. Chim.* [2] **1**, 330 (1864).

Orthorhombic crystals; d 6.875; mp > 400°; 2.13 g dissolve in 100 g water at 9.3°, 2.4 g at 12°, 10.86 g at 100°. Insol in alcohol and ethyl ether.

9403. Thallium Selenide. Thallous selenide. $SeTl_2$; mol wt 487.73. Se 16.19%, Tl 83.81%. Tl_2Se. Prepared by the action of hydrogen selenide on a soln of thallous carbonate: Kuhlmann, *Bull. Soc. Chim.* [2] **1**, 330 (1864).

Dark gray plates with a metallic luster. mp 340° (Kuhlmann); mp 338° [Palabon, *Compt. Rend.* **145**, 118 (1907); **173**, 142 (1921)]. Insol in water and acids.

9404. Thallium Sesquioxide. Thallic oxide; thallium peroxide. O_3Tl_2; mol wt 456.76. O 10.51%, Tl 89.49%. Tl_2O_3.

Brown powder. d 9.65. mp 717°. Insol in water. Dec by HCl with evolution of chlorine and by H_2SO_4 with evolution of oxygen.

9405. Thallium Sulfate. Thallous sulfate; Eccothal. O_4STl_2; mol wt 504.83. O 12.68%, S 6.35%, Tl 80.97%. Tl_2SO_4.

White, rhomboid prisms. *Poisonous!* d 6.77; mp 632°. Soly in 100 ml water at 0°: 2.70 g; at 20°: 4.87 g; at 100°: 18.45 g. LD_{50} orally in rats: 25 mg/kg, E. W. Schafer, *Toxicol. Appl. Pharmacol.* **21**, 315 (1972).

Caution: Ingestion causes G.I. colic, vomiting trembling, convulsions, paralysis, dyspnea, collapse, death, *cf. Clinical Toxicology of Commercial Products,* R. E. Gosselin *et al.,* Eds. (Williams & Wilkins, Baltimore, 4th ed., 1976) section III, pp 307-311. *See also* Thallium.
USE: As rat poison, as ant bait and as a reagent in analytical chemistry.

9406. Thallium Sulfide. Thallous sulfide. STl_2; mol wt 440.83. S 7.27%, Tl 92.73%. Tl_2S.

Bluish-black, cryst powder. d 8.39. mp 448.5°. Almost insol in water, alkali hydoxides, sulfides or cyanides. Sol in mineral acids.

9407. Thallium Trifluoride. Thallic fluoride. F_3Tl; mol wt 261.38. F 21.81%, Tl 78.19%. TlF_3. Prepd from Tl_2O_3 and F_2: Hannebohn, Klemm, *Z. Anorg. Allgem. Chem.* **229**, 343 (1936); Kwasnik in *Handbook of Preparative Inorganic Chemistry,* **Vol. 1**, G. Brauer, Ed. (Academic Press, New York, 2nd ed., 1963) pp 230-231.

Orthorhombic crystals. Very sensitive to moisture. d_4^{25} 8.36. d 8.65 also reported: Hebecker, Hoppe, *Naturwiss.* **53**, 104 (1966). mp 550°. Dec on heating in air, but may be melted in an atm of fluorine. Instantly dec by water, forming thallium hydroxide and a dark brown sediment. May be stored in sealed quartz tubes.

9408. Thaumatin. Talin. Sweet-tasting basic protein extracted from the fruit of the tropical plant, *Thaumatococcus danielli* Benth., *Marantaceae,* found in Western Africa from Sierra Leone to Zaire, in Sudan and Uganda. Composed of five different forms, thaumatins I, II, III, b and c; thaumatins I and II predominate. All are nearly 100,000 times sweeter than sucrose and have molecular weights of about 22,000. Isoln and characterization: H. van der Wel, K. Loeve, *Eur. J. Biochem.* **31**, 221 (1972). Extraction process: J. Higgenbotham, U.S. pat. **4,011,206** (1976 to Tate and Lyle Ltd.). Electrophysical study of effects on taste receptors: Brouwer *et al., Acta Physiol. Scand.* **89**, 550 (1973). Spectrometric investigation: O. Korver *et al., Eur. J. Biochem.* **35**, 554 (1973). Studies on the primary structure of thaumatin I: R. B. Iyengar *et al., ibid.* **96**, 193 (1979). Thaumatins I and II consist of almost identical sequences of 207 amino acids. Thaumatin I crystallizes in two different forms: H. van der Wel *et al., FEBS Letters* **56**, 316 (1975). The crystal structure shows two distinct regions with the amino acids either in sheets or in complex loops. Crystal structure: A. M. de Vos *et al., Proc. Nat. Acad. Sci. USA* **82**, 1406 (1985). Cloning and expression in *E. coli* of the structural gene of thaumatin II: L. Edens *et al., Gene* **18**, 1 (1982); C. T. Verrips, **Eur. pat. Appls. 54,330** and **54,331** (both 1982 to Unilever). Cloning of the natural gene: A. M. Ledeboer *et al., Gene* **30**, 23 (1984). *Reviews:* R. Cagan, *Science* **181**, 32 (1973); H. van der Wel, *Chem. & Ind. (London)* **1983**, 19.

Intensely sweet taste, licorice aftertaste. Strongly cationic, isoelectric pt greater than or equal to 11.7. uv max: 278 nm (pH 5.6); 283, 290 nm (pH 13.0). About 750-1600 times sweeter than sucrose on a wt basis; 30,000-100,000 times on a molar basis. Threshold values are near $10^{-4}\%$. The proteins lose sweetness on heating, on splitting of disulfide bridges and also at pHs < 2.5 which points to the importance of the tertiary structure for the sweetness. *See* Korver *et al., loc. cit.*
USE: Potential low-calorie sweetener.

9409. Theaflavine. *[2S-[2α(2S*,3S*),3α]]-1,8-Bis(3,4-dihydro-3,5,7-trihydroxy-2H-1-benzopyran-2-yl)-3,4,6-trihydroxy-5H-benzocyclohepten-5-one.* $C_{29}H_{24}O_{12}$; mol wt 564.50. C 61.70%, H 4.29%, O 34.01%. From black tea extracts: Roberts, Myers, *J. Sci. Food Agr.* **10**, 176 (1959). Structure: Takino *et al., Tetrahedron Letters* **1965**, 4019. Configuration: Brown *et al., ibid.* **1966**, 1193.

Crystals from water, dec 237-240°. Absorption max (ethanol): 216, 271, 384, 470 nm (ϵ 35500, 19500, 8700, 3600). Nonaacetate, $C_{47}H_{42}O_{21}$, crystals, mp 167-168°. uv max (alc): 211, 250, 314, 353 nm (ϵ 25100, 13200, 7100, 4700).

9410. Thearubigins. Weakly acidic class of orange-brown phenolic pigments produced during the fermentation step of tea manufacture. Accounts for as much as 20% of dry weight of black tea; in combination with theaflavine, *q.v.*, determines the color, brightness and body of a tea infusion. Isolation: E. A. H. Roberts *et al., J. Sci. Food Agric.* **8,** 72 (1957); E. A. H. Roberts, *ibid.* **9,** 212 (1958). Effects on tea characteristics: R. F. Smith, *ibid.* **19,** 530 (1968); M. Hazarika *et al., ibid.* **35,** 1208 (1984). Structure studies: A. G. Brown *et al., Phytochemistry* **8,** 2333 (1969); J. E. Berkowitz *et al., ibid.* **10,** 2271 (1971); D. J. Cattell, H. E. Nursten, *ibid.* **15,** 1967 (1976). HPLC analysis: B. L. Wedzicha, T. J. Donovan, *J. Chromatog.* **478,** 217 (1989); R. G. Bailey *et al., ibid.* **542,** 115 (1991).

9411. Thebaine. *(5α)-6,7,8,14-Tetrahydro-4,5-epoxy-3,6-dimethoxy-17-methylmorphinan;* paramorphine. $C_{19}H_{21}$-NO_3; mol wt 311.38. C 73.29%, H 6.80%, N 4.50%, O 15.41%. From opium, which contains from 0.3-1.5% depending on its origin. Discussion of structure and bibliography: Small, Lutz, *Chemistry of the Opium Alkaloids, U.S. Public Health Reports,* **Suppl. No. 103,** Washington (1932); Small, Browning, *J. Org. Chem.* **3,** 618 (1938); Cherbuliez, Araqui, *Helv. Chim. Acta* **26,** 2251 (1943); Ghosh, Robinson, *J. Chem. Soc.* **1944,** 506; K. W. Bentley, *The Chemistry of the Morphine Alkaloids* (Oxford, 1954) p 184 sqq. Config: Kalvoda *et al., Helv. Chim. Acta* **38,** 1847 (1955). Syntheses: Rapoport *et al., J. Am. Chem. Soc.* **89,** 1942 (1967); Schwartz, Mami, *ibid.* **97,** 1239 (1975); Barber, Rapoport, *J. Med. Chem.* **18,** 1074 (1975). Absorption spectrum: Csokán, *Z. Anal. Chem.* **124,** 344 (1942). Toxicity data: Eddy, *J. Pharmacol. Exp. Ther.* **66,** 182 (1939).

Orthorhombic, rectangular plates by sublimation at 170-180° under atmospheric pressure and a 1 mm distance. mp 193° (rapid heating). $[\alpha]_D^{15}$ −219° (p = 2 in alc); $[\alpha]_D^{23}$ −230° (p = 5 in chloroform). pK at 15° = 6.05. pH of satd water soln 7.6. One gram dissolves in 1460 ml water at 15° [Kolthoff, *Biochem. Z.* **162,** 336 (1925)], in about 15 ml hot alcohol or 13 ml chloroform, about 200 ml ether, 25 ml benzene, 12 ml pyridine; not very sol in petr ether.

Hydrochloride monohydrate, $C_{19}H_{21}NO_3 \cdot HCl \cdot H_2O$, orthorhombic prisms from alcohol, $[\alpha]_D^{23}$ −164° (p = 2). Sol in about 12 parts water, in alcohol; pH of 0.05 molar soln 4.95. LD_{50} s.c. in rabbits: 14 mg/kg (Eddy).

Oxalate hexahydrate, $2C_{19}H_{21}NO_3 \cdot C_2H_2O_4 \cdot 6H_2O$, prisms, sol in about 10 parts water, in alcohol. Almost insol in ether.

Binoxalate monohydrate, $C_{19}H_{21}NO_3 \cdot C_2H_2O_4 \cdot H_2O$, prisms, sol in 45 parts water.

Bitartrate monohydrate, $C_{19}H_{21}NO_3 \cdot C_4H_6O_6 \cdot H_2O$, prisms, sol in 130 parts water, quite sol in hot water, hot alcohol.

Salicylate, $C_{19}H_{21}NO_3 \cdot C_7H_6O_3$, crystals, sol in 750 parts water.

Note: This is a controlled substance (opiate) listed in the U.S. Code of Federal Regulations, Title 21 Part 1308.12 (1995).

9412. Thebainone. *7,8-Didehydro-4-hydroxy-3-methoxy-17-methylmorphinan-6-one;* thebainone-A. $C_{18}H_{21}NO_3$; mol wt 299.37. C 72.22%, H 7.07%, N 4.68%, O 16.03%. Prepn from thebaine, codeinone or β-ethylthiocodide: Morris, Small, *J. Am. Chem. Soc.* **56,** 2159 (1934). Earlier references and discussion of structure: Small, Lutz, *Chemistry of the Opium Alkaloids, U.S. Public Health Reports* **Suppl. No. 103,** Washington (1932). About anomalies in nomenclature and difference from metathebainone *see* Henry, *Plant Alkaloids* (London, 1939) p 249. Description of all thebainones: K. W. Bentley, *The Chemistry of the Morphine Alkaloids* (Oxford, 1954) p 219.

Crystals from ethyl acetate, mp 146°. $[\alpha]_D^{28}$ −47° (c = 1.16 in 95% alc). One gram dissolves in 250 ml water, about 120 ml boiling water. Sol in chloroform, benzene, acetone; sparingly sol in ether, alcohol, methanol.

Sesquihydrate, crystals from water, mp 90°.

Methanolate, crystals from methanol, mp 118°.

Hydrochloride, $C_{18}H_{21}NO_3 \cdot HCl$, crystals from alcohol, mp 256° (turns red at mp). $[\alpha]_D^{30}$ −25° (c = 1.63).

Hydriodide, $C_{18}H_{21}NO_3 \cdot HI$, crystals, mp 165°, solidifies and remelts 260°.

Methiodide, mp 251°.

9413. Thenaldine. *1-Methyl-N-phenyl-N-(2-thienylmethyl)-4-piperidinamine; 1-methyl-4-N-2-thenylanilinopiperidine;* 1-methyl-4-amino-*N*-phenyl-*N*-(2-thenyl)piperidine; thenophenopiperidine; 1-methyl-4-[phenyl-(2-thenyl)amino]piperidine; thenalidine; Sandostene. $C_{17}H_{22}N_2S$; mol wt 286.44. C 71.28%, H 7.74%, N 9.78%, S 11.19%. Prepn: Stoll, Bourquin, U.S. pats. **2,717,251** and **2,757,175** (1955 and 1956 to Sandoz).

$bp_{0.02}$ 158-160°, mp 95-97°.
Tartrate, crystals, mp 170-172°.
THERAP CAT: Tartrate as antihistaminic; antipruritic.

9414. Thenium Closylate. *N,N-Dimethyl-N-(2-phenoxyethyl)-2-thiophenemethanaminium salt with 4-chlorobenzenesulfonic acid (1:1); dimethyl(2-phenoxyethyl)-2-thenylammonium p-chlorobenzenesulfonate;* 611C55; Bancaris; Canopar. $C_{21}H_{24}ClNO_4S_2$; mol wt 454.01. C 55.56%, H 5.33%, Cl 7.81%, N 3.09%, O 14.10%, S 14.13%. Prepn: Copp, **Brit.** pat. **864,885** (1961 to Wellcome Found.). Anthelmintic activity: Burrows, Lillis, *Am. J. Vet. Res.* **23,** 77 (1962).

Crystals from isopropanol + ether, mp 159-160°. Soly in water at 20°: 0.6% w/v.

THERAP CAT (VET): Anthelmintic.

9415. 3-Thenoic Acid. *3-Thiophenecarboxylic acid; β-thiophenic acid.* $C_5H_4O_2S$; mol wt 128.15. C 46.86%, H 3.15%, O 24.97%, S 25.02%. Prepd by the oxidation of 3-thenaldehyde with silver oxide: Campaigne, LeSuer, *Org. Syn.* **33**, 94 (1953).

Crystals from water, mp 137-138°. Soly in water at 25°: 0.43 g/100 g. pKa 6.23. Volatile with steam.

9416. Thenyldiamine. *N,N-Dimethyl-N'-2-pyridinyl-N'-(3-thienylmethyl)-1,2-ethanediamine; 2-[(2-dimethylamino-ethyl)-3-thenylamino]pyridine; N,N-dimethyl-N'-(α-pyridyl)-N'-(3-methylthienyl)ethylenediamine; N-(α-pyridyl)-N-(β-thenyl)-N',N'-dimethylethylenediamine; N-(2-dimethylami-noethyl)-N-2-pyridyl-3-thenylamine; dethylandiamine;* Win-2848; Thenfadil. $C_{14}H_{19}N_3S$; mol wt 261.39. C 64.33%, H 7.33%, N 16.08%, S 12.27%. Prepn: Campaigne, LeSuer, *J. Am. Chem. Soc.* **71**, 333 (1949). Toxicity study: Hoppe, Lands, *J. Pharmacol. Exp. Ther.* **97**, 371 (1949).

Free base, liquid. $bp_{1.0}$ 169-172°. n_D^{20} 1.5915. Hydrochloride, $C_{14}H_{19}N_3S.HCl$, crystals from methanol, mp 169.5-170°. Bitter taste. Sol in water up to 20%. Slightly sol in alc. pH of 1% aq soln 6.5. LD_{50} orally in rats: 525 mg/kg (Hoppe, Lands).

THERAP CAT: Hydrochloride as antihistaminic.

9417. Theobroma Oil. Cacao butter; cocoa butter. From roasted seeds of *Theobroma cacao* L., *Sterculiaceae. Constit.* Chiefly glycerides of stearic, palmitic, oleic, arachidic, and linoleic acids.

Yellowish-white solid; brittle below 25°; chocolate odor and taste; d_{25}^{100} 0.858-0.864; mp 30-35°; n_D^{40} 1.4537-1.4578. Sapon. no. 188-195. Iodine no. 35-40. Insoluble in water. Slightly sol in alcohol; sol in boiling abs alc; very sol in chloroform, ether, benzene, petr ether.

USE: Lubricant in massage; base for suppositories and ointments. Manuf chocolate, toilet soaps, creams, etc.

9418. Theobromine. *3,7-Dihydro-3,7-dimethyl-1H-purine-2,6-dione; 3,7-dimethylxanthine.* $C_7H_8N_4O_2$; mol wt 180.17. C 46.67%, H 4.48%, N 31.10%, O 17.76%. The principal alkaloid of the cacao bean which contains 1.5-3% of the base. Also present in cola nuts and in tea. Usually extracted from the hull of cacao beans which contains 0.7-1.2%. Extraction process: Schwyzer, *Die Fabrikation Pharmazeutischer und Chemisch-Technischer Produkte* (Berlin, 1931). Synthesis starting with 3-methyluric acid: Fischer, Ach, *Ber.* **31**, 1980 (1898); *cf.* Gebner, Krebs, *J. Gen. Chem. USSR* **16**, 179-186 (1946). Comparison with theophylline, *q.v.*, of metabolism in humans: D. J. Birkett *et al., Drug Metabol. Dispos.* **13**, 725 (1985). Pharmacokinetics: A. Lelo *et al., Brit. J. Clin. Pharmacol.* **22**, 177 (1986). Bronchodilator effect in asthma: F. E. R. Simons *et al., J. Allerg. Clin. Immunol.* **76**, 703 (1985).

Monoclinic needles (lamellar twinning on 001) from water, mp 357°. Sublimes 290-295°. Kb at 18°: 1.3×10^{-14}; Ka 0.9×10^{-10}. Absorption spectrum: Hartley, *J. Chem. Soc.* **87**, 1803, 1810 (1905). One gram dissolves in about 2000 ml water, 150 ml boiling water, 2220 ml 95% alcohol; sol in the fixed alkali hydroxides, concd acids, in about 22 parts of 20% aq tribasic sodium phosphate soln; moderately sol in ammonia. Almost insol in benzene, ether, chloroform, carbon tetrachloride. Forms salts which are dec by water, and compds with bases which are more stable.

Calcium salicylate, **Theocalcin, Calcium Diuretin.** Double salt or mixture of calcium theobromine and calcium salicylate. Contains no less than 44% theobromine. White, amorphous powder. Slightly saline taste. Partly sol in water. Aq solns are alkaline to phenolphthalein.

Sodium acetate, **Thesodate.** Equimolar mixture of sodium theobromine and sodium acetate, containing $1H_2O$. Theobromine 59.6%, anhyd sodium acetate 27.1%. White, odorless or almost odorless, hygroscopic powder. Absorbs CO_2 from the air becoming incompletely sol. Very sol in water, sparingly sol in cold alcohol; the solns are strongly alkaline. pH about 10. *Keep tightly closed.*

Sodium salicylate, **Diuretin.** Equimolar mixture of sodium theobromine and sodium salicylate, containing $1H_2O$. Theobromine 47.3%, sodium salicylate 42.1%. White, odorless or almost odorless, hygroscopic powder. Absorbs CO_2 from the air and becomes incompletely soluble. Soluble in 1 part water; slightly sol in alcohol; the solns are strongly alkaline. pH about 10.

THERAP CAT: Diuretic, bronchodilator, cardiotonic.

THERAP CAT (VET): Diuretic, myocardial stimulant, vasodilator.

9419. 1-Theobromineacetic Acid. *2,3,6,7-Tetrahydro-3,7-dimethyl-2,6-dioxo-1H-purine-1-acetic acid; 3,6-dihydro-3,7-dimethyl-2,6-dioxo-1(2H)-purineacetic acid; 3,6-dihydro-2,6-diketo-3,7-dimethyl-1(2)-purineacetic acid.* $C_9H_{10}N_4O_4$; mol wt 238.20. C 45.38%, H 4.23%, N 23.52%, O 26.87%. Prepd from sodium theobromine + chloroacetic acid: E. Merck *et al.*, Ger. pat. 352,980, *C.A.* **17**, 1307[1] (1923).

Crystals, mp 260°.
Sodium salt, $C_9H_9N_4NaO_4$, *sodium theobromine acetate,* **Técarine.** Crystalline powder, freely sol in water, forming a neutral soln.

THERAP CAT: Bronchodilator.

9420. Theofibrate. *2-(4-Chlorophenoxy)-2-methylpropanoic acid 2-(1,2,3,6-tetrahydro-1,3-dimethyl-2,6-dioxo-7H-purin-7-yl)ethyl ester; 2-(p-chlorophenoxy)-2-methylpropionic acid ester with 7-(2-hydroxyethyl)theophylline; 1-(theophyllin-7-yl)ethyl 2-(p-chlorophenoxy)isobutyrate; etofylline clofibrate;* ML-1024; Duolip. $C_{19}H_{21}ClN_4O_5$; mol wt 420.85. C 54.23%, H 5.03%, Cl 8.42%, N 13.31%, O 19.01%. Deriv of clofibric acid, *q.v.*, with antilipemic, antithrombotic, and platelet-aggregation inhibitory activity: Prepn: G. Metz, M. Specker, Ger. pat. 2,308,826; *eidem,* U.S. pat. 3,984,413 (1974, 1976 both to Merckle); *eidem, Arzneimittel-Forsch.* **25**, 1686 (1975). Hypolipemic activity: G. Metz *et al., ibid.* **27**, 1173 (1977). Series of articles on chemistry, biopharmaceutic evaluation, pharmacology, toxicology, clinical studies: *ibid.* **30**, 2013-2074 (1980).

Colorless crystals from ethanol, mp 133-135°. Practically insol in water at pH 2-7.4 and in cold alcohols. Sol in acetone, chloroform, hot alcohols. LD$_{50}$ in mice, rats, dogs (g/kg): 11.7, 17.0, > 10.0 orally (Metz *et al.*).
THERAP CAT: Antihyperlipoproteinemic.

9421. Theophylline. *3,7-Dihydro-1,3-dimethyl-1H-purine-2,6-dione;* 1,3-dimethylxanthine; Accurbron; Aerobin; Aerolate; Afonilum; Armophylline; Austyn; Bilordyl; Bronchoretard; Bronkodyl; Cétraphylline; Duraphyllin; Diffumal; Elixophyllin; Etheophyl; Euphyllin; Euphylong; LaBID; Lasma; Nuelin; Physpan; Pro-Vent; PulmiDur; Pulmo-Timelets; Respbid; Slo-Bid; Slo-Phyllin; Solosin; Talotren; Teosona; Theobid; Theoclear; Theochron; Theo-Dur; Theolair; Theon; Theophyl; Theograd; Theostat; Theovent; Unifyl; Uniphyl; Uniphyllin; Xanthium. C$_7$H$_8$N$_4$O$_2$; mol wt 180.17. C 46.67%, H 4.48%, N 31.10%, O 17.76%. Xanthine derivative with diuretic, cardiac stimulant and smooth muscle relaxant activities; isomeric with theobromine, *q.v.* Small amounts occur in tea. Synthesis starting with dimethylurea and ethyl cyanoacetate: Traube, *Ber.* 33, 3035 (1900); Grinberg, *J. Appl. Chem. USSR* 13, 1461 (1940); Gebner, Krebs, *J. Gen. Chem. USSR* 16, 179 (1946). Bioavailability: K. Svedmyr *et al., Allergy* 37, 111 (1982). Comprehensive description: J. L. Cohen, *Anal. Profiles Drug Subs.* 4, 466-493 (1975). Symposium on clinical experience in asthma and allergy: *Am. J. Med.* 79(6A), 1-78 (1985); on pharmacology, bioavailability, pharmacokinetics and efficacy in obstructive pulmonary disease: *J. Allerg. Clin. Immunol.* 78(4 Part 2), 669-824 (1986). Review of clinical toxicology: J. D. Truwit, *Crit. Care Clinics* 7, 639-657 (1991).

Monohydrate, thin monoclinic tablets from water, mp 270-274°. Bitter taste. pKa (25°): 8.77; pKb: 13.5, 11.5. One gram dissolves in 120 ml water, 80 ml alcohol, about 110 ml chloroform. Sol in hot water, in alkali hydroxides, ammonia, dil HCl or HNO$_3$; sparingly sol in ether. uv max (0.1N NaOH): 274 nm.
Ethanolamine, C$_9$H$_{15}$N$_5$O$_3$, *Monotheamin.* White, crystalline, nonhygroscopic powder contg 75% anhydr theophylline and 25% ethanolamine. Sol in water.
Isopropanolamine, C$_{10}$H$_{17}$N$_5$O$_3$, *Oxyphyllin.* Prepn: Greenbaum, *Am. J. Pharm.* 109, 550 (1937). Prisms from alcohol. Very sol in water. pH of aq solns 9.20.
Lysine salt, *Paidomal.*
Sodium acetate, *Theocin.* Equimolar mixture of theophylline sodium and sodium acetate, containing 1H$_2$O. White, odorless, crystalline powder; bitter taste. Sol in about 25 parts water. Insol in alcohol, chloroform or ether.
Sodium glycinate, *Biophylline, Englate, Glytheonate, Panophylline.* Prepn: Krantz *et al., J. Am. Pharm. Assoc.* 36, 248 (1947). Equimolar mixture of theophylline sodium and glycine. Darkens at 180°. Soly in water 18% w/v. Saturated soln has a pH of 8.7-9.1, d^{20} 1.05, and is stable to carbon dioxide.

THERAP CAT: Bronchodilator.
THERAP CAT (VET): Bronchodilator.

9422. Thermolysin. Bacillus thermoproteolyticus neutral proteinase; E.C. 3.4.24.4. Proteolytic enzyme of mol wt 37,500 that hydrolyzes protein bonds on the *N*-terminal side of hydrophobic amino acid residues. Contains a zinc atom essential for activity and four Ca^{2+} ions essential for thermal and conformational stability. Isoln from *Bacillus thermoproteolyticus:* S. Endo, *J. Ferment. Technol.* 40, 346 (1962). Properties and amino acid composition: Y. Ohta *et al., J. Biol. Chem.* 241, 5919 (1966). Site of enzymatic hydrolysis: Y. Ohta, Y. Ogura, *J. Biochem.* 58, 607 (1965). Substrate specificity studies: H. Matsubara *et al., Biochem. Biophys. Res. Commun.* 21, 242 (1965); 24, 427 (1966); K. Morihara, H. Tsuzuki, *Biochim. Biophys. Acta* 118, 215 (1966). Stability studies: Y. Ohta, *J. Biol. Chem.* 242, 509 (1967). Inhibition studies: H. Matsubara *et al., Biochem. Biophys. Res. Commun.* 34, 719 (1969); J. Murphy *et al., Arch. Biochem. Biophys.* 202, 405 (1980). Purification: H. Matsubara, *Methods Enzymol.* 19, 642 (1970). Structure studies: K. Titani *et al., Nature* 238, 35 (1972); B. W. Matthews *et al., ibid.* 37, 41; P. M. Colman *et al., J. Mol. Biol.* 70, 701 (1972). Function of the metal ions: J. Feder *et al., Biochemistry* 10, 4552 (1971); G. Voordouw, R. S. Roche, *ibid.* 14, 4667 (1975); R. S. Roche, G. Voordouw, *CRC Crit. Rev. Biochem.* 5, 1 (1978). Effect of the histidyl residue on the mechanism of action: S. Blumberg *et al., Isr. J. Chem.* 12, 643 (1974); M. K. Pangburn, K. A. Walsch, *Biochemistry* 14, 4050 (1975). *Review: Experientia* 26, Suppl., 31-59 (1976).
Crystals. uv max: 280 nm (ϵ 66,300). Optimum pH 7.0-8.5; stable at pH 6.0-9.0. The refrigerated lyophilized enzyme is stable for months; frozen enzyme soln can be kept for weeks without significant loss of activity. Not deactivated at 65°, but loses half of its activity upon heating at 80° for 1 hr.
USE: In studies of protein sequences.

9423. Thermorubin. Antibiotic substances isolated from *Thermoactinomyces antibioticus:* Craveri *et al., Clin. Med.* 71, 511 (1964); *eidem,* Ger. pat. 1,180,891; *eidem,* U.S. pat. 3,300,379 (1964, 1967 both to Lepetit). Structure of thermorubin A, the major component: Moppett *et al., J. Am. Chem. Soc.* 94, 3269 (1972); revised structure: F. Johnson *et al., ibid.* 102, 5580 (1980). Inhibition of bacterial protein synthesis: G. Pirali *et al., Biochim. Biophys. Acta* 366, 310 (1974). Studies on mechanism of action: F. Lin, A. Wishnia, *Biochemistry* 21, 477, 484 (1982).

Thermorubin A

Orange-red rosettes and needles from chloroform, darkens 190°, chars 300° without melting. [α]$_D^{25}$ −14° (c = 0.4 in dioxane). Absorption max [dioxane-cyclohexane (1:1)]: 300, 328, 414, 430 nm (E$_{1cm}^{1\%}$ 1041, 1066, 313, 332). Sol in dioxane, pyridine, tetrahydrofuran, DMF, DMSO, concd alkalies, concd H$_2$SO$_4$, glacial acetic acid; slightly sol in methanol, ethanol, butanol, ethyl acetate, acetone, chloroform, benzene, cyclohexane. Practically insol in water, ether, petr ether, hexane. LD$_{50}$ i.p. in mice: 300 mg/kg (Craveri, 1967).
Diacetate, crystals from ethyl acetate. Absorption max [0.1N HCl-dioxane (1:1)]: 337, 436, 462 nm (E$_{1cm}^{1\%}$ 1500, 113, 162).
Thermorubin A, C$_{32}$H$_{24}$O$_{12}$, dihydrate, orange-red rosettes from ethyl acetate. Absorption max (ethanol): 250, 300, 328, 435 nm (ϵ 33700, 54640, 49980, 16300). pKa values: 4.7, 7.0, 9.0. Easily tautomerizes in soln. Thermally labile, decomposes rapidly in soln at temps > 60°.

9424. Thevetin A. *(3β,5β)-3-[(O-β-D-Glucopyranosyl-(1→6)-O-D-glucopyranosyl-(1→4)-6-deoxy-3-O-methyl-α-L-*

glucopyranosyl)oxy]-14-hydroxy-19-oxocard-20(22)-enolide.
$C_{42}H_{64}O_{19}$; mol wt 872.96. C 57.79%, H 7.39%, O 34.82%.
Glycoside isolated from *Thevetia neriifolia* Juss., *Apocynaceae:* Bloch *et al., Helv. Chim. Acta* **43**, 652 (1960). Separation from thevetin B: Delalande, Baisse, **U.S.** pats. **3,030,-355; 3,043,829** (1962). Structure: K. Tori *et al., Tetrahedron Letters* **1977**, 717.

Crystals from water. mp 208-210°. $[\alpha]_D^{24}$ −72.0° ±1.5° (c = 1.48 in methanol).
Acetyl deriv, $C_{58}H_{80}O_{27}$, crystals from methanol + ether, mp 143-149°. $[\alpha]_D^{26}$ −54.2° ±1° (c = 1.86 in chloroform).

9425. Thevetose. *6-Deoxy-3-O-methylglucose; 3-O-methylglucomethylose; quinovose 3-methyl ether.* $C_7H_{14}O_5$; mol wt 178.19. C 47.19%, H 7.92%, O 44.90%. Isoln of L-thevetose from *Thevetia nereifolia* Juss. and *Tanghinia venenifera* Poir., *Apocynaceae:* Frèrejacque, *Compt. Rend.* **221**, 645 (1945); Frèrejacque, Hasenfratz, *ibid.* **222**, 815 (1946); Sigg *et al., Helv. Chim. Acta.* **38**, 166 (1955); from thevetin: Helfenberger, Reichstein, *ibid.* **31**, 1470 (1948); from *Bowiea volubilis* Harv., *Liliaceae:* Tschesche, Dölberg, *Ber.* **90**, 2378 (1957). Synthesis: Blindenbacher, Reichstein, *Helv. Chim. Acta* **31**, 1669 (1948). Isoln of D-thevetose from *Adenium honghcl* D.C., *Apocynaceae:* Frèrejacque, *Compt. Rend.* **230**, 127 (1950). Rare instance of two optical isomers of the same sugar being present in the vegetable kingdom.

α-L-Thevetose

L-Form, needles from acetone + ether, mp 128-130°. $[\alpha]_D^{19}$ −66° → −36.9° (c = 1.46 in water).
β-L-Triacetate, $C_{13}H_{20}O_8$, crystals, mp 118-119°. $[\alpha]_D^{18}$ −7.5° (c = 1.12 in acetone).
L-Thevetosazone, crystals, mp 136-139°. $[\alpha]_D^{24}$ +110.8° → +40.3°.
D-Form, crystals, mp 116°. $[\alpha]_D^{20}$ +84° → +33°.
β-D-Triacetate, $C_{13}H_{20}O_8$, crystals, mp 121°. $[\alpha]_D^{20}$ +6° (acetone).
α-D-Triacetate, crystals, mp 105°. $[\alpha]_D^{20}$ +122° (acetone).
D-Thevetosazone, crystals, mp 123-128°. $[\alpha]_D^{20}$ −117.8° → −26.4°.

9426. Thiabendazole. *2-(4-Thiazolyl)-1H-benzimidazole;* 4-(2-benzimidazolyl)thiazole; MK-360; Omnizole; Thiaben; Thibenzole; Bovizole; Eprofil; Equizole; Mintezol; Top Form Wormer; Mertect; Lombristop; Minzolum; Nemapan; Polival; TBZ; Tecto. $C_{10}H_7N_3S$; mol wt 201.25. C 59.68%, H 3.51%, N 20.88%, S 15.93%. Prepd by the reaction of 4-thiazolecarboxamide with *o*-phenylenediamine in polyphosphoric acid: H. D. Brown *et al., J. Am. Chem. Soc.* **83**, 1764 (1961); L. H. Sarett, H. D. Brown, **U.S.** pat. **3,017,415** (1962 to Merck & Co.). Synthesis of labeled thiabendazole: D. J. Tocco *et al., J. Med. Chem.* **7**, 399 (1964). Alternate route of synthesis: V. J. Grenda *et al., J. Org. Chem.* **30**, 259 (1965). Anthelmintic props: H. D. Brown *et al., loc. cit.;* K. C. Kates *et al., J. Parasitol.* **57**, 356 (1971). Fungicidal props: H. J. Robinson *et al., J. Invest. Dermatol.* **42**, 479 (1966). Systemic props in plants: D. C. Erwin *et al., Phytopathology* **58**, 860 (1968). Toxicity: H. J. Robinson *et al., Toxicol. Appl. Pharmacol.* **7**, 53 (1965).

Residue analysis: IUPAC Appl. Chem. Div., *Pure Appl. Chem.* **52**, 2567 (1980). Comprehensive description: V. K. Kapoor, *Anal. Profiles Drug Subs.* **16**, 611-639 (1986).

Colorless crystals, mp 304-305°. uv max (methanol): 298 nm (ε 23330). Fluorescence max in acid soln: 370 nm (310 nm excitation). Max solv in water at pH 2.2: 3.84%. Soluble in DMF, DMSO. Slightly soluble in alcohols, esters, chlorinated hydrocarbons. LD_{50} in mice, rats, rabbits (g/kg): 3.6, 3.1, > 3.8 orally (Robinson).
Hypophosphite, **Arbotect.** Amber liquid. d^{25} 1.103.
USE: Systemic fungicide for spoilage control of citrus fruit; for treatment and prevention of Dutch elm disease in trees; for control of fungal diseases of seed potatoes.
THERAP CAT: Anthelmintic (Nematodes).
THERAP CAT (VET): Anthelmintic, fungicide.

9427. Thiacetazone. *N-[4-[[(Aminothioxomethyl)hydrazono]methyl]phenyl]acetamide; 2-[[4-(acetylamino)phenyl]methylene]hydrazinecarbothioamide;* 4′-formylacetanilide thiosemicarbazone; *p*-acetamidobenzaldehyde thiosemicarbazone; *p*-acetylaminobenzaldehyde thiosemicarbazone; *p*-acetaminobenzylidenethiosemicarbazone; amithiozone; thibone; Myrizone; Neustab; Panrone; Seroden; Tebethion; Thiocarbazil; Thioparamizone; Tibione; Tiobicina. $C_{10}H_{12}$-N_4OS; mol wt 236.30. C 50.83%, H 5.12%, N 23.71%, O 6.77%, S 13.57%. Prepn: G. Domagk *et al., Naturwiss.* **33**, 315 (1946); Behnisch *et al., Angew. Chem.* **60A**, 113 (1948); Chabrier, Cattelain, *Bull. Soc. Chim. France* **1950M**, 52; Das, Mukherjee, *J. Am. Chem. Soc.* **75**, 1241 (1953). Toxicity data: Bavin *et al., J. Pharm. Pharmacol.* **2**, 764 (1950). HPLC determn in body fluids: P. J. Jenner, *J. Chromatog.* **276**, 463 (1983). Review of clinical experience: H. C. Hinshaw, W. McDermott, *Am. Rev. Tuberculosis* **61**, 145-157 (1950); of clinical safety and efficacy: P. Nunn *et al., Trans. Roy. Soc. Trop. Med. Hyg.* **87**, 578 (1993).

Minute, pale yellow crystals from abs alc. Bitter taste. Darkens on exposure to light. Dec 225-230°. uv max (ethanol): 328 nm, extinction 0.580. Sol in hot alc; very sparingly sol in cold alc. Insol in water, acetone, benzene, carbon tetrachloride, chloroform, carbon disulfide, petr ether. Practically insol in the other common organic solvents except glycols. Soly in propylene glycol ∼1%. LD_{50} s.c. in mice: 1-2 g/kg (Bavin).
THERAP CAT: Antibacterial (tuberculostatic).

9428. Thialbarbital. *5-(2-Cyclohexen-1-yl)dihydro-5-(2-propenyl)-2-thioxo-4,6(1H,5H)-pyrimidinedione; 5-allyl-5-(2-cyclohexen-1-yl)-2-thiobarbituric acid; thialbarbitone; Kemithal.* C_{13}-$H_{16}N_2O_2S$; mol wt 264.35. C 59.07%, H 6.10%, N 10.60%, O 12.10%, S 12.13%. Prepn: Volwiler, Tabern, **U.S.** pat. **2,153,730** (1939 to Abbott).

Crystals, mp 148-150°.
Sodium salt, $C_{13}H_{15}N_2NaO_2S$, pale yellow crystals from methanol, mp 130-132°. Very sol in water, alc. Practically insol in benzene, ether. pH of 2.5% soln in water: 10.5.

Note: This is a controlled substance (depressant) listed in the U.S. Code of Federal Regulations, Title 21 Part 1308.13 (1995).

THERAP CAT: Anesthetic (intravenous).

THERAP CAT (VET): Anesthetic.

9429. Thiambutene. *N,N-Diethyl-1-methyl-3,3-di-2-thienylallylamine;* 3-diethylamino-1,1-di(2'-thienyl)but-1-ene; diethylthiambutene; 191C49; NIH-4185. $C_{16}H_{21}NS_2$; mol wt 291.48. C 65.93%, H 7.26%, N 4.81%, S 22.00%. Prepn: Adamson, *J. Chem. Soc.* **1950,** 885; U.S. pat. 2,561,899 (1951 to Burroughs Wellcome). Toxicity data: J. S. McKenzie, N. R. Beechey, *Arch. Int. Pharmacodyn. Ther.* **135,** 376 (1962). Use: Hayes, *Vet. Rec.* **83,** 528 (1968). Evaluation as analgesic and preanesthetic agent: W. D. Harbison *et al., Aust. Vet. J.* **50,** 543 (1974). HPLC determn: I. Jane *et al., J. Chromatog.* **323,** 191 (1985).

bp$_{0.03}$ 122-128°. LD$_{50}$ i.p. in mice: 90 mg/kg (McKenzie, Beechey).

Hydrochloride, $C_{16}H_{21}NS_2 \cdot HCl$, *Themalon.* Crystals, mp 152-153°.

Note: This is a controlled substance (opiate) listed in the U.S. Code of Federal Regulations, Title 21 Part 1308.11 (1995).

THERAP CAT (VET): Analgesic (narcotic).

9430. Thiamine. *3-[(4-Amino-2-methyl-5-pyrimidinyl)methyl]-5-(2-hydroxyethyl)-4-methylthiazolium chloride;* vitamin B$_1$; aneurin; thiamine monochloride; thiaminium chloride. $C_{12}H_{17}ClN_4OS$; mol wt 300.81. C 47.91%, H 5.70%, Cl 11.79%, N 18.63%, O 5.32%, S 10.66%. Essential nutrient required for carbohydrate metabolism; also involved in nerve function. Biosynthesized by microorganisms and plants. Dietary sources include whole grains, meat products, vegetables, milk, legumes and fruit. Also present in rice husks and yeast. Converted *in vivo* to thiamine diphosphate, a coenzyme in the decarboxylation of α-keto acids. Chronic deficiency may lead to neurological impairment, beriberi, Wernicke-Korsakoff syndrome. Isoln from rice bran: B. C. P. Jansen, W. F. Donath, *Chem. Weekblad* **23,** 201 (1926). Structure: R. R. Williams, *J. Am. Chem. Soc.* **58,** 1063 (1936); R. R. Williams, J. K. Cline, *ibid.* 1504; R. R. Williams *et al., ibid.* **59,** 526 (1937). Review of syntheses: Knobloch in H. Vogel, *Chemie und Technik der Vitamine* vol. II (Stuttgart, 1953) pp 1-128. Toxicity data: D. Winter *et al., Int. Z. Vitaminforsch.* **37,** 82 (1967). HPLC determn in foods, pharmaceuticals, body tissues: T. Kawaski, *Methods Enzymol.* **122,** 15 (1986); in plasma and pharmacokinetics: H. Mascher, C. Kikuta, *J. Pharm. Sci.* **82,** 56 (1993). Review of bioavailability, absorption, and role in nutrition: F. L. Iber *et al., Am. J. Clin. Nutr.* **36,** 1067-1082 (1982). *Reviews:* "Thiamin—Twenty Years of Progress", *Ann. N.Y. Acad. Sci.* **378,** H. Z. Sable, C. J. Grubier, Eds. (1982) 470 pp; "Thiamin, Vitamin B$_1$, Aneurin" in *Vitamins,* W. Friedrich, Ed. (de Gruyter, Berlin, 1988) pp 339-401.

Hydrochloride, $C_{12}H_{17}ClN_4OS \cdot HCl$, *thiamine chloride hydrochloride, thiamine dichloride, Benerva, Betabion, Betalin S, Betaxin, Bewon, Metabolin, Vitaneurin.* Comprehensive description: K. A. M. Al-Rashood *et al., Anal. Profiles Drug Subs.* **18,** 413-458 (1989). Monoclinic plates in rosette-like clusters. Slight thiazole odor. Bitter taste. dec

248°. One gram dissolves in $\sim$1 ml water, 18 ml glycerol, 100 ml 95% alcohol, 315 ml abs alcohol; more sol in methanol. Sol in propylene glycol. Practically insol in ether, benzene, hexane, chloroform. pH of a 1% w/v soln in water 3.13; pH of a 0.1% w/v soln in water 3.58. On exposure to air of average humidity, the vitamin absorbs an amount of water corresponding to nearly one mol, forming a hydrate. LD$_{50}$ in mice (mg/kg): 89.2 i.v.; 8224 orally (Winter).

Mononitrate, $C_{12}H_{17}N_5O_4S$. Prepn: R. J. Turner, G. J. Schmitt, U.S. pat. 2,844,579 (1958 to Am. Cyanamid). Crystals, mp 196-200° (dec). Practically nonhygroscopic. pKa 4.8. Soly in water (g/100 ml): 2.7 (25°); $\sim$30 (100°). pH of 2% aq soln 6.5 to 7.1. More stable than the hydrochloride; suitable for enrichment of flours and feeds, multivitamin prepns.

THERAP CAT: Vitamin (enzyme cofactor).

THERAP CAT (VET): Vitamin (enzyme cofactor).

9431. Thiamine Diphosphate. *3-[(4-Amino-2-methyl-5-pyrimidinyl)methyl]-4-methyl-5-(4,6,6-trihydroxy-3,5-dioxa-4,6-diphosphahex-1-yl)thiazolium chloride P,P'-dioxide; thiamine trihydrogen pyrophosphate (ester);* cocarboxylase; thiamine pyrophosphate; thiamine diphosphate ester chloride; TDP; TPP; Berolase; Bivitasi; Cocalose; Cocarvit; Nutrase; Pyrolase. $C_{12}H_{19}ClN_4O_7P_2S$; mol wt 460.77. C 31.28%, H 4.16%, Cl 7.69%, N 12.16%, O 24.31%, P 13.44%, S 6.96%. Coenzyme required for the decarboxylation of α-keto acids. Most abundant form of thiamine in animal tissues. Identification as cofactor and enzymatic synthesis: K. Lohmann, P. Schuster, *Biochem. Z.* **294,** 183 (1937). Chemical synthesis: Weijlard, Tauber, *J. Am. Chem. Soc.* **60,** 2263 (1938); Weil-Malherbe, *Biochem. J.* **34,** 980 (1940); Weijlard, *J. Am. Chem. Soc.* **63,** 1160 (1941); Karrer, Viscontini, *Helv. Chim. Acta* **29,** 711 (1946). LC determn in blood: J. W. I. Brunnekreeft *et al., J. Chromatog.* **491,** 89 (1989). Role in pathogenesis of Wernicke-Korsakoff Syndrome: R. F. Butterworth *et al., Alcohol. Clin. Exp. Res.* **17,** 1084 (1993). Review of enzyme activity: J. Ullrich *et al., Vitam. Horm. (New York)* **28,** 365 (1970); of mechanism of action: R. Kluger, *Chem. Rev.* **87,** 863-876 (1987); M. Louloudi, N. Hadjiliadis, *Coord. Chem. Rev.* **135,** 429-468 (1994).

Monohydrate, crystals from alc contg some HCl, dec 240-244°. mp 238-240° from abs ethanol. uv max: 242 nm. Soluble in water. pH of 0.3% soln 2.23. The dry substance is very stable.

Free ester tetrahydrate, $C_{12}H_{18}N_4O_7P_2S \cdot 4H_2O$. Prepn: Wenz *et al.,* U.S. pat. 2,991,284 (1961 to E. Merck). dec 220-225°.

Thiamine monophosphate, $C_{12}H_{18}ClN_4O_4PS$, *3-[(4-amino-2-methyl-5-pyrimidinyl)methyl]-4-methyl-5-[2-(phosphono-oxy)ethyl]thiazolium chloride, thiamine monophosphate chloride.* Prepd by sulfuric acid hydrolysis of thiamine diphosphate: Lohmann, Schuster, *loc. cit.* Monohydrate; prisms from dil acetone or water, mp $\sim$200°. Freely sol in water.

THERAP CAT: Vitamin (enzyme cofactor).

9432. Thiamine Disulfide. *N,N'-[Dithiobis[2-(2-hydroxyethyl)-1-methyl-2,1-ethenediyl]]bis[N-[(4-amino-2-methyl-5-pyrimidinyl)methyl]formamide];* aneurin disulfide; vitamin B$_1$ disulfide; Aktivin; Neolamin. $C_{24}H_{34}N_8O_4S_2$; mol wt 562.72. C 51.23%, H 6.09%, N 19.91%, O 11.37%, S 11.40%. Isoln: Zima, Williams, *Ber.* **73,** 941 (1940). Prepn: Warnat, U.S. pat. 2,458,453 (1949 to Hoffmann-La Roche); Matsukawa, Hirano, *J. Pharm. Soc. Japan* **73,** 379 (1953); Hirano, Iwatsu, *ibid.* **73,** 1115 (1953); Kawasaki, *ibid.* **76,** 706 (1956). Bioavailability: N. Aoyagi *et al., Chem. Pharm. Bull.* **34,** 281 (1986).

(Z,Z)-isomer

When anhydr, mp 177° with intense yellow coloration. Very sparingly sol in benzene, acetone, ether, ethanol; freely sol if the crystals contain solvated acetone or water. Soly in water at 37° (mg/ml): 37.7 (pH 1.2); 3.0 (pH 3); 1.33 (pH 5); 0.568 (pH 7.2).

Hydrochloride, crystals from methanol + water, dec 231°.

Phosphate, $C_{24}H_{36}N_8O_{10}P_2S_2$, *thiamine monophosphate disulfide, Biotinin, Vitamogen.*

O,O-Diisobutyrate, $C_{32}H_{46}N_8O_6S_2$, *O-isobutyrylthiamine disulfide, bisibutiamine, sulbutiamine, Arcalion, Neodaian, Vitaberin.*

THERAP CAT: Vitamin (enzyme cofactor).

9433. Thiamine Triphosphate. *3-[(4-Amino-2-methyl-5-pyrimidinyl)methyl]-4-methyl-5-(4,6,8,8-tetrahydroxy-3,5,7-trioxa-4,6,8-triphosphaoct-1-yl)thiazolium inner salt P,P',P''-trioxide;* thiamine triphosphoric acid ester; TTP. $C_{12}H_{19}N_4$-$O_{10}P_3S$; mol wt 504.29. C 28.58%, H 3.80%, N 11.11%, O 31.73%, P 18.43%, S 6.36%. One of the bioactive forms of thiamine; concentrated in neurons and in excitable tissues such as skeletal muscle. Believed to have a noncofactor role in nerve function. Prepn: L. Velluz *et al., Bull. Soc. Chim. France* **15**, 871 (1948); Roux *et al., Bull. Soc. Chim. Biol.* **30**, 592 (1948); M. Viscontini *et al., Helv. Chim. Acta* **32**, 1478 (1949). Relation to thiamine diphosphate: L. Velluz *et al., J. Biol. Chem.* **180**, 1137 (1949). Identification in rat liver: A. Rossi-Fanelli *et al., Science* **116**, 711 (1952). Biosynthesis by baker's yeast: K.-H. Kiessling, *Nature* **172**, 1187 (1953). Enzymatic synthesis: J. R. Cooper, K. Nishino, *Methods Enzymol.* **122**, 24 (1986). Chromatographic dctermn in brain extracts: J. R. Cooper, T. Matsuda, *ibid.* 20. Review of metabolism, tissue distribution and potential neurophysiological role: J. R. Cooper, J. H. Pincus, *Neurochem. Res.* **4**, 223-239 (1979); L. Bettendorff, *Metabol. Brain Dis.* **9**, 183-209 (1994).

Hemihydrate, very hygroscopic, rhomb-shaped microcrystals from water-alcohol-acetone, dec 228-232°. Freely sol in water. Insol in the usual organic solvents.

9434. Thiamiprine. *6-[(1-Methyl-4-nitro-1H-imidazol-5-yl)thio]-1H-purin-2-amine; 2-amino-6-[(1-methyl-4-nitroimidazol-5-yl)thio]purine;* 2-amino-6-[1'-methyl-4'-nitro-5'-imidazolyl)mercaptopurine; Guaneran. $C_9H_8N_8O_2S$; mol wt 292.28. C 36.98%, H 2.76%, N 38.34%, O 10.95%, S 10.97%. Prepd from thioguanine and 1-methyl-4-nitro-5-chloroimidazole: Hitchings, Elion, U.S. pat. **3,056,785** (1962 to Burroughs Wellcome).

Crystals, dec slowly > 200°. uv max: 320 nm at pH 1; 315 nm at pH 11.

THERAP CAT: Antineoplastic.

9435. Thiamorpholine. *Thiomorpholine;* tetrahydro-1,-4-thiazine; 1,4-thiazan; parathiazan. C_4H_9NS; mol wt 103.19. C 46.56%, H 8.79%, N 13.57%, S 31.08%. Prepd by heating β,β'-dichlorodiethyl sulfide with alcoholic ammonia under pressure: Davies, *J. Chem. Soc.* **117**, 298, 306 (1920).

Mobile liquid. Strong odor resembling that of piperidine. Absorbs CO_2 from air. bp$_{758}$ 169°; bp$_{743}$ 166-167°. Volatile with steam. Miscible with water and many organic liquids.

Hydrochloride, $C_4H_9NS.HCl$, hygroscopic needles from alcohol + ether, dec 163°. Freely sol in water.

Picrolonate, $C_4H_9NS.C_{10}H_8N_4O_5$, deep orange prisms from alcohol, dec 242° (darkens at 210°).

9436. Thiamphenicol. *[R-(R*,R*)]-2,2-Dichloro-N-[2-hydroxy-1-(hydroxymethyl)-2-[4-(methylsulfonyl)phenyl]ethyl]acetamide;* D-*d-threo*-2-dichloroacetamido-1-(4-methylsulfonyl)phenyl-1,3-propanediol; dextrosulphenidol; Win 5063-2; 8053CB; Hyrazin; Igralin; Neomyson; Rigelon; Thiamcol; Thionicol; Thiophenicol; Urfamycine; Urophenil. $C_{12}H_{15}Cl_2NO_5S$; mol wt 356.23. C 40.46%, H 4.24%, Cl 19.90%, N 3.93%, O 22.46%, S 9.00%. Synthesis: Cutler *et al., J. Am. Chem. Soc.* **74**, 5475 (1952); Suter *et al., ibid.* **75**, 4330 (1953); Suter, U.S. pats. **2,759,927; 2,759,970-1, 2, 6** (1956 to Sterling Drugs); Brit. pat. **770,277** (1957 to Parke, Davis).

Crystals, mp 164.3-166.3°. $[\alpha]_D^{25}$ +12.9° (ethanol). uv max (95% ethanol): 224, 266, 274 nm (ϵ 13,700, 800, 700). Appreciably sol in water. Soluble in alcohol.

DL-Form, *racephenidol, racephenicol, Dexawin, Thiocymetin.*

THERAP CAT: Antibacterial.

THERAP CAT (VET): D-Form as antimicrobial; DL-form in control of fowl cholera.

9437. Thiamylal. *Dihydro-5-(1-methylbutyl)-5-(2-propenyl)-2-thioxo-4,6(1H,5H)-pyrimidinedione; 5-allyl-5-(1-methylbutyl)-2-thiobarbituric acid;* thioseconal. $C_{12}H_{18}N_2$-O_2S; mol wt 254.35. C 56.67%, H 7.13%, N 11.01%, O 12.58%, S 12.61%. Prepn: **Brit.** pat. **613,704** (1948 to Lilly); Abe *et al., J. Pharm. Soc. Japan* **75**, 891 (1955); Izumi, Nakanishi, **Japan.** pat. **8785('56)** (to Yoshitomi); U.S. pat. **2,876,225** (1959 to Abbott).

Crystals from dil ethanol, mp 132-133°.

Sodium salt, $C_{12}H_{17}N_2NaO_2S$, *Surital.*

Caution: May be habit forming. This is a controlled substance (depressant) listed in the U.S. Code of Federal Regulations, Title 21 Parts 329.1 and 1308.13 (1995).

THERAP CAT: Anesthetic (intravenous).
THERAP CAT (VET): Anesthetic (intravenous).

9438. Thianaphthene. *Benzo[b]thiophene;* benzothiofuran. C_8H_6S; mol wt 134.20. C 71.60%, H 4.51%, S 23.89%. Occurs in lignite tar. Catalytic synthesis from styrene and hydrogen sulfide: Moore, Greensfelder, *J. Am. Chem. Soc.* **69,** 2008 (1947); from ethyl benzene and hydrogen sulfide: Hansch, Hawthorne, *ibid.* **70,** 2495 (1948).

Leaflets. Odor similar to that of naphthalene. mp 32°; bp_{760} 221°; bp_{20} 103-105°. Volatile with steam. Sol in the usual organic solvents.

USE: Manuf of pharmaceuticals, thioindigo.

9439. Thiarubrines. $C_{13}H_8S_2$; mol wt 228.34. C 68.38%, H 3.53%, S 28.09%. Class of dark red bioactive dithiacyclohexadiene polyacetylenes found in various *Asteraceae*; A and B forms are most common. Used medicinally in Africa: J. M. Watt, M. G. Breyer-Brandwijk, *The Medicinal and Poisonous Plants of Southern and Eastern Africa,* (E. & S. Livingstone Ltd., Edinburgh, 2nd edition, 1962) pp 202-205; leaves are consumed by chimpanzees possibly as a pharmaceutic: E. Rodriguez *et al., Experientia* **41,** 419 (1985). Isoln: F. Bohlmann, K.-M. Kleine, *Ber.* **98,** 3081 (1965); and separation by HPLC: R. A. Norton *et al., Phytochemistry* **24,** 356 (1985); from *Ambrosia chamissonis:* S. Ellis *et al., ibid.* **33,** 224 (1993). NMR spectra and biosynthetic study of A: M. L. Gomez-Barrios *et al., ibid.* **31,** 2703 (1992). Synthesis of A: M. Koreeda, W. Yang, *J. Am. Chem. Soc.* **116,** 10793, (1994); of B: E. Block *et al., ibid.* **9403.** Antibiotic properties of A: G. H. N. Towers *et al., Planta Med.* **51,** 225 (1985); photo-enhancement of activity: G. P. Constabel, G. H. N Towers, *ibid.* **55,** 35 (1989). Antiviral properties of A: J. B. Hudson *et al., ibid* **52,** 51 (1986); photo-enhancement of activity against HIV: J. B. Hudson *et al., Photochem. Photobiol.* **57,** 675 (1993).

Thiarubrine A

Thiarubrin(e) A, 3-(5-hexene-1,3-diynyl)-6-(1-propynyl)-1,-2-dithiin. Red liquid. λmax (EtOH): 490 nm. uv max (EtOH): 345, 233 nm.
Thiarubrin(e)B, 3-(3-buten-1-ynyl)-6-(1,3-pentadiynyl)-1,2-dithiin. λmax (EtOH): 490 nm. uv max (EtOH): 345, 243 nm.

9440. Thiazesim. *5-[2-(Dimethylamino)ethyl]-2,3-dihydro-2-phenyl-1,5-benzothiazepin-4(5H)-one;* thiazenone; tiazesim. $C_{19}H_{22}N_2OS$; mol wt 326.46. C 69.90%, H 6.79%, N 8.58%, O 4.90%, S 9.82%. Prepn of the hydrochloride: Krapcho *et al., J. Med. Chem.* **6**(5), 544 (1963); Krapcho, U.S. pat. 3,075,967 (1963 to Olin Mathieson). Metabolism: J. Dreyfuss *et al., J. Pharm. Sci.* **57,** 1497, 1505 (1968).

Hydrochloride, $C_{19}H_{22}N_2OS.HCl$, *SQ-10496, Altinil.* Crystals from acetonitrile, mp 222-224°.

THERAP CAT: Antidepressant.

9441. Thiazinamium Methylsulfate. *N,N,N,α-Tetramethyl-10H-phenothiazine-10-ethanaminium methyl sulfate; trimethyl(1-methyl-2-phenothiazin-10-ylethyl)ammonium methyl sulfate;* trimethyl[1-methyl-2-(10-phenothiazinyl)ethyl]ammonium methyl sulfate; N-[β-(10-phenothiazinyl)propyl]trimethylammonium methyl sulfate; RP-3554; Multergan; Padisal. $C_{19}H_{26}N_2O_4S_2$; mol wt 410.56. C 55.59%, H 6.38%, N 6.82%, O 15.59%, S 15.62%. Prepd by treating promethazine with dimethyl sulfate: **Brit.** pat. 641,452 (1950 to Rhône-Poulenc). Crystal structure: P. Marsau, Y. Cam, *Acta Crystallogr.* **B29,** 980 (1973). Determn in body fluids: J. H. G. Jonkman *et al., J. Pharm. Pharmacol.* **27,** 849 (1975). Bioavailability: *idem et al., Clin. Pharmacol. Ther.* **21,** 457 (1977). Pharmacokinetics: *idem et al., Arzneimittel-Forsch.* **23,** 223 (1983); *idem et al., Int. J. Clin. Pharmacol. Ther. Toxicol.* **21,** 454 (1983).

Crystals, mp 206-210° (some dec). Discolors on exposure to light. Soly in water (25°): ~10%. Freely sol in abs ethanol; sparingly sol in acetone. Practically insol in ether, benzene.

THERAP CAT: Antihistaminic.

9442. Thiazole. C_3H_3NS; mol wt 85.13. C 42.33%, H 3.55%, N 16.45%, S 37.67%. First described as the pyridine of the thiophene series: A. Hantzsch, J. Weber, *Ber.* **20,** 3118 (1887); A. Hantzsch, *Ann.* **249,** 1 (1888). Book: *The Chemistry of Heterocyclic Compounds,* A. Weissberger, E. C. Taylor, Eds., vol. **34** (New York, Wiley, 1979).

Colorless or pale yellow liquid. Characteristic foul odor; d^{17} ~1.20; bp 115-118°. Slightly sol in water; sol in many organic solvents. Forms compds with auric, mercuric, and platinic chlorides.

9443. Thiazolinobutazone. *4-Butyl-1,2-diphenyl-3,5-pyrazolidinedione compd with 4,5-dihydro-2-thiazolamine (1:1);* 2-thiazoline-2-ammonium 4-*n*-butyl-1,2-diphenyl-3,5-pyrazolidinedionate; phenylbutazone 2-amino-2-thiazoline salt; TZB; LAS-11871; Fordonal. $C_{22}H_{26}N_4O_2S$; mol wt 410.54. C 64.36%, H 6.38%, N 13.65%, O 7.79%, S 7.81%. A heterocyclic derivative of phenylbutazone, *q.v.* Prepn: J. Moragues *et al., Arzneimittel-Forsch.* **24,** 1785 (1974). Pharmacology: M. Márquez, D. J. Roberts, *ibid.* 1786, 1790; M. Colombo *et al., ibid.* **26,** 1347 (1976).

White crystals, mp 164-166°. LD_{50} orally in rats, mice: 1425, 1650 mg/kg (Colombo).

THERAP CAT: Anti-inflammatory.

9444. Thiazolsulfone. *5-[(4-Aminophenyl)sulfonyl]-2-thiazolamine; 2-amino-5-sulfanilylthiazole;* 4-aminophenyl-2'-aminothiazolyl-5'-sulfone; thiazosulfone; thiazolesulfone;

Promizole. $C_9H_9N_3O_2S_2$; mol wt 255.32. C 42.34%, H 3.55%, N 16.46%, O 12.53% S 25.12%. Prepd from the reaction products of *p*-nitrobenzenesulfonyl chloride and 2 mols of 2-aminothiazole: Bambas, *J. Am. Chem. Soc.* **67**, 671 (1945); Bambas, U.S. pat. **2,389,126** (1945 to Parke, Davis).

Fine needles from alc, mp 219-221° (some decompn). Sol in water at pH 6.5 to the extent of 30-40 mg per 100 ml at 28-30°. Freely sol in acetone, dioxane, 70% alc, dil acids. Moderately sol in abs alcohol, ethyl acetate and ether. It is a very weak acid, dissolves in 10% alkali (decompn) and forms an alkali metal salt at pH 10.

THERAP CAT: Antibacterial.

9445. Thiazol Yellow G. *2,2'-(1-Triazene-1,3-diyldi-4,-1-phenylene)bis[6-methyl-7-benzothiazolesulfonic acid] disodium salt; C.I. Direct Yellow 9;* 2,2'-[(diazoamino)di-*p*-phenylene]bis[6-methyl-7-benzothiazolesulfonic acid] disodium salt; C.I. 19540; Chlorazol Yellow 2G; Clayton yellow; Diazamine Golden Yellow T; Titan yellow. $C_{28}H_{19}$-$N_5Na_2O_6S_4$; mol wt 695.73. C 48.34%, H 2.75%, N 10.07%, Na 6.61%, O 13.80% S 18.44%. Prepn from 2-(*p*-amino-phenyl)-6-methyl-7-benzothiazolesulfonic acid ("dehydro-thio-*p*-toluidinesulfonic acid"): *Beilstein* vol. **27**, 2nd suppl, 509; F. J. Welcher, *Organic Analytical Reagents* vol. **4** (Van Nostrand, New York, 1948) p 391; H. King *et al.*, *Analyst* **92**, 695 (1967). *See also: Colour Index* vol. **2** (3rd ed., 1971) p 2010; H. J. Conn's *Biological Stains*, R. D. Lillie, Ed. (Williams & Wilkins, Baltimore, 9th ed., 1977) p 371.

Yellowish-brown powder. Sol in water, or alc (yellow solns), NaOH (reddish-yellow soln), H_2SO_4 (brownish-yellow soln). *Protect from light.*

USE: Dyeing cotton, viscose rayon, natural silk; as biological stain; as an analytical reagent, most often for the determination of Mg; as pH indicator, yellow 11.0 to red 13.0; as fluorescent dye for microscopy.

9446. Thibenzazoline. *1,3-Dihydro-1,3-bis(hydroxymethyl)-2H-benzimidazole-2-thione;* 1,3-bis(hydroxymethyl)-2-benzimidazolinethione; 2-mercaptobenzimidazole-1,3-dimethylol; Thyrcocordon. $C_9H_{10}N_2O_2S$; mol wt 210.26. C 51.41%, H 4.79%, N 13.32%, O 15.22% S 15.25%. Prepn: Monti, Venturi, *Gazz. Chim. Ital.* **76**, 365 (1946).

Very bitter. mp 160-162°. Soluble dil in alkalies.
THERAP CAT: Antihyperthyroid.

9447. Thidiazuron. *N-Phenyl-N'-1,2,3-thiadiazol-5-ylurea;* 5-(*N*-phenylcarbamoylamino)-1,2,3-thiadiazole; thiadiazuron; TDZ; SN-49537; Dropp. $C_9H_8N_4OS$; mol wt 220.25. C 49.08%, H 3.66%, N 25.44%, O 7.26% S 14.56%. Non-purine containing urea derivative with cytokinin activity. Prepn: H. Schulz, F. Arndt, Ger. pat. **2,214,632**; *eidem*, U.S. pat. **3,883,547** (1973, 1975 both to Schering AG). Field trial in cotton: A. R. Hopkins, R. F. Moore, *J. Econ. Entomol.* **73**, 768 (1980). Effect of temperature and adjuvants on activity in cotton: C. E. Snipes, G. D. Wills, *Weed Sci.* **42**, 13 (1994). Review of effect on shoot forma-

tion and axillary shoot proliferation: C.-Y. Lu, *In Vitro Cell. Dev. Biol.* **29P**, 92-96 (1993); on woody plant tissue culture: C. A. Huetteman, J. E. Preece, *Plant Cell, Tissue Organ Cult.* **33**, 105-119 (1993).

Crystals from isopropanol, mp 217° (dec). Very sol in DMSO, DMF, acetone, cyclohexanone, isophorone. Slightly sol in aliphatic and aromatic hydrocarbons and water.

USE: Cotton defoliant; plant growth regulator in tissue culture.

9448. Thienamycin. *[5R-[5α,6α(R*)]]-3-[(2-Aminoethyl)thio]-6-(1-hydroxyethyl)-7-oxo-1-azabicyclo[3.2.0]hept-2-ene-2-carboxylic acid.* $C_{11}H_{16}N_2O_4S$; mol wt 272.33. C 48.52%, H 5.92%, N 10.29%, O 23.50% S 11.77%. The first member of a family of des-thia-carbapenem nucleus antibiotics having a thioethylamine side-chain on the enamine portion of the fused 5-membered ring. Produced by *Streptomyces cattleya*: J. S. Kahan *et al.*, U.S. pat. **3,950,357** (1976 to Merck & Co.). Discovery, isoln, taxonomy, physical properties: *eidem*, *J. Antibiot.* **32**, 1 (1979). Structure, absolute configuration: G. Albers-Schönberg *et al.*, *J. Am. Chem. Soc.* **100**, 6491 (1978). Synthesis of the (±)-form: D. B. R. Johnston *et al.*, *ibid.* 313; F. A. Bouffard *et al.*, *J. Org. Chem.* **45**, 1130 (1980); S. M. Schmitt *et al.*, *ibid.* 1135, 1142; D. G. Melillo *et al.*, *Tetrahedron Letters* **21**, 2783 (1980); *eidem*, *ibid.* **22**, 913 (1981); M. Shiozaki, T. Hiraoka, *Tetrahedron* **38**, 3457 (1982). Alternate prepn of key intermediate: J. D. Buynak *et al.*, *Chem. Commun.* **1986**, 941. Stereocontrolled synthesis of the naturally occurring (+)-form: T. Salzmann *et al.*, *J. Am. Chem. Soc.* **102**, 6161 (1980); S. Karady *et al.*, *ibid.* **103**, 6765 (1981). Alternate synthetic routes: S. T. Hodgson *et al.*, *Tetrahedron* **41**, 5871 (1985); T. Iimori, M. Shibasaki, *Tetrahedron Letters* **26**, 1523 (1985); T. Chiba, T. Nakai, *ibid.* 4647; D. J. Hart, D. C. Ha, *ibid.* 5493; T. Kametani *et al.*, *J. Org. Chem.* **50**, 2327 (1985); H. Maruyama, T. Hiraoka, *ibid.* **51**, 399 (1986); D. G. Melillo *et al.*, *ibid.* 1498. Continuous production in immobilized cell systems: E. J. Arcuri *et al.*, *Biotechnol. Bioeng.* **28**, 842 (1986). Biosynthesis: J. M. Williamson *et al.*, *J. Biol. Chem.* **260**, 4637 (1985). *In vitro* antibacterial activity: F. P. Tally *et al.*, *Antimicrob. Ag. Chemother.* **14**, 436 (1978); S. S. Weaver *et al.*, *ibid.* **15**, 518 (1979). Evaluation of *in vitro* and *in vivo* activities: H. Kropp *et al.*, *Antimicrob. Ag. Chemother* **17**, 993 (1980). Comparative study vs gram-positive and gram-negative aerobic and anaerobic species and β-lactamase stability: H. C. Neu, P. Labthavikul, *ibid.* **21**, 180 (1982). *N*-Acyl derivatives as anti-inflammatory agents: J. B. Doherty *et al.*, U.S. pat. **4,465,687** (1984 to Merck & Co). Pharmacokinetics, bacteriological efficacy: P. Patamasucon, G. H. McCracken, *Antimicrob. Ag. Chemother.* **21**, 390 (1982). Review of early syntheses of the carbapenems: T. Kametani, *Heterocycles* **17**, 463-506 (1982).

White hygroscopic solid. $[\alpha]_D^{27}$ +82.7° (c = 1.0 in water). uv max (water, pH 4-8): 296.5 nm (ε 7900); (pH 2): 309 nm; (pH 12): 300.5 nm. Freely sol in water, sparingly sol in methanol. In dilute soln, stability is optimal between pH 6-7, declining with unusual rapidity above that range. Susceptible to inactivation by dilute solns of hydroxylamine and cysteine.

N-Formimidoylthienamycin monohydrate, *see* imipenem.

9449. Thiethylperazine. *2-(Ethylthio)-10-[3-(4-methyl-1-piperazinyl)propyl]phenothiazine;* 3-ethylmercapto-10-(1'-methylpiperazinyl-4'-propyl)phenothiazine. $C_{22}H_{29}N_3S_2$;

mol wt 399.62.   C 66.12%, H 7.31%, N 10.51%, S 16.05%.
Prepn: Bourquin *et al., Helv. Chim. Acta* **41**, 1072 (1958).

Crystals from acetone, mp 62-64°.   bp$_{0.01}$ 227°.
Dimaleate, $C_{22}H_{29}N_3S_2 \cdot 2C_4H_4O_4$, *GS-95, NSC-130044, Torecan Maleate, Toresten, Tresten.* Crystals from methanol, dec 188-190°.
Dihydrochloride, $C_{22}H_{29}N_3S_2 \cdot 2HCl$, crystals from ethanol, mp 214-216°.
Dimalate, $C_{22}H_{29}N_3S_2 \cdot 2C_4H_6O_5$, crystals from ethanol, mp 139°.
THERAP CAT: Antiemetic.

9450.   Thimerfonate Sodium. *Ethyl(4-mercaptoben-zenesulfonato-S^4)mercury sodium salt;* sodium *p*-ethylmer-curithiophenylsulfonate; ethyl[(*p*-sulfophenyl)thio]mercury sodium salt; Sulfo-Merthiolate. $C_8H_9HgNaO_3S_2$; mol wt 440.87.   C 21.80%, H 2.06%, Hg 45.50%, Na 5.21%, O 10.89%, S 14.55%.   Prepn: Waldo, *J. Am. Chem. Soc.* **53**, 992 (1931); Kharasch, U.S. pat. 1,672,615.

Powder, very sol in water. Forms a stable solution.
THERAP CAT: Anti-infective (topical).

9451.   Thimerosal. *Ethyl[2-mercaptobenzoato(2—)-O, S]mercurate(1—) sodium; [(o-carboxyphenyl)thio]ethylmer-cury sodium salt;* sodium ethylmercurithiosalicylate; thio-mersalate; mercurothiolate; Merthiolate; Merzonin; Vita-septol. $C_9H_9HgNaO_2S$; mol wt 404.82.   C 26.70%, H 2.24%, Hg 49.55%, Na 5.68%, O 7.90%, S 7.92%.   Prepd by reacting ethylmercuric chloride (or ethylmercuric hydrox-ide) with thiosalicylic acid: Kharasch, U.S. pat. 1,672,615 (1928); Trikojus, *Nature* **158**, 472 (1946); Swirska *et al., Przemysl Chem.* **39**, 371 (1960), *C.A.* **55**, 3507a (1961). Tox-icity: Mason *et al., Clin. Toxicol.* **4**, 185 (1971).

Cream-colored, crystalline powder. Stable in air, but not in sunlight. One gram dissolves in about 1 ml water, in about 8 ml alcohol. Practically insol in ether and benzene. Stabilization of solns with EDTA: Davisson, U.S. pat. **2,864,844** (1958 to Lilly). pH of 1% aq soln: 6.7. LD$_{50}$ s.c. in rats: 98 mg/kg (Mason).
USE: Pharmaceutic aid (preservative).
THERAP CAT: Anti-infective.
THERAP CAT (VET): Antibacterial, antifungal (topical).

9452.   Thioacetaldehyde. *2,4,6-Trimethyl-s-trithiane;* trithioacetaldehyde. $C_6H_{12}S_3$; mol wt 180.36.   C 39.96%, H 6.71%, S 53.34%. Occurs in α- and β-forms. Prepn: Bau-mann, Fromm, *Ber.* **22**, 2600 (1889); Fromm, Engler, *Ber.* **58**, 1916 (1925). Molecular structure: Hassel, Viervoll, *Acta Chem. Scand.* **1**, 164 (1947). Stereochemistry: Schön-berg, Barakat, *J. Chem. Soc.* **1947**, 693.

α-Form, monoclinic plates, mp 101°.
β-Form, needles from acetone, mp 126°.

9453.   Thioacetamide. *Ethanethioamide.* C_2H_5NS; mol wt 75.13.   C 31.97%, H 6.71%, N 18.64%, S 42.68%. CH$_3$-CSNH$_2$.   Prepd by heating ammonium acetate and alumi-num sulfide: Kindler, Finndorf, *Ber.* **54**, 1080 (1921); from acetonitrile + H_2S: Kindler, *Ann.* **431**, 203 (1923); from acetamide + K_3PS_4: Schultz, Ranke, *Arch. Pharm.* **294**, 82 (1961). Toxicity data: Ambrose *et al., J. Ind. Hyg. Toxicol.* **31**, 158 (1949). Review of uses: H. F. Walton, *Thioacet-amide as Analytical Reagent,* Arapahoe Chemicals, Inc., Boulder, Colorado.
Crystals from benzene, mp 113-114°. Slight odor of mer-captans. uv max (water): 210, 261, 318 nm (log ε 3.66, 4.08, 1.8). Soly in water at 25° 16.3 g/100 ml; in ethanol 26.4 g/100 g. Sparingly sol in ether. MLD orally in rats: 200 mg/kg (Ambrose).
Note: This substance may reasonably be anticipated to be a carcinogen: *Seventh Annual Report on Carcinogens* (PB95-109781, 1994) p 379.
USE: Substitute for H$_2$S in laboratory qualitative analyses.

9454.   Thioacetic Acid. *Ethanethioic acid; thiolacetic acid;* thiacetic acid. C_2H_4OS; mol wt 76.12.   C 31.56%, H 5.30%, O 21.02%, S 42.13%. CH$_3$COSH.   Prepn by distil-ling glacial acetic acid with phosphorus pentasulfide: Keku-lé, *Ann.* **90**, 309 (1854); from acetic anhydride and hydrogen sulfide: Ellingboe, *Org. Syn.* **31**, 105 (1951).
Yellow, fuming liquid; pungent odor. d$_4^{10}$ 1.075; bp 93°. Not solidified at −17°. Sol in water, particularly hot; very sol in alcohol.

9455.   Thiobarbital. *5,5-Diethyldihydro-2-thioxo-4,6-(1H,5H)-pyrimidinedione; 5,5-diethyl-2-thiobarbituric acid;* Ibition. $C_8H_{12}N_2O_2S$; mol wt 200.26.   C 47.98%, H 6.04%, N 13.99%, O 15.98%, S 16.01%. Prepn: Fischer, Dilthey, *Ann.* **335**, 350 (1904); Carrington, *J. Chem. Soc.* **1944**, 124. Metabolism in man: Bush *et al., J. Pharmacol. Exp. Ther.* **134**, 110 (1961).

Pale yellow needles from water, mp 180°. Sol in about 88 parts hot water, in ethanol, chloroform, ether, acetone, am-monia, alkalies. Sparingly sol in toluene. Practically insol in benzene.
THERAP CAT: Antihyperthyroid.

9456.   Thiobenzyl Alcohol. *Benzenemethanethiol;* benz-yl mercaptan; α-toluenethiol. C_7H_8S; mol wt 124.21.   C 67.69%, H 6.49%, S 25.82%. $C_6H_5CH_2SH$.
Colorless liquid; odor of leek; d^{20} 1.058; bp 194-195°. Oxidizes in air to dibenzyl disulfide.
Caution: Can cause mild irritation to mucous mem-branes.

9457.   Thiobutabarbital. *5-Ethyldihydro-5-(1-methyl-propyl)-2-thioxo-4,6(1H,5H)-pyrimidinedione; 5-sec-butyl-5-ethyl-2-thiobarbituric acid; 5-ethyl-5-(1-methylpropyl)-2-thiobarbituric acid;* thibutabarbital. $C_{10}H_{16}N_2O_2S$; mol wt 228.32.   C 52.61%, H 7.06%, N 12.27%, O 14.02%, S 14.04%. Prepn: Volwiler, Tabern, U.S. pats. 2,153,729; 2,153,731 (both 1939 to Abbott).

Crystals, mp 163-165°.

Sodium salt, $C_{10}H_{15}N_2NaO_2S$, **Inactin, Brevinarcon.** Crystalline, slightly hygroscopic solid. Readily sol in water.

Note: This is a controlled substance (depressant) listed in the U.S. Code of Federal Regulations, Title 21 Part 1308.13 (1995).

THERAP CAT: Anesthetic (intravenous).

9458. Thiocarbamizine. *2,2'-[[[4-[(Aminocarbonyl)-amino]phenyl]arsinidene]bis(thio)]bis[benzoic acid]; (p-urei-dophenylarsylenedithio)di-o-benzoic acid; (p-ureidobenzene-arsylenedithio)di-o-benzoic acid; 4-carbamidophenyl bis[o-carboxyphenylthio]arsenite; bis[o-carboxyphenylmercapto]-(p-ureidophenyl)arsine; thiocarbamisin.* $C_{21}H_{17}AsN_2O_5S_2$; mol wt 516.43. C 48.84%, H 3.32%, As 14.51%, N 5.42%, O 15.49%, S 12.42%. Prepd by heating o-HSC$_6$H$_4$COOH and Carbarsone oxide in aq alkali, purified by repeated precipitation with acid and dissolving in alkali: Rohrmann, **U.S. pat. 2,516,831** (1950 to Lilly). *Compare* Arsenamide, Thiocarbarsone, Carbarsone.

White, cryst powder. Sparingly sol in water, alcohols. Insol in acids. Freely sol in dil alkali [*see* Arsenamide].

THERAP CAT: Antiamebic.

9459. Thiocarbarsone. *2,2'-[[[4-[(Aminocarbonyl)ami-no]phenyl]arsinidene]bis(thio)]bis[acetic acid]; (p-ureido-phenylarsylenedithio)diacetic acid; 4-carbamidophenyl bis-[carboxymethylthio]arsenite; bis[carboxymethylmercapto]-(p-ureidophenyl)arsine.* $C_{11}H_{13}AsN_2O_5S_2$; mol wt 392.29. C 33.68%, H 3.34%, As 19.10%, N 7.14%, O 20.39%, S 16.35%. Prepd by heating HSCH$_2$COOH and Carbarsone oxide in aq alkali, purified by repeated precipitation with acid and dissolving in alkali: Rohrmann, **U.S. pat. 2,516,831** (1950 to Lilly). *Compare* Arsenamide.

White, cryst powder. Sparingly sol in water, alcohols. Insol in acids. Freely sol in dil alkali [*see* Arsenamide].

THERAP CAT: Antiamebic.

9460. Thiocolchicine. *(S)-N-[5,6,7,9-Tetrahydro-1,2,3-trimethoxy-10-(methylthio)-9-oxobenzo[a]heptalen-7-yl]acet-amide.* $C_{22}H_{25}NO_5S$; mol wt 415.51. C 63.59%, H 6.06%, N 3.37%, O 19.25%, S 7.72%. Prepn: Velluz, Muller, *Bull. Soc. Chim. France* **1954,** 755; Muller, Velluz, U.S. pat. **2,820,029** (1958 to UCLAF). Prepn of 2-glucoside analog: Velluz, Muller, *Bull. Soc. Chim. France* **1955,** 194; Muller, Velluz, *loc. cit.*

Yellow cubic crystals from ethyl acetate, mp 192-194°. $[\alpha]_D^{20}$ −221. Sol in ethanol, acetone, chloroform. Practically insol in water, ether.

2-Glucoside analog, $C_{27}H_{33}NO_{10}S$, **thiocolchicoside, 2-de-methoxy-2-glucosidoxythiocolchicine, Coltramyl, Coltrax, Miorel, Musco-Ril.** Crystals from ethanol + 1N NaOH, dec 220°. $[\alpha]_D$ −609° (water), −240° (ethanol).

THERAP CAT: Thiocolchicoside as muscle relaxant (skele-tal).

9461. Thiocresol. *Methylbenzenethiol;* tolylmercaptan; toluenethiol. C_7H_8S; mol wt 124.21. C 67.69%, H 6.49%, S 25.82%. CH$_3$C$_6$H$_4$SH.

All three isomers boil at ~195°. Insol in water. Sol in alcohol or ether.

o-Thiocresol, mp 15°; volatile with steam.

m-Thiocresol, liquid, mp below −20°.

p-Thiocresol, leaflets, mp 43-44°.

THERAP CAT: Antiseptic; dermatologic.

9462. Thioctic Acid. *1,2-Dithiolane-3-pentanoic acid; 1,2-dithiolane-3-valeric acid; α-lipoic acid; 6,8-thioctic acid; 5-(1,2-dithiolan-3-yl)valeric acid; 5-[3-(1,2-dithiolanyl)]pen-tanoic acid; δ-[3-(1,2-dithiacyclopentyl)]pentanoic acid; pro-togen A; acetate replacing factor; pyruvate oxidation factor; Biletan; Thioctacid; Thioctan; Tioctan.* $C_8H_{14}O_2S_2$; mol wt 206.33. C 46.57%, H 6.84%, O 15.51%, S 31.08%. Growth factor for many bacteria and protozoa; prosthetic group, coenzyme, or substrate in plants, microorganisms, and ani-mal tissues. Isoln of naturally occurring d-form: L. J. Reed *et al., Science* **114,** 93 (1951); *eidem, J. Am. Chem. Soc.* **75,** 1267 (1953); Patterson *et al., ibid.* **76,** 1823 (1954). Syn-theses of dl-form: Bullock *et al., ibid.* **74,** 1868, 3455 (1952); Hornberger *et al., ibid.* 2382; Reed, U.S. pats. **2,980,716** and **3,049,549** (1961, 1962 to Res. Corp.); Lewis, Raphael, *J. Chem. Soc.* **1962,** 4263; Ose *et al.,* U.S. pat. **3,223,712** (1965 to Yamanouchi); J. Tsuji *et al., J. Org. Chem.* **43,** 3606 (1978). Biosynthesis via linoleic acid: J. P. Carreau in *Methods in Enzymology* vol. **62,** D. McCormick, L. Wright, Eds. (Academic Press, New York, 1974) pp 152-158. Enan-tioselective synthesis of d-form: P. C. Bulmanpage *et al., Chem. Commun.* **1986,** 1408. Clinical study in treatment of Wilson's disease: S. F. Gomes da Costa, *Arzneimittel-For-sch.* **20,** 1210 (1970). Use in treatment of mushroom poi-soning: R. Plotzker *et al., Am. J. Med. Sci.* **283,** 79 (1982); J. P. Hanrahan, M. A. Gordon, *J. Am. Med. Assoc.* **251,** 1057 (1984). *Reviews:* Wagner, Folkers, *Vitamins and Coenzymes* (Interscience, New York, 1964) pp 244-263; Schmidt *et al., Angew. Chem. Int. Ed.* **4,** 846 (1965); Sch-midt *et al., Advan. Enzymol. Relat. Areas Mol. Biol.* **32,** 423 (1969).

d-Form

Sodium salt, $C_8H_{13}NaO_2S_2$, white powder, sol in water. pH of aq solns about 7.4.

d-Form, crystals by vacuum sublimation (at 85-90° and 25 microns). mp 46-48° (microblock). $[\alpha]_D^{23}$ +104° (c = 0.88 in benzene). uv max (methanol): 333 nm (ε 150). pKa 5.4. Practically insol in water. Sol in fat solvents.

dl-Form, yellow needles from cyclohexane, mp 60-61°. bp 160-165°. uv spectrum: Calvin, *Fed. Proc.* **13,** 703 (1954).

Practically insol in water. Sol in fat solvents. Forms a water-soluble sodium salt.

l-Form, crystals from cyclohexane, mp 45-47.5° (microblock). $[\alpha]_D^{23}$ −113° (c = 1.88 in benzene). uv max (methanol): 330 nm (ϵ 140).

Ethylenediamine, *Tioctidasi.*

THERAP CAT: Treatment of liver disease; antidote to poisonous mushrooms (*Amanita* species).

9463. Thiocyanate Sodium. Sodium sulfocyanate; sodium rhodanide; sodium thiocyanate. CNNaS; mol wt 81.07. C 14.81%, N 17.28%, Na 28.36%, S 39.55%. NaSCN. Contains at least 98% NaSCN. Toxicity study: Anderson, Chen, *J. Am. Pharm. Assoc.* **29**, 152 (1940).

Colorless or white, deliquesc crystals. mp about 300°. Sol in about 0.6 part water; freely sol in alcohol, acetone. When dissolved in water the temp is considerably lowered. The solns are neutral. *Keep well closed.* LD_{50} orally in rats: 764 mg/kg; i.v. in mice: 484 mg/kg (Anderson, Chen).

USE: Manuf other thiocyanates, especially organic.

9464. Thiocyanic Acid. Hydrogen thiocyanate; Rhodanwasserstoffsäure (German). CHNS; mol wt 59.09. C 20.33%, H 1.71%, N 23.70%, S 54.26%. HSCN. Thiocyanic acid is believed to be a tautomeric mixture of HSCN and HNCS (*isothiocyanic acid*): Beard, Dailey, *J. Chem. Phys.* **18**, 1437 (1950). Conveniently prepd in the laboratory from KNCS + KHSO$_4$: Birckenbach, Bucher, *Ber.* **73**, 1153 (1940). Dil aq solns may be prepd from ammonium thiocyanate by the action of ion exchange resins: Klement, *Z. Anorg. Allgem. Chem.* **260**, 268 (1949). Toxicology of organic thiocyanates: Y. Yokoi, *Japan. J. Pharmacol.* **3**, 99 (1954), *C.A.* **48**, 13965f (1954).

Colorless gas or white solid depending upon degree of polymerization. Freely sol in water. Very strong acid. A 5% aq soln may be kept refrigerated for several weeks. Also sol in some organic solvents.

Ethyl ester, C$_3$H$_5$NS, *ethyl thiocyanate, ethyl sulfocyanate.* Liquid. d$_4^{23}$ 1.007. bp 146°. n$_D^{15}$ 1.4684. Insol in water. Misc with alcohol, ether. MLD in mice (mg/kg): 52 orally; 39.1 s.c.; 18.3 i.p.; 6 i.v. (Yokoi).

9465. Thiodicarb. *N,N′-[Thiobis[(methylimino)carbonyloxy]]bisethanimidothioic acid dimethyl ester;* bis-[*O*-(1-methylthioethylimino)-*N*-methylcarbamic acid]-*N,N′*-sulfide; *N,N′*-bis[1-methylthioacetaldehyde *O*-(*N*-methylcarbamoyl)oxime]sulfide; *O*-[[*N*-[*N′*-(1-methylthioethylideneiminooxycarbonyl)-*N′*-methylaminosulfenyl]-*N*-methylcarbamoyl]]-*S*-methylacetohydroximate; dicarbosulf; bismethomyl thioether; UC-51762; CGA-45156; Larvin. C$_{10}$H$_{18}$N$_4$O$_4$S$_3$; mol wt 354.48. C 33.88%, H 5.12%, N 15.81%, O 18.05%, S 27.14%. Prepn: Neth. pat. Appl. **7,508,197** (1976 to Ciba-Geigy); T. D. J. D'Silva, Ger. pat. **2,654,331**; *eidem*, U.S. pat. **4,382,957** (1977, 1983 both to Union Carbide). Activity: A. A. Sousa *et al.*, *J. Econ. Entomol.* **70**, 803 (1977). Field trials: E. P. Pieters, D. L. Pitts, *J. Ga. Entomol. Soc.* **15**, 207 (1980); J. R. Bradley, Jr., A. M. Agnello, *J. Econ. Entomol.* **81**, 706 (1988).

Crystals, mp 173-174°. LD_{50} in rats (mg/kg): 160 orally; > 1600 dermally (Sousa).

USE: Insecticide.

9466. 2,2′-Thiodiethanol. Thiodiglycol; thiodiethylene glycol; bis(hydroxyethyl)sulfide. C$_4$H$_{10}$O$_2$S; mol wt 122.19. C 39.32%, H 8.25%, O 26.19%, S 26.24%. HOCH$_2$CH$_2$SCH$_2$CH$_2$OH. Prepd from ethylene oxide and hydrogen sulfide: Chichibabin, Bestuzher, *Compt. Rend.* **200**, 242 (1935); Nenitzescu, Scarlatescu, *Ber.* **68**, 587 (1935); Headlee, cited in G. O. Curme, F. Johnston, *Glycols* (Reinhold, New York, 1952) p 103.

Liquid. d$_4^{20}$ 1.1824. mp −16°. bp$_{14}$ 168°. n$_D^{20}$ 1.519. Flash pt, open cup: 320°F (160°C). Misc with water, alc. Slightly sol in ether.

9467. Thiodiglycolic Acid. *2,2′-Thiobis[acetic acid];* dimethylsulfide-α,α′-dicarboxylic acid; mercaptodiacetic acid. C$_4$H$_6$O$_4$S; mol wt 150.16. C 32.00%, H 4.03%, O 42.62%, S 21.36%. HOOCCH$_2$SCH$_2$COOH. Prepn from sodium chloroacetate and hydrogen sulfide: Loven, *Ber.* **27**, 3059 (1894).

Crystals from water. mp 129°. Sol in water, alcohol.

USE: Detection of copper, lead, mercury, silver: Dubsky *et al.*, *C.A.* **34**, 6185 (1940).

9468. 3,3′-Thiodipropionic Acid. *3,3′-Thiobis[propanoic acid];* β,β-thiodipropionic acid; thiodihydracrylic acid; diethyl sulfide 2,2′-dicarboxylic acid. C$_6$H$_{10}$O$_4$S; mol wt 178.21. C 40.44%, H 5.66%, O 35.91%, S 17.99%. Prepn: Brit. pat. **571,628** (1945 to Am. Cyanamid); Gresham, Shaver, U.S. pat. **2,449,992** (1948 to B. F. Goodrich).

Nacreous leaflets from hot water, mp 134°. K at 25° = 7.8 × 10^{-5}. One gram dissolves in 26.9 ml water at 26°. Freely sol in hot water, alcohol, acetone.

USE: Antioxidant for soap products and polymers of ethylene. In plasticizers and lubricants. Proposed for edible fats, oils, other foods.

9469. Thioformamide. *Methanethioamide.* CH$_3$NS; mol wt 61.11. C 19.66%, H 4.95%, N 22.92%, S 52.47%. HCSNH$_2$. Prepd by the action of phosphorus pentasulfide on formamide: Willstätter, Wirth, *Ber.* **42**, 1911 (1909). Synthesis: Erlenmeyer, Menzi, *Helv. Chim. Acta* **31**, 2071 (1948); Schmitz, U.S. pat. **2,682,558** (1954 to du Pont); R. Tull, L. M. Weinstock, *Angew. Chem. Int. Ed.* **8**, 278 (1969). Spectroscopic studies: R. Sugisaki *et al.*, *J. Mol. Spectrosc.* **49**, 241 (1974); R. H. Judge *et al.*, *Can. J. Chem.* **65**, 2100 (1987). Use in synthesis of thiazoles: T. J. Cousineau, J. A. Secrist III, *J. Org. Chem.* **44**, 4351 (1979); H. Fukatsu *et al.*, *Heterocycles* **29**, 1517 (1989).

Prisms from ethyl acetate or from ether + petr ether, mp 26-27°. uv max (CH$_3$OH): 263 nm (ϵ 12500). Turns yellow and dec to a dark-colored mass. May be stored as sphere in abs ether, preferably over P$_2$O$_5$. Sol in water (with considerable cooling). Freely sol in tetrahydrofuran, alcohol, ether, acetone, ethyl acetate; sparingly sol in cold chloroform, benzene, petr ether, carbon disulfide. The decompn products are hydrogen cyanide, hydrogen sulfide, ammonia, and some solid, amorphous, sulfur-contg compounds.

Monohydrate, yellow oil.

USE: In the synthesis of the thiazole containing compds.

9470. 5-Thio-D-glucose. *5-Thio-α-D-glucopyranose;* α-D-glucothiopyranose. C$_6$H$_{12}$O$_5$S; mol wt 196.22. C 36.73%, H 6.16%, O 40.77%, S 16.34%. The nearest analog of normal D-glucose available to date. Thought to interfere with cellular transport systems utilizing D-glucose due to this structural similarity. The first chemical other than a hormone or alkylating agent which can interfere reversibly with spermatogenesis and which also has been effective against malignant cultured cells. Prepn: M. S. Feather, R. L. Whistler, *Tetrahedron Letters* **1962**, 667; Rowell, Whistler, *J. Org. Chem.* **31**, 1514 (1966); U. G. Nayak, R. L. Whistler, *ibid.* **34**, 97 (1969); Abd El-Rahman, Whistler, *Org. Prep. Proced. Int.* **5**, 245 (1973); H. Driguez, B. Henrissat, *Tetrahedron Letters* **22**, 5061 (1981). Pharmacology: D. J. Hoffman, R. L. Whistler, *Biochemistry* **7**, 4479 (1968); R. L. Whistler, W. C. Lake, *Biochem. J.* **130**, 919 (1972). Radioprotective effect on tissue *in vivo*: V. L. Schuman *et al.*, *Int. J. Radiat. Oncol. Biol. Phys.* **8**, 589 (1982). Brief review: *Science* **186**, 431 (1974).

Crystals from methanol, mp 135-136°. $[\alpha]_D^{20}$ +188° (c = 1.56 in water). LD_{50} i.p. in mice: 5.5 g/kg (Schuman).
USE: Tool for examination of D-glucose biochemistry.

9471. Thioglycerol. *3-Mercapto-1,2-propanediol; α-monothioglycerol; thioglycerin.* $C_3H_8O_2S$; mol wt 108.16. C 33.31%, H 7.46%, O 29.58%, S 29.65%. $HOCH_2CH(OH)$-CH_2SH. Prepd by heating glycerol-α-monochlorohydrin with KHS in alcoholic solution. *Ref:* Sutton, *J. Am. Med. Assoc.* **104**, 2168 (1935).
Yellowish, very viscous liq; slight sulfidic odor. d 1.295. Slightly sol in water; miscible with alcohol. Insol in ether.
THERAP CAT: Vulnerary.
THERAP CAT (VET): Has been used in 1:5000 concn as a vulnerary.

9472. Thioglycolic Acid. *Mercaptoacetic acid; thioglycollic acid.* $C_2H_4O_2S$; mol wt 92.12. C 26.08%, H 4.38%, O 34.74%, S 34.81%. $HSCH_2COOH$. Prepd by the action of sodium sulfhydrate on sodium chloroacetate; by electrolysis of dithioglycollic acid (from sodium sulfide and sodium chloroacetate). Toxicity study: Deichmann, Mergard, *J. Ind. Hyg. Toxicol.* **30**, 373 (1948).
Liquid, strong, unpleasant odor. Readily oxidized by air. d 1.325. mp −16.5°. bp_{29} 123°. bp_{15} 108°. bp_5 96°. K_1 at 25° = 2.1 × 10^{-4}; K_2 = 2.1 × 10^{-11}. Miscible with water, alcohol, ether, chloroform, benzene, and many other organic solvents. LD_{50} orally in rats: 0.15 ml/kg (Deichmann, Mergard).
Caution: Can cause severe burns and blistering of skin.
USE: Sensitive reagent for iron, molybdenum, silver, tin. With ferric iron a blue color appears, and when an alkali hydroxide is added to a soln contg ferrous salts and thioglycolic acid, a yellow precipitate forms. Used in the manuf of thioglycolates. The ammonium and sodium salts are commonly used for cold waving and the calcium salt is a depilatory. The sodium salt also is used in bacteriology in the prepn of thioglycolate media.

9473. Thioguanine. *2-Amino-1,7-dihydro-6H-purine-6-thione; 2-aminopurine-6-thiol;* Lanvis. $C_5H_5N_5S$; mol wt 167.19. C 35.92%, H 3.01%, N 41.89%, S 19.18%. Prepn: Elion, Hitchings, *J. Am. Chem. Soc.* **77**, 1676 (1955); Hitchings, Elion, U.S. pats. **2,697,709; 2,800,473; 2,884,667;** Hitchings *et al.*, U.S. pats. **3,019,224; 3,132,144** (1954, 1957, 1959, 1962, and 1964, all to Burroughs Welcome).

Needles from water, mp > 360°.
THERAP CAT: Antineoplastic.

9474. Thioguanosine. 2-Amino-6-mercapto-9-β-D-ribofuranosylpurine. $C_{10}H_{13}N_5O_4S$; mol wt 299.31. C 40.13%, H 4.38%, N 23.40%, O 21.38%, S 10.71%. Prepn: Fox *et al., J. Am. Chem. Soc.* **80**, 1669 (1948).

Hemihydrate, tiny tapered prisms from water, dec 224-227°. $[\alpha]_D^{22}$ −64° (c = 1.3 in 0.1N NaOH). uv max (pH 4-6): 257, 342 nm (ε 8820, 24800). pKa 8.33.
USE: In cancer research, *compare* 6-Mercaptopurine and Nebularine.

9475. Thiokol®. *Polysulfide rubber;* thiorubber. Polysulfide polymers prepd from dihaloalkanes and sodium polysulfide. Reviews of prepn, chemistry and applications: Berenbaum in *High Polymers,* H. Mark *et al.,* Eds., **vol. 13** entitled *Polyethers* part **3**, N. G. Gaylord, Ed. (Interscience, New York, 1962) pp 43-114; Panek, *ibid.* pp 115-224; Berenbaum in *Encyclopedia of Polymer Science and Technology* **vol. 11** (Interscience, New York, 1969) pp 425-447.

Thiokol A

Thiokol A, Ethanite, Perduren. The first commercial polysulfide polymer, prepd from ethylene dichloride and sodium polysulfide. Sulfur content 84%; d about 1.6. Mixes with natural rubber. Cured polymer retains unpleasant odor; irritating fumes evolve during manuf. Stable to the usual organic solvents and dil mineral acids. Unstable to alkalies and oxidizing substances. Of low tensile strength and abrasion resistance. Not recommended where tropic or arctic climates prevail.
Thiokol FA. Prepd from ethylene dichloride, dichlorodiethyl formal and sodium polysulfide. Sulfur content 47%; d 1.34. No odor. Excellent solvent resistant characteristics but not as good as Thiokol A. Low temperature flexibility to −50°F.
USE: In rubber and resin manuf. As lining for flexible oil pipes and self-sealing gasoline tanks.

9476. Thiolactic Acid. *2-Mercaptopropanoic acid;* 2-mercaptopropionic acid; 2-thiolpropionic acid; α-mercaptopropanoic acid. $C_3H_6O_2S$; mol wt 106.15. C 33.95%, H 5.70%, O 30.15%, S 30.21%. $CH_3CH(SH)COOH$. First prepd from α-chloropropionic acid and potassium hydrosulfide: Schacht, *Ann.* **129**, 3 (1864); prepn from α-chloropropionic acid and sodium thiosulfate: Martin, U.S. pat. **2,413,361** (1946 to Martin Labs). From sodium sulfide, sulfur and α-bromopropionic acid: Dumesnil, U.S. pat. **2,985,557** (1961 to Frank E. Jonas). *Review:* Reid, *Organic Chemistry of Bivalent Sulfur* **vol. I** (Chemical Publ. Co., New York, 1958) p 449.
Oil, disagreeable odor. Solidif on cooling, crystals mp about 10°. d_4^{15} 1.220. bp_{16} 117°. n_D^{16} 1.4823. Miscible with water, alcohol, ether, acetone.
Copper salt, $C_3H_5CuO_2S$, yellow precipitate. Practically insol in water and in dil HNO_3.
Barium salt, $Ba(C_3H_5O_2S)_2$, rubbery compd, very sol in water. Practically insol in alcohol.
Mercury salt, $Hg(C_3H_5O_2S)_2$, small shiny platelets, sparingly sol in water, very sol in alcohol.
Sodium salt pentahydrate, $C_3H_5NaO_2S.5H_2O$, crystals. Dec about 250°. Sol in water, methanol, ethanol. Less sol in propanol.
Calcium salt decahydrate, $Ca(C_3H_5O_2S)_2.10H_2O$, crystals. Dec 270°. Sol in water, methanol, ethanol. Less sol in propanol.
Platinum salt, $Pt(C_3H_5O_2S)_2$, greenish-yellow precipitate. Practically insol in water and dil acids. Sol in sodium hydroxide and sodium carbonate.
Silver salt, $C_3H_5AgO_2S$, yellow precipitate. Practically insol in water and dil HNO_3.
USE: In depilatory and hair waving prepns.

9477. 2-Thiolhistidine. *α-Amino-2,3-dihydro-2-thioxo-1H-imidazole-4-propanoic acid; 2-mercaptohistidine;* α-amino-2-mercapto-4-imidazolepropionic acid. $C_6H_9N_3O_2S$; mol wt 187.22. C 38.49%, H 4.85%, N 22.44%, O 17.09%, S 17.13%. Prepn from histidine: Ashley, Harington, *J. Chem. Soc.* **1930**, 2586; from aspartic acid: Harington, Overhoff, *Biochem. J.* **27**, 338 (1933). Alternate procedures: Hegedüs, *Helv. Chim. Acta* **38**, 22 (1955); Marei, Raphael, *J. Chem. Soc.* **1958**, 2624. *Review:* J. P. Greenstein, M. Winitz, *Chemistry of the Amino Acids,* **vol. 3** (John Wiley, New York, 1961) pp 2671-2675.

L-form

L-Form, plates from water, darkens at 290°, not melted at 310°. $[\alpha]_D^{20}$ −10° (c = 2 in N HCl). pK_1^1 1.84; pK_2^1 8.47; pK_3^1 11.4. Readily sol in hot water, sparingly sol in water at room temp. Tends to form supersaturated solns. Practically insol in alcohol, other organic solvents.

Dihydrochloride, $C_6H_9N_3O_2$.2HCl, large prisms, dec 197-199°. Freely sol in water, sparingly sol in alcohol.

9478. Thiolutin. *6-(Acetamido)-4-methyl-1,2-dithiolo-[4,3-b]pyrrol-5(4H)-one; N-(4,5-dihydro-4-methyl-5-oxo-1,2-dithiolo[4,3-b]pyrrol-6-yl)acetamide; 3-acetamido-5-methyl-pyrrolin-4-one[4,3-d]-1,2-dithiole; acetopyrrothine.* $C_8H_8-N_2O_2S_2$; mol wt 228.30. C 42.09%, H 3.53%, N 12.27%, O 14.02%, S 28.09%. Antibiotic isolated from several strains of *Streptomyces albus*: F. W. Tanner, Jr. *et al.*, 118th Am. Chem. Soc. Meet. (Chicago, Sept. 1950), *Abstracts of Papers*, p 18A; F. W. Tanner, Jr. *et al.*, U.S. pat. 2,689,854 (1954 to Pfizer). Characterization, similarity to aureothricin, *q.v.*: W. D. Celmer *et al.*, *J. Am. Chem. Soc.* **74**, 6304 (1952). Structure: W. D. Celmer, I. A. Solomons, *ibid.* **77**, 2861 (1955). Total synthesis: U. Schmidt, F. Geiger, *Ann.* **664**, 168 (1963); K. Hagio, N. Yoneda, *Bull. Chem. Soc. Japan* **47**, 1484 (1974). Bactericidal, protozoicidal, fungicidal properties: H. Seneca *et al.*, *Antibiot. & Chemother.* **2**, 357 (1952). Mode of action: A. Jimenez *et al.*, *Antimicrob. Ag. Chemother.* **3**, 729 (1973). Inhibition of microbiological growth in beer: J. B. Bockelmann, F. B. Strandskov, U.S. pat. 2,798,811 (1957 to Schaefer Brewing Co.). Activity against soil borne pathogens: P. R. Deb, B. K. Dutta, *Curr. Sci. (India)* **53**, 659 (1984). As allergy inhibitor: P. Stahl *et al.*, Ger. pat. 3,434,562 (1986 to Boehringer, Mannheim), *C.A.* **105**, 54609k (1986).

Brilliant yellow needles from *n*-butanol, dec 273-276°. Sublimes 200°/0.1 mm. uv max (methanol): 250, 311, 388 nm (ε 6300, 5700, 11,000) (Celmer, Solomons). Sparingly sol in water (210 mg/l); more sol in methanol, ethanol, chloroform, acetone (1% solns in acetone have been prepd), glacial acetic acid, methyl isobutyl ketone. Less sol in ether, benzene, hexane. Stable in acid and neutral solns, decomp in alkaline soln. Effective vs gram-positive and gram-negative bacteria, fungi and ameboid parasites. LD_0 in mice (mg/kg): 5-10 s.c.; 10 orally; LD_{50} in mice (mg/kg): 25 s.c.; 25 orally (Seneca).

9479. Thiomalic Acid. *Mercaptobutanedioic acid;* mercaptosuccinic acid. $C_4H_6O_4S$; mol wt 150.16. C 32.00%, H 4.03%, O 42.62%, S 21.36%. HOOCCH(SH)CH$_2$COOH.

White crystals; sulfidic odor. mp 149-150°. Soly in water at 40° about 50 g/100 ml; in ethanol at 25° about 50 g/100 ml. Also sol in acetone. Moderately sol in ether. Practically insol in benzene. The aq soln gives a transitory blue color with ferric chloride.

9480. Thionalide. *2-Mercapto-N-2-naphthalenylacet-amide;* thioglycollic-β-aminonaphthalide. $C_{12}H_{11}NOS$; mol wt 217.29. C 66.33%, H 5.10%, N 6.45%, O 7.36%, S 14.76%.

White to ivory-colored needles. mp 111-112°. Insol in water. Freely sol in most organic solvents.

USE: As a reagent for copper, mercury, silver, thallium, and bismuth.

9481. Thionaphthene-2-carboxylic Acid. *Benzo[b]thio-phene-2-carboxylic acid;* 2-benzothiophenecarboxylic acid; TNCA; BL-5583. $C_9H_6O_2S$; mol wt 178.21. C 60.66%, H 3.39%, O 17.96%, S 17.99%. Prepn: R. Weissgerber, O. Kruber, *Ber.* **53**, 1551 (1920); *eidem*, Ger. pat. 341,837 (1921), *Frdl.* **13**, 279 (1921); D. A. Shirley, M. D. Cameron, *J. Am. Chem. Soc.* **72**, 2788 (1950); T. Higa, A. J. Krubsack, *J. Org. Chem.* **41**, 3399 (1976). Use as anti-osteoporotic agent: C. M. Samour, J. A. Vida, U.S. pat. 4,101,668 (1978 to Bristol-Myers). Effect on total skeletal calcium in mice: J. C. Robin *et al.*, *J. Med.* **11**, 15 (1980). Mechanism of action study: J. C. Robin *et al.*, *Calcif. Tissue Int.* **36**, 194 (1984). Hypocalcemic effect: A. J. Johannesson *et al.*, *Endocrinology* **117**, 1508 (1985).

Prisms from alcohol or needles from water, mp 236° (Weissgerber). Also reported as crystals from benzene-ethyl acetate, mp 236-238° (Higa).

9482. Thionazin. *Phosphorothioic acid O,O-diethyl O-pyrazinyl ester; O,O-diethyl-O-(2-pyrazinyl) phosphorothio-ate;* ethyl pyrazinyl phosphorothioate; EN-18133; ENT-25580; American Cyanamid 18133; Cynem; Nemafos; Zino-phos. $C_8H_{13}N_2O_3PS$; mol wt 248.24. C 38.71%, H 5.28%, N 11.28%, O 19.34%, P 12.48%, S 12.92%. Prepn: Dixon *et al.*, U.S. pat. 2,918,468; Gordon, U.S. pat. 2,938,831; Miller, Forbes, U.S. pat. 3,091,614; Gagliardi, U.S. pat. 3,340,-262 (1959, 1960, 1963, 1967 all to Am. Cyanamid). Toxicity study: T. B. Gaines, *Toxicol. Appl. Pharmacol.* **14**, 515 (1969).

Amber liquid, mp −1.7°. bp 80°. n_D^{25} 1.5131. Vapor pressure at 30°: 3 × 10^{-3} mm Hg. Slightly sol in water; misc with most organic solvents. LD_{50} in female, male rats (mg/kg): 3.5, 6.4 orally; 11, 17 dermally (Gaines).

Caution: Cholinesterase inhibitor. See Parathion.

USE: Nematocide; insecticide.

9483. Thionine. *3,7-Diaminophenothiazin-5-ium chlo-ride;* Lauth's violet; C.I. 52000. $C_{12}H_{10}ClN_3S$; mol wt 263.75. C 54.65%, H 3.82%, Cl 13.44%, N 15.93%, S 12.16%. Prepn from *p*-phenylenediamine: Lauth, *Compt. Rend.* **82**, 1441 (1876); *Bull. Soc. Chim. France* [2] **26**, 422 (1876); Bernthsen, Ger. pat. 25,150 (1883 to BASF), *Frdl.* **1**, 253; Loiseleur, *Ann. Inst. Pasteur* **86**, 262 (1954); Loiseleur, Petit, *Compt. Rend.* **250**, 2573 (1960); Balestic, Magat, *C.A.* **55**, 23542g (1961). *See also: Colour Index* vol. **4** (3rd ed., 1971) p 4469.

Blackish-green glistening needles. Absorption max (water): 602.5 nm. Difficultly sol in cold, easily sol in hot water, giving first a blue, then a violet soln. The soln becomes somewhat bluer when HCl is added. Brownish-red precipitate with NaOH. Yellowish-green soln with concd H_2SO_4 which upon dilution turns blue, then violet on further dilution.

USE: For general nuclear staining; for counting bacteria in milk; as antioxidant for linseed oil. Has been used as catalyst in condensation of 1,3-butadienyl ethyl ether with maleic anhydride.

9484. Thionyl Bromide. Br_2OS; mol wt 207.87. Br 76.88%, O 7.70%, S 15.43%. $SOBr_2$. Prepd from thionyl chloride and hydrogen bromide: Besson, *Compt. Rend.* **122**, 320 (1896); **123**, 884 (1896); Hibbert, Pullman, *Inorg. Syn.* **1**, 113 (1939).

Orange-yellow liquid. d_4^{20} 2.688. mp $-52°$. bp_{773} 138°; bp_{20} 48°. Somewhat less stable than thionyl chloride. Dec slowly on standing in glass-stoppered bottles. Hydrolyzed by water. Miscible with benzene, chloroform, carbon tetrachloride.

Caution: Highly irritating to skin, eyes, mucous membranes, respiratory tract.

9485. Thionyl Chloride. Sulfurous oxychloride. Cl_2OS; mol wt 118.97. Cl 59.60%, O 13.45%, S 26.95%. $SOCl_2$. Usually prepd by the oxidation of sulfur dichloride with sulfur trioxide: Michaelis, *Ann.* **274**, 173 (1893); Edwards, U.S. pat. **2,362,057** (1944 to Hooker Electrochemical); Fehér in *Handbook of Preparative Inorganic Chemistry*, vol. 1, G. Brauer, Ed. (Academic Press, New York, 2nd ed., 1963) pp 382-383; Macaluso, "Sulfur Compounds" in Kirk-Othmer *Encyclopedia of Chemical Technology*, vol. 19 (Interscience, New York, 2nd ed., 1969) pp 398-401. Purification by distilling from quinoline and boiled linseed oil: Martin, Fieser, *Org. Syn.* coll. vol. II, 570 (1943). Alternate methods of purification: Cottle, *J. Am. Chem. Soc.* **68**, 1380 (1946); Kunkel; Rosenberg, Flaxman, U.S. pats. **3,155,457**; **3,156,529** (both 1964 to Hooker Chem.); Friedman, Wetter, *J. Chem. Soc. (A)* **1967**, 36.

Colorless to pale yellow or reddish, fuming, refractive liquid. Suffocating odor. d_4^0 1.676; d_4^{10} 1.655; d_4^{20} 1.638. mp $-104.5°$. bp_{760} 76°; $bp_{96.6}$ 20°. n_D^{20} 1.517. Dec when heated above 140° forming Cl_2, SO_2, and S_2Cl_2. Hydrolyzed by water forming SO_2 and HCl. Miscible with benzene, chloroform, carbon tetrachloride.

Caution: Vapors and liquid are strongly irritating and corrosive to skin, mucous membranes and eyes.

USE: For making acyl chlorides, to replace OH or SH groups with chlorine atoms; reacts with Grignard reagents to form the corresp sulfoxides. Review of use in organic synthesis: J. S. Pizey, *Synthetic Reagents*, vol. 1 (John Wiley, New York, 1974) pp 321-357.

9486. Thionyl Fluoride. Thionyl difluoride. F_2OS; mol wt 86.06. F 44.15%, O 18.59%, S 37.26%. SOF_2. Usually prepd by the action of antimony trifluoride on thionyl chloride in the presence of antimony pentafluoride: Booth, Mericola, *J. Am. Chem. Soc.* **62**, 640 (1940); Smith, Muetterties, *Inorg. Syn.* **6**, 162 (1960). May also be prepared by the action of other fluorides on thionyl chloride: Kemmitt, Sharp, *Advan. Fluorine Chem.* **4**, 228-229 (1965).

Colorless gas. Suffocating odor. Does not attack glass. Stored in compressed form in steel cylinders. mp $-129.5°$. bp $-43.8°$. Crit temp $+88°$. d (liq; $-100°$) 1.780. d (solid; $-183°$) 2.095. Trouton constant 22.6. Very slowly hydrolyzed by cold water. Sol in ether, benzene.

Caution: Highly irritating to eyes, respiratory tract.

9487. Thiopental Sodium. *5-Ethyldihydro-5-(1-methylbutyl)-2-thioxo-4,6(1H,5H)-pyrimidinedione monosodium salt; 5-ethyl-5-(1-methylbutyl)-2-thiobarbituric acid sodium salt;* thiobumal sodium; penthiobarbital sodium; thiopentone sodium; thionembutal; Intraval Sodium; Nesdonal Sodium; Pentothal Sodium; Trapanal. $C_{11}H_{17}N_2NaO_2S$; mol wt 264.32. C 49.98%, H 6.48%, N 10.60%, Na 8.70%, O 12.11%, S 12.13%. Prepn: U.S. pats. **2,153,729** (1939); **2,876,225** (1959). Prepn of nonhygroscopic crystals: Hartop, U.S. pat. **3,109,001** (1963 to Abbott). Acute toxicity: Christensen, Lee, *Toxicol. Appl. Pharmacol.* **26**, 495 (1973).

Comprehensive description: M. J. McLeish, *Anal. Profiles Drug Subs. Excip.* **21**, 535-572 (1992).

Yellowish-white, hygroscopic powder. Alliaceous, garlic-like odor. Sol in water, alcohol. Insol in ether, benzene, petr ether. Aq solns are alkaline to litmus. Solns dec on standing; on boiling precipitation occurs. LD_{50} in mice (mg/kg): 149 i.p.; 78 i.v. (Christensen, Lee).

Caution: May be habit forming. This is a controlled substance (depressant) listed in the U.S. Code of Federal Regulations, Title 21 Parts 329.1 and 1308.13 (1995).

THERAP CAT: Anesthetic (intravenous).

THERAP CAT (VET): Short-acting anesthetic.

9488. Thiopeptin. Thiofeed. Sulfur containing peptide antibiotic complex produced by *Streptomyces tateyamensis* no. 7906: Miyairi *et al.*, **Ger. pat. 1,929,355** (1969 to Fujisawa), *C.A.* **72**, 88921w (1970). Consists of the a and b series of thiopeptins A_1, A_2, A_3, A_4, and B (major component). Structurally similar to thiostrepton, *q.v.* Biological and chemical studies: Miyairi *et al.*, *Antimicrob. Ag. Chemother.* **1**, 192 (1972). Characterization of thiopeptins B: Miyairi *et al.*, *J. Antibiot.* **23**, 113 (1970); structural studies: Muramatsu *et al.*, *ibid.* **25**, 537 (1972); **30**, 383 (1977). Total structures of components: O. D. Hensens, G. Albers-Schönberg, *Tetrahedron Letters* **1978**, 3649.

Thiopeptins B, faint yellow crystals, mp 219-222° (dec). $[\alpha]_D^{23}$ $-80°$ (c = 1 in chloroform). uv spectrum (methanol): shoulders at 230-250, 295, 305 nm. Sol in dioxane, DMSO, DMF, pyridine, chloroform. Insol in ether, benzene, n-hexane, petr ether, water. Fairly sol in methanol, acetone, ethyl acetate.

THERAP CAT (VET): Antibiotic feed additive.

9489. Thiophanate. *[1,2-Phenylenebis(iminocarbonothioyl)]biscarbamic acid diethyl ester; 4,4'-o-phenylenebis[3-thioallophanic acid]diethyl ester;* 1,2-bis(3-ethoxycarbonyl-2-thioureido)benzene; Cercobin; Topsin; Nemafax. $C_{14}H_{18}N_4O_4S_2$; mol wt 370.45. C 45.39%, H 4.90%, N 15.12%, O 17.28%, S 17.31%. Prepn: T. Noguchi *et al.*, **Ger. pat. 1,806,123**; *eidem*, U.S. pat. **4,020,095** (1969, 1977 both to Nippon Soda Co.). Photodegradation: H. Buchenauer *et al.*, *Pestic. Sci.* **5**, 343 (1973). Activity vs nematode infection in ruminants: Eichler, *Brit. Vet. J.* **129**, 533 (1973). Toxicology in animals: *idem, ibid.* **130**, 570 (1974).

Colorless plates from acetone, dec 194°. LD_{50} orally in mice and rats: > 15 g/kg (Eichler).

*O,O-*Dimethyl analog, $C_{12}H_{14}N_4O_4S_2$, *thiophanate-methyl*, Cercobin-M, Topsin-M. Prepn: T. Noguchi *et al., loc. cit.* Persistence in soil: J. R. Fleeker *et al., J. Agr. Food Chem.* **22**, 592 (1974). Pharmacology, toxicity: Y. Hashimoto *et al., Toxicol. Appl. Pharmacol.* **23**, 606, 616 (1972). Colorless prisms, mp 181.5-182.5°. Sol in acetone, methanol, chloroform, acetonitrile; slightly sol in other organics. Practically insol in water. LD_{50} in rats, mice, guinea pigs, rabbits (g/kg): 3.40, 6.64, 3.64, 2.27 orally (Hashimoto).

USE: Systemic fungicide.

THERAP CAT (VET): Anthelmintic (nematodes).

9490. Thiophene. Thiofuran; thiofurfuran; thiole; thiotetrole; divinylene sulfide. C_4H_4S; mol wt 84.14. C 57.10%, H 4.79%, S 38.11%. Found in coal tar, in coal gas, and in technical benzene: V. Meyer, *Ber.* **16**, 1471 (1883); **17**, 2642 (1884). Made available in commercial quantities by a process utilizing the dehydrogenation of butane with sulfur as the dehydrogenating agent, followed by cyclization with sulfur to form the thiophene ring: Rasmussen, Ray, *Chem. Inds.* **60**, 593, 620 (1947). Laboratory prepn by heating sodium succinate with phosphorus trisulfide: R. Phillips, *Org. Syn.* **coll. vol. II**, 578 (1943). Also prepd by passing ethylene or acetylene into boiling sulfur; or by passing acetylene and hydrogen sulfide over hot bauxite or nickel hydroxide: U.S. pat. **1,421,743**, *C.A.* **16**, 3093 (1922). *Review:* B. Buchholz in Kirk-Othmer *Encyclopedia of Chemical Technology* vol. 22 (Wiley-Interscience, New York, 3rd ed., 1983) pp 965-973.

Liquid. Slight aromatic odor resembling that of benzene. d_4^0 1.0873; d_4^{25} 1.0573; d_4^{50} 1.0285. mp $-38.3°$. bp_{760} 84.4°; bp_{400} 64.7°; bp_{200} 46.5°; bp_{100} 30.5°; bp_{60} 20.1°; bp_{40} +12.5°; bp_{20} 0.0°; bp_{10} $-10.9°$; bp_5 $-20.8°$; n_D^{25} 1.52684. Absorption spectrum: Purvis, *J. Chem. Soc.* **97**, 1653, 1656 (1910). Insol in water; miscible with most organic solvents. May be heated to 850° without decompn.

USE: Solvent similar to benzene, but suitable for lower and higher temps; manuf of resins from thiophene-phenol mixtures and formaldehyde; manuf of dyes and pharmaceuticals.

9491. 2-Thiophenecarboxylic Acid. *2-Thenoic acid.* $C_5H_4O_2S$; mol wt 128.15. C 46.86%, H 3.15%, O 24.97%, S 25.02%. Prepn: Voerman, *Rec. Trav. Chim.* **26**, 293 (1907); Sy, de Malleray, *Bull. Soc. Chim. France* **1963**, 1276; Gross *et al.*, *Ber.* **96**, 1382 (1963).

Needles from water, mp 128.5°. Very sol in ether, alc, hot water; moderately sol in $CHCl_3$; slightly sol in petr ether.

Sodium salt, $C_5H_3NaO_2S$, *sodium 2-thiophenecarboxylate, sodium 2-thenoate, Trophires.*

USE: Sodium salt as lubricating-grease thickener, Morway, Kolfenbach, U.S. pat. **2,576,031** (1951 to Standard Oil).

9492. Thiophenol. *Benzenethiol;* phenylmercaptan. C_6H_6S; mol wt 110.18. C 65.41%, H 5.49%, S 29.10%. C_6H_5SH. Prepd by the reduction of benzenesulfonyl chloride with zinc dust in sulfuric acid: Adams, Marvel, *Org. Syn.* **1**, 71 (1921).

Liquid. Repulsive, penetrating, garlic-like odor, esp when impure. d_4^{25} 1.0728; bp_{760} 168.3°; bp_{100} 103.6°; bp_{50} 86.2°; bp_{20} 69.7°; $bp_{1.0}$ 18.6°; n_D^{25} 1.58603. Heat of fusion 24.90 cal/g; spec heat at 25°: 0.3829; entropy at 25°: 52.6. Insol in water. Very sol in alc; miscible with ether, benzene, CS_2. Feebly acidic. Oxidizes in air, esp when dissolved in alcoholic ammonia, forming diphenyl disulfide, $C_6H_5SSC_6H_5$. The hydrogen of the SH group is easily replaced by metals.

9493. Thiopropazate. *4-[3-(2-Chlorophenothiazin-10-yl)propyl]-1-piperazineethanol acetate;* 2-chloro-10-[3-[1-(2-acetoxyethyl)-4-piperazinyl]propyl]phenothiazine; 10-[3-[1-(2-acetoxyethyl)-4-piperazinyl]propyl]-2-chlorophenothiazine; *N*-(β-acetoxyethyl)-*N'*-[γ-(2'-chloro-10'-phenothiazinyl)propyl]piperazine; 1-(2-acetoxyethyl)-4-[3-(2-chloro-10-phenothiazinyl)propyl]piperazine. $C_{23}H_{28}ClN_3O_2S$; mol wt 446.01. C 61.94%, H 6.33%, Cl 7.95%, N 9.42%, O 7.17%, S 7.19%. Prepd from 2-chloro-10-(γ-chloropropyl)phenothiazine and piperazine in butanone followed by treatment with β-bromoethyl acetate in toluene: Cusic, U.S. pat. **2,766,235** (1956); Anderson *et al.*, *Arzneimittel-Forsch.* **12**, 937 (1962).

Free base, $bp_{0.1}$ 214-218°. Sol in ether.
Dihydrochloride, $C_{23}H_{28}ClN_3O_2S.2HCl$, *Dartal, Dartalan.* Crystals from 95% ethanol, dec 223-229°. Freely sol in water; much less sol in alc, chloroform; almost insol in ether.
Dimaleate, $C_{23}H_{28}ClN_3O_2S.2C_4H_4O_4$, crystals from methanol + ethanol, mp 167-169°.

THERAP CAT: Antipsychotic.

9494. Thioproperazine. *N,N-Dimethyl-10-[3-(4-methyl-1-piperazinyl)propyl]phenothiazine-2-sulfonamide;* 3-dimethylsulfamoyl-10-[3-(4-methylpiperazino)propyl]phenothiazine; 2-dimethylsulfamoyl-10-[3'-(4''-piperazino)propyl]phenothiazine; thioperazine; RP-7843; SKF-5883. $C_{22}H_{30}N_4O_2S_2$; mol wt 446.64. C 59.16%, H 6.77%, N 12.54%, O 7.16%, S 14.36%. Prepn: **Brit. pat. 814,512** (1959 to Rhône-Poulenc).

Crystals, mp 140°.
Fumarate, $C_{22}H_{30}N_4O_2S_2.C_4H_4O_4$, crystals, mp 182°.
Dimethanesulfonate, $C_{22}H_{30}N_4O_2S_2.2CH_3SO_3H$, *thioproperazine mesylate, Majeptil, Vontil.*

THERAP CAT: Antipsychotic; antiemetic.

9495. Thioquinox. *1,3-Dithiolo[4,5-b]quinoxaline-2-thione; trithiocarbonic acid cyclic 2,3-quinoxalinediyl ester;* 2,3-quinoxalinedithiol cyclic trithiocarbonate; trithiocarbonic acid cyclic ester with 2,3-quinoxalinedithiol; 2-thio-1,3-dithiolo[4,5-b]quinoxaline; chinothionat; quinothionate; Eradex; Readex. $C_9H_4N_2S_3$; mol wt 236.34. C 45.74%, H 1.71%, N 11.85%, S 40.70%. Prepn starting with 2,3-dimercaptoquinoxaline and thiophosgene: Sasse *et al.*, U.S. pat. **3,141,886** (1964 to Bayer); **Ger. pat. 1,088,965** (1960); **Belg. pat. 580,478** (1958). Toxicity study: Sasse, *Hoefchen-Briefe* **13**, 197 (1960), *C.A.* **56**, 1801h (1962).

Brownish to yellow powder, mp 180°. The commercial product begins to melt at 165° but does not decompose until the temp exceeds 200°. Vapor pressure 1×10^{-7} mm Hg at 20°. Practically insol in water. Slightly sol in acetone, methanol, anhydr ethanol, petr ether, kerosene. LD_{50} orally in rats: 3.4 g/kg (Sasse).

USE: Acaricide, fungicide.

9496. Thioredoxin. Ubiquitous, small, hydrogen carrier protein that participates in a wide variety of biochemical reactions, e.g., ribonucleotide reduction, methionine sulfoxide, sulfate and disulfide reduction, phosphate transfer reactions. Thioredoxin is oxidized from a dithiol to a disulfide during ribonucleotide reduction, and regenerated by reduced triphosphopyridine nucleotide (TPNH) and the flavoprotein thioredoxin reductase: Laurent *et al.*, *J. Biol. Chem.* **239**, 3436 (1964). Prepn from *Escherichia coli*: Williams *et al.*, *ibid.* **242**, 5226 (1967). Complete amino acid sequence: A. Holmgren, *Eur. J. Biochem.* **6**, 475 (1968). Purification and characterization: A. Holmgren *et al., J. Biol. Chem.* **256,** 3118 (1981). Prepn of bacteriophage T4-induced thioredoxin: Berglund, Sjoberg, *ibid.* **245,** 6030 (1970). Amino acid sequence: Sjoberg, Holmgren, *ibid.* **247,** 8063 (1972). The T4 thioredoxin shows no structural homology with thioredoxin from *E. coli.* Purification, characterization and amino acid sequence of thioredoxin from *Corynebacterium nephridii:* M. Meng, H. P. C. Hogenkamp, *ibid.* **256,** 9174 (1981).

9497. Thioridazine. *10-[2-(1-Methyl-2-piperidinyl)ethyl]-2-(methylthio)-10H-phenothiazine;* 2-methylmercapto-10-[2-(*N*-methyl-2-piperidyl)ethyl]phenothiazine; 3-methylmercapto-*N*-[2'-(*N'*-methyl-2-piperidyl)ethyl]phenothiazine. $C_{21}H_{26}N_2S_2$; mol wt 370.58. C 68.06%, H 7.07%, N 7.56%, S 17.31%. Dopamine receptor blocker; parent compound of sulforidazine and mesoridazine, *q.q.v.* Prepn: Bourquin *et al., Helv. Chim. Acta* **41,** 1072 (1958). Toxicity: E. I. Goldenthal, *Toxicol. Appl. Pharmacol.* **18,** 185 (1971). Effect on dopaminergic function: D. M. Niedzwiecki *et al., J. Pharmacol. Exp. Ther.* **228,** 636 (1984); C. D. Kilts *et al., ibid* **231,** 334 (1984). GLC determn in plasma: E. C. Dinovo *et al., J. Pharm. Sci.* **65,** 667 (1976). HPLC sepn of isomers: R. Whelpton *et al., J. Chromatog.* **426,** 223 (1988). Clinical evaluation: R. Axelsson *et al., Curr. Ther. Res.* **21,** 587 (1977). Comprehensive description: E. M. Abdel-Moety, K. A. Al-Rashood in *Analytical Profiles of Drug Substances* **vol. 18,** K. Florey, Ed. (Academic Press, New York, 1989) pp 459-525.

Crystals from acetone, mp 72-74°. $bp_{0.02}$ 230°. uv max (95% ethanol): 263, 314 nm (ε 38172, 4595); (0.1*N* HCl): 230, 263 nm (ε 20939, 45954); (0.1 *N* NaOH): 313 nm (ε 5226). pKa 9.5. Sol in alcohol (1 in 6), chloroform (1 in 0.81), ether (1 in 3); freely sol in dehydrated alcohol. Insol in water. LD_{50} orally in rats: 995 ±39 mg/kg (Goldenthal).

Hydrochloride, $C_{21}H_{26}N_2S_2$·HCl, *TP-21, Aldazine, Mellaril, Melleretten, Melleril, Mallorol, Novoridazine, Orsanil, Ridazin, Stalleril.* Crystals from acetone, mp 158-160°. uv max (water): 262, 310 nm (ε 41842, 3215); (95% ethanol): 264, 310 nm (ε 41598, 3256); (0.1*N* HCl): 264, 305 nm (ε 42371, 5495); (0.1*N* NaOH): 263 nm (ε 18392). Sol in water (1 in 9); freely sol in ethanol (1 in 10), methanol, chloroform (1 in 5). Insol in ether.

THERAP CAT: Antipsychotic.

9498. Thiosalicylic Acid. *2-Mercaptobenzoic acid; o*-sulfhydrylbenzoic acid. $C_7H_6O_2S$; mol wt 154.19. C 54.53%, H 3.92%, O 20.75%, S 20.80%. Prepd by heating *o*-halogenated benzoic acids with an alkaline hydrosulfide in the presence of copper: *Ger. pat.* **189,200** (1906); by reduction of dithiosalicylic acid: Claasz, *Ber.* **45,** 2427 (1912); *Ger. pat.* **205,450;** C. F. H. Allen, D. D. MacKay, *Org. Syn. coll. vol.* **II,** 580 (1943).

Sulfur-yellow flakes, plates, needles from glacial acetic acid or alcohol. Softens at 158°. mp 164-165°. Sublimes. Slightly sol in hot water; freely in glacial acetic acid, alc. Exposure to air yields dithiosalicylic acid, which is produced also when an alcoholic soln of thiosalicylic acid

comes in contact with $FeCl_3$ (resulting in a transitory blue color).

Sodium salt, $C_7H_5NaO_2S$, *Jecto-Sal, Thiocyl.* Crystals, sol in water.

USE: Manuf thioindigo dyes, reagent for determn of iron.

9499. Thiosemicarbazide. *Hydrazinecarbothioamide.* CH_5N_3S; mol wt 91.14. C 13.18%, H 5.53%, N 46.11%, S 35.18%. $NH_2CSNHNH_2$.

White, cryst powder. mp 182-184°. Sol in water or alc. LD_{50} orally in adult Norway rats: 13 mg/kg, Dieke, *Proc. Soc. Exp. Biol. Med.* **70,** 688 (1949).

USE: As a reagent for detection of metals.

9500. Thiosinamine. *(2-Propenyl)thiourea;* allyl thiourea; allylthiocarbamide; Aminosin; Rhodalline. $C_4H_8N_2S$; mol wt 116.19. C 41.35%, H 6.94%, N 24.11%, S 27.60%. $CH_2=CHCH_2NHCSNH_2$. Made by warming a mixture of equal parts of allyl mustard oil and abs alcohol with an equal amount of 30% ammonia. Growth regulating activity: Karanov, Vasilev, *Izv. Inst. Fiziol. Rast. Bulg. Akad. Nauk* **16,** 167 (1970), *C.A.* **73,** 119454y (1970). Mechanism of action studies as an egg suppressive agent in schistosomiasis, A. B. Machado *et al., J. Parasitol.* **56,** 392 (1970). *In vivo* and *in vitro* effects: I. Popiel, D. A. Erasmus, *Trans. Roy. Soc. Trop. Med. Hyg.* **75,** 287 (1981).

White crystals; bitter taste; slight garlic odor; d 1.22; mp 78°. Sol in about 30 parts water; sol in alcohol, slightly sol in ether; insol in benzene. LD s.c. in rats: 850 mg/kg.

THERAP CAT (VET): Has been used to minimize scar tissue.

9501. 1-Thiosorbitol. *1-Thio-D-glucitol.* $C_6H_{14}O_5S$; mol wt 198.24. C 36.35%, H 7.12%, O 40.35%, S 16.18%. $CH_2OH(CHOH)_4CH_2SH.$ Prepd by heating an aq soln of glucose with a catalyst, sulfur, and hydrogen at 130° and 1500 psi for 3 hrs: Farlow *et al., J. Am. Chem. Soc.* **70,** 1392 (1948).

Crystals from abs alcohol, mp 92-93°. Nonhygroscopic. $[\alpha]_D^{27} -1.9°$ (c = 2). Readily sol in water, pyridine, ethylene glycol, formamide; practically insol in benzene, petr ether, carbon tetrachloride; carbon disulfide. Solubilities at 20° in g/100 ml solvent: 1.7 in abs alcohol; 1.2 in dioxane; 0.016 in ethyl ether; 0.016 in trichloroethylene; 0.010 in acetone. Strong reducing agent.

USE: Suggested as a polymer stabilizer; anticorrosion agent for pickling baths; in plating baths.

9502. Thiostrepton. Gargon; Thiactin; Bryamycin. $C_{72}H_{85}N_{19}O_{18}S_5$; mol wt 1664.91. C 51.94%, H 5.15%, N 15.98%, O 17.30%, S 9.63%. Polypeptide antibiotic contg sulfur. Produced by *Streptomyces azureus* isolated from New Mexican soil: Pagano *et al., Antibiot. Ann.* **1955-56,**

554; Vandeputte, Dutcher, *ibid.* 560; Brit. pat. **795,570** (1958); Donovick *et al.;* Platt, U.S. pats. **2,982,689; 2,982,-698** (1961 all to Olin Mathieson). Structure studies: Bodanszky *et al., J. Am. Chem. Soc.* **84,** 2003 (1962); **86,** 2478 (1964); Anderson *et al., Nature* **225,** 233 (1970). Total structure determn: K. Tori *et al., Tetrahedron Letters* **1976,** 185; *eidem, J. Antibiot.* **32,** 1072 (1979). ^{13}C-NMR study: *eidem, ibid.* **34,** 124 (1981). Mode of action as an inhibitor of protein synthesis: Cannon, Burns, *FEBS Letters* **18,** 1 (1971); Cundliffe, *Biochem. Biophys. Res. Commun.* **44,** 912 (1971). Identity with thiactin: Bodanszky *et al., J. Antibiot.* **16A,** 76 (1963). Comprehensive description: K. Florey, Ed. in *Analytical Profiles of Drug Substances* vol. **7** (Academic Press, New York, 1978) pp 423-444. *Review:* Pestka, Bodley, in *Antibiotics* vol. **3,** J. W. Corcoran, F. E. Hahn, Eds. (Springer-Verlag, New York, 1975) pp 551-573; E. Cundliffe, *ibid.* vol. **5**(pt. 1), F. E. Hahn, Ed. (1979) pp 329-343.

Crystals from chloroform + methanol, dec 246-256°. $[\alpha]_D^{23}$ −98.5° (glacial acetic acid); −61° (dioxane); −20° (pyridine). Sol in chloroform, dioxane, pyridine, glacial acetic acid, DMF. Practically insol in water, the lower alcohols, nonpolar organic solvents, dil aq acid or base. Dissolved by methanolic acid or base with decompn. Stable in the presence of gastric and intestinal juices and urine. No uv maxima but shows characteristic shoulders at 225, 250, 280 nm ($E_{1cm}^{1\%}$ 520, 380, 255).

Hemisuccinate, mp 200-220°, forms a water-soluble potassium salt. Prepn: Bodanszky, Fried, U.S. pat. **3,181,995** (1965 to Olin Mathieson).

THERAP CAT: Antibacterial.

THERAP CAT (VET): Has been used as an antibacterial.

9503. Thiothixene. *N,N-Dimethyl-9-[3-(4-methyl-1-piperazinyl)propylidene]thioxanthene-2-sulfonamide; cis-9-[3-(4-methyl-1-piperazinyl)propylidene]-2-(dimethylsulfonamido)thioxanthene; tiotixene;* Navane; Orbinamon. $C_{23}H_{29}N_3O_2S_2$; mol wt 443.63. C 62.27%, H 6.59%, N 9.47%, O 7.21%, S 14.46%. Prepn of *cis/trans*-isomer mixture and sepn of isomers: Bloom, Muren, Belg. pat. **647,066** corresp to U.S. pat. **3,310,553** (1964, 1967 both to Pfizer); Muren, Bloom, *J. Med. Chem.* **13,** 17 (1970). Only *cis*-isomer exhibits therapeutic activity. Structure studies: J. P. Schaefer, *Chem. Commun.* **1967,** 743. Pharmacological studies: A. Weissman, *Psychopharmacologia* **12,** 142 (1968). HPLC determn in serum: C. Dilger *et al., Arzneimittel-Forsch.* **38,** 1522 (1988). Book: T. A. Ban, *Psychopharmacology of Thiothixene* (Raven Press, New York, 1978) 485 pp. Comprehensive description: D. K. Wyatt, L. T. Grady in *Analytical Profiles of Drug Substances* vol. **18,** K. Florey, Ed. (Academic Press, New York, 1989) pp 527-565.

cis-form

cis-Isomer, crystals, mp 147.5-149°. uv max (methanol): 228, 260, 310 nm (log ε 4.6, 4.3, 3.9). Has greater pharmacologic activity than *trans*-isomer. LD_{50} in mice, rats (mg/kg): 100, 55 i.p. (Weissman).

trans-Isomer, mp 123-124.5°. uv max (methanol): 229, 252, 301 nm (log ε 4.5, 4.2, 3.9). LD_{50} i.p. in mice: 235 mg/kg (Weissman).

cis-trans-Isomer mixture, crystals, mp 114-118°.

Dimaleate, mp 158-160.5°.

Dioxalate, mp 229°.

THERAP CAT: Antipsychotic.

9504. 2-Thiouracil. *2,3-Dihydro-2-thioxo-4(1H)-pyrimidinone;* 2-mercapto-4-hydroxypyrimidine; 4-hydroxy-2(1H)-pyrimidinethione; 2-mercapto-4(1H)-pyrimidinone; 6-hydroxy-2-mercaptopyrimidine; 2-mercapto-4-pyrimidone; Deracil. $C_4H_4N_2OS$; mol wt 128.15. C 37.49%, H 3.15%, N

21.86%, O 12.48%, S 25.02%. Occurs in seeds of *Brassica, Cruciferae:* Purves, *Brit. J. Exp. Pathol.* **22,** 241 (1941). Prepd by condensing ethyl formylacetate with thiourea: Wheeler, Liddle, *Am. Chem. J.* **40,** 550 (1908). Antithyroid activity results from its interference with the iodination of thyroxine precursors: *see* Maloof, Soodak, *Pharmacol. Rev.* **15,** 72-79 (1963). Inhibition of nucleic acid metabolism: Cardeilhac, *Proc. Soc. Exp. Biol. Med.* **125,** 692 (1967). Toxicology: K. K. Carroll, R. L. Noble, *J. Pharmacol. Exp. Ther.* **97,** 478 (1949).

Minute, bitter crystals, no definite mp. Very slightly soluble in water (1:2000); practically insol in alcohol, ether, acids. Readily sol in alkaline solns. LD_{100} i.p. in rats: 1500 mg/kg (Carroll, Noble).

THERAP CAT: Treatment of hyperthyroidism; angina pectoris; congestive heart failure.

THERAP CAT (VET): Thyroid depressant. In hyperthyroidism and to promote fattening.

9505. Thiourea. Thiocarbamide. CH_4N_2S; mol wt 76.12. C 15.78%, H 5.30%, N 36.80%, S 42.12%. H_2NCS-NH_2. Made by fusing ammonium thiocyanate: Powers and Powers, Mitchell, U.S. pat. **2,552,584; 2,560,596** (both 1951 to Koppers); by treating cyanamide with hydrogen sulfide: Robin, Jr., U.S. pat. **2,173,067** (1940 to Am. Cyanamid); Lewis, U.S. pat. **2,393,917** (1946 to Monsanto); Van de Kamp, U.S. pat. **2,357,149** (1944 to Merck & Co.). Toxicity data: Dieke *et al., J. Pharmacol. Exp. Ther.* **90,** 262 (1947).

Crystals, mp 176-178°. d 1.405. Soluble in 11 parts water, in alcohol; sparingly sol in ether. Neutral reaction. Forms addition compds with metallic salts. LD_{50} orally in wild Norway rats: 1830 mg/kg (Dieke).

Caution: Chronic administration in rats has resulted in hepatic tumors, bone marrow depression and goiters: *Clinical Toxicology of Commercial Products,* R. E. Gosselin *et al.,* Eds. (Williams & Wilkins, Baltimore, 5th ed., 1984) Section II, p 350. This substance may reasonably be anticipated to be a carcinogen: *Seventh Annual Report on Carcinogens* (PB95-109781, 1994) p 381.

USE: Photographic fixing agent and to remove stains from negatives; manuf resins; vulcanization accelerator; a reagent for bismuth, selenite ions.

9506. Thioxanthene. Thiaxanthene; dibenzothiopyran; dibenzopenthiophene; diphenylenemethane sulfide. $C_{13}H_{10}$-S; mol wt 198.29. C 78.75%, H 5.08%, S 16.17%. Conveniently obtained from thioxanthone in 74% yield by reduction wth lithium aluminum hydride: Mustafa, Hilmy, *J. Chem. Soc.* **1952,** 1345.

Needles or rods from alcohol + chloroform. Sublimes easily. mp 128°. bp_{730} 340°. Dipole moment: 1.44 D. Freely sol in chloroform. Moderately sol in alcohol, ether.

9507. Thioxanthone. *Thioxanthen-9-one;* thiaxanthone; 9-oxothioxanthene. $C_{13}H_8OS$; mol wt 212.27. C 73.56%, H 3.80%, O 7.54%, S 15.11%. Prepd by the interaction of diphenyl sulfide and phosgene with aluminum chloride as catalyst: Szmant *et al., J. Org. Chem.* **18,** 745 (1953).

Yellow needles from chloroform, mp 211°. bp_{715} 273°. Freely sol in benzene, chloroform, carbon disulfide, hot glacial acetic acid. Slightly sol in alc. Practically insol in water, alkaline solns. Sol in concd sulfuric acid, giving a yellow soln with strong green fluoresence in visible light.

9508. THIP. *4,5,6,7-Tetrahydroisoxazolo[5,4-c]pyridin-3(2H)-one;* 4,5,6,7-tetrahydroisoxazolo[5,4-c]pyridin-3-ol; gaboxadol. $C_6H_8N_2O_2$; mol wt 140.14. C 51.42%, H 5.75%, N 19.99%, O 22.83%. Structural analog of muscimol, *q.v.*, with potent GABA agonist activity. Prepn: P. Krogsgaard-Larsen, *Acta Chem. Scand.* B **31**, 584 (1977); idem, **Japan. Kokai** 79 36290 corresp to U.S. pat. **4,278,676** (1979, 1981 both to H. Lundbeck & Co.). GABA agonist effects: P. Krogsgaard-Larsen et al., *Nature* **268**, 53 (1977). Antinociceptive effects: D. A. Kendall et al., *J. Pharmacol. Exp. Ther.* **220**, 482 (1982). HPLC determn: S. M. Madsen, *J. Chromatog.* **238**, 509 (1982). Muscle relaxant activity: R. C. Hill, **Belg. pat.** 890,136 (1982 to Sandoz), *C.A.* **97**, 11847k (1982). ³H-THIP binding study: E. Falch, P. Krogsgaard-Larsen, *J. Neurochem.* **38**, 1123 (1982). Pharmacodynamics and potential therapeutic uses: A. V. Christensen et al., *Pharm. Weekbl., Sci. Ed.* **4**, 145 (1982). Effects in tardive dyskinesia: S. Korsgaard et al., *Arch. Gen. Psychiatry* **39**, 1017 (1982).

Zwitterionic. Colorless cryst mp 242-244° (dec). uv max (methanol): 212 nm (log ε 3.64). pKa (water, 25°): 4.44 ±0.03; 8.48 ±0.04.

Hydrobromide, $C_6H_9BrN_2O_2$. Faintly reddish cryst from methanol-ether, mp 162-163° (dec). LD_{50} in mice (mg/kg): 80 i.v., 145 i.p.; > 320 orally (U.S. patent).

USE: As a molecular probe to study GABA receptors.

9509. Thiphenamil. *α-Phenylbenzeneethanethioic acid S-[2-(diethylamino)ethyl]ester; diphenylthioacetic acid S-(2-diethylaminoethyl) ester;* diphenylthiolacetic acid 2-diethylaminoethyl ester; S-[2-(diethylamino)ethyl]diphenylthioacetate; β-diethylaminoethyl diphenylthioacetate. $C_{20}H_{25}NOS$; mol wt 327.49. C 73.35%, H 7.69%, N 4.28%, O 4.89%, S 9.79%. Prepn: Richardson, U.S. pat. **2,390,555** (1945 to Wm. P. Poythress); Clinton, U.S. pat. **2,510,773** (1950 to Sterling Drug); Richardson, *J. Pharm. Sci.* **55**, 1316 (1966).

Hydrochloride, $C_{20}H_{25}NOS \cdot HCl$, *Thiphen, Trocinate.* Rosettes of tiny needles from benzene + petr ether, large prisms from abs ethanol + ethyl acetate, mp 129-130°. Sol in water. Aq solns are about neutral to litmus.

THERAP CAT: Anticholinergic; relaxant (smooth muscle).

9510. Thiram. *Tetramethylthioperoxydicarbonic diamide; bis(dimethylthiocarbamoyl) disulfide;* bis(dimethylthiocarbamyl) disulfide; tetramethylthiuram disulfide; TMTD; ENT-987; SQ-1489; NSC-1771; Thiurad; Thylate; Fernasan; Nomersan; Rezifilm; Pomarsol; Tersan; Tuads; Arasan. $C_6H_{12}N_2S_4$; mol wt 240.44. C 29.97%, H 5.03%, N 11.65%, S 53.35%. Prepn: v. Braun, Stechele, *Ber.* **35**, 820 (1902); **36**, 2280 (1903); Romani, *C.A.* **16**, 854 (1922); Cummings, Simmons, *Ind. Eng. Chem.* **20**, 1173 (1928). Acute toxicity: Gaines, *Toxicol. Appl. Pharmacol.* **14**, 515 (1969). Review of carcinogenic risk: *IARC Monographs* **12**, 225 (1976).

Crystals from chloroform + alcohol, mp 155-156° (commercial grades, mp 146°). d 1.29. Insol in water, dil caustic, gasoline. Soly in alcohol and in ether less than 0.2%; soly in acetone 1.2%; in benzene 2.5%. Sol in chloroform, acetone. LD_{50} orally in rats: 640 mg/kg (Gaines).

Caution: Potential symptoms of overexposure are irritation of mucous membranes; dermatitis; with ethanol consumption: flushing, erythema, pruritus, urticaria, headache, nausea, vomiting, diarrhea, weakness, dizziness and dyspnea. *See NIOSH Pocket Guide to Chemical Hazards* (DHHS/NIOSH 90-117, 1990) p 212. A potent skin sensitizer. Toxicity greater in presence of fats, oils, fat solvents. *See Clinical Toxicology of Commercial Products,* R. E. Gosselin et al., Eds. (Williams & Wilkins, Baltimore, 5th ed., 1984) section III, pp 383-386.

USE: Rubber accelerator; vulcanizer; seed disinfectant; fungicide; bacteriostat in soap; animal repellent.

THERAP CAT: Antiseptic.

9511. Thomas Phosphate. Thomas flour; calcium phosphate, tetrabasic. $Ca_3(PO_4)_2 \cdot CaO$ also represented by $4CaO.P_2O_5$. A byproduct of the Thomas steel process used in Europe: Iron ore of the "Minette" type from Alsace and the Rhineland contains about 2% phosphates which are combined with calcium and separated during the refining process.

USE: Agricultural fertilizer.

9512. Thonzonium Bromide. *N-[2-[[(4-Methoxyphenyl)methyl]-2-pyrimidinylamino]ethyl]-N,N-dimethyl-1-hexadecanaminium bromide; hexadecyl[2-[(p-methoxybenzyl)-2-pyrimidinylamino]ethyl]dimethylammonium bromide;* cetyl-[2-[(p-methoxybenzyl)-2-pyrimidinylamino]ethyl]dimethylammonium bromide; N-(2-pyrimidyl)-N-(p-methoxybenzyl)-N',N'-dimethyl-N'-cetyl-N'-bromoethylenediamine; Thonzide. $C_{32}H_{55}BrN_4O$; mol wt 591.72. C 64.96%, H 9.37%, Br 13.50%, N 9.47%, O 2.70%. Prepn: Fand, U.S. pat. **2,742,470** (1956 to Nepera Chem.).

Crystals, mp 91-92°.
USE: Detergent.

9513. Thonzylamine Hydrochloride. *N-[(4-Methoxyphenyl)methyl]-N',N'-dimethyl-N-2-pyrimidinyl-1,2-ethanediamine monohydrochloride; 2-[(2-dimethylaminoethyl)(p-methoxybenzyl)amino]pyrimidine hydrochloride;* N,N-dimethyl-N'-(p-methoxybenzyl)-N'-(2-pyrimidyl)ethylenediamine hydrochloride; Anahist; Neohetramine hydrochloride; Tonamil. $C_{16}H_{23}ClN_4O$; mol wt 322.84. C 59.53%, H 7.18%, Cl 10.98%, N 17.35%, O 4.96%. Prepd by treating the sodium salt of 2-(p-methoxybenzyl)aminopyrimidine with N,N-dimethyl-2-chloroethylamine: Friedman, Tolstoouhov, U.S. pat. **2,465,865** (1949). Toxicity study: Reinhard, Seudi, *Proc. Soc. Exp. Biol. Med.* **66**, 512 (1947).

Crystals, mp 173-176° (free base, oily liq, bp$_{2.2}$ 185-187°). Freely sol in water; sol in alcohol, chloroform. Practically insol in ether. pH 5.1-5.7 (2% aq solution). LD$_{50}$ orally in guinea pigs: 493 mg (base)/kg (Reinhard, Seudi).
THERAP CAT: Antihistaminic.

9514. Thorium. Th; at. wt 232.0381 (characteristic terrestrial isotopic composition); at. no. 90; valence 4. No stable nuclides. Natural isotopes (mass numbers): 232 ($\sim$100%, T$_{1/2}$ 1.40 × 10^{10} years, longest-lived known isotope, α-decay); 228 (natural decay product of ^{232}Th); 227, 231 (natural decay products of ^{235}U); 230, 234 (natural decay products of ^{238}U); other isotopes: 213-226, 229, 233, 235, 236. ^{232}Th found in nature undergoes natural decay by the emission of α-, β-, and γ-rays, eventually forming ^{208}Pb. Occurs in the minerals thorite, thorianite, orangite, yttrocrasite; in monazite sand (principal ore of commercial significance); present in the earth's crust 8-15 ppm. Isoln from thorite (ThSiO$_4$): J. J. Berzelius, *K. Sven. Ventenskapsakad. Handl.* **9**, 1 (1829); *Pogg. Ann.* **16**, 385 (1829). Prepn of metal: *idem, Ann. Physik* **16**, 385 (1829); Marden, *Trans. Electrochem. Soc.* **66**, 39 (1934). Monographs: F. L. Cuthbert, *Thorium Production Technology* (Addison-Wesley, Reading, Mass., 1958); L. Grainger, *Uranium and Thorium* (Pitman, London, 1958). *Reviews: Mellor's* **vol. 7**, 174-253 (1930); "The Actinides," in *Comprehensive Inorganic Chemistry* **vol. 5**, J. C. Bailar, Jr., *et al.*, Eds. (Pergamon Press, Oxford, 1973) pp 1-715; L. I. Katzin in Kirk-Othmer *Encyclopedia of Chemical Technology* **vol. 22** (Wiley-Interscience, New York, 3rd ed., 1983) pp 989-1002; L. I. Katzin, D. C. Sonnenberger in *The Chemistry of the Actinide Elements* **vol. 1**, J. J. Katz *et al.*, Eds. (Chapman and Hall, New York, 1986) pp 41-101. Review of prepn and purification: J. C Spirlet *et al., Adv. Inorg. Chem.* **31**, 1-40 (1987). Review of toxicology and health effects: *Toxicological Profiles for Thorium* (PB91-180448, 1990) 174 pp; *Patty's Industrial Hygiene and Toxicology* **vol. 2C**, G. D. Clayton, F. E. Clayton, Eds. (Wiley-Interscience, New York, 4th ed., 1994) pp 2249-2258.
Grayish-white, lustrous, radioactive metal; somewhat ductile and malleable. Two alloptropic phases: face-centered cubic α-form, d^{25} 11.724, transforms to β-form at 1360°; body-centered cubic β-form transforms to liquid at mp 1750° (Katzin, Sonnenberger). Also reported as mp 1690° (Cuthbert). bp $\sim$3800°. Heat of fusion < 19.2 kJ/mol. Heat of vaporization $\sim$586 kJ/mol. Heat capacity (25°) 27.32 J/mol.K. Darkens on prolonged exposure to air. Finely divided metal is pyrophoric in air. HCl attacks metal vigorously, leaving up to 25% as an undissolved residue. Nitric acid passivates metal. Dilute HF and H$_2$SO$_4$ and concentrated H$_3$PO$_4$ and HClO$_4$ attack thorium slowly, with evolution of H. Metal not attacked by alkali hydroxides.
USE: As fuel in nuclear reactors, as source of fissionable ^{233}U. In manuf incandescent gas-light mantles, welding electrodes, ceramics. As hardener in Mg alloys; for filament coatings in incandescent lamps and vacuum tubes; as chemical catalyst.

9515. Thorium Chloride. Thorium tetrachloride. Cl$_4$-Th; mol wt 373.85. Cl 37.93%, Th 62.07%. ThCl$_4$. It may contain from 7 to 9 mols H$_2$O. Prepn of the octahydrate: Kremer, *J. Am. Chem. Soc.* **64**, 1009 (1942).
White, odorless crystals; d 4.59; mp 770°. bp 921°. Sol in water, alcohol. Sol in ethylenediamine.

9516. Thorium Iodide. Thorium tetraiodide. I$_4$Th; mol wt 739.66. I 68.63%, Th 31.37%. ThI$_4$. Prepd by passing iodine vapors in an inert gas over the metal at elevated temp: Allen, Yost, *J. Chem. Phys.* **22**, 855 (1954). *See also* Cuthbert, *Thorium Production Technology* (Addison-Wesley, Reading, Mass., 1958).
Pale yellow crystals, mp 566°. bp 837°. Dec by light or heat. When heated with Th gives ThI$_3$.

9517. Thorium Nitrate. N$_4$O$_{12}$Th; mol wt 480.06. N 11.67%, O 39.99%, Th 48.34%. Th(NO$_3$)$_4$. Commercial form described as tetrahydrate; also crystallizes with 6 and 12 mols H$_2$O. Obtained from monazite: Pearce *et al., Inorg. Syn.* **II**, 38 (1946).

Tetrahydrate, white, slightly deliquesc crystals. Very sol in water, alcohol, the soln having an acid reaction. *Keep well closed.*
USE: Thorium nitrate with 1% cerium nitrate constitutes the usual impregnating liq for incandescent mantles. As a reagent for determination of fluorine.

9518. Thorium Oxide. Thorium dioxide; thoria. O$_2$-Th; mol wt 264.04. O 12.12%, Th 87.88%. ThO$_2$. *Review:* Keller in *Comprehensive Inorganic Chemistry* **vol. 5**, J. C. Bailar, Jr. *et al.*, Eds. (Pergamon Press, Oxford, 1973) pp 221-223.
White, heavy, infusible, cryst powder; when heated is incandescent. d 10.0. mp 3390°. Insol in water or alkalies. Sol in acids with difficulty.
Suspension contg 20-25% ThO$_2$, *Thorotrast.* Reviews of tumor induction by Thorotrast: Looney, *Am. J. Roentgenol., Radium Ther. Nucl. Med.* **83**, 163-185; Smoron, Battifora, *Cancer* **30**, 1252-1259 (1972).
Note: Thorium oxide has been listed as a known carcinogen: *Seventh Annual Report on Carcinogens* (PB95-109781, 1994) p 69.
THERAP CAT: Has been used as diagnostic aid (radiopaque medium).

9519. Thorium Sulfate. O$_8$S$_2$Th; mol wt 424.17. O 30.18%, S 15.12%, Th 54.70%. Th(SO$_4$)$_2$. Crystallizes with 4, 6, 8 and 9 mols H$_2$O.
Nonahydrate, colorless or white, monoclinic crystals; dec when strongly heated; d 2.8. Sol in about 70 parts cold, in about 15 parts hot water.

9520. Thorium Tetracyanoplatinate(II). *Thorium(4+) tetrakis(cyano-C)platinate(2—) (1:2);* platinous thorium cyanide; thorium platinocyanide. C$_8$N$_8$Pt$_2$Th; mol wt 830.34. C 11.57%, N 13.49%, Pt 46.99%, Th 27.94%. Th[Pt(CN)$_4$]$_2$.
Hexadecahydrate, yellow crystals. Sparingly sol in water.
USE: For fluorescent screens, *see* Barium Platinous Cyanide.

9521. Thozalinone. *2-(Dimethylamino)-5-phenyl-4(5H)-oxazolone; 2-(dimethylamino)-5-phenyl-2-oxazolin-4-one;* 5-phenyl-2-(dimethylamino)-2-oxazolin-4-one; tozalinone; Stimsen. C$_{11}$H$_{12}$N$_2$O$_2$; mol wt 204.23. C 64.69%, H 5.92%, N 13.72%, O 15.67%. Prepn: Hardy, Jr. *et al.*, U.S. pat. **3,037,990** (1962 to Am. Cyanamid). Pharmacological studies: B. M. Bernstein, C. N. Latimer, *Psychopharmacologia* **12**, 338 (1968); U. H. Lindberg, J. Pedersen, *Acta Pharm. Suecica* **5**, 15 (1968). In treatment of Parkinsonism: W. D. Gray, C. E. Edward, U.S. pat. **3,665,075** (1972 to Am. Cyanamid).

Crystals from water, mp 133-136°.
THERAP CAT: Antidepressant.

9522. Threonine. L-*Threonine;* Thr; T; 2-amino-3-hydroxybutyric acid; α-amino-β-hydroxybutyric acid; [R-(R*,-S*)]-2-amino-3-hydroxybutanoic acid. C$_4$H$_9$NO$_3$; mol wt 119.12. C 40.33%, H 7.62%, N 11.76%, O 40.29%. An essential amino acid for human development; last of the common amino acids to be discovered. Identified in oat protein: S. B. Schryver, H. W. Buston, *Proc. Roy. Soc. London* **99B**, 476 (1925/6); isolated from blood fibrin: W. C. Rose, *J. Biol. Chem.* **109**, 77 (1935). Structure: C. E. Meyer, W. C. Rose, *ibid.* **115**, 721 (1936). Early chemistry and biochemistry: *Amino Acids and Proteins*, D. M. Greenberg, Ed. (Charles C. Thomas, Springfield, IL, 1951) 950 pp., *passim*; J. P. Greenstein, M. Winitz, *Chemistry of the Amino Acids* **vols 1-3** (John Wiley and Sons, Inc., New York, 1961) pp. 2238-2258, *passim*. GC/MS metabolism study: O. Ballevre *et al., Anal. Biochem.* **193**, 212 (1991). Clinical evaluation in amyotrophic lateral sclerosis: J. B. Roufs, *Med. Hypotheses* **34**, 20 (1991); O. Blin *et al., J. Neurol.* **239**, 79 (1992). Review of microbial production:

Daoust, *Develop. Ind. Microbiol.* **7**, 41 (1966); R. Bhatta-charyya *et al., Hind. Antibiot. Bull.* **30**, 54-65 (1988). Review of biosynthesis: G. N. Cohen in *Amino Acids, Biotechnol. Ser.* **3**, 147-171 (1983); of regulation of biosynthesis: G. Galili, *Plant Cell* **7**, 899-906 (1995).

Crystals, dec 255-257°. $[\alpha]_D^{26}$ $-28.3°$ (c = 1.09). pK_1' 2.15; pK_2' 9.12. Freely sol in water. Insol in common neutral solvents.

DL-Form hemihydrate, orthorhombic crystals, dec 229-230°.

9523. D-Threose. *[S-(R*,S*)]-2,3,4-Trihydroxybutanal.* $C_4H_8O_4$; mol wt 120.11. C 40.00%, H 6.71%, O 53.28%. From calcium D-xylonate by oxidation with H_2O_2: Ruff, *Ber.* **34**, 1370 (1901). Improved procedure using strontium D-xylonate and ferric acetate catalyst: Hockett, *J. Am. Chem. Soc.* **57**, 2260, 2265 (1935). From tetraacetyl-D-xylononitrile: Maquenne, *Compt. Rend.* **130**, 1403 (1900); *Ann. Chim.* [7], **24**, 404 (1901); Bonner, Roth, *J. Am. Chem. Soc.* **81**, 5454 (1959); from monobenzylidene-D-arabitol: Steiger, Reichstein, *Helv. Chim. Acta* **19**, 1016 (1939); from 1,1-diethylsulfonyl-D-*threo*-3,4,5-trihydroxypent-1-ene: Hough, Taylor, *J. Chem. Soc.* **1955**, 1212; from D-galactose: Perlin, Brice, *Can. J. Chem.* **34**, 541 (1956). Synthesis of DL-threose: Lake, Glattfeld, *J. Am. Chem. Soc.* **66**, 1091 (1944); Schmid, Grob, *Helv. Chim. Acta* **32**, 77 (1949); Sonogashira, Nakagawa, *Bull. Chem. Soc. Japan* **45**, 2616 (1972).

Syrup. Shows mutarotation. Final $[\alpha]_D^{20}$ $-12.3°$ (20 min, c = 4). Very sol in water; slightly in alcohol. Practically insol in ether, petr ether.

Phenylosazone $C_{16}H_{18}N_4O_2$, dec 164-165°. Identical with D-erythrose phenylosazone.

Triacetate, $C_{10}H_{14}O_7$, prisms from abs ethanol, mp 117-118°. $[\alpha]_D^{25}$ $+34.4°$ (c = 2 in chloroform). Soluble in hot water, chloroform, acetone, ethyl acetate; sparingly sol in abs alcohol, methanol, ether.

9524. L-Threose. $C_4H_8O_4$; mol wt 120.11. C 40.00%, H 6.71%, O 53.28%. By degradation of L-xylose through the oxime and tetraacetyl-L-xylononitrile to L-threose diacetamide: Hockett *et al., J. Am. Chem. Soc.* **60**, 278 (1938); from D-glucitol: Hutson, Weigel, *J. Chem. Soc.* **1961**, 1546.

Not isolated. Shows mutarotation; final $[\alpha]_D^{20}$ $+13.2°$ (c = 4.5). Sol in water.

Phenylosazone, $C_{16}H_{18}N_4O_2$, dec 162°.

9525. Thrombin. Blood-coagulation factor IIa; factor IIa; fibrinogenase; E.C. 3.4.21.5. Multifunctional serine protease generated at the site of vascular injury from prothrombin *q.v.* Key enzyme in the coagulation cascade; converts fibrinogen into fibrin and activates factor XIII, *q.v.*, which cross-links and stabilizes the fibrin polymer. Enhances inflammation and tissue repair via receptor-mediated cellular responses; activates platelets and endothelial cells and stimulates proliferation of fibroblasts and smooth muscle cells. The mature enzyme is glycosylated and consists of two polypeptide chains, designated A and B, covalently linked by a disulfide bridge. Exists in several forms, the largest and most active being the 39 kDa α-thrombin. The

human α-thrombin A chain contains 36 amino acids; the B chain contains 259 residues. Purification by conversion from bovine prothrombin: W. H. Seegers *et al., J. Biol. Chem.* **126**, 91 (1938); W. H. Seegers, D. A. McGinty *ibid.* **146**, 511 (1942); H. P. Smith, U.S. pat. **2,398,077** (1946 to Parke, Davis). Sequence of bovine A chain: S. Magnussen, *Biochem. J.* **110**, 25P (1968). Primary structure of human α-thrombin: R. J. Butkowski *et al., J. Biol. Chem.* **252**, 4942 (1977). Review of structure: M. T. Stubbs, W. Bode, *Thromb. Res.* **69**, 1-58 (1993). Review of cloning, structure and function of thrombin receptors: S. R. Coughlin *et al., Cold Spring Harb. Symp. Quant. Biol.* **57**, 149-154 (1992); C. Tapparelli *et al., Trends Pharmacol. Sci.* **14**, 426-428 (1993). Review of role in hemostasis: J. W. Fenton *et al., Hematol. Oncol. Clin. North Am.* **7**, 1107-1119 (1993); in cell adhesion and proliferation: R. Bar-Shavit *et al., Am. J. Resp. Cell Mol.* **6**, 123-130 (1992); in inflammation and healing and effects in RA: R. Morris *et al., Ann. Rheum. Dis.* **53**, 72-79 (1994).

Topical thrombin, Thrombinar, Thrombogen, Thrombostat. A standardized prepn of bovine thrombin. Review of clinical experience: R. T. Tidrick *et al., Surgery* **14**, 191-196 (1943); in combination with fibrin: M. Brennan, *Blood Rev.* **5**, 240-244 (1991). White powder. Sol in water, isotonic saline.

THERAP CAT: Hemostatic (local).

9526. Thrombomodulin. Thrombin-binding protein present on the surface of vascular endothelial cells that modulates the pro- vs anti- coagulant activities of thrombin, *q.v.* Alters the macromolecular specificity of thrombin, decreasing its ability to catalyze clot formation and converting it to a potent activator of protein C, *q.v.* Also expressed by platelets, synovial cells, and certain tumor cells. Identification: C. T. Esmon, W. G. Owen, *Proc. Nat. Acad. Sci. USA* **78**, 2249 (1981). Isoln from rabbit lung: N. L. Esmon *et al., J. Biol. Chem.* **257**, 859 (1982); from human placenta: H. H. Salem *et al., ibid.* **259** 12246 (1984). Human thrombomodulin is a single-chain glycoprotein containing 559 amino acids; mol wt ~75 kDa. A soluble, truncated form has been isolated from human plasma and urine: H. Ishii, P. W. Majerus, *J. Clin. Invest.* **76**, 2178 (1985). Review: N. L. Esmon, *Prog. Hemost. Thromb.* **9**, 29-55 (1989); of structure and function: W. A. Dittman, P. W. Majerus, *Blood* **75**, 329-336 (1990); of tissue distribution and expression by tumor cells: L. M. Fink *et al., Int. J. Dev. Biol.* **37**, 221-226 (1993).

9527. Thromboplastin. *Blood-coagulation factor III;* Factor III; cytozyme; thrombokinase; thrombokinin; tissue factor; tissue thromboplastin; zymoplastic substance; thrombostop; Tachostyptan. An integral membrane glycoprotein which, in the presence of Ca^{2+} ions, initiates coagulation by augmenting the proteolytic attack of factor VII on factors IX and X, *q.q.v.*: Nemerson, *Biochemistry* **5**, 601 (1966). Prepn from cattle brains: Hess, *J. Am. Med. Assoc.* **66**, 558 (1916). From rabbit brains or lung tissue: Singher, Swart, U.S. pat. **2,842,480** (1958 to Ortho); *see also* U.S. pats. **2,847,347** and **2,847,350** (1958 to Ortho). Purification of bovine tissue factor: R. Bach *et al., J. Biol. Chem.* **256**, 8324 (1981). The tissue thromboplastin activity seems to be brought about by a lipoprotein comprising a combination of phospholipids: phosphatidyl ethanolamine (PE), phosphatidyl serine (PS), and phosphatidyl choline (PC): R. B. Hunter, *et al., Fibrinogen and Fibrin Turnover of Clotting Factors* (Schattauer, Stuttgart, 1963) pp 429-454. Removal of phospholipids from tissue factor results in loss of coagulant activity as well as solubilizing the protein component. Purification and characterization of the protein component: Nemerson, Pitlick, *Biochemistry* **9**, 5100 (1970); of tissue factor apoprotein: F. A. Pitlick, Y. Nemerson, *Methods Enzymol.* **45B**, 37-48 (1976). Located on the plasma membrane of endothelial cells readily available for complexing with clotting factors: Zeldis *et al., Science* **175**, 766 (1972). Synthesis by cultured cells: Zacharski, McIntyre, *J. Med.* **4**, 118 (1973). Possibly a vitamin K dependent clotting factor: L. R. Zacharski *et al., Ann. N.Y. Acad. Sci.* **370**, 311 (1981). Use as clinical reagent: F. D. Ziegler *et al., Ann. Clin. Lab.*

Sci. **11**, 202 (1981); A. L. Suchman, P. F. Griner, *Ann. Int. Med.* **104**, 810 (1986).
THERAP CAT: Hemostatic.

9528. Thrombopoietin. Thrombocytopoiesis-stimulating factor; TSF; TPO; c-Mpl ligand; megakaryocyte potentiating factor. Glycopeptide hormone which is the major physiological regulator of platelet production. Promotes the maturation of megakaryocytes and increases platelet size and number. Produced in the kidney; detectable in plasma and urine of thrombocytopenic animals and humans. Identification of humoral factor that increases platelet levels: E. Kelemen *et al., Acta Haematol. (Basel)* **20**, 350 (1958). Partial purification from rabbit plasma: B. L. Evatt *et al., J. Lab. Clin. Med.* **83**, 364 (1974); from human plasma: A. Vannucchi *et al., Leukemia* **2**, 236 (1988). Production by cultured human embryonic kidney cells: T. P. McDonald *et al., J. Lab. Clin. Med.* **85**, 59 (1975). Purification: *idem et al., Exp. Hematol.* **17**, 865 (1989). Review of biological activity and properties: *idem, Am. J. Pediatr. Hematol. Oncol.* **14**, 8-21 (1992). Recombinant TPO acts both as a proliferative and a maturation factor for megakaryocytes. Cloning and characterization of human TPO: F. J. de Sauvage *et al., Nature* **369**, 533 (1994); of murine: S. Lok *et al., ibid.* 565; K. Kaushansky *et al., ibid.* 568; F. Wendling *et al., ibid.* 571.

9529. Thrombospondin. Glycoprotein G; thrombin-sensitive protein; TSP. A major glycoprotein constituent of human platelet α-granules that is released in response to platelet activation by α-thrombin, and plays an important role in mediating platelet aggregation. Discovery and initial isoln and properties: N. L. Baenziger *et al., Proc. Nat. Acad. Sci. USA* **68**, 240 (1971); *idem, J. Biol. Chem.* **247**, 2723 (1972). Thrombospondin was initially thought to be a single polypeptide of mol wt about 190,000, but has been shown to be a disulfide-linked trimer of mol wt 450,000: I. Hagen, *Biochim. Biophys. Acta* **392**, 242 (1975); D. R. Phillips, P. P. Agin, *J. Biol. Chem.* **252**, 2121 (1977); J. W. Lawler *et al., Thromb. Haemostasis* **37**, 355 (1977); *idem, J. Biol. Chem.* **253**, 8609 (1978); S. S. Margossian *et al., ibid.* **256**, 7495 (1981). Synthesis and secretion by cells in culture: G. J. Raugi *et al., J. Cell Biol.* **95**, 351 (1982). Quantitative analysis: J. N. George *et al., J. Lab. Clin. Med.* **92**, 430 (1978). The binding of secreted thrombospondin to platelet membranes is dependent on Ca^{2+}: D. R. Phillips *et al., J. Biol. Chem.* **255**, 11629 (1980). Evidence for calcium-sensitive structure: J. Lawler *et al., ibid.* **257**, 12257 (1982). Radioimmunoassay: S. D. Saglio, H. S. Slayter, *Blood* **59**, 162 (1982). Identity of thrombospondin with the endogenous lectin secreted by activated platelets: E. A. Jaffee *et al., Nature* **295**, 246 (1982). *See also* Fibrinogen.
Partial specific volume: 0.714 ml/g. Intrinsic viscosity: 40 ml/g in buffered saline at pH 7.6, 20°.

9530. Thromboxanes. Compounds derived from prostaglandin endoperoxides that cause platelet aggregation, contraction of arteries and other biological effects. Found in platelets, leucocytes, lung tissue, spleen, kidney, and umbilical artery. They are important mediators of the actions of polyunsaturated fatty acids transformed by cyclooxygenase. Discovery and structure of thromboxane B_2 (originally referred to as PHD): M. Hamberg, B. Samuelsson, *Proc. Nat. Acad. Sci. USA* **71**, 3400 (1974). Discovery and structure of thromboxane A_2 and identity with the unstable component of RCS (rabbit aorta contracting substance) *see* P. J. Piper, J. R. Vane, *Nature* **223**, 29 (1969); M. Hamberg *et al., Proc. Nat. Acad. Sci. USA* **72**, 2994 (1975). Biosynthesis and biological properties: P. Needleman *et al., Science* **193**, 163 (1976). Physiological review: J. R. Vane *et al., Int. Rev. Exp. Pathol.* **23**, 161-207 (1982). *Reviews:* B. Samuelsson in *Organic Chemistry*, A. T. Blomquist, H. H. Wasserman, Eds. vol. 36, entitled "Prostaglandin Research", P. Crabbe, Ed. (Academic Press, New York, 1977) pp 17-46; E. Granström *et al., Advan. Prostaglandin Thromboxane Leukotriene Res.* **10**, 15-58 (1982); L. J. Roberts *et al., ibid.* 211-225. Books: *Advances in Prostaglandin and Thromboxane Research* vols. **1-8**, B. Samuelsson, R. Paoletti, Eds. (Raven Press, New York, 1976-1980); *New Synthetic Routes to Prostaglandins and Thromboxanes*, S. M. Roberts, F. Scheinmann, Eds. (Academic Press, New York, 1982).

Thromboxane A_2

Thromboxane B_2

Thromboxane A_2, $C_{20}H_{32}O_5$, *(5Z,9α,11α,13E,15S)-9,11-epoxy-15-hydroxythromboxa-5,13-dien-1-oic acid, [1S-[1α,3α-(1E,3R*)4β(Z),5α]]-7-[3-(3-hydroxy-1-octenyl)-2,6-dioxabicyclo[3.1.1]hept-4-yl]-5-heptenoic acid, TXA₂.* Highly unstable, biologically active bicyclic oxitane-oxane compound derived from the endoperoxide *PGG_2* and rapidly converted to thromboxane B_2 by addition of water. Formed by incubation of arachidonic acid or PGG_2 with washed platelets. It induces irreversible platelet aggregation and causes contraction of the isolated rabbit aorta and release of serotonin and ADP from platelets in platelet-rich plasma. Synthesis of stable analogs: E. J. Corey *et al., Tetrahedron Letters* **1980**, 137; K. M. Massey, G. M. Bundy, *ibid.* 445; S. Ohuchida *et al., J. Am. Chem. Soc.* **103**, 4597 (1981); V. N. Kale, D. L. J. Clive, *J. Org. Chem.* **49**, 1554 (1984). Synthesis of biologically active unstable analogs: S. S. Bhagwat *et al., Tetrahedron Letters* **1985**, 1955. Total synthesis and structure of thromboxane A2: *eidem, Nature* **315**, 511 (1985). Formation and effects in human platelets: J. Svensson *et al., Acta Physiol. Scand.* **98**, 285 (1970). Biological half-life: 32 ±2 sec at 37°.

Thromboxane B_2, $C_{20}H_{34}O_6$, *(5Z,9α,13E,15S)-9,11,15-trihydroxythromboxa-5,13-dien-1-oic acid, [2R-[2α(1E,3S*),-3β(Z),4β,6α]]-7-[tetrahydro-4,6-dihydroxy-2-(3-hydroxy-1-octenyl)-2H-pyran-3-yl]-5-heptenoic acid, TXB₂, PHD.* A stable metabolite of thromboxane A_2 in platelets, initially considered biologically inactive. It is released during anaphylaxis in isolated guinea pig lungs and has been isolated from guinea pig brain homogenates and carrageenin-induced granuloma. TXB₂ has also been reported as possessing chemotactic properties. Total synthesis: N. A. Nelson, R. W. Jackson, *Tetrahedron Letters* **1976**, 3275; R. C. Kelly *et al., ibid.* 3279; from a prostaglandin $F_{2α}$ derivative: W. P. Schneider, R. A. Morge, *ibid.* 3283; stereospecific synthesis from D-glucose: S. Hanessian, P. Lavallee, *Can. J. Chem.* **55**, 562 (1977); E. J. Corey *et al., Tetrahedron Letters* **1977**, 1625; S. Hanessian, P. Lavallee, *Can. J. Chem.* **59**, 870 (1981). Metabolism: L. J. Roberts *et al., J. Biol. Chem.* **252**, 7415 (1966). Biological properties: J. R. Boot *et al., J. Physiol.* **257**, 47P (1976); L. S. Wolfe *et al., Biochem. Biophys. Res. Commun.* **70**, 907 (1976); W.-C. Chang *et al., Prostaglandins* **13**, 3 (1977). Plates from ethyl acetate/ether/petr ether, mp 95-96°. $[\alpha]_D^{25}$ +57.4° (c = 0.26 in ethyl acetate).

9531. Thuja. Arbor vitae; yellow cedar; false white cedar; tree of life. Dried, leafy, young twigs of *Thuja occidentalis* L., Cupressaceae. **Habit.** North America; cultivated in Europe. **Constit.** Thujone, fenchone, thujetic acid, tannin, pinipicrin.

9532. Thujic Acid. *5,5-Dimethyl-1,3,6-cycloheptatriene-1-carboxylic acid;* 4,4-dimethylcyclohepta-2,5,7-trienecarboxylic acid; dehydroperillic acid. $C_{10}H_{12}O_2$; mol wt 164.20. C 73.15%, H 7.37%, O 19.49%. Antibiotic substance from heartwood of *Thuja plicata* D. Don, Cupressaceae (Western red cedar). Isoln: Anderson, Sherrard, *J. Am. Chem. Soc.* **55**, 3813 (1933); Erdtman, Gripenberg, *Acta Chem. Scand.* **2**, 625 (1948). Structure: Gripenberg, *ibid.* **3**, 1137 (1949); **5**, 995 (1951); **10**, 487 (1956); Davis, Tulinsky, *Tetrahedron Letters* **1962**, 839.

Crystals from petr ether, mp 88-89°. uv max: 220, 280 nm (log ϵ 4.3, 3.7); min: 240 nm (log ϵ 3.0).

Methyl ester, $C_{11}H_{14}O_2$, crystals, mp 34.5-35° or liq with pleasant odor, bp_{14} 112-113°. d_4^{22} 1.0225. n_D^{22} 1.5130.

Hexahydrothujic acid, $C_{10}H_{18}O_2$, liq, bp_{14} 150-152°. n_D^{20} 1.4671.

9533. Thujone. *4-Methyl-1-(1-methylethyl)bicyclo-[3.1.0]hexan-3-one; 3-thujanone.* $C_{10}H_{16}O$; mol wt 152.24. C 78.90%, H 10.59%, O 10.51%. A constituent of many essential oils; present in thuja, etc. Equilibrium mixture contains 33% α-thujone and 67% β-thujone: Eastman, Winn, *J. Am. Chem. Soc.* **82**, 5908 (1960). α- and β-Thujones differ only in the stereochemistry of the 4-methyl group. Conformation: Hach *et al.*, *Tetrahedron Letters* **1970**, 3175. Chemistry: J. P. Kutney *et al.*, *Bioorg. Chem.* **7**, 289 (1978); *eidem*, *Can. J. Chem.* **57**, 3145 (1979); **58**, 2641 (1980). Toxicity study: K. C. Rice, R. S. Wilson, *J. Med. Chem.* **19**, 1054 (1976). *Review:* J. L. Simonsen, *The Terpenes* vol. **II** (University Press, Cambridge, 1949) pp 32-52.

β-Thujone

Colorless or almost colorless liquid. uv max (isooctane): 300 nm (ϵ 23). Practically insol in water. Sol in alc and many other organic solvents. LD_{50} s.c. in mice: 134.2 mg/kg (Rice, Wilson).

l-Form, α-*thujone*. bp_{17} 83.8-84.1°. d_4^{25} 0.9109. n_D^{15} 1.4490. $[\alpha]_D^{20}$ −19.2°. LD_{50} s.c. in mice: 87.5 mg/kg (Rice, Wilson).

d-Form, *d*-*isothujone, β-thujone*. bp_{17} 85.7-86.2°. d_4^{25} 0.9135. n_D^{25} 1.4500. $[\alpha]_D^{15}$ +72.5°. LD_{50} s.c. in mice: 442.2 mg/kg (Rice, Wilson).

Caution: Ingestion may cause convulsions.

9534. Thujopsene. *1,1a,4,4a,5,6,7,8-Octahydro-2,4a,8,-8-tetramethylcyclopropa[d]naphthalene;* widdrene. $C_{15}H_{24}$; mol wt 204.36. C 88.16%, H 11.84%. From wood oil of the Japanese Hiba tree, *Thujopsis dolobrata* Sieb, and Zucc., *Cupressaceae:* Yano, *J. Soc. Chem. Ind. Japan* **16**, 443 (1913); Uchida, *ibid* **31**, 501 (1928). Identity with widdrene: Erdtman, Thomas, *Acta Chem. Scand.* **12**, 267 (1958). Structure: Norin, *ibid.* **15**, 1676 (1961). Stereochemistry: Sisido *et al.*, *J. Org. Chem.* **26**, 1964 (1961); Norin *Acta Chem. Scand.* **17**, 738 (1963). Synthesis: Dauben, Ashcraft, *J. Am. Chem. Soc.* **85**, 3673 (1963); Büchi, White, *ibid.* **86**, 2884 (1964).

Liquid. bp_{10} 120°. $[\alpha]_D$ −110° (c = 2 in chloroform). n_D^{25} 1.5031. d^{24} 0.932. uv max (alc): 212 nm (ϵ 4680).

9535. Thulium. Tm; at. wt 168.93421; at. no. 69; valences 2, 3. A rare earth metal of the yttrium group; member of the lanthanide series. Naturally occurring isotope (mass number): 169; known artificial radioactive isotopes: 147; 148; 150-168; 170-176. Estimated abundance in earth's crust: 0.2-0.5 ppm. Found in small quantities in euxenite, ytterspar, sipylite, gadolinite, and other rare earth minerals. Discovered in crude erbium oxide: Cleve, *Compt. Rend.* **89**, 478, 521, 708 (1879). Obtained in a state of high purity by fractional crystn of its bromide: James, *J. Am. Chem. Soc.* **32**, 517 (1910); **33**, 1332 (1911). Sepn from other rare earths by ion exchange: Spedding *et al., ibid.* **76**, 2557 (1954). Toxicity study: Haley, *J. Pharm. Sci.* **54**, 663 (1965). Review of prepn, properties and compds: *The Rare Earths,* F. H. Spedding, A. H. Daane, Eds. (Krieger, Huntington, N.Y., 1971, reprint of 1961 ed.) 641 pp; Hulet, Bode, "Separation Chemistry of the Lanthanides and Transplutonium Actinides" in *MTP Int. Rev. Sci.: Inorg. Chem., Ser. One,* vol. 7, K. W. Bagnall, Ed. (University Park Press, Baltimore, 1972) pp 1-45; Moeller, "The Lanthanides" in *Comprehensive Inorganic Chemistry* vol. 4, J. C. Bailar, Jr. *et al.,* Eds. (Pergamon Press, Oxford, 1973) pp 1-101; F. H. Spedding in Kirk Othmer *Encyclopedia of Chemical Technology* vol. 19 (John Wiley & Sons, New York, 3rd ed., 1982) pp 833-854; *Chemistry of the Elements,* N. N. Greenwood, A. Earnshaw, Eds. (Pergamon Press, New York, 1984) pp 1423-1449. Brief review of properties: G. T. Seaborg, *Radiochim. Acta* **61**, 115-122 (1993).

Silvery-white, easily worked metal. Hexagonal close-packed crystals, d 9.3208. mp 1545°. bp 1725°; also reported as bp 1950° (Spedding, 1982). Heat of fusion: 16.84 kJ/mol. Heat of sublimation (25°): 232.2 kJ/mol. Solns of thulium salts show a characteristic absorption spectrum: Exner, Haschek, cited in *Mellor's* vol. V, 698 (1929).

Oxide, O_3Tm_2, *thulia*. Dense powder of greenish-white color. Prepd by igniting the oxalate; dissolves slowly in strong acids; exhibits a reddish glow on gentle heating.

Hydroxide, $Tm(OH)_3$, white precipitate.

Chloride heptahydrate, $TmCl_3.7H_2O$, deliquesc crystals. Sol in water, in alcohol. LD_{50} in mice: 485 mg/kg i.p.; 6.25 g/kg orally (Haley).

Sulfate octahydrate, $Tm_2(SO_4)_3.8H_2O$, obtained by ppting an aq soln of thulium chloride and sulfuric acid with alc.

Oxalate hexahydrate, $Tm_2(C_2O_4)_3.6H_2O$, greenish-white precipitate. Sol in aq alkali oxalates with formation of double oxalates.

9536. Thurfyl Nicotinate. *Nicotinic acid tetrahydrofurfuryl ester;* tetrahydrofurfuryl nicotinate; nicotafuryl; Trafuril. $C_{11}H_{13}NO_3$; mol wt 207.23. C 63.76%, H 6.32%, N 6.76%, O 23.16%. Prepn: Hartmann, Merz, U.S. pat. **2,485,152** (1949 to Ciba). Brief review: A. A. Fisher, *Cutis* **51**, 225-227 (1993).

Oil, $bp_{0.25}$ 114-116°. Sol in water, oil.

THERAP CAT: Rubefacient.

9537. Thyme. Dried leaves and flowering tops of *Thymus vulgaris* L., *Labiatae.* Habit. Southern Europe; cultivated in gardens. *Constit.* Volatile oil; tannin, gum.

USE: As a seasoning in foods.

9538. Thymidine. 1-(2-Deoxy-β-D-ribofuranosyl)-5-methyluracil; thymine-2-desoxyriboside. $C_{10}H_{14}N_2O_5$; mol wt 242.23. C 49.58%, H 5.83%, N 11.56%, O 33.03%. Constituent of deoxyribonucleic acid, *q.v.* Isoln from thymonucleic acid: Levene, London, *J. Biol. Chem.* **83**, 793 (1929). Structure: Levene, Tipson, *ibid.* **109**, 623 (1935). Conformation: Lemieux, *Can. J. Chem.* **39**, 116 (1961); Tollin *et al., Nature* **217**, 1148 (1968). Prepn of thymidine-3'-phosphate and of thymidine-5'-phosphate: Tener, *J. Am. Chem. Soc.* **83**, 165 (1961). *Review: Basic Principles in Nucleic Acid Chemistry,* vol. 1, P. O. P. Ts'o, Ed. (Academic Press, New York, 1974) *passim. See also* Nucleic Acids.

Z. Liu *et al., ibid.* **27**, 1144 (1981). Toxicity: P. M. Jenner *et al., Food Cosmet. Toxicol.* **2**, 327 (1964).

Rosettes of needles from ethyl acetate. mp 185°. Yields a sublimate of thymine when heated. $[\alpha]_D^{25}$ +30.6° (c = 1.029). uv max (pH 7.2): 206.5, 267 nm ($\epsilon \times 10^{-3}$ 9.8, 9.7), Voet *et al., Biopolymers* **1**, 193 (1963). Sol in water, methanol, hot alcohol, hot acetone, hot ethyl acetate, pyridine, glacial acetic acid; sparingly sol in hot chloroform.

Monotrityl thymidine, $C_{29}H_{28}N_2O_5$, prepd by the action of triphenylmethyl chloride on thymidine in pyridine. mp 125°. $[\alpha]_D^{24}$ +11.4° (c = 1.01 in acetone).

9539. Thymine. *5-Methyl-2,4(1H,3H)-pyrimidinedione;* 5-methyluracil; 2,4-dihydroxy-5-methylpyrimidine. C_5H_6-N_2O_2; mol wt 126.11. C 47.62%, H 4.80%, N 22.21%, O 25.37%. A pyrimidine derivative; constituent of nucleic acids. Originally isolated from thymus nucleic acid: Levene, *Z. Physiol. Chem.* **39**, 4 (1903). Prepn by heating 2-ethylmercapto-4-hydroxy-5-methylpyrimidine: Wheeler, Merriam, *Am. Chem. J.* **29**, 478 (1903); **43**, 29 (1910). From methylcyanacetylurea by catalytic reduction: Bergmann, Johnson, *J. Am. Chem. Soc.* **55**, 1733 (1933). From β-methylmalic acid: Scherp, *J. Am. Chem. Soc.* **68**, 912 (1946). Crystal structure of monohydrate: Gerdil, *Acta Cryst.* **14**, 333 (1961). *Review:* Ts'o, "Bases, Nucleosides and Nucleotides" in *Basic Principles in Nucleic Acid Chemistry* vol. 1, P. O. P. Ts'o, Ed. (Academic Press, New York, 1974) pp 453-584. *See also* Nucleic Acids.

Dendritic or star-shaped plates from water, sometimes short needles. Sublimes in platelets. Dec 335-337° (Kofler stage). Weak acid, pK at 25° = 9.94. uv max (pH 7.0): 205, 264.5 nm ($\epsilon \times 10^{-3}$ 9.5, 7.9). Sol in hot water; slightly sol in cold water (4 g/l at 25°). Somewhat sol in alc; sparingly sol in ether; readily sol in alkalies with formation of salts. Oxidation yields urea, ethanal, pyruvic acid, formic acid. Hydrazine reacts with thymine forming urea and 4-methylpyrazolone. Thymine forms a silver salt which is sol in excess ammonia. Its mercuric and lead salts are insol.

Thymine-2-desoxyriboside, *see* Thymidine.

USE: In biochemical research.

9540. Thymol. *5-Methyl-2-(1-methylethyl)phenol;* 5-methyl-2-isopropyl-1-phenol; 1-methyl-3-hydroxy-4-isopropylbenzene; 3-*p*-cymenol; 3-hydroxy-*p*-cymene; thyme camphor; *m*-thymol. $C_{10}H_{14}O$; mol wt 150.22. C 79.96%, H 9.39%, O 10.65%. Isolated by Neumann in 1719. Obtained from the essential oil of *Thymus vulgaris* L. and *Monarda punctata* L., Labiatae: Arppe, *Ann.* **58**, 41 (1846); Meyer, *Pharm. Ztg.* **81**, 192, 205 (1936). Also occurs in other volatile oils. Produced synthetically from *p*-cymene, piperitone, or *m*-cresol: Austerweil, **Brit.** pat. **221,227** (1923); Jennen, Verdroncken, *Compt. Rend.* **245**, 183 (1957); Bottoms, U.S. pat. **2,840,616** (1958 to Natl. Cylinder Gas). Bactericidal activity: J. M. Schaffer, F. W. Tilley, *J. Bacteriol.* **14**, 259 (1927). Mold elimination on surfaces: O. W. Richards, K. J. Hawley, *J. Chem. Ed.* **16**, 6 (1939). In vitro antifungal activity: H. B. Myers, *J. Am. Med. Assoc.* **89**, 1834 (1927). Effectiveness as antifungal preservative: M. Dersarkissian, M. Goodberry: *Studies Conserv.* **25**, 28 (1980). Use as clinical preservative: T. Z. Liu, *Clin. Chem.* **25**, 336 (1979); T.

Crystals, mp 51.5°. bp ~233°. Appreciably volatile at 100°; volatilizes in water vapors. Characteristic odor; pungent, somewhat caustic taste. d_4^{25} 0.9699. n_D^{20} 1.5227; n_D^{25} 1.5204. One gram dissolves in ~1000 ml water, 1 ml alcohol, 0.7 ml chloroform, 1.5 ml ether, 1.7 ml olive oil at 25°. Sol in glacial acetic acid, oils, fixed alkali hydroxides. LD_{50} orally in rats: 980 mg/kg (Jenner).

Incompat: Acetanilide, antipyrine, camphor, monobromated camphor, chloral hydrate, menthol, quinine sulfate, salol, urethane, spirit nitrous ether; in triturations because of liquefaction.

Acetate, $C_{12}H_{16}O_2$, *acetylthymol, thymyl acetate.* Yellowish, oily liq; thymol odor. d^0 1.009. bp 243.5-245.5°. Practically insol in water. Miscible with alcohol, benzene chloroform, ether.

Caution: Mild irritant.

Carbonate, $C_{21}H_{26}O_3$, white crystals; thymol odor; volatilizes with steam. mp 49°. Insol in water, acids, alkalies. Sol in hot alcohol, chloroform, ether, carbon tetrachloride.

USE: For destroying mold; preserving documents, art objects and urine. Stabilizer (antioxidant) for trichloroethylene, halothane.

THERAP CAT: Antiseptic (topical); anthelmintic (Nematodes).

THERAP CAT (VET): Has been used as anthelmintic, and as an antiseptic, external and internal.

9541. Thymol Blue. *4,4'-(3H-2,1-Benzoxathiol-3-ylidene)bis[5-methyl-2-(1-methylethyl)phenol] S,S-dioxide;* α-hydroxy-α,α-bis(5-hydroxycarvacryl)-o-toluenesulfonic acid γ-sultone; thymolsulfonephthalein. $C_{27}H_{30}O_5S$; mol wt 466.60. C 69.50%, H 6.48%, O 17.14%, S 6.87%.

Brownish-green, cryst powder; characteristic odor. Insol in water. Sol in alcohol, dil alkali solns.

USE: As as acid-base indicator; pH: red 1.2 to yellow 2.8; also yellow 8.0 to blue 9.6.

9542. Thymolphthalein. *3,3-Bis[4-hydroxy-2-methyl-5-(1-methylethyl)phenyl]-1(3H)-isobenzofuranone; 5',5''-di-isopropyl-2',2''-dimethylphenolphthalein.* $C_{28}H_{30}O_4$; mol wt 430.54. C 78.11%, H 7.02%, O 14.86%. Obtained by heating phthalic anhydride with thymol at 110° in the presence of stannic chloride.

Needles, mp ~253°. Insol in water; sol in alcohol, acetone; also sol in dil alkalies with a blue color, in H_2SO_4 with a carmine-red color.

USE: As pH indicator: colorless 9.3 to blue 10.5. Also as reagent for blood after decolorizing the alkaline soln by boiling with zinc dust.

9543. Thymomodulin. Leucotrofina. Cell-free thymic hormone preparation extracted from calf thymus by acid hydrolysis. Composed of a mixture of biologically active acidic peptides of mol wt <10,000. Modulates the maturation of T-cells. Prepn of crude extract from calf thymus: B.

Brunetti, E. Pini, U.S. pat. 3,657,417 (1972 to Ellem). Electrophoretic characterization: C. Secchi *et al., Riv. Eur. Sci. Med. Farmacol.* 4, 499 (1982), *C.A.* 101, 21878m (1984). Biological activity and comparison with other thymic hormones: J. J. Twomey, N. M. Kouttab, *Cell. Immunol.* 72, 186 (1982). Purification and pharmacodynamics: *eidem, Drugs Exptl. Clin. Res.* 10, 921 (1984). Antileukopenic activity in myelodepressed cancer patients: Z. Uray *et al., Aggressologie* 21, 215 (1980); C. Gallo Curcio *et al., Int. J. Immunother.* 2, 189 (1986). Use in prophylaxis of infantile asthma: R. Genova, A. Guerra, *Ped. Med. Chir.* 5, 395 (1983). Preliminary study in AIDS: G. Valesini *et al., Eur. J. Cancer Clin. Oncol.* 22, 531 (1986).
THERAP CAT: Immunoregulator.

9544. Thymopentin. *N-(N-(N-(N²-L-arginyl-L-lysyl)-L-α-aspartyl)-L-valyl)-L-tyrosine;* thymopoietin pentapeptide; TP-5; ORF-15244; Immunox; Sintomodulina; Timunox. $C_{30}H_{49}N_9O_9$; mol wt 679.77. C 53.01%, H 7.27%, N 18.54%, O 21.18%. Thymic hormone analog corresponding to residues 32-36 of thymopoietin, *q.v.,* which exhibits the full biological activity of the natural hormone. Synthesis: G. Goldstein *et al., Science* 204, 1309 (1979); *eidem,* U.S. pat. 4,190,646 (1980 to Sloan-Kettering). Bioavailability: T. Audhya, G. Goldstein, *Int. J. Pept. Protein Res.* 22, 187 (1983). Pharmacology: K. Bolla *et al., Int. J. Clin. Pharmacol. Res.* 4, 431 (1984). Comparison of biological activity with thymopoietin and splenin: T. Audhya *et al., Proc. Nat. Acad. Sci. USA* 81, 2847 (1984). Clinical study in treatment of primary immunodeficiencies: F. Aitui *et al., Lancet* 1, 551 (1983); in AIDS: N. Clumeck *et al., Int. J. Clin. Pharmacol. Res.* 4, 459 (1984); in rheumatoid arthritis: M. G. Malaise *et al., Lancet* 1, 832 (1985). *Review:* E. A. Boyse, *Surv. Immunol. Res.* 4, 6-10 (1985).

Arg–Lys–Asp–Val–Tyr

THERAP CAT: Immunoregulator.

9545. Thymopoietin. Thymin (formerly); TP. A polypeptide thymic hormone which selectively induces T-cell maturation from prothymocytes and inhibits B-cell differentiation. mol wt ~5500 daltons. First detected as a factor causing impaired neuromuscular transmission in myasthenia gravis, a neuromuscular disease associated with thymic abnormalities: G. Goldstein, S. Wittingham, *Lancet* 2, 315 (1966); G. Goldstein, *ibid.* 2, 119 (1968). Isoln from bovine thymus as two closely homologous forms, thymopoietins I and II: *idem, Nature* 247, 11 (1974); *idem, Ann. N.Y. Acad. Sci.* 249, 177 (1975); *idem,* U.S. pat. 4,077,949 (1978 to Sloan-Kettering). Bovine TPs contain 49 amino acids; I differs from II only in residues 1, 2, and 43. The active site corresponds to residues 32-36. Human TP contains 48 amino acids. Human and bovine forms are highly homologous and are identical at the active site. Proposed sequence of bTP-II: D. H. Schlesinger, G. Goldstein, *Cell* 5, 361 (1975). Revised sequence of II, sequence of I, isoln and sequence of splenin, *q.v.,* a thymopoietin-like peptide found in spleen: T. Audhya *et al., Biochemistry* 20, 6195 (1981). Isoln and sequence of human thymopoietin: T. Audhya *et al., Proc. Nat. Acad. Sci. USA* 84, 3545 (1987). First synthesis of a bioactive fragment: D. H. Schlesinger, *Cell* 5, 367 (1975). Identification of active site and synthesis of thymopentin, *q.v.,* the pentapeptide corresponding to the active site: G. Goldstein *et al., Science* 204, 1309 (1979). Total synthesis of originally proposed amino acid sequence of II: M. Fujino *et al., Chem. Pharm. Bull.* 25, 1486 (1977); of revised sequence: T. Abiko, H. Sekino, *ibid.* 35, 2016 (1987). Effect on T-cell differentiation: R. S. Basch, G. Goldstein, *Proc. Nat. Acad. Sci. USA* 71, 1474 (1974); M. P. Scheid *et al., Ann. N.Y. Acad. Sci.* 249, 531 (1975). Comparison of bioactivities of thymopoietin and splenin: T. Audhya *et al., Proc. Nat. Acad. Sci. USA* 81, 2847 (1984). Review of thymopoietin and other thymic hormones: J. F. Bach, *J. Immunopharmacol.* 1, 277-310 (1979); of thymopoietin and immunoregulation: G. Goldstein, C. Lau in *Polypeptide Hormones,* R. F. Beers, E. G. Bassett, Eds. (Raven Press, New York, 1980) pp 459-467; E. A. Boyse, *Surv. Immunol. Res.* 4, 6-10 (1985).

9546. Thymosin. Cell free, lymphocytopoietic hormone and biologically active thymic factor, comprising a family of acidic, heat-stable polypeptides, that plays an important part in the development of T-cells (thymus-dependent lymphocytes) and immunological functions in animals and man. Extraction from calf thymus and partial purification: A. L. Goldstein *et al., Proc. Nat. Acad. Sci. USA* 56, 1010 (1966). Purification and characterization of thymosins and isoln of one of the active components, *thymosin fraction 5: eidem, ibid.* 69, 1800 (1972). Improved isoln and properties: J. A. Hooper *et al., Ann. N.Y. Acad. Sci.* 249, 125 (1975). Thymosin fraction 5 is a mixture of small polypeptides ranging in mol wt from ~1000 to 15,000, it is carbohydrate and lipid-free and contains no nucleotides or unusual amino acids. It is an immunopotentiating agent that can act in place of the thymus gland in immuno-deprived or thymus-deprived patients to restore some immune functions. Nomenclature of the polypeptide fragments of fraction 5 is based on isoelectric points: α-region below 5.0, β-region 5.0-7.0, and γ-region above 7.0. Isoln and preliminary structural determination of *thymosin* α_1, a biologically active, 28-residue polypeptide component in the highly acidic region of fraction 5: A. L. Goldstein *et al., Proc. Nat. Acad. Sci. USA* 74, 725 (1977). Confirmation of primary structure: T. Michalewsky *et al., Int. J. Peptide Protein Res.* 21, 93 (1983). Isoln, characterization, biological activities, amino acid sequence analysis of α_1 and *polypeptide* β_1, a biologically inactive fragment: T. L. K. Low *et al., J. Biol. Chem.* 254, 981 (1979); T. L. K. Low, A. L. Goldstein, *ibid.* 987. Synthesis of α_1 by solution methods: C. Birr, U. Stollenwerk, *Angew. Chem. Int. Ed.* 18, 394 (1979); S. S. Wang *et al., J. Am. Chem. Soc.* 101, 253 (1979). Automated solid phase synthesis: *eidem, Int. J. Peptide Protein Res.* 15, 1 (1980). *Thymosin* β_4 is a biologically active fragment that appears to act on stem cells to form prothymocytes. Extrathymic occurrence of β_4: E. Hannappel *et al., Proc. Nat. Acad. Sci. USA* 79, 2172 (1982). Amino acid sequence of β_4: T. L. K. Low *et al., ibid.* 78, 1162 (1981); T. L. K. Low, A. L. Goldstein, *J. Biol. Chem.* 257, 1000 (1982). Automated solid phase synthesis of β_4: S. S. Wang *et al., Int. J. Peptide Protein Res.* 18, 413 (1981); T. L. K. Low *et al., Biochemistry* 22, 733 (1983). In vitro synthesis of β_4 by rat spleen mRNA: A. W. Filipowicz, B. L. Horecker, *Proc. Nat. Acad. Sci. USA* 80, 1811 (1983). Biosynthesis of β_4 by spleen cells: G. Xu *et al., ibid.* 79, 4006 (1982). Thymosin β_4 (and fraction 5) have been shown to stimulate secretion of luteinizing hormone-releasing factor (LH-RH, *q.v.):* R. W. Rebar *et al., Science* 214, 669 (1981). Isoln of *thymosin* β_8 and *thymosin* β_9: E. Hannappel *et al., Proc. Nat. Acad. Sci. USA* 79, 1708 (1982). Synthesis of β_8: T. Abiko, H. Sekino, *Chem. Pharm. Bull.* 31, 1320 (1983). Synthesis of β_9: *eidem, J. Appl. Biochem.* 4, 449 (1982). For specific immunological activities of thymosin *see* Zisblatt *et al., Proc. Nat. Acad. Sci. USA* 66, 1170 (1970); Goldstein *et al., Radiat. Res.* 41, 579 (1970); Goldstein *et al., J. Immunol.* 104, 359 (1970); Bach *et al., Proc. Nat. Acad. Sci. USA* 68, 2734 (1971); A. L. Goldstein *et al.,* in *Progress in Cancer Research and Therapy,* vol. 2, M. A. Chirigos, Ed. (Raven Press, New York, 1977) pp 241-254. Evidence for mediation of thymosin fraction 5 effects on lymphocytes by prostaglandin E_2, *q.v.:* G. R. Garaci *et al., Science* 220, 1163 (1983). Clinical studies: D. W. Wara *et al., N. Engl. J. Med.* 292, 70 (1975); L. A. Schafer *et al., Ann. N.Y. Acad. Sci.* 277, 609 (1976); J. Costanzi *et al., ibid.* 332, 148 (1979). Review of clinical studies: G. D. Marshall *et al., Recent Results Cancer Res.* 75, 100-105 (1980). *Reviews:* A. L. Goldstein *et al., Recent Progr. Horm. Res.* 26, 505-538 (1970); Robey, "Thymosin Reviewed" in *Thymic Hormones,* T. D. Luckey, Ed. (University Park Press, Baltimore, 1973) pp 159-16; A. L. Goldstein, *Trans. Am. Clin. Climatol. Assoc.* 88, 79-94 (1976); A. L. Goldstein *et al., Recent Progr. Horm. Res.* 37, 369-416 (1981).

Soluble in 5% perchloric acid. Activity remains relatively stable at 100°.

9547. Thymostatin. Calf thymic extract which markedly inhibits *in vitro* and *in vivo* the incorporation of labelled nucleosides into the DNA and RNA of lymphocytes. Also inhibited incorporation of labelled nucleosides *in vitro* into other cell types both lymphoid and non-lymphoid, but did not affect incorporation *in vitro* of labelled amino acids into cellular protein. A carbohydrate containing peptide with a particle size < 2000. Prepn and properties: Goldstein *et al.*, *Proc. Nat. Acad. Sci. USA* **57**, 821 (1967). Relatively heat stable; insol in chloroform but sol in chloroform-methanol (2:1).

9548. Thymostimulin. TP-1. Biological response modifier; partially purified extract of calf thymus composed of a mixture of bovine thymic peptides. Extraction and purification: G. Bergesi, R. Falchetti, *Folia Allergol. Immunol. Clin.* **24**, 204 (1977). Pharmacology and biological properties: R. Falchetti *et al.*, *Drugs Exp. Clin. Res.* **3**, 39 (1977). Clinical studies in combination with antimicrobial therapy: P. Periti *et al.*, *J. Chemother.* **5**, 37 (1993); as an adjunct to cancer chemotherapy: M. Federico *et al.*, *Am J. Clin Oncol.* **18**, 8 (1995); with zidovudine, *q.v.*, vs HIV infection: G. Barbaro *et al. Curr. Ther. Res.* **56**, 369 (1995). Review of pharmacology and therapeutic use: K. L. Dechant, H. M. Bryson, *Clin. Immunother.* **1**, 378-398 (1994).
THERAP CAT: Immunomodulator.

9549. *o*-**Thymotic Acid.** *2-Hydroxy-6-methyl-3-(1-methylethyl)benzoic acid; 3-hydroxy-2-p-cymenecarboxylic acid; o*-thymotinic acid; 6-methyl-3-isopropylsalicylic acid. $C_{11}H_{14}$-O_3; mol wt 194.23. C 68.02%, H 7.27%, O 24.71%. Prepn from thymol: Kolbe, Lautemann, *Ann.* **115**, 205 (1861); Spallino, Provenzal, *Gazz. Chim. Ital.* **39** (II), 326 (1909); Royer *et al.*, *Bull. Soc. Chim. France* **1955**, 1421.

Monoclinic prismatic needles from water, mp 127°. Volatile with steam. One gram dissolves in 10 liters of water at 20°. Sol in alc, ether, chloroform, benzene, petr ether.
Sodium salt, $C_{11}H_{13}NaO_3$, white cryst mass, freely sol in water.
Amide, $C_{11}H_{15}NO_2$, needles from alc, mp 137°, dec 205°. Sol in org solvents and in aq solns of sodium carbonate.
Acetonyl ester, $C_{14}H_{18}O_4$, needles from alcohol, mp 75°. Sol in alcohol.
The methyl and ethyl esters appear to be liquids at room temperature.
USE: As salicylic acid and salicylates.

9550. Thymyl N-Isoamylcarbamate. *Isoamylcarbamic acid thymyl ester;* isopropyl-*m*-cresyl ester of isoamylcarbamic acid; Egressin. $C_{16}H_{25}NO_2$; mol wt 263.38. C 72.97%, H 9.57%, N 5.32%, O 12.15%. Prepd by interaction of isoamylamine and thymyl chloroformate: Zima, v. Werder, U.S. pat. **2,524,185** (1949 to E. Merck).

Needles from petr ether, mp 57°. Practically insol in water (< 1:50,000). Upon alkaline saponification, it is split into isoamylamine and thymol (along with CO_2).
THERAP CAT: Anthelmintic (Nematodes).

9551. Thyroid. Tiroidina; Thyradin; Thyrocrine. Thyroid gland of domesticated animals that are used as food by man, freed from connective tissue and fat, dried and powdered. Contains not less than 0.17% and not over 0.23% iodine in thyroid combination. 1 part ≃ 5 parts fresh gland. Chemistry and physiology: Rawson *et al.*, *The Hor-*

mones **vol. III** (Academic Press, New York, 1955) pp 433-519.
Yellowish powder; slight meat-like odor; saline taste.
THERAP CAT: Thyroid hormone.
THERAP CAT (VET): In myxedema. Has been used in obesity, renal insufficiency, chronic skin conditions and to increase spermatogenesis, libido, lactation.

9552. Thyroidin. Iodothyrin. A dried extract of thyroid diluted with milk sugar or other suitable diluent, equal in potency to official thyroid.
THERAP CAT: Thyroid hormone.

9553. Thyropropic Acid. *4-(4-Hydroxy-3-iodophenoxy)-3,5-diiodohydrocinnamic acid;* 3,3',5-triiodothyropropionic acid; β-[4-(4-hydroxy-3-iodophenoxy)-3,5-diiodophenyl]-propionic acid; Birodan. $C_{15}H_{11}I_3O_4$; mol wt 635.96. C 28.33%, H 1.74%, I 59.86%, O 10.06%. Thyroid hormone analog. Prepd by iodination of 3,5-diiodothyropropionic acid: Tomita, Lardy, *J. Biol. Chem.* **219**, 595 (1956).

Crystals from abs ethanol, mp 200°.
THERAP CAT: Antihyperlipoproteinemic.

9554. Thyroprotein. *Thyroactive protein;* Protamone-D. Iodinated casein. Use in feed supplements: Kohler, *The Use of Thyroactive Protein in Dairy Cattle Feeding,* The Texas Nutrition Conference Oct. **1953**, Abstracts of papers, pp 5-7. Prepn: Reineke *et al.*, *J. Biol. Chem.* **143**, 285 (1942); **147**, 115 (1943); Turner, Reineke, U.S. pats. **2,329,-445, 2,379,842, 2,385,117, 2,478,065** (1944, 1945, 1945, 1949 to Am. Dairies & Quaker Oats); Whitmoyer, Moore, U.S. pat. **2,382,193** (1945 to Whitmoyer Labs.); West, Van Bruggen, U.S. pats. **2,642,426, 2,709,671** (1953, 1955 to Feed Prods).
USE: Feed supplement, *Fed. Regist.* **45**, 41360 (1980).

9555. Thyroxine. *O-(4-Hydroxy-3,5-diiodophenyl)-3,5-diiodotyrosine; 3-[4-(4-hydroxy-3,5-diiodophenoxy)-3,5-diiodophenyl]alanine;* β-[(3,5-diiodo-4-hydroxyphenoxy)-3,5-diiodophenyl]alanine; 3,5,3',5'-tetraiodothyronine. $C_{15}H_{11}$-I_4NO_4; mol wt 776.87. C 23.19%, H 1.43%, I 65.34%, N 1.80%, O 8.24%. An amino acid of the thyroid gland, exerts a stimulating effect on metabolism. Its natural occurrence in the free form has not been demonstrated. It may be a cleavage product of thyroglobulin. The L-form is twice as active physiologically as the racemic product, the D-form has very little activity. Obtained from thyroid glands of animals: Kendall, *J. Am. Med. Assoc.* **64**, 2042 (1915); *J. Biol. Chem.* **39**, 125 (1919). Structure: Harington, *Biochem. J.* **20**, 293, 300 (1926). Synthesis: Harington, Barger, *ibid.* **21**, 169 (1927); Canzanelli, *et al., ibid.* **29**, 1617 (1935); Ginger, Anthony, U.S. pat. **2,889,363;** Anthony, Ginger, U.S. pat. **2,889,364** (both 1959 to Baxter Labs.). *See also* R. Pitt-Rivers, J. R. Tata, *The Thyroid Hormones* (Pergamon Press, 1959). Commercial syntheses: Nahm, Siedel in *Forschung in Hoechst* (Jubiläumsjahr 1963) p 663; Nahm, Siedel, *Ber.* **96**, 1 (1963).

L-form

DL-Thyroxine, needle-like crystals. Dec 231-233°. Insol in water, in alcohol, and in the other usual organic solvents, but in the presence of mineral acids or alkalies it dissolves in alcohol; sol in solns of the alkali hydroxides and in hot solns of the alkali carbonates. When alkali hydroxide solns of thyroxine are satd with sodium chloride, the sodium salt of thyroxine separates.

L-Thyroxine, crystals, dec 235-236°. $[\alpha]_{546}^{25}$ —3.2° (0.66 g in 6.07 g of 0.5N NaOH and 13.03 g alc); $[\alpha]_D^{20}$ —4.4° (3% in 0.13N NaOH in 70% EtOH).

L-Thyroxine Sodium Salt. *See* Levothyroxine Sodium.

D-Thyroxine, *Debetrol*. Crystals, dec 237°. $[\alpha]_{546}^{21}$ +2.97° (0.74 g in 6 g of 0.5N NaOH and 14 g of alcohol).

D-Thyroxine Sodium Salt, $C_{15}H_{10}I_4NNaO_4$, *dextrothyroxine sodium, Biotirmone, Choloxin, Detyroxin, Dethyrona, Dextroid, Dynothel, Eulipos.*

THERAP CAT: L-Form as thyroid hormone. D-Form as antihyperlipoproteinemic.

9556. Tiadenol. *2,2'-[1,10-Decanediylbis(thio)]bisethanol; 2,2'-(decamethylenedithio)diethanol;* 1,10-bis(2-hydroxyethylthio)decane; LL-1558; Delipid; Eulip; Fonlipol; Tiaden; Tiaterol. $C_{14}H_{30}O_2S_2$; mol wt 294.52. C 57.09%, H 10.27%, O 10.86%, S 21.77%. HOCH$_2$CH$_2$S(CH$_2$)$_{10}$SCH$_2$-CH$_2$OH. Prepn: Williams, Cossar, U.S. pat. **3,021,215** (1962 to Eastman Kodak); Lafon, **Ger.** pat. **2,038,836** (1971 to Orsymonde), *C.A.* **75**, 35528d (1971). Series of articles on pharmacology, metabolism, and toxicology: *Therapie* **27**, 395-444 (1972).

Crystals, mp 69.5°. uv max (ethanol): 212 nm. Soluble in ethanol, chloroform. Practically insol in water.

THERAP CAT: Antihyperlipoproteinemic.

9557. Tiagabine. *(R)-1-[4,4-Bis(3-methyl-2-thienyl)-3-butenyl]-3-piperidinecarboxylic acid;* (—)-(R)-1-[4,4-bis(3-methyl-2-thienyl)-3-butenyl]nipecotic acid; TGB; NO-328; NO-05-0328; NNC-05-0328; A-70569. $C_{20}H_{25}NO_2S_2$; mol wt 375.56. C 63.96%, H 6.71%, N 3.73%, O 8.52%, S 17.08%. GABA uptake inhibitor. Prepn: F. C. Groenvald, C. Braestrup, **PCT Int.** pat. **Appl. 87 00,171** (1987 to Novo); F. C. Gronvald, C. Braestrup, U.S. pat. **5,010,090** (1991 to Novo Nordisk); K. E. Andersen *et al., J. Med. Chem.* **36**, 1716 (1993). Pharmacology: C. L. Faingold *et al., Exp. Neurology* **126**, 225 (1994). HPLC determn in human plasma: L. E. Gustavson, S. Chu, *J. Chromatog.* **574**, 313 (1992). Clinical evaluation as add-on therapy: J. Bauer *et al., J. Epilepsy* **8**, 83 (1995). Review of pharmacology: W. J. Giardina, *ibid.* **7**, 161-166 (1994). *Reviews:* H. Mengel, *Epilepsia* **35**, Suppl. 5, S81-S84 (1994); L. Gram, *ibid.* S85-S87.

Hydrochloride, $C_{20}H_{25}NO_2S_2$·HCl, white to off-white, odorless crystalline powder, mp 192° (dec) (Mengel). $[\alpha]_D^{20}$ —11°. pKa$_1$ 3.3; pKa$_2$ 9.4. Partition coefficient (octanol-water, pH 7.4): 39.3. Soly in water: 3%. Practically insol in hexane.

THERAP CAT: Anticonvulsant.

9558. Tiamenidine. *N-(2-Chloro-4-methyl-3-thienyl)-4,5-dihydro-1H-imidazol-2-amine;* 2-[(2-chloro-4-methyl-3-thienyl)amino]-2-imidazoline; 2-chloro-4-methyl-3-(2'-imidazolin-2'-ylamino)thiophene; thiamenidine. $C_8H_{10}ClN_3S$; mol wt 215.71. C 44.55%, H 4.67%, Cl 16.44%, N 19.48%, S 14.87%. A thiophene analog of clonidine, *q.v.* Prepn: R. Rippel *et al.,* **Ger.** pat. **1,941,761;** *eidem,* U.S. pat. **3,758,476** (1971, 1973, both to Hoechst). Structural studies: J. M. Leger, *Compt. Rend. Ser. C* **289**, 93 (1979); A. Carpy *et al., Mol. Pharmacol.* **21**, 400 (1981). GC mass spectrometry determn in human plasma: T. A. Bryce, J. L. Burrows, *Biomed. Mass Spectrom.* **6**, 27 (1979). Pharmacology: E. Lindner, J. Kaiser, *Arch. Int. Pharmacol.* **211**, 305 (1974). Activity as α-adrenoceptor agonist: A. G. Roach *et al., J. Pharmacol. Exp. Ther.* **227**, 421 (1983).

Crystals from isopropanol + petr ether, mp 152°.

Hydrochloride, $C_8H_{10}ClN_3S$·HCl, *HOE-440, Sundralen.* Crystals from isopropanol + petr ether, mp 228-229°. LD$_{50}$ in rats, mice (mg/kg): 40, 45 i.v.; in mice: 170 s.c., 400 orally (Lindner, Kaiser).

THERAP CAT: Antihypertensive.

9559. Tiamulin. *[3aS-(3aα,4β,5α,6α,8β,9α,9aβ,10S*)]-[[2-(Diethylamino)ethyl)thio]acetic acid 6-ethenyldecahydro-5-hydroxy-4,6,9,10-tetramethyl-1-oxo-3a,9-propano-3aH-cyclopentacycloocten-8-yl ester;* 14-desoxy-14-[(2-diethylaminoethyl)mercaptoacetoxy]mutilin; thiamutilin; tiamutin; SQ-14055. $C_{28}H_{47}NO_4S$; mol wt 493.75. C 68.11%, H 9.59%, N 2.84%, O 12.96%, S 6.49%. Derivative of pleuromutilin, *q.v.* Prepn: H. Egger, **Ger.** pat. **2,248,237** corresp to U.S. pat. **3,919,290** (1973, 1975 both to Sandoz); H. Egger, H. Reinshagen, *J. Antibiot.* **29**, 915 (1976). Biosynthesis: F. Knauseder, E. Brandl, *ibid.* 125. *In vitro* activity: H. Werner *et al., ibid.* **31**, 756 (1978). Metabolism: J. Dreyfuss *et al., ibid.* **32**, 496 (1979). Effect vs *Mycoplasma:* C. O. Baughn *et al., Avian Dis.* **22**, 620 (1978); R. F. Goodwin, *Vet. Rec.* **104**, 194 (1979). Mechanism of action: G. Högenauer in *Antibiotics* vol. 5(pt. 1), F. E. Hahn, Ed. (Springer-Verlag, New York, 1979) pp 344-360. Treatment of swine dysentery: M. D. Anderson, *Vet. Med. Small Anim. Clin.* **78**, 98 (1983).

Fumarate, $C_{32}H_{51}NO_8S$, *81723 hfu, SQ-22947, Denagard, Dynamutilin, Tiamutin.* Crystals from acetone, mp 147-148° (after stirring in ethyl acetate and drying at 60° and 80° overnight).

THERAP CAT (VET): Antibacterial.

9560. Tianeptine. *7-[(3-Chloro-6,11-dihydro-6-methyl-dibenzo[c,f][1,2]thiazepin-11-yl)amino]heptanoic acid S,S-dioxide.* $C_{21}H_{25}ClN_2O_4S$; mol wt 436.96. C 57.72%, H 5.77%, Cl 8.11%, N 6.41%, O 14.65%, S 7.34%. Tricyclic compound with psychostimulant, anti-ulcer and anti-emetic properties. Prepn: C. Malen *et al.,* **Ger.** pat. **2,011,806** corresp to U.S. pat. **3,758,528** (1970, 1973 both to Sci. Union et Cie-Soc. Franc. Rech. Med.). Neuropharmacology study: C. Malen, J. Poignant, *Experientia* **28**, 811 (1972).

Sodium salt, $C_{21}H_{24}ClN_2NaO_4S$, *S-1574, Stablon.* Solid, mp 180°.

THERAP CAT: Antidepressant.

9561. Tiapride. *N-[2-(Diethylamino)ethyl]-2-methoxy-5-(methylsulfonyl)benzamide;* N-[2-(diethylamino)ethyl]-5-(methylsulfonyl)-o-anisamide; thiapride; FLC-1374. $C_{15}H_{24}$-N_2O_4S; mol wt 328.43. C 54.86%, H 7.37%, N 8.53%, O

19.49%, S 9.76%. Dopamine receptor antagonist structurally related to sulpiride, *q.v.* Prepn: G. Bulteau *et al.,* Ger. pat. **2,327,192; Brit.** pat. **1,394,563** (1973, 1975 both to Soc. Etudes Sci. Ind. L'Ile de France). Crystal structure: C. Houttemane *et al., Acta Cryst.* **C39,** 585 (1983). Dopamine receptor binding studies: T. Arima *et al., Japan. J. Pharmacol.* **41,** 419 (1986). Pharmacokinetics: E. Rey *et al., Int. J. Clin. Pharmacol. Ther. Toxicol.* **20,** 62 (1982). Clinical evaluation in neuroleptic-induced dyskinesia: W. Greil *et al., Neuropsychobiology* **14,** 17 (1985); in alcohol withdrawal syndrome: R. Agricola *et al., J. Int. Med. Res.* **10,** 160 (1982); G. K. Shaw *et al., Brit. J. Psychiatry* **150,** 164 (1987).

Crystals, mp 123-125°.
Hydrochloride, $C_{15}H_{24}N_2O_4S.HCl$, *Gramalil, Italprid, Luxoben, Sereprile, Tiapridal, Tiapridex.*
THERAP CAT: Antidyskinetic.

9562. Tiaprofenic Acid. *5-Benzoyl-α-methyl-2-thiopheneacetic acid;* α-methyl-5-benzoyl-2-thienylacetic acid; FC-3001; RU-15060; Suralgan; Surgam. $C_{14}H_{12}O_3S$; mol wt 260.31. C 64.60%, H 4.65%, O 18.44%, S 12.32%. Nonsteroidal anti-inflammatory. Prepn: F. Clémence, O. Le Martret, Ger. pat. **2,055,264** and **Fr.** pat. **2,112,111** (1971 and 1972 to Roussel-UCLAF), *C.A.* **75,** 63597u (1971) and **78,** 97473c (1973); F. Clémence *et al., Eur. J. Med. Chem. - Chem. Ther.* **9,** 390 (1974). Pharmacology: *eidem, ibid.;* H. Fujimura *et al., Oyo Yakuri* **9,** 715 (1975), *C.A.* **83,** 188321w (1975). Review of pharmacology and efficacy in rheumatic diseases and pain control: E. M. Sorkin, R. N. Brogden, *Drugs* **29,** 208-235 (1985).

mp 96° (isopropyl ether).
THERAP CAT: Anti-inflammatory.

9563. Tiaprost. *7-[3,5-Dihydroxy-2-[3-hydroxy-4-(3-thienyloxy)-1-butenyl]cyclopentyl]-5-heptenoic acid;* (15-*R,S*)-16-(3-thienyloxy)-ω-tetranor-PGF$_{2α}$. $C_{20}H_{28}O_6S$; mol wt 396.50. C 60.58%, H 7.12%, O 24.21%, S 8.09%. Analog of prostaglandin F$_{2α}$, *q.v.* Prepn: W. Bartmann *et al.,* Ger. pat. **2,524,955** corresp to U.S. pat. **4,258,053** (1977, 1981 both to Hoechst); *eidem, Prostaglandins* **17,** 301 (1979). Efficacy in bovine endometritis: W. Bentele *et al., Tierarztliche Umschau* **35,** 676, 678, 683 (1980). Pharmacological effects: W. v. Rechenberg *et al., Blue Book for the Vet. Profession* (Hoechst AG) **30,** 417 (1981). Administration in cattle: R. Humble, *ibid.* 425.

Tromethamine salt, $C_{24}H_{39}NO_9S$, *tiaprost trometamol, Iliren.*
THERAP CAT (VET): Luteolytic.

9564. Tiaramide. *4-[(5-Chloro-2-oxo-3(2H)-benzothiazolyl)acetyl]-1-piperazineethanol;* 5-chloro-3-[4-(2-hydroxyethyl)-1-piperazinyl]carbonylmethyl-2-benzothiazolinone; tialamide. $C_{15}H_{18}ClN_3O_3S$; mol wt 355.85. C 50.63%, H 5.10%, Cl 9.96%, N 11.81%, O 13.49%, S 9.01%. Prepn: Umio *et al.,* **Japan.** pat. **71 15,302,** *C.A.* **75,** 36127j (1971);

eidem, **Japan.** pat. **71 18,752,** *C.A.* **75,** 63824r (1971); Umio, **U.S.** pat. **3,661,921** (1971, 1971, 1972 all to Fujisawa). Pharmacology: Takashima *et al., Arzneimittel-Forsch.* **22,** 711 (1972); Tsurumi *et al., ibid.* 716, 724. Metabolism: Noda *et al., ibid.* 732. Toxicity studies: Watanabe *et al., ibid.* **23,** 504 (1973). Mode of antiasthmatic action: G. C. Folco *et al., Pharmacol. Res. Commun.* **11,** 703 (1979).

pKa 6.2.
Hydrochloride, $C_{15}H_{18}ClN_3O_3S.HCl$, *NTA-194, FK-1160, Solantal.* White, odorless, bitter tasting crystalline powder, mp 159-161°. pH (10% aqueous soln): 3.4-3.7. Very soluble in water; slightly sol in organic solvents. LD$_{50}$ in male mice, rats (mg/kg): 178, 203 i.v.; 298, 540 i.p.; 564, 3600 orally (Watanabe).
THERAP CAT: Antiasthmatic; anti-inflammatory.

9565. Tibezonium Iodide. *N,N-Diethyl-N-methyl-2-[[4-[4-(phenylthio)phenyl]-3H-1,5-benzodiazepin-2-yl]thio]-ethanaminium iodide;* diethylmethyl[2-[[4-[p-(phenylthio)-phenyl]-3H-1,5-benzodiazepin-2-yl]thio]ethyl]ammonium iodide; 2-[β-(N-diethylamino)ethylthio]-4-(p-phenylthio)-phenyl-3H-1,5-benzodiazepine methiodide; thiabenzazonium iodide; Rec-15-0691; Antoral. $C_{28}H_{32}IN_3S_2$; mol wt 601.62. C 55.90%, H 5.36%, I 21.09%, N 6.98%, S 10.66%. 1,5-Benzodiazepine deriv with bactericidal activity. Prepn: D. Nardi *et al.,* **Swiss** pat. **555,347** (1974 to Recordati), *C.A.* **82,** 43480 (1975). Synthesis, physical characteristics, antibacterial activity: *eidem, Experientia* **31,** 440 (1975); *eidem, Farmaco Ed. Sci.* **30,** 248 (1975). Structure activity study: C. Greico *et al., ibid.* **32,** 909 (1977). Antimicrobial activity: M. Veronese *et al., Chemotherapy* **23,** 90 (1977). Toxicity study: D. Nardi *et al., Experientia* **31,** 440 (1975).

Crystals from isopropanol, mp 162°. LD$_{50}$ in mice, rats (mg/kg): 9000, > 10000 orally; 42, 35 i.p. (Nardi).
THERAP CAT: Antibacterial.

9566. Tibolone. *(7α,17α)-17-Hydroxy-7-methyl-19-nor-pregn-5(10)-en-20-yn-3-one;* 7α-methyl-17α-ethynyl-17β-hydroxy-19-norandrost-5(10)-en-3-one; 7α-methyl-17α-ethynyl-17β-hydroxyestr-5(10)-en-3-one; Org-OD-14; Livial. $C_{21}H_{28}O_2$; mol wt 312.45. C 80.73%, H 9.03%, O 10.24%. Synthetic steroid with weak estrogenic, androgenic and progestogenic activity. Prepn: **Neth.** pat. **Appl. 6,406,797;** H. P. de Jongh, N. P. van Vliet, **U.S.** pat. **3,340,279** (1965, 1967 both to Organon). Improved process: M. S. de Winter, E. A. Harryvan, **U.S.** pat. **3,475,465** (1969 to Organon). Endocrinological profile: J. de Visser *et al., Arzneimittel-Forsch.* **34,** 1010 (1984). Series of articles on pharmacology and clinical efficacy in post-menopausal women: *Maturitas* Suppl 1, 1-72 (1987). Clinical effect in osteroporosis: P. Geusens *et al., ibid.* **13,** 155 (1991).

Crystals, mp 165-169°.
THERAP CAT: In treatment of menopausal syndrome.

9567. Ticarbodine. *2,6-Dimethyl-N-[3-(trifluoromethyl)phenyl]-1-piperidinecarbothioamide;* α,α,α-*trifluoro-2,6-dimethylthio-1-piperidinecarboxy-m-toluidide;* EL-974; Tribodine. $C_{15}H_{19}F_3N_2S$; mol wt 316.39. C 56.94%, H 6.05%, F 18.01%, N 8.85%, S 10.13%. Deriv of thiourea, *q.v.* Prepn: H. D. Porter, H. M. Taylor, U.S. pat. **3,659,012** (1972 to Lilly). Anthelmintic efficacy in dogs: R. J. Boisvenue *et al., Am. J. Vet. Res.* **33**, 709 (1972); G. F. Slonka *et al., ibid.* 1075; D. K. Hass, J. A. Collins, *Proc. Helminthol. Soc. Wash.* **43**, 135 (1976).

Light yellow cryst, mp 123-126°.
THERAP CAT (VET): Anthelmintic.

9568. Ticarcillin. *[2S-[2α,5α,6β(S*)]]-6-[(Carboxy-3-thienylacetyl)amino]-3,3-dimethyl-7-oxo-4-thia-1-azabicyclo[3.2.0]heptane-2-carboxylic acid; N-(2-carboxy-3,3-dimethyl-7-oxo-4-thia-1-azabicyclo[3.2.0]hept-6-yl)-3-thiophenemalonamic acid;* 6-[D(−)-α-carboxy-3-thienylacetamido]penicillanic acid; α-carboxy-3-thienylmethylpenicillin. $C_{15}H_{16}N_2O_6S_2$; mol wt 384.43. C 46.87%, H 4.20%, N 7.29%, O 24.97%, S 16.68%. Broad spectrum semi-synthetic antibiotic related to penicillin. Prepn: **Belg.** pat. **646,991**; E. G. Brain, J. H. Nayler, U.S. pat. **3,282,926** (1964, 1966 to Beecham Group Ltd.). *In vitro* studies: H. C. Neu, E. B. Winshell, *Antimicrob. Ag. Chemother.* **1970**, 385; R. Sutherland *et al., ibid.* 390; N. J. Legakis, J. Papavassiliou, *J. Antibiot.* **28**, 912 (1975). *In vivo* studies: P. Acred *et al., Antimicrob. Ag. Chemother.* **1970**, 396. Absorption and excretion: R. Sutherland, P. J. Wise, *ibid.* 402. Clinical pharmacology: V. Rodriguez *et al., Antimicrob. Ag. Chemother.* **4**, 31 (1973); R. D. Libke *et al., Clin. Pharmacol. Ther.* **17**, 441 (1975). Review of pharmacology and therapeutic efficacy: R. N. Brogden *et al., Drugs* **20**, 325-352 (1980).

Disodium salt, $C_{15}H_{14}N_2Na_2O_6S_2$, *BRL-2288, Aerugipen, Monapen, Ticar, Ticarpen, Ticillin.* Creamy-white hygroscopic non-crystalline powder. Readily sol in water (> 100 g/100 ml water) giving a clear soln with pH between 6.0 and 8.0. Aq solns are relatively stable; acid solns relatively unstable (Sutherland).
THERAP CAT: Antibacterial.
THERAP CAT (VET): Antibacterial.

9569. Ticlopidine. *5-[(2-Chlorophenyl)methyl]-4,5,6,7-tetrahydrothieno[3,2-c]pyridine;* 5-(o-chlorobenzyl)-4,5,6,7-tetrahydrothieno[3,2-c]pyridine. $C_{14}H_{14}ClNS$; mol wt 263.79. C 63.75%, H 5.35%, Cl 13.44%, N 5.31%, S 12.16%. Platelet aggregation inhibitor. Prepn: **Ger.** pat. **2,404,308**; A. R. J. Castaigne, U.S. pat. **4,051,141** (1974, 1977 both to Cent. Etudes Ind. Pharm.); E. Braye, U.S. pat. **4,127,580**

(1978 to Parcor). Metabolism: P. Godard *et al., Eur. J. Drug Metab. Pharmacokinet.* **3**, 67 (1978); *eidem, ibid.* **4**, 133 (1979); A. Tuong *et al., ibid.* **6**, 91 (1981). Mode of action: G. Leblondel, P. Allain, *Biochem. Pharmacol.* **27**, 2099 (1978); J. R. O'Brien *et al., Thromb. Res.* **13**, 245 (1978); J. J. Bruno, *ibid.* **1983**, Suppl. 4, 59. Pharmacology: A. Akashi *et al., Arzneimittel-Forsch.* **30**, 409, 415 (1980). Clinical studies: J. J. Thebault *et al., J. Int. Med. Res.* **5**, 405 (1977); C. Lecrubier *et al., Therapie* **32**, 189 (1977); T. Katsumura *et al., Angiology* **33**, 357 (1982). Review of pharmacodynamics, pharmacokinetics and therapeutic use: E. Saltiel, A. Ward, *Drugs* **34**, 222-262 (1987). Comprehensive description: F. J. Al-Shammary, N. A. A. Mian in *Analytical Profiles of Drug Substances and Excipients* vol. **21**, H. G. Brittain, Ed. (Academic Press, San Diego, 1992) pp 573-609.

Hydrochloride, $C_{14}H_{14}ClNS.HCl$, *4-C-32, 53-32 C, Anagregal, Caudaline, Panaldine, Ticlid, Ticlodix, Ticlodone, Ticlosin, Tiklid.* Crystals from ethanol, mp 190°. uv max (water): 214, 268, 295 nm ($A_{1cm}^{1\%}$ 303.8, 13.14, 2). pKa 7.64. Almost sol in water; sol in 95% alcohol, methanol, chloroform. Insol in ether. LD_{50} in mice (mg/kg/24 hrs): 55 i.v.; > 300 orally (Castaigne).
THERAP CAT: Antithrombotic.

9570. Ticrynafen. *[2,3-Dichloro-4-(2-thienylcarbonyl)phenoxy]acetic acid; [2,3-dichloro-4-(2-thenoyl)phenoxy]acetic acid;* [2,3-dichloro-4-(2-thiophenecarbonyl)phenoxy]acetic acid; tienilic acid; thienylic acid; ANP-3624; CE-3624; SKF-62698; Diflurex; Selacryn. $C_{13}H_8Cl_2O_4S$; mol wt 331.18. C 47.15%, H 2.43%, Cl 21.41%, O 19.32%, S 9.68%. A heterocyclic derivative of phenoxyacetic acid. Prepn: J. Godfroid, J. Thuillier, **Ger.** pat. **2,048,372** corresp to **U.S.** pat. **3,758,506** (1971, 1973, both to C.E.R.P.H.A.) and **Fr.** pat. **2,115,042** (1972 to C.E.R.P.H.A.). Synthesis and pharmacology: G. Thuillier *et al., Eur. J. Med. Chem.* **9**, 625 (1974). Pharmacokinetics in healthy volunteers: A. L. Kerremans *et al., Eur. J. Clin. Pharmacol.* **22**, 515 (1982). Comparative study in hypertensive patients: B. T. Emmerson *et al., ibid.* 203. Hepatotoxicity study: J. W. Manier *et al., Am. J. Gastroenterol.* **77**, 401 (1982).

Crystals from 50% ethanol, mp 148-149°; also reported as mp 157°. LD_{50} i.v., orally in mice: 225, 1275 mg/kg, **U.S.** pat. **3,758,506**.
THERAP CAT: Diuretic; uricosuric; antihypertensive.

9571. Tiemonium Iodide. *4-[3-Hydroxy-3-phenyl-3-(2-thienyl)propyl]-4-methylmorpholinium iodide;* N-methyl-N-[3-hydroxy-3-phenyl-3-(α-thienyl)propyl]morpholinium iodide; Visceralgina. $C_{18}H_{24}INO_2S$; mol wt 445.36. C 48.54%, H 5.43%, I 28.49%, N 3.15%, O 7.18%, S 7.20%. Prepn: **Brit.** pat. **953,386** (1964 to C.E.R.M.).

Solid, mp 189-191°.
THERAP CAT: Anticholinergic; antispasmodic.

9572. Tigemonam. *[S-(Z)]-[[[1-(2-Amino-4-thiazolyl)-2-[[2,2-dimethyl-4-oxo-1-(sulfooxy)-3-azetidinyl]amino]-2-*

oxoethylidene]amino]oxy]acetic acid; [[[(Z)-(2-amino-4-thi-azolyl)][(3S)-1-hydroxy-2,2-dimethyl-4-oxo-3-azetidinyl]car-bamoyl]methylene]amino]oxy]acetic acid hydrogen sulfate (ester). $C_{12}H_{15}N_5O_9S_2$; mol wt 437.41. C 32.95%, H 3.46%, N 16.01%, O 32.92%, S 14.66%. Orally active synthetic monosulfactam, structurally similar to the monobactam aztreonam, *q.v.* Prepn: C. Yoshida *et al., J. Antibiot.* **38,** 1536 (1985); W. A. Slusarchyk *et al.,* **Belg. pat. 904,121;** W. A. Slusarchyk, W. H. Koster, U.S. pat. **4,638,061** (1986, 1987 both to Squibb). Antibacterial spectrum and β-lactamase stability *in vitro:* S. K. Tanaka *et al., Antimicrob. Ag. Chemother.* **31,** 219 (1987); N.-X. Chin, H. C. Neu, *ibid.* **32,** 84 (1988). Antibacterial activity and pharmacokinetics in animals: J. M. Clark *et al., ibid.* **31,** 226 (1987).

Disodium salt, $C_{12}H_{13}N_5Na_2O_9S_2$, *SQ-30213.* mp 140-145° (dec).

Dicholine salt, $C_{22}H_{41}N_2O_{11}S_2$, *SQ-30836, Tigemen.*
THERAP CAT: Antibacterial.

9573. Tiglic Acid. *(E)-2-Methyl-2-butenoic acid; (E)-2-methylcrotonic acid; trans-2,3-dimethylacrylic acid.* $C_5H_8O_2$; mol wt 100.12. C 59.98%, H 8.05%, O 31.96%. The stable isomer of angelic acid. Found as glyceride in croton oil, as butyl ester in the oil of the Roman camomile, *Anthemis nobilis* L., *Compositae,* and as geranyl tiglate in oil of geranium. Is formed during the charcoaling of maple wood. Formation by the intestinal roundworm, *Ascaris lumbricoides:* Bueding, *J. Biol. Chem.* **202,** 505 (1953). Has been found in crude sodium penicillin: Cram, Tishler, *J. Am. Chem. Soc.* **70,** 4238 (1948). Sepn by partition chromatography: Bueding, *loc. cit.* Synthesis from 2-hydroxy-2-methylbutyronitrile: Crawford, *J. Soc. Chem. Ind. (London)* **64,** 231 (1945). Review and bibliography: Buckles *et al., Chem. Rev.* **55,** 659-677 (1955).

Triclinic plates, rods from water. Spicy odor. *Vesicant.* d 0.972. mp 63.5-64°. bp_{760} 198.5°; $bp_{11.5}$ 95.0-96°. Volatile with steam. n_D^{81} 1.4342. K at 25° = 9.6 × 10^{-6}. uv max (H_2O): 216-217 nm (ε 10700). Molar heat of combustion 635.1 kcal. Sparingly sol in cold water; freely sol in hot water. Sol in alcohol, ether.

Calcium salt trihydrate, $Ca(C_5H_7O_2)_2 \cdot 3H_2O$, leaflets. Much less sol in water than calcium angelate: 100 parts of aq soln satd at 17° contains 6.05 parts of anhydr calcium tiglate.

Amide, C_5H_9NO, crystals, mp 75-76°.

Methyl ester, $C_6H_{10}O_2$, liquid; d_4^{20} 0.9498; bp_{766} 139.6°; n_D^{20} 1.4370.

Ethyl ester, $C_7H_{12}O_2$, liquid; $d_4^{19.5}$ 0.9247; bp_{752} 156°; bp_{11} 55.5°. n_D^{20} 1.4350. Heat of formn at constant vol: 953.2 kcal, at constant pressure: 954.4 kcal.

Geranyl ester, $C_{15}H_{24}O_2$, liquid; pleasant odor; d_{15}^{15} 0.9279; bp_7 149-151°.

USE: The esters in perfumes and flavoring agents. The free acid as a breaker of emulsions.

9574. Tigloidine. *2-Methyl-2-butenoic acid [1α,3α(E),-5α]- 8-methyl-8-azabicyclo[3.2.1]oct-3-yl ester; (E)-1αH,5αH-tropan-3β-ol 2-methylcrotonate;* tiglylpseudotropine; 3β-tigloyloxytropane; tiglic acid ester with pseudotropine. $C_{13}H_{21}NO_2$; mol wt 223.32. C 69.92%, H 9.48%, N 6.27%, O 14.33%. Isoln from *Duboisia myoporoides* R. Br., *Solanaceae* and prepn from tropine and tigloyl chloride: Barger *et al., J. Chem. Soc.* **1937,** 1820; isoln from *Datura innoxia* Miller,

Solanaceae: Evans, Wellendorf, *ibid.* **1959,** 1406. Pharmacology: Sanghvi *et al., Eur. J. Pharmacol.* **4,** 246 (1968).

Hydrobromide, $C_{13}H_{21}NO_2 \cdot HBr$, *Tiglyssin.* mp 234-235°. Soluble in chloroform.
THERAP CAT: Antispasmodic.

9575. Tigogenin. *(3β,5α,25R)-Spirostan-3-ol.* $C_{27}H_{44}$-O_3; mol wt 416.64. C 77.84%, H 10.64%, O 11.52%. The aglycon of tigonin. Usually separated from crude gitogenin obtained from leaves of *Digitalis lanata* Ehrh., *Scrophulariaceae:* Windhaus, *Z. Physiol. Chem.* **150,** 205 (1925); Jacobs, Fleck, *J. Biol. Chem.* **88,** 545 (1930). From the sisal plant *Agave sisalana* L., *Amaryllidaceae:* Rubin, U.S. pats. **2,991,282** (1961); **3,303,187** (1967). Differs from smilagenin *(q.v.)* in the configuration at C-5. Structure: Marker, Rohrmann, *J. Am. Chem. Soc.* **62,** 898 (1940). Structure and synthesis: Mazur, Sondheimer, *ibid.* **81,** 3161 (1959); Caglioti, Magi, *Tetrahedron* **19,** 1127 (1963).

Crystals from dil methanol, mp 203°. $[\alpha]_D^{20}$ −62°. More sol in acetone, in ether, and esp in petr ether than gitogenin. Pptd by digitonin.

Acetyltigogenin, $C_{29}H_{46}O_4$, crystals from methanol, mp 206°. $[\alpha]_D^{18}$ −74.4°.

9576. Tigonin. $C_{56}H_{92}O_{27}$; mol wt 1197.33. C 56.18%, H 7.74%, O 36.08%. Isoln from *Digitalis lanata* Ehrh., *Scrophulariaceae:* Tschesche, *Ber.* **69,** 1665 (1936). Also occurs in *D. purpurea* L., *Scrophulariaceae,* in *Chlorogalum pomeridianum* (DC.) Kunth, *Liliaceae:* Liang, Noller, *J. Am. Chem. Soc.* **57,** 525 (1935). Built from 2 glucose, 2 galactose, 1 xylose, and 1 tigogenin unit.

Hygroscopic, amorphous flakes from 95% alc. After drying at 118° over P_2O_5 *in vacuo* sinters at 220°. mp ~260° *(in vacuo).* Sol in water. Not pptd by ether from water (difference from digitonin); amorphous ppt with amyl alcohol.

Cholesterol tigonide, an equimolar addition compd, long needles from methanol, dec above 200°.

9577. Tiliacorine. *(1α,1′α)-6′,7-Didemethoxy-6′,7-epoxyrodiasine.* $C_{36}H_{36}N_2O_5$; mol wt 576.69. C 74.98%, H 6.29%, N 4.86%, O 13.87%. From bark of *Tiliarcora acuminata* Miers, and *T. racemosa* Colebr., *Menispermaceae.* Isoln: Van Itallie, Steenhauer, *Pharm. Weekbl.* **59,** 1381 (1922). Structure: Anjaneyulu *et al., Chem. & Ind. (London)* **1959,** 1119; Rao, Row, *J. Org. Chem.* **25,** 981 (1960). Revised structure: Anjaneyula *et al., J. Sci. Ind. Res. (India)* **21B,** 602 (1962); *eidem, Tetrahedron* **25,** 3091 (1969); M. Shamma, J. E. Foy, *J. Org. Chem.* **41,** 1293 (1976). Abs config and biosynthesis: D. S. Bhakuni *et al., Chem. Commun.* **1978,** 226.

Bitter crystals, mp 271-272°. $[\alpha]_D$ +105.3°. uv max: 295, 265 nm (log ϵ 3.91, 3.48). Sol in alcohol, benzene, chloroform, ether.

9578. Tilidine. *trans*-(±)-*2-(Dimethylamino)-1-phenyl-3-cyclohexene-1-carboxylic acid ethyl ester;* ethyl 2-(dimethylamino)-1-phenyl-3-cyclohexene-1-carboxylate; 3-*trans*-dimethylamino-4-phenyl-4-*trans*-carbethoxy-Δ^1-cyclohexene; $C_{17}H_{23}NO_2$; mol wt 273.38. C 74.69%, H 8.48%, N 5.12%, O 11.71%. Prepn: **S. Afr. pat. 66 06,476; Brit.** pat. **1,120,-186;** G. Satzinger, **U.S.** pat. **3,557,127** (1967, 1968, 1971, all to Warner-Lambert). Synthesis of 6-C^{14} hydrochloride: K.-O. Vollmer, F. W. Koss, *Arzneimittel-Forsch.* **20,** 990 (1970). Pharmacology and toxicity data: M. Hermann *et al., ibid.* 977, 983. Metabolism: K.-O. Vollmer, H. Achenbach, *ibid.* **24,** 1237 (1974). HPLC determn in pharmaceutics: D. Zivanov-Stakic *et al., Farmaco* **44,** 759 (1989). Clinical trial in comparison with nefopam, *q.v.*: J. Abeloos *et al., Acta Anaesthesiol. Belg.* **34,** 283 (1983); in combination with naloxone, *q.v.*: H. Van Cauwenberge *et al., Int. J. Clin. Pharmacol. Res.* **12,** 1 (1992).

bp$_{0.01}$ 95.5-96°.
Hydrochloride hemihydrate, $C_{17}H_{23}NO_2 \cdot HCl \cdot \frac{1}{2}H_2O$, *Gö-1261C, W-5759A, Lucayan, Valoron.* Crystals, mp 125°. Easily sol in water. LD$_{50}$ (7 day) in mice, rats (mg/kg): 437.0, 417.7 i.g.; 490.0, 400.0 s.c.; 52.0, 74.1 i.v. (Herrmann).
Hydrochloride, crystals from ethyl acetate-methyl ethyl ketone, mp 159°.
cis-(±)-Form, bp$_{0.01}$ 97.5-98°.
cis-(±)-Hydrochloride, $C_{17}H_{23}NO_2 \cdot HCl \cdot 1\frac{1}{2}H_2O$, crystals from ethyl acetate-methyl ethyl ketone, mp 84°.
Note: This is a controlled substance (opiate) listed in the U.S. Code of Federal Regulations, Title 21 Part 1308.11 (1995).
THERAP CAT: Analgesic (narcotic).

9579. Tilisolol. *4-[3-[(1,1-Dimethylethyl)amino]-2-hydroxypropoxy]-2-methyl-1(2H)-isoquinolinone;* 4-[3-(tert-butylamino)-2-hydroxypropoxy]-*N*-methylisocarbostyril. $C_{17}H_{24}NO_3$; mol wt 304.39. C 67.08%, H 7.95%, N 9.20%, O 15.77%. β-Adrenergic blocker. Prepn: H. Fukushima, Y. Suzuki, **Ger.** pat. **2,631,080;** *eidem,* **U.S.** pat. **4,129,565** (1977, 1978 both to Nisshin). Antiarrhythmic activity: K. Hashimoto *et al., J. Pharmacol. Exp. Ther.* **223,** 801 (1982); and pharmacological profile: Y. Nakagawa *et al., Arzneimittel-Forsch.* **34,** 194 (1984). Receptor binding study: T. Nagatomo *et al., Japan J. Pharmacol.* **34,** 249 (1984). Vasodilating activity in humans: T. Imaizumi *et al., Arzneimittel-Forsch.* **38,** 1342 (1988). HPLC determn in plasma: K. Yonezawa *et al., J. Chromatog.* **339,** 219 (1985). Toxicity data: S. Kawase *et al., Oyo Yakuri* **34,** 499 (1987), *C.A.* **108,** 161293m (1988).

Hydrochloride, $C_{17}H_{24}N_2O_3 \cdot HCl$, *N-696, Selecal.* White crystals from ethyl acetate, mp 203-205°. LD$_{50}$ in male, female mice and rats (mg/kg): 1393, 1290, 145, 188 orally; 1219, 1245, 176, 169 s.c.; 578, 557, 39.5, 29.2 i.p.; 74.3, 104.7, 75.8, 38.1 i.v. (Kawase).
THERAP CAT: Antiarrhythmic; antihypertensive.

9580. Tilmicosin. 4^A-*O-De(2,6-dideoxy-3-C-methyl-α-L-ribo-hexopyranosyl)-20-deoxo-20-(3,5-dimethyl-1-piperidinyl)-tylosin;* 20-deoxo-20-(3,5-dimethylpiperidin-1-yl)desmycosin; EL-870; LY-177370; Micotil. $C_{46}H_{80}N_2O_{13}$; mol wt 869.15. C 63.57%, H 9.28%, N 3.22%, O 23.93%. Macrolide antibiotic; structurally related to tylosin, *q.v.* Prepd as 85:15 mixture of *cis/trans* isomers: M. Debono, H. A. Kirst, **Eur.** pat. **Appl. 103,465;** *eidem,* **U.S.** pat. **4,820,695** (1984, 1989 both to Lilly); M. Debono *et al., J. Antibiot.* **42,** 1253 (1989). *In vitro* antibacterial activity: E. E. Ose, *ibid.* **40,** 190 (1987). Metabolism and tissue residue studies in cattle: A. L. Donoho *et al., ACS Symposium Series* **503,** 158-167 (1992). Trial in treatment of calf pneumonia: E. E. Ose, L. V. Tonkinson, *Vet. Rec.* **123,** 367 (1988). Prophylactic efficacy in bovine respiratory tract disease: D. W. Morck *et al., J. Am. Vet. Med. Assoc.* **202,** 273 (1993).

cis-form

uv max: 283 nm (ϵ 22643). pKa' (66% DMF): 7.4, 8.5. $[\alpha]_D^{23}$ +12.75° (c = 0.010004 in CHCl$_3$, 5 cm).
THERAP CAT (VET): Antibacterial.

9581. Tilorone. *2,7-Bis[2-(diethylamino)ethoxy]-9H-fluoren-9-one;* bis-DEAE-fluorenone. $C_{25}H_{34}N_2O_3$; mol wt 410.56. C 73.14%, H 8.35%, N 6.82%, O 11.69%. The first recognized synthetic, small mol wt compound that is an orally active interferon inducer: Krueger, Mayer, *Science* **169,** 1213 sqq. (1970). Prepn: Fleming *et al.,* **Ger.** pat. **1,964,761** corresp to **U.S.** pat. **3,592,819** (1970, 1971 to Richardson Merrell); E. R. Andrews *et al., J. Med. Chem.* **17,** 882 (1974); H. M. Burke, M. M. Joullie, *ibid.* **21,** 1084 (1978). Mechanism of action: De Clercq, Merigan, *J. Infec. Dis.* **123,** 190 (1971). Toxicity studies: Kaufman *et al., Proc. Soc. Exp. Biol. Med.* **137,** 357 (1971). *Review:* P. Chandra *et al.,* "Tilorone Hydrochloride" in *Antibiotics* **vol. 5**(pt. 2), F. E. Hahn, Ed. (Springer-Verlag, New York, 1979) pp 385-413.

Dihydrochloride, $C_{25}H_{34}N_2O_3 \cdot 2HCl$, crystals from butanone-methanol, mp 235-237°. uv max (water): 269 nm (E$_{1cm}^{1\%}$ 1600). LD$_{50}$ in mice, rats (single dose): 959, 852 mg/kg orally; 145, 244 mg/kg i.p. (Krueger, Mayer).

9582. Tiludronic Acid. *[[(4-Chlorophenyl)thio]methylene]bisphosphonic acid;* ACPMD; Cl-TMBP; ME-3737; SR-41319; Skelid. $C_7H_9ClO_6P_2S$; mol wt 318.61. C 26.39%, H 2.85%, Cl 11.13%, O 30.13%, P 19.44%, S 10.06%. Prepn and anti-inflammatory activity: J. C. Breliere *et al.,* **Eur. pat. Appl. 100,718;** *eidem,* **U.S. pat. 4,876,248** (1984, 1989 both to Sanofi). HPLC determn in plasma and urine: J.-P. Fels *et al., J. Chromatog.* **430,** 73 (1988). Determn of dissociation constants: M. Bonnery *et al., Bull. Soc. Chim. Fr.* **1988,** 49. Clinical evaluation in Paget's disease: M. Audran *et al., Clin. Rheumatol.* **8,** 71 (1989); in reduction of postmenopausal bone loss: J. Y. Reginster *et al., Lancet* **2,** 1469 (1989).

pK_1 10.85; pK_2 6.90; pK_3 2.95; pK_4 1.30.
Di-*tert*-butylamine salt, mp 253° (dec).

THERAP CAT: In treatment of Paget's disease and osteoporosis.

9583. Timepidium Bromide. *3-(Di-2-thienylmethylene)-5-methoxy-1,1-dimethylpiperidinium bromide;* SA-504; Mepidium; Sesden. $C_{17}H_{22}BrNOS_2$; mol wt 400.40. C 51.00%, H 5.54%, Br 19.96%, N 3.50%, O 4.00%, S 16.02%. Prepn: M. Kawazu *et al.,* **Ger. pat. 2,128,808** corresp to U.S. pat. **3,764,607** (1971, 1973 both to Tanabe). Pharmacological properties: H. Tamaki *et al., Japan. J. Pharmacol.* **22,** 685 (1972), *C.A.* **78,** 52812 (1973). Metabolism: J. Sugihara, N. Taga, *Radioisotopes* **26,** 238 (1977). Absorption, distribution, excretion: M. Yoshikawa, *Oyo Yakuri* **14,** 179 (1977), *C.A.* **88,** 163772 (1978). Teratological study: Y. Fujisawa *et al., ibid.* **7,** 1293 (1973), *C.A.* **81,** 99413 (1974). Chronic toxicity study: K. Doi *et al., ibid.* **13,** 851 (1977), *C.A.* **88,** 32014 (1978).

Colorless cryst from acetone/ether, mp 198-200°.
THERAP CAT: Anticholinergic.

9584. Timiperone. *4-[4-(2,3-Dihydro-2-thioxo-1H-benzimidazol-1-yl)-1-piperidinyl]-1-(4-fluorophenyl)-1-butanone;* 1-[1-[3-(4-fluorobenzoyl)propyl]-4-piperidyl]-2-mercaptobenzimidazole; 4-fluoro-4-[4-(2-thioxo-1-benzimidazolinyl)piperidino]butyrophenone; 1-[1-[3-(4-fluorobenzoyl)propyl]-4-piperidyl]-2,3-dihydrobenzimidazole-2-thione; DD-3480; Tolopelon. $C_{22}H_{24}FN_3OS$; mol wt 397.52. C 66.47%, H 6.09%, F 4.78%, N 10.57%, O 4.02%, S 8.07%. Butyrophenone derivative with neuroleptic activity. Prepn: M. Sato *et al.,* **Japan. Kokai 75 84,578;** K. Ueno *et al.,* U.S. pat. **3,963,727** (1975, 1976 both to Daiichi); M. Sato *et al., J. Med. Chem.* **21,** 1116 (1978). Improved synthesis: M. Sato, M. Arimoto, *Chem. Pharm. Bull.* **30,** 719 (1982). Pharmacology: T. Yamasaki *et al., Japan. J. Pharmacol.* **27** (Suppl.), 124P (1977); *eidem, Arzneimittel-Forsch.* **31,** 701, 707 (1981); T. Shibuya *et al., Int. J. Clin. Pharmacol. Ther. Toxicol.* **20,** 251 (1982). Pharmacokinetics and metabolism: H. Tachizawa *et al., Drug. Metabol. Dispos.* **9,** 442 (1981); K. Sudo *et al., Xenobiotica* **11,** 685 (1981). Multicenter controlled clinical trials in schizophrenia: R. Takahashi *et al., J. Int. Med. Res.* **10,** 257 (1982); T. Kariya *et al., ibid.* **11,** 66 (1983).

Crystals from acetone, mp 201-203°. uv max (ethanol): 226.5, 246, 309 nm. Slightly sol in water. LD_{50} in male rats, mice (mg/kg): 232, 478 orally (Yamasaki).

THERAP CAT: Antipsychotic.

9585. Timolol. *(S)-1-[(1,1-Dimethylethyl)amino]-3-[[4-(4-morpholinyl)-1,2,5-thiadiazol-3-yl]oxy]-2-propanol;* S-(−)-3-(3-tert-butylamino-2-hydroxypropoxy)-4-morpholino-1,2,-5-thiadiazole; (−)-3-morpholino-4-(3-tert-butylamino-2-hydroxypropoxy)-1,2,5-thiadiazole. $C_{13}H_{24}N_4O_3S$; mol wt 316.42. C 49.35%, H 7.64%, N 17.71%, O 15.17%, S 10.13%. β-Adrenergic blocker. Prepn: B. K. Wasson, **Ger. pat. 1,925,956;** *idem,* U.S. pat. **3,655,663** (1969, 1972 both to Frosst). Manufacturing process: L. M. Weinstock *et al.,* **Ger. pat. 1,925,955;** *eidem,* U.S. pat. **3,657,237** (1970, 1972 both to Frosst). Synthesis and activity data: Wasson *et al., J. Med. Chem.* **15,** 651 (1972). Pharmacology: Franciosa *et al., Clin. Pharmacol. Ther.* **13,** 138 (1972); Ulrych *et al., ibid.* 232. Review of efficacy in glaucoma: R. C. Heel *et al., Drugs* **17,** 38-55 (1979). Clinical evaluation in hypertension: B. A. Rofman *et al., Hypertension* **2,** 643 (1980). Multicenter study of effect in myocardial infarction: *N. Engl. J. Med.* **304,** 801 (1981); of efficacy in limiting infarct size: M. Sederholm *et al., ibid.* **310,** 9 (1984). Comprehensive description: D. J. Mazzo, A. E. Loper in *Analytical Profiles of Drug Substances* vol. 16, K. Florey, Ed. (Academic Press New York, 1987) pp 641-692.

Hydrogen maleate salt, $C_{13}H_{24}N_4O_3S \cdot C_4H_4O_4$, *MK-950, Aquanil, Betim, Blocadren, Proflax, Temserin, Tenopt, Timacar, Timacor, Timoptic, Timoptol.* White crystals from ethanol, mp 201.5-202.5°. $[\alpha]_{405}^{24}$ −12.0° (c = 5 in 1*N* HCl), $[\alpha]_D^{25}$ −4.2°. uv max (0.1*N* HCl): 294 nm ($A_{1cm}^{1\%}$ 200). Sol in water, ethanol, methanol; sparingly sol in chloroform; very slightly sol in cyclohexane. Practically insol in isooctane, ether. Stable in soln up to pH 12.

(±)-Free base, crystalline solid from isopropyl ether, mp 71.5-72.5°.

THERAP CAT: Antihypertensive; antiarrhythmic (class II); antianginal; antiglaucoma agent.

9586. Timonacic. *4-Thiazolidinecarboxylic acid;* ATC; norgamen; thioproline; NSC-25855; Detoxepa; Hepalidine; Heparegen; Tiazolidin. $C_4H_7NO_2S$; mol wt 133.17. C 36.08%, H 5.30%, N 10.52%, O 24.03%, S 24.08%. Prepd from DL- or L-cysteine and formaldehyde. Prepn of *l*-form: Ratner, Clarke, *J. Am. Chem. Soc.* **59,** 200 (1937); of *dl*- and *l*-forms: Werner *et al., Helv. Chim. Acta* **30,** 432 (1947); **Fr. pat. M3184** (1965 to Sogespar S.A.), *C.A.* **63,** 12980h (1965). Proton magnetic resonance and conformation: Martin, Mathur, *J. Am. Chem. Soc.* **87,** 1065 (1965). Series of articles on pharmacology and toxicology: *Gazz. Med. Ital.* **131,** 251-286 (1972). Toxicity data: Bertrand, Piton, *ibid.* 265. Induces reverse transformation of tumor cells to normal. Clinical evaluation in cancer: A. Brugarolas, M. Gosalvez, *Lancet* **1,** 68 (1980).

dl-Form, crystals, mp 195°. LD$_{50}$ orally in mice: 400 mg/kg (Bertrand, Piton).

l-Form, crystals from water, dec 196-197°. Readily sol in hot water, acid, alkali; sparingly sol in cold water. Practically insol in alcohol.

THERAP CAT: Hepatoprotectant.

9587. Tin. Sn; at. wt 118.69; at. no. 50; valence 2, 4. Ten naturally occurring isotopes: 112 (0.95%); 114 (0.65%); 115 (0.34%); 116 (14.24%); 117 (7.57%); 118 (24.01%); 119 (8.59%); 120 (32.97%); 122 (4.71%); 124 (5.98%); artificial, radioactive isotopes: 108-111; 113; 121; 123; 125-132. Found in cassiterite, stannite, and tealite. Occurrence in earth's crust: 6×10^{-4}%. The metal of commerce is about 99.8% pure. Prepn of high purity tin: Baralis, Marone, *Met. Ital.* **59**, 494 (1967), *C.A.* **67**, 119613a (1967). Physical properties: Kirshenbaum, Cahill, *J. Inorg. Nucl. Chem.* **25**, 232 (1963). *Monograph:* C. L. Mantell, *Tin: Its Mining, Production, Technology and Applications* (Reinhold, New York, 1949). *Reviews:* Abel in *Comprehensive Inorganic Chemistry* vol. 2, J. C. Bailar, Jr. *et al.*, Eds. (Pergamon Press, Oxford, 1973) pp 43-104; W. Germain *et al.*, in Kirk-Othmer *Encyclopedia of Chemical Technology* vol. 23 (Wiley-Interscience, New York, 3rd ed., 1983) pp 18-42.

Almost silver-white, lustrous, soft, very malleable and ductile metal; only slightly tenacious; easily powdered. When being bent, emits the crackling "tin cry". Brittle at 200°. At −40° crumbles to gray amorphous powder ("gray tin"), slowly changing back above 20° to white tin. Available in the form of bars, foil, powder, shot, etc. Stable in air, but when in powder form it oxidizes, esp in presence of moisture. d 7.31. mp 231.9° bp 2507° (2780°K). Specific heat (25°) 0.053 cal/g/°C. Brinell hardness 2.9. Reacts slowly with cold dil HCl or dil HNO$_3$, hot dil H$_2$SO$_4$; readily with concd HCl, aqua regia; very slowly attacked by acetic acid; slowly attacked by cold, more readily by hot caustic alkali; concd HNO$_3$ converts it into insol metastannic acid.

Caution: Potential symptoms of overexposure to inorganic tin compounds are irritation of eyes and skin. Potential symptoms of overexposure to organic tin compounds are headache, vertigo; eye irritation; psychoneurologic disturbances; sore throat, cough; abdominal pain, vomiting; urine retention; paresis, focal anesthesia; skin burns; pruritis. *See NIOSH Pocket Guide to Chemical Hazards* (DHHS/NIOSH 90-117, 1990) p 212-215. *See also Patty's Industrial Hygiene and Toxicology* vol. 2A, G. D. Clayton, F. E. Clayton, Eds. (Wiley-Interscience, New York, 3rd ed., 1981) pp 1940-1968.

USE: Chiefly for tin-plating, soldering alloys, babbitt and type metals, manuf tin salts, collapsible tubes.

9588. Tinidazole. *1-[2-(Ethylsulfonyl)ethyl]-2-methyl-5-nitro-1H-imidazole;* ethyl[2-(2-methyl-5-nitro-1-imidazolyl)ethyl]sulfone; CP-12574; Fasigin; Fasigyn; Pletil; Simplotan; Sorquetan; Tricolam; Trimonase. C$_8$H$_{13}$N$_3$O$_4$S; mol wt 247.28. C 38.86%, H 5.30%, N 16.99%, O 25.88%, S 12.97%. Prepn: K. Butler, U.S. pat. **3,376,311** (1968 to Pfizer); M. W. Miller *et al., J. Med. Chem.* **13**, 849 (1970). Series of articles on prepn, antiprotozoal activity and pharmacokinetics: *Antimicrob. Ag. Chemother.* **1969**, 257-270. Review of antiprotozoal activity: P. R. Sawyer *et al., Drugs* **11**, 423-440 (1976); of antibacterial activity, pharmacology and therapeutic efficacy vs anaerobes: A. A. Carmine *et al., ibid.* **24**, 85-117 (1982). Symposium on clinical experience

in anaerobic infection: *J. Antimicrob. Chemother.* **10**, Suppl. A, 1-184 (1982). Clinical trial in *Helicobacter pylori* positive gastritis: G. Oderda *et al., Gut* **33**, 1328 (1992).

Colorless crystals from benzene, mp 127-128°. LD$_{50}$ in mice (mg/kg): > 3600 orally; > 2000 i.p. (Miller).

THERAP CAT: Antiprotozoal (Trichomonas, Giardia); antiamebic; antibacterial.

9589. Tinofedrine. *α-[1-[(3,3-Di-3-thienyl-2-propenyl)-amino]ethyl]benzenemethanol;* α-[1-[(3,3-di-3-thienylallyl)-amino]ethyl]benzyl alcohol. C$_{20}$H$_{21}$NOS$_2$; mol wt 355.52. C 67.57%, H 5.95%, N 3.94%, O 4.50%, S 18.04%. Prepn: K. Thiele, K. Posselt, **Ger. pat. 1,921,453** (1970 to Degussa), *C.A.* **74**, 76320c (1971); K. Posselt, H. Offermanns, **Ger. pat. 2,150,977** corresp to U.S. pat. **3,767,675** (1972, 1973 both to Degussa); K. Thiele *et al., Arzneimittel-Forsch.* **28**, 2047 (1978). Absolute configuration determined by circular dichroism: J. Engle *et al., Chem. Ztg.* **105**, 85 (1981). Series of articles on pharmacology, hemodynamics, pharmacokinetics, metabolism: *Arzneimittel-Forsch.* **28**, 1335-1367 (1978). Synthesis of radioactive tinofedrine: A. Saus, K. Posselt, *ibid.* **30**, 917 (1980). Spectroscopic study: K. Thiele *et al., ibid.* 1057. EEG study in geriatric patients: B. Saletu, P. Anderer, *ibid.* 1218. General effects on heart and blood circulation: F. Stroman, K. Thiemer, *ibid.* **31**, 1892 (1981).

Hydrochloride, C$_{20}$H$_{21}$NOS$_2$.HCl, *D-8955, Novocebrin.* Cryst from isopropanol, mp 226-229° (dec). LD$_{50}$ in mice, rats (mg/kg): 1890, 6600 orally, 20.15, 14.00 i.v., P. St. Janiak, *Drugs of the Future* **4**, 286 (1979).

THERAP CAT: Vasodilator (cerebral).

9590. Tinoridine. *2-Amino-4,5,6,7-tetrahydro-6-(phenylmethyl)thieno[2,3-c]pyridine-3-carboxylic acid ethyl ester;* 2-amino-6-benzyl-4,5,6,7-tetrahydrothieno[2,3-c]pyridine-3-carboxylic acid ethyl ester; 2-amino-3-ethoxycarbonyl-6-benzyl-4,5,6,7-tetrahydrothieno[2,3-c]pyridine; ethyl 2-amino-6-benzyl-4,5,6,7-tetrahydrothieno[2,3-c]pyridine-3-carboxylate; Y-3642. C$_{17}$H$_{20}$N$_2$O$_2$S; mol wt 316.42. C 64.53%, H 6.37%, N 8.85%, O 10.11%, S 10.13%. Prepn: M. Nakanishi *et al.,* **Ger. pat. 1,812,404;** *eidem,* **U.S. pat. 3,563,997** (1969, 1971 both to Yoshitomi). Pharmacological studies: M. Nakanishi *et al., J. Pharm. Soc. Japan* **90**, 272-334 (1970), *C.A.* **73**, 54504t-54510s (1970); *eidem, Arzneimittel-Forsch.* **20**, 998, 1004 (1970); *Yakuri to Chiro* **2**, 1028 (1974). Protective effect vs. carbon tetrachloride hepatotoxicity: H. Yasuda *et al., Toxicol. Appl. Pharmacol.* **52**, 407 (1980).

Slightly sol in H$_2$O, mp 112-113°. LD$_{50}$ in mice, rats (mg/kg): 5400, > 10200 orally; 1600, 1250 i.p. (Nakanishi, 1970).

Hydrochloride, C$_{17}$H$_{20}$N$_2$O$_2$S.HCl, *Dimaten, Nonflamin.* Yellowish white to yellow powder, mp 234-235° (dec). Practically odorless with a bitter taste. Slightly sol in methanol; very slightly sol in water, ether, acetone, benzene. LD$_{50}$ orally in mice: 1601 mg/kg (Nakanishi, 1974).

THERAP CAT: Analgesic; antipyretic; anti-inflammatory.

9591. Tin Phosphides. Several stoichiometries reported: *Mellor's* vol. **VIII**, 847-849 (1931). Crystal structure of Sn$_4$P$_3$ and discussion of other phases: Olofsson, *Acta Chem. Scand.* **21**, 1659 (1967); **24**, 1153 (1970).

USE: Tin phosphide with a phosphorus content > 10% used in manuf of phosphor bronze.

Consult the Name Index before using this section.

9592. Tinzaparin. Innohep; Logiparin. Low molecular weight heparin prepared by enzymatic depolymerization of porcine mucosal heparin with heparinase from *Flavobacterium heparinum.* Approx mol wt 4900 Da. Manufacturing process: J. I. Nielsen, P. B. Ostergaard, **Eur.** pat. **Appl. 244,236**; J. I. Nielsen, **U.S.** pat. **5,106,734** (1987, 1992 both to Novo). Pharmacokinetics: P. C. Pedersen *et al., Thromb. Res.* **61**, 477 (1991). Multicenter clinical trial: A. Liezorovicz *et al., Brit. J. Surg.* **78**, 412 (1991).
    THERAP CAT: Antithrombotic.

9593. Tiocarlide. *N,N'-[4-(3-Methylbutoxy)phenyl]thiourea; 4,4'-bis(isopentyloxy)thiocarbanilide;* 4,4'-di(isoamyloxy)thiocarbanilide; 1,3-bis(*p*-isoamyloxyphenyl)-2-thiourea; thiocarlide; DATC; Datanil; Disocarban; Isoxyl. $C_{23}H_{32}N_2O_2S$; mol wt 400.59. C 68.96%, H 8.05%, N 6.99%, O 7.99%, S 8.00%. Prepn: Buu-Hoi, Xuong, *Compt. Rend.* **237**, 498 (1953); Huebner, Scholz, U.S. pat. **2,703,815** (1955 to Ciba).

    Crystals from ethanol, mp 134-145°.
    THERAP CAT: Antibacterial (tuberculostatic).

9594. Tioclomarol. *3-[3-(4-Chlorophenyl)-1-(5-chloro-2-thienyl)-3-hydroxypropyl]-4-hydroxy-2H-1-benzopyran-2-one; 3-[5-chloro-α-(p-chloro-β-hydroxyphenethyl)-2-thenyl]-4-hydroxycoumarin;* Apegmone. $C_{22}H_{16}Cl_2O_4S$; mol wt 447.34. C 59.07%, H 3.61%, Cl 15.85%, O 14.31%, S 7.17%. Slow-acting anticoagulant, related structurally to warfarin, *q.v.* Prepn: E. Boschetti *et al.,* **S. Afr.** pat. **67 07,267** corresp to **U.S.** pat. **3,574,234** (1968, 1971 both to Lipha). Synthesis and anticoagulant activity: *eidem, Chim. Ther.* **7**, 20 (1972).

    White cryst from methanol, mp 104°.
    THERAP CAT: Anticoagulant.

9595. Tioconazole. *1-[2-[(2-Chloro-3-thienyl)methoxy]-2-(2,4-dichlorophenyl)ethyl]-1H-imidazole;* 1-[2,4-dichloro-β-[(2-chloro-3-thienyl)oxy]phenethyl]imidazole; UK-20349; Fungibacid; Gyno-Trosyd; Trosyd; Trosyl; Vagistat; Zoniden. $C_{16}H_{13}Cl_3N_2OS$; mol wt 387.72. C 49.57%, H 3.38%, Cl 27.43%, N 7.23%, O 4.13%, S 8.27%. Antimycotic imidazole derivative. Prepn: G. E. Gymer, **Belg.** pat. **841,309**; *idem,* **U.S.** pat. **4,062,966** (1976, 1977 both to Pfizer). Antifungal spectrum: S. Jevons, *Antimicrob. Ag. Chemother.* **15**, 597 (1979); F. C. Odds, *J. Antimicrob. Chemother.* **6**, 749 (1980). Pharmacology: M. S. Marriott *et al., Dermatologica* **166**, Suppl. 1, 1 (1983). Clinical trial in dermatomycosis: Y. M. Clayton *et al., Clin. Exp. Dermatol.* **7**, 543 (1982). Series of articles on pharmacology and clinical efficacy in gynecological use: *Gynak. Rundsch.* **23**, Suppl. 1, 1-60 (1983).

Hydrochloride, $C_{16}H_{13}Cl_3N_2OS.HCl$, crystals, mp 168-170°.
    THERAP CAT: Antifungal (topical).

9596. Tiomesterone. *(1α,7α,17β)-1,7-Bis(acetylthio)-17-hydroxy-17-methylandrost-4-en-3-one; 17β-hydroxy-1α,7α-dimercapto-17-methylandrost-4-en-3-one 1,7-diacetate;* 1α,7α-bis(acetylthio)-17α-methyltestosterone; 1α,7α-bis(acetylmercapto)-17α-methyltestosterone; thiomesterone; STA-307; Emdabol; Emdabolin; Protabol. $C_{24}H_{34}O_4S_2$; mol wt 450.66. C 63.96%, H 7.60%, O 14.20%, S 14.23%. Prepn: Krämer, *Ber.* **96**, 2803 (1963); Brueckner *et al.,* **U.S.** pat. **3,087,942** (1963 to E. Merck).

    Crystals, mp 205-206°. $[α]_D$ −66° (dioxane). uv max: 237.5 nm (ε 19800).
    THERAP CAT: Androgen.

9597. Tiopronin. *N-(2-Mercapto-1-oxopropyl)glycine;* N-(2-mercaptopropionyl)glycine; α-mercaptopropionylglycine; Acadione; Capen; Epatiol; Mucolysin; Thiola; Thiosol. $C_5H_9NO_3S$; mol wt 163.20. C 36.80%, H 5.56%, N 8.58%, O 29.41%, S 19.65%. Prepn: Mita *et al.,* **U.S.** pat. **3,246,-025** (1966 to Santen). Antidotal effects in mice: H. Fujimura *et al., Nippon Yakurigaku Zasshi* **60**, 278 (1964), *C.A.* **62**, 972h (1965). Anti-inflammatory pharmacology in animals: F. Capasso, *Agents Actions* **11**, 741 (1981). Clinical trial in chronic hepatitis: F. Ichida *et al., J. Int. Med. Res.* **10**, 325 (1982). Clinical comparison with D-penicillamine in rheumatoid arthritis: B. Amor *et al., Arthritis Rheum.* **25**, 698 (1982); G. Pasero *et al., ibid.* 923.

    Crystals from ethyl acetate, mp 95-97°. LD_{50} i.v. in mice: 2.1 g/kg (Fujimura).
    THERAP CAT: Antidote (heavy metal poisoning); hepatoprotectant; mucolytic.

9598. Tiotropium Bromide. *(1α,2β,4β,5α,7β)-7-[(Hydroxydi-2-thienylacetyl)oxy]-9,9-dimethyl-3-oxa-9-azoniatricyclo[3.3.1.0²,⁴]nonane bromide;* 6β,7β-epoxy-3β-hydroxy-8-methyl-1αH,5αH-tropanium bromide, di-2-thienylglycolate; Ba-679 BR; Ba-679. $C_{19}H_{22}BrNO_4S_2$; mol wt 472.42. C 48.31%, H 4.69%, Br 16.91%, N 2.96%, O 13.55%, S 13.58%. Muscarinic receptor antagonist. Prepn: R. Banholzer *et al.,* **Eur.** pat. **Appl. 418,716** (1991 to Boehringer Ingelheim), *C.A.* **116**, 20937c (1992). Pharmacology: E.-B. Haddad *et al., Mol. Pharmacol.* **45**, 899 (1994); T. Takahashi *et al., Am. J. Respir. Crit. Care Med.* **150**, 1640 (1994). Clinical evaluation: F. P. V. Maesen *et al., Eur. Respir. J.* **6**, 1031 (1993). *Review:* P. J. Barnes *et al., Life Sci.* **56**, 853-859 (1995).

mp 218-220°.
THERAP CAT: Bronchodilator.

9599. Tioxidazole. *(6-Propoxy-2-benzothiazolyl)car-bamic acid methyl ester;* methyl 6-propoxy-2-benzothiazolyl-carbamate; Sch-21480; Tiox. $C_{12}H_{14}N_2O_3S$; mol wt 266.32. C 54.12%, H 5.30%, N 10.52%, O 18.02%, S 12.04%. Structurally related to benzimidazole anthelmintics. Prepn and anthelmintic activity: M. M. Nafissi-Varchei, **Belg. pat.** **840,945**; U.S. pat. **4,006,242** (1976, 1977 both to Schering). Physical data, activity against roundworms: E. Panitz *et al.,* *Experientia* **34**, 733 (1978). Anthelmintic effect against gastrointestinal parasites in naturally infected horses: J. H. Drudge *et al., Am. J. Vet. Res.* **41**, 1383 (1980); E. T. Lyons *et al., ibid.* **42**, 1048 (1981).

White odorless solid, mp 178-180°. Insoluble in water. Slightly sol in organic solvents.
THERAP CAT (VET): Anthelmintic for horses.

9600. Tioxolone. *6-Hydroxy-1,3-benzoxathiol-2-one;* 6-hydroxy-2-oxo-1,3-benzoxathiole; thioxolone; Camyna; Stepin. $C_7H_4O_3S$; mol wt 168.17. C 49.99%, H 2.40%, O 28.54%, S 19.07%. Prepd from resorcinol and ammonium thiocyanate in presence of copper sulfate: Werner, **U.S. pat.** **2,332,418** (1943 to Winthrop); Urushibara, Koga, *Bull. Chem. Soc. Japan* **29**, 419 (1956); Berg, Fiedler, **U.S. pat.** **2,886,488** (1959 to Thomae). Formerly believed to be the 4-hydroxy isomer: Fiedler, *Ber.* **95**, 1771 (1962).

Crystals from water, mp 160°. Practically insol in water. Sol in ethanol, isopropanol, propylene glycol, ether, benzene, toluene. Hydrolyzed by alkali.
Combination with hydrocortisone, *Psoil.*
THERAP CAT: Antiseborrheic; antiacne.

9601. Tipepidine. *3-(Di-2-thienylmethylene)-1-methyl-piperidine;* 1-methyl-3-piperidylidenedi(2-thienyl)methane; tipedine; AT-327; CR-662. $C_{15}H_{17}NS_2$; mol wt 275.44. C 65.41%, H 6.22%, N 5.09%, S 23.28%. Prepn: Okumura *et al., Tanabe Seiyaku Kenkyu Nempo* **3**, 30 (1958), *C.A.* **53**, 10214 (1959); Ponomarev, Martem'yanova, *USSR* pat. **176,903** (1965), *C.A.* **64**, 12648b (1966). Synthesis: *eidem, Khim. Geterosikl. Soedin.* **1967**, 174, *C.A.* **67**, 73501g (1967). Prepn of deriv: **Brit.** pat. **924,544** (1963); Yamamoto, Yoshikawa, **Japan.** pat. **17,988('62)** (both to Tanabe Seiyaku), *C.A.* **59**, 11443f, 11446b (1963). Pharmacology and clinical efficacy: Higaki *et al., Tanabe Seiyaku Kenkyu Nempo* **4**, 35 (1959), *C.A.* **54**, 3725h (1960); Kase *et al., Chem. Pharm. Bull.* **7**, 372 (1959). Metabolism: Watanabe *et al., Yakuga-ku Zasshi* **89**, 29 (1969); Sasaki *et al., ibid.* 345.

Yellow crystals from petr ether, mp 64-65°. $bp_{4.5}$ 178-184°. LD_{50} in mice (mg/kg): 294 i.p.; 308 i.m.; 867 orally (Higaki).
Citrate monohydrate, $C_{15}H_{17}NS_2 \cdot C_6H_8O_7 \cdot H_2O$, *bithiodine.* Yellow crystals, mp 138-139°. Sol in water, ethanol, propylene glycol.

Hibenzate, $C_{29}H_{27}NO_4S_2$, *Asverin, Sotal.* mp 187-190°.
THERAP CAT: Antitussive.

9602. Tiquizium Bromide. *trans-3-(Di-2-thienylmethyl-ene)octahydro-5-methyl-2H-quinolizinium bromide;* 3-(di-2-thienylmethylene)-5-methyl-*trans*-quinolizidinium bromide; HSR-902; HS-902; Thiaton. $C_{19}H_{24}BrNS_2$; mol wt 410.44. C 55.60%, H 5.89%, Br 19.47%, N 3.41%, S 15.63%. Quaternary ammonium salt with anticholinergic activity. Prepn: H. Kato *et al.,* **Belg. pat.** **866,988** (1978 to Hokuriku), *C.A.* **90**, 151996p (1979); H. Kato, E. Koshinaka, **U.S. pat.** **4,205,074** (1980 to Hokuriku). Synthesis and anticholinergic activity: E. Koshinaka *et al., Chem. Pharm. Bull.* **27**, 1454 (1979). Pharmacological comparison with other antispasmodics: S. Kubo *et al., Japan. J. Pharmacol.* **30**, 103P (1980). Pharmacology: M. Yamazaki *et al., Oyo Yakuri* **23**, 417, 423 (1981), *C.A.* **97**, 66016u, 66017v (1982). HPLC determn in biological fluids: T. Yamada *et al., Yaku-gaku Zasshi* **103**, 1319 (1983), *C.A.* **100**, 131957y (1984).

Needles from methanol-acetone, mp 278-281° (dec).
THERAP CAT: Antispasmodic.

9603. Tiratricol. *[4-(4-Hydroxy-3-iodophenoxy)-3,5-di-iodophenyl]acetic acid;* 3,3',5-triiodo-4-(4-hydroxyphenoxy)-phenylacetic acid; 3,5-diiodo-4-(3-iodo-4-hydroxyphenoxy)-phenylacetic acid; 3,3',5-triiodothyroacetic acid; Triac; Triacana. $C_{14}H_9I_3O_4$; mol wt 621.94. C 27.04%, H 1.46%, I 61.21%, O 10.29%. Thyroid hormone analog. Prepd by iodination of 4-(4'-hydroxyphenoxy)-3,5-diiodophenylacetic acid: Wilkinson, *Biochem. J.* **63**, 601 (1956); Hems, **Brit.** pat. **803,149** (1958 to Glaxo); Wilkinson, **Brit.** pat. **805,761** (1958 to Nat. Res. Dev. Corp.); Meltzer *et al., J. Org. Chem.* **26**, 1418 (1961). Clinical evaluation in thyroidectomy: S. I. Sherman, P. W. Ladenson, *J. Clin. Endocrinol. Metab.* **75**, 901 (1992).

Needles from methanol+ water, mp 65°, resolidifies 110°, mp 180-183°.
THERAP CAT: Antihypothyroid.

9604. Tirilazad. *(16α)-21-[4-(2,6-di-1-pyrrolidinyl-4-pyrimidinyl)-1-piperazinyl]-16-methylpregna-1,4,9(11)-triene-3,20-dione.* $C_{38}H_{52}N_6O_2$; mol wt 624.87. C 73.04%, H 8.39%, N 13.45%, O 5.12%. Member of a novel class of nonglucocorticoid, 21-aminosteroid antioxidants known as *lazaroids* which inhibit lipid peroxidation and have cyto-protectant activity. Prepn: J. M. McCall *et al.,* **PCT Int.** pat. **Appl. 87 01,706;** *eidem,* **U.S. pat.** **5,175,281** (1987, 1992 both to Upjohn). Inhibition of iron-dependent lipid peroxidation *in vitro:* J. M. Braughler *et al., J. Biol. Chem.* **262**, 10438 (1987). Effects on experimental head injury in mice: E. D. Hall *et al., J. Neurosurg.* **68**, 456 (1988); in post-traumatic spinal cord ischemia in cats: E. D. Hall, *ibid.* 462. HPLC determn in plasma: J. W. Cox, R. H. Pullen, *J. Chromatog.* **424**, 293 (1988). Clinical pharmacokinetics: J. C. Fleishaker *et al., J. Clin. Pharmacol.* **33**, 175, 182 (1993). Veterinary evaluation in endotoxemia of neonatal calves: M. L. Rose, S. D. Semrad, *Am. J. Vet. Res.* **53**, 2305 (1992). Review: E. D. Hall, *Ann. Neurol.* **32**, S137-S142 (1992).

Monomethanesulfonate monohydrate, $C_{38}H_{52}N_6O_2$.-$CH_3SO_3H.H_2O$, *tirilazad mesylate, U-74006F, Freedox.* mp 181-185° (dec). uv max: 234, 285 nm (ϵ 52000, 17000).

THERAP CAT: Lipid peroxidation inhibitor.

9605. Tirofiban. *N-(Butylsulfonyl)-O-[4-(4-piperidin-yl)butyl]-L-tyrosine; N*-(butylsulfonyl)-4-[4-(4-piperidyl)but-oxy]-L-phenylalanine; 2-*S*-(*n*-butylsulfonylamino)-3[4-(pi-peridin-4-yl)butyloxyphenyl]propionic acid. $C_{22}H_{36}N_2O_5S$; mol wt 440.60. C 59.97%, H 8.24%, N 6.36%, O 18.16%, S 7.28%. Specific nonpeptide platelet fibrinogen receptor (GPIIb/IIIa) antagonist. Prepn: M. S. Egbertson *et al.,* **Eur.** pat. **Appl. 478,363** (1992 to Merck); J. Y. L. Chung *et al., Tetrahedron* **49,** 5767 (1993). Physicochemical properties: J. A. McCauley *et al., J. Phys. D: Appl. Phys.* **26,** B85 (1993). Pharmacology: J. J. Lynch *et al., J. Pharmacol. Exp. Ther.* **272,** 20 (1995). Clinical pharmacokinetics and pharmacodynamics: J. S. Barrett *et al., Clin. Pharmacol. Ther.* **56,** 377 (1994).

White solid, mp 223-225°.
Hydrochloride monohydrate, $C_{22}H_{36}N_2O_5S.HCl.H_2O$, *MK-383, Aggrastat.* Solid, mp 131-132°. $[\alpha]_D^{25}$ — 14.4° (c = 0.92 in methanol).

THERAP CAT: Antithrombotic; in treatment of unstable angina.

9606. Tiron®. *4,5-Dihydroxy-1,3-benzenedisulfonic acid disodium salt;* 1,2-dihydroxybenzene-3,5-disulfonic acid disodium salt; disodium-1,2-dihydroxybenzene-3,5-disul-fonate; sodium catechol disulfonate; sodium pyrocatechol-2,4-disulfonate; disodium pyrocatechol-3,5-disulfonate. C_6-$H_4Na_2O_8S_2$; mol wt 314.20. C 22.94%, H 1.28%, Na 14.63%, O 40.74%, S 20.41%. Prepd by heating 5-bromo-4-hydroxy-*m*-benzenedisulfonic acid with sodium hydroxide in water and acidifying product with hydrochloric acid: Fukayama *et al.,* **Japan.** pat. 4327('52) (to Sanwa Pure Chem.), *C.A.* **48,** 5215c (1954).

Crystals, nonhygroscopic. Very freely sol in water. Pro-duces water-sol, colored compds with metal salts. With ferric chloride the resulting complex yields a deep blue soln at pH below 5. Titanium salts give an orange color, copper salts produce a green-yellow, and hexavalent molybdenum a canary-yellow color.

USE: Colorimetric reagent for iron, manganese, titanium, molybdenum: Yoe, Jones, *Ind. Eng. Chem., Anal. Ed.* **16,** 111 (1944). Zirconium complexes: Intorre, Martell, *J. Am. Chem. Soc.* **82,** 358 (1960).

9607. Tiropramide. *α-(Benzoylamino)-4-[2-(diethyl-amino)ethoxy]-N,N-dipropylbenzenepropanamide;* DL-*α*-benz-amido-*p*-[2-(diethylamino)ethoxy]-*N,N*-dipropylhydrocin-namamide; *O*-(2-diethylaminoethyl)-*N*-benzoyl-DL-tyrosyl-

di-*n*-propylamide; CR-605. $C_{28}H_{41}N_3O_3$; mol wt 467.65. C 71.91%, H 8.84%, N 8.99%, O 10.26%. Smooth muscle relaxant; deriv of tyrosine. Prepn: F. Makovec *et al.,* **Ger.** pat. **2,503,992;** *eidem,* **U.S.** pat. **4,004,008** (1975, 1977 both to Rotta). Activity on gastrointestinal motility: P. Senin *et al., Gastrointestinal Motility in Health and Disease,* H. L. Duthie, Ed. (University Park Press, Baltimore, 1978) pp 417-427. Mechanism of action: R. R. Vidal y Plana *et al., J. Pharm. Pharmacol.* **33,** 19 (1981).

Cryst from petr ether, mp 65-67°. LD_{50} in rats: 33.9 mg/kg i.v. (Makovec).
Hydrochloride, $C_{28}H_{41}N_3O_3.HCl$, *Alfospas, Maiorad.*

THERAP CAT: Antispasmodic.

9608. Tissue Plasminogen Activator. Fibrinokinase; extrinsic plasminogen activator; t-PA; TPA. Mol wt ~70,000. Serine protease catalyzing the enzymatic conver-sion of plasminogen to plasmin, *q.q.v.,* through the hydroly-sis of a single Arg-Val bond. Component of the mammalian fibrinolytic system responsible for the specific activation of plasminogen associated with fibrin clots. Distribution of t-PA in various organs differs widely and appears to be related to the degree of tissue vascularization. Identification of plasminogen-activating ability of tissue fragments: T. Astrup, P. M. Permin, *Nature* **159,** 681 (1947); H. J. Tag-non, M. L. Petermann, *Proc. Soc. Exp. Biol. Med.* **70,** 359 (1949). Extraction of soluble activator with potassium thio-cyanate: T. Astrup, A. Stage, *Nature* **170,** 929 (1952). Localization of plasminogen activator in vascular endothe-lial cells: A. J. Todd, *J. Pathol. Bacteriol.* **78,** 281 (1959). Tissue distribution and physiological significance: T. As-trup, *Fed. Proc.* **25,** 42 (1966). Purification: F. Bachmann *et al., Biochemistry* **3,** 1578 (1964); P. Kok, T. Astrup, *ibid.* **8,** 79 (1969); D. C. Rijken *et al., Biochim. Biophys. Acta* **580,** 140 (1979). Immunological identity with plasminogen acti-vators from vascular wall and circulating blood: *eidem, Thromb. Res.* **18,** 815 (1980). Differentiation from uroki-nase: O. Matsuo *et al., Thromb. Haemostasis* **45,** 225 (1981). Purification and characterization of human melanoma t-PA: D. C. Rijken, D. Collen, *J. Biol. Chem.* **256,** 7035 (1981). Cloning and expression of cDNA for human t-PA: D. Pen-nica *et al., Nature* **301,** 214 (1983). t-PA exists in 2 forms. The native, single chain form (t-PA I) can be broken by proteolytic cleavage into 2 polypeptides connected by a di-sulfide bond (t-PA II): P. Wallen *et al., Biochim. Biophys. Acta* **719,** 318 (1982). Conversion of t-PA I to t-PA II dur-ing fibrinolysis and comparison of biological activity: D. C. Rijken *et al., J. Biol. Chem.* **257,** 2920 (1982).

Activity is markedly enhanced by fibrin, primarily as a result of specific binding of the activator to the fibrin clot: S. M. Camiolo *et al., Proc. Soc. Exp. Biol. Med.* **138,** 277 (1971); B. Wiman, D. Collen, *Nature* **272,** 549 (1978). Fi-brin-bound t-PA exhibits greater affinity for plasminogen than does non-bound activator, localizing fibrinolysis at the clot site: M. Hoylaerts *et al., J. Biol. Chem.* **257,** 2912 (1982). Experimental thrombolysis in rabbits: O. Matsuo *et al., Nature* **291,** 590 (1981); in humans: W. Weimar *et al., Lancet* **2,** 1018 (1981). Comparative clinical trials of recom-binant t-PA in myocardial infarction: The Thrombolysis in Myocardial Infarction (TIMI) Study Group, *N. Engl. J. Med.* **312,** 932 (1985); M. Verstraete *et al., Lancet* **1,** 842 (1985); in acute pulmonary embolism: S. Z. Goldhaber *et al., Lancet* **2,** 293 (1988). Review of the fibrinolytic system: D. Collen, *Thromb. Haemostasis* **43,** 77-89 (1980). Review of pharmacology and bioactivity: D. Collen, R. Lijnen, *Arteriosclerosis* **4,** 579-585 (1984); of pharmacology and clinical efficacy: S. D. Rogers *et al.*, *Pharmacotherapy* **7**, 111-121 (1987); D. Collen *et al.*, *Drugs* **38**, 346-388 (1989).

Practically insol in solns of low ionic strength at neutral pH. Sol in 1.6M KSCN at neutral pH; 0.3M K-acetate at pH 4.2.

Recombinant human t-PA, *alteplase, duteplase, rt-PA, Actilyse, Actiplas, Activase, Prolysis, Tiplagen*. Two-chain form produced by recombinant DNA technology. Amino acid sequence identical to the naturally occurring form.

THERAP CAT: Thrombolytic.

9609. Titanic(IV) Acid. Orthotitanic acid; titanic hydroxide. $Ti(OH)_4.xH_2O$. The H_2O content depends upon the method and time of drying.

White, amorphous powder. Insol in water; when freshly pptd it is sol in dil HCl or H_2SO_4 and in boiling caustic alkali solns; after drying it is almost insol.

9610. Titanium. Ti; at. wt 47.88; at. no. 22; valence 2, 3, 4 (mostly tetravalent). Five natural isotopes (mass numbers): 48 (73.94%); 46 (7.93%); 47 (7.28%); 49 (5.51%); 50 (5.34%). Artificial isotopes: 43-45; 51. Ninth most abundant element in earth's crust; 0.63% by wt. Occurs as the oxide in the minerals rutile, ilmenite, perovskite, anatase, or octahedrite and brookite; other minerals include *sphene* or *titanite* ($CaTiSiO_5$) and *benitoite* ($BaTiSi_3O_9$). Discovered by Gregor in 1789; investigated and named by Klaproth in 1795; isolated by Berzelius in 1825. Prepn: de Boer, *Ind. Eng. Chem.* **19**, 1256 (1927); Fast, *Z. Anorg. Chem.* **241**, 42 (1939); Ehrlich in *Handbook of Preparative Inorganic Chemistry* vol. **2**, G. Brauer, Ed. (Academic Press, New York, 2nd ed., 1965) pp 1161-1172. Reviews: *Gmelin's, Titanium* (8th ed.) **41** (1951); Everhart, *Titanium and Titanium Alloys* (Reinhold, New York, 1954); Brophy *et al.*, *Titanium Bibliography* 1900-1951 + suppl (Washington, 1954); McQuillan & McQuillan, *Titanium* (Butterworth's, London, 1956); Barksdale, *Titanium, Its Occurrence, Chemistry and Technology* (Ronald Press, New York, 2nd ed, 1966); Clark, "Titanium" in *Comprehensive Inorganic Chemistry* vol. **3**, J. C. Bailar, Jr. *et al.*, Eds. (Pergamon Press, Oxford, 1973) pp 355-417.

Dark gray, lustrous metal. Dimorphic: α-form: hexagonal structure below 882.5°; β-form: body-centered cubic crystals above 882.5°. Brittle when cold; malleable when hot. Ductile only when free of oxygen; traces of oxygen or nitrogen increase strength. mp 1677°. bp 3277°. Calculated d (α- form): 4.506 (25°); (β-form): 4.400 (900°). Specific heat (25°): 5.98 cal/g-atom/°C. Decomposes steam at 700-800°. Combines with oxygen at a red heat. Can burn in an atm of oxygen under certain conditions: *Chem. & Eng. News* **36**, 36 (Aug. 4, 1958). Attacked by acids only on heating; oxidized by nitric acid to the dioxide. Reacts with fluorine at 150°, with chlorine at 300°, with bromine at 360°, with iodine above 360°. Combines with nitrogen at 800°. Forms alloys with aluminum, chromium, cobalt, copper, iron, lead, nickel, tin.

USE: As alloy with copper and iron in titanium bronze; as addition to steel to impart great tensile strength; to aluminum to impart resistance to attack by salt solns and by organic acids; to remove traces of oxygen and nitrogen from incandescent lamps. Surgical aid (fracture fixation).

9611. Titanium Dichloride. Cl_2Ti; mol wt 118.77. Cl 59.70%, Ti 40.30%. $TiCl_2$. Prepn: Clifton, McWood, *J. Phys. Chem.* **60**, 311 (1956); Ehrlich *et al.*, *Z. Anorg. Allgem. Chem.* **292**, 139 (1957).

Black crystals. mp 1035°. d 3.13. Burns like tinder when heated in air; dec by water. Sol in alcohol. Practically insol in chloroform, ether, carbon disulfide. *Keep dry and well closed.*

9612. Titanium Dioxide. Unitane; C.I. Pigment White 6; C.I. 77891. O_2Ti; mol wt 79.87. O 40.07%, Ti 59.93%. TiO_2. Found in nature as the minerals *rutile* (tetragonal), *anatase* or *octahedrite* (tetragonal), *brookite* (orthorhombic), *ilmenite* ($FeTiO_3$), and *perovskite* ($CaTiO_3$). May be prepd by direct combination of titanium and oxygen; by treatment of titanium salts in aq soln; by the reaction of volatile, inorganic titanium compds with oxygen; by oxidation or hydrolysis of organic compds of titanium. Industrial prepn from ilmenite or rutile: Faith, Keyes & Clark's *Industrial*

Chemicals, F. A. Lowenheim, M. K. Moran, Eds. (Wiley-Interscience, New York, 4th ed., 1975) pp 814-821. Prepn of synthetic rutile: Merker, **U.S.** pat. **2,760,874** (1956 to National Lead). Prepn of spectroscopically pure material by dissolving titanium in an ammoniacal soln of 90% H_2O_2: Czanderna *et al.*, *J. Am. Chem. Soc.* **79**, 5407 (1957). Comprehensive description: H. G. Brittain *et al.* in *Analytical Profiles of Drug Substances and Excipients* vol. **21**, H. G. Brittain, Ed. (Academic Press, San Diego, 1992) pp 659-691.

White powder, mp 1855°. d (rutile): 4.23; (anatase): 3.90; (brookite): 4.13. Insol in water, HCl, HNO_3, or dil H_2SO_4. Sol in hot concd H_2SO_4, in HF. The reactivity depends on a previous heat treatment; prolonged heating produces a less sol material. Also made sol by fusion with potassium bisulfate or with alkali hydroxides or carbonates to form alkali titanates. Possesses perhaps the greatest hiding power of all inorganic white pigments. *Titania* is a name applied to large TiO_2 crystals (translucent water-white or with yellowish cast) suitable for use in jewelry. These crystals have a refractive index (2.7) higher than diamonds (2.4), but lack the hardness of diamonds. When substantially pure, a massive single crystal (boule) of rutile has the properties of a precious gem with a very light straw color and with reflectance, refraction and brilliance measuring greater than those of a diamond.

Note: The rutile structure is common among metal fluorides and oxides of the type MF_2 and MO_2.

Caution: Potential symptom of overexposure is slight lung fibrosis. Potential occupational carcinogen. *See NIOSH Pocket Guide to Chemical Hazards* (DHHS/NIOSH 90-117, 1990) p 214.

USE: Airfloated ilmenite is used for titanium pigment manuf. Rutile sand is suitable for welding-rod-coating materials, as ceramic colorant, as source of titanium metal. As color in the food industry. Anatase titanium dioxide is used for welding-rod-coatings, acid resistant vitreous enamels, in specification paints, exterior white house paints, acetate rayon, white interior air-dry and baked enamels and lacquers, inks and plastics, for paper filling and coating, in water paints, tanners' leather finishes, shoe whiteners, and ceramics. High opacity and tinting values are claimed for rutile-like pigments.

THERAP CAT: Protectant (topical); ultraviolet sunscreen.

9613. Titanium Hydride. H_2Ti; mol wt 49.88. H 4.04%, Ti 95.96%. TiH_2. Prepd by the reduction of titanium oxide with calcium hydride in the presence of hydrogen above 600°: Alexander, **U.S.** pat. **2,427,338** (1947).

Gray-black metallic powder. Stable in air. Dissociation appreciable at 450°.

USE: Additive in powder metallurgy; getter for oxygen and nitrogen in electronic tubes; wetting agent for ceramic to metal seals; source of pure hydrogen.

9614. Titanium Isopropylate. Isopropyl titanate(IV). $C_{12}H_{28}O_4Ti$; mol wt 284.22. C 50.71%, H 9.93%, O 22.52%, Ti 16.84%. $Ti[OCH(CH_3)_2]_4$. Prepn: Bradley *et al.*, *J. Chem. Soc.* **1952**, 5020.

Liquid. Fumes in air. mp $\sim$20°: R. Feld, P. L. Cowe, *The Organic Chemistry of Titanium* (Butterworths, Washington, 1965) p 36. bp_{760} 220°; bp_{10} 104°. d_4^{20} 0.9711. Apparent viscosity 2.11 cp at 25°. Dielectric constant: 3.64 at 62 kilocycles. Dec rapidly in water. Sol in anhydr ethanol, ether, benzene, chloroform. *Precautions must be taken in handling to exclude water from solvents and diluents.*

USE: Polymerization catalyst.

9615. Titanium Sesquisulfate. Titanous sulfate. $O_{12}S_3Ti_2$; mol wt 383.92. O 50.01%, S 25.06%, Ti 24.94%. $Ti_2(SO_4)_3$. (Crystallizes also as a hydrate with $8H_2O$).

Green, cryst powder. Insol in water, alc, concd H_2SO_4. Sol in dil HCl or dil H_2SO_4, giving violet-colored solns.

USE: Same as titanium trichloride.

9616. Titanium Sulfate. *Titanyl sulfate;* titanoxy sulfate. O_5STi; mol wt 159.93. O 50.02%, S 20.05%, Ti 29.93%. $TiOSO_4$. Prepn: Hayek, Engelbrecht, *Monatsh.* **80**, 640 (1949).

White or slightly yellow powder; dec by water.

Note: The name titanyl sulfate is sometimes applied to $Ti(SO_4)_2$.

USE: As mordant in dyeing.

THERAP CAT: Dermatologic.

9617. Titanium Tetrabromide. Br_4Ti; mol wt 367.48. Br 86.97%, Ti 13.03%. $TiBr_4$. Prepn: Olsen, Ryan, *J. Am. Chem. Soc.* **54**, 2215 (1932).

Amber-yellow or orange, very hygroscopic crystals. mp 28.25°. bp 233.45°. d^{20} 3.25. Also reported as 2.6: Duppa, *Proc. Roy. Soc. London* **8**, 42 (1857). Dec by water; sol in abs ether, abs alcohol. *Keep tightly closed.*

9618. Titanium Tetrachloride. Cl_4Ti; mol wt 189.68. Cl 74.76%, Ti 25.24%. $TiCl_4$. Purification: Baxter *et al., J. Am. Chem. Soc.* **45**, 1228 (1923); **48**, 3117 (1926). Survey of preparative methods: Ehrlich in *Handbook of Preparative Inorganic Chemistry*, Vol. 2, G. Brauer, Ed. (Academic Press, New York, 2nd ed., 1965) pp 195-199.

Colorless liquid; penetrating acid odor; absorbs moisture from the air and evolves dense white fumes. d 1.726. mp −24.1°. bp 136.4°. Sol in cold water, alcohol; dec by hot water. *Keep tightly closed.*

Caution: Irritant to eyes, respiratory tract.

USE: Formerly used with potassium bitartrate as a mordant in textile industry, and with dyewoods in dyeing leather; also as smoke-producing screen with ammonia; manuf iridescent glass and artificial pearls.

9619. Titanium Tetrafluoride. F_4Ti; mol wt 123.86. F 61.35%, Ti 38.65%. TiF_4. Prepd by the reaction of hydrogen fluoride with titanium tetrachloride: Ruff, Ipsen, *Ber.* **36**, 1777 (1903); Ruff, Plato, *ibid.* **37**, 673 (1904); by the action of fluorine on the metal or the dioxide: Haendler *et al., J. Am. Chem. Soc.* **76**, 2177 (1954).

Powdery white mass. d_4^{20} 2.798. Sublimes at 284°. mp >400°. Hisses on contact with water. Very hygroscopic; the dihydrate may be crystallized from aqueous soln (also obtained from solns of TiO_2 in HF). Aq solns hydrolyze slowly, and the existence of an oxyfluoride, $TiOF_2$, has been established. Also sol in alcohol and pyridine from which the compds $TiF_4.2EtOH$ and $TiF_4.C_5H_5N$ have been isolated. Insol in ether. Dry ammonia is absorbed at room temp to form $TiF_4.4NH_3$, but at 120° $TiF_4.2NH_3$ is the stable phase. This is sol in water and is sufficiently stable to sublime. TiF_4 is also sol in phosphorus oxychloride, but at 30° a reaction occurs and POF_3 is evolved. *See also* Eméleus in *Fluorine Chemistry* vol. I, J. H. Simons, Ed. (Academic Press, New York, 1950) p 47.

9620. Titanium Trichloride. Titanous chloride. Cl_3Ti; mol wt 154.23. Cl 68.96%, Ti 31.04%. $TiCl_3$. Prepn by reduction of the tetrachloride: Ingraham *et al., Inorg. Syn.* **6**, 52 (1960); Sherfey, *ibid.* 57.

Dark red-violet, unstable, deliquesc crystals; dec on heating above 500°. The dry powder is pyrophoric in air. Very reactive and readily dissociated by moisture in air. Sol in water, exothermic process; sol in alcohol. Practically insol in ether. *Keep tightly closed.*

USE: A powerful reducing agent. Reduces nitrate to ammonia; when boiled with aq SO_2, sulfur is separated; hence is used as an aq soln for estimation of nitro groups, ferric ions, per-salts, etc. Removes stains, etc. (stripper) in laundering.

9621. Titanocene Dichloride. *Dichlorobis(η^5-2,4-cyclopentadien-1-yl)titanium; dichlorodi-π-cyclopentadienyltitanium;* biscyclopentadienyltitanium(IV) dichloride. $C_{10}H_{10}Cl_2$-Ti; mol wt 248.96. C 48.24%, H 4.05%, Cl 28.48%, Ti 19.23%. $(C_5H_5)_2TiCl_2$. Also Cp_2TiCl_2. Prepn: Wilkinson, Birmingham, *J. Am. Chem. Soc.* **76**, 4281 (1954). Structure: Clearfield *et al., Can. J. Chem.* **53**, 1622 (1975).

Bright red acicular crystals from toluene. mp 289 ± 2°. d 1.60. Moderately sol in toluene, chloroform, alcohol, other hydroxylic solvents; sparingly sol in water, petr ether, benzene, ether, carbon disulfide, carbon tetrachloride.

USE: Catalyst; with aluminum alkyls as Ziegler-Natta polymerization catalyst.

9622. Titin. Connectin. Giant, elastic protein found in vertebrate skeletal and cardiac muscles; comprises ~10% of the myofibrillar mass. One of the largest polypeptides known; mol wt: ~3 × 10⁶ Da. Structure consists of repeats of 2 motifs, each with approximately 100 amino acid residues, organized into a super-repeating pattern. Enriched in proline (8-9 mole %) which contributes to its flexibility. Integral component of the cytoskeletal lattice of muscle cells (*see also* Nebulin). Thought to act as a "molecular ruler" that regulates the assembly of myosin and other thick filament proteins; connects the ends of the thick filament with the Z-line. Also responsible for the generation of resting tension in stretched muscle fibers. Identification in rabbit muscle: K. Maruyama *et al., Nature* **262**, 58 (1976). Characterization and localization in myofibrils: K. Wang *et al., Proc. Nat. Acad. Sci. USA* **76**, 3698 (1979). Purification methods: K. Wang, *Methods Enzymol.* **85B**, 264 (1982). Distribution and organization in the cytoskeletal matrix of striated muscle: idem, *Adv. Exp. Med. Biol.* **170**, 285-305 (1984). Quantitative determn in muscle fibers by gel electrophoresis: H. L. M. Granzier, K. Wang, *Electrophoresis* **14**, 56 (1993). *Reviews:* J. Trinick, *Trends Biochem. Sci.* **19**, 405-409 (1994); K. Maruyama, *Biophys. Chem.* **50**, 73-85 (1994).

9623. Tixocortol. *(11β)-11,17-Dihydroxy-21-mercaptopregn-4-ene-3,20-dione;* 11β,17α-dihydroxy-21-thio-3,20-dioxo-4-pregnene. $C_{21}H_{30}O_4S$; mol wt 378.53. C 66.63%, H 7.99%, O 16.91%, S 8.47%. Synthesis and biological activity: S. S. Simons *et al., J. Steroid Biochem.* **13**, 311 (1980). Use in fluorescent chemoaffinity labeling: *eidem, Biochemistry* **18**, 4915 (1979). Prepn of the 21-pivalate: D. R. Torossian *et al.,* Ger. pat. **2,357,778**; *eidem,* U.S. pat. **4,014,909** (1974, 1977 both to Jouveinal). Series of articles on pharmacology, *in vitro* and *in vivo* activity of the pivalate: *Arzneimittel-Forsch.* **31**, 453-469 (1981).

Fine white solid, dec 220-221°. uv max (95% ethanol): 241 nm (ϵ 1.65×10⁴).

21-Pivalate, $C_{26}H_{38}O_5S$, *(11β)-21-[(2,2-dimethyl-1-oxopropyl)thio]-11,17-dihydroxypregn-4-ene-3,20-dione, JO-1016, Pivalone, Rectovalone, Tiovalon.* Crystals from ethanol, mp 195-200°. $[\alpha]_D^{20}$ +145° (c = 1 in dioxane). uv max (methanol): 229 nm (log ϵ 4.259).

USE: Free thiol in chemoaffinity labeling.

THERAP CAT: Anti-inflammatory.

9624. Tizanidine. *5-Chloro-N-(4,5-dihydro-1H-imidazol-2-yl)-2,1,3-benzothiadiazol-4-amine;* 5-chloro-4-(2-imidazolin-2-ylamino)-2,1,3-benzothiadiazole; DS-103-282; Sirdalud; Ternelin. $C_9H_8ClN_5S$; mol wt 253.71. C 42.61%, H 3.18%, Cl 13.97%, N 27.60%, S 12.64%. α_2-Adrenergic agonist; centrally active myotonolytic. Prepn and antitremor activity: P. Neumann, Neth. pat. Appl. **7,306,228**; *eidem,* U.S. pat. **3,843,668** (1973, 1974 both to Wander-Sandoz). Pharmacology: A. C. Sayers *et al., Arzneimittel-Forsch.* **30**, 793 (1980); and mechanism of action studies: L. Turski *et al., Brain Res.* **379**, 367 (1986); H. Ono *et al., Gen. Pharmacol.* **17**, 137 (1986). Pharmacokinetics: V. Heazlewood *et al., Eur. J. Clin. Pharmacol.* **25**, 65 (1983). Clinical evaluation in spasticity of multiple sclerosis: P. M. Newman *et al.,* *ibid.* **23**, 31 (1982); M. C. Hoogstraten *et al., Acta Neurol. Scand.* **77**, 224 (1988).

Crystals from methanol, mp 221-223°. LD$_{50}$ orally in mice: 235 mg/kg (Sayers).

THERAP CAT: Muscle relaxant (skeletal).

9625. TMD. *Decahydro-1,1,4a-trimethyl-2-naphthalenol;* 1,1,10-trimethyl-*trans*-2-decalol; 4,4,10β-trimethyl-*trans*-decal-3β-ol. C$_{13}$H$_{24}$O; mol wt 196.33. C 79.53%, H 12.32%, O 8.15%. Prepn: B. Gaspert *et al., J. Chem. Soc.* **1958**, 624; C. Djerassi, D. Marshall, *J. Am. Chem. Soc.* **80**, 3986 (1958); T. G. Halsall *et al., J. Chem. Soc.* **1959**, 2798; S. L. Mukherjee, P. C. Dutta, *ibid.* **1960**, 67. Inhibitor of cholesterol biosynthesis: J. A. Nelson *et al., J. Am. Chem. Soc.* **100**, 4900 (1978); T.-Y. Chang *et al., J. Biol. Chem.* **254**, 11258 (1979).

Crystals, mp 60-65° (subl). Sol in alcohol; slightly sol in water.

d-Form, crystals, mp 87-88° (subl). [α]$_D^{26}$ +15.9° (CHCl$_3$).
l-Form, crystals, mp 87-88° (subl). [α]$_D^{26}$ − 16.9° (CHCl$_3$).

9626. TNAZ. *1,3,3-Trinitroazetidine.* C$_3$H$_4$N$_4$O$_6$; mol wt 192.09. C 18.76%, H 2.10%, N 29.17%, O 49.98%. Melt-castable nitramine explosive. Synthesis and crystal structure: T. G. Archibald *et al., J. Org. Chem.* **55**, 2920 (1990). Alternate synthesis: A. R. Katritzky *et al., J. Heterocyclic Chem.* **31**, 271 (1994); T. Axenrod *et al., J. Org. Chem.* **60**, 1959 (1995). Decomposition studies: P. Politzer, J. M. Seminario, *Chem. Phys. Letters* **207**, 27 (1993). Thermolysis: J. C. Oxley *et al., J. Phys. Chem.* **98**, 7004 (1994). *Review:* S. Borman, *Chem. & Eng. News* **72**, 18-22 (Jan. 17, 1994).

Explosive. White crystals from CCl$_4$, mp 100°-101°. d 1.84 g/cm^3.

USE: Munitions.

9627. Tobacco Mosaic Virus. TMV. Most thoroughly investigated and best known virus. A ribonucleoprotein contg 95% protein and 5% ribonucleic acid. Molecular weight is 39.0 × 10^6. TMV is a rod-shaped particle, ~3000 Å long, consisting of 2130 polypeptide chains arranged in some regular order to form a tubular molecule containing a core of RNA. The polypeptide chains consist of 158 amino acid residues each; their mol wt is 17,533. The RNA is responsible for the infectivity of the virus. The protein functions to protect the RNA and determine the macroscopic properties of the virus such as morphology and serological specificity. Isoln from diseased tobacco plants: Stanley, *Science* **81**, 644 (1935). Prepn and characterization of essentially uniform TMV particles: Boedtker, Simmons, *J. Am. Chem. Soc.* **80**, 2550 (1958); Knight, *Biochem. Prepns.* **9**, 132 (1962). Sequential arrangement of the 158 amino acid residues of the protein subunit: Tsugita *et al., Proc. Nat. Acad. Sci. USA* **46**, 1463 (1960). Synthesis of TMV-RNA by cell free extracts from infected tobacco leaves: Kim, Wildman, *Biochem. Biophys. Res. Commun.* **8**, 394 (1962); Cochran, *Chem. & Eng. News* **40**, 64 (Sept. 17, 1962). *Reviews:* Anderer, *Advan. Protein Chem.* **18**, 1 (1963); Caspar, *ibid.* **37**; Klug, Caspar, *Advan. Virus Res.* **7**, 233-277 (1960); Lauffer, Stevens, *ibid.* **13**, 1 (1968); Reddi, *ibid.* **17**, 51 (1972); L. Hirth, K. E. Richards, *ibid.* **26**, 145-199 (1981).

USE: Popular tool for studying the correlation between chemical structure and biological function.

9628. Tobramycin. *O-3-Amino-3-deoxy-α-D-glucopyranosyl-(1→6)-O-[2,6-diamino-2,3,6-trideoxy-α-D-ribo-hexopyranosyl-(1→4)]-2-deoxy-D-streptamine;* 4-[2,6-diamino-2,3,6-trideoxy-α-D-glycopyranosyl]-6-[3-amino-3-deoxy-α-D-glycopyranosyl]-2-deoxystreptamine; nebramycin factor 6; NF 6; Gernebcin; Tobracin; Tobradistin; Tobralex; Tobramaxin; Tobrex. C$_{18}$H$_{37}$N$_5$O$_9$; mol wt 467.52. C 46.24%, H 7.98%, N 14.98%, O 30.80%. Single factor antibiotic comprising about 10% of *nebramycin*, (formerly called *tenebrimycin, tenemycin*), the aminoglycosidic antibiotic complex produced by *Streptomyces tenebrarius*. See also apramycin. Series of articles on isolation, separation and evaluation of the nebramycin complex: Stark *et al.,* Higgens, Kastner; Thompson, Presti; Wick, Welles, *Antimicrob. Ag. Chemother.* **1967**, 314-348. *Review:* Stark *et al., Folia Microbiol. (Prague)* **16**, 205-217 (1971). Tobramycin is structurally similar to the kanamycins and the gentamicins. Elucidation of structure: Koch, Rhoades, *Antimicrob. Ag. Chemother.* **1970**, 309. Synthesis: Y. Takagi *et al., Bull. Chem. Soc. Japan* **49**, 3649 (1976); M. Tanabe *et al., Tetrahedron Letters* **1977**, 3607. Activity studies: Black, Griffith; Preston, Wick, *ibid.* **314**, 322; Meyers, Hirschman, *J. Clin. Pharmacol.* **12**, 313, 321 (1972). Toxicology: Welles *et al., Toxicol. Appl. Pharmacol.* **22**, 332 (1972); De Rosa *et al., J. Int. Med. Res.* **2**, 100 (1974).

Basic, water-sol substance. [α]$_D$ 128°. LD$_{50}$ in mice, rats (mg/kg): 441, 969 s.c. (Welles).

Sulfate, *Nebcin, Nebicina, Obracin, Tobra.*

THERAP CAT: Antibacterial.

9629. Tocainide. *2-Amino-N-(2,6-dimethylphenyl)propanamide;* 2-aminopropiono-2',6'-xylidide. C$_{11}$H$_{16}$N$_2$O; mol wt 192.26. C 68.72%, H 8.39%, N 14.57%, O 8.32%. Antiarrhythmic agent related to lidocaine, *q.v.* Prepn: R. N. Boys *et al., Ger. pat.* **2,235,745** (1973 to Astra), *C.A.* **78**, 140411 (1973); and resolution of isomers: E. W. Byrnes *et al., J. Med. Chem.* **22**, 1171 (1979). *In vitro* effects on muscle contractions: R. Dengler, R. Rüdel, *Arzneimittel-Forsch.* **29**, 270 (1979). Inhibition of leucocyte locomotion in dogs: G. J. Stewart *et al., Lab. Invest.* **42**, 302 (1980). Biotransformation in humans: A. T. Elvin *et al., J. Pharm. Sci.* **69**, 47 (1980). Kinetics: C. Groffner, *Clin. Pharmacol. Ther.* **27**, 64 (1980). Use in refractory ventricular arrhythmias: D. M. Roden *et al., Am. Heart J.* **100**, 15 (1980). HPLC resolution of enantiomers: K. M. McErlane, G. K. Pillai, *J. Chromatog.* **274**, 129 (1983); determn in plasma: A. J. Sedman, J. Gal, *ibid.* **306**, 155 (1984). Pharmacokinetics of enantiomers: A. H. Thomson *et al., Brit. J. Pharmacol.* **21**, 149 (1986). *In vitro* and *in vivo* pharmacodynamics of enantiomers: A. J. Block *et al., J. Cardiovasc. Pharmacol.* **11**, 216 (1988). Review of pharmacology and therapeutic efficacy: B. Holmes *et al., Drugs* **26**, 93-123 (1983); D. M. Roden, R. L. Woosley, *N. Engl. J. Med.* **315**, 41-45 (1986).

(±)-Form hydrochloride, C$_{11}$H$_{16}$N$_2$O.HCl, *W-36095, Taquidil, Tonocard, Xylotocan.* Crystals from ethanol/ether, mp 246-247°.

R-(−)-Form hydrochloride, crystals from ethanol-diethyl ether, mp 265-266°. [α]$_D$ −42.16° (c = 2.63 in methanol).

S-(+)-Form hydrochloride, crystals from ethanol-diethyl ether, mp 264.5°. [α]_D +42.35° (c = 2.63 in methanol).
THERAP CAT: Antiarrhythmic (class IB).

9630. Tocamphyl. *1,2,2-Trimethyl-1,3-cyclopentanedicarboxylic acid 1-[1-(4-methylphenyl)ethyl] ester, compd with 2,2'-iminobis[ethanol] (1:1); camphoric acid 1-(p,α-dimethylbenzyl) ester, compd with 2,2'-imidodiethanol (1:1); p,α-dimethylbenzyl camphorate diethanolamine salt; p-toluoylmethylcarbinolmono-d-camphoric acid ester diethanolamine salt; methyl p-tolylcarbinol camphorate diethanolamine salt; p-tolylmethylcarbinol camphoric acid ester diethanolamine salt; diethanolamine p-tolylmethylcarbinol camphorate; diethanolamine d-methyltoluylcarbinol camphorate; Biliphorine; Hepatoxane; Syncuma.* C_{23}H_{37}NO_6; mol wt 423.55. C 65.22%, H 8.81%, N 3.31%, O 22.66%. Prepn: **Swiss.** pat. 211,203 (1940 to Chemiewerk Homburg); *Chem. Zentr.* **1941**, I, 2972.

Crystals, sol in water.
THERAP CAT: Choleretic.

9631. Tocol. *3,4-Dihydro-2-methyl-2-(4,8,12-trimethyltridecyl)-2H-1-benzopyran-6-ol; 2-methyl-2-(4,8,12-trimethyltridecyl)-6-chromanol; 2-methyl-2-phytyl-6-chromanol; 6-hydroxy-2-methyl-2-phytylchroman; 2-methyl-2-phytyl-6-hydroxychroman.* C_{26}H_{44}O_2; mol wt 388.63. C 80.35%, H 11.41%, O 8.23%. Synthesis by the condensation of hydroquinone and phytol in the presence of anhydr formic acid: Pendse, Karrer, *Helv. Chim. Acta* **40**, 1837 (1957). Antioxidant activity of tocol and its methyl derivs: Olcott, van der Veen, *Lipids* **3**, 331 (1968).

Colorless, viscous oil. bp_{0.001} 165-175°.
Acetate, C_{28}H_{46}O_3, viscous oil. bp_{0.001} 180-185°.
USE: Antioxidant.

9632. β-Tocopherol. *3,4-Dihydro-2,5,8-trimethyl-2-(4,8,12-trimethyltridecyl)-2H-1-benzopyran-6-ol; 2,5,8-trimethyl-2-(4,8,12-trimethyltridecyl)-6-chromanol; 5,8-dimethyltocol; cumotocopherol; neotocopherol; p-xylotocopherol.* C_{28}H_{48}O_2; mol wt 416.69. C 80.71%, H 11.61%, O 7.68%. Accompanies vitamin E (α-tocopherol) and γ-tocopherol from natural sources. Is biologically less active than α-tocopherol. May be separated by fractional crystn of the allophanates: Emerson *et al., Science* **83**, 421 (1936); *J. Biol. Chem.* **113**, 319 (1936); Baxter *et al., J. Am. Chem. Soc.* **65**, 918 (1943).

Pale yellow, viscous oil. bp_{0.1} 200-210°. [α]_D^{20} +6.37°. uv max: 297 nm (E_{1cm}^{1%} 86.4). Insol in water. Freely sol in oils, fats, acetone, alcohol, chloroform, ether, other fat solvents. Very stable to heat and alkalies. Slowly oxidized by atmospheric oxygen, rapidly by ferric and silver salts. Gradually darkens on exposure to light. Not pptd by digitonin.
Allophanate, C_{30}H_{50}N_2O_4, crystals, mp 138-139°. [α]_D^{18} +6.7° (chloroform).
Acetate, C_{30}H_{50}O_3, pale yellow, oily liquid. bp_{0.3} 215-220°. Insol in water. Freely sol in acetone, chloroform, ether. Less readily sol in alcohol.

THERAP CAT: *See* Vitamin E.
THERAP CAT (VET): In "stiff lamb disease".

9633. γ-Tocopherol. *3,4-Dihydro-2,7,8-trimethyl-2-(4,8,12-trimethyltridecyl)-2H-1-benzopyran-6-ol; 2,7,8-trimethyl-2-(4,8,12-trimethyltridecyl)-6-chromanol; 7,8-dimethyltocol; o-xylotocopherol.* C_{28}H_{48}O_2; mol wt 416.69. C 80.71%, H 11.61%, O 7.68%. Accompanies vitamin E (α-tocopherol) and β-tocopherol from natural sources. Is biologically less active than α-tocopherol. May be separated by fractional crystn of the allophanates: Emerson *et al., Science* **83**, 421 (1936); *J. Biol. Chem.* **113**, 319 (1936); Baxter *et al., J. Am. Chem. Soc.* **65**, 918 (1943).

Pale yellow, viscous oil. Has been crystallized at −30°. bp_{0.1} 200-210°. [α]_D^{20} −2.4° (alc). uv max: 298 nm (E_{1cm}^{1%} 92.8). Insol in water. Freely sol in oils, fats, acetone, alcohol, chloroform, ether, other fat solvents. Very stable to heat and alkalies. Slowly oxidized by atmospheric oxygen, rapidly by ferric and silver salts. Gradually darkens on exposure to light. Not pptd by digitonin.
Allophanate, C_{30}H_{50}N_2O_4, crystals, mp 136-138°. [α]_D^{18} +3.4° (chloroform).
Acetate, C_{30}H_{50}O_3, pale yellow, oily liquid. bp_{0.3} 215-220°. Insol in water. Freely sol in acetone, chloroform, ether. Less readily sol in alcohol.
THERAP CAT: *See* Vitamin E.
THERAP CAT (VET): *See* Vitamin E.

9634. δ-Tocopherol. *[2R-[2R*(4R*,8R*)]]-3,4-Dihydro-2,8-dimethyl-2-(4,8,12-trimethyltridecyl)-2H-1-benzopyran-6-ol; 8-methyltocol.* C_{27}H_{46}O_2; mol wt 402.66. C 80.54%, H 11.51%, O 7.95%. Appears to be a rather common member of the vitamin E complex. Ingredient of Mixed Tocopherols Concentrate, N.F. It was found to constitute approx 30% of the mixed tocopherols in soybean oil, 5% of those in wheat germ oil, and there is evidence of its occurrence in cottonseed and peanut oils. Claimed to be the most potent antioxidant of the tocopherols. It has only one hundredth of the activity of natural α-tocopherol in the Evans resorption sterility test for vitamin E. Isoln from soybean oil: Stern *et al., J. Am. Chem. Soc.* **69**, 869 (1947). Synthesis: Green *et al., J. Chem. Soc.* **1959**, 3374; **Brit.** pat. **900,085** (1961 to Hoffmann-La Roche).

Pale yellow, viscous oil. [α]_{546}^{25} +3.4° (c = 15.5 in alc); [α]_{546}^{25} +1.1° (c = 10.9 in benzene). uv max: 298 nm (E_{1cm}^{1%} 91.2). Solubilities similar to γ-tocopherol, *q.v.*
THERAP CAT: *See* Vitamin E.

9635. ε-Tocopherol. *[R-(E,E)]-3,4-Dihydro-2,5,8-trimethyl-2-(4,8,12-trimethyl-3,7,11-tridecatrienyl)-2H-1-benzopyran-6-ol; 2,5,8-trimethyl-2-(4,8,12-trimethyltrideca-3,7,11-trienyl)chroman-6-ol; 5-methyltocol.* C_{28}H_{42}O_2; mol wt 410.64. C 81.90%, H 10.31%, O 7.79%. Isoln from wheat germ oil and from bran: Eggitt, Ward, *J. Sci. Food Agr.* **4**, 569 (1953); Eggitt, Norris, *ibid.* **6**, 689 (1955); 7, 496 (1956). Structure: Green *et al., J. Chem. Soc.* **1959**, 3362; *Chem. & Ind. (London)* **1960**, 73; McHale *et al., J. Chem. Soc.* **1963**, 784. Synthesis of all-*trans*-ε-tocopherol: Schudel *et al., Helv. Chim. Acta* **46**, 2517 (1963).

Pale yellow oil. uv max (ethanol): 296 nm ($E_{1cm}^{1\%}$ 87).

4-Phenylazobenzoate, $C_{41}H_{50}N_2O_3$, orange crystals from isopropanol, mp 70-71°.

9636. ζ₁-Tocopherol. *3,4-Dihydro-2,5,7,8-tetramethyl-2-(4,8,12-trimethyl-3,7,11-tridecatrienyl)-2H-1-benzopyran-6-ol; 2,5,7,8-tetramethyl-2-(4,8,12-trimethyl-3,7,11-tridecatrien-yl)-6-chromanol; 5,7,8-trimethyltocotrien-3',7',11'-ol.* $C_{29}H_{44}O_2$; mol wt 424.67. C 82.02%, H 10.44%, O 7.54%. Isoln from wheat bran: Green *et al., J. Sci. Food Agr.* **6**, 274 (1955); Green *et al., Chem. & Ind. (London)* **1960**, 73. Structure: Green *et al., J. Chem. Soc.* **1959**, 3362. Attempt at synthesis: McHale *et al., ibid.* **1963**, 784. *Review:* M. Kofler *et al.,* "Physicochemical Properties and Assay of the Tocopherols" in R. S. Harris, I. G. Wood, *Vitam. Horm.* (New York) **20**, 407-439 (1962). Synthesis of all-*trans*-ζ₁-tocopherol: Schudel *et al., Helv. Chim. Acta* **46**, 2517 (1963).

all trans-form

uv max (ethanol): 292.5 nm ($E_{1cm}^{1\%}$ 91).

9637. ζ₂-Tocopherol. *3,4-Dihydro-2,5,7-trimethyl-2-(4,8,12-trimethyltridecyl)-2H-1-benzopyran-6-ol; 2,5,7-trimethyl-2-(4,8,12-trimethyltridecyl)-6-chromanol; 5,7-dimethyltocol.* $C_{28}H_{48}O_2$; mol wt 416.69. C 80.71%, H 11.61%, O 7.68%. Isoln from rice: Green, Marcinkiewicz, *Nature* **177**, 86 (1956). Structure: Green *et al., J. Chem. Soc.* **1959**, 3362. Synthesis: Karrer, Fritsche, *Helv. Chim. Acta* **21**, 1234 (1938); Bergel *et al., J. Chem. Soc.* **1938**, 1382; McHale *et al., ibid.* **1958**, 1600.

Pale yellow oil. Crystallizes in prisms at −4°. uv max (ethanol): 292 nm ($E_{1cm}^{1\%}$ 83.0).

4-Phenylazobenzoate, $C_{41}H_{56}N_2O_3$, crystals from 2-propanol, mp 61°.

3',5'-Dinitrophenylurethane, $C_{35}H_{51}N_3O_7$, plates from 2-propanol + water, mp 65°.

9638. η-Tocopherol. *3,4-Dihydro-2,7-dimethyl-2-(4,8,-12-trimethyltridecyl)-2H-1-benzopyran-6-ol; 2,7-dimethyl-2-(4,8,12-trimethyltridecyl)-6-chromanol; 7-methyltocol.* $C_{27}H_{46}O_2$; mol wt 402.66. C 80.54%, H 11.51%, O 7.95%. Isolation from rice: Green, Marcinkiewicz, *Nature* **177**, 86 (1956). Synthesis: McHale *et al., J. Chem. Soc.* **1958**, 1600; Green *et al., ibid.* **1959**, 3374; Marcinkiewicz *et al., ibid.* 3377.

Pale yellow oil. uv max (ethanol): 298 nm ($E_{1cm}^{1\%}$ 104.1).

4-Phenylazobenzoate, $C_{40}H_{54}N_2O_3$, crystals from 2-propanol, mp 55-56°.

3',5'-Dinitrophenylurethane, $C_{34}H_{49}N_3O_7$, crystals from ethanol + water, mp 115-117°.

p-Nitrophenylurethane, $C_{34}H_{50}N_2O_5$, crystals from methanol, mp 112°.

9639. Tocoretinate. *[2R*(4R*,8R*)]-(±)-Retinoic acid 3,4-dihydro-2,5,7,8-tetramethyl-2-(4,8,12-trimethyltridecyl)-2H-1-benzopyran-6-yl ester; (±)-(2R*)-2,5,7,8-tetramethyl-2-[(4R*,8R*)-4,8,12-trimethyltridecyl]-6-chromanyl retinoate;* tretinoin tocoferil; DL-α-tocopheryl retinoate; L-300; N-021; Olcenon. $C_{49}H_{76}O_3$; mol wt 713.14. C 82.53%, H 10.74%, O 6.73%. Wound healing agent that stimulates proliferation of normal skin fibroblasts. Prepn: H. Fukawa, K. Tanaka, **Japan. Kokai 73 00,469**; *eidem,* **U.S. pat. 3,878,202** (1973, 1975 both to Nisshin Flour Milling Co.). Pharmacology: K. Sakyo *et al., Oyo Yakuri* **43**, 111 (1992), *C.A.* **116**, 228223z (1992). Pharmacokinetics: T. Nakazawa *et al., ibid.* 205, *C.A.* **117**, 19823 (1993). Mechanism of action study: N. Kawamura *et al., Dig. Dis. Sci.* **39**, 2191 (1994). Clinical evaluation in skin ulcers: T. Doi, *Skin Res.* **36**, 384 (1994); K. Nakagawa *et al., ibid.* 209. Acute toxicity study: Y. Harada *et al., Oyo Yakuri* **43**, 227 (1992), *C.A.* **117**, 20721 (1992).

Light yellow oil. uv max (ethanol): 365 nm ($E_{1cm}^{1\%}$ 642). LD₅₀ in mice (mg/kg): > 1000 i.v.; > 2000 orally (Fukawa, 1975).

THERAP CAT: Vulnerary.

9640. Todralazine. *2-(1-Phthalazinyl)hydrazinecarboxylic acid ethyl ester; 3-(1-phthalazinyl)carbazic acid ethyl ester; N¹-carbethoxy-N²-phthalazinehydrazine; carboethoxyphthalazinohydrazine; ecarazine.* $C_{11}H_{12}N_4O_2$; mol wt 232.24. C 56.89%, H 5.21%, N 24.12%, O 13.78%. Prepn: S. Biniecki *et al., Bull. Acad. Polon. Sci. Ser. Sci.* **6**, 227 (1958), *C.A.* **52**, 18424g (1958); *eidem,* **Belg. pat. 647,722**; *eidem,* **U.S. pat. 3,591,588** (1964, 1971 both to Polfa). Spectrofluorometric determn in plasma: A. Ishii, T. Deguchi, *Chem. Pharm. Bull.* **26**, 2241 (1978). Pharmacological study: M. Filczewski, E. Boguka, *Pol. J. Pharmacol. Pharm.* **31**, 127 (1979), *C.A.* **91**, 204508 (1979). Absorption, distribution, excretion in rats and humans: A. Ishii *et al., Oyo Yakuri* **18**, 61 (1979), *C.A.* **92**, 104012 (1980). Clinical study: W. Reiterer, H. Czitober, *Arzneimittel-Forsch.* **27**, 2163 (1977). Toxicity study: F. Parravincini *et al., Farmaco Ed. Sci.* **34**, 299 (1979).

Hydrochloride, $C_{11}H_{12}N_4O_2 \cdot HCl$, *CEPH, BT-621, Apiracohl, Aperdor, Apride, Atapren, Binazin, Illcut, Propat.* LD₅₀ i.p. in mice: 500 mg/kg (Parravincini).

THERAP CAT: Antihypertensive.

9641. Tofenacin. *N-Methyl-2-[(2-methylphenyl)phenylmethoxy]ethanamine; N-methyl-2-[(o-methyl-α-phenylbenzyl)oxy]ethylamine;* N-demethylorphenadrine; N-methylaminoethyl 2-methylbenzhydryl ether; phenyl-(o-tolylmethyl) methylaminoethyl ether; [2-(phenyl-o-tolylmethoxy)ethyl]-methylamine; N-methyl-2-(2-methylbenzhydryloxy)ethylamine; N-methyl-2-[α-(2-tolylbenzyl)oxy]ethylamine. $C_{17}H_{21}NO$; mol wt 255.36. C 79.96%, H 8.29%, N 5.49%, O 6.27%. Metabolite of orphenadrine, *q.v.,* formed by N-demethylation. Prepn: **Belg. pat. 628,167**; Harms, **U.S. pat.**

3,407,258 (1963, 1968 both to Brocades-Stheeman & Pharmacia). Resolution of isomers: van der Stelt *et al., Arzneimittel-Forsch.* **19**, 2010 (1969). Physical and chemical properties: Doorenbos *et al., Pharm. Weekblad* **101**, 525 (1966). Pharmacology: Funcke *et al., Arch. Int. Pharmacodyn.* **177**, 28 (1969); Den Besten *et al., Arzneimittel-Forsch.* **20**, 538 (1970). Metabolic studies: Hespe, Prins, *Eur. J. Pharmacol.* **8**, 119 (1969); Hespe, Kafoe, *ibid.* **13**, 113 (1970). Clinical studies: Bram, Shanmuganathan, *Curr. Ther. Res.* **13**, 625 (1971).

Liquid, $bp_{0.7}$ 139-143°.
Hydrochloride, $C_{17}H_{21}NO.HCl$, *BS-7331, Elamol, Tofacine*. White to off-white crystals, mp 143-147° (Doorenbos); 147-148° (Harms). Freely soluble in water, methanol, ethanol, chloroform. Slightly sol in acetone. Practically insol in ethyl ether. LD_{50} in mice (mg/kg): 182 orally, 82 s.c., 58 i.p., 36.5 i.v.; in rats: 72 i.p.; in guinea pigs: 92 s.c. (Funcke).
THERAP CAT: Antidepressant.

9642. Tofisopam. *1-(3,4-Dimethoxyphenyl)-5-ethyl-7,8-dimethoxy-4-methyl-5H-2,3-benzodiazepine; EGYT 341; Grandaxin; Seriel.* $C_{22}H_{26}N_2O_4$; mol wt 382.46. C 69.09%, H 6.85%, N 7.32%, O 16.73%. The first 5H-2,3-benzodiazepine. Prepn: J. Korosi *et al., Hung.* pat. **155,572** (1969 to Pharm. Res. Inst.), *C.A.* **70**, 115026a (1969); Brit. pat. **1,202,579** corresp to U.S. pat. **3,736,315** (1970, 1973 both to EGYT); J. Korosi, T. Lang, *Ber.* **107**, 3883 (1974). Synthesis and conformation: *eidem, Ther. Hung.* **23**, 132 (1975). FT ^{13}C NMR study: A. Neszmelyi *et al., Ber.* **107**, 3894 (1974). Pharmacology: L. Petocz, I. Kosoczky, *Ther. Hung.* **23**, 134 (1975). Human pharmacokinetics: S. Ronai *et al., ibid.* 139. Comparative efficacy: H. L. Goldberg, R. J. Finnerty, *Am. J. Psychiatry* **136**, 196 (1979).

Colorless to light cream cryst powder from isopropyl alcohol, mp 156-157°. uv max (methanol): 310, 272, 239 nm (ϵ 16100, 11200, 26300).
THERAP CAT: Anxiolytic.

9643. Tolan. *1,1'-(1,2-Ethynediyl)bisbenzene; diphenylacetylene; diphenylethyne.* $C_{14}H_{10}$; mol wt 178.23. C 94.34%, H 5.66%. $C_6H_5C{\equiv}CC_6H_5$. Prepd by the oxidation of benzil dihydrazone with mercuric oxide: Schlenk, Bergmann, *Ann.* **463**, 76 (1928); by dehydrohalogenation of stilbene dibromide: Söderbäck, *Ann.* **443**, 161 (1925); Smith, Hoehn, *J. Am. Chem. Soc.* **63**, 1180 (1941); Smith, Falkof, *Org. Syn. coll. vol. III*, 350 (1955). Improved procedure: L. F. Fieser, *Experiments in Organic Chemistry* (Boston, 3rd ed., 1955) p 181.
Monoclinic, pseudo-rhombic rods or large spears from 95% ethanol. mp 60-61° (also reported as 62.5°). bp_{760} 300°; bp_{19} 170°. Dipole moment 0.3. Specific heat at 20°: 0.297. uv max: 216, 221, 269, 272, 279, 288, 297 nm (ϵ 20600, 20300, 23450, 25200, 33000, 23250, 29400). Insol in water. Freely sol in ether, hot alcohol.

9644. Tolazamide. *N-[[(Hexahydro-1H-azepin-1-yl)-amino]carbonyl]-4-methylbenzenesulfonamide; 1-(hexahydro-1H-azepin-1-yl)-3-(p-tolylsulfonyl)urea; N-(p-toluenesulfonyl)-N'-hexamethyleniminourea; tolazolamide; U-17835; Diabewas; Norglycin; Tolanase; Tolinase.* $C_{14}H_{21}N_3O_3S$; mol wt 311.41. C 54.00%, H 6.80%, N 13.49%, O 15.41%, S 10.30%. Prepn: Wright, Willette, *J. Med. Pharm. Chem.* **5**, 815 (1962); Wright, **Brit.** pat. **887,886** (1962 to Upjohn). Mode of action study: Marshall *et al., Metab. Clin. Exp.* **19**, 1046 (1970). Clinical experience and review of literature: Balodimos, Marble, *Curr. Ther. Res.* **13**, 6-12 (1971).

Crystals, mp 170-173°.
THERAP CAT: Antidiabetic.

9645. Tolazoline. *4,5-Dihydro-2-(phenylmethyl)-1H-imidazole; 2-benzyl-2-imidazoline; 2-benzyl-2-iminazoline; benzazoline; 2-benzyl-4,5-imidazoline; phenylmethylimidazoline.* $C_{10}H_{12}N_2$; mol wt 160.22. C 74.97%, H 7.55%, N 17.48%. α-Adrenergic blocker. Prepn: Sonn, U.S. pat. **2,161,938** (1939 to Ciba). HPLC determn in serum: L. M. L. Todesco *et al., Ther. Drug Monit.* **9**, 78 (1987). Review of pharmacology and clinical use: R. M. Ward, *Clin. Perinatol.* **11**, 703-713 (1984).

Hydrochloride, $C_{10}H_{12}N_2.HCl$, *Lambral, Priscol, Priscoline, Vaso-Dilatan.* Bitter crystals, mp 174°. Freely sol in water, alcohol; sol in chloroform; very slightly sol in ether, ethyl acetate. pH of 2.5% soln 4.9-5.3.
THERAP CAT: Vasodilator (peripheral).

9646. Tolbutamide. *N-[(Butylamino)carbonyl]-4-methylbenzenesulfonamide; 1-butyl-3-(p-tolylsulfonyl)urea; tolylsulfonylbutylurea; 3-(p-tolyl-4-sulfonyl)-1-butylurea; N-n-butyl-N'-tosylurea; N'-4-methylbenzenesulfonyl-N''-butylurea; N-(sulfonyl-p-methylbenzene)-N'-n-butylurea; D-860; U-2043; Artosin; Diaben; Diasulfon; Dolipol; Glyconon; Ipoglicone; Mobenol; Orabet; Orinase; Oterben; Pramidex; Rastinon; Tolbusal.* $C_{12}H_{18}N_2O_3S$; mol wt 270.35. C 53.31%, H 6.71%, N 10.36%, O 17.75%, S 11.86%. Description: Ehrhart, *Naturwiss.* **43**, 93 (1956). Prepn: **Brit.** pat. **808,071**; Aumüller, Herr, **Ger.** pat. **1,066,575** (both 1959 to Hoechst); Ruschig *et al.*, **U.S.** pat. **2,968,158** (1961 to Upjohn). Comprehensive description: W. F. Beyer, E. H. Jensen, *Anal. Profiles Drug Subs.* **3**, 513-543 (1974).

Crystals, mp 128.5-129.5°.
Sodium salt, $C_{12}H_{17}N_2NaO_3S$, mp 130-133°. Tetrahydrate, mp 41-43°.
THERAP CAT: Antidiabetic.
THERAP CAT (VET): Hypoglycemic agent.

9647. Tolciclate. *Methyl(3-methylphenyl)carbamothioic acid O-(1,2,3,4-tetrahydro-1,4-methanonaphthalen-6-yl) ester; O-(1,4-methano-1,2,3,4-tetrahydro-6-naphthyl)-N-methyl-N-(m-tolyl)thiocarbamate; KC-9147; Fungifos; Kilmicen; Tolmicen.* $C_{20}H_{21}NOS$; mol wt 323.46. C 74.27%, H 6.54%, N 4.33%, O 4.95%, S 9.91%. Topical antimycotic agent with high liposolubility. Prepn: P. Melloni *et al.*, **Ger.** pat. **2,313,845** corresp to U.S. pat. **3,855,263** (1973,

1974 both to Carlo Erba). *In vitro* and *in vivo* study: I. deCarneri *et al., Arzneimittel-Forsch.* **26**, 769 (1976). Antimycotic studies: A. Bianchi *et al., Antimicrob. Ag. Chemother.* **12**, 429 (1977). Clinical studies in dermatomycosis: L. C. Cucé *et al., J. Int. Med. Res.* **8**, 144 (1980); C. Intini *et al., Pharmatherapeutica* **2**, 439 (1980).

White cryst powder from isopropanol, mp 92-94°. Practically insol in water. Soly (mg/ml): 14.9 in *n*-hexane; 23.9 in *n*-octanol. LD$_{50}$ in mice, rats, dogs (mg/kg): 4000, 6000, 5000 orally (deCarneri).

THERAP CAT: Antifungal.

9648. Tolcyclamide. *N-[(Cyclohexylamino)carbonyl]-4-methylbenzenesulfonamide;* 1-cyclohexyl-3-*p*-tolylsulfonylurea; tolhexamide; glycyclamide; cyclamide; K-386; Diaboral. C$_{14}$H$_{20}$N$_2$O$_3$S; mol wt 296.39. C 56.73%, H 6.80%, N 9.45%, O 16.19%, S 10.82%. Prepn: Logemann, Artini, *Ber.* **90**, 2527 (1957).

Crystals from trichloroethylene, mp 174-176°.
THERAP CAT: Antidiabetic.

9649. Toldimfos Sodium. *(4-Dimethylamino-o-tolyl)-phosphonous acid sodium salt;* sodium (4-dimethylamino-o-tolyl)phosphonate; *p*-dimethylamino-*o*-toluenephosphonous acid sodium salt; Foston; Tonofosfan. C$_9$H$_{13}$NNaO$_2$P; mol wt 221.17. C 48.88%, H 5.92%, N 6.33%, Na 10.39%, O 14.47%, P 14.00%. Prepd from *N,N*-dimethyl-*m*-toluidine and phosphorus trichloride: Benda, Schmidt, **Ger.** pat. **397,813** (1924 to Cassella), *Frdl.* **14**, 1409.

Trihydrate, scales, needles, or prisms from alc. Freely sol in cold water, hot alcohol.
THERAP CAT (VET): Phosphorus source

9650. Tolfenamic Acid. *2-[(3-Chloro-2-methylphenyl)-amino]benzoic acid;* N-(3-chloro-*o*-tolyl)anthranilic acid; N-(2-methyl-3-chlorophenyl)anthranilic acid; GEA-6414; Clotam; Tolfedine; Tolfine. C$_{14}$H$_{12}$ClNO$_2$; mol wt 261.71. C 64.25%, H 4.62%, Cl 13.55%, N 5.35%, O 12.23%. Deriv of anthranilic acid, related structurally to mefenamic and flufenamic acids, *q.q.v.* Prepn: **Neth.** pat. **Appl. 6,600,251** (1966 to Gea A/S), *C.A.* **66**, 2377 (1967); R. A. Scherrer, F. W. Short, **U.S.** pat. **3,313,848** (1967 to Parke, Davis). Inhibition of prostaglandin biosynthesis: I. B. Linden *et al., Scand. J. Rheumatol.* **5**, 129 (1976). HPLC determn: F. Nielsen-Kudsk, *Acta Pharmacol. Toxicol.* **47**, 267 (1980). Metabolism: T. Kuninaka *et al., Yakugaku Zasshi* **101**, 232 (1981), *C.A.* **95**, 168 (1981). Human pharmacokinetics: P. Pentikaeinen *et al., Eur. J. Clin. Pharmacol.* **19**, 359 (1981). Pharmacological studies: S. Yamashita *et al., Toho Igakkai Zasshi* **28**, 76-105 (1981), *C.A.* **95**, 16183, 180846 (1981). Clinical study: V. Rejholec *et al., Scand. J. Rheumatol.* **Suppl. 33**, 50 (1980); **Suppl. 36**, 1 (1980).

Crystals from abs ethanol, mp 207-207.5°.
THERAP CAT: Anti-inflammatory; analgesic.
THERAP CAT (VET): Anti-inflammatory.

9651. o-Tolidine. *3,3'-Dimethyl-[1,1'-biphenyl]-4,4'-diamine; 3,3'-dimethylbenzidine;* 4,4'-diamino-3,3'-dimethylbiphenyl. C$_{14}$H$_{16}$N$_2$; mol wt 212.29. C 79.21%, H 7.60%, N 13.20%. Made by alkaline reduction of *o*-nitrotoluene with zinc, and subsequent rearrangement of the *o*-hydrazotoluene formed, by boiling with HCl: Van Loon, *Chem. Weekbl.* **5**, 689 (1907). *See also* Schultz *et al., Ann.* **352**, 111 (1907). Crystal and molecular structure: Chawdhury *et al., Acta Crystallogr. Sect. B*, **24**, 1222 (1968). Metabolism: Dieteren, *Arch. Environ. Health* **12**, 30 (1966). Carcinogenic activity: Pliss, Zebenzhinskii, *J. Nat. Cancer Inst.* **45**, 283 (1970).

White to reddish crystals or cryst powder. mp 129-131°. Slightly sol in water; sol in alcohol, ether, dil acids. *Keep well closed and protected from light.*

Sulfate, C$_{14}$H$_{16}$N$_2$.H$_2$SO$_4$, white to gray mass. Slightly sol in water, alcohol; sol in dil acids.

Note: *o*-Tolidine may reasonably be anticipated to be a carcinogen: *Seventh Annual Report on Carcinogens* (PB95-109781, 1994) p 178.

USE: Manuf dyes; also as very sensitive reagent for gold (1:10 million detectable), and for free chlorine in water.

9652. Tolindate. *Methyl (3-methylphenyl)carbamothioic acid O-(2,3-dihydro-1H-inden-5-yl) ester; m,N-dimethylthiocarbanilic acid O-5-indanyl ester; O-(5-indanyl) m,N-dimethylthiocarbanilate;* Dalnate. C$_{18}$H$_{19}$NOS; mol wt 297.42. C 72.69%, H 6.44%, N 4.71%, O 5.38%, S 10.78%. Prepared by treating 5-indanyl thionochloroformate with *N*-methyl-*m*-toluidine: Elpern, Youlus, **U.S.** pat. **3,509,200** (1970 to USV).

Crystals, mp 94-95°.
THERAP CAT: Antifungal.

9653. Toliprolol. *1-[(1-Methylethyl)amino]-3-(3-methylphenoxy)-2-propanol; 1-(isopropylamino)-3-(m-tolyloxy)-2-propanol;* 1-(3-methylphenoxy)-3-(isopropylamino)-2-propanol; 1-(3-methylphenoxy)-2-hydroxy-3-isopropylaminopropane; MHIP. C$_{13}$H$_{21}$NO$_2$; mol wt 223.32. C 69.92%, H 9.48%, N 6.27%, O 14.33%. β-Adrenergic blocker. Prepn: **Neth.** pat. **Appl. 6,409,883**; H. Koppe *et al.,* **U.S.** pat. **3,459,782** (1965, 1969 both to Boehringer, Ing.); **Neth.** pat. **Appl. 6,410,522**; R. Howe, **U.S.** pat. **3,432,545** (1965, 1969 both to I.C.I.). The (−)-isomer is the more potent adrenergic β-receptor antagonist. Resolution of isomers: Howe, Rao, *J. Med. Chem.* **11**, 1118 (1968). Structure-activity studies: Crowther *et al., ibid.* **12**, 638 (1969); Somani, Laddu, *Eur. J. Pharmacol.* **14**, 209 (1971). Metabolism: Stock, Westermann, *Biochem. Pharmacol.* **14**, 227 (1965). Review of pharmacology and clinical data: Marmo *et al., Clin. Ter.* **62**, 11-51, 117-163 (1972).

Crystals from ethyl acetate-petr ether, mp 75-76°. Also reported as mp 79°.

Hydrochloride, $C_{13}H_{22}NO_2 \cdot HCl$, *ICI-45763, Ko-592, Doberol, Sinorytmal.* Crystals from ethanol-ether, mp 120-121°. uv max (water): 270 nm ($E_{1cm}^{1\%}$ 49.8).

THERAP CAT: Antianginal, antihypertensive.

9654. Tollens Reagent. A solution prepd from equal amounts of 10% silver nitrate and 10% sodium hydroxide solutions to which enough dilute ammonia solution has been added to dissolve the precipitated silver oxide. Tollens reagent oxidizes aldehydes to the corresponding acids; during the reaction the silver, bound in form of a complex, is reduced to metallic silver and forms a characteristic silver mirror. *Refs:* B. Tollens, *Ber.* **15**, 1635 (1882); W. Ponndorf, *ibid.* **64B**, 1913 (1937); S. Siggia, E. Segel, *Anal. Chem.* **25**, 640 (1953); J. M. Kolthoff, P. J. Elving, *Treatise on Analytical Chemistry* vol. **13** (New York, 1966) p 183.

Caution: Tollens reagent should always be prepared freshly; old, opaque or "dried out" solutions are explosive, H. Waldmann, *Chimia* **13**, 297 (1959).

USE: Reagent in characterization of sugars, aldehydes, hydrazides. As oxidizing agent.

9655. Tolmetin. *1-Methyl-5-(4-methylbenzoyl)-1H-pyrrole-2-acetic acid; 1-methyl-5-p-toluoylpyrrole-2-acetic acid; 5-(p-toluoyl)-1-methylpyrrole-2-acetic acid;* McN-2559. $C_{15}H_{15}NO_3$; mol wt 257.29. C 70.02%, H 5.88%, N 5.44%, O 18.66%. Prepn: Carson, **Fr.** pat. **1,574,570** (1969 to McNeil Labs.), *C.A.* **72**, 100498y (1969). Pharmacology: Carson *et al., J. Med. Chem.* **14**, 646 (1971); S. Wong *et al., J. Pharmacol. Exp. Ther.* **185**, 127 (1973). *Review:* S. Wong in *Pharmacological and Biochemical Properties of Drug Substances* vol. **1**, M. E. Goldberg, Ed. (Am. Pharm. Assoc., Washington, DC, 1977) pp 233-255. Review of pharmacology and therapeutic efficacy: R. N. Brogden *et al., Drugs* **15**, 429-450 (1978).

Crystals from acetonitrile, mp 155-157° (dec).

Sodium salt dihydrate, $C_{15}H_{14}NNaO_3 \cdot 2H_2O$, *McN-2559-21-98, Reutol, Tolectin, Tolmene.*

THERAP CAT: Anti-inflammatory.

9656. Tolnaftate. *Methyl(3-methylphenyl)carbamothioic acid O-2-naphthalenyl ester; m,N-dimethylthiocarbanilic acid O-2-naphthyl ester; O-2-naphthyl m,N-dimethylthiocarbanilate; 2-naphthyl N-methyl-N-(3-tolyl)thionocarbamate; naphthiomate T;* Sch-10144; *Aftate; Chinofungin; Fungistop; Hi-Alarzin; Sporiline; Timoped; Tinactin; Tinaderm; Tonoftal.* $C_{19}H_{17}NOS$; mol wt 307.42. C 74.23%, H 5.57%, N 4.56%, O 5.20%, S 10.43%. Prepn: **Fr.** pat. **1,337,797;** Miyazaki *et al.,* **U.S.** pat. **3,334,126** (1963, 1967 to Japan Soda); Noguchi *et al., J. Pharm. Soc. Japan* **88**, 335 (1968). Pharmacology and toxicology: Noguchi *et al., Antimicrob. Ag. Chemother.* **1962**, 259; Hashimoto *et al., Toxicol. Appl. Pharmacol.* **8**, 380 (1966); Noguchi *et al., ibid.* 368. Clinical comparison with undecylenic acid: J. F. Fuerst *et al., Cutis* **25**, 544 (1980); F. Battistini *et al., Int. J. Dermatol.* **22**, 388 (1983). Mode of action study: M. P. Gupta *et al., J. Vet. Med. Mycol.* **29**, 45 (1991). LC determn in pharmaceuticals: A. K. Dash, *J. Pharm. Biomed. Anal.* **11**, 847 (1993).

Crystals from alcohol, mp 110.5-111.5°. Practically insol in water. Sparingly sol in methanol, ethanol. Sol in chloroform (1:1.5), acetone (1:7), CCl_4 (1:9). LD_{50} in mice, rats (g/kg): > 10, > 6 orally; > 6, > 4 s.c. (Hashimoto).

THERAP CAT: Antifungal.

THERAP CAT (VET): Antifungal.

9657. Tolonidine. *N-(2-Chloro-4-methylphenyl)-4,5-dihydro-1H-imidazol-2-amine; 2-(2-chloro-p-toluidino)-2-imidazoline;* 2-(2-chloro-4-methylphenyl)amino-1,3-diazacyclopentene-2. $C_{10}H_{12}ClN_3$; mol wt 209.68. C 57.28%, H 5.77%, Cl 16.91%, N 20.04%. Orally active antihypertensive agent, related structurally to clonidine, *q.v.* Prepn: **Neth.** pat. **Appl. 6,411,516;** U.S. pats. **3,236,857; 3,454,701** (1965, 1966, 1969 all to Boehringer, Ing.). Synthesis and hypotensive activity: P. B. M. Timmermans *et al., Rec. Trav. Chim.* **97**, 51 (1978). Pharmacological properties: C. Cosnier *et al., Arzneimittel-Forsch.* **25**, 1557, 1802, 1926 (1975). Structure-activity study: P. B. M. Timmermans, P.A. Van Zwieten, *J. Med. Chem.* **20**, 1636 (1977). Chromatographic study: P. B. M. Timmermans *et al., J. Chromatog.* **144**, 215 (1977). Quantum chemical studies: *eidem, Arzneimittel-Forsch.* **27**, 2266 (1977). Radioimmunoassay disposition study: B. Jarrott, S. Spector, *J. Pharmacol. Exp. Ther.* **207**, 195 (1978). Toxicity study: D. Cosnier *et al., Arzneimittel-Forsch.* **25**, 1926 (1975).

Crystals from benzene/petr ether, mp 148-150°. pKa 9.41.

Nitrate, $C_{10}H_{12}ClN_3 \cdot HNO_3$, *CERM-10137, Euctan.* Crystals, mp 162-164°. LD_{50} in male mice, rats (mg/kg): 160, 420 orally; 21.25, 42 i.v. (Cosnier).

THERAP CAT: Antihypertensive.

9658. Tolonium Chloride. *3-Amino-7-(dimethylamino)-2-methylphenothiazin-5-ium chloride; 3-amino-7-dimethylamino-2-methylphenazathionium chloride;* blutene chloride; toluidine blue O; dimethyltoluthionine chloride; C.I. Basic Blue 17; C.I. 52040; Klot; Tolazul. $C_{15}H_{16}ClN_3S$; mol wt 305.83. C 58.91%, H 5.27%, Cl 11.59%, N 13.74%, S 10.48%. Prepd from dimethyl-*p*-phenylenediamine, sodium thiosulfate, and *o*-toluidine: Dändliker, Bernthsen, U.S. pat. **416,055** (1888 to BASF). Prepn of hemostatic compositions contg tolonium chloride: D. A. Hoff, U.S. pat. **2,809,913** (1957 to Warren-Teed). Prepn of clear, colorless, stable, isotonic solns of purified leucotoluidine blue O by reducing toluidine blue O with sodium hydrosulfite at pH 2.5-3.5: B. March, E. E. Moore, U.S. pat. **2,571,593** (1951 to Abbott). Clinical studies in bleeding disorders: J. Allen *et al., Surg. Gynecol. Obstet.* **89**, 692 (1949). Clinical use for parathyroid identification during thyroidectomy: R. M. Yeager, E. T. Krementz, *Ann. Surg.* **169**, 829 (1969). Acute and chronic toxicity study: T. J. Haley, F. Stolarsky, *Stanford Med. Bull.* **9**, 96 (1951). Review of therapeutic and diagnostic use: A. Mashberg, *J. Am. Dent. Assoc.* **106**, 319-323 (1983). *See also Colour Index* vol. **4** (3rd ed., 1971) p 4471; H. J. Conn's *Biological Stains,* R. D. Lillie, Ed. (Williams & Wilkins, Baltimore, 9th ed., 1977) p 428.

Dark green powder. Sol in water (3.82 g/100 ml), giving a blue to violet soln; sol in alc (0.57 g/100 ml), giving a blue soln. Absorption max (water): 640.4 nm. LD$_{50}$ in mice, rats, rabbits (mg/kg): 27.56, 28.93, 13.44 i.v. (Haley, Storlarsky).

USE: Direct dyeing, printing of wool, silk. Biological stain.

THERAP CAT: Hemostatic. Diagnostic aid (oral carcinoma).

9659. Toloxatone. *5-(Hydroxymethyl)-3-(3-methylphenyl)-2-oxazolidinone; 5-(hydroxymethyl)-3-m-tolyl-2-oxazolidinone;* MD-69276; Humoryl; Perenum. C$_{11}$H$_{13}$NO$_3$; mol wt 207.23. C 63.76%, H 6.32%, N 6.76%, O 23.16%. Reversible monoamine oxidase type A inhibitor. Prepn: C. Fauran *et al.,* **Ger. pat. 2,012,120;** *eidem,* **U.S. pat. 3,655,687** (1970, 1972 both to Delalande); *eidem, Chim. Ther.* **8,** 324 (1973). Pharmacology: G. Raynaud *et al., ibid.* 328. Psychopharmacological profile: J.-P. Kan *et al., Eur. J. Med. Chem.* **12,** 13 (1977); H. Giono-Barber *et al., Arzneimittel-Forsch.* **27,** 1188 (1977). Pharmacokinetics: M. S. Benedetti *et al., ibid.* **32,** 276 (1982). Metabolism: A. Malnoe, M. S. Benedetti, *Xenobiotica* **9,** 281 (1979). GLC determn in plasma: S. Vajta *et al., J. Chromatog.* **274,** 139 (1983). Clinical evaluation in panic disorder: G. Perna *et al., J. Clin. Psychopharmacol.* **14,** 414 (1994).

Crystals from isopropyl alcohol, mp 76°. LD$_{50}$ orally in mice (mg/kg): 1850 (Fauran); also reported as 1500 (Raynaud).

THERAP CAT: Antidepressant.

9660. Tolperisone. *2-Methyl-1-(4-methylphenyl)-3-(1-piperidinyl)-1-propanone; 2,4'-dimethyl-3-piperidinopropiophenone;* 1-piperidino-2-methyl-3-(p-tolyl)-3-propanone; 2-methyl-3-piperidino-1-p-tolylpropan-1-one; mydetone. C$_{16}$-H$_{23}$NO; mol wt 245.36. C 78.32%, H 9.45%, N 5.71%, O 6.52%. Prepn: Nádor *et al.,* **Hung. pat. 144,997** (1956); Yokoyama *et al.,* **Japan. pat. 20,390('65)** (to Eisai). Pharmacology and toxicity: J. Porszasz *et al., Acta Physiol. Acad. Sci. Hung.* **18,** 149 (1960); *eidem, Arzneimittel-Forsch.* **11,** 257 (1961).

Hydrochloride, C$_{16}$H$_{23}$NO.HCl, *N-553, Abbsa, Atmosgen, Arantoick, Besnoline, Isocalm, Kineorl, Menopatol, Metosomin, Minacalm, Muscalm, Mydocalm, Naismeritin, Tolisartine.* Crystals from methyl ethyl ketone, mp 176-177°. LD$_{50}$ s.c. in mice: 620 mg/kg (Porszasz, 1961).

THERAP CAT: Muscle relaxant (skeletal).

9661. Tolpovidone ^{131}I. 1-(ω-p-Iodo-^{131}I-benzylpolyethylene)-2-pyrrolidinone; iodinated p-toluidine polyvinylpyrrolidone; radio-iodinated (^{131}I) tolpovidone; Raovin.

USE: Differential diagnosis of source of hypoalbuminemia.

9662. Tolpropamine. *N,N,4-Trimethyl-γ-phenylbenzenepropanamine; N,N-dimethyl-3-phenyl-3-p-tolylpropylamine;* 3-dimethylamino-1-phenyl-1-p-tolylpropane. C$_{18}$-

H$_{23}$N; mol wt 253.39. C 85.32%, H 9.15%, N 5.53%. Prepn: Bockmühl, Stein, **Ger. pat. 925,468** (1955 to Hoechst); Klosa, *J. Prakt. Chem.* **34,** 312 (1966). Pharmacology: Sendrail, Gleizes, *Therapie* **15,** 119 (1960).

Hydrochloride, C$_{18}$H$_{23}$N.HCl, *Pragman Gelee.* mp 182-184°.

THERAP CAT: Topical antihistaminic, antipruritic.

9663. Tolrestat. *N-[[6-Methoxy-5-(trifluoromethyl)-1-naphthalenyl]thioxomethyl]-N-methylglycine;* tolrestatin; AY-27773; Alredase; Lorestat. C$_{16}$H$_{14}$F$_3$NO$_3$S; mol wt 357.35. C 53.78%, H 3.95%, F 15.95%, N 3.92%, O 13.43%, S 8.97%. Orally active aldose reductase inhibitor. Prepn and pharmacology: K. Sestanj *et al., Eur. pat. Appl.* **59,596** (1982 to Ayerst); *ibid., J. Med. Chem.* **27,** 255 (1984). Prevention of cataracts in galactosemic rats: N. Simard-Duquesne *et al., Proc. Soc. Exp. Biol. Med.* **178,** 599 (1985). Kinetics and metabolism in man: D. R. Hicks *et al., Clin. Pharmacol. Ther.* **36,** 493 (1984). Ultraviolet and HPLC determn in serum: D. R. Hicks, M. Kraml, *Ther. Drug Monit.* **6,** 328 (1984). Effect on erythrocyte sorbitol levels in diabetic patients: P. Raskin *et al., Clin. Pharmacol. Ther.* **38,** 625 (1985).

mp 164-165°.
Methyl ester, C$_{17}$H$_{16}$F$_3$NO$_3$S, mp 109-110°.
THERAP CAT: Treatment of diabetic neuropathy.

9664. Toltrazuril. *1-Methyl-3-[3-methyl-4-[4-[(trifluoromethyl)thio]phenoxy]phenyl]-1,3,5-triazine-2,4,6(1H,3H,-5H)-trione;* 1-methyl-3-[4-[p-[(trifluoromethyl)thio]phenoxy]-m-tolyl]-s-triazine-2,4,6(1H,3H,5H)-trione; Bay Vi 9142; Baycox. C$_{18}$H$_{14}$F$_3$N$_3$O$_4$S; mol wt 425.39. C 50.82%, H 3.32%, F 13.40%, N 9.88%, O 15.04%, S 7.54%. Triazinetrione anticoccidial. General prepn: **Belg. pat. 826,900;** J. H. Reisdorff *et al.,* **U.S. pat. 3,966,725** (1975, 1976 both to Bayer). Prepn and use as animal growth promotant: **Belg. pat. 866,389;** A. Haberkorn *et al.,* **U.S. pat. 4,219,552** (1978, 1980 both to Bayer). Series of articles on efficacy vs coccidia in chickens: E. Kutzer *et al., Wien. Tierärztl. Monatssch.* **72,** 321-340 (1985). Field trial in sheep: B. Gjerde, O. Helle, *Acta Vet. Scand.* **27,** 124 (1986); in chickens: H. D. Chapman, *J. Comp. Path.* **97,** 21 (1987).

mp 194°.
THERAP CAT (VET): Coccidiostat.

9665. *o*-Tolualdehyde. *2-Methylbenzaldehyde;* o-toluylaldehyde. C$_8$H$_8$O; mol wt 120.15. C 79.97%, H 6.71%, O

13.32%. Prepd by reacting nitropropane with *o*-xylyl bromide in the presence of sodium ethanoate: Hass, Bender, *Org. Syn.* **30,** 99 (1950).

Liquid. d_4^{19} 1.0386. bp_{760} 200-202°; bp_{15} 94-96°; bp_6 68-72°. n_D^{25} 1.5430; n_D^{19} 1.549; n_α^{19} 1.5423; n_β^{19} 1.5650; n_γ^{19} 1.5798.

9666. *o*-Toluamide. *2-Methylbenzamide.* C_8H_9NO; mol wt 135.17. C 71.09%, H 6.71%, N 10.36%, O 11.84%. Prepd by reacting *o*-tolunitrile, hydrogen peroxide, 95% alcohol, and sodium hydroxide at 40-50°: Noller, *Org. Syn.* coll. vol. **II,** 586 (1943). Also prepared by reacting the nitrile with boron fluoride in dil acetic acid: Hauser, Hoffenberg, *J. Org. Chem.* **20,** 1448 (1955).

Crystals from water, mp 144-145°. Very sol in alcohol, hot water, concd HCl, less sol in ether. Sparingly sol in benzene. Practically insol in cold water. *Explosive.* Keep away from open flame.

9667. Toluene. *Methylbenzene;* toluol; phenylmethane; Methacide. C_7H_8; mol wt 92.14. C 91.25%, H 8.75%. Obtained mainly from tar oil. Review of mfg processes: Faith, Keyes & Clark's *Industrial Chemicals,* F. A. Lowenheim, M. K. Moran, Eds. (Wiley-Interscience, New York, 4th ed., 1975) pp 822-830. Solubility: F. P. Schwarz, *Anal. Chem.* **52,** 10 (1980). Myelotoxic potential: L. Greenburg *et al., J. Am. Med. Assoc.* **118,** 573 (1942). Comparison with benzene of effects on hematopoiesis and bone marrow metabolism: H. W. Gerard, *AMA Arch. Ind. Health* **13,** 468 (1956). Acute toxicity: H. F. Smyth *et al., Am. Ind. Hyg. Assoc. J.* **30,** 470 (1969). Evaluation of chronic occupational exposure: H. Tahti *et al., Int. Arch. Occup. Environ. Health* **48,** 61 (1981). *Review:* M. C. Hoff in Kirk-Othmer *Encyclopedia of Chemical Technology* vol. 23 (Wiley-Interscience, New York, 3rd ed., 1983) pp 246-273. Review of reproductive toxicity: J. M. Donald *et al., Env. Health Persp.* **94,** 237-244 (1991); of toxicology and human exposure: *Toxicological Profile for Toluene* (PB90-198904, 1989) 125 pp.

Flammable, refractive liq; benzene-like odor. d_4^{20} 0.866. mp −95°. bp 110.6°. n_D^{20} 1.4967. Flash pt, closed cup: 40°F (4.4°C). Soly in water at 23.5°C (w/w): 0.067%. Very slightly sol in water; misc with alc, chloroform, ether, acetone, glacial acetic acid, carbon disulfide. LD_{50} orally in rats: 7.53 g/kg (Smyth).
Caution: Readily absorbed by inhalation, ingestion and somewhat by skin contact. Direct contact may cause severe dermatitis due to drying and defatting action. May present lung aspiration hazard if ingested. Potential symptoms of acute overexposure by inhalation may include local irritation; CNS excitation and depression. Low concentrations may result in transitory mild upper respiratory tract irritation, mild eye irritation, lacrimation, metallic taste, slight nausea, hilarity, lassitude, drowsiness and impaired balance. High concentrations may cause paresthesia, vision disturbances, dizziness, nausea, headache, narcosis and collapse; death from respiratory failure or sudden ventricular fibrillation. Chronic overexposure by inhalation has been associat-

ed with hepatotoxicity and nephrotoxicity. Syndromes following chronic inhalation involve severe muscle weakness, cardiac arrhythmias, gastrointestinal and neuropsychiatric complaints. *See Patty's Industrial Hygiene and Toxicology* vol. **2B,** G. D. Clayton, F. E. Clayton, Eds. (Wiley-Interscience, New York, 4th ed., 1994) pp 1326-1332; *Clinical Toxicology of Commercial Products,* R. E. Gosselin *et al.,* Eds. (Williams & Wilkins, Baltimore, 5th ed., 1984) Section II, p 153, Section III, p 397-404.
USE: In manuf benzoic acid, benzaldehyde, explosives, dyes, and many other organic compds; as a solvent for paints, lacquers, gums, resins; thinner for inks, perfumes, dyes; in the extraction of various principles from plants; as gasoline additive.

9668. Toluene 2,4-Diisocyanate. *2,4-Diisocyanato-1-methylbenzene;* 2,4-diisocyanatotoluene; 2,4-tolylene diisocyanate; TDI; Nacconate 100. $C_9H_6N_2O_2$; mol wt 174.16. C 62.07%, H 3.47%, N 16.09%, O 18.37%. Usually prepd from toluene-2,4-diamine and phosgene. *Review:* Astle, *Industrial Organic Nitrogen Compounds* (New York, 1961) pp 284-313; Faith, Keyes & Clark's *Industrial Chemicals,* F. A. Lowenheim, M. K. Moran, Eds. (Wiley-Interscience, New York, 4th ed., 1975) pp 831-835.

Liquid at room temperature. Sharp, pungent odor. mp 19.5-21.5°. d_4^{20} liq 1.2244. bp_{760} 251°; bp_{11} 126°. Darkens on exposure to sunlight. Reacts with water with evolution of carbon dioxide. Flash pt, open cup: 132° (270°F). Misc with alcohol (decompn), diglycol monomethyl ether, ether, acetone, carbon tetrachloride, benzene, chlorobenzene, kerosene, olive oil. Concd alkaline compds such as NaOH or *tert*-amines may cause run-away polymerization.
Caution: Potential symptoms of overexposure are irritation of nose and throat; choking, paroxysmal cough; chest pain, retrosternal soreness; nausea, vomiting and abdominal pain; bronchial spasm, pulmonary edema; dyspnea, asthma; conjunctivitis, lacrimation; dermatitis, skin sensitization. *See NIOSH Pocket Guide to Chemical Hazards* (DHHS/ NIOSH 90-117, 1990) p 214. *See also Clinical Toxicology of Commercial Products,* R. E. Gosselin *et al.,* Eds. (Williams & Wilkins, Baltimore, 5th ed., 1984) Section II, p 414. This substance may reasonably be anticipated to be a carcinogen: *Seventh Annual Report on Carcinogens* (PB95-109781, 1994) p 383.
USE: In the manuf of polyurethane foams and other elastomers.

9669. Toluene-3,4-dithiol. 1,2-Dimercapto-4-methylbenzene; "dithiol". $C_7H_8S_2$; mol wt 156.27. C 53.80%, H 5.16%, S 41.04%. Prepd from toluene-3,4-disulfonyl chloride with tin and hydrochloric acid: Mills, Clark, *J. Chem. Soc.* **1936,** 178.

Crystals, mp 31°. bp_{84} 185-187°. Sol in benzene, in aq alkali hydroxide solns.
USE: For the detection of bismuth, molybdenum, rhenium, tin, tungsten, see Bickford *et al., J. Am. Pharm. Assoc., Sci. Ed.* **37,** 255 (1948).

9670. *p*-Toluenesulfinic Acid. $C_7H_8O_2S$; mol wt 156.21. C 53.82%, H 5.16%, O 20.49%, S 20.53%. CH_3-$C_6H_4SO_2H$. Prepd by reduction of *p*-toluenesulfonyl chloride with zinc dust: Whitmore, Hamilton, *Org. Syn.* **2,** 89

(1922). Because the sulfinic acid is difficult to dry without partial conversion to the sulfonic acid, the sodium salt, $CH_3C_6H_4SO_2Na.2H_2O$, is usually prepd. The free sulfinic acid is then obtained as needed by dissolving the sodium salt in cold water and carefully acidifying the soln with the exact amt of HCl needed.

Long, rhombic plates or needles from water. mp 85°. Freely sol in alc, ether; sparingly sol in water, hot benzene.

9671. p-Toluenesulfonic Acid. *4-Methylbenzenesulfonic acid;* tosic acid. $C_7H_8O_3S$; mol wt 172.20. C 48.82%, H 4.68%, O 27.87%, S 18.62%. $CH_3C_6H_4SO_3H$. Prepd by sulfonation of toluene with 96-100% H_2SO_4; when carried out at 75° the compn of the reaction product is 75% para-, 19% ortho- and 6% meta-toluenesulfonic acid. Convenient lab prepn: L. F. Fieser, *Experiments in Organic Chemistry,* (Boston, 3rd ed., 1955) p 144. The separation of toluene from petroleum fractions can be accomplished by sulfonation with H_2SO_4 at 60°.

Monoclinic leaflets or prisms. Also reported as crystallizing with $1H_2O$ or $4H_2O$. When anhydrous, mp 106-107°. Metastable form, mp 38°. bp_{20} 140°. $bp_{0.1}$ 185-187°. Freely sol in water, about 67 g/100 ml. Sol in alc and ether.

Sodium salt, $C_7H_7NaO_3S$, orthorhombic plates, very sol in water.

Caution: Highly irritating to skin, mucous membranes.

USE: In dye chemistry; in manuf of oral antidiabetic drugs.

9672. p-Toluenesulfonyl Chloride. Tosyl chloride. $C_7H_7ClO_2S$; mol wt 190.65. C 44.10%, H 3.70%, Cl 18.60%, O 16.78%, S 16.82%. $CH_3C_6H_4SO_2Cl$. Made by treating toluene with chlorosulfonic acid.

Crystals, mp 69-71°. bp_{15} 146°. Insol in water; freely sol in alcohol, benzene, ether.

9673. Toluic Acid. *Methylbenzoic acid.* $C_8H_8O_2$; mol wt 136.15. C 70.57%, H 5.92%, O 23.50%. Prepn of m- and o-forms by oxidation of corresponding xylene: Toland, U.S. pat. **2,903,480** (1959 to California Res. Corp.); Hay et al., J. Org. Chem. **25**, 616 (1960). Prepn of p-form by reaction of p-tolyldiazonium tetrafluoroborate with nickel carbonyl and acetic acid: Clark, Cookson, *J. Chem. Soc.* **1962**, 686; by oxidation of p-xylene: Taves, U.S. pat. **3,030,413** (1962 to Hercules Powder). Manuf of p-form from toluene: Braunworth, U.S. pat. **3,046,305** (1962 to Pure Oil).

m-Toluic acid, prisms from water, mp 111-113°, bp 263°. Sublimes. Sol in 1170 parts water at 15°, 60 parts boiling water; very sol in alcohol, ether.

o-Toluic acid, crystals, mp 107-108°; bp 258-260°; volatile with steam. Slightly sol in cold water; sol in 35 parts boiling water; very sol in alcohol.

p-Toluic acid, crystals, mp 180-181°; bp 274-275°. Sparingly sol in hot water; very sol in alcohol, ether, methanol.

9674. Toluidine. C_7H_9N; mol wt 107.16. C 78.46%, H 8.47%, N 13.07%. Prepn: J. S. Muspratt, A. W. Hofmann, *Ann.* **54**, 1 (1845); of each isomer: F. Beilstein, A. Kuhlberg, ibid. **156**, 66 (1870); P. Kovacic, J. L. Foote, J. Am. Chem. Soc. **83**, 743 (1961); P. Kovacic et al., ibid. **84**, 759 (1962). Toxicity data: H. F. Smyth, Am. Ind. Hyg. Assoc. J. **23**, 95 (1962). GC determn in urine: K. El-Bayoumy et al., Cancer Res. **46**, 6064 (1986).

m-Toluidine, 3-methylbenzamine, 3-aminotoluene, 3-methylaniline. Liquid, mp ∼ −50°. bp 203-204°. d_{25}^{25} 0.990. n_D^{22} 1.5711. Slightly sol in water; sol in alcohol, ether, dil acids.

o-Toluidine, 2-methylbenzamine, 2-aminotoluene, 2-methylaniline. Light yellow liquid becoming reddish brown on exposure to air and light. bp 200-202°. d_{20}^{20} 1.008. n_D^{20} 1.5688. Flash pt, closed cup: 185°F (85°C). Slightly sol in water; sol in alcohol, ether, dil acids. Keep well closed and protected from light. LD_{50} orally in rats: 0.94 g/kg (Smyth).

p-Toluidine, 4-methylbenzamine, 4-aminotoluene, 4-methylaniline. Lustrous plates or leaflets, mp 44-45°. bp 200-201°. d_{20}^4 1.046. n_D^{59} 1.5532. Flash pt, closed cup: 188°F (86°C). Sol in about 135 parts water; freely sol in alcohol, ether, acetone, methanol, carbon disulfide, oils, dil acids.

Caution: Potential symptoms of overexposure to o-toluidine are anoxia, headache, cyanosis; weakness, dizziness, drowsiness; microhematuria; eye burns; dermatitis. See *NIOSH Pocket Guide to Chemical Hazards* (DHHS/NIOSH 90-117, 1990) p 216. o-Toluidine and its hydrochloride may reasonably be anticipated to be carcinogens: *Seventh Annual Report on Carcinogens* (PB95-109781, 1994) p 387.

USE: Manufacture of various dyes and other organic chemicals. o-Isomer also in printing textiles blue black; making colors fast to acids. p-Isomer also as a reagent for lignin, nitrite, phloroglucinol.

9675. o-Tolunitrile. *2-Methylbenzonitrile; 2-methylbenzenecarbonitrile;* o-cyanotoluene; o-methylbenzonitrile. C_8H_7N; mol wt 117.15. C 82.02%, H 6.02%, N 11.96%. $CH_3C_6H_4CN$. Prepd from o-toluidine by diazotization in HCl soln and treatment of the diazonium chloride with potassium cuprocyanide: Herb, *Ann.* **258**, 9 (1890); Clarke, Read, *Org. Syn. coll.* vol. I (2nd ed., 1941) p 514.

Liquid; d_4^{20} 0.9955; d_4^{45} 0.9737; d_4^{75} 0.9481; mp −13°; bp_{760} 205.2°; bp_{100} 135°; bp_{40} 110°; bp_{20} 93°; bp_{10} 77.9°; bp_5 64°; $bp_{1.0}$ 36.7°; n_D^{23} 1.52720. Absorption spectrum. Daly, Ewbank, *J. Chem. Soc.* **87**, 1357 (1905); Purvis, ibid. **107**, 503 (1915). Insol in water. Miscible with alc, ether.

9676. p-Tolunitrile. Prepd from p-toluidine in the manner described for o-tolunitrile.

Needles from alc. d_4^{30} 0.9785; d_4^{45} 0.9640; d_4^{60} 0.9512; d_4^{75} 0.9390; mp 29.5°; bp_{760} 217.6°; bp_{100} 145.2°; bp_{60} 130°; bp_{40} 109.5°; bp_{20} 101.7°; bp_{10} 85.8°; bp_5 71.3°; $bp_{1.0}$ 42.5°. Absorption spectrum: Baly, Ewbank, *J. Chem. Soc.* **87**, 1357 (1905); Purvis, ibid. **107**, 503 (1915). Insol in water. Very sol in alcohol, ether.

9677. 2-(p-Toluyl)benzoic Acid. p-Toluyl-o-benzoic acid; 4′-methylbenzophenone-2-carboxylic acid. $C_{15}H_{12}O_3$; mol wt 240.26. C 74.99%, H 5.03%, O 19.98%. Prepd from phthalic anhydride and toluene in the presence of aluminum chloride: Friedel, Crafts, *Ann. Chim. Phys.* [6] **14**, 447 (1888); Fieser, *Org. Syn.* **4**, 73 (1925).

Monohydrate, triclinic pinacoidal crystals from alcohol. Sweet taste. Becomes anhydr at 100°, then melts at 146°. Slightly sol in boiling water; freely sol in alcohol, benzene, ether, acetone, boiling toluene.

Methyl ester, $C_{16}H_{14}O_3$, plates from methanol, mp 53°. Soluble in alcohol, benzene.

9678. Toluylene Blue. *N-[4-[(2,4-Diamino-5-methylphenyl)imino]-2,5-cyclohexadien-1-ylidene]-N-methylmethanaminium chloride; [4-(4,6-diamino-m-tolyl)imino-2,5-cyclohexadien-1-ylidene]dimethylammonium chloride;* C.I. 49410. $C_{15}H_{19}ClN_4$; mol wt 290.80. C 61.96%, H 6.59%, Cl 12.19%, N 19.27%. Prepd by condensation of p-nitrosodimethylaniline with 2,4-diaminotoluylene: Witt, *Ber.* **12**, 933 (1879); **Ger. pat. 15,272;** *Frdl.* **1,** 274. Also formed by irradiating a mixture of dimethyl-p-phenylenediamine and m-toluylenediamine: Loiseleur, *Compt. Rend.* **237**, 461 (1953). See also *Colour Index* vol. 4 (3rd ed., 1971) p 4443.

Monohydrate, prismatic, copper-brown, shiny crystals. Gives blue soln with cold water, alc, acetic acid.

USE: Biological stain.

9679. Tolycaine. *2-(2-Diethylaminoacetamido)-m-toluic acid methyl ester; 2-methyl-6-carbomethoxy-N-diethylaminoacetanilide; methyl 2-diethylaminoacetamido-m-toluate; 3-methyl-2-diethylaminoacetylaminobenzoic acid methyl ester.* $C_{15}H_{22}N_2O_3$; mol wt 278.35. C 64.73%, H 7.97%, N 10.06%, O 17.24%. Prepn: Hiltmann et al., Ger. pat. **1,018,070** (1957 to Bayer); U.S. pat. **2,921,077** (1960 to Schenley).

Oil, bp_5 190-192°.

Hydrochloride, $C_{15}H_{22}N_2O_3 \cdot HCl$, *Baycain.* Crystals, mp 139-140.5°.

THERAP CAT: Anesthetic (local).

9680. Tolylhydrazine. $C_7H_{10}N_2$; mol wt 122.17. C 68.82%, H 8.25%, N 22.93%. Prepd by stannous chloride reduction of the diazonium salt of the corresponding toluidine: Hunsberger et al., J. Org. Chem. **21**, 394 (1956).

m-Tolylhydrazine. Oily liquid. bp_{760} 243°, bp_{16} 132-134°. d_{15}^{15} 1.061-1.062, d_4^{20} 1.057-1.058. Insol in water; sol in alc, chloroform, ether.

Nitrate, $C_7H_{10}N_2 \cdot HNO_3$, needles, mp 145-147°.

o-Tolylhydrazine. Needles, mp 56-59°. Sparingly sol in water; sol in alcohol, chloroform, ether, slightly in cold petr ether.

Nitrate, leaflets, mp 98-100°. Very sol in water, alcohol; insol in ether.

p-Tolylhydrazine. Rhombic bipyramids, mp 61° or 65-66°. bp 240-244° with slight decompn. Sparingly sol in water; sol in alcohol, benzene, ether.

Nitrate, leaflets, mp 152-153°.

USE: *o*-Tolylhydrazine as reagent for galactose.

9681. p-Tolylsulfonylmethylnitrosamide. *N,4-Dimethyl-N-nitrosobenzenesulfonamide; N-methyl-N-nitroso-p-toluenesulfonamide; Diazald.* $C_8H_{10}N_2O_3S$; mol wt 214.25. C 44.85%, H 4.70%, N 13.08%, O 22.40%, S 14.97%. Prepd by the action of nitrous acid on *p*-tolylsulfonylmethylamide: Ger. pat. **224,388** (1910 to Bayer); Frdl. **10**, 1216; Chem. Zentr. **1910, II**, 609; Takizawa, J. Pharm. Soc. Japan **70**, 490 (1950); de Boer, Backer, Rec. Trav. Chim. **73**, 229 (1954); Org. Syn. **34**, 96 (1954).

Yellow crystals from benzene + petr ether, mp 62°. Stable in an ordinary brown bottle for several years. A white

coating (formed of *p*-tolylsulfonylmethylamide) does no harm. Insol in water. Sol in ether, petr ether, benzene, chloroform, carbon tetrachloride. Yields diazomethane on treatment with alkali.

USE: In the laboratory prepn of diazomethane. Directions for use: de Boer, Backer, Rec. Trav. Chim. **73**, 232 (1954).

9682. Tomatidine. *(3β,5α,22β,25S)-Spirosolan-3-ol; 5α-tomatidan-3β-ol; 5α,20β_F,22α_F,25β_F,27-azaspirostan-3β-ol.* $C_{27}H_{45}NO_2$; mol wt 415.66. C 78.02%, H 10.91%, N 3.37%, O 7.70%. By hydrolysis of tomatine: Kuhn et al., Ber. **83**, 448 (1950). Isoln from the roots of Rutgers tomato plant [*Lycopersicon esculentum* Mill., cultivar. "Rutgers"]: Brink, Folkers, J. Am. Chem. Soc. **73**, 4018 (1951); Fontaine et al., ibid. 878; Sato et al., ibid. 880; Kuhn, Low, U.S. pat. **2,770,-618** (1956 to Amer. Home Prod.). Structure: Sato et al., J. Org. Chem. **25**, 783 (1960); Schreiber, Adams, Experientia **17**, 13 (1961). Synthesis: Uhle, Moore, J. Am. Chem. Soc. **76**, 6412 (1954); Uhle, ibid. **83**, 1460 (1961); Kessar et al., Tetrahedron **27**, 2869 (1971).

Plates from ethyl acetate, mp 202-206°. $[\alpha]_D^{25}$ +8° (chloroform).

Hydrochloride, $C_{27}H_{45}NO_2 \cdot HCl$, crystals from abs ethanol, mp 265-270°. $[\alpha]_D^{25}$ — 5° (methanol).

9683. Tomatine. *(3β,5α,22β,25S)-Spirosolan-3-yl O-β-D-glucopyranosyl-(1→2)-O-[β-D-xylopyranosyl-(1→3)]-O-β-D-glucopyranosyl-(1→4)-β-D-galactopyranoside;* lycopersicin. $C_{50}H_{83}NO_{21}$; mol wt 1034.20. C 58.07%, H 8.09%, N 1.35%, O 32.49%. Occurs in the extract of leaves of wild tomato plants: Fontaine et al., Arch. Biochem. **18**, 467 (1948); Kuhn, Low, Ber. **81**, 552 (1948); Kuhn et al., ibid. **83**, 448 (1950); Bognar, Makleit, Pharmazie **11**, 376 (1956). Yields on partial hydrolysis, besides α-tomatine, the main constituent, $β_1$-, $β_2$-, γ- and δ-tomatine: Kuhn et al., Ber. **90**, 203 (1957). α-Tomatine consists of one mol tomatidine linked to a tetrasaccharide composed of 2 mols D-glucose, 1 mol D-xylose and 1 mol D-galactose: Kuhn et al., Angew. Chem. **68**, 212 (1956). Proposed as an alternate precipitant to digitonin: Schultz, Sander, Z. Physiol. Chem. **308**, 122 (1957). Structure: Reichstein, ibid. **74**, 887 (1962). Toxicity study: Wilson et al., Toxicol. Appl. Pharmacol. **3**, 39 (1961).

Needles from methanol, mp 263-268°. $[\alpha]_D^{20}$ — 18° (c = 0.55 in pyridine). Sol in ethanol, methanol, dioxane, propylene glycol. Practically insol in water, ether, petr ether. Stable to strong alkali but hydrolyzed by acids to produce cryst tomatidine and a soln rich in reducing sugars. Has been found to inhibit the growth of various fungi and bacteria. LD orally in rats: 900-1000 mg/kg (Wilson).

USE: Precipitating agent for steroids.

9684. Tomudex®. *N-[[5-[[(1,4-Dihydro-2-methyl-4-oxo-6-quinazolinyl)methyl]methylamino]-2-thienyl]carbonyl]-L-glutamic acid;* N-(5-[N-(3,4-dihydro-2-methyl-4-oxoquinazolin-6-ylmethyl)-N-methylamino]-2-thenoyl)-L-glutamic acid; ICI-D-1694; ZD-1694. $C_{21}H_{22}N_4O_6S$; mol wt 458.50. C 55.01%, H 4.84%, N 12.22%, O 20.94%, S 6.99%. Folate-based inhibitor of thymidylate synthase; rapidly and extensively metabolized to its more potent polyglutamate derivatives. Prepn: L. R. Hughes, **Eur. pat. Appl. 239,362** (1987 to ICI; Natl. Res. Dev.); *idem,* **U.S. pat. 4,992,550** (1991 to ICI); P. R. Marsham *et al., J. Med. Chem.* **34,** 1594 (1991). Pharmacokinetics: D. I. Jodrell *et al., Cancer Chemother. Pharmacol.* **28,** 331 (1991). HPLC determn of metabolites in cultured cells: W. Gibson *et al., Biochem. Pharmacol.* **45,** 863 (1993). Review of clinical toxicology: S. J. Clarke *et al., Adv. Exp. Med. Biol.* **338,** 601-604 (1993); of mechanism of action: A. L. Jackman *et al., ibid.* **339,** 265-276. Historical development, pharmacology, and clinical evaluation: S. J. Clarke *et al., ibid.* 277-287; A. L. Jackman *et al., Eur. J. Cancer* **31A,** 1277-1282 (1995).

Pale yellow powder. Sol in water.
Monohydrate, $C_{21}H_{22}N_4O_6S.H_2O$, mp 180-184°.
THERAP CAT: Antineoplastic.

9685. Tonin. β-Angiotensin I converting enzyme (formerly). A converting enzyme that differs from renin, *q.v.,* in its ability to form angiotensin II directly from angiotensinogen by cleaving the Phe-His bond; it can also convert angiotensin I to angiotensin II. Its presence was discovered in rat submaxillary glands: R. Boucher *et al., Hypertension 72,* J. Genest, E. Koiw, Eds. (Springer-Verlag, New York, 1972) p 512; *idem, Circ. Res.* **Suppl. I,** 203 (1974). Purification and characterization: S. Demassieux *et al., Can. J. Biochem.* **54,** 788 (1976). Crystal data: K. Hayakawa *et al., J. Mol. Biol.* **123,** 107 (1978). Radioimmunoassay: J. Gutkowska *et al., Can. J. Biochem.* **56,** 769 (1978). Purification by affinity chromatography: M. Ikeda *et al., Hypertension 3,* 81 (1981); by gel permeation and HPLC: C. Lazure *et al., Anal. Biochem.* **125,** 406 (1982). Isoln using chromatofocusing: E. S. P. Cheng, B. J. Morris, *ibid.* **126,** 295 (1982). N-Terminal amino acid sequence of rat tonin: N. G. Seidah *et al., Can. J. Biochem.* **56,** 920 (1978). Substrate specificity studies: *eidem, Proc. Am. Peptide Symp. 6th,* E. Gross, J. Meienhofer, Eds. (Pierce Chem. Co., Rockford, Ill., 1979) p 921; M. Chretien *et al., FEBS Letters* **113,** 173 (1980). Formation of angiotensin II by tonin from partially purified human angiotensinogen: C. Grise *et al., Can. J. Biochem.* **59,** 250 (1981). Pressor effect in anephric animals: E. L. Schiffrin *et al., Can. J. Physiol. Pharmacol.* **59,** 864 (1981). Sequence homologies between tonin and other peptides: C. Lazure *et al., Nature* **292,** 383 (1981). Immunohistochemical study: T. B. Oerstavik *et al., J. Histochem. Cytochem.* **30,** 1123 (1982). Role as renin activator: J. Gutkowska *et al., Can. J. Biochem.* **60,** 843 (1982). Role in exptl hypertension: R. Garcia *et al., Hypertension, Int. Symp., 3rd,* H. Villarreal, Ed. (Wiley, New York, 1981) p 79. *Reviews:* R. Boucher *et al., Circ. Res.* **Suppl. II,** 26-29 (1977); R. Boucher, J. Genest, *Endocrine Functions of the Brain,* M. Motta, Ed. (Raven Press, New York, 1980) pp 373-384; J. Genest, *Heterogeneity of Renin and Renin Substrate,* M. P. Sambhi, Ed. (Elsevier, New York, 1981) pp 11-24.

Mol wt determn is 31,400 by gel filtration and 28,700 by sedimentation equilibrium. Activity is not affected by pepstatin. Can be incubated at 20° for 150 min, between pH 3.4-8 without significant loss of enzymatic activity; loses 15% of its original activity at pH 2.8. After 5 min of incubation at 100°, 60-65% activity remains.

9686. Topiramate. *2,3:4,5-Bis-O-(1-methylethylidene)-β-D-fructopyranose sulfamate;* 2,3:4,5-di-O-isopropylidene-β-D-fructopyranose sulfamate; McN-4853; RWJ-17021-000; Topamax. $C_{12}H_{21}NO_8S$; mol wt 339.37. C 42.47%, H 6.24%, N 4.13%, O 37.72%, S 9.45%. Sulfamate substituted monosaccharide; structurally distinct antiepileptic agent. Prepn: B. E. Maryanoff, J. F. Gardocki, **U.S. pat. 4,513,-006** (1985 to McNeil); and anticonvulsant activity: B. E. Maryanoff *et al., J. Med. Chem.* **30,** 880 (1987). GC determn in plasma: M. L. Holland *et al., J. Chromatog.* **433,** 276 (1988). Comparative pharmacokinetics: M. Bialer, *Clin. Pharmacokinet.* **24,** 441 (1993).

Crystals from ethyl acetate + hexane, mp 125-126°. $[\alpha]_D^{23}$ −34.0° (c = 0.4 in methanol).
THERAP CAT: Anticonvulsant.

9687. Topotecan. *(S)-10-[(Dimethylamino)methyl]-4-ethyl-4,9-dihydroxy-1H-pyrano[3',4':6,7]indolizino[1,2-b]quinoline-3,14(4H,12H)-dione;* 9-[(dimethylamino)methyl]-10-hydroxy-(20S)-camptothecin; hycamptamine; SKF-104864. $C_{23}H_{23}N_3O_5$; mol wt 421.45. C 65.55%, H 5.50%, N 9.97%, O 18.98%. DNA topoisomerase I inhibitor; semisynthetic analog of camptothecin, *q.v.* Prepn: J. C. Boehm *et al.,* **Eur. pat. Appl. 321,122** (1989 to SmithKline Beckman), *C.A.* **112,** 7778z (1990); W. D. Kingsbury *et al., J. Med. Chem.* **34,** 98 (1991). HPLC determn in plasma: J. H. Beijnen *et al., J. Pharm. Biomed. Anal.* **8,** 789 (1990). Clinical pharmacology: E. K. Rowinsky *et al., J. Clin. Oncol.* **10,** 647 (1992).

Monohydrochloride, $C_{23}H_{23}N_3O_5.HCl$, NSC-609669, SKF-104864A. Soluble in water up to 1 mg/ml.
THERAP CAT: Antineoplastic.

9688. Toremifene. *(Z)-2-[4-(4-Chloro-1,2-diphenyl-1-butenyl)phenoxy]-N,N-dimethylethanamine;* (Z)-4-chloro-1,2-diphenyl-1-[4-[2-(N,N-dimethylamino)ethoxy]phenyl]-1-butene. $C_{26}H_{28}ClNO$; mol wt 405.97. C 76.92%, H 6.95%, Cl 8.73%, N 3.45%, O 3.94%. Nonsteroidal antiestrogen structurally similar to tamoxifen, *q.v.* Prepn: R. J. Toivola *et al.,* **Eur. pat. Appl. 95,875; U.S. pat. 4,696,949** (1983, 1987 both to Farmos). Pharmacology: S. Kallio *et al., Cancer Chemother. Pharmacol.* **17,** 103 (1986). Antitumor effects *in vitro* and *in vivo:* L. Kangas *et al., ibid.* 109. HPLC determn in plasma: W. M. Holleran *et al., Anal. Letters* **20,** 871 (1987). Clinical evaluation in advanced breast cancer: R. Valavaara *et al., Eur. J. Cancer Clin. Oncol.* **24,** 785 (1988). *Review:* L. Kangas, *Prog. Cancer Res. Ther.* **35,** *Proc. 3rd Int. Congr. Hormones Cancer, Hamburg, 1987,* 374-377 (1988).

mp 108-110°.
Citrate, $C_{26}H_{28}ClNO.C_6H_8O_7$, *FC-1157a*, *Fareston.* mp 160-162°.
THERAP CAT: Antiestrogen; antineoplastic.

9689. Toril Oil. From the fruit of *Torilis anthriscus* (L.) Gmel., *Umbelliferae.* A Japanese folk remedy for ascaris. It is relatively non-toxic for higher animals and very toxic for Lumbricus, leech and ascaris.

9690. Torsemide. *N-[[(1-Methylethyl)amino]carbonyl]-4-[(3-methylphenyl)amino]-3-pyridinesulfonamide;* 1-isopropyl-3-[(4-m-toluidino-3-pyridyl)sulfonyl]urea; 3-isopropylcarbamylsulfonamido-4-(3'-methylphenyl)aminopyridine; torasemide; AC-4464; BM-02015; JDL-464; Demadex; Toradiur; Torem; Unat. $C_{16}H_{20}N_4O_3S$; mol wt 348.43. C 55.16%, H 5.79%, N 16.08%, O 13.78%, S 9.20%. Sulfonylurea loop diuretic. Prepn: J. E. DeLarge *et al.*, **Ger.** pat. 2,516,025; *eidem,* U.S. pat. 4,018,929 (1975, 1977 both to A. Christiaens, S.A.); J. DeLarge, C. L. Lapiere, *Ann. Pharm. Fr.* **36**, 369 (1978). Pharmacokinetics in humans: M. Lesne *et al., Int. J. Clin. Pharmacol. Ther. Toxicol.* **20**, 382 (1982). Preliminary evaluation in acute heart failure: R. Stroobandt *et al., Arch. Int. Pharmacodyn.* **260**, 151 (1982). Clinical pharmacology: D. C. Brater *et al., Clin. Pharmacol. Ther.* **42**, 187 (1987). Series of articles on pharmacology, mode of action and renal effects in animals: *Arzneimittel-Forsch.* **35**, 1520-1541 (1985); on pharmacology, pharmacokinetics and clinical studies: *Eur. J. Clin. Pharmacol.* **31**, Suppl., 1-55 (1986); *Arzneimittel-Forsch.* **38**, 143-214 (1988).

mp 163-164°. pKa 6.44.
THERAP CAT: Diuretic.

9691. Torularhodin. *3',4'-Didehydro-β,ψ-caroten-16'-oic acid.* $C_{40}H_{52}O_2$; mol wt 564.85. C 85.06%, H 9.28%, O 5.66%. Carotenoid pigment found in *Torula rubra* and *Rhodotorula mucilaginosa* yeasts. Isoln: Karrer, Rutschmann, *Helv. Chim. Acta* **26**, 2109 (1943). Structure and synthesis: Isler *et al., ibid.* **42**, 864 (1959).

Fine dark purple needles from methanol + ether or toluene, mp 210-212° (vac, some decompn). Absorption max in CS_2: 582, 541, 502 nm; in methanol: 529, 493, 460 nm. Freely sol in carbon disulfide, chloroform, pyridine; less sol in ether, benzene, hot ethanol; sparingly sol in methanol. Practically insol in petr ether.
Methyl ester, $C_{41}H_{54}O_2$, dark red needles from benzene + methanol, mp 172-173°.

9692. Tosufloxacin. *7-(3-Amino-1-pyrrolidinyl)-1-(2,4-difluorophenyl)-6-fluoro-1,4-dihydro-4-oxo-1,8-naphthyridine-3-carboxylic acid;* A-61827. $C_{19}H_{15}F_3N_4O_3$; mol wt

404.35. C 56.44%, H 3.74%, F 14.10%, N 13.86%, O 11.87%. Trifluorinated quinolone antibacterial. Prepn: H. Narita *et al.,* **Ger.** pat. 3,514,076, *C.A.* **104**, 129888r (1986); **Belg.** pat. 904,086, *C.A.* **105**, 208850w (1986); Y. Todo *et al.,* **U.S.** pat. 4,704,459 (1985, 1985, 1987 all to Toyama); and activity: D. T. W. Chu *et al., J. Med. Chem.* **29**, 2363 (1986); H. Narita *et al., Yakugaku Zasshi* **106**, 802 (1986), *C.A.* **106**, 196291v (1987). *In vitro* activity studies of the base: P. B. Fernandes *et al., Antimicrob. Ag. Chemother.* **32**, 27 (1988); and *in vivo* animal studies of the toluenesulfonate: M. Takahata *et al., J. Antimicrob. Chemother.* **22**, 143 (1988). Series of articles on antibacterial activity and clinical evaluation of the toluenesulfonate: *Chemotherapy (Tokyo)* **36**, Suppl. 9, 1-1538 (1988).

Hydrochloride, $C_{19}H_{15}F_3N_4O_3.HCl$, *A-60969.* Crystals from conc HCl-ethanol (1:3), mp 247-250° (dec).
Toluenesulfonic acid salt monohydrate, $C_{19}H_{15}F_3N_4O_3.-C_7H_8O_3S.H_2O$, *tosufloxacin tosilate, A-64730, T-3262, Ozex, Tosuxacin.* mp 258-260°.
THERAP CAT: Antibacterial.

9693. Toxaphene. Chlorinated camphene; camphechlor; polychlorocamphene; Hercules 3956; Alltox; Geniphene; Motox; Phenacide; Phenatox; Strobane-T; Toxakil. A very complex, but reproducible mixture of at least 177 C_{10} polychloro derivs., having an approx overall empirical formula of $C_{10}H_{10}Cl_8$. Produced by the chlorination of camphene to 67-69% chlorine by weight and made up of compds of $C_{10}H_8Cl_{10}$, $C_{10}H_{18-n}Cl_n$ (mostly polychlorobornanes) and $C_{10}H_{16-n}Cl_n$ (polychlorobornenes and/or polychlorotricyclenes) with n = 6 to 9. Prepn: Buntin, U.S. pat. 2,565,471 (1951 to Hercules Powder). Isoln of components in crystalline form: Casida *et al., Science* **183**, 520 (1974); *eidem, J. Agr. Food Chem.* **22**, 939 (1974). Acute toxicity data: T. B. Gaines, *Toxicol. Appl. Pharmacol.* **14**, 515 (1969). Mutagenicity studies: N. K. Hooper *et al., Science* **205**, 591 (1979). Livestock toxicity and tissue residues: L. Penumarthy *et al., Vet. Toxicol.* **18**, 60 (1976). Reviews: Liebmann *et al., Arch. Pflanzenschutz* **7**, 131-150 (1971); F. Korte *et al., Pure Appl. Chem.* **51**, 1583-1601 (1979); M. A. Saleh, *Rev. Environ. Contam. Toxicol.* **118**, 1-85 (1990).
Yellow waxy solid, mp 65-90°. Pleasant piney odor. Vapor pressure at 20°: 3×10^{-7} mm Hg. d^{25} 1.630. Log partition coefficient in octanol/water 6.44. Dehydrochlorinates in the presence of alkali, prolonged exposure to sunlight, and at temps about 155°. Soly in water: 3 mg/l. Freely sol in aromatic hydrocarbons. Corrosive to iron. LD_{50} in male, female rats (mg/kg): 90, 80 orally; 1075, 780 dermally (Gaines).
Caution: Potential symptoms of overexposure are nausea, confusion, agitation, tremors, convulsions and unconsciousness; dry, red skin. See *NIOSH Pocket Guide to Chemical Hazards* (DHHS/NIOSH 90-117, 1990) p 62. *See also Clinical Toxicology of Commercial Products,* R. E. Gosselin *et al.,* Eds. (Williams & Wilkins, Baltimore, 5th ed., 1984) Section III, pp 386-387. This substance may reasonably be anticipated to be a carcinogen: *Seventh Annual Report on Carcinogens* (PB95-109781, 1994) p 390.
USE: Insecticide. *Compare* Strobane. Not recommended for use in dairy barns or on milking animals (Penumarthy).

9694. Toxiferine I. *C-Toxiferine I.* $[C_{40}H_{46}N_4O_2]^{2+}$. From calabash curare: Schmid, Karrer, *Helv. Chim. Acta* **30**, 1162 (1947); from *Strychnos toxifera* Schomb., *Loganiaceae:* Wieland *et al., Ann.* **547**, 156 (1941); King, *J. Chem. Soc.* **1949**, 3263. Identity with *toxiferine V* and *toxiferine*

XI: Battersby *et al., ibid.* **1960,** 1848. Structure: Arnold *et al., Helv. Chim. Acta* **44,** 620 (1961). Synthesis: Berlage *et al., ibid.* **42,** 394 (1959); Grdinic *et al., J. Am. Chem. Soc.* **86,** 3357 (1964). Pharmacokinetics: P. G. Waser, J. Reller, *Agents Actions* **2,** 170 (1972). ^{13}C-NMR study: E. Wenkert *et al., J. Org. Chem.* **43,** 1099 (1978).

Dichloride, $C_{40}H_{46}Cl_2N_4O_2$, crystals. $[\alpha]_D^{22}$ −546° (c = 0.30). uv max (ethanol): 292 nm (log ε 4.62). Sol in water.

9695. Toxoflavin. *1,6-Dimethylpyrimido[5,4-e]-1,2,4-triazine-5,7(1H,6H)-dione;* 1,6-dimethyl-5,7-dioxo-1,5,6,7-tetrahydropyrimido[5,4-*e*]-*as*-triazine; xanthothricin. C_7H_7-N_5O_2; mol wt 193.17. C 43.53%, H 3.65%, N 36.26%, O 16.57%. Highly toxic antibiotic from cultures of *Pseudomonas cocovenenans* which also produces bongkrekic acid, *q.v.* Isoln: van Veen, Mertens, *Rec. Trav. Chim.* **53,** 257, 398 (1934); Machlowitz *et al., Antibiot. & Chemother.* **4,** 259 (1954). Structure: Van Damm *et al., Rec. Trav. Chim.* **79,** 255 (1960). Synthesis: Daves *et al., J. Am. Chem. Soc.* **83,** 3904 (1961); **84,** 1724 (1962); Yoneda *et al., Tetrahedron Letters* **1971,** 851. Mode of action: Latuasan, Berends, *Biochim. Biophys. Acta* **52,** 502 (1961). Biosynthesis: Levenberg, Linton, *J. Biol. Chem.* **241,** 846 (1966).

Bright yellow platelets from propanol, dec 172-173°. uv max: 257.5, 394 nm (ε 16400, 2500). Acts as a pH indicator with a sharp loss of color at pH 10.5, but is destroyed by alkali. Sol in water, chloroform, ethyl acetate, ethanol. LD_{50} in mice: 1.7 mg/kg i.v.; 8.4 mg/kg orally.

9696. Toxohormone. Name given in 1948 by Nakahara and Fukuoka to a factor produced by living cancer cells and released into the circulation to produce decreases in liver catalase activity, tryptophan pyrrolase activity, liver ferritin and plasma iron. *See* Nakahara, Fukuoka, *Japan. J. Cancer Res.* **40,** 45 (1949). This tumor-specific concept has been questioned: Greenfield, Meister, *J. Nat. Cancer Inst.* **11,** 997 (1951); Olivares *et al., Science* **157,** 327 (1967); Kampschmidt, Upchurch, *Proc. Soc. Exp. Med. Biol.* **127,** 632 (1968). Described as a polypeptide of low mol wt having 30-40 amino acid residues of which 12-13 are different, and with a high content of glycine, glutamic acid, aspartic acid, alanine and leucine. Amino acid and lipid composition of a highly purified toxohormone prepd from human malignant tissue: Yunoki, Griffin, *Cancer Res.* **21,** 537 (1961); Masuno *et al., ibid.* **41,** 284 (1981). *Reviews:* W. Nakahara, F. Fukuoka, *Chemistry of Cancer Toxin Toxohormone* (C. C. Thomas, Springfield, Illinois, 1961) 75 pp; Nakahara, *Methods Cancer Res.* **2,** 203-237 (1967); Olivares, Kampschmidt in *Oncology, Proc. Int. Cancer Congr., 10th, Houston 1970* **3,** 158-170 (1971); S. Fujii, *Gann Monogr. Cancer Res.* **24,** 215-222 (1979).

Thermostable, non-heat-coagulable, water-sol and alcohol precipitable.

9697. Toxopyrimidine. *4-Amino-2-methyl-5-pyrimidinemethanol;* 6-amino-5-hydroxymethyl-2-methylpyrimidine; 4-amino-5-hydroxymethyl-2-methylpyrimidine; pyramin; pyramine. $C_6H_9N_3O$; mol wt 139.16. C 51.79%, H 6.52%, N 30.20%, O 11.50%. A metabolite of thiamine. Prepd from 4-amino-5-aminomethyl-2-methylpyrimidine dihydrochloride or ethyl 4-amino-2-methyl-5-pyrimidinecarboxylate: Dornow, Petsch, *Ber.* **86,** 1404 (1953); DiBella, Hennessy, *J. Org. Chem.* **26,** 2017 (1961).

Needles from water, mp 193-198°; crystals from methanol + ether, mp 198-200°. Sublimes at 0.01 mm between 155-170°.

9698. Toyocamycin. *4-Amino-7-β-D-ribofuranosyl-7H-pyrrolo[2,3-d]pyrimidine-5-carbonitrile;* 4-amino-5-cyano-7-(D-ribofuranosyl)-7*H*-pyrrolo[2,3-*d*]pyrimidine; uramycin B; vengicide; antibiotic 1037; E-212. $C_{12}H_{13}N_5O_4$; mol wt 291.27. C 49.48%, H 4.50%, N 24.04%, O 21.97%. Antibiotic substance extracted from the culture filtrate and mycelium of *Streptomyces toyocaensis*. Isoln: Nishimura *et al., J. Antibiot.* **9A,** 60 (1956). Structure: Ohkuma, *ibid.* **14A,** 343 (1961). Synthesis of aglycone: Taylor, Hendess, *J. Am. Chem. Soc.* **86,** 951 (1964). Total synthesis: Tolman *et al., ibid.* **90,** 524 (1968); **91,** 2102 (1969). Biosynthesis: Uematsu, Suhadolnik, *Biochemistry* **9,** 1260 (1970); *eidem, J. Biol. Chem.* **245,** 4365 (1970). Crystal and molecular structure: P. Prusiner, M. Sundaralingam, *Acta Crystallogr.* **B34,** 517 (1978).

Fine needles from methanol or acetone, mp 243°. Recrystallization from water yields the hydrate, $C_{12}H_{13}N_5O_4 \cdot H_2O$, mp 239-243°. $[\alpha]_D^{16}$ −45.7° (c = 1.05 in 0.1*N* HCl). uv max (H_2O): 230, 277 nm ($E_{1cm}^{1\%}$ 400, 548). Soluble in acetic acid, acidic solns. Moderately sol in methanol, ethanol, acetone, dioxane, butanol, water, ether. Practically insol in chloroform, ethyl acetate, petr ether. LD_{100} s.c. in mice: 10-20 mg/kg (Nishimura).

9699. Tralkoxydim. *2-[1-(Ethoxyimino)propyl]-3-hydroxy-5-(2,4,6-trimethylphenyl)-2-cyclohexen-1-one;* 2-[1-(ethoxyimino)propyl]-3-hydroxy-5-mesitylcyclohex-2-en-1-one; PP-604; Achieve; Grasp. $C_{20}H_{27}NO_3$; mol wt 329.44. C 72.92%, H 8.26%, N 4.25%, O 14.57%. Cereal selective post-emergent herbicide. Prepn: R. B. Warner *et al., Eur. pat.* **Appl. 80,301;** *eidem,* U.S. pat. **4,717,418** (1983, 1988 both to ICI). Physical properties and herbicidal activity: R. B. Warner *et al., Proc. Brit. Crop Prot. Conf. - Weeds* **1987,** 19. Field trials on grass weeds: J. Rola, *ibid.* 363; P. B. Sutton *et al., ibid.* 389. Mechanism of action study: J. Secor, C. Cseke, *Plant Physiol.* **86,** 10 (1988).

White crystalline solid, mp 106°. Vapor pressure at 20°: 4×10^{-10} kPa. Soly at 20° (mg/l): water 6 at pH 6.5, 5 at pH 5.0; at 24° (g/l): hexane 18; toluene 213; dichloromethane > 500; methanol 25; acetone 89; ethyl acetate 110. LD_{50} orally in male and female rats, male and female mice, male rabbit (mg/kg): 1324, 934, 1231, 1100, 519; dermally in male and female rats: > 2000, > 2000 mg/kg (Warner 1987).

USE: Post-emergent herbicide.

9700. Tralomethrin. *2,2-Dimethyl-3-(1,2,2,2-tetrabromoethyl)cyclopropanecarboxylic acid cyano(3-phenoxyphenyl)methyl ester;* (S)-α-cyano-3-phenoxybenzyl (1R)-cis-2,2-dimethyl-3-[(RS)-1,2,2,2-tetrabromoethyl]cyclopropanecarboxylate; RU-25474; OMS-3048; Saga; Scout; Tracker; Tralate; Tralox. $C_{22}H_{19}Br_4NO_3$; mol wt 665.01. C 39.73%, H 2.88%, Br 48.06%, N 2.11%, O 7.22%. Synthetic pyrethroid consisting of two active diastereomers whose absolute configurations differ at the monobrominated carbon atom. Prepn: **Belg. pat. 873,201;** J. Martel *et al.,* **U.S. pat. 4,279,835** (1979, 1981 both to Roussel-Uclaf). Insecticidal activity: M. Benoit *et al., Pest. Biochem. Physiol.* **26,** 284 (1986). Mechanism of action study: M. Roche *et al., ibid.* **24,** 306 (1985). HPLC determn in water, sediment and fish tissue: J. Mao *et al., J. Agric. Food Chem.* **41,** 596 (1993). Field trials: D. D. Amalraj *et al., Indian J. Malariol.* **28,** 141 (1991); W. R. Halliday *et al., J. Agric. Entomol.* **9,** 145 (1992).

USE: Insecticide.

9701. Tramadol. *trans-(±)-2-[(Dimethylamino)methyl]-1-(3-methoxyphenyl)cyclohexanol;* E-265; CG-315E; U-26225A. $C_{16}H_{25}NO_2$; mol wt 263.38. C 72.97%, H 9.57%, N 5.32%, O 12.15%. Prepn: **Brit. pat. 997,399;** K. Flick, E. Frankus, **U.S. pat. 3,652,589** (1965, 1972 both to Grünenthal); K. Flick *et al., Arzneimittel-Forsch.* **28,** 107 (1978). Series of articles on pharmacology and clinical studies: *ibid.* 114-219. Toxicology: F. Lagler *et al., ibid.* 164. Mechanism of action study: R. B. Raffa *et al., J. Pharmacol. Exp. Ther.* **267,** 331 (1993). HPLC determn in urine: B. Elsing, G. Blaschke, *J. Chromatog.* **612,** 223 (1993). Symposium on pharmacology and clinical experience: *Drugs* **47,** Suppl. 1, 1-46 (1994).

Hydrochloride, $C_{16}H_{25}NO_2 \cdot HCl$, *Crispin, Tramal, Ultram, Zydol.* White crystals, mp 180-181°. Sol in water. LD_{50} in mice, rats (mg/kg): 350, 228 orally; 200, 286 s.c. (Lagler).
THERAP CAT: Analgesic.

9702. Tramazoline. *4,5-Dihydro-N-(5,6,7,8-tetrahydro-1-naphthalenyl)-1H-imidazol-2-amine; 2-[(5,6,7,8-tetrahydro-1-naphthyl)amino]-2-imidazoline.* $C_{13}H_{17}N_3$; mol wt 215.30. C 72.52%, H 7.96%, N 19.52%. α-Adrenergic agonist. Prepn: Berg, **Ger. pats. 1,191,381; 1,195,323** (both 1965 to

Thomae), *C.A.* **63,** 8373c; 13274d (1965). Pharmacology and toxicity data: R. Engelhorn, H. Klupp, *Arzneimittel-Forsch.* **12,** 971 (1962). Activity studies: Sachsenröder *et al., ibid.* **22,** 392 (1972).

Crystals from isopropanol, mp 142-143°.
Hydrochloride monohydrate, $C_{13}H_{17}N_3 \cdot HCl \cdot H_2O$, *KB-227, Biciron, Ellatun, Rhinaspray, Rhinogutt, Rhinospray, Rinogutt, Towk.* Crystals from alc + ether or acetone + ether, mp 172-174°. Sol in water. LD_{50} orally in mice: 195 mg/kg (Engelhorn, Klupp).
THERAP CAT: Decongestant.

9703. Trandolapril. *[2S-[1[R*(R*)],2α,3aα,7aβ]]-1-[2-[[1-(Ethoxycarbonyl)-3-phenylpropyl]amino]-1-oxopropyl]-octahydro-1H-indole-2-carboxylic acid;* (3aR,7aS)-1-[N-[1(S)-(ethoxycarbonyl)-3-phenylpropyl]-(S)-alanyl]octahydro-indole-2(S)-carboxylic acid; (2S,3aR,7aS)-1-[(S)-N-[(S)-1-carboxy-3-phenylpropyl]alanyl]hexahydro-2-indolinecarboxylic acid 1-ethyl ester; RU-44570; Odrik; Gopten. $C_{24}H_{34}N_2O_5$; mol wt 430.54. C 66.95%, H 7.96%, N 6.51%, O 18.58%. Angiotensin converting enzyme (ACE) inhibitor. Prepn: H. Urbach *et al.,* **Eur. pat. Appl. 84,164;** *eidem,* **U.S. pat. 4,933,361** (1983, 1990 both to Hoechst). Enzyme inhibition and pharmacology: N. L. Brown *et al., Eur. J. Pharmacol.* **148,** 79 (1988). Clinical pharmacology: F. De Ponti *et al., Eur. J. Clin. Pharmacol.* **40,** 149 (1991). Clinical trial in prevention of death after myocardial infarction: L. Kober *et al., N. Engl. J. Med.* **333,** 1670 (1995). Series of articles on pharmacology and clinical trials: *Am. J. Hypertension* **8,** 63S-74S (1995).

Diacid, $C_{22}H_{30}N_2O_5$, *trandolaprilat, RU-44403.*
THERAP CAT: Antihypertensive.

9704. Tranexamic Acid. *trans-4-(Aminomethyl)cyclohexanecarboxylic acid;* AMCHA; RP-18429; Anvitoff; Cyklokapron; Emorhalt; Exacyl; Frenolyse; Hexapromin; Hexatron; Rikavarin; Spiramin; Tranex; Tranexan; Transamin; Trasamlon; Ugurol. $C_8H_{15}NO_2$; mol wt 157.21. C 61.12%, H 9.62%, N 8.91%, O 20.35%. Prepn: A. Einhorn, C. Ladisch, *Ann.* **310,** 194 (1900); M. Levine, R. Sedlecky, *J. Org. Chem.* **24,** 115 (1959); **Neth. pat. Appl. 6,503,605;** T. Naito *et al.,* **U.S. pat. 3,499,925** (1965, 1970 to both Daiichi Seiyaku and Mitsubishi Chem.). Pharmacology: Andersson *et al., Scand. J. Haematol.* **2,** 230 (1965), *C.A.* **64,** 2608c (1966). Resoln of isomers and antiplasmin activity: M. Shimizu *et al., Chem. Pharm. Bull.* **16,** 357 (1968), T. Naito *et al., ibid.,* 728. Controlled clinical study in treatment of acute upper gastrointestinal tract bleeding: D. Barer *et al., N. Engl. J. Med.* **308,** 1571 (1983). Toxicity data: B. Melander *et al., Acta Pharmacol. Toxicol.* **22,** 340 (1965). *Review:* Kjellman, Schannong, *Nord. Med.* **83,** 166 (1970).

mp 386-392° (dec). Soly in water: about 1 g/6 ml. Very slightly sol in alcohol, ether. Practically insol in most other organic solvents. Chemically stable; not hygroscopic. LD_{50} in mice, rats (mg/kg): 1500, 1200 i.v. (Melander). *cis*-form, mp 236-238° (dec).

THERAP CAT: Hemostatic.

9705. Tranilast. *2-[[3-(3,4-Dimethoxyphenyl)-1-oxo-2-propenyl]amino]benzoic acid; N-(3',4'-dimethoxycinnamoyl)anthranilic acid; N-5'; Rizaben.* $C_{18}H_{17}NO_5$; mol wt 327.34. C 66.05%, H 5.23%, N 4.28%, O 24.44%. Orally active anti-allergic agent. Prepn: K. Harita *et al.*, **Ger. pat.** **2,402,398;** *idem*, U.S. pat. **3,940,422** (1974, 1976 both to Kissei). Pharmacological properties: H. Azuma *et al.*, *Brit. J. Pharmacol.* **58,** 483 (1976). Mechanism of action study: Y. Iijima *et al.*, *Biochem. Biophys. Res. Commun.* **93,** 912 (1980). Clinical study in pediatric bronchial asthma: H. Shioda, *Allergy* **34,** 213 (1979). Toxicity studies: M. Nakazawa *et al.*, *Oyo Yakuri* **12,** 385, 407 (1976), *C.A.* **88,** 115327-8 (1978). Series of articles on teratogenicity tests: *Iyakuhin Kenkyu* **9,** 148-193 (1978), *C.A.* **88,** 130930f, 146300m-303q (1978).

Crystals from chloroform, mp 211-213°. LD_{50} in male, female mice, male, female rats (mg/kg): 780, 680, 1600, 1100 orally; 410, 385, 405, 395 i.p.; 2630, 2820, 3630, 3060 s.c. (Nakazawa, p 385).

THERAP CAT: Antiallergic.

9706. Transferrins. A group of homologous non-heme, iron-binding glycoproteins of approx mol wts of 76,000-81,000. They are widely distributed in a variety of physiological fluids and cells, esp in the sera of most vertebrates, in egg whites and in mammalian milk, tears and leukocytes. They are involved in iron transport to developing red cells for hemoglobin synthesis. Each protein molecule specifically binds with two Fe^{3+} ions to form salmon-pink complexes; bicarbonate or carbonate ions are involved in the formation of these colored complexes. Reviews of isoln, properties and biological functions: Feeney, Komatsu, *Struct. Bonding* **1,** 149-206 (1966); Aisen, "The Transferrins" in *Inorganic Biochemistry*, vol. 1, G. L. Eichhorn, Ed. (Elsevier, New York, 1973) pp 280-305; Bezkorovainy, Zschocke, *Arzneimittel-Forsch.* **24,** 476-485, 726-737 (1974); P. Aisen, A. Leibman, *Bioinorganic Chemistry* II, K. N. Raymond, Ed. (A.C.S., Washington, 1977) pp 104-126; P. Aisen, I. Listowsky, *Ann. Rev. Biochem.* **49,** 357-393 (1980). Review of transferrin receptors: R. Newman *et al.*, *Trends Biochem. Sci.* **7,** 397-399 (1982).

Serum transferrin, β_1-metal-combining protein, siderophilin. Commonly called transferrin. Structure of carbohydrate moiety: Jamieson *et al.*, *J. Biol. Chem.* **246,** 3686 (1971). Composed of two homologous domains, each containing a binding site for metal ions. The sites are similar, but not identical, in their metal-binding properties. X-ray studies: L. J. DeLucas *et al.*, *J. Mol. Biol.* **123,** 285 (1978). Resolution of the two sites by Eu(III) excitation spectroscopy: P. B. O'Hara, R. Bersohn, *Biochemistry* **21,** 5269 (1982). N-Terminal amino acid sequence of human serum transferrin: M.-H. Metz-Boutigue *et al.*, *Biochim. Biophys. Acta* **670,** 243 (1981). Absorption max of human serum Fe^{3+}-transferrin: about 465 nm ($E_{1cm}^{1\%}$ 0.57); uv max: 280 nm ($E_{1cm}^{1\%}$ 14.3).

Conalbumin, ovotransferrin, Diarconal. Isolated from egg white; distinguished from ovalbumin by its lower thermal coagulation point: Osborne, Campbell, *J. Am. Chem. Soc.* **22,** 422 (1900). Sepn from other egg-white proteins: Longworth *et al.*, *ibid.* **62,** 2580 (1940). Primary structure of hen ovotransferrin: J. Williams *et al.*, *Eur. J. Biochem.* **122,** (1982). Purification, characterization and function of the iron-binding fragments: W.-W. Keung *et al.*, *J. Biol. Chem.* **257,** 1177, 1184 (1982). Antibacterial activity: P. Valenti *et al.*, *Antimicrob. Ag. Chemother.* **21,** 840 (1982). Absorption max of Fe^{3+}-complex: 470 nm ($E_{1cm}^{1\%}$ 0.62).

Lactoferrin, lactotransferrin. Important component of the human milk bacteriostatic system; also found in human and bovine tear proteins. Isoln from human whey by a single chromatographic step: L. Bläckberg, O. Hernell, *FEBS Letters* **109,** 180 (1980). Sequential purification of lactoferrin, lysozyme, and secretory IgA from human milk: M. Boesman-Finkelstein, R. A. Finkelstein, *ibid.* **144,** 1 (1982). Partial C-terminal amino acid sequence of human lactoferrin: M.-H. Metz-Boutigue *et al.*, *ibid.* **142,** 107 (1982).

9707. Transforming Growth Factor-β. TGF-β. Family of multifunctional cytokines that regulate cellular differentiation, motility and growth. Also regulate the synthesis and deposition of the extracellular matrix. Involved in various physiological processes including embryogenesis, immunoregulation, bone remodeling and wound healing. Secreted by virtually all cell types as a biologically inactive (latent) form which is stored at the cell surface and in the extracellular matrix. The mature, active cytokine is a homodimer; mol wt 25 kDa. At least 5 isoforms have been identified. The three mammalian isoforms (TGF-β_1, β_2, and β_3) exhibit 70-80% sequence homology, bind to the same receptors and exert similar biological effects. TGF-β_1 is the most abundant. Biological effects are mediated by binding to membrane receptors that exist on virtually all cells. Released from degranulating platelets at the site of a wound, TGF-β initiates angiogenesis and collagen synthesis. Acts as a chemoattractant and activator of macrophages and fibroblasts. Dysregulation is implicated in the pathogenesis of fibrotic diseases. Inhibits the growth of most epithelial and lymphoid cells. Escape from this normal control mechanism is implicated in carcinogenic transformation. Initial description: J. E. DeLarco, G. J. Todaro, *Proc. Nat. Acad. Sci. USA* **75,** 4001 (1978). Crystal structure of TGF-β_2: S. Daopin *et al.*, *Science* **257,** 369 (1992). Review of physiological actions: A. B. Roberts, M. B. Sporn, *Growth Factors* **8,** 1-9 (1993); of latent forms, binding proteins and receptors: K. Miyazono *et al.*, *ibid.*, 11-22. Proposed mechanisms for tumor cell transformation: M. J. Newman, *Cancer Metast. Rev.* **12,** 239-254 (1993). Role in embryogenesis: N. L. McCartney-Francis, S. M. Wahl, *J. Leukocyte Biol.* **55,** 401-409 (1994); in tissue fibrosis: W. A. Border, N. A. Noble, *New Engl. J. Med.* **331,** 1286-1292 (1994); in kidney disease: K. Sharma, F. N. Ziyadeh, *Am. J. Physiol.* **266,** F829-F842 (1994). Review of potential clinical applications in oncology: J. Kekow, G. J. Wiedemann, *Int. J. Oncol.* **7,** 177-182 (1995).

9708. Tranylcypromine. *trans-(±)-2-Phenylcyclopropanamine*; SKF-*trans*-385. $C_9H_{11}N$; mol wt 133.19. C 81.16%, H 8.32%, N 10.52%. Monoamine oxidase inhibitor. Prepn: Burger, Yost, *J. Am. Chem. Soc.* **70,** 2198 (1948); R. E. Tedeschi, U.S. pat. **2,997,422** (1961 to SK & F).

Liquid, $bp_{1.5-1.6}$ 79-80°.
Hydrochloride, $C_9H_{11}N \cdot HCl$, crystals from ethyl acetate + ether, mp 164-166°.
Sulfate, $(C_9H_{11}N)_2 \cdot H_2SO_4$, *Parnate, Tylciprine.* Crystals, sol in water; very slightly sol in alcohol, ether. Practically insol in chloroform.

THERAP CAT: Antidepressant.

9709. Trapidil. *N,N-Diethyl-5-methyl-[1,2,4]triazolo[1,-5-a]pyrimidin-7-amine; 7-diethylamino-5-methyl-s-triazolo-[1,5-a]pyrimidine*; trapymin; AR-12008; Avantrin; Rocornal. $C_{10}H_{15}N_5$; mol wt 205.26. C 58.52%, H 7.37%, N 34.12%. First triazolopyrimidine registered as a drug. Prepn: E. Tenor *et al.*, **East Ger.** pat. **55,956** (1967), *C.A.* **67,** 90830f (1967); E. Tenor, R. Ludwig, *Pharmazie* **26,** 534 (1971). Physical and chemical properties: S. Pfeifer *et al.*, *ibid.* 539. Pharmacology and toxicology: H. Fuller *et al.*, *ibid.* 554. Metabolic studies: S. Pfeifer *et al.*, *ibid.* 549; *ibid.* **27,** 752 (1972); I. Bornschein *et al.*, *ibid.* **33,** 51 (1978). Mechanism of action: K. Satoh *et al.*, *Arzneimittel-Forsch.*

30, 1264 (1980). Antiarrhythmic activities in rabbits: M. Sakanashi *et al., ibid.* 33, 215 (1983). Clinical hemodynamic effects: M. Di Donato *et al., ibid.* 35, 1295 (1985). GC determn in biological fluids: A. Marzo *et al., ibid.* 37, 947 (1987).

White to yellowish, odorless and bitter crystalline powder, mp 98-99.4° (Pfeifer); 102-104° from heptane (Tenor). Eutectic temp of mixture with azobenzene: 48°. Very sol in water, $1N$ sulfuric acid, 10% ammonium hydroxide; easily sol in methanol, isopropanol, *n*-butanol, chloroform and benzene; sol in ether. Practically insol in hexane, heptane. $pK_s = 2.79$. uv max (methanol): 222, 270, 307 nm (log ϵ 4.28, 3.83, 4.28). Very stable except under extremely alk conditions. LD_{50} in mice, rats (mg/kg): 115, 76 i.v.; 380, 235 orally; 155, 100 i.p.; 132, 100 s.c. (Fuller).

Hydrochloride, $C_{10}H_{15}N_5.HCl$, mp 212°.

THERAP CAT: Vasodilator (coronary).

9710. Traumatic Acid. *2-Dodecenedioic acid;* 1-decene-1,10-dicarboxylic acid. $C_{12}H_{20}O_4$; mol wt 228.29. C 63.14%, H 8.83%, O 28.03%. A wound hormone of plants. Isoln from pods of green beans: English *et al., Proc. Nat. Acad. Sci. USA* 25, 323 (1939). The naturally occurring traumatic acid is the *trans*-form. Synthesis of the *trans*-form: *eidem, J. Am. Chem. Soc.* 61, 3434 (1939); U.S. pats. 2,339,259 (1944); 2,391,824 (1945); Truscheit, Eiter, *Ann.* 658, 86 (1962); Dolezal, *Coll. Czech. Chem. Commun.* 35, 1932 (1970); Schreurs *et al., Rec. Trav. Chim. Pays-Bas Belg.* 90, 1331 (1971); Prakasa Rao, Nayak, *Synthesis* 1975, 608; J. H. Babler, R. K. Moy, *Syn. Commun.* 9, 669 (1979). Synthesis of the *cis*-form: Lauer, Gensler, *J. Am. Chem. Soc.* 67, 1171 (1945).

HOOC~~~~~~~~~COOH

trans-form

trans-Form, crystals from alc, acetone, or 1,2-dimethoxy-ethane, mp 166-167°. $bp_{0.001}$ 150-160°. Very sparingly sol in water; sol in alc, ether, benzene, chloroform.

cis-Form, crystals from ethyl acetate and petr ether, mp 67-68°.

9711. Traxanox. *9-Chloro-7-(1H-tetrazol-5-yl)-5H-[1]benzopyrano[2,3-b]pyridin-5-one;* 9-chloro-7-(5-1H-tetrazol-yl)-5-oxo-5H-[1]benzopyrano[2,3-b]pyridine. $C_{13}H_6ClN_5O_2$; mol wt 299.68. C 52.10%, H 2.02%, Cl 11.83%, N 23.37%, O 10.68%. Antiallergic agent which selectively inhibits release of allergic mediators from mast cells. Prepn: T. Oe, M. Tsuruda, Ger. pat. 2,521,980; *eidem,* U.S. pat. 4,085,111 (1975, 1978 both to Yoshitomi). Antianaphylactic activity in animals: K. Goto *et al., Japan. J. Pharmacol.* 30, 537 (1980). Mechanism of action study: K. Goto *et al., Int. Arch. Allergy Appl. Immunol.* 68, 332 (1982). Clinical pharmacology and pharmacokinetics: A. Ebihara *et al., Arzneimittel-Forsch.* 37, 1388 (1987). Metabolism in humans: M. Tateno *et al., Yakuri to Chiryo* 16, 3251 (1988), *C.A.* 110, 18040n (1989).

mp > 300°.

Sodium salt pentahydrate, $C_{13}H_5ClN_5NaO_2.5H_2O$, *Y-12141, Clearnal.*

THERAP CAT: Antiallergic; antiasthmatic.

9712. Trazodone. *2-[3-[4-(3-Chlorophenyl)-1-piperazin-yl]propyl]-1,2,4-triazolo[4,3-a]pyridin-3(2H)-one.* $C_{19}H_{22}-ClN_5O$; mol wt 371.87. C 61.37%, H 5.96%, Cl 9.53%, N 18.83%, O 4.30%. Prepn: Palazzo, Silvestrini, U.S. pat. **3,381,009** (1968 to Angelini Francesco). Pharmacology: Catanese, Lisciani, *Boll. Chim. Farm.* 109, 369 (1970); B. Silvestrini, E. Quadri, *Eur. J. Pharmacol.* 12, 231 (1970). Analytical data: Baiocchi *et al., Arzneimittel-Forsch.* 23, 400 (1973). Crystal structure: J. P. Fillers, S. W. Hawkinson, *Acta Crystallogr.* B35, 498 (1979). Review of pharmacological properties and therapeutic use: R. N. Brogden *et al., Drugs* 21, 401-429 (1981); B. Silvestrini *et al.,* in *Pharmacological and Biochemical Properties of Drug Substances* vol. 3, M. E. Goldberg, Ed. (Am. Pharm. Assoc., Washington, DC, 1981) pp 94-119. Comprehensive description: D. Gorecki, R. Verbeeck, *Anal. Profiles Drug Subs.* 16, 693-729 (1986). Symposium on clinical efficacy: *Psychopharmacology* 95, Suppl., 1-56 (1988).

Crystals, mp 86-87°. Also reported as mp 96° (Baiocchi). pKa (50% ethanol): 6.14.

Hydrochloride, $C_{19}H_{22}ClN_5O.HCl$, *AF-1161, Bimaran, Desyrel, Molipaxin, Pragmazone, Thombran, Tombran, Trazolan, Trittico.* White, odorless plates from ethanol, mp 223°. Soluble in chloroform; sparingly sol in water, ethanol, methanol. Practically insol in common organic solvents. uv max (water): 211, 246, 274, 312 nm (ϵ 50100, 11730, 3840, 3840). LD_{50} i.v. in mice: 96 mg/kg (Silvestrini, Quadri).

THERAP CAT: Antidepressant.

9713. Trehalose. *α-D-Glucopyranosyl-α-D-glucopyranoside;* α,α-trehalose; natural trehalose; mycose; (α-D-gluco-sido)-α-D-glucoside. $C_{12}H_{22}O_{11}$; mol wt 342.30. C 42.11%, H 6.48%, O 51.41%. About 23-30% is found in trehala manna, the cocoons of a parasitic beetle (*Larinus* species), found on *Echinops pesicus*. Trehalose also occurs in fungi, e.g., *Amanita muscaria.* Extraction procedure: Harding, *Chem. Zentr.* 1924, I, 2017. Isoln from compressed baker's yeast: Stewart *et al., J. Am. Chem. Soc.* 72, 2059 (1950). Structure: Schlubach, Maurer, *Ber.* 58, 1179 (1925); Bredereck, *ibid.* 63, 959 (1930); Hassid, Ballou in *The Carbohydrates,* W. Pigman, Ed. (Academic Press, New York, 1957) p 507. Synthesis: Haworth, Hickenbottom, *J. Chem. Soc.* **1931,** 2847; Lemieux, Bauer, *Can. J. Chem.* 32, 340 (1954); A. A. Pavia *et al., Carbohyd. Res.* 79, 79 (1980). *Review:* C. K. Lee, *Dev. Food Carbohyd.* 2, 1-89 (1980).

Dihydrate, orthorhombic, bisphenoidal crystals from dil alcohol. Sweet taste. mp 96.5-97.5°. The water of crystn escapes around 130°. Anhydrous trehalose melts at 203°. $[\alpha]_D^{20}$ +178° (c = 7 of the dihydrate). Sol in water, hot alcohol. Insol in ether. Does not reduce Fehling's soln. Is fermented by yeast. Is not split by α-glucosidase. Acid hydrolysis gives 2 mols D-glucose.

9714. Tremetone. *1-[2,3-Dihydro-2-(1-methylethenyl)-5-benzofuranyl]ethanone; 2,3-dihydro-2-isopropenyl-5-benzofuranyl methyl ketone;* 2-isopropenyl-2,3-dihydro-5-acetylbenzofuran. $C_{13}H_{14}O_2$; mol wt 202.25. C 77.20%, H 6.98%, O 15.82%. Principal ketone suspected of being the active toxin of *Eupatorium urticaefolium* Reichard, *Compositae* (white snakeroot). Isoln and structure: Bonner, DeGraw, *Tetrahedron* **18**, 1295 (1962). Synthesis of dihydrotremetone: DeGraw, Bonner, *ibid.* 1311. Synthesis of racemic tremetone: DeGraw *et al.*, *ibid.* **19**, 19 (1963); Bohlmann, Buehmann, *Ber.* **105**, 863 (1972). Abs config: Bonner *et al.*, *Tetrahedron* **20**, 1419 (1964). *Review:* Christensen, *Econ. Bot.* **19**, 293-300 (1965).

Liquid. $[\alpha]_D^{28}$ −59.6° (c = 5.52 in absolute ethanol). n_D^{25} 1.5658. d_4^{28} 1.080. uv max (ethanol): 227, 280, 285 nm (ε 11,950, 12,600, 12,300).

Dihydrotremetone, $C_{13}H_{16}O_2$, *(R)-1-[2,3-dihydro-2-(1-methylethyl)-5-benzofuranyl]ethanone*. Liquid. bp 216-221°. $[\alpha]_D^{25}$ −47.0° (c = 1.78 in abs ethanol). uv max (ethanol): 231, 279 nm (ε 39,500; 18,800).

9715. Tremorine. *1,1'-(2-Butyne-1,4-diyl)bispyrrolidine; 1,1'-(2-butynylene)dipyrrolidine;* 1,4-dipyrrolidino-2-butyne. $C_{12}H_{20}N_2$; mol wt 192.30. C 74.95%, H 10.48%, N 14.57%. Prepn from pyrrolidine + 1,4-dichloro-2-butyne: Maier, **Ger.** pat **896,810** (1953 to BASF); Reppe *et al.*, *Ann.* **596**, 79 (1955); Biel, DiPierro, *J. Am. Chem. Soc.* **80**, 4609 (1958). Review of chemical and biological studies: Karlen, *Acta Pharm. Suecica* **1970**, 169-200; *see also* Kolla, Obvintseva, *Farmakol. Toksikol.* **36**, 736-745 (1973).

Liquid. $bp_{2.5}$ 116-116.5°; bp $_{0.1}$ 93-95°.
Methiodide, $C_{12}H_{20}N_2 \cdot 2CH_3I$, crystals, mp 239-240°.
USE: To produce exptl parkinsonism: Everett *et al.*, *Science* **124**, 79 (1956).

9716. Trenbolone. *(17β)-17-Hydroxyestra-4,9,11-trien-3-one;* 4,9,11-estratrien-17β-ol-3-one; 17β-hydroxy-19-norandrosta-4,9,11-trien-3-one; 19-norandrosta-4,9,11-trien-17β-ol-3-one; trienbolone; trienolone. $C_{18}H_{22}O_2$; mol wt 270.37. C 79.96%, H 8.20%, O 11.84%. Prepn of base: Velluz *et al.*, *C. R. Acad. Sci.* **257**, 569 (1963); Heller *et al.*, *Steroids* **10**, 211 (1967). Base and 17-acetate: **Fr.** pat. **M1958** and **Brit.** pat. **1,035,683** (1963 and 1966 to Roussel-UCLAF), *C.A.* **60**, 3039h (1964); **65**, 17027c (1966). 17-Cyclohexylmethylcarbonate: Nedelec, Costerousse, **Fr.** pat. **M5979** (1968 to Roussel-UCLAF), *C.A.* **71**, 50356g (1969). Pharmacology of the acetate: Krüskemper *et al.*, *Arzneimittel-Forsch.* **17**, 449 (1967). Animal studies: Beranger, Malterre, *C. R. Soc. Biol.* **162**, 1157 (1968); Best, *Vet. Rec.* **91**, 624 (1972).

Crystals, mp 186°. $[\alpha]_D^{20}$ +19° (c = 0.45 in ethanol). Also reported as mp 183-186° from acetone-water. uv max: 239, 340.5 nm (ε 5260, 28000) (Heller).

Acetate, $C_{20}H_{24}O_3$, *(17β)-17-acetyloxyestra-4,9,11-trien-3-one, 17β-acetoxy-3-oxoestra-4,9,11-triene, 17β-acetoxyestra-4,9,11-trien-3-one, Finaplix.* Crystals, mp 96-97°. $[\alpha]_D^{20}$ +36.8° (c = 0.37 in methanol).

Cyclohexylmethylcarbonate, $C_{26}H_{34}O_4$, *(17β)-17-[(cyclohexylmethoxy)carbonyl]oxyestra-4,9,11-trien-3-one, trenbolone hexahydrobenzylcarbonate, Parabolan.* Crystals from cyclohexane-petr ether, mp 90-95°. $[\alpha]_D^{20}$ +41.6° (c = 0.5 in ethanol).

Note: This is a controlled substance (anabolic steroid) in the U.S. Code of Federal Regulations, Title 21 Part 1308.13, as defined in Part 1308.02 (1995).

THERAP CAT: Anabolic.
THERAP CAT (VET): Anabolic.

9717. Trengestone. *(9β,10α)-6-Chloropregna-1,4,6-triene-3,20-dione;* 6-chloro-1,6-didehydroretroprogesterone; 6-chloro-1,6-bisdehydroretroprogesterone; Ro-4-8347; Retroid. $C_{21}H_{25}ClO_2$; mol wt 344.88. C 73.14%, H 7.31%, Cl 10.28%, O 9.28%. A member of a class of hormonally active steroids termed "retrosteroids", which are characterized by a (9β, 10α)-configuration in contrast to the usual (9α,10β)-configuration of steroids. Prepn: Threadgold, **Belg.** pat. **652,597** corresp to Reerink *et al.*, U.S. pat. **3,422,122** (1965, 1969 to Phillips). Activity studies: Sadovsky *et al.*, *Gynecol. Invest.* **1**, 319 (1970); Kalra *et al.*, *J. Endocrinol.* **51**, 675 (1971). Metabolism: Breuer *et al.*, *Acta Endocrinol. (Copenhagen)* **74**, 127 (1973); Dixon *et al.*, *Steroids* **22**, 35 (1973).

Crystals from acetone, mp 208-209° (dec). uv max: 229, 253, 302 nm (ε 11500, 10520, 10650).

THERAP CAT: Progestogen.

9718. Trepibutone. *2,4,5-Triethoxy-γ-oxobenzenebutanoic acid;* 3-(2,4,5-triethoxybenzoyl)propionic acid; AA-149; Supacal. $C_{16}H_{22}O_6$; mol wt 310.35. C 61.92%, H 7.15%, O 30.93%. Prepn: T. Murata *et al.*, **Ger.** pat. **2,244,324** corresp to U.S. pat. **3,943,169** (1973, 1976 both to Takeda). Properties and stabilities: M. Mitani *et al.*, *Takeda Kenkyusho Ho* **36**, 206 (1977), *C.A.* **88**, 126284f (1978). Pharmacokinetics: *eidem, ibid.* 215, *C.A.* **88**, 115067m (1978). Biotransformation: T. Kobayashi *et al.*, *Xenobiotica* **8**, 535 (1978). Metabolism studies: S. Tanayama *et al.*, *ibid.* 365, 377. Spasmolytic action in dogs: H. Satoh *et al.*, *Eur. J. Pharmacol.* **48**, 309 (1978). Mechanism of choleretic action: *eidem, ibid.* 125. Toxicity studies: S. Sato *et al.*, *Takeda Kenkyusho Ho* **36**, 263 (1977), *C.A.* **88**, 11535x (1978). Toxicity data: *Japan. Med. Gaz.* **18**(6), 13 (1981).

Colorless needles from aq ethanol or plates from aq acetone, mp 150-151°. Stable to heat, humidity, indoor diffused sunlight. Aq solns heated to 100° for 10 hr showed no degradation. LD_{50} in male mice, rats (mg/kg): 1340, 2450 orally, 530, 410 i.p. *(Japan. Med. Gaz.)*.

THERAP CAT: Choleretic; antispasmodic.

9719. Tretoquinol. *(±)-1,2,3,4-Tetrahydro-1-[(3,4,5-trimethoxyphenyl)methyl]-6,7-isoquinolinediol;* 1-(3',4',5'-trimethoxybenzyl)-6,7-dihydroxy-1,2,3,4-tetrahydroisoquinoline; trimethoquinol. $C_{19}H_{23}NO_5$; mol wt 345.40. C 66.07%, H 6.71%, N 4.06%, O 23.16%. Synthesis of *dl*-form: Yamato *et al.*, *Tetrahedron, Suppl.* **8**, 129 (1966). Pharmacology: Fogelman, Grundy, *Brit. J. Pharmacol.* **38**, 416 (1970). Metabolic studies: Meshi *et al.*, *Biochem. Pharma-*

col. **19**, 2937 (1970); Satoh *et al., Chem. Pharm. Bull.* **19**, 667 (1971).

dl-Form hydrochloride, pale yellow crystals from methanol + ether, decomp 224.5-226°.

l-Form hydrochloride, $C_{19}H_{23}NO_5 \cdot HCl$, *AQ-110, Inolin, Vems.* Pale yellow crystals, freely sol in water; sol in alcohol.

THERAP CAT: The *l*-form hydrochloride as bronchodilator.

9720. TRH. *5-Oxo-L-prolyl-L-histidyl-L-prolinamide; thyrotropin-releasing factor;* thyrotropin releasing hormone; TRF; protirelin; lopremone (rescinded USAN); TSH-releasing factor; pyroglutamylhistidylprolinamide; thyroliberin; Antepan; Stimu-TSH; Thypinone; Thyrefact. $C_{16}H_{22}N_6O_4$; mol wt 362.39. C 53.03%, H 6.12%, N 23.19%, O 17.66%. A hypothalamic neurohormone which stimulates the release and synthesis of TSH, *q.v.*, from the anterior pituitary via the hypophyseal portal system; the first of the hypothalamic regulatory hormones to be isolated, characterized, and synthesized. Isoln from bovine hypothalami: Schreiber *et al., Experiencia* **18**, 338 (1962); Schally *et al., Endocrinology* **78**, 726 (1966). Purification from ovine hypothalami: Guillemin *et al., C.R. Acad. Sci., Ser. D* **262**, 2278 (1966). Isoln from porcine hypothalami and properties: Schally *et al., Biochem. Biophys. Res. Commun.* **25**, 165 (1966); Schally *et al., J. Biol. Chem.* **244**, 4077 (1969). Structural studies: Folkers *et al., Biochem. Biophys. Res. Commun.* **37**, 123 (1969); Burgus *et al., C.R. Acad. Sci., Ser. D.* **269**, 226 (1969). Identity of isolated TRH with synthetic tripeptide: Burgus *et al., ibid.* 1870; Bowers *et al., Endocrinology* **86**, 573, 1143 (1970). Solubility: Burgus *et al., Experientia* **23**, 417 (1967). Synthesis: Flouret, *J. Med. Chem.* **13**, 843 (1970); Chang *et al., ibid.* **14**, 481 (1971); E. Gross *et al., Angew. Chem., Int. Ed.* **12**, 664 (1972); P. G. Pietta *et al., J. Org. Chem.* **39**, 44 (1974). Review of synthetic methods: Rivier, *Methods Enzymol.* **37**, 408 (1975). TRH is also believed to induce the secretion of the pituitary lactogenic hormone, prolactin, *q.v.:* Tasjian *et al., Biochem. Biophys. Res. Commun.* **43**, 516 (1971); Bowers *et al., ibid.* **51**, 512 (1973). It has been shown to block and reverse leukotriene-induced hypotension in the unanesthetized guinea pig: W. E. Lux *et al., Nature* **302**, 822 (1983). Clinical studies: Hershman, *N. Engl. J. Med.* **290**, 886 (1974). Reviews of TRH and other hypothalamic releasing hormones: Schally *et al., Recent Progr. Horm. Res.* **24**, 497 (1968); Burgus, Guillemin, *Ann. Rev. Biochem.* **39**, 499 (1970); *Polypeptide Hormones,* R. F. Beers. E. G. Bassett, Eds. (Raven Press, New York, 1980) pp 165-278.

pyroGlu–His–ProNH₂

Purified TRH is partially sol in chloroform, highly sol in absolute methanol. Completely insol in pyridine. Inactivated by diazotized sulfanilic acid (Pauly reagent) and by plasma, serum, or whole blood *in vitro.* Resists inactivation by proteolytic enzymes.

Tartrate, $C_{16}H_{22}N_6O_4 \cdot xC_4H_6O_6$, *Irtonin, Xantium.*

THERAP CAT: Prohormone.

9721. Triacetin. *1,2,3-Propanetriol triacetate;* glyceryl triacetate; triacetyl glycerine; Enzactin; Fungacetin. $C_9H_{14}O_6$; mol wt 218.21. C 49.54%, H 6.47%, O 43.99%. $C_3H_5(OCOCH_3)_3$. Prepd by acetylation of glycerol: Dunbar, Bolstad, *J. Org. Chem.* **21**, 1041 (1956); by reaction of oxygen with a liquid phase mixture of allyl acetate and acetic acid using a bromide as catalyst: Keith, U.S. pat. **2,911,437** (1959 to Sinclair Refining). As an antifungal: **Brit.** pat.

845,029 (1960 to Wisc. Alumni Res. Found.). Acute toxicity: A. Wretlind, *Acta Physiol. Scand.* **40**, 338 (1957).

Colorless, somewhat oily liquid having a slight, fatty odor and a bitter taste. d_4^{25} 1.1562, d_4^{20} 1.1596, d_{20}^{20} 1.163. mp −78°. bp 258-260°, bp_{40} 172°. n_D^{20} 1.4307. Sol in 14 parts water. Miscible with alcohol, ether, chloroform. Slightly sol in carbon disulfide. LD_{50} i.v. in mice: 1600 ±81 mg/kg (Wretlind).

USE: As fixative in perfumery; solvent in manuf celluloid, photographic films. Technical triacetin (a mixture of mono-, di-, and small quantities of triacetin) as a solvent for basic dyes, particularly indulines, and tannin in dyeing.

THERAP CAT: Antifungal (topical).

9722. 1-Triacontanol. Melissyl alcohol; myricyl alcohol; 1-hydroxytriacontane. $C_{30}H_{62}O$; mol wt 438.82. C 82.11%, H 14.24%, O 3.65%. $CH_3(CH_2)_{28}CH_2OH$. Present in plant cuticle waxes and in beeswax as the palmitate. Prepn: Robinson, *J. Chem. Soc.* **1934**, 1545; K. Maruyama *et al., J. Org. Chem.* **45**, 737 (1980). Use as a plant growth regulator: S. K. Ries *et al., Science* **195**, 1339 (1977); S. K. Ries, **Belg.** pat. **854,587** corresp to U.S. pat. **4,150,970** (1977, 1979 to Michigan State Univ.). Use of colloidally dispersed triacontanol in plant growth enhancement studies: R. G. Laughlin *et al., Science* **219**, 1219 (1983).

Crystals. d_{95} 0.777. mp 87°. Practically insol in water; freely sol in benzene, ether; very slightly sol in cold, more sol in hot alcohol.

USE: Plant growth regulator.

9723. Triadimefon. *1-(4-Chlorophenoxy)-3,3-dimethyl-1-(1H-1,2,4-triazol-1-yl)-2-butanone;* BAY MEB 6447; Bayleton. $C_{14}H_{16}ClN_3O_2$; mol wt 293.75. C 57.24%, H 5.49%, Cl 12.07%, N 14.30%, O 10.89%. Systemic fungicide active against mildews and rusts of grains, fruits, vegetables and ornamentals. Prepn and activity: W. Meiser *et al.,* **Ger.** pat. **2,201,063** corresp to U.S. pat. **3,912,752** (1973, 1975 to Bayer). Fungal metabolism: M. Gasztonyi, *Pestic. Sci.* **12**, 433 (1981). *Reviews:* F. Michel, P. Pourcharesse, *Def. Veg.* **31**, 97-109 (1977); T. J. Martin, D. B. Morris, *Pflanzenschutz-Nachr.* **32**, 31-79 (1979).

Crystals, mp 82°. Soly in water at 20°: 260 mg/l. Moderately sol in most organic solvents except aliphatics. LD_{50} in male, female rats (mg/kg): 568, 363 orally (Michel, Pourcharesse).

USE: Systemic agricultural fungicide.

9724. Triadimenol. *β-(4-Chlorophenoxy)-α-(1,1-dimethylethyl)-1H-1,2,4-triazole-1-ethanol;* 1-(4-chlorophenoxy)-3,3-dimethyl-1-(1H-1,2,4-triazol-1-yl)butan-2-ol; BAY KWG 0519; Bayfidan; Baytan; Spinnaker; Summit. $C_{14}H_{18}ClN_3O_2$; mol wt 295.77. C 56.85%, H 6.13%, Cl 11.99%, N 14.21%, O 10.82%. Agricultural fungicide systemically active against powdery mildews and rusts of grains. Commercial product is mixture of 2 diastereoisomers. Prepn: **Belg.** pat. **814,831**; W. Kramer *et al.,* U.S. pat. **3,952,002** (1974, 1976 both to Bayer AG). Comprehensive description: P. E. Frohberger, *Pflanzenschutz-Nachr.* **31**, 11 (1978). Field trials in cereal diseases: J. Trägner-Born, T. van den Boom, *ibid.* 25. Inhibits fungal ergosterol biosynthesis: H. Buchenauer, *Pestic. Sci.* **9**, 507 (1978). Active fungal metabolite of triadimefon, *q.v.:* M. Gasztonyi, *ibid.* **12**, 433 (1981). GC separation of enantiomers: T. Clark, A. H. B. Deas, *J. Chromatog.* **329**, 181 (1985).

Crystals, mp 112-117°. Slight, non-specific odor. Soly in water at 20°: 0.012 g/100 g. Soluble in alcohol, ketones. LD$_{50}$ in male, female rats (mg/kg): 1161, 1105 orally; >5000 dermally, 24-hr; LD$_{50}$ in quail: >10000 mg/kg (Frohberger).

USE: Systemic agricultural fungicide; cereal seed protectant.

9725. Triafur. *2-Amino-5-(5-nitro-2-furyl)-1,3,4-thiadiazole;* 2-(5-nitro-2-furyl)-5-amino-1,3,4-thiadiazole; Furidiazina. C$_6$H$_4$N$_4$O$_3$S; mol wt 212.19. C 33.96%, H 1.90%, N 26.40%, O 22.62%, S 15.11%. Prepn: Skagius *et al., Acta Chem. Scand.* **14,** 1054 (1960); Skagius, **Brit.** pat. 852,795 (1960 to Pharmacia); Sherman, *J. Org. Chem.* **26,** 88 (1961). Bacteriostatic and fungistatic activity: Jeney, Zsolnai, *Zentralbl. Bakteriol., Parasitenk., Infektionskr., Hyg., Abt. I Orig.* **204,** 430 (1967).

Yellow cryst from DMF + water or acetic acid, mp 280°.

9726. Triallate. *Bis(1-methylethyl)carbamothioic acid S-(2,3,3-trichloro-2-propenyl) ester; diisopropylthiocarbamic acid S-(2,3,3-trichloroallyl) ester;* 2,3,3-trichloro-2-propene-1-thiol diisopropylcarbamate; *S*-2,3,3-trichloroallyl diisopropylthiocarbamate; *S*-(2,3,3-trichloro-2-propenyl) bis(1-methylethyl)carbamothioate; CP-23426; Avadex BW; Far-Go. C$_{10}$H$_{16}$Cl$_3$NOS; mol wt 304.67. C 39.42%, H 5.29%, Cl 34.91%, N 4.60%, O 5.25%, S 10.52%. Prepn: M. W. Harman, J. J. D'Amico, **U.S.** pat. **3,330,821** (1967 to Monsanto). Herbicidal activity: R. Grover *et al., Weed Res.* **19,** 363 (1979). Mutagenic evaluation *in vitro* (Ames Test): F. De Lorenzo *et al., Cancer Res.* **38,** 13 (1978); G. R. Douglas *et al., Mutat. Res.* **85,** 45 (1981). Soil persistence: A. E. Smith, B. J. Hayden, *Bull. Environm. Contam. Toxicol.* **29,** 240 (1982). Field studies for pre-emergent use: T. G. Reeves, C. L. Touhey, *Aust. J. Exp. Agr. Anim. Husb.* **12,** 55 (1972); E. M. Randall, R. H. Jarvis, *Exp. Husb.* **38,** 32 (1982); post-emergent use: R. P. Garnett, *Aspects Appl. Biol.* **13,** 73 (1980). GC determn in soils: A. E. Smith, *J. Chromatog.* **97,** 103 (1974). HPLC determn in soils: A. Pena Heras, F. Sanchez-Rasero, *ibid.* **358,** 302 (1986).

USE: Herbicide.

9727. Triamcinolone. *(11β, 16α)-9-Fluoro-11,16,17,21-tetrahydroxypregna-1,4-diene-3,20-dione;* Δ^1-9α-fluoro-16α-hydroxyhydrocortisone; 9α-fluoro-16α-hydroxyprednisolone; Δ^1-16α-hydroxy-9α-fluorohydrocortisone; 16α-hydroxy-9α-fluoroprednisolone; CL-19823; Aristocort; Cinolone; Kenacort; Ledercort (tabl.); Omcilon; Tricortale; Volon. C$_{21}$H$_{27}$FO$_6$; mol wt 394.44. C 63.95%, H 6.90%, F 4.82%, O 24.34%. Prepn: Bernstein *et al., J. Am. Chem. Soc.* **78,** 5693 (1956); **81,** 1689 (1959); Thoma *et al., ibid.* **79,** 4818 (1957); Bernstein *et al.,* Allen *et al.,* **U.S.** pats. **2,789,-118; 3,021,347** (1957, 1962, both to Am. Cyanamid). Comprehensive description: K. Florey, *Anal. Profiles Drug Subs.* **1,** 367-396, 423-442 (1972); D. H. Sieh, *ibid.* **11,** 593-614, 651-661 (1982).

Crystals, mp 269-271°. mp also reported as 260-262.5°. [α]$_D^{25}$ +75° (acetone). uv max: 238 nm (ε 15800).

16,21-Diacetate, C$_{25}$H$_{31}$FO$_8$, *(11β,16α)-16,21-bis(acetyloxy)-9-fluoro-11,17-dihydroxypregna-1,4-diene-3,20-dione, 16α,21-diacetoxy-9α-fluoro-11β,17α-dihydroxy-1,4-pregnadiene-3,20-dione,* Cenocort, CINO-40, Tracilon. Solvated crystals, mp 186-188° (with effervescence, mp 235° after drying). [α]$_D^{25}$ +22° (chloroform). uv max: 239 nm (ε 15200).

THERAP CAT: Glucocorticoid.

THERAP CAT (VET): Glucocorticoid.

9728. Triamcinolone Acetonide. *(11β,16α)-9-Fluoro-11,21-dihydroxy-16,17-[1-methylethylidenebis(oxy)]pregna-1,4-diene-3,20-dione; 9α-fluoro-11β,16α,17,21-tetrahydroxypregna-1,4-diene-3,20-dione cyclic 16,17-acetal with acetone;* 9α-fluoro-16α-hydroxyprednisolone acetonide; triamcinolone 16α,17-acetonide; 9α-fluoro-11β,21-dihydroxy-16α,-17α-isopropylidenedioxy-1,4-pregnadiene-3,20-dione; 9α-fluoro-16α,17-isopropylidenedioxyprednisolone; Adcortyl; Azmacort; Delphicort; Extracort; Ftorocort; Kenacort-A; Kenalog; Ledercort Cream; Nasacort; Respicort; Rineton; Solodelf; Tramacin; Triam; Tricinolon; Vetalog; Volon A; Volonimat. C$_{24}$H$_{31}$FO$_6$; mol wt 434.50. C 66.34%, H 7.19%, F 4.37%, O 22.09%. Prepd by stirring a suspension of triamcinolone in acetone in the presence of a trace of perchloric acid: Fried *et al., J. Am. Chem. Soc.* **80,** 2338 (1958); Bernstein *et al., ibid.* **81,** 1689 (1959); Bernstein, Allen, **U.S.** pat. **2,990,401** (1961 to Am. Cyanamid). Alternate synthesis using 2,3-dibromo-5,6-dicyanoquinone: Hydorn, **U.S.** pat. **3,035,050** (1962 to Olin Mathieson). Clinical trial in chronic asthma: I. L. Bernstein *et al., Chest* **81,** 20 (1982). Comprehensive description: K. Florey, *Anal. Profiles Drug Subs.* **1,** 397-421 (1972); D. H. Sieh, *ibid.* **11,** 615-649 (1982).

Crystals, mp 292-294°. [α]$_D^{23}$ +109° (c = 0.75 in chloroform). uv max (abs alc.): 238 nm (ε 14600). Sparingly sol in methanol, acetone, ethyl acetate.

21-Acetate, crystals, mp 268-270°. [α]$_D^{23}$ +92° (c = 0.59 in chloroform).

21-Disodium phosphate, C$_{24}$H$_{30}$FNa$_2$O$_9$P, *Aristosol.*

21-Hemisuccinate, C$_{28}$H$_{35}$FO$_9$, *Solutedarol.*

THERAP CAT: Glucocorticoid; antiasthmatic (inhalant); antiallergic (nasal).

THERAP CAT (VET): Glucocorticoid.

9729. Triamcinolone Benetonide. *(11β,16α)-21-[3-(Benzoylamino)-2-methyl-1-oxopropoxy]-9-fluoro-11-hydroxy-16,17-[(1-methylethylidene)bis(oxy)]pregna-1,4-diene-3,20-dione; 9-fluoro-11β,16α,17,21-tetrahydroxypregna-1,4-diene-3,20-dione cyclic 16,17-acetal with acetone 21-ester with N-benzoyl-2-methyl-β-alanine;* 9α-fluoro-16α-hydroxyprednisolone 16α,17α-acetonide 21-(β-benzoylamino)isobutyrate; triamcinolone acetonide β-benzoylaminoisobutyrate; TBI; Tibicorten. C$_{35}$H$_{42}$FNO$_8$; mol wt 623.72. C 67.40%, H 6.79%, F 3.05%, N 2.25%, O 20.52%. Prepn: C. Cavazza *et al.,* **Ger.** pat. **2,047,218;** *eidem,* **U.S.** pat. **3,749,712** (1971, 1973 both to Sigma-Tau). Pharmacology: E. T. Ordonez,

Arzneimittel-Forsch. **21,** 248 (1971). Percutaneous absorption by rats and rabbits: W. H. Down *et al., Toxicol. Letters* **1,** 95 (1977). Clinical study: D. J. Tazelaar, *J. Int. Med. Res.* **5,** 338 (1977). HPLC analysis: S. Muck *et al., Boll. Chim. Farm.* **120,** 240 (1981). For structure *see* Triamcinolone Acetonide.

Crystalline powder, mp 203-207°. $[\alpha]_D^{20}$ +96° ±3° (c = 1 in ethanol). Sol in methanol, acetone, ethanol, dioxane, pyridine, DMF, chloroform. Insol in water.

THERAP CAT: Glucocorticoid; anti-inflammatory (topical).

9730. Triamcinolone Hexacetonide. *(11β,16α)-21-(3,3-dimethyl-1-oxobutoxy)-9-fluoro-11-hydroxy-16,17-[(1-methylethylidene)bis(oxy)]pregna-1,4-diene-3,20-dione; 9-fluoro-11β,16α,17,21-tetrahydroxypregna-1,4-diene-3,20-dione cyclic 16,17-acetal with acetone, 21-(3,3-dimethylbutyrate);* 21-*tert*-butylacetate-9α-fluoro-11β-hydroxy-16α,17α-(isopropylidenedioxy)pregna-1,4-diene-3,20-dione; 21-(3,3-dimethylbutyryloxy)-9α-fluoro-11β-hydroxy-16α,17α-(isopropylidenedioxy)pregna-1,4-diene-3,20-dione; triamcinolone acetonide *tert*-butyl acetate; TATBA; CL-34433; Aristospan; Hexatrione; Lederlon; Lederspan. $C_{30}H_{41}FO_7$; mol wt 532.65. C 67.65%, H 7.76%, F 3.57%, O 21.03%. The hexacetonide ester of the potent glucocorticoid, triamcinolone, *q.v.* Prepn of syringeable suspension: Nash, Naeger, U.S. pat. **3,457,348** (1969 to Am. Cyanamid). Anti-inflammatory activity in rabbits: I. M. Hunneyball, *Agents Actions* **11,** 490 (1981). Early clinical studies: Bilka, *Minnesota Med.* **50,** 483 (1967); Layman, Peterson, *ibid.* 669. Clinical studies of intra-articular therapy in arthritis: R. C. Allen *et al., Arthritis Rheum.* **29,** 997 (1986); M. Talke, *Fortschr. Med.* **104,** 742 (1986). Toxicity study: Tonelli, *Steroids* **8,** 857 (1966). Comprehensive description: V. Zbinovsky, G. P. Chrekian, *Anal. Profiles Drug Subs.* **6,** 579-595 (1977). For structure *see* Triamcinolone Acetonide.

Fine, white, needle-like crystals, mp 295-296° (dec), also reported as mp 271-272° (dec). uv max (ethanol): 238 nm (ε 15500). $[\alpha]_D^{25}$ +90°±2 (c = 1.13% in chloroform). Soly in g/100 ml at 25°: chloroform and dimethylacetamide > 5; ethyl acetate 0.77, methanol 0.59, diethyl carbonate 0.50, glycerin 0.42, propylene glycol 0.13; absolute alcohol 0.03; water 0.0004.

THERAP CAT: Anti-inflammatory.

9731. Triamterene. *6-Phenyl-2,4,7-pteridinetriamine; 2,4,7-triamino-6-phenylpteridine;* 6-phenyl-2,4,7-triaminopteridine; ademin(e); pterofen; pterophene; NSC-77625; SKF-8542; Dyren; Dyrenium; Dytac; Jatropur; Teriam; Triteren; Urocaudal. $C_{12}H_{11}N_7$; mol wt 253.27. C 56.91%, H 4.38%, N 38.71%. Prepn: Spickett, Timmis, *J. Chem. Soc.* **1954,** 2887; Pachter, *J. Org. Chem.* **28,** 1191 (1963); Weinstock, Wiebelhaus, U.S. pat. **3,081,230** (1963 to SK & F); Osdene *et al., J. Med. Chem.* **10,** 431 (1967). *Review: Ther. Triamteren, Wiener Symp.* **1966,** K. Fellinger, Ed. (Georg Thieme, Vienna, 1967).

Yellow plates from butanol, mp 316° (Spickett, Timmis); also reported as crystals from DMF, mp 327° (Osdene *et al.*). uv max (4.5% formic acid): 356 nm (ε 21000).

THERAP CAT: Diuretic.

9732. Triasulfuron. *2-(2-Chloroethoxy)-N-[[(4-methoxy-6-methyl-1,3,5-triazin-2-yl)amino]carbonyl]benzenesulfonamide;* 1-[2-(2-chloroethoxy)phenylsulfonyl]-3-(4-methoxy-6-methyl-1,3,5-triazin-2-yl)urea; CGA-131036; Amber; Logran. $C_{14}H_{16}ClN_5O_5S$; mol wt 401.83. C 41.85%, H 4.01%, Cl 8.82%, N 17.43%, O 19.91%, S 7.98%. Sulfonylurea herbicide for broadleaf weeds. Prepn: W. Meyer, W. Föry, *Eur.* pat. **Appl. 44,808;** *eidem,* U.S. pat. **4,514,212** (1982, 1985 both to Ciba-Geigy). Comprehensive description: J. Amrein, H. R. Gerber, *Proc. Brit. Crop Prot. Conf. - Weeds* **1985,** 55-62. Uptake and metabolism in wheat: A. M. Meyer, F. Müller, *Proc. Brighton Crop Prot. Conf. -*

Weeds **1989,** 441. Behavior in soil: F. K. Oppong, G. R. Sagar, *Weed Res.* **32,** 157, 167 (1992). Immunoassay in soil: J. A. Schlaeppi *et al., J. Agric. Food Chem.* **42,** 1914 (1994).

White crystals, mp 186°. Vapor pressure (20°): 7.5 × 10^{-13} mm Hg. Soly in water (20°): 1.5 g/l (pH 7). LD_{50} in rats (mg/kg): > 5000 orally; > 2000 dermally; LC_{50} (4 hr) in rats: > 5185 mg/m³ by inhalation (Amrein, Gerber).

USE: Herbicide.

9733. Triaziquone. *2,3,5-Tris(1-aziridinyl)-2,5-cyclohexadiene-1,4-dione; 2,3,5-tris(1-aziridinyl)-p-benzoquinone;* 2,3,5-tris(aziridino)-1,4-benzoquinone; 2,3,5-tris(ethyleneimino)benzoquinone; Bayer 3231; Trenimon. $C_{12}H_{13}N_3O_2$; mol wt 231.25. C 62.33%, H 5.67%, N 18.17%, O 13.84%. Prepn: Gauss, Domagk, U.S. pat. **2,976,279** (1961 to Schenley).

Purple acicular crystals from ethyl acetate, mp 162.5-163°. Sparingly sol in cold water; sol in acetone, benzene, chloroform, ethyl acetate, methanol and warm acetic acid. Active against *Entamoeba histolytica*.

THERAP CAT: Antineoplastic.

9734. Triazolam. *8-Chloro-6-(2-chlorophenyl)-1-methyl-4H-[1,2,4]triazolo[4,3-a][1,4]benzodiazepine; 8-chloro-6-(o-chlorophenyl)-1-methyl-4H-s-triazolo[4,3-a][1,4]benzodiazepine;* clorazolam; U-33030; Halcion; Novodorm; Songar. $C_{17}H_{12}Cl_2N_4$; mol wt 343.21. C 59.49%, H 3.52%, Cl 20.66%, N 16.32%. Prepn: J. B. Hester, Ger. pat. **2,012,190;** *eidem,* U.S. pat. **3,701,782** (1970, 1972 both to Upjohn); J. B. Hester *et al., J. Med. Chem.* **14,** 1078 (1971). Pharmacology: T. Furukawa *et al., Igaku Kenkyu* **45,** 285 (1975), *C.A.* **84,** 130354p. Metabolism: F. S. Eberts, *Drug Metab. Dispos.* **5,** 547 (1977). HPLC determn in urine: T. Inoue, S. Suzuki, *J. Chromatog.* **422,** 197 (1987). Clinical studies: K. K. Okawa, G. S. Allens, *J. Int. Med. Res.* **6,** 343 (1978); A. J. Bowen, *ibid.* 337; A. J. Puech *et al., Therapie* **33,** 287 (1978). Toxicity: *Pharm. Weekblad.* **113,** 725 (1978). Review of pharmacology and therapeutic efficacy: G. E. Pakes *et al., Drugs* **22,** 81-110 (1981).

Tan crystals from 2-propanol, mp 233-235°. LD_{50} in mice, rats (mg/kg): > 100, > 5000 orally *(Pharm. Weekblad.).*

Note: This is a controlled substance (depressant) listed in the U.S. Code of Federal Regulations, Title 21 Part 1308.14 (1995).

THERAP CAT: Sedative, hypnotic.

9735. 1*H*-1,2,4-Triazole. Pyrrodiazole. $C_2H_3N_3$; mol wt 69.07. C 34.78%, H 4.38%, N 60.84%. Prepn starting with thiosemicarbazide and formic acid: Ainsworth, *Org. Syn.* **40**, 99 (1960).

Needles from ethanol + benzene, mp 120-121°. bp_{760} 260° (decompn). Appreciably sol in water, alcohol.

Hydrochloride, $C_2H_3N_3 \cdot HCl$, platelets, dec 169°.

9736. Triazophos. Phosphorothioic acid *O,O*-diethyl *O*-(*1-phenyl-1H-1,2,4-triazol-3-yl)ester;* 1-phenyl-3-(*O,O*-diethylthionophosphoryl)-1,2,4-triazole; HOE-2960; Hostathion. $C_{12}H_{16}N_3O_3PS$; mol wt 313.32. C 46.00%, H 5.15%, N 13.41%, O 15.32%, P 9.89%, S 10.23%. Cholinesterase inhibitor. Prepn: O. Scherer, H Mildenberger, **S. Afr.** pat. **68 03,471;** *eidem,* **U.S.** pat. **3,686,200** (1968, 1972 both to Hoechst). Properties and analytical methods: W. G. Thier *et al. Anal. Methods Pest. Plant Growth Regul.* **10**, 127 (1978). Degradation studies: K.-D. Bock *et al., Environ. Quality Safety, Suppl. 3 (Pesticides)* 833 (1975). GC determn: H.-J. Stan, D. Mrowetz, *J. Chromatog.* **279**, 173 (1983). Soil adsorption studies: G. S. Rao *et al., Indian J. Agric. Sci.* **61**, 674 (1991). Field studies as insecticide and acaricide: S. J. B. Hay, *Proc. 6th Brit. Crop Prot. Council - Insect. Fungic.* 597 (1972). Nematicidal activity: A. Sirohi, Siyanand, *Indian J. Nematol.* **22**, 134 (1992).

Light brown oil, mp 273-278°K. Vapor pressure 2.9 × 10^{-6} mbar (303°K). d 1.247. Soly at 296°K: 39 ppm in water. Soly at 298°K (g/100 ml): *n*-hexane 0.7, toluene 30, ethyl acetate 30, acetone 30, ethanol 30. LD_{50} in rats (mg/kg): 82 orally, 1100 dermally (Thier). LD_{50} in rats (mg/kg): 107 i.p.; in rabbits (mg/kg): 280 dermally (Hay). USE: Pesticide.

9737. Triazoxide. *7-Chloro-3-(1H-imidazol-1-yl)-1,2,4-benzotriazine 1-oxide;* 3-(imidazol-1-yl)-7-chloro-benzo-1,2,-4-triazine-1-oxide; BAY SAS 9244. $C_{10}H_6ClN_5O$; mol wt 247.64. C 48.50%, H 2.44%, Cl 14.32%, N 28.28%, O 6.46%. Prepn: K. Sasse *et al.,* **Ger.** pat. **2,802,488;** *eidem,* **U.S.** pat. **4,239,760** (1979, 1980 both to Bayer). GC determn in water: R. Brennecke, K. Vogeler, *Pflanzenschutz-Nachr.* **37**, 46 (1984). HPLC determn of adsorption coeff on soil: W. Kördel *et al., Sci. Total Environ.* **162**, 119 (1995). Physicochemical properties, toxicology, and fungicidal activity: S. Dutzmann, *Brit. Crop Prot. Council Monogr.* **57**, 85 (1994). Field trials in combination with tebuconazole, *q.v.:* A. Wainwright *et al., ibid.* 103.

Light yellow crystals, mp 182° (Dutzmann); also reported as mp 172° (Sasse). Vapor pressure at 25°: 5.2 × 10^{-12} Pa. Soly at 20° (g/l): water 0.03; *n*-hexane < 1; dichloromethane 50-100; 2-propanol 2-5; toluene 20-50. LD_{50} in rats

(mg/kg): > 5000 dermally; 100-200 orally. LC_{50} (4hr) in rats (mg/l): 0.8-3.2 by inhalation (Dutzmann).

USE: Fungicide.

9738. Tribenoside. *Ethyl 3,5,6-tris-O-(phenylmethyl)-D-glucofuranoside;* Ba-21401; Alven; Flebosan; Glyvenol; Hemocuron; Venex. $C_{29}H_{34}O_6$; mol wt 478.59. C 72.78%, H 7.16%, O 20.06%. Prepn: Druey, Huber, **U.S.** pat. **3,157,634** (1964 to Ciba). Pharmacology: Lecomte, *C.R. Soc. Biol.* **163**, 1469 (1969); Helfer, Jaques, *Pharmacology* **5**, 23 (1971). Increase in capillary resistance in patients with rheumatoid arthritis: W. C. Dick *et al., Ann. Rheum. Dis.* **28**, 187 (1969). GLC determn in plasma: A. Sioufi, F. Pommier, *J. Pharm. Sci.* **69**, 167 (1980). Excretion as hippuric acid: *eidem, Eur. J. Drug Metab. Pharmacokinet.* **7**, 223 (1982). Review of pharmacology: R. Jaques, *ibid.* **15**, 445-460 (1977).

$bp_{1.2}$ 270-280°. $[\alpha]_D^{26}$ +8° (chloroform).

THERAP CAT: Sclerosing agent.

9739. Tribromoacetic Acid. $C_2HBr_3O_2$; mol wt 296.74. C 8.10%, H 0.34%, Br 80.78%, O 10.78%. CBr_3COOH. Prepn: Müller, **U.S.** pat. **2,057,964** (1936 to Winthrop Chem.); A. M. Kovalevskaya, S. A. Shkylar, *Zh. Org. Khim.* **1**, 1540 (1965). Study of far IR spectrum: G. Statz, E. Lippert, *Ber. Bunsenges. Phys. Chem.* **71**, 673 (1967).

Monoclinic prisms, mp 129-135°. bp 245° (dec). Sol in water, alcohol, ether; slightly sol in petr ether. Dec in boiling water to bromoform.

USE: Catalyst for polymerization; as brominating agent: W. J. Szczepek, *Polish J. Chem.* **55**, 709 (1981).

9740. 2,4,6-Tribromoaniline. Aniline tribromide. $C_6H_4Br_3N$; mol wt 329.82. C 21.85%, H 1.22%, Br 72.68%, N 4.25%. Prepd by controlled bromination of aniline: Suthers *et al., J. Org. Chem.* **27**, 447 (1962).

Needles; d 2.35; mp 120-122°; bp 300°. Insol in water. Sol in hot alc, chloroform, ether; slightly sol in cold alcohol.

9741. Tribromo-*tert*-butyl Alcohol. *1,1,1-Tribromo-2-methyl-2-propanol;* acetone-bromoform; brometone. $C_4H_7Br_3O$; mol wt 310.81. C 15.46%, H 2.27%, Br 77.12%, O 5.15%. $(CH_3)_2C(OH)CBr_3$. Prepd from acetone and bromoform in the presence of sodium or potassium hydroxide: Aldrich, *J. Am. Chem. Soc.* **33**, 386 (1911); Viehe *et al., Ber.* **96**, 426 (1963).

Crystals from dil alcohol, mp 167-176° (Aldrich), from ether, mp 169° (Viehe). Camphor-like odor and taste. Volatilizes in air. Slightly sol in water; sol in alcohol, ether.

Caution: This substance may be habit forming and is listed in the U.S. Code of Federal Regulations, Title 21 Part 329.1 (1995).

USE: As modifier in polymerization of vinyl chloride, Seymour, **U.S.** pat. **2,716,112** (1955 to U.S. Rubber).

9742. 2,4,6-Tribromo-*m*-cresol. *2,4,6-Tribromo-3-methylphenol;* 2,4,6-tribromo-3-hydroxytoluene; Micatex. $C_7H_5Br_3O$; mol wt 344.83. C 24.38%, H 1.46%, Br 69.52%, O 4.64%. Prepn by bromination of *m*-cresol: Biilmann, Rimbert, *Bull. Soc. Chim. France* [4] **33**, 1473 (1923).

Crystals from 50% aq alcohol, mp 84°.
THERAP CAT: Antifungal (topical).

9743. Tribromoethanol. Tribromoethyl alcohol. C_2-H_3Br_3O; mol wt 282.76. C 8.50%, H 1.07%, Br 84.78%, O 5.66%. Br_3CCH_2OH. Prepd by the reduction of bromal with aluminum isopropylate. See U.S. pats. 1,572,742 (1926); 1,725,054 (1929); 1,882,944 (1932). Acute toxicity: R. R. Burtner, G. Lehmann, *J. Pharmacol. Exp. Ther.* **63,** 183 (1938).

Crystals, mp 79-82°, decompn starting at 70°. bp_{10} 92-93°. Slight ethereal or aromatic odor and taste. Sol in 40 parts water at 40°; sol in alcohol, ether, benzene, warm petr ether; very sol in amylene hydrate. The aq and alcoholic solns dec on exposure to light; in aq soln dibromoacetaldehyde and HBr are formed, both of which are strongly irritant. *Protect from light.* LD_{50} orally in rats: 1.09 g/kg (Burtner, Lehmann).

Solution in amylene hydrate, *bromethol, Avertin.*
THERAP CAT: Has been used as a basal anesthetic.
THERAP CAT (VET): Has been used as a general anesthetic and narcotic.

9744. 2,4,6-Tribromophenol. Bromol. $C_6H_3Br_3O$; mol wt 330.80. C 21.79%, H 0.91%, Br 72.46%, O 4.84%. Prepd by controlled bromination of phenol: Konecny, *J. Am. Chem. Soc.* **76,** 4993 (1954). Toxicity study: E. F. Stohlman, *Pub. Health Reports* **66,** 1303 (1951).

Long crystals; d 2.55; mp 94-96°; bp 244°. Sublimable. Sol in 14,000 parts water at 15°; sol in alcohol, chloroform, ether, glycerol oils. LD_{50} orally in rats: < 2000 mg/kg (Stohlman).

9745. 1,2,3-Tribromopropane. *sym*-Tribromopropane; glycerol tribromohydrin; tribromohydrin; allyl tribromide. $C_3H_5Br_3$; mol wt 280.78. C 12.83%, H 1.79%, Br 85.37%. $BrCH_2CHBrCH_2Br$. Prepd by the addn of bromine to allyl bromide in carbon tetrachloride: Perkin, Simonsen, *J. Chem. Soc.* **87,** 859 (1905); Johnson, McEwen, *Org. Syn.* **5,** 99 (1925). By gentle heating of propylene bromide with bromine in the presence of iron wire: Kronstein, *Ber.* **54,** 7 (1921); Tapley, Giesy, *J. Am. Pharm. Assoc.* **15,** 173 (1926); by reaction of bromotrichloromethane with allyl bromide initiated by γ-rays: Heiba, Anderson, *J. Am. Chem. Soc.* **79,** 4940 (1957).

Liquid; d^{23} 2.436; mp 16.5°; bp_{760} 220°; bp_{200} 170°; bp_{100} 148°; bp_{60} 134°; bp_{40} 123°; bp_{20} 106°; bp_{10} 90°; bp_5 76°; $bp_{1.0}$ 47.5°; n_D^{18} 1.58436. Insol in water; sol in alcohol, ether, chloroform.
USE: As a nematocide: Youngson, Goring, U.S. pat. 3,003,914 (1959 to Dow).

9746. Tribromosilane. Tribromomonosilane; silicobromoform. Br_3HSi; mol wt 268.81. Br 89.18%, H 0.37%, Si 10.45%. $SiHBr_3$. Prepd by passing hydrogen bromide over heated silicon or a silicide such as copper silicide: Schumb, Young, *Inorg. Syn.* **1,** 38 (1939); Schenk in *Handbook of Preparative Inorganic Chemistry* vol. **1,** G. Brauer, Ed. (Academic Press, New York, 2nd ed., 1963) pp 692-694.
Mobile liquid. Spontaneously flammable when the contact surface with air is large. d_4^{17} 2.7. mp −73.5°. bp_{760} 111.8°. The vapor pressure at 0° is given as 8.8 mm Hg.

Dipole moment 0.79. Hydrolyzed by water with the formation of silicoformic anhydride, $H_2Si_2O_3$ and HBr. Sol in chlorinated hydrocarbons.

9747. Tribromsalan. *3,5-Dibromo-N-(4-bromophenyl)-2-hydroxybenzamide; 3,4',5-tribromosalicylanilide;* TBS; Temasept IV; Tuasol 100. $C_{13}H_8Br_3NO_2$; mol wt 449.92. C 34.70%, H 1.79%, Br 53.28%, N 3.11%, O 7.11%. Preparation: **Brit. pat.** 840,366 (1960 to Unilever); Lemaire *et al., J. Pharm. Sci.* **50,** 831 (1961); Lamberti; Schramm, Lemaire; Schramm, U.S. pats. 2,967,885; 3,064,048; 3,057,920 (1961, 1962 and 1962, all to Lever Bros.); Majewski, U.S. pat. 3,254,121 (1966 to Dow).

Crystals, mp 227-228°. Practically insol in water. Sol in hot acetone; very sol in DMF.
USE: Bacteriostat in detergents. Banned by FDA from use in cosmetics.

9748. Tributylamine. *N,N-Dibutyl-1-butanamine.* C_{12}-$H_{27}N$; mol wt 185.35. C 77.76%, H 14.68%, N 7.56%. $(CH_3CH_2CH_2CH_2)_3N$. Prepd by vapor phase alkylation of ammonia with butanol: Lemon, Myerly, U.S. pat. 3,022,-349 (1962 to Union Carbide Corp.).
Hygroscopic liquid; characteristic odor. d_{20}^{20} 0.7782; bp 216-217°. Sparingly sol in water; very sol in alcohol, ether. *Keep well closed.*
Caution: Causes CNS stimulation, skin irritation, sensitization.

9749. Tributyl Phosphate. *Phosphoric acid tributyl ester.* $C_{12}H_{27}O_4P$; mol wt 266.32. C 54.12%, H 10.22%, O 24.03%, P 11.63%. $(C_4H_9)_3PO_4$. Prepd by reaction of $POCl_3$ with butyl alcohol: Pianfetti, Janey, U.S. pat. 3,020,303 (1962 to FMC). Toxicity study: Smyth, Carpenter, *J. Ind. Hyg. Toxicol.* **26,** 269 (1944).
Colorless, odorless liq; d_{25}^{25} 0.976; bp 289° with decompn; bp_{27} 177-178°; mp below −80°. Flash pt 146°. n_D^{25} 1.4215. One ml dissolves in about 165 ml water; miscible with usual organic solvents. LD_{50} orally in rats: 3.0 g/kg (Smyth, Carpenter).
Caution: Potential symptoms of overexposure are irritation of eyes, respiratory system and skin; headache; nausea. *See NIOSH Pocket Guide to Chemical Hazards* (DHHS/NIOSH 90-117, 1990) p 216.
USE: Plasticizer for cellulose esters, lacquers, plastics, and vinyl resins.

9750. Tributyrin. *Butanoic acid 1,2,3-propanetriyl ester;* glyceryl tributyrate. $C_{15}H_{26}O_6$; mol wt 302.37. C 59.58%, H 8.67%, O 31.75%. $(C_3H_7COO)_3C_3H_5$. Prepd by esterification of glycerol with excess butyric acid: Weatherby *et al., J. Am. Chem. Soc.* **47,** 2249 (1925).
Oily liq; bitter taste; d_4^{20} 1.032; mp −75°; bp_{760} 305-310°. bp_{15} about 190°; n_D^{20} 1.4358. Insol in water. Very sol in alc, ether.

9751. Tricaine. *3-Aminobenzoic acid ethyl ester methanesulfonate;* ethyl *m*-aminobenzoate methanesulfonate; MS-222; Finquel. $C_{10}H_{15}NO_5S$; mol wt 261.30. C 45.97%, H 5.79%, N 5.36%, O 30.62%, S 12.27%. Prepn: Billeter *et al.,* U.S. pat. 1,678,317 (1928 to Sandoz); Ger. pat. 454,698 (1927); *Frdl.* **16,** 2427 (1931). Pharmacology in fish: Jolly *et al., Vet. Rec.* **91,** 424 (1972).

Fine needles from alcohol + ethyl acetate, mp 149-150°. Soly in water at 20°: 1 g/0.8 ml. Also given as 1:9. Slightly acid reaction. Aq solns stable to boiling.

THERAP CAT (VET): Anesthetic for fish.

9752. Tricarballylic Acid. *1,2,3-Propanetricarboxylic acid;* β-carboxyglutaric acid. $C_6H_8O_6$; mol wt 176.13. C 40.92%, H 4.58%, O 54.50%. HOOCCH$_2$CH(COOH)-CH$_2$COOH. The calcium salt is found in maple-sugar sand, and on the walls of evaporators in beet sugar manuf. Prepared by hydrolysis of ethyl propane-1,1,2,3-tetracarboxylate: Clarke, Murray, *Org. Syn. coll. vol.* **I** (2nd ed., 1941) p 523; from aconitic acid by reduction with sodium amalgam: Fittig, *Ann.* **314,** 15 (1901); by hydrolysis of 1-cyano-2,3-dicarbomethoxypropane, obtained by reaction of dimethyl itaconate with HCN in the presence of KCN: Bavley, Tate, U.S. pat. **2,992,268** (1958 to Pfizer).

Large, orthorhombic prisms from water or ether, mp 166°. K$_1$ at 30° = 3.25 × 10^{-4}; K$_2$ = 2.65 × 10^{-5}; K$_3$ = 1.48 × 10^{-6}. At 18° 50 g dissolve in 100 ml water and 0.9 g dissolve in 100 ml ether. Quite sol in alc. Absorption spectrum: Bielecki, Henri, *Ber.* **46,** 2596 (1913). The trisodium salt is neutral to litmus.

Trimethyl ester, $C_9H_{14}O_6$, d$_4^{20}$ 1.1822; bp$_{13}$ 150°; n$_D^{20}$ 1.4398.

9753. Trichlorfon. *(2,2,2-Trichloro-1-hydroxyethyl)-phosphonic acid dimethyl ester;* O,O-dimethyl-1-hydroxy-2,-2,2-trichloroethylphosphonate; O,O-dimethyl 2,2,2-trichloro-1-hydroxyethylphosphonate; chlorofos; metrifonate; trichlorphene; Bayer L 1359; Cekufon; Combot; Danex; Dipterex; Dylox; Neguvon; Proxol; Tugon. $C_4H_8Cl_3O_4P$; mol wt 257.44. C 18.66%, H 3.13%, Cl 41.31%, O 24.86%, P 12.03%. Cholinesterase inhibitor. Prepn: Lorenz, U.S. pat. **2,701,225** (1955 to Bayer); Barthel *et al., J. Am. Chem. Soc.* **76,** 4186 (1954); Lorenz *et al., ibid.* **77,** 2554 (1955). Activity: R. L. Metcalf *et al., J. Econ. Entomol.* **52,** 44 (1959). Identification of metabolites: C. Lange, *Z. Chem.* **20,** 446 (1980). Toxicity study: T. B. Gaines, *Toxicol. Appl. Pharmacol.* **14,** 515 (1969). Review of carcinogenic risk: *IARC Monographs* **30,** 207-231 (1983).

White crystals, mp 83-84°. d$_4^{20}$ 1.73. n$_D^{20}$ 1.3439. Vapor pressure at 20°: 7.8 × 10^{-6} mm Hg. Soly at 25°: water 15.4 g/100 ml; chloroform 75 g/100 mg; ether 17 g/100 ml; benzene 15.2 g/100 ml. Very slightly sol in pentane and hexane. Dec by alkali. LD$_{50}$ in male, female rats (mg/kg): 630, 560 orally (Gaines).

Butanoic acid ester, $C_8H_{14}Cl_3O_5P$, *2,2,2-trichloro-1-(dimethoxyphosphinoyl)ethyl butyrate, butonate, T-113.* Prepn: Arthur, Casida, *J. Agr. Food Chem.* **6,** 360 (1958); *eidem,* U.S. pat. **2,911,435** (1959 to Wisconsin Alumni Res. Found.). Mode of action studies: Knowles, Casida, *J. Agr. Food Chem.* **14,** 566 (1966). Liquid. bp$_{0.5}$ 129°.

USE: Insecticide.

THERAP CAT (VET): Anthelmintic (nematodes); ectoparasiticide.

9754. Trichlormethiazide. *6-Chloro-3-(dichloromethyl)-3,4-dihydro-2H-1,2,4-benzothiadiazine-7-sulfonamide 1,1-dioxide;* 6-chloro-3-dichloromethyl-7-sulfamyl-3,4-dihydro-1,2,4-benzothiadiazine 1,1-dioxide; 3-dichloromethyl-6-chloro-7-sulfamyl-3,4-dihydro-1,2,4-benzothiadiazine 1,1-dioxide; 3-dichloromethylhydrochlorothiazide; hydrotrichlorothiazide; trichloromethiazide; Achletin; Anatran; Anistadin; Aponorin; Carvacron; Diurese; Esmarin; Fluitran; Flutra; Intromene; Kubacron; Metahydrin; Naqua; Salurin; Tachionin; Tolcasone; Triflumen. $C_8H_8Cl_3N_3O_4S_2$; mol wt 380.66. C 25.24%, H 2.12%, Cl 27.94%, N 11.04%, O 16.81%, S 16.85%. Prepn: deStevens *et al., Experientia* **16,** 113 (1960); Sherlock *et al., ibid.* 184. Toxicity: E. I. Goldenthal, *Toxicol. Appl. Pharmacol.* **18,** 185 (1971).

Crystals from methanol + acetone + water, dec 266-273°, also reported as 248-250°. Soly (mg/ml) at 25°: water 0.8; ethanol 21; methanol 60. LD$_{50}$ orally in rats: > 20000 mg/kg (Goldenthal).

THERAP CAT: Diuretic; antihypertensive.

THERAP CAT (VET): Diuretic.

9755. Trichloroacetaldehyde. Chloral; anhydr chloral. C_2HCl_3O; mol wt 147.39. C 16.30%, H 0.68%, Cl 72.16%, O 10.86%. Cl$_3$CCHO. Prepn by chlorinating alcohol, treating with H$_2$SO$_4$ and then distilling: Liebig, *Ann.* **1,** 189 (1832); Personne, *J. Pharm. Chim.* **[4] 10,** 350 (1869); **11,** 205 (1870); Brochet, *Bull. Soc. Chim. France* [3] **17,** 228 (1897); *Ann. Chim.* [7] **10,** 332 (1897); Trillat, *Bull. Soc. Chim. France* [3] **17,** 230 (1897); Besson, *Ger. pat.* **133,021** (1902); *Chem. Zentr.* **1902,** II, 553; Ohse, **Ger.** pat. **734,723** (1943), *C.A.* **38,** 3671 (1944). Prepn by chlorination of a mixture of alcohol and acetaldehyde: Société d'Electrochimie, **Fr.** pat. **612,904;** *Chimie et Industrie* **21,** 567 (1929). Prepn from chloral hydrate by azeotropic distn: Mahoney, Pierson, U.S. pat. **2,584,036** (1952 to Merck & Co.). From hypochlorous acid and trichloroethylene: Stevens *et al.,* U.S. pat. **2,759,978** (1956 to Columbia-Southern). In the laboratory anhydr chloral may be quickly obtained by shaking pharmaceutical grade chloral hydrate with concd H$_2$SO$_4$, separating the two layers, and distilling: *cf.* Gattermann, Wieland, *Praxis des Organischen Chemikers* (de Gruyter, Berlin, 40th ed., 1961) p 334. *See also* Chloral Hydrate.

Oily liquid. Pungent, irritating odor; d$_4^{20}$ 1.510; d$_4^{25}$ 1.5050. mp −57.5°. bp$_{760}$ 97.8°. n$_D^{20}$ 1.45572; n$_{He}^{21.4}$ 1.45412. Freely sol in water forming chloral hydrate. Sol in alcohol forming chloral alcoholate; sol in ether. Polymerizes under the influence of light and in presence of sulfuric acid forming a white solid trimer called *metachloral.*

Caution: This substance may be habit forming and is listed in the U.S. Code of Federal Regulations, Title 21 Part 329.1 (1995).

USE: Manuf chloral hydrate, DDT.

9756. Trichloroacetic Acid. TCA. $C_2HCl_3O_2$; mol wt 163.39. C 14.70%, H 0.62%, Cl 65.10%, O 19.58%. CCl$_3$-COOH. Prepd by oxidation of chloral hydrate with nitric acid: Parkes, Hollingshead, *Chem. & Ind. (London)* **1954,** 222. Manuf by chlorination of acetic acid: Eaker, U.S. pat. **2,832,803** (1958 to Monsanto). Toxicity study: G. W. Bailey, J. L. White, *Residue Rev.* **10,** 97 (1965).

Very deliquesc crystals; slight characteristic odor. d$_4^{61}$ 1.629; mp 57-58°; bp 196-197°. Sol in 0.1 part water; very sol in alcohol, ether; dec by heating with caustic alkalies into chloroform and alkali carbonate. Its aq soln is very acid. pH of 0.1 molar aq soln 1.2. *Keep tightly closed in a cool place.* Storage of trichloroacetic acid solns in water of less than 30% strength is not recommended. Decompn products are chloroform, hydrochloric acid, carbon dioxide and carbon monoxide. LD$_{50}$ orally in rats: 5000 mg/kg (Bailey, White).

Ethyl ester, $C_4H_5Cl_3O_2$, *ethyl trichloroacetate.* Clear liq; odor resembling menthol. d$_4^{20}$ 1.383. bp 168°. n$_D^{20}$ 1.4507. Insol in water. Misc with alcohol, ether.

Sodium salt, $C_2Cl_3NaO_2$, *sodium trichloroacetate, Konesta, Varitox.* Yellow deliquesc powder, mp > 300°. Soly in water at 25°: 1.2 kg/l. Sol in ethanol.

Caution: Very corrosive! When sufficient penetration has occurred, the parts should be irrigated with sodium carbonate soln.

USE: As a decalcifier and fixative in microscopy; also as a precipitant of protein. As herbicide.

THERAP CAT: Caustic.
THERAP CAT (VET): Caustic, vesicant.

9757. Trichloroacetonitrile. Trichloromethylnitrile; Tritox. C_2Cl_3N; mol wt 144.39. C 16.64%, Cl 73.66%, N 9.70%. CCl_3CN. Prepd from ethyl trichloroacetate and aq ammonia, and physical and thermodynamic properties: Davies, Jenkin, *J. Chem. Soc.* **1954**, 2374; by action of phosphorus pentoxide on trichloroacetamide: Carpenter, *J. Org. Chem.* **27**, 2085 (1962). Manuf by reaction of methylnitrile, HCl and chlorine gas: Käbisch, U.S. pat. **2,745,868** (1956 to Degussa). Toxicity study: H. F. Smyth *et al.*, *Am. Ind. Hyg. Assoc. J.* **23**, 95 (1962).

Liquid, bp_{760} 85.7°; d_4^{25} 1.4403, d_4^{35} 1.4223; $n_D^{20.0}$ 1.4409, $n_D^{27.0}$ 1.4375. LD_{50} orally in rats: 0.25 g/kg (Smyth).

Caution: A strong irritant to eyes, skin.
USE: Insecticide.

9758. 2,4,6-Trichloroanisole. Tyrene. $C_7H_5Cl_3O$; mol wt 211.47. C 39.76%, H 2.38%, Cl 50.29%, O 7.57%. Prepd by reaction of 2,4,6-trichlorophenol with dimethyl sulfate: Kohn, Heller, *Monatsh.* **46**, 91 (1925).

Monoclinic needles from alc, mp 60°, $bp_{738.2}$ 240°, bp_{28} 132°. Faint odor similar to that of acetophenone. Sublimes slowly at room temp. Volatile with steam. Practically insol in water. Sol in methanol, dioxane, benzene, cyclohexanone.

USE: Formerly as a dye assistant for polyester fibers.

9759. 1,2,3-Trichlorobenzene. *vic*-Trichlorobenzene. $C_6H_3Cl_3$; mol wt 181.45. C 39.72%, H 1.67%, Cl 58.62%. Prepared from 3,4,5-trichloroaniline by diazotization: Cohen, Hartley, *J. Chem. Soc.* **87**, 1365 (1905); Holleman, *Rec. Trav. Chim.* **37**, 196 (1918); from 2,3,4-trichloroaniline by diazotization: Dadien *et al.*, *Monatsh.* **61**, 431 (1932); from 2,3,4-trichloroaniline by treatment with ethyl nitrite: Beilstein, Kurbatow, *Ann.* **192**, 234 (1878).

Platelets from alc, d 1.69. mp 52.6°. bp 221°. n_D^{19} 1.5776. Flash pt 113° (235.4°F). Volatile with steam. Insol in water. Sparingly sol in alc. Freely sol in benzene, carbon disulfide.

Caution: Irritating to eyes, mucous membranes.

USE: A commercial grade (mixture of isomeric trichlorobenzenes) is used to combat termites.

9760. 1,2,4-Trichlorobenzene. *unsym*-Trichlorobenzene. $C_6H_3Cl_3$; mol wt 181.45. C 39.72%, H 1.67%, Cl 58.62%. Prepd from 2,4-dichloroaniline or 2,5-dichloroaniline or 3,4-dichloroaniline by diazotization and treatment with Cu_2Cl_2: Beilstein, Kurbatow, *Ann.* **192**, 230 (1878); van der Lande, *Rec. Trav. Chim.* **51**, 104, 110 (1932); from 1,3-diaminobenzene by tetrazotization and treatment with Cu_2Cl_2: Cohn, Fischer, *Monatsh.* **21**, 278 (1900).

Liquid; mp 17°; d_{25}^{25} 1.4634; bp 213°; n_D^{25} 1.5524. Flash pt 110° (230°F). Volatile with steam. Insol in water. Sparingly sol in alc. Miscible with ether, benzene, petr ether, carbon disulfide.

9761. 1,3,5-Trichlorobenzene. *sym*-Trichlorobenzene. $C_6H_3Cl_3$; mol wt 181.45. C 39.72%, H 1.67%, Cl 58.62%. Prepd from 2,4,6-trichloraniline by diazotization and treatment with alcohol: Jackson, Lamar, *Am. Chem. J.* **18**, 667 (1896); Backer, van der Baan, *Rec. Trav. Chim.* **56**, 1177 (1937).

Crystals; mp 63.4°; bp 208.4°. n_D^{19} 1.5662. Flash pt 107° (224.6°F). Volatile with steam. Insol in water. Sparingly sol in alc. Freely sol in ether, benzene, petr ether, carbon disulfide, glacial acetic acid.

9762. α,α,β-Trichloro-*n*-butyraldehyde. *2,2,3-Trichlorobutanal;* butylchloral; anhydr butylchloral; butyrchloral; crotonchloral. $C_4H_5Cl_3O$; mol wt 175.44. C 27.38%, H 2.87%, Cl 60.62%, O 9.12%. $CH_3CHClCCl_2CHO$. Prepd by the action of chlorine on acetaldehyde: Krämer, Pinner, *Ber.* **3**, 883 (1870); Pinner, *Ann.* **179**, 24 (1875); on paraldehyde: *idem;* on crotonaldehyde after saturation with gaseous HCl: High, U.S. pat. **2,280,290** (1942); Brown, Plump, U.S. pat. **2,351,000** (1944). From crotonaldehyde and chlorine: Ropp *et al.*, *Org. Syn.* coll. vol. IV, 130 (1963). *See also* Butylchloral Hydrate.

Oily liquid. Pungent, disagreeable odor; d_4^{20} 1.3956; bp_{760} 164.5-165.5°; bp_{25} 57-60°. n_D^{20} 1.47554. Freely sol in water forming butylchloral hydrate; sol in alcohol forming an alcoholate; sol in ether. Polymerizes.

9763. 2,3,6-Trichloro-*p*-cresol. *2,3,6-Trichloro-4-methylphenol;* 2,3,5-trichloro-4-hydroxytoluene; 2,3,5-trichloro-4-hydroxy-1-methylbenzene. Prepared by chlorination of *p*-cresol: Zincke *et al.*, *Ann.* **328**, 268 (1903).

Long needles from glacial acetic acid or petr ether. mp 66-67°. Slightly sol in water; freely sol in the usual organic solvents, and in dil alkali and soda solns.

9764. 2,4,6-Trichloro-*m*-cresol. *2,4,6-Trichloro-3-methylphenol;* 2,4,6-trichloro-3-hydroxytoluene. $C_7H_5Cl_3O$; mol wt 211.47. C 39.76%, H 2.38%, Cl 50.29%, O 7.57%. Prepd by chlorination of *m*-cresol, chlorination of thymol in the presence of iron, or by action of concd sulfuric acid on 2,4,4-trichloro-3-methyl-6-isopropyl-$\Delta^{2,5}$-cyclohexadienone: Crowther, McCombie, *J. Chem. Soc.* **103**, 536 (1913).

Needles or plates from water or petr ether. mp 45-47°; bp 265°, bp_{14} 142-144°. Volatile with steam. Slightly sol in water, benzene, petr ether, glacial acetic acid. Freely sol in methanol, alc, chloroform, xylene, ether and in alkali solns.

9765. 4,5,6-Trichloro-*o*-cresol. *2,3,4-Trichloro-6-methylphenol;* 3,4,5-trichloro-2-hydroxytoluene; 1-methyl-2-hydroxy-3,4,5-trichlorobenzene. Prepd by chlorination of 1-methyl-2-hydroxy-4-chlorobenzene: Kohn, Syreia, *J. Am. Chem. Soc.* **70**, 3950 (1948).

Needles from petr ether, mp 77°. Slightly soluble in water; freely sol in the customary organic solvents and in caustic alkali solns.

9766. 1,1,1-Trichloroethane. Methylchloroform; Chlorothene. $C_2H_3Cl_3$; mol wt 133.40. C 18.01%, H 2.27%, Cl 79.73%. CH_3CCl_3. Prepd by the action of chlorine on 1,1-

dichloroethane: Sutton, *Proc. Roy. Soc.* **A133,** 673 (1931); by the catalytic addition of HCl to 1,1-dichloroethylene: **Ger.** pat. **523,436** (1931); **U.S.** pat. **2,209,000** (Nutting, Huscher, 1940). Review of mfg processes: Faith, Keyes & Clark's *Industrial Chemicals,* F. A. Lowenheim, M. K. Moran, Eds. (Wiley-Interscience, New York, 4th ed., 1975) pp 836-843.

Liquid. Nonflammable. mp $-32.5°$. d_4^{20} 1.3376. bp_{760} 74.1°. n_D^{20} 1.43838. Insoluble in water. Absorbs some water. Sol in acetone, benzene, carbon tetrachloride, methanol, ether.

Caution: Potential symptoms of overexposure are headache, lassitude, CNS depression, poor equilibrium; irritation of eyes; dermatitis; cardiac arrhythmias. *See NIOSH Pocket Guide to Chemical Hazards* (DHHS/NIOSH 90-117, 1990) p 148.

USE: In cold type metal cleaning, also in cleaning plastic molds.

9767. 1,1,2-Trichloroethane. Vinyl trichloride. C_2H_3-Cl_3; mol wt 133.40. C 18.01%, H 2.27%, Cl 79.73%. CH_2-$ClCHCl_2$. Prepd by catalytic chlorination of ethane or ethylene: Joseph, **U.S.** pat. **2,752,401** and Pye, **U.S.** pat. **2,752,402** (both 1956 to Dow); Reynolds, **U.S.** pat. **2,783,-286** (1957 to Olin Mathieson). Toxicity data: H. F. Smyth *et al., Am. Ind. Hyg. Assoc. J.* **30,** 470 (1969).

Nonflammable liquid; pleasant odor; d_4^{20} 1.4416. mp $-35°$. bp 113-114°. n_D^{20} 1.4711. Insol in water. Misc with alcohol, ether, and many other organic liquids. LD_{50} orally in rats: 0.58 ml/kg (Smyth).

Caution: Potential symptoms of overexposure are irritation of nose and eyes; central nervous system depression; liver and kidney damage. Potential occupational carcinogen. *See NIOSH Pocket Guide to Chemical Hazards* (DHHS/NIOSH 90-117, 1990) p 216.

USE: Solvent for fats, waxes, natural resins, alkaloids.

9768. 2,2,2-Trichloroethanol. Trichloroethyl alcohol. $C_2H_3Cl_3O$; mol wt 149.40. C 16.08%, H 2.02%, Cl 71.19%, O 10.71%. CCl_3CH_2OH. Prepd by reduction of the corresponding ester, acid chloride, or acid with lithium aluminum hydride: Sroog *et al., J. Am. Chem. Soc.* **71,** 1710 (1949). Manufacture by reduction of chloral hydrate with an amine borane: Chamberlain, Schechter, **U.S.** pat. **2,898,379** (1959 to Callery Chem.).

Hygroscopic liquid, ethereal odor. At low temps it crystallizes in rhombic tablets. mp at 18°; bp 151-153°; d_{20}^{20} 1.55. Sol in about 12 parts water; miscible with alcohol or ether. pH of aq soln is 5-6, but on prolonged contact with water some free acid is formed. *Keep well closed and protected from light.* LD_{50} orally in rats: 600 mg/kg, *Handbook of Toxicology* **vol. 1,** W. S. Spector, Ed. (Saunders, Philadelphia, 1955) pp 302-303.

THERAP CAT: Sedative; hypnotic.

9769. Trichloroethylene. *Trichloroethene;* ethinyl trichloride; Tri-Clene; Trilene; Trichloren; Algylen; Trimar; Triline; Trethylene; Westrosol; Chlorylen; Gemalgene; Germalgene. C_2HCl_3; mol wt 131.39. C 18.28%, H 0.77%, Cl 80.95%. $CCl_2=CHCl$. Prepn from *sym*-tetrachloroethane by elimination of HCl (by boiling with lime): **Ger.** pat. **171,900.** By passing tetrachloroethane vapor over $CaCl_2$ catalyst at 300°: **Ger.** pat. **263,457;** without catalyst at 450-470°: **Brit.** pat. **575,530** (1946 to du Pont). Physical properties: E. W. McGovern, *Ind. Eng. Chem.* **35,** 1230 (1943); S. A. Mumford, J. W. C. Phillips, *J. Chem. Soc.* **1950,** 75. Review of mfg processes: S. A. Miller, *Chem. Process Eng.* **47,** 268 (1966); Faith, Keyes & Clark's *Industrial Chemicals,* F. A. Lowenheim, M. K. Moran, Eds. (Wiley-Interscience, New York, 4th ed., 1975) pp 844-848. Toxicity data: H. S. Smyth *et al., Am. Ind. Hyg. Assoc. J.* **30,** 470 (1969). Found to induce hepatocellular carcinomas in National Cancer Institute tests on mice: *Chem. & Eng. News* **54,** 4 (Apr. 5, 1976). Review of carcinogenicity studies: *IARC Monographs* **20,** 545-572 (1979). *See also:* B. L. Van Duuren, *Environ. Res.* **49,** 333 (1989). Review of use as anesthetic: J. V. Farman, *Brit. J. Anaesthesia* **53,** Suppl. 3, 3S-9S (1981). Review of metabolism, toxicity and carcinogenicity: J. V. Bruckner *et al., Crit. Rev. Toxicol.* **20,** 31-50 (1989).

Nonflammable, mobile liquid. Characteristic odor resembling that of chloroform. *Use with adequate ventilation.* $d_4^{?}$

1.4904; d_4^{15} 1.4695; d_4^{20} 1.4642; d_4^{25} 1.4559. Vapor density: 4.53 (air = 1.00). mp $-84.8°$. bp_{760} 86.9°; bp_{400} 67.0°; bp_{200} 48.0°; bp_{100} 31.4°; bp_{60} 20.0°; bp_{20} $-1.0°$; bp_5 $-22.8°$; $bp_{1.0}$ $-43.8°$; n_D^{17} 1.47914; n_D^{20} 1.4775; n_D^{25} 1.45560. Soly in water (25°): 0.11 g/100 g. Misc with ether, alcohol, chloroform. Dissolves most fixed and volatile oils. Slowly dec (with formn of HCl) by light in the presence of moisture. Preserve in sealed, light-resistant containers; avoid prolonged exposure to excessive heat. Medicinal trichloroethylene may contain thymol as a preservative; industrial grades may contain stabilizers such as triethylamine. LD_{50} orally in rats: 4.92 ml/kg; LC (4 hrs) in rats: 8000 ppm (Smyth).

Caution: Potential symptoms of overexposure are headache, vertigo; visual disturbance, tremors, somnolence, nausea and vomiting; irritation of eyes; dermatitis; cardiac arrhythmias, paresthesia. Potential occupational carcinogen. *See NIOSH Pocket Guide to Chemical Hazards* (DHHS/NIOSH 90-117, 1990) p 216. *See also* Patty's *Industrial Hygiene and Toxicology* **vol. 2B,** G. D. Clayton, F. E. Clayton, Eds. (Wiley-Interscience, 3rd ed., 1982) pp 3553-3560; *Clinical Toxicology of Commercial Products,* R. E. Gosselin *et al.,* Eds. (Williams & Wilkins, Baltimore, 5th ed., 1984) Section II, pp 165-166.

USE: Solvent for fats, waxes, resins, oils, rubber, paints, and varnishes. Solvent for cellulose esters and ethers. Used for solvent extraction in many industries. In degreasing, in dry cleaning. In the manuf of organic chemicals, pharmaceuticals, such as chloroacetic acid.

THERAP CAT: Anesthetic (inhalation).

THERAP CAT (VET): Anesthetic (inhalation).

9770. Trichlorofluoromethane. Trichloromonofluoromethane; fluorotrichloromethane; fluorocarbon 11; FC 11; Freon 11; Frigen 11; Arcton 11. CCl_3F; mol wt 137.37. C 8.74%, Cl 77.43%, F 13.83%. Prepn: Henne, *Organic Reactions* **2,** 64 (1944). Manuf: Faith, Keyes & Clark's *Industrial Chemicals,* F. A. Lowenheim, M. K. Moran, Eds. (Wiley-Interscience, New York, 4th ed., 1975) pp 325-330.

Liquid at temps below 23.7°. Faint ethereal odor. Nonflammable. $d_4^{17.2}$ 1.494; d_{gas}^{25} 5.04 (air = 1). mp $-111°$. bp_{760} 23.7°; bp_{400} $+6.8°$; bp_{200} $-9.1°$; bp_{100} $-23.0°$; bp_{60} $-32.3°$; bp_{40} $-39.0°$; bp_{20} $-49.7°$; bp_{10} $-59.0°$; bp_5 $-67.6°$; $bp_{1.0}$ $-84.3°$. Crit temp 198°; crit press. 43.2 atm (635 lb/sq inch, abs). $n_D^{18.5}$ 1.3865. Dipole moment 0.45. Practically insol in water. Sol in alcohol, ether, other, organic solvents.

Caution: Potential symptoms of overexposure are incoordination, tremors; dermatitis; frostbite; cardiac arrhythmias, cardiac arrest. *See NIOSH Pocket Guide to Chemical Hazards* (DHHS/NIOSH 90-117, 1990) p 116. *See also* Patty's *Industrial Hygiene and Toxicology* **vol. 2B,** G. D. Clayton, F. E. Clayton, Eds. (Wiley-Interscience, New York, 3rd ed., 1981) p 3073-3091.

Note: Consult latest Government regulations on use as aerosol propellant.

USE: In refrigeration machinery requiring a refrigerant effective at negative pressures. As aerosol propellant.

9771. 3,4,6-Trichloro-2-nitrophenol. 2-Nitro-3,4,6-trichlorophenol; 2,4,5-trichloro-6-nitrophenol; Dowlap. C_6H_2-Cl_3NO_3; mol wt 242.44. C 29.72%, H 0.83%, Cl 43.87%, N 5.78%, O 19.80%. Prepd by dissolving 2,4,5-trichlorophenol in glacial acetic acid and treating with concd nitric acid: Kohn, Fink, *Monatsh.* **58,** 73 (1931); Harrison *et al., J. Chem. Soc.* **1943,** 235.

Pale yellow crystals from petr ether, mp 92-93°.

USE: To combat the sea lamprey, an eel-like fish which attacks trout, especially in the Great Lakes region.

9772. 2,4,5-Trichlorophenol. Collunosol; Dowicide 2. $C_6H_3Cl_3O$; mol wt 197.45. C 36.50%, H 1.53%, Cl 53.87%,

O 8.10%. Prepd by treating 1,2,4,5-tetrachlorobenzene with methanolic NaOH in autoclave at 160° for several hrs: Harrison *et al., J. Chem. Soc.* **1943**, 235; Agfa, **Ger.** pat. **411,052** (1925); *Chem. Zentr.* **1925, I,** 2411. Toxicity data: Deichmann, *Fed. Proc.* **2,** 76 (1943).

Needles from alcohol or ligroin. Strong phenolic odor. mp 67°. Sublimes. bp_{746} 248°. bp_{760} 253°. Weak monobasic acid. K at 25° = 4.3 × 10⁻⁸. Soly (g/100 g of solvent at 25°): acetone 615; benzene 163; carbon tetrachloride 51; ether 525; denatured alcohol formula 30, 525; methanol 615; liquid petrolatum (at 50°) 56; soybean oil 79; toluene 122; water < 0.2. LD_{50} orally in rats: 0.82 g/kg (Deichmann).

Sodium salt sesquihydrate, *Dowicide B.* Flakes [prepd according to **U.S.** pat. **1,991,329** (1935 to Dow)]. Solubility (g/100 g solvent at 25°): acetone 163; denatured alcohol formula 30, 186; ethylene glycol 33; methanol 241; water 113. pH of satd aq soln 11.0-13.0.

USE: Fungicide, bactericide.

9773. 2,4,6-Trichlorophenol. Dowicide 2S; Omal. $C_6H_3Cl_3O$; mol wt 197.45. C 36.50%, H 1.53%, Cl 53.87%, O 8.10%. Prepd by direct chlorination of phenol: Tiessens, *Rec. Trav. Chim.* **50,** 115 (1931); Chulkov *et al., Org. Chem. Ind. USSR* **3,** 97 (1937); *Chem. Zentr.* **1938, I,** 1419; *C.A.* **31,** 4967 (1937). Prepn of sodium salt monohydrate: Hunter, Seyfried, *J. Am. Chem. Soc.* **43,** 154 (1921).

Crystals from ligroin. Strong phenolic odor. Volatile with steam, but not from alkaline soln. d 1.4901. mp 69°. bp_{760} 246°. Soly (g/100 g of solvent): Acetone 525; benzene 113; carbon tetrachloride 37; diacetone alcohol 335; ether 354; denatured alcohol formula 30, 400; methanol 525; pine oil 163; Stoddard solvent 16; toluene 100; turpentine 37; water < 0.1.

Sodium salt monohydrate, flaky crystals. Freely sol in water, alcohol, ether, acetone. pH of satd aq soln 11.0-13.0.

Note: 2,4,6-Trichlorophenol may reasonably be anticipated to be a carcinogen: *Seventh Annual Report on Carcinogens* (PB95-109781, 1994) p 392.

USE: Fungicide, bactericide, preservative.

9774. 1,1,1-Trichloro-2-propanol. 1,1,1-Trichloroisopropyl alcohol; trichloroisopropanol; Isopral. $C_3H_5Cl_3O$; mol wt 163.43. C 22.05%, H 3.08%, Cl 65.08%, O 9.79%. $CCl_3CH(OH)CH_3$. Prepd by reaction of chloral and methylmagnesium bromide: Kharasch *et al., J. Am. Chem. Soc.* **63,** 2305 (1941). Toxicity study: Burtner, Lehmann, *J. Pharmacol. Exp. Ther.* **63,** 183 (1938).

Monoclinic crystals; camphor-like odor; pungent taste; mp 50°; bp 161-162°. Sublimes at ordinary temp. Sol in about 35 parts water; freely sol in alcohol or ether. *Keep well closed in a cool place.* LD_{50} orally in rats: 1 g/kg (Burtner, Lehmann).

9775. 3′,4′,5-Trichlorosalicylanilide. 3,4-Dichloroanilide of 5-chlorosalicylic acid; 5-chlorosalicylic acid 3′,4′-dichloroanilide; Anobial. $C_{13}H_8Cl_3NO_2$; mol wt 316.57. C 49.32%, H 2.55%, Cl 33.60%, N 4.42%, O 10.11%. Prepn: Bindler, Model, **U.S.** pat. **2,703,332** (1955 to Geigy).

Crystals from chlorobenzene mp 246-248°. Forms a water sol sodium salt.

USE: As skin antiseptic and deodorant in soaps and cosmetics.

9776. Trichlorosilane. Trichloromonosilane; silicochloroform. Cl_3HSi; mol wt 135.45. Cl 78.52%, H 0.74%, Si 20.73%. $SiHCl_3$. Prepd from Si and HCl or from SiH_4 and HCl in presence of $AlCl_3$: Gattermann, *Ber.* **22,** 190 (1889); Ruff, Albert, *Ber.* **38,** 2226 (1905); Stock, Zeidler, *Ber.* **56,** 986 (1923); Schenk in *Handbook of Preparative Inorganic Chemistry* **vol. 1,** G. Brauer, Ed. (Academic Press, New York, 2nd ed., 1963) pp 691-692. Toxicity study: H. F. Smyth *et al., J. Ind. Hyg. Toxicol.* **31,** 60 (1949).

Volatile, mobile liquid. Fumes in air. Supports combustion. d_0^0 1.3830; d_2^{20} 1.3417; d_4^{25} 1.3313. mp −126.5°. bp_{760} 31.8°, also reported as 36.5°. Viscosity in cp: 0.397 at 0°; 0.332 at 20°; 0.316 at 25°. n_D^{20} 1.4020; n_D^{25} 1.3983. Dec by water. Sol in benzene, carbon disulfide, chloroform, carbon tetrachloride. Dipole moment 0.97. LD_{50} orally in rats: 1.03 g/kg (Smyth).

USE: In organic synthesis.

9777. 2,2′,2″-Trichlorotriethylamine. *2-Chloro-N,N-bis(2-chloroethyl)ethanamine;* tris(β-chloroethyl)amine; RA_2; Nitrogen Lost; HN3. $C_6H_{12}Cl_3N$; mol wt 204.53. C 35.24%, H 5.91%, Cl 52.00%, N 6.85%. $N(CH_2CH_2Cl)_3$. Prepd by the action of thionyl chloride on triethanolamine: Ward, *J. Am. Chem. Soc.* **57,** 914 (1935); **U.S.** pat. **2,072,-348** (1937 to Hercules Powder). Improved procedure: Contardi, Dumontel, *Chim. Ind. (Milan)* **29,** 169 (1947), *C.A.* **42,** 5849 (1948); Witten in Kirk-Othmer, *Encyclopedia of Chemical Technology* **vol. 7** (Interscience, New York, 1951) p 130; Wilson, Tishler, *J. Am. Chem. Soc.* **73,** 3635 (1951).

Mobile liquid. Faint odor of fish + soap. *Vesicant, necrotizing irritant. Never use without appropriate gas mask.* Volatility at 25° = 0.120 mg/l. d_4^{25} 1.2347, mp −4°. bp_{15} 144°. n_D^{25} 1.4925. Very slightly sol in water. Miscible with dimethylformamide, carbon disulfide, carbon tetrachloride, many other organic solvents and oils. The undiluted liq dec on standing and forms polymeric quaternary ammonium salts which are insol in the free base.

Hydrochloride, $C_6H_{12}Cl_3N.HCl,$ *trimustine, trichlormethine, Trillekamin.* Crystals, mp 130-131°. Freely sol in water, sol in alcohol.

THERAP CAT: Hydrochloride as antineoplastic.

9778. Trichlorourethan. *2,2,2-Trichloroethanol carbamate;* trichloroethyl urethan; 2,2,2-trichloroethyl carbamate; carbamic acid trichloroethyl ester. $C_3H_4Cl_3NO_2$; mol wt 192.43. C 18.73%, H 2.10%, Cl 55.27%, N 7.28%, O 16.63%. $NH_2COOCH_2CCl_3$. Prepd from 2,2,2-trichloroethyl alcohol and carbamoyl chloride in ether: Willstätter, Duisberg, *Ber.* **56,** 2283 (1923).

Needles from petr ether, mp 64-65°. Soluble in about 100 parts of water. Very sol in chloroform, alcohol, ether. Sol in benzene. Sparingly sol in petr ether.

9779. Trichodermin. *12,13-Epoxytrichothec-9-en-4-ol acetate;* WG-696. $C_{17}H_{24}O_4$; mol wt 292.38. C 69.84%, H 8.27%, O 21.89%. Antifungal metabolite from *Trichoderma viride* ND8: **Neth. pat. Appl. 302,527** (1964 to Loevens); *C.A.* **62,** 1050f (1965); also isolated from *Myrothecium roridum.* Structure: Godtfredsen, Vangedal, *Proc. Chem. Soc.* **1964,** 188; Gutzwiller *et al., Helv. Chim. Acta* **47,** 2234 (1964); Godtfredsen, Vangedal, *Acta Chem. Scand.* **19,** 1088 (1965); Abrahamsson, Nilsson, *ibid.* **20,** 1044 (1966). Total synthesis: Colvin *et al., Chem. Commun.* **1971,** 858; *eidem, J. Chem. Soc. Perkin Trans. I* **1973,** 1989. Activity studies: Yamamoto *et al., Takeda Kenkyusho Nempo* **28,** 69 (1969), *C.A.* **72,** 76058g (1970).

Crystals from pentane at −70°. mp 46°, 58-60° (Colvin). bp$_{0.05}$ 110-112°. $[\alpha]_D^{20}$ −11° (chloroform). uv max (ethanol): 205 nm (ϵ 2400). Sparingly soluble in water; soluble in all common organic solvents. Alkaline hydrolysis yields *trichodermol (Roridan C)*, crystals, mp 124-125°, *see also* verrucarrins. LD$_{50}$ in mice (mg/kg): 500-1000 s.c.; > 1000 orally (Neth. pat.).

9780. Trichosanthin. GLQ223. Active component of the traditional Chinese medicine, Tian Hua Fen (Radix Trichosanthis), known since 300 A.D. and used in gynecological disorders and abortion. Basic polypeptide of 234 amino acids, M$_r$ 26,000 Da, isolated from the roots of *Trichosanthes kirilowii* Maxim, *Cucurbitaceae*. Purification, characterization, and amino acid sequence: W. Yu *et al.*, *Pure Appl. Chem.* **58**, 789 (1986); J. M. Maraganore *et al.*, *J. Biol. Chem.* **262**, 11628 (1987). Conformation: S. Kubota *et al.*, *Biochim. Biophys. Acta* **871**, 101 (1986); and homology with ricin, *q.v.: eidem, Int. J. Peptide Protein Res.* **30**, 646 (1987). Inhibition of protein synthesis: H. W. Yeung *et al.*, *ibid.* **31**, 265 (1988). Inhibition of HIV replication: M. S. McGrath *et al.*, *Proc. Nat. Acad. Sci. USA* **86**, 2844 (1989). Abortifacient activity in animals: M. C. Chang *et al.*, *Contraception* **19**, 175 (1979); S. K. Saksena *et al.*, *ibid.* **20**, 367 (1979); I. F. Lau *et al.*, *ibid.* **21**, 77 (1980). Clinical trial as abortifacient: K. F. Cheng, *Obstet. Gynecol.* **59**, 494 (1982).
Crystals. pI 9.4.
THERAP CAT: Abortifacient.

9781. Trichostatin(s). Antibiotic A-300; A-300-I. Antifungal antibiotic isolated from metabolites of *Streptomyces hygroscopicus*. It is composed of trichostatins A (major), B, the ferric chelate of trichostatin A, and C, the first glycosyl hydroxamate from a natural source. Isoln of A and B: N. Tsuji *et al.*, **Japan. Kokai 74 14,691,** *C.A.* **81,** 48547h (1974). Structural elucidation of A and B: *eidem, J. Antibiot.* **29,** 1 (1976). Isoln and structure of C: N. Tsuji, M. Kobayashi, *ibid.* **31,** 939 (1978).

Trichostatin A

Trichostatin A, C$_{17}$H$_{22}$N$_2$O$_3$, *[R-(E,E)]-7-[4-(dimethylamino)phenyl]-N-hydroxy-4,6-dimethyl-7-oxo-2,4-heptadienamide.* Crystals from ethyl acetate, mp 150-151°. $[\alpha]_D^{20.5}$ +62.8° (±1.1°) (c = 1.007 in ethanol). uv max (ethanol): 252, 265, 341 nm (E$_{1cm}^{1\%}$ 531, 582, 648). Sol in lower alcohols, sparingly sol in chloroform, ethyl acetate, acetone, benzene.
Trichostatin B, C$_{51}$H$_{63}$FeN$_6$O$_9$, *tris[7-[4-(dimethylamino)phenyl]-N-hydroxy-4,6-dimethyl-7-oxo-2,4-heptadienamidato-O^{N}O^J]iron.* Trihydrate, dark reddish purple prisms from methanol, mp 192° (dec). uv max (ethanol): 253, 277, 341, 450 nm (E$_{1cm}^{1\%}$ 624, 651, 918, 50).
Trichostatin C, C$_{23}$H$_{32}$N$_2$O$_8$, *7-[4-(dimethylamino)phenyl]-N-(β-D-glucopyranosyloxy)-4,6-dimethyl-7-oxo-2,4-heptadienamide.* Colorless prisms from methanol, mp 171-173°. $[\alpha]_D^{24}$ +50.5° (±0.9°) (c = 0.987 in methanol). uv max (methanol): 268, 344 nm (ϵ 14600; 14300).

9782. Trichothecin. *[4β(Z)]-12,13-Epoxy-4-[(1-oxo-2-butenyl)oxy]trichothec-9-en-8-one; 12,13-epoxy-4-hydroxytrichothec-9-en-8-one crotonate.* C$_{19}$H$_{24}$O$_5$; mol wt 332.40. C 68.66%, H 7.28%, O 24.07%. Mycotoxin with antibiotic activity. Produced by *Trichothecium roseum*: Freeman, Morrison, *Nature* **162**, 30 (1948). Activity: G. G. Freeman, *J.*

Gen. Microbiol. **12,** 213 (1955). Structure: Godtfredsen, Vangedal, *Proc. Chem. Soc.* **1964,** 188; Gutzwiller *et al.*, *Helv. Chim. Acta* **47,** 2234 (1964). Biogenesis: Jones, Lowe, *J. Chem. Soc.* **1960,** 3959. Biosynthetic studies: Achilladelis *et al.*, *J. Chem. Soc. Perkin Trans. I* **1972,** 1425; Machida, Nozoe, *Tetrahedron Letters* **1972,** 1969; *eidem, Tetrahedron* **28,** 5113 (1972).

Slender needles from petrol ether, mp 118°. $[\alpha]_D^{18}$ +44° (c = 1 in CHCl$_3$). uv max (hexane): 217 nm (ϵ 18000); (methanol): 215 nm (ϵ 19000). Slightly sol in water (400 mg/l at 25°). Freely sol in most organic solvents. Aq solns are stable at pH 1-10 for at least 48 hrs at 20°. At pH 12 the antifungal activity is destroyed rapidly. Aq solns at pH 7 can be maintained at 100° for at least one hour without loss of activity. LD$_{50}$ i.v. in mice: ∼300 mg/kg (Freeman).

9783. Tricine. *N-[2-Hydroxy-1,1-bis(hydroxymethyl)-ethyl]glycine; N-[tris(hydroxymethyl)methyl]glycine.* C$_6$H$_{13}$-NO$_5$; mol wt 179.17. C 40.22%, H 7.31%, N 7.82%, O 44.65%. One of the zwitterionic amino acids known as "Good" buffers; active in the pH range 6-8.5. Prepn: N. E. Good, *Arch. Biochem. Biophys.* **96,** 653 (1962); and characterization: N. E. Good *et al.*, *Biochemistry* **5,** 467 (1966). Temperature effects on pK: R. N. Roy *et al.*, *J. Am. Chem. Soc.* **95,** 8231 (1973); M. L. Soni, R. C. Kapoor, *Int. J. Quant. Chem.* **20,** 385 (1981); and on pH: R. N. Roy *et al.*, *Cryobiology* **22,** 589 (1985). Scavenger of hydroxyl radicals: M. Hicks, J. M. Gebicki, *FEBS* **199,** 92 (1986); B. Halliwell *et al.*, *Anal. Biochem.* **165,** 215 (1987). Use as buffer: R. S. Gardner, *J. Cell. Biol.* **42,** 320 (1969); R. S. Spendlove *et al.*, *Proc. Soc. Exp. Biol. Med.* **137,** 258 (1971); H. Schaegger, G. Von Jagow, *Anal. Biochem.* **166,** 368 (1987).

Crystals from alcohol/water, mp 187°. pKa$_1$ ∼2.3; pKa$_2$ (0.1M): 0°, 8.6; 20°, 8.15; 37°, 7.8. pKa$_2$ (20°): 8.15 (0.2M); 8.15 (0.01M). ΔpKa/°C −0.021. Saturated aqueous soln is 0.8M at 0°.
USE: Biological buffer.

9784. Triclabendazole. *5-Chloro-6-(2,3-dichlorophenoxy)-2-methylthio-1H-benzimidazole;* CGA-89317; Fasinex. C$_{14}$H$_9$Cl$_3$N$_2$OS; mol wt 359.66. C 46.75%, H 2.52%, Cl 29.57%, N 7.79%, O 4.45%, S 8.92%. Benzimidazole derivative effective against liver fluke. Prepn: J.-J. Gallay *et al.*, **Belg. pat. 865,872** corresp to **U.S. pat. 4,197,307** (1978, 1980 both to Ciba-Geigy). Efficacy against immature and mature *Fasciola hepatica* in sheep, goats: K. Wolff *et al.*, *Vet. Parasitol.* **13,** 145 (1983); in sheep: J. C. Boray *et al.*, *Vet. Rec.* **113,** 315 (1983). Effective also against *F. gigantica* in sheep: N. Güralp, R. Tinar, *J. Helminthol.* **58,** 113 (1984). Brief review: J. Eckert *et al.*, *Biol. Muench. Tieraerztl. Wochenschr.* **97,** 349 (1984).

Crystals, mp 175-176°.
THERAP CAT (VET): Anthelmintic (fasciola).

9785. Triclobisonium Chloride. *N,N,N',N'-Tetramethyl-N,N'-bis[1-methyl-3-(2,2,6-trimethylcyclohexyl)propyl]-1,6-*

hexanediaminium dichloride; hexamethylenebis[dimethyl[1-methyl-3-(2,2,6-trimethylcyclohexyl)propyl]ammonium chloride]; *N,N'*-bis[1-methyl-3-(2,2,6-trimethylcyclohexyl)-propyl]-*N,N'*-dimethyl-1,6-hexanediamine bis(methochloride); Ro-5-0810/1; Triburon. $C_{36}H_{74}Cl_2N_2$; mol wt 605.90. C 71.36%, H 12.31%, Cl 11.70%, N 4.62%. Obtained as the hemihydrate. Prepn: Goldberg, Teitel, U.S. pat. **3,064,052** (1962 to Hoffmann-La Roche). Clinical application: Edelson *et al., Antibiot. Ann.* **1958-1959,** 110; Robinson, Harmon, *ibid.* 113. Comprehensive description: B. C. Rudy, B. Z. Senkowski, *Anal. Profiles Drug Subs.* **2,** 507-521 (1973).

White cryst powder, mp 243-253° (dec). Sol in water, chloroform, alcohol.
THERAP CAT: Antiseptic.

9786. Triclocarban. *N-(4-Chlorophenyl)-N'-(3,4-dichlorophenyl)urea; 3,4,4'-trichlorocarbanilide;* 1-(3',4'-dichlorophenyl)-3-(4'-chlorophenyl)urea; TCC; Cutisan; Nobacter; Solubacter. $C_{13}H_9Cl_3N_2O$; mol wt 315.59. C 49.48%, H 2.87%, Cl 33.70%, N 8.88%, O 5.07%. Prepn from 3,4-dichloroaniline and 4-chlorophenyl isocyanate: Beaver *et al., J. Am. Chem. Soc.* **79,** 1236 (1957); Beaver, Stoffel, U.S. pat. **2,818,390** (1957 to Monsanto); **Brit.** pat. **769,273** (1957).

Fine white plates, mp 255.2-256°.
USE: Bacteriostat and antiseptic in soaps and other cleansing compositions.
THERAP CAT: Antiseptic, disinfectant.

9787. Triclofenol Piperazine. *2,4,5-Trichlorophenol compd with piperazine (2:1);* bis(2,4,5-trichlorophenol) piperazine; piperazine salt of bis(2,4,5-trichlorophenol); CI-416; Ranestol (obsolete). $C_{16}H_{16}Cl_6N_2O_2$; mol wt 481.03. C 39.95%, H 3.35%, Cl 44.22%, N 5.82%, O 6.65%. Prepn: Short, Elslager, U.S. pat. **2,980,681** (1961 to Parke, Davis); *J. Med. Pharm. Chem.* **5,** 642 (1962).

Crystals, mp 109-110°.
THERAP CAT: Anthelmintic (Nematodes).

9788. Triclofos. *2,2,2-Trichloroethanol dihydrogen phosphate;* trichloroethyl phosphate. $C_2H_4Cl_3O_4P$; mol wt 229.38. C 10.47%, H 1.76%, Cl 46.37%, O 27.90%, P 13.50%. Prepn: Hems *et al., Brit. Med. J.* **1,** 1834 (1962).

Monosodium salt, $C_2H_3Cl_3NaO_4P$, *Sch-10159, Trichloryl, Tricloryl, Triclos.* Sol in water. LD_{50} orally in mice: 1.4 g/kg (Hems).
THERAP CAT: Sedative, hypnotic.

9789. Triclopyr. *[(3,5,6-Trichloro-2-pyridinyl)oxy]acetic acid;* Dowco 233; Garlon; Timbrel. $C_7H_4Cl_3NO_3$; mol

wt 256.47. C 32.78%, H 1.57%, Cl 41.47%, N 5.46%, O 18.71%. Selective post-emergence herbicide. Prepn: L. D. Markley, U.S. pat. **3,862,952** (1975 to Dow). Activity: B. C. Byrd *et al., Proc. South. Weed Sci. Soc.* **28,** 251 (1975). Acute toxicity: E. E. Kenaga, *Down Earth* **35,** 25 (1979).

Fluffy solid, mp 148-150°. Vapor pressure at 25°: 1.26×10^{-6} mm Hg. pKa 2.68. Subject to photolysis. Soly in water at 25°: 440 mg/l. Soly at 25° (g/kg): acetone 989; 1-octanol 307. LD_{50} orally in rats: 713 mg/kg (Kenaga).
USE: Herbicide.

9790. Triclosan. *5-Chloro-2-(2,4-dichlorophenoxy)phenol;* 2,4,4'-trichloro-2'-hydroxydiphenyl ether; CH-3635; Aquasept; Gamophen; Irgasan DP 300; Sapoderm; Ster-Zac. $C_{12}H_7Cl_3O_2$; mol wt 289.54. C 49.78%, H 2.44%, Cl 36.73%, O 11.05%. Prepn: **Neth.** pat. **Appl. 6,401,526** corresp to E. Model, J. Bindler, U.S. pat. **3,506,720** (1964, 1970 to Geigy). Physical and bacteriostatic properties: C. A. Savage, *Drug Cosmet. Ind.* **109**(3), 36, 161 (1971). Metabolism: J. G. Black *et al., Toxicology* **3,** 33 (1975). Toxicology: F. L. Lyman, T. Furia, *Ind. Med. Surg.* **38**(2), 64 (1969), *C.A.* **71,** 89601h (1969). Use as disinfectant and textile preservative: E. Model, J. Bindler, U.S. pat. **3,629,477** (1971 to Geigy).

White to off-white crystalline powder, slight, faintly aromatic odor. mp 54-57.3°. Vapor pressure (20°C) 4×10^{-6} mm Hg. pKa 7.9. Insol in water. Readily sol in alkaline solns and many organic solvents.
USE: Bacteriostat and preservative for cosmetic and detergent prepns.
THERAP CAT: Antiseptic, disinfectant.

9791. Tricromyl. *3-Methyl-4H-1-benzopyran-4-one; 3-methylchromone;* 3-methyl-γ-benzopyrone; Cromonalgina. $C_{10}H_8O_2$; mol wt 160.17. C 74.99%, H 5.03%, O 19.98%. Based on an ancient Egyptian drug now termed *bezr el khelda.* Prepn from *o*-hydroxypropiophenone: Clerc-Bory *et al., Bull. Soc. Chim. France* **1955,** 1083; Mentzer, Meunier, **Fr.** pat. **980,785** (1951 to Lab. Franc. Chimiother.); Mentzer, U.S. pat. **2,769,015** (1956 one-half to Laroche-Navarron).

Crystals from ethanol, mp 68°. uv max (alc): 304 nm.
THERAP CAT: Antispasmodic; vasodilator (coronary).

9792. Tridecylbenzene. *1-Phenyltridecane;* Detergent Alkylate #5; Tridane. $C_{19}H_{32}$; mol wt 260.46. C 87.62%, H 12.38%. $C_6H_5(CH_2)_{12}CH_3$. Prepn: Ziegler *et al., Ann.* **511,** 13 (1934). Manufacture: Williamson, Bieneman, U.S. pat. **3,207,800** (1965 to Continental Oil).
Liquid, bp 346°, bp_{10} 188-189.5°. mp 10°. d_4^{20} 0.8550, d_4^{25} 1.8515. n_D^{20} 1.4821, n_D^{25} 1.4800. Forms stable foams in the presence of fat.
USE: In manuf of detergents and surface-active agents. Can be sulfonated.

9793. Tridemorph. *2,6-Dimethyl-4-tridecylmorpholine.* N-tridecyl-2,6-dimethylmorpholine; 2,6-dimethyl-4-tridecyltetrahydro-1,4-oxazine. $C_{19}H_{39}NO$; mol wt 297.52. C 76.70%, H 13.21%, N 4.71%, O 5.38%. Prepn: W. Sanne *et* al., **Belg.** pat. **614,214;** *eidem,* U.S. pat. **3,468,885** (1962, 1969 both to BASF); K.-H. König *et al., Angew. Chem. Int. Ed. Engl.* **4,** 336 (1965). Structure determn of two diastereomers: D. Kost, E. Gurfinkel, *J. Chromatog.* **108,** 207 (1975). Fungicidal activity: J. Kradel *et al., Proc. 5th Brit. Insectic. Fungic. Conf.,* 16 (1969); *in vitro* activity comparison of tridemorph and commercial formulation Calixin: A. Kerkenaar, A. A. Sijpesteijn, *Pest. Biochem. Physiol.* **12,** 124 (1979). Uptake and distribution in barley: E.-H. Pommer *et al., Proc. 5th Brit. Insectic. Fungic. Conf.,* 347 (1969); R. H. Waring, M. S. Wolfe, *Pestic. Sci.* **6,** 169 (1975). Behavior in soil: S. Otto, N. Drescher, *Proc. 7th Brit. Insectic. Fungic. Conf.,* 57 (1973). Metabolic fate in rats: D. R. Hawkins *et al., Pestic. Sci.* **5,** 535 (1974); of Calixin: R. H. Waring, *Xenobiotica* **4,** 717 (1974). Mode of action: A. Kerkenaar *et al., Pest. Biochem. Physiol.* **12,** 195 (1979). Comparison of toxicity of tridemorph and Calixin: J. Merkle *et al., Teratology* **29,** 259 (1984). Field trials: P. Lakshmanan, S. Mohan, *Pesticides* **22,** 27 (1988).

Oil, bp$_{0.7}$ 130-133°; bp$_{1.3}$ 139-142°. n_D^{25} 1.4568. Commercial product, *Calixin.* Reaction mixture of C$_{11}$-C$_{14}$ 4-alkyl-2,6-dimethylmorpholine homologs containing 60-70% of tridemorph.
USE: Systemic agricultural fungicide.

9794. Tridihexethyl Iodide. γ-*Cyclohexyl-N,N,N-triethyl-γ-hydroxybenzenepropanaminium iodide;* (3-cyclohexyl-3-hydroxy-3-phenylpropyl)triethylammonium iodide; 3 diethylamino-1-cyclohexyl-1-phenyl-1-propanol ethiodide; 3-diethylamino-1-phenyl-1-cyclohexyl-1-propanol ethiodide; α-(2-diethylaminoethyl)-α-phenylcyclohexanemethanol ethiodide; propethonum iodide; tridihexethide; 921 C; Claviton. C$_{21}$H$_{36}$INO; mol wt 445.43. C 56.63%, H 8.15%, I 28.49%, N 3.14%, O 3.59%. Prepn: Lobby, U.S. pat. **2,913,494** (1959 to Am. Cyanamid).

Bitter crystals, mp 179-184°. Solubility in water at 25°: 1.1 g/100 ml. Freely sol in alc, chloroform. Very slightly sol in ether. pH of a 1% aq soln 5.5-7.
Note: The name *Pathilon* previously used for the iodide is now used for *tridihexethyl chloride.*
THERAP CAT: Anticholinergic.

9795. Tridiphane. *2-(3,5-Dichlorophenyl)-2-(2,2,2-trichloroethyl)oxirane;* Dowco 356; Tandem. C$_{10}$H$_7$Cl$_5$O; mol wt 320.43. C 37.48%, H 2.20%, Cl 55.32%, O 4.99%. Prepn: L. D. Markley, E. J. Norton, Ger. pat. **2,519,073;** *eidem,* U.S. pat. **4,211,549** (1975, 1980 both to Dow). Herbicidal activity: J. A. Jagschitz, *Proc. Ann. Meet. Northeast Weed Sci. Soc.* **39,** 274 (1985); D. B. Vitolo *et al., ibid.* **40,** 272 (1986). Field trial in combination with atrazine, *q.v.:* P. J. Dryden, M. J. Watson, *Proc. 38th N.Z. Weed and Pest Control Conf.* **1985,** 191. Mechanism of action: G. L. Lamoureux, D. G. Rusness, *Pestic. Biochem. Physiol.* **26,** 323 (1986). Metabolism in mouse: J. Magdalou, B. D. Hammock, *Toxicol. Appl. Pharmacol.* **91,** 439 (1987). Toxicological studies: T. R. Hanley *et al., Fundam. Appl. Toxicol.* **8,** 179 (1987); J. A. John-Greene *et al., Toxicology* **43,** 325 (1987).

Yellow oil, n_D^{25} 1.5720. LD$_{50}$ for technical grade (89-91% pure) in male, female mice, male, female rats (mg/kg): ~1200, ~740, 1700-2300, 1500-1900 orally (Hanley).
USE: Herbicide.

9796. Trientine. *N,N′-Bis(2-aminoethyl)-1,2-ethanediamine; triethylenetetramine;* 1,8-diamino-3,6-diazaoctane; 3,6-diazaoctane-1,8-diamine; 1,4,7,10-tetraazadecane; trien; TETA; TECZA. C$_6$H$_{18}$N$_4$; mol wt 146.24. C 49.28%, H 12.41%, N 38.31%. (NH$_2$CH$_2$CH$_2$NHCH$_2$—)$_2$. Prepn: A. W. von Hoffmann, *Ber.* **23,** 3711 (1890); J. van Alphen, *Rec. Trav. Chim.* **55,** 412 (1936); G. D. Jones *et al., J. Org. Chem.* **9,** 125 (1944); of the dihydrochloride: H. B. Dixon *et al., Lancet* **1,** 853 (1972). Effects on the metabolism of copper in rats: F. W. Sunderman *et al., Toxicol. Appl. Pharmacol.* **38,** 177 (1976); H. Harders *et al., Arzneimittel-Forsch.* **30,** 254 (1980). Use of the dihydrochloride in the treatment of Wilson's disease: J. M. Walshe, *Prog. Clin. Biol. Res.* **34,** 271 (1979); R. H. Haslam *et al., Dev. Pharmacol. Ther.* **1,** 318 (1980); J. M. Walshe, *Lancet* **1,** 643 (1982). Review of chemistry, physical properties and industrial uses: R. D. Spitz, "Diamines and Higher Amines, Aliphatic" in Kirk-Othmer *Encyclopedia of Chemical Technology* **vol. 7** (Wiley-Interscience, New York, 3rd ed., 1979) pp 580-602. Orphan drug developed by Merck. *Review:* J. M. Walshe, *Orphan Drugs,* F. E. Karch, Ed. (Marcel Dekker, New York, 1982) pp 57-71.
Oily liquid, mp 12°. bp 266-267°. n_D^{20} 1.4971; d^{15} 0.9817. Flash pt 290°F (143°C). pH 14. Sol in water, alc. *Corrosive!* LD$_{50}$ orally in rats: 2.5 g/kg (Spitz).
Dihydrochloride, C$_6$H$_{18}$N$_4$.2HCl, *Cuprid, Syprine.* Hygroscopic crystals, mp 115-118°.
USE: As thermosetting resin; epoxy curing agent; lubricating oil additive; analytical reagent for Cu, Ni.
THERAP CAT: Chelating agent; Wilson's disease treatment.

9797. Trietazine. *6-Chloro-N,N,N′-triethyl-1,3,5-triazine-2,4-diamine;* 2-chloro-4-diethylamino-6-ethylamino-s-triazine; 2-ethylamino-4-diethylamino-6-chloro-s-triazine. G-27901; NC-1667; Aventox; Gesafloc; Remtal. C$_9$H$_{16}$-ClN$_5$; mol wt 229.71. C 47.06%, H 7.02%, Cl 15.43%, N 30.49%. Prepn: Pearlman, Banks, *J. Am. Chem. Soc.* **70,** 3726 (1948). Toxicity study: G. W. Bailey, J. L. White, *Residue Rev.* **10,** 97 (1965).

Crystals from propanol, mp 100-102°. Soly at 25°: 20 ppm in water; 17% in acetone; 20% in benzene; > 50% in chloroform; 10% in dioxane; 3% in ethanol. LD$_{50}$ orally in rats: 1750 mg/kg (Bailey, White).
USE: Herbicide.

9798. Triethanolamine. *2,2′,2″-Nitrilotrisethanol;* trihydroxytriethylamine; tris(hydroxyethyl)amine; triethylolamine; trolamine. C$_6$H$_{15}$NO$_3$; mol wt 149.19. C 48.30%, H 10.13%, N 9.39%, O 32.17%. N(CH$_2$CH$_2$OH)$_3$. Produced along with mono- and diethanolamine by ammonolysis of ethylene oxide. *See the refs under Ethanolamine.* Monograph: E. J. Fischer, *Triäthanolamin und andere Alkanolamine* (Heidelberg, 4th ed., 1954).
Very hygroscopic, viscous liq. Slight ammoniacal odor. Turns brown on exposure to air and light. d$_4^{20}$ 1.1242; d$_4^{60}$ 1.0985. One gallon weighs 9.37 lbs. mp 21.57°. bp$_{760}$

335.4°. *See:* McDonald *et al., J. Chem. Eng. Data* **4**, 311 (1959). Viscosity (centipoise) at 25°: 590.5; viscosity at 60°: 65.7. Strong base. K at 25° = 3.15 × 10^{-10}. pH of 0.1N aq soln 10.5. n_D^{20} 1.4852. Flash pt 365°F. Miscible with water, methanol, acetone. Soly at 25° in benzene, 4.2%; in ether, 1.6%; in carbon tetrachloride, 0.4%; in *n*-heptane, <0.1%.

Hydrochloride, crystals from ethanol, mp 177°.

Salicylate, *Mobisyl, Myoflex.*

USE: Intermediate in the manuf of surface active agents, textile specialties, waxes, polishes, herbicides, petroleum demulsifiers, toilet goods, cement additives, cutting oils. In making emulsions with mineral and vegetable oils, paraffin and waxes. Solvent for casein, shellac, dyes; manuf synthetic resins; increasing the penetration of organic liquids into wood and paper. In the production of lubricants for the textile industry. Pharmaceutic aid (alkalizer).

THERAP CAT: Analgesic.

9799. Triethylamine. *N,N-Diethylethanamine.* C_6H_{15}-N; mol wt 101.19. C 71.22%, H 14.94%, N 13.84%. $(C_2H_5)_3N$. Prepd by reaction of *N,N*-diethylacetamide with lithium aluminum hydride: Uffer, Schlittler, *Helv. Chim. Acta* **31**, 1397 (1948). Manuf by vapor phase alkylation of ammonia with ethanol: Lemon, Myerly, U.S. pat. **3,022,349** (1962 to Union Carbide). Toxicity study: H. F. Smyth *et al., Arch. Ind. Hyg. Occup. Med.* **4**, 119 (1951).

Liquid; strong ammoniacal odor; d_4^{25} 0.7255; mp -115°; bp 89-90°; n_D^{20} 1.4003. Flash pt, closed cup: 20°F (-6°C). Slightly sol in water above 18.7°; misc with alcohol, ether, also with water below 18.7°. *Keep well closed.* LD$_{50}$ orally in rats: 0.46 g/kg (Smyth).

Hydrochloride, C_6H_{15}N.HCl, crystals from alcohol, mp 253-254°, sublimes at 245°; d 1.069. Sol in 0.7 parts water; sol in alcohol, chloroform; very slightly sol in benzene. Practically insol in ether.

Caution: Potential symptoms of overexposure to triethylamine are irritation of eyes, respiratory system and skin. *See NIOSH Pocket Guide to Chemical Hazards* (DHHS/NIOSH 90-117, 1990) p 218.

USE: In the prepn of quaternary ammonium compds.

9800. Triethylammonium Formate. $C_7H_{17}NO_2$; mol wt 147.22. C 57.11%, H 11.64%, N 9.51%, O 21.74%. $(C_2H_5)_3N.HOOCH$. Prepd by neutralizing 50% formic acid with triethylamine and evaporating the resulting soln on a steambath for twelve hours at 20 mm: Alexander, Wildman, *J. Am. Chem. Soc.* **70**, 1187 (1948).

Light brown syrup. Very sol in water. Sol in alc.

USE: In syntheses employing the Leuckart reaction.

9801. Triethylenediamine. *1,4-Diazabicyclo[2.2.2]-octane;* Dabco. $C_6H_{12}N_2$; mol wt 112.17. C 64.24%, H 10.78%, N 24.97%. Prepn: Krause *et al.,* Brit. pat. **871,754** (1958 to Houdry Process).

Crystals, extremely hygroscopic. Sublimes readily at room temp. mp 158°. bp 174°. pKa$_1$ 3.0; pKa$_2$ 8.7. Soly (g/100 g at 25°) in water: 45; acetone: 13; benzene: 51; ethanol: 77; methyl ethyl ketone: 26.1.

USE: Catalyst in making urethane foams.

9802. Triethylene Glycol. *2,2'-[1,2-Ethanediylbis-(oxy)]bisethanol;* 2,2'-ethylenedioxybis(ethanol). $C_6H_{14}O_4$; mol wt 150.17. C 47.99%, H 9.40%, O 42.62%. Prepd from ethylene oxide and ethylene glycol in the presence of sulfuric acid: Matignon *et al., Bull. Soc. Chim.* [5] **1**, 1308 (1934). Manuf by forming ether-ester of HOCH$_2$COOH with glycol and then hydrogenating: Gresham, U.S. pat. **2,654,786** (1953 to du Pont). Toxicity data: Stenger *et al., Arzneimittel-Forsch.* **18**, 1536 (1968).

Colorless, hygroscopic, practically odorless liquid. d_4^{15} 1.1274; bp 285°; bp$_{14}$ 165°. mp -7.2°. Viscosity at 20°: 47.8 cp; n_D^{15} 1.4578. Misc with water, alcohol, benzene, toluene; sparingly sol in ether. Practically insol in petr ether. LD$_{50}$ in mice, rats (g/kg): 21, 15-22 orally; 7.3-9.5, 11.7 i.v. (Stenger).

USE: In various plastics to increase pliability; in air disinfection.

9803. Triethylenemelamine. *2,4,6-Tris(1-aziridinyl)-1,-3,5-triazine;* 2,4,6-tris(ethylenimino)-*s*-triazine; 2,4,6-triethylenimino-1,3,5-triazine; triethanomelamine; tretamine; triamelin; TEM; NSC-9706; Persistol. $C_9H_{12}N_6$; mol wt 204.23. C 52.93%, H 5.92%, N 41.15%. Prepd from ethylenimine and cyanuric chloride: Bestian, *Ann.* **566**, 210 (1950); Wystrach, Kaiser, U.S. pat. **2,520,619** (1950 to Am. Cyanamid); Kaiser, Schaefer, U.S. pat. **2,653,934** (1953); Wystrach *et al., J. Am. Chem. Soc.* **77**, 5915 (1955). Toxicology: F. S. Philips, J. B. Thiersch, *J. Pharmacol. Exp. Ther.* **100**, 398 (1950). Review of carcinogenic risk: *IARC Monographs* **9**, 95-105 (1975); of comparative mutagenicity: C. Ramel, *Environ. Sci. Res.* **24**, 943-976 (1981).

Minute crystals from chloroform, dec 139°. Soly (w/w) at 26°: in water 40%, chloroform 28.1%, methylene chloride 19.7%, methanol 12.5%, acetone 10.6%, dioxane 9.6%, ethanol 7.7%, benzene 5.6%, dimethyl Cellosolve 4.8%, methyl ethyl ketone 4.7%, ethyl acetate 4.5%, carbon tetrachloride 3.6%. Stable at reduced temps; ampuled aq solns stored at 4° are stable for about 3 months. At room temp, aq solns polymerize. LD$_{50}$ in mice, rats (mg/kg): 2.8, 1.0 i.p.; 15, 13 orally (Philips).

USE: Manuf of resinous products, textile finishing agents. Insect sterilant. Research tool used as positive control for mutagenicity assays.

THERAP CAT: Antineoplastic.

9804. Triethylenephosphoramide. *1,1',1''-Phosphinyl-idynetrisaziridine; tris(1-aziridinyl)phosphine oxide;* phosphoric acid triethyleneimide; aphoxide; APO; TEPA. $C_6H_{12}N_3OP$; mol wt 173.15. C 41.62%, H 6.99%, N 24.27%, O 9.24%, P 17.89%. Prepn: Bestian, *Ann.* **566**, 231 (1950). Toxicity study: Gaines, *Toxicol. Appl. Pharmacol.* **14**, 515 (1969).

Crystals, mp 41°. bp$_{23}$ 90-91°. Extremely sol in water. Very sol in alcohol, ether, acetone. LD$_{50}$ orally in male rats: 37 mg/kg (Gaines).

USE: Insect chemosterilant; in dyeing, creaseproofing and flameproofing textiles; stabilizer for polymers; in photographic emulsion hardening.

THERAP CAT: Antineoplastic.

9805. Triethylenethiophosphoramide. *1,1',1''-Phosphinothioylidynetrisaziridine; tris(1-aziridinyl)phosphine sulfide;* thiotepa; Tespamin; Tifosyl. $C_6H_{12}N_3PS$; mol wt 189.22. C 38.09%, H 6.39%, N 22.21%, P 16.37%, S 16.95%. Preparation: Kuh, Seeger, U.S. pat. **2,670,347** (1954 to Am. Cyanamid); Saijo, Endo, Japan. pat. **55 218** (1955 to Semimoto Chem.). Toxicity data: Scherf *et al., Arzneimittel-Forsch.* **20,** 1467 (1970). Review of carcinogenicity studies: *IARC Monographs* **9,** 85-94 (1975).

Crystals from pentane or ether, mp 51.5°. Soly in water at 25°: 19 g/100 ml. Freely sol in alc; sol in benzene, ether, chloroform. LD_{50} i.v. in rats: 15 mg/kg (Scherf).

Note: This substance may reasonably be anticipated to be a carcinogen: *Seventh Annual Report on Carcinogens* (PB95-109781, 1994) p 394.

USE: Insect sterilant.
THERAP CAT: Antineoplastic.

9806. Triethyl Phosphate. Ethyl phosphate. $C_6H_{15}O_4$-P; mol wt 182.16. C 39.56%, H 8.30%, O 35.13%, P 17.00%. $(C_2H_5)_3PO_4$. Prepd from tetraethyl hypophosphate with ethanol in the presence of aluminum ethoxide: Mukaiyama *et al., J. Org. Chem.* **27,** 1815 (1962). Manuf by treating triethyl phosphite with diethyl hydrogen phosphate: McCall, Coover, U.S. pat. **2,960,529** (1960 to Kodak).

Liquid; d^{19} 1.0725; bp 215-216°; bp_{10} 90-95°; n_D^{17} 1.4067. Sol in water with some decompn; sol in alcohol, ether.

USE: As ethylating agent; formation of polyesters which are used as insecticides.

9807. Triethyl Phosphine. $C_6H_{15}P$; mol wt 118.16. C 60.99%, H 12.80%, P 26.21%. $(C_2H_5)_3P$. Prepd from ethyllithium and aluminum trichloride: Screttas, Isbell, *J. Org. Chem.* **27,** 2573 (1962). Manuf from white phosphorus, ethylene and hydrogen at elevated pressures: Oppegard, U.S. pat. **2,687,437** (1954 to du Pont).

Colorless liquid; odor of hyacinths; d_4^{15} 0.800; bp_{744} 127-128°. Practically insol in water. Miscible with alcohol, ether.

USE: In organic syntheses.

9808. Trifenmorph. *4-(Triphenylmethyl)morpholine; 4-tritylmorpholine;* WL-8008; Frescon. $C_{23}H_{23}NO$; mol wt 329.44. C 83.85%, H 7.04%, N 4.25%, O 4.86%. Prepn: Adams *et al.,* U.S. pat. **3,577,413** (1971 to Shell). Lab tests as molluscicide: Webbe, Sturrock, *Ann. Trop. Med. Parasitol.* **58,** 234 (1964); *C.A.* **61,** 15286g (1964). Field trials and toxicity: Boyce *et al., Nature* **210,** 1140 (1966). Degradn: K. I. Beynon *et al., Pestic. Sci.* **3,** 689 (1972). Neurophysiological action: R. B. Morton, D. R. Gardner, *Experientia* **32,** 611 (1976).

Crystalline solid, mp 174-176°, resolidifies and melts again at 185-187°. Vapor pressure at 20°: 1.4×10^{-7} mm Hg. Soly in water at 20°: 0.02 mg/l. Soly at 20° (g/l): carbon tetrachloride 300; chloroform 450; tetrachloroethylene 255. Stable to heat and alkali but hydrolyzed by mild acidic conditions. LD_{50} orally in rats: 1200-2000 mg/kg (Boyce).

USE: As molluscicide (against snails carrying bilharzia flukes).

9809. Triflumuron. *2-Chloro-N-[[[4-(trifluoromethoxy)phenyl]amino]carbonyl]benzamide; N-(2-chlorobenzoyl)-N'-[4-(trifluoromethoxy)phenyl]urea;* trifluron; SIR-8514; BAY SIR 8514; Alsystin; Baycidal; Starycide. $C_{15}H_{10}ClF_3$-N_2O_3; mol wt 358.70. C 50.23%, H 2.81%, Cl 9.88%, F 15.89%, N 7.81%, O 13.38%. Arthropod insecticide which inhibits chitin biosynthesis: Prepn: W. Sirrenberg *et al.,* **Ger.** pat. **2,601,780** corresp to U.S. pat. **4,139,636** (1977, 1979 both to Bayer AG); **Belg.** pat. **867,046** corresp to R.

H. Rigterink, U.S. pat. **4,170,657** (1978, 1979 both to Dow Chemical). Insecticidal activity: C. H. Schaefer *et al., J. Econ. Entomol.* **71,** 427 (1978); S. C. Chang, *ibid.* **72,** 479 (1979). Stability under field conditions: C. H. Schaefer, E. F. Dupras, *J. Agr. Food Chem.* **27,** 1031 (1979). Comprehensive description: G. Zoebelein *et al., Z. Angew. Entomol.* **89,** 289-297 (1980).

Crystals, mp 198° (Sirrenberg). Also reported as mp 188-190° (Rigterink). Slightly toxic to fish. Safe to bees. LD_{50} in rats, mice (mg/kg): > 5000, > 5000 ip., s.c., orally (Zoebelein).

USE: Insecticide (larvicide).

9810. Trifluomeprazine. *N,N,β-Trimethyl-2-(trifluoromethyl)-10H-phenothiazine-10-propanamine; 10-[3-(dimethylamino)-2-methylpropyl]-2-(trifluoromethyl)phenothiazine;* RP-7746. $C_{19}H_{21}F_3N_2S$; mol wt 366.45. C 62.28%, H 5.78%, F 15.55%, N 7.64%, S 8.75%. Prepn: **Brit.** pat. **813,861** corresp to G. E. Ullyot, U.S. pat. **2,921,069** (1959, 1962 both to SK&F). Metabolism study: T. L. Flanagan *et al., J. Pharm. Sci.* **51,** 996 (1962). Chromatographic studies: M. W. Anders, G. J. Mannering, *J. Chromatog.* **7,** 258 (1962); R. J. Warren *et al., J. Pharm Sci.* **55,** 144 (1966); B. B. Wheals, *J. Chromatog.* **187,** 65 (1980). Pharmacological studies: R. C. Kelsey, G. J. Frishmuth, *Arch Int. Pharmacodyn. Ther.* **173,** 44 (1968); E. W. Baur, *J. Pharmacol. Exp. Ther.* **177,** 219 (1971).

uv max (95% ethanol): 308, 258 nm (log ε 3.60, 4.55). Maleate, $C_{19}H_{21}F_3N_2S.C_4H_4O_4$, *Nortran.*
THERAP CAT (VET): Tranquilizer.

9811. Trifluoperazine. *10-[3-(4-Methyl-1-piperazinyl)-propyl]-2-(trifluoromethyl)-10H-phenothiazine; 2-trifluoromethyl-10-[3'-(1-methyl-4-piperazinyl)propyl]phenothiazine.* $C_{21}H_{24}F_3N_3S$; mol wt 407.50. C 61.90%, H 5.94%, F 13.99%, N 10.31%, S 7.87%. Prepn: Craig *et al., J. Org. Chem.* **22,** 709 (1957); **Brit.** pat. **813,861;** G. E. Ullyot, U.S. pat. **2,921,069** (1959 and 1960 to SK & F). Metabolism: T. L. Flanagan *et al., J. Pharm. Sci.* **51,** 996 (1962); C. L. Huang, K. G. Bhansali, *ibid.* **57,** 1511 (1968). Toxicity: P. J. Fowler *et al., Arzneimittel-Forsch.* **27,** 866 (1977). Comprehensive description: A. Post *et al., Anal. Profiles Drug Subs.* **9,** 543-581 (1980).

$bp_{0.6}$ 202-210°. uv max (ethanol): 258, 307.5 nm (log ε 4.50, 3.50). LD_{50} orally in rats, mice: 542.7, 424.0 mg/ml (Fowler).

Dihydrochloride, $C_{21}H_{24}F_3N_3S.2HCl$, *triftazin, triphthasine, Eskazinyl, Eskazine, Jatroneural, Modalina, Stelazine, Terfluzine.* Cream colored fine powder from abs alc, mp 242-243°. Hygroscopic. Freely sol in water. Insol in dil base, ether, benzene. pK_1 3.9, pK_2 8.1. pH of 5% aq soln 2.2.

THERAP CAT: Antipsychotic.

9812. Trifluoracetic Acid. Perfluoroacetic acid. $C_2HF_3O_2$; mol wt 114.02. C 21.07%, H 0.88%, F 49.99%, O 28.06%. CF_3COOH. Prepn: F. Swarts, *Bull. Sci. Acad. Roy. Belg.* **8,** 343 (1922); by oxidation of fluorine olefins: A. L. Henne, U.S. pat. **2,371,757** (1945 to du Pont); A. L. Henne *et al., J. Am. Chem. Soc.* **67,** 918 (1945). Improved prepn: A. L. Henne, P. Trott, *ibid.* **69,** 1820 (1947). Photochemical prepn: R. N. Hazeldine, F. Nyman, *J. Chem. Soc.* **1959,** 387; *eidem, ibid.* 420. Toxicity and metabolism: M. M. Airaksinen, T. Tammisto, *Ann. Med. Exp. Biol. Fenn.* **46,** 242 (1968). *Review:* J. B. Milne in *Chem. Non-Aqueous Solvents* **vol. 5B,** J. J. Lagowski, Ed. (Academic Press, New York, 1978) pp 1-52.

Liquid, sharp biting odor. bp 72.4°. mp -15.4°. d^{20} 1.5351. Miscible with ether, acetone, ethanol, benzene, CCl_4, hexane. Strong, non-oxidizing acid. pka 0.3. LD_{50} i.v. in mice: 1200 mg/kg (Airaksinen, Tammisto).

USE: In organic synthesis; dissolves protein when mixed with liquid SO_2.

9813. Trifluperidol. *1-(4-Fluorophenyl)-4-[4-hydroxy-4-[3-(trifluoromethyl)phenyl]-1-piperidinyl]-1-butanone; 4'-fluoro-4-[4-hydroxy-4-(α,α,α-trifluoro-m-tolyl)piperidino]-butyrophenone; p-fluoro-4-[4'-hydroxy-4'-(3''-trifluoromethyl)phenyl]piperidinobutyrophenone;* 1-(3'-*p*-fluorobenzoylpropyl)-4-hydroxy-4-(3''-trifluoromethylphenyl)piperidine; ω-[4-hydroxy-4-(*m*-trifluoromethylphenyl)piperidino]-*p*-fluorobutyrophenone; flumoperone; $C_{22}H_{23}F_4NO_2$; mol wt 409.42. C 64.54%, H 5.66%, F 18.56%, N 3.42%, O 7.82%. Prepn: P. A. J. Janssen, **Brit. pat. 895,309;** *idem,* **U.S. pat. 3,438,991** (1962, 1969 both to Janssen). Pharmacology: *idem, Arzneimittel-Forsch.* **11,** 932 (1961).

Hydrochloride, $C_{22}H_{23}F_4NO_2$.HCl, *R-2498, Psicoperidol, Psychoperidol, Triperidol.* Crystals from acetone, mp 200.5-201.3°. Sol in water. LD_{50} in rats (mg/kg): 14 i.v.; 70 s.c. (Janssen, 1961).

THERAP CAT: Antipsychotic.

9814. Triflupromazine. *N,N-Dimethyl-2-(trifluoromethyl)-10H-phenothiazine-10-propanamine; 10-[3-(dimethylamino)propyl]-2-trifluoromethylphenothiazine;* 2-trifluoromethyl-10-(γ-dimethylaminopropyl)phenothiazine; fluopromazine. $C_{18}H_{19}F_3N_2S$; mol wt 352.42. C 61.35%, H 5.43%, F 16.17%, N 7.95%, S 9.10%. Prepn: Yale *et al., J. Am. Chem. Soc.* **79,** 4375 (1957); **Brit. pat. 813,861** (1959 to SK & F). Comprehensive description of the hydrochloride: K. Florey, *Anal. Profiles Drug Subs.* **2,** 523-550 (1973).

Liquid, $bp_{0.7}$ 176°; $bp_{0.4}$ 162-164°. n_D^{23} 1.5780.

Hydrochloride, $C_{18}H_{19}F_3N_2S$.HCl, *Adazine, Fluorofen, Psyquil, Siquil, Vespral, Vesprin, Vetame.* Crystals from xylene, dec 173-174°. uv max: 255, 305 nm ($E_{1cm}^{1\%}$ 700, 90). Sol in water, ethanol, acetone. pH of 2% aq soln 4.1, if the pH is raised to 6.4, pptn of the free base results.

THERAP CAT: Antipsychotic.
THERAP CAT (VET): Tranquilizer.

9815. Trifluralin. *2,6-Dinitro-N,N-dipropyl-4-(trifluoromethyl)benzenamine; α,α,α-trifluoro-2,6-dinitro-N,N-dipropyl-p-toluidine;* 2,6-dinitro-*N,N*-dipropyl-α,α,α-trifluoro-*p*-toluidine; 2,6-dinitro-*N,N*-dipropyl-4-trifluoromethylaniline; *N,N*-dipropyl-2,6-dinitro-4-trifluoromethylaniline; L-36352; Lilly 36352; Olitref; Treflan; Tri-4; Triflurex. $C_{13}H_{16}F_3N_3O_4$; mol wt 335.28. C 46.57%, H 4.81%, F 17.00%, N 12.53%, O 19.09%. Pre-emergence herbicide. Prepn: **Brit. pat. 917,253,** *C.A.* **59,** 9889b (1963); Soper, **U.S. pat. 3,403,180** (1963; 1968, both to Lilly). Photodecomposition: E. Leitis, D. G. Crosby, *J. Agr. Food Chem.* **22,** 842 (1974). Soil degradation: T. Golab *et al., ibid.* **27,** 163 (1979). Toxicity data: Goldenthal, *Toxicol. Appl. Pharmacol.* **18,** 185 (1971). Carcinogenicity study: J. F. Robens, *Vet. Human Toxicol.* **22,** 328 (1980).

Yellow crystals, mp 46-47°. $bp_{4.2}$ 139-140°. Slightly sol in water (0.0024 g/100 ml); freely sol in acetone, Stoddard solvent, xylene. LD_{50} orally in rats: 500 mg/kg (Goldenthal).

USE: Herbicide.

9816. Trifluridine. *α,α,α-Trifluorothymidine; 2'-deoxy-5-(trifluoromethyl)uridine;* 5-(trifluoromethyl)-2'-deoxyuridine; F3TDR; NSC-75520; TFT Thilo; Virophta; Viroptic. $C_{10}H_{11}F_3N_2O_5$; mol wt 296.20. C 40.55%, H 3.74%, F 19.24%, N 9.46%, O 27.01%. Prepn: C. Heidelberger *et al., J. Am. Chem. Soc.* **84,** 3597 (1962); *eidem, J. Med. Chem.* **7,** 1 (1964); C. Heidelberger, **U.S. pat. 3,201,387** (1965 to U.S. Dept. HEW). Crystal structure: A. H. Tench, *Diss. Abstr. Int. B* **33,** 3587 (1973). NMR study: R. J. Cushley *et al., J. Am. Chem. Soc.* **90,** 709 (1968). Metabolism: D. L. Dexter *et al., Cancer Res.* **32,** 247 (1972); W. J. O'Brien, H. F. Edelhauser, *Invest. Ophthalmol. Vis. Sci.* **16,** 1093 (1977). Pharmacodynamics: B. L. Wigdahl, J. R. Parkhurst, *Antimicrob. Ag. Chemother.* **14,** 470 (1978); G. J. Smith *et al., Biochem. Biophys. Res. Commun.* **83,** 1538 (1978). Teratogenicity study: M. Itoi *et al., Arch. Ophthalmol.* **93,** 46 (1975). Cytotoxicity and mutagenicity study: E. Huberman, C. Heidelberger, *Mutat. Res.* **14,** 130 (1972). Clinical studies: H. E. Kaufman, *Invest. Ophthalmol. Vis. Sci.* **17,** 941 (1978); R. A. Hyndiuk *et al., Arch. Ophthalmol.* **96,** 1839 (1978). Review of mechanism of antiviral activity: C. Heidelberger, *Ann. N.Y. Acad. Sci.* **255,** 317 (1975). Review of pharmacology and therapeutic use: A. A. Carmine *et al., Drugs* **23,** 329-353 (1982).

Cryst from ethyl acetate, mp 186-189°. uv max (0.1*N* HCl): 260 nm (ε 9960); (0.1*N* NaOH): 260 nm (ε 6590).

THERAP CAT: Antiviral (ophthalmic).

9817. Triflusal. *2-(Acetyloxy)-4-(trifluoromethyl)benzoic acid; α,α,α-trifluoro-2,4-cresotic acid acetate;* acetyl-4-trifluoromethylsalicylic acid; UR-1501; Disgren. $C_{10}H_7F_3O_4$; mol wt 248.16. C 48.40%, H 2.84%, F 22.97%, O 25.79%. Analog of aspirin, *q.v.,* with an inhibitory effect on platelet aggregation. Prepn: M. Hauptschein, **U.S. pat. 3,019,253** (1962 to Pennsalt); E. F. Barra, A. C. M. Boga, **Ger. pat.**

2,641,556 corresp to U.S. pat. **4,096,252** (1977, 1978 both to Uriach). Inhibition of platelet aggregation in man and rat: J. Garcia-Rafanell, J. Morell, *Therapie* **32**, 337 (1977). Effect on blood coagulation: M. Rutllant *et al., Curr. Ther. Res.* **22**, 510 (1977). Pharmacokinetics: V. Rimbau *et al., Arch. Farmacol. Toxicol.* **7**, 11 (1981). Clinical studies: R. M. Masso *et al., Curr. Ther. Res.* **25**, 791 (1979); E. Sala-Planell *et al., Angiologia* **33**, 71 (1981).

White cryst solid from petr ether/ether, mp 120-122° (upon slow heating); 110-112° (upon quick heating). Misc with ethanol. Practically insol in water. LD$_{50}$ in mice, rats (mg/kg): 437, 402 orally; 380, 217 i.p., *RTECS* **Vol. I**, R. J. Lewis, R. L. Tatken, Eds. (1980) p 546.

THERAP CAT: Antithrombotic.

9818. Trifolium. Meadow clover; red clover; purple clover; cow clover. Dried inflorescence of *Trifolium pratense* L., *Leguminosae. Habit.* Europe, Asia, Northern Africa; naturalized in U.S. *Constit.* Tannin, resins, fat, trifolianol, trifoliin.

THERAP CAT: Antispasmodic; expectorant.

9819. Triforine. *N,N'-[1,4-Piperazinediylbis(2,2,2-trichloroethylidene)]bisformamide;* 1,4-di(2,2,2-trichloro-1-formamidoethyl)piperazine; 1,4-bis(1-formamido-2,2,2-trichloroethyl)piperazine; Cela W-524; CME-74770; Basforin; Funginex; Saprol. C$_{10}$H$_{14}$Cl$_6$N$_4$O$_2$; mol wt 434.96. C 27.61%, H 3.24%, Cl 48.90%, N 12.88%, O 7.36%. Prepn: W. Ost *et al.,* **Ger.** pat. **1,901,421** corresp to U.S. pat. **3,595,916** (1969, 1971, both to Boehringer, Ing.).

White crystals, mp 155°. Vapor pressure at 25° = 2×10^{-7} mm Hg. Soly in water at 20°: 27-29 ppm. Sol in CMF, DMSO, *N*-methylpyrrolidone; moderately sol in tetrahydrofuran. Insol in acetone, benzene, carbon tetrachloride, chloroform, methylene chloride, petroleum ether. Dec to chloral and piperazine salts by conc H$_2$SO$_4$ and HCl; slowly dec to chloroform and piperazine by conc alk. Half life in the soil ∼3 weeks. Low toxicity to fish and bees.

USE: Fungicide.

9820. Triglyme. *2,5,8,11-Tetraoxadodecane;* triethylene glycol dimethyl ether. C$_8$H$_{18}$O$_4$; mol wt 178.23. C 53.91%, H 10.18%, O 35.91%. CH$_3$O(CH$_2$)$_2$O(CH$_2$)$_2$O(CH$_2$)$_2$OCH$_3$. Prepd by high pressure hydrogenation of 1,1,2,2-tetrakis(2-methoxyethoxy)ethane: McNamee, MacDowell, U.S. pat. **2,425,042** (1947 to Carbide & Carbon Chem.).

Liquid. d$_4^{20}$ 0.990. Flash pt 111°. mp −45°. bp$_{760}$ 216°; bp$_{10}$ 103.5°; bp$_{0.9}$ 20°. n$_D^{20}$ 1.4233. Miscible with water, hydrocarbon solvents.

USE: Solvent.

9821. Trigonellamide Chloride. *3-(Aminocarbonyl)-1-methylpyridinium chloride;* N^1-methylnicotinamide chloride; 1-methylpyridine-3-carboxylic acid amide chloride; nicotinamide chloromethylate; nicotinamide methyl chloride. C$_7$-H$_9$ClN$_2$O; mol wt 172.61. C 48.71%, H 5.26%, Cl 20.54%, N 16.23%, O 9.27%. One of the principal excretion products of the metabolism of nicotinic acid in man, dog, and rat. Coenzyme action: Warburg, Christian, *Biochem. Z.* **287**, 291 (1936). Synthesis by refluxing nicotinamide with methyl iodide in methanol, then shaking the nicotinamide methiodide with AgCl: Karrer *et al., Helv. Chim. Acta* **19**,

826 (1936); and isolation from urine: Huff, Perizweig, *J. Biol. Chem.* **150**, 395 (1943). Differentiation from NAD: Carpenter, Kodicek, *Biochem. J.* **46**, 421 (1950). *In vitro* metabolism: G. S. Johnson, *Eur. J. Biochem.* **112**, 635 (1980); H. Hoshino *et al., Biochim. Biophys. Acta* **801**, 250 (1984). HPLC determn in urine: M. A. Kutnink *et al., J. Liq. Chromatog.* **7**, 969 (1984).

Crystals from methanol. Dec 240°. Moderately sol in water. More sol in alcohol, butanol, isobutanol. Insol in amyl alcohol, octyl alcohol, benzene, chlorobenzene, chloroform. Destroyed upon boiling the aq soln, more rapidly in the presence of alkali. At room temp it is destroyed in alkaline soln. Reacts with ketones in aq alkaline soln to produce a greenish-blue fluorescence; on acidification the fluorescence changes to blue, and is intensified by heating.

9822. Trigonelline. *3-Carboxy-1-methylpyridinium inner salt;* nicotinic acid *N*-methylbetaine; coffearine; caffearine; gynesine; trigenolline. C$_7$H$_7$NO$_2$; mol wt 137.14. C 61.31%, H 5.14%, N 10.21%, O 23.33%. In seeds of *Trigonella foenumgraecum* L., *Leguminosae,* in coffee beans, in seeds of *Strophanthus* spp, *Apocynaceae* and of *Cannabis sativa* L., *Moraceae,* in seeds of many other plants; also in sea urchin, *Arabacia pustulosa,* and in jellyfish, *Velella spirans.* Excreted in urine after taking nicotinic acid: Ackermann, *Z. Biol.* **59**, 17 (1912). Isoln from normal urine: Linnewah, Renwein, *Z. Physiol. Chem.* **207**, 48 (1932); **209**, 110 (1932). Syntheses: Turnau, *Monatsh.* **26**, 551 (1905); Sarett *et al., J. Biol. Chem.* **135**, 483 (1940); Green, Tong, *J. Am. Chem. Soc.* **78**, 4896 (1956); Kosower, Patton, *J. Org. Chem.* **26**, 1318 (1961). Toxicity study: Brazda, Coulson, *Proc. Soc. Exp. Biol. Med.* **62**, 19 (1946).

Monohydrate, crystals from ethanol, mp 230-233°. Salty taste. Very sol in water; sol in alcohol. Practically insol in ether, chloroform. LD$_{50}$ s.c. in rats: 5.0 g/kg (Brazda, Coulson).

Hydrochloride, C$_7$H$_7$NO$_2$.HCl, crystals from 90% alcohol, mp 258-259°. Very sol in water; slightly in alcohol. Practically insol in ether, benzene.

9823. Trihexyphenidyl Hydrochloride. *α-Cyclohexyl-α-phenyl-1-piperidinepropanol hydrochloride;* 3-(1-piperidyl)-1-cyclohexyl-1-phenyl-1-propanol hydrochloride; 1-phenyl-1-cyclohexyl-3-piperidyl-1-propanol hydrochloride; benzhexol chloride; Aparkane; Artane; Broflex; Cyclodol; Pacitane; Paralest; Pargitan; Parkinane; Parkopan; Peragit; Pipanol; Sedrena; Tremin; Triphedinon; Triphenidyl; Tsiklodol. C$_{20}$H$_{32}$ClNO; mol wt 337.93. C 71.09%, H 9.54%, Cl 10.49%, N 4.14%, O 4.73%. Synthesis: Denton *et al., J. Am. Chem. Soc.* **71**, 2053 (1949); Adamson, Wilkinson, U.S. pat. **2,682,543** (1954 to Burroughs Wellcome); Denton, U.S. pat. **2,716,121** (1955 to Am. Cyanamid). Resolution into isomers: Adamson, Duffin, **Brit.** pat. **750,156** (1956 to Wellcome Found.).

• HCl

Crystals, dec 258.5°. Free base, mp 114.3-115.0°. Soly (g/100 ml): water at 25°, 1.0; alcohol 6; chloroform 5. More sol in methanol; very slightly sol in ether, benzene. pH of a 1% aq soln 5.5-6.0.

l-Form, crystals from isopropyl alc, mp 264°. $[\alpha]_D^{20}$ −30° (c = 0.4 in chloroform). Free base, mp 112-113°. $[\alpha]_D^{20}$ −25° (c = 0.4 in ethanol).

THERAP CAT: Anticholinergic; antiparkinsonian.

9824. Trihydrazine Dihydriodide. Hydrazine dihydriodide. $H_{14}I_2N_6$; mol wt 351.96. H 4.01%, I 72.11%, N 23.88%. $(N_2H_5^+I^-)_2.N_2H_4$. Prepared from hydrazine (85-100%) and hydriodic acid in ethanol at pH 8-9: Curtius, Schulz, *J. Prakt. Chem.* **42**, 540 (1890); Gilbert, Decius, *J. Am. Chem. Soc.* **80**, 3871 (1958).

Large, white, biaxial needles from abs ethanol, mp 90-92°. Crystals stored in a glass-stoppered bottle at room temp showed no deterioration over a six month period. Freely sol in water with formation of hydrazine monoiodide and hydriodic acid. The hydrazine is very strongly held in the dry crystals. Exposure in a vacuum still at 25 microns and room temperature for 5½ hours showed no loss in weight. The neutral hydrazine molecule is probably hydrogen-bonded to the $N_2H_5^+$ ions.

9825. Trillium. Beth root; Indian balm; ground lily; birthroot. Dried rhizome of *Trillium erectum* L. and other spp of *Trillium, Liliaceae. Habit.* Canada, south to Tennessee and Missouri; also Japan. *Constit.* Trilline, fixed oil, tannin.

THERAP CAT: Formerly in metrorrhagia, menorrhagia, various types of hemorrhage; astringent in diarrhea.

9826. Trilobine. *(1'α)-6',7-Epoxy-6,12'-dimethoxy-2'-methyloxyacanthan.* $C_{35}H_{34}N_2O_5$; mol wt 562.67. C 74.71%, H 6.09%, N 4.98%, O 14.22%. From root of *Cocculus trilobus* DC. and *C. sarmentosus* Diels, *Menispermaceae:* Tomita *et al., J. Pharm. Soc. Japan* **48**, 83 (1928); **50**, 127 (1930); **62**, 468, 481 (1942). Structure: Kondo, Tomita, *Ann.* **497**, 104 (1932); Inubushi, Nomura, *Tetrahedron Letters* **1962**, 1133. Total synthesis of trilobine and isotrilobine: Y. Inubushi *et al., Chem. Pharm. Bull.* **25**, 1636 (1977).

Trilobine   R = H
Isotrilobine   R = CH₃

Crystals from benzene, mp 237°. $[\alpha]_D$ +307° (chloroform). Practically insol in water. Sparingly sol in alcohol, acetone, ether. Sol in chloroform.

N-Methyl deriv., $C_{36}H_{36}N_2O_5$, *isotrilobine, homotrilobine, N-methyltrilobine.* Crystals from acetone, mp 215°. $[\alpha]_D$ +317° (chloroform).

9827. Trilostane. *(4α,5α,17β)-4,5-Epoxy-3,17-dihydroxyandrost-2-ene-2-carbonitrile; 4,5-epoxy-17-hydroxy-3-oxoandrostane-2-carbonitrile; 2α-cyano-4α,5α-epoxyandrostan-*

17β-ol-3-one; Win-24540; Desopan; Modrastane; Modrenal. $C_{20}H_{27}NO_3$; mol wt 329.44. C 72.92%, H 8.26%, N 4.25%, O 14.57%. Prepn: R. O. Clinton, A. J. Manson, U.S. pat. **3,296,255** (1967 to Sterling); H. C. Neumann *et al., J. Med. Chem.* **13**, 948 (1970). Inhibition of steroid biosynthesis: G. O. Potts *et al., Steroids* **32**, 257 (1978). Disposition in animals: J. F. Baker *et al., Arch. Int. Pharmacodyn. Ther.* **243**, 4 (1980). Metabolism in rats: Y. Mori *et al., Chem. Pharm. Bull.* **29**, 2646 (1981); in humans: D. T. Robinson *et al., J. Steroid Biochem.* **21**, 601 (1984). HPLC determn in plasma: P. Powles *et al., J. Chromatog.* **311**, 434 (1984); and of major metabolite, *17-ketotrilostane:* R. R. Brown *et al., ibid.* **339**, 440 (1985). Clinical study in Cushing's syndrome: P. Komanicky *et al., J. Clin. Endocrinol. Metab.* **47**, 1042 (1978). Clinical trials in breast cancer: C. G. Beardwell *et al., Cancer Chemother. Pharmacol.* **10**, 158 (1983); C. J. Williams *et al., Cancer Treat. Rep.* **71**, 1197 (1987).

Tan crystals from pyridine/dioxane, mp 257.8-270° (dec). $[\alpha]_D^{25}$ +137.4° (c = 1 in pyridine). uv max (ethanol): 252 nm (ε 8300).

THERAP CAT: Adrenocortical suppressant; in treatment of breast cancer.

9828. Trimazosin. *4-(4-Amino-6,7,8-trimethoxy-2-quinazolinyl)-1-piperazinecarboxylic acid 2-hydroxy-2-methylpropyl ester.* $C_{20}H_{29}N_5O_6$; mol wt 435.48. C 55.16%, H 6.71%, N 16.08%, O 22.04%. Prepn: H. J. Hess, **Ger.** pat. **2,120,495** corresp to U.S. pat. **3,669,968** (1971, 1972 both to Pfizer). Pharmacological study in rats: J. P. Buyniski *et al., Clin. Exp. Hypertens.* **2**, 1039 (1980), *C.A.* **94**, 76779x (1981). Cardiocirculatory effects in congestive heart failure: N. A. Awan *et al., Am. J. Cardiol.* **44**, 126 (1979). Long-term therapeutic use: K. T. Weber *et al., N. Engl. J. Med.* **303**, 242 (1980).

White cryst from chloroform/diisopropyl ether, mp 158-159°.

Hydrochloride monohydrate, $C_{20}H_{29}N_5O_6.HCl.H_2O$, *CP-19106-1, Cardovar, Supres.* White cryst, mp 166-169° (dec).

THERAP CAT: Antihypertensive.

9829. Trimebutine. *3,4,5-Trimethoxybenzoic acid 2-(dimethylamino)-2-phenylbutyl ester.* $C_{22}H_{29}NO_5$; mol wt 387.48. C 68.20%, H 7.54%, N 3.61%, O 20.65%. Anticholinergic. Prepn by esterification: C. P. J. Roux, D. R. Torossian, **Fr.** pat. **1,344,455**, *C.A.* **60**, 15777g (1964); by transesterification: D. R. Torossian, G. G. Aubard, **Ger.** pat. **1,151,716** corresp to **Brit.** pat. **1,342,547** (1964, 1972, 1974 all to Jouveinal). Clinical trials in treatment of spastic colon: K. Lüttecke, *J. Int. Med. Res.* **6**, 86 (1978); M. G. Moshal, M. Herron, *ibid.* **7**, 231 (1979). Effect on gastrointestinal tract contractions in dogs: K. Yamada *et al., Japan. J. Pharmacol.* **33**, 301 (1983); on spontaneous contraction of guinea pig colon: H. Takenaga *et al., ibid.* **34**, 177 (1984). HPLC determn in plasma: A. Astier, A. M. Deutsch, *J. Chromatog.* **224**, 149 (1981).

Crystals from ethanol, mp 78-80°C. Soluble in methylene chloride.

Maleate, $C_{22}H_{29}NO_5.C_4H_4O_4$, **TM-906, Cerekinon, Debridat, Digerent, Foldox, Polibutin, Spabucol, Trimedat.** Crystals from water, mp 105-106°.

THERAP CAT: Antispasmodic.

9830. Trimecaine. *2-Diethylamino-2',4',6'-trimethylacetanilide; N-sym*-trimethylphenyldiethylaminoacetamide; 2-diethylaminoacetyl-2',4',6'-trimethylanilide; Mesocaine; Mesidicaine; Mesokain. $C_{15}H_{24}N_2O$; mol wt 248.37. C 72.54%, H 9.74%, N 11.28%, O 6.44%. Prepn: Löfgren, Lundqvist, U.S. pat. **2,441,498** (1948 to Astra); Borovansky *et al., J. Am. Pharm. Assoc.* **48,** 402 (1959).

Crystals, mp 44°. bp$_6$ 187°; bp$_{0.6}$ 154-155°.
Hydrochloride, $C_{15}H_{24}N_2O.HCl$, crystals from acetone, mp 140°. LD$_{50}$ s.c. in mice: 295 mg/kg.

THERAP CAT: Hydrochloride as local anesthetic.

9831. Trimedlure. *4(or 5)-Chloro-2-methylcyclohexanecarboxylic acid, 1,1-dimethylethyl ester; tert*-butyl 4(or 5)-chloro-2-methylcyclohexanecarboxylate. $C_{12}H_{21}ClO_2$; mol wt 232.75. C 61.93%, H 9.09%, Cl 15.23%, O 13.75%. Substituted cyclohexane developed as attractant for the Mediterranean fruit fly, or medfly, *Ceratitis capitata* (Weidemann). Prepn and physical props: M. Beroza *et al., J. Agr. Food Chem.* **9,** 361 (1961). Commercial product consists mainly of the four isomers having the methyl and ester substituents on the ring in a *trans* configuration. Separation and identification of isomers: T. P. McGovern, M. Beroza, *J. Org. Chem.* **31,** 1472 (1966). Extension of activity by fixatives: *eidem, J. Econ. Entomol.* **60,** 379 (1967). Controlled release: S. Nakagawa *et al., ibid.* **72,** 625 (1979). Repellent effect of high concns: *eidem, ibid.* **64,** 762 (1971). Rate of loss from wicks: J. R. King, P. J. Landolt, *ibid.* **77,** 221 (1984). Toxicity studies: M. Beroza *et al., Toxicol. Appl. Pharmacol.* **31,** 421 (1975).

trans-form

Oil, bp 107-113°. n_D^{20} 1.460. LD$_{50}$ in rats (mg/kg): 4556 ±1136 orally; in rabbits (mg/kg): > 2025 dermally. LC$_{50}$ in rainbow trout, bluegill sunfish (ppm): 11.5, 14.7 (24 hr) (Beroza).

USE: In traps for monitoring medfly infestations.

9832. Trimellitic Acid. *1,2,4-Benzenetricarboxylic acid;* 1,2,4-tricarboxybenzene. $C_9H_6O_6$; mol wt 210.14. C 51.44%, H 2.88%, O 45.68%. Obtained by oxidation of coal with nitric acid: Grosskinsky, *Glückauf* **88,** 376 (1952), *C.A.* **46,** 7731 (1952); by oxidation of β-indancarboxylic acid with nitric acid: Braun *et al., Ber.* **53,** 1160 (1920). Manuf by oxidation of pseudocumene: Backlund, U.S. pat. **3,009,-953** (1961 to Union Oil of Calif.).

Crystals from acetic acid or dilute alcohol. mp 218-220° (also reported 229-234° dec). Solubilities at 25° in g/100 g solvent: carbon tetrachloride 0.004; ligroin 0.03; mixed xylenes 0.006; dimethylformamide 31.3; ethyl acetate 1.7; acetone 7.9; water 2.1; ethanol 25.3. Practically insol in chloroform, benzene, carbon disulfide.

USE: Intermediate in the prepn of resins, plasticizers, dyes, inks, adhesives.

9833. Trimellitic Anhydride. *1,3-Dihydro-1,3-dioxo-5-isobenzofurancarboxylic acid;* trimellitic acid 1,2-anhydride; anhydrotrimellitic acid; 1,3-dioxo-5-phthalancarboxylic acid. $C_9H_4O_5$; mol wt 192.13. C 56.26%, H 2.10%, O 41.64%. Prepd by subliming trimellitic acid above its mp: Alder, Dortmann, *Ber.* **85,** 556 (1952); by heating crude trimellitic acid with V_2O_5: McKinnis, U.S. pat. **2,998,431** (1959 to Union Oil of Calif.).

Crystals, mp 161-163.5°. bp$_{14}$ 240-245°. Solubilities at 25° in g/100 g solvent: carbon tetrachloride 0.002; ligroin 0.06; mixed xylenes 0.4; dimethylformamide 15.5; acetone 49.6; ethyl acetate 21.6.

USE: In the preparation of resins, adhesives, polymers, dyes, printing inks.

9834. Trimeprazine. *N,N,β-Trimethyl-10H-phenothiazine-10-propanamine;* 10-[3-(dimethylamino)-2-methylpropyl]phenothiazine; 10-(2-methyl-3-dimethylaminopropyl)-phenothiazine; alimemazine; methylpromazine; Bayer 1219; RP-6549. $C_{18}H_{22}N_2S$; mol wt 298.45. C 72.44%, H 7.43%, N 9.39%, S 10.74%. Prepn: Jacob, Robert, U.S. pat. **2,837,518** (1958 to Rhône-Poulenc). Metabolism studies: Robinson, *J. Pharm. Pharmacol.* **18,** 19 (1966).

Crystals, mp 68°. bp$_{0.3}$ 150-175°.
Tartrate, $(C_{18}H_{22}N_2S)_2.C_4H_6O_6$, **Panectyl, Repeltin, Temaril, Theralene, Vallergan.** Crystals. Sol in water; slightly sol in alcohol.

THERAP CAT: Antipruritic.
THERAP CAT (VET): Antipruritic.

9835. Trimetazidine. *1-[(2,3,4-Trimethoxyphenyl)methyl]piperazine.* $C_{14}H_{22}N_2O_3$; mol wt 266.34. C 63.14%, H 8.33%, N 10.52%, O 18.02%. Coronary vasodilator. Prepn: Regnier, Canevari, Fr. pat. **1,302,958** (1962 to Sci. Union et Cie, Soc. Franç. Rech. Med.); Fr. pat. **M805;** J. Servier, U.S. pat. **3,262,852** (1961, 1966 to Biofarma). Pharmacology: Fujita, *Japan. J. Pharmacol.* **17,** 19 (1967); Nagata *et al., ibid.* **19,** 628 (1969); **21,** 337 (1971). GC-MS determn in biological fluids: L. Fay *et al., J. Chromatog.* **490,** 198 (1989). Review of pharmacology, toxicology and clinical studies: C. Harpey *et al., Cardiovasc. Drug Rev.* **6,** 292-312 (1989).

bp$_2$ 200-205°.

Dihydrochloride, C$_{14}$H$_{22}$N$_2$O$_3$.2HCl, *Kyurinett, Vastarel F, Yoshimilon*. Crystals, mp 225-228°. LD$_{50}$ in male, female mice, rats (mg/kg): 91, 107, 124, 124 i.v.; 264, 245, 327, 288 i.p.; 528, 608, 1147, 987 orally (Harpey).

THERAP CAT: Antianginal.

9836. Trimethadione. *3,5,5-Trimethyl-2,4-oxazolidinedione;* 3,5,5-trimethyl-2,4-dioxooxazolidine; troxidone; Absentol; Epidione; Petidon; Ptimal; Tridione. C$_6$H$_9$NO$_3$; mol wt 143.14. C 50.35%, H 6.34%, N 9.79%, O 33.53%. Synthesis: Spielman, *J. Am. Chem. Soc.* **66**, 1244 (1944); U.S. pat. **2,575,692** (1951 to Abbott); Davies, Hook, U.S. pat. **2,559,011** (1951 to Brit. Schering). Metabolism to dimethadione, *q.v.:* T. C. Butler, *J. Pharmacol. Exp. Ther.* **108**, 11 (1953); T. C. Butler, W. J. Waddell, *ibid.* **110**, 241 (1954). Anticonvulsant activity: H. Ferngren, *Acta Pharmacol. Toxicol.* **26**, 177 (1968); H. H. Frey, *ibid.* **27**, 295 (1969). Pharmacodynamics and metabolism in rats: D. O. Thueson *et al., Epilepsia* **15**, 563 (1974). Evaluation in teratogenesis: J. German *et al., Teratology* **3**, 349 (1970); A. B. Rifkind, *Toxicol. Appl. Pharmacol.* **30**, 452 (1974). GC determn in serum: E. Tanaka, S. Misawa, *J. Chromatog.* **413**, 376 (1987); LC determn in serum: M. Okamoto *et al., Chromatographia* **23**, 325 (1987). Use in treatment of petit mal epilepsy: S. Livingston *et al., J. Am. Med. Assoc.* **194**, 227 (1965). Use in dissolution of pancreatic stones in humans and dogs: A. Noda *et al., Lancet* **2**, 351 (1984); *eidem, Gastroenterology* **93**, 1002 (1987).

Granules, crystals, mp 46-46.5°. bp$_5$ 78-80°. Slight camphor-like odor. Burning, faintly bitter taste. Soly in water about 5%; increased by the addition of urethan. Freely sol in alcohol, benzene, chloroform, ether. Practically insol in petr ether. The pH of a 5% soln is ~6.0.

THERAP CAT: Anticonvulsant.

THERAP CAT (VET): Anticonvulsant.

9837. Trimethaphan Camsylate. *Decahydro-2-oxo-1,3-bis(phenylmethyl)thieno[1',2':1,2]thieno[3,4-d]imidazol-5-ium salt with (1S)-7,7-dimethyl-2-oxobicyclo[2.2.1]heptane-1-methanesulfonic acid (1:1);* trimetaphan camphorsulfonate; methioplegium; trimethaphan camphorsulfonate; trimethaphen camphorsulfonate; 4,6-dibenzyl-5-oxo-1-thia-4,6-diazatricyclo[6.3.0.0$^{3.7}$]undecanium (+)-β-camphorsulfonate; *d*-3,4-(1',3'-dibenzyl-2'-ketoimidazolido)-1,2-trimethylene-thiophanium *d*-camphorsulfonate; Nu-2222; Arfonad. C$_{32}$H$_{40}$N$_2$O$_5$S$_2$; mol wt 596.81. C 64.40%, H 6.76%, N 4.69%, O 13.40%, S 10.75%. Available from the Roche synthesis of biotin: Randall *et al., J. Pharmacol. Exp. Ther.* **97**, 48 (1949); Scurr, Wyman, *Lancet* **266**, 338 (Feb. 13, 1954). May be prepd by treating (+)-3,4-(1,3-dibenzylureylene)-tetrahydro-2-oxothiophene with 3-ethoxypropylmagnesium

bromide, followed by reduction, ring closure, and conversion into the (+)-β-camphorsulfonate. Comprehensive description: K. W. Blessel *et al., Anal. Profiles Drug Subs.* **3**, 545-564 (1974).

Bitter crystals, dec ~245°. [α]$_D^{20}$ +22.0° (c = 4 in water). One gram dissolves in less than 5 ml water and in less than 2 ml alcohol. Slightly sol in acetone, ether; pH of a 1% aq soln is 5.0-6.0.

THERAP CAT: Antihypertensive.

9838. Trimethidinium Methosulfate. *1,3,8,8-Tetramethyl-3-[3-(trimethylammonio)propyl]-3-azoniabicyclo-[3.2.1]octane bis(methyl sulfate); 3-[3-(dimethylamino)propyl]-1,3,8,8-tetramethyl-3-azoniabicyclo[3.2.1]octane methyl sulfate methosulfate;* N-(γ-trimethylammoniumpropyl)-N-methylcamphidinium methyl sulfate; camphidonium; Wy-1395; HA-106; Ostensin. C$_{19}$H$_{42}$N$_2$O$_8$S$_2$; mol wt 490.68. C 46.51%, H 8.63%, N 5.71%, O 26.09%, S 13.07%. Prepd from 1-(3-dimethylaminopropyl)-3,4,4-trimethyl-3,5-ethylenepiperidine and methyl sulfate: Schmidt, Ger. pat. **1,086,703** (1960 to Thomae), *C.A.* **55**, 27379i (1961). Ganglion blocking agent.

Crystals, mp 192-193°.

THERAP CAT: Antihypertensive.

9839. Trimethobenzamide. *N-[(2-Dimethylaminoethoxy)benzyl]-3,4,5-trimethoxybenzamide;* 4-(2-dimethylaminoethoxy)-N-(3,4,5-trimethoxybenzoyl)benzylamine. C$_{21}$H$_{28}$N$_2$O$_5$; mol wt 388.46. C 64.93%, H 7.27%, N 7.21%, O 20.59%. Prepn: Goldberg, Teitel, U.S. pat. **2,879,293** (1959 to Hoffmann-La Roche). Pharmacology: K. W. Blessel *et al., Anal. Profiles Drug Subs.* **2**, 551-570 (1973).

Hydrochloride, C$_{21}$H$_{28}$N$_2$O$_5$.HCl, *Ro-2-9578, Anaus, Tigan, Xametina*. Crystals, mp 187.5-190°. Freely sol in water (approx soly at 25° > 50%). A 5% aq soln is stable to autoclaving at 120° for 20 min at pH 3-7.

THERAP CAT: Antiemetic.

9840. Trimethoprim. *5-[(3,4,5-Trimethoxyphenyl)methyl]-2,4-pyrimidinediamine; 2,4-diamino-5-(3,4,5-trimethoxybenzyl)pyrimidine;* Instalac; Monotrim; Proloprim; Syraprim; Tiempe; Trimanyl; Trimogal; Trimopan; Trimpex; Uretrim; Wellcoprim. C$_{14}$H$_{18}$N$_4$O$_3$; mol wt 290.32. C 57.92%, H 6.25%, N 19.30%, O 16.53%. Prepn from guanidine and β-ethoxy-3,4,5-trimethoxybenzylbenzalnitrile: Stenbuck, Hood, U.S. pat. **3,049,544** (1962 to Burroughs Wellcome); Hoffer, U.S. pat. **3,341,541** (1967 to Hoffmann-La Roche). Improved synthesis: B. Roth *et al., J. Med. Chem.* **23**, 379, 535 (1980). Toxicity data: Yamamoto *et al., Chemotherapy (Tokyo)* **21**, 187 (1973). *Review:* Burchall in *Antibiotics* **vol. 3**, J. W. Corcoran, F. E. Hahn, Eds. (Springer-Verlag, New York, 1975) pp 304-320. Comprehensive description: G. J. Manius, *Anal. Profiles Drug Subs.* **7**, 445-475 (1978). Review of antibacterial activity, pharmacokinetics and therapeutic use: R. N. Brogden *et al., Drugs* **23**, 405-430 (1982).

White to cream, bitter crystalline powder, mp 199-203°. Soly in g/100 ml at 25°: DMAC 13.86; benzyl alcohol 7.29; propylene glycol 2.57; chloroform 1.82; methanol 1.21; water 0.04; ether 0.003; benzene 0.002. pKa 6.6. LD_{50} orally in mice: 7000 mg/kg (Yamamoto).

Note: See Sulfamethoxazole, Sulfadiazine, Sulfametrole, Sulfamoxole, and Sulfalene for lists of trade names of mixtures with Trimethoprim.

THERAP CAT: Antibacterial.

THERAP CAT (VET): Antibacterial.

9841. Trimethylamine. *N,N-Dimethylmethanamine.* C_3H_9N; mol wt 59.11. C 60.96%, H 15.35%, N 23.70%. $(CH_3)_3N$. Together with other amines it is a degradation product of nitrogenous plant and animal substances. It is formed during the distillation of sugar beet residues which contain betaine. In conjugated form it is widely distributed in animal tissue and especially in fish. It is converted to the free tertiary amine during putrefaction. It has been detected in menstrual blood, and in urine which was stored at room temp. Prepn from paraformaldehyde and ammonium chloride: Adams, Brown, *Org. Syn.* **1,** 75 (1921); **coll. vol. I,** 2nd ed., p 528; by the action of formaldehyde and formic acid on ammonia: Sommelet, Ferrand, *Bull. Soc. Chim.* [4] **35,** 446 (1924). Physical properties: J. G. Aston *et al., J. Am. Chem. Soc.* **66,** 1171 (1944).

Gas. Pungent, fishy, ammoniacal odor, saline taste. Liquefiable by pressure at ordinary temp or by condensation. mp −117.08°. bp_{760} 2.87°; bp_{747} 3.2-3.8°. d_4^0 0.6709. Strong base. pK_b (25°): 4.13. Readily absorbed by water, alcohol with which it is miscible; also sol in ether, benzene, toluene, xylene, ethylbenzene, chloroform. Sold as 25% water soln or as liquefied gas.

Hydrochloride, $C_3H_9N.HCl$, *trimethylammonium chloride.* Prepn: *Org. Syn.* **coll. vol. I,** 2nd ed., p 531. Monoclinic deliquesc crystals from alc. Odor less intense than that of basc. Dec 277-278°. Sinters and sublimes at 200°. Sol in water, alcohol; moderately sol in chloroform. Insol in ether. *Keep well closed.*

Oxide, C_3H_9NO, *trimethyloxamine.* Needles with $2H_2O$ from water. Anhydr at 96°. mp 257°. Sol in water, methanol. Strong alkaline reaction.

USE: In the manuf of quaternary ammonium compds; as insect attractant; as warning agent for natural gas.

9842. Trimethyl Borate. *Boric acid trimethyl ester;* methyl borate. $C_3H_9BO_3$; mol wt 103.91. C 34.68%, H 8.73%, B 10.40%, O 46.19%. $B(OCH_3)_3$. Prepn from pyridine-boron trichloride complex: Gerrard, Lappert, *Chem. & Ind. (London)* **1952,** 53; from methanol and boric oxide, borax or boric acid: Schlesinger, *J. Am. Chem. Soc.* **75,** 213 (1953); from methyl orthosilicates and boron halide: Wiberg, Krüerke, *Z. Naturforsch.* **8b,** 608 (1953); from boric acid and methanol: Steinberg, Hunter, *Ind. Eng. Chem.* **49,** 174 (1957). Several mfg processes: U.S. pats. **2,689,259; 2,884,439; 2,937,195** (to Callery Chem.); U.S. pats. **2,880,-227** and **2,884,440** (to Olin Mathieson); U.S. pat. **2,855,427** (to Am. Potash & Chem.); U.S. pat. **2,739,979** (to USAEC). Acute toxicity: H. F. Smyth *et al., Am. Ind. Hyg. Assoc. J.* **23,** 95 (1962).

Liquid, d 0.915. bp 67-68°. Flash pt 29° (84.2°F). Miscible with tetrahydrofuran, ether, isopropylamine, hexane, methanol, Nujol and other organic liquids. Stable in the absence of moisture, but hydrolyzes in the presence of water to methanol and boric acid. Forms an azeotrope with methanol: 70% $B(OCH_3)_3$ +30% methanol, d 0.87; bp 52- 54°; flash pt 34° (93.2°F). LD_{50} orally in rats: 6.14 ml/kg (Smyth).

USE: As solvent for waxes, resins, oils; catalyst in the manuf of ketones; analysis of paint and varnish ingredients;

as neutron detector gas in the presence of a scintillation counter; as a promoter of diborane reactions.

9843. N,N-1-Trimethyl-3,3-diphenylpropylamine. *N,-N,α-Trimethyl-γ-phenylbenzenepropanamine;* 4,4-diphenyl-2-dimethylaminobutane; 3-dimethylamino-1,1-diphenylbutane; Recipavrin. $C_{18}H_{23}N$; mol wt 253.39. C 85.32%, H 9.15%, N 5.53%. Prepn: Perrine, *J. Org. Chem.* **18,** 898 (1953).

l-Form hydrochloride, $C_{18}H_{23}N.HCl$, rods from acetone, mp 180-182°. $[\alpha]_D^{20}$ −43.3° (c = 1.04 in water).

d-Form hydrochloride, $C_{18}H_{23}N.HCl$, mp 179-181°. $[\alpha]_D^{20}$ +43.1° (c = 0.53 in water).

Nitrate, crystals from water, mp 118-120°.

THERAP CAT: Antispasmodic.

9844. Trimethylene Bromide. *1,3-Dibromopropane;* α,-γ-dibromopropane; ω,ω'-dibromopropane; trimethylene dibromide. $C_3H_6Br_2$; mol wt 201.89. C 17.85%, H 3.00%, Br 79.16%. $BrCH_2CH_2CH_2Br$. Prepd by the action of hydrobromic acid on trimethylene glycol in the presence of sulfuric acid: Kamm, Marvel, *Org. Syn.* **coll. vol. I,** p 30 (1941); *cf.* Derick, Hess, *J. Am. Chem. Soc.* **40,** 545 (1918); Norris, Mulliken, *ibid.* **42,** 2096 (1920): Kamm, Newcomb, *ibid.* **43,** 2229 (1921). From trimethylene glycol and PBr_3: Bogert, Slocum, *ibid.* **46,** 765 (1924).

Colorless liquid; d_4^{25} 1.9712; bp_{760} 167° (mp −36°); n_D^{15} 1.5249. Slightly sol in water (1.68 g/l at 30°); sol in alc, ether. Upon prolonged heating trimethylene bromide dec and part of it is converted to propylene bromide (1,2-dibromopropane). Boiling with water yields trimethylene glycol.

9845. Trimethylene Glycol. *1,3-Propanediol.* $C_3H_8O_2$; mol wt 76.10. C 47.35%, H 10.60%, O 42.05%. $HOCH_2$-CH_2CH_2OH. Prepd by reduction of ethyl glycidate with lithium aluminum hydride: Walborsky, Colombini, *J. Org. Chem.* **27,** 2387 (1962).

Colorless to pale yellow, very viscid, sweet liquid. d_4^{20} 1.0597; bp 210-212°; n_D^{20} 1.4398. Miscible with water, alc.

9846. Trimethylene Oxide. *Oxetane;* 1,3-epoxypropane. C_3H_6O; mol wt 58.08. C 62.04%, H 10.41%, O 27.55%. Prepd by dropwise addition of 3-chloropropyl acetate to hot potassium hydroxide soln: Noller, *Org. Syn.* **29,** 92 (1949); modified procedure: Searles, *J. Am. Chem. Soc.* **73,** 124 (1951).

Oil, agreeable aromatic odor. d_0^0 0.8975; d_4^{25} 0.8930. bp_{750} 48°; bp_{736} 45-46°. n_D^{25} 1.3895; n_D^{23} 1.3905. Reacts with Grignard reagents and organolithium compds to give, after hydrolysis, 3-substituted propanols.

9847. Trimethyl Isopropyl Butanamide. *N,2,3-Trimethyl-2-(1-methylethyl)butanamide;* 2-isopropyl-N,2,3-dimethylbutanamide; WS23. $C_{10}H_{21}NO$; mol wt 171.28. C 70.12%, H 12.36%, N 8.18%, O 9.34%. Prepn: D. G. Rowsell *et al.,* Ger. pat. **2,317,538;** *eidem,* U.S. pat. **4,296,255** (1973, 1981 both to Wilkinson Sword). Structure-activity study: H. R. Watson *et al., J. Soc. Cosmet. Chem.* **29,** 185 (1978). Brief review: M. A. Parrish, *Mfg. Chem.* **58,** 31-32 (February, 1987).

Colorless solid, mp 58-61° (Rowsell); also reported as white, crystalline solid, melting range 63° (Parrish). bp$_{0.35mm}$ 83-85°.

USE: Physiological coolant in foods, beverages, tobacco products, toiletries, cosmetics and pharmaceuticals.

9848. 2,4,6-Trimethylpyridine. γ-Collidine; *sym*-collidine; 2,4,6-collidine; α,γ,α′-collidine. C$_8$H$_{11}$N; mol wt 121.18. C 79.29%, H 9.15%, N 11.56%. Found in small amts in coal tar, in shale oil. Produced commercially from coal tar to some extent; also produced by the Hantzsch pyridine synthesis: Hantzsch, *Ann.* **215**, 1 (1882); Mosher in *Heterocyclic Compounds* vol. 1, R. C. Elderfield, Ed., (John Wiley, New York, 1950) pp 462-472; from acetone and ammonia: Mosher in Kirk-Othmer *Encyclopedia of Chemical Technology* vol. **11** (New York, 1953) p 287, *cf.* Dürkopf, *Ber.* **21**, 2713 (1888); from paraldehyde, acetone, ammonium acetate, and ammonia water: **Ger.** pat. **349,267;** *Frdl.* **14**, 539; from 3,5-dimethyl-2-cyclohexen-1-one, ammonium acetate, and ammonia water: Frank, Meikle, *J. Am. Chem. Soc.* **72**, 4184 (1950).

Liquid. Aromatic odor. d$_4^{16.4}$ 0.9191; d$_4^{22.1}$ 0.9166 (commercial grade: d$_{15.5}^{15.5}$ 0.920-0.935, approx 7.74 lbs/gal). mp −46°. bp$_{762}$ 170.5°; bp$_{760}$ 171°; bp$_{31}$ 65°; bp$_{2.7}$ 10°. n$_D^{25}$ 1.4959; n$_D^{22.1}$ 1.49770. pKa (25°): 6.69. Flash pt 136°F. Dielectric constant at 22° = 6.6 (λ = 70 cm). More sol in cold water than in hot water: 20.8 g/100 ml at 6°; 3.5 g/100 ml at 20°; 1.8 g/100 ml at 100°. Miscible with ether. Sol in methanol, ethanol, chloroform, benzene, toluene, dil acids.

9849. Trimethylsilyl Triflate. *Trifluoromethanesulfonic acid trimethylsilyl ester;* trimethylsilyltrifluoromethane sulfonate. C$_4$H$_9$F$_3$O$_3$SSi; mol wt 222.26. C 21.62%, H 4.08%, F 25.64%, O 21.60%, S 14.43%, Si 12.64%. CF$_3$SO$_2$-OSi(CH$_3$)$_3$. Prepn: M. Schmeisser *et al., Ber.* **103**, 868 (1970); H. W. Roesky, H. H. Giere, *Z. Naturforsch.* **25B**, 773 (1970); D. Haebich, F. Effenberger, *Synthesis* **1978**, 755; T. Morita *et al., ibid.* **1981**, 745. As catalytic reagent: H. Vorbrueggen, K. Krolikiewicz, *Angew. Chem. Int. Ed.* **14**, 421 (1975); as silylating agent: G. A. Olah *et al., J. Org. Chem.* **46**, 5212 (1981).

Liquid, bp$_{760}$ 140°, bp$_{10}$ 36.5°. n$_D^{20}$ 1.3630. Fumes in air; sensitive to atm moisture.

USE: Catalyst and silylating agent for organic syntheses.

9850. Trimetozine. *4-(3,4,5-Trimethoxybenzoyl)morpholine;* N-(3,4,5-trimethoxybenzoyl)tetrahydro-1,4-oxazine; V-7; Opalène; Trioxazine. C$_{14}$H$_{19}$NO$_5$; mol wt 281.31. C 59.78%, H 6.81%, N 4.98%, O 28.44%. Prepd from morpholine and 3,4,5-trimethoxybenzoyl chloride: **Brit.** pat. **872,350** (1961 to Egyesült Gyogyszer és Tápszergyár); Pettit *et al., J. Med. Pharm. Chem.* **5**, 800 (1962).

Crystals, mp 120-122°. Slightly sol in water, alcohol.

THERAP CAT: Anxiolytic.

9851. Trimetrexate. *5-Methyl-6-[[(3,4,5-trimethoxyphenyl)amino]methyl]-2,4-quinazolinediamine;* 2,4-diamino-5-methyl-6-[(3,4,5-trimethoxyanilino)methyl]quinazoline; TMQ; NSC-249008; JB-11; CI-898. C$_{19}$H$_{23}$N$_5$O$_3$; mol wt 369.42. C 61.77%, H 6.28%, N 18.96%, O 12.99%. Lipophilic dihydrofolate reductase inhibitor structurally related to methotrexate, *q.v.,* with antimicrobial and antitumor activity. Prepn: E. F. Elslager, L. M. Werbel, **Brit.** pat. **1,345,502** (1974 to Parke, Davis); E. F. Elslager *et al., J. Med. Chem.* **26**, 1753 (1983); of water soluble salts: N. L. Colbry, **Eur.** pat. **Appl. 51,415;** *idem,* U.S. pat. **4,376,858** (1982, 1983 both to Warner-Lambert). *In vitro* antifolate activity and *in vivo* antitumor effect: J. R. Bertino *et al., Biochem. Pharmacol.* **28**, 1983 (1979). Pharmacology: E. C. Weir *et al., Cancer Res.* **42**, 1696 (1982); R. C. Jackson *et al., Advan. Enzyme Regul.* **22**, 187 (1984). GC-MS determn in human plasma: P. L. Stetson, W. D. Ensminger, *J. Chromatog.* **383**, 69 (1986). Clinical evaluation with leucovorin vs *Pneumocystis carinii* in patients with AIDS: C. J. Allegra *et al., N. Engl. J. Med.* **317**, 978 (1987). Review of pharmacology and clinical efficacy: J. T. Lin, J. R. Bertino, *J. Clin. Oncol.* **5**, 2032-2040 (1987); P. J. O'Dwyer *et al., NCI Monographs* **5**, 105-109 (1987).

Monoacetate monohydrate, C$_{19}$H$_{23}$N$_5$O$_3$.C$_2$H$_4$O$_2$.H$_2$O, crystals from aqueous acetic acid, mp 215-217°. Poorly sol in water. LD$_{50}$ i.p. in mice: 175 mg/kg (Jackson).

D-Glucuronate, C$_{19}$H$_{23}$N$_5$O$_3$.C$_6$H$_{10}$O$_7$, NSC-352122, *Neutrexin, Oncotrex.* Tan colored solid. Soly in water: > 50 mg/ml.

THERAP CAT: Antineoplastic; antipneumocystis.

9852. Trimipramine. *10,11-Dihydro-N,N,β-trimethyl-5H-dibenz[b,f]azepine-5-propanamine; 5-[3-(dimethylamino)-2-methylpropyl]-10,11-dihydro-5H-dibenz[b,f]azepine;* 5-(3-dimethylamino-2-methylpropyl)iminodibenzyl; trimepramine; trimeproprimine; RP-7162; Sapilent. C$_{20}$H$_{26}$N$_2$; mol wt 294.44. C 81.59%, H 8.90%, N 9.51%. Prepn: Jacob, Messer, *Compt. Rend.* **252**, 2117 (1961). Toxicity studies: Okamoto, *C.A.* **72**, 77299y (1970). Chemistry: Bever, Bredenstein, *Deut. Apoth.-Ztg.* **113**, 1562 (1973). Comprehensive description of the maleate: A. A. Al-Badr, *Anal. Profiles Drug Subs.* **12**, 683-712 (1983).

Crystals, mp 45°.

Maleate, C$_{20}$H$_{26}$N$_2$.C$_4$H$_4$O$_4$, *Stangyl (tabl), Surmontil (tabl).* Crystals, mp 142°. Sol in chloroform; slightly sol in water, ethanol. Practically insol in ether.

Methanesulfonate, *Stangyl (amp), Surmontil (amp).*

THERAP CAT: Antidepressant.

9853. Trimoprostil. *(5Z,11α,13E,15R)-15-Hydroxy-11,-16,16-trimethyl-9-oxoprosta-5,13-dien-1-oic acid;* (Z)-7-[(1R,-2R,3R)-2-[(E)-(3R)-3-hydroxy-4,4-dimethyl-1-octenyl]-3-methyl-5-oxocyclopentyl]-5-heptenoic acid; nat-11R,16,16-trimethyl-15R-hydroxy-9-oxoprosta-*cis*-5-*trans*-13-dienoic acid; 11R,16,16-trimethyl-(11-desoxyprostaglandin E$_2$); 11-deoxy-11α,16,16-trimethyl-PGE$_2$; TM-PGE$_2$; Ro-21-6937;

Ulstar. $C_{23}H_{38}O_4$; mol wt 378.55. C 72.98%, H 10.12%, O 16.91%. Synthetic prostaglandin E_2 analog with antisecretory activity. Prepn: G. W. Holland *et al.*, **Ger.** pat. **2,437,-622**; *eidem*, **U.S.** pats. **4,052,446; 4,190,587** (1975, 1977, 1980 all to Hoffmann-La Roche). Pharmacology: D. E. Wilson, S. L. Winter, *Prostaglandins* **16**, 127 (1978); S. P. Lee *et al.*, *Eur. J. Clin. Invest.* **17**, 1 (1987). Effect on bicarbonate secretion: M. Feldman, *J. Clin. Invest.* **72**, 295 (1983); on inhibition of gastric acid secretion: R. J. Wills *et al.*, *Clin. Pharmacol. Ther.* **37**, 113 (1985). Metabolism: S. J. Kolis *et al.*, *Drug Metab. Dispos.* **14**, 465 (1986). Clinical pharmacokinetics: R. J. Wills *et al.*, *J. Clin. Pharmacol.* **26**, 48 (1986). Clinical evaluation: H. G. Dammann *et al.*, *Arzneimittel-Forsch.* **36**, 500 (1986). Multicenter clinical comparison with cimetidine, *q.v.*: K. D. Bardhan *et al.*, *Scand. J. Gastroenterol.* **23**, 134 (1988); with aldioxa, *q.v.*: A. Ishimori *et al.*, *Acta Ther.* **15**, 27 (1989). Toxicity study: M. Shimizu *et al.*, *Shin'yaku to Rinsho* **35**, 2199 (1986), *C.A.* **106**, 150180e (1987).

Colorless oil. $[\alpha]_D$ −51.54° (c = 1 in $CHCl_3$). LD_{50} in mice, rats (mg/kg): 41, 23 orally; 70, 21 i.p.; 68, 29 s.c. (Shimizu).

THERAP CAT: Antiulcerative.

9854. Trimyristin. Myristin; glyceryl trimyristate. $C_{45}H_{86}O_6$; mol wt 723.17. C 74.74%, H 11.99%, O 13.27%. Occurs in many vegetable fats and oils, notably in coconut oil and nutmeg butter.

White to yellowish-gray solid; d_4^{60} 0.885; mp 56-57°; n_D^{60} 1.4429. Insol in H_2O; sol in alc, benzene, chloroform, ether.

9855. Trinexapac-ethyl. *4-(Cyclopropylhydroxymethyl-ene)-3,5-dioxocyclohexanecarboxylic acid ethyl ester*; cimectacarb; CGA-163935; Moddus; Primo; Vision. $C_{13}H_{16}O_5$; mol wt 252.27. C 61.90%, H 6.39%, O 31.71%. Gibberellin biosynthesis inhibitor. Prepn: H.-G. Brunner, **Eur.** pat. **Appl. 126,713**; *idem*, **U.S.** pat. **4,693,745** (1984, 1987 both to Ciba-Geigy). Comprehensive description: E. Kerber *et al.*, *Proc. Brighton Crop Prot. Conf - Weeds* **1989**, 83-88. Field trials in cereals and oilseed rape: J. Amrein *et al.*, *ibid.* 89; in turfgrass: D. B. Vitolo *et al.*, *Proc. Plant Growth Regul. Soc. Am.* **18**, 136 (1991). Mechanism of action: R. Adams *et al.*, *Curr. Plant Sci. Biotechnol. Agric.* **13**, 818 (1992).

mp 36°. n_D^{30} 1.5350. pKa 4.7. Vapor pressure (20°): 1.6 × 10^{-3} Pa. Soly in water at 20° (g/l): 27 (pH 7); 5 (pH 5). Soly at 20° (g/ml): methanol > 1; acetonitrile > 1; cyclo-hexanone > 1; isopropanol 0.9; *n*-octanol 0.18; hexane 0.035. Partition coefficient (*n*-octanol/water at pH 3): 2.1. LD_{50} in rats (mg/kg): 4460 orally; >4000 dermally (Kerber).

USE: Plant growth regulator.

9856. sym-Trinitrobenzene. *1,3,5-Trinitrobenzene*; benzite. $C_6H_3N_3O_6$; mol wt 213.11. C 33.82%, H 1.42%, N 19.72%, O 45.05%. Prepared by decarboxylation of trinitrobenzoic acid, obtained by oxidation of TNT: Clarke, Hartman, *Org. Syn.* **2**, 93 (1922); by the action of alkali on 2,4,6-trinitrobenzaldehyde: Secareanu, *Bull. Soc. Chim.* **51**, 591 (1932).

Orthorhombic bipyramidal plates from glacial acetic acid. d_4^{20} 1.76; mp 122.5°; d_4^{152} 1.4775. Can be sublimed by careful heating, explodes when heated rapidly. Absorption spectrum: Hatzsch, Picton, *Ber.* **42**, 2121 (1909). Soly (g/100 g solvent): Water 0.035; benzene 6.2; methanol 4.9; alcohol 1.9; ether 1.5; carbon disulfide 0.25; petr ether 0.05. Freely sol in dil Na_2SO_3 soln. Trinitrobenzene is dimorphous, the other (rare) form melts at 61°.

USE: Explosive, less sensitive to impact than TNT but more powerful and brisant: Robertson, *J. Chem. Soc.* **119**, 8 (1921); van Duin, *Rec. Trav. Chim.* **39**, 687 (1920).

9857. 2,4,6-Trinitrobenzoic Acid. *sym*-Trinitrobenzoic acid. $C_7H_3N_3O_8$; mol wt 257.12. C 32.70%, H 1.18%, N 16.34%, O 49.78%. Prepd by chromic acid oxidation of 2,4,6-trinitrotoluene: Clarke, Hartman, *Org. Syn.* **2**, 95 (1922); Kastens, Kaplan, *Ind. Eng. Chem.* **42**, 402 (1950).

Orthorhombic crystals from water, mp 228.7°. Sublimes with decompn forming CO_2 and trinitrobenzene. Soly at 25°: 2.05% (w/w) in water, 26.6% in alcohol, 14.7% in ether. Also sol in acetone, methanol; slightly sol in benzene.

9858. 2,4,7-Trinitrofluorenone. *2,4,7-Trinitro-9H-flu-oren-9-one*; TNF. $C_{13}H_5N_3O_7$; mol wt 315.20. C 49.54%, H 1.60%, N 13.33%, O 35.53%. Prepn by nitration of fluoren-one: Schmidt, Bauer, *Ber.* **38**, 3758 (1905); Orchin *et al.*, *J. Am. Chem. Soc.* **69**, 1225 (1947); by nitration of 2,5-dinitrofluorenone: Ray, Francis, *J. Org. Chem.* **8**, 58 (1943). Structure: Bell, *J. Chem. Soc.* **1928**, 1990. Crystal structure: D. L. Dorset *et al.*, *Acta Crystallogr.* **B28**, 3122 (1972); H. L. Ammon, *ibid.* **B29**, 2314 (1973). Carcinogenicity study: C. Huggins, N. C. Yang, *Science* **137**, 257 (1962). Evaluation of employee exposure to TNF in workplace using HPLC: M. J. Seymour, *J. Chromatog.* **236**, 530 (1982); R. E. McCullen, A. N. Sanghvi, *ASTM Spec. Tech. Publ.* **786**, 26 (1982).

Yellow needles from acetic acid, mp 175.2-176°.

USE: In photocopiers; forms charge-transfer complexes with aromatic hydrocarbons and amines.

9859. Trinitromethane. Nitroform. CHN_3O_6; mol wt 151.04. C 7.95%, H 0.67%, N 27.82%, O 63.56%. $CH(NO_2)_3$. Prepn from tetranitromethane and $K_4[Fe(CN)_6]$ in aq soln: Chattaway, Harrison, *J. Chem. Soc.* **109**, 171 (1916); by nitration of acetylene with nitric acid: Hager, *Ind. Eng. Chem.* **41**, 2168 (1949).

Crystals, mp 15° (the unstable *aci*-form, mp 50°). d_4^{25} (liq) 1.469. Heat of combustion 746 cal/g. Dipole moment 2.61 (benzene). Dec above 25°. Explodes when heated rapidly. Sol in water, giving an intensely yellow soln, although the dry crystals are pure white.

Potassium salt, $CK(NO_2)_3$, moderately stable crystals. Soly in water at 0°: 16.7 g/100 ml; at 60°: 193.8 g/100 ml. Soly in ethanol: 5.29 g/l.

Caution: Slightly irritating to eyes, mucous membranes.

USE: In the manuf of explosives and propellants.

9860. 2,4,6-Trinitrotoluene. TNT; α-trinitrotoluol; *sym*-trinitrotoluene; 1-methyl-2,4,6-trinitrobenzene; trotyl; tolit; trilit. $C_7H_5N_3O_6$; mol wt 227.13. C 37.02%, H 2.22%, N 18.50%, O 42.26%. Prepn by nitration of toluene with mixed acid ($HNO_3 + H_2SO_4$) in three steps or by continuous flow according to the Schmid-Meissner and Biazi processes: Swift, Tittensor, *J. Soc. Chem. Ind.* **59**, 92 (1940); Johnston in *McGraw-Hill Encyclopedia of Science and Technology* **9**, 104 (1960). Physical constants and applications: Lothrop, Handrick, *Chem. Rev.* **44**, 419-445 (1949); M. A. Cook, *The Science of High Explosives,* ACS monograph no. 139 (Reinhold, New York, 1958).

Monoclinic rhombohedra from alcohol. The commercial crystals (needles) are yellow. mp 80.1°. d_4^{20} 1.654. Burns at 295° when not confined. Can be distilled under reduced pressure. Vapors are toxic. Dipole moment 1.37. Very sparingly sol in water: About 0.01% at 25°, one gram dissolves in 700 ml of boiling water. Sol in acetone, benzene. Less sol than 2,4,6-trinitrophenol in alcohol, ether, carbon disulfide. Reacts vigorously with reducing agents.

Caution: Potential symptoms of overexposure are liver damage, jaundice; cyanosis; sneezing; coughing, sore throat; peripheral neuropathy, muscle pain; kidney damage; cataract; sensitization dermatitis; leukocytosis; anemia; cardiac irregularities. *See NIOSH Pocket Guide to Chemical Hazards* (DHHS/NIOSH 90-117, 1990) p 220.

USE: High explosive. Must be detonated by a high velocity initiator such as nitramine or by efficient concussion.

9861. Triolein. *9-Octadecenoic acid 1,2,3-propanetriyl ester;* olein; glyceryl trioleate. $C_{57}H_{104}O_6$; mol wt 885.45. C 77.32%, H 11.84%, O 10.84%. One of the chief constituents of nondrying oils and fats. From Palestine olive oil: Hilditch, Madison, *J. Soc. Chem. Ind.* **60**, 258 (1941); from cacao butter: Meara, *J. Chem. Soc.* **1949**, 2154. Prepn by esterification of oleic acid: Wheeler *et al., J. Biol. Chem.* **132**, 687 (1940); Swicklik *et al., J. Am. Oil Chem. Soc.* **32**, 69 (1955). Synthesis: Serebrennikova *et al., Dokl. Akad. Nauk SSSR* **140**, 1083 (1961).

Colorless to yellowish, oily liquid; tasteless, odorless. d_4^{15} 0.915. mp $-4°$ to $-5°$. bp_{15} 235-240°. n_D^{20} 1.4676; n_D^{60} 1.4561. Practically insol in water; sol in chloroform, ether, carbon tetrachloride; slightly sol in alcohol.

9862. Triostins. A *"quinoxaline antibiotic complex"* similar to echinomycin, *q.v.* Powerful, selective inhibitor of nucleic acid synthesis *in vitro.* Isoln of triostin C from *Streptomyces* S-2-210 resembling *S. aureus:* J. Shoji, K. Katagiri, *J. Antibiot.* **14A**, 335 (1961). Isoln of triostins A and B: H. Otsuka, J. Shoji, *ibid.* **19A**, 128 (1966). Structure of C: *eidem, Tetrahedron* **21**, 2931 (1965); of minor components: *eidem, ibid.* **23**, 1535 (1967); of A: H. Otsuka *et al., J. Antibiot.* **29**, 107 (1976). Biosynthesis: T. Yoshida, K. Katagiri, *Biochemistry* **8**, 2645 (1969). Synthesis of A:

P. K. Chakravarty, R. K. Olsen, *Tetrahedron Letters* **1978**, 1613; M. Shin *et al., Pept. Chem., 18th* **1980**, 207, *C.A.* **95**, 98304 (1981). Conformation of A in soln: J. R. Kalman *et al., J. Chem. Soc. Perkin Trans. I* **1979**, 1313. Review of chemistry and biochemistry: M. J. Waring in *Antibiotics* **vol. 5**(pt. 2), F. E. Hahn, Ed. (Springer-Verlag, New York, 1979) pp 173-194.

Triostin A   R = CH$_3$
Triostin C   R = CH(CH$_3$)$_2$

Needles from chloroform + methanol.

Triostin A, $C_{50}H_{62}N_{12}O_{12}S_2$, mp 245-248° (dec). $[\alpha]_D^{25}$ $-157°$ (c = 0.97 in chloroform). uv max (methanol): 243, 320 nm (log ϵ 4.75, 4.11).

Triostin C, $C_{54}H_{70}N_{12}O_{12}S_2$, mp > 260° (dec). $[\alpha]_D^{24}$ $-143.9°$ (c = 1.2 in chloroform). uv max (methanol): 243, 320 nm (log ϵ 4.87, 4.13).

9863. s-Trioxane. *1,3,5-Trioxane;* 1,3,5-trioxacyclohexane; metaformaldehyde; trioxymethylene. $C_3H_6O_3$; mol wt 90.08. C 40.00%, H 6.71%, O 53.28%. *Review:* Walker, Carlisle, *Chem. & Eng. News.* **21**, 1250 (1943).

Stable, cyclic trimer of formaldehyde possessing characteristic chloroform-like odor. Crystalline solid, mp 64°, bp_{759} 114.5° without dec. Sublimes readily. Easily sol in water (17.2 g/100 ml at 18°, 21.1 g/100 ml at 25°), alcohols, ketones, ether, acetone, chlorinated and aromatic hydrocarbons, and other organic solvents. Only slightly sol in pentane, petroleum ether, and lower paraffins. d^{65} 1.17. Flash pt 45°. On distillation with water, forms an azeotrope boiling at 91.4° containing approx 70% trioxane by wt. Slowly depolymerized by strong acids in aq soln, but inert to alkalies. In non-aqueous systems, readily converted to monomeric formaldehyde by small concentrations of strong acids at a rate determined by the acid concentration.

9864. Trioxsalen. *2,5,9-Trimethyl-7H-furo[3,2-g][1]-benzopyran-7-one;* 6-hydroxy-β,2,7-trimethyl-5-benzofuran-acrylic acid δ-lactone; 4,5',8-trimethylpsoralen; NSC-71047; Trisoralen. $C_{14}H_{12}O_3$; mol wt 228.25. C 73.67%, H 5.30%, O 21.03%. Synthetic trimethyl psoralen deriv. Prepn: K. D. Kaufman, *J. Org. Chem.* **26**, 117 (1961); U.S. pat. **3,201,-421** (1965). Protective effect vs UV-B erythema: P. G. Agache, L. Coupez, *Arch. Dermatol. Res.* **268**, 85 (1980). Clinical study: N. Vaatainen *et al., Clin. Exp. Dermatol.* **6**, 133 (1981). Comprehensive description: M. M. A. Hassan, M. A. Loutfy, *Anal. Profiles Drug Subs.* **10**, 705-727 (1981). *See also* Psoralen, Methoxsalen.

Prisms from chloroform, mp 234.5-235°. uv max (methanol): 250, 295, 335 nm (log ε 4.35, 3.99, 3.80). Sol in methylene chloride. Slightly sol in alc, chloroform. Practically insol in water.

THERAP CAT: Pigmentation agent (photosensitizer).

9865. Tripalmitin. *Hexadecanoic acid 1,2,3-propanetriyl ester;* palmitin; glyceryl tripalmitate. $C_{51}H_{98}O_6$; mol wt 807.34. C 75.87%, H 12.24%, O 11.89%. $(C_{15}H_{31}COO)_3$-C_3H_5. Occurs in fats. Prepd from glycerol and palmitic acid in the presence of Twitchell reagent: Ozaki, *Biochem. Z.* **177**, 159 (1926); or in the presence of trifluoroacetic anhydride: Bourne *et al., J. Chem. Soc.* **1949**, 2976.

Needles from ether, mp 66° (occurs also in lower-melting, unstable forms). bp 310-320°; d_4^{70} 0.8730; d_4^{80} 0.8663. n_D^{80} 1.43807. Saponification value 208.5. Insol in water. Practically insol in alcohol (0.0043 parts/100 parts of abs alcohol at 21°). Freely sol in ether, benzene, chloroform.

9866. Tripamide. *(3aα,4α,7α,7aα)-3-(Aminosulfonyl)-4-chloro-N-(octahydro-4,7-methano-2H-isoindol-2-yl)benzamide;* 4-chloro-*N*-(*endo*-hexahydro-4,7-methanoisoindolin-2-yl)-3-sulfamoylbenzamide; toripamide; ADR-033; E-614; Normanal. $C_{16}H_{20}ClN_3O_3S$; mol wt 369.87. C 51.96%, H 5.45%, Cl 9.59%, N 11.36%, O 12.98%, S 8.67%. Sulfonamide with diuretic and peripheral vasodilator activity. Prepn: H. Hamano *et al.,* **Japan.** pat. 73 05,585 (1973 to Eisai), *C.A.* **78**, 136070r (1973). Synthesis of 14C-tripamide: T. Nakamura *et al., J. Label. Comp. Radiopharm.* **14**, 191 (1978). Pharmacological study: T. Satoh *et al., Oyo Yakuri* **21**, 607 (1981), *C.A.* **96**, 28414u (1982). Metabolism: T. Horie *et al., Xenobiotica* **11**, 197, 693 (1981).

Colorless needles.
THERAP CAT: Antihypertensive; diuretic.

9867. Triparanol. *4-Chloro-α-[4-[2-(diethylamino)ethoxy]phenyl]-α-(4-methylphenyl)benzeneethanol;* 1-[*p*-(2-diethylaminoethoxy)phenyl]-1-(*p*-tolyl)-2-(*p*-chlorophenyl)ethanol; 2-(*p*-chlorophenyl)-1-[*p*-(2-diethylaminoethoxy)phenyl]-1-(*p*-tolyl)ethanol; MER-29; Trianel; Hipocolestina; Triparin; Acosterina; Mctasclene; Diticyl; Drenaren; Clotrox; Tropalin; Trikosterol; Valip; Verdiana; Metasqualene; Sclane. $C_{27}H_{32}ClNO_2$; mol wt 438.01. C 74.04%, H 7.36%, Cl 8.09%, N 3.20%, O 7.31%. Prepn: Allen, Palopoli *et al.,* U.S. pat. 2,914,562 (1959 to Wm. S. Merrell).

Crystals, mp 102-104°. Sol in alcohol; slightly sol in olive oil; practically insol in water.

Caution: Withdrawn from market because of association with formation of irreversible cataracts; *see* Laughlin, Carey, *J. Am. Med. Assoc.* **181**, 339 (1962).

THERAP CAT: Antilipemic.

9868. Tripelennamine. *N,N-Dimethyl-N'-(phenylmethyl)-N'-2-pyridinyl-1,2-ethanediamine; 2-[benzyl(2-dimethylaminoethyl)amino]pyridine; N-benzyl-N',N'-dimethyl-N-(2-pyridyl)ethylenediamine; N,N-dimethyl-N'-benzyl-N'-(α-pyridyl)ethylenediamine;* β-dimethylaminoethyl-2-pyridylbenzylamine; β-dimethylaminoethyl-2-pyridylaminotoluene. $C_{16}H_{21}N_3$; mol wt 255.36. C 75.26%, H 8.29%, N 16.46%. Prepn: C. P. Huttrer *et al., J. Am. Chem. Soc.* **68**, 1999 (1946); C. Djerassi *et al.,* U.S. pat. **2,406,594** (1946 to Ciba); R. J. Horclois, U.S. pat. **2,502,151** (1950 to Rhône-Poulenc). Crystal structure: M. Parvez, *Acta Crystallogr.* **C43**, 1408 (1987). Toxicity studies: D. P. Waller *et al., Clin. Toxicol.* **16**, 17 (1980). Comprehensive description: H. G. Piskorik, *Anal. Profiles Drug Subs.* **14**, 108-133 (1985).

Yellow oil. Amine odor. $bp_{0.1}$ 138-142°, $bp_{1.7}$ 185-190°, bp_{20} 193-205°. n_D^{25} 1.5759-1.5765. Miscible with water.

Hydrochloride, $C_{16}H_{21}N_3 \cdot HCl$, *Dehistin, Azaron, Pyribenzamine, PBZ, Vetibenzamina.* Crystals from ethyl acetate + methanol, mp 192-193°. Bitter taste, produces temporary numbness of the tongue. uv max (water): 244, 305 (ε 14470, 4780). One gram dissolves in 0.77 ml water, in 6 ml alcohol, in 6 ml chloroform, in about 350 ml acetone. Practically insol in benzene, ether, ethyl acetate. About neutral to litmus. pH of aq soln contg 25 mg/ml: 6.71; 50 mg/ml: 6.67; 100 mg/ml: 5.56. LD_{50} in mice (mg/kg): 47 i.p. (Walker).

Citrate, $C_{16}H_{21}N_3 \cdot C_6H_8O_7$, crystals, mp 106-110°. Less bitter than the hydrochloride. Freely sol in water, alcohol. Very slightly sol in ether. Practically insol in benzene, chloroform. 1% aq soln has a pH of 4.25.

THERAP CAT: Antihistaminic.
THERAP CAT (VET): Antihistaminic.

9869. Triphenylcarbinol. *Triphenylmethanol;* tritanol. $C_{19}H_{16}O$; mol wt 260.34. C 87.66%, H 6.19%, O 6.15%. $(C_6H_5)_3COH$. Prepd by the action of phenylmagnesium bromide on benzophenone: Acree, *Ber.* **37**, 2755 (1904); Peters *et al., J. Am. Chem. Soc.* **47**, 452 (1925); Dubsky, Jacot-Guillarmod, *Helv. Chim. Acta* **53**, 1965 (1970); by the action of potassium permanganate on triphenylfluoro- or triphenylchloromethane: Blicke, *J. Am. Chem. Soc.* **46**, 1518 (1924). Convenient lab prepn from methyl benzoate and phenylmagnesium bromide: L. F. Fieser, *Experiments in Organic Chemistry* (Boston, 3rd ed., 1955) p 77.

Trigonal crystals from benzene; d_4^0 1.199; mp 164.2°. Distills between 360 and 380° without decompn. Absorption spectrum in alc: Orndorff *et al., J. Am. Chem. Soc.* **49**, 1543 (1927). Insol in water, petr ether. Easily sol in alc, ether, benzene. Sol in concd H_2SO_4 with an intensely yellow color, in glacial acetic acid without color.

9870. Triphenylene. 9,10-Benzphenanthrene; 1,2,3,4-dibenznaphthalene; isochrysene. $C_{18}H_{12}$; mol wt 228.29. C 94.70%, H 5.30%. Occurs in coal tar: Kaffer, *Ber.* **68**, 1812 (1935). Synthesis from cyclohexanone: Mannich, *Ber.* **40**, 163 (1907); Nenitzescu, Curcaneanu, *Ber.* **70**, 346 (1937); from 9-phenanthrylmagnesium bromide and succinic anhydride: Bergmann, *J. Am. Chem. Soc.* **59**, 1441 (1937); from 1,2,3,4-tetrahydrophenanthrene: Bachmann, Struve, *J. Org. Chem.* **4**, 472 (1937); by reaction of 3 mols *o*-bromoiodobenzene in the presence of lithium: Heaney, Millar, *Org. Syn.* **40**, 105 (1960).

Long needles from alc or chloroform. Sublimes; d 1.302; mp 199°; bp 425°. Absorption spectrum: Clar, Lombardi, *Ber.* **65**, 1414 (1932). Solns have blue fluorescence. Does not react with maleic anhydride.

9871. Triphenylmethane. *1,1′,1″-Methylidenetris[benzene];* Tritan. $C_{19}H_{16}$; mol wt 244.34. C 93.40%, H 6.60%. $(C_6H_5)_3CH$. Prepd by ether reduction of triphenylchloromethane obtained by reaction of carbon tetrachloride and benzene in the presence of aluminum chloride: Norris, *Org. Syn.* **4**, 81 (1925); from diphenylchloromethane and phenylmagnesium bromide: Sayles, Kharasch, *J. Org. Chem.* **26**, 4210 (1961).

Orthorhombic pyramidal, solvated crystals containing one mol benzene, mp 78.2°, dries on exposure to air. When dry, mp 93.4° (stable form; there are 2 metastable forms). d_4^{100} 1.0134; bp_{760} 360°; bp_{200} 239.7°; bp_{100} 228°; bp_{60} 221°; bp_{40} 215.5°; bp_{20} 206.8°; bp_{10} 197°; bp_5 188°; $bp_{1.0}$ 170°. n_D^{100} 1.59546. Absorption spectrum: Orndorff, *J. Am. Chem. Soc.* **49**, 1543 (1927); Anderson, *ibid.* **50**, 209 (1928); **51**, 1890 (1929). Very sol in ether, hot alcohol, chloroform. Sol in petr ether, benzene, CS_2. Slightly sol in glacial acetic acid. Soly (parts/100 parts satd soln (w/w) at 30°) 48.6 in chloroform; 12.5 in hexane; 53 in CS_2; 7.24 parts in benzene (19°).

9872. Triphenyl Phosphate. $C_{18}H_{15}O_4P$; mol wt 326.29. C 66.26%, H 4.63%, O 19.61%, P 9.49%. $(C_6H_5O)_3PO$. Prepd from P_2O_5 and phenol: Prahl, U.S. pat. **2,805,240** (1957); by reaction of triethyl phosphite with chloramine-T: Cadogan, Moulden, *J. Chem. Soc.* **1961**, 3079. Toxicology: Sutton *et al.*, *Arch. Environ. Health* **1**, 33 (1960).

Nonflammable needles; mp 49-50°; bp_{11} 245°. Insol in water. Sol in benzene, chloroform, ether, acetone, moderately sol in alcohol.

Caution: Potential symptoms of overexposure are minor changes in blood enzymes. *See NIOSH Pocket Guide to Chemical Hazards* (DHHS/NIOSH 90-117, 1990) p 220.

USE: Noncombustible substitute for camphor in celluloid; rendering acetylcellulose, nitrocellulose, airplane "dope", etc., stable and fireproof; impregnating roofing paper; plasticizer in lacquers and varnishes.

9873. Triphenylphosphine. $C_{18}H_{15}P$; mol wt 262.29. C 82.43%, H 5.76%, P 11.81%. $(C_6H_5)_3P$. Prepd from phenylmagnesium bromide and phosphorus trichloride: Pfeiffer, Pietsch, *Ber.* **37**, 4621 (1904); Sauvage, *Compt. Rend.* **139**, 675 (1904); Dodonon, Medox, *Ber.* **61**, 910 (1928); Denney *et al.*, *J. Am. Chem. Soc.* **83**, 1729 (1961).

Odorless monoclinic platelets or prisms from ether, mp 80.5°. bp > 360° (in inert gas). d_4^{25} 1.194; d_4^{80}(liq) 1.075. For more physical constants *see* Forward *et al.*, *J. Chem. Soc.* **1949**, 5121. Is triboluminescent. Freely sol in ether; sol in benzene, chloroform, glacial acetic acid; less sol in alcohol; practically insol in water.

USE: In organic synthesis; polymerization initiator.

9874. Triphenyltetrazolium Chloride. 2,3,5-Triphenyl-2H-tetrazolium chloride; red tetrazolium; TPTZ; TTC; RT; VitaStain. $C_{19}H_{15}ClN_4$; mol wt 334.81. C 68.16%, H 4.52%, Cl 10.59%, N 16.73%. Prepn: H. v. Pechmann, P. Runge, *Ber.* **27**, 2920 (1894); Atkinson *et al.*, *Science* **111**, 385 (1950); R. Kuhn, D. Jerchel, *Ber.* **74**, 945 (1941); W. Ried, *Angew. Chem.* **14**, 391 (1952); R. Price, *J. Chem. Soc. (A)* **1971**, 3379. Review on the significance of tetrazolium salts as indicators in biochemical and biological oxidation-reduction processes: W. Ried, *loc. cit.*

Solvated nearly colorless needles from alcohol or chloroform, dry at 105°. Turns yellow on exposure to light. Dec 243°. Sol in water, alcohol, acetone; insol in ether. Oxidizes aldoses and ketoses, as well as other α-ketols, and is thereby reduced to a water-insoluble, deep red pigment, a triphenylformazan. LD_{50} i.v. in mice: 5600 μg/kg, *RTECS* **Vol. II**, R. J. Lewis, R. L. Tatken, Eds. (1980) p 690.

USE: In analytical chemistry as a sensitive reagent for reducing sugars, and to distinguish between α-ketols and simple aldehydes. For staining plant and animal tissue. Germination indicator in testing the ability of seeds to germinate. If the seed is alive, its embryonic tissue hydrogenates the colorless tetrazolium salt to a deep red, insol triphenylformazan: G. Lakon, *Ber. Dtsch. Bot. Ges.* **60**, 299, 434 (1942). In histochemical studies. In determination of antibiotics; of dehydrogenases.

9875. Triphenyltin Hydroxide. *Hydroxytriphenylstannane; hydroxytriphenyltin;* fentin hydroxide; fenolovo; DuTer. $C_{18}H_{16}OSn$; mol wt 367.03. C 58.90%, H 4.39%, O 4.36%, Sn 32.34%. $(C_6H_5)_3SnOH$. Prepn by alkaline hydrolysis of $(C_6H_5)_3SnCl$: Kushlefsky *et al.*, *Inorg. Chem.* **2**, 187 (1963); of $(C_6H_5)_3SnI$: Poller, *J. Inorg. Nucl. Chem.* **24**, 593 (1963). GC determn in vegetable crops: H. H. Van den Broek *et al.*, *Analyst (London)* **113**, 1237 (1988). Review of analytical methods: A. Van Rossum *et al.*, *Anal. Methods Pestic. Plant Growth Regul.* **11**, 227-246 (1980). *Reviews:* Ingham, *Chem. Rev.* **60**, 459-539 (1960); R. Bock, *Residue Rev.* **79**, 1-270 (1981).

Crystals. mp 122-123.5°. Thermally dec to $(C_6H_5)_4Sn$, $(C_6H_5)_2SnO$ and H_2O. Slightly sol in alcohol, toluene. Practically insol in water.

Acetate, $C_{20}H_{18}O_2Sn$, *triphenyltin acetate, acetoxytriphenyltin, fentin acetate, Brestan.* Prepn from acetic acid and $(C_6H_5)_3SnOH$: van der Kirk, Luijten, *J. Appl. Chem. (London)* **6**, 49 (1956); and $(C_6H_5)_3SnH$: Weber, Becker, *J. Org. Chem.* **27**, 1258 (1962). Small needles, mp 122-124°. Sol in ether, slightly sol in alcohol, benzene.

Note: The EPA has determined that triphenyltin hydroxide is teratogenic in rats: *Fed. Reg.* **50**, 1107 (1985).

USE: Antifeeding compounds for insect pest control; nonsystemic fungicide.

9876. Tripoli. Finely granulated, white or gray, porous siliceous rock. Used as abrasive and paint filler. The original tripoli, mined near Tripoli in Northern Africa was a diatomaceous earth, but this definition is no longer correct. Tripoli grains are an amorphous form of SiO_2, soft, porous, and free from sharp cutting edges, making it especially suitable as an ingredient of metal polishes. Also used in oil well drilling muds.

9877. Triprolidine. *(E)-2-[1-(4-Methylphenyl)-3-(1-pyrrolidinyl)-1-propenyl]pyridine; trans-2-[3-(1-pyrrolidinyl)-1-p-tolylpropenyl]pyridine; trans-1-(2-pyridyl)-3-pyrrolidino-1-p-tolylprop-1-ene; trans-1-(4-methylphenyl)-1-(2-pyridyl)-3-pyrrolidinoprop-1-ene.* $C_{19}H_{22}N_2$; mol wt 278.40. C 81.97%, H 7.97%, N 10.06%. Histamine H_1-receptor antagonist. Prepn: Adamson, U.S. pats. **2,712,020; 2,712,023** (both 1955 to Burroughs Wellcome); Adamson *et al.*, *J. Chem. Soc.* **1958**, 312. Structure-activity studies: Ison, Casy, *J. Pharm. Pharmacol.* **23**, 848 (1971). Crystal and molecular structure: James, Williams, *Can. J. Chem.* **52**, 1880 (1974). Pharmacokinetics and antihistaminic effects in humans: K. J. Simons *et al.*, *J. Allergy Clin. Immunol.* **77**, 326 (1986). Comprehensive description: S. A. Benezra, C.-H. Yang, *Anal. Profiles Drug Subs.* **8**, 509-528 (1979).

Crystals from light petr, mp 59-61°. uv max (ethanol): 236, 285 nm (ε 15300, 6800).

Hydrochloride monohydrate, $C_{19}H_{22}N_2 \cdot HCl \cdot H_2O$, 295C51, Actidil, Actidilon, Pro-Actidil, Pro-Entra, Venen. Crystals from water, mp 116-118°. uv max (ethanol): 235, 283 nm (ε 15000, 7400). Moderately sol in water, ethanol, methanol.

Oxalate, $C_{19}H_{22}N_2 \cdot C_2H_2O_4$, crystals from methanol, dec 173-174°. uv max (ethanol): 233, 283 nm (ε 16200, 8200). THERAP CAT: Antihistaminic.

9878. Triptorelin. *6-D-Tryptophanluteinizing hormone-releasing factor (pig);* 6-D-tryptophan-LH-RH; D-trp⁶-LHRH; D-Trp⁶LRH; D-trp⁶-gonadorelin; détryptoréline; AY-25650; Wy-42462; Wy-42422. $C_{64}H_{82}N_{18}O_{13}$; mol wt 1311.47. C 58.61%, H 6.30%, N 19.22%, O 15.86%. Synthetic peptide agonist analog of LH-RH, *q.v.* Prepn: A. V. Schally, D. H. Coy, Ger. pat. 2,625,843; *eidem,* U.S. pat. 4,010,125 (1976, 1977); D. H. Coy *et al., J. Med. Chem.* 19, 423 (1976). Comparison with LH-RH of *in vitro* activity: D. H. Coy *et al., Biochem. Biophys. Res. Commun.* 67, 576 (1975). Pharmacokinetics and metabolism in humans: J. L. Barron *et al., J. Clin. Endocrinol. Metab.* 54, 1169 (1982). HPLC analysis: D. C. Serti *et al., J. Liq. Chromatog.* 4, 1135 (1981). RIA in human serum: M. Mason-Garcia *et al., Proc. Nat. Acad. Sci. USA* 82, 1547 (1985). Clinical trial in prostatic carcinoma: H. Parmar *et al., Lancet* 2, 1201 (1985); for *in vitro* fertilization: A. Hazout *et al., Fertil. Steril.* 59, 596 (1993).

5–oxoPro–His–Trp–Ser–Tyr–D-Trp–Leu–Arg–Pro–GlyNH₂

Fluffy, white solid. $[\alpha]_D^{23} -58.8°$ (c = 0.33 in acetic acid). Acetate, $C_{64}H_{82}N_{18}O_{13} \cdot C_2H_4O_2$, *Decapeptyl.* THERAP CAT: Gonad-stimulating principle; antineoplastic (hormonal).

9879. Triptycene. *9,10-Dihydro-9,10-o-benzenoanthracene.* $C_{20}H_{14}$; mol wt 254.33. C 94.45%, H 5.55%. Synthesis by three different methods: Bartlett *et al., J. Am. Chem. Soc.* 64, 2649 (1942); Wittig, *Org. Syn.* 39, 75 (1959); Friedman, Logullo, *J. Am. Chem. Soc.* 85, 1549 (1963).

Crystals from cyclohexane or methylcyclohexane, mp 253-254°.

9880. 2,4,6-Tripyridyl-s-triazine. *2,4,6-Tri-2-pyridyl-s-triazine;* 2,4,6-tripyridyl-1,3,5-triazine; tripyridyltriazine; TPTZ. $C_{18}H_{12}N_6$; mol wt 312.33. C 69.22%, H 3.87%, N 26.91%. Synthesis: Case, Kroft, *J. Am. Chem. Soc.* 81, 905 (1959). Preparation: Schaefer, U.S. pat. 3,294,798 (1966 to Am. Cyanamid). Thermal stability data: Johns *et al., J. Chem. Eng. Data* 7, 227 (1962). *Review:* "2,4,6-Tripyridyl-s-triazine" in Diehl *et al., The Iron Reagents* (The G. Frederick Smith Chem. Co., Columbus, Ohio, 1965) pp 41-56.

Crystals, mp 210-220° (Schaefer); trihydrate from aqueous ethanol, mp 244-245° (Case, Kroft). Reacts with ferrous ions to yield intense violet color over pH range 3.4-5.8. Absorption max Fe(TPTZ)₂²⁺ (water): 593 nm (ε 22600), Collins *et al., Anal. Chem.* 31, 1862 (1959). USE: Reagent for the spectrophotometric determn of iron.

9881. Tris-BP. *2,3-Dibromo-1-propanol phosphate(3:1);* phosphoric acid tris(2,3-dibromopropyl) ester; tris(2,3-dibromopropyl) phosphate; Apex 462-5; Flammex AP; Flammex T 23P; Firemaster LV-T 23P; Firemaster T 23P; T 23P; Fyrol HB 32. $C_9H_{15}Br_6O_4P$; mol wt 697.61. C 15.50%, H 2.17%, Br 68.72%, O 9.17%, P 4.44%. Prepn: G. E. Walter, I. Hornstein, U.S. pat. 2,574,515 (1951 to Glenn. L. Martin Co.); D. E. Overbeek, R. C. Nametz, U.S. pat. 3,046,297 (1962 to Michigan Chem. Co.); R. W. Rimmer, U.S. pat. 3,223,755 (1965 to duPont). Use in flameproofing: W. D. Paist, N. Van Gorder, U.S. pat. 2,662,834 (1953 to Celanese). Mutagenicity studies: M. J. Prival *et al., Science* 195, 76 (1977); A. Nakamura *et al., Mutat. Res.* 66, 373 (1979). Carcinogenicity studies: B. L. Van Duuren *et al., Cancer Res.* 38, 3236 (1978); G. Reznik *et al., J. Nat. Cancer Inst.* 63, 205 (1979). Review of toxicology: F. A. Daniher, *Proc. Symp. Text. Flammability* 4, 126-143 (1976). *Review:* A. Blum, B. N. Ames, *Science* 195, 17 (1977).

Viscous liquid. LD₅₀ orally in rats: > 5.0 g/kg (Daniher).

Note: This substance may reasonably be anticipated to be a carcinogen: *Seventh Annual Report on Carcinogens* (PB95-109781, 1994) p 396.

USE: Flame retardant. Formerly used in children's sleepwear.

9882. Tris(ethylenediamine)cadmium Dihydroxide. Tris(ethylenediamine)cadmium hydroxide; tri(en)cadmium hydroxide; Cadoxen. $C_6H_{26}CdN_6O_2$; mol wt 326.72. C 22.06%, H 8.02%, Cd 34.41%, N 25.72%, O 9.79%. [Cd(H₂NCH₂CH₂NH₂)₃](OH)₂. Prepared by shaking a given amount of cadmium oxide in 10 times its wt of 30% aq ethylenediamine soln for 15 minutes and centrifuging; the supernatant liquor is the product: Jayme, Neuschäfer, *Naturwiss.* 44, 62 (1957). The soln contains about 4.5% Cd (w/w) and dissolves about 3% (w/w) cellulose, giving a clear, highly viscous soln.

USE: Solvent for cellulose: Jayme, Ger. pat. 1,079,318 (1960 to E. Merck), *C.A.* 55, 18107d (1961); solvent for sulfite pulps.

9883. Tris(hydroxymethyl)nitromethane. *2-(Hydroxymethyl)-2-nitro-1,3-propanediol;* 2-nitro-2-(hydroxymethyl)-1,3-propanediol; trimethylolnitromethane. $C_4H_9NO_5$; mol wt 151.12. C 31.79%, H 6.00%, N 9.27%, O 52.94%. Prepn from trioxymethylene and nitromethane: Boileau, *Mém. Poudres* 35, Annexe 7-76 (1953).

Crystals from ethyl acetate + benzene, mp 214° (pure); mp 180° (usual laboratory product); mp 175-176° (tech). Soly in water at 20°: 220 g/100 ml. Freely sol in alcohols. Sparingly sol in benzene, other hydrocarbons. pH of $0.1M$ aq soln 4.5.

Caution: Irritating to skin, mucous membranes.

USE: Bactericide for inanimate objects. To inhibit bacterial growth in circulating industrial water systems, cutting oils, nonprotein glues and sizings.

9884. Trisilane. Trisilicopropane; trisilicon octahydride; silicopropane; trisilicane. H_8Si_3; mol wt 92.32. H 8.73%, Si 91.27%. Si_3H_8. Obtained by separation of mixed silanes prepared from magnesium silicide and hydrochloric acid: Stock, Somiesky, *Ber.* **49**, 111 (1916); **54B**, 524 (1921); **56B**, 247 (1923); Culbertson, **U.S.** pat. **2,551,571** (1951 to Union Carbide); prepared by conversion of silane to higher silanes in an ozonizer type of electric discharge: Spanier, MacDiarmid, *Inorg. Chem.* **1**, 432 (1962).

Liquid. mp −117.4°; bp 52.9°; d^0 0.743; vapor pressure 95.5 mm Hg at 0°. Much less stable than silane or disilane. Detonates in air. Dec in water. Reacts vigorously with CCl_4 and $CHCl_3$.

9885. Tristearin. *Octadecanoic acid 1,2,3-propanetriyl ester;* stearin; glyceryl tristearate. $C_{57}H_{110}O_6$; mol wt 891.50. C 76.80%, H 12.44%, O 10.77%. Present in many animal and vegetable fats, especially the hard ones like cacao butter and tallow. Prepd from stearic acid and glycerol in the presence of Al_2O_3: Ingram, **Brit.** pat. **663,566** (1951 to I.C.I.); by catalytic hydrogenation of many oils: *Bailey's Industrial Oil and Fat Products* (Wiley, New York, 3rd ed., 1964) pp 881-882.

White powder; d_4^{80} 0.862; mp ~55°; on further heating solidifies and melts again at 72°. n_D^{80} 1.4385. Insol in water. Sol in benzene, chloroform, hot alcohol; almost insol in cold alcohol, ether, petr ether.

USE: In textile sizes. Formerly in making candles.

9886. Tris(trimethylsilyl)silane. *1,1,1,3,3,3-Hexamethyl-2-(trimethylsilyl)trisilane.* $C_9H_{28}Si_4$; mol wt 248.66. C 43.47%, H 11.35%, Si 45.18%. $((CH_3)_3Si)_3SiH$. Radical based mediator. Prepn: H. Gilman *et al., Chem. & Ind. (London)* **1965**, 848; H. Bürger, W. Kilian, *J. Organometal. Chem.* **18**, 299 (1969). Si-H bond dissociation energies: J. M. Kanabus-Kaminska *et al., J. Am. Chem. Soc.* **109**, 5268 (1987). Identification as reducing agent: C. Chatgilialoglu *et al., J. Org. Chem.* **53**, 3641 (1988). Use as a reducing agent: M. Ballestri *et al., ibid.* **56**, 678 (1991); D. H. R. Barton *et al., Tetrahedron Letters* **33**, 6629 (1992); in carbonylation: I. Ryu *et al., Synlett.* **1993**, 143.

Liquid, bp_{8mm} 80°-83°.

USE: Reducing and hydrosilyating agent.

9887. Tristriphenylphosphine Rhodium Carbonyl Hydride. *Carbonylhydrotris(triphenylphosphine)rhodium;* hydridocarbonyltris(triphenylphosphine)rhodium(I). $C_{55}H_{46}$-OP_3Rh; mol wt 918.80. C 71.90%, H 5.05%, O 1.74%, P 10.11%, Rh 11.20%. $HRh(CO)[P(C_6H_5)_3]_3$. Prepn: Bath, Vaska, *J. Am. Chem. Soc.* **85**, 3500 (1963); Levison, Robinson, *J. Chem. Soc. (A)* **1970**, 2947; Ahmad *et al., J. Chem. Soc. Dalton Trans.* **1972**, 843; Ahmad *et al., Inorg. Syn.* **15**, 45 (1974). Structure: LaPlaca, Ibers, *J. Am. Chem. Soc.* **85**, 3501 (1963); *eidem, Acta Crystallogr.* **18**, 511 (1965). Chemistry: Evans *et al., J. Chem. Soc. (A)* **1968**, 2660; O'Connor, Wilkinson, *ibid.* 2665.

Yellow microcrystals, d 1.33. mp 120-122° in air; 172-174° under nitrogen. Moderately sol in benzene, chloroform, dichloromethane; sparingly sol in cyclohexane; insol in light petroleum. Dissociates extensively in organic solvents.

USE: Catalyst.

9888. Trithiocarbonic Acid. *Carbonotrithioic acid;* dihydrogen thiocarbonate. CH_2S_3; mol wt 110.22. C 10.90%,

H 1.83%, S 87.27%. Prepared by the treatment of $BaCS_3$ with ice-cold 10% HCl: v. Halban *et al., Z. Elektrochem.* **29**, 445 (1923); Mills, Robinson, *J. Chem. Soc.* **1928**, 2326; Gattow, Krebs, *Angew. Chem.* **74**, 29 (1962).

Highly refractive red oil. mp −26.9°. bp +57.8°. d_4^{20} 1.483; d_4^{25} 1.476. n_D^{20} 1.8225. Dec by water, alcohol. Addition of sulfur produces tetrathiocarbonic acid CH_2S_4.

9889. Trithiozine. *4-[Thioxo(3,4,5-trimethoxyphenyl)-methyl]morpholine;* 4-(3,4,5-trimethoxythiobenzoyl)morpholine; sulmetozine (rescinded INN); tritiozine; ISF-2001; Tresanil. $C_{14}H_{19}NO_4S$; mol wt 297.38. C 56.55%, H 6.44%, N 4.71%, O 21.52%, S 10.78%. Non-anticholinergic gastric secretion inhibitor. Prepn: G. Pifferi, **Ger.** pat. **2,102,246**; *idem,* **U.S.** pat. **3,862,138** (1972, 1975 both to ISF). Synthesis and pharmacology: G. Pifferi *et al., Chim. Ther.* **8**, 462 (1973). HPLC determn in plasma and urine: T. Crolla *et al., J. Chromatog.* **222**, 257 (1981). Clinical studies: M. Elakovic *et al., J. Int. Med. Res.* **8**, 347 (1980); U. Marini *et al., Clin. Ter.* **92**, 399 (1980). Evaluation of gastric acid suppression: K. Gibinski, *Curr. Med. Res. Opin.* **7**, 516 (1981).

Pale yellow solid from ethanol, mp 141-143°. LD_{50} i.p. in mice: 2000 mg/kg (Pifferi, 1973).

THERAP CAT: Antisecretory (gastric).

9890. Triticum. Agropyrum; couch grass; dog grass; graminis; quick grass. Dried rhizome and roots of *Agropyron repens* L., Beauv., *Gramineae,* gathered in spring. *Habit.* Europe, Northern Asia; naturalized in the U.S. *Constit.* Triticin, glucose, mannite, inosite.

9891. Tritium. Triterium. T or $_1^3H$. at. wt 3.016. Exists in the diatomic state; mol wt 6.032. Naturally occurring radioactive isotope of hydrogen. Under normal conditions the total atmospheric content of molecular T_2 gas is only 11 g. Half-life 12.26 years, low β-emitter. First prepd by the bombardment of deuterophosphoric acid with fast deuterons: M. L. E. Oliphant *et al., Proc. Roy. Soc.* **A144**, 692 (1934); T. W. Bonner, *Phys. Rev.* **53**, 711 (1938). Produced commercially from ^6Li by slow neutron bombardment: $_3^6Li + _0^1n \rightarrow _1^3H + _2^4He$. Details of process producing tritium by neutron irradiation of lithium fluoride: Jenks *et al.,* **U.S.** pat. **3,079,317** (1963). *Reviews:* E. A. Evans, *Tritium and Its Compounds* (Butterworth, London, 1966) 441 pp; Mackay, Dove, "Deuterium and Tritium" in *Comprehensive Inorganic Chemistry* vol. 1, J. C. Bailar Jr. *et al.,* Eds. (Pergamon Press, Oxford, 1973) pp 77-116; J. J. Katz in Kirk-Othmer *Encyclopedia of Chemical Technology* vol. 7 (Wiley-Interscience, New York, 3rd ed., 1979) pp 554-564.

Gas, having the properties of hydrogen. mp −254.54° (20.62°K) at 162 mm (triple point). bp −248.12° (25.04°K). Crit temp −232.56°. Crit press. 18.317 atm. Molar density of liquid: 45.35 moles/l (20.62°K).

USE: In fusion-based thermonuclear weapons (hydrogen bombs): energy is released by deuteron bombardment according to the reaction: $_1^3H + _1^2H \rightarrow _2^4He + _0^1n + 18$ meV. Widely used as a radioactive tracer in chemical, biochemical and biological research.

9892. Tritolyl Phosphate. *Phosphoric acid tris(methylphenyl) ester;* tricresyl phosphate; TCP; PX-917; Celluflex 179; Kronitex TCP; Lindol. $C_{21}H_{21}O_4P$; mol wt 368.37. C 68.47%, H 5.75%, O 17.37%, P 8.41%. A mixture of isomeric tritolyl phosphates, usually excluding the very toxic *ortho*-isomer as much as possible. Prepd from cresol and phosphoric oxychloride, phosphoric acid or pentachloride: Prahl, **U.S.** pat. **2,805,240** (1957); Bondy, Gumb, **Brit.** pat. **890,642** (1962 to Coalite and Chem. Prod.); Faith, Keyes & Clark's *Industrial Chemicals,* F. A. Lowenheim, M. K.

Consult the Name Index before using this section.

Moran, Eds. (Wiley-Interscience, New York, 4th ed., 1975) pp 849-853.

Oily, flame resistant liquid. d_{25}^{25} 1.16; bp_{10} ~265°. Pour point −28°. n_D^{25} 1.55. Flash pt (closed cup) 410°. Insol in water (<0.002% at 85°). Sp heat 0.38. Misc with all the common organic solvents and thinners, linseed oil, china wood oil, castor oil.

USE: As plasticizer in vinyl plastics manuf, as flame-retardant, solvent for nitrocellulose, in cellulosic molding compositions, as additive to extreme pressure lubricants, as a nonflammable fluid in hydraulic systems, as lead scavenger in gasoline: Yust, Bame, U.S. pat. **2,889,212** (1959 to Shell); to sterilize certain surgical instruments.

9893. Tri-*o*-tolyl Phosphate. *Phosphoric acid tris(2-methylphenyl) ester;* tri-o-cresyl phosphate. $C_{21}H_{21}O_4P$; mol wt 368.37. C 68.47%, H 5.75%, O 17.37%, P 8.41%.

Colorless or pale yellow liquid. *Poisonous!* mp 25.6°. bp ~410° with slight decompn. Flash pt 225°. d 1.1955. n 1.5575. Sparingly sol in water, slightly sol in alcohol, sol in ether.

Caution: Potential symptoms of overexposure are gastrointestinal disturbance; peripheral neuropathy; cramps in calves, paresthesia in feet or hands; weak feet, wrist drop and paralysis. *See NIOSH Pocket Guide to Chemical Hazards* (DHHS/NIOSH 90-117, 1990) p 220. Ingestion may cause acute nausea, vomiting, diarrhea. Following a latent period, polyneuritis progressing to paralysis of extremities has been seen in severe poisoning. *See Clinical Toxicology of Commercial Products,* R. E. Gosselin *et al.,* Eds. (Williams & Wilkins, Baltimore, 5th ed., 1984) Section III, pp 388-393; *Patty's Industrial Hygiene and Toxicology* vol. **2D,** G. D. Clayton, F. E. Clayton, Eds. (John Wiley & Sons, New York, 4th ed., 1994) pp 3063-3085.

USE: As plasticizer in lacquers and varnishes.

9894. Tritoqualine. *7-Amino-4,5,6-triethoxy-3-(5,6,7,8-tetrahydro-4-methoxy-6-methyl-1,3-dioxolo[4,5-g]isoquinolin-5-yl)-1(3H)-isobenzofuranone; 7-amino-4,5,6-triethoxy-3-(5,-6,7,8-tetrahydro-4-methoxy-6-methyl-1,3-dioxolo[4,5-g]isoquinolin-5-yl)phthalide;* tritocaline; L-554; Hypostamine; Inhibostamin; Livalfa. $C_{26}H_{32}N_2O_8$; mol wt 500.55. C 62.39%, H 6.44%, N 5.60%, O 25.57%. Prepn: **Fr. pat. 1,295,309** (1962 to Lab. de Recherches Biol. Laborec). Activity: Hahn *et al., Arzneimittel-Forsch.* **20,** 1490 (1970).

Crystals, mp 183°.
THERAP CAT: Antihistaminic.

9895. Triuret. *Diimidotricarbonic diamide;* 1,3-dicarbamylurea; carbonyldiurea. $C_3H_6N_4O_3$; mol wt 146.11. C 24.66%, H 4.14%, N 38.35%, O 32.85%. H₂NCONHCO-NHCONH₂. Prepared by the action of phosgene on urea: Schiff, *Ann.* **291,** 374 (1896); Blair, *J. Am. Chem. Soc.* **48,** 101 (1926); Haworth, Mann, *J. Chem. Soc.* **1943,** 603; Werner, Gray, *Sci. Proc. Roy. Dublin Soc.* **24,** 111 (1946).

Crystals from ammonia water, dec 233°. Freely sol in liq ammonia. Forms a mono- and dipotassium salt, see Blair, *loc. cit.*

9896. Troclosene Potassium. *1,3-Dichloro-1,3,5-triazine-2,4,6(1H,3H,5H)-trione potassium salt;* 3,5-dichlorotetrahydro-2,4,6-trioxo-s-triazin-1(2H)-yl potassium; potassium troclosene; potassium dichloroisocyanurate; ACL-59. $C_3Cl_2KN_3O_3$; mol wt 236.06. C 15.26%, Cl 30.04%, K 16.56%, N 17.80%, O 20.33%. Structure studies and prepn: Petterson *et al., J. Org. Chem.* **25,** 1595 (1960). Prepn from trisodium isocyanurate and gaseous Cl: Symes *et al.,* U.S. pats. **3,035,056/7** (1962 to Monsanto).

USE: Useful source of available Cl in solid bleach and detergent formulations.
THERAP CAT: Anti-infective (topical).

9897. Trofosfamide. *N,N,3-Tris(2-chloroethyl)tetrahydro-2H-1,3,2-oxazaphosphorin-2-amine 2-oxide; 2-[bis(2-chloroethyl)amino]-3-(2-chloroethyl)tetrahydro-1,3,2-oxazaphosphorine 2-oxide; N,N,N'-tris(2-chloroethyl)-N',O-propylene phosphoric acid ester diamide;* trilophosphamide; trophosphamide; NSC-109723; Z-4828; Ixoten. $C_9H_{18}Cl_3N_2O_2P$; mol wt 323.59. C 33.41%, H 5.61%, Cl 32.87%, N 8.66%, O 9.89%, P 9.57%. The 3-(2-chloroethyl) deriv. of cyclophosphamide, *q.v.* Prepn: **Brit. pat. 1,188,159,** *C.A.* **73,** 44892d (1970); Arnold *et al.,* **Ger. pat. 2,107,936** [*C.A.* **77,** 152238m (1972)] (1970, 1972, both to Asta-Werke); K. Pankiewicz *et al., J. Am. Chem. Soc.* **101,** 7712 (1979). Pharmacology: N. Brock, *Int. Congr. Chemother., Proc. 5th,* K. H. Spitzy, H. Haschek, Eds. (Verlag Wiener Med. Akad., Vienna, 1967) II(1), pp 155-161; J. Potel, N. Brock, *Arzneimittel-Forsch.* **21,** 1250 (1971); N. Brock, J. Potel, *ibid.* **24,** 1149 (1974); Harrison, Fuquay, *Proc. Soc. Exp. Biol. Med.* **139,** 957 (1972). Crystal and molecular structure: A. Perales, S. Garcia-Blanco, *Acta Crystallogr.* **33B,** 1939 (1977).

Crystals from ether, mp 50-51°. $[\alpha]_D^{25}$ −28.6° (c = 2 in CH_3OH). LD_{50} i.p. in mice: 212 mg/kg (Brock, Potel).
THERAP CAT: Antineoplastic.

9898. Troglitazone. *5-[[4-[(3,4-Dihydro-6-hydroxy-2,5,-7,8-tetramethyl-2H-1-benzopyran-2-yl)methoxy]phenyl]methyl]-2,4-thiazolidinedione;* (±)-5-[4-[(6-hydroxy-2,5,7,8-tetramethylchroman-2-yl)methoxy]benzyl]-2,4-thiazolidinedione; romglizone; CS-045; CI-991. $C_{24}H_{27}NO_5S$; mol wt 441.55. C 65.28%, H 6.16%, N 3.17%, O 18.12%, S 7.26%. Oral hypoglycemic agent which improves insulin sensitivity and decreases hepatic glucose production. Prepn: **Japan. Kokai 85 51,189;** T. Yoshioka *et al.,* **U.S. pat. 4,572,912** (1985, 1986 both to Sankyo); T. Yoshioka *et al., J. Med. Chem.* **32,** 421 (1989). Mechanism of action studies: T. P. Ciaraldi *et al., Metabolism* **39,** 1056 (1990); M. Kellerer *et al., Diabetes* **43,** 447 (1994). Clinical evaluation: T. Kuzuya *et al., Diabetes Res. Clin. Pract.* **11,** 147 (1991). Clinical metabolic effects: S. L. Suter *et al., Diabetes Care* **15,** 193 (1992).

Crystals from benzene-acetone, mp 184-186°.
THERAP CAT: Antidiabetic.

9899. Troleandomycin. *Oleandomycin triacetate ester;* triacetyloleandomycin; NSC-108166; Cyclamycin; Wytrion; Evramycin; TAO; Triocetin. $C_{41}H_{67}NO_{15}$; mol wt 813.98. C 60.50%, H 8.30%, N 1.72%, O 29.48%. For structure *see* Oleandomycin. Semi-synthetic macrolide antibiotic. Prepd from oleandomycin: Celmer *et al., Antibiot. Ann.* **1957-1958**, p 476; **Brit.** pat. **877,730** (1958 to Pfizer); M. Khristov, N. Petkov, *Farmatsiya* **27**, 1 (1977), *C.A.* **90**, 23458c (1979). *Review:* S. Ross, *Antimicrobial Therapy,* B. M. Kagan, Ed. (Saunders, Philadelphia, 1970) pp 134-144.
Crystals from isopropanol. Practically tasteless. Dec 176°. $[\alpha]_D^{25}$ −23° (methanol). pKa 6.6. Soly in water <0.1 g/100 ml.
THERAP CAT: Antibacterial.

9900. Trolnitrate Phosphate. *2,2′2″-Nitrilotrisethanol trinitrate (ester) phosphate (1:2) (salt);* triethanolamine trinitrate biphosphate; trinitrotriethanolamine diphosphate; tri(2-nitroxyethyl)amine biphosphate; tris(2-nitroxyethyl)-amine biphosphate; Angitrit; Bentonyl; Duronitrin; Metamine; Nitretamin; Nitroduran; Ortin; Praenitron; Vasomed. $C_6H_{18}N_4O_{17}P_2$; mol wt 480.17. C 15.01%, H 3.78%, N 11.67%, O 56.64%, P 12.90%. Coronary vasodilator. Prepd by the nitration of triethanolamine followed by precipitation with phosphoric acid: Metadiet, **Fr.** pat. **984,523** (1951); Junkmann *et al.,* **Ger.** pat. **830,955** (1952 to Schering).

Crystals, mp 107-109°.
THERAP CAT: Antianginal.

9901. Tromantadine. *2-[2-(Dimethylamino)ethoxy]-N-tricyclo[3.3.1.1³,⁷]dec-1-ylacetamide; N-1-adamantyl-N-[2-(dimethylamino)ethoxy]acetamide;* 1-(dimethylaminoethoxy-acetamido)adamantane. $C_{16}H_{28}N_2O_2$; mol wt 280.41. C 68.53%, H 10.06%, N 9.99%, O 11.41%. Prepn: Scherm, Peteri, **Ger.** pat. **1,941,218** (1971 to Merz), *C.A.* **74**, 99516k (1971); May, Peteri, *Arzneimittel-Forsch.* **23**, 718 (1973). Chemistry and toxicology: Peteri, Sterner, *ibid.* 577.

Hydrochloride, $C_{16}H_{28}N_2O_2$.HCl, *D-41, Viru-Merz, Viruserol.* Crystals, mp 157-158°. LD_{50} orally in rats: 630 mg/kg; i.v. in mice: 71.0 mg/kg (Peteri, Sterner).
THERAP CAT: Antiviral.

9902. Tromethamine. *2-Amino-2-hydroxymethyl-1,3-propanediol;* trimethylol aminomethane; tris(hydroxymethyl)aminomethane; trisamine; tris buffer; trometamol; tromethane; TRIS; THAM; Talatrol; Tris Amino; Trizma. $C_4H_{11}NO_3$; mol wt 121.14. C 39.66%, H 9.15%, N 11.56%, O 39.62%. May be prepd by reduction or catalytic hydrogenation of the corresp nitro compd. Prepn of similar compds: Hass, Vanderbilt, **U.S.** pat. **2,174,242** (1940); Johnson, De-

gering, *J. Org. Chem.* **8**, 7 (1943); Boileau, *Mém. Poudres* **35**, Annexe 7-76 (1953). Prepn by electrolytic reduction: Mc-Millan, **U.S.** pat. **2,485,982** (1949 to Comm. Solvents Corp.). Titrimetric standard: Whitehead, *J. Chem. Ed.* **36**, 297 (1959). Monograph: *Ann. New York Acad. Sci.* **92**, Art. 2, pp 333-812 (June 17, 1961). Crystal structure: R. Rudman *et al., Science* **200**, 531 (1978). GC determn in plasma: H. Hulshoff, H. B. Kostenbauder, *J. Chromatog.* **145**, 155 (1978). Pharmacokinetics in rabbits: H. Brasch, H. Iven, *Arch. Int. Pharmacodyn.* **254**, 4 (1981). Use as a biological buffer: R. A. Durst, B. R. Staples, *Clin. Chem.* **18** 206 (1972); S. P. Fling, D. S. Gregerson, *Anal. Biochem.* **155**, 83 (1986); T. Higa, D. M. Desiderio, *ibid.* **173**, 463 (1988). Interaction with hydroxyl radicals: M. Hicks, J. M. Gebicki, *FEBS Letters* **199**, 92 (1986). *Review:* Nahas, *Pharmacol. Rev.* **14**, 447-472 (1962).

Crystalline mass, mp 171-172°. bp_{10} 219-220°. Weak, monoacidic base: pKb (25°): 5.91. pKa (20°): 8.3. pKa (37°): 7.82. pH of 0.1 molar aq soln 10.4. Aq solns do not absorb CO_2 from the air. Soly at 25° (mg/ml) in water: 550; ethylene glycol: 79.1; methanol: 26; anhydr ethanol: 14.6; 95% ethanol: 22.0; DMF: 14; acetone: 2.0; ethyl acetate: 0.5; olive oil: 0.4; cyclohexane: 0.1; chloroform: 0.05; carbon tetrachloride: <0.05.
USE: In the synthesis of surface-active agents, vulcanization accelerators, pharmaceuticals. As emulsifying agent for cosmetic creams and lotions, mineral oil and paraffin wax emulsions, leather dressings, textile specialties, polishes, cleaning compds, so-called soluble oils. Absorbent for acidic gases. Biological buffer.
THERAP CAT: Alkalizer.

9903. Tropacine. *endo-α-Phenylbenzeneacetic acid 8-methyl-8-azabicyclo[3.2.1]oct-3-yl ester; 1αH,5αH-tropan-3α-ol diphenylacetate;* diphenylacetic acid 3α-tropanyl ester; 3α-tropanyl diphenylacetate; tropine diphenylacetate. $C_{22}H_{25}NO_2$; mol wt 335.45. C 78.77%, H 7.51%, N 4.18%, O 9.54%. Prepd from tropine and diphenylacetyl chloride: **Swiss** pat. **202,181** (1939 to Ciba), *C.A.* **33**, 8922⁸ (1939); Friess *et al., Toxicol. Appl. Pharmacol.* **2**, 574 (1960).

Hydrochloride, crystals from chloroform + ether, mp 217-218°.
THERAP CAT: Anticholinergic.

9904. Tropacocaine. *exo-8-Methyl-8-azabicyclo[3.2.1]-octan-3-ol benzoate (ester); 1αH,5αH-tropan-3β-ol benzoate;* benzoylpseudotropeine; benzoyl-ψ-tropeine; pseudotropine benzoate; ψ-tropine benzoate; tropacaine. $C_{15}H_{19}NO_2$; mol wt 245.32. C 73.44%, H 7.81%, N 5.71%, O 13.04%. From Javanese coca leaves. Prepared by heating pseudotropine with water and benzoic anhydride: Wilstätter, *Ber.* **29**, 943 (1896). Stereochemistry: Beyerman *et al., Rec. Trav. Chim.* **75**, 1445 (1956).

Plates, tablets, mp 49°. Distills *in vacuo* without dec. pK at 15° = 4.32; K = 1.9×10^{-5}; pH of 0.06M soln 8.4. Freely sol in alc, ether, chloroform, benzene, petr ether, dil acids; slightly sol in water. MLD i.v. in rats: 15-20 mg/kg, Hirschfelder, *Physiol. Rev.* **12**, 262 (1932).

Hydrochloride, $C_{15}H_{19}NO_2 \cdot HCl$, strongly refractive prisms from alcohol, dec 283°. Sol in water, slightly in abs alc, practically insol in ether. Aq solns are stable to boiling water for ~20 minutes. pH of 0.1M soln 5.8.

9905. Tropaeolin O. *4-[(2,4-Dihydroxyphenyl)azo]benzenesulfonic acid monosodium salt; C.I. Acid Orange 6;* sodium *p-(2,4-dihydroxyphenylazo)benzenesulfonate;* sodium azoresorcinolsulfanilate; C.I. Food Yellow 8; C.I. 14270; Resorcinol Yellow; Yellow T; Tropeolin O; Tropaeolin R; Chrysoine; Gold Yellow. $C_{12}H_9N_2NaO_5S$; mol wt 316.27. C 45.57%, H 2.87%, N 8.86%, Na 7.27%, O 25.29%, S 10.14%. Prepd from resorcinol and a *p*-sulfobenzenediazonium salt, followed by reaction with an inorganic sodium salt: Sisley, *Bull. Soc. Chim. France* [3] **25**, 869 (1901). *See also: Colour Index* vol. 4 (3rd ed., 1971) p 4064. Brief review: H. J. Conn's *Biological Stains*, R. D. Lillie, Ed. (Williams & Wilkins, Baltimore, 9th ed., 1977) p 106.

Brown powder. Sol in water with reddish-yellow color; sol in alcohol.
USE: As indicator. pH: yellow 11, orange-brown 12.7. Occasionally used as plasma stain.

9906. Tropaeolin OO. *4-[[4-(Phenylamino)phenyl]azo]benzenesulfonic acid monosodium salt; C.I. Acid Orange 5; p-[(p-anilinophenyl)azo]benzenesulfonic acid sodium salt;* sodium *p-[(p-anilinophenyl)azo]benzenesulfonate;* Diphenylamine Orange; Orange GS; Orange N; Orange IV; Fast Yellow; Acid Yellow D; C.I. 13080. $C_{18}H_{14}N_3NaO_3S$; mol wt 375.38. C 57.59%, H 3.76%, N 11.19%, Na 6.12%, O 12.79%, S 8.54%. Prepd from diphenylamine and *p*-sulfobenzenediazonium chloride, followed by reaction with NaOH: Witt, *Ber.* **12**, 258 (1879); *Colour Index* vol. 4 (3rd ed., 1971) p 4045.

Orange-yellow scales or yellow powder. Sol in water.
USE: As indicator. pH: red 1.4 to yellow 2.6.

9907. Tropane. *8-Methyl-8-azabicyclo[3.2.1]octane; 1αH,5αH-tropane;* 2,3-dihydro-8-methylnortropidine. $C_8H_{15}N$; mol wt 125.21. C 76.74%, H 12.07%, N 11.19%. Structure and synthesis: Ladenburg, *Ber.* **16**, 1408 (1883); Merling, *Ber.* **25**, 3124 (1892); Willstätter, Iglauer, *Ber.* **30**, 721 (1897); **33**, 1170 (1900); *eidem, Ann.* **317**, 315, 350 (1901); Robinson, *J. Chem. Soc.* **111**, 762, 876 (1917); Hess, *Ber.* **51**, 1007 (1918); Schöpf, *Angew. Chem.* **50**, 779, 797 (1937); Ruggli, Maeder, *Helv. Chim. Acta* **27**, 436 (1944); Keagle, Hartung, *J. Am. Chem. Soc.* **68**, 1608 (1946).

Liquid; bp 163-169°; d_{15}^{15} 0.9259. Sparingly sol in water, the soly decreasing with an increase in temp. Tropane and water will mix provided the volume of tropane is in excess.

9908. Tropenzile. *(3-endo, 6-exo)-α-Hydroxy-α-phenyl-benzeneacetic acid 6-methoxy-8-methyl-8-azabicyclo[3.2.1]oct-3-yl ester; 6-methoxy-1αH,5αH-tropan-3α-ol benzilate;* 6-methoxytropine benzilate; benzilic acid 6-methoxytropine ester. $C_{23}H_{27}NO_4$; mol wt 381.47. C 72.42%, H 7.13%, N 3.67%, O 16.78%. Prepn: Stoll *et al., Helv. Chim. Acta* **38**, 571 (1955); **Swiss** pat. **325,296** (1957 to Sandoz).

Needles from benzene + petr ether, mp 99-101°.
Hydrochloride, $C_{23}H_{27}NO_4 \cdot HCl$, mp 146-148°.
Methyl bromide, $C_{23}H_{27}NO_4 \cdot CH_3Br$, *tropenzilium bromide.* Ingredient of *Palerol.*
THERAP CAT: Antispasmodic.

9909. Tropesin. *(±)-1-(4-Chlorobenzoyl)-5-methoxy-2-methyl-1H-indole-3-acetic acid 2-carboxy-2-phenylethyl ester; d,l-2-phenyl-3-[1-(4-chlorobenzoyl)-5-methoxy-2-methylindole-3-acetoxy]propionic acid; 1-p-chlorobenzoyl-5-methoxy-2-methyl-3-indolylacetic acid 2-phenyl-2-carboxyethyl ester;* Repanidal. $C_{28}H_{24}ClNO_6$; mol wt 505.95. C 66.47%, H 4.78%, Cl 7.01%, N 2.77%, O 18.97%. Tropic acid ester of indomethacin, *q.v.* Prepn: L. Fisnerova *et al.,* **Ger.** pat. **2,727,629**; *eidem,* **U.S.** pat. **4,136,194** (1978, 1979 both to Spofa). Clinical pharmacokinetics: I. Janku *et al., Drugs Exp. Clin. Res.* **IX**, 407 (1983). Review of pharmacology and clinical experience: J. Grimová *et al., Drugs Today* **27**, 391-400 (1991).

Crystals from nitromethane, mp 130-132°. Also reported as mp 127-129°; 128-130°. LD_{50} in female mice, rats (mg/kg): 190, 140 orally (Fisnerova, 1979).
THERAP CAT: Anti-inflammatory; analgesic.

9910. Tropic Acid. *α-(Hydroxymethyl)benzeneacetic acid;* 2-phenylhydracrylic acid; *α-phenyl-β-hydroxypropionic acid;* tropaic acid; tropeic acid. $C_9H_{10}O_3$; mol wt 166.18. C 65.05%, H 6.07%, O 28.88%. Degradation product of tropane alkaloids, esp atropine: Lossen, *Ann.* **133**, 351, 370 (1865). Resolution of isomers: McKenzie, Wood, *J. Chem. Soc.* **115**, 828 (1919). Absolute config of isomers: Fodor, Csepreghy, *ibid.* **1961**, 3222; Watson, Youngson, *J. Chem. Soc. Perkin Trans. I* **1972**, 1597. Prepn: Sletzinger,

Paulsen, U.S. pat. **2,390,278** (1945 to Merck & Co.); Blicke, U.S. pat. **2,716,650** (1955 to U. of Michigan); **Ger.** pat. **923,426** (1955 to Sterling Drug). Biosynthetic studies: Louden, Leete, *J. Am. Chem. Soc.* **84**, 1510, 4507 (1962).

(±)-Form, needles or plates from water or benzene, mp 118°. K at 25° = 7.5×10^{-5}. Absorption spectrum: Dobbie, Fox, *J. Chem. Soc.* **103**, 1194 (1913). 1 gram dissolves in 50 ml water; freely sol in boiling water; sol in alcohol, ether, slightly in benzene; practically insol in petr ether.
(+)-Form, mp 107°. $[\alpha]_D^{20}$ +72° (c = 0.5 in water).
(−)-Form, mp 126-128°. $[\alpha]_D^{20}$ −72° (c = 0.5 in water).

9911. Tropicamide. *N-Ethyl-α-(hydroxymethyl)-N-(4-pyridinylmethyl)benzeneacetamide; N-ethyl-2-phenyl-N-(4-pyridylmethyl)hydracrylamide; N*-ethyl-*N*-(γ-picolyl)tropamide; Mydriacyl; Mydriaticum. $C_{17}H_{20}N_2O_2$; mol wt 284.36. C 71.81%, H 7.09%, N 9.85%, O 11.25%. Prepn: Rey-Bellet, Spiegelberg, U.S. pat. **2,726,245** (1955 to Hoffmann-La Roche). Comprehensive description: K. W. Blessel *et al.*, *Anal. Profiles Drug Subs.* **3**, 565-580 (1974).

Crystals, mp 96-97°. uv max (0.025 mg/ml in 0.1*N* HCl): 254 nm (ε 5.1×10^3).
THERAP CAT: Anticholinergic (ophthalmic).

9912. Tropine. *endo-8-Methyl-8-azabicyclo[3.2.1]octan-3-ol; 1αH,5αH-tropan-3α-ol;* 2,3-dihydro-3α-hydroxy-8-methylnortropidine; 2,3-dihydro-3α-hydroxytropidine. $C_8H_{15}NO$; mol wt 141.21. C 68.04%, H 10.71%, N 9.92%, O 11.33%. Prepd by reaction of tropidine and HBr, followed by hydrolysis and separation of isomers, tropine and pseudotropine: Ladenburg, *Ber.* **35**, 1159 (1902); by hydrogenation of tropinone in the presence of Raney nickel: Van de Kamp, Sletzinger, U.S. pat. **2,366,760** (1945 to Merck & Co.); Stoll *et al.*, U.S. pat. **2,746,976** (1956 to Sandoz). Yield of tropine and pseudotropine under varying conditions of reduction: Beckett *et al.*, *Tetrahedron* **6**, 319 (1959). Separation of tropine and pseudotropine by gold salt formation: Ladenburg, *loc. cit.*; by fractional distillation under reduced pressure: Friess *et al.*, *Toxicol. Appl. Pharmacol.* **2**, 574 (1960). Stereochemistry: Beyerman *et al.*, *Rec. Trav. Chim.* **75**, 1445 (1956).

Hygroscopic plates from ether, mp 63°. bp 233°. pK at 15° = 3.80. pH of 0.05 molar soln 11.5. Freely sol in water and alcohol; sol in ether and chloroform.
Note: Esters of tropine are known as *tropeines.*

9913. Tropine Benzylate. *endo-α-Hydroxy-α-phenylbenzeneacetic acid 8-methyl-8-azabicyclo[3.2.1]oct-3-yl ester; 1αH,5αH-tropan-3α-ol benzilate;* benzilic acid 3α-tropanyl ester; glykin; BAT; BTE; BETE. $C_{22}H_{25}NO_3$; mol wt 351.45. C 75.19%, H 7.17%, N 3.99%, O 13.66%. Prepn: Hromatka *et al.*, *Monatsh.* **83**, 1321 (1952).

Prisms from ether or benzene, mp 152-153°.
Hydrochloride, $C_{22}H_{25}NO_3.HCl$, mp 239-240°.

9914. Tropisetron. *endo-1H-Indole-3-carboxylic acid 8-methyl-8-azabicyclo[3.2.1]oct-3-yl ester; 3α-tropanyl-1H-indole-3-carboxylic acid ester; 1αH,5αH-tropan-3α-yl indole-3-carboxylate;* ICS-205-930. $C_{17}H_{20}N_2O_2$; mol wt 284.36. C 71.81%, H 7.09%, N 9.85%, O 11.25%. Specific serotonin (5-HT₃) receptor antagonist. Prepn: P. Donatsch *et al.*, **Ger.** pat. **Appl.** **3,322,574** (1983 to Sandoz); P. Donatsch *et al.*, U.S. pat. **4,789,673** (1988). Pharmacology: B. P. Richardson *et al.*, *Nature* **316**, 126 (1985); F. M. Williams *et al.*, *J. Cardiovasc. Pharmacol.* **7**, 550 (1985). Receptor binding studies: C. Waeber *et al.*, *Neuroscience* **31**, 393 (1989). Clinical pharmacokinetics: V. Fischer *et al.*, *Drug Metab. Dispos.* **20**, 603 (1992). Series of articles on antiemetic efficacy in chemotherapy and radiotherapy: *Drugs* **43**, Suppl. 3, 6-39 (1992). Review of pharmacology: C. Seynaeve *et al.*, *Anti-Cancer Drugs* **2**, 343-355 (1991); and toxicology: K. Kutz, *Ann. Oncol.* **4**, Suppl. 3, S15-S18 (1993).

mp 201-202° (from methylene chloride/ethyl acetate).
Monohydrochloride, $C_{17}H_{20}N_2O_2.HCl$, *Navoban, Novaban.* mp 283-285° (dec).
THERAP CAT: Antiemetic.

9915. Tropomyosins. Fibrous proteins involved in the regulation of muscle relaxation. They are present in all forms of striated and smooth muscles and probably in nonmuscle cells as well. Native tropomyosin consisting of two proteins, tropomyosin and *troponin*, is the Ca^{2+}-sensitive regulatory protein that controls the interaction between actin and myosin necessary for the production of force in muscle. In all skeletal tissues, there are two forms of tropomyosin chains, designated α and β. Their ratio depends on the muscle source. Troponin consists of three subunits, troponin T (the tropomyosin binding subunit), troponin I (the actomyosin ATPase inhibitory subunit) and troponin C (the Ca^{2+}-binding subunit). All three subunits, in addition to tropomyosin, are responsible for the native tropomyosin activity. Isoln of tropomyosin from skeletal muscle and cardiac muscle: K. Bailey, *Nature* **157**, 368 (1946); *idem, Biochem. J.* **43**, 271 (1948). Structure studies: R. S. Hodges, L. B. Smillie, *Biochem. Biophys. Res. Commun.* **41**, 987 (1970). Amino acid sequence studies: J. Sodek *et al.*, *Proc. Nat. Acad. Sci. USA* **69**, 3800 (1972). X-ray crystal structure of troponin C: M. Sundaralingam *et al.*, *Science* **227**, 945 (1985). *Reviews:* C. E. Bodwell, K. Laki, in *Contractile Proteins and Muscle*, K. Laki, Ed. (Dekker, New York, 1971); W. F. Harrington in *The Proteins*, **vol. 4**, H. Neurath, R. L. Hill, Eds. (Academic Press, New York, 1979) pp 317-327. Comprehensive review of isolation, preparation, identification and role in the contractile process: *Methods Enzymol.* **vol. 85**, Part B, entitled "Structural and Contractile

Proteins", D. W. Frederiksen, L. W. Cunningham, Eds. (Academic Press, New York, 1982) 774 pp.

9916. Tropylium Bromide. *Cycloheptatrienylium bromide;* cycloheptatrienocarbonium bromide. C_7H_7Br; mol wt 171.04. C 49.16%, H 4.13%, Br 46.72%. Prepd by bromination of 1,3,5-cycloheptatriene in carbon tetrachloride, followed by removal of the CCl_4 and heating the residue *in vacuo* for several days: von Doering, Knox, *J. Am. Chem. Soc.* **79**, 352 (1957); King, Stone, *Inorg. Syn.* **7**, 99 (1963).

Yellow prisms from ethanol, mp 203°. Freely sol in water. Practically insol in ether.

9917. Trospectomycin. *[2R-(2α,4aβ,5aβ,6β,7β,8β,9α,-9aα,10aβ)]-2-Butyldecahydro-4a,7,9-trihydroxy-6,8-bis(methylamino)-4H-pyrano[2,3-b][1,4]benzodioxin-4-one;* 6'-n-propylspectinomycin. $C_{17}H_{30}N_2O_7$; mol wt 374.43. C 54.53%, H 8.08%, N 7.48%, O 29.91%. Semi-synthetic analog of spectinomycin, *q.v.* Prepn: D. R. White, **Ger.** pat. **3,308,-196**; *idem*, **U.S.** pat. **4,532,336** (1983, 1985 both to Upjohn); D. R. White *et al.*, *J. Antibiotics* **36**, 339 (1983); P. M. Herrinton *et al.*, *J. Org. Chem.* **58**, 678 (1993). *In vitro* activity: G. E. Zurenko *et al.*, *Antimicrob. Ag. Chemother.* **32**, 216 (1988). Mode of action: E. M. Veringa *et al.*, *Int. J. Med. Microbiol.* **271**, 311 (1989). Pharmacokinetic and tolerance study: G. R. Peters *et al.*, *Int. J. Clin. Pharmacol. Ther. Toxicol.* **28**, 361 (1990). HPLC determn in plasma and serum: R. J. Simmonds *et al.*, *J. Liq. Chromatog.* **13**, 1125 (1990). Clinical study in gonorrhea: T. F. Mroczkowski *et al.*, *Drugs Exptl. Clin. Res.* **19**, 41 (1993).

Sulfate pentahydrate, $C_{17}H_{30}N_2O_7 \cdot H_2SO_4 \cdot 5H_2O$, *U-63366F, Spexil.* Crystals from acetone/water.
THERAP CAT: Antibacterial.

9918. Trospium Chloride. *endo-3-[(Hydroxydiphenylacetyl)oxy]spiro[8-azoniabicyclo[3.2.1]octane-8,1'-pyrrolidinium] chloride; 3α-hydroxyspiro[1αH,5αH-nortropane-8,1'-pyrrolidinium] chloride benzilate;* azoniaspiro(3α-benziloyloxynortro-pane-8,1'-pyrrolidine) chloride; azoniaspiro(3α-diphenylglycoloyloxynortropan-8,1'-pyrrolidine) chloride; 3α-benziloyloxyspiro(nortropane-8,1'-pyrrolidinium) chloride; Relaspium; Spasmex. $C_{25}H_{30}ClNO_3$; mol wt 427.97. C 70.16%, H 7.07%, Cl 8.28%, N 3.27%, O 11.22%. Tropine derivative with anticholinergic activity. Prepn: **Neth.** pat. **Appl. 6,402,155**; R. Pfleger *et al.*, **U.S.** pat. **3,480,626** (1964, 1969 both to Pfleger); H. Bertholdt *et al.*, *Arzneimittel-Forsch.* **17**, 719 (1967). Pharmacology and toxicology in animals: H. Antweiler *et al.*, *ibid.* **16**, 1581 (1966). Inhibition of gastric motility and acid secretion in humans: G. Lux, P. Frühmorgen, *Fortschr. Med.* **96**, 2113 (1978). Fluorimetric determn in plasma and urine: G. Schladitz-Keil *et al.*, *J. Chromatog.* **345**, 99 (1985). Bioavailability: *eidem, Arzneimittel-Forsch.* **36**, 984 (1986).

Crystals from ethanol-ether, mp 255-257° (dec). LD_{50} in mice (mg/kg): 12.3 i.v. (Antweiler).
THERAP CAT: Antispasmodic.

9919. Trovafloxacin. *(1α,5α,6α)-7-(6-Amino-3-azabicyclo[3.1.0]hex-3-yl)-1-(2,4-difluorophenyl)-6-fluoro-1,4-dihydro-4-oxo-1,8-naphthyridine-3-carboxylic acid;* CP-99219. $C_{20}H_{15}F_3N_4O_3$; mol wt 416.36. C 57.70%, H 3.63%, F 13.69%, N 13.46%, O 11.53%. Fluorinated quinolone antibacterial. Prepn: K. E. Brighty, **U.S.** pat. **5,164,402** (1992 to Pfizer). Antibacterial spectrum *in vitro:* B. B. Gooding, R. N. Jones, *Antimicrob. Ag. Chemother.* **37**, 349 (1993); H. C. Neu, N.-X. Chin, *ibid.* **38**, 2615 (1994).

Hydrochloride, $C_{20}H_{15}F_3N_4O_3 \cdot HCl$. Pale yellow crystals from acetonitrile/methanol, mp 246° (dec). Monomethanesulfonate, $C_{20}H_{15}F_3N_4O_3 \cdot CH_4O_3S$, *trovafloxacin mesylate,* CP-99219-27.
THERAP CAT: Antibacterial.

9920. Troxerutin. *2-[3,4-Bis(2-hydroxyethoxy)phenyl]-3-[[6-O-(6-deoxy-α-L-mannopyranosyl)-β-D-glucopyranosyl]oxy]-5-hydroxy-7-(2-hydroxyethoxy)-4H-1-benzopyran-4-one;* 7,3',4'-tris[O-(2-hydroxyethyl)]rutin; trioxyethylrutin; tri-(hydroxyethyl)rutoside; Posorutin; Ruven; Vastribil; Veinamitol; Veniten. $C_{33}H_{42}O_{19}$; mol wt 742.69. C 53.37%, H 5.70%, O 40.93%. The principal component of a mixture, the *O-(β-hydroxyethyl)rutosides,* which also contains mono-, di-, tetra- and other trihydroxyethyl derivs of rutin, *q.v.* The mixture is prepd by the hydroxyethylation of the phenolic groups of rutin with glycochlorohydrin in alk medium: J. Favre, **Swiss** pat. **349,614** (1957); *see also* **Brit.** pat. **833,-174** (1960 to Zyma), *C.A.* **54**, 21135i (1960). Isolation and identification of major components of the mixture: P. Courbat *et al.*, *Helv. Chim. Acta* **49**, 1203, 1420 (1966). Prepn of troxerutin: P. J. Courbat, **U.S.** pat. **3,420,815** (1969 to Zyma). Metabolism in man: A. M. Hackett *et al.*, *Arzneimittel-Forsch.* **26**, 925 (1976).

Yellow powder, mp 181°. Sol in water, glycerol, propylene glycol. Practically insol in cold ethanol, methanol (forms alcoholate), ether, benzene, chloroform.

O-(β-Hydroxyethyl)rutinosides (mixture), *HR, Paroven, Relvene, Varemoid, Venoruton.* Yellow powder, mp 156°. Sol in water, methanol, glycerol, propylene glycol. Practically insol in cold ethanol (forms alcoholate), ether, benzene, chloroform.

THERAP CAT: Treatment of venous disorders.

9921. Troxipide. (±)-*3,4,5-Trimethoxy-N-3-piperidinylbenzamide;* 3-(3,4,5-trimethoxybenzamido)piperidine; KU-54; Aplace. $C_{15}H_{22}N_2O_4$; mol wt 294.35. C 61.21%, H 7.53%, N 9.52%, O 21.74%. Prepn: Belg. pat. **736,840**; T. Irikura *et al.,* U.S. pat. **3,647,805** (1969, 1972 both to Kyorin). Anti-ulcer activity: T. Irikura, K. Kasuga, *J. Med. Chem.* **14**, 357 (1971). Pharmacokinetics in man: T. Irikura *et al., Iyakuhin Kenkyu* **12**, 971 (1981), *C.A.* **96**, 79377s (1982). Effect on gastric mucosa in rats: Y. Abe *et al., Nippon Yakurigaku Zasshi* **83**, 317 (1984), *C.A.* **101**, 456g (1984); on glucosamine synthetase: Y. Abe *et al., Oyo Yakuri* **27**, 521 (1984), *C.A.* **101**, 17001c (1984). Metabolism in rats: K. Tagaki *et al., ibid.* 1151, *C.A.* **101**, 103555t (1984); K. Tagaki, K. Endo, *ibid.* 1167, *C.A.* **101**, 103556u (1984). Toxicity: T. Irikura *et al., Kiso to Rinsho* **12**, 3422 (1978).

Needles from acetonitrile, mp 179-181.5°. Sol in ethanol. LD_{50} in male, female rats, male, female mice (mg/kg): 500, 2100, 2200, 2000 orally; >4150, >4150, 1600, 1550 s.c.; 340, 340, 300, 305 i.p. (Irikura, 1978).

Hydrochloride hemihydrate, $C_{15}H_{22}N_2O_4 \cdot HCl \cdot \frac{1}{2}H_2O$, needles from acetonitrile, mp 206-209°. Sol in ethanol.

THERAP CAT: Antiulcerative.

9922. Truxillic Acid. *2,4-Diphenyl-1,3-cyclobutanedicarboxylic acid.* $C_{18}H_{16}O_4$; mol wt 296.32. C 72.96%, H 5.44%, O 21.60%. Cinnamic acid polymers obtained from the minor alkaloids of cocaine: Liebermann, *Ber.* **21**, 2342 (1888). Five stereoisomers have been obtained: α-, γ-, ε-, peri- and epi-isomers. Stereochemical configurations: Stoermer, Bacher, *Ber.* **57B**, 15-23 (1924).

	a	b	c	d	e	f
α-Isomer	COOH	H	H	C_6H_5	H	COOH
γ-Isomer	COOH	H	H	C_6H_5	COOH	H
ε-Isomer	H	COOH	C_6H_5	H	H	COOH
peri-Isomer	COOH	H	C_6H_5	H	COOH	H
epi-Isomer	COOH	H	C_6H_5	H	H	COOH

α-Isomer (d = C_6H_5; b,c,e = H; a,f = COOH), *γ-isatropaic acid, cocaic acid.* Prepd by irradiation of cinnamic acid in water: White, Dunathan, *J. Am. Chem. Soc.* **78**, 6055 (1956). Crystals from acetic acid, mp 284-285°. Sparingly sol in boiling water, in ether, benzene, carbon disulfide; sol in hot glacial acetic acid, hot alc; sparingly sol in acetone.

γ-Isomer (d = C_6H_5; b,c,f = H; a,e = COOH), *ε-isatropaic acid.* γ-Truxillic anhydride, obtained by heating α-truxillic acid with acetic anhydride and sodium acetate, was heated with alkali and the free acid pptd with HCl: Liebermann, *Ber.* **22**, 124 (1889). Needles from dil alc, mp 228°. Very slightly sol in hot water; sol in ether.

ε-Isomer (c = C_6H_5; a,d,e = H; b,f = COOH), *β-cocaic acid.* Prepd by fusion of α-truxillic acid with KOH: Hesse, *Ann.* **271**, 180 (1892). Needles from ether, mp 192°. Freely sol in glacial acetic acid, abs alcohol, chloroform; less sol in benzene. Practically insol in ligroin.

peri-Isomer (c = C_6H_5; b,d,f = H; a,e = COOH), *η-truxillic acid.* Prepn: γ-Truxillic anhydride heated at low pressure forms the peri-anhydride, converted to the acid by warming with alcoholic KOH: Stoermer, Bacher, *loc. cit.* Crystals from benzene + ligroin, mp 266° (effervescence). Soluble in alcohol. Practically insol in ether, benzene.

epi-Isomer (c = C_6H_5; b,d,e = H; a,f = COOH). Prepd by boiling peri-truxillic acid with excess 10% NaOH: Stoermer, Bacher, *loc. cit.* Crystals from dil alc, mp 285-287°. By melting or heating with acetic anhydride in a tube, ε-isomer forms. Practically insol in ether, benzene.

9923. Trypan Blue. *3,3'-[(3,3'-Dimethyl[1,1'-biphenyl]-4,4'-diyl)bis(azo)]bis[5-amino-4-hydroxy-2,7-naphthalenedisulfonic acid] tetrasodium salt;* C.I. Direct Blue 14; C.I. 23850; 3,3'-[(3,3'-dimethyl-4,4'-biphenylene)bis(azo)]bis(5-amino-4-hydroxy-2,7-naphthalenedisulfonic acid) tetrasodium salt; tetrasodium 3,3'-[(3,3'-dimethyl-4,4'-biphenylene)bis(azo)]bis(5-amino-4-hydroxy-2,7-naphthalenedisulfonate); sodium ditolyl-diazobis-8-amino-1-naphthol-3,6-disulfonate; Benzamine Blue; Diamine Blue; Benzo Blue; Congo Blue; Dianil Blue; Naphthylamine Blue; Niagara Blue. $C_{34}H_{24}N_6Na_4O_{14}S_4$; mol wt 960.82. C 42.50%, H 2.52%, N 8.75%, Na 9.57%, O 23.31%, S 13.35%. Prepd by coupling diazotized o-tolidine with 5-amino-4-hydroxy-2,7-naphthalenedisulfonic acid in sodium carbonate soln: Lewers, Lowry, *Ind. Eng. Chem.* **17**, 1289-1290 (1925); *Colour Index vol. 4* (3rd ed., 1971) p 4198. Toxicity study: Anderson *et al., Proc. Soc. Exp. Biol. Med.* **31**, 825 (1934). Review of teratogenicity studies: R. L. Cahen, *Clin. Pharmacol. Ther.* **5**, 480 (1964).

Bluish-gray powder. Sol in water forming a deep blue soln with violet tinge. Almost insol in alcohol. LD_{100} i.v. in rats: 300 mg/kg (Anderson).

USE: Biological stain.

9924. Trypan Red. *4,4'-[(3-Sulfo[1,1'-biphenyl]-4,4'-diyl)bis(azo)]bis[3-amino-2,7-naphthalenedisulfonic acid] pentasodium salt;* 4,4'-[(3-sulfo-4,4'-biphenylene)bis(azo)]bis-(3-amino-2,7-naphthalenedisulfonic acid) pentasodium salt; pentasodium 4,4'-[(3-sulfo-4,4'-biphenylene)bis(azo)]bis(3-amino-2,7-naphthalenedisulfonate); C.I. 22850. $C_{32}H_{19}N_6Na_5O_{15}S_5$; mol wt 1002.81. C 38.33%, H 1.91%, N 8.38%, Na 11.46%, O 23.93%, S 15.99%. Prepd by coupling diazotized 4,4'-diaminobiphenyl-3-sulfonic acid with sodium 2-amino-3,6-naphthalenedisulfonate: Krauss, *J. Am. Chem. Soc.* **36**, 961 (1914); *Colour Index vol. 4* (3rd ed., 1971) p 4181.

Reddish-brown powder. Sol in water. Practically insol in alcohol.

USE: Biological stain.

THERAP CAT: Antiprotozoal (Trypanosoma).

THERAP CAT (VET): Has been used as a trypanocide.

9925. Tryparsamide. *[4-[(2-Amino-2-oxoethyl)amino]phenyl]arsonic acid monosodium salt;* N-(carbamoylmethyl)arsanilic acid monosodium salt; monosodium N-phenylglycinamide-p-arsonate; Glyphenarsine; Tryparsone; Tryponarsyl; Trypothane. $C_8H_{10}AsN_2NaO_4$; mol wt 296.09. C 32.45%, H 3.40%, As 25.30%, N 9.46%, Na 7.76%, O 21.61%. Prepd by heating a soln of arsanilic acid in aq NaOH, Na_2CO_3 and chloroacetamide: Jacobs, Heidelberger, *J. Am. Chem. Soc.* **41,** 1590 (1919); *Org. Syn.* **8,** 100 (1928). U.S. pats. **1,280,119; 1,280,124; 1,280,126** (Sept. 24, 1918); **2,465,308** (1949).

Hemihydrate, platelets, slowly affected by light, stable to air. One gram dissolves in about 2 ml water. Slightly sol in alcohol. Insol in ether, chloroform. pH of 1:20 aq soln 6.5. *Keep tryparsamide in tight containers, preferably at a temp not above 20° and protected from light.*

THERAP CAT: Antiprotozoal (Trypanosoma).

9926. Trypsin. Parenzyme; Parenzymol; Tryptar; Trypure. Mol wt 24,000. Proteolytic enzyme formed in the small intestine by the action of a peptidase, enterokinase, on the pancreatic cell product, trypsinogen. Acts on lysyl and arginyl bonds of peptide chains and hydrolyzes even esters and amides. *Reviews:* Desnuelle, "Trypsin" in *The Enzymes* vol. **4,** P. D. Boyer *et al.,* Eds. (Academic Press, New York, 2nd ed., 1960) pp 119-132; Keil, *ibid.* vol. **3** (3rd ed., 1971) pp 250-275; Inagami, "Trypsin" in *Proteins, Structure and Function* vol. **1,** M. Funatsu *et al.,* Eds. (Kodansha, Tokyo, Wiley, New York, 1972) pp 1-83.

Yellow to grayish-yellow powder or crystals. Stable indefinitely in dry form at room temp. Sol in water. Practically insol in alcohol or glycerol. Readily sol in Sorensen's sodium phosphate buffer soln. Acts optimally at pH values between 7 and 9. Solns lose 75% of their potency within 3 hrs at room temp. Prepn of stabilized trypsin compositions contg partially hydrolyzed gelatin: Sullivan, Martin, U.S. pat. **2,930,736** (1960 to National Drug).

THERAP CAT: Enzyme (proteolytic).
THERAP CAT (VET): Enzyme (proteolytic).

9927. Tryptamine. *1H-Indole-3-ethanamine; 3-(2-aminoethyl)indole;* 2-(3-indolyl)ethylamine. $C_{10}H_{12}N_2$; mol wt 160.22. C 74.97%, H 7.55%, N 17.48%. Occurs in plants. Synthesis starting with nitroethylene and indole: Noland, Hartman, *J. Am. Chem. Soc.* **76,** 3227 (1954). Alternate routes: Thesing, Schulde, *Ber.* **85,** 324 (1952); Jackson, Smith, *J. Chem. Soc.* **1965,** 3498; Tacconi, *Farmaco Ed. Sci.* **20,** 902 (1965); S. Takano *et al., Heterocycles* **6,** 1167 (1977); I. Fleming, M. Woolias, *J. Chem. Soc. Perkin Trans. I* **1979,** 829. X-ray structure determn: Wakahara *et al., Tetrahedron Letters* **1970,** 4999. *Review* of tryptamine syntheses: J. E. Saxton in R. H. F. Manske, *The Alkaloids,* vol. VIII (1965) pp 8-10.

Needles from petr ether, mp 118°. uv max (ethanol): 222, 282, 290 nm (log ε 4.56, 3.78, 3.71). Sol in ethanol, acetone. Practically insol in water, ether, benzene, chloroform.

Hydrochloride, $C_{10}H_{12}N_2$.HCl, needles from ethanol + ethyl acetate, mp 248°. uv max (95% ethanol): 221, 275, 281, 290 nm (log ε 4.52, 3.73, 3.75, 3.69).

9928. Tryptazan. *α-Amino-1H-indazole-3-propanoic acid; α-amino-1H-indazole-3-propionic acid;* 3-indazolealanine. $C_{10}H_{11}N_3O_2$; mol wt 205.22. C 58.53%, H 5.40%, N 20.48%, O 15.59%. Synthesis from isatin: Snyder *et al., J. Am. Chem. Soc.* **74,** 2009 (1952).

Hemihydrate, crystals from water, darkens at 240°, dec 267-268°. pKa_1 2.20; pKa_2 8.95. Soly in water (g/100 ml): 0.8 at 29°; 12 at 100°; in 95% ethanol: 0.1 at 29°.

9929. Tryptophan. *L-Tryptophan;* Trp; W; *(S)-α-amino-1H-indole-3-propanoic acid; l-α-aminoindole-3-propionic acid; l-α-amino-3-indolepropionic acid;* 2-amino-3-indolylpropanoic acid; *l-β-3-indolylalanine;* Ardeytropin; Kalma; Optimax WV; Pacitron; Sedanoct; Trofan; Tryptan. $C_{11}H_{12}N_2O_2$; mol wt 204.23. C 64.69%, H 5.92%, N 13.72%, O 15.67%. An essential amino acid for human development; precursor of serotonin, *q.v.* Isoln from casein: F. G. Hopkins, S. W. Cole, *J. Physiol.* **27,** 418 (1902). Structure: A. Ellinger, *Ber.* **39,** 2515 (1906); A. Ellinger, A. C. Flamand, *Ber.* **40,** 3029 (1907). Early chemistry and biochemistry: *Amino Acids and Proteins,* D. M. Greenberg, Ed. (Charles C. Thomas, Springfield, IL, 1951) 950 pp., *passim;* J. P. Greenstein, M. Winitz, *Chemistry of the Amino Acids* vols 1-3 (John Wiley and Sons, Inc., New York, 1961) pp. 2316-2347, *passim.* Intrinsic fluorescent/phosphorescent moiety in proteins; used in characterizing structure and conformational changes: E. A. Burstein *et al., Photochem. Photobiol.* **18,** 263 (1973); C. Pokalsky et al., *J. Biol. Chem.* **270,** 3809 (1995); review of phosphorescence: S. Papp, J. M. Vanderkooi, *Photochem. Photobiol.* **49,** 775-784 (1989). Review of microbial production: T. K. Maiti, S. P. Chatterjee, *Hind. Antibiot. Bull.* **33,** 26-61 (1991). Review of biosynthesis: I. P. Crawford, G. V. Stauffer, *Ann. Rev. Biochem.* **49,** 163-195 (1980); of nutrition and metabolism: J. C. Peters, *Adv. Exp. Med. Biol.* **294,** 345-358 (1991). Review of use in depression: S. N. Young, *J. Psychiatr. Neurosci.* **16,** 241-246 (1991); in neuropsychiatric disorders: R. Sandyk, *Int. J. Neurosci.* **67,** 127-144 (1992). Review as toxic agent in eosinophilia-myalgia syndrome (EMS): D. S. Milburn, C. W. Myers, *DICP Ann. Pharmacother.* **25,** 1259 (1991); L. D. Kaufman, R. M. Philen, *Drug Safety* **8,** 89-98 (1993).

Leaflets or plates from dil alc, dec 289° (rapid heating). $[\alpha]_D^{23}$ −31.5° (c = 1); $[\alpha]_D^{20}$ +2.4° (0.5N HCl); $[\alpha]_D^{20}$ +0.15° (c = 2.43 in 0.5N NaOH). pK_1 2.38; pK_2 9.39. Soly in water (g/l): 8.23 at 0°; 10.57 at 20°; 11.36 at 25°; 17.06 at 50°; 27.95 at 75°; 49.87 at 100°. Sol in hot alcohol, in alkali hydroxides. Insol in chloroform.

Hydrochloride, $C_{11}H_{12}N_2O_2$.HCl, needles from methanol, dec 251°.

USE: Probe for studying protein structure and dynamics.

THERAP CAT: In treatment of depression, schizophrenia and other neuropsychiatric disorders.

9930. Tryptophol. *3-Indoleethanol;* 3-ω-hydroxyethylindole; 2-(3-indolyl)ethyl alcohol; 2-indolyl(3)-ethanol; *β-indolylethyl alcohol;* 3-β-hydroxyethylindole. $C_{10}H_{11}NO$; mol wt 161.20. C 74.51%, H 6.88%, N 8.69%, O 9.92%. Prepd by treatment of indolemagnesium bromide with ethylene oxide: Oddo, Cambieri, *Gazz. Chim. Ital.* **60,** 19 (1939); Snyder, Pilgrim, *J. Am. Chem. Soc.* **70,** 1962 (1948); by Bouveault-Blanc reduction of 3-indoleacetic ester: Jackson, *J. Biol. Chem.* **88,** 659 (1930); by treatment of indolemagnesium iodide with ethylene chlorohydrin: Majima, Hoshino, *Ber.* **58,** 2042 (1925); from gramine: Snyder, Pilgrim, *J. Am. Chem. Soc.* **70,** 3770 (1948). By lithium aluminum hydride reduction of methyl 3-indoleglycolate: Speeter, Anthony, U.S. pat. **3,076,814** (1963 to Upjohn).

Platelets from ether + petr ether, mp 59°. bp$_{2.0}$ 174°. Slightly sol in water. Sol in methanol, ethanol, ether, acetone, chloroform, ethyl acetate, glacial acetic acid. Moderately sol in benzene, amyl alcohol, hot carbon disulfide. Sparingly sol in petr ether.

9931. TSH. *Thyrotropin;* thyrotropic hormone; thyreotrophic hormone; thyroid-stimulating hormone; TTH; Dermathycin; Thytropar. Member of the glycoprotein family of hormones, produced by the anterior lobe of the pituitary gland. Appears to be essential for the correct functioning of the thyroid gland. Stimulates the production of thyroxine, thereby raising the basal metabolic rate. Massive doses of vitamin A inhibit the secretion of thyrotropic hormone. Like HCG, LH, and FH, *q.q.v.,* TSH is composed of α and β subunits, *cf.* T. H. Liao, J. G. Pierce, *J. Biol. Chem.* **245,** 3275 (1970); J. G. Pierce, T. H. Liao, *Fed. Proc.* **29,** 600 (1970). Amino acid sequence of bovine TSH: Shome *et al., J. Biol. Chem.* **246,** 833 (1971); Liao, Pierce, *ibid.* 850; of human α and β subunits: J. G. Pierce *et al., Recent Progr. Horm. Res.* **27,** 165 (1971). Diagnostic use: C. W. H. Havard, M. Boss, *Brit. Med. J.* **3,** 678 (1974). *Reviews:* Salter, *Hormones* **2,** 301-349 (1950); Sonnenberg, *Vitam. Horm. (New York)* **16,** 205-261 (1958); J. G. Pierce, T. F. Parsons, *Ann. Rev. Biochem.* **50,** 465-495 (1981). *Monograph:* S. C. Werner, *Thyrotropin* (C. C. Thomas, Springfield, Ill., 1963) 392 pp.
Behaves like a globulin with a broad isoelectric zone. Biological activity is destroyed by heating and by proteolysis with pepsin, trypsin, and chymotrypsin. Also inactivated by oxidizing agents, such as potassium permanganate and elemental iodine.
THERAP CAT: Thyrotropic hormone; diagnostic aid (thyroid function).
THERAP CAT (VET): Thyrotropic hormone.

9932. Tsuduranine. *(R)-5,6,6a,7-Tetrahydro-1,2-dimethoxy-4H-dibenzo[de,g]quinolin-10-ol; 1,2-dimethoxy-6aβ-noraporphin-10-ol;* tuduranine. $C_{18}H_{19}NO_3$; mol wt 297.35. C 72.71%, H 6.44%, N 4.71%, O 16.14%. In sinomenine mother liquors. Isoln: Gotô, *Ann.* **521,** 175 (1935). Structure: Gotô, Shishido, *Ann.* **539,** 262 (1939). Synthesis of *dl*-form: Narayanaswami *et al., Indian J. Chem.* **7,** 945 (1969).

Difficult to crystallize. Minute needles from slowly evaporating ether, mp about 125° (softens at 105°), or 204° depending on cryst form. $[\alpha]_D^{20}$ −127.5° (c = 0.855 in ethanol). Freely sol in the usual organic solvents.
Hydrochloride, shiny scales from water, dec 286°. $[\alpha]_D^{15}$ −148° (c = 0.88 in water + methanol). Sparingly sol in water.

9933. T-2 Toxin. *(3α,4β,8α)-12,13-Epoxytrichothec-9-ene-3,4,8,15-tetrol 4,15-diacetate 8-(3-methylbutanoate); 3α-hydroxy-4β,15-diacetoxy-8α-(3-methylbutyryloxy)-12,13-epoxy-Δ^9-tricothecene; 8α-(3-methylbutyryloxy)-4β,15-diacetoxyscirp-9-en-3α-ol;* fusariotoxin T-2; insariotoxin; mycotoxin T-2; NSC-138780. $C_{24}H_{34}O_9$; mol wt 466.53. C 61.79%, H 7.35%, O 30.87%. *Tricothecene mycotoxin* isolated from *Fusarium tricinctum:* J. R. Bamburg *et al., Tetrahedron* **24,** 3329 (1968). Physicochemical data: A. E. Pohland *et al., Pure Appl. Chem.* **54,** 2119 (1982). Synthesis: M. C. Wani *et al., J. Org. Chem.* **52,** 3468 (1987). Biosyn-

thetic study: F. Van Middlesworth *et al., J. Org. Chem.* **55,** 1237 (1990). Toxicology studies: W. F. O. Marasas *et al., Toxicol. Appl. Pharmacol.* **15,** 471 (1969); H. B. Schiefer, D. S. Hancock, *ibid.* **76,** 464 (1984); D. A. Creasia *et al., Fund. Appl. Toxicol.* **14,** 54 (1990). Implicated as a chemical warfare agent in Southeast Asia with nivalenol, *q.v.:* N. Wade, *Science* **214,** 34 (1981); R. T. Rosen, J. D. Rosen, *Biomed. Mass Spectrom.* **9,** 443 (1982). *Review: Developments in Food Science* vol. **4,** Y. Ueno, Ed., entitled "Trichothecenes: Chemical, Biological and Toxicological Aspects" (Kodansha Ltd. and Elsevier, New York, 1983) 310 pp. Review of pharmacokinetics and metabolism: B. Yagen, M. Bialer, *Drug Metab. Rev.* **25,** 281-323 (1993).

Crystals, mp 151-152°. $[\alpha]_D^{26}$ +15° (c = 2.58 in ethanol). Freely soluble in ethyl alcohol, ethyl acetate, chloroform, DMSO and other organic solvents; slightly soluble in petroleum ether; very slightly soluble in water. LD$_{50}$ orally in female rats: 4.0 mg/kg (Marasas). LD$_{50}$ (mg/kg) in mice: 5.2 i.p., 4.2 i.v.; in rats: 7.0 intragastric, 0.9-1.3 i.p., 0.9 i.v., 2.0 s.c.; in guinea pigs: 3.0-4.0 orally, 5.3 intragastric, 1.0 i.m., 1.0-2.0 i.v., 1.0-2.0 s.c.; in pigs: 5.0 orally, 3.0 i.v. (Yagen, Bailer).
Caution: May be highly irritating to skin and mucous membranes. Direct contact may cause extensive inflammation and tissue necrosis (Marasas). Topical exposure has lead to systemic toxicity and death in experimental animals (Schiefer, Hancock).

9934. Tuaminoheptane. *2-Heptanamine; 1-methylhexylamine;* 2-aminoheptane; Tuamine. $C_7H_{17}N$; mol wt 115.22. C 72.97%, H 14.87%, N 12.16%. $CH_3(CH_2)_4CH(CH_3)NH_2$. α-Adrenergic agonist; topical vasoconstrictor. Prepd by heating 2-bromoheptane on steam bath with alcoholic ammonia in pressure tube: Clarke, *J. Am. Chem. Soc.* **21,** 1027 (1899); by hydrogenation of methyl amyl ketone and ammonia in the presence of Raney nickel: Norton *et al., J. Org. Chem.* **19,** 1054 (1954).
Volatile liq; d_2^{25} 0.7600-0.7660; bp$_{760}$ 142-144°; n_D^{25} 1.4150-1.4200. Slightly soluble in water. pH of 1% aq soln 11.45. Freely sol in alcohol, ether, petr ether, chloroform, benzene.
Hydrochloride, $C_7H_{17}N.HCl$, needles, mp 133°. Soluble in water. pH about 5.4.
Sulfate, $2C_7H_{17}N.H_2SO_4$, *Heptedrine.* Crystals, readily sol in water. The pH of a 1% soln is about 5.4.
THERAP CAT: Nasal decongestant.

9935. Tuberactinomycin. Polypeptide antibiotic mixture produced by *Streptomyces griseoverticillatus* var *tuberacticus:* Nagata *et al., J. Antibiot.* **21,** 681 (1968); *eidem,* U.S. pat. **3,639,580** (1972). Composed of tuberactinomycins A, B, N, and O. Structures: Yoshioka *et al., Tetrahedron Letters* **1971,** 2043; Wakamiya *et al., Bull. Chem. Soc. Japan* **46,** 949 (1973); Wakamiya, Shiba, *J. Antibiot.* **27,** 900 (1974); *eidem,* **28,** 292 (1975). Total synthesis of tuberac-

Tuberactinomycins				
	A	B	N	O
R_1	OH	H	OH	H
R_2	OH	OH	H	H

Consult the Name Index before using this section.

tinomycin O: T. Teshima *et al., Tetrahedron Letters* **1976**, 2343; *eidem, J. Antibiot.* **30**, 1073 (1977). Chemical studies on tuberactinomycin: T. Wakamiya *et al., Heterocycles* **15**, 999 (1981).

Hydrochloride, white solid, mp 244-264° (dec). $[\alpha]_D^{25}$ −31.5° (c = 1 in water). uv maxima in water: 268 nm ($E_{1cm}^{1\%}$ 330); in 1*N* HCl: 268.5 nm ($E_{1cm}^{1\%}$ 313); in 0.1*N* NaOH: 285 nm ($E_{1cm}^{1\%}$ 206.5). Sol in water; weakly sol in methanol. Practically insol in ethanol, pyridine, ether, chloroform, dioxane; insol in acetone, benzene. pKa$_1$ 7.2; pKa$_2$ 10.3.

Tuberactinomycin A, $C_{25}H_{43}N_{13}O_{11}$, *1-[(3R,4R)-4-hydroxy-L-3,6-diaminohexanoic acid]viomycin.*

Tuberactinomycin B, see Viomycin.

Tuberactinomycin N, see Enviomycin.

Tuberactinomycin O, $C_{25}H_{43}N_{13}O_9$, *(R)-6-[L-2-(2-amino-1,4,5,6-tetrahydro-4-pyrimidinyl)glycine]viomycin.*

THERAP CAT: Antibacterial (tuberculostatic).

9936. Tubercidin. *7-β-D-Ribofuranosyl-7H-pyrrolo[2,3-d]pyrimidin-4-amine; 4-amino-7-β-D-ribofuranosyl-7H-pyrrolo[2,3-d]pyrimidine;* 7-deazaadenosine; sparsamycin A; U-10071. $C_{11}H_{14}N_4O_4$; mol wt 266.26. C 49.62%, H 5.30%, N 21.04%, O 24.04%. Antibiotic substance produced in the culture broth of *Streptomyces tuberidicus.* Isoln: Anzai *et al., J. Antibiot.* **10A**, 201 (1957). Structure: Susuki, Marumo, *ibid.* **14A**, 34 (1961). Total synthesis: Tolman *et al., J. Am. Chem. Soc.* **91**, 2102 (1969). Crystal structure: Stroud, *Acta Crystallogr.* **29B**, 690 (1973); Abola, Sundaralingham, *ibid.* 697.

Needles from water, dec 247-248°. $[\alpha]_D^{17}$ −67° (50% acetic acid). uv max (0.01*N* NaOH): 270 nm (ε 12,100). Sol in acidic and alkaline soln. One gram dissolves in 330 ml water, 200 ml methanol, 2000 ml ethanol. Practically insol in acetone, ethyl acetate, chloroform, benzene, petr ether. LD$_{50}$ i.v. in mice: 45 mg/kg (Anzai).

THERAP CAT: Antifungal; antibacterial (tuberculostatic); antineoplastic.

9937. Tuberculin. A filtrate from triturated *Mycobacteria tuberculosis.* **Old Tuberculin** is the culture filtrate prepd by boiling and then filtering the *M. tuberculosis* culture. Purified protein derivative is a more refined tuberculin prepd by precipitating the culture filtrate with ammonium sulfate. Purification and dermal reactivity: F. B. Seibert, J. T. Glenn, *Am. Rev. Tuberc.* **44**, 9 (1941). Review of tuberculin skin test: G. W. Comstock *et al., Am. Rev. Resp. Dis.* **124**, 356 (1981).

Purified Protein Derivative, PPD, Aplisol, Tubersol.

THERAP CAT: Diagnostic aid (tuberculosis).

9938. Tuberin. *N-[2-(4-Methoxyphenyl)ethenyl]formamide; N-trans-(p-Methoxystyryl)formamide; N-formyl trans-p-methoxystyrylamine.* $C_{10}H_{11}NO_2$; mol wt 177.20. C 67.78%, H 6.26%, N 7.90%, O 18.06%. Antitubercular antibiotic isolated from the broth filtrate of *Streptomyces amakusaensis:* Ohkuma *et al., J. Antibiot.* **15A**, 115 (1962); Sumiki *et al., Japan.* pat. **64** 7399 (1964 to Inst. Phys. & Chem. Res.), *C.A.* **62**, 8355g (1965). Structure and synthesis: Anzai *et al., J. Antibiot.* **15A**, 110, 117, 123 (1962). Alternate synthesis: I. J. Massey, I. T. Harrison, *Chem. Ind. (London)* **1977**, 920. Biosynthetic studies: K. M. Cable *et al., J. Chem. Soc., Perkin Trans. I* **1987**, 1593.

Prisms from benzene, mp 132-133°. Stable in weakly acidic or weakly alkaline solns. uv max (methanol): 219, 285 nm ($E_{1cm}^{1\%}$ 870, 1710). Sol in the lower alcohols, ethyl acetate, acetone; moderately sol in carbon tetrachloride, chloroform; sparingly sol in water, benzene. Practically insol in petr ether.

Dihydrotuberin, $C_{10}H_{13}NO_2$, liquid. n_D^{19} 1.5349.

9939. Tubocurarine Chloride. *7',12'-Dihydroxy-6,6'-dimethoxy-2,2',2'-trimethyltubocuraranium chloride hydrochloride.* $C_{37}H_{42}Cl_2N_2O_6$; mol wt 681.66. C 65.20%, H 6.21%, Cl 10.40%, N 4.11%, O 14.08%. Isoln of *d*-form from *Chondodendron tomentosum* R. & P., *Menispermaceae:* Dutcher, *J. Am. Chem. Soc.* **68**, 419 (1946). Purification: *ibid.* **74**, 2221 (1952). Isoln of *l*-form from *Ch. tomentosum:* King, *Nature* **158**, 515 (1946); *J. Chem. Soc.* **1947**, 936. Structure: King, *ibid.* **1948**, 265. Stereochemistry: Hultin, *Acta Chem. Scand.* **17**, 753 (1963). Revised structure: Everett *et al., Chem. Commun.* **1970**, 1020; Codding, James, *ibid.* **1972**, 1174. Synthesis of *d*-form: J. Naghaway, T. O. Soine, *J. Pharm. Sci.* **68**, 655 (1979). Synthesis of *dl*-form: Veronin *et al., Doklady Akad. Nauk SSSR* **122**, 77 (1958), *C.A.* **53**, 1345f (1959). Pharmacology and toxicology: R. D. Sofia, L. C. Knobloch, *Toxicol. Appl. Pharmacol.* **28**, 227 (1974); C. A. Winter, J. T. Lehman, *J. Pharmacol. Exp. Ther.* **100**, 489 (1950). Toxicity data: H. Rosen *et al., Proc. Soc. Exp. Biol. Med.* **120**, 511 (1965). Use as diagnostic aid for myasthenia gravis: F. F. Foldes *et al., J. Am. Med. Assoc.* **203**, 133 (1968). Comprehensive description: C. Papastephanou in *Analytical Profiles of Drug Substances* vol. 7, K. Florey, Ed. (Academic Press, New York, 1978) pp 477-500.

(+)-Tubocurarine chloride

d-Form, *Jexin, Tubadil, Delacurarine, Curarin-HAF, Tubarine, Intocostrin.* Hexagonal and pentagonal microplatelets from water. Can exist in the form of various hydrates. The anhydrous material (dec 274-275°) takes up water in moist atm until it reaches the pentahydrate stage, dec ~270°. $[\alpha]_D^{22}$ +190° (c = 0.5); $[\alpha]_D^{23}$ +219° (c = 0.785 in methanol). Corresp calcd values for the anhydr salt: $[\alpha]_D^{22}$ +190°; $[\alpha]_D^{23}$ +245° (methanol). uv max (H$_2$O): 280 nm ($E_{1cm}^{1\%}$ 118). Soly (25°): ~50 mg/ml water; but supersatd solns are formed readily (Dutcher). Soly also reported as ~1 g/40 ml water; ~1 g/75 ml ethanol. Presence of 1.0*N* HCl diminishes soly by about one-third. Also sol in methanol. Insol in pyridine, chloroform, benzene, acetone, ether. LD$_{50}$ in mice, rats (mg/kg): 33.2, 27.8 orally (Rosen).

l-Form pentahydrate, needles from water, mp 268° (effervescence). $[\alpha]_D^{20}$ −258° (c = 0.38) for the anhydr salt.

Dimethyl ether, $C_{39}H_{46}Cl_2N_2O_6$, *d*-tubocurarine chloride dimethyl ether, Mecostrin chloride. Crystals, dec about 236° (effervescence). $[\alpha]_D^{25}$ +185° to +195° (c = 0.5 in H$_2$O). uv max (acidified 0.005% aq soln): 280 nm (A$_{1cm}^{1\%}$ ~89). Soluble in water, dil sodium hydroxide. Sparingly sol in alc, in dil HCl, chloroform. Practically insol in benzene, ether.

Caution: Large doses and overdoses may cause respiratory paralysis and hypotension.

THERAP CAT: The *d*-form and dimethyl ether as muscle relaxants (skeletal). *d*-Form as diagnostic aid (myasthenia gravis).

THERAP CAT (VET): Muscle relaxant.

9940. Tubulin. Colchicine-binding protein. The subunit protein of *microtubules*, which are large protein assemblies that play an important role in eukaryotic cell form determination and dynamics. Microtubules have the general structure of long hollow cylinders within which 13 protofilaments of tubulin are arranged in a parallel manner to the cylinder axis. The axial arrangement of the protofilaments with respect to each other results in the appearance of a helical structure, and the *in vitro* microtubule assembly process generally follows the laws of helical protein polymerization, cf. *Thermodynamics of the Polymerization of Proteins*, F. Oosawa, S. Asakura, Eds. (Academic Press, New York, 1975). Tubulin is an asymmetric dimer consisting of two nearly identical molecules, *α-tubulin* and *β-tubulin*, each having mol wts of about 55,000. The two molecules can be separated due to differences in electrophoretic mobilities. Isoln from mammalian brain using colchicine binding: R. C. Weisenberg *et al., Biochemistry* **7**, 4466 (1968). Discovery of conditions for microtubule assembly *in vitro:* R. C. Weisenberg, *Science* **177**, 1104 (1972); G. G. Borisy, J. B. Olmstead, *ibid.* 1196. Prepn of large quantities of brain tubulin through successive assembly-disassembly cycles: M. L. Shelanski *et al., Proc. Nat. Acad. Sci. USA* **70**, 765 (1973). Purification of tubulin from rat pancreas: J. F. Launay *et al., Biochem. Biophys. Res. Commun.* **111**, 253 (1983). Structure of two human α-tubulin genes: C. Wilde *et al., Proc. Nat. Acad. Sci. USA* **79**, 96 (1982). Structure and arrangement of protofilaments in microtubules and tubulin sheets: B. F. McEwen, *Diss. Abstr. B* **43**, 942 (1982). Series of articles on prepn, isoln, and purification of tubulin from various sources: *Methods Enzymol.* **85**, Pt. B, 376-417 (1982). Reviews: J. A. Snyder, J. R. McIntosh, *Ann. Rev. Biochem.* **45**, 699-720 (1976); S. N. Timasheff, L. M. Grisham, *ibid.* **49**, 565-591 (1980); M. F. Carlier, *Mol. Cell. Biochem.* **47**, 97-113 (1982).

Purified calf brain tubulin retains many of its *in vivo* biochemical characteristics, such as the ability to self-assemble into microtubules and the response of the assembly reaction to inhibitory effects of cold temperature, Ca^{2+}, and antimicrotubule agents, *e.g.* vinblastine and colchicine, *q.q.v.* uv max (PG buffer): 278 nm (ϵ 1.33 ml mg⁻¹ cm⁻¹).

9941. Tuftsin. N^2-*[1-(N^2-L-Threonyl-L-lysyl)-L-prolyl]*-L-*arginine.* $C_{21}H_{40}N_8O_6$; mol wt 500.60. C 50.39%, H 8.05%, N 22.38%, O 19.18%. A naturally occurring tetrapeptide having a variety of immunopotentiating properties, especially stimulation and enhancement of phagocytosis. It also exhibits antitumor and antibacterial activity and has been shown to possess chemotactic, migration-enhancing, and mitogenic properties for leukocytes. Discovered during research on the physiological role of cytophilic gamma-globulin: V. A. Najjar, K. Nishioka, *Nature* **228**, 672 (1970). Produced in the spleen; present in mammalian blood in the gamma globulin fraction as part of the larger molecule *leucokinin.* Isoln and characterization: K. Nishioka *et al., Biochim. Biophys. Acta* **310**, 217 (1973); V. A. Najjar, U.S. pat. 3,778,426 (1973 to Research Corp.). Solid phase synthesis: K. Nishioka *et al., Biochim. Biophys. Res. Commun.* **47**, 172 (1972); *eidem, Biochim. Biophys. Acta* **310**, 230 (1973). Synthesis by fragment condensation: J. Vicar *et al., Coll. Czech. Chem. Commun.* **41**, 3467 (1976); by liquid phase method: S. Nozaki *et al., Bull. Chem. Soc. Japan* **50**, 422 (1977). ¹³C-NMR and circular dichroism studies: I. Z. Siemion *et al., Eur. J. Biochem.* **112**, 339 (1980). Conformational studies have provided conflicting evidence on the structure, cf. M. Blumenstein *et al., Biochemistry* **18**, 4247 (1979). Specific receptors on macrophages, monocytes, and granulocytes are thought to mediate the biological activity of tuftsin: A. Constantopoulos, V. A. Najjar, *J. Biol. Chem.* **248**, 3819 (1973); R. M. G. Nair *et al., Immunochemistry* **15**, 901 (1978); Z. Bar-Shavit *et al., Biochem. Biophys. Res. Commun.* **94**, 1445 (1980). Its physiological significance has been shown in patients in whom tuftsin deficiency has resulted in a human syndrome with increased incidence of severe infections: *Macrophages and Lymphocytes, Part A*, M.

R. Escobar, H. Friedman, Eds. (Plenum Press, New York, 1980) pp 131-147; *Lymphokine Reports*, E. Pick, Ed. (Academic Press, New York, 1980) pp 157-159; V. A. Najjar, *Med. Biol.* **59**, 134 (1981). General biological proprties: V. A. Najjar, *Mol. Cell. Biochem.* **41**, 1 (1981). Antitumor activity: K. Nishioka *et al., ibid.* 13. Bactericidal activity: J. Martinez, F. Winternitz, *ibid.* 123. Analogs: F. Z. Siemion, *ibid.* 99. Reviews: V. A. Najjar, *Exp. Cell. Biol.* **46**, 114-126 (1978); *eidem, Adv. Exp. Med. Biol.* **121A**, 131-147 (1980); K. Nishioka *et al., Life Sci.* **28**, 1081-1090 (1981); V. A. Najjar, *Mol. Cell. Biochem.* **41**, 3 (1981). Conference proceedings: *Ann. N.Y. Acad. Sci.* **419**, entitled "Antineoplastic, Immunogenic and Other Effects of the Tetrapeptide Tuftsin: a Natural Macrophage Activator", V. A. Najjar, M. Fridkin, Eds. (1983) pp 1-273.

9942. Tulobuterol. 2-Chloro-α-*[[(1,1-dimethylethyl)-amino]methyl]benzenemethanol;* α-*[(tert-*butylamino)methyl]-*o*-chlorobenzyl alcohol. $C_{12}H_{18}ClNO$; mol wt 227.73. C 63.29%, H 7.97%, Cl 15.57%, N 6.15%, O 7.03%. A β-adrenergic receptor agonist, related structurally to terbutaline, *q.v.* Prepn: H. Kato, S. Kurata, *Ger.* pat. 2,244,737 (1973 to Hokuriku), *C.A.* **78**, 147538a (1973). Pharmacology: S. Kubo *et al., Arzneimittel-Forsch.* **25**, 1028 (1975); *eidem, ibid.* **27**, 1433 (1977); I. Uesaka *et al., ibid.* 1439. Metabolism: T. Fujiihashi *et al., Oyo Yakuri* **18**, 347 (1979), *C.A.* **92**, 121781 (1980); K. Matsumura *et al., Yakugaku Zasshi* **101**, 198 (1981), *C.A.* **94**, 167404 (1981). Determn of tulobuterol and its metabolites in human urine: *eidem, J. Chromatog.* **222**, 53 (1981). Toxicological studies: S. Kubo *et al., Oyo Yakuri* **13**, 197, 317 (1977), *C.A.* **88**, 83591-2 (1978).

Crystals, mp 89-91°. LD_{50} in male mice, rats, rabbits (mg/kg): 305, 850, 563 orally; 170, 417, 164 s.c. (Kubo, 1975).

Hydrochloride, $C_{12}H_{18}ClNO.HCl$, *C-78, Atenos, Berachin, Brelomax, Bremax, Hokunalin, Respacal.* White crystalline powder, mp 161-163°.

THERAP CAT: Bronchodilator.

9943. Tumor Necrosis Factor. TNF. Cytokine produced by activated macrophages as part of the cellular immune response. Originally characterized by its selective hemorrhagic necrosis of tumor cells. Identification in the sera of endotoxin treated mice previously sensitized with bacillus Calmette-Guerin (BCG): E. A. Carswell *et al., Proc. Nat. Acad. Sci. USA* **72**, 3666 (1975). Partial purification: S. Green *et al., ibid.* **73**, 381 (1976). Preliminary characterization of murine TNF: T. Haranaka, N. Satomi, *Japan. J. Exp. Med.* **51**, 191 (1981); F. C. Kull, P. Cuatrecasas, *J. Immunol.* **126**, 1279 (1981); of rabbit TNF: M. R. Ruff, G. E. Gifford, *ibid.* **125**, 1671 (1980); N. Matthews *et al., Brit. J. Cancer* **42**, 416 (1980). Identification of macrophages as cellular source of TNF: D. N. Männel *et al., Infect. Immun.* **30**, 523 (1980); N. Santomi *et al., Japan. J. Exp. Med.* **51**, 317 (1981). Differentiation from interferon: N. Bloksma *et al., Cancer Immunol. Immunother.* **14**, 41 (1982). Cytotoxic effect on the malaria parasite *Plasmodium falciparum:* C. G. Haidaris *et al., Infect. Immun.* **42**, 385 (1983); A. O. Wozencraft *et al., ibid.* **43**, 664 (1984). Activity against transplanted human and murine tumors in mice: T. Haranaka *et al., Int. J. Cancer* **34**, 263 (1984). Human TNF is a trimer of 3 identical subunits with 157 amino acid residues and mol wt 17,350 Da. Cloning and expression of cDNA for human TNF in *E. coli:* D. Pennica *et al., Nature* **312**, 724 (1984); T. Shirai *et al., ibid.* **313**, 803 (1985); A. M. Wang *et al., Science* **228**, 149 (1985). Identity with *cachectin:* B. Buetler, A. Cerami, *Nature* **320**, 584 (1986). Structure: E. Y. Jones *et al., ibid.* **338**, 225 (1989). Preliminary evaluation with interferon-γ in metastatic melanoma: S. Retsas *et al., Brit. Med. J.* **298**, 1290 (1989).

Consult the Name Index before using this section.

Reviews: M. R. Ruff, G. E. Gifford, *Lymphokines* **2**, 235 (1981); L. J. Old, *Sci. Am.* **258**(5), 59-60, 70-75 (1988).

Note: Lymphotoxin, a TNF-like factor produced by lymphocytes has been referred to as *TNF-β.*

9944. Tung Oil. China wood oil. A drying oil from seeds of *Aleurites cordata* Steud., *Euphorbiaceae,* indigenous to China and Japan, but now grown also in Florida. Chief fatty acid component is eleostearic acid. Unlike linseed and soybean oils, it need not be refined.

Pale yellow liquid; characteristic disagreeable odor. On long keeping, or on heating for a short time at 300°, polymerizes to a stiff jelly. d 0.936-0.943. Iodine no. 163-171. Sapon no. 190-197. Sol in chloroform, ether, carbon disulfide, oils; the polymerized product is practically insol in the usual organic solvents.

USE: Manuf quick-drying wood varnishes, linoleum, and floor cloth; in India rubber substitutes, insulating masses; for waterproofing paper and other tissues.

9945. Tungsten. Wolfram. W; at. wt 183.85; at. no. 74; valences 6, 5, 4, 3, 2. Naturally occurring isotopes: 180 (0.135%); 182 (26.4%); 183 (14.4%); 184 (30.6%); 186 (28.4%); artificial radioactive isotopes: 173-179; 181; 185; 187-189. Discovered by C. W. Scheele in 1781, isolated in 1783 by J. J. and F. de Elhuyar. One of the rarer metals, comprises about 1.5 ppm of the earth's crust. Chief ores are *wolframite* [(Fe,Mn)WO$_4$] and scheelite (CaWO$_4$). Found chiefly in China, Malaya, Mexico, Alaska, South America and Portugal. Scheelite ores mined in the U.S. carry from 0.4-1.0% WO$_3$. Description of isoln processes: K. C. Li, C. Y. Wang, *Tungsten*, A.C.S. Monograph Series no. 94 (Reinhold, New York, 3rd ed., 1955) pp 113-269; G. D. Rieck, *Tungsten and Its Compounds* (Pergamon Press, New York, 1967) 154 pp. *Reviews:* Parish, *Advan. Inorg. Chem. Radiochem.* **9**, 315-354 (1966); Rollinson, "Chromium, Molybdenum and Tungsten" in *Comprehensive Inorganic Chemistry* vol. 3, J. C. Bailar, Jr. *et al.,* Eds. (Pergamon Press, Oxford, 1973) pp 623-624, 742-769.

Steel-gray to tin-white metal; body centered cubic structure. d$_4^{20}$ 18.7-19.3; depends on extent of working. Hardness 6.5-7.5. mp 3410°. bp$_{760}$ 5900°. Spec heat (20°): 0.032 cal/g/°C. Heat of fusion 44 cal/g. Heat of vaporization 1150 cal/g. Electrical resistivity (20°): 5.5 µohm-cm. Stable in dry air at ordinary temps, forms the trioxide at red heat. Not attacked by water, but oxidized to the dioxide by steam. Very stable to acids, attacked only superficially by concd nitric acid or aqua regia. Powdered tungsten can be pyrophoric under the right conditions. Slowly sol in fused potassium hydroxide or sodium carbonate in presence of air; sol in a fused mixture of NaOH and nitrate. Attacked by fluorine at room temp; by chlorine at 250-300° giving the hexachloride in absence of air, and the trioxide and oxychloride in the presence of air. LD$_{50}$ i.p. in rats: 5 g/kg, *Toxic Substances List,* H. E. Christensen, Ed. (1973) p 965.

USE: To increase hardness, toughness, elasticity, and tensile strength of steel; manuf alloys; manuf filaments for incandescent lamps and in electron tubes; in contact points for automotive, telegraph, radio and television apparatus; in phonograph needles. *Tungsten carbides* (W$_2$C, WC) used in rock drills, metal-cutting tools, wire-drawing dies. WC used as catalyst instead of platinum: Bennett *et al., Science* **184**, 563 (1974).

9946. Tungsten Hexafluoride. F$_6$W; mol wt 297.83. F 38.27%, W 61.73%. WF$_6$. Prepd by direct fluorination of powdered tungsten. Can be purified by distillation, preferably under pressure: Ruff, Ascher, *Z. Anorg. Allgem. Chem.* **196**, 413 (1931); Henkel, Klemm, *ibid.* **222**, 68 (1935); Marchi, *Inorg. Syn.* **3**, 181 (1950); from ClF and tungsten: Pitts, Jache, *Inorg. Chem.* **7**, 1661 (1968).

Colorless gas or pale yellow liquid. Orthorhombic, deliquescent crystals when solid. d$_{liq}^{15}$ 3.441. mp +2.3°. bp 17.5°. Soly in anhydr HF: 3.14 moles/1000 g HF, Frlec, Hyman, *Inorg. Chem.* **6**, 1596 (1967). May be stored in glass pressure ampuls.

9947. Tungsten Trioxide. Tungstic anhydride. O$_3$W; mol wt 231.84. O 20.70%, W 79.30%. WO$_3$. Prepn from sodium tungstate: Hein, Herzog, in *Handbook of Prepara-*

tive Inorganic Chemistry vol. 2, G. Brauer, Ed. (Academic Press, New York, 2nd ed., 1965) pp 1423-1424.

Canary yellow, heavy powder; dark orange when heated, regaining the original color on cooling. Insol in water. Sol in caustic alkalies; very slightly sol in acids.

USE: Manuf tungstates which are used for x-ray screens and for fireproofing fabrics.

9948. Tungstic(VI) Acid. H$_2$O$_4$W; mol wt 249.85. H 0.81%, O 25.61%, W 73.58%. H$_2$WO$_4$. Prepn: Morley, *J. Chem. Soc.* **1930,** 1990.

Yellow or greenish-yellow powder. Insol in water and acid except hydrofluoric acid. Slowly sol in solns of caustic alkalies. When freshly pptd from a soluble tungstate it contains 1 mol H$_2$O, and is appreciably sol in water.

9949. Tunicamycin. A family of nucleoside antibiotics produced by *Streptomyces lysosuperificus.* Isoln, characterization: A. Takatsuki *et al., J. Antibiot.* **24**, 215 (1971). Biological properties: A. Takatsuki, G. Tamura, *ibid.* 224. Effect on microorganisms: A. Takatsuki *et al., ibid.* **25**, 75 (1972). Tunicamycin is produced as a mixture of at least 10 homologous antibiotics, the main components being tunicamycins V, VII, II and X, also referred to respectively as A, B, C, D. They contain uracil, *N*-acetylglycosamine, an 11-carbon aminodialdose called *tunicamine,* and a fatty acid linked to the amino group of tunicamine. The homologs differ in their fatty acid components, which vary in degree of saturation and chain length and branching. Structural elucidation: A. Takatsuki *et al., Agr. Biol. Chem.* **41**, 2307 (1977); T. Ito *et al., ibid.* **44**, 695 (1980). HPLC sepn of components: W. C. Mahoney, D. Duskin, *J. Biol. Chem.* **254**, 6572 (1979). Approaches to synthesis: Y. Fukuda *et al., Bull. Chem. Soc. Japan* **55**, 880 (1982). Tunicamycin has been shown to interfere with glycoprotein synthesis in yeast and mammalian systems: A. Takatsuki *et al., Agr. Biol. Chem.* **39**, 2089 (1975); S. Kuo, J. O. Lampen, *Arch. Biochem. Biophys.* **172**, 574 (1976). Effect on epidermal glycoprotein and glycosaminoglycan synthesis *in vitro:* I. A. King, A. Tabiowo, *Biochem. J.* **198**, 331 (1981). Enhancement of antiviral and anticellular activity of interferon, *q.v.:* R. K. Maheshwari *et al., Science* **219**, 1339 (1983). *Review:* A. D. Elbein, *Trends Biochem. Sci.* **6**, 219-221 (1981). Book: *Tunicamycin,* G. Tamura, Ed. (Japan Sci. Soc., Tokyo, 1982) 220 pp. *See also* Streptovirudin.

Tunicamycins II, V, VII, X
(n = 8, 9, 10, 11)

White cryst powder, mp 234-235° (dec). [α]$_D^{20}$ +52° (c = 0.5 in pyridine). uv max (methanol): 205, 260 nm (E$_{1cm}^{1\%}$ 230, 110). Sol in alk water, pyridine, hot methanol. Slightly sol in ethanol, butanol. Practically insol in acetone, ethyl acetate, chloroform, benzene, acidic water. When dissolved in water at 100° and held for 30 min, stable at neutral and alk pH, unstable at acidic pH.

USE: As a tool in studying glycoproteins in a wide variety of biological systems.

9950. Tunichrome B-1. *(E,Z)-3,5-Dihydroxy-L-tyrosyl-α,β-didehydro-3,5-dihydroxy-N-[2-(3,4,5-trihydroxyphenyl)-ethenyl]tyrosinamide;* TB-1. C$_{26}$H$_{25}$N$_3$O$_{11}$; mol wt 555.50. C 56.22%, H 4.54%, N 7.56%, O 31.68%. Major component of a group of polyphenolic yellow blood pigments isolated from the sea squirt (tunicate) *Ascidia nigra* (Linnaeus). The *tunichromes* are strong biological reducing agents involved in the selective accumulation of vanadium by *A. nigra*. Isoln of crude tunichrome mixture from the blood cells of *A. nigra*, action in converting vanadium(V) to vanadium(III): I. G. Macara *et al.*, *Biochem. J.* **181**, 457 (1979). Isoln of tunichrome B-1 by HPLC under anaerobic conditions, characterization and structure determn by mass spec, nmr: R. C. Bruening *et al.*, *J. Am. Chem. Soc.* **107**, 5298 (1985). Brief account of the role of tunichromes in studies into the biochemical nature of vanadium and other trace metals: R. J. Seltzer, *Chem. & Eng. News* **63**, 67 (Sept. 16, 1985). *Review:* R. C. Bruening *et al.*, *J. Nat. Prod.* **49**, 193 (1986).

Yellow solid, sol in methanol. uv max (methanol): 210, 340 nm (ε 68000, 19600).

9951. Turanose. *3-O-α-D-Glucopyranosyl-D-fructose;* 3-(α-D-glucosido)-D-fructose. $C_{12}H_{22}O_{11}$; mol wt 342.30. C 42.11%, H 6.48%, O 51.41%. Prepd from melezitose: Tanret, *Compt. Rend.* **142**, 1424 (1906); Bridel, Aagard, *Bull. Soc. Chim. Biol.* **9**, 884 (1927); Hudson, Pacsu, *J. Am. Chem. Soc.* **52**, 2522 (1930); Pacsu in *Methods in Carbohydrate Chemistry* vol. **I** (Academic Press, New York, 1962) p 353. Structure: Isbell, Pigman, *J. Res. N.B.S.* **20**, 787 (1938); Isbell, *ibid.* **26**, 35 (1941); Hudson, *J. Org. Chem.* **9**, 117, 470 (1944); Hassid, Ballou in *The Carbohydrates*, W. Pigman, Ed. (Academic Press, New York, 1957) p 508. Crystal structure: A. Neuman *et al.*, *Acta Crystallogr.* **B34**, 242 (1978).

Nonhygroscopic prisms from water + alc. Very sweet taste. Dec 157°. Shows mutarotation. $[\alpha]_D^{20}$ +27.3° → +75.8° (c = 4 in water). Freely sol in water, methanol. One gram dissolves in 19 ml of 95% alc. Acid hydrolysis yields 1 mol D-glucose and 1 mol D-fructose. Reducing power about ½ of D-glucose.

9952. Turicine. *(2R-cis)-2-Carboxy-4-hydroxy-1,1-dimethylpyrrolidinium inner salt; d-N,N-dimethyl-4-hydroxypyrrolidine-2-carboxylic acid betaine; d-4-hydroxyproline betaine; d-4-hydroxystachydrine; allohydroxy-D-proline betaine.* $C_7H_{13}NO_3$; mol wt 159.19. C 52.82%, H 8.23%, N 8.80%, O 30.15%. Isolated from *Stachys (betonica) officinalis* L., Trev., and *Stachys sylvatica* L., Wald-Ziest, *Labiatae:* Küng, Trier, *Z. Physiol. Chem.* **85**, 209 (1913); Guggenheim, *Die Biogenen Amine* (S. Karger, New York, 1951) p 247. Stereoisomeric with betonicine. Prepd by the reaction of allohydroxy-D-proline with methyl iodide and silver oxide: Patchett, Witkop, *J. Am. Chem. Soc.* **79**, 185 (1957).

Crystals from water, ethanol and ethyl acetate, mp 259-260°. $[\alpha]_D^{20}$ +37.8° (c = 1 in water).

Monohydrate, efflorescent prisms from water. Sweet taste. Less sol in water than betonicine; slightly sol in alcohol. Neutral to litmus.

Hydrochloride, $C_7H_{13}NO_3$·HCl, needles from abs alcohol, dec 222°. $[\alpha]_D^{24}$ +25° (c = 7).

Aurichloride, $C_7H_{13}NO_3$·HAuCl₄, scaly clusters from water, mp 232°.

Platinichloride monohydrate, $(C_7H_{13}NO_3)_2$·H_2PtCl_6·H_2O, dec 223°.

L-*Turicine, allohydroxy-L-proline betaine*, crystals, mp 252-254°. $[\alpha]_D^{20}$ −39.0° (c = 1 in water). *Ref:* Friess *et al.*, *J. Am. Chem. Soc.* **79**, 459 (1957).

9953. Turkey-Red Oil. Sulfated castor oil; red oil. Prepd by treating castor oil with 15-30% H_2SO_4 at 25-30° for several hrs, followed by washing and neutralizing with NaOH soln. The product is an anion-active wetting agent. The commercial mfg process actually esterifies the 12-hydroxyl group in ricinoleic acid (the main constit of castor oil) thus producing a true sulfate ester, and not a sulfonate.

USE: In finishing cotton and linen; to obtain bright, clear colors in dyeing fabrics.

9954. Turks Island Salt. A salt mixture used in biological laboratories to make artificial sea water. When the natural product is not available, the following recipe may be used: sodium chloride (NaCl) 2815 g; potassium chloride (KCl) 67 g; magnesium chloride ($MgCl_2$·$6H_2O$) 551 g; magnesium sulfate ($MgSO_4$·$7H_2O$) 692 g. Dissolve each salt (technical grade) separately and dilute the combined solns to 95 l; then add calcium chloride ($CaCl_2$) 145 g dissolved in 5 l H_2O. *Ref:* Abderhalden, *Handbuch der Biologischen Arbeitsmethoden*, Abt. IX, Teil 1, 2. Hälfte, Band 2, p 1101.

9955. Turmeric. *Curcuma longa;* tumeric; saffron Indian. From rhizome of *Curcuma longa* Linn. (*C. domestica* Valeton), *Zingiberaceae*. Habit. India, China, East Indies. Constit. Yellow coloring matter (curcumin), *p,p*-dihydroxydicinnamoylmethane, *p*-hydroxycinnamoylferuloylmethane, *p,α*-dimethylbenzyl alcohol, 1-methyl-4-acetyl-1-cyclohexene, turmerone, α-phellandrene, sabinene, zingiberene, cineol, borneol, caprylic acid. Isoln of curcumin from turmeric: Janaki, Bose, *J. Indian Chem. Soc.* **44**, 985 (1967).

Turmeric has an aromatic pepper-like but somewhat bitter taste and gives curry dishes their characteristic yellowish color.

USE: Condiment (as curry powder), color for ointments.

9956. ar-Turmerone. *(S)-2-Methyl-6-(4-methylphenyl)-2-hepten-4-one; 2-methyl-6-p-tolyl-2-hepten-4-one.* $C_{15}H_{20}O$; mol wt 216.32. C 83.29%, H 9.32%, O 7.40%. Isoln of naturally occurring (+)-form from Curcuma oil *(Curcuma longa* Linn., *Zingiberaceae):* H. Rupe *et al.*, *Helv. Chim. Acta* **17**, 372 (1934). Structure: H. Rupe, A. Gassmann, *ibid.* **19**, 569 (1936). Abs config: V. K. Honwad, A. S. Rao, *Tetrahedron* **20**, 2921 (1964). Total synthesis: A. I. Meyers, R. K. Smith, *Tetrahedron Letters* **1979**, 2749; T. Sato *et al.*, *ibid.* **1980**, 3377.

Oil. bp₁₀ 159-160°. $[\alpha]_D^{20}$ +82.21° (Rupe, Gassman); also reported as $[\alpha]_D^{22}$ +59.9° (c = 4.5 in hexane) (Sato). d_4^{20} 0.9634.

(±)-form. Synthesis: J. Colonge, J. Chambion, *C. R. Acad. Sci.* **222**, 557 (1946); T.-L. Ho, *Syn. Commun.* **11**, 579 (1981). n_D^{21} 1.5218.

9957. Turpentine. Gum thus. Oleoresin from *Pinus palustris* Mill. and from other species of *Pinus*, *Pinaceae*.

Yellowish, opaque, sticky masses; characteristic odor and taste. Insol in water. Sol in alc, chloroform, ether, glacial acetic acid.

Note: "Turpentine" also is used to designate "oil of turpentine".

Caution: Absorbed through skin, lungs, intestine. Irritates skin, mucous membranes. Causes skin eruption, G.I. irritation, delirium, ataxia, kidney damage, coma. Inhalation causes palpitation, dizziness, nervous disturbances, chest pain, bronchitis, nephritis. Benign skin tumors from chronic contact, cf. *Clinical Toxicology of Commercial Products*, R. E. Gosselin *et al.*, Eds. (Williams & Wilkins, Baltimore, 4th ed., 1976) section III, pp 315-317.

USE: As source of turpentine oil; also as constituent of stimulating ointments. Industrially, as an insecticide, solvent for waxes, in production of synthetic camphor; in shoe, stove, furniture, etc., polishes.

THERAP CAT: Rubefacient; counterirritant.

THERAP CAT (VET): Internally: antiseptic, carminative, expectorant. Externally: counterirritant, rubefacient.

9958. Tutin. *[1aS-(1aα,1bβ,2β,5β,6α,6aβ,7β,7aα,8S*)]-Hexahydro-1b,6-dihydroxy-6a-methyl-8-(1-methylethenyl)-spiro(2,5-methano-7H-oxireno[3,4]cyclopent[1,2-d]oxepin-7,-2'-oxiran)-3(2H)-one.* $C_{15}H_{18}O_6$; mol wt 294.30. C 61.22%, H 6.16%, O 32.62%. Poisonous constituent of *Coriaria ruscifolia* L. or *C. japonica* A. Gray, (tutu) *Coriariaceae*. Naturally occurring form is (+)-tutin. Isoln: Easterfield, Aston, *J. Chem. Soc.* **79**, 125 (1901). Structure: Johns, Markham, *ibid.* **1961**, 3006; Craven, *Nature* **1963** (1963); Mackay, Mathieson, *Tetrahedron Letters* **1963**, 1399. Absolute configuration: Okuda, Yoshida, *ibid.* **1965**, 2137. Biosynthesis: A. Corbella *et al.*, *Chem. Commun.* **1969**, 634. NMR analysis: J. W. Blunt *et al.*, *Aust. J. Chem.* **32**, 1339 (1979). Stereospecific synthesis: K. Watamatsu *et al.*, *Tetrahedron* **42**, 5551 (1986). Determn in toxic honey: J. L. Love *et al.*, *N.Z. J. Technol.* **2**, 179 (1986). Effect on central nervous system: D. R. Curtis *et al.*, *Brain Res.* **63**, 419 (1973); A. Nistri *et al.*, *Lancet* **1**, 996 (1974). Toxicity data: C. H. Jarboe *et al.*, *J. Med. Chem.* **11**, 729 (1968).

White, odorless crystals. mp 209-212°; also reported as mp 204-205° (Wakamatsu). $[\alpha]_D^{20}$ +9.25° (alc); $[\alpha]_D^{17}$ +13.9° (c = 0.75, methanol). Sol in water, alcohol, ether. LD_{50} i.p. in female mice: 3.0 mg/kg (Jarboe).

Diacetate, $C_{19}H_{22}O_8$, crystals from ethanol-water, mp 197-201°.

Caution: Poisoning from ingestion of *C. ruscifolia* has been reported and is due to presence of tutin. Symptoms include giddiness, stupor, coma, delirium, convulsions.

9959. Twistane. *Tricyclo[4.4.0.0^{3,8}]decane.* $C_{10}H_{16}$; mol wt 136.24. C 88.16%, H 11.84%. Synthesis: Whitlock, *J. Am. Chem. Soc.* **84**, 3412 (1962); Gautier, Deslongchamps, *Can. J. Chem.* **45**, 297 (1967); Whitlock, Siefken, *J. Am. Chem. Soc.* **90**, 4929 (1968). Absolute configuration: Keiichi *et al.*, *Tetrahedron Letters* **1968**, 5467.

Crystals, mp 163-164.8°.

9960. Tybamate. *Butylcarbamic acid 2-[[(aminocarbonyl)oxy]methyl]-2-methylpentyl ester; carbamate of 2-(hydroxymethyl)-2-methylpentyl ester of butylcarbamic acid; N-butyl-2-methyl-2-propyl-1,3-propanediol dicarbamate; 2-methyl-2-propyltrimethylene butylcarbamate carbamate;* Nospan; Solacen; Tybatran. $C_{13}H_{26}N_2O_4$; mol wt 274.36. C 56.91%, H 9.55%, N 10.21%, O 23.33%. Prepd from 2-methyl-2-propyl-3-hydroxypropyl carbamate + butyl isocyanate: Berger, Ludwig, U.S. pat. **2,937,119** (1960 to Carter Prod.). Comprehensive description: P. Reisberg *et al.*, *Anal. Profiles Drug Subs.* **4**, 494-515 (1975).

Crystals from 1,1,2-trichloroethane + hexane (1:2), mp 49-51°, $bp_{0.06}$ 150-152°.

THERAP CAT: Anxiolytic.

9961. Tylocrebrine. *9,11,12,13,13a,14-Hexahydro-2,3,-5,6-tetramethoxydibenzo[f,h]pyrrolo[1,2-b]isoquinoline; 2,3,-5,6-tetramethoxyphenanthro[9,10:6',7']indolizidine.* $C_{24}H_{27}NO_4$; mol wt 393.48. C 73.26%, H 6.92%, N 3.56%, O 16.26%. Phenanthroindolizidine alkaloid; isomeric with tylophorine (*q.v.*). Isoln of *l*-isomer from *Tylophora crebriflora* S. T. Blake, *Asclepiadaceae*: E. Gellert *et al.*, *J. Chem. Soc.* **1962**, 1008; K. V. Rao *et al.*, *J. Pharm. Sci.* **59**, 1501 (1970). Antileukemic activity: E. Gellert, R. Rudzats, *J. Med. Chem.* **7**, 361 (1964). *d*-Isomer isolated from *Ficus septica*, *Moraceae*. Synthesis of the *dl*-form and structure: E. Gellert *et al.*, *loc. cit.*; B. Chauncy, E. Gellert, *Austr. J. Chem.* **23**, 2503 (1970). Inhibits protein synthesis. G. R. Donaldson *et al.*, *Biochem. Biophys. Res. Commun.* **31**, 104 (1968); E. Battaner, D. Vasquez, *Biochim. Biophys. Acta* **254**, 316 (1971). Mode of action studies: M. T. Huang, A. P. Grollman, *Mol. Pharmacol.* **8**, 538 (1972). *Review:* R. S. Gupta, *Antibiotics (N.Y.)* **6**, 47 (1983).

dl-Form, needles from chloroform + methanol, mp 219-221°.

l-Form, crystals from methanol, dec 218-220°. uv max: 263, 342, 360 nm (log ε 4.81, 3.25, 3.09). $[\alpha]_D^{24}$ −45° (c = 0.74 in chloroform). pKa (50% aq ethanol): 6.7.

d-Form, crystals from methanol, mp 220-222°. $[\alpha]_D^{22}$ +20.5°.

9962. Tylophorine. *(S)-9,11,12,13,13a,14-Hexahydro-2,3,6,7-tetramethoxydibenzo[f,h]pyrrolo[1,2-b]isoquinoline; 2,3,6,7-tetramethoxyphenanthro[9,10:6',7']indolizidine.* $C_{24}H_{27}NO_4$; mol wt 393.48. C 73.26%, H 6.92%, N 3.56%, O 16.26%. Major alkaloid from *Tylophora asthmatica* Wight et Arn., *Asclepiadaceae*. Also found in other *Asclepiadaceae*, *Moraceae*, *Urticaceae* and *Lauraceae*. Naturally occurring form originally isolated and reported to be levorotatory; later corrected to dextrorotatory. Isoln: A. N. Ratnagiriswaran, K. Venkatachalam, *Indian J. Med. Res.* **22**, 433 (1935); R. N. Chopra *et al.*, *Arch. Pharm.* **275**, 236 (1937); T. R. Govindachari *et al.*, *J. Chem. Soc.* **1954**, 2801. Structure: *eidem*, *Tetrahedron* **9**, 53 (1960). Absolute configuration: *eidem*, *J. Chem. Soc. Perkin Trans. I* **1974**, 1161. Synthesis of the *dl*-form: *eidem*, *Chem. & Ind. (London)* **1960**, 664; N. A. Khatri *et al.*, *J. Am. Chem. Soc.* **103**, 6387

(1981). Synthesis and verification of dextrorotation of naturally occurring form: T. F. Buckley 3rd, H. Rapoport, *J. Org. Chem.* **48**, 4222 (1983); J. E. Nordlander, F. G. Njoroge, *ibid.* **52**, 1627 (1987). Stereoselective synthesis of *l*-form: M. Ihara et al., *Tetrahedron Letters* **29**, 4135 (1988). Biosynthesis: Mulchandani et al., *Phytochemistry* **8**, 1931 (1969); *ibid* **10**, 1047 (1971); D. S. Bhakuni, V. K. Mangla, *Tetrahedron* **37**, 401 (1981). Pharmacology: C. Gopalakrishnan et al., *Indian J. Med. Res.* **69**, 513 (1979); **71**, 940 (1980).

Crystals, dec 282-284°. $[\alpha]_D^{23}$ +15° (c = 0.7 in chloroform); $[\alpha]_D^{21}$ +73° (c = 0.7 in chloroform). uv max in ethanol: 257, 286, 339, 356 nm (log ϵ 4.7, 4.42, 3.28, 3.19). Unstable in solutions, decomposition with yellowing sets in promptly accompanied by decreasing rotatory strength.
l-Form, colorless crystals, dec 286-287°. $[\alpha]_D^{27}$ −11.6° (c = 1.07 in chloroform). uv max 255, 290, 340, 352 nm (log ϵ 4.74, 4.49, 3.30, 2.93). Sol in chloroform; slightly sol in abs alc, ether, cold benzene. Practically insol in water.
dl-Form, crystals from chloroform + ethanol, mp 292°.

9963. Tylosin. Tylan; Tylon. $C_{46}H_{77}NO_{17}$; mol wt 916.11. C 60.31%, H 8.47%, N 1.53%, O 29.69%. Macrolide antibiotic isolated from a strain of *Streptomyces fradiae* found in soil from Thailand: Hamill et al., *Antibiot. & Chemother.* **11**, 328 (1961); *eidem,* U.S. pat. **3,178,341** (1965 to Lilly). Prodn in batch and chemostat cultures: P. P. Gray, S. Bhuwapathanapun, *Biotech. Bioeng.* **22**, 1785 (1980). Partial structure: Morin, Gorman, *Tetrahedron Letters* **1964**, 2339. Structure: Morin et al., *ibid.* **1970**, 4737; Achenbach et al., *Ber.* **108**, 2481 (1975). Configurational study: S. Omura et al., *Tetrahedron Letters* **1977**, 1045. Abs config: *eidem, J. Antibiot.* **33**, 915 (1980); N. D. Jones et al., *ibid.* **35**, 420 (1982). Synthesis of *tylonolide*, the aglycone: S. Masamune et al., *J. Am. Chem. Soc.* **98**, 7874 (1976); K. Tatsuta et al., *Tetrahedron Letters* **22**, 3997 (1981). Relationship of ribosomal binding and antibacterial properties: J. W. Corcoran et al., *J. Antibiot.* **30**, 1012 (1977). Biosynthesis studies: E. T. Seno et al., *Antimicrob. Ag. Chemother.* **11**, 455 (1977); S. Omura et al., *J. Antibiot.* **31**, 254 (1978).

Crystals from water, mp 128-132°. $[\alpha]_D^{25}$ −46° (c = 2 in methanol). uv max: 282 nm ($E_{1cm}^{1\%}$ 245). Soly in water at 25°: 5 mg/ml. Sol in lower alcohols, esters and ketones, in chlorinated hydrocarbons, benzene, ether. Solns are stable at pH 4-9; at pH <4 another active compd, *desmycosin* is formed.
Hydrochloride, $C_{46}H_{77}NO_{17} \cdot HCl$, crystals from ethanol + ether, mp 141-145°.
THERAP CAT (VET): Antibacterial.

9964. Tyloxapol. *4-(1,1,3,3-Tetramethylbutyl)phenol polymer with formaldehyde and oxirane;* oxyethylated tertiary octylphenol formaldehyde polymer; oxyethylated tertiary octylphenol-polymethylene polymer; *p*-isooctylpolyoxyethylenephenol formaldehyde polymer; tyloxypal; Alevaire; Superinone; Triton A-20; Triton WR-1339. Nonionic detergent with surface-tension-reducing properties. Prepn: Bock, Rainey, U.S. pat. **2,454,541** (1948 to Rohm & Haas); Cornforth et al., *Nature* **168**, 150 (1951).
Freely sol in water; sol in benzene, toluene, chloroform, carbon tetrachloride, carbon disulfide, acetic acid. Alkaline pH. Oxidized by metals. Aq solns grow molds in contact with air.
THERAP CAT: Mucolytic.

9965. Tymazoline. *4,5-Dihydro-2-[[5-methyl-2-(1-methylethyl)phenoxy]methyl]-1H-imidazole;* 2-[(thymyloxy)methyl]-2-imidazoline; 2-(thymyloxymethyl)glyoxalidine; 2-[(p-mentha-1,3,5-trien-2-yloxy)methyl]-2-imidazoline; Pernazene. $C_{14}H_{20}N_2O$; mol wt 232.33. C 72.38%, H 8.68%, N 12.06%, O 6.89%. Prepd from an alkyl thymyloxyacetimidate and ethylenediamine: Sonn, U.S. pat. **2,149,473** (1939 to Ciba); Djerassi, Scholz, *J. Am. Chem. Soc.* **69**, 1688 (1947). Pharmacology: Pham-Huu-Chanh et al., *Therapie* **24**, 797 (1969).

Hydrochloride, $C_{14}H_{20}N_2O \cdot HCl$, crystals, mp 215-217° (Sonn); from water or methyl ethyl ketone + ethanol, mp 223.5-225° (Djerassi, Scholz).
THERAP CAT: Nasal decongestant.

9966. Tyramine. *4-(2-Aminoethyl)phenol;* 4-hydroxyphenethylamine; tyrosamine; 2-*p*-hydroxyphenylethylamine; *p*-β-aminoethylphenol; α-(4-hydroxyphenyl)-β-aminoethane. $C_8H_{11}NO$; mol wt 137.18. C 70.04%, H 8.08%, N 10.21%, O 11.66%. Decarboxylation product of tyrosine. Found in mistletoes, putrefied animal tissue, ripe cheese, ergot. Synthesis: Barger, *J. Chem. Soc.* **95**, 1127 (1909); Waser, *Helv. Chim. Acta* **8**, 766 (1925); Buck, *J. Am. Chem. Soc.* **55**, 3389 (1933). Crystal and molecular structure: A. Podder et al., *Acta Crystallogr.* **B35**, 649 (1979).

Crystals from benzene or alcohol, mp 164-165°. bp_{25} 205-207°; bp_2 166°. Alkaline reaction. One gram dissolves in 95 ml water at 15°; in 10 ml boiling alcohol. Sparingly sol in benzene, xylene.
Hydrochloride, $C_8H_{11}NO \cdot HCl$, *Mydrial, Uteramin.* Crystals from alcohol + ether, mp 269°. Sol in water with neutral reaction.
THERAP CAT: Adrenergic.

9967. Tyrocidine. Peptide antibiotic mixture produced by *Bacillus brevis:* major constituent of tyrothricin, *q.v.* Isoln: Hotchkiss et al., *J. Biol. Chem.* **141**, 155, 163 (1941); Moses, U.S. pat. **3,265,572** (1966 to Penick). Separated into the three components, tyrocidines A, B, and C. Review of chemistry and biosynthesis: E. Katz, A. L. Demain, *Bacteriol. Rev.* **41**, 449-474 (1977). Regulatory role in bacterial sporogenesis: H. Ristow et al., *Nature* **280**, 165 (1979); *eidem, Eur. J. Biochem.* **129**, 395 (1982); W. Pschorn et al., *ibid.* 403. Structure-activity relationship: W. Danders et al., *Antimicrob. Ag. Chemother.* **22**, 785 (1982).

Tyrocidine A

Hydrochloride mixture, **Brevicidin, Rapicidin,** rods or needles from methanol, dec 240°. $[\alpha]_D^{20}$ −101° (c = 1.2 in 95% alc). Soluble in 95% alc, acetic acid, pyridine; slightly sol in water, acetone, abs alcohol. Practically insol in ether, chloroform, hydrocarbons.

Tyrocidine A, $C_{66}H_{87}N_{13}O_{13}$. Separation from the tyrocidine mixture: Battersby, Craig, *J. Am. Chem. Soc.* **74,** 4019, 4023 (1952). Structure: Paladine, Craig, *ibid.* **76,** 688 (1954). Synthesis: Ohno *et al., Bull. Chem. Soc. Japan* **39,** 1738 (1966); K. Okamoto *et al., ibid.* **50,** 231 (1977).

Tyrocidine A hydrochloride, crystals from methanol + water, mp 240-242°. $[\alpha]_D^{25}$ −111° (c = 1.37 in 50% alc). Freely sol in aq methanol or alc; slightly sol in methanol, ethanol. Practically insol in chloroform, acetone, ether.

Tyrocidine B, $C_{68}H_{88}N_{14}O_{13}$. Purification and amino acid sequence determination: King, Craig, *J. Am. Chem. Soc.* **77,** 6624, 6627 (1955). Synthesis: Kuromizu, Izumiya, *Experientia* **26,** 587 (1970). Possesses the same structure as tyrocidine A except that L-tryptophan replaces the L-phenylalanine. Crystals from methanol + isopropyl ether.

Tyrocidine B hydrochloride pentahydrate, crystals, mp 236-237°. $[\alpha]_D$ −93.0° (c = 0.5 in methanol).

Tyrocidine C, $C_{70}H_{89}N_{15}O_{13}$. Separation from the tyrocidine mixture: Ruttenberg *et al., Biochemistry* **4,** 11 (1965). Possesses same structure as tyrocidine B except that D-tryptophan replaces D-phenylalanine attached to L-asparagine. Synthesis: Kuromizu, Izumiya, *Tetrahedron Letters* **1970,** 1471.

THERAP CAT: Antibacterial.

9968. Tyropanoate Sodium. *α-Ethyl-2,4,6-triiodo-3-[(1-oxobutyl)amino]benzenepropanoic acid monosodium salt; 3-butyramido-α-ethyl-2,4,6-triiodohydrocinnamic acid sodium salt;* α-ethyl-β-(2,4,6-triiodo-3-butyramidophenyl)propionic acid sodium salt; sodium tyropanoate; Win-8851-2; Bilopac; Bilopaque; Lumopaque; Tyropaque. $C_{15}H_{17}I_3NNaO_3$; mol wt 663.01. C 27.17%, H 2.58%, I 57.42%, N 2.11%, Na 3.47%, O 7.24%. Prepn: S. Archer, J. O. Hoppe, U.S. pat. **2,895,988** (1959 to Sterling Drug); and toxicity: J. O. Hoppe *et al., J. Med. Chem.* **13,** 997 (1970).

Colorless solid from ethyl ether, mp 208-210°. Soluble in water. LD_{50} i.v. in mice: 720 mg/kg (Hoppe).

Free acid, $C_{15}H_{18}I_3NO_3$, crystals from ethyl oxyacetate, mp 182-184°.

THERAP CAT: Diagnostic aid (radiopaque medium—cholecystographic).

9969. Tyrosinase. A copper-containing enzyme widely distributed in plants, animals, and man. Catalyzes the hydroxylation of tyrosine in the liver and in melanin-forming cells to 3,4-dihydroxyphenylalanine (dopa). Causes the cut surface of many fruits and plants to darken. The enzyme, as isolated from the common edible mushroom or potato, is characterized by its ability to catalyze aerobic oxidation of both monohydric and *o*-dihydric phenols. These activities are commonly referred to as **cresolase (monophenolase, monophenoloxidase)** and **catecholase (o-dihydric phenolase)** activities. Since monophenolase activity is frequently lost during purification of the enzyme, the name **polyphenolase (polyphenoloxidase)** is preferred by some workers for prepns which have mainly *o*-dihydric phenolase activity. *Ref:* C. R. Dawson, W. B. Tarpley "Copper Oxidases" in J. B. Sumner, K. Myrbäck, *The Enzymes* vol. II, (Academic Press, New York, 1951) pp 456-483. Prepn from potato peels: Kubowitz, *Biochem. Z.* **299,** 32 (1938). Extraction from mushrooms: Cohen, Lerner, U.S. pat. **2,956,929** (1960 to Gillette). Separation of α-, β-, γ-, and δ-tyrosinases of mushroom tyrosinase: Bouchilloux *et al., J. Biol. Chem.* **238,** 1699 (1963). Isoln and properties of crystalline tyrosinase from *Neurospora:* Fling *et al., ibid.,* p 2045. Isoln

and properties of β-tyrosinase: Kumagai *et al., J. Biol. Chem.* **245,** 1767 (1970).

THERAP CAT: Antihypertensive.

9970. Tyrosine. *L-Tyrosine;* Tyr; Y; β-(*p*-hydroxyphenyl)alanine; α-amino-*p*-hydroxyhydrocinnamic acid; (*S*)-α-amino-4-hydroxybenzenepropanoic acid. $C_9H_{11}NO_3$; mol wt 181.19. C 59.66%, H 6.12%, N 7.73%, O 26.49%. Nonessential amino acid for human development; precursor for the synthesis of thyroid hormones and select neurotransmitters, such as dopamine and norepinephrine. May be considered essential by the brain. Name derived from the Greek "tyros" for "cheese" from which it was first identified and isolated: J. Liebig, *Ann.* **57,** 127 (1846); *idem, ibid.* **62,** 257 (1847). Early chemistry and biochemistry: *Amino Acids and Proteins,* D. M. Greenberg, Ed. (Charles C. Thomas, Springfield, IL, 1951) 950 pp., *passim;* J. P. Greenstein, M. Winitz, *Chemistry of the Amino Acids* vol 1-3 (John Wiley and Sons, Inc., New York, 1961) pp. 2348-2367, *passim.* Synthesis of labeled form: Y. Watanabe *et al., Acta Radiol.* **376** (Suppl.), 110 (1991). Intrinsic fluorophore in proteins; used in characterizing structure and conformational changes: V. Giancotti *et al., Biochim. Biophys. Acta* **624,** 60 (1980); S. T. Ferreira *et al., Biophys. J.* **66,** 1185 (1994); B. Kierdaszuk *et al., Photochem. Photobiol.* **61,** 319 (1995). Review of determn in blood: E. Robins, *Meth. Biochem. Anal.* **17,** 287-309 (1969). Review of neurotransmitter synthesis: C. J. Gibson, *Retina* **2,** 332-340 (1982). Review of toxicity: C. Laberge *et al., Adv. Exp. Med. Biol.* **206,** 209-221 (1986). Review of post-translational phosphorylation: S. Atherton-Fessler *et al., Semin. Cell Biol.* **4,** 433-442 (1993); sulfation: C. Niehrs *et al., Chemico-Biolog. Interact.* **92,** 257-271 (1994).

Fine silky needles, dec 342-344° (closed capillary, bath preheated to 280°, rapid heating). d 1.456. pK_1' 2.20; pK_2' 9.11; pK_3' 10.07. $[\alpha]_D$ −10.6° (c = 4 in *N* HCl); $[\alpha]_D^{18}$ −13.2° (c = 4 in 3*N* NaOH). Soly in water (g/100 g): 0.02 at 0°; 0.045 at 25°; 0.105 at 50°; 0.244 at 75°; 0.565 at 100°. Insol in common neutral solvents, such as abs alcohol, ether, acetone, except water. Sol in alkaline solns.

DL-Form, synthetic product. Stout needles. Dec 316°. Soly in water (g/100 g): 0.0147 at 0°; 0.351 at 25°; 0.0836 at 50°.

D-Form. Crystals. Dec 310-314°. $[\alpha]_D^{25}$ +10.3° (c = 4 in *N* HCl). Soly in water (g/100 g): 0.196 at 0°; 0.1052 at 50°.

USE: Probe for studying protein structure and dynamics.

9971. *m*-Tyrosine. *3-Hydroxy-L-phenylalanine;* α-amino-3-hydroxyhydrocinnamic acid; metatyrosine. $C_9H_{11}NO_3$; mol wt 181.19. C 59.66%, H 6.12%, N 7.73%, O 26.49%. A possible precursor of catecholamines: Sourkes *et al., Nature* **189,** 577 (1961). An intermediate in an alternate pathway for the biosynthesis of catecholamines, where with the existing hydroxylating enzymes *m*-hydroxylation of phenylalanine to *m*-tyrosine occurs before *p*-hydroxylation (forming dopa) and is followed by subsequent decarboxylation to dopamine. Formation *in vitro* of dopa from L-*m*-tyrosine: Tong *et al., Biochem. Biophys. Res. Commun.* **43,** 819 (1971); *in vivo:* Hollunger, Persson, *Acta Pharmacol. Toxicol.* **34,** 391 (1974). Biosynthesis and metabolism studies: D'Iorio *et al., Advan. Neurol.* **5,** 265 (1974). Has also been isolated from a plant source, *Euphorbia myrsinites* L. Euphorbiaceae: Mothes *et al., Z. Naturforsch.* **19b,** 1161 (1964). *m*-Tyrosine has the ability to cross the blood-brain barrier and is decarboxylated to *m*-tyramine which stimulates dopamine receptors, presumably accounting for the demonstrated pharmacological effects of *m*-tyrosine. Pharmacological studies: Carlsson, Lindqvist, *Eur. J. Pharmacol.* **2,** 187 (1967); Rubenson, *J. Pharm. Pharmacol.* **23,** 228, 412 (1971); Sandler *et al., Nature* **229,** 414 (1971); Ungerstedt *et al., Eur. J. Pharmacol.* **21,** 230 (1973). Crystal and molecular structure: Byrkjedal *et al., Acta Chem. Scand.* **28B,** 750 (1974).

mp 267-270° (dec). $[\alpha]_D^{22}$ — 14.5° (70% ethanol); $[\alpha]_D^{22}$ +8.9° (70% ethanol, 2N HCl).

9972. Tyrothricin. "Dubos crude crystals"; Coltirot; Martricin; Hydrotricine; Dermotricine; Tyroderm; Solutricine; Tyri 10. Polypeptide antibiotic mixture extracted from cultures of soil bacilli belonging to the *Tyrothrix* group of bacteria. Usually obtained by extracting acidified cultures of *Bacillus brevis* with alc and precipitating with NaCl soln. Surface, submerged, and aerated cultures are used. Tyrothricin contains from 10 to 20% gramicidin and from 40 to 60% tyrocidine. Solubility data: Weiss *et al.*, *Antibiot. & Chemother.* **7**, 374 (1957). Function in the sporulation process: Sarkar, Paulus, *Nature New Biol.* **239**, 228 (1972). Toxicity data: H. Sous *et al.*, *Arzneimittel-Forsch.* **7**, 98 (1957).

Gray to brown powder, decomp 215-220°. Has 2-3% absorbed or combined moisture. Stable to light, air and temps up to 50°. Practically insol in water. Sol in alcohol (about 28 mg/ml), in methanol and in propylene glycol. One part of a 2% alcoholic soln contg 10% formaldehyde soln U.S.P. dissolves in 49 parts of water without cloudiness. Tyrothricin should not be dissolved in ether or acetone as only the gramicidin component is soluble. Soly in mg/ml at about 28°:   water 2.1; isopropanol 5.6; benzene 0.30; isooctane 0.042; carbon tetrachloride 0.455; ethyl acetate 2.65; acetone 6.8; ether 3.25; dioxane 11.1; chloroform 1.6. Aqueous suspensions may be attacked by microorganisms. Solutions and emulsions should not be sterilized by heat. Incompatible with alkalies and strong acids. Tyrothricin gives a positive biuret and Hopkins-Cole test. LD_{50} in mice (mg/kg): > 1500 s.c.; 100 i.p.; > 3000 orally (Sous).

THERAP CAT: Topical antibacterial.

THERAP CAT (VET): Topical antibacterial.

U

9973. Ubenimex. *[S-(R*,S*)]-N-(3-Amino-2-hydroxy-1-oxo-4-phenylbutyl)-L-leucine; [(2S,3R)-3-amino-2-hydroxy-4-phenylbutanoyl]-L-leucine;* NK-421; Bestatin. $C_{16}H_{24}N_2O_4$; mol wt 308.38. C 62.32%, H 7.84%, N 9.08%, O 20.75%. Dipeptide antitumor antibiotic produced by *Streptomyces olivoreticuli.* Competitive inhibitor of aminopeptidase B and leucine aminopeptidase. Prepn from fermentation broth: H. Umezawa *et al.,* Ger. pat. **2,528,984;** *eidem,* U.S. pats. **4,029,547; 4,052,449** (1976, 1977, 1977 all to Microbiochem. Res. Found., Japan); *eidem, J. Antibiot.* **29,** 97 (1976). Structure: H. Suda *et al., ibid.* 100. Syntheses: *eidem, ibid.* 600; R. Nishizawa *et al., ibid.* **36,** 695 (1983). Synthesis of stereoisomers and structure-activity study: *eidem, J. Med. Chem.* **20,** 510 (1977). Stereocontrolled synthesis: S. Kobayashi *et al., Tetrahedron Letters* **25,** 5079 (1984). Crystal structure: J. S. Ricci *et al., J. Org. Chem.* **47,** 3063 (1982). Cell surface binding studies: H. Umezawa *et al., J. Antibiot.* **29,** 857 (1976); W. E. Müller *et al., Int. J. Immunopharmacol.* **4,** 393 (1982). Mitogenic effect on lymphocytes *in vitro:* M. Ishizuka *et al., J. Antibiot.* **33,** 653 (1980). Immunopotentiating and antitumor activities in mice: M. Ishizuka *et al., ibid.* 642. Acute toxicity: T. Sakakibara *et al., Japan. J. Antibiot.* **36,** 2971 (1983), *C.A.* **100,** 132296u (1984). Clinical evaluation in bladder cancer: H. Blomgren *et al., Biomed. Pharmacother.* **38,** 143 (1984); in acute nonlymphocytic leukemia: K. Ota *et al., Cancer Immunol. Immunother.* **23,** 5 (1986). Review of bioactivity and clinical studies: H. Umezawa, *Drugs Exptl. Clin. Res.* **10,** 519-531 (1984).

Colorless needles, mp 233-236°. $[\alpha]_D^{20}$ −15.5° (c = 1.0 in 1N HCl). pKa 8.1, 3.1. uv max: 241.5, 248, 253, 258, 264.5, 268 nm ($E_{1cm}^{1\%}$ 3.8, 4.0, 5.0, 6.0, 4.6, 2.7). Sol in acetic acid, DMSO, methanol. Less sol in water. Insol in ethyl acetate, benzene, hexane, chloroform. LD_{50} (14 day) in mice, rats (g/kg): 1.3-1.9, 1.9-2.1 s.c.; 0.19, 0.78-0.90 i.p.; >4.0, >2.0 orally (Sakakibara).

THERAP CAT: Immunomodulator; antineoplastic.

9974. Ubiquinones. Coenzymes Q; mitoquinones; SA; Q-275. A group of lipid-soluble benzoquinones involved in electron transport in mitochondria, *i.e.,* in the oxidation of succinate or reduced nicotine adenine dinucleotide (NADH) via the cytochrome system. Occurs in the majority of aerobic organisms, from bacteria to higher plants and animals. Ubiquinone structures, analogous to the menaquinones, *q.v.,* are based on the 2,3-dimethoxy-5-methylbenzoquinone nucleus with a variable terpenoid side chain contg one to twelve mono-unsaturated *trans*-isoprenoid units with 10 units being the most common in animals. According to the existing dual system of nomenclature the compds can be described as: coenzyme Q_n in which n = 1-12, or ubiquinone(x) in which x designates the total number of carbon atoms in the side chain and can be any multiple of 5. Differences in properties are due to the difference in length of the side chain. Naturally occurring members are the coenzymes Q_6-Q_{10}. The entire series has been prepd synthetically. Recent syntheses of ubiquinone 50: S. Terao *et al., J. Org. Chem.* **44,** 868 (1979); Y. Naruta, *ibid.* **45,** 4097 (1980); K. Sato *et al., Chem. Commun.* **1982,** 153. *Reviews: Ciba Foundation Symposium on Quinones in Electron Transport,* Wolstenholme, O'Connor, Eds. (Churchill, London, 1961) 453 pp; Wagner, Folkers, *Vitamins and Coenzymes* (Interscience, New York, 1964) pp 435-468; Crane in *Progr. Chem. Fats Lipids,* vol. 7(2), Holman, Ed. (Pergamon Press, 1964) pp 267-289; *Methods Enzymol.* **18C,** 135-237 (1971); R. A. Morton in *The Vitamins* vol. V, W. H. Sebrell, R. S. Harris, Eds. (Academic Press, New York, 2nd ed., 1972) pp 355-391. Series of books: *Biomedical and*

Clinical Aspects of Coenzyme Q vols. **1-3,** K. Folkers, Y. Yamamura, Eds. (Elsevier, New York, 1977, 1980, 1981).

Ubiquinone 50

Ubiquinone 50, $C_{59}H_{90}O_4$, *coenzyme Q_{10}, coenzyme Q-199, ubidecarenone, ubiquinone 10, NSC-140865, Adelir, Caomet, Decafar, Decorenone, Dymion, Heartcin, Inokiton, Iuvacor, Mitocor, Neuquinon, Taidecanone, Ubifactor, Ubiquasar, Ubisan, Ubivis, Ubiten, Udekinon.*

Ubichromenol 50, cyclic isomer of ubiquinone 50.

THERAP CAT: Cardiotonic.

9975. Ubiquitin. ATP-dependent proteolytic factor; APF-1; Ub; ubiquitous immunopoietic polypeptide; UBIP. Ubiquitous polypeptide, mol wt 8500 daltons, so named due to its presence in all eukaryotes including plants; widely distributed in the organism. Highly conserved sequence of 76 amino acids is identical in a wide variety of sources including humans, fish, and insects. Participates in diverse cellular functions, such as protein degradation, chromatin structure, and heat shock, by conjugation to other proteins through its carboxyl terminus. Isoln: G. Goldstein *et al., Proc. Nat. Acad. Sci. USA* **72,** 11 (1975); as APF-1 and recognition of role as marker in protein degradation: A. Ciechanover *et al., Biochem. Biophys. Res. Commun.* **81,** 1100 (1978); A. Ciechanover *et al., Proc. Nat. Acad. Sci. USA* **77,** 1365 (1980); A. Hershko *et al., ibid.* 1783. Amino acid sequence: D. H. Schlesinger *et al., Biochemistry* **14,** 2214 (1975). Identity of APF-1 with ubiquitin and revised amino acid sequence: A. Ciechanover *et al., J. Biol. Chem.* **255,** 7525 (1980); K. D. Wilkinson *et al., ibid.* 7529; K. D. Wilkinson, T. K. Audhya, *ibid.* **256,** 9235 (1981). Structural studies: W. J. Cook *et al., J. Mol. Biol.* **130,** 353 (1979); S. Vijay-Kumar *et al., J. Biol. Chem.* **262,** 6396 (1987); P. L. Weber *et al., Biochemistry* **26,** 7282 (1987). Role of chromatin structure: D. C. Watson *et al., Nature* **276,** 196 (1978); R. D. Mueller *et al., J. Biol. Chem.* **260,** 5147 (1985); in receptor structure: M. Siegelman *et al., Science* **231,** 823 (1986). Identity as heat shock protein: U. Bond, M. J. Schlesinger, *Mol. Cell. Biol.* **5,** 949 (1985). Role as marker for protein degradation in polymeric form: A. Hershko, H. Heller, *Biochem. Biophys. Res. Commun.* **128,** 1079 (1985); V. Chau *et al., Science* **243,** 1576 (1989). Enzymic determn: I. A. Rose, J. V. B. Warms, *Proc. Nat. Acad. Sci. USA* **84,** 1477 (1987). Brief review of functions and mechanisms: D. Finley, A. Varshavsky, *Trends Biochem. Sci.* **10,** 343 (1985). Review of protein degradation: M. Rechsteiner, *Ann. Rev. Cell. Biol.* **3,** 1-30 (1987); A. Hershko, *J. Biol. Chem.* **263,** 15327-15240 (1988). *Review:* K. D. Wilkinson, *Anti-Cancer Drug Des.* **2,** 211-229 (1987). Book: *Ubiquitin,* M. Rechsteiner, Ed. (Plenum Press, New York, 1988).

9976. Uglow Black Silver. A sand coated with silver and manganese. Used in the sterilization of large amounts of water: Uglow, Gan, Z. *Hyg. Infektionskrankh.* **117,** 488 (1935); Granata, *Ann. Igiene* **56,** 284 (1946); *Chimie & Industrie* **58,** 148 (1947).

9977. Uintahite. Gilsonite; uintaite. A variety of asphalt from the Uinta (or Uintah) valley, near Fort Duchesne, Utah: E. S. Dana, *A System of Mineralogy* (John Wiley, New York, 6th ed., 1901) p 1020.

Black, lustrous masses; streak and powder a rich brown. Nonconductor of electricity. Burns easily with a brilliant flame. Sol in heavy lubricating petroleum, warm oil of turpentine, and alc.

USE: Potential raw material for gasoline and coke. In the making of asphalt floor tile, storage battery boxes, inks, paints, and insulation for hot underground piping.

9978. Ujothion. *5-Benzyldihydro-6-thioxo-2H-1,3,5-thiadiazine-3(4H)-acetic acid.* $C_{12}H_{14}N_2O_2S_2$; mol wt

282.39. C 51.04%, H 5.00%, N 9.92%, O 11.33%, S 22.71%. Prepn: Rieche *et al.*, *Arch. Pharm.* **296**, 770 (1963); Schade *et al.*, **East Ger.** pat. **31,793** (1964), *C.A.* **63**, 16371a (1965).

Crystals, mp 152°.
THERAP CAT: Antifungal.

9979. Uliginosins. Antibiotics isolated from *Hypericum uliginosum* HBK, a woody herb found in Mexico and Central America. Isoln: Taylor, Brooker, *Lloydia* **32**, 217 (1969). Structure: Parker, Johnson, *J. Am. Chem. Soc.* **90**, 4716 (1968); Parker *et al.*, *ibid.* 4723. Synthesis of uliginosin A and dihydrouliginosin B: Meikle, Stevens, *Chem. Commun.* **1972**, 123; *eidem, J. Chem. Soc., Perkin Trans. I* **1978**, 1303. Synthesis of uliginosin B: *eidem, Tetrahedron Letters* **1972**, 4787; *eidem, J. Chem. Soc., Perkin Trans. I* **1979**, 2563.

Uliginosin A

Uliginosin B

Uliginosin A. $C_{28}H_{36}O_8$. *3,5-dihydroxy-4,4-dimethyl-2-(2-methyl-1-oxopropyl)-6-[[2,4,6-trihydroxy-3-(3-methyl-2-butenyl)-5-(2-methyl-1-oxopropyl)phenyl]methyl]-2,5-cyclohexadien-1-one.* Pale yellow crystals, mp 160.5-161.5° from acetonitrile-chloroform (4:1). uv max (cyclohexane): 229, 293 nm (ε 31500, 25000).

Uliginosin B. $C_{28}H_{34}O_8$. *2-[[5,7-dihydroxy-2,2-dimethyl-8-(2-methyl-1-oxopropyl)-2H-1-benzopyran-6-yl]methyl]-3,5-dihydroxy-4,4-dimethyl-6-(2-methyl-1-oxopropyl)-2,5-cyclohexadien-1-one.* Pale yellow crystals, mp 139.5-142.0° from nitromethane. uv max (cyclohexane): 230, 270 nm (ε 34000, 37000).

9980. Ultramarine. C.I. Pigment Blue 29; C.I. 77007. A blue pigment occurring naturally as the mineral lapis lazuli. Made by igniting a mixture of kaolin, Na_2CO_3 (or Na_2SO_4), S and carbon. The resulting aluminosulfosilicates resemble zeolites structurally and have the approx formula $Na_7Al_6Si_6O_{24}S_3$. *See: Colour Index* **vol. 4** (3rd ed., 1971) p 4653.

Blue lumps or powder. Insol in water; readily dec by acids, even carbonic acid, with liberation of H_2S.

USE: As a pigment in calico printing, wall paper, mottled soap; bluing in laundry use; for coloring tiles, cements, rubber, but is now largely replaced by coal tar dyes; as a color in food.

9981. Umbelliferone. *7-Hydroxy-2H-1-benzopyran-2-one; 7-hydroxycoumarin;* hydrangin; skimmetin. $C_9H_6O_3$; mol wt 162.14. C 66.67%, H 3.73%, O 29.60%. The aglucon of skimmin. Present in many plants. Obtained by distillation of resins from umbelliferae: Zwenger, *Ann.* **115**,

1, 15 (1860). Prepn: Bert, *Compt. Rend.* **214**, 230 (1942); Austerweil, *ibid.* **248**, 1810 (1959); Dressler, Reabe, U.S. pat. **3,503,996** (1970 to Koppers). Main product of metabolism of coumarin in man: Schilling *et al.*, *Nature* **221**, 664 (1969). Metabolism studies of umbelliferone: Indahl, Scheline, *Xenobiotica* **1**, 13 (1971). Use in brain intracellular pH measurements: T. M. Sundt, R.E. Anderson, *Brain Res.* **186**, 355 (1980); *eidem, J. Neurophysiol.* **44**, 60 (1980). Use in fluorescent immunoassays: S. G. Thompson, J. F. Burd, *Antimicrob. Ag. Chemother.* **18**, 264 (1980); T. M. Li *et al.*, *Anal. Biochem.* **118**, 102 (1981).

Needles from water, mp 225-228°. Develops odor of coumarin on heating. Sublimes. Absorption spectrum: Sen, Bagchi, *J. Org. Chem.* **24**, 316 (1959). One gram dissolves in ~100 ml boiling water; freely sol in alcohol, chloroform, acetic acid; sol in dil alkalies; sparingly sol in ether. Solns show blue fluorescence.

USE: In sunscreen lotions and creams; as intracellular and pH sensitive fluorescent indicator and blood-brain barrier probe.

9982. Undecoylium Chloride. Acylcolaminoformylmethylpyridinium chloride. The acyl group is a mixture containing 8-14 carbon atoms, but principally a 10-11 carbon atom chain. $[C_5H_5N^+\text{-}CONHCH_2CH_2OOC(CH_2)_n\text{-}CH_3]Cl^-$, *where* n = 6-12.

Complex with iodine, *Virac.* $[C_5H_5N^+\text{-}CONHCH_2CH_2O\text{-}OC(CH_2)_nCH_3]Cl^-.I_2$. Freely sol in water.

THERAP CAT: Antiseptic (topical).

9983. Undecylenic Acid. *10-Undecenoic acid; 10-hendecenoic acid;* 9-undecylenic acid; Declid; Renselin; Sevinon. $C_{11}H_{20}O_2$; mol wt 184.28. C 71.70%, H 10.94%, O 17.36%. $CH_2{=}CH(CH_2)_8CO_2H$. Occurs in sweat. Obtained by pyrolysis of ricinoleic acid. Prepn by vacuum distn of castor oil: Krafft, *Ber.* **10**, 2035 (1877); Perkins, Cruz, *J. Am. Chem. Soc.* **49**, 1073 (1927), found that distn at 400° under a pressure of 50 mm produced a distillate composed of ~40% heptaldehyde and 20% undecylenic acid. Toxicology: G. W. Newell *et al.*, *J. Invest. Dermatol.* **13**, 145 (1949).

Liquid or crystals. Odor suggestive of perspiration. d_4^{24} (vac) 0.9072; d_4^{25} 0.9102; d_4^{45} 0.8993; $d_4^{79.9}$ (vac) 0.8653. mp 24.5°; bp_{760} 275° (dec); bp_{182} 232-235°; bp_{130} 230-235°; bp_{100} 213.5°; bp_{90} 198-200°; bp_{15} 168.3°; $bp_{1.0}$ 131°. n_D^{25} 1.4486. Neutralization value 304.5; iodine value 137.8. Insol in water. Sol in alcohol, chloroform, ether. LD_{50} in mice (g/kg): 8.15 orally; 0.960 i.p. (Newell).

Zinc salt, $C_{22}H_{38}O_4Zn$, *zinc undecylenate.* Amorphous white powder, mp 115-116°. Resembles zinc stearate in appearance and physical properties. Can be prepd by dissolving zinc oxide in dil undecylenic acid and concentrating the solution.

Methyl ester, $C_{12}H_{22}O_2$, liq; d_4^{15} 0.889; bp_{760} 248°; bp_{10} 124°; mp $-27.5°$; n_D^{20} 1.43928; n_D^{25} 1.43727. Sol in alcohol, chloroform, petr ether, oils.

THERAP CAT: Antifungal (topical).
THERAP CAT (VET): Antifungal (topical).

9984. Unoprostone. *[1R-[1α(Z),2β,3α,5α]]-7-[3,5-Dihydroxy-2-(3-oxodecyl)cyclopentyl]-5-heptenoic acid;* 13,14-dihydro-15-keto-20-ethyl-PGF$_{2α}$. $C_{22}H_{38}O_5$; mol wt 382.54. C 69.08%, H 10.01%, O 20.91%. Prepn: R. Ueno *et al.*, **Eur.** pat. **289,349**; *eidem,* **U.S.** pat. **5,221,763** (1988, 1993 both to Ueno). Pharmacological characterization: Y. Goh, J. Kishino, *Japan. J. Ophthalmol.* **38**, 236 (1994). **Mechanism of action study:** M. Sakurai *et al.*, *ibid.* **37**, 252 (1993). Clinical evaluation in glaucoma and ocular hypertension: I. Azuma *et al.*, *ibid.* 514.

Isopropyl ester, $C_{25}H_{44}O_5$, **UF-021, Rescula.**
THERAP CAT: Antiglaucoma; in treatment of ocular hypertension.

9985. Uracil. *2,4(1H,3H)-Pyrimidinedione;* 2,4-dioxopyrimidine; 2,4-pyrimidinediol; 4-hydroxy-2(1*H*)-pyrimidinone; 2-hydroxy-4(1*H*)-pyrimidinone; 2-hydroxy-4(3*H*)-pyrimidinone. $C_4H_4N_2O_2$; mol wt 112.09. C 42.86%, H 3.60%, N 24.99%, O 28.55%. Obtained by hydrolysis of nucleic acids, *cf.* Levene, Bass, *Nucleic Acids* (New York, 1931). Alternate routes of formation: Johnson, *J. Am. Chem. Soc.* **63**, 263 (1941); Fox, Harada, *Science* **133**, 1923 (1961); Takemoto, Yamamoto, *Synthesis* **1971**, 154. Several desmotropic forms. Crystal structure: Stewart, *Acta Cryst.* **23**, 1102 (1967). *Review:* Ts'o "Bases, Nucleosides and Nucleotides" in *Basic Principles in Nucleic Acid Chemistry* **vol. 1**, P. O. P Ts'o, Ed. (Academic Press, New York, 1974) pp 453-584. *See also* Nucleic Acids.

Needles from water, mp 335° with effervescence. uv max (pH 7.0): 202.5, 259.5 nm ($\epsilon \times 10^{-3}$ 9.2, 8.2). Freely sol in hot water; sparingly in cold water (100 parts of water at 25° dissolves 0.358 part of uracil). Almost insol in alc, ether; sol in ammonia water and in other alkalies. pK = 9.45.
USE: In biochemical research.

9986. Uracil Mustard. *5-[Bis(2-chloroethyl)amino]-2,-4-(1H,3H)-pyrimidinedione;* 5-[bis(2-chloroethyl)amino]uracil; 2,6-dihydroxy-5-bis[2-chloroethyl]aminopyramidine; 5-[di(β-chloroethyl)amino]uracil; uramustine; demethyldopan; desmethyldopan; NSC-34462; U-8344. $C_8H_{11}Cl_2N_3O_2$; mol wt 252.10. C 38.12%, H 4.40%, Cl 28.13%, N 16.67%, O 12.69%. Synthesis from 5-aminouracil: Lyttle, Petering, *J. Am. Chem. Soc.* **80**, 6459 (1958); Lyttle, U.S. pat. **2,969,364** (1961 to Upjohn). Teratology study: S. Chaube, M. L. Murphy, "The Teratogenic Effects of the Recent Drugs Active in Cancer Chemotherapy" in *Advances in Teratology* **3**, D. H. M. Woollam, Ed. (Academic Press, New York, 1968) pp 181-237.

Crystals from methanol + water, dec 206°. Sparingly sol in water. uv max (0.01*N* H_2SO_4 in 95% ethanol): 257 nm (ϵ 5675). LD$_{50}$ i.p. in rats: ~1.25-2.5 mg/kg (Chaube, Murphy).
THERAP CAT: Antineoplastic.

9987. Uramil. *5-Amino-2,4,6(1H,3H,5H)-pyrimidinetrione;* 5-aminobarbituric acid; dialuramide; 5-amino-2,4,6-pyrimidinetriol; Murexan. $C_4H_5N_3O_3$; mol wt 143.10. C 33.57%, H 3.52%, N 29.36%, O 33.54%. Prepd by boiling alloxantin with ammonium chloride: Wöhler, Liebig, *Ann.* **26**, 241 (1838); by reduction of 5-nitrobarbituric acid with tin and HCl: Hartman, Sheppard, *Org. Syn.* **coll. vol. II**, 617 (1943); from barbituric acid, sodium nitrite and sodium hydrosulfite: Koppel, Robins, *J. Am. Chem. Soc.* **80**, 2751 (1958).

Crystals. Discolors in the air. Does not melt below 400°. Insol in cold water, in ether, chloroform; sparingly sol in hot water; sol in solns of alkali hydroxides, in ammonia and in sulfuric acid. Forms a slightly sol lead salt.

9988. Uramite®. A polymerized urea-formaldehyde prepn, contg 38% nitrogen. Prepn of a fertilizer contg 40% nitrogen from urea and formaldehyde: Darden, U.S. pat. **2,766,283** (1956 to du Pont).
Yellow, granular material of small particle size. Essentially insol in water.
USE: Nitrogen fertilizer for turfs and ornamentals.

9989. Uranediol. *(3β,5α,17α,17aβ)-17-Methyl-D-homoandrostane-3,17a-diol.* $C_{21}H_{36}O_2$; mol wt 320.52. C 78.70%, H 11.32%, O 9.98%. From pregnant mares' urine: Marker *et al., J. Am. Chem. Soc.* **60**, 210, 1061, 1561, 2719 (1938); Klyne, *Biochem. J.* **43**, 611 (1948); Brooks *et al., ibid.* **51**, 694 (1952). Structure: Klyne, *Nature* **166**, 559 (1950). Formation by acid hydrolysis of 5α-pregnane-3β,-20β-diol: Hirschmann, Williams, *J. Biol. Chem.* **238**, 2305 (1963). Configuration: Hirschmann *et al., J. Org. Chem.* **31**, 375 (1966).

Needles from aq ethanol, mp 216-219°. Sublimes at 180° and 0.06-0.1 mm. $[\alpha]_D^{15}$ +3.7° (c = 1.8 in chloroform).
Diacetate, $C_{25}H_{40}O_4$, plates from aq methanol, mp 159.5-160.5°. Sublimes at 190° and 0.1 mm. $[\alpha]_D^{20}$ −30.4° (c = 1.4 in chloroform); $[\alpha]_D^{20}$ −29.6° (ethanol).
Dibenzoate, $C_{35}H_{44}O_4$, leaflets from chloroform + methanol, mp 209-210°. $[\alpha]_D^{29}$ +18.6° (c = 1.05 in chloroform). uv max: 230, 272 nm (log ϵ 4.43, 3.26).

9990. Uranium. U; at. wt 238.0289 (characteristic naturally occurring isotopic mixture); at. no. 92; valence 6, 5, 4, 3. No stable nuclides; three naturally occurring isotopes (mass numbers): 238, $T_{1/2}$ 4.47 × 10⁹ years (99.275%); 235, $T_{1/2}$ 7.04 × 10⁸ years (0.718%); 234, $T_{1/2}$ 2.46 × 10⁵ years (0.005%); twelve artificial isotopes: 226-233; 236; 237; 239; 240. Occurrence in the earth's crust 2.1 ppm. Mined as uranium ore; main ores of commercial interest are carnotite, pitchblende, tobernite and autunite. Commercially important mines located in Elliot Lake—Blind River area in Canada, Rand gold fields in South Africa, Colorado and Utah in U.S., in Australia and France. Discovery from pitchblende: M. H. Klaproth, *Chem. Ann.* **II**, 387 (1789). Prepn of metal: E. Péligot, *C. R. Acad. Sci.* **12**, 735 (1841); *idem, Ann. Chim. Phys.* **5**, 5 (1842). Flowsheet and details of prepn of pure uranium metal: *Chem. Eng.* **62**, no. 10, 113 (1955); Spedding *et al., U.S. pat.* **2,852,364** (1958 to U.S.A. E.C.). *Reviews:* Mellor's **vol. XII**, 1-138 (1932); C. D. Harrington, A. R. Ruehle, *Uranium Production Technology* (Van Nostrand, Princeton, 1959); E. H. P. Cordfunke, *The Chemistry of Uranium* (Elsevier, New York, 1969) 250 pp; several authors in *Handb. Exp. Pharmakol.* **36**, 3-306 (1973); "The Actinides," in *Comprehensive Inorganic Chemistry* **vol. 5**, J. C. Bailar, Jr., *et al.*, Eds. (Pergamon Press, Oxford, 1973) *passim*; F. Weigel in Kirk-Othmer *Encyclopedia of Chemical Technology* **vol. 23** (Wiley-Interscience, New York, 3rd ed., 1983) pp 502-547; *idem* in *The Chemistry of the Actinide Elements* **vol. 1**, J. J. Katz *et al.*, Eds. (Chapman and Hall, New York, 1986) pp 169-442; J. C. Spirlet *et* *al., Adv. Inorg. Chem.* **31**, 1-40 (1987). Review of toxicology and health effects: *Toxicological Profile for Uranium* (PB91-180471, 1990) 205 pp.

Silver-white, lustrous, radioactive metal; malleable and ductile. Tarnishes rapidly in air, forming a layer of dark-colored oxide. Three allotropic modifications: orthorhombic α-form, d 19.07, transforms to β-form at 667.8° ±1.3°; tetragonal, β-form, d 18.11, transforms to γ-form at 774.9° ±1.6°; body-centered cubic γ-form, d 18.06, transforms to liquid at mp. mp 1132.8° ±0.8°. Heat of vaporization 446.7 kJ/mol; heat of fusion 19.7 kJ/mol; heat of sublimation 487.9 kJ/mol. Finely divided U metal and some U compounds may ignite spontaneously in air or oxygen. Rapidly soluble in aqueous HCl. Non-oxidizing acids, such as sulfuric, phosphoric and hydrofluoric, react only very slowly with U; nitric acid dissolves massive U at a moderate rate. Dissolution of finely divided U in nitric acid may approach explosive violence. Uranium metal is inert to alkalies.

Caution: Uranium is a chemical hazard as well as a radiological hazard. Direct contact with U metal or insoluble U compds may cause dermatitis. Radiation hazard is caused by the direct emission of α-particle radiation and by α-particles emitted from radon gas and its particulate daughters formed during the natural decay of U. Uranium and U compds are potential occupational carcinogens. *See* NIOSH *Pocket Guide to Chemical Hazards* (DHHS/NIOSH 90-117, 1990) p 222; *Patty's Industrial Hygiene and Toxicology* vol. **2C**, G. D. Clayton, F. E. Clayton, Eds. (Wiley-Interscience, New York, 4th ed., 1994) pp 2297-2317.

USE: [235]U in nuclear power reactors and nuclear weapons. Uranium depleted of [235]U to manuf armor-piercing ammunition, in inertial guidance devices and gyro compasses, as a counterweight for missile reentry vehicles, as radiation shielding material, and x-ray targets.

9991. Uranium Dioxide. Uranous oxide; black uranium oxide. O_2U; mol wt 270.03. O 11.85%, U 88.15%. UO_2. Occurs in nature as the minerals *uraninite* or *pitchblende*. Review of uranium oxides: Keller in *Comprehensive Inorganic Chemistry* vol. **5**, J. C. Bailar, Jr. *et al.*, Eds. (Pergamon Press, Oxford, 1973) pp 224-233.

Brown to black powder, or cubic crystals. d 10.97. mp 2865°. When obtained by heating the urano-uranic oxide or the oxalate in hydrogen, it is brown or copper-red in color and is pyrophoric. Insol in water, dil acids; sol in concd acids.

USE: Nuclear fuel.

9992. Uranium Hexafluoride. F_6U; mol wt 352.02. F 32.38%, U 67.62%. UF_6. Prepd by the action of fluorine on uranium metal or carbide; on uranium pentachloride; on uranium tetrafluoride; on triuranium octaoxide in the presence of carbon. It appears that all uranium compds when heated with fluorine to a sufficiently high temp give UF_6: National Nuclear Energy Series **VIII-5**, J. J. Katz, E. Rabinowitch, *The Chemistry of Uranium*, part **I** (New York, 1951) pp 396-449.

Volatile, white monoclinic cryst solid. $d_4^{20.7}$ 5.09. d^{70} (liq) 3.595. mp 64.8°. Sublimes at 56.5°. Critical temp 230.2°, crit pressure 45.5 atm: Oliver *et al.*, *J. Am. Chem. Soc.* **75**, 2827 (1953). Reacts vigorously with water, forming mainly UO_2F_2 and HF. Sol in liq chlorine and bromine. Dissolves in nitrobenzene to give a dark red soln fuming in air. Sol in carbon tetrachloride, chloroform, and *sym*-tetrachloroethane, $Cl_2CHCHCl_2$, which forms the most stable soln, extensive reaction occurring only after several days at room temp. Sol in fluorocarbons (C_6F_{12} or C_7F_{16}) without reaction. Best handled in copper apparatus.

9993. Uranium Peroxide. O_4U; mol wt 302.03. O 21.19%, U 78.81%. UO_4. Exists as a di- or tetrahydrate. Prepd from $UO_2(NO_3)_2.6H_2O$: Watt *et al.*, *J. Am. Chem. Soc.* **72**, 3341 (1950); Boggs, El-Chehabi, *ibid.* **79**, 4258 (1957); Sato, *J. Appl. Chem.* **13**, 361 (1963).

Dihydrate, hygroscopic, pale yellow crystals. Dec slowly in the temp range 90-195° to form a different peroxide U_2O_7, an orange, hygroscopic solid which dec to UO_3 and oxygen on contact with water, and to uranyl salts and oxygen on contact with acids. At temps of 200° and above, the dec of $UO_4.2H_2O$ leads to UO_3 without the formation of U_2O_7.

9994. Uranium Tetrachloride. Cl_4U; mol wt 379.84. Cl 37.33%, U 62.67%. UCl_4. Prepd from uranium(VI) oxide dihydrate and hexachloropropene: Hermann, Suttle, *Inorg. Syn.* **5**, 143 (1957).

Dark green octahedral crystals (tetragonal symmetry). Oxidizes in air and dec on contact with water. Should be stored in sealed ampuls. d_4^{25} 4.725. mp 590°, bp_{760} 791°. Heat of formation (solid) 250.9 kcal/mol at 0°. Freely sol in water (dec). Also sol in polar organic solvents. Insol in nonpolar solvents such as hydrocarbons and ethyl ether.

9995. Uranium Tetrafluoride. F_4U; mol wt 314.02. F 24.20%, U 75.80%. UF_4. Prepd from uranyl oxide and dichlorodifluoromethane: Booth *et al.*, *J. Am. Chem. Soc.* **68**, 1969 (1946); also prepd by the action of hydrogen fluoride on uranium dioxide or by addition of hydrofluoric acid to a U(IV) soln: Eméleus in *Fluorine Chemistry*, vol. **I**, J. H. Simons, Ed. (Academic Press, New York, 1950) p 59.

Monoclinic green crystals, mp > 1100°. Changes to U_3O_8 when heated in air. Insol in water. Sol in concd acids and alkalies with decompn.

Hemipentahydrate, orthorhombic green crystals, insol in water.

9996. Uranium Trichloride. Cl_3U; mol wt 344.39. Cl 30.88%, U 69.12%. UCl_3. Prepd by hydrogen reduction of uranium(IV) chloride: Suttle, *Inorg. Syn.* **5**, 145 (1957).

Dark purple crystals. Less hygroscopic than uranium(IV) chloride. d 5.51. mp 842°. Heat of formation at 25° 212.0 kcal/mol. Freely sol in water, giving a purple soln, which evolves hydrogen and turns green because of oxidation to uranium(IV). Less sol in polar organic solvents than uranium(IV) chloride and insol in nonpolar solvents.

9997. Uranium Trioxide. Uranic oxide; red uranium oxide. O_3U; mol wt 286.03. O 16.78%, U 83.22%. UO_3. Review of uranium oxides: Keller in *Comprehensive Inorganic Chemistry* vol. **5**, J. C. Bailar, Jr. *et al.*, Eds. (Pergamon Press, New York, 1973) pp 224-233.

Red or brownish-yellow powder. d 7.29. Insol in water; sol in acids.

9998. Uranyl Acetate. $C_4H_6O_6U$; mol wt 388.12. C 12.38%, H 1.56%, O 24.73%, U 61.33%. $UO_2(C_2H_3O_2)_2$.

Dihydrate, yellow, cryst powder; slight acetic odor. d 2.89. Sol in 10 parts of water, usually incompletely, due to the presence of basic salt. Freely sol in water acidulated with acetic acid, slightly in alcohol.

USE: As a reagent for pptn of sodium; in dry copying inks, and as activator in bacterial oxidation processes.

9999. Uranyl Chloride. Uranium dioxydichloride. Cl_2O_2U; mol wt 340.93. Cl 20.80%, O 9.39%, U 69.82%. UO_2Cl_2. Prepd by the reaction of O_2 with uranium(IV) chloride at 300-350°: Leary, Suttle, *Inorg. Syn.* **5**, 148 (1957).

Bright yellow crystals. Orthorhombic system. Very hygroscopic. Appreciably volatile above 775°. Dec *in vacuo* above 450° yielding chlorine and UO_2 plus U_3O_8. Very freely sol in water. Forms cryst hydrates. Sol in polar organic solvents such as acetone and alcohol, but insol in less polar solvents such as benzene. Aq solns are unstable.

10000. Uranyl Nitrate. N_2O_8U; mol wt 394.04. N 7.11%, O 32.48%, U 60.41%. $UO_2(NO_3)_2$.

Hexahydrate, yellow crystals; greenish luster by reflected light. When shaken, rubbed, or crushed, the crystals show remarkable triboluminescence with occasional detonations. d 2.807; mp 60°. Sol in ~1.5 parts of water, freely in alcohol, ether. The aq soln is acid. Solns of uranium nitrate in ether should not be allowed to stand in sunlight as explosion may occur.

USE: As intensifier in photography; manuf uranium glaze, decorating porcelain; also as reagent in analytical chemistry.

10001. Uranyl Phosphate. HO_6PU; mol wt 366.01. H 0.28%, O 26.23%, P 8.46%, U 65.03%. UO_2HPO_4.

Tetrahydrate, yellow, microcryst powder. Insol in water; sol in acids.

10002. Uranyl Sulfate. O_6SU; mol wt 366.09. O 26.22%, S 8.76%, U 65.02%. UO_2SO_4.

Trihydrate, lemon-yellow, cryst mass. d 3.28. Sol in ~5 parts water, 25 parts alcohol.

10003. Urapidil. *6-[[3-[4-(2-Methoxyphenyl)-1-piperazinyl]propyl]amino]-1,3-dimethyl-2,4(1H,3H)-pyrimidinedione;* 6-[[3-[4-(o-methoxyphenyl)-1-piperazinyl]propyl]-amino]-1,3-dimethyluracil; B-66256; Ebrantil; Eupressyl; Mediatensyl; Uraprene. $C_{20}H_{29}N_5O_3$; mol wt 387.48. C 62.00%, H 7.54%, N 18.07%, O 12.39%. α_1-Adrenergic antagonist; deriv of uracil, *q.v.* Prepn: W. Pruesse *et al.*, **Ger.** pat. **1,942,405;** *eidem,* U.S. pat. **3,957,786** (1971, 1976 both to Byk Gulden); K. Klemm *et al., Arzneimittel-Forsch.* **27,** 1895 (1977). Series of articles on pharmacology, pharmacokinetics, metabolism: *ibid.* 1898-1932. Toxicity data: J. Koenig *et al., ibid.* 1919. Mode of action: M. Eltze, *Eur. J. Pharmacol.* **59,** 1 (1979); H. R. Kaplan, R. D. Smith, *Fed. Proc.* **40,** 2268 (1981). Hemodynamic responses in man: G. G. Belz *et al., Clin. Pharmacol. Ther.* **37,** 48 (1985). Clinical trials: A. Barankay *et al., Arzneimittel-Forsch.* **31,** 849 (1981); H. Liebau *et al., J. Hypertens.* **4,** Suppl. 6, S141 (1986). Series of articles on clinical efficacy in hypertension: *Drugs* **35,** Suppl. 6, 147-192 (1988).

Crystals from water, mp 156-158°. uv max (methanol): 237, 268 nm (ϵ 1.10×10^4, 2.67×10^4). pKa 7.10. LD_{50} in male mice, rats (mg/kg): 750, 550 orally; 260, 145 i.v. (Koenig).

THERAP CAT: Antihypertensive.

10004. Urazole. *1,2,4-Triazolidine-3,5-dione; bicarbamimide;* 1H-1,2,4-triazole-3,5(2H,4H)-dione; hydrazodicarbonimide; 3,5-diketotriazolidine. $C_2H_3N_3O_2$; mol wt 101.06. C 23.77%, H 2.99%, N 41.58%, O 31.66%. Prepd by heating biuret with hydrazine hydrate at 108°: Stolle, Krauch, *J. Prakt. Chem.* [2] **88,** 313 (1913). From methyl or ethyl allophanate: Gordon, Audrieth, *Inorg. Syn.* **5,** 52 (1957).

Leaflets from water, dec 249-250°. Weak acid, Ka: 1.6×10^{-6}. pH of satd aqueous soln at 25° = 3.15. Soly in water (g/100 ml): 2.83 (0°); 23.7 (65°). Sparingly sol in alc, practically insol in ether. Sol in concd HCl from which it crystallizes without change. Forms ammonium and sodium salts.

10005. Urea. Carbamide; carbonyldiamide; Aquacare; Aquadrate; Basodexan; Hyanit; Keratinamin; Nutraplus; Onychomal; Pastaron; Ureaphil; Ureophil; Urepearl. CH_4N_2O; mol wt 60.06. C 20.00%, H 6.71%, N 46.65%, O 26.64%. H_2NCONH_2. Product of protein metabolism. From cyanamide by hydrolysis, from CO_2 by ammonolysis: One mol of carbon dioxide reacts with two mols of ammonia to yield ammonium carbamate. The carbamate then dec to form urea and water: Wöhler, *Ann. Physik* **12,** 253 (1828). Continuous process: **Brit.** pat. **844,110** (1960 to Inventa, AG), *C.A.* **55,** 7298b (1961). Review of commercial processes: Tonn, Jr., *Chem. Eng.* **62,** no. 10, 186-190 (1955); Moore, Parks, *Chem. & Eng. News* **37,** 84-90 (Nov. 23, 1959); Faith, Keyes & Clark's *Industrial Chemicals,* F. A. Lowenheim, M. K. Moran, Eds. (Wiley-Interscience, New York, 4th ed., 1975) pp 854-861. Prepn from ammonia, carbon monoxide and sulfur in methanol: Applegath, U.S. pat. **2,857,430** (1958 to Monsanto); Franz, Ap-

plegath, *J. Org. Chem.* **26,** 3304, 3306, 3309 (1961). Comprehensive review: G. T. Y. Chao, *Urea, Its Properties and Manufacture* (Chao's Institute, W. Covina, Calif., 1967) 508 pp.

Tetragonal prisms. Develops odor of NH_3. Cooling, saline taste. mp 132.7°. On further heating it decomposes to biuret, NH_3, and cyanuric acid. d_4^{18} 1.32; d_4^{18} of water solns (w/w): 10% 1.027; 20% 1.054; 50% 1.145. pH of a 10% water soln: 7.2. One gram dissolves in 1 ml water, 10 ml 95% alc, 1 ml boiling 95% alc, 20 ml abs alc, 6 ml methanol, 2 ml glycerol. Almost insol in chloroform, ether. Sol in concd HCl. Water solns dec on heating, giving off some NH_3. Pure urea should not give the biuret reaction unless heated above the melting point. In practice all reagent grade urea gives a positive biuret reaction. *See* Biuret.

USE: Fertilizer, because of easily available high nitrogen content. In animal feeds. Is reacted with aldehydes to make resins and plastics. Is condensed with malonic ester to form barbituric acid. Used extensively in the paper industry to soften cellulose. In ammoniated dentifrices.

THERAP CAT: Diuretic.

THERAP CAT (VET): Nutritional factor (partial source of dietary nitrogen in ruminants). To enhance action of sulfonamides. Has beeen used to promote healing in infected wounds, septic metritis. Diuretic. Antiseptic.

10006. Urea Hydrochloride. CH_5ClN_2O; mol wt 96.52. C 12.44%, H 5.22%, Cl 36.73%, N 29.02%, O 16.58%. CO-$(NH_2)_2$.HCl.

White to faintly yellow, deliquesc crystals. Dec at 145°. Sol in water. The aq soln is acid. *Keep well closed.*

10007. Urea Hydrogen Peroxide. Carbamide peroxide; hydrogen peroxide carbamide; Debrox; Exterol; Gly-Oxide; Hyperol; Ortizon (obsolete); Perhydrit; Perhydrol-Urea. $CH_6N_2O_3$; mol wt 94.07. C 12.77%, H 6.43%, N 29.78%, O 51.02%. $CO(NH_2)_2$.H_2O_2. Usually contains 34-35% H_2O_2.

White crystals or cryst powder. Dec in air into urea, oxygen, and water. Sol in 2.5 parts water; partly dec by alcohol and ether into H_2O_2 and urea.

USE: For extemporaneous prepn of H_2O_2.

THERAP CAT: Disinfectant; cerumenolytic.

10008. Urea Nitrate. Acidogen nitrate. $CH_5N_3O_4$; mol wt 123.07. C 9.76%, H 4.10%, N 34.14%, O 52.00%. $CO(NH_2)_2$.HNO_3.

White, odorless leaflets. mp 152° with decompn. Has tendency to explode, especially when rubbed while heating. Sparingly sol in cold water or in HNO_3; sol in hot water or alcohol. The aq soln is acid.

Caution: Mildly irritating to skin, mucous membranes.

10009. Urease. Urea amidohydrolase; Urastrat. Nickel-dependent metalloenzyme which hydrolyzes urea to ammonium carbonate. First enzyme crytallized; mol wt approx 489,000 Da. Isolation from jack bean, *Canavalia ensiformis* and crystallization: J. B. Sumner, *J. Biol. Chem.* **69,** 435 (1926). Early structure-activity studies: G. Gorin *et al., Biochemistry* **1,** 911 (1962); C. J. Bailey, D. Boulter, *Biochem. J.* **113,** 669 (1969). Identification as nickel containing: N. E. Dixon *et al., J. Am. Chem. Soc.* **97,** 4131 (1975). Crystal structure: E. Jabri *et al., Science* **268,** 998 (1995). *Review:* J. E. Varner, "Urease" in *The Enzymes* **vol. 4,** P. D. Boyer *et al.,* Eds. (Academic Press, New York, 2nd ed., 1960) pp 247-256; F. J. Reithel, *ibid.* **vol. 4** (3rd ed., 1971) pp 1-21.

Colorless, octahedral crystals. Soluble in water, in dil alkali and dil ammonia. uv max: 278.5 nm. Isoelectric point pH 5.0-5.1; also reported as pH 4.8 (Reithel).

USE: Clinical reagent in determination of urea in body fluid.

10010. Urea Stibamine. Carbostibamide. No definitely known and agreed structure. Originally reported to consist of urea and p-aminophenylstibinic acid with the formula $C_7H_{12}N_3O_4Sb$: Brahmachari, *Ind. J. Med. Res.* **10,** 492 (1922). Later suggested that the substance was identical with ammonium p-carbamidophenylstibinate, NH₂CONH-$C_6H_4SbO(OH)ONH_4$: Niyogi, *J. Ind. Chem. Soc.* **5,** 285 (1928); Ghosh *et al., Ind. J. Med. Res.* **16,** 461 (1928), and Gray *et al., Proc. Roy. Soc.* **B108,** 54 (1931), produced evidence that urea stibamine was not a single substance and attempted to prove that the active principle was a disubstituted urea: *sym*-diphenylcarbamido-4,4-distibinic acid. The antimony content of commercial samples seems to vary between 39 and 42%: Guha *et al., Nature* **151**, 108 (1943). Prepd by the addition of urea to a warm aq suspension of stibanilic acid: Datta *et al., Sci. Cult.* **11**, 385 (1946), *C.A.* **41**, 105 (1947); Christiansen, Green, U.S. pat. **2,488,268** (1949 to Squibb).

White powder. Sol in water. Partially sol in alc, ether.

THERAP CAT: Anthelmintic (Nematodes, Schistosoma); antiprotozoal (Leishmania).

10011. Uredepa. *[Bis(1-aziridinyl)phosphinyl]carbamic acid ethyl ester;* ethyl [bis(1-aziridinyl)phosphinyl]carbamate; ethyl *N*-[bis(ethyleneimido)phosphoro]carbamate; bis-(ethylenimido)phosphorylurethane; urethimine; AB-100; NSC-37095; Avinar. $C_7H_{14}N_3O_3P$; mol wt 219.18. C 38.36%, H 6.44%, N 19.17%, O 21.90%, P 14.13%. Outline of prepn: Bardos *et al., Nature* **183**, 399 (1959); Bardos, Papanastassiou, U.S. pat. **3,201,313** (1965 to Armour Pharm.).

Crystals from benzene + cyclohexane, mp 88-90°. Readily sol in water but dec in aq soln.

USE: Has been used experimentally as an insect chemosterilant.

THERAP CAT: Antineoplastic.

10012. Urena. A bast fiber from the plant *Urena lobata* Dill. ex L., *Malvaceae.* The plant is tropical and subtropical and grows wild, sometimes becoming a noxious weed in South and Middle America, Asia, Indonesia, the Philippines, Madagascar, and Africa. It is cultivated to a small extent in Brazil, India, and Madagascar, and to a much larger extent in the former Belgian Congo and French Equatorial Africa. Harvested when in full bloom (pink flowers). The fiber is creamy white, lustrous, soft and flexible, comparable to jute. The presence of steroids in the rest of the plant has been demonstrated. *Ref:* E. E. Stout, *Introduction to Textiles* (John Wiley, New York, 1960) pp 56-57.

10013. Urethan. *Carbamic acid ethyl ester; ethyl carbamate;* urethane; ethyl urethan. $C_3H_7NO_2$; mol wt 89.09. C 40.44%, H 7.92%, N 15.72%, O 35.92%. $NH_2COOC_2H_5$. Prepd by heating urea with alcohol under pressure; by warming urea nitrate with alcohol and sodium nitrite. Toxicity data: K. J. Franklin, *J. Pharmacol. Exp. Ther.* **42**, 1 (1931). Review of carcinogenic action and metabolism: Mirvish in *Advan. Cancer Res.* **11**, 1-42 (1968).

Crystals, mp 48-50°. Cooling saline taste, d 1.1. bp 182-184°. Sublimes readily at 103° and 54 mm pressure. One gram dissolves in 0.5 ml water, 0.8 ml alcohol, 0.9 ml chloroform, 1.5 ml ether, 2.5 ml glycerol, 32 ml olive oil. The aq soln is neutral. MLD i.p. in mice: 2.1-2.2 g/kg (Franklin).

Note: This substance may reasonably be anticipated to be a carcinogen: *Seventh Annual Report on Carcinogens* (PB95-109781, 1994) p 398.

USE: Molten urethan is a good solvent for various organic materials. Intermediate in organic synthesis. In the prepn and modification of amino resins. As solubilizer and cosolvent for pesticides, fumigants.

THERAP CAT: Antineoplastic.

THERAP CAT (VET): Anesthetic.

10014. Uric Acid. *7,9-Dihydro-1H-purine-2,6,8(3H)-trione;* 8-hydroxyxanthine; purine-2,6,8-triol; purine-2,6,8(1*H*,-3*H*,9*H*)-trione; 2,6,8-trioxypurine. $C_5H_4N_4O_3$; mol wt 168.11. C 35.72%, H 2.40%, N 33.33%, O 28.55%. Discovered by Scheele and independently by Bergman in 1776. It forms the chief end-product of the nitrogenous metabolism of birds and of scaly reptiles and is found in their excrement; present in the urine of all carnivorous animals. Prepn from urea: Bills *et al., J. Org. Chem.* **27**, 4633 (1962). Role

in biological processes: Bishop, Talbott, *Pharmacol. Revs.* **5**, 231 (1953).

White, odorless, tasteless crystals; dec by heat without melting and with evolution of HCN. d 1.89. One gram dissolves in about 15,000 parts cold water, about 2000 parts boiling water; sol in glycerol, in solns of alkali hydroxides, their carbonates, sodium acetate and sodium phosphate. Insol in alcohol, ether. Gives murexide reaction.

10015. Uricase. *Urate oxidase; uric oxidase;* urico-oxidase; uriKoxidase. An enzyme responsible for the oxidative scission of the purine skeleton, and therefore of key importance to the catabolism of nitrogenous compds in general in organisms that do not eliminate intact uric acid or some other purine. All mammals except the primates (including man) are uricolytic organisms. Porcine liver or bovine kidney is the starting material for various purified prepns of uricase. Isoln from porcine liver: Holmberg, *Biochem. J.* **33**, 1901 (1939); Miller *et al., J. Biol. Chem.* **216**, 625 (1955); Robbins, Grant, U.S. pat. **2,878,161** (1959 to Armour). *Review:* Mahler, "Uricase" in *The Enzymes,* vol. **8**, P. D. Boyer *et al.,* Eds. (Academic Press, New York, 1963) pp 285-296.

Pale, brownish-green crystals or shiny, transparent, striated plates. Practically insol in water. Slightly sol in buffered alkali solns. Solns at pH 7.5-10.5 are relatively stable. Shows unusually high absorption in the region of 330-350 nm (for highly purified uricase: $A_{276}^{1\%} = 11.3$, $A_{330}^{1\%} = 2.0$, both in 1% Na_2CO_3; $A_{280}/A_{330} = 5.6$). Isoelec. pt. pH 6.3. Copper content of the enzyme (specific activity 120-125) equals ~0.56 mg/g of enzyme protein. There appears to be a satisfactory correlation between copper content and activity of the purified enzyme. The enzyme is sensitive to cyanide ion; the presence of $10^{-4}M$ KCN inhibits its activity.

USE: In the determination of serum and urine uric acid.

10016. Uridine. 1-β-D-Ribofuranosyluracil; uracil riboside. $C_9H_{12}N_2O_6$; mol wt 244.20. C 44.27%, H 4.95%, N 11.47%, O 39.31%. Nucleoside; widely distributed in nature. Prepd by hydrolysis of yeast nucleic acid with weak alkali, *cf.* Levene, Bass, *Nucleic Acids* (New York, 1931). Improved isolns: Harris, Thomas, *J. Chem. Soc.* **1948**, 1936; Elmore, *ibid.* **1950**, 2084; Lorine, Ploeser, *J. Biol. Chem.* **178**, 439 (1949). Crystal structure: Green *et al., Chem. Commun.* **1971**, 53. *Review: Basic Principles in Nucleic Acid Chemistry* vol. **1**, P. O. P. Ts'o, Ed. (Academic Press, New York, 1974) *passim. See also* Nucleic Acids.

Needles from dil alc, mp 165°. $[\alpha]_D^{20}$ +4° (c = 2). uv max (pH 7.3): 261, 205 nm (ε × 10^{-3} 10.1, 9.8), Voet *et al., Biopolymers* **1**, 193 (1963). Sol in water. Upon prolonged refluxing with HCl, furfurol is formed.

10017. Uridine 5'-Diphosphate. *Uridine 5'-(trihydrogen diphosphate);* UDP; uridine 5'-pyrophosphate; uridine-5-pyrophosphoric acid. $C_9H_{14}N_2O_{12}P_2$; mol wt 404.16. C 26.75%, H 3.49%, N 6.93%, O 47.50%, P 15.33%. Can be isolated from calf's liver, thymus, and yeast. The commercial product is derived from yeast. Pentose nucleic acids (isolated from yeast) are digested with rattlesnake venom (freed of 5'-monoesterase) and the nucleotides are separated by chromatography: Cohn, Volkin, *Arch. Biochem. Biophys.* **35**, 465 (1952); *J. Biol. Chem.* **203**, 319 (1953). For alternate procedures *see* the refs under Uridine Diphosphate Glucose. Syntheses: Chambers, *J. Am. Chem. Soc.* **81**, 3032 (1959); Moffatt, Khorana, *ibid.* **83**, 649 (1961).

Lithium salt monohydrate, $C_9H_{12}N_2O_{12}P_2Li_2.H_2O$, crystals, sol in water.

Sodium salt trihydrate, $C_9H_{12}N_2O_{12}P_2Na_2.3H_3O$, crystals. pKa'_1 6.5; pKa'_2 9.4. uv max (pH 7): 262 nm; (pH 11): 261 nm. Freely sol in water.

10018. Uridine Diphosphate Glucose. *Uridine 5'-(trihydrogen diphosphate) mono-α-D-glucopyranosyl ester;* UDPG; uridine 5'-pyrophosphate glucose ester; uridine-5'-diphosphoglucose; co-waldenase; co-galactoisomerase. $C_{15}H_{24}N_2O_{17}P_2$; mol wt 566.31. C 31.81%, H 4.27%, N 4.95%, O 48.03%, P 10.94%. The coenzyme of the galactowaldenase system which catalyzes the conversion of galactose-1-phosphate into glucose-1-phosphate. Isoln from baker's yeast: Caputto *et al., J. Biol. Chem.* **184**, 333 (1950). Also present in animal tissue. Synthesis: Michelson, Todd, *J. Chem. Soc.* **1956**, 3459; Moffatt, Khorana, *J. Am. Chem. Soc.* **80**, 3756 (1958). *Reviews:* Leloir, Cardini in *The Enzymes* vol. 2A, P. D. Boyer *et al.*, Eds. (Academic Press, New York, 2nd ed., 1960) pp 39-61; A. M. Michelson, *The Chemistry of Nucleosides and Nucleotides* (Academic Press, New York, 1963) pp 153-250; D. W. Hutchison, *Nucleotides and Coenzymes* (John Wiley, New York, 1964) pp 36-82.

Isolated as the barium or calcium salt, white powder, sol in water.

Sodium salt dihydrate, $C_{15}H_{23}N_2O_{17}P_2Na.2H_2O$, white powder. uv max (pH 2.0): 262 nm. Freely sol in water.

10019. Uridine 5'-Triphosphate. *Uridine 5'-(tetrahydrogen triphosphate);* UTP; Uteplex. $C_9H_{15}N_2O_{15}P_3$; mol wt 484.14. C 22.33%, H 3.12%, N 5.79%, O 49.57%, P 19.19%. Pyrimidine analog of ATP. Isoln from rabbit muscle: Lipton *et al., J. Am. Chem. Soc.* **75**, 5450 (1953). Synthesis: Kenner *et al., J. Chem. Soc.* **1954**, 2288; Hall, Khorana, *J. Am. Chem. Soc.* **76**, 5056 (1954). Prepd enzymatically from uridine diphosphate or from ribonucleic acid.

Barium salt tetrahydrate,$(C_9H_{12}N_2O_{15}P_3)_2Ba_3.4H_2O$, white powder.

Trisodium salt dihydrate, $C_9H_{12}N_2Na_3O_{15}P_3.2H_2O$, *Utipina*. White powder, sol in water, very sparingly sol in alcohol. pKa_1 6.6; pKa_2 9.5. uv max (pH 7): 262 nm (a_M 10,000); (pH 11): 261 nm (a_M 8100).

10020. 5'-Uridylic Acid. Uridine 5'-phosphoric acid; uridine 5'-monophosphate; UMP. $C_9H_{13}N_2O_9P$; mol wt 324.18. C 33.34%, H 4.04%, N 8.64%, O 44.42%, P 9.55%. Nucleotide; widely distributed in nature. Synthesis by phosphorylation of 2',3'-O-benzylidene uridine with diphenyl phosphorochloridate: Brown *et al., J. Chem. Soc.* **1950**, 408; Smith, *Biochem. Prepn.* **8**, 130 (1961). Monograph on the synthesis of nucleotides: G. R. Pettit, *Synthetic Nucleotides* vol. 1 (Van Nostrand Reinhold, New York, 1972) 252 pp. Crystal structure of hydrated barium salt: Shefter, Trueblood, *Acta Cryst.* **18**, 1067 (1965). *Reviews: see* Uridine; Nucleic Acids.

Free acid: pKa_1' 6.4, pKa_2' 9.5. uv max (pH 7.0): 262 nm (a_M 10,000).

Disodium salt dihydrate, $C_9H_{11}N_2Na_2O_9P.2H_2O$, white powder, characteristic meaty taste. uv max (0.1M HCl): 262 nm (ε 10,000). Soly in water at 20°: about 41 g/100 ml H_2O.

10021. Urinastatin. *Bikunin trypsin inhibitor;* mingin; urinary trypsin inhibitor; UTI; ulinastatin; Miraclid. Acid-stable glycoprotein (pI 2.8), mol wt ~67,000 by gel filtration. Consists of a single polypeptide chain of 147 amino acid residues containing two Kunitz-type protease inhibitor domains, and two carbohydrate side chains. Normally present in serum and urine. Isoln from human urine: N. R. Shulman, *J. Biol. Chem.* **213**, 655 (1955); and purification: G. J. Proksch, J. I. Routh, *J. Lab. Clin. Med.* **79**, 491 (1972). Characterization: M. Balduyck *et al., Eur. J. Biochem.* **158**, 417 (1986). Protease inhibition: H. Sumi, N. Toki, *Proc. Soc. Exp. Biol. Med.* **167** 530 (1981). Determn by enzyme immunoassay: N. Nishino *et al., Haemostasis* **19**, 112 (1989). Biodistribution and clinical kinetics: B.-M. Jönsson-Berling, K. Ohlsson, *Scand. J. Clin. Lab. Invest.* **51**, 549 (1991). Protective effect in exptl pancreatitis: T. Hirano, T. Manabe, *Arch. Surg.* **128**, 1322 (1993). Clinical evaluation vs ischemic myocardial injury: S. Shimai *et al., Japan. Circ. J.* **53**, 1144 (1989); vs postoperative hyperamylasemia: D. Korenaga *et al., Eur. Surg. Res.* **23**, 214 (1991). Receptor binding studies: H. Kobayashi *et al., J. Biol. Chem.* **269**, 20642 (1994). Review of early literature: H. J. Faarvang, *Scand. J. Clin. Lab. Invest.* **17**, Suppl. 83, 11-77 (1965). Review of characterization as subunit of inter-α-trypsin inhibitor (inter-α-inhibitor): W. Gebhard, K. Hochstrasser in *Proteinase Inhibitors,* A. J. Barrett, G. Salvesen, Eds. (Elsevier, Amsterdam, 1986) pp 389-401; W. Gebhard *et al., Biol. Chem. Hoppe-Seyler* **371**, Suppl., 13-22 (1990).

THERAP CAT: Protease inhibitor.

10022. Urobilins. Bile pigments which occur in feces; and in urine. More specifically described as dipyrrylmethene compds joined at the two α-pyrrolyl positions by methylene bridges to either pyrrolenone or pyrrolidone rings. Reviews of occurrence, proposed structures, stereoisomerism, and interrelationships of *i-urobilin, d-urobilin, stercobilin, urobilin IXα,* and *d-urobilin IXα:* A. H. Jackson *et al., Nature* **209**, 581 (1966); C. H. Gray, D. C. Nicholson, *Medicine* **46**, 83 (1967); W. Rüdiger, *Fortschr. Chem. Org. Naturst.* **29**, 60-139 (1971). Structure of stercobilin and *d-*urobilin: C. H. Gray, D. C. Nicholson, *J. Chem. Soc.* **1958**,

3085. Revised structure of *d*-urobilin: S. D. Killilea, P. O'Carra, *Biochem. J.* **129**, 1179 (1972). Stereochemistry: W. J. Cole *et al., J. Chem. Soc.* **1965**, 4085. Spectral study: W. J. Cole *et al., J. Chem. Soc. (C)* **1966**, 1321. Synthesis of optically active stercobilin: H. Plieninger, J. Ruppert, *Ann.* **736**, 43 (1970); of optically active urobilins: H. Plieninger *et al., ibid.*, 62. Abs config of stercobilin: H. Brockmann, Jr. *et al., Proc. Nat. Acad. Sci.* **68**, 2141 (1971).

(–)-Stercobilin

USE: Spot-test reagent for detection of copper, mercury, zinc: F. J. Welcher, *Organic Analytical Reagents,* vol. **IV** (Van Nostrand, New York, 1949).

10023. Urochloralic Acid. *2,2,2-Trichloroethyl β-D-glucopyranosiduronic acid; 2,2,2-trichloroethyl β-D-glucosiduronic acid; β,β,β-trichloroethyl-D-glucuronide.* $C_8H_{11}Cl_3$-O_7; mol wt 325.53. C 29.52%, H 3.41%, Cl 32.67%, O 34.40%. Isolated from urine of man and dog after ingestion of chloral hydrate: von Mering, Musculus, *Ber.* **8**, 663 (1875); from urine of calves fed trichloroethylene: Seto, Schultze, *J. Am. Chem. Soc.* **78**, 1616 (1956).

Needles, mp 142°. $[\alpha]_D^{27}$ —50.0° (c = 0.7). Freely soluble in water, alcohol. One gram dissolves in 234 ml of anhydr ether at 20°.

10024. Urokinase. Win-22005; Abbokinase; Actosolv; Breokinase; Persolv; Purochin; Ukidan; Uronase; Win-Kinase. Serine protease which activates plasminogen to plasmin; present in mammalian blood and urine. Produced as prourokinase, *q.v.*, and converted to active form by plasmin or kallikrein, *q.q.v.* Description of fibrinolytic activity: J. R. B. Williams, *Brit. J. Exp. Pathol.* **32**, 530 (1951). Isolation from human urine and activity: T. Astrup, I. Sterndorff, *Proc. Soc. Exp. Biol. Med.* **81**, 675 (1952); G. W. Sobel *et al., Am. J. Physiol.* **171**, 768 (1952). Species specificity and distribution in mammals: S. R. Mohler *et al., Am. J. Physiol.* **192**, 186 (1958). Isoln from human male urine: H. O. Singher, L. Zuckerman, U.S. pats. **2,961,382** and **2,989,440** (1960, 1961 to Ortho); N. O. Kjeldgaard, J. Ploug, U.S. pat. **2,983,647** (1961 to Lövens Kemiske Fabrik); J. Doczi, U.S. pat. **3,081,236** (1963 to Warner-Lambert). Prepn of crystalline form: A. Lesuk *et al., Science* **147**, 880 (1965). Two variants of bioactive urokinase, high molecular weight (HMW-UK, ~50000 Da) and low molecular weight (LMW-UK, ~30000 Da) have been identified: W. F. White *et al., Biochemistry* **5**, 2160 (1966). Both are disulfide-linked dimers consisting of a heavy chain (B) and a light chain (A). HMW-UK is converted to LMW-UK by proteolytic cleavage of the A-chain to form the A_1-chain. Structural characterization: M. Nobuhara *et al., J. Biochem.* **90**, 225 (1981). Structure and amino acid sequences: W. A. Günzler *et al., Z. Physiol. Chem.* **363**, 133 (1982); G. J. Steffens *et al., ibid.* 1043; W. A. Günzler *et al., ibid.* 1155. Expression of gene coding for human urokinase in *E. coli:* B. Ratzkin *et al., Proc. Nat. Acad. Sci. USA* **78**, 3313 (1981). Clinical evaluation in pulmonary embolism: P. Petitpretz *et al., Circulation* **70**,

861 (1984). Double-blind multicenter clinical trial in coronary thrombosis following myocardial infarction: H. Kambara *et al., Japan. Circ. J.* **51**, 1072 (1987). Clinical applications of urokinase-treated tubing: T. Ohshiro *et al., Methods Enzymol.* **137**, 529 (1988). Series of articles on plasminogen activation, fibrinolysis and clinical efficacy: *Proc. Serono Symp., Thrombosis and Urokinase* **9**, 1-257 (1977). *Review:* F. Duckert, *Handb. Exp. Pharmacol.* **46**, 209-237 (1978).

THERAP CAT: Thrombolytic.

10025. Urothion. *2-Amino-7-(1,2-dihydroxyethyl)-6-(methylthio)thieno[3,2-g]pteridin-4(3H)-one.* $C_{11}H_{11}N_5O_3S_2$; mol wt 325.37. C 40.61%, H 3.41%, N 21.52%, O 14.75%, S 19.71%. Constituent of normal human urine: Koschara, *Z. Physiol. Chem.* **277**, 284 (1943). Structure: Tschesche *et al., Ber.* **88**, 1251 (1955). Revised structure: Goto *et al., Tetrahedron Letters* **1967**, 4507; *eidem, J. Biochem. (Tokyo)* **65**, 611 (1969). Synthesis: Sakurai, Goto, *Tetrahedron Letters* **1968**, 2941; *eidem, J. Biochem. (Tokyo)* **65**, 755 (1969).

Orange-colored clusters of crystals, not melted at 360°. Shows mutarotation: $[\alpha]_D^{20}$ —20° (after 15 hrs in 0.05*N* NaOH). Soly in water at pH 6.6 about 1:10,000. More sol in alkaline soln. Also sol in acids. Practically insol in the usual organic solvents.

10026. Ursodiol. *(3α,5β,7β)-3,7-Dihydroxycholan-24-oic acid;* 17β-(1-methyl-3-carboxypropyl)etiocholane-3α,7β-diol; 3α,7β-dioxycholanic acid; ursodeoxycholic acid; Actigall; Arsacol; Cholit-Ursan; Delursan; Desol; Destolit; Deursil; Litursol; Lyeton; Peptarom; Solutrat; Urdes; Ursacol; Urso; Ursobilin; Urschol; Ursodamor; Ursofalk; Ursolvan. $C_{24}H_{40}O_4$; mol wt 392.58. C 73.43%, H 10.27%, O 16.30%. Epimeric with chenodiol, *q.v.*, with respect to the hydroxyl group at C_7. Found in bear bile (combined with taurine). Isoln: Shoda, *J. Biochem. (Japan)* **7**, 505 (1927). Structure: Kaziro, *Z. Physiol. Chem.* **185**, 151 (1929); **197**, 206 (1931); Iwasaki, *ibid.* **244**, 181 (1936). Toxicity data: M. Ardenne, P. G. Reitnauer, *Arzneimittel-Forsch.* **20**, 323 (1970). Effect on cholesterol and bile acid metabolism: G. S. Tint *et al., Gastroenterology* **91**, 1007 (1986). Clinical trial in primary biliary cirrhosis: R. E. Poupon *et al., N. Engl. J. Med.* **324**, 1548 (1991). Review of pharmacology, toxicology, efficacy: A. Ward *et al., Drugs* **27**, 95-131 (1984). Brief review of clinical effects and comparison with chenodiol: H. Fromm, *Gastroenterology* **87**, 229-233 (1984).

Bitter plates from alc. mp 203°. $[\alpha]_D^{20}$ +57° (c = 2 in abs ethanol). Freely sol in ethanol, glacial acetic acid; slightly sol in chloroform, sparingly sol in ether. Practically insol in water. LD_{50} in mice (g/kg): 0.1 i.v. (Ardenne, Reitnauer); in rats, mice (mg/kg): 2000, 6000 s.c.; 1000, 1200 i.p., 310, 260 i.v. (Ward).

Diformate, $C_{26}H_{40}O_6$, crystals, mp 170°.
Diacetate, $C_{28}H_{44}O_6$, crystals, mp 98-102°.
THERAP CAT: Anticholelithogenic.

10027. Ursolic Acid. *(3β)-3-Hydroxyurs-12-en-28-oic acid;* urson; prunol; micromerol; malol. $C_{30}H_{48}O_3$; mol wt 456.71. C 78.90%, H 10.59%, O 10.51%. In leaves and berries of *Arctostaphylos uva-ursi* (L.) Spreng (bearberry), of *Vaccinium macrocarpon* Ait. (cranberry), *Rhododendron hymenanthes* Makino, *Ericaceae*. In the protective wax-like coating of apples, pears, prunes, and other fruits. Isoln from apple peelings: Sando, *J. Biol. Chem.* **56**, 457 (1923). Structure: Ruzicka *et al.*, *Helv. Chim. Acta* **28**, 199 (1945); Zurcher *et al.*, *ibid.* **37**, 2145 (1954). Conversion from α-amyrin: Boar *et al.*, *J. Chem. Soc. (C)* **1970**, 678. Chemistry: Mezzetti *et al.*, *Planta Med.* **20**, 244 (1971).

Large, lustrous prisms from abs alcohol, fine hair-like needles from dil alcohol, mp 285-288°. $[\alpha]_D^{21}$ +67.5° (c = 1 in *N* alc KOH). Soly at 15°: One part dissolves in 88 parts methanol, 178 alcohol (35 boiling alcohol), 140 ether, 388 chloroform, 1675 carbon disulfide. Moderately sol in acetone. Sol in hot glacial acetic acid and in 2% alcoholic NaOH. Insol in water and petr ether.

Acetate, $C_{32}H_{50}O_4$, mp 289-290°. $[\alpha]_D$ +62.3° (c = 1.15 in chloroform).

Methyl ester, $C_{31}H_{50}O_3$, mp 171°. $[\alpha]_D^{20}$ +58° (c = 1.2 in pyridine).

Methyl ester acetate, $C_{33}H_{52}O_4$, mp 246-247°.

USE: As emulsifying agent in pharmaceuticals, foods.

10028. Urushiol. Main constituent of the irritant oil of poison ivy, *Toxicodendron radicans* (L.) Kuntze, poison oak, *T. diversilobum*, the Asiatic lacquer tree, *T. verniciferum* D.C. and other plants of the genus *Toxicodendron*, *Anacardiaceae*. Yoshida, *J. Chem. Soc.* **43**, 472 (1883); Hill *et al.*, *J. Am. Chem. Soc.* **56**, 2736 (1934). It is a mixture of several compds which are derivatives of catechol with unsaturated C_{15} or C_{17} side chains and which, upon hydrogenation, yield the same 3-pentadecylcatechol, *q.v.* or 3-heptadecylcatechol: Majima *et al.*, *Ber.* **55**, 172 (1922); Corbett, Billets, *J. Pharm. Sci.* **64**, 1715 (1975). Urushiol from *T. verniciferum* is a mixture of I-IV and urushiol from *T. radicans* is a mixture of I, II, III, and V: Symes, Dawson, *J. Am. Chem. Soc.* **76**, 2959 (1954); Sunthankar, Dawson, *ibid.* **76**, 5070 (1954). Separation of components: Markiewitz, Dawson, *J. Org. Chem.* **30**, 1610 (1965). Structural characterization of urushiol from *T. diversilobum*: Corbett, Billets, *loc. cit.* Separation of urushiol congeners: C. Y. Ma *et al.*, *J. Chromatog.* **200**, 163 (1980). Isoln of pure urushiol from poison ivy or oak extracts and separation of congeners by reversed phase chromatography: M. A. El Sohly *et al.*, *J. Nat. Prod.* **45**, 532 (1982). Allergenic properties of congeners: E. S. Watson *et al.*, *J. Pharm. Sci.* **70**, 785 (1981). Review of extraction and compositional analysis: J. H. P. Tyman, *Chem. Soc. Rev.* **8**, 499 (1979).

I	R = (CH₂)₇CH₃
II	R = CH=CH(CH₂)₅CH₃
III	R = CH=CHCH₂CH=CH(CH₂)₂CH₃
IV	R = CH=CHCH₂CH=CHCH=CHCH₃
V	R = CH=CHCH₂CH=CHCH₂CH=CH₂

Pale yellow liq; $d_4^{21.5}$ 0.9687; bp 200-210°. Soluble in alcohol, ether, benzene. Moderately sol in petr ether.

Caution: An extremely active allergen causing skin reactions similar to poison ivy.

THERAP CAT: Antiallergic (hyposensitization therapy).

10029. Urylon. A synthetic fiber, condensation product of nonamethylenediamine and urea. Prepn of such polyurea polymers: **Brit. pat. 900,787** (1962 to Toyo Koatsu Ind.).

Fiber softens at 205-210°, mp at about 237°. Sp gr 1.07. Loses not more than 10% of its strength after 10 hrs of exposure to 40% sulfuric acid or 40% caustic soda at room temp. Also loses strength on prolonged exposure to light or to a temp of 150°. Hydrolyzes at 130° in the presence of steam.

USE: For fishing nets and lines, knitted fabric, and blends of staple fiber with wool and cotton.

10030. Uscharidin. *[2α(2S,3R,6R),3β,5α]-14-Hydroxy-19-oxo-2,3-[(tetrahydro-3-hydroxy-6-methyl-4-oxo-2H-pyran-3,2-diyl)bis(oxy)]card-20(22)-enolide.* $C_{29}H_{38}O_9$; mol wt 530.62. C 65.64%, H 7.22%, O 27.14%. African arrow poison and cardiac glycoside produced by *Calotropis procera* R. Br., *Asclepiadaceae*: Hesse *et al.*, *Ann.* **566**, 130 (1950). Prepn by hydrolysis of uscharin with 2*N* H_2SO_4 in methanol: Hesse *et al.*, *Ann.* **537**, 67 (1938); by treatment of voruscharin with mercuric chloride: Hesse, Ludwig, *Ann.* **632**, 158 (1960). Structure: Crout *et al.*, *J. Chem. Soc.* **1964**, 2187. Revised structure: Brüschweiler *et al.*, *Helv. Chim. Acta* **52**, 2276 (1969). Stereochemistry: H. T. A. Cheung, T. R. Watson, *J. Chem. Soc. Perkin Trans. I* **1980**, 2162.

Rhombic plates, decomp 290°. $[\alpha]_D^{20}$ +38° (c = 0.9 in methanol).

10031. Usnic Acid. *2,6-Diacetyl-7,9-dihydroxy-8,9b-dimethyl-1,3(2H,9bH)-dibenzofurandione;* usninic acid; usnein; Usniacin. $C_{18}H_{16}O_7$; mol wt 344.32. C 62.79%, H 4.68%, O 32.53%. Antibacterial substance found in lichens. Isoln from varieties of *Usnea barbata* (L.) Wigg., *Usneaceae*: Rochleder, Heldt, *Ann.* **48**, 11 (1843); Widman, *Ann.* **310**, 230 (1900); **324**, 139 (1902). Isoln from *Ramalina reticulata:* Marshak, *Public Health Reports* **62**, 3 (1947); Stark *et al.*, *J. Am. Chem. Soc.* **72**, 1819 (1950). Occurs in nature in both the *d*- and *l*-forms as well as a racemic mixture. Structure: Curd, Robertson, *J. Chem. Soc.* **1937**, 894; Schöpf, Ross, *Ann.* **546**, (1941); Barton, Brunn, *J. Chem. Soc.* **1953**, 603. Resolution of (±)-usnic acid: Dean *et al.*, *ibid.* 1250. Synthesis: Barton *et al.*, *ibid.* **1956**, 530. Biosynthesis *in vitro:* Penttila, Fales, *Chem. Commun.* **1966**, 656. Absolute configuration of (+)-form: S. Huneck *et al.*, *Tetrahedron Letters* **22**, 351 (1981). Toxicity study: *Antibiotics* I, D. Gottlieb, P. Shaw, Eds. (Springer Verlag, New York, 1967) p 611.

Yellow orthorhombic prisms from acetone, mp 204°. $[\alpha]_D^{16}$ +509.4° (c = 0.697 in chloroform). Monobasic acid. Soly at 25° (g/100 ml): water <0.01; acetone 0.77; ethyl acetate 0.88; ethanol 0.02; methyl Cellosolve 0.22; ethyl Cellosolve 0.32; furfural 7.32; furfuryl alcohol 1.21. LD_{50} i.v. in mice: 25 mg/kg (Gottlieb, Shaw).

Sodium salt dihydrate, $C_{18}H_{15}NaO_7.2H_2O$, pale yellow, silky needles. Slightly sol in water.

10032. Ustilagic Acid. Ustizeain B. $C_{37}H_{62-66}O_{17}$; mol wt about 780. A mixture of partially acylated derivatives of a di-D-glucosyldihydroxyhexadecanoic acid. Antibiotic substance produced by the corn smut fungus *Ustilago zeae*: Haskins, *Can. J. Res.* **28C**, 213 (1950); Lemieux *et al.*, *Can. J. Chem.* **29**, 409, 415 (1951); Reed, Holder, *Can. J. Med. Sci.* **31**, 505 (1953); Haskins, U.S. pat. **2,698,843** (1955 to National Research Council, Canada). Biosynthesis: Boothroyd *et al.*, *Can. J. Biochem. Physiol.* **33**, 289 (1955). Production by submerged culture: Lemieux, Can. pat. **600,121** (1960 to National Research Council, Canada).

Long crystals from ether, mp 146-147°. $[\alpha]_D^{23}$ +7° (pyridine). Freely sol in methanol, pyridine, 2,3-butanediol, 1,2-propanediol. Sparingly sol in ethanol, butanol, acetone. Insol in water, glycerol, ethyl acetate, ether, benzene, petr ether. Shows *in vitro* activity against *Cryptococcus neoformans*, *Candida albicans*, and some saprophytic fungi. Practically ineffective in rabbits and mice suffering from fungus diseases.

10033. Utrophin. Dystrophin-related protein; DRP; DMDL protein. Autosomal homolog of dystrophin, *q.v.*, the protein affected by mutation in Duchenne muscular dystrophy (DMD). Cytoplasmic protein; mol wt 395 kDa. Localized in the neuromuscular junctions of normal adult muscle. Widely distributed in a variety of other tissues including brain, stomach, kidney, spleen, liver, and lung. The name is derived from its ubiquitous pattern of expression. Like dystrophin, utrophin binds to actin and to a transmembrane glycoprotein complex, known as dystrophin-associated proteins (DAPs), to link the cytoskeleton with the extracellular matrix. Thought to partially compensate for the lack of dystrophin in DMD patients. Identification in mouse skeletal muscle: D. R. Love *et al.*, *Nature* **339**, 55 (1989). Primary structure: J. M. Tinsley *et al.*, *ibid.* **360**, 591 (1992). Association with DAPs: K. Matsumura *et al.*, *ibid*, 588; J. M. Tinsley *et al.*, *Proc. Nat. Acad. Sci. USA* **91**, 8307 (1994). Analysis of actin-binding: S. J. Winder *et al.*, *J. Cell Sci.* **108**, 63 (1995). Expression in DMD patients: Y. Mizuno *et al.*, *J. Neurol. Sci.* **119**, 43 (1993). Review and comparison with dystrophin: J. M. Tinsley *et al.*, *Curr. Opin. Genet. Devel.* **3**, 484-490 (1993); D. R. Love *et al.*, *Neuromusc. Disord.* **3**, 5-21 (1993); J. M. Tinsley, K. E. Davies, *ibid.* 537-539.

10034. Uva Ursi. Bearberry. Dried leaves of *Arctostaphylos uva-ursi* (L.), Spreng., *Ericaceae*. *Habit.* Northern Europe, North America, Asia. *Constit.* Volatile oil, arbutin, quercetin, ericolin, methylarbutin, myricetin, tannin (6-7%); ursolic, gallic and malic acids.

THERAP CAT: Antiseptic (urinary).

THERAP CAT (VET): Has been used as a diuretic.

10035. Uzarin. *(3β,5α)-3-[(6-O-β-D-Glucopyranosyl-β-D-glucopyranosyl)oxy]-14-hydroxycard-20(22)-enolide.* $C_{35}H_{54}O_{14}$; mol wt 698.81. C 60.16%, H 7.79%, O 32.05%. Isoln from the dried root of a *Gomphocarpus* sp, *Asclepiadaceae*: Windaus, Haack, *Ber.* **63**, 1377 (1930); Tschesche, Brathge, *ibid.* **85**, 1042 (1952); Schmid *et al.*, *Helv. Chim. Acta* **42**, 72 (1959). Yields uzarigenin by enzymic cleavage. Structure: Tschesche, Bohle, *Ber.* **68**, 2252 (1935). Structure of uzarigenin: Rangaswami, Reichstein, *Helv. Chim. Acta* **32**, 939 (1949); Russel *et al.*, *ibid.* **44**, 1320 (1961). Unlike all other known cardiac glycosides, has the A/B-*trans* configuration: L. F. Fieser, M. Fieser, *Steroids* (Reinhold, New York, 1959) pp 762-763. *Review:* Heusser, *Fortschr. Chem. Org. Naturst.* **7**, 101 (1950).

β-D-glucose–β-D-glucose—O   Uzarigenin

Prisms from pyridine + water, mp 266-270°; stout needles from methanol + ether, mp 206-208°. $[\alpha]_D^{20}$ −27° (c = 1.075 in pyridine); $[\alpha]_D^{19}$ −1.4° (c = 0.85 in methanol). uv max: 217 nm (log ∈ 4.23). Sol in pyridine, hot methyl Cellosolve; sparingly sol in water. Practically insol in ether, chloroform, acetone.

Uzarigenin, $C_{23}H_{34}O_4$, *(3β,5α)-3,14-dihydroxycard-20(22)-enolide, odorigenin.* Synthesis: Stache *et al., Ann.* **726**, 136 (1969); Kamano *et al., J. Org. Chem.* **39**, 2319 (1974); *eidem, J. Chem. Soc. Perkin Trans. I* **1975**, 1972. Crystals from methanol, mp 240-256°. $[\alpha]_D^{20}$ +10.5° (c = 1.056 in alc).

3-O-Acetyluzarigenin, $C_{25}H_{36}O_5$, hexagonal plates from methanol, mp 262-266°. $[\alpha]_D^{22}$ +4.6° (c = 1.09 in chloroform).

THERAP CAT: Antidiarrheal.

V

10036. Vaccenic Acid. *trans-11-Octadecenoic acid.* $C_{18}H_{34}O_2$; mol wt 282.47. C 76.54%, H 12.13%, O 11.33%. $CH_3(CH_2)_5CH=CH(CH_2)_9COOH$. Neutralization equivalent 282.5; iodine no. 89.9. Found in butterfat and in other animal fats. Growth-promoting factor for rats. Isoln: Bertram, *Biochem. Z.* **197**, 433 (1928). Synthesis: Böeseken, Hoagland, *Rec. Trav. Chim.* **46**, 632 (1927); Ahmad *et al., J. Am. Chem. Soc.* **70**, 3391 (1948). Configuration and ir spectrum: Rao, Daubert, *ibid.* 1102.

Platelets from acetone, mp 43-44°; n_D^{60} 1.4439; n_D^{70} 1.4402. Methyl ester, $C_{19}H_{36}O_2$, bp$_3$ 172-173°.

10037. Vacciniin. *β-D-Glucopyranose 6-benzoate;* 6-benzoyl-D-glucose. $C_{13}H_{16}O_7$; mol wt 284.27. C 54.93%, H 5.67%, O 39.40%. From the berries of *Vaccinium vitisidaea* L., *V. macrocarpum* Ait., and *V. oxycoccus* L., *Ericaceae:* Griebel, *Z. Untersuch. Nahr.-u. Genussm.* **19**, 241 (1910); Brigl, Zerrweck, *Z. Physiol. Chem.* **229**, 117 (1934). Prepn and chromatography: Bock *et al., Nahrung* **3**, 1036 (1960).

Crystalline powder from water, mp 120-123°. Bitter taste. $[\alpha]_D^{20}$ +47.4° (c = 3.13 in methanol). Sol in water, alcohol, acetone, methanol; sparingly sol in chloroform, benzene. Practically insol in ether, petr ether.

Phenylhydrazone, yellow cryst powder from ethanol + water, mp 134-135°.

10038. Valacidin. Proposed $C_{26}H_{24}N_4O_8$. Antibiotic substance produced by a *Streptomyces* (NRRL 2675) similar to *S. lavendulae:* Bromer, McGuire, U.S. pat. **2,970,943** (1961 to Lilly).

Reddish-brown solid. pKa' 7.0. uv max (methanol): 375, 294, 246 nm (ε 16800, 27100, 40400). Sol in alkaline aq solns and in most polar solvents. Slightly sol in lower alcohols. Substantially insol in most nonpolar organic solvents and in aq acid solns. Stable in aq solns up to pH 9.0.

USE: Preserving agent. In embalming fluids (compatible with formaldehyde). To preserve biological specimens.

10039. Valacyclovir. *L-Valine 2-[(2-amino-1,6-dihydro-6-oxo-9H-purin-9-yl)methoxy]ethyl ester;* L-valine ester with 9-[(2-hydroxyethoxy)methyl]guanine; valacyclovir; ValACV. $C_{13}H_{20}N_6O_4$, mol wt 324.34. C 48.14%, H 6.22%, N 25.91%, O 19.73%. L-Valine ester prodrug of acyclovir, *q.v.* Prepn: T. A. Krenitsky *et al.,* Eur. pat. **Appl. 308,065**; L. M. Beauchamp, U.S. pat. **4,957,924** (1989, 1990 both to Wellcome). Evaluation as prodrug: L. M. Beauchamp *et al., Antiviral Chem. Chemother.* **3**, 157 (1992). Clinical pharmacokinetics: S. Weller *et al., Clin. Pharmacol. Ther.* **54**, 595 (1993).

Hydrochloride, *256U, BW-256U87, BW-256, Valtrex.* Crystalline solid, occurs as hydrate. uv max (water): 252.8 nm (ε 8530). Soly in water: 174 mg/ml.

THERAP CAT: Antiviral.

10040. n-Valeraldehyde. *Pentanal;* valeral; valeric aldehyde. $C_5H_{10}O$; mol wt 86.13. C 69.72%, H 11.70%, O 18.58%. $CH_3(CH_2)_3CHO$. Prepn: Lieben, Rossi, *Ann.* **159**, 70 (1871); Olsen, U.S. pat. **2,548,171** (1951 to GAF); Sisti *et al., J. Org. Chem.* **27**, 279 (1962). Toxicity study: H. F. Smyth *et al., Am. Ind. Hyg. Assoc. J.* **30**, 470 (1969).

Liquid, bp 102-103°. d_4^0 0.8095. n_D^{20} 1.3944. Very slightly sol in water; miscible with many organic solvents. LD_{50} orally in rats: 5.66 ml/kg (Smyth).

Caution: A mild irritant.

USE: In flavoring compds, resin chemistry, rubber accelerators.

10041. Valerian. Dried rhizome and roots of *Valeriana officinalis* L., *Valerianaceae. Habit.* Europe, Northern Asia; naturalized in eastern U.S. *Constit.* Volatile oil (~1%); valerine (valerianin) and chatinine (alkaloids); valeric, formic and malic acids; tannin, resin.

Component of *Valerbé.*

THERAP CAT: Sedative.

10042. Valeric Acid, Normal. *Pentanoic acid;* valerianic acid; propylacetic acid. $C_5H_{10}O_2$; mol wt 102.13. C 58.80%, H 9.87%, O 31.33%. $CH_3(CH_2)_3COOH$. Obtained by decompn of n-propylmalonic acid: Fürth, *Monatsh.* **9**, 308 (1888); from n-butyl chloride: Gilman, Kirby, *Org. Syn.* coll. vol. I, 363 (2nd ed., 1941). Industrial synthesis by oxidation of amyl alcohol or by fermentation processes. Toxicity study: L. Orö, A. Wretlind, *Acta Pharmacol. Toxicol.* **18**, 141 (1961).

Colorless liquid; unpleasant odor; d_4^{20} 0.939; mp −34.5°; bp 186-187°; bp$_{23}$ 96°; n_D^{20} 1.4086. Sol in 30 parts water; freely sol in alcohol, ether. LD_{50} i.v. in mice: 1290±53 mg/kg (Orö, Wretlind).

Ethyl ester, $C_7H_{14}O_2$, *ethyl n-valerate.* Liquid. d_4^{20} 0.877. bp 145-146°. n_D^{20} 1.3732. Insol in water. Misc with alcohol, ether.

USE: Intermediate in perfumery.

10043. Valeronitrile. *Pentanenitrile;* 1-butyl cyanide; 1-cyanobutane. C_5H_9N; mol wt 83.13. C 72.24%, H 10.91%, N 16.85%. $CH_3(CH_2)_3CN$. Prepn: A. Lieben, A. Rossi, *Ann.* **158**, 137 (1871); W. Kantlehner *et al., ibid.* **1980**, 389; H. G. Thomas, H. D. Greyn, *Synthesis* **1990**, 129. Acute toxicity and metabolism: H. Tanii, K. Hashimoto, *Arch. Toxicol.* **55**, 47 (1984). Comparative toxicity of aliphatic nitriles: M. A. Wallig *et al., Food Chem. Toxicol.* **26**, 149 (1988).

bp$_{15}$ 45-47°. bp$_{739.3}$ 140.4°. bp 141. d^0 0.8164. n_D^{20} 1.3962. Log P (n-octanol/water): 0.94. LD_{50} orally in male mice: 2.297 mmol/kg (Tanii).

USE: Solvent.

10044. Valethamate Bromide. *N,N-Diethyl-N-methyl-2-[(3-methyl-1-oxo-2-phenylpentyl)oxy]ethanaminium bromide; 3-methyl-2-phenylvaleric acid diethyl(3-hydroxyethyl)methylammonium bromide ester;* 2-phenyl-3-methylvaleric acid β-(diethylamino)ethyl ester bromomethylate; 3-methyl-2-phenylvaleric acid 2-diethylaminoethyl ester methyl bromide; 2-diethylaminoethyl 2-phenyl-3-methylvalerate methyl bromide; diethyl(2-hydroxyethyl)methylammonium 3-methyl-2-phenylvalerate bromide; Resitan; Epidosin; Murel. $C_{19}H_{32}BrNO_2$; mol wt 386.37. C 59.06%, H 8.35%, Br 20.68%, N 3.63%, O 8.28%. Prepn: Stühmer, Funke, **Ger.** pats. **969,245; 971,136; 1,112,989** (all 1958 to Kali-Chemie);

Martin, Habicht, **Ger.** pat. **1,091,124** and **U.S.** pat. **2,987,-517** (1960, 1961 to Cilag-Chemie).

Crystals from ethanol + ether or acetone, mp 100-101°. Freely sol in water, very sol in alcohol. Practically insol in ether. Aq solns are stable to storage; 0.6% ampuled solns showed no loss after one year at room temp.

THERAP CAT: Anticholinergic.

10045. Validamycins. Antibiotic complex produced by *Streptomyces hygroscopicus* var *limoneus.* Consists of validamycins A (major component), B, C, D, E, and F. Isoln, characterization, and biological properties of validamycins A and B: Iwasa *et al., J. Antibiot.* **24,** 107, 119 (1971). Isoln and characterization of validamycins C, D, E, F: Horii *et al., ibid.* **25,** 48 (1972). Manuf: Horii *et al.,* **Japan.** pat. **39,697('72)** (to Takeda), *C.A.* **78,** 122657a (1973). Structure of validamycin A: Horii, Kameda, *Chem. Commun.* **1972,** 747; revised structure: T. Suami *et al., J. Antibiot.* **33,** 98 (1980). Bioassay methods: Iwasa *et al., ibid.* **24,** 114 (1971). Total synthesis of (±)-validoxylamines A and B, constituents of validamycins: S. Ogawa *et al., J. Org. Chem.* **49,** 2594 (1984). Total synthesis of validamycin B: S. Ogawa, Y. Miyamoto, *Chem. Commun.* **1987,** 1843; of validamycin A: *eidem, Chem. Letters* **1988,** 889.

Validamycin A

Validamycin A, $C_{20}H_{35}NO_{13}$, *[1S-(1α,4α,5β,6α)]-1,5,6-Trideoxy-4-O-β-D-glucopyranosyl-5-(hydroxymethyl)-1-[[4,5,6-trihydroxy-3-(hydroxymethyl)-2-cyclohexen-1-yl]amino]-D-chiro-inositol, Validacin, Valimon.* Colorless hydrophilic powder. Does not show sharp mp; softens at 100°, dec at 135°. $[\alpha]_D^{24}$ +110° (c = 1 in water or pyridine), 92° (c = 1 in DMF). pKa 6.0. Sol in water, methanol, DMF, DMSO; sparingly sol in ethanol, acetone. Insol in ethyl acetate, diethyl ether.

Hydrochloride, $C_{20}H_{35}NO_{13}$·HCl, crystalline powder, mp 95° (dec). $[\alpha]_D^{22}$ +49° (c = 1 in water). Sol in water, alcohols. Insol in acetone, diethyl ether.

USE: Fungicide.

10046. Valine. *L-Valine;* Val; V; 2-aminoisovaleric acid; 2-amino-3-methylbutyric acid; α-aminoisovaleric acid; (*S*)-2-amino-3-methylbutanoic acid. $C_5H_{11}NO_2$; mol wt 117.15. C 51.26%, H 9.46%, N 11.96%, O 27.31%. An essential amino acid for human development. Identified from organ extracts in 1856 by von Group-Besanez; isolated from proteins in 1879 by Schützenberger who proposed that it was aminovaleric acid. Structure confirmation: E. Fischer, *Ber.* **39,** 2320 (1906). Early chemistry and biochemistry: *Amino Acids and Proteins,* D. M. Greenberg, Ed. (Charles C. Thomas, Springfield, IL, 1951) 950 pp., *passim;* J. P. Greenstein, M. Winitz, *Chemistry of the Amino Acids* vols **1-3** (John Wiley and Sons, Inc., New York, 1961) pp. 2368-2380, *passim.* Conformation study: R. H. Yun, J. Hermans, *Protein Eng.* **4,** 761 (1991). Nutritional assessment study: V. R. Young *et al., J. Nutr.* **102,** 1159 (1972); in parenteral nutrition: P. Reiderer *et al., Nutr. Metab.* **24,** 209 (1980); in hemodialysis patients: G. A. Young *et al., Kidney Int.* **21,** 492 (1982). Brief review of metabolism: P. Kamoun, *Trends Biochem. Sci.* **17,** 175-176 (1992).

Leaflets from water + alcohol, mp 315° (closed capillary). d 1.230. Sublimes. $[M]_D$ +33.1° (*5N* HCl); +72.6° (glacial acetic acid); $[\alpha]_D^{26}$ +13.9° (p = 0.9); $[\alpha]_D^{23}$ +22.9° (c = 0.8 in 20% HCl). Soly in water at 0°: 83.4 g/l; at 25°: 88.5 g/l; at 50°: 96.2 g/l; at 65°: 102.4 g/l. Insol in common neutral solvents.

DL-Form, sublimes without melting at ordinary speed of heating. Dec 298° (closed capillary, very rapid heating). One part dissolves in 11.7 parts of water at 15°, in 14.1 parts of water at 25°. Insol in common neutral solvents.

10047. Valinomycin. $C_{54}H_{90}N_6O_{18}$; mol wt 1111.34. C 58.36%, H 8.16%, N 7.56%, O 25.91%. Cyclododecadepsipeptide ionophore antibiotic produced by *Streptomyces fulvissimus* and related to the enniatins, *q.v.* Composed of 3 moles each of L-valine, D-α-hydroxyisovaleric acid, D-valine, and L-lactic acid linked alternately to form a 36-membered ring: Brockmann *et al., Ber.* **88,** 57 (1955); *Ann.* **603,** 216 (1957). Structural studies: Shemyakin *et al., Tetrahedron Letters* **1963,** 351; *Tetrahedron* **19,** 995 (1963). Proposed structure: Brockmann *et al., Naturwiss.* **50,** 689 (1963). Structure and synthesis: Shemyakin *et al., Tetrahedron Letters* **1963,** 1921. Solid phase synthesis: Gisin *et al., J. Am. Chem. Soc.* **91,** 2691 (1969); Losse, Klengel, *Tetrahedron* **27,** 1423 (1971). Biosynthesis: Smirnova *et al., C.A.* **73,** 97347m (1970); Ristow *et al., FEBS Letters* **42,** 127 (1974). Conformation: Ivanov *et al., Biochem. Biophys. Res. Commun.* **34,** 803 (1969); Onishi, Urry, *ibid.* **36,** 194 (1969); Duax *et al., Science* **176,** 911 (1972). *Review:* Y. A. Ovchinnokov, V. T. Ivanov, "The Cyclic Peptides: Structure, Conformation, and Function", in *The Proteins,* **vol.** V, H. Neurath, R. L. Hill, Eds. (Academic Press, New York, 3rd ed., 1982) pp 563-573.

Shiny rectangular platelets from dibutyl ether, mp 190° (hot stage). $[\alpha]_D^{20}$ +31.0° (c = 1.6 in benzene). Neutral reaction. Practically insol in water. Freely sol in petr ether, ether, benzene, chloroform, glacial acetic, butyl acetate, acetone. Active *in vitro* against *Mycobacterium tuberculosis.*

USE: Insecticide, nematocide, Patterson, Wright, **U.S.** pat. **3,520,973** (1970 to Am. Cyanamid).

10048. Valnoctamide. *2-Ethyl-3-methylpentanamide; 2-ethyl-3-methylvaleramide; α-ethyl-β-methylvaleramide; valmethamide;* McN-X-181; Axiquel; Nirvanil. $C_8H_{17}NO$; mol wt 143.23. C 67.09%, H 11.96%, N 9.78%, O 11.17%. Prepn by hydrolysis of 5-ethyl-5-(1-methylpropyl)barbituric acid: Freifelder *et al., J. Org. Chem.* **26,** 203 (1961).

Crystals, mp 113.5-114°.
THERAP CAT: Tranquilizer.

10049. Valproic Acid. *2-Propylpentanoic acid; 2-propyl-valeric acid;* di-*n*-propylacetic acid; Mylproin. $C_8H_{16}O_2$; mol wt 144.21. C 66.63%, H 11.18%, O 22.19%. $(CH_3CH_2-CH_2)_2CHCOOH$. Prepn: Oberreit, *Ber.* **29**, 1998 (1896); Keil, *Z. Physiol. Chem.* **282**, 137 (1947); Wiemann, Thuan, *Bull. Soc. Chim. France* **1958**, 199. Pharmacology: Meunier *et al., Therapie* **18**, 435 (1963); Lebreton *et al., ibid.* **19**, 451 (1964); Carraz *et al., ibid.* **20**, 419 (1965). Activity of deriv: Meunier, U.S. pat. **3,325,361** (1967 to Chemetron); *idem,* Fr. pat. CAM 244 (1969 to Lab. Berthier), *C.A.* **74**, 146380y (1971). Mechanism of action: B. Meldrum, *Brain Res. Bull.* **5**, Suppl. 2, 579 (1980). Review of early clinical studies: Carraz *et al., Encephale* **54**, 458 (1965). Use in infantile spasms: D. S. Bachman, *Arch. Neurol.* **39**, 49 (1982). Toxicity data: Jenner *et al., Food Cosmet. Toxicol.* **2**, 327 (1964). Discussion of potential teratogenicity: R. W. Hurd *et al., Lancet* **1**, 181 (1983). Review of pharmacology, therapeutic indications, and adverse reactions: E. M. Rimmer, A. Richens, *Pharmacother.* **5**, 171-184 (1985). Comprehensive description: Z. L. Chang in *Analytical Profiles of Drug Substances* vol. 8, K. Florey, Ed. (Academic Press, New York, 1979) pp 529-556. *Reviews:* R. M. Pinder *et al., Drugs* **13**, 81-123 (1977); A. T. Dren *et al.,* in *Pharmacological and Biochemical Properties of Drug Substances* vol. 2, M. E. Goldberg, Ed. (Am. Pharm. Assoc., Washington, DC, 1979) pp 58-97.

Colorless liquid, bp_{14} 120-121°; bp_{20} 128-130°. $n_D^{24.5}$ 1.425. d_{25} 0.904. pKa 4.6. Very slightly sol in water. LD_{50} orally in rats: 670 mg/kg (Jenner).

Sodium salt (1:1), $C_8H_{15}NaO_2$, *sodium valproate, DPA sodium, Convulex, Depakene, Depakin, Dépakine, Epilim, Ergenyl, Eurekene, Labazene, Leptilan, Orfiril, Valcote.* White crystals. pKa 4.8. Hygroscopic. LD_{50} orally in mice: 1700 mg/kg (Meunier).

Sodium salt (2:1), $C_{16}H_{31}NaO_4$, *sodium hydrogen bis(2-propylpentanoate), divalproex sodium, Abbott 50711, Depakote.*

Magnesium salt, $C_{16}H_{30}MgO_4$, *Logical.*
THERAP CAT: Anticonvulsant; antiepileptic.

10050. Valpromide. *2-Propylpentanamide; 2-propylvaleramide;* 2-propylpentamide; dipropylacetamide; Dépamide. $C_8H_{17}NO$; mol wt 143.23. C 67.09%, H 11.96%, N 9.78%, O 11.17%. $(CH_3CH_2CH_2)_2CHCONH_2$. Method of prepn: Fischer, Dilthey, *Ber.* **35**, 853 (1902); Wiemann, Thuan, *Compt. Rend.* **212**, 105 (1941); Benoit-Guyod *et al., ibid.* **1965**, 1660. Patents: Meunier, U.S. pats. **3,301,754** and **3,325,-361** (both 1967 to Chemetron). Pharmacology: Carraz *et al., Therapie* **19**, 468 (1964); Eymard, Werbenac, *ibid.* **27**, 11 (1972).

White, odorless and bitter crystalline powder, mp 125-126°. Practically insol in water.
THERAP CAT: Anticonvulsant.

10051. Valsartan. *N-(1-Oxopentyl)-N-[[2'-(1H-tetrazol-5-yl)[1,1'-biphenyl]-4-yl]methyl]-L-valine;* N-[p-(o-1H-tetrazol-5-ylphenyl)benzyl]-N-valeryl-L-valine; (S)-N-(1-carboxy-2-methylprop-1-yl)-N-pentanoyl-N-[2'-(1H-tetrazol-5-yl)-biphenyl-4-ylmethyl]amine; CGP-48933. $C_{24}H_{29}N_5O_3$; mol wt 435.53. C 66.19%, H 6.71%, N 16.08%, O 11.02%. Nonpeptide angiotensin II AT_1-receptor antagonist. Prepn: P. Bühlmayer *et al., Eur.* pat. Appl. **443,983**; *eidem,* U.S. pat. **5,399,578** (1991, 1995 both to Ciba Geigy); *idem et al., Bioorg. Med. Chem. Letters* **4**, 29 (1994). Pharmacological profile: L. Criscione *et al., Brit. J. Pharmacol.* **110**, 761 (1993). HPLC determn in human plasma: A. Sioufi *et al., J. Liq. Chromatog.* **17**, 2179 (1994). Clinical pharmacology: P. Müller *et al., Eur. J. Clin. Pharmacol.* **47**, 231 (1994).

Crystals from diisopropyl ether, mp 116-117°.
THERAP CAT: Antihypertensive.

10052. Vamicamide. *(R*,R*)-(±)-α-[2-(Dimethyl-amino)propyl]-α-phenyl-2-pyridineacetamide;* 4-dimethyl-amino-2-phenyl-2-(2-pyridyl)valeramide; FK-176. $C_{18}H_{23}-N_3O$; mol wt 297.40. C 72.70%, H 7.80%, N 14.13%, O 5.38%. Prepn: I. Ueda *et al., Eur.* pat. Appl. **105,458**; *eidem,* U.S. pat. **4,564,621** (1984, 1986 both to Fujisawa); H. Oyasu *et al., J. Med. Chem.* **37**, 1378 (1994). Comparative pharmacology: *idem et al., Arzneimittel-Forsch.* **44**, 1242 (1994); T. Yamamoto *et al., Drug Dev. Res.* **34**, 9 (1995). HPLC determn of enantiomers in serum and urine: A. Suzuki *et al., J. Chromatog.* **617**, 279 (1993). Clinical pharmacokinetics: M. Terakawa, N. Iyota, *Yakubutsu Dotai* **9**, 799 (1994), *C.A.* **122**, 204469g (1995).

Crystals from 40% aq ethanol, mp 156-157° (Oyasu, *J. Med. Chem.*); also reported as crystals from diethyl ether, mp 132-134° (Ueda, 1986).

(−)-(2R,4R)-Form, prisms from ethyl acetate/isopropyl ether, mp 104°. $[\alpha]_D$ −103.1° (c = 0.5 in methanol).

(+)-(2S,4S)-Form, prisms from ethyl acetate/isopropyl ether (1:4), mp 103-104°. $[\alpha]_D$ +103.3° (c = 0.5 in methanol).
THERAP CAT: Anticholinergic.

10053. Vanadium. V; at. wt 50.9415; at. no. 23; valencies 2, 3, 4, 5. Two naturally occurring isotopes: ^{51}V (99.75%); ^{50}V (0.25%); the latter is radioactive: $T_{1/2}$ 6 × 10^{15} years. Artificial isotopes: 46-49; 52-54. Abundance in earth's crust: 0.01% by wt. Widespread in nature; over 65 minerals known including *patronite* (polysulfide), *vanadinite* $(9PbO.3V_2O_5.PbCl_2)$, *roscoelite* $[2K_2O.2Al_2O_3.(Mg,Fe)O.-3V_2O_5.10SiO_2.4H_2O]$ and *carnotite* $(K_2O.2U_2O_3.V_2O_5.-3H_2O)$. [Carnotite is also an imporant source of uranium.] Discovered by Selström in 1830; prepd by Roscoe in 1869. Prepn: Prandtl, Manz, *Z. Anorg. Allgem. Chem.* **79**, 209 (1912); Marden, Rich, *J. Ind. Eng. Chem.* **19**, 786 (1927); McKechnie, Seybolt, *J. Electrochem. Soc.* **97**, 311 (1950); Gregory, Lilliendahl, *ibid.* **98**, 395 (1951); *Handbook of Preparative Inorganic Chemistry* vol. 2, G. Brauer, Ed. (Academic Press, New York, 2nd ed., 1965) pp 1252-1255. Review: Clark, "Vanadium" in *Comprehensive Inorganic Chemistry* vol. 3, J. C. Bailar, Jr. *et al.,* Eds. (Pergamon Press, Oxford, 1973) pp 491-551.

Light gray or white lustrous powder, fused hard lumps or body-centered cubic crystals; not tarnished in air and not appreciably affected by moisture at ordinary temp. mp 1917°. $d^{18.7}$ 6.11. Sp heat (20-100°) 0.12 cal/g/°C. Electrical resistivity 24.8 microhms/cm. Insol in water. Not attacked by hot or cold hydrochloric acid, by cold sulfuric acid. Reacts with hot sulfuric acid, hydrofluoric acid, nitric acid, aqua regia. Not attacked by bromine water, or by aq alkalies. The metal precipitates gold, silver and platinum from their salts; reduces mercuric salts to mercurous, ferric salts to ferrous.

USE: Manuf rust-resistant vanadium steel.

10054. Vanadium Carbonyl. Vanadium hexacarbonyl. C_6O_6V; mol wt 219.00. C 32.91%, O 43.83%, V 23.26%. $V(CO)_6$. Prepn: Natta et al., C.A. **54**, 16252 (1960); Ercoli et al., J. Am. Chem. Soc. **82**, 2966 (1960); Hileman in Preparative Inorganic Reactions **1**, 107 (1964).

Blue-green pyrophoric crystals. Sensitive to air. Should be stored under nitrogen. Dec 60-70° in a nitrogen-filled tube.

10055. Vanadium Pentafluoride. F_5V; mol wt 145.93. F 65.09%, V 34.91%. VF_5. Prepd from vanadium tetrafluoride by thermal disproportionation at 650° in N_2 current according to the eq $2VF_4 \rightarrow VF_5 + VF_3$: Ruff, Lickfett, Ber. **44**, 2548 (1911); Kwasnik in Handbook of Preparative Inorganic Chemistry, vol. 1, G. Brauer, Ed. (Academic Press, New York, 2nd ed., 1963) p 253; from the elements: Trevorrow, et al., J. Am. Chem. Soc. **79**, 5167 (1957); Clark, Emeleus, J. Chem. Soc. **1957**, 2119. Review of transition metal pentafluorides: Peacock, Advan. Fluorine Chem. **7**, 113-145 (1973).

Liquid. mp 19.5°; bp 47.9°. $d_4^{19.5}$ 2.502. Heat of vaporization 10.60 kcal/mole. Appreciable vapor pressure at room temp. Hydrolyzed by water, dil alkali. Soly in anhydrous HF: 3.3 moles/l. Freely sol in alcohol, chloroform, acetone, ligroin. Insol in carbon disulfide. Dec toluene and ether. Etches glass slowly at room temp; more rapidly in presence of moisture with formation of yellow color. May be stored in sealed vessels made from iron, nickel, copper, platinum.

10056. Vanadium Pentoxide. Vanadic anhydride. O_5V_2; mol wt 181.88. O 43.98%, V 56.02%. V_2O_5. Prepd by heating ammonium metavanadate, NH_4VO_3.

Yellow to rust-brown orthorhombic crystals; d 3.35; mp 690°. Loses oxygen reversibly in the region 700-1125°. One gram dissolves in ~125 ml water; sol in concd acids, forming red to yellow solns; sol in alkalies, forming vanadates; insol in alcohol. Its acid solutions are reduced by SO_2, Zn + HCl, and by evaporation with HCl.

Caution: Reported to be a respiratory irritant and to cause skin pallor, greenish-black tongue, chest pain, cough, dyspnea, palpitation, lung changes. When ingested, causes G.I. disturbances. May also cause a papular skin rash: Zenz et al., Arch. Environ. Health **5**, 542 (1962); E. Browning, Toxicity of Industrial Metals (Appleton-Century-Crofts, New York, 2nd ed., 1969) pp 340-347.

USE: As catalyst in the oxidation of SO_2 to SO_3, alcohol to acetaldehyde, etc.; for the manuf of yellow glass; inhibiting ultraviolet light transmission in glass; depolarizer; as developer in photography; in form of ammonium vanadate as mordant in dyeing and printing fabrics and in manuf of aniline black.

10057. Vanadium Tetrafluoride. F_4V; mol wt 126.94. F 59.87%, V 40.13%. VF_4. Prepd from VCl_4 and HF or from the elements: Ruff, Lickfett, Ber. **44**, 2539 (1911); Cavell, Clark, J. Chem. Soc. **1962**, 2692; Kwasnik in Handbook of Preparative Inorganic Chemistry, vol. 1, G. Brauer, Ed. (Academic Press, New York, 2nd ed., 1963) pp 252-253.

Bright lime-green powder; very hygroscopic; hydrolyzes to form a brown powder and eventually a blue paste. d 3.15. Sublimes at 100-120° in a vacuum. Also disproportionates to VF_3 and VF_5 at 100-120° in a vacuum. Freely sol in water with blue color, in acetone deep green, in glacial acetic acid blue green. Sparingly sol in alc, chloroform, SO_2Cl_2. May be stored in sealed iron or copper vessels.

10058. Vanadium Trifluoride. F_3V; mol wt 107.94. F 52.80%, V 47.20%. VF_3. Prepd from VCl_3 and HF: Ruff, Lickfett, Ber. **44**, 2539 (1911); from VCl_2 and HF: Emeleus, Gutmann, J. Chem. Soc. **1949**, 2979; by thermal decompn in an inert atm of $(NH_4)_3VF_6$: Sturm, Sheridan, Inorg. Syn. **7**, 87 (1963).

Greenish-yellow powder. d 3.363. mp approx 1406°. Sublimes at bright red heat. Almost insol in water, alcohol, acetone, ethyl acetate, acetic anhydride, glacial acetic acid, toluene, carbon tetrachloride, chloroform, carbon disulfide.

Black color with NaOH soln. Aq solns have strong reducing properties.

Trihydrate, dark green rhombohedral crystals, obtained by crystallizing vanadium trioxide dissolved in hydrofluoric acid, or by electrolytic reduction of a soln of vanadium pentoxide in aq HF. Loses one molecule of water at 100°. Dissolves in water to some extent, forming autocomplexes.

10059. Vanadium Trioxide. Vanadium sesquioxide; vanadic oxide. O_3V_2; mol wt 149.88. O 32.02%, V 67.98%. V_2O_3. Prepd by reduction of V_2O_5 with hydrogen or carbon monoxide.

Black powder. On exposure to air it is gradually converted into indigo-blue crystals of V_2O_4. d 4.87; mp 1940°. Insol in water; difficultly sol in acids.

USE: Catalyst, e.g., when making ethanol from ethylene.

10060. Vanadium Trisulfate. $O_{12}S_3V_2$; mol wt 390.07. O 49.22%, S 24.66%, V 26.12%. $V_2(SO_4)_3$. Prepd by the reduction of V_2O_5 in sulfuric acid with elemental sulfur: Auger, Compt. Rend. **173**, 306 (1921); Riveng, Bull. Soc. Chim. France **4**, 1697 (1937); Claunch, Jones, Inorg. Syn. **7**, 92 (1963).

Lemon-yellow powder. When heated in vacuum at or slightly below 410°, it dec into $VOSO_4$ and SO_2. Stable in dry air. Upon exposure to moist air for several weeks, a green hydrate forms. Very slowly sol in water at room temp, somewhat faster in boiling water; practically insol in concd sulfuric acid; sol in dil and concd nitric acid. Powerful reducing agent.

10061. Vanadium Trisulfide. Vanadium sesquisulfide. S_3V_2; mol wt 198.08. S 48.56%, V 51.44%. V_2S_3.

Greenish-black powder; d 4.7. Dec when heated. Insol in water, cold HCl, dil H_2SO_4; sol in hot HCl, hot dil H_2SO_4, HNO_3.

10062. Vanadyl Dichloride. Vanadium oxydichloride. Cl_2OV; mol wt 137.85. Cl 51.44%, O 11.61%, V 36.96%. $VOCl_2$. Usually contains some water. Prepd according to the eq $V_2O_5 + 3VCl_3 + VOCl_3 \rightarrow 6VOCl_2$: Funk, Weiss, Z. Anorg. Allgem. Chem. **295**, 327 (1958); Oppermann, ibid. **351**, 113 (1967).

Green, very deliquesc tabular crystals. d 2.88. Disproportionates at 384° to VOCl and $VOCl_3$. Slowly dec by water. Sol in abs alcohol, glacial acetic acid. Keep tightly closed.

USE: Has been used as mordant in printing fabrics.

10063. Vanadyl Sulfate. Vanadium oxysulfate. O_5SV; mol wt 163.00. O 49.08%, S 19.67%, V 31.25%. $VOSO_4$. Dihydrate, blue, cryst powder. Sol in water.

USE: Dihydrate as mordant in dyeing and printing textiles; manuf colored glass; for blue and green glazes on pottery.

10064. Vanadyl Trichloride. Trichlorooxovanadium; vanadium oxytrichloride. Cl_3OV; mol wt 173.30. Cl 61.37%, O 9.23%, V 29.40%. $VOCl_3$. Prepn by the action of dry Cl_2 on V_2O_3 or V_2O_5: Brown, Griffitts, Inorg. Syn. **1**, 106 (1939); **4**, 80 (1953); Oppermann, Z. Anorg. Allgem. Chem. **351**, 113 (1967). Toxicity data: Smyth et al., Am. Ind. Hyg. Assoc. J. **30**, 470 (1969).

Yellow liquid, emitting red fumes on exposure to moist air. d 1.84. bp 126-127°. mp −77°. Dec in presence of moisture into vanadic acid and HCl. When a small quantity of water is added, it becomes thick and almost blood-red because of formation of vanadic acid. Keep tightly closed. LD_{50} orally in rats: 0.14 g/kg (Smyth).

10065. Vanaspati. Manuf by hydrogenation of vegetable oils, peanut, cottonseed and sesame in the presence of activated-Ni catalyst: Gupta, Indian Chem. Engr. **3**, 163 (1961), C.A. **56**, 15895h (1962).

Product mp 33-37°; max free fatty acid (as oleic) 0.25%.

USE: Hydrogenated shortening. Used in the Orient.

10066. Vancomycin. Vancocin. $C_{66}H_{75}Cl_2N_9O_{24}$; mol wt 1449.27. C 54.70%, H 5.22%, Cl 4.89%, N 8.70%, O 26.50%. Amphoteric glycopeptide antibiotic substance produced by Streptomyces orientalis from Indonesian and Indian soil which inhibits bacterial mucopeptide biosynthesis by formation of complexes. Isoln: M. H. McCormick et al.,

Antibiot. Ann. **1955-56,** 606; **U.S.** pat. **3,067,099** (1962 to Lilly). Purif: H. M. Higgins *et al., Antibiot. Ann.* **1957-58,** 906. Structure studies: F. J. Marshall, *J. Med. Chem.* **8,** 18 (1965); W. D. Weringa *et al., J. Chem. Soc. Perkin Trans. I* **1972,** 443; K. A. Smith *et al., ibid.* **1974,** 2369. Total structure determination by x-ray analysis: G. M. Sheldrick *et al., Nature* **271,** 223 (1978). Revised configuration: M. P. Williamson, D. H. Williams, *J. Am. Chem. Soc.* **103,** 6580 (1981). Further revision of vancomycin structure: C. M. Harris *et al., ibid.* **105,** 6915 (1983). Complex formation: M. Nieto, H. P. Perkins, *Biochem. J.* **123,** 789 (1971). Mode of binding: D. H. Williams, J. R. Kalman, *J. Am. Chem. Soc.* **99,** 2768 (1977). Toxicology: R. C. Anderson *et al., Antibiot. Ann.* **1956-57,** 75. Series of articles on antimicrobial activity, pharmacokinetics and clinical usage: *J. Antimicrob. Chemother.* **14,** Suppl. D, 1-109 (1984). Review of clinical studies: J. W. Lightbown, *Experimental Chemotherapy* vol. III, R. J. Schitzer, F. Hawking, Eds. (Assoc. Press, London, 1964) pp 278-289. *General review:* Jordan, Reynolds, in *Antibiotics* vol. 3, J. W. Corcoran, F. E. Hahn, Eds. (Springer-Verlag, New York, 1975) pp 704-718.

Monohydrochloride, $C_{66}H_{75}Cl_2N_9O_{24} \cdot HCl$, *Lyphocin, Vancor.* White solid; uv max (H_2O): 282 nm $(E^{1\%}_{1cm}$ 40). Soly in water: >100 mg/ml. Moderately sol in dil methanol. Insol in the higher alcohols, acetone, ether. Low concns of urea increase the soly in neutral aq solns. Ammonium sulfate and sodium chloride ppt the antibiotic from acidic solns. LD_{50} in mice (mg/kg): 489 i.v.; 1734 i.p.; 5000 s.c. and orally (Anderson).

THERAP CAT: Antibacterial.

10067. Vanilla. Cured, full-grown, unripe fruit of *Vanilla planifolia* Andr., *Orchidaceae. Habit.* Mexico, West Indies, Reunion, Mauritius, Seychelles. *Constit.* 2-2.75% vanillin, ~4% resin; vanillic acid, ~10% sugar.

USE: In manuf of confectionery and in various bakery products; perfumery; flavor for beverages; pharmaceutic aid (flavor).

10068. Vanillic Acid. *4-Hydroxy-3-methoxybenzoic acid.* $C_8H_8O_4$; mol wt 168.15. C 57.14%, H 4.80%, O 38.06%. Obtained from vanillin by oxidation with silver oxide or by controlled caustic fusion: Pearl, *Org. Syn.* **30,** 101 (1950).

White, odorless needles. mp 210°. Sublimes undecomposed. Sol in 860 parts water; very sol in alc; sol in ether. Not colored by $FeCl_3$. Its salts are freely sol in water.

10069. Vanillin. *4-Hydroxy-3-methoxybenzaldehyde;* methylprotocatechuic aldehyde; vanillic aldehyde; 3-methoxy-4-hydroxybenzaldehyde. $C_8H_8O_3$; mol wt 152.15. C 63.15%, H 5.30%, O 31.55%. Occurs naturally in vanilla, in

potato parings, in Siam benzoin, etc.; made synthetically from eugenol or guaiacol; also from the waste (lignin) of the wood pulp industry: Sörensen, Mehlum, U.S. pat. **2,752,-394** (1956). Production from lignosulfonic acid compds: Craig, Logan, U.S. pats. **3,054,659** and **3,054,825** (both 1962 to Ontario Paper Co.). Toxicity: P. M. Jenner *et al., Food Cosmet. Toxicol.* **2,** 327 (1964).

White or very slightly yellow needles; pleasant aromatic vanilla odor and taste. Slowly oxidizes somewhat on exposure to moist air. Affected by light. d 1.056; mp 80-81° (81-83°); bp 285°; bp_{15} 170°. One gram dissolves in 100 ml water, 16 ml water at 80°, ~20 ml glycerol; freely sol in alc, chloroform, ether, carbon disulfide, glacial acetic acid, pyridine; also sol in oils and aq solns of alkali hydroxides. Solns are acid to litmus. *Preserve in tight, light-resistant containers.* LD_{50} orally in rats: 1580 mg/kg (Jenner).

USE: Pharmaceutic aid (flavor). As a flavoring agent in confectionery, beverages, foods; in perfumery. One part vanillin equals 400 parts vanilla pods; in manuf liqueurs, 2.5-3 parts vanillin replace 500 parts tincture vanilla. Also as reagent in analytical chemistry.

10070. Vanilmandelic Acid. *α,4-Dihydroxy-3-methoxybenzeneacetic acid;* 3-methoxy-4-hydroxymandelic acid; 4-hydroxy-3-methoxymandelic acid; VMA. $C_9H_{10}O_5$; mol wt 198.18. C 54.55%, H 5.09%, O 40.37%. Catecholamine metabolite. Urine levels elevated in various pathologies. Misnamed *vanillinemandelic acid* and *vanillylmandelic acid.* Prepn from vanillin cyanohydrin: Gardner, Hibbert, *J. Am. Chem. Soc.* **66,** 608 (1944). Improved procedure: Shaw *et al., J. Org. Chem.* **23,** 30 (1958); E. F. Recondo, H. Rinderknecht, *ibid.* **25,** 2248 (1960); I. Goodman *et al., Biochem. Prep.* **13,** 75 (1971). Resolution: Armstrong *et al., Biochim. Biophys. Acta* **25,** 422 (1957). Determination in urine: T. C. Stewart, J. A. Freeman, *Vanilmandelic Acid & Catecholamine Determinations* (Am. Soc. Clin. Pathol., Chicago, 1976) pp 1-81.

DL-Form, scales from ether + benzene, dec 131-133°; also reported as mp 134-135° (Goodman). uv max (0.1*N* HCl): 230, 279 nm (ε 6320, 2810); (0.1*N* NaOH): 247, 285, 345 nm (ε 6860, 3960, 630). Readily resinifies on heating or on prolonged exposure to air. Freely sol in water, acetone. Mod sol in ether, acetonitrile. Sparingly sol in benzene. Mass spectral data: T. R. Sharp, *Org. Mass Spectrom.* **15,** 381 (1980).

L-Form, crystals, dec 152°. $[\alpha]_D^{22}$ +128° (c = 0.7).

D-Form, crystals, dec 152°. $[\alpha]_D^{23}$ −131°.

10071. Vanitiolide. *4-[(4-Hydroxy-3-methoxyphenyl)-thiomethyl]morpholine;* 4-(thiovanilloyl)morpholine; 4-hydroxy-3-methoxybenzoic acid morpholide thione; 2-hydroxy-5-(morpholinothiocarbonyl)anisole; 4-[4-hydroxy(thio-m-anisoyl)]morpholine; 2-methoxy-4-(morpholinothiocarbonyl)phenol; 4-(morpholinothiocarbonyl)guaiacol; Bildux. $C_{12}H_{15}NO_3S$; mol wt 253.32. C 56.90%, H 5.97%, N 5.53%, O 18.95%, S 12.66%. Prepn of similar compds, 4-[o- and p-methoxy(thiobenzoyl)]morpholine and 4-(thiosalicyloyl)-

morpholine: Chabrier *et al.*, *Bull. Soc. Chim. France* **1950**, 1167.

THERAP CAT: Choleretic.

10072. Vasicine. *1,2,3,9-Tetrahydropyrrolo[2,1-b]quin-azolin-3-ol;* peganine. $C_{11}H_{12}N_2O$; mol wt 188.23. C 70.19%, H 6.43%, N 14.88%, O 8.50%. Isoln from *Adhatoda vasica* Nees, *Acanthaceae:* Hooper, *Pharm. J.* **18**, 841 (1888); Sen, Ghose, *J. Indian Chem. Soc.* **1**, 315 (1924); Mehta *et al.*, *J. Org. Chem.* **28**, 445 (1963). Isoln from *Peganum harmala* L., *Zygophyllaceae:* Späth, Nikawitz, *Ber.* **67**, 45 (1934); Späth, Kuffner, *ibid.* **67**, 868 (1934). Structure and synthesis: Späth *et al.*, *ibid.* **68**, 699 (1935); Späth, Platzer, *ibid.* **69**, 255 (1936). Synthesis of *dl*-vasicine: Southwick, Casanova, *J. Am. Chem. Soc.* **80**, 1168 (1958). Reviews: Späth, *Monatsh.* **72**, 115 (1938); H. T. Openshaw, "The Quinazoline Alkaloids" in *The Alkaloids* vol. III, R. H. F. Manske, H. L. Holmes, Eds. (Academic Press, New York, 1953) pp 101-118; Ray, *J. Indian Chem. Soc.* **35**, 697 (1958).

dl-Form, needles from alc. mp 210°. Sublimes in high vacuum. Sol in acetone, alcohol, chloroform; slightly sol in water, ether, benzene.

l-Form, needles from alc, mp 212°. $[\alpha]_D^{14}$ $-254°$ (c = 2.4 in $CHCl_3$); $[\alpha]_D^{14}$ $-62°$ (c = 2.4 in alc). In dil HCl this alkaloid is dextrorotatory.

Hydrochloride dihydrate, needles, mp 208° (dry).
Hydriodide dihydrate, needles, mp 195° (dry).
Methiodide, needles from methanol, mp 187°.
Acetylvasicine, $C_{11}H_{11}N_2OCOCH_3$, crystals, mp 123°, $bp_{0.01}$ 230-240°.

10073. Vasopressin. Antidiuretic hormone; beta-hypophamine; β-hypophamine; Leiormone; Pitressin; Tonephin; Vasophysin. The water-soluble pressor principle prepd by synthesis or obtained from the posterior lobe of the pituitary of healthy domestic animals used for food by man: *U.S.P.* XVIII, 770. Separation from oxytocin: Kamm *et al.*, *J. Am. Chem. Soc.* **50**, 573 (1928). Purification: Turner *et al.*, *J. Biol. Chem.* **191**, 21 (1951). Two vasopressins, differing only in the amino acid at position 8, have been isolated: arginine vasopressin, from human, beef, chicken, horse and sheep pituitaries, and lysine vasopressin (lypressin, *q.v.*), from porcine pituitaries. *Review:* E. Schröder, K. Lübke, *The Peptides* vol. II (Academic Press, New York, 1966) pp 336-350; C. R. W. Edwards, "Vasopressin" in *Hormones in Blood* vol. 2, C. H. Gray, V. H. T. James, Eds. (Academic Press, London, 3rd ed., 1979) pp 423-450. Book: *Antidiuretic Hormone* vol. 4, M. L. Forsling, Ed. (Eden Press, Quebec, 1980) 165 pp.

Cys–Tyr–Phe–Gln–Asn–Cys–Pro–Arg–GlyNH₂

Arginine Vasopressin

Arginine vasopressin, $C_{46}H_{65}N_{15}O_{12}S_2$, argipressin, rindervasopressin. Structure: du Vigneaud *et al.*, *J. Am. Chem. Soc.* **75**, 4880 (1953). Synthesis: du Vigneaud *et al.*, *ibid.* **80**, 3355 (1958); Bodanszky *et al.*, *ibid.* **86**, 4452 (1964); Meienhofer *et al.*, *ibid.* **92**, 7199 (1970); Jones *et al.*, *J. Org. Chem.* **38**, 2865 (1973). Fever suppressing activity: K. E. Cooper *et al.*, *J. Physiol. (London)* **295**, 33 (1979). Effect on female sexual behavior in rats: P. Södersten *et al.*, *Nature* **301**, 608 (1983).

Ornithine vasopressin, *see* ornipressin.

THERAP CAT: Antidiuretic and vasopressor hormone; hemostatic.

THERAP CAT (VET): Antidiuretic hormone; in deficiency conditions.

10074. Veatchine. $C_{22}H_{33}NO_2$; mol wt 343.51. C 76.92%, H 9.68%, N 4.08%, O 9.32%. From bark of *Garrya veatchii* Kellogg, *Garryaceae*, where it occurs together with garryine and other alkaloids: Oneto, *J. Am. Pharm. Assoc.* **35**, 204 (1946); Wiesner *et al.*, *Can. J. Chem.* **30**, 608 (1952). Structure: Wiesner *et al.*, *J. Am. Chem. Soc.* **76**, 6068 (1954); Pelletier, *ibid.* **82**, 2398 (1960). Stereochemistry: Vorbrüggen, Djerassi, *Tetrahedron Letters* **1961**, 119; *J. Am. Chem. Soc.* **84**, 2990 (1962). Coexistence of epimers found in crystal and molecular structure: S. W. Pelletier *et al.*, *ibid.* **100**, 7976 (1978). ¹³C-NMR study of epimers: N. V. Mody, S. W. Pelletier, *Tetrahedron* **34**, 2421 (1978). Racemic syntheses and resolution: Nagata *et al.*, *ibid.* **86**, 929 (1964), **89**, 1499 (1967); Guthrie *et al.*, *Coll. Czech. Chem. Commun.* **31**, 602 (1966). Total synthesis of optically active form: Wiesner *et al.*, *Experientia* **26**, 471 (1970).

Crystals from dil acetone, mp 119-120°. Bitter taste. pH 11.5. $[\alpha]_D^{27.5}$ $-69.01°$ (c = 1.06 in ethanol). Readily sol in water, ethanol.

Hydrochloride, $C_{22}H_{33}NO_2.HCl$, crystals from abs ethanol + ether, dec 267-271°. Sol in water.

10075. Vecuronium Bromide. *1-[(2β,3α,5α,16β,17β)-3,-17-Bis(acetyloxy)-2-(1-piperidinyl)androstan-16-yl]-1-methyl-piperidinium bromide;* NC-45; Org-NC-45; Musculax; Norcuron. $C_{34}H_{57}BrN_2O_4$; mol wt 637.74. C 64.03%, H 9.01%, Br 12.53%, N 4.39%, O 10.04%. A non-depolarizing neuromuscular blocking agent and homolog of pancuronium bromide, *q.v.* Prepn: W. R. Buckett *et al.*, *J. Med. Chem.* **16**, 1116 (1973); I. C. Carlyle *et al.*, *Eur. pat.* **Appl. 8824** (1980 to Akzo), *C.A.* **93**, 155850 (1980). Series of articles on pharmacology, pharmacokinetics, pharmacodynamics, clinical studies: *Brit. J. Anaesth.* **52**, Suppl. 1, 1S-72S (1980). Clinical pharmacology: M. R. Fahey *et al.*, *Anesthesiology* **55**, 6 (1981). Comparison with pancuronium bromide: S. L. Son *et al.*, *ibid.* 12.

Crystals, mp 227-229°. LD_{50} in mice: 0.061 mg/kg i.v. (Buckett).

THERAP CAT: Muscle relaxant (skeletal).

10076. Vellosimine. *Sarpagan-17-al;* velosimine. $C_{19}H_{20}N_2O$; mol wt 292.38. C 78.05%, H 6.89%, N 9.58%, O 5.47%. From bark of *Geissosperum vellosii* Allem., *Apocynaceae:* Rapoport *et al.*, *J. Am. Chem. Soc.* **80**, 1601 (1958). Structure: Rapoport, Moore, *J. Org. Chem.* **27**, 2981 (1962); Ohashi *et al.*, *Tetrahedron* **19**, 2241 (1963).

Crystals from methanol, mp 305-306°. Can be sublimed at 180-100° (0.01 mm). $[\alpha]_D^{26}$ +48°. uv max (ethanol): 280, 289 nm (ϵ 8000, 6430).

10077. Velnacrine. (±)-*9-Amino-1,2,3,4-tetrahydro-1-acridinol.* $C_{13}H_{14}N_2O$; mol wt 214.27. C 72.87%, H 6.59%, N 13.07%, O 7.47%. Cognition enhancing acetylcholinesterase inhibitor; major metabolite of tacrine, *q.v..* Prepn: G. M. Shutske, F. A. Pierrat, **Eur. pat. Appl.** 179,383; *eidem,* **U.S. pat 4,631,286** (both 1986 to Hoechst-Roussel); G. M. Shutske *et al., J. Med. Chem* **31,** 1278 (1988); *eidem, ibid.* **32,** 1805 (1989). Clinical pharmacokinetics: S. K. Puri *et al., J. Clin. Pharmacol.* **30,** 948 (1990). Pharmacology: M. F. M. Braga *et al., Brit. J. Pharmacol.* **102,** 909 (1991). Structure-activity study: D. A. Butterfield, A. Rangachari, *Biochem. Biophys. Res. Commun.* **185,** 596 (1992). HPLC determn in biological fluids: R. S. Hsu *et al., J. Chromatog.* **494,** 347 (1989). Clinical evaluation in Alzheimer's disease: M. F. Murphy *et al., Ann. N.Y. Acad Sci.* **640,** 253 (1991).

Powder, mp 245°.
Maleate, $C_{13}H_{14}N_2O.C_4H_4O_4$, *HP-029, Mentane.* Crystals from methanol and ethyl acetate, mp 171-173°. LD_{50} in mice (mg/kg): 136 orally; 162 i.p. (Shutske 1988, 1989).
THERAP CAT: Nootropic.

10078. Venice Turpentine. Larch turpentine. Oleoresin from *Larix decidua* Mill. (*L. europaea* Lam. & DC.), *Pinaceae. Habit.* Middle and Southern Europe. *Constit.* Volatile oil, resin. Use in light microscopy: D. A. Johansen, *Plant Microtechnique* (McGraw Hill, New York, 1940) pp 115-116.
Yellow, sometimes greenish, limpid, tenacious, thick liquid; pleasant, aromatic odor; hot, pungent, somewhat bitter taste. Becomes hard and brittle on prolonged exposure to air. Insol in water; sol in glacial acetic acid, amyl alcohol, acetone, caustic alkalies; slowly but freely sol in alc.
USE: As clearing agent and mounting medium for light microscopy.

10079. Venlafaxine. (±)-*1-[2-(Dimethylamino)-1-(4-methoxyphenyl)ethyl]cyclohexanol;* (±)-1-[α-[(dimethylamino)methyl]-*p*-methoxybenzyl]cyclohexanol; *N,N*-dimethyl-2-(1-hydroxycyclohexyl)-2-(4-methoxyphenyl)ethylamine; venlafaxine; $C_{17}H_{27}NO_2$; mol wt 277.41. C 73.61%, H 9.81%, N 5.05%, O 11.53%. Serotonin noradrenaline reuptake inhibitor (SNRI). Prepn: G. E. M. Husbands *et al.,* **Eur. pat. Appl.** 112,669; **U.S. pat. 4,535,186** (1984, 1985 both to Am. Home Prods.); and resolution of isomers: J. P. Yardley *et al., J. Med. Chem.* **33,** 2899 (1990). Receptor binding studies: E A. Muth *et al., Biochem. Pharmacol.* **35,** 4493 (1986). HPLC determn in biological fluids: D. R. Hicks *et al., Ther. Drug. Monitor.* **16,** 100 (1994). Clinical pharmacokinetics: K. J. Klamerus *et al., J. Clin. Pharmacol.* **32,** 716 (1992). Clinical trial: E. Schweizer *et al., J. Clin. Psychopharmacol.* **11,** 233 (1991). Review of pharmacology and clinical efficacy: S. A. Montgomery, *J. Clin. Psychiatry* **54,** 119-126 (1993).

Hydrochloride, $C_{17}H_{27}NO_2.HCl$, *Wy-45030, Effexor.* White to off-white crystalline solid from methanol/ethyl acetate, mp 215-217°. Soly (mg/ml): 572 water. Partition coefficient (octanol/water): 0.43.
(+)-Form, crystals from ethyl acetate, mp 102-104°. $[\alpha]_D^{25}$ +27.6° (c = 1.07 in 95% ethanol).
(+)-Form hydrochloride, *Wy-45655.* Crystals from methanol/ether, mp 240-240.5°. $[\alpha]_D^{25}$ −4.7° (c = 0.945 in ethanol).
(−)-Form, crystals from ethyl acetate, mp 102-104°. $[\alpha]_D^{25}$ −27.1° (c = 1.04 in 95% ethanol).
(−)-Form hydrochloride, *Wy-45651.* Crystals from methanol/ether, mp 240-240.5°. $[\alpha]_D^{25}$ +4.6° (c = 1.0 in ethanol).
THERAP CAT: Antidepressant.

10080. Venturicidins. Antifungal antibiotics isolated from *Streptomyces aureofaciens* strains. Preliminary isolation work: Rhodes *et al., Nature* **192,** 952 (1962). Isoln of venturicidins A and B and activity studies: Brufani *et al., Helv. Chim. Acta* **51,** 1293 (1968); *see also* Langcake *et al., Biochem. Soc. Trans.* **2,** 202 (1974). Final structures: Brufani *et al., Experientia* **27,** 604 (1971); *eidem, Helv. Chim. Acta* **55,** 2329 (1972).

Venturicidin A   R = NH₂CO
Venturicidin B   R = H

Venturicidin A. $C_{41}H_{67}NO_{11}$, *venturicidin B 3′-carbamate.* Needles from chloroform-petr ether, mp 145-147°. Also reported as mp 140-142°. $[\alpha]_D$ +119° (c = 0.5 in chloroform). uv max (alcohol): 206, 247 (shoulder), ~300 nm (shoulder) (log ϵ 3.80, 2.23, 2.08).
Venturicidin B. $C_{40}H_{66}O_{10}$, *(3-decarbamoyloxy)-3-hydroxyventuricidin A.* Amorphous white powder from chloroform-ether, mp 168-170°. Also reported as mp 145-149° (ethyl acetate-petr ether). $[\alpha]_D$ +100° (c = 0.847 in chloroform).

10081. Veralipride. *5-(Aminosulfonyl)-2,3-dimethoxy-N-[[1-(2-propenyl)-2-pyrrolidinyl]methyl]benzamide; N-[(1-allyl-2-pyrrolidinyl)methyl]-5-sulfamoyl-o-veratramide; LIR-1660;* Agréal; Agradil; Veralipril. $C_{17}H_{25}N_3O_5S$; mol wt 383.47. C 53.25%, H 6.57%, N 10.96%, O 20.86%, S 8.36%. Prepn: **Neth. pat. Appl.** 7,707,982 (1978 to Soc. Etud.-Sci. et Ind. de l'Ile-de-France), *C.A.* **89,** 30768 (1978). Pharmacological studies: P. Bouyard *et al., Sem. Hop.* **56,** 1475 (1980); J. C. Czyba, *ibid.* 1483. Clinical studies: R. Renaud, J. Macler, *ibid.* **57**, 353 (1981); S. Angeli, P. Fougère, *ibid.* **58**, 111 (1982).

THERAP CAT: Treatment of menopausal disorders.

10082. Veralkamine. *(3β,16β,17α,22α)-17-Methyl-18-nor-16,28-secosolanida-5,12-diene-3,16-diol; 17-methyl-20α-((2S,5S)-5-methyl-2-piperidyl)-18-nor-17α-pregna-5,12-diene-3β,16β-diol; (22S:25S)-12,26-epimino-17β-methyl-18-nor-cholesta-5,12-diene-3β,16β-diol; (17S:12S:25S)-22,26-epimino-18(13→17)-abeo-cholesta-5,12-diene-3β,16β-diol;* veralcamine. $C_{27}H_{43}NO_2$; mol wt 413.64. C 78.40%, H 10.48%, N 3.39%, O 7.74%. Steroidal alkaloid isolated from *Veratrum album* sp. *lobelianium* (Bernh.) Suessenguth, *Liliaceae:* Tomko *et al., Pharm. Zentralhalle* **99**, 373 (1960), *C.A.* **55**, 2013e. Structure studies: Tomko *et al., Coll. Czech. Chem. Commun.* **27**, 1404 (1962). Complete structure: Tomko *et al., Tetrahedron Letters* **1967**, 3907; *eidem, Tetrahedron* **24**, 4865 (1968); Hoehne *et al., ibid.* 4875.

Crystals from ethanol, mp 119-123° and 165-169°; $[\alpha]_D^{24}$ −84.1° ±3 (c = 0.533 in $CHCl_3$).
N,O,O-Triacetate, mp 152-154°. $[\alpha]_D^{27}$ −8.0° ($CHCl_3$).
N-Monoacetate, mp 191-193°. $[\alpha]_D^{23}$ −79.1° ($CHCl_3$).

10083. Verapamil. *α-[3-[[2-(3,4-Dimethoxyphenyl)eth-yl]methylamino]propyl]-3,4-dimethoxy-α-(1-methylethyl)-benzeneacetonitrile; 5-[(3,4-dimethoxyphenethyl)methylami-no]-2-(3,4-dimethoxyphenyl)-2-isopropylvaleronitrile; α-iso-propyl-α-[(N-methyl-N-homoveratryl)-γ-aminopropyl]-3,4-dimethoxyphenylacetonitrile;* iproveratril; D-365. $C_{27}H_{38}$-N_2O_4; mol wt 454.61. C 71.34%, H 8.43%, N 6.16%, O 14.08%. Coronary vasodilator with calcium blocking activity. Prepn: Belg. pat. 615,861; Dengel, U.S. pat. 3,261,859 (1962, 1966 both to Knoll). Physical and chemical data: Appel, *Arzneimittel-Forsch.* **12**, 562 (1962). Synthesis and absolute configuration of enantiomers: Ramuz, *Helv. Chim. Acta* **58**, 2050 (1975). Stereospecific synthesis of enantiomers: L. J. Theodore, W. L. Nelson, *J. Org. Chem.* **52**, 1309 (1987). Pharmacology: Haas, Hartfelder, *ibid.* 549; Schlepper, Witzleb, *ibid.* 559. Metabolism: McIlhenny, *J. Med. Chem.* **14**, 1178 (1971). HPLC determn in plasma: C. Horne *et al., Arzneimittel-Forsch.* **37**, 956 (1987). Series of articles on clinical studies: *ibid.* **20**, 1277-1336 (1970). Comparative study in variant angina: D. D. Waters *et al., Am. J. Cardiol.* **47**, 179 (1981); in arrhythmias: B. N. Singh *et al., Drugs* **25**, 125 (1983). Symposium on pharmacology and clinical efficacy in hypertension: *Am. J. Cardiol.* **57**, 1D-107D (1986). Comprehensive description: Z. L. Chang, *Anal. Profiles Drug Subs.* **17**, 643-674 (1988). *Reviews:* S. H. Baky, E. B. Kirsten, in *Pharmacological and Biochemical Properties of Drug Substances* vol. 3, M. E. Goldberg, Ed. (Am. Pharm. Assoc., Washington, DC, 1981) pp 226-261; D. J. Triggle, V. C. Swamy, *Circ. Res.* **52**(2), Pt. 2, 117-128 (1983).

Viscous, pale yellow oil, $bp_{0.01}$ 243-246°. n_D^{25} 1.5448. Practically insol in water. Sparingly sol in hexane; sol in benzene, ether. Freely sol in the lower alcohols, acetone, ethyl acetate, chloroform.
Hydrochloride, $C_{27}H_{38}N_2O_4$·HCl, *Arpamyl, Berkatens, Calan, Cardiagutt, Cardibeltin, Cordilox, Dignover, Drosteakard, Geangin, Isoptin, Quasar, Securon, Univer, Vasolan, Veracim, Veramex, Veraptin, Verelan, Verexamil.* Crystals, dec 138.5-140.5° (corr). Soly (21°): 7 g/100 g water (pH = 4.24). pH of 0.1% aq soln: 5.25. uv max: 232, 278 nm. Sparingly sol in chloroform. Sol in ethanol, isopropanol, acetone, ethyl acetate; freely sol in methanol, DMF. Soly (mg/ml): water 83, ethanol (200 proof) 26, propylene glycol 93, ethanol (190 proof) >100, methanol >100, 2-propanol 4.6, ethyl acetate 1.0, DMF >100, methylene chloride >100, hexane 0.001. pKa 8.6. LD_{50} in rats, mice (mg/kg): 16, 8 i.v. (Baky, Kirsten).
THERAP CAT: Antianginal; antiarrhythmic (class IV).

10084. Veratraldehyde. *3,4-Dimethoxybenzaldehyde; 3,-4-dimethoxybenzenecarbonal;* veratric aldehyde; protocate-chualdehyde dimethyl ether. $C_9H_{10}O_3$; mol wt 166.18. C 65.05%, H 6.07%, O 28.88%. Prepd by methylation of vanillin: Kostanecki, Tambor, *Ber.* **39**, 4022 (1906); Buck, *Org. Syn.* **13**, 102 (1933); Alt, U.S. pat. 3,007,968 (1957 to Monsanto); by oxidation of veratryl alcohol with chromium(VI) oxide-pyridine complex: Holum, *J. Org. Chem.* **26**, 4814 (1961).

Needles from ether, petr ether, toluene, or carbon tetra-chloride. Odor of vanilla beans; mp 42-43°; bp_{760} 281°; bp_{53} 201°; bp_{10} 155°. Slightly sol in hot water; freely sol in alcohol and ether. Solns are oxidized to veratric acid under the influence of light.

10085. Veratramine. *(3β,23β)-14,15,16,17-Tetradehydro-veratraman-3,23-diol.* $C_{27}H_{39}NO_2$; mol wt 409.61. C 79.17%, H 9.60%, N 3.42%, O 7.81%. Secondary base from *Veratrum grandiflorum* (Maxim.) Loes. f., and from *V. viride* Ait., *Liliaceae.* Isoln and structure: Saito, *Bull. Chem. Soc. Japan* **15**, 22 (1940); Jacobs, Craig, *J. Biol. Chem.* **160**, 555 (1945); Jacobs, Sato, *ibid.* **181**, 55 (1949); **191**, 71 (1951); Tamm, Wintersteiner, *J. Am. Chem. Soc.* **74**, 3842 (1952); Wintersteiner, *Festschrift Arthur Stoll* (Birkhäuser-Verlag, Basel) pp 166-176. Total synthesis: Masamune *et al., J. Am. Chem. Soc.* **89**, 4521 (1967); Johnson *et al., ibid.* 4523; Masamune *et al., Tetrahedron* **27**, 3369 (1971); Kutney *et al., Can. J. Chem.* **53**, 1796 (1975). Stereochemistry: Sicher, Tichy, *Tetrahedron Letters* **1959**(12), 6 (1959); Kataoka, *Chem. & Ind. (London)* **1961**, 512; Bailey *et al., Tetrahedron Letters* **1963**, 555. Revised stereochemistry: Scott *et al., ibid.* **1967**, 2381; Kupchan, Suffness, *J. Am. Chem. Soc.* **90**, 2730 (1968); Sprague *et al., Tetrahedron* **27**, 4857 (1971). *See also:* Veratrum Viride.

Crystals, mp 206-207°. Slightly sol in water. Sol in methanol, alcohol. Precipitated by digitonin. $[\alpha]_D^{25}$ −71.8° (c = 1.21); $[\alpha]_D^{25}$ −70° (c = 1.56 in methanol). uv max: 268 nm. Dihydroveratramine, crystals, mp 192.5-194°. $[\alpha]_D^{25}$ +26° (c = 1.26 in acetic acid).

10086. Veratric Acid. *3,4-Dimethoxybenzoic acid;* dimethylprotocatechuic acid. $C_9H_{10}O_4$; mol wt 182.18. C 59.34%, H 5.53%, O 35.13%. Isolated from seed of *Schoenocaulon officinale* (Schlecht. & Cham.) A. Gray (*Sabadilla officinarum* Brandt). Prepn: Arthur, Ng, *J. Chem. Soc.* **1959**, 3094.

Monohydrate, odorless crystals. At 100° becomes anhydr; mp 180-181° when anhydr. Sublimes in rhombic crystals. Sol in 2150 parts cold, 165 parts boiling water; very sol in alcohol or ether. Its barium salt is but slightly sol in water.

10087. Veratridine. *(3β,4α,16β)-4,9-Epoxycevane-3,4,-12,14,16,17,20-heptol 3-(3,4-dimethoxybenzoate);* 3-veratroylveracevine. $C_{36}H_{51}NO_{11}$; mol wt 673.80. C 64.17%, H 7.63%, N 2.08%, O 26.12%. From seed of *Schoenocaulon officinale* (Schlecht. & Cham.) A. Gray and also from the rhizome of *Veratrum album* L. *Liliaceae.* Can be isolated from commercial veratrine as the sparingly sol nitrate: Blount, *J. Chem. Soc.* **1935**, 122; Vejdelek *et al., Chem. Listy* **50**, 603 (1956); *Coll. Czech. Chem. Commun.* **22**, 98 (1957). Purification and properties: L. C. McKinney *et al., Anal. Biochem.* **153**, 33 (1986). Toxicity studies: O. Krayer *et al., J. Pharmacol. Exp. Ther.* **82**, 167 (1944); K. Tanaka, *ibid.* **113**, 89 (1955); Swiss, Bauer, *Proc. Soc. Exp. Biol. Med.* **76**, 847 (1951). Review of chemistry and structure of veratridine and other *Veratrum* alkaloids: Kupchan, By, in *The Alkaloids* vol. **10**, R. H. F. Manske, Ed. (Academic Press, New York, 1968) pp 193-285.

Yellowish-white, amorphous powder. Tenaciously retains water. mp 180° (after drying at 130°). $[\alpha]_D^{20}$ +8.0° (ethanol). pKa 9.54 ±0.02. Insol in water. Slightly sol in ether. LD_{50} in mice (mg/kg): 1.35 i.p. (Swiss, Bauer); 0.42 i.v. (Krayer); 6.3 s.c. (Tanaka).
Nitrate, amorphous powder, sparingly sol in water.
Sulfate, slender needles, very hygroscopic.

Perchlorate, long thin needles from water, mp 259-260° (after drying at 120° *in vacuo*).

10088. Veratrine (Mixture). This veratrine is a mixture of the alkaloids cevadine, veratridine, cevadilline, sabadine, cevine from seeds of *Schoenocaulon officinale* (Schlecht & Cham.) A. Gray, *Liliaceae.*
White or grayish-white powder. mp 145-155°. One gram dissolves in about 1800 ml water, 1000 ml boiling water, 2.8 ml alcohol, 0.7 ml chloroform, 4.2 ml ether, 80 ml olive oil; freely sol in dil acids, benzene, amyl alcohol, slightly in glycerol; insol in petr ether.
Caution: Poisonous! Exceedingly irritating to mucous membranes, causing violent sneezing when inhaled.
THERAP CAT: Topical counterirritant.
THERAP CAT (VET): Has been used as emetic, ruminatoric.

10089. Veratrole. *1,2-Dimethoxybenzene; o-dimethoxybenzene;* pyrocatechol dimethyl ether. $C_8H_{10}O_2$; mol wt 138.17. C 69.55%, H 7.30%, O 23.16%. Prepd by methylation of pyrocatechol: Ullmann, *Ann.* **327**, 104 (1903); Drahowzal, Klamann, *Monatsh.* **82**, 588 (1951). Toxicity study: P. M. Jenner *et al., Food Cosmet. Toxicol.* **2**, 327 (1964).

Crystals or liquid; d_{25}^{25} 1.084; mp 22-23°; bp 206-207°. Slightly sol in water; sol in alcohol, ether, fatty oils. LD_{50} in rats, mice (mg/kg): 1360, 2020 orally (Jenner).

10090. Veratrum viride. American hellebore; green hellebore; American veratrum; Indian poke. Dried rhizome and roots of *Veratrum viride* Ait., *Liliaceae. Habit.* North America. *Constit.* The alkaloids jervine, pseudojervine, rubijervine, cevadine, germitrine, germidine, veratralbine, veratroidine. *See* Seiferle *et al., J. Econ. Entomol.* **35**, 35 (1942); Fried *et al., J. Am. Chem. Soc.* **72**, 4621 (1950). *Review:* Kupchan, *J. Pharm. Sci.* **50**, 273-287 (1961).
Mixture of the alkaloids of *Veratrum viride, alkavervir, Veriloid.*
Partially purified prepn of the ester alkaloids, *Vergitryl.*
THERAP CAT: Antihypertensive.
THERAP CAT (VET): Has been used as a circulatory depressant, stomachic, emetic, parasiticide.

10091. Verazide. *4-Pyridinecarboxylic acid [(3,4-dimethoxyphenyl)methylene]hydrazide; isonicotinic acid veratrylidenehydrazide;* 1-isonicotinoyl-2-veratrylidenehydrazine; 3,4-dimethoxybenzal isonicotinoylhydrazone. $C_{15}H_{15}N_3O_3$; mol wt 285.30. C 63.15%, H 5.30%, N 14.73%, O 16.82%. Prepd from isonicotinylhydrazine and veratraldehyde: Fox, Gibas, *J. Org. Chem.* **18**, 983 (1953).

Crystals from 2-propanol, mp 189-190°.
THERAP CAT: Antibacterial (tuberculostatic).

10092. Verbascose. *β-D-Fructofuranosyl O-α-D-galactopyranosyl-(1→6)-O-α-D-galactopyranosyl(1→6)-O-α-D-galactopyranosyl-(1→6)-α-D-glucopyranoside; O-α-D-galactopyranosyl-(1→6)-[O-α-D-galactopyranosyl-(1→6)-]_2-O-α-D-glucopyranosyl-(1→2)-β-D-fructofuranoside.* $C_{30}H_{52}O_{26}$; mol wt 828.73. C 43.48%, H 6.32%, O 50.20%. Oligosaccharide isolated from roots of mullein *Verbascum thapsus* L., *Scrophulariaceae:* Bourquelot, Bridel, *Compt. Rend.* **151**, 760 (1910); from birch wood, *Betula verrucosa* Ehrh., *Betulaceae:* Lindberg, Selleby, *Acta Chem. Scand.* **12**, 1512 (1958). Structure: Wickström *et al., Bull. Soc. Chim. France* **1956**, 827. *Review:* D. French "The Raffinose Family of Oligosaccharides" in *Advan. Carbohyd. Chem.* **9**, 180-181 (1954).

Needles, mp 219-220°, mp 253°. $[\alpha]_D^{20}$ +170° (water); $[\alpha]_D^{25}$ +146° (c = 2.1 in water).

10093. Verbenalin. *[1S-(1α,4aα,7α,7aα)]-1-(β-D-Gluco-pyranosyloxy)-1,4a,5,6,7,7a-hexahydro-7-methyl-5-oxocyclo-penta[c]pyran-4-carboxylic acid methyl ester;* cornin. $C_{17}H_{24}O_{10}$; mol wt 388.37. C 52.58%, H 6.23%, O 41.20%. A glycoside isolated from *Verbena officinalis* L., *Verbenaceae, Cornus florida* L., *Cornaceae. Refs:* Bourdier, *J. Pharm. Chim.* **27**(6), 49, 101 (1908); Holste, *Arch. Exp. Path. Pharmakol.* **101**, 46 (1924); Miller, *J. Am. Pharm. Assoc.* **17**, 744 (1928). Identity with cornin: Reichert, *Arch. Pharm.* **273**, 357 (1935). Structure: Büchi, Manning, *Tetrahedron Letters* **1960**, 5 (1960); *eidem, Tetrahedron* **18**, 1049 (1962). Synthesis of *verbenalol,* the aglycone: P. Callant *et al., ibid.* **37**, 2085 (1981).

Bitter needles, mp 182-183°. $[\alpha]_D^{25}$ −173° (water). uv max: 238, 290 nm (ε 9600, 105) in ethanol. Freely sol in water, slightly in alcohol, ethyl acetate, acetone. Practically insol in chloroform, ether.

10094. d-Verbenone. *(1R)-4,6,6-Trimethylbicyclo[3.1.1]-hept-3-en-2-one; (1R,5R)-(+)-2-pinen-4-one.* $C_{10}H_{14}O$; mol wt 150.22. C 79.96%, H 9.39%, O 10.65%. A constituent of Spanish verbena oil (from *Verbena triphylla* L., *Verbenaceae*): Kerschbaum, *Ber.* **33**, 885 (1900). Prepn from α-pinene: Bain, Gary, U.S. pat. **2,911,442** (1959 to Glidden). Structure: Blumann, Zeitschel, *Ber.* **46**, 1178 (1913); Weinhaus, Schumm, *Ann.* **439**, 20 (1924). Absolute configuration: Hurst, Whitham, *J. Chem. Soc.* **1960**, 2864. *Review:* J. L. Simonsen, *The Terpenes* vol. II (University Press, Cambridge, 2nd ed., 1949) p 232-239.

Oil. Characteristic odor. mp 6.5°. bp_{12} 102-105°; bp_{760} 227-228°. d^{20} 0.9780. n_D^{18} 1.4957. $[\alpha]_D^{18}$ +249.62°. uv max (ethanol): 253 nm (ε 6730). Practically insol in water. Miscible in all proportions with the usual organic solvents. LD in mice: 250 mg/kg i.p. (*Toxic Substances List,* 1972 Edition, H. E. Christensen, Ed.).

10095. Vermiculite. Hydrated magnesium-aluminum-iron silicate. Typical analysis of Montana vermiculite ore: SiO_2 38.64%, MgO 22.68%, Al_2O_3 14.94%, Fe_2O_3 9.29%, K_2O 7.84%, CaO 1.23%, Cr_2O_3 0.29%, Mn_3O_4 0.11%, Cl 0.28%.
Monoclinic crystals, pseudo-hexagonal characteristics. Soft, resilient, glabrous feel. Dissolves in hot concd H_2SO_4. Has base exchange capacities. Sold as the ore or in "expanded" form. The expanded grades can absorb large quantities of liquids and remain free flowing. The ore is not sol in weak acids or bases, but expanded grades may be attacked. Vermiculites should be classed as montmorillonoids: Rustum Roy, *Chem. & Eng. News* **32**, 4842 (1954).
Note: A food grade vermiculite is known as *Verxite.*

USE: As catalyst; loose fill insulation, filler and packing material.

10096. Vernamycin B. Group of peptide lactones produced by *Streptomyces loidensis:* Donovick *et al.,* U.S. pat. **2,990,325** (1961 to Olin Mathieson). Antimicrobial antibiotic structurally similar to virginiamycin S_1: Bodanszky, Ondetti, *Antimicrob. Ag. Chemother.* **1963**, 360. Identity with pristinamycins: Preud'homme *et al., Bull. Soc. Chim. France* **1968**, 585. Nomenclature: Crooy, De Neys, *J. Antibiot.* **25**, 371 (1972). For structure see Virginiamycin S_1.
Crystals, dec 130-135°. $[\alpha]_D^{25}$ −72°. uv max (methanol): 230 nm ($E_{1cm}^{1\%}$ 590). Sol in methanol, ethanol, isopropanol, butanol, acetone, methyl ethyl ketone, ethyl acetate, amyl acetate, ether, dioxane, dimethylformamide, benzene, acetic acid; slightly sol in water, hexane.
Vernamycin $B_α$, see Mikamycin B.
Vernamycin $B_β$, 4-[N-methyl-4-(methylamino)-L-phenyl-alanine]virginiamycin S_1, ostreogrycin B_2, pristinamycin I_B.
Vernamycin $B_γ$, 2-D-alanine-4-[4-(dimethylamino)-N-methyl-L-phenylalanine]virginiamycin S_1, ostreogrycin B_1, pristinamycin I_C.
Vernamycin $B_δ$, 2-D-alanine-4-[N-methyl-4-(methylamino)-L-phenylalanine]virginiamycin S_1.

10097. Vernolate. *Dipropylcarbamothioic acid S-propyl ester; dipropyl thiocarbamic acid S-propyl ester;* S-propyl dipropylthiocarbamate; R-1607; Vernam. $C_{10}H_{21}NOS$; mol wt 203.35. C 59.07%, H 10.41%, N 6.89%, O 7.87%, S 15.77%. Selective soil incorporated herbicide. Prepn: **Brit. pat. 808,753;** H. Tilles, J. Antognini, U.S. pat. **2,913,327** (both 1959 to Stauffer); H. Tilles, *J. Am. Chem. Soc.* **81**, 714 (1959). Toxicity study: G. W. Bailey, J. L. White, *Residue Rev.* **10**, 97 (1965).

Clear liquid, bp_{30} 149-150°. Vapor pressure at 25°: 1.04 × 10^{-2} mm Hg. d_4^{30} 0.9440; n_D^{30} 1.4736. Soly in water at 25°: 107 mg/l. Misc with most organic solvents. LD_{50} orally in rats: 1780 mg/kg (Bailey, White).
USE: Herbicide.

10098. Vernolepin. *[3aR-(3aα,4α,5aα,9aα,9bβ)]-5a-Ethenyloctahydro-4-hydroxy-3,9-bis(methylene)-2H-furo[2,3-f][2]benzopyran-2,8(3H)-dione.* $C_{15}H_{16}O_5$; mol wt 276.29. C 65.21%, H 5.84%, O 28.95%. The first recognized naturally occurring elemanolide dilactone; a sesquiterpene lactone with tumor inhibiting properties. Isoln from *Vernonia hymenolepis* A. Rich, *Compositae* and structure: S. M. Kupchan *et al., J. Am. Chem. Soc.* **90**, 3596 (1968); structural elucidation: *eidem, J. Org. Chem.* **34**, 3903 (1969). Biological properties: S. M. Kupchan *et al., J. Med. Chem.* **14**, 1147 (1971). Approach to stereosynthesis of (+)-form: H. Iio *et al., Tetrahedron* **35**, 941 (1979); K. Kondo *et al. Tetrahedron Letters* **34**, 4219 (1993). Total synthesis of (±)-form: P. A. Grieco *et al., J. Am. Chem. Soc.* **98**, 1612 (1976); F. Zutterman *et al., Tetrahedron* **35**, 2389 (1979).

Crystals from chloroform-petr ether, mp 181-182°. $[\alpha]_D^{28}$ +72° (c = 1.04 in acetone). uv max (methanol): 208 nm (ε 20300).
(±)-Form, crystals from chloroform, mp 210-211°.

10099. Vernolic Acid. *11-(3-Pentyloxiranyl)-9-undecen-oic acid; 12,13-epoxyoleic acid; cis-12,13-epoxyoctadec-cis-9-*

enoic acid; octadec-cis-12,13-epoxy-cis-9-enoic acid. $C_{18}H_{32}$-O_3; mol wt 296.45. C 72.93%, H 10.88%, O 16.19%. Naturally occurring as the (+) and (−) forms. Principal seed-oil acid from *Vernonia anthelmintica* Willd. (*Serratula anthelmintica* Roxb., *Conyza anthelmintica* Linn.) *Compositae*. Isoln and structure: Gunstone, *J. Chem. Soc.* **1954**, 1611; from *Plectranthus spp.*: C. M. J. Daulatabad, A. M. Mirajkur, *J. Chem. Tech. Biotechnol.* **45**, 143 (1989). Synthesis and absorption spectrum: Osbond, *ibid.* **1961**, 5270; *idem*, **Brit.** pat **909,354** (1962 to Roche). Enantiospecific synthesis of (−)-form: C. A. Moustakis *et al.*, *Tetrahedron Letters* **27**, 303 (1986).

(+)-form

(−)-form, *[2R-[2α(Z),3α]]-11-(3-pentyloxiranyl)-9-undecenoic acid, (12R,13S)-(−)-12,13-epoxyoleic acid.* Crystals from acetone at −25°, mp 30-31°. Liq at room temp. $[α]_D$ −8°.

(+)-form, *[2S-[2α(Z),3α]]-11-(3-pentyloxiranyl)-9-undecenoic acid, (12S,13R)-(+)-12,13-epoxyoleic acid.*

10100. Verrucarins. Macrocyclic tricothecane derivs which are secondary metabolites of the soil fungi *Myrothecium verrucaria* (Albertini et Schweinitz) Ditmar ex Fries; they are characterized by antibiotic, antifungal, and cytostatic activity. Verrucarins are triesters of the sesquiterpene alcohol *verrucarol* and closely related to the *roridins*, which are diesters of the same alcohol. Isoln of verrucarins A (major), B, C, D, E, F, G: E. Haerri *et al.*, *Helv. Chim. Acta* **45**, 840 (1962). Isoln of A: Symth, Kraskin, U.S. pat. **3,087,859** (1963 to Rohm & Haas); Vittimberga, *J. Org. Chem.* **28**, 1786 (1963). Identity with muconomycin A: Vittimberga, Vittimberga, *ibid.* **30**, 746 (1965). Structure of A: J. Gutzwiller, C. Tamm, *Helv. Chim. Acta* **48**, 157 (1965). Stereochemistry: W. Zürcher *et al.*, *ibid.* 840; A. T. McPhail, G. T. Sim, *J. Chem. Soc. (C)* **1966**, 1394. Synthesis of A: W. C. Still, H. Ohmizu, *J. Org. Chem.* **46**, 5242 (1981). Isoln of B: W. Loeffler *et al.*, **Belg.** pat. **627,002** (1963 to Sandoz), *C.A.* **60**, 10485h (1964). Structure of B: J. Gutzwiller, C. Tamm, *Helv. Chim. Acta* **48**, 177 (1965). Toxicological studies of A and B: Guarino *et al.*, *Biotechnol. Bioeng.* **10**, 457 (1968); of A: Mortimer *et al.*, *Res. Vet. Sci.* **12**, 508 (1971). Biosynthetic studies of A and B: B. Müller, C. Tamm, *Helv. Chim. Acta* **58**, 483 (1975); G. A. Cordell, *Chem. Rev.* **76**, 425 (1976). Isoln of J (muconomycin B): Vittimberga, Vittimberga, *loc. cit.*; of H and J: B. Boehner *et al.*, *Helv. Chim. Acta* **48**, 1079 (1965). Structure of J: E. Fetz *et al.*, *ibid.* 1669. Partial synthesis of tetrahydroverrucarin J and revised structure of J: W. Breitenstein, C. Tamm, *ibid.* **61**, 1975 (1978). Isoln and structure of L: *eidem, ibid.* **60**, 1522 (1977). Isoln of L: B. B. Jarvis *et al.*, *J. Antibiot.* **34**, 121 (1981). Total synthesis of verrucarol: R. H. Schlessinger, R. A. Nugent, *J. Am. Chem. Soc.* **104**, 1116 (1982). Toxicity study: E. M. Rüsch, H. Stählin, *Arzneimittel-Forsch.* **15**, 893 (1965). Comprehensive review of the verrucarins and roridins: C. Tamm, *Fortschr. Chem. Org. Naturst.* **31**, 64-117 (1974). For a description of the stereochemistry of the roridins *see* B. R. Jarvis *et al.*, *J. Nat. Prod.* **45**, 440 (1982).

Verrucarin A

Verrucarin A, $C_{27}H_{34}O_9$, *muconomycin A*. Colorless rectangular plates from ether/acetone, mp > 360° (dec). $[α]_D^{23}$ +206° (c = 1.012 in chloroform); uv max (ethanol): 260 nm (log ε 4.25). LD_{50} in mice, rats, rabbits (mg/kg): 1.5, 0.87, 0.54 i.v. (Rüsch, Stählin).

Verrucarin B, $C_{27}H_{32}O_9$, *(2'S,3R')-2'-deoxy-2',3'-epoxyverrucarin A.* Colorless needles from acetone/ether, decomp > 330°. $[α]_D^{23}$ +147° (c = 1.066 in benzene). $[α]_D^{22}$ +101° ±1.5° (c = 1.416 in dioxane). uv max (methanol): 258.5 nm (log ε 4.37).

Verrucarin J, $C_{27}H_{32}O_8$, *(2'E)-2',3'-didehydro-2'-deoxyverrucarin A, muconomycin B.* Crystals from ether, dec > 235°. $[α]_D^{22}$ +54° (benzene). $[α]_D^{22}$ +20° ± 2° (c = 1.011 in chloroform). uv max: 261, 220.5 nm (ε 22000, 21600).

Verrucarin K, $C_{27}H_{34}O_8$, *12,13-deepoxy-12,13-didehydroverrucarin A.* Crystals from methylene chloride/ether, dec > 320°. $[α]_D^{23}$ +218° ±2° (c = 0.58 in chloroform). uv max (ethanol): 259 nm (log ε 4.19).

Caution: Verrucarins are extremely toxic; can cause severe local irritation and inflammation of the skin.

10101. Versalide®. *1-(3-Ethyl-5,6,7,8-tetrahydro-5,5,8,8-tetramethyl-2-naphthalenyl)ethanone; 3'-ethyl-5',6',7',8'-tetrahydro-5',5',8',8'-tetramethyl-2'-acetonaphthone.* $C_{18}H_{26}O$; mol wt 258.40. C 83.67%, H 10.14%, O 6.19%. Prepn: Davidson, Lusskin, U.S. pat. **3,045,047** (1962 to Trubek Lab.). GC determn in fragrances: H. H. Wisneski *et al.*, *J. Assoc. Offic. Anal. Chem.* **65**, 598 (1982). Toxicity: K. R. Butterworth, P. L. Mason, *Food Cosmet. Toxicol.* **19**, 753 (1981).

Crystals, mp 46.5°. bp_2 130°. Soluble in alc.
USE: As musk for perfumes, cosmetics, soaps.

10102. Versen-Ol®. *N-[2-[Bis(carboxymethyl)amino]ethyl]-N-(2-hydroxyethyl)glycine trisodium salt; N-(carboxymethyl)-N'-(2-hydroxyethyl)-N,N'-ethylenediglycine trisodium salt; N-hydroxyethylethylenediaminetriacetic acid trisodium salt; trisodium N-hydroxyethylethylenediaminetriacetate.* $C_{10}H_{15}N_2Na_3O_7$; mol wt 344.21. C 34.89%, H 4.39%, N 8.14%, Na 20.04%, O 32.54%. Prepn: Young, U.S. pat. **2,811,550** (1957 to Refined Prod.); Kroll, Dexter, U.S. pat. **2,845,457** (1958 to Geigy).

USE: As chelating agent for trivalent iron from pH 7 to pH 10. Forms strong 1:1 ferric chelates. The iron chelates formed will then chelate other heavy metal or alkaline earth ions.

10103. Verticillins. Three antibiotics, verticillins A, B and C, produced by a species of *Verticillium* (strain TM-759), an imperfect fungus isolated from a basidiocarp of *Coltricia cinnamomea (Polystictus cinnamomeus).* Isoln, IR, NMR spectra of verticillin A: Katagiri *et al.*, *J. Antibiot.* **23**, 420 (1970); *see also* Chepenko *et al.*, *C. A.* **78**, 156654n (1973). Structure: Minato *et al.*, *Chem. Commun.* **1971**, 44. Structure of verticillins A, B and C and absolute configuration of verticillins A and B: *eidem, J. Chem. Soc. Perkin Trans. I* **1973**, 1819. Synthetic studies: Häusler, Schmidt, *Ber.* **107**, 2804 (1974); Schmidt *et al.*, *ibid.* 2816.

Verticillin A

Verticillin A, $C_{30}H_{28}N_6O_6S_4$. Pale yellow plates from chloroform, $C_{30}H_{28}N_6O_6S_4 \cdot CHCl_3$, mp 199-213° (dec); pale yellow needles from pyridine, $C_{30}H_{28}N_6O_6S_4 \cdot 2/3 C_5H_5N$, mp 202-217° (dec); pale yellow amorphous powder from tetrahydrofuran, mp 203-214° (dec). $[\alpha]_D$ +703.7° (c = 0.422 in dioxane). uv max (dioxane): 306 nm (ε 5960). LD_{50} i.p. in mice: 7.6 mg/kg (Katagiri).

Verticillin B, $C_{30}H_{28}N_6O_7S_4$. The mono-3-hydroxymethyl analog of verticillin A. Pale yellow prisms from chloroform, mp 230-233° (dec). $[\alpha]_D^{21}$ +704.7° (c = 0.493 in dioxane). uv max (dioxane): 306 nm (ε 5600).

Verticillin C, $C_{30}H_{28}N_6O_7S_5$. Differs from verticillin B by having a trisulfide rather than disulfide bridge in one of the two dioxopiperazine rings. Pale yellow amorphous powder from methanol-water, mp 230-235° (dec). $[\alpha]_D$ +765.0° (c = 0.506 in dioxane). uv max (dioxane): 303 nm (ε 5500).

10104. Verticine. *(3β,5α,6α)-Cevane-3,6,20-triol;* peimine. $C_{27}H_{45}NO_3$; mol wt 431.66. C 75.13%, H 10.51%, N 3.24%, O 11.12%. A member of the hexacyclic *Ceveratrum* group of alkaloids. Isoln from *Fritillaria verticillata* Willd. var. *Thunbergii* Baker, *Liliaceae:* Fukuda: *Sci. Repts. Tohoku Univ.,* Ser. A, **18,** 323 (1929); Morimoto, Kimata, *Chem. Pharm. Bull. (Tokyo)* **8,** 302 (1960). Also isolated from *F. roylei* Hook: Chou, Chen, *Chinese J. Physiol.* **6,** 265 (1932), *C.A.* **26,** 5703 (1932). Structure: Ito *et al., ibid.* **11,** 1337 (1963). Abs config: S. Ito *et al., Tetrahedron Letters* **1968,** 5373. Total synthesis: J. P. Kutney *et al., J. Am. Chem. Soc.* **99,** 964 (1977).

Needles from ethanol, mp 223-224°. $[\alpha]_D^{16}$ −19.4° (ethanol); $[\alpha]_D^{17}$ −20° (CHCl₃). pK'a 9.5. uv max (ethanol + HCl): 215 nm (ε 10).

Hydrochloride, $C_{27}H_{45}NO_3 \cdot HCl$, prisms from water, dec 291-294°. $[\alpha]_D^{18}$ −18.5°.

Perchlorate, $C_{27}H_{45}NO_3 \cdot HClO_4$, prisms from water, dec 273°. $[\alpha]_D$ −15° (c = 0.05).

Methiodide, $C_{27}H_{45}NO_3 \cdot CH_3I$, crystals from butanol + ether, dec 205-210°. $[\alpha]_D^{18}$ −14.0° (water). Sol in water, methanol, ethanol; practically insol in chloroform, benzene, acetone.

3-Glucoside, $C_{33}H_{55}NO_8$, *peiminoside.* Pale brown powder. Sol in water, methanol, alc; sparingly sol in benzene.

10105. Vesnarinone. *1-(3,4-Dimethoxybenzoyl)-4-(1,2,3,-4-tetrahydro-2-oxo-6-quinolinyl)piperazine;* 3,4-dihydro-6-[4-(3,4-dimethoxybenzoyl)-1-piperazinyl]-2(1H)-quinolinone; 6-[4-(3,4-dimethoxybenzoyl)-1-piperazinyl]-3,4-dihydrocarbostyril; 1-(1,2,3,4-tetrahydro-2-oxo-6-quinolyl)-4-veratroylpiperazine; piteranometozine; OPC-8212; Arkin. $C_{22}H_{25}N_3O_4$; mol wt 395.46. C 66.82%, H 6.37%, N 10.63%, O 16.18%. Positive inotropic agent. Prepn: Y. H. Yang *et al.,* Belg. pat. 890,942; M. Tominaga *et al.,* U.S. pat. 4,415,572 (1982, 1983 both to Otsuka); M. Tominaga *et al., Chem. Pharm. Bull.* **32,** 2100 (1984). Physicochemical properties: T. Shimizu *et al., Arzneimittel-Forsch.* **34,** 334 (1984). Series of articles on mechanism of action and pharmacology: *ibid.* 342-402. Effects on canine arrhythmias: K. Hashimoto, H. Mitsuhashi, *Brit. J. Pharmacol.* **88,** 915 (1986). Metabolism: G. Miyamoto *et al., Xenobiotica* **18,** 1143 (1988). Clinical pharmacokinetics: A. Ohnishi, T. Ishizaki, *J. Clin. Pharmacol.* **28,** 719 (1988). HPLC determn in plasma and urine: G. Miyamoto *et al., J. Chromatog.* **338,** 450 (1985). Clinical evaluations in congestive heart failure: H. Asanoi *et al., J. Am. Coll. Cardiol.* **9,** 865 (1987); A. M. Feldman *et al., Am. Heart J.* **116,** 771 (1988).

Colorless granules from ethanol-chloroform, mp 238.1-239.5° (Tominaga); also reported as odorless, tasteless, pale yellow crystalline powder, mp 238.1-239.8° (Shimizu). uv max (methanol, ethanol, chloroform): 271 nm (ε = 2.51 × 10⁴, 2.52 × 10⁴, 2.30 × 10⁴, c = 10 μg/ml). % solubility (25°): glacial acetic acid 18.68; chloroform 14.19; benzyl alcohol 5.913; N-methyl-2-pyrrolidone 3.407; DMSO 2.509; 60% sulfolane 1.229; DMF 1.179; dioxane 0.1653; methanol 0.1151; acetone 0.06389; ethanol 0.04005; water 0.002086. pKa 2.86.

THERAP CAT: Cardiotonic.

10106. Vetivones. $C_{15}H_{22}O$; mol wt 218.34. C 82.52%, H 10.16%, O 7.33%. Ketonic components of vetiver oil; from roots of *Vetiveria zizanioides (Andropogon muricatus* Retz., *Gramineae).* Two isomers, the α-form having the stronger odor. Isoln and structure: Pfau, Plattner, *Helv. Chim. Acta* **22,** 640 (1939); Naves, Perrottet, *ibid.* **24,** 3 (1941).

α-Vetivone   β-Vetivone

α-Vetivone. *(4R-cis)-4,4a,5,6,7,8-Hexahydro-4,4a-dimethyl-6-(1-methylethylidene)-2(3H)-naphthalenone;* 4βH,5α-eremophila-1(10),7(11)-dien-2-one; isonootkatone. Structure: Endo, DeMayo, *Chem. Commun.* **1967,** 89; Marshall, Andersen, *Tetrahedron Letters* **1967,** 1611. Total syntheses of the racemate: Marshall *et al., Chem. Commun.* **1967,** 753; Marshall, Warne, *J. Org. Chem.* **36,** 178 (1971); Vandergen *et al., Rec. Trav. Chim.* **90,** 1034, 1045 (1971); Dastur, *J. Am. Chem. Soc.* **96,** 2605 (1974). Crystals from pentane, strong agreeable odor. mp 51.5° (Naves, Perrottet); mp of racemate 30-35° (Marshall, Andersen). n_D^{20} 1.5384. $[\alpha]_D^{20}$ +248° (chloroform). d_4^{20} 1.003 (liquid, supercooled). uv max (ethanol): 233 nm (ε 13200).

β-*Vetivone.* *(5R-cis)-6,10-Dimethyl-2-(1-methylethylidene)-spiro[4.5]dec-6-en-8-one; 2-isopropylidene-6,10-dimethylspiro-[4.5]dec-6-en-8-one.* Structure: Marshall *et al., J. Am. Chem. Soc.* **89**, 2748, 2750 (1967). Syntheses of racemate: Marshall, Johnson, *Chem. Commun.* **1968**, 391; *eidem, J. Org. Chem.* **35**, 192 (1970); K. Uneyama *et al., Chem. Letters* **1977**, 493; S. Torii *et al., Bull. Chem. Soc. Japan* **51**, 3590 (1978); E. Wenkert *et al., J. Am. Chem. Soc.* **100**, 1267 (1978). Stereospecific synthesis of (−)-β-vetivone: Deighton *et al., Chem. Commun.* **1975**, 662; M. Asaoka *et al., Chem. Letters* **1988**, 1225. Crystals from pentane, mp 44-46°. d_4^{20} 1.000 (liquid, supercooled). d_4^{45} 0.9804. $[\alpha]_D^{20}$ −38.9° (c = 10 in alcohol). n_D^{20} 1.5309 (liquid, supercooled), n_D^{45} 1.5216.
USE: Of potential interest in the perfume industry.

10107. Vetrabutine. α-*(3,4-Dimethoxyphenyl)-N,N-dimethylbenzenebutanamine; N,N-dimethyl-α-(3-phenylpropyl)veratrylamine; 1-(3,4-dimethoxyphenyl)-1-dimethylamino-4-phenylbutane; 3,4-dimethoxy-N,N-dimethyl-α-(3-phenylpropyl)benzylamine; dimophebumine.* $C_{20}H_{27}NO_2$; mol wt 313.44. C 76.64%, H 8.68%, N 4.47%, O 10.21%. Prepn: **Brit.** pat. **802,723** (1958 to Thomae).

bp$_{0.1}$ 166-168°.
Hydrochloride, $C_{20}H_{27}NO_2$.HCl, *SP-281, Monzal, Monzaldon.* Solid, mp 146-148°.
THERAP CAT: Uterine relaxant.
THERAP CAT (VET): Uterine relaxant.

10108. Viburnum opulus. Cramp bark; high bush cranberry; cranberry tree; water elder; squaw bush; snowball bush. Dried bark of *Viburnum opulus* L., var. *americanum* Ait., *Caprifoliaceae. Habit.* Europe, Asia, northern North America, south to Pennsylvania. *Constit.* Viburnin, bitter resin, tannin, sugar; citric, malic, oxalic, and valeric acids.
THERAP CAT: Antispasmodic (uterine).

10109. Viburnum prunifolium. Black hawk; sweet viburnum; sheep berry; stag bush; sloe-leaved viburnum. Dried root bark of *Viburnum prunifolium* L., *Caprifoliaceae. Habit.* U.S. *Constit.* Viburnin, bitter resin, tannin; citric, malic, oxalic, and valeric acids.
THERAP CAT: Antispasmodic (uterine).

10110. Vicianin. Mandelonitrile vicianoside. $C_{19}H_{25}$-NO_{10}; mol wt 427.41. C 53.39%, H 5.90%, N 3.28%, O 37.43%. In seeds of *Vicia angustifolia* Clos, and *V. sativa* L., *Leguminosae.* Isoln: Bertrand, *Compt. Rend.* **143**, 832 (1906); *Bull. Soc. Chim.* [4] **1**, 151 (1907); Bertrand, Weisweiller, *Compt. Rend.* **147**, 252 (1908); **151**, 325 (1910); *Bull. Soc. Chim.* [4] **9**, 147 (1911). Preparation of hexaacetyl-β-vicianoside of (−)-mandelonitrile, believed to be vicianin hexaacetate: Chaudhury, Robertson, *J. Chem. Soc.* **1949**, 2054.

Monohydrate, needles from water. When anhydr mp 160° (copper block). $[\alpha]_D^{18}$ −21° (satd aq soln). Sol in hot water, slightly in cold water, in alcohol; practically insol in chloroform, benzene, petr ether.

10111. Vicianose. *6-O-α-L-Arabinopyranosyl-D-glucose; 6-(α-L-arabinosido)-D-glucose.* $C_{11}H_{20}O_{10}$; mol wt 312.27. C 42.31%, H 6.46%, O 51.24%. By enzymatic hydrolysis of vicianin, gein, and violutin: Bertrand, Weisweiller, *Compt. Rend.* **150**, 181 (1910); *Bull. Soc. Chim.* [4] **9**, 38 (1911).

Synthesis: Helferich, Bredereck, *Ann.* **465**, 166 (1928); Wallenfels, Beck, *Ann.* **630**, 46 (1960).

Needles from 75% alc, dec 210°. Weak sweetish taste. Shows mutarotation. $[\alpha]_D^{20}$ +63° → 40° (22 hrs, c = 8). Very sol in water, practically insol in abs alc. Reduces Fehling's soln somewhat more than maltose.

10112. Vicine. *2,6-Diamino-5-(β-D-glucopyranosyloxy)-4(1H)-pyrimidinone; 2,4-diamino-6-oxypyrimidine-5-(β-D-glucopyranoside); divicine 5-glucoside; vicioside; divicine-β-glucoside.* $C_{10}H_{16}N_4O_7$; mol wt 304.26. C 39.48%, H 5.30%, N 18.41%, O 36.81%. Isoln from vetch seeds *(Vicia sativa* L., *Leguminosae):* Ritthausen, *Ber.* **9**, 301 (1876); Levene, *J. Biol. Chem.* **18**, 306 (1914); Levene, Senior, *ibid.* **25**, 611 (1916); Herissey, Cheymol, *Bull. Soc. Chim. Biol.* **13**, 29 (1931); Gmelin, Hasenmaier, *Arzneimittel-Forsch.* **7**, 755 (1957). Structure: Bendich, *Trans. N.Y. Acad. Sci.* **15**, 58 (1952); Bendich, Clements, *Biochim. Biophys. Acta* **12**, 462 (1953).

Needles from water, dec 243-244°. $[\alpha]_D^{26}$ −11.7° (c = 3.9 in 0.2N NaOH). uv max (0.1N HCl): 274 nm (ε 16400); (0.1N NaOH): 269, 248, 235, 230 nm (ε 9520, 4360, 5180, 5090). One gram dissolves in about 100 ml water; slightly sol in alcohol; readily sol in dil acids and alkalies.
Note: Another *vicin*, a rhamnosidoglucoside from vetch flowers, is described by Karrer, Widmer, *Helv. Chim. Acta* **10**, 67 (1927). It has the formula: $C_{21}H_{21}O_{12}.2H_2O$.

10113. Vidarabine. *9-β-D-Arabinofuranosyl-9H-purine-6-amine monohydrate; 9-β-D-arabinofuranosyladenine monohydrate;* arabinosyladenine; adenine arabinoside; spongoadenosine; ara-A; CI-673; Arasena-A; Vira-A. $C_{10}H_{13}N_5$-$O_4.H_2O$; mol wt 285.26. C 42.11%, H 5.30%, N 24.55%, O 28.04%. Purine nucleoside first synthesized as a potential anticancer agent: Lee *et al., J. Am. Chem. Soc.* **82**, 2648 (1960); Reist *et al., J. Org. Chem.* **27**, 3274 (1962); Glaudemans, Fletcher, *ibid.* **28**, 3004 (1963); Reist *et al., ibid.* **29**, 3725 (1964). Production by culturing a strain of *Streptomyces antibioticus:* **Brit.** pat. **1,159,290** (1969 to Parke, Davis), *C.A.* **71**, 79757z (1969). Crystal and molecular structure: Bunick, Voet, *Acta Crystallogr.* **30B**, 1641 (1974). Metabolism and distribution studies: J. J. Brink, G. A. LePage, *Cancer Res.* **24**, 1042 (1964). *In vitro* antiviral activity: J. de Rudder, M. Privat de Garilhe, *Antimicrob. Ag. Chemother.* **1965**, 578. Series of articles on *in vitro* and *in vivo* antiviral activity: *ibid.* **1968**, 136-179. Comparative clinical evaluation with acyclovir, *q.v.*, in varicella-zoster infection: J. L. Vildé *et al., J. Med. Virol.* **20**, 127 (1986); in herpes simplex encephalitis: R. J. Whitley *et al., Infection* **15**, Suppl. 1, S 3 (1987). Pharmacokinetics in children: T. C. Shope *et al., J. Infect. Dis.* **148**, 721 (1983). HPLC determn in plasma and urine: W. P. McCann *et al., Antimicrob. Ag. Chemother.* **28**, 265 (1985). Toxicity data: S. M. Kurtz *et al., ibid.* **1968**, 180. Book: *Adenine Arabinoside: An Antiviral Agent,* D. Paven-Langston *et al.,* Eds. (Raven Press, New York, 1975) xviii + 425 pp. *Reviews:* C. E. Cass in *Antibiotics* vol. 5(pt. 2), F. E. Hahn, Ed. (Springer-Verlag, New York, 1979) pp 85-109; R. A. Buchanan, F. Hess, *Pharmacol. Ther.* **8**, 143-171 (1980). Comprehensive description: W. Hong *et al.,* in *Analytical Profiles of Drug*

Substances vol. **15**, K. Florey, Ed. (Academic Press, New York, 1986) pp 647-672.

Crystals from water, mp 257.0-257.5° (0.4 H_2O). $[\alpha]_D^{27}$ −5° (c = 0.25). uv max (pH 1): 257.5 nm (ε 12700); pH 7: 259 nm (ε 13400); pH 13: 259 nm (ε 14000). LD_{50} in mice (mg/kg): 4677 i.p.; > 7950 orally (Kurtz).

THERAP CAT: Antiviral.

10114. Vigabatrin. *4-Amino-5-hexenoic acid;* γ-vinyl-γ-aminobutyric acid; gamma-vinyl GABA; γ-vinyl GABA; GVG; MDL-71754; RMI-71754; Sabril. $C_6H_{11}NO_2$; mol wt 129.16. C 55.80%, H 8.58%, N 10.84%, O 24.77%. Irreversible inhibitor of γ-aminobutyric acid transaminase, the enzyme responsible for the degradation of the neurotransmitter γ-aminobutyric acid (GABA). Prepn: B. W. Metcalf, M. Jung, U.S. pat. **3,960,927** (1976 to Richardson-Merrell); and *in vitro* enzyme inactivation: B. Lippert *et al., Eur. J. Biochem.* **74**, 441 (1977). Mechanism of action study: P. J. Schechter *et al., Eur. J. Pharmacol.* **45**, 319 (1977). Anticonvulsant activity and toxicity studies: W. Löscher, *Neuropharmacol.* **21**, 803 (1982). HPLC determn in plasma and urine: J. A. Smithers *et al., J. Chromatog.* **341**, 232 (1985). The *S*(+)-enantiomer is the pharmacologically active form. Pharmacokinetics of enantiomers in humans: K. D. Haegele, P. J. Schechter, *Clin. Pharmacol. Ther.* **40**, 581 (1986). Clinical studies in treatment resistant epilepsy: C. A. Tassinari *et al., Arch. Neurol.* **44**, 907 (1987); T. R. Browne *et al., Neurology* **37**, 184 (1987). Series of articles on clinical use in adult and childhood epilepsy: *J. Child Neurol.* **6**, Suppl. 2, S3-S69 (1991). Reviews of early literature and mechanism of action: M. J. Iadarola, K. Gale, *Mol. Cell. Biochem.* **39**, 305-330 (1981); of pharmacology and toxicology: E. J. Hammond, B. J. Wilder, *Clin. Neuropharmacol.* **8**, 1-12 (1985). *Review:* S. M. Grant, R. C. Heel, *Drugs* **41**, 889-926 (1991).

Crystals from acetone/water, mp 209°. Freely sol in water. LD_{50} i.p. in mice: > 2500 mg/kg (Löscher).

THERAP CAT: Anticonvulsant.

10115. Villikinin. An intestinal villi-stimulating material extracted from canine intestinal mucosa and human plasma: Kokas *et al., Fed. Proc.* **22**, no. 2, part I, 225 (Mar.-Apr. 1963). *Review:* Grossman, *Physiol. Rev.* **30**, 33 (1950).

Heat stable, sol in water, dil acids, alcohol, acetone.

10116. Viloxazine. *2-[(2-Ethoxyphenoxy)methyl]morpholine;* 2-(2-ethoxyphenoxymethyl)tetrahydro-1,4-oxazine; ICI-58834. $C_{13}H_{19}NO_3$; mol wt 237.30. C 65.80%, H 8.07%, N 5.90%, O 20.23%. Prepn: K. B. Mallion *et al.,* **Brit. pat. 1,138,405;** *idem,* U.S. pat. **3,714,161** (1969, 1973 both to I.C.I.). Manufacturing process: S. A. Lee, **Brit. pat. 1,260,886;** *idem,* U.S. pat. **3,712,890** (1972, 1973 both to I.C.I.). Pharmacology: Mallion *et al., Nature* **238**, 157 (1972); *see also* Bereen: *Lancet* **1**, 379 (1973). Toxicity: R. D. Brosnan *et al., J. Int. Med. Res.* **4**, 83 (1976). Clinical evaluation in narcolepsy: C. Guilleminault *et al., Sleep* **9**, 275 (1986). Review of pharmacology and therapeutic efficacy: R. M. Pinder *et al., Drugs* **13**, 401 (1977).

Hydrochloride, $C_{13}H_{19}NO_3.HCl$, *Catatrol, Vivalan, Vicilan, Vivarint.* mp 185-186°. LD_{50} in mice (mg/kg): 1000 orally, 60 i.v. (Brosnan).

THERAP CAT: Antidepressant.

10117. Viminol. α-[[Bis(1-methylpropyl)amino]methyl]-1-[(2-chlorophenyl)methyl]-1H-pyrrole-2-methanol; 1-(o-chlorobenzyl)-α-[(di-sec-butylamino)methyl]pyrrole-2-methanol; 1-[α-(N-o-chlorobenzyl)pyrryl]-2-di-*sec*-butylaminoethanol; diviminol; Z-424. $C_{21}H_{31}ClN_2O$; mol wt 362.94. C 69.50%, H 8.61%, Cl 9.77%, N 7.72%, O 4.41%. Prepn: U. M. Teotino, D. Della Bella, **S. Afr. pat. 67 02,732,** *eidem,* U.S. pat. **3,539,589** (1970 to Whitefin Holding). Synthesis and metabolic fate of ^{14}C-viminol: Cameron *et al., Arzneimittel-Forsch.* **23**, 708 (1973). Viminol possesses 3 asymmetric carbon atoms and thus is composed of 6 stereoisomers. Role of stereoisomers in determining specific pharmacological effects: D. Della Bella *et al., Nature New Biol.* **241**, 282 (1973). Additional pharmacology data: M. Babbini *et al., Eur. J. Pharmacol.* **23**, 137 (1973). *Review:* D. Della Bella, *Boll. Chim. Farm.* **111**, 5-19 (1972).

$bp_{0.1 mm}$ 160-165°. *p*-Hydroxybenzoate, $C_{21}H_{31}ClN_2O.C_7H_6O_3$, *Dividol.* mp 128-130° (dec). LD_{50} i.p. in rats: 205.9 mg/kg (Babbini).

THERAP CAT: Analgesic.

10118. Vinbarbital Sodium. *5-Ethyl-5-(1-methyl-1-butenyl)-2,4,6(1H,3H,5H)-pyrimidinetrione sodium salt;* 5-ethyl-5-(1-methyl-1-butenyl)barbituric acid sodium salt; sodium 5-ethyl-5-(1-methyl-1-butenyl)barbiturate; Delvinal Sodium. $C_{11}H_{15}N_2NaO_3$; mol wt 246.24. C 53.66%, H 6.14%, N 11.38%, Na 9.34%, O 19.49%. Prepn: Cope, Hancock, *J. Am. Chem. Soc.* **61**, 776 (1939); Cope, U.S. pat. **2,187,703** (1940 to Sharp & Dohme). Pharmacology and toxicity: J. P. Hendrix, *J. Pharmacol. Exp. Ther.* **68**, 22 (1940).

Hygroscopic crystals, bitter taste. Slightly sol in ether and chloroform. A 1% aq soln is alkaline to phenolphthalein and has a pH between 8.5 and 9.5. Unbuffered aq solns are unstable. LD_{50} orally in rats: 130 mg/kg (Hendrix).

Free acid, $C_{11}H_{15}N_2O_3$, mp 161-163°. *Caution:* May be habit forming. This is a controlled substance (depressant) listed in the U.S. Code of Federal Regulations, Title 21 Parts 329.1 and 1308.13 (1995).

THERAP CAT: Sedative, hypnotic.

THERAP CAT (VET): Sedative, hypnotic.

10119. Vinblastine. *Vincaleukoblastine;* VLB; 29060-LE. $C_{46}H_{58}N_4O_9$; mol wt 810.99. C 68.13%, H 7.21%, N 6.91%, O 17.76%. Antitumor alkaloid isolated from *Vinca rosea* Linn., *Apocynaceae:* Noble *et al., Ann. N.Y. Acad. Sci.* **76**, Art. 3, 882-894 (1958); Gorman *et al., J. Am. Chem. Soc.* **81**, 4745, 4754 (1959); Beer *et al.,* U.S. pat. **3,097,137** (1963 to Can. Pats. Dev.); Svoboda, U.S. pat. **3,225,030** (1965 to Lilly). Structural studies: Neuss *et al., J. Am. Chem. Soc.*

86, 1440 (1964). Structure and stereochemistry: Moncrief, Lipscomb, *ibid.* **87**, 4963 (1965). Partial synthesis of vinblastine-type alkaloids: Potier *et al., Chem. Commun.* **1975**, 670. Studies on total synthesis: J. P. Kutney, *Lloydia* **40**, 107 (1977). Biosynthesis from catharanthine and vindoline, *q.v.:* P. Mangeney *et al., J. Am. Chem. Soc.* **101**, 2243 (1979). ^{13}C-NMR spectral analysis: Wenkert *et al., Helv. Chim. Acta* **58**, 1560 (1975). Pharmacology: S. M. Sieber *et al., Cancer Treat. Rep.* **60**, 1127 (1976). Toxicology: C. Lu, M. Meistrich, *Cancer Res.* **39**, 3575 (1979). Comprehensive description: J. H. Burns in *Analytical Profiles of Drug Substances* vol. 1, K. Florey, Ed. (Academic Press, New York, 1972) pp 443-462; F. J. Muhtadi, A. F. A. Afify in *Analytical Profiles of Drug Substances and Excipients* vol. 21, H. G. Brittain, Ed. (Academic Press, San Diego, 1992) pp 611-658.

Solvated needles from methanol, mp 211-216°. $[\alpha]_D^{26}$ +42° (chloroform). uv max (ethanol): 214, 259 nm (log ϵ 4.73, 4.21). Practically insol in water, petr ether. Sol in alcohols, acetone, ethyl acetate, CHCl$_3$.

Sulfate, C$_{46}$H$_{58}$N$_4$O$_9$.H$_2$SO$_4$, *Exal, Velban, Velbe.* Crystals, mp 284-285°. $[\alpha]_D^{26}$ −28° (c = 1.01 in methanol). pKa$_1$ 5.4; pKa$_2$ 7.4. uv max (methanol): 212, 262, 284, 292 nm (log ϵ 4.75, 4.28, 4.22, 4.18). Very slightly sol in ethanol. Practically insol in ether. One part is sol in 10 parts of water, 50 parts of chloroform. LD$_{50}$ i.v. in mice: 9.5 mg/kg (Lu, Meistrich).

Dihydrochloride dihydrate, C$_{46}$H$_{58}$N$_4$O$_9$.2HCl.2H$_2$O, crystals, dec 244-246°.

THERAP CAT: Antineoplastic.

10120. Vincamine. *(3α,14β,16α)-14,15-Dihydro-14-hydroxyeburnamenine-14-carboxylic acid methyl ester;* 13a-ethyl-2,3,5,6,12,13,13a,13b-octahydro-12-hydroxy-1H-indolo[3,2,1-de]pyrido[3,2,1-ij][1,5]naphthyridine-12-carboxylic acid methyl ester; Angiopac; Arteriovinca; Devincan; Equipur; Novicet; Ocu-Vinc; Oxygeron; Perval; Pervincamine; Pervone; Sostenil; Tripervan; Vincadar; Vincafarm; Vincafolina; Vincafor; Vincagil; Vincalen; Vincamidol; Vincapront; Vincasaunier; Vincimax; Vinodrel Retard; Vraap. C$_{21}$H$_{26}$N$_2$O$_3$; mol wt 354.45. C 71.16%, H 7.39%, N 7.90%, O 13.54%. Major indole alkaloid of *Vinca minor* L., *Apocynaceae* occurring naturally in the *d*-form: E. Schlittler, A. Furlenmeier, *Helv. Chim. Acta* **36**, 2017 (1953); Pailer, Belohlav, *Monatsh.* **85**, 1055 (1954); King *et al., J. Chem. Soc.* **1955**, 4206; J. Trojanek *et al., Coll. Czech. Chem. Commun.* **26**, 867 (1961). Isoln from *Tabernaemontana rigida* Miers, *Apocynaceae:* M. P. Cava *et al., J. Org. Chem.* **33**, 1055 (1968). Structure: J. Trojanek *et al., Tetrahedron Letters* **1961**, 702; J. Mokry *et al., ibid.* **1962**, 433; O. Clauder, *ibid.* 1147; Plat *et al., Bull. Soc. Chim. France* **1962**, 1082. Abs config: J. Trojanek *et al., Chem. & Ind. (London)* **1965**, 1261; Blaha *et al., Coll. Czech. Chem. Commun.* **33**, 3833 (1968). X-ray determn of molecular structure: H. P. Weber, T. J. Petcher, *J. Chem. Soc. Perkin Trans. II* **1973**, 2001. Total synthesis: C. Szantay *et al., Tetrahedron Letters* **1973**, 191; *eidem, Tetrahedron* **33**, 1803 (1977); P. Pfäffli *et al., Helv. Chim. Acta* **58**, 1131 (1975); W. Oppolzer *et al., ibid.* **60**, 1801 (1977); of *dl*-form: M. E. Kuehne, *J. Am. Chem. Soc.* **86**, 2946 (1964); J. L. Herrmann *et al., ibid.* **96**, 3702 (1974); K. H. Gibson, J. E. Saxton, *Tetrahedron* **33**,

833 (1977); K. Irie, Y. Ban, *Heterocycles* **18**, 255 (1982); T. R. Govindachari, S. Rajeswari, *Indian J. Chem.* **22B**, 531 (1983). Synthesis of stereoisomers: J. Warnant *et al., Ger. pat.* **2,115,718**; *eidem,* U.S. pat. **3,770,724** (1971, 1973 both to Roussel-UCLAF); G. Rossey, A. Wick, *J. Org. Chem.* **47**, 4725 (1982). HPLC determn: A. Amato *et al., J. Chromatog.* **270**, 387 (1983). Toxicity data: L. Szporny, K. Szász, *Arch. Exp. Pathol. Pharmakol.* **236**, 296 (1959). Hypotensive activity in dogs and man: Z. Szabo, Z. Nagy, *Arzneimittel-Forsch.* **10**, 811 (1960). Clinical pharmacology: C. C. Lim *et al., Brit. J. Clin. Pharmacol.* **9**, 100 (1980). Clinical pharmacokinetics: H. Millart, D. Lamiable *et al., Int. J. Clin. Pharmacol. Ther. Toxicol.* **21**, 581 (1983). Brief review: P. Cook, I. James, *N. Engl. J. Med.* **305**, 1562 (1981).

Yellow crystals from acetone or methanol, mp 232-233°. $[\alpha]_D^{23}$ +41° (in pyridine). uv max: 225, 278 nm (log ϵ 4.14, 3.61). LD$_{50}$ in mice (mg/kg): 75 i.v.; > 1000 s.c. (Szporny, Szász); 1000 orally (Nagy).

Hydrochloride, C$_{21}$H$_{26}$N$_2$O$_3$.HCl, *Cetal, Esberidin.*

(±)-Form, crystals from methanol, mp 228-229° (Gibson, Saxton).

Note: Other alkaloids found in the "vincamine fraction" of *Vinca minor* L. are: *vincine, vincaminine* and *vincinine:* Holubek *et al., Tetrahedron Letters* **1963**, 897.

THERAP CAT: Vasodilator.

10121. Vincetoxin. Cynanchin. C$_{50}$H$_{82}$O$_{20}$; mol wt 1003.19. C 59.86%, H 8.24%, O 31.90%. C$_{46}$H$_{70}$O$_{16}$-(OCH$_3$)$_4$. From roots of *Vincetoxicum officinale* Moench., *Asclepiadaceae:* Tanret, *Compt. Rend.* **100**, 277 (1885); Kubler, *Arch. Pharm.* **246**, 660 (1908); Gager, Zechner, *ibid.* **276**, 431 (1938). Constitution: Korte, *Ber.* **88**, 1527 (1955); Korte, Ripphahn, *Ann.* **621**, 58 (1959). Known in two forms: water-soluble and water-insoluble.

The water-soluble form is a light yellow, amorphous powder, mp 132° (dec). $[\alpha]_D$ −50°. Sol in alcohol, chloroform; insol in ether.

The water-insoluble form is an amorphous powder mp 59°. Sol in alcohol, chloroform, ether.

10122. Vinclozolin. *3-(3,5-Dichlorophenyl)-5-ethenyl-5-methyl-2,4-oxazolidinedione;* 3-(3,5-dichlorophenyl)-5-methyl-5-vinyloxazolidine-2,4-dione; BAS-352F; Ronilan; Vorlan. C$_{12}$H$_9$Cl$_2$NO$_3$; mol wt 286.11. C 50.38%, H 3.17%, Cl 24.78%, N 4.90%, O 16.78%. Prepn: D. Mangold *et al., Ger. pat.* **2,207,576** (1973 to BASF), *C.A.* **79**, 137120q (1973). Activity: C. Hess, F. Locher, *Proc. Brit. Insectic. Fungic. Conf.* **2**, 693 (1975). Comparative mechanism of action: A. C. Pappas, D. J. Fisher, *Pestic. Sci.* **10**, 239 (1979).

Crystalline solid, mp 108°. Soly in water at 20°: 1 g/l. Soly in (g/kg): acetone 435; benzene 146; chloroform 319; ethyl acetate 253. Slowly hydrolyzed in alkaline soln. LD$_{50}$ orally in rats: 10 g/kg (Hess, Locher).

Caution: Causes eye and skin irritation; potential skin sensitizer (Mallinckrodt Technical Data Sheet).

USE: Agricultural fungicide.

10123. Vinconate. *(±)-3-Ethyl-2,3,3a,4-tetrahydro-1H-indolo[3,2,1-de][1,5]naphthyridine-6-carboxylic acid methyl ester;* chanodesethylapovincamine. C$_{18}$H$_{20}$N$_2$O$_2$; mol wt

296.37. C 72.95%, H 6.80%, N 9.45%, O 10.80%. Synthetic hexahydrocanthane alkaloid; vincamine analog. Prepn: **Belg. pat. 870,887**; J. A. A. Hannart, **U.S. pat. 4,200,638** (1979, 1980 both to Omnium Chim.). Pharmacology: K. Ninomiya et al., *Iyakuhin Kenkyu* **23**, 620 (1992), *C.A.* **118**, 32823 (1992). Pharmacokinetics: M. Kinbara, Y. Satou, *ibid.* 275, *C.A.* **117**, 123938 (1992). Nootropic activity: H. Kinoshita et al., *Res. Commun. Chem. Pathol. Pharmacol.* **84**, 175 (1994). Clinical pharmacology: B. Saletu et al., *Arch. Gerontol. Geriatr.* **3**, 127 (1984). Acute toxicity study: N. Nishimura et al., *Yakuri to Chiryo* **20**, Suppl. 8, S1995 (1992), *C.A.* **118**, 52241 (1992).

mp 120-121°.
Monohydrochloride, $C_{18}H_{20}N_2O_2$·HCl, *OC-340, OM-853.* mp 194-195°. LD_{50} in male, female mice, rats (mg/kg): 947, 699, 2582, 2348 orally; 127, 131, 112, 117 i.v.; 4073, 3446, >6000, >6000 s.c. (Nishimura).
THERAP CAT: Nootropic.

10124. Vincristine. *22-Oxovincaleukoblastine; leurocristine;* VCR; LCR. $C_{46}H_{56}N_4O_{10}$; mol wt 824.97. C 66.97%, H 6.84%, N 6.79%, O 19.39%. Antitumor alkaloid isolated from *Vinca rosea* Linn. (*Catharanthus roseus* G. Don), *Apocynaceae:* Svoboda, *Lloydia* **24**, 173 (1961). Structure: Neuss et al., *J. Am. Chem. Soc.* **86**, 1440 (1964); Moncrief, Lipscomb, *ibid.* **87**, 4963 (1965). Pharmacology: R. H. Adamson et al., *Arch. Int. Pharmacodyn. Ther.* **157**, 299 (1965); S. M. Sieber et al., *Cancer Treat. Rep.* **60**, 127 (1976). Prepn and pharmacology of [³H]vincristine: Owellen, Donigian, *J. Med. Chem.* **15**, 894 (1972). Symposium on vincristine: *Cancer Chemother. Rep.* **52**, 453-535 (1968). Biosynthesis from catharanthine, q.v. and vindoline, q.v.: A. Rahman et al., *Tetrahedron Letters* **1976**, 2351; P. Mangeney et al., *J. Am. Chem. Soc.* **101**, 2243 (1979). Comprehensive description of the sulfate: J. H. Burns, *Anal. Prof. Drug Subs.* **1**, 463-480 (1972). For structure *see* Vinblastine, R = CHO.
Blades from methanol, mp 218-220°. $[\alpha]_D^{25}$ +17°; $[\alpha]_D^{25}$ +26.2° (ethylene chloride). pKa: 5.0, 7.4 in 33% DMF. uv max (ethanol): 220, 255, 296 nm (log a_m 4.65, 4.21, 4.18). LD_{50} i.p. in mice: 5.2 mg/kg (Adamson).
Sulfate, $C_{46}H_{56}N_4O_{10}$·H_2SO_4, *Kyocristine, Oncovin, Vincosid, Vincrex.* Crystals from ethanol.
THERAP CAT: Antineoplastic.

10125. Vindesine. *3-(Aminocarbonyl)-O⁴-deacetyl-3-de-(methoxycarbonyl)vincaleukoblastine;* desacetylvinblastine amide; VDS; Compound 112531; NSC-245467. $C_{43}H_{55}$-N_5O_7; mol wt 753.94. C 68.50%, H 7.35%, N 9.29%, O 14.85%. Synthetic deriv of the dimeric *Catharanthus* alkaloid vinblastine, q.v. Prepn: G. J. Cullinan, K. Gerzon, **Ger. pat. 2,415,980** (1974 to Lilly), *C.A.* **82**, 72191r (1975); *see also* C. J. Burnett et al., *J. Med. Chem.* **21**, 88 (1978). ¹³C-NMR spectroscopy: D. E. Dorman, J. W. Paschal, *Org. Magn. Reson.* **8**, 413 (1976). Antitumor activity in rodents: M. J. Sweeney et al., *Cancer Res.* **38**, 2886 (1978). Pharmacokinetics: R. J. Owellen et al., *ibid.* **37**, 2603 (1977). Use in acute lymphocytic leukemia of childhood: W. Krivit et al., *Cancer Chemother. Pharmacol.* **2**, 267 (1979). Brief review of preclinical and early clinical data: R. W. Dyke et al., *ibid.* 229. Toxicology: G. C. Todd et al., *Toxicol. Environ. Health* **1**, 843 (1976); R. J. Owellen et al., *Biochem. Pharmacol.* **26**, 1213 (1977).

Crystals from ethanol-methanol, mp 230-232°. $[\alpha]_D^{25}$ +39.4° (c = 1.0 in methanol). uv max (methanol): 214, 266, 288, 296 nm (ε 53400, 17450, 13950, 12500). pKa' (DMF 66%) 5.39, 7.36; (H_2O) 6.04, 7.67.
Sulfate salt, $C_{43}H_{55}N_5O_7$·H_2SO_4, *LY-099094, Eldisine, Fildesin.* Amorphous solid from ethanol-isopropyl alcohol, mp >250° (solvated form). LD_{50} in mice, rats (mg/kg): 6.3 ± 0.6, 2.0 ± 0.2 i.v. (Todd); in mice: 8.8 ± 2.5 i.p. (Owellen, *Biochem. Pharmacol.*).
THERAP CAT: Antineoplastic.

10126. Vindoline. *(2β,3β,4β,5α,12β,19α)-4-(Acetyloxy)-6,7-didehydro-3-hydroxy-16-methoxy-1-methylaspidospermidine-3-carboxylic acid methyl ester.* $C_{25}H_{32}N_2O_6$; mol wt 456.54. C 65.77%, H 7.06%, N 6.14%, O 21.03%. Major alkaloid from the leaves of *Vinca rosea* Linn. (*Catharanthus roseus* G. Don), *Apocynaceae:* Gorman et al., *J. Am. Pharm. Assoc.* **48**, 256 (1959); Svoboda et al., *ibid.* 659; Moza, Trojánek, *Coll. Czech. Chem. Commun.* **28**, 1419 (1963). Structure: Gorman et al., *J. Am. Chem. Soc.* **84**, 1058 (1962); Neuss, *Bull. Soc. Chim. France* **1963**, 1509. Stereochemistry: Moncrief, Lipscomb, *J. Am. Chem. Soc.* **87**, 4963 (1965). Review of chemistry: Neuss et al., *Advan. Chemother.* **1**, 133 (1964). Total synthesis of (±)-vindoline: Ando et al., *J. Am. Chem. Soc.* **97**, 6880 (1975); Y. Ban et al., *Tetrahedron Letters* **1978**, 151; J. P. Kutney et al., *J. Am. Chem. Soc.* **100**, 4220 (1978). Lacks physiological activity alone but is contained as the pentacyclic moiety in the antineoplastic agents vinblastine and vincristine, q.q.v.

Vindoline	R = OCH_3
Demethoxyvindoline	R = H

Needles from acetone + petr ether, mp 164-165°; prisms, mp 174-175°; $[\alpha]_D^{20}$ −18° (chloroform); pKa 5.5 in 66% DMF (Moza, Trojánek). Also reported as crystals, mp 154-155°; $[\alpha]_D^{27}$ +42° (chloroform) (Gorman). uv max (ethanol): 212, 250, 304 nm (log ε 4.49, 3.74, 3.57).
Hydrochloride, $C_{25}H_{32}N_2O_6$·HCl, crystals from acetone, mp 161-164°.
Demethoxyvindoline, $C_{24}H_{30}N_2O_5$, *vindorosine, vindolidine.* Isoln: Moza, Trojánek, *loc. cit.* Structure: *eidem, Coll. Czech. Chem. Commun.* **28**, 1427 (1963). Needles from benzene + petr ether, mp 167°. $[\alpha]_D^{16}$ −31° (chloroform). uv max (methanol): 250, 302 nm (log ε 3.98, 3.52).

10127. Vinorelbine. *3',4'-Didehydro-4'-deoxy-C'-norvincaleukoblastine;* nor-5'-anhydrovinblastine; NVBKW-2307. $C_{45}H_{54}N_4O_8$; mol wt 778.95. C 69.39%, H 6.99%, N 7.19%, O 16.43%. Semi-synthetic vinca alkaloid; structurally related to vinblastine, *q.v.* Prepn: **Japan. Kokai 80 31,096**; N. Langlois *et al.*, U.S. pat. **4,307,100** (1980, 1981 both to Agence Nat. Valorisation Recherche); P. Mangeney *et al.*, *Tetrahedron* **35**, 2175 (1979). Pharmacology: G. Mathé, P. Reizenstein, *Cancer Letters* **27**, 285 (1985). Clinical pharmacokinetics: R. Rahmani *et al.*, *Cancer Res.* **47**, 5796 (1987). HPLC determn in biological fluids: F. Jehl *et al.*, *J. Chromatog.* **525**, 225 (1990). Symposium: *Seminars in Oncology* **16**, Suppl. 4, 1-45 (1989). *Review:* M. Marty *et al.*, *Nouv. Rev. Fr. Hematol.* **31**, 77-84 (1989).

Residue. $[\alpha]_D^{20}$ +52.4° (c = 0.3 in CHCl$_3$). uv max (ethanol): 215, 268, 282, 293, 310 nm (ϵ 3700, 11000, 9500, 7600, 4400).

Ditartrate, C$_{45}$H$_{54}$N$_4$O$_8$.2C$_4$H$_6$O$_6$, *Eunades*, *Navelbine*. Yellow-white amorphous powder. Sol in water and ethanol.

THERAP CAT: Antineoplastic.

10128. Vinpocetine. *(3α,16α)-Eburnamenine-14-carboxylic acid ethyl ester;* 3α,16α-apovincaminic acid ethyl ester; ethyl apovincamin-22-oate; RGH-4405; Cavinton; Ceractin. C$_{22}$H$_{26}$N$_2$O$_2$; mol wt 350.46. C 75.40%, H 7.48%, N 7.99%, O 9.13%. Deriv of vincamine, *q.v.*, with vasodilating activity. Prepn: C. Lörincz *et al.*, Ger. pat. **2,253,750** corresp to U.S. pat. **4,035,370** (1973, 1977 both to Gedeon Richter); *eidem, Arzneimittel-Forsch.* **26**, 1907 (1976). Series of articles on pharmacology, biochemistry, metabolism, pharmacokinetics, clinical studies: *ibid.* 1908-1989. Toxicity studies: E. Pálosi, L. Szporny, *ibid.* 1926; E. Cholnoky, L. I. Dömök, *ibid.* 1939. HPLC studies: G. Szepesi, M. Gazdag, *J. Chromatog.* **205**, 57, 341 (1981). Evaluation of effectiveness as antimotion drug: E. I. Matsnev, D. Bodo, *Aviat. Space Environ. Med.* **55**, 281 (1984). One-pot synthesis from vincamine: Y. Kuge *et al.*, *Synth. Commun.* **24**, 759 (1994).

Crystals from benzene, mp 147-153° (dec). $[\alpha]_D^{20}$ +114° (c = 1 in pyridine). uv max (96% ethanol): 229, 275, 315 nm (log ϵ 4.45, 4.08, 3.85). LD$_{50}$ in mice, rats (mg/kg): 534, 503 orally; 240, 133.8 i.p.; 58.7, 42.6 i.v. (Pálosi, Szporny), also reported as 161.2 mg/kg i.p. in mice (Cholnoky, Dömök).

THERAP CAT: Vasodilator (cerebral).

10129. Vintiamol. *N-[(4-Amino-2-methyl-5-pyrimidinyl)methyl]-N-[4-hydroxy-1-methyl-2-[(3-oxo-3-phenyl-1-propenyl)thio]-1-butenyl]formamide;* N-[(4-amino-2-methyl-5-pyrimidinyl)methyl]-N-[2-[(2-benzoylvinyl)thio]-4-hydroxy-1-methyl-1-butenyl]formamide; S-(benzoylvinyl)thiamine. C$_{21}$H$_{24}$N$_4$O$_3$S; mol wt 412.51. C 61.15%, H 5.86%, N 13.58%, O 11.64%, S 7.77%. Prepn and sepn of isomers: Fusco, Tenconi, *Farmaco Ed. Sci.* **20**, 866 (1965).

trans-isomer

Higher melting isomer: crystals from ethanol, mp 202-204°, very sparingly sol in chloroform. Lower melting isomer: separated from mixture of both isomers by its greater soly in chloroform; crystals from ethanol contg 1 molecule of ethanol, mp 100-103°.

THERAP CAT: Vitamin (enzyme cofactor).

10130. Vinyl Acetate. *Acetic acid ethenyl ester; acetic acid vinyl ester.* C$_4$H$_6$O$_2$; mol wt 86.09. C 55.81%, H 7.02%, O 37.17%. CH$_3$COOCH=CH$_2$. Prepn: Schnizer, U.S. pat. **2,859,241** (1958 to Celanese); Sharp, Steitz, U.S. pat. **2,860,-159** (1958 to Pan Am. Petroleum); Foster, Tobler, *J. Am. Chem. Soc.* **83**, 851 (1961). Toxicity data: Smyth, Carpenter, *J. Ind. Hyg. Toxicol.* **30**, 63 (1948). Study of chronic human exposure to vinyl acetate: Deese, Joyner, *Am. Ind. Hyg. Assoc. J.* **30**, 449 (1969). *Reviews:* Leonard "Vinyl Acetate" in *Vinyl and Diene Monomers* (part 1), E. C. Leonard, Ed. (Wiley-Interscience, New York, 1970) pp 263-328; Faith, Keyes & Clark's *Industrial Chemicals*, F. A. Lowenheim, M. K. Moran, Eds (Wiley-Interscience, New York, 4th ed., 1975) pp 862-867; W. Daniels, "Poly(Vinyl Acetate)" in Kirk-Othmer *Encyclopedia of Chemical Technology* vol. 23 (Wiley-Interscience, New York, 3rd ed., 1983) pp 817-847.

Liquid. Polymerizes in light to a colorless, transparent mass. bp 72.7°. Two reported melting pts: $-100°$; $-93°$ (Daniels). d$_4^{20}$ 0.932. Flash pt, closed cup: 18°F ($-8°$C). Soly in water (20°): 1 g/50 ml. Misc with alc, ether. LD$_{50}$ orally in rats: 2.92 g/kg (Smyth, Carpenter).

USE: In polymerized form for plastic masses, films and lacquers.

10131. Vinylbital. *5-Ethenyl-5-(1-methylbutyl)-2,4,6-(1H,3H,5H)-pyrimidinetrione; 5-(1-methylbutyl)-5-vinylbarbituric acid; 5-vinyl-5-(1-methylbutyl)barbituric acid; butyvinal; Speda; Optanox.* C$_{11}$H$_{16}$N$_2$O$_3$; mol wt 224.26. C 58.91%, H 7.19%, N 12.49%, O 21.40%. Prepn: Seefelder, *Festschr. Carl Wurster* **1960**, 71, *C.A.* **56**, 9928d (1962).

Crystals, mp 90-91.5°.

Note: This is a controlled substance (depressant) listed in the U.S. Code of Federal Regulations, Title 21 Part 1308.13 (1995).

THERAP CAT: Sedative, hypnotic.

10132. Vinyl Chloride. *Chloroethylene.* C$_2$H$_3$Cl; mol wt 62.50. C 38.44%, H 4.84%, Cl 56.73%. CH$_2$=CHCl. Prepd from ethylene dichloride and alcoholic potassium: Regnault, *Ann.* **14**, 22 (1835); by halogenation of ethylene: Miller, Jenks, U.S. pat. **2,896,000** (1959 to National Distillers and Chemical Corp.). Review of mfg processes: Faith, Keyes & Clark's *Industrial Chemicals*, F. A. Lowenheim, M. K. Moran, Eds. (Wiley-Interscience, New York, 4th ed., 1975) pp 868-873. Acute toxicity study: L. Prodan *et al.*, *Ann. N.Y. Acad. Sci.* **246**, 154 (1975). Series of articles on toxicology and "vinyl chloride disease": *ibid*, 1-337. Review of carcinogenic risk: *IARC Monographs* **19**, 377-437 (1979). *Review:* in Kirk-Othmer *Encyclopedia of Chemical Technology* vol. 12 (John Wiley & Sons, 3rd ed., 1983) pp 865-885.

Colorless gas; liquefies in a freezing mixture. Polymerizes in light or in presence of catalyst. mp $-153.8°$. bp $-13.37°$. d$_4^{20}$ 0.9106. n_D^{20} 1.3700. Vapor pressure at 20°:

2530 mm Hg. *Flammable!* Flash pt, closed cup: −78° (−112°F). Sol in alc, ether, carbon tetrachloride, benzene. Slightly sol in water. LD_{50} in mice, rats, guinea pigs, rabbits (mg/l): 293.75, 390, 595, 295 by inhalation (Prodan).

Caution: Potential symptoms of overexposure are weakness; abdominal pain, gastrointestinal bleeding; hepatomegaly; pallor or cyanosis of extremities. See *NIOSH Pocket Guide to Chemical Hazards* (DHHS/NIOSH 90-117, 1990) p 224. See also *Patty's Industrial Hygiene and Toxicology* vol. 2E, G. D. Clayton, F. E. Clayton, Eds. (John Wiley & Sons, Inc., New York, 4th ed., 1994) pp 4169-4177. This substance has been listed as a known carcinogen: *Seventh Annual Report on Carcinogens* (PB95-109781, 1994) p 70.

USE: In the plastics industry; as refrigerant; in organic syntheses.

10133. Vinyl Ether. *1,1'-Oxybisethene;* divinyl ether; divinyl oxide; ethenyloxyethene; Vinethene; Vinesthene. C_4H_6O; mol wt 70.09. C 68.55%, H 8.63%, O 22.83%. $CH_2=CHOCH=CH_2$. Prepn: Major, Ruigh, U.S. pat. 2,021,872 (1935 to Merck & Co.); Mittag, Smidt, U.S. pat. 2,832,807 (1958 to Consortium für electrochem. Ind. GmbH).

Very volatile, flammable liq; characteristic odor. Dec on exposure to light or to acid fumes, forming acetaldehyde, and polymerizes to a solid glass-like mass. d_{20}^{20} 0.774; d_4^{20} 0.773. bp 28.4°. n_D^{20} 1.3989. Crit temp 183° (calc). Critical press. ∼30 atm. Specific heat ∼0.53 cal/g. Flash point (closed cup) < −30° (< −22°F). Explosives limits (% by vol in air), lower: 1.7, upper: 27.0. Autoignition temp 360° (680°F). 0.53 g dissolves in 100 g water at 37°. Oil-water soly coefficient at 37°: 41. Miscible with alcohol, ether, oils and other organic solvents. Pharmaceutical product contains not less than 96.0% and not more than 97.0% vinyl ether, the remainder consisting of dehydrated alcohol. It may contain not more than 0.025% of suitable preservative. *U.S.P.* XXI (1985). *Keep protected from light and acid fumes, in a cool place, and away from an open flame.* Lethal concn for mice in air: 51,233 ppm.

THERAP CAT: Anesthetic (inhalation).

10134. Vinylidene Chloride. *1,1-Dichloroethene; 1,1-dichloroethylene; asym-*dichloroethylene. $C_2H_2Cl_2$; mol wt 96.94. C 24.78%, H 2.08%, Cl 73.14%. $CH_2=CCl_2$. Prepn from ethylene chloride: Reilly, U.S. pat. 2,140,548 (1938 to Dow). By dehydrochlorination of 1,1,2-trichloroethane: Conrad, Gould, U.S. pat. 2,989,570 (1961 to Ethyl Corp.).

Liquid. Mild, sweet odor resembling that of chloroform. d_4^{20} 1.2129. mp −122.5°. bp_{760} 31.7°. n_D^{20} 1.4249. Flash pt −15°. Practically insol in water. Sol in organic solvents. At temps > 0° and especially in the presence of oxygen or other suitable catalysts polymerizes to a plastic. Several inhibitors to preserve the monomer have been invented. Uncontrolled polymerization may lead to explosive reaction products with oxygen or ozone: Reinhardt, *Chem. Eng. News* 25, 2136 (1947).

Caution: Irritant to skin, mucous membranes; narcotic in high concentrations; has caused liver, kidney injury in exptl animals: Irish in *Industrial Hygiene and Toxicology* vol. 2, F. A. Patty, Ed. (Interscience, New York, 2nd ed., 1962) pp 1305-1307; Prendergast *et al.*, *Toxicol. Appl. Pharmacol.* 10, 270 (1967).

USE: Intermediate in the production of "vinylidene polymer plastics" such as *Saran* and *Velon*.

10135. Violacein. *(E)-3-[1,2-Dihydro-5-(5-hydroxy-1H-indol-3-yl)-2-oxo-3H-pyrrol-3-ylidene]-1,3-dihydro-2H-indol-2-one; 3-[2-(5-hydroxyindol-3-yl)-5-oxo-2-pyrrolin-4-ylidene]-2-indolinone; 3-[2-(5-hydroxyindol-3-yl)-5-oxopyrrolin-4-ylidene]oxindole; 5-(5-hydroxy-3-indolyl)-3-(3-oxindolylidene)-2-oxopyrroline.* $C_{20}H_{13}N_3O_3$; mol wt 343.34. C 69.97%, H 3.82%, N 12.24%, O 13.98%. Pigment isolated from *Chromobacterium violaceum (Bacillus violaceus):* Tobie, *J. Bact.* 29, 223 (1935); Strong, *Science* 100, 287 (1944); Ballantine *et al.*, *J. Chem. Soc.* 1958, 755. Structure and synthesis: Ballantine *et al.*, *ibid.* 1960, 2292. *Review:* DeMoss in *Antibiotics* vol. 2, D. Gottlieb, P. D. Shaw, Eds. (Springer, New York, 1967).

Purplish-black needles, prisms. Practically insol in water; slightly sol in alcohol; moderately sol in dioxane.

10136. Violaxanthin. *5,6:5',6'-Diepoxy-5,5',6,6'-tetrahydro-β,β-carotene-3,3'-diol;* zeaxanthin diepoxide. $C_{40}H_{56}O_4$; mol wt 600.88. C 79.96%, H 9.39%, O 10.65%. Widely distributed carotenoid pigment. Formed in plants from zeaxanthin. Isoln from yellow pansies *(Viola tricolor):* Kuhn, Winterstein, *Ber.* 64, 326 (1931). Structure: Karrer *et al.*, *Helv. Chim. Acta* 14, 1044 (1931); 16, 977 (1933); 19, 1024 (1936); 27, 1684 (1944). Partial synthesis: Karrer, Jucker, *ibid.* 28, 300 (1945). Abs config: Bartlett *et al.*, *J. Chem. Soc. (C)* 1969, 2527. Isoln from *Viola tricolor* and configuration of the 15-*cis*-isomer: P. Molnar, J. Szabolcs, *Phytochemistry* 19, 623 (1980).

Orange prisms from methanol, reddish-brown acicular crystals from CS_2. mp 200°. $[\alpha]_{Cd}^{20}$ +35° (c = 0.08 in chloroform). Absorption max (in alc): 471.5, 442.5, 417.5 nm. Sol in alcohol, methanol, carbon disulfide, ether; almost insol in petr ether.

15-*cis*-Isomer, irregular yellow plates from benzene/petrol, mp 109°. uv max (benzene): 479, 448, 423, 337 nm (log ∊ 4.91, 4.98, 4.83, 4.77). Friction causes crystals to form orange prisms.

10137. Viologen. A term coined by Michaelis to designate the chlorides of certain quaternary bases derived from γ,γ'-dipyridyl. Presently used as dichlorides, dibromides and diiodides. Prepn: Michaelis, *Biochem. Z.* 250, 564 (1932); Michaelis, Hill, *J. Am. Chem. Soc.* 55, 1481 (1933); *J. Gen. Physiol.* 16, 859-873 (1933). Viologens are useful as oxidation-reduction indicators because their potential range is very negative. In contrast to other redox indicators, the color is exhibited by the reduced form, whereas usually the oxidized form is the colored one and secondly, the redox potential of these substances is independent of pH. Review of electrochemistry: C. L. Bird, A. T. Kuhn, *Chem. Soc. Rev.* 10, 49-82 (1981).

Ethyl viologen, 1,1-diethyl-4,4'-bipyridinium, N,N'-diethyl-γ,γ'-dipyridylium. Normal potential at 30°: −0.449 volts.

Benzyl viologen, 1,1'-bis(phenylmethyl)-4,4'-bipyridinium, N,N'-dibenzyl-γ,γ'-dipyridylium. Normal potential at 30°: −0.359 volts.

Betaine viologen, N,N'-dibetaine-γ,γ'-dipyridylium. Normal potential at 30°: −0.444 volts.

Dimethyl analog, see Paraquat.

USE: As biological oxidation-reduction indicators.

10138. Violuric Acid. *2,4,5,6(1H,3H)-Pyrimidinetetrone 5-oxime; alloxan 5-oxime; 5-isonitrosobarbituric acid; 5-(hydroxyimino)barbituric acid.* $C_4H_3N_3O_4$; mol wt 157.09. C 30.58%, H 1.92%, N 26.75%, O 40.74%. Known to exist as a mixture of keto-enol tautomers. Prepd from alloxan and hydroxylamine: Cresole, *Ber.* 16, 1133 (1883); Guinchard, *Ber.* 32, 1723 (1899).

Orthorhombic crystals, dec 240-241°. pK 4.7. Sparingly sol in water to a violet-colored soln; sol in alcohol. With $FeCl_3$ produces blue color.

USE: Analytical reagent for chromatographic separation of cations. Forms chelates: Leermakers, Hoffman, *J. Am. Chem. Soc.* **80**, 5663 (1958).

10139. Viomycin. Celiomycin; florimycin; tuberactinomycin B. $C_{25}H_{43}N_{13}O_{10}$; mol wt 685.70. C 43.79%, H 6.32%, N 26.56%, O 23.33%. Polypeptide antibiotic produced by various *Streptomyces* species including *S. puniceus, S. floridae, S. vinaceus.* Isoln: A. C. Finlay *et al., Am. Rev. Tuberc.* **63**, 1 (1951); Bartz *et al., ibid.* 4. Production: Marsh *et al.,* U.S. pat. **2,633,445** (1953 to Ciba); Freaney, U.S. pat. **2,828,245** (1958 to C.S.C.). Purification and chemical characterization: T. Kitagawa *et al., Chem. Pharm. Bull.* **20**, 2176 (1972). Proposed structure: Yoshioka *et al., Tetrahedron Letters* **1971**, 2043. Alternate structure: B. W. Bycroft *et al., ibid.* **1968**, 5901; *eidem, Experientia* **27**, 501 (1971); *eidem, J. Chem. Soc., Perkin Trans. I* **1972**, 820, 827. Confirmed structure: Noda *et al., J. Antibiot.* **25**, 427 (1972); T. Kitagawa *et al., Chem. Pharm. Bull.* **20**, 2215 (1972). X-ray crystallography: B. W. Bycroft, *Chem. Commun.* **1972**, 660. Biosynthesis: J. H. Carter II *et al., Biochemistry* **13**, 1221, 1227 (1974). Toxicological study: H. Keller *et al., Arzneimittel-Forsch.* **6**, 61 (1956). Enzyme immunoassay for determn in body fluids: T. Kitagawa *et al., Chem. Pharm. Bull.* **30**, 2487 (1982). For structure *see* Tuberactinomycin.

Sulfate, *Viocin.* Hygroscopic plates, mp 266° (dec). $[\alpha]_D^{18}$ −29.5° (c = 1 in H_2O). uv max (water or 0.1N HCl; 0.1N $NaOH$): 268 nm (log ϵ 4.4); 285 nm (log ϵ 4.2). LD_{50} in mice (mg/kg): 240 i.v. (Finlay), 1750 s.c. (Keller). Sol in water. Relatively insol in most organic solvents. Solns adjusted to pH 5-6 are quite stable. Soly data: Weiss *et al., Antibiot. & Chemother.* **7**, 374 (1957).

Pantothenate sulfate, $C_{25}H_{43}N_{13}O_{10}\cdot C_9H_{17}NO_5\cdot H_2SO_4$, *Vionactan, Viothenat.* Prepn: H. Keller, H. Mückter, Ger. pat. **1,011,800** (1957 to Grünenthal), *C.A.* **53**, 13519c (1959). Crystals, mp 242° (dec).

Hydrochloride, hygroscopic plates, mp 270° (dec). $[\alpha]_D^{18}$ −16.6° (c = 1 in H_2O). uv max (water; 0.1N HCl; 0.1N $NaOH$): 268 nm (log ϵ 4.5); 268 nm (log ϵ 4.4); 285 nm (log ϵ 4.3).

THERAP CAT: Antibacterial (tuberculostatic).

10140. VIP. *Vasoactive intestinal polypeptide;* vasoactive intestinal peptide. Neuroactive polypeptide gastrointestinal hormone, related to glucagon and secretin, *q.q.v.* It exhibits a wide variety of biological activities, including relaxation of systemic and vascular smooth muscle and stimulation of the exocrine pancreas, of secretion of insulin and of cyclic-AMP formation in the small intestine. Initially isolated from porcine upper intestinal wall: S. I. Said, V. Mutt, *Science* **169**, 1217 (1970); *eidem, Eur. J. Biochem.* **28**, 199 (1972). Porcine VIP is a 28-residue peptide having an *N*-terminal seryl sequence: M. Bodanszky *et al., Proc. Nat. Acad. Sci. USA* **70**, 382 (1973); V. Mutt, S. I. Said, *Eur. J. Biochem.* **42**, 581 (1974). Synthesis: M. Bodanszky *et al., J. Am. Chem. Soc.* **96**, 4973 (1974); *see also* S. I. Said *et al.,* U.S. pat. **3,862,927** (1975). Solid-phase synthesis: D. H. Coy, J. Gardner, *Int. J. Peptide Protein Res.* **15**, 73 (1980); R. Colombo, *Experientia* **38**, 773 (1982). Isoln of chicken VIP: A. Nilsson, *FEBS Letters* **47**, 284 (1974). Amino acid sequence: *idem, ibid.* **60**, 322 (1975). Synthesis of C-terminal fragment: M. Bodanszky *et al., Bioorg. Chem.* **5**, 339 (1976). Solid-phase synthesis: R. Colombo, *Int. J. Peptide Protein Res.* **19**, 71 (1982). Multiple immunoreactive forms of VIP have also been found in human colonic mucosa: R. Dimaline, G. J. Dockray, *Gastroenterology* **75**, 387 (1978). The presence of a common form of VIP in brain and gastrointestinal tract has been demonstrated by radioimmunoassay and radioreceptor

assay: J. Besson *et al., Acta Endocrinol.* **87**, 799 (1978). Although the precise effect of VIP on neuronal function is not completely known, it is believed to play an important role in neurotransmission, *cf.* M. G. Bryant *et al., Lancet* **1**, 991 (1976); Y. Matsuzaki *et al., Science* **210**, 1252 (1980). Review of possible functions as a neural peptide: S. I. Said *et al., Advan. Biochem. Psychopharmacol.* **22**, 75-82 (1980). Book: *Vasoactive Intestinal Peptide,* S. I. Said, Ed. (Raven Press, New York, 1982) 528 pp.

```
 1
His–Ser–Asp–Ala–Val–Phe–Thr–Asp–Asn–Tyr–Thr–Arg–Leu–Arg
                                                         |
 28
Asn–Leu–Ile–Ser–Asn–Leu–Tyr–Lys–Lys–Val–Ala–Met–Gln–Lys
```

Porcine VIP

10141. Viquidil. *(3R-cis)-3-(3-Ethenyl-4-piperidinyl)-1-(6-methoxy-4-quinolinyl)-1-propanone; quinicine;* 1-(6-methoxy-4-quinolyl)-3-(3-vinyl-4-piperidyl)-1-propanone; chinicine; mequiverine; quinotoxine; quinotoxol; LM-192. $C_{20}H_{24}N_2O_2$; mol wt 324.42. C 74.05%, H 7.46%, N 8.63%, O 9.86%. An isomer of quinine; occurs naturally as the *d*-form. Present in small quantities in cinchona barks; formed by heating quinine with glycerol at 180°: Howard, *J. Chem. Soc.* **24**, 61 (1871); **25**, 101 (1872); Miller, Rohde, *Ber.* **33**, 3214 (1900); Howard, Chick, *Pharm. J.* **99**, 143 (1917). Conversion to quinine: Rabe, Kindler, *Ber.* **51**, 466 (1918). Partial synthesis: Prostenik, Prelog, *Helv. Chim. Acta* **26**, 1965 (1943). Total synthesis: Woodward, Doering, *J. Am. Chem. Soc.* **66**, 849 (1944); **67**, 860 (1945); U.S. pat. **2,500,-444** (1950 to Polaroid); Grethe *et al., Helv. Chim. Acta* **56**, 1485 (1973). Review of synthesis, chemistry and pharmacology: Quevauviller *et al., Ann. Pharm. Fr.* **24**, 39 (1966). Metabolic studies: Guérémy *et al., Arzneimittel-Forsch.* **22**, 1336 (1972); Uzan *et al., ibid.* 1341.

Yellow viscous oil, $[\alpha]_D$ +43°. Slightly sol in water; freely sol in alcohol, chloroform, ether.

Hydrochloride, $C_{20}H_{24}N_2O_2\cdot HCl$, *Desclidium, Permiran.* Yellow, odorless and bitter tasting powder, mp 184 ±4°. uv max (chloroform): 246, 355 nm. Sol in alc; sparingly sol in water. Practically insol in acetone.

THERAP CAT: Vasodilator; antiarrhythmic.

10142. Virginiamycin. Staphylomycin; virgimycin; antibiotic no. 899; SKF-7988; Eskalin V; Stafac; Staphylomycine. Antibiotic complex produced by a *Streptomyces* related to *S. virginiae:* Somer, Van Dijck, *Antibiot. & Chemother.* **5**, 632 (1955). A mixture of two principal components, virginiamycin M_1 (factor M_1) and virginiamycin S_1 (factor S). Separation of factors M_1 and S: Vanderhaeghe *et al., ibid.* **7**, 606 (1957); of the complex into six components: Gosselinckx, Parmentier, Chomatog. Sym., *2nd Brussels* **1962**, 181; of factor S into three components: Vanderhaeghe *et al., Tetrahedron Letters* **1971**, 2687. The product of commerce contains ~75% of fraction M_1, 5% of fraction S. Structure of virginiamycin S_1: Vanderhaeghe, Parmentier, *J. Am. Chem. Soc.* **82**, 4414 (1960); of virginiamycin M_1: Delpierre *et al., Tetrahedron Letters* **1966**, 369; *eidem, J. Chem. Soc. (C)* **1966**, 1653; Kingston *et al., ibid.* 1669. Nomenclature: Crooy, De Neys, *J. Antibiot.* **25**, 371 (1972). Prepn of derivatives of virginiamycin S_1: G. Janssen *et al., J. Antibiot.* **30**, 141 (1977). Biosynthesis of virginiamycin M_1: D. G. I. Kingston, M. X. Kolpak, *J. Am. Chem. Soc.* **102**, 5964 (1980).

Virginiamycin S₁

Virginiamycin M₁

Antibiotic complex, amorphous, white powder. Dec 138-140°. $[\alpha]_D^{20}$ −124° (c = 0.5 in ethanol). Sparingly sol in water and dil acid. Dissolves in aq alkali (above pH 9.5) with rapid inactivation. Sol in methanol, ethanol, acetone, ethyl acetate, chloroform, benzene. Practically insol in ligroin.

Virginiamycin S₁, $C_{43}H_{49}N_7O_{10}$, *staphylomycin S*. Crystals from methanol, mp 240-242°. $[\alpha]_D^{20}$ −28° (c = 1 in ethanol). uv max (ethanol): 305 nm (log ε 3.85). Approx solys (percent): ether 0.1; methanol 0.5; ethanol 1.5; benzene 2.5; acetone and ethyl acetate 3; dioxane 4. Very sol in chloroform, DMF. Practically insol in water, petr ether.

Virginiamycin M₁, $C_{28}H_{35}N_3O_7$, *mikamycin A, ostreogrycin A, pristinamycin II_A, staphylomycin M₁, streptogramin A, vernamycin A*. Tan powder from acetone + petr ether, dec 165-167°. $[\alpha]_D$ −190° ± 2° (c = 0.5 in ethanol). uv max (methanol): 216 nm ($E_{1cm}^{1\%}$ 582). Approx solys (percent): ether, 0.1; benzene 0.3; ethyl acetate 0.5; acetone 2; methanol and ethanol 4; dioxane and tetrahydrofuran 5. Very sol in chloroform, DMF. Practically insol in water, petr ether.

THERAP CAT: Antibacterial.

THERAP CAT (VET): Antibacterial; food additive.

10143. **Viridicatin.** *3-Hydroxy-4-phenyl-2(1H)-quinolin-one; 3-hydroxy-4-phenylcarbostyril;* 2,3-dihydroxy-4-phenyl-quinoline. $C_{15}H_{11}NO_2$; mol wt 237.26. C 75.94%, H 4.67%, N 5.90%, O 13.49%. Antibiotic substance from the mycelium of *Penicillium viridicatum* Westling, the chief mold on stored corn: Cunningham, Freeman, *Biochem. J.* **53**, 328 (1953); from various strains of *P. cyclopium* Westling: Bracken *et al., ibid.* **57**, 587 (1954); from *P. puberulum* Bainier: Austin, Meyers, *J. Chem. Soc.* **1964**, 1197. Biosynthesis: Luckner, *Tetrahedron Letters* **1962**, 1035; Luckner, Mothes, *Arch. Pharm.* **296**, 18 (1963); Framm *et al., Eur. J. Biochem.* **37**, 78 (1973). Synthesis: Eistert, Selzer, *Z. Naturforsch.* **17b**, 202 (1962). Mass spectra: Luckner *et al., Tetrahedron* **25**, 2575 (1969). Antibacterial activity and inhibitory effect on plant growth: Taniguchi, Satomura, *Agr. Biol. Chem.* **34**, 506 (1970).

Lustrous needles from methanol or ethanol, mp 268°. Sublimes unchanged in high vacuum at 160-170°. Very weak acid. uv and ir curves: Cunningham, Freeman, *loc. cit.* Practically insol in water, aq NaHCO₃. Sparingly sol in cold organic solvents, cold concd HCl, dil mineral acids; sol in cold, aq 2N KOH, glacial acetic acid.

Sodium salt, $C_{15}H_{10}NNaO_2$, prismatic needles from aq NaOH, dec 260-265°. Sol in water.

Monoacetylviridicatin, $C_{17}H_{13}NO_3$, crystals from aq ethanol, mp 200-201°. Dec slowly on standing.

O,O-Dimethylviridicatin, $C_{17}H_{15}NO_2$, plates from ethanol, mp 86-87°.

O,N-Dimethylviridicatin, $C_{17}H_{15}NO_2$, plates from ethanol, mp 197-198°.

10144. **Viridin.** *1β-Hydroxy-2β-methoxy-18-norandrosta-5,8,11,13-tetraeno[6,5,4-bc]furan-3,7,17-trione.* $C_{20}H_{16}O_6$; mol wt 352.34. C 68.18%, H 4.58%, O 27.25%. Antibiotic substance from *Gliocladium virens:* Grove *et al., J. Chem. Soc.* **1965**, 3803. Previously reported isolation from *Trichoderma viride:* Brian, McGowan, *Nature* **156**, 144 (1945); Brian *et al., Ann. Appl. Biol.* **33**, 190 (1946). Epimerizes to β-viridin where methoxyl group is equatorial. Separation of the two isomers: Vischer *et al., Nature* **165**, 528 (1950). Structure: Grove *et al., Chem. Commun.* **1965**, 343. Configuration: *eidem, J. Chem. Soc. (C)* **1966**, 743; Neidle, Hursthouse, *J. Chem. Soc. Perkin Trans. II* **1972**, 760. Biosynthesis: Blight *et al., Chem. Commun.* **1968**, 1117; Grove, *J. Chem. Soc. (C)* **1969**, 549.

Prisms from benzene, mp 245° (dec). $[\alpha]_D^{19}$ −224°. uv max: 242, 300 nm (log ε 4.49, 4.22). Needles from acetone, mp 222-224° (dec). Prisms from glacial acetic acid, mp 200-205° (dec). Sol in water, chloroform; sparingly sol in carbon disulfide, carbon tetrachloride. Practically insol in ether, camphor. Aq solns lose activity rapidly unless acidified to pH 3.

β-Isomer, prisms from benzene, mp 240-245° (dec). $[\alpha]_D^{16}$ −23°. uv max: 243, 300 nm (log ε 4.45, 4.25).

THERAP CAT: Antifungal.

10145. **Viscose.** A viscous orange-red aq soln of sodium cellulose xanthogenate obtained by dissolving wood pulp cellulose in sodium hydroxide soln and treating with carbon disulfide. Manuf: Bachlott, U.S. pat. **2,855,321** (1958 to du Pont); Von Kohorn, U.S. pat. **2,985,647** (1961).

USE: Intermediate in the manuf of rayon and cellophane.

10146. **Viscosin.** $C_{54}H_{95}N_9O_{16}$; mol wt 1126.40. C 57.58%, H 8.50%, N 11.19%, O 22.73%. Antibiotic substance produced by *Pseudomonas viscosa.* Isoln: Kochi *et al., Bact. Proc.* **1951**, 29; Groupé *et al., Proc. Soc. Exptl. Biol. Med.* **78**, 354 (1951); Ohno *et al., J. Agr. Chem. Soc. Japan* **27**, 665 (1953); *C.A.* **49**, 3012 (1955); Toki, Ohno, *C.A.* **53**, 243b (1959). Synthesis studies: Fukuda, Bani, *C.A.* **56**, 10265i (1962). Structure: Toki, Ohno, *loc. cit.* Revised structure: Hiramoto *et al., Tetrahedron Letters* **1970**, 1087;

eidem, Chem. Pharm. Bull. **19**, 1308, 1315 (1971). Shows antitubercular activity.

D-CH₃(CH₂)₆CHCH₂C—Leu—D-Glu—D-allo-Thr—D-Val—Leu—D-Ser
| |
OH O   └─Ile—D-Ser—Leu─┘

Powder, dec 270-273°. $[\alpha]_D^{29}$ —168.3° (ethanol). Stable to heat. Practically insol in water. Sol in methanol, ethanol, ether, acetone, alkaline phosphate buffer solns.

10147. Visnadine. *[9R-[9α(R*),10α]]-2-Methylbutanoic acid 10-(acetyloxy)-9,10-dihydro-8,8-dimethyl-2-oxo-2H,8H-benzo[1,2-b:3,4-b']dipyran-9-yl ester; 2-Methylbutyric acid 9-ester with 9,10-dihydro-9,10-dihydroxy-8,8-dimethyl-2H,8H-benzo[1,2-b:3,4-b']dipyran-2-one acetate;* 8,8-dimethyl-2-oxo-9,10-dihydro-2H,8H-benzo[1,2-b:3,4-b']dipyran-9,10-diyl-10-acetate-9-(α-methylbutyrate); 3,4,5-trihydroxy-2,2-dimethyl-6-chromanacrylic acid δ-lactone 4-acetate 3-(2-methylbutyrate); 4'-acetoxy-3'-(α-methylbutyryloxy)-2',2'-dimethyldihydropyrano(7,8:6',5')coumarin; 3-(α-methylbutyryloxy)-4-acetoxy-3,4-dihydroseseline; Cardine; Carduben; Vibeline; Visnamine. $C_{21}H_{24}O_7$; mol wt 388.42. C 64.94%, H 6.23%, O 28.83%. Isoln from seeds of *Ammi visnaga* L., *Umbelliferae* (Bishop's weed): Smith *et al., Science* **115**, 520 (1952); *eidem, J. Am. Chem. Soc.* **79**, 3534 (1957); Smith, Haber; Smith, Hosansky, U.S. pats. **2,816,118; 2,980,699** (1957, 1961 both to Penick); Le Men, U.S. pat. **2,995,574** (1961 to Lab. Roger Bellon); Baddar *et al., J. Chem. Soc.* **1963**, 4522. Synthesis of (±)-form: Shanbhag *et al., Tetrahedron* **21**, 3591 (1965). Pharmacology and toxicology: Nkondi *et al., Therapie* **21**, 1267 (1966); Erbring *et al., Arzneimittel-Forsch.* **17**, 283 (1967); Eyraud, Aurousseau, *ibid.* **23**, 201 (1973).

Needles from light petr, or ether + hexane, mp 85-88°. $[\alpha]_D^{20}$ +9.2° (alc), $[\alpha]_D^{30}$ +42.5° (c = 2 in dioxane). Slightly sol in water; quite sol in ethanol, methanol. Very sol in chloroform, acetone, ether, benzene, DMF. LD₅₀ in mice (mg/kg): 2240 orally; > 370 s.c. (Erbring).

(±)-Form, crystals from petr ether, mp 150-152° (Shanbhag).

THERAP CAT: Vasodilator (coronary).

10148. Visnagin. *4-Methoxy-7-methyl-5H-furo[3,2-g][1]-benzopyran-5-one;* 5-methoxy-2-methylfuranochromone; Visnacorin. $C_{13}H_{10}O_4$; mol wt 230.22. C 67.82%, H 4.38%, O 27.80%. A constituent of *Ammi visnaga* Lam., *Umbelliferae*. Isoln and structure: Späth, Gruber, *Ber.* **74**, 1492 (1941). Synthesis: Aneja *et al., Tetrahedron* **3**, 230 (1958); Badawi, Fayez, *Tetrahedron Letters* **1967**, 1029.

Thread-like needles from water. mp 142-145°. Very slightly sol in water. Sparingly sol in alc. Freely sol in chloroform.

10149. Vital Red. *3-Amino-4-[[4'-[(2-amino-6-sulfo-1-naphthalenyl)azo]-3,3'-dimethyl[1,1'-biphenyl]-4-yl]azo]-2,7-naphthalenedisulfonic acid trisodium salt; C.I. Direct Red 34;* C.I. 23570; brilliant vital red; Brilliant Congo R; Vital Red Evans; ditolyldiazo-3,6-disulfo-β-naphthylamine-β-naphthylamine-6-sulfonic acid sodium salt. $C_{34}H_{25}N_6Na_3O_9S_3$; mol wt 826.77. C 49.39%, H 3.05%, N 10.16%, Na 8.34%, O 17.42%, S 11.64%. Prepd by diazotization of o-tolidine and coupling the resulting tetrazotolidine first with 2-naphthylamine-3,6-disulfonic acid (amino R salt) and next with 2-naphthylamine-6-sulfonic acid (Brönner acid): Palkin, Evans, *J. Am. Chem. Soc.* **47**, 429 (1925); **Ger.** pat. **41,095;** *Frdl.* **1**, 476; *Colour Index* vol. **4** (3rd ed., 1971) p 4191.

Cherry-red crystals from water. Absorption max: 500 nm (aq soln). Sol in water giving a brownish-red soln. Gives blue solution with concd HCl. Slightly sol in alc.

USE: For blood volume determinations.

10150. Vitamin A. *Retinol; all-trans-retinol;* 3,7-dimethyl-9-(2,6,6-trimethyl-1-cyclohexen-1-yl)-2,4,6,8-nonatetraen-1-ol; anti-infective vitamin; antixerophthalmic vitamin; axerophthol; biosterol; lard-factor; oleovitamin A; ophthalamin (obsolete); vitamin A₁; vitamin A alcohol; Acon; Afaxin; Agiolan; Alphalin; Anatola; Aoral; Apexol; Apostavit; Atav; Avibon; Avita; Avitol; Axerol; "Dohyfral" A; Epiteliol; Nio-A-Let; Prepalin; Testavol; Vaflol; Vi-Alpha; Vitpex; Vogan; Vogan-Neu. $C_{20}H_{30}O$; mol wt 286.46. C 83.86%, H 10.56%, O 5.59%. Occurs in animals (not in plants); carotenoids are converted into vitamin A by the liver. Extracted from fish liver oils where it occurs mostly in esterified form. Structure: Karrer *et al., Helv. Chim. Acta* **14**, 1036, 1431; Karrer, Morf, *ibid.* **16**, 625 (1933); Heilbron *et al., Biochem. J.* **26**, 1178, 1194 (1932). Stereochemistry: *see* L. Zechmeister, *Chem. Rev.* **34**, 267 (1944). Identity of lard-factor and vitamin A: Ames, Harris, *Science* **120**, 391 (1954). Total synthesis of vitamin A from β-ionone and a propargyl halide: Eiter, Truscheit, U.S. pat. **3,060,229** (1962 to Bayer); by conversion of retinal: Wendler *et al., J. Am. Chem. Soc.* **72**, 234 (1950); Klein, Kapp, U.S. pat. **2,972,634** (1961 to Nopco); T. Mukaiyama, A. Ishida, *Chem. Letters* **1975**, 1201. Stereospecific synthesis of all *(E)*-form: P. S. Manchand *et al., Helv. Chim. Acta* **59**, 567 (1976); A. Fischli *et al., ibid.* 397; G. Cardillo *et al., J. Chem. Soc., Perkin Trans. I* **1979**, 1729. Toxicology: J. J. Kamm, *J. Am. Acad. Dermatol.* **6**, 652 (1982). Comprehensive monograph of the chemistry, physics, physiology of vitamin A and its provitamins: W. H. Sebrell, R. S. Harris, *The Vitamins,* Vol. I (Academic Press, New York, 2nd ed., 1967) 570 pp. Book: *The Retinoids,* Vol. 1-2, M. B. Sporn *et al.,* Eds. (Academic Press, New York, 1984). *See also:* Neovitamin A.

Yellow prisms from propylene oxide or petr ether. Solvated crystals from more polar solvents, such as methanol or ethyl formate. mp 62-64°. Distills at 120-125° at 5×10^{-3} mm pressure. n_D^{22} 1.6410 (calculated from refractive indexes of 20-70% solns in mineral oil). uv max: 324-325 nm ($E_{1cm}^{1\%}$ 1835): Baxter, Robeson, *J. Am. Chem. Soc.* **64**, 2407 (1942). Practically insol in water or glycerol. Sol in abs alcohol, methanol, chloroform, ether, fats and oils. Ultraviolet light inactivates vitamin A and its solns which exhibit a characteristic green fluorescence. The free alcohol is sensitive to air-oxidation, but oil solns of it are quite stable. Esters of vitamin A are more stable to oxidation. LD₅₀ (10 day) in mice (mg/kg): 1510 i.p.; 2570 orally (Kamm).

Acetate, $C_{22}H_{32}O_2$, pale yellow prismatic crystals from methanol, mp 57-58°. uv max (ethanol): 326 nm ($E_{1cm}^{1\%}$ 1550). Biopotency 2.904×10^6 I.U./g. LD₅₀ (10 day) orally in mice: 4100 mg/kg (Kamm).

Palmitate, $C_{36}H_{60}O_2$, *Arovit, Optovit-A,* is the ester preponderant in fish liver oils. Amorphous or cryst. mp 28-29°. uv max (ethanol): 325-328 nm ($E_{1cm}^{1\%}$ 975). Biopotency 1.817×10^6 I.U./g. LD$_{50}$ (10 day) in mice, rats (mg/kg): 6060, 7910 orally (Kamm).

Note: The U.S.P. unit of vitamin A is equal to 0.30 microgram of vitamin A equivalent to 0.344 microgram of vitamin A acetate. Observed biopotency 3.33×10^6 I.U./g.

THERAP CAT: Vitamin (antixerophthalmic).

THERAP CAT (VET): Nutritional factor.

10151. Vitamin A$_2$. *3,4-Didehydroretinol;* retinol$_2$; dehydroretinol. $C_{20}H_{28}O$; mol wt 284.44. C 84.45%, H 9.92%, O 5.62%. Combines with opsin to form the visual pigment, porphyropsin, *q.v.,* in fresh-water fish. A mixture of stereoisomers. Isoln from pike liver oils: Shantz, *Science* **108,** 417 (1948). *See also* Shantz, Brinkman, *J. Biol. Chem.* **183,** 467 (1950); Farrar *et al., J. Chem. Soc.* **1952,** 503. Synthesis and characteristics of stereoisomers: Schwieter, *Chimia* **14,** 362 (1960); *Helv. Chim. Acta* **45,** 517, 528, 541, 548 (1962). *Review: The Vitamins,* **vol. 1,** W. H. Sebrell, R. S. Harris, Eds. (Academic Press, New York, 2nd ed., 1967) *passim.*

Stereoisomeric mixture, golden yellow oil. Readily affected by oxygen. uv max (ethanol): 351, 287 nm ($E_{1cm}^{1\%}$ 1460, 820). Has about 40% of the bioactivity of cryst vitamin A$_1$. Stereoisomers: All-*trans*, mp 17-19°; 13-*cis*, mp 73-74°; 9-*cis*, mp 77-79°; 9,13-di-*cis*, mp $< -30°$.

10152. Vitamin B$_{12}$. Cyanocobalamin; 5,6-dimethylbenzimidazolyl cyanocobamide; cobinamide cyanide phosphate 3'-ester with 5,6-dimethyl-1-α-D-ribofuranosylbenzimidazole inner salt; LLD factor; *Lactobacillus lactis* Dorner factor; extrinsic factor; antipernicious anemia principle; Anacobin; Antipernicin; Bedodeka; Bedoz; Behepan; Berubi; Berubigen; Betalin-12; Betolvex; Cobalin; Crystamine; Cykobemin; Cytacon; Cytamen; Cytobion; Docémine; Docibin; Docigram; Docivit; Dodex; Fresmin; Macrabin; Millevit; Redisol; Rubesol; Rubramin PC; Sytobex; Vibalt; Vitarubin. $C_{63}H_{88}CoN_{14}O_{14}P$; mol wt 1355.38. C 55.83%, H 6.54%, Co 4.35%, N 14.47%, O 16.53%, P 2.29%. Prototype of the family of naturally occurring cobalt coordination compds known as *corrinoids.* Analogs of vitamin B$_{12}$ which differ only in the β-ligand of the cobalt are termed *cobalamins.* Synthesized almost exclusively by bacteria. Dietary sources include fish, meat, liver, and dairy products; plants have little or no cobalamins. Also found in soil and water, the richest sources being activated sewage sludge (*see* Milorganite) or manure. Converted by the body into its bioactive forms, methylcobalamin and cobamamide, *q.q.v.,* which serve as enzyme cofactors. Severe deficiency may result in megaloblastic anemia and/or neurological impairment. Isoln from mammalian liver: E. L. Rickes *et al., Science* **107,** 396 (1948); from cultures of *Streptomyces griseus*: *eidem, ibid.* **108,** 634 (1948); E. L. Rickes, T. R. Wood, U.S. pat. **2,563,794** (1951 Merck & Co.). Fermentation process using *Pseudomonas denitrificans*: R. A. Long, U.S. pat. **3,018,225** (1962 to Merck & Co.). Purification from sewage sludge: P. J. Van Melle, U.S. pat. **3,057,851** (1962 to Armour). Structure: D. C. Hodgkin *et al., Nature* **176,** 325 (1955); R. Bonnett *et al., ibid.* 328. X-ray structure analysis: D. C. Hodgkin, *Fortschr. Chem. Org. Naturst.* **15,** 167-227 (1958). Stereochemistry: Stora, *Bull. Soc. Chim. France* **1959,** 1421. Total synthesis: R. B. Woodward, *Pure Appl. Chem.* **33,** 145 (1973). Nomenclature: IUPAC rules, *ibid.* **48,** 497 (1976). Gastrointestinal absorption is dependent on Castle's intrinsic factor, *q.v.,* and *haptocorrin* (R-protein); transport of the absorbed vitamin is mediated by *transcobalamins*: B. Seetharam, D. H. Alpers, *Ann. Rev. Nutr.* **2,** 343-369 (1982); R. Gräsbeck, *Clin. Biochem.* **17,** 99-107 (1984). Role of Co—C bond in B$_{12}$-dependent reactions: J. M. Pratt, *Pure Appl. Chem.* **65,** 1513 (1993). Review of biosynthesis: A. I. Scott, *Angew. Chem. Int. Ed.* **32,** 1223-

1243 (1993). Comprehensive description: J. Kirschbaum in *Analytical Profiles of Drug Substances* vol. **10,** K. Florey, Ed. (Academic Press, New York, 1981) pp 183-288. Book: *B$_{12}$* **vols. 1 and 2,** D. Dolphin, Ed. (Wiley-Interscience, New York, 1982) 672 and 506 pp. *Reviews:* B. T. Golding in *Comprehensive Organic Chemistry* vol. **5,** E. Haslam, Ed. (Pergamon, New York, 1979) pp 549-584; "Vitamin B$_{12}$" in *Vitamins,* W. Friedrich, Ed. (de Gruyter, Berlin, 1988) pp 837-928.

Hygroscopic, dark red crystals. When exposed to air, may absorb ~12% water. The hydrated crystals are stable to air. Darkens at 210-220°. Not melted at 300°. $[\alpha]_{656}^{23}$ $-59° \pm 9°$ (dil aq soln). Absorption max (water): 278, 361, 550 nm ($A_{1cm}^{1\%}$ 115, 204, 64). Odorless and tasteless. One gram dissolves in ~80 ml water. Aq solns are neutral, maximum stability in the pH range 4.5-5. Solns in this pH range can be autoclaved for 20 min at 120°. Soluble in alc. Insol in acetone, CHCl$_3$, ether.

Zinc tannate complex, *cyanocobalamin zinc tannate, Depinar, Bevitam.* Long acting injectable vitamin B$_{12}$ prepn. Prepn: Thompson, U.S. pat. **2,920,015** (1960 to Armour). Comparative clinical study: A. Killander, I. Werner, *Acta Hematol.* **40,** 305 (1968). Practically insol in water. On reconstitution forms a colloidal suspension.

THERAP CAT: Vitamin (hematopoietic).

THERAP CAT (VET): Vitamin (hematopoietic).

10153. Vitamin B$_{12}$, Radioactive. Radioactive cyanocobalamin; radiocyanocobalamin. Prepn of ^{60}Co labelled compound: Chaiet *et al., Science* **111,** 601 (1950). Properties identical with unlabelled vitamin B$_{12}$, except for the presence of radioactive cobalt. Evaluation of dual-isotope Schilling test for pernicious anemia: L. S. Zuckier, L. R. Chervu, *J. Nucl. Med.* **25,** 1032 (1984).

^{60}Co Vitamin B$_{12}$, *Racobalamin-60, Rubratope-60,* T$_{1/2}$ 5.27 years, emitting beta and gamma rays.

^{58}Co Vitamin B$_{12}$, T$_{1/2}$ 70.8 days, emitting beta, gamma and x-rays.

^{57}Co Vitamin B$_{12}$, *Rubratope-57,* T$_{1/2}$ 271 days, emitting gamma and x-rays.

Combination of ^{57}Co and ^{58}Co labelled Vitamin B$_{12}$, *Dicopac.*

THERAP CAT: Diagnostic aid (pernicious anemia).

10154. Vitamin B$_{12r}$. *Vitamin B$_{12}$-Co(II);* cob(II)alamin; 5,6-dimethyl-1-(3-O-phosphono-α-D-ribofuranosyl)-1H-benzimidazole monoester with cobinamide-Co(II) inner salt. $C_{62}H_{88}CoN_{13}O_{14}P$; mol wt 1329.37. C 56.02%, H 6.67%, Co 4.43%, N 13.70%, O 16.85%, P 2.33%. A one-electron reduction product of vitamin B$_{12}$ containing a divalent cobalt atom. Undergoes a further one-electron reduction to vitamin B$_{12s}$. Prepn: H. Diehl, R. Murie, *Iowa State Coll. J. Sci.* **26,** 555 (1952); R. O. Brady, H. A. Barker, *Biochem.*

Biophys. Res. Commun. **4**, 373 (1961); H.-U. Blaser, J. Halpern, *J. Am. Chem. Soc.* **102**, 1684 (1980). Crystal structure: B. Kräutler *et al., ibid.* **111**, 8936 (1989). Valence of reduced forms of B$_{12}$: H. A. O. Hill *et al., Chem. & Ind. (London)* **1964**, 197. Review of chemistry of B$_{12r}$, B$_{12s}$ and analogs: R. Bonnett, *Chem. Rev.* **63**, 573-605 (1963); of prepns: D. Dolphin, *Methods Enzymol.* **18**, (Pt. C), 34-52 (1971). Role in enzymatic reactions: J. M. Pratt, *Pure Appl. Chem.* **65**, 1513 (1993).

Dark brown amorphous solid. Oxidized by air to aquocobalamin. Absorption max (H$_2$O): 474, 404, 312, 289, 260 nm (log ε 3.99, 3.88, 4.46, 4.29, 4.25).

Vitamin B$_{12s}$, C$_{62}$H$_{89}$CoN$_{13}$O$_{14}$P, cobinamide-Co(I) dihydrogen phosphate (ester) 3'-ester with 5,6-dimethyl-1-α-D-ribofuranosyl-1H-benzimidazole, cob(I)alamin, hydridocobalamin. Obtained as a gray-green product in solution.

10155. Vitamin D$_1$. C$_{56}$H$_{88}$O$_2$; mol wt 793.31. C 84.79%, H 11.18%, O 4.03%. C$_{28}$H$_{44}$O.C$_{28}$H$_{44}$O. A 1:1 molecular compd of lumisterol and vitamin D$_2$. Prepn: Windaus *et al., Ann.* **489**, 252 (1931); **493**, 259 (1932).

Crystals from acetone, mp 124-125°. Sublimes in high vacuum at 135°. Dec 180°. Soly at 16° (g/ml): 0.037 of acetone; 0.024 of petr ether (bp 35-40°); 0.020 of methanol. uv max (0.02% sol in petr ether): 265 nm (ε 1.56). [α]$_D^{20}$ +140.5° (8.9 mg in 2 ml acetone), +140.5° (ethanol), +127° (chloroform).

10156. Vitamin D$_2$. *(3β,5Z,7E,22E)-9,10-Secoergosta-5,-7,10(19),22-tetraen-3-ol;* calciferol; ergocalciferol; oleovitamin D$_2$; activated ergosterol; viosterol; Condol; Decaps; Dee-Ron; Deltalin; De-Rat Concentrate; Deratol; Detalup; Diactol; Divit Urto; Drisdol; D-Tracetten; Ergorone; Ertron; Fortodyl; Hi-Deratol; Infron; Metadee; Mina D$_2$; Mulsiferol; Mykostin; Ostelin; Radiostol; Radsterin; Shock-Ferol; Sterogyl; Uvesterol-D; Vio-D. C$_{28}$H$_{44}$O; mol wt 396.66. C 84.79%, H 11.18%, O 4.03%. The synthetic form of vitamin D. Prepd from ergosterol by uv irradiation in a suitable solvent: Askew *et al., Proc. Roy. Soc. London* **B109**, 488 (1931/32). The best wavelengths for production of vitamin D$_2$ seem to be from 275 to 300 nm: Knudson, Benford, *J. Biol. Chem.* **124**, 287 (1938); Bunker *et al., J. Am. Chem. Soc.* **62**, 508 (1940). Prepn by electron bombardment of ergosterol using longer waves: **Austrian pat. 119,-210.** Some U.S. pats. covering the manuf of vitamin D$_2$ are **1,902,785** (1933); **2,030,792** (1936); also **1,680,818** (1928) and **1,871,136** (1932). Direct total synthesis: B. Lythgoe *et al., Tetrahedron Letters* **1977**, 3685. Discussion of structural problems: Papers by Askew and Windaus *et al.,* beginning in 1930. Stereochemistry: Crowfoot, Dunitz, *Nature* **162**, 608 (1948); *Chem. & Ind. (London)* **1957**, 1149. General review: Inhoffen, *Angew. Chem.* **72**, 875 (1960). A polar, biologically active metabolite of vitamin D$_2$, *25-hydroxyergocalciferol,* which is ~1.5 times more active in curing rickets in rats, has been isolated from pig plasma. *See* Suda *et al., Biochem. Biophys. Res. Commun.* **35**, 182 (1969); *eidem, Biochemistry* **8**, 3515 (1969).

Prisms from acetone, mp 115-118°. Sublimes in very high vacuum (0.0006 mm) without dec. [α]$_D^{25}$ +82.6° (c = 3 in acetone); [α]$_D^{20}$ +102.5° (alcohol); [α]$_D^{20}$ +52° (CHCl$_3$). uv max (hexane): 264.5 nm (E$_{1cm}^{1\%}$ 458.9 ± 7.5). Not precipitated by digitonin (diff from ergosterol). Insol in water. Sol in the usual organic solvents. 1 ml acetone dissolves 0.0695 g

at 7°. Slightly sol in vegetable oils. Commercial solns are usually made with propylene glycol or sesame oil. Vitamin D$_2$ crystals have a potency of 40 units of vitamin D (U.S.P.) per μg. The behavior in storage is described under vitamin D$_3$. *See also* Huber, Barlow, *J. Biol. Chem.* **149**, 125 (1943) for stability information.

USE: Rodenticide.

THERAP CAT: Vitamin (antirachitic).

THERAP CAT (VET): Nutritional factor (antirachitic). Low activity in poultry.

10157. Vitamin D$_3$. *(3β,5Z,7E)-9,10-Secocholesta-5,7,10-(19)-trien-3-ol;* activated 7-dehydrocholesterol; oleovitamin D$_3$; cholecalciferol; colecalciferol; CC; Duphafral D$_3$ 1000; Delsterol; Deparal; Ebivit; Micro-Dee; Neo Dohyfral D$_3$; Provitina; Ricketon; Trivitan; D$_3$-Vicotrat; Vi-De-3-hydrosol; Vigantol; Vigorsan. C$_{27}$H$_{44}$O; mol wt 384.65. C 84.31%, H 11.53%, O 4.16%. The vitamin that mediates intestinal calcium absorption, bone calcium metabolism and probably, muscle activity. It usually acts as a hormone precursor as it requires two stages of metabolism, first to 25-hydroxycholecalciferol, *q.v.,* and then to 1α,25-dihydroxycholecalciferol, *q.v.,* before reaching actual hormonal form. Occurs in and is isolated from fish liver oils. Methods of sepn include chromatography, molecular distillation, esterification and fractionation of the esters. Prepd by irradiation of its provitamin 7-dehydrocholesterol, *q.v.:* Windaus *et al., Z. Physiol. Chem.* **241**, 100 (1936); Windaus *et al., Ann.* **533**, 118 (1938); Akhtar, Gibbons, *Tetrahedron Letters* **1965**, 509. Direct total synthesis: B. Lythgoe *et al., Tetrahedron Letters* **1977**, 3685. Laser photochemical production: V. Malatesta *et al., J. Am. Chem. Soc.* **103**, 6781 (1981). General review: Inhoffen, *Angew. Chem.* **72**, 875-881 (1960). Review of metabolism: Haussler, Rasmussen, *J. Biol. Chem.* **247**, 2328-2335 (1972); *Nature* **245**, 180-182 (1973).

Fine needles from dilute acetone, mp 84-85°. [α]$_D^{20}$ +84.8° (c = 1.6 in acetone); [α]$_D^{20}$ +51.9° (c = 1.6 in chloroform). uv max (alcohol or hexane): 264.5 nm. (E$_{1cm}^{1\%}$ 450-490): Huber *et al., J. Am. Chem. Soc.* **67**, 609 (1945). Not precipitated by digitonin (diff from 7-dehydrocholesterol). Practically insol in water; sol in the usual organic solvents; slightly sol in vegetable oils. Oxidized and inactivated by moist air within a few days. Deterioration of pure cryst vitamin D$_3$ is negligible after storage of 1 year in amber evacuated ampuls at refrigerator temps; vitamin D$_2$ may be kept for 9 months under the same conditions. Additional stability information: Huber, Barlow, *J. Biol. Chem.* **149**, 125 (1942). Generally vitamin D$_3$ is considered more stable than vitamin D$_2$.

Note: Vitamin D$_3$ is approx as effective as vitamin D$_2$ in the human and in the rat. It is also fully active in chicks. Vitamin D$_2$ is only 1-2 percent as potent for the chick as vitamin D$_3$. Because of this difference it is important that poultry feeds are supplemented with vitamin D$_3$ rather than D$_2$. One unit (U.S.P. or international) is defined as the activity of 0.025 γ of vitamin D$_3$ contained in the U.S.P. vitamin D reference standard.

THERAP CAT: Vitamin (antirachitic).

THERAP CAT (VET): Nutritional factor (antirachitic).

10158. Vitamin D$_4$. *(3β,5Z,7E)-9,10-Secoergosta-5,7,10-(19)-trien-3-ol;* 22:23-dihydrovitamin D$_2$; 22,23-dihydroergocalciferol. C$_{28}$H$_{46}$O; mol wt 398.67. C 84.36%, H

11.63%, O 4.01%. Prepd from 22:23-dihydroergosterol by irradiation with light of the magnesium arc: Windaus, Trautmann, *Z. Physiol. Chem.* **247**, 185 (1937). Synthesis: P. J. Kocienski *et al., J. Chem. Soc., Perkin Trans. I* **1979**, 1290.

Platelets from dil acetone, mp 96-98°. Originally given as mp 107-108°, *see* Windaus, Guntzel, *Ann.* **538**, 122 (1939). $[\alpha]_D^{18}$ +89.3° (c = 0.47 in acetone). uv max: 265 nm. Not precipitated by digitonin. Practically insol in water. Sol in the usual organic solvents except petr ether; slightly sol in vegetable oils.

10159. Vitamin E. *[2R-2R*(4R*,8R*)]-3,4-Dihydro-2,-5,7,8-tetramethyl-2-(4,8,12-trimethyltridecyl)-2H-1-benzo-pyran-6-ol; 2,5,7,8-tetramethyl-2-(4',8',12'-trimethyltridecyl)-6-chromanol;* α-tocopherol; (+)-α-tocopherol; 5,7,8-tri-methyltocol; antisterility vitamin. $C_{29}H_{50}O_2$; mol wt 430.71. C 80.87%, H 11.70%, O 7.43%. Found largely in plant materials. Present in highest concns (0.1-0.3%) in wheat germ, corn, sunflower seed, rapeseed, soybean oils, alfalfa and lettuce. Natural α-tocopherol is usually found with β-and γ-tocopherols, *q.q.v.* Isoln from wheat germ: Evans *et al., J. Biol. Chem.* **113**, 319 (1936). Structure: Fernholz, *J. Am. Chem. Soc.* **59**, 1154 (1937); **60**, 700 (1938). Synthesis: Karrer *et al., Helv. Chim. Acta* **21**, 520, 820 (1938); Bergel *et al., J. Chem. Soc.* **1938**, 1382; Smith *et al., Science* **88**, 37 (1938); Smith, Sprung, *J. Am. Chem. Soc.* **65**, 1276 (1943). Recent syntheses: N. Cohen *et al., Helv. Chim. Acta* **61**, 837 (1978); *eidem, J. Am. Chem. Soc.* **101**, 6710 (1979); R. Barn-er, M. Schmid, *Helv. Chim. Acta* **62**, 2384 (1979). Abs config of natural α-tocopherol: Mayer *et al., ibid.* **46**, 963 (1963). Stereoselective synthesis of the side chain: C. H. Heathcock, E. T. Jarvi, *Tetrahedron Letters* **23**, 2825 (1982). Review of industrial processes: Rubel, *Vitamin E Manufacture* (Noyes Dev. Corp., Park Ridge, N.J., 1969). *Reviews: The Vitamins* Vol. **5**, W. H. Sebrell, R. S. Harris, Eds. (Academic Press, New York, 1972) pp 165-317; J. M. Bieri, P. M. Farrell, *Vitam. Horm. (New York)* **34**, 31-75 (1976). Book: *Ann. N.Y. Acad. Sci.* **393**, entitled "Vitamin E: Biochemical, Hematological and Clinical Aspects", B. Lubin, L. J. Machlin, Eds. (1982) 506 pp. Review of medical uses: J. G. Bieri *et al., N. Engl. J. Med.* **308**, 1063-1071 (1983).

$[\alpha]_{5461}^{25}$ −3.0° (benzene); $[\alpha]_{5461}^{25}$ +0.32° (alc).
Succinate, *Vitamin E acid succinate.* Prepn: Demole *et al., Helv. Chim. Acta* **22**, 65 (1939); McArthur, Watson, *Can. Chem. Process Inds.* **23**, 350 (1939); Baxter *et al., J. Am. Chem. Soc.* **65**, 918 (1943). Needles from petr ether, mp 76-77°. uv max (ethanol): 286 nm ($E_{1cm}^{1\%}$ 38.5). Practically insol in water.
Nicotinate, $C_{35}H_{53}NO_3$, *Hijuven, Juvela Nicotinate, Ren-ascin.*
*dl-*Form, slightly viscous, pale yellow oil. Natural α-tocopherol has been crystallized, mp 2.5°-3.5°. d_4^{25} 0.950;

bp$_{0.1}$ 200-220°; n_D^{25} 1.5045. uv max: 294 nm ($E_{1cm}^{1\%}$ 71). Practically insol in water. Freely sol in oils, fats, acetone, alcohol, chloroform, ether, other fat solvents. Stable to heat and alkalies in the absence of oxygen. Not affected by acids up to 100°. Slowly oxidized by atm oxygen, rapidly by ferric and silver salts. Gradually darkens on exposure to light.
USE: As an antioxidant in vegetable oils and shortening.
THERAP CAT: Treatment of vitamin E deficiency.
THERAP CAT (VET): Nutritional factor. Interrelationship with selenium. (Prevents muscle degeneration, also encephalomalaria and exudative diathesis.) Has been used to promote fertility.

10160. Vitamin E Acetate. *[2R*(4R*,8R*)]-3,4-Dihy-dro-2,5,7,8-tetramethyl-2-(4,8,12-trimethyltridecyl)-2H-1-ben-zopyran-6-ol acetate; 2,5,7,8-tetramethyl-2-(4,8,12-trimethyl-tridecyl)-6-chromanol acetate;* α-tocopherol acetate; α-to-copheryl acetate; $C_{31}H_{52}O_3$; mol wt 472.75. C 78.76%, H 11.09%, O 10.15%. Prepn from *dl-*α-tocopherol and acetic anhydride: Surmatis, Weber, U.S. pat. **2,723,278** (1955 to Hoffmann-La Roche). Prepn of *d-* and *l-*forms: Robeson, Nelan, *J. Am. Chem. Soc.* **84**, 3196 (1962). Stereoselective synthesis: K.-K. Chan *et al., J. Org. Chem.* **43**, 3435 (1978). Total synthesis of all eight stereoisomers: N. Cohen *et al., Helv. Chim. Acta* **64**, 1158 (1981). Comprehensive description: B. C. Rudy, B. Z. Senkowski, *Anal. Profiles Drug Subs.* **3**, 111-126 (1974).

*dl-*Form, *Detulin, Ephynal, Eprolin, Epsilan-M, Eusovit, E-Vimin, Evion, Juvela, OptoVit-E, Toco 500, Vitagutt.* Pale yellow, viscous liquid. mp −27.5°. $d_4^{21.3}$ 0.9533. bp$_{0.01}$ 184°; bp$_{0.025}$ 194°; bp$_{0.3}$ 224°. n_D^{20} 1.4950-1.4972. uv max (cyclo-hexane): 285.5 nm. Practically insol in water. Freely sol in acetone, chloroform, ether. Less readily sol in alc. Unlike the free vitamins, the acetate is practically unaffected by the oxidizing influence of air, light, and ultraviolet light.
*d-*Form, *[2R-[2R*(4R*,8R*)]]-3,4-dihydro-2,5,7,8-tetra-methyl-2-(4,8,12-trimethyltridecyl)-2H-1-benzopyran-6-ol ace-tate, E-Vicotrat, Spondyvit, Tocopherex.* Crystals, mp 28°. $[\alpha]_D^{25}$ +0.25° (c = 10 in chloroform); $[\alpha]_D^{25}$ +3.2° (in ethanol).
*l-*Form, crystals, mp 23°. $[\alpha]_D^{25}$ −2.0° (in ethanol).
Note: The international unit of vitamin E is equal to one mg of standard *dl-*α-tocopheryl acetate. The *d-*form is more active: 1 mg = 1.36 I.U. *l-*α-Tocopheryl acetate has 42% of the activity of *d-*α-tocopheryl acetate in the rat hemolysis test. Based on this activity a potency ratio of 1.4:1.0 for *d-*α-tocopheryl acetate compared to *dl-*α-tocopheryl acetate has been established.
THERAP CAT: Vitamin.
THERAP CAT (VET): Vitamin.

10161. Vitamin K. General term referring to a group of naphthoquinone derivatives required for the bioactivation of proteins involved in hemostasis. The designation "K" was derived from the German "Koagulationsvitamin." Vitamin K compds are classified into 3 groups: *phylloquinone* (K_1), *q.v.,* found in green plants; *menaquinones* (K_2), *q.v.,* prima-rily produced by intestinal bacteria; and *menadione* (K_3), *q.v.,* and derivatives which are synthetic, lipid soluble com-pounds. Reduced *in vivo* to *dihydrovitamin K* (KH_2) which serves as a coenzyme in the conversion of glutamic acid residues to γ-carboxyglutamic acid (Gla), *q.v.,* in the post-translational modification of blood coagulation factors II, VII, IX and X, *q.q.v.,* and the anticoagulant proteins C and S. Other Gla-containing proteins, such as the bone matrix protein, osteocalcin, *q.v.,* have been identified in a wide vari-ety of tissues. This γ-carboxylation is accompanied by the oxidation of KH_2 to *vitamin K epoxide* which is then recy-cled back to vitamin K. Discovery: H. Dam, *Biochem. Z.* **215**, 475 (1929); **220**, 158 (1930); *Nature* **135**, 652 (1935). Historical survey: H. Dam, *Vit. Horm.* **24**, 295-306 (1966). Menadione and phylloquinone are metabolized by animals to menaquinone-4: C. Martius, H. O. Esser, *Biochem. Z.* **331**, 1 (1958); H. H. W. Thijssen, M. J. Drittij-Reijnders, *Brit. J. Nutr.* **72**, 415 (1994). Comprehensive review: W. Friedrich in *Vitamins* (de Gruyter, New York, 1988) pp 285-338. Review of metabolism and role in human nutrition: M. J. Shearer, *Blood Rev.* **6**, 92-104 (1992); J. W. Suttie, *J. Am. Diet. Assoc.* **92**, 585-590 (1992); of mechanism of action: P. Dowd *et al., Nat. Prod. Rep.* **11**, 251-264 (1994); of pharmacology and therapeutic use: J. A. Thorp *et al., Drugs* **49**, 376-387 (1995). Physiological review: M. J. Shearer, *Lancet* **345**, 229-234 (1995).

10162. Vitamin K₅. *4-Amino-2-methyl-1-naphthalenol; 4-amino-2-methyl-1-naphthol;* 1-hydroxy-2-methyl-4-amino-naphthalene; 2-methyl-4-amino-1-hydroxynaphthalene; 2-methyl-4-amino-1-naphthol; 3-methyl-4-hydroxy-1-naphthylamine; Synkamin. $C_{11}H_{11}NO$; mol wt 173.21. C 76.28%, H 6.40%, N 8.09%, O 9.24%. Prepd from 2-methylnaphthalene or from menadione: Sah, *Rec. Trav. Chim.* **59**, 458 (1940); **60**, 373 (1941); Veldstra, Wiardi, *ibid.* **61**, 547 (1942); **62**, 75 (1943). Use as food preservative: Yang, Steele, *Food Technol.* **12**, 501 (1958).

Hydrochloride, $C_{11}H_{11}NO \cdot HCl$, *Kayvisyn.* Needles from dil HCl, darkens at 262°, dec 280-282°. Turns pink to dark violet on exposure to air and light. Freely sol in water. Slightly sol in alc. Insol in ether.
N-Acetyl analog, $C_{13}H_{13}NO_2$, *4-acetamido-2-methyl-1-naphthol, K-Vitrat.* Needles, mp 208-210°.
USE: Food preservative.
THERAP CAT: Vitamin (prothrombogenic).

10163. Vitamin(s) L. Factors presumably necessary for lactation. Vitamin L₁: See *o-Aminobenzoic Acid*. Vitamin L₂ was shown to be 7-[tetrahydro-3,4-dihydroxy-5-(methylmercaptomethyl)-2-furyl]adenine: Nakahara *et al., Sci. Papers Inst. Phys. Chem. Research (Tokyo)* **40**, 433 (1943), *C.A.* **41**, 6317c (1947). Manuf: Sato *et al., Japan.* pats. 7244('62) and **16,015** ('69), *C.A.* **59**, 1746f (1963) and *C.A.* **77**, 102146j (1972).

Vitamin L₂

10164. Vitamin T. Tegotin; termitin; torutilin; factor T; vitamin T Goetsch; Goetsch's vitamin; Temina. A complex of growth-promoting substances, originally obtained from termites: Goetsch, *Oesterr. Zool. Z.* **1**, 58-85 (1946). Also obtainable from roaches, yeasts and fungi. The isoln procedure has never been described. Isoln of "vitamin T complexes" from yeast: Koch *et al., Ger.* pat. **1,000,962** (1957 to Aschaffenburger Zellstoffwerke), *C.A.* **54**, 18898d (1960). Vitamin T may be a mixture of known vitamins and growth-promoting factors: Koch *et al., Naturwiss.* **38**, 339 (1951). Polemic by Barkow, Goetsch, *ibid.* **42**, 346 (1955).

10165. Vitamin U. *(S)-(3-Amino-3-carboxypropyl)-dimethyl sulfonium chloride;* methylmethioninesulfonium chloride; MMSC; Cabagin-U; Epadyn-U; Vitas-U. C_6H_{14}-$ClNO_2S$; mol wt 199.70. C 36.09%, H 7.07%, Cl 17.75%, N

7.01%, O 16.02%, S 16.06%. The anti-ulcer vitamin reported in cabbage leaves and other green vegetables: G. Cheney, *California Med.* **77**, 248 (1952). Activity: V. Z. Szabo, G. Vargha, *Arzneimittel-Forsch.* **10**, 23 (1960); K. Seri *et al., ibid.* **29**, 1517 (1979). Prepn: Y. Kanai, Y. Kawamura, **Japan.** pat. 4757('62) (to Nippon Kayaku), *C.A.* **58**, 12672e (1963); H. Wagner, **Ger.** pat. **1,239,697** (1967 to Degussa), *C.A.* **67**, 73866e (1967).

Needles, mp 134° (dec). LD_{50} i.v. in mice: 2760 mg/kg, I. V. Zaikonnikova, L. G. Urazaeva, *C.A.* **80**, 10388y (1974). Bromide analog, $C_6H_{14}BrNO_2S$, *methylmethioninesulfonium bromide, Ardésyl.* mp 139° (dec).
THERAP CAT: Treatment of gastric disorders.

10166. Vitride®. *Dihydrobis(2-methoxyethanolato-O,-O')aluminate(1—) sodium;* sodium bis(2-methoxyethoxy)-aluminum hydride. $C_6H_{16}AlNaO_4$; mol wt 202.16. C 35.65%, H 7.98%, Al 13.35%, Na 11.37%, O 31.66%. NaAl-$H_2(OCH_2CH_2OCH_3)_2$. Prepn: Casensky *et al.*, **U.S.** pat. **3,507,895** (1970). Properties: Vit, *Eastman Organic Chemical Bulletin* **42**(3), 1 (1970).
Highly viscous liquid at room temp. Miscible in a great variety of inert solvents (e.g., aromatic hydrocarbons and ethers). Immiscible in paraffinic hydrocarbons. Thermally stable at 200°C; thermal decomposition starts at 205°C and is vigorous.
USE: Reducing agent.

10167. Vitronectin. Serum spreading factor; S-protein. Multifunctional adhesive glycoprotein found in serum and in various tissues. Has binding sites for integrins, collagen, heparin, complement components, and perforin, *q.q.v.* Thought to play a regulatory role in hemostasis, wound healing, tissue remodeling and cancer. Synthesized by hepatocytes as a single chain polypeptide, mol wt ~78 kDa. Binds readily to proteoglycans of the extracellular matrix and is deposited in normal loose connective tissues, in dermal elastin fibers, and in the vascular wall. Also synthesized by platelets, macrophages and megakaryocytes. Released by activated platelets in different multimeric forms. Promotes cellular attachment, spreading and migration of a wide variety of cell types and serves as a matrix-associated regulator of blood coagulation. Identification: R. Holmes, *J. Cell Biol.* **32**, 297 (1967). Isoln and comparison with fibronectin, *q.v.*: D. Barnes *et al., J. Supramol. Struct.* **14**, 47 (1980). Purification and tissue distribution: E. G. Hayman *et al., Proc. Nat. Acad. Sci. USA* **80**, 4003 (1983). Amino acid sequence: S. Suzuki *et al., EMBO J.* **4**, 2519 (1985). Identity with S-protein: K. T. Preissner *et al., Biochem. Biophys. Res. Commun.* **134**, 951 (1986); B. R. Tomasini, D. F. Mosher, *Blood* **68**, 737 (1986). Review of structure and biological function in hemostasis: K. T. Preissner, *Ann. Rev. Cell Biol.* **7**, 275-310 (1991); K. T. Preissner, D. Jenne, *Thromb. Haemost.* **66**, 123-132; 189-194 (1991). Review of vitronectin receptors and role in cell growth and differentiation: B. Felding-Habermann, D. A. Cheresh, *Curr. Opin. Cell Biol.* **5**, 864-868 (1993).

10168. Voacamine. *12-Methoxy-13-[(3α)-17-methoxy-17-oxovobasan-3-yl]ibogamine-18-carboxylic acid methyl ester;* voacangidine. $C_{43}H_{52}N_4O_5$; mol wt 704.91. C 73.27%, H 7.44%, N 7.95%, O 11.35%. Bisindole alkaloid found in *Voacanga africana* Stapf., *V. thouarsii* R. & Sch., var *obtusa* (K. Sch.) Pichon and *V. schweinfurthii* Staph., *Apocynaceae*: Janot, Goutarel, *Compt. Rend.* **240**, 1719 (1955); Rao, *J. Org. Chem.* **23**, 1455 (1958); Janot, Goutarel, **U.S.** pat. **2,823,204** (1958 to Lab. Gobey). Identity with voacanginine: LeBarre, Gillo, *Bull. Acad. Roy. Med. Belg.* **20**, 194 (1956). Cleaved by acid catalysis to voacangine: Winkler, *Naturwiss.* **48**, 694 (1961). Structure: Goutarel *et al., Compt. Rend.* **243**, 1670 (1956); Percheron, *Ann. Chim. (Paris)* **4**, 303 (1959); Büchi *et al., J. Am. Chem. Soc.* **85**,

1893 (1963). *Review:* Gorman *et al.* in *The Alkaloids* **1**, J. E. Saxton, Ed. (The Chem. Soc., London, 1971) pp 242-249.

voacangine

Prisms from acetone + methanol, dec 223°. $[\alpha]_D^{20}$ −52° (chloroform). uv max: 225, 295 nm (log ϵ 4.72, 4.28). Sol in chloroform, acetone; slightly sol in methanol, ethanol.

Voacangine, $C_{22}H_{28}N_2O_3$, *carbomethoxyibogaine, 12-methoxyibogamine-18-carboxylic acid methyl ester.* Structure: Bartlett *et al., J. Am. Chem. Soc.* **80**, 126 (1958). Prismatic needles from ethanol, mp 136-137°. Sublimes$_{0.01}$ 135°. $[\alpha]_D^{20}$ −42° (c = 1.26 in chloroform). pKa 7.4 (40% aq methanol); pKa 5.73 (33% DMF). uv max (methanol): 225, 287 300 nm (log ϵ 4.43, 3.97, 3.93). Freely sol in acetone, chloroform; slightly sol in methanol, ethanol.

10169. Vodka. Little water. Defined as ethyl alcohol obtained from grain or potatoes, filtered through charcoal, and diluted with distilled water. Manuf by "spirit flow" process: Jacobs, U.S. pat. **2,946,687** (1960 to Heublein).

10170. Vomicine. *4-Hydroxy-19-methyl-16,19-secostrychnidine-10,16-dione;* 12-hydroxy-*N*-methylstrychnine. $C_{22}H_{24}N_2O_4$; mol wt 380.44. C 69.46%, H 6.36%, N 7.36%, O 16.82%. From seed of *Strychnos nux vomica* L., *Loganiaceae:* Wieland, Oertel, *Ann.* **469**, 193 (1929). Structure: Huisgen *et al., ibid.* **573**, 121 (1951). Synthesis: Rosenmund, *Angew. Chem.* **75**, 1127 (1963). *Reviews:* R. Robinson in *Progress in Organic Chemistry* vol. I (Butterworths, London, 1952) pp 2-21; J. B. Hendrickson in *The Alkaloids* vol. VI, R. H. F. Manske, H. L. Holmes, Eds. (Academic Press, New York, 1960) pp 195-204.

Hexagonal prisms from acetone, mp 284°. $[\alpha]_D^{22}$ +80° (c = 0.5 in alc). Weak, mono-acidic base forming salts with an acid reaction. Freely sol in chloroform, sol in hot alcohol, acetone, slightly in ether, ethyl acetate.

Hydrochloride, $C_{22}H_{24}N_2O_4 \cdot HCl$, prisms from water, dec 245°. uv max (ethanol): 222, 263, 297 nm (log ϵ 4.27, 3.77, 3.52). Sparingly sol in water.

Methyl ether, $C_{23}H_{26}N_2O_4$, needles from alcohol, dec 290°. $[\alpha]_D^{20}$ +16° (c = 0.5 in alcohol).

10171. Vomitoxin. *12,13-Epoxy-3,7,15-trihydroxytrichothec-9-en-8-one;* deoxynivalenol; dehydronivalenol. $C_{15}H_{20}O_6$; mol wt 296.32. C 60.80%, H 6.80%, O 32.40%. Isoln of the trichothecene mycotoxin from *Fusarium roseum* and structure: N. Morooka *et al., J. Food Hyg. Soc. Japan* **13**, 368 (1972); T. Yoshizawa, N. Morooka, *Agr. Biol. Chem.* **37**, 2933 (1973). Isoln from *F. graminearum:* R. F. Vesonder *et al., Appl. Microbiol.* **26**, 1008 (1973); *eidem, Appl. Environ. Microbiol.* **31**, 280 (1976). Emetic and refusal activity in swine: D. M. Forsyth *et al., ibid.* **34**, 547 (1977). HPLC analysis: G. A. Bennett *et al., J. Am. Oil Chem. Soc.* **58**, 1002A (1981). Implicated as a chemical warfare agent with nivalenol, *q.v.* in Southeast Asia: N. Wade, *Science* **214**, 34 (1981).

Fine needles from ethyl acetate + petr ether, mp 151-153°. $[\alpha]_D^{25}$ +6.35° (c = 0.07 in ethanol). uv max (ethanol): 218 nm (ϵ 4500). LD$_{50}$ i.p. in male, female mice (mg/kg): 70.0, 76.7 (Yoshizawa, Morooka).

10172. VX. *Methylphosphonothioic acid S-[2-[bis(1-methylethyl)amino]ethyl] O-ethyl ester; O*-ethyl *S*-[2-(diisopropylamino)ethyl]methylphosphonothioate; Tx 60. $C_{11}H_{26}NO_2PS$; mol wt 267.37. C 49.41%, H 9.80%, N 5.24%, O 11.97%, P 11.58%, S 11.99%. Nerve gas. Prepn: R. V. Ley, G. L. Sainsbury, **Brit. pat. 1,346,409;** A. W. Wardrop, C. Stratford, **Brit. pat. 1,346,410** (both 1974 to U.K. Sec. of State for Defence), *C.A.* **81**, 4068y, 4069z (1974); S. R. Eckhaus *etal.*, U.S. pat. **3,911,059** (1975 to U.S.A. Sec. for Army). Activity studies in humans: F. R. Sidell, W. A. Groff, *Toxicol. Appl. Pharmacol.* **27**, 241 (1974); F. N. Craig *et al., J. Invest. Dermatol.* **68**, 357 (1977). Toxicity study: J. J. Gordon, L. Leadbeater, *Toxicol. Appl. Pharmacol.* **40**, 109 (1977).

Nonvolatile, odorless liquid. pKa' 7.9. LD$_{50}$ s.c. in rabbits: 15.4 μg/kg (Gordon, Leadbeater).
Caution: Cholinesterase inhibitor; more potent than sarin, *q.v.*

USE: Chemical warfare agent.

W

10173. Warburganal. *[1S-(1α,4aα,8aβ)]-1,4,4a,5,6,7,8,-*
8a-Octahydro-1-hydroxy-5,5,8a-trimethyl-1,2-naphthalene-
dicarboxaldehyde. $C_{15}H_{22}O_3$; mol wt 250.34. C 71.97%, H
8.86%, O 19.17%. Drimane sesquiterpene with antifeedant
activity against the African army worm. Biological activity
includes plant growth regulation, cytotoxic, antimicrobial
and molluscicidal properties. Isoln from *Warburgia ugan-*
densis, Canellaceae and structure: I. Kubo *et al., Chem.*
Commun. **1976**, 1013. Relationship between structure and
antifeedant activity: K. Nakanishi, I. Kubo, *Isr. J. Chem.*
16, 28 (1977). Synthesis of (±)-warburganal: S. P. Tanis,
K. Nakanishi, *J. Am. Chem. Soc.* **101**, 4398 (1979); T.
Nakata *et al., ibid.* 4400; A. S. Kende, T. J. Blacklock,
Tetrahedron Letters **1980**, 3119; P. A. Wender, S. L. Eck,
ibid. **1982**, 1871; D. M. Hollinshead *et al., J. Chem. Soc.*
Perkin Trans. I **1983**, 1579. *See also:* **Japan. Kokai** 80
136,238, and **80 136,240** (both 1980 to Inst. Phys. Chem.
Res.); **Japan. Kokai** 81 43,236 (1981 to Suntory Ltd.);
Japan. Kokai 83 38,232 (1983 to Teikoku Zoki). Synthesis
of (−)-warburganal: H. Okawara *et al., Tetrahedron Letters*
1982, 1087.

mp 98-99°. uv max (methanol): 224 nm (ε 6300). $[α]_D^{21}$
−260° (c = 0.350 in $CHCl_3$).

10174. Warfarin. *4-Hydroxy-3-(3-oxo-1-phenylbutyl)-*
2H-1-benzopyran-2-one; 3-(α-acetonylbenzyl)-4-hydroxy-
coumarin; 1-(4'-hydroxy-3'-coumarinyl)-1-phenyl-3-buta-
none; 3-α-phenyl-β-acetylethyl-4-hydroxycoumarin; compound 42; WARF compound 42; Co-Rax; Rodex. $C_{19}H_{16}$-
O_4; mol wt 308.33. C 74.01%, H 5.23%, O 20.76%. The
commercial product is the racemic mixture; the $S(−)$-form
is more active than the R-isomer. Prepd by the Michael
condensation of benzylidene-acetone with 4-hydroxy-
coumarin: Stahmann *et al., U.S. pat.* 2,427,578 (1947);
Schroeder, Link, *U.S. pat.* 2,765,321 (1956 to Wisconsin
Alumni Res. Found.); Link, *U.S. pat.* 2,777,859 (1957).
Resolution and abs configuration: West *et al., J. Am.*
Chem. Soc. **83**, 2676 (1961); Preis, *Dissertation Abstr.* **18**, 793
(1958); Preis *et al., U.S. pat.* 3,239,529 (1966 to Wisconsin
Alumni Res. Found.). Mechanism of action: Bell *et al.,*
Biochemistry **11**, 1959 (1972). Conformation in soln: E. J.
Valente *et al., J. Med. Chem.* **20**, 1849 (1977); **21**, 141, 231
(1978). Human metabolism: R. J. Lewis, W. F. Trager,
Ann. N.Y. Acad. Sci. **179**, 205 (1971). Stereospecific HPLC
determn in plasma: C. Banfield, M. Rowland, *J. Pharm.*
Sci. **72**, 921 (1983). Antimetastatic effect in lung cancer: L.
R. Zacharski *et al., Cancer* **53**, 2046 (1984); in rat adenocar-
cinoma: B. L. Neubauer *et al., J. Urol.* **135**, 163 (1986).
Toxicity studies: E. C. Hagan, J. L. Radomski, *J. Am.*
Pharm. Assoc.-Sci. Ed. **42**, 379 (1953); N. Back *et al., Phar-*
macol. Res. Commun. **10**, 445 (1978). Review of therapeutic
uses: J. V. Lloyd, *Med. J. Aust.* **142**, 197-201 (1985); and
pharmacology: J. Hirsh *et al., Chest* **102**, Suppl., 312S-326S

(1992). Comprehensive description: S. A. Babhair *et al.,*
Anal. Profiles Drug Subs. **14**, 423-452 (1985).
Crystals from alc, mp 161°. uv max (water, pH 10): 308
nm (ε 13610). Soluble in acetone, dioxane. Moderately sol
in methanol, ethanol, isopropanol, some oils. Freely sol in
alkaline aq solns (forms a water-soluble sodium salt). Practically insol in water, benzene, cyclohexane, Skellysolves A
and B. Warfarin has an acidic enol which forms metallic
salts and an acetate, mp 117-118°, and a ketone which forms
an oxime, mp 182-183° and a 2,4-dinitrophenylhydrazone,
mp 215-216°.
Sodium salt, $C_{19}H_{15}NaO_4$, *Coumadin, Marevan, Panwar-*
fin, Prothromadin, Tintorane, Warfilone, Waran. Slightly bitter, crystalline powder. Discolored by light. Very sol
in water; freely sol in alcohol; very slightly sol in chloroform, ether. LD_{50} in male rats, female rats, mice, rabbits
(mg/kg): 323, 58, 374, ∼800 orally (Hagen); also reported
as LD_{50} in male, female rats (mg/kg): 100.3, 8.7 orally
(Back).
Potassium salt, $C_{19}H_{15}KO_4$, *Athrombin-K.*
Compound with 2-(dimethylamino)ethanol, $C_{23}H_{27}NO_5$,
warfarin-deanol, MD-6134, Adoisine.
Caution: Potential symptoms of overexposure are hematuria, back pain; hematoma of arms and legs; epistaxis,
bleeding lips and mucous membrane hemorrhage; abdominal pain, vomiting and fecal blood; petechial rash; abnormal
hematologic indices. *See NIOSH Pocket Guide to Chemical*
Hazards (DHHS/NIOSH 90-117, 1990) p 224. *See also*
Clinical Toxicology of Commercial Products, R. E. Gosselin
et al., Eds. (Williams & Wilkins, Baltimore, 5th ed., 1984)
Section III, pp 395-397.

USE: Rodenticide.

THERAP CAT: Anticoagulant.

10175. Water. Hydrogen oxide. H_2O; mol wt 18.02. H
11.19%, O 88.81%. *Reviews:* N. E. Dorsey, *Properties of*
Ordinary Water-Substance, A.C.S. Monograph Series no. **81**,
(Reinhold, New York, 1940) 673 pp; D. Eisenberg, W.
Kauzmann, *The Structure and Properties of Water* (Oxford
University Press, New York, 1969) 296 pp; Ebsworth *et al.,*
in *Comprehensive Inorganic Chemistry* vol. 2, J. C. Bailar, Jr.
et al., Eds. (Pergamon Press, Oxford, 1973) pp 741-747.
Pyrogen-free water (water for injection) is distilled water
rendered free of fever-producing proteins (bacteria and their
metabolic products). Method of prepn: Ishizuka *et al., C.A.*
49, 15177 (1955).
Liquid. Temp of max density 3.98°. $d^{3.98}$ 1.000000 g/ml
(0.999972 g/cc). d_4^{25} 0.997. d^0 (ice) 0.917 g/cc; d_4^0 (liq)
0.999868. Density tables: Bigg, *Brit. J. Appl. Phys.* **18**, 521
(1967); Kell, *J. Chem. Eng. Data* **12**, 66 (1967). Expands on
freezing. mp 0°. bp 100°. One liter satd vapor weighs
0.5974 g at 100° and 760 mm. Crit temp 374.2°; crit pressure 218 atm. Sp. heat (liq; 14°) 1.000 cal/g/°C. Latent
heat of fusion: 1.436 kcal/mole. Latent heat of vaporization: 9.717 kcal/mole. n_D^{20} 1.3330. Dielectric const (0°)
87.740. Dipole moment (25°) in benzene 1.76; in dioxane
1.86. Ionization const for pure water only: K (25°)
1.008×10^{-14}; at moderate concn of solutes (e.g. 1.0*M*
KOH): K (25°) 0.971×10^{-14}. The most universal solvent
known.

10176. Water Gas. Blue gas. Obtained by blowing
steam through incandescent coke. *Composition:* 6% CO_2;
42% CO; 51% H_2; 1% N_2.
Caution: Asphyxiant. USE: In the manuf of ammonia as
source of hydrogen. *Cf.* Producer Gas.

10177. Watermelon. Arbuse. *Citrullus vulgaris* Schrad.,
Cucurbitaceae, cultivated in hot and temperate zones the
world over. Contains diuretic principles: Bliss *et al., Am. J.*
Pharm. **105**, 53 (1933); Roby *et al., ibid.* **111**, 68 (1939).

10178. Wheat Germ Oil. Cav-Ecol; Myopone; Dena-
mone. Obtained by hydraulic expression or solvent extraction of wheat germ which constitutes ∼2% of a wheat grain,
the seed of *Triticum aestivum* L. (*T. sativum* Lam., *T. vul-*
gare Vill.), *Gramineae.* Constit. (of the oil): Linoleic acid
44.1%, oleic acid 30.0%, satd acids 15.1%, linolenic acid
10.8%, unsaponifiable matter 4.7%. The unsaponifiable
matter contains vitamin E-active tocopherols (reported as
0.5% of the oil and as 2 international vitamin E units per

gram of the oil), sitosterols, dihydrositosterol and other cryst alcohols and phospholipids. *Review:* E. W. Eckey, *Vegetable Fats and Oils* (Reinhold, New York, 1954) pp 291-293.

Bland yellow oil resembling corn oil. d_{25}^{25} 0.925-0.933. n_D^{40} 1.469-1.478. Acid value 6-20. Saponif value 179-190. Iodine value 115-129. Thiocyanogen value 80-85. Hydroxyl value 10-48. Reichert-Meissl value 0.3-1.4. Polenske value 0.4-2.1. Hehner value 76-95. Miscible with chloroform, ether, benzene, petr ether. Slightly sol in alc.

USE: Nutritional supplement. Source of natural vitamin E and unsatd fatty acids (vitamin F).

THERAP CAT (VET): Dietary source of vitamin E.

10179. Whisky. Whiskey. A liquid produced by distillation of the fermented mash of malted cereal grains, which has been stored in wood containers for not less than 4 years. Straight whisky contains 47-53% abs alcohol by vol, 0.05-0.16% acid calculated as acetic acid, and 0.038-0.15% esters as ethyl acetate. Whisky marked 100 proof contains 50% ethanol (v/v). It also contains small quantities of other natural constituents (congeners), which vary according to the grain used, method of fermentation, etc., and which are largely responsible for the characteristic aroma and flavor.

Blended whisky contains at least 40% ethanol (v/v) and is made from at least 20% of 100 proof straight whisky mixed with neutral spirits, *q.v.* The most popular blends are made from 65 vols neutral spirits and 35 vols straight whisky.

Light to deep amber liquid; characteristic odor and taste. d 0.923-0.935 at 25°.

THERAP CAT: Sedative, vasodilator (peripheral).

10180. White Pine. Deal pine; Northern pine; Weymouth pine. The dried inner bark of *Pinus strobus* L., *Pinaceae*. *Constit.* Coniferin glycoside, coniferyl alcohol, tannin, oleoresin, volatile oils.

Weak yellowish-orange to light yellowish-brown when powdered, with a slightly terebinthinate odor and a slightly mucilaginous, sweet, then bitter and astringent taste.

THERAP CAT: The bark as expectorant.

10181. Wieland-Gumlich Aldehyde. *(17R)-19,20-Didehydro-17,18-epoxycuran-17-ol; 1-deacetyldiaboline;* caracurine VII. $C_{19}H_{22}N_2O_2$; mol wt 310.40. C 73.52%, H 7.14%, N 9.03%, O 10.31%. Decompn product of isonitrosostrychnine: Wieland, Kaziro, *Ann.* **506**, 60 (1933). Isoln from *Strychnos toxifera* Benth., and *S. subcordata* Spruce, *Loganiaceae:* Asmis *et al., Helv. Chim. Acta* **37**, 1983 (1954); Penna *et al., Gazz. Chim. Ital.* **87**, 1163 (1957). Identity with caracurine VII: Bernauer *et al., Helv. Chim. Acta* **41**, 1405 (1958). Structure: Deyrup *et al., ibid.* **45**, 2266 (1962). Chemistry of the degradation from strychnine, *q.v.:* Hymon *et al., ibid.* **52**, 1564-1602 (1969). Synthesis: L. Szabo, O. Clauder, *Acta Chim. Acad. Sci. Hung.* **95**, 85 (1977). ^{13}C-NMR study: E. Wenkert *et al., J. Org. Chem.* **43**, 1099 (1978).

Crystals from acetone + methanol, decomp 213-214°. $[\alpha]_D^{22}$ −133.8° (c = 0.52 in methanol). Sol in methanol, ethanol, chloroform; slightly sol in acetone, ethyl acetate.

Hydrochloride hemihydrate, crystals from ethanol, mp >300°. uv max (water): 240, 290 nm (log ε 3.80, 3.40). Sol in water, ethanol, methanol, warm chloroform; slightly sol in acetone, chloroform.

10182. Wild Cherry. Wild black cherry bark. Dried stem bark of *Prunus serotina* Ehrh., *Rosaceae,* collected in autumn. *Habit.* North America. *Constit.* Prunasin; the enzyme emulsin capable of hydrolyzing prunasin to benz-

aldehyde, glucose, and hydrocyanic acid; benzoic, trimethyl-gallic and *p*-coumaric acids; tannin, volatile oil.

USE: Flavoring in foods. Pharmaceutic aid (flavor).

10183. Wildfire Toxin. *N-[2-Amino-4-(3-hydroxy-2-oxo-3-azetidinyl)-1-oxobutyl]-L-threonine.* $C_{11}H_{19}N_3O_6$; mol wt 289.29. C 45.67%, H 6.62%, N 14.53%, O 33.18%. Isoln from *Pseudomonas tabaci,* the bacterium responsible for wildfire disease of tobacco: Woolley *et al., J. Biol. Chem.* **197**, 409 (1952). Highly toxic to a variety of organisms including bacteria, algae, higher plants and animals: Sinden *et al., Toxicol. Appl. Pharmacol.* **14**, 82 (1969). Mode of action: Lovrekovich *et al., Nature* **197**, 917; **198**, 710 (1963). Structural studies: Woolley *et al., J. Biol. Chem.* **215**, 485 (1955); Stewart, *J. Am. Chem. Soc.* **83**, 435 (1961). Revised structure: *idem, Nature* **229**, 174 (1971).

Colorless, very hygroscopic substance. Very soluble in water; readily sol in methanol. Moderately sol in ethanol; sparingly sol in *n*-butanol and higher alcohols. Practically insol in acetone, ethyl acetate, and less polar organic liquids. Shows marked instability. Slowly loses potency in aq soln at pH 6. Complete loss of activity at these conditions and 0° in one month. Inactivation more rapid in methanol or when pH is shifted to either side of 6, esp in alkaline range.

10184. Withaferin A. *(4β,5β,6β,22R)-5,6-Epoxy-4,22,27-trihydroxy-1-oxoergosta-2,24-dien-26-oic acid δ-lactone; 4β,-27-dihydroxy-1-oxo-5β,6β-epoxywitha-2,24-dienolide.* $C_{28}H_{38}O_6$; mol wt 470.61. C 71.46%, H 8.14%, O 20.40%. Isolated from the leaves of *Withania somnifera* Dun., *Solanaceae:* Lavie, Yarden, *J. Chem. Soc.* **1962**, 2925; Kirson *et al., Tetrahedron* **26**, 2209 (1970); Subramanian, Sethi, *Indian J. Pharm.* **32**, 16 (1970); from roots of *Withania coagulans:* eidem, *Curr. Sci.* **38**, 267 (1969); from *Acnistus arborescens* (L.): Kupchan *et al., J. Org. Chem.* **34**, 3858 (1969). Structure: Lavie *et al., J. Chem. Soc.* **1965**, 7517; Kupchan, *loc. cit.* Crystal structure: McPhail, Sim, *J. Chem. Soc. (C)* **1968**, 962. Antitumor activity studies: B. Shohat *et al., Cancer Chemother. Rep.* **51**, 271 (1967); B. Shohat, H. Joshua, *Int. J. Cancer* **8**, 487 (1971). Synthetic studies: M. Hirayama *et al., J. Chem. Soc. Perkin Trans. I* **1981**, 88. Stereoselective synthesis: *eidem, Tetrahedron Letters* **23**, 4725 (1982).

White prisms from acetone-petroleum ether, mp 252-253° (Kupchan). Also reported as mp 243-245° (ethyl acetate). $[\alpha]_D^{28}$ +125° (c = 1.30 in CHCl$_3$). uv max (ethanol): 214, 335 nm (ε 17300, 165).

10185. Woodruff. Woodward herb; petit muguet; Waldmeister (German). Leaves of *Asperula odorata* L., *Rubiaceae.* *Habit.* Europe, Siberia, North Africa, Australia. *Constit.* Coumarin, tannin, asperuloside, fatty oil, essential oil, bitter principle. Discussion of its fragrance, prepn of *asperule absolute,* use of *Asperula* in various fragrant formulations: *Perfumery Essent. Oil Record,* **1963,** (June) p 382.

USE: Flavoring agent; the fresh leaves in flavoring May wine, the dried leaves in sachets.

10186. Woodward's Reagent K. *2-Ethyl-5-(3-sulfophenyl)isoxazolium inner salt;* N-ethyl-5-phenylisoxazolium-3'-sulfonate. $C_{11}H_{11}NO_4S$; mol wt 253.28. C 52.16%, H 4.38%, N 5.53%, O 25.27%, S 12.66%. Prepd by reacting 5-phenylisoxazole with chlorosulfonic acid, followed by alkylation with triethyloxonium fluoroborate, and acid hydrolysis: Woodward *et al., J. Am. Chem. Soc.* **83,** 1010 (1961).

Crystals, dec 207-208°. Stable. Non-hygroscopic.
USE: Peptide-bond former.

10187. Worenine. *5,6-Dihydro-14-methylbis[1,3]benzodioxolo[5,6-a:5',6'-g]quinolizinium;* 13-methyl-ψ-coptisine. $[C_{20}H_{16}NO_4]^+$. Protoberberine alkaloid isolated from root of *Coptis japonica* Makino, *Ranunculaceae:* Kitasato, *J. Pharm. Soc. Japan* No. **542,** 48, 315 (1927); *Acta Phytochim.* **3,** 175 (1927), *C.A.* **22,** 1779-1780 (1928); in *Coptis chinensis* Franch., *Ranunculaceae:* Schramm, Wej-ds, *Pharmazie* **14,** 405 (1959). Synthesis of 13-methyl-ψ-coptisine and its probable identity with worenine: T. R. Govindachari *et al., Indian J. Chem.* **9,** 1313 (1971).

Chloride, $C_{20}H_{16}ClNO_4$, yellow needles from water, mp 288-289°.
Iodide, $C_{20}H_{16}INO_4$, crystals from ethanol, mp 297-299°.

10188. Wortmannin. *[1S-(1α,6bα,9aβ,11α,11bβ)]-11-(Acetyloxy)-1,6b,7,8,9a,10,11,11b-octahydro-1-(methoxymethyl)-9a,11b-dimethyl-3H-furo[4,3,2-de]indeno[4,5-h]-2-benzopyran-3,6,9-trione.* $C_{23}H_{24}O_8$; mol wt 428.44. C 64.48%, H 5.65%, O 29.87%. Antifungal antibiotic from *Penicillium wortmanni* Klocker: Brian *et al., Trans. Brit. Mycol. Soc.* **40,** 365 (1957). Similar to viridin, *q.v.* Structure: MacMillan *et al., Chem. Commun.* **1968,** 613; *eidem, J. Chem. Soc. Perkin Trans. I* **1972,** 2898. Absolute stereochemistry: Petcher *et al., Chem. Commun.* **1972,** 1061; MacMillan *et al., ibid.* 1063.

Neutral solid, mp 240°. Unstable in aq solns pH 3-8.

X

10189. Xamoterol. (±)-*N*-[2-[[2-Hydroxy-3-(4-hydroxy-phenoxy)propyl]amino]ethyl]-4-morpholinecarboxamide; 1-(4-hydroxyphenoxy)-3-[2-(4-morpholinocarboxamido)ethyl-amino]-2-propanol. $C_{16}H_{25}N_3O_5$; mol wt 339.39. C 56.62%, H 7.42%, N 12.38%, O 23.57%. β_1-Adrenoceptor partial agonist. Prepn: **Belg. pat. 867,376;** B. G. Main, J. J. Barlow, **U.S. pat. 4,143,140** (1978, 1979 both to ICI); prepn and adrenoceptor stimulant activity: J. J. Barlow *et al., J. Med. Chem.* **24,** 315 (1981). Cardiovascular activity: A. Nuttall, H. M. Snow, *Brit. J. Pharmacol.* **77,** 381 (1982). Hemodynamics, cardioselectivity and kinetics: G. Jennings *et al., Clin. Pharmacol. Ther.* **35,** 594 (1984). Hemodynamic effects: J.-M. R. Detry *et al., Eur. Heart J.* **4,** 584 (1983); H. Sato *et al., Circulation* **75,** 213 (1987). Metabolism: T. R. Marten *et al., Drug Metab. Dispos.* **12,** 652 (1984). HPLC determn in biological fluids: C. J. Oddie *et al., J. Chromatog.* **308,** 370 (1984). Clinical evaluations: F. L. Tseu *et al., Brit. Heart J.* **56,** 469 (1986); L. Barrios *et al., Eur. J. Clin. Pharmacol.* **29,** 667 (1986). Clinical comparison with atenolol, *q.v.,* in asthmatic patients: J. W. J. Lammers *et al., Brit. J. Clin. Pharmacol.* **22,** 595 (1986).

Hemifumarate, $(C_{16}H_{25}N_3O_5)_2 \cdot C_4H_4O_4$, *ICI-118587, Carwin, Corwin, Xamtol.* Crystals from ethanol, mp 168-169° (dec).

THERAP CAT: Cardiotonic.

10190. Xanomeline. 3-[4-(Hexyloxy)-1,2,5-thiadiazol-3-yl]-1,2,5,6-tetrahydro-1-methylpyridine. $C_{14}H_{23}N_3OS$; mol wt 281.42. C 59.75%, H 8.24%, N 14.93%, O 5.69%, S 11.39%. Selective muscarinic M_1-receptor agonist. Prepn: P. Sauerberg, P. H. Olesen, **Eur. pat. Appl. 384,288** (1990 to Ferrosan); *eidem,* **U.S. pat. 5,043,345** (1991 to Novo Nordisk); *eidem et al., J. Med. Chem.* **35,** 2274 (1992). Prepn of crystalline tartrate: L. M. Osborne *et al.,* **PCT Int. pat. Appl. 94 29,303** (1994 to Novo Nordisk). Muscarinic receptor binding study: H. E. Shannon *et al., J. Pharmacol. Exp. Ther.* **269,** 271 (1994). Pharmacology: F. P. Bymaster *et al., ibid.* 282. HPLC determn in plasma: C. L. Hamilton *et al., J. Chromatog.* **613,** 365 (1993).

Oxalate, $C_{14}H_{23}N_3OS \cdot C_2H_2O_4$. Crystals from acetone, mp 148°.
(+)-L-Hydrogen tartrate, $C_{14}H_{23}N_3OS \cdot C_4H_6O_6$, *xanomeline tartrate, LY-246708, NNC-11-0232, Lomeron, Memcor.* Crystals from 2-propanol, mp 95.5°.

THERAP CAT: Cholinergic; nootropic.

10191. Xanthan Gum. Polysaccharide B-1459; Keltrol F; Kelzan. Mol wt > 10^6. Polysaccharide gum produced by the bacterium *Xathomonas campestris:* Jeanes *et al., J. Polymer Sci.* **5,** 519 (1961). Composed of D-glucosyl, D-mannosyl and D-glucosyluronic acid residues and differing proportions of *O*-acetyl and pyruvic acid acetal. The primary structure consists of a cellulose backbone with trisaccharide side chains, the repeating unit being a pentasaccharide. Structural studies: Sloneker *et al., Can. J. Chem.* **40,** 2066, 2188 (1962); **42,** 1261 (1964); P. E. Jansson *et al., Carbohyd. Res.* **45,** 275 (1975); L. D. Melton *et al., ibid.* **46,**

245 (1976); E. R. Morris *et al., J. Mol. Biol.* **110,** 1 (1977); G. Holzwarth, E. B. Prestridge, *Science* **197,** 757 (1977). Secondary and tertiary structure in solutions and gels: D. A. Rees, E. J. Welsh, *Angew. Chem. Int. Ed.* **16,** 214 (1977). *Review:* McNeely, Kang in *Industrial Gums,* R. L. Whistler, Ed. (Academic Press, New York, 2nd ed., 1973) pp 486-497.

Cream-colored, odorless, free-flowing powder. Dissolves readily in water with stirring to give highly viscous solns at low concns. Forms strong films on evaporation of aq solns. Resistant to heat degradation. Aq solns are highly pseudoplastic.

USE: In foods, non-foods, and cosmetics as stabilizer and emulsifying agent. Potential use in chemically enhanced oil recovery.

10192. Xanthatin. *[3aR-(3aα,7β,8aβ)]-3,3a,4,7,8,8a-Hexahydro-7-methyl-3-methylene-6-(3-oxo-1-butenyl)-2H-cyclohepta[b]furan-2-one.* $C_{15}H_{18}O_3$; mol wt 246.31. C 73.15%, H 7.37%, O 19.49%. Antimicrobial agent from *Xanthium strumarium* L., sens. lat.; *X. pennsylvanicum* Wallr., *Compositae; X. riparium.* Isoln: J. E. Little *et al., Arch. Biochem.* **27,** 247 (1950); T. A. Geissman *et al., J. Am. Chem. Soc.* **76,** 685 (1954). Structure: P. G. Deuel, T. A. Geissman, *ibid.* **79,** 3778 (1957). Evaluation as repellent of blue mussel: A. Harada *et al., Agric. Biol. Chem.* **49,** 1887 (1985).

Flat needles from methanol or ethanol, mp 114.5-115°. uv max: 275, 213 nm (ε 22800, 7300). $[\alpha]_D^{30}$ − 20° (ethanol). Sol in ether, acetone, alcohol; slightly sol in hot water (neutral pH). Practically insol in petr ether, 5% NaOH, 5% HCl.

10193. Xanthine. 3,7-Dihydro-1H-purine-2,6-dione; 2,6-(1H,3H)-purinedione; 2,6-dioxopurine. $C_5H_4N_4O_2$; mol wt 152.11. C 39.48%, H 2.65%, N 36.83%, O 21.04%. Occurs in animal organs, yeast, potatoes, coffee beans, tea. First isolated from urinary bladder stones: *Beilstein* **26,** 447 (1937). Prepd by treating a sulfuric acid soln of guanine with sodium nitrite: Fischer, *Ann.* **215,** 309 (1882). Several other syntheses, *cf.* Levene, Bass, *Nucleic Acids* (New York, 1931).

Scales, plates from water. Dec on heating without melting and with partial sublimation. Kb at 40° = 6.09×10^{-14}; Ka at 40° = 1.19×10^{-10}. Absorption spectrum: Kalckar, *J. Biol. Chem.* **167,** 429 (1947). One gram dissolves in 14.5 liters of water at 16°, in 1.4 liters boiling water; less sol in alcohol. Sol in mineral acids, freely sol in NH_4OH and in NaOH solns.

1,3-Dimethylxanthine, *see* Theophylline.
3,7-Dimethylxanthine, *see* Theobromine.
1,3,7-Trimethylxanthine, *see* Caffeine.

10194. Xanthinol Niacinate. *3-Pyridinecarboxylic acid compd with 3,7-dihydro-7-[2-hydroxy-3-[(2-hydroxyethyl)methylamino]propyl]-1,3-dimethyl-1H-purine-2,6-dione (1:1);* 7-[2-hydroxy-3-[(2-hydroxyethyl)methylamino]propyl]theophylline, compd with nicotinic acid; 7-[3-[N-(2-hydroxyethyl)-amino]-2-hydroxypropyl]theophylline nicotinate; xanthinol nicotinate; SK-331-A; Angiomin; Complamin; Sadamin; Xavin. $C_{19}H_{26}N_6O_6$; mol wt 434.45. C 52.53%, H 6.03%, N 19.34%, O 22.10%. Prepn: Bestian, **Ger. pat. 1,102,750** (1961 to Wuelfing), *C.A.* **56,** 11602i (1962). Alternate synthesis: Korbonits *et al., Acta Pharm. Hung.* **38,** 98 (1968).

Pharmacology: Hemmer, Diezemann, *Arzneimittel-Forsch.* **12**, 672 (1962); Lennartz, *ibid.* 675; Stamm, *ibid.* 679.

Crystals, mp 180°. Freely sol in water.
THERAP CAT: Vasodilator (peripheral).

10195. Xanthocillin. Xantocillin; Brevicid. Antibiotic complex produced by *Penicillium notatum* Westling: Rothe, *Pharmazie* **5**, 190 (1950); Barwald, **Brit. pat. 898,498** (1962 to Arzneimittelwerk VEB), *C.A.* **57**, 9013a (1962). Consists of at least three antibiotics, xanthocillins X, Y$_1$ and Y$_2$; the first being the major component (about 70%).

Xanthocillin X	R = H	R' = H
Xanthocillin Y$_1$	R = OH	R' = H
Xanthocillin Y$_2$	R = OH	R' = OH

Xanthocillin X, C$_{18}$H$_{12}$N$_2$O$_2$, *4,4'-(2,3-diisocyano-1,3-buta-diene-1,4-diyl)bisphenol, bis(p-hydroxybenzylidene)ethylene isocyanide, 1,4-bis(p-hydroxyphenyl)-2,3-diisonitrilo-1,3-buta-diene.* Structure: Hagedorn, Tönjes, *Pharmazie* **11**, 409 (1956); **12**, 567 (1957); Hagedorn *et al.*, *Ber.* **93**, 1584 (1960). Synthesis: Hagedorn, Eholzer, *Angew. Chem.* **74**, 215 (1962). Biosynthetic studies: Achenbach, Grisebach, *Z. Naturforsch.* **20b**, 137 (1965); Achenbach, König, *Ber.* **105**, 784 (1972). Crystal structure: D. Britton *et al.*, *Cryst. Struct. Commun.* **10**, 1497 (1981). Studies on antiviral activity: Takatsuki *et al.*, *J. Antibiot.* **21**, 671 (1968); *eidem, ibid.* **22**, 151 (1969); Kitahara, *ibid.* **34**, 1556 (1981). Clusters of yellow needles from alc; yellow rhombs from ethyl acetate. Chars at about 210°. Sol (up to 1%) in alc, ether, acetone, dioxane. Freely sol in aqueous alkaline solns. Practically insol in water, petr ether, benzene, chloroform. Forms a water-sol dipotassium salt, the concd solns of which have a pH of 10.5.
Dimethyl ether, C$_{20}$H$_{16}$N$_2$O$_2$, yellow needles from dioxane + alcohol, dec 181°.
Xanthocillin Y$_1$, C$_{18}$H$_{12}$N$_2$O$_3$, *4-[4-(4-hydroxyphenyl)-2,3-diisocyano-1,3-butadienyl]-1,2-benzenediol.*
Xanthocillin Y$_2$, C$_{18}$H$_{12}$N$_2$O$_4$, *4,4'-(2,3-diisocyano-1,3-butadiene-1,4-diyl)bis-1,2-benzenediol.* Structure of xanthocillins Y$_1$ and Y$_2$: Achenbach *et al.*, *Ber.* **105**, 3061 (1972).
USE: In feed supplements.
THERAP CAT: Antibacterial.

10196. Xanthone. *9H-Xanthen-9-one;* 9-oxoxanthene; diphenylene ketone oxide; dibenzo-γ-pyrone; Genicide. C$_{13}$H$_8$O$_2$; mol wt 196.21. C 79.58%, H 4.11%, O 16.31%. Prepd by heating phenyl salicylate: Graebe, *Ann.* **254**, 279 (1889); Holleman, *Org. Syn.* coll. vol. I (2nd ed., 1941) p 552.

Polymorphic needles from alcohol, mp 174°. bp$_{730}$ 351°. Sparingly volatile with steam. Dipole moment 3.0. Solubility: 0.55 g/100 ml of cold alcohol; 6.71 g/100 ml of boiling alcohol. Freely sol in chloroform. Slightly sol in hot water,

ether, petr ether, benzene, toluene, xylene. Dissolves in concd H$_2$SO$_4$ to a yellow soln with pale blue fluorescence.
Oxime, *xanthoxime*, mp 161°.
USE: In the prepn of xanthydrol. Ovicide for codling moth eggs; less effective as larvicide.

10197. Xanthophyll. *β,ε-Carotene-3,3'-diol;* lutein; vegetable lutein; vegetable luteol; Bo-Xan. C$_{40}$H$_{56}$O$_2$; mol wt 568.88. C 84.45%, H 9.92%, O 5.62%. One of the most widespread carotenoid alcohols in nature. Originally isolated from egg yolk, also isolated by chromatography from nettles, algae, and petals of many yellow flowers. Occurs also in colored feathers of birds: Volker, *Z. Physiol. Chem.* **288**, 20 (1951). Extraction from petals of *Tagetes patula* L., *Compositae:* Karrer *et al.*, *Helv. Chim. Acta* **30**, 531 (1947). Occurs together with zeaxanthin, *q.v.* Dipalmitate occurs in *Helenium autumnale* L., *Compositae* and other flowers: Kuhn, Winterstein, *Naturwiss.* **18**, 754 (1930). Conversion to zeaxanthin with sodium alcoholate: Karrer, Jucker, *ibid.* 266. Does not possess vitamin A potency: Schumacher *et al., Poultry Sci.* **23**, 529 (1944). Stereochemistry: Zechmeister, *Chem. Rev.* **34**, 267 (1944). Structure: Karrer, *Helv. Chim Acta* **34**, 2160 (1951). Abs config: Goodfellow *et al., Chem. Commun.* **1970**, 1578; Buchecker *et al., Chimia* **25**, 192 (1971); *eidem, Helv. Chim. Acta* **57**, 631 (1974). Synthesis: H. Mayer, A. Rüttimann, *ibid.* **63**, 1451 (1980). Sepn and determn of configurational isomers: A. Rüttiman *et al., J. High Resolut. Chromatog. Chromatog. Commun.* **6**, 612 (1983). *Reviews:* Zechmeister, *Carotinoide* (Berlin, 1934); Mayer, *The Chemistry of Natural Coloring Matters* (New York, 1943); Karrer, Jucker, *Carotenoids* (New York, 1950).

Yellow prisms with metallic luster from ether + methanol, mp 190° (corr), (a higher mp indicates impure material). Also reported as mp 183° [Buchecker (1974)]. [α]$_{Cd}^{18}$ +165° (c = 0.7 in benzene). Absorption max (dioxane): 481, 453, 429, 333, 268 nm (ε 142000, 152000, 100000, 15500, 35000). Insol in water, sol in fats and in fat solvents. More sol in boiling methanol (1:700) than zeaxanthin.
Dipalmitate, C$_{72}$H$_{116}$O$_4$, *helenien, Adaptinol.* Red needles from alcohol, mp 92°.

10198. Xanthopterin. *2-Amino-1,5-dihydro-4,6-pteridinedione; 2-amino-4,6-pteridinediol;* 2-amino-4,6-dihydroxypteridine; 2-amino-4,6-dihydroxypyrimido[4,5-b]pyrazine. C$_6$H$_5$N$_5$O$_2$; mol wt 179.14. C 40.23%, H 2.81%, N 39.09%, O 17.86%. Pigment first found in wings of butterflies. Widely distributed in insects and in animals. Has been separated from urine and from the crab. Isoln and structure: Schöpf, Becker, *Ann.* **507**, 266 (1933); **524**, 55, 126 (1936); Schöpf, Kottler, *Ann.* **539**, 128 (1939); Wieland, Purrmann, *Ann.* **544**, 163 (1940). *See also* Fukushima, Shiota, *J. Biol. Chem.* **247**, 4549 (1972). Synthesis: Purrmann, *Ann.* **546**, 98 (1940); **548**, 284 (1941); Koschara, *Z. Physiol. Chem.* **277**, 159 (1943); Totter, *J. Biol. Chem.* **154**, 105 (1944); Elion *et al., J. Am. Chem. Soc.* **71**, 741 (1949); Koschara, **Ger. pat. 859,471** (1952 to Bayer), *C.A.* **52**, 10222f (1958); Stuart, Wood, *J. Chem. Soc.* **1963**, 4186. Facile synthesis: Taylor, Jacobi, *J. Am. Chem. Soc.* **95**, 4455 (1973); Taylor *et al., J. Org. Chem.* **40**, 2341 (1975). Exhibits tumor inhibitory properties.

Monohydrate, orange-yellow crystals, sinters around 360°, decomp above 410°. Practically insol in water. Freely sol in dil NH$_4$OH and NaOH giving yellow solns and in 2N HCl giving colorless solns. uv max at pH 11: 255, 390 nm (E$_{1cm}^{1\%}$ 0.92, 0.355). Can be converted by yeast into folic acid.

10199. Xanthosine. Xanthine riboside; 9-β-D-ribofuranosyl xanthine; 9-β-D-ribofuranosyl-9H-purine-2,6-diol; 9-β-D-ribofuranosyl-9H-purine-2,6-(1H,3H)-dione. $C_{10}H_{12}N_4O_6$; mol wt 284.23. C 42.26%, H 4.26%, N 19.71%, O 33.77%. Prepd from guanosine by treatment with sodium nitrite and acetic acid: Levene, Jacobs, *Ber.* **43**, 3163 (1910); Gulland, Macrae, *J. Chem. Soc.* **1933**, 662. Synthesis: Howard *et al.*, *ibid.* **1949**, 232. Biosynthesis: Magasanik, Brooke, *J. Biol. Chem.* **206**, 83 (1954); Korn, Buchanan, *ibid.* **217**, 183 (1955); Bolis *et al., ibid.* **219**, 917 (1956).

Dihydrate, long prisms from water. Anhydr felted clusters (warts) from alc. Dec on heating, no distinct melting range. $[\alpha]_D^{30}$ −51.2° (p = 8 in 0.3N NaOH). uv max: 253 nm (ε 8.790). a_M (molar absorbancy): 11.4 × 10³ at 248.5 nm in water at pH 8.0. Sparingly sol in cold water; freely sol in hot water; sol in hot dil alcohol. Easily hydrolyzed by mineral acids.

10200. Xanthoxyletin. *5-Methoxy-8,8-dimethyl-2H,8H-benzo[1,2-b:5,4-b']dipyran-2-one;* 7-hydroxy-5-methoxy-2,2-dimethyl-2H-1-benzopyran-6-acrylic acid δ-lactone; xanthoxylin N; xanthoxyloin. $C_{15}H_{14}O_4$; mol wt 258.27. C 69.76%, H 5.46%, O 24.78%. Found in *Xanthoxylum americanum* Mill., *Melicope ternata* Forst., *M. mantelli* Buch., *Halifordia scleroxyla* F. Muell., *Rutaceae; Chloroxylon swietenia* DC., *Meliaceae.* Isoln: Staples, *Am. J. Pharm.* **1829**, 123; Witte, *Arch. Pharm.* **212**, 283 (1878); Lloyd, *Am. J. Pharm.* **1890**, 229; Gordin, *J. Am. Chem. Soc.* **28**, 1649 (1906); Bell *et al., J. Chem. Soc.* **1936**, 627; Robertson, Subramaniam, *ibid.* **1937**, 286; Dieterle, Kruta, *Arch. Pharm.* **275**, 45 (1937); Briggs, Locker, *J. Chem. Soc.* **1951**, 3131; King *et al., ibid.* **1954**, 1392; Hegarty, Lahey, *Aust. J. Chem.* **9**, 120 (1956); Cambie, *J. Chem. Soc.* **1960**, 2376. Synthesis: Joshi, Kamat, *Tetrahedron Letters* **1966**, 5767.

Elongated prisms from methanol, mp 132-133°. uv max: 277, 269, 322, 347 nm (log ε 4.28, 4.32, 3.99, 4.05 in ethanol). Very easily sol in benzene, chloroform, hot alcohol; sol in acetone; sparingly sol in ether. Sol in 49 parts cold acetone and 25,000 parts cold water.

10201. Xanthoxylin. *1-(2-Hydroxy-4,6-dimethoxyphenyl)ethanone; 2'-hydroxy-4',6'-dimethoxyacetophenone;* phloracetophenone 4,6-dimethyl ether; 2,4-dimethoxy-6-hydroxyacetophenone; brevifolin. $C_{10}H_{12}O_4$; mol wt 196.20. C 61.22%, H 6.16%, O 32.62%. Found in *Xanthoxylum piperitum* DC., *X. alatum* Roxb., *Rutaceae, Artemisia brevifolia* Wallich, *Compositae, Hippomane mancinella* L., *Sapium sebiferum* Roxb., *Euphorbiaceae.* Isoln: Stenhouse, *Ann.* **89**, 251 (1854); **104**, 236 (1857); Semmler, Schossberger, *Ber.* **44**, 2885 (1911); Smith, Smith, *Pharm. J.* **119**, 688 (1927); **123**, 604, 611 (1929); Schaeffer *et al., J. Am. Pharm. Assoc.* **43**, 43 (1954); Chu *et al., C.A.* **53**, 10532g (1959). Synthesis: Kostanecki, Lambor, *Ber.* **32**, 2260 (1899); Canter *et al., J. Chem. Soc.* **1931**, 1245; Belton *et al., Sci. Proc. Roy. Dublin Soc.* **25**, 19 (1949); Schmid, Bolleter, *Helv. Chim. Acta* **33**, 917 (1950); MacKenzie *et al., J. Chem. Soc.* **1950**, 2965; Dean, Robertson, *ibid.* **1953**, 1244; Kawano, *Chem. & Ind. (London)* **1959**, 368.

Crystals from ethanol, mp 81-83°. Practically insol in water. Sol in alcohol, ether.

10202. Xanthurenic Acid. *4,8-Dihydroxy-2-quinolinecarboxylic acid; 4,8-dihydroxyquinaldic acid.* $C_{10}H_7NO_4$; mol wt 205.17. C 58.54%, H 3.44%, N 6.83%, O 31.19%. Excreted by pyridoxine-deficient animals after the ingestion of tryptophan. Isoln from the urine of albino rats fed almost exclusively on fibrin: Musajo, *Gazz. Chim. Ital.* **67**, 165 (1937). Synthesis from ethyl oxalacetate and o-anisidine: Musajo, Minchilli, *Ber.* **74**, 1842 (1941); Mebane, Oroshnik, *J. Am. Chem. Soc.* **73**, 3520 (1951); Furst, Olsen, *J. Org. Chem.* **16**, 412 (1951).

Sulfur-yellow crystals, mp 286°. uv max (water): 243, 342 nm (ε 30000, 6500). Insol in water. Sol in aq alkali hydroxides and carbonates (yellow solns) and in hot dil HCl. It gives a red color with Millon reagent, intense ruby-red with alkali diazobenzenesulfonates, and intense green with $FeSO_4$. When the acid is dissolved in very dil $NaHCO_3$, the green color with $FeSO_4$ is still visible at a diln of 1:200,000.

Sodium salt, $C_{10}H_6NNaO_4 \cdot nH_2O$, yellow crystals, anhydr at 170°. Slightly sol in water.

Methyl ester, $C_{11}H_9NO_4$, yellow crystals, dec 262°. Sol in NH_4OH, alkali hydroxides and carbonates.

10203. Xanthyletin. *8,8-Dimethyl-2H,8H-benzo[1,2-b:5,-4-b']dipyran-2-one;* 7-hydroxy-2,2-dimethyl-2H-1-benzopyran-6-acrylic acid δ-lactone; 2,2-dimethylchromenocoumarin. $C_{14}H_{12}O_3$; mol wt 228.25. C 73.67%, H 5.30%, O 21.03%. Linear pyranocoumarin found in *Xanthoxylum americanum* Mill., *Luvanga scandens* Ham., *Citrus acida* Roxb., *Rutaceae; Chloroxylon swietenia* DC., *Meliaceae.* Isoln: Bell, Robertson, *J. Chem. Soc.* **1936**, 1828; Bell *et al., ibid.* **1937**, 1542; Späth *et al., Ber.* **72**, 1450 (1939); Bose, Mookerjee, *J. Indian Chem. Soc.* **21**, 181 (1944); Mookerjee, *ibid.* **23**, 41 (1946); King *et al., J. Chem. Soc.* **1954**, 1392. Synthesis: Späth, Hillel, *Ber.* **72**, 2093 (1939); Steck, *Can. J. Chem.* **49**, 2297 (1971); P. Waykole *et al., Indian J. Chem.* **19B**, 238 (1980); V. K. Ahluwalia *et al., Monatsh.* **112**, 119 (1981); J. Banerji *et al., Indian J. Chem.* **21B**, 496 (1982).

Elongated flat prisms from methanol, mp 130-131°. $bp_{0.1}$ 140-145°. uv max: 266, 348 nm (log ε 4.34, 4.15).

10204. Xenazoic Acid. *4-[(2-[1,1'-Biphenyl]-4-yl-1-ethoxy-2-oxoethyl)amino]benzoic acid; p-[(α-ethoxy-p-phenylphenacyl)amino]benzoic acid; p-[N-(α-ethoxy-β-keto-β-parabiphenyl)ethylamino]benzoic acid; p-(α-ethoxy-p-phenylphenacylamido)benzoic acid; xenalamine; CV-58903; SKF-8318; Xenovis.* $C_{23}H_{21}NO_4$; mol wt 375.42. C 73.58%, H 5.64%, N 3.73%, O 17.05%. Prepn: Cavallini *et al., J. Med. Pharm. Chem.* **2**, 99 (1960).

Crystals from ethanol, mp 192° (dec).
THERAP CAT: Antiviral.

10205. Xenbucin. *α-Ethyl-[1,1'-biphenyl]-4-acetic acid; 2-(4-biphenylyl)butyric acid; α-(4-biphenylyl)butyric acid; α-(4-diphenylyl)butyric acid; α-(p-xenyl)butyric acid; 4-diphenylylethylacetic acid; 4-biphenylylethylacetic acid; ethyl-p-xenylacetic acid; α-ethyl-p-phenyl-α-toluic acid;* MG-1559; Maggioni 1559; Liosol; Liposana. $C_{16}H_{16}O_2$; mol wt 240.30. C 79.97%, H 6.71%, O 13.32%. Prepn: Blicke, Grier, *J. Am. Chem. Soc.* **65**, 1725 (1943); Cavallini, Massarani, *Farmaco Ed. Sci.* **11**, 167 (1956). Fibrinolytic activity: R. J. Gryglewski, M. Eckstein, *Nature* **214**, 626 (1967).

Crystals from glacial acetic acid, mp 123-125°. Soluble in ethanol, methanol, acetone, carbon tetrachloride, benzene, ether, dimethylformamide. Practically insol in water.

trans-4-Phenylcyclohexylamine salt (1:1), $C_{28}H_{33}NO_2$, *butixirate,* MG-5771, *Flectar.* Prepn, analgesic and anti-inflammatory properties: G. Cantarelli *et al., Farmaco Ed. Sci.* **24**, 140 (1969); **Brit. pat. 1,168,542** (1969 to Maggioni), *C.A.* **72**, 43197 (1970). Therapeutic effectiveness: P. Croce, *Clin. Ter.* **50**, 241 (1969). Crystals, mp 224-227°. LD_{50} i.p. in mice: 183 μM/kg (Cantarelli).

THERAP CAT: Free acid as antihyperlipoproteinemic; *trans*-4-phenylcyclohexylamine salt as analgesic, anti-inflammatory.

10206. Xenon. Xe; at. wt 131.29; at. no. 54. Valences 2, 4, 6, 8. Group VIIIA(18), also known as Group 0. A noble gas characterized by an electronic structure in which the outer *p* subshell is entirely filled. Naturally occurring stable isotopes (mass numbers): 124 (0.10%); 126 (0.09%); 128 (1.91%); 129 (26.4%); 130 (4.1%); 131 (21.2%); 132 (26.9%); 134 (10.4%); 136 (8.9%); known artificial, radioactive isotopes: 110-123; 125; 127; 133; 135; 137-145. Discovered in the final residues obtained after evaporating liq air: Ramsay, Travers, *Proc. Roy. Soc.* **63** [A], 405 (1898). Occurs frequently in gases evolved from thermal springs; concentration in air: 0.087 ppm by vol. Obtained commercially from the atmosphere by distillation-liquefaction process. Extraction from liq air residues: Allen, Moore, *J. Am. Chem. Soc.* **53**, 2512 (1931). *Xenon platinum hexafluoride* was the first reported xenon compound: N. Bartlett, *Proc. Chem. Soc.* **1962**, 218. Teratogenicity study: G. A. Lane *et al., Science* **210**, 899 (1980). Review of biology, chemistry and anesthetic properties: R. M. Featherstone, C. A. Muelbaecher, *Pharmacol. Rev.* **15**, 97 (1963). Review of diagnostic use of radioactive compds for pulmonary studies: F. Fazio, P. Wollman, *Clin. Physiol.* **1**, 323 (1981); for cerebral blood flow: H. Yonas *et al., Adv. Tech. Stand. Neurosurg.* **15**, 3 (1987). Reviews of chemistry and compds: *Noble-Gas Compounds,* H. H. Hyman, Ed. (Univ. Chicago Press, Chicago, 1963) 404 pp; J. H. Holloway, *Noble-Gas Chemistry* (Methuen, London, 1968) 213 pp; Sladky, "Noble Gases" in *MTP Int. Rev. Sci.: Inorg. Chem., Ser. One* vol. 3, V. Gutman, Ed. (Butterworths, London, 1972) p 1-52; Cockett, Smith, "The Monatomic Gases" in *Comprehensive Inorganic Chemistry* vol. 1, J. C. Bailar, Jr. *et al.,* Eds. (Pergamon Press, Oxford, 1973) pp 139-211; Bartlett, Sladky, *ibid.,* pp. 213-330; S.-C. Hwang, W. R. Weltmer, Jr. in Kirk Othmer *Encyclopedia of Chemical Technology* vol. 13 (John Wiley & Sons, 4th ed., 1995) pp 1-38; G. J. Schrobilgen, J. M. Whalen, *ibid.* pp 38-53; *Chemistry of the Elements,* N. N.

Greenwood, A. Earnshaw, Eds. (Pergamon Press, New York, 1984) pp 1042-1059.

Colorless, odorless, tasteless, relatively inert, monatomic gas; will form compds with highly electronegative elements such as O, F, Cl. Soly of gas in water (20°): 108.1 cm³/kg water. Triple point temp 161.35°K, press 81.66 kPa. Critical temp 289.74°K, critical press 5840 kPa, critical d 1100 kg/m³. Gas: d⁰ (101.3 kPa) 5.8971 kg/m³, d (normal bp) 11 kg/m³. Liquid: normal bp 165.02°K, d (normal bp) 3057 kg/m³, d (triple pt) 3084 kg/m³, heat of vaporization (normal bp) 12.640 kJ/mol. Solid: d (triple pt) 3540 kg/m³, heat of vaporization (triple pt) 15.1 kJ/mol, heat of fusion (triple pt) 2.3 kJ/mol. Solid form exists as face-centered cubic crystals at normal pressure. Spectrum: Collie, *Proc. Roy. Soc.* **97** [A], 349 (1920). Emission spectra: T. Jacksier, R. M. Barnes, *Appl. Spectrosc.* **48**, 65 (1994).

Xenon-133, Xeneisol Xe 133.

Hydrate, Xe.xH₂O, mp 24°, and a deuterate, Xe.6D₂O, have been prepd: R. de Forcrand, *Compt. Rend.* **176**, 355 (1923); **181**, 15 (1925); Tamman, Krige, *Z. Anorg. Chem.* **146**, 179 (1925).

Xenon difluoride, F₂Xe. Prepd from the elements: Weeks *et al., J. Am. Chem. Soc.* **84**, 4612 (1962); Hoppe *et al., Z. Anorg. Allgem. Chem.* **324**, 214 (1963). Colorless crystals; d 4.32. Triple pt. 129.03°: Schreiner *et al., J. Phys. Chem.* **72**, 1162 (1968). Sublimes without decompn. Soly in water at 0°: 25 g/l.

Xenon tetrafluoride, F₄Xe. First prepd by direct combination of the elements at 6 atm and 400°: Claassen *et al., J. Am. Chem. Soc.* **84**, 3593 (1962). Colorless crystals; d 4.04. Triple pt 117.10°: Schreiner *et al., loc. cit.* Sublimes without decompn. Hydrolyzes to form Xe, O₂, HF and XeO₃.

Xenon hexafluoride, F₆Xe. Laboratory prepn: Chernick *et al., Inorg. Syn.* **8**, 258 (1966). Colorless solid; greenish-yellow vapor; vapor press. about 30 mm at 25°. mp 49.48°; bp 75.57°; d²⁴·⁴(solid) 3.411; d⁵⁵·²(liq) 3.173: Schreiner *et al., J. Chem. Phys.* **51**, 4838 (1969). Hydrolyzed by water to form XeOF₄ and XeO₃. More powerful oxidizing and fluorinating agent than XeF₂ and XeF₄. Cannot be stored in glass or quartz containers.

Xenon trioxide, O₃Xe. Prepn: Williamson, Koch, *Science* **139**, 1046 (1963); Jaselskis *et al., J. Am. Chem. Soc.* **88**, 2149 (1966). *Powerful explosive,* formed when XeF₄ and XeF₆ are hydrolyzed. Colorless, hygroscopic solid; d 4.55. Aqueous solns, *"xenic acid",* may be prepd in concns >2M.

Caution: Can act as a simple asphyxiant by displacing air. See: *Matheson Gas Data Book* (Matheson Co., Inc., 4th ed., East Rutherford, NJ, 1966) pp 499-500.

USE: Gas in lamps designed to resemble natural daylight; in lamps of extremely high brilliance. Isotopes in leak detection systems for nuclear reactors.

THERAP CAT: Anesthetic (inhalation). Xenon¹³³ as diagnostic aid (radioactive imaging agent).

10207. p-Xenylcarbimide. *4-Isocyanato-1,1'-biphenyl; isocyanic acid 4-biphenylyl ester; p-diphenyl isocyanate.* $C_{13}H_9NO$; mol wt 195.22. C 79.98%, H 4.65%, N 7.17%, O 8.20%. C₆H₅C₆H₄NCO. Prepd from the corresponding amine and phosgene: Kaplan, *J. Chem. Eng. Data* **6**, 272 (1961).

Needles, mp 56°; bp 283° (decomp), bp₂·₃ 137-140°. Very sol in ether.

USE: As a reagent for the identification of hydroxy compounds, with which it gives crystalline derivatives of definite melting point.

10208. Xenytropium Bromide. *endo-(±)-8-([1,1'-Biphenyl]-4-ylmethyl)-3-(3-hydroxy-2-phenyl-1-oxopropoxy)-8-methyl-8-azoniabicyclo[3.2.1]octane bromide; 3α-hydroxy-8-(p-phenylbenzyl)-1αH,5αH-tropanium bromide (±)-tropate; 8-p-phenylbenzylatropinium bromide; p-biphenylmethyl-(dl-tropyl-α-tropinium)bromide; 4-diphenylmethyl-dl-tropeyltropinium bromide;* N-399; Gastropin; Gastripon. $C_{30}H_{34}BrNO_3$; mol wt 536.51. C 67.16%, H 6.39%, Br 14.89%, N 2.61%, O 8.95%. Anticholinergic. Prepn: Nador, Gyermek, **U.S. pat. 2,833,773** (1958 to Licencia-Budapest).

Crystals from ethanol, dec 220-222°.
THERAP CAT: Antispasmodic.

10209. Xibenolol. (±)-*1-[(Dimethylethyl)amino]-3-(2,3-dimethylphenoxy)-2-propanol;* 3-*(tert*-butylamino)-3-(2′,3′-dimethylphenoxy)-2-propanol; (±)-1-*(tert*-butylamino)-(2,3-xylyloxy)-2-propanol. $C_{15}H_{25}NO_2$; mol wt 251.37. C 71.67%, H 10.02%, N 5.57%, O 12.73%. Non-selective β-adrenergic blocker. Prepn: W. Kunz *et al.,* **Ger. pat. 1,236,523** (1967 to Sanol-Arzneimittel), *C.A.* **67,** 64046k (1967); K. Tsukamoto *et al.,* U.S. pat. **4,018,824** (1977 to Teikoku Hormone). Pharmacology in dogs: N. Himori *et al., Arch. Int. Pharmacol.* **220,** 4 (1976); *eidem, Arch. Pharmacol.* **316,** 19 (1986). Metabolism in animals: S. Honma, A. Kambekawa, *Chem. Pharm. Bull.* **23,** 1045 (1975). Stereoselective metabolism and pharmacokinetics in man: S. Honma *et al., ibid.* **33,** 760 (1985). Toxicology: T. Usui *et al., Yakuri to Chiryo* **12,** Suppl. 6, 969, 987 (1984); *C.A.* **102,** 125280x, 125281y (1985).

Crystals from ether, mp 57°. $bp_{0.7mm}$ 134-136°. uv max (ethanol): 271.2, 274, 279.3 nm (ε 1.08×10^3, 1.07×10^3, 1.11×10^3).
Hydrochloride, $C_{15}H_{25}NO_2 \cdot HCl$, *D-32, Selapin, Rhythminal, Rythminal.* mp 135-137°.
THERAP CAT: Antiarrhythmic.

10210. Xibornol. *6-Isobornyl-3,4-xylenol;* 6-isobornyl-3,4-dimethylphenol; 6-(2-isobornyl)-3,4-xylen-1-ol; 3,4-dimethyl-6-isobornylphenol; *Nanbacine.* $C_{18}H_{26}O$; mol wt 258.40. C 83.67%, H 10.14%, O 6.19%. Syntheses: Moldovanskaya *et al., C.A.* **66,** 76163p (1967); Starkov, Glushkova, *J. Appl. Chem. USSR* **40,** 209 (1967), *C.A.* **67,** 72960u (1967), *C.A.* **71,** 70225s (1969). Prepn of cryst substance: **Brit. pat. 1,206,774** and Mardiguian, Fournier, **Ger. Offen. pat. 2,032,170** (1970 and 1971 to MARPHA Soc. d'Etudes et d'Exploitation de Marques), *C.A.* **74,** 13299g, 88171t (1971). Activity study: Capponi, *Bull. Soc. Pathol. Exot.* **62,** 658 (1969), *C.A.* **73,** 118883a (1970).

Crystals from petr ether, mp 94-96°, bp_3 165-168°. Also reported as very viscous, pale yellow liquid; bp_9 185-189°. d_4^{20} 1.0240. n_D^{20} 1.5382 (Starkov, Glushkova).
USE: Rubber antioxidant.
THERAP CAT: Antibacterial.

10211. Ximoprofen. *4-[3-(Hydroxyimino)cyclohexyl]-α-methylbenzeneacetic acid;* p-(3-oxocyclohexyl)hydratropic acid oxime; *XIFAM;* 2-[4-(3-oximinocyclohexyl)phenyl]-propionic acid; *13832-JL.* $C_{15}H_{19}NO_3$; mol wt 261.32. C 68.94%, H 7.33%, N 5.36%, O 18.37%. Nonsteroidal anti-inflammatory. Prepn: J. G. Maillard, **Ger. pat. 2,442,910;**

idem, U.S. pat. **3,935,255** (1975, 1976 both to Lab. Jacques Logeais). Metabolism: B. C. Mayo *et al., Xenobiotica* **20,** 233 (1990). Clinical pharmacokinetics: I. W. Taylor *et al., Brit. J. Clin. Pharmacol.* **32,** 242 (1991). GC determn in plasma: I. W. Taylor, L. F. Chasseaud, *J. Chromatog.* **495,** 275 (1989). HPLC determn in urine: G. R. Morris *et al., ibid* **530,** 377 (1990). Clinical evaluation: M. Dougados *et al., J. Rheumatol.* **16,** 1167 (1989).

Crystals from water/methanol, mp 178°.
THERAP CAT: Anti-inflammatory.

10212. Xipamide. *4-Chloro-5-sulfamoyl-2′,6′-salicyloxylidide;* 4-chloro-2′,6′-dimethyl-5-sulfamoylsalicylanilide; 4-chloro-5-sulfamylsalicyloyl-2′,6′-dimethylanilide; *Bei-1293; Aquaphor; Chronexan; Diurexan; Lumitens.* $C_{15}H_{15}ClN_2O_4S$; mol wt 354.81. C 50.78%, H 4.26%, Cl 9.99%, N 7.90%, O 18.04%, S 9.04%. Salicylic acid derivative. Prepn: **Neth. pat. Appl. 6,607,680;** Liebenow, U.S. pat. **3,567,777** (1966, 1970 both to P. Beiersdorf). Clinical studies: O. Hammer, U. Dembowski, *Med. Klin.* **64,** 1862 (1969); R. Fischer, A. Lenhartz, *Med. Welt* **1970,** 270. Review of pharmacology and therapeutic efficacy: B. N. C. Prichard, R. N. Brogden, *Drugs* **30,** 313-332 (1985).

Crystals from methanol-water, mp 256°.
THERAP CAT: Diuretic. Antihypertensive.

10213. Xylazine. *N-(2,6-Dimethylphenyl)-5,6-dihydro-4H-1,3-thiazin-2-amine;* 5,6-dihydro-2-(2,6-xylidino)-4H-1,3-thiazine; 2-(2,6-dimethylphenylamino)-4H-5,6-dihydro-1,3-thiazine; *Bay 1470; Bay Va 1470; Wh-7286.* $C_{12}H_{16}N_2S$; mol wt 220.34. C 65.41%, H 7.32%, N 12.71%, S 14.55%. Prepn: O. Behner *et al.,* **Belg. pat. 634,552,** *C.A.* **61,** 4369k (1964); *eidem,* **Ger. pat. 1,173,475,** *C.A.* **61,** 13323c (1964); *eidem,* U.S. pat. **3,235,550** (1964, 1964, 1966 all to Bayer). Pharmacology: G. Sagner *et al., Deut. Tieraerztl. Wochenschr.* **75,** 565 (1968); Kroneberg, Schlossman, *Arch. Pharmakol. Exp. Pathol.* **268,** 348 (1971). Metabolic studies: Duhm *et al., Berlin. München. Tieraerztl. Wochenschr.* **82,** 104 (1969). Veterinary clinical studies: Clarke, Hall, *Vet. Rec.* **85,** 512 (1969); Burns, McMullan, *Vet. Med.* **67,** 77 (1972).

Colorless, almost tasteless crystals from benzene-ligroin, mp 140-142°. Also reported as mp 136-139°. Soluble in dilute acids, benzene, acetone, chloroform; sparingly sol in petr ether. Practically insol in water, alkalies. LD_{50} in mice (mg/kg): 43 i.v.; 121 s.c.; 240 orally; LD_{50} in rats (mg/kg): 130 orally (Sagner).
Hydrochloride, $C_{12}H_{16}N_2S \cdot HCl$, *Narcoxyl, Rompun, Xylapan, Xylasol.*
THERAP CAT (VET): Sedative, analgesic, muscle relaxant.

10214. Xylene. *Dimethylbenzene;* xylol. C_8H_{10}; mol wt 106.17. C 90.51%, H 9.49%. $C_6H_4(CH_3)_2$. First isolated from a crude wood distillate: Cahours, *Compt. Rend.* **30,** 319 (1850). Obtained from coal tar: Fittig, *Ann.* **153,** 265

(1870). The xylene of commerce is a mixture of the three isomers *o-*, *m-* and *p*-xylene, the *m*-isomer predominating. Manuf from pseudocumene: Seubold, U.S. pat. **2,960,545** (1960 to Union Oil); by catalytic isomerization of a hydrocarbon fraction: Berger, U.S. pat. **3,078,318** (1963 to Universal Oil Prod.). Separation of isomers by clathration: Schaeffer, U.S. pat. **3,029,300** (1962 to Union Oil). Use as clearing agent: K. Kubota, *J. Polym. Sci.* **5**, 1179 (1967); J. B. Matthews, *J. Clin. Pathol.* **34**, 103 (1981). Toxicity data: H. F. Smyth *et al., Am. Ind. Hyg. Assoc. J.* **23**, 95 (1962). Series of articles on toxicity: *Environ. Health Perspect.* **101**, Suppl. 6, 115-149 (1993). Review of mfg processes: Faith, Keyes & Clark's *Industrial Chemicals*, F. A. Lowenheim, M. K. Moran, Eds. (Wiley-Interscience, New York, 4th ed., 1975) pp 874-881; of toxicology and human exposure: *Toxicological Profile for Total Xylenes* (PB91-181552, 1990) 191 pp.

Mobile, flammable liquid; d ∼0.86; bp 137-140°. Flash pt 29°. Practically insol in water. Miscible with abs alcohol, ether, and many other organic liquids.

m-Xylene, colorless liquid; d_4^{15} 0.8684; mp −47.4°; bp 139.3°; n_D^{20} 1.4973. Flash pt, closed cup: 77°F (25°C). Insol in water. Miscible with alcohol, ether, and many other organic solvents. LD$_{50}$ orally in rats: 7.71 ml/kg (Smyth).

o-Xylene, colorless liquid; d_4^{20} 0.8801; mp −25°; bp 144°; n_D^{20} 1.5058. Flash pt, closed cup: 63°F (17°C). Insol in water. Miscible with alc, ether.

p-Xylene, colorless plates or prisms at low temp; d_4^{20} 0.86104; mp 13-14°; bp 137-138°; n_D^{20} 1.49575: Thorne *et al., Ind. Eng. Chem. Anal. Ed.* **17**, 481 (1945). Flash pt, closed cup: 77°F (25°C). Insol in water. Sol in alcohol, ether, organic solvents.

Caution: Potential symptoms of acute overexposure by inhalation to *m-*, *o-*, or *p*-isomers are flushing and reddening of the face, a feeling of increased heat due to dilation of superficial blood vessels, disturbed vision, dizziness, tremors, salivation, cardiac stress, drowsiness, incoordination and staggering gait, CNS depression, confusion, coma. Symptoms of chronic inhalation exposure may include respiratory irritation, CNS excitation followed by CNS depression, paresthesia, tremors, apprehension, impaired memory, weakness, nervous irritation, vertigo, headache, anorexia, nausea, flatulence, anemia, mucosal hemorrhage. Overexposure by ingestion may cause severe GI distress. Aspiration into lungs may cause chemical pneumonitis, pulmonary edema, hemorrhage. Direct contact may cause eye irritation, conjuctivitis, corneal burns; skin irritation and dermatitis due to defatting action. See *Patty's Industrial Hygiene and Toxicology* **vol. 2B**, G. D. Clayton, F. E. Clayton, Eds. (Wiley-Interscience, New York, 4th ed., 1994) p 1332-1339; *NIOSH Pocket Guide to Chemical Hazards* (DHHS/NIOSH 90-117, 1990) p 226.

USE: As solvent; raw material for production of benzoic acid, phthalic anhydride, isophthalic and terephthalic acids as well as their dimethyl esters used in the manufacture of polyester fibers; manuf dyes and other organics; sterilizing catgut; with Canada balsam as oil-immersion in microscopy; clearing agent in microscope technique.

10215. Xylenol. *Dimethylphenol.* $C_8H_{10}O$; mol wt 122.17. C 78.65%, H 8.25%, O 13.10%. $(CH_3)_2C_6H_3OH$. Constituent of "cresylic acid." There are 6 isomers of xylenol. They are only slightly sol in water but freely sol in alcohol, chloroform, ether, benzene, etc.; also sol in NaOH soln. Prepn and physical properties: R. J. L. Andon *et al., J. Chem. Soc.* **1960**, 5246.

2,3-Dimethylphenol, *vic-o-xylenol.* Needles from water or dil alc, mp 75°, bp 218°: Thöl, *Ber.* **18**, 2561 (1885); also reported as mp 72.57° ±0.02°, bp$_{760}$ 216.87° ±0.001° (Andon).

2,4-Dimethylphenol, *as-m-xylenol.* Crystals, bp$_{766}$ 211.5°, mp 25.4-26°: Jacobsen, *Ber.* **11**, 17 (1878); **18**, 3463 (1885); also reported as mp 24.54° ±0.01°, bp$_{760}$ 210.931° ±0.001° (Andon).

2,5-Dimethylphenol, *p-xylenol.* Crystals from alcohol + ether, mp 74.5°, bp$_{762}$ 211.5° (Jacobsen); bp 213.5°: Würtz, *Ann.* **147**, 372 (1868); also reported as mp 74.85° ±0.02°, bp$_{760}$ 211.132° ±0.002° (Andon).

2,6-Dimethylphenol, *vic-m-xylenol.* Needles, mp 49°, bp 203°: Gattermann, *Ann.* **357**, 313 (1907); also reported as mp 45.62° ±0.01°, bp$_{760}$ 201.030° ±0.001° (Andon).

3,4-Dimethylphenol, *as-o-xylenol.* Needles from water, mp 62.5°, bp 225°: Jacobsen, *Ber.* **17**, 159 (1884); also reported as mp 65.11° ±0.01°, bp$_{760}$ 226.947° ±0.001° (Andon).

3,5-Dimethylphenol, *sym-m-xylenol.* Needles from water, mp 64°, bp 219.5°: Thöl, *Ber.* **18**, 359 (1885); also reported as mp 63.27° ±0.02°, bp$_{760}$ 221.962° ±0.003° (Andon). LD$_{50}$ orally in mice: 620 mg/kg: W. A. McOmie *et al., J. Am. Pharm. Assoc.* **38**, 366 (1949).

USE: For the prepn of coal tar disinfectants; manuf of artificial resins.

10216. Xylenol Blue. *4,4'-(3H-2,1-Benzoxathiol-3-ylidene)bis[2,5-dimethylphenol] S,S-dioxide; p*-xylenolsulfonephthalein; 1,4-dimethyl-5-hydroxybenzenesulfonephthalein; *α,4,4'-trihydroxy-2,5,2',5'-tetramethyltriphenylmethane-2''*-sulfonic acid γ-sultone. $C_{23}H_{22}O_5S$; mol wt 410.49. C 67.30%, H 5.40%, O 19.49%, S 7.81%. Prepd from *o*-sulfobenzoic acid dichloride or *o*-sulfobenzoic acid anhydride and *p*-xylenol in the presence of zinc chloride: Cohen, *Biochem. J.* **16**, 31 (1922).

Brown crystals from alcohol.
USE: Indicator, used in 0.02% soln: pH 1.2 red, 2.8 yellow, 9.6 blue.

10217. Xylidine. *ar,ar-Dimethylbenzenamine;* dimethylaniline; aminodimethylbenzene. $C_8H_{11}N$; mol wt 121.18. C 79.29%, H 9.15%, N 11.56%. $(CH_3)_2C_6H_3NH_2$. There are six isomeric xylidines. Prepd by reduction of corresponding nitro-compounds: Allchin, U.S. pat. **1,867,962** (to I.C.I.); physical properties of all six isomers also given: Birch *et al., J. Am. Chem. Soc.* **71**, 1362 (1949); van Loon *et al., Rec. Trav. Chim.* **79**, 977 (1960); Bergmann, Berkovic, *J. Org. Chem.* **26**, 919 (1961).

All except *o*-4-xylidine are liquids above 20°. d 0.97-0.99, and bp 213-226°. They are sparingly sol in water, sol in alcohol and form more or less sol salts with the strong mineral acids.

Caution: Potential symptoms of overexposure are anoxia, cyanosis; lung, liver and kidney damage. See *NIOSH Pocket Guide to Chemical Hazards* (DHHS/NIOSH 90-117, 1990) p 226.

USE: Chiefly in the manuf of dyes.

10218. Xylitol. *xylo*-Pentane-1,2,3,4,5-pentol; xylite; Eutrit; Kannit; Klinit; Kylit; Newtol; Torch; Xyliton. $C_5H_{12}O_5$; mol wt 152.15. C 39.47%, H 7.95%, O 52.58%. Intermediate in metabolism of D-glucose through glucuronate cycle in livers. Prepd by reduction of xylose: G. Bertrand, *Bull. Soc. Chim. France* [3] **5**, 555 (1891); E. Fischer, R. Stahel, *Ber.* **24**, 538 (1891). Prepn of metastable crystals: M. L. Wolfrom, E. J. Kohn, *J. Am. Chem. Soc.* **64**, 1739 (1942); of stable form: J. F. Carson *et al., ibid.* **65**, 1777 (1943). Crystal structure: H. S. Kim, G. A. Jeffrey, *Acta Crystallogr.* **25B**, 2607 (1969). Use in prevention of dental caries: E. Grunberg *et al., Int. J. Vit. Nutr. Res.* **43**, 227 (1973); A. Scheinin, K. K. Makinen, **Ger.** pat. **2,606,533** (1976 to Hoffmann-La Roche), *C.A.* **85**, 149140h (1976). Acute toxicity: S. Salminen *et al., Toxicol. Lett.* **18**, Suppl. 1, 37 (1983). Reviews of toxicity, metabolism and use as dietary additive: *International Symposium on Metabolism, Physiology and Clinical Uses of Pentoses and Pentitols*, B. L. Horecker *et al.*, Eds. (Springer-Verlag, New York, 1969) 408 pp; *Sugars in Nutrition*, H. L. Sipple, K. W. McNutt, Eds. (Academic Press, New York, 1974) *passim*; G. E. Demetrakopoulos, H. Amos, *World Rev. Nutr. Diet* **32**, 96-122 (1978); R. Ylikahri, *Adv. Food Res.* **25**, 159-180 (1979).

Book: *Xylitol*, J. N. Counsel, Ed. (Applied Science, London, 1978) 191 pp.

$$CH_2OH - HC-OH - HO-CH - HC-OH - CH_2OH$$

Stable form: orthorhombic needles from THF, prisms from ethanol; mp 93-94.5°; d 1.52. Metastable form: colorless, monoclinic, lath-shaped crystals from anhydrous methanol; hygroscopic; mp 61-61.5°. Soly of stable form (g/100 g soln): abs methanol 6.0; abs ethanol 1.2; water 64.2. Relative sweetness equal to sucrose. LD_{50} orally in mice: approx 22 g/kg (Salminen).

USE: As oral and intravenous nutrient; in anticaries preparations.

10219. Xylometazoline. *2-[[4-(1,1-Dimethylethyl)-2,6-dimethylphenyl]methyl]-4,5-dihydro-1H-imidazole; 2-(4-tert-butyl-2,6-dimethylbenzyl)-2-imidazoline.* $C_{16}H_{24}N_2$; mol wt 244.38. C 78.64%, H 9.90%, N 11.46%. α-Adrenergic agonist. Prepn: Hüni, U.S. pat. **2,868,802** (1959 to Ciba). Pharmacological studies: Morimoto, Tanaka, *C.A.* **72**, 20437n (1970).

mp 131-133°.
Hydrochloride, $C_{16}H_{24}N_2$.HCl, *Neo-Rinoleina, Novorin, Olynth, Otriven, Otrivin, Otrix, Therapin, Xymelin.* Soly in water: up to 3%; also sol in methanol, ethanol. Practically insol in ether, benzene.

THERAP CAT: Decongestant (nasal).

10220. Xylose. *D-Xylose; wood sugar;* Xylomed; Xylo-Pfan. $C_5H_{10}O_5$; mol wt 150.13. C 40.00%, H 6.71%, O 53.28%. Widely distributed in plant materials, especially in wood (maple, cherry), in straw, in hulls. Not found in free state, but in form of xylan, a polysaccharide built from D-xylose units and occurring in association with cellulose. Xylose occurs also as part of glycosides. Isoln from corn cobs by boiling with 8% H_2SO_4: Monroe, *J. Am. Chem. Soc.* **41**, 1002 (1919). Peanut shells and cottonseed hulls also are practical sources of xylose: Ling, Nanji, *J. Chem. Soc.* **1923**, 620. Configuration: Hudson, Yanovsky, *J. Am. Chem. Soc.* **39**, 1029 (1917); Haworth, *Nature* **116**, 430 (1925). Review on history, constitution and prepn: Harding, *Sugar* **24**, 14 (1922).

α-D-Xylose

Monoclinic needles or prisms. Very sweet taste. mp 144-145° (Wheeler, Tollens, *Ann.* **254**, 309); mp 153-154° (Hébert, *Compt. Rend.* **110**, 970). d_4^{20} 1.525. Shows mutarotation. $[\alpha]_D^{20}$ +92° → +18.6° (16 hrs c = 10). One gram dissolves in 0.8 ml water. Sol in pyridine, hot alcohol. pKa (18°): 12.14. Reduces warm Fehling's soln. Upon heating with water in closed tube to 140° or by boiling with dil H_2SO_4, furfurol is formed.

USE: In tanning, dyeing, and as a diabetic food.
THERAP CAT: Diagnostic aid (intestinal function).

10221. Xylulose. *threo-Pentulose.* $C_5H_{10}O_5$; mol wt 150.13. C 40.00%, H 6.71%, O 53.28%. L-Form has been found in the urine of humans with pentosuria. Prepn of DL-form: Gascoigne, *Chem. & Ind. (London)* **1959**, 402; of D-

form: Mendicino, *J. Am. Chem. Soc.* **82**, 4975 (1960); of L-form: Wolfrom, Bennett, *J. Org. Chem.* **30**, 458 (1965). Isoln of DL-form from the acid hydrolysate of bagasse hemicellulose: Banerjee *et al., Sci. Cult. (Calcutta)* **27**, 498 (1961), *C.A.* **56**, 11682d (1962). Enzymic prepn of L-form: Hough, Jones, *Chem. & Ind. (London)* **1952**, 907; *eidem, J. Chem. Soc.* **1952**, 4047. Formation of L-form in normal humans and guinea pigs, and its utilization by guinea-pig liver prepns: Touster *et al., J. Am. Chem. Soc.* **76**, 5005 (1954). *Reviews: The Carbohydrates*, W. Pigman, Ed. (Academic Press, New York, 1957) pp 80, 86-87, 759, 795; *Methods in Carbohydrate Chemistry* vol. 1, R. L. Whistler, M. L. Wolfrom, Eds. (Academic Press, New York, 1962) pp 94-101.

$$CH_2OH - C=O - HC-OH - HO-CH - CH_2OH$$

L-isomer

D-Isomer, syrup. $[\alpha]_D^{18}$ −33° (c = 2.5).
D-Isomer *p*-bromophenylhydrazone, $C_{11}H_{15}BrN_2O_4$, pale yellow crystals from abs ethanol + water, mp 128-129°. $[\alpha]_D^{20}$ +24° (15 min) → −31° (7 days, in pyridine).
L-Isomer, syrup. $[\alpha]_D^{21}$ +31°.
L-Isomer *p*-bromophenylhydrazone, yellow plates from dil alc, mp 128°. $[\alpha]_D^{20}$ −20° (10 min) → +22° (5 hrs, c = 0.5 in ethanol).

10222. 1-Xylylazo-2-naphthol. *1-[(2,4-Dimethylphenyl)-azo]-2-naphthalenol; C.I. Solvent Orange 7;* 1-(2,4-xylylazo)-2-naphthol; FD & C Red no. 32; Oil Red XO; Ext. D & C Red no. 14; C.I. 12140. $C_{18}H_{16}N_2O$; mol wt 276.34. C 78.24%, H 5.84%, N 10.14%, O 5.79%. Once reported as 2,5-xylylazo deriv. Prepd by coupling diazotized *m*-xylidene with 2-naphthol: J. M. Tedder, *J. Chem. Soc.* **1957**, 4003; R. B. Smyth, G. G. McKeown, *J. Chromatog.* **5**, 395 (1961). Metabolism: J. L. Radomski, *J. Pharm. Exp. Ther.* **136**, 378 (1962).

Red needles, mp 166°. Insol in water; sol in ethanol, acetone, benzene.

Caution: Delisted for use in foods, drugs, and cosmetics by the FDA.

10223. Xylyl Bromide. C_8H_9Br; mol wt 185.06. C 51.92%, H 4.90%, Br 43.18%. Prepn of *m*-isomer: Wenner, *J. Org. Chem.* **17**, 523 (1952); of *o*-isomer: Dev, *J. Indian Chem. Soc.* **32**, 403 (1955); of *p*-isomer: Cockburn *et al., J. Chem. Soc.* **1960**, 3340.

m-Xylyl bromide, 1-(bromomethyl)-3-methylbenzene, α-bromo-m-xylene, m-methylbenzyl bromide, ω-bromo-m-xylene. Liquid, bp 212-215° with slight dec. d^{23} 1.371. Practically insol in water; sol in alcohol, ether.
o-Xylyl bromide, prisms, mp 21°. bp 223-234°, bp_{742} 216-217°, bp_{15} 102°. n_D^{20} 1.5730. d^{23} 1.381. Practically insol in water. Sol in alcohol, ether.

p-Xylyl bromide, needles from alcohol, mp 38°. bp$_{740}$ 218-220°, bp$_{15}$ 120°. d 1.324. Practically insol in water. Very sol in chloroform, hot ether.

Caution: Powerful lacrimator.

USE: In organic syntheses; in war-gas formulations.

10224. Xylyl Chloride. C_8H_9Cl; mol wt 140.61. C 68.34%, H 6.45%, Cl 25.21%. Prepn of *o*-isomer: Rabjohn, *J. Am. Chem. Soc.* **76**, 5479 (1954); of *m*-isomer: van Zanten, Nauta, *Rec. Trav. Chim.* **79**, 1211 (1960); of *p*-isomer: Newman, George, *J. Org. Chem.* **26**, 4306 (1961).

m-Xylyl chloride, 1-(chloromethyl)-3-methylbenzene, α-chloro-m-xylene, m-methylbenzyl chloride, ω-chloro-m-xylene. Liquid, bp 195-196°. bp$_{10}$ 90-92°. d^{20} 1.064. Practically insol in water; miscible with abs alcohol, ether.

o-Xylyl chloride, liquid, bp 195-203°, bp$_{25}$ 95-96°. n_D^{20} 1.5391. Practically insol in water; miscible with abs alcohol, ether.

p-Xylyl chloride, fuming liquid, characteristic odor. bp 200-202°, bp$_1$ 48-50°. n_D^{17} 1.5360. Practically insol in water; miscible with abs alcohol, ether.

Caution: Powerful lacrimator.

Y

10225. Yam, Mexican. Giant yam; cabeza de negro. *Dioscorea macrostachya* Benth., *Dioscoreaceae* or one of ~600 other *Dioscoreae*, native to subtropical countries. The underground tubers weigh as much as 90 lbs. and contain steroidal sapogenins. Account of nature, origins, cultivation and utilization of the useful members of the *Dioscoreaceae:* D. G. Coursey, *Yams* (Longmans, London, 1967) 230 pp. *See also* Dioscorea and Barbasco.
USE: In the partial synthesis of hormones having a steroid structure.

10226. Yangonin. *(E)-4-Methoxy-6-[2-(4-methoxyphenyl)ethenyl]-2H-pyran-2-one; 4-methoxy-6-(p-methoxystyryl)-2H-pyran-2-one;* 5-hydroxy-3-methoxy-7-(p-methoxyphenyl)-2,4,6-heptatrienoic acid δ-lactone; 4-methoxy-6-[β-(p-anisyl)vinyl]-α-pyrone; 6-(p-methoxystyryl)-4-methoxy-α-pyrone. $C_{15}H_{14}O_4$; mol wt 258.27. C 69.76%, H 5.46%, O 24.78%. From root of *Piper methysticum* Forst., *Piperaceae* (kava): Winzhermer, *Arch. Pharm.* **246**, 338 (1908); Borsche, Gerhardt, *Ber.* **47**, 2902 (1914). From *Ranunculus quelpaertensis* Nakai, *Ranunculaceae:* Shibata *et al., Bull. Chem. Soc. Japan* **45**, 930 (1972). Structure: Chmielewska *et al., Tetrahedron* **4**, 36 (1958); Bu'Lock, Smith, *J. Chem. Soc.* **1960**, 502. Synthesis: Harris, Combs, *J. Org. Chem.* **33**, 2399 (1968); R. Bacardit, *J. Heterocycl. Chem.* **19**, 157 (1982). Molecular and crystal structure: Engel, Nowacki, *Z. Kristallogr., Kristallgeometrie, Kristallphys., Kristallchem.* **134**, 180 (1971). Activity studies: Kretzschmar *et al., Arch. Int. Pharmacodyn. Ther.* **180**, 471 (1969).

Crystals from methanol, mp 155-157°. uv max (ethanol): 360 nm (log ε 4.33). Practically insol in water; sol in hot alcohol, glacial acetic acid, ethyl acetate, acetone; slightly sol in benzene, ether.
Dihydroyangonin, $C_{15}H_{16}O_4$, needles from alc, mp 106-107°. uv max (methanol): 274, 228 nm (log ε 3.82, 3.71).

10227. Yeast. The moist, living cells of a fungus or fungi whose usual and dominant growth form is unicellular. The term yeast is not one with an exact botanical meaning, *see* Henrici's *Molds, Yeasts and Actinomycetes* (New York, 1947) p 264 sqq. Normally produces alcoholic fermentation in fluids contg sugar. Moist yeast is usually combined with a starchy or absorbent base and comes on the market in the form of white or yellowish-white, soft, easily broken masses of a characteristic, slightly sour odor. Description of a modern process of manuf: Schultz, Swift, U.S. pat. **2,717,-837** (1955 to Standard Brands).
The dried yeast of pharmacopoeias consists of the dry cells of any suitable strain of *Saccharomyces cerevisiae* Meyen, *Saccharomytaceae* or *Candida utilis* (Hanneberg) Lodder and Kreger-Van Rij, *Cryptococcaceae* (torula yeast), usually obtained as a by-product from the brewing of beer made from an extract of cereal grains and hops. The yeast cells are washed free of beer and dried, and may be debittered. These yeasts are commonly known as "Brewer's Dried Yeast" and "Debittered Brewer's Dried Yeast." Dried yeast may be obtained also by growing suitable strains of yeast, using media other than those required for the production of beer, and under appropriate environmental conditions. The yeast thus obtained is commonly known as "Primary Dried Yeast."
Dried yeast that fulfills pharmacopeal requirements contains not less than 40% protein, and, in each gram, the equivalent of not less than 0.12 mg of thiamine hydrochloride, 0.04 mg of riboflavin, and 0.25 mg of nicotinic acid. It

occurs as yellowish-white to weak yellowish-orange flakes, granules or powder, having an odor indicative of the type. It is inactive in fermenting power.
USE: Moist yeast in baking bread, brewing; producing alcohol by fermentation of sugar, molasses, and cereals; as a source of vitamins. Dried yeast as a source of vitamins; in baking.
THERAP CAT: Source of protein and vitamin B complex.
THERAP CAT (VET): Dietary source of B vitamins.

10228. Yellow AB. *1-Phenylazo-2-naphthalenamine; C.I. Solvent Yellow 5;* FD & C Yellow no. 3; Ext. D & C Yellow no. 9; C.I. 11380. $C_{16}H_{13}N_3$; mol wt 247.30. C 77.71%, H 5.30%, N 16.99%. Prepd from diazotized aniline and β-naphthylamine: Lawson, *Ber.* **18**, 796 (1885); *Colour Index* vol. 4 (3rd ed., 1971) p 4021. Toxicity studies: Allmark *et al., J. Pharm. Pharmacol.* **7**, 591 (1955); Hansen *et al., Toxicol. Appl. Pharmacol.* **5**, 16 (1963). Formerly used in the U.S. for coloring oleomargarine.

Red platelets from abs alc, mp 102-104°. Practically insol in water. Sol in alcohol, carbon tetrachloride, acetic acid, vegetable oils. An alcoholic soln becomes redder with HCl addition, and is unaltered by NaOH addition.
Caution: Delisted for internal use in 1959: *Fed. Reg.* **24**, 883 (1959), *C.A.* **53**, 12505a. Delisted also for external use in 1960. The FDA has declared this color unsafe due to the impurity 2-naphthylamine, *q.v.*, a proven carcinogen.

10229. Yellow OB. *1-[(2-Methylphenyl)azo]-2-naphthalenamine; C.I. Solvent Yellow 6;* 1-o-tolylazo-2-naphthylamine; FD & C Yellow no. 4; Ext. D & C Yellow no. 10; C.I. 11390. $C_{17}H_{15}N_3$; mol wt 261.33. C 78.13%, H 5.79%, N 16.08%. Prepd from diazotized o-toluidine and β-naphthylamine: Krüss, *Z. Physik. Chem.* **51**, 270 (1905); Norman, *J. Chem. Soc.* **101**, 1913 (1912); Fischer, *J. Prakt. Chem.* **104**, 102 (1922); Hodgson, Foster, *J. Chem. Soc.* **1942**, 30. Toxicity studies: Allmark *et al., J. Pharm. Pharmacol.* **7**, 591 (1955); Hansen *et al., Toxicol. Appl. Pharmacol.* **5**, 16 (1963). Formerly used in the U.S. for coloring oleomargarine.

Deep red crystals from alc. Also reported as bright yellow needles (Krüss). mp 125-126° (Fischer). Practically insol in water. Sol in alc, ether, benzene, carbon tetrachloride, vegetable oils, glacial acetic acid. An alcoholic soln becomes redder with HCl addition, but is unaltered by NaOH addition.
Caution: Delisted for internal use in 1959: *Fed. Reg.* **24**, 883 (1959), *C.A.* **53**, 12505a (1959). Delisted also for external use in 1960. The FDA has declared this color unsafe due to the impurity 2-naphthylamine, *q.v.*, a proven carcinogen.

10230. Yellow Phenolphthalein. In the mfg process of phenolphthalein, *q.v.*, a stage is reached where certain by-products formed in the synthesis have not yet been removed: Ebert, *et al.*, U.S. pat. **1,711,048**, *C.A.* **23**, 2990 (1929); Hubacher, U.S. at. **1,940,495**, *C.A.* **28**, 1366 (1934); U.S. pat. **2,192,485**, *C.A.* **34**, 4396 (1940). These are yellow to brownish-yellow bodies mixed with the white phenolphthalein and imparting a characteristic yellow color to the product. Compds isolated from a sample prepd according to

U.S. pat. **2,168,346** are: white phenolphthalein 93.16%, fluoran 0.32%, isophenolphthalein 0.08%, white compd mp 250° 0.04%, 2-(4-hydroxybenzoyl)benzoic acid 0.10%. Total compds isolated: 93.70%. Review of chemistry and pharmacology: Hubacher, Doernberg, *J. Am. Pharm. Assoc., Sci. Ed.* **37**, 261 (1948).

The commercially available yellow phenolphthalein is a powder of moderate yellow color. d_4^{20} 1.290-1.296. mp 255-260°. One gram dissolves in 12 ml alcohol, in 102 ml ether. Solns show a slight greenish fluorescence. Found to be 2.5 times more active as a laxative in rhesus monkeys than phenolphthalein.

THERAP CAT: Cathartic.

10231. Yig. *Iron yttrium oxide;* yttrium iron garnet. $Fe_5O_{12}Y_3$; mol wt 737.94. Fe 37.84%, O 26.02%, Y 36.14%. $Y_3Fe_5O_{12}$. Man-made mineral. Prepn by molten salt technique: Nielsen, *Electronics*, Jan. 31, **1964**, pp 44-45. Advances in manufacture and applications: Gundlach, *ibid.*, Oct. 14, **1968**, pp 104-118.

Hard, brittle, garnet-type crystals. Hardness about equal to that of quartz. mp > 1040°. The melt is easily amenable to doping and crystal nucleation by epitaxy. Crystals may be oriented along the simple axes (100), (110), and (111) using the natural (110) face of the crystal.

USE: In microwave tunable filters and limiters.

10232. Ylangene. *1,3-Dimethyl-8-(1-methylethyl)tricyclo[4.4.0.0²,⁷]dec-3-ene; (1S,2R,6R,7R,8S)-(+)-8-isopropyl-1,3-dimethyltricyclo[4.4.0.0²,⁷]dec-3-ene;* α-ylangene. $C_{15}H_{24}$; mol wt 204.36. C 88.16%, H 11.84%. From ylang-ylang oil: Heraut, Dimitrov, *Chem. Listy* **46**, 432 (1952); from oil of birch buds: Holub *et al., Coll. Czech. Chem. Commun.* **24**, 3730 (1959); from oil of *Juniperus oxycedrus* Linn., *Cupressaceae:* Motl *et al., ibid.* **25**, 1656 (1960). Structure: Motl *et al., Chem. & Ind. (London)* **1963**, 1759; Hunter, Brogden, *J. Org. Chem.* **29**, 982 (1964); Motl *et al., Tetrahedron Letters* **1965**, 451. Stereoisomer of copaene, *q.v.* Total synthesis of (±)-form: Heathcock *et al., J. Am. Chem. Soc.* **89**, 4133 (1967); Corey, Watt, *ibid.* **95**, 2303 (1973).

Oil. d_4^{20} 0.9091. n_D^{20} 1.4934. $[\alpha]_D^{20}$ +15.4°.

10233. Ylang-Ylang Oil. Cananga oil. Volatile oil distilled in Madagascar, Reunion Island, Comoro Islands and the Philippine Islands from freshly picked flowers of *Cananga odorata* Hook & Thoms., *Anonaceae.* The first distillate yields the so-called "ylang-ylang extra" and is the finest oil. The first, second and third fractions follow, the first two having insignificant commercial value. *Constit.* Geraniol and linalool esters of acetic and benzoic acids; *p*-cresol methyl ether, cadinene, a sesquiterpene, a phenol.

Light yellow, very fragrant liquid. d_{20}^{20} 0.930-0.950. Rotation: −27° to −50° in 100-mm tube.

USE: In delicate perfumes.

10234. Ylides. Compounds in which an atom from group V or VI of the periodic table, bearing a positive charge, is connected to a carbon atom carrying an unshared pair of electrons. *Ref:* A. William Johnson, *Ylid Chemistry* (Academic Press, New York, 1966); B. M. Trost, L. S. Melvin, *Sulfur Ylides* (Academic Press, New York, 1975); J. March, *Advanced Organic Chemistry* (John Wiley & Sons, New York, 4th ed., 1992) p 39.

10235. Yogurt. Yoghurt; joghurt; kisselo-mleko; mazun; leben raib; dahi. Fermented whole milk. Produced by evaporating milk to ~50% of its volume and maintaining at 50°C for 12 hrs after adding "maya" a mixture of *Lactobacillus bulgaricus*, L. *acidophilus*, and *Streptococcus lactis*. Used as food. Contains antibacterial principles. Differentiation of yogurt and *clabber* as produced in Kentucky: *J. Am. Med. Assoc.* **209**, 778 (1969).

10236. Yohimbine. *(16α,17α)-17-Hydroxyyohimban-16-carboxylic acid methyl ester;* quebrachine; corynine; aphrodine. $C_{21}H_{26}N_2O_3$; mol wt 354.45. C 71.16%, H 7.39%, N 7.90%, O 13.54%. Indole alkaloid with α_2-adrenergic blocking activity. Found in *Corynanthe johimbe* K. Schum., *Rubiaceae* and related trees, also in *Rauwolfia serpentina* (L.) Benth., *Apocynaceae:* Raymond-Hamet, *J. Pharm. Chim.* **19**, 209 (1934); Hofmann, *Helv. Chim. Acta* **37**, 849 (1954); Stoll, Jucker, *Ullmanns Encyklopädie der Technischen Chemie*, vol. 3 (Munich, 3rd ed., 1953) p 266; Bader *et al., J. Am. Chem. Soc.* **76**, 1695 (1954). Structure: Witkop, *Ann.* **554**, 83 (1943); Clemo, Swan, *J. Chem. Soc.* **1946**, 617. Stereochemistry: Janot *et al., Bull. Soc. Chim. France* **1952**, 1085; Godfredsen, Vandegal, *Acta Chim. Scand.* **10**, 1414 (1956); Van Tamelen *et al., J. Am. Chem. Soc.* **78**, 4628 (1956); Ban, Yonemitsu, *Tetrahedron* **20**, 2877 (1964). Synthesis: Van Tamelen *et al., J. Am. Chem. Soc.* **80**, 5006 (1958); Liljegren, Potts, *J. Org. Chem.* **27**, 377 (1962). Total synthesis: Van Tamelen *et al., J. Am. Chem. Soc.* **91**, 7315 (1969); Stork, Guthikonda, *ibid.* **94**, 5109 (1972); T. Kametani *et al., Chem. Pharm. Bull.* **24**, 2500 (1976); E. Wenkert *et al., J. Am. Chem. Soc.* **100**, 4894 (1978); **101**, 5370 (1979); **104**, 2244 (1982); I. Ninomiya *et al., Heterocycles* **14**, 631 (1980). Pharmacokinetics in humans: J. A. Owen *et al., Eur. J. Clin. Pharmacol.* **32**, 577 (1987). Clinical studies in impotence: K. Reid *et al., Lancet* **2**, 421 (1987); A. Morales *et al., J. Urol.* **137**, 1168 (1987). Review of pharmacology and use in molecular studies of α_2-adrenoreceptor: M. R. Goldberg, D. Robertson, *Pharmacol. Rev.* **35**, 143-180 (1987). Comprehensive description: A. G. Mekkawi, A. A. Al-Badr in *Analytical Profiles of Drug Substances* vol. 16, K. Florey, Ed. (Academic Press, New York, 1986) pp 731-768.

Orthorhombic needles from dil alc, mp 234°. Also mp 235-237°. $[\alpha]_D^{20}$ +50.9° to +62.2° (ethanol); $[\alpha]_D^{20}$ +108° (pyridine); $[\alpha]_{546}^{20}$ +129° (c = 0.5 in pyridine). uv max (methanol): 226, 280, 291 nm (log ε 4.56, 3.88, 3.80). Sparingly sol in water. Sol in alcohol, chloroform, hot benzene; moderately sol in ether.

Hydrochloride, $C_{21}H_{26}N_2O_3$.HCl, *Aphrodyne, Yocon, Yohimex, Yohydrol.* Orthorhombic plates, prisms from alc; dec 302°. $[\alpha]_D^{22}$ +105° (H_2O). Sol in ~120 ml water, 400 ml alc. The aq soln is about neutral.

USE: Pharmacological probe for the study of α_2-adrenoceptor.

THERAP CAT: α-Adrenergic blocker; mydriatic.

THERAP CAT (VET): Adrenergic blocking agent. Has been used as an aphrodisiac.

10237. allo-Yohimbine. *17-Hydroxyyohimban-16-carboxylic acid methyl ester;* alloyohimbine; dihydroyohimbine. $C_{21}H_{26}N_2O_3$; mol wt 354.45. C 71.16%, H 7.39%, N 7.90%, O 13.54%. From bark of *Corynanthe johimbe* K. Schum., *Rubiaceae:* Warnat, *Ber.* **59**, 2388 (1926); Hahn, Brandenberg, *ibid.* **60**, 699 (1927). Identity with dihydroyohimbine: Heinemann, *ibid.* **67**, 15 (1934). Structure and stereochemistry: LeHir *et al., Bull. Soc. Chim. France* **1953**, 1027; Janot *et al., ibid.* **1961**, 637. Revised structure and total synthesis: Töke *et al., J. Org. Chem.* **38**, 2496 (1973). Total synthesis: E. Wenkert *et al., J. Am. Chem. Soc.* **101**, 5370

(1979). *Review:* A. Chatterjee *et al.*, "Rauwolfia Alkaloids" in Zechmeister, *Progress in the Chemistry of Organic Natural Products* vol. XIII (Springer Verlag, Vienna, 1956) pp 354-356.

Trihydrate, needles from 50% alcohol, mp 98-99°. After drying at 100° and 0.01 mm Hg for 24 hrs, the anhydrous form melts at 135-140°. $[\alpha]_D^{19}$ +84° (c = 0.40 in pyridine). uv max (methanol): 225, 280, 290 nm (log ϵ 4.52, 3.91, 3.74). Sol in ethanol, methanol, pyridine, dil acids; slightly sol in benzene. Practically insol in water.

10238. **α-Yohimbine.** *17α-Hydroxy-20α-yohimban-16β-carboxylic acid methyl ester;* corynanthidine; isoyohimbine; mesoyohimbine; rauwolscine. $C_{21}H_{26}N_2O_3$; mol wt 354.45. C 71.16%, H 7.39%, N 7.90%, O 13.54%. From bark of *Corynanthe johimbe* K. Schum., *Rubiaceae:* Hahn, Brandenburg, *Ber.* **60**, 669 (1927); Wilbaut, van Gastel, *Rec. Trav. Chim.* **54**, 88 (1935); from *Rauwolfia canescens* L., *Apocynaceae:* Mookerjee, *J. Indian Chem. Soc.* **18**, 33 (1941), *C.A.* **35**, 7967 (1941); Stoll *et al.*, *Helv. Chim. Acta* **38**, 270 (1955). Identity with corynanthidine: Janot, Goutarel, *Bull. Soc. Chim. France* **1946**, 535. Identity with isoyohimbine: Heinemann, *Ber.* **60**, 15 (1934). Identity with rauwolscine: Chatterjee *et al.*, *Chem. & Ind. (London)* **1954**, 491. Structure and stereochemistry: Le Hir *et al.*, *Bull. Soc. Chim. France* **1953**, 1027; Wenkert, Liu, *Experientia* **11**, 302 (1955); Aldrich *et al.*, *J. Am. Chem. Soc.* **81**, 2481 (1959); Janot *et al.*, *Bull. Soc. Chim France* **1961**, 637. Total synthesis: Töke *et al.*, *J. Org. Chem.* **38**, 2496 (1973).

Crystals, mp 243-244°. $[\alpha]_D^{19}$ −18° (pyridine); $[\alpha]_D^{19}$ −27° (abs alcohol). pKa 6.34. uv max (methanol): 227, 281 nm (log ϵ 4.50, 3.93). Sol in warm methanol, ethanol; slightly sol in ether, benzene; practically insol in petr ether, water. Hydrochloride, $C_{21}H_{26}N_2O_3$·HCl, crystals, mp 288°. $[\alpha]_D^{20}$ +55.5° (water).

10239. **Ytterbium.** Yb; at. wt 173.04; at. no. 70; valences 2, 3. A rare earth metal of the yttrium group; member of the lanthanide series. Naturally occurring isotopes (mass numbers): 168 (0.13%); 170 (3.05%); 171 (14.3%); 172 (21.9%); 173 (16.12%); 174 (31.8%); 176 (12.7%); known artificial radioactive isotopes: 152-167; 169; 175; 177; 178. Estimated abundance in earth's crust: 2.66-3.1 ppm. Occurs in the rare earth minerals xenotime, ytterbite (gadolinite), monazite. Discovered independently: Urbain, *Compt. Rend.* **145**, 759 (1907); and called *aldebaranium:* von Welsbach, *Monatsh.* **29**, 181 (1908); **34**, 1713 (1913). Sepn and purification: Urbain, *Congress of Applied Chemistry* [X] **94** (1909); Prandtl, *Z. Anorg. Chem.* **238**, 321 (1938); Spedding *et al.*, *J. Am. Chem. Soc.* **74**, 2783 (1952); **76**, 2557 (1954). Spectrum: Exner, Haschek *et al.*, cited by Mellor, *A Comprehensive Treatise on Inorganic and Theoretical Chemistry* **5**, 706 (1929). Toxicity study: Haley, *J. Pharm. Sci.* **54**, 663 (1965). Reviews of prepn, properties and compds: *The Rare Earths,* F. H. Spedding, A. H. Daane, Eds. (Krieger, Huntington, N.Y., 1971, reprint of 1961 ed.) 641 pp; Hulet,

Bode, "Separation Chemistry of the Lanthanides and Transplutonium Actinides", in *MTP Int. Rev. Sci.: Inorg. Chem.*, *Ser. One*, vol. 7, K. W. Bagnall, Ed. (University Park Press, Baltimore, 1972) pp 1-45; Moeller, "The Lanthanides", in *Comprehensive Inorganic Chemistry*, vol. 4, J. C. Bailar Jr. *et al.*, Eds. (Pergamon Press, Oxford, 1973) pp 1-101; F. H. Spedding in Kirk Othmer *Encyclopedia of Chemical Technology* vol. 19 (John Wiley & Sons, New York, 3rd ed., 1982) pp 833-854; *Chemistry of the Elements*, N. N. Greenwood, A. Earnshaw, Eds. (Pergamon Press, New York, 1984) pp 1423-1449. Brief review of properties: G. T. Seaborg, *Radiochim. Acta* **61**, 115-122 (1993).

Silvery, ductile metal. Crystalline forms: face-centered cubic α-form, d 6.977, transforms to β-form at 798°; body-centered cubic β-form, d 6.54, exists at > 798°. mp 819°. bp 1196°. Heat of fusion: 7.657 kJ/mol. Heat of sublimation (25°): 152.1 kJ/mol. Forms both di- and trivalent salts.

Oxide, O_3Yb_2, *ytterbia.* Colorless mass, sol in dil acids.
Hydroxide, $Yb(OH)_3$, colorless gelatinous precipitate.
Chloride, $YbCl_3$, hexahydrate, deliquescent crystals, d 2.575; mp 150-155°. LD_{50} in mice: 395 mg/kg i.p.; 6.7 g/kg orally (Haley).
Nitrate, $Yb(NO_3)_3$, tetrahydrate, transparent hygroscopic prisms from concd nitric acid. LD_{50} (hexahydrate) in rats: 255 mg/kg i.p.; 3.1 g/kg orally (Haley).
Sulfate, $Yb_2(SO_4)_3$, octahydrate, lustrous colorless crystals, soly decreases with rise in temp.

10240. **Yttrium.** Y; at. wt 88.90585; at. no. 39; valence 3. Group IIIb(3). A rare earth metal. Naturally occurring isotope (mass number): 89; known artificial radioactive isotopes: 80-88; 90-100. Estimated abundance in earth's crust: 28.1-31 ppm. Natural sources: xenotime, fergusonite, samarskite, yttrialite, gadolinite, and other rare earth minerals. Discovered in 1794 by Gadolin; separated as yttria in 1843 by Mosander; named by Ekeberg. Sepn by fractional precipitation: Bonardi, James, *J. Am. Chem. Soc.* **37**, 2642 (1915); Willand, James, *ibid.* **38**, 1198 (1916); Wichers *et al.*, *ibid.* **40**, 1615 (1918); by ion exchange: Spedding *et al.*, *ibid.* **69**, 2812 (1947); Mayer, Freiling, *ibid.* **75**, 5647 (1953). Toxicity study: Cochran *et al.*, *Arch. Ind. Hyg. Occup. Med.* **1**, 637 (1950). Reviews of prepn, properties and compds: *The Rare Earths,* F. H. Spedding, A. H. Daane, Eds. (Krieger, Huntington, N.Y., 1971, reprint of 1961 ed.) 641 pp; Vickery, "Scandium, Yttrium and Lanthanum", in *Comprehensive Inorganic Chemistry*, vol. 3, J. C. Bailar Jr. *et al.*, Eds. (Pergamon Press, Oxford, 1973) pp 329-353; Moeller, "The Lanthanides", *ibid.* vol. 4, pp 1-101; F. H. Spedding in Kirk Othmer *Encyclopedia of Chemical Technology* vol. 19 (John Wiley & Sons, New York, 3rd ed., 1982) pp 833-854; *Chemistry of the Elements*, N. N. Greenwood, A. Earnshaw, Eds. (Pergamon Press, New York, 1984) pp 1102-1110, 1423-1449.

Iron-gray, lustrous powder; darkens on exposure to light. Forms hexagonal close-packed crystals. d 4.4689. mp 1522°. bp 3338°. Heat of fusion: 11.43 kJ/mol. Heat of sublimation (25°): 424.7 kJ/mol. E°(aq) Y^{3+}/Y −2.37 V (calc). Oxidizes on heating in air or oxygen; dec cold water slowly, boiling water rapidly.

Oxide, O_3Y_2, *yttria.* White powder, body-centered cubic structure, d 5.03. Obtained by igniting yttrium or its salts. Sol in dil acids; readily absorbs ammonia from the air; displaces ammonia from ammonium salts. LD_{50} i.p. in rats: 500 mg/kg (Cochran).
Hydroxide, $Y(OH)_3$, white gelatinous ppt, dries to a white powder which absorbs CO_2 from the air, obtained by the action of ammonium or alkali hydroxides on a soln of an yttrium salt.
Chloride, YCl_3, hexahydrate, colorless, deliquesc crystals. Sol in water, alc. Anhydr chloride obtained by heating in a stream of HCl.
Carbonate, $Y_2(CO_3)_3$, trihydrate, white to reddish-white powder, prepd by hydrolysis of yttrium trichloroacetate. Insol in water. Sol in dil mineral acids.
Sulfate, $Y_2(SO_4)_3$, octahydrate, monoclinic crystals. Sol in concd sulfuric acid with formation of $Y(HSO_4)_3$. Soly in water decreases with increase in temp. Forms double salts with alkali sulfates.

Nitrate, $Y(NO_3)_3$, hexahydrate, deliquesc crystals, sol in water, produces basic nitrates on partial decompn. LD_{50} i.p. in rats: 350 mg/kg (Cochran).

USE: Yttrium doped with rare earths as phosphors for color television receivers. Oxide for mantles in gas and acetylene lights. Chloride in prepn of pure metal. Yttrium aluminum garnets (YAGS) in lasers and for making artificial diamonds.

Z

10241. Zafirlukast. *[3-[[2-Methoxy-4-[[[(2-methylphen-yl)sulfonyl]amino]carbonyl]phenyl]methyl]-1-methyl-1H-indol-5-yl]carbamic acid cyclopentyl ester;* cyclopentyl 3-[2-methoxy-4-[(o-tolylsulfonyl)carbamoyl]benzyl]-1-methylin-dole-5-carbamate; *N*-[4-[5-(cyclopentyloxycarbonyl)amino-1-methylindol-3-ylmethyl]-3-methoxybenzoyl]-2-methylben-zenesulfonamide; ICI-204219; Accolate. $C_{31}H_{33}N_3O_6S$; mol wt 575.69. C 64.68%, H 5.78%, N 7.30%, O 16.68%, S 5.57%. Leukotriene D_4 antagonist. Prepn: F. J. Brown *et al.,* **Eur.** pat. **Appl. 199,543;** P. R. Bernstein *et al.,* **U.S.** pat. **4,859,692** (1986, 1989 both to ICI); V. G. Matassa *et al., J. Med. Chem.* **33,** 1781 (1990). Preclinical pharmacology: R. D. Krell *et al., Am. Rev. Respir. Dis.* **141,** 978 (1990). Effect on allergen-induced bronchoconstriction: S. R. Findlay *et al., J. Allergy Clin. Immunol.* **89,** 1040 (1992); K. M. O'Shaughnessy *et al., Am. Rev. Respir. Dis.* **147,** 1431 (1993); on exercise-induced asthma: H. K. Makker *et al., ibid.* 1413.

White solid from methanol, mp 138-140°.
THERAP CAT: Antiasthmatic.

10242. Zalcitabine. *2′,3′-Dideoxycytidine;* dideoxycyti-dine; ddCyd; Hivid. $C_9H_{13}N_3O_3$; mol wt 211.22. C 51.18%, H 6.20%, N 19.89%, O 22.72%. Pyrimidine nucleoside ana-log with antiviral activity. Prepn: J. P. Horwitz *et al., J. Org. Chem.* **32,** 817 (1967); R. Marumoto, M. Honjo, *Chem. Pharm. Bull.* **22,** 128 (1974). Prepn and structure-activity study: T.-S. Lin *et al., J. Med. Chem.* **30,** 440 (1987). *In vitro* inhibition of viral DNA synthesis: P. C. van der Vliet, M. M. Kwant, *Biochemistry* **20,** 2628 (1981); of HIV-1 virus replication: H. Mitsuya, S. Broder, *Proc. Nat. Acad. Sci. USA* **83,** 1911 (1986). Phosphorylation by human cells: D. A. Cooney *et al., Biochem. Pharmacol.* **35,** 2065 (1986). Metabolism by human cells: M. C. Starnes, Y. Cheng, *J. Biol. Chem.* **262,** 988 (1987). Potential application in ac-quired immunodeficiency syndrome (AIDS): E. De Clercq, *J. Med. Chem.* **29,** 1561 (1986).

Crystals from ethanol + benzene, mp 215-217° (Horwitz); also reported as mp 209-210° (Lin). $[\alpha]_D^{25}$ +81° (c = 0.635 in water). uv max in 0.1*N* HCl: 280 nm (ε 17720); in 0.1*N* NaOH: 270 nm (ε 8410).
THERAP CAT: Antiviral.

10243. Zaldaride. *1,3-Dihydro-1-[1-[(4-methyl-4H,6H-pyrrolo[1,2-a][4,1]benzoxazepin-4-yl)methyl]-4-piperidinyl]-2H-benzimidazol-2-one.* $C_{26}H_{28}N_4O_2$; mol wt 428.53. C 72.87%, H 6.59%, N 13.07%, O 7.47%. Non-specific ion channel and calmodulin blocker. Prepn: J. W. F. Wasley, J. Norman, **Eur.** pat. **Appl. 233,483;** *eidem,* **U.S.** pat. **4,758,-559** (1987, 1988 both to Ciba-Geigy). Synthesis: S. Boyer *et al., J. Heterocyclic Chem.* **25,** 1003 (1988). Calmodulin inhibition: J. A. Norman *et al., Mol. Pharmacol.* **31,** 535 (1987); and ion channel inhibition: R. Neuhaus, B. F. X. Reber, *Eur. J. Pharmacol.* **226,** 183 (1992). Inhibition of gastric acid secretion: E. W. Black *et al., J. Pharmacol. Exp.*

Ther. **248,** 208 (1988); of secretory diarrhea in mice and rats: J. E. Shook *et al., ibid.* **251,** 247 (1989). LC determn in human plasma: D. Chollet, M. Salanon, *J. Chromatog.* **593,** 73 (1992). Clinical trial in travelers' diarrhea: H. L. DuPont *et al., Gastroenterology* **104,** 709 (1993).

Crystals, mp 173°-175° as hydrate.
Maleate, $C_{26}H_{28}N_4O_2 \cdot C_4H_4O_4$, *CGS-9343B, ZY-17617B.* Crystals, mp 189°-190°.
THERAP CAT: Antidiarrheal.

10244. Zaltoprofen. *(±)-10,11-Dihydro-α-methyl-10-oxodibenzo[b,f]thiepin-2-acetic acid;* 2-(10,11-dihydro-10-oxodibenzo[b,f]thiepin-2-yl)propionic acid; CN-100; Sore-ton; Peon. $C_{17}H_{14}O_3S$; mol wt 298.36. C 68.44%, H 4.73%, O 16.09%, S 10.75%. Anti-inflammatory activity resides in (*S*)-enantiomer. Prepn: Y. Fujimoto, S. Yamabe, **Ger.** pat. **2,941,869;** *eidem,* **U.S.** pat. **4,247,706** (1980, 1981 both to Nippon Chemiphar). Series of articles on pharmacology: *Arzneimittel-Forsch.* **36,** 1796-1822 (1986). Resolution of racemate: M. Yamamoto *et al., Chirality* **2,** 280 (1990). HPLC determn in human plasma and urine; pharmacoki-netics: J. Oshima *et al., J. Chromatog.* **414,** 381 (1987).

Pale yellow crystals from benzene/*n*-hexane, mp 130.5-131.5°; also reported as mp 131-133°. Tasteless, odorless. Freely sol in acetone, chloroform; sol in methanol; slightly sol in ethanol, benzene. Practically insol in water, cyclohex-ane.
(*S*)-form, crystals from ethylendichloride, mp 129.5-131°. $[\alpha]_D^{25}$ +32.4° (c = 1 in CHCl$_3$).
THERAP CAT: Anti-inflammatory.

10245. Zatebradine. *3-[3-[[2-(3,4-Dimethoxyphenyl)eth-yl]methylamino]propyl]-1,3,4,5-tetrahydro-7,8-dimethoxy-2H-3-benzazepin-2-one;* 1-(7,8-dimethoxy-1,3,4,5-tetrahydro-2H-3-benzazepin-2-on-3-yl)-3-[*N*-methyl-*N*-[2-(3,4-dimeth-oxyphenyl)ethyl]amino]propane. $C_{26}H_{36}N_2O_5$; mol wt 456.58. C 68.40%, H 7.95%, N 6.14%, O 17.52%. Specific bradycardic agent; sinus node inhibitor. Prepn: M. Reiffen *et al.,* **Eur.** pat. **Appl. 65,229;** *eidem,* **U.S.** pat. **4,490,369** (1982, 1984 both to Thomae); and structure-activity study: M. Reiffen *et al., J. Med. Chem.* **33,** 1496 (1990). Pharma-cology: W. Kobinger, C. Lillie, *Eur. J. Pharmacol.* **104,** 9 (1984). Mechanism of action study: M. Goethals *et al., Circulation* **88,** 2389 (1993). Pharmacokinetics: W. Roth *et al., J. Pharm. Sci.* **82,** 99 (1993). GC-MS determn in plasma: J. Schmid *et al., J. Chromatog.* **658,** 93 (1994).

Hydrochloride, $C_{26}H_{36}N_2O_5 \cdot$HCl, *UL-FS-49.* Two crystal-line modifications, mp 188° and mp 168°. Sol in water.
THERAP CAT: Antianginal.

10246. Zearalenone. *[S-(E)]-3,4,5,6,9,10-Hexahydro-14,-16-dihydroxy-3-methyl-1H-2-benzoxacyclotetradecin-1,7(8H)-dione;* 6-(10-hydroxy-6-oxo-*trans*-1-undecenyl)-β-resorcylic acid lactone; Compd F-2; FES. $C_{18}H_{22}O_5$; mol wt 318.37. C 67.91%, H 6.97%, O 25.13%. One of a group of compounds known under the more general name of *RALs* or *resorcylic acid lactones*. Isoln of *l*-form from the mycelia of the fungus *Gibberella zeae (Fusarium graminearum):* Stob *et al., Nature* **196**, 1318 (1962); Andrews, Stob, U.S. pat. 3,196,019 (1965 to Purdue Res. Found.); Hodge *et al.,* U.S. pats. 3,239,341; 3,239,342 (both 1966 to Commercial Solvents). Structure: Urry *et al., Tetrahedron Letters* 1966, 3109. Synthesis of *dl*-form: Taub *et al., Chem. Commun.* 1967, 225; Neth. pat. 6,812,148 (1968 to Merck & Co.); Vlattas *et al., J. Org. Chem.* **33**, 4176 (1968); T. Takahashi *et al., J. Am. Chem. Soc.* **101**, 5072 (1979); eidem, *Tetrahedron Letters* **22**, 1363 (1981). Total synthesis and abs config: Taub *et al., Tetrahedron* **24**, 2443 (1968); Girotra, Wendler, U.S. pat. 3,551,455 (1970 to Merck & Co.). Physicochemical data: A. E. Pohland *et al., Pure Appl. Chem.* **54**, 2219 (1982).

Crystals, mp 164-165°. $[\alpha]_{546}^{25}$ −170.5° (c = 1.0 in CH$_3$OH). uv max (CH$_3$OH): 236, 274, 316 nm (ε 29700; 13900; 6020). Sol in aq alkali, ether, benzene, alcohols. Practically insol in water.
Diacetate, C$_{22}$H$_{26}$O$_7$, crystals, mp 123-125°.
dl-Form, crystals, mp 187-189°.
THERAP CAT (VET): Anabolic.

10247. Zeatin. *(E)-2-Methyl-4-(1H-purin-6-ylamino)-2-buten-1-ol; trans-zeatin.* C$_{10}$H$_{13}$N$_5$O; mol wt 219.25. C 54.78%, H 5.98%, N 31.94%, O 7.30%. Naturally occurring plant growth hormone; cytokinin originally isolated from sweet corn kernels, *Zea mays* L. Gramineae. Isoln and structure determn: D. S. Letham *et al., Proc. Chem. Soc.* **1964**, 230. Synthesis: G. Shaw, D. V. Wilson, *ibid.* 231; G. Shaw *et al., J. Chem. Soc. (C)* **1966**, 921; J. Corse, J. Kuhnle, *Synthesis* 1972, 618; G. M. Gray, *ibid.* **1983**, 488; idem, *Eur. pat. Appl.* 86,454 (1983 to J. T. Baker). Inhibition of mitochondrial function: C. O. Miller, *Plant Physiol.* **69**, 1274 (1982); translocation in soybean explants: L. Noodén, D. S. Letham, *J. Plant Growth Regul.* **2**, 265 (1984). Reviews: D. S. Letham, *Ann. Rev. Plant Physiol.* **18**, 349-363 (1967); D. S. Letham, L. M. S. Palni, *ibid.* **34**, 163-197 (1983).

Crystals from water, mp 207-208°. uv max in 0.1*M* HCl: 207, 275 nm (ε 14500, 14650); at pH 7.2: 212, 270 nm (ε 17050, 16150); in 0.1*M* NaOH: 220, 276 nm (ε 15900, 14650).

10248. Zeaxanthin. *β,β-Carotene-3,3′-diol; all trans-β-carotene-3,3′-diol; (3R,3′R)-dihydroxy-β-carotene; zeaxanthol; anchovyxanthin.* C$_{40}$H$_{56}$O$_2$; mol wt 568.88. C 84.45%, H 9.92%, O 5.62%. One of the most widespread carotenoid alcohols in nature. Occurs together and is isomeric with xanthophyll, *q.v.* It is the pigment of yellow corn *Zea mays* L., Gramineae. Isoln from yellow corn grits: Karrer *et al., Helv. Chim. Acta* **13**, 268 (1930). Isoln by chromatography: Kuhn, Grundmann, *Ber.* **67**, 596 (1934). Isoln from algae, bacteria: A. J. Aasen *et al., Acta Chem. Scand.* **26**, 404 (1972). Isoln of dipalmitate from *Physalis* petal: Kuhn *et al., Ber.* **63**, 1489 (1930). Does not possess vitamin A potency: Schumacher *et al., Poultry Sci.* **23**, 529 (1944). Stereochemistry: Zechmeister, *Chem. Rev.* **34**, 267 (1944). Abs config of natural zeaxanthin: T. E. de Ville *et al., Chem. Commun.* **1969**, 1311; J. R. Hlubucek *et al., J. Chem. Soc. Perkin Trans. I* **1974**, 848. Total synthesis: Isler *et al.,*

Helv. Chim. Acta **39**, 2041 (1956); **40**, 456 (1957); Loeber *et al., J. Chem. Soc. (C)* **1971**, 404; A. Rüttimann, H. Mayer, *Helv. Chim. Acta* **63**, 1456 (1980); P. R. Ellis *et al., ibid.* **64**, 1092 (1980); E. Widmer *et al., ibid.* **65**, 944, 958 (1982). Sepn of configurational isomers by HPLC: A. Rüttimann *et al., J. High Resolut. Chromatog. Chromatog. Commun.* **6**, 612 (1983). First isoln of enantiomeric and *meso*-zeaxanthin in nature: T. Maoka *et al., Comp. Biochem. Physiol.* **83B**, 121 (1986). Mfg procedure: Isler *et al.,* U.S. pat. 2,917,539 (1959 to Hoffmann-La Roche). Reviews: Zechmeister, *Carotinoide* (Berlin, 1934); Mayer, *The Chemistry of Natural Coloring Matters* (New York, 1943); Karrer, Jucker, *Carotenoids* (New York, 1950).

Yellow rhombic plates with steel-blue metallic luster from ethanol, mp 207° (Zechmeister); mp 215.5° (Kuhn). Optically inactive. Absorption bands (ethanol) 483, 451.5 nm. Practically insol in water. Slightly sol in petr ether, methanol; sol in carbon disulfide, benzene, chloroform, carbon tetrachloride, pyridine, ethyl acetate; sol in glacial acetic acid upon addition of hexane. Less sol in boiling methanol (1:1550) than xanthophyll.
Dipalmitate, C$_{72}$H$_{116}$O$_4$, *physalien, physalin.* Fine yellow or red needles from benzene, mp 97°.

10249. Zein. A prolamine; an alcohol-soluble protein present in amounts of 2.5-10% (dry basis) in corn *(Zea mays* L., *Gramineae).* The greater part of zein has a mol wt of 38,000. Does not contain lysine or tryptophan. Extracted commercially from gluten meal with dil isopropanol: Swallen, *Ind. Eng. Chem.* **33**, 394 (1941). Improved method: Carter, Reck, Ger. pat. 2,002,337 (1971 to Nutrilite Prod.), *C.A.* **75**, 117341b (1971). Review: Mossé, *Ann. Physiol. Végétale* **3**, 105 (1961).
White to slightly yellow powder. d 1.226. When completely dry may be heated to 200° without visible signs of decompn. Soluble in aq alcohols, the glycols, the ethyl ether of ethylene glycol, furfuryl alcohol, tetrahydrofurfuryl alcohol, aq alkaline solns of pH 11.5 (or greater). Tends to become denatured when in soln and becomes insol. Insol in water, acetone; but readily sol in acetone-water mixtures between the limits of 60-80% acetone by volume. Insol in anhydr alcohols except methanol.
USE: Manuf plastics, paper coatings, adhesives, substitutes for shellac, laminated board, in solid color printing, films, edible coatings for foodstuffs.

10250. Zeolites. Crystalline, hydrated alkali-aluminum silicates of the general formula M$_{2/n}$O.Al$_2$O$_3$.ySiO$_2$.wH$_2$O where M represents a group IA or IIA element, *n* is the cation valence, *y* is 2 or greater and *w* is the number of water molecules contained in the channels or interconnected voids within the zeolite. The cations are mobile and capable of undergoing ion exchange. Zeolites occur naturally in sedimentary and volcanic rocks, altered basalts, ores, clay deposits. Some 40 known zeolite minerals and a great number of synthetic zeolites are available commercially. Ref: Milton, U.S. pat. 2,882,243 (1959 to Union Carbide). Reviews: D. W. Breck, R. A. Anderson, "Molecular Sieves" in Kirk-Othmer *Encyclopedia of Chemical Technology* vol. 15 (Wiley-Interscience, New York, 3rd ed., 1981) pp 638-669; P. L. Layman, *Chem. & Eng. News* **60**, 10 (Sept. 27, 1982); J. Haggin, *ibid.* **60**, 9 (Dec. 13, 1982). Books: *Molecular Sieve Zeolites - I, II,* Advances in Chemistry Series **101, 102** (A.C.S., Washington, 1971) 526 pp, 459 pp; *Molecular Sieves,* Advances in Chemistry Series **121** (A.C.S., Washington, 1973) 634 pp; D. W. Breck, *Zeolite Molecular Sieves* (John Wiley, New York, 1974) 771 pp; *Natural Zeolites: Occurrence, Properties, Use,* L. B. Sand, F. A. Mumpton, Eds. (Pergamon, Oxford, 1978). Studies on fibrous zeolites as possible environmental hazards: A. N. Rohl *et al., Science* **216**, 518 (1982); Y. I. Baris, *Arch. Environ. Health* **37**, 177 (1982). Use as microscopic chemical reactor vessels: R. Pool, *Science* **263**, 1698 (1994).

USE: As molecular sieves, filters, adsorbents, catalysts, drying agents, cation exchangers, dispersing agents, detergent builders.

10251. Zeranol. *3,4,5,6,7,8,9,10,11,12-Decahydro-7,14,-16-trihydroxy-3-methyl-1H-2-benzoxacyclotetradecin-1-one;* 6-(6,10-dihydroxyundecyl)-β-resorcylic acid μ-lactone; zearalanol; MK-188; P-1496; Ralgro; Ralabol; Ralone. $C_{18}H_{26}O_5$; mol wt 322.40. C 67.06%, H 8.13%, O 24.81%. Prepn from zearalenone, *q.v.:* Hodge *et al., U.S.* pat. **3,239,345** (1956 to Commercial Solvents). Metabolism in steers: Sharp, Dyer, *J. Anim. Sci.* **34,** 176 (1972).

Isopropyl alcohol-water mixtures yield two crystalline diastereoisomers. The more soluble has mp 146-148°; $[\alpha]_D^{25}$ ~+39° (methanol). The less soluble has mp 178-180°; $[\alpha]_D^{25}$ ~+46° (methanol). uv max (methanol): 218, 265, 304 nm.
THERAP CAT (VET): Anabolic.

10252. Zidovudine. *3'-Azido-3'-deoxythymidine;* azidothymidine; AZT; BW-A509U; Aztec; Retrovir. $C_{10}H_{13}N_5$-O_4; mol wt 267.24. C 44.94%, H 4.90%, O 23.95%. Pyrimidine nucleoside analog; reverse transcriptase inhibitor. Prepn: J. P. Horwitz *et al., J. Org. Chem.* **29,** 2076 (1964); R. P. Glinski *et al., ibid.* **38,** 4299 (1973). Total synthesis: C. K. Chu *et al., Tetrahedron Letters* **29,** 5349 (1988). *In vitro* antiviral, antimetabolite and antineoplastic activity: E. De Clercq *et al., Biochem. Pharmacol.* **29,** 1849 (1980). *In vitro* activity against HIV-1 virus: H. Mitsuya *et al., Proc. Nat. Acad. Sci. USA* **82,** 7096 (1985). Preliminary clinical trial in AIDS: R. Yarchoan *et al., Lancet* **1,** 575 (1986). Clinical pharmacokinetics: R. W. Klecker *et al., Clin. Pharmacol. Ther.* **41,** 407 (1987). Use in treatment of AIDS and AIDS-related complex: J. L. Rideout *et al.,* **Ger.** pat. **3,608,606** (1986 to Wellcome Found.); *eidem, U.S.* pat. **4,724,232** (1988 to Burroughs Wellcome). Toxicology studies: K. M. Ayers, *Am. J. Med.* **85,** Suppl. 2A, 186 (1988). Symposium on clinical experience: *J. Infect.* **18,** Suppl. 1, 1-101 (1989). Clinical trials of reduced dose in AIDS: M. A. Fischl *et al., New Engl. J. Med.* **323,** 1009 (1990); A. C. Collier *et al., ibid.* 1015. Comprehensive description: M. L. Sethi in *Analytical Profiles of Drug Substances* vol. **20,** K. Florey Ed. (Academic Press, New York, 1991) pp 729-765.

Needles from petr ether, mp 106-112° (Horwitz). Also reported as crystals from water, mp 120-122° (Glinski). $[\alpha]_D^{25}$ +99° (c = 0.5 in water). Soly in water (25°C): 25 mg/ml. uv max (water): 266.5 nm (ε 11650). LD_{50} in male, female mice, male, female rats (mg/kg): 3568, 3062, 3084, 3683 orally; > 750 i.v. (all species) (Ayers).
THERAP CAT: Antiviral.

10253. Zileuton. *N-(1-Benzo[b]thien-2-ylethyl)-N-hydroxyurea;* (±)-N-hydroxy-N-(1-benzo[b]thien-2-ylethyl)-urea; A-64077; Abbott 64077; Leutrol. $C_{11}H_{12}N_2O_2S$; mol wt 236.29. C 55.91%, H 5.12%, N 11.86%, O 13.54%, S 13.57%. Inhibitor of 5-lipoxygenase, the initial enzyme in the biosynthesis of leukotrienes from arachidonic acid. Prepn: J. B. Summers, Jr. *et al.,* **Eur.** pat. **Appl. 279,263;**

eidem, U.S. pat. **4,873,259** (1988, 1989 both to Abbott). Chromatographic resolution of enantiomers: S. B. Thomas *et al., J. Chromatog.* **623,** 390 (1992). Inhibition of 5-lipoxygenase: G. W. Carter *et al., J. Pharmacol. Exp. Ther.* **256,** 929 (1991). Pharmacokinetics: P. Rubin *et al., Agents Actions Suppl.* **35,** 103 (1991). Clinical evaluation in rheumatoid arthritis: M. E. Weinblatt *et al., J. Rheumatol.* **19,** 1537 (1992). Clinical trial in asthma: E. Israel *et al., Ann. Intern. Med.* **119,** 1059 (1993). Review of development: R. L. Bell *et al., Int. J. Immunopharmacol.* **14,** 505-510 (1992).

Crystals, mp 157-158°.
THERAP CAT: Antiasthmatic; anti-inflammatory.

10254. Zimeldine. *(Z)-3-(4-Bromophenyl)-N,N-dimethyl-3-(3-pyridinyl)-2-propen-1-amine;* (Z)-3-(4'-bromophenyl)-3-(3''-pyridyl)dimethylallylamine; zimelidine; cis-H-102/09. $C_{16}H_{17}BrN_2$; mol wt 317.23. C 60.58%, H 5.40%, Br 25.19%, N 8.83%. Antidepressant that inhibits membranal 5-hydroxytryptamine uptake. Prepn: P. B. Berntsson *et al.,* **S. Afr.** pat. **72 01,503;** *eidem,* **U.S.** pat. **3,928,369** (1972, 1975 both to AB Hassle). X-ray crystallographic study: S. Abrahamsson *et al., Acta Chem. Scand.* **A30,** 609 (1976). *In vivo* study in rats and humans: S. B. Ross *et al., Life Sci.* **19,** 205 (1976). Pharmacokinetics: D. Brown, *Eur. J. Clin. Pharmacol.* **17,** 111 (1980). Mode of action: K. Fuxe *et al., Neurosci. Letter* **13,** 307 (1979). Metabolism: J. Lundström *et al., Arzneimittel-Forsch.* **31,** 486 (1981). Clinical studies: A. Georgotas *et al., Am. J. Psychiatry* **139,** 1057 (1982); E. Syvalahti *et al., J. Int. Med. Res.* **10,** 250 (1982). Profile of antidepressant action: S. A. Montgomery *et al., Advan. Biochem. Psychopharmacol.* **32,** 35 (1982). Reversal of ethanol-induced memory impairment in human subjects: H. Weingartner *et al., Science* **221,** 472 (1983). Review of pharmacology and therapeutic efficacy: R. C. Heel *et al., Drugs* **24,** 169-206 (1982). Symposium on pharmacology, pharmacokinetics, metabolism, clinical studies: *Brit. J. Clin. Pract.* **1982,** Suppl. 19, 1-122.

Dihydrochloride monohydrate, $C_{16}H_{17}BrN_2 \cdot 2HCl \cdot H_2O$, *Normud, Zelmid.* Crystals, mp 193°.
THERAP CAT: Antidepressant.

10255. Zinc. Zn; at. wt 65.38; at. no. 30; valence 2. Group 2b element. Abundance in earth's crust: 0.02% by wt. Natural isotopes: 64 (48.89%); 66 (27.81%); 68 (18.57%); 67 (4.11%); 70 (0.62%); eight radioactive isotopes and two isomers. Occurs in smithsonite or zinc spar, sphalerite or zinc blende, zincite, willemite, *franklinite,* [(Zn,Mn,-Fe)O.(Fe.Mn$_2$)O$_3$] or *gahnite* (ZnAl$_2$O$_4$). Has been known since very early times. Commercial forms: ingots; lumps; sheets; wire; shot; strips; sticks; granules; mossy; powder (dust). Prepn: Gowland, Bannister, *Metallurgy of Non-Ferrous Metals* (Griffin, London, 1930); *Zinc Production, Properties and Uses* (Zinc Development Association, London, 1968). *Reviews:* Zinc, C. H. Mathewson, Ed., A.C.S. Monograph Series no. **142** (Reinhold, New York, 1959) 721 pp; Schlechter, Thompson, "Zinc and Zinc Alloys" in Kirk-Othmer, *Encyclopedia of Chemical Technology,* **vol. 22** (Interscience, New York, 2nd ed., 1970) pp 555-603; Aylett, "Group IIB" in *Comprehensive Inorganic Chemistry,* **vol. 3,** J. C. Bailar, Jr. *et al.,* Eds. (Pergamon Press, Oxford, 1973) pp 187-328.

Bluish-white, lustrous metal; distorted hexagonal close-packed structure; stable in dry air; becomes covered with a white coating of basic carbonate on exposure to moist air. mp 419.5°. bp 908°. d^{25} 7.14. Heat capacity at constant pressure (25°): 6.07 cal/mole deg. Mohs' hardness 2.5. When heated to 100-150° becomes malleable, at 210° becomes brittle and pulverizable. Burns in air with a bluish-green flame. Loses electrons in aqueous systems to form Zn^{2+} E° (aq) Zn/Zn^{2+} 0.763 V. Slowly attacked by H_2SO_4 or HCl; oxidizing agents or metal ions, e.g. Cu^{2+}, Ni^{2+}, Co^{2+}, accelerate the process. Reacts slowly with ammonia water and acetic acid; rapidly with HNO_3. Reacts with alkali hydroxides to form *"zincates"*, ZnO_2^{2-}, which are actually hydroxo complexes such as $Zn(OH)_3^-$; $Zn(OH)_4^{2-}$, $[Zn(OH)_4(H_2O)_2]^{2-}$.

Caution: Inhalation of fumes may result in sweet taste, throat dryness, cough, weakness, generalized aching, chills, fever, nausea, vomiting. Zinc chloride fumes have caused injury to mucous membranes and skin irritation. Ingestion of sol salts may cause nausea, vomiting, purging. *See* E. Browning, *Toxicity of Industrial Metals* (Appleton-Century-Crofts, New York, 2nd ed., 1969) pp 348-355.

USE: Galvanizing sheet iron; as ingredient of alloys such as bronze, brass, Babbitt metal, German silver, and special alloys for die-casting; as a protective coating for other metals to prevent corrosion; for electrical apparatus, especially dry cell batteries, household utensils, castings, printing plates, building materials, railroad car linings, automotive equipment; as reducing agent in organic chemistry; for deoxidizing bronze; extracting gold by the cyanide process, purifying fats for soaps; bleaching bone glue; manuf sodium hydrosulfite; insulin zinc salts; as reagent in analytical chemistry, e.g., in the Marsh and Gutzeit test for arsenic; as a reducer in the determination of iron. It is a nutritional trace element.

10256. Zinc Acetate. $C_4H_6O_4Zn$; mol wt 183.48. C 26.18%, H 3.30%, O 34.88%, Zn 35.64%. $Zn(C_2H_3O_2)_2$. Prepn of anhydr salt from zinc nitrate and acetic anhydride: Späth, *Monatsh.* **33**, 235 (1912). Clinical evaluations in Wilson's disease: G. M. Hill *et al.*, *Hepatology* **7**, 522 (1987); G. J. Brewer *et al.*, *J. Lab. Clin. Med.* **109**, 526 (1987). Toxicity: H. F. Smyth *et al.*, *Am. Ind. Hyg. Assoc. J.* **30**, 470 (1969).

Dihydrate, crystallizes from dil acetic acid; faint, acetous odor; astringent taste; slightly efflorescent. d 1.735; mp 237°. One gram dissolves in 2.3 ml water, 1.6 ml boiling water, 30 ml alcohol, about 1 ml boiling alcohol. The aq soln is neutral or slightly acid to litmus; pH about 5-6. *Keep in well-closed containers. Incompat.* with zinc salts in general: Acacia, alkalies and their carbonates, oxalates, phosphates, sulfides, lime water, vegetable astringent decoctions and infusions. LD_{50} orally in rats: 2.46 g/kg (Smyth).

USE: Preserving wood; as mordant in dyeing; manuf glazes for painting on porcelain. As a reagent in testing for albumin, tannin, urobilin, phosphate, blood.

THERAP CAT: Styptic, astringent. Formerly used as an emetic.

THERAP CAT (VET): Antiseptic, astringent, protective (topical). Has been used as an emetic.

10257. Zinc Bacitracin. *Bacitracin zinc complex;* bacitracin zinc salt; Baciferm. Contains about 7% Zn. Prepd by the action of zinc salts on bacitracin broth: Anker *et al.*, *J. Bacteriol.* **55**, 249 (1948); Hodge, Lafferty as cited in U.S. pat. 2,803,584 (1957 to CSC). The potency is usually between 50 and 60 units/mg of bacitracin activity: Gross, *Drug Cosmet. Ind.* **75**, 612 (1954); Gross *et al.*, *J. Am. Pharm. Assoc., Sci. Ed.* **45**, 447 (1956).

Creamy powder contg 1-4% H_2O. Less bitter than bacitracin. Soly in water at 25° = 0.23-0.45% (w/w). Solubilities determined by Weiss *et al.*, *Antibiot. & Chemother.* **7**, 374 (1957) in mg/ml at about 28°: water 5.1; methanol 6.55; ethanol 2.0; isopropanol 0.16; ethyl acetate 1.3; chloroform 0.01; petr ether 0.025. More stable than bacitracin at room and elevated temps. Zinc bacitracin can be used in formulations requiring heat processing.

USE: In ointments, tablets, implantation pellets, suppositories, and troches, either alone or in combination with other antibiotics or therapeutic agents: Hodge, Lafferty, *loc. cit.*

THERAP CAT: Antibacterial.

THERAP CAT (VET): Antimicrobial.

10258. Zinc Bromide. Br_2Zn; mol wt 225.20. Br 70.96%, Zn 29.04%. $ZnBr_2$. Usually contains at least 97% $ZnBr_2$, the remainder being chiefly water.

Very hygroscopic, granular powder; sharp, metallic taste. d 4.22; mp 394°; bp 697° with partial decompn. One gram dissolves in 0.25 ml water, 0.5 ml 90% alcohol; sol in ether, solns of alkali hydroxides. The aq soln is acid to litmus; pH about 4. *Keep tightly closed.*

USE: Making silver bromide collodion emulsions for photography; in the shielding of viewing windows for nuclear reactions.

10259. Zinc Caprylate. *Octanoic acid zinc salt.* $C_{16}H_{30}O_4Zn$; mol wt 351.80. C 54.63%, H 8.60%, O 18.19%, Zn 18.59%. $Zn(C_8H_{15}O_2)_2$. Prepd from ammonium caprylate and zinc sulfate: van Renesse, *Ann.* **171**, 380 (1874).

Lustrous scales from alc, mp 136°. Sparingly sol in boiling water; moderately sol in boiling alcohol. *Keep well closed.* Dec in moist atm giving off caprylic acid.

USE: As fungicide like zinc propionate.

10260. Zinc Carbonate. CO_3Zn; mol wt 125.40. C 9.58%, O 38.28%, Zn 52.15%. $ZnCO_3$. Occurs in nature as the minerals *smithsonite, zincspar*. Prepn: Hüttig *et al.*, *Monatsh.* **72**, 31 (1939).

Rhombohedral structure. Solubility in water at 15° 0.001 g/100 g; sol in dil acids, alkalies, solns of NH_4^+ salts. Basic carbonate, *zinc carbonate hydroxide, zinc subcarbonate.* Variable composition, usually characterized as $3Zn(OH)_2 \cdot 2ZnCO_3$. Occurs as the mineral *hydrozincite*. Reagent specification: 70% ZnO minimum.

USE: As pigment; manuf of porcelains, pottery, rubber.

THERAP CAT: Astringent, topical antiseptic.

THERAP CAT (VET): Astringent, antiseptic, protective (topical). Also used in rations to prevent Zn deficiency diseases.

10261. Zinc Chloride. Butter of zinc. Cl_2Zn; mol wt 136.30. Cl 52.02%, Zn 47.98%. $ZnCl_2$. Usually contains at least 95% $ZnCl_2$; remainder is chiefly water and oxychloride. Toxicity: Bruner, *Fed. Proc.* **9**, 260 (1950).

White, odorless, very deliquesc granules, or fused pieces or rods. d^{25} 2.907; mp ~290°; bp 732°. Solubility in H_2O: 432 g/100 g (25°); 614 g/100 g (100°). One gram dissolves in 0.25 ml of 2% HCl, in 1.3 ml alcohol, 2 ml glycerol; freely sol in acetone. With much water some zinc oxychloride is formed. The aq soln is acid to litmus; pH about 4. *Keep tightly closed.* LD i.v. in rats: 60-90 mg/kg (Bruner).

Caution: Potential symptoms of overexposure to fumes are conjunctivitis; irritation of nose and throat; coughing, copious sputum; dyspnea, chest pain, pulmonary edema and bronchopneumonia; pulmonary fibrosis, cor pulmonale; fever; cyanosis; tachypnea; skin burns; irritation of skin and eyes. *See NIOSH Pocket Guide to Chemical Hazards* (DHHS/NIOSH 90-117, 1990) p 226.

USE: Deodorant, disinfecting and embalming material; alone or with phenol and other antiseptics for preserving railway ties; fireproofing lumber; with ammonium chloride as flux for soldering; etching metals; manuf parchment paper, artificial silk, dyes, activated carbon, cold-water glues, vulcanized fiber; browning steel, galvanizing iron, copper-plating iron; in magnesia cements; petroleum oil refining; cement for metals and for facing stone; mordant in printing and dyeing textiles; carbonizing woolen goods; producing crepe and crimping fabrics; mercerizing cotton; sizing and weighting fabrics; vulcanizing rubber; solvent for cellulose; preserving anatomical specimens; in microscopy for separating silk, wool, and plant fibers; as dehydrating agent in chemical syntheses. Dentin desensitizer.

THERAP CAT: Astringent.

THERAP CAT (VET): Antiseptic, astringent. Has been used in ulcers, fistulas, pododermatitis.

10262. Zinc Chromate(VI) Hydroxide. Zinc yellow; buttercup yellow; C.I. Pigment Yellow 36. A basic salt of somewhat variable composition. Approx $Zn_2CrO_4(OH)_2$.

Hydrate, yellow, odorless, fine powder. Slightly sol in water; sol in dil acids, including acetic acid.

USE: As pigment in paints, varnishes, oil colors, linoleum, rubber, etc.

10263. Zinc Citrate. *2-Hydroxy-1,2,3-propanetricarboxylic acid zinc salt.* $C_{12}H_{10}O_{14}Zn_3$; mol wt 574.37. C 25.09%, H 1.75%, O 39.00%, Zn 34.15%. $Zn_3(C_6H_5O_7)_2$. Prepd from zinc carbonate and citric acid: Heldt, *Ann.* **47**, 157 (1843).
Dihydrate, odorless powder. Slightly sol in water; sol in dil mineral acids, in alkali hydroxides.
USE: In toothpaste and mouthwash.

10264. Zinc Cyanide. C_2N_2Zn; mol wt 117.43. C 20.46%, N 23.86%, Zn 55.69%. $Zn(CN)_2$. Usually contains about 85% zinc cyanide, some water and oxide.
White powder. *Poison!* Insol in water; sol in solns of alkali cyanides or hydroxides; not appreciably attacked by organic acids, but readily dec by dil mineral acid with evolution of hydrogen cyanide.
USE: Electroplating; removing NH_3 from producer gas.

10265. Zinc Fluoride. F_2Zn; mol wt 103.39. F 36.75%, Zn 63.25%. ZnF_2. Prepd from $ZnCO_3$ and HF: Ruff, *Die Chemie des Fluors* (Berlin, 1920) p 36; Emeleus in *Fluorine Chemistry* vol. **1**, J. H. Simons, Ed. (Academic Press, New York, 1950) p 38; Kwasnik in *Handbook of Preparative Inorganic Chemistry* vol. **1**, G. Brauer, Ed. (Academic Press, New York, 2nd ed., 1963) p 242.
Tetragonal needles (rutile lattice) or white cryst mass. d^{25} 5.00: Haendler *et al., J. Am. Chem. Soc.* **74**, 3167 (1952). mp 872°. bp 1500°. Soly in water: 5×10^{-5} moles/l, Kwasnik, *loc. cit.* Slightly sol in aq HF, more sol in HCl, HNO_3, NH_4OH. Zinc fluoride used for fluorinations should be slightly hydrated.
Tetrahydrate, rhombohedral crystals, becomes anhydr at 100°. Soly in water: 1.516 g/100 ml. May be stored in glass bottles.
USE: In the fluorination of organic compds, manuf phosphors for fluorescent electric lights, glazes and enamels for porcelain, preserving wood, in electroplating baths.

10266. Zinc Formate. $C_2H_2O_4Zn$; mol wt 155.43. C 15.46%, H 1.30%, O 41.18%, Zn 42.07%. $Zn(HCOO)_2$. Forms dihydrate readily. Prepd from zinc carbonate and formic acid: Kendall, Adler, *J. Am. Chem. Soc.* **43**, 1470 (1921).
Dihydrate, crystals. d 2.207. Solubility in H_2O (20°): 5.2 g/100 g. Practically insol in alc.

10267. Zinc Hexafluorosilicate. Zinc fluosilicate; zinc silicofluoride. F_6SiZn; mol wt 207.47. F 54.94%, Si 13.54%, Zn 31.52%. $ZnSiF_6$.
Hexahydrate, white crystals. Soluble in water. pH of 1% aqueous solution 3.2.
USE: Mothproofing agent; laundry sour; hardener for concrete.

10268. Zinc Iodate. I_2O_6Zn; mol wt 415.20. I 61.13%, O 23.12%, Zn 15.75%. $Zn(IO_3)_2$. May contain 1-2 mols of water.
White, crystalline powder. Soluble in 115 parts cold, 77 parts hot water.
THERAP CAT: Topical antiseptic.

10269. Zinc Iodide. I_2Zn; mol wt 319.20. I 79.51%, Zn 20.49%. ZnI_2. Usually contains at least 98% ZnI_2.
White or almost white, odorless, hygroscopic, granular powder; sharp, saline taste; becomes brown on exposure to air and light, due to liberation of iodine. d^{25} 4.74; mp ~446°; bp ~625° with decompn. One gram dissolves in 0.3 ml water, 2 ml boiling water, 2 ml glycerol; freely sol in alcohol, ether. The aq soln is acid to litmus; pH about 5. *Keep well closed and protected from light.*
THERAP CAT: Topical antiseptic, astringent.

10270. Zinc Iodide-Starch. Trommsdorff's starch. A soln is prepd by heating 4 parts starch, 20 $ZnCl_2$, and 2 ZnI_2 with 1 liter water. Deteriorates with age and acquires a blue color.
USE: For detecting nitrites, free Cl, and other oxidizing agents.

10271. Zinc Lactate. $C_6H_{10}O_6Zn$; mol wt 243.53. C 29.59%, H 4.14%, O 39.42%, Zn 26.85%. $Zn(C_3H_5O_3)_2$.

Prepd from zinc carbonate and lactic acid: Pederson *et al., J. Biol. Chem.* **68**, 151 (1926).
Trihydrate, crystals. Soluble in 60 parts cold, 6 parts boiling water.

10272. Zinc Meta-arsenite. ZMA. As_2O_4Zn; mol wt 279.23. As 53.66%, O 22.92%, Zn 23.42%. $Zn(AsO_2)_2$.
White powder. Sol in acids.
USE: Wood preservative, insecticide.

10273. Zinc Nitrate. N_2O_6Zn; mol wt 189.40. N 14.79%, O 50.68%, Zn 34.52%.
Hexahydrate, colorless, odorless crystals; d 2.065; mp ~36°. Sol in about 0.5 part water, freely in alcohol. The aq soln is acid to litmus; pH of 5% aq soln 5.1. *Keep well closed and in a cool place.* Also available in the form of fused pieces. In this form it contains only about 20% water. Technical flake usually contains 25.6% H_2O.
USE: As a mordant in dyeing.

10274. Zinc Nitride. N_2Zn_3; mol wt 224.18. N 12.50%, Zn 87.50%. Zn_3N_2. Prepd by heating zinc amide, $Zn(NH_2)_2$, to 330° or by heating zinc dust to 600° in a stream of ammonia: Juza *et al., Z. Anorg. Chem.* **239**, 273 (1938).
Blackish-gray, cryst solid.

10275. Zinc Nitrite. N_2O_4Zn; mol wt 157.40. N 17.80%, O 40.66%, Zn 41.54%. $Zn(NO_2)_2$. Prepd by treating sodium nitrite with zinc sulfate in alcohol: Ephraim, Bolle, *Ber.* **48**, 643 (1915).
Hydrolyzes so quickly that it cannnot be prepd from water, from which only basic salts will separate.

10276. Zinc Oleate. *9-Octadecenoic acid zinc salt.* $C_{36}H_{66}O_4Zn$; mol wt 628.31. C 68.82%, H 10.59%, O 10.19%, Zn 10.41%. $(C_{17}H_{33}CO_2)_2Zn$. Usually contains small amounts of the zinc salts of other fatty acids. Prepn: Grabner, *Monatsh.* **42**, 287 (1921).
White, dry, greasy powder. Insol in water; sol in alcohol, ether, carbon disulfide, benzene, petr ether.

10277. Zinc Ortho-arsenate. $As_2O_8Zn_3$; mol wt 474.01. As 31.61%, O 27.00%, Zn 41.39%. $Zn_3(AsO_4)_2$. Occurs in nature as an octahydrate *koettigite*, $Zn_3(AsO_4)_2.8H_2O$, and as a basic arsenate *adamite*, $Zn_3(AsO_4)_2.Zn(OH)_2$.
White, odorless powder. *Poison!* Insol in water; sol in acids, alkali hydroxides, ammonia.

10278. Zinc Oxalate. C_2O_4Zn; mol wt 153.41. C 15.66%, O 41.72%, Zn 42.62%. ZnC_2O_4. Prepn: Chatterjee, Dhar, *J. Phys. Chem.* **28**, 1020 (1924).
Dihydrate, powder. Very slightly sol in water; sol in dil mineral acids, ammonia.

10279. Zinc Oxide. Flowers of zinc; philosopher's wool; zinc white; C.I. Pigment White 4; C.I. 77947. OZn; mol wt 81.39. O 19.66%, Zn 80.34%. ZnO. Occurs as the mineral *zincite*. Prepd by vaporization of metallic zinc and oxidation of the vapors with preheated air (French process); also from franklinite, (American process) or from zinc sulfide: Faith, Keyes & Clark's *Industrial Chemicals*, F. A. Lowenheim, M. K. Moran, Eds. (Wiley-Interscience, New York, 4th ed., 1975) pp 882-888. Purification: Depew, U.S. pat. **2,372,367** (1945 to American Zinc, Lead & Smelting). The medicinal grade contains 99.5% or more ZnO; technical grades contain 90-99% ZnO and a few tenths of 1% of lead. *See also: Colour Index* vol. **4** (3rd ed., 1971) p 4687.
White or yellowish-white, odorless powder. Hexagonal crystals: d 5.67. Also reported as d_4^{20} 5.607. Sublimes at normal pressure. n_D 2.0041, 2.0203. American process zinc oxide pH 6.95. French process zinc oxide pH 7.37. Practically insol in water. Sol in dil acetic or mineral acids, ammonia, ammonium carbonate, fixed alkali hydroxide solns.
Caution: Potential symptoms of overexposure to fumes are sweet, metallic taste; dry throat, cough; chills, fever; tight chest, dyspnea, rales, reduced pulmonary function; headache; blurred vision; muscle cramps, low back pain; nausea, vomiting; fatigue, lassitude, malaise. *See NIOSH Pocket Guide to Chemical Hazards* (DHHS/NIOSH 90-117, 1990) p 228.
USE: As pigment in white paints instead of lead carbonate; in cosmetics, driers, quick-setting cements; with syrupy

phosphoric acid or $ZnCl_2$ in dental cements; manuf opaque glass and certain types of transparent glass; manuf enamels, automobile tires, white glue, matches, white printing inks, porcelains, zinc green; as a reagent in analytical chemistry; in electrostatic copying paper; as flame retardant; in electronics as semiconductor.

THERAP CAT: Astringent, protectant (topical).

THERAP CAT (VET): Antiseptic, astringent, protective (topical).

10280. Zinc Perchlorate. Cl_2O_8Zn; mol wt 264.29. Cl 26.83%, O 48.43%, Zn 24.74%. $Zn(ClO_4)_2$.

Hexahydrate, deliquesc crystals, mp 106°. Freely sol in water. Sol in alcohol. Also forms a tetrahydrate.

10281. Zinc Permanganate. Mn_2O_8Zn; mol wt 303.26. Mn 36.23%, O 42.21%, Zn 21.56%. $Zn(MnO_4)_2$. The article of commerce is about 95% pure.

Hexahydrate, violet-brown or almost black, deliquesc crystals; similar to potassium permanganate in appearance. Deteriorates on exposure to light and air. Sol in 3 parts water; dec by alc. *Keep well closed and protected from light.*

THERAP CAT: Antiseptic; astringent.

10282. Zinc Peroxide. ZPO; zinc superoxide. O_2Zn; mol wt 97.39. O 32.86%, Zn 67.14%. ZnO_2. The peroxide of commerce contains 50-60% ZnO_2, the remainder is ZnO.

White to yellowish-white, odorless powder. Dec above 150°. Insol in, but gradually dec by, water. Sol in dil acids, liberating hydrogen peroxide.

USE: Accelerator in rubber compounding; curing agent for synthetic elastomers. Deodorant for wounds and skin diseases.

THERAP CAT: Antiseptic (topical), astringent.

10283. Zinc *p*-Phenolsulfonate. *p-Hydroxybenzenesulfonic acid zinc salt;* 1-phenol-4-sulfonic acid zinc salt; zinc *p*-hydroxybenzenesulfonate; zinc sulfocarbolate; zinc sulfophenate; Phenozin. $C_{12}H_{10}O_8S_2Zn$; mol wt 411.73. C 35.01%, H 2.45%, O 31.09%, S 15.58%, Zn 15.88%. $Zn(HOC_6H_4SO_3)_2$. It is at least 99.5% pure. Prepn: Rojahn, *Deut. Apoth.-Ztg.* **50,** 1095 (1935), *C.A.* **31,** 6816⁸ (1937).

Octahydrate, crystals or cryst powder; odorless. Effloresces in dry air; loses all its H_2O at about 120°. One gram dissolves in 1.6 ml water, 0.4 ml boiling water, 1.8 ml alcohol. The aq soln is acid to litmus; pH about 4.

USE: In insecticide formulations.

THERAP CAT: Astringent.

THERAP CAT (VET): Has been used as an intestinal antiseptic, and externally to promote healing of ulcers, slowly granulating wounds.

10284. Zinc Phosphate. $O_8P_2Zn_3$; mol wt 386.11. O 33.15%, P 16.04%, Zn 50.81%. $Zn_3(PO_4)_2$. The tetrahydrate occurs in nature as the mineral *hopeite;* the article of commerce is about 98% pure.

Tetrahydrate, white, odorless powder. Insol in water or alcohol. Sol in dil mineral acids, acetic acid, ammonia, and in alkali hydroxide solns.

USE: In dental cements.

10285. Zinc Phosphide. P_2Zn_3; mol wt 258.12. P 24.00%, Zn 76.00%. Zn_3P_2. Usually prepd from the elements.

Dark gray tetragonal crystals, lustrous or dull powder; faint phosphorus odor. d 4.55. When strongly heated with the exclusion of air it melts and finally sublimes. The mp has been given as 420° and the bp as 1100°. When kept dry it is quite stable. Insol in water, alcohol. Sol in benzene, carbon disulfide; reacts with HCl, H_2SO_4 with evolution of spontaneously flammable phosphine. Reacts violently with concd H_2SO_4, HNO_3 and other oxidizing agents. *Keep dry.* LD_{50} orally in rats: 40.5-46.7 mg/kg, *Handbook of Toxicology,* **vol. 1,** W. S. Spector, Ed. (Saunders, Philadelphia, 1956) pp 316-317.

USE: In rat and field mice poison preparations.

10286. Zinc Propionate. $C_6H_{10}O_4Zn$; mol wt 211.53. C 34.07%, H 4.76%, O 30.25%, Zn 30.91%. $Zn(C_3H_5O_2)_2$. Prepd by dissolving zinc oxide in dil propionic acid and concentrating the soln: Gaze, *Arch. Pharm.* **229,** 488 (1891).

Clusters of plates, tablets; crystallizes also in needles as the monohydrate. The soly of the anhydr salt is 32% (w/w) in water at 15°; 2.8% in alc at 15°; 17.2% in boiling alc. *Keep well closed.* Dec in moist atm, giving off propionic acid.

USE: As fungicide on adhesive tape to reduce plaster irritation caused by molds, fungi, and bacterial action.

THERAP CAT: Antifungal (topical).

10287. Zinc Pyrophosphate. *Diphosphoric acid, zinc salt (1:2).* $O_7P_2Zn_2$; mol wt 304.72. O 36.75%, P 20.33%, Zn 42.92%. $Zn_2P_2O_7$.

White, crystalline powder. d^{23} 3.75. Insoluble in water. Soluble in dilute mineral acids.

10288. Zinc Salicylate. *2-Hydroxybenzoic acid zinc salt.* $C_{14}H_{10}O_6Zn$; mol wt 339.62. C 49.51%, H 2.97%, O 28.27%, Zn 19.25%. $Zn[C_6H_4(OH)COO]_2$. Prepd from sodium salicylate and zinc sulfate: Clark, Kao, *J. Am. Chem. Soc.* **70,** 2151 (1948).

Trihydrate, needles or cryst powder. Sol in water, alc. The aq soln is practically neutral to litmus.

THERAP CAT: Antiseptic; astringent.

10289. Zinc Selenate. O_4SeZn; mol wt 208.35. O 30.72%, Se 37.90%, Zn 31.39%. $ZnSeO_4$. Heptahydrate prepd by the action of selenic acid on zinc carbonate: Banks, *J. Chem. Soc.* **1934,** 1010.

Pentahydrate, triclinic crystals. Dec above 50°. d_4^{20} 2.591. Sol in water.

Hexahydrate, tetragonal crystals. d 2.325.

10290. Zinc Selenide. SeZn; mol wt 144.35. Se 54.70%, Zn 45.30%. ZnSe. Prepd by mixing a soln of a zinc salt and potassium selenide: Berzelius, cited in *Mellor's Comprehensive Treatise on Inorganic and Theoretical Chemistry* **10,** 776 (1930); by passing selenium vapor over zinc heated *in vacuo:* Moser, Doctor, *Z. Anorg. Chem.* **118,** 284 (1921); by passing H_2Se into a methanol soln of zinc acetate; Nitsche, U.S. pat. **2,805,917** (1957 to du Pont); from $ZnSe.N_2H_4$ (zinc selenide hydrazinate): Conn *et al.,* U.S. pat. **3,014,779** (1961 to Merck & Co.).

Yellow, cubic crystals. d_4^{15} 5.42. mp >1100°. Dec in air. Insol in water. Dec in dil nitric acid.

USE: Commercial phosphor.

10291. Zinc Silicate. Zinc orthosilicate. O_4SiZn_2; mol wt 222.86. O 28.72%, Si 12.60%, Zn 58.68%. Zn_2SiO_4. Occurs in nature as the mineral *willemite.* Prepd by heating the proper amounts of ZnO and SiO_2 at about 1200°.

White powder, insol in water or dil acids.

USE: In television screens.

10292. Zinc Stearate. *Octadecanoic acid zinc salt.* $C_{36}H_{70}O_4Zn$; mol wt 632.34. C 68.38%, H 11.16%, O 10.12%, Zn 10.34%. $Zn(C_{18}H_{35}O_2)_2$. Usually occurs as a mixture of the zinc salts of stearic and palmitic acids, and usually with some excess of zinc oxide. Contains 13.5-15% ZnO. Prepd from stearic acid and zinc chloride: Vold, Hattiangdi, *Ind. Eng. Chem.* **41,** 2311 (1949).

Fine, soft, bulky powder; slight characteristic odor; neutral reaction. Repels water. mp about 120°. Insol in water, alcohol, ether. Sol in benzene; dec by dil acids.

USE: In tablet manuf; in cosmetic and pharmaceutical powders and ointments; as a flatting and sanding agent in lacquers; as a drying lubricant and dusting agent for rubber; as a plastic mold releasing agent; as a waterproofing agent for concrete, rock wool, paper, textiles.

THERAP CAT (VET): Antiseptic, astringent, protective (topical).

10293. Zinc Sulfate. White vitriol; zinc vitriol; Keratol; Optraex; Solvezink; Solvazinc; Zincaps; Zincate; Zincomed; Z Span. O_4SZn; mol wt 161.45. O 39.64%, S 19.86%, Zn 40.50%. $ZnSO_4$. Prepn and physical properties: *Gmelin's, Zink* (8th ed.) **32,** 936-960 (1956). Effects of oral zinc sulfate in acne treatment: G. Michaelsson *et al., Arch. Dermatol.* **113,** 31 (1977). Clinical evaluations in Wilson's Disease: T. U. Hoogenraad, C. J. A. Van den Hamer, *Acta Neurol. Scand.* **67,** 356 (1983); T. U. Hoogenraad *et al., J. Neurolog. Sci.* **77,** 137 (1987).

Monohydrate, *dried zinc sulfate.* Powder or granules. Loses H_2O above 238°. Sol in water. Practically insol in alcohol.

Heptahydrate, *Op-Thal-Zin, Redeema.* Odorless crystals or granules or powder; astringent taste. Efflorescent in dry air. d 1.97. mp 100°. At 280° loses all H_2O; dec above 500°. One gram dissolves in 0.6 ml water, 2.5 ml glycerol. Insol in alcohol. The aq soln is acid to litmus; pH about 4.5. *Keep well closed.*

Note: The monohydrate does not cake as the heptahydrate does, and hence is more convenient for use during the warm season and in warm climates.

Caution: Irritating to skin, mucous membranes: H. E. Stokinger, *Patty's Industrial Hygiene and Toxicology* 2A, G. D. Clayton, F. E. Clayton, Eds. (John Wiley & Sons, New York, 1981) pp 2033-2049.

USE: As mordant in calico-printing; preserving wood and skins; with hypochlorite for bleaching paper; manuf lithopone and other zinc salts; clarifying glue; electrodepositing Zn; also as reagent in analytical chemistry.

THERAP CAT: Ophthalmic astringent. Zinc supplement.

THERAP CAT (VET): Astringent. Has been used as an emetic.

10294. Zinc Sulfide. Zinc Blende. SZn; mol wt 97.46. S 32.90%, Zn 67.10%. ZnS. Occurs in nature as the minerals *wurtzite* (hexagonal, d 4.087) and *sphalerite* (cubic, d 4.102). Precipitated zinc sulfide of commerce usually contains 15-20% water of hydration. The dried precipitate may have been heated to 725° in the absence of air to obtain substantial conversion to wurtzite, the form preferred by the pigment industry.

White to grayish-white or yellowish powder. When contg water, it slowly oxidizes in air to sulfate. Insol in water, alkalies. Sol in dil mineral acids.

USE: Pigment for paints, oilcloths, linoleum, leather, dental rubber, etc., especially in the form of lithopone; mixed with ZnO as "mineral white." Anhyd zinc sulfide is used in x-ray screens and with a trace of a radium or mesothorium salt in luminous dials of watches; also television screens.

10295. Zinc Tannate. Sal barnit. Compd of ZnO and tannin in variable proportions. Prepd from an aq tannin soln and zinc hydroxide in excess ammonium hydroxide plus an ammonium salt: **Ger. pat. 479,229** (1926 to Lab. Reumella Adolf Boas), *C.A.* **23**, 4778[1] (1929).

Yellowish, odorless powder. Practically insol in water, alcohol. Sol in dil acids.

THERAP CAT: Astringent; antiseptic.

10296. Zinc Tartrate. $C_4H_4O_6Zn$; mol wt 213.46. C 22.51%, H 1.89%, O 44.97%, Zn 30.63%. $ZnC_4H_4O_6$. Prepd from potassium tartrate and zinc chloride. Prepn and solubility in water: Cantoni, Zachoder, *Bull. Soc. Chim.* [3] **33**, 747 (1905).

Dihydrate, cryst powder.

10297. Zinc Telluride. TeZn; mol wt 192.99. Te 66.12%, Zn 33.88%. ZnTe. Prepd by fusing Zn and Te: Braithwaite, *Proc. Phys. Soc. (London)* **64B**, 274 (1951); Bube, *Proc. IRE* **43**, 1836 (1955), *see also* Dennis, Anderson, *J. Am. Chem. Soc.* **36**, 882 (1914). From zinc oxide and powdered tellurium in aq alkaline medium: Montignie, *Bull. Soc. Chim. (France)* **1947**, 750.

Gray or brownish-red powder. Ruby-red crystals (cubic system) by sublimation. Stable in dry air. d_4^{15} 6.34. mp 1239°. Forms a monohydrate. Prolonged contact with water or dil HCl yields H_2 and H_2Te.

USE: In semiconductor research, as photoconductor.

10298. Zinc Thiocyanate. Zinc sulfocyanate. $C_2N_2S_2$-Zn; mol wt 181.56. C 13.23%, N 15.43%, S 35.32%, Zn 36.02%. $Zn(SCN)_2$.

White, deliquesc crystals. Sol in water, alcohol. The aq soln is only slightly acid to litmus. *Keep well closed and protected from light.*

USE: To assist in textile dyeing.

10299. Zinc Valerate. *Pentanoic acid zinc salt.* $C_{10}H_{18}$-O_4Zn; mol wt 267.64. C 44.88%, H 6.78%, O 23.91%, Zn 24.43%. $Zn(C_5H_9O_2)_2$. The article of commerce is usually

basic. Prepd from valeric acid and zinc hydrate: Lieben, Rossi, *Ann.* **159**, 58 (1871).

Dihydrate, lustrous scales or powder; valerian odor; sweetish taste. Gradually dec on exposure to air. One gram dissolves in 70 ml water or in 22 ml alcohol when free from basic salt; dec by acid. *Keep tightly closed and in cool place.*

10300. Zineb. *[[1,2-Ethanediylbis[carbamodithioato]]-(2—)]zinc; [ethylenebis(dithiocarbamato)]zinc;* zinc ethylenebis(dithiocarbamate); ethylenebis(dithiocarbamic acid) zinc salt; ENT-14874; Parzate C; Lonacol; Dithane Z-78; Tiesene; Tritoftorol. $(C_4H_6N_2S_4Zn)_n$. Polymeric salt of ethylenebisdithiocarbamic acid; related to maneb and mancozeb, *q.q.v.* Prepd from a zinc salt and nabam, *q.v.:* C. B. Luginbuhl, U.S. pat. **2,690,448** (1954 to du Pont). Fungicidal activity is due to degradation products, principally *ethylenethiuram monosulfide:* R. A. Ludwig *et al., Can. J. Bot.* **33**, 42 (1955). Activity against potato blight: B. K. De, S. B. Chattopadhyay, *Pesticide* **18**, 52 (1984). Decomposition of zineb to ethylenebisdicarbamate and ethylene thiourea, *q.v.,* in tomatoes: B. D. Ripley, D. F. Cox, *J. Agr. Food Chem.* **26**, 1137 (1978). HPLC determn: K. Gustafsson *et al., ibid.* **29**, 729 (1981). Toxicity data: T. B. Gaines, *Toxicol. Appl. Pharmacol.* **14**, 515 (1969). Aquatic toxicology: C. J. Van Leeuwen *et al., Aquat. Toxicol.* **7**, 145 (1985).

Powder or crystals from chloroform + alcohol. Practically insol in water. The powder spreads easily on water, also forms aq suspensions. Sol in carbon disulfide, pyridine. LD_{50} orally in rats: > 5000 mg/kg (Gaines).

Caution: Irritation of skin and mucous membranes has been reported: *Clinical Toxicology of Commercial Products,* R. E. Gosselin *et al.,* Eds. (Williams & Wilkins, Baltimore, 5th Ed., 1984) Section II, p 313.

USE: Agricultural fungicide.

10301. Zingerone. *4-(4-Hydroxy-3-methoxyphenyl)-2-butanone;* (4-hydroxy-3-methoxyphenyl)ethyl methyl ketone; vanillylacetone; zingherone; zingiberone. $C_{11}H_{14}O_3$; mol wt 194.23. C 68.02%, H 7.27%, O 24.71%. Isolated from ginger root or prepd from vanillin and acetone followed by catalytic hydrogenation: Nomura, *J. Chem. Soc.* **111**, 769 (1917); *idem,* U.S. pats. **1,263,796** (1918); **1,306,710** (1919); Cotton, U.S. pat. **2,381,210** (1945 to Penn. Coal Prod.); K. Banno, T. Mukaiyama, *Bull. Chem. Soc. Japan* **49**, 1453 (1976).

Crystals from acetone, petr ether, ether + petr ether, mp 40-41°. bp_{14} 187-188°. Sparingly sol in water, petr ether; sol in ether, dil alkalies.

10302. Zinostatin. *Neocarzinostatin;* neocarcinostatin; NCS; NSC-69856; Neocarzinostatin K. An antitumor acidic antibiotic consisting of 2 components: a protein of mol wt 10,700 containing 109 amino acid residues of 18 kinds and a labile non-protein chromophore which possesses the full biological activity of NCS. Isoln from the culture filtrate of *Streptomyces carcinostaticus* var. F-41: N. Ishida *et al., J. Antibiot.* **18A**, 68 (1965). Characterization: H. Maeda *et al., ibid.* **19A**, 253 (1966). Prodn: N. Ishida *et al.,* U.S. pat. **3,334,022** (1967 to Empire Corp.). Amino acid sequence: J. Meienhofer *et al., Science* **178**, 875 (1972). Review of protein structural studies: *eidem, Progress in Peptide Research* vol. **II**, S. Lande, Ed. (Gordon and Breach, New York, 1972) pp 295-306. Spectral characterization and separation of the chromophore: M. A. Napier *et al., Biochem. Biophys. Res. Commun.* **89**, 635 (1979). Partial structure of the chromophore: G. Albers-Schönberg et al., ibid. **95**, 1351 (1980). Roles of chromophore and apo-protein in biological activity: L. S. Kappen et al., Proc. Nat. Acad. Sci. USA **77**, 1970 (1980). Molecular basis of action: I. H. Goldberg et al., Mol. Biol. Biochem. Biophys. **32**, 308 (1980). Absorption, distribution, excretion: K. Toriyama et al., Japan. J. Antibiot. **28**, 64 (1975). Review of pharmacology: T. A. Beerman et al., in Advances in Enzyme Regulation vol. **14**, G. Weber, Ed. (Pergamon Press, New York, 1976) pp 207-228.

White powder, mp 260° (dec). uv max (water): 278 nm ($E_{1cm}^{1\%}$ 15). Isoelectric pt: pH 3.26 [Japan. Med. Gaz. **14**(3), 11 (1977)]. Sol in water. Insol in most common org solvents. See M. Kohno et al., Japan. J. Antibiot. **27**, 707 (1974) for stability of aq solns. LD$_{50}$ in mice (mg/kg): 1050 orally, 0.96 i.v. (Toriyama).

THERAP CAT: Antineoplastic.

10303. Zipeprol. 4-(2-Methoxy-2-phenylethyl)-α-(methoxyphenylmethyl)-1-piperazineethanol; α-(α-methoxybenzyl)-4-(β-methoxyphenethyl)-1-piperazineethanol; 1-(2-hydroxy-3-methoxy-3-phenylpropyl)-4-(2-methoxy-2-phenylethyl)-piperazine; 1-(2-methoxy-2-phenylethyl)-4-(2-hydroxy-3-methoxy-3-phenylpropyl)piperazine. $C_{23}H_{32}N_2O_3$; mol wt 384.52. C 71.84%, H 8.39%, N 7.29%, O 12.48%. Prepn: Mauvernay et al., **Ger. pat. 2,109,366** corresp to U.S. pat. **3,718,650** (1971, 1973 both to Mauvernay). Pharmacology: G. Rispat et al., Arzneimittel-Forsch. **26**, 523 (1976); D. Cosnier et al., ibid. 848.

Crystals from abs ethanol, mp 83°.

Dihydrochloride, $C_{23}H_{32}N_2O_3 \cdot 2HCl$, CERM-3024, Antituxil-Z, Citizeta, Mirsol, Respilene, Respirase, Zitoxil. Crystals from abs ethanol, mp 231°. Very stable in aq soln. LD$_{50}$ orally in mice: 301 mg/kg (Rispat).

THERAP CAT: Antitussive.

10304. Ziprasidone. 5-[2-[4-(1,2-Benzisothiazol-3-yl)-1-piperazinyl]ethyl]-6-chloro-1,3-dihydro-2H-indol-2-one; 5-(2-(4-(1,2-benzisothiazol-3-yl)piperazinyl)ethyl)-6-chlorooxindole; CP-88059. $C_{21}H_{21}ClN_4OS$; mol wt 412.94. C 61.08%, H 5.13%, Cl 8.59%, N 13.57%, O 3.87%, S 7.77%. Dopamine D_2/serotonin 5-HT$_2$ antagonist. Prepn: J. A. Lowe III, A. A. Nagel, **Eur. pat. Appl. 281,309**; eidem, **U.S. pat. 4,831,031** (1988, 1989 both to Pfizer). Clinical pharmacology: C. J. Bench et al., Psychopharmacology **112**, 308 (1993). HPLC determn in serum: J. S. Janiszewski et al., J. Chromatog. B **668**, 133 (1995). Brief review of comparative pharmacology: J. M. Goldstein, Exp. Opin. Invest. Drugs **4**, 291-298 (1995).

Hydrochloride monohydrate, $C_{21}H_{21}ClN_4OS \cdot HCl \cdot H_2O$, CP-88059-1. Also occurs as the hemihydrate, mp > 300°.

THERAP CAT: Antipsychotic.

10305. Ziram. (T-4)-Bis(dimethylcarbamodithioato-S,S')zinc; bis(dimethyldithiocarbamato)zinc; zinc dimethyldithiocarbamate; dimethyldithiocarbamic acid zinc salt; zinc bis(dimethylthiocarbamoyl) disulfide; methyl cymate; Methasan; Zimate; Zirberk; Karbam White; Corozate; Fuclasin; Fuklasin; Zerlate. $C_6H_{12}N_2S_4Zn$; mol wt 305.83. C 23.56%,

H 3.95%, N 9.16%, S 41.94%, Zn 21.38%. Prepd from zinc oxide, dimethylamine, and carbon disulfide: Olin, Deger, **U.S. pat. 2,492,314** (1949 to Sharples Chemicals). Crystal structure: Klug, Acta Cryst. **21**, 536 (1966). Toxicity study: Hodge et al., J. Pharmacol. Exp. Ther. **118**, 174 (1956).

Crystals from hot chloroform + alcohol, mp 250°. Can form a flammable dust. d$_4^{25}$ 1.66. Practically insol in water. Soly per 100 ml of solvent at 25°: < 0.2 g, alcohol; < 0.5 g, acetone; < 0.5 g, benzene; < 0.2 g, carbon tetrachloride, more sol in chloroform; < 0.2 g, ether; 0.5 g, naphtha. Sol in dil caustic solns. LD$_{50}$ orally in rats: 1.4 g/kg (Hodge).

Caution: Irritant to skin and mucous membranes.

USE: Rubber vulcanization accelerator; agricultural fungicide.

10306. Zirconium. Zr; at. wt 91.22; at. no. 40; valence 4; also 3. Five naturally occurring isotopes: 90 (51.46%); 91 (11.23%); 92 (17.11%); 94 (17.40%); 96 (2.80%); artificial radioactive isotopes: 81-89, 93, 95, 97-99. Occurrence in earth's crust: 0.023%. Occurs in the minerals zircon, malacon, baddeleyite, zirkelite, eudialyte; frequently found in the rare-earth minerals; in monazite sand. Discovered by Klaproth in 1789; prepd by Berzelius in 1824. Prepn: Fast, Z. Anorg. Chem. **239**, 145 (1938); purification of zirconium by ion exchange columns: Ayres, J. Am. Chem. Soc. **69**, 1879 (1947). Sepn of zirconium and hafnium: Fischer et al., Angew. Chem. Int. Ed. **5**, 15 (1966). Reviews of zirconium and its compds: W. B. Blumenthal, The Chemical Behavior of Zirconium (Van Nostrand, Princeton, 1958); Gmelin's Handb. Anorg. Chem., Zirconium (8th ed.) **42**, (1958) 448 pp; Larsen, "Zirconium and Hafnium Chemistry" in Advan. Inorg. Chem. Radiochem. **13**, 1-333 (1970); Bradley, Thornton, "Zirconium and Hafnium" in Comprehensive Inorganic Chemistry, vol. 3, J. C. Bailar, Jr. et al., Eds. (Pergamon Press, Oxford, 1973) pp 419-490.

Bluish-black, amorphous powder or grayish-white lustrous metal (platelets or flakes) of hexagonal lattice below 865°, body-centered cubic above 865°, mp 1857°; bp 3577°. d 6.5. Brinnell hardness: 85. Can absorb up to 10 atoms per cent of oxygen or nitrogen. Reacts with hydrofluoric acid, aqua regia, hot phosphoric acid. Not attacked by cold, very slightly attacked by hot, concd sulfuric or hydrochloric acid; not attacked by nitric acid. Attacked by fused potassium hydroxide or nitrate. On prolonged heating the compact form combines with oxygen, nitrogen, carbon, and the halogens. The powder form has a very low ignition temp and is very explosive when mixed with oxidizing agents.

Caution: Zirconium and its salts generally have low systemic toxicity. A granulomatous disease of the skin, particularly in the axilla, has been reported in users of a deodorant contg sodium zirconium lactate: see E. Browning, Toxicity of Industrial Metals (Appleton-Century-Crofts, New York, 2nd ed., 1969) pp 356-360. Consult latest Government regulations on use in aerosol antiperspirants.

USE: Pure zirconium (hafnium-free) is a valuable structural material for atomic reactors because of its low nuclear cross-section and high corrosion and heat resistance. Because of hafnium's high neutron absorption characteristics, it must be removed from zirconium which is to be used in nuclear reactors; removal unnecessary for other commercial purposes. As an ingredient of priming or explosive mixtures; flashlight powders; as deoxidizer in metallurgy; as "getter" in vacuum tubes; in constructing rayon spinnerets in lamp filaments, flash bulbs.

10307. Zirconium Chloride. Zirconium tetrachloride. Cl_4Zr; mol wt 233.03. Cl 60.85%, Zr 39.15%. ZrCl$_4$. Lewis acid. In large-scale prepns zirconium oxide is converted to the carbide, which is chlorinated to yield the tetrachloride: Kroll et al., J. Electrochem. Soc. **94**, 1 (1948). Process not easily adaptable to lab use. Lab prepn based on the equation ZrO$_2$ + 2CCl$_4$ → ZrCl$_4$ + 2COCl$_2$: Hummers et al., Inorg. Syn. **4**, 121 (1953).

Lustrous monoclinic crystals: Krebs, *Angew. Chem. Int. Ed.* **8**, 146 (1969); idem, *Z. Anorg. Allgem. Chem.* **378**, 263 (1970). Tetrahedral symmetry in gas phase with the Zr-Cl distance of 2.33 Å. Extremely hygroscopic, forms HCl vapor and gives off fumes in moist air. d 2.803. Sublimes at 331°. mp 437° under its own pressure which is about 25 atm at this temp. Decomposed by water to form $ZrOCl_2$ and HCl; sol in alcohol, ether. LD_{50} in mice, rats: 655, 1688 mg/kg orally, *Toxic Substances List*, H. E. Christensen, Ed. (1972) p. 547.

USE: Friedel-Crafts catalyst. Component of Ziegler-type catalysts in the condensation of ethylene. Starting material in the synthesis of a number of organic derivs of zirconium, such as alkoxides and zircocene. The alkoxides have been shown to be of value in the curing of silicone plastic films. The alkoxyzirconium carboxylates are said to be useful in the water-repellent treatment of textiles and other fibrous materials. *Review:* Blumenthal, *J. Chem. Ed.* **39**, 604-610 (1962).

10308. Zirconium Fluoride. Zirconium tetrafluoride. F_4Zr; mol wt 167.22. F 45.45%, Zr 54.55%. ZrF_4. Prepd by thermal decompn of $(NH_4)_2ZrF_6$: v. Hevesy, Dullenkamp, *Z. Anorg. Allgem. Chem.* **221**, 161 (1934); according to the equation $ZrCl_4 + 4HF \rightarrow ZrF_4 + 4HCl$: Wolter, *Chemiker-Ztg.* **51**, 607 (1908); from zirconium oxide and fluorine: Haendler *et al.*, *J. Am. Chem. Soc.* **76**, 2177 (1954). Strongly refractive, crystalline mass (monoclinic system). d^{16} 4.6. Sublimes above 600°. Solubility in water (20°): 1.32 g/100 ml. Does not react with water; forms stable trihydrate. Freely soluble in hydrofluoric acid. LD_{50} in mice: 98 mg/kg orally: *Toxic Substances List*, H. E. Christensen, Ed. (1972) p 547.

10309. Zirconium Hydride. Ideal composition: ZrH_2. Prepd by the reduction of zirconium oxide with calcium hydride in the presence of hydrogen above 600°: Alexander, U.S. pat. **2,427,339** (1947 to Metal Hydride).
Gray-black metallic powder. Stable, no reaction with water.
USE: Powerful reducing agent in acid solution or at high temps; hydrogenation catalyst; in the vacuum tube industry and powder metallurgy.

10310. Zirconium Hydroxide. H_4O_4Zr; mol wt 159.25. H 2.53%, O 40.19%, Zr 57.28%. $Zr(OH)_4$. Proposed structures of freshly pptd and aged compds, cyclic tetramers: Zaitsev, *Russ. J. Inorg. Chem.* **11**, 900 (1966).
White, bulky, amorphous powder; d 3.25. Insol in water; sol in mineral acids when freshly pptd; less sol when aged. It colors turmeric paper brown.
USE: In the pigment, dye, and glass industries.

10311. Zirconium Iodide. Zirconium tetraiodide. I_4Zr; mol wt 598.84. I 84.77%, Zr 15.23%. ZrI_4. Prepd from the elements at a furnace temp of 450°: Eberly, *Inorg. Syn.* **7**, 52 (1963).
Orange-colored, crystalline solid. Sublimes at 431°. mp 499° (elevated pressure). Fumes heavily in air. Dissolves in water with the liberation of steam. Heat of formation: 90 kcal/mol at 25°. Magnetic susceptibility -0.238×10^{-6} c.g.s. electromagnetic units.

10312. Zirconium Nitrate. $N_4O_{12}Zr$; mol wt 339.24. N 16.52%, O 56.59%, Zr 26.89%. $Zr(NO_3)_4$. Prepn according to the equation $ZrCl_4 + 4N_2O_5 \rightarrow Zr(NO_3)_4 + 4NO_2Cl$: Field, Hardy, *Proc. Chem. Soc.* **1962**, 76. Pentahydrate obtained from strong nitric acid. The zirconium nitrate of commerce is usually somewhat basic.
Pentahydrate, white, very hygroscopic crystals or white pieces or scales. Very sol in water; sol in alcohol. The aq soln is acid to litmus. *Keep tightly closed.*

10313. Zirconium Oxide. Zirconia; zirconium dioxide; zirconic anhydride. O_2Zr; mol wt 123.22. O 25.97%, Zr 74.03%. ZrO_2. Occurs in nature as the mineral *baddeleyite.* Prepn: Clark, Reynolds, *Ind. Eng. Chem.* **29**, 711 (1937); Henderson, Higbie, *J. Am. Chem. Soc.* **76**, 5878 (1954).
White, heavy, amorphous, odorless, tasteless powder or monoclinic crystals. Also forms tetragonal crystals above ∼1100°; cubic above ∼1900°. d 5.85. mp 2680°. bp 4300°. Practically insol in water; slightly sol in HCl, HNO_3; slowly

sol in HF; dissolves on heating with a mixt of 2 parts H_2SO_4 and 1 part H_2O. Fusion with sodium carbonate results in the formation of sodium zirconate which dec hydrolytically with water to form sodium hydroxide and practically insol zirconium hydroxide.
USE: Instead of lime for the oxyhydrogen light; with earths of the yttrium group in incandescent lighting (Nernst lamps); as pigment, abrasive; manuf enamels, white glass, refractory crucibles, and furnace linings.
THERAP CAT: Dermatologic.

10314. Zirconium Silicate. Zirconium orthosilicate. O_4SiZr; mol wt 183.31. O 34.91%, Si 15.32%, Zr 49.77%. $ZrSiO_4$. Occurs in nature as the mineral *zircon.* Widely scattered in all kinds of rocks and in beach sand (South Carolina, Northeast Florida). Usually separated from sand by electrostatic and electromagnetic elutriation, yielding 99% pure $ZrSiO_4$. Prepn from SiO_2 and ZrO_2 in an arc furnace: Curtis, Sowman, *J. Am. Ceram. Soc.* **36**, 195 (1953); can also be made from solns of zirconium salts and sodium silicate.
Tetragonal, bipyramidal crystals. Should be colorless. Colors are from impurities and radioactive bombardment, often removable by calcination. Dissociates to ZrO_2 and SiO_2 when heated above 1540°. Recombines when cooled slowly, but forms a mixture of monoclinic ZrO_2 and vitreous SiO_2 when rapidly quenched. Undissociated (α-form) $ZrSiO_4$ has a high sp gr, 4.7, and a high birefringence; the dissociated, γ-form has a lower sp gr, 3.9-4.0, and a low birefringence. Hardness of α-form about the same as that of quartz (6-7.5). Very inert chemically. Unaffected by aq reagents.
USE: In refractories, ceramics, glazes, cements, coatings for casting molds, polishing materials, gem stones, catalyst in alkyl and alkenyl hydrocarbon manuf, in fritted glass filters, as stabilizer in silicone rubbers. In Europe in cosmetic creams and powders.

10315. Zirconium Sulfate. Disulfatozirconic acid. O_8S_2Zr; mol wt 283.35. O 45.17%, S 22.63%, Zr 32.19%. $Zr(SO_4)_2$. Tetrahydrate usually prepd by treating zirconium oxychloride with hot concd H_2SO_4: Blumenthal, *J. Chem. Ed.* **39**, 604 (1962). Toxicity study: Cochran *et al.*, *Arch. Ind. Hyg. Occup. Med.* **1**, 637 (1950).
Tetrahydrate, crystalline solid. Converted to monohydrate at 100°; to anhydr form at 380°. Soly in water (18°): 52.5 g/100 g of soln. A soln at room temp deposits a solid on standing. The more dilute the soln, the more rapid the deposition. The composition of the solid is $4ZrO_2\cdot3SO_3$-$15H_2O$, known as *Hauser's salt:* Hauser, Herzfeld, *Z. Anorg. Allgem. Chem.* **67**, 369 (1910). This pptn does not take place if the soln has been heated above 64°. LD_{50} in rats: 3.5 g/kg orally; 175 mg/kg i.p. (Cochran).
USE: Catalyst support; precipitation of amino acids and proteins; in the tanning industry.

10316. Zirconyl Acetate. Diacetatozirconic acid. Approx formula: $Zr(OH)_2(C_2H_3O_2)_2$. Obtained as aq soln by adding acetic acid to an aq slurry of carbonated hydrous zirconia: Blumenthal, *J. Chem. Ed.* **39**, 604 (1962). Toxicity data: Cochran *et al.*, *Arch. Ind. Hyg. Occup. Med.* **1**, 637 (1950).
Available as 22% ZrO_2 soln. d 1.46. mp −7°. pH 3.8-4.2. Stable at room temp. Also available as 13% ZrO_2 soln. d 1.20. pH 3.3-4.0. Undergoes exchange with anion exchange resins, but not with cation exchangers. Evaporation of the solns under reduced pressure yields the solid compd. It is resinous or glue-like in appearance and amorphous under x-rays. Its properties indicate it to be a highly polymerized product. The powdered solid readily dissolves when added slowly to rapidly swirling water. On heating, the solns yield a solid hydrolyzate, which is a more basic acetate with the compn $ZrOOHC_2H_3O_2$. When the mixt is allowed to cool and stand at room temp, this ppt slowly redissolves. LD_{50} in rats: 4.1 g/kg orally; 300 mg/kg i.p. (Cochran).
USE: Precipitating agent for gelatin and starch on textiles and paper; water-repellent for textiles (especially in combination with silicones).

10317. Zirconyl Chloride. *Dichlorooxozirconium;* zirconium oxychloride; basic zirconium chloride. Cl_2OZr; mol wt 178.13. Cl 39.81%, O 8.98%, Zr 51.21%. $ZrOCl_2$. Prepn from $ZrCl_4$ and Cl_2O: Dehnicke, Meyer, *Z. Anorg. Allgem. Chem.* **331**, 121 (1964). Octahydrate conveniently prepd by crystn of an aq soln of zirconium chloride: v. Siemens, Zander, *Wissenschaftl. Veröffentl. Siemens* **2**, 484 (1922); Blumenthal, *J. Chem. Ed.* **39**, 607 (1962). Toxicity study: Cochran *et al., Arch. Ind. Hyg. Occup. Med.* **1**, 637 (1950).

Octahydrate, tetragonal crystals from water. Crystals contain tetramers of the form $[Zr_4(OH)_8(H_2O)_{16}]^{8+}$; d 1.91: Mak, *Can. J. Chem.* **46**, 3491 (1968). Freely sol in water, alcohol. The pH is about equal to that of HCl of the same molarity. LD_{50} in rats: 400 mg/kg i.p.; 3.5 g/kg orally (Cochran).

USE: To make other zirconium compds; to precipitate acid dyes; to prepare high quality pigment toners; to improve the properties of color lakes. Aq $ZrOCl_2$ solns have considerable solvent action on sparingly sol sulfates, such as calcium sulfate. The free acid in $ZrOCl_2$ solns may be neutralized, and a sol compd $ZrOOHCl.xH_2O$ can be recovered. It is highly polymerized in soln and amorphous in the solid state; has been used in the prepn of body deodorants and antiperspirant preparations.

10318. Zoapatanol. *[2S-[2α(S*),3β,6E]]-9-[3-Hydroxy-6-(2-hydroxyethylidene)-2-methyl-2-oxepanyl]-2,6-dimethyl-2-nonen-5-one.* $C_{20}H_{34}O_4$; mol wt 338.49. C 70.97%, H 10.12%, O 18.91%. Oxepane diterpenoid isolated from the leaves of the zoapatle plant, *Montanoa tomentosa, Compositeae,* which has been used by Mexican women to prepare "tea" to induce menses and labor. Isoln and structure: M. P. Wachter, R. M. Kanojia, U.S. pat. **4,086,358** (1978 to Ortho); S. D. Levine *et al., J. Am. Chem. Soc.* **101**, 3404 (1979); R. M. Kanojia *et al., J. Org. Chem.* **47**, 1310 (1982). Spasmogenic effects: J. B. Smith *et al., Life Sci.* **28**, 2743 (1981). Total syntheses of (±)-form: R. Chen, D. A. Rowand, *J. Am. Chem. Soc.* **102**, 6609 (1980); K. C. Nicolau *et al., ibid.* 6611; V. V. Kane, D. L. Doyle, *Tetrahedron Letters* **22**, 3027, 3031 (1981). Mass spec: C. J. Shaw, *Org. Mass Spectrom.* **16**, 281 (1981). ^{13}C-NMR study: M. L. Cotter, *Org. Magn. Res.* **17**, 14 (1981).

Pale yellow oil.

10319. Zolamine. *N-[(4-Methoxyphenyl)methyl]-N',N'-dimethyl-N-2-thiazolyl-1,2-ethanediamine; 2-[(2-dimethylaminoethyl)(p-methoxybenzyl)amino]thiazole; N',N'-dimethyl-N-(2-thiazolyl)-N-(p-methoxybenzyl)ethylenediamine.* $C_{15}H_{21}N_3OS$; mol wt 291.42. C 61.82%, H 7.26%, N 14.42%, O 5.49%, S 11.00%. This is an isostere of pyrilamine, *q.v.,* and is prepd accordingly: Shigeya Saijo, *J. Pharm. Soc. Japan* **72**, 1009 (1952).

Oily liquid. Putrid odor. bp$_7$ 217-219°.
Hydrochloride, $C_{15}H_{21}N_3OS.HCl$, *194-B, WL-291.* Odorless crystals from alc. Slightly bitter taste. mp 167.5-167.8°. Sol in water. Ingredient of *Otodyne* (obsolete).
THERAP CAT: Antihistaminic; anesthetic (local).

10320. Zolimidine. *2-[4-(Methylsulfonyl)phenyl]imidazo[1,2-a]pyridine;* zoliridine; Solimidin. $C_{14}H_{12}N_2O_2S$; mol wt 272.33. C 61.75%, H 4.44%, N 10.29%, O 11.75%, S 11.77%. Non-anticholinergic gastroprotective agent. Prepn: **Brit. pat. 991,589;** L. Almirante *et al.,* **U.S. pat. 3,318,880** (1965, 1967 both to Selvi); *eidem, J. Med. Chem.*

8, 305 (1965). Metabolism: *eidem, Farmaco Ed. Sci.* **29**, 941 (1974). Series of articles on pharmacology: *Panminerva Med.* **16**, 301-359 (1974). Pharmacokinetics: E. Schraven, D. Trottnow, *Arzneimittel-Forsch.* **26**, 213 (1976). Comparative study of effects on rat gastric mucosa: A. P. Green *et al., J. Pharm. Pharmacol.* **33**, 348 (1981). Clinical study: A. Materia *et al., Clin. Ter.* **97**, 183 (1981). Toxicity study: L. Almirante *et al., J. Med. Chem.* **8**, 305 (1965).

Crystals, mp 242-244°. LD_{50} orally in rats: 3710 mg/kg (Almirante, 1967).
Hydrochloride monohydrate, LD_{50} i.p. in mice: 800 mg/kg (Almirante, 1965).
THERAP CAT: Antiulcerative.

10321. Zolpidem. *N,N,6-Trimethyl-2-(4-methylphenyl)-imidazo[1,2-a]pyridine-3-acetamide; N,N,6-trimethyl-2-p-tolylimidazo[1,2-a]pyridine-3-acetamide;* SL-80.0750. $C_{19}H_{21}N_3O$; mol wt 307.40. C 74.24%, H 6.89%, N 13.67%, O 5.20%. Selective benzodiazepine receptor agonist not related chemically to benzodiazepines. Prepn: J. P. Kaplan, P. George, Eur. pat. Appl. 50,563; *eidem,* U.S. pat. **4,382,938** (1982, 1983 both to Synthelabo). Neuropharmacology: S. Arbilla *et al., Arch. Pharmacol.* **330**, 248 (1985); H. Depoortere *et al., J. Pharmacol. Exp. Ther.* **237**, 649 (1986). Neurochemical profile: B. Scatton *et al., ibid.* 659. Binding study in rat brain: S. Arbilla *et al., Eur. J. Pharmacol.* **130**, 257 (1986). HPLC determn in plasma: P. Guinebault *et al., J. Chromatog.* **383**, 206 (1986). Clinical evaluation of hypnotic activity: A. N. Nicholson, P. A. Pascoe, *Brit. J. Clin. Pharmacol.* **21**, 205 (1986). Evaluation as pre-anesthetic medication: J. N. Cashman *et al., ibid.* **24**, 85 (1987).

mp 196°. pKa 6.2.
L-(+)-Hemitartrate, $(C_{19}H_{21}N_3O)_2.C_4H_6O_6,$ *SL-80.0750-23N, Ambien, Ivadal, Niotal, Stilnoct, Stilnox.* Soly in water (20°): 23 mg/ml.
Note: This is a controlled substance (depressant) listed in the U.S. Code of Federal Regulations, Title 21 Part 1308.14 (1995).
THERAP CAT: Hypnotic.

10322. Zomepirac. *5-(4-Chlorobenzoyl)-1,4-dimethyl-1H-pyrrole-2-acetic acid;* 1,4-dimethyl-5-(p-chlorobenzoyl)-pyrrole-2-acetic acid. $C_{15}H_{14}ClNO_3$; mol wt 291.73. C 61.76%, H 4.84%, Cl 12.15%, N 4.80%, O 16.45%. Prepn: J. R. Carson, Ger. pat. 2,102,746; *idem,* U.S. pat. **3,752,826** (1971, 1973 both to McNeil); J. R. Carson, S. Wong, *J. Med. Chem.* **16**, 172 (1973). Pharmacology: R. Sofia *et al., Pharmacol. Res. Commun.* **11**, 179 (1979); P. O'Neill *et al., J. Pharmacol. Exp. Ther.* **209**, 366 (1979). Determn in plasma by HPLC: K.-T. Ng, T. Snyderman, *J. Chromatog.* **178**, 241 (1979). Metabolism: J. M. Grindel *et al., Drug Metab. Dispos.* **8**, 343 (1980); W. N. Wu *et al., ibid.* 349. Pharmacokinetics: R. K. Nayak *et al., Clin. Pharmacol. Ther.* **27**, 395 (1980). Series of articles on pharmacology, kinetics, clinical studies: *J. Clin. Pharmacol.* **20**, 213-424 (1980). Preclinical narcotic abuse liability evaluation: J. H. Woods *et al., Arzneimittel-Forsch.* **33**, 218 (1983). Multicenter clinical study in painful conditions: C. E. Steele, W. L. Jefferson, *Curr. Med. Res. Opin.* **8**, 382 (1983). Review of pharmacology and therapeutic efficacy: P. A. Morley *et al., Drugs* **23**, 250-275 (1982). Comprehensive description: M. Zinic *et al.* in *Analytical Profiles of Drug Substances* vol. **15**, K. Florey, Ed. (Academic Press, New York, 1986) pp 673-698.

White cryst from 2-propanol, mp 178-179°.
Sodium salt dihydrate, $C_{15}H_{13}ClNNaO_3 \cdot 2H_2O$, *Zomax*, *Zomaxin*, *Zopirac*. Crystals from isopropanol + water, mp 295-296°.
THERAP CAT: Analgesic; anti-inflammatory.

10323. Zonisamide. *1,2-Benzisoxazole-3-methanesulfon-amide;* 3-(sulfamoylmethyl)-1,2-benzisoxazole; AD-810; CI-912; Aleviatin; Exceglan; Excegram. $C_8H_8N_2O_3S$; mol wt 212.23. C 45.28%, H 3.80%, N 13.20%, O 22.62%, S 15.11%. Prepn and anticonvulsant activity: H. Uno *et al.*, **Japan Kokai 78 77,057**; *eidem*, U.S. pat. **4,172,896** (1978, 1979 both to Dainippon); *eidem*, *J. Med. Chem.* **22**, 180 (1979). Anticonvulsant activity and neurotoxic effects: Y. Masuda *et al.*, *Epilepsia* **20**, 623 (1979). Mechanism of action study: T. Ito *et al.*, *Arzneimittel-Forsch.* **30**, 603 (1980). Pharmacology and toxicity: Y. Masuda *et al., ibid.* 477. Pharmacological effects on CNS: M. Hori *et al., ibid.* **37**, 1124 (1987). Pharmacology: K. Nakatsuji *et al., ibid.* 1131. Pharmacokinetics: T. Ito *et al., ibid.* **32**, 1581 (1982); J. G. Wagner *et al., Ther. Drug Monit.* **6**, 277 (1984). Absorption and distribution: K. Matsumoto *et al., Arzneimittel-Forsch.* **33**, 961 (1983); E. M. Cornford, K. P. Landon, *Ther. Drug Monit.* **7**, 247 (1985). HPLC determn in serum: U. Juergens, *J. Chromatog.* **385** 233 (1987). Clinical evaluations in partial seizures: J. C. Sackellares *et al., Epilepsia* **26**, 206 (1985); comparison with carbamazepine, *q.v.* in epilepsy: A. J. Wilensky *et al., ibid.* 212. Brief review: E. J. Hammond *et al., Gen. Pharmacol.* **18**, 303 (1987).

Tasteless, odorless, white needles from ethyl acetate, mp 160-163° (Uno); also reported as mp 162-166° (Masuda, 1980). Sparingly sol in water, chloroform, *n*-hexane. Sol in methanol, ethanol, ethyl acetate, and acetic acid. LD_{50} in mice, rats (mg/kg): 1892, 2001 orally; 1273, 2569 s.c.; 699, 733 i.p.; 604, 748 i.v. (Masuda, 1980).
THERAP CAT: Anti-epileptic.

10324. Zopiclone. *4-Methyl-1-piperazinecarboxylic acid 6-(5-chloro-2-pyridinyl)-6,7-dihydro-7-oxo-5H-pyrrolo[3,4-b]-pyrazin-5-yl ester;* 6-(5-chloropyrid-2-yl)-5-(4-methylpiper-azin-1-yl)carbonyloxy-7-oxo-6,7-dihydro-5*H*-pyrrolo[3,4-b]-pyrazine; RP-27267; Amoban; Amovane; Imovance; Imovane; Sopivan; Zimovane. $C_{17}H_{17}ClN_6O_3$; mol wt 388.81. C 52.52%, H 4.41%, Cl 9.12%, N 21.61%, O 12.34%. The first of a family of non-benzodiazepines showing a pharmacological profile similar to chlordiazepoxide, *q.v.*, and nitrazepam, *q.v.* Prepn: C. Cotrel *et al.*, **Ger. pat. 2,300,491**; *eidem*, U.S. pat. **3,862,149** (1973, 1975 both to Rhone-Poulenc); C. Jeanmart, C. Cotrel, *Compt. Rend. Ser. C* **287**, 377 (1978). *In vitro* and *in vivo* inhibition of benzodiazepine binding: J. C. Blanchard *et al., Life Sci.* **24**, 2417 (1979); *see also* P. H. Wu *et al., ibid.* **28**, 1023 (1981). Clinical study in sleep disorders: R. Duriez *et al., Therapie* **34**, 317 (1979). Comparative study with benzodiazepines: E. Wickström, K. E. Giercksky, *Eur. J. Clin. Pharmacol.* **17**, 93 (1980). Experimental and clinical pharmacology, clinical trials including sleep laboratory studies and investigations of dependence inducing potential: *Pharmacology* **27**, Suppl. 2, 1-250 (1983). Series of articles on pharmacokinetics, pharmacology and efficacy in insomnia: *Sleep* **10**, Suppl 1, 1-79 (1987); on comparative clinical studies with triazolam, *q.v.*, in insomnia: *Int. Clin. Psychopharmacol.* **5**, Suppl. 2, 21-55 (1990). Review of pharmacology and therapeutic efficacy: K. L. Goa, R. C. Heel, *Drugs* **32**, 48-65 (1986).

Crystals from acetonitrile/diisopropyl ether (1:1), mp 178°.
THERAP CAT: Sedative; hypnotic.

10325. Zopolrestat. *3,4-Dihydro-4-oxo-3-[[5-(trifluoro-methyl)-2-benzothiazolyl]methyl]-1-phthalazineacetic acid;* 2-[4-oxo-3-[5-(trifluoromethyl)benzothiazol-2-ylmethyl]-3,4-dihydrophthalazin-1-yl]acetic acid; CP-73850. $C_{19}H_{12}F_3N_3-O_3S$; mol wt 419.38. C 54.42%, H 2.88%, F 13.59%, N 10.02%, O 11.44%, S 7.65%. Aldose reductase inhibitor. Prepn: B. L. Mylari *et al.*, **Eur. pat. Appl. 222,576**; E. R. Larson, B. L. Mylari, U.S. pat. **4,939,140** (1987, 1990 both to Pfizer); B. L. Mylari *et al. J. Med. Chem.* **34**, 108 (1991). Pharmacology: B. Tesfamariam *et al., J. Cardiovasc. Pharmacol.* **21**, 205 (1993); B. Tesfamariam *et al., Am. J. Physiol* **265**, H1189 (1993). Clinical pharmacokinetics: P. B. Inskeep *et al., J. Clin. Pharmacol.* **34**, 760 (1994).

Crystals, mp 197-198°. pKa (dioxane/water): 5.46 (1:1); 6.38 (2:1). log *P* (*n*-octanol/water): 3.43.
THERAP CAT: Treatment of diabetic complications.

10326. Zorubicin. *(2S-cis)-Benzoic acid [1-[4-[(3-amino-2,3,6-trideoxy-α-L-lyxo-hexopyranosyl)oxy]-1,2,3,4,6,11-hexa-hydro-2,5,12-trihydroxy-7-methoxy-6,11-dioxo-2-napthacen-yl]ethylidene]hydrazide;* benzoic acid hydrazide 3-hydrazone with daunorubicin; RP-22050. $C_{34}H_{35}N_3O_{10}$; mol wt 645.67. C 63.25%, H 5.46%, N 6.51%, O 24.78%. Semi-synthetic antibiotic related to daunorubicin, *q.v.* Prepn: G. Jolles, **Ger. pat. 2,327,211** (1974 to Rhone-Poulenc), *C.A.* **82**, 171381x (1975). Biological activity: R. Maral *et al., Compt. Rend. Ser. D* **275**, 301 (1972); R. Maral, *Cancer Chemother. Pharmacol.* **2**, 31 (1979). Distribution and metabolism in mice: R. Baurain *et al., ibid.* 37. Mechanism of action: G. P. Sartiano *et al., J. Antibiot.* **32**, 1038 (1979). Acute cardiovascular effects in dogs: E. H. Herman, R. S. Young, *Cancer Treat. Rep.* **63**, 1771 (1979). Clinical study in breast cancer: J. N. Ingle, *ibid.* 1701.

Hydrochloride, $C_{34}H_{35}N_3O_{10} \cdot HCl$, NSC-164011, *Rubidazone*. Red-orange crystalline powder from ethanol. $[\alpha]_D^{20}$ −50° (c = 0.2 in water). uv max (methanol): 232.5, 253,

480, 495 nm (ϵ 40225, 35300, 10480, 10300). LD_{50} in mice (mg/kg): 13.66 s.c., 4.42 i.p., 8.50 i.v. (Maral, 1972).

THERAP CAT: Antineoplastic.

10327. Zotepine. *2-[(8-Chlorodibenzo[b,f]thiepin-10-yl)-oxy]-N,N-dimethylethanamine;* 2-chloro-11-(2-dimethylaminoethoxy)dibenzo[b,f]thiepine; Lodopin; Nipolept. $C_{18}H_{18}$-ClNOS; mol wt 331.87. C 65.15%, H 5.47%, Cl 10.68%, N 4.22%, O 4.82%, S 9.66%. A tricyclic enol-ether compound with psychotropic and neurotropic activity. Prepn: S. Umio *et al.,* **Ger. pat. 1,907,670**; *eidem,* U.S. pat. **3,704,245** (1969, 1972 both to Fujisawa); I. Ueda *et al., Chem. Pharm. Bull.* **26,** 3058 (1978). Pharmacological study: S. Uchida *et al., Arzneimittel-Forsch.* **29,** 1588 (1979). Pharmacokinetics and metabolism: K. Noda *et al., ibid.* 1595. Toxicological and teratological studies: K. Fukuhara *et al., ibid.* 1600.

Crystals from cyclohexane, mp 90-91°. uv max (95% ethanol): 266 nm. LD_{50} in male mice, rats (mg/kg): 108, 458 orally; 43.3, 39.7 i.v.; 40.0, 97.0 i.p.; 84.9, 2080 s.c. (Fukuhara).

THERAP CAT: Antipsychotic.

10328. Zoxazolamine. *5-Chloro-2-benzoxazolamine;* 2-amino-5-chlorobenzoxazole; McN-485; Deflexol; Flexilon; Flexin; Zoxamin; Zoxine. $C_7H_5ClN_2O$; mol wt 168.58. C 49.87%, H 2.99%, Cl 21.03%, N 16.62%, O 9.49%. Prepd by reacting 2-amino-5-chlorophenol with ethanolic cyanogen bromide: Nagana *et al., J. Am. Chem. Soc.* **75,** 2770 (1953); Marsh, Sam, U.S. pats. **2,780,633** (1957); **2,890,985** (1959). From 4-chloro-2-hydroxyphenylthiourea: Sam, U.S. pat. **2,978,458** (1961 to McNeil Labs.); from 2,5-dichlorobenzoxazole: Button *et al.,* U.S. pat. **2,969,370** (1961 to Dow). *Review:* C. K. Cain, A. P. Roszkowski, "Benzoxazoles, Benzothiazoles and Benzimidazoles" in *Medicinal Chemistry: A Series of Monographs* **4,** G. DeStevens, Ed. (Academic Press, New York, 1964) pp 325-357.

Plates from benzene, mp 184-185°. uv max (methanol): 244, 285 nm. Slightly sol in water; sol in alcohol, propylene glycol. LD_{50} in mice, rats (mg/kg): 376, 102 i.p.; 678, 730 orally (Cain, Roszkowski).

Hydrochloride, $C_7H_5ClN_2O.HCl$, needles, mp 229° (dec).

Hydrobromide, $C_7H_5ClN_2O.HBr$, plates, mp 240° (dec).

2-Acetylzoxazolamine, $C_9H_7ClN_2O_2$, crystals, mp 212-214°. LD_{50} i.p. in mice: 1500 mg/kg (Cain, Roszkowski).

THERAP CAT: Muscle relaxant (skeletal); uricosuric.

10329. Zygadenine. *4α,9-Epoxycevane-3β,4,14,15α,16β,-20-hexol.* $C_{27}H_{43}NO_7$; mol wt 493.64. C 65.69%, H 8.78%, N 2.84%, O 22.69%. Occurs together with germine in various species of *Zygadenus* (death camas) and *Veratrum (Liliaceae):* Heyl *et al., J. Am. Chem. Soc.* **35,** 258 (1913); Heyl, Herr, *ibid.* **71,** 1751 (1949); Kupchan, Deliwala, *ibid.* **75,** 1025 (1953); Klohs *et al., ibid.* 4925; Stoll, Seebeck, *Helv. Chim. Acta* **36,** 1570 (1953); Kupchan *et al., J. Am. Chem. Soc.* **77,** 689, 755 (1955). Structure: Kupchan, *ibid.* **78,** 3546 (1956); **81,** 1925 (1959).

Fine needles from benzene. mp 214-216°. $[\alpha]_D^{20}$ −48.4° (c = 1.26 in chloroform). Not precipitated by digitonin. Soluble in alcohol, benzene, chloroform.

Hydrochloride, $C_{27}H_{43}NO_7.HCl$, crystals, dec 231-234°.

Sulfate, $2(C_{27}H_{43}NO_7).H_2SO_4$, crystals, dec 237-242°.

10330. Zymosan. Anticomplementary factor. Crude yeast cell wall prepns consisting chiefly of protein-carbohydrate complexes. Prepn from whole yeast cells: Pillemer, Ecker, *J. Biol. Chem.* **137,** 139 (1941); Pillemer *et al., J. Exp. Med.* **103,** 1 (1956); from yeast cell walls: Z. Holan, *Folia Microbiol.* **25,** 501 (1980). Composition of zymosan: F. J. DiCarlo, J. V. Fiore, *Science* **127,** 756 (1958). Enhances the specific defenses of an organism through the activation of the properdin system, *q.v.* Activity studies: Brade *et al., J. Immunol.* **111,** 1389 (1973); **112,** 1115 (1974). Induces aggregation and a release reaction in human platelet-rich plasma: Zucker, Grant, *ibid.* 1219. *Review:* Fitzpatrick, DiCarlo, *Ann. N.Y. Acad. Sci.* **118,** art. 4, 233 (1964).

Light gray powder. Practically insol in water, but readily disperses to give a homogeneous suspension.

USE: In the assay of properdin.

MISCELLANEOUS TABLES

ALPHABETICAL LIST OF TABLES

Glossary

Listed below are brief definitions for selected terms used in the monographs and/or frequently encountered within the references cited.

abzyme Antibody that can act like an enzyme in a specific chemical reaction; catalytic antibody.

adjuvant Substance administered with an antigen to augment its immunogenicity.

adrenergic System of autonomic innervation activated by epinephrine, norepinephrine or dopamine; particularly referring to substances that mimic the actions of the sympathetic nervous system.

agonist Receptor-specific activator; also a substance that mimics the effects of an endogenous regulatory compound.

alkaloid A nitrogenous, basic, usually bitter, substance found in plants. Term also applied to synthetic structural analogs.

allele One of the alternative forms of a gene present at a single site in a chromosome that determines certain inherited characteristics.

allomone Chemical messenger used for communication between different species. Causes a reaction in the other species which is favorable to the organism secreting the substance.

aminoglycoside Carbohydrate typically consisting of 3 or 4 rings which has primary or secondary amines as ring substituents. Classification of antibiotics derived from various species of *Streptomyces,* which interfere with the function of bacterial ribosomes, e.g. streptomycin, gentamicin.

angiogenesis The development and formation of new blood vessels; postulated that tumor growth is preceded by an increase in new capillaries which converge upon the tumor.

antagonist Receptor-specific blocker; also a compound devoid of intrinsic regulatory activity which blocks or diminishes the action of an agonist.

anti Stereodescriptor restricted to bicyclo[X.Y.Z]anes in which the Z bridge reference group is oriented toward the Y bridge. Analogous to *trans* in double-bonded nitrogen systems such as oximes.

antibiotic Substance produced and secreted by various microorganisms which in low concentrations is toxic to other species.

antibody Glycoproteins (immunoglobulins) produced in vertebrates by B-cells as a response to antigenic challenge. Exist in many different forms, each with a different binding site that specifically recognizes the antigen that induced its formation.

antigen Immunogen; an agent that elicits an immune response including the production of an antibody; also considered to be any substance to which an antibody can bind.

antisense The DNA strand which is not transcribed; therefore, it has the same oligonucleotide sequence as the mRNA which codes for the protein.

apoptosis A regulated series of events leading to the death and removal of a cell in response to specific developmental or physiological signals; also known as "programmed cell death".

autosomal Pertaining to any chromosome that is not a sex chromosome.

bacteriocin A protein produced by bacteria that is toxic to another closely related strain.

biological response modifiers Agents that alter the immune response such as interleukins, interferons, colony-stimulating factors, and tumor necrosis factor.

bioremediation The biodegradation of organic pollutants for the purpose of restoring or reclaiming soil, air, water, or sludge.

biosensor An automated analytical device based on a specific biomolecular event; the result is measured optically or electronically.

cachexia The wasting of bodily tissue and malnutrition produced by chronic diseases.

carcinoma A malignant tumor arising from epithelial cells.

cell adhesion The ability of cells to become and to remain associated with like cells to form tissues.

chelate The cyclic structure formed by unshared electrons on neighboring atoms complexing with a metal ion; used as a sequestrant in metal poisoning.

chemoprevention Use of pharmacological agents or physiological nutrients to slow or to halt the carcinogenic process.

chemoselectivity Pertains to the preferential reaction of a reagent with one functional group of a multifunctional molecule.

chromophore The group of atoms in a molecule which absorbs light and is responsible for the color of the molecule.

cis Relative stereodescriptor limited to cyclic systems with two stereocenters. Configuration where the two groups of higher priority are on the same side of the reference plane. For alternate definition, see *zusammen.*

clinical trial, Phase I Safety testing and pharmacological profiling in humans; usually involves 20-80 healthy volunteers and takes about one year.

clinical trial, Phase II Effectiveness testing in volunteer patients; usually involves 100-300 volunteers and takes about two years.

clinical trial, Phase III Extensive clinical testing in humans to determine efficacy and to identify adverse reactions; usually involves 1000-3000 patients and lasts about three years.

clinical trial, Phase IV Additional post-marketing studies to evaluate long term effects.

cloning Proliferation of a single cell to a large colony; replication from a single gene introduced into a host cell.

coenzyme Nonprotein molecule that is essential for the activity of an enzyme; functions as a cofactor of an enzyme.

combinatorial chemistry Process for the simultaneous synthesis of all possible products (matrix compound synthesis) in a single reaction vessel from a given set of reactants.

corticosteroid   A 21-carbon steroid hormone produced by the adrenal cortex in response to the release of corticotropin (ACTH) or angiotensin.

cytochromes   Family of colored proteins, related by the presence of a porphyrin metal (usually Fe), which undergo rapid oxidation and reduction in organisms.

cytokines   Diverse group of soluble, nonantibody proteins, secreted by a variety of cell types, which modulate the functional activities of individual cells by interaction with specific cell surface receptors, e.g. interferon, interleukin.

cytomegalovirus   Subgroup of herpes virus that causes disease in immunocompromised hosts.

dative bond   Semipolar linkage between two atoms where one atom contributes both electrons; also referred to as a coordinate covalent bond.

denaturation   Unfolding of a protein or DNA, by action of certain physical or chemical procedures resulting in loss of native conformation.

domain   A region in a protein that can be structural or functional.

eicosanoids   Collective term applied to substances such as prostaglandins, thromboxanes, leukotrienes and lipoxins, which are all derived from arachidonic acid; play an important role in inflammation.

enantioselectivity   Pertains to a stereoselective reaction in which the potential products are mirror images.

endo   Stereodescriptor used only for ring positions on [X.Y.Z.]anes in which X≥Y>Z>0 and the group of higher priority is on the opposite side of the ring reference plane to the Z-bridgehead.

endocytosis   Process in which the cell membrane engulfs extracellular particles to form coated vesicles interior to the cell.

endogenous   Developing or originating within the organism.

endosome   Vesicle, derived from the cell membrane, that can fuse with lysosomes.

endotoxins   Heat-stable lipopolysaccharide-protein complexes contained in cell walls of gram-negative bacteria including non-infectious gram-negatives; released only when the cells are disrupted.

entactogen   Non-hallucinogenic psychoactive compound which is purported to facilitate communication and introspective states. Derived from the Greek roots of *en* for within or inside, *gen* to produce or originate and the Latin root *tactus* for touch.

enterotoxin   Exotoxin produced by enterobacteria, which, when ingested or produced within the intestine, affects its function, inducing nausea, diarrhea, vomiting, etc.

entgegen **(E)**   German for opposite. Relative stereodescriptor for configuration about double bonds; used when groups of higher priority are on different sides of the vertical reference plane.

enzyme   A protein, specifically produced, which catalyzes biochemical reactions.

enzyme-linked immunosorbent assay (ELISA)   An immunoassay in which an enzyme linked to an antibody is used to detect the presence of an antigen.

eosinophils   A class of granular white blood cells (granulocytes) identified by their affinity for eosin dye; contain a variety of enzymes that help inactivate chemical mediators such as histamine.

ergot   The hardened web-like mass of threads of *Claviceps purpurea* developed on rye attacked by the fungus. Ergot alkaloids are used as oxytocics and in treatment of migraine.

erythema   Redness of the skin produced by congestion of the capillaries.

excitotoxin   Neurotoxic analogs of glutamic acid which mimic its excitatory effects on neurons of the CNS.

exo   Stereodescriptor used only for ring positions on [X.Y.Z.]anes in which X≥Y>Z>0 and the group of higher priority is on the same side of the ring reference plane to the Z-bridgehead.

exocytosis   Process of expelling cellular material from the interior to the exterior of the cell.

exogenous   Developed or originating outside the organism.

exon   One of the regions in the discontinuous genes of higher organisms that codes for amino acids. The exons are transcribed to form messenger RNA, which is translated to protein.

exotoxins   Disease-producing proteins secreted by species of certain bacteria into surrounding medium.

extracellular matrix   The non-cellular portion of tissues, located between and adjacent to cells, which is largely filled by a complex mixture of secreted proteins and glycoproteins; critical in maintaining tissue integrity.

flavoprotein   Flavin dependent enzymes which normally contain FAD or FMN as tightly bound coenzymes.

geminal *(gem)*   Indicates that the two substituents in a di-substituted compound are on the same carbon atom.

gene fusion   The altering of the coding sequence of a gene by joining part of it to the coding sequence of a different gene to produce a novel hybrid gene.

genome   The genetic constitution of an organism; entire DNA sequence content of an organism.

glucocorticoid   A 21-carbon steroid hormone secreted by adrenal cortex. Acts on carbohydrate, lipid and protein metabolism and inhibits the release of corticotropin (ACTH).

glycoprotein   Protein modified by the covalent attachment of one or more sugar molecules.

hapten   Small nonantigenic molecule which becomes an antigen when combined with a larger carrier molecule.

hybridoma   Cell line produced by fusing an antibody-secreting cell with a myeloma cell; used to produce specific monoclonal antibodies in great quantity.

hypertrophy   Abnormal enlargement of an organ or tissue, not due to tumor formation; caused by the increase in size of its cellular components.

idiotype The segment of an immunoglobulin molecule that determines its antigenic specificity.

immunity, active Immunity where the formation of a specific antibody and increased resistance results from stimulation of the host's tissues by an antigen.

immunity, passive Immunity acquired from transfer of preformed protective substances, as in the injection of immune serum.

immunogenicity The ability to elicit an immune response within an organism.

immunophilins High affinity, cytoplasmic receptors that bind immunosuppressives such as cyclosporin and rapamycin.

immunosensors Biosensors wherein antibodies are the biomolecular component used to recognize their appropriate antigens.

immunotoxins Molecules that comprise a cytotoxic agent linked to an antibody specific for target cell antigens.

interleukins Cytokines, typically produced by lymphocytes, that regulate immune and inflammatory responses.

intron Non-coding segment of DNA located within a eukaryotic gene separating two of the coding sequences (exons). Removed by RNA splicing before the formation of the mature mRNA.

kairomones Chemical messengers used for communication between different species which are either nonadaptive or actually detrimental to the organisms producing them.

lantobiotics Polycyclic bacteriocins consisting of the thioether amino acids lanthionine and beta-methyllanthionine.

lipocortins (calpactins) Phospholipase-inhibiting proteins acting in the first step of the eicosanoid synthesis pathway.

lipoprotein Conjugated protein containing a lipid or group of lipids, such as a glyceride, a phospholipid, cholesterol, or a combination of these.

lymph The yellowish, slightly basic liquid which is derived from tissue fluids, collected throughout the body and transported in lymph vessels to the circulatory system.

lymphocyte Immunologically competent cells occurring in lymphatic tissue, spleen and blood, which recognize antigens. Classified according to their origin and function into B-lymphocytes (bursa-equivalent) and T-lymphocytes (thymus-dependent).

lymphokine Nonimmunoglobulin protein synthesized primarily by T-lymphocytes that modulates an immune response, e.g. interleukin-2, γ-interferon.

lysosomes Cytoplasmic, subcellular particles containing enzymes; involved in intracellular digestive processes. Present in practically all animal cells; especially prominent in phagocytic cells.

major histocompatibility complex Large, multigene cluster that encodes for the cell surface glycoproteins which are essential in the cellular immune response to distinguish ''self'' and ''non-self''.

meso compound Pertaining to a compound which contains equal numbers of enantiomeric groups which are identically linked. Compound having at least two chiral centers, yet no optical rotation because it contains a plane of symmetry.

mesomerism The existence of chemical structures having the same molecular arrangements that differ only in the location of electrons; the actual electronic distribution is not represented by any one structure, but is a hybrid of contributing structures.

metalloprotein Conjugated protein which contains one or more metal cations.

monoclonal Arising from a single clone of cells; identical progeny of a single cell.

morphogens Soluble substances that control the embryonic differentiation of cells and tissues.

mucins Viscous glycoprotein constituents of saliva, gastric juice, etc., which are lubricants or protectants in body cavities or on body surfaces.

multidrug resistance The resistance of tumor cells to varied chemotherapeutic agents mediated by the expression of a transmembrane pump, P-glycoprotein, which actively pumps the drugs out of the cells, lowering the intracellular concentration and cytotoxic effectiveness.

mutagenesis Process of genetic mutation.

mutein Name applied to a protein arising as a result of a mutation.

mycotoxins Fungal toxins that are harmful to other organisms.

naphthenes Five and six carbon cycloparaffins and their alkyl derivatives found in crude petroleum; sometimes includes polycyclics found in higher-boiling fractions.

neuroleptic Agent that pharmacologically manages the manifestations of a psychotic disorder, affecting cognition and behavior; also called major tranquilizer or antipsychotic.

neurotrophic factors Polypeptides that support the growth and survival of specific populations of neurons.

nootropic Agent having positive effects on organically impaired cognition or nervous system function; from the Greek *noos* or *nous* for mind and *-tropic* for nourishing, stimulating.

nor Originally the abbreviation for *Nitrogen ohne Radikal*. Currently a subtractive prefix for stereoparents indicating ring contraction or the loss of a methyl group from the carbon skeleton.

oncogene A gene that induces uncontrolled cell proliferation and has acquired the ability under certain circumstances to transform normal cells into cancer cells.

penaldic acids *N*-Acylaminomalonaldehydric acids which are found in the degradation products of the penicillins.

peroxidases Enzymes of the oxidoreductase class which catalyze oxidation of substrates by hydrogen peroxide.

pheromones Chemical substances used for communication between individual organisms of the same species. Perceived primarily by the olfactory sense and to a lesser extent the gustatory sense. From the Greek *pherein* to transfer, and *hormon* to excite or stimulate.

phytosterols Sterols obtained from higher plants; differ from sterols occurring in animals by having C_1 or C_2 residues at C-24 and/or a double bond at C-22.

polymerase Type of enzyme which catalyzes polymerization reactions, especially of nucleotides to polynucleotides.

polymerase chain reaction (PCR) An enzyme-based technique for producing large quantities of specific DNA fragments.

polymorphism The quality of existing in several different forms.

primary structure (protein) Amino acid sequence.

prodrug A drug precursor; a substance which is inactive by itself and which must undergo chemical conversion by metabolic processes to become an active pharmacological agent.

progestogen Substance which induces effects similar to those of progesterone.

protamines Basic proteins that occur combined with nucleic acid in the sperm of fish.

protein sequence database Database containing experimentally determined amino acid and/or DNA sequences of known proteins.

protein translocation Movement of proteins across membranes into cell organelles, such as ribosomes or mitochondria.

pyrogens Fever-producing substances; agents which produce a febrile response when administered parenterally to man and certain animals.

rational drug design A computer-aided approach in which tools of organic chemistry, such as molecular modeling and x-ray crystallography, are utilized to produce pharmacologically active molecules.

receptor Protein or protein complex that binds a specific ligand such as a neurotransmitter, hormone, lymphokine, etc., producing a secondary biological signal within the cell.

rectus (R) — (right) Opposite of (*S*). Absolute term describing the spatial arrangement about an asymmetric carbon when the substituent of lowest priority lies below the reference plane and the priority of the remaining substituents decreases in a clockwise direction.

regioselectivity Pertains to the preferential formation of one structural isomer when two or more are possible.

ribosome Organelle found in all living cells; macromolecular assembly of proteins and rRNA which serves as the site for protein synthesis.

roentgen (r) A unit of radiation. That quantity of χ or γ radiation which produces one esu of charge in one cubic centimeter of air under standard conditions.

saponins Foaming glycoside of sterols or triterpenes that are widely distributed in plants. Causes cell lysis.

sarcoma Malignant tumor arising from connective tissue or muscle cells.

secondary structure (protein) Residue by residue conformation of the backbone of the polypeptide chain. Typical conformations include the α helices, β sheets and random coils; given rise to by the regular hydrogen-bond interactions within contiguous stretches of polypeptide chain.

serotype Classification of bacterial subtypes based upon variations in their constituent antigens.

sinister (S) — (left) Opposite of (*R*). Absolute term describing the spatial arrangement about an asymmetric carbon when the substituent of lowest priority lies below the reference plane and the priority of the remaining substituents decreases in a counterclockwise direction.

stenosis Contraction or narrowing of a duct or canal.

stereoselectivity Pertains to the preferential formation of one stereoisomer when two or more are possible.

stereospecific reaction A reaction where starting materials, which differ only in their configuration, are converted to stereoisomerically distinct products. Applies when the reaction processes are constrained to proceed in a stereochemically defined manner.

steroidogenesis Biosynthesis of steroids, as by adrenal glands or gonads.

surfactants Agents that alter the physical and chemical properties of surfaces such as synthetic detergents and soap.

synergistic The joint action of agents wherein their combined effect is greater than the sum of their individual effects.

syn Stereodescriptor restricted to bicyclo[X.Y.Z]anes in which the Z bridge reference group is oriented toward the X bridge; analogous to *cis* in double-bonded nitrogen systems such as oximes.

tar acids Phenols obtained from coal tar distillates or synthesized from coal tar hydrocarbons.

teratogen Agent that causes abnormal fetal development.

terpene Hydrocarbon derived from essential oils, resins, etc., containing multiples of the C_5 isoprene unit as its carbon skeleton. Monoterpene is $C_{10}H_{16}$.

tertiary structure (protein) Three dimensional conformation of the polypeptide chain after folding.

toxoid A toxin which has been treated by a chemical agent or by heat to destroy its toxic properties while retaining its ability to stimulate antibody production.

trans Relative stereodescriptor limited to cyclic systems with two stereocenters. Configuration where the two groups of highest priority are on opposite sides of the reference plane.

transcription factor A protein that binds to a specific DNA sequence and influences the expression of neighboring genes.

transduction Transfer of genetic material from one cell to another by a viral vector.

transfection Introduction of exogenous DNA (usually viral or bacterial) into mammalian cells or bacterial cells.

transferrin Blood plasma protein which transports iron as Fe^{3+} in two iron binding sites; an anion is bound with each of the Fe^{3+}.

transgenic Containing artificially introduced DNA as a result of experimental splicing of a segment of DNA from one genome into the germ line of another.

transposon Small DNA segment that is able to move from one location to another within the genome; associated with antibiotic resistance in bacteria.

trophic Pertaining to nutrition.

tropism A growth response elicited by an external stimulus.

vicinal (*vic*) Indicates that two substituents in a disubstituted compound are on adjacent carbon atoms.

zusammen (Z) German for together. Relative stereodescriptor for configuration about double bonds; used when groups of higher priority are on the same side of the vertical reference plane.

zwitterion Dipolar ion in which positive and negative charges are physically separated.

Common Heterocyclic Ring Systems

O-containing rings

| furan | oxirane | 2*H*-pyran | 4*H*-pyran | 2*H*-chromene | benzo[*b*]furan |

S-containing rings

thiophene   benzo[*b*]thiophene   parathiazine

N-containing rings

pyrrole   pyrrolidine   pyrazole   pyrazolidine   imidazole   imidazoline   imidazolidine

pyridine   piperidine   pyrazine   piperazine   pyrimidine

indole   purine   benzimidazole   quinoline

Other

phenothiazine   morpholine

Abbreviated Terms Used by the U.S. Adopted Names (USAN) Council for Radicals and Adducts

Abbreviation	Chemical Name
aceturate	*N*-acetylglycinate
acistrate	2′-acetate (ester) and octadecanoate (salt)
axetil	1-acetoxyethyl
besylate	benzenesulfonate
buteprate	(1-oxobutoxy) (ester) and (1-oxopropoxy) (ester)
camsylate	camphorsulfonate
caproate	hexanoate
closylate	*p*-chlorobenzenesulfonate
cyclotate	4-methylbicyclo[2.2.2]oct-2-ene-1-carboxylate
cypionate	cyclopentanepropionate
dapropate	*N,N*-dimethyl-β-alanine
diolamine	diethanolamine
edamine	ethylenediamine
edetate*	ethylenediaminetetraacetate
edisylate	1,2-ethanedisulfonate
enanthate	heptanoate
epolamine	1-pyrrolidineethanol
erbumine	2-methyl-2-propanamine
estolate	propanoate and dodecyl sulfate (salt)
esylate	ethanesulfonate
etabonate	(ethoxycarbonyl)oxy
fostedate	tetradecyl hydrogen phosphate
gluceptate	glucoheptonate
hybenzate	*o*-(4-hydroxybenzoyl)benzoate
hyclate	monohydrochloride, hemiethanolate, hemihydrate
isethionate	2-hydroxyethanesulfonate
meglumine	*N*-methylglucamine
mesylate	methanesulfonate
mofetil	2-(4-morpholinyl)ethyl
napsylate	2-naphthalenesulfonate
olamine	ethanolamine
pamoate	4,4′-methylenebis[3-hydroxy-2-naphthoate]
pendetide	N^6-[*N*-[2-[[2-[bis(carboxymethyl)amino]ethyl](carboxymethyl)amino]ethyl]-*N*-(carboxymethyl)glycyl]-N^2-(*N*-glycyl-L-tyrosyl-L-lysine-tyrosyl)
phenpropionate	3-phenylpropionate
pivalate	trimethylacetate
pivoxetil	1-(2-methoxy-2-methyl-1-oxopropoxy)ethyl
pivoxil	(2,2-dimethyl-1-oxopropoxy)methyl
proxetil	1-[(isopropoxycarbonyl)oxy]ethyl
salnacedin	*N*-acetyl-L-cysteine 2-hydroxybenzoate
tebutate	*tert*-butyl acetate
tosylate	*p*-toluenesulfonate
triflutate	trifluoroacetate
trolamine	triethanolamine
xinafoate	1-hydroxy-2-naphthalenecarboxylate

*The sodium salts are named as follows: edetate monosodium (one Na ion); edetate disodium (two Na ions); edetate trisodium (three Na ions); edetate sodium (four Na ions).

International Nonproprietary Names (INN) for Radicals and Groups Proposed for Pharmaceutical Substances by the World Health Organization (WHO)

Proposed Name	Chemical Name
aceturate	*N*-acetylglycinate
acistrate	2′-acetate, stearate (salt)
acoxil	acetoxymethyl
amsonate	4,4′-diaminostilbene-2,2′-disulfonate
axetil	1-acetoxyethyl
besilate	benzenesulfonate
bezomil	(benzoyloxy)methyl
buciclate	*trans*-4-butylcyclohexanecarboxylate
bunapsilate	3,7-di-*tert*-butyl-1,5-naphthalenedisulfonate
buteprate	butyrate propionate
camsilate	camphorsulfonate
carbesilate	*p*-carboxybenzenesulfonate
ciclotate	4-methylbicyclo[2.2.2]oct-2-ene-1-carboxylate

Proposed Name	Chemical Name
cilexetil	(*RS*)-1-[[(cyclohexyloxy)carbonyl]oxy]ethyl
cipionate	cyclopentanepropionate
closilate	*p*-chlorbenzenesulfonate
crobefate	(±)-(*E*)-6-hydroxy-4′-methoxy-3-(*p*-methoxybenzylidene)flavanone, phosphate, ion(2−)
cromacate	[(6-hydroxy-4-methyl-2-oxo-2*H*-1-benzopyran-7-yl)oxy]acetate
cromesilate	6,7-dihydroxycoumarin-4-methanesulfonate
dapropate	*N*,*N*-dimethyl-β-alanine
deanil	2-(dimethylamino)ethyl
decil	decyl
dibudinate	2,6-di-*tert*-butyl-1,5-naphthalenedisulfonate
dibunate	2,6-di-*tert*-butyl-1-naphthalenesulfonate
digolil	2-(2-hydroxyethoxy)ethyl
diolamine	diethanolamine
docosil	docosyl
dofosfate	octadecyl hydrogen phosphate
edamine	ethylene diamine
edisilate	1,2-ethanedisulfonate
embonate	4,4′-methylenebis(3-hydroxy-2-naphthoate)
enantate	heptanoate
epolamine	1-pyrrolidineethanol
erbumine	*tert*-butylamine
esilate	ethanesulfonate
estolate	propionate lauryl sulfate
etabonate	ethyl carbonate
farnesil	(2*E*,6*E*)-3,7,11-trimethyl-2,6,10-dodecatrienyl
fendizoate	*o*-[(2′-hydroxy-4-biphenylyl)carbonyl]benzoate
fostedate	tetradecyl hydrogen phosphate
gluceptate	glucoheptonate
hibenzate	*o*-(4-hydroxybenzoyl)benzoate
hyclate	monohydrochloride hemiethanolate hemihydrate
isetionate	2-hydroxyethanesulfonate
lauril	*n*-dodecyl
laurilsulfate	*n*-dodecylsulfate
megallate	3,4,5-trimethoxybenzoate
mesilate	methanesulfonate
metembonate	3-methoxy-2-naphthoate
metilsulfate	methylsulfate
mofetil	2-morpholinoethyl
napadisilate	1,5-naphthalenedisulfonate
napsilate	2-naphthalenesulfonate
octil	octyl
olamine	ethanolamine
oxoglurate	2-oxoglutarate
pendetide	N^6-[[*N*-[2-[[2-[bis(carboxymethyl)amino]ethyl](carboxymethyl)amino]ethyl]-*N*-(carboxymethyl)glycyl]-N^2-(*N*-glycyl-L-tyrosyl)-L-lysine
pivalate	trimethylacetate
pivetil	1-hydoxyethyl pivalate (ester)
pivoxetil	1-(2-methoxy-2-methyl-1-oxopropoxy)ethyl
pivoxil	(pivaloyloxy)methyl
proxetil	1-[(isopropoxycarbonyl)oxy]ethyl
steaglate	stearoyl-glycolate
tebutate	*tert*-butyl acetate
tenoate	2-thiophenecarboxylate
teoclate	8-chlorotheophyllinate
teprosilate	1,2,3,6-tetrahydro-1,3-dimethyl-2,6-dioxopurine-7-propanesulfonate
tofesilate	1,2,3,6-tetrahydro-1,3-dimethyl-2,6-dioxopurine-7-ethanesulfonate
tosilate	*p*-toluenesulfonate
triclofenate	2,4,5-trichlorophenolate
triflutate	trifluoroacetate
trolamine	triethanolamine
troxundate	[2-(2-ethoxyethoxy)ethoxy]acetate
xinafoate	1-hydroxy-2-naphthoate

Cancer Chemotherapy Drug Regimens

Listed below are selected acronyms for combination cancer chemotherapy regimens comprising substances in *The Merck Index*.

Acronym	Drug regimens
ABP	doxorubicin + bleomycin + prednisone
ABVD	doxorubicin + bleomycin + vinblastine + dacarbazine
AC	doxorubicin + cyclophosphamide
ACVB	doxorubicin + cyclophosphamide + vindesine + bleomycin
ADIC	doxorubicin + dacarbazine
APO	doxorubicin + prednisone + vincristine + 6-mercaptopurine + asparaginase + methotrexate
AV	doxorubicin + vincristine
BACOP	bleomycin + doxorubicin + cyclophosphamide + vincristine + prednisone
BAPP	bleomycin + doxorubicin + cisplatin + prednisone
BCD	methotrexate + doxorubicin + cisplatin
BCVPP	carmustine + cyclophosphamide + vinblastine + procarbazine + prednisone
BEP	bleomycin + etoposide + cisplatin
BMP	bleomycin + methotrexate + cisplatin
BOLD	bleomycin + vincristine + lomustine + dacarbazine
CA	cyclophosphamide + doxorubicin
CAF	cyclophosphamide + doxorubicin + fluorouracil
CAMF	cyclophosphamide + doxorubicin + methotrexate + fluorouracil
CAP	cyclophosphamide + doxorubicin + cisplatin
CAP-BOP	cyclophosphamide + doxorubicin + procarbazine + bleomycin + vincristine + prednisone
CAV	cyclophosphamide + doxorubicin + vincristine
CAVE	cyclophosphamide + doxorubicin + vincristine + etoposide
CAVEP	cyclophosphamide + doxorubicin + vincristine + etoposide + cisplatin
CBV	cyclophosphamide + carmustine + etoposide
CC	carboplatin + cyclophosphamide
CFP	cyclophosphamide + fluorouracil + prednisone
CFPMV	cyclophosphamide + fluorouracil + prednisone + methotrexate + vincristine
CFPT	cyclophosphamide + fluorouracil + prednisone + tamoxifen
CHAD	cyclophosphamide + hexamethylmelamine + doxorubicin + cisplatin
CHAMOCA	cyclophosphamide + hydroxyurea + dactinomycin + methotrexate + vincristine + doxorubicin
CHAP-5	cyclophosphamide + hexamethylmelamine + doxorubicin + cisplatin
CHF	cyclophosphamide + hexamethylmelamine + fluorouracil
ChlVPP	chlorambucil + vinblastine + procarbazine + prednisone
CHO	cyclophosphamide + doxorubicin + vincristine
CHOP	cyclophosphamide + doxorubicin + vincristine + prednisone
CHOP-B	cyclophosphamide + doxorubicin + vincristine + prednisone + bleomycin
CMF	cyclophosphamide + methotrexate + fluorouracil
CMFP	cyclophosphamide + methotrexate + fluorouracil + prednisone
CMFVP	cyclophosphamide + methotrexate + fluorouracil + vincristine + prednisone
C-MOPP	cyclophosphamide + mechlorethamine + vincristine + procarbazine + prednisone
CMV	cisplatin + methotrexate + vinblastine
COAP	cyclophosphamide + vincristine + cytarabine + prednisolone
CODE	cisplatin + vincristine + doxorubicin + etoposide
COMLA	cyclophosphamide + vincristine + methotrexate* + cytarabine
COMP	cyclophosphamide + vincristine + methotrexate + prednisone
COP	cyclophosphamide + vincristine + prednisone
COP-BLAM	cyclophosphamide + vincristine + prednisone + bleomycin + doxorubicin + procarbazine
COPP	cyclophosphamide + vincristine + prednisone + procarbazine
CVF	cyclophosphamide + vincristine + fluorouracil
CVP	cyclophosphamide + vincristine + prednisone
CYVADIC	cyclophosphamide + vincristine + doxorubicin + dacarbazine
DICEP	cyclophosphamide + etoposide + cisplatin
EAP	etoposide + doxorubicin + cisplatin
EFP	etoposide + fluorouracil + cisplatin
ELF	etoposide + leucovorin + fluorouracil
EMA-CO	etoposide + methotrexate + dactinomycin + cyclophosphamide + vincristine
ESHAP	etoposide + methylprednisolone + cytarabine + cisplatin
FA	fluorouracil + doxorubicin
FAC	fluorouracil + doxorubicin + cyclophosphamide
FAM	fluorouracil + doxorubicin + mitomycin C
FAMTX	fluorouracil + doxorubicin + methotrexate
FAP	fluorouracil + doxorubicin + cisplatin
FEB	fluorouracil + epirubicin + carmustine
FUVAC	fluorouracil + vinblastine + doxorubicin + cyclophosphamide
HAD	hexamethylmelamine + doxorubicin + cisplatin
H-CAP	hexamethylmelamine + cyclophosphamide + doxorubicin + cisplatin

Cancer Chemotherapy Drug Regimens (Continued)

Acronym	Drug regimens
Hexa-CAF	hexamethylmelamine + cyclophosphamide + methotrexate + fluorouracil
ICE	ifosfamide + carboplatin + etoposide
LOPP	chlorambucil + vincristine + procarbazine + prednisone
LSA$_2$-L$_2$	cyclophosphamide + vincristine + prednisone + daunorubicin + methotrexate + cytarabine + thioguanine + colaspase + hydroxyurea + carmustine
MAC	methotrexate + dactinomycin + chlorambucil
MACC	methotrexate + doxorubicin + cyclophosphamide + lomustine
MACOP-B	methotrexate* + doxorubicin + cyclophosphamide + vincristine + prednisone + bleomycin
M-BACOD	methotrexate + bleomycin + doxorubicin + cyclophosphamide + vincristine + dexamethasone
MBD	methotrexate + bleomycin + cisplatin
MCF	mitoxantrone + cyclophosphamide + fluorouracil
MINE	mesna + ifosfamide + mitoxantrone + etoposide
MIP	mitomycin + ifosfamide + cisplatin
MOP	mechlorethamine + vincristine + procarbazine
MOPP	mechlorethamine + vincristine + procarbazine + prednisone
M-VAC	methotrexate + vinblastine + doxorubicin + cisplatin
MVP	mitomycin + vindesine + cisplatin
MVPP	mechlorethamine + vinblastine + procarbazine + prednisone
PAC	cisplatin + doxorubicin + cyclophosphamide
PC	cisplatin + cyclophosphamide
PCV	procarbazine + lomustine + vincristine
PE	cisplatin + etoposide
PEB	cisplatin + etoposide + bleomycin
PMF	cisplatin + mitomycin C + fluorouracil
ProMACE	prednisone + methotrexate + doxorubicin + cyclophosphamide + etoposide
ProMACE-CytaBOM	prednisone + methotrexate + doxorubicin + cyclophosphamide + etoposide + cytarabine + bleomycin + vincristine + methotrexate*
ProMACE-MOPP	prednisone + methotrexate* + doxorubicin + cyclophosphamide + etoposide + mechlorethamine + vincristine + procarbazine + prednisone
PVB	cisplatin + vinblastine + bleomycin
SMF	streptozocin + mitomycin + fluorouracil
VAB-6	vinblastine + dactinomycin + bleomycin + cisplatin + cyclophosphamide
VAC	vincristine + dactinomycin + cyclophosphamide
VAD	vincristine + doxorubicin + dexamethasone
VAMP	vincristine + prednisone + methotrexate + 6-mercaptopurine
VAP-cyclo	vincristine + doxorubicin + prednisolone + cyclophosphamide
VIP	vindesine + ifosfamide + cisplatin
VMF	etoposide + methotrexate + fluorouracil
VP	vindesine + cisplatin

*with folinic acid (leucovorin) rescue

Code Letters Used by Companies for Experimental Substances

A	Abbott Laboratories, USA	CH	Chinoin, Hungary
	Ajinomoto Co., Ltd., Japan	CHF	Chiesi Farmaceutici S.p.A., Italy
	Asta Medica AG, Germany	CHX	Chemex Pharm., USA
	Astra AB, Sweden	CI	Parke-Davis & Co., USA
	Eli Lilly & Co., USA	CK	Schering AG, Germany
	Ohmeda Pharm. Prods. Div., Inc., USA	CL	American Cyanamid Co., USA
	Takeda Chem. Ind., Ltd., Japan		Chemie Linz AG, Austria
AA	Takeda Chem. Ind. Ltd., Japan		Cutter Biologicals, USA
AB	Armour Pharmaceutical Co., USA	CLY	Yoshitomi Pharm. Ind. Ltd., Japan
	Dainippon Pharmaceutical Co., Japan	CM	Sanofi S.A., France
ABT	Abbott Laboratories, USA	CN	Parke-Davis & Co., USA
AC	A. Christiaens S.A., Belgium	CP	Continental Pharma S.A., Belgium
	Ajinomoto Co., Ltd., Japan		Monsanto Chemical Co., USA
	American Cyanamid Co., USA		Pfizer, Inc., USA
	Hercules Inc., USA	CR	Rotta Farmaceutici S.p.A., Italy
ACL	Monsanto Chemical Co., USA	CRD	Monsanto Chemical Co., USA
ACP	Amchem Products, Inc., USA		Tanabe Seiyaku Co., Ltd., Japan
AD	Dainippon Pharm. Co., Ltd., Japan	CRL	Labs. Lafon, France
Ad	Boehringer Ingelheim KG, Germany	CS	Sankyo Co., Ltd., Japan
ADR	Adria Labs., Inc., USA	CV	Takeda Chem. Ind. Ltd., Japan
AF	Angelini Francesco S.p.A., Italy	CY	Sanofi Winthrop, France
AG	Agouron Pharmaceuticals, USA		
	Takeda Chem. Ind., Ltd., Japan	D	Asta Medica AG, Germany
AGN	Allergan, Inc., USA		Chemiewerke Homburg, Germany
AH	Allen & Hanburys, Ltd., UK		Daiichi Pharmaceutical Co., Ltd., Japan
AHR	A. H. Robins Co., Inc., USA		Knoll AG, Germany
ALO	Alcon Labs., Inc., USA		Ortho Pharm. Corp., USA
AM	Kyorin Pharm. Co., Ltd., Japan	DA	Istituto De Angeli S.p.A., Italy
AMA	Takeda Chem. Ind., Ltd., Japan	DAN	Labs. Delagrange, France
AMI	Advanced Magnetics, Inc., USA	DD	Daiichi Pharmaceutical Co. Ltd., Japan
AMR	Affiliated Medical Research, Inc., USA	DE	Santen Pharm. Co. Ltd., Japan
AN	Rhône-Poulenc Agrochimie S.A., France	DF	Dompè Farmaceutici S.p.A., Italy
ANP	Anphar-Rolland, France	DH	Hoechst Marion Roussel, USA
ASL	DuPont Merck Pharmaceutical Co., USA	DJ	Daiichi Pharmaceutical Co., Ltd., Japan
AW	Sandoz AG, Switzerland	DL	Gruppo Lepetit S.p.A., Italy
AY	Wyeth-Ayerst Labs., USA	DMP	DuPont Merck Pharmaceutical Co., USA
		DN	DowElanco, USA
B	Bayer AG, Germany	DO	Uniroyal, Inc., USA
	Bracco S.p.A., Italy	DPX	E. I. du Pont de Nemours & Co., USA
Ba	Ciba-Geigy AG, Switzerland	DS	Sandoz Ltd., Switzerland
BAS	BASF Group, Germany	DU	Solvay-Duphar BV, Netherlands
BAX	Baxter Healthcare Corp., USA	DUP	E. I. duPont de Nemours & Co., USA
BAY, Bay	Bayer AG, Germany		
BB	British Biotech plc, UK	E	Bayer AG, Germany
BC	Bristol-Myers Squibb Co., USA		Eisai Co. Ltd., Japan
	Sandoz Ltd., Switzerland		Laboratorios Dr. Esteve S.A., Spain
BDH	BDH Pharmaceuticals, UK	EF	Efamol Ltd., UK
BI	Boehringer Ingelheim KG, Germany	EGIS	EGIS Pharmaceutical Works, Hungary
BL	Bristol-Myers Squibb Co., USA	EGYT	EGIS Pharmaceutical Works, Hungary
BM	Boehringer Mannheim GmbH, Germany	EI	American Cyanamid Co., USA
BMS	Bristol-Myers Squibb Co., USA	EL	DowElanco, USA
BMY	Bristol-Myers Squibb Co., USA	EM	Evans Medical Ltd., UK
BN	Beaufour-Ipsen, France	EMD	E. Merck AG, Germany
BO	Bottu SA, France	EN	Endo Labs., USA
BOF	Beiersdorf AG, Germany	ES	Eprova Ltd., Switzerland
BP	Bioprojet Pharma, France	EU	Procter & Gamble Pharm., Inc., USA
BRL	SmithKline Beecham plc, UK	EX	Lakeside Labs., Inc., USA
BTS	The Boots Co. plc, UK	EXP	E. I. duPont de Nemours & Co., USA
BW	Glaxo Wellcome plc, UK		
BY	Byk Gulden Lomberg Chemische Fabrik GmbH, Germany	F	Fujirebio, Inc., Japan
			Pierre Fabré SA, France
			Procter & Gamble Pharm., Inc., USA
C	Lab. Cassenne, France	FB	Bayer AG, Germany
	Ciba-Geigy AG, Switzerland	FBA	Bayer AG, Germany
	Cilag AG, Switzerland	FC	Orion Corporation, Finland
CC	Cilag AG, Switzerland	FCE	Farmitalia Carlo Erba s.r.l., Italy
CCI	Glaxo Wellcome plc, UK	FG	AB Ferrosan, Sweden
CERM	Centre Europeen de Recherches Mauvernay, France	FH	Boehringer Ingelheim KG, Germany
		FI	Farmitalia Carlo Erba s.r.l., Italy
CG	Glaxo Wellcome plc, UK	FI	Ferrer Internacional S.A., Spain
	Chemie Grunenthal GmbH, Germany	FID	Fidia S.p.A., Italy
CGA	Ciba-Geigy AG, Switzerland	FK	Fujisawa Pharmaceutical Co., Ltd., Japan
CGP	Ciba-Geigy AG, Switzerland	FL	Chinoin, Hungary
CGS	Ciba-Geigy AG, Switzerland	FLC	Labs. Delagrange, France

FMC	FMC Corp., USA
FPL	Fisons plc, UK
FR	Fujisawa Pharm. Co. Ltd., Japan
FUT	Torii & Co. Ltd., Japan
FW	Rohm & Haas Co., USA
G	Ciba-Geigy AG, Switzerland
	Givaudan Corp., USA
GC	Hopkins Agric. Chem. Co., USA
GEA	GEA A/S, Denmark
GI	Glaxo Wellcome plc, UK
GL	Genelabs Technologies, Inc., USA
GM	Glycomed, Inc., USA
G	Gdecke AG, Germany
GP	Ciba-Geigy AG, Switzerland
	Gensia, Inc., USA
GR	Glaxo Wellcome plc, UK
GS	Ciba-Geigy AG, Switzerland
	Gilead Sciences, USA
	Pfizer, Inc., USA
GX	Glaxo Wellcome plc, UK
H	Hässle Lakemedel AB, Sweden
	Bracco S.p.A., Italy
HB	Hoechst AG, Germany
HC	Sandoz AG, Switzerland
	Toyo Jozo Co. Ltd., Japan
HE	Helopharm KG, Germany
HF	Sandoz AG, Switzerland
IIGP	H. G. Pars Labs., USA
HH	Hommel GmbH, Switzerland
HN	Hafslund Nycomed AS, Norway
HOE	Hoechst AG, Germany
HPEK	Paul B. Elder Co., USA
HQ	Tanabe Seiyaku Co. Ltd., Japan
HR	Hoechst Marion Roussel, USA
HS	Sandoz AG, Switzerland
HSR	Hokuriku Seiyaku Co. Ltd., Japan
I	Farmitalia Carlo Erba s.r.l., Italy
	Ohmeda Pharm. Prods. Div., Inc., USA
IBI	Istituto Biochimico Italiano S.p.A., Italy
ICI	Zeneca Pharmaceuticals, UK
ICN	ICN Pharmaceuticals, USA
ID	Sumitomo Pharmaceutical Co., Ltd., Japan
IMI	Farmitalia Carlo Erba s.r.l., Italy
INF	Parke-Davis & Co., USA
ISF	ISF S.p.A., Italy
ITF	Italfarmaco S.p.A., Italy
JB	Lakeside Labs., USA
JDL	A. Christiaens S.A., Belgium
JM	Johnson Matthey, Inc., USA
JO	Jouveinal Labs., France
JTP	Japan Tobacco Industry Corp., Japan
K	Farmitalia Carlo Erba s.r.l., Italy
	Kali-Chemie Pharma GmbH, Germany
	Klinge Pharma GmbH, Germany
	Kowa Company Ltd., Japan
KB	Kanebo Pharmaceuticals Ltd., Japan
KC	Kali-Chemie Pharma GmbH, Germany
KD	Klinge Pharma GmbH, Germany
KF	Dr. Karl Thomae GmbH, Germany
Ko	Boehringer Ingelheim KG, Germany
KP	Kaken Pharmaceutical Co., Ltd., Japan
KRN	Kirin Brewery Co. Ltd., Japan
KT	Kyowa Hakko Kogyo Co. Ltd., Japan
KU	Kyorin Pharm. Co. Ltd., Japan
KW	Kyowa Hakko Kogyo Co., Ltd., Japan
KWD	Draco Lakemedel AB, Sweden

L	Alfa Wassermann S.p.A., Italy
	Gruppo Lepetit S.p.A., Italy
	Lab. Labaz, France
	Lederle Labs., USA
	Merck & Co., Inc., USA
	Sandoz AG, Switzerland
LA	Hoffmann-La Roche, Inc., USA
	Lipha Santé, France
	Lacer SA, Spain
LAC	AB Bofors, Sweden
LAS	Almirall Laboratorios S.A., Spain
LB	Hoechst AG, Germany
LCB	Lipha Santé, France
LD	Lab. Dausse, France
LG	Lab. Guidotti S.p.A., Italy
	Lifegroup, Italy
LGD	Ligand Pharmaceuticals, USA
LJC	Lederle Labs., USA
LK	Lek Pharmaceuticals, Slovenia
LL	Lab. Lafon, France
	Lederle Labs., USA
LM	Farmitalia Carlo Erba s.r.l., Italy
	Lipha S.A., France
	Pharmuka S.F., France
LR	Istituto Luso Farmaco D'Italia S.p.A., Italy
LS	AB Leo, Sweden
	Lipha Santé, France
	Rhône-Poulenc Agrochimie S.A., France
LT	Sandoz AG, Switzerland
Lu	H. Lundbeck A/S, Denmark
LW	Luitpold Pharma GmbH, Germany
LY	Eli Lilly & Co., USA
M	Farmoplant, Italy
	Imperial Chemical Industries Ltd., UK
	Mochida Pharm. Co., Ltd., Japan
	Reckitt & Colman Ltd., UK
M & B	May & Baker Ltd., UK
MBR	3M Pharmaceuticals, USA
MC	Mobil Chemical Co., USA
MCI	Mitsubishi Kasei Corp., Japan
McN	McNeil Labs., Inc., USA
MD	Lab. Delalande, France
MDL	Hoechst Marion Roussel, USA
ME	Meiji Seika Kaisha Ltd., Japan
Me	Boehringer Ingelheim KG, Germany
MEN	Menarini Industrie Farmaceutiche Riunite s.r.l., Italy
MER	Hoechst Marion Roussel, USA
MF	Boehringer Mannheim GmbH, Germany
	Mediolanum Farmaceutici S.p.A., Italy
MJ	Mead Johnson Pharmaceuticals, USA
MK	Merck & Co., Inc., USA
MKC	Mitsubishi Kasei Corp., Japan
ML	Merckle GmbH, Germany
MO	Abbott Laboratories, USA
MON	Monsanto Chemical Co., USA
MP	Mallinckrodt Inc., USA
MPI	Matrix Pharmaceutical, USA
MPV	Orion Corporation, Finland
MRL	Hoechst Marion Roussel, USA
MS	Mitsui Pharm., Inc., Japan
MT	Meiji Seika Kaisha Ltd., Japan
MY	Mitsubishi Kasei Corp., Japan
MZ	Chinoin, Hungary
N	H. Lundbeck A/S, Denmark
	Nelson Res. & Dev., USA
	Nisshin Pharmaceutical Co. Ltd., Japan
	Stauffer Chemical Co., USA
	UCB S.A., Belgium
NA	Dr. Karl Thomae GmbH, Germany
	Organon Co., USA
NAT	National Drug Co., USA

NC	Fisons plc, UK	RS	Syntex Labs., Inc., USA
	Nippon Chemiphar Co., Ltd., Japan	RU	Roussel-UCLAF SA, France
	Nissan Chem. Ind. Ltd., Japan	RV	Knoll AG, Germany
	Sandoz AG, Switzerland		Ravizza S.p.A., Italy
	Warner-Lambert Pharm. Co., USA	RWJ	R. W. Johnson Pharm. Res. Inst., USA
NCI	Nissan Chem. Ind. Ltd., Japan	RX	Reckitt & Colman Ltd., UK
NDR	National Drug Co., USA		
NE	Procter & Gamble Pharm., Inc., USA	S	Boehringer Ingelheim KG, Germany
NF	Procter & Gamble Pharm., Inc., USA		Hoechst AG, Germany
NIH	Natl. Inst. Health, USA		Lab. Servier, France
NIK	Nikken Chemical Co., Ltd., Japan		Shionogi & Co., Ltd., Japan
NK	Nippon Kayaku Co., Ltd., Japan		Simes S.p.A., Italy
NKK	Nihon Nohyaku Co. Ltd., Japan		Solco Basel AG, Switzerland
NKT	Nippon Kayaku Co. Ltd., Japan		Sumitomo Pharmaceutical Co., Ltd., Japan
NNC	Novo-Nordisk A/S, Denmark	SA	Siegfried Pharma AG, Switzerland
NO	Novo-Nordisk A/S, Denmark	SaH	Sandoz AG, Switzerland
NP	Nippon Soda Co., Ltd., Japan	SAN	Sandoz AG, Switzerland
NS	NeuroSearch A/S, Denmark	SAS	Schiapparelli S.p.A., Italy
	Nippon Shinyaku Co., Ltd., Japan	SB	SmithKline Beecham plc, UK
NSC	Natl. Cancer Inst., USA	SC	G. D. Searle & Co., USA
NSP	Nippon Soda Co. Ltd., Japan	SCE	Takeda Chem. Ind., Ltd., Japan
NTA	Fujisawa Pharmaceutical Co., Ltd., Japan	Sch	Schering-Plough Corp., USA
NZ	Nissan Chem. Ind. Ltd., Japan	SD	Lab. Diamant, France
			Santen Pharm. Co., Ltd., Japan
OK	Otsuka Pharm. Co. Ltd., Japan		Shell Oil Co., USA
ONO	Ono Pharm. Co. Ltd., Japan		Siegfried Pharma AG, Switzerland
OP	Ono Pharm. Co. Ltd., Japan		Simes S.p.A., Italy
OPC	Otsuka Pharm. Co. Ltd., Japan	SDZ	Sandoz AG, Switzerland
OR	Orion Corporation, Finland	SE	Lab. Servier, France
ORF	Ortho Pharm. Co., USA	SF	Meiji Seika Kaisha Ltd., Japan
ORG	Organon, Netherlands		Proter S.p.A., Italy
			Sandoz Inc., USA
P	Asta Medica AG, Germany	SG	Chugai Pharmaceutical Co., Japan
	Bayer AG, Germany	SGD	Siegfried Pharma AG, Switzerland
	Dainippon Pharmaceutical Co., Japan	SH	Schering AG, Germany
	Farmitalia Carlo Erba s.r.l., Italy	SKF	SmithKline Beecham plc, UK
	Leo Pharm, Denmark	SL	Synthelabo SA, France
	Pfizer, Inc., USA	SM	Greenwich Pharm., Inc., USA
	Pierrel S.p.A., Italy		Sumitomo Pharm. Co., Ltd., Japan
	UCB S.A., Belgium	SMS	Sandoz AG, Germany
PA	Pfizer, Inc., USA	SN	Sandoz AG, Germany
PAA	Parke-Davis & Co., USA		Schering AG, Germany
PB	Dr. Karl Thomae GmbH, Germany	SP	Shaman Pharmaceuticals, USA
PC	Pitman-Moore Co., USA	SPA-S	Societa Prodotti Antibiotici S.p.A., Italy
	ProCyte Corp., USA		
	Sumitomo Chemical Co., Japan	SPC	Sphinx Pharmaceutical Corp., USA
PD	Parke-Davis & Co., USA	SPE	Unilabo, France
PF	Lab. Prodes S.A., Spain	SQ	Bristol-Myers Squibb Co., USA
PH	Solvay Duphar B.V., Netherlands	SR	Chevron Chemical Co., USA
Ph	Pharmacia & Upjohn, Inc., USA		Sanofi SA, France
PN	Sandoz AG, Switzerland	SRG	Schering-Plough, USA
PP	Eisai Co., Ltd., Japan	ST	Boehringer Ingelheim KG, Germany
	Zeneca Agrochemicals, UK		Chemie Linz AG, Austria
PR	Lab. Prodes S.A., Spain		Sigma Tau Industrie Farmaceutiche Riunite S.p.A., Italy
PSC	Sandoz AG, Switzerland		
		Su	Ciba-Geigy AG, Switzerland
R	3M Pharmaceuticals, USA	SUN	Suntory Pharmaceutical, Japan
	Cilag AG, Switzerland	SW	Sankyo Co., Ltd., Japan
	Janssen Pharmaceutica N.V., Belgium		
	Lab. Roussel, France	T	Takeda Chem. Ind. Ltd., Japan
	Zeneca Agrochemicals, UK		Toyama Chem. Co. Ltd., Japan
RB	Lab. Roger Bellon, France	TA	Tanabe Seiyaku Co., Ltd., Japan
RC	Lab. Robert & Carriere, France	TAK	Takeda Chem. Ind., Ltd., Japan
RD	Lab. Anphar-Rolland, France	TAP	TAP Pharmaceuticals, USA
	The Boots Co. plc, UK	TC	Chinoin, Hungary
Rec	Recordati S.p.A., Italy	TCV	Takeda Chem. Ind., Ltd., Japan
RG	Rhône-Poulenc S.A., France	TH	Theraplix S.A., France
	Salutas Fahlberg-List Pharma GmbH, Germany	Th	Boehringer Ingelheim KG, Germany
		TI	Takeda Chemical Ind. Co., Ltd., Japan
RGH	Gedeon Richter, Ltd., Hungary	TJ	Tsumura Corp., Japan
RH	Rohm & Haas Co., USA	TMS	Toyo Jozo Co. Ltd., Japan
RMI	Hoechst Marion Roussel, USA	TN	Taiyo Pharm. Ind. Co., Ltd., Japan
Ro	Hoffmann-La Roche Ltd., Switzerland	TO	Torii & Co., Ltd., Japan
RP	Rhône-Poulenc, S.A., France	TPN	Sandoz AG, Germany
RPR	Rhône-Poulenc Rorer Pharm., Inc., USA	TR	Miles, Inc., USA

Code Letters Used by Companies for Experimental Substances (Continued)

TS	Taisho Seiyaku KK, Japan		WG	3M Pharmaceuticals, USA
TSAA	Takeda Chem. Ind. Ltd., Japan		WHR	William H. Rorer, USA
TVX	Troponwerke Dinklage & Co., Germany		Win	Sanofi Winthrop Inc., USA
TZU	Teikoku Hormone Mfg. Co. Ltd., Japan		WL	Wyeth-Ayerst Labs., USA
			WR	Walter Reed Army Hospital, USA
U	Pharmacia & Upjohn, Inc., USA		WY	Wyeth-Ayerst Labs., USA
UC	Union Carbide Corp., USA			
UCB	UCB SA, Belgium		X	Simes S.p.A., Italy
UK	Pfizer, Inc., USA		XR	Xenova Group plc, UK
UP	UPSA, France		XZ	Pfizer, Inc., USA
UR	J. Uriach & Cia S.A., Spain			
			Y	Yoshitomi Pharmaceutical Ind., Japan
VC	Mobil Chemical Co., USA		YM	Yamanouchi Pharm. Co. Ltd., Japan
VCS	Velsicol Chemical Corp., USA		YTR	Taiho Pharmaceutical Co. Ltd., Japan
W	Carter-Wallace, Inc., USA		Z	Zambon Group S.p.A., Italy
	Sandoz AG, Switzerland			Zeria Pharm. Co., Ltd., Japan
	Warner-Lambert Co., USA		ZD	Zeneca Pharmaceuticals, UK
WAC	Wassermann SEEFT S.A., Spain		ZK	Schering AG, Germany
WAL	Boehringer Ingelheim KG, Germany		ZM	Zeneca Pharmaceuticals, UK
WAS	Wassermann-SEEFT S.A., Spain		ZY	Zyma SA, Switzerland
WAY	Wyeth-Ayerst Labs., USA			

Company Register

The following table lists the full company name and location corresponding to the abbreviations given in the Name Index for companies associated with trademarks.

Aaciphar	Aaciphar (subsidiary of Akzo Nobel), Brussels, Belgium
Aandersen	Aandersen Farmaceutisk Institut S.p.A., Rome, Italy
Abbott	Abbott Labs., Abbott Park, Illinois
ABC	Istituto Biologico Chemioterapico ABC S.p.A., Torino, Italy
Abello	Abello S.A. (subsidiary of Merck & Co., Inc.), Madrid, Spain
AB Leo	AB Leo (subsidiary of Pharmacia AB), Helsingborg, Sweden
Adria	Pharmacia Adria (formerly Adria Labs., Inc.) (subsidiary of Pharmacia AB), Columbus, Ohio
Adrian-Marinier	Labs. Doms-Adrian (formerly Labs. Adrian-Marinier), Courbevoie, France
Adroka	Adroka AG, Allschwil, Switzerland
AFI	A/S Farmaceutisk Industri, Oslo, Norway
AGIPS	AGIPS Pharmaceutici s.r.l., Rapallo, Italy
Agpharm	Agpharm AG, Lucerne, Switzerland
Air Prods. Chem.	Air Products and Chemicals, Inc., Allentown, Pennsylvania
Ajinomoto	Ajinomoto Co., Ltd., Tokyo, Japan
Akorn	Akorn, Inc., Abita Springs, Louisiana
Akzo	Akzo Nobel N.V., Arnhem, Netherlands
Albert-Roussel	Albert-Roussel Pharma GmbH (subsidiary of Hoechst AG), Wiesbaden, Germany
Alcon	Alcon Labs., Inc. (subsidiary of Nestle S.A.), Fort Worth, Texas
Alconox	Alconox Inc., New York, New York
Alcon-Thilo	Alcon Pharma GmbH, Freiburg, Germany
Aldrich	Aldrich Chemical Co., Inc. (subsidiary of Sigma-Aldrich Corp.), Milwaukee, Wisconsin
Alfa	Alfa Wassermann S.p.A., Bologna, Italy
Allard	Lab. Allard S.A. (subsidiary of Bristol-Myers Squibb Co.), Paris, France
Allen & Hanburys	Allen & Hanburys Ltd. (subsidiary of Glaxo Wellcome), Uxbridge, Middlesex, UK
Allergan	Allergan Inc., Irvine, California
Allergopharma	Allergopharma Joachim Ganzer KG, Reinbeck, Germany
Allied	Allied Signal, Inc. (formerly Allied Chemical Co.), Morristown, New Jersey
Almirall	Almirall S.A. Laboratorios, Barcelona, Spain
Alza	Alza Corp., Palo Alto, California
Amchem	Rhône-Poulenc Agrochemie (formerly Amchem Products Inc.), Lyons, France
Am. Cyanamid	American Cyanamid Co. (division of Am. Home Prod. Corp.), Wayne, New Jersey
Am. Home	American Home Products Corp., Madison, New Jersey
Amino	Amino AG, Neuenhof, Switzerland
Amsa	Amsa, s.r.l., Berberino di Mugello, Italy
Am. Urologicals	American Urologicals, Inc., Pembroke Pines, Florida
ANA	Labs. Roche-Nicholas S.A. (formerly Labs. Nicholas-Ana) (subsidiary of Hoffmann-La Roche), Gaillard, France
Anaquest	Ohmeda, Inc. (formerly Anaquest, Inc.) (subsidiary of BOC Group), Liberty Corner, New Jersey
Anasco	Anasco GmbH (subsidiary of Boehringer, Ing.), Ingelheim, Germany
Andreu	Labs. Dr. Andreu, S.A. (subsidiary of Hoffmann-La Roche), Barcelona, Spain
Andromaco	Labs. Andromaco, S.A., Madrid, Spain
Angelini Francesco	Aziende Chimiche Riunite Angelini Francesco S.p.a., Rome, Italy
Ankerpharm	Ankerpharm GmbH, Rudolstadt, Germany
Anphar-Rolland	Lab. Anphar-Rolland (division of Lipha Santé), Evry, France
Antibioticos	Antibioticos S.A. (subsidiary of Montedison), Madrid, Spain
Apotex	Apotex, Inc., Weston, Ontario, Canada
Apothecon	Apothecon Products (division of Bristol-Myers Squibb Co.), Princeton, New Jersey
APS	Berk Pharmaceuticals (formerly Approved Prescription Services Ltd.), Eastbourne, East Sussex, UK
Ardeypharm	Ardeypharm GmbH Pharmazeutische Fabrik, Herdecke, Germany
Ariston	Qumíca Ariston S.A., Buenos Aires, Argentina
Armour Pharm.	Armour Pharmaceutical Co. (subsidiary of Rhône-Poulenc Rorer), Collegeville, Pennsylvania
Armstrong	Armstrong Labs. S.A.C.I.F., Buenos Aires, Argentina
Arnolds	Arnolds Veterinary Products Ltd., Harlescott, Shrewsbury, UK
Aron	Aron-Médicia (division of Lipha Santé), Lyon, France
Artesan	Artesan Pharma GmbH, Luchow, Germany
Arun	Arun Products Ltd., Arundel, W. Sussex, UK
Arzneimittelwerk Dresden	Arzneimittelwerk Dresden GmbH (subsidiary of Asta Medica AG), Radebeul, Germany
Asahi	Asahi Chemical Industry Co., Ltd., Tokyo, Japan
Asche	Asche AG (subsidiary of Schering AG), Hamburg, Germany
Ascher	B. F. Ascher & Co., Inc., Lenexa, Kansas
Ashland	Ashland Chemical Co. (division of Ashland Oil Inc.), Dublin, Ohio
Assia	Assia Chemical Labs., Ltd., Tel Aviv, Israel
Asta	Asta Medica AG (subsidiary of Degussa AG), Frankfurt, Germany
Astier	Lab. Astier, Paris, France
Astra	Astra AB, Sodertalje, Sweden
Astra-Simes	Astra-Simes S.p.A., Milan, Italy
Athenstaedt	Athenstaedt GmbH & Co. KG, Bremen, Germany
Ausonia	Ausonia Farmaceutici, Pomezia (Rome), Italy
A.V.P.	A.V.P. Pharmaceuticals, Inc., Clarence, New York

Ayerst	Wyeth-Ayerst Laboratories (division of Am. Home Prod. Corp.), Philadelphia, Pennsylvania
Azupharma	Azupharma GmbH, Gerlingen, Germany
Bago	Labs. Bago S.A., Buenos Aires, Argentina
Bailly	Lab. A. Bailly-SPEAB, Paris, France
Baldacci	Labs. Baldacci, S.p.A., Pisa, Italy
Banyu	Banyu Pharmaceutical Co., Ltd., Tokyo, Japan
Barnes-Hind	Pilkington Barnes-Hind, Inc. (subsidiary of Pilkington Brothers plc), Sunnyvale, California
BASF	BASF Group (formerly Badische Anilin- und Soda-Fabrik) Ludwigshafen, Germany
Basotherm	Basotherm GmbH (subsidiary of Boehringer Ingelheim KG), Biberach an der Riss, Germany
Bastian	Bastian-Werk GmbH, Munich, Germany
Bausch & Lomb	Bausch & Lomb Pharmaceuticals, Inc., Tampa, Florida
Baxter	Baxter International, Inc. (formerly Baxter Travenol Labs.), Deerfield, Illinois
Bayer	Bayer AG, Leverkusen, Germany
Bayropharm	Bayropharm GmbH (subsidiary of Bayer AG), Leverkusen, Germany
Beecham	SmithKline Beecham plc (formerly Beecham Research Labs.), Brentford, Middlesex, UK
Behringwerke	Behringwerke AG (subsidiary of Hoechst AG), Marburg, Germany
Beiersdorf	Beiersdorf AG, Hamburg, Germany
Bellon	Labs. Roger Bellon (subsidiary of Rhône-Poulenc Rorer), Neuilly-sur-Seine, France
Belupo	Belupo, Zagreb, Yugoslavia
Bencard	Bencard Pharmaceuticals Ltd. (subsidiary of SmithKline Beecham), Brentford, Middlesex, UK
Bene-Arzneimittel	Bene-Arzneimittel GmbH, Munich, Germany
Bengué	Bengué & Co., Ltd., Maidenhead, Berks., UK
Benvegna	Neoterapici Benvegna s.r.l., Palermo, Italy
Benzon	Benzon Pharma A/S (subsidiary of Hafslund Nycomed), Copenhagen, Denmark
Bergamon	Bergamon S.p.A., Rome, Italy
Berk	Berk Pharmaceuticals Ltd. (subsidiary of Rhône-Poulenc), Eastbourne, East Sussex, UK
Berlex	Berlex Labs. Inc. (subsidiary of Schering AG), Wayne, New Jersey
Berlin-Chemie	Berlin-Chemie, Berlin, Germany
Bernabo	Bernabo & Cia S.A.C.I.I.F.A., Buenos Aires, Argentina
Besins-Iscovesco	Labs. Besins-Iscovesco, Paris, France
Beta	Lab. Biologico Chemioterapico Beta s.r.l., Brescia, Italy
Beytout	Labs. Beytout, Saint-Mandé, France
Biochemie	Biochemie GmbH (subsidiary of Sandoz), Vienna, Austria
Biocodex	Lab. Biocodex, Montrouge, France
Bioindustria	Bioindustria Farmaceutici S.p.A., Novi Ligure, Italy
Biokema	Biokema S.A., Crissier, Switzerland
Biologici	Biologici Italia Lab. s.r.l., Novate Milanese, Italy
Biopharma	Biopharma (Labs. Biopharmaceutiques de France) (subsidiary of Servier), Neuilly-sur-Seine, France
Biorex	Biorex Labs, Enfield, Middlesex, UK
Biosedra	Lab. Biosedra (subsidiary of Fresenius AG), Louviers, France
Biosintetica	Biosintetica, Sao Paulo, Brazil
Bock	Bock Pharmacal Co., St. Louis, Missouri
Boehringer, Ing.	Boehringer Ingelheim KG (subsidiary of C. H. Boehringer Sohn), Ingelheim, Germany
Boehringer, Mann.	Boehringer Mannheim GmbH, Mannheim, Germany
Bofors	AB Bofors, Nobel-Pharma, Molndal, Sweden
B.O.I.	Labs. B.O.I., Barcelona, Spain
Bolar	Circa Pharm. Inc. (formerly Bolar Pharmaceutical Co., Inc.), Copiague, New York
Bonomelli	Bonomelli S.p.A., Dolzago, Italy
Boots	The Boots Co. plc, Nottingham, UK
Bottu	Bottu S.A. (subsidiary of Rhône-Poulenc), Antony, France
Bouchara	Lab. du Dr. E. Bouchara, Levallois-Perret, France
Bracco	Bracco S.p.A., Milan, Italy
Braun	B. Braun Melsungen AG, Melsungen, Germany
Brenner-Efeka	Brenner-Efeka Pharma GmbH (subsidiary of Am. Home Prod. Corp.), Münster, Germany
Breon	Sterling Drug Inc. (formerly Breon Labs., Inc.), New York, New York
Bristol	Bristol-Myers Squibb Co. (formerly Bristol Labs.), Princeton, New Jersey
Bristol-Myers Squibb	Bristol-Myers Squibb Co., Princeton, New Jersey
Britannia	Britannia Pharmaceuticals Ltd., Redhill, Surrey, UK
Brocades	Yamanouchi Pharmaceutical Co., Ltd. (formerly Brocades Ltd.), Tokyo, Japan
Brocchieri	Stabilimento Chimico Farmaceutico dr. L. Brocchieri s.r.l., Rome, Italy
Bruneau	Lab. Bruneau, Boulogne, France
Brunton	Brunton Chemists Ltd., Sunderland, Durham, UK
Bruschettini	Bruschettini s.r.l., Genoa, Italy
Burroughs Wellcome	Glaxo Wellcome, plc, London, UK
Byk-Gulden	Byk-Gulden Lomberg Chem. Fabrik, GmbH, Konstanz, Germany
Caber	Farmaceutici Caber s.r.l., Pisa, Italy
Calbiochem	Calbiochem, San Diego, California
Calgon	Calgon Vestal Labs. (division of Bristol-Myers Squibb), St. Louis, Missouri
Calmic	Calmic, Berkhamsted, UK
Carnrick	Carnrick Labs., Inc., Cedar Knolls, New Jersey
Carrion	Labs. Carrion (subsidiary of Synthélabo), Courbevoie, France

Carter-Wallace	Carter-Wallace, Inc., New York, New York
Casasco	Labs. Casasco S.A.I.C., Buenos Aires, Argentina
Cascan	Cascan GmbH & Co., KG (subsidiary of Glaxo Wellcome), Wiesbaden, Germany
Cassella-Riedel	Cassella-Riedel Pharma GmbH (subsidiary of Hoechst AG), Frankfurt, Germany
Cassenne	Lab. Cassenne (subsidiary of Roussel-UCLAF), Paris, France
Cehasol	Cehasol Chem.-Pharm. Produkte GmbH, Vienna, Austria
Cenci	H. R. Cenci Labs., Inc., Fresno, California
Centerchem	Centerchem Pharmaceuticals, Inc., Stamford, Connecticut
Central Pharm.	Central Pharmaceuticals, Inc., Seymour, Indiana
Century	Century Pharmaceuticals, Inc., Indianapolis, Indiana
C.E.P.A.	Compania Espanola de Penicilina y Antibioticos, S.L., Madrid, Spain
CFPI	CFPI, Gennevilliers, France
Chassot	Chassot AG, Belp, Switzerland
Chauvin	Labs. Chauvin S.A., Montpellier, France
Chefaro	Deutsche Chefaro Pharma GmbH (subsidiary of Akzo Pharma B.V.), Waltrop, Germany
Chemagro	Chemagro Agricultural Chemicals Div. (division of Miles), Kansas City, Missouri
Chemie Linz	Chemie Linz AG, Linz, Austria
Chemil	Chemil Farmaceutici s.r.l., Milan, Italy
Chemipharm	Chemipharm GmbH & Co., Chem.-Pharm. Fabrik KG, Cologne, Germany
Chemofux	Chemofux GmbH, Vienna, Austria
Chephasaar	Chephasaar, Chem.-Pharm. Fabrik GmbH, St. Ingelbert, Germany
Chesebrough	Chesebrough-Ponds USA (subsidiary of Unilever, Inc.), Greenwich, Connecticut
Chevron	Chevron Chemical Co., San Ramon, California
Chibret	Labs. Merck Sharp & Dohme-Chibret S.A. (subsidiary of Merck & Co., Inc.), Paris, France
Chiesi	Chiesi Farmaceutici S.p.A., Parma, Italy
Chinoin	Chinoin Gyogyszer es Vegyeszeti Termekek Gyara, Budapest, Hungary
Chinosolfabrik	Chinosolfabrik (subsidiary of Riedel-de Haen AG), Seelze, Germany
Christiaens	A. Christiaens S.A., Brussels, Belgium
Chugai	Chugai Pharmaceutical Co., Ltd., Tokyo, Japan
Ciba	Ciba-Geigy AG (formerly Ciba AG), Basel, Switzerland
Ciba-Geigy	Ciba-Geigy AG, Basel, Switzerland
Cidan	Labs. Cidan, Benicarlo (Castellon), Spain
Cilag	Cilag AG (subsidiary of Johnson & Johnson), Schaffhausen, Switzerland
Circle	Circle Pharmaceuticals, Inc., Indianapolis, Indiana
Cleary	W. A. Cleary Chemical Corp., Somerset, New Jersey
Coli	Farmaceutici Coli s.r.l., Rome, Italy
Collins	L. D. Collins & Co., Ltd., Hertfordshire, UK
Combe	Combe Inc., White Plains, New York
Comm. Solvents	Commercial Solvents Corp., Northbrook, Illinois
Conoco	Conoco Inc. (division of DuPont), Houston, Texas
Consolidated Chem.	Consolidated Chemicals Ltd., Wrexham, Clywd., UK
Consolidated Midland	Consolidated Midland Corp., Brewster, New York
Cont. Pharma	Continental Pharma S.A. (subsidiary of Monsanto), Brussels, Belgium
Cooper	Cooper Labs., Inc., Parsippany, New Jersey
CooperVision	CooperVision Inc., Fairport, New York
Coop. Farm.	Cooperative Farmaceutica Societa Coop., a.r.l., Milan, Italy
Corvi	Camillo Corvi S.p.A.(subsidiary of Roussel-UCLAF), Piacenzi, Italy
Cosmopharma	Cosmopharma N.V., Zwanenburg, Netherlands
Cox Pharm	A.H. Cox & Co. Ltd., Barnstaple, UK
CP Pharm.	CP Pharmaceuticals, Ltd., Wrexham, Clywd., UK
Crinos	Crinos Industria Farmacobiologica S.p.A., Como Villaguardia, Italy
Critikon	Critikon Inc. (subsidiary of Johnson & Johnson), Tampa, Florida
Croda	Croda, Inc., Parsippany, New Jersey
Crookes	Crookes Healthcare Ltd. (subsidiary of The Boots Co.), Nottingham, UK
Crosara	Lab. Farmaco Biologico Crosara S.p.A., Rome, Italy
Crystal Chem.	Crystal Chemical Inter-America, Houston, Texas
CT	C.T.-Lab. Farmaceutico s.r.l., San Remo, Italy
Cusi	Laboratorios Cusi S.A., El Masnou (Barcelona), Spain
Cutter	Cutter Biologicals (division of Miles Labs.), Emeryville, California
C-Vet	C-Vet Ltd., Leyland, Lancashire, UK
Dagra	Dagra B.V., Diemen, Netherlands
Daiichi	Daiichi Pharmaceutical Co., Ltd., Tokyo, Japan
Dainippon	Dainippon Pharmaceutical Co., Ltd., Osaka, Japan
Damor	Farmaceutici Damor S.p.A., Naples, Italy
Daniels	Daniels Pharmaceuticals, Inc., St. Petersburg, Florida
Dausse	Labs. Dausse (subsidiary of Synthélabo), Paris, France
Davis & Geck	Davis & Geck (division of American Cyanamid Co.), Wayne, New Jersey
DDSA	DDSA Pharmaceuticals, London, UK
De Angeli	Istituto de Angeli S.p.A., Milan, Italy
Debat	Lab. Debat, Garches, France
Deglaude	Théranol-Deglaude, Bagneux, France
Degussa	Degussa AG, Frankfurt, Germany
Delagrange	Labs. Delagrange (subsidiary of Synthélabo), Chilly-Mazarin, France
Delalande	Lab. Delalande (subsidiary of Synthélabo), Courbevoie, France
Delalande Isnardi	Lab. Delalande Isnardi S.p.A., Imperia, Italy

Delandale	Lorex Synthélabo, Ltd. (formerly Delandale Labs.) (subsidiary of L'Oreal S.A.), Marlow, Bucks, UK
Dermik	Dermik Labs., Inc. (subsidiary of Rhône-Poulenc Rorer), Collegeville, Pennsylvania
Desitin	Desitin Arzneimittel GmbH, Hamburg, Germany
Dexo	Labs. Dexo, Nanterre, France
Diagnostic Data	DDI Pharm. Inc. (formerly Diagnostic Data, Inc.), Mountain View, California
Diamant	Labs. Diamant (subsidiary of Roussel-UCLAF), Puteaux, France
Diamond	Diamond Shamrock Inc., San Antonio, Texas
Dieckmann	Dieckmann Arzneimittel GmbH (subsidiary of Merck & Co., Inc.), Munich, Germany
Dista	Dista Products, Ltd. (division of Eli Lilly & Co.), Liverpool, UK
Doak	Doak Pharmacal Co., Inc., Westbury, New York
Doetsch, Grether	Doetsch, Grether & Cie AG, Basel, Switzerland
Dolder	Dolder AG, Basel, Switzerland
Dolorgiet	Dolorgiet GmbH & Co. KG, St. Augustin/Bonn, Germany
Dompé	Dompé Farmaceutici S.p.A., Milan, Italy
Dorsch	Dorsch GmbH, Gräfelfing, Germany
Dorsey	Dorsey Labs. (division of Sandoz Pharmaceuticals Corp.), East Hanover, New Jersey
Dow	The Dow Chemical Co., Midland, Michigan
DowElanco	DowElanco, Indianapolis, Indiana
Draco	Draco Lakemedel AB (subsidiary of AB Astra), Lund, Sweden
Dreluso	Dreluso Pharmazeutika, Dr. Elten & Sohn GmbH, Hessisch Oldendorf, Germany
Dukron	Dukron Italiana s.r.l., Campoverde (Latina), Italy
Dulcis	Allergan France S.A. (formerly Laboratoire Dulcis) (subsidiary of Allergan, Inc.), Mougins, France
Dumex	Dumex A/S, Copenhagen, Denmark
Duncan Farm	Duncan Farmaceutici S.p.A., Verona, Italy
Duncan, Flockhart	Duncan, Flockhart & Co., Ltd. (subsidiary of Glaxo Wellcome), Uxbridge, Middlesex, UK
Duphar	Solvay Duphar B.V. (subsidiary of Solvay S.A.), Weesp, Netherlands
Du Pont	E. I. du Pont de Nemours & Co., Inc., Wilmington, Delaware
DuPont Merck	DuPont Merck Pharmaceutical Co., Wilmington, Delaware
Durachemie	Durachemie GmbH & Co. KG (subsidiary of Cyanamid GmbH), Wolfratshausen, Germany
Eastman-Kodak	Eastman-Kodak Co., Rochester, New York
Eaton	Procter & Gamble Pharm. Inc. (formerly Norwich Eaton Pharmaceuticals), Norwich, New York
Ebewe	Ebewe Arzneimittelfabrik (subsidiary of BASF), Unterach am Attersee, Austria
EGIS	Egis Pharmaceutical Works, Budapest, Hungary
EGYT	Egis Pharmaceutical Works (formerly Egyesult Gyogyszer es Tapszer Gyar), Budapest, Hungary
Eisai	Eisai Co., Ltd., Tokyo, Japan
Elan	Elan Pharma Ltd., Westmeath, Ireland
Elanco	Elanco Animal Health Co. (subsidiary of Eli Lilly & Co.), Indianapolis, Indiana
Elder	SPI Pharm. Inc. (formerly Paul B. Elder Co., Inc.) (subsidiary of ICN Pharm. Inc.), Costa Mesa, California
E.L.E.A.	Labs. Elea, S.A.C.I.F. y A., Buenos Aires, Argentina
Elmu	Elmuquimica Farmaceutica S.L. (Especialidades Latinas Medicamentos Universales S.A.), Madrid, Spain
E. Merck	E. Merck AG, Darmstadt, Germany
Emery	Emery Industries, Inc., Mauldin, South Carolina
Endo	Endo Laboratories (subsidiary of Du Pont Merck Pharm.), Wilmington, Delaware
Engelhard	Engelhard Corp., Iselin, New Jersey
Engelhard KG	Karl Engelhard, Fabrik Pharmazeutischer Praparate, Frankfurt, Germany
Erba	Farmitalia Carlo Erba s.r.l. (subsidiary of Pharmacia), Milan, Italy
Erco	Ercopharm A/S, Kvistgard, Denmark
Errekappa	Errekappa Euroterapici S.p.A., Milan, Italy
Espe	Espe Fabrik Dental Medizin GmbH & Co. KG, Seefeld, Germany
Esseti	Esseti s.a.s. Laboratorio Chimico Farmaco Biologico, S. Giorgi-Cremano, Italy
Essex	Essex Pharma (subsidiary of Schering-Plough), Munich, Germany
Esteve	Labs. Dr. Esteve, S.A., Barcelona, Spain
Ethicon	Ethicon, Inc. (subsidiary of Johnson & Johnson), Somerville, New Jersey
Europharma	Europharma, S.A., Madrid, Spain
Eutherapie	Lab. Eutherapie (subsidiary of Servier), Neuilly-sur-Seine, France
Evans	Evans Healthcare Ltd. (subsidiary of Medeva plc), Leatherhead, Surrey, UK
Exa	Exa Impex Ltda., Buenos Aires, Argentina
Fabre	Pierre Fabre S.A., Castres, France
Falk	Dr. Falk Pharma GmbH, Freiburg, Germany
Farillon	Farillon Ltd., Romford, Essex, UK
Farmacologico	Lab. Farmacologico Milanese s.r.l., Caronno Pertusella, Italy
Farmades	Farmades S.p.A., Rome, Italy
Farmasa	Farmasa, Sao Paulo, Brazil
Farmetrusca	Farmetrusca s.a.s. di C. Pini e C., Tavarnuzze, Italy
Farmigea	Farmigea S.p.A.-Industria Chimico Farmaceutica, Pisa, Italy
Farmila	Farmila-Farmaceutici Milano s.r.l., Milan, Italy
Farmitalia	Farmitalia Carlo Erba s.r.l. (subsidiary of Pharmacia), Milan, Italy
Farmos	Orion-Farmos, Espoo, Finland
Fatol	Fatol Arzneimittel GmbH, Schiffweiler, Germany

Faulding	F. H. Faulding & Co. Pty Ltd., Salisbury South, SA, Australia
Fawns & McAllen	Fawns & McAllen Pty, Ltd., Clayton, Victoria, Australia
FBA Pharmaceuticals	FBA Pharmaceuticals, Inc., New York, New York
Fermenta	Fermenta Animal Health Co., Kansas City, Missouri
Ferndale	Ferndale Labs., Inc., Ferndale, Michigan
Ferrer	Ferrer Internacional S.A., Barcelona, Spain
Ferring	Ferring AB, Malmo, Sweden
Ferrosan	AB Ferrosan (subsidiary of Novo Nordisk), Malmo, Sweden
Fidia	Fidia S.p.A. (Farmaceutici Italiani Derivati Industrialie Affini S.p.A.), Albano Terme, Italy
FIRMA	FIRMA, Fabbr. Ital. Ritrov. Medic. Aff. S.p.A., Florence, Italy
Firmenich	Firmenich S.A., Geneva, Switzerland
Fisons	Fisons plc, Pharmaceuticals Div., Coalville, Leicestershire, UK
FMC	FMC Corp., Chicago, Illinois
Forest	Forest Pharmaceuticals, Inc., St. Louis, Missouri
Formenti	Prodotti Formenti s.r.l., Milan, Italy
Fort Dodge	Fort Dodge Labs. (subsidiary of Am. Home Prod. Corp.), Fort Dodge, Iowa
Foscama	Biomedica Foscama Industria Chimico-Farmaceutica S.p.A., Rome, Italy
Fougera	E. Fougera & Co. (division of Altana, Inc.), Melville, New York
Fournier	Laboratoires Fournier S.A., Dijon, France
Francia	Francia Farmaceutici Industria Farmaco Biologica s.r.l., Milan, Italy
Fresenius	Fresenius AG, Bad Homburg, Germany
Frosst	Merck Frosst Canada (subsidiary of Merck & Co., Inc.), Montreal, Canada
Fujinaga	Fujinaga Pharmaceutical Co., Ltd., Tokyo, Japan
Fujisawa	Fujisawa Pharmaceutical Co., Ltd., Osaka, Japan
Fumouze	Labs. Fumouze, S.A. (subsidiary of Carter-Wallace), Clichy, France
Funai	Merrell-Dow Funai K.K., Osaka, Japan
Fuso	Fuso Pharmaceutical Industries, Ltd., Osaka, Japan
Gador	Gador S.A., Buenos Aires, Argentina
GAF	International Specialty Products, Inc. (formerly GAF Corp.), Wayne, New Jersey
Galen	Galen Ltd., Portadown, Co., Armagh, Northern Ireland
Galenica	Galenica Vertretungen AG, Bern, Switzerland
Galenika	Galenika, Zemun, Yugoslavia
Galenus	Galenus Mannheim GmbH (subsidiary of Boehringer Mannheim GmbH), Mannheim, Germany
Gallier	Labs. Gallier S.A., Aubervilliers, France
Galma	Galma Pharmaceutical Chemical Factory, Tel Aviv, Israel
Gea	Gea A/S (subsidiary of Bristol-Myers Squibb), Frederiksberg, Denmark
Gebro	Gebro Broschek GmbH, Tirol, Austria
Gedeon Richter	Chemical Works of Gedeon Richter Ltd., Budapest, Hungary
Geigy	Ciba-Geigy (formerly J. R. Geigy AG), Basel, Switzerland
Geistlich	Geistlich-Pharma, Wolhusen, Switzerland
Genentech	Genentech, Inc. (subsidiary of Hoffmann-La Roche), South San Francisco, California
Geneva	Geneva Pharmaceuticals, Inc., Broomfield, Colorado
Genevrier	Labs. Genevrier, Sophia Antipolis, France
Gentili	Istituto Gentili S.p.A., Pisa, Italy
Gerot	Gerot-Pharmazeutika, GmbH, Vienna, Austria
Geymonat	Geymonat S.p.A., Anagni, Italy
Ghimas	Ghimas S.p.A., Casalecchio di Reno, Italy
Gibipharma	Gibipharma S.p.A., Pero (Milano), Italy
Gist-Brocades	Yamanouchi Europe (formerly Gist-Brocades N.V.), Leiderdorp, Netherlands
Giuliani	Giuliani S.p.A., Milan, Italy
Giulini	Giulini Pharma GmbH (subsidiary of Kali-Chemie), Hannover, Germany
Givaudan	Givaudan-Roure Intl. S.A. (subsidiary of Hoffmann-La Roche), Vernier, Switzerland
Glaxo	Glaxo Wellcome plc, London, UK
Glenbrook	Glenbrook Labs. (division of Sterling Drug Inc.), New York, New York
Glenwood	Glenwood Inc., Tenafly, New Jersey
Globopharm	Globopharm AG, Küsnacht, Switzerland
Glyco	Glyco Chemicals, Inc., Greenwich, Connecticut
Gobbi-Novag	Labs. Gobbi-Novag S.A., Buenos Aires, Argentina
Godecke	Gödecke AG (subsidiary of Warner-Lambert), Berlin, Germany
GP	GP Labs. (division of Pfizer), West Ryde, N.S.W., Australia
Gray	Gray Pharmaceutical Co., Norwalk, Connecticut
Green Cross	The Green Cross Corp., Osaka, Japan
Greenwich	Greenwich Pharmaceuticals, Inc., Fort Washington, Pennsylvania
Grelan	Grelan Pharmaceutical Co., Ltd., Tokyo, Japan
Grossman	Labs. Grossman S.A., Mexico City, Mexico
Grunenthal	Grünenthal GmbH, Stolberg, Germany
Guardian	Guardian Laboratories (division of United-Guardian, Inc.), Smithtown, New York
Guerbet	Lab. Guerbet, Roissy-Charles-de-Gaulle, France
Guidotti	Lab. Guidotti, S.p.A., Pisa, Italy
Halsey	Halsey Drug Co., Brooklyn, New York
Hamilton	Hamilton Pharm. Pty. Ltd., Adelaide, Australia
Hassle	Hässle Lakemedel AB (subsidiary of Astra AB), Molndal, Sweden
Hauck	Roberts/Hauck Pharm. Corp. (subsidiary of Roberts Pharm. Corp), Eatontown, New Jersey
Hausmann	Hausmann AG, St. Gallen, Switzerland

Haver	Haver Labs. (division of Mobay), Shawnee, Kansas
Hefa	Hefa Pharma Vertriebs Arzneimittel GmbH & Co. KG, Werne, Germany
Heilit	Heilit Arzneimittel GmbH, Reinbeck, Germany
Heilmittelwerke	Heilmittelwerke Wien GmbH (subsidiary of Hafslund Nycomed), Vienna, Austria
Hek	Hek Pharma GmbH, Quarnbeck, Germany
Helopharm	Helopharm W. Petrik GmbH & Co. KG, Berlin, Germany
Henkel	Henkel KGaA, Dusseldorf, Germany
Hennig	Hennig Arzneimittel GmbH & Co. KG, Florsheim am Main, Germany
Henning	Henning Berlin GmbH, Berlin, Germany
Herbert	Herbert Pharma GmbH, Wiesbaden-Bierstadt, Germany
Herbrand	Dr. Herbrand KG Chem.-Pharm. Werk, Gengenback, Germany
Hercules	Hercules Inc., Wilmington, Delaware
Hermal	Hermal Kurt Herrmann (subsidiary of E. Merck), Reinbek, Germany
Hermes	Hermes Arzneimittel GmbH, Grosshesselohe/Munich, Germany
Heterochemical	HeteroChemical Corp., Valley Stream, New York
Heumann	Heumann Pharma GmbH, Nurenberg, Germany
Heyl	Heyl Chem.-Pharm. Fabrik GmbH & Co. KG, Berlin, Germany
Hishiyama	Hishiyama Pharmaceutical Co., Ltd., Osaka, Japan
Hobein	Dr. Hobein & Co., GmbH Arzneimittel Nachf., Meckenheim, Germany
Hoechst	Hoechst AG, Frankfurt/Main, Germany
Hoechst Marion Roussel	Hoechst Marion Roussel Inc., Kansas City, Kansas
Hoei	Hoei Pharmaceutical Co., Ltd., Osaka, Japan
Hokuriku	Hokuriku Seiyaku Co., Ltd., Tokyo, Japan
Holzinger	Holzinger, Dr. et Mr. L., pharm. Erzeugungs- u. Vertriebsgesellschaft mbH, Vienna Austria
Homburg	Asta Medica AG (formerly Chemiewerk Homburg) (subsidiary of Degussa), Frankfurt, Germany
Hommel	Chemische Werke Hommel GmbH, Mullheim/Baden, Germany
Hopkins	Hopkins Agricultural Chemical Co., Madison, Wisconsin
Hor-Fer-Vit	Hor-Fer-Vit Pharmazeutische GmbH, Oldenburg, Germany
Horii	Horii Pharmaceutical Ind., Ltd., Osaka, Japan
Horita	Horita Yakuhin Gosei K.K., Nagoya, Japan
Hormon-Chemie	Hormon-Chemie GmbH, Munich, Germany
Hormosan	Hormosan-Kwizda GmbH, Frankfurt/Main, Germany
Horner	Frank W. Horner, Inc. (subsidiary of Carter-Wallace), Ville Mont-Royal, Quebec, Canada
Hosbon	Labs. Hosbon, S.A., Sao Paulo, Brazil
Houde	Lab. Houdé (subsidiary of Roussel-UCLAF), Puteaux, France
Hough, Hoseason	Hough, Hoseason & Co., Ltd., Manchester, UK
Hoyer	Hoyer GmbH & Co., Pharmazeutische Präparate, Neuss, Germany
HW & D	Becton Dickinson Microbiology Systems (formerly Hynson, Westcott & Dunning, Inc.), Cockeysville, Maryland
Hyland	Baxter-Hyland Labs. (division of Baxter Healthcare Corp.), Glendale, California
Hyrex	Hyrex Pharmaceuticals, Memphis, Tennessee
IBI	Istituto Biochimico Italiano G. Lorenzini S.p.A., Milan, Italy
Ibirn	Istituto Bioterapico Nazionale s.r.l., Rome, Italy
IBP	Istituto Biochimico Pavese s.p.a., Pavia, Italy
Ichthyol	Ichthyol-Gesellschaft Cordes, Hermanni & Co., Hamburg, Germany
ICI	Imperial Chemical Industries, Ltd., London, UK
ICN	ICN Pharmaceuticals Inc., Costa Mesa, California
IFCI	IFCI (Industria Farmaceutica Cosmetica Italiana S.p.A.), Bologna, Italy
Ika	Ikapharm, Ramat-Gan, Israel
IMC	Mallinckrodt Group, Inc. (formerly International Minerals & Chemical Corp.), St. Louis, Missouri,
Inava	Labs. Inava, Departement Médical de Pierre Fabre Médicament, Bologna, France
Indian Health	Indian Health Institute and Lab., Ltd., Calcutta, India
Innothera	Innothéra, Arcueil, France
Inpharzam	Inpharzam S.A. (subsidiary of Zambon), Brussels, Belgium
Intersan	Intersan, Institut für pharmazeutische und klinische Forschung GmbH, Ettlingen, Germany
Intersint	Intersint Italiana s.r.l. (subsidiary of Fisons), Rome, Italy
Intervet	Intervet UK Ltd. (subsidiary of Akzo Nobel), Cambridge, UK
INTES	INTES (Industria Terapeutica Splendore), Casoria, Italy
Inverni	Inverni Della Beffa S.p.A., Milan, Italy
Ipsen	Institut de Produits de Synthese et d'Extraction Naturelle, Paris, France
IRBI	IRBI S.p.A., Rome, Italy
Iromedica	Iromedica AG, St. Gallen, Switzerland
Isagro	Isagro S.p.A., Milan, Italy
Isei	Isei Co., Ltd., Yamagata, Japan
ISF	ISF S.p.A., Milan, Italy
ISI	Istituto Sierovaccinogeno Italiano S.p.A., Castelvecchio Pascoli, Italy
Isis	Isis Pharma GmbH (subsidiary of Schwarz Pharma), Zwickau, Germany
ISK Biosci	ISK Biosciences Corp., Mentor, Ohio
ISM	Istituto Sieroterapico Milanese, Milan, Italy
Isnardi	Pietro Isnardi & C., S.p.A., Imperia Oneglia, Italy
Isola-Ibi	Isola-Ibi, Genova-Quarto, Italy
I.S.O.M.	I.S.O.M. S.p.A., Milan, Italy
Ist. Chim. Inter	Istituto Chimico Internazionale di Dr. Giuseppe Rendre s.r.l., Rome, Italy

Italchimici	Italchimici S.p.A. (subsidiary of Fisons), Rome, Italy
Italfarmaco	Italfarmaco S.p.A., Milan, Italy
Ital Suisse	Ital Suisse Co. s.a.s. di Giancarlo Ceroni & C., Casarile, Italy
Itherapia	Itherapia Pharmavertrieb GmbH, Munich, Germany
Iwaki	Iwaki Seiyaku Co., Ltd., Tokyo, Japan
J & J	Johnson & Johnson, New Brunswick, New Jersey
Janssen	Janssen Pharmaceutica N.V. (subsidiary of Johnson & Johnson), Beerse, Belgium
Jenapharm	Jenapharm GmbH, Jena, Germany
Jouveinal	Labs. Jouveinal, Fresnes, France
Jugoremedija	Jugoremedija, Belgrade, Yugoslavia
Kabi	Kabi Pharmacia AG (subsidiary of Pharmacia AB), Dubendorf, Sweden
KabiVitrum	KabiVitrum Ltd., Uxbridge, Middlesex, UK
Kade	Dr. Kade Pharmazeutische Fabrik GmbH, Berlin, Germany
Kaigai	Kaigai Seiyaku Co., Ltd., Tokyo, Japan
Kaken	Kaken Pharmaceutical Co., Ltd., Chiba, Japan
Kali-Chemie	Kali-Chemie AG (subsidiary of Solvay S.A.), Hannover, Germany
Kanebo	Kanebo Pharmaceuticals, Ltd., Tokyo, Japan
Karlspharma	Karlspharma pharmazeutische Produkte GmbH, Karlsruhe, Germany
Katwijk	Katwijk Farma B.V., Katwijk aan Zee, Netherlands
Kayaku	Kayaku Co., Ltd., Tokyo, Japan
Kelco	Kelco Company (subsidiary of Monsanto), San Diego, California
Key	Key Pharmaceuticals, Inc. (subsidiary of Schering-Plough), Kenilworth, New Jersey
Kissei	Kissei Pharmaceutical Co., Ltd., Matsumoto, Japan
Klinge	Klinge Pharma GmbH (subsidiary of Fujisawa Pharm. Co., Ltd.), Munich, Germany
Klinge-Nattermann	Klinge-Nattermann Puren GmbH, Munich, Germany
Knoll	Knoll AG (subsidiary of BASF), Ludwigshafen, Germany
Kodama	Kodama Ltd., Tokyo, Japan
Kohler	Dr. Franz Kohler Chemie GmbH, Alsbach-Hähnlein, Germany
Kotobuki	Kotobuki Seiyaku, Nagano, Japan
Kowa	Kowa Company Ltd., Nagoya, Japan
Kowa Yakuhin	Kowa Yakuhin Kogyo, Tokyo, Japan
Kremers-Urban	Kremers-Urban Co., Milwaukee, Wisconsin
Kreussler	Chemische Fabrik Kreussler & Co. GmbH, Wiesbaden, Germany
Krewel	Krewel-Werke GmbH, Eitorf, Germany
Kureha	Kureha Kagaku, Tokyo, Japan
Kwizda	F. J. Kwizda Chemische Fabrik, Vienna, Austria
Kyorin	Kyorin Pharmaceutical Co., Ltd., Tokyo, Japan
Kyoritsu	Kyoritsu Pharmaceutical Co., Ltd., Nara, Japan
Kyowa	Kyowa Hakko Kogyo Co., Ltd., Tokyo, Japan
Kyowa Yakuhin	Kyowa Yakuhin, Osaka, Japan
L.A.B.	Laboratories for Applied Biology, Ltd., London, UK
Labatec	Labatec-Pharma S.A., Geneva, Switzerland
Labaz	Lab. Labaz (subsidiary of Sanofi Winthrop), Paris, France
Labcatal	Lab. Labcatal, Montrouge, France
Labinca	Laboratorios Labinca S.A., Buenos Aires, Argentina
Labopharma	Labopharma, Chem.-Pharm. Fabrik, GmbH, Berlin, Germany
Lacer	Lacer S.A., Barcelona, Spain
Lafare	Lab. Farmaceutico Reggiano s.r.l., Ercolana Resina, Italy
Lafi	Lafi, Sao Paulo, Brazil
Lafon	Labs. Lafon, Maisons-Alfort, France
Lagap	Lagap Pharmaceuticals Ltd., Bordon, Hants., UK
Lakeside	Lakeside Labs. (subsidiary of Merrell Dow), Cincinnati, Ohio
Lampugnani	Lampugnani Farmaceutici S.p.A., Milan, Italy
Lannett	The Lannett Co., Inc., Philadelphia, Pennsylvania
Lappe	Produpharm Lappe, Bergish Gladbach, Germany
Laroche Navarron	Lab. Syntex (formerly Laroche Navarron), Puteaux, France
Lasa	Lasa Labs., Barcelona, Spain
Laser	Laser, Inc., Crown Point, Indiana
Latema	Labs. de Therapeutique Moderne, LTM, Suresnes, France
Lazar	Dr. Lazar y Cia.S.A.Q.eI., Buenos Aires, Argentina
Leciva	Leciva, Prague, Czech Republic
Lederle	Lederle Labs. (subsidiary of American Cyanamid Co.), Wayne, New Jersey
Lefrancq	Lab. Soekami Lefrancq, St-Ouen, France
Lek	Lek, Ljubljana, Yugoslavia
Lemmon	Lemmon Co. (subsidiary of Teva Pharm. Ind. Ltd.), Sellersville, Pennsylvania
Lemoine	Labs. Lemoine, Courbevoie, France
Lenza	Farmaceutici Lenza s.r.l., Naples, Italy
Leo Pharm	Leo Pharmaceutic Products, Ballerup, Denmark
Lepetit	Gruppo Lepetit S.p.a. (subsidiary of Hoechst Marion Roussel), Milan, Italy
Lichtenstein	Lichtenstein Pharmaceutica GmbH, Koblenz-Kesselheim, Germany
Lifepharma	Lifepharma s.r.l., Milan, Italy
Lilly	Eli Lilly & Co., Indianapolis, Indiana
Lindopharm	Lindopharm GmbH, Hilden, Germany
Lipha	Lipha Santé S.A., Lyon, France
Lipo	Lipo Chemicals, Inc., Paterson, New Jersey

Lisapharma	Lisapharma S.p.A., Como, Italy
Locatelli	Farmaceutici Locatelli s.r.l., Rome, Italy
Logeais	Labs. Jacques Logeais, Issy-les-Moulineaux, France
Lotus	Lotus Biochem. Corp., Radford, Virginia
Lucien	Lab. Therabel Lucien Pharma, Neuilly-sur-Seine, France
Luitpold	Luitpold Pharma GmbH, Munich, Germany
Lundbeck	H. Lundbeck A/S, Copenhagen, Denmark
Lusofarmaco	Istituto Luso Farmaco d'Italia S.p.A., Milan, Italy
Lyocentre	Lab. Lyocentre (subsidiary of Akzo Nobel), Aurillac, France
Lysoform	Lysoform Dr. Hans Rosemann GmbH, Berlin, Germany
Lyssia	Lyssia GmbH (subsidiary of Kali-Chemie), Hannover, Germany
3M	Minnesota Mining & Manufacturing Co., St. Paul, Minnesota
3M Pharma	3M Pharmaceuticals, St. Paul, Minnesota
M & B	Rhône-Poulenc Rorer (formerly May & Baker, Ltd.), Eastbourne, E. Sussex, UK
Macfarlan Smith	Macfarlan Smith Ltd., Edinburgh, UK
Mack, Illert.	Heinrich Mack Nachf., Chem.-Pharm. Fabrik (subsidiary of Pfizer Inc.), Illertissen, Germany
Madaus	Madaus AG, Cologne, Germany
Maggioni	Maggioni-Winthrop Farmaceutici S.p.A., Milan, Italy
Magis	Lab. Magis Farmaceutici S.p.A., Brescia, Italy
Makara	Makara GmbH, Dusseldorf, Germany
Makhteshim-Agan	Makhteshim-Agan, Beer-Sheva, Israel
Malesci	Malesci Istituto Farmacobiologico, S.p.A., Florence, Italy
Mallard	Mallard, Inc., Detroit, Michigan
Mallinckrodt	Mallinckrodt, Inc., St. Louis, Missouri
Manetti Roberts	L. Manetti & H. Roberts & C. per Azioni, Florence, Italy
Mann	Dr. Gerhard Mann, Chem.-Pharm. Fabrik GmbH (subsidiary of Bausch & Lomb), Berlin, Germany
Marion	Hoechst Marion Roussel (formerly Marion Labs., Inc.), Kansas City, Missouri
Martinet	Lab. Martinet, Paris, France
Maruishi	Maruishi Pharmaceutical Co., Ltd., Osaka, Japan
Maruko	Maruko Seiyaku Co., Ltd., Nagoya, Japan
Master	Master Pharma s.r.l., Parma, Italy
Maurer	Maurer-Pharma GmbH, Saarbrücken, Germany
Maurry Biol.	H. E. Maurry Biological Co., Inc., Los Angeles, California
Max Ritter	Max Ritter Pharma S.A., Gland, Switzerland
Mayrand	Mayrand Inc., Winston-Salem, North Carolina
Mazer	Mazer Chemicals, Inc., Gurnee, Illinois
McGaw	McGaw, Inc. (subsidiary of IVAX Corp.), Irvine, California
McNeil	McNeil Lab., Inc. (subsidiary of Johnson & Johnson), Fort Washington, Pennsylvania
MCP	MCP Pharmaceuticals Ltd., Livingston, West Lothian, UK
Mead Johnson	Mead Johnson Pharmaceuticals (subsidiary of Bristol-Myers Squibb), Princeton, New Jersey
Mect	Mect Corp, Tokyo, Japan
Medac	Medac Gesellschaft für klinische Spezialpräparate mbH, Hamburg, Germany
Medial	3M Pharma (formerly Labs. Medial-Riker), Pithiviers, France
Medica	Medica B.V., Hertogenbosch, Netherlands
Medice	Medice, Chem.-Pharm. Fabrik Pütter GmbH & Co. KG, Iserlohn, Germany
Medichemie	Medichemie AG, Ettingen, Switzerland
Medici	Lab. Farm. Dr. Medici s.r.l., Rome, Italy
Medicia	Aron-Médicia (division of Lipha Santé), Lyon, France
Medici Domus	Medici Domus s.r.l., Vittore-Olona, Italy
Medinova	Medinova AG, Zurich, Switzerland
Mediolanum	Mediolanum Farmaceutici S.p.A., Milan, Italy
Medi-Physics	Medi-Physics, Inc. (subsidiary of Amersham International plc), Arlington Heights, Illinois
Medo	Medo Pharmaceuticals Ltd. (subsidiary of Schwarz Pharma AG), London, UK
Medosan	Medosan s.r.l. Industrie Biochimiche Riunite, Cecchina, Italy
Meiji	Meiji Seika Kaisha, Ltd., Tokyo, Japan
Meiji Yakuhin	Meiji Yakuhin, K.K., Tokyo, Japan
Mekos	Mekos A/S, Hovik, Norway
Menarini	A. Menarini Industrie Farmaceutiche Riunite s.r.l., Florence, Italy
Mendelejeff	Stabilimento chimico Farmaceutico Mendelejeff s.r.l., Rome, Italy
Mentholatum	Mentholatum Co., Buffalo, New York
Mepha	Mepha Pharma AG, Aesch, Switzerland
Mepros	Mepros, Bladel, Netherlands
Merck & Co.	Merck & Co., Inc., Whitehouse Station, New Jersey
Merck-Clevenot	Labs. Merck-Clevenot (subsidiary of E. Merck AG), Nogent-sur-Marne, France
Merckle	Merckle GmbH, Blaubeuren, Germany
Merieux	Institut Merieux (subsidiary of Rhône-Poulenc), Lyon, France
Merrell	Hoechst Marion Roussel (formerly Marion Merrell Dow Pharmaceuticals), Kansas City, Missouri
Merz	Merz & Co., GmbH & Co., Frankfurt/Main, Germany
Meuselbach	Meuselbach Pharma GmbH, Meuselbach, Germany
Meyer	Meyer Labs., Inc., Fort Lauderdale, Florida
Midy	Sterling-Midy, Paris, France

Mikasa	Mikasa Seiyaku Co., Ltd., Tokyo, Japan
Miles	Miles, Inc. (subsidiary of Bayer AG), Westhaven, Connecticut
Millot-Solac	Sanofi Winthrop (formerly Labs. Millot-Solac), Gentilly, France
Minden	Minden Pharma GmbH (subsidiary of Knoll AG), Minden, Germany
Miquel	Laboratorios Miquel S.A. (subsidiary of Otsuka Pharm. Co., Ltd.), Barcelona, Spain
Misemer	Misemer Pharmaceuticals, Inc., Springfield, Missouri
Mission Pharmacal	Mission Pharmacal Co., San Antonio, Texas
Mitsubishi	Mitsubishi Kasei Corp., Tokyo, Japan
Mitsui	Mitsui Pharmaceuticals, Inc., Tokyo, Japan
Mobil	Mobil Oil Corp., Fairfax, Virginia
Mochida	Mochida Pharmaceutical Co., Ltd., Tokyo, Japan
Mohan	Mohan Medicine Research Inst., Tokyo, Japan
Molimin	Molimin Arzneimittel GmbH, Hallstadt/Ofr., Germany
Molteni	Molteni L. & C. dei F. Ili Alitti Societa Esercizio S.p.A., Scandicci, Italy
Monsanto	Monsanto Chemical Co., St. Louis, Missouri
Morishita	Morishita Pharmaceutical Co., Ltd., Osaka, Japan
Morton-Norwich	Morton International, Inc. (formerly Morton-Norwich Products, Inc.), Chicago, Illinois
MSD	MSD Sharp & Dohme GmbH (subsidiary of Merck & Co., Inc.), Haar, Germany
Much	Much Pharma AG (subsidiary of Am. Home Prod. Corp.), Haan, Germany
Mulli	Dr. Mulli Pharma GmbH & Co., KG, Neuenburg, Germany
Mundipharma	Mundipharma GmbH, Limburg/Lahn, Germany
Nagase	Nagase Iyaku, Osaka, Japan
Napp	Napp Labs. Ltd., Cambridge, UK
Nativelle	Procter & Gamble Pharm. France (formerly Lab. Nativelle S.A.), Neuilly-sur-Seine, France
Nattermann	A. Nattermann & Cie, GmbH (subsidiary of Rhône-Poulenc), Cologne, Germany
NEGMA	Labs. NEGMA, Toussus-le-Noble, France
Nelson	Nelson Research & Development, Irvine, California
Neos	Neos Donner KG, Berlin, Germany
Newport	Newport Pharmaceuticals, Inc., Newport Beach, California
Nichiiko	Nihon Iyakuhin Kogyo Co., Ltd., Toyama, Japan
Nicholas	Nicholas Labs., Ltd. (subsidiary of Hoffmann-La Roche), Welwyn Garden City, UK
Nigy	Lab. Nigy, Paris, France
Nikken	Nikken Chemicals Co., Ltd., Tokyo, Japan
Nion	Nion Corp., El Monte, California
Nippon Chemiphar	Nippon Chemiphar Co., Ltd., Tokyo, Japan
Nippon Kayaku	Nippon Kayaku Co., Ltd., Tokyo, Japan
Nippon Shinyaku	Nippon Shinyaku Co., Ltd., Kyoto, Japan
Nippon Shoji	Nippon Shoji Kaisha, Ltd., Osaka, Japan
Nippon Soda	Nippon Soda Co., Ltd., Tokyo, Japan
Nippon Zoki	Nippon Zoki Pharmaceutical Co., Ltd., Osaka, Japan
Nissan	Nissan Kagaku Kogyo K.K., Tokyo, Japan
Nisshin	Nisshin Pharmaceutical Co., Ltd., Osaka, Japan
Nissui	Nissui Seiyaku Co., Ltd., Tokyo, Japan
Nor-Am	AgrEvo GmbH (formerly Nor-Am Chemical Co.), Wilmington, Delaware
Norbrook	Norbrook Labs. Ltd., Kidderminster, Worcs., UK
Norden	Norden Labs. Inc. (subsidiary of SmithKline Beecham), Lincoln, Nebraska
Nordic	Nordic Labs., Quebec, Canada
Nordmark	Nordmark Arzneimittel GmbH (subsidiary of Knoll AG), Uetersen/Holstein, Germany
Norgine	Norgine GmbH, Marburg, Germany
Norton	H. N. Norton & Co., Ltd., Harlow, Essex, UK
Norwich	Procter & Gamble Pharmaceuticals, Inc. (formerly Norwich-Eaton Pharmaceuticals), Cincinnati, Ohio
Nourypharma	Nourypharma B.V. (subsidiary of Akzo Nobel), Oss, Netherlands
Novo	Novo Nordisk A/S, Bagsvaerd, Denmark
Novocol	Novocol Chemical Mfg. Co., Inc., Hicksville, New York
Nuovo Cons.	Nuovo Consorzio Sanitario Nazionale di Malizia Dr. Paolo, Rome, Italy
Nycomed	Hafslund Nycomed A/S, Oslo, Norway
Nyegaard	Nyegaard & Co., A/S, Oslo, Norway
Oberval	Lab. Oberval (division of Lipha Santé), Lyon, France
Ohmeda	Ohmeda Pharm. Prods. Div., Liberty Corner, New Jersey
Ohta	Ohta Pharmaceutical Co., Ltd., Tokyo, Japan
Olin	Olin Corp., Stamford, Connecticut
OM	Lab. OM, S.A., Geneva, Switzerland
Ono	Ono Pharmaceutical Co., Ltd., Osaka, Japan
Opfermann	Opfermann Arzneimittel GmbH, Wiehl, Germany
OPW	Oranienburger Pharmawerk GmbH, Oranienburg, Germany
Organon	Organon International B.V. (subsidiary of Akzo Nobel), Oss, Netherlands
Orion	Orion Corporation, Espoo, Finland
Ortega	Ortega Pharm. Co., Jacksonville, Florida
Ortho	Ortho-McNeil Pharmaceutical Corp. (subsidiary of Johnson & Johnson), Raritan, New Jersey
Otsuka	Otsuka Pharmaceutical Co., Ltd., Tokyo, Japan
OTW	Organotherapeutische Werke GmbH, Ettlingen, Germany
Paines & Byrne	Paines & Byrne Ltd., Byfleet, Surrey, UK
Panthox & Burck	Panthox & Burck Istituto Biochimico Italo Svizzero S.p.A., Milan, Italy

Parke, Davis	Parke, Davis & Co. (division of Warner-Lambert Pharmaceutical Co.), Morris Plains, New Jersey
Parmed	Parmed Pharmaceuticals, Niagara Falls, New York
Penick	Penick Corp., Newark, New Jersey
Penn	Penn Pharmaceuticals Ltd., High Wycombe, Bucks, UK
Pennwalt	Elf Atochem North America, Inc. (formerly Pennwalt Corp.) (subsidiary of Elf Aquitaine), Philadelphia, Pennsylvania
Penreco	Penreco (division of Pennzoil Products Co.), Karns City, Pennsylvania
Permamed	Permamed AG, Basel, Switzerland
Person & Covey	Person & Covey, Inc., Glendale, California
Pfizer	Pfizer Inc., New York, New York
Pfleger	Dr. R. Pfleger Chemische Fabrik GmbH, Bamberg, Germany
Pharmachim	Pharmachim, Sofia, Bulgaria
Pharmacia	Pharmacia & Upjohn, Inc., London, UK
Pharma Investi	Pharma Investi S.A., Madrid, Spain
Pharmascience	Laboratoires Pharmascience, Courbevoie, France
Pharma Selz	Pharma Selz GmbH, Florsheim/Dalsheim, Germany
Pharma-Stern	Pharma-Stern GmbH, Wedel/Holstein, Germany
Pharma Wernigerode	Pharma Wernigerode GmbH, Wernigerode, Germany
Pharmax	Pharmax Ltd., Bexley, Kent, UK
Pharm. Mfg.	Pharmaceutical Manufacturing Company, Bolton, Lanc., UK
Pharmuka	Labs. Roger Bellon (formerly Pharmuka S.F.) (subsidiary of Rhône-Poulenc Rorer), Genevilliers, France
Piam	Vecchi & C. Piam di G. Assereto, Genoa, Italy
Pickles	J. Pickles & Sons, Knaresborough, N. Yorks, UK
Pierrel	Pierrel, S.p.A., Milan, Italy
Pirri	Istituto Pirri s.r.l., Milan, Italy
Pitman-Moore	Pitman-Moore Inc. (subsidiary of Mallinckrodt), Mundelein, Illinois
Plantorgan	Plantorgan GmbH & Co. OHG, Bad Zwischenahn, Germany
Plant Protection	Zeneca Agrochemicals (formerly Plant Protection Ltd.), Haslemere, Surrey, UK
Plessner	The Paul Plessner Co., St. Petersburg, Florida
Pliva	Pliva, Zagreb, Yugoslavia
Plough	Plough USA (subsidiary of Schering-Plough), Memphis, Tennessee
Pohl	G. Pohl-Boskamp GmbH & Co., Hohenlocksted, Germany
Polfa	Polfa, Warsaw, Poland
Poli	Poli Industria Chimica, S.p.A., Milan, Italy
Polifarma	Polifarma S.p.A., Rome, Italy
Poythress	Poythress Labs., Inc., Richmond, Virginia
Procter & Gamble	Procter & Gamble, Cincinnati, Ohio
Prodes	Laboratorios Prodes, S.A., Barcelona, Spain
Promedica	Lab. Promedica, La-Chaussée-St-Victor, France
Promonta	Chemische Fabrik Promonta GmbH, Hamburg, Germany
Protea	Protea Pharmaceuticals (subsidiary of Fisons), Pennant Hills, NSW, Australia
Proter	Proter S.p.A., Milan, Italy
Protina	Protina Chemische GmbH, Ismaning, Germany
Provet	Provet AG, Lyssach, Switzerland
Pulitzer	Pulitzer Italiana S.p.A., Rome, Italy
Purdue Frederick	The Purdue Frederick Co., Inc., Norwalk, Connecticut
Puropharma	Puropharma s.r.l., Milan, Italy
PVO Intl	Stepan Co. (formerly PVO International, Inc.), Maywood, New Jersey
Quinoderm	Quinoderm Ltd., Oldham, Lanc., UK
Rachelle	Rachelle Labs., Inc., Culver, Indiana
Radiumpharma	Radiumpharma s.r.l., Milan, Italy
RAFA	RAFA Labs., Ltd., Jerusalem, Israel
Ralston-Purina	Purina Mills, Inc., St. Louis, Missouri
R.A.N.	R.A.N. Novesia Arzneimittel GmbH, Neuss, Germany
Ratiopharm	Ratiopharm GmbH & Co., Ulm/Donautal, Germany
Ravensberg	Ravensberg GmbH Chemische Fabrik, Konstanz, Germany
Ravizza	Ravizza S.p.A. (subsidiary of Knoll AG), Muggio (Milan), Italy
Reckitt & Colman	Reckitt & Colman Ltd., N. Humberside, UK
Recordati	Recordati Industria Chimica e Farmaceutica S.p.A., Milan, Italy
Redel	Julius Redel, Cesra-Arzneimittelfabrik GmbH & Co., Baden-Baden, Germany
Reed & Carnrick	Reed & Carnrick (division of Block Drug, Inc.), Jersey City, New Jersey
Reid-Provident	Reid-Rowell, Inc. (formerly Reid-Provident Labs., Inc.), Marietta, Georgia
Reiss	Dr. Rudolf Reiss Chemische Werke GmbH & Co. KG, Cologne, Germany
Rentschler	Dr. Rentschler Arzneimittel GmbH & Co., Laupheim, Germany
Rexall	Rexall Group, Boca Raton, Florida
Rhein-Pharma	Rhein-Pharma GmbH, Plankstadt, Germany
Rhone-Poulenc	Rhône-Poulenc S.A., Courbevoie, France
Riker	3M Pharmaceuticals (formerly Riker Labs.) (subsidiary of the 3M Company), St. Paul, Minnesota
Riom	Riom Labs.-CERM, Riom, France
Ripari-Gero	Istituto Farmaco Biologico Ripari-Gero, S.p.A., Monteriggioni, Italy
Ritsert	Dr. E. Ritsert, GmbH & Co., KG, Eberbach, Germany
Robeco	Robeco Chemicals, Inc., New York, New York
Robert et Carriére	Lab. Robert et Carriére (subsidiary of Synthélabo), Paris, France

Roberts	Roberts Pharmaceutical Corp., Eatontown, New Jersey
Robins	A. H. Robins Co., Inc. (subsidiary of Am. Home Prod. Corp.), Richmond, Virginia
Robisch	Chemische Fabrik G. Robisch GmbH, Munich, Germany
Robugen	Robugen GmbH Pharmazeutische Fabrik, Eslingen, Germany
Roche	Hoffmann-La Roche Ltd., Basel, Switzerland
Rodleben	Rodleben Pharma GmbH, Rodleben, Germany
Roemmers	Roemmers S.A.I.C.F., Buenos Aires, Argentina
Roerig	Roerig (division of Pfizer Inc.), New York, New York
Roger	Labs. Roger S.A., Barcelona, Spain
Roger Bellon	Labs. Roger Bellon (subsidiary of Rhône-Poulenc Rorer), Neuilly-sur-Seine, France
Rohm	Rohm Pharma GmbH, Darmstadt, Germany
Rohm & Haas	Rohm & Haas Co., Philadelphia, Pennsylvania
Roland	Roland Arzneimittel GmbH (subsidiary of Byk-Gulden), Hamburg, Germany
Rolland	Albert Rolland S.A. (subsidiary of E. Merck), Paris, France
Rorer	Rhône-Poulenc Rorer, Inc. (subsidiary of Rhône-Poulenc), Collegeville, Pennsylvania
Rosa-Phytopharma	Lab. Rosa-Phytopharma S.A., Paris, France
Rosemont	Rosemont Pharm. Ltd., Leeds UK
Ross	Ross Labs. (division of Abbott Labs.), Columbus, Ohio
Roter	Roter B.V., Hilversum, Netherlands
Rotta	Rotta Farmaceutici S.p.A. (subsidiary of Rhône-Poulenc S.A.), Milan, Italy
Roussel	Lab. Roussel (subsidiary of Roussel-UCLAF), Paris, France
Roussel-UCLAF	Roussel-UCLAF S.A. (subsidiary of Hoechst), Paris, France
Roux-Ocefa	Labs. Roux-Ocefa S.A., Buenos Aires, Argentina
Rowa	Rowa Ltd., Bantry, Co. Cork, Ireland
Rowell	Rowell Labs., Inc., Baudette, Minnesota
Roxane	Roxane Labs., Inc. (subsidiary of Boehringer, Ing.), Columbus, Ohio
R. P. Drugs	R. P. Drugs Ltd., Leeds, Yorkshire, UK
RPR	Rhône-Poulenc Rorer, Inc. (subsidiary of Rhône-Poulenc S.A.), Collegeville, Pennsylvania
Rybar	Rybar Labs., Ltd., Andover, Hants, UK
Rystan	Rystan Co., Inc. (subsidiary of Carnrick Labs., Inc.), Cedar Knolls, New Jersey
Sagitta	Sagitta Arzneimittel GmbH, Feldkirchen-Westerham, Germany
Salsbury	Salsbury Labs. Inc. (subsidiary of Solvay S.A.), Charles City, Iowa
Salus	Salus Researches S.p.A., Rome, Italy
Salutas Fahlberg-List	Salutas Fahlberg-List Pharma GmbH, Magdeburg, Germany
S.A.L.V.A.T.	S.A.L.V.A.T., S.A., Laboratorios, Barcelona, Spain
Samil	Samil S.p.A. (subsidiary of Sandoz), Rome, Italy
Sana	San-a Yakuhin Kogyo Co., Ltd., Tokyo, Japan
San Carlo	San Carlo Farmaceutici S.p.A., Rome, Italy
Sandoz	Sandoz Ltd., Basel, Switzerland
Sanko	Sanko Seiyaku Kogyo Co., Ltd., Tokyo, Japan
Sankyo	Sankyo Co., Ltd., Tokyo, Japan
Sankyo Zoki	Sankyo Zoki Co., Tokyo, Japan
Sanofi	Sanofi S.A. (subsidiary of Elf Aquitaine), Paris, France
Sanofi Winthrop	Sanofi Winthrop, Inc. (division of Sanofi S.A.), New York, New York
Sanol	Sanol GmbH (subsidiary of Schwarz Pharma), Monheim, Germany
Sanorania	Sanorania GmbH (subsidiary of Upjohn), Berlin, Germany
Sanraku	Sanraku Inc., Tokyo, Japan
Santen	Santen Pharmaceutical Co., Ltd., Osaka, Japan
Sanwa	Sanwa Kagaku Kenkyusho Co., Ltd., Nagoya, Japan
Sanzen	Sanzen Pharmaceutical Co., Ltd., Ohmiya, Japan
Sapos	Sapos S.A., Geneva, Switzerland
Sarm	Sarm s.r.l., Rome Italy
Sarva	Sarva-Syntex S.A., Brussels, Belgium
Sato	Sato Pharmaceutical Co., Ltd., Tokyo, Japan
Sato Yakuhin	Sato Yakuhin Kogyo Co., Ltd., Kashiwara, Japan
Sauba	Lab. Sauba, Montreuil, France
Saunier-Daguin	Labs. Saunier-Daguin, Cléry-St-André, France
Sauter	Labs. Sauter S.A. (subsidiary of Hoffmann-La Roche), Geneva, Switzerland
Savage	Savage Labs. (division of Altana), Melville, New York
Savoma	Savoma Medicinali S.p.A., Parma, Italy
Sawai	Sawai Pharmaceutical Co., Ltd., Osaka, Japan
Scarium	Scarium AG, Lucerne, Switzerland
Schaper & Brummer	Schaper & Brummer GmbH & Co., KG, Salzgitter (Ringelheim), Germany
Schering	Schering-Plough Corp., Kenilworth, New Jersey
Schering AG	Schering AG, Berlin, Germany
Scheurich	E. Scheurich, Pharmwerk GmbH, Appenweier, Germany
Schiapparelli	Schiapparelli Searle s.r.l., Milan, Italy
Schülke & Mayr	Schülke & Mayr GmbH, Norderstedt, Germany
Schwabe	Dr. Schwabe GmbH, Munich, Germany
Schwarz	Schwarz Pharma AG, Monheim, Germany
Schwarzhaupt	Kommanditgesellschaft Schwarzhaupt GmbH & Co., Cologne, Germany
Sclavo	Sclavo S.p.A., Siena, Italy
Scott	O. M. Scott and Sons Co., Marysville, Ohio
Searle	G. D. Searle & Co. (subsidiary of Monsanto), Skokie, Illinois
Seiko	Seikoeiyo Yakuhin K.K., Osaka, Japan
Senju	Senju Pharmaceutical Co., Ltd., Osaka, Japan

Septa	Septa Chemifarma, Barcelona, Spain
Serono	Ares-Serono NV, Geneva, Switzerland
Serpero	Serpero, Industria Galenica Milanese S.p.A., Milan, Italy
Servier	Lab. Servier, Gidy, France
Servipharm	Servipharm Ltd. (subsidiary of Ciba-Geigy), Basel, Switzerland
Shell	Shell Oil Co., Houston, Texas
Shionogi	Shionogi & Co., Ltd., Osaka, Japan
Showa Shinyaku	Showa Yakka, Tokyo, Japan
Showa Yakuhin	Showa Yakuhin Kako Co., Ltd., Tokyo, Japan
Sidus	Instituto Sidus I.C.S.A., Buenos Aires, Argentina
SIFI	Societa Industria Farmaceutica Italiana S.p.A., Catania, Italy
Sigma	Sigma Chemical Co. (subsidiary of Sigma-Aldrich Corp.), St. Louis, Missouri
Sigma-Tau	Sigma-Tau Industrie Farmaceutiche Riunite S.p.A., Rome, Italy
Sigurta	Sigurta s.r.l., Milan, Italy
Silva Araújo Roussel	Labs. Silva Araújo Roussel S.A., Rio de Janeiro, Brazil
SIMES	Simes S.p.A., Milan, Italy
Sinclair	Sinclair Pharmaceuticals Ltd., Godalming, Surrey, UK
Sintopharma	Labs. Sintopharma S.A., Sao Paulo, Brazil
Sintyal	Sintyal S.A., Buenos Aires, Argentina
Sis-Ter	Sistemi Terapeutici S.p.A., Palazzo Pignano, Italy
SIT	Specialita Igienico Terapeutiche s.r.l., Mede (Pavia), Italy
S&K	S&K Pharma, Schumann u. Kohl GmbH, Perl, Germany
SK & F	SmithKline Beecham plc (formerly Smith Kline & French Labs.), Brentford, Middlesex, UK
SKB	SmithKline Beecham plc, Brentford, Middlesex, UK
Smith & Nephew	Smith & Nephew Pharmaceuticals Ltd., Romford, Essex, UK
Sobio	Lab. Sobio (subsidiary of SmithKline Beecham), Paris, France
Solvay	Solvay S.A., Brussels, Belgium
SPA	Societa Prodotti Antibiotici S.p.A., Milan, Italy
Specia	Specia (formerly Societe Parisienne d'Expansion Chimique), (subsidiary of Rhône-Poulenc), Paris, France
Speywood	Speywood Labs., Ltd., Maidenhead, Berks, UK
Spitzner	W. Spitzner, Arzneimittelfabrik GmbH, Ettlingen, Germany
Spofa	Spofa, Prague, Czech Republic
Spret-Mauchant	Labs. Roger Bellon (formerly Spret-Mauchant) (subsidiary of Rhône-Poulenc Rorer), Neuilly-sur-Seine, France
Squibb	Bristol-Myers Squibb Co. (formerly E. R. Squibb & Sons), Princeton, New Jersey
Squibb-Heyden	Squibb-von Heyden GmbH (subsidiary of Bristol-Myers Squibb), Munich, Germany
SS Pharm.	SS Pharmaceutical Co., Ltd., Tokyo, Japan
Stada	Stada Arzneimittel AG, Bad Vilbel, Germany
Stafford-Miller	Stafford-Miller Ltd., Welwyn Garden City, Herts., UK
Standard Oil	BP America (formerly Standard Oil Co.), Cleveland, Ohio
Stauffer	Rhône-Poulenc Basic Chemicals Co. (formerly Stauffer Chemical Co.), Shelton, Connecticut
Steiner	Steiner & Co., Deutsche Arzneimittel Gesellschaft, Berlin, Germany
Stepan	Stepan Co., Northfield, Illinois
Sterling-Winthrop	Sanofi Winthrop, Inc. (formerly Sterling Winthrop, Inc.) (subsidiary of Sanofi S.A.), New York, New York
Sterwin	Sterwin Labs., Inc. (division of Pitman-Moore, Inc.), Millsboro, Delaware
Stiefel	Stiefel Labs., Inc., Coral Gables, Florida
Stockli	Dr. J. Stöckli, Netstal, Switzerland
Streuli	G. Streuli & Co. AG, Uznach, Switzerland
Stricker	Werner Stricker AG, Zollikofen, Switzerland
Stroder	Ist. Farmaco Biologico Stroder s.r.l., Florence, Italy
Stroschein	Pharma Stroschein GmbH, Hamburg, Germany
Stuart	Stuart Pharmaceuticals (division of Zeneca, Inc.), Wilmington, Delaware
Sudmedica	Sudmedica GmbH Chemisch-pharmazeutische Fabrik, Munich, Germany
Sumitomo	Sumitomo Pharmaceuticals Co., Ltd., Osaka, Japan
Syncro	Syncro Argentina S.A.Q.I.C.I.F., Buenos Aires, Argentina
Syntetic	A/S Syntetic, Brabrand, Denmark
Syntex	Syntex Labs., Inc. (subsidiary of Hoffmann-La Roche), Palo Alto, California
Synthelabo	Synthélabo S.A. (subsidiary of L'Oreal S.A.), Paris, France
Szabo	Szabo-Multipharma S.A., Buenos Aires, Argentina
TAD	TAD Pharmazeutisches Werk GmbH, Cuxhaven, Germany
Taiho	Taiho Pharmaceutical Co., Ltd., Tokyo, Japan
Taisho	Taisho Pharmaceutical Co., Ltd., Tokyo, Japan
Taiyo	Taiyo Pharmaceutical Industry Co., Ltd., Nagoya, Japan
Takata	Takata Pharmaceutical Co., Ltd., Tokyo, Japan
Takeda	Takeda Chem. Ind. Ltd., Osaka, Japan
Tanabe Seiyaku	Tanabe Seiyaku Co., Ltd., Osaka, Japan
TAP	TAP Pharmaceuticals Inc. (subsidiary of Takeda), Deerfield, Illinois
Taro	Taro Pharm USA, Hawthorne, New York
Tatsumi	Tatsumi Kagaku Co., Ltd., Kanazawa, Japan
Teikoku Kagaku	Teikoku Chemical Ind. Co., Ltd., Osaka, Japan
Teikoku Zoki	Teikoku Hormone Mfg. Co., Ltd., Osaka, Japan
Teisan	Teisan Seiyaku K.K., Tokyo, Japan

Temmler	Temmler Pharma GmbH (subsidiary of Asta Medica AG), Marburg, Germany
Tenneco	Tenneco Inc., Houston, Texas
Teva	Teva Pharmaceutical Industries Ltd., Petah Tikva, Israel
Thames	Thames Labs., Ltd., London, UK
Theranol	Theranol Deglaude, Bagneux, France
Theraplix	Theraplix S.A. (subsidiary of Rhône-Poulenc), Paris, France
Thiemann	Thiemann Arzneimittel GmbH (subsidiary of Akzo Nobel), Waltrop, Germany
Thomae	Dr. Karl Thomae GmbH (subsidiary of Boehringer Ingelheim), Biberach, Germany
Tiber	Tiber Prodotti Chimico Biologici s.r.l., Rome, Italy
Tillotts	Tillotts Labs., Henlow, Bedfordshire, UK
Toa Eiyo	Toa Eiyo Co., Ltd., Tokyo, Japan
Tobishi	Tobishi Pharmaceutical Co., Ltd., Tokyo, Japan
Tokyo Tanabe	Tokyo Tanabe Seiyaku Co., Ltd., Tokyo, Japan
Torii	Torii & Co., Ltd., Tokyo, Japan
Tosi	Dr. A. Tosi Farmaceutici s.r.l., Novara, Italy
Tosse	E. Tosse & Co. mbH (subsidiary of Byk-Gulden), Hamburg, Germany
Towa Seiyaku	Towa Seiyaku, Tokyo, Japan
Towa Yakuhin	Towa Yakuhin Co., Ltd., Kadoma, Japan
Toyama	Toyama Chemical Co., Ltd., Tokyo, Japan
Toyo Jozo	Toyo Jozo Co., Ltd., Tokyo, Japan
Toyo Pharmar	Toyo Pharmar Co., Ltd., Toyama, Japan
Travenol	Baxter International, Inc. (formerly Baxter Travenol Labs., Inc.), Deerfield, Illinois
Trommsdorff	H. Trommsdorff GmbH & Co. Arzneimittel (subsidiary of Ferrer), Alsdorf, Germany
Tropenwerke	Tropenwerke GmbH & Co. KG (subsidiary of Bayer AG), Koln, Germany
Tsuruhara	Tsuruhara Pharmaceutical Co., Ltd., Ikeda, Japan
Tubi Lux	Tubi Lux Farma, Rome, Italy
UCB	UCB, S.A. (formerly Union Chimique Belge), Brussels, Belgium
Ulmer	The Ulmer Pharmacal Co., Minneapolis, Minnesota
Unifa	Unifa S.A.Q.el., Buenos Aires, Argentina
Unimed	Unimed Pharmaceutical, Inc., Buffalo Grove, Illinois
Union Carbide	Union Carbide Corp., Danbury, Connecticut
Uniroyal	Uniroyal Chemical Co., Inc., Middlebury, Connecticut
Univet	Univet, Ltd., Bichester, Oxon, UK
UOP	UOP (formerly Universal Oil and Petroleum, Inc.) (unit of Union Carbide), Des Plaines, Illinois
Upjohn	Pharmacia & Upjohn, Inc., London, UK
UPSA	UPSA (subsidiary of Bristol-Myers Squibb), Rueil-Malmaison, France
Upsher-Smith	Upsher-Smith Labs., Inc., Minneapolis, Minnesota
Uriach	J. Uriach & Cia S.A., Barcelona, Spain
Ursapharm	Ursapharm Arzneimittel GmbH, Saarbrücken, Germany
Usafarma	Usafarma S.A., Sao Paulo, Brazil
U.S. Ethicals	Rhône-Poulenc Rorer (formerly U.S. Ethicals, Inc.), Collegeville, Pennsylvania
USV	USV Pharmaceuticals Corp., Tarrytown, New York
Vaillant-Defresne	Lab. Vaillant-Defresne, Courbevois, France
Valeas	Valeas S.p.A., Milan, Italy
Valpan	Lab. Valpan, Le Mee-sur-Seine, France
Vanderbilt	R. T. Vanderbilt Co., Inc., Norwalk, Connecticut
Van Dyk	Van Dyk & Company, Inc., Belleville, New Jersey
Vangard	Vangard Labs., Glasgow, Kentucky
Velsicol	Velsicol Chemical Corp., Rosemont, Illinois
Verla-Pharm	Verla-Pharm. Arzneimittelfabrik, Tutzing, Germany
Veterinaria	Veterinaria AG, Zurich, Switzerland
Veyron-Froment	Lab. Veyron et Froment, Marseille, France
Vifor	Vifor S.A., Geneva, Switzerland
Villela	Mauricio Villela S.A., Rio de Janeiro, Brazil
Villette	Lab. Villette (Dpt. des Labs. Pharmascience), Courbevoie, France
Vineland	Vineland Chemical Co., Vineland, New Jersey
Virbach	Virbach AG, Ksnacht, Switzerland
Vita	Vita Farmaceutici S.p.A., Turin, Italy
Vitacain	Vitacain Pharmaceutical Co., Ltd., Moriguchi, Japan
Viti	Marco Viti Farmaceutici S.p.A., Mozzate (CO), Italy
Voigt	Dr. med. Hans Voigt GmbH, Limburg/Lahn, Germany
Von Boch	Von Boch Arzneimittel s.r.l. Istituto Farmacobiologico, Rome, Italy
Wacker	Wacker-Chemie GmbH (subsidiary of Hoechst), Munich, Germany
Wakamoto	Wakamoto Pharmaceutical Co., Ltd., Tokyo, Japan
Walker	Walker Corp. & Co., Inc., Syracuse, New York
Wallace	Wallace Mfg. Chemists, Ltd., London, UK
Wallace Labs.	Wallace Labs. (division of Carter-Wallace, Inc.), Cranbury, New Jersey
Wampole	Wampole Labs. (division of Carter-Wallace, Inc.), Cranbury, New Jersey
Wander	Sandoz-Wander Pharma AG (subsidiary of Sandoz), Bern, Switzerland
Warner-Chilcott	Warner-Chilcott Labs. (subsidiary of Warner-Lambert Co.), Morris Plains, New Jersey
Warner-Lambert	Warner-Lambert Co., Morris Plains, New Jersey
Warner Wellcome	Lab. Warner Wellcome, Courbevoie, France
Wassermann	Sociedad Espanola de Espec. Farmaco-Terapeuticas Wassermann, S.A., Barcelona, Spain
W.B. Pharm.	Boehringer Ing., Ltd., Bracknell, Burks, UK

Weimer	Weimer Pharma GmbH, Rastatt, Germany
Welcker-Lyster	Welcker-Lyster Ltd., Montreal, Canada
Westwood-Squibb	Westwood-Squibb Pharmaceuticals, Inc. (subsidiary of Bristol-Myers Squibb Co.), Buffalo, New York
Whitby	Whitby Inc. (subsidiary of UCB), Richmond, Virginia
Whitehall	Whitehall Labs. (subsidiary of Am. Home Prod. Corp.), Madison, New Jersey
Willows	Willows Francis Vet Ltd., Crawley, W. Sussex, UK
Will-Pharma	Will-Pharma S.A., Brussels, Belgium
Winthrop	Sanofi Winthrop Inc. (formerly Winthrop Pharmaceuticals) (subsidiary of Sanofi S.A.), New York, New York
Winzer	Dr. Winzer, Chem.-Pharm. Fabrik GmbH, Konstanz, Germany
Wisconsin Pharmacal	Wisconsin Pharmacal Co. Inc., Jackson, Wisconsin
Woelfer	Otto A. H. Woelfer GmbH, Bovenau, Germany
Woelm	Woelm Pharma GmbH & Co. (subsidiary of Rhône-Poulenc), Eschwege, Germany
Wolff	Dr. August Wolff GmbH & Co. Arzneimittel, Bielefeld, Germany
Wyeth	Wyeth-Ayerst Labs. (division of Am. Home Prod. Corp.), Philadelphia, Pennsylvania
Xttrium	Xttrium Labs., Inc., Chicago, Illinois
Yamagata	Yamagata Seiyaku Ltd., Yamagata, Japan
Yamanouchi	Yamanouchi Pharmaceutical Co., Ltd., Tokyo, Japan
Yoshitomi	Yoshitomi Pharmaceutical Ind. Ltd., Osaka, Japan
Zambeletti	Dr. L. Zambeletti S.p.A. (subsidiary of SmithKline Beecham), Milan, Italy
Zambon	Zambon Group S.p.A., Vicenza, Italy
Zdravlje	Zdravlje, Leskovac, Yugoslavia
Zeneca	Zeneca Pharmaceutics, Cheshire, UK
Zeneca Agro	Zeneca Agrochemicals, Haslemere, Surrey, UK
Zeria	Zeria Pharmaceutical Co., Ltd., Tokyo, Japan
Zoecon	Zoecon Corp. (subsidiary of Sandoz), Dallas, Texas
Zoja	Zoja Giorgio S.p.A., Milan, Italy
Zyma	Zyma S.A. (subsidiary of Ciba-Geigy), Nyon, Switzerland

Numerical Prefixes Commonly Used in Forming Chemical Names

Numeral	Prefix	Numeral	Prefix	Numeral	Prefix
$\frac{1}{2}$	hemi-	13	trideca-	28	octacosa-
1	mono-	14	tetradeca-	29	nonacosa-
$1\frac{1}{2}$	sesqui-	15	pentadeca-	30	triaconta-
2	di-, bi-	16	hexadeca-	40	tetraconta-
$2\frac{1}{2}$	hemipenta-	17	heptadeca-	50	pentaconta-
3	tri-	18	octadeca-	60	hexaconta-
4	tetra-	19	nonadeca-	70	heptaconta-
5	penta-	20	eicosa-	80	octaconta-
6	hexa-	21	heneicosa-	90	nonaconta-
7	hepta-	22	docosa-	100	hecta-
8	octa-	23	tricosa-	101	henhecta-
9	ennea-, nona-	24	tetracosa-	102	dohecta-
10	deca-	25	pentacosa-	110	decahecta-
11	hendeca-, undeca-	26	hexacosa-	120	eicosahecta-
12	dodeca-	27	heptacosa-	200	dicta-

Alchemical Symbols Used in Biology and Botany

Symbol	Meaning
☉	Sun; gold; annual plant.
☉ ☉	Biennial plant.
○ or ☽	Moon; silver; *sometimes* female sex.
☿	Mercury.
♀	Venus; copper; *female sex.*
♁	Earth; terra; *sometimes* male sex.
♁ ♀	Having male and female flowers separate.
♁ — ♀	Having male and female flowers on the same plant.
♁ : ♀	Having male and female flowers on different plants.
♂	Mars; iron; *male sex.*
♂	Conjunction; mating; mated.
♃	Jupiter; tin; perennial herb.
♄	Saturn; lead.
🜍	Sulfur.
♒	Arsenic.

Prescription Notation

The units of quantities written on prescriptions in the Apothecaries System are usually designated by symbols. Occasionally abbreviations are used. The following list provides these symbols and abbreviations.

Apothecaries' Weight

Symbol	Abbreviation	Unit
—	gr.	grain
Ə	sc.	scruple
ℨ	dr.	dram
℥	oz.	ounce
℔	lb.	pound

Apothecaries' Volume

Symbol	Abbreviation	Unit
♏	min.	minim
f ℨ	fl. dr.	fluid dram
f ℥	fl. oz.	fluid ounce
O.	pt.	pint
Cong.	gal.	gallon

Roman numerals (lower case) are always used following the symbol to designate the number of units required, but, if the abbreviation is used, arabic numerals are used and precede the appreviation. For example:

ℨ iv   but   4 dr.

For less than one unit, one-half may be designated by ss following the symbol, but other fractions must be designated by arabic numeral fractions. For example:

ℨ ss   but   gr. $\frac{1}{8}$

Roman numerals are also usually used to designate the number of dosage forms required. For example:

Caps. No. xiv

Latin Terms

Abbreviation	Expected form	Meaning	Abbreviation	Expanded form	Meaning
a.	ante	before	d.	dexter	right
a., aur.	auris	ear	d.	dies	a day
aa.	ana	of each	d.	dosis	a dose
a.c.	ante cibos	before meals	da.	da	give
ad	ad	to, up to	dent., d.	dentur	let be given, give
add.	adde, addantur, addendus, addendo	add, let them be added, to be added, by adding	det.	detur	let it be given
			dieb. alt.	diebus alternis	on alternate days
			dieb. secund.	diebus secundis	every second day
ad lib.	ad libitum	at pleasure	dil.	dilue	dilute
adm.	admove	apply	disp.	dispensa, dispensetur	dispense
ad man. med.	ad manus medici	(to be delivered) into the hands of the (pre-scribing) physician	div.	divide	divide
			div. in par. aeq.	dividatur in partes aequales	divide in equal parts
ad satur.	ad saturandum	to saturation			
agit.	agita	shake			
alb.	albus	white	d.t.d.	dentur tales doses	give such doses
alt. hor.	alternis horis	every other hour	dulc.	dulcis	sweet
A.M.	ante meridiem	before noon	dur.	durus	hard
ampl.	amplus	large	e.m.p.	ex modo prescripto	after the manner prescribed
ampul.	ampulla	ampul, ampule			
applicand.	applicandus	to be applied	et	et	and
aq.	aqua	water	ex	ex	out of
aq. bull.	aqua bulliens	boiling water	ext.	extractum	an extract
aq. dest.	aqua destillata	distilled water	f., ft.	fac, fiat, fiant	make, let be made
aq. ferv.	aqua fervens	hot water	ferv.	fervens	hot
aq. frig.	aqua frigida	cold water	filt.	filtra	filter (imperative form)
b.	bis	twice	fl.	fluidus	fluid
ben.	bene	well	flav.	flavus	yellow
bib.	bibe	drink	fldext.	fluidextractum	fluidextract
b.i.d.	bis in die	twice a day	fort.	fortis	strong
bol.	bolus	a large pill	frig.	frigidus	cold
brevis	brevis	short	garg.	gargarisma	a gargle
bull.	bulliens, bulliat, bulliant	boiling, let boil	Gm.	gramma	gram
			gr.	granum	grain
c̄	cum	with	gran.	granulatus	granulated
C	centum	a hundred	gtt.	gutta	a drop
cap.	capiat	let the patient take	h.s.	hora somni	at the hour of sleep, at bedtime
caps.	capsula	a capsule			
cerat.	ceratum	wax ointment	hydrarg.	hydrargyrum.	mercury
chart.	charta	paper, a powder in paper	i.c.	inter cibos	between meals
			juxt.	juxta	near
chart. cerat.	charta cerata	waxed paper, parchment paper	Kal.	Kalium	potassium
			l.a.	lege artis	according to the art
chirurg.	chirurgicalis	surgical	laev.	laevus	left
chord chirurg.	chorda chirurgicalis	surgical "catgut"	lb.	libra	pound
			lev.	levis	light
cito disp!	cito dispensetur!	let it be dispensed quickly	liq.	liquor	a liquor, a solution
			lot.	lotio	lotion
coch. amp.	cochleare amplum	a tablespoonful	m.	mane	in the morning
			M.	misce	mix
coch. mag.	cochleare magnum	a tablespoonful	m.	mitte	send
			mag.	magnus	large
coch. med.	cochleare medium	a dessertspoonful	m. dict.	more dicto	as directed
			min.	minimum	minim
coch. mod.	cochleare modicum	a dessertspoonful	mist.	mistura	mixture
			mixt.	mixtura	mixture
coch. parv.	cochleare parvum	a teaspoonful	m.p.	modo praescripto	in the manner prescribed
col.	cola	strain (imperative form)	m.t.d.	mitte tales doses	send such doses
			n.	naris	nostril
colet.	coletur	let it be strained	Natr.	Natrium	sodium
collun.	collunarium	a nose wash	nebul.	nebula	a spray
collut.	collutorium	a mouth wash	n. et m.	nocte maneque	night and morning
collyr.	collyrium	an eye wash	nig.	niger	black
comp.	compositus	compound, compounded	no.	numero	number
			noct.	nocte	at night
cong.	congius	a gallon	non rep.	non repetatur	do not repeat
consperg.	consperge, conspergetur	dust, sprinkle	O.	Octarius	a pint
			o.d.	oculus dexter	right eye
cont.	contra	against	o.l.	oculus laevus	left eye
contus.	contusus	bruised	omn. hor.	omni hora	at every hour
coq.	coque, coquatur	boil, let it boil	omn. man.	omni mane	on every morning

Latin Terms

Abbreviation	Expected form	Meaning	Abbreviation	Expanded form	Meaning
opt.	optimus	best	s.a.	secundum artem	according to art
o.s.	oculus sinister	left eye	sat.	saturatus	saturated
o.u.	oculi uterque	both eyes	scat.	scatula	box
p. ae.	partes aequales	equal parts	scat. orig.	scatula originalis	original package, manufacturer's package (and label)
parv.	parvus	small			
p.c.	post cibos	after meals			
phial.	phiala	bottle	sic.	siccus	dried
Plumb.	Plumbum	lead	s.n.	secundum naturam	according to nature
P.M.	post meridiem	after noon			
p.o.	per os	by mouth	sol.	solubilis	soluble
pond.	ponderosus	heavy	sol.	solutio	solution
ppt.	praecipitatus	precipitated	solv.	solve	dissolve
p.r.n.	pro re nata	as occasion arises, as needed	s.o.s.	si opus sit	if there is need
			spir.	spiritus	spirit
pro rect.	pro recto	rectal	spiss.	spissus	dried
pt.	perstetur	let it be continued	ss.	semis	one half
pulv.	pulvis, pulveres	powder, powders	stat.	statim	immediately
q., qq.	quodque, quaeque	each, every	suc.	succus	juice
			sum.	sume, sumendus	take, to be taken
q.i.d.	quater in die	four times a day	S.V.R.	spiritus vini rectificatus	alcohol
qq. hor.	quaque hora	every hour			
Q.R.	quantum rectum	the quantity is correct	syr.	syrupus	syrup
			tab.	tabella	tablet
q.s.	quantum sufficiat, quantum satis	a sufficient quantity	tal.	talis, tales, talia	such
			ter.	tere	rub
			t.i.d.	ter in die	three times a day
quot. op. sit	quoties opus sit	as often as necessary	tinct., tr.	tinctura	tincture
q.v.	quantum voleris	as much as you wish	trit.	tritura	triturate
℞	recipe	you take	ult.	ultime	lastly
recen.	recens	fresh	unct.	unctus	smeared
rect.	rectificatus	rectified	ung., ungt.	unguentum	ointment
ren. sem.	renovetur semel	shall be renewed (only) once	ust.	ustus	burnt
			ut dict.	ut dictum	as directed
rept.	repetatur	let it be repeated	v.	vel	or
rub.	ruber	red	vesp.	vesper	evening
S., Sig.	signa, signetur	write, it shall be written (as instruction to the patient)	vir.	viridis, viride	green
			vol.	volatilis, volatile	volatile

Greek Alphabet

Name of letter	Capital	Lower case	Trans-literation	Name of letter	Capital	Lower case	Transliteration
alpha	A	α	a	nu	N	ν	n
beta	B	6 or β	b	xi	Ξ	ξ	x
gamma	Γ	γ	g	omicron	O	o	o short
delta	Δ	δ	d	pi	Π	π	p
epsilon	E	ϵ	e short	rho	P	ρ	r
zeta	Z	ζ	z	sigma	Σ	σ or ς	s
eta	H	η	e long	tau	T	τ	t
theta	Θ	θ	th	upsilon	Υ	ν	y
iota	I	ι	i	phi	Φ	ϕ or φ	f
kappa	K	κ	k, c	chi	X	χ	ch as in German echt
lambda	Λ	λ	l	psi	Ψ	ψ	ps
mu	M	μ	m	omega	Ω	ω	o long

Russian Alphabet

Cyrillic print	Trans-literation	Pronunciation	Cyrillic print	Trans-literation	Pronunciation
А а	a	*a* in far	С с	s	*s* in say
Б б	b	*b*	Т т	t	*t*
В в	v	*v*	У у	u	*oo* in boot
Г г	g (h)	*g* in gay	Ф ф	f	*f*
Д д	d	*d*	Х х	kh	like German *ch*
Е е	e	*e* in fell; also *ye* in yell	Ц ц	t̂s	*ts* in hoots
Ж ж	zh	*z* in azure	Ч ч	ch	*ch* in church
З з	z	*z* in zeal	Ш ш	sh	*sh*
И и	i	*i* in meet	Щ щ	shch	*shch*, as in fre*sh ch*eese
Й й	ĭ	*y* in boy	Ъ ъ	"	mute*
К к	k	*k*	Ы ы	y	*y* in rhythm (hard)
Л л	l	*l*	Ь ь	'	mute (softens preceding consonant)
М м	m	*m*	Э э	ė	*e* in met
Н н	n	*n*	Ю ю	îu	*u* in union
О о	o	*o* in or	Я я	îa	*ya* in yard
П п	p	*p*			
Р р	r	*r*			

* Hard sign; used to separate a consonant from a soft vowel especially in foreign words; frequently replaced by an apostrophe.

Roman Numerals

I	II	III	IV	V	VI	VII	VIII	IX	X
1	2	3	4	5	6	7	8	9	10
XX	XXX	XL	L	LX	LXX	LXXX	XC	IC	C
20	30	40	50	60	70	80	90	99	100
CC	CCC	CD	D	DC	DCC	DCCC	CM	XM	M
200	300	400	500	600	700	800	900	990	1000

Chemical Terms Translator

English	French	German	Russian	Japanese
acetic acid	acide acétique	Essigsäure	уксусная кислота	酢酸
acetone	acétone	Aceton	ацетон	アセトン
acidic	acide	sauer	кислотный	酸性の
alkaline	alcalin	alkalisch	щелочной	アルカリ性の
amorphous	amorphe	amorph	аморфный	無定形の
anhydrous	anhydre	wasserfrei	безводный	無水の
aqueous	aqueux	wäßrig	водный	含水の
boiling point	point d'ébullition	Siedepunkt	точка кипения	沸点
chloroform	chloroforme	Chloroform	хлороформ	クロロフォルム
chromatography	chromatographie	Chromatographie	хроматография	クロマトグラフ
cold	froid	kalt	холодный	冷たい
colorless	incolore	farblos	бесцветный	無色の
compound	composé	Verbindung	соединение	化合物
condensation	condensation	Kondensation	конденсация	濃縮
cool	refroidir	kühlen	охлаждать	冷却する
corrosive	corrosif	korrosiv; ätzend	коррозионный	腐食性の
crude	brut	roh	неочищенный	粗製の
crystal	cristal	Kristall	кристалл	結晶
crystalline	cristallin	kristalline	кристаллический	結晶の
decomposition	décomposition	Zersetzung	разложение	分解
dehydration	déshydratation	Dehydratisierung	дегидратация	脱水
deliquescense	déliquescence	Zerfließlichkeit	расплывание	潮解
density	densité	Dichte	плотность	密度
dextrorotatory	dextrogyre	rechtsdrehend	правовращающий	右旋性の
dilute (adjective)	dilué	verdünnt	разбавленный	希釈した
dilute (verb)	diluer	verdünnen	разбавлять	希釈する
dissolve	dissoudre	lösen	растворять	溶解する
distill	distiller	destillieren	дистиллировать	蒸留する
dry	sécher	trocknen	сушить	乾燥する
easily soluble	facilement soluble	leicht löslich	легкорастворимый	溶解しやすい
empirical formula	formule empirique	Summenformel	эмпирическая формула	実験式
esterification	estérification	Veresterung	эстерификация	エステル化
ethanol	éthanol	Äthanol	этанол	エタノール
ether	éther	Äther	простой эфир	エーテル
evaporation	évaporation	Verdunstung	выпаривание	蒸発
explosive	explosif	explosiv	взрывчатый	爆発性の
flammable	inflammable	entflammbar	воспламеняемый	発火性の
flash point	point d'éclair	Flammpunkt	температура вспышки	引火点
fraction	fraction	Fraktion	фракция	留分、画分
freezing point	point de congélation	Gefrierpunkt	точка замерзания	凝固点
gas	gaz	Gas	газ	気体
heat	chauffer	erhitzen	нагревать	加熱する
hot	chaud	heiß	горячий	熱い
hydrochloric acid	acide chlorhydrique	Salzsäure	соляная кислота	塩酸
hygroscopic	hygroscopique	hygroskopisch	гигроскопический	吸湿性の
immiscible	immiscible	unmischbar	несмешивающийся	混和しにくい
impure	impur	unrein	нечистый	不純な
IR (infrared)	IR (infrarouge)	IR (infrarot)	ИК (инфракрасный)	赤外
insoluble	insoluble	unlöslich	нерастворимый	不溶性の
isolation	isolation	Isolierung	выделение	単離
isomer	isomère	Isomer	изомер	アイソマー
levorotatory	lévogyre	linksdrehend	левовращающий	左旋性の

English	French	German	Russian	Japanese
light	lumière	Licht	свет	光
liquid	liquide	Flüssigkeit	жидкость	液体
lustrous	lustré	glänzend	блестящий	光沢のある
MS (mass spectro-metry)	spectrométrie de masse	MS (Massenspektrometrie)	МС (масс-спектрометрия)	質量分析
maximum	maximum	Maximum	максимум	最大
melting point	point de fusion	Schmelzpunkt	точка плавления	融点
minimum	minimum	Minimum	минимум	最小
miscible	miscible	mischbar	смешивающийся	混和しやすい
mix	mélanger	mischen	смешивать	混ぜる
mixture	mélange	Mischung	смесь	混合物
mole	mole	Mol	моль	モル
molecular weight	poids moléculaire	Molekulargewicht	молекулярный вес	分子量
nitric acid	acide nitrique	Salpetersäure	азотная кислота	硝酸
NMR (nuclear mag-netic resonance)	résonance magnétique nucléaire	NMR (kernmagnetische Resonanz)	ЯМР (ядерный магнитный резонанс)	ＮＭＲ（核磁気共鳴）
normal (N)	normal (N)	normal (N)	нормальный (о растворе)	規定（濃度単位）
odorless	inodore	geruchlos	без запаха	無臭の
oil	huile	Öl	нефть	油
optical rotation	rotation optique	optische Drehung	оптическое вращение	旋光
phase	phase	Phase	фаза	相
powder	poudre	Pulver	порошок	粉末
practically insoluble	pratiquement insoluble	nahezu unlöslich	практически нерастворимый	事実上不溶の
precipitate (noun)	précipité	Präzipitat	осадок	析出物
precipitate (verb)	précipiter	präzipitieren	осаждать	析出する
preparation	préparation	Präparat	препарат	調製
pure	pur	rein	чистый	純粋な
racemate	racémate	Razemat	рацемат	ラセミ体
recrystallize	recristalliser	umkristallisieren	перекристаллизовывать	再結晶する
residue	résidu	Rückstand	остаток	残留物
resolution	résolution	Auflösung	разрешающая способность	解像度
saturated	saturé	gesättigt	насыщенный	飽和した
slightly soluble	légèrement soluble	schwer löslich	слегка растворимый	僅かに溶解する
sodium hydroxide	hydroxyde de sodium	Natriumhydroxid	едкий натр	水酸化ナトリウム
solid	solide	Feststoff	твердое вещество	固体
solubility	solubilité	Löslichkeit	растворимость	溶解性
soluble	soluble	löslich	растворимый	可溶性の
solution	solution	Lösung	раствор	溶液
solvent	solvant	Lösungsmittel	растворитель	溶媒
somewhat soluble	quelque peu soluble	mäßig löslich	частично растворимый	いく分溶解する
sparingly soluble	difficilement soluble	wenig löslich	слаборастворимый	ほんの少し溶解する
sublime	sublimer	sublimieren	сублимировать	昇華する
synthesis	synthèse	Synthese	синтез	合成
uv (ultraviolet)	uv (ultraviolet)	UV (ultraviolett)	УФ (ультрафиолетовый)	紫外
vapor pressure	pression de vapeur	Dampfdruck	упругость пара	蒸気圧
very soluble	très soluble	sehr leicht löslich	высокорастворимый	よく溶解する
viscous	visqueux	viskos	вязкий	粘性の
volatile	volatil	flüchtig	летучий	揮発性の
warm	réchauffer	erwärmen	подогревать	温める
water	eau	Wasser	вода	水
yield	rendement	Ausbeute	выход	収率

Boiling Points of Solvents

The following tables list boiling points which have been extracted from pertinent monographs of *The Merck Index*. Only those solvents whose boiling points are reported at 1 atmospheric pressure (760 mm Hg) have been included, except where noted. Table 1 lists compounds arranged in ascending order by boiling point (°C). Table 2 lists the solvents alphabetically.

Table 1

Solvent	Boiling Point	Solvent	Boiling Point
Trimethylamine	2.87	Carbon tetrachloride	76.7
Cryofluorane	4.1	Ethyl acetate	77
Dimethylamine	7	Acrylonitrile	77.3
Neopentane	9.5	*n*-Butylamine	78
Ethylene oxide	10.7	*n*-Butyl chloride	78.5
Ethyl chloride	12.3	Ethyl alcohol	78.5
Acetaldehyde	21	Methyl ethyl ketone	79.6
Trichlorofluoromethane	23.7	Hexafluorobenzene	80
1-Pentene	30.1	Benzene	80.1
Furan	31.36	Cyclohexane	80.7
Methyl formate	31.5	Propyl formate	81-82
Vinylidene chloride	31.7	Acetonitrile	81.6
Isopropylamine	33-34	1,2-Dimethoxyethane	82-83
Isoprene	34.067	*tert*-Butyl alcohol	82.41
Propylene oxide	34.23	Isopropyl alcohol	82.5
Ethyl ether	34.6	Cyclohexene	83
Ethanethiol	34.7-35.04	Ethylene dichloride	83-84
Isopropyl chloride	35-36	Diisopropylamine	84
2-Pentene (*trans*-form)	35.85	Thiophene	84.4
Pentane	36.1	Fluorobenzene	84.73
Methyl sulfide	36.2	*tert*-Amyl methyl ether	86.3
2-Pentene (*cis*-form)	37.0	Trichloroethylene	86.7
Amylene	37.5-38.5	Propylidene chloride	87
Ethyl bromide	38.2	Tetrahydropyran	88
Methylene chloride	39.75	Bromodichloromethane	88.4-88.6
Methylal	41.6	Pyrrolidine	88.5-89
Methyl iodide	42.5	Isopropyl acetate	89
Allyl chloride	44-45	Isopropyl iodide	89-90
tert-Butylamine	44-46	Triethylamine	89-90
Propyl chloride	46-47	Propyl ether	89-91
Carbon disulfide	46.5	Methacrylonitrile	90.3
Propylamine	48-49	*sec*-Butyl bromide (*dl*-form)	91.2
Isoflurane	48.5	Isobutyl bromide	91.5
Propionaldehyde	49	Ethyl sulfide	92
Cyclopentane	49.3	Methyl isopropyl ketone	94.2
tert-Butyl chloride	51.0	Ethyl chloroformate	95
Acrolein	52.5	Propylene dichloride	95-96
Ethyl formate	53-54	Allyl alcohol	96-97
Allylamine	55-58	Methylene bromide	97
Methyl *tert*-butyl ether	55.2	Propionitrile	97.2
Diethylamine	55.5	*n*-Propyl alcohol	97.2
Ethylenimine	56-57	Isobutyl formate	98
Acetone	56.5	*n*-Butyl mercaptan	98.4
Methyl acetate	56.9	*n*-Heptane	98.4
Ethylidene chloride	57.3	Ethyl propionate	99
Isopropyl bromide	59-60	Isooctane	99.3
Acetylene dichloride (*cis*-form)	60	Ethyl acrylate	99.4
Chloroform	61-62	*sec*-Butyl alcohol (*dl*-form)	99.5
sec-Butylamine (all isomers)	63	Water	100
3-Chloro-1-butene (*dl*-form)	63.9-64.2	Formic acid	100.5
Isobutyraldehyde	64	Dioxane	101.1
Methanol	64.7	Nitromethane	101.2
Tetrahydrofuran	66	*n*-Butyl bromide	101.3
Trimethyl borate	67-68	Diethyl ketone	101.5
sec-Butyl chloride (*dl*-form)	68	Propyl acetate	101.6
Isobutylamine	68-69	Methyl propyl ketone	102
Isobutyl chloride	68-69	Propyl iodide	102-103
Isopropyl ether	68-69	Acetal	102.7
1-Chloro-2-methylpropene	68.1	Benzotrifluoride	103.46
n-Hexane	69	Allyl acetate	104
Propyl bromide	71	*n*-Amylamine	104
Ethyl iodide	72	Isobutyronitrile	104
Vinyl acetate	72-73	Crotonaldehyde	104.0
Trifluoroacetic acid	72.4	Phosphorus oxychloride	105.8
tert-Butyl bromide	73.3	Piperidine	106
1,1,1-Trichloroethane	74.1	Amyl chloride	107.8
Butyraldehyde	74.8	Isobutyl alcohol	108

Table 1 (Continued)

Solvent	Boiling Point	Solvent	Boiling Point
n-Dipropylamine	110	2,6-Lutidine	144
Toluene	110.6	o-Xylene	144
sec-Butyl acetate (dl-form)	112-113	Methyl lactate	144-145
1,1,2-Trichloroethane	113-114	Ethyl chloroacetate	144-146
o-Fluorotoluene	114	Methyl Cellosolve® acetate	145
Neopentyl alcohol	114	γ-Picoline	145
Nitroethane	114-115	Styrene	145-146
Propargyl alcohol	114-115	Tetrachloroethane	146.5
Pyridine	115-116	Isobutyl isobutyrate	147
N-Methylmorpholine	115.4	Allyl isothiocyanate	148-154
3-Pentanol	115.6	Bromoform	149-150
m-Fluorotoluene	116	2,2,2-Trichloroethanol	151-153
p-Fluorotoluene	116	2-Heptanone	151.5
sec-Butyl acetate (d-form)	116-117	Cumene	152-153
Ethylenediamine	116-117	Isobutyric acid	152-155
n-Butyl alcohol	117-118	N,N-Dimethylformamide	153
Isopropylacetone	117-118	Ethyl lactate	154
Butyronitrile	117.5	α-Pinene (all isomers)	155-156
Epichlorohydrin	117.9	Anisole	155.5
Acetic acid glacial	118	Cyclohexanone	155.6
Isobutyl acetate	118	2-Ethoxyethyl acetate	156
Allyl cyanide	119	Bromobenzene	156.2
Methyl isothiocyanate	119	1-Hexanol	157
Propylenediamine	119-120	Ethyl amyl ketone	157-162
2-Pentanol	119.3	2-Heptanol (dl-form)	158-160
Chloroacetone	119.7	o-Chlorotoluene	158.97
Isobutyl iodide	120	n-Dibutylamine	159-160
Ethyl butyrate	120-121	n-Propylbenzene	159.2
Isoamyl bromide	120-121	Cyclohexanol	161
2-Nitropropane	120.3	Pentachloroethane	161-162
Caprylene	121	m-Chlorotoluene	161.75
Tetrachloroethylene	121	Furfural	161.8
Crotyl alcohol (trans-form)	121.2	p-Chlorotoluene	161.99
Chlorodibromomethane	121.3-121.8	Diglyme	162
Crotyl alcohol (cis-form)	123.6	β-Pinene (l-form)	162-163
n-Butyl acetate	125-126	Methacrylic acid	163
Octane	125.6	Methyl oxalate	163-164
Bromine trifluoride	125.75	N,N-Dimethylacetamide	163-165
Ethyl carbonate	126	Butyric acid	163.5
Hexamethyldisilazane	126	β-Pinene (d-form)	164-166
sec-Propylene chlorohydrin	126-127	Mesitylene	164.7
Methyl butyl ketone	127	2-Amino-2-methyl-1-propanol	165
2-Methyl-1-butanol	128	n-Butyl n-butyrate	165
α-Picoline	128-129	Pyruvic acid	165 (dec)
Ethylene chlorohydrin	128-130	β-Pinene (dl-form)	165-166
Morpholine	128.9	Propionic anhydride	167.0
1H-Pyrrole	129.8	Diacetone alcohol	167.9
Mesityl oxide	130	Isocrotonic acid	168-169
Ethyl isothiocyanate	130-132	Thiophenol	168.3
Methyl chloroacetate	130-132	tert-Butylbenzene	168.5
Methyl thiocyanate	130-133	Methyl acetoacetate	169-171 (dec)
n-Butyl iodide	130.4	Furfuryl alcohol	170
Isopentyl alcohol	130.5	Isobutylbenzene	170.5
Cyclopentanone	130.6	Ethanolamine	170.8
Chlorobenzene	131-132	2,4,6-Trimethylpyridine	171
Ethylene dibromide	131-132	Butyl Cellosolve®	171-172
1-Nitropropane	131.6	Phenetole	171-173
Cyclohexylamine	134.5	Isoamyl ether	172
2-Ethoxyethanol	135	Hexyl methyl ketone	172-173
Ethyl isovalerate	135	m-Dichlorobenzene	173
Ethylbenzene	136.25	sec-Butylbenzene	173.5
p-Xylene	137-138	p-Dichlorobenzene	174.12
1-Pentanol	137.5	1,3-Dichloro-2-propanol	174.3
Acetic anhydride	139	Isovaleric acid	175-177
m-Xylene	139.3	Limonene (d-form)	175.5-176[a]
Propylene dibromide	140-142	Limonene (dl,l-forms)	175.5-176.5[a]
2,3-Dibromopropene	140-143	Eucalyptol	176-177
Acetylacetone	140.5	Tetramethylurea	176.5
Valeronitrile	141	Methyl carbamate	177
Acrylic acid	141.0	p-Cymene	177.10
Propionic acid	141.1	sym-Dichloroethyl ether	178
Isoamyl acetate	142	Tetrahydrofurfuryl alcohol	178
n-Butyl ether	142-143	2-Octanol (dl-form)	178.5

Table 1 (Continued)

Solvent	Boiling Point	Solvent	Boiling Point
Benzaldehyde	179	o-Chloroaniline	208.84
Benzyl chloride	179	Nitrobenzene	210-211
Isoamyl butyrate	179	Trimethylene glycol	210-212
2,3-Butylene glycol (d-form)	179-182	Formamide	210.5 (dec)
N-Methylformamide	180-185	2,4-Xylenol	210.931
o-Dichlorobenzene	180.5	Ethyl benzoate	211-213
Ethyl acetoacetate	180.8	2,5-Xylenol	211.132
Methylene iodide	181	Benzyl acetate	213
n-Butyl sulfide	182	1,2,4-Trichlorobenzene	213
Phenol	182	Triethyl phosphate	215-216
Urethan	182-184	Isophorone	215.3
n-Butylbenzene	183.1	Triglyme	216
2-Ethyl-1-hexanol	184-185	Tributylamine	216-217
Aniline	184-186	2,3-Xylenol	216.87
Benzylamine	185	Naphthalene	217.9
Crotonic acid	185.0	Carbitol® acetate	218.5
Diethyl oxalate	185.7	Diethyl maleate	219.5[b]
Benzyl ethyl ether	186	1,3-Dimethyl-2-imidazolidinone	220[c]
Valeric acid, normal	186-187	1,2,3-Tribromopropane	220
n-Amyl ether	186.75	Methyl salicylate	220-224
Hexachloroethane	186.8	Benzotrichloride	220.8
Propylene glycol (l-form)	187-189	3,5-Xylenol	221.692
Decalin® (trans-form)	187.25	Acetamide	222
Acetonylacetone	188	3,4-Xylenol	226.947
Diethyl Carbitol®	188	Ethylene cyanohydrin	228 (dec)
Iodobenzene	188-189	Butyl Carbitol®	230.4
Propylene glycol (dl-form)	188.2	Methyl nonyl ketone	231.5-232.5[d]
Dimethyl sulfoxide	189	m-Nitrotoluene	231.9
Dipropylene glycol monomethyl ether	189.6	Benzyl cyanide	233.5
Ethylene glycol diacetate	190-191	Quinoline	237.7
Benzonitrile	190.7	Dimethyl sulfone	238
o-Cresol	191-192	1,5-Pentanediol	239
Isoamyl isovalerate	191-194	Caprylic acid	239.7
Glycerol formal	191-195	Isoquinoline	243.25
Succinyl chloride	192-193	Diethylene glycol	244-245
N,N-Dimethylaniline	192-194	Ethohexadiol	244.2
Methyl Carbitol®	193	2-Pyrrolidone	245
1-Octanol	194-195	2-Phenoxyethanol	245.2
Phenyl acetate	195-196	Cinnamaldehyde	246.0 (dec)
Decalin® (cis-form)	195.7	n-Butyl benzoate	250
Carbitol®	196	Eugenol	255
Salicylaldehyde	196-197	Triacetin	258-260
Trichloroacetic acid	196-197	Diacetin	259
Ethylene glycol	197.6	Phenyl ether	259
Hexylene glycol	198	1-Chloronaphthalene	259.3
Ethyl malonate	198-199	Isoamyl benzoate	260-262
Methyl benzoate	198-200	2,5-Tetrahydrofurandimethanol	265
Octyl acetate	199	Succinonitrile	265-267
Butyric anhydride	199.4-201.4	Triethylenetetramine	266-267
p-Toluidine	200-201	Diethanolamine	268.8
o-Toluidine	200-202	Ethyl cinnamate	271
Pine oil	200-220	Isoamyl salicylate	274-278
2,6-Xylenol	201.03	Tetraglyme	275.3
p-Cresol	201.8	o-Nitroanisole	277
Acetophenone	202	2,4-Dimethylsulfolane	280-281 (dec)
m-Cresol	202	1-Bromonaphthalene	281.1
1-Methylpyrrolidone	202	Dimethyl phthalate	283.7
Benzyl formate	203	Sulfolane	285
m-Toluidine	203-204	Triethylene glycol	285
Butyrolactone	204	Glycerol	290.0 (dec)
Camphor	204	Diethyl phthalate	295
Benzyl alcohol	204.7	Benzyl ether	295-298 (dec)
Benzal chloride	205	Phenyl carbonate	302-306
n-Caproic acid	205	Benzyl benzoate	323-324
Veratrole	206-207	Triethanolamine	335.4
Ethyl cyanoacetate	206.0	n-Butyl phthalate	340
Tetralin®	207.2	Butyl stearate	343
1,3-Butylene glycol	207.5	Methyl abietate	360-365 (dec)
Bromoacetic acid	208		

Table 2

Solvent	Boiling Point	Solvent	Boiling Point
Acetal	102.7	*n*-Butyl ether	142-143
Acetaldehyde	21	*n*-Butyl iodide	130.4
Acetamide	222	*n*-Butyl mercaptan	98.4
Acetic acid glacial	118	*n*-Butyl phthalate	340
Acetic anhydride	139	Butyl stearate	343
Acetone	56.5	*n*-Butyl sulfide	182
Acetonitrile	81.6	Butyraldehyde	74.8
Acetonylacetone	188	Butyric acid	163.5
Acetophenone	202	Butyric anhydride	199.4-201.4
Acetylacetone	140.5	Butyrolactone	204
Acetylene dichloride (*cis*-form)	60	Butyronitrile	117.5
Acrolein	52.5	Camphor	204
Acrylic acid	141.0	*n*-Caproic acid	205
Acrylonitrile	77.3	Caprylene	121
Allyl acetate	104	Caprylic acid	239.7
Allyl alcohol	96-97	Carbitol®	196
Allylamine	55-58	Carbitol® acetate	218.5
Allyl chloride	44-45	Carbon disulfide	46.5
Allyl cyanide	119	Carbon tetrachloride	76.7
Allyl isothiocyanate	148-154	Chloroacetone	119.7
2-Amino-2-methyl-1-propanol	165	*o*-Chloroaniline	208.84
n-Amylamine	104	Chlorobenzene	131-132
Amyl chloride	107.8	3-Chloro-1-butene (*dl*-form)	63.9-64.2
Amylene	37.5-38.5	Chlorodibromomethane	121.3-121.8
n-Amyl ether	186.75	Chloroform	61-62
tert-Amyl methyl ether	86.3	1-Chloro-2-methylpropene	68.1
Aniline	184-186	1-Chloronaphthalene	259.3
Anisole	155.5	*m*-Chlorotoluene	161.75
Benzal chloride	205	*o*-Chlorotoluene	158.97
Benzaldehyde	179	*p*-Chlorotoluene	161.99
Benzene	80.1	Cinnamaldehyde	246.0 (dec)
Benzonitrile	190.7	*m*-Cresol	202
Benzotrichloride	220.8	*o*-Cresol	191-192
Benzotrifluoride	103.46	*p*-Cresol	201.8
Benzyl acetate	213	Crotonaldehyde	104.0
Benzyl alcohol	204.7	Crotonic acid	185.0
Benzylamine	185	Crotyl alcohol (*cis*-form)	123.6
Benzyl benzoate	323-324	Crotyl alcohol (*trans*-form)	121.2
Benzyl chloride	179	Cryofluorane	4.1
Benzyl cyanide	233.5	Cumene	152-153
Benzyl ether	295-298 (dec)	Cyclohexane	80.7
Benzyl ethyl ether	186	Cyclohexanol	161
Benzyl formate	203	Cyclohexanone	155.6
Bromine trifluoride	125.75	Cyclohexene	83
Bromoacetic acid	208	Cyclohexylamine	134.5
Bromobenzene	156.2	Cyclopentane	49.3
Bromodichloromethane	88.4-88.6	Cyclopentanone	130.6
Bromoform	149-150	*p*-Cymene	177.10
1-Bromonaphthalene	281.1	Decalin® (*cis*-form)	195.7
n-Butyl acetate	125-126	Decalin® (*trans*-form)	187.25
sec-Butyl acetate (*d*-form)	116-117	Diacetin	259
sec-Butyl acetate (*dl*-form)	112-113	Diacetone alcohol	167.9
n-Butyl alcohol	117-118	2,3-Dibromopropene	140-143
sec-Butyl alcohol (*dl*-form)	99.5	*n*-Dibutylamine	159-160
tert-Butyl alcohol	82.41	*m*-Dichlorobenzene	173
n-Butylamine	78	*o*-Dichlorobenzene	180.5
sec-Butylamine (all isomers)	63	*p*-Dichlorobenzene	174.12
tert-Butylamine	44-46	*sym*-Dichloroethyl ether	178
n-Butylbenzene	183.1	1,3-Dichloro-2-propanol	174.3
sec-Butylbenzene	173.5	Diethanolamine	268.8
tert-Butylbenzene	168.5	Diethylamine	55.5
n-Butyl benzoate	250	Diethyl Carbitol®	188
n-Butyl bromide	101.3	Diethylene glycol	244-245
sec-Butyl bromide (*dl*-form)	91.2	Diethyl ketone	101.5
tert-Butyl bromide	73.3	Diethyl maleate	219.5[b]
n-Butyl *n*-butyrate	165	Diethyl oxalate	185.7
Butyl Carbitol®	230.4	Diethyl phthalate	295
Butyl Cellosolve®	171-172	Diglyme	162
n-Butyl chloride	78.5	Diisopropylamine	84
sec-Butyl chloride (*dl*-form)	68	1,2-Dimethoxyethane	82-83
tert-Butyl chloride	51.0	*N,N*-Dimethylacetamide	163-165
1,3-Butylene glycol	207.5	Dimethylamine	7
2,3-Butylene glycol (*d*-form)	179-182	*N,N*-Dimethylaniline	192-194

Table 2 (Continued)

Solvent	Boiling Point	Solvent	Boiling Point
N,N-Dimethylformamide	153	Iodobenzene	188-189
1,3-Dimethyl-2-imidazolidinone	220[c]	Isoamyl acetate	142
Dimethyl phthalate	283.7	Isoamyl benzoate	260-262
2,4-Dimethylsulfolane	280-281 (dec)	Isoamyl bromide	120-121
Dimethyl sulfone	238	Isoamyl butyrate	179
Dimethyl sulfoxide	189	Isoamyl ether	172
Dioxane	101.1	Isoamyl isovalerate	191-194
n-Dipropylamine	110	Isoamyl salicylate	274-278
Dipropylene glycol monomethyl ether	189.6	Isobutyl acetate	118
Epichlorohydrin	117.9	Isobutyl alcohol	108
Ethanethiol	34.7-35.04	Isobutylamine	68-69
Ethanolamine	170.8	Isobutylbenzene	170.5
Ethohexadiol	244.2	Isobutyl bromide	91.5
2-Ethoxyethanol	135	Isobutyl chloride	68-69
2-Ethoxyethyl acetate	156	Isobutyl formate	98
Ethyl acetate	77	Isobutyl iodide	120
Ethyl acetoacetate	180.8	Isobutyl isobutyrate	147
Ethyl acrylate	99.4	Isobutyraldehyde	64
Ethyl alcohol	78.5	Isobutyric acid	152-155
Ethyl amyl ketone	157-162	Isobutyronitrile	104
Ethylbenzene	136.25	Isocrotonic acid	168-169
Ethyl benzoate	211-213	Isoflurane	48.5
Ethyl bromide	38.2	Isooctane	99.3
Ethyl butyrate	120-121	Isopentyl alcohol	130.5
Ethyl carbonate	126	Isophorone	215.3
Ethyl chloride	12.3	Isoprene	34.067
Ethyl chloroacetate	144-146	Isopropyl acetate	89
Ethyl chloroformate	95	Isopropylacetone	117-118
Ethyl cinnamate	271	Isopropyl alcohol	82.5
Ethyl cyanoacetate	206.0	IIsopropylamine	33-34
Ethylene chlorohydrin	128-130	Isopropyl bromide	59-60
Ethylene cyanohydrin	228 (dec)	Isopropyl chloride	35-36
Ethylenediamine	116-117	Isopropyl ether	68-69
Ethylene dibromide	131-132	Isopropyl iodide	89-90
Ethylene dichloride	83-84	Isoquinoline	243.25
Ethylene glycol	197.6	Isovaleric acid	175-177
Ethylene glycol diacetate	190-191	Limonene (*d*-form)	175.5-176[a]
Ethylene oxide	10.7	Limonene (*dl,l*-forms)	175.5-176.5[a]
Ethylenimine	56-57	2,6-Lutidine	144
Ethyl ether	34.6	Mesitylene	164.7
Ethyl formate	53-54	Mesityl oxide	130
2-Ethyl-1-hexanol	184-185	Methacrylic acid	163
Ethylidene chloride	57.3	Methacrylonitrile	90.3
Ethyl iodide	72	Methanol	64.7
Ethyl isothiocyanate	130-132	Methyl abietate	360-365 (dec)
Ethyl isovalerate	135	Methyl acetate	56.9
Ethyl lactate	154	Methyl acetoacetate	169-171 (dec)
Ethyl malonate	198-199	Methylal	41.6
Ethyl propionate	99	Methyl benzoate	198-200
Ethyl sulfide	92	2-Methyl-1-butanol	128
Eucalyptol	176-177	Methyl *tert*-butyl ether	55.2
Eugenol	255	Methyl butyl ketone	127
Fluorobenzene	84.73	Methyl carbamate	177
m-Fluorotoluene	116	Methyl Carbitol®	193
o-Fluorotoluene	114	Methyl Cellosolve® acetate	145
p-Fluorotoluene	116	Methyl chloroacetate	130-132
Formamide	210.5 (dec)	Methylene bromide	97
Formic acid	100.5	Methylene chloride	39.75
Furan	31.36	Methylene iodide	181
Furfural	161.8	Methyl ethyl ketone	79.6
Furfuryl alcohol	170	*N*-Methylformamide	180-185
Glycerol	290.0 (dec)	Methyl formate	31.5
Glycerol formal	191-195	Methyl iodide	42.5
n-Heptane	98.4	Methyl isopropyl ketone	94.2
2-Heptanol (*dl*-form)	158-160	Methyl isothiocyanate	119
2-Heptanone	151.5	Methyl lactate	144-145
Hexachloroethane	186.8	*N*-Methylmorpholine	115.4
Hexafluorobenzene	80	Methyl nonyl ketone	231.5-232.5[d]
Hexamethyldisilazane	126	Methyl oxalate	163-164
n-Hexane	69	Methyl propyl ketone	102
1-Hexanol	157	1-Methylpyrrolidone	202
Hexylene glycol	198	Methyl salicylate	220-224
Hexyl methyl ketone	172-173	Methyl sulfide	36.2

Table 2 (Continued)

Solvent	Boiling Point	Solvent	Boiling Point
Methyl thiocyanate	130-133	1*H*-Pyrrole	129.8
Morpholine	128.9	Pyrrolidine	88.5-89
Naphthalene	217.9	2-Pyrrolidone	245
Neopentane	9.5	Pyruvic acid	165 (dec)
Neopentyl alcohol	114	Quinoline	237.7
o-Nitroanisole	277	Salicylaldehyde	196-197
Nitrobenzene	210-211	Styrene	145-146
Nitroethane	114-115	Succinonitrile	265-267
Nitromethane	101.2	Succinyl chloride	192-193
1-Nitropropane	131.6	Sulfolane	285
2-Nitropropane	120.3	Tetrachloroethane	146.5
m-Nitrotoluene	231.9	Tetrachloroethylene	121
Octane	125.6	Tetraglyme	275.3
1-Octanol	194-195	Tetrahydrofuran	66
2-Octanol (*dl*-form)	178.5	2,5-Tetrahydrofurandimethanol	265
Octyl acetate	199	Tetrahydrofurfuryl alcohol	178
Pentachloroethane	161-162	Tetrahydropyran	88
Pentane	36.1	Tetralin®	207.2
1,5-Pentanediol	239	Tetramethylurea	176.5
1-Pentanol	137.5	Thiophene	84.4
2-Pentanol	119.3	Thiophenol	168.3
3-Pentanol	115.6	Toluene	110.6
1-Pentene	30.1	*m*-Toluidine	203-204
2-Pentene (*cis*-form)	37.0	*o*-Toluidine	200-202
2-Pentene (*trans*-form)	35.85	*p*-Toluidine	200-201
Phenetole	171-173	Triacetin	258-260
Phenol	182	1,2,3-Tribromopropane	220
2-Phenoxyethanol	245.2	Tributylamine	216-217
Phenyl acetate	195-196	Trichloroacetic acid	196-197
Phenyl carbonate	302-306	1,2,4-Trichlorobenzene	213
Phenyl ether	259	1,1,1-Trichloroethane	74.1
Phosphorus oxychloride	105.8	1,1,2-Trichloroethane	113-114
α-Picoline	128-129	2,2,2-Trichloroethanol	151-153
γ-Picoline	145	Trichloroethylene	86.7
α-Pinene (all isomers)	155-156	Trichlorofluoromethane	23.7
β-Pinene (*d*-form)	164-166	Triethanolamine	335.4
β-Pinene (*dl*-form)	165-166	Triethylamine	89-90
β-Pinene (*l*-form)	162-163	Triethylene glycol	285
Pine oil	200-220	Triethylenetetramine	266-267
Piperidine	106	Triethyl phosphate	215-216
Propargyl alcohol	114-115	Trifluoracetic acid	72.4
Propionaldehyde	49	Triglyme	216
Propionic acid	141.1	Trimethylamine	2.87
Propionic anhydride	167.0	Trimethyl borate	67-68
Propionitrile	97.2	Trimethylene glycol	210-212
Propyl acetate	101.6	2,4,6-Trimethylpyridine	171
n-Propyl alcohol	97.2	Urethan	182-184
Propylamine	48-49	Valeric acid, normal	186-187
n-Propylbenzene	159.2	Valeronitrile	141
Propyl bromide	71	Veratrole	206-207
Propyl chloride	46-47	Vinyl acetate	72-73
sec-Propylene chlorohydrin	126-127	Vinylidene chloride	31.7
Propylenediamine	119-120	Water	100
Propylene dibromide	140-142	*m*-Xylene	139.3
Propylene dichloride	95-96	*o*-Xylene	144
Propylene glycol (*dl*-form)	188.2	*p*-Xylene	137-138
Propylene glycol (*l*-form)	187-189	2,3-Xylenol	216.87
Propylene oxide	34.23	2,4-Xylenol	210.931
Propyl ether	89-91	2,5-Xylenol	211.132
Propyl formate	81-82	2,6-Xylenol	201.03
Propylidene chloride	87	3,4-Xylenol	226.947
Propyl iodide	102-103	3,5-Xylenol	221.692
Pyridine	115-116		

[a]bp measured at 763 mm Hg.
[b]bp measured at 758 mm Hg.
[c]bp measured at 754 mm Hg.
[d]bp measured at 761 mm Hg.

Density of Liquids

(Applicable to the usual purity of the compounds listed)

Liquid	lbs/gal	kg/l	Liquid	lbs/gal	kg/l
acetal	6.9	0.83	butyl butyrate	7.2	0.86
acetic acid, glacial	8.7	1.04	butyl Carbitol	8.0	0.96
acetic anhydride	9.0	1.08	butyl Cellosolve	7.5	0.90
acetone	6.6	0.79	n-butyl chloride	7.4	0.89
acetonitrile	6.5	0.78	butyl citrate	8.7	1.04
acetophenone	8.6	1.03	1,3-butyleneglycol	8.3	0.99
acetyl chloride	9.2	1.10	n-butyl ether	6.4	0.77
acetylene dichloride	10.7	1.28	butyl ethyl ether	6.3	0.75
acetylene tetrachloride	13.3	1.59	butyl lactate	8.1	0.97
acrylonitrile	6.6	0.79	butyl oxalate	8.3	0.99
alcohol	6.8	0.81	n-butyl phthalate	8.7	1.04
alcohol, anhydrous	6.6	0.79	butyl propionate	7.2	0.86
aldehol	6.9	0.83	butyl stearate	7.2	0.86
aldol	9.2	1.10	n-butyraldehyde	6.8	0.81
allyl alcohol	7.1	0.85	butyric acid	8.0	0.96
allyl bromide	11.6	1.39	butyric anhydride	8.1	0.97
allyl chloride	7.7	0.92	n-caproic acid	7.7	0.92
aluminum subacetate soln	8.7	1.04	caprylic acid	7.6	0.91
ammonia solution, diluted (10%)	8.0	0.96	caraway oil	7.5	0.90
ammonia solution, strong (28%)	7.5	0.90	Carbitol	8.2	0.98
amyl acetate	7.3	0.87	Carbitol acetate	8.5	1.02
n-amyl alcohol	6.8	0.81	carbon disulfide	10.5	1.26
amyl alcohol, secondary	6.8	0.81	carbon tetrachloride	13.3	1.59
amyl alcohol, tertiary	6.7	0.80	cardamom oil	7.8	0.93
amylamine	6.3	0.75	carvacrol	8.1	0.97
amylene hydrate	6.7	0.80	carvone	8.0	0.96
amyl mercaptan	7.1	0.85	castor oil	8.0	0.96
amyl nitrate	7.1	0.85	Cellosolve	7.8	0.93
amyl oleate	7.2	0.86	Cellosolve acetate	8.1	0.97
p-tert-amylphenol	7.7	0.92	chenopodium oil	8.0	0.96
amyl phthalate	8.5	1.02	chlordane	13.4	1.61
amyl propionate	7.3	0.87	o-chloroaniline	10.1	1.21
amyl salicylate	8.8	1.05	chlorobenzene	9.2	1.10
anethole	8.2	0.98	β-chloroethyl acetate	9.6	1.15
aniline	8.5	1.02	chloroethylbenzene	8.7	1.04
anise oil	8.2	0.98	chloroform	12.3	1.47
o-anisidine	9.1	1.09	β-chlorophenetole	9.6	1.15
anisole	8.3	0.99	o-chlorophenol	10.3	1.23
aromatic ammonia spirit	7.5	0.90	chloropicrin	13.8	1.65
balsam canada	8.3	0.99	chlorosulfonic acid	14.7	1.76
bay oil	8.1	0.97	chlorotoluene	9.0	1.08
benzal chloride	10.5	1.26	cinnamon oil	8.8	1.05
benzaldehyde	8.7	1.04	citronellal	7.1	0.85
benzene	7.3	0.87	citronella oil	7.5	0.90
benzotrichloride	11.5	1.38	clove oil	8.7	1.04
benzoyl chloride	10.2	1.22	coconut oil	7.7	0.92
benzyl acetate	8.8	1.05	cod liver oil	7.7	0.92
benzyl alcohol	8.7	1.04	collodion	6.4	0.77
benzyl benzoate	9.3	1.11	coriander oil	7.2	0.86
benzyl Cellosolve	8.9	1.07	corn oil	7.6	0.91
benzyl chloride	9.2	1.10	cottonseed oil	7.6	0.91
benzyl ether	8.7	1.04	creosote	9.0	1.08
bergamot oil	7.3	0.87	cresol	8.6	1.03
bitter almond oil	8.7	1.04	m-cresol	8.6	1.03
bitter orange oil	7.1	0.85	o-cresol	8.7	1.04
bornyl acetate	8.2	0.98	p-cresol	8.6	1.03
bromine	26.0	3.11	crotonaldehyde	7.2	0.86
bromobenzene	12.5	1.50	crotonyl alcohol	7.3	0.87
bromoethylbenzene	11.6	1.39	cubeb oil	7.6	0.91
bromoform	24.0	2.88	cumene	7.2	0.86
o-bromotoluene	11.9	1.43	cyclohexane	6.5	0.78
p-bromotoluene	11.7	1.40	cyclohexanol	8.0	0.96
n-butyl acetate	7.3	0.87	cyclohexanone	7.9	0.95
butyl acetate, secondary	7.2	0.86	cyclohexene	6.7	0.80
butyl acetyl ricinoleate	7.8	0.93	cyclohexyl glycolate	8.6	1.03
n-butyl alcohol	6.8	0.81	cyclopentane	6.2	0.74
butyl alcohol, secondary	6.7	0.80	cyclopentanol	7.9	0.95
butyl alcohol, tertiary	6.6	0.79	cyclopentanone	7.9	0.95
butylamine, tertiary	5.8	0.69	cyclopentene	6.4	0.77
n-butyl bromide	10.6	1.27	m-cymene	7.2	0.86

Density of Liquids (Continued)

Liquid	lbs/gal	kg/l	Liquid	lbs/gal	kg/l
o-cymene	7.3	0.87	ethylene glycol	9.3	1.11
p-cymene	7.1	0.85	ethylene glycol diacetate	9.2	1.10
decalin	7.5	0.90	ethylene glycol monacetate	9.2	1.10
n-decyl alcohol	6.9	0.83	ethyl formate	7.7	0.92
diacetone alcohol	7.8	0.93	ethyl iodide	16.1	1.93
diallyl ether	6.7	0.80	ethyl isobutyrate	7.2	0.86
dibutylamine	6.4	0.77	ethyl isopropyl ether	6.2	0.74
dibutyl phthalate	8.7	1.04	ethyl isovalerate	7.2	0.86
dichloroacetic acid	13.1	1.57	ethyl lactate	8.6	1.03
o-dichlorobenzene	10.9	1.31	ethyl maleate	8.9	1.07
sym-dichloroethyl ether	10.2	1.22	ethyl malonate	8.8	1.05
dichloroisopropyl ether	9.3	1.11	ethyl methyl ether	6.0	0.72
diethanolamine	9.1	1.09	ethyl nitrite	7.5	0.90
diethylamine	5.9	0.71	ethyl nitrite spirit	6.9	0.83
2-diethylaminoethanol	7.2	0.86	ethyl oxalate	9.0	1.08
diethylaniline	7.8	0.93	ethylphenylethanolamine	8.7	1.04
diethyl carbinol	6.8	0.81	ethyl phenyl ketone	8.5	1.02
diethyl Carbitol	7.6	0.91	ethyl phthalate	9.4	1.13
diethylene glycol	9.3	1.11	ethyl propionate	7.4	0.89
diethylene glycol monolaurate	8.0	0.96	2-ethyl-3-propylacrolein	7.1	0.85
diethylformamide	7.5	0.90	ethyl n-propyl ether	6.1	0.73
diethyl ketone	6.8	0.81	ethyl propyl ketone	6.8	0.81
diethyl phthalate	9.4	1.13	ethyl salicylate	9.4	1.13
diethyl sulfate	9.8	1.17	ethyl silicate	7.8	0.93
diglycol chlorohydrin	9.7	1.16	ethylsulfuric acid	11.0	1.32
diglycol laurate	8.1	0.97	ethyl trichloroacetate	11.5	1.38
diglycol oleate	7.7	0.92	ethyl urethan	8.2	0.98
diisobutyl ketone	6.7	0.80	eucalyptol	7.7	0.92
diisopropanolamine	8.3	0.90	eucalyptus oil	7.6	0.91
diisopropylamine	6.0	0.72	eugenol	8.9	1.07
dimercaprol	10.3	1.23	expressed almond oil	7.6	0.91
dimethoxytetraglycol	8.4	1.01	fennel oil	8.0	0.96
dimethylamine	5.7	0.68	ferric chloride soln	11.0	1.32
dimethylaniline	8.0	0.96	ferric subsulfate soln	12.9	1.55
dimethylformamide	7.9	0.95	ferrous iodide syrup	11.4	1.37
dimethyl furan	7.4	0.89	formaldehyde soln	9.0	1.08
dimethyl glyoxal	8.3	0.99	formamide	9.5	1.14
dimethyl phthalate	9.9	1.19	formic acid, 90%	10.0	1.20
dimethyl sulfate	11.1	1.33	formic acid, 25%	8.8	1.05
dioctylamine	6.7	0.80	furan	7.8	0.93
1,4-dioxane	8.6	1.03	furfural	9.7	1.16
dioxolane	8.9	1.07	furfuryl alcohol	9.4	1.13
diphenylamine	9.7	1.16	geranial	7.4	0.89
dipropylamine	6.1	0.73	geraniol	7.4	0.89
dipropylene glycol	8.6	1.03	glycerin	10.4	1.25
dipropyl ketone	6.8	0.81	glycerophosphoric acid	13.2	1.58
dwarf pine needle oil	7.2	0.66	glyceryl laurate	8.2	0.98
epichlorohydrin	9.9	1.19	halibut liver oil	7.7	0.92
ethanolamine	8.5	1.02	heptadecanol	7.1	0.85
ether (at 20°)	6.0	0.72	2,4-heptadiene	6.1	0.73
ethohexadiol	7.8	0.93	n-heptane	5.7	0.68
ethyl acetate	7.5	0.90	heptyl ether	6.8	0.81
ethyl acetoacetate	8.6	1.03	1,5-hexadiene	5.7	0.68
ethylamine	5.9	0.71	2,4-hexadiene	5.9	0.71
m-ethylaniline	8.0	0.96	hexaethyltetraphosphate	10.8	1.29
ethyl benzoate	8.8	1.05	n-hexaldehyde	6.9	0.83
ethyl bromide	11.9	1.43	n-hexyl alcohol	6.8	0.81
2-ethylbutyl alcohol	6.9	0.83	n-hexyl methyl ketone	6.8	0.81
ethyl butyl ketone	6.8	0.81	hydrazine	8.4	1.01
2-ethylbutyraldehyde	6.8	0.81	hydriodic acid, 47%	12.5	1.50
ethyl n-butyrate	7.3	0.87	hydrobromic acid, 40%	11.5	1.38
2-ethylbutyric acid	7.7	0.92	hydrochloric acid, 36%	9.8	1.17
ethyl caproate	7.5	0.90	hydrocyanic acid, 97%	5.8	0.69
ethyl carbonate	8.1	0.97	hydrofluoric acid, 55%	9.8	1.17
ethyl chloroacetate	9.7	1.16	hydrofluosilicic acid, 30%	10.6	1.27
ethyl chloroformate	9.5	1.14	hydrogen peroxide, anhyd	12.2	1.46
ethyl cinnamate	8.8	1.05	hydrogen peroxide, 30%	9.2	1.10
ethyl citrate	9.5	1.14	hydroxybutyraldehyde	9.2	1.10
ethylene chlorohydrin	10.1	1.21	hypophosphorous acid, 30%	9.4	1.13
ethylenediamine, anhyd	7.5	0.90	indalone	9.0	1.08
ethylene dibromide	18.2	2.18	indan	8.0	0.96
ethylene dichloride	10.5	1.26	β-ionone	7.9	0.95

Liquid	lbs/gal	kg/l	Liquid	lbs/gal	kg/l
isoamyl acetate	7.3	0.87	methyl isobutyrate	7.4	0.89
isoamyl alcohol	6.8	0.81	methyl lactate	9.0	1.08
isoamyl benzoate	8.3	0.99	methyl phthalate	9.9	1.19
isoamyl bromide	10.2	1.22	methyl propyl ether	6.1	0.73
isoamyl butyrate	7.1	0.85	methyl propyl ketone	6.7	0.80
isoamyl chloride	7.4	0.89	methyl salicylate	9.9	1.19
isoamyl ether	6.5	0.78	monacetin	10.1	1.21
isoamyl isovalerate	7.1	0.85	α-monochlorohydrin	11.1	1.33
isoamyl nitrite	7.3	0.87	monoisopropanolamine	8.2	0.98
isoamyl phthalate	8.6	1.03	morpholine	8.3	0.99
isoamyl salicylate	8.9	1.07	mustard oil	7.6	0.91
isobutenyl chloride	7.7	0.92	myristica oil	7.5	0.90
isobutyl acetate	7.3	0.87	neatsfoot oil	7.6	0.91
isobutyl alcohol	6.8	0.81	nitric acid (68%)	11.7	1.40
isobutylamine	6.0	0.72	nitric acid, fuming	12.5	1.50
isobutyl benzoate	8.3	0.99	nitrobenzene	10.0	1.20
isobutyl bromide	10.5	1.26	nitroglycerin	13.3	1.59
isobutyl n-butyrate	7.2	0.86	nitromethane	9.4	1.13
isobutyl chloride	7.4	0.89	m-nitrotoluene	9.6	1.15
isobutyl ether	6.3	0.75	o-nitrotoluene	9.7	1.16
isobutyl formate	7.3	0.87	n-nonane	5.9	0.71
isobutyl nitrite	7.2	0.86	octanal	6.8	0.81
isobutyl propionate	7.4	0.89	octane	5.9	0.71
isobutyric acid	7.9	0.95	octyl acetate	7.3	0.87
isoeugenol	9.1	1.09	octyl alcohol	6.9	0.83
isoprene	5.7	0.68	octyl amine	6.6	0.79
isopropyl acetate	7.3	0.87	oleic acid	7.5	0.90
isopropyl alcohol	6.5	0.78	olive oil	7.6	0.91
isopropylamine	5.8	0.69	orange flower oil	7.2	0.86
isopropylbenzene	7.2	0.86	orange oil	7.0	0.84
isopropyl bromide	10.9	1.31	paraldehyde	8.3	0.99
isopropyl chloride	7.2	0.86	peanut oil	7.6	0.91
isopropyl ether	6.0	0.72	pennyroyal oil, American	7.8	0.93
isovaleric acid	7.8	0.93	pentachloroethane	14.0	1.68
isovaleronitrile	6.5	0.78	pentane	5.2	0.62
lactic acid, 85–90%	10.1	1.21	peppermint oil	7.5	0.90
lavender oil	7.4	0.89	perchloric acid, 70%	14.0	1.68
lead subacetate soln	10.4	1.25	persic oil	7.6	0.91
lemon oil	7.1	0.85	peruvian balsam	9.7	1.16
lemon grass oil	7.5	0.90	petroleum benzin	5.5	0.66
d-limonene	7.0	0.84	phenetole	8.0	0.96
linoleic acid	7.5	0.90	phenylacetaldehyde	8.6	1.03
linseed oil	7.7	0.92	phenyl acetate	8.9	1.07
liquefied phenol	8.9	1.07	phenyl Cellosolve	9.2	1.10
liquid petrolatum	7.4	0.89	phenylethanolamine	9.0	1.08
liquid petrolatum, light	7.1	0.85	phenyl ethyl acetate	8.6	1.03
l-menthol	7.4	0.89	phenylethyl alcohol	8.5	1.02
mercury	112.9	13.53	phosphoric acid, 85%	14.2	1.70
mesityl oxide	7.1	0.85	o-phthalyl chloride	11.7	1.40
methyl acetate	7.7	0.92	3-picoline	8.0	0.96
methyl acetoacetate	8.9	1.07	pimenta oil	8.6	1.03
methylal	7.1	0.85	dl-pinene	7.1	0.85
methyl alcohol	6.6	0.79	pine needle oil	7.4	0.89
methyl amyl acetate	7.1	0.85	pine oil	7.8	0.93
methyl amyl alcohol	6.7	0.80	piperidine	7.2	0.86
methyl amyl carbinol	6.8	0.81	polyethylene glycol 300	9.4	1.13
methyl-n-amyl ketone	6.8	0.81	polyethylene glycol 400	9.4	1.13
N-methylaniline	8.2	0.98	polysorbate 80	9.0	1.08
methyl butyl ketone	6.9	0.83	poppy oil	7.7	0.92
methyl Carbitol	8.6	1.03	potassium hydroxide soln, 50%	12.7	1.52
methyl Cellosolve	8.0	0.96	propionaldehyde	6.7	0.80
methyl Cellosolve acetate	8.4	1.01	propionic acid	8.3	0.99
methyl cyclohexanone	7.7	0.92	propionic anhydride	8.3	0.99
methylene bromide	20.8	2.49	propionitrile	6.5	0.78
methylene chloride	11.1	1.33	propyl acetate	7.4	0.89
methylene iodide	27.7	3.32	n-propyl alcohol	6.7	0.80
methyl ethyl ketone	6.7	0.80	propyl chloride	7.4	0.89
methyl formate	8.1	0.97	propylene chlorohydrin	9.2	1.10
methyl hexyl ketone	6.8	0.81	propylenediamine	7.3	0.87
methyl-3-hydroxybutyrate	8.8	1.05	propylene dichloride	9.7	1.16
methyl iodide	18.8	2.25	propylene glycol	8.6	1.03
methyl isobutyl ketone	6.7	0.80	propylene oxide	6.9	0.83

Density of Liquids (Continued)

Liquid	lbs/gal	kg/l	Liquid	lbs/gal	kg/l
propyl ether	6.1	0.73	thyme oil	7.7	0.92
pyridine	8.2	0.98	toluene	7.2	0.86
quinoline	9.1	1.09	*m*-toluidine	8.2	0.98
rapeseed oil	7.6	0.91	*o*-toluidine	8.4	1.01
rosemary oil	7.5	0.90	triacetin	9.7	1.16
rose oil	7.1	0.85	tributylamine	6.5	0.78
safrol	9.2	1.10	tributyl citrate	8.7	1.04
salicylaldehyde	9.7	1.16	tributyl phosphate	8.1	0.97
santal oil	8.1	0.97	1,1,2-trichloroethane	12.0	1.44
sassafras oil	8.9	1.07	trichloroethylene	12.2	1.46
sesame oil	7.6	0.91	triethanolamine	9.4	1.13
sodium carbonate soln, 10%	9.1	1.09	triethylamine	6.0	0.72
sodium hydroxide soln, 50%	12.8	1.53	triethyleneglycol	9.4	1.13
sorbitol	11.0	1.32	triethylphosphate	8.9	1.07
sorbitol soln	10.4	1.25	trimethylene glycol	8.8	1.05
soya oil	7.7	0.92	trimethyl phosphate	10.1	1.21
spearmint oil	7.7	0.92	triolein	7.6	0.91
sperm oil	7.3	0.87	tung oil	7.8	0.93
styrene	7.6	0.91	turpentine oil	7.2	0.86
succinaldehyde	8.9	1.07	turpentine oil, rectified	7.2	0.86
sulfuric acid, 96%	15.3	1.83	undecane	6.2	0.74
sulfurous acid, 6%	8.5	1.02	1-undecanol	6.9	0.83
syrup	11.0	1.32	undecylemic acid	7.6	0.91
terebene	7.2	0.86	valeric acid	7.9	0.95
terpineol	7.8	0.93	valeronitrile	6.7	0.80
tetrachloroethylene	13.6	1.63	vinyl acetate	7.8	0.93
tetradecanol	7.0	0.84	vinyl ether	6.4	0.77
tetraethylene glycol	9.4	1.13	vitamin K_1	8.1	0.97
tetrahydrofuran	7.4	0.89	water	8.3	0.99
tetrahydrofurfuryl alcohol	8.8	1.05	whiskey	7.7	0.92
tetralin	8.1	0.97	*m*-xylene	7.2	0.86
thioglycolic acid	11.0	1.32	*o*-xylene	7.5	0.90
thionyl chloride	13.6	1.63	ylang-ylang oil	7.8	0.93
thiophene	8.9	1.07			

Percentage Solution Table

Although the making of a solution of given percentage strength is a simple matter when both the substance and the solvent are *weighed,* the preparation of a *definite volume* of solution of given percentage strength is not so easy, because the displacement volume of the substance is not known. Hence, it is practically impossible to determine the ultimate volume readily produced by a given *weight* of substance and a definite *volume* of solvent. For weight-in-volume solutions of low concentration, the margin of error is relatively low but as the concentration increases the error increases also. Accurate strength solutions of higher concentration are best made by weighing.

However, weight-in-volume solutions are prescribed by the U.S.P. to be used in compounding prescriptions whenever gases or solids are dissolved in liquids, since most physicians have in mind a certain weight of substance in a definite volume of solution. Such solutions are defined by the U.S.P. as "the number of grams of an active constituent in 100 milliliters of solution" regardless of whether water or some other liquid is the vehicle.

For such weight-in-volume solutions, the following tables will afford a ready means of ascertaining the quantities (by weight) of substance required to prepare varying volumes of definite w/v percentage strength. The calculations are based on 1 fl oz (480 minims) of water = 455 grains (round number) in the Apothecaries table and 100 ml of water = 100 g in the Metric table.

APOTHECARIES

Strength of solution (% w/v)	$\frac{1}{4}$ fl oz	$\frac{1}{2}$ fl oz	1 fl oz	2 fl oz	3 fl oz	4 fl oz	6 fl oz	8 fl oz	10 fl oz	12 fl oz	16 fl oz
					Grains of substance required per given volume of solution						
0.01	0.0114	0.023	0.046	0.091	0.137	0.182	0.273	0.364	0.455	0.546	0.728
0.02	0.023	0.046	0.091	0.182	0.273	0.364	0.546	0.728	0.910	1.092	1.456
0.04	0.046	0.091	0.182	0.364	0.546	0.728	1.092	1.456	1.820	2.184	2.912
0.05	0.057	0.114	0.2275	0.455	0.683	0.910	1.365	1.820	2.275	2.730	3.640
0.1	0.114	0.228	0.455	0.910	1.365	1.820	2.730	3.640	4.55	5.46	7.28
0.2	0.2275	0.455	0.910	1.820	2.730	3.640	5.460	7.28	9.10	10.92	14.56
0.25	0.284	0.569	1.138	2.275	3.413	4.55	6.83	9.10	11.38	13.65	18.20
0.5	0.569	1.138	2.275	4.55	6.83	9.10	13.65	18.20	22.75	27.3	36.4
1	1.138	2.275	4.55	9.10	13.65	18.2	27.3	36.4	45.5	54.6	72.8
2	2.275	4.55	9.10	18.20	27.30	36.4	54.6	72.8	91.0	109.2	145.6
3	3.413	6.83	13.65	27.30	40.95	54.6	81.9	109.2	136.5	163.8	218.4
4	4.55	9.10	18.20	36.40	54.60	72.8	109.2	145.6	182.0	218.4	291.2
5	5.69	11.37	22.75	45.50	68.25	91.0	136.5	182.0	227.5	273	364
10	11.38	22.75	45.50	91.0	136.50	182	273	364	455	546	728
15	17.06	34.13	68.25	136.5	204.75	273	409.5	546	682.5	819	1092
20	22.75	45.50	91.0	182	273	364	546	728	910	1092	1456
25	28.44	56.90	113.75	227.5	341.25	455	682.5	910	1137.5	1365	1820
30	34.13	68.25	136.5	273	409.5	546	819	1092	1365	1638	2184
40	45.5	91.0	182	364	546	728	1092	1456	1820	2184	2912

METRIC

Strength of solution (% w/v)	10 ml	15 ml	25 ml	30 ml	60 ml	90 ml	100 ml	120 ml	150 ml	200 ml	500 ml	1000 ml
					Grams of substance required per given volume of solution							
0.01	0.001	0.0015	0.0025	0.003	0.006	0.009	0.01	0.012	0.015	0.02	0.05	0.1
0.02	0.002	0.003	0.005	0.006	0.012	0.018	0.02	0.024	0.03	0.04	0.1	0.2
0.05	0.005	0.008	0.013	0.015	0.03	0.045	0.05	0.06	0.075	0.1	0.25	0.5
0.1	0.01	0.015	0.025	0.03	0.06	0.09	0.1	0.12	0.15	0.2	0.5	1.0
0.2	0.02	0.03	0.05	0.06	0.12	0.18	0.2	0.24	0.30	0.4	1.0	2.0
0.25	0.025	0.038	0.063	0.075	0.15	0.225	0.25	0.3	0.375	0.5	1.25	2.5
0.5	0.05	0.075	0.125	0.15	0.3	0.45	0.5	0.6	0.75	1.0	2.5	5.0
1	0.1	0.15	0.25	0.3	0.6	0.9	1.0	1.2	1.5	2.0	5.0	10.0
1.5	0.15	0.225	0.375	0.45	0.9	1.35	1.5	1.8	2.25	3.0	7.5	15.0
2	0.2	0.3	0.5	0.6	1.2	1.8	2.0	2.4	3.0	4.0	10.0	20.0
3	0.3	0.45	0.75	0.9	1.8	2.7	3.0	3.6	4.5	6.0	15.0	30.0
4	0.4	0.6	1.0	1.2	2.4	3.6	4.0	4.8	6.0	8.0	20.0	40.0
5	0.5	0.75	1.25	1.5	3.0	4.5	5.0	6.0	7.5	10.0	25.0	50.0
10	1.0	1.5	2.5	3.0	6.0	9.0	10.0	12.0	15.0	20.0	50.0	100.0
15	1.5	2.25	3.75	4.5	9.0	13.5	15.0	18.0	22.5	30.0	75.0	150.0
20	2.0	3.0	5.0	6.0	12.0	18.0	20.0	24.0	30.0	40.0	100.0	200.0
25	2.5	3.75	6.25	7.5	15.0	22.5	25.0	30.0	37.5	50.0	125.0	250.0
40	4.0	6.0	10.0	12.0	24.0	36.0	40.0	48.0	60.0	80.0	200.0	400.0

Isotonic Solutions

The following table provides data for adjusting the osmolality of aqueous solutions, so that they will be isosmotic with normal saline solution (0.9% NaCl) and therefore isotonic with blood and tears. This can be achieved by using either the freezing point depression method or the sodium chloride equivalent method. The freezing point depression values have been determined experimentally and published. (Refs. 1-6) NaCl equivalent values have been calculated from these data.

The top values listed for each chemical are NaCl equivalents. The second values, in italics, are freezing point depression values in degrees Celsius. The percentage concentration (w/v) at isotonicity is given in bold face in the last column.

Using the freezing point depression method an isotonic solution is prepared by matching the freezing point depression value of 0.9% NaCl which is accepted as 0.52. Freezing point depression values are additive in this concentration range; therefore, to make a 2% Hepes buffer solution isotonic with tears, the calculation is as follows:

Freezing point depression of isotonic solution   0.52
Freezing point depression of 2% Hepes from table   0.163
    Difference to be supplied by NaCl   0.357
    0.9% NaCl has a freezing point depression of 0.52,
    x% NaCl has a freezing point depression of 0.357.
Therefore $x/0.9 = 0.357/0.52$
            $x = 0.618$

Thus 0.618 g/100ml NaCl combined with 2 g/100ml Hepes produces an isotonic solution. Note: If more than one component is in solution the freezing point depressions of each must be added together before the difference value is obtained.

Using the NaCl equivalents method to prepare a solution involves: Multiplying the number of grams of each component corrected for volume by its NaCl equivalent from the table at the nearest listed concentration. For example to make 60 ml of 1% boric acid isotonic: Add the NaCl equivalents of each solution component.

       60 ml of 1% boric acid soln (0.6g boric acid x 0.5g NaCl equiv.) = 0.3g

Calculate the number of grams of NaCl required for the volume being used. (60 ml need 0.54 g) Subtract the NaCl equivalents of the components from the total NaCl required. (0.54 - 0.3 = 0.24g) The difference is the amount of NaCl which must be added to that specific volume. Note: If the difference is less than zero then the solution is already hypertonic and cannot be adjusted without altering the concentration of the components.

Sodium Chloride Equivalents and Freezing Point Depressions (°C)
for Certain Concentrations (w/v) of Solution

Chemical	Concentration of Solution, NaCl Equivalents					At Isosmotic Concentration	
	0.5%	1%	2%	3%	5%		
Acetrizoate methylglucam	0.09	0.08	0.08	0.08	0.08	0.07	**12.12%**
	0.024°	*0.04°*	*0.09°*	*0.13°*	*0.22°*	*0.52°*	**12.12%**
Acetrizoate sodium	0.10	0.10	0.10	0.10	0.10	0.09	**9.64%**
	0.027°	*0.05°*	*0.10°*	*0.16°*	*0.27°*	*0.52°*	**9.64%**
Acetylcysteine	0.20	0.20	0.20	0.20	—	0.20	**4.58%**
	0.055°	*0.11°*	*0.22°*	*0.34°*	—	*0.52°*	**4.58%**
Acetylsulfanilamide sodium	0.24	0.23	0.23	0.23	—	0.23	**3.85%**
	0.066°	*0.13°*	*0.26°*	*0.40°*	—	*0.52°*	**3.85%**
Acriflavine	0.10	0.10	0.09	0.09	—	—	
	0.025°	*0.05°*	*0.10°*	*0.15°*	—	—	
Adenosine phosphate	0.50	0.41	—	—	—	—	
	0.140°	*0.23°*	—	—	—	—	
Adrenalone hydrochloride	0.30	0.27	0.24	0.22	—	0.21	**4.24%**
	0.086°	*0.15°*	*0.27°*	*0.38°*	—	*0.52°*	**4.24%**
Alcohol	0.65	0.65	—	—	—	0.65	**1.39%**
	0.188°	*0.37°*	—	—	—	*0.52°*	**1.39%**
Alcohol, dehydrate	0.70	0.70	—	—	—	0.70	**1.28%**
	0.203°	*0.40°*	—	—	—	*0.52°*	**1.28%**
Alum, potassium	0.20	0.18	0.16	0.15	0.15	0.14	**6.35%**
	0.054°	*0.10°*	*0.18°*	*0.26°*	*0.41°*	*0.52°*	**6.35%**
Aminacrine hydrochloride	0.20	0.17	—	—	—	—	
	0.052°	*0.09°*	—	—	—	—	
Aminoacetic acid	0.42	0.41	0.41	—	—	0.41	**2.20%**
	0.119°	*0.23°*	*0.47°*	—	—	*0.52°*	**2.20%**
Aminocaproic acid	0.26	0.26	0.26	0.26	—	0.26	**3.52%**
	0.075°	*0.14°*	*0.29°*	*0.44°*	—	*0.52°*	**3.52%**

Chemical	Concentration of Solution, NaCl Equivalents						
	0.5%	1%	2%	3%	5%	At Isosmotic Concentration	
p-Aminohippuric acid	0.13 0.035°	0.13 0.07°	— —	— —	— —	— —	
p-Aminosalicylate sodium	0.30 0.086°	0.29 0.16°	0.29 0.32°	0.28 0.47°	— —	0.27 0.52°	**3.27%** **3.27%**
Ammonium carbonate	0.70 0.202°	0.70 0.40°	— —	— —	— —	0.70 0.52°	**1.29%** **1.29%**
Ammonium chloride	1.10 0.315°	— —	— —	— —	— —	1.07 0.52°	**0.84%** **0.84%**
Ammonium lactate	0.33 0.093°	0.33 0.18°	0.33 0.37°	— —	— —	0.33 0.52°	**2.76%** **2.76%**
Ammonium nitrate	0.69 0.200°	0.69 0.40°	— —	— —	— —	0.69 0.52	**1.30%** **1.30%**
Ammonium phosphate, dibasic	0.58 0.165°	0.55 0.31°	— —	— —	— —	0.51 0.52°	**1.76%** **1.76%**
Ammonium sulfate	0.55 0.158°	0.55 0.31°	— —	— —	— —	0.54 0.52°	**1.68%** **1.68%**
Antimony potassium tartrate	0.22 0.065°	0.18 0.10°	0.15 0.17°	0.13 0.23°	0.10 0.33°	— —	
Antipyrine	0.18 0.050°	0.17 0.09°	0.16 0.17°	0.14 0.25°	0.14 0.39°	0.13 0.52°	**6.81%** **6.81%**
Arecoline hydrobromide	0.030 0.084°	0.27 0.15°	0.25 0.28°	0.24 0.41°	— —	0.23 0.52°	**3.88%** **3.88%**
Arginine glutamate	0.17 0.048°	0.17 0.09°	0.17 0.19°	0.17 0.29°	0.17 0.48°	0.17 0.52°	**5.37%** **5.37%**
L-Arginine hydrochloride	0.31 0.087°	0.30 0.17°	0.28 0.32°	0.27 0.46°	— —	0.26 0.52°	**3.43%** **3.43%**
Arsenic trioxide	0.30 0.085°	0.30 0.16°	— —	— —	— —	— —	
Ascorbic acid	0.20 0.053°	0.18 0.10°	0.18 0.20°	0.18 0.31°	0.18 0.51°	0.18 0.52°	**5.94%** **5.94%**
Atropine methylnitrate	0.20 0.055°	0.18 0.10°	0.16 0.18°	0.15 0.26°	0.14 0.41°	0.14 0.52°	**6.52%** **6.52%**
Atropine sulfate	0.14 0.039°	0.13 0.07°	0.12 0.13°	0.11 0.19°	0.11 0.31°	0.10 0.52°	**8.85%** **8.85%**
Aurothioglucose	0.03 0.007°	0.03 0.01°	0.03 0.02°	0.03 0.04°	0.03 0.07°	— —	
Barbital sodium	0.32 0.087°	0.30 0.17°	0.29 0.33°	0.29 0.50°	— —	0.29 0.52°	**3.12%** **3.12%**
Benzalkonium chloride	0.18 0.048°	0.16 0.09°	0.15 0.17°	0.14 0.24°	0.13 0.38°	— —	
Benzethonium chloride	0.08 0.022°	0.05 0.02°	0.03 0.03°	0.02 0.04°	0.02 0.05°	— —	
Benzyl alcohol	0.18 0.049°	0.17 0.09°	0.16 0.18°	0.15 0.26°	— —	— —	
Bismuth potassium tartrate	0.10 0.033°	0.09 0.05°	0.07 0.08°	0.06 0.10°	0.05 0.14°	— —	
Bismuth sodium tartrate	0.14 0.041°	0.13 0.07°	0.13 0.13°	0.12 0.19°	0.11 0.31°	0.10 0.52°	**8.91%** **8.91%**

Chemical	Concentration of Solution, NaCl Equivalents					At Isosmotic Concentration	
	0.5%	1%	2%	3%	5%		
Boric acid	0.52	0.50	—	—	—	0.47	1.9%
	0.146°	*0.28°*	—	—	—	*0.52°*	1.9%
Bromodiphenhydramine hydrochloride	0.20	0.17	0.14	0.10	0.07	—	
	0.067°	*0.10°*	*0.16°*	*0.18°*	*0.20°*	—	
Butabarbital sodium	0.27	0.27	0.27	0.27	—	0.27	3.33%
	0.078°	*0.15°*	*0.31°*	*0.47°*	—	*0.52°*	3.33%
Caffeine	0.08	0.08	—	—	—	—	
	0.025°	*0.04°*	—	—	—	—	
Calcium aminosalicylate	0.30	0.27	0.23	0.21	—	—	
	0.091°	*0.15°*	*0.26°*	*0.36°*	—	—	
Calcium chloride dihydrate	0.50	0.51	—	—	—	0.53	1.70%
	0.145°	*0.29°*	—	—	—	*0.52°*	1.70%
Calcium chloride hexahydrate	0.34	0.35	0.36	—	—	0.36	2.5%
	0.097°	*0.20°*	*0.41°*	—	—	*0.52°*	2.5%
Calcium chloride, anhydrous	0.70	0.70	—	—	—	0.70	1.29%
	0.206°	*0.40°*	—	—	—	*0.52°*	1.29%
Calcium disodium edetate	0.21	0.21	0.21	0.20	—	0.20	4.50%
	0.061°	*0.12°*	*0.24°*	*0.35°*	—	*0.52°*	4.50%
Calcium gluconate	0.18	0.16	0.15	0.14	—	—	
	0.050°	*0.09°*	*0.16°*	*0.23°*	—	—	
Calcium lactate	0.26	0.23	0.22	0.21	—	0.20	4.5%
	0.073°	*0.13°*	*0.25°*	*0.37°*	—	*0.52°*	4.5%
Chiniofon	0.14	0.13	0.12	0.11	—	—	
	0.039°	*0.07°*	*0.13°*	*0.20°*	—	—	
Chloramine-T	0.24	0.23	0.22	0.22	—	0.22	4.1%
	0.064°	*0.12°*	*0.25°*	*0.38°*	—	*0.52°*	4.1%
Chlorobutanol, hydrated	0.24	—	—	—	—	—	—
	0.071°	—	—	—	—	—	—
Chlorophyll	0.14	0.10	0.08	0.06	0.05	—	
	0.037°	*0.05°*	*0.08°*	*0.11°*	*0.15°*	—	
Citric acid	0.18	0.18	0.17	0.17	0.16	0.16	5.52%
	0.050°	*0.09°*	*0.19°*	*0.28°*	*0.47°*	*0.52°*	5.52%
Congo red	0.05	0.05	0.05	0.05	0.05	—	
	0.015°	*0.03°*	*0.05°*	*0.09°*	*0.15°*	—	
Cromolyn sodium	0.16	0.14	0.11	0.09	0.05		
	0.046°	*0.08°*	*0.12°*	*0.14°*	*0.15°*		
Cupric sulfate	0.20	0.18	0.16	0.15	0.14	0.13	6.85%
	0.054°	*0.09°*	*0.17°*	*0.25°*	*0.39°*	*0.52°*	6.85%
Cupric sulfate, anhydrous	0.30	0.27	0.25	0.23	—	0.22	4.09%
	0.084°	*0.15°*	*0.28°*	*0.39°*	—	*0.52°*	4.09%
Cyclophosphamide	0.10	0.10	0.10	—	—	—	
	0.031°	*0.06°*	*0.12°*	—	—	—	
Dextrose	0.16	0.16	0.16	0.16	0.16	0.16	5.51%
	0.045°	*0.09°*	*0.18°*	*0.27°*	*0.47°*	*0.52°*	5.51%
Dextrose, anhydrous	0.18	0.18	0.18	0.18	0.18	0.18	5.05%
	0.050°	*0.10°*	*0.20°*	*0.31°*	*0.51°*	*0.52°*	5.05%
Diatrizoate sodium	0.10	0.09	0.09	0.09	0.09	0.09	10.55%
	0.025°	*0.04°*	*0.09°*	*0.14°*	*0.24°*	*0.52°*	10.55%

Isotonic Solutions (Continued)

Chemical	Concentration of Solution, NaCl Equivalents					At Isosmotic Concentration	
	0.5%	1%	2%	3%	5%		
Diethanolamine	0.31 *0.089°*	0.31 *0.17°*	0.31 *0.35°*	— —	— —	0.31 *0.52°*	**2.90%** **2.90%**
Dimethyl sulfoxide	0.42 *0.122°*	0.42 *0.24°*	0.42 *0.48°*	— —	— —	0.42 *0.52°*	**2.16%** **2.16%**
Disodium edetate	0.24 *0.070°*	0.23 *0.13°*	0.22 *0.24°*	0.21 *0.36°*	— —	0.20 *0.52°*	**4.44%** **4.44%**
Dopamine hydrochloride	0.30 *0.085°*	0.30 *0.17°*	0.29 *0.33°*	0.29 *0.50°*	— —	0.29 *0.52°*	**3.11%** **3.11%**
Ephedrine hydrochloride	0.32 *0.087°*	0.30 *0.16°*	0.29 *0.33°*	0.28 *0.48°*	— —	0.28 *0.52°*	**3.2%** **3.2%**
Ephedrine lactate	0.28 *0.075°*	0.26 *0.14°*	0.25 *0.28°*	0.24 *0.42°*	— —	0.24 *0.52°*	**3.72%** **3.72%**
Ephedrine sulfate	0.24 *0.070°*	0.23 *0.13°*	0.22 *0.24°*	0.20 *0.35°*	— —	0.20 *0.52°*	**4.54%** **4.54%**
Epinephrine bitartrate	0.18 *0.050°*	0.18 *0.09°*	0.17 *0.19°*	0.16 *0.28°*	0.16 *0.45°*	0.16 *0.52°*	**5.7%** **5.7%**
Epinephrine hydrochloride	0.30 *0.088°*	0.29 *0.16°*	0.27 *0.31°*	0.26 *0.45°*	— —	0.26 *0.52°*	**3.47%** **3.47%**
Ergonovine maleate	0.20 *0.055°*	0.16 *0.08°*	0.13 *0.14°*	— —	— —	— —	
Ethylenediamine	0.46 *0.130°*	0.44 *0.25°*	0.43 *0.50°*	— —	— —	— —	
Evans blue	0.06 *0.017°*	0.06 *0.03°*	0.06 *0.06°*	0.05 *0.09°*	0.05 *0.14°*	—	
Ferric ammonium citrate, green	0.18 *0.054°*	0.17 *0.09°*	0.16 *0.17°*	0.15 *0.25°*	0.14 *0.39°*	—	
Ferric cacodylate	0.10 *0.023°*	0.09 *0.04°*	0.08 *0.09°*	— —	— —	—	
Ferrous gluconate	0.16 *0.048°*	0.15 *0.08°*	0.14 *0.15°*	0.12 *0.21°*	0.11 *0.33°*	—	
Ferrous lactate	0.22 *0.062°*	0.21 *0.12°*	0.21 *0.23°*	— —	— —	—	
Fluorescein sodium	0.36 *0.099°*	0.31 *0.18°*	0.29 *0.33°*	0.27 *0.47°*	— —	0.27 *0.52°*	**3.34%** **3.34%**
Fluorouracil	0.16 *0.045°*	0.13 *0.07°*	— —	— —	— —	—	
Folinic acid-SF calcium	0.06 *0.013°*	0.05 *0.02°*	0.05 *0.05°*	0.04 *0.07°*	0.04 *0.12°*	—	
D-Fructose	0.18 *0.050°*	0.18 *0.10°*	0.18 *0.20°*	0.18 *0.31°*	0.18 *0.51°*	0.18 *0.52°*	**5.05%** **5.05%**
Galactose, anhydrous	0.18 *0.053°*	0.18 *0.10°*	0.18 *0.21°*	0.18 *0.31°*	— —	0.18 *0.52°*	**4.92%** **4.92%**
Gentamicin sulfate	0.05 *0.015°*	0.05 *0.03°*	0.05 *0.06°*	0.05 *0.09°*	0.05 *0.15°*	—	
Glucoheptonate calcium	0.12 *0.037°*	0.12 *0.06°*	0.11 *0.12°*	0.10 *0.17°*	0.10 *0.27°*	—	
Glucosulfone sodium	0.18 *0.049°*	0.16 *0.08°*	0.14 *0.16°*	0.13 *0.23°*	0.13 *0.36°*	—	

Chemical	Concentration of Solution, NaCl Equivalents						
	0.5%	1%	2%	3%	5%	At Isosmotic Concentration	
D-Glucuronic acid	0.20	0.20	0.19	0.19	0.18	0.18	**5.02%**
	0.061°	*0.11°*	*0.22°*	*0.32°*	*0.51°*	*0.52°*	**5.02%**
L-Glutamic acid	0.25	0.25	0.25	—	—	—	
	0.070°	*0.14°*	*0.29°*	—	—	—	
Glycerin	0.36	0.35	0.35	—	—	0.35	**2.6%**
	0.104°	*0.20°*	*0.40°*	—	—	*0.52°*	**2.6%**
Glycine	0.41	0.41	0.41	—	—	0.41	**2.19%**
	0.118°	*0.23°*	*0.47°*	—	—	*0.52°*	**2.19%**
Gold sodium thiomalate	0.10	0.10	0.10	0.09	0.09	—	
	0.032°	*0.06°*	*0.11°*	*0.15°*	*0.25°*	—	
Guanidine hydrochloride	0.72	0.65	—	—	—	0.61	**1.47%**
	0.208°	*0.37°*	—	—	—	*0.52°*	**1.47%**
Heparin sodium	0.07	0.07	0.07	0.07	0.07	0.07	**12.2%**
	0.021°	*0.04°*	*0.08°*	*0.12°*	*0.21°*	*0.52°*	**12.2%**
Hepes	0.17	0.16	0.14	0.14	0.13	0.13	**6.80%**
	0.048°	*0.08°*	*0.16°*	*0.23°*	*0.38°*	*0.52°*	**6.80%**
Histamine dihydrochloride	0.40	0.40	0.40	—	—	0.40	**2.24%**
	0.115°	*0.23°*	*0.46°*	—	—	*0.52°*	**2.24%**
Histamine phosphate	0.28	0.25	0.24	0.23	—	0.22	**4.1%**
	0.080°	*0.14°*	*0.27°*	*0.39°*	—	*0.52°*	**4.1%**
Histidine monohydrochloride	0.30	0.29	0.28	0.26	—	—	
	0.082°	*0.16°*	*0.31°*	*0.46°*	—	—	
Hyaluronidase	0.01	0.01	0.01	0.01	0.01	—	
	0.004°	*0.00°*	*0.01°*	*0.02°*	*0.03°*	—	
Indigotindisulfonate sodium	0.30	0.30	—	—	—	—	
	0.085°	*0.17°*	—	—	—	—	
o-Iodohippurate sodium	0.16	0.16	0.16	0.15	0.15	0.15	**5.92%**
	0.047°	*0.09°*	*0.18°*	*0.26°*	*0.44°*	*0.52°*	**5.92%**
Iodophthalein sodium	0.20	0.17	0.14	0.12	0.11	0.09	**9.58%**
	0.055°	*0.09°*	*0.15°*	*0.21°*	*0.31°*	*0.52°*	**9.58%**
Iodopyracet	0.12	0.11	0.11	0.11	0.10	0.10	**9.21%**
	0.036°	*0.06°*	*0.12°*	*0.18°*	*0.29°*	*0.52°*	**9.21%**
Iodopyracet diethylamine	0.14	0.12	0.12	0.11	0.11	0.10	**8.73%**
	0.035°	*0.06°*	*0.13°*	*0.19°*	*0.30°*	*0.52°*	**8.73%**
Iopamidol	0.03	0.03	0.03	0.03	0.03	—	
	0.008°	*0.01°*	*0.03°*	*0.05°*	*0.08°*	—	
Isopropyl Alcohol	0.53	0.53	—	—	—	0.53	**1.71%**
	0.153°	*0.30°*	—	—	—	*0.52°*	**1.71%**
Kanamycin sulfate	0.08	0.07	0.07	0.07	0.07	—	
	0.021°	*0.04°*	*0.08°*	*0.12°*	*0.21°*	—	
Lactic acid	0.44	0.41	0.39	—	—	0.39	**2.3%**
	0.124°	*0.23°*	*0.45°*	—	—	*0.52°*	**2.3%**
Lactose	0.06	0.07	0.08	0.08	0.09	0.09	**9.75%**
	0.019°	*0.04°*	*0.08°*	*0.13°*	*0.24°*	*0.52°*	**9.75%**
Lithium carbonate	1.06	—	—	—	—	0.98	**0.92%**
	0.303°	—	—	—	—	*0.52°*	**0.92%**
Lithium chloride	1.40	—	—	—	—	1.40	**0.65%**
	0.405°	—	—	—	—	*0.52°*	**0.65%**

Chemical	Concentration of Solution, NaCl Equivalents						
	0.5%	1%	2%	3%	5%	At Isosmotic Concentration	
Lyapolate sodium	0.10 *0.025°*	0.09 *0.05°*	0.09 *0.10°*	0.09 *0.15°*	0.09 *0.26°*	0.09 *0.52°*	**9.96%** **9.96%**
Magnesium chloride	0.48 *0.136°*	0.45 *0.26°*	0.45 *0.51°*	— —	— —	0.45 *0.52°*	**2.02%** **2.02%**
Magnesium sulfate	0.18 *0.049°*	0.17 *0.09°*	0.16 *0.17°*	0.15 *0.26°*	0.15 *0.41°*	0.14 *0.52°*	**6.3%** **6.3%**
Magnesium sulfate, anhydrous	0.34 *0.093°*	0.32 *0.18°*	0.30 *0.34°*	0.29 *0.49°*	— —	0.28 *0.52°*	**3.18%** **3.18%**
Mannitol	0.16 *0.047°*	0.17 *0.09°*	0.17 *0.20°*	0.17 *0.30°*	0.18 *0.51°*	0.18 *0.52°*	**5.07%** **5.07%**
Merbromin	0.16 *0.044°*	0.14 *0.08°*	0.12 *0.13°*	0.11 *0.18°*	0.09 *0.27°*	— —	
Mercuric cyanide	0.16 *0.047°*	0.15 *0.08°*	0.15 *0.16°*	0.14 *0.23°*	0.13 *0.38°*	— —	
Mercury bichloride	0.14 *0.038°*	0.13 *0.07°*	0.12 *0.14°*	0.12 *0.20°*	0.10 *0.33°*	— —	
Methacholine bromide	0.29 *0.087°*	0.28 *0.16°*	0.26 *0.29°*	0.24 *0.42°*	— —	0.24 *0.52*	**3.77%** **3.77%**
Methacholine chloride	0.34 *0.099°*	0.32 *0.18°*	0.30 *0.38°*	0.28 *0.49°*	— —	0.28 *0.52°*	**3.21%** **3.21%**
Methenamine	0.22 *0.061°*	0.23 *0.12°*	0.24 *0.27°*	0.24 *0.41°*	— —	0.24 *0.52°*	**3.68%** **3.68%**
Methiodal sodium	0.24 *0.068°*	0.24 *0.13°*	0.24 *0.27°*	0.24 *0.41°*	— —	0.24 *0.52°*	**3.81%** **3.81%**
Methionine	0.32 *0.091°*	0.28 *0.16°*	0.25 *0.28°*	— —	— —	— —	
Methylergonovine maleate	0.10 *0.028°*	0.10 *0.05°*	— —	— —	— —	— —	
N-Methylglucamine	0.20 *0.057°*	0.20 *0.11°*	0.18 *0.21°*	0.18 *0.31°*	0.18 *0.51°*	0.18 *0.52°*	**5.02%** **5.02%**
Metrizamide	0.04 *0.010°*	0.04 *0.02°*	0.03 *0.04°*	0.03 *0.06°*	— —	— —	
Monoethanolamine	0.53 *0.154°*	0.53 *0.30°*	— —	— —	— —	0.53 *0.52°*	**1.70%** **1.70%**
Morphine hydrochloride	0.16 *0.044°*	0.15 *0.08°*	0.15 *0.16°*	0.14 *0.24°*	— —	— —	
Morphine nitrate	0.22 *0.061°*	0.19 *0.10°*	0.16 *0.18°*	0.15 *0.25°*	— —	— —	
Morphine sulfate	0.16 *0.046°*	0.14 *0.07°*	0.12 *0.13°*	0.11 *0.17°*	0.09 *0.25°*	— —	
Neoarsphenamine	0.42 *0.116°*	0.40 *0.22°*	0.39 *0.44°*	— —	— —	0.39 *0.52°*	**2.32%** **2.32%**
Neomycin sulfate	0.14 *0.041°*	0.12 *0.06°*	0.10 *0.11°*	0.09 *0.15°*	0.08 *0.22°*	— —	
Nicotinamide	0.30 *0.083°*	0.26 *0.14°*	0.23 *0.26°*	0.21 *0.37°*	— —	0.20 *0.52°*	**4.49%** **4.49%**
Nicotinic acid	0.26 *0.074°*	0.25 *0.14°*	— —	— —	— —	— —	

Isotonic Solutions (Continued)

Chemical	Concentration of Solution, NaCl Equivalents						
	0.5%	1%	2%	3%	5%	At Isosmotic Concentration	
Papaverine hydrochloride	0.10 0.028°	0.10 0.06°	0.10 0.12°	— —	— —	— —	
Paraldehyde	0.25 0.071°	0.25 0.14°	0.25 0.28°	0.25 0.43°	— —	0.25 0.52	**3.65%** **3.65%**
Phenobarbital sodium	0.24 0.069°	0.24 0.13°	0.23 0.26°	0.23 0.39°	— —	0.23 0.52°	**3.95%** **3.95%**
Phenol	0.38 0.104°	0.35 0.19°	0.33 0.38°	— —	— —	0.32 0.52°	**2.8%** **2.8%**
Phentolamine mesylate	0.18 0.052°	0.17 0.09°	0.16 0.17°	0.14 0.24°	0.13 0.36°	0.11 0.52°	**8.23%** **8.23%**
Phenylethyl alcohol	0.25 0.070°	0.25 0.14°	0.25 0.28°	— —	— —	— —	
Polyethylene glycol 300	0.12 0.034°	0.12 0.06°	0.12 0.14°	0.12 0.21°	0.13 0.37°	0.13 0.52°	**6.73%** **6.73%**
Polyethylene glycol 400	0.08 0.022°	0.08 0.04°	0.09 0.09°	0.09 0.15°	0.09 0.27°	0.11 0.52°	**8.50%** **8.50%**
Polyethlene glycol 1500	0.06 0.015°	0.06 0.03°	0.07 0.07°	0.07 0.12°	0.07 0.21°	0.09 0.52°	**10.00%** **10.00%**
Polyethylene glycol 1540	0.02 0.005°	0.02 0.01°	0.02 0.02°	0.03 0.04°	0.03 0.09°	— —	
Polyethylene glycol 4000	0.02 0.004°	0.02 0.00°	0.02 0.02°	0.02 0.03°	0.02 0.06°	— —	
Polymyxin B sulfate	0.10 0.033°	0.09 0.04°	0.07 0.07°	0.06 0.09°	0.04 0.13°	— —	
Polysorbate 80	0.02 0.005°	0.02 0.01°	0.02 0.02°	0.02 0.03°	0.02 0.05°	— —	
Polyvinyl alcohol (99% hydrolyzed)	0.02 0.004°	0.02 0.00°	0.02 0.02°	0.02 0.03°	0.03 0.07°	— —	
Potassium acetate	0.59 0.172°	0.59 0.34°	— —	— —	— —	0.59 0.52°	**1.53%** **1.53%**
Potassium chlorate	0.50 0.140°	0.49 0.27°	— —	— —	— —	0.48 0.52°	**1.88%** **1.88%**
Potassium chloride	0.76 0.219°	0.76 0.43°	— —	— —	— —	0.76 0.52°	**1.19%** **1.19%**
Potassium iodide	0.34 0.104°	0.34 0.20°	0.34 0.40°	— —	— —	0.34 0.52°	**2.59%** **2.59%**
Potassium nitrate	0.58 0.163°	0.56 0.32°	— —	— —	— —	0.56 0.52°	**1.62%** **1.62%**
Potassium permanganate	0.39 0.112°	0.39 0.22°	0.39 0.44°	— —	— —	— —	
Potassium phosphate, anhydrous	0.50 0.140°	0.46 0.26°	0.43 0.49°	— —	— —	0.43 0.52°	**2.11%** **2.11%**
Potassium phosphate, monobasic	0.48 0.133°	0.44 0.25°	0.42 0.48°	— —	— —	0.41 0.52°	**2.18%** **2.18%**
Potassium sorbate	0.44 0.125°	0.41 0.23°	0.40 0.46°	— —	— —	0.40 0.52°	**2.23%** **2.23%**
Potassium sulfate	0.46 0.132°	0.44 0.25°	0.43 0.49°	— —	— —	0.43 0.52°	**2.11%** **2.11%**

Chemical	Concentration of Solution, NaCl Equivalents					At Isosmotic Concentration	
	0.5%	1%	2%	3%	5%		
Potassium thiocyanate	0.61 *0.180°*	0.59 *0.34°*	— —	— —	— —	0.59 *0.52°*	**1.52%** **1.52%**
Povidone	0.01 *0.004°*	0.01 *0.00°*	0.01 *0.01°*	0.10 *0.01°*	0.01 *0.03°*	— —	
Propylene glycol	0.44 *0.126°*	0.43 *0.25°*	0.43 *0.49°*	— —	— —	0.43 *0.52°*	**2.10%** **2.10%**
Pyridoxine hydrochloride	0.41 *0.118°*	0.36 *0.20°*	0.32 *0.36°*	0.29 *0.51°*	— —	— —	
Resorcinol	0.28 *0.082°*	0.28 *0.16°*	0.28 *0.31°*	0.27 *0.47°*	— —	0.27 *0.52°*	**3.3%** **3.3%**
Riboflavin phosphate, sodium	0.08 *0.022°*	0.08 *0.04°*	0.08 *0.09°*	0.08 *0.15°*	— —	— —	
Rose bengal	0.08 *0.020°*	0.07 *0.04°*	0.07 *0.08°*	0.07 *0.12°*	0.07 *0.19°*	0.06 *0.52°*	**14.9%** **14.9%**
Rose bengal B	0.08 *0.022°*	0.08 *0.04°*	0.08 *0.08°*	0.08 *0.15°*	0.08 *0.21°*	— —	
Silver nitrate	0.33 *0.095°*	0.33 *0.19°*	0.33 *0.38°*	— —	— —	0.33 *0.52°*	**2.74%** **2.74%**
Silver protein, mild	0.17 *0.047°*	0.17 *0.09°*	0.17 *0.18°*	0.17 *0.28°*	0.16 *0.47°*	0.16 *0.52°*	**5.51%** **5.51%**
Silver protein, strong	0.12 *0.033°*	0.08 *0.04°*	0.06 *0.06°*	0.05 *0.08°*	0.04 *0.10°*	— —	
Sodium acetate	0.47 *0.136°*	0.46 *0.26°*	0.45 *0.51°*	— —	— —	0.45 *0.52°*	**2.03%** **2.03%**
Sodium acetate, anhydrous	0.08 *0.226°*	0.77 *0.44°*	— —	— —	— —	0.76 *0.52°*	**1.18%** **1.18%**
Sodium antimonyl tartrate	0.14 *0.039°*	0.13 *0.07°*	0.13 *0.14°*	0.12 *0.20°*	0.12 *0.33°*	0.11 *0.52°*	**7.9%** **7.9%**
Sodium arsenate, dibasic	0.26 *0.074°*	0.25 *0.14°*	0.25 *0.27°*	0.24 *0.41°*	— —	0.24 *0.52*	**3.83%** **3.83%**
Sodium ascorbate	0.34 *0.097°*	0.32 *0.18°*	0.30 *0.35°*	— —	— —	0.30 *0.52°*	**2.99%** **2.99%**
Sodium benzoate	0.40 *0.116°*	0.40 *0.23°*	0.40 *0.46°*	— —	— —	0.40 *0.52°*	**2.25%** **2.25%**
Sodium bicarbonate	0.68 *0.197°*	0.65 *0.38°*	— —	— —	— —	0.65 *0.52°*	**1.39%** **1.39%**
Sodium biphosphate monohydrate	0.45 *0.128°*	0.43 *0.24°*	0.41 *0.47°*	— —	— —	0.41 *0.52°*	**2.21%** **2.21%**
Sodium biphosphate dihydrate	0.40 *0.109°*	0.36 *0.20°*	0.34 *0.38°*	— —	— —	0.32 *0.52°*	**2.77%** **2.77%**
Sodium biphosphate, anhydrous	0.50 *0.142°*	0.46 *0.26°*	0.43 *0.49°*	— —	— —	0.43 *0.52°*	**2.1%** **2.1%**
Sodium bismuth thioglycolate	0.20 *0.055°*	0.19 *0.10°*	0.18 *0.20°*	0.18 *0.30°*	0.17 *0.49°*	0.17 *0.52°*	**5.29%** **5.29%**
Sodium bisulfite	0.64 *0.186°*	0.61 *0.35°*	— —	— —	— —	0.60 *0.52°*	**1.5%** **1.5%**
Sodium borate	0.48 *0.137°*	0.42 *0.24°*	0.37 *0.42°*	— —	— —	0.35 *0.52°*	**2.6%** **2.6%**

Chemical	Concentration of Solution, NaCl Equivalents					At Isosmotic Concentration	
	0.5%	1%	2%	3%	5%		
Sodium bromide	0.58	0.58	—	—	—	0.57	**1.6%**
	0.166°	*0.32°*	—	—	—	*0.52°*	**1.6%**
Sodium cacodylate	0.38	0.32	0.30	0.28	—	0.27	**3.3%**
	0.104°	*0.18°*	*0.33°*	*0.48°*	—	*0.52°*	**3.3%**
Sodium carbonate, anhydrous	0.74	0.70	—	—	—	0.68	**1.32%**
	0.214°	*0.40°*	—	—	—	*0.52°*	**1.32%**
Sodium carbonate, monohydrated	0.64	0.60	—	—	—	0.58	**1.56%**
	0.183°	*0.34°*	—	—	—	*0.52°*	**1.56%**
Sodium carboxymethyl cellulose	0.03	0.03	—	—	—	—	—
	0.007°	*0.01°*	—	—	—	—	
Sodium chloride	1.00	—	—	—	—	1.00	**0.9%**
	0.289°	—	—	—	—	*0.52°*	**0.9%**
Sodium citrate	0.32	0.31	0.30	0.30	—	0.30	**3.02%**
	0.091°	*0.17°*	*0.34°*	*0.51°*	—	*0.52°*	**3.02%**
Sodium colistimethate	0.16	0.15	0.14	0.14	0.13	0.13	**6.85%**
	0.045°	*0.08°*	*0.16°*	*0.23°*	*0.38°*	*0.52°*	**6.85%**
Sodium folate	0.14	0.12	0.11	0.10	—	—	
	0.040°	*0.06°*	*0.12°*	*0.16°*	—	—	
Sodium hypophosphite	0.68	0.61	—	—	—	—	
	0.190°	*0.35°*	—	—	—	—	
Sodium iodide	0.41	0.39	0.39	—	—	0.38	**2.37%**
	0.113°	*0.22°*	*0.44°*	—	—	*0.52°*	**2.37%**
Sodium lactate	0.58	0.55	—	—	—	0.52	**1.72%**
	0.164°	*0.31°*	—	—	—	*0.52°*	**1.72%**
Sodium lauryl sulfate	0.10	0.08	0.07	0.05	—	—	—
	0.029°	*0.04°*	*0.06°*	*0.08°*	—	—	
Sodium metabisulfite	0.70	0.67	—	—	—	0.65	**1.38%**
	0.206°	*0.38°*	—	—	—	*0.52°*	**1.38%**
Sodium nitrate	0.74	0.68	—	—	—	0.66	**1.36%**
	0.214°	*0.39°*	—	—	—	*0.52*	**1.36%**
Sodium nitrite	0.86	0.84	—	—	—	0.83	**1.08%**
	0.248°	*0.48°*	—	—	—	*0.52*	**1.08%**
Sodium nitroferricyanide	0.30	0.29	0.28	0.28	—	0.27	**3.30%**
	0.086°	*0.16°*	*0.32°*	*0.47°*	—	*0.52°*	**3.30%**
Sodium phosphate	0.30	0.29	0.28	0.27	—	0.27	**3.33%**
	0.086°	*0.16°*	*0.32°*	*0.47°*	—	*0.52°*	**3.33%**
Sodium phosphate, dibasic dihydrate	044	0.42	0.41	—	—	0.40	**2.23%**
	0.127°	*0.24°*	*0.47°*	—	—	*0.52°*	**2.23%**
Sodium phosphate, dibasic dodecahydrate	0.24	0.22	0.21	0.21	—	0.20	**4.45%**
	0.064°	*0.12°*	*0.24°*	*0.35°*	—	*0.52°*	**4.45%**
Sodium phosphate, exsiccated	0.56	0.53	—	—	—	0.51	**1.75%**
	0.159°	*0.30°*	—	—	—	*0.52°*	**1.75%**
Sodium propionate	0.62	0.61	—	—	—	0.61	**1.47%**
	0.177°	*0.35°*	—	—	—	*0.52°*	**1.47%**
Sodium ricinoleate	0.10	0.10	0.10	0.09	0.09	—	
	0.033°	*0.06°*	*0.11°*	*0.16°*	*0.25°*	—	
Sodium salicylate	0.38	0.36	0.36	—	—	0.36	**2.53%**
	0.106°	*0.20°*	*0.41°*	—	—	*0.52°*	**2.53%**

Chemical	Concentration of Solution, NaCl Equivalents					At Isosmotic Concentration	
	0.5%	1%	2%	3%	5%		
Sodium succinate	0.32	0.32	0.31	—	—	0.31	**2.90%**
	0.092°	0.18°	0.36°	—	—	0.52°	**2.90%**
Sodium sulfate	0.28	0.26	0.25	0.23	—	0.23	**3.95%**
	0.079°	0.14°	0.28°	0.40°	—	0.52°	**3.95%**
Sodium sulfate, anhydrous	0.58	0.54	—	—	—	0.51	**1.78%**
	0.165°	0.30°	—	—	—	0.52°	**1.78%**
Sodium sulfite, exsiccated	0.72	0.65	—	—	—	—	
	0.204°	0.37°	—	—	—	—	
Sodium tartrate	0.33	0.33	0.33	—	—	0.33	**2.72%**
	0.098°	0.19°	0.38°	—	—	0.52°	**2.72%**
Sodium thiosulfate	0.32	0.31	0.31	—	—	0.30	**2.98%**
	0.092°	0.18°	0.35°	—	—	0.52°	**2.98%**
Sorbitol hemihydrate	0.16	0.16	0.16	0.16	0.16	0.16	**5.48%**
	0.045°	0.09°	0.19°	0.28°	0.48°	0.52°	**5.48%**
Streptomycin calcium chloride	0.20	0.20	0.19	0.19	0.18	0.18	**5.0%**
	0.057°	0.11°	0.21°	0.32°	0.52°	0.52°	**5.0%**
Streptomycin hydrochloride	0.18	0.17	0.17	0.16	0.16	—	
	0.050°	0.09°	0.19°	0.28°	0.46°	—	
Streptomycin sulfate	0.08	0.07	0.07	0.06	0.06	—	
	0.020°	0.03°	0.07°	0.10°	0.17°	—	
Strychnine hydrochloride	0.20	0.18	0.14	—	—	—	
	0.060°	0.09°	0.16°	—	—	—	
Strychnine nitrate	0.12	0.12	—	—	—	—	
	0.035°	0.06°	—	—	—	—	
Sucrose	0.08	0.08	0.09	0.09	0.09	0.10	**9.25%**
	0.023°	0.04°	0.09°	0.15°	0.26°	0.52°	**9.25%**
Sulfobromophthalein sodium	0.07	0.06	0.05	0.05	0.04	—	
	0.019°	0.03°	0.06°	0.08°	0.12°	—	
Tannic acid	0.03	0.03	0.03	0.03	0.03	—	
	0.009°	0.01°	0.03°	0.05°	0.08°	—	
Tartaric acid	0.26	0.25	0.24	0.23	—	0.23	**3.9%**
	0.075°	0.14°	0.27°	0.40°	—	0.52°	**3.9%**
Tetraethylammonium bromide	0.36	0.33	0.30	0.28	—	0.28	**3.17%**
	0.098°	0.18°	0.34°	0.49°	—	0.52°	**3.17%**
Tetraethylammonium chloride	0.36	0.34	0.33	—	—	0.33	**2.67%**
	0.100°	0.19°	0.38°	—	—	0.52°	**2.67%**
Thiamine hydrochloride	0.26	0.25	0.23	0.22	—	0.21	**4.24%**
	0.074°	0.13°	0.26°	0.37°	—	0.52°	**4.24%**
Thiocyanate sodium	0.71	0.17	—	—	—	0.17	**1.27%**
	0.205°	0.41°	—	—	—	0.52°	**1.27%**
Thio-TEPA	0.16	0.16	0.16	0.16	0.16	0.16	**5.67%**
	0.045°	0.09°	0.18°	0.27°	0.46°	0.52°	**5.67%**
Tribromoethanol	0.06	0.05	0.05	—	—	—	
	0.015°	0.03°	0.05°	—	—	—	
Triethanolamine	0.20	0.21	0.22	0.22	—	0.22	**4.05%**
	0.058°	0.12°	0.25°	0.38°	—	0.52°	**4.05%**
Trisodium edetate, monohydrate	0.29	0.29	0.28	0.27	—	0.27	**3.31%**
	0.079°	0.15°	0.31°	0.47°	—	0.52°	**3.31%**

Chemical	Concentration of Solution, NaCl Equivalents					At Isosmotic Concentration	
	0.5%	1%	2%	3%	5%		
Tromethamine	0.26 *0.075°*	0.26 *0.15°*	0.26 *0.30°*	0.26 *0.45°*	— —	0.26 *0.52°*	**3.41%** **3.41%**
Trypan blue	0.26 *0.075°*	0.26 *0.15°*	— —	— —	— —	— —	
Tubocurarine chloride	0.14 *0.042°*	0.13 *0.07°*	0.11 *0.12°*	0.10 *0.17°*	0.09 *0.26°*	— —	
Urea	0.55 *0.158°*	0.53 *0.30°*	— —	— —	— —	0.52 *0.52°*	**1.73%** **1.73%**
Urethan	0.31 *0.089°*	0.31 *0.17°*	0.31 *0.35°*	— —	— —	0.31 *0.52°*	**2.93%** **2.93%**
Uridine	0.12 *0.035°*	0.12 *0.06°*	0.12 *0.13°*	0.12 *0.20°*	0.12 *0.33°*	0.11 *0.52°*	**8.18%** **8.18%**
Warfarin sodium	0.18 *0.049°*	0.17 *0.09°*	0.16 *0.18°*	0.15 *0.26°*	0.15 *0.43°*	0.15 *0.52°*	**6.10%** **6.10%**
Zinc chloride	0.66 *0.190°*	0.61 *0.35°*	— —	— —	— —	— —	
Zinc p-phenosulfonate	0.18 *0.053°*	0.18 *0.10°*	0.18 *0.19°*	0.17 *0.29°*	0.17 *0.48°*	— —	
Zinc sulfanilate	0.22 *0.066°*	0.21 *0.12°*	0.20 *0.22°*	0.19 *0.32°*	0.18 *0.51°*	— —	
Zinc sulfate	0.16 *0.045°*	0.15 *0.08°*	0.14 *0.15°*	0.13 *0.22°*	0.12 *0.35°*	0.12 *0.52°*	**7.65%** **7.65%**
Zinc sulfate (dried)	0.24 *0.072°*	0.23 *0.136°*	0.22 *0.252°*	0.21 *0.362°*	— —	0.20 *0.52°*	**4.52%** **4.52%**

References
1. E. R. Hammarlund, K. Pedersen-Bjergaard, *J. Am. Pharm. Assoc., Sci. Ed.* **47,** 107 (1958).
2. E. R. Hammarlund *et al., J. Pharm. Sci.* **54,** 160 (1965).
3. E. R. Hammarlund, G. L. Van Pevenage, *ibid.* 55, 1448 (1996).
4. W. E. Fassett *et al., ibid.* **58,** 1540 (1969).
5. C. Sapp *et al., ibid.* **64,** 1884 (1975).
6. E. R. Hammarlund, *ibid.* **70,** 1161 (1981).
7. A. R. Gennaro (Ed)., *Remington's Pharmaceutical Sciences* (Mack Publ. Co., Easton, PA., 17th ed., 1985) pp 1455-1472.
8. E. R. Hammarlund, *J. Pharm. Sci.* **78,** 519 (1989).

Indicators

The following tables list indicators commonly used in acid-base titrimetry for visible end-point determinations. Table 1 lists individual compounds which exhibit a color change over a range of pH values. Table 2 lists mixtures of compounds which exhibit a very sharp color change at a specific pH.

Table 1

Indicator	Chemical name	Acid color[1]	pH range	Basic color[1]	Preparation of stock solution[2]
methyl violet	mixture of tetra-, penta-, and hexamethyl-p-rosaniline hydrochloride salts	Y	0.15-3.2	V	0.01-0.05% in water
cresol red (acid range)	o-cresolsulfonephthalein	R	0.2-1.8	Y	Dissolve 100 mg in 2.65 ml 0.1N NaOH, dil to 100 ml with water
metanil yellow	4′-anilinoazobenzene-m-sulfonic acid, sodium salt	R	1.2-2.3	Y	0.01-0.1% in water
m-cresol purple (acid range)	m-cresolsulfonephthalein	R	1.2-2.8	Y	Dissolve 100 mg in 2.65 ml 0.1N NaOH, dil to 100 ml with water
xylenol blue (acid range)	p-xylenolsulfonephthalein	R	1.2-2.8	Y	Dissolve 40 mg in 0.98 ml 0.1M NaOH, dil to 100 ml with water
thymol blue (acid range)	thymolsulfonephthalein	R	1.2-2.8	Y	Dissolve 100 mg in 2.15 ml 0.1N NaOH, dil to 100 ml with water
tropeolin OO	4′-phenylaminoazobenzene-4-sulfonic acid, sodium salt	R	1.3-3.2	Y	1% in water; 0.04% in 50% aq alc[3]
quinaldine red	2-(p-dimethylaminostyryl)-quinoline ethiodide	C	1.4-3.2	R	0.1% in alc or 90% aq ethanol[3]
α-dinitrophenol	2,4-dinitrophenol	C	2.0-4.7	Y	0.1% in 70% aq alc[3]
methyl yellow; dimethyl yellow	p-dimethylaminoazobenzene	R	2.9-4.0	Y	0.1% in 90% aq alc[3]
bromophenol blue	3,3′,5,5′-tetrabromophenol-sulfonephthalein	Y	3.0-4.6	Pu	Dissolve 100 mg in 1.5 ml 0.1N NaOH, dil to 100 ml with water
tetrabromophenol blue	tetrabromophenol tetrabromo-sulfonephthalein	Y	3.0-4.6	B	Dissolve 100 mg in 1.01 ml 0.1N NaOH, dil to 100 ml with water
bromochlorophenol blue	3,3′-dibromo-5,5′-dichloro-phenolsulfonephthalein	Y	3.0-4.6	Pu	Dissolve 40 mg in 0.69 ml 0.1M NaOH, dil to 100 ml with water
Congo red	diphenyl-4,4′-bis-(2-azo-1-naphthylamine)sulfonic acid, sodium salt	B	3.0-5.2	R	0.1% in water
methyl orange	4′-dimethylaminoazobenzene-4-sulfonic acid, sodium salt	R	3.1-4.4	O	0.04-0.1% in water
p-ethoxychrysoidine hydrochloride	4-ethoxy-2′,4′-diaminoazo-benzene hydrochloride	R	3.5-5.5	Y	0.1% in 90% aq alc[3]; 0.2% in alc
naphthyl red	α-naphthylaminoazobenzene	R	3.7-5.0	Y	0.1% in ethanol or 70% aq alc[3]
alizarin sodium sulfonate	dihydroxyanthraquinone sodium sulfonate	Y	3.7-5.2	V	0.1% in water
bromocresol green	3,3′,5,5′-tetrabromo-m-cresolsulfonephthalein	Y	3.8-5.4	B	Dissolve 100 mg in 1.45 ml 0.1N NaOH, dil to 100 ml with water
γ-dinitrophenol	2,5-dinitrophenol	C	4.0-5.8	Y	0.1% in 70% aq alc[3]
methyl red	4′-dimethylaminoazobenzene-2-carboxylic acid	R	4.4-6.2	Y	Dissolve 100 mg in 3.7 ml 0.1N NaOH, dil to 100 ml with water; 0.1% in alc
lacmoid		R	4.4-6.4	B	0.2% in alc
chlorophenol red	3,3′-dichlorophenolsulfone-phthalein	Y	4.8-6.4	Pu	Dissolve 100 mg in 2.35 ml 0.1N NaOH, dil to 100 ml with water
benzoyl auramine G		V	5-5.6	Y	0.25% in methanol
bromocresol purple	5,5′-dibromo-o-cresolsulfone-phthalein	Y	5.2-6.8	Pu	Dissolve 100 mg in 1.85 ml 0.1N NaOH, dil to 100 ml with water

Table 1 (Continued)

Indicator	Chemical name	Acid color[1]	pH range	Basic color[1]	Preparation of stock solution[2]
bromophenol red	3,3′-dibromophenolsulfone-phthalein	Y	5.2-6.8	R	Dissolve 40 mg in 0.78 ml 0.1*M* NaOH, dil to 100 ml with water
p-nitrophenol		C	5.6-7.6	Y	0.1% in water; 0.2% in alc
bromothymol blue	3,3′-dibromothymolsulfone-phthalein	Y	6.0-7.6	B	Dissolve 100 mg in 1.6 ml 0.1*N* NaOH, dil to 100 ml with water
phenol red	phenolsulfonephthalein	Y	6.4-8.2	R	Dissolve 100 mg in 2.85 ml 0.1*N* NaOH, dil to 100 ml with water
rosolic acid; aurin	*p*-quinonemono(bis-4-oxyphenylmethide)	Y	6.6-8.0	R	0.2% in 50% aq alc[3]
neutral red	3-amino-6-dimethylamino-2-methylphenazinium chloride	R	6.8-8.0	Y	0.1% in 70% aq alc[3]
quinoline blue	cyanine	C	7.0-8.0	V	1% in alc
cresol red (basic range)	*o*-cresolsulfonephthalein	Y	7.0-8.8	R	Dissolve 100 mg in 2.65 ml 0.1*N* NaOH, dil to 100 ml with water
α-naphtholphthalein		C	7.3-8.7	G-B	0.1% in alc or 70% aq alc[3]
tropeolin OOO; α-naphthol orange	α-naphtholazobenzene-*p*-sulfonic acid, sodium salt	Y	7.4-8.9	P	0.1% in water
m-cresol purple (basic range)	*m*-cresolsulfonephthalein	Y	7.4-9.0	Pu	Dissolve 100 mg in 2.65 ml 0.1*N* NaOH, dil to 100 ml with water
ethyl bis(2,4-dinitro-phenyl)acetate		C	7.5-9.1	B	satd soln in 1:1 acetone-alc
thymol blue (basic range)	thymolsulfonephthalein	Y	8.0-9.6	B	Dissolve 100 mg in 2.15 ml 0.1*N* NaOH, dil to 100 ml with water
xylenol blue (basic range)	*p*-xylenolsulfonephthalein	Y	8.0-9.6	B	Dissolve 40 mg in 0.98 ml 0.1*M* NaOH, dil to 100 ml with water
o-cresolphthalein	di-*o*-cresolphthalide	C	8.2-9.8	R-V	0.04% in 50% aq alc[3]
phenolphthalein	di-*p*-dioxydiphenylphthalide	C	8.2-9.8	P	0.1% in alc or 60% aq alc[3]
thymolphthalein	dithymolphthalide	C	9.3-10.5	B	0.1% in 80% aq alc[3]
α-naphtholbenzein	dimethylphenolphthalein	Br	9.8-11.0	G-B	0.1% in alc
alizarin yellow GG; salicyl yellow	3′-nitro-4-oxyazobenzene-3-carboxylic acid, sodium salt	Y	10.0-12.1	Br-Y	0.1% in ethanol or water
alizarin yellow R	4′-nitro-4-oxyazobenzene-3-carboxylic acid, sodium salt	Y	10.0-12.1	R	0.1% in water
Nile blue	diethylaminonaphthophenazoxonium sulfate	B	10.1-11.1	R	0.1 % in water
nitramine	2,4,6-trinitrophenylmethyl-nitramine	C	10.8-13.0	Br	0.1% in 70% aq alc[3]
tropeolin O	2′,4′-dioxyazobenzene-4-sulfonic acid, sodium salt	Y	11.0-12.7	R-Br	0.1% in water
Poirrier blue C4B	triphenylrosaniline sulfonic acid, sodium or potassium salt	B	11.0-13.0	Pu	0.1% in water
indigo carmine	indigotin-5,5′-disulfonic acid, disodium salt	B	11.5-14.0	Y	0.25% in 50% aq ethanol[3]
trinitrobenzene		C	12.0-14.0	O	0.1% in 70% aq alc[3]

Table 2

Composition of indicator stock solution[2]	Solvent	Acid color[1]	Transition pH	Basic color[1]
dimethyl yellow, 0.05% + methylene blue, 0.05%	alc	B-V	3.25	G
methyl orange, 0.1% + indigo carmine, 0.25%	aq	V	4.1	G
bromocresol green, sodium salt, 0.05% + methyl orange, 0.01%	aq	O	4.3	B-G
bromocresol green, 0.075% + methyl red, 0.05%	alc	wine-R	5.1	G
methyl red, 0.1% + methylene blue, 0.05%	alc	R-V	5.4	G
bromocresol green, sodium salt, 0.05% + chlorophenol red, sodium salt, 0.05%	aq	Y-G	6.1	B-V
bromocresol purple, sodium salt, 0.05% + bromothymol blue, sodium salt, 0.05%	aq	Y	6.7	V-B
neutral red, 0.05% + methylene blue, 0.05%	alc	V-B	7.0	G
bromothymol blue, sodium salt, 0.05% + phenol red, sodium salt, 0.05%	aq	Y	7.5	V
cresol red, sodium salt, 0.025% + thymol blue, sodium salt, 0.075%	aq	Y	8.3	V
phenolphthalein, 0.075% + α-naphtholphthalein, 0.025%	alc	P	8.9	V
phenolphthalein, 0.075% + thymol blue, 0.025%	50% alc	Y	9.0	V
phenolphthalein, 0.05% + thymolphthalein, 0.05%	alc	C	9.9	V
phenolphthalein, 0.033% + Nile blue, 0.133%	alc	B	10.0	R

[1]The indicator colors are abbreviated as follows: B, blue; Br, brown; C, colorless; G, green; O, orange; P, pink; Pu, purple; R, red; V, violet; and Y, yellow.
[2]One to five drops of the indicator stock soln should be added to every 10 ml soln to be titrated.
[3]When preparing the aq alc solns, it is important to dissolve the indicator in alc first and then dil with water to the indicated conc. For example, to make a 0.1% soln in 70% aq alc, dissolve 100 mg indicator in 70 ml alc, then add 30 ml water.

References

I. M. Kolthoff, *Acid-Base Indicators* (Macmillan, New York, 1937).
I. M. Kolthoff, V. A. Stenger, *Volumetric Analysis* **vols. I, II** (Interscience, New York, 2nd ed., 1942).
R. G. Bates, *Determination of pH* (Wiley & Sons, New York, 1954).
E. Bishop, *Indicators* (Pergamon Press, Oxford, 1972).
F. J. Green, *The Sigma-Aldrich Handbook of Stains, Dyes, and Indicators* (Aldrich, Milwaukee, Wisconsin, 1990).

Amino Acid Abbreviations and pKa Values

NAME	ABBREVIATIONS		pKa VALUES		
	3 Letter	1 Letter	—COOH	—NH$_3^+$	R group
Alanine	Ala	A	2.34	9.69	
Arginine	Arg	R	2.17	9.04	12.48
Asparagine	Asn	N	2.01	8.8	
Aspartic Acid	Asp	D	1.89	9.60	3.65
Cysteine	Cys	C	1.96	8.18	10.29
Glutamine	Gln	Q	2.17	9.13	
Glutamic Acid	Glu	E	2.19	9.67	4.25
Glycine	Gly	G	2.34	9.6	
Histidine	His	H	1.8	9.17	6.00
Isoleucine	Ile	I	2.35	9.68	
Leucine	Leu	L	2.36	9.60	
Lysine	Lys	K	2.18	8.95	10.52
Methionine	Met	M	2.28	9.2	
Phenylalanine	Phe	F	1.83	9.12	
Proline	Pro	P	1.99	10.96	
Serine	Ser	S	2.21	9.15	13.6
Threonine	Thr	T	2.11	9.62	13.6
Tryptophan	Trp	W	2.38	9.39	
Tyrosine	Tyr	Y	2.2	9.11	10.06
Valine	Val	V	2.32	9.61	

Saturated Solutions

The following table provides the data for making saturated solutions of the substance listed at the temperature designated. Data are provided for making saturated solutions by weight (g of substance per 100 g of saturated solution) and by volume (g of substance per 100 ml of saturated solution and the ml of water required to make such a solution).

To make one *fluid ounce* of a saturated solution: multiply the grams of substance per 100 ml of saturated solution by 4.55 to obtain the number of grains required, by 0.01039 to obtain the number of avoirdupois ounces, by 0.00947 to obtain the number of apothecaries (Troy) ounces; also multiply the ml of water by 16.23 to obtain the number of minims, or divide by 100 to obtain the number of fluid ounces.

To make one *fluid dram:* multiply the grams of substance per 100 ml of saturated solution by 0.5682 to obtain the number of grains required; also multiply the ml of water by 0.60 to obtain the number of minims required.

Substance	Formula	Temp, °C	g/100 g satd soln	g/100 ml satd soln	ml water/ 100 ml satd soln	Specific gravity
acetanilide	$C_6H_5NHCOCH_3$	25	0.54	0.54	99.2	0.997
p-acetophenetidin	$C_6H_4(OC_2H_5)NHCH_3CO$	25	0.0766	0.0766	99.92	1.00
p-acetotoluide	$CH_3CONHC_6H_4CH_3$	25	0.12	0.12	99.7	0.9979
alanine	$CH_3CH(NH_2)COOH$	25	14.1	14.7	89.5	1.042
aluminum ammonium sulfate	$Al_2(SO_4)_3(NH_4)_2SO_4.24H_2O$	25	12.4	13	92	1.05
aluminum chloride hydrated	$AlCl_3.6H_2O$	25	55.5	75	60	1.35
aluminum fluoride	$Al_2F_6.5H_2O$	20	0.499	0.5015	100.0	1.0051
aluminum potassium sulfate	$AlK(SO_4)_2$	25	6.62	7.02	99.1	1.061
aluminum sulfate	$Al_2(SO_4)_3.18H_2O$	25	48.8	63	66 555	1.29
o-aminobenzoic acid	$C_6H_4NH_2COOH$	25	0.52	0.519	99.4	0.999
DL-α-amino-n-butyric acid	$CH_3CH_2CH(NH_2)COOH$	25	17.8	18.6	86.2	1.046
DL-α-aminoisobutyric acid	$(CH_3)_2C(NH_2)COOH$	25	13.3	13.7	89.5	1.031
ammonium arsenate	$NH_4H_2AsO_4$	20	32.7	40.2	83.0	1.228
ammonium benzoate	$NH_4C_7H_5O_2$	25	18.6	19.4	84.7	1.040
ammonium bromide	NH_4Br	15	41.7	53.8	75.2	1.290
ammonium carbonate		25	20	22	88	1.10
ammonium chloride	NH_4Cl	15	26.3	28.3	79.3	1.075
ammonium citrate, dibasic	$(NH_4)_2HC_6H_5O_7$	25	48.7	60.5	61.5	1.22
ammonium dichromate	$(NH_4)_2Cr_2O_7$	25	27.9	33	85	1.18
ammonium iodide	NH_4I	25	64.5	106.2	58.3	1.646
ammonium molybdate	$(NH_4)_6Mo_7O_{24}.4H_2O$	25	30.6	39	88	1.27
ammonium nitrate	NH_4NO_3	25	68.3	90.2	41.8	1.320
ammonium oxalate	$(NH_4)_2C_2O_4.H_2O$	25	4.95	5.06	97.0	1.019
ammonium perchlorate	NH_4ClO_4	25	21.1	23.7	88.7	1.123
ammonium periodate	NH_4IO_4	16	2.63	2.68	99.2	1.018
ammonium persulfate	$(NH_4)_2S_2O_8$	25	42.7	53	71	1.24
ammonium phosphate, dibasic	$(NH_4)_2HPO_4$	14.5	56.2	75.5	58.8	1.343
ammonium phosphate, monobasic	$NH_4H_2PO_4$	25	28.4	33	83	1.16
ammonium salicylate	$NH_4C_7H_5O_3$	25	50.8	58.2	56.4	1.145
ammonium silicofluoride	$(NH_4)_2SiF_6$	17.5	15.7	17.2	92.3	1.095
ammonium sulfate	$(NH_4)_2SO_4$	20	42.6	53.1	71.7	1.248
ammonium sulfite	$(NH_4)_2SO_3.H_2O$	25	39.3	47.3	73.2	1.204
ammonium thiocyanate	NH_4CNS	25	62.2	71	43	1.14
amyl alcohol	$C_5H_{11}OH$	25	2.61	2.60	96.9	0.995
aniline	$C_6H_5NH_2$	22	3.61	3.61	96.2	0.998
aniline hydrochloride	$C_6H_5NH_2.HCl$	25	49	54	56	1.10
aniline sulfate	$(C_6H_5NH_2)_2.H_2SO_4$	25	5.88	6	96	1.02
L-asparagine	$NH_2COCH_2CH(NH_2)COOH$	25	2.44	2.46	98.2	1.007
barium bromide	$BaBr_2$	20	51	87.2	83.8	1.710
barium chlorate	$Ba(ClO_3)_2$	25	28.5	36.8	92.6	1.294
barium chloride	$BaCl_2$	20	26.3	33.4	93.8	1.27
barium iodide	$BaI_2.7\frac{1}{2}H_2O$	25	68.8	157.0	71.1	2.277
barium nitrate	$Ba(NO_3)_2$	25	9.4	10.2	97.9	1.080
barium nitrite	$Ba(NO_2)_2$	17	40	59.6	89.4	1.490
barium perchlorate	$Ba(ClO_4)_2$	25	75.3	145.8	47.8	1.936
benzamide	$C_6H_5CONH_2$	25	1.33	1.33	98.6	0.999
benzoic acid	$C_7H_6O_2$	25	0.367	0.367	99.63	1.00
beryllium sulfate	$BeSO_4.4H_2O$	25	28.7	37.3	93.0	1.301
boric acid	H_3BO_3	25	4.99	5.1	97	1.02
n-butyl alcohol	$CH_3(CH_2)_2CH_2OH$	25	79.7	67.3	17.1	0.845
cadmium bromide	$CdBr_2.4H_2O$	25	52.9	94.0	83.9	1.775
cadmium chlorate	$Cd(ClO_3)_2.12H_2O$	18	76.4	174.5	54.0	2.284

Substance	Formula	Temp, °C	g/100 g satd soln	g/100 ml satd soln	ml water/ 100 ml satd soln	Specific gravity
cadmium chloride	$CdCl_2.2\frac{1}{2}H_2O$	25	54.7	97.2	80.8	1.778
cadmium iodide	CdI_2	20	45.9	73.0	86.3	1.590
cadmium sulfate	$3(CdSO_4).8H_2O$	25	43.4	70.3	91.8	1.619
calcium bromide	$CaBr_2$	20	58.8	107.2	75.0	1.82
calcium chlorate	$Ca(ClO_3)_2.2H_2O$	18	64.0	110.7	62.3	1.729
calcium chloride	$CaCl_2.6H_2O$	25	46.1	67.8	79.2	1.47
calcium chromate	$CaCrO_4.2H_2O$	18	14.3	16.4	98.7	1.149
calcium ferrocyanide	$Ca_2Fe(CN)_6$	25	36.5	49.6	86.2	1.357
calcium iodide	CaI_2	20	67.6	143.8	69.0	2.125
calcium lactate	$Ca(C_3H_5O_3)_2.5H_2O$	25	4.95	5	96	1.01
calcium nitrite	$Ca((NO_2)_2.4H_2O$	18	45.8	65.7	77.8	1.427
calcium sulfate	$CaSO_4.2H_2O$	25	0.208	0.208	99.70	0.999
camphoric acid	$C_8H_{14}(COOH)_2$	25	0.754	0.754	99.246	1.00
carbon disulfide	CS_2	22	0.173	0.173	99.63	0.998
cerium nitrate	$Ce(NO_3)_3.6H_2O$	25	63.7	119.9	68.2	1.880
cesium bromide	$CsBr$	21.4	53.1	89.8	79.5	1.693
cesium chloride	$CsCl$	25	65.7	126.3	65.9	1.923
cesium iodide	CsI	22.8	48.0	74.1	80.5	1.545
cesium nitrate	$CsNO_3$	25	21.9	26.1	92.9	1.187
cesium perchlorate	$CsClO_4$	25	2.01	2.03	99.0	1.010
cesium periodate	$CsIO_4$	15	2.10	2.13	99.5	1.017
cesium sulfate	Cs_2SO_4	25	64.5	129.8	71.7	2.013
chloral hydrate	$CCl_3CHO.H_2O$	25	79.4	120	31	1.51
chloroform	$CHCl_3$	29.4	0.703	0.705	99.57	1.0028
chromic oxide	CrO_3	18	62.5	106.3	64.0	1.703
chromium potassium sulfate	$Cr_2K_2(SO_4)_4.24H_2O$	25	19.6	22	90	1.12
citric acid	$(CH_2)_2COH(COOH)_3.H_2O$	25	67.5	88.6	42.7	1.311
cobalt chlorate	$Co(ClO_3)_2$	18	64.2	119.3	66.5	1.857
cobalt nitrate	$Co(NO_3)_2$	18	49.7	78.2	79.1	1.572
cobalt perchlorate	$Co(ClO_4)_2$	26	71.8	113.5	44.7	1.581
cupric ammonium chloride	$CuCl_2.2NH_4Cl.2H_2O$	25	30.3	35.5	82	1.17
cupric ammonium sulfate	$CuSO_4.(NH_4)_2SO_4$	19	15.3	17.3	96.0	1.131
cupric bromide	$CuBr_2$	25	55.8	102.5	81.2	1.84
cupric chlorate	$Cu(ClO_3)_2$	18	62.2	105.2	64.1	1.692
cupric chloride	$CuCl_2.2H_2O$	25	53.3	80	70	1.50
cupric nitrate	$Cu(NO_3)_2.6H_2O$	20	56.0	94.5	74.3	1.688
cupric selenate	$CuSeO_4$	21.2	14.7	17.2	99.4	1.165
cupric sulfate	$CuSO_4.5H_2O$	25	18.5	22.3	98.7	1.211
dextrose	$C_6H_{12}O_6.H_2O$	25	49.5	59	60	1.19
ether	$(C_2H_5)_2O$	22	5.45	5.34	93.0	0.985
ethyl acetate	$CH_3COOC_2H_5$	25	7.47	7.44	92.1	0.996
ferric ammonium citrate		25	67.7	97	46	1.43
ferric ammonium oxalate	$Fe(NH_4)_3(C_2O_4)_3.3H_2O$	25	51.5	65	61	1.26
ferric ammonium sulfate	$FeSO_4.(NH_4)_2SO_4$	16.5	19.1	22.4	94.3	1.165
ferric chloride	$FeCl_3$	25	73.1	131.1	48.3	1.793
ferric nitrate	$Fe(NO_3)_3$	25	46.8	70.2	79.8	1.50
ferric perchlorate	$Fe(ClO_4)_3.10H_2O$	25	79.9	132.1	33.2	1.656
ferrous sulfate	$FeSO_4.7H_2O$	25	42.1	52.8	72.7	1.255
gallic acid	$C_6H_2(OH)_3COOH.H_2O$	25	1.15	1.15	99.05	1.002
D-glutamic acid	$C_5H_9O_4N$	25	0.86	0.86	99.15	1.0002
glycine	NH_2CH_2COOH	25	20.0	21.7	86.8	1.083
hydroquinone	$C_6H_4(OH)_2$	20	6.7	6.78	94.4	1.012
m-hydroxybenzoic acid	$C_6H_4OHCOOH$	25	0.975	0.975	99.03	1.000
lactose	$C_{12}H_{22}O_{11}.H_2O$	25	15.9	17	90	1.07
lead acetate	$Pb(C_2H_3O_2)_2$	25	36.5	49.0	85.1	1.340
lead bromide	$PbBr_2$	25	0.97	0.98	99.6	1.006
lead chlorate	$Pb(ClO_3)_2$	18	60.2	117.0	77.3	1.944
lead chloride	$PbCl_2$	25	1.07	1.08	99.6	1.007
lead iodide	PbI_2	25	0.08	0.08	99.7	0.998
lead nitrate	$Pb(NO_3)_2$	25	37.1	53.6	91.0	1.445
DL-leucine	$C_6H_{13}O_2N$	25	0.976	0.975	98.9	0.999
L-leucine	$C_6H_{13}O_2N$	25	2.24	2.24	97.85	1.0012
lithium benzoate	$LiC_7H_5O_2$	25	27.7	30.4	79.6	1.100
lithium bromate	$LiBrO_3$	18	60.4	110.5	72.5	1.830

Saturated Solutions (Continued)

Substance	Formula	Temp, °C	g/100 g satd soln	g/100 ml satd soln	ml water/100 ml satd soln	Specific gravity
lithium carbonate	Li_2CO_3	15	1.36	1.38	100.0	1.014
lithium chloride	$LiCl.H_2O$	25	45.9	59.5	70.2	1.296
lithium citrate	$Li_3C_6H_5O_7$	25	31.8	38.6	82.8	1.213
lithium dichromate	$Li_2Cr_2O_7.H_2O$	18	52.6	82.9	74.8	1.574
lithium fluoride	LiF	18	0.27	0.27	99.9	1.002
lithium formate	$LiCHO_2$	18	27.9	31.8	80.4	1.140
lithium iodate	$LiIO_3$	18	44.6	69.9	86.8	1.566
lithium nitrate	$LiNO_3$	19	48.9	64.5	67.5	1.318
lithium perchlorate	$LiClO_4.3H_2O$	25	37.5	47.6	79.5	1.269
lithium salicylate	$LiC_7H_5O_3$	25	52.7	63.6	57.1	1.206
lithium sulfate	$Li_2SO_4.H_2O$	25	27.2	33	88.5	1.21
magnesium bromide	$MgBr_2.6H_2O$	18	50.1	83.1	82.8	1.655
magnesium chlorate	$Mg(ClO_3)_2$	18	56.3	90.0	69.7	1.594
magnesium chloride	$MgCl_2.6H_2O$	25	62.5	79	47.5	1.26
magnesium chromate	$MgCr_2O_4.7H_2O$	18	42.0	59.7	82.5	1.422
magnesium dichromate	$MgCrO_7.5H_2O$	25	81.0	138.8	32.6	1.712
magnesium iodate	$Mg(IO_3)_2.4H_2O$	18	6.44	6.95	100.8	1.078
magnesium iodide	$MgI_2.8H_2O$	18	59.7	114.0	77.1	1.909
magnesium molybdate	$MgMoO_4$	25	15.9	18.4	97.4	1.159
magnesium nitrate	$Mg(NO_3)_2.6H_2O$	25	42.1	58.6	80.5	1.388
magnesium perchlorate	$Mg(ClO_4)_2.6H_2O$	25	49.9	73.6	73.9	1.472
magnesium selenate	$MgSeO_4$	20	35.3	50.8	93.0	1.440
magnesium sulfate	$MgSO_4.7H_2O$	25	55.3	72	58.5	1.30
manganese chloride	$MnCl_2$	25	43.6	63.2	82.0	1.449
manganese nitrate	$Mn(NO_3)_2.6H_2O$	18	57.3	93.2	69.2	1.624
manganese silicofluoride	$MnSiF_6$	17.5	37.7	54.5	90.1	1.446
manganese sulfate	$MnSO_4$	25	39.4	59.1	90.8	1.499
mercuric acetate	$Hg(C_2H_3O_2)_2$	25	30.2	38	88	1.26
mercuric bromide	$HgBr_2$	25	0.609	0.610	99.6	1.0023
mercury bichloride	$HgCl_2$	25	6.6	6.96	98.5	1.054
methylene blue	$C_{16}H_{18}N_3ClS.3H_2O$	25	4.25	4.3	97	1.01
methyl salicylate	$C_6H_4OHCOOCH_3$	25	0.12	0.12	99.88	1.00
monochloracetic acid	$CH_2ClCOOH$	25	78.8	105	28	1.33
β-naphthalenesulfonic acid	$C_{10}H_7SO_3H$	30	56.9	67.9	51.4	1.193
nickel ammonium sulfate	$NiSO_4(NH_4)_2SO_4.6H_2O$	25	9.0	9.5	96	1.05
nickel chlorate	$Ni(ClO_3)_2$	18	56.7	94.2	72.0	1.658
nickel chlorate	$Ni(ClO_3)_2.6H_2O$	18	64.5	107.2	59.1	1.661
nickel nitrate	$Ni(NO_3)_2.6H_2O$	25	77	122	36	1.58
nickel perchlorate	$Ni(ClO_4)_2$	26	70.8	112.2	46.4	1.584
nickel perchlorate	$Ni(ClO_4)_2.9H_2O$	18	52.4	82.7	75.1	1.576
nickel sulfate	$NiSO_4.6H_2O$	25	47.3	64	71	1.35
DL-norleucine	$C_6H_{13}NO_2$	25	1.13	1.13	98.97	0.999
oxalic acid	$H_2C_2O_4.2H_2O$	25	9.81	10.3	94.2	1.044
phenol	C_6H_5OH	20	6.1	6.14	94.5	1.0057
β-phenylalanine	$C_6H_5CH_2CH(NH_2)COOH$	25	2.88	2.89	97.5	1.0035
m-phenylenediamine	$C_6H_8N_2$	20	23.1	23.8	79.3	1.032
p-phenylenediamine	$C_6H_8N_2$	20	3.69	3.70	96.67	1.0038
phenyl salicylate	$C_6H_4OHCOOC_6H_5$	25	0.015	0.015	99.84	0.999
phenyl thiourea	$CS(NH_2)NHC_6H_5$	25	0.24	0.24	99.6	0.998
phosphomolybdic acid	$20MoO_3.2H_3PO_4.48H_2O$	25	74.3	135	46	1.81
phosphotungstic acid	Approx. $20WO_3.2H_3PO_4.25H_2O$	25	71.4	160	64	2.24
potassium acetate	$KC_2H_3O_2$	25	68.7	97.1	44.3	1.413
potassium antimony tartrate	$KSbOC_4H_4O_6$	25	7.64	8.02	96.9	1.049
potassium bicarbonate	$KHCO_3$	25	26.6	31.6	87.5	1.188
potassium bitartrate	$KC_4H_5O_6$	25	0.65	0.65	99.3	0.999
potassium bromate	$KBrO_3$	25	7.53	7.89	97.5	1.054
potassium bromide	KBr	25	40.6	56.0	82.0	1.380
potassium carbonate	$K_2CO_3.1\frac{1}{2}H_2O$	25	52.9	82.2	73.5	1.559
potassium chlorate	$KClO_3$	25	8.0	8.41	96.6	1.051
potassium chloride	KCl	25	26.5	31.2	86.8	1.178
potassium chromate	K_2CrO_4	25	39.4	54.1	83.7	1.381
potassium citrate	$K_3C_6H_5O_7$	25	60.91	92.1	59.2	1.514
potassium dichromate	$K_2Cr_2O_7$	25	13.0	14.2	95.0	1.092
potassium ferricyanide	$K_3Fe(CN)_6$	22	32.1	38.1	80.8	1.187

Substance	Formula	Temp, °C	g/100 g satd soln	g/100 ml satd soln	ml water/ 100 ml satd soln	Specific gravity
potassium ferrocyanide	$K_4Fe(CN)_6$	25	24.0	28.2	89.2	1.173
potassium fluoride	$KF.2H_2O$	18	48.0	72.0	78.0	1.500
potassium formate	$KCHO_2$	18	76.8	120.6	36.4	1.571
potassium hydroxide	KOH	15	51.7	79.2	74.2	1.536
potassium iodate	KIO_3	25	8.40	8.99	98.0	1.071
potassium iodide	KI	25	59.8	103.2	69.1	1.721
potassium meta-antimonate	$KSbO_3$	18	2.73	2.81	99.7	1.025
potassium nitrate	KNO_3	25	28.0	33.4	86.0	1.193
potassium nitrite	KNO_2	20	74.3	121.5	42.3	1.649
potassium oxalate	$K_2C_2O_4.H_2O$	25	28.3	34	86	1.20
potassium perchlorate	$KClO_4$	25	2.68	2.72	99.0	1.014
potassium periodate	KIO_4	13	0.658	0.661	99.83	1.005
potassium permanganate	$KMnO_4$	25	7.10	7.43	97.3	1.046
potassium sodium tartrate	$KNaC_4H_4O_6.4H_2O$	25	39.71	51.9	78.8	1.308
potassium stannate	K_2SnO_3	15.5	42.7	69.2	92.9	1.620
potassium sulfate	K_2SO_4	25	10.83	11.8	96.9	1.086
quinine salicylate	$C_{20}H_{24}N_2O_2.C_6H_4(OH)COOH.2H_2O$	25	0.065	0.065	99.84	0.999
resorcinol	$C_6H_4(OH)_2$	25	58.8	67.2	47.2	1.142
rubidium bromate	$RbBrO_3$	16	2.15	2.18	99.4	1.016
rubidium bromide	$RbBr$	25	52.7	85.6	76.9	1.625
rubidium chloride	$RbCl$	25	48.6	72.8	77.1	1.50
rubidium iodate	$RbIO_3$	15.6	2.72	2.78	99.5	1.022
rubidium iodide	RbI	24.3	63.6	117.7	67.3	1.850
rubidium nitrate	$RbNO_3$	25	40.1	55.0	82.4	1.375
rubidium perchlorate	$RbClO_4$	25	1.88	1.90	99.3	1.012
rubidium periodate	$RbIO_4$	16	0.645	0.648	99.85	1.0052
rubidium sulfate	Rb_2SO_4	25	33.8	45.6	89.7	1.354
silicotungstic acid	$H_4SiW_{12}O_{40}$	18	90.6	258	26.8	2.843
silver acetate	$Ag(C_2H_3O_2)$	25	1.10	1.11	99.40	1.0047
silver bromate	$AgBrO_3$	25	0.204	0.2037	99.65	0.9985
silver fluoride	$AgF.2H_2O$	15.8	64.5	168.4	92.7	2.61
silver nitrate	$AgNO_3$	25	71.5	164	65.5	2.29
silver perchlorate	$AgClO_4.H_2O$	25	84.5	237.1	43.5	2.806
silver acetate	$NaC_2H_3O_2$	25	33.6	40.5	80.0	1.205
sodium ammonium sulfate	$NaNH_4SO_4$	15	25.2	29.6	87.9	1.174
sodium arsenate	$Na_3AsO_4.12H_2O$	17	21.1	23.5	88.0	1.119
sodium benzenesulfonate	$NaC_6H_5SO_3$	25	16.4	17.6	90.1	1.076
sodium benzoate	$NaC_7H_5O_2$	25	36.0	41.5	73.9	1.152
sodium bicarbonate	$NaHCO_3$	15	8.28	8.80	97.6	1.061
sodium bisulfate	$NaHSO_4.H_2O$	25	59	87	60	1.47
sodium bromide	$NaBr.2H_2O$	25	48.6	75.0	79.4	1.542
sodium carbonate	$Na_2CO_3.10H_2O$	25	22.6	28.1	96.5	1.242
sodium chlorate	$NaClO_3$	25	51.7	74.3	69.6	1.440
sodium chloride	$NaCl$	25	26.5	31.7	88.1	1.198
sodium chromate	Na_2CrO_4	18	40.1	57.4	85.7	1.430
sodium citrate	$Na_3C_6H_5O_7.5H_2O$	25	48.1	61.2	66.0	1.272
sodium dichromate	$Na_2Cr_2O_7$	18	63.9	111.4	63.0	1.743
sodium ferrocyanide	$Na_4Fe(CN)_6$	25	17.1	19.4	93.9	1.131
sodium fluoride	NaF	25	3.98	4.14	99.7	1.038
sodium formate	$NaCHO_2$	18	44.7	58.9	73.0	1.316
sodium hydroxide	$NaOH$	25	50.8	77	74	1.51
sodium hypophosphite	NaH_2PO_2	16	52.1	72.4	66.6	1.386
sodium iodate	$NaIO_3.H_2O$	25	8.57	9.21	98.5	1.075
sodium iodide	NaI	25	64.8	124.3	67.7	1.919
sodium molybdate	Na_2MoO_4	18	39.4	56.6	87.0	1.435
sodium nitrate	$NaNO_3$	25	47.9	66.7	72.5	1.391
sodium nitrite	$NaNO_2$	20	45.8	62.3	73.8	1.359
sodium oxalate	$Na_2(CO_2)_2$	25	3.48	3.58	99.1	1.025
sodium paratungstate	$(Na_2O)_3(WO_3)_7.16H_2O$	0	26.7	35.2	96.5	1.316
sodium perchlorate	$NaClO_4$	25	67.8	114.1	54.1	1.683
sodium periodate	$NaIO_4.3H_2O$	25	12.6	13.9	96.2	1.103
sodium phenolsulfonate	$C_6H_4(OH)SO_3Na$	25	16.1	17.4	90.5	1.079
sodium phosphate dibasic	Na_2HPO_4	17	4.2	4.4	99.9	1.043
sodium phosphate tribasic	Na_3PO_4	14	9.5	10.5	99.8	1.103

Substance	Formula	Temp, °C	g/100 g satd soln	g/100 ml satd soln	ml water/ 100 ml satd soln	Specific gravity
sodium pyrophosphate	$Na_2H_2P_2O_7.6H_2O$	25	13.0	14.4	95.8	1.104
sodium salicylate	$NaC_7H_5O_3$	25	53.6	67.0	58.0	1.248
sodium selenate	Na_2SeO_4	18	29.0	38.1	93.4	1.313
sodium silicofluoride	$NaSiF_6$	20	0.773	0.737	99.76	1.0054
sodium sulfate	Na_2SO_4	25	21.8	26.4	94.5	1.208
sodium sulfate	$Na_2SO_4.10H_2O$	25	27.7	33.3	87.0	1.207
sodium sulfide	$Na_2S.9H_2O$	25	52.3	63	57	1.20
sodium sulfite, anhydrous	Na_2SO_3	25	23	28.5	95.5	1.24
sodium thiocyanate	$NaCNS$	25	62.9	87	51	1.38
sodium thiosulfate	$Na_2S_2O_3.5H_2O$	25	66.8	93	46	1.39
sodium tungstate	$Na_2WO_4.10H_2O$	18	42.0	66.1	91.3	1.573
stannous chloride	$SnCl_2$	15	72.9	133.1	49.5	1.827
strontium chlorate	$Sr(ClO_3)_2$	18	63.6	117.0	67.0	1.839
strontium chloride	$SrCl_2.6H_2O$	15	33.4	45.5	90.7	1.36
strontium iodide	$SrI_2.6H_2O$	20	64.0	137.8	77.5	2.15
strontium nitrate	$St(NO_3)_2$	25	44.2	65.3	82.5	1.477
strontium nitrite	$Sr(NO_2)_2$	19	39.3	56.8	87.8	1.445
strontium perchlorate	$Sr(ClO_4)_2$	25	75.6	158.5	50.8	2.084
strontium salicylate	$Sr(C_7H_5O_3)_2$	25	4.58	4.68	97.5	1.019
succinic acid	$(CH_2)_2(COOH)_2$	25	7.67	7.82	94.5	1.021
succinimide	$(CH_2CO)_2NH.H_2O$	25	30.6	32.7	74.2	1.067
sucrose	$C_{12}H_{22}O_{11}$	25	67.89	90.9	43.0	1.340
tartaric acid	$C_2H_2(OH)_2(COOH)_2$	15	58.5	76.9	54.7	1.31
tetraethyl ammonium iodide	$N(C_2H_5)_4I$	25	32.9	36.2	74.0	1.102
tetramethyl ammonium iodide	$N(CH_3)_4I$	25	5.51	5.60	96.1	1.016
thallium chloride	$TlCl$	25	0.40	0.40	99.6	1.0005
thallium nitrate	$TlNO_3$	25	10.4	11.4	98.0	1.093
thallium nitrite	$TlNO_2$	25	32.1	43.7	92.5	1.360
thallium perchlorate	$TlClO_4$	25	13.5	15.2	97.1	1.122
thallium sulfate	Tl_2SO_4	25	5.48	5.74	99.0	1.047
trichloroacetic acid	CCl_3COOH	25	92.3	149.6	12.41	1.615
uranyl chloride	UO_2Cl_2	18	76.2	208.5	65.2	2.736
uranyl nitrate	$UO_2(NO_3)_2.6H_2O$	25	68.9	120	54.5	1.74
urea	$(NH_2)_2CO$	25	53.8	62	53.5	1.15
urea phosphate	$CO(NH_2)_2.H_3PO_4$	24.5	52.4	66.1	60.1	1.26
urethan	$NH_2CO_2C_2H_5$	25	82.8	88.8	18.5	1.073
D-valine	$(CH_3)_2CHCH(NH_2)COOH$	25	8.14	8.26	93.3	1.015
DL-valine	$(CH_3)_2CHCH(NH_2)COOH$	25	6.61	6.68	94.5	1.012
zinc acetate	$Zn(C_2H_3O_2)_2$	25	25.7	30.0	86.5	1.165
zinc benzenesulfonate	$Zn(C_6H_5SO_3)_2$	25	29.5	34.9	83.4	1.182
zinc chlorate	$Zn(ClO_3)_2$	18	65.0	124.4	67.0	1.914
zinc chloride	$ZnCl_2$	25	67.5	128	61	1.89
zinc iodide	ZnI_2	18	81.2	221.3	51.2	2.725
zinc phenolsulfonate	$(C_6H_5OSO_3)_2Zn.8H_2O$	25	39.8	47.3	71.5	1.185
zinc selenate	$ZnSeO_4$	22	37.8	58.9	97.0	1.559
zinc silicofluoride	$ZnSiF_6.6H_2O$	20	32.9	47.2	96.3	1.434
zinc sulfate	$ZnSO_4.7H_2O$	25	36.7	54.6	94.7	1.492
zinc valerate	$Zn(C_5H_9O_2)_2$	25	1.27	1.27	98.8	1.001

COOLING MIXTURES

Substance	Quantity of substance	Quantity of water	Temp of resulting mixtures in °C	Approx temp, °F
ammonium nitrate	100	94	−4.0	25
sodium acetate	85	100	−4.7	23
sodium nitrate	75	100	−5.3	22.5
sodium thiosulfate cryst	110	100	−8.0	18
calcium chloride, 6H$_2$O	100	246 (ice)	−9.0	16
sodium chloride	36	100	−10.0	14
ammonium nitrate	45	100 (ice)	−16.8	1.5
sodium nitrate	50	100 (ice)	−17.8	0
ammonium thiocyanate	133	100	−18.0	0
sodium chloride	33	100 (ice)	−21.3	−6
calcium chloride, 6H$_2$O	100	123 (ice)	−21.5	−6.5
sodium bromide	66	100 (ice)	−28	−18
magnesium chloride cryst	85	100 (ice)	−34	−29
sulfuric acid (66.1% H$_2$SO$_4$)	100	109.7 (snow)	−37.0	−34.6
calcium chloride, 6H$_2$O	100	81 (ice)	−40.3	−40
calcium chloride, 6H$_2$O	100	70 (ice)	−55	−67
alcohol at 4°C with solid carbon dioxide			−72	−98
chloroform with solid carbon dioxide			−77	−106
acetone with solid carbon dioxide			−86	−123
ether with solid carbon dioxide			−100	−148
solid carbon dioxide (dry ice) sublimes at			−78.5	−109

THERMOMETRIC EQUIVALENTS

Temperature Scales

Symbol	Designation	Zero point	Freezing point of water	Boiling point of water at standard atm press.
°C	degree Celsius or Centigrade	freezing point of water	0°C	100°C
°K	degree Kelvin or absolute temperature in degrees Centigrades	absolute zero	273.15°K	373.15°K
°F	degree Fahrenheit	−17.8°C	+32°F	212°F
°Rank	degree Rankine or absolute temperature in degrees Fahrenheit	absolute zero	491.4°Rank	671.4°Rank

The following formulas may be used to convert temperatures from one scale to another:

Temperature given in	Temperature wanted in			
	°C	°K	°F	°Rank
°C	C	C + 273.15	1.8C + 32	1.8C + 491.4
°K	K − 273.15	K	1.8K − 459.4	1.8K
°F	0.556F − 17.8	0.556F + 255.3	F	F + 459.4
°Rank	0.556Rank − 273.1	0.556Rank	Rank − 459.4	Rank

International System of Units (SI)

SI Basic and Supplementary Units

Name	Symbol	Physical Quantity
SI BASE UNITS		
meter	m	length
kilogram	kg	mass
second	s	time
ampere	A	electric current
kelvin	K	thermodynamic temperature
mole	mol	amount of substance
candela	cd	luminous intensity
SI SUPPLEMENTARY UNITS		
radian	rad	plane angle
steradian	sr	solid angle

SI Derived Units with Special Names

Formula	Symbol	Special Name	Physical Quantity
$1 \, kg \cdot m/s^2$	$= 1 \, N$	1 newton	force
$1 \, N/m^2$	$= 1 \, Pa$	1 pascal	pressure or stress
$1 \, N \cdot m$	$= 1 \, J$	1 joule	work, energy, or quantity of heat
$1 \, J/s$	$= 1 \, W$	1 watt	power or radiant energy flux
$1 \, W/A$	$= 1 \, V$	1 volt	electric potential, potential difference, or electromotive foce
$1 \, A/V$	$= 1 \, S$	1 siemens	electric conductance
$1 \, V/A$	$= 1 \, \Omega$	1 ohm	electric resistance
$1 \, A \cdot s$	$= 1 \, C$	1 coulomb	quantity of electricity or electric charge
$1 \, C/V$	$= 1 \, F$	1 farad	electric capacitance
$1 \, V \cdot s$	$= 1 \, Wb$	1 weber	magnetic flux
$1 \, Wb/A$	$= 1 \, H$	1 henry	inductance
$1 \, Wb/m^2$	$= 1 \, T$	1 tesla	magnetic flux density or magnetic induction
$1 \, cd \cdot sr$	$= 1 \, lm$	1 lumen	luminous flux
$1 \, lm/m^2$	$= 1 \, lx$	1 lux	illuminance
$1 \, J/kg$	$= 1 \, Gy$	1 gray	absorbed dose (of ionizing radiation)
1 (disintegration)/s	$= 1 \, Bq$	1 becquerel	activity (of a radionuclide)
1 (cycle)/s	$= 1 \, Hz$	1 hertz	frequency (of a periodic phenomenon)
$K - 273$	$= \, °C$	degree Celsius	temperature

SI Prefixes

Factor	Prefix	Symbol	Factor	Prefix	Symbol
10^{-24}	yocto	y	10	deca	da
10^{-21}	zepto	z	10^2	hecto	h
10^{-18}	atto	a	10^3	kilo	k
10^{-15}	femto	f	10^6	mega	M
10^{-12}	pico	p	10^9	giga	G
10^{-9}	nano	n	10^{12}	tera	T
10^{-6}	micro	μ	10^{15}	peta	P
10^{-3}	milli	m	10^{18}	exa	E
10^{-2}	centi	c	10^{21}	zetta	Z
10^{-1}	deci	d	10^{24}	yotta	Y

References
G. G. Stoner, *Textile Research Journal* 46, 623, 850 (1976).
Federal Register 47, 8399 (1982).

Universal Conversion Factors

To convert units of one system into another, find the given units in the table and multiply by the appropriate conversion factor. For example:

To convert 26.5 centimeters into inches, find centimeters in the table, read across to inches and multiply by 0.3937.

$$26.5 \text{ cm} \times \frac{0.3937 \text{ inch}}{\text{cm}} = 10.4 \text{ inches}$$

To convert 8.75 inches into centimeters, find inches in the table, read across to centimeters and multiply by 2.540.

$$8.75 \text{ inches} \times \frac{2.540 \text{ cm}}{\text{inch}} = 22.2 \text{ cm}$$

TO CONVERT	INTO	MULTIPLY BY	TO CONVERT	INTO	MULTIPLY BY
A			baryes	dynes/sq cm	1.000
abcoulombs	statcoulombs	2.998×10^{10}	bolts (US cloth)	meters	36.576
acres	sq chains (Gunter's)	10.	Btu	liter—atmosphere	10.409
acres	sq rods	160.	Btu	ergs	1.0550×10^{10}
acres	square links		Btu	foot-lb	778.3
	(Gunter's)	1×10^5	Btu	gram-calories	252.0
acres	hectares or		Btu	horsepower-hr	3.931×10^{-4}
	sq hectometers	0.4047	Btu	joules	1,054.8
acres	sq ft	43,560.0	Btu	kilogram-calories	0.2520
acres	sq meters	4,047.	Btu	kilogram-meters	107.5
acres	sq miles	1.562×10^{-3}	Btu	kilowatt-hr	2.928×10^{-4}
acres	sq yd	4,840.	Btu/hr	foot-lb/sec	0.2162
acre-feet	cu ft	43,560.0	Btu/hr	gram-cal/sec	0.0700
acre-feet	gallons	3.259×10^5	Btu/hr	horsepower-hr	3.929×10^{-4}
amperes/sq cm	amps/sq in	6.452	Btu/hr	watts	0.2931
amperes/sq cm	amps/sq meter	10^4	Btu/min	foot-lb/sec	12.96
amperes/sq in	amps/sq cm	0.1550	Btu/min	horsepower	0.02356
amperes/sq in	amps/sq meter	1,550.0	Btu/min	kilowatts	0.01757
amperes/sq meter	amps/sq cm	10^{-4}	Btu/min	watts	17.57
amperes/sq meter	amps/sq in	6.452×10^{-4}	Btu/sq ft/min	watts/sq in	0.1221
ampere-hours	coulombs	3,600.0	bucket (Br dry)	cubic cm	1.818×10^4
ampere-hours	faradays	0.03731	bushels	cu ft	1.2445
ampere-turns	gilberts	1.257	bushels	cu in	2,150.42
ampere-turns/cm	amp-turns/in	2.540	bushels	cu meters	0.03524
ampere-turns/cm	amp-turns/meter	100.0	bushels	liters	35.24
ampere-turns/cm	gilberts/cm	1.257	bushels	pecks	4.0
ampere-turns/in	amp-turns/cm	0.3937	bushels	pints (dry)	64.0
ampere-turns/in	amp-turns/meter	39.37	bushels	quarts (dry)	32.0
ampere-turns/in	gilberts/cm	0.4950			
ampere-turns/meter	amp-turns/cm	0.01			
ampere-turns/meter	amp-turns/in	0.0254	**C**		
ampere-turns/meter	gilberts/cm	0.01257	calories, gram		
angstrom unit	inches	3937×10^{-9}	(mean)	Btu (mean)	3.9685×10^{-3}
angstrom unit	meters	1×10^{-10}	candle/sq cm	lamberts	3.142
angstrom unit	microns or (mu)	1×10^{-4}	candle/sq inch	lamberts	0.4870
ares	acres (US)	0.02471	centares (centiares)	sq meters	1.0
ares	sq yd	119.60	centigrade	Fahrenheit	$1.8 C° + 32$
ares	acres	0.02471	centigrams	grams	0.01
ares	sq meters	100.0	centiliters	ounces (fl)	0.3382
astronomical units	kilometers	1.495×10^8	centiliters	cu in	0.6103
atmospheres	ton/sq in	0.007348	centiliters	drams	2.705
atmospheres	cm of mercury	76.0	centiliters	liters	0.01
atmospheres	mm of mercury	760.0	centimeters	feet	3.281×10^{-2}
atmospheres	torrs	760.0	centimeters	inches	0.3937
atmospheres	ft of water (at 4°C)	33.90	centimeters	kilometers	10^{-5}
atmospheres	in of mercury (at 0°C)	29.92	centimeters	meters	0.01
atmospheres	kg/sq cm	1.0333	centimeters	miles	6.214×10^{-6}
atmospheres	kg/sq meter	10,332.	centimeters	millimeters	10.0
atmospheres	lb/sq in	14.70	centimeters	mils	393.7
atmospheres	tons/sq ft	1.058	centimeters	yards	1.094×10^{-2}
			centimeter-dynes	cm-grams	1.020×10^{-3}
B			centimeter-dynes	meter-kg	1.020×10^{-8}
barrels (US, dry)	cu in	7056.	centimeter-dynes	lb-ft	7.376×10^{-8}
barrels (US. dry)	quarts (dry)	105.0	centimeter-grams	cm-dynes	980.7
barrels (US, liq)	gallons	31.5	centimeter-grams	meter-kg	10^{-5}
barrels (oil)	gallons (oil)	42.0	centimeter-grams	lb-ft	7.233×10^{-5}
bars	atmospheres	0.9869	centimeters of		
bars	dynes/sq cm	10^6	mercury	atmospheres	0.01316
bars	kg/sq meter	1.020×10^4	centimeters of		
bars	lb/sq ft	2,089.	mercury	ft of water	0.4461
bars	lb/sq in	14.50			

TO CONVERT	INTO	MULTIPLY BY	TO CONVERT	INTO	MULTIPLY BY
centimeters of mercury	kg/sq meter	136.0	cubic inches	bushels	4.650×10^{-4}
centimeters of mercury	lb/sq ft	27.85	cubic inches	cu cm	16.39
centimeters of mercury	lb/sq in	0.1934	cubic inches	cu ft	5.787×10^{-4}
centimeters/sec	ft/min	1.1969	cubic inches	cu meters	1.639×10^{-5}
centimeters/sec	ft/sec	0.03281	cubic inches	cu yards	2.143×10^{-5}
centimeters/sec	kilometers/hr	0.036	cubic inches	drams (fl)	4.4329
centimeters/sec	knots	0.1943	cubic inches	gallons (US)	4.329×10^{-3}
centimeters/sec	meters/min	0.6	cubic inches	gallons (UK)	3.605×10^{-3}
centimeters/sec	miles/hr	0.02237	cubic inches	gills	0.1385
centimeters/sec	miles/min	3.728×10^{-4}	cubic inches	liters	0.01639
centimeters/sec/sec	ft/sec/sec	0.03281	cubic inches	milliliters	16.39
centimeters/sec/sec	km/hr/sec	0.036	cubic inches	mil-ft	1.061×10^{5}
centimeters/sec/sec	meters/sec/sec	0.01	cubic inches	minims	265.974
centimeters/sec/sec	miles/hr/sec	0.02237	cubic inches	ounces (fl)	0.5541
chain	inches	792.00	cubic inches	pecks	1.860×10^{-3}
chain	meters	20.12	cubic inches	pints (dry)	0.0298
chains (surveyors' or Gunter's)	yards	22.00	cubic inches	pints (liq)	0.03463
circular mils	sq cm	5.067×10^{-6}	cubic inches	quarts (dry)	0.0149
circular mils	sq mils	0.7854	cubic inches	quarts (liq)	0.01732
circular mils	sq in	7.854×10^{-7}	cubic meters	bushels	28.38
circumference	radians	6.283	cubic meters	cu cm	10^{6}
cords	cord ft	8	cubic meters	cu ft	35.31
cord feet	cu ft	16	cubic meters	cu in	61,023.74
coulomb	statcoulombs	2.998×10^{9}	cubic meters	cu yards	1.308
coulombs	faradays	1.036×10^{-5}	cubic meters	gallons (US)	264.2
coulombs/sq cm	coulombs/sq in	64.52	cubic meters	gallons (UK)	220.0
coulombs/sq cm	coulombs/sq meter	10^{4}	cubic meters	liters	1,000.0
coulombs/sq in	coulombs/sq cm	0.1550	cubic meters	milliliters	10^{6}
coulombs/sq in	coulombs/sq meter	1,550.	cubic meters	pecks	113.51
coulombs/sq meter	coulombs/sq cm	10^{-4}	cubic meters	pints (dry)	1,816.166
coulombs/sq/meter	coulombs/sq in	6.452×10^{-4}	cubic meters	pints (liq)	2,113.0
cubic centimeters	cu ft	3.531×10^{-5}	cubic meters	quarts (dry)	908.083
cubic centimeters	cu in	0.06102	cubic meters	quarts (liq)	1,057
cubic centimeters	cu meters	10^{-6}	cubic yards	cu cm	7.646×10^{5}
cubic centimeters	cu yards	1.308×10^{-4}	cubic yards	cu ft	27.0
cubic centimeters	drams (fl)	0.2705	cubic yards	cu in	46,656.0
cubic centimeters	gallons (US)	2.642×10^{-4}	cubic yards	cu meters	0.7646
cubic centimeters	gills	8.454×10^{-3}	cubic yards	gallons (US)	202.0
cubic centimeters	liters	0.001	cubic yards	gallons (UK)	168.2
cubic centimeters	milliliters	1.0	cubic yards	liters	764.6
cubic centimeters	minims	16.231	cubic yards	milliliters	7.646×10^{5}
cubic centimeters	ounces (fl)	0.0338	cubic yards	pints (liq)	1,615.9
cubic centimeters	pints (liq)	2.113×10^{-3}	cubic yards	quarts (liq)	807.9
cubic centimeters	quarts (liq)	1.057×10^{-3}	cubic yards/min	cubic ft/sec	0.45
cubic feet	bushels	0.8036	cubic yards/min	gallons/sec	3.367
cubic feet	cu cm	28,316.85	cubic yards/min	liters/sec	12.74
cubic feet	cu in	1,728.0			
cubic feet	cu meters	0.02832		**D**	
cubic feet	cu yards	0.03704	Dalton	gram	1.650×10^{-24}
cubic feet	drams (fl)	7660.05	days	seconds	86,400.0
cubic feet	gallons (US)	7.48052	decigrams	grams	0.1
cubic feet	gallons (UK)	6.229	deciliters	liters	0.1
cubic feet	gills	239.38	decimeters	meters	0.1
cubic feet	liters	28.32	degrees (angle)	quadrants	0.01111
cubic feet	milliliters	28,316.85	degrees (angle)	radians	0.01745
cubic feet	minims	459,603.1	degrees (angle)	seconds	3,600.0
cubic feet	ounces (fl)	957.51	degrees/sec	radians/sec	0.01745
cubic feet	pecks	3.2143	degrees/sec	revolutions/min	0.1667
cubic feet	pints (dry)	51.428	degrees/sec	revolutions/sec	2.778×10^{-3}
cubic feet	pints (liq)	59.84	dekagrams	grams	10.0
cubic feet	quarts (dry)	25.714	dekaliters	liters	10.0
cubic feet	quarts (liq)	29.92	dekameters	meters	10.0
cubic feet/min	cu cm/sec	472.0	drams (avdp)	drams (apoth)	0.4557
cubic feet/min	gallons/sec	0.1247	drams (avdp)	grains	27.3437
cubic feet/min	liters/sec	0.4720	drams (avdp)	grams	1.7718
cubic feet/min	lb of water/min	62.43	drams (avdp)	kilograms	1.7718×10^{-3}
cubic feet/sec	million gals/day	0.646317	drams (avdp)	milligrams	1771.85
cubic feet/sec	gallons/min	448.831	drams (avdp)	ounces (apoth or troy)	0.0570
			drams (avdp)	ounces (avdp)	0.0625
			drams (avdp)	pennyweights	1.139

TO CONVERT	INTO	MULTIPLY BY
drams (avdp)	pounds (apoth or troy)	4.747×10^{-3}
drams (avdp)	pounds (avdp)	3.906×10^{-3}
drams (avdp)	scruples	1.367
drams (apoth)	drams (avdp)	2.1943
drams (apoth)	grains	60.0
drams (apoth)	grams	3.8879
drams (apoth)	kilograms	3.888×10^{-3}
drams (apoth)	milligrams	3887.93
drams (apoth)	ounces (apoth or troy)	0.125
drams (apoth)	ounces (avdp)	0.1371429
drams (apoth)	pennyweights	2.5
drams (apoth)	pounds (apoth or troy)	0.0104
drams (apoth)	pounds (avdp)	8.571×10^{-3}
drams (apoth)	scruples	3.0
drams (fl)	cu cm	3.6967
drams (fl)	cu ft	1.3055×10^{-4}
drams (fl)	cu in	0.2256
drams (fl)	gallons (US)	9.7656×10^{-4}
drams (fl)	gills	0.03125
drams (fl)	liters	3.6967×10^{-3}
drams (fl)	milliliters	3.6967
drams (fl)	minims	60.0
drams (fl)	ounces (fl)	0.125
drams (fl)	pints (liq)	7.8125×10^{-3}
drams (fl)	quarts (liq)	3.9063×10^{-3}
dynes	grams	1.020×10^{-3}
dynes	joules/cm	10^{-7}
dynes	joules/meter (newtons)	10^{-5}
dynes	kilograms	1.020×10^{-6}
dynes	poundals	7.233×10^{-5}
dynes	pounds	2.248×10^{-6}
dynes/cm	ergs/sq mm	.01
dynes/sq cm	atmospheres	9.869×10^{-7}
dynes/sq cm	bars	10^{-6}
dynes/sq cm	in of mercury at 0°C	2.953×10^{-5}
dynes/sq cm	in of water at 4°C	4.015×10^{-4}

E

TO CONVERT	INTO	MULTIPLY BY
ell	cm	114.30
ell	inches	45.
em, pica	inches	.167
em, pica	cm	.4233
ergs	Btu	9.480×10^{-11}
ergs	dyne-centimeters	1.0
ergs	foot-lb	7.367×10^{-8}
ergs	gram-calories	0.2389×10^{-7}
ergs	gram-cm	1.020×10^{-3}
ergs	horsepower-hr	3.7250×10^{-14}
ergs	joules	10^{-7}
ergs	kg-calories	2.389×10^{-11}
ergs	kg-meters	1.020×10^{-8}
ergs	kilowatt-hr	0.2778×10^{-13}
ergs	watt-hr	0.2778×10^{-10}
ergs/sec	Btu/min	5.688×10^{-6}
ergs/sec	dyne-cm/sec	1.000
ergs/sec	ft-lb/min	4.427×10^{-6}
ergs/sec	ft-lb/sec	7.3756×10^{-8}
ergs/sec	horsepower	1.341×10^{-10}
ergs/sec	kg-calories/min	1.433×10^{-9}
ergs/sec	kilowatts	10^{-10}

F

TO CONVERT	INTO	MULTIPLY BY
Fahrenheit	centigrade	0.556 F° − 17.8
farads	microfarads	10^{6}
faradays/sec	ampere (abs)	9.6500×10^{4}
faradays	ampere-hr	26.80

TO CONVERT	INTO	MULTIPLY BY
faradays	coulombs	9.649×10^{4}
fathoms	meters	1.828804
fathoms	feet	6.0
feet	centimeters	30.48
feet	inches	12.0
feet	kilometers	3.048×10^{-4}
feet	meters	0.3048
feet	miles (naut)	1.645×10^{-4}
feet	miles (stat)	1.894×10^{-4}
feet	millimeters	304.8
feet	mils	1.2×10^{4}
feet	yards	0.333
feet of water	atmospheres	0.02950
feet of water	in of mercury	0.8826
feet of water	kg/sq cm	0.03048
feet of water	kg/sq meter	304.8
feet of water	lb/sq ft	62.43
feet of water	lb/sq in	0.4335
feet/min	cm/sec	0.5080
feet/min	feet/sec	0.01667
feet/min	km/hr	0.01829
feet/min	meters/min	0.3048
feet/min	miles/hr	0.01136
feet/sec	cm/sec	30.48
feet/sec	km/hr	1.097
feet/sec	knots	0.5921
feet/sec	meters/min	18.29
feet/sec	miles/hr	0.6818
feet/sec	miles/min	0.01136
feet/sec/sec	cm/sec/sec	30.48
feet/sec/sec	km/hr/sec	1.097
feet/sec/sec	meters/sec/sec	0.3048
feet/sec/sec	miles/hr/sec	0.6818
feet/100 feet	percent grade	1.0
foot-candles	lumens/sq meter	10.764
foot-pounds	Btu	1.286×10^{-3}
foot-pounds	ergs	1.356×10^{7}
foot-pounds	gram-calories	0.3238
foot-pounds	hp-hr	5.050×10^{-7}
foot-pounds	joules	1.356
foot-pounds	kg-calories	3.24×10^{-4}
foot-pounds	kg-meters	0.1383
foot-pounds	kilowatt-hr	3.766×10^{-7}
foot-pounds/min	Btu/min	1.286×10^{-3}
foot-pounds/min	foot-lb/sec	0.01667
foot-pounds/min	horsepower	3.030×10^{-5}
foot-pounds/min	kg-calories/min	3.24×10^{-4}
foot-pounds/min	kilowatts	2.260×10^{-5}
foot-pounds/sec	Btu/hr	4.6263
foot-pounds/sec	Btu/min	0.07717
foot-pounds/sec	horsepower	1.818×10^{-3}
foot-pounds/sec	kg-calories/min	0.01945
foot-pounds/sec	kilowatts	1.356×10^{-3}
furlongs	miles (US)	0.125
furlongs	rods	40.0
furlongs	feet	660.0

G

TO CONVERT	INTO	MULTIPLY BY
gallons (US)	cu cm	3,785.0
gallons (US)	cu ft	0.1337
gallons (US)	cu in	231.0
gallons (US)	cu meters	3.785×10^{-3}
gallons (US)	cu yards	4.951×10^{-3}
gallons (US)	drams (fl)	1024.0
gallons (US)	gallons (UK)	0.83267
gallons (US)	gills	32.0
gallons (US)	liters	3.785
gallons (US)	milliliters	3,785.0
gallons (US)	minims	61,440.0
gallons (US)	ounces (fl)	128.0
gallons (US)	pints (liq)	8.0

TO CONVERT	INTO	MULTIPLY BY	TO CONVERT	INTO	MULTIPLY BY
gallons (US)	quarts (liq)	4.0	grams	tons (short)	1.102×10^{-6}
gallons (UK)	cu ft	0.1605	grams/cm	lb/in	5.600×10^{-3}
gallons (UK)	cu in	277.4	grams/cu cm	lb/cu ft	62.43
gallons (UK)	cu meters	4.546×10^{-3}	grams/cu cm	lb/cu in	0.03613
gallons (UK)	cu yards	5.946×10^{-3}	gramc/cu cm	lb/mil-ft	3.405×10^{-7}
gallons (UK)	gallons (US)	1.20095	grams/liter	grains/gal	58.417
gallons (UK)	liters	4.546	grams/liter	lb/1,000 gal	8.345
gallons of water	lb of water	8.3453	grams/liter	lb/cu ft	0.062427
gallons/min	cu ft/sec	2.228×10^{-3}	grams/liter	parts/million	1,000.0
gallons/min	liters/sec	0.06308	grams/sq cm	lb/sq ft	2.0481
gallons/min	cu ft/hr	8.0208	gram-calories	Btu	3.9683×10^{-3}
gauss	lines/sq in	6.452	gram-calories	ergs	4.1868×10^{7}
gauss	webers/sq cm	10^{-8}	gram-calories	ft-lb	3.0880
gauss	webers/sq in	6.452×10^{-8}	gram-calories	horsepower-hr	1.5596×10^{-6}
gauss	webers/sq meter	10^{-4}	gram-calories	kilowatt-hr	1.1630×10^{-6}
gilberts	ampere-turns	0.7958	gram-calories	watt-hr	1.1630×10^{-3}
gilberts/cm	amp-turns/cm	0.7958	gram-calories/sec	Btu/hr	14.286
gilberts/cm	amp-turns/in	2.021	gram-centimeters	Btu	9.297×10^{-8}
gilberts/cm	amp-turns/meter	79.58	gram-centimeters	ergs	980.7
gills	cu cm	118.2941	gram-centimeters	joules	9.807×10^{-5}
gills	cu ft	4.1775×10^{-3}	gram-centimeters	kg-cal	2.343×10^{-8}
gills	cu in	7.21875	gram-centimeters	kg-meters	10^{-5}
gills	drams (fl)	32.0			
gills	gallons (US)	0.03125			
gills	liters	0.1183	**H**		
gills	milliliters	118.2941	hand	cm	10.16
gills	minims	1920.0	hectares	acres	2.471
gills	ounces (fl)	4.0	hectares	sq ft	1.076×10^{5}
gills	pints (liq)	0.25	hectograms	grams	100.0
gills	quarts (liq)	0.125	hectoliters	liters	100.0
gills (UK)	cu cm	142.07	hectometers	meters	100.0
grade	radian	0.01571	hectowatts	watts	100.0
grains	drams (apoth)	0.0167	henries	millihenries	1,000.0
grains	drams (avdp)	0.03657143	hogsheads (Br)	cu ft	10.114
grains	grams	0.0648	hogsheads (US)	cu ft	8.42184
grains	kilograms	6.479×10^{-5}	hogsheads (US)	gallons (US)	63
grains	milligrams	64.799	horsepower	Btu/min	42.44
grains	ounces (apoth or troy)	2.083×10^{-3}	horsepower	ft-lb/min	33,000
			horsepower	ft-lb/sec	550.0
grains	ounces (avdp)	2.286×10^{-3}	horsepower	horsepower (metric)	1.014
grains	pennyweights	0.04167	horsepower	kg-calories/min	10.68
grains	pounds (apoth or troy)	1.736×10^{-4}	horsepower	kilowatts	0.7457
			horsepower	watts	745.7
grains	pounds (avdp)	1.423×10^{-4}	horsepower (boiler)	Btu/hr	33,479.
grains	scruples	0.05	horsepower (boiler)	kilowatts	9.803
grains (troy)	grains (avdp)	1.0	horsepower (metric)	horsepower	0.9863
grains (troy)	grams	0.06480	horsepower (metric)	ft-lb/sec	542.5
grains (troy)	ounces (avdp)	2.286×10^{-3}	horsepower-hours	Btu	2,547.
grains (troy)	pennyweight (troy)	0.04167	horsepower-hours	ergs	2.6845×10^{13}
grains/US gal	parts/million	17.118	horsepower-hours	ft-lb	1.98×10^{6}
grains/US gal	lb/million gal	142.86	horsepower-hours	gm-cal	641,190.
grains/UK gal	parts/million	14.286	horsepower-hours	joules	2.684×10^{6}
grams	drams (apoth)	0.2572	horsepower-hours	kg-calories	641.1
grams	drams (avdp)	0.5644	horsepower-hours	kg-meters	2.737×10^{5}
grams	dynes	980.7	horsepower-hours	kilowatt-hr	0.7457
grams	grains	15.43	hours	days	4.167×10^{-2}
grams	joules/cm	9.807×10^{-5}	hours	weeks	5.952×10^{-3}
grams	joules/meter (newtons)	9.807×10^{-3}	hundredweights (long)	cwt (short)	1.12
grams	kilograms	0.001	hundredweights (long)	kilograms	50.802
grams	milligrams	1,000.0			
grams	ounces (apoth or troy)	0.03215	hundredweights (long)	ounces	1792.0
grams	ounces (avdp)	0.03527	hundredweights (long)	pounds	112.
grams	pennyweights	0.643			
grams	pounds (apoth or troy)	2.679×10^{-3}	hundredweights (long)	slugs	3.4811
grams	pounds (avdp)	2.205×10^{-3}	hundredweights (long)	tons (long)	0.05
grams	poundals	0.07093			
grams	scruples	0.7716	hundredweights (long)	tons (short)	0.056
grams	slugs	6.852×10^{-5}			

TO CONVERT	INTO	MULTIPLY BY	TO CONVERT	INTO	MULTIPLY BY
hundredweights (short)	cwt (short)	0.8929	kilograms	joules/meters (newtons)	9.807
hundredweights (short)	kilograms	45.359	kilograms	milligrams	10^6
hundredweights (short)	ounces	1600.0	kilograms	ounces (apoth or troy)	32.151
hundredweights (short)	pounds	100.0	kilograms	ounces (avdp)	35.274
			kilograms	pennyweights	643.015
hundredweights (short)	slugs	3.1081	kilograms	pounds (apoth or troy)	2.679
hundredweights (short)	tons (metric)	0.0453592	kilograms	pounds (avdp)	2.205
			kilograms	poundals	70.93
hundredweights (short)	tons (long)	0.0446429	kilograms	scruples	771.62
			kilograms	slugs	0.0685
hundredweights (short)	tons (short)	0.05	kilograms	tons (metric)	0.001
			kilograms	tons (long)	9.842×10^{-4}
I			kilograms	tons (short)	1.102×10^{-3}
			kilograms/cu meter	gm/cu cm	0.001
inches	centimeters	2.540	kilograms/cu meter	lb/cu ft	0.06243
inches	feet	0.0833	kilograms/cu meter	lb/cu in	3.613×10^{-5}
inches	meters	2.540×10^{-2}	kilograms/cu meter	lb/mil-foot	3.405×10^{-10}
inches	miles	1.578×10^{-5}	kilograms/meter	lb/ft	0.6720
inches	miles (naut)	1.3715×10^{-5}	kilograms/sq cm	dynes	980,665.
inches	millimeters	25.40	kilograms/sq cm	atmospheres	0.9678
inches	mils	1,000.0	kilograms/sq cm	ft of water	32.81
inches	yards	2.778×10^{-2}	kilograms/sq cm	in of mercury	28.96
inches of mercury	atmospheres	0.03342	kilograms/sq cm	lb/sq ft	2,048.
inches of mercury	ft of water	1.133	kilograms/sq cm	lb/sq in	14.22
inches of mercury	kg/sq cm	0.03453	kilograms/sq meter	atmospheres	9.678×10^{-5}
inches of mercury	kg/sq meter	345.3	kilograms/sq meter	bars	9.807×10^{-5}
inches of mercury	lb/sq ft	70.73	kilograms/sq meter	ft of water	3.281×10^{-3}
inches of mercury	lb/sq in	0.4912	kilograms/sq meter	in of mercury	2.896×10^{-3}
inches of water (at 4°C)	atmospheres	2.458×10^{-3}	kilograms/sq meter	lb/sq ft	0.2048
inches of water (at 4°C)	in of mercury	0.07355	kilograms/sq meter	lb/sq in	1.422×10^{-3}
inches of water (at 4°C)	kg/sq cm	2.540×10^{-3}	kilograms/sq mm	kg/sq meter	10^6
			kilograms-calories	Btu	3.968
inches of water (at 4°C)	oz/sq in	0.5781	kilogram-calories	ft-lb	3,088.
inches of water (at 4°C)	lb/sq ft	5.204	kilogram-calories	hp-hr	1.560×10^{-3}
			kilogram-calories	joules	4,186
inches of water (at 4°C)	lb/sq in	0.03613	kilogram-calories	kg-meters	426.9
international ampere	ampere (abs)	.9998	kilogram-calories	kilojoules	4.186
international volt	volts (abs)	1.0003	kilogram-calories	kilowatt-hr	1.163×10^{-3}
international volt	joules	9.654×10^4	kilogram-meters	Btu	9.294×10^{-3}
			kilogram-meters	ergs	9.804×10^7
J			kilogram-meters	ft-lb	7.233
joules	Btu	9.480×10^{-4}	kilogram-meters	joules	9.804
joules	ergs	10^7	kilogram-meters	kg-calories	2.342×10^{-3}
joules	ft-lb	0.7376	kilogram-meters	kilowatt-hr	2.723×10^{-6}
joules	kg-calories	2.389×10^{-4}	kilolines	maxwells	1,000.0
joules	kg-meters	0.1020	kiloliters	liters	1,000.0
joules	watt-hr	2.778×10^{-4}	kilometers	centimeters	10^5
joules/cm	grams	1.020×10^4	kilometers	feet	3,281.
joules/cm	dynes	10^7	kilometers	inches	3.937×10^4
joules/cm	joules/meter (newtons)	100.0	kilometers	meters	1,000.0
			kilometers	miles	0.6214
joules/cm	poundals	723.3	kilometers	millimeters	10^6
joules/cm	pounds	22.48	kilometers	yards	1,094.
			kilometers/hr	cm/sec	27.78
K			kilometers/hr	ft/min	54.68
kilograms	cwt (long)	0.0197	kilometers/hr	ft/sec	0.9113
kilograms	cwt (short)	0.022	kilometers/hr	knots	0.5396
kilograms	drams (apoth)	257.21	kilometers/hr	meters/min	16.67
kilograms	drams (avdp)	564.38	kilometers/hr	miles/hr	0.6214
kilograms	dynes	980,665.	kilometers/hr/sec	cm/sec/sec	27.78
kilograms	grains	15,432.36	kilometers/hr/sec	ft/sec/sec	0.9113
kilograms	grams	1000.0	kilometers/hr/sec	meters/sec/sec	0.2778
kilograms	joules/cm	0.09807	kilometers/hr/sec	miles/hr/sec	0.6214
			kilowatts	Btu/min	56.92
			kilowatts	foot-lb/min	4.426×10^4
			kilowatts	foot-lb/sec	737.6
			kilowatts	horsepower	1.341
			kilowatts	kg-calories/min	14.34

TO CONVERT	INTO	MULTIPLY BY	TO CONVERT	INTO	MULTIPLY BY
kilowatts	watts	1,000.0	meters	millimeters	1,000.0
kilowatt-hours	Btu	3,413.	meters	rods	0.1988
kilowatt-hours	ergs	3.600×10^{13}	meters	yards	1.094
kilowatt-hours	ft-lb	2.655×10^{6}	meters	varas	1.179
kilowatt-hours	gram-calories	859,850.	meters/min	cm/sec	1.667
kilowatt-hours	horsepower-hr	1.341	meters/min	ft/min	3.281
kilowatt-hours	joules	3.6×10^{6}	meters/min	ft/sec	0.05468
kilowatt-hours	kg-calories	860.5	meters/min	km/hr	0.06
kilowatt-hours	kg-meters	3.671×10^{5}	meters/min	knots	0.03238
kilowatt-hours	lb of water evaporated from and at 212°F.	3.53	meters/min	miles/hr	0.03728
			meters/sec	feet/min	196.8
kilowatt-hours	lb of water raised from 62° to 212°F.	22.75	meters/sec	feet/sec	3.281
knots	ft/hr	6,080.	meters/sec	kilometers/hr	3.6
knots	kilometers/hr	1.8532	meters/sec	kilometers/min	0.06
knots	nautical miles/hr	1.0	meters/sec	miles/hr	2.237
knots	statute miles/hr	1.151	meters/sec	miles/min	0.03728
knots	yd/hr	2,027.	meters/sec/sec	cm/sec/sec	100.0
knots	ft/sec	1.689	meters/sec/sec	ft/sec/sec	3.281
			meters/sec/sec	km/hr/sec	3.6
L			meters/sec/sec	miles/hr/sec	2.237
league	miles (approx.)	3.0	meter-kilograms	cm-dynes	9.807×10^{7}
light year	miles	5.9×10^{12}	meter-kilograms	cm-gm	10^{5}
light year	kilometers	9.46091×10^{12}	meter-kilograms	lb-ft	7.233
lines/sq cm	gauss	1.0	microfarad	farads	10^{-6}
lines/sq in	gauss	0.1550	micrograms	grams	10^{-6}
lines/sq in	webers/sq cm	1.550×10^{-9}	microhms	megohms	10^{-12}
lines/sq in	webers/sq in	10^{-8}	microhms	ohms	10^{-6}
lines/sq in	webers/sq meter	1.550×10^{-5}	microliters	liters	10^{-6}
links(engineer's)	inches	12.0	microns	meters	1×10^{-6}
links(surveyor's)	inches	7.92	miles (naut)	feet	6,080.27
liters	bushels	0.02838	miles (naut)	inches	7.2913×10^{4}
liters	cu cm	1,000.0	miles (naut)	kilometers	1.853
liters	cu ft	0.03531	miles (naut)	meters	1,853.
liters	cu in	61.024	miles (naut)	miles (stat)	1.1516
liters	cu meters	0.001	miles (naut)	yards	2,027.
liters	cu yards	1.308×10^{-3}	miles (statute)	centimeters	1.609×10^{5}
liters	drams (fl)	270.512	miles (statute)	feet	5,280.
liters	gallons (US)	0.2642	miles (statute)	inches	6.336×10^{4}
liters	gallons (UK)	0.220	miles (statute)	kilometers	1.609
liters	gills	8.454	miles (statute)	meters	1,609.
liters	milliliters	1,000.	miles (statute)	miles (naut)	0.8684
liters	minims	16,230.73	miles (statute)	rods	320.
liters	ounces (fl)	33.814	miles (statute)	yards	1,760.
liters	pecks	0.1135	miles/hr	cm/sec	44.70
liters	pints (dry)	1.8162	miles/hr	ft/min	88.
liters	pints (liq)	2.113	miles/hr	ft/sec	1.467
liters	quarts (dry)	0.9081	miles/hr	km/hr	1.609
liters	quarts (liq)	1.057	miles/hr	km/min	0.02682
liters/min	cu ft/sec	5.886×10^{-4}	miles/hr	knots	0.8684
liters/min	gal/sec	4.403×10^{-3}	miles/hr	meters/min	26.82
lumens/sq ft	foot-candles	1.0	miles/hr	miles/min	0.1667
lumen	spherical candle power	.07958	miles/hr/sec	cm/sec/sec	44.70
			miles/hr/sec	ft/sec/sec	1.467
lumen	watt	.001496	miles/hr/sec	km/hr/sec	1.609
lumen/sq ft	lumen/sq meter	10.76	miles/hr/sec	meters/sec/sec	0.4470
lux	foot-candles	0.0929	miles/min	cm/sec	2,682.
			miles/min	ft/sec	88.
M			miles/min	km/min	1.609
maxwells	kilolines	0.001	miles/min	knots/min	0.8684
maxwells	webers	10^{-8}	miles/min	miles/hr	60.0
megalines	maxwells	10^{6}	mil-feet	cu in	9.425×10^{-6}
megohms	microhms	10^{12}	milliers	kilograms	1,000.
megohms	ohms	10^{6}	milligrams	drams (apoth)	2.572×10^{-4}
meters	centimeters	100.0	milligrams	drams (avdp)	5.644×10^{-4}
meters	feet	3.281	milligrams	grains	0.01543236
meters	inches	39.37	milligrams	grams	0.001
meters	kilometers	0.001	milligrams	kilograms	10^{-6}
meters	miles (naut)	5.396×10^{-4}	milligrams	ounces (apoth or troy)	3.215×10^{-5}
meters	miles (stat)	6.214×10^{-4}	milligrams	ounces (avdp)	3.527×10^{-5}
			milligrams	pennyweights	6.43×10^{-4}

TO CONVERT	INTO	MULTIPLY BY	TO CONVERT	INTO	MULTIPLY BY
milligrams	pounds (apoth or troy)	2.679×10^{-6}	ounces (avdp)	milligrams	28,349.5
milligrams	pounds (avdp)	2.2046×10^{-6}	ounces (avdp)	ounces (apoth or troy)	0.9115
milligrams	scruples	7.7162×10^{-4}	ounces (avdp)	pennyweights	18.23
milligrams/liter	parts/million	1.0	ounces (avdp)	pounds (apoth or troy)	0.0759
millihenries	henries	0.001	ounces (avdp)	pounds (avdp)	0.0625
milliliters	cu cm	1.0	ounces (avdp)	scruples	21.875
milliliters	cu ft	3.531×10^{-5}	ounces (avdp)	slugs	1.9426×10^{-3}
milliliters	cu in	0.06102	ounces (avdp)	tons (long)	2.790×10^{-5}
milliliters	drams (fl)	0.2705	ounces (avdp)	tons (metric)	2.835×10^{-5}
milliliters	gallons (US)	2.642×10^{-4}	ounces (avdp)	tons (short)	3.125×10^{-5}
milliliters	gills	8.454×10^{-3}	ounces (apoth or troy)	drams (apoth)	8.0
milliliters	liters	0.001	ounces (apoth or troy)	drams (avdp)	17.554
milliliters	minims	16.231	ounces (apoth or troy)	grains	480.0
milliliters	ounces (fl)	0.0338	ounces (apoth or troy)	grams	31.103481
milliliters	pints (liq)	2.113×10^{-3}	ounces (apoth or troy)	kilograms	0.0311
milliliters	quarts (liq)	1.057×10^{-3}	ounces (apoth or troy)	milligrams	31,103.48
millimeters	centimeters	0.1	ounces (apoth or troy)	ounces (avdp)	1.09714
millimeters	feet	3.281×10^{-3}	ounces (apoth or troy)	pennyweights	20.0
millimeters	inches	0.03937	ounces (apoth or troy)	pounds (apoth or troy)	0.08333
millimeters	kilometers	10^{-6}	ounces (apoth or troy)	pounds (avdp)	0.0686
millimeters	meters	0.001	ounces (apoth or troy)	scruples	24.0
millimeters	miles	6.214×10^{-7}	ounces (fl)	cu cm	29.5735
millimeters	mils	39.37	ounces (fl)	cu ft	1.0444×10^{-3}
millimeters	yards	1.094×10^{-3}	ounces (fl)	cu in	1.805
millimeters of Hg	atmospheres	1.316×10^{-3}	ounces (fl)	drams (fl)	8.0
millimeters of Hg	torr	1.0	ounces (fl)	gallons (US)	7.8125×10^{-3}
millimicrons	meters	1×10^{-9}	ounces (fl)	gills	0.25
million gals/day	cu ft/sec	1.54723	ounces (fl)	liters	0.02957
mils	centimeters	2.540×10^{-3}	ounces (fl)	milliliters	29.5735
mils	feet	8.333×10^{-5}	ounces (fl)	minims	480.0
mils	inches	0.001	ounces (fl)	pints (liq)	0.0625
mils	kilometers	2.540×10^{-8}	ounces (fl)	quarts (liq)	0.03125
mils	yards	2.778×10^{-5}	ounces/sq in	dynes/sq cm	4309
miner's inches	cu ft/min	1.5	ounces/sq in	pounds/sq in	0.0625
minims (Br)	cu cm	0.059192			
minims	cu cm	0.061612		**P**	
minims	cu ft	2.176×10^{-6}	parsecs	miles	19×10^{12}
minims	cu in	3.7598×10^{-3}	parsecs	kilometers	3.084×10^{13}
minims	drams (fl)	0.0167	parts/million	grains/US gal	0.0584
minims	gallons (US)	1.628×10^{-5}	parts/million	grains/UK gal	0.07016
minims	gills	5.208×10^{-4}	parts/million	lb/million gal	8.345
minims	liters	6.161×10^{-5}	pecks (Br)	cu in	554.6
minims	milliliters	0.061612	pecks (Br)	liters	9.091901
minims	ounces (fl)	0.0021	pecks	bushels	0.25
minims	pints (liq)	1.302×10^{-4}	pecks	cu ft	0.3111
minims (liq)	quarts (liq)	6.51×10^{-5}	pecks	cu in	537.605
minutes (angles)	degrees	0.01667	pecks	cu meters	8.8098×10^{-3}
minutes (angles)	quadrants	1.852×10^{-4}	pecks	liters	8.8098
minutes (angles)	radians	2.909×10^{-4}	pecks	pints (dry)	16.0
minutes (angles)	seconds	60.0	pecks	quarts (dry)	8.0
myriagrams	kilograms	10.0	pennyweights	drams (apoth)	0.4
myriameters	kilometers	10.0	pennyweights	drams (avdp)	0.8777
myriawatts	kilowatts	10.0	pennyweights	grains	24.0
			pennyweights	grams	1.55517
	N		pennyweights	kilograms	1.555×10^{-3}
nepers	decibels	8.686	pennyweights	milligrams	1,555.17
newtons	dynes	1×10^{5}	pennyweights	ounces (apoth or troy)	0.05
	O				
ohm (Int)	ohm (abs)	1.0005			
ohms	megohms	10^{-6}			
ohms	microhms	10^{6}			
ounces (avdp)	cwt (long)	5.5804×10^{-4}			
ounces (avdp)	cwt (short)	6.25×10^{-4}			
ounces (avdp)	drams (apoth)	7.292			
ounces (avdp)	drams (avdp)	16.0			
ounces (avdp)	grains	437.5			
ounces (avdp)	grams	28.349527			
ounces (avdp)	kilograms	0.0283			

TO CONVERT	INTO	MULTIPLY BY	TO CONVERT	INTO	MULTIPLY BY
pennyweights	ounces (avdp)	0.0549	pounds (apoth or troy)	ounces (apoth or troy)	12.0
pennyweights	pounds (apoth or troy)	4.1667×10^{-3}	pounds (apoth or troy)	ounces (avdp)	13.1657
pennyweights	pounds (avdp)	3.428×10^{-3}	pounds (apoth or troy)	pennyweights	240.0
pennyweights	scruples	1.2	pounds (apoth or troy)	pounds (avdp)	0.822857
pints (dry)	bushels	0.0156	pounds (apoth or troy)	scruples	288.0
pints (dry)	cu ft	0.0194	pounds (apoth or troy)	tons (long)	3.6753×10^{-4}
pints (dry)	cu in	33.60	pounds (apoth or troy)	tons (metric)	3.7324×10^{-4}
pints (dry)	cu meters	5.506×10^{-4}	pounds (apoth or troy)	tons (short)	4.1143×10^{-4}
pints (dry)	liters	0.5506	pounds of water	cu ft	0.01602
pints (dry)	pecks	0.0625	pounds of water	cu in	27.68
pints (dry)	quarts (dry)	0.5	pounds of water	gallons	0.1198
pints (liq)	cu cm	473.2	pounds of water/min	cu ft/sec	2.670×10^{-4}
pints (liq)	cu ft	0.01671	pound-feet	cm-dynes	1.356×10^{7}
pints (liq)	cu in	28.875	pound-feet	cm-gm	13,825.
pints (liq)	cu meters	4.732×10^{-4}	pound-feet	meter-kg	0.1383
pints (liq)	cu yards	6.189×10^{-4}	pounds/cu ft	gm/cu cm	0.01602
pints (liq)	drams (fl)	128.0	pounds/cu ft	kg/cu meter	16.02
pints (liq)	gallons (US)	0.125	pounds/cu ft	lb/cu in	5.787×10^{-4}
pints (liq)	gills	4.0	pounds/cu ft	lb/mil-foot	5.456×10^{-9}
pints (liq)	liters	0.4732	pounds/cu in	gm/cu cm	27.68
pints (liq)	milliliters	473.2	pounds/cu in	kg/cu meter	2.768×10^{4}
pints (liq)	minims	7680.0	pounds/cu in	lb/cu ft	1,728.
pints (liq)	ounces (fl)	16.0	pounds/cu in	lb/mil-foot	9.425×10^{-6}
pints (liq)	quarts (liq)	0.5	pounds/ft	kg/meter	1.488
Planck's quantum	erg-sec	6.624×10^{-27}	pounds/in	gm/cm	178.6
poise	gm/cm sec	1.00	pounds/mil-foot	gm/cu cm	2.306×10^{6}
poundals	dynes	13,826.	pounds/sq ft	atmospheres	4.725×10^{-4}
poundals	grams	14.10	pounds/sq ft	ft of water	0.01602
poundals	joules/cm	1.383×10^{-3}	pounds/sq ft	in of mercury	0.01414
poundals	joules/meter (newtons)	0.1383	pounds/sq ft	kg/sq meter	4.882
poundals	kilograms	0.01410	pounds/sq ft	lb/sq in	6.944×10^{-3}
poundals	pounds	0.03108	pounds/sq in	atmospheres	0.06804
pounds (avdp)	cwt (long)	8.929×10^{-3}	pounds/sq in	ft of water	2.307
pounds (avdp)	cwt (short)	0.01	pounds/sq in	in of mercury	2.036
pounds (avdp)	drams (apoth)	116.67	pounds/sq in	kg/sq meter	703.1
pounds (avdp)	drams (avdp)	256.0	pounds/sq in	lb/sq ft	144.0
pounds (avdp)	dynes	44.4823×10^{4}			
pounds (avdp)	grains	7,000.0		**Q**	
pounds (avdp)	grams	453.5924	quadrants (angle)	degrees	90.0
pounds (avdp)	joules/cm	0.04448	quadrants (angle)	minutes	5,400.0
pounds (avdp)	joules/meter (newtons)	4.448	quadrants (angle)	radians	1.571
pounds (avdp)	kilograms	0.4536	quadrants (angle)	seconds	3.24×10^{5}
pounds (avdp)	milligrams	453,592.37	quarts (dry)	bushels	0.0313
pounds (avdp)	ounces (apoth or troy)	14.5833	quarts (dry)	cu ft	0.0389
pounds (avdp)	ounces (avdp)	16.0	quarts (dry)	cu in	67.20
pounds (avdp)	pennyweights	291.667	quarts (dry)	cu meters	1.101×10^{-3}
pounds (avdp)	poundals	32.17	quarts (dry)	liters	1.1012
pounds (avdp)	pounds (apoth or troy)	1.21528	quarts (dry)	pecks	0.125
pounds (avdp)	scruples	350.0	quarts (dry)	pints (dry)	2.0
pounds (avdp)	slugs	3.108×10^{-2}	quarts (liq)	cu cm	946.4
pounds (avdp)	tons (long)	4.464×10^{-4}	quarts (liq)	cu ft	0.03342
pounds (avdp)	tons (metric)	4.536×10^{-4}	quarts (liq)	cu in	57.75
pounds (avdp)	tons (short)	5.0×10^{-4}	quarts (liq)	cu meters	9.464×10^{-4}
pounds (apoth or troy)	drams (apoth)	96.0	quarts (liq)	cu yards	1.238×10^{-3}
pounds (apoth or troy)	drams (avdp)	210.65	quarts (liq)	drams (fl)	256.0
pounds (apoth or troy)	grains	5,760.0	quarts (liq)	gallons (US)	0.25
pounds (apoth or troy)	grams	373.2417	quarts (liq)	gills	8.0
pounds (apoth or troy)	kilograms	0.3732	quarts (liq)	liters	0.9464
pounds (apoth or troy)	milligrams	373,241.72	quarts (liq)	milliliters	946.4
			quarts (liq)	minims	15,360.0
			quarts (liq)	ounces (fl)	32.0
			quarts (liq)	pints (liq)	2.0

TO CONVERT	INTO	MULTIPLY BY	TO CONVERT	INTO	MULTIPLY BY
R			square feet	sq miles	3.587×10^{-8}
radians	degrees	57.30	square feet	sq mm	9.290×10^{4}
radians	minutes	3,438	square feet	sq yd	0.1111
radians	quadrants	0.6366	square inches	circular mils	1.273×10^{6}
radians	seconds	2.063×10^{5}	square inches	sq cm	6.452
radians/sec	degrees/sec	57.30	square inches	sq ft	6.944×10^{-3}
radians/sec	rev/min	9.549	square inches	sq mm	645.2
radians/sec	rev/sec	0.1592	square inches	sq mils	10^{6}
radians/sec/sec	rev/min/min	573.0	square inches	sq yd	7.716×10^{-4}
radians/sec/sec	rev/min/sec	9.549	square kilometers	acres	247.1
radians/sec/sec	rev/sec/sec	0.1592	square kilometers	sq cm	10^{10}
revolutions	degrees	360.0	square kilometers	sq ft	1.076×10^{7}
revolutions	quadrants	4.0	square kilometers	sq in	1.550×10^{9}
revolutions	radians	6.283	square kilometers	sq meters	10^{6}
revolutions/min	degrees/sec	6.0	square kilometers	sq miles	0.3861
revolutions/min	radians/sec	0.1047	square kilometers	sq yd	1.196×10^{6}
revolutions/min	rev/sec	0.01667	square meters	acres	2.471×10^{-4}
revolutions/min/min	radians/sec/sec	1.745×10^{-3}	square meters	sq cm	10^{4}
revolutions/min/min	rev/min/sec	0.01667	square meters	sq ft	10.76
revolutions/min/min	rev/sec/sec	2.778×10^{-4}	square meters	sq in	1,550.
revolutions/sec	degrees/sec	360.0	square meters	sq miles	3.861×10^{-7}
revolutions/sec	radians/sec	6.283	square meters	sq mm	10^{6}
revolutions/sec	rev/min	60.0	square meters	sq yd	1.196
revolutions/sec/sec	radians/sec/sec	6.283	square miles	acres	640.0
revolutions/sec/sec	rev/min/min	3,600.0	square miles	sq ft	2.788×10^{7}
revolutions/sec/sec	rev/min/sec	60.0	square miles	sq km	2.590
rods	chain (Gunter's)	0.25	square miles	sq meters	2.590×10^{6}
rods	meters	5.029	square miles	sq yd	3.098×10^{6}
rods (surveyors'			square millimeters	circular mils	1,973
measure)	yards	5.5	square millimeters	sq cm	0.01
rods	feet	16.5	square millimeters	sq ft	1.076×10^{-5}
			square millimeters	sq in	1.550×10^{-3}
S			square mils	circular mils	1.273
scruples	drams (apoth)	0.3333	square mils	sq cm	6.452×10^{-6}
scruples	drams (avdp)	0.7314	square mils	sq in	10^{-6}
scruples	grains	20.0	square yards	acres	2.066×10^{-4}
scruples	grams	1.296	square yards	sq cm	8,361
scruples	kilograms	1.296×10^{-3}	square yards	sq ft	9.0
scruples	milligrams	1295.97	square yards	sq in	1,296.
scruples	ounces (apoth or		square yards	sq meters	0.8361
	troy)	0.0417	square yards	sq miles	3.228×10^{-7}
scruples	ounces (avdp)	0.0457	square yards	sq mm	8.361×10^{5}
scruples	pennyweights	0.8333			
scruples	pounds (apoth or				
	troy)	3.472×10^{-3}	**T**		
scruples	pounds (avdp)	2.857×10^{-3}	temperature (°C)	absolute tempera-	
seconds (angle)	degrees	2.778×10^{-4}	+273	ture (°C)	1.0
seconds (angle)	minutes	0.01667	temperature (°C)		
seconds (angle)	quadrants	3.087×10^{-6}	+17.78	temperature (°F)	1.8
seconds (angle)	radians	4.848×10^{-6}	temperature (°F)	absolute tempera-	
slugs	cwt (long)	0.2873	+460	ture (°F)	1.0
slugs	cwt (short)	0.3217	temperature (°F)		
slugs	grams	1.459×10^{4}	−32	temperature (°C)	5/9
slugs	kilograms	14.59	tons (long)	cwt (long)	20.
slugs	ounces (avdp)	514.79	tons (long)	cwt (short)	22.4
slugs	pounds (avdp)	32.17	tons (long)	kilograms	1,016.
slugs	tons (long)	1.436×10^{-2}	tons (long)	ounces (avdp)	35,840.0
slugs	tons (short)	1.609×10^{-2}	tons (long)	pounds (avdp)	2,240.0
sphere	steradians	12.57	tons (long)	slugs	69.621
square centimeters	circular mils	1.973×10^{5}	tons (long)	tons (metric)	1.0160
square centimeters	sq ft	1.076×10^{-3}	tons (long)	tons (short)	1.120
square centimeters	sq in	0.1550	tons (metric)	cwt (short)	22.046
square centimeters	sq meters	0.0001	tons (metric)	kilograms	1,000.0
square centimeters	sq miles	3.861×10^{-11}	tons (metric)	ounces (avdp)	35,273.96
square centimeters	sq mm	100.0	tons (metric)	pounds (avdp)	2,205.
square centimeters	sq yd	1.196×10^{-4}	tons (metric)	tons (long)	0.9842
square feet	acres	2.296×10^{-5}	tons (metric)	tons (short)	1.1023
square feet	circular mils	1.833×10^{8}	tons (short)	cwt (long)	17.857
square feet	sq cm	929.0	tons (short)	cwt (short)	20.0
square feet	sq in	144.0	tons (short)	grams	9.072×10^{5}
square feet	sq meters	0.09290	tons (short)	kilograms	907.1847

TO CONVERT	INTO	MULTIPLY BY	TO CONVERT	INTO	MULTIPLY BY
tons (short)	ounces (apoth or troy)	29,166.66	watts (abs)	Btu (mean)/min	0.056884
tons (short)	ounces (avdp)	32,000.0	watts (abs)	joules/sec	1.
tons (short)	pounds (apoth or troy)	2,430.56	watt-hours	Btu	3.413
			watt-hours	ergs	3.60×10^{10}
tons (short)	pounds (avdp)	2,000.0	watt-hours	ft-lb	2,656.
tons (short)	slugs	62.16	watt-hours	gm-cal	859.85
tons (short)	tons (long)	0.89286	watt-hours	horsepower-hr	1.341×10^{-3}
tons (short)	tons (metric)	0.9072	watt-hours	kg-cal	0.8605
tons (short)/sq ft	kg/sq meter	9,765.	watt-hours	kg-meters	367.2
tons (short)/sq ft	lb/sq in	2,000.	watt-hours	kilowatt-hr	0.001
tons of water/24 hr	lb of water/hr	83.333	watt (int)	watt (abs)	1.0002
tons of water/24 hr	gal/min	0.16643	webers	maxwells	10^8
tons of water/24 hr	cu ft/hr	1.3349	webers	kilolines	10^5
torr	mm of mercury	1.0	webers/sq in	gauss	1.550×10^7
torr	atmospheres	1.316×10^{-3}	webers/sq in	lines/sq in	10^8
			webers/sq in	webers/sq cm	0.1550
	V		webers/sq in	webers/sq meter	1,550.
volt/inch	volt/cm	0.39370	webers/sq meter	gauss	10^4
volt (abs)	statvolts	0.003336	webers/sq meter	lines/sq in	6.452×10^4
			webers/sq meter	webers/sq cm	10^{-4}
	W		webers/sq meter	webers/sq in	6.452×10^{-4}
watts	Btu/hr	3.4129		**Y**	
watts	Btu/min	0.05688	yards	centimeters	91.44
watts	ergs/sec	10^7	yards	feet	3.0
watts	ft-lb/min	44.27	yards	inches	36.0
watts	ft-lb/sec	0.7378	yards	kilometers	9.144×10^{-4}
watts	horsepower	1.341×10^{-3}	yards	meters	0.9144
watts	horsepower (metric)	1.360×10^{-3}	yards	miles (naut)	4.934×10^{-4}
watts	kg-calories/min	0.01433	yards	miles (stat)	5.682×10^{-4}
watts	kilowatts	0.001	yards	millimeters	914.4

Table of Radioactive Isotopes

The following table is based principally on the *Table of Isotopes,* C. M. Lederer, V. S. Shirley, Eds. (Wiley-Interscience, New York, 7th ed., 1978) and the *Table of Radioactive Isotopes,* V. S. Shirley, Ed. (Wiley-Interscience, New York, 8th ed., 1986). Additional references are cited in the monographs for each element in the monograph section of *The Merck Index.*

Columns 1, 2 and 3 list, respectively, the atomic number (Z), the name of the element and the mass number (A) of the isotope. An asterisk preceding the mass number indicates that the radioisotope occurs in nature. An m following the mass number indicates a metastable isotope. Customary designations for naturally occurring radioisotopes are given in parentheses after the mass number.

The total half-life of the isotope is listed in column 4. The half-life reported represents the most accurate value as published in the *Table of Radioactive Isotopes.* Except in the case of the transuranium elements, isotopic modifications for which the mass number assignment is doubtful are omitted. Generally, only those isotopes with a half-life of 60 minutes or longer are included in the table. Exceptions are made in the following instances:

1. Radioisotopes already used as biological tracers
2. Short-lived products, including nuclear isomers, of radioactive parents with half-lives of 60 minutes or greater
3. Naturally occurring radioisotopes
4. Isotopes of the transuranium elements

The following abbreviations for time units are employed: y = years, d = days, h = hours, min = minutes, s = seconds, ms = milliseconds, μs = microseconds, ns = nanoseconds.

In the last column are the principal modes of disintegration and energies of the radiations in million electronvolts (MeV). Symbols used to represent the various modes of decay are:

α	alpha particle emission	K	electron capture
β⁻	beta particle	IT	isomeric transition
β⁺	positron	e⁻	internal conversion electron
γ	gamma ray	x	x-ray
		S.F.	spontaneous fission

When x-radiation can be associated with a specific element, the symbol of this element precedes the x designation, as As-x. The energies of these radiations are enclosed in parentheses following the appropriate symbol for a given type of radiation. For β^- and β^+, values of E_{max} are listed. An attempt was made to arrange the modes of disintegration and energies in order of decreasing importance. Radiation types and energies of minor importance are omitted unless useful for identification purposes. Annihilation radiation associated with β^+ emission is not explicitly indicated. The reader should consult the literature cited for detailed decay schemes.

Z	Element	A	Half-life	Radiation (MeV)
1	Hydrogen	3	12.33y	β^-(0.0186)
4	Beryllium	7	53.29d	K, γ(0.478)
		10	1.6×10^6y	β^-(0.555), no γ
6	Carbon	11	20.39min	β^+(0.961), no γ
		14	5730y	β^-(0.156), no γ
7	Nitrogen	13	9.965min	β^+(1.190), no γ
9	Fluorine	18	1.8295h	β^+(0.635), K, O-x
11	Sodium	22	2.602y	β^+(0.545, 1.83), K, Ne-x, γ(1.275)
		24	14.659h	β^-(1.39), γ(2.75, 1.37)
12	Magnesium	28	20.90h	β^-(0.46), γ(1.34, 0.94, 0.40, 0.031)
13	Aluminum	26	7.2×10^5y	β^+(1.16), K, Mg-x, γ(1.81, 1.12)
14	Silicon	31	2.622h	β^-(1.48), γ(1.27)
		32	104y	β^-(0.213), no γ
15	Phosphorus	32	14.282d	β^-(1.71), no γ
		33	25.34d	β^-(0.25), no γ
16	Sulfur	35	87.51d	β^-(0.167)
		38	2.84h	β^-(1.0, 3.0), γ(1.94)
17	Chlorine	36	3.01×10^5y	β^-(0.71), K, S-x, no γ
		38	37.24min	β^-(4.81, 1.11, 2.77), γ(2.17, 1.60)
		39	55.6min	β^-(1.91, 2.18, 3.45), γ(1.27, 0.25, 1.52)
18	Argon	37	35.04d	K, Cl-x, no γ
		41	1.827h	β^-(1.20, 2.49), γ(1.29)
		42	32.9y	β^-

Table of Radioactive Isotopes (Continued)

Z	Element	A	Half-life	Radiation (MeV)
19	Potassium	*40	1.277×10^9y	β^-(1.34), K, Ar-x, γ(1.46)
		42	12.360h	β^-(3.52, 1.97), γ(1.52, 0.31)
		43	22.3h	β^-(0.83, 0.46, 1.22, 1.82), γ(0.618, 0.373, 0.39, 0.59, 0.22)
20	Calcium	41	1.03×10^5y	K, K-x
		45	163.8d	β^-(0.255)
		47	4.536d	β^-(1.98, 0.67), γ(1.30, 0.81, 0.49)
21	Scandium	43	3.891h	β^+(1.2, 0.82, 0.39), K, γ(0.373)
		44m	2.442d	IT, Sc-x, γ(0.271), e^-
		44	3.927h	β^+(1.47), K, γ(1.16)
		46	83.83d	β^-(0.357), Ti-x, γ(1.12, 0.889)
		47	3.341d	β^-(0.439, 0.60), γ(0.159)
		48	1.821d	β^-(0.65), γ(1.31, 1.04, 0.984)
22	Titanium	44	47.3y	K, γ(0.068, 0.078)
		45	3.080h	β^+(1.044), K, Sc-x
23	Vanadium	48	15.976d	β^+(0.696), K, Ti-x, γ(1.31, 0.983)
		49	330d	K, Ti-x, no γ
24	Chromium	48	21.56h	K, V-x, γ(0.308, 0.116), no β^+
		51	27.704d	K, V-x, γ(0.32)
25	Manganese	52	5.591d	K, β^+(0.574), Cr-x, γ(0.744, 0.935, 1.434)
		53	3.74×10^6y	K, Cr-x, no γ
		54	312.20d	K, Cr-x, γ(0.835), no β^+, no β^-
		56	2.5785h	β^-(2.84, 1.03, 0.72), γ(0.847, 1.81, 2.11)
26	Iron	52	8.275h	β^+(0.804), K, Mn-x, γ(0.169)
		55	2.73y	K, Mn-x, no γ
		59	44.496d	β^-(0.273, 0.475), γ(1.10, 1.29, 0.192)
		60	$\sim 1 \times 10^5$y	β^-
27	Cobalt	55	17.53h	β^+(1.04, 1.50), K, Fe-x, γ(0.932, 0.480, 1.41)
		56	77.7d	K, β^+(1.46), Fe-x, γ(0.847, 1.04, 1.24, 1.77, 2.60, 3.26, 2.02)
		57	271.77d	K, Fe-x, γ(0.136, 0.122, 0.014)
		58m	9.15h	IT, Co-x, γ(0.025), no β^+
		58	70.916d	K, β^+(0.474), Fe-x, γ(0.811)
		60	5.271y	β^-(0.318), γ(1.173, 1.332)
		61	1.650h	β^-(1.22), γ(0.067)
28	Nickel	56	6.10d	K, Co-x, γ(0.158, 0.270, 0.480, 0.75, 0.812, 1.56), no β^+
		57	1.503d	K, β^+(0.85, 0.72), Co-x, γ(1.378, 0.127, 1.76)
		59	7.5×10^4y	K, Co-x, no γ
		63	100.1y	β^-(0.066), no γ
		65	2.520h	β^-(2.14, 0.65, 1.02), γ(1.48, 0.367, 1.12)
		66	2.275d	β^-(0.20), no γ
29	Copper	61	3.408h	β^+(1.21), K, Ni-x, γ(0.28, 0.66)
		64	12.701h	K, β^-(0.571), β^+(0.657), Ni-x, γ(1.34)
		67	2.580d	β^-(0.395, 0.484, 0.577), γ(0.185, 0.092)
30	Zinc	62	9.26h	K, β^+(0.66), Cu-x, γ(0.041, 0.25, 0.51, 0.60)
		65	244.1d	K, β^+(0.325), γ(1.12), Cu-x
		69m	13.76h	IT, Zn-x, γ(0.439), e^-
		69	55.6min	β^-(0.90)
		72	1.938d	β^-(0.3), γ(0.016, 0.145, 0.192)
31	Gallium	66	9.49h	β^+(4.15), K, Zn-x, γ(1.04. 2.75)
		67	3.261d	K, Zn-x, γ(0.093, 0.184, 0.300, 0.393), e^-, no β^+
		68	1.135h	β^+(1.9), K, Zn-x, γ(1.08)
		72	14.10h	β^-(0.64, 0.96, 1.51, 2.53, 3.17), γ(0.835, 0.63, 2.20, 2.50)
		73	4.87h	β^-(1.20), γ(0.297, 0.534, 0.326)
32	Germanium	66	2.26h	K, β^+(1.02), Ga-x, γ(0.382, 0.044, 0.065, 0.109, 0.190, 0.246, 0.273, 0.338, 0.471)

Z	Element	A	Half-life	Radiation (MeV)
		68	270.8d	K, Ga-x
		69	1.627d	K, β^+(1.22, 0.61), γ(1.11, 0.573, 0.872)
		71	11.15d	K, Ga-x
		73m	499ms	IT, γ(0.053, 0.0133), e^-
		75	1.3797h	β^-(1.19, 0.98, 0.92, 0.72), γ(0.265, 0.199)
		77	11.30h	β^-(2.20, 1.38, 0.71), γ(0.264, 0.211, 0.216, 0.416, 0.714)
		78	1.467h	β^-(0.71), γ(0.277)
33	Arsenic	71	2.70d	K, β^+(0.81), Ge-x, γ(0.175, 0.023), e^-
		72	1.083d	β^+(3.323, 2.490), K, Ge-x, γ(0.834)
		73	80.30d	K, γ(0.054, 0.0133), Ge-x, e^-
		74	17.78d	K, β^-(1.35, 0.72), β^+(1.53, 0.94), Ge-x, γ(0.596, 0.635)
		76	1.097d	β^-(2.97, 2.41, 1.79), γ(0.559, 0.657)
		77	1.6179d	β^-(0.68), γ(0.239, 0.250, 0.088)
		78	1.512h	β^-(4.42, 3.70, 3.00, 2.50), γ(0.62, 0.70, 0.83, 1.31)
34	Selenium	72	8.40d	K, As-x, γ(0.046), e^-
		73	7.15h	β^+(1.32), K, As-x, γ(0.361, 0.067)
		75	119.77d	K, γ(0.265, 0.136, 0.280, 0.121, 0.401), As-x
		79	$<6.5 \times 10^4$y	β^-(0.16), no γ
35	Bromine	75	1.62h	β^+(1.74, 1.59, 1.45, 1.34, 1.13), K, γ(0.286)
		76	16.2h	β^+(3.98, 3.44), K, Se-x, γ(0.559, 0.657, 1.216)
		77	2.37650d	K, β^+(0.34), Se-x, γ(0.24, 0.30, 0.52, 0.58, 0.82)
		80m	4.42h	IT, Br-x, γ(0.037, 0.049), e^-
		80	17.68min	β^-(1.997, 1.38), K, β^+(0.85), Se-x, γ(0.616, 0.667)
		82	1.4708d	β^-(0.444), γ(0.554, 0.619, 0.698, 0.777, 0.828, 1.04, 1.32, 1.48), no K, no β^+
		83	2.39h	β^-(0.925), γ(0.530)
36	Krypton	76	14.8h	K, γ(0.252), no β^+
		77	1.24h	β^+(1.875, 1.700, 1.550), K, Br-x, γ(0.130)
		79	1.460d	K, β^+(0.613), Br-x, γ(0.044–1.34)
		81m	13s	IT, Kr-x, γ(0.19)
		81	2.10×10^5y	K, Br-x, γ(0.276)
		83m	1.86h	IT, Kr-x, γ(0.032, 0.009)
		85m	4.480h	β^-(0.84), IT, Kr-x, γ(0.51, 0.305)
		85	10.72y	β^-(0.67), γ(0.517)
		87	1.272h	β^-(3.8, 1.3), γ(0.403, 2.55, 0.85)
		88	2.84h	β^-(0.52, 2.8, 0.9), γ2.392, 0.028, 0.166, 0.196, 0.362)
37	Rubidium	81	4.58h	K, β^+(1.05), Kr-x, γ(0.446)
		82m	6.472h	K, β^+(0.800), γ(0.776, 0.554)
		82	1.273min	β^+(3.15), K, Kr-x, γ(0.776)
		83	86.2d	K, Kr-x, γ(0.521, 0.530, 0.553, 0.009, 0.79), no β^+
		84	32.87d	K, β^+(1.66, 0.78), β^-(0.89), Kr-x, γ(0.882)
		86	18.66d	β^-(1.78, 0.71), γ(1.08)
	*87	87	4.80×10^{10}y	β^-(0.272), no γ
		88	17.8min	β^-(5.080, 3.240, 2.350), γ(1.836, 0.898, 2.678)
38	Strontium	82	25.55d	K, Rb-x, no γ
		83	1.3504d	K, β^+(1.227), Rb-x, γ(0.763)
		85m	1.1258h	IT, K, Rb-x, Sr-x, γ(0.150, 0.231), e^-, no β^+
		85	64.84d	K, Rb-x, γ(0.514)
		87m	2.795h	IT, γ(0.388), e^-
		89	50.55d	β^-(1.488), γ(0.01 with Y^{89m})
		90	28.5y	β^-(0.546), no γ
		91	9.52h	β^-(2.67, 1.36, 1.09), γ(1.024, 0.556, 0.750)
		92	2.71h	β^-(0.55, 1.5), γ(1.384)
39	Yttrium	85m	4.86h	β^+(2.24), K, Sr-x, γ(0.767, 0.232, 2.124), no IT
		85	2.68h	β^+(1.58, 1.15), K, Sr-x, γ(0.504, 0.232), no IT
		86	14.74h	K, β^+(1.04, 1.25, 1.60, 2.02, 2.34, 3.15), Sr-x, γ(1.077)
		87m	12.9h	IT, β^+?(1.150), Y-x, γ(0.381)

Z	Element	A	Half-life	Radiation (MeV)
		87	3.346d	K, β^+?(0.7), Sr-x, γ(0.485, 0.39)
		88	106.61d	K, β^+(0.76), Sr-x, γ(0.898, 1.84)
		90	2.671d	β^-(2.288), γ(2.186)
		91m	49.71min	IT, Y-x, γ(0.556)
		91	58.51d	β^-(1.543), γ(1.21)
		92	3.54h	β^-(3.64), γ(0.448, 0.560, 0.934, 1.40, 1.85)
		93	10.25h	β^-(2.89), γ(0.267, 0.95, 1.92)
40	Zirconium	86	16.5h	K, Y-x, γ(0.243), no β^+
		87	1.733h	β^+(2.260), K, Y-x, γ(1.228)
		88	83.4d	K, Y-x, γ(0.394)
		89	3.268d	K, β^+(0.897), Y-x, γ(0.909)
		90m	809.2ms	IT, Zr-x, γ(2.319)
		93	1.53×10^6y	β^-(0.060)
		95	64.02d	β^-(0.360, 0.396), γ(0.724, 0.757)
		97	16.90h	β^-(1.91), γ(0.743)
41	Niobium	89	2.03h	β^+(3.320), γ(1.627)
		90m	18.82s	IT, Nb-x, γ(0.122)
		90	14.60h	β^+(1.50), K, Zr-x, γ(0.141, 2.319)
		91m	62d	IT, Nb-x, γ(0.1045, 1.2), K
		91	680y	K, Zr-x
		92m	10.15d	K, γ(0.934)
		93m	13.6y	IT, Nb-x, γ(0.0304)
		94	2.03×10^4y	β^-(0.473), γ(0.703, 0.871)
		95m	3.61d	IT, β^-(1.160), Nb-x, γ(0.204, 0.236), e$^-$
		95	34.97d	β^-(0.160), γ(0.766)
		96	23.35h	β^-(0.748, 0.500), γ(0.778, 1.091)
		97m	1.00min	IT, γ(0.743)
		97	1.202h	β^-(1.267), γ(0.658)
42	Molybdenum	90	5.67h	K, β^+(1.085), Nb-x, γ(0.122, 0.257)
		93m	6.85h	IT, Mo-x, γ(0.264, 0.685, 1.477), e$^-$, no β^+
		93	3500y	K, Nb-x
		99	2.7477d	β^-(1.214), Tc-x, γ(0.041, 0.181, 0.740, 0.780)
43	Technetium	93	2.75h	K, β^+(0.807), Mo-x, γ(1.363, 1.520)
		95m	61d	K, IT, β^+(0.71), Mo-x, γ(0.204, 0.582, 0.835)
		95	20.0h	K, Mo-x, γ(0.766), no β^+
		96	4.28d	K, Mo-x, γ(0.778, 0.813, 0.850), no β^+
		97m	90.5d	IT, Tc-x, γ(0.096)
		97	2.6×10^6y	K, Mo-x
		98	4.2×10^6y	β^-(0.397), γ(0.652, 0.745)
		99m	6.006h	IT, Tc-x, γ(0.141, 0.002), e$^-$
		99	2.13×10^5y	β^-(0.292)
44	Ruthenium	95	1.64h	K, β^+(1.200, 0.910), Tc-x, γ(0.340, 0.625, 1.09)
		97	2.88d	K, Tc-x, γ(0.216, 0.324), no β^+
		103	39.254d	β^-(0.12, 0.22), γ(0.497)
		105	4.44h	β^-(1.187, 1.109), γ(0.316, 0.469, 0.676, 0.724)
		106	1.020y	β^-(0.0392), no γ
45	Rhodium	99	16d	K, β^+(1.100), Ru-x, γ(0.090, 0.353, 0.528)
		99m	4.7h	K, β^+(0.74), Ru-x, γ(0.090, 0.277, 0.341, 0.618, 1.261)
		100	20.8h	K, β^+(2.62, 2.07, 1.26), Ru-x, γ(0.446, 0.540, 1.107, 2.376)
		101m	4.34d	K, IT, Ru-x, Rh-x, γ(0.307)
		101	3.3y	K, γ(0.127, 0.198, 0.325)
		102m	207d	K, β^-(1.15), β^+(1.29, 0.82), Ru-x, γ(0.475)
		102	~2.9y	K, Ru-x, γ(0.418, 0.475, 0.632, 0.698, 0.768, 1.05, 1.11), no β^+
		103m	56.12min	IT, Rh-x, γ(0.040), e$^-$
		105m	45s	IT, Rh-x, γ(0.129), e$^-$
		105	1.4733d	β^-(0.568, 0.249), γ(0.306, 0.319)
		106m	2.17h	β^-(0.925, 0.730), γ(0.451, 0.512, 0.717, 0.749)
		106	29.80s	β^-(3.53, 3.1, 2.4), γ(0.512, 0.622, 1.128)
46	Palladium	100	3.63d	K, Rh-x, γ(0.0748, 0.0840, 0.126)
		101	8.47h	K, β^+(0.776, 0.488), Rh-x, γ(0.270, 0.296)
		103	16.97d	K, Rh-x, γ(0.040, 0.357, 0.497)

Z	Element	A	Half-life	Radiation (MeV)
		107	6.5×10^6y	β^-(0.04)
		109	13.7h	β^-(1.028), Ag-x, γ(0.088, 0.311, 0.636), e$^-$
		111m	5.5h	IT, β^-, Pd-x, γ(0.391)
		111	23.4min	β^-(2.10), γ(0.060, 0.071, 0.377, 0.547, 0.580, 0.623, 0.651)
		112	21.04h	β^-(0.28), γ(0.0185)
47	Silver	103	1.095h	K, β^+(1.680, 1.3, 1.15), Pd-x, γ(0.12, 0.15, 0.24, 0.27, 1.01)
		104	1.153h	β^+(0.99), K, Pd-x, γ(0.556, 0.768)
		105	41.29d	K, β^+(0.325), Pd-x, γ(0.064, 0.280, 0.344, 0.443)
		106m	8.46d	K, Pd-x, γ(0.451, 0.512, 0.717, 1.046), no β^+
		107m	44.3s	IT, Ag-x, γ(0.93), e$^-$
		108m	127y	K, IT, Pd-x, Ag-x, γ(0.434, 0.614, 0.723)
		109m	39.6s	IT, Ag-x, γ(0.088), e$^-$
		110m	249.76d	β^-(0.087, 0.53), IT, γ(0.658, 0.885, 0.937)
		110	24.6s	β^-(2.891, 2.22), K, γ(0.658)
		111m	1.080min	IT, γ(0.060)
		111	7.45d	β^-(1.04, 0.69), γ(0.342)
		112	3.14h	β^-(3.94, 3.35, 1.96), γ(0.617)
		113	5.37h	β^-(2.03), γ(0.298)
48	Cadmium	107	6.50h	K, β^+(0.302), Ag-x, γ(0.093, 0.796, 0.829)
		109	1.2665y	K, Ag-x, γ(0.088)
		113m	13.7y	β^-(0.590), IT, γ(0.264)
		115m	44.6d	β^-(1.62), γ(0.934, 1.29, 0.485), no IT
		115	2.228d	β^-(1.11, 0.58), In-x, γ(0.336, 0.528)
		117m	3.36h	β^-(0.715, 0.415), In-x, γ(0.564, 0.860, 1.029, 1.066, 1.235, 1.443, 1.997)
		117	2.49h	β^-(2.220, 1.80, 1.29, 0.67), In-x, γ(0.273, 0.344, 0.434)
49	Indium	109	4.2h	K, β^+(0.79), Cd-x, γ(0.203), e$^-$
		110	4.9h	K, Cd-x, γ(0.937, 0.884, 0.658), no β^+, no IT
		110	1.152h	β^+(2.25), K, Cd-x, γ(0.658)
		111	2.807d	K, Cd-x, γ(0.172, 0.247)
		113m	1.658h	IT, In-x, γ(0.393), e$^-$, no K
		114m	49.51d	IT, K, In-x, γ(0.192, 0.558, 0.725)
		114	1.1983min	β^-(1.99, 0.67), K, β^+(0.40), Cd-x, γ(1.30)
		115m	4.486h	IT, β^-(0.83), In-x, γ(0.336), e$^-$
		*115	4.41×10^{14}y	β^-(0.495), no γ
		117m	1.942h	β^-(1.77, 1.62), IT, In-x, γ(0.159)
		117	43.8min	β^-(0.74), Sn-x, γ(0.159, 0.553)
50	Tin	110	4.11h	K, In-x, γ(0.283)
		113	115.09d	K, In-x, γ(0.392)
		117m	13.61d	IT, Sn-x, γ(0.159), e$^-$
		119m	293.0d	IT, Sn-x, γ(0.065, 0.024), e$^-$
		121m	55y	β^-(0.354), Sb-x, γ(0.037)
		121	1.1275d	β^-(0.383), no γ
		123	129.2d	β^-(1.42), γ(1.030, 1.089)
		125	9.64d	β^-(2.35, 0.40), γ(0.823, 0.916, 1.067, 1.089)
		127	2.10h	β^-(1.45), γ(1.114)
51	Antimony	115	32.1min	K, β^+(1.51), Sn-x, γ(0.497)
		116m	1.005h	K, β^+(1.16), Sn-x, γ(0.100, 0.136, 0.407, 0.543, 0.973, 1.072, 1.294)
		117	2.80h	K, β^+(0.57), Sn-x, γ(0.159, 0.862)
		118m	5.00h	K, Sn-x, γ(0.041, 0.254, 1.05, 1.23), β^+(?)
		118	3.6min	K, β^+(2.7), Sn-x, γ(1.23)
		119	1.587d	K, Sn-x, γ(0.024), e$^-$
		120	5.76d	K, Sn-x, γ(0.090, 0.200, 1.03, 1.17), no β^+, no IT
		122	2.70d	β^-(1.414, 1.980, 0.723), K, Sn-x, γ(0.564)
		124	60.20d	β^-(2.301), γ(0.603, 1.69, 0.722), no K
		125	2.73y	β^-(0.621, 0.444, 0.302), γ(0.428, 0.464, 0.601, 0.636)
		126	12.4d	β^-(1.9), γ(0.666, 0.695, 0.721)
		127	3.85d	β^-(1.493, 1.244), γ(0.473, 0.686, 0.784)
		128	9.01h	β^-(2.30), γ(0.314, 0.53, 0.64, 0.75)
		129	4.40h	β^-(1.80, 1.53, 1.38, 0.62, 0.52), γ(0.545, 0.813, 0.915)

Z	Element	A	Half-life	Radiation (MeV)
52	Tellurium	116	2.49h	K, Sb-x, γ(0.094), β^+(?)
		117	1.03h	K, β^+(1.75), Sb-x, γ(0.720)
		118	6.00d	K, Sb-x, no γ
		119m	4.69d	K, Sb-x, γ(0.153, 0.270, 1.22), no β^+, no IT
		119	16.05h	K, β^+(0.627), Sb-x, γ(0.644)
		121m	154d	IT, K, Te-x, Sb-x, γ(0.212)
		121	16.8d	K, Sb-x, γ(0.573, 0.508), no β^+
		123m	119.7d	IT, Te-x, γ(0.0885, 0.159), e$^-$
		123	1.3×10^{13}y	K, Sb-x
		125m	58d	IT, Te-x, γ(0.110, 0.035), e$^-$
		127m	109d	IT, Te-x, γ(0.058, 0.089), β^-
		127	9.35h	β^-(0.70), I-x, γ(0.360, 0.418)
		129m	33.6d	IT, β^-(1.60, 0.91), Te-x, γ(0.696, 0.730)
		129	1.160h	β^-(1.453, 0.989, 0.69, 0.29), I-x, γ(0.460, 0.487)
		131m	1.25d	β^-(2.46, 0.57, 0.42), IT, Te-x, I-x, γ(0.150, 0.774, 0.794)
		131	25.0min	β^-(2.14, 1.69, 1.35), I-x, γ(0.150, 0.453)
		132	3.26d	β^-(0.22), I-x, γ(0.228)
53	Iodine	120	1.35h	K, β^+(4.595, 4.030, 3.400, 2.490, 1.54), Te-x, γ(0.560, 1.523)
		121	2.12h	K, β^+(1.2), Te-x, γ(0.212, 0.532)
		122	3.62min	β^+(3.1), K, Te-x, γ(0.564)
		123	13.2h	K, Te-x, γ(0.159), no β^+
		124	4.18d	K, β^+(2.14, 1.53), Te-x, γ(0.603)
		125	60.14d	K, Tc-x, γ(0.035)
		126	13.02d	K, β^-(1.25, 0.865, 0.385), β^+(1.13), Te-x, (0.389)
		129	1.57×10^7y	β^-(0.150), Xe-x, γ(0.038)
		130	12.36h	β^-(0.62, 1.04, 1.7), γ(0.418, 0.539, 0.669, 0.739)
		131	8.040d	β^-(0.607, 0.336), Xe-x, γ(0.080, 0.284, 0.364, 0.637, 0.723)
		132	2.284h	β^-(2.16, 1.61, 1.22, 1.04, 0.80), γ(0.667, 0.773)
		133	20.8h	β^-(1.27, 0.94, 0.7), γ(0.53, 0.87)
		135	6.55h	β^-(1.32, 0.87), γ(0.527, 1.132, 1.260, 1.458, 1.678)
54	Xenon	122	20.1h	K, I-x, γ(0.149, 0.350)
		123	2.08h	K, β^+(1.51), I-x, γ(0.149, 0.178)
		125	16.9h	K, I-x, γ(0.055, 0.188, 0.242)
		127m	1.153min	IT, Xe-x, γ(0.125, 0.173)
		127	36.41d	K, I-x, γ(0.058, 0.145, 0.172, 0.203, 0.375)
		129m	8.89d	IT, Xe-x, γ(0.040, 0.197), e$^-$
		131m	11.9d	IT, Xe-x, γ(0.164), e$^-$
		133m	2.19d	IT, Xe-x, γ(0.233), e$^-$
		133	5.245d	β^-(0.346), Cs-x, γ(0.081)
		135m	15.65min	IT, Xe-x, γ(0.527)
		135	9.104h	β^-(0.905), γ(0.250, 0.61)
55	Cesium	126	1.64min	β^+(3.4), K, Xe-x, γ(0.389)
		127	6.25h	K, β^+(1.063, 0.685), Xe-x, γ(0.125, 0.287, 0.411, 0.462, 0.587)
		128	3.62min	β^+(2.88, 2.44), K, Xe-x, γ(0.443)
		129	1.3358d	K, Xe-x, γ(0.372, 0.411, 0.549)
		131	9.69d	K, Xe-x
		132	6.475d	K, β^+(0.40), β^-, Xe-x, γ(0.668)
		134m	2.91h	IT, β^-(0.55), Cs-x, γ(0.127)
		134	2.062y	β^-(0.658, 0.089), γ(0.605, 0.796)
		135	3.0×10^6y	β^-(0.21), no γ
		136	13.16d	β^-(0.341, 0.657), Ba-x, γ(0.818, 1.048)
		137	30.0y	β^-(0.514, 1.18), Ba-x, γ(0.662)
56	Barium	126	1.67h	K, γ(0.218, 0.234, 0.241, 0.258)
		128	2.43d	K, Cs-x, γ(0.273)
		129(129m)	2.23(2.17)h	K, β^+(1.43, 1.24, 1.0), Cs-x, γ(0.129, 0.182, 0.21, 0.22, 1.45)
		131	11.8d	K, Xe-x, γ(0.124, 0.216, 0.373, 0.496), no β^+
		133m	1.621d	IT, Ba-x, γ(0.276, 0.012), e$^-$
		133	10.54y	K, Cs-x, γ(0.080, 0.276, 0.302, 0.356, 0.382), e$^-$
		135m	1.196d	IT, Ba-x, γ(0.268)
		137m	2.552min	IT, Ba-x, γ(0.662)

Z	Element	A	Half-life	Radiation (MeV)
		139	1.41h	β^-(2.38, 2.23,0.95), La-x, γ(0.166, 1.421)
		140	12.75d	β^-(1.02, 0.83, 0.59, 0.46), La-x, γ(0.537)
57	Lanthanum	131	59min	K, β^+(1.94, 1.42, 0.70), Ba-x, γ(0.418)
		132	4.8h	β^+(3.66, 3.20, 2.62), K, Ba-x, γ(0.465)
		133	3.91h	K, β^+(1.2), Ba-x, γ(0.279, 0.290, 0.302, 0.618, 0.633)
		134	6.45min	β^+(2.67), K, Ba-x, γ(0.563, 0.605, 1.555)
		135	19.48h	K, Ba-x, γ(0.481, 0.588, 0.87), no β^+
		136	9.87min	K, β^+(1.85), Ba-x, γ(0.819)
		137	6×10^4y	K, Ba-x, no γ
		*138	1.06×10^{11}y	K, β^-(0.205), Ba-x, γ(0.788, 1.436)
		140	1.68d	β^-(2.164, 1.680, 1.365, 1.150, 0.857, 0.510), γ(0.487, 1.596)
		141	3.93h	β^-(2.43, 0.9), γ(1.354)
		142	1.54h	β^-(4.52, 3.85, 2.98, 2.31, 2.11, 1.98, 1.79, 1.23, 0.87), γ(0.641, 2.398)
58	Cerium	132	3.51h	K, γ(0.182)
		133	5.40h	K, β^+(1.3) La-x, γ(0.058, 0.131, 0.477, 0.510)
		134	3.16d	K, La-x
		135	17.8h	K, β^+(0.80), La-x, γ(0..266, 0.300, 0.518, 0.607)
		137m	1.43d	IT, K, Ce-x, γ(0.1.69, 0.254, 0.762, 0.825), e^-
		137	9.0h	K, La-x, γ(0.477), e^-
		139	137.66d	K, La-x, γ(0.166)
		141	32.50d	β^-(0.444, 0.582), Pr-x, γ(0.145)
		143	1.38d	β^-(1.40, 1.125, 0.74), Pr-x, γ(0.293), e^-
		144	284.9d	β^-(0.316), Pr-x, γ(0.080, 0.134)
59	Praseodymium	136	13.1min	K, β^+(3.00, 2.56), Ce-x, γ(0.540, 0.552)
		137	1.28h	K, β^+(1.7), Ce-x, γ(0.837)
		138m	2.1h	K, β^+(1.65), Ce-x, γ(0.302, 0.789, 1.038)
		139	4.41h	K, β^+(1.09), Ce-x, γ(0.255, 1.347, 1.376, 1.631)
		140	3.39min	K, β^+(2.366), Ce-x, γ(0.307, 1.596)
		142	19.13h	β^-(2.164), γ(1.576)
		143	13.58d	β^-(0.932)
		144	17.28min	β^-(2.99, 2.3, 0.81), γ(0.697, 1.49, 2.19)
		145	5.98h	β^-(1.8), γ(0.072, 0.676, 0.748, 0.921, 0.979, 1,051, 1.150)
		146	24.15min	β^-(4.1, 3.6, 2.8, 2.1), γ(0.454)
60	Neodymium	139m	5.5h	K, IT, β^+(1.170), Nd-x, Pr-x γ(0.114, 0.231, 0.708, 0.738, 0.982)
		140	3.37d	K, Pr-x, no γ
		141	2.49h	K, β^+(0.79), Pr-x, γ(0.145, 1.127, 1.147, 1.293)
		*144	2.1×10^{15}y	α(1.83)
		147	10.98d	β^-(0.810, 0.369), γ(0.091, 0.531), e^-
		149	1.73h	β^-(1.55, 1.425, 1.130, 1.025), Pm-x, γ(0.030, 0.114, 0.655)
61	Promethium	142	40.5s	β^+(3.80), K, Nd-x, γ(1.576)
		143	265d	K, Nd-x, γ(0.742), no β^+
		144	363d	K, Nd-x,γ(0.477, 0.618, 0.696), no β^+
		145	17.7y	K, Nd-x, γ(0.067, 0.072), e^-
		146	5.53y	K, β^-(0.795), Nd-x, γ(0.453, 0.75)
		147	2.6234y	β^-(0.224), γ(0.122)
		148m	41.29d	β^-(0.69, 0.50, 0.40), IT, Pm-x, Sm-x, γ(0.550, 0.630)
		148	5.37d	β^-(2.48, 1.93,1.02), γ(0.551, 1.47)
		149	2.211d	β^-(1.064, 0.784), γ(0.286)
		150	2.68h	β^-(3.260, 2.930, 2.100, 1.600), γ(0.344)
		151	1.1833d	β^-(1.19, 1.13, 1.05, 0.98, 0.84, 0.73, 0.50, 0.35), Sm-x, γ(0.168, 0.275, 0.340)
62	Samarium	142	1.2082h	K, β^+(1.03), Pm-x
		145	340d	K, Pm-x, γ(0.061), e^-
		146	1.03×10^8y	α(2.55)
		*147	1.06×10^{11}y	α(2.23)
		151	90y	β^-(0.076), γ(0.022), e^-
		153	1.946d	β^-(0.81, 0.71, 0.64), Eu-x, γ(0.103, 0.070), e^-
		156	9.4h	β^-(0.715, 0.43), Eu-x, γ(0.0226–0.291)

Z	Element	A	Half-life	Radiation (MeV)
63	Europium	145	5.93d	K, β^+(1.67, 0.78), Sm-x, γ(0.656, 0.894, 1.66)
		146	4.59d	K, β^+(2.11, 1.47, 0.80), Sm-x, γ(0.633, 0.634, 0.747)
		147	24d	K, β^+, α(2.91, 0.002%), Sm-x
		148	54.5d	K, β^+(0.92), α(2.63, 9×10^{-7}%), Sm-x, γ (0.550, 0.630)
		149	93.1d	K, Sm-x, γ(0.277, 0.328)
		150	12.62h	β^-(1.01), K, β^+(1.24), Sm-x, γ(0.334, 0.407)
		150	35.8y	K, Sm-x, γ(0.334, 0.439, 0.584)
		$152m_1$	9.32h	β^-(1.89, 1.55, 0.56), K, β^+(0.89, 0.77), γ(0.122, 0.344, 0.841, 0.963)
		$152m_2$	1.600h	IT, Eu-x, γ(0.090), e^-, no β
		152	13.33y	K, β^-(1.492, 0.690, 0.360), β^+(0.727, 0.479), Gd-x, Sm-x, γ(0.122, 0.344, 0.964, 1.086, 1.112, 1.408)
		154	8.8y	β^-(0.843, 0.579, 0.274), Gd-x, γ(0.123, 1.005, 1.274)
		155	4.96y	β^-(0.152, 0.252), Gd-x, γ(0.087, 0.105)
		156	15.19d	β^-(2.460, 1.215, 0.495, 0.300), Gd-x, γ(0.089), 0.812, 1.230, 1.242)
		157	15.15h	β^-(1.28), Gd-x, γ(0.064, 0.373)
		158	46.0min	β^-(3.40, 2.43), γ(0.079, 0.977)
64	Gadolinium	146	48.27d	K, β^+(0.350), Eu-x, γ(0.115, 0.155)
		147	1.588d	K, β^+(1.250, 0.970), Eu-x, γ(0.229, 0.370, 0.396)
		148	75y	α(3.18)
		149	9.25d	K, α(3.018; 0.00046%), Eu-x, γ(0.150)
		150	1.79×10^6y	α(2.73)
		151	120d	K, α(2.60; 8×10^{-7}%), Eu-x, γ(0.154, 0.175, 0.243)
		*152	1.08×10^{14}y	α(2.4)
		153	241.6d	K, Eu-x, γ(0.070, 0.097, 0.103)
		159	18.56h	β^-(0.95, 0.89, 0.60), Tb-x, γ(0.363)
65	Terbium	148	100h	K, β^+(4.6), Gd-x, γ(0.489, 0.783)
		149	4.15h	K, α(3.967), β^+(1.78), Gd-x, γ(0.165, 0.352, 0.388, 0.652, 0.817, 0.853)
		150	3.27h	K, β^+(3.70, 3.04), α(3.492), Gd-x, γ(0.496, 0.638)
		151	17.6h	K, α(3.41; 0.0095%), Gd-x, γ(0.108, 0.252, 0.287, 0.479)
		152	17.5h	K, β^+(2.800, 2.456, 1.890), Gd-x, γ(0.344)
		153	2.34d	K, β^+(0.570, 0.520, 0.460, 0.339), Gd-x, γ(0.212)
		154	21.4d	K, β^+, Gd-x, γ(0.123, 1.274, 1.291, 1.997, 2.064, 2.187)
		$154m_1$	9.0h	K, IT, Gd-x, γ(0.123, 0.248)
		$154m_2$	22.6h	K, IT, γ(0.226, 0.347, 0.427, 1.420)
		155	5.32d	K, Gd-x, γ(0.105)
		156m	5.0h	IT, K, β^+(2.640), γ(0.0884)
		156	5.34d	K, Gd-x, γ(0.089, 0.199, 0.534, 1.222)
		157	150y	K, Gd-x, γ(0.054)
		158	150y	K, β^-(0.853), Gd-x, γ(0.944, 0.962)
		160	72.3d	β^-(1.76, 0.87, 0.57, 0.45), Dy-x, γ(0.299, 0.879, 0.966)
		161	6.91d	β^-(0.590, 0.520, 0.460), Dy-x, γ(0.075)
66	Dysprosium	152	2.38h	K, α(3.63), Tb-x, γ(0.257)
		153	6.4h	K, β^+(0.980, 0.670), α(3.464), Tb-x, γ(0.081, 0.099, 0.214, 0.254)
		154m	13h	α(3.37)
		154	2.9×10^6y	α(2.872)
		155	10.0h	K, β^+(1.08, 0.85), Tb-x, γ(0.227)
		157	8.1h	K, Tb-x, γ(0.326), no β^+
		159	144.4d	K, Tb-x, γ(0.058)
		165	2.334d	β^-(1.305, 1.215), Ho-x, γ(0.095, 0.362)
		166	3.400d	β^-(0.402, 0.48), Ho-x, γ(0.082)
67	Holmium	160m	5.02h	IT, K, β^+(1.9, 1.01, 0.57, 0.31), Dy-x, γ(0.060, 0.087, 0.197, 0.645, 0.728, 0.879, 0.966), includes daughter radiation from Ho^{160}
		160	25.6min	K, γ(see Ho^{160m})
		161	2.48h	K, Dy-x, γ(0.026)
		162m	1.133h	IT, K, Ho-x, Dy-x, γ(0.081, 0.185, 1.220)

Z	Element	A	Half-life	Radiation (MeV)
		162	15min	K, β^+(1.200, 1.120), Dy-x, γ(0.080, 1.319)
		166m	1200y	β^-($\leq$0.065), Er-x, γ(0.184, 0.712, 0.810)
		166	1.117d	β^-(1.85, 1.78), Er-x, γ(0.081)
		167	3.1h	β^-(0.970, 0.610, 0.315), Er-x, γ(0.347)
68	Erbium	160	1.191d	K, Ho-x, no γ
		161	3.24h	K, β^+(0.820), Ho-x, γ(0.211, 0.826)
		163	1.250h	K, β^+(0.188), Ho-x, γ(0.298, 0.436, 1.113)
		165	10.36h	K, Ho-x, no γ
		167m	2.28s	IT, Er-x, γ(0.208)
		169	9.40d	β^-(0.34), γ(0.008)
		171	7.52h	β^-(1.49, 1.06, 0.58), Tm-x, γ(0.112, 0.296, 0.308)
		172	2.054d	β^-(0.356), Tm-x, γ(0.408, 0.610)
69	Thulium	162	21.7min	K, β^+(3.82, 2.3, 0.90), Er-x, γ(0.102, 0.227, 0.799)
		163	1.81h	K, β^+(1.05, 0.71), Er-x, γ(0.104)
		165	1.2525d	K, β^+(0.30), Er-x, γ(0.054, 0.248, 0.807)
		166	7.70h	K, β^+(1.928, 1.219), Er-x, γ(0.081, 0.084, 0.184, 0.705, 2.052)
		167	9.24d	K, Er-x, γ(0.208, 0.532)
		168	93.1d	K, β^-(?), Er-x, γ(0.184, 0.198, 0.447, 0.816)
		170	128.6d	β^-(0.968, 0.884), Yb-x, γ(0.084)
		171	1.92y	β^-(0.097), Yb-x, γ(0.067)
		172	2.650d	β^-(1.870, 1.790), Yb-x, γ(1.094, 1.387, 1.466)
		173	8.24d	β^-(0.89, 1.3), Yb-x, γ(0.399)
70	Ytterbium	164	1.26h	K, Tm-x, γ(0.097)
		166	2.362d	K, Tm-x, γ(0.082)
		169	32.022d	K, Tm-x, γ(0.063, 0.110, 0.131, 0.177, 0.198, 0.308)
		175	4.19d	β^-(0.466, 0.073), Lu-x, γ(0.114, 0.283, 0.396)
		177	1.9h	β^-(1.40), Lu-x, γ(1.080)
71	Lutetium	169	1.4192d	K, β^+(1.271), Yb-x, γ(0.191, 0.960)
		170	2.00d	K, β^+(2.39), Yb-x, γ(0.985, 1.054, 1.226, 1.280, 1.365, 2.042, 2.126)
		171	8.24d	K, Yb-x, γ(0.740)
		172	6.70d	K, Yb-x, γ(0.181, 0.901, 1.094), no β^+
		173	1.37y	K, Yb-x, γ(0.0787, 0.272)
		174m	142d	IT, K, Lu-x, γ(0.045, 0.067), no β^-
		174	3.31y	K, β^+(0.380), Yb-x, γ(0.0765, 1.24)
		176m	3.635h	β^-(1.31, 1.22), Hf-x, γ(0.0884), no IT
		*176	3.59 $\times$ 10^{10}y	β^-(0.565), Hf-x, γ(0.089, 0.203, 0.306), no K
		177m	160.9d	β^-, IT, Lu-x, Hf-x, γ(0.105, 0.208, 0.229, 0.379)
		177	6.71d	β^-(0.497, 0.385, 0.175), Hf-x, γ(0.113, 0.208)
		179	4.59h	β^-(1.35), γ(0.214)
72	Hafnium	169	1.87h	K, β^+(1.3), Lu-x, γ(0.115)
		170	16.01h	K, Lu-x, γ(0.165, 0.573)
		171	12.1h	K, Lu-x, γ(0.122, 1.072)
		172	1.87y	K, Lu-x, γ(0.024, 0.082, 0.125), no α
		173	23.6h	K, Lu-x, γ(0.124, 0.297)
		*174	2.0 $\times$ 10^{15}y	α(2.50)
		175	70d	K, Lu-x, γ(0.343)
		180m	5.519h	IT, Hf-x, γ(0.058, 0.215, 0.333, 0.444)
		181	42.39d	β^-(0.408), Ta-x, γ(0.133, 0.346, 0.482)
		182	9.0 $\times$ 10^{6}y	β^-, γ(0.271)
73	Tantalum	173	3.65h	K, β^+(2.48), Hf-x, γ(0.070, 0.090, 0.172)
		174	1.18h	K, β^+(2.525), Hf-x, γ(0.206)
		175	10.5h	K, β^+, Hf-x, γ(0.082, 0.207, 0.349)
		176	8.08h	K, β^+, Hf-x, γ(0.088, 1.159)
		177	2.358d	K, Hf-x, γ(0.113, 0.208)
		178	9.31min	K, β^+(0.89, 0.80), Hf-x, γ(1.106, 1.341, 1.351, 1.403)
		178	2.45h	K, Hf-x, γ(0.089, 0.213, 0.326, 0.426)
		179	1.79y	K, Hf-x, no γ
		180m	8.152h	K, β^-(0.71, 0.61), Hf-x, γ(0.093, 0.102)
		182	115.0d	β^-(1.713, 1.470), W-x, γ(1.121, 1.221)
		183	5.1d	β^-(0.615, 0.470), W-x, γ(0.108, 0.246, 0.354)
		184	8.7h	β^-(1.19, 1.45, 2.64), W-x, γ(0.111, 0.253, 0.318, 0.414, 0.921)

Z	Element	A	Half-life	Radiation (MeV)
74	Tungsten	176	2.5h	K, Ta-x, γ(0.061, 0.095)
		177	2.21h	K, Ta-x, γ(0.030, 0.427, 1.036)
		178	21.5d	K, Ta-x
		181	121.2d	K, Ta-x, γ(0.0062, 0.136, 0.152)
		185	75.1d	β^-(0.433), γ(0.125)
		187	23.9h	β^-(1.312, 0.622), Re-x, γ(0.072, 0.134, 0.480)
		188	69.4d	β^-(0.349), Re-x, γ(0.227, 0.290)
75	Rhenium	181	19.9h	K, W-x, γ(0.361, 0.366)
		182	12.7h	K, β^+(1.74), W-x, γ(0.100, 1.121, 1.189, 1.222)
		182	2.667d	K, W-x, γ(0.100, 0.131, 0.169, 0.229, 1.121, 1.222, 1.231), no β^+
		183	70.0d	K, W-x, γ(0.046, 0.162, 0.292)
		184m	165d	IT, K, Re-x, W-x, γ(0.111, 0.161, 0.217, 0.253, 0.921)
		184	38.0d	K, W-x, γ(0.111, 0.792, 0.895, 0.903)
		186	3.777d	β^-(107, 0.933), K, W-x, Os-x, γ(0.137, 0.632, 0.768)
		*187	4.6×10^{10}y	β^-(0.0026)
		188	16.98h	β^-(2.12, 196), Os-x, γ(0.155)
		189	1.013d	β^-(1.00), Os-x, γ(0.031, 0.696, 0.217, 0.219, 0.245)
76	Osmium	182	21.6h	K, Re-x, γ(0.180, 0.263, 0.510)
		183m	9.9h	K, IT, Os-x, γ(1.102, 1.108)
		183	13.0h	K, Re-x, γ(0.382)
		185	93.6d	K, Re-x, γ(0.646, 0.875), no β^+
		189m	5.8h	IT, Os-x, γ(0.0308), e$^-$
		190m	9.9min	IT, Os-x, γ(0.187, 0.361, 0.616), e$^-$
		191m	13.10h	IT, Os-x, γ(0.074)
		191	15.4d	β^-(0.143), Ir-x, γ(0.129), e$^-$
		193	1.271d	β^-(1.13, 1.06, 0.99, 0.85, 0.67), Ir-x, γ(0.073, 0.139, 0.460)
		194	6.0y	β^-(0.053), Ir-x, γ(0.043)
77	Iridium	183	57min	K, Os-x, γ(0.088, 0.229)
		184	3.02h	K, β^+(2.9, 2.5), Os-x, γ(0.120, 0.264, 0.390)
		185	14.0h	K, Os-x, γ(0.037, 0.060, 0.254)
		186	15.8h	K, β^+(1.94), Os-x, γ(0.137, 0.297, 0.434)
		187	10.5h	K, Os-x, γ(0.026, 0.065, 0.913)
		188	1.729d	K, β^+(1.66), Os-x, γ(0.155, 0.478)
		189	13.2d	K, Os-x, γ(0.031, 0.070, 0.245)
		190m$_1$	1.2h	IT, Ir-x, γ(0.0263)
		190m$_2$	3.2h	K, IT, Os-x, Ir-x, γ(0.187, 0.361, 0.503, 0.617)
		190	11.78d	K, Os-x, γ(0.187, 0.371, 0.519, 0.558, 0.569, 0.605), no β^+
		192	73.831d	β^-(0.672, 0.536), K, Os-x, Pt-x, γ(0.296, 0.308, 0.316, 0.468)
		193m	10.60d	IT, Ir-x, γ(0.0802)
		194	19.15h	β^-(2.24, 1.92, 1.62, 0.98), γ(0.328)
		195	2.8h	β^-(1.15, 0.98), Pt-x, γ(0.10, 0.13)
		196m	1.40h	β^-(1.16), γ(0.356, 0.393, 0.447, 0.521, 0.647), no IT
78	Platinum	186	2.0h	K, α(4.23; 1.4×10^{-4}%), Ir-x, γ(0.690)
		187	2.35h	K, Ir-x, γ(0.110, 0.285)
		188	10.2d	K, α(3.93; 3×10^{-5}%), Ir-x, γ(0.188, 0.195)
		189	10.89h	K, β^+(0.885, 0.479), Ir-x, γ(0.094, 0.244, 0.545, 0.569, 0.608, 0.721)
		*190	6.0×10^{11}y	α(3.18)
		191	2.9d	K, Ir-x, γ(0.082, 0.360, 0.409)
		*192	$>6.0 \times 10^{16}$y	α(2.6)
		193m	4.33d	IT, Pt-x, γ(0.0126, 0.136)
		193	50y	K, Ir-x, no γ
		195m	4.02d	IT, Pt-x, γ(0.0311, 0.0991, 0.130)
		197m	1.573h	IT, β^-, Pt-x, γ(0.279, 0.346)
		197	18.3h	β^-(0.719, 0.642, 0.451), Au-x, γ(0.077, 0.191)
		200	12.5h	β^-, γ(0.076)

Z	Element	A	Half-life	Radiation (MeV)
79	Gold	191	3.18h	K, Pt-x, γ(0.278, 0.284, 0.400, 0.586, 0.674)
		192	4.94h	K, β^+(2.492, 2.192), Pt-x, γ(0.316)
		193	17.65h	K, Pt-x, γ(0.256), no α
		194	1.646d	K, β^+(1.49), Pt-x, γ(0.328)
		195	186.09d	K, Pt-x, γ(0.099)
		196m$_2$	9.7h	IT, Au-x, γ(0.020, 0.085, 0.148, 0.175, 0.188)
		196	6.183d	K, β^-(0.259), Pt-x, γ(0.333, 0.356, 0.426)
		197m	7.8s	IT, Au-x, γ(0.279)
		198	2.6935d	β^-(1.371), γ(0.412, 0.676, 1.09), no K
		199	3.139d	β^-(0.296, 0.250, 0.462) Hg-x, γ(0.158, 0.208)
80	Mercury	191	49min	K, Au-x, γ(0.252)
		192	4.85h	K, Au-x, γ(0.157, 0.275)
		193m	11.8h	K, IT, β^+(1.17), Hg-x, Au-x, γ(0.258, 0.408)
		193	3.80h	K, Au-x, γ(0.038)
		194	520y	K, Au-x, no γ
		195m	1.73d	K, IT, Hg-x, Au-x, γ(0.262, 0.560)
		195	9.5h	K, Au-x, γ(0.061, 0.780)
		197m	23.8h	IT, K, Hg-x, γ(0.134), e$^-$
		197	2.6725d	K, Au-x, γ(0.077)
		203	46.60d	β^-(0.21), γ(0.279)
81	Thallium	195	1.16h	K, β^+(1.8), Hg-x, γ(0.242, 0.279, 0.564, 0.885, 1.363)
		196m	1.41h	K, IT, Hg-x, γ(0.426, 0.635, 0.695)
		196	1.84h	K, Hg-x, γ(0.426, 0.611, 0.635, 1.496)
		197	2.84h	K, β^+(1.2), Hg-x, γ(0.152, 0.426)
		198m	1.87h	K, IT, Hg-x, Tl-x, γ(0.412, 0.587, 0.637)
		198	5.3h	K, β^+(2.4), Hg-x, γ(0.412)
		199	7.42h	K, Hg-x, γ(0.158, 0.208, 0.247, 0.455), no β^+
		200	1.087d	K, β^+(1.44, 107), Hg-x, γ(0.579, 0.828)
		201	3.046d	K, Hg-x, γ(0.135, 0.167)
		202	12.23d	K, Hg-x, γ(0.439, 0.52, 0.96)
		204	3.78y	β^-(0.763), K, Hg-x
		*206(RaE')	4.20min	β^-(1.53)
		*207(AcC'')	4.77min	β^-(1.43), γ(0.90)
		*208(ThC'')	3.053min	β^-(1.80, 1.52, 1.29), γ(2.61, 0.583)
		209	2.20min	β^-(1.18), Pb-x, γ(0.12, 0.45, 1.56)
		*210(RaC'')	1.30min	β^-(2.3, 1.9, 1.3), γ(0.296, 0.795)
82	Lead	198	2.4h	K, Tl-x, γ(0.117, 0.173, 0.290), no α
		199	1.50h	K, β^+(2.8), Tl-x, γ(0.353, 0.367, 0.721, 1.658)
		200	21.5h	K, Tl-x, γ(0.148)
		201m	1.02min	IT, Pb-x, γ(0.629)
		201	9.33h	K, β^+(0.55), Tl-x, γ(0.331, 0.361)
		202m	3.62h	IT, K, Tl-x, Pb-x, γ(0.422, 0.787, 0.961)
		202	5.3×10^4y	K, Tl-x
		203	2.169d	K, Tl-x, γ(0.279)
		204m	1.120h	IT, Pb-x, γ(0.375, 0.899, 0.912)
		205	1.9×10^7y	K, Tl-x
		207m	796ms	IT, γ(0.570, 1.06)
		209	3.253h	β^-(0.645), no γ
		*210(RaD)	22.3y	β^-(0.017), α(3.72; 1.7×10^{-6}%); γ(0.0465)
		*211(AcB)	36.1min	β^-(1.36, 0.53), γ(0.405, 0.427, 0.832)
		*212(ThB)	10.64h	β^-(0.569, 0.331), Bi-x, γ(0.239)
		*214(RaB)	26.8min	β^-(0.59, 0.65, 1.03), γ(0.352)
83	Bismuth	201m	59.1min	K, α(5.240), Pb-x
		201	1.80h	K, Pb-x, γ(0.629 with Pb^{201m})
		202	1.72h	K, β^+, Pb-x, γ(0.422, 0.961)
		203	11.76h	K, β^+(1.35, 0.74), Pb-x, γ(0.820)
		204	11.22h	K, Pb-x, γ(0.899, 0.984), no β^+
		205	15.31d	K, β^+(0.990), Pb-x, γ(0.703, 0.988, 1.764)
		206	6.243d	K, β^+(0.98), Pb-x, γ(0.516, 0.537, 0.803, 0.881, 1.719)
		207	32.2y	K, Pb-x, γ(0.570, 1.06, 1.77)
		208	3.68×10^5y	K, Pb-x, γ(2.61), no β^+
		210m	3.00×10^6y	α(4.96, 4.92, 4.57), γ(0.266, 0.305), no β^-
		*210(RaE)	5.013d	β^-(1.16), α(4.69, 4.65)
		*211(AcC)	2.14min	α(6.62, 6.28), β^-, γ(0.351)

Z	Element	A	Half-life	Radiation (MeV)
		*212(ThC)	1.0092h	β^-(2.25), α(6.05, 6.09), Tl-x, γ(0.727)
		213	45.59min	β^-(1.420, 1.018), α(5.87, 5.55), γ(0.439)
		*214(RaC)	19.9min	β^-(3.26, 1.88, 1.51, 1.0, 0.4), α(5.512, 5.448), γ(0.609)
		215	7.6min	β^-
84	Polonium	204	3.53h	K, α(5.38), γ(0.137, 0.270, 0.534, 0.884, 1.016)
		205	1.80h	K, α(5.220), γ(0.849, 0.872, 1.001)
		206	8.8d	K, α(5.22), Bi-x, γ(0.011, 0.286, 0.511, 0.807, 1.032)
		207	5.83h	K, α(5.11), β^+(1.14, 0.89), Pb-x, γ(0.743, 0.912, 0.992)
		208	2.898y	α(5.11), K, Bi-x, γ(0.292, 0.571, 0.603, 0.862)
		209	102y	α(4.88), K, Bi-x, γ(0.26)
		*210(RaF)	138.376d	α(5.30), γ(0.803)
		*211(AcC′)	516ms	α(7.45, 6.89), γ(0.56, 0.88)
		*212(ThC′)	298ns	α(8.78)
		213	4.2μs	α(8.38)
		*214(RaC′)	163.69μs	α(7.69), γ(0.799)
		*215(AcA)	1.780ms	α(7.38), β^-
		*216(ThA)	150ms	α(6.78)
		*218(RaA)	3.11min	α(6.00)
85	Astatine	207	1.80h	K, α(5.76), γ(0.588, 0.814)
		208	1.63h	K, α(5.641), Po-x, γ(0.177, 0.660, 0.685)
		209	5.41h	K, α(5.65), Po-x, γ(0.545, 0.782, 0.790)
		210	8.1h	K, α(5.52, 5.44, 5.36), Po-x, γ(0.245, 1.181, 1.483)
		211	7.214h	K, α(5.87), Po-x, γ(0.67)
		215	100μs	α(8.03)
		216	300μs	α(7.80)
		217	32.3ms	α(7.07)
		218	1.6s	α(6.70, 6.65)
		219	54s	α(6.28), β^-
86	Radon	210	2.4h	α(6.04), K, γ(0.073, 0.233, 0.458, 0.649)
		211	14.6h	K, α(5.85, 5.78, 5.62), At-x, γ(0.442, 0.674, 0.678, 0.947)
		218	35ms	α(7.13), γ(0.609)
		*219(An)	3.96s	α(6.82, 6.55, 6.42), Po-x, γ(0.272, 0.401)
		*220(Tn)	55.6s	α(6.29), γ(0.550)
		*222(Rn)	3.825d	α(5.49), γ(0.510)
87	Francium	212	20.0min	K, α(6.41, 6.38, 6.26)
		220	27.4s	α(6.68, 6.64, 6.58), β^-, γ(0.045, 0.106, 0.154, 0.162)
		221	4.9min	α(6.34, 6.12), At-x, γ(0.218)
		222	14.4min	β^-(1.78), α
		*223(AcK)	21.8min	β^-(1.15), α(5.34), Ra-x, γ(0.050, 0.080, 0.234)
88	Radium	222	38.0s	α(6.56), γ(0.324)
		*223 (AcX)	11.43d	α(5.74, 5.71, 5.60, 5.54), Rn-x, γ(0.154, 0.270, 0.324)
		*224(ThX)	3.66d	α(5.68, 5.45), Rn-x, γ(0.241)
		225	14.8d	β^-(0.32), Ac-x, γ(0.040), no α
		*226(Ra)	1600y	α(4.78, 4.60), Rn-x, γ(0.187)
		*228(MsTh$_1$)	5.75y	β^-(0.039, 0.015), γ(0.135), no α
89	Actinium	224	2.9h	K, α(6.20, 6.14, 6.04), Ra-x, γ(0.133, 0.217)
		225	10.0d	α(5.83, 5.79, 5.64), Fr-x, γ(0.099, 0.150, 0.187)
		226	1.2d	β^-(1.105, 0.885), K, α(5.399), Th-x, Ra-x, γ(0.158, 0.185, 0.230, 0.253)
		*227(Ac)	21.77y	β^-(0.046), α(4.95, 4.94), Th-x, γ(0.086, 0.100, 0.160)
		*228(MsTh$_2$)	6.13h	β^-(2.18, 1.85, 1.7, 1.11, 0.64; 0.45), Th-x, γ(0.339, 0.911, 0.969)
		229	1.045h	β^-(1.09), γ(0.135, 0.146, 0.165, 0.262)
90	Thorium	226	31min	α(6.34, 6.23), Ra-x, γ(0.111, 0.242)
		*227(RaAc)	18.718d	α(6.04, 5.98, 5.76), Ra-x, γ(0.050, 0.236, 0.256)
		*228(RaTh)	1.913y	α(5.43, 5.34), Ra-x, γ(0.084, 0.132, 0.167, 0.214)

Z	Element	A	Half-life	Radiation (MeV)
		229	7340y	α(4.90, 4.84, 4.81), Ra-x, γ(0.137, 0.20)
		*230(Io)	7.54×10^4y	α(4.68, 4.62), Ra-x, γ(0.068)
		*231(UY)	1.0633d	β^-(0.302, 0.218, 0.138), Pa-x, γ(0.026, 0.084)
		*232(Th)	1.405×10^{10}y	α(4.01, 3.95), γ(0.059)
		*234(UX$_1$)	24.10d	β^-(0.199, 0.104), Pa-x, γ(0.029, 0.063, 0.092, 0.093, 0.113)
91	Protactinium	228	22h	K, α(6.11, 6.08, 6.03, 5.80), Th-x, γ(0.410, 0.463, 0.965, 0.969)
		229	1.4d	K, α(5.67, 5.62, 5.58, 5.54), Th-x
		230	17.4d	K, β^-(0.509), α(5.34, 5.33, 5.31, 5.30), Th-x, γ(0.444, 0.455, 0.899, 0.919, 0.952)
		*231(Pa)	3.276×10^4y	α(5.06, 5.02, 5.01, 4.95, 4.73), Ac-x, γ(0.027, 0.284, 0.300, 0.303)
		232	1.31d	β^-(0.32, 1.19, 1.30), U-x, γ(0.150, 0.894, 0.969)
		233	27.0d	β^-(0.257, 0.15, 0.568), U-x, γ(0.312)
		*234m(UX$_2$)	1.17min	β^-(2.29), IT, U-x, γ(0.765, 1.00)
		*234(UZ)	6.70h	β^-(0.68, 0.51, 0.28), U-x, γ(0.131, 0.570)
92	Uranium	230	20.8d	α(5.89, 5.82), Th-x, γ(0.072, 0.154, 0.231)
		231	4.2d	K, α(5.46), Pa-x, γ(0.026, 0.084)
		232	68.9y	α(5.32, 5.27), Pa-x, γ(0.058, 0.129)
		233	1.592×10^5y	α(4.82, 4.78), Th-x, γ(0.029, 0.042, 0.055, 0.097, 0.317)
		*234(U$_{II}$)	2.454×10^5y	α(4.77, 4.72), Th-x, γ(0.053, 0.121), e$^-$
		235m	26min	IT
		*235(AcU)	7.037×10^8y	α(4.40, 4.37, 4.22), Th-x, γ(0.186)
		236	2.342×10^7y	α(4.49, 4.44), γ(0.049), e$^-$
		237	6.75d	β^-(0.248, 0.09), γ(0.060, 0.208)
		*238(U$_1$)	4.468×10^9y	α(4.20, 4.15), γ(0.050), e$^-$
		239	23.47min	β^-(1.21, 1.29), Np-x, γ(0.044, 0.075)
		240	14.1h	β^-(0.36), Np-x, γ(0.044)
93	Neptunium	231	48.8min	K, α(6.28), γ(0.264, 0.348, 0.371), no β^+
		232	14.7min	K, U-x, γ(0.282, 0.327, 0.864), no β^+
		233	36.2min	K, α(5.54), U-x, γ(0.299, 0.312)
		234	4.4d	K, β^+(0.8), U-x, γ(1.528, 1.559, 1.602)
		235	1.085y	K, α(5.02), U-x, γ(0.026, 0.084)
		236	22.5h	K, β^-(0.537), U-x, γ(0.045, 0.643, 0.688)
		236	1.550×10^5y	K, β^-, γ(0.160)
		237	2.140×10^6y	α(4.79, 4.77), Pa-x, γ(0.029, 0.086)
		238	2.117d	β^-(1.24, 0.28, 0.25), γ(0.984, 1.026, 1.029)
		239	2.355d	β^-(0.437, 0.332, 0.393, 0.713), Pu-x, γ(0.106, 0.228, 0.278)
		240m	7.22min	β^-(2.18, 1.60, 1.30), γ(0.555, 0.597)
		240	1.032h	β^-(0.89), γ(0.448, 0.566, 0.974)
		241	13.9min	β^-(1.4), γ(0.135, 0.175), no α
		241	3.4h	β^-
94	Plutonium	232	34.1min	K, α(6.60, 6.54)
		233	20.9min	K, α(6.30), γ(0.235, 0.500, 0.535, 0.688, 1.004)
		234	8.8h	K, α(6.20, 6.15), Np-x, no γ
		235	25.3min	K, α(5.86), Np-x, γ(0.049)
		236	2.851y	α(5.77, 5.72), U-x, γ(0.047, 0.110, 0.165), e$^-$
		237m	180ms	IT, Pu-x, γ(0.145), e$^-$
		237	45.17d	K, α(5.37, 5.66), Np-x, γ(0.033, 0.044, 0.060)
		238	87.74y	α(5.50, 5.46), U-x, γ(0.0435, 0.0998, 0.153)
		239	2.411×10^4y	α(5.16, 5.14, 5.11), U-x, γ(0.052, 0.129)
		240	6563y	α(5.17, 5.12), U-x, γ(0.045)
		241	14.4y	β^-(0.021), α(4.90, 4.85), U-x, γ(0.149)
		242	3.763×10^5y	α(4.90, 4.86), γ(0.045, 0.103)
		243	4.956h	β^-(0.579, 0.490), Am-x, γ(0.084)
		244	8.26×10^7y	α(4.59, 4.55), S.F.
		245	10.5h	β^-(1.210, 0.930), γ(0.327)
		246	10.85d	β^-(0.15, 0.33), Am-x, γ(0.044, 0.180, 0.224)
95	Americium	237	1.217h	K, α(6.04), γ(0.280)
		238	1.63h	K, α(5.94), Pu-x, γ(0.561, 0.919)
		239	11.9h	K, α(5.78), Pu-x, γ(0.228, 0.278)
		240	2.117d	K, α(5.38, 5.34), Pu-x, γ(0.889, 0.988)

Z	Element	A	Half-life	Radiation (MeV)
		241	432.7y	α(5.49, 5.44), Np-x, γ(0.060), e$^-$
		242m	141y	IT, α(5.21), Am-x, Np-x, γ(0.0486), e$^-$
		242	16.01h	β^-(0.625, 0.667), K, Pu-x, Cm-x, γ(0.042, 0.045)
		243	7380y	α(5.28, 5.23), Np-x, γ(0.075)
		244m	~26min	β^-(1.50), K, Cm-x
		244	10.1h	β^-(0.387), Cm-x, γ(0.744, 0.898)
		245	2.05h	β^-(0.91), Cm-x, γ(0.253), e$^-$
		246	25.0min	β^-(1.31, 1.60, 2.10), Cm-x, γ(0.799, 1.062, 1.079)
96	Curium	238	2.4h	K, α(6.52)
		239	~3h	K, Am-x, γ(0.188), no α
		240	27d	α(6.29, 6.25)
		241	32.8d	K, α(5.94), Am-x, γ(0.015, 0.472)
		242	162.94d	α(6.12, 6.07), Pu-x, γ(0.561, 0.605)
		243	28.5y	α(5.79, 5.74), K, Pu-x, γ(0.210, 0.228, 0.278), e$^-$
		244	18.11y	α(5.81, 5.77), γ(0.043)
		245	8500y	α(5.36, 5.31), Pu-x, γ(0.133, 0.174)
		246	4730y	α(5.39, 5.34), γ(0.045)
		247	1.56×10^7y	α(4.87, 5.27), γ(0.402)
		248	3.40×10^5y	α(5.08, 5.04), S.F.
		249	1.0692h	β^-(0.86), γ(0.560, 0.634)
		250	$<1.3 \times 10^4$y	S.F.
97	Berkelium	243	4.5h	K, α(6.76, 6.72, 6.57, 6.54, 6.21), Cm-x, γ(0.755, 0.84, 0.946)
		244	4.35h	K, α(6.67, 6.62), Cm-x, γ(0.218, 0.892)
		245	4.94h	K, α(6.35, 6.31, 6.15, 6.12), Cm-x, γ(0.253, 0.39)
		246	1.80d	K, Cm-x, γ(0.799)
		247	1380y	α(5.53, 5.69, 5.71), Am-x, γ(0.084, 0.27)
		248	23.7h	β^-(0.65), K, Cm-x
		249	320d	β^-(0.125), α(5.42), γ(0.327)
		250	3.217h	β^-(0.73, 1.76), Cf-x, γ(0.990, 1.03)
		251	56min	β^-
98	Californium	240	1.06min	α(7.59)
		241	3.8min	α(7.34)
		242	3.5min	α(7.39, 7.35)
		243	10.7min	K, α(7.06, 7.17)
		244	19.4min	α(7.22, 7.18)
		245	43.6min	K, α(7.14)
		246	1.487d	α(6.76, 6.72), Cm-x
		247	3.11h	K, α(6.30), Bk-x, γ(0.294, 0.418, 0.448)
		248	334d	α(6.26), S.F.
		249	350.6y	α(5.81), Cm-x, γ(0.388, 0.333, 0.253)
		250	13.08y	α(6.03, 5.99), γ(0.0429), S.F.
		251	898y	α(5.85, 5.67), Cm-x, γ(0.18)
		252	2.645y	α(6.12, 6.08), Cm-x, γ(0.043), S.F.
		253	17.81d	β^-(0.27), α(5.98)
		254	60.5d	S.F., α(5.84)
99	Einsteinium	243	21s	α(7.89)
		244	37s	K, α(7.57)
		245	1.33min	K, α(7.73)
		246	7.7min	K, α(7.36)
		247	4.7min	K, α(7.31)
		248	27min	K, α(6.87)
		249	1.703h	K, α(6.77), γ(0.379, 0.813)
		250	8.6h	K, γ(0.303, 0.829), no α
		251	1.38d	K, α(6.49), γ(0.153, 0.178)
		252	1.291y	α(6.63, 6.56), K, Bk-x, γ(0.139, 0.785)
		253	20.4d	α(6.64), Bk-x, γ(0.387, 0.389, 0.429)
		254m	1.638d	β^-(0.48, 1.13), α(6.38), K, Fm-x, γ(0.648, 0.692)
		254	275.7d	α(6.44), Bk-x, γ(0.034, 0.036, 0.043), no β^-
		255	38.3d	β^-, α(6.30), S.F.
100	Fermium	244	3.7ms	S.F.
		245	4.2s	α(8.15)
		246	1.1s	α(8.24), S.F.
		248	36s	α(7.87, 7.83), S.F.

Z	Element	A	Half-life	Radiation (MeV)
		249	2.6min	α(7.53)
		250	30min	α(7.44)
		251	5.30h	K, α(6.83), γ(0.406, 0.453, 0.881)
		252	1.0579d	α(7.04)
		253	3.00d	K, α(6.94, 6.67)
		254	3.240h	α(7.20, 7.16), γ(0.042, 0.094), S.F.
		255	20.07h	α(7.02), γ(0.204)
		256	2.627h	S.F., α(6.92)
		257	100.5d	α(6.52), Cf-x, γ(0.115, 0.241), S.F.
101	Mendelevium	248	7s	K, α(8.36, 8.32)
		249	24s	K, α(8.03)
		250	52s	K, α(7.82, 7.75)
		251	4.0min	K, α(7.55)
		252	2.3min	K
		254	10min	K, no α
		255	27min	K, α(7.33)
		256	1.27h	K, α(7.22)
		257	5.2h	K, α(7.07)
		258	55d	α(6.79, 6.73)
102	Nobelium	251	800ms	α(8.68, 8.60)
		252	2.30s	α(8.42, 8.37), S.F.
		253	1.7min	α(8.01)
		254	55s	α(8.10)
		255	3.1min	α(8.11), K
		256	3.3s	α(8.43), S.F.
		257	25s	α(8.32, 8.27, 8.22)
		258	1.2ms	S.F.
		259	1.00h	α(7.49, 7.52), K
103	Lawrencium	255	22s	α(8.43, 8.37), no K
		256	28s	α(8.52, 8.47, 8.43, 8.39)
		257	646ms	α(8.86, 8.80)
		258	4.3s	α(8.62, 8.59)
		259	5.4s	α(8.46)
		260	3.0min	α(8.04)
104	Unnilquadium	257	3.8s	α(9.00, 8.95, 8.78, 8.70)
		258	13ms	S.F.
		259	3.4s	α(8.77, 8.86)
		260	21ms	S.F.
		261	1.08min	α(8.30)
105	Unnilpentium	260	1.52s	α(9.12, 9.07, 9.04), S.F.
		261	1.8s	α(8.93), S.F.
		262	34s	α(8.45, 8.66)
106	Unnilhexium	261	~260ms	α(9.560)
		263	800ms	α(9.25, 9.06)
107	Unnilseptium	262	~115ms	α(9.704)
		262	4.7ms	α(10.376)
108	Unniloctium	265	~1.8ms	α(10.360)
109	Unnilennium	266	~3.5ms	α(11.100)

Radioactive Isotopes Used in Medical Diagnosis and Therapy

Name and Symbol	Principal Nuclear Properties	Form	Use	Mode of Administration
Americium ^{241}Am	Half-life 432.7y α(5.49, 5.44) γ(0.060)	Encapsulated source	Diagnostic: Radiation source for bone mineral analyzer Therapeutic: Intracavitary radiation source in treatment of malignancies	External irradiation Intracavitary irradiation
Calcium ^{47}Ca	Half-life 4.536d β^-(0.67, 1.98) γ(1.297)	Calcium chloride	Diagnostic: Calcium metabolism studies	i.v.
Cesium ^{137}Cs Daughter ^{137m}Ba	Half-life 30.0y β^-(1.176, 0.514) Half-life 2.552min γ(0.662)	Cesium chloride or cesium sulfate (encased in needles or applicator cells)	Therapeutic: Teletherapy source or intracavitary or interstitial radiation source in treatment of malignancies	External, intracavitary or interstitial irradiation
Chromium ^{51}Cr	Half-life 27.704d K; γ(0.32)	Chromic chloride	Diagnostic: Determn of serum protein loss into the gastrointestinal tract	i.v.
		Chromium disodium edetate	Diagnostic: Determn of glomerular filtration rate	i.v.
		Labeled human serum albumin	Diagnostic: Placenta localization; gastrointestinal protein loss	i.v.
		Sodium chromate-labeled red blood cells	Diagnostic: Determn of red cell volume or mass; red cell survival time; evaluation of blood loss; spleen imaging; placenta localization	i.v.
Cobalt ^{60}Co	Half-life 5.271y β^-(0.318, 1.48) γ(1.173, 1.332)	Metallic cobalt	Therapeutic: Teletherapy source or intracavitary or interstitial radiation source in treatment of malignancies	External, intracavitary or interstitial irradiation
		Radioactive vitamin B$_{12}$	Diagnostic: In Schilling test–for absence of intrinsic factor (pernicious anemia) or other defects of intestinal vitamin B$_{12}$ absorption	Orally
^{57}Co	Half-life 271.77d K; γ(0.122)	Radioactive vitamin B$_{12}$	Diagnostic: In Schilling test–for absense of intrinsic factor (pernicious anemia) or other defects of intestinal vitamin B$_{12}$ absorption	Orally
^{58}Co	Half-life 71.916d K; β^+(0.48) γ(0.811)	Radioactive vitamin B$_{12}$	Diagnostic: In Schilling test–for absence of intrinsic factor (pernicious anemia) or other defects of intestinal vitamin B$_{12}$ absorption	Orally
Copper ^{64}Cu	Half-life 12.701h β^-(0.571) β^+(0.657) γ(1.34)	Copper versenate Copper acetate	Diagnostic: Brain scan Diagnostic: Study of Wilson's disease	i.v. Orally or i.v.
Fluorine ^{18}F	Half-life 1.8295h β^+(0.635)	Fludeoxyglucose (FDG)	Diagnostic: Functional brain imaging	i.v.
		Sodium fluoride (reactor produced)	Diagnostic: Bone scan	Orally or i.v.
Gadolinium ^{153}Gd	Half-life 241.6d K; γ(0.70, 0.097, 0.103)	Sealed source	Diagnostic: Radiation source for bone mineral analyzer	External irradiation
Gallium ^{67}Ga	Half-life 3.261d K; γ(0.093, 0.184, 0.300, 0.393)	Gallium citrate	Diagnostic: Detection of neoplastic and inflammatory lesions; tumor seeking agent	i.v.

Radioactive Isotopes Used in Medical Diagnosis and Therapy (Continued)

Name and Symbol	Principal Nuclear Properties	Form	Use	Mode of Administration
Gold ^{198}Au	Half-life 2.6935d β^-(1.371, 0.962) γ(0.412)	Colloidal gold	Diagnostic: Liver imaging	i.v.
		Colloidal gold or seeds	Therapeutic: Treatment of wide-spread abdominal carcinomatosis with ascites; carcinomatosis of pleura with effusion; lymphomas; interstitially in metastatic tumors	Intraperitoneal or intrapleural irradiation
Indium ^{113m}In	Half-life 1.658h γ(0.393)	Indium-colloid	Diagnostic: Liver and spleen imaging	i.v.
		Indium-Fe(OH)$_3$	Diagnostic: Pulmonary perfusion imaging; cardiac output	i.v.
		Indium-labeled red blood cells	Diagnostic: Determn of blood volume	i.v.
		Indium-pentetate (DTPA)	Diagnostic: Brain scan; renal function studies	i.v.
		Indium-transferrin	Diagnostic: Static cardiovascular blood pool imaging; hepatic and placenta blood pool imaging; placenta localization	i.v.
^{111}In	Half-life 2.807d K; γ(0.172, 0.247)	Indium bleomycin	Diagnostic: Tumor detection	i.v.
		Indium chloride	Diagnostic: Hematopoietic bone marrow imaging; tumor detection	i.v.
		Indium oxyquinoline (oxine) - labeled leukocytes	Diagnostic: Detection of abscesses, infections and inflammation	i.v.
		Indium oxyquinoline (oxine) - labeled platelets	Diagnostic: Detection of deep vein thrombosis; cardiac thrombosis; renal transplant rejection	i.v.
		Indium oxyquinoline (oxine) - labeled red blood cells	Diagnostic: Detection of gastro-intestinal bleeding	i.v.
		Indium-pentetate (DTPA)	Diagnostic: Cisternography	Intrathecal, intracisternal or intraventricular
			Diagnostic: Gastric emptying studies Cardiac output; renal scintigraphy	Orally i.v.
		Indium-pentetreotide	Diagnostic: Neuroendocrine tumor detection	i.v.
		Indium-satumomab pendetide	Diagnostic: Tumor detection	i.v.
Iodine ^{131}I	Half-life 8.040d β^-(0.607, 0.81, 0.336) γ(0.080, 0.284, 0.364, 0.637, 0.723)	Diiodofluorescein	Diagnostic: Brain scan	i.v.
		Iobenguane (MIBG)	Diagnostic: Adrenomedullary imaging and tumor detection	i.v.
			Therapeutic: Antineoplastic (radiation source) in treatment of neuroendocrine tumors	
		Iodinated fats and fatty acids e.g. oleic acid, triolein	Diagnostic: Pancreatic function; intestinal fat absorption	Orally
		Iodinated fibrinogen	Diagnostic: *In vitro* determn of fibrinolytic enzymes	*In vitro*

Name and Symbol	Principal Nuclear Properties	Form	Use	Mode of Administration
Iodine ^{131}I (Cont'd)		Iodinated human serum albumin (IHSA)	Diagnostic: Plasma volume determn; peripheral vascular flow; cardiac output; circulation time; cerebral vascular flow	i.v.
			Diagnostic: Brain scan; plancenta localization	i.v.
			Diagnostic: Cisternography	Intrathecal
		Iodinated human serum albumin (macroaggregated)	Diagnostic: Pulmonary perfusion imaging	i.v.
		Iodinated human serum albumin (microaggregated)	Diagnostic: Hepatic blood pool imaging	i.v.
		Iodinated levothyroxine	Diagnostic: Metabolic study of endogenous thyroxine	i.v.
			Diagnostic: *In vitro* determn of thyroid function	*In vitro*
		Iodinated liothyronine	Diagnostic: *In vitro* determn of thyroid function	*In vitro*
		Iodinated povidone	Diagnostic: Protein-loss enteropathy	i.v.
		Iodinated rose bengal	Diagnostic: Liver function–hepatic excretion studies	i.v.
		Iodohippurate sodium	Diagnostic: Determn of renal function, renal blood flow, urinary tract obstruction; renal imaging	i.v.
		Sodium iodide	Diagnostic: Thyroid function studies; thyroid imaging	Orally or i.v.
			Therapeutic: Hyperthyroidism; cancer of the thyroid	Orally or i.v.
^{125}I	Half-life 60.14d K; γ(0.035)	Iodinated fats or fatty acids	Diagnostic: Pancreatic function; intestinal fat absorption	Orally
		Iodinated fibrinogen	Diagnostic: Localization of deep vein thrombosis; study of fibrinogen metabolism	i.v.
			Diagnostic: *In vitro* determn of fibrinolytic enzymes	*In vitro*
		Iodinated human serum albumin (IHSA)	Diagnostic: Determn of blood or plasma volume; circulation time; cardiac output	i.v.
		Iodinated levothyroxine	Diagnostic: Metabolic study of endogenous thyroxine	i.v.
			Diagnostic: *In vitro* determn of thyroid function	*In vitro*
		Iodinated liothyronine	Diagnostic: *In vitro* determn of thyroid function	*In vitro*
		Iodinated povidone	Diagnostic: Protein-loss enteropathy	i.v.
		Iodinated rose bengal	Diagnostic: Liver function–hepatic excretion studies	i.v.
		Sealed source	Diagnostic: Radiation source for bone mineral analyzer	External irradiation
		Sodium iodide	Diagnostic: Thyroid function studies; thyroid imaging	Orally or i.v.

Name and Symbol	Principal Nuclear Properties	Form	Use	Mode of Administration
^{123}I	Half-life 13.2h K; γ(0.159)	Iobenguane (MIBG)	Diagnostic: Adrenomedullary imaging and tumor detection Therapeutic: Antineoplastic (radiation source) in treatment of neuroendocrine tumors	i.v.
		Iodohippurate sodium	Diagnostic: Determn of renal function, renal blood flow, urinary obstruction; renal imaging	i.v.
		Iofetamine hydrochloride (IMP)	Diagnostic: Cerebral imaging	i.v.
		Sodium iodide	Diagnostic: Thyroid function studies; thyroid imaging	Orally or i.v.
Iridium ^{192}Ir	Half-life 73.831d β^-(0.67) γ(0.296, 0.308, 0.317, 0.468, 0.589, 0.604, 0.612)	Seed encased in nylon ribbon	Therapeutic: Interstitial treatment of tumors	Interstitial irradiation
Iron ^{59}Fe	Half-life 44.496d β^-(0.273, 0.475) γ(1.095, 1.292)	Ferric chloride Ferrous citrate Ferrous sulfate	Diagnostic: Ferrokinetics	i.v.
		Labeled red blood cells	Diagnostic: Red cell maturation studies	i.v.
^{55}Fe	Half-life 2.73y K	Labeled red blood cells	Diagnostic: Red cell maturation studies	i.v.
Krypton ^{85}Kr	Half-life 10.72y β^-(0.67) γ(0.517	Gas	Diagnostic: Cardiac abnormalities; skeletal muscle, coronary or cerebral blood flow	Intramuscular or intra-arterial injection
^{81m}Kr	Half-life 13s γ(0.19)	Gas	Diagnostic: Lung ventilation studies	Inhalation
Lead RaD(^{210}Pb) Daughter RaE(^{210}Bi)	Half-life 22.3y β^-(0.017) γ(0.047) Half-life 5.013d β^-(1.16)	Beta ray applicator	Therapeutic: See Strontium(^{90}Sr)	External irradiation
Mercury ^{197}Hg	Half-life 2.6725d K; γ(0.077)	Chlormerodrin	Diagnostic: Brain scan; renal imaging	i.v.
		Merisoprol	Diagnostic: Determn of renal function	i.v.
^{203}Hg	Half-life 46.60d β^-(0.214) γ(0.279)	Chlormerodrin	Diagnostic: Brain scan; renal imaging	i.v.
Phosphorous ^{32}P	Half-life 14.282d β^-(1.71)	Chromic phosphate	Therapeutic: Neoplastic suppressant; treatment for peritoneal or pleural effusions caused by metastatic disease	Intrapleural or intraperitoneal
		Labeled red blood cells	Diagnostic: Blood volume determn	i.v.
		Sodium phosphate	Diagnostic: Study of peripheral vascular disease; localization of ocular, brain and skin tumors; study of breast carcinomas Therapeutic: Polycythemia vera; chronic myelocytic leukemia; chronic lymphocytic leukemia; skeletal metastases; neoplastic suppressant	Orally or i.v. Orally or i.v.

Name and Symbol	Principal Nuclear Properties	Form	Use	Mode of Administration
Potassium ^{43}K	Half-life 22.3h β^-(0.83, 0.46, 1.22, 1.82) γ(0.618, 0.373, 0.39, 0.59, 0.22)	Potassium chloride	Diagnostic: Myocardial scan; determn of total exchangeable potassium	i.v.
^{42}K	Half-life 12.360h β^-(3.52) γ(1.524)	Potassium carbonate	Diagnostic: Localization of brain tumors; determn of intracellular fluid space	Orally or i.v.
		Potassium chloride	Diagnostic: Tumor detection; renal blood flow; determn of total exchangeable potassium	i.v.
Radium ^{226}Ra	Half-life 1600y α(4.78, 4.60) γ(0.187)	Radium bromide; α and β particles filtered by platinum	Therapeutic: Treatment of malignancies such as cancer of uterine cervix and fundus, oral pharynx, urinary bladder, skin and metastatic cancer of lymph nodes	Interstitial irradiation
Radon (Radium Emanation) ^{222}Rn (Daughter of ^{226}Ra)	Half-life 3.825d α(5.49) γ(0.510)	Gaseous radon; α and β particles filtered by 0.3 mm of gold	Therapeutic: See Radium (^{226}Ra)	Interstitial irradiation
Ruthenium ^{106}Ru	Half-life 1.020y β^-(0.039)	Beta ray applicator	Therapeutic: See Strontium(^{90}Sr)	External irradiation
Daughter (^{106}Rh)	Half-life 29.80s β^-(3.53, 3.1, 2.4) γ(0.512), 0.622, 1.128)			
Selenium ^{75}Se	Half-life 119.77d K; γ(0.265, 0.136, 0.121, 0.280, 0.401)	Selenomethionine	Diagnostic: Imaging of pancreas and parathyroid glands	i.v.
Sodium ^{24}Na	Half-life 14.659h β^-(1.389, 4.17) γ(1.369, 2.754)	Sodium chloride	Diagnostic: Determn of circulation times, sodium space, total exchangeable sodium	i.v.
^{22}Na	Half-life 2.602y β^+(0.545, 1.83) K; γ(1.275)	Sodium chloride	Diagnostic: Determn of sodium space and total exchangeable sodium	i.v.
Strontium ^{85}Sr	Half-life 64.84d K; γ(0.514)	Strontium chloride Strontium nitrate	Diagnostic: Bone imaging	i.v.
^{87m}Sr	Half-life 2.795h γ(0.388)			
^{90}Sr	Half-life 28.5y β^-(0.546)	Beta ray applicator	Therapeutic: Treatment of benign conditions of eye such as pterygia, traumatic corneal ulceration, corneal scars, vernal conjunctivitis, hemangioma of eyelid, vascularization of cornea and in preparation for a corneal transplant	External irradiation
Daughter ^{90}Y	Half-life 2.671d β^-(2.288) γ(2.186)			
Sulfur ^{35}S	Half-life 87.51d β^-(0.167)	Sodium sulfate	Diagnostic: Determn of extracellular fluid volume	i.v.

Name and Symbol	Principal Nuclear Properties	Form	Use	Mode of Administration
Technetium ^{99m}Tc	Half-life 6.006h IT, γ(0.141)	Sodium pertechnetate	Diagnostic: Brain imaging. Cerebral angiography; thyroid imaging; salivary gland imaging; placenta localization; blood pool imaging; gastric mucosa imaging; cardiac function studies; renal blood flow studies	Orally or i.v. i.v.
			Urinary bladder imaging	Urethral catheterization
			Nasolacrimal drainage system imaging	Direct instillation
		Sodium pertechnetate labeled red blood cells	Diagnostic: Determn of red blood cell volume, short-term survival studies	i.v.
			In vitro compatibility studies	In vitro
		Tc-albumin	Diagnostic: Blood pool imaging; cardiovascular studies; placenta localization; determn of blood or plasma volumes	i.v.
		Tc-albumin (aggregated)	Diagnostic: Pulmonary perfusion imaging	i.v.
		Tc-albumin (microaggregated)	Diagnostic: Liver imaging	i.v.
		Tc-bicisate	Diagnostic: Brain imaging	i.v.
		Tc-butedronate (DPD)	Diagnostic: Brain imaging	i.v.
		Tc-disofenin (DISIDA)	Diagnostic: Hepatobiliary imaging	i.v.
		Tc-etidronate (EHDP)	Diagnostic: Bone imaging	i.v.
		Tc-exametazime (HM-PAO)	Diagnostic: Cerebral perfusion imaging	i.v.
		Tc-gluceptate	Diagnostic: Brain imaging; renal imaging; assess renal and brain perfusion	i.v.
		Tc-labeled red blood cells	Diagnostic: Determn of red cell volume; short-term red cell survival studies	i.v.
		Tc-lidofenin (HIDA)	Diagnostic: Hepatobiliary imaging	i.v.
		Tc-mebrofenin	Diagnostic: Hepatobiliary imaging	i.v.
		Tc-medronate (MDP)	Diagnostic: Bone imaging	i.v.
		Tc-mertiatide (MAG3)	Diagnostic: Renal imaging	i.v.
		Tc-oxidronate (HDP)	Diagnostic: Bone imaging	i.v.
		Tc-pamidronate (ADP)	Diagnostic: Bone imaging	i.v.
		Tc-pentetate (DTPA)	Diagnostic: Brain imaging; renal imaging; assess renal and brain perfusion; estimate glomerular filtration rate	i.v.
			Lung ventilation studies	Inhalation
		Tc-polyphosphates	Diagnostic: Bone imaging; myocardial imaging; blood pool imaging; detection of gastrointestinal bleeding	i.v.

Name and Symbol	Principal Nuclear Properties	Form	Use	Mode of Administration
Technetium ^{99m}Tc (Cont'd)		Tc-pyrophosphate	Diagnostic: Bone imaging; cardiac imaging; blood pool imaging; detection of gastrointestinal bleeding	i.v.
		Tc-sestamibi (HEXAMIBI)	Diagnostic: Myocardial perfusion imaging	i.v.
		Tc-succimer (DMSA)	Diagnostic: Renal imaging	i.v.
		Tc-sulfur colloid	Diagnostic: Liver, spleen and bone marrow imaging Esophageal transit studies; gastro-esophageal reflux scintigraphy; determn of pulmonary aspiration of gastric contents; detection of intrapulmonary and lower gastrointestinal bleeding Lung ventilation imaging	i.v. Orally Inhalation
		Tc-teboroxime	Diagnostic: Myocardial perfusion imaging	i.v.
		Tc-tetrofosmin	Diagnostic: Myocardial perfusion imaging	i.v.
Thallium ^{201}Tl	Half-life 3.046d K; γ(0.135, 0.167)	Thallous chloride	Diagnostic: Myocardial perfusion imaging; localization of sites of parathyroid hyperactivity	i.v.
Xenon ^{133}Xe	Half-life 5.245d β^-(0.346) γ(0.081)	Gas	Diagnostic: Pulmonary perfusion imaging; cerebral blood flow; lung ventilation studies	Inhalation
		Gas in saline solution	Diagnostic: Pulmonary perfusion imaging; regional blood flow; lung ventilation studies	Intra-arterial or intramuscular injection
Xenon ^{127}Xe	Half-life 36.41d K; γ(0.172, 0.203)	Gas	Diagnostic: Pulmonary perfusion imaging; lung ventilation studies	Inhalation
Ytterbium ^{169}Yb	Half-life 32.022d K; γ(0.063, 0.100, 0.131, 0.177, 0.198, 0.308)	Yb-pentetate (DTPA)	Diagnostic: Cisternography; brain scan	Intrathecal instillation or i.v.

References
1. S. Baum, R. Bramlet, *Basic Nuclear Medicine* (Appleton-Century-Crofts, New York, 1975).
2. C. Behrns *et al., Atomic Medicine* (The Williams & Wilkins Co., Baltimore, 5th ed., 1969).
3. H. M. Chilton, R. L. Witcofski, *Nuclear Pharmacy* (Lea & Febiger, Philadelphia, 1986).
4. *CRC Handbook of Radioactive Nuclides,* Y. Wang, Ed. (The Chemical Rubber Co., Cleveland, Ohio, 1969).
5. *Diagnostic and Investigational Uses of Radiolabeled Blood Elements,* R. J. Davey, M. E. Wallace, Eds. (Am. Assoc. of Blood Banks, Arlington, Va., 1987).
6. *Martindale,* J. E. F. Reynolds, Ed. (Pharmaceutical Press, London, 29th ed., 1989) pp. 1377-1382.
7. *Physician's Desk Reference for Radiology and Nuclear Medicine,* L. M. Freeman, M. D. Blaufox, Editorial Consultants (Medical Economics Co., Oradell, N.J., 4th ed., 1974/75, 5th ed., 1975/76).
8. *Remington's Pharmaceutical Sciences,* A. R. Gennaro, Ed. (Mack Publ. Co., Easton, PA, 17th ed., 1985), pp. 471-499.
9. D. P. Swanson *et al., Pharmaceuticals in Medical Imaging* (Macmillan Publishing Co., Inc., New York, 1990).

Table of Minerals

Minerals are naturally occurring, homogeneous solids formed in the Earth's crust by the inorganic forces of nature, and possess definite, but generally not fixed, chemical compositions, physical characteristics, and highly ordered atomic arrangements. This table contains a group of common minerals selected for their relative importance and frequency of occurrence. Several physical properties commonly used to identify minerals have been included for each entry. The information provided here should be used with the understanding that there is a certain amount of variation in physical properties from specimen to specimen. For more information about the material presented here, consult the references cited at the end of the table.

Streak is the color of a finely powdered mineral obtained by rubbing the mineral on a streak plate, which is an unglazed piece of porcelain.

Since minerals are frequently polymorphic, the crystal system reported in the table is that of the most commonly observed form. The term amorphous (amor) is used to describe noncrystalline minerals that lack an ordered atomic arrangement. The crystal systems have been abbreviated in the table as follows:

hex	hexagonal	ortho	orthorhombic
iso	isometric	tet	tetragonal
mono	monoclinic	trig	trigonal

Hardness, which is the resistance of a smooth surface of a mineral to scratching, is expressed in terms of the Mohs' Hardness Scale. This relative scale spans from 1-9 and is defined as follows:

<2.5	can be scratched by a fingernail
>2.5-3	cannot be scratched by a fingernail; can be scratched by a copper cent
>3-5.5	cannot be scratched by a copper cent; can be scratched by a knife
>5.5-<7	cannot be scratched by a knife; can be scratched by quartz
>7	cannot be scratched by quartz

Mineral	Composition	Color	Streak	Crystal System	Hardness	Index of Refraction	Specific Gravity
Acanthite	Ag_2S	gray-black	black	mono	2-2.5	-	7.3
Actinolite	$Ca_2(Mg,Fe)_5Si_8O_{22}(OH)_2$	green	white	mono	6	1.64	3.0-3.4
Aegirine	$NaFeSi_2O_6$	dark green, brown, black	white	mono	6	1.82	3.6
Allanite	$(Ca,Ce)_3(Al,Fe)_2Si_3O_{12}(OH)$	brown to pitch-black	colorless	mono	5.5-6	1.70-1.81	3.5-4.2
Almandine	$Fe_3Al_2Si_3O_{12}$	red to black	white	iso	7-7.5	1.83	4.0-4.3
Alunite	$KAl_3(SO_4)_2(OH)_6$	white, red	colorless	hex	4	1.57	2.6-2.8
Amblygonite	$LiAlFPO_4$	white, pale green, blue	colorless	trig	6	1.60	3.0-3.1
Analcime	$NaAlSi_2O_6.H_2O$	colorless, white	colorless	iso	5-5.5	1.48-1.49	2.27
Anatase	TiO_2	brown	white	tet	5.5-6	2.6	3.9
Andalusite	Al_2SiO_5	reddish-brown, white, gray	colorless	ortho	7.5	1.64	3.16-3.20
Andradite	$Ca_3Fe_2Si_3O_{12}$	yellow, green, brown, black	white	iso	7-7.5	1.89	3.8-3.9
Anglesite	$PbSO_4$	colorless, white	colorless	ortho	3	1.88	6.2-6.4
Anhydrite	$CaSO_4$	colorless	colorless	ortho	3-3.5	1.58	2.89-2.98
Anthophyllite	$(Mg,Fe)_7Si_8O_{22}(OH)_2$	white, gray	colorless	ortho	5.5-6	1.61-1.71	2.85-3.2
Antimony	Sb	white to gray	gray	hex	3-3.5	-	6.7
Antlerite	$Cu_3(SO_4)(OH)_4$	dark emerald-green	light green	ortho	3.5-4	1.74	3.88
Apatite	$Ca_5(PO_4)_3(F,Cl,OH)$	green, blue, violet, brown, colorless	colorless	hex	5	1.63	3.15-3.20
Apophyllite	$KCa_4(Si_4O_{10})_2F.8H_2O$	colorless, white	colorless	tet	4.5-5	1.54	2.3-2.4
Aragonite	$CaCO_3$	colorless, white	colorless	ortho	3.5-4	1.68	2.95
Argentite	Ag_2S	black	dark gray-black	iso	2-2.5	-	7.2-7.4
Arsenic	As	tin-white	gray	hex	3.5	-	5.7

Mineral	Composition	Color	Streak	Crystal System	Hardness	Index of Refraction	Specific Gravity
Arsenopyrite	FeAsS	silver-white	black	mono	5.5-6	-	6.0-6.2
Atacamite	$Cu_2Cl(OH)_3$	emerald-green	light green	ortho	3-3.5	1.86	3.75-3.77
Augite	$(Ca,Na)(Mg,Fe,Al)$ $(Si,Al)_2O_6$	black, dark green	white	mono	6	1.67-1.73	3.25-3.55
Autunite	$Ca(UO_2)_2(PO_4)_2.10\text{-}12H_2O$	lemon-yellow	yellow	tet	2-2.5	1.58	3.1-3.2
Axinite	$(Ca,Fe,Mn)_3Al_2(BO_3)$ $(Si_4O_{12})(OH)$	clove-brown	colorless	trig	6.5-7	1.69	3.27-3.35
Azurite	$Cu_3(CO_3)_2(OH)_2$	azure-blue	light blue	mono	3.5-4	1.76	3.77
Barite	$BaSO_4$	colorless, white	colorless	ortho	3-3.5	1.64	4.5
Bauxite	A mixture of Al hydroxides	yellow, brown, gray, white	colorless	-	1-3	-	2.0-2.55
Beryl	$Be_3Al_2(Si_6O_{18})$	bluish-green, yellow, colorless	colorless	hex	7.5-8	1.57-1.61	2.65-2.8
Biotite	$K(Mg,Fe)_3(AlSi_3O_{10})$ $(OH)_2$	dark brown, green to black	colorless	mono	2.5-3	1.61-1.70	2.95-3
Bismuth	Bi	silver-white	silver-white	hex	2-2.5	-	9.8
Boehmite	$\gamma\text{-}AlO(OH)$	white, yellow, brown	white	ortho	3	1.65	3.0-3.1
Borax	$Na_2B_4O_5(OH)_4.8H_2O$	colorless, white	colorless	mono	2-2.5	1.47	1.71
Bornite	Cu_5FeS_4	brownish bronze	gray-black	tet	3	-	5.1
Boulangerite	$Pb_5Sb_4S_{11}$	bluish-lead-gray	brownish-gray	mono	2.5-3	-	6.23
Bournonite	$PbCuSbS_3$	steel-gray to black	steel-gray to black	ortho	2.5-3	-	5.8-5.9
Braunite	$3Mn_2O_3.MnSiO_3$	gray to black	gray to black	tet	6-6.5	-	4.7-4.8
Brochantite	$Cu_4SO_4(OH)_6$	green	green	mono	3.5-4	1.78	4.0
Brookite	TiO_2	brown to black	white to gray	ortho	5.5-6	2.6	4.1
Brucite	$Mg(OH)_2$	white, gray, green	colorless	hex	2.5	1.57	2.39
Bustamite	$(Mn,Ca,Fe)SiO_3$			trig	5.5-6.5	1.67-1.70	3.3-3.4
Calaverite	$AuTe_2$	tin-white to brass yellow	yellowish to green-gray	mono	2.5	-	9.4
Calcite	$CaCO_3$	colorless, white	colorless	hex	3	1.66	2.71
Cancrinite	$Na_6Ca_2(AlSiO_4)_6$ $(CO_3)_2$	yellow, white, pink	white	hex	6	1.52	2.4-2.5
Carnallite	$KMgCl_3.6H_2O$	colorless, white	white	ortho	2.5	1.48	1.60
Carnotite	$K_2(UO_2)_2$ $(VO_4)_2.3 H_2O$	yellow	yellow	mono	2	1.93	4-5
Cassiterite	SnO_2	brown to black	white	tet	6-7	2.00	6.8-7.1
Celestite	$SrSO_4$	colorless, white, pale blue	colorless	ortho	3-3.5	1.62	3.95-3.97
Cerussite	$PbCO_3$	colorless, white	colorless	ortho	3-3.5	2.08	6.55
Chabazite	$Ca_2Al_2Si_4O_{12}.6H_2O$	white, flesh-red	colorless	hex	4-5	1.48	2.05-2.15
Chalcanthite	$CuSO_4.5H_2O$	blue	pale blue	trig	2.5	1.54	2.28
Chalcocite	Cu_2S	steel-gray	gray-black	ortho	2.5-3	-	5.7

Table of Minerals (Continued)

Mineral	Composition	Color	Streak	Crystal System	Hardness	Index of Refraction	Specific Gravity
Chalcopyrite	$CuFeS_2$	brass-yellow	greenish-black	tet	3.5-4	-	4.1-4.3
Chlorargyrite	AgCl	pearl-gray, colorless	colorless	iso	2-3	2.07	~5.5
Chlorite	$Mg,Fe_3(Si,Al)_4O_{10}$ $(OH)_2.(Mg,Fe)_3(OH)_6$	green of various shades	colorless	mono	2-2.5	1.57-1.67	2.6-2.9
Chloritoid	$(Fe,Mg,Mn)_2(Al,Fe)$ $Al_3O_2(SiO_4)_2(OH)_4$	grassy green	white	mono	6.5	1.72-1.73	3.58-3.61
Chondrodite	$Mg_5(SiO_4)_2(F,OH)_2$	light yellow, brown	colorless	mono	6-6.5	1.60-1.63	3.1-3.2
Chromite	$FeCr_2O_4$	iron-black to brownish black	dark brown	iso	5.5	2.16	4.6
Chrysoberyl	$BeAl_2O_4$	yellowish to green	colorless	ortho	8.5	1.75	3.65-3.8
Chrysocolla	$(Cu,Al)_2H_2Si_2O_5$ $(OH)_4.nH_2O$	light green to turquoise-blue	light blue	amor	2-4	~4	2.0-2.4
Cinnabar	HgS	red to vermillion	bright red	hex	2-2.5	2.81	8.1
Clinozoisite	$Ca_2Al_3OSi_3O_{12}(OH)$	grayish-white, green	colorless	mono	6-6.5	1.67-1.72	3.25-3.37
Cobaltite	CoAsS	silver-white	gray-black	iso	5.5-6	-	6.3
Colemanite	$CaB_3O_4(OH)_3.H_2O$	colorless, white	colorless	mono	4-4.5	1.59	2.42
Columbite	$(Fe,Mn)Nb_2O_6$	black	dark brown to black	ortho	6	-	5.3-7.3
Copper	Cu	copper-red	copper-red	iso	2.5-3	-	8.9
Cordierite	$(Mg,Fe)_2Al_4Si_5$ $O_{18}.nH_2O$	blue, gray, brown	colorless	ortho	7-7.5	1.53-1.57	2.60-2.66
Corundum	Al_2O_3	gray, blue, pink, brown	white	trig	9	1.77	4.0
Covellite	CuS	indigo-blue to black	lead-gray to black	hex	1.5-2	-	4.6
Cristobalite	SiO_2	colorless, white	colorless	iso	7	1.48	2.32
Crocoite	$PbCrO_4$	red	orange-yellow	mono	2.5-3	2.36	5.9-6.1
Cryolite	Na_3AlF_6	colorless, white	colorless	mono	2.5	1.34	2.95-3.0
Cummingtonite	$(Fe,Mg,Mn)_7Si_8O_{22}$ $(OH)_2$	brown	white	mono	6	1.66-1.68	3.2-3.6
Cuprite	Cu_2O	red	brownish-red	iso	3.5-4	-	6.0
Danburite	$CaB_2(SiO_4)_2$	-	-	ortho	7	1.63	2.97-3.02
Datolite	$CaB(SiO_4)(OH)$	colorless, pale green	colorless	mono	5-5.5	1.65	2.8-3.0
Diamond	C	colorless, yellow, black	colorless	iso	10	2.42	3.5
Diaspore	AlO(OH)	white, gray	colorless	ortho	6.5-7	1.72	3.35-3.45
Diopside	$Ca(Mg,Fe)Si_2O_6$	white, green	white	mono	6	1.67	3.25-3.40
Dioptase	$CuSiO_2(OH)_2$	green		hex	5	1.65	3.3
Dolomite	$CaMg(CO_3)_2$	colorless, white, pink	colorless	hex	3.5-4	1.68	2.85
Enargite	Cu_3AsS_4	gray-black	gray-black	ortho	3	-	4.4
Enstatite	$MgSiO_3$	gray-brown, green, brown	colorless	ortho	5.5	1.65	3.2-3.5

Mineral	Composition	Color	Streak	Crystal System	Hardness	Index of Refraction	Specific Gravity
Epidote	$Ca_2(Al,Fe)Al_2(SiO_4)_3(OH)$	yellowish to blackish green	colorless	mono	6-7	1.72-1.78	3.35-3.45
Epsomite	$MgSO_4.7H_2O$	white	white	ortho	2	1.46	1.67
Erythrite	$Co_3(AsO_4)_2.8H_2O$	red to pink	pink	mono	1.5-2.5	1.66	3.06
Euclase	$BeAl(SiO_4)(OH)$	-	-	mono	7.5	1.66	3.1
Eucryptite	$LiAlSiO_4$	-	-	hex	-	1.55	2.67
Fayalite	Fe_2SiO_4	-	-	ortho	6.5	1.86	4.39
Fluorite	CaF_2	colorless, violet, green	colorless	iso	4	1.43	3.18
Franklinite	$(Fe,Zn,Mn)(Fe,Mn)_2O_4$	iron-black	dark brown	iso	6	-	5.15
Gahnite	$ZnAl_2O_4$	dark green	colorless	iso	7.5-8	1.80	4.55
Galena	PbS	lead-gray	lead-gray	iso	2.5	-	7.6
Garnet	$(Ca,Mg,Fe,Mn)_3(Al,Fe,Cr)_2(SiO_4)_3$	brown to red	-	iso	6.5-7.5	1.71-1.88	3.5-4.3
Gersdorffite	$NiAsS$	silver-white	gray-black	iso	5.5	-	6.3
Gibbsite	$Al(OH)_3$	white, red, brown	white	mono	2.5-3.5	1.57	2.3-2.4
Glauconite	$(K,Na)(Fe,Mg,Al)_2(Si,Al)_4O_{10}(OH)_2$	dark green to black	pale green	mono	2	1.62	2.5-2.8
Glaucophane	$Na_2Mg_3Al_2Si_8O_{22}(OH)_2$	blue, blue-black	white	mono	6	1.62-1.67	3.0-3.3
Goethite	$\alpha\text{-FeO(OH)}$	yellow-brown to dark brown	yellow-brown	ortho	5-5.5	2.39	4.37
Gold	Au	gold-yellow	gold-yellow	iso	2.5-3	-	15.0-19.3
Graphite	C	steel-gray to iron-black	black	hex	1-1.5	-	2.23
Grossular	$Ca_3Al_2Si_3O_{12}$	white, green, brown	white	iso	7-7.5	1.73	3.6
Grunerite	$Fe_7Si_8O_{22}(OH)_2$	light brown	colorless	mono	6	1.71	3.6
Gypsum	$CaSO_4.2H_2O$	colorless, white, gray	colorless	mono	2	1.52	2.32
Halite	$NaCl$	colorless, white	colorless	iso	2.5	1.54	2.1-2.3
Hausmannite	Mn_3O_4	brownish black	brown	tet	5.5-6	-	4.84
Hedenbergite	$CaFeSi_2O_6$	green to black	white	mono	5-6	1.73	3.40-3.55
Hematite	Fe_2O_3	red to vermillion	red-brown	hex	5.5-6.5	-	5.2
Hemimorphite	$Zn_4(Si_2O_7)(OH)_2.H_2O$	white	colorless	ortho	4.5-5	1.62	3.4-3.5
Heulandite	$CaAl_2Si_7O_{18}.6H_2O$	colorless, white	colorless	mono	3.5-4	1.48	2.18-2.20
Hornblende	$NaCa_2(Mg,Fe,Al)_5(Al,Si)_8O_{22}(OH)_2$	green, brown, black	white	mono	5-6	1.62-1.72	3.0-3.4
Humite	$(Mg,Fe)_7(SiO_4)_3(F,OH)_2$	yellow, brownish red	white	ortho	6-6.5	1.64	3.15-3.35
Hypersthene	$(Mg,Fe)SiO_3$	brown, black	white	ortho	5-6	1.68-1.73	3.4-4.0
Ilmenite	$FeTiO_3$	black	black	hex	5.5-6	-	4.7
Iron	Fe	gray	gray	iso	4	-	7.8
Jadeite	$NaAlSi_2O_6$	green, white	colorless	mono	6.5-7	1.66	3.3-3.5

Mineral	Composition	Color	Streak	Crystal System	Hardness	Index of Refraction	Specific Gravity
Jamesonite	$Pb_4FeSb_6S_{14}$	steel-gray to gray-black	steel-gray to gray-black	mono	2-3	-	5.5-6.0
Jarosite	$KFe_3(SO_4)_2(OH)_6$	brown	yellow	trig	3	1.82	2.9-3.3
Kainite	$KMg(Cl,SO_4).2\ 3/4\ H_2O$	-	-	mono	3	1.51	2.1
Kaolinite	$Al_2Si_2O_5(OH)_4$	white	colorless	trig	2-2.5	1.55	2.6-2.63
Kernite	$Na_2B_4O_6(OH)_2.3H_2O$	colorless, white	colorless	mono	3	1.47	1.95
Kyanite	Al_2SiO_5	blue	colorless	trig	5-7	1.72	3.56-3.66
Laumontite	$CaAl_2Si_4O_{12}.4H_2O$	white	white	mono	4	1.52	2.25-2.30
Lawsonite	$CaAl_2(Si_2O_7)(OH)_2.H_2O$	colorless, gray, pink	colorless	ortho	8	1.67	3.09
Lazulite	$(Mg,Fe)Al_2(PO_4)_2(OH)_2$	azure-blue	colorless	mono	5-5.5	1.64	3.0-3.1
Lazurite	$(Na,Ca)_8(AlSiO_4)_6(SO_4,S,Cl)_2$	azure-blue, greenish-blue	colorless	iso	5-5.5	1.50	2.4-2.45
Lepidolite	$K(Li,Al)_{2-3}(AlSi_3O_{10})(O,OH,F)_2$	pink to grayish white	colorless	mono	2.5-4	1.55-1.59	2.8-3.0
Leucite	$KAlSi_2O_6$	gray, white	colorless	iso	5.5-6	1.51	2.45-2.50
Lithiophilite	$Li(Mn,Fe)PO_4$	bluish-gray	colorless	ortho	4.5-5	1.67	3.5
Magnesite	$MgCO_3$	white, yellow	colorless	hex	3.5-5	1.70	3.0-3.2
Magnetite	Fe_3O_4	black	black	iso	6	-	5.18
Malachite	$Cu_2CO_3(OH)_2$	bright green	light green	mono	3.5-4	1.88	3.9-4.03
Manganite	$MnO(OH)$	steel-gray to iron-black	dark brown	mono	4	-	4.3
Marcasite	FeS_2	pale yellow	grayish-black	iso	6-6.5	-	4.9
Margarite	$CaAl_2(Al_2Si_2O_{10})(OH)_2$	pink, gray, white	colorless	mono	3.5-5	1.65	3.0-3.1
Melanterite	$FeSO_4.7H_2O$	pale green	white	mono	2	1.48	1.90
Melilite	$Ca_2Al(AlSiO_7) - Ca_2Mg(Si_2O_7)$	nearly colorless, reddish brown, gray	white	tet	5-6	-	2.94-3.05
Microcline	$KAlSi_3O_8$	colorless, white	colorless	trig	6	1.53	2.54-2.56
Microlite	$Ca_2Ta_2O_6(O,OH,F)$	yellow, brown, black	yellow to brown	iso	5.5	1.92-1.99	5.48-5.56
Millerite	NiS	brass-yellow	greenish-black	hex	3-3.5	-	5.5
Mimetite	$Pb_5(AsO_4)_3Cl$	white, yellow	white	hex	3.5-4	2.1-2.2	7.2-7.3
Molybdenite	MoS_2	lead-gray	grayish-black	hex	1-1.5	-	4.7
Monazite	$(Ce,La,Y,Th)PO_4$	yellowish-brown	colorless	mono	5-5.5	1.79	5.0-5.3
Monticellite	$CaMgSiO_4$			ortho	5	1.65	3.2
Montmorillonite	$(Al,Mg)_2Si_4O_{10}(OH)_2.nH_2O$	white, gray, green-gray	white	mono	2	1.50-1.64	2.0-2.7
Muscovite	$KAl_2(AlSi_3)O_{10}(OH)_2$	colorless, pale tints	colorless	mono	2-2.5	1.60	2.76-2.88
Natrolite	$Na_2Al_2Si_3O_{10}.2H_2O$	colorless, white	colorless	ortho	5-5.5	1.48	2.25
Nepheline	$(Na,K)AlSiO_4$	colorless, gray, brown	colorless	hex	5.5-6	1.54	2.55-2.65

Mineral	Composition	Color	Streak	Crystal System	Hardness	Index of Refraction	Specific Gravity
Nickel Skutterudite	$(Ni,Co)As_3$	silver-white	black	iso	5.5	-	6.1-6.9
Nickeline	NiAs	pale copper-red	brown-black	hex	5-5.5	-	7.8
Niter	KNO_3	white	colorless	ortho	2	1.50	2.09-2.14
Nitratite	$NaNO_3$	colorless, white	colorless	hex	1-2	1.59	2.29
Olivine	$(Mg,Fe)_2SiO_4$	olive to grayish-green	colorless	ortho	6.5-7	1.69	3.27-4.37
Opal	$SiO_2.nH_2O$	colorless, white	colorless	amor	5-6	1.44	1.9-2.2
Orpiment	As_2S_3	lemon-yellow	yellow	mono	1.5-2	2.8	3.49
Orthoclase	$KAlSi_3O_8$	colorless, white	colorless	mono	6	1.52	2.54-2.56
Pectolite	$Ca_2NaH(SiO_3)_3$	white	colorless	trig	5	1.60	2.7-2.8
Pentlandite	$(Fe,Ni)_9S_8$	yellowish-bronze	bronze-brown	iso	3.5-4	-	4.6-5.0
Periclase	MgO	white	white	iso	5	1.73	3.58
Phenacite	Be_2SiO_4	white, colorless	colorless	hex	7.5-8	1.65	2.97-3.0
Phlogopite	$KMg_3(AlSi_3O_{10})(OH)_2$	pale brown	colorless	mono	2.5-3	1.56-1.64	2.86
Plagioclase	$(Na,Ca)(Al,Si)_4O_8$	colorless, white, gray	colorless	trig	6	1.53-1.59	2.62
Platinum	Pt	steel-gray	steel-gray	iso	4-4.5	-	14-19
Prehnite	$Ca_2Al(AlSi_3O_{10})(OH)_2$	apple-green, white	colorless	ortho	6-6.5	1.63	2.8-2.95
Proustite	Ag_3AsS_3	ruby-red	bright red	hex	2-2.5	3.09	5.55
Pumpellyite	$Ca_2MgAl_2(SiO_4)(Si_2O_7)(OH)_2.H_2O$	bluish green to nearly white	white	mono	5.5-6	-	3.18-3.23
Pyrargyrite	Ag_3SbS_3	red	red	hex	2.5	3.08	5.85
Pyrite	FeS_2	pale brass-yellow	brownish-black	iso	6-6.5	-	5.0
Pyrochlore	$(Ca,Na)_2(Nb,Ta)_2O_6(OH,F)$	yellow, brown, black	white	iso	5-5.5	-	4.2-6.4
Pyrolusite	MnO_2	iron-black	black	tet	1-2	-	4.7
Pyromorphite	$Pb_5(PO_4)_3Cl$	green, brown, yellow, gray	colorless	hex	3.5-4	2.06	6.5-7.1
Pyrope	$Mg_3Al_2Si_3O_{12}$	red to black	white	iso	7-7.5	1.71	3.6
Pyrophyllite	$Al_2Si_4O_{10}(OH)_2$	white, gray, green	colorless	mono	1-2	1.59	2.8-2.9
Pyrrhotite	$Fe_{1-x}S$	brownish-bronze	black	mono	4	-	4.6
Quartz	SiO_2	colorless, white, various colors	colorless	hex	7	1.54	2.65
Realgar	AsS	red	red to orange	mono	1.5-2	2.60	3.48
Rhodochrosite	$MnCO_3$	pink, rose-red, brown	colorless	hex	3.5-4.5	1.82	3.45-3.6
Rhodonite	$MnSiO_3$	pink, brown	colorless	trig	5.5-6	1.73-1.75	3.58-3.70
Riebeckite	$Na_2(Mg,Fe)_3Fe_2Si_8O_{22}(OH)_2$	dark blue, black	blue-gray	mono	6	1.66-1.71	3.3-3.6
Romanechite	$BaMn_9O_{16}(OH)_4$	black	black	ortho	5-6	-	3.7-4.7

Mineral	Composition	Color	Streak	Crystal System	Hardness	Index of Refraction	Specific Gravity
Rutile	TiO_2	reddish-brown	pale brown	tet	6-6.5	2.61	4.18-4.25
Scapolite	$3NaAlSi_3O_8.NaCl - 3CaAlSi_2O_8.CaCO_3$	white, pink, gray	colorless	tet	5-6	1.55-1.60	2.65-2.74
Scheelite	$CaWO_4$	white, yellow, brown	colorless	tet	4.5-5	1.92	5.9-6.1
Serpentine	$Mg_3Si_2O_5(OH)_4$	shades of green	colorless	mono	2-5	1.55	2.3-2.66
Shattuckite	$Cu_5(SiO_3)_4(OH)_2$	blue		ortho	-	1.78	3.8
Siderite	$FeCO_3$	light to dark brown	yellow, brown	hex	3.5-4	1.88	3.83-3.88
Sillimanite	Al_2SiO_5	white, brown, gray	colorless	ortho	6-7	1.66	3.23
Silver	Ag	silver-white	silver-white	iso	2.5-3	-	10.5
Skutterudite	$(Co,Ni)As_3$	silver-white	black	iso	5.5-6	-	6.1-6.9
Smithsonite	$ZnCO_3$	green, blue, white	colorless	hex	5	1.85	4.35-4.40
Sodalite	$Na_8(AlSiO_4)_6Cl_2$	blue, white	colorless	iso	5.5-6	1.48	2.15-2.3
Sperrylite	$PtAs_2$	tin-white	black	iso	6-7	-	10.6
Spessartine	$Mn_3Al_2Si_3O_{12}$	orange, red, brown	white	iso	7-7.5	1.80	4.0-4.2
Sphalerite	ZnS	yellow-brown to black	white to yellow-brown	iso	3.5	2.37	3.9-4.1
Spinel	$MgAl_2O_4$	red, black, blue, green, brown	colorless	iso	8	1.72	3.6-4.0
Spodumene	$LiAlSi_2O_6$	white, gray, pink, green	colorless	mono	6.5-7	1.67	3.15-3.20
Staurolite	$Fe_2Al_9O_6(SiO_4)_4(O,OH)_2$	red-brown	colorless	ortho	7-7.5	1.75	3.65-3.75
Stibnite	Sb_2S_3	lead-gray to black	lead-gray to black	ortho	2	-	4.5
Stilbite	$NaCa_2Al_5Si_{13}O_{36}.14H_2O$	white	colorless	mono	3.5-4	1.50	2.1-2.2
Stilpnomelane	$K(Fe,Mg)_6Si_8Al(O,OH)_{27}.2-4H_2O$	black, yellowish to reddish brown	white	mono	3-4	1.58-1.74	2.59-2.96
Stromeyerite	$(Ag,Cu)_2S$	steel-gray	-	ortho	2.5-3	-	6.2-6.3
Strontianite	$SrCO_3$	colorless, white	colorless	ortho	3.5-4	1.67	3.7
Sulfur	S	yellow	colorless	ortho	1.5-2.5	2.04	2.05-2.09
Sylvanite	$(Au,Ag)Te_2$	tin-white	gray	mono	2	-	8-8.2
Sylvite	KCl	colorless, white	colorless	iso	2	1.49	1.99
Talc	$Mg_3Si_4O_{10}(OH)_2$	white, green, gray	colorless	mono	1	1.59	2.7-2.8
Tantalite	$(Fe,Mn)Ta_2O_6$	black	dark brown to black	ortho	6	-	5.3-7.3
Tennantite	$Cu_{12}As_4S_{13}$	black	brown to black	iso	3-4.5	-	4.6-5.1
Tetrahedrite	$Cu_{12}Sb_4S_{13}$	grayish black	black	iso	3-4.5	-	4.7-5.0
Thorite	$ThSiO_4$	brown to black	white	tet	5	1.8	5.3
Titanite	$CaTiO(SiO_4)$	brown, green, yellow	colorless	mono	5-5.5	1.91	3.4-3.55

Table of Minerals (Continued)

Mineral	Composition	Color	Streak	Crystal System	Hardness	Index of Refraction	Specific Gravity
Topaz	$Al_2SiO_4(F,OH)_2$	colorless, yellow, blue	colorless	ortho	8	1.61-1.63	3.4-3.6
Torbernite	$Cu(UO_2)_2(PO_4)_2.8-12H_2O$	emerald-green to apple-green	green	tet	2-2.5	1.59	3.2-3.7
Tourmaline	$(Na,Ca)(Li,Mg,Al)(Al,Fe,Mn)_6(BO_3)_3(Si_6O_{18})(OH)_4$	black, green, brown, blue, pink	colorless	hex	7-7.5	1.64-1.68	3.0-3.25
Tremolite	$Ca_2Mg_5Si_8O_{22}(OH)_2$	white	white	mono	6	1.61	3.0
Tridymite	SiO_2	colorless	colorless	hex	7	1.47	2.26
Triphylite	$Li(Fe,Mn)PO_4$	bluish-gray	colorless	ortho	4.5-5	1.69	3.42-3.56
Turquoise	$CuAl_6(PO_4)_4(OH)_8.5H_2O$	blue, bluish-green	colorless	trig	6	1.62	2.6-2.8
Ulexite	$NaCaB_5O_6(OH)_6.5 H_2O$	white	colorless	trig	1-2.5	1.50	1.96
Uraninite	UO_2	black	brownish-black	iso	5.5	-	9.0-9.7
Uvarovite	$Ca_3Cr_2Si_3O_{12}$	green	white	iso	7.5	1.87	3.90
Vanadinite	$Pb_5(VO_4)_3Cl$	ruby-red, brown, yellow	colorless	hex	3	2.25-2.42	6.7-7.1
Vermiculite	$(Mg,Fe,Al)_3(Al,Si)_4O_{10}(OH)_2.4H_2O$	yellow, brown, green	white	mono	1.5	1.55-1.58	2.4
Vesuvianite	$Ca_{10}(Mg,Fe)_2Al_4Si_9O_{34}(OH)_4$	green, brown, yellow, blue	colorless	tet	6.5	1.70-1.75	3.35-4.45
Vivianite	$Fe_3(PO_4)_2.8H_2O$	blue, blue-green	blue	mono	1-1.5	1.60	2.7
Wavellite	$Al_3(PO_4)_2(OH)_3.5 H_2O$	green, white, brown	colorless	ortho	3.5-4	1.54	2.33
Willemite	Zn_2SiO_4	yellow-green, white, brown	colorless	hex	5.5	1.69	3.9-4.2
Witherite	$BaCO_3$	colorless, white	colorless	ortho	3.5	1.68	4.3
Wolframite	$(Fe,Mn)WO_4$	brown to black	black to brown	mono	4-4.5	-	7.0-7.5
Wollastonite	$CaSiO_3$	colorless, white	colorless	trig	5-5.5	1.63	2.8-2.9
Wulfenite	$PbMoO_4$	yellow, red	colorless	tet	3	2.40	6.8
Wurtzite	ZnS	brownish black	brown	hex	3.5-4	2.35	4.09
Xenotime	YPO_4	yellowish to reddish brown	white	tet	4-5	-	4.4-5.1
Zincite	ZnO	red to orange-yellow	orange-yellow	hex	4-4.5	2.01	5.68
Zircon	$ZrSiO_4$	brown, gray, colorless	colorless	tet	7.5	1.92-1.96	4.68
Zoisite	$Ca_2Al_3Si_3O_{12}(OH)$	gray, green, pink	white	ortho	7	1.69	3.3

References

1. L. G. Berry *et al., Mineralogy: Concepts, Descriptions, Determinations* (W. H. Freeman and Co., San Francisco, 2nd ed., 1983).
2. A. M. Clark, *Hey's Mineral Index: Mineral Species, Varieties and Synonyms* (Chapman & Hall, London, 3rd ed., 1993).
3. C. Klein, C. S. Hurlbut, Jr., *Manual of Mineralogy* (John Wiley & Sons, New York, 20th ed., 1985).
4. P. G. Read, *Dictionary of Gemmology* (Butterworth, Heinemann, Oxford, 2nd ed., 1988).

ORGANIC NAME REACTIONS

Organic Name Reactions

The Organic Name Reactions (ONR) section is intended to serve the professional chemist and student by describing organic chemical reactions which have come to be recognized and referred to by name within the chemistry community. Since this section last appeared in *The Merck Index, Tenth Edition* (1983), many new name reactions have surfaced in the chemical literature. A select group has been chosen for addition to this section. Each reaction description is designed to be informative and representative of the pertinent literature; however, it is not meant to be comprehensive. The descriptions are composed of the following: (1) name(s) associated with the reaction, (2) the original and/or primary contributor(s) connected with the discovery and/or development of the reaction, (3) a concise description of the transformation, emphasizing such reaction parameters as starting materials, reagents, products, intermediates, mechanism, application, and/or reaction conditions, (4) a reaction scheme, (5) key references, and (6) cross references to other ONR based on commonalities. Included in the back of this section is an index which also lists supplementary terms.

Abbreviations

Ac	acetyl	E	electrophile
BBN	borabicyclo[3.3.1]nonane	ee	enantiomeric excess
BINAP	2,2'-bis(diphenylphosphino)-1,1'-binaphthyl	Et	ethyl
		EWG	electron withdrawing group
BOC	*t*-butyloxycarbonyl	HA	protic acid
Bu	butyl	HMPT	hexamethylphosphoric triamide
cat	catalytic	LDA	lithium diisopropylamide
Cp	cyclopentadienide	LHMDS	lithium hexamethyldisilazide
dba	dibenzylideneacetone	Me	methyl
DCC	dicyclohexylcarbodiimide	NuH	nucleophile
DEAD	diethylazadicarboxylate	Ph	phenyl
DME	dimethylether	Pr	propyl
dppf	dichloro[1,1'-bis(diphenyl-phosphino)ferrocene]	salen	*N,N'*-ethylenebis(salicylideneimine)
		Tf	trifluoromethanesulfonyl
dppp	1,3-bis(diphenylphosphino)propane	Ts	*p*-toluenesulfonyl

1. Acetoacetic Ester Synthesis

Base-catalyzed alkylation or arylation of β-ketoesters. Subsequent mild hydrolysis and decarboxylation yield substituted acetones. Alternately, treatment with concentrated base produces substituted esters:

$$CH_3COCH_2COOCH_2CH_3 \xrightarrow[\text{2. RX}]{\text{1. base}} CH_3COCHCOOCH_2CH_3$$

with R below the CH, then branching:

$$\xrightarrow[\text{dil OH}^-]{\text{H}^+ \text{ or}} CH_3COCH_2R$$

$$\xrightarrow[\text{conc OH}^-]{} CH_3COOH + RCH_2COOCH_2CH_3$$

Synthetic applications: R. Kluger, M. Brandl, *J. Org. Chem.* **51**, 3964 (1986); T. Yamamitsu *et al.*, *Perkin Trans. I* **1989**, 1811.

2. Acyloin Condensation

L. Bouveault, R. Loquin, *Compt. Rend.* **140**, 1593 (1905).

Reductive coupling of esters by sodium to yield acyloins (α-hydroxyketones). Yields are greatly improved in the presence of trimethylchlorosilane:

K. T. Finley, *Chem. Rev.* **64**, 573 (1964); K. Ziegler, *Houben-Weyl* **4/2**, 729-822 (1955); S. M. McElvain, *Org. React.* **4**, 256 (1948); J. J. Bloomfield *et al.*, *ibid.* **23**, 259 (1976); R. Brettle, *Comp. Org. Syn.* **3**, 613-632 (1991). *Cf.* Benzoin Condensation.

3. Akabori Amino Acid Reactions

S. Akabori, *J. Chem. Soc. Japan* **52**, 606 (1931); *Ber.* **66**, 143, 151 (1933); *J. Chem. Soc.* **64**, 608 (1943).

1. Formation of aldehydes by oxidative decomposition of α-amino acids when heated with sugars according to the equation:

$$\underset{\text{RCHCOOH}}{\overset{NH_2}{|}} \xrightarrow[\text{sugar}]{O_2} RCHO + CO_2 + NH_3$$

2. Reduction of α-amino acids and esters by sodium amalgam and ethanolic hydrogen chloride to the corresponding α-amino aldehydes:

$$\underset{\text{RCHCOOEt}}{\overset{NH_2 \cdot HCl}{|}} \xrightarrow[\text{HCl/EtOH}]{Na(Hg)} \underset{\text{RCHCHO}}{\overset{NH_2 \cdot HCl}{|}}$$

3. Formation of alkamines by heating mixtures of aromatic aldehydes and amino acids. No reaction was observed with tertiary amino groups.

E. Takagi *et al.*, *J. Pharm. Soc. Japan* **71**, 648 (1951); **72**, 812 (1952); A. Lawson, H. V. Morley, *J. Chem. Soc.* **1955**, 1695; A. Lawson, *ibid.* **1956**, 307; K. Dose, *Ber.* **90**, 1251 (1957); V. N. Belikov *et al.*, *Izv. Akad. Nauk SSSR, Ser. Khim.* **1969**, 2536.

4. Aldol Reaction (Condensation)

R. Kane, *Ann. Phys. Chem., Ser. 2*, **44**, 475 (1838); *idem, J. Prakt. Chem.* **15**, 129 (1838).

Traditionally, it is the acid- or base-catalyzed condensation of one carbonyl compound with the enolate/enol of another, which may or may not be the same, to generate a β-hydroxy carbonyl compound – an aldol. The method is compromised by self-condensation, polycondensation, generation of regioisomeric enols/enolates, and dehydration of the aldol followed by Michael addition, *q.v.* The development of methods for the preparation and use of preformed enolates or enol derivatives, that dictate specific carbon-carbon bond formation, have revolutionized the coupling of carbonyl compounds:

MX = LDA, LHMDS, CH_3MgBr, Bu_2BOTf/NR_3, 9-BBNCl/NR_3, LDA + $ZnCl_2$, LDA + $(C_5H_5)_2ZrCl_2$, LDA + Ti(O-i Pr)$_3$Cl
Lewis acid = $TiCl_4$, $SnCl_4$, $AlCl_3$, $BF_3 \cdot O(CH_2CH_3)_2$, $ZnCl_2$

Historical perspective: C. H. Heathcock, *Comp. Org. Syn.* **2**, 133-179 (1991). General review: T. Mukaiyama, *Org. React.* **28**, 203-331 (1982). Application of lithium and magnesium enolates: C. H. Heathcock, *Comp. Org. Syn.* **2**, 181-238 (1991); of boron enolates: B. M. Kim *et al., ibid.* 239-275; of transition metal enolates: I. Paterson, *ibid.* 301-319. Stereoselective reactions of ester and thioester enolates: M. Braun, H. Sacha, *J. Prakt. Chem.* **335**, 653-668 (1993). Review of asymmetric methodology: A. S. Franklin, I. Paterson, *Contemp. Org. Syn.* **1**, 317-338 (1994). *Cf.* Claisen-Schmidt Condensation; Henry Reaction; Ivanov Reaction; Knoevenagel Condensation; Reformatsky Reaction; Robinson Annulation.

5. Algar-Flynn-Oyamada Reaction

J. Algar, J. P. Flynn, *Proc. Roy. Irish Acad.* **42B**, 1 (1934); B. Oyamada, *J. Chem. Soc. Japan* **55**, 1256 (1934).

Alkaline hydrogen peroxide oxidation of *o*-hydroxyphenyl styryl ketones (chalcones) to flavonols *via* the intermediate dihydroflavonols:

T. S. Wheeler, *Record Chem. Progr.* **18**, 133 (1957); W. P. Cullen *et al., J. Chem. Soc. C* **1971**, 2848. Mechanism: T. R. Gormley, W. I. O'Sullivan, *Tetrahedron* **29**, 369 (1973). *Cf.* Auwers Synthesis.

6. Allan-Robinson Reaction

J. Allan, R. Robinson, *J. Chem. Soc.* **125**, 2192 (1924).

Preparation of flavones or isoflavones by condensing *o*-hydroxyaryl ketones with anhydrides of aromatic acids and their sodium salts:

S. F. Dyke *et al., J. Org. Chem.* **26**, 2453 (1961); Seshandri in *The Chemistry of Flavonoid Compounds*, T. A. Geissman, Ed. (New York, 1962) p 182; Gripenberg in *ibid.* p 411; W. Rahman, K. T. Nasim, *J. Org. Chem.* **27**, 4215 (1962); D. L. Dreyer *et al., Tetrahedron* **20**, 2977 (1964). *Cf.* Baker-Venkataraman Rearrangement; Kostanecki Acylation.

7. Allylic Rearrangements

L. Claisen, *Ber.* **45**, 3157 (1912).

Migration of a carbon-carbon double bond in a three carbon (allylic) system on treatment with nucleophiles under S_N1 conditions (or under S_N2 conditions when the nucleophilic attack takes place at the γ-carbon):

Reviews: J. R. DeWolfe, W. G. Young, *Chem. Rev.* **56**, 753 (1956); W. G. Young, *J. Chem. Ed.* **39**, 455 (1962); P. delaMare in *Molecular Rearrangements* Part 1, P. de Mayo, Ed. (Wiley-Interscience, New York, 1963) pp 27-110; K. Mackenzie in *The Chemistry of Alkenes*, S. Patai, Ed. (Interscience, New York, 1964) pp 436-453; R. H. DeWolfe, W. G. Young in *ibid.* pp 681-738; J. March, *Advanced Organic Chemistry* (Wiley-Interscience, New York, 4th ed., 1992) pp 327-330.

8. Amadori Rearrangement

M. Amadori, *Atti Accad. Nazl. Lincei* **2**(6), 337 (1925), *C.A.* **20,** 902 (1926); *ibid.* **9**(6), 68, 226 (1929), *C.A.* **23,** 3211, 3443 (1929).

Conversion of *N*-glycosides of aldoses to *N*-glycosides of the corresponding ketoses by acid or base catalysis:

J. E. Hodge, *Advan. Carbohyd. Chem.* **10,** 169 (1955); R. U. Lemieux in *Molecular Rearrangements* Part 2, P. de Mayo, Ed. (Wiley-Interscience, New York, 1964) p 753. ^{13}C-NMR studies: W. Funcke, *Ann.* **1978,** 2099. *Review:* K. Maruoka, H. Yamamoto, *Comp. Org. Syn.* **6,** 789-791 (1991).

9. Arens-van Dorp Synthesis; Isler Modification

D. A. van Dorp, J. F. Arens, *Nature* **160,** 189 (1947); J. F. Arens *et al., Rec. Trav. Chim.* **68,** 604, 609 (1949); O. Isler *et al., Helv. Chim. Acta* **39,** 259 (1956).

The preparation of alkoxyethynyl alcohols from ketones and ethoxyacetylene. In the **Isler modification** the tedious preparation of ethoxyacetylene is obviated by treating β-chlorovinyl ether with lithium amide to yield lithium ethoxy-acetylene, which is then condensed with the ketone:

H. Heusser *et al., Helv. Chim. Acta* **33,** 370 (1950); J. F. Arens, *Advan. Org. Chem.* **2,** 117-212 (1960); H. Meerwein, *Houben-Weyl* **6/3,** 189 (1965). *Cf.* Favorskii-Babayan Synthesis; Nef Synthesis.

10. Arndt-Eistert Synthesis

F. Arndt, B. Eistert, *Ber.* **68,** 200 (1935).

Procedure for converting an acid to its next higher homolog:

Application to carbohydrate synthesis: J. Dyong *et al., Ber.* **112,** 161 (1979); to unsaturated diazoketones: T. Hudlicky, J. P. Sheth, *Tetrahedron Letters* **1979,** 2667. Alternative reagent for diazomethane: T. Aoyama, T. Shioiri, *ibid.* **21,** 4461 (1980). *Reviews:* W. E. Bachmann, W. S. Struve, *Org. React.* **1,** 38 (1942); B. Eistert in *Newer Methods in Preparative Organic Chemistry* **vol. 1** (Interscience, New York, 1948) pp 513-570; G. B. Gill, *Comp. Org. Syn.* **3,** 888-889 (1991). *Cf.* Wolff Rearrangement.

11. Auwers Synthesis

K. v. Auwers *et al., Ber.* **41,** 4233 (1908); **48,** 85 (1915); **49,** 809 (1916); K. v. Auwers, P. Pohl, *Ann.* **405,** 243 (1914).

Expansion of coumarones to flavonols by treatment of 2-bromo-2-(α-bromobenzyl)coumarones with alcoholic alkali:

T. H. Minton, H. Stephen, *J. Chem. Soc.* **121,** 1598 (1922); J. Kalff, R. Robinson, *ibid.* **127,** 1968 (1925); B. H. Ingham *et al., ibid.* **1931,** 895; B. G. Acharya *et al., ibid.* **1940,** 817; S. Wawzonek, *Heterocyclic Compounds* **2,** 245 (1951). *Cf.* Algar-Flynn-Oyamada Reaction.

12. Baeyer-Drewson Indigo Synthesis

A. Baeyer, V. Drewson, *Ber.* **15,** 2856 (1882).

Formation of indigos by an aldol reaction, *q.v.,* of *o*-nitrobenzaldehydes to acetone, pyruvic acid or acetaldehyde. Of interest mainly as a method of protecting *o*-nitrobenzaldehydes:

K. Venkataraman, *Chemistry of Synthetic Dyes* **2,** 1008 (New York, 1952); M. Sainsbury, *Rodd's Chemistry of Carbon Compounds* **IVB,** 346, 353 (1977).

13. Baeyer-Villiger Reaction
A. Baeyer, V. Villiger, *Ber.* **32,** 3625 (1899); **33,** 858 (1900).
The oxidation of ketones to esters or lactones by peracids:

Reviews: P. A. S. Smith in *Molecular Rearrangements* Part 1, P. de Mayo, Ed. (Wiley-Interscience, New York, 1963) pp 577-591; J. B. Lee, B. C. Uff, *Quart. Rev. Chem. Soc.* **21,** 429-457 (1967); C. H. Hassall, *Org. React.* **9,** 73 (1957); G. R. Krow, *ibid.* **43,** 251-798 (1993); *idem, Comp. Org. Syn.* **7,** 671-688 (1991).

14. Baker-Venkataraman Rearrangement
W. Baker, *J. Chem. Soc.* **1933,** 1381; H. S. Mahal, K. Venkataraman, *ibid.* **1934,** 1767.
Base-catalyzed rearrangement of *o*-acyloxyketones to β-diketones, important intermediates in the synthesis of chromones and flavones:

K. Venkataraman in Zechmeister, *Progress in the Chemistry of Organic Natural Products* **vol. 17** (1959) p 2; Gripenberg in *The Chemistry of Flavonoid Compounds,* Geissman, Ed. (New York, 1962) p 410; P. K. Jain *et al., Synthesis* **1982,** 221; H. Henecka, *Houben-Weyl* **7/2a,** 531 (1973); **8,** 580 (1952). *Cf.* Allan-Robinson Reaction; Kostanecki Acylation.

15. Bamberger Rearrangement
E. Bamberger, *Ber.* **27,** 1347, 1548 (1894).
Intermolecular rearrangement of *N*-phenylhydroxylamines in aqueous acid to give the corresponding 4-aminophenols:

Early review: H. J. Shine, *Aromatic Rearrangements* (Elsevier, New York, 1967) pp 182-190. Kinetic and mechanistic study: G. Kohnstam *et al., Perkin Trans. II* **1984,** 423. Synthetic application: D. Johnston, D. Elder, *J. Labelled Compd. Radiopharm.* **25,** 1315 (1988). *Cf.* Hofmann-Martius Rearrangement.

16. Bamford-Stevens Reaction; Shapiro Reaction
W. R. Bamford, T. S. Stevens, *J. Chem. Soc.* **1952,** 4735.
Formation of olefins by base-promoted decomposition of *p*-toluenesulfonylhydrazones of aldehydes and ketones:

The formation of unrearranged alkenes, generally the less substituted isomers, by treatment of ketone derived *p*-toluenesulfonylhydrazones with alkyl lithium reagents is known as the **Shapiro reaction:** R. H. Shapiro, M. J. Heath, *J. Am. Chem. Soc.* **89,** 5734 (1967). Use of *N,N*-diethylaminosulfonylhydrazones: J. Kang *et al., Bull. Korean Chem. Soc.* **13,** 192 (1992).
Silicon directing effect: T. K. Sarkar, B. K. Ghorai, *Chem. Commun.* **1992,** 1184. *Reviews:* R. H. Shapiro, *Org.*

React. **23,** 405-507 (1976); R. M. Adlington, A. G. M. Barrett, *Accts. Chem. Res.* **16,** 55-59 (1983); K. Maruka, H. Ya-mamoto, *Comp. Org. Syn.* **6,** 776-779 (1991); A. R. Chamberlin, D. J. Sall, *ibid.* **8,** 944-949.

17. Barbier(-type) Reaction
P. Barbier, *C. R. Acad. Sci.* **128,** 110 (1899).
One-step procedure for the preparation of alcohols from organic halides and aldehydes or ketones:

$$R\!-\!X \quad + \quad \text{(ketone)} \quad \xrightarrow{M} \quad \text{(alcohol)} \qquad \begin{array}{l} M = Mg, Li, Sm(II), Zn \\ R = alkyl, aryl, vinyl, allyl \\ X = Cl, Br, I \end{array}$$

Review of mechanistic studies of Sm-mediated coupling: D. P. Curran *et al., Synlett* **1992,** 943-961. Book: C. Blomberg, *The Barbier Reaction and Related One-Step Processes,* K. Hafner *et al.,* Eds. (Springer-Verlag, New York, 1993) 183 pp. Zn-promoted coupling: F. Hong *et al., Chem. Commun.* **1994,** 289. Sm-mediated coupling: M. Kunishima *et al., Chem. Pharm. Bull.* **42,** 2190 (1994). Comparison with Ni(0) insertion chemistry for intramolecular cyclization: M. Kihara *et al., Tetrahedron* **48,** 67 (1992). *Cf.* Grignard Reaction.

18. Barbier-Wieland Degradation
H. Wieland, *Ber.* **45,** 484 (1912); P. Barbier, R. Locquin, *Compt. Rend.* **156,** 1443 (1913).
Stepwise carboxylic acid degradation of aliphatic acids (particularly in sterol side chains) to the next lower homolog. The ester is converted to a tertiary alcohol that is dehydrated with acetic anhydride, and the olefin oxidized with chromic acid to a lower homologous carboxylic acid:

$$RCH_2COOCH_3 \xrightarrow[\text{2. HX}]{\text{1. 2PhMgX}} RCH_2COHPh_2 \xrightarrow{Ac_2O} RCH\!=\!CHPh_2 \xrightarrow{CrO_3} RCOOH + Ph_2CO$$

H. Wieland *et al., Z. Physiol. Chem.* **161,** 80 (1926); C. W. Shoppee, *Ann. Repts.* (Chem. Soc., London) **44,** 184 (1947); W. Baker *et al., J. Chem. Soc.* **1958,** 1007; J. R. Dias, R. Ramachandra, *Tetrahedron Letters* **1976,** 3685. *Cf.* Krafft Degradation; Miescher Degradation.

19. Bart Reaction; Scheller Modification
H. Bart, **Ger.** pat. **250,264** (1910); **254,092** (1910); **264,924** (1910); **268,172** (1912); *Ann.* **429,** 55 (1922); F. Scheller, **Brit.** pat. **261,026;** A. W. Ruddy *et al., J. Am. Chem. Soc.* **64,** 828 (1942).
Formation of aromatic arsonic acids by treating aromatic diazonium compounds with alkali arsenites in the presence of cupric salts or powdered silver or copper; in the **Scheller modification** primary aromatic amines are diazotized in the presence of arsenious chloride and a trace of cuprous chloride:

$$PhN_2^+Cl^- + Na_3AsO_3 \longrightarrow PhAsO_3Na_2 + NaCl + N_2$$

$$PhNH_2 + HNO_2 + AsCl_3 + (H_2O) \longrightarrow PhAsO_3H_2 + N_2 + 3HCl$$

The modified Bart reaction can be applied to the formation of arylstibonic acids:

$$ArNH_2 + HNO_2 + SbCl_3 + (H_2O) \longrightarrow ArSbO_3H_2 + N_2 + 3HCl$$

C. F. Hamilton, J. F. Morgan, *Org. React.* **2,** 415 (1944); G. O. Doak, H. G. Steinman, *J. Am. Chem. Soc.* **68,** 1987 (1946); K. H. Saunders, *Aromatic Diazo-Compounds and Their Technical Applications* (London, 1949) p 330; W. A. Cowdry, D. S. Davies, *Quart. Rev.* **6,** 363 (1952).

20. Barton Deoxygenation (Barton-McCombie Reaction)
D. H. R. Barton, S. W. McCombie, *Perkin Trans. I* **1975,** 1574.
Deoxygenation of alcohols *via* their thiocarbonyl derivatives which undergo free radical scission upon treatment with tri-*n*-butyltin hydride:

$$\underset{R^1}{\overset{S}{\big\|}}_{O\!-\!R^2} \xrightarrow{n\text{-}Bu_3SnH} \left[\underset{R^1}{\overset{S\!-\!Snn\text{-}Bu_3}{\big|}}_{O\!-\!R^2} \longrightarrow \underset{R^1}{\overset{S\!-\!Snn\text{-}Bu_3}{\big\|}}_{O} + \cdot R^2 \right] \xrightarrow{n\text{-}Bu_3SnH} R^2H$$

$$R^1 = H, CH_3, SCH_3, Ph, OPh, imidazolyl; R^2 = alkyl$$

Mechanistic study: J. E. Forbes, S. Z. Zard, *Tetrahedron Letters* **30,** 4367 (1989). *Review:* M. Pereyre *et al., Tin in Organic Synthesis* (Butterworths, Boston, 1987) pp 84-96. Review of methodological improvements, particularly the replacement of tri-*n*-butyltin hydride with silicon hydrides: C. Chatgilialoglu, C. Ferreri, *Res. Chem. Intermed.* **19,** 755-775 (1993).

21. Barton Olefin Synthesis (Barton-Kellogg Reaction)

D. H. R. Barton *et al., Chem. Commun.* **1970,** 1226; R. M. Kellogg, S. Wassenaar, *Tetrahedron Letters* **1970,** 1987; R. M. Kellogg *et al., ibid.* 4689.

Olefin synthesis by two-fold extrusion of nitrogen and sulfur from a Δ³-1,3,4-thiadiazoline intermediate. Particularly applicable to the synthesis of moderately hindered *tetra*-substituted ethylenes:

R = Bu, Ph, OCH₂CH₃, N(CH₂CH₃)₃

Scope and limitations: D. H. R. Barton *et al., Perkin Trans. I* **1974,** 1794. Synthetic applications: A. P. Schaap, G. R. Faler, *J. Org. Chem.* **38,** 3061 (1973); L. K. Bee *et al., ibid.* **40,** 2212 (1975); M. D. Bachi *et al., Tetrahedron Letters* **1978,** 4167; J. E. McMurry *et al., J. Am. Chem. Soc.* **106,** 5018 (1984). *Cf.* McMurry Reaction.

22. Barton Reaction

D. H. R. Barton *et al., J. Am. Chem. Soc.* **82,** 2640 (1960); **83,** 4076 (1961).

Conversion of a nitrite ester to a γ-oximino alcohol by photolysis involving the homolytic cleavage of a nitrogen-oxygen bond followed by hydrogen abstraction:

M. Akhtar, *Advan. Photochem.* **2,** 263 (1964); R. H. Hesse, *Advan. Free Radical Chem.* **3,** 83 (1969); J. Kalvoda, *Angew. Chem. Int. Ed.* **8,** 525 (1964). Mechanism: D. H. R. Barton *et al., Perkin Trans. I* **1979,** 1159.

23. Baudisch Reaction

O. Baudisch *et al., Naturwiss.* **27,** 768, 769 (1939); *Science* **92,** 336 (1940); *J. Am. Chem. Soc.* **63,** 622 (1941).

Synthesis of *o*-nitrosophenols from benzene or substituted benzenes, hydroxylamine and hydrogen peroxide in the presence of copper salts:

K. Maruyama *et al., Tetrahedron Letters* **1966,** 5889; *J. Org. Chem.* **32,** 2516 (1967); *Bull. Chem. Soc. Japan* **44,** 3120 (1971); W. Seidenfaden, *Houben-Weyl* **10/1,** 1025, 1027 (1971).

24. Baylis-Hillman Reaction

A. B. Baylis, M. E. D. Hillman, **Ger.** pat. **2,155,113;** *eidem,* **U.S.** pat. **3,743,669** (1972, 1973 both to Celanese).

Coupling of activated vinyl systems with aldehydes, catalyzed by 1,4-diazabicyclo[2.2.2]octane (DABCO), to yield α-hydroxyalkylated or -arylated products:

EWG = COOR, COR, CN, SO₂R, CONR₂

Scope and limitations/mechanistic studies: Y. Fort *et al., Tetrahedron* **48,** 6371 (1992); E. L. M. van Rozendaal *et al., ibid.* **49,** 6931 (1993). Rate enhancement study: J. Augé *et al., Tetrahedron Letters* **35,** 7947 (1994). Use of chiral auxillary: S. E. Drewes *et al., Syn. Commun.* **23,** 1215 (1993). Synthetic applications: *idem et al., ibid.* 2807; P. Perlmutter, T. D. McCarthy, *Aust. J. Chem.* **46,** 253 (1993). *Review:* S. E. Drewes, G. H. P. Roos, *Tetrahedron* **44,** 4653-4670 (1988).

25. Béchamp Reduction

A. J. Béchamp, *Ann. Chim. Phys.* **42**(3), 186 (1854).

Reduction of aromatic nitro compounds to the corresponding amines by iron, ferrous salts or iron catalysts in aqueous acid:

$$ArNO_2 + 2Fe + 6HCl \longrightarrow ArNH_2 + 2H_2O + 2FeCl_3$$

J. Werner, *Ind. Eng. Chem.* **40**, 1575 (1948); **41**, 1841 (1949); S. Yagi *et al.*, *Bull. Chem. Soc. Japan* **29**, 194 (1956); A. Courtin, *Helv. Chim. Acta* **62**, 2280 (1980). *Reviews:* C. S. Hamilton, J. F. Morgan, *Org. React.* **2**, 428 (1944); R. Schröter, *Houben-Weyl* **11/1**, 394-409 (1957).

26. Beckmann Rearrangement; Beckmann Fragmentation

E. Beckmann, *Ber.* **19**, 988 (1886).

Acid-mediated isomerization of oximes to amides. Oximes of cyclic ketones give ring enlargements:

Certain oximes, particularly those having a quaternary carbon *anti* to the hydroxyl, are likely to undergo the **Beckmann fragmentation** to form nitriles instead of amides:

Application to steroidal oximes: P. Catsoulacos, D. Catsoulacos, *J. Heterocyclic Chem.* **30**, 1 (1993). *Reviews:* L. G. Donaruma, W. Z. Heldt, *Org. React.* **11**, 1-156 (1960); R. E. Gawley, *ibid.* **35**, 1-420 (1988); C. G. McCarty in *The Chemistry of the Carbon-Nitrogen Double Bond*, S. Patai, Ed. (Interscience, New York, 1970) pp 408-439; J. R. Hauske, *Comp. Org. Syn.* **1**, 98-100 (1991); K. Maruoka, H. Yamamoto, *ibid.* **6**, 763-775; D. Craig, *ibid.* **7**, 689-702. *Cf.* Schmidt Reaction; Tiemann Rearrangement.

27. Bénary Reaction

E. Bénary, *Ber.* **63**, 1573 (1930); **64**, 2543 (1931).

Action of Grignard reagents on enamino ketones or aldehydes yields β substituted α,β-unsaturated ketones or aldehydes:

T. Cuvigny, H. Normant, *Bull. Soc. Chim. France* **1960**, 515. Use of lithio derivatives instead of Grignard reagents: C. Jutz, *Ber.* **91**, 1867 (1958). Mechanism: A. Pasteur *et al.*, *Bull. Soc. Chim. France* **1965**, 2328.

28. Benkeser Reduction

R. A. Benkeser *et al.*, *J. Am. Chem. Soc.* **74**, 5699 (1952); **77**, 3230 (1955).

Reduction of aromatic and olefinic compounds with lithium or calcium and low molecular weight amines to monounsaturated olefins, as well as the fully reduced products. The extent of reduction and selectivity can be controlled by varying the reaction conditions:

$$R = CH_3, CH_2CH_3, CH_2CH_2CH_3, CH_2CH_2NH_2$$

Selectivity study: R. A. Benkeser *et al*, *Tetrahedron Letters* no.16, 1 (1960). Comparative review: E. M. Kaiser, *Synthesis* **1972**, 391-415 *passim*. Scope and limitations: R. A. Benkeser *et al.*, *J. Org. Chem.* **48**, 2796 (1983). Synthetic applications: C. Eaborn *et al.*, *Perkin Trans. I* **1975**, 475; R. Eckrich, D. Kuck, *Synlett* **1993**, 344. *Cf.* Birch Reduction.

29. Benzidine Rearrangement; Semidine Rearrangement

A. W. Hofmann, *Proc. Roy. Soc. London* **12**, 576 (1863); P. Jacobson *et al.*, *Ber.* **26**, 688 (1893).

The acid-catalyzed rearrangement of hydrazobenzenes to 4,4'-diaminobiphenyls. If the hydrazobenzene contains a *para* substituent, the product is a *p*-aminodiphenylamine (**Semidine rearrangement**):

D. L. H. Williams, *Comprehensive Chemical Kinetics* **vol. 13,** C. H. Bamford, C. F. H. Tipper, Eds. (Elsevier, New York, 1972) pp 437-448; R. A. Cox, E. Buncel, *The Chemistry of Hydrazo, Azo and Azoxy Groups,* pt. 2, S. Patai, Ed. (Wiley, New York, 1975) pp 775-807; H. J. Shine *et al., J. Am. Chem. Soc.* **103,** 955 (1981).

30. Benzilic Acid Rearrangement (Benzil-Benzilic Acid Rearrangement)
J. Liebig, *Ann.* **25,** 27 (1838); N. Zinin, *ibid.* **31,** 329 (1939).

Base-induced rearrangement of benzil to benzylic acid *via* phenyl group migration. More commonly perceived to include the migrations of other groups in α-dicarbonyl compounds:

$$C_6H_5-\overset{O}{\overset{||}{C}}-\overset{O}{\overset{||}{C}}-C_6H_5 \xrightarrow{KOH} C_6H_5-\overset{C_6H_5}{\underset{OH}{\overset{|}{C}}}-COO^-K^+$$

Reviews: S. Selman, J. F. Eastham, *Quart. Rev.* **14,** 221 (1960); D. J. Cram, *Fundamentals of Carbanion Chemistry* (Academic Press, 1965) pp 238-243; G. B. Gill, *Comp. Org. Syn.* **3,** 821-838 (1991).

31. Benzoin Condensation
A. J. Lapworth, *J. Chem. Soc.* **83,** 995 (1903); **85,** 1206 (1904).

Cyanide-catalyzed condensation of aromatic aldehydes to give benzoins (acyloins):

$$ArCHO + Ar'CHO \xrightarrow{KCN} ArCHOHCOAr'$$

H. Staudinger, *Ber.* **46,** 3530, 3535 (1913); W. S. Ide, J. S. Buck, *Org. React.* **4,** 269 (1948); H. Herlinger, *Houben-Weyl* **7/2a,** 653 (1973); A. Hassner, K. M. L. Rai, *Comp. Org. Syn.* **1,** 541-577 (1991). *Cf.* Acyloin Condensation.

32. Bergius Process
F. Bergius, *Gas World* **58,** 490 (1913); **Brit.** pat. **18,232** (1914).

Formation of petroleum-like hydrocarbons by hydrogenation of coal at high temperatures and pressures (*e.g.,* 450°C and 300 atm) with or without catalysts; production of toluene by subjecting aromatic naphthas to cracking temperatures at 100 atm with a low partial pressure of hydrogen in the presence of a catalyst. B. T. Brooks, *The Chemistry of the Nonbenzenoid Hydrocarbons* (New York, 1950) p 115; McGraw-Hill *Encyclopedia of Science and Technology* **vol. 2** (New York, 1960) p 166; R. M. Baldwin in Kirk-Othmer *Encyclopedia of Chemical Technology* **vol. 6** (Wiley, New York, 4th ed., 1993) p 569. *Cf.* Fischer-Tropsch Syntheses.

33. Bergman Reaction
R. R. Jones, R. G. Bergman, *J. Am. Chem. Soc.* **94,** 660 (1972); R. G. Bergman, *Accts. Chem. Res.* **6,** 25 (1973).

The cyclizaton of enediynes to generate 1,4-benzenoid diradicals:

Application to ring annulation: J. W. Grissom *et al., Tetrahedron* **50,** 4635 (1994). Kinetic study: *idem et al., J. Org. Chem.* **59,** 5833 (1994). Reaction energetics: E. Kraka, D. Cremer, *J. Am. Chem. Soc.* **116,** 4929 (1994). Reviews of enediyne chemistry and its application to the development of antitumor agents: K. C. Nicolaou *et al., Proc. Natl. Acad. Sci. USA* **90,** 5881-5888 (1993); K. Nicolaou, *Chem. Brit.* **1994,** 33-36, 41.

34. Bergmann Azlactone Peptide Synthesis
M. Bergmann *et al., Ann.* **449,** 277 (1926).

Conversion of an acetylated amino acid and an aldehyde into an azlactone with an alkylene side chain, reaction with a second amino acid with ring opening and formation of an acylated unsaturated dipeptide, followed by catalytic hydrogenation and hydrolysis to the dipeptide:

$$C_6H_5HC \underset{N \quad O}{\diagdown} + \begin{array}{c} COOH \\ | \\ CH(NH_2) \\ | \\ CH_2 \\ | \\ CH_2COOH \end{array} \longrightarrow C_6H_5HC = \underset{\underset{NHCOCH_3}{|}}{\overset{\overset{COOH}{|}}{C}} CONHCHCH_2CH_2COOH \xrightarrow{H_2}$$

(with CH$_3$ on the oxazolone ring)

$$C_6H_5CH_2\underset{\underset{NHCOCH_3}{|}}{CH}CONHCHCH_2CH_2COOH \xrightarrow{HCl} C_6H_5CH_2\underset{\underset{NH_2}{|}}{CH}CONHCHCH_2CH_2COOH \quad \overset{COOH}{|}$$

J. S. Fruton, *Advan. Protein Chem.* **V,** 15 (1949); S. Archer in *Amino Acids and Proteins,* D. M. Greenberg, Ed. (Thomas, Springfield, Ill., 1951) p 181; H. D. Springall, *The Structural Chemistry of Proteins* (New York, 1954) p 29; E. Baltazzi, *Quart. Rev. (London)* **10,** 235 (1956). *Cf.* Erlenmeyer-Plöchl Azlactone and Amino Acid Synthesis.

35. Bergmann Degradation
M. Bergmann, *Science* **79,** 439 (1934).

Stepwise degradation of polypeptides involving benzoylation, conversion to azides and treatment of the azides with benzyl alcohol; this treatment yields, *via* rearrangement to isocyanates, carbobenzoxy compounds which undergo catalytic hydrogenation and hydrolysis to the amide of the degraded peptide:

$$R'CONHCHRCON_3 \xrightarrow{C_6H_5CH_2OH} R'CONHCHRNHCOOCH_2C_6H_5 \xrightarrow[H_2O]{H_2}$$

$$R'CONH_2 + RCHO + CO_2 + C_6H_5CH_3 + NH_3$$

M. Bergmann, L. Zervas, *J. Biol. Chem.* **113,** 341 (1936); H. D. Springall, *The Structural Chemistry of Proteins* (New York, 1954) p 321. *Cf.* Curtius Rearrangement.

36. Bergmann-Zervas Carbobenzoxy Method
M. Bergmann, L. Zervas, *Ber.* **65,** 1192 (1932).

Formation of the *N*-carbobenzoxy derivative of an amino acid for use in peptide synthesis and liberation of the amino group at an appropriate stage of synthesis by hydrogenolysis of the labile carbon-oxygen bond:

$$C_6H_5CH_2OCOCl + H_2NCHRCOOH \longrightarrow C_6H_5CH_2OCONHCHRCOOH \xrightarrow[H_2NCHR'COOCH_3]{Ac_2O}$$

$$C_6H_5CH_2OCONHCHRCONHCHR'COOCH_3 \xrightarrow{H_2} C_6H_5CH_3 + HOOCNHCHRCONHCHR'COOCH_3 \xrightarrow{-CO_2}$$

$$H_2NCHRCONHCHR'COOCH_3$$

C. L. A. Schmidt, *The Chemistry of the Amino Acids and Proteins* (Thomas, Springfield, Ill., 1944) p 262; S. Archer in *Amino Acids and Proteins,* D. M. Greenberg, Ed. (Charles C. Thomas, Springfield, Ill., 1951) p 177; G. W. Kenner, *J. Chem. Soc.* **1956,** 3689; T. W. Greene, *Protective Groups in Organic Synthesis* (Wiley, New York, 1981) p 239. *Cf.* Fischer Peptide Synthesis.

37. Bernthsen Acridine Synthesis
A. Bernthsen, *Ann.* **192,** 1 (1878); **224,** 1 (1884).

Formation of 5-substituted acridines by heating diarylamines in organic acids or anhydrides, usually in the presence of zinc chloride:

A. Albert, *The Acridines* (London, 1951) p 67; A. Albert, *Heterocyclic Compounds* **4,** 502 (1952); N. P. Buu-Hoi *et al., J. Chem. Soc.* **1955,** 1082; R. M. Acheson in *The Chemistry of Heterocyclic Compounds,* A. Weissberger, Ed., *Acridines* (Interscience, New York, 1956) pp 19-25; F. D. Popp, *J. Org. Chem.* **27,** 2658 (1962).

38. Betti Reaction
M. Betti, *Gazz. Chim. Ital.* **30 II,** 301 (1900); **33 II,** 2 (1903); F. Pirrone, *ibid.* **66,** 518 (1936); **67,** 529 (1937).
   The reaction of aromatic aldehydes, primary aromatic or heterocyclic amines and phenols leading to α-aminobenzylphenols:

$$R—NH_2 \quad + \quad R'—CHO \quad +$$

   J. P. Phillips, *Chem. Rev.* **56,** 286 (1956); J. P. Phillips, E. M. Barrall, *J. Org. Chem.* **21,** 692 (1956). *Review:* J. P. Phillips, Leach, *Trans. Kentucky Acad. Sci.* **24**(3-4), 95 (1964). *Cf.* Mannich Reaction.

39. Biginelli Reaction
P. Biginelli, *Ber.* **24,** 1317, 2962 (1891); **26,** 447 (1893).
   Synthesis of tetrahydropyrimidinones by the acid-catalyzed condensation of an aldehyde, a β-keto ester and urea:

$$CH_3COCH_2COOC_2H_5 \quad + \quad RCHO \quad + \quad H_2NCONH_2 \longrightarrow$$

   H. E. Zaugg, W. B. Martin, *Org. React.* **14,** 88 (1965); D. J. Brown, *The Pyrimidines* (Wiley, New York, 1962) p 440; *ibid.,* Suppl. I, **1970,** p 326; F. Sweet, Y. Fissekis, *J. Am. Chem. Soc.* **95,** 8741 (1973).

40. Birch Reduction
A. J. Birch, *J. Chem. Soc.* **1944,** 430; **1945,** 809; **1946,** 593; **1947,** 102, 1642; **1949,** 2531.
   Reduction of aromatic rings by means of alkali metals in liquid ammonia to give mainly unconjugated dihydro derivatives:

R = electron-donating   R = electron-withdrawing

   Reviews: A. J. Birch, H. Smith, *Quart. Rev. (London)* **12,** 17 (1958); D. Caine, *Org. React.* **23,** 1-258 (1976); P. W. Rabideau, Z. Marcinow, *ibid.* **42,** 1-334 (1992); J. M. Hook, L. N. Mander, *Nat. Prod. Rep.* **3,** 35-85 (1986); L. N. Mander, *Comp. Org. Syn.* **8,** 489-521 (1991). *Cf.* Benkeser Reduction.

41. Bischler-Möhlau Indole Synthesis
A. Bischler *et al., Ber.* **25,** 2860 (1892); **26,** 1336 (1893); R. Möhlau, *ibid.* **14,** 171 (1881); **15,** 2480 (1882); E. Fischer, T. Schmitt, *ibid.* **21,** 1071 (1888).
   Formation of 2-substituted indoles by heating ω-halogeno- or ω-hydroxy-ketones with excess aniline *via* cyclization of the intermediate 2-arylaminoketone:

   P. L. Julian *et al., Heterocyclic Compounds* **3,** 22 (1952); R. J. Sundberg, *The Chemistry of Indoles* (Academic Press, New York, 1970) p 164; R. K. Brown in *The Chemistry of Heterocyclic Compounds,* A. Weissberger, E. C. Taylor, Eds., *Indoles, Part I,* W. J. Houlihan, Ed. (Wiley, New York, 1972) p 317.

42. Bischler-Napieralski Reaction
A. Bischler, B. Napieralski, *Ber.* **26,** 1903 (1893).
   Cyclodehydration of β-phenethylamides to 3,4-dihydroisoquinoline derivatives by means of condensing agents such as phosphorous pentoxide or zinc chloride:

W. M. Whaley, T. R. Govindachari, *Org. React.* **6,** 74 (1951); T. Kametani *et al., Tetrahedron* **27,** 5367 (1971); G. Fodor *et al., Angew. Chem. Int. Ed.* **11,** 919 (1972); G. Fodor, S. Nagubandi, *Tetrahedron* **36,** 1279 (1980); *eidem, Heterocycles* **15,** 165 (1981). Review of enantioselective modifications: M. O. Rozwadowska, *ibid.* **39,** 903-931 (1994). *Cf.* Bradsher Reaction; Pechmann Condensation; Pictet-Gams Isoquinoline Synthesis; Pictet-Hubert Reaction; Skraup Reaction.

43. Blaise Ketone Synthesis; Blaise-Maire Reaction
E. E. Blaise, A. Koehler, *Bull. Soc. Chim.* [4] **7,** 215 (1910); E. E. Blaise, M. Maire, *Compt. Rend.* **145,** 73 (1907); E. E. Blaise, *Bull. Soc. Chim.* [4] **9,** 1 (1911).

Formation of ketones by treatment of acid halides with organozinc compounds; the use of β-hydroxy carbonyl chlorides to give β-hydroxy ketones, convertible into α,β-unsaturated ketones in boiling dilute sulfuric acid, is known as the **Blaise-Maire reaction:**

$$RZnCl + R'COCl \longrightarrow RCOR' + ZnCl_2$$

$$AcOCH_2CHRCOCl + R'ZnCl \longrightarrow AcOCH_2CHRCOR' \longrightarrow HOCH_2CHRCOR' \longrightarrow H_2C{=}CRCOR'$$

J. Cason, *Chem. Rev.* **40,** 17 (1947); D. A. Shirley, *Org. React.* **8,** 29 (1954).

44. Blaise Reaction
E. E. Blaise, *Compt. Rend.* **132,** 478 (1901).

Formation of β-oxoesters by treatment of α-bromocarboxylic esters with zinc in the presence of nitriles. The intermediate organozinc compound reacts with the nitrile and the complex is hydrolyzed with 30% potassium hydroxide:

$$R'R''CBrCOOC_2H_5 + Zn \longrightarrow R'R''C(ZnBr)COOC_2H_5 \xrightarrow{RCN}$$

$$R'R''C(CR{=}NZnBr)COOC_2H_5 \xrightarrow{aq KOH} RCOCR'R''COOC_2H_5 + NH_3 + Zn(OH)Br$$

A. Horeau, J. Jacques, *Bull. Soc. Chim.* **1947,** Mem. 58; J. Cason *et al., J. Org. Chem.* **18,** 1594 (1953); H. Henecka, *Houben-Weyl* **7/2a,** 518 (1973); K. Nützel, *ibid.* **13/2a,** 829. *Cf.* Reformatsky Reaction.

45. Blanc (Chloromethylation) Reaction
G. Blanc, *Bull. Soc. Chim. France* [4], **33,** 313 (1923).

Introduction of the chloromethyl group into aromatic rings on treatment with formaldehyde and hydrogen chloride in the presence of zinc chloride:

$$ArH + CH_2O + HCl \xrightarrow{ZnCl_2} ArCH_2Cl$$

Reviews: R. C. Fuson, C. H. McKeever, *Org. React.* **1,** 63 (1942); G. Olah, W. S. Tolgyesi in *Friedel-Crafts and Related Reactions* **vol. II,** Part 2, G. Olah, Ed. (Interscience, New York, 1963) pp 659-784. *Cf.* Quelet Reaction.

46. Blanc Reaction-Blanc Rule
H. G. Blanc, *Compt. Rend.* **144,** 1356 (1907).

Cyclization of dicarboxylic acids on heating with acetic anhydride to give either cyclic anhydrides or ketones depending on the respective positions of the carboxyl groups: 1,4- and 1,5-diacids yield anhydrides, while diacids in which the carboxy groups are in 1,6 or further removed positions yield ketones:

H. Kwart, K. King in *The Chemistry of Carboxylic Acids and Esters,* J. Patai, Ed. (Interscience, London, 1969) p 362; K. D. Bode, *Houben-Weyl* **7/2,** 640 (1973). *Cf.* Ruzicka Large Ring Synthesis.

47. Bodroux-Chichibabin Aldehyde Synthesis

F. Bodroux, *Compt. Rend.* **138,** 92 (1904); A. E. Chichibabin, *Ber.* **37,** 186, 850 (1904).

Formation of aldehydes by treatment of orthoformates with Grignard reagents:

$$HC(OC_2H_5)_3 + RMgX \longrightarrow RCH(OC_2H_5)_2 \xrightarrow[H_2O]{H^+} RCHO$$

L. I. Smith *et al., J. Org. Chem.* **6,** 437, 489 (1941); H. W. Post, *The Chemistry of the Aliphatic Orthoesters* (New York, 1943) p 96; H. Meerwein, *Houben-Weyl* **6/3,** 243 (1965); R. H. DeWolfe, *Carboxylic Orthoacid Derivatives* (Academic Press, New York, 1970) p 224. *Cf.* Bouveault Aldehyde Synthesis.

48. Bodroux Reaction

F. Bodroux, *Bull. Soc. Chim. France* **33,** 831 (1905), **35,** 519 (1906); **1,** 912 (1907); *Compt. Rend.* **138,** 1427 (1904); **140,** 1108 (1905); **142,** 401 (1906).

Formation of substituted amides by reaction of a simple aliphatic or aromatic ester with an aminomagnesium halide obtained by treatment of a primary or secondary amine with a Grignard reagent at room temperature:

$$RCOOR' + IMgNHR'' \longrightarrow IMgOCR(OR')NHR'' \xrightarrow{IMgNHR''} RCONHR''$$

H. L. Bassett, C. R. Thomas, *J. Chem. Soc.* **1954,** 1188; K. Nützel, *Houben-Weyl* **13/2a,** 278 (1973).

49. Bogert-Cook Synthesis

M. T. Bogert, *Science* **77,** 289 (1933); J. W. Cook, C. L. Hewett, *J. Chem. Soc.* **1933,** 1098.

Condensation of β-phenylethylmagnesium bromide with cyclohexanones followed by cyclodehydration of the tertiary alcohol with concentrated sulfuric acid with formation of octahydrophenanthrene derivatives and a small amount of spiran:

L. F. Fieser, M. Fieser, *Natural Products Related to Phenanthrene* (New York, 1949) p 90. For a general approach to the synthesis of phenanthrenoid compounds, *see* D. A. Evans *et al., J. Am. Chem. Soc.* **99,** 7083 (1977).

50. Bohn-Schmidt Reaction

R. Bohn, **Ger. pat.** **46,654** (1889); R. E. Schmidt, **Ger. pat.** **60,855** (1891).

Hydroxylation of anthraquinones containing at least one hydroxyl group by treatment with fuming sulfuric acid or sulfuric acid and boric acid in the presence of a catalyst such as mercury:

Reviews: Fieser, Fieser, *Organic Chemistry* (New York, 1956) p 903; M. Phillips, *Chem. Rev.* **6,** 168 (1929). Studies and proposed mechanism: J. Winkler, W. Jenny, *Helv. Chim. Acta* **48,** 119 (1965).

51. Boord Olefin Synthesis

L. C. Swallen, C. E. Boord, *J. Am. Chem. Soc.* **52,** 651 (1930); **53,** 1505 (1931); **55,** 3293 (1933); H. B. Dykstra *et al., ibid.* **52,** 3396 (1930).

Regiospecific synthesis of olefins from aldehydes and Grignard reagents:

C. Niemann, C. D. Wagner, *J. Org. Chem.* **7**, 227 (1942); P. Bandart, *Bull. Soc. Chim.* **11**, 336 (1944); L. Crombie, *Quart. Rev. (London)* **6**, 131 (1952); M. Schlosser, *Houben-Weyl* **5/1b**, 213 (1972).

52. Borsche-Drechsel Cyclization

E. Drechsel, *J. Prakt. Chem.* **38**(2), 69 (1858); W. Borsche, M. Feise, *Ber.* **20**, 378 (1904).

Formation of carbazole by acid-catalyzed rearrangement of cyclohexanone phenylhydrazone to tetrahydrocarbazole followed by oxidation:

N. Campbell, B. M. Barclay, *Chem. Rev.* **40**, 361 (1947); W. Freudenberg, *Heterocyclic Compounds* **3**, 298 (1952); P. Bruck, *J. Org. Chem.* **35**, 2222 (1970). *Cf.* Bucherer Carbazole Synthesis; Fischer Indole Synthesis; Piloty-Robinson Synthesis.

53. Bouveault Aldehyde Synthesis

L. Bouveault, *Bull. Soc. Chim. France* **31**, 1306, 1322 (1904).

Action of Grignard reagents on *N,N*-disubstituted formamides yields aldehydes:

L. I. Smith, J. Nichols, *J. Org. Chem.* **6**, 489 (1941); J. Sicé *J. Am. Chem. Soc.* **75**, 3697 (1953); E. R. H. Jones *et al.*, *J. Chem. Soc.* **1958**, 1054. Use of lithio derivatives instead of Grignard reagents: E. A. Evans, *Chem. & Ind. (London)* **1957**, 1596. *Cf.* Bodroux-Chichibabin Aldehyde Synthesis.

54. Bouveault-Blanc Reduction

L. Bouveault, G. Blanc, *Compt. Rend.* **136**, 1676 (1903); *Bull. Soc. Chim. France* [3] **31**, 666 (1904).

Formation of alcohols by reduction of esters with sodium and an alcohol:

$$RCOOCH_3 \xrightarrow[R'OH]{Na} RCH_2OH + CH_3OH$$

H. O. House, *Modern Synthetic Reactions* (W. A. Benjamin, Menlo Park, California, 2nd ed., 1972) p 150.

55. Boyland-Sims Oxidation

E. Boyland *et al., J. Chem. Soc.* **1953**, 3623; E. Boyland, P. Sims, *ibid.* **1954**, 980.

Alkaline persulfate oxidation of aromatic amines to yield predominantly the *o*-amino aryl sulfates. Acid-catalyzed hydrolysis generates the *o*-hydroxy aryl amines:

Regioselectivity/mechanistic study: E. J. Behrman, *J. Org. Chem.* **57**, 2266 (1992). *Review: idem, Org. React.* **35**, 421-511 (1988). *Cf.* Elbs Persulfate Oxidation.

56. Bradsher Reaction

C. K. Bradsher, *J. Am. Chem. Soc.* **62**, 486 (1940).

Acid-catalyzed cyclodehydration of *o*-acyldiarylmethanes to anthracene derivatives:

Extension to an *o*-acyldiaryl ether: H. Ishibashi *et al., Tetrahedron* **50**, 10215 (1994). *Review:* C. K. Bradsher, *Chem. Rev.* **87**, 1277-1297 (1987). *Cf.* Bischler-Napieralski Reaction.

57. Brook Rearrangement

A. G. Brook, *J. Am. Chem. Soc.* **80,** 1886 (1958); *idem et al., ibid.* **81,** 981 (1959).

Base-catalyzed silicon migration from carbon to oxygen in α-, β- and γ-silyl alcohols, yielding silyl ethers:

$$base = NR_3, NaOH, Na, Na/K, NaH$$
$$n = 1,2,3$$

Early review: A. G. Brook, *Accts. Chem. Res.* **7,** 77-84 (1974). Synthetic applications: H. J. Reich *et al., J. Am. Chem. Soc.* **112,** 5609 (1990); K. Takeda *et al., Synlett* **1993,** 841; I. Fleming, U. Ghosh *et al., Perkin Trans. I* **1994,** 257.

58. Bucherer-Bergs Reaction

H. T. Bucherer, H. T. Fischbeck, *J. Prakt. Chem.* **140,** 69 (1934); H. T. Bucherer, W. Steiner, *ibid.* 291; H. Bergs, **Ger.** pat. **566,094** (1929).

Preparation of hydantoin from carbonyl compounds by reaction with potassium cyanide and ammonium carbonate, or from the corresponding cyanohydrin and ammonium carbonate:

Application to the synthesis of α-amino acids: E. Ware, *Chem. Rev.* **46,** 422 (1950); H. Wieland *et al., Houben Weyl* **11/2,** 371 (1958); Y. Maki *et al., Chem. Pharm. Bull.* **21,** 2466 (1973); A. Commeyras *et al., Tetrahedron* **36,** 2649 (1980); G. Grunewald *et al., J. Med. Chem.* **23,** 754 (1980). *Cf.* Strecker Amino Acid Synthesis; Urech Cyanohydrin Method.

59. Bucherer Carbazole Synthesis

H. T. Bucherer, F. Seyde, *J. Prakt. Chem.* **77**(2), 403 (1908).

Formation of carbazoles from naphthols or naphthylamines, aryl hydrazines and sodium bisulfite:

Reviews: N. L. Drake, *Org. React.* **1,** 114 (1942); E. Enders, *Houben-Weyl* **10/2,** 250 (1967). *Cf.* Borsche-Drechsel Cyclization.

60. Bucherer Reaction

H. T. Bucherer, *J. Prakt. Chem.* [2] **69,** 49 (1904); R. Lepetit, *Bull. Soc. Ind. Mulhouse* **1903,** 326.

Reversible formation of β-naphthylamine from β-naphthol and aqueous ammonium sulfite or bisulfite *via* intermediate formation of tetralonesulfonic and tetraloneiminosulfonic acids:

N. L. Drake, *Org. React.* **1,** 105 (1942); H. Seeboth, *Angew. Chem. Int. Ed.* **6,** 307 (1967); M. S. Gibson in *The Chemistry of the Amino Group,* S. Patai, Ed. (Interscience, London, 1968) p 37; Z. Allan *et al., Tetrahedron Letters* **1969,** 4855.

61. Buchner-Curtius-Schlotterbeck Reaction

E. Buchner, T. Curtius, *Ber.* **18,** 2371 (1885); F. Schlotterbeck, *Ber.* **40,** 479 (1907); **42,** 2559 (1909).

Formation of ketones from aldehydes and aliphatic diazo compounds; ethylene oxides may also be formed:

$$RCHN_2 + R'CHO \longrightarrow RCH_2COR' + N_2$$

B. Eistert in *Newer Methods of Preparative Organic Chemistry,* English Ed. (New York, 1948) p 521; C. D. Gutsche, *Org. React.* **8,** 364 (1954); J. B. Bastus, *Tetrahedron Letters* **1963,** 955.

62. Buchner Method of Ring Enlargement

E. Buchner, *Ber.* **29,** 106 (1896); E. Buchner, K. Schottenhammer, *Ber.* **53,** 865 (1920).

Diazoacetic acid ester reacts with benzene and homologs to give the corresponding esters of noncaradienic acid, transformed at high temperatures to derivatives of cycloheptatriene and phenylacetic acid:

W. von F. Doering, L. H. Knox, *J. Am. Chem. Soc.* **79,** 352 (1957); W. Kirmse, *Carbene Chemistry* (Academic Press, New York, 2nd ed., 1971); A. F. Noels *et al., J. Org. Chem.* **46,** 873 (1981). *Cf.* Pfau-Plattner Azulene Synthesis.

63. Camps Quinoline Synthesis

R. Camps, *Ber.* **22,** 3228 (1899); *Arch. Pharm.* **237,** 659 (1899); **239,** 591 (1901).

Formation of hydroxyquinolines from *o*-acylaminoacetophenones in alcoholic sodium hydroxide. The relative proportions of the isomeric products are mainly determined by the acyl residue on the amino nitrogen:

R. H. F. Manske, *Chem. Rev.* **30,** 127 (1942); B. Witkop *et al., J. Am. Chem. Soc.* **73,** 2641 (1951); J. Bornstein *et al., ibid.* **76,** 2760 (1954); R. C. Elderfield, *Heterocyclic Compounds* **4,** 60 (1952).

64. Cannizzaro Reaction

S. Cannizzaro, *Ann.* **88,** 129 (1853); K. List, H. Limpricht, *Ann.* **90,** 180 (1854).

Base-catalyzed dismutation of aromatic or aliphatic aldehydes with no α-hydrogen into the corresponding acids and alcohols. When the aldehydes are not identical the reaction is called the "crossed Cannizarro reaction":

$$2RCHO \xrightarrow{\text{NaOH}} RCOO^-Na^+ + RCH_2OH$$

T. A. Geissman, *Org. React.* **2,** 94 (1944); F. P. B. Van der Maeden *et al., Rec. Trav. Chim. Pays-Bas* **91**(2), 221 (1972); C. G. Swain *et al., J. Am. Chem. Soc.* **101,** 3576 (1979); R. S. McDonald, C. E. Sibley, *Can. J. Chem.* **59,** 1061 (1981). *Cf.* Meerwein-Ponndorf-Verley Reduction; Oppenauer Oxidation; Tishchenko Reaction.

65. Carroll Rearrangement

M. F. Carroll, *J. Chem. Soc.* **1940,** 704; **1941,** 507; W. Kimel, A. C. Cope, *J. Am. Chem. Soc.* **65,** 1992 (1943).

Preparation of γ,δ-unsaturated ketones by base-catalyzed reaction of allylic alcohols with β-ketoesters or thermal rearrangement of allyl acetoacetates:

Detailed experimental: S. R. Wilson, C. E. Augelli, *Org. Syn.* **68,** 210 (1990). Synthetic applications: A. V. Echavarren *et al., Tetrahedron Letters* **32,** 6421 (1991); N. Ouvrard *et al., ibid.* **34,** 1149 (1993). *Cf.* Claisen Rearrangement.

66. Castro-Stephens Coupling (Stephens-Castro Coupling, Castro Reaction)

C. E. Castro, R. D. Stephens, *J. Org. Chem.* **28,** 2163 (1963); R. D. Stephens, C. E. Castro, *ibid.* 3313; A. M. Sladkov *et al., Bull. Acad. Sci. USSR, Div. Chem. Sci.* **1963,** 2043.

The coupling of cuprous acetylides with aryl halides to yield arylacetylenes:

$$ArX + Cu-C{\equiv}C-R \xrightarrow[\Delta]{\text{pyridine}} Ar-C{\equiv}C-R$$

$$X = I, Br, Cl$$
$$R = alkyl, aryl, vinyl$$

Synthetic applications: J. D. Kinder *et al., Synlett* **1993,** 149; E. Defranq *et al., Helv. Chim. Acta* **76,** 425 (1993). Early reviews: G. H. Posner, *Org. React.* **22,** 253-400 *passim* (1975); A. M. Sladkov, I. R. Gol'ding, *Russ. Chem. Rev.* **48,** 868-896 (1979). *Cf.* Glaser Coupling.

67. Chapman Rearrangement

O. Mumm *et al., Ber.* **48,** 379 (1915); A. W. Chapman, *J. Chem. Soc.* **127,** 1992 (1925); **1927,** 174; **1929,** 569.

Thermal rearrangement of aryl imidates to *N,N*-diaryl amides:

$$\underset{\underset{OAr''}{|}}{Ar'-C}=NAr''' \xrightarrow{\Delta} Ar'-\underset{\underset{O}{\|}}{C}-N\underset{Ar'''}{\overset{Ar''}{<}}$$

J. W. Schulenberg, S. Archer, *Org. React.* **14,** 1 (1965); C. G. McCarty in *The Chemistry of the Carbon-Nitrogen Double Bond,* S. Patai, Ed. (Interscience, New York, 1970) p 439; C. G. McCarty, L. A. Garner in *The Chemistry of Amidines and Imidates,* S. Patai, Ed. (Interscience, New York, 1975) p 189.

68. Chichibabin Pyridine Synthesis

A. E. Chichibabin, *J. Russ. Phys. Chem. Soc.* **37,** 1229 (1906); *J. Prakt. Chem.* **107,** 122 (1924).

Condensation of carbonyl compounds with ammonia or amines under pressure to form pyridine derivatives; the reaction is reversible and produces different pyridine derivatives along with byproducts:

$$3RCH_2CHO + NH_3 \longrightarrow$$

M. M. Sprung, *Chem. Rev.* **26,** 301 (1940); R. L. Frank, R. P. Seven, *J. Am. Chem. Soc.* **71,** 2629 (1949); H. S. Mosher, *Heterocyclic Compounds* **1,** 456 (1950); J. A. Gautier, J. Renault, *Bull. Soc. Chim. France* **1955,** 588; C. P. Farley, E. L. Eliel, *J. Am. Chem. Soc.* **78,** 3477 (1956); A. T. Soldatenkov, *Zh. Org. Khim.* **16,** 188 (1980). *Cf.* Hantzsch (Dihydro)Pyridine Synthesis; Kröhnke Pyridine Synthesis.

69. Chichibabin Reaction

A. E. Chichibabin, O. A. Zeide, *J. Russ. Phys. Chem. Soc.* **46,** 1216 (1914), *C.A.* **9,** 1901 (1915).

Amination of pyridines and other heterocyclic nitrogen compounds with alkali-metal amides:

$$\text{pyridine} + NaNH_2 \xrightarrow{100-200°} \text{2-aminopyridine}$$

M. T. Leffler, *Org. React.* **1,** 19 (1942); H. S. Mosher, *Heterocyclic Compounds* **1,** 405 (1950); A. F. Pozharskii *et al., Russ. Chem. Rev.* **47,** 1042 (1978); H. J. W. van den Haak *et al., J. Org. Chem.* **46,** 2134 (1981). *Cf.* Kröhnke Pyridine Synthesis.

70. Chugaev Reaction

L. Chugaev (Tschugaeff), *Ber.* **32,** 3332 (1899).

Formation of olefins from alcohols without rearrangement through pyrolysis of the corresponding xanthates *via cis*-elimination:

$$RCH_2CH_2OH + NaOH + CS_2 \longrightarrow RCH_2CH_2OCSSNa \xrightarrow{CH_3I}$$

$$RCH_2CH_2OCSSCH_3 \xrightarrow{\Delta} RCH=CH_2 + COS + CH_3SH$$

C. H. DePuy, R. W. King, *Chem. Rev.* **60,** 444 (1960); H. R. Nace, *Org. React.* **12,** 57 (1962); K. Harano, T. Taguchi, *Chem. Pharm. Bull. Japan* **20,** 2357 (1972). *Cf.* Cope Elimination Reaction.

71. Ciamician-Dennstedt Rearrangement

G. L. Ciamician, M. Dennstedt, *Ber.* **14,** 1153 (1881).

Expansion of the pyrrole ring by heating with chloroform or other halogeno compounds in alkaline solution. The intermediate dichlorocarbene, by addition to the pyrrole, forms an unstable dihalogenocyclopropane which rearranges to a 3-halogenopyridine:

A. H. Corwin, *Heterocyclic Compounds* **1,** 309 (1950); H. S. Mosher, *ibid.* 475; P. S. Skell, R. S. Sandler, *J. Am. Chem. Soc.* **80,** 2024 (1958); E. Vogel, *Angew. Chem.* **72,** 8 (1960).

72. Claisen Condensation (Acetoacetic Ester Condensation)

L. Claisen, O. Lowman, *Ber.* **20,** 651 (1887).

Base-catalyzed condensation of an ester containing an α-hydrogen atom with a molecule of the same ester or a different one to give β-keto esters:

$$CH_3COOC_2H_5 + CH_3COOC_2H_5 \xrightarrow{C_2H_5O^-} CH_3COCH_2COOC_2H_5 + C_2H_5OH$$

C. R. Hauser, B. E. Hudson, *Org. React.* **1,** 266-322 (1942); H. O. House, *Modern Synthetic Reactions* (W. A. Benjamin, Menlo Park, California, 2nd ed., 1972) pp 734-746; J. F. Garst, *J. Chem. Ed.* **56,** 721 (1979); J. E. Bartmess *et al., J. Am. Chem. Soc.* **103,** 1338 (1981); B. R. Davis, P. J. Garratt, *Comp. Org. Syn.* **2,** 795-805 (1991). *Cf.* Dieckmann Reaction.

73. Claisen Rearrangement; Eschenmoser-Claisen Rearrangement; Johnson-Claisen Rearrangement; Ireland-Claisen Rearrangement

L. Claisen, *Ber.* **45,** 3157 (1912), L. Claisen, E. Tietze, *ibid.* **58,** 275 (1925); **59,** 2344 (1926).

Highly stereoselective [3,3]-sigmatropic rearrangement of allyl vinyl or allyl aryl ethers to yield γ,δ-unsaturated carbonyl compounds or *o*-allyl substituted phenols, respectively:

When R' = NR₂, the reaction is referred to as the **Eschenmoser-Claisen rearrangement**: A. E. Wick *et al., Helv. Chim. Acta* **47,** 2425 (1964); M. Lautens *et al., Tetrahedron Letters* **31,** 5829 (1990); B. Coates *et al., ibid.* **32,** 4199 (1991).

When R' = OR, the reaction is referred to as the **Johnson-Claisen rearrangement**: W. S. Johnson *et al., J. Am. Chem. Soc.* **92,** 741 (1970); R. Bao *et al., Synlett* **1992,** 217; D. Basavaiah, S. Pandiaraju, *Tetrahedron Letters* **36,** 757 (1995).

When R' = OSiR₃ or OLi, the reaction is referred to as the **Ireland-Claisen rearrangement**: R. E. Ireland, R. H. Mueller, *J. Am. Chem. Soc.* **94,** 5897 (1972); R. E. Ireland *et al., J. Org. Chem.* **56,** 650 (1991); *idem et al., ibid.* 3572; K. Hattori, H. Yamamoto, *Tetrahedron* **50,** 3099 (1994).

Inclusive reviews: S. J. Rhoads, N. R. Raulins, *Org. React.* **22,** 1-252 (1975); F. E. Ziegler, *Chem. Rev.* **88,** 1423-1452 (1988); P. Wipf, *Comp. Org. Syn.* **5,** 827-873 (1991). *Cf.* Carroll Rearrangement; Cope Rearrangement; Overman Rearrangement.

74. Claisen-Schmidt Condensation

L. Claisen, A. Claparède, *Ber.* **14,** 2460 (1881); J. G. Schmidt, *ibid.* 1459.

Condensation of an aromatic aldehyde with an aliphatic aldehyde or ketone in the presence of a relatively strong base (hydroxide or alkoxide ion) to form an α,β-unsaturated aldehyde or ketone:

A. T. Nielsen, W. J. Houlihan, *Org. React.* **16,** 1 (1968); H. O. House, *Modern Synthetic Reactions* (W. A. Benjamin, Menlo Park, California, 2nd ed., 1972) pp 632-639; J. A. Fine, P. Pulaski, *J. Org. Chem.* **38,** 1747 (1973). *Cf.* Aldol Reaction.

75. Clemmensen Reduction

E. Clemmensen, *Ber.* **46,** 1837 (1913); **47,** 51, 681 (1914).

Reduction of carbonyl groups of aldehydes and ketones to methylene groups with zinc amalgam and hydrochloric acid:

$$RCOR' \xrightarrow[HCl]{Zn(Hg)} RCH_2R'$$

E. L. Martin, *Org. React.* **1,** 155 (1942); M. Smith in *Reduction,* R. L. Augustine, Ed. (M. Dekker, New York, 1968) pp 95-170; W. Reusch, *ibid.* pp 186-194; J. G. St. C. Buchanan, P. D. Woodgate, *Quart. Rev.* **23,** 522 (1969); D. Straschewski, *Angew. Chem.* **71,** 726 (1959); E. Vedejs, *Org. React.* **22,** 401 (1975); S. Yamamura, S. Nishiyama, *Comp. Org. Syn.* **8,** 309-313 (1991). *Cf.* Haworth Phenanthrene Synthesis; Wolff-Kishner Reduction.

76. Combes Quinoline Synthesis

A. Combes, *Bull. Soc. Chim. France* **49,** 89 (1888).

Formation of quinolines by condensation of β-diketones with primary arylamines followed by acid-catalyzed ring closure of the intermediate Schiff base:

W. S. Johnson, F. J. Matthews, *J. Am. Chem. Soc.* **66,** 210 (1944); F. W. Bergstrom, *Chem. Rev.* **35,** 156 (1944); J. C. Perche *et al., Perkin Trans. I* **1972,** 260; J. Born, *J. Org. Chem.* **37,** 3952 (1972). *Cf.* Conrad-Limpach Reaction; Doebner Reaction.

77. Conrad-Limpach Reaction

M. Conrad, L. Limpach, *Ber.* **20,** 944 (1887); **24,** 2990 (1891).

Thermal condensation of arylamines with β-ketoesters followed by cyclization of the intermediate Schiff bases to 4-hydroxyquinolines:

R. H. Manske, *Chem. Rev.* **30,** 121 (1942); R. H. Reitsema, *ibid.* **43,** 47 (1948); H. Henecka, *Chemie der Beta Dicarbonylverbindungen* (Berlin, 1950) p 307; R. C. Elderfield, *Heterocyclic Compounds* **4,** 30 (1952); J. M. Barker *et al., J. Chem. Res. (S)* **1980,** 4. *Cf.* Combes Quinoline Synthesis; Doebner Reaction.

78. Cope Elimination Reaction

A. C. Cope *et al., J. Am. Chem. Soc.* **71,** 3929 (1949); *idem et al., ibid.* **75,** 3212 (1953).

Formation of an olefin and a hydroxylamine by pyrolysis of an amine oxide:

Early reviews: C. H. DePuy, R. W. King, *Chem. Rev.* **60,** 448 (1960); A. C. Cope, E. R. Trumbull, *Org. React.* **11,** 317-493 *passim* (1960). Synthetic application: E. Tojo *et al., Heterocycles* **27,** 2367 (1988). Mechanistic study: R. D. Bach, M. L. Braden, *J. Org. Chem.* **56,** 7194 (1991). Methods development: A. D. Woolhouse, *J. Heterocyclic Chem.* **30,** 873 (1993). Synthetic applications of the reverse reaction (retro-Cope elimination): E. Ciganek, *J. Org. Chem.* **55,** 3007 (1990); M. B. Gravestock *et al., Chem. Commun.* **1993,** 169. *Cf.* Chugaev Reaction; Hofmann Degradation.

79. Cope Rearrangement; Oxy-Cope Rearrangement

A. C. Cope *et al., J. Am. Chem. Soc.* **62,** 441 (1940).

Highly stereoselective [3,3]-sigmatropic rearrangement of 1,5-dienes; "all-carbon" equivalent of the Claisen rearrangement, *q.v.*:

When R = OH, the transformation is referred to as the **oxy-Cope rearrangement:** J. Berson, M. Jones, *ibid.* **86,** 5019 (1964).

Reviews: S. J. Rhodds, N. R. Raulins, *Org. React.* **22,** 1-252 (1975); S. R. Wilson, *ibid.* **43,** 93-250 *passim* (1993); R. K. Hill, *Comp. Org. Syn.* **5,** 785-826 (1991). Review of hetero-Cope rearrangements: S. Blechert, *Synthesis* **1989,** 71-82. Brief review of synthetic applications: K. Durairaj, *Curr. Sci.* **66,** 917-922 (1994).

80. Corey-Kim Oxidation

E. J Corey, C. U. Kim, *J. Am. Chem. Soc.* **94,** 7586 (1972).

Oxidation of primary and secondary alcohols *via* the intermediacy of their alkoxysulfonium salts. Upon the addition of base, the salt rearranges intramolecularly to aldehydes and ketones, respectively:

Application to the synthesis of α-hydroxy ketones: E. J. Corey, C. U. Kim, *Tetrahedron Letters* **1974,** 287; of 1,3-dicarbonyl compounds: S. Katayama *et al., Synthesis* **1988,** 178. *Cf.* Pfitzner-Moffatt Oxidation; Swern Oxidation.

81. Corey-Winter Olefin Synthesis

E. J. Corey, R. A. E. Winter, *J. Am. Chem. Soc.* **85,** 2677 (1963).

Synthesis of olefins from 1,2-diols and thiocarbonyldiimidazole. Treatment of the intermediate cyclic thionocarbonate with trimethylphosphite yields the olefin by *cis*-elimination:

M. Tichy, J. Sicher, *Tetrahedron Letters* **1969,** 4609; T. L. Nagabhushan, *Can. J. Chem.* **48,** 383 (1970); J. Daub *et al., Tetrahedron Letters* **1972,** 4435; S. Hanessian *et al., ibid.* **1978,** 737; I. Dyong *et al., Ber.* **112,** 1849 (1979); E. J. Corey, P. B. Hopkiss, *Tetrahedron Letters* **23,** 1797 (1982).

82. Cornforth Rearrangement

J. W. Cornforth, *The Chemistry of Penicillin* (Princeton University Press, New Jersey, 1949) p 700.

Thermal rearrangement of 4-carbonyl substituted oxazoles to their isomeric oxazoles *via* the postulated dicarbonyl nitrile ylides:

Mechanistic study: M. J. S. Dewar, I. J. Turchi, *J. Am. Chem. Soc.* **96,** 6148 (1974). Scope and limitations: *eidem, J. Org. Chem.* **40,** 1521 (1975). Extension to the synthesis of 5-aminothiazoles: S. L. Corrao *et al., ibid.* **55,** 4484 (1990). Synthetic application: G. L'abbé *et al., Perkin Trans. I* **1993,** 2259.

83. Craig Method

L. C. Craig, *J. Am. Chem. Soc.* **56,** 231 (1934).

Introduction of a halogen into the α-position of aminopyridines by treatment with sodium nitrite in hydrohalic acid, followed by warming:

$$\text{pyridine-NH}_2 \xrightarrow[\text{NO}_2^-]{\text{HX}} \text{pyridine-X}$$

H. S. Mosher, *Heterocyclic Compounds* **1,** 515, 555 (1950); H. E. Mertel in *The Chemistry of Heterocyclic Compounds,* A. Weissberger, Ed., *Pyridine and its Derivatives Part Two,* E. Klingsberg, Ed. (Interscience, New York, 1961) p 334.

84. Criegee Reaction

R. Criegee, *Ber.* **64,** 260 (1931).

Oxidative cleavage of vicinal glycols by lead tetraacetate:

$$\underset{\overset{|}{R}}{\overset{\overset{OH}{|}}{R'-C}}-\underset{\overset{|}{R'''}}{\overset{\overset{OH}{|}}{C-R''}} + Pb(OAc)_4 \longrightarrow R-\overset{O}{\overset{||}{C}}-R' + R''-\overset{O}{\overset{||}{C}}-R''' + Pb(OAc)_2 + 2AcOH$$

Reviews: R. Criegee in *Newer Methods of Preparative Organic Chemistry* **vol. 1** (Interscience, New York, 1948) pp 12-20; C. A. Bunton in *Oxidation in Organic Chemistry* **vol. 1,** K. Wiberg, Ed. (Academic Press, New York, 1965) pp 367-407; H. O. House, *Modern Synthetic Reactions* (W. A. Benjamin, Menlo Park, California, 2nd ed., 1972) pp 359-387; A. S. Perlin, *Advan. Carbohyd. Chem.* **14,** 9 (1959); K. W. Bentley in *Elucidation of Organic Structures by Physical and Chemical Methods* pt. 2, K. W. Bentley, G. W. Kirby, Eds. (Wiley, New York, 2nd ed., 1973) pp 169-177. *Cf.* Malaprade Reaction.

85. Curtius Rearrangement; Curtius Reaction

T. Curtius, *Ber.* **23,** 3023 (1890); *idem, J. Prakt. Chem.* [2] **50,** 275 (1894).

Formation of isocyanates by thermal decomposition of acyl azides:

$$RCON_3 \xrightarrow{\Delta} R-N=C=O$$

The stepwise conversion of a carboxylic acid to an amine having one fewer carbon unit, *via* the azide and isocyanate, is referred to as the **Curtius reaction**:

$$RCOOH \longrightarrow RCON_3 \longrightarrow RNCO \longrightarrow RNH_2$$

Synthetic applications: R. Lo Scalzo *et al., Gazz. Chim. Ital.* **118,** 819 (1988); N. De Kimpe *et al., J. Org. Chem.* **59,** 8215 (1994). *Reviews:* P. A. S. Smith, *Org. React.* **3,** 337-449 (1946); J. H. Saunders, R. J. Slocombe, *Chem. Rev.* **43,** 205 (1948); D. V. Banthorpe in *The Chemistry of the Azido Group,* S. Patai, Ed. (Interscience, New York, 1971) pp 397-405; T. Shioiri, *Comp. Org. Syn.* **6,** 795-828 (1991). *Cf.* Bergmann Degradation; Hofmann Reaction; Lossen Rearrangement; Schmidt Reaction.

86. Dakin Reaction

H. D. Dakin, *Am. Chem. J.* **42,** 477 (1909).

Replacement of the formyl or acetyl groups in phenolic aldehydes or ketones by a hydroxyl group by means of hydrogen peroxide:

$$\text{(4-hydroxybenzaldehyde)} \xrightarrow[\text{NaOH}]{\text{H}_2\text{O}_2} \text{(hydroquinone)} + HCOOH$$

J. E. Leffler, *Chem. Rev.* **45,** 385 (1949); W. M. Schubert, R. R. Kintner in *The Chemistry of the Carbonyl Group,* S. Patai, Ed. (Interscience, New York, 1966) p 749; M. B. Hocking, J. H. Ong, *Can. J. Chem.* **55,** 102 (1977); M. B. Hocking *et al., ibid.* **56,** 2646 (1978).

87. Dakin-West Reaction

H. D. Dakin, R. West, *J. Biol. Chem.* **78,** 91, 745, 757 (1928).

Reaction of α-amino acids with acetic anhydride in the presence of base to give α-acetamido ketones. The reaction occurs *via* the intermediate azlactone:

$$\underset{NH_2}{RCHCOOH} \xrightarrow{Ac_2O} \underset{NHCOCH_3}{RCHCOCH_3}$$

J. W. Cornforth, *Heterocyclic Compounds* **5**, 345 (1957). *Mechanism:* R. Knorr, R. Huisgen, *Ber.* **103**, 2598 (1970); W. Steglich, G. Hofle, *ibid.* **104**, 3644 (1971); G. Hofle *et al., ibid.* **105**, 1718 (1972); N. Allinger *et al., J. Org. Chem.* **39**, 1730 (1974); I. Charles *et al., Perkin Trans. 1* **1980**, 1139.

88. Darzens Condensation (Darzens-Claisen Reaction, Glycidic Ester Condensation)
G. Darzens, *Compt. Rend.* **139**, 1214 (1904); **141**, 766 (1905); **142**, 214 (1906).

Formation of α,β-epoxy esters (glycidic esters) by the condensation of aldehydes or ketones with esters of α-haloacids; the corresponding thermally unstable glycidic acids yield aldehydes or ketones on decarboxylation:

M. S. Newman, B. J. Magerlein, *Org. React.* **5**, 413 (1949); M. Ballester, *Chem. Rev.* **55**, 283 (1955); H. O. House, *Modern Synthetic Reactions* (W. Benjamin, Menlo Park, California, 2nd ed., 1972) pp 666-671; R. F. Borch, *Tetrahedron Letters* **1972**, 3761; S. Colonna *et al., Perkin Trans. 1* **1978**, 8; T. Rosen, *Comp. Org. Syn.* **2**, 409-439 (1991).

89. Darzens-Nenitzescu Synthesis of Ketones
G. Darzens, *Compt. Rend.* **150**, 707 (1910); C. D. Nenitzescu, I. P. Cantuniari, *Ann.* **510**, 269 (1934); C. D. Nenitzescu, C. Cioranescu, *Ber.* **69**, 1820 (1936).

Acylation of olefins with acid chlorides or anhydrides catalyzed by Lewis acids. When performed in the presence of a saturated hydrocarbon the product is the saturated ketone:

G. A. Olah, *Friedel-Crafts and Related Reactions* **vol. 1** (Interscience, New York, 1963) p 129; C. D. Nenitzescu, A. T. Balaban, *ibid.* **vol. 3,** Part 2, 1069 (1964); L. Ötvös *et al., Acta Chimica Acad. Sci. Hungaricae* **71**(2), 193 (1972); H. O. House, *Modern Synthetic Reactions* (W. A. Benjamin, Menlo Park, California, 2nd ed., 1972) p 786; J. K. Groves, *Chem. Soc. Rev.* **1**, 73 (1972). *Cf.* Friedel-Crafts Reaction; Nencki Reaction; Nenitzescu Reductive Acylation.

90. Darzens Synthesis of Tetralin Derivatives
G. Darzens, *Compt. Rend.* **183**, 748 (1926).

Cyclization of α-benzyl-α-allylacetic acid type compounds by moderate heating in concentrated sulfuric acid to yield tetralin derivatives:

E. Bergmann, *Chem. Rev.* **29**, 536 (1941); J. N. Chatterjea *et al., Indian J. Chem.* **20B**, 264 (1981).

91. Delépine Reaction
M. Delépine, *Compt. Rend.* **120**, 501 (1895); **124**, 292 (1897).

Preparation of primary amines by reaction of alkyl halides with hexamethylenetetramine followed by acid hydrolysis of the formed quaternary salts:

$$RCH_2X + C_6H_{12}N_4 \longrightarrow [RCH_2C_6H_{12}N_4]^+X^- \xrightarrow{HX} RCH_2NH_3^+X^-$$

S. J. Angyal, *Org. React.* **8,** 197 (1954); Y. Basace *et al., Bull. Soc. Chim. France* **1971,** 1468; S. N. Quessy *et al., Perkin Trans. I* **1979,** 512.

92. de Mayo Reaction
P. de Mayo *et al., Proc. Chem. Soc. London* **1962,** 119; P. de Mayo, H. Takeshita, *Can. J. Chem.* **41,** 440 (1963).
Synthesis of 1,5-diketones by photoaddition of enol derivatives of 1,3-diketones to olefins, followed by a retro-aldol reaction, *q.v.*:

P. de Mayo, *Accts. Chem. Res.* **4,** 49 (1971); H. Meier, *Houben-Weyl* **4/5b,** 924 (1975); W. Oppolzer, *Pure Appl. Chem.* **53,** 1189 (1981).

93. Demjanov Rearrangement
N. J. Demjanov, M. Lushnikov, *J. Russ. Phys. Chem. Soc.* **35,** 26 (1903); *Chem. Zentr.* **1903,** I, 828.
Deamination of primary amines by diazotization to give rearranged alcohols. In the case of alicyclic amines, ring enlargement or contraction occurs:

P. A. S. Smith, D. R. Baer, *Org. React.* **11,** 157 (1960); H. Stetter, P. Goebel, *Ber.* **96,** 550 (1963); R. Kotani, *J. Org. Chem.* **30,** 350 (1965); V. Dave *et al., Can. J. Chem.* **57,** 1557 (1979); R. K. Murray, Jr., T. M. Ford, *J. Org. Chem.* **44,** 3504 (1979); D. Fattori *et al., Tetrahedron* **49,** 1649 (1993). *Cf.* Tiffeneau-Demjanov Rearrangement; Wagner-Meerwein Rearrangement.

94. Dess-Martin Oxidation
D. B. Dess, J. C. Martin, *J. Org. Chem.* **48,** 4155 (1983).
Mild oxidation of primary and secondary alcohols to aldehydes and ketones, respectively, employing the triace-toxyperiodinane (the "Dess-Martin Periodinane" reagent):

Scope and limitations of fluoroalkyl-substituted carbinols as substrates: R. J. Linderman, D. M. Graves, *J. Org. Chem.* **54,** 661 (1989). Methods development: D. B. Dess, J. C. Martin, *J. Am. Chem. Soc.* **113,** 7277 (1991). Application to the synthesis of 2'- and 3'-ketonucleosides: V. Samano, M. J. Robins, *J. Org. Chem.* **55,** 5186 (1990); of substituted oxazoles: P. Wipf, C. P. Miller, *ibid.* **58,** 3604 (1993). Preparation of the periodinane: R. E. Ireland, L. Liu, *ibid.* 2899.

95. D-Homo Rearrangement of Steroids
L. Ruzicka, H. Meldahl, *Helv. Chim. Acta* **21,** 1760 (1938); **22,** 421 (1939).
Originally discovered in 17β-hydroxy-20-ketosteroids, but thoroughly studied in the 17α-hydroxy-20-keto series, this reaction involves an acid- or base-catalyzed acyloin rearrangement which yields a 6-membered D-ring:

R. B. Turner, *J. Am. Chem. Soc.* **75**, 3484 (1953); D. K. Fukushima *et al., ibid.* **77**, 6585 (1955); N. L. Wendler *et al., ibid.* **78**, 5027 (1956); I. Elphimoff-Felkin, *Bull. Soc. Chim. France* **1956**, 1845; N. L. Wendler *et al., Tetrahedron* **11**, 163 (1960). *Review:* N. L. Wendler in *Molecular Rearrangements* Part 2, P. de Mayo, Ed. (Wiley-Interscience, New York, 1964) p 1114.

96. Dieckmann Reaction

W. Dieckmann, *Ber.* **27**, 102, 965 (1894); **33**, 595, 2670 (1900); *Ann.* **317**, 51, 93 (1901).

Base-catalyzed cyclization of dicarboxylic acid esters to give β-ketoesters, the intramolecular equivalent of the Claisen condensation, *q.v.*:

J. P. Schaefer, J. J. Bloomfield, *Org. React.* **15**, 1-203 (1967); H. O. House, *Modern Synthetic Reactions* (W. A. Benjamin, Menlo Park, California, 2nd ed., 1972) pp 740-743; H. Kwart, K. Sing in *The Chemistry of Carboxylic Acids and Esters,* S. Patai, Ed. (Interscience, New York, 1969) p 341; B. R. Davis, P. J. Garrett, *Comp. Org. Syn.* **2**, 806-829 (1991). *Cf.* Gabriel-Colman Rearrangement.

97. Diels-Alder Reaction

O. Diels, K. Alder, *Ann.* **460**, 98 (1928); **470**, 62 (1929); *Ber.* **62**, 2081, 2087 (1929).

The 1,4-addition of the double bond of a dienophile to a conjugated diene to generate a six-membered ring, such that up to four new stereocenters may be created simultaneously. The [4+2]-cycloaddition usually occurs with high regio- and stereoselectivity:

diene   dienophile

Heteroatomic analogs of the diene (e.g., CHR=CR—CR=O, O=CR—CR=O, and RN=CR—CR=NR) and dienophile (e.g., RN=NR, R₂C=NR, and RN=O) may also serve as reactants.

Early reviews: M. C. Kloetzel, *Org. React.* **4**, 1-59 (1948); H. L. Holmes *ibid.* 60-173; L. W. Butz, A. W. Rytina, *ibid.* **5**, 136-192 (1949). Intermolecular reactions: W. Oppolzer, *Comp. Org. Syn.* **5**, 315-399 (1991). Intramolecular reactions: E. Ciganek, *Org. React.* **32**, 1-374 (1984); W. R. Roush, *Comp. Org. Syn.* **5**, 513-550 (1991). Use of heterodienophiles: S. M. Weinreb, *ibid.* 401-449. Use of nitroso dienophiles: J. Streith, A. DeFoin, *Synthesis* **1994**, 1107-1117. Use of heterodienes: D. L. Boger, *ibid,* 451-512. Review of diastereoselectivity: J. M. Coxon *et al.,* "Diastereofacial Selectivity in the Diels-Alder Reaction" in *Advances in Detailed Reaction Mechanisms* **3**, 131-166 (1994); T. Oh, M. Reilly, *Org. Prep. Proceed. Int.* **26**, 131-158 (1994); H. Waldmann, *Synthesis* **1994**, 535-551. *Cf.* Wagner-Jauregg Reaction.

98. Dienone-Phenol Rearrangement

K. von Auwers, K. Ziegler, *Ann.* **425**, 217 (1921).

Transformation of a 4,4-disubstituted cyclohexadienone into a 3,4-disubstituted phenol upon acid treatment:

Reviews: C. J. Collins, J. F. Eastham in *The Chemistry of the Carbonyl Group,* S. Patai, Ed. (Interscience, New York, 1966) pp 775-778; A. J. Waring, *Adv. Alicyclic Chem.* **1,** 207 (1967); B. Miller in *Mechanisms of Molecular Migrations* vol. **1,** B. S. Thyagarajan, Ed. (Interscience, New York, 1968) pp 275-285; B. Miller, *Accts. Chem. Res.* **8,** 277 (1975).

99. Dimroth Rearrangement
O. Dimroth, *Ann.* **364,** 183 (1909); **459,** 39 (1927).

Rearrangement of *N*-alkylated or arylated iminoheterocycles to the corresponding alkylamino or arylamino heterocycles:

D. J. Brown, J. S. Harper in *Pteridine Chemistry,* W. Pfleiderer, E. C. Taylor, Eds. (Macmillan, New York, 1964) p 219; D. J. Brown in *Mechanisms of Molecular Migrations* vol. **1,** B. S. Thyagarajan, Ed. (Wiley-Interscience, New York, 1968) p 209; D. J. Brown in *The Pyrimidines* Suppl. I (Interscience, New York, 1970) p 287; A. Albert, *Perkin Trans. I* **1973,** 2659; D. J. Brown, K. Lenega, *ibid.* **1974,** 372; D. J. Brown, T. Nagamatsu, *Aust. J. Chem.* **32,** 1585 (1979).

100. Doebner-Miller Reaction; Beyer Method for Quinolines
O. Doebner, W. v. Miller, *Ber.* **16,** 2464 (1883).

Acid-catalyzed synthesis of quinolines from primary aromatic amines and α,β-unsaturated carbonyl compounds. When the latter are prepared *in situ* from two molecules of aldehyde or an aldehyde and methyl ketone, the reaction is known as the **Beyer method for quinolines:**

F. W. Bergström, *Chem. Rev.* **35,** 153 (1944); Y. Ogata *et al., J. Chem. Soc. B* **1969,** 805; G. A. Dauphinee, T. P. Forrest, *J. Chem. Soc. D* **1969,** 327; *Can J. Chem.* **56,** 632 (1978); C. M. Leir, *J. Org. Chem.* **42,** 911 (1977). *Cf.* Gould-Jacobs Reaction; Knorr Quinoline Synthesis.

101. Doebner Reaction
O. Doebner, *Ann.* **242,** 265 (1887); *Ber.* **20,** 277 (1887); **27,** 352, 2020 (1894).

Formation of substituted cinchoninic acids from aromatic amines on heating with aldehydes and pyruvic acid:

F. W. Bergström, *Chem. Rev.* **35,** 156 (1944); R. C. Elderfield, *Heterocyclic Compounds* **4,** 25 (1952); C. Centini, *Rev. Soe. Venez. Quim.* **7**(5), 265 (1970), *C.A.* **74,** 76301x (1971); G. E. Gream, A. K. Serelis, *Aust. J. Chem.* **31,** 863 (1978). *Cf.* Combes Quinoline Synthesis; Conrad-Limpach Reaction.

102. Doering-LaFlamme Allene Synthesis
W. von E. Doering, P. M. LaFlamme, *Tetrahedron* **2,** 75 (1958); **U.S.** pat. **2,933,544** (1960).

Treatment of an olefin with bromoform and an alkoxide to yield the 1,1-dibromocyclopropane which reacts with an active metal to produce an allene:

Reviews: M. Murray, *Houben-Weyl* **5/2a,** 985 (1977); V. Nair, *Comp. Org. Syn.* **4,** 1009-1012 (1991).

103. Dötz Reaction

K. H. Dötz, *Angew. Chem. Int. Ed.* **14,** 644 (1975).

Three component cyclization of an aromatic or vinylic alkoxy pentacarbonyl chromium carbene complex, an alkyne, and carbon monoxide, generating a Cr(CO)₃ coordinated phenol:

Solvent effects: K. S. Chan *et al., J. Organometal. Chem.* **334,** 9 (1987). Methods development: S. Chamberlin *et al., Tetrahedron* **49,** 5531 (1993); S. Chamberlin, W. D. Wulff, *J. Org. Chem.* **59,** 3047 (1994). Synthetic applications: W. D. Wulff *et al., J. Am. Chem. Soc.* **110,** 7419 (1988); D. L. Boger, I. C. Jacobson, *J. Org. Chem.* **56,** 2115 (1991). *Review:* K. H. Dötz, *New J. Chem.* **14,** 433-445 (1990).

104. Duff Reaction

J. C. Duff, E. J. Bills, *J. Chem. Soc.* **1932,** 1987; **1934,** 1305; **1941,** 547; **1945,** 276.

ortho-Formylation of phenols or *para*-formylation of aromatic amines with hexamethylenetetramine in the presence of an acidic catalyst:

L. N. Ferguson, *Chem. Rev.* **38,** 230 (1946); Y. Ogata, F. Sugiura, *Tetrahedron* **24,** 5001 (1968); F. Wada *et al., Bull. Chem. Soc. Japan* **53,** 1473 (1980). *Cf.* Reimer-Tiemann Reaction.

105. Dutt-Wormall Reaction

P. K. Dutt, H. R. Whitehead, A. Wormall, *J. Chem. Soc.* **119,** 2088 (1921); P. K. Dutt, *ibid.* **125,** 1463 (1924).

Preparation of diazoaminosulfinates by reaction of diazonium salts with aryl- or alkylsulfonamides followed by alkaline hydrolysis to yield the corresponding sulfinic acid of the sulfonamide, and the azide:

$$\text{ArN}_2\text{Cl} \xrightarrow{\text{H}_2\text{NSO}_2\text{R}} \text{ArN}\!=\!\text{NNHSO}_2\text{R} \xrightarrow{\text{HO}^-} \text{ArN}_3 + \text{HO}_2\text{SR}$$

H. Bretschneider, H. Rager, *Monatsh.* **81,** 970 (1950); I. G. Laing, *Rodd's Chemistry of Carbon Compounds* **IIIC,** 107 (1973); C. Grundmann, *Houben-Weyl* **10/3,** 808 (1965).

106. Edman Degradation

P. Edman, *Acta Chem. Scand.* **4,** 283 (1950).

Cyclic degradation of peptides based on the reaction of phenylisothiocyanate with the free amino group of the *N*-terminal residue such that amino acids are removed one at a time and identified as their phenylthiohydantoin derivatives:

S. Bösze *et al., J. Chromatog. A* **668,** 345 (1994). *Reviews:* R. A. Laursen *et al., Methods Biochem. Anal.* **26,** 201-284 (1980); R. L. Heinrikson, "The Edman Degradation in Protein Sequence Analysis" in *Biochemical and Biophysi-*

cal Studies of Proteins and Nucleic Acids, T.-B. Lo *et al.,* Eds. (Elsevier, New York, 1984) pp 285-302; K.-K. Han *et al., Int. J. Biochem.* **17,** 429-445 (1985); C. G. Fields *et al., Peptide Res.* **6,** 39-47 (1993).

107. Ehrlich-Sachs Reaction
P. Ehrlich, F. Sachs, *Ber.* **32,** 2341 (1899).

Formation of *N*-phenyl imines by the base-catalyzed condensation of compounds containing active methylene groups with aromatic nitroso compounds; nitrones also may be formed:

$$C_6H_5CH_2CN + C_6H_5NO \longrightarrow \underset{C_6H_5}{\overset{NC}{\diagup}}=NC_6H_5 + H_2O$$

F. Barrow, F. J. Thorneycroft, *J. Chem. Soc.* **1939,** 769; A. McGookin, *J. Appl. Chem.* **5,** 65 (1955); F. Bell, *J. Chem. Soc.* **1957,** 516; D. M. W. Anderson, F. Bell, *ibid.* **1959,** 3708; D. M. W. Anderson, J. L. Duncan, *ibid.* **1961,** 1631; W. Seidenfaden, *Houben-Weyl* **10/1,** 1079 (1971).

108. Einhorn-Brunner Reaction
A. Einhorn *et al., Ann.* **343,** 229 (1905); K. Brunner, *Ber.* **47,** 2671 (1914); *Monatsh.* **36,** 509 (1915).

Formation of substituted 1,2,4-triazoles by acid-catalyzed condensation of hydrazines or semicarbazides with diacyl-amines:

M. R. Atkinson, J. B. Polya, *J. Chem. Soc.* **1952,** 3418; **1954,** 141, 3319; Theilheimer, *Synthetic Methods* **9,** No. 449 (1955); K. T. Potts, *Chem. Rev.* **61,** 103 (1961). *Cf.* Pellizzari Reaction.

109. Elbs Persulfate Oxidation
K. Elbs, *J. Prakt. Chem.* **48,** 179 (1893).

Hydroxylation of phenols to predominately *p*-diphenols by potassium persulfate in alkaline solution:

$$\text{C}_6\text{H}_5\text{—OH} \xrightarrow[\text{OH}^-]{\text{K}_2\text{S}_2\text{O}_8} \text{HO—C}_6\text{H}_4\text{—OH}$$

S. M. Sethna, *Chem. Rev.* **49,** 91 (1951); J. B. Lee, B. C. Uff, *Quart. Rev.* **21,** 453 (1967); E. J. Behrman, *Org. React.* **35,** 421-511 (1988). *Cf.* Boyland-Sims Oxidation.

110. Elbs Reaction
K. Elbs, E. Larsen, *Ber.* **17,** 2847 (1884).

Formation of anthracenes by intramolecular condensation of diaryl ketones containing a methyl or methylene substituent adjacent to the carbonyl group:

L. F. Fieser, *Org. React.* **1,** 129 (1942); G. N. Badger, B. J. Christie, *J. Chem. Soc.* **1956,** 3435; N. P. Buu-Hoi, D. Lavit, *Rec. Trav. Chim.* **76,** 419 (1957); M. Newman, V. K. Khanna, *J. Org. Chem.* **45,** 242 (1980).

111. Emde Degradation
H. Emde, *Ber.* **42,** 2590 (1909); *Ann.* **391,** 88 (1912).

Modification of the Hofmann degradation, *q.v.,* method for reductive cleavage of the carbon-nitrogen bond by treatment of an alcoholic or aqueous solution of a quaternary ammonium halide with sodium amalgam. Also used as a catalytic method with palladium and platinum catalysts. The method succeeds with ring compounds not degraded by the Hofmann procedure:

Reviews: A. Birch, *Quart Rev.* **4**, 78 (1950); *Org. React.* **7**, 143, 278 (1953); F. Möller, *Houben-Weyl* **11/1**, 973 (1955); Z. Spialter, J. A. Pappalardo, *Acyclic Aliphatic Tertiary Amines* (Macmillan, New York, 1965) pp 79-81; K. W. Bentley in *Elucidation of Organic Structures by Physical and Chemical Methods Pt. 2,* K. W. Bentley, G. W. Kirby, Eds. (Wiley, New York, 2nd ed., 1973) pp 296-298.

112. Emmert Reaction
B. Emmert, E. Asendorf, *Ber.* **72**, 1188 (1939); B. Emmert, E. Pirot, *ibid.* **74**, 714 (1941).

Formation of 2-pyridyldialkylcarbinols by condensation of ketones with pyridine or its homologs in the presence of aluminum or magnesium amalgam:

C. H. Tilford *et al., J. Am. Chem. Soc.* **70**, 4001 (1948); H. L. Lochti *et al., ibid.* **75**, 4477 (1953); R. Abramovitch, R. Vinutha, *J. Chem. Soc. C* **1969**, 2104; C. A. Russell *et al., J. Chem. Soc. D* **1970**, 1406; R. Tschesche, W. Führer, *Ber.* **111**, 3502 (1978).

113. Ene Reaction (Alder-Ene Reaction); Conia Reaction
K. Alder *et al., Ber.* **76**, 27 (1943).

The addition of an alkene having an allylic hydrogen (ene) to a compound containing a multiple bond (enophile) to form a new bond between two unsaturated termini, with an allylic shift of the ene double bond, and transfer of the allylic hydrogen to the enophile. The mechanism is related to that of the Diels-Alder reaction, *q.v.*:

enophile – carbonyl and thiocarbonyl compounds, imines, alkenes, alkynes
Lewis acid = BF$_3$ • O(CH$_2$CH$_3$)$_2$, SnCl$_4$, Al(CH$_2$CH$_3$)Cl$_2$, Al(CH$_3$)$_2$Cl

Lewis acid-promoted cyclization of 5-hexenals: J. A. Marshall, *Chemtracts-Org. Chem.* **5**, 1-7 (1992). Review of alkenes as enophiles: B. B. Snider, *Comp. Org. Syn.* **5**, 1-27 (1991). Review of carbonyl compounds as enophiles: *idem, ibid.* **2**, 527-561; in conjunction with asymmetric synthesis: K. Mikami, M. Shimizu, *Chem. Rev.* **92**, 1021-1050 (1992); K. Mikami *et al., Synlett* **1992**, 255-265.

The intramolecular Ene reaction of unsaturated ketones, in which the carbonyl functionality serves as the ene component, *via* its tautomer, and the olefinic moiety serves as the enophile, is known as the **Conia reaction**:

F. Rouessac *et al., Tetrahedron Letters* **1965**, 3319. *Review:* J. M. Conia, P. Le Perchec, *Synthesis* **1975**, 1-19.

114. Erlenmeyer-Plöchl Azlactone and Amino Acid Synthesis
E. Erlenmeyer, *Ann.* **275**, 1 (1893); J. Plöchl, *Ber.* **17**, 1616 (1884).

Formation of azlactones by intramolecular condensation of acylglycines in the presence of acetic anhydride. The reaction of azlactones with carbonyl compounds, followed by hydrolysis and reduction, yields the amino acid; drastic hydrolysis gives the α-oxo acid:

C. L. A. Schmidt, *The Chemistry of the Amino Acids and Proteins* (Springfield, Ill., 1944) p 54; H. E. Carter, *Org. React.* **3**, 198 (1946); M. Crawford, W. T. Little, *J. Chem. Soc.* **1959**, 729; W. Steglich, *Fortschr. Chem. Forsch.* **12**, 84 (1969). *Cf.* Bergmann Azlactone Peptide Synthesis; Perkin Reaction.

115. Eschenmoser Fragmentation (Eschenmoser-Tanabe Fragmentation)

A. Eschenmoser *et al.*, *Helv. Chim. Acta* **50**, 708 (1967); J. Schreiber *et al.*, *ibid.* 2101; M. Tanabe *et al.*, *Tetrahedron Letters* **1967**, 3943.

Cleavage of α,β-epoxyketones under mild conditions, *via* sulfonylhydrazone intermediates, to yield acetylenic and carbonyl compounds:

R⁵ = *p*-toluene, 2,4,6-trimethylbenzene, 2,4-dinitrobenzene
optional catalyst = C_6H_5N, $NaHCO_3$, Na_2CO_3, silica gel

Early review: D. Felix *et al.*, *Helv. Chim. Acta* **54**, 2896-2912 (1971). Synthetic applications: C. B. Reese, H. P. Sanders, *Synthesis* **1981**, 276; W. Dai, J. A. Katzenellenbogen, *J. Org. Chem.* **58**, 1900 (1993); A. Abad *et al.*, *Synlett* **1991**, 787. *Cf.* Grob Fragmentation; Wharton Reaction.

116. Étard Reaction

A. L. Étard, *Compt. Rend.* **90**, 534 (1880); *Ann. Chim. Phys.* **22**, 218 (1881).

Oxidation of an arylmethyl group to an aldehyde by treatment with chromyl chloride:

$$C_6H_5CH_3 \xrightarrow{CrO_2Cl_2} C_6H_5CHO$$

W. H. Hartford, M. Darrin, *Chem. Rev.* **58**, 1 (1958); H. O. House, *Modern Synthetic Reactions* (W. A. Benjamin, Menlo Park, California, 2nd ed., 1972) p 289; C. D. Nenitzescu *et al.*, *Rev. Roum. Chim.* **14**, 1543, 1553 (1969); I. I. Schiketanz *et al.*, *ibid.* **22**, 1097 (1977).

117. Favorskii-Babayan Synthesis

A. Favorskii, *J. Russ. Phys. Chem. Soc.* **37**, 643 (1905); *Chem. Zentr.* **1905**, **II**, 1018; A. Babayan *et al.*, *J. Gen. Chem. (USSR)* **9**, 1631 (1939), *C.A.* **34**, 2788 (1940).

Synthesis of acetylenic alcohols from ketones and terminal acetylenes in the presence of anhydrous alkali:

A. W. Johnson, *The Chemistry of Acetylenic Compounds* vol. 1 (London, 1946) p 14; R. A. Raphael, *Acetylenic Compounds in Organic Synthesis* (New York, 1955) p 10; M. F. Shostakovskii *et al.*, *Zh. Org. Khim.* **4**, 1747 (1968), *C.A.* **70**, 19507n (1969); A. V. Shchelkunov *et al.*, *ibid.* **6**, 930 (1970), *C.A.* **73**, 34609c (1970); Y. M. Vilenchik *et al.*, *ibid.* **14**, 447 (1978), *C.A.* **88**, 169530a (1978). *Cf.* Arens-van Dorp Synthesis; Nef Synthesis.

118. Favorskii Rearrangement; Wallach Degradation

A. E. Favorskii, *J. Prakt. Chem.* **88**(2), 658 (1913); O. Wallach, *Ann.* **414**, 296 (1918).

Base-catalyzed rearrangement of α-haloketones to acids or esters. The rearrangement of α,α'-dibromocyclohexanones to 1-hydroxycyclopentanecarboxylic acids, followed by oxidation to the ketones is known as the **Wallach degradation**:

Detailed experimental procedure: D. W. Goheen, W. R. Vaughan, *Org. Syn.* **coll. vol. 4,** 594 (1963). Application to the synthesis of carboxylic acids: T. Satoh *et al., Bull. Chem. Soc. Japan* **66,** 2339 (1993). Applications to asymmetric synthesis: *idem et al., Tetrahedron Letters* **34,** 4823 (1993); E. Lee, C. H. Yoon, *Chem. Commun.* **1994,** 479. *Reviews:* A. S. Kende, *Org. React.* **11,** 261-316 (1960); P. J. Chenier, *J. Chem. Ed.* **55,** 286 (1978); A. Baretta, B. Waegill, "A Survey of Favorskii Rearrangement Mechanisms" in *Reactive Intermediates,* R. A. Abramovitch, Ed. (Plenum Press, New York, 1982) pp 527-585; J. Mann, *Comp. Org. Syn.* **3,** 839-859 (1991).

119. Feist-Bénary Synthesis

F. Feist, *Ber.* **35,** 1537, 1545 (1902); E. Bénary, *Ber.* **44,** 489, 493 (1911).

Formation of furans from α-halogenated ketones or ethers and 1,3-dicarbonyl compounds in the presence of pyridine. When ammonia is used as the condensing agent pyrrole derivatives are always formed as secondary products:

T. Reichstein, H. Zschokke, *Helv. Chim. Acta* **14,** 1270 (1931); **15,** 268, 1105, 1112 (1932); R. C. Elderfield, T. N. Dodd, *Heterocyclic Compounds* **1,** 132 (1950); J. Kagan, K. C. Mattes, *J. Org. Chem.* **45,** 1524 (1980). Alternative substrate: R. C. Cambie *et al., Syn. Commun.* **20,** 1923 (1990). *Cf.* Hantzsch Pyrrole Synthesis.

120. Fenton Reaction

H. J. H. Fenton, *Proc. Chem. Soc.* **9,** 113 (1893); *J. Chem. Soc.* **65,** 899 (1894).

Oxidation of α-hydroxy acids with hydrogen peroxide and ferrous salts (Fenton's reagent) to α-keto acids or of 1,2-glycols to hydroxy aldehydes:

W. A. Waters in *Organic Chemistry* **vol. 4,** H. Gilman, Ed. (Wiley, New York, 1953) p 1157; G. Sosnovsky, D. Rawlinson in *Organic Peroxides* **vol. 2,** D. Swern, Ed. (Interscience, New York, 1970) pp 269-336; C. Walling, *Accts. Chem. Res.* **8,** 125 (1975); C. Walling, K. Amarnath, *J. Am. Chem. Soc.* **104,** 1185 (1982); T. Tezuka *et al., ibid.* **103,** 3045 (1981).

121. Ferrier Rearrangement

R. J. Ferrier, *Perkin Trans. I* **1979,** 1455.

The stereochemically controlled conversion of hex-5-enopyranosides into cyclohexanones (inosose derivatives), catalyzed by mercury(II) salts, such that the 5-hydroxyl and the 3-substituent of the product are predominantly in a *trans* relationship:

Stereochemical/mechanistic study: A. S. Machado *et al., Carbohyd. Res.* **233,** C5 (1992). Scope and limitations: N. Chida *et al., Bull. Chem. Soc. Japan* **64,** 2118 (1991). Synthetic applications: D. H. R. Barton *et al., Tetrahedron* **46,** 215 (1990); R. Chretien *et al., Nat. Prod. Letters* **2,** 69 (1993).

122. Finkelstein Reaction

H. Finkelstein, *Ber.* **43,** 1528 (1910).

Reaction of alkyl halides with sodium iodide in acetone:

$$RBr + NaI \longrightarrow RI + NaBr$$

C. K. Ingold, *Structure and Mechanism in Organic Chemistry* (Cornell Univ. Press, London, 2nd ed., 1969) p 435; J. Hayami *et al., Tetrahedron Letters* **1973,** 385; S. Samaan, F. Rolla, *Phosphorus and Sulfur* **4,** 145 (1978); W. B. Smith, G. D. Branum, *Tetrahedron Letters* **22,** 2055 (1981).

123. Fischer-Hepp Rearrangement

O. Fischer, E. Hepp, *Ber.* **19,** 2991 (1886).

Rearrangement of secondary aromatic nitrosamines to *p*-nitrosoarylamines:

H. J. Shine, *Aromatic Rearrangements* (Elsevier, New York, 1967) p 231; D. L. H. Williams in *Comprehensive Chemical Kinetics* **vol. 13** (1972) p 454; S. Johal *et al., Perkin Trans. II* **1980,** 165.

124. Fischer Indole Synthesis

E. Fischer, F. Jourdan, *Ber.* **16,** 2241 (1883); E. Fischer, O. Hess, *ibid.* **17,** 559 (1884).

Formation of indoles on heating aryl hydrazones of aldehydes or ketones in the presence of catalysts such as Lewis or proton acids:

B. Robinson, *Chem. Rev.* **63,** 373 (1963); **69,** 227 (1969); W. A. Remers, R. K. Brown in *The Chemistry of Heterocyclic Compounds,* A. Weissberger, Ed., *Indoles, Part I,* W. J. Houlihan, Ed. (Wiley, New York, 1972) pp 232-317; I. I. Grandberg, V. I. Sorokin, *Russ. Chem. Rev.* **43,** 115 (1974); H. Ishii, *Accts. Chem. Res.* **14,** 275 (1981); B. Robinson, *The Fischer Indole Synthesis* (Wiley, New York, 1982) 923 pp. *Cf.* Borsche-Drechsel Cyclization; Piloty-Robinson Synthesis.

125. Fischer Oxazole Synthesis

E. Fischer, *Ber.* **29,** 205 (1896).

Condensation of equimolar amounts of aldehyde cyanohydrins and aromatic aldehydes in dry ether in the presence of dry hydrochloric acid:

R. H. Wiley, *Chem. Rev.* **37,** 410 (1945); J. W. Cornforth, R. H. Cornforth, *J. Chem. Soc.* **1949,** 1028 ; J. W. Cornforth, *Heterocyclic Compounds* **5,** 309 (1957); T. Onaka, *Tetrahedron Letters* **1971,** 4391.

126. Fischer Peptide Synthesis

E. Fischer, *Ber.* **36,** 2982 (1903).

Formation of polypeptides by treatment of an α-chloro or α-bromo acyl chloride with an amino acid ester, hydrolysis to the acid and conversion to a new acid chloride which is again condensed with a second amino acid ester, and so on. The terminal chloride is finally converted to an amino group with ammonia:

C. L. A. Schmidt, *The Chemistry of the Amino Acids and Proteins* (Thomas, Springfield, Ill., 1944) p 257; B. Rockland in *Amino Acids and Proteins,* D. M. Greenberg, Ed. (Charles C. Thomas, Springfield, Ill., 1951) p 232; H. D. Springall, *The Structural Chemistry of Proteins* (New York, 1954) p 24. *Cf.* Bergmann-Zervas Carbobenzoxy Method.

127. Fischer Phenylhydrazine Synthesis
E. Fischer, *Ber.* **8,** 589 (1875).

Formation of arylhydrazines by reduction of diazo compounds with excess sodium sulfite and hydrolysis of the substituted hydrazine sulfonic acid salt with hydrochloric acid. The process is a standard industrial method for production of arylhydrazines:

$$ArN_2^+ + NaSO_3^- \longrightarrow ArN{=}NSO_3Na \xrightarrow{NaHSO_3} ArN(SO_3Na)NHSO_3Na \xrightarrow{HCl} ArNHNH_2$$

G. H. Colemann, *Org. Syn.* **coll. vol. I,** 432 (1932); K. H. Saunders, *The Aromatic Diazo-Compounds and Their Technical Applications* (London, 1949) p 183; R. Huisgen, R. Lux, *Ber.* **93,** 540 (1960).

128. Fischer Phenylhydrazone and Osazone Reaction
E. Fischer, *Ber.* **17,** 579 (1884).

Formation of phenylhydrazones and osazones by heating sugars with phenylhydrazine in dilute acetic acid:

$$\overset{\displaystyle CHO}{\underset{\displaystyle HCOH}{|}} \xrightarrow{C_6H_5NHNH_2} \overset{\displaystyle HC{=}NNHC_6H_5}{\underset{\displaystyle HCOH}{|}} \longrightarrow \overset{\displaystyle HCNHNHC_6H_5}{\underset{\displaystyle COH}{\|}} \xrightarrow{-C_6H_5NH_2}$$

$$\overset{\displaystyle HC{=}NH}{\underset{\displaystyle CO}{|}} \xrightarrow{C_6H_5NHNH_2} \overset{\displaystyle HC{=}NNHC_6H_5}{\underset{\displaystyle C{=}NNHC_6H_5}{|}}$$

E. G. V. Percival, *Advan. Carbohyd. Chem.* **3,** 23 (1948); F. Micheel, *Chemie der Zucker und Polysaccharide* (Leipzig, 1956) p 54; W. Pigman, *The Carbohydrates* **1957,** 452, 455; H. Simon *et al., Fortschr. Chem. Forsch.* **14,** 451 (1970).

129. Fischer-Speier Esterification Method
E. Fischer, A. Speier, *Ber.* **28,** 3252 (1895)

Esterification of acids by refluxing with excess alcohol in the presence of hydrogen chloride or other acid catalysts:

$$CH_3COOH + CH_3CH_2OH \xrightarrow{3\% HCl} CH_3COOCH_2CH_3 + H_2O$$

E. D. Hughes, *Quart. Rev.* **2,** 110 (1948); A. J. Kirby in *Comprehensive Chemical Kinetics* **vol. 10,** C. H. Bamford, C. F. H. Tipper, Eds. (Elsevier, New York, 1972) p 57; E. K. Euranto in *The Chemistry of Carboxylic Acids and Esters,* S. Patai, Ed. (Interscience, New York, 1969) p 505.

130. Fischer-Tropsch Syntheses; Synthol Process; Oxo Synthesis
F. Fischer, H. Tropsch, *Ber.* **56,** 2428 (1923).

Synthesis of hydrocarbons, aliphatic alcohols, aldehydes, and ketones by the catalytic hydrogenation of carbon monoxide using enriched synthesis gas from passage of steam over heated coke. The ratio of products varies with conditions. The high pressure **Synthol process** gives mainly oxygenated products and addition of olefins in the presence of cobalt catalyst (**Oxo synthesis**) produces aldehydes. Normal pressure synthesis leads mainly to petroleum-like hydrocarbons.

C. Masters, *Adv. Organomet. Chem.* **17,** 61 (1979); C. K. Rofer-DePoorter, *Chem. Rev.* **31,** 447 (1981); W. A. Herrmann, *Angew. Chem. Int. Ed.* **21,** 117 (1982). *Cf.* Bergius Process; Oxo Process.

131. Flood Reaction
E. A. Flood, *J. Am. Chem. Soc.* **55,** 1735 (1933).

Formation of trialkylsilyl halides from hexaalkyldisiloxanes using concentrated sulfuric acid in the presence of ammonium chloride or fluoride, or by treatment of the intermediate silane sulfates with hydrogen chloride in the presence of ammonium sulfate:

$$R_3SiOSiR_3 \xrightarrow[H_2SO_4]{NH_4Cl} R_3SiCl$$

$$R_3SiOSiR_3 \xrightarrow{H_2SO_4} (R_3Si)_2SO_2 \xrightarrow[(NH_4)_2SO_4]{HCl} R_3SiCl$$

H. W. Post, *Silicones and Other Organic Silicone Compounds* (New York, 1949) p 64; E. G. Rochow *et al., The Chemistry of Organometallic Compounds* (New York, 1957) p 158, 159. Synthetic applications: L. Birkofer, O. Stuhl, *Top. Curr. Chem.* **88,** 33 (1980).

132. Forster Reaction
M. O. J. Forster, *J. Chem. Soc.* **75,** 934 (1899); H. Decker, P. Becker, *Ann.* **395,** 362 (1913).

Formation of secondary amines by condensation of a primary amine with an aldehyde, addition of alkyl halide to the Schiff base, and subsequent hydrolysis:

$$RNH_2 + C_6H_5CHO \longrightarrow RN{=}CHC_6H_5 \xrightarrow{R'Cl} [RR'N{=}CHC_6H_5]^+Cl^- \xrightarrow{H_2O} RR'NH + C_6H_5CHO$$

H. Glaser, *Houben-Weyl* **11/1,** 108 (1957); F. Möller, *ibid.* p 956.

133. Forster Reaction
M. O. J. Forster, *J. Chem. Soc.* **107,** 260 (1915).

Formation of diazoketones from α-oximinoketones by reaction with chloramine:

M. P. Cava, R. L. Litle, *Chem. & Ind. (London)* **1957,** 367; W. Kirmse *et al., Angew. Chem.* **69,** 106 (1957). Mechanism: J. Meinwald *et al., J. Am. Chem. Soc.* **81,** 4751 (1959). Review and applications: F. Weygand, H. J. Bestmann, *Angew. Chem.* **72,** 535 (1960); W. Rundel, *ibid.* **74,** 469 (1962). Application to steroids: M. P. Cava, B. R. Vogt, *J. Org. Chem.* **30,** 3776 (1965).

134. Franchimont Reaction
A. P. N. Franchimont, *Ber.* **5,** 1048 (1872).

Carboxylic acid dimerization to 1,2-dicarboxylic acids by treating α-bromocarboxylic acids with potassium cyanide followed by hydrolysis and decarboxylation:

N. Zelinsky, *Ber.* **21,** 3160 (1888); O. Poppe, *ibid.* **23,** 113 (1890); R. C. Fuson *et al., J. Am. Chem. Soc.* **51,** 1536 (1929); **52,** 4074 (1930); **60,** 1237 (1938); H. N. Rydon, *J. Chem. Soc.* **1936,** 593; H. Henecka, *Chemie der Beta-Dicarbonylverbindungen* (Berlin, 1950) p 176.

135. Frankland-Duppa Reaction
E. Frankland, *Ann.* **126,** 109 (1863); E. Frankland, B. F. Duppa, *Ann.* **135,** 25 (1865).

Formation of α-hydroxycarboxylic esters by reaction of dialkyl oxalates with alkyl halides in the presence of zinc, or amalgamated zinc, and acid:

$$ROOC{-}COOR + 2R'I + 2Zn + 2HCl \longrightarrow R'R'C(OH)COOR + ROH + ZnI_2 + ZnCl_2$$

E. Krause, A. von Grosse, *Die Chemie der metallorganischen Verbindungen* (Berlin, 1937) p 225; K. Nützel, *Houben-Weyl* **13/2a,** 741 (1973).

136. Frankland Synthesis
E. Frankland, *Ann.* **71,** 213 (1849); **85,** 3641 (1853).

Synthesis of zinc dialkyls from alkyl halides and zinc:

$$2C_2H_5I + 2Zn \longrightarrow (C_2H_5)_2Zn + ZnI_2$$

Reviews: K. Nützel, *Houben-Weyl* **13/2a,** 570 (1973); C. R. Noller, *Org. Syn.* **coll. vol. II,** 184 (1943).

137. Freund Reaction; Gustavson Reaction; Hass Cyclopropane Process
A. Freund, *Monatsh.* **3,** 625 (1882); G. Gustavson, *J. Prakt. Chem.* [2] **36,** 300 (1887); H. B. Hass *et al., Ind. Eng. Chem.* **28,** 1178 (1936).

Formation of alicyclic hydrocarbons by the action of sodium (Freund reaction) or zinc (Gustavson reaction) on open chain dihalo compounds; 1,3-dichloropropane derived from the chlorination of propane obtained from natural gas is cyclized in the Hass cyclopropane process by treating with zinc dust in aqueous alcohol in the presence of catalytic sodium iodide:

H. Gilman, *Organic Chemistry* **I** (New York, 1943) p 74; J. D. Bartleson *et al., J. Am. Chem. Soc.* **68,** 2513 (1946); R. N. Shortsidge *et al., ibid.* **70,** 946 (1948); B. T. Brooks, *The Chemistry of the Nonbenzenoid Hydrocarbons* (New York, 1950) p 88; H. F. Ebel, A. Lüttringhaus, *Houben-Weyl* **13/1,** 492 (1970).

138. Friedel-Crafts Reaction
C. Friedel, J. M. Crafts, *Compt. Rend.* **84,** 1392, 1450 (1877).
The alkylation or acylation of aromatic compounds catalyzed by aluminum chloride or other Lewis acids:

RCOX = acyl halides, anhydrides
RX = alkyl halides, alkenes, alkynes, alcohols

Reviews: C. C. Price, *Org. React.* **3,** 1 (1946); G. A. Olah, *Friedel-Crafts and Related Reactions* **vols. 1-4** (Interscience, New York, 1963-1965); J. K. Groves, *Chem. Soc. Rev.* **1,** 73 (1972); H. Heaney, *Comp. Org. Syn.* **2,** 733-752, 753-768 (1991); **3,** 293-339. Aliphatic version: S. C. Eyley, *ibid.* **2,** 707-731. *Cf.* Darzens-Nenitzescu Synthesis of Ketones; Haworth Phenanthrene Synthesis; Nencki Reaction.

139. Friedlaender Synthesis
P. Friedlaender, *Ber.* **15,** 2572 (1882); P. Friedlaender, C. F. Gohring, *ibid.* **16,** 1833 (1883).
Base-catalyzed condensation of 2-aminobenzaldehydes with ketones to form quinoline derivatives:

Reviews: R. H. Manske, *Chem. Rev.* **30,** 124 (1942); F. W. Bergstrom, *ibid.* **35,** 151 (1944); C. C. Cheng, S. J. Yan, *Org. React.* **28,** 37 (1982). Application to carbocyclic ketones: G. Kempter *et al., Ber.* **97,** 16 (1964); to 5- and 6-membered cyclic ketones containing O, S, or N: G. Kempter, S. Hirschberg, *ibid.* **98,** 419 (1965); K. Rao *et al., J. Heterocyclic Chem.* **16,** 1241 (1979). *Cf.* Niementowski Quinoline Synthesis; Pfitzinger Reaction.

140. Fries Rearrangement
K. Fries, G. Fink, *Ber.* **41,** 4271 (1908); K. Fries, W. Pfaffendorf, *ibid.* **43,** 212 (1910).
Rearrangement of phenolic esters to *o*- and/or *p*-phenolic ketones on heating with aluminum chloride or other Lewis acid catalysts:

and/or

A. H. Blatt, *Org. React.* **1,** 342 (1942); A. Gerecs in *Friedel-Crafts and Related Reactions* vol. **3,** Part 1, G. Olah, Ed. (Interscience, New York, 1964) pp 499-533; F. R. Jensen, G. Goldman in *ibid.* Part 2, p 1349; H. Kwart, K. Sing in *The Chemistry of Carboxylic Acids and Esters,* S. Patai, Ed. (Interscience, New York, 1969) pp 341-373; R. Martin, *Bull. Soc. Chim. France* **1979,** 373; R. Martin *et al., Monatsh.* **81,** 111 (1980). Photo-Fries rearrangement: J. C. Anderson, C. B. Reese, *Proc. Chem. Soc.* **1960,** 217; D. Bellus *et al., Chem. Rev.* **67,** 599 (1967); D. Bellus, *Advan. Photochem.* **8,** 109 (1971); D. Veierov *et al., Perkin Trans. II* **1980,** 1659; D. J. Crouse *et al., J. Org. Chem.* **46,** 374 (1981).

141. Fritsch-Buttenberg-Wiechell Rearrangement
P. Fritsch, *Ann.* **279,** 319 (1894); W. P. Buttenberg, *ibid.* 327; H. Wiechell, *ibid.* 332.
The rearrangement of 1,1-diaryl-2-haloethylenes to diaryl acetylenes with strong bases:

$$Ar'—C\equiv C—Ar$$

G. Köbrich, *Angew. Chem. Int. Ed.* **4,** 49 (1965); G. Köbrich, P. Buck in *Acetylenes,* H. G. Viehe, Ed. (Marcel Dekker, New York, 1969) pp 117, 131; G. Köbrich, *Angew. Chem. Int. Ed.* **11,** 473 (1972); P. J. Stang, D. P. Fox, *J. Org. Chem.* **43,** 364 (1978).

142. Fujimoto-Belleau Reaction

C. I. Fujimoto, *J. Am. Chem. Soc.* **73,** 1856 (1951); B. Belleau, *ibid.* 5441.

Synthesis of cyclic α-substituted α,β-unsaturated ketones from enol lactones and Grignard reagents prepared from primary halides:

Review: J. Weill-Raynal, *Synthesis* **1969,** 49.

143. Gabriel-Colman Rearrangement (Phthalimidoacetic Ester → Isoquinoline Rearrangement, Gabriel Isoquinoline Synthesis)

S. Gabriel, J. Colman, *Ber.* **33,** 980, 996, 2630 (1900); **35,** 2421 (1902).

Formation of isoquinoline derivatives by the action of sodium ethoxide on phthalimidoacetic esters:

C. F. H. Allen, *Chem. Rev.* **47,** 284 (1950); H. Henecka, *Houben-Weyl* **8,** 578 (1952); J. H. M. Hill, *J. Org. Chem.* **30,** 620 (1965). *Cf.* Dieckmann Reaction.

144. Gabriel Ethylenimine Method (Gabriel-Marckwald Ethylenimine Synthesis)

S. Gabriel *et al., Ber.* **21,** 1049 (1888); W. Marckwald *et al., ibid.* **32,** 2036 (1899); **33,** 764 (1900); **34,** 3544 (1901).

Formation of ethylenimines (aziridines) by elimination of hydrogen halides from aliphatic vicinal haloamines with alkali. The method can be extended to the preparation of five- and six-membered-ring amines:

O. C. Dermer, G. E. Ham, *Ethylenimine and Other Aziridines* (Academic Press, New York, 1969) pp 1-59.

145. Gabriel Synthesis

S. Gabriel, *Ber.* **20,** 2224 (1887).

Conversion of alkyl halides to primary amines by treatment with potassium phthalimide and subsequent hydrolysis:

M. S. Gibson, R. W. Bradshaw, *Angew. Chem. Int. Ed.* **7,** 919 (1968); B. Dietrich *et al., J. Am. Chem. Soc.* **103,** 1282 (1981); O. Mitsunobu, *Comp. Org. Syn.* **6,** 79-85 (1991).

146. Gattermann Aldehyde Synthesis

L. Gattermann, *Ber.* **31,** 1149 (1898); *Ann.* **313,** (1907).

Preparation of phenolic aldehydes, phenol ethers or heterocyclic compounds by treatment of the aromatic substrate with hydrogen cyanide and hydrogen chloride in the presence of Lewis acid catalysts:

W. E. Truce, *Org. React.* **9**, 37 (1957); E. Baltazzi, L. I. Krimen, *Chem. Rev.* **63**, 526 (1963). *Cf.* Houben-Hoesch Reaction; Reimer-Tiemann Reaction.

147. Gattermann-Koch Reaction

L. Gattermann, J. A. Koch, *Ber.* **30**, 1622 (1897); L. Gattermann, *Ann.* **347**, 347 (1906).

Formylation of benzene, alkylbenzenes or polycyclic aromatic hydrocarbons with carbon monoxide and hydrogen chloride in the presence of aluminum chloride at high pressure. Addition of cuprous chloride allows the reaction to proceed at atmospheric pressure:

N. N. Crounse, *Org. React.* **5**, 290 (1949); G. A. Olah, S. J. Kuhn in *Friedel-Crafts and Related Reactions* **vol. 3**, Part 2, G. Olah, Ed. (Interscience, New York, 1964) pp 1153-1156; L. Toniolo, M. Graziani, *J. Organometal. Chem.* **194**, 221 (1980).

148. Glaser Coupling; Eglinton Reaction; Cadiot-Chodkiewicz Coupling

C. Glaser, *Ber.* **2**, 422 (1869).

Homocoupling of terminal alkynes catalyzed by cuprous salts in the presence of an oxidant and ammonium chloride:

Synthetic applications: F. M. Menger *et al., J. Am. Chem. Soc.* **115**, 6600 (1993); L. Guo *et al., Chem. Commun.* **1994**, 243.

This coupling may also be effected by cupric salts in pyridine and is often referred to as the **Eglinton reaction**. It is particularly applicable to cyclizations: G. Eglinton, A. R. Galbraith, *Chem. & Ind. (London)* **1956**, 737; N. Hébert *et al., J. Org. Chem.* **57**, 1777 (1992).

Heterocoupling may be accomplished *via* the **Cadiot-Chodkiewicz coupling** of terminal alkynes with haloalkynes, catalyzed by cuprous salts in the presence of aliphatic amines:

W. Chodkiewicz *et al., Compt. Rend.* **245**, 322 (1957); B. N. Ghose, *Syn. React. Inorg. Met.-Org. Chem.* **24**, 29 (1994).

Inclusive reviews: P. Cadiot, W. Chodkiewicz, "Couplings of Acetylenes" in *Chemistry of Acetylenes,* H. G. Viehe, Ed. (Marcel Dekker, New York, 1969) pp 597-647; K. Sonogashira, *Comp. Org. Syn.* **3**, 551-561. *Cf.* Castro-Stephens Coupling; Ullmann Reaction.

149. Gomberg-Bachmann Reaction

M. Gomberg, W. E. Bachmann, *J. Am. Chem. Soc.* **46**, 2339 (1924).

Formation of diaryl compounds from aryl diazonium salts and aromatic compounds in the presence of alkali:

W. E. Bachmann, R. A. Hoffman, *Org. React.* **2**, 224 (1944); O. C. Dermer, M. T. Edmison, *Chem. Rev.* **57**, 77 (1957); D. H. Hey, *Advan. Free-Radical Chem.* **2**, 47 (1966); D. E. Rosenberg *et al., Tetrahedron Letters* **21**, 4141 (1980). For intramolecular version, *see* Pschorr Reaction.

150. Gomberg Free Radical Reaction

M. Gomberg, *J. Am. Chem. Soc.* **22**, 757 (1900).

Formation of free radicals by abstraction of the halogen from triarylmethyl halides with metals:

$$2(C_6H_5)_3CCl + Zn \longrightarrow 2(C_6H_5)_3C^{\bullet} + ZnCl_2$$

A. R. Forrester *et al.*, in *Organic Chemistry of Stable Free Radicals* (Academic Press, New York, 1968); Scholle, Rozantsev, *Russ. Chem. Rev.* **42**, 1101 (1973); J. M. McBride, *Tetrahedron* **30**, 2009 (1974).

151. Gould-Jacobs Reaction

R. G. Gould, W. A. Jacobs, *J. Am. Chem. Soc.* **61**, 2890 (1939).

Synthesis of 4-hydroxyquinolines from anilines and diethyl ethoxymalonate *via* cyclization of the intermediate anilinomethylenemalonate followed by hydrolysis and decarboxylation:

R. H. Reitsema, *Chem. Rev.* **43**, 53 (1948); R. C. Elderfield, *Heterocyclic Compounds* **4**, 38 (1952); C. C. Price, R. N. Roberts, *Org. Syn.* **coll. vol. III,** 272 (New York, 1955); D. G. Markees, L. S. Schwab, *Helv. Chim. Acta* **55**, 1319 (1972); R. Albrecht, G. A. Hoyer, *Ber.* **105**, 3118 (1972); J. M. Barker *et al., J. Chem. Res. (S)* **1980**, 4. *Cf.* Doebner-Miller Reaction; Knorr Quinoline Synthesis.

152. Graebe-Ullmann Synthesis

C. Graebe, F. Ullmann, *Ann.* **291**, 16 (1896); F. Ullmann, *ibid.* **332**, 82 (1904).

Formation of carbazoles by the action of nitrous acid on 2-aminodiphenylamines, followed by decomposition of the resulting benzotriazoles:

O. Bremer, *Ann.* **514**, 279 (1934); S. H. Tucker *et al., J. Chem. Soc.* **1942**, 500; N. Campbell, B. Barclay, *Chem. Rev.* **40**, 360 (1947); C. C. Colser *et al., J. Chem. Soc.* **1951**, 110; B. W. Ashton, H. Suschitzky, *ibid.* **1957**, 4559; R. A. Abramovitch, I. D. Spenser, *Advan. Heterocyclic Chem.* **3**, 128 (1964).

153. Griess Diazo Reaction; Witt and Knoevenagel Diazotization Methods

P. Griess, *Ann.* **106**, 123 (1858); **121**, 257 (1862); E. Knoevenagel, *Ber.* **23**, 2994 (1890); O. N. Witt, *ibid.* **42**, 2953 (1909).

Formation of aromatic diazonium salts from primary aromatic amines and nitrous acid or other nitrosating agents:

$$ArNH_2 + NaNO_2 + 2HX \longrightarrow ArN_2^+X^- + NaX + 2H_2O$$

$$2ArNH_2 + N_2O_3 + 2HNO_3 + H_2O \longrightarrow 2ArN_2^+NO_3^- + 4H_2O \text{ (Griess reaction)}$$

$$2ArNH_2 + Na_2S_2O_5 + 4HNO_3 \longrightarrow 2ArN_2^+NO_3^- + Na_2S_2O_7 + 4H_2O \text{ (Witt method)}$$

$$ArNH_2 + RONO + HX \longrightarrow ArN_2^+X^- + ROH + H_2O \text{ (Knoevenagel method)}$$

N. Kornblum, *Org. React.* **2**, 264 (1944); W. A. Cowdry, D. S. Davies, *Quart. Rev.* **6**, 358 (1952); Ridd, *ibid.* **15**, 418 (1961); B. I. Belov, V. V. Kozlov, *Russ. Chem. Rev.* **32**, 59 (1963); K. Schank in *The Chemistry of Diazonium and Diazo Groups,* S. Patai, Ed. (Wiley, New York, 1978) p 645; J. B. Fox, Jr., *Anal. Chem.* **51**, 1493 (1979).

154. Grignard Degradation

W. Steinkopf *et al., Ann.* **512**, 136 (1934); **543**, 128 (1940).

Stepwise dehalogenation of a polyhalo compound through its Grignard reagent which on treatment with water yields a product containing one halogen atom less:

V. Grignard, *Compt. Rend.* **130,** 1322 (1900); F. F. Blicke, *Heterocyclic Compounds* **1,** 222 (1950); K. Nützel, *Houben-Weyl* **13/2a,** 128 (1973).

155. Grignard Reaction
V. Grignard, *Compt. Rend.* **130,** 1322 (1900).

Traditionally, it is the addition of organomagnesium compounds (Grignard reagents) to carbonyl compounds to generate alcohols. A more modern interpretation extends the scope of the reaction to include the addition of Grignard reagents to a wide variety of electrophilic substrates:

Early review: D. A. Shirley, *Org. React.* **8,** 28-58 (1954). Preparation of Grignard reagents: Y. H. Lai, *Synthesis* **1981,** 585-604. Mechanistic study: K. Maruyama, T. Katagiri, *J. Phys. Org. Chem.* **2,** 205 (1989). Review of stereoselective addition of carbonyl compounds: D. M. Huryn, *Comp. Org. Syn.* **1,** 49-75 (1991). General review: G. S. Silverman, P. E. Rakita in Kirk-Othmer *Encyclopedia of Chemical Technology* **vol. 12** (Wiley-Interscience, New York, 4th ed., 1994) pp 768-786. *Cf.* Barbier(-type) Reaction.

156. Grob Fragmentation
C. A. Grob, W. Baumann, *Helv. Chim. Acta* **38,** 594 (1955).

Carbon-carbon bond cleavage primarily *via* a concerted process involving a five atom system:

The intramolecular version is useful for the preparation of medium-size rings:

C. A. Grob, *Angew. Chem. Int. Ed.* **8,** 535-546 (1969); L. Birladeanu *et al., J. Am. Chem. Soc.* **92,** 6387 (1970); S. Schreiber, *ibid.* **102,** 6163 (1980); J. R. Mahajan, H. C. deAraújo, *Synthesis* **1981,** 49; M. Ochiai *et al., J. Org. Chem.* **54,** 4832 (1989); X. L. Armesto *et al., ibid.* **59,** 4659 (1994); P. Weyerstahl, H. Marschall, *Comp. Org. Syn.* **6,** 1044-1065 (1991). *Cf.* Eschenmoser Fragmentation; Wharton Reaction.

157. Grundmann Aldehyde Synthesis
C. Grundmann, *Ann.* **524,** 31 (1936).

Transformation of an acid into an aldehyde of the same chain length by conversion of the acid chloride, *via* the diazo ketone, to the acetoxy ketone, reduction with aluminum isopropoxide and hydrolysis to the glycol, and cleavage with lead tetraacetete:

E. Mosetting, *Org. React.* **8,** 225 (1954); O. Bayer, *Houben-Weyl* **7/1,** 239 (1954); H. K. Mangold, *J. Org. Chem.* **24,** 405 (1959). *Cf.* Sonn-Müller Method; Stephen Aldehyde Synthesis.

158. Guareschi-Thorpe Condensation
I. Guareschi, *Mem. Reale Accad. Sci. Torino* **II,** 46, 7, 11, 25 (1896); H. Baron, F. G. P. Renfry, J. F. Thorpe, *J. Chem. Soc.* **85,** 1726 (1904).

Synthesis of pyridine derivatives by condensation of cyanoacetic ester with acetoacetic ester in the presence of ammonia. In a second type of synthesis a mixture of cyanoacetic ester and a ketone is treated with alcoholic ammonia:

C. Hollins, *The Synthesis of Nitrogen Ring Compounds* (New York, 1924) p 197; V. Migrdichian, *The Chemistry of Organic Cyanogen Compounds* (New York, 1947) p 322; H. S. Mosher, *Heterocyclic Compounds* **1**, 466 (1950). *Cf.* Hantzsch (Dihydro)Pyridine Synthesis; Kröhnke Pyridine Synthesis.

159. Guerbet Reaction
M. Guerbet, *Compt. Rend.* **128**, 511 (1899).

Condensation of alcohols at high temperature and pressure in the presence of sodium alkoxide or copper by a dehydrogenation, aldol condensation, *q.v.,* and hydrogenation sequence:

H. Machemer, *Angew. Chem.* **64**, 213 (1952); S. Veibel, J. T. Nielsen, *Tetrahedron* **23**, 1723 (1967); G. Gregorio *et al., J. Organometal. Chem.* **37**, 385 (1972); E. Klein *et al., Ann.* **1973**, 1004.

160. Gutknecht Pyrazine Synthesis
H. Gutknecht, *Ber.* **12**, 2290 (1879); **13**, 1116 (1880).

Cyclization of α-amino ketones, produced by reduction of isonitroso ketones to yield the dihydropyrazines which are dehydrogenated with mercury(I) oxide or copper(II) sulfate, or sometimes with atmospheric oxygen:

I. J. Krems, P. E. Spoerri, *Chem. Rev.* **40**, 291 (1947); Y. T. Pratt, *Heterocyclic Compounds* **6**, 379, 385 (1957).

161. Haller-Bauer Reaction
A. Haller, E. Bauer, *Compt. Rend.* **148**, 70, 127 (1909); **149**, 5 (1909).

Cleavage of non-enolizable ketones with sodium amide; frequently applied to ketones, resulting in the formation of trisubstituted acetic acids:

K. E. Hamlin, A. W. Weston, *Org. React.* **9**, 1 (1957); H. M. Walborsky *et al., J. Org. Chem.* **36**, 2937 (1971); E. M. Kaiser, C. O. Warner, *Synthesis* **1975**, 395.

162. Hammick Reaction
P. Dyson, D. L. Hammick, *J. Chem. Soc.* **1937**, 1724.

Decarboxylation of α-picolinic or related acids in the presence of carbonyl compounds accompanied by the formation of a new carbon-carbon bond:

D. L. Hammick *et al., J. Chem. Soc.* **1939**, 809; **1949**, 659; N. H. Cantwell, E. V. Brown, *J. Am. Chem. Soc.* **75**, 1489 (1953); M. J. Betts, B. R. Brown, *J. Chem. Soc.* **1967**, 1730; E. V. Brown, M. B. Shambhu, *J. Org. Chem.* **36**, 2002 (1971).

163. Hantzsch (Dihydro)Pyridine Synthesis

A. Hantzsch, *Ann.* **215,** 1, 72 (1882); *Ber.* **18,** 1744 (1885); **19,** 289 (1886).

Synthesis of dihydropyridines by condensation of two moles of a β-dicarbonyl compound with one mole of an aldehyde in the presence of ammonia. Dehydrogenation is accomplished with an oxidizing agent:

H. S. Mosher, *Heterocyclic Compounds* **1,** 462 (1950); R. M. Kellog *et al., J. Org. Chem.* **45,** 2854 (1980); Y. Watanabe *et al., Synthesis* **1983,** 761. Substitution of α-cyano ketones for β-dicarbonyl compounds: S. Marchalín, J. Kuthan, *Coll. Czech. Chem. Commun.* **50,** 1962 (1985). Mechanistic study: A. R. Katritzky *et al., Tetrahedron* **42,** 5729 (1986). Extension to the synthesis of unsymmetrical dihydropyridines: J. B. Sainani *et al., Indian J. Chem.* **34B,** 17 (1995). *Cf.* Chichibabin Pyridine Synthesis; Guareschi-Thorpe Condensation; Kröhnke Pyridine Synthesis.

164. Hantzsch Pyrrole Synthesis

A. Hantzsch, *Ber.* **23,** 1474 (1890).

Formation of pyrrole derivatives from α-chloromethyl ketones, β-keto esters and ammonia or amines:

R. Elderfield, T. N. Dodd, Jr., *Heterocyclic Compounds* **1,** 132 (1950); A. H. Corwin, *ibid.* 290; M. W. Roomi, S. F. Mac-Donald, *Can. J. Chem.* **48,** 1689 (1970). *Cf.* Feist-Bénary Synthesis; Knorr Pyrrole Synthesis; Paal Knorr Pyrrole Synthesis.

165. Harries Ozonide Reaction (Ozonolysis)

C. Harries, *Ann.* **343,** 311 (1905).

Treatment of olefins with ozone as a method of cleaving olefinic linkages. On hydrolysis or catalytic hydrogenation the initially formed ozonide yields two molecules of carbonyl compounds:

Reviews: P. S. Bailey, *Chem. Rev.* **58,** 925 (1958); L. J. Chinn, *Selection of Oxidants in Synthesis: Oxidation at the Carbon Atom* (Dekker, New York, 1971) pp 151-160; K. W. Bentley in *Elucidation of Organic Structures by Physical and Chemical Methods,* part 2, K. W. Bentley, G. W. Kirby, Eds. (Wiley, New York, 2nd ed., 1973) pp 150-169; P. S. Bailey, *Ozonation in Organic Chemistry* **vols. 1, 2** (Academic Press, New York, 1978, 1982). Mechanism: R. Criegee, *Record Chem. Progr.* (Kresge-Hooker Sci. Lab.) **18,** 111 (1957); R. W. Murray, *Accts. Chem. Res.* **1,** 313 (1968).

166. Haworth Methylation

W. N. Haworth, *J. Chem. Soc.* **107,** 13 (1915).

Formation of methylated methyl glycosides from monosaccharides with dimethyl sulfate and 30% sodium hydroxide. The glycosidic methyl group is hydrolyzed with acid to yield the free methylated sugar:

W. N. Haworth, H. Machemer, *J. Chem. Soc.* **1932,** 2270; C. C. Barker *et al., ibid.* **1946,** 783; E. J. Bourne, S. Peat, *Advan. Carbohyd. Chem.* **5,** 146 (1950); W. Pigman, *The Carbohydrates* **1957,** 369. *Cf.* Purdie Methylation.

167. Haworth Phenanthrene Synthesis

R. D. Haworth, *J. Chem. Soc.* **1932,** 1125, 2717; *idem et al., ibid.* 1784, 2248, 2720; **1934,** 454.

Preparation of phenanthrenes from naphthalenes *via* a series of steps including a Friedel-Crafts acylation and two Clemmensen or Wolff-Kishner reductions, *q.q.v.*:

E. Berliner, *Org. React.* **5,** 229 (1949); I. Agranat, Y. S. Shih, *Synthesis* **1974,** 865; R. Menicagli, O. Piccolo, *J. Org. Chem.* **45,** 2581 (1980). *Cf.* Friedel-Crafts Reaction.

168. Hayashi Rearrangement

M. Hayashi, *J. Chem. Soc.* **1927,** 2516; **1930,** 1513, 1520, 1524.

Rearrangement of *o*-benzoylbenzoic acids in the presence of sulfuric acid or phosphorous pentoxide:

J. W. Cook, *J. Chem. Soc.* **1932,** 1472; M. Hayashi *et al., Bull. Chem. Soc. Japan* **11,** 184 (1936); R. B. Sandin, L. F. Fieser, *J. Am. Chem. Soc.* **62,** 3098 (1940); R. B. Sandin *et al., ibid.* **78,** 3817 (1956); R. Goncalves *et al., J. Org. Chem.* **17,** 705 (1952); S. Cristol, M. L. Caspar, *ibid.* **33,** 2020 (1968).

169. Heck Reaction

R. F. Heck, J. P. Nolley, Jr., *J. Org. Chem.* **37,** 2320 (1972).

Stereospecific palladium-catalyzed coupling of alkenes with organic halides or triflates lacking *sp³*-hybridized β-hydrogens:

R⁴ = aryl, alkenyl, benzyl
X = I, Br, OSO₂CF₃
[Pd] = Pd(OCOCH₃)₂, PdCl₂, Pd(dba)₃, Pd(PPh₃)₄
L = PAr₃, dppp, binap
base = N(CH₂CH₃)₃, K₂CO₃, NaOCOCH₃

Variation of reaction parameters in the context of the asymmetric synthesis of (+)-vernolepin: K. Ohrai *et al., J. Am. Chem. Soc.* **116,** 11737 (1994). Review of intramolecular reactions in natural product synthesis: L. E. Overman, *Pure Appl. Chem.* **66,** 1423-1430 (1994). *Reviews:* R. F. Heck, *Org. React.* **27,** 345-390 (1982); A. de Meijere, F. E. Meyer, *Angew. Chem. Int. Ed.* **33,** 2379-2411 (1994); W. Cabri, I. Candiani, *Accts. Chem. Res.* **28,** 2-7 (1995). *Cf.* Stille Coupling; Suzuki Coupling.

170. Helferich Method

B. Helferich, E. Schmitz-Hillebrecht, *Ber.* **66,** 378 (1933).

Glycosidation of an acetylated sugar by heating with a phenol in the presence of zinc chloride or *p*-toluenesulfonic acid as catalyst:

W. W. Pigman, R. M. Goepp, *Chemistry of the Carbohydrates* (New York, 1948) p 194; W. W. Pigman, *The Carbohydrates* (New York, 1957) p 198; B. Helferich, J. Zirner, *Ber.* **96,** 385 (1963); A. Piskala *et al., Nucleic Acid Chem.* **1,** 455 (1978).

171. Hell-Volhard-Zelinsky Reaction

C. Hell, *Ber.* **14,** 891 (1881); J. Volhard, *Ann.* **242,** 141 (1887); N. Zelinsky, *Ber.* **20,** 2026 (1887).

α-Halogenation of carboxylic acids with halogen and phosphorus, presumably involving the enol form of the intermediate acyl halide:

$$RCH_2COOH \xrightarrow{Br_2, P} RCH_2COBr \xrightarrow{Br_2} \underset{Br}{RCH-COBr} \xrightarrow{H_2O} \underset{Br}{R-CH-COOH}$$

N. O. V. Sonntag, *Chem. Rev.* **52,** 237 (1953); H. J. Harwood, *ibid.* **62,** 102 (1962); H. Kwart, E. V. Scalzi, *ibid.* **86,** 5496 (1964); A. R. Sexton *et al., J. Am. Chem. Soc.* **91,** 7098 (1969).

172. Henkel Reaction (Raecke Process, Henkel Process)

B. Raecke, **Ger.** pats. **936,036** (1952) and **958,920** (1952) to Henkel & Co.

Industrial scale thermal rearrangement or disproportionation of alkaline salts of aromatic acids to symmetrical diacids in the presence of cadmium or other metallic salts:

Review: B. Raecke, *Angew. Chem.* **70,** 1 (1958); Y. Ogata *et al., J. Org. Chem.* **25,** 2082 (1960); E. McNelis, *ibid.* **30,** 1209 (1965); J. Szammer, L. Otvos, *Radiochem. Radioanal. Lett.* **45,** 359 (1980).

173. Henry Reaction

L. Henry, *Compt. Rend.* **120,** 1265 (1895); J. Kamlet, **U.S.** pat. **2,151,517** (1939).

Base-catalyzed aldol-type condensation, *q.v.*, of nitroalkanes with aldehydes or ketones:

Application to sugars: R. Fernández *et al., Carbohyd. Res.* **247,** 239 (1993). Reagent-controlled asymmetric induction: H. Sasai *et al., Tetrahedron Letters* **34,** 855 (1993); R. Chinchilla *et al., Tetrahedron: Asymmetry* **5,** 1393 (1994). *Reviews:* J. C. Snowden, *Advan. Carbohyd. Chem.* **6,** 291-318 (1951); H. H. Baer, L. Urbas in *The Chemistry of the Nitro and Nitroso Groups,* Pt. 2, H. Feuer, Ed. (Interscience, New York, 1970) pp 117-130; G. Rosini, *Comp. Org. Syn.* **2,** 321-340 (1991). *Cf.* Knoevenagel Condensation.

174. Herz Reaction

R. Herz, **Ger.** pat. **360,690** (1914 to Cassella & Co.); **U.S.** pat. **1,637,023** (1928); **U.S.** pat. **1,699,432** (1929).

Formation of *o*-aminothiophenols by heating aromatic amines with excess sulfur monochloride. The initial products are thiazothionium halides (Herz compounds) which will undergo chlorination if the position *para* to the amino group is unsubstituted:

W. K. Warburton, *Chem. Rev.* **57,** 1011 (1957); L. D. Huestis *et al., J. Org. Chem.* **30,** 2763 (1965); P. Hope, L. A. Wiles, *J. Chem. Soc. C* **1967,** 1642; B. K. Strelets, L. S. Efros, *Zh. Org. Khim.* **1969,** 153; S. W. Schneller, *Int. J. Sulfur Chem.* **8,** 579 (1976).

175. Hilbert-Johnson Reaction

T. B. Johnson, G. E. Hilbert, *Science* **69,** 579 (1929); G. E. Hilbert, T. B. Johnson, *J. Am. Chem. Soc.* **52,** 2001, 4489 (1930).

Reaction of 2,4-dialkoxypyrimidines with halogenoses to yield pyrimidine nucleosides:

W. Zorbach, *Methods Carbohyd. Chem.* **6,** 445 (1972); T. Ueda, H. Ohtsuka, *Chem. Pharm. Bull.* **21,** 1451, 1530 (1973); H. Vorbrüggen, B. Bennua, *Ber.* **114,** 1279 (1981). *Review:* J. Pliml, M. Prystas, *Advan. Heterocyclic Chem.* **8,** 115 (1967). *Cf.* Vorbrüggen Glycosylation.

176. Hinsberg Oxindole Synthesis

O. Hinsberg, *Ber.* **21,** 110 (1888); **25,** 2545 (1892); **41,** 1367 (1908).

Formation of oxindoles from secondary aryl amines and the sodium bisulfite addition compound of glyoxal; primary aryl amines give glycine or glycinamide derivatives:

O. Hinsberg, J. Rosenzweig, *Ber.* **27,** 3253 (1894); C. Hollins, *Synthesis of Nitrogen Ring Compounds* (London, 1924) p 112; H. Burton, *J. Chem. Soc.* **1932,** 546; P. L. Julian *et al., Heterocyclic Compounds* **3,** 139 (1952). *Cf.* Stollé Synthesis.

177. Hinsberg Sulfone Synthesis

O. Hinsberg, *Ber.* **27,** 3259 (1894); **28,** 1315 (1895).

Formation of sulfonylquinol derivatives by addition of quinones to cold dilute aqueous solutions of sulfinic acids:

R. M. Scribner, *J. Org. Chem.* **31,** 3671 (1966); H. Ulrich *et al., Houben-Weyl* **7/3a,** 661 (1977). *Cf.* Thiele Reaction.

178. Hinsberg Synthesis of Thiophene Derivatives

O. Hinsberg, *Ber.* **43**, 901 (1910).

Formation of thiophene carboxylic acids from α-diketones and dialkyl thiodiacetates:

H. Wynberg, D. J. Zwanenburg, *J. Org. Chem.* **29**, 1919 (1964); H. Wynberg, H. J. Kooreman, *J. Am. Chem. Soc.* **87**, 1739 (1965); A. Birch, D. A. Crombie, *Chem. Ind.* **1971**, 177; D. J. Chadwick *et al., Perkin Trans. I* **1972**, 2079.

179. Hoch-Campbell Aziridine Synthesis

J. Hoch, *Compt. Rend.* **198**, 1865 (1934); K. N. Campbell, J. F. McKenna, *J. Org. Chem.* **4**, 198 (1939).

Formation of aziridines by treatment of ketoximes with Grignard reagents and subsequent hydrolysis of the organometallic complex:

K. N. Campbell *et al., J. Org. Chem.* **8**, 99, 103 (1943); **9**, 184 (1944); J. P. Freeman, *Chem. Rev.* **73**, 283 (1973); O. C. Dermer, G. E. Ham, *Ethylenimine and Other Aziridines* (Academic Press, New York, 1969) pp 65-68; G. Alvernhe, A. Laurent, *J. Chem. Res. (S)* **1978**, 28.

180. Hofmann Degradation (Exhaustive Methylation)

A. W. Hofmann, *Ber.* **14**, 659 (1881).

Formation of an olefin and a tertiary amine by pyrolysis of a quaternary ammonium hydroxide:

A. C. Cope, E. R. Trumbull, *Org. React.* **11**, 317-493 *passim* (1960). K. W. Bentley in *Elucidation of Organic Structures by Physical and Chemical Methods,* Pt. 2, K. W. Bentley, G. W. Kirby, Eds. (Wiley, New York, 2nd ed., 1973) pp 255-289; S. Huenig *et al., Ann.* **1979**, 1925; S. Nomoto *et al., Tetrahedron Letters* **1979**, 521; M. J. Minch *et al., J. Org. Chem.* **43**, 31 (1978). *Cf.* Cope Elimination Reaction; Emde Degradation.

181. Hofmann Isonitrile Synthesis (Carbylamine Reaction)

A. W. Hofmann, *Ann.* **146**, 107 (1868); *Ber.* **3**, 767 (1870).

Formation of isonitriles by the reaction of primary amines with chloroform in the presence of alkali; the odor of the isocyanide is a test for a primary amine:

$$C_2H_5NH_2 + CHCl_3 + 3NaOH \longrightarrow C_2H_5NC + 3NaCl + 3H_2O$$

P. A. S. Smith, N. W. Kalenda, *J. Org. Chem.* **23**, 1599 (1958); M. B. Frankel *et al., Tetrahedron Letters* **1959**, 5; H. L. Jackson, B. C. McKusick, *Org. Syn.* **coll. vol. IV**, 438 (1963).

182. Hofmann-Löffler-Freytag Reaction

A. W. Hofmann, *Ber.* **16**, 558 (1883); **18**, 5, 109 (1885); K. Löffler, C. Freytag, *ibid.* **42**, 3427 (1909).

Formation of pyrrolidines or piperidines by thermal or photochemical decomposition of protonated *N*-haloamines:

$$RCH_2(CH_2)_nNClR' \xrightarrow[\Delta \text{ or } h\nu]{H^+} RCHCl(CH_2)_nNH_2R'^+ \longrightarrow \quad n = 3 \text{ or } 4$$

M. E. Wolff, *Chem. Rev.* **63**, 55 (1963); E. J. Corey, W. R. Hertler, *J. Am. Chem. Soc.* **82**, 1657 (1960); R. Furstoss *et al., Tetrahedron Letters* **1970**, 1263; S. Titouani *et al., Tetrahedron* **36**, 2961 (1980).

183. Hofmann-Martius Rearrangement
A. W. Hofmann, C. A. Martius, *Ber.* **4**, 742 (1871); A. W. Hofmann, *ibid.* **5**, 720 (1872).
Thermal conversion of *N*-alkylaniline hydrohalides to *o*- and *p*-alkylanilines:

H. Hart, J. R. Kosak, *J. Org. Chem.* **27**, 116 (1962); Y. Ogata *et al., Tetrahedron* **20**, 2717 (1964); *J. Org. Chem.* **35**, 1642 (1970); G. F. Grillot in *Mechanisms of Molecular Migration* **vol. 3**, B. S. Thyagarajan, Ed. (Wiley, New York, 1971) p 237; W. Czuba, *Pol. J. Chem.* **52**, 1403 (1978). *Cf.* Bamberger Rearrangement.

184. Hofmann Reaction
A. W. Hofmann, *Ber.* **14**, 2725 (1881).
Conversion of primary carboxylic amides to primary amines with one fewer carbon atom upon treatment with sodium hypobromite (or sodium hydroxide and bromine) *via* the intermediate isocyanate:

$$RCONH_2 \xrightarrow[\text{NaOH}]{\text{Br}_2} [R-N=C=O] \xrightarrow{\text{H}_2\text{O}} RNH_2$$

Early review: E. S. Wallis, J. F. Lane, *Org. React.* **3**, 267-306 (1949). Alternative reagents/strategies: S. Kajigaeshi *et al., Chem. Letters* **1989**, 463; S. Jew *et al., Arch. Pharm. Res.* **15**, 333 (1992); D. S. Rane, M. M. Sharma, *J. Chem. Tech. Biotechnol.* **59**, 271-277 (1994). *Review:* T. Shioiri, *Comp. Org. Syn.* **6**, 800-806 (1991). *Cf.* Curtius Rearrangement; Lossen Rearrangement; Schmidt Reaction; Weerman Degradation.

185. Hofmann-Sand Reactions
K. A. Hofmann, J. Sand, *Ber.* **33**, 1340, 1353 (1900).
Olefin mercuration with mercuric salts (halides, acetates, nitrates, or sulfates) in aqueous solution. In alcoholic solutions the accelerated reaction produces alkoxyalkyl compounds:

$$H_2C=CH_2 + HgX_2 + NaOH \xrightarrow{-\text{NaX}} HOCH_2CH_2HgX \longrightarrow XHgCH_2CH_2OCH_2CH_2HgX$$

$$H_2C=CH_2 + Hg(OCOCH_3)_2 + ROH \longrightarrow ROCH_2CH_2HgOCOCH_3 + CH_3COOH$$

J. Sand, *Ber.* **34**, 1385, 2906, 2910 (1901); *Ann.* **329**, 135 (1903); J. Chatt, *Chem. Rev.* **48**, 7 (1951); E. R. Rochow *et al., Chemistry of Organometallic Compounds* (New York, 1957) p 109; W. Kitching, *Organomet. Chem. Rev.* **3**, 35 (1968); K. P. Geller, H. Straub, *Houben-Weyl* **13/2b**, 130 (1974).

186. Hooker Reaction
S. C. Hooker, *J. Am. Chem. Soc.* **58**, 1174 (1936).
Oxidation of 2-hydroxy-3-alkyl-1,4-quinones with dilute alkaline permanganate with shortening of the alkyl side chain by a methylene group and simultaneous exchange of hydroxyl and alkyl or alkenyl group positions:

S. C. Hooker, A. Steyermark, *J. Am. Chem. Soc.* **58**, 1179 (1936); L. F. Fieser, M. Fieser, *ibid.* **70**, 3215 (1948); L. F. Fieser, A. R. Bader, *ibid.* **73**, 681 (1951); L. F. Fieser, M. Fieser, *Advanced Organic Chemistry* (New York, 1961) p 870.

187. Houben-Fischer Synthesis
J. Houben, W. Fischer, *J. Prakt. Chem.* [2] **123**, 89, 262, 313 (1929).
Formation of aromatic nitriles by basic hydrolysis of trichloromethyl aryl ketimines. Acidic hydrolysis yields ketones:

$$C_6H_6 + Cl_3CCN \longrightarrow \underset{C_6H_5}{\overset{Cl_3C}{>}}=NH \cdot HCl$$

$$\xrightarrow{KOH} PhCN + CHCl_3 + KCl + H_2O$$

$$\xrightarrow{H^+/H_2O} PhCOCCl_3 + NH_4Cl$$

J. Houben, W. Fischer, *Ber.* **63**, 2464 (1930); **64**, 240, 2636, 2645 (1931); **66**, 339 (1933); D. T. Mowry, *Chem. Rev.* **42**, 221 (1948); P. E. Spoerri, A. S. DuBois, *Org. React.* **5**, 390 (1949); G. Hesse, *Houben-Weyl* **4/2** 103 (1955); W. Ruske in *Friedel-Crafts and Related Reactions* **vol. III,** Part 1, G. A. Olah, Ed. (Interscience, New York, 1964) p 407. *Cf.* Houben-Hoesch Reaction.

188. Houben-Hoesch Reaction
K. Hoesch, *Ber.* **48**, 1122 (1915); J. Houben, *ibid.* **59**, 2878 (1926).

Synthesis of acylphenols from phenols or phenolic ethers by the action of organic nitriles in the presence of hydrochloric acid and aluminum chloride as catalyst:

Reviews: P. E. Spoerri, A. S. DuBois, *Org. React.* **5**, 387 (1949); Thomas, *Anhydrous Aluminum Chloride in Organic Chemistry* (New York, 1941) p 504; W. Ruske in *Friedel-Crafts and Related Reactions* **vol. III,** Part 1, G. A. Olah, Ed. (Interscience, New York, 1964) p 383. *Cf.* Gatterman Aldehyde Synthesis; Houben-Fischer Synthesis.

189. Houdry Cracking Process
E. Houdry, **U.S. pats. 1,957,648** and **1,957,649** (1934).

Decomposition of petroleum or heavy petroleum fractions into more useful lower boiling materials by heating at 500° and 30 psi, over a silica-alumina-manganese oxide catalyst.

E. Houdry *et al.*, *Oil Gas J.* **37**, 40 (1938); A. N. Sachanen, *Chemical Constituents of Petroleum* (New York, 1945) p 260; V. Haensel, M. J. Sterba, *Ind. Eng. Chem.* **40**, 1662 (1948); Kirk-Othmer *Encyclopedia of Chemical Technology* **4**, 323, 357 (New York, 1979); E. Boye, *Chemiker-Ztg.* **81**, 341 (1957); S. Gussow *et al.*, *Oil Gas J.* **78**, 96 (1980).

190. Hunsdiecker Reaction (Borodine Reaction)
C. Hunsdiecker *et al.*, **U.S. pat. 2,176,181** (1939); H. Hunsdiecker, C. Hunsdiecker, *Ber.* **75**, 291 (1942); A. Borodine, *Ann.* **119**, 121 (1861).

Synthesis of organic halides by thermal decarboxylation of silver salts of carboxylic acids in the presence of halogens:

$$RCOOAg + X_2 \xrightarrow{\Delta} RX + CO_2 + AgX$$

R. G. Johnson, R. K. Ingham, *Chem. Rev.* **56**, 219 (1956); C. V. Wilson, *Org. React.* **9**, 341 (1957); S. J. Cristol, W. C. Firth, Jr., *J. Org. Chem.* **26**, 280 (1961); F. F. Knapp, Jr., *Steroids* **33**, 245 (1979); A. I. Meyers, M. P. Fleming, *J. Org. Chem.* **44**, 3405 (1979). *Cf.* Kochi Reaction; Simonini Reaction.

191. Hydroboration Reaction
H. C. Brown, B. C. Subba Rao, *J. Am. Chem. Soc.* **78**, 5694 (1956); *J. Org. Chem.* **22**, 1135, 1136 (1957).

Addition of boron hydrides to alkenes, allenes, and alkynes to form organoboranes, such that boron adds to the less substituted carbon. Attack usually takes place on the less hindered side in a *cis* fashion:

Diastereofacial and regioselectivity study: B. W. Gung *et al.*, *Syn. Commun.* **24**, 167 (1994). Methods development for asymmetric synthesis: U. P. Dhokte, H. C. Brown, *Tetrahedron Letters* **35**, 4715 (1994). Application to hydration: G. Zweifel, H. C. Brown, *Org. React.* **13**, 1-54 (1963). General reviews: H. O. House, *Modern Synthetic Reactions* (W. A. Benjamin, Menlo Park, California, 2nd ed., 1972) pp 106-130; K. Smith, A. Pelter, *Comp. Org. Syn.* **8**, 703-731 (1991). Reviews of asymmetric synthesis: H. C. Brown, *Tetrahedron* **37**, 3547-3587 (1981); K. Burgess, M. J. Ohlmeyer, *Adv. Chem. Ser.* **230**, 163-177 (1992). *Cf.* Suzuki Coupling.

192. Ivanov Reaction

D. Ivanov, A. Spassoff, *Bull. Soc. Chim. France* **49,** 19, 375 (1931); D. Ivanov *et al., ibid.* **51,** 1321, 1325, 1331 (1932). The addition of enediolates of aryl acetic acids (Ivanov reagents) to electrophiles, particularly carbonyl compounds:

Early reviews: B. Blagoev, D. Ivanov, *Synthesis* **1970,** 615; D. Ivanov *et al., ibid.* **1975,** 83. Synthetic application: Y. A. Zhdanov *et al., Carbohyd. Res.* **29,** 274 (1973). Kinetic and mechanistic study: J. Toullec *et al., J. Org. Chem.* **50,** 2563 (1985). *Cf.* Aldol Reaction; Knoevenagel Condensation.

193. Jacobsen Epoxidation

W. Zhang *et al., J. Am. Chem. Soc.* **112,** 2801 (1990); E. N. Jacobsen *et al., ibid.* **113,** 7063 (1991).

Chiral (salen)manganese(III)-catalyzed asymmetric epoxidation of alkenes. Enantio- and diastereoselectivity depend strongly on the nature of the substrate:

Methods development: E. N. Jacobsen *et al., Tetrahedron* **50,** 4323 (1994); S. Chang *et al., J. Am. Chem. Soc.* **116,** 6937 (1994); B. D. Brandes, E. N. Jacobsen, *J. Org. Chem.* **59,** 4378 (1994). Large-scale preparation of ligand: J. F. Larrow *et al., ibid.* 1939. *Review:* E. N. Jacobsen, "Asymmetric Catalytic Epoxidation of Unfunctionalized Olefins" in *Catalytic Asymmetric Synthesis,* I. Ojima, Ed. (VCH, New York, 1993) pp 159-202. For parallel studies, *see* N. Hosoya *et al., Synlett* **1993,** 641; H. Sasaki *et al., ibid.* **1994,** 356.

194. Jacobsen Rearrangement

O. Jacobsen, *Ber.* **19,** 1209 (1886); **20,** 901 (1887).

Reaction of polymethylbenzenes with concentrated sulfuric acid to give rearranged polymethylbenzenesulfonic acids. Under identical conditions halogenated polymethylbenzenes undergo disproportionation:

L. I. Smith, *Org. React.* **1**, 370 (1942); H. Suzuki *et al., Bull. Chem. Soc. Japan* **36**, 1642 (1963); A. Koeberg-Telder, H. Cerfontain, *Perkin Trans. II* **1977**, 717; M. Nakada *et al., Bull. Chem. Soc. Japan* **52**, 3671 (1979).

195. Janovsky Reaction

J. V. Janovsky, L. Erb, *Ber.* **19**, 2155 (1886).

Reaction of aldehydes and ketones containing α-methylene groups with *m*-dinitrobenzenes in the presence of a strong base resulting in the formation of an intense purple coloration, used for the detection of carbonyl compounds:

Reviews: Akatsuka, *J. Pharm. Soc. Japan* **80**, 389 (1960); Foster, Mackie, *Tetrahedron* **18**, 1131 (1962); Pollitt, Saunders, *J. Chem. Soc.* **1965**, 4615; M. Kimura *et al, Chem. Pharm. Bull. Japan* **17**, 531 (1969); K. Kohashi *et al., ibid.* **25**, 50 (1977). *Cf.* Zimmermann Reaction.

196. Japp-Klingemann Reaction

F. R. Japp, F. Klingemann, *Ann.* **247**, 190 (1888); *Ber.* **20**, 2942, 3284, 3398 (1887).

Formation of hydrazones by coupling of aryldiazonium salts with active methylene compounds in which at least one of the activating groups is acyl or carboxyl. This group usually cleaves during the process:

Review: R. R. Philips, *Org. React.* **10**, 143 (1959); H. C. Yao, P. Resnick, *J. Am. Chem. Soc.* **84**, 3504 (1962); M. O. Lozinskii, A. A. Gershkovich, *ibid.* **8**, 785 (1972); A. Kozikowski, W. C. Floyd, *Tetrahedron Letters* **1978**, 19.

197. Jones Oxidation

K. Bowden *et al., J. Chem. Soc.* **1946**, 39.

The oxidation of primary and secondary alcohols to acids and ketones, respectively, in the presence of chromic acid, aqueous sulfuric acid, and acetone. Isolated multiple bonds are not disturbed under these conditions:

P. Bladon *et al., J. Chem. Soc.* **1951**, 2402; E. R. H. Jones *et al., ibid.* **1953**, 457, 2548, 3019; C. Djerassi *et al., J. Org. Chem.* **21**, 1547 (1956); R. N. Warriner *et al., Aust. J. Chem.* **31**, 1113 (1978); S. V. Ley, A. Madin, *Comp. Org. Syn.* **7**, 253-256 (1991). *Cf.* Sarett Oxidation.

198. Jourdan-Ullmann-Goldberg Synthesis

F. Jourdan, *Ber.* **18**, 1444 (1885); F. Ullmann, *ibid.* **36**, 2382 (1903); I. Goldberg, *ibid.* **39**, 1691 (1906); **40**, 4541 (1907).

Synthesis of substituted diphenylamines, useful as intermediates in the synthesis of acridones:

(Jourdan)

(Ullmann)

(Goldberg)

Reviews: R. M. Acheson, *Acridines* (Interscience, New York, 1956) p 148; Schulenberg, Archer, *Org. React.* **14,** 19 (1965).

199. Julia Olefination

M. Julia, J.-M. Paris, *Tetrahedron Letters* **1973,** 4833.

The formation of predominantly *trans*-olefins *via* the addition of phenyl sulfones to aldehydes or ketones, followed by alcohol functionalization and subsequent reductive elimination with sodium amalgam:

R^3 = $COCH_3$, X = $OCOCH_3$; R^3 = COPh, SO_2CH_3, X = Cl

Reviews: P. Kocienski, *Phosphorus and Sulfur* **24,** 97-127 (1985); S. E. Kelly, *Comp. Org. Syn.* **1,** 792-806 (1991).

200. Kendall-Mattox Reaction

V. R. Mattox, E. C. Kendall, *J. Am. Chem. Soc.* **70,** 882 (1948); **72,** 2290 (1950); *J. Biol. Chem.* **188,** 287 (1951); E. C. Kendall, W. F. McGuckin, *J. Am. Chem. Soc.* **74,** 5811 (1952).

Formation of a conjugated ketone from an α-bromoketone *via* a phenylhydrazone or semicarbazone:

C. Djerassi, *J. Am. Chem. Soc.* **71,** 1003 (1949); B. A. Koechlin *et al., J. Biol. Chem.* **184,** 393 (1950); N. L. Wendler *et al., J. Am. Chem. Soc.* **73,** 3818 (1951); J. J. Beereboom *et al., ibid.* **75,** 3500 (1953); C. R. Engel, *ibid.* **78,** 4727 (1956); E. W. Warnhoff, *J. Org. Chem.* **28,** 887 (1963).

201. Kiliani-Fischer Synthesis

H. Kiliani, *Ber.* **18,** 3066 (1885); E. Fischer, *ibid.* **22,** 2204 (1889).

Extension of the carbon atom chain of aldoses by treatment with cyanide. Hydrolysis of the cyanohydrins followed by reduction of the lactone yields the homologous aldose:

ONR-48

$$\underset{\text{HCOH}}{\overset{\text{HC}=\text{O}}{|}} \xrightarrow{\text{HCN}} \underset{\text{HCOH}}{\overset{\overset{\text{CN}}{|}}{\underset{|}{\text{HC}-\text{OH}}}} \xrightarrow{H_2O} \underset{\text{HCOH}}{\overset{\text{O}=\text{C}}{\underset{|}{\text{HCOH}}}}O \xrightarrow{\text{NaHg}} \underset{\text{HCOH}}{\overset{\text{HC}=\text{O}}{\underset{|}{\text{HCOH}}}}$$

Reviews: C. S. Hudson, *Advan. Carbohyd. Chem.* **1,** 2 (1945); T. Moury, *Chem. Rev.* **42,** 239 (1948); L. Hough, A. C. Richardson, *The Carbohydrates* **1A,** 118 (1972); R. Kuhn, P. Klesse, *Ber.* **91,** 1989 (1958); R. Varma, D. French, *Carbohyd. Res.* **25,** 71 (1972); R. Blazer, T. W. Whalen, *J. Am. Chem. Soc.* **102,** 5082 (1980). *Cf.* Urech Cyanohydrin Method.

202. Kishner Cyclopropane Synthesis
N. M. Kishner, A. Zavadovskii, *J. Russ. Phys. Chem. Soc.* **43,** 1132 (1911).

Formation of cyclopropane derivatives by decomposition of pyrazolines formed by reacting α,β-unsaturated ketones or aldehydes with hydrazine:

L. I. Smith, E. R. Rogier, *J. Am. Chem. Soc.* **73,** 3840 (1951); G. S. Hammond, R. W. Todd, *ibid.* **76,** 4081 (1954); T. L. Jacobs, *Heterocyclic Compounds* **5,** 109 (1957). Mechanistic aspects of pyrazoline decomposition to cyclopropanes: R. G. Bergman in *Free Radicals* vol. **1,** J. Kochi, Ed. (Wiley, New York, 1973) p 191; R. J. Crawford, M. Ohno, *Can. J. Chem.* **52,** 3134 (1974); R. J. Crawford, H. Tokunaga, *ibid.* 4033; J. A. Berson in *Rearrangements in Ground and Excited States* vol. **1,** P. de Mayo, Ed. (Academic Press, New York, 1980) p 326.

203. Knoevenagel Condensation; Doebner Modification
E. Knoevenagel *Ber.* **31,** 2596 (1898); O. Doebner, *Ber.* **33,** 2140 (1900).

Condensation of aldehydes or ketones with active methylene compounds in the presence of ammonia or amines; the use of malonic acid and pyridine is known as the **Doebner modification**:

$$RCHO + H_2C(COOR)_2 \xrightarrow{\text{base}} RCH=C(COOR)_2 \xrightarrow[-CO_2]{HO^-} RCH=CH-COOH$$

Early reviews: J. R. Johnson, *Org. React.* **1,** 210 (1942); G. Jones, *ibid.* **15,** 204 (1967); H. O. House, *Modern Synthetic Reactions* (W. A. Benjamin, Menlo Park, California, 2nd ed., 1972) pp 646-653. Catalytic methodology: D. Brillon, G. Suavé, *J. Org. Chem.* **57,** 1838 (1992); D. Prajapati, J. S. Sandhu, *Chem. Letters* **1992,** 1945; *eidem, Perkin Trans. I* **1993,** 739; Y. Lin *et al., J. Organometal. Chem.* **448,** 215 (1993). Development of enantioselective methods: L. F. Tietze, P. Saling, *Chirality* **5,** 329 (1993). Application to the synthesis of indole alkaloids: L. F. Tietze *et al., Synthesis* **1994,** 1185. *Review:* L. F. Tietze, U. Beifuss, *Comp. Org. Syn.* **2,** 341-394 (1991). *Cf.* Aldol Reaction; Henry Reaction; Ivanov Reaction.

204. Knoop-Oesterlin Amino Acid Synthesis
F. Knoop, H. Oesterlin, *Z. Physiol. Chem.* **148,** 294 (1925).

Preparation of α-amino acids by catalytic hydrogenation of α-oxo acids in aqueous ammonia in the presence of platinum, palladium or Raney nickel catalysts, probably *via* an unstable iminocarboxylate ion intermediate:

$$RCOCOO^- + NH_3 \longrightarrow \underset{}{\overset{NH}{\overset{\|}{RCCOO^-}}} \xrightarrow[Pd]{H_2} \underset{}{\overset{NH_2}{\overset{|}{RCHCOOH}}}$$

H. R. V. Arnstein, R. Bentley, *Quart. Rev.* **4,** 186 (1950); S. Nakamura, K. Ashida, *J. Agr. Chem. Soc. Japan* **24,** 185 (1950-1951); T. Wieland *et al., Houben-Weyl* **11/2,** 311, 482 (1958); C. W. Huffman, W. G. Skelly, *Chem. Rev.* **63,** 632 (1963).

205. Knorr Pyrazole Synthesis
L. Knorr, *Ber.* **16,** 2587 (1883).

Formation of pyrazole derivatives from hydrazines, hydrazides, semicarbazides, and aminoguanidines by condensation with 1,3-dicarbonyl compounds; substituted hydrazines yield two structurally isomeric pyrazoles:

$$R'\text{-}CO\text{-}CH(R)\text{-}CO\text{-}R'' + NH_2NHAr \longrightarrow \text{(pyrazole)} + \text{(pyrazole)} + 2H_2O$$

T. J. Jacobs, *Heterocyclic Compounds* **5,** 46 (1957); M. H. Palmer, *Structure and Reactions of Heterocyclic Compounds* (Arnold, London, 1967) pp 378-385. *Cf.* Pechmann Pyrazole Synthesis.

206. Knorr Pyrrole Synthesis

L. Knorr, *Ber.* **17,** 1635 (1884); *Ann.* **236,** 290 (1886); L. Knorr, H. Lange, *Ber.* **35,** 2998 (1902).

Formation of pyrrole derivatives by condensation of α-amino ketones as such or generated *in situ* from isonitrosoketones with carbonyl compounds containing active α-methylene groups:

A. H. Corwin, *Heterocyclic Compounds* **1,** 287 (1950); H. Fischer, *Org. Syn.* **coll. vol. III,** 573 (1955); S. Hauptmann, M. Martin, *Z. Chem.* **8,** 333 (1968); A. J. Castro *et al., J. Org. Chem.* **35,** 2815 (1970); Y. Tamura *et al., Chem. & Ind. (London)* **1971,** 767; H. Rapoport, J. Harbuck, *J. Org. Chem.* **36,** 853 (1971). *Cf.* Hantzsch Pyrrole Synthesis; Paal-Knorr Pyrrole Synthesis.

207. Knorr Quinoline Synthesis

L. Knorr, *Ann.* **236,** 69 (1886); **245,** 357, 378 (1888).

Formation of α-hydroxyquinolines from β-ketoesters and arylamines above 100°. The intermediate anilide undergoes cyclization by dehydration with concentrated sulfuric acid:

F. W. Bergstrom, *Chem. Rev.* **35,** 157 (1944); C. R. Hauser, G. A. Reynolds, *J. Am. Chem. Soc.* **70,** 2402 (1948); *Org. Syn.* **coll. vol. III,** 593 (1955); R. C. Elderfield, *Heterocyclic Compounds* **4,** 30 (1952); A. J. Hodgkinson, B. Staskum, *J. Org. Chem.* **34,** 1709 (1969). *Cf.* Doebner-Miller Reaction; Gould-Jacobs Reaction.

208. Koch-Haaf Carboxylations

H. Koch, *Brennstoff Chem.* **36,** 321 (1955); H. Koch, W. Haaf, *Ann.* **618,** 251 (1958).

Formation of tertiary carboxylic acids by treating alcohols with carbon monoxide in strong acid:

H. Langhals *et al., Tetrahedron Letters* **22,** 2365 (1981). *Reviews:* K. E. Möller, *Brennstoff Chem.* **47,** 10 (1966); Y. T. Eidus *et al., Russ. Chem. Rev.* **42,** 199 (1973); H. Bahrmann, "Koch Reactions" in *New Syntheses with Carbon Monoxide,* J. Falbe, Ed. (Springer-Verlag, New York, 1980) pp 372-413.

Extension to olefins:

$$-C=C- + CO + H_2O \xrightarrow{H^+} -\underset{|}{C}-\underset{|}{C}-$$

G. Olah, J. Olah in *Friedel-Crafts and Related Reactions* **vol. 3,** Part 2, G. A. Olah, Ed. (Interscience, New York, 1964) pp 1272-1296; C. W. Bird, *Chem. Rev.* **62,** 283 (1962).

209. Kochi Reaction
J. K. Kochi, *J. Am. Chem. Soc.* **87,** 2500 (1965).

Synthesis of organic chlorides by decarboxylation of carboxylic acids in the presence of lead tetraacetate and lithium chloride:

$$RCOOH + Pb(OAc)_4 \longrightarrow RCOOPb(OAc)_3 + AcOH \xrightarrow[\Delta]{LiCl} RCl + CO_2$$

R. A. Sheldon, J. K. Kochi, *Org. React.* **19,** 279 (1972); M. Mannier, J. P. Aycard, *Can. J. Chem.* **57,** 1257 (1979). *Cf.* Hunsdiecker Reaction.

210. Koenigs-Knorr Synthesis
W. Koenigs, E. Knorr, *Ber.* **34,** 957 (1901).

Formation of glycosides from acetylated glycosyl halides and alcohols or phenols in the presence of silver or mercury salts. The reaction proceeds with inversion of configuration:

Reviews: Evans *et al., Advan. Carbohyd. Chem.* **6,** 41-52 (1951); K. Igarashi, *ibid.* **34,** 243 (1977); H. M. Flowers, *Methods Carbohyd. Chem.* **6,** 474-480 (1972); R. R. Schmidt, *Comp. Org. Syn.* **6,** 33-64 (1991).

211. Kolbe Electrolytic Synthesis; Crum Brown-Walker Reaction
H. Kolbe, *Ann.* **69,** 257 (1849).

Formation of symmetrical dimers by the electrolysis of carboxylates (decarboxylative dimerization). The coupling of two distinct carboxylates yields unsymmetrical products:

$$2RCOO^- \xrightarrow{electrolysis} R-R + 2CO_2$$

The dimerization of half-esters is known as the **Crum Brown-Walker reaction**: A. Crum Brown, J. Walker, *ibid.* **261,** 107 (1891).

$$2ROOCCH_2COO^- \xrightarrow{electrolysis} ROOCCH_2CH_2COOR + 2CO_2$$

Reviews: B. C. L. Weedon, *Quart. Rev.* **6,** 380 (1952); A. K. Vijh, B. E. Conway, *Chem. Rev.* **67,** 623 (1967); L. Eberson in *Organic Electrochemistry,* M. M. Baizer, Ed. (M. Dekker, New York, 1973) pp 469-507; H. J. Schäfer, *Comp. Org. Syn.* **3,** 633-658 (1991).

212. Kolbe-Schmitt Reaction
H. Kolbe, *Ann.* **113,** 125 (1860); R. Schmitt, *J. Prakt. Chem.* [2] **31,** 397 (1885).

Formation of aromatic hydroxy acids by carboxylation of phenolates, mostly in the *ortho* position, by carbon dioxide:

Reviews: A. S. Lindsey, H. Jeskey, *Chem. Rev.* **57,** 583 (1957); D. C. Ayres, *Carbanions in Synthesis* **1966,** 168-173; J. L. Hales *et al., J. Chem. Soc.* **1954,** 3145; J. March, *Advanced Organic Chemistry* (Wiley-Interscience, New York, 4th ed., 1992) p 546.

213. Kostanecki Acylation
S. von Kostanecki, A. Rozycki, *Ber.* **34,** 102 (1901).

Formation of chromones or coumarins by acylation of *o*-hydroxyaryl ketones with aliphatic acid anhydrides, followed by cyclization:

W. Baker, *J. Chem. Soc.* **1933,** 1381; C. R. Hauser, *Org. React.* **8,** 91 (1954); T. Szell *et al., Tetrahedron* **25,** 715 (1969); *idem et al., Helv. Chim. Acta* **52,** 2636 (1969); Y. A. Shaikh, K. N. Trivedi, *J. Indian Chem. Soc.* **49,** 599, 713 (1972). *Cf.* Allan-Robinson Reaction; Baker-Venkataraman Rearrangement.

214. Krafft Degradation

F. Krafft, *Ber.* **12,** 1664 (1879).

Conversion of carboxylic acids, especially of high molecular weight, into the next lower homolog by dry distillation of the alkaline earth salt with the corresponding acetate, followed by chromic acid oxidation of the methyl ketone:

$$(RCH_2COO)_2M + (AcO)_2M \longrightarrow RCH_2COCH_3 \xrightarrow{CrO_3} RCOOH$$

F. C. Whitmore, *Organic Chemistry* (New York, 1951) p 255; F. Klages, *Lehrbuch der organischen Chemie* **I** (Berlin, 1952) pp 262, 266, 368. *Cf.* Barbier-Wieland Degradation.

215. Krapcho Decarbalkoxylation

A. P. Krapcho *et al., Tetrahedron Letters* **1967,** 215.

The decarbalkoxylation of malonate esters, β-keto esters, α-cyano esters and α-sulfonyl esters in dipolar aprotic solvents, at high temperatures, in the presence of water and/or salt, to yield esters, ketones, nitriles and sulfonyl derivatives, respectively:

$R^1 = CH_3, CH_2CH_3$
EWG = COOR, COR, CN, SO$_2$R
solvent (possibly wet) = DMSO, DMF, HMPT
MX = LiCl, NaCl, LiI, NaCN, KCN, *n*-Bu$_4$NOAc

Scope and limitations: A. P. Krapcho *et al., J. Org. Chem.* **43,** 138 (1978). Mechanistic studies: A. M. Bernard *et al., Tetrahedron* **46,** 3929 (1990); P. J. Gilligan, P. J. Krenitsky, *Tetrahedron Letters* **35,** 3441 (1994). Review of synthetic applications: A. P. Krapcho, *Synthesis* **1982,** 805-822, 893-914.

216. Kröhnke Oxidation

F. Kröhnke *et al., Ber.* **69,** 2006 (1936); **71,** 2583 (1938); **72,** 440 (1939).

Transformation of activated halides into aldehydes *via* their pyridinium salts, which yield nitrones upon treatment with *p*-nitrosodimethylaniline. Aldehydes or ketones are generated upon hydrolysis:

R = Ar, R'CO, R'CH=CH

A. A. Goldberg, H. A. Walker, *J. Chem. Soc.* **1954,** 2540; F. Kröhnke, *Angew. Chem. Int. Ed.* **2,** 380 (1963); A. Markovac *et al., Heterocyclic Chem.* **14,** 19 (1977); I. Maeba *et al., Perkin Trans. I* **1991,** 939; S. N. Kilenyi, *Comp. Org. Syn.* **7,** 657-659 (1991). *Cf.* Sommelet Reaction.

217. Kröhnke Pyridine Synthesis

W. Zecher, F. Kröhnke, *Ber.* **94**, 690, 698 (1961); *eidem, Angew. Chem. Int. Ed.* **1**, 626 (1962).

1,4-Michael addition, *q.v.*, of α-pyridinium methyl ketone salts to α,β-unsaturated ketones, generating the 1,5-dicarbonyl compounds which undergo ammonium acetate-promoted ring closure, to yield substituted pyridines:

Early review: F. Kröhnke, *Synthesis* **1976**, 1-24. Synthetic applications: J. N. Chatterjea *et al., Indian J. Chem.* **15B**, 430 (1977); G. R. Newkome *et al., J. Org. Chem.* **51**, 850 (1986); P. Lhoták, A. Kurfürst, *Coll. Czech. Chem. Commun.* **57**, 1937 (1992). *Cf.* Chichibabin Pyridine Synthesis; Guareschi-Thorpe Condensation; Hantzsh (Dihydro)Pyridine Synthesis.

218. Kucherov Reaction

M. Kucherov, *Ber.* **14**, 1540 (1881).

Hydration of acetylenic hydrocarbons with dilute sulfuric acid in the presence of mercuric sulfate or boron trifluoride as catalyst:

Reviews: A. D. Petrov, *Usp. Khim.* **21**, 250 (1952); M. Miocque *et al., Ann. Chim. (Paris)* **8**, 157 (1963); M. M. Khan, A. E. Martell, *Homogeneous Catalysis by Metal Complexes* vol. **2** (Academic Press, New York, 1974) p 1974; B. S. Krupin, A. A. Petrov, *J. Gen. Chem. USSR* **33**, 3799 (1963); W. L. Budde, R. E. Dessy, *Tetrahedron Letters* **1963**, 651; *J. Am. Chem. Soc.* **85**, 3964 (1963); K. G. Golodova, S. I. Yakimovich, *Zh. Org. Khim.* **8**, 2015 (1972). Extension to allenes: A. V. Fedorova, A. A. Petrov, *J. Gen. Chem. USSR* **32**, 1740 (1962).

219. Kuhn-Winterstein Reaction

R. Kuhn, A. Winterstein, *Helv. Chim. Acta* **11**, 87 (1928).

Conversion of 1,2-glycols into *trans* olefins by reaction with diphosphotetraiodide (P_2I_4) or other halogenated reagents. This reaction is useful in the preparation of polyenes:

Kuhn *et al., Ber.* **71**, 1510 (1938); **84**, 566 (1961); **88**, 309 (1955); Inhoffen *et al., Ann.* **684**, 24 (1965).

220. Ladenburg Rearrangement

A. Ladenburg, *Ber.* **16**, 410 (1883); *Ann.* **247**, 1 (1888).

Thermal rearrangement of an alkyl- or benzylpyridinium halide to an alkyl- or benzylpyridine:

J. H. Brewster, E. L. Eliel, *Org. React.* **7**, 135 (1953); L. E. Tenenbau in *Pyridine and Its Derivatives*, Pt. 2, E. Klingsberg, Ed. (Interscience, New York, 1961) p 163.

221. Lebedev Process

S. V. Lebedev, *Zh. Obshch. Khim.* **3**, 698 (1933).

Formation of butadiene from ethanol by catalytic pyrolysis. The catalysts used are mixtures of silicates and aluminum and zinc oxides:

$$2CH_3CH_2OH \longrightarrow H_2C{=}CHCH{=}CH_2 + H_2 + 2H_2O$$

S. V. Lebedev, **Fr.** pat. **665,917** (1928); **Brit.** pat. **331,482** (1929); **U.S.S.R.** pat. **24,393** (1931); C. Ellis, *The Chemistry of Petroleum Derivatives* **II** (New York, 1937) p 173; G. Egloff, G. Hulla, *Chem. Rev.* **36,** 67 (1945); Y. A. Gorin, *Zh. Obshch. Khim.* **20,** 1596 (1950); Kirk-Othmer *Encyclopedia of Chemical Technology* **vol. 4** (New York, 3rd ed., 1978) p 322.

222. Lehmstedt-Tanasescu Reaction
K. Lehmstedt, *Ber.* **65,** 834 (1932); I. Tanasescu, *Bull. Soc. Chim. France* **41,** 528 (1927).

Preparation of acridones (and 10-hydroxyacridones) from *o*-nitrobenzaldehyde and a halobenzene in the presence of concentrated sulfuric acid containing nitrous acid as catalyst:

I. Tanasescu, Z. Frenkel, *ibid.* **1960,** 693. Mechanism: Silberg, Frenkel, *Rev. Roumaine Chim.* **10,** 1035 (1965).

223. Letts Nitrile Synthesis
E. A. Letts, *Ber.* **5,** 669 (1872).

Formation of nitriles by heating aromatic carboxylic acids with metal thiocyanates:

$$RCOOH + KSCN \xrightarrow{\Delta} RCN + CO_2 + KHS$$

G. Krüss, *Ber.* **17,** 1766 (1884); E. E. Reid, *Am. Chem. J.* **43,** 162 (1910); G. D. van Epps, E. E. Reid, *J. Am. Chem. Soc.* **38,** 2120 (1916); D. T. Mowry, *Chem. Rev.* **42,** 264 (1948); F. Klages, *Lehrbuch der organischen Chemie* **I** (Berlin, 1959) p 362.

224. Leuckart (Leukart) Reaction; Leuckart-Wallach Reaction; Eschweiler-Clarke Reaction
R. Leuckart, *Ber.* **18,** 2341 (1885).

Reductive alkylation of ammonium (or amine) salts of formic acid or formamides by aldehydes or ketones:

When the reaction is performed in the presence of excess formic acid it is referred to as the **Leuckart-Wallach reaction:** O. Wallach, *Ann.* **272,** 99 (1892). Application to steroids: W. E. Solomons, N. J. Doorenbos, *J. Pharm. Sci.* **63,** 19 (1974); A. M. Bellini *et al., Steroids* **56,** 395 (1991).

The reductive methylation of primary or secondary amines employing formaldehyde and formic acid is known as the **Eschweiler-Clarke reaction**: W. Eschweiler, *Ber.* **38,** 880 (1905); H. T. Clarke *et al., J. Am. Chem. Soc.* **55,** 4571 (1933). Synthetic applications: E. Farkas, C. J. Sunman, *J. Org. Chem.* **50,** 1110 (1985); J. Casanova, P. Devi, *Syn. Commun.* **23,** 245 (1993).

Early reviews: M. L. Moore, *Org. React.* **5,** 301-330 (1949); F. Möller, R. Schröter, *Houben-Weyl* **11/1,** 648-664 (1957). Synthesis of *N,N'*-dimethylamines: R. D. Bach, *J. Org. Chem.* **33,** 1647 (1968). Application to deoxybenzoins: M. J. Villa *et al., Heterocycles* **24,** 1943 (1986). Mechanistic study: P. I. Awachie, V. C. Agwada, *Tetrahedron* **46,** 1899 (1990). Optimized procedure: R. Carlson *et al., Acta Chem. Scand.* **47,** 1046 (1993).

225. Leuckart Thiophenol Reaction
R. Leuckart, *J. Prakt. Chem.* [2] **41,** 179 (1890).

Decomposition of diazoxanthates, by warming gently in faintly acidic cuprous media, to the corresponding aryl xanthates which afford aryl thiols on alkaline hydrolysis and aryl thioethers on warming:

D. S. Tarbell, D. K. Fukushima, *Org. Syn.* **coll. vol. III,** 809 (1955); K. H. Saunders, *The Aromatic Diazo-Compounds and Their Technical Applications* (London, 1949) p 325; D. S. Tarbell, M. A. McCall, *J. Am. Chem. Soc.* **74,** 48 (1952); A. R. Forrester, J. L. Wardell, *Rodd's Chemistry of Carbon Compounds* **IIIA,** 422 (1971); A. Schöberl, A. Wagner, *Houben-Weyl* **9,** 12 (1955).

226. Lieben Iodoform Reaction (Haloform Reaction)

A. Lieben, *Ann.* (Suppl.) **7,** 218 (1870).

Cleavage of methyl ketones with halogens (mostly iodine) and base to carboxylic acids and haloform:

$$RCOCH_3 + NaOI \longrightarrow RCOCl_3 \xrightarrow{NaOH} CHI_3 + RCOO^-Na^+$$

R. C. Fuson, B. A. Bull, *Chem. Rev.* **15,** 275 (1934); R. N. Seelye, T. A. Turney, *J. Chem. Ed.* **36,** 572 (1959); J. March, *Advanced Organic Chemistry* (Wiley-Interscience, New York, 4th ed., 1992) p 632; H. O. House, *Modern Synthetic Reactions* (W. A. Benjamin, Menlo Park, California, 2nd ed., 1972) pp 464-465.

227. Lobry de Bruyn-van Ekenstein Transformation

C. A. Lobry de Bruyn, *Rec. Trav. Chim.* **14,** 150 (1895); C. A. Lobry de Bruyn, W. A. van Ekenstein, *ibid.* 195, 203; **16,** 262 (1897).

Isomerization of carbohydrates in alkaline media, considered to embrace both epimerization of aldoses and ketoses and aldose-ketose interconversion:

Reviews: Evans, *Chem. Rev.* **31,** 544 (1942); Sattler, *Advan. Carbohyd. Chem.* **3,** 113 (1948); Pigman, *The Carbohydrates* (Academic Press, New York, 1957) p 60; Speck, *Advan. Carbohyd. Chem.* **13,** 63 (1958); Schaffer, *J. Org. Chem.* **29,** 1473 (1964); M. H. Johansson, O. Samuelson, *Chem. Scr.* **9,** 151 (1976).

228. Lossen Rearrangement

W. Lossen, *Ann.* **161,** 347 (1872); **175,** 271, 313 (1874).

Conversion of a hydroxamic acid to an isocyanate *via* the intermediacy of its *O*-acyl, sulfonyl, or phosphoryl derivative. In the presence of amines, ureas are formed; in the presence of water, amines containing one less carbon than the starting material, are generated:

Early reviews: H. L. Yale, *Chem. Rev.* **33,** 209 (1943); L. Bauer, O. Exner, *Angew. Chem. Int. Ed.* **13,** 376 (1974). Application to the structural analysis of peptides: J. T. Capecchi *et al., J. Org. Chem.* **48,** 2014 (1983). Investigation of reaction conditions leading to the formation of ureas: J. Pihuleac, L. Bauer, *Synthesis* **1989,** 61. Extention to *N*-phosphinoylhydroxylamines: J. Fawcett *et al., Chem. Commun.* **1992,** 227. *Review:* T. Shiori, *Comp. Org. Syn.* **6,** 821-825 (1991). *Cf.* Curtius Rearrangement; Hofmann Reaction; Schmidt Reaction.

229. McFadyen-Stevens Reaction

J. S. McFadyen, T. S. Stevens, *J. Chem. Soc.* **1936,** 584.

Base-catalyzed thermal decomposition of acylbenzenesulfonylhydrazines to aldehydes:

$$R-\overset{O}{\underset{}{C}}-NHNHSO_2Ar \xrightarrow[\text{glycol, }160°]{Na_2CO_3} R-\overset{O}{\underset{}{C}}-H$$

E. Mosettig, *Org. React.* **8**, 232-240 (1954); S. Siddappa, G. A. Bhat, *J. Chem. Soc. C* **1971**, 178; S. B. Matin *et al., J. Org. Chem.* **39**, 2285 (1974); M. Nair, H. Shechter, *Chem. Commun.* **1978**, 793. Alternative hydrazide reagent: C. C. Dudman *et al., Tetrahedron Letters* **1980**, 4645.

230. McLafferty Rearrangement
F. W. McLafferty, *Anal. Chem.* **31**, 82 (1959).

Electron-impact-induced cleavage of carbonyl compounds having a hydrogen in the γ-position, to an enolic fragment and an olefin:

D. G. I. Kingston *et al., Chem. Rev.* **74**, 215 (1974); K. Biemann, *Mass Spectrometry* (New York, 1962) p 119; Djerassi *et al., J. Am. Chem. Soc.* **87**, 817 (1965); **91**, 2069 (1969); **94**, 473 (1972); M. J. Lacey *et al., Org. Mass Spectrom.* **5**, 1391 (1971); G. Eadon, *J. Am. Chem. Soc.* **94**, 8938 (1972); F. Turecek, V. Hanus, *Org. Mass Spectrom.* **15**, 8 (1980). *Cf.* Norrish Type Cleavage.

231. McMurry Reaction
J. E. McMurry, M. P. Fleming, *J. Am. Chem. Soc.* **96**, 4708 (1974); S. Tyrlik, I. Wolochowicz, *Bull. Soc. Chim. France* **1973**, 2147; T. Mukaiyama *et al., Chem. Letters* **1973**, 1041.

Deoxygenative coupling of carbonyl compounds to alkenes induced by low-valent titanium:

$$\overset{}{\underset{}{}}C=O \xrightarrow[\text{reagent}]{\text{Ti(0) or Ti(I)}} \overset{}{\underset{}{}}C=C\overset{}{\underset{}{}}$$

Ti(0) or Ti(I) reagent is formed *in situ* from $TiCl_3(DME)_{1.5}$ + Zn/Cu or $TiCl_3$ + $LiAlH_4$

Synthetic application: A. Fürstner, D. N. Jumbam, *Tetrahedron* **48**, 5991 (1992). *Reviews:* J. E. McMurry, *Chem. Rev.* **89**, 1513-1524 (1989); G. M. Robertson, *Comp. Org. Syn.* **3**, 583-595 (1991). *Cf.* Barton Olefin Synthesis.

232. Madelung Synthesis
W. Madelung, *Ber.* **45**, 1128 (1912).

Formation of indole derivatives by intramolecular cyclization of an *N*-(2-alkylphenyl)alkanamide by a strong base at high temperature:

R. K. Brown in *The Chemistry of Heterocyclic Compounds,* A. Weissberger, Ed., *Indoles, Part I,* W. J. Houlihan, Ed. (Wiley, New York, 1972) pp 385-396; W. J. Houlihan *et al., J. Org. Chem.* **46**, 4511, 4515 (1981).

233. Maillard Reaction ("Browning" Reaction)
L. C. Maillard, *Compt. Rend.* **154**, 66 (1912); *Ann. Chim.* **9**, 5, 258 (1916).

The reaction of amino groups of amino acids, peptides or proteins with the "glycosidic" hydroxyl group of sugars ultimately resulting in the formation of brown pigments.

G. P. Ellis, *Advan. Carbohyd. Chem.* **14**, 63 (1959); E. F. L. J. Anet, *ibid.* **19**, 181 (1964); H. Simon, A. Kraus, *Fortschr. Chem. Forsch.* **14**, 431 (1970); H. E. Nursten, *Food Chem.* **6**, 263 (1981); M. C. DeFigueiredo Toledo, P. A. Bobbio, *An. Acad. Bras. Cienc. (Brazil)* **53**, 707-712 (1981); *Maillard Reaction* in Food Proc Int. Symposium, Uddevalla, Sweden, 1979, C. Erickson, Ed. (Pergamon Press, New York, 1982).

234. Malaprade Reaction (Periodic Acid Oxidation)
L. Malaprade, *Bull. Soc. Chim. France* [4] **43**, 683 (1928); *Compt. Rend.* **186**, 382 (1928).

Compounds containing two hydroxyl groups, or a hydroxyl and an amino group, attached to adjacent carbon atoms, undergo cleavage of the carbon-carbon bond when treated with periodic acid to yield aldehydes:

$$RCHOHCHOHR' + HIO_4 \longrightarrow RCHO + R'CHO + H_2O + HIO_3$$

$$RCHOHCHNH_2R' + HIO_4 \longrightarrow RCHO + R'CHO + NH_3 + HIO_3$$

H. O. House, *Modern Synthetic Reactions* (W. A. Benjamin, Menlo Park, California, 2nd ed., 1972) pp 353-359; K. W. Bentley in *Elucidation of Organic Structures by Physical and Chemical Methods,* Pt. 2, K. W. Bentley, G. W. Kirby, Eds. (Wiley, New York, 2nd ed., 1973) pp 177-185. *Cf.* Criegee Reaction.

235. Malonic Ester Syntheses

Syntheses based on the strongly activated methylene group of malonic esters which on reaction with sodium ethoxide forms a resonance-stabilized ion that can be alkylated or acylated. After hydrolysis, the free alkylmalonic acids readily decarboxylate to mono- or disubstituted monocarboxylic acids:

$$H_5C_2OOCCH_2COOC_2H_5 \xrightarrow[\text{2. RX}]{\text{1. NaOC}_2\text{H}_5} H_5C_2OOCCHRCOOC_2H_5 \xrightarrow{\text{HCl}} RCH_2COOH$$

H. O. House, *Modern Synthetic Reactions* (W. A. Benjamin, Menlo Park, California, 2nd ed., 1972) pp 510-518, 756-761. Use of crown ethers as catalysts: D. H. Hunter, R. A. Perry, *Synthesis* **1977,** 37. *Cf.* Perkin Alicyclic Synthesis.

236. Mannich Reaction

C. Mannich, W. Krosche, *Arch. Pharm.* **250,** 647 (1912).

Reaction of compounds having an active hydrogen with non-enolizable aldehydes and ammonia or primary or secondary amines to give aminomethylated products (Mannich bases):

$$(CH_3)_2NH + HCHO + CH_3COCH_3 \longrightarrow (CH_3)_2NCH_2CH_2COCH_3 + H_2O$$

Early reviews: F. F. Blicke, *Org. React.* **1,** 303 (1942); H. O. House, *Modern Synthetic Reactions* (W. A. Benjamin, Menlo Park, California, 2nd ed., 1972) pp 654-660. Development of diastereoselective methods: P. C. B. Page *et al., J. Org. Chem.* **58,** 6902 (1993). *p*-Substituted phenols as substrates: D. A. Leigh, P. Linnane, *Tetrahedron Letters* **34,** 5639 (1993). Application to the synthesis of vinylphosphonates: H. Krawezyk, *Syn. Commun.* **24,** 2263 (1994). *Reviews:* M. Tramontini, L. Angiolini, *Tetrahedron* **46,** 1791-1837 (1990); E. F. Kleinman, *Comp. Org. Syn.* **2,** 893-951 (1991); H. Heane, *ibid.* 953-973; L. E. Overman, D. J. Ricca, *ibid.* 1007-1046. *Cf.* Betti Reaction; Robinson-Schöpf Reaction.

237. Marschalk Reaction

C. Marschalk *et al., Bull. Soc. Chim. France* **3,** 1545 (1936).

Sodium dithionite reduction of 1-hydroxy- or aminoanthraquinones to their leuco-forms, followed by condensation with aldehydes to yield the 2-alkylated anthraquinones. 2-Hydroxyanthraquinones yield 1-alkylated products:

X = OH, Y = O; X = NH$_2$, Y = NH

Scope and limitations. K. Krohn, W. Baltus, *Tetrahedron* **44,** 49 (1988). Synthetic applications: F. Suzuki *et al., J. Am. Chem. Soc.* **100,** 2272 (1978); L. M. Harwood *et al., Can. J. Chem.* **62,** 1922 (1984); M. T. Furlong *et al., Syn. Commun.* **20,** 2691 (1990); N. R. Ayyangar *et al., Indian J. Chem.* **31B,** 3 (1992).

238. Martinet Dioxindole Synthesis

A. Guyot, J. Martinet, *Compt. Rend.* **156,** 1625 (1913).

Formation of derivatives of dioxindole from esters of mesoxalic acid and aromatic amines or amino quinolines:

J. Martinet, *ibid.* **166,** 851, 998 (1918); *Ann. Chim.* [9] **11,** 85 (1919); W. Langenbeck *et al., Ann.* **499,** 201 (1932); **512,** 276 (1934); W. C. Sumpter, *Chem. Rev.* **37,** 472 (1945); P. L. Julian *et al., Heterocyclic Compounds* **3,** 239 (1952).

239. Meerwein Arylation

H. Meerwein *et al., J. Prakt. Chem.* **152,** 237 (1939).

Arylation of olefins by treatment with diazonium salts in the presence of cupric salts:

$$Z-C=C- \xrightarrow[\text{CuCl}_2]{\text{ArN}_2^+\text{Cl}^-} Z-C=C-$$

$$Z = C=C, C=O, Ar, CN, H$$

Reviews: C. S. Rondestvedt, Jr., *Org. React.* **11,** 189 (1960); *ibid.* **24,** 225-259 (1976); S. Oae *et al., Bull. Chem. Soc. Japan* **53,** 1065 (1980). *Cf.* Pschorr Reaction.

240. Meerwein-Ponndorf-Verley Reduction

H. Meerwein, R. Schmidt, *Ann.* **444,** 221 (1925); W. Ponndorf, *Angew. Chem.* **39,** 138 (1926); A. Verley, *Bull. Soc. Chim. France* **37,** 537, 871 (1925).

Reduction of aldehydes or ketones to the corresponding alcohols with aluminum alkoxides (the reverse of the Oppenauer oxidation, *q.v.*):

$$\underset{O}{R-C-R'} + \underset{OH}{H_3C-CH-CH_3} \xrightarrow{Al[OCH(CH_3)_2]_3} \underset{OH}{R-CH-R'} + \underset{O}{H_3C-C-CH_3}$$

A. L. Wilds, *Org. React.* **2,** 178 (1944); H. O. House, *Modern Synthetic Reactions* (W. A. Benjamin, Menlo Park, California, 2nd ed., 1972) pp 68, 154, 155; R. Monicagli *et al., J. Organometal. Chem.* **50,** C15-C17 (1973); J. Hutton, *Syn. Commun.* **9,** 483 (1979); L. Horner, U. B. Kaps, *Ann.* **1980,** 192; R. M. Kellogg, *Comp. Org. Syn.* **8,** 88-91 (1991). *Cf.* Cannizzaro Reaction; Tischenko Reaction.

241. Meisenheimer Rearrangements

J. Meisenheimer, *Ber.* **52,** 1667 (1919).

Formation of *O, N, N*-trisubstituted hydroxylamines from tertiary amine oxides *via* [1,2]-R group migration, or [2,3]-sigmatropic rearrangement when R' = allyl:

[1,2]-Rearrangements: N. Castagnoli, Jr. *et al., Tetrahedron* **26,** 4319 (1970); J. B. Bremner *et al., Aust. J. Chem.* **41,** 293 (1988); R. Yoneda *et al., Tetrahedron Letters* **35,** 3749 (1994). *Cf.* Stevens Rearrangement; [1,2]-Wittig Rearrangement.

[2,3]-Rearrangements: V. Rautenstrauch, *Helv. Chim. Acta* **56,** 2492 (1973); Y. Yamamato *et al., J. Org. Chem.* **41,** 303 (1976); T. Kurihara *et al., Chem. Pharm. Bull.* **42,** 475 (1994). *Cf.* Mislow-Evans Rearrangement; Sommelet-Hauser Rearrangement; [2,3]-Wittig Rearrangement.

Application of both rearrangements: *idem et al., ibid.* 31.

242. Menschutkin Reaction

N. Menschutkin, *Z. Physik. Chem.* **5,** 589 (1890); **6,** 41 (1890).

Reaction of tertiary amines with alkyl halides to form quaternary salts:

$$R_3N + R'X \longrightarrow R_3R'N^+X^-$$

C. K. Ingold, *Structure and Mechanism in Organic Chemistry* (Cornell Univ. Press, New York, 2nd ed., 1969) p 435; M. H. Abraham, *Progr. Phys. Org. Chem.* **11,** 1 (1974); E. M. Arnett, R. Reich, *J. Am. Chem. Soc.* **102,** 5892 (1980); J. W. Viers, J. C. Schug, *ibid.* **104,** 850 (1982).

243. Merrifield Solid-Phase Peptide Synthesis

R. B. Merrifield, *J. Am. Chem. Soc.* **85,** 2149 (1963).

Synthesis of long peptides involving the following steps: (1) attachment of the C-terminal amino acid to an insoluble polymeric support resin, P, (2) elongation of the peptide chain, and (3) cleavage of the peptide from the resin:

Elongation step:

$$\xrightarrow[\text{2. NEt}_3]{\text{1. CF}_3\text{COOH}}$$

(steps repeated)

Book: C. Birr, *Aspects of Merrifield Peptide Synthesis,* K. Hafner *et al.,* Eds. (Springer-Verlag, New York, 1978) 102 pp. Autobiographical account: B. Merrifield, *Life During a Golden Age of Peptide Chemistry,* J. I. Seeman, Ed. (ACS, Washington, D.C., 1993) pp 54-118. Method for monitoring synthesis: B. D. Larsen *et al., J. Am. Chem. Soc.* **115,** 6247 (1993). Application: D. D. Smith *et al., Int. J. Peptide Protein Res.* **44,** 183 (1994). Brief review: *Biologically Active Peptides* **vol. 1,** W. V. Williams, D. B. Weiner, Eds. (Technomic, Lancaster, Pennsylvania, 1993) pp 115-121, 144-147.

244. Meyer Reaction
G. Meyer, *Ber.* **16,** 1439 (1883).

Preparation of alkylstannonic acids by reacting alkali stannite with an alkyl iodide. When applied to alkali arsenites or plumbites the reaction yields alkylarsonic and alkylplumbonic acids, respectively:

$$\text{Na}_2\text{SnO}_2 + \text{RX} \longrightarrow \text{RSnO}_2\text{Na} + \text{NaX}$$

$$\text{Na}_3\text{AsO}_3 + \text{RX} \longrightarrow \text{RAsO}_3\text{Na}_2 + \text{NaX}$$

W. R. Cullen, *Advan. Organometal. Chem.* **4,** 148 (1966).

245. Meyer-Schuster Rearrangement; Rupe Rearrangement
K. H. Meyer, K. Schuster, *Ber.* **55,** 819 (1922); H. Rupe, E. Kambli, *Helv. Chim. Acta* **9,** 672 (1926).

Acid-catalyzed rearrangement of secondary and tertiary α-acetylenic alcohols to α,β-unsaturated carbonyl compounds: aldehydes when the acetylenic group is terminal, ketones when it is internal:

$$\underset{\text{R}_2\text{CC}}{\overset{\text{OH}}{|}}\equiv\text{CR}' \xrightarrow{\text{H}^+} \text{R}_2\text{C}=\text{CHCOR}'$$

The conversion of tertiary alkylacetylenic carbinols with a terminal acetylenic group to predominantly α,β-unsaturated ketones and not the expected aldehydes, is referred to as the **Rupe rearrangement**:

$$\text{RCH}_2-\underset{\underset{\text{R}'}{|}}{\overset{\overset{\text{OH}}{|}}{\text{C}}}-\text{C}\equiv\text{CH} \xrightarrow[\Delta]{\text{HCOOH}} \text{R}-\text{CH}=\text{CR}'\text{COCH}_3$$

Reviews: R. Heilmann, R. Glenat, *Ann. Chim. (Paris)* **8,** 178 (1963); S. A. Vartanyan, S. O. Babanyan, *Russ. Chem. Rev.* **36,** 670 (1967); S. Swaminathan, K. V. Narayanan, *Chem. Rev.* **71,** 429 (1971); M. Edens *et al., J. Org. Chem.* **42,** 3403 (1977); K. E. Stevens, P. Yates, *Chem. Commun.* **1980,** 990. Titanium/copper or silver based catalysis: P. Chabardes, *Tetrahedron Letters* **29,** 6253 (1988).

246. Meyer Synthesis (Victor Meyer Synthesis)
V. Meyer, O. Stuber, *Ber.* **5,** 203 (1872).

Formation of aliphatic nitrites and nitro derivatives by the reaction of aliphatic halides with metal nitrites:

$$\text{RX} + \text{MNO}_2 \longrightarrow \text{RONO} + \text{RNO}_2 + \text{MX}$$

R. B. Reynolds, H. Adkins, *J. Am. Chem. Soc.* **51,** 279 (1929). *Reviews:* H. B. Hass, E. F. Riley, *Chem. Rev.* **32,** 373 (1943); N. Kornblum, *Org. React.* **12,** 101-156 (1962). Application to the synthesis of α,ω-dinitroalkanes: J. K. Stille, E. D. Vessel, *J. Org. Chem.* **25,** 478 (1960); G. Leston, *Org. Syn.* **4,** 368 (1963).

247. Meyers Aldehyde Synthesis
A. I. Meyers *et al., J. Am. Chem. Soc.* **91,** 763 (1969); *eidem, J. Org. Chem.* **38,** 36 (1973).

Synthesis of aldehydes from alkylhalides and 2-lithiomethyltetrahydro-3-oxazine:

J. March, *Advanced Organic Chemistry* (Wiley-Interscience, New York, 4th ed., 1992) pp 478-479; A. I. Meyers *et al., J. Org. Chem.* **46,** 783 (1981).

248. Michael Reaction (Addition, Condensation)
A. Michael, *J. Prakt. Chem.* [2] **35,** 349 (1887).

Base-promoted conjugate addition of carbon nucleophiles (donors) to activated unsaturated systems (acceptors):

donor = malonates, cyanoacetates, acetoacetates, carboxylic esters, ketones, aldehydes, nitriles, nitro compounds, sulfones
acceptor = α,β-unsaturated ketones, esters, aldehydes, amides, carboxylic acids, nitriles
sulfoxides, sulfones, nitro compounds, phosphonates, phosphoranes
base = $NaOCH_2CH_3$, $NH(CH_2CH_3)_2$, KOH, $KOC(CH_3)_3$, $N(CH_2CH_3)_3$, NaH, BuLi, LDA

Reviews: E. D. Bergmann *et al., Org. React.* **10,** 179-555 (1959); H. O. House, *Modern Synthetic Reactions* (W. A. Benjamin, Menlo Park, California, 2nd ed., 1972) pp 595-623; M. E. Jung, *Comp. Org. Syn.* **4,** 1-67 (1991). Extension to organometallic nucleophiles: V. J. Lee, *ibid.* 69-137, 139-168; J. A. Kozlowski, ibid. 169-198. Reviews of stereoselective synthesis: H.-G. Schmalz, *ibid.* 199-236; D. A. Oare, C. H. Heathcock, *Top. Stereochem.* **20,** 87-170 (1991); J. d'Angelo *et al., Tetrahedron Asymmetry* **3,** 459-505 (1992). *Cf.* Nagata Hydrocyanation; Robinson Annulation.

249. Michaelis-Arbuzov Reaction
A. Michaelis, R. Kaehne, *Ber.* **31,** 1048 (1898); A. E. Arbuzov, *J. Russ. Phys. Chem. Soc.* **38,** 687 (1906); *Chem. Zentr.* **1906,** II, 1639.

Formation of monoalkylphosphonic esters from alkyl halides and trialkyl phosphites, *via* the intermediate phosphonium salt:

$$P(OR)_3 + R'X \longrightarrow [(RO)_3PR']X \xrightarrow{\Delta} (RO)_2P(O)R'$$

G. M. Kosolapoff, *Org. React.* **6,** 276 (1951); K. Sasse, *Houben-Weyl* **12/1,** 433 (1963); B. A. Arbuzov, *Pure Appl. Chem.* **9,** 307 (1964); D. Redmore, *Chem. Rev.* **71,** 317 (1971); G. Bauer, G. Haegele, *Angew. Chem. Int. Ed.* **16,** 477 (1977); A. K. Bhattacharya, G. Thyagarajan, *Chem. Rev.* **81,** 415 (1981).

250. Miescher Degradation
C. Meystre *et al., Helv. Chim. Acta* **27,** 1815 (1944).

Adaptation of the Barbier-Wieland degradation, *q.v.*, to permit simultaneous elimination of three carbon atoms, as in degradation of the bile acid side chain to the methyl ketone stage. Conversion of the methyl ester of the bile acid to the tertiary alcohol, followed by dehydration, bromination, dehydrohalogenation and oxidation of the diene yields the chain-shortened ketone:

$$RR'CHCH_2CH_2COOCH_3 \xrightarrow{PhMgBr} RR'CHCH_2CH_2C(OH)Ph_2 \xrightarrow{-H_2O} RR'CHCH_2CH=CPh_2 \xrightarrow{(CH_2CO)_2NBr}$$

$$RR'CHCHBrCH=CPh_2 \xrightarrow{-HBr} RR'C=CH-CH=CPh_2 \xrightarrow{CrO_3} RCOR' + OHCCH=CPh_2$$

C. W. Shoppee, *Ann. Repts.* (Chem. Soc. London) **44,** 184 (1947); F. S. Spring, *J. Chem. Soc.* **1950,** 3355; A. Wettstein, G. Anner, *Experientia* **1954,** 407; C. J. W. Brooks, *Rodd's Chemistry of Carbon Compounds* **IID,** 26 (1970); P. G. Marshall, *ibid.* 233, 253, 323.

251. Mignonac Reaction
G. Mignonac, *Compt. Rend.* **172,** 223 (1921).

Formation of amines by catalytic hydrogenation of aldehydes or ketones in liquid ammonia and absolute ethanol in the presence of a nickel catalyst:

$$RCOR' + NH_3 + H_2 \xrightarrow{\text{Ni}} RR'CHNH_2 + H_2O$$

F. Randvere, *Anales farm. bioquim. (Buenos Aires)* **18**, 81 (1948); *Houben-Weyl* **4/2**, 51 (1955).

252. Milas Hydroxylation of Olefins

N. A. Milas *et al., J. Am. Chem. Soc.* **58**, 1302 (1936); **59**, 543, 2342, 2345 (1937); **61**, 1844 (1939); **62**, 1841 (1940).

Formation of *cis*-glycols by reaction of alkenes with hydrogen peroxide and either ultraviolet light or a catalytic amount of osmium, vanadium, or chromium oxide:

F. D. Gunstone, *Advan. Org. Chem.* **1**, 115 (1960); P. N. Rylander, *Organic Syntheses with Noble Metal Catalysts* (Academic Press, New York, 1973) p 60. *Cf.* Sharpless Dihydroxylation.

253. Mislow-Evans Rearrangement

P. Bickart *et al., J. Am. Chem. Soc.* **90**, 4869 (1968); D. A. Evans *et al., ibid.* **93**, 4956 (1971).

[2,3]-Sigmatropic rearrangement of allylic sulfoxides to allylic sulfenates which are captured by thiophiles to generate the allylic alcohols, thereby effecting the 1,3-transposition of sulfoxide and alcohol functions. The reverse process is accomplished by treating the alcohol with arylsulfenyl chloride, followed by thermal rearrangement of the sulfenate to generate the allylic sulfoxide:

thiophile = $P(OCH_3)_3$, $(CH_3CH_2)_2NH$, piperidine, PhS$^-$

Early review. D. A. Evans, G. C. Andrews, *Accts. Chem. Res.* **7**, 147 (1974). Acid-catalyzed modification: Y. Masaki *et al., Chem. Pharm. Bull.* **33**, 2531 (1985). Synthetic applications: H. J. Reich, S. Wollowitz, *J. Am. Chem. Soc.* **104**, 7051 (1982); G. H. Posner *et al., J. Org. Chem.* **52**, 4836 (1987); A. Padwa *et al., ibid.* **56**, 4252 (1991). *Cf.* Meisenheimer Rearrangements; [2,3]-Wittig Rearrangement.

254. Mitsunobu Reaction

O. Mitsunobu *et al., Bull. Chem. Soc. Japan* **40**, 935 (1967); O. Mitsunobu, Y. Yamada, *ibid.* 2380.

Condensation of alcohols and acidic components (NuH) on treatment with dialkyl azodicarboxylates and trialkyl- or triarylphosphines occurring primarily with inversion of configuration *via* the proposed intermediary oxyphosphonium salts:

NuH = phosphoric mono- and diesters, carboxylic acids, phenols, imides, oximes, hydroxymates, heterocycles, thiols, thioamides, β-keto esters

Methods development: J. E. Macor, J. M. Wehner, *Heterocycles* **35**, 349 (1993); R. S. Coleman, E. B. Grant, *Tetrahedron Letters* **35**, 8341 (1994); R. F. C. Brown *et al., Tetrahedron* **50**, 5469 (1994). Synthetic applications: M. L. Edwards *et al., ibid.* 5579; M. A. Poelert *et al., Rec. Trav. Chim. Pays-Bas* **113**, 355 (1994). *Reviews:* O. Mitsunobu, *Synthesis* **1981**, 1-28; B. R. Castro, *Org. React.* **29**, 1-162 (1983); D. L. Hughes, *ibid.* **42**, 335-656 (1992).

255. Nagata Hydrocyanation

W. Nagata *et al., Tetrahedron Letters* **1962,** 461.

Alkylaluminum-mediated 1,4-addition of hydrogen cyanide to α,β-unsaturated carbonyl compounds:

Early review: W. Nagata, M. Yoshioka, *Org. React.* **25,** 255-476 (1977). Synthetic application: T. F. Gallagher, J. L. Adams, *J. Org. Chem.* **57,** 3347 (1992). *Cf.* Michael Reaction.

256. Nametkin Rearrangement

S. S. Nametkin, *Ann.* **432,** 207 (1923).

A special case of carbonium ion rearrangement in camphene hydrochloride derivatives involving the migration of a methyl group:

H. Henecka, *Houben-Weyl* **4/2,** 16 (1955); P. S. Moervs *et al., J. Am. Chem. Soc.* **100,** 260 (1978). *Cf.* Retropinacol Rearrangement; Wagner-Meerwein Rearrangement.

257. Nazarov Cyclization Reaction

I. N. Nazarov *et al., Izv. Akad. Nauk S.S.S.R., Otd. Khim. Nauk* **1942,** 200.

Protic or Lewis acid-catalyzed electrocyclic ring closure of divinyl ketones, or their equivalents, to yield 2-cyclopentenones:

Silicon-directed cyclizations to α-methylenecyclopentanones: H. T. Kang *et al., Tetrahedron Letters* **33,** 3495 (1992). *Reviews:* S. E. Denmark, *Comp. Org. Syn.* **5,** 751-784 (1991); K. L. Habermas *et al., Org. React.* **45,** 1-158 (1994).

258. Neber Rearrangement

P. W. Neber, A. v. Friedolsheim, *Ann.* **449,** 109 (1926); P. W. Neber, G. Huh, *ibid.* **515,** 283 (1935).

Formation of α-amino ketones by treatment of sulfonic esters of ketoximes with potassium ethoxide, followed by hydrolysis:

Reviews: C. O'Brien, *Chem. Rev.* **64,** 81 (1964); C. G. McCarty in *The Chemistry of the Carbon-Nitrogen Double Bond,* S. Patai, Ed. (Interscience, New York, 1970) p 447; Y. Tamura *et al., Synthesis* **1973,** 215; R. F. Parcell, J. C. Sanchez, *J. Org. Chem.* **46,** 5229 (1981); K. Maruoka, H. Yamamoto, *Comp. Org. Syn.* **6,** 786-789 (1991).

259. Nef Reaction

J. U. Nef, *Ann.* **280,** 263 (1894).

Formation of aldehydes and ketones from primary and secondary nitroalkanes, respectively, by treatment of their salts with sulfuric acid:

Reviews: P. Salomaa in *The Chemistry of the Carbonyl Group,* S. Patai, Ed. (Interscience, N.Y., 1966) pp 177-210; H. W. Pinnick, *Org. React.* **38,** 655-792 (1990); D. S. Grierson, H.-P. Husson, *Comp. Org. Syn.* **6,** 937-944 (1991).

260. Nef Synthesis
J. U. Nef, *Ann.* **308,** 281 (1899).
　Addition of sodium acetylides to aldehydes and ketones to yield acetylenic carbinols; occasionally and erroneously referred to as the Nef reaction, *q.v.*:

$$RCOR' + NaC\equiv CH + H_2O \longrightarrow RR'C(OH)C\equiv CH + NaOH$$

　Farbenfabriken Bayer, **Ger.** pats. **280,226; 285,770** (1913); J. H. Saunders, *Org. Syn.* **20,** 40 (1940); A. W. Johnson, *The Chemistry of the Acetylenic Compounds* (London, 1946) p 11; C. D. Hurd, W. D. McPhee, *J. Am. Chem. Soc.* **69,** 239 (1947); W. Oroschnik, A. O. Mebane, *ibid.* **71,** 2062 (1949); R. A. Raphael, *Acetylenic Compounds in Organic Synthesis* (London, 1955) p 10. *Cf.* Arens-van Dorp Synthesis; Favorskii-Babayan Synthesis.

261. Nencki Reaction
M. Nencki, N. Sieber, *J. Prakt. Chem.* (2) **23,** 147 (1881).
　The ring acylation of phenols with acids in the presence of zinc chloride, or the modification of the Friedel-Crafts reaction, *q.v.*, by substitution of ferric chloride for aluminum chloride.
　M. Nencki, W. Schmid, *ibid.* 546; M. Nencki, *ibid.* **25,** 273 (1882); U. S. Chiema, K. Venkataraman, *J. Chem. Soc.* **1932,** 918; C. W. Schellhammer, *Houben-Weyl* **7/2a,** 284 (1973). *Cf.* Darzens-Nenitzescu Synthesis of Ketones.

262. Nenitzescu Indole Synthesis
C. D. Nenitzescu, *Bull. Soc. Chim. Romania* **11,** 37 (1929).
　Synthesis of 5-hydroxyindole derivatives by condensation of *p*-benzoquinone with β-aminocrotonic esters:

Reviews: R. K. Brown in *The Chemistry of Heterocyclic Compounds,* A. Weissenberger, E. C. Taylor, Eds., *Indoles Part Two,* W. J. Houlihan, Ed. (Wiley, New York, 1972) p 413; G. R. Allen, Jr., *Org. React.* **20,** 337 (1973); U. Kuecklander, W. Huehnermann, *Arch. Pharm.* **312,** 515 (1979); J. L. Bernier, J. P. Henichart, *J. Org. Chem.* **46,** 4197 (1981).

263. Nenitzescu Reductive Acylation
C. D. Nenitzescu, E. Cioranescu, *Ber.* **69,** 1820 (1936).
　Hydrogenative acylation of cycloolefins with acid chlorides in the presence of aluminum chloride; with five- and six-membered rings no change in ring size occurs but with seven-membered rings rearrangement takes place with formation of a cyclohexane derivative:

　C. Nenitzescu, C. N. Ionescu, *Ann.* **491,** 189 (1931); C. D. Nenitzescu, J. P. Cantuniari, *ibid.* **510,** 269 (1934); C. D. Nenitzescu, I. Chicos, *Ber.* **68,** 1584 (1935); C. A. Thomas, *Anhydrous Aluminum Chloride in Organic Chemistry* (New York, 1941) p 759; S. L. Friess, R. Pinson, *J. Am. Chem. Soc.* **73,** 3512 (1951); Olah, *Friedel-Crafts and Related Reactions* **vol. III,** Part 2 (New York, 1964) p 1069. *Cf.* Darzens-Nenitzescu Synthesis of Ketones.

264. Nicholas Reaction
R. F. Lockwood, K. M. Nicholas, *Tetrahedron Letters* **1977,** 4163.
　The reaction of dicobalthexacarbonyl-stabilized propargyl cations with nucleophiles, followed by oxidative demetalation to yield propargylated products:

$$R^1—C≡C—\underset{R^4}{\overset{OR^2}{\underset{|}{\overset{|}{C}}}}—R^3 \quad \xrightarrow[\text{2. H}^+ \text{ or Lewis acid}]{\text{1. Co}_2(CO)_8} \quad \left[R^1—C≡C—\underset{\underset{Co_2(CO)_6}{|}}{\overset{+}{C}}\underset{R^4}{\overset{R^3}{<}} \right] \quad \xrightarrow[\text{2. [O]}]{\text{1. NuH}} \quad R^1—C≡C—\underset{R^4}{\overset{Nu}{\underset{|}{C}}}{<}^{R^3}$$

R^1 = H, SiR$_3$, alkyl, aryl; R^2 = H, CH$_3$

H$^+$ = HBF$_4 \cdot$O(CH$_3$)$_2$, CF$_3$COOH; Lewis acid = BF$_3 \cdot$OEt$_2$, EtAlCl$_2$, TiCl$_4$, Bu$_2$BOSO$_2$CF$_3$

NuH = electron rich aromatics, β-dicarbonyl compounds, ketones, enolates, allyl silanes, hydride, amines, enamines

[O] = Fe(NO$_3$)$_3$, Ce(NH$_4$)$_2$(NO$_3$)$_6$, (CH$_3$)$_3$NO

Stereochemical study: S. L. Schreiber *et al., J. Am. Chem. Soc.* **109,** 5749 (1987). Scope and limitations: K. D. Roth, *Synlett* **1992,** 435; K. D. Roth, U. Müller, *Tetrahedron Letters* **34,** 2919 (1993). Synthetic applications: P. A. Jacobi, W. Zheng, *ibid.* 2581, 2585; E. Tyrrell *et al., Synlett* **1993,** 769. *Review:* K. M. Nicholas, *Accts. Chem. Res.* **20,** 207-214 (1987).

265. Niementowski Quinazoline Synthesis

S. v. Niementowski, *J. Prakt. Chem.* [2] **51,** 564 (1895).

Formation of 4-oxo-3,4-dihydroquinazolines by cyclization of the reaction products of anthranilic acid and amides:

Reviews: T. A. Williamson, *Heterocyclic Compounds* **6,** 331 (1957); W. L. F. Armarego, *Advan. Heterocyclic Chem.* **1,** 253 (1963); E. Cuny *et al., Tetrahedron Letters* **21,** 3029 (1980).

266. Niementowski Quinoline Synthesis

S. v. Niementowski, *Ber.* **27,** 1394 (1894); **28,** 2809 (1895); **38,** 2044 (1905); **40,** 4285 (1907).

Formation of γ-hydroxyquinoline derivatives from anthranilic acids and carbonyl compounds:

R. H. Manske, *Chem. Rev.* **30,** 127 (1942); T. A. Williamson, *Heterocyclic Compounds* **6,** 331 (1957); W. L. F. Armarego, *Quinazolines* (Interscience, New York, 1967) p 74; E. Cuny *et al., Tetrahedron Letters* **21,** 3029 (1980). *Cf.* Friedlaender Synthesis; Pfitzinger Reaction.

267. Nierenstein Reaction

D. A. Clibbens, M. Nierenstein, *J. Chem. Soc.* **107,** 1491 (1915).

Formation of ω-chloroacetophenones by reaction of diazomethane in dry ether with aroyl chlorides. Coumaranones are obtained if an *ortho*-hydroxy group is present:

$$ArCOCl + CH_2N_2 \longrightarrow ArCOCH_2Cl + N_2$$

W. E. Bachman, W. S. Struve, *Org. React.* **1,** 38 (1942); Y. Miyahara, *J. Heterocycl. Chem.* **16,** 1147 (1979).

268. Norrish Type Cleavage

R. G. W. Norrish, C. H. Bamford, *Nature* **138,** 1016 (1936); **140,** 195 (1937).

Norrish Type I Cleavage: Homolytic cleavage of aldehydes and ketones originating from their excited nπ* state. Synthetically useful for the ring cleavage of cyclic ketones:

Norrish Type II Cleavage: Reaction originating from the nπ* excited state of aldehydes and ketones that involves intramolecular γ-hydrogen abstraction followed by cleavage of the resulting diradical to an olefin and an enol which tautomerizes to the carbonyl compound:

P. J. Wagner, *Accts. Chem. Res.* **4,** 168 (1971); J. D. Coyle, H. A. J. Carless, *Chem. Soc. Rev.* **1,** 465 (1972); O. L. Chapman, D. S. Weiss, *Org. Photochem.* **3,** 197 (1973); J. March, *Advanced Organic Chemistry* (Wiley-Interscience, New York, 4th ed., 1992) p 242. *Cf.* McLafferty Rearrangement.

269. Noyori Hydrogenation

T. Ikariya *et al., Chem. Commun.* **1985,** 922; R. Noyori *et al., J. Am. Chem. Soc.* **108,** 7117 (1986).

Homogeneous asymmetric catalytic hydrogenation of olefinic and carbonyl bonds mediated by enantiopure ruthenium(II) BINAP complexes. The substrates must have coordinating functionalities in neighboring positions which serve as directing groups during the transformation:

Detailed experimental procedure: M. Kitamura *et al., Org. Syn.* **71,** 1 (1993). Methods development for enamide substrates: *idem et al., J. Org. Chem.* **59,** 297 (1994). Development and use of arene substituted BINAP catalysts: K. Mashima *et al., ibid.* 3064. *Reviews:* H. Takaya *et al., Adv. Chem. Ser.* **230,** 123-142 (1992); *idem et al.,* "Asymmetric Hydrogenation" in *Catalytic Asymmetric Synthesis,* I. Ojima, Ed. (VCH, New York, 1993) pp 1-39; R. Noyori, *Asymmetric Catalysis in Organic Synthesis* (John Wiley & Sons, New York, 1994) pp 16-94.

270. Oppenauer Oxidation

R. V. Oppenauer, *Rec. Trav. Chim.* **56,** 137 (1937).

The aluminum alkoxide-catalyzed oxidation of a secondary alcohol to the corresponding ketone (the reverse of the Meerwein-Ponndorf-Verley reduction, *q.v.*):

$$RCHOHR' + CH_3COCH_3 \underset{}{\overset{Al(OC_3H_7)_3}{\rightleftharpoons}} RCOR' + CH_3CHOHCH_3$$

T. Beresin in *Newer Methods of Preparative Organic Chemistry,* English Ed. (Interscience, New York, 1948) p 125; C. Djerassi, *Org. React.* **6,** 207 (1951); L. Horner, U. B. Kaps, *Ann.* **1980,** 192. *Cf.* Cannizzaro Reaction.

271. Overman Rearrangement

L. E. Overman, *J. Am. Chem. Soc.* **96,** 597 (1974); **98,** 2901 (1976).

Formal [3,3]-sigmatropic rearrangement of the trichloroacetimidate of allylic alcohols to allylic trichloroacetamides, thereby transposing the hydroxyl and amino functions with good chirality transfer:

Early review: L. E. Overman, *Accts. Chem. Res.* **13**, 218-224 (1980). Synthetic applications: M. Isobe *et al., Tetrahedron Letters* **31**, 3327 (1990); T. Allmendinger *et al., ibid.* 7301; J. Gonda *et al., Synthesis* **1993**, 729. Use of a chiral template and mechanistic studies: T. Eguchi *et al., Tetrahedron* **49**, 4527 (1993). *Cf.* Claisen Rearrangement.

272. Oxo Process (Hydroformylation Reaction)

O. Roelen, **U.S.** pat. **2,327,066** (1943); R. H. Hasek (Eastman), *Org. Chem. Bull.* **27**, No. 1 (1955).

Formation of alcohols from olefins, carbon monoxide and hydrogen in the liquid phase in the presence of catalysts (metallic cobalt compounds such as Raney cobalt or cobalt carbonyls) at 115-190° and high pressures (100-200 atmospheres) in a Fischer-Tropsch-type reaction, *q.v.* The process is sometimes carried out in two stages, the initial stage giving largely aldehydes which are then reduced to the alcohols.

B. Cornils, "Hydroformylation. Oxo Synthesis, Roelen Reaction" in *New Syntheses with Carbon Monoxide,* J. Falbe, Ed. (Springer-Verlag, Berlin, 1980) pp 1-225. Reppe modification (olefin + CO + H_2O + $Fe(CO)_5$): R. Massoudi *et al., J. Am. Chem. Soc.* **109**, 7428 (1987).

273. Paal-Knorr Pyrrole Synthesis

C. Paal, *Ber.* **18**, 367 (1885); L. Knorr, *ibid.* 299.

Formation of pyrroles by heating 1,4-dicarbonyl compounds with ammonia or primary amines in a sealed tube:

H. Fischer, H. Orth, *Die Chemie des Pyrrols* **1** (Leipzig, 1934) p 34; D. M. Young, C. F. H. Allen, *Org. Syn.* **16**, 25 (1936); A. H. Corwin, *Heterocyclic Compounds* **1**, 290 (1950); N. P. Buu-Hoi *et al., J. Org. Chem.* **20**, 639, 850 (1955); H. H. Wassermann *et al., Tetrahedron* **32**, 1863 (1976). *Cf.* Hantzsch Pyrrole Synthesis; Knorr Pyrrole Synthesis.

274. Parham Cyclization

W. E. Parham *et al., J. Org. Chem.* **40**, 2394 (1975).

Four- to seven-membered ring annulation of aryl bromides bearing *ortho* side chains having an electrophilic moiety, accomplished by halogen-metal exchange and subsequent nucleophilic ring closure:

E = COOH, CONR_2, epoxide, CH_2Br, CH_2Cl, OCONR_2, NCHAr, CONRCOCH_2R, POPh_2

Synthetic applications: G. J. Quallich *et al., ibid.* **57**, 761 (1992); M. R. Paleo *et al., ibid.* 2029; A. Couture *et al., Chem. Commun.* **1994**, 1329. *Review:* W. E. Parham, C. K. Bradsher, *Accts. Chem. Res.* **15**, 300-305 (1982).

275. Passerini Reaction

M. Passerini, *Gazz. Chim. Ital.* **51**, 126, 181 (1921).

Formation of α-hydroxycarboxamides on treatment of an isonitrile with a carboxylic acid and an aldehyde or ketone:

Synthetic applications: J. R. Falck, S. Manna, *Tetrahedron Letters* **22**, 619 (1981); R. Bossio *et al., Synthesis* **1993**, 783. Modifications: W. E. Lumma, *J. Org. Chem.* **46**, 3668 (1981); T. Carofiglio *et al., Organometallics* **12**, 2726 (1993). *Reviews:* I. Ugi, *Angew. Chem. Int. Ed.* **1**, 8 (1962); I. Ugi *et al.* in *Isonitrile Chemistry,* I. Ugi, Ed. (Academic Press, New York, 1971) pp 133-143; I. Ugi *et al., Comp. Org. Syn.* **2**, 1083-1087 (1991).

276. Paterno-Büchi Reaction

E. Paterno, G. Chieffi, *Gazz. Chim. Ital.* **39**, 341 (1909); G. Büchi *et al., J. Am. Chem. Soc.* **76**, 4327 (1954).

Formation of oxetanes by photochemical cycloaddition of carbonyl compounds to olefins:

D. R. Arnold, *Advan. Photochem.* **6**, 301 (1968); G. Jones, II, *Org. Photochem.* **5**, 1 (1981); S. C. Freilich, K. S. Peters, *J. Am. Chem. Soc.* **103**, 6255 (1981); J. A. Porco, Jr., S. L. Schreiber, *Comp. Org. Syn.* **5**, 151-192 (1991).

277. Pauson-Khand Reaction

I. U. Khand *et al., Perkin Trans I* **1973**, 977.

The formal [2+2+1] cycloaddition of an alkene, alkyne, and carbon monoxide to form cyclopentenones:

Use of a chiral auxillary: X. Verdaguer *et al., J. Am. Chem. Soc.* **116**, 2153 (1994); V. Bernardes *et al., J. Org. Chem.* **60**, 6670 (1995). Catalytic version: N. Jeong *et al., J. Am. Chem. Soc.* **116**, 3159 (1994). *Reviews:* N. E. Schore, *Org. React.* **40**, 1-90 (1991); *idem, Comp. Org. Syn.* **5**, 1037-1064 (1991).

278. Payne Rearrangement

G. B. Payne, *J. Org. Chem.* **27**, 3819 (1962).

Base-promoted isomerization of 2,3-epoxyalcohols:

In conjunction with nucleophilic ring opening: T. Katsuki *et al., ibid.* **47**, 1373 (1982); C. H. Behrens *et al., ibid.* **50**, 5687 (1985); P. C. B. Page *et al., Perkin Trans. I* **1990**, 1375; T. Konosu *et al., Chem. Pharm. Bull.* **40**, 562 (1992).

279. Pechmann Condensation

H. v. Pechmann, C. Duisberg, *Ber.* **16**, 2119 (1883).

Synthesis of coumarins by condensation of phenols with β-keto esters in the presence of Lewis acid catalysts:

Early reviews: S. Sethna, *Chem. Rev.* **36**, 10 (1945); S. Sethna, R. Phadke, *Org. React.* **7**, 1 (1953). T. Kappe, E. Ziegler, *Org. Prep. Proced.* **1**, 61 (1969); T. Kappe, C. Mayer, *Synthesis* **1981**, 524; A. G. Osborne, *Tetrahedron* **37**, 2021 (1981). *Cf.* Bischler-Napieralski Reaction; Simonis Chromone Cyclization.

280. Pechmann Pyrazole Synthesis

H. v. Pechmann, *Ber.* **31**, 2950 (1898).

Formation of pyrazoles from acetylenes and diazomethane. The analogous addition of diazoacetic esters to the triple bond yields pyrazolecarboxylic acid derivatives:

R. A. Raphael, *Acetylenic Compounds in Organic Synthesis* (London, 1955) p 179; T. L. Jacobs, *Heterocyclic Compounds* **5**, 70 (1957); B. Eistert *et al., Houben-Weyl* **10/4**, 840 (1968). *Cf.* Knorr Pyrazole Synthesis.

281. Pellizzari Reaction

G. Pellizzari, *Gazz. Chim. Ital.* **41**, II, 20 (1911).

Formation of substituted 1,2,4-triazoles by the condensation of amides and acyl hydrazines. When the acyl groups of the amide and the acylhydrazine are different, interchange of acyl groups may occur with formation of a mixture of triazoles:

$$R\text{—}C(\text{OH})(\text{NH}) + HN(C_6H_5)\text{—}N\text{=}C(\text{OH})\text{—}R \xrightarrow{\Delta} \text{triazole} + 2H_2O$$

M. R. Atkinson, J. B. Polya, *J. Chem. Soc.* **1952**, 3418; P. Karrer, *Organic Chemistry* (New York, 4th ed., 1950) p 802; C. W. Bird, C. K. Wong, *Tetrahedron Letters* **1974**, 1251. *Cf.* Einhorn-Brunner Reaction.

282. Pelouze Synthesis
J. Pelouze, *Ann.* **10,** 249 (1834).

Formation of nitriles from alkali cyanides by alkylation with alkyl sulfates or alkyl phosphates:

$$ROSO_2OK + KCN \longrightarrow RCN + K_2SO_4$$

V. Migrdichian, *Chemistry of Organic Cyanogen Compounds* (New York, 1947) p 6; D. T. Mowry, *Chem. Rev.* **42,** 192 (1948); P. Kurtz, *Houben-Weyl* **8,** 306 (1952).

283. Perkin Alicyclic Synthesis
W. H. Perkin, Jr., *Ber.* **16,** 1793 (1883).

Synthesis of alicyclic compounds from α,ω-dihaloalkanes and compounds containing active methylene groups in the presence of sodium ethoxide:

H. O. House, *Modern Synthetic Reactions* (W. A. Benjamin, Inc., Menlo Park, California, 2nd ed., 1972) pp 492-570. *Cf.* Malonic Ester Syntheses.

284. Perkin Reaction
W. H. Perkin, *J. Chem. Soc.* **21,** 53, 181 (1868); **31,** 388 (1877).

Formation of α,β-unsaturated carboxylic acids by aldol condensation, *q.v.,* of aromatic aldehydes and acid anhydrides in the presence of an alkali salt of the acid:

Reviews: J. R. Johnson, *Org. React.* **1,** 210 (1942); H. O. House, *Modern Synthetic Reactions* (W. A. Benjamin, Menlo Park, California, 2nd ed., 1972) pp 660-663; N. Poonia *et al., Bull. Chem. Soc. Japan* **53,** 3338 (1980); T. Rosen, *Comp. Org. Syn.* **2,** 395-408 (1991). *Cf.* Erlenmeyer-Plöchl Azlactone and Amino Acid Synthesis; Stobbe Condensation.

285. Perkin Rearrangement (Coumarin-Benzofuran Ring Contraction)
W. H. Perkin, *J. Chem. Soc.* **23,** 368 (1870).

Formation of benzofuran-2-carboxylic acids and benzofurans by heating 3-halocoumarins with alkali:

R. C. Elderfield, V. B. Meyer, *Heterocyclic Compounds* **2,** 2, 5 (1951).

286. Perkow Reaction

W. Perkow et al., Naturwiss. **39**, 353 (1952).

Formation of enol phosphates on treatment of α-halocarbonyl compounds with trialkyl phosphites:

F. W. Lichtenthaler, *Chem. Rev.* **61**, 607 (1961); B. Miller in *Topics in Phosphorus Chemistry* **vol. 2**, M. Grayson, E. J. Griffith, Eds. (John Wiley, New York, 1965) p 178; K. Sasse, *Houben-Weyl* **12/1**, 423 (1963); A. J. Kirby, S. G. Warren, *The Organic Chemistry of Phosphorus* (Elsevier, Amsterdam, 1967) p 123; B. A. Arbuzow, *Pure Appl. Chem.* **9**, 306 (1964); I. J. Borowitz et al., *J. Org. Chem.* **38**, 1713 (1973); T. Winkler, W. L. Bencze, *Helv. Chim. Acta* **63**, 402 (1980); M. Sekine et al., *J. Org. Chem.* **46**, 4030 (1981).

287. Peterson Reaction (Olefination)

D. J. Peterson, *J. Org. Chem.* **33**, 780 (1968).

Reaction of α-silyl carbanions with carbonyl compounds yielding β-silylalkoxides which undergo instantaneous elimination to afford olefins:

L. Birkofer, Ö. Stiehl, *Top. Curr. Chem.* **88**, 58 (1980); E. Colvin, *Silicon in Organic Synthesis* (Butterworth, London, 1981) p 143; D. J. Ager, *Synthesis* **1984**, 384-398; *idem, Org. React.* **38**, 1-223 (1990); S. E. Kelly, *Comp. Org. Syn.* **1**, 731-737, 782-783 (1991). *Cf.* Tebbe Olefination; Wittig Reaction.

288. Petrenko-Kritschenko Piperidone Synthesis

P. Petrenko-Kritschenko et al., *Ber.* **39**, 1358 (1906); **40**, 2882 (1907); **41**, 1692 (1908); **42**, 2020, 3683 (1909).

Formation of piperidones *via* cyclization of two moles of aldehyde and one mole each of acetonedicarboxylic ester and ammonia or a primary amine:

R. Robinson, *J. Chem. Soc.* **111**, 762, 876 (1917); C. Mannich, O. Hieronimus, *Ber.* **75**, 49 (1942); H. S. Mosher, *Heterocyclic Compounds* **1**, 659 (New York, 1950). *Cf.* Robinson-Schöpf Reaction.

289. Pfau-Plattner Azulene Synthesis

A. St. Pfau, P. A. Plattner, *Helv. Chim. Acta* **22**, 202 (1939).

Formation of azulenes by ring enlargement of indanes on addition of diazoacetic ester, hydrolysis, dehydrogenation and decarboxylation of the resulting acid:

P. A. Plattner *et al., ibid.* **23,** 907 (1940); **24,** 483 (1941); **25,** 590 (1942); B. Eistert, *Newer Methods of Preparative Organic Chemistry* (Interscience, New York, 1948) p 555; D. H. Reid, *Chem. Soc. Spec. Publ.* **12,** 69 (1958); K. Hafner, *Angew. Chem.* **70,** 419 (1958). *Cf.* Buchner Method of Ring Enlargement.

290. Pfitzinger Reaction

W. Pfitzinger, *J. Prakt. Chem.* [2] **33,** 100 (1886); **38,** 582 (1888).

Formation of quinoline-4-carboxylic acids by condensation of isatic acids from isatin with α-methylene carbonyl compounds; subsequent decarboxylation yields quinolines:

C. Hollins, *The Synthesis of Nitrogen Ring Compounds* (London, 1924) p 286; R. H. Manske, *Chem. Rev.* **30,** 126 (1942); F. W. Bergstrom, *ibid.* **35,** 152 (1944); R. C. Elderfield, *Heterocyclic Compounds* **4,** 47 (1952); N. P. Buu-Hoi *et al., Bull. Soc. Chim. France* **1968,** 2476; M. H. Palmer, P. S. McIntyre, *J. Chem. Soc. B* **1969,** 539. *Cf.* Friedlaender Synthesis; Niementowski Quinoline Synthesis.

291. Pfitzner-Moffatt Oxidation (Moffatt Oxidation)

K. E. Pfitzner, J. G. Moffatt, *J. Am. Chem. Soc.* **85,** 3027 (1963).

Mild oxidation of primary and secondary alcohols, promoted by dicyclohexylcarbodiimide activation of dimethyl sulfoxide, evidently involving the alkoxysulfonium ylides, which rearrange intramolecularly to generate aldehydes and ketones, respectively:

Reviews: J. G. Moffatt, "Sulfoxide-Carbodiimide and Related Oxidations" in *Oxidation* vol. 2, R. L. Augustine, D. J. Trecker, Eds. (Dekker, New York, 1971) pp 1-64; T. T. Tidwell, *Org. React.* **39,** 297-572 *passim* (1990). T. V. Lee, *Comp. Org. Syn.* **7,** 291-303 *passim* (1991). *Cf.* Corey-Kim Oxidation; Swern Oxidation.

292. Pictet-Gams Isoquinoline Synthesis

A. Pictet, A. Gams, *Ber.* **43,** 2384 (1910).

Formation of isoquinolines by cyclization of acylated aminomethyl phenyl carbinols or their ethers with phosphorus pentoxide in toluene or xylene:

Reviews: W. M. Whaley, T. R. Govindachari, *Org. React.* **6**, 151 (1951); W. Y. Gensler, *Heterocyclic Compounds* **4**, 361 (1952); W. Herz, L. Tsai, *J. Am. Chem. Soc.* **77**, 3529 (1955); A. A. Bindra *et al., Tetrahedron Letters* **1968**, 2677; N. Ardabilchi *et al., Perkin Trans. I* **1979**, 539. *Cf.* Bischler-Napieralski Reaction.

293. Pictet-Hubert Reaction; Morgan-Walls Reaction

A. Pictet, A. Hubert, *Ber.* **29**, 1182 (1896); C. T. Morgan, L. P. Walls, *J. Chem. Soc.* **1931**, 2447; **1932**, 2225.

Phenanthridine cyclization by dehydrative ring closure of acyl-*o*-aminobiphenyls on heating with zinc chloride at 250-300° (Pictet-Hubert), or with phosphorus oxychloride in boiling nitrobenzene (Morgan-Walls):

L. P. Walls, *J. Chem. Soc.* **1945**, 294; J. Cymerman, W. F. Short, *ibid.* **1949**, 703; R. S. Theobald, K. Schofield, *Chem. Rev.* **46**, 175 (1950); L. P. Walls, *Heterocyclic Compounds* **4**, 574 (1952); J. Eisch, H. Gilman, *Chem. Rev.* **57**, 525 (1957); N. Campbell, *Chemistry of Carbon Compounds* **IVA**, 691 (1957). *Cf.* Bischler-Napieralski Reaction.

294. Pictet-Spengler Isoquinoline Synthesis

A. Pictet, T. Spengler, *Ber.* **44**, 2030 (1911).

Formation of tetrahydroisoquinoline derivatives by condensation of β-arylethylamines with carbonyl compounds and cyclization of the Schiff bases formed:

Reviews: W. M. Whaley, T. R. Govindachari, *Org. React.* **6**, 151 (1951); R. A. Abramovitch, I. D. Spenser, *Advan. Heterocyclic Chem.* **3**, 79 (1964); K. Stuart, R. Woo-Ming, *Heterocycles* **3**, 223 (1975); D. Soerens *et al., J. Org. Chem.* **44**, 535 (1979); H. Ernst *et al., Ber.* **114**, 1894 (1981). Stereochemical study: E. Dominguez *et al., Tetrahedron* **43**, 1943 (1987). Review of enantioselective modifications: M. D. Rozwadowski, *Heterocycles* **39**, 903-931 (1994).

295. Piloty-Robinson Synthesis

O. Piloty, *Ber.* **43**, 489 (1910); G. M. Robinson, R. Robinson, *J. Chem. Soc.* **43**, 639 (1918).

Formation of pyrroles by heating azines of enolizable ketones with acid catalysts, usually zinc chloride or hydrogen chloride:

Review: N. V. Sidgwick, *Organic Chemistry of Nitrogen Compounds* (Oxford, 3rd ed., 1966) pp 619-641; H. Posvic *et al., J. Org. Chem.* **39**, 2575 (1974). *Cf.* Borsche-Drechsel Cyclization; Fischer Indole Synthesis.

296. Pinacol Rearrangement

R. Fittig, *Ann.* **114**, 54 (1860).

Acid-catalyzed rearrangement of vicinal diols to aldehydes or ketones:

Reviews: C. J. Collins, *Quart. Rev.* **14**, 357 (1960); C. J. Collins, J. F. Eastham in S. Patai, Ed., *Chemistry of the Carbonyl Group* (Interscience, New York, 1966) pp 762-767; B. Rickborn, *Comp. Org. Syn.* **3**, 721-732 (1991). *Cf.* Tiffeneau-Demjanov Rearrangement.

297. Pinner Reaction (Amidine and Ortho Ester Synthesis)

A. Pinner, F. Klein, *Ber.* **10,** 1889 (1877); **11,** 4, 1475 (1878); **16,** 352, 1643 (1883).

Formation of imino esters (alkyl imidates) by addition of dry hydrogen chloride to a mixture of a nitrile and an alcohol. Treatment of alkyl imidates with ammonia or primary or secondary amines affords amidines, while treatment with alcohols yields ortho-esters:

Reviews: R. Roger, D. Neilson, *Chem. Rev.* **61,** 179 (1961); E. N. Zil'berman, *Russ. Chem. Rev.* **31,** 615 (1962); P. L. Compagnon, M. Moeque, *Ann. Chim. (Paris)* **5,** 23 (1970); B. Decroix *et al., J. Chem. Res.* **1978,** 134.

298. Pinner Triazine Synthesis

A. Pinner, *Ber.* **23,** 2919 (1890).

Preparation of 2-hydroxy-4,6-diaryl-*s*-triazines by reaction of aryl amidines and phosgene. The reaction may be extended to halogenated aliphatic amidines:

A. Pinner, *Ber.* **25,** 1414 (1892); **28,** 483 (1895); J. Ephraim, *ibid.* **26,** 2226 (1893); P. Flatow, *ibid.* **30,** 2006 (1897); T. Rappaport, *ibid.* **34,** 1990 (1901); H. Schroeder, C. Grundmann, *J. Am. Chem. Soc.* **78,** 2447 (1956); E. M. Smolin, L. Rapoport, *The Chemistry of Heterocyclic Compounds*, A. Weissberger, Ed., *s-Triazines and Derivatives* (Interscience, New York, 1959) p 186.

299. Piria Reaction

R. Piria, *Ann.* **78,** 31 (1851).

Formation of arylsulfamic acids or sulfonation products or both by refluxing aromatic nitro compounds with a metal sulfite and boiling the mixture with dilute acid to yield the amines and sulfamic acids:

J. F. Bunnett, R. E. Zahler, *Chem. Rev.* **49,** 398 (1951); R. Schroter, *Houben-Weyl* **11/1,** 457 (1957); R. Budziarek, *Chem. & Ind. (London)* **1978,** 583.

300. Polonovski Reaction; Potier-Polonovski Reaction

M. Polonovski, M. Polonovski, *Bull. Soc. Chim. France* **41,** 1190 (1927).

Rearrangement of tertiary amine oxides upon treatment with acetic anhydride or acetyl chloride, in which one of the alkyl groups attached to the nitrogen is cleaved, generating the *N,N*-disubstituted acetamide and aldehyde:

Reviews: A. R. Katritzky, J. N. Lagowski, *Chemistry of Heterocyclic N-Oxides* (Academic Press, New York, 1971) p 279, 362; D. Grierson, *Org. React.* **39**, 85-295 (1990); D. S. Grierson, H.-P. Husson, *Comp. Org. Syn.* **6**, 909-924 (1991).

The reaction proceeds *via* an iminium ion intermediate which becomes the stable reaction product when trifluoroacetic anhydride is employed. This modified procedure is commonly referred to as the **Potier-Polonovski reaction**: A. Cave *et al., Tetrahedron* **23**, 4681 (1967); T. Tamminen *et al., ibid.* **45**, 2683 (1989); R. J. Sundberg *et al., ibid.* **48**, 277 (1992).

301. Pomeranz-Fritsch Reaction; Schlittler-Müller Modification

C. Pomeranz, *Monatsh.* **14**, 116 (1893); P. Fritsch, *Ber.* **26**, 419 (1893); E. Schlittler, J. Müller, *Helv. Chim. Acta* **31**, 914, 1119 (1948).

Formation of isoquinolines by the acid-catalyzed cyclization of benzalaminoacetals prepared from aromatic aldehydes and aminoacetal; in the **Schlittler-Müller modification** the starting materials are benzyl amines and glyoxal semiacetal:

M. J. Bevis *et al., Tetrahedron* 27, 1253 (1971); E. V. Brown, *J. Org. Chem.* **42**, 3208 (1977); R. Hirsenkorn, *Tetrahedron Letters* **32**, 1775 (1991). *Reviews:* W. J. Gensler, *Org. React.* **6**, 191 (1951); *idem, Heterocyclic Compounds* **4**, 368 (1952); J. M. Bobbit, A. J. Bourque, *Heterocycles* **25**, 601-614 (1987).

302. Ponzio Reaction

G. Ponzio, *Gazz. Chim. Ital.* **27, I**, 171 (1897).

Formation of dinitrophenylmethanes from benzaldoximes by oxidation with nitrogen dioxide in ether:

J. L. Riebsomer, *Chem. Rev.* **36**, 183 (1945); L. F. Fieser, W. von E. Doering, *J. Am. Chem. Soc.* **68**, 2252 (1946); L. F. Fieser, M. Fieser, *Reagents for Organic Synthesis* (New York, 1967) p 325; H. G. Padeken *et al.*, Houben-Weyl **10/1**, 113 (1971). Improved procedure: H. Suzuki *et al., Bull. Chem. Soc. Japan* **61**, 2929 (1988).

303. Prévost Reaction

C. Prévost, *Compt. Rend.* **196**, 1129 (1933), *C.A.* **27**, 3195 (1933).

Hydroxylation of olefins with iodine and silver benzoate in an anhydrous solvent to give *trans*-glycols:

Reviews: C. V. Wilson, *Org. React.* **9,** 350 (1957); F. D. Gunstone, *Advan. Org. Chem.* **1,** 117 (1960); H. O. House, *Modern Synthetic Reactions* (W. A. Benjamin, Menlo Park, California, 2nd ed., 1972) p 438; S. Amin *et al., J. Org. Chem.* **46,** 2573 (1981). *Cf.* Woodward *cis*-Hydroxylation.

304. Prilezhaev (Prileschajew) Reaction
N. Prilezhaev, *Ber.* **42,** 4811 (1909).

Formation of epoxides by the reaction of alkenes with peracids:

$$RHC{=}CHR' + C_6H_5CO_2OH \longrightarrow \underset{R \quad \quad R'}{\triangle O} + C_6H_5COOH$$

Reviews: D. Swern, *Chem. Rev.* **45,** 16 (1949); *Org. React.* **7,** 378 (1953); H. O. House, *Modern Synthetic Reactions* (W. A. Benjamin, Menlo Park, California, 2nd ed., 1972) pp 302-319; D. I. Metelitra, *Russ. Chem. Rev.* **41,** 807 (1972); D. Schnurgfeil, *Z. Chem.* **20,** 445 (1980).

305. Prins Reaction
H. J. Prins, *Chem. Weekblad* **16,** 64, 1072, 1510 (1919), *C.A.* **13,** 3155 (1919).

Acid-catalyzed addition of olefins to formaldehyde to give 1,3-diols, allylic alcohols or *meta*-dioxanes:

Reviews: R. Arundale, L. A. Mikeska, *Chem. Rev.* **51,** 505 (1952); V. I. Isagulyants *et al., Russ. Chem. Rev.* **1968,** 17; C. W. Roberts in *Friedel-Crafts and Related Reactions* **vol. 2,** Part 2, G. A. Olah, Ed. (Interscience, 1964) pp 1175-1210; D. R. Adams, S. P. Bhatnagar, *Synthesis* **1977,** 661; K. H. Schulte-Elte *et al., Helv. Chim. Acta* **62,** 2673 (1979); R. El Gharbi *et al., Synthesis* **1981,** 361; B. B. Snider, *Comp. Org. Syn.* **2,** 527-561 (1991).

306. Pschorr Reaction
R. Pschorr, *Ber.* **29,** 496 (1896).

Synthesis of phenanthrene derivatives from diazotized α-aryl-*o*-aminocinnamic acids by intramolecular arylation:

Reviews: P. H. Leake, *Chem. Rev.* **56,** 27 (1956); D. F. De Tar, *Org. React.* **9,** 409 (1957); R. A. Abramovitch, *Advan. Free Rad. Chem.* **2,** 88 (1967); T. Kametani, K. Fukumoto, *J. Heterocyclic Chem.* **8,** 341 (1971); S. Foldeak, *Tetrahedron* **27,** 3465 (1971); T. S. Kametani *et al., ibid.* **27,** 5367 (1971); F. F. Gadallah *et al., J. Org. Chem.* **38,** 2386 (1973); S. M. Kupchan *et al., ibid.* 405; G. Daidone *et al., J. Heterocyclic Chem.* **17,** 1409 (1980). *Cf.* Gomberg-Bachman Reaction; Meerwein Arylation.

307. Pummerer Rearrangement
R. Pummerer, *Ber.* **43,** 1401 (1910).

Rearrangement of sulfoxides to α-acyloxythioethers in the presence of acyclic anhydrides. When nucleophiles other than those derived from the anhydride are present, different functionalization is achieved:

Diastereoselectivity in the preparation of 4-phenyl-4-butanolides: H. Su *et al., Bull. Chem. Soc. Japan* **66,** 2603 (1993). Application to vinyl sulfoxides (the additive Pummerer reaction): D. Craig, K. Daniels, *Tetrahedron* **49,** 11263 (1993). Regiospecific cyclization: G. Majumdar, D. Mal, *Indian J. Chem.* **33B,** 700 (1994). Asymmetric synthesis: Y. Kita *et al., Tetrahedron Letters* **35,** 3575 (1994). *Reviews:* O. DeLucchi *et al., Org. React.* **40,** 157-405 (1991); D. S. Grierson, H.-P. Husson, *Comp. Org. Syn.* **6,** 924-937 (1991).

308. Purdie Methylation (Irvine-Purdie Methylation)

T. Purdie, J. C. Irvine, *J. Chem. Soc.* **83,** 1021 (1903).

Exhaustive methylation of a methyl glycoside by repeated treatment with methyl iodide and silver oxide, followed by hydrolysis of the pentamethyl ether with dilute acid to yield the anomeric hydroxyl group:

C. C. Barker *et al., ibid.* **1946,** 753; E. J. Bourne, S. Peat, *Advan. Carbohyd. Chem.* **5,** 146 (1950); W. Pigman, *The Carbohydrates* (New York, 1957) p 370; P. V. Kovac *et al., Carbohyd. Res.* **58,** 327 (1977). *Cf.* Haworth Methylation.

309. Quelet Reaction

R. Quelet, *Compt. Rend.* **195,** 155 (1932).

Passage of dry hydrochloric acid through a solution in ligroin of a phenolic ether and an aliphatic aldehyde in the presence or absence of a dehydration catalyst to yield α-chloroalkyl derivatives by substitution in the *para* position to the ether group or in the *ortho* position in *para* substituted phenolic ethers:

R. Quelet, *Compt. Rend.* **196,** 1411 (1933); **198,** 102 (1934); **199,** 150 (1934); **202,** 956 (1936); *Bull. Soc. Chim. France* **7,** 196 (1940); U. Neda, R. Oda, *J. Soc. Chem. Ind. Japan* **47,** 565 (1944). *Cf.* Blanc (Chloromethylation) Reaction.

310. Ramberg-Bäcklund Reaction

L. Ramberg, B. Bäcklund, *Arkiv Kemi Mineral Geol.* **13A**(27), 50 (1940), *C.A.* **34,** 4725[5] (1940).

Reaction of α-halo sulfones with strong bases to yield alkenes:

Reviews: L. A. Paquette, *Accts. Chem. Res.* **1,** 209-216 (1968); F. G. Bordwell, *ibid.* **3,** 28 (1970); L. Paquette, *Org. React.* **25,** 1 (1977); G. D. Hartman, R. D. Hartman, *Synthesis* **1982,** 504; J. M. Clough, *Comp. Org. Syn.* **3,** 861-886 (1991).

311. Raschig Phenol Process

F. Raschig, **Fr.** pat. **698,341** (1930), *C.A.* **25,** 3012 (1931).

Commercial process for the production of phenol by the hydrolysis of chlorobenzene, produced by the chlorination of benzene with hydrochloric acid and air:

$$C_6H_6 + HCl + 1/2 O_2 \text{ (air)} \xrightarrow[\text{Cu-Fe}]{230°} C_6H_5Cl + H_2O \xrightarrow[\text{SiO}_2]{425°} C_6H_5OH + HCl$$

W. H. Prahl, **U.S.** pat. **1,963,761** (1934); **2,156,402** (1939); J. A. Kent, *Riegel's Industrial Chemistry* (New York, 1962) p 339; W. L. Faith, D. B. Keyes, R. L. Clark, *Industrial Chemistry* (New York, 3rd ed., 1965) p 586; R. N. Shreve, *Chemical Process Industries* (New York, 3rd ed., 1967) p 105; Kirk-Othmer *Encyclopedia of Chemical Technology* **vol. 17** (New York, 1982) p 378.

312. Reed Reaction

C. F. Reed, **U.S.** pat. **2,046,090** (1933); **U.S.** pat. **2,174,110** (1934); **U.S.** pat. **2,174,492** (1938).

Photochemical sulfonation of paraffins and cycloparaffins by sulfur dioxide and chlorine under irradiation with ultraviolet light:

$$RH + SO_2 + Cl_2 \xrightarrow{h\nu} RSO_2Cl + HCl$$

F. Asinger *et al., Ber.* **75**, 34, 42, 344 (1942); J. H. Helberger *et al., Ann.* **562,** 23 (1949); H. Eckoldt, *Houben-Weyl* **9,** 407-427 (1955); A. Schönberg, *Präparative organische Photochemie* (Berlin, 1958) p 201; G. Sosnovsky, *Free Radical Reactions in Preparative Organic Chemistry* (New York, 1964) p 105.

313. Reformatsky (Reformatskii) Reaction

S. Reformatskii, *Ber.* **20,** 1210 (1887); *J. Russ. Phys. Chem. Soc.* **22,** 44 (1890).

Condensation of aldehydes or ketones with organozinc derivatives of α-halo esters to yield β-hydroxy esters:

Early reviews: R. L. Shriner, *Org. React.* **1,** 1 (1942); H. O. House, *Modern Synthetic Reactions* (W. A. Benjamin, Menlo Park, California, 2nd ed., 1972) pp 671-682; M. W. Rathke, *Org. React.* **22,** 423 (1975). Use of thiocarbonyl electrophiles: M. Chandrasekharam *et al., Tetrahedron Letters* **34,** 6439 (1993). Application to the synthesis of β-keto esters: C. Kashima *et al., J. Org. Chem.* **58,** 793 (1993). Asymmetric synthesis: D. Pini *et al., Tetrahedron: Asymmetry* **5,** 1875 (1994). Methods development for the synthesis of β-lactones: H. Schick *et al., Tetrahedron* **51,** 2939 (1995). *Reviews:* A. Fürstner, *Synthesis* **1989,** 571-590; M. W. Rathke, P. Weipert, *Comp. Org. Syn.* **2,** 277-299 (1991). *Cf.* Aldol Reaction; Blaise Reaction.

314. Reimer-Tiemann Reaction

K. Reimer, F. Tiemann, *Ber.* **9,** 824, 1268, 1285 (1876).

Formation of phenolic aldehydes from phenols, chloroform and alkali:

Review: H. Wynberg, *Chem. Rev.* **60,** 169 (1960); H. Wynberg, E. W. Meijer, *Org. React.* **28,** 2 (1982); H. Wynberg, *Comp. Org. Syn.* **2,** 769-775 (1991). *Cf.* Duff Reaction; Gattermann Aldehyde Synthesis.

315. Reissert Indole Synthesis

A. Reissert, *Ber.* **30,** 1030 (1897).

Condensation of an *o*-nitrotoluene with oxalic ester, reduction to the amine, and cyclization to the indole:

W. O. Kermack *et al., J. Chem. Soc.* **119,** 1602 (1921); P. C. Julian *et al., Heterocyclic Compounds* **3,** 18 (1962); J. G. Cannon *et al., J. Med. Chem.* **24,** 238 (1981).

316. Reissert Reaction (Grosheintz-Fischer-Reissert Aldehyde Synthesis)

A. Reissert, *Ber.* **38,** 1603, 3415 (1905); J. M. Grosheintz, H. O. L. Fischer, *J. Am. Chem. Soc.* **63,** 2021 (1941).

Formation of 1-acyl-2-cyano-1,2-dihydroquinoline derivatives (Reissert compounds) by reaction of acid chlorides with quinoline and potassium cyanide; hydrolysis of these compounds yields aldehydes and quinaldic acid:

Reviews: E. Mosettig, *Org. React.* **8,** 220 (1954); W. E. McEwen, R. L. Cobb, *Chem. Rev.* **55,** 511 (1955); F. D. Popp, *Advan. Heterocycl. Chem.* **9,** 1 (1968); *idem, ibid.* **24,** 187 (1979); *idem, Bull. Soc. Chim. Belg.* **90,** 609 (1981); *idem* in *The Chemistry of Heterocyclic Compounds* **vol. 32,** Part 2, G. Jones, Ed. (Wiley, New York, 1982) p 353.

317. Reppe Chemistry

The term designates that phase of acetylene chemistry involving the use of acetylene at high pressures in the presence of suitable catalysts to carry out the fundamental reactions of vinylation, ethynylation, cyclopolymerization and carbonylation as developed from 1928 onward by Walter Reppe and associates in the I. G. Farbenindustrie laboratories in Ludwigshafen:

$$H_2C=CHOR \xleftarrow{\text{ROH}} HC\equiv CH \xrightarrow{\text{RSH}} H_2C=CHSR$$

$$RR'CO + HC\equiv CH \longrightarrow RR'C(OH)C\equiv CH \xrightarrow{\text{RR'CO}} RR'C(OH)C\equiv CC(OH)RR'$$

$$4HC\equiv CH \longrightarrow \text{cyclooctatetraene}$$

$$4HC\equiv CH + 4C_2H_4OH + Ni(CO)_4 + 2HCl \longrightarrow 4H_2C=CHCOOC_2H_5 + NiCl_2 + H_2$$

J. W. Copenhaver, M. H. Bigelow, *Acetylene and Carbon Monoxide Chemistry* (New York, 1949) p 246; W. Reppe, *Acetylene Chemistry,* U.S. Dept. Commerce PB 18852-S (1949); *Neue Entwicklungen auf dem Gebiet des Acetylens und Kohlenoxyds* (Berlin 1949); H. Kröper, *Houben-Weyl* **4/2,** 413-422 (1955); D. W. F. Hardie, *Acetylene, Manufacture and Uses* (New York, 1965) p 67; L. F. Fieser, M. Fieser, *Reagents for Organic Synthesis* (New York, 1967) pp 61, 183, 185, 190, 519, 720, 722, 723. Review of carbonylations: A. Mullen, "Carbonylations Catalyzed by Metal Carbonyls-Reppe Reactions" in *New Syntheses with Carbon Monoxide,* J. Falbe, Ed. (Springer-Verlag, Berlin, 1980) pp 243-308. Mechanistic study of cyclooctatetraene synthesis: R. E. Colborn, K. P. C. Vollhardt, *J. Am. Chem. Soc.* **108,** 5470 (1986); C. J. Lawrie *et al., Organometallics* **8,** 2274 (1989).

318. Retro-Diels-Alder Reaction
Thermal dissociation of Diels-Alder adducts, occurring most readily when one or both fragments are particularly stable:

Early review: H. Kwart, K. King, *Chem. Rev.* **68,** 415-447 (1968). Acceleration by alkoxide substituent: O. Papies, W. Grimme, *Tetrahedron Letters* **21,** 2799 (1980). Application to the synthesis of enethiols: Y. Vallée *et al., Syn. Commun.* **23,** 1267 (1993); of cyclopentadienyl ligands: B. Y. Lee *et al., J. Am. Chem. Soc.* **116,** 2163 (1994). Role in structure elucidation *via* mass spectrometry: F. Turecek, V. Hanus, *Mass Spectrom. Rev.* **3,** 85-152 (1984). Application to natural product synthesis: A. Ichihara, *Synthesis* **1987,** 207-222; R. W. Sweger, A. W. Czarnik, *Comp. Org. Syn.* **5,** 551-592 (1991).

319. Retropinacol Rearrangement
N. Zelinsky, J. Zelikow, *Ber.* **34,** 3249 (1901).
Conversion of an alcohol to the rearranged olefin on treatment with acid:

Application to sterols: W. F. Johns, *J. Org. Chem.* **26,** 4583 (1961); L. M. Harrison, P. V. Fennessey, *J. Steroid. Biochem.* **36,** 407 (1990); to cyclohexanols: W. Hueckel, S. K. Gupte, *Ann.* **685,** 105 (1965), *C. A.* **63,** 11283d (1965). In conjuction with ring expansion: T. Kimura *et al., J. Org. Chem.* **43,** 1247 (1978). *Cf.* Nametkin Rearrangement; Wagner-Meerwein Rearrangement.

320. Reverdin Reaction
F. Reverdin, *Ber.* **29,** 997, 2595 (1896).
Migration of iodine during nitration of iodophenolic ethers:

F. Reverdin, *Bull. Soc. Chim. France* [4] **1,** 618 (1907); G. M. Robinson, *J. Chem. Soc.* **109,** 1078 (1916); D. V. Nightingale, *Chem. Rev.* **40,** 128 (1947); M. J. S. Dewar, *Electronic Theory of Organic Chemistry* (Oxford, 1949) p 232.

321. Riehm Quinoline Synthesis
P. Riehm *et al., Ber.* **18,** 2245 (1885); **19,** 1394 (1886); *idem, Ann.* **238,** 9 (1887).
Formation of quinoline derivatives by prolonged heating of arylamine hydrochlorides with ketones with or without use of aluminum chloride or phosphorus pentachloride:

$$\text{[benzene ring with NH}_2 \cdot \text{HCl]} + 2CH_3COCH_3 \xrightarrow[\text{AlCl}_3]{\Delta} \text{[quinoline with CH}_3 \text{ groups]} + CH_4 + 2H_2O$$

E. Knoevenagel *et al., Ann.* **55,** 1923, 1934 (1922); **56,** 2414 (1923); C. Hollins, *The Synthesis of Nitrogen Ring Compounds* (London, 1924) p 263; D. J. Craig, *J. Am. Chem. Soc.* **60,** 1458 (1938); R. C. Elderfield, J. R. McCarthy, *ibid.* **73,** 975 (1951); R. C. Elderfield, *Heterocyclic Compounds* **4,** 16 (1952).

322. Riemschneider Thiocarbamate Synthesis
R. Riemschneider, F. Wojahn, *Pharmazie* **4,** 460 (1949); *Chim. Ind. (Paris)* **64,** 99 (1950); *Pharm. Zentralhalle* **89,** 118 (1950).

The action of concentrated sulfuric acid followed by treatment with ice-water serves to transform arylthiocyanates into the corresponding thiocarbamates:

$$ArSCN \xrightarrow[\text{H}_2\text{O}]{\text{conc H}_2\text{SO}_4} ArS-C{\overset{OH}{\underset{NH}{}}} \rightleftharpoons ArSCONH_2$$

R. Riemschneider, *Chim. Ind. (Milan)* **33,** 483 (1951); *idem et al., J. Am. Chem. Soc.* **73,** 5905 (1951); R. Riemschneider, G. Orlick, *Angew. Chem.* **64,** 420 (1952); R. Riemschneider, *Chim. Ind. (Milan)* **34,** 353 (1952); *idem, Z. Naturforsch.* **7b,** 277 (1952); R. Riemschneider, G. Orlick, *Monatsh.* **84,** 316 (1953); K. Schmidt, P. Kolleck-Bös, *J. Am. Chem. Soc.* **75,** 6067 (1953).

323. Riley Oxidations (Selenium Dioxide Oxidation)
H. L. Riley *et al., J. Chem. Soc.* **1932,** 1875.

Oxidations of organic compounds with selenium dioxide; *e.g.,* the oxidation of active methylene groups to carbonyl groups:

$$CH_3CH_2CHO + SeO_2 \longrightarrow CH_3COCHO + H_2O + Se$$

N. Rabjohn, *Org. React.* **5,** 331 (1949); *Oxidation,* E. N. Trachtenberg, R. L. Augustine, Eds. (Marcel Dekker, New York, 1969) pp 119-187; H. O. House, *Modern Synthetic Reactions* (W. A. Benjamin, Menlo Park, California, 2nd ed., 1972) pp 407-411.

324. Ritter Reaction
J. J. Ritter, P. P. Minieri, *J. Am. Chem. Soc.* **70,** 4045 (1948); J. J. Ritter, J. Kalish, *ibid.* 4048.

Synthesis of amides from nitriles and alcohols or alkenes in strongly acidic media:

$$\text{[(H}_3\text{C)}_2\text{C=CH}_2] + R-C{\equiv}N \xrightarrow{H^+} \left[(H_3C)_3-\overset{+}{N}{\equiv}C-R \right] \xrightarrow{H_2O} (H_3C)_3-\text{NH-CO-R}$$

Reviews: L. I. Krimen, D. J. Cota, *Org. React.* **17,** 213-325 (1969); R. C. Larock, W. W. Leong, *Comp. Org. Syn.* **4,** 292-294 (1991); R. Bishop, *ibid.* **6,** 261-300 (1991). Synthetic applications: S. Top, G. Jaouen, *J. Org. Chem.* **46,** 78 (1981); D. M. Fink, R. C. Effland, *Syn. Commun.* **24,** 2793 (1994); W. M. Samaniego *et al., Tetrahedron Letters* **35,** 6967 (1994).

325. Robinson Annulation
W. S. Rapson, R. Robinson, *J. Chem. Soc.* **1935,** 1285.

Formation of six-membered ring α,β-unsaturated ketones by the addition of cyclohexanones to methyl vinyl ketone (or simple derivatives of methyl vinyl ketone) or its equivalents, followed by an intramolecular aldol condensation, *q.v.*:

$$\text{[cyclohexanone with R]} + \text{[MVK]} \xrightarrow{\text{base}} \text{[intermediate]} \longrightarrow \text{[bicyclic alcohol]} \xrightarrow{-H_2O} \text{[octalone]}$$

Early review: R. E. Gawley, *Synthesis* **1976,** 777-794. Improved methodology: T. Sato *et al., Tetrahedron Letters* **31,** 1581 (1990). Stereochemical study: C. Nussbaumer, *Helv. Chim. Acta* **73,** 1621 (1990). Synthetic applications: R. V. Bonnert *et al., Perkin Trans. I* **1991,** 1225; S. Kim, P. L. Fuchs, *J. Am. Chem. Soc.* **115,** 5934 (1993). *Cf.* Michael Reaction; Wichterle Reaction.

326. Robinson-Schöpf Reaction

R. Robinson, *J. Chem. Soc.* **111**, 762, 876 (1917); C. Schöpf, *Angew. Chem.* **50**, 779, 797 (1937).
Synthesis of tropinones from a dialdehyde, methylamine and acetonedicarboxylic acid:

K. Alder *et al., Ann.* **601**, 147 (1956); R. D. Guthrie, J. F. McCarthy, *J. Chem. Soc. C* **1967**, 62; R. V. Stevens, A. W. M. Lee, *J. Am. Chem. Soc.* **101**, 7032 (1979); M. Langlois *et al., Syn. Commun.* **22**, 3115 (1992). *Cf.* Mannich Reaction; Petrenko-Kritschenko Piperidone Synthesis.

327. Rosenmund Reduction

K. W. Rosenmund, *Ber.* **51**, 585 (1918); K. W. Rosenmund, F. Zetzsche, *ibid.* **54**, 425 (1921).
Catalytic reduction of acid chlorides to aldehydes. To prevent further hydrogenation a poison is added to the catalyst:

$$RCOCl + H_2 \xrightarrow{\text{Pd-BaSO}_4} RCHO + HCl$$

Reviews: E. Mosettig, R. Mozingo, *Org. React.* **4**, 362 (1948); A. Rachlin *et al., Org. Syn.* **51**, 8 (1971); J. A. Peters, H. Van Bekkum, *Rec. Trav. Chim.* **100**, 21 (1981). Investigation of reaction parameters: W. F. Maier *et al., J. Am. Chem. Soc.* **108**, 2608 (1986). Modified procedure applied to the synthesis of esters: V. V. Grushin, H. Alper, *J. Org. Chem.* **56**, 5159 (1991).

328. Rosenmund-von Braun Synthesis

K. W. Rosenmund, E. Struck, *Ber.* **52**, 1749 (1916); J. von Braun, G. Manz, *Ann.* **488**, 111 (1931).
Conversion of aryl halides to aromatic nitriles in the presence of cuprous cyanide:

$$ArX + CuCN \xrightarrow{250°} ArCN + CuX$$

Reviews: D. T. Mowry, *Chem. Rev.* **42**, 207 (1948); J. E. Callen *et al., Org. Syn.* **3**, 212 (1955); M. S. Newman, *ibid.* 631; K. Takagi *et al., Bull. Chem. Soc. Japan* **48**, 3298 (1975); P. Bouyssou *et al., J. Heterocyclic Chem.* **29**, 895 (1992).

329. Rothemund Reaction

P. Rothemund, *J. Am. Chem. Soc.* **57**, 2010 (1935); **61**, 2912 (1939).
Preparation of *meso*-tetrasubstituted porphyrins by condensation of pyrrole with an aldehyde:

Ball *et al., J. Am. Chem. Soc.* **68**, 2278 (1946); Thomas, Martell, *ibid.* **78**, 1335 (1956). Mechanism: Badger *et al., Aust. J. Chem.* **17**, 1028 (1964); R. G. Little, *J. Heterocyclic Chem.* **18**, 833 (1981).

330. Rubottom Oxidation

A. G. Brook, D. M. Macrae, *J. Organometal. Chem.* **77**, C19 (1974); A. Hassner *et al., J. Org. Chem.* **40**, 3427 (1975); G. M. Rubottom *et al., Tetrahedron Letters* **1974**, 4319.
Oxidation of enolsilanes with *m*-chloroperbenzoic acid (*m*-CPBA) to afford α-hydroxy ketones:

Synthetic applications: R. Gleiter *et al., J. Org. Chem.* **57**, 252 (1992); C. R. Johnson *et al., J. Am. Chem. Soc.* **114**, 9414 (1992); M. T. Crimmins *et al., ibid.* **115**, 3146 (1993).

331. Ruff-Fenton Degradation
O. Ruff, *Ber.* **31**, 1573 (1898); **32**, 550 (1899); H. J. H. Fenton, *Proc. Chem. Soc.* **9**, 113 (1893).
Shortening of the carbon chain of sugars by the oxidation of aldonic acids (as calcium salts) with hydrogen peroxide and ferric salts:

$$RCH(OH)COOH + H_2O_2 \xrightarrow{Fe^{3+}} RCHO + CO_2 + 2H_2O$$

W. Pigman, *The Carbohydrates* (Academic Press, New York, 1957) p 118; H. S. Isbell, M. A. Salam, *Carbohyd. Res.* **90**, 123 (1981).

332. Ruzicka Large Ring Synthesis
L. Ruzicka *et al., Helv. Chim. Acta* **9**, 249, 339, 389, 499 (1926).
Formation of large ring alicyclic ketones from dicarboxylic acids by thermal decomposition of salts with metals of the second and fourth groups of the periodic table (Ca, Th, Ce):

L. Ruzicka, *Chem. & Ind. (London)* **54**, 2 (1935); H. Gilman, *Organic Chemistry* **vol. 1** (New York, 1943) p 78; K. Ziegler, *Houben-Weyl* **4/2**, 754 (1955). *Cf.* Blanc Reaction.

333. Sabatier-Senderens Reduction
P. Sabatier, J. B. Senderens, *Compt. Rend.* **128**, 1173 (1899).
Catalytic hydrogenation of organic compounds in the vapor phase by passage over hot, finely divided nickel (the oldest of all hydrogenation methods).
E. B. Maxted in *Handbuch der Katalyse* **vol. 7**, G. M. Schwab, Ed. (Vienna, 1943) p 624; H. Roth *et al., Houben-Weyl* **2**, 288 (1953); G. Schiller, *ibid.* **4/2**, 284 (1955); H. O. House, *Modern Synthetic Reactions* (W. A. Benjamin, Menlo Park, California, 2nd ed., 1972) Chapter 1.

334. Sakurai Reaction (Hosomi-Sakurai Reaction)
A. Hosomi, H. Sakurai, *Tetrahedron Letters* **1976**, 1295; A. Hosomi *et al., Chem. Letters* **1976**, 941.
Lewis acid-promoted nucleophilic addition of allylic silanes to carbon electrophiles accompanied by regiospecific transposition of the allylic moiety:

E = aldehydes, ketones, enones, acid chlorides, acetals, ketals, epoxides, iminium salts
Lewis acid = $TiCl_4$, $AlCl_3$, $BF_3 \cdot O(CH_2CH_3)_2$, $SnCl_4$, $(CH_3CH_2)_2AlCl$, cat $(CH_3)_3SiOSO_2CF_3$

Synthetic applications: I. E. Markó, D. J. Bayston, *Tetrahedron Letters* **34**, 6595 (1993); H. Hioki *et al., ibid.* 6131. [$TiCp_2(OSO_2CF_3)_2$] as catalyst: T. K. Hollis *et al., ibid.* 4309. *Reviews:* I. Fleming *et al., Org. React.* **37**, 57-575 (1989); Y. Yamamoto, N. Sasaki, "The Stereochemistry of the Sakurai Reaction" in *Stereochemistry of Organometallic and Inorganic Compounds* **vol. 3**, I. Bernal, Ed. (Elsevier, New York, 1989) pp 363-437; I. Fleming, *Comp. Org. Syn.* **2**, 563-593 (1991).

335. Sandmeyer Diphenylurea Isatin Synthesis
T. Sandmeyer, *Z. Farb. Textile Chem.* **2**, 129 (1903).
Formation of a cyanoformamidine by treatment of a symmetrical diphenylthiourea with potassium cyanide in alcohol containing lead carbonate, reduction with ammonium sulfide and ring-closure with concentrated sulfuric acid to isatin-2-anil; also formed smoothly by ring closure of the cyanoformamidine with aluminum chloride in benzene or carbon disulfide:

J. R. Geigy & Co., **Ger.** pat. **115,169, 116,563** (1900); *Friedländer* **6**, 574, 575 (1900-1902); A. Reissert, *Ber.* **37,** 3708 (1904); G. Schultz *et al., J. Prakt. Chem.* [2] **74,** 74, 76 (1906); C. Hollins, *The Synthesis of Nitrogen Ring Compounds* (London, 1924) p 102; C. S. Marvel, G. S. Hiers, *Org. Syn.* **coll. vol. I,** 327 (1943); P. L. Julian *et al., Heterocyclic Compounds* **3,** 207 (1952); O. Bayer, W. Eckert, *Houben-Weyl* **7/4,** 11 (1968). *Cf.* Sandmeyer Isonitrosoacetanilide Isatin Synthesis.

336. Sandmeyer Isonitrosoacetanilide Isatin Synthesis

T. Sandmeyer, *Helv. Chim. Acta* **2,** 234 (1919).

Formation of isonitrosoacetodiphenylamidine by condensation of chloral hydrate, hydroxylamine and aniline, cyclization with concentrated sulfuric acid, and quantitative hydrolysis to isatin on dilution:

J. Martinet, P. Cousset, *Compt. Rend.* **172,** 1234 (1921); C. Hollins, *The Synthesis of Nitrogen Ring Compounds* (London, 1924) p 103; C. S. Marvel, G. S. Hiers, *Org. Syn.* **coll. vol. I,** 327 (1943); P. L. Julian *et al., Heterocyclic Compounds* **3,** 208 (1952); F. E. Sheibley, J. S. McNulty, *J. Org. Chem.* **21,** 171 (1956); O. Bayer, W. Eckert, *Houben-Weyl* **7/4,** 14 (1968). *Cf.* Sandmeyer Diphenylurea Isatin Synthesis.

337. Sandmeyer Reaction; Gattermann Reaction; Körner-Contardi Reaction

T. Sandmeyer, *Ber.* **17,** 1633, 2650 (1884); L. Gattermann, *Ber.* **23,** 1218 (1890); G. Körner, A. Contardi, *Atti Accad. nazl. Lincei* **23 II,** 464 (1914), *C.A.* **9,** 1478 (1915).

Substitution of diazonium groups in aromatic compounds by halo or cyano groups in the presence of cuprous salts (Sandmeyer reaction), copper powder and hydrochloric or hydrobromic acid (Gattermann reaction) or cupric salts (Körner-Contardi reaction):

$$C_6H_5N_2^+X^- \xrightarrow{\text{CuCl}} C_6H_5Cl$$

Early reviews: H. H. Hodgson, *Chem. Rev.* **40,** 251-277 (1947); W. A. Coudrey, D. S. Davies, *Quart. Rev.* **6,** 358-379 (1952); A. Roedig, *Houben-Weyl* **5/4,** 438 (1960); R. Stroh, *ibid.* **5/3,** 846 (1962). Direct conversion of aryl amines to aryl halides: M. P. Doyle, *J. Org. Chem.* **42,** 2426 (1977). Mechanistic studies: J. K. Kochi, *J. Am. Chem. Soc.* **79,** 2942 (1957); C. Galli, *Perkin Trans. II* **1984,** 897. Synthetic application: C. Corral *et al., Heterocycles* **23,** 1431 (1985). Improved methodology: N. Suzuki *et al., Perkin Trans. I* **1987,** 645.

338. Sarett Oxidation; Collins Oxidation

G. I. Poos, G. E. Arth, R. E. Beyler, L. H. Sarett, *J. Am. Chem. Soc.* **75,** 422 (1953).

Oxidation of primary and secondary alcohols to aldehydes (and/or carboxylic acids) and ketones by means of CrO_3-pyridine complex:

$$\underset{\text{OH}}{\overset{\text{H}}{>}C< } \xrightarrow[C_5H_5N]{CrO_3} >C=O$$

J. R. Holum, *J. Org. Chem.* **26,** 4814 (1961); E. J. Kris, *Chem. & Ind. (London)* **1961,** 1834; V. I. Stenberg, R. J. Perkins, *J. Org. Chem.* **28,** 323 (1963); P. G. Gassman, P. G. Pape, *J. Org. Chem.* **29,** 160 (1964); H. O. House, *Modern Synthetic Reactions* (W. A. Benjamin, Menlo Park, California, 2nd ed., 1972) pp 264-273. Mechanistic studies: F. Hasan, J. Rocek, *J. Am. Chem. Soc.* **97,** 1444, 3762 (1975).

The **Collins oxidation** is characterized by a modified procedure (dichloromethane as solvent) that reliably oxidizes primary alcohols to aldehydes: J. C. Collins, *Tetrahedron Letters* **1968,** 3363; J. C. Collins, W. W. Hess, *Org. Syn.* **52,** 5 (1972); R. W. Ratcliffe, *ibid.* **55,** 84 (1976). *Cf.* Jones Oxidation.

339. Schiemann Reaction (Balz-Schiemann Reaction)

G. Balz, G. Schiemann, *Ber.* **60,** 1186 (1927).

Formation of diazonium fluoroborates by diazotization of aromatic amines in the presence of fluoroborates, followed by their thermal decomposition to aryl fluorides:

$$ArNH_2 + HNO_2 + HBF_4 \longrightarrow ArN_2^+ BF_4^- \xrightarrow{\Delta} ArF + N_2 + BF_3$$

Reviews: A. Roe, *Org. React.* **5,** 193 (1949); H. Suschitzky, *Advan. Fluorine Chem.* **4,** 1 (1965); T. K. Al'sing, E. G. Sochilin, *Zh. Org. Khim.* **7,** 530 (1971); R. Bartsch *et al., J. Am. Chem. Soc.* **98,** 6753 (1976); H. G. O. Becker, G. Israel, *J. Prakt. Chem.* **321,** 579 (1979).

340. Schmidt Reaction

R. F. Schmidt, *Ber.* **57,** 704 (1924).

Acid-catalyzed addition of hydrazoic acid to carboxylic acids, aldehydes and ketones to give amines, nitriles and amides, respectively. Tertiary alcohols and substituted alkenes yield imines upon treatment with hydrazoic acid:

$$RCOOH \xrightarrow[H_2SO_4]{HN_3} RNH_2$$

$$RCHO \xrightarrow[H_2SO_4]{HN_3} RCN + RNHCHO$$

$$RCOR \xrightarrow[H_2SO_4]{HN_3} RCONHR$$

$$R_3COH \xrightarrow[H_2SO_4]{HN_3} R_2C{=}NR$$

$$R_2C{=}CR_2 \xrightarrow[H_2SO_4]{HN_3} R_2CHC{=}NR \atop \qquad\quad R$$

Early reviews: H. Wolff, *Org. React.* **3,** 307-336 (1946); P. A. S. Smith in *Molecular Rearrangements* Part 1, P. de Mayo, Ed. (Wiley-Interscience, New York, 1963) pp 507-558; D. V. Banthorpe, *The Chemistry of the Azido Group,* S. Patai, Ed. (Interscience, New York, 1971) pp 405-421; G. I. Koldobskii, *Russ. Chem. Rev.* **47,** 1084 (1978). Application to cyclic ketones: A. Lévai *et al., Heterocycles* **34,** 1523 (1992); J.-Y. Mérour *et al., J. Heterocyclic Chem.* **31,** 87 (1994); to alcohols and alkenes: W. H. Pearson *et al., J. Am. Chem. Soc.* **115,** 10183 (1993). Extension to dialkyl acylphosphonates: M. Sprecher, D. Kost, *ibid.* **116,** 1016 (1994). *Review:* T. Shioiri, *Comp. Org. Syn.* **6,** 817-821 (1991). *Cf.* Beckmann Rearrangement; Curtius Rearrangement; Hofmann Reaction; Lossen Rearrangement.

341. Scholl Reaction

R. Scholl, C. Seer, *Ann.* **394,** 111 (1912).

Coupling of aromatic molecules by treatment with Lewis acid catalysts:

$$2ArH \xrightarrow[H^+]{AlCl_3} Ar{-}Ar + H_2$$

Review: C. D. Nenitzescu, A. T. Balaban in *Friedel-Crafts and Related Reactions,* **vol. 2,** part 2, G. Olah, Ed. (Wiley, New York, 1964) pp 979-1048; G. A. Clowes, *J. Chem. Soc. C* **1968,** 2519; A. C. Buchanan *et al., J. Am. Chem. Soc.* **102,** 5262 (1980).

342. Schöllkopf Bis-Lactim Amino Acid Synthesis

U. Schöllkopf *et al., Angew. Chem. Int. Ed.* **18,** 863 (1979); **20,** 798 (1981).

Asymmetric amino acid synthesis *via* diastereoselective alkylation of the lithiated bis-lactim ether (derived from L-Val and Gly or Ala) by an electrophile. Subsequent acid hydrolysis liberates L-Val-OCH$_3$ and the *(R)*-α-substituted amino acid ester. When the bis-lactim is generated from D-Val, the *(S)*-enantiomer forms:

R = H, CH$_3$
E = alkyl halides, aldehydes, ketones, thioketones, acid chlorides, epoxides, acrylates

Synthetic applications: S. Kotha, A. Kuki, *Chem. Commun.* **1992,** 404; M. S. Allen *et al., Syn. Commun.* **22,** 2077 (1992). Isotopic labeling: N. R. Thomas, D. Gani, *Tetrahedron* **47,** 497 (1991). *Reviews:* U. Schöllkopf, *Top. Curr. Chem.* **109,** 65-84 (1983); *idem, Pure Appl. Chem.* **55,** 1799-1806 (1983); R. M. Williams, *Synthesis of Optically Active α-Amino Acids* (Pergamon, New York, 1989) pp 1-33.

343. Schotten-Baumann Reaction

C. Schotten, *Ber.* **17**, 2544 (1884); E. Baumann, *ibid.* **19**, 3218 (1886).

Acylation of alcohols or amines with acid chlorides in aqueous alkaline solution:

$$ROH \; + \; \underset{O}{\overset{R'}{\underset{\|}{C}}}X \quad \xrightarrow{\text{NaOH}} \quad \underset{O}{\overset{R'}{\underset{\|}{C}}}OR$$

Review: N. O. V. Sonntag, *Chem. Rev.* **52**, 272-273 (1953). Synthetic applications: M. Tsuchiya *et al., Bull. Chem. Soc. Japan* **42**, 1756 (1969); G. I. Georg, *Bioorg. Med. Chem. Letters* **4**, 335 (1994).

344. Semmler-Wolff Reaction (Wolff-Semmler Aromatization, Wolff Aromatization)

W. Semmler, *Ber.* **25**, 3352 (1892); L. Wolff, *Ann.* **322**, 351 (1902).

Rearrangement of α,β-unsaturated cyclohexenyl ketoximes into aromatic amines under acidic conditions:

Review: R. T. Conley, S. Ghosh in *Mechanisms of Molecular Migrations* **vol. 4**, B. S. Thyagarajan, Ed., (Interscience, New York, 1971) p 251; M. I. El-Sheikh, J. M. Cook, *J. Org. Chem.* **45**, 2585 (1980); Y. Tamura *et al., Synthesis* **1980**, 483.

345. Serini Reaction

A. Serini *et al., Ber.* **72**, 391 (1939).

Zinc-promoted rearrangement of 17-hydroxy-20-acetoxysterol derivatives into C-20 ketones; the reaction is applicable to other cyclic, as well as open-chain alcohols:

Reviews: C. W. Shoppe, *Chimia* **4**, 418 (1948); L. F. Fieser, M. Fieser, *Steroids* (Reinhold Publishing Corp., New York, 1959) p 628; N. L. Wendler in *Molecular Rearrangements* Part 2, P. de Mayo, Ed. (Wiley-Interscience, New York, 1964) p 1038; E. Ghera, *Chem. Commun.* **1968**, 1639; *J. Org. Chem.* **35**, 660 (1970).

346. Sharpless Dihydroxylation

E. N. Jacobsen *et al., J. Am. Chem. Soc.* **110**, 1968 (1988).

Osmium-catalyzed asymmetric *cis*-dihydroxylation of olefins:

A premix of the four reagent components is commercially available. The composition containing (DHQD)$_2$-PHAL is termed AD-mix-β; the composition containing (DHQ)$_2$-PHAL is termed AD-mix-α. (DHQD)$_2$-PHAL = 1,4-bis(9-O-dihydroquinidine)phthalazine; (DHQ)$_2$-PHAL = 1,4-bis(9-O-dihydroquinine)phthalazine. R_L = largest substituent; R_M = medium-sized substituent; R_S = smallest substituent.

Note: The scheme shown is an empirical mnemonic indicating olefin orientation and face selectivity. It is not to be considered an absolute predictor of new diol configurations.

Allyl and vinyl silanes as substrates: A. R. Bassindale *et al., Perkin Trans. I* **1994**, 1061. Chemoselective dihydroxylation of a polyene: S. C. Sinha, E. Keinan, *J. Org. Chem.* **59**, 949 (1994). *Reviews:* R. A. Johnson, K. B. Sharpless, "Catalytic Asymmetric Dihydroxylation" in *Catalytic Asymmetric Synthesis*, I. Ojima, Ed. (VCH, New York, 1993) pp 227-272; H. C. Kolb *et al., Chem. Rev.* **94**, 2483-2547 (1994). *Cf.* Milas Hydroxylation of Olefins.

347. Sharpless Epoxidation

T. Katsuki, K. B. Sharpless, *J. Am. Chem. Soc.* **102,** 5974 (1980).

Titanium-catalyzed asymmetric epoxidation of allylic alcohols employing titanium alkoxide, an optically active tartrate ester and an alkyl hydroperoxide. A high degree of enantiomeric purity is attainable having predictable absolute stereochemistry:

D-(-)-diethyl tartrate (unnatural)

$(CH_3)_3COOH$, $Ti(O\text{-}i\text{-}Pr)_4$
CH_2Cl_2

70-90% yields,
>90% ee

L-(+)-diethyl tartrate (natural)

Note: The asterisk at a chiral center denotes a preponderance of either the *R* or *S* configuration.

Mechanistic studies: S. S. Woodward *et al., J. Am. Chem. Soc.* **113,** 106 (1991); M. G. Finn, K. B. Sharpless, *ibid.* 113. Methods development for the synthesis of enantiopure allylic alcohols: D. C. Dittmer *et al., J. Org. Chem.* **58,** 718 (1993). Alkenylsilanols as substrates: T. H. Chan *et al., Can. J. Chem.* **71,** 60 (1993). *Reviews:* R. A. Johnson, K. B. Sharpless, *Comp. Org. Syn.* **7,** 389-436 (1991); E. Höft, *Top. Curr. Chem.* **164,** 63-77 (1993).

348. Sharpless Oxyamination

K. B. Sharpless *et al., J. Am. Chem. Soc.* **97,** 2305 (1975).

Osmium-mediated *cis*-addition of nitrogen and oxygen moieties to mono-, di- and tri-substituted olefins to yield vicinal amino or amido alcohols:

OsO_3NR^3 (stoichiometric)

Chloramine-T • $3H_2O$, cat OsO_4

$R^4OCON(Na)Cl$, $Hg(NO_3)_2$
$(CH_3CH_2)_4NOCOCH_3$, cat OsO_4

$R^3 = C(CH_3)_3$, 1-adamantyl
$R^4 = CH_2CH_3$, $C(CH_3)_3$, CH_2Ph
$R^5 = R^3$, Ts, $COOR^4$

Methods development in the context of taxol synthesis: L. Mangatal *et al., Tetrahedron* **45,** 4177 (1989). Synthetic applications: S. K. Dubey, E. E. Knaus, *Can. J. Chem.* **61,** 565 (1983); M. Lemaire *et al., Synlett* **1990,** 615. Brief review: *Organic Syntheses by Oxidation with Metal Compounds,* W. J. Mijs, C. R. H. I. de Jonge, Eds. (Plenum Press, New York, 1986) pp 642-645.

349. Simmons-Smith Reaction

H. E. Simmons, R. D. Smith, *J. Am. Chem. Soc.* **80,** 5323 (1958).

Stereospecific synthesis of cyclopropanes by treatment of olefins with methylene iodide and zinc-copper couple:

$+ CH_2I_2 + Zn(Cu) \xrightarrow{(C_2H_5)_2O} \quad + ZnI_2 + Cu$

Reviews: H. E. Simmons *et al., Org. React.* **20,** 1 (1973); C. Girard, J. M. Conia, *J. Chem. Res. (S)* **1978,** 182; W. Ratier *et al., ibid.* 179; A. Sele *et al., Helv. Chim. Acta* **62,** 866 (1979); J. Joska, J. Fajkos, *Coll. Czech. Chem. Commun.* **46,** 2751 (1981).

350. Simonini Reaction

A. Simonini, *Monatsh.* **13,** 320 (1892); **14,** 81 (1893).

The preparation of aliphatic esters by the reaction of the silver salt of a carboxylic acid with iodine:

$$2RCOOAg + I_2 \xrightarrow{\Delta} RCOOR + 2AgI + CO_2$$

H. Wieland, F. G. Fischer, *Ann.* **446,** 49 (1926); J. Kleinberg, *Chem. Rev.* **40,** 381 (1947); R. G. Johnson, R. K. Ingham, *ibid.* **56,** 259 (1956); C. V. Wilson, *Org. React.* **9,** 332 (1957); N. J. Bunce, M. Hadley, *Can. J. Chem.* **54,** 2612 (1976). *Cf.* Hunsdiecker Reaction.

351. Simonis Chromone Cyclization

E. Petschek, H. Simonis, *Ber.* **46,** 2014 (1913).

Formation of chromones from phenol and β-keto esters in the presence of phosphorus pentoxide, phosphorus oxychloride or sulfuric acid. Coumarins may also form (Pechmann condensation, *q.v.*):

Reviews: S. M. Sethna, N. M. Shah, *Chem. Rev.* **36,** 14 (1945); S. M. Sethna, R. Phadke, *Org. React.* **7,** 15 (1953); R. N. Lacey, *J. Chem. Soc.* **1954,** 854; O. Dann, G. Illing, *Ann.* **605,** 158 (1957); S. F. Tan, *Aust. J. Chem.* **25,** 1367 (1972).

352. Skraup Reaction

Z. H. Skraup, *Ber.* **13,** 2086 (1880).

Synthesis of quinolines from aromatic amines, glycerol, an oxidizing agent and sulfuric acid:

Early review: R. H. F. Manske, M. Kulka, *Org. React.* **7,** 80-99 (1953). G. M. Badger *et al., Aust. J. Chem.* **16,** 814, 828 (1963); M. Wahren, *Tetrahedron* **20,** 2773 (1964); E. B. Mullock *et al., J. Chem. Soc. C* **1970,** 829; N. P. Buu-Hoi *et al., Perkin Trans. I,* **1972,** 260, 263. *Cf.* Bischler-Napieralski Reaction.

353. Smiles Rearrangement; Truce-Smiles Rearrangement

A. A. Levi *et al., J. Chem. Soc.* **1931,** 3264; W. J. Evans, S. Smiles, *ibid.* **1935,** 181; **1936,** 329.

Intramolecular nucleophilic aromatic substitution in alkaline solution resulting in the migration of an aromatic system from one heteroatom to another. The two-carbon unit joining X and Y is usually part of an aromatic ring but may also be aliphatic:

X = S, SO, SO$_2$, O, COO
YH = OH, NHR, SH, CH$_2$R, CONHR
Z = NO$_2$, SO$_2$R

The conversion of *o*-methyldiaryl sulfones to *o*-benzylbenzenesulfinic acids is referred to as the **Truce-Smiles rearrangement**: W. E. Truce *et al., J. Am. Chem. Soc.* **80,** 3625 (1958); G. P. Crowther, C. R. Hauser, *J. Org. Chem.* **33,** 2228 (1968).

Early reviews: J. F. Bunnett, R. E. Zahler, *Chem. Rev.* **49,** 362 (1951); H. J. Shine, *Aromatic Rearrangements* (Elsevier, New York, 1967) pp 307-316; W. E. Truce *et al., Org. React.* **18,** 99-215 (1970). Conversion of phenols to anilines: I. G. C. Coutts, M. R. Southcott, *Perkin Trans. I* **1990,** 767. Kinetic study: K. Bowden, P. R. Williams, *Perkin Trans. II* **1991,** 215. Methods development for aliphatic substrates: M. Sako *et al., Chem. Pharm. Bull.* **42,** 806 (1994). Application to the synthesis of phenothiazines: S. K. Mukherjee *et al., Pharmazie* **49,** 453 (1994); J. Mukesh *et al., ibid.* 689.

354. Sommelet-Hauser Rearrangement

M. Sommelet, *Compt. Rend.* **205,** 56 (1937).

Rearrangement of benzyl quaternary ammonium salts to *ortho* substituted benzyldialkylamines on treatment with alkali metal amides:

$$\text{C}_6\text{H}_5\text{CH}_2\overset{+}{\text{N}}(\text{CH}_3)_3 \quad \xrightarrow{\text{NH}_2^-} \quad \text{o-CH}_3\text{C}_6\text{H}_4\text{CH}_2\text{N}(\text{CH}_3)_2$$

Early reviews: H. E. Zimmerman in *Molecular Rearrangements* Part 1, P. de Mayo, Ed. (Wiley-Interscience, New York, 1963) pp 382-391; S. H. Pine, *Org. React.* **18**, 403-464 (1970). Extension to sulfur ylides: M. Yamamoto *et al.*, *Bull. Chem. Soc. Japan* **62**, 958 (1989); H. Ishibashi *et al.*, *Chem. Pharm. Bull.* **39**, 2878 (1991). Effects of aromatic substitution: T. Tanaka *et al.*, *ibid.* **40**, 518 (1992). Selectivity studies (Sommelet-Hauser rearrangement vs Stevens rearrangement, *q.v.*): T. Kitano *et al.*, *Perkin Trans. I* **1992**, 2851; T. Tanaka *et al.*, *J. Org. Chem.* **57**, 5034 (1992). *Cf.* Meisenheimer Rearrangements; [2,3]-Wittig Rearrangement.

355. Sommelet Reaction
M. Sommelet, *Compt. Rend.* **157**, 852 (1913); *Bull. Soc. Chim. France* [4] **23**, 95 (1918).

Preparation of aldehydes from aralkyl halides by treatment with hexamethylenetetramine to yield the quaternary salt, followed by mild hydrolysis:

$$\text{ArCH}_2\text{X} + \text{C}_6\text{H}_{12}\text{N}_4 \longrightarrow [\text{ArCH}_2-\text{C}_6\text{H}_{12}\text{N}_4]^+\text{X}^- \xrightarrow[\text{H}_2\text{O}]{\Delta} \text{ArCHO}$$

Early reviews: S. J. Angyal, *Org. React.* **8**, 197-217 (1954); Bayer, *Houben-Weyl* **7/1**, 194 (1954). Synthetic applications: S. Miyano *et al.*, *Bull. Chem. Soc. Japan* **59**, 3285 (1986); D. Evans *et al.*, *Heterocycles* **26**, 1569 (1987). *Cf.* Kröhnke Oxidation.

356. Sonn-Müller Method
A. Sonn, E. Müller, *Ber.* **52**, 1927 (1919).

Reaction sequence employed to convert aromatic anilides to aldehydes. Treatment of the anilide with phosphorus pentachloride generates the imidoyl chloride, which is reduced to the imine with a mixture of stannous chloride and hydrochloric acid. Subsequent hydrolysis yields the aldehyde:

$$\text{ArCONHPh} \xrightarrow{\text{PCl}_5} \underset{\text{Cl}}{\overset{\text{Ar}}{\text{C}}}=\text{NPh} \xrightarrow[\text{HCl}]{\text{SnCl}_2} \text{ArCH}=\text{NPh} \xrightarrow{\text{H}_2\text{O}} \text{ArCHO}$$

T. Reichstein, H. Zschokke, *Helv. Chim. Acta* **15**, 1105 (1932); W. E. Bachmann, *J. Am. Chem. Soc.* **57**, 1381 (1935); T. S. Work, *J. Chem. Soc.* **1942**, 429; L. N. Ferguson, *Chem. Rev.* **38**, 244 (1946); E. Mosettig, *Org. React.* **8**, 240 (1954); L. F. Fieser, M. Fieser, *Advanced Organic Chemistry* (New York, 1961) p 832. *Cf.* Grundmann Aldehyde Synthesis; Stephen Aldehyde Synthesis.

357. Staudinger Reaction
H. Staudinger, J. Meyer, *Helv. Chim. Acta* **2**, 635 (1919).

Synthesis of phosphazo compounds by the reaction of tertiary phosphines with organic azides:

$$\text{R}_3\text{P} + \text{N}_3\text{X} \longrightarrow \text{R}_3\text{P}=\text{N}-\text{N}=\text{N}-\text{X} \xrightarrow{-\text{N}_2} \text{R}_3\text{P}=\text{N}-\text{X}$$
$$\text{phosphazide}$$

Review: Y. G. Gololobov *et al.*, *Tetrahedron* **37**, 437 (1981).

358. Stephen Aldehyde Synthesis
H. Stephen, *J. Chem. Soc.* **127**, 1874 (1925); T. Stephen, H. Stephen, *ibid.* **1956**, 4695.

Reaction sequence employed to convert nitriles to aldehydes. Treatment of the nitrile with a mixture of stannous chloride and hydrochloric acid yields the imine salt complex which is subsequently hydrolyzed to the aldehyde. Practically applied only to aromatic aldehydes:

$$\text{RCN} \xrightarrow[\text{HCl}]{\text{SnCl}_2} (\text{RCH}=\text{NH}\cdot\text{HCl})_2\,\text{SnCl}_4 \xrightarrow{\text{H}_2\text{O}} \text{RCHO}$$

L. N. Ferguson, *Chem. Rev.* **38**, 243 (1946); E. Mosettig, *Org. React.* **8**, 246 (1954); O. Bayer, *Houben-Weyl* **7/1**, 299 (1954); E. N. Zilberman, P. S. Pyryalova, *J. Gen. Chem. U.S.S.R.* (Engl. trans.) **33**, 3348 (1963); C. G. Stuckwisch, *J. Org. Chem.* **37**, 318 (1972). *Cf.* Grundmann Aldehyde Synthesis; Sonn-Müller Method.

359. Stevens Rearrangement

T. S. Stevens *et al., J. Chem. Soc.* **1928,** 3193; **1930,** 2107, 2119; **1932,** 55, 1926, 1932.

Migration of an alkyl group from a sulfonium or quaternary ammonium salt to an adjacent carbanionic center on treatment with strong base. The product is a rearranged tertiary amine or sulfide:

$$Y = NR \text{ or } S$$

Early reviews: H. E. Zimmerman in *Molecular Rearrangements* Part 1, P. de Mayo, Ed. (Wiley-Interscience, New York, 1963) pp 345-406; D. J. Cram, *Fundamentals of Carbanion Chemistry* (Academic Press, New York, 1965) pp 223-229; S. M. Pine, *Org. React.* **18,** 403-464 (1970). Selectivity studies (Stevens rearrangement vs Sommelet-Hauser rearrangement, *q.v.*): T. Kitano *et al., Perkin Trans. I* **1992,** 2851; T. Tanaka *et al., J. Org. Chem.* **57,** 5034 (1992). *Review:* I. E. Markó, *Comp. Org. Syn.* **3,** 913-932 (1991). *Cf.* Meisenheimer Rearrangements; [1,2]-Wittig Rearrangement.

360. Stieglitz Rearrangement

J. Stieglitz, P. N. Leech, *Ber.* **46,** 2147 (1913); *J. Am. Chem. Soc.* **36,** 272 (1914).

Rearrangement of trityl hydroxylamines to Schiff bases on treatment with phosphorus pentachloride:

$$Ar_3CNHOH \xrightarrow{PCl_5} Ar_2C{=}NAr + POCl_3 + 2HCl$$

Reviews: P. A. S. Smith in *Molecular Rearrangements* Part 1, P. de Mayo, Ed. (Wiley-Interscience, New York, 1963) p 479; *Trans. N.Y. Acad. Sci.* **31,** 504 (1969); N. Koga, J. P. Anselme, *Tetrahedron letters* **1969,** 4773; R. V. Hoffman, D. J. Poclker, *J. Org. Chem.* **44,** 2364 (1979).

361. Stille Coupling

M. Kosugi *et al., Chem. Letters* **1977,** 301 (1977); D. Milstein, J. K. Stille, *J. Am. Chem. Soc.* **100,** 3636 (1978).

Palladium-catalyzed cross coupling reaction of organostannanes with organic halides, acetates or perfluorinated sulfonates lacking a *sp³*-hybridized β-hydrogen:

$$R_3Sn{-}R^1 + R^2{-}X \xrightarrow{[Pd]} R^1{-}R^2 + R_3SnX$$

$$[Pd] = Pd(PPh_3)_4, PhCl I_2Pd(PPh_3)_2Cl$$
$$R^1 = \text{alkynyl, alkenyl, aryl, allyl, benzyl, alkyl}$$
$$R^2 = \text{acyl, alkenyl, allyl, benzyl, aryl}$$
$$X = Cl, Br, I, OCOCH_3, OSO_2(C_nF_{2n+1}), n = 0,1,4$$

Allylic acetates as substrates: L. Del Valle *et al., J. Org. Chem.* **55,** 3019 (1990). Effect of additives: S. Gronowitz *et al., J. Organometal. Chem.* **460,** 127 (1993); V. Farina *et al., J. Org. Chem.* **59,** 5905 (1994). Synthesis of α-methylene lactones: R. M. Adlington *et al., Perkin Trans. I* **1994,** 1697. Solid-phase synthesis of 1,4-benzodiazepines: M. J. Plunkett, J. A. Ellman, *J. Am. Chem. Soc.* **117,** 3306 (1995). *Reviews:* J. K. Stille, *Angew. Chem. Int. Ed.* **25,** 508-524 (1986); M. Pereyre *et al., Tin in Organic Synthesis* (Butterworths, Boston, 1987) pp 185-207 *passim.* Review of synthetic applications: T. N. Mitchell, *Synthesis* **1992,** 803-815. *Cf.* Heck Reaction; Suzuki Coupling.

362. Stobbe Condensation

H. Stobbe, *Ber.* **26,** 2312 (1893); *Ann.* **282,** 280 (1894).

Condensation of aldehydes or ketones with diethyl succinate in the presence of a strong base to form monoesters of α-alkylidene (or arylidene) succinic acids:

Reviews: W. S. Johnson, G. H. Daub, *Org. React.* **6**, 1 (1951); H. O. House, *Modern Synthetic Reactions* (W. A. Benjamin, Menlo Park, California, 2nd ed., 1972) pp 663-666; R. J. Hart, H. G. Heller, *Perkin Trans. I* **1972**, 1321; N. R. El-Rayyes, *J. Prakt. Chem.* **315**, 295 (1973); V. B. Bagos *et al., Helv. Chim. Acta* **62**, 90 (1979). *Cf.* Perkin Reaction.

363. Stollé Synthesis

R. Stollé, *Ber.* **46**, 3915 (1913); **47**, 2120 (1914); *J. Prakt. Chem.* **105**, 137 (1923); **128**, 1 (1930).

Formation of indole derivatives by the reaction of arylamines with α-haloacid chlorides or oxalyl chloride, followed by cyclization of the resulting amides with aluminum chloride:

W. C. Sumpter, *Chem. Rev.* **34**, 396 (1944); **37**, 446 (1945); P. L. Julian *et al., Heterocyclic Compounds* **3**, 142, 209 (1952); A. H. Beckett *et al., Tetrahedron* **24**, 6093 (1968). *Cf.* Hinsberg Oxindole Synthesis.

364. Stork Enamine Reaction

G. Stork *et al., J. Am. Chem. Soc.* **76**, 2029 (1954); G. Stork, H. Landesman, *ibid.* **78**, 5128 (1956).

Synthesis of α-alkyl or α-acyl carbonyl compounds from enamines and alkyl or acyl halides:

Reviews: J. Szmuszkovicz, *Adv. Org. Chem.* **4**, 1 (1963); A. G. Cook, Ed., *Enamines* (Marcel Dekker, New York, 1969); H. O. House, *Modern Synthetic Reactions* (W. A. Benjamin, Menlo Park, California, 2nd ed., 1972) pp 570-580, 766-772; S. F. Dyke, *Chemistry of Enamines* (Cambridge University Press, New York, 1973); P. W. Hickmott, *Tetrahedron* **38**, 1975 (1982).

365. Strecker Amino Acid Synthesis

A. Strecker, *Ann.* **75**, 27 (1850); **91**, 349 (1854).

Synthesis of α-amino acids by reaction of aldehydes with ammonia and hydrogen cyanide followed by hydrolysis of the resulting α-aminonitriles. Safer, milder, and more selective reaction conditions have been developed, especially in regard to asymmetric synthesis. The scope of the reaction has been extended to include primary and secondary amines:

Reviews: J. P. Greenstein, M. Winitz, *Chemistry of the Amino Acids* **vol. 3** (New York, 1961) pp 698-700; G. C. Barrett, *Chemistry and Biochemistry of the Amino Acids* (Chapman and Hall, New York, 1985) pp 251, 261. Asymmetric synthesis using enantiopure sulfinimines: F. A. Davis *et al., Tetrahedron Letters* **35**, 9351 (1994). Review of stereoselective synthesis: R. O. Duthaler, *Tetrahedron* **50**, 1539-1650 *passim* (1994). *Cf.* Bucherer-Bergs Reaction.

366. Strecker Degradation

A. Strecker, *Ann.* **123,** 363 (1862).

Reaction of an α-amino acid with a carbonyl compound in aqueous solution or suspension to give carbon dioxide and an aldehyde or ketone containing one less carbon atom. Inorganic oxidizing agents can also be used to bring about the transformation:

Mechanism: Schönberg *et al., J. Chem. Soc.* **1948,** 176. *Reviews:* Schönberg, Moubacher, *Chem. Rev.* **50,** 261 (1952); N. van Chuyen *et al., Agr. Biol. Chem.* **36,** 1199 (1972), *C.A.* **77,** 114855e (1972) Photo-promoted hypochlorite oxidation: Y. Ogata *et al., Bull. Chem. Soc. Japan* **54,** 2057 (1981).

367. Strecker Sulfite Alkylation

A. Strecker, *Ann.* **148,** 90 (1868).

Formation of alkyl sulfonates by reaction of alkyl halides with alkali or ammonium sulfites in aqueous solution in the presence of iodide:

$$RX + M_2SO_3 \xrightarrow{\text{NaI}} RSO_3M + MX$$

A. Collmann, *ibid.* 101; W. Hemilian, *ibid.* **168,** 145 (1873); *Ber.* **6,** 562 (1873); **Swiss.** pats. **105,845; 104,907** (1925); F. C. Wagner, E. E. Reid, *J. Am. Chem. Soc.* **53,** 3409 (1931); C. Weygand, *Organic Preparations* (New York, 1945) p 306; M. Quaedvlieg, *Houben-Weyl* **9,** 372 (1955).

368. Suzuki Coupling

N. Miyaura *et al., Tetrahedron Letters* **1979,** 3437; N. Miyaura, A. Suzuki, *Chem. Commun.* **1979,** 866.

Palladium-catalyzed cross coupling of organic halides or perfluorinated sulfonates with organoboron derivatives proceeding with high stereo- and regioselectivity:

$$R^1{-}BY_2 \ + \ R^2{-}X \ \xrightarrow[\text{base}]{[Pd]} \ R^1{-}R^2$$

$$BY_2 = B(OR)_2, \ 9{-}BBN, \ B(CHCH_3CH(CH_3)_2)_2$$
$$X = I, \ Br, \ Cl, \ OSO_2(C_nF_{2n+1}), \ n = 0,1,4$$
$$R^1 = \text{aryl, alkenyl, alkyl}$$
$$R^2 = \text{aryl, alkenyl, alkynyl, benzyl, allyl, alkyl}$$
$$[Pd] = Pd(PPh_3)_4, \ Pd(dppf)Cl_2$$
$$\text{base} = Na_2CO_3, \ NaOCH_2CH_3, \ TlOH, \ N(CH_2CH_3)_3, \ K_3PO_4$$

Competition with Heck reaction, *q.v.,* when using an alkenyl boronate ester: A. R. Hunt *et al., Tetrahedron Letters* **34,** 3599 (1993). Alternative palladium catalysts: G. Marck *et al., ibid.* **35,** 3277 (1994); T. I. Wallow, B. M. Novak, *J. Org. Chem.* **59,** 5034 (1994). *Reviews:* A. Suzuki, *Pure Appl. Chem.* **63,** 419-422 (1991); A. R. Martin, Y. Yang, *Acta Chem. Scand.* **47,** 221-230 (1993). *Cf.* Hydroboration Reaction; Stille Coupling.

369. Swarts Reaction

F. Swarts, *Bull. Acad. Roy. Belg.* **24,** 309 (1892).

Fluorination of organic polyhalides with antimony trifluoride (or zinc and mercury fluorides) in the presence of a trace of a pentavalent antimony salt:

$$ClHC{=}CClCHCl_2 + SbF_3 \xrightarrow{SbCl_5} ClHC{=}CClCF_3 + SbCl_3$$

A. L. Henne, *Org. React.* **2,** 49 (1944); M. Hudlicky, *Chemistry of Organic Fluorine Compounds* (MacMillan, New York, 1962) pp 93-98.

370. Swern Oxidation (Moffatt-Swern Oxidation)

K. Omura, D. Swern, *Tetrahedron* **34,** 1651 (1978).

Mild oxidation of primary and secondary alcohols, promoted by oxalyl chloride activation of dimethyl sulfoxide, evidently involving the dimethyl alkoxysulfonium salts. Upon the addition of base, the intermediates rearrange intramolecularly to generate aldehydes or ketones, respectively:

Reactivity/selectivity studies: M. Marx, T. T. Tidwell, *J. Org. Chem.* **49**, 788 (1984). *Reviews:* A. J. Mancuso, D. Swern, *Synthesis* **1981**, 165-185 *passim;* T. T. Tidwell, *Org. React.* **39**, 297-572 *passim* (1990). *Cf.* Corey-Kim Oxidation; Pfitzner-Moffatt Oxidation.

371. Tafel Rearrangement
J. Tafel, H. Hahl, *Ber.* **40**, 3312 (1907).

Rearrangement of the carbon skeleton of substituted acetoacetic esters to hydrocarbons with the same number of carbon atoms by electrolytic reduction at a lead cathode in alcoholic sulfuric acid:

$$CH_3COCRR'COOC_2H_5 \nearrow CH_3CH_2CRR'CH_3 \quad \text{(normal)}$$
$$\searrow CH_3CH_2CH_2CHRR' \quad \text{(rearranged)}$$

J. Tafel, W. Jürgen, *Ber.* **42**, 2548 (1909); J. Tafel, *ibid.* **45**, 437 (1912); C. J. Brockman, *Electro-organic Chemistry* (New York, 1926) p 321; H. Stenzl, F. Fichter, *Helv. Chim. Acta* **17**, 669 (1934); **19**, 392 (1936); **20**, 846 (1937); F. Asinger, H. H. Vogel, *Houben-Weyl* **5/1a**, 280, 471 (1970).

372. Tebbe Olefination (Methylenation)
F. N. Tebbe *et al., J. Am. Chem. Soc.* **100**, 3611 (1978); S. H. Pine *et al., ibid.* **102**, 3270 (1980).

Exchange of the oxygen atom of a carbonyl function for the methylene group of the proposed titanium carbene complex (the Tebbe reagent) to yield terminal alkenes:

$$Y = H, R, OR, NR_2$$

Comparative study with Wittig reaction, *q.v.:* S. H. Pine *et al., Synthesis* **1991**, 165. *Reviews:* K. A. Brown-Wensley *et al., Pure Appl. Chem.* **55**, 1733-1744 (1983); S. E. Kelly, *Comp. Org. Syn.* **1**, 743-746 (1991); S. H. Pine, *Org. React.* **43**, 1-91 (1993). *Cf.* Peterson Reaction.

373. Thiele Reaction (Thiele-Winter Acetoxylation)
J. Thiele, *Ber.* **31**, 1247 (1898).

Formation of triacetoxy aromatic compounds by the reaction of quinones with acetic anhydride catalyzed by sulfuric acid or boron trifluoride:

Review: J. F. W. McOmie, J. N. Blatchly, *Org. React.* **19**, 199 (1972). J. M. Blatchly *et al., Perkin Trans. I* **1972**, 2286; J. F. W. McOmie, S. A. Saleh, *ibid.* **1974**, 384; M. Hirama, S. Ito, *Chem. Letters* **1977**, 627. *Cf.* Hinsberg Sulfone Synthesis.

374. Thorpe Reaction
H. Baron, F. G. P. Remfry, Y. F. Thorpe, *J. Chem. Soc.* **85**, 1726 (1904); K. Ziegler *et al., Ann.* **504**, 94 (1933).

Base-catalyzed self-condensation of nitriles to yield imines which tautomerize to enamines:

Reviews: J. P. Schaefer, J. J. Bloomfield, *Org. React.* **15,** 1 (1967); H. O. House, *Modern Synthetic Reactions* (W. Benjamin, Menlo Park, California, 2nd ed., 1972) p 742; E. C. Taylor, A. McKillop, *Chemistry of Enaminonitriles and o-Aminonitriles* (Wiley-Interscience, N.Y., 1970) pp 1-58; *eidem, Advan. Org. Chem.* **7,** 1 (1970). *Cf.* Ziegler Method.

375. Tiemann Rearrangement

F. Tiemann, *Ber.* **24,** 4162 (1891).

Rearrangement of amide oximes (available from nitriles and hydroxylamine) to monosubstituted ureas by treatment with benzenesulfonyl chloride and water:

$$RC\overset{\overset{\displaystyle NOH}{\|}}{-}NH_2 \xrightarrow{C_6H_5SO_2Cl} RC\overset{\overset{\displaystyle NOSO_2C_6H_5}{\|}}{-}NH_2 \xrightarrow{H_2O} RHN\overset{\overset{\displaystyle O}{\|}}{-}C-NH_2$$

P. A. S. Smith, *Org. React.* **3,** 366 (1946); M. W. Partridge, H. A. Turner, *J. Pharm. Pharmacol.* **5,** 103 (1953); R. F. Plapinger, O. O. Owens, *J. Org. Chem.* **21,** 1186 (1956); J. Garapon *et al., Tetrahedron Letters* **1970,** 4905. *Cf.* Beckmann Rearrangement.

376. Tiffeneau-Demjanov Rearrangement

M. Tiffeneau *et al., Compt. Rend.* **205,** 54 (1937).

Rearrangement of β-amino alcohols upon diazotization with nitrous acid to give carbonyl compounds. Cyclic alcohols yield ring expanded or contracted products:

Reviews: P. A. S. Smith, D. R. Baer, *Org. React.* **11,** 157-188 (1960); H. Metzger, *Houben-Weyl* **10/4,** 233 (1968); D. J. Coveney, *Comp. Org. Syn.* **3,** 781-782 (1991). W. E. Parham, C. S. Roosevelt, *J. Org. Chem.* **37,** 1975 (1972); D. Fattori *et al., Tetrahedron* **49,** 1649 (1993). *Cf.* Demjanov Rearrangement; Pinacol Rearrangement.

377. Tishchenko Reaction

L. Claisen, *Ber.* **20,** 646 (1887); V. Tishchenko, *J. Russ. Phys. Chem. Soc.* **38,** 355, 482, 540, 547 (1906); *Chem. Zentr.* **1906 II,** 1309, 1552, 1555, 1556.

Formation of esters from aldehydes by an oxidation-reduction process in the presence of aluminum or sodium alkoxides:

$$2RCHO \xrightarrow{Al(OC_2H_5)_3} RCOOCH_2R$$

O. Kamm. W. F. Kamm, *Org. Syn.* **coll. vol. I,** 104 (1941); Y. Ogata, A. Kawasaki, *Tetrahedron* **25,** 929, 2845 (1969); P. R. Stapp, *J. Org. Chem.* **38,** 1433 (1973); G. Fouquet *et al., Ann.* **1979,** 1591. *Reviews:* L. Cichon, *Wiad. Chem.* **20,** 641, 783 (1966), *C.A.* **66,** 54672b, 94408b (1967). *Cf.* Cannizzaro Reaction; Meerwein-Ponndorf-Verley Reduction.

378. Traube Purine Synthesis

W. Traube, *Ber.* **33,** 1371, 3035 (1900).

Preparation of 4,5-diaminopyrimidines by introduction of the amino group into the 5-position of 4-amino-6-hydroxy- or 4,6-diaminopyrimidines by nitrosation and ammonium sulfide reduction, followed by ring closure with formic acid or chlorocarbonic ester:

J. H. Davidson, *The Nucleic Acids* **I** (New York, 1955) p 131; A. R. Katritzky, *Quart. Rev.* **10,** 397 (1956); *idem, Rev. Pure Appl. Chem.* **11,** 178 (1961); J. H. Lister, *Purines* (Wiley, New York, 1971) pp 31-90.

379. Trost Allylation (Tsuji-Trost Reaction)

J. Tsuji *et al., Tetrahedron Letters* **1965,** 4387; B. M. Trost, T. J. Fullerton, *J. Am. Chem. Soc.* **95,** 292 (1973).

Palladium-catalyzed allylation of nucleophiles proceeding in an S$_N$2 or S$_N$2' fashion depending on the catalyst, nucleophile, and substituents on the substrate:

$$\text{/\\/}X \xrightarrow[\text{Base}]{[Pd]^0, \text{NuH}} \text{/\\/}Nu$$

NuH = malonates, β-diketones, β-keto esters, enamines, β-keto sulfones, bis-sulfones

X = Br, Cl, OCOOR, SO$_2$R, OCOR, OCONR$_2$, OPO(OR)$_2$, NO$_2$

[Pd]^0 = Pd(PPh$_3$)$_4$, Pd$_2$(dba)$_3$ • CHCl$_3$, + PPh$_3$, Pd(OCOCH$_3$)$_2$ + PPh$_3$

Base = NaH, if necessary

Scope and limitations under neutral conditions: J. Tsuji *et al., J. Org. Chem.* **50,** 1523 (1985). Application to the synthesis of polyprenoids: E. Keinan, D. Eren, *Pure Appl. Chem.* **60,** 89 (1988). Review of intramolecular applications: B. M. Trost, *Adv. Chem. Ser.* **230,** 463-478 (1992). *Review:* C. G. Frost *et al., Tetrahedron: Asymmetry* **3,** 1089-1122 (1992).

380. Tscherniac-Einhorn Reaction

J. Tscherniac, **Ger.** pat. **134,979;** A. Einhorn *et al., Ann.* **343,** 207 (1905); **361,** 113 (1908).

Introduction of the amidomethyl group into aromatic rings or activated methylene groups in the presence of sulfuric acid:

$$\text{RCONHCH}_2\text{OH} + \langle \bigcirc \rangle \longrightarrow \text{RCONHCH}_2 - \langle \bigcirc \rangle$$

Reviews: R. Schröter, *Houben-Weyl* **11/1,** 795 (1957); Hellman *Angew. Chem.* **69,** 463 (1957); H. E. Zaugg, W. B. Martin, *Org. React.* **14,** 52 (1965). C. D. Nenitzescu, I. G. Dinulescu, *Rev. Chim.* **2,** 47 (1954), *C.A.* **50,** 15445b (1956); H. E. Zaugg *et al., J. Org. Chem.* **34,** 11, 14 (1969); K. Bott, *Ber.* **106,** 2513 (1973); A. R. Mitchell *et al., Tetrahedron Letters* **1976,** 3795.

381. Twitchell Process

E. Twitchell, **U.S.** pats. **601,603; 628,503** (1898); **Ger.** pats. **365,522; 385,074.**

Commercial process for splitting fats to glycerol and fatty acids by heating the sulfuric-acid-washed fat 20-48 hours in an open tank with steam in a mixture of 25-50% water, 0.5% sulfuric acid and 0.75-1.25% Twitchell reagent (sulfonated petroleum products):

$$\begin{array}{l} \text{CH}_2\text{OOCR} \\ | \\ \text{CHOOCR} \\ | \\ \text{CH}_2\text{OOCR} \end{array} + 3\text{H}_2\text{O} \longrightarrow \begin{array}{l} \text{CH}_2\text{OH} \\ | \\ \text{CHOH} \\ | \\ \text{CH}_2\text{OH} \end{array} + 3\text{RCOOH}$$

E. Twitchell, *J. Am. Chem. Soc.* **22,** 22 (1900); **28,** 196 (1906); J. W. Lawrie, *Glycerol and the Glycols* (New York, 1928) p 32; R. B. Trusler, *J. Oil & Fat Ind.* **8,** 141 (1931); A. F. Bailey, *Industrial Oil and Fat Products* (New York, 1945) p 668; C. J. Marsel, H. D. Allen, *Chem. Eng.* **54**(6), 104 (1947); V. Mills, H. K. McClain, *Ind. Eng. Chem.* **41,** 1982 (1949); L. Lascaray, *J. Am. Oil Chemists Soc.* **29,** 362 (1952); Faith, Keyes & Clark's *Industrial Chemicals* (Wiley-Interscience, New York, 4th ed., 1975) p 431.

382. Ugi Reaction (Four-Component Condensation, 4CC)

I. Ugi, *Angew. Chem. Int. Ed.* **1,** 8 (1962).

The α-addition of an iminium ion and the conjugate base of a carboxylic acid to an isocyanide, followed by spontaneous rearrangement of the α-adduct to yield an α-aminocarboxamide derivative. Carbonyl compounds and amines, or their condensation products, serve as precursors to the iminium ion. The nature of the product depends primarily on the acid component:

When four discrete reactants are used, the reaction is often referred to as the four-component condensation (4CC). Diastereoselective methods development: H. Kunz *et al., Synthesis* **1991,** 1039; M. Goebel, I. Ugi, *ibid.* 1095. *Reviews:* I. Ugi, *Proc. Estonian Acad. Sci. Chem.* **40,** 1-13 (1991); I. Ugi *et al., Comp. Org. Syn.* **2,** 1083-1109 (1991).

383. Ullmann Reaction

F. Ullmann, *Ann.* **332,** 38 (1904); F. Ullmann, P. Sponagel, *Ber.* **38,** 2211 (1905).

Copper-mediated coupling of aryl halides. Biaryl ether synthesis is similarly accomplished with aryl halides and phenols:

$$2 \; C_6H_5{-}I \xrightarrow{Cu} C_6H_5{-}C_6H_5 + CuI_2$$

P. E. Fanta, *Chem. Rev.* **38**, 139 (1946); **64**, 613 (1964); A. A. Moroz, M. S. Shvartsberg, *Russ. Chem. Rev.* **43**, 679 (1974); P. E. Fanta, *Synthesis* **1974**, 9; M. F. Semmelhack *et al., J. Am. Chem. Soc.* **103**, 6460 (1981); D. W. Knight, *Comp. Org. Syn.* **3**, 499-507 (1991). *Cf.* Glaser Coupling.

384. Urech Cyanohydrin Method; Ultee Cyanohydrin Method

F. Urech, *Ann.* **164**, 225 (1872); A. J. Ultee, *Rec. Trav. Chim.* **28**, 1 (1909).

Cyanohydrin formation by addition of alkali cyanide to the carbonyl group in the presence of acetic acid (Urech method) or by reaction of the carbonyl compound with anhydrous hydrogen cyanide in the presence of a basic catalyst (Ultee cyanohydrin method):

$$RR'CO + KCN + CH_3COOH \longrightarrow RR'C(OH)CN + CH_3COO^-K^+$$

$$RR'CO + HCN \xrightarrow{HO^-} RR'C(OH)CN$$

A. J. Ultee, *Ber.* **39**, 1856 (1906); *Rec. Trav. Chim.* **28**, 248, 257 (1909); K. N. Welch, G. R. Clemo, *J. Chem. Soc.* **1928**, 2629; H. R. Dittmar, *U.S.* pat. **2,101,823** (1937); V. Migrdichian, *The Chemistry of Organic Cyanogen Compounds* (New York, 1947) p 173; D. T. Mowry, *Chem. Rev.* **42**, 231 (1948); P. Kurz, *Houben-Weyl* **8**, 274 (1952); R. F. B. Cox, R. T. Stormont, *Org. Syn.* **coll. vol. 2**, 7 (1955). *Cf.* Bucherer-Bergs Reaction; Kiliani-Fischer Synthesis.

385. Urech Hydantoin Synthesis

F. Urech, *Ann.* **165**, 99 (1873).

Formation of hydantoins from α-amino acids by treatment with potassium cyanate in aqueous solution and heating the salt of the intermediate hydantoic acid with 25% hydrochloric acid:

H. D. Dakin, *Am. Chem. J.* **44**, 48 (1910); T. B. Johnson, *J. Am. Chem. Soc.* **35**, 780 (1913); W. J. Boyd, W. Robson, *Biochem. J.* **29**, 542, 546, 2256 (1935); E. Ware, *Chem. Rev.* **46**, 407 (1950); M. Sainsbury, R. S. Theobald, *Rodd's Chemistry of Carbon Compounds* **IVC,** 185 (1986).

386. Vilsmeier-Haack Reaction

A. Vilsmeier, A. Haack, *Ber.* **60**, 119 (1927).

Formylation of activated aromatic or heterocyclic compounds with disubstituted formamides and phosphorus oxychloride:

Reviews: M. R. de Maheas, *Bull. Soc. Chim. France* **1962**, 1989; W. G. Jackson *et al., J. Am. Chem. Soc.* **103**, 533 (1981); C. Jutz, *Advan. Org. Chem.* **9**, 225-342 (1976); O. Meth-Cohn, S. P. Stanforth, *Comp. Org. Syn.* **2**, 777-794 (1991).

387. Voight Amination

K. Voight, *J. Prakt. Chem.* [2] **34**, 1 (1886).

Amination of benzoins with amines in the presence of phosphorus pentoxide or hydrochloric acid:

H. H. Strain, *J. Am. Chem. Soc.* **51,** 269 (1929); R. M. Cowper, T. S. Stevens, *J. Chem. Soc.* **1940,** 347; P. L. Julian *et al., J. Am. Chem. Soc.* **67,** 1203 (1945); R. E. Lutz *et al., ibid.* **70,** 2016 (1948); I. A. Kaye *et al., ibid.* **75,** 746 (1953); J. Iwao *et al., J. Pharm. Soc. Japan* **74,** 551 (1954); R. E. Lutz, J. W. Baker, *J. Org. Chem.* **21,** 49 (1956).

388. Volhard-Erdmann Cyclization
J. Volhard, H. Erdmann, *Ber.* **18,** 454 (1885).

Synthesis of alkyl and aryl thiophenes by cyclization of disodium succinate or other 1,4-difunctional compounds (γ-oxo acids, 1,4-diketones, chloroacetyl-substituted esters) with phosphorus heptasulfide:

L. H. Friedburg, *J. Am. Chem. Soc.* **12,** 83 (1890); *J. Chem. Soc.* **58,** 1400 (1890); R. Phillips, *Org. Syn.* **coll. vol. II,** 578 (1943); F. F. Blicke, *Heterocyclic Compounds* **1,** 212 (1950); D. E. Wolf, K. Folkers, *Org. React.* **4,** 412 (1951); R. F. Feldkamp, B. F. Tullar, *Org. Syn.* **coll. vol. IV,** 671 (1963).

389. von Braun Amide Degradation
H. von Pechmann, *Ber.* **33,** 611 (1900); J. von Braun, *ibid.* **37,** 3210 (1904).

$$\text{RCONHR'} \xrightarrow{\text{PCl}_5} \text{RCN} + \text{R'Cl}$$

Mechanistic study: B. A. Phillips *et al., Tetrahedron* **29,** 3309 (1973). Application to *N-t*-butylamides: R. B. Perni, G. W. Gribble, *Org. Prep. Proceed. Int.* **15,** 297 (1983).

390. von Braun Reaction
J. von Braun, *Ber.* **40,** 3914 (1907); **42,** 2219 (1909); **44,** 1250 (1911).

Reaction of tertiary amines with cyanogen bromide to form disubstituted cyanamides and an alkyl halide:

$$\text{R}_2\text{NR'} + \text{BrCN} \longrightarrow \text{R}_2\text{NCN} + \text{R'Br}$$

Mechanistic correlation with Ritter, Bischler-Napieralski, Beckmann and Schmidt reactions, *q.q.v.*: G. Fodor, S. Nagubandi, *Tetrahedron* **36,** 1279 (1980). Synthetic applications: S. Siddiqui *et al., Z. Naturforsch.* **37b,** 1481 (1982); *idem et al., Pakistan J. Sci. Ind. Res.* **30,** 163 (1987). *Reviews:* H. A. Hageman, *Org. React.* **7,** 198-262 (1953); J. H. Cooley, E. J. Evain, *Synthesis* **1989,** 1-7.

391. von Richter (Cinnoline) Synthesis
V. von Richter, *Ber.* **16,** 677 (1883).

Formation of cinnoline derivatives by diazotization of *o*-aminoarylpropiolic acids or *o*-aminoarylacetylenes followed by hydration and cyclization:

M. Busch, M. Klett, *Ber.* **25,** 2847 (1892); N. J. Leonard, *Chem. Rev.* **37,** 270 (1945); K. Schofield, J. C. E. Simpson, *J. Chem. Soc.* **1945,** 512; K. Schofield, T. Swain, *ibid.* **1949,** 2393; J. C. E. Simpson, *Condensed Pyridazine and Pyrazine Rings* (New York, 1953) p 16; G. R. Ramage, J. K. Landquist, *Chemistry of Carbon Compounds* **IVB,** 1217 (1959). *Cf.* Widman-Stoermer Synthesis.

392. von Richter Reaction
V. von Richter, *Ber.* **4,** 21, 459, 553 (1871).

Carboxylation of substituted aromatic nitro compounds with ethanolic potassium cyanide at 120-270°, the carboxyl group entering a position *ortho* to that previously occupied by the eliminated nitro group:

J. F. Bunnett, *Quart. Rev.* **12**, 15 (1958); D. Samuel, *J. Chem. Soc.* **1960**, 1318; J. Sauer, R. Huisgen, *Angew. Chem.* **72**, 314 (1960); M. Rosenblum, *J. Am. Chem. Soc.* **82**, 3796 (1960); E. Cullen, P. L'Ecuyer, *Can. J. Chem.* **39**, 144, 154, 382, 862 (1961); E. F. Ullman, E. A. Bartkus, *Chem. & Ind. (London)* **1962**, 93; K. M. Ibne-Rasa, E. Koubak, *J. Org. Chem.* **28**, 3240 (1963).

393. Vorbrüggen Glycosylation

U. Niedballa, H. Vorbrüggen, *Angew. Chem. Int. Ed.* **9**, 461 (1970).

The reaction of silylated heterocyclic bases with peracylated sugars in the presence of Lewis acids to yield natural β-nucleosides. If the sugar lacks a 2α-acyloxy substituent, an anomeric mixture forms:

R = COCH₃, COPh
Lewis acid = BF₃ • O(CH₂CH₃)₂, SnCl₄, (CH₃)₃SiOSO₂CF₃, (CH₃)₃SiClO₄, TiCl₄, AlCl₃

Scope and limitations: H. Vorbrüggen *et al., Ber.* **114**, 1234 (1981). Mechanistic study: H. Vorbrüggen, G. Höfle, *ibid.* 1256. Synthetic applications: U. Niedballa, H. Vorbrüggen, *J. Org. Chem.* **39**, 3654, 3660, 3664, 3668, 3672 (1974); R. O. Dempcy, E. B. Skibo, *ibid.* **56**, 776 (1991); S. H. Kawai, G. Just, *Nucleosides Nucleotides* **10**, 1485 (1991). *Cf.* Hilbert-Johnson Reaction.

394. Wacker Oxidation

J. Smidt *et al., Angew. Chem. Int. Ed.* **1**, 176 (1959).

The oxidation of ethylene to acetaldehyde employing palladium chloride and cupric chloride as catalysts and molecular oxygen as oxidant. The reaction has been extensively developed for the oxidation of terminal alkenes to methyl ketones:

Application to hydroxy-α,β-unsaturated esters: S. X. Auclair *et al., Tetrahedron Letters* **33**, 7739 (1992). Use of a multicomponent catalytic system: E. Monflier *et al., ibid.* **36**, 387 (1995). Synthetic applications: M. Romero *et al., ibid.* **35**, 3255 (1994); L. A. Paquette, X. Wang, *J. Org. Chem.* **59**, 2052 (1994). *Reviews:* L. S. Hegedus, *Comp. Org. Syn.* **4**, 552-559 (1991); J. Tsuji, *ibid.* **6**, 449-468.

395. Wagner-Jauregg Reaction

T. Wagner-Jauregg, *Ber.* **63**, 3213 (1930); *Ann.* **491**, 1 (1931).

Addition of maleic anhydride to diarylethylenes with formation of *bis* adducts which can be converted to aromatic ring systems:

F. Bergmann *et al., J. Am. Chem. Soc.* **69,** 1773, 1777, 1779 (1947); K. Alder in *Newer Methods of Preparative Organic Chemistry,* English Ed. (Interscience, New York, 1948) p 425; M. C. Kloetzel, *Org. React.* **4,** 32 (1948). *Cf.* Diels-Alder Reaction.

396. Wagner-Meerwein Rearrangement

G. Wagner, *J. Russ. Phys. Chem. Soc.* **31,** 690 (1899); H. Meerwein, *Ann.* **405,** 129 (1914).

Carbon-to-carbon migration of alkyl, aryl or hydride ions. The original example is the acid-catalyzed rearrangement of camphene hydrochloride to isobornyl chloride:

C. Le Drian, P. Vogel, *Helv. Chim. Acta* **70,** 1703 (1987); M. Asaoka, H. Takei, *Tetrahedron Letters* **28,** 6343 (1987); L. U. Román *et al., J. Org. Chem.* **56,** 1938 (1991). Review of applications to alcohols: Y. Pocker in *Molecular Rearrangements* Part 1, P. de Mayo, Ed. (Wiley-Interscience, New York, 1963) pp 6-15; to bicyclic systems: J. Berson, *ibid.* 111-231; to terpenes: J. F. King, P. de Mayo, *ibid.* 813-840; to alkaloids: E. W. Warnhof, *ibid.* 842-879; to steroids: N. L. Wendler, *ibid.* 1020-1028. *Reviews:* R. L. Cargill *et al., Accts. Chem. Res.* **7,** 106-113 (1974); H. Hogeveen, E. M. G. A. Van Kruchten, *Top. Curr. Chem.* **80,** 89-124 (1979); J. R. Hanson, *Comp. Org. Syn.* **3,** 705-719 (1991). *Cf.* Demjanov Rearrangement; Nametkin Rearrangement; Retropinacol Rearrangement.

397. Walden Inversion

P. Walden, *Ber.* **28,** 1287, 2766 (1895).

Inversion of configuration of a chiral center in bimolecular nucleophilic substitution (S$_N$2) reactions:

H. A. Bent, *Chem. Rev.* **68,** 587 (1968); D. P. G. Harmon, *J. Chem. Ed.* **47,** 398 (1970); L. Kryger *et al., Acta Chem. Scand.* **26,** 2339, 2349 (1972); C. W. Shoppee, J. Nemorin, *Perkin Trans. I* **1973,** 542; K.-C. To *et al., J. Chem. Phys.* **74,** 1499 (1981).

398. Wallach Rearrangement

O. Wallach, L. Belli, *Ber.* **13,** 525 (1880).

Acid-catalyzed rearrangement of azoxybenzenes to *p*-hydroxyazobenzenes:

Reviews: K. H. Schündehütte, *Houben-Weyl* **10/3,** 771-773 (1965); E. Buncel in *Mechanisms of Molecular Migrations* **vol. 1,** B. S. Thyagarajan, Ed. (Interscience, New York, 1968) p 61; R. A. Cox, E. Buncel in *The Chemistry of Hydrazo, Azo and Azoxy Groups,* pt. 2, S. Patai, Ed. (Wiley, New York, 1975) pp 808-837; J. Yamamoto *et al., Tetrahedron* **36,** 3177 (1980).

399. Weerman Degradation

R. A. Weerman, *Rec. Trav. Chim.* **37,** 1, 16 (1918).

Formation of an aldose with one less carbon atom from an aldonic acid by a Hofmann-type reaction, *q.v.,* of the corresponding amide. This is a general reaction of α-hydroxy carboxylic acids:

W. N. Haworth *et al., J. Chem. Soc.* **1934,** 1722; **1938,** 1975; E. S. Wallis, J. F. Lane, *Org. React.* **3,** 275 (1946); J. C. Sowden in *The Carbohydrates*, W. Pigman, Ed. (New York, 1957) p 120; L. F. Fieser, M. Fieser, *Advanced Organic Chemistry* (New York, 1961) p 945.

400. Weiss Reaction

U. Weiss, J. M. Edwards, *Tetrahedron Letters* **1968,** 4885.

Reaction of 1,2-dicarbonyl compounds with 3-oxoglutarates to yield *cis*-bicyclo[3.3.0]octane-3,7-dione or [n.3.3]propellanedione (n > 2) tetracarboxylates. Subsequent acid-catalyzed hydrolysis and decarboxylation yield the respective 2,4,6,8-unsubstituted diones:

$E = COOCH_3$; R^1 or $R^2 = H$, alkyl, aryl or R^1—$R^2 = (CH_2)_n$, n > 2

Experimental procedure: S. H. Bertz *et al., Org. Syn.* **coll. vol. VII,** 50 (1990). Review of synthetic applications: A. K. Gupta *et al., Tetrahedron* **47,** 3665-3710 (1991). *Review:* H.-U. Reissig, "The Weiss Reaction" in *Organic Synthesis Highlights*, J. Mulzer *et al.,* Eds. (VCH, New York, 1991) pp 121-125.

401. Wessely-Moser Rearrangement

F. Wessely, G. H. Moser, *Monatsh.* **56,** 97 (1930).

Rearrangement of flavones and flavanones possessing a 5-hydroxyl group, through fission of the heterocyclic ring and reclosure of the intermediate diaroylmethanes in the alternate direction:

alternate substrates: chromones, isoflavones, flavonols, xanthones, furanochromones

Reviews: Wheeler, *Record Chem. Progr.* **18,** 133 (1957); T. R. Seshadri, *Tetrahedron* **6,** 169 (1959), H. D. Locksley, *Fortschr. Chem. Org. Naturst.* **30,** 292 (1973).

402. Westphalen-Lettré Rearrangement

T. Westphalen, *Ber.* **48,** 1064 (1915); H. Lettré, M. Müller, *ibid.* **70,** 1947 (1937).

Dehydration of 5-hydroxycholesterol derivatives accompanied by C-10 to C-5 methyl migration in compounds with a β-substituent at C-6:

Early review: N. L. Wendler in *Molecular Rearrangements* Part 2, P. de Mayo, Ed. (Wiley-Interscience, New York, 1964) p 1027. A. T. Rowland, *J. Org. Chem.* **29,** 222 (1964); J. W. Blunt *et al., Tetrahedron* **21,** 1567 (1965); K. Kieslich, G. Schulz, *Ann.* **726,** 152 (1969); B. Marples, J. G. L. Jones, *J. Chem. Soc. C* **1970,** 2273; J. Wicha, *Tetrahedron Letters* **1972,** 2877; P. Kocovsky *et al., Coll. Czech. Chem. Commun.* **44,** 234 (1979).

403. Wharton Reaction

P. S. Wharton, D. H. Bohlen, *J. Org. Chem.* **26,** 3615 (1961); P. S. Wharton, *ibid.* 4781.

Reduction of α,β-epoxy ketones by hydrazine to allylic alcohols:

Improved procedure: C. Dupuy, J. L. Luche, *Tetrahedron* **45,** 3437 (1989). Synthetic applications: S. Takano *et al., Synlett* **1991,** 636; T. Yoshimitsu *et al., Synthesis* **1994,** 1029. *Review:* A. R. Chamberlin, D. J. Sall, *Comp. Org. Syn.* **8,** 927-929 (1991). *Cf.* Eschenmoser Fragmentation; Grob Fragmentation.

404. Whiting Reaction
P. Nayler, M. C. Whiting, *J. Chem. Soc.* **1954,** 4006.

Alkynediols are reduced by lithium aluminum hydride in ether or tertiary amines to dienes:

$$\underset{\substack{| \quad\quad | \\ \text{OH} \quad \text{OH}}}{\text{RCHC}\equiv\text{CCHR}} \xrightarrow{\text{LiAlH}_4} \text{RCH}=\text{CH}-\text{CH}=\text{CHR'}$$

R. A. Raphael, *Acetylene Compounds in Organic Synthesis* (New York, 1955) p 114; O. Isler *et al., Helv. Chim. Acta* **39,** 454 (1956); L. F. Fieser, M. Fieser, *Reagents for Organic Synthesis* (New York, 1967) p 385.

405. Wichterle Reaction
O. Wichterle *et al., Coll. Czech. Chem. Commun.* **13,** 300 (1948).

Modification of the Robinson annulation, *q.v.,* in which 1,3-dichloro-*cis*-2-butene is used instead of methyl vinyl ketone:

M. Kobayashi, T. Matsumoto, *Chem. Lett.* **1973,** 957; H. Yoshioka *et al., Tetrahedron Letters* **1979,** 3489. *Review:* M. Hudlicky, *Coll. Czech. Chem. Commun.* **58,** 2229-2244 (1993).

406. Widman-Stoermer Synthesis
O. Widman, *Ber.* **17,** 722 (1884); R. Stoermer, H. Fincke, *Ber.* **42,** 3115 (1909).

Synthesis of cinnolines by cyclization of diazotized *o*-aminoarylethylenes at room temperature:

N. J. Leonard, *Chem. Rev.* **37,** 270 (1945); J. C. E. Simpson, *Condensed Pyridazine and Pyrazine Rings* (New York, 1953) p 6; T. L. Jacobs, *Heterocyclic Compounds* **6,** 137 (1957); G. R. Ramage, J. K. Landquist, *Chemistry of Carbon Compounds* **IVB,** 1217 (1959). *Cf.* von Richter (Cinnoline) Synthesis.

407. Willgerodt-Kindler Reaction
C. Willgerodt, *Ber.* **20,** 2467 (1887); **21,** 534 (1888); K. Kindler, *Ann.* **431,** 193 (1923).

Conversion of aryl alkyl ketones to amides and/or the ammonium salts of the corresponding acids by aqueous ammonium polysulfide or by sulfur and a primary or secondary amine:

$$\text{ArCO(CH}_2)_n\text{CH}_3 \xrightarrow[\text{H}_2\text{O}]{\text{(NH}_4)_2\text{S}_x} \text{Ar(CH}_2)_{n+1}-\text{CONH}_2 + \text{Ar(CH}_2)_{n+1}-\text{COO}^-\text{NH}_4{}^+$$

Reviews: M. Carmack, M. A. Spielman, *Org. React.* **3,** 83 (1946); R. Wegler *et al., Newer Methods of Preparative Organic Chemistry* **vol. 3** (Academic Press, New York, 1964) pp 1-51; E. E. Campaigne in *The Chemistry of the Carbonyl Group,* S. Patai, Ed. (Wiley, New York, 1966) p 954; A. L. J. Beckwith, *The Chemistry of Amides,* J. Zabicky, Ed. (Interscience, London, 1970) pp 145-147; S. W. Schneller, *Int. J. Sulfur Chem.* **8,** 591 (1976).

408. Williamson Synthesis
A. W. Williamson, *J. Chem. Soc.* **4,** 229 (1852).

Synthesis of ethers by alkylation of alkoxides with alkyl halides or alkyl sulfates:

$$\text{RO}^-\text{Na}^+ + \text{R'X} \longrightarrow \text{ROR'} + \text{NaX}$$

$$\text{RO}^-\text{Na}^+ + \text{R'}_2\text{SO}_4 \longrightarrow \text{ROR'} + \text{R'SO}_4{}^-\text{Na}^+$$

Reviews: O. C. Dermer, *Chem. Rev.* **14**, 409 (1934); H. Feuer, J. Hooz in *The Chemistry of the Ether Linkage,* S. Patai, Ed. (Wiley, New York, 1967) pp 446-460; H. O. Kalinowski *et al., Ber.* **114**, 477 (1981); J. March, *Advanced Organic Chemistry* (Wiley-Interscience, New York, 4th ed., 1992) p 386.

409. Wittig Reaction; Horner Reaction; Horner-Wadsworth-Emmons Reaction

G. Wittig, U. Schöllkopf, *Ber.* **87**, 1318 (1954); G. Wittig, W. Haag, *ibid.* **88**, 1654 (1955).

Alkene formation from carbonyl compounds and phosphonium ylides, proceeding primarily through the proposed betaine and/or oxaphosphetane intermediates. The stereoselectivity can be controlled by the choice of ylide, carbonyl compound, and reaction conditions:

When the ylide is replaced with a phosphine oxide carbanion, the reaction is referred to as the **Horner reaction:** L. Horner *et al., Ber.* **91**, 61 (1958); *idem et al., ibid.* **92**, 2499 (1959).

When the ylide is replaced with a phosphonate carbanion, the reaction is referred to as the **Horner-Emmons-Wadsworth reaction:** W. S. Wadsworth, Jr., W. D. Emmons, *J. Am. Chem. Soc.* **83**, 1733 (1961).

Application to the synthesis of β,γ-unsaturated amides: T. Janecki *et al., Tetrahedron* **51**, 1721 (1995). *Reviews:* A. Maercker, *Org. React.* **14**, 270-490 (1965); K. P. C. Vollhardt, *Synthesis* **1975**, 765-780; W. S. Wadsworth, Jr., *Org. React.* **25**, 73-253 (1977); I. Gosney, A. G. Rowley in *Organophosphorus Reagents in Organic Synthesis,* J. I. G. Cadogan, Ed. (Academic Press, New York, 1979) pp 17-153; B. E. Maryanoff, A. B. Reitz, *Chem. Rev.* **89**, 863-927 (1989); S. E. Kelly, *Comp. Org. Syn.* **1**, 755-782 (1991). Reviews of mechanistic studies: W. E. McEwen *et al., ACS Symposium Series* **486**, 149-161 (1992); E. Vedejs, M. J. Peterson, *Top. Stereochem.* **21**, 1-157 (1994). *Cf.* Peterson Reaction; Tebbe Reaction.

410. [1,2]-Wittig Rearrangement

G. Wittig, L. Löhmann, *Ann.* **550**, 260 (1942); G. Wittig, *Experientia* **14**, 389 (1958).

Rearrangement of ethers with alkyl lithiums to yield alcohols *via* a [1,2]-shift:

Reviews: H. E. Zimmerman in *Molecular Rearrangements* Part 1, P. de Mayo, Ed. (Wiley-Interscience, New York, 1963) p 372-377; L. Brandsma, J. F. Arens in *Chemistry of the Ether Linkage,* S. Patai, Ed. (Interscience, New York, 1967) pp 570-580; U. Schöllkopf, *Angew. Chem.* **82**, 795 (1970); A. R. Lepley, A. G. Giumanini in *Mechanisms of Molecular Migrations* **vol. 3,** B. S. Thyagarajan, Ed. (Interscience, New York, 1971); U. Schöllkopf, *Ind. Chim. Belg.* **36**, 1057 (1971); G. Tennant, *Ann. Rep. Progr. Chem. Sec. B* **68**, 241 (1972); R. W. Hoffmann, *Angew. Chem.* **91**, 625 (1979); *idem, Nachr. Chem. Tech. Lab.* **30**, 483 (1982). *Cf.* Meisenheimer Rearrangements; Stevens Rearrangement.

411. [2,3]-Wittig Rearrangement

J. Cast *et al., J. Chem. Soc.* **1960**, 3521; U. Schöllkopf, K. Fellenberger, *Ber.* **698**, 80 (1966); Y. Makisumi, S. Notzumoto, *Tetrahedron Letters* **1966**, 6393.

[2,3]-Sigmatropic rearrangement of the conjugate bases of allylic ethers with high regioselectivity. The stereoselectivity is highly dependent on the the nature of the substrate:

Y = alkynyl, alkenyl, Ph, COR, CN
base = LDA, *n*-BuLi, NaNH₂

Methods development for ring contractions generating enediynes: H. Audrain *et al., Tetrahedron* **50**, 1469 (1994). Review of stereoselectivity: K. Mikami, T. Nakai, *Synthesis* **1994**, 594. *Reviews:* J. A. Marshall, *Comp. Org. Syn.* **3**, 975-1014 (1991); T. Nakai, K. Mikami, *Org. React.* **46**, 105-209 (1994). *Cf.* Meisenheimer Rearrangements; Mislow-Evans Rearrangement; Sommelet-Hauser Rearrangement.

412. Wohl Degradation; Zemplén Modification

A. Wohl, *Ber.* **26**, 730 (1893); **32**, 3666 (1899); G. Zemplén, *Ber.* **59**, 1254, 2402 (1926).

Method for the conversion of an aldose into an aldose with one less carbon atom by the reversal of the cyanohydrin synthesis. In the Wohl method the nitrile group is eliminated by treatment with ammoniacal silver oxide; in the **Zemplén modification** sodium alkoxide is used in the elimination of the nitrile:

$$
\begin{array}{ccccccccc}
\underset{\substack{| \\ \text{HCOH} \\ | \\ (\text{HCOH})_3 \\ | \\ \text{CH}_2\text{OH}}}{\text{HC}=\text{O}} & \xrightarrow{\text{NH}_2\text{OH}} & \underset{\substack{| \\ \text{HCOH} \\ | \\ (\text{HCOH})_3 \\ | \\ \text{CH}_2\text{OH}}}{\text{HC}=\text{NOH}} & \xrightarrow[\text{NaOAc}]{\text{Ac}_2\text{O}} & \underset{\substack{| \\ \text{HCOAc} \\ | \\ (\text{HCOAc})_3 \\ | \\ \text{CH}_2\text{OAc}}}{\text{CN}} & \xrightarrow[\text{NH}_3]{\text{AgOH}} & \underset{\substack{| \\ (\text{HCOH})_3 \\ | \\ \text{CH}_2\text{OH}}}{\text{HC}(\text{NHCOCH}_3)_2} & \xrightarrow{\text{H}_2\text{SO}_4} & \underset{\substack{| \\ (\text{HCOH})_3 \\ | \\ \text{CH}_2\text{OH}}}{\text{CHO}}
\end{array}
$$

Reviews: V. Deulofeu, *Advan. Carbohyd. Chem.* **4**, 129, 138 (1949); R. Bognár *et al., Ann.* **680**, 118 (1964); W. W. Wendall, *Tetrahedron Letters* **1970**, 3439; L. Hough, A. C. Richardson, *The Carbohydrates* **1A**, 128 (1972).

413. Wohl-Ziegler Reaction

A. Wohl, *Ber.* **52**, 51 (1919); K. Ziegler *et al., Ann.* **551**, 30 (1942).

Allylic bromination of olefins with *N*-bromosuccinimide. Peroxides or ultraviolet light are used as initiators:

Reviews: C. Djerassi, *Chem. Rev.* **43**, 271 (1948); L. Horner, E. M. Winkelman, *Angew. Chem.* **71**, 349 (1959); S. S. Novikov *et al., Russ. Chem. Rev.* **31**, 671 (1962); A. Nechvatal, *Adv. Free-Radical Chem.* **4**, 175-201 (1972).

414. Wolff-Kishner Reduction; Huang-Minlon Modification

N. Kishner, *J. Russ. Phys. Chem. Soc.* **43**, 582 (1911), *C.A.* **6**, 347 (1912); L. Wolff, *Ann.* **394**, 86 (1912); Huang-Minlon, *J. Am. Chem. Soc.* **68**, 2487 (1946).

Reduction of carbonyl compounds on heating with hydrazine hydrate and a base. In the **Huang-Minlon modification** diethylene glycol is used as a solvent:

$$
\text{RCOR}' \xrightarrow[\text{KOH}]{\text{H}_2\text{NNH}_2} [\text{RR}'\text{C}=\text{NNH}_2] \longrightarrow \text{RCH}_2\text{R}' + \text{N}_2
$$

Reviews: D. Todd, *Org. React.* **4**, 378 (1948); H. H. Szmant, *Angew. Chem. Int. Ed.* **7**, 120 (1968); F. Asinger, H. H. Vogel, *Houben-Weyl* **5/1a**, 251, 456 (1970); H. Balli, *ibid.* **5/1b**, 629 (1972); R. O. Hutchins, M. K. Hutchins, *Comp. Org. Syn.* **8**, 327-343 (1991). *Cf.* Clemmensen Reduction; Haworth Phenanthrene Synthesis.

415. Wolff Rearrangement

L. Wolff, *Ann.* **394**, 25 (1912).

Rearrangement of diazoketones to ketenes by action of heat, light or some metallic catalysts. The rearrangement is the key step in the Arndt-Eistert synthesis, *q.v.*:

$$
\text{RCOCHN}_2 \xrightarrow{-\text{N}_2} \text{RCH}=\text{C}=\text{O} \begin{array}{c} \xrightarrow{\text{R}'\text{OH}} \text{RCH}_2\text{COOR}' \\ \xrightarrow{\text{R}'\text{NH}_2} \text{RCH}_2\text{CONHR}' \end{array}
$$

Reviews: P. A. S. Smith in P. de Mayo, *Molecular Rearrangements* Part 1, P. de Mayo, Ed. (Wiley-Interscience, New York, 1963) pp 528-550, 558-568; W. Kirmse, *Carbene Chemistry* (Academic Press, New York, 2nd ed., 1971) pp 475-492; H. Meier, K. P. Zeller, *Angew. Chem. Int. Ed.* **14**, 32 (1975); M. Torres, *Pure Appl. Chem.* **52**, 1623 (1980); C. B. Gill, *Comp. Org. Syn.* **3**, 887-912 (1991).

416. Wolffenstein-Böters Reaction

O. Böters, R. Wolffenstein, **Ger.** pat. **194,883** (1906); **Fr.** pat. **380,121** (1907); **Brit.** pat. **17,521** (1907); **U.S.** pat. **923,761** (1909).

Simultaneous oxidation and nitration of aromatic compounds to nitrophenols with nitric acid or the higher oxides of nitrogen in the presence of a mercury salt as catalyst:

R. Wolffenstein, O. Böters, *Ber.* **46**, 586 (1913); F. H. Westheimer *et al., J. Am. Chem. Soc.* **69**, 773 (1947); M. Carmack *et al., ibid.* 785; E. E. Aristoff *et al., Ind. Eng. Chem.* **40**, 1281 (1948); W. Seidenfaden, W. Pawellek, *Houben-Weyl* **10/1**, 815 (1971).

417. Woodward *cis*-Hydroxylation

R. B. Woodward, **U.S.** pat. **2,687,435** (1954); R. B. Woodward, F. V. Brutcher, *J. Am. Chem. Soc.* **80**, 209 (1958).

The hydroxylation of an olefin with iodine and silver acetate in wet acetic acid to give *cis*-glycols:

L. B. Barkley, M. W. Farrar, *J. Am. Chem. Soc.* **76**, 5014 (1954); W. S. Knowles, Q. E. Thompson, *ibid.* **79**, 3212 (1957); W. F. Forbes, R. Shelton, *J. Org. Chem.* **24**, 436 (1959); F. D. Gunstone, *Advan. Org. Chem.* **1**, 117 (1960). Application to steroids: L. Mangoni, V. Dovinola, *Tetrahedron Letters* No. 60, 5235 (1969); P. Kocovsky, V. Cerny, *Coll. Czech. Chem. Commun.* **42**, 163 (1977). Modification: L. Mangoni *et al., Gazz. Chim. Ital.* **105**, 377 (1975). *Cf.* Prévost Reaction.

418. Wurtz-Fittig Reaction

B. Tollens, R. Fittig, *Ann.* **131**, 303 (1864); R. Fittig, J. König, *Ann.* **144**, 277 (1867).

Formation of alkylated aromatic hydrocarbons on treatment of a mixture of an alkyl halide and an aryl halide with sodium:

T. L. Kwa, C. Boelhouwer, *Tetrahedron* **25**, 5771 (1969); B. J. Wakefield, *Comp. Organometal. Chem.* **7**, 45 (1982).

419. Wurtz Reaction

A. Wurtz, *Ann. Chim. Phys.* [3] **44**, 275 (1855); *Ann.* **96**, 364 (1855).

Coupling of two sp^3-carbon centers by treating alkyl halides with sodium:

$$2RX + 2Na \longrightarrow R\!-\!R + 2NaX$$

J. L. Wardell, *Comp. Organometal. Chem.* **1**, 52 (1982); W. E. Lindsell, *ibid.* 193; B. J. Wakefield, *ibid.* **7**, 45; D. C. Billington, *Comp. Org. Syn.* **3**, 413-423 (1991).

420. Ziegler Method (Thorpe-Ziegler Method)

K. Ziegler *et al., Ann.* **504**, 94 (1933).

Cyclization of dinitriles at high dilution in dialkyl ether in the presence of ether-soluble metal alkylanilide and hydrolysis of the resultant imino-nitrile with formation of macrocyclic ketones in good yields:

K. Ziegler *et al., Ann.* **511**, 1 (1933); **512**, 164; **513**, 43 (1934); *idem, Ber.* **67**, 139 (1934); *idem, et al., Ann.* **528**, 114, 143 (1937); R. C. Fuson in *Organic Chemistry* vol. I, H. Gilman, Ed. (New York, 1943) p 89; V. Migrdichian, *The Chemistry of Organic Cyanogen Compounds* (New York, 1947) p 288; K. Ziegler, *Houben-Weyl* **4/2**, 758 (1955). *Cf.* Thorpe Reaction.

421. Ziegler-Natta Polymerization

K. Ziegler *et al., Angew. Chem.* **67**, 426, 541 (1955); G. Natta, *ibid.* **68**, 393 (1956).

Polymerization of vinyl monomers under mild conditions using Lewis acid catalysts to give a stereoregulated, or tactic, polymer.

K. Ziegler, *ibid.* **71**, 623 (1959); **72**, 829 (1960); N. G. Gaylord, H. F. Mark, *Linear and Stereoregulated Addition Polymers* (Interscience, New York, 1959) p 571; C. L. Arcus in *Progress in Stereochemistry* vol. 3, P. B. D. de la Mare, W. Klyne, Eds. (Butterworth Inc., Washington, D.C., 1962) pp 269-288; M. N. Berger *et al., Adv. Catalysis* **19**, 211 (1969); T. Keii, *Kinetics of Ziegler-Natta Polymerization* (Halsted Press, New York, 1973) pp 129-162; A. W. Langer, Jr., *Ann. N.Y. Acad. Sci.* **295**, 110 (1977); *Developments in Polymerization* vol. 2, R. N. Haward, Ed. (Burgess-Intl., Philadelphia, 1979) pp 81-148; H. J. Sinn, W. Kaminsky, *Advan. Organomet. Chem.* **18**, 207 (1980); D. M. P. Mingos, *Comp. Organometal. Chem.* **3**, 72-75 (1982); P. D. Gavens *et al., ibid.* 475-547.

422. Zimmermann Reaction

W. Zimmermann, *Z. Physiol. Chem.* **233**, 257 (1935).

The reaction that occurs between methylene ketones and aromatic polynitro compounds in the presence of alkali.

When applied to 17-oxosteroids, the colored compounds formed can be used for the quantitative determination of 17-oxosteroids:

W. Zimmerman *et al., ibid.* **289,** 91 (1952); *idem, ibid.* **300,** 141 (1955). Studies on mechanism: Neunhoffer *et al., ibid.* **323,** 116 (1961); Foster, Mackie, *Tetrahedron* **18,** 1131 (1962); H. Hoffmeister, C. Rufer, *Ber.* **98,** 2376 (1965); B. T. Rudd, O. M. Galal, *Proc. Assoc. Clin. Biochem.* **4,** 175 (1967); C. S. Feldkamp *et al., Microchem. J.* **22,** 201 (1977). *Cf.* Janovsky Reaction.

423. Zincke Disulfide Cleavage
T. Zincke, *Ber.* **44,** 769 (1911).

Formation of sulfenyl halides by three essentially similar methods involving the action of chlorine or bromine on aryl disulfides, thiophenols, or arylbenzyl sulfides:

$$ArSSAr + X_2 \longrightarrow 2ArSX$$

$$ArSH + X_2 \longrightarrow ArSX + HX$$

$$ArSCH_2C_6H_5 + 2X_2 \longrightarrow ArSX + C_6H_5CHX_2 + HX$$

T. Zincke *et al., Ber.* **45,** 471 (1912); **51,** 751 (1918); *Ann.* **391,** 55 (1912); **400,** 1 (1913); **406,** 103 (1914); **416,** 86 (1918); M. H. Hubacher, *Org. Syn.* **coll. col. II,** 455 (1943); N. Kharasch *et al., Chem. Rev.* **39,** 283 (1946); A. Schöberl, A. Wagner, *Houben-Weyl* **9,** 268 (1955); E. Kühle, *Synthesis* **1970,** 561.

424. Zincke Nitration
T. Zincke, *J. Prakt. Chem.* **61,** 561 (1900).

Replacement of *ortho-* or *para*-bromine or iodine atoms (but not fluorine or chlorine atoms) in phenols by a nitro group on treatment with nitrous acid or a nitrite in acetic acid:

L. C. Raiford, W. Heyl, *Am. Chem. J.* **43,** 393 (1910); **44,** 209 (1911); H. H. Hodgson, J. Nixon, *J. Chem. Soc.* **1932,** 273; L. C. Raiford, G. R. Miller, *J. Am. Chem. Soc.* **55,** 2125 (1933); L. C. Raiford, A. L. LeRosen, *ibid.* **66,** 1872 (1944); W. Seidenfaden, D. Pawellek, *Houben-Weyl* **10/1,** 821 (1971).

425. Zincke-Suhl Reaction
T. Zincke, R. Suhl, *Ber.* **39,** 4148 (1906).

Phenol-dienone rearrangement of *p*-cresols by addition of carbon tetrachloride in the presence of aluminum chloride with formation of 4-methyl-4-trichloromethylcyclohexa-2,5-dienone:

M. S. Newman, A. G. Pinkus, *J. Org. Chem.* **19,** 978, 985, 992, 997 (1954); M. S. Newman, L. L. Wood, Jr., *J. Am. Chem. Soc.* **81,** 6450 (1959); G. A. Olah, *Friedel-Crafts and Related Reactions* **vol. I** (Interscience, New York, 1963) p 128.

INDEX

Boldface numbers refer to the primary entry.

Organic Name Reactions

CHEMICAL ABSTRACTS SERVICE REGISTRY NUMBERS

CHEMICAL ABSTRACTS SERVICE REGISTRY NUMBERS

The following tables provide the Chemical Abstracts Service (CAS) registry numbers for the monograph title compounds and selected derivatives. The latter are indicated by an asterisk following the name. Two formats are employed: the first table is ordered alphabetically by compound name; the second table is ordered numerically by CAS registry number.

CHEMICAL ABSTRACTS REGISTRY NUMBERS-I

Abamectin [71751-41-2]
Abecarnil [111841-85-1]
Abietic Acid [514-10-3]
Abikoviromycin [31774-33-1]
Abrin [1393-62-0]
Abrine [526-31-8]
Abscisic Acid [21293-29-8]
Absinthin [1362-42-1]
Acacetin [480-44-4]
Acacia [9000-01-5]
Acacic Acid [1962-14-7]
Acadesine [2627-69-2]
Acamprosate Calcium [77337-73-6]
Acarbose [56180-94-0]
Acebutolol [37517-30-9]
Acebutolol Hydrochloride*
   [34381-68-5]
Acecainide [32795-44-1]
Acecainide Hydrochloride*
   [34118-92-8]
Acecarbromal [77-66-7]
Aceclofenac [89796-99-6]
Acedapsone [77-46-3]
Acediasulfone [80-03-5]
Acediasulfone Sodium Salt*
   [127-60-6]
Acefylline [652-37-9]
Acefylline Sodium Salt* [837-27-4]
Aceglatone [642-83-1]
Aceglutamide [2490-97-3]
Aceglutamide Aluminum Complex*
   [12607-92-0]
Acemetacin [53164-05-9]
Acenaphthene [83-32-9]
Acenocoumarol [152-72-7]
Aceperone [807-31-8]
Acephate [30560-19-1]
Acepromazine [61-00-7]
Acepromazine Maleate* [3598-37-6]
Acerin [8001-08-9]
Acesulfame [33665-90-6]
Acesulfame Potassium Salt*
   [55589-62-3]
Acetal [105-57-7]
Acetaldehyde [75-07-0]
Acetaldehyde Ammonia [75-39-8]
Acetaldehyde Sodium Bisulfite
   [918-04-7]
Acetaldoxime [107-29-9]
Acetamide [60-35-5]
Acetamidine Hydrochloride
   [124-42-5]
ε-Acetamidocaproic Acid [57-08-9]
ε-Acetamidocaproic Acid Sodium
   Salt* [7234-48-2]
ε-Acetamidocaproic Acid Zinc Salt*
   [70020-71-2]
Acetamidoeugenol [305-13-5]
Acetaminophen [103-90-2]
Acetaminosalol [118-57-0]
Acetanilide [103-84-4]
p-Acetanisidine [51-66-1]
Acetarsone [97-44-9]
Acetarsone Diethylamine Salt*
   [534-33-8]
Acetatocobalamin* [22465-48-1]
Acetazolamide [59-66-5]
Acetazolamide Sodium Salt*
   [1424-27-7]

Acetiamine [299-89-8]
Acetiamine Hydrochloride*
   [28008-04-0]
Acetic Acid Glacial [64-19-7]
Acetic Anhydride [108-24-7]
Acetimidoquinone [50700-49-7]
Acetoacetanilide [102-01-2]
Acetoacetic Acid [541-50-4]
Acetobromglucose [572-09-8]
Acetochlor [34256-82-1]
Acetohexamide [968-81-0]
Acetohydroxamic Acid [546-88-3]
Acetoin [513-86-0]
Acetomeroctol [584-18-9]
Acetone [67-64-1]
Acetone Cyanohydrin [75-86-5]
Acetonedicarboxylic Acid [542-05-2]
Acetone Sodium Bisulfite [540-92-1]
Acetonitrile [75-05-8]
Acetonylacetone [110-13-4]
Acetophenazine [2751-68-0]
Acetophenazine Dimaleate*
   [5714-00-1]
Acetophenone [98-86-2]
Acetorphan [81110-73-8]
Acetosulfone Sodium [128-12-1]
m-Acetotoluide* [537-92-8]
o-Acetotoluide* [120-66-1]
p-Acetotoluide* [103-89-9]
Acetoxime [127-06-0]
Acetoxolone [6277-14-1]
Acetoxolone Aluminium Salt*
   [29728-34-5]
21-Acetoxypregnenolone [566-78-9]
Acetozone [644-31-5]
Acetrizoate Sodium [129-63-5]
Acetulan [8028-98-6]
Aceturic Acid [543-24-8]
Acetylacetone [123-54-6]
N-Acetyl-L-aspartic Acid* [997-55-7]
Acetyl Bromide [506-96-7]
α-Acetylbutyrolactone [517-23-7]
Acetylcarnitine [14992-62-2]
Acetylcarnitine L-Form Hydrochlo-
   ride* [5080-50-2]
Acetyl Chloride [75-36-5]
Acetylcholine Bromide [66-23-9]
Acetylcholine Chloride [60-31-1]
Acetylcysteine [616-91-1]
Acetyldigitoxins α-Form* [1111-39-3]
α-Acetyldigoxin* [5511-98-8]
β-Acetyldigoxin* [5355-48-6]
Acetylene [74-86-2]
Acetylene Dibromide [540-49-8]
Acetylene Dichloride [540-59-0]
Acetyleneurea [496-46-8]
N-Acetylglucosamine* [7512-17-6]
Acetyl Iodide [507-02-8]
Acetylleucine Monoethanolamine
   [149-90-6]
N-Acetylmethionine [65-82-7]
5-Acetyl-2-methoxybenzaldehyde
   [531-99-7]
N-Acetylmuramic Acid* [10597-89-4]
Acetyl Nitrate [591-09-3]
N-Acetylpenicillamine [15537-71-0]
Acetylpheneturide [13402-08-9]
Acetylsalicylsalicylic Acid [530-75-6]

N^4-Acetylsulfamethoxazole*
   [21312-10-7]
Acetyl Sulfamethoxypyrazine
   [3590-05-4]
N^4-Acetylsulfanilamide [121-61-9]
N-Acetylsulfanilic Acid [121-62-0]
N-Acetylsulfanilyl Chloride [121-60-8]
Acetyl Sulfisoxazole* [80-74-0]
Acetyltannic Acid [1397-74-6]
Acid Fuchsin [3244-88-0]
Acidic Fibroblast Growth Factor*
   [106096-92-8]
Acid Violet 7B [5905-34-0]
Acifluorfen [50594-66-6]
Acifran [72420-38-3]
Acifran (−)-Form* [77103-92-5]
Acifran (+)-Form* [77103-91-4]
Acipimox [51037-30-0]
Acitretin [55079-83-9]
Aclacinomycin A* [57576-44-0]
Aclacinomycin B* [57596-79-9]
Aclatonium Napadisilate [55077-30-0]
Aconine [509-20-6]
Aconitic Acid [499-12-7]
Aconitine [302-27-2]
Acranil [1684-42-0]
Acridine [260-94-6]
Acriflavine [8048-52-0]
Acriflavine Dihydrochloride*
   [8063-24-9]
Acrinathrin [101007-06-1]
Acrisorcin [7527-91-5]
Acrivastine [87848-99-5]
Acrolein [107-02-8]
Acrylamide [79-06-1]
Acrylic Acid [79-10-7]
Acrylonitrile [107-13-1]
Actarit [18699-02-0]
ACTH [9002-60-2]
Actinium [7440-34-8]
Actinobolin [24397-89-5]
Actinodaphnine [517-69-1]
Actinomycetin [1402-37-5]
Actinomycin F_1 [1402-44-4]
Actinoquinol [15301-40-3]
Actinoquinol Sodium Salt*
   [7246-07-3]
Actinorhodine [1397-77-9]
Actiphenol [526-02-3]
ACV [32467-88-2]
Acyclovir [59277-89-3]
Adamantane [281-23-2]
Adenine [73-24-5]
A-Denopterin [25663-23-4]
Adenosine [58-61-7]
Adenosine Diphosphate [58-64-0]
Adenosine Triphosphate [56-65-5]
S-Adenosylmethionine [29908-03-0]
3′-Adenylic Acid [84-21-9]
5′-Adenylic Acid [61-19-8]
Adinazolam [37115-32-5]
Adinazolam Monomethanesulfonate*
   [57938-82-6]
Adiphenine Hydrochloride [50-42-0]
Adiphenine Methyl Bromide*
   [6113-04-8]
Adipic Acid [124-04-9]
Adipsin [104118-48-1]
Adlumidine [550-49-2]

Adlumine [524-46-9]
Adonitol [488-81-3]
Adonitoxin [17651-61-5]
Adrafinil [63547-13-7]
Adrenalone [99-45-6]
Adrenalone Hydrochloride* [62-13-5]
Adrenochrome [54-06-8]
Adrenochrome Monosemicarbazone*
  [69-81-8]
Adrenochrome Oxime Sesquihydrate*
  [6055-73-8]
Adrenochrome Thiosemicarbazone*
  [113185-69-6]
Adrenoglomerulotropin [1210-56-6]
Adrenolutin [642-75-1]
Adrenosterone [382-45-6]
AET [56-10-0]
Affinin [25394-57-4]
Aflatoxin B₁* [1162-65-8]
Aflatoxin B₂* [7220-81-7]
Afloqualone [56287-74-2]
Agar [9002-18-0]
Agaricic Acid [666-99-9]
Agaritine [2757-90-6]
Agmatine [306-60-5]
Agmatine Sulfate* [2482-00-0]
Agroclavine [548-42-5]
Agrocybin [544-44-5]
Ahistan [518-61-6]
AICAR [3031-94-5]
AIR [25635-88-5]
Ajacine [509-17-1]
Ajaconine [545-61-9]
Ajmaline [4360-12-7]
Ajoene [92285-01-3]
Ajugarin-I* [62640-05-5]
Ajugarin-II* [62640-06-6]
Ajugarin-III* [62640-07-7]
Ajugarin-IV* [82225-47-6]
Ajugarin-V* [82231-14-9]
Aklomide [3011-89-0]
Akuammicine [639-43-0]
Akuammine [3512-87-6]
AL 721 [99751-63-0]
Alacepril [74258-86-9]
Alachlor [15972-60-8]
Alafosfalin [60668-24-8]
Alanine [56-41-7]
DL-Alanine* [302-72-7]
β-Alanine [107-95-9]
Alanine D-Form* [338-69-2]
L-Alanosine [5854-93-3]
Alantolactone [546-43-0]
Alazopeptin [1397-84-8]
Albaspidin [58409-52-2]
Albendazole [54965-21-8]
Albizziin [1483-07-4]
Albofungin [37895-35-5]
Albomycin [1414-39-7]
Alborixin [57760-36-8]
Albuterol [18559-94-9]
Albuterol Sulfate* [51022-70-9]
Albutoin [830-89-7]
Alcian Blue [12040-44-7]
Alclofenac [22131-79-9]
Alclofenac Monoethanolamine
  Complex* [22131-80-2]
Alclometasone [67452-97-5]
Alclometasone 17,21-Dipropionate*
  [66734-13-2]
Alcuronium [23214-96-2]
Alcuronium Dichloride* [15180-03-7]
Aldicarb [116-06-3]
Aldioxa [5579-81-7]
Aldol [107-89-1]
Aldosterone [52-39-1]
Aldrin [309-00-2]
Alendronic Acid [66376-36-1]
Alendronic Acid Monosodium Salt
  Trihydrate* [121268-17-5]

Aleuritic Acid [533-87-9]
Aleuritic Acid Ethyl Ester*
  [6003-09-4]
Aleuritic Acid Hydrazide*
  [6003-10-7]
Alexidine [22573-93-9]
Alexitol Sodium [66813-51-2]
Alfadolone Acetate [23930-37-2]
Alfaprostol [74176-31-1]
Alfaxalone [23930-19-0]
Alfaxalone Mixture with Alfadolone
  Acetate* [8067-82-1]
Alfentanil [71195-58-9]
Alfentanil Hydrochloride Mono-
  hydrate* [70879-28-6]
Alfuzosin [81403-80-7]
Alfuzosin Hydrochloride*
  [81403-68-1]
Algestone [595-77-7]
Algestone Acetophenide [24356-94-3]
Algestone Cyclic Acetal with
  Acetone* [4968-09-6]
Algin [9005-38-3]
Alginic Acid [9005-32-7]
Alginic Acid Calcium Salt*
  [9005-35-0]
Alginic Acid Potassium Salt*
  [9005-36-1]
Alglucerase [143003-46-7]
Alibendol [26750-81-2]
Alinidine [33178-86-8]
Alitame [80863-62-3]
Alizapride [59338-93-1]
Alizapride Hydrochloride*
  [59338-87-3]
Alizarin [72-48-0]
Alizarin Cyanine Green F [4403-90-1]
Alizarin Blue [568-02-5]
Alizarine Orange [568-93-4]
Alizarine Yellow R [2243-76-7]
Alkannin [517-88-4]
Alkofanone [7527-94-8]
Allantoin [97-59-6]
Allenolic Acid [553-39-9]
Allethrin I* [584-79-2]
Allethrin II* [497-92-7]
Allicin [539-86-6]
Allidochlor [93-71-0]
Alliin [556-27-4]
Allobarbital [52-43-7]
Allocholesterol [517-10-2]
Alloclamide [5486-77-1]
Allocryptopine [485-91-6]
Allocupreide Sodium [5965-40-2]
Allopregnane [641-85-0]
Allopregnane-3α,20α-diol [566-58-5]
Allopregnane-3α,20β-diol [566-57-4]
Allopregnane-3β,20α-diol [566-56-3]
Allopregnane-3β,20β-diol [516-53-0]
Allopregnane-3β,21-diol-11,20-dione
  [566-02-9]
Allopregnane-3β,17α-diol-20-one
  [570-54-7]
3,20-Allopregnanedione [566-65-4]
Allopregnane-3β,11β,17α,20β,21-pen-
  tol [516-39-2]
Allopregnane-3β,17α,20β,21-tetrol
  [566-41-6]
Allopregnane-3α,11β,17α,21-tetrol-
  20-one [302-91-0]
Allopregnane-3β,11β,17α,21-tetrol-
  20-one [651-43-4]
Allopregnane-3β,17α,20α-triol
  [570-50-3]
Allopregnane-3β,17α,20β-triol
  [520-86-5]
Allopregnane-3β,17α,21-triol-11,20-
  dione [516-45-0]
Allopregnane-3β,11β,21-triol-20-one
  [516-16-5]

Allopregnane-3β,17α,21-triol-20-one
  [516-47-2]
Allopregnan-3α-ol-20-one [516-54-1]
Allopregnan-3β-ol-20-one [516-55-2]
Allopregnan-20α-ol-3-one [516-59-6]
Allopregnan-20β-ol-3-one [516-58-5]
Allopurinol [315-30-0]
D-Allose [2595-97-3]
Allotetrahydrocortisone [547-77-3]
Alloxan [50-71-5]
Alloxantin [76-24-4]
Allura Red AC [25956-17-6]
Allyl Acetate [591-87-7]
Allyl Alcohol [107-18-6]
Allylamine [107-11-9]
Allyl Bromide [106-95-6]
Allyl Chloride [107-05-1]
Allyl Cyanide [109-75-1]
Allylestrenol [432-60-0]
Allyl Ether [557-40-4]
Allyl Ethyl Ether [557-31-3]
Allyl Iodide [556-56-9]
Allyl Isothiocyanate [57-06-7]
Allylprodine [25384-17-2]
Allyl Sulfide [592-88-1]
Allylurea [557-11-9]
Almagate [66827-12-1]
Aminoprofen [39718-89-3]
Almitrine [27469-53-0]
Almitrine Dimethanesulfonate*
  [29608-49-9]
Aloe-Emodin [481-72-1]
Aloeresin A* [74545-79-2]
Aloin [5133-19-7]
Aloxidone [526-35-2]
Aloxiprin [9014-67-9]
Alphaprodine [77-20-3]
Alphaprodine dl-Form Hydro-
  chloride* [561-78-4]
Alpidem [82626-01-5]
Alpiropride [81982-32-3]
Alprazolam [28981-97-7]
Alprenolol [13655-52-2]
Alprenolol Hydrochloride*
  [13707-88-5]
Alsactide [34765-96-3]
Alseroxylon* [8001-95-4]
Alstonidine [25394-75-6]
Alstonine [642-18-2]
Althiazide [5588-16-9]
Altrenogest [850-52-2]
Altretamine [645-05-6]
D-Altrose [1990-29-0]
Aluminon [569-58-4]
Aluminum [7429-90-5]
Aluminum Acetate Solution
  [8006-13-1]
Aluminum Ammonium Sulfate
  [7784-25-0]
Aluminum Antimonide [25152-52-7]
Aluminum Benzoate [555-32-8]
Aluminum Bis(acetylsalicylate)
  [23413-80-1]
Aluminum Borate [11121-16-7]
Aluminum Boroformate [8011-44-7]
Aluminum Borohydride [16962-07-5]
Aluminum Bromide [7727-15-3]
Aluminum tert-Butoxide [556-91-2]
Aluminum Calcium Hydride
  [16941-10-9]
Aluminum Carbide [1299-86-1]
Aluminum Cesium Sulfate
  [14284-36-7]
Aluminum Chlorate [15477-33-5]
Aluminum Chlorate Nonahydrate*
  [7784-15-8]
Aluminum Chloride [7446-70-0]
Aluminum Chloride Hexahydrate*
  [7784-13-6]
Aluminum Diacetate [142-03-0]

* Denotes derivative of title compound

Aluminum Ethoxide [555-75-9]
Aluminum Fluoride [7784-18-1]
Aluminum Hexafluorosilicate [17099-70-6]
Aluminum Hexaurea Sulfate Tri-iodide [15304-14-0]
Aluminum Hydride [7784-21-6]
Aluminum Hydroxide [21645-51-2]
Aluminum Hydroxychloride [1327-41-9]
Aluminum Hypophosphite [7784-22-7]
Aluminum Iodide [7784-23-8]
Aluminum Isopropoxide [555-31-7]
Aluminum Lactate [18917-91-4]
Aluminum Lithium Hydride [16853-85-3]
Aluminum Magnesium Silicate [12511-31-8]
Aluminum β-Naphtholdisulfonate [1300-81-8]
Aluminum Nicotinate [1976-28-9]
Aluminum Nitrate [13473-90-0]
Aluminum Nitride [24304-00-5]
Aluminum Oleate [688-37-9]
Aluminum Oxalate [814-87-9]
Aluminum Oxide [1344-28-1]
Aluminum Palmitate [555-35-1]
Aluminum Phosphate [7784-30-7]
Aluminum Phosphide [20859-73-8]
Aluminum Potassium Sulfate [10043-67-1]
Aluminum Rubidium Sulfate [13530-57-9]
Aluminum Selenide [1302-82-5]
Aluminum Silicate [12141-46-7]
Aluminum Sodium Sulfate [10102-71-3]
Aluminum Stearate [637-12-7]
Aluminum Subacetate Solution [8000-61-1]
Aluminum Sulfate [10043-01-3]
Aluminum Sulfide [1302-81-4]
Aluminum Tartrate [815-78-1]
Aluminum Thiocyanate [538-17-0]
Aluminum Zinc Sulfate [22992-10-5]
Alverine [150-59-4]
Alverine Citrate* [5560-59-8]
Amanitin [11030-71-0]
α-Amanitin* [23109-05-9]
β-Amanitin* [21150-22-1]
Amanozine [537-17-7]
Amantadine [768-94-5]
Amantadine Hydrochloride* [665-66-7]
Amantanium Bromide [58158-77-3]
Amaranth (Dye) [915-67-3]
Amarogentin [21018-84-8]
Amarolide [29913-86-8]
Ambazone [539-21-9]
Ambenonium Chloride [115-79-7]
Amber [9000-02-6]
Ambergris [8038-65-1]
Amberlite [9079-25-8]
Ambrosin [509-93-3]
Ambroxol [18683-91-5]
Ambroxol Hydrochloride* [23828-92-4]
Ambucaine [119-29-9]
Ambucaine Hydrochloride* [5870-25-7]
Ambucetamide [519-88-0]
Ambuphylline [5634-34-4]
Ambuside [3754-19-6]
Ambutonium Bromide [115-51-5]
Amcinonide [51022-69-6]
Amdinocillin [32887-01-7]
Amdinocillin Pivoxil [32886-97-8]
Amdinocillin Pivoxil Hydrochloride* [32887-03-9]
Americium [7440-35-9]

Ametryn [834-12-8]
Amezinium Methyl Sulfate [30578-37-1]
Amfenac [51579-82-9]
Amfenac Sodium Salt Monohydrate* [61618-27-7]
Amicarbalide [3459-96-9]
Amicetin [17650-86-1]
Amicibone [23271-63-8]
Amicoumacin A [78654-44-1]
Amidephrine [3354-67-4]
Amidephrine Monomethanesulfonate* [1421-68-7]
Amidinomycin [3572-60-9]
Amidochlor [40164-67-8]
Amido-G-Acid [86-65-7]
Amidomycin [552-33-0]
Amido-R-Acid [92-28-4]
Amifostine [20537-88-6]
Amifostine Monohydrate* [63717-27-1]
Amifostine Trihydrate* [112901-68-5]
Amikacin [37517-28-5]
Amikacin Sulfate* [39831-55-5]
Amiloride [2609-46-3]
Amiloride Hydrochloride Dihydrate* [17440-83-4]
Aminacrine [90-45-9]
Aminacrine Hydrochloride* [134-50-9]
Amine 220 [95-38-5]
Amineptine [57574-09-1]
Amineptine Hydrochloride* [30272-08-3]
Aminitrozole [140-40-9]
p-Aminoacetanilide [122-80-5]
Aminoacetonitrile [540-61-4]
Aminoacetonitrile Hydrochloride* [6011-14-9]
Aminoacetonitrile Sulfate (1:1)* [151-63-3]
Aminoacetonitrile Sulfate (2:1)* [5466-22-8]
D-Amino Acid Oxidase [9000-88-0]
L-Amino Acid Oxidase [9000-89-9]
α-Aminoadipic Acid [542-32-5]
1-Aminoanthraquinone [82-45-1]
1-Aminoanthraquinone-2-carboxylic Acid [82-24-6]
p-Aminoazobenzene [60-09-3]
o-Aminoazotoluene [97-56-3]
m-Aminobenzoic Acid [99-05-8]
o-Aminobenzoic Acid [118-92-3]
p-Aminobenzoic Acid [150-13-0]
p-Aminobenzoic Acid Diethylamine Salt* [6018-84-4]
2-Aminobenzothiazole [136-95-8]
6-Aminobenzothiazole [533-30-2]
N-(p-Aminobenzoyl)glutamic Acid [4271-30-1]
2-Amino-1-butanol [96-20-8]
α-Aminobutyric Acid [80-60-4]
β-Aminobutyric Acid [541-48-0]
γ-Aminobutyric Acid [56-12-2]
ε-Aminocaproic Acid [60-32-2]
Aminocarb [2032-59-9]
7-Aminocephalosporanic Acid [957-68-6]
4-Amino-4′-chlorodiphenyl [135-68-2]
Aminochlorthenoxazin [3567-76-8]
2-Amino-4,6-dichlorophenol [527-62-8]
1-[(2-Aminoethyl)amino]-2-propanol [123-84-2]
2-Amino-2-ethyl-1,3-propanediol [115-70-8]
Aminoglutethimide [125-84-8]
Aminoguanidine [79-17-4]
Aminoguanidine Hydrochloride* [1937-19-5]

p-Aminohippuric Acid [61-78-9]
p-Aminohippuric Acid Sodium Salt* [94-16-6]
3-Amino-4-hydroxybutyric Acid [589-44-6]
4-Amino-3-hydroxybutyric Acid [352-21-6]
α-Aminoisobutyric Acid [62-57-7]
δ-Aminolevulinic Acid [106-60-5]
β-Amino-α-methylphenethyl Alcohol [52500-61-5]
2-Amino-2-methyl-1,3-propanediol [115-69-5]
2-Amino-2-methyl-1-propanol [124-68-5]
2-Amino-4-methylthiazole [1603-91-4]
Aminometradine [642-44-4]
3-Amino-2-naphthoic Acid [5959-52-4]
4-Amino-1-naphthol [2834-90-4]
1-Amino-2-naphthol-4-sulfonic Acid [116-63-2]
1-Amino-2-naphthol-6-sulfonic Acid [5639-34-9]
6-Aminonicotinic Acid [3167-49-5]
2-Amino-5-nitrothiazole [121-66-4]
6-Aminopenicillanic Acid [551-16-6]
Aminopentamide [60-46-8]
m-Aminophenol [591-27-5]
o-Aminophenol [95-55-6]
p-Aminophenol [123-30-8]
p-Aminophenylacetic Acid [1197-55-3]
4-Amino-3-phenylbutyric Acid [1078-21-3]
Aminophylline [317-34-0]
2-Amino-4-picoline [695-34-1]
2-Amino-4-picoline Camphor-sulfonate* [12261-97-1]
2-Amino-4-picoline Hydrochloride* [2403-84-1]
Aminopromazine [58-37-7]
Aminopromazine Fumarate* [3688-62-8]
2-Aminopropanol [78-91-1]
3-Aminopropionitrile [151-18-8]
p-Aminopropiophenone [70-69-9]
α-(α-Aminopropyl)benzyl Alcohol [5897-76-7]
Aminopropylon [3690-04-8]
Aminopterin [54-62-6]
α-Aminopyridine [504-29-0]
β-Aminopyridine* [462-08-8]
Aminopyrine [58-15-1]
Aminopyrine Bicamphorate* [94442-12-3]
Aminopyrine Hydrochloride* [6170-29-2]
Aminopyrine Salicylate* [603-57-6]
Aminoquinuride [3811-56-1]
Aminorex [2207-50-3]
Aminorex Fumarate* [13425-22-4]
p-Aminosalicylic Acid [65-49-6]
p-Aminosalicylic Acid Calcium Salt* [133-15-3]
p-Aminosalicylic Acid Hydrazide [6946-29-8]
p-Aminosalicylic Acid Potassium Salt* [133-09-5]
p-Aminosalicylic Acid Sodium Salt Dihydrate* [6018-19-5]
4-Amino-2-sulfobenzoic Acid [527-76-4]
2-Aminothiazole [96-50-4]
2-Amino-1,1,3-tricyanopropene [868-54-2]
Amioca [9037-22-3]
Amiodarone [1951-25-3]
Amiodarone Hydrochloride* [19774-82-4]
Amiphenazole [490-55-1]
Amiprilose [56824-20-5]

Amiprilose Hydrochloride*
[60414-06-4]
Amisometradine [550-28-7]
Amisulpride [71675-85-9]
Amiton [78-53-5]
Amitraz [33089-61-1]
Amitriptyline [50-48-6]
Amitriptyline Hydrochloride*
[549-18-8]
Amitriptylinoxide [4317-14-0]
Amitrole [61-82-5]
Amixetrine [24622-72-8]
Amixetrine Hydrochloride*
[24622-52-4]
Amlexanox [68302-57-8]
Amlodipine [88150-42-9]
Amlodipine (±)-Form Maleate*
[88150-47-4]
Ammonia [7664-41-7]
Ammoniacum [9000-03-7]
Ammonium Acetate [631-61-8]
Ammonium Acetate Solution*
[8013-61-4]
Ammonium Benzoate [1863-63-4]
Ammonium Bicarbonate [1066-33-7]
Ammonium Bifluoride [1341-49-7]
Ammonium Bimalate [5972-71-4]
Ammonium Binoxalate [5972-72-5]
Ammonium Bisulfate [7803-63-6]
Ammonium Bisulfide [12124-99-1]
Ammonium Bisulfite [10192-30-0]
Ammonium Bitartrate [3095-65-6]
Ammonium Borate [12007-58-8]
Ammonium Bromide [12124-97-9]
Ammonium Caprylate [5972-76-9]
Ammonium Carbamate [1111-78-0]
Ammonium Carbonate [10361-29-2]
Ammonium Ceric Nitrate
[16774-21-3]
Ammonium Cerous Sulfate
[21995-38-0]
Ammonium Chloride [12125-02-9]
Ammonium Chromate(VI) [7788-98-9]
Ammonium Chromic Sulfate
[13548-43-1]
Ammonium Citrate, Dibasic
[3012-65-5]
Ammonium Cobaltous Phosphate
[14590-13-7]
Ammonium Cobaltous Sulfate
[13596-46-8]
Ammonium Cupric Chloride
[15610-76-1]
Ammonium Dichromate(VI)
[7789-09-5]
Ammonium Dithiocarbamate
[513-74-6]
Ammonium Ferric Chromate
[7789-08-4]
Ammonium Ferric Citrate [1185-57-5]
Ammonium Ferric Oxalate
[14221-47-7]
Ammonium Ferric Sulfate
[10138-04-2]
Ammonium Ferricyanide [14221-48-8]
Ammonium Ferrocyanide
[14481-29-9]
Ammonium Ferrous Sulfate
[10045-89-3]
Ammonium Fluoride [12125-01-8]
Ammonium Formate [540-69-2]
Ammonium Hexafluoroaluminate
[7784-19-2]
Ammonium Hexafluorogallate
[14639-94-2]
Ammonium Hexafluorophosphate
[16941-11-0]
Ammonium Hexafluorosilicate
[16919-19-0]

Ammonium Hypophosphite
[7803-65-8]
Ammonium Iodide [12027-06-4]
Ammonium Lactate [52003-58-4]
Ammonium Magnesium Chloride
[39733-35-2]
Ammonium Mandelate [530-31-4]
Ammonium Mercuric Chloride
[33445-15-7]
Ammonium Molybdate(VI)
[12027-67-7]
Ammonium Nickel Sulfate
[15699-18-0]
Ammonium Nitrate [6484-52-2]
Ammonium Nitroferricyanide
[14402-70-1]
Ammonium Oleate [544-60-5]
Ammonium Osmium Chloride
[12125-08-5]
Ammonium Oxalate [1113-38-8]
Ammonium Palmitate [593-26-0]
Ammonium Pentachlorozincate
[14639-98-6]
Ammonium Perchlorate [7790-98-9]
Ammonium Peroxydisulfate
[7727-54-0]
Ammonium Phosphate, Dibasic
[7783-28-0]
Ammonium Phosphate, Monobasic
[7722-76-1]
Ammonium Phosphite [51503-61-8]
Ammonium Phosphomolybdate
[12026-66-3]
Ammonium Phosphotungstate
[1311-90-6]
Ammonium Picrate [131-74-8]
Ammonium Platinic Chloride
[16919-58-7]
Ammonium Platinous Chloride
[13820-41-2]
Ammonium Salicylate [528-94-9]
Ammonium Selenate [7783-21-3]
Ammonium Selenite [7783-19-9]
Ammonium Sodium Phosphate
[13011-54-6]
Ammonium Stearate [1002-89-7]
Ammonium Sulfamate [7773-06-0]
Ammonium Sulfate [7783-20-2]
Ammonium Sulfide [12135-76-1]
Ammonium Sulfite [10196-04-0]
Ammonium Tetrachloroaluminate
[7784-14-7]
Ammonium Tetrachlorozincate
[14639-97-5]
Ammonium Thiocyanate [1762-95-4]
Ammonium Thiosulfate [7783-18-8]
Ammonium Titanium Oxalate
[10580-02-6]
Ammonium Tungstate(VI)
[11120-25-5]
Ammonium Uranate(VI) [7783-22-4]
Ammonium Uranium Carbonate
[18077-75-5]
Ammonium Uranium Fluoride
[18433-40-4]
Ammonium Valerate [42739-38-8]
Ammonium Vanadate(V) [7803-55-6]
Ammonium Zirconyl Carbonate
[12616-24-9]
Amobarbital [57-43-2]
Amobarbital Sodium Salt* [64-43-7]
Amocarzine [36590-19-9]
Amodiaquin [86-42-0]
Amodiaquin Dihydrochloride
Dihydrate* [6398-98-7]
Amolanone [76-65-3]
Amolanone Hydrochloride*
[6009-67-2]
Amoproxan [22661-76-3]

Amoproxan Hydrochloride*
[22661-96-7]
Amorolfine [78613-35-1]
Amorolfine Hydrochloride*
[78613-38-4]
Amoscanate [26328-53-0]
Amosulalol [85320-68-9]
Amosulalol Monohydrochloride*
[93633-92-2]
Amotriphene [5585-64-8]
Amoxapine [14028-44-5]
Amoxicillin [26787-78-0]
Amoxicillin Trihydrate* [61336-70-7]
Ampelopsin [27200-12-0]
Amperozide [75558-90-6]
Amphecloral [5581-35-1]
Amphenidone [134-37-2]
Amphenone B [2686-47-7]
Amphetamine [300-62-9]
Amphetamine d-Form Tannate*
[1407-85-8]
Amphetamine Phosphate* [139-10-6]
Amphetamine Sulfate* [60-13-9]
Amphetaminil [17590-01-1]
Amphomycin [1402-82-0]
Amphotalide [1673-06-9]
Amphotericin B [1397-89-3]
Ampicillin [69-53-4]
Ampicillin Sodium Salt* [69-52-3]
Ampicillin Trihydrate* [7177-48-2]
Ampiroxicam [99464-64-9]
Amprolium [121-25-5]
Amprotropine Phosphate [134-53-2]
Ampyrone [83-07-8]
Amrinone [60719-84-8]
Amsacrine [51264-14-3]
Amsonic Acid [81-11-8]
Amtolmetin Guacil [87344-06-7]
Amygdalin [29883-15-6]
n-Amylamine [110-58-7]
Amylase [9000-92-4]
α-Amylase (Bacterial)* [9000-85-5]
α-Amylase (Porcine)* [9000-90-2]
β-Amylase (Sweet Potato)*
[9000-91-3]
Amylbenzene [538-68-1]
d-Amyl Bromide [534-00-9]
n-Amyl Bromide [110-53-2]
tert-Amyl Bromide [507-36-8]
n-Amyl Butyrate [540-18-1]
n-Amyl Caproate [540-07-8]
Amyl Chloride [543-59-9]
6-n-Amyl-m-cresol [53043-14-4]
Amylene [513-35-9]
Amylene Dichloride [507-45-9]
n-Amyl Ether [693-65-2]
tert-Amyl Isovalerate [542-37-0]
n-Amyl Mercaptan [110-66-7]
tert-Amyl Methyl Ether [994-05-8]
Amylocaine Hydrochloride [532-59-2]
Amylpenicillin Sodium [575-47-3]
α-Amyrin [638-95-9]
β-Amyrin [559-70-6]
Anabasine [494-52-0]
Anabsinthin [6903-12-4]
Anagestone [2740-52-5]
Anagestone Acetate* [3137-73-3]
Anagrelide [68475-42-3]
Anagrelide Hydrochloride*
[58579-51-4]
Anagyrine [486-89-5]
Anakinra* [143090-92-0]
Anastrozole [120511-73-1]
Anatabine [581-49-7]
Anazolene Sodium [3861-73-2]
Ancitabine [31698-14-3]
Ancitabine Hydrochloride*
[10212-25-6]
Ancrod [9046-56-4]
Ancymidol [12771-68-5]

Denotes derivative of title compound

Andrographolide [5508-58-7]
Androisoxazole [360-66-7]
Androstane [438-22-2]
Androstane-3β,11β-diol-17-one [514-17-0]
Androstenediol [521-17-5]
Androstenediol 3-Acetate* [1639-43-6]
Androstenediol 17-Acetate* [5937-72-4]
Androstenediol 3-Acetate-17-benzoate* [5953-63-9]
Androstenediol Diacetate* [2099-26-5]
Androstenediol Dipropionate* [2297-30-5]
4-Androstene-3,17-dione [63-05-8]
(3α,5α)-Androst-16-en-3-ol [1153-51-1]
Androsterone [53-41-8]
Anemonin [508-44-1]
Anethole [104-46-1]
Anethole Trithione [532-11-6]
Angelic Acid [565-63-9]
Angelica Lactone [1333-38-6]
Angiotensin [1407-47-2]
Angiotensin Amide* [53-73-6]
Anhalamine [643-60-7]
Anhalonidine [17627-77-9]
Anhalonine [519-04-0]
Anilazine [101-05-3]
Anileridine [144-14-9]
Anileridine Dihydrochloride* [126-12-5]
Aniline [62-53-3]
Aniline Mustard [553-27-5]
Anilinephthalein [509-77-3]
1-Anilino-8-naphthalenesulfonate [82-76-8]
Aniracetam [72432-10-1]
p-Anisaldehyde [123-11-5]
Anise Alcohol [105-13-5]
p-Anisic Acid [100-09-4]
Anisindione [117-37-3]
Anisole [100-66-3]
Anisomycin [22862-76-6]
Anisotropine Methylbromide [80-50-2]
o-(p-Anisoyl)benzoic Acid [1151-15-1]
p-Anisoyl Chloride [100-07-2]
Anistreplase [81669-57-0]
Annatto [1393-63-1]
Annotinine [559-49-9]
p-Anol [539-12-8]
Anot [3572-44-9]
Anserine [584-85-0]
Antazoline [91-75-8]
Antazoline Hydrochloride* [2508-72-7]
Antazoline Phosphate* [154-68-7]
Antheridiol [22263-79-2]
Anthiolimine [305-97-5]
Anthracene [120-12-7]
Anthragallol [602-64-2]
Anthralin [1143-38-0]
Anthralin Triacetate* [16203-97-7]
Anthramycin [4803-27-4]
Anthranol [529-86-2]
Anthraquinone [84-65-1]
Anthrarobin [577-33-3]
Anthrarufin [117-12-4]
Anthrimide [82-22-4]
Anthrone [90-44-8]
α-Antiarin [23605-05-2]
Antimony [7440-36-0]
Antimony Chloride Oxide [7791-08-4]
Antimony Dichlorotrifluoride [7791-16-4]
Antimony Pentachloride [7647-18-9]
Antimony Pentafluoride [7783-70-2]

Antimony Pentasulfide [1315-04-4]
Antimony Pentoxide [1314-60-9]
Antimony Potassium Oxalate [5965-33-3]
Antimony Potassium Tartrate [28300-74-5]
Antimony Sodium Tartrate [34521-09-0]
Antimony Sodium Thioglycollate [539-54-8]
Antimony Sulfate [7446-32-4]
Antimony Thioglycollamide [6533-78-4]
Antimony Tribromide [7789-61-9]
Antimony Trichloride [10025-91-9]
Antimony Trifluoride [7783-56-4]
Antimony Triiodide [7790-44-5]
Antimony Trioxide [1309-64-4]
Antimony Triselenide [1315-05-5]
Antimony Trisulfide [1345-04-6]
Antimycin A₁ [642-15-9]
Antimycin A₃ [522-70-3]
Antipyrine [60-80-0]
Antipyrine Acetylsalicylate* [569-84-6]
Antipyrine Mandelate* [603-64-5]
Antipyrine Methylethylglycolate* [5794-16-1]
Antipyrine Salicylacetate* [603-59-8]
Antipyrine Salicylate [520-07-0]
α₁-Antitrypsin [9041-92-3]
Antrafenine [55300-29-3]
ANTU [86-88-4]
Apafant [105219-56-5]
Apalcillin [63469-19-2]
Apalcillin Sodium Salt* [58795-03-2]
Apamin [24345-16-2]
Apazone [13539-59-8]
Apazone Dihydrate* [22304-30-9]
Aphidicolin [38966-21-1]
Apholate [52-46-0]
Aphylline [577-37-7]
Apicycline [15599-51-6]
Apigenin [520-36-5]
Apigetrin [578-74-5]
Apiin [26544-34-3]
Apiole (Dill) [484-31-1]
Apiole (Parsley) [523-80-8]
Apiose [639-97-4]
Aplasmomycin [61230-25-9]
Apoatropine [500-55-0]
Apocodeine [641-36-1]
Apocynin [498-02-2]
Apo-β-erythroidine [478-85-3]
Apomorphine [58-00-4]
Apomorphine Hydrochloride* [41372-20-7]
Apomorphine Methylbromide* [602-81-3]
Aporeine [2030-53-7]
Apraclonidine [66711-21-5]
Apraclonidine Dihydrochloride* [73217-88-6]
Apraclonidine Hydrochloride* [73218-79-8]
Apramycin [37321-09-8]
Aprindine [37640-71-4]
Aprindine Hydrochloride* [33237-74-0]
Aprobarbital [77-02-1]
Aprobarbital Sodium Salt* [125-88-2]
Apronalide [528-92-7]
Aprotinin [9087-70-1]
Apyrase [9000-95-7]
Aquacobalamin* [13422-52-1]
Arabinose [147-81-9]
Arabitol [2152-56-9]
D-Araboflavin [5978-87-0]
Arachidic Acid [506-30-9]
Arachidonic Acid [506-32-1]

Aramite [140-57-8]
Aranidipine [86780-90-7]
Araroba [1393-64-2]
Arbaprostil [55028-70-1]
Arbekacin [51025-85-5]
Arborescin [6831-14-7]
Arbutamine [128470-16-6]
Arbutamine Hydrochloride* [125251-66-3]
Arbutin [497-76-7]
Arecaidine [499-04-7]
Arecoline [63-75-2]
Arecoline Hydrobromide* [300-08-3]
Arecoline p-Stibonobenzoic Acid* [17162-36-6]
Argatroban [74863-84-6]
Argatroban Monohydrate* [141396-28-3]
Arginine [74-79-3]
Arginine Glutamate [4320-30-3]
Arginine Hydrochloride* [1119-34-2]
Argol [8007-14-5]
Argon [7440-37-1]
Aricine [482-91-7]
Aristolochic Acid [313-67-7]
Armepavine [524-20-9]
Armstrong's Acid [81-04-9]
Arogenic Acid [53078-86-7]
Arotinolol [68377-92-4]
Arotinolol Hydrochloride* [68377-91-3]
Arprinocid [55779-18-5]
Arsacetin [618-22-4]
Arsanilic Acid [98-50-0]
Arsenamide [531-72-6]
Arsenic [7440-38-2]
Arsenic Acid [7778-39-4]
Arsenic Disulfide [1303-32-8]
Arsenic Hemiselenide [1303-35-1]
Arsenic Pentafluoride [7784-36-3]
Arsenic Pentaselenide [1303-37-3]
Arsenic Pentasulfide [1303-34-0]
Arsenic Pentoxide [1303-28-2]
Arsenic Tribromide [7784-33-0]
Arsenic Trichloride [7784-34-1]
Arsenic Trifluoride [7784-35-2]
Arsenic Triiodide [7784-45-4]
Arsenic Trioxide [1327-53-3]
Arsenic Triselenide [1303-36-2]
Arsenic Trisulfide [1303-33-9]
Arsenoacetic Acid [544-27-4]
Arsine [7784-42-1]
Arsonoacetic Acid [107-38-0]
Arsphenamine [139-93-5]
Arsthinol [119-96-0]
Arteether [75887-54-6]
Artemether [71963-77-4]
Artemisin [481-05-0]
Artemisinin [63968-64-9]
Artesunate [88495-63-0]
Asafetida [9000-04-8]
Asaprol [516-18-7]
Asarinin [133-04-0]
Ascaridole [512-85-6]
Ascorbic Acid [50-81-7]
Ascorbigen [8075-98-7]
Asiaticoside [16830-15-2]
Asoxime Chloride [34433-31-3]
L-Asparaginase [9015-68-3]
Asparagine [70-47-3]
Asparagine D-Form* [2058-58-4]
Aspartame [22839-47-0]
Aspartic Acid [56-84-8]
Aspartic Acid D-Form* [1783-96-6]
Aspergillic Acid [490-02-8]
Asperlicin [93413-04-8]
Asperlicin B* [93413-08-2]
Asperlicin C* [93413-06-0]
Asperlicin D* [93413-07-1]
Asperlicin E* [93413-05-9]

Denotes derivative of title compound   **REG-5**

Asperuloside [14259-45-1]
Aspidin [584-28-1]
Aspidinol [519-40-4]
Aspidospermine [466-49-9]
Aspirin [50-78-2]
Aspirin Guaiacol Ester* [55482-89-8]
Aspoxicillin [63358-49-6]
Astacin [514-76-1]
Astatine [7440-68-8]
Astaxanthin [472-61-7]
Astemizole [68844-77-9]
Atenolol [29122-68-7]
Athamantin [1892-56-4]
Atipamezole [104054-27-5]
Atipamezole Hydrochloride*
  [104075-48-1]
Atisine [466-43-3]
Atorvastatin [134523-00-5]
Atorvastatin Calcium Salt*
  [134523-03-8]
Atorvastatin Lactone* [125995-03-1]
Atorvastatin Sodium Salt*
  [134523-01-6]
Atovaquone [95233-18-4]
Atractyloside [17754-44-8]
Atracurium Besylate [64228-81-5]
Atracurium Besylate 1R-cis,1R'-cis
  Form* [96946-42-8]
Atranorin [479-20-9]
Atrazine [1912-24-9]
Atrial Natriuretic Factor [85637-73-6]
Atrolactamide [2019-68-3]
Atrolactic Acid [515-30-0]
Atropic Acid [492-38-6]
Atropine [51-55-8]
Atropine Methylnitrate* [52-88-0]
Atropine N-Oxide [4438-22-6]
Atropine N-Oxide Hydrochloride*
  [4574-60-1]
Aucubin [479-98-1]
Auranofin [34031-32-8]
Aurantiogliocladin [483-54-5]
Aurantiogliocladin Dihydro Deriv*
  [4373-40-4]
Aurantiogliocladin Quinhydrone
  Deriv* [4516-93-2]
Aureothin [2825-00-5]
Aureothricin [574-95-8]
Aurin [603-45-2]
Aurodox [12704-90-4]
Aurothioglucose [12192-57-3]
Aurothioglycanide [16925-51-2]
Avilamycin [11051-71-1]
Avoparcin [37332-99-3]
Azacitidine [320-67-2]
Azacosterol [313-05-3]
Azacosterol Dihydrochloride*
  [1249-84-9]
Azacyclonol [115-46-8]
Azacyclonol Hydrochloride*
  [1798-50-1]
Azadirachtin [11141-17-6]
Azafrin [507-61-9]
8-Azaguanine [134-58-7]
Azamethonium Bromide [306-53-6]
Azanidazole [62973-76-6]
Azaperone [1649-18-9]
Azaserine [115-02-6]
Azasetron [123040-69-7]
Azasetron Hydrochloride*
  [123040-16-4]
Azatadine [3964-81-6]
Azatadine Dimaleate* [3978-86-7]
Azathioprine [446-86-6]
Azathioprine Sodium Salt*
  [55774-33-9]
6-Azathymine [932-53-6]
6-Azauridine [54-25-1]
6-Azauridine 2',3',5'-Triacetate*
  [2169-64-4]

Azelaic Acid [123-99-9]
Azelastine [58581-89-8]
Azelastine Monohydrochloride*
  [79307-93-0]
2-Azetidinecarboxylic Acid
  [2517-04-6]
Azidamfenicol [13838-08-9]
Azidocillin [17243-38-8]
Azidocillin Potassium Salt*
  [22647-32-1]
Azidocillin Sodium Salt* [35334-12-4]
Azimilide [149908-53-2]
Azimilide Dihydrochloride*
  [149888-94-8]
Azinphos-methyl [86-50-0]
Azintamide [1830-32-6]
Azithromycin [83905-01-5]
Azlocillin [37091-66-0]
Azlocillin Sodium Salt* [37091-65-9]
Azobenzene [103-33-3]
2,2'-Azobisisobutyronitrile [78-67-1]
Azodicarbonamide [123-77-3]
Azolitmin [1395-18-2]
Azomycin [527-73-1]
Azosemide [27589-33-9]
Azoxybenzene [495-48-7]
Azoxybenzene trans-Form*
  [20972-43-4]
Aztreonam [78110-38-0]
Azulene [275-51-4]
Azulene Sodium Sulfonate*
  [51873-93-9]
Azure A [531-53-3]
Azure B [531-55-5]
Azure C [531-57-7]
Bacampicillin [50972-17-3]
Bacampicillin Hydrochloride*
  [37661-08-8]
Bacilysin [29393-20-2]
Bacimethrin [3690-12-8]
Bacitracin [1405-87-4]
Bacitracin Methylenedisalicylic Acid
  [1405-88-5]
Baclofen [1134-47-0]
Badische Acid [86-60-2]
Baicalein [491-67-8]
Bakankosin [1398-17-0]
Bambermycins [11015-37-5]
Bambuterol [81732-65-2]
Bambuterol Hydrochloride*
  [81732-46-9]
Bamethan [3703-79-5]
Bamethan Sulfate* [5716-20-1]
Bamifylline [2016-63-9]
Bamifylline Hydrochloride*
  [20684-06-4]
Bamipine [4945-47-5]
Bamipine Dihydrochloride*
  [61732-85-2]
Baptigenin [5908-63-4]
Barban [101-27-9]
Barbital [57-44-3]
Barbital Sodium Salt* [144-02-5]
Barbituric Acid [67-52-7]
Barium [7440-39-3]
Barium Acetate [543-80-6]
Barium Benzenesulfonate [515-72-0]
Barium Bromate [13967-90-3]
Barium Bromide [10553-31-8]
Barium Carbonate [513-77-9]
Barium Chlorate [13477-00-4]
Barium Chloride [10361-37-2]
Barium Chromate(VI) [10294-40-3]
Barium Cyanide [542-62-1]
Barium Dithionate [13845-17-5]
Barium Ferrocyanide [13821-06-2]
Barium Fluoride [7787-32-8]
Barium Formate [541-43-5]
Barium Hexafluorosilicate
  [17125-80-3]

Barium Hydroxide [17194-00-2]
Barium Hypophosphite [14871-79-5]
Barium Iodate [10567-69-8]
Barium Iodide [13718-50-8]
Barium Manganate(VI) [7787-35-1]
Barium Mercuric Iodide [10048-99-4]
Barium Nitrate [10022-31-8]
Barium Nitrite [13465-94-6]
Barium Oxalate [516-02-9]
Barium Oxide [1304-28-5]
Barium Perchlorate [13465-95-7]
Barium Permanganate [7787-36-2]
Barium Peroxide [1304-29-6]
Barium Phosphate, Dibasic
  [10048-98-3]
Barium Platinous Cyanide [562-81-2]
Barium Selenide [1304-39-8]
Barium Silicide [1304-40-1]
Barium Sulfate [7727-43-7]
Barium Sulfide [21109-95-5]
Barium Sulfide, Black [8011-62-9]
Barium Sulfite [7787-39-5]
Barium Thiocyanate [2092-17-3]
Barium Thiosulfate [35112-53-9]
Barium Titanate(IV) [12047-27-7]
Barium Uranium Oxide [10380-31-1]
Barnidipine [104713-75-9]
Barnidipine Hydrochloride*
  [104757-53-1]
Barthrin [70-43-9]
Basic Aluminum Carbonate Gel
  [1339-92-0]
Basic Fibroblast Growth Factor*
  [106096-93-9]
Basic Lead Carbonate [1344-36-1]
Batimastat [130370-60-4]
Batrachotoxin [23509-16-2]
Batrachotoxinin A* [19457-37-5]
Batroxobin [9039-61-6]
Batyl Alcohol [544-62-7]
BBB* [4482-83-1]
Bebeerine [477-60-1]
Becanthone [15351-04-9]
Becanthone Hydrochloride*
  [5591-22-0]
Beclamide [501-68-8]
Beclobrate [55937-99-0]
Beclomethasone [4419-39-0]
Beclomethasone Dipropionate*
  [5534-09-8]
Beclotiamine [13471-78-8]
Beeswax [8012-89-3]
Befunolol [39552-01-7]
Befunolol (R)-(+)-Form Hydro-
  chloride* [66685-79-8]
Befunolol (S)-(−)-Form Hydro-
  chloride* [66717-59-7]
Befunolol Hydrochloride*
  [39543-79-8]
Behenic Acid [112-85-6]
Belladonnine [510-25-8]
Belladonnine Bisulfate Monohydrate*
  [5908-89-4]
Bemegride [64-65-3]
Benactyzine [302-40-9]
Benactyzine Hydrochloride* [57-37-4]
Benactyzine Methobromide*
  [3166-62-9]
Benalaxyl [71626-11-4]
Benapryzine [22487-42-9]
Benapryzine Hydrochloride*
  [3202-55-9]
Benazepril [86541-75-5]
Benazeprilat* [86541-78-8]
Benazepril Hydrochloride*
  [86541-74-4]
Bencyclane [2179-37-5]
Bencyclane Fumarate* [14286-84-1]
Bendazac [20187-55-7]
Bendazol [621-72-7]

*Denotes derivative of title compound

Bendazol Hydrochloride* [1212-48-2]
Bendiocarb [22781-23-3]
Bendroflumethiazide [73-48-3]
Benexate Hydrochloride [78718-25-9]
Benexate Hydrochloride Compd with β-Cyclodextrin (1:1)* [91574-91-3]
Benfluorex [23602-78-0]
Benfluorex Hydrochloride* [23642-66-2]
Benfluralin [1861-40-1]
Benfotiamine [22457-89-2]
Benfuracarb [82560-54-1]
Benfurodil Hemisuccinate [3447-95-8]
Benidipine [105979-17-7]
Benidipine (±)-α-Form Hydrochloride* [91599-74-5]
Benmoxine [7654-03-7]
Benomyl [17804-35-2]
Benorylate [5003-48-5]
Benoxaprofen [67434-14-4]
Benoxinate [99-43-4]
Benoxinate Hydrochloride* [5987-82-6]
Benperidol [2062-84-2]
Benproperine [2156-27-6]
Benproperine Pamoate* [64238-92-2]
Benproperine Trihydrogen Phosphate* [19428-14-9]
Benserazide [322-35-0]
Benserazide Combination with Levodopa* [37270-69-2]
Benserazide Hydrochloride* [14919-77-8]
Bentazon [25057-89-0]
Bentiromide [37106-97-1]
Bentonite [1302-78-9]
β-Benzalbutyramide [7236-47-7]
Benzal Chloride [98-87-3]
Benzaldehyde [100-52-7]
Benzamide [55-21-0]
Benzanilide [93-98-1]
1,2-Benzanthracene [56-55-3]
Benzanthrone [82-05-3]
Benzarone [1477-19-6]
Benzathine [140-28-3]
Benzbromarone [3562-84-3]
Benzene [71-43-2]
Benzenearsonic Acid [98-05-5]
Benzeneboronic Acid [98-80-6]
Benzenestibonic Acid [535-46-6]
Benzenesulfonic Acid [98-11-3]
Benzenesulfonic Anhydride [512-35-6]
Benzenesulfonyl Chloride [98-09-9]
1,2,4-Benzenetriol [533-73-3]
Benzestrol [85-95-0]
Benzethonium Chloride [121-54-0]
Benzetimide [14051-33-3]
Benzetimide Hydrochloride* [5633-14-7]
Benzhydrylamine [91-00-9]
Benzidine [92-87-5]
Benzil [134-81-6]
Benzil Dioxime [23873-81-6]
Benzilic Acid [76-93-7]
Benzilonium Bromide [1050-48-2]
Benzimidazole [51-17-2]
2-Benzimidazolethiol [583-39-1]
Benziodarone [68-90-6]
Benznidazole [22994-85-0]
Benzo Azurine G [2429-71-2]
Benzocaine [94-09-7]
Benzoctamine [17243-39-9]
Benzoctamine Hydrochloride* [10085-81-1]
Benzodepa [1980-45-6]
Benzofuran [271-89-6]
Benzoguanamine [91-76-9]
Benzohydrol [91-01-0]
Benzoic Acid [65-85-0]
Benzoic Anhydride [93-97-0]

Benzoin [119-53-9]
Benzoin Oxime [441-38-3]
6,7-Benzomorphan [575-19-9]
Benzonatate [104-31-4]
Benzonitrile [100-47-0]
Benzophenone [119-61-9]
Benzophenone-6 [131-54-4]
Benzopinacol [464-72-2]
Benzopurpurine 4B [992-59-6]
1,2-Benzopyran [254-04-6]
Benzo[a]pyrene [50-32-8]
Benzo[e]pyrene [192-97-2]
Benzo[f]quinoline [85-02-9]
Benzoquinonium Chloride [311-09-1]
Benzoresorcinol [131-56-6]
Benzothiazole [95-16-9]
1H-Benzotriazole [95-14-7]
Benzotrichloride [98-07-7]
Benzotrifluoride [98-08-8]
Benzoxiquine [86-75-9]
Benzoxonium Chloride [19379-90-9]
Benzoyl Chloride [98-88-4]
Benzoylecgonine [519-09-5]
Benzoyl Isothiocyanate [532-55-8]
Benzoylpas [13898-58-3]
Benzoylpas Calcium Salt Pentahydrate* [5631-00-5]
Benzoylpas Sodium Salt* [537-20-2]
Benzoyl Peroxide [94-36-0]
3,4-Benzphenanthrene [195-19-7]
Benzphetamine [156-08-1]
Benzphetamine Hydrochloride* [5411-22-3]
Benzpiperylon [53-89-4]
Benzpyrinium Bromide [587-46-2]
Benzquinamide [63-12-7]
Benzthiazide [91-33-8]
Benztropine Mesylate [132-17-2]
Benzydamine [642-72-8]
Benzydamine Hydrochloride* [132-69-4]
Benzyl Acetate [140-11-4]
Benzyl Alcohol [100-51-6]
Benzylamine [100-46-9]
Benzylaniline [103-32-2]
Benzyl Benzoate [120-51-4]
Benzyl Bromide [100-39-0]
Benzyl Chloride [100-44-7]
Benzyl Cinnamate [103-41-3]
Benzyl Cyanide [140-29-4]
Benzyl Ether [103-50-4]
Benzyl Ethyl Ether [539-30-0]
Benzyl Formate [104-57-4]
Benzyl Fumarate [538-64-7]
Benzylhydrochlorothiazide [1824-50-6]
Benzylideneacetone [122-57-6]
Benzylideneaniline [538-51-2]
Benzylimidobis(p-methoxyphenyl)methane [524-96-9]
Benzyl Methyl Ether [538-86-3]
Benzylmorphine [14297-87-1]
Benzylpenicillinic Acid [61-33-6]
Benzylpenicillin Sodium [69-57-8]
o-Benzylphenol [28994-41-4]
p-Benzylphenol [101-53-1]
Benzyl Salicylate [118-58-1]
Benzylsulfamide [104-22-3]
Benzyl Sulfide [538-74-9]
p-(Benzylsulfonamido)benzoic Acid [536-95-8]
S-Benzylthiuronium Chloride [538-28-3]
Benzylurea [538-32-9]
Bephenium [7181-73-9]
Bephenium Hydroxynaphthoate* [3818-50-6]
Bepridil [64706-54-3]
Bepridil Hydrochloride Monohydrate* [74764-40-2]

Beractant [108778-82-1]
Beraprost [88430-50-6]
Beraprost Sodium Salt* [88475-69-8]
Berbamine [478-61-5]
Berberine [2086-83-1]
Berbine [483-49-8]
Bergapten(e) [484-20-8]
Bergenin [477-90-7]
Berkelium [7440-40-6]
Bermoprofen [78499-27-1]
Berninamycin [58798-97-3]
Beryllium [7440-41-7]
Beryllium Acetate [543-81-7]
Beryllium Acetate, Basic [1332-52-1]
Beryllium Acetylacetonate [10210-64-7]
Beryllium Borohydride [17440-85-6]
Beryllium Bromide [7787-46-4]
Beryllium Carbide [506-66-1]
Beryllium Chloride [7787-47-5]
Beryllium Fluoride [7787-49-7]
Beryllium Formate [1111-71-3]
Beryllium Hydride [7787-52-2]
Beryllium Hydroxide [13327-32-7]
Beryllium Iodide [7787-53-3]
Beryllium Nitrate [13597-99-4]
Beryllium Nitride [1304-54-7]
Beryllium Oxide [1304-56-9]
Beryllium Perchlorate [13597-95-0]
Beryllium Potassium Fluoride [7787-50-0]
Beryllium Potassium Sulfate [53684-48-3]
Beryllium Selenate [10039-31-3]
Beryllium Sodium Fluoride [13871-27-7]
Beryllium Sulfate [13510-49-1]
Besipirdine [119257-34-0]
Besipirdine Hydrochloride* [130953-69-4]
Besipirdine Maleate* [119257-40-8]
Betahistine [5638-76-6]
Betahistine Dihydrochloride* [5579-84-0]
Betaine [107-43-7]
Betaine Hydrochloride* [590-46-5]
Betaine Monohydrate* [590-47-6]
Betaine Sodium Aspartate* [93227-64-6]
Betamethasone [378-44-9]
Betamethasone 21-Acetate* [987-24-6]
Betamethasone 17-Benzoate* [22298-29-9]
Betamethasone 17,21-Dipropionate* [5593-20-4]
Betamethasone 21-Phosphate Disodium Salt* [151-73-5]
Betamethasone 17-Valerate* [2152-44-5]
Betamipron [3440-28-6]
Betasine [3734-24-5]
Betaxolol [63659-18-7]
Betaxolol Hydrochloride* [63659-19-8]
Betazole [105-20-4]
Betazole Dihydrochloride* [138-92-1]
Bethanechol Chloride [590-63-6]
Bethanidine [55-73-2]
Bethanidine Sulfate* [114-85-2]
Betonicine [515-25-3]
Betoxycaine [3818-62-0]
Betoxycaine Monohydrochloride* [5003-47-4]
Betulin [473-98-3]
Bevantolol [59170-23-9]
Bevantolol Hydrochloride* [42864-78-8]
Bevonium Methyl Sulfate [5205-82-3]
Bezafibrate [41859-67-0]

Bezitramide [15301-48-1]
Bialamicol [493-75-4]
Bialamicol Dihydrochloride*
 [3624-96-2]
Biapenem [120410-24-4]
Bibenzonium Bromide [15585-70-3]
Bibenzyl [103-29-7]
Bibrocathol [6915-57-7]
Bicalutamide [90357-06-5]
Bicine [150-25-4]
Bicozamycin [38129-37-2]
Bicuculline [485-49-4]
Bidisomide [103810-45-3] (unspeci-
 fied)
Bidisomide [116078-65-0] (±)
Bietamiverine [479-81-2]
Bietamiverine Dihydrochloride*
 [2691-46-5]
Bietamiverine Hydrochloride*
 [1477-10-7]
Bietanautine [6888-11-5]
Bietaserpine [53-18-9]
Bietaserpine Bitartrate* [1111-44-0]
Bifemelane [90293-01-9]
Bifemelane Hydrochloride*
 [62232-46-6]
Bifenox [42576-02-3]
Bifenthrin [82657-04-3]
Bifidus Factor [9007-03-8]
Bifluranol [34633-34-6]
Bifonazole [60628-96-8]
Biguanide [56-03-1]
Bikhaconitine [6078-26-8]
Bilirubin [635-65-4]
Biliverdine [114-25-0]
Binapacryl [485-31-4]
Binedaline [60662-16-0]
Binedaline Hydrochloride*
 [57647-35-5]
Binifibrate [69047-39-8]
Biocytin [576-19-2]
Biopterin [22150-76-1]
Bioresmethrin [28434-01-7]
Biotin [58-85-5]
Biotin l-Sulfoxide [3376-83-8]
Biperiden [514-65-8]
Biperiden Hydrochloride* [1235-82-1]
Biphenamine [3572-52-9]
Biphenamine Hydrochloride*
 [5560-62-3]
p-Biphenylamine [92-67-1]
2,4'-Biphenyldiamine [492-17-1]
Bipiperidyl Mustard [6802-93-3]
α-Bisabolol [515-69-5]
α-Bisabolol [25428-43-7] (±)-form
α-Bisabolol (−)-Epi-form*
 [78148-59-1]
α-Bisabolol (+)-Epi-form*
 [76738-75-5]
α-Bisabolol (−)-Form* [23089-26-1]
α-Bisabolol (+)-Form* [23178-88-3]
Bisacodyl [603-50-9]
Bisacodyl Complex with Tannin*
 [1336-29-4]
Bis(4-amino-1-anthraquinonyl)amine
 [128-87-0]
Bisantrene [78186-34-2]
Bisantrene Dihydrochloride*
 [71439-68-4]
2,5-Bis(1-aziridinyl)-3,6-bis(2-meth-
 oxyethyl)-1,4-benzoquinone
 [800-24-8]
Bisbentiamine [2667-89-2]
Bis(p-chlorophenoxy)methane
 [555-89-5]
Bisdequalinium Chloride [52951-36-7]
Bis(p-dimethylaminobenzylidene)benz-
 idine [6001-51-0]
Bis(1,2-dimethylpropyl)borane
 [1069-54-1]

Bis(2-ethylhexyl) Phthalate [117-81-7]
Bis(2-ethylhexyl) Sebacate [122-62-3]
Bismark Brown R [8005-78-5]
Bismark Brown Y [10114-58-6]
Bis(1-methylamyl) Sodium Sulfosuc-
 cinate [6001-97-4]
Bis[methylthio]methane [1618-26-4]
Bismuth [7440-69-9]
Bismuth Aluminate [12284-76-3]
Bismuth Bromide [7787-58-8]
Bismuth Bromide Oxide [7787-57-7]
Bismuth Butylthiolaurate
 [53897-25-9]
Bismuth Chloride [7787-60-2]
Bismuth Chloride Oxide [7787-59-9]
Bismuth Ethyl Camphorate
 [52951-37-8]
Bismuth Fluoride [7787-61-3]
Bismuth Hydroxide [10361-43-0]
Bismuthine [18288-22-7]
Bismuth Iodide [7787-64-6]
Bismuth Iodide Oxide [7787-63-5]
Bismuth Iodosubgallate [138-58-9]
Bismuth Nitrate [10361-44-1]
Bismuth Oleate [52951-38-9]
Bismuth Oxalate [6591-55-5]
Bismuth Oxide [1304-76-3]
Bismuth Pentafluoride [7787-62-4]
Bismuth Phosphate [10049-01-1]
Bismuth Potassium Iodide
 [41944-01-8]
Bismuth Potassium Tartrate
 [5798-41-4]
Bismuth Selenide [12068-69-8]
Bismuth Sodium Iodide [53778-50-0]
Bismuth Sodium Tartrate
 [31586-77-3]
Bismuth Sodium Triglycollamate
 [5798-43-6]
Bismuth Subacetate [5142-76-7]
Bismuth Subcarbonate [5892-10-4]
Bismuth Subgallate [22650-86-8]
Bismuth Subnitrate [1304-85-4]
Bismuth Subsalicylate [14882-18-9]
Bismuth Sulfate [7787-68-0]
Bismuth Sulfide [1345-07-9]
Bismuth Telluride [1304-82-1]
Bismuth Tetroxide [12048-50-9]
Bismuth Tribromophenate [5175-83-7]
Bismuth Valerate, Basic [5798-47-0]
Bis(1-naphthylmethyl)amine
 [5798-49-2]
Bisoprolol [66722-44-9]
Bisoprolol Hemifumarate*
 [104344-23-2]
Bisoxatin Acetate [14008-48-1]
Bisphenol A [80-05-7]
Bisphenol B [77-40-7]
1,4-Bis(trichloromethyl)benzene
 [68-36-0]
Bis(triphenylphosphine)dicarbonyl-
 nickel [13007-90-4]
Bitertanol [55179-31-2]
Bithionol [97-18-7]
Bithionol Sodium Salt* [6385-58-6]
Bitolterol [30392-40-6]
Bitolterol Methanesulfonate*
 [30392-41-7]
Bitoscanate [4044-65-9]
Biuret [108-19-0]
Bixin [6983-79-5]
Blancophor R [2606-93-1]
Blasticidin S [2079-00-7]
Bleomycins Sulfate* [9041-93-4]
Bolandiol [19793-20-5]
Bolandiol Dipropionate* [1986-53-4]
Bolasterone [1605-89-6]
Boldenone [846-48-0]
Boldenone 10-Undecenoate*
 [13103-34-9]

Boldine [476-70-0]
Bombesin [31362-50-2]
Bomyl [122-10-1]
Bongkrekic Acid [11076-19-0]
Bopindolol [62658-63-3]
Bopindolol Maleate* [62658-64-4]
Bopindolol Malonate* [82857-38-3]
Borazine [6569-51-3]
Boric Acid [10043-35-3]
Boric Anhydride [1303-86-2]
Borneol [507-70-0]
Bornyl Acetate [76-49-3]
d-Bornyl α-Bromoisovalerate
 [52964-40-6]
Bornyl Chloride [464-41-5]
d-Bornyl Isovalerate [53022-14-3]
Bornyl Salicylate [560-88-3]
Boromycin [34524-20-4]
Boron [7440-42-8]
Boron Carbide [12069-32-8]
Boron Monoxide [12505-77-0]
Boron Nitride [10043-11-5]
Boron Tribromide [10294-33-4]
Boron Trichloride [10294-34-5]
Boron Trifluoride [7637-07-2]
Boron Trifluoride Etherate [109-63-7]
Bostrycoidin [4589-33-7]
β-Boswellic Acid [631-69-6]
Bottromycin [1393-68-6]
Botulin A* [93384-43-1]
Bradycor [140661-97-8]
Bradykinin [58-82-2]
Brallobarbital [561-86-4]
Brassidic Acid [506-33-2]
Brassinolide [72962-43-7]
Brazilin [474-07-7]
Brefeldin A [20350-15-6]
Brequinar [96187-53-0]
Brequinar Sodium Salt* [96201-88-6]
Bretylium Tosylate [61-75-6]
Brilliant Blue FCF [3844-45-9]
Brilliant Green [633-03-4]
Brimonidine [59803-98-4]
Brimonidine D-Tartrate* [70359-46-5]
Brodifacoum [56073-10-0]
Brodimoprim [56518-41-3]
Bromacil [314-40-9]
Bromadiolone [28772-56-7]
Bromal [115-17-3]
Bromal Hydrate [507-42-6]
Bromazepam [1812-30-2]
Bromcresol Green [76-60-8]
Bromcresol Purple [115-40-2]
Bromelain [9001-00-7]
Bromethalin [63333-35-7]
Bromfenac [91714-94-2]
Bromfenac Monosodium Salt*
 [91714-93-1]
Bromfenac Monosodium Salt
 Sesquihydrate* [120638-55-3]
Bromhexine [3572-43-8]
Bromhexine Hydrochloride*
 [611-75-6]
Bromic Acid [7789-31-3]
Bromindione [1146-98-1]
Bromine [7726-95-6]
Bromine Pentafluoride [7789-30-2]
Bromine Trifluoride [7787-71-5]
Bromisovalum [496-67-3]
N-Bromoacetamide [79-15-2]
p-Bromoacetanilide [103-88-8]
Bromoacetic Acid [79-08-3]
Bromoacetone [598-31-2]
p-Bromoacetophenone [99-90-1]
ω-Bromoacetophenone [70-11-1]
p-Bromoaniline [106-40-1]
5-Bromoanthranilic Acid [5794-88-7]
Bromobenzene [108-86-1]
p-Bromobenzenesulfonyl Chloride
 [98-58-8]

Denotes derivative of title compound

p-Bromobenzoic Acid [586-76-5]
p-Bromobenzyl Bromide [589-15-1]
p-Bromobenzyl Chloride [589-17-3]
p-Bromobenzyl Chloroformate
 [5798-78-7]
α-Bromobenzyl Cyanide [5798-79-8]
α-Bromobutyric Acid [80-58-0]
3-Bromo-d-camphor [76-29-9]
α-Bromo-n-caproic Acid [616-05-7]
Bromocriptine [25614-03-3]
Bromocriptine Methanesulfonate*
 [22260-51-1]
Bromodichloromethane [75-27-4]
Bromodiphenhydramine [118-23-0]
Bromodiphenhydramine Hydrochloride* [1808-12-4]
Bromofenofos [21466-07-9]
Bromoform [75-25-2]
α-Bromoisobutyric Acid [2052-01-9]
α-Bromoisovaleric Acid [565-74-2]
β-Bromoisovaleric Acid [5798-88-9]
Bromolysergide [478-84-2]
p-Bromomandelic Acid [6940-50-7]
1-Bromonaphthalene [90-11-9]
2-Bromonaphthalene [580-13-2]
p-Bromophenacyl Bromide [99-73-0]
m-Bromophenol* [591-20-8]
o-Bromophenol* [95-56-7]
p-Bromophenol* [106-41-2]
p-Bromophenylhydrazine [589-21-9]
p-Bromophenyl Isocyanate
 [2493-02-9]
Bromophos [2104-96-3]
Bromopride [4093-35-0]
Bromopride Hydrochloride*
 [52423-56-0]
β-Bromopropionic Acid [590-92-1]
Bromopropylate [18181-80-1]
Bromosalicylchloranilide [3679-64-9]
5-Bromosalicylhydroxamic Acid
 [5798-94-7]
5-Bromosalicylic Acid Acetate
 [1503-53-3]
Bromosaligenin [2316-64-5]
Bromosuccinic Acid [923-06-8]
N-Bromosuccinimide [128-08-5]
5-Bromouracil [51-20-7]
Bromoxynil [1689-84-5]
Bromperidol [10457-90-6]
Brompheniramine [86-22-6]
Brompheniramine d-Form* [132-21-8]
Brompheniramine d-Form Maleate*
 [2391-03-9]
Brompheniramine Maleate*
 [980-71-2]
Bromphenol Blue [115-39-9]
Bromthymol Blue [76-59-5]
Bronopol [52-51-7]
Broparoestrol [479-68-5]
Brotizolam [57801-81-7]
Brovincamine [57475-17-9]
Brovincamine Hydrogen Fumarate*
 [84964-12-5]
Broxyquinoline [521-74-4]
Bruceantin [41451-75-6]
Brucine [357-57-3]
Bryostatin 1* [83314-01-6]
Bucetin [1083-57-4]
Bucillamine [65002-17-7]
Buckminsterfullerene [99685-96-8]
Bucladesine [362-74-3]
Bucladesine Barium Salt*
 [18837-96-2]
Bucladesine Sodium Salt*
 [16980-89-5]
Buclizine [82-95-1]
Buclizine Dihydrochloride* [129-74-8]
Buclosamide [575-74-6]
Buclosamide Combination with
 Salicylic Acid* [75199-98-3]

Bucloxic Acid [32808-51-8]
Bucloxic Acid Calcium Salt*
 [32808-53-0]
Bucolome [841-73-6]
Bucrylate [1069-55-2]
Bucumolol [58409-59-9] (unspecified)
Bucumolol [80750-21-6] (±)-form
Bucumolol Hydrochloride*
 [30073-40-6] (unspecified)
Bucumolol Hydrochloride*
 [36556-75-9] (±)-form
Budesonide [51333-22-3]
Budipine [57982-78-2]
Budipine Hydrochloride*
 [63661-61-0]
Budralazine [36798-79-5]
Bufalin [465-21-4]
Bufencarb [8065-36-9]
Bufeniode [22103-14-6]
Bufetolol [53684-49-4]
Bufetolol Hydrochloride*
 [35108-88-4]
Bufexamac [2438-72-4]
Buflomedil [55837-25-7]
Buflomedil Hydrochloride*
 [35543-24-9]
Bufogenin B [465-19-0]
Buformin [692-13-7]
Buformin Hydrochloride* [1190-53-0]
Bufotalin [471-95-4]
Bufotenine [487-93-4]
Bufotoxin [464-81-3]
Bufuralol [54340-62-4]
Bufuralol Hydrochloride*
 [59652-29-8]
Bufuralol (−)-Hydrochloride*
 [57704-10-6]
Bufuralol (+)-Hydrochloride*
 [57704-11-7]
Bulan [117-26-0]
Bulbocapnine [298-45-3]
Bumadizon [3583-64-0]
Bumadizon Calcium Salt Hemihy-
 drate* [69365-73-7]
Bumetanide [28395-03-1]
Bunaftine [32421-46-8]
Bunamidine [3748-77-4]
Bunamidine Hydrochloride*
 [1055-55-6]
Bunamiodyl Sodium [1923-76-8]
Bunazosin [80755-51-7]
Bunazosin Hydrochloride*
 [52712-76-2]
Bunitrolol [34915-68-9]
Bunitrolol Hydrochloride*
 [23093-74-5]
Buparvaquone [88426-33-9]
Buphanamine [6793-24-4]
Buphanitine [4673-18-1]
Bupirimate [41483-43-6]
Bupivacaine [2180-92-9]
Bupivacaine Hydrochloride
 Monohydrate* [14252-80-3]
Bupranolol [14556-46-8]
Bupranolol Hydrochloride*
 [15148-80-8]
Buprenorphine [52485-79-7]
Buprenorphine Hydrochloride*
 [53152-21-9]
Bupropion [34911-55-2]
Bupropion Hydrochloride*
 [31677-93-7]
Buquinolate [5486-03-3]
Buramate [4663-83-6]
Burgundy Mixture [11125-96-5]
Buserelin [57982-77-1]
Buserelin Monoacetate* [68630-75-1]
Buspirone [36505-84-7]
Buspirone Hydrochloride*
 [33386-08-2]

Busulfan [55-98-1]
Butabarbital Sodium [143-81-7]
Butacaine [149-16-6]
Butacaine Sulfate* [149-15-5]
Butacetin [2109-73-1]
Butachlor [23184-66-9]
1,3-Butadiene [106-99-0]
Butalamine [22131-35-7]
Butalamine Hydrochloride*
 [56974-46-0]
Butalbital [77-26-9]
Butallylonal [1142-70-7]
Butallylonal Sodium Salt* [3486-86-0]
Butamben [94-25-7]
Butamirate [18109-80-3]
Butamirate Citrate* [18109-81-4]
Butamisole [54400-59-8]
Butamisole Hydrochloride*
 [54400-62-3]
Butane [106-97-8]
Butanilicaine [3785-21-5]
Butaperazine [653-03-2]
Butaperazine Dimaleate* [1063-55-4]
Butaverine [55837-14-4]
Butaverine Hydrochloride*
 [17824-89-4]
Butazolamide [16790-49-1]
Butedronic Acid [51395-42-7]
Butedronic Acid Tetrasodium Salt*
 [97772-98-0]
Butenafine [101828-21-1]
Butenafine Hydrochloride*
 [101827-46-7]
1-Butene [106-98-9]
2-Butene [107-01-7]
Butethal [77-28-1]
Butethamate [14007-64-8]
Butethamate Citrate* [3639-12-1]
Butethamine [2090-89-3]
Butethamine Hydrochloride*
 [553-68-4]
Butethamine meta-Isomer Hydro-
 chloride* [553-58-2]
Buthalital* [468-65-5]
Buthalital Sodium [510-90-7]
Buthiazide [2043-38-1]
Buthiobate [51308-54-4]
Buthionine Sulfoximine [5072-26-4]
Butibufen [55837-18-5]
Butidrine Hydrochloride [1506-12-3]
Butirosin [12772-35-9]
Butirosin A* [34291-02-6]
Butirosin B* [34291-03-7]
Butirosin Sulfate Dihydrate*
 [51022-98-1]
Butobendine [55769-65-8]
Butobendine Dihydrochloride*
 [55769-64-7]
Butoconazole [64872-76-0]
Butoconazole Nitrate* [64872-77-1]
Butoctamide [32838-26-9]
Butoctamide Hydrogen Succinate*
 [32838-28-1]
Butoctamide Hydrogen Succinate
 Calcium Salt* [32266-82-3]
Butofilolol [64552-17-6]
Butofilolol Maleate* [88606-96-6]
Butonate* [126-22-7]
Butopyronoxyl [532-34-3]
Butorphanol [42408-82-2]
Butorphanol Tartrate* [58786-99-5]
Butoxycaine [3772-43-8]
Butoxycaine Hydrochloride*
 [2350-32-5]
Butralin [33629-47-9]
Butriptyline [35941-65-2]
Butriptyline Hydrochloride*
 [5585-73-9]
Butropium Bromide [29025-14-7]
n-Butyl Acetate [123-86-4]

sec-Butyl Acetate [105-46-4]
tert-Butyl Acetate [540-88-5]
tert-Butylacetic Acid [1070-83-3]
n-Butyl Acrylate [141-32-2]
n-Butyl Alcohol [71-36-3]
sec-Butyl Alcohol [78-92-2]
tert-Butyl Alcohol [75-65-0]
n-Butylamine [109-73-9]
sec-Butylamine [13952-84-6]
tert-Butylamine [75-64-9]
Butylate [2008-41-5]
Butylated Hydroxyanisole [25013-16-5]
Butylated Hydroxytoluene [128-37-0]
n-Butylbenzene [104-51-8]
sec-Butylbenzene [135-98-8]
tert-Butylbenzene [98-06-6]
n-Butyl Benzoate [136-60-7]
n-Butyl Bromide [109-65-9]
sec-Butyl Bromide [78-76-2]
tert-Butyl Bromide [507-19-7]
n-Butyl n-Butyrate [109-21-7]
Butyl Carbitol [112-34-5]
n-Butyl Carbonate [542-52-9]
Butyl Cellosolve [111-76-2]
n-Butyl Chloride [109-69-3]
sec-Butyl Chloride [78-86-4]
tert-Butyl Chloride [507-20-0]
tert-Butyl Chloroacetate [107-59-5]
Butyl Citrate [77-94-1]
α-Butylene Dibromide [533-98-2]
1,3-Butylene Glycol [107-88-0]
2,3-Butylene Glycol [513-85-9]
n-Butyl Ether [142-96-1]
tert-Butyl Hydroperoxide [75-91-2]
tert-Butyl Hypochlorite [507-40-4]
Butylidene Chloride [541-33-3]
n-Butyl Iodide [542-69-8]
sec-Butyl Iodide [513-48-4]
Butyl Isocyanate [111-36-4]
n-Butylmalonic Acid [534-59-8]
n-Butyl Mercaptan [109-79-5]
sec-Butyl Mercaptan [513-53-1]
tert-Butyl Mercaptan [75-66-1]
n-Butylmercuric Chloride [543-63-5]
1-Butyl-3-metanilylurea [4618-41-1]
Butyl Methoxydibenzoylmethane [70356-09-1]
n-Butyl Nitrite [544-16-1]
tert-Butyl Nitrite [540-80-7]
Butylparaben [94-26-8]
p-tert-Butylphenol [98-54-4]
4-tert-Butylphenyl Salicylate [87-18-3]
n-Butyl Phthalate [84-74-2]
n-Butyl Propionate [590-01-2]
N-Butylscopolammonium Bromide [149-64-4]
Butyl Stearate [123-95-5]
n-Butyl Sulfide [544-40-1]
Butyraldehyde [123-72-8]
n-Butyramide [541-35-5]
Butyric Acid [107-92-6]
Butyric Anhydride [106-31-0]
Butyroin [496-77-5]
Butyrolactone [96-48-0]
Butyronitrile [109-74-0]
n-Butyryl Chloride [141-75-3]
Buzepide [3691-21-2]
Buzepide Methiodide* [15351-05-0]
Cabenegrin A-I* [84297-59-6]
Cabenegrin A-II* [84297-60-9]
Cabergoline [81409-90-7]
Cabergoline Diphosphate* [85329-89-1]
Cacodyl [471-35-2]
Cacodylic Acid [75-60-5]
Cacotheline [561-20-6]
Cactinomycin [8052-16-2]
Cadalene [483-78-3]
Cadaverine [462-94-2]

Cadmium [7440-43-9]
Cadmium Acetate [543-90-8]
Cadmium Bromide [7789-42-6]
Cadmium Carbonate [513-78-0]
Cadmium Chloride [10108-64-2]
Cadmium Cyanide [542-83-6]
Cadmium Fluoride [7790-79-6]
Cadmium Hydroxide [21041-95-2]
Cadmium Iodide [7790-80-9]
Cadmium Nitrate [10325-94-7]
Cadmium Oxide [1306-19-0]
Cadmium Potassium Cyanide [14402-75-6]
Cadmium Salicylate [19010-79-8]
Cadmium Selenate [13814-62-5]
Cadmium Selenide [1306-24-7]
Cadmium Succinate [141-00-4]
Cadmium Sulfate [10124-36-4]
Cadmium Sulfide [1306-23-6]
Cadmium Telluride [1306-25-8]
Cadmium Tungstate(VI) [7790-85-4]
Cadralazine [64241-34-5]
Cadusafos [95465-99-9]
Cafaminol [30924-31-3]
Cafestol [469-83-0]
Caffeic Acid [331-39-5]
Caffeine [58-08-2]
Calcifediol [19356-17-3]
Calcimycin [52665-69-7]
Calcipotriene [112965-21-6]
Calcitonin [9007-12-9]
Calcitonin, human synthetic* [21215-62-3]
Calcitonin, porcine* [12321-44-7]
Calcitonin, salmon synthetic* [47931-85-1]
Calcitriol [32222-06-3]
Calcium [7440-70-2]
Calcium Acetate [62-54-4]
Calcium Acetylsalicylate [69-46-5]
Calcium Acetylsalicylate Complex with Urea* [5749-67-7]
Calcium Aluminosilicate [1327-39-5]
Calcium Arsenate [7778-44-1]
Calcium Arsenite [52740-16-6]
Calcium Ascorbate [5743-27-1]
Calcium 3-Aurothio-2-propanol-1-sulfonate [5743-29-3]
Calcium Bisulfite, Solution [13780-03-5]
Calcium Borate [12007-56-6]
Calcium Borogluconate [5743-34-0]
Calcium Bromide [7789-41-5]
Calcium Bromolactobionate [33659-28-8]
Calcium N-Carbamoylaspartate [16649-79-9]
Calcium N-Carbamoylaspartate D-Form* [16649-82-4]
Calcium N-Carbamoylaspartate DL-Form* [30905-63-6]
Calcium Carbide [75-20-7]
Calcium Carbonate [471-34-1]
Calcium Chlorate [10137-74-3]
Calcium Chloride [10043-52-4]
Calcium Chromate(VI) [13765-19-0]
Calcium Citrate [813-94-5]
Calcium Cyanamide [156-62-7]
Calcium Cyanamide Citrated [8013-88-5]
Calcium Cyanide [592-01-8]
Calcium Cyclamate [139-06-0]
Calcium Dichromate(VI) [14307-33-6]
Calcium 2-Ethylbutanoate [136-91-4]
Calcium Ferrous Citrate [53684-61-0]
Calcium Fluoride [7789-75-5]
Calcium Fluorophosphate [7789-74-4]
Calcium Formate [544-17-2]
Calcium Gluconate [299-28-5]

Calcium Glycerophosphate [27214-00-2]
Calcium Hexafluorosilicate [16925-39-6]
Calcium Hydride [7789-78-8]
Calcium Hydroxide [1305-62-0]
Calcium Hypochlorite [7778-54-3]
Calcium Hypophosphite [7789-79-9]
Calcium Iodate [7789-80-2]
Calcium Iodide [10102-68-8]
Calcium Iodobehenate [1319-91-1]
Calcium Iodostearate [1301-16-2]
Calcium Lactate [814-80-2]
Calcium Levulinate [591-64-0]
Calcium Magnesium Acetate [76123-46-1]
Calcium Mesoxalate [21085-60-9]
Calcium Methionate [819-17-0]
Calcium Molybdate(VI) [7789-82-4]
Calcium Nitrate [10124-37-5]
Calcium Nitrite [13780-06-8]
Calcium Oleate [142-17-6]
Calcium Oxalate [563-72-4]
Calcium Oxide [1305-78-8]
Calcium Palmitate [542-42-7]
Calcium Permanganate [10118-76-0]
Calcium Peroxide [1305-79-9]
Calcium Phenolsulfonate [127-83-3]
Calcium Phenoxide [5793-84-0]
Calcium Phosphate, Dibasic [7757-93-9]
Calcium Phosphate, Monobasic [7758-23-8]
Calcium Phosphate, Tribasic [7758-87-4]
Calcium Phosphide [1305-99-3]
Calcium Phosphite [21056-98-4]
Calcium Polycarbophil [9003-97-8]
Calcium Propionate [4075-81-4]
Calcium Pyrophosphate [35405-51-7]
Calcium D-Saccharate [5793-88-4]
Calcium Selenide [1305-84-6]
Calcium Silicate [1344-95-2]
Calcium Stearate [1592-23-0]
Calcium Succinate [140-99-8]
Calcium Sulfate [7778-18-9]
Calcium Sulfide [20548-54-3]
Calcium Sulfite [10257-55-3]
Calcium Tartrate [3164-34-9]
Calcium Thiocyanate [2092-16-2]
Calcium Thioglycollate [814-71-1]
Calcium Thiosulfate [10124-41-1]
Calcium Tungstate(VI) [7790-75-2]
Caldariomycin [465-61-2]
C-Calebassine [7257-29-6]
Californium [7440-71-3]
Calmagite [3147-14-6]
Calmodulin [77107-46-1]
Calomelol [8011-82-3]
Calotropin [1986-70-5]
Calusterone [17021-26-0]
Calycanthine [595-05-1]
Camazepam [36104-80-0]
Cambendazole [26097-80-3]
Camostat [59721-28-7]
Camostat Monomethanesulfonate* [59721-29-8]
Campesterol [474-62-4]
Camphene [79-92-5]
d-Camphocarboxylic Acid [18530-30-8]
d-Camphocarboxylic Acid Ammonium Salt* [6799-26-4]
d-Camphocarboxylic Acid Basic Bismuth Salt* [4154-53-4]
Camphor [76-22-2]
Camphoric Acid [5394-83-2]
d-Camphorsulfonic Acid [3144-16-9]
Camphotamide [4876-45-3]
Camptothecin [7689-03-4]

Denotes derivative of title compound

Camylofine [54-30-8]
Camylofine Dihydrochloride*
[5892-41-1]
Canadine [522-97-4]
Canadine d-Form* [2086-96-6]
Canadine dl-Form* [29074-38-2]
Canadine l-Form* [5096-57-1]
Canavanine [543-38-4]
Candelilla Wax [8006-44-8]
Candesartan [139481-59-7]
Candesartan Cilexetil* [145040-37-5]
Candicidin [1403-17-4]
Candidin [1405-90-9]
Cannabidiol [13956-29-1]
Cannabinol [521-35-7]
Canrenone [976-71-6]
Canrenone Free Acid Potassium Salt*
[2181-04-6]
Cantharidin [56-25-7]
Canthaxanthin [514-78-3]
Capobenic Acid [21434-91-3]
Capobenic Acid Sodium Salt*
[27276-25-1]
Caprenin [138184-95-9]
Capreomycin [11003-38-6]
Capreomycin Disulfate* [1405-36-3]
Capreomycin IA* [37280-35-6]
Capreomycin IB* [33490-33-4]
n-Capric Acid [334-48-5]
n-Caproic Acid [142-62-1]
Caproic Aldehyde [66-25-1]
Caprolactam [105-60-2]
Caproyl Chloride [142-61-0]
Caprylene [111-66-0]
Caprylic Acid [124-07-2]
Caprylic Aldehyde [124-13-0]
CAPS [1135-40-6]
Capsaicin [404-86-4]
Capsanthin [465-42-9]
Captafol [2425-06-1]
Captan [133-06-2]
Captodiamine [486-17-9]
Captodiamine Hydrochloride*
[904-04-1]
Captopril [62571-86-2]
Captopril Combination with Hydro-
chlorothiazide* [110075-07-5]
Capuride [5579-13-5]
Caramiphen [77-22-5]
Caramiphen Ethanedisulfonate*
[125-86-0]
Caramiphen Hydrochloride*
[125-85-9]
Carazolol [57775-29-8]
Carbachol [51-83-2]
Carbadox [6804-07-5]
Carbamazepine [298-46-4]
Carbamyl Chloride [463-72-9]
Carbanilic Acid [501-82-6]
Carbanilide [102-07-8]
Carbarsone [121-59-5]
Carbaryl [63-25-2]
Carbazochrome Salicylate
[13051-01-9]
Carbazochrome Sodium Sulfonate
[51460-26-5]
Carbazole [86-74-8]
9-Carbazoleacetic Acid [524-80-1]
Carbendazim [10605-21-7]
Carbenicillin [4697-36-3]
Carbenicillin Disodium Salt*
[4800-94-6]
Carbenicillin Phenyl Sodium*
[21649-57-0]
Carbenoxolone [5697-56-3]
Carbenoxolone Disodium Salt*
[7421-40-1]
Carbetapentane [77-23-6]
Carbetapentane Citrate* [23142-01-0]
Carbetocin [37025-55-1]

Carbic Anhydride [129-64-6]
Carbidopa [28860-95-9]
Carbidopa Combination with
Levodopa* [57308-51-7]
Carbimazole [22232-54-8]
Carbinoxamine [486-16-8]
Carbinoxamine l-Form d-Tartrate*
[49746-00-1]
Carbinoxamine Maleate* [3505-38-2]
Carbiphene [15687-16-8]
Carbiphene Hydrochloride*
[467-22-1]
Carbitol [111-90-0]
Carbobenzoxy Chloride [501-53-1]
Carbocloral [541-79-7]
Carbocysteine [638-23-3]
Carbofuran [1563-66-2]
Carbohydrazide [497-18-7]
γ-Carboline [244-69-9]
Carbomycin A* [4564-87-8]
Carbomycin B* [21238-30-2]
Carbon [7440-44-0]
Carbon Dioxide [124-38-9]
Carbon Diselenide [506-80-9]
Carbon Disulfide [75-15-0]
Carbonic Anhydrase [9001-03-0]
Carbon Monoxide [630-08-0]
Carbon Nitride [143334-20-7]
Carbon Suboxide [504-64-3]
Carbon Tetrachloride [56-23-5]
Carbon Tetrafluoride [75-73-0]
Carbon Tetraiodide [507-25-5]
N,N'-Carbonyldiimidazole [530-62-1]
Carbonyl Fluoride [353-50-4]
Carbophenothion [786-19-6]
Carboplatin [41575-94-4]
Carboprost [35700-23-3]
Carboprost Methyl Ester*
[35700-21-1]
Carboprost Tromethamine Salt*
[58551-69-2]
Carboquone [24279-91-2]
Carbostyril [59-31-4]
Carboxin [5234-68-4]
β-Carboxyaspartic Acid [75898-26-9]
γ-Carboxyglutamic Acid [53861-57-7]
Carboxymethylcellulose Sodium
[9004-32-4]
Carboxypolymethylene [9007-20-9]
Carbromal [77-65-6]
Carbubarb [960-05-4]
Carbutamide [339-43-5]
Carbuterol [34866-47-2]
Carbuterol Hydrochloride*
[34866-46-1]
Cardiotoxin [11061-96-4]
3-Carene [13466-78-9]
Carfimate [3567-38-2]
Cargutocin [33605-67-3]
Carindacillin [35531-88-5]
Carindacillin Sodium Salt*
[26605-69-6]
Carisoprodol [78-44-4]
Carlsbad Salt Artificial [8007-49-6]
Carminic Acid [1260-17-9]
Carmofur [61422-45-5]
Carmoxirole [98323-83-2]
Carmoxirole Hydrochloride*
[115092-85-8]
Carmustine [154-93-8]
Carnauba Wax [8015-86-9]
Carnegine [490-53-9]
Carnidazole [42116-76-7]
Carnitine [461-06-3]
Carnitine D-Form* [541-14-0]
Carnitine DL-Form* [406-76-8]
Carnitine L-Form* [541-15-1]
Carnitine D-Form Hydrochloride*
[10017-44-4]

Carnitine DL-Form Hydrochloride*
[461-05-2]
Carnitine L-Form Hydrochloride*
[6645-46-1]
Carnosine [305-84-0]
Caro's Acid [7722-86-3]
α-Carotene [7488-99-5]
β-Carotene [7235-40-7]
γ-Carotene [472-93-5]
δ-Carotene [472-92-4]
Carotol [465-28-1]
Caroverine [23465-76-1]
Caroxazone [18464-39-6]
Carpaine [3463-92-1]
Carpetimycin A* [76025-73-5]
Carpetimycin B* [76094-36-5]
Carphenazine [2622-30-2]
Carphenazine Dimaleate* [2975-34-0]
Carpipramine [5942-95-0]
Carpipramine Dihydrochloride
Monohydrate* [7075-03-8]
Carprofen [53716-49-7]
Carpronium Chloride [13254-33-6]
Carrageenan [9000-07-1]
Carsalam [2037-95-8]
Cartap [15263-53-3]
Carteolol [51781-06-7]
Carteolol Hydrochloride*
[51781-21-6]
Carthamin [36338-96-2]
Carticaine [23964-58-1]
Carticaine Hydrochloride*
[23964-57-0]
Carubicin [50935-04-1]
Carubicin Hydrochloride*
[52794-97-5]
Carumonam [87638-04-8]
Carumonam Disodium Salt*
[86832-68-0]
Carvacrol [499-75-2]
Carvedilol [72956-09-3]
Carvone [99-49-0]
Caryophyllene [87-44-5]
Carzenide [138-41-0]
Carzinophilin [1403-28-7]
Cascarillin [10118-56-6]
Casein [9000-71-9]
Casimiroedine [5853-02-1]
Casimiroin [477-89-4]
Cassaidine [26296-41-3]
Cassaine [468-76-8]
Cassamine [471-71-6]
Cassella's Acid F [494-44-0]
Castanospermine [79831-76-8]
Catalase [9001-05-2]
Catalposide [6736-85-2]
Catechin [154-23-4]
Catechin dl-Form* [7295-85-4]
Catechin l-Form* [18829-70-4]
Catharanthine [2468-21-5]
Cathinone [71031-15-7]
Caulophylline [486-86-2]
2C-B [66142-81-2]
2C-B Hydrochloride* [56281-37-9]
Ceanothic Acid [21302-79-4]
Cedrol [77-53-2]
Cefaclor [70356-03-5]
Cefadroxil [66592-87-8]
Cefamandole [34444-01-4]
Cefamandole Nafate* [42540-40-9]
Cefatrizine [51627-14-6]
Cefatrizine Compd with Propylene
Glycol* [64217-62-5]
Cefazedone [56187-47-4]
Cefazedone Sodium Salt*
[63521-15-3]
Cefazolin [25953-19-9]
Cefazolin Sodium Salt* [27164-46-1]
Cefbuperazone [76610-84-9]

Cefbuperazone Sodium Salt* [76648-01-6]
Cefcapene Pivoxil [105889-45-0]
Cefcapene Pivoxil Free Acid* [135889-00-8]
Cefclidin [105239-91-6]
Cefdinir [91832-40-5]
Cefditoren [104145-95-1]
Cefditoren Pivaloyloxymethyl Ester* [117467-28-4]
Cefepime [88040-23-7]
Cefepime Hydrochloride Mono-hydrate* [123171-59-5]
Cefetamet [65052-63-3]
Cefetamet Pivaloyloxymethyl Ester* [65243-33-6]
Cefixime [79350-37-1]
Cefmenoxime [65085-01-0]
Cefmenoxime Hydrochloride (syn-Iso-mer)* [75738-58-8]
Cefmenoxime Sodium Salt (syn-Isom-er)* [65085-02-1]
Cefmetazole [56796-20-4]
Cefmetazole Sodium Salt* [56796-39-5]
Cefminox [84305-41-9]
Cefodizime [69739-16-8]
Cefodizime Disodium Salt* [86329-79-5]
Cefonicid [61270-58-4]
Cefonicid Disodium Salt* [61270-78-8]
Cefoperazone [62893-19-0]
Cefoperazone Sodium Salt* [62893-20-3]
Ceforanide [60925-61-3]
Ceforanide Sodium Salt* [61336-41-2]
Cefotaxime [63527-52-6]
Cefotaxime Sodium Salt (syn-Isomer)* [64485-93-4]
Cefotetan [69712-56-7]
Cefotetan Disodium Salt* [74356-00-6]
Cefotiam [61622-34-2]
Cefotiam Dihydrochloride* [66309-69-1]
Cefotiam Hexetil Hydrochloride* [95789-30-3]
Cefoxitin [35607-66-0]
Cefoxitin Sodium Salt* [33564-30-6]
Cefozopran [113359-04-9]
Cefpimizole [84880-03-5]
Cefpimizole Monosodium Salt* [85287-61-2]
Cefpiramide [70797-11-4]
Cefpiramide Sodium Salt* [74849-93-7]
Cefpirome [84957-29-9]
Cefpirome Sulfate* [98753-19-6]
Cefpodoxime Proxetil [87239-81-4]
Cefprozil [92665-29-7] (anhydrous)
Cefprozil [121123-17-9] (monohy-drate)
Cefprozil E-Form* [92676-86-3]
Cefprozil Z-Form* [121412-77-9]
Cefroxadine [51762-05-1]
Cefsulodin [62587-73-9]
Cefsulodin Sodium Salt* [52152-93-9]
Ceftazidime [72558-82-8]
Ceftazidime Pentahydrate* [78439-06-2]
Cefteram [82547-58-8]
Cefteram Pivaloyloxymethyl Ester* [82547-81-7]
Ceftezole [26973-24-0]
Ceftezole Sodium Salt* [41136-22-5]
Ceftibuten [97519-39-6]
Ceftibuten Dihydrate* [118081-34-8]
Ceftiofur [80370-57-6]

Ceftiofur Monosodium Salt* [104010-37-9]
Ceftizoxime [68401-81-0]
Ceftizoxime Sodium Salt* [68401-82-1]
Ceftriaxone [73384-59-5]
Ceftriaxone Disodium Salt Hemi-heptahydrate* [104376-79-6]
Cefuroxime [55268-75-2]
Cefuroxime Sodium Salt* [56238-63-2]
Cefuzonam [82219-78-1]
Celesticetin [2520-21-0]
Celestin Blue [1562-90-9]
Celiprolol [56980-93-9]
Celiprolol Hydrochloride* [57470-78-7]
Cellobiose [528-50-7]
Cellocidin [543-21-5]
Celluloid [8050-88-2]
Cellulose [9004-34-6]
Cellulose Ethyl Hydroxyethyl Ether [9004-58-4]
Centaurein [35595-03-0]
Centchroman [78994-24-8]
Centchroman Hydrochloride* [51023-56-4]
Cephacetrile Sodium [23239-41-0]
Cephaeline [483-17-0]
Cephalexin [15686-71-2]
Cephalexin Monohydrate* [23325-78-2]
Cephalexin Monohydrochloride Monohydrate* [105879-42-3]
Cephalexin Sodium Salt* [38932-40-0]
Cephaloglycin [3577-01-3]
Cephaloglycin Dihydrate* [22202-75-1]
Cephalonic Acid [18456-04-7]
Cephaloridine [50-59-9]
Cephalosporin C [61-24-5]
Cephalosporin P_1 [28393-42-2]
Cephalothin [153-61-7]
Cephalothin Sodium Salt* [58-71-9]
Cephapirin Sodium [24356-60-3]
Cepharanthine [481-49-2]
Cephradine [38821-53-3]
Cephradine Dihydrate* [31828-50-9]
Cephradine Monohydrate* [75975-70-1]
Ceresin [8001-75-0]
Ceric Fluoride [10060-10-3]
Ceric Oxide [1306-38-3]
Ceric Sulfate [13590-82-4]
Cerium [7440-45-1]
Ceronapril [111223-26-8]
Cerous Bromide [14457-87-5]
Cerous Carbonate [537-01-9]
Cerous Chloride [7790-86-5]
Cerous Fluoride [7758-88-5]
Cerous Iodide [7790-87-6]
Cerous Nitrate [10108-73-3]
Cerous Oxalate [139-42-4]
Cerous Sulfate [13454-94-9]
Cerulenin [17397-89-6]
Ceruletide [17650-98-5]
Ceruletide Diethylamine Salt* [71247-25-1]
Ceruloplasmin [9031-37-2]
Cervicarcin [18700-78-2]
Cesium [7440-46-2]
Cesium Bromide [7787-69-1]
Cesium Carbonate [534-17-8]
Cesium Chloride [7647-17-8]
Cesium Hydroxide [21351-79-1]
Cesium Iodide [7789-17-5]
Cesium Nitrate [7789-18-6]
Cesium Sulfate [10294-54-9]
Cetalkonium Chloride [122-18-9]

Cetamolol [34919-98-7]
Cetamolol Hydrochloride* [77590-95-5]
Cethexonium Bromide [1794-74-7]
Cetiedil [14176-10-4]
Cetiedil Citrate* [16286-69-4]
Cetirizine [83881-51-0]
Cetirizine Dihydrochloride* [83881-52-1]
Cetotiamine [137-76-8]
Cetoxime [25394-78-9]
Cetoxime Hydrochloride* [22204-29-1]
Cetraric Acid [489-49-6]
Cetraxate [34675-84-8]
Cetraxate Hydrochloride* [27724-96-5]
Cetrimonium Bromide [57-09-0]
Cetrimonium Stearate [124-23-2]
Cetrimonium Tosylate* [138-32-9]
Cetyl Alcohol [36653-82-4]
Cetyldimethylethylammonium Bro-mide [124-03-8]
Cetyl Lactate [35274-05-6]
Cetyl Palmitate [540-10-3]
Cetylpyridinium Chloride [123-03-5]
Cetylpyridinium Chloride Mono-hydrate* [6004-24-6]
Cevadine [62-59-9]
Cevine [124-98-1]
Chalcomycin [20283-48-1]
Chalcone [94-41-7]
Chalcopyrite [1308-56-1]
D-Chalcose [3150-28-5]
Chamazulene [529-05-5]
Chanoclavine [2390-99-0]
CHAPS [75621-03-3]
CHAPS Hydroxy Analog* [82473-24-3]
Chartreusin [6377-18-0]
Charybdotoxin [95751-30-7]
Chaulmoogric Acid [29106-32-9]
Chaulmoogric Acid Ethyl Ester* [623-32-5]
Chaulmoogric Acid Methyl Ester* [24828-59-9]
Chavicine [495-91-0]
Chavicol [501-92-8]
Cheirolin [505-34-0]
Chelerythrine [34316-15-9]
Chelidonic Acid [99-32-1]
Chelidonine (±)-Form* [20267-87-2]
Chelidonine (−)-Form* [88200-01-5]
Chelidonine (+)-Form* [476-32-4]
Chenodiol [474-25-9]
CHES [103-47-9]
Chimaphilin [482-70-2]
Chimonanthine [5545-89-1]
Chimyl Alcohol [506-03-6]
Chitin [1398-61-4]
Chlophedianol [791-35-5]
Chlophedianol Hydrochloride* [511-13-7]
Chloracetyl Chloride [79-04-9]
Chloracizine [800-22-6]
Chloral Alcoholate [515-83-3]
Chloral Ammonia [507-47-1]
Chloral Betaine [2218-68-0]
Chloral Formamide [515-82-2]
Chloral Hydrate [302-17-0]
α-Chloralose [15879-93-3]
Chloramben [133-90-4]
Chlorambucil [305-03-3]
Chloramine-B [127-52-6]
Chloramine-T [127-65-1]
Chloraminophenamide [121-30-2]
Chloramphenicol [56-75-7]
Chloramphenicol Monosuccinate Ar-ginine Salt* [34327-18-9]

Denotes derivative of title compound

Chloramphenicol Monosuccinate Sodium Salt* [982-57-0]
Chloramphenicol Palmitate* [530-43-8]
Chloramphenicol Pantothenate Calcium Complex (4:1)* [31342-36-6]
Chloranil [118-75-2]
Chloranilic Acid [87-88-7]
Chlorazanil [500-42-5]
Chlorazanil Hydrochloride* [2019-25-2]
Chlorbenside [103-17-3]
Chlorbenzoxamine [522-18-9]
Chlorbenzoxamine Dihydrochloride* [5576-62-5]
Chlorbetamide [97-27-8]
Chlorbicyclen [2550-75-6]
Chlorbicyclen trans-Isomer* [26382-17-2]
Chlorcyclizine [82-93-9]
Chlorcyclizine Hydrochloride* [1620-21-9]
Chlordan(e) [57-74-9]
Chlordantoin [5588-20-5]
Chlordecone [143-50-0]
Chlordiazepoxide [58-25-3]
Chlordiazepoxide Hydrochloride* [438-41-5]
Chlordimeform [6164-98-3]
Chlorendic Anhydride [115-27-5]
Chlorfenac [85-34-7]
Chlorfenethol [80-06-8]
Chlorfenvinphos [470-90-6]
Chlorguanide [500-92-5]
Chlorguanide Hydrochloride* [637-32-1]
Chlorhexadol [3563-58-4]
Chlorhexidine [55-56-1]
Chlorhexidine Diacetate* [56-95-1]
Chlorhexidine D-Digluconate* [18472-51-0]
Chlorhexidine Dihydrochloride* [3697-42-5]
Chloric Acid [7790-93-4]
Chlorimuron Ethyl [90982-32-4]
Chlorinated Lime [7778-54-3]
Chlorindanol [145-94-8]
Chlorine [7782-50-5]
Chlorine Dioxide [10049-04-4]
Chlorine Heptoxide [10294-48-1]
Chlorine Monofluoride [7790-89-8]
Chlorine Monoxide [7791-21-1]
Chlorine Trifluoride [7790-91-2]
Chlorisondamine Chloride [69-27-2]
Chlormadinone Acetate [302-22-7]
Chlormadinone Acetate Mixture with Ethinyl Estradiol* [37301-55-6]
Chlormadinone Acetate Mixture with Mestranol* [8065-91-6]
Chlormequat Chloride [999-81-5]
Chlormerodrin [62-37-3]
Chlormezanone [80-77-3]
Chlormidazole [3689-76-7]
Chlormidazole Hydrochloride* [74298-63-8]
Chlornaphazine [494-03-1]
Chlornitrofen [1836-77-7]
Chloroacetaldehyde [107-20-0]
Chloroacetamide [79-07-2]
Chloroacetic Acid [79-11-8]
Chloroacetic Anhydride [541-88-8]
Chloroacetone [78-95-5]
p-Chloroacetophenone [99-91-2]
ω-Chloroacetophenone [532-27-4]
Chloroacetyl Isocyanate [4461-30-7]
N-(3-Chloroallyl)hexaminium Chloride [4080-31-3]
m-Chloroaniline* [108-42-9]
o-Chloroaniline* [95-51-2]
p-Chloroaniline* [106-47-8]

Chloroarsenol [151-07-5]
Chloroazodin [502-98-7]
Chlorobenzene [108-90-7]
p-Chlorobenzenesulfonic Acid [98-66-8]
Chlorobenzilate [510-15-6]
m-Chlorobenzoic Acid [535-80-8]
o-Chlorobenzoic Acid [118-91-2]
p-Chlorobenzoic Acid [74-11-3]
o-Chlorobenzylidenemalononitrile [2698-41-1]
p-Chlorobenzylpseudothiuronium Chloride [544-47-8]
Chlorobutanol [57-15-8]
1-Chloro-2-butene [591-97-9]
3-Chloro-1-butene [563-52-0]
3-Chloro-d-camphor [508-29-2]
4-Chloro-m-cresol [59-50-7]
Chlorocyanohydrin [513-96-2]
Chlorodibromomethane [124-48-1]
1-Chloro-2,4-dinitrobenzene [97-00-7]
2-Chloro-1,3-dinitrobenzene [606-21-3]
β-Chloroethyl Acetate [542-58-5]
N-(2-Chloroethyl)dibenzylamine Hydrochloride [55-43-6]
2-Chloroethyl Vinyl Ether [110-75-8]
Chloroform [67-66-3]
Chlorogenic Acid [327-97-9]
Chlorogenin [562-34-5]
1-Chlorohexane [544-10-5]
α-Chlorohydrin [96-24-2]
Chloromethyl Methyl Ether [107-30-2]
1-Chloro-2-methyl-1-propene [513-37-1]
3-Chloro-2-methyl-1-propene [563-47-3]
1-Chloronaphthalene [90-13-1]
2-Chloronaphthalene [91-58-7]
Chlorophacinone [3691-35-8]
p-Chlorophenacyl Bromide [536-38-9]
Chlorophyll [1406-65-1]
Chloropicrin [76-06-2]
Chloroprednisone [52080-57-6]
Chloroprednisone 21-Acetate* [14066-79-6]
Chloroprocaine Hydrochloride [3858-89-7]
β-Chloropropionic Acid [107-94-8]
β-Chloropropionitrile [542-76-7]
6-Chloropurine [87-42-3]
Chloropyramine [59-32-5]
Chloropyramine Hydrochloride* [6170-42-9]
Chloroquine [54-05-7]
Chloroquine Diphosphate* [50-63-5]
Chloroselenic Acid [7790-95-6]
N-Chlorosuccinimide [128-09-6]
Chlorosulfonic Acid [7790-94-5]
Chlorothalonil [1897-45-6]
Chlorothen [148-65-2]
Chlorothen Citrate* [148-64-1]
Chlorothiazide [58-94-6]
Chlorothiazide Sodium Salt* [7085-44-1]
Chlorothricin [34707-92-1]
Chlorothymol [89-68-9]
m-Chlorotoluene* [108-41-8]
o-Chlorotoluene* [95-49-8]
p-Chlorotoluene* [106-43-4]
Chlorotrianisene [569-57-3]
Chlorotris(triphenylphosphine)rhodium [14694-95-2]
Chloroxine [773-76-2]
Chloroxylenol [88-04-0]
Chlorozotocin [54749-90-5]
Chlorphenesin [104-29-0]
Chlorphenesin Carbamate [886-74-8]
Chlorpheniramine [132-22-9]

Chlorpheniramine d-Form* [25523-97-1]
Chlorpheniramine d-Form Maleate* [2438-32-6]
Chlorpheniramine Maleate* [113-92-8]
Chlorphenoxamide [3576-64-5]
Chlorphenoxamine [77-38-3]
Chlorphenoxamine Hydrochloride* [562-09-4]
Chlorphentermine [461-78-9]
Chlorphentermine Hydrochloride* [151-06-4]
Chlorproethazine [84-01-5]
Chlorproethazine Hydrochloride* [4611-02-3]
Chlorproguanil [537-21-3]
Chlorproguanil Hydrochloride* [15537-76-5]
Chlorpromazine [50-53-3]
Chlorpromazine Hydrochloride* [69-09-0]
Chlorpropamide [94-20-2]
Chlorpropham [101-21-3]
Chlorprothixene [113-59-7]
Chlorpyrifos [2921-88-2]
Chlorpyrifos O,O-Dimethyl Analog* [5598-13-0]
Chlorquinaldol [72-80-0]
Chlorsulfuron [64902-72-3]
Chlortetracycline [57-62-5]
Chlortetracycline Hydrochloride* [64-72-2]
Chlorthalidone [77-36-1]
Chlorthenoxazin(e) [132-89-8]
Chlorthion [500-28-7]
Chlorzoxazone [95-25-0]
Cholane [548-98-1]
Cholanic Acid [25312-65-6]
Cholanthrene [479-23-2]
Cholecystokinin [9011-97-6]
(5α)-Cholestane [481-21-0]
Cholestanol [80-97-7]
Cholesterol [57-88-5]
Cholestyramine Resin [11041-12-6]
Cholic Acid [81-25-4]
Choline [62-49-7]
Choline Alfoscerate [28319-77-9]
Choline Chloride [67-48-1]
Choline Dehydrocholate [4201-78-9]
Choline Dihydrogen Citrate [77-91-8]
Choline Esterase [9001-08-5]
Choline Salicylate [2016-36-6]
Choline Theophyllinate [4499-40-5]
Chondrillasterol [481-17-4]
Chondrocurine [477-58-7]
Chondrofoline [31944-97-5]
Chondroitin Sulfate [9007-28-5]
Chondroitin 4-Sulfate* [24967-93-9]
Chondroitin 6-Sulfate* [25322-46-7]
Chondroitin 4-Sulfate Disodium Salt* [39455-18-0]
Chondrosine [499-14-9]
Chonemorphine [4282-07-9]
Chorismic Acid [617-12-9]
Chromic Acetate [1066-30-4]
Chromic Bromide [10031-25-1]
Chromic Carbonate [29689-14-3]
Chromic Chloride [10025-73-7]
Chromic Fluoride [7788-97-8]
Chromic Formate [27115-36-2]
Chromic Hydroxide [1308-14-1]
Chromic Nitrate [13548-38-4]
Chromic Oxide [1308-38-9]
Chromic Phosphate [7789-04-0]
Chromic Phosphate Radioactive Form* [24381-60-0]
Chromic Potassium Oxalate [14217-01-7]

Denotes derivative of title compound

Chromic Potassium Sulfate [10141-00-1]
Chromic Sulfate [10101-53-8]
Chromium [7440-47-3]
Chromium Carbonyl [13007-92-6]
Chromium Dioxide [12018-01-8]
Chromium Picolinate [14639-25-9]
Chromium Tetrafluoride [10049-11-3]
Chromium Trioxide [1333-82-0]
Chromocarb [4940-39-0]
Chromocarb Diethylamine* [23915-80-2]
Chromomycin A$_3$* [7059-24-7]
Chromonar [804-10-4]
Chromonar Hydrochloride* [655-35-6]
Chromotrope 2B [548-80-1]
Chromotropic Acid [148-25-4]
Chromous Acetate [628-52-4]
Chromous Bromide [10049-25-9]
Chromous Chloride [10049-05-5]
Chromous Fluoride [10049-10-2]
Chromous Formate [4493-37-2]
Chromous Oxalate [814-90-4]
Chromous Sulfate [13825-86-0]
Chromyl Chloride [14977-61-8]
Chromyl Fluoride [7788-96-7]
Chrysamminic Acid [517-92-0]
Chrysanthemaxanthin [27780-11-6]
Chrysanthemic Acid [10453-89-1]
Chrysanthemic Acid dl-cis-Form* [2935-23-1]
Chrysanthemic Acid dl-trans-Form* [705-16-8]
Chrysanthemic Acid d-trans-Form* [4638-92-0]
Chrysanthemic Acid l-trans-Form* [2259-14-5]
Chrysanthenone [473-06-3]
Chrysarobin [491-59-8]
6-Chrysenamine [2642-98-0]
Chrysene [218-01-9]
Chrysergonic Acid [53092-91-4]
Chrysin [480-40-0]
Chrysoidine [532-82-1]
Chrysoidine Citrate* [5909-04-6]
Chrysoidine Free Base* [495-54-5]
Chrysophanic Acid [481-74-3]
Chymopapain [9001-09-6]
α-Chymotrypsin* [9004-07-3]
Ciafos [2636-26-2]
Cichoriin [531-58-8]
Cicletanine [89943-82-8]
Cicletanine Hydrochloride* [82747-56-6]
Ciclonicate [53449-58-4]
Ciclopirox [29342-05-0]
Ciclopirox Ethanolamine Salt (1:1)* [41621-49-2]
Ciclosidomine [66564-16-7]
Ciclosidomine Hydrochloride* [26209-07-4]
Cicrotoic Acid [25229-42-9]
Cicutoxin [505-75-9]
Cidofovir [113852-37-2]
Cifenline [53267-01-9]
Cifenline Succinate* [100678-32-8]
Cilastatin [82009-34-5]
Cilastatin Sodium Salt* [81129-83-1]
Cilazapril [92077-78-6]
Ciliary Neurotrophic Factor (Human)* [133423-39-9]
Cilnidipine [102106-21-8] (unspecified)
Cilnidipine [132203-70-4] (±)
Cilostazol [73963-72-1]
Cimaterol [54239-37-1]
Cimetidine [51481-61-9]
Cimetidine Hydrochloride* [70059-30-2]

Cimetropium Bromide [51598-60-8]
Cimigenol [3779-59-7]
Cinametic Acid [35703-32-3]
Cinchomeronic Acid [490-11-9]
Cinchonamine [482-28-0]
Cinchonidine [485-71-2]
Cinchonine [118-10-5]
Cinchophen [132-60-5]
Cinchophen Allyl Ester* [524-34-5]
Cinchophen Hydrochloride* [132-58-1]
Cinchophen Sodium Salt* [5949-18-8]
Cinchotoxine [69-24-9]
Cinepazet [23887-41-4]
Cinepazet Maleate* [50679-07-7]
Cinepazide [23887-46-9]
Cinepazide Maleate* [26328-04-1]
Cinerin I* [25402-06-6]
Cinerin II* [121-20-0]
Cinitapride [66564-14-5]
Cinitapride Tartrate* [96623-56-2]
Cinmetacin [20168-99-4]
Cinmethylin [87818-31-3]
Cinnabarine [146-90-7]
Cinnamaldehyde [104-55-2]
Cinnamedrine [90-86-8]
Cinnamic Acid [621-82-9]
Cinnamoyl Chloride [102-92-1]
Cinnamoylcocaine [521-67-5]
O-Cinnamoyltaxicin-I* [11034-45-0]
O-Cinnamoyltaxicin-II triacetate* [3835-52-7]
O-Cinnamoyltaxicin-I triacetate* [13452-36-3]
Cinnamyl Alcohol [104-54-1]
Cinnamyl Anthranilate [87-29-6]
Cinnamyl Cinnamate [122-69-0]
Cinnarizine [298-57-7]
Cinnoline [253-66-7]
Cinobufotalin [1108-68-5]
Cinolazepam [75696-02-5]
Cinoxacin [28657-80-9]
Cinoxate [104-28-9]
Cinromide [58473-74-8]
Cioteronel [89672-11-7]
Ciprofibrate [52214-84-3]
Ciprofloxacin [85721-33-1]
Ciprofloxacin Monohydrochloride Monohydrate* [86393-32-0]
Ciramadol [63269-31-8]
Ciramadol Hydrochloride* [63323-46-4]
Cisapride [81098-60-4]
Cisplatin [15663-27-1]
Citalopram [59729-33-8]
Citalopram Hydrobromide* [59729-32-7]
Citicoline [987-78-0]
Citicoline Sodium Salt* [33818-15-4]
Citiolone [1195-16-0]
Citraconic Acid [498-23-7]
Citral [5392-40-5]
Citramalic Acid [2306-22-1]
β-Citraurin [650-69-1]
Citrazinic Acid [99-11-6]
Citric Acid [77-92-9]
Citrinin [518-75-2]
Citromycetin [478-60-4]
Citronellal [106-23-0]
α-Citronellal* [141-26-4]
β-Citronellol [106-22-9]
Citrulline [372-75-8]
Citrulline Hydrochloride* [34312-10-2]
Citrulline Malate (Salt)* [54940-97-5]
Citrullol [1390-93-8]
Citrus Red 2 [6358-53-8]
Civetone [542-46-1]
Cladribine [4291-63-8]
Clanobutin [30544-61-7]

Clarase [9001-11-0]
Clarithromycin [81103-11-9]
Clavulanic Acid [58001-44-8]
Clavulanic Acid Combination of Potassium Salt with Ticarcillin Disodium* [116876-37-0]
Clazuril [101831-36-1]
Clebopride [55905-53-8]
Clebopride Malate* [57645-91-7]
Clemastine [15686-51-8]
Clemastine Hydrogen Fumarate* [14976-57-9]
Clemizole [442-52-4]
Clemizole Hydrochloride* [1163-36-6]
Clenbuterol [37148-27-9]
Clenbuterol Hydrochloride* [21898-19-1]
Clentiazem [96125-53-0]
Clentiazem Maleate* [96128-92-6]
1,6-Cleve's Acid [119-79-9]
1,7-Cleve's Acid [119-28-8]
Clidanac [28968-07-2]
Clidinium Bromide [3485-62-9]
Clinafloxacin [105956-97-6]
Clinafloxacin Hydrochloride* [105956-99-8]
Clindamycin [18323-44-9]
Clindamycin Hydrochloride Monohydrate* [58207-19-5]
Clinofibrate [30299-08-2]
Clioxanide [14437-41-3]
Clobazam [22316-47-8]
Clobenfurol [3611-72-1]
Clobenoside [29899-95-4]
Clobenzepam [1159-93-9]
Clobenzepam Hydrochloride* [2726-03-6]
Clobenzorex [13364-32-4]
Clobenzorex Hydrochloride* [5843-53-8]
Clobenztropine [5627-46-3]
Clobetasol [25122-41-2]
Clobetasol 17-Propionate* [25122-46-7]
Clobetasone [54063-32-0]
Clobetasone 17-Butyrate* [25122-57-0]
Clobutinol [14860-49-2]
Clobutinol Hydrochloride* [1215-83-4]
Clobuzarit [22494-47-9]
Clocapramine [47739-98-0]
Clocapramine Dihydrochloride Monohydrate* [60789-62-0]
Clocinizine [298-55-5]
Cloconazole [77175-51-0]
Cloconazole Monohydrochloride* [77174-66-4]
Clocortolone [4828-27-7]
Clocortolone 21-Acetate* [4258-85-9]
Clocortolone 21-Pivalate* [34097-16-0]
Clodinafop-Propargyl [105512-06-9]
Clodronic Acid [10596-23-3]
Clodronic Acid Disodium Salt* [22560-50-5]
Clofazimine [2030-63-9]
Clofenamide [671-95-4]
Clofentezine [74115-24-5]
Clofibrate [637-07-0]
Clofibrate Mixture with Androsterone* [8075-95-4]
Clofibric Acid [882-09-7]
Clofibric Acid Basic Aluminum Salt* [24818-79-9]
Clofibric Acid Magnesium Salt* [14613-30-0]
Clofibric Acid Pyridoxine Salt* [29952-87-2]
Cloflucarban [369-77-7]

* Denotes derivative of title compound

Clofoctol [37693-01-9]
Cloforex [14261-75-7]
Clomacran [5310-55-4]
Clomacran Phosphate* [22199-46-8]
Clomestrone [4091-75-2]
Clometacin [25803-14-9]
Clomethiazole [533-45-9]
Clomethiazole Ethanedisulfonate* [1867-58-9]
Clomethiazole Hydrochloride* [6001-74-7]
Clometocillin [1926-49-4]
Clometocillin Sodium Salt* [54530-86-8]
Clomiphene [911-45-5]
Clomiphene Citrate* [50-41-9]
Clomipramine [303-49-1]
Clomipramine Hydrochloride* [17321-77-6]
Clomocycline [1181-54-0]
Clonazepam [1622-61-3]
Clonidine [4205-90-7]
Clonidine Hydrochloride* [4205-91-8]
Clonitazene [3861-76-5]
Clonitrate [2612-33-1]
Clonixin [17737-65-4]
Clonixin Lysine Salt* [55837-30-4]
Clopamide [636-54-4]
Clopamide Mixture with Pindolol* [71789-16-7]
Clopenthixol [982-24-1]
Clopenthixol Dihydrochloride* [633-59-0]
Clopenthixol cis(Z)-Form* [53772-83-1]
Clopenthixol cis(Z)-Form Dihydrochloride* [58045-23-1]
Cloperastine [3703-76-2]
Cloperastine Hydrochloride* [14984-68-0]
Clopidogrel [113665-84-2]
Clopidogrel Hydrogen Sulfate* [135046-48-9]
Clopidol [2971-90-6]
Clopirac [42779-82-8]
Cloprednol [5251-34-3]
Cloprostenol [40665-92-7]
Cloprostenol Sodium Salt* [55028-72-3]
Clopyralid [1702-17-6]
Cloquintocet-Mexyl [99607-70-2]
Cloranolol [39563-28-5]
Cloranolol Hydrochloride* [54247-25-5]
Clorazepic Acid [23887-31-2]
Clorazepic Acid Dipotassium Salt* [57109-90-7]
Clorazepic Acid Monopotassium Salt* [5991-71-9]
Clorexolone [2127-01-7]
Cloricromen [68206-94-0]
Cloricromen Hydrochloride* [74697-28-2]
Clorindione [1146-99-2]
Clorophene [120-32-1]
Clorprenaline [3811-25-4]
Clorprenaline Hydrochloride Monohydrate* [5588-22-7]
Clorsulon [60200-06-8]
Clortermine [10389-73-8]
Clortermine Hydrochloride* [10389-72-7]
Closantel [57808-65-8]
Clospirazine [24527-27-3]
Clospirazine Hydrochloride* [27007-85-8]
Clostebol [1093-58-9]
Clostebol Acetate* [855-19-6]
Clothiapine [2058-52-8]
Clotiazepam [33671-46-4]

Clotrimazole [23593-75-1]
Cloxacillin [61-72-3]
Cloxacillin Benzathine Salt* [23736-58-5]
Cloxacillin Sodium Monohydrate* [7081-44-9]
Cloxazolam [24166-13-0]
Cloxotestosterone [53608-96-1]
Cloxotestosterone Acetate* [13867-82-8]
Cloxyquin [130-16-5]
Clozapine [5786-21-0]
Clupeine [9007-31-2]
Cnicin [24394-09-0]
Cobalt [7440-48-4]
Cobaltic Acetate [917-69-1]
Cobaltic-Cobaltous Oxide [1308-06-1]
Cobaltic Fluoride [10026-18-3]
Cobaltic Oxide Monohydrate [12016-80-7]
Cobaltic Potassium Nitrite [13782-01-9]
Cobaltous Acetate [71-48-7]
Cobaltous Arsenate [7785-24-2]
Cobaltous Bromide [7789-43-7]
Cobaltous Carbonate [513-79-1]
Cobaltous Chloride [7646-79-9]
Cobaltous Chromate(III) [13455-25-9]
Cobaltous Cyanide [542-84-7]
Cobaltous Fluoride [10026-17-2]
Cobaltous Formate [544-18-3]
Cobaltous Hydroxide [21041-93-0]
Cobaltous Iodide [15238-00-3]
Cobaltous Nitrate [10141-05-6]
Cobaltous Oxalate [814-89-1]
Cobaltous Oxide [1307-96-6]
Cobaltous Phosphate [13455-36-2]
Cobaltous Potassium Sulfate [13596-22-0]
Cobaltous Sulfate [10124-43-3]
Cobaltous Sulfide [1317-42-6]
Cobaltous Thiocyanate [3017-60-5]
Cobamamide [13870-90-1]
Cobrotoxin [12584-83-7]
Cocaethylene [529-38-4]
Cocaine [50-36-2]
Coclaurine [486-39-5]
Codamine [21040-59-5]
Codeine [76-57-3]
Codeine Acetate* [5913-71-3]
Codeine Hydrobromide* [125-25-7]
Codeine Hydrochloride* [1422-07-7]
Codeine Methyl Bromide [125-27-9]
Codeine N-Oxide [3688-65-1]
Codeine Phosphate [52-28-8]
Codeine Salicylate* [6020-73-1]
Codeine Sulfate [1420-53-7]
Coenzyme A [85-61-0]
Coherin [9044-70-6]
Colchiceine [477-27-0]
Colchicine [64-86-8]
Colestipol [26658-42-4]
Colestipol Hydrochloride* [37296-80-3]
Colforsin [66575-29-9]
Colfosceril Palmitate [63-89-8]
Colistin [1066-17-7]
Colistin Sodium Methanesulfonate* [8068-28-8]
Colistin Sulfate* [1264-72-8]
Collagenase [9001-12-1]
Collinomycin [27267-69-2]
Colloidal Bismuth Subcitrate [57644-54-9]
Colocynthin [1398-78-3]
4-221-Colony-stimulating Factor 1 (Human Clone p3ACSF-69 Protein Moiety Reduced)* [117276-75-2]
Colostrokinin [37203-40-0]
Colpormon [1247-71-8]

α-Colubrine* [509-44-4]
β-Colubrine* [509-36-4]
Columbamine [3621-36-1]
Columbin [546-97-4]
Concanavalin A [11028-71-0]
Condurangin [1401-98-5]
Conessine [546-06-5]
Conessine Dihydrobromide* [5913-82-6]
Congo Red [573-58-0]
Congressane [2292-79-7]
Conhydrine [495-20-5]
β-Coniceine [538-90-9]
γ-Coniceine [1604-01-9]
Coniferin [531-29-3]
Coniferyl Alcohol [458-35-5]
Coniine [458-88-8]
Coniine Hydrobromide* [637-49-0]
Coniine Hydrochloride* [555-92-0]
Conquinamine [464-86-8]
Convallamarogenin [16683-27-5]
Convallatoxin [508-75-8]
Convicine [19286-37-4]
COP [26591-04-8]
COP 1 [28704-27-0]
Copaene [3856-25-5]
Copaiba [8001-61-4]
Copal [9000-14-0]
Coparaffinate [8001-60-3]
Copper [7440-50-8]
Copper Phthalocyanine [147-14-8]
Coproergostane [511-21-7]
Coprogen [31418-71-0]
Coprostane [481-20-9]
Coprosterol [360-68-9]
Coptine [1391-14-6]
Coptisine [3486-66-6]
Cord Factor [61512-20-7]
Cordycepin [73-03-0]
Coriamyrtin [2571-86-0]
Coroxon [321-54-0]
Corticosterone [50-22-6]
Cortisone [53-06-5]
Cortisone 21-Acetate* [50-04-4]
Cortisone 21-Cyclopentanepropionate* [509-00-2]
Cortivazol [1110-40-3]
Cortol [516-38-1]
Cortolone [516-42-7]
Cortolone 3,20,21-Triacetate* [2638-46-2]
Corybulbine [518-77-4]
Corycavamine [521-85-7]
Corycavidine [521-93-7]
Corydaldine [493-49-2]
Corydaline [518-69-4]
Corydine [476-69-7]
Corynantheine [18904-54-6]
Corynanthine [483-10-3]
Corypalmine [6018-40-2]
Corytuberine [517-56-6]
Cosyntropin [16960-16-0]
Cotarnine [82-54-2]
Cotarnine Chloride* [10018-19-6]
Cotarnine Hydrochloride* [36647-02-6]
Cotarnine Phthalate* [6190-36-9]
Cotinine [486-56-6]
Cotinine Fumarate* [5695-98-7]
Cotoin [479-21-0]
Coumachlor [81-82-3]
Coumafuryl [117-52-2]
Coumalic Acid [500-05-0]
Coumaphos [56-72-4]
Coumaran [496-16-2]
p-Coumaric Acid [7400-08-0]
Coumarilic Acid [496-41-3]
Coumarin [91-64-5]
Coumarin-3-carboxylic Acid [531-81-7]

Coumestrol [479-13-0]
Coumetarol [4366-18-1]
Coumingine [26241-81-6]
Coumithoate [572-48-5]
Creatine [57-00-1]
Creatinine [60-27-5]
Creolin [12751-04-1]
Creosol [93-51-6]
Creosote, Coal Tar [8001-58-9]
Creosote, Wood [8021-39-4]
Creosote, Wood, Calcium deriv.*
 [8001-83-0]
Cresol [1319-77-3]
m-Cresol* [108-39-4]
o-Cresol* [95-48-7]
p-Cresol* [106-44-5]
o-Cresolphthalein [596-27-0]
Cresol Red [1733-12-6]
m-Cresotic Acid [50-85-1]
o-Cresotic Acid [83-40-9]
p-Cresotic Acid [89-56-5]
m-Cresyl Acetate [122-46-3]
o-Cresyl Acetate [533-18-6]
CRF [9015-71-8]
Crilanomer [37291-07-9]
Crimidine [535-89-7]
Croceic Acid [132-57-0]
Crocetin [27876-94-4]
Crocetin Di-gentiobiose Ester*
 [42553-65-1]
Cromakalim* [94470-67-4]
Cromolyn [16110-51-3]
Cromolyn Disodium Salt*
 [15826-37-6]
Cropropamide [633-47-6]
Cropropamide Combination with
 Crotethamide* [8015-51-8]
Crotamine [37196-57-9]
Crotamiton [483-63-6]
Crotethamide [6168-76-9]
Crotonaldehyde [4170-30-3]
Crotonic Acid [107-93-7]
Crotoxin [9007-40-3]
Crotoxyphos [7700-17-6]
Crotyl Alcohol [6117-91-5]
CRP [9007-41-4]
Crufomate [299-86-5]
Cryofluorane [76-14-2]
Cryolite [15096-52-3]
Cryptopine [482-74-6]
Cryptoxanthin [472-70-8]
Cubane [277-10-1]
Cubebin [18423-69-3]
Cucurbitacin B* [6199-67-3]
Cucurbitacin E* [18444-66-1]
Cumene [98-82-8]
Cumic Acid [536-66-3]
Cumic Alcohol [536-60-7]
Cumidine [99-88-7]
Cuminaldehyde [122-03-2]
Cuminaldehyde Thiosemicarbazone*
 [3811-20-9]
Cupferron [135-20-6]
Cupreine [524-63-0]
Cupric Acetate [142-71-2]
Cupric Acetate, Basic [52503-64-7]
Cupric Acetoarsenite [12002-03-8]
Cupric Arsenite [10290-12-7]
Cupric Borate [39290-85-2]
Cupric Bromide [7789-45-9]
Cupric Butyrate [540-16-9]
Cupric Carbonate, Basic [12069-69-1]
Cupric Chlorate [14721-21-2]
Cupric Chloride [7447-39-4]
Cupric Chromate(VI) [13548-42-0]
Cupric Chromite [12018-10-9]
Cupric Citrate [866-82-0]
Cupric Ferrocyanide [13601-13-3]
Cupric Fluoride [7789-19-7]
Cupric Formate [544-19-4]

Cupric Gluconate [527-09-3]
Cupric Glycinate [13479-54-4]
Cupric Hexafluorosilicate
 [12062-24-7]
Cupric Hydroxide [20427-59-2]
Cupric Nitrate [3251-23-8]
Cupric Oleate [1120-44-1]
Cupric Oxalate [814-91-5]
Cupric Oxide [1317-38-0]
Cupric Perchlorate [13770-18-8]
Cupric p-Phenolsulfonate [547-56-8]
Cupric Phosphate [7798-23-4]
Cupric Salicylate [16048-96-7]
Cupric Selenate [15123-69-0]
Cupric Selenide [1317-41-5]
Cupric Selenite [10214-40-1]
Cupric Stearate [660-60-6]
Cupric Sulfate [7758-98-7]
Cupric Sulfate, Basic [1332-14-5]
Cupric Sulfide [1317-40-4]
Cupric Tartrate [815-82-7]
Cupric Tungstate(VI) [13587-35-4]
Cuprobam [7076-63-3]
Cuprous Acetate [598-54-9]
Cuprous Bromide [7787-70-4]
Cuprous Chloride [7758-89-6]
Cuprous Cyanide [544-92-3]
Cuprous Iodide [7681-65-4]
Cuprous Mercuric Iodide [13876-85-2]
Cuprous Oxide [1317-39-1]
Cuprous Potassium Cyanide
 [13682-73-0]
Cuprous Selenide [20405-64-5]
Cuprous Sulfide [22205-45-4]
Cuprous Sulfite [13982-53-1]
Cuprous Thiocyanate [1111-67-7]
Cuproxoline [13007-93-7]
Curare [8063-06-7]
C-Curarine I [7168-64-1]
C-Curarine III [6866-93-9]
Curcumin [458-37-7]
Curine [436-05-5]
Curium [7440-51-9]
Curvularin [10140-70-2]
Cuscohygrine [454-14-8]
Cuspareine [442-33-1]
Cusparine [529-92-0]
Cyacetacide [140-87-4]
Cyamelide [462-02-2]
Cyamemazine [3546-03-0]
Cyanamide [420-04-2]
Cyanazine [21725-46-2]
Cyanic Acid [420-05-3]
Cyanidin Chloride [528-58-5]
Cyanoacetamide [107-91-5]
Cyanoacetic Acid [372-09-8]
Cyanofenphos [13067-93-1]
Cyanogen [460-19-5]
Cyanogen Azide [764-05-6]
Cyanogen Bromide [506-68-3]
Cyanogen Chloride [506-77-4]
Cyanogen Iodide [506-78-5]
Cyanuric Acid [108-80-5]
Cycasin [14901-08-7]
Cyclacillin [3485-14-1]
Cyclamic Acid [100-88-9]
Cyclandelate [456-59-7]
Cyclarbamate [5779-54-4]
Cyclazocine [3572-80-3]
Cyclethrin [97-11-0]
Cyclexanone [15301-52-7]
Cyclexedrine [532-52-5]
Cyclexedrine Hydrochloride*
 [64011-61-6]
Cyclic AMP [60-92-4]
Cyclic GMP [7665-99-8]
Cyclizine [82-92-8]
Cyclizine Hydrochloride* [303-25-3]
Cyclobarbital [52-31-3]

Cyclobarbital Calcium Salt*
 [143-76-0]
Cyclobendazole [31431-43-3]
Cyclobenzaprine [303-53-7]
Cyclobenzaprine Hydrochloride*
 [6202-23-9]
Cyclobutane [287-23-0]
Cyclobutyrol [512-16-3]
Cyclobutyrol Sodium Salt*
 [1130-23-0]
Cyclobuxine D [2241-90-9]
Cyclocumarol [518-20-7]
Cyclodrine [52109-93-0]
Cyclodrine Hydrochloride*
 [78853-39-1]
Cyclofenil [2624-43-3]
Cycloguanil [516-21-2]
Cycloguanil Pamoate* [609-78-9]
Cycloheptanone [502-42-1]
Cyclohexane [110-82-7]
Cyclohexanecarboxylic Acid [98-89-5]
Cyclohexanol [108-93-0]
Cyclohexanone [108-94-1]
Cyclohexene [110-83-8]
Cycloheximide [66-81-9]
Cyclohexylamine [108-91-8]
Cyclohexyl Bromide [108-85-0]
Cyclohexylcarbinol [100-49-2]
Cyclohexyl Chloride [542-18-7]
2-Cyclohexyl-4,6-dinitrophenol
 [131-89-5]
Cycloleucine [52-52-8]
Cyclomethycaine [139-62-8]
Cyclomethycaine Hydrochloride*
 [537-61-1]
Cyclonite [121-82-4]
Cyclonium Iodide [6577-41-9]
Cyclopentadiene [542-92-7]
Cyclopentamine [102-45-4]
Cyclopentamine Hydrochloride*
 [3459-06-1]
Cyclopentane [287-92-3]
Cyclopentanol [96-41-3]
Cyclopentanone [120-92-3]
1,2-Cyclopentenophenanthrene
 [482-66-6]
Cyclopenthiazide [742-20-1]
Cyclopentobarbital [76-68-6]
Cyclopentobarbital Sodium Salt*
 [302-34-1]
Cyclopentolate [512-15-2]
Cyclopentolate Hydrochloride*
 [5870-29-1]
Cyclophosphamide [50-18-0] (anhy-
 drous form)
Cyclophosphamide [6055-19-2]
Cyclopropane [75-19-4]
Cyclopropyl Methyl Ether [540-47-6]
Cyclorphan [4163-15-9]
Cycloserine [68-41-7]
Cyclosporin A* [59865-13-3]
Cyclosporin B* [63775-95-1]
Cyclosporin C* [59787-61-0]
Cyclosporin D* [63775-96-2]
Cyclosporin G* [74436-00-3]
Cyclothiazide [2259-96-3]
Cyclovalone [579-23-7]
Cycothiamin(e) [6092-18-8]
Cycrimine Hydrochloride [126-02-3]
Cyfluthrin [68359-37-5]
Cyfluthrin (1R,3R,αR)-Form*
 [85649-12-3]
Cyfluthrin (1R,3R,αS)-Form*
 [85649-15-6]
Cyfluthrin (1R,3S,αS)-Form*
 [85649-19-0]
Cyhalothrin [68085-85-8]
Cyheptamide [7199-29-3]
Cyhexatin [13121-70-5]

Denotes derivative of title compound

Cymarin [508-77-0]
Cymarose [579-04-4]
Cymene [25155-15-1]
Cymiazole [61676-87-7]
Cynanchogenin [6870-10-6]
Cynarin(e) [30964-13-7]
Cypermethrin [52315-07-8]
Cyphenothrin [39515-40-7]
Cyprenorphine [4406-22-8]
Cyprenorphine Hydrochloride* [16550-22-4]
Cyproconazole [94361-06-5]
Cyprodinil [121552-61-2]
Cyproheptadine [129-03-3]
Cyproheptadine Hydrochloride Monohydrate* [6032-06-0]
Cyproheptadine Hydrochloride Sesquihydrate* [41354-29-4]
Cyproquinate [19485-08-6]
Cyproterone [2098-66-0]
Cyproterone Acetate* [427-51-0]
Cyromazine [66215-27-8]
Cystamine [51-85-4]
L-Cystathionine [56-88-2]
Cysteamine [60-23-1]
Cysteic Acid [13100-82-8]
Cysteine [52-90-4]
Cysteine Hydrochloride* [52-89-1]
Cystine [56-89-3]
meso-Cystine* [6020-39-9]
L-Cystine S,S-dioxide* [30452-69-8]
L-Cystine S,S'-dioxide* [533-77-7]
Cystine D-Form* [349-46-2]
Cystine DL-Form* [923-32-0]
Cys-Tyr-Cys-interferon γ* [98059-18-8]
Cytarabine [147-94-4]
Cythioate [115-93-5]
Cytidine [65-46-3]
2'-Cytidylic Acid [85-94-9]
3'-Cytidylic Acid [84-52-6]
Cytisine [485-35-8]
Cytochalasin B* [14930-96-2]
Cytochrome c [9007-43-6]
Cytohemin [19554-22-4]
Cytolipin H [4682-48-8]
Cytosine [71-30-7]
2,4-D [94-75-7]
Dacarbazine [4342-03-4]
Dactinomycin [50-76-0]
Daidzein [486-66-8]
Dalapon [75-99-0]
Daltroban [79094-20-5]
Damar [9000-16-2]
Damascenine [483-64-7]
Daminozide [1596-84-5]
Danazol [17230-88-5]
Danofloxacin [112398-08-0]
Danofloxacin Monomethane-sulfonate* [119478-55-6]
Dansyl Chloride [605-65-2]
Danthron [117-10-2]
Dantrolene [7261-97-4]
Dantrolene Sodium Salt Hemi-heptahydrate* [24868-20-0]
Daphnandrine [1183-76-2]
Daphnetin [486-35-1]
Daphnin [486-55-5]
Daphnoline [479-36-7]
Dapiprazole [72822-12-9]
Dapiprazole Monohydrochloride* [72822-13-0]
Dapsone [80-08-0]
Darvan [9003-24-1]
Datiscetin [480-15-9]
Daucol [887-08-1]
Daunorubicin [20830-81-3]
Daunorubicin Hydrochloride* [23541-50-6]
Dauricine [524-17-4]

Dazomet [533-74-4]
2,4-DB [94-82-6]
DBHBT [63147-28-4]
DBMC [497-39-2]
DCPA [1861-32-1]
DDD (Analytical) [6088-51-3]
DDT [50-29-3]
Deaminooxytocin [113-78-0]
Deanol [108-01-0]
Deanol Aceglumate* [3342-61-8]
Deanol Acetamidobenzoate* [3635-74-3]
Deanol Bitartrate* [5988-51-2]
Debrisoquin [1131-64-2]
Debrisoquin Sulfate* [581-88-4]
Decaborane(14) [17702-41-9]
Decalin [91-17-8]
Decamethonium Bromide [541-22-0]
Decamethylcyclopentasiloxane [541-02-6]
Decamethylene Glycol [112-47-0]
Decamethyltetrasiloxane [141-62-8]
Dechlorane Plus [13560-89-9]
Decimemide [14817-09-5]
Decoquinate [18507-89-6]
n-Decyl Alcohol [112-30-1]
Deet [134-62-3]
Defensins [103220-14-0]
Deferoxamine [70-51-9]
Deferoxamine Hydrochloride* [1950-39-6]
Deferoxamine Methanesulfonate* [138-14-7]
Defibrotide [83712-60-1]
Deflazacort [14484-47-0]
Defosfamide [3733-81-1]
Deguelin [522-17-8]
Dehydroacetic Acid [520-45-6]
Dehydroascorbic Acid [490-83-5]
(3β)-7-Dehydrocholesterol [434-16-2]
Dehydrocholic Acid [81-23-2]
Dehydrocholic Acid Sodium Salt* [145-41-5]
11-Dehydrocorticosterone [72-23-1]
Dehydroemetine [4914-30-1]
Dehydroemetine (±)-Form Dihydro-chloride* [14358-43-1]
Dehydroergosterol [516-85-8]
3-Dehydroretinal [472-87-7]
7-Dehydrositosterol [521-04-0]
Delapril [83435-66-9]
Delapril Hydrochloride* [83435-67-0]
Delavirdine [136817-59-9]
Delavirdine Mesylate* [147221-93-0]
Delcosine* [545-56-2]
Delmadinone Acetate [13698-49-2]
Delphinidin [528-53-0]
Delphinine [561-07-9]
Delsoline [509-18-2]
Deltamethrin [52918-63-5]
Demanyl Phosphate [6909-62-2]
Demecarium Bromide [56-94-0]
Demeclocycline [127-33-3]
Demeclocycline Hydrochloride* [64-73-3]
Demecolcine [477-30-5]
Demegestone [10116-22-0]
Demethoxyvindoline* [5231-60-7]
Demeton [8065-48-3]
Demexiptiline [24701-51-7]
Demexiptiline Hydrochloride* [18059-99-9]
Denatonium Benzoate [3734-33-6]
Denopamine [71771-90-9]
Denopterin [22006-84-4]
6-Deoxy-L-ascorbic Acid [528-81-4]
Deoxycholic Acid [83-44-3]
Deoxycorticosterone [64-85-7]
Deoxycorticosterone Acetate [56-47-3]

Deoxydihydrostreptomycin [26086-49-7]
Deoxyepinephrine [501-15-5]
2-Deoxy-D-glucose [154-17-6]
1-Deoxynojirimycin [19130-96-2]
Deoxyribonuclease I [9003-98-9]
Deoxyribonuclease I Recombinant Human Form* [143831-71-4]
D-2-Deoxyribose [533-67-5]
2-Deoxystreptamine [2037-48-1]
Deoxyuridine [951-78-0]
(±)-Deprenyl* [2323-36-6]
Deptropine [604-51-3]
Deptropine Citrate* [2169-75-7]
Dequalinium Chloride [522-51-0]
Dermatan Sulfate* [24967-94-0]
Dermostatin [11120-15-3]
Desacetoxymatricarin* [17946-87-1]
Desaspidin BB [114-43-2]
Desatrine [6746-01-6]
Deserpidine [131-01-1]
Deserpidine Hydrochloride* [6033-69-8]
Deserpidine 10-Methoxy Deriv* [865-04-3]
Deserpidine Oxalate* [119506-01-3]
Desflurane [57041-67-5]
Desipramine [50-47-5]
Desipramine Hydrochloride* [58-28-6]
Desirudin* [120993-53-5]
Deslanoside [17598-65-1]
Deslorelin [57773-65-6]
Desmopressin [16679-58-6]
Desmopressin Monoacetate* [62288-83-9]
Desmosterol [313-04-2]
Desogestrel [54024-22-5]
Desogestrel Mixture with Ethinyl Estradiol* [71138-35-7]
Desomorphine [427-00-9]
Desonide [638-94-8]
Desosamine [5779-39-5]
Desoximetasone [382-67-2]
6-Desoxy-D-glucosamine [6018-53-7]
11-Desoxy-17-hydroxycorticosterone [152-58-9]
Desthiobiotin [533-48-2]
Destomycin A [14918-35-5]
Detaxtran [9015-73-0]
Detaxtran Hydrochloride* [9064-91-9]
Detomidine [76631-46-4]
Detoxin C₁* [74717-53-6]
Detoxin D₁* [37878-19-6]
Deuterium [7782-39-0]
Deuterium Oxide [7789-20-0]
Devarda's Metal [8049-11-4]
Dexamethasone [50-02-2]
Dexamethasone 21-Acetate* [1177-87-3]
Dexamethasone 21-Phosphate Diso-dium Salt* [2392-39-4]
Dexetimide [21888-98-2]
Dexetimide Hydrochloride* [21888-96-0]
Dexfenfluramine* [3239-44-9]
Dexfenfluramine Hydrochloride* [3239-45-0]
Dexmedetomidine* [113775-47-6]
Dexpanthenol [81-13-0]
Dextran [9004-54-0]
Dextranase [9025-70-1]
Dextranomer [56087-11-7]
Dextran Sulfate Sodium [9011-18-1]
Dextri-Maltose [8006-91-5]
Dextrin [9004-53-9]
Dextroamphetamine Sulfate [51-63-8]
Dextromethorphan Hydrobromide* [125-69-9]

Denotes derivative of title compound

Dextromoramide [357-56-2]
Dezocine [53648-55-8]
DFDD [603-55-4]
DFDT [475-26-3]
Dhurrin [499-20-7]
Diaboline [509-40-0]
Diacerein [13739-02-1]
Diacetazotol [83-63-6]
Diacetin [25395-31-7]
Diacetonamine [625-04-7]
Diacetone Acrylamide [2873-97-4]
Diacetone Alcohol [123-42-2]
Diacetoneglucose [582-52-5]
Diacetyl [431-03-8]
Diacetyldihydromorphine [509-71-7]
Diacetylmorphine [561-27-3]
Dialifor [10311-84-9]
Diallate [2303-16-4]
Diallylamine [124-02-7]
Diallylcyanamide [538-08-9]
Diamfenetide [36141-82-9]
2,4-Diaminoanisole [615-05-4]
2,4-Diaminoanisole Sulfate*
  [39156-41-7]
p-Diaminoazobenzene [538-41-0]
3,5-Diaminobenzoic Acid [535-87-5]
2,6-Diamino-2′-butyloxy-3,5′-azopyri-
  dine [617-19-6]
4,4′-Diaminodiphenylamine [537-65-5]
p,p′-Diaminodiphenylmethane
  [101-77-9]
2,4-Diamino-6-hydroxypyrimidine
  [56-06-4]
2,3-Diaminophenazine [655-86-7]
2,4-Diaminophenol [95-86-3]
2,3-Diaminopropionic Acid [515-94-6]
2,6-Diaminopurine [1904-98-9]
3,4-Diaminopyridine [54-96-6]
Diamond [7782-40-3]
Diamond Ink [8006-35-7]
Diampromide [552-25-0]
Diamthazole Dihydrochloride
  [136-96-9]
Diamyl Sodium Sulfosuccinate
  [922-80-5]
1,2-Dianilinoethane [150-61-8]
Dianisidine [119-90-4]
Diathymosulfone [5964-62-5]
Diatretyne I [53318-35-7]
Diatretyne II [3625-68-1]
Diatrizoate Sodium [737-31-5]
Diaveridine [5355-16-8]
Diazepam [439-14-5]
Diazinon [333-41-5]
Diaziquone [57998-68-2]
Diazoacetic Ester [623-73-4]
Diazoaminobenzene [136-35-6]
p-Diazobenzenesulfonic Acid
  [305-80-6]
Diazomethane [334-88-3]
6-Diazo-5-oxo-L-norleucine [157-03-9]
5-Diazouracil [2435-76-9]
Diazoxide [364-98-7]
Dibekacin [34493-98-6]
Dibekacin Sulfate* [58580-55-5]
Dibenzalacetone [538-58-9]
1,2:5,6-Dibenzanthracene [53-70-3]
Dibenzepin [4498-32-2]
Dibenzepin Hydrochloride*
  [315-80-0]
Dibenzoylmethane [120-46-7]
2,3:6,7-Dibenzphenanthrene
  [222-93-5]
Dibenzylamine [103-49-1]
Dibenzyl Chlorophosphonate
  [538-37-4]
Dibenzyl Disulfide [150-60-7]
Dibenzyl Phosphite [17176-77-1]
Diborane(6) [19287-45-7]
Diboron Tetrachloride [13701-67-2]

Diboron Tetrahydroxide [13675-18-8]
9,10-Dibromoanthracene [523-27-3]
p-Dibromobenzene [106-37-6]
α,α′-Dibromo-d-camphor [514-12-5]
Dibromochloropropane [96-12-8]
1,2-Dibromo-2,4-dicyanobutane
  [35691-65-7]
4′,5′-Dibromofluorescein [596-03-2]
Dibromogallic Acid [602-92-6]
3,5-Dibromo-4-hydroxybenzenesul-
  fonic Acid [4232-99-9]
Dibromopropamidine [496-00-4]
Dibromopropamidine Isethionate*
  [614-87-9]
2,3-Dibromopropene [513-31-5]
2,6-Dibromoquinone-4-chlorimide
  [537-45-1]
3,5-Dibromosalicylaldehyde [90-59-5]
3,5-Dibromosalicylic Acid [3147-55-5]
2,3-Dibromosuccinic Acid [526-78-3]
3,5-Dibromo-L-tyrosine [300-38-9]
Dibromsalicil [523-88-6]
Dibucaine Hydrochloride [61-12-1]
Dibutoline Sulfate [532-49-0]
n-Dibutylamine [111-92-2]
Di-tert-butyl Ether [6163-66-2]
Di-tert-butyl Malonate [541-16-2]
2,6-Di-tert-butylpyridine [585-48-8]
Dibutyl Succinate [141-03-7]
Di-tert-butyl Succinate [926-26-1]
Dibutyltin Dilaurate [77-58-7]
Dicamba [1918-00-9]
Dicapthon [2463-84-5]
Dicentrine [517-66-8]
Dichlobenil [1194-65-6]
Dichlofenthion [97-17-6]
Dichlofluanid [1085-98-9]
Dichlone [117-80-6]
Dichloralphenazone [480-30-8]
Dichloramine T [473-34-7]
Dichlorisone [7008-26-6]
Dichlorisone 21-Acetate* [79-61-8]
Dichloroacetic Acid [79-43-6]
1,1-Dichloroacetone [513-88-2]
1,3-Dichloroacetone [534-07-6]
2,2-Dichloroacetyl Chloride [79-36-7]
3,4-Dichloroaniline [95-76-1]
m-Dichlorobenzene [541-73-1]
o-Dichlorobenzene [95-50-1]
p-Dichlorobenzene [106-46-7]
2,2′-Dichlorobenzidine [84-68-4]
3,3′-Dichlorobenzidine [91-94-1]
3,3′-Dichlorobenzidine Dihydrochloride* [612-83-9]
Dichlorobenzyl Alcohol [1777-82-8]
1,1-Dichloro-2,2-bis(p-chlorophenyl)-
  ethane [72-54-8]
Dichloro(2-chlorovinyl)arsine
  [541-25-3]
2,3-Dichloro-5,6-dicyanobenzoquinone
  [84-58-2]
Dichlorodifluoromethane [75-71-8]
1,3-Dichloro-5,5-dimethylhydantoin
  [118-52-5]
sym-Dichloroethyl Ether [111-44-4]
4′,5′-Dichlorofluorescein [2320-96-9]
2,6-Dichloroindophenol Sodium
  [620-45-1]
sym-Dichloromethyl Ether [542-88-1]
Dichlorophen [97-23-4]
Dichlorophenarsine Hydrochloride
  [536-29-8]
2,4-Dichlorophenol [120-83-2]
2,6-Dichlorophenol [87-65-0]
1,3-Dichloro-2-propanol [96-23-1]
1,3-Dichloropropene [542-75-6]
Dichloroxylenol [133-53-9]
Dichlorphenamide [120-97-8]
Dichlorprop [120-36-5]
Dichlorvos [62-73-7]

Diclazuril [101831-37-2]
Diclobutrazol [75736-33-3]
Diclofenac Sodium [15307-79-6]
Diclofop-Methyl [51338-27-3]
Dicloxacillin [3116-76-5]
Dicloxacillin Sodium Salt Mono-
  hydrate* [13412-64-1]
Dicobalt Octacarbonyl [10210-68-1]
Dicofol [115-32-2]
Dicrotophos [141-66-2]
Dicryl [2164-09-2]
Dictamnine [484-29-7]
Dicumarol [66-76-2]
Dicyanine [52260-69-2]
Dicyanodiamide [461-58-5]
Dicyanodiamidine Sulfate [591-01-5]
9-Dicyanomethylene-2,4,7-trinitro-
  fluorene [1172-02-7]
Dicyclohexylamine [101-83-7]
Dicyclohexylcarbodiimide [538-75-0]
Dicyclomine Hydrochloride [67-92-5]
Didanosine [69655-05-6]
Didecyldimethylammonium Chloride
  [7173-51-5]
Didemnin A* [77327-04-9]
Didemnin B* [77327-05-0]
Didemnin C* [77327-06-1]
Dideoxyadenosine [4097-22-7]
Dieldrin [60-57-1]
Dienestrol [84-17-3]
Dienestrol Diacetate* [84-19-5]
Dienochlor [2227-17-0]
Diethadione [702-54-5]
Diethanolamine [111-42-2]
Diethazine [60-91-3]
Diethazine Hydrochloride* [341-70-8]
Diethofencarb [87130-20-9]
Diethylacetic Acid [88-09-5]
Diethylamine [109-89-7]
2-Diethylaminoethanol [100-37-8]
Diethylaniline [91-66-7]
N,N-Diethylbenzhydrylamine
  [519-72-2]
Diethylbromoacetamide [511-70-6]
Diethylcarbamazine [90-89-1]
Diethylcarbamazine Citrate (Hydro-
  gen Citrate)* [1642-54-2]
N,N′-Diethylcarbanilide [85-98-3]
Diethyl Carbitol [112-36-7]
Diethylene Glycol [111-46-6]
Diethylene Glycol Monolaurate
  [141-20-8]
Diethyl Ketone [96-22-0]
Diethylmagnesium [557-18-6]
Diethyl Maleate [141-05-9]
Diethylmalonic Acid [510-20-3]
Diethyl Oxalate [95-92-1]
Diethylpropion [90-84-6]
Diethylpropion Hydrochloride*
  [134-80-5]
Diethylsilane [542-91-6]
Diethylstilbestrol [56-53-1]
Diethylstilbestrol Dipropionate*
  [130-80-3]
Diethyl Sulfate [64-67-5]
Diethylzinc [557-20-0]
Difemerine [80387-96-8]
Difemerine Hydrochloride*
  [70280-88-5]
Difenamizole [20170-20-1]
Difenoconazole [119446-68-3]
Difenoxin [28782-42-5]
Difenpiramide [51484-40-3]
Difenzoquat [49866-87-7]
Difenzoquat Methyl Sulfate*
  [43222-48-6]
Diflorasone [2557-49-5]
Diflorasone Diacetate* [33564-31-7]
Difloxacin [98106-17-3]

Denotes derivative of title compound

Difloxacin Monohydrochloride*
[91296-86-5]
Diflubenzuron [35367-38-5]
Diflucortolone [2607-06-9]
Diflucortolone 21-Valerate*
[59198-70-8]
Diflunisal [22494-42-4]
2,4-Difluoroaniline [367-25-9]
p-Difluorobenzene [540-36-3]
4,4'-Difluorodiphenyl [398-23-2]
Difluprednate [23674-86-4]
Digallic Acid [536-08-3]
Digalogenin [6877-35-6]
Diginatigenin [559-57-9]
Diginatin [52589-12-5]
Diginin [467-53-8]
Digitalin [752-61-4]
Digitalose [4481-08-7]
Digitogenin [511-34-2]
Digitonin [11024-24-1]
Digitoxigenin [143-62-4]
Digitoxin [71-63-6]
Digitoxose [527-52-6]
Diglyme [111-96-6]
Digoxigenin [1672-46-4]
Digoxin [20830-75-5]
Dihexyverine [561-77-3]
Dihexyverine Hydrochloride*
[5588-25-0]
Dihydralazine [484-23-1]
Dihydralazine Sulfate* [7327-87-9]
Dihydrazine Sulfate [13464-80-7]
Dihydrocodeine [125-28-0]
Dihydrocodeinone Enol Acetate
[466-90-0]
Dihydrocodeinone Enol Acetate
Hydrochloride* [20236-82-2]
Dihydroequilin [3563-27-7]
Dihydroequilin 17α-Form* [651-55-8]
Dihydroergocristine Methanesul-
fonate* [24730-10-7]
Dihydroergotamine [511-12-6]
Dihydroergotamine Methanesul-
fonate* [6190-39-2]
Dihydro-β-erythroidine [23255-54-1]
Dihydroisocodeine [795-38-0]
Dihydrokawain* [587-63-3]
Dihydromorphine [509-60-4]
Dihydroresorcinol [504-02-9]
Dihydrostreptomycin [128-46-1]
Dihydrostreptomycin Pantothenate*
[3563-84-6]
Dihydrostreptomycin Sesquisulfate*
[5490-27-7]
Dihydrostreptomycin Trihydrochlo-
ride* [6533-54-6]
Dihydrotachysterol [67-96-9]
Dihydrothebaine [561-25-1]
Dihydroxyacetone [96-26-4]
Dihydroxyaluminum Acetylsalicylate
[53230-06-1]
Dihydroxyaluminum Aminoacetate
[13682-92-3]
Dihydroxyaluminum Sodium Car-
bonate [539-68-4]
Dihydroxymaleic Acid [526-84-1]
9,10-Dihydroxystearic Acid [120-87-6]
Dihydroxytartaric Acid [76-30-2]
2,4-Diiodoaniline [533-70-0]
4',5'-Diiodofluorescein [38577-97-8]
3,5-Diiodosalicylic Acid [133-91-5]
3,5-Diiodothyronine [534-51-0]
3,5-Diiodotyrosine [66-02-4]
Diisoamylamine [544-00-3]
Diisobutyl Sodium Sulfosuccinate
[127-39-9]
Diisopromine [5966-41-6]
Diisopromine Hydrochloride*
[24358-65-4]
Diisopropylamine [108-18-9]

Diisopropylamine Dichloroacetate
[660-27-5]
Diisopropyl Paraoxon [3254-66-8]
Dikegulac [18467-77-1]
Dilazep [35898-87-4]
Dilazep Dihydrochloride*
[20153-98-4]
Dilevalol [75659-07-3]
Dilevalol Hydrochloride*
[75659-08-4]
Diloxanide [579-38-4]
Diloxanide Furoate* [3736-81-0]
Diltiazem [42399-41-7]
Diltiazem d-cis Form Hydrochloride*
[33286-22-5]
Dimecrotic Acid [7706-67-4]
Dimecrotic Acid Magnesium Salt*
[54283-65-7]
Dimefline [1165-48-6]
Dimefline Hydrochloride* [2740-04-7]
Dimefox [115-26-4]
Dimemorfan [36309-01-0]
Dimemorfan Phosphate* [36304-84-4]
Dimenhydrinate [523-87-5]
Dimenoxadol [509-78-4]
Dimepheptanol [545-90-4]
Dimercaprol [59-52-9]
2,3-Dimercapto-1-propanesulfonic
Acid [74-61-3]
2,3-Dimercapto-1-propanesulfonic
Acid Sodium Salt* [4076-02-2]
Dimestrol [130-79-0]
Dimetacrine [4757-55-5]
Dimetacrine Tartrate* [3759-07-7]
Dimetan [122-15-6]
Dimethadione [695-53-4]
Dimethazan [519-30-2]
Dimethazone [81777-89-1]
Dimethenamid [87674-68-8]
Dimethicone [9006-65-9]
Dimethindene [5636-83-9]
Dimethindene Maleate* [3614-69-5]
Dimethirimol [5221-53-4]
Dimethisoquin [86-80-6]
Dimethisoquin Hydrochloride*
[2773-92-4]
Dimethisterone [79-64-1]
Dimethisterone Mixture with Ethinyl
Estradiol* [8015-19-8]
Dimethoate [60-51-5]
Dimethocaine [94-15-5]
Dimethomorph [110488-70-5]
Dimethoxanate [477-93-0]
Dimethoxane [828-00-2]
1,2-Dimethoxyethane [110-71-4]
2,6-Dimethoxyquinone [530-55-2]
Dimethylacetal [534-15-6]
N,N-Dimethylacetamide [127-19-5]
Dimethylamine [124-40-3]
p-Dimethylaminoazobenzene [60-11-7]
p-Dimethylaminobenzaldehyde
[100-10-7]
p-Dimethylaminobenzalrhodanine
[536-17-4]
4-(Dimethylamino)benzoic Acid
[619-84-1]
4-(Dimethylamino)benzoic Acid 2-
Ethylhexyl Ester* [21245-02-3]
4-(Dimethylamino)benzoic Acid 3-
Methylbutyl Ester* [21245-01-2]
p-Dimethylaminobenzophenone
[530-44-9]
N,N-Dimethylaniline [121-69-7]
9,10-Dimethyl-1,2-benzanthracene
[57-97-6]
5,6-Dimethylbenzimidazole [582-60-5]
p,α-Dimethylbenzyl Alcohol
[536-50-5]
2,3-Dimethyl-1,3-butadiene [513-81-5]
Dimethylcadmium [506-82-1]

Dimethyl Carbate [39589-98-5]
N,N'-Dimethylcyclobuxine D*
[4282-16-0]
5,5-Dimethyl-1,3-cyclohexanedione
[126-81-8]
N,N-Dimethylformamide [68-12-2]
N,N-Dimethylglycine [1118-68-9]
N,N-Dimethylglycine Hydrazide Hy-
drochloride [539-64-0]
Dimethylglyoxime [95-45-4]
1,1-Dimethylhydrazine [57-14-7]
1,2-Dimethylhydrazine [540-73-8]
1,3-Dimethyl-2-imidazolidinone
[80-73-9]
Dimethylmercury [593-74-8]
Dimethyl(methylene)ammonium
Iodide [33797-51-2]
N,N-Dimethyl-1-naphthylamine
[86-56-6]
Dimethylolpropionic Acid [4767-03-7]
Dimethyl-p-phenylenediamine
[99-98-9]
Dimethyl Phthalate [131-11-3]
Dimethyl Sulfate [77-78-1]
2,4-Dimethylsulfolane [1003-78-7]
Dimethyl Sulfone [67-71-0]
Dimethyl Sulfoxide [67-68-5]
Dimethylthiambutene [524-84-5]
N,N'-Dimethylthiourea [534-13-4]
N,N-Dimethyltryptamine [61-50-7]
Dimethylzinc [544-97-8]
Dimetilan [644-64-4]
Dimetofrine [22950-29-4]
Dimetofrine Hydrochloride*
[22775-12-8]
Dimetridazole [551-92-8]
Diminazene Aceturate [908-54-3]
Dimorpholamine [119-48-2]
Dimoxyline [147-27-3]
β,β'-Dinaphthylamine [532-18-3]
Diniconazole [83657-24-3]
Diniconazole (—)-Form* [83657-18-5]
Diniconazole (+)-Form* [83657-19-6]
Dinitolmide [148-01-6]
2,4-Dinitroaniline [97-02-9]
2,6-Dinitroaniline [606-22-4]
2,4-Dinitrobenzaldehyde [528-75-6]
2,4-Dinitrobenzenesulfenyl Chloride
[528-76-7]
3,4-Dinitrobenzoic Acid [528-45-0]
3,5-Dinitrobenzoic Acid [99-34-3]
3,5-Dinitrobenzoyl Chloride [99-33-2]
4,4'-Dinitrocarbanilide [587-90-6]
Dinitrocresol [534-52-1]
3,7-Dinitro-5-oxophenothiazine
[574-81-2]
2,4-Dinitrophenol [51-28-5]
2,5-Dinitrophenol [329-71-5]
2,6-Dinitrophenol [573-56-8]
2,4-Dinitrophenylhydrazine [119-26-6]
2,4-Dinitroresorcinol [519-44-8]
5,5'-Dinitrosalicil [528-10-9]
Dinobuton [973-21-7]
Dinocap [39300-45-3]
Dinoseb [88-85-7]
Dinosterol [58670-63-6]
Dinsed [96-62-8]
Diopterin [6807-82-5]
Dioscin [19057-60-4]
Dioscorine [3329-91-7]
Diosgenin [512-04-9]
Diosmetin [520-34-3]
Diosmin [520-27-4]
Diosphenol [490-03-9]
Dioxadrol [6495-46-1]
Dioxadrol d-Form Hydrochloride*
[1162-15-8]
Dioxadrol Hydrochloride*
[3666-69-1]
Dioxane [123-91-1]

Dioxaphetyl Butyrate [467-86-7]
Dioxathion [78-34-2]
Dioxethedrine [497-75-6]
Dioxybenzone [131-53-3]
Dioxypyramidon [519-65-3]
2,5-Di-*tert*-pentylhydroquinone [79-74-3]
Diperodon Hydrochloride [537-12-2]
Diphemanil Methylsulfate [62-97-5]
Diphemethoxidine [13862-07-2]
Diphenadione [82-66-6]
Diphenamid [957-51-7]
Diphenane [101-71-3]
Diphencyprone [886-38-4]
Diphenhydramine [58-73-1]
Diphenhydramine Compd with *p*-Sulfonamide Benzoate* [60539-06-2]
Diphenhydramine Hydrochloride* [147-24-0]
Diphenic Acid [482-05-3]
Diphenidol [972-02-1]
Diphenidol Hydrochloride* [3254-89-5]
Diphenolic Acid [126-00-1]
Diphenoxylate [915-30-0]
Diphenoxylate Hydrochloride* [3810-80-8]
Diphenyl [92-52-4]
Diphenylacetamide [519-87-9]
Diphenylacetic Acid [117-34-0]
Diphenylamine [122-39-4]
Diphenylamine-2,2'-dicarboxylic Acid [579-92-0]
N,N'-Diphenylbenzidine [531-91-9]
1,4-Diphenyl-1,3-butadiene [886-65-7]
sym-Diphenylcarbazide [140-22-7]
Diphenylcarbazone [538-62-5]
1,1-Diphenylethene [530-48-3]
N-(1,2-Diphenylethyl)nicotinamide [553-06-0]
1,3-Diphenylguanidine [102-06-7]
1,1-Diphenylhydrazine [530-50-7]
Diphenylketene [525-06-4]
Diphenylmagnesium [555-54-4]
Diphenylmethane [101-81-5]
N,N'-Diphenyl-*p*-phenylenediamine [74-31-7]
1,1-Diphenyl-2-picrylhydrazyl (Free Radical) [1898-66-4]
Diphenylpyraline [147-20-6]
Diphenylpyraline 8-Chloro-theophyllinate* [606-90-6]
Diphenylpyraline Hydrobromide* [5807-87-4]
Diphenylpyraline Hydrochloride* [132-18-3]
sym-Diphenylpyrophosphorodiamidic Acid [541-04-8]
Diphenyl Sulfone [127-63-9]
sym-Diphenylthiourea [102-08-9]
Diphetarsone [515-76-4]
Diphosgene [503-38-8]
Dipicrylamine [131-73-7]
Dipin [738-99-8]
Dipipanone [467-83-4]
Dipipanone Hydrochloride* [856-87-1]
Dipiproverine [117-30-6]
Dipiproverine Dihydrochloride* [2404-18-4]
Dipivefrin [52365-63-6]
Dipivefrin Hydrochloride* [64019-93-8]
Diploicin [527-93-5]
Diponium Bromide [2001-81-2]
Diprenorphine [14357-78-9]
Dipropalin [1918-08-7]
Dipropetryn [4147-51-7]
n-Dipropylamine [142-84-7]

Dipropylene Glycol Monomethyl Ether [34590-94-8]
Dipropyl Ketone [123-19-3]
5,5-Dipropyl-2,4-oxazolidinedione [512-12-9]
Dipyridamole [58-32-2]
α,α'-Dipyridyl [366-18-7]
γ,γ'-Dipyridyl [553-26-4]
Dipyrocetyl [486-79-3]
Dipyrone [5907-38-0]
Dipyrone Magnesium Salt* [6150-97-6]
Diquat Dibromide [85-00-7]
1,3-Di-6-quinolylurea [532-05-8]
Diresorcinol [531-02-2]
Dirithromycin [62013-04-1]
Disilane [1590-87-0]
Disodium Dihydrogen Hypophosphate [7782-95-8]
Disodium Phenyl Phosphate [3279-54-7]
Disofenin [65717-97-7]
Disophenol [305-85-1]
Disopyramide [3737-09-5]
Disopyramide Phosphate* [22059-60-5]
Disparlure [29804-22-6] (2S-cis)
Disparlure [54910-51-9] (2S-cis)
Distigmine Bromide [15876-67-2]
Disulfamide [671-88-5]
Disulfiram [97-77-8]
Disulfoton [298-04-4]
Disul-sodium [136-78-7]
Ditazol [18471-20-0]
Dithianone [3347-22-6]
Dithiazanine Iodide [514-73-8]
2,2'-Dithiobis[benzothiazole] [120-78-5]
2,4-Dithiobiuret [541-53-7]
4,4'-Dithiodimorpholine [103-34-4]
3,3-Dithiodipyridine Dihydrochloride [538-45-4]
Dithiopyr [97886-45-8]
Dithiosalicylic Acid [527-89-9]
1,4-Dithiothreitol [3483-12-3]
Dithizone [60-10-6]
Ditiocarb Sodium [148-18-5]
1,1-Di-*p*-tolylethane [530-45-0]
1,2-Di-*p*-tolylethane [538-39-6]
p-Ditolylmercury [537-64-4]
Diuron [330-54-1]
Divicine [32267-39-3]
Dixanthogen [502-55-6]
Dixyrazine [2470-73-7]
Dizocilpine [77086-21-6]
Dizocilpine (−)-Form* [77086-19-2]
Dizocilpine Maleate* [77086-22-7]
Djenkolic Acid [498-59-9]
DMPA [299-85-4]
DMPO [3317-61-1]
Dobesilate Calcium [20123-80-2]
Dobutamine [34368-04-2]
Dobutamine Hydrochloride* [49745-95-1]
Docarpamine [74639-40-0]
Docetaxel [114977-28-5]
1-Docosanol* [661-19-8]
Docusate Calcium [128-49-4]
Docusate Potassium* [7491-09-0]
Docusate Sodium [577-11-7]
Dodecahedrane [4493-23-6]
Dodecamethylcyclohexasiloxane [540-97-6]
Dodecamethylpentasiloxane [141-63-9]
1-Dodecanol [112-53-8]
[S-(all-E)]-3-(1,3,5,7,9-Dodecapentaenyloxy)-1,2-propanediol [91423-46-0]
Dodecarbonium Chloride [100-95-8]

Dodemorph [1593-77-7]
Dodine [2439-10-3]
Dofetilide [115256-11-6]
Doisynolic Acid [109784-48-7]
Dolasetron [115956-12-2]
Dolasetron Methanesulfonate* [115956-13-3]
Domesticine [476-71-1]
Domiodol [61869-07-6]
Domiphen Bromide [538-71-6]
Domoic Acid [14277-97-5]
Domperidone [57808-66-9]
Donovan's Solution [8012-54-2]
Dopa D-Form* [5796-17-8]
Dopa DL-Form* [63-84-3]
Dopamine [51-61-6]
Dopamine Hydrochloride* [62-31-7]
Dopan [520-09-2]
Dopastin [37134-80-8]
Dopexamine [86197-47-9]
Dopexamine Dihydrochloride* [86484-91-5]
Doramectin [117704-25-3]
Dorzolamide [120279-96-1]
Dorzolamide Hydrochloride* [130693-82-2]
Dothiepin [113-53-1]
Dothiepin Hydrochloride* [897-15-4]
Dowicide 9 [13347-42-7]
Doxacurium Chloride [106819-53-8]
Doxapram [309-29-5]
Doxapram Hydrochloride Monohydrate* [7081-53-0]
Doxazosin [74191-85-8]
Doxazosin Monomethanesulfonate* [77883-43-3]
Doxefazepam [40762-15-0]
Doxenitoin [3254-93-1]
Doxepin [1668-19-5]
Doxepin *cis*-Form Hydrochloride* [25127-31-5]
Doxepin *trans*-Form Hydrochloride* [3607-18-9]
Doxepin Hydrochloride* [1229-29-4]
Doxifluridine [3094-09-5]
Doxofylline [69975-86-6]
Doxorubicin [23214-92-8]
Doxorubicin Hydrochloride* [25316-40-9]
Doxycycline [564-25-0] (anhydrous)
Doxycycline [17086-28-1] (monohydrate)
Doxycycline Hyclate* [24390-14-5]
Doxylamine [469-21-6]
Doxylamine Succinate* [562-10-7]
Drazoxolon [5707-69-7]
Drimenin [2326-89-8]
Drocarbil* [900-77-6]
Drofenine [1679-76-1]
Drofenine Hydrochloride* [548-66-3]
Droloxifene [82413-20-5]
Droloxifene Citrate* [97752-20-0]
Drometrizole [2440-22-4]
Dromostanolone [58-19-5]
Dromostanolone Propionate* [521-12-0]
Droperidol [548-73-2]
Droprenilamine [57653-27-7]
Droprenilamine Hydrochloride* [59182-63-7]
Dropropizine [17692-31-8]
Dropropizine S-Form* [99291-24-4]
Drosophilin A [484-67-3]
Drospirenone [67392-87-4]
Drotebanol [3176-03-2]
Droxicam [90101-16-9]
Droxidopa [23651-95-8]
DSIP [62568-57-4]
DTBP [110-05-4]
Dulcin [150-69-6]

Denotes derivative of title compound

Duloxetine [116539-59-4]
Duloxetine Hydrochloride* [136434-34-9]
Durapatite [1306-06-5]
Durene [95-93-2]
Durohydroquinone [527-18-4]
Duroquinone [527-17-3]
Dyclonine [586-60-7]
Dyclonine Hydrochloride* [536-43-6]
Dydrogesterone [152-62-5]
Dymanthine [124-28-7]
Dymanthine Hydrochloride* [1613-17-8]
Dymixal [8013-96-5]
Dynorphin [74913-18-1]
Dyphylline [479-18-5]
Dypnone [495-45-4]
Dysprosium [7429-91-6]
E-2020 [120011-70-3]
Ebastine [90729-43-4]
Ebimar [9013-42-7]
Ebrotidine [100981-43-9]
Eburnamonine [474-00-0]
Eburnamonine (±)-Form* [2580-88-3]
Eburnamonine (—)-Form* [4880-88-0]
Eburnamonine (—)-Form Phosphate* [94134-60-8]
Ecabet [33159-27-2]
Ecabet Sodium Salt* [86408-72-2]
Ecadotril* [112573-73-6]
Ecdysone* [3604-87-3]
Ecgonidine [484-93-5]
Ecgonine [481-37-8]
Ecgonine l-Form Monohydrate* [5796-30-5]
Ecgonine Hydrochloride* [5796-31-6]
Echinenone [432-68-8]
Echinochrome A [517-82-8]
Echinomycin [512-64-1]
Echinopsine [83-54-5]
Echinuline [1859-87-6]
Echitamine [6871-44-9]
Echothiophate Iodide [513-10-0]
Econazole [27220-47-9]
Econazole Nitrate* [24169-02-6] (unspecified)
Econazole Nitrate* [68797-31-9] (±)
ECTEOLA-Cellulose [9015-13-8]
Ectylurea [95-04-5]
Edatrexate [80576-83-6]
Edestin [9007-57-2]
Edetate Calcium Disodium [62-33-9]
Edetate Disodium [139-33-3]
Edetate Sodium [64-02-8]
Edetate Trisodium [150-38-9]
Edetic Acid [60-00-4]
Edetic Acid Dipotassium Salt Dihydrate* [25102-12-9]
Edifenphos [17109-49-8]
Edoxudine [15176-29-1]
Edrophonium Chloride [116-38-1]
EEDQ [16357-59-8]
Eflornithine [67037-37-0]
Eflornithine Hydrochloride Monohydrate* [96020-91-6]
Efloxate [119-41-5]
Efonidipine [111011-63-3]
Efonidipine Hydrochloride* [111011-53-1]
Efonidipine Hydrochloride Ethanol* [111011-76-8]
Efrotomycin [56592-32-6]
EGCG [989-51-5]
Eicosamethylnonasiloxane [2652-13-3]
(all-Z)-5,8,11,14,17-Eicosapentaenoic Acid [10417-94-4]
(all-Z)-5,8,11,14,17-Eicosapentaenoic Acid Ethyl Ester* [86227-47-6]
Einsteinium [7429-92-7]
Elaidic Acid [112-79-8]

Elaiomycin [23315-05-1]
Elastase [9004-06-2]
Elcatonin [60731-46-6]
Eledoisin [69-25-0]
Eledoisin Trifluoroacetate* [10129-92-7]
Element 104 [53850-36-5]
Element 105 [53850-35-4]
Element 106 [54038-81-2]
Element 107 [54037-14-8]
Element 108 [54037-57-9]
Element 109 [54038-01-6]
Elenolide [24582-91-0]
Elgodipine [119413-55-7] (±)
Elgodipine [121488-95-7] (unspecified)
Elgodipine Hydrochloride* [121489-04-1]
Ellagic Acid [476-66-4]
Ellipticine [519-23-3]
Ellipticinium Acetate [58337-35-2]
Elliptone [478-10-4]
Eltoprazine [98224-03-4]
Eltoprazine Hydrochloride* [98206-09-8]
Elymoclavine [548-43-6]
Emamectin [121124-29-6] (major component)
Emamectin [121424-52-0] (minor component)
Emamectin [137335-79-6]
Emamectin Benzoate* [137512-74-4]
Embelin [550-24-3]
Embramine [3565-72-8]
Embramine Hydrochloride* [13977-28-1]
Emedastine [87233-61-2]
Emedastine Difumarate* [87233-62-3]
Emepronium Bromide [3614-30-0]
Emetamine [483-19-2]
Emetine [483-18-1]
Emetine Dihydrochloride* [316-42-7]
Emitefur [110690-43-2]
Emodin [518-82-1]
Emorfazone [38957-41-4]
Emylcamate [78-28-4]
Enalapril [75847-73-3]
Enalaprilat [76420-72-9]
Enalapril Maleate* [76095-16-4]
Enallylpropymal [1861-21-8]
Enallylpropymal Sodium Salt* [59005-68-4]
Enanthotoxin [20311-78-8]
Encainide [66778-36-7]
Encainide Hydrochloride* [66794-74-9]
Enciprazine [68576-86-3]
Enciprazine Dihydrochloride* [68576-88-5]
Endiandric Acid A* [74591-03-0]
Endiandric Acid B* [76060-33-8]
Endiandric Acid C* [76060-34-9]
Endobenzyline Bromide [52080-56-5]
Endosulfan [115-29-7]
Endothall [145-73-3]
Endothelin [117399-94-7]
Endralazine [39715-02-1]
Endralazine Methanesulfonate* [65322-72-7]
Endrin [72-20-8]
Enduracidin [11115-82-5]
Enduracidin A Hydrochloride* [33368-20-6]
Enduracidin B Hydrochloride* [34765-98-5]
Enfenamic Acid [23049-93-6]
Enflurane [13838-16-9]
Enilconazole [35554-44-0]
Enocitabine [55726-47-1]
Enoxacin [74011-58-8]

Enoxacin Sesquihydrate* [84294-96-2]
Enoximone [77671-31-9]
Enoxolone [471-53-4]
Enprostil [73121-56-9]
Enrofloxacin [93106-60-6]
Enterobactin [28384-96-5]
Enterogastrone [9007-67-4]
Enterolactone [78473-71-9]
Enteromycin [3552-16-7]
Entprol [102-60-3]
Enviomycin [33103-22-9]
Enviroxime [72301-79-2]
Eosine I Bluish [548-24-3]
Eosine Yellowish—(YS) [17372-87-1]
Epalrestat [82159-09-9]
Epanolol [86880-51-5]
Eperisone [64840-90-0]
Eperisone Hydrochloride* [56839-43-1]
Ephedrine dl-Form* [90-81-3]
Ephedrine l-Form* [299-42-3]
Ephedrine dl-Form Hydrochloride* [134-71-4]
Ephedrine l-Form Hydrochloride* [50-98-6]
Ephedrine l-Form Sulfate* [134-72-5]
Epiandrosterone [481-29-8]
Epibatidine [140111-52-0]
Epichlorohydrin [106-89-8]
Epicholestanol [516-95-0]
Epicholesterol [474-77-1]
Epicillin [26774-90-3]
Epicinchonidine* [550-54-9]
Epicinchonine* [485-70-1]
16-Epiestriol [547-81-9]
Epimestrol [7004-98-0]
Epinastine [80012-43-7]
Epinastine Hydrobromide* [127786-29-2]
Epinastine Hydrochloride* [108929-04-0]
Epinephrine [51-43-4]
Epinephrine Bitartrate* [51-42-3]
Epinephrine dl-Form* [329-65-7]
Epinephrine dl-Form Hydrochloride* [329-63-5]
Epinephrine Hydrochloride* [55-31-2]
Epinephryl Borate* [5579-16-8]
Epiquinidine [572-59-8]
Epiquinine [572-60-1]
Epirizole [18694-40-1]
Epirubicin [56420-45-2]
Epirubicin Hydrochloride* [56390-09-1]
Epithiazide [1764-85-8]
Epitiostanol [2363-58-8]
EPN [2104-64-5]
Epoetin Alfa* [113427-24-0]
Epoetin Beta* [122312-54-3]
Epostane [80471-63-2]
Epoxiconazole [106325-08-0] (undefined)
Epoxiconazole [133855-98-8]
Eprazinone [10402-90-1]
Eprazinone Dihydrochloride* [10402-53-6]
Eprinomectin [123997-26-2]
Eprinomectin Component B$_{1a}$* [133305-88-1]
Eprinomectin Component B$_{1b}$* [133305-89-2]
Epristeride [119169-78-7]
Eprosartan [133040-01-4]
Eprosartan Mesylate* [144143-96-4]
Eprozinol [32665-36-4]
Eprozinol Dihydrochloride* [27588-43-8]
Epsiprantel [98123-83-2]

Eptastigmine [101246-68-8]
Eptastigmine Tartrate* [121652-76-4]
Eptazocine [72522-13-5]
EPTC [759-94-4]
Equilenin [517-09-9]
Equilin [474-86-2]
Equine Chorionic Gonadotropin*
  [9002-70-4]
Equol [531-95-3]
Erbium [7440-52-0]
Erbium Sulfate [13478-49-4]
Erbon [136-25-4]
Erdin [10088-68-3]
Erdin (+)-Form* [3405-51-4]
Erdin (+)-Form Methyl Ester*
  [427-63-4]
Erdmann's Salt [13600-89-0]
Erdosteine [84611-23-4]
Ergocornine [564-36-3]
Ergocorninine [564-37-4]
Ergocristine [511-08-0]
Ergocristinine [511-07-9]
α-Ergocryptine* [511-09-1]
β-Ergocryptine* [20315-46-2]
Ergocryptinine [511-10-4]
Ergoflavin [3101-51-7]
Ergoloid Mesylates [8067-24-1]
Ergometrinine [479-00-5]
Ergonovine [60-79-7]
Ergonovine Maleate* [129-51-1]
Ergosine [561-94-4]
Ergostane [511-20-6]
Ergostanol [6538-02-9]
α-Ergostenol [632-32-6]
β-Ergostenol [632-31-5]
γ-Ergostenol [516-78-9]
Ergosterol [57-87-4]
Ergotamine [113-15-5]
Ergotamine Tartrate* [379-79-3]
Ergotaminine [639-81-6]
Ergothioneine [497-30-3]
Ergotinine [8006-08-4]
Ergotoxine [8006-25-5]
Eriochrome Black T [1787-61-7]
Eriodictyol [552-58-9]
Eritadenine [23918-98-1]
Erucic Acid [112-86-7]
Erythritol [149-32-6]
Erythritol Anhydride [564-00-1]
Erythrityl Tetranitrate [7297-25-8]
Erythrocentaurin [50276-98-7]
α-Erythroidine [466-80-8]
β-Erythroidine [466-81-9]
Erythromycin [114-07-8]
Erythromycin Acistrate [96128-89-1]
Erythromycin Estolate [3521-62-8]
Erythromycin Ethylsuccinate*
  [41342-53-4]
Erythromycin Glucoheptonate
  [23067-13-2]
Erythromycin Lactobionate
  [3847-29-8]
Erythromycin Propionate [134-36-1]
Erythromycin Stearate [643-22-1]
Erythrophlamine [511-00-2]
Erythrophleine [36150-73-9]
Erythropoietin [11096-26-7]
Erythropterin [7449-03-8]
D-Erythrose [583-50-6]
L-Erythrose [533-49-3]
D-Erythrose 4-Phosphate [585-18-2]
Erythrosine [16423-68-0]
L-Erythrulose [533-50-6]
Esaprazole [64204-55-3]
Escin [6805-41-0]
α-Escin* [66795-86-6]
β-Escin* [11072-93-8]
Esculetin [305-01-1]
Esculin [531-75-9]
Eseridine [25573-43-7]

Eseridine Salicylate* [5995-96-0]
Esmolol [81147-92-4]
Esmolol Hydrochloride* [81161-17-3]
Essential Fatty Acids [11006-87-4]
Estazolam [29975-16-4]
Estradiol [50-28-2]
α-Estradiol [57-91-0]
Estradiol Benzoate* [50-50-0]
Estradiol Cypionate* [313-06-4]
Estradiol Dipropionate* [113-38-2]
Estradiol Enanthate* [4956-37-0]
Estradiol Undecylate* [3571-53-7]
Estradiol 17-Valerate* [979-32-8]
Estragole [140-67-0]
Estramustine [2998-57-4]
Estramustine 17-(Dihydrogenphos-
  phate) Disodium Salt* [52205-73-9]
Estriol [50-27-1]
Estriol 16,17-Bis(sodium Hemisuc-
  cinate)* [113-22-4]
Estrone [53-16-7]
Estrone Acetate* [901-93-9]
Estrone dl-Form* [19973-76-3]
Estrone Methyl Ether* [1091-94-7]
Estrone Propionate* [975-64-4]
Estrone Sulfate* [481-97-0]
Estrone Sulfate Piperazine Salt*
  [7280-37-7]
Etafedrine [7681-79-0]
Etafedrine Hydrochloride*
  [5591-29-7]
Etafenone [90-54-0]
Etamiphyllin [314-35-2]
Etamiphyllin Heparinate*
  [59547-58-9]
Etamiphyllin Hydrochloride*
  [17140-68-0]
Etamycin [299-20-7]
Etanidazole [22668-01-5]
Etaqualone [7432-25-9]
Eterobarb [27511-99-5]
Etersalate [62992-61-4]
Ethacridine [442-16-0]
Ethacridine Lactate Monohydrate*
  [6402-23-9]
Ethacrynic Acid [58-54-8]
Ethacrynic Acid Sodium Salt*
  [6500-81-8]
Ethadione [520-77-4]
Ethalfluralin [55283-68-6]
Ethambutol [74-55-5]
Ethambutol Dihydrochloride*
  [1070-11-7]
Ethamivan [304-84-7]
Ethamsylate [2624-44-4]
Ethane [74-84-0]
Ethanearsonic Acid [507-32-4]
Ethanearsonic Acid Disodium Salt*
  [5982-55-8]
1,2-Ethanedisulfonic Acid [110-04-3]
1,2-Ethanedithiol [540-63-6]
Ethanethiol [75-08-1]
Ethanolamine [141-43-5]
Ethanolamine Oleate* [2272-11-9]
Ethaverine [486-47-5]
Ethaverine Hydrochloride* [985-13-7]
Ethchlorvynol [113-18-8]
Ethebenecid [1213-06-5]
Ethenzamide [938-73-8]
Ethephon [16672-87-0]
Ethiazide [1824-58-4]
Ethinamate [126-52-3]
Ethinyl Estradiol [57-63-6]
Ethiodized Oil [8008-53-5]
Ethion [563-12-2]
Ethionamide [536-33-4]
Ethionine [13073-35-3]
Ethionine DL-Form* [67-21-0]
Ethiozin [64529-56-2]
Ethirimol [23947-60-6]

Ethisterone [434-03-7]
Ethofumesate [26225-79-6]
Ethoheptazine [77-15-6]
Ethoheptazine Citrate* [6700-56-7]
Ethohexadiol [94-96-2]
Ethopabate [59-06-3]
Ethoprop [13194-48-4]
Ethopropazine [522-00-9]
Ethopropazine Hydrochloride*
  [1094-08-2]
Ethosuximide [77-67-8]
Ethotoin [86-35-1]
Ethoxazene [94-10-0]
Ethoxazene Hydrochloride*
  [2313-87-3]
2-Ethoxyethanol [110-80-5]
2-Ethoxyethyl Acetate [111-15-9]
2-Ethoxynaphthalene [93-18-5]
Ethoxyquin [91-53-2]
Ethoxzolamide [452-35-7]
Ethybenztropine [524-83-4]
Ethybenztropine Hydrobromide*
  [24815-25-6]
Ethybenztropine Hydrochloride*
  [26598-44-7]
Ethyl Acetate [141-78-6]
Ethyl Acetoacetate [141-97-9]
Ethyl Acrylate [140-88-5]
Ethyl Alcohol [64-17-5]
Ethylamine [75-04-7]
Ethylamine Hydriodide* [506-58-1]
Ethylamine Hydrochloride*
  [557-66-4]
Ethylamine Oleate* [39664-27-2]
N-Ethylamphetamine [457-87-4]
Ethyl Amyl Ketone [106-68-3]
Ethylaniline [103-69-5]
Ethylbenzene [100-41-4]
Ethylbenzhydramine [642-58-0]
Ethylbenzhydramine Hydrochloride*
  [86-24-8]
Ethylbenzhydramine Methyl Iodide*
  [5982-52-5]
Ethyl Benzoate [93-89-0]
Ethyl Benzoylacetate [94-02-0]
α-Ethylbenzyl Alcohol [93-54-9]
Ethylbenzylaniline [92-59-1]
Ethyl Biscoumacetate [548-00-5]
Ethyl Bromide [74-96-4]
Ethyl α-Bromopropionate [535-11-5]
Ethyl tert-Butyl Ether [637-92-3]
Ethyl Butyrate [105-54-4]
Ethyl Caprate [110-38-3]
Ethyl Caproate [123-66-0]
Ethyl Caprylate [106-32-1]
Ethyl β-Carboline-3-carboxylate
  [74214-62-3]
Ethyl Carbonate [105-58-8]
Ethyl Cellulose [9004-57-3]
Ethyl Chloride [75-00-3]
Ethyl Chloroacetate [105-39-5]
Ethyl Chloroformate [541-41-3]
Ethyl α-Chloropropionate [535-13-7]
Ethyl Cyanoacetate [105-56-6]
Ethyl Dibunate [5560-69-0]
Ethyl Diethylmalonate [77-25-8]
Ethyldimethyl-9-octadecenylam-
  monium Bromide [6458-13-5]
Ethylene [74-85-1]
Ethylene Bromohydrin [540-51-2]
Ethylene Chlorohydrin [107-07-3]
Ethylene Cyanohydrin [109-78-4]
Ethylenediamine [107-15-3]
Ethylene Dibromide [106-93-4]
Ethylene Dichloride [107-06-2]
Ethylene Glycol [107-21-1]
Ethylene Glycol Diacetate [111-55-7]
Ethylene Glycol Dinitrate [628-96-6]
Ethylene Glycol Monoacetate
  [542-59-6]

Denotes derivative of title compound

Ethylene Oxide [75-21-8]
Ethylene Thiourea [96-45-7]
Ethylenimine [151-56-4]
Ethylestrenol [965-90-2]
Ethyl Ether [60-29-7]
Ethyl Formate [109-94-4]
2-Ethyl-1-hexanol [104-76-7]
Ethylhydrocupreine [522-60-1]
Ethylhydrocupreine Hydrochloride*
  [3413-58-9]
Ethylidene Chloride [75-34-3]
Ethylidene Diacetate [542-10-9]
Ethylidene Dicoumarol [1821-16-5]
Ethyl Iodide [75-03-6]
Ethyl Isobutyrate [97-62-1]
Ethyl Isothiocyanate [542-85-8]
Ethyl Isovalerate [108-64-5]
Ethyl Lactate [97-64-3]
Ethyl Laurate [106-33-2]
Ethyl Levulinate [539-88-8]
Ethyl Linoleate [544-35-4]
Ethyl Loflazepate [29177-84-2]
N-Ethylmaleimide [128-53-0]
Ethyl Malonate [105-53-3]
Ethyl Menthane Carboxamide
  [39711-79-0]
Ethylmercuric Chloride [107-27-7]
N-(Ethylmercuri)-p-toluenesulfonani-
  lide [517-16-8]
Ethyl Methanesulfonate [62-50-0]
Ethyl Methyl Ether [540-67-0]
Ethylmethylthiambutene [441-61-2]
Ethylmorphine [76-58-4]
Ethylmorphine Hydrochloride
  Dihydrate* [6746-59-4]
Ethylmorphine Methiodide*
  [6696-59-9]
Ethyl Nitrite [109-95-5]
N-Ethyl-N-nitrosourea [759-73-9]
Ethylnorepinephrine [536-24-3]
Ethylnorepinephrine Hydrochloride*
  [3198-07-0]
Ethyl Oenanthate [106-30-9]
Ethyl Oxalacetate [108-56-5]
Ethylparaben [120-47-8]
Ethyl Pelargonate [123-29-5]
Ethyl Phenylacetate [101-97-3]
Ethyl Phosphorochloridite [589-57-1]
3-Ethyl-4-picoline [529-21-5]
4-Ethyl-2-picoline [536-88-9]
5-Ethyl-2-picoline [104-90-5]
1-Ethyl-3-piperidinol [13444-24-1]
Ethyl Propionate [105-37-3]
3-Ethylpyridine [536-78-7]
4-Ethylpyridine [536-75-4]
Ethyl Salicylate [118-61-6]
Ethyl Silicate [78-10-4]
Ethylstibamine [1338-98-3]
Ethyl Sulfate [540-82-9]
Ethyl Sulfide [352-93-2]
2-(Ethylsulfonyl)ethanol [513-12-2]
Ethyl Tartrate [87-91-2]
Ethyl Tartrate, Acid [608-89-9]
2-(Ethylthio)ethanol [110-77-0]
Ethyl p-Toluenesulfonate [80-40-0]
Ethyl Vanillin [121-32-4]
Ethynodiol [1231-93-2]
Ethynodiol Diacetate* [297-76-7]
Ethynylbenzene [536-74-3]
Etidocaine [36637-18-0]
Etidocaine Hydrochloride*
  [36637-19-1]
Etidronic Acid [2809-21-4]
Etidronic Acid Disodium Salt*
  [7414-83-7]
Etifelmin [341-00-4]
Etifelmin Hydrochloride* [1146-95-8]
Etifoxine [21715-46-8]
Etifoxine Hydrochloride*
  [56776-32-0]

Etilefrin [709-55-7]
Etilefrin dl-Form* [10128-36-6]
Etilefrin dl-Form Hydrochloride*
  [534-87-2]
Etiocholane [438-23-3]
Etiocholanic Acid [438-08-4]
Etiocobalamin [27792-36-5]
Etioporphyrin [26608-34-4]
Etiproston [59619-81-7]
Etiroxate [17365-01-4]
Etiroxate Hydrochloride*
  [55327-22-5]
Etisazol [7716-60-1]
Etizolam [40054-69-1]
Etodolac [41340-25-4]
Etodroxizine [17692-34-1]
Etodroxizine Dimaleate* [56335-21-8]
Etofenamate [30544-47-9]
Etofibrate [31637-97-5]
Etofylline [519-37-9]
Etofylline Combination with
  Theophylline* [71990-71-1]
Etofylline Nicotinate [13425-39-3]
Etoglucid [1954-28-5]
Etomidate [33125-97-2]
Etomidoline [21590-92-1]
Etonitazene [911-65-9]
Etoperidone [52942-31-1]
Etoperidone Hydrochloride*
  [57775-22-1]
Etoposide [33419-42-0]
Etorphine [14521-96-1]
Etoxadrol [28189-85-7]
Etoxadrol (+)-Hydrochloride*
  [23239-37-4]
Etozolin [73-09-6]
Etretinate [54350-48-0]
Etrimfos [38260-54-7]
Etryptamine [2235-90-7]
Etryptamine Acetate* [118-68-3]
Etymemazine [523-54-6]
Etymemazine Hydrochloride*
  [3737-33-5]
β-Eucaine [500-34-5]
Eucalyptol [470-82-6]
Eucalyptus Gum [9000-20-8]
Eucatropine [100-91-4]
Eugenol [97-53-0]
Euparin [532-48-9]
Eupatorin [855-96-9]
Euprocin [1301-42-4]
Europium [7440-53-1]
Evan's Blue [314-13-6]
Evodiamine [518-17-2]
Exalamide [53370-90-4]
Exaltolide [106-02-5]
Exametazime [105613-48-7]
Exifone [52479-85-3]
Exiproben [26281-69-6]
Exiproben Sodium Salt* [3478-44-2]
Exosurf [99732-49-7]
Factor V [9001-24-5]
Factor VII [9001-25-6]
Factor VIII [9001-27-8]
Factor IX [9001-28-9]
Factor X [9001-29-0]
Factor XI [9013-55-2]
Factor XII [9001-30-3]
Factor XIII [9013-56-3]
Fadrozole [102676-47-1]
Fadrozole Hydrochloride*
  [102676-31-3] (unspecified)
Fadrozole Hydrochloride*
  [102676-96-0] (±)
Fagarine [524-15-2]
Famciclovir [104227-87-4]
Famotidine [76824-35-6]
Famphur [52-85-7]
Fampridine [504-24-5]
Fantofarone [114432-13-2]

α-Farnesene [502-61-4]
β-Farnesene [18794-84-8]
Farnesol [4602-84-0]
Fast Green FCF [2353-45-9]
Fasudil [103745-39-7]
Fasudil Hydrochloride* [105628-07-7]
Fazadinium Bromide [49564-56-9]
Febantel [58306-30-2]
Febarbamate [13246-02-1]
Febrifugine [24159-07-7]
Febuprol [3102-00-9]
Feclemine [3590-16-7]
Feclemine Dihydrochloride*
  [115097-93-3]
Fedotozine [123618-00-8]
Fedotozine D(—)-Tartrate*
  [133267-27-3]
Felbamate [25451-15-4]
Felbinac [5728-52-9]
Felbinac Ethyl Ester* [14062-23-8]
Felinine [471-09-0]
Felodipine [72509-76-3]
Felypressin [56-59-7]
Femoxetine [59859-58-4]
Femoxetine Hydrochloride*
  [56222-04-9]
Fenadiazole [1008-65-7]
Fenalamide [4551-59-1]
Fenalcomine [34616-39-2]
Fenalcomine Hydrochloride*
  [34535-83-6]
Fenamiphos [22224-92-6]
Fenapanil [61019-78-1]
Fenarimol [60168-88-9]
Fenbendazole [43210-67-9]
Fenbenicillin [1926-48-3]
Fenbenicillin Potassium Salt*
  [1177-30-6]
Fenbuconazole [114369-43-6]
Fenbufen [36330-85-5]
Fenbutatin Oxide [13356-08-6]
Fenbutrazate [4378-36-3]
Fenbutrazate Hydrochloride*
  [6474-85-7]
Fenbutrazate Hydrochloride Mixture
  with Phenmetrazine Theoclate*
  [8004-38-4]
Fencamfamine [1209-98-9]
Fencamfamine Hydrochloride*
  [2240-14-4]
Fencamine [28947-50-4]
Fencamine Hydrochloride*
  [63918-50-3]
d-Fenchone [4695-62-9]
Fencibutirol [5977-10-6]
Fenclozic Acid [17969-20-9]
Fendiline [13042-18-7]
Fendiline Hydrochloride*
  [13636-18-5]
Fendosal [53597-27-6]
Fenethazine [522-24-7]
Fenethylline [3736-08-1]
Fenethylline Hydrochloride*
  [1892-80-4]
Fenfluramine [458-24-2]
Fenfluramine l-Form* [37577-24-5]
Fenfluramine l-Form Hydrochloride*
  [3616-78-2]
Fenfluramine Hydrochloride*
  [404-82-0]
Fenipentol [583-03-9]
Fenitrothion [122-14-5]
Fenofibrate [49562-28-9]
Fenoldopam [67227-56-9]
Fenoldopam Monomethanesulfonate*
  [67227-57-0]
Fenoprofen [31879-05-7]
Fenoprofen Calcium Salt Dihydrate*
  [53746-45-5]
Fenoterol [13392-18-2]

Fenoterol Hydrobromide* [1944-12-3]
Fenoterol Hydrochloride* [1944-10-1]
Fenoverine [37561-27-6]
Fenoxaprop-ethyl [82110-72-3]
Fenoxazoline [4846-91-7]
Fenoxazoline Hydrochloride*
 [21370-21-8]
Fenoxedil [54063-40-0]
Fenoxedil Hydrochloride*
 [27471-60-9]
Fenoxycarb [72490-01-8]
Fenozolone [15302-16-6]
Fenpentadiol [15687-18-0]
Fenpiclonil [74738-17-3]
Fenpiprane [3540-95-2]
Fenpiprane Hydrochloride*
 [3329-14-4]
Fenpiverinium Bromide [125-60-0]
Fenpropathrin [39515-41-8]
Fenpropidin [67306-00-7]
Fenpropimorph [67306-03-0]
Fenpropimorph cis-Form*
 [67564-91-4]
Fenproporex [15686-61-0]
Fenproporex Hydrochloride*
 [18305-29-8]
Fenprostalene [69381-94-8]
Fenpyroximate [134098-61-6]
Fenquizone [20287-37-0]
Fenquizone Monopotassium Salt*
 [52246-40-9]
Fenretinide [65646-68-6]
Fenspiride [5053-06-5]
Fenspiride Hydrochloride*
 [5053-08-7]
Fensulfothion [115-90-2]
Fentanyl [437-38-7]
Fentanyl Citrate* [990-73-8]
Fenthion [55-38-9]
Fentiazac [18046-21-4]
Fenticlor [97-24-5]
Fenticonazole [72479-26-6]
Fenticonazole Mononitrate*
 [73151-29-8]
Fentonium Bromide [5868-06-4]
Fenuron [101-42-8]
Fenvalerate [51630-58-1]
Fenvalerate (S,S)-Isomer*
 [66230-04-4]
Fepradinol [63075-47-8]
Feprazone [30748-29-9]
Ferbam [14484-64-1]
Fermium [7440-72-4]
Ferric Acetate, Basic [10450-55-2]
Ferric Albuminate [8001-11-4]
Ferric and Ammonium Acetate Solu-
tion [8006-27-7]
Ferric Bromide [10031-26-2]
Ferric Chloride [7705-08-0]
Ferric Chromate(VI) [10294-52-7]
Ferric Citrate [2338-05-8]
Ferric Ferrocyanide [14038-43-8]
Ferric Fluoride [7783-50-8]
Ferric Formate [555-76-0]
Ferric Hydroxide [20344-49-4]
Ferric Hypophosphite [7783-84-8]
Ferriclate Calcium Sodium
 [34150-62-4]
Ferric Nitrate [10421-48-4]
Ferric Oxide [1309-37-1]
Ferric Oxide, Saccharated [8047-67-4]
Ferric Phosphate [10045-86-0]
Ferric Pyrophosphate [10058-44-3]
Ferric Sodium Edetate [15708-41-5]
Ferric Sodium Pyrophosphate
 [10045-87-1]
Ferric Subsulfate Solution [1310-45-8]
Ferric Sulfate [10028-22-5]
Ferric Thiocyanate [4119-52-2]
Ferrite [1317-54-0]

Ferritin [9007-73-2]
Ferrocene [102-54-5]
Ferrocholinate [1336-80-7]
Ferroglycine Sulfate [17169-60-7]
Ferrosoferric Oxide [1317-61-9]
Ferrous Bromide [7789-46-0]
Ferrous Carbonate Mass [8030-35-1]
Ferrous Carbonate Saccharated
 [8001-10-3]
Ferrous Chloride [7758-94-3]
Ferrous Citrate [23383-11-1]
Ferrous Fluoride [7789-28-8]
Ferrous Fumarate [141-01-5]
Ferrous Gluconate [299-29-6]
Ferrous Hydroxide [18624-44-7]
Ferrous Iodide [7783-86-0]
Ferrous Lactate [5905-52-2]
Ferrous Oxalate [516-03-0]
Ferrous Oxide [1345-25-1]
Ferrous Phosphate [14940-41-1]
Ferrous Phosphide [1310-43-6]
Ferrous Selenide [1310-32-3]
Ferrous Succinate [10030-90-7]
Ferrous Sulfate [7720-78-7]
Ferrous Sulfate Heptahydrate*
 [7782-63-0]
Ferrous Sulfate Monohydrate*
 [17375-41-6]
Ferrous Sulfide [1317-37-9]
Ferrous Thiocyanate [6010-09-9]
Fertilysin [1477-57-2]
Fertirelin [38234-21-8]
Fertirelin Monoacetate* [66002-66-2]
Ferulic Acid [1135-24-6]
Fervenulin [483-57-8]
Fexofenadine [83799-24-0]
Fialuridine [69123-98-4]
Fibrin [9001-31-4]
Fibrinogen [9001-32-5]
Fibroblast Growth Factor
 [62031-54-3]
Fibroin [9007-76-5]
Fichtelite [2221-95-6]
Ficin [9001-33-6]
Filgrastim* [121181-53-1]
Filicinic Acid [2065-00-1]
Filipin [11078-21-0]
Filipin III* [480-49-9]
Finasteride [98319-26-7]
Fipexide [34161-24-5]
Fipexide Hydrochloride* [34161-23-4]
Fipronil [120068-37-3]
Firefly Luciferin [2591-17-5]
Fisetin [528-48-3]
Flavaspidic Acid [114-42-1]
Flavine-Adenine Dinucleotide
 [146-14-5]
Flavone [525-82-6]
Flavopereirine [486-18-0]
Flavoxanthin [512-29-8]
Flavoxate [15301-69-6]
Flavoxate Hydrochloride* [3717-88-2]
Flavoxate Succinate* [28782-19-6]
Flecainide [54143-55-4] (unspecified)
Flecainide [99495-87-1] (±)
Flecainide Monoacetate* [54143-56-5]
Fleroxacin [79660-72-3]
Flesinoxan [98206-10-1]
Flesinoxan Hydrochloride*
 [98205-89-1]
Flindersine [523-64-8]
Floctafenine [23779-99-9]
Flomoxef [99665-00-6]
Flomoxef Sodium Salt* [92823-03-5]
Flopropione [2295-58-1]
Florantyrone [519-95-9]
Floredil [53731-36-5]
Floredil Hydrochloride* [30116-80-4]
Florfenicol [73231-34-2]
Flosequinan [76568-02-0]

Floxacillin [5250-39-5]
Floxacillin Sodium Monohydrate*
 [34214-51-2]
Floxuridine [50-91-9]
Fluacizine [30223-48-4]
Fluanisone [1480-19-9]
Fluazacort [19888-56-3]
Fluazifop-butyl [69806-50-4]
Fluazinam [79622-59-6]
Flubendazole [31430-15-6]
Flubenzimine [37893-02-0]
Fluchloralin [33245-39-5]
Flucloronide [3693-39-8]
Fluconazole [86386-73-4]
Flucycloxuron [94050-52-9] (E-iso-
mer)
Flucycloxuron [94050-53-0] (Z-iso-
mer)
Flucycloxuron [113036-88-7]
Flucythrinate [70124-77-5]
Flucytosine [2022-85-7]
Fludarabine [21679-14-1]
Fludarabine 5'-Monophosphate*
 [75607-67-9]
Fludeoxyglucose F^{18} [105851-17-0]
Fludiazepam [3900-31-0]
Fludioxonil [131341-86-1]
Fludrocortisone [127-31-1]
Fludrocortisone 21-Acetate*
 [514-36-3]
Flufenamic Acid [530-78-9]
Flufenamic Acid Aluminum Salt*
 [61891-34-7]
Flufenamic Acid Butyl Ester*
 [67330-25-0]
Fluindione [957-56-2]
Flumazenil [78755-81-4]
Flumecinol [56430-99-0]
Flumedroxone Acetate [987-18-8]
Flumequine [42835-25-6]
Flumethasone [2135-17-3]
Flumethasone 21-Pivalate*
 [2002-29-1]
Flumethiazide [148-56-1]
Flumethrin [69770-45-2]
Flumetramide [7125-73-7]
Flumetsulam [98967-40-9]
Flumiclorac [87547-04-4]
Flumiclorac-pentyl* [87546-18-7]
Flunarizine [52468-60-7]
Flunarizine Dihydrochloride*
 [30484-77-6]
Flunisolide [3385-03-3]
Flunisolide 21-Acetate* [4533-89-5]
Flunitrazepam [1622-62-4]
Flunixin [38677-85-9]
Flunixin Meglumine Salt*
 [42461-84-7]
Flunoxaprofen [66934-18-7]
Fluoboric Acid [16872-11-0]
Fluocinolone Acetonide [67-73-2]
Fluocinonide [356-12-7]
Fluocortin Butyl [41767-29-7]
Fluocortolone [152-97-6]
Fluocortolone 21-Hexanoate*
 [303-40-2]
Fluometuron [2164-17-2]
9H-Fluorene [86-73-7]
9H-Fluorene-2,7-diamine [525-64-4]
N-2-Fluorenylacetamide [53-96-3]
Fluorescamine [38183-12-9]
Fluorescein [2321-07-5]
Fluorescein Disodium Salt* [518-47-8]
Fluorescin [518-44-5]
Fluoresone [2924-67-6]
Fluoridamid [47000-92-0]
Fluorine [7782-41-4]
Fluorine Dioxide [7783-44-0]
Fluorine Monoxide [7783-41-7]

Denotes derivative of title compound

Fluorine Nitrate [7789-26-6]
Fluorine Perchlorate [10049-03-3]
Fluoroacetamide [640-19-7]
Fluoroacetic Acid [144-49-0]
3-Fluoro-D-alanine [35455-20-0]
p-Fluoroaniline [371-40-4]
Fluorobenzene [462-06-6]
p-Fluorobenzoic Acid [456-22-4]
1-Fluoro-2,4-dinitrobenzene [70-34-8]
Fluoroform [75-46-7]
Fluoromethane [593-53-3]
Fluorometholone [426-13-1]
Fluorometholone 17-Acetate*
   [3801-06-7]
5-Fluoroorotic Acid [703-95-7]
p-Fluorophenylacetic Acid [405-50-5]
Fluorosalan [4776-06-1]
Fluorosulfonic Acid [7789-21-1]
m-Fluorotoluene* [352-70-5]
o-Fluorotoluene* [95-52-3]
p-Fluorotoluene* [352-32-9]
Fluorouracil [51-21-8]
Fluosilicic Acid [16961-83-4]
Fluosol DA [75216-20-5]
Fluoxetine [54910-89-3]
Fluoxetine Hydrochloride*
   [59333-67-4]
Fluoxymesterone [76-43-7]
Flupentixol [2709-56-0]
Flupentixol Decanoate* [30909-51-4]
Flupentixol Dihydrochloride*
   [2413-38-9]
Fluperolone Acetate [2119-75-7]
Fluphenazine [69-23-8]
Fluphenazine Decanoate* [5002-47-1]
Fluphenazine Dihydrochloride*
   [146-56-5]
Fluphenazine Enanthate* [2746-81-8]
Flupirtine [56995-20-1]
Flupirtine Maleate* [75507-68-5]
Fluprednidene Acetate [1255-35-2]
Fluprednisolone [53-34-9]
Fluproquazone [40507-23-1]
Fluprostenol [40666-16-8]
Fluprostenol Sodium Salt*
   [55028-71-2]
Flurandrenolide [1524-88-5]
Flurazepam [17617-23-1]
Flurazepam Dihydrochloride*
   [1172-18-5]
Flurbiprofen [5104-49-4]
Flurogestone Acetate [2529-45-5]
Flurothyl [333-36-8]
Fluroxene [406-90-6]
Fluroxypyr [69377-81-7]
Fluroxypyr 1-Methylheptyl Ester*
   [81406-37-3]
Flurprimidol [56425-91-3]
Flusilazole [85509-19-9]
Fluspirilene [1841-19-6]
Flutamide [13311-84-7]
Flutazolam [27060-91-9]
Fluticasone Propionate [80474-14-2]
Flutoprazepam [25967-29-7]
Flutriafol [76674-21-0]
Flutrimazole [119006-77-8]
Flutropium Bromide [63516-07-4]
Fluvalinate [69409-94-5]
Fluvastatin [93957-54-1]
Fluvastatin Sodium Salt* [93957-55-2]
Fluvoxamine [54739-18-3]
Fluvoxamine Maleate* [61718-82-9]
Folescutol [15687-22-6]
Folescutol Hydrochloride*
   [36002-19-4]
Folic Acid [59-30-3]
Folinic Acid [58-05-9]
Folinic Acid Calcium Salt Penta-
   hydrate* [6035-45-6]

Folinic Acid l-Form Calcium Salt*
   [80433-71-2]
Folpet [133-07-3]
Fomepizole [7554-65-6]
Fominoben [18053-31-1]
Fominoben Hydrochloride*
   [24600-36-0]
Fomocaine [17692-39-6]
Fonazine [7456-24-8]
Fonazine Methanesulfonate*
   [7455-39-2]
Fonofos [944-22-9]
Formaldehyde [50-00-0]
Formaldehyde Sodium Bisulfite
   [870-72-4]
Formamide [75-12-7]
Formanilide [103-70-8]
Formebolone [2454-11-7]
Formestane [566-48-3]
Formic Acid [64-18-6]
Formicin [625-51-4]
Formocortal [2825-60-7]
Formononetin [485-72-3]
Formosulfathiazole* [12041-72-4]
Formoterol [73573-87-2]
Formothion [2540-82-1]
Formyl Fluoride [1493-02-3]
N^2-Formylsulfisomidine [795-13-1]
Fortimicin A* [55779-06-1]
Fortimicin B* [54783-95-8]
Foscarnet Sodium [63585-09-1]
Fosetyl Al [39148-24-8]
Fosfestrol [522-40-7]
Fosfestrol Disodium Salt* [5965-09-3]
Fosfestrol Tetrasodium Salt*
   [4719-75-9]
Fosfomycin [23155-02-4]
Fosfomycin Tromethanol*
   [78964-85-9]
Fosfosal [6064-83-1]
Fosinopril [98048-97-6]
Fosinopril Diacid* [95399-71-6]
Fosinopril Sodium Salt* [88889-14-9]
Fospirate [5598-52-7]
Fosthietan [21548-32-3]
Fotemustine [92118-27-9]
Francium [7440-73-5]
Frangulin [60529-33-1]
Frangulin A* [521-62-0]
Frangulin B* [14101-04-3]
Fraxetin [574-84-5]
Fraxin [524-30-1]
Fredericamycin A [80455-68-1]
Frenolicin [10023-07-1]
Frequentin [29119-03-7]
Friedelin [559-74-0]
Fructose [57-48-7]
DL-Fructose [6035-50-3]
Fructose-1,6-diphosphate [488-69-7]
Fructose Ferric Form* [12286-76-9]
Fructose-6-phosphate [643-13-0]
FSH [9002-68-0]
Fucosamine [24724-90-1]
D-Fucose [3615-37-0]
L-Fucose [2438-80-4]
Fucosterol [17605-67-3]
Fucoxanthin [3351-86-8]
Fulvoplumierin [20867-01-0]
Fumagillin [23110-15-8]
Fumaric Acid [110-17-8]
Fumigatin [484-89-9]
Fumonisin B_1 [116355-83-0]
Fungichromin [6834-98-6]
Fungisterol [53260-54-1]
Funtumine [474-45-3]
Furaltadone [139-91-3]
Furan [110-00-9]
2-Furanacrylic Acid [539-47-9]
2-Furanacrylonitrile [7187-01-1]
Furazabol [1239-29-8]

Furazolidone [67-45-8]
Furazolium Chloride [5118-17-2]
Furcellaran [9000-21-9]
Furethidine [2385-81-1]
Furfural [98-01-1]
Furfuryl Alcohol [98-00-0]
5-Furfuryl-5-isopropylbarbituric Acid
   [1146-21-0]
Furfurylmethylamphetamine
   [13445-60-8]
Furfurylmethylamphetamine d-Form*
   [3776-93-0]
Furfurylmethylamphetamine dl-Form*
   [14817-79-9]
Furfurylmethylamphetamine d-Form
   Cyclohexylsulfamate* [3776-92-9]
Furfurylmethylamphetamine dl-Form
   Cyclohexylsulfamate* [14611-84-8]
α-Furildioxime [522-27-0]
2-Furoic Acid [88-14-2]
Furonazide [3460-67-1]
Furosemide [54-31-9]
Furoyl Chloride [1300-32-9]
Fursultiamine [804-30-8]
Furtrethonium [7618-86-2]
Furtrethonium Iodide* [541-64-0]
Fusafungine [1393-87-9]
Fusaric Acid [536-69-6]
Fusarubin [1702-77-8]
Fuscin [83-85-2]
Fusidic Acid [6990-06-3]
Fusidic Acid Sodium Salt* [751-94-0]
Fustin [20725-03-5]
Fyrol FR-2 [13674-87-8]
Gabapentin [60142-96-3]
Gabexate [39492-01-8]
Gabexate Monomethanesulfonate*
   [56974-61-9]
Gadodiamide [131410-48-5]
Gadodiamide Injection* [131410-51-0]
Gadolinium [7440-54-2]
Gadopentetic Acid [80529-93-7]
Gadopentetic Acid Dimeglumine Salt*
   [86050-77-3]
Gadoteridol [120066-54-8]
Galactaric Acid [526-99-8]
Galactitol [608-66-2]
Galactoflavin [5735-19-3]
D-Galactosamine [7535-00-4]
D-Galactose [59-23-4]
D-Galactose Microparticulate Form*
   [90881-70-2]
D-Galactose Transpulmonary Micro-
   particulate Form* [144046-30-0]
D-Galacturonic Acid [685-73-4]
Galangin [548-83-4]
Galanthamine [357-70-0]
Galanthamine Hydrobromide*
   [1953-04-4]
Galegine [543-83-9]
Galipine [525-68-8]
Gallacetophenone [528-21-2]
Gallamine Triethiodide [65-29-2]
Gallein [2103-64-2]
Gallic Acid [149-91-7]
Gallic Acid Methyl Ester* [99-24-1]
Gallium [7440-55-3]
Gallium Arsenide [1303-00-0]
Gallium Citrate* [41183-64-6]
Gallium Nitrate* [13494-90-1]
Gallium Phosphide [12063-98-8]
Gallium Trifluoride [7783-51-9]
Gallocyanine [1562-85-2]
Gallopamil [16662-47-8]
Gallopamil d-Form Hydrochloride*
   [38176-09-9]
Gallopamil l-Form Hydrochloride*
   [36222-39-6]
Gallopamil Hydrochloride*
   [16662-46-7]

Gamabufotalin [465-11-2]
Gamboge [9000-25-3]
Gambogic Acid [2752-65-0]
Ganciclovir [82410-32-0]
Ganciclovir Sodium Salt*
  [107910-75-8]
Ganglefene [299-61-6]
Ganglioside G$_{m1}$* [37758-47-7]
Gardenin A* [21187-73-5]
Gardenin B* [2798-20-1]
Gardenin C* [29550-05-8]
Gardenin D* [29202-00-4]
Gardenin E* [29550-07-0]
Gardol® [137-16-6]
Garryine [561-51-3]
Gaultherin [490-67-5]
Gefarnate [51-77-4]
Geissoschizoline [18397-07-4]
Geissospermine [427-01-0]
Gelatin [9000-70-8]
Gellan Gum [71010-52-1]
Gelsemine [509-15-9]
Gemcitabine [95058-81-4]
Gemcitabine Hydrochloride*
  [122111-03-9]
Gemeprost [64318-79-2]
Gemfibrozil [25812-30-0]
Genistein [446-72-0]
Genisteine-Alkaloid [446-95-7]
Genite [97-16-5]
Gentamicin [1403-66-3]
Gentamicin A* [13291-74-2]
Gentamicin C$_1$* [25876-10-2]
Gentamicin C$_{1a}$* [26098-04-4]
Gentamicin C$_2$* [25876-11-3]
Gentamicin C Complex Sulfate*
  [1405-41-0]
Gentianine [439-89-4]
Gentian Violet [548-62-9]
Gentiobiose [554-91-6]
Gentiopicrin [20831-76-9]
Gentisic Acid [490-79-9]
Gentisic Acid Sodium Salt*
  [4955-90-2]
Gentisin [437-50-3]
Gentisyl Alcohol [495-08-9]
Gentrogenin [427-28-1]
Geosmin [19700-21-1]
Gepefrine [18840-47-6]
Gepefrine Tartrate* [60763-48-6]
Gephyrotoxin [55893-12-4]
Geraniol [106-24-1]
Geranylhydroquinone [10457-66-6]
German Chamomile Oil* [8002-66-2]
Germane [7782-65-2]
Germanium [7440-56-4]
Germanium Dichloride [10060-11-4]
Germanium Dioxide [1310-53-8]
Germanium Tetrachloride
  [10038-98-9]
Germanium Tetrafluoride [7783-58-6]
Germine [508-65-6]
Gestodene [60282-87-3]
Gestodene Mixture with Ethinyl
  Estradiol* [109852-02-0]
Gestonorone Caproate [1253-28-7]
Gestrinone [16320-04-0]
Ghatti Gum [9000-28-6]
Gibberellic Acid [77-06-5]
Gibbs Reagent [101-38-2]
Gigantine [32829-58-6]
[6]-Gingerol [23513-14-6]
Ginkgolide A* [15291-75-5]
Ginkgolide B* [15291-77-7]
Ginkgolide C* [15291-76-6]
Giractide [24870-04-0]
Giractide Hexaacetate Salt*
  [29365-11-5]
Gitalin [1405-76-1]
Gitogenin [511-96-6]

F-Gitonin [28591-01-7]
Gitoxigenin [545-26-6]
Gitoxin [4562-36-1]
Gitoxin Pentaacetate* [7242-04-8]
Gladiolic Acid [478-05-7]
Glafenine [3820-67-5]
Glaucarubin [1448-23-3]
Glaucine [475-81-0]
Gliadin [9007-90-3]
Glibornuride [26944-48-9]
Gliclazide [21187-98-4]
Glimepiride [93479-97-1]
Gliotoxin [67-99-2]
Glipizide [29094-61-9]
Gliquidone [33342-05-1]
Glisoxepid [25046-79-1]
Glucagon [9007-92-5]
Glucametacin [52443-21-7]
Glucamine [488-43-7]
D-Glucaric Acid [87-73-0]
D-Glucoascorbic Acid [528-88-1]
Glucofrangulin [52731-38-1]
α-Glucogallin [53318-36-8]
β-Glucogallin [13405-60-2]
Glucoheptonic Acid [87-74-1]
Glucoheptonic Acid Sodium Salt*
  [13007-85-7]
Gluconic Acid [526-95-4]
Gluconic Acid Magnesium Salt
  Dihydrate* [59625-89-7]
Gluconolactone [90-80-2]
Glucosamine [3416-24-8]
Glucosamine α-Form* [28905-11-5]
Glucosamine β-Form* [28905-10-4]
Glucosamine Sulfate Salt*
  [29031-19-4]
Glucose [50-99-7]
Glucose Oxidase [9001-37-0]
Glucose-6-phosphate [56-73-5]
α-Glucose-1-phosphate [59-56-3]
α-Glucose-1-phosphate Barium Salt
  Trihydrate* [6056-70-8]
α-Glucose-1-phosphate Calcium Salt*
  [119119-85-6]
α-Glucose-1-phosphate Dipotassium
  Salt Dihydrate* [5996-14-5]
Glucosulfone Sodium [554-18-7]
N^4-β-D-Glucosylsulfanilamide
  [53274-53-6]
Glucovanillin [494-08-6]
D-Glucuronic Acid [6556-12-3]
β-Glucuronidase [9001-45-0]
D-Glucuronolactone [32449-92-6]
Glutamic Acid [56-86-0]
L-Glutamic Acid 5-Ethyl Ester
  [1119-33-1]
Glutamic Acid D-Form* [6893-26-1]
Glutamic Acid DL-Form* [617-65-2]
Glutamic Acid Hydrochloride*
  [138-15-8]
Glutamine [56-85-9]
Glutamine DL-Form* [585-21-7]
Glutaraldehyde [111-30-8]
Glutaric Acid [110-94-1]
Glutaronitrile [544-13-8]
Glutathione [70-18-8]
Glutathione Disulfide* [27025-41-8]
Glutethimide [77-21-4]
Glyburide [10238-21-8]
Glybuthiazol(e) [535-65-9]
Glybuzole [1492-02-0]
Glycarsamide [144-87-6]
Glyceraldehyde [367-47-5]
Glyceraldehyde 3-Phosphate
  [142-10-9]
Glyceric Acid [473-81-4]
Glycerol [56-81-5]
Glycerol 1-Octanoate* [502-54-5]
Glycerophosphoric Acid [57-03-4]
Glyceryl p-Aminobenzoate [136-44-7]

Glyceryl Iodide [554-10-9]
Glyceryl Monostearate [31566-31-1]
Glycidol [556-52-5]
Glycine [56-40-6]
Glycine Sulfate [513-29-1]
Glycinin [9007-93-6]
Glycobiarsol [116-49-4]
Glycocholic Acid [475-31-0]
Glycocyamine [352-97-6]
Glycogen [9005-79-2]
Glycol Dilaurate [624-04-4]
Glycolic Acid [79-14-1]
Glycol Salicylate [87-28-5]
Glyconiazide [3691-74-5]
Glycopyrrolate [596-51-0]
Glycosine [6873-15-0]
N-Glycylglycine [556-50-3]
Glycyrrhizic Acid [1405-86-3]
Glyhexamide [451-71-8]
Glymidine [339-44-6]
Glymidine Sodium Salt* [3459-20-9]
Glyodin [556-22-9]
Glyoxal [107-22-2]
Glyoxal-Sodium Bisulfite [517-21-5]
Glyoxylic Acid [298-12-4]
Glyphosate [1071-83-6]
Glyphosine [2439-99-8]
Glypinamide [1228-19-9]
Goitrin [500-12-9]
Gold [7440-57-5]
Gold, Explosive [1332-12-3]
Gold Monochloride [10294-29-8]
Gold Monocyanide [506-65-0]
Gold Monoiodide [10294-31-2]
Gold Monosulfide [1303-60-2]
Gold Monoxide [1303-57-7]
Gold, Radioactive, Colloidal
  [10043-49-9]
Gold Selenate [10294-32-3]
Gold Selenide [1303-62-4]
Gold Sodium Thiomalate [12244-57-4]
Gold Sodium Thiosulfate [10233-88-2]
Gold Stannate [1345-24-0]
Gold Tribromide [10294-28-7]
Gold Tribromide, Acid [17083-68-0]
Gold Trichloride [13453-07-1]
Gold Trichloride, Acid [16903-35-8]
Gold Tricyanide [535-37-5]
Gold Trihydroxide [1303-52-2]
Gold Trioxide [1303-58-8]
Gold Trisulfide [1303-61-3]
Goserelin [65807-02-5]
Gossyplure [50933-33-0]
Gossyplure (Z,E)-Form* [53042-79-8]
Gossyplure (Z,Z)-Form* [52207-99-5]
Gossypol [303-45-7]
Gougerotin [2096-42-6]
Gramicidin S [113-73-5]
Gramine [87-52-5]
Granaticin [19879-06-2]
Granaticin B* [19879-03-9]
Grandisol [26532-22-9]
Granisetron [109889-09-0]
Granisetron Hydrochloride*
  [107007-99-8]
Granulocyte Colony-Stimulating
  Factor [143011-72-7]
Granulocyte-Macrophage Colony-
  Stimulating Factor [83869-56-1]
Graphite [7782-42-5]
Graphite Fluoride [11113-63-6]
Graphitic Acid [1399-57-1]
Gratiogenin [7067-16-5]
Gratioside [53762-90-6]
Gravitol(e) [6006-09-3]
Grayanotoxin I* [4720-09-6]
Grayanotoxin II* [4678-44-8]
Grayanotoxin III* [4678-45-9]
Grepafloxacin [119914-60-2] (un-
  specified)

Denotes derivative of title compound

Grepafloxacin [146863-02-7] (±)
Grepafloxacin Hydrochloride*
[161967-81-3] (±)
Grindelic Acid [1438-57-9]
Grisein [1391-82-8]
Griseofulvin [126-07-8]
Griseoviridin [53216-90-3]
Guaiac [9000-29-7]
Guaiacol [90-05-1]
Guaiacol Benzoate [531-37-3]
Guaiacol Carbonate [553-17-3]
Guaiacol Phenylacetate* [4112-89-4]
Guaiacol Phosphate [563-03-1]
Guaiacol Phosphate Sodium Salt*
[5978-58-5]
Guaiacol Valerate [531-39-5]
Guaiapate [852-42-6]
Guaiazulene [489-84-9]
Guaiazulene 3-Sulfonate Sodium
Salt* [6223-35-4]
Guaiazulene Trinitrobenzene Deriv*
[4968-29-0]
Guaifenesin [93-14-1]
Guaiol [489-86-1]
Guaithylline* [5634-38-8]
Guamecycline [16545-11-2]
Guamecycline Dihydrochloride*
[13040-98-7]
Guanabenz [5051-62-7]
Guanabenz Monoacetate*
[23256-50-0]
Guanacline [1463-28-1]
Guanacline Monosulfate* [1562-71-6]
Guanadrel [40580-59-4]
Guanadrel Sulfate* [22195-34-2]
Guanazodine [32059-15-7]
Guanethidine [55-65-2]
Guanethidine Monosulfate*
[645-43-2]
Guanethidine Sulfate* [60-02-6]
Guanfacine [29110-47-2]
Guanfacine Hydrochloride*
[29110-48-3]
Guanidine [113-00-8]
Guanidinium Aluminum Sulfate
Hexahydrate [10199-21-0]
Guanine [73-40-5]
Guanochlor [5001-32-1]
Guanochlor Sulfate* [551-48-4]
Guanosine [118-00-3]
Guanoxabenz [24047-25-4]
Guanoxabenz Hydrochloride*
[23256-40-8]
Guanoxan [2165-19-7]
Guanoxan Sulfate* [5714-04-5]
3'-Guanylic Acid [117-68-0]
5'-Guanylic Acid [85-32-5]
Guaran [9000-30-0]
Guar Gum [9000-30-0]
Guinea Green B [4680-78-8]
D-Gulonic Acid [20246-33-7]
L-Gulonic Acid [526-97-6]
D-Gulose [4205-23-6]
L-Gulose [6027-89-0]
Gum Benzoin [9000-05-9]
Gum Tragacanth [9000-65-1]
Gusperimus [104317-84-2]
Gusperimus (—)-Form Trihydro-
chloride* [84937-45-1]
Gusperimus Trihydrochloride*
[85468-01-5]
Gutta-Percha [9000-32-2]
Guvacine [498-96-4]
Gymnemic Acid [1399-64-0]
Gypsogenin [639-14-5]
Hachimycin [1394-02-1]
Hadacidin [689-13-4]
Hafnium [7440-58-6]
Halazepam [23092-17-3]
Halazone [80-13-7]

Halcinonide [3093-35-4]
Halethazole [15599-36-7]
Halimide® [19014-05-2]
Halobetasol Propionate [66852-54-8]
Halofantrine [69756-53-2]
Halofantrine Hydrochloride*
[36167-63-2]
Halofuginone [55837-20-2]
Halofuginone Hydrobromide*
[64924-67-0]
Halometasone [50629-82-8]
Haloperidol [52-86-8]
Haloperidol Decanoate* [74050-97-8]
Haloperidol Hydrochloride*
[1511-16-6]
Halopredone Acetate [57781-14-3]
Haloprogin [777-11-7]
Halopropane [679-84-5]
Halostachine [495-42-1]
Halothane [151-67-7]
Haloxazolam [59128-97-1]
Haloxon [321-55-1]
Halquinol [8067-69-4]
Hamamelitannin [469-32-9]
Hamamelose [4573-78-8]
Hamycin [1403-71-0]
Haplophytine [16625-20-0]
Harmaline [304-21-2]
Harmalol [525-57-5]
Harman [486-84-0]
Harmine [442-51-3]
Hasubanonine [1805-85-2]
HCS [11085-36-2]
Hecogenin [467-55-0]
Hectorite [12173-47-6]
Hederagenin [465-99-6]
α-Hederin [27013-91-8]
Helenalin [6754-13-8]
Helenynolic Acid [7309-58-2]
Helicin [618-65-5]
Heliosupine [32728-78-2]
Helium [7440-59-7]
Hellebrin [13289-18-4]
Helminthosporal [723-61-5]
Helminthosporol [1619-29-0]
Helveticoside [630-64-8]
Helvolic Acid [29400-42-8]
Hematein [475-25-2]
Hematin [15489-90-4]
Hematoporphyrin [14459-29-1]
Hematoporphyrin Hydrochloride
Monohydrate* [6033-08-5]
Hematoxylin [517-28-2]
Heme [14875-96-8]
Hementin [87041-58-5]
Hemicholinium [16478-59-4]
Hemicholinium Dibromide*
[312-45-8]
Hemi-Dewar Biphenyl [4655-30-5]
Hemin [16009-13-5]
Hemipyocyanine [528-71-2]
Hemisulfur Mustard [693-30-1]
Heparamine [33260-52-9]
Heparin [9005-49-6]
Heparin Calcium Salt* [37270-89-6]
Heparin Magnesium Salt*
[54479-70-8]
Heparin Potassium Salt* [9005-48-5]
Heparin Sodium Salt* [9041-08-1]
Hepaxanthin [512-39-0]
HEPES [7365-45-9]
Hepronicate [7237-81-2]
Heptabarbital [509-86-4]
Heptachlor [76-44-8]
Heptaminol [372-66-7]
Heptaminol 5'-Adenylate*
[57249-13-5]
Heptaminol Hydrochloride*
[543-15-7]
Heptanal [111-71-7]

Heptanal Sodium Bisulfite
[13495-04-0]
n-Heptane [142-82-5]
Heptanoic Acid [111-14-8]
1-Heptanol [111-70-6]
2-Heptanol [543-49-7]
2-Heptanone [110-43-0]
Heptenophos [23560-59-0]
Heptoxime [530-97-2]
D-manno-Heptulose [3615-44-9]
Hercynine [534-30-5]
Herquinone [26871-30-7]
Hesperetin [520-33-2]
Hesperidin [520-26-3]
Hetacillin [3511-16-8]
Hetacillin Potassium Salt*
[5321-32-4]
Hetastarch [9004-62-0]
5-HETE [70608-72-9]
Heteronium Bromide [7247-57-6]
Hetolin [2390-22-9]
Hexaaminecobalt Trichloride
[10534-89-1]
Hexaborane(10) [23777-80-2]
Hexacarbacholine Bromide [306-41-2]
Hexachlorobenzene [118-74-1]
Hexachloroethane [67-72-1]
Hexachlorophene [70-30-4]
Hexacyclonate Sodium [7009-49-6]
Hexadecyl 3-Hydroxy-2-naphthoate
[531-84-0]
Hexadimethrine Bromide
[28728-55-4]
Hexaflumuron [86479-06-3]
Hexafluorenium Bromide [317-52-2]
Hexafluorobenzene [392-56-3]
Hexalure [23192-42-9]
Hexamethonium [60-26-4]
Hexamethonium Bromide* [55-97-0]
Hexamethyldisilazane [999-97-3]
Hexamethylene Glycol [629-11-8]
Hexamethylolmelamine [531-18-0]
Hexamidine [3811-75-4]
Hexamidine Isethionate* [659-40-5]
n-Hexane [110-54-3]
1,6-Hexanediamine [124-09-4]
Hexanitrohexaazaisowurtzitane
[135285-90-4]
1-Hexanol [111-27-3]
Hexapropymate [358-52-1]
Hexazinone [51235-04-2]
Hexazole [4671-03-8]
Hexedine [5980-31-4]
3-Hexen-1-ol [544-12-7]
Hexestrol [84-16-2]
Hexestrol Bis(β-diethylaminoethyl
Ether) [2691-45-4]
Hexestrol Bis(β-diethylaminoethyl
ether) Dihydrochloride* [69-14-7]
Hexestrol Diphosphate* [14188-82-0]
Hexethal Sodium [144-00-3]
Hexetidine [141-94-6]
Hexobarbital [56-29-1]
Hexobarbital Sodium Salt* [50-09-9]
Hexobendine [54-03-5]
Hexobendine Dihydrochloride*
[50-62-4]
Hexocyclium Methyl Sulfate
[115-63-9]
Hexoprenaline [3215-70-1]
Hexoprenaline Dihydrochloride*
[4323-43-7]
Hexoprenaline Sulfate* [32266-10-7]
Hexylcaine Hydrochloride [532-76-3]
2-Hexyldecanoic Acid [25354-97-6]
2-Hexyldecanoic Acid Sodium Salt*
[536-37-8]
Hexylene Glycol [107-41-5]
Hexyl Methyl Ketone [111-13-7]
4-Hexylresorcinol [136-77-6]

Denotes derivative of title compound

Hexythiazox [78587-05-0]
Hippuric Acid [495-69-2]
Hirsutic Acid C [3650-17-7]
Hirudin [8001-27-2]
Histaminase [9001-53-0]
Histamine [51-45-6]
Histamine Phosphate* [51-74-1]
Histapyrrodine [493-80-1]
Histapyrrodine Hydrochloride*
  [6113-17-3]
Histidine [71-00-1]
Histrelin [76712-82-8]
HMPA [680-31-9]
Holarrhenine [561-22-8]
Holmium [7440-60-0]
Holomycin [488-04-0]
Homarine [445-30-7]
Homatropine [87-00-3]
Homatropine DL-Hydrobromide*
  [51-56-9]
Homatropine DL-Methylbromide*
  [80-49-9]
Homidium [3546-21-2]
Homidium Bromide* [1239-45-8]
Homobatrachotoxin* [23509-17-3]
Homocamfin [535-86-4]
Homochelidonine [476-33-5]
Homochlorcyclizine [848-53-3]
Homochlorcyclizine Dihydrochloride*
  [1982-36-1]
Homochlorcyclizine Hydrochloride*
  [32059-24-8]
Homocysteine [6027-13-0]
Homocystine [462-10-2]
Homoeriodictyol [446-71-9]
Homofenazine [3833-99-6]
Homofenazine Difumarate*
  [3093-71-8]
Homofenazine Dihydrochloride*
  [1256-01-5]
Homogentisic Acid [451-13-8]
Homosalate [118-56-9]
Homoserine [672-15-1]
Homovanillic Acid [306-08-1]
HON [4439-84-3]
Honokiol [35354-74-6]
Hopantenic Acid [18679-90-8]
Hordenine [539-15-1]
HPA-23 [89899-81-0]
HQNO [341-88-8]
Human Chorionic Gonadotropin*
  [9002-61-3]
Human Growth Hormone*
  [12629-01-5]
Humulene [6753-98-6]
β-Humulene* [116-04-1]
Humulon [26472-41-3]
Huperzine A [102518-79-6]
Hyalobiuronic Acid [499-15-0]
Hyaluronic Acid [9004-61-9]
Hyaluronic Acid Sodium Salt*
  [9067-32-7]
Hyaluronidases [9001-54-1]
Hycanthone [3105-97-3]
Hycanthone Mesylate* [23255-93-8]
Hydantoin [461-72-3]
Hydnocarpic Acid [459-67-6]
Hydracarbazine [3614-47-9]
Hydracrylic Acid [503-66-2]
Hydralazine [86-54-4]
Hydralazine Hydrochloride*
  [304-20-1]
Hydrallostane [516-41-6]
Hydramethylnon [67485-29-4]
Hydramitrazine [13957-36-3]
Hydramitrazine d-Tartrate*
  [20423-87-4]
Hydrargaphen [14235-86-0]
Hydrastine [118-08-1]

Hydrastine Hydrochloride*
  [5936-28-7]
Hydrastinine [6592-85-4]
Hydrazine [302-01-2]
Hydrazine Dihydrochloride*
  [5341-61-7]
Hydrazine Hydrate [7803-57-8]
Hydrazine Sulfate [10034-93-2]
Hydrazine Tartrate [634-62-8]
4-Hydrazinobenzenesulfonic Acid
  [98-71-5]
2-Hydrazinoethanol [109-84-2]
Hydrazoic Acid [7782-79-8]
Hydrindantin [5103-42-4]
Hydriodic Acid [10034-85-2]
Hydrobenzoin [492-70-6]
Hydrobromic Acid [10035-10-6]
Hydrocarbostyril [553-03-7]
Hydrochloric Acid [7647-01-0]
Hydrochlorothiazide [58-93-5]
Hydrocinchonidine [485-64-3]
Hydrocinchonine [485-65-4]
Hydrocinnamic Acid [501-52-0]
Hydrocodone [125-29-1]
Hydrocodone Bitartrate Hemi-
  pentahydrate* [34195-34-1]
Hydrocortamate [76-47-1]
Hydrocortamate Hydrochloride*
  [125-03-1]
Hydrocortisone [50-23-7]
Hydrocortisone 21-Acetate* [50-03-3]
Hydrocortisone 17-Butyrate*
  [13609-67-1]
Hydrocortisone 21-Phosphate Di-
  sodium Salt* [6000-74-4]
Hydrocortisone 21-Sodium
  Succinate* [125-04-2]
Hydrocortisone 17-Valerate*
  [57524-89-7]
Hydrocotarnine [550-10-7]
Hydroflumethiazide [135-09-1]
Hydrofluoric Acid [7664-39-3]
Hydrofuramide [494-47-3]
Hydrogen [1333-74-0]
Hydrogen Bromide [10035-10-6]
Hydrogen Chloride [7647-01-0]
Hydrogen Cyanide [74-90-8]
Hydrogen Fluoride [7664-39-3]
Hydrogen Iodide [10034-85-2]
Hydrogen Peroxide [7722-84-1]
Hydrogen Selenide [7783-07-5]
Hydrogen Sulfide [7783-06-4]
Hydrogen Telluride [7783-09-7]
Hydrogen Tetracarbonylferrate(II)
  [17440-90-3]
Hydrohydrastinine [494-55-3]
Hydromorphone [466-99-9]
Hydromorphone Hydrochloride*
  [71-68-1]
Hydrone [8049-21-6]
Hydroorotic Acid [155-54-4]
Hydroprene [41096-46-2] (unspecified)
Hydroprene [41205-09-8] (±)
Hydroprene (S)-Form* [65733-18-8]
Hydroquinidine [1435-55-8]
Hydroquinidine Hydrochloride*
  [1476-98-8]
Hydroquinine [522-66-7]
Hydroquinone [123-31-9]
Hydrotalcite* [12304-65-3]
Hydroxocobalamin [13422-51-0]
Hydroxyamphetamine [1518-86-1]
Hydroxyamphetamine Hydro-
  bromide* [306-21-8]
p-Hydroxybenzaldehyde [123-08-0]
p-Hydroxybenzoic Acid [99-96-7]
α-Hydroxybenzylphosphinic Acid
  [52705-43-8]
4'-Hydroxybutyranilide [101-91-7]
γ-Hydroxybutyrate [591-81-1]

γ-Hydroxybutyrate Sodium Salt*
  [502-85-2]
β-Hydroxybutyric Acid [300-85-6]
3-Hydroxycamphor [10373-81-6]
Hydroxychloroquine [118-42-3]
Hydroxychloroquine Sulfate*
  [747-36-4]
1α-Hydroxycholecalciferol
  [41294-56-8]
24-Hydroxycholesterol [474-73-7]
25-Hydroxycholesterol [2140-46-7]
Hydroxycodeinone [508-54-3]
Hydroxydione Sodium [53-10-1]
20-Hydroxyecdysone* [5289-74-7]
p-Hydroxyephedrine [365-26-4]
N-Hydroxyethylpromethazine Chloride [2090-54-2]
O-(β-Hydroxyethyl)rutinosides (mixture)* [55965-63-4]
Hydroxyglutamic Acid [533-62-0]
8-Hydroxy-7-iodo-5-quinolinesulfonic
  Acid [547-91-1]
8-Hydroxy-7-iodo-5-quinolinesulfonic
  Acid Compd with Chloroquine
  (2:1)* [7270-12-4]
4-Hydroxyisophthalic Acid [636-46-4]
Hydroxylamine [7803-49-8]
Hydroxylamine Hydrochloride*
  [5470-11-1]
Hydroxylamine Sulfate* [10039-54-0]
Hydroxylupanine [15358-48-2]
3-Hydroxymethyl-β-carboline*
  [65474-79-5]
1-(Hydroxymethyl)-5,5-dimethylhy-
  dantoin [116-25-6]
17-Hydroxy-16-methylene-Δ⁶-proges-
  terone [10087-54-4]
17-Hydroxy-16-methylene-Δ⁶-proges-
  terone 17-Acetate* [805-84-5]
5-(Hydroxymethyl)-2-furaldehyde
  [67-47-0]
N-(Hydroxymethyl)nicotinamide
  [3569-99-1]
1-Hydroxy-2-naphthoic Acid [86-48-6]
3-Hydroxy-2-naphthoic Acid [92-70-6]
Hydroxypethidine [468-56-4]
Hydroxyphenamate [50-19-1]
N-(4-Hydroxyphenyl)glycine
  [122-87-2]
17α-Hydroxyprogesterone [68-96-2]
17α-Hydroxyprogesterone Caproate*
  [630-56-8]
Hydroxyproline [51-35-4]
Hydroxyproline cis-Form* [618-27-9]
Hydroxypropyl Cellulose [9004-64-2]
Hydroxypropyl Methylcellulose
  [9004-65-3]
8-Hydroxyquinoline [148-24-3]
8-Hydroxyquinoline Aluminum
  Sulfate* [153-77-5]
8-Hydroxyquinoline Sulfate*
  [134-31-6]
8-Hydroxy-5-quinolinesulfonic Acid
  [84-88-8]
Hydroxystilbamidine [495-99-8]
Hydroxystilbamidine Isethionate*
  [533-22-2]
Hydroxystreptomycin [6835-00-3]
Hydroxytetracaine [490-98-2]
5-Hydroxytryptophan [56-69-9]
5-Hydroxytryptophan D-Form*
  [4350-07-6]
5-Hydroxytryptophan DL-Form*
  [114-03-4]
5-Hydroxytryptophan L-Form*
  [4350-09-8]
Hydroxyurea [127-07-1]
Hydroxyzine [68-88-2]
Hydroxyzine Dihydrochloride*
  [2192-20-3]

Denotes derivative of title compound

Hydroxyzine Pamoate* [10246-75-0]
Hyenanchin [3484-46-6]
Hygrine [496-49-1]
Hygromycin [6379-56-2]
Hygromycin B [31282-04-9]
Hygrophylline [3573-82-8]
Hymecromone [90-33-5]
Hymecromone O,O-Diethyl Phosphorothioate [299-45-6]
Hymexazol [10004-44-1]
Hyodeoxycholic Acid [83-49-8]
Hyoscyamine [101-31-5]
Hyoscyamine Hydrobromide* [306-03-6]
Hyoscyamine Sulfate Dihydrate* [6835-16-1]
Hypalon [9008-08-6]
Hypaphorine [487-58-1]
Hypericin [548-04-9]
Hypochlorous Acid [7790-92-3]
Hypoglycine A [156-56-9]
Hypoglycine B [502-37-4]
Hypophosphoric Acid [7803-60-3]
Hypophosphorous Acid [6303-21-5]
Hypoxanthine [68-94-0]
Hypusine [34994-11-1]
Hypusine Dihydrochloride* [82310-93-8]
Ibafloxacin [91618-36-9]
Ibogaine [83-74-9]
Ibopamine [66195-31-1]
Ibopamine Hydrochloride* [75011-65-3]
Ibotenic Acid [2552-55-8]
Ibudilast [50847-11-5]
Ibufenac [1553-60-2]
Ibuprofen [15687-27-1]
Ibuprofen S-Form L-Lysine Salt Monohydrate* [141505-32-0]
Ibuprofen Methylglucamine Salt* [135861-34-6]
Ibuproxam [53648-05-8]
Ibutilide [122647-31-8]
Ibutilide Fumarate* [122647-32-9]
Icariin* [489-32-7]
Ichthammol [8029-68-3]
Ichthyopterin [490-58-4]
Idarubicin [58957-92-9]
Idarubicin Hydrochloride* [57852-57-0]
Idebenone [58186-27-9]
Idose [2152-76-3]
Idoxuridine [54-42-2]
Idrocilamide [6961-46-2]
Ifenprodil [23210-56-2]
Ifosfamide [3778-73-2]
Illudin M* [1146-04-9]
Illudin S* [1149-99-1]
Iloprost [78919-13-8]
Imazamethabenz [81405-85-8]
Imazapyr [81334-34-1]
Imazapyr Isopropylamine Salt* [81510-83-0]
Imazaquin [81335-37-7]
Imazethapyr [81335-77-5]
Imazethapyr Ammonium Salt* [101917-66-2]
Imiclopazine [7224-08-0]
Imiclopazine Dihydrochloride* [7414-95-1]
Imidacloprid [105827-78-9]
Imidapril [89371-37-9]
Imidaprilat* [89371-44-8]
Imidapril Monohydrochloride* [89396-94-1]
Imidazole [288-32-4]
Imidazole Salicylate [36364-49-5]
2-Imidazolidinone [120-93-4]
Imidocarb [27885-92-3]

Imidocarb Dihydrochloride* [5318-76-3]
Iminodiacetic Acid [142-73-4]
4,4'-Iminodicyclohexanecarboxylic Acid [53555-68-3]
Imipenem [74431-23-5]
Imipenem Combination with Cilastatin Sodium* [85960-17-4]
Imipramine [50-49-7]
Imipramine Hydrochloride* [113-52-0]
Imipramine N-Oxide [6829-98-7]
Imipramine N-Oxide Hydrochloride* [20438-98-6]
Imiquimod [99011-02-6]
Imolamine [318-23-0]
Imolamine Hydrochloride* [15823-89-9]
Imperatorin [482-44-0]
Imperialine [18059-10-4]
Improsulfan [13425-98-4]
Improsulfan Hydrochloride* [3458-22-8]
Improsulfan Tosylate* [32784-82-0]
Inaperisone [99323-21-4] (unspecified)
Inaperisone [125676-39-3] (±)
Inaperisone Hydrochloride* [90878-85-6] (unspecified)
Inaperisone Hydrochloride* [118230-97-0] (±)
Indaconitine [4491-19-4]
Indalpine [63758-79-2]
Indan [496-11-7]
Indanazoline [40507-78-6]
Indanazoline Hydrochloride* [40507-80-0]
Indanthrene [81-77-6]
Indapamide [26807-65-8]
1H-Indazole [271-44-3]
Indecainide [74517-78-5]
Indecainide Hydrochloride* [73681-12-6]
Indeloxazine Hydrochloride [65043-22-3]
Indene [95-13-6]
Indenolol [60607-68-3]
Indenolol Hydrochloride* [81789-85-7]
Indican (Metabolic Indican) [487-94-5]
Indican (Plant Indican) [487-60-5]
Indigo [482-89-3]
Indigo Carmine [860-22-0]
Indinavir [150378-17-9]
Indinavir Sulfate* [157810-81-6]
Indium [7440-74-6]
Indium Antimonide [1312-41-0]
Indium Arsenide [1303-11-3]
Indium Gallium Aluminium Phosphide [108424-49-3]
Indium Gallium Aluminium Phosphide [108730-13-8]
Indium 111In Satumomab Pendetide* [138955-27-8]
Indium Oxide [1312-43-2]
Indium Phosphide [22398-80-7]
Indium Selenide [1312-42-1]
Indium Sulfate [13464-82-9]
Indium Telluride [1312-45-4]
Indium Trichloride [10025-82-8]
Indium Trifluoride [7783-52-0]
Indobufen [63610-08-2]
Indocyanine Green [3599-32-4]
Indole [120-72-9]
Indoleacetic Acid [87-51-4]
Indolebutyric Acid [133-32-4]
Indolmycin [21200-24-8]
3-Indolylacetone [1201-26-9]
Indomethacin [53-86-1]

Indomethacin Sodium Salt Trihydrate* [74252-25-8]
Indoprofen [31842-01-0]
Indoprofen (−)-Form* [53086-14-9]
Indoprofen (+)-Form* [53086-13-8]
Indoramin [26844-12-2]
Indoramin Hydrochloride* [38821-52-2]
Indospicine [16377-00-7]
Inosine [58-63-9]
Inosine Pranobex [36703-88-5]
Inosinic Acid [131-99-7]
Inositol [87-89-8]
Inositol Monophosphate* [573-35-3]
Inositol Niacinate [6556-11-2]
Insularine [549-07-5]
Insulin [9004-10-8]
Insulinase [9013-83-6]
Insulin (Bovine)* [11070-73-8]
Insulin (Human)* [11061-68-0]
Insulin (Porcine)* [12584-58-6]
Integrelin [157630-07-4]
Interferon-γ [82115-62-6]
Interferon Alfa-2a* [76543-88-9]
Interferon Alfa-2b* [99210-65-8]
Interferon Beta-1b* [145155-23-3]
Interferon Gamma-1b* [98059-61-1]
Interleukin-2 [85898-30-2]
Interleukin-2 (Recombinant Human)* [110942-02-4]
Inulin [9005-80-5]
Invertase [9001-57-4]
Invert Sugar [8013-17-0]
Iobenguane [80663-95-2]
Iobenzamic Acid [3115-05-7]
Iocarmic Acid [10397-75-8]
Iocarmic Acid Di-N-methylglucamine Salt* [54605-45-7]
Iocetamic Acid [16034-77-8]
Iodamide [440-58-4]
Iodamide N-Methyl-D-glucamine Salt* [18656-21-8]
Iodic Acid [7782-68-5]
Iodinated Glycerol [5634-39-5]
Iodine [7553-56-2]
Iodine Colloidal [7553-56-2]
Iodine Heptafluoride [16921-96-3]
Iodine Monobromide [7789-33-5]
Iodine Monochloride [7790-99-0]
Iodine Pentafluoride [7783-66-6]
Iodine Pentoxide [12029-98-0]
Iodine Trichloride [865-44-1]
Iodinin [68-81-5]
Iodipamide [606-17-7]
Iodipamide Bis[N-methylglucamine] Salt* [3521-84-4]
Iodipamide Disodium Salt* [2618-26-0]
Iodixanol [92339-11-2]
Iodized Oil [8001-40-9]
Iodoacetic Acid [64-69-7]
Iodoalphionic Acid [577-91-3]
p-Iodoaniline [540-37-4]
o-Iodoanisole [529-28-2]
Iodobenzene [591-50-4]
o-Iodobenzoic Acid [88-67-5]
Iodochlorhydroxyquin [130-26-7]
Iodofenphos [18181-70-9]
Iodoform [75-47-8]
o-Iodohippurate Sodium [133-17-5]
o-Iodophenol [533-58-4]
p-Iodophenol [540-38-5]
Iodophthalein Sodium [2217-44-9]
Iodophthalein Sodium Free Acid* [386-17-4]
Iodopsin [1415-94-7]
Iodopyracet [300-37-8]
Iodopyrrole [87-58-1]
Iodoquinol [83-73-8]
N-Iodosuccinimide [516-12-1]

3-Iodotyrosine [70-78-0]
Iofetamine ^{123}I [75917-92-9]
Iofetamine ^{123}I Hydrochloride*
[85068-76-4]
Ioglycamic Acid [2618-25-9]
Ioglycamic Acid Meglumine Salt*
[14317-18-1]
Iohexol [66108-95-0]
Iomeglamic Acid [25827-76-3]
Iopamidol [60166-93-0]
Iopanoic Acid [96-83-3]
Iopanoic Acid d-Form* [17879-96-8]
Iopanoic Acid l-Form* [17879-97-9]
Iopanoic Acid Sodium Salt*
[2497-78-1]
Iopentol [89797-00-2]
Iophendylate [99-79-6]
Iophenoxic Acid [96-84-4]
Iopromide [73334-07-3]
Iopronic Acid [41473-08-9]
Iopydol [5579-92-0]
Iopydone [5579-93-1]
Iothalamic Acid [2276-90-6]
Iothalamic Acid ^{131}I Labeled Sodium
Salt* [15845-98-4]
Iothalamic Acid N-Methylglucamine
Salt* [13087-53-1]
Iothalamic Acid Sodium Salt*
[1225-20-3]
Iothion [534-08-7]
Iotrolan [79770-24-4]
Ioversol [87771-40-2]
Ioxaglate Meglumine mixture with
Ioxaglate Sodium* [76820-74-1]
Ioxaglic Acid [59017-64-0]
Ioxilan [107793-72-6]
Ipodate [5587-89-3]
Ipodate Calcium Salt* [1151-11-7]
Ipodate Sodium Salt* [1221-56-3]
Ipratropium Bromide [22254-24-6]
Ipriflavone [35212-22-7]
Iprindole [5560-72-5]
Iprindole Hydrochloride*
[20432-64-8]
Iproclozide [3544-35-2]
Iprodione [36734-19-7]
Iproniazid [54-92-2]
Iproniazid Phosphate* [305-33-9]
Ipronidazole [14885-29-1]
Ipsapirone [95847-70-4]
Ipsapirone Hydrochloride*
[92589-98-5]
Irbesartan [138402-11-6]
Iridium [7439-88-5]
Iridium Hexafluoride [7783-75-7]
Iridium Sesquioxide [1312-46-5]
Iridium Trichloride [10025-83-9]
Iridomyrmecin [485-43-8]
Irigenin [548-76-5]
Irinotecan [97682-44-5]
Irinotecan Hydrochloride*
[100286-90-6]
Irinotecan Hydrochloride Trihydrate*
[136572-09-3]
Irisolone [3301-68-6]
Iron [7439-89-6]
α-Irone [79-69-6]
β-Irone [79-70-9]
γ-Irone [79-68-5]
Iron Pentacarbonyl [13463-40-6]
Iron Sorbitex [1338-16-5]
Irsogladine [57381-26-7]
Irsogladine Maleate* [84504-69-8]
Isanic Acid [506-25-2]
Isatide [464-73-3]
Isatin [91-56-5]
Isatropic Acid [596-56-5]
Isaxonine [4214-72-6]
Isaxonine Phosphate* [65606-21-5]
Isazofos [42509-80-8]

Isbogrel [89667-40-3]
Isepamicin [58152-03-7]
Isepamicin Sulfate* [67814-76-0]
Isethionic Acid [107-36-8]
Isoaminile [77-51-0]
Isoaminile Citrate* [28416-66-2]
Isoaminile Cyclamate* [10075-36-2]
Isoamyl Acetate [123-92-2]
Isoamylamine [107-85-7]
Isoamyl Benzoate [94-46-2]
Isoamyl Bromide [107-82-4]
Isoamyl Butyrate [106-27-4]
Isoamyl Chloride [107-84-6]
Isoamyl Cyanide [542-54-1]
Isoamyl Ether [544-01-4]
Isoamyl Formate [110-45-2]
Isoamyl Iodide [541-28-6]
Isoamyl Isovalerate [659-70-1]
Isoamyl Nitrate [543-87-3]
Isoamyl Nitrite [110-46-3]
Isoamyl Phthalate [605-50-5]
Isoamyl Salicylate [87-20-7]
Isoapo-β-erythroidine [2581-07-9]
Isoascorbic Acid [89-65-6]
Isobenzan [297-78-9]
Isoborneol [124-76-5]
Isobornyl Thiocyanoacetate
[115-31-1]
Isobutol* [41663-50-7]
Isobutyl Acetate [110-19-0]
Isobutyl Alcohol [78-83-1]
Isobutylamine [78-81-9]
Isobutyl p-Aminobenzoate [94-14-4]
Isobutylbenzene [538-93-2]
Isobutyl Bromide [78-77-3]
Isobutyl n-Butyrate [539-90-2]
Isobutyl Carbamate [543-28-2]
Isobutyl Chloride [513-36-0]
Isobutyl Chlorocarbonate [543-27-1]
Isobutylene [115-11-7]
Isobutyl Ether [628-55-7]
Isobutyl Formate [542-55-2]
Isobutyl Iodide [513-38-2]
Isobutyl Isobutyrate [97-85-8]
Isobutyl Isovalerate [589-59-3]
Isobutyl Mercaptan [513-44-0]
Isobutyl Nitrate [543-29-3]
Isobutyl Nitrite [542-56-3]
Isobutyl Propionate [540-42-1]
Isobutyl Stearate [646-13-9]
Isobutyl Sulfide [592-65-4]
Isobutyl Thiocyanate [591-84-4]
Isobutyl Urethane [539-89-9]
Isobutyraldehyde [78-84-2]
Isobutyric Acid [79-31-2]
Isobutyronitrile [78-82-0]
Isocarboxazid [59-63-2]
Isocaryophyllene* [118-65-0]
Isochondrodendrine [477-62-3]
Isocil [314-42-1]
Isocinchomeronic Acid [100-26-5]
Isoconazole [27523-40-6]
Isoconazole Nitrate* [24168-96-5]
Isocorybulbine [22672-74-8]
Isocorydine [475-67-2]
Isocorypalmine [53447-14-6]
Isocrotonic Acid [503-64-0]
Isocyanic Acid [75-13-8]
Isodurene [527-53-7]
Isoestradiol [517-04-4]
8-Isoestrone [517-06-0]
Isoetharine [530-08-5]
Isoetharine Hydrochloride*
[2576-92-3]
Isoeugenol [97-54-1]
Isoeugenol cis-Form* [5912-86-7]
Isoeugenol trans-Form* [5932-68-3]
Isofenphos [25311-71-1]
Isofezolac [50270-33-2]
Isoflavone [574-12-9]

Isoflupredone [338-95-4]
Isoflupredone 21-Acetate* [338-98-7]
Isoflurane [26675-46-7]
Isoflurophate [55-91-4]
L-Isoglutamine [636-65-7]
Isoladol [530-34-7]
Isolan [119-38-0]
Isoleucine [73-32-5]
α-Isolupanine* [486-87-3]
Isolysergic Acid [478-95-5]
Isomaltol [3420-59-5]
Isometamidium Chloride [34301-55-8]
Isomethadol [25117-79-7]
Isomethadone [466-40-0]
Isomethadone d-Form* [26594-41-2]
Isomethadone dl-Form* [116836-09-0]
Isomethadone l-Form* [561-10-4]
Isomethadone d-Form Hydrochloride*
[63814-06-2]
Isomethadone l-Form Hydrochloride*
[7487-81-2]
Isometheptene [503-01-5]
Isometheptene Hydrochloride*
[6168-86-1]
Isoniazid [54-85-3]
Isoniazid 4-Aminosalicylate*
[2066-89-9]
Isoniazid Methanesulfonate*
[13447-95-5]
Isoniazid Methanesulfonate Calcium*
[6059-26-3]
Isoniazid Methanesulfonate Sodium*
[3804-89-5]
Isonicotinic Acid [55-22-1]
Isonicotinic Acid Diethylamide
[530-40-5]
Isonipecotic Acid [498-94-2]
Isonitrosoacetone [306-44-5]
Isonitrosoacetophenone [532-54-7]
Isonixin [57021-61-1]
Isooctane [540-84-1]
Isooctyl Alcohol [26952-21-6]
Isopelletierine* [539-00-4]
Isopentyl Alcohol [123-51-3]
Isophorone [78-59-1]
Isophthalic Acid [121-91-5]
Isophytol [505-32-8]
Isopilosine [491-88-3]
Isopimaric Acid [5835-26-7]
Isoprene [78-79-5]
Isopromethazine [303-14-0]
Isopropalin [33820-53-0]
Isopropamide Iodide [71-81-8]
Isopropenyl Acetate [108-22-5]
4-(5-Isopropenyl-2-methyl-1-cyclopen-
ten-1-yl)-2-butanone [87-45-6]
Isopropyl Acetate [108-21-4]
Isopropyl Acetoacetate [542-08-5]
Isopropylacetone [108-10-1]
Isopropyl Alcohol [67-63-0]
Isopropylamine [75-31-0]
Isopropyl Bromide [75-26-3]
Isopropyl Chloride [75-29-6]
Isopropyl Ether [108-20-3]
Isopropylidene Glycerol [100-79-8]
Isopropyl Iodide [75-30-9]
Isopropyl Myristate [110-27-0]
Isopropyl Nitrite [541-42-4]
Isoproterenol [7683-59-2]
Isoproterenol dl-Form Sulfate
Dihydrate* [6700-39-6]
Isoproturon [34123-59-6]
Isopyrocalciferol [474-70-4]
Isoquassin [21293-20-9]
Isoquercitrin [21637-25-2]
Isoquinoline [119-65-3]
Isorubijervine [468-45-1]
Isosafrole [120-58-1]
Isosorbide [652-67-5]
Isosorbide Dinitrate [87-33-2]

Denotes derivative of title compound

Isosorbide-5-mononitrate* [16051-77-7]
β-Isosparteine [24915-04-6]
Isothebaine [568-21-8]
Isothipendyl [482-15-5]
Isothipendyl Hydrochloride* [1225-60-1]
Isovaleraldehyde [590-86-3]
Isovaleramide [541-46-8]
Isovaleric Acid [503-74-2]
Isovaleryl Chloride [108-12-3]
Isovaleryl Diethylamide [533-32-4]
2-Isovalerylindane-1,3-dione [83-28-3]
Isovaline [595-39-1]
Isovaline D-Form* [3059-97-0]
Isovaline L-Form* [595-40-4]
Isoxaben [82558-50-7]
Isoxepac [55453-87-7]
Isoxicam [34552-84-6]
Isoxsuprine [395-28-8]
Isoxsuprine Hydrochloride* [579-56-6]
Isradipine [75695-93-1]
Itaconic Acid [97-65-4]
Itraconazole [84625-61-6]
Itramin Tosylate [13445-63-1]
Ivermectin [70288-86-7]
Janus Green B [2869-83-2]
Japan Wax [8001-39-6]
Japonilure [64726-91-6]
Jasmolin I* [4466-14-2]
Jasmolin II* [1172-63-0]
Jasmone [488-10-8]
Jatrorrhizine [3621-38-3]
Javanicin [476-45-9]
Jervine [469-59-0]
Jesaconitine [16298-90-1]
Josamycin [16846-24-5]
Josamycin Propionate* [51016-68-3]
Juglone [481-39-0]
Julocrotine [492-87-5]
Justicidin A* [25001-57-4]
Justicidin B* [17951-19-8]
Juvenile Hormone I* [13804-51-8]
Juvenile Hormone II* [34218-61-6]
Juvenile Hormone III* [22963-93-5]
Kaempferol [520-18-3]
Kainic Acid [487-79-6]
Kallidin [342-10-9]
Kallikrein [9001-01-8]
Kanamycin [8063-07-8]
Kanamycin A* [59-01-8]
Kanamycin A Sulfate* [25389-94-0]
Kanamycin B* [4696-76-8]
Karanjin [521-88-0]
Karaya Gum [9000-36-6]
Karsil [2533-89-3]
Kawain [500-64-1]
Kawain (±)-Form* [3155-48-4]
Kebuzone [853-34-9]
Keratinase [9025-41-6]
Kermesic Acid [18499-92-8]
Kerosene [8008-20-6]
Ketamine [6740-88-1]
Ketamine Hydrochloride* [1867-66-9]
Ketanserin [74050-98-9]
Ketanserin Tartrate* [83846-83-7]
Ketazolam [27223-35-4]
Ketene [463-51-4]
Kethoxal [27762-78-3]
Kethoxal Bis(thiosemicarbazone)* [2507-91-7]
Ketipic Acid [533-76-6]
Ketobemidone [469-79-4]
Ketobemidone Hydrochloride* [5965-49-1]
Ketoconazole [65277-42-1]
α-Ketoglutaric Acid [328-50-7]
2-Keto-L-gulonic Acid [526-98-7]
Ketoprofen [22071-15-4]

Ketoprofen Lysine Salt* [57469-78-0]
11-Ketoprogesterone [516-15-4]
Ketorolac [74103-06-3]
Ketorolac (±)-Form Tromethamine Salt* [74103-07-4]
Ketotifen [34580-13-7]
Ketotifen Fumarate* [34580-14-8]
Khellin [82-02-0]
Khellol Glucoside [17226-75-4]
Kinetin [525-79-1]
Kino [8052-27-5]
Kinoprene [42588-37-4]
Kistrin [127829-86-1]
Kitol [4626-00-0]
Kojic Acid [501-30-4]
Kopsine [559-48-8]
Krebiozen [9008-19-9]
Krypton [7439-90-9]
Krypton Difluoride* [13773-81-4]
Kurchessine [6869-45-0]
Kurcholessine [6869-47-2]
Kyanmethin [461-98-3]
Kynurenic Acid [492-27-3]
Kynurenine [343-65-7]
Labetalol [36894-69-6]
Labetalol Hydrochloride* [32780-64-6]
Lachesine [1164-38-1]
Lacidipine [103890-78-4]
Lactaroviolin [85-33-6]
Lactate Dehydrogenase [9001-60-9]
D-Lactic Acid [10326-41-7]
DL-Lactic Acid [598-82-3]
L-Lactic Acid [79-33-4]
Lactic Acid Lactate [617-57-2]
Lactitol [585-86-4]
Lactitol Dihydrate* [81025-03-8]
Lactitol Monohydrate* [81025-04-9]
Lactobacillic Acid [19625-10-6]
Lactobionic Acid [96-82-2]
p-Lactophenetide [539-08-2]
Lactose [63-42-3]
Lactucin [1891-29-8]
Lactulose [4618-18-2]
Laidlomycin [56283-74-0]
Laidlomycin Propionate Potassium* [84799-02-0]
Lamifiban [144412-49-7]
Lamifiban Trifluoroacetate Salt* [144412-50-0]
Laminaran [9008-22-4]
Lamivudine [134678-17-4]
Lamotrigine [84057-84-1]
Lanatoside A* [17575-20-1]
Lanatoside B* [17575-21-2]
Lanatoside C* [17575-22-3]
Lanatoside D* [11030-31-2]
Lankamycin [30042-37-6]
Lanoconazole [101530-10-3]
Lanosterol [79-63-0]
Lansoprazole [103577-45-3]
Lanthanum [7439-91-0]
Lanthionine [922-55-4]
Lapachol [84-79-7]
Lappaconitine [32854-75-4]
Lapyrium Chloride [6272-74-8]
Lasalocid A [25999-31-9]
Laserpitin [7067-12-1]
Lasiocarpine [303-34-4]
Latanoprost [130209-82-4]
Latrunculin A* [76343-93-6]
Latrunculin B* [76343-94-7]
Laudanidine [301-21-3]
Laudanine [85-64-3]
Laudanosine [2688-77-9]
Laudexium Methyl Sulfate [3253-60-9]
Laureline [81-38-9]
Laureline (±)-Form* [3749-97-1]
Laureline (S)-Form* [65981-49-9]

Laurepukine [34029-94-2]
Lauric Acid [143-07-7]
Laurocapram [59227-89-3]
Lauroguadine [135-43-3]
Laurolinium Acetate [146-37-2]
Laurotetanine [128-76-7]
3-O-Lauroylpyridoxol Diacetate [1562-13-6]
Lauryl Bromide [143-15-7]
Lawesson's Reagent [19172-47-5]
Lawrencium [22537-19-5]
Lawsone [83-72-7]
Lazabemide [103878-84-8]
Lazabemide Monohydrochloride* [103878-83-7]
Lazurite [1302-83-6]
Lead [7439-92-1]
Lead Acetate [301-04-2]
Lead Antimonate(V) [13510-89-9]
Lead Arsenate [10102-48-4]
Lead Arsenite [10031-13-7]
Lead Azide [13424-46-9]
Lead Borate [10214-39-8]
Lead Bromate [34018-28-5]
Lead Bromide [10031-22-8]
Lead Butyrate [819-73-8]
Lead Chlorate [10294-47-0]
Lead Chloride [7758-95-4]
Lead Chromate(VI) [7758-97-6]
Lead Chromate(VI) Oxide [18454-12-1]
Lead Dioxide [1309-60-0]
Lead Fluoride [7783-46-2]
Lead Formate [811-54-1]
Lead Hexafluorosilicate [25808-74-6]
Lead Hydroxide [1311-11-1]
Lead Hypophosphite [10294-58-3]
Lead Iodide [10101-63-0]
Lead Lactate [18917-82-3]
Lead Molybdate(VI) [10190-55-3]
Lead Monoxide [1317-36-8]
Lead Nitrate [10099-74-8]
Lead Oleate [1120-46-3]
Lead Oxalate [814-93-7]
Lead Phosphate [7446-27-7]
Lead Selenate [7446-15-3]
Lead Selenite [7488-51-9]
Lead Sesquioxide [1314-27-8]
Lead Sodium Thiosulfate [10101-94-7]
Lead Stearate [1072-35-1]
Lead Subacetate [1335-32-6]
Lead Sulfate [7446-14-2]
Lead Sulfide [1314-87-0]
Lead Telluride [1314-91-6]
Lead Tetraacetate [546-67-8]
Lead Tetrafluoride [7783-59-7]
Lead Tetroxide [1314-41-6]
Lead Thiocyanate [592-87-0]
Lead Vanadate(V) [10099-79-3]
Lecithin [8002-43-5]
Ledol [577-27-5]
Lefetamine [7262-75-1]
Lefetamine Hydrochloride* [14148-99-3]
Leiopyrrole [5633-16-9]
Leiopyrrole Hydrochloride* [14435-78-0]
Lenacil [2164-08-1]
Lenampicillin [86273-18-9]
Lenampicillin Hydrochloride* [80734-02-7]
Lenograstim* [135968-09-1]
Lenthionine [292-46-6]
Lentinan [37339-90-5]
Leonurine [24697-74-3]
Lepidine [491-35-0]
Leptodactyline [13957-33-0]
Leptophos [21609-90-5]
Lercanidipine [100427-26-7]

Lercanidipine Hydrochloride*
[132866-11-6]
Lesopitron [132449-46-8]
Lesopitron Dihydrochloride*
[132449-89-9]
Lethane 60 [301-11-1]
Lethane 384 [112-56-1]
Letosteine [53943-88-7]
Letrozole [112809-51-5]
Leucine [61-90-5]
Leucinocaine Mesylate [135-44-4]
Leucocyanidin [480-17-1]
Leucodrin [14225-07-1]
Leucoglycodrin [13190-88-0]
Leucomycins [1392-21-8]
Leucopterin [492-11-5]
Leukotriene A$_4$* [72059-45-1]
Leukotriene B$_4$* [71160-24-2]
Leukotriene C$_4$* [72025-60-6]
Leukotriene D$_4$* [73836-78-9]
Leukotriene E$_4$* [75715-89-8]
Leuprolide [53714-56-0]
Leuprolide Acetate* [74381-53-6]
Levallorphan [152-02-3]
Levallorphan Tartrate* [71-82-9]
Levamisole [14769-73-4]
Levamisole Hydrochloride*
[16595-80-5]
Levcromakalim [94535-50-9]
Levobunolol [47141-42-4]
Levobunolol Hydrochloride*
[27912-14-7]
Levocabastine [79516-68-0]
Levocabastine Hydrochloride*
[79547-78-7]
Levodopa [59-92-7]
Levomepate [428-07-9]
Levomethadyl Acetate [1477-40-3]
Levomethadyl Acetate Hydro-
chloride* [43033-72-3]
Levomethorphan Hydrobromide*
[125-68-8]
Levonorgestrel* [797-63-7]
Levophacetoperane [24558-01-8]
Levophacetoperane Hydrochloride*
[23257-56-9]
Levopimaric Acid [79-54-9]
Levopropoxyphene [2338-37-6]
Levopropoxyphene Hydrochloride*
[1596-70-9]
Levopropoxyphene 2-Naphthalene-
sulfonate* [5714-90-9]
Levorphanol [77-07-6]
Levorphanol d-Form* [125-73-5]
Levorphanol dl-Form* [297-90-5]
Levorphanol dl-Form Hydrobromide*
[5985-35-3]
Levorphanol Tartrate Dihydrate*
[5985-38-6]
Levosulpiride* [23672-07-3]
Levothyroxine Sodium [55-03-8]
Levulinic Acid [123-76-2]
LH [9002-67-9]
LH-RH [9034-40-6]
LH-RH Hydrochloride* [51952-41-1]
Lichenin [1402-10-4]
Lidamidine [66871-56-5]
Lidamidine Hydrochloride*
[65009-35-0]
Lidocaine [137-58-6]
Lidocaine Hydrochloride Mono-
hydrate* [6108-05-0]
Lidofenin [59160-29-1]
Lidoflazine [3416-26-0]
Light Green SF Yellowish [5141-20-8]
Lignin [9005-53-2]
Lignoceric Acid [557-59-5]
Ligroin [8032-32-4]
Limaprost [74397-12-9]
Limettin [487-06-9]

Limonene [138-86-3]
Limonin [1180-71-8]
Linalool [78-70-6]
Linalyl Acetate [115-95-7]
Linamarin [554-35-8]
Linarin [480-36-4]
Linatine [10139-06-7]
Lincomycin [154-21-2]
Lincomycin Hydrochloride Mono-
hydrate* [7179-49-9]
Lindane [58-89-9]
Lindlar Catalyst [53092-86-7]
Lineatin [65035-34-9]
Linoleic Acid [60-33-3]
Linoleic Acid Aluminum Salt*
[645-17-0]
Linoleic Acid Cyclohexylamide*
[3207-50-9]
Linolenic Acid [463-40-1]
γ-Linolenic Acid [506-26-3]
Linuron [330-55-2]
Liothyronine [6893-02-3]
Liothyronine Hydrochloride*
[6138-47-2]
Liothyronine Sodium Salt* [55-06-1]
Lipase [9001-62-1]
Lipoprotein Lipase [9004-02-8]
Lipotropic Hormone [9035-55-6]
Lisinopril [83915-83-7]
Lisuride [18016-80-3]
Lisuride Maleate* [19875-60-6]
Lithium [7439-93-2]
Lithium Acetate [546-89-4]
Lithium Acetylsalicylate [552-98-7]
Lithium Amide [7782-89-0]
Lithium Benzoate [553-54-8]
Lithium Bitartrate [868-16-6]
Lithium Borate [1303-94-2]
Lithium Borohydride [16949-15-8]
Lithium Bromide [7550-35-8]
Lithium Carbonate [554-13-2]
Lithium Chloride [7447-41-8]
Lithium Chromate(VI) [14307-35-8]
Lithium Citrate [919-16-4]
Lithium Dichromate(VI) [13843-81-7]
Lithium Fluoride [7789-24-4]
Lithium Formate [556-63-8]
Lithium Hydride [7580-67-8]
Lithium Hydroxide [1310-65-2]
Lithium Iodide [10377-51-2]
Lithium Nitrate [7790-69-4]
Lithium Oxalate [553-91-3]
Lithium Oxide [12057-24-8]
Lithium Perchlorate [7791-03-9]
Lithium Selenate [15593-52-9]
Lithium Selenite [15593-51-8]
Lithium Silicate [10102-24-6]
Lithium Sulfate [10377-48-7]
Lithium Tetracyanoplatinate(II)
[14402-73-4]
Lithocholic Acid [434-13-9]
Lithopone [1345-05-7]
Litmocidin [1392-27-4]
Lobelanidine [552-72-7]
Lobelanine [579-21-5]
Lobeline [90-69-7]
Lobeline (±)-Form* [134-65-6]
Lobeline Hydrochloride* [134-63-4]
Lobeline Sulfate* [134-64-5]
Lobenzarit [63329-53-3]
Lobenzarit Disodium Salt*
[64808-48-6]
Lochnericine [72058-36-7]
Lochneridine [5980-01-8]
Lodoxamide [53882-12-5]
Lodoxamide Diethyl Ester*
[53882-13-6]
Lodoxamide Tromethamine Salt*
[63610-09-3]
Lofentanil [61380-40-3]

Lofentanil Oxalate* [61380-41-4]
Lofepramine [23047-25-8]
Lofepramine Hydrochloride*
[26786-32-3]
Lofexidine [31036-80-3]
Lofexidine Hydrochloride*
[21498-08-8]
Loflucarban [790-69-2]
Loganin [18524-94-2]
Loline [25161-91-5]
Lomefloxacin [98079-51-7]
Lomefloxacin Monohydrochloride*
[98079-52-8]
Lomerizine [101477-55-8]
Lomerizine Dihydrochloride*
[101477-54-7]
Lomustine [13010-47-4]
Lonapalene [91431-42-4]
Lonazolac [53808-88-1]
Lonazolac Calcium Salt* [75821-71-5]
Longifolene [475-20-7]
Lonidamine [50264-69-2]
Lonomycins Lonomycin A*
[58785-63-0]
Lonomycins Lonomycin A Sodium
Salt* [58845-80-0]
Looplure [14959-86-5]
Loperamide [53179-11-6]
Loperamide Hydrochloride*
[34552-83-5]
Lophophorine [17627-78-0]
Lophotoxin [78697-56-0]
Loprazolam [61197-73-7]
Loprazolam Methanesulfonate*
[70111-54-5]
Loprinone [106730-54-5]
Loracarbef [121961-22-6]
Lorajmine [47562-08-3]
Lorajmine Hydrochloride*
[40819-93-0]
Loratadine [79794-75-5]
Lorazepam [846-49-1]
Lorcainide [59729-31-6]
Lorcainide Hydrochloride*
[58934-46-6]
Lormetazepam [848-75-9]
Lornoxicam [70374-39-9]
Losartan [114798-26-4]
Losartan Monopotassium Salt*
[124750-99-8]
Loteprednol Etabonate [82034-46-6]
Lotrifen [66535-86-2]
Lovastatin [75330-75-5]
Loxapine [1977-10-2]
Loxapine Succinate* [27833-64-3]
Loxiglumide [107097-80-3]
Loxoprofen [68767-14-6]
Loxoprofen Sodium Salt*
[80382-23-6]
Lucanthone Hydrochloride [548-57-2]
Lucensomycin [13058-67-8]
Lucifer Yellow CH [67769-47-5]
Lufenuron [103055-07-8]
Lumazine [487-21-8]
Lumichrome [1086-80-2]
Lumiflavine [1088-56-8]
Luminol [521-31-3]
Lumisterol [474-69-1]
Lunacridine [83-58-9]
Lunacridine (±)-Form* [52306-35-1]
Lunacrine [82-40-6]
Lunacrine dl-Form* [95723-59-4]
Lunasine [6901-22-0]
Lunasine (−)-Form Perchlorate*
[121543-68-8]
Lunine [518-60-5]
Lunularic Acid [23255-59-6]
Lupanine d-Form* [550-90-3]
Lupanine dl-Form* [4356-43-8]
Lupanine l-Form* [486-88-4]

*Denotes derivative of title compound

Lupeol [545-47-1]
Lupinine [486-70-4]
Luprostiol [73523-00-9]
Lupulon [468-28-0]
Luteolin [491-70-3]
Lutetium [7439-94-3]
2,6-Lutidine [108-48-5]
Lututrin [1407-04-1]
Lyapolate Sodium [25053-27-4]
Lycoctonine [26000-17-9]
Lycoctonine N-Succinylanthranilic
  Acid Ester* [25867-19-0]
Lycodine [20316-18-1]
Lycofawcine [3175-90-4]
Lycomarasmine [7611-43-0]
Lycopene [502-65-8]
Lycophyll [19891-75-9]
Lycopodine [466-61-5]
Lycoramine [21133-52-8]
Lycorine [476-28-8]
Lycoxanthin [19891-74-8]
Lymecycline [992-21-2]
Lynestrenol [52-76-6]
Lynestrenol Mixture with Ethinyl
  Estradiol* [8064-76-4]
Lynestrenol Mixture with Mestranol*
  [8015-14-3]
Lypressin [50-57-7]
Lysalbinic Acid [9006-58-0]
Lysergamide [478-94-4]
Lysergic Acid [82-58-6]
Lysergide [50-37-3]
Lysidine [534-26-9]
Lysine [56-87-1]
Lysine Acetylsalicylate [62952-06-1]
L-Lysine L-Glutamate [5408-52-6]
L-Lysine Monohydrochloride*
  [657-27-2]
Lysostaphin [9011-93-2]
Lysozyme [9001-63-2]
Lysozyme Hydrochloride*
  [9066-59-5]
Lyxoflavine [13123-37-0]
D-Lyxose [1114-34-7]
Mabuterol [56341-08-3]
Mabuterol d-Form Hydrochloride*
  [95656-54-5]
Mabuterol dl-Form Hydrochloride*
  [95656-48-7]
Mabuterol l-Form Hydrochloride*
  [95656-55-6]
Maclurin [519-34-6]
Macromerine [2970-95-8]
Macrophage Colony-Stimulating Fac-
  tor [81627-83-0]
Macusine A* [6801-39-4]
Macusine B* [6792-07-0]
Macusine C* [6801-17-8]
Maddrell's Salt [10361-03-2]
MADU [840-50-6]
Maduramicin [84878-61-5]
Mafenide [138-39-6]
Mafenide Acetate* [13009-99-9]
Mafenide Hydrochloride* [138-37-4]
Mafenide Propionate* [12001-72-8]
Magaldrate [1317-26-6]
Magenta I [632-99-5]
Magnesium [7439-95-4]
Magnesium Acetate [142-72-3]
Magnesium Acetylsalicylate
  [132-49-0]
Magnesium Amide [7803-54-5]
Magnesium Benzoate [553-70-8]
Magnesium Bisulfate [10028-26-9]
Magnesium Borate [13703-82-7]
Magnesium Bromate [7789-36-8]
Magnesium Bromide [7789-48-2]
Magnesium Carbonate Hydroxide
  [39409-82-0]
Magnesium Chlorate [10326-21-3]

Magnesium Chloride [7786-30-3]
Magnesium Citrate [3344-18-1]
Magnesium Citrate, Dibasic
  [144-23-0]
Magnesium Fluoride [7783-40-6]
Magnesium Formate [557-39-1]
Magnesium Germanide [1310-52-7]
Magnesium Hexafluorosilicate
  [16949-65-8]
Magnesium Hydride [60616-74-2]
Magnesium Hydroxide [1309-42-8]
Magnesium Iodide [10377-58-9]
Magnesium Lactate [18917-93-6]
Magnesium Mandelate [18937-33-2]
Magnesium Metasilicate*
  [13776-74-4]
Magnesium Nitrate [10377-60-3]
Magnesium Oleate [1555-53-9]
Magnesium Orthosilicate*
  [10034-94-3]
Magnesium Oxalate [547-66-0]
Magnesium Oxide [1309-48-4]
Magnesium Perborate [14635-87-1]
Magnesium Perchlorate [10034-81-8]
Magnesium Permanganate
  [10377-62-5]
Magnesium Peroxide [14452-57-4]
Magnesium Phosphate, Dibasic
  [7757-86-0]
Magnesium Phosphate, Monobasic
  [13092-66-5]
Magnesium Phosphate, Tribasic
  [7757-87-1]
Magnesium Potassium Selenate
  [28041-84-1]
Magnesium Pyrophosphate
  [13446-24-7]
Magnesium Salicylate [18917-89-0]
Magnesium Selenate [14986-91-5]
Magnesium Selenide [1313-04-8]
Magnesium Selenite [15593-61-0]
Magnesium Silicide [22831-39-6]
Magnesium Stannide [1313-08-2]
Magnesium Stearate [557-04-0]
Magnesium Sulfate [7487-88-9]
Magnesium Sulfate Heptahydrate*
  [10034-99-8]
Magnesium Sulfate Trihydrate*
  [15320-30-6]
Magnesium Sulfite [7757-88-2]
Magnesium Thiocyanate [306-61-6]
Magnesium Thiosulfate [10124-53-5]
Magnesium Trisilicate* [14987-04-3]
Magneson [74-39-5]
Magnoflorine [2141-09-5]
Magnoline [6859-66-1]
Magnolol [528-43-8]
Malachite Green [569-64-2]
Malathion [121-75-5]
Maleamic Acid [557-24-4]
Maleanilic Acid [555-59-9]
Maleic Acid [110-16-7]
Maleic Anhydride [108-31-6]
Maleic Hydrazide [123-33-1]
Maleuric Acid [105-61-3]
Malic Acid [6915-15-7]
Malic Acid D-(+)-Form* [636-61-3]
Malic Acid DL-Form* [617-48-1]
Malic Acid L-(—)-Form* [97-67-6]
Malonic Acid [141-82-2]
Malononitrile [109-77-3]
Malotilate [59937-28-9]
Maltol [118-71-8]
Maltose [69-79-4]
Malvidin Chloride [643-84-5]
Mancozeb [8018-01-7]
Mandelic Acid [90-64-2]
Mandelic Acid Calcium Salt*
  [134-95-2]

Mandelic Acid Isoamyl Ester
  [5421-04-5]
Mandelic Acid Sodium Salt*
  [114-21-6]
Mandelonitrile [532-28-5]
Mandelonitrile Glucoside [138-53-4]
Maneb [12427-38-2]
Manganese [7439-96-5]
Manganese Acetate [638-38-0]
Manganese Borate [12228-91-0]
Manganese Bromide [13446-03-2]
Manganese Carbonate [598-62-9]
Manganese Carbonyl [10170-69-1]
Manganese Chloride [7773-01-5]
Manganese Difluoride [7782-64-1]
Manganese Dioxide [1313-13-9]
Manganese Hypophosphite
  [10043-84-2]
Manganese Iodide [7790-33-2]
Manganese Nitrate [10377-66-9]
Manganese Oleate [23250-73-9]
Manganese Oxalate [640-67-5]
Manganese Oxide [1317-35-7]
Manganese Phosphate, Dibasic
  [7782-76-5]
Manganese Pyrophosphate
  [53731-35-4]
Manganese Selenide [1313-22-0]
Manganese Sesquioxide [1317-34-6]
Manganese Silicate [7759-00-4]
Manganese Sulfate [7785-87-7]
Manganese Sulfide [18820-29-6]
Manganese Trifluoride [7783-53-1]
Mangostin [6147-11-1]
Manidipine [89226-50-6]
Manidipine Dihydrochloride*
  [89226-75-5]
Mannitol [69-65-8]
Mannitol Hexanitrate [15825-70-4]
Mannomustine [551-74-6]
D-Mannose [3458-28-4]
Maprotiline [10262-69-8]
Maprotiline Hydrochloride*
  [10347-81-6]
Margaric Acid [506-12-7]
Marrubiin [465-92-9]
Matricarin [5989-43-5]
Matrine [519-02-8]
Maytansine [35846-53-8]
Mazindol [22232-71-9]
Mazipredone [13085-08-0]
Mazipredone Hydrochloride*
  [60-39-9]
MCPA [94-74-6]
MCPA Sodium Salt* [3653-48-3]
MDA [4764-17-4]
MDA Hydrochloride* [6292-91-7]
MDE [14089-52-2] (unspecified)
MDE [82801-81-8] (±)
MDE Hydrochloride* [74341-78-9]
MDMA [42542-10-9] (unspecified)
MDMA [69610-10-2] (±)
MDMA Hydrochloride* [64057-70-1]
  (unspecified)
MDMA Hydrochloride* [92279-84-0]
  (±)
Mebendazole [31431-39-7]
Mebeverine [3625-06-7]
Mebeverine Hydrochloride*
  [2753-45-9]
Mebhydroline [524-81-2]
Mebhydroline 1,5-Naphthalene-
  disulfonate Salt* [6153-33-9]
Mebiquine [23910-07-8]
MeBmt [59865-23-5]
Mebrofenin [78266-06-5]
Mebutamate [64-55-1]
Mecamylamine [60-40-2]
Mecamylamine Hydrochloride*
  [826-39-1]

Denotes derivative of title compound

Mechlorethamine [51-75-2]
Mechlorethamine Hydrochloride* [55-86-7]
Mechlorethamine Oxide Hydrochloride [302-70-5]
Meclizine [569-65-3]
Meclizine Dihydrochloride Monohydrate* [31884-77-2]
Meclocycline [2013-58-3]
Meclocycline 5-Sulfosalicylate* [73816-42-9]
Meclofenamic Acid [644-62-2]
Meclofenoxate [51-68-3]
Meclofenoxate Hydrochloride* [3685-84-5]
Mecloqualone [340-57-8]
Mecloralurea [1954-79-6]
Mecloxamine [5668-06-4]
Meconic Acid [497-59-6]
Meconin [569-31-3]
Mecoprop [7085-19-0]
Mecrylate [137-05-3]
Mecysteine Hydrochloride [18598-63-5]
Medazepam [2898-12-6]
Medazepam Hydrochloride* [2898-11-5]
Medetomidine [86347-14-0]
Medetomidine Hydrochloride* [86347-15-1]
Medibazine [53-31-6]
Medibazine Dihydrochloride* [96588-03-3]
Medicagol [1983-72-8]
Medicarpin [32383-76-9]
Medifoxamine [32359-34-5]
Medifoxamine Fumarate* [16604-45-8]
Medmain [576-11-4]
Medrogestone [977-79-7]
Medronic Acid [1984-15-2]
Medroxyprogesterone [520-85-4]
Medroxyprogesterone 17-Acetate* [71-58-9]
Medrylamine [524-99-2]
Medrysone [2668-66-8]
Medullipin I [97621-73-3]
Medullipin II* [122392-03-4]
Mefenamic Acid [61-68-7]
Mefenorex [17243-57-1]
Mefenorex Hydrochloride* [5586-87-8]
Mefexamide [1227-61-8]
Mefloquine Hydrochloride [51773-92-3]
Mefloquine Hydrochloride Combination with Sulfadoxine and Pyrimethamine* [69191-18-0]
Mefluidide [53780-34-0]
Mefruside [7195-27-9]
Megestrol Acetate [595-33-5]
Megestrol Acetate Mixture with Ethinyl Estradiol* [8064-66-2]
Megestrol Acetate Mixture with Mestranol* [8064-51-5]
Meglumine Acetrizoate [22154-43-4]
Meglumine Diatrizoate [131-49-7]
Meglumine Diatrizoate Mixture with Sodium Diatrizoate* [8064-12-8]
Meglutol [503-49-1]
Melamine [108-78-1]
Melanostatin [9083-38-9]
Melarsoprol [494-79-1]
Melatonin [73-31-4]
Meldrum's Acid [2033-24-1]
Melengestrol [5633-18-1]
Melengestrol Acetate* [2919-66-6]
Melezitose [597-12-6]
Melibiose [585-99-9]
Melilotoside [618-67-5]

Melinamide [14417-88-0]
Melinonine A [6801-41-8]
Melitracen [5118-29-6]
Melitracen Hydrochloride* [10563-70-9]
Melittin [20449-79-0]
Mellitic Acid [517-60-2]
Meloxicam [71125-38-7]
Melperone [3575-80-2]
Melperone Hydrochloride* [1622-79-3]
Melphalan [148-82-3]
Memantine [19982-08-2]
Memantine Hydrochloride* [41100-52-1]
Menadiol [481-85-6]
Menadiol Diacetate* [573-20-6]
Menadiol Dibutyrate* [53370-44-8]
Menadiol Sodium Diphosphate* [131-13-5]
Menadione [58-27-5]
Menadione Dimethylpyrimidinol Bisulfite* [14451-99-1]
Menadione Sodium Bisulfite* [130-37-0]
Menadione Sodium Bisulfite* [6147-37-1] (Trihydrate)
Menadoxime [573-01-3]
Menaquinone 4* [863-61-6]
Menaquinone 6* [84-81-1]
Menaquinone 7* [2124-57-4]
Menazon [78-57-9]
Menbutone [3562-99-0]
Menbutone Magnesium Salt* [16643-66-6]
Mendelevium [7440-11-1]
Menogaril [71628-96-1]
Menthol [89-78-1]
l-Menthone [14073-97-3]
Menthyl Acetate [89-48-5]
Menthyl Borate [53370-45-9]
Menthyl Salicylate [89-46-3]
Menthyl Valerate [89-47-4]
Meobentine [46464-11-3]
Meobentine Sulfate* [58503-79-0]
Meparfynol [77-75-8]
Meparfynol Carbamate* [302-66-9]
Mepartricin [11121-32-7]
Mepartricin A* [62534-68-3]
Mepartricin B* [62534-69-4]
Mepazine [60-89-9]
Mepenzolate Bromide [76-90-4]
Meperidine [57-42-1]
Meperidine Hydrochloride* [50-13-5]
Mephenesin [59-47-2]
Mephenesin Carbamate* [533-06-2]
Mephenoxalone [70-07-5]
Mephentermine [100-92-5]
Mephentermine Sulfate Dihydrate* [6190-60-9]
Mephenytoin [50-12-4]
Mephobarbital [115-38-8]
Mephosfolan [950-10-7]
Mepindolol [23694-81-7]
Mepindolol Sulfate Salt* [56396-94-2]
Mepiprazole [20326-12-9]
Mepiprazole Dihydrochloride* [20344-15-4]
Mepiquat Chloride [24307-26-4]
Mepitiostane [21362-69-6]
Mepivacaine [96-88-8]
Mepivacaine Hydrochloride* [1722-62-9]
Mepixanox [17854-59-0]
Meprednisone [1247-42-3]
Meprobamate [57-53-4]
Meprylcaine [495-70-5]
Meprylcaine Hydrochloride* [956-03-6]
Meptazinol [54340-58-8]

Meptazinol Hydrochloride* [59263-76-2]
Mequitazine [29216-28-2]
Meralein Sodium [4386-35-0]
Meralluride [8069-64-5]
Merbromin [129-16-8]
Mercamphamide [127-50-4]
Mercamphamide Mixture of Sodium Salt with Theophylline* [8012-34-8]
Mercamphamide Sodium Salt* [124-82-3]
2-Mercaptobenzothiazole [149-30-4]
2-Mercaptoethanol [60-24-2]
Mercaptomerin Sodium [21259-76-7]
6-Mercaptopurine [50-44-2]
Mercufenol Chloride [90-03-9]
Mercumallylic Acid [86-36-2]
Mercumallylic Acid Sodium Salt Compd with Theophylline* [8018-15-3]
Mercuric Acetate [1600-27-7]
Mercuric Arsenate [7784-37-4]
Mercuric Benzoate [583-15-3]
Mercuric Bromide [7789-47-1]
Mercuric Chloride [7487-94-7]
Mercuric Chloride, Ammoniated [10124-48-8]
Mercuric Cyanide [592-04-1]
Mercuric Dichromate(VI) [7789-10-8]
Mercuric Fluoride [7783-39-3]
Mercuric Iodate [7783-32-6]
Mercuric Iodide, Red [7774-29-0]
Mercuric Nitrate [10045-94-0]
Mercuric Oleate [1191-80-6]
Mercuric Oxide, Red [21908-53-2]
Mercuric Oxide, Yellow [21908-53-2]
Mercuric Oxycyanide [1335-31-5]
Mercuric Salicylate [5970-32-1]
Mercuric Sodium p-Phenolsulfonate [535-55-7]
Mercuric Stearate [645-99-8]
Mercuric Subsulfate [1312-03-4]
Mercuric Succinimide [584-43-0]
Mercuric Sulfate [7783-35-9]
Mercuric Sulfide, Black [1344-48-5]
Mercuric Sulfide, Red [1344-48-5]
Mercuric Thiocyanate [592-85-8]
Mercurophen [52486-78-9]
Mercurous Acetate [631-60-7]
Mercurous Bromide [10031-18-2]
Mercurous Chlorate [10294-44-7]
Mercurous Chloride [10112-91-1]
Mercurous Fluoride [13967-25-4]
Mercurous Iodide [15385-57-6]
Mercurous Nitrate [10415-75-5]
Mercurous Oxide [15829-53-5]
Mercurous Sulfate [7783-36-0]
Mercury [7439-97-6]
Merisoprol Hg 197 [5579-94-2]
Meropenem [96036-03-2] (anhydrous)
Meropenem [119478-56-7] (trihydrate)
Merphyrin [15375-94-7]
Mersalyl [492-18-2]
Mesaconic Acid [498-24-8]
Mesalamine [89-57-6]
Mescaline [54-04-6]
Mesembrine [24880-43-1]
Mesitylene [108-67-8]
Mesityl Oxide [141-79-7]
Mesna [19767-45-4]
Mesoridazine [5588-33-0]
Mesoridazine Benzenesulfonate* [32672-69-8]
Mesoxalic Acid [473-90-5]
Mesquite Gum [9000-47-9]
Mestanolone [521-11-9]
Mesterolone [1424-00-6]
Mestilbol [18839-90-2]
Mestranol [72-33-3]
Mesulfen [135-58-0]

Denotes derivative of title compound

Metabutoxycaine [3624-87-1]
Metabutoxycaine Hydrochloride*
[550-01-6]
Metachrome Yellow [584-42-9]
Metaclazepam [84031-17-4]
Metaclazepam Hydrochloride*
[61802-93-5]
[2.2]Metacyclophane [2319-97-3]
Metalaxyl [57837-19-1]
Metaldehyde [9002-91-9]
Metamfepramone [15351-09-4]
Metamitron [41394-05-2]
Metampicillin [6489-97-0]
Metampicillin Sodium Salt*
[6489-61-8]
Metanephrine [5001-33-2]
Metanilic Acid [121-47-1]
Metanil Yellow [587-98-4]
Metaphanine [1805-86-3]
Metapramine [21730-16-5]
Metaproterenol [586-06-1]
Metaproterenol Sulfate* [5874-97-5]
Metaraminol [54-49-9]
Metaraminol Bitartrate (Hydrogen L-
Tartrate)* [33402-03-8]
Metaraminol Hydrochloride*
[5967-52-2]
Metaraminol Oxalate Dihydrate*
[5967-53-3]
Metaxalone [1665-48-1]
Metazocine [3734-52-9]
Metcaraphen [561-79-5]
Metcaraphen Hydrochloride*
[1950-31-8]
Meteloidine [526-13-6]
Metepa [57-39-6]
Metergoline [17692-51-2]
Metescufylline [15518-82-8]
Metformin [657-24-9]
Metformin p-Chlorophenoxyacetate
(Salt)* [25672-33-7]
Metformin Embonate* [34461-22-8]
Metformin Hydrochloride*
[1115-70-4]
Methabenzthiazuron [18691-97-9]
Methacholine Chloride [62-51-1]
Methacrifos [62610-77-9]
Methacrylic Acid [79-41-4]
Methacrylic Acid Methyl Ester*
[80-62-6]
Methacrylonitrile [126-98-7]
Methacycline [914-00-1]
Methacycline Hydrochloride*
[3963-95-9]
Methadone Hydrochloride
[1095-90-5]
Methadone Hydrochloride dl-Form*
[125-56-4]
Methadone Hydrochloride l-Form*
[5967-73-7]
Methadyl Acetate [509-74-0]
Methafurylene [531-06-6]
Methafurylene Fumarate* [5429-41-4]
Methallatal [115-56-0]
Methallenestril [517-18-0]
Methallibure [926-93-2]
Methamidophos [10265-92-6]
Methamphetamine [537-46-2]
Methamphetamine Hydrochloride*
[51-57-0]
Metham Sodium [137-42-8]
Methandriol [521-10-8]
Methandriol Diacetate* [2061-86-1]
Methandriol Dipropionate*
[3593-85-9]
Methandrostenolone [72-63-9]
Methane [74-82-8]
Methanearsonic Acid [124-58-3]
Methanesulfonic Acid [75-75-2]
Methanesulfonyl Chloride [124-63-0]

Methanethiol [74-93-1]
Methanol [67-56-1]
Methantheline Bromide [53-46-3]
Methaphenilene [493-78-7]
Methaphenilene Hydrochloride*
[7084-07-3]
Methapyrilene [91-80-5]
Methapyrilene Fumarate*
[33032-12-1]
Methapyrilene Hydrochloride*
[135-23-9]
Methaqualone [72-44-6]
Methaqualone Hydrochloride*
[340-56-7]
Metharbital [50-11-3]
Methargen [53370-43-7]
Methazolamide [554-57-4]
Methazole [20354-26-1]
Methcathinone [5650-44-2] (unspec-
ified)
Methcathinone [28521-94-0] (±)
Methcathinone Hydrochloride*
[2576-11-6] (unspecified)
Methcathinone Hydrochloride*
[49656-78-2] (±)
Methdilazine [1982-37-2]
Methdilazine Hydrochloride*
[1229-35-2]
Methenamine [100-97-0]
Methenamine Allyl Iodide
[36895-62-2]
Methenamine Anhydromethylene-
citrate [6190-43-8]
Methenamine Hippurate [5714-73-8]
Methenamine Mandelate [587-23-5]
Methenamine Salicylate [620-34-8]
Methenamine Sulfosalicylate
[20480-93-7]
Methenamine Tetraiodine
[12001-65-9]
Methenolone [153-00-4]
Methenolone 17-Acetate* [434-05-9]
Methenolone 17-Enanthate*
[303-42-4]
Methestrol [130-73-4]
Methetoin [5696-06-0]
Methicillin Sodium [132-92-3]
Methidathion [950-37-8]
Methimazole [60-56-0]
Methiocarb [2032-65-7]
Methiodal Sodium [126-31-8]
Methionic Acid [503-40-2]
Methionine [63-68-3]
Methionine D-Form* [348-67-4]
Methionine DL-Form* [59-51-8]
Methionine Hydroxy Analog
[583-91-5]
Methionyl Human Growth Hormone*
[82030-87-3]
Methioprim [588-36-3]
Methiotriazamine [6043-86-3]
Methisazone [1910-68-5]
Methitural [467-43-6]
Methitural Sodium Salt* [730-68-7]
Methixene [4969-02-2]
Methixene Hydrochloride Mono-
hydrate* [7081-40-5]
Methocarbamol [532-03-6]
Methohexital Sodium [22151-68-4]
Methomyl [16752-77-5]
Methoprene [40596-69-8] (unspecified)
Methoprene [41205-06-5] (±)
Methopterin [2410-93-7]
Methotrexate [59-05-2]
Methotrexate Disodium Salt*
[7413-34-5]
Methotrexate Monohydrate*
[6745-93-3]
Methotrimeprazine [60-99-1]

Methotrimeprazine Maleate*
[7104-38-3]
Methoxamine [390-28-3]
Methoxamine Hydrochloride*
[61-16-5]
Methoxsalen [298-81-7]
Methoxyamine [67-62-9]
Methoxychlor [72-43-5]
Methoxyethylbenzeneboronic Acid
(—)-S-Form* [166191-23-7]
Methoxyethylbenzeneboronic Acid
(+)-R-Form [159752-39-3]
Methoxyflurane [76-38-0]
10-Methoxyharmalan [3589-73-9]
3-Methoxy-4-hydroxyphenylglycol
[534-82-7]
2-(Methoxymethyl)-5-nitrofuran
[586-84-5]
2-Methoxynaphthalene [93-04-9]
Methoxyphenamine [93-30-1]
Methoxyphenamine Hydrochloride*
[5588-10-3]
3-(o-Methoxyphenyl)-2-phenylacrylic
Acid [25333-25-9]
3-(o-Methoxyphenyl)-2-phenylacrylic
Acid Magnesium Salt* [14222-61-8]
N-(p-Methoxyphenyl)-p-phenylenedi-
amine [101-64-4]
Methoxypromazine [61-01-8]
Methoxypromazine Maleate*
[3403-42-7]
6-Methoxy-α-tetralone [1078-19-9]
1-Methoxy-3-(trimethylsilyloxy)-1,3-
butadiene [54125-02-9] (E)-form
1-Methoxy-3-(trimethylsilyloxy)-1,3-
butadiene [59414-23-2] (undefined)
5-Methoxytryptamine [608-07-1]
Methscopolamine Bromide [155-41-9]
Methsuximide [77-41-8]
Methyclothiazide [135-07-9]
Methyl Abietate [127-25-3]
N-Methylacetanilide [579-10-2]
Methyl Acetate [79-20-9]
Methyl Acetoacetate [105-45-3]
Methyl Acetylsalicylate [580-02-9]
Methyl Acrylate [96-33-3]
Methylal [109-87-5]
Methyl Allyl Trisulfide [34135-85-8]
Methylamine [74-89-5]
2-Methylaminoethanol [109-83-1]
p-Methylaminophenol Sulfate
[55-55-0]
Methylaniline [100-61-8]
Methyl Anthranilate [134-20-3]
2-Methylanthraquinone [84-54-8]
Methylarbutin [6032-32-2]
3-Methylarsacetin [25384-21-8]
Methylbenzethonium Chloride
[25155-18-4]
Methyl Benzoate [93-58-3]
Methyl Benzoylsalicylate [610-60-6]
α-Methylbenzylamine [98-84-0]
Methyl Blue [28983-56-4]
Methyl Bromide [74-83-9]
2-Methyl-1-butanol [137-32-6]
3-Methyl-2-butanol [598-75-4]
Methyl tert-Butyl Ether [1634-04-4]
Methyl Butyl Ketone [591-78-6]
2-Methyl-3-butyn-2-ol [115-19-5]
Methyl Butyrate [623-42-7]
Methyl Carbamate [598-55-0]
Methyl Carbitol [111-77-3]
Methyl Carbonate [616-38-6]
Methyl Cellosolve [109-86-4]
Methyl Cellosolve Acetate [110-49-6]
Methylcellulose [9004-67-5]
Methyl Chloride [74-87-3]
Methyl Chloroacetate [96-34-4]
Methyl Chlorocarbonate [79-22-1]
3-Methylcholanthrene [56-49-5]

Methylcobalamin [13422-55-4]
Methylconiine [553-75-3]
3'-Methyl-1,2-cyclopentenophenan-
threne [549-38-2]
5-Methylcytosine [554-01-8]
Methyl Demeton [8022-00-2]
β-Methyldigoxin* [30685-43-9]
p-Methyldiphenhydramine
[19804-27-4]
Methyldiphenylamine [552-82-9]
Methyldopa [555-30-6]
3-O-Methyldopa [300-48-1]
Methyldopa Ethyl Ester Hydro-
chloride* [2508-79-4]
Methyleneaminoacetonitrile
[109-82-0]
4,4'-Methylenebis[2-chloroaniline]
[101-14-4]
Methylene Blue [61-73-4]
Methylene Bromide [74-95-3]
α-Methylene Butyrolactone
[547-65-9]
Methylene Chloride [75-09-2]
Methylenedigallic Acid [552-21-6]
5,5'-Methylenedisalicylic Acid
[122-25-8]
Methylene Iodide [75-11-6]
Methylenomycin A* [52775-76-5]
Methylenomycin B* [52775-77-6]
N-Methylephedrine [552-79-4]
N-Methylepinephrine [554-99-4]
N-Methylepinephrine DL-Form*
[6032-14-0]
Methylergonovine [113-42-8]
Methylergonovine Maleate*
[57432-61-8]
Methylergonovine Tartrate*
[6209-37-6]
Methyl Ether [115-10-6]
Methyl Ethyl Ketone [78-93-3]
α-Methylfentanyl [79704-88-4]
α-Methylfentanyl Hydrochloride*
[1443-44-3]
Methyl Fluorosulfonate [421-20-5]
N-Methylformamide [123-39-7]
Methyl Formate [107-31-3]
N-Methylglucamine [6284-40-8]
N-Methylglucamine Antimonate*
[133-51-7]
N-Methyl-α-L-glucosamine
[42852-95-9]
α-Methylglucoside [97-30-3]
Methyl Green [14855-76-6]
Methylguanidine [471-29-4]
Methylhexaneamine [105-41-9]
Methylhydrazine [60-34-4]
Methyl Iodide [74-88-4]
Methyl Isobutyrate [547-63-7]
Methyl Isocyanate [624-83-9]
Methyl Isopropyl Ketone [563-80-4]
Methyl Isothiocyanate [556-61-6]
Methyl Isovalerate [556-24-1]
Methyl Lactate [547-64-8]
Methyl Linoleate [112-63-0]
Methyl Malonate [108-59-8]
Methyl Methanesulfonate [66-27-3]
3-Methyl-6,7-methylenedioxy-1-
piperonylisoquinoline [550-73-2]
Methyl N-Methylnipecotate
[1690-72-8]
Methyl N-Methylnipecotate
Methobromide* [60296-01-7]
N-Methylmyosmine [525-74-6]
Methyl Nicotinate [93-60-7]
Methyl Nitrate [598-58-3]
Methyl p-Nitrobenzenesulfonate
[6214-20-6]
N-Methyl-N'-nitro-N-nitrosoguanidine
[70-25-7]
Methyl Nonyl Ketone [112-12-9]

Methyl Orange [547-58-0]
Methyl Oxalate [553-90-2]
Methylparaben [99-76-3]
Methyl Parathion [298-00-0]
5-Methyl-5-(3-phenanthryl)hydantoin
[3784-92-7]
5-Methyl-5-(3-phenanthryl)hydantoin
Sodium Salt* [74449-59-5]
N-Methylphenazonium Methosulfate
[299-11-6]
Methylphenidate [113-45-1]
Methylphenidate Hydrochloride*
[298-59-9]
3-Methyl-5-phenylhydantoin
[6846-11-3]
3'-Methylphthalanilic Acid [85-72-3]
Methylprednisolone [83-43-2]
Methylprednisolone Aceponate*
[86401-95-8]
Methylprednisolone 21-Acetate*
[53-36-1]
Methylprednisolone 21-Phosphate
Disodium Salt* [5015-36-1]
Methylprednisolone 21-Succinate
Sodium Salt* [2375-03-3]
p-(2-Methylpropenyl)phenol Acetate
[53370-48-2]
Methyl Propionate [554-12-1]
Methyl Propyl Ether [557-17-5]
Methyl Propyl Ketone [107-87-9]
5-Methylpyrazole-3-carboxylic Acid
[402-61-9]
Methyl Pyridyl Ketone [350-03-8]
Methyl 4-Pyridyl Ketone Thiosemi-
carbazone [3115-21-7]
Methyl 4-Pyridyl Ketone Thiosemi-
carbazone Hydrochloride*
[2260-13-1]
1-Methylpyrrolidone [872-50-4]
N-Methylpyrroline [554-15-4]
Methyl Red [493-52-7]
Methyl Salicylate [119-36-8]
4'-(Methylsulfamoyl)sulfanilanilide
[547-53-5]
Methyl Sulfate [75-93-4]
Methyl Sulfide [75-18-3]
N-Methyltaurine [107-68-6]
17-Methyltestosterone [58-18-4]
17α-Methyltestosterone 3-Cyclopent-
yl Enol Ether [67-81-2]
4-Methyl-5-thiazoleethanol [137-00-8]
Methyl Thiocyanate [556-64-9]
Methylthiouracil [56-04-2]
Methyltrienolone [965-93-5]
α-Methyl-m-tyrosine [305-96-4]
6-Methyluracil [626-48-2]
Methyl Vinyl Ketone [78-94-4]
Methymycin [497-72-3]
Methyprylon [125-64-4]
Methysergide [361-37-5]
Methysergide Hydrogen Maleate*
[129-49-7]
Methysticin [495-85-2]
Methysticin (±)-Dihydromethysticin*
[3155-57-5]
Metiazinic Acid [13993-65-2]
Meticrane [1084-65-7]
Metipranolol [22664-55-7]
Metipranolol Hydrochloride*
[36592-77-5]
Metitepine [20229-30-5]
Metitepine Maleate* [19728-88-2]
Metizoline [17692-22-7]
Metizoline Hydrochloride*
[5090-37-9]
Metobromuron [3060-89-7]
Metochalcone [18493-30-6]
Metoclopramide [364-62-5]
Metoclopramide Monohydrochloride
Monohydrate* [54143-57-6]

Metocurine Iodide [7601-55-0]
Metofenazate [388-51-2]
Metofenazate Difumarate* [522-23-6]
Metofoline [2154-02-1]
Metolachlor [51218-45-2]
Metolazone [17560-51-9]
Metomidate [5377-20-8]
Metopimazine [14008-44-7]
Metopon [143-52-2]
Metopon Hydrochloride* [124-92-5]
Metoprolol [37350-58-6]
Metoprolol Succinate* [98418-47-4]
Metoprolol Tartrate* [56392-17-7]
Metoquinone [622-91-3]
Metoserpate [1178-28-5]
Metoserpate Hydrochloride*
[1178-29-6]
Metralindole [54188-38-4]
Metralindole Hydrochloride*
[53734-79-5]
Metribuzin [21087-64-9]
Metrizamide [31112-62-6]
Metrizoic Acid [1949-45-7]
Metrizoic Acid Sodium Salt*
[7225-61-8]
Metronidazole [443-48-1]
Metronidazole Hydrochloride*
[69198-10-3]
Metron S [13946-02-6]
Metsulfuron Methyl [74223-64-6]
Meturedepa [1661-29-6]
Metyrapone [54-36-4]
Metyridine [114-91-0]
Metyrosine [672-87-7]
Metyrosine DL-Form* [620-30-4]
Mevaldic Acid [541-07-1]
Mevalonic Acid [150-97-0]
Mevastatin [73573-88-3]
Mevinphos [7786-34-7]
Mexacarbate [315-18-4]
Mexazolam [31868-18-5]
Mexenone [1641-17-4]
Mexicain [52500-59-1]
Mexiletine [31828-71-4]
Mexiletine Hydrochloride*
[5370-01-4]
Mezlocillin [51481-65-3]
Mezlocillin Sodium Salt Mono-
hydrate* [59798-30-0]
Mianserin [24219-97-4]
Mianserin Hydrochloride*
[21535-47-7]
Mibefradil [116644-53-2]
Mibefradil Dihydrochloride*
[116666-63-8]
Mibolerone [3704-09-4]
Miboplatin [103775-75-3]
Michler's Base [101-61-1]
Michler's Ketone [90-94-8]
Miconazole [22916-47-8]
Miconazole Nitrate* [22832-87-7]
Micranthine [36104-64-0]
Micrococcin P [1392-46-7]
Micronomicin [52093-21-7]
Midazolam [59467-70-8]
Midazolam Hydrochloride*
[59467-96-8]
Midazolam Maleate* [59467-94-6]
Midecamycin A₁* [35457-80-8]
Midecamycin A₃* [36025-69-1]
Midodrine [42794-76-3]
Midodrine Hydrochloride*
[3092-17-9]
Mifepristone [84371-65-3]
MIF-I* [2002-44-0]
Miglitol [72432-03-2]
Milbemycin β₁* [51596-16-8]
Milbemycin A₃* [51596-10-2]
Milbemycin A₄* [51596-11-3]
Milbemycin D* [77855-81-3]

Denotes derivative of title compound

Milbemycin Oxime [129496-10-2]
Milbemycins [51570-36-6]
Mildiomycin [67527-71-3]
Milnacipran [92623-85-3]
Milnacipran Hydrochloride*
[101152-94-7]
Milorganite [8049-99-8]
Miloxacin [37065-29-5]
Milrinone [78415-72-2]
Milrinone Lactate* [100286-97-3]
Miltefosine [58066-85-6]
Mimosine [500-44-7]
Minaprine [25905-77-5]
Minaprine Dihydrochloride*
[25953-17-7]
Mineral Spirits [64475-85-0]
Minocycline [10118-90-8]
Minocycline Hydrochloride*
[13614-98-7]
Minoxidil [38304-91-5]
Miokamycin [55881-07-7]
Mipafox [371-86-8]
Mirex [2385-85-5]
Mirimostim* [121547-04-4]
Mirtazepine [61337-67-5]
Misch Metal [8049-20-5]
Misoprostol [59122-46-2]
Mitobronitol [488-41-5]
Mitoguazone [459-86-9]
Mitolactol [10318-26-0]
Mitomycin A* [4055-39-4]
Mitomycin B* [4055-40-7]
Mitomycin C* [50-07-7]
Mitomycins [1404-00-8]
Mitotane [53-19-0]
Mitoxantrone [65271-80-9]
Mitoxantrone Dihydrochloride*
[70476-82-3]
Mitragynine [4098-40-2]
Mivacurium Chloride [106861-44-3]
Mizoribine [50924-49-7]
MMT [12108-13-3]
Mocimycin [50935-71-2]
Moclobemide [71320-77-9]
Modafinil [68693-11-8]
Mofebutazone [2210-63-1]
Mofegiline [119386-96-8]
Mofegiline Hydrochloride*
[120635-25-8]
Mofezolac [78967-07-4]
Molgramostim* [99283-10-0]
Molindone [7416-34-4]
Molindone Hydrochloride*
[15622-65-8]
Molsidomine [25717-80-0]
Molybdenum [7439-98-7]
Molybdenum Disulfide [1317-33-5]
Molybdenum Hexafluoride
[7783-77-9]
Molybdenum Sesquioxide [1313-29-7]
Molybdenum Trioxide [1313-27-5]
Molybdic(VI) Acid [7782-91-4]
Mometasone Furoate [83919-23-7]
Monacetin [26446-35-5]
Monardein Chloride [1329-77-7]
Monensin [17090-79-8]
Monoamine Oxidase [9001-66-5]
Monobenzone [103-16-2]
Monobutyrin [557-25-5]
Monobutyrin (R)-Form* [5309-42-2]
Monocrotaline [315-22-0]
Monocrotophos [6923-22-4]
Monorden [12772-57-5]
Monosodium Glutamate [142-47-2]
Monotropein [5945-50-6]
Montan Wax [8002-53-7]
Montelukast [158966-92-8]
Montelukast Monosodium Salt*
[151767-02-1]
Montmorillonite [1318-93-0]

Monuron [150-68-5]
Moperone [1050-79-9]
Moperone Hydrochloride*
[3871-82-7]
Mopidamol [13665-88-8]
Moprolol [5741-22-0]
Moprolol l-Form* [77164-20-6]
Moprolol l-Form Hydrochloride*
[113482-87-4]
Moprolol Hydrochloride*
[27058-84-0]
MOPS [1132-61-2]
Moquizone [19395-58-5]
Moquizone Hydrochloride*
[19395-78-9]
Morantel [20574-50-9]
Morantel Tartrate* [26155-31-7]
Morazone [6536-18-1]
Morazone Hydrochloride*
[50321-35-2]
Morclofone [31848-01-8]
Moricizine [31883-05-3]
Moricizine Hydrochloride*
[29560-58-5]
Morin [480-16-0]
Morindin [60450-21-7]
Moroxydine [3731-59-7]
Moroxydine Hydrochloride*
[3160-91-6]
Morphazinamide [952-54-5]
Morphazinamide Hydrochloride*
[1473-73-0]
Morphenol [519-56-2]
Morpheridine [469-81-8]
Morphinan [468-10-0]
Morphine [57-27-2]
Morphine Hydrochloride [52-26-6]
Morphine Mucate* [596-19-0]
Morphine Sulfate [64-31-3]
Morphine Sulfate Pentahydrate*
[6211-15-0]
Morpholine [110-91-8]
Morpholine Salicylate* [147-90-0]
7-Morpholinomethyltheophylline
[5089-89-4]
Morphothebaine [478-53-5]
Mosapramine [89419-40-9]
Mosapramine Dihydrochloride*
[98043-60-8]
Motilin [52906-92-0]
Motretinide [56281-36-8]
Moveltipril [85856-54-8]
Moveltipril Calcium Salt*
[85921-53-5]
Moxalactam [64952-97-2]
Moxalactam Disodium Salt*
[64953-12-4]
Moxastine [3572-74-5]
Moxaverine [10539-19-2]
Moxaverine Hydrochloride*
[1163-37-7]
Moxestrol [34816-55-2]
Moxidectin [113507-06-5]
Moxisylyte [54-32-0]
Moxisylyte Hydrochloride*
[964-52-3]
Moxonidine [75438-57-2]
Moxonidine Hydrochloride*
[75438-58-3]
MPTP [28289-54-5]
Mucochloric Acid [87-56-9]
Mucochloric Anhydride [4412-09-3]
Muconic Acid [505-70-4]
Mupirocin [12650-69-0]
Muramic Acid [1114-41-6]
Muramyl Dipeptide [53678-77-6]
Murexide [3051-09-0]
Murexine [20284-40-6]
Muscalure [27519-02-4]
Muscarine [300-54-9]

Muscazone [2255-39-2]
Muscimol [2763-96-4]
Muscone [541-91-3]
Mustard Gas [505-60-2]
Muzolimine [55294-15-0]
Mycaminose [519-21-1]
Mycarose [6032-92-4]
Mycelianamide [22775-52-6]
Myclobutanil [88671-89-0]
Mycobacidin [539-35-5]
Mycobacillin [18524-67-9]
Mycomycin [544-51-4]
Mycophenolic Acid [24280-93-1]
Mycosamine [527-38-8]
Myoral [645-74-9]
β-Myrcene [123-35-3]
Myricetin [529-44-2]
Myristic Acid [544-63-8]
Myristicin [607-91-0]
Myristyl Alcohol [112-72-1]
Myristyltrimethylammonium Bromide
[1119-97-7]
Myrophine [467-18-5]
Myrrh [9000-45-7]
Myrtecaine [7712-50-7]
Myrtol [8002-55-9]
Mytatrienediol [5108-94-1]
Myxin [13925-12-7]
Nabam [142-59-6]
Nabilone [51022-71-0]
Nabumetone [42924-53-8]
NAD [53-84-9]
NADH* [58-68-4]
Nadifloxacin [124858-35-1]
Nadolol [42200-33-9]
Nadoxolol [54063-51-3]
Nadoxolol Hydrochloride*
[35991-93-6]
NADP [53-59-8]
NADPH* [53-57-6]
Naepaine [2188-67-2]
Nafamostat [81525-10-2]
Nafamostat Dimethanesulfonate*
[82956-11-4]
Nafarelin [76932-56-4]
Nafarelin Acetate Hydrate*
[86220-42-0]
Nafcillin Sodium [985-16-0]
Nafiverine [5061-22-3]
Nafronyl [31329-57-4]
Nafronyl Acid Oxalate* [3200-06-4]
Naftalofos [1491-41-4]
Naftifine [65472-88-0]
Naftifine Hydrochloride*
[65473-14-5]
Naftopidil [57149-07-2] (unspecified)
Naftopidil [132295-16-0] (±)
Naftopidil Dihydrochloride*
[57149-08-3]
Nalbuphine [20594-83-6]
Nalbuphine Hydrochloride*
[23277-43-2]
Naled [300-76-5]
Nalidixic Acid [389-08-2]
Nalmefene [55096-26-9]
Nalorphine [62-67-9]
Nalorphine Hydrobromide*
[1041-90-3]
Nalorphine Hydrochloride* [57-29-4]
Naloxone [465-65-6]
Naloxone Hydrochloride* [357-08-4]
(Dihydrate)
Naloxone Hydrochloride*
[51481-60-8] (Dihydrate)
Naltrexone [16590-41-3]
Naltrexone Hydrochloride*
[16676-29-2]
Nandinine [572-76-9]
Nandrolone [434-22-0]

Nandrolone Cyclohexanecarboxylate* [18470-94-5]
Nandrolone Cyclohexanepropionate* [912-57-2]
Nandrolone Decanoate* [360-70-3]
Nandrolone Dodecanoate* [26490-31-3]
Nandrolone Furylpropionate* [7642-64-0]
Nandrolone p-Hexyloxyphenylpropionate* [52279-57-9]
Nandrolone Phenpropionate* [62-90-8]
Nandrolone Propionate* [7207-92-3]
Napalm [8031-21-8]
Napelline [5008-52-6]
Naphazoline [835-31-4]
Naphazoline Hydrochloride* [550-99-2]
Naphthacene [92-24-0]
Naphthalene [91-20-3]
1-Naphthaleneacetic Acid [86-87-3]
1,8-Naphthalenediamine [479-27-6]
1,6-Naphthalenedisulfonic Acid [525-37-1]
2,6-Naphthalenedisulfonic Acid [581-75-9]
2,7-Naphthalenedisulfonic Acid [92-41-1]
1-Naphthalenesulfonic Acid [85-47-2]
2-Naphthalenesulfonic Acid [120-18-3]
1-Naphthalenethiol [529-36-2]
2-Naphthalenethiol [91-60-1]
Naphthalic Acid [518-05-8]
1-Naphthoic Acid [86-55-5]
2-Naphthoic Acid [93-09-4]
1-Naphthol [90-15-3]
2-Naphthol [135-19-3]
1-Naphthol-8-amino-3,6-di-sulfonic Acid [90-20-0]
1-Naphthol-4,8-disulfonic Acid [117-56-6]
2-Naphthol-3,6-disulfonic Acid [148-75-4]
2-Naphthol-6,8-disulfonic Acid [118-32-1]
α-Naphtholphthalein [596-01-0]
2-Naphthol Sodium Salt* [875-83-2]
1-Naphthol-2-sulfonic Acid [567-18-0]
1-Naphthol-4-sulfonic Acid [84-87-7]
2-Naphthol-6-sulfonic Acid [93-01-6]
Naphthol Yellow S [846-70-8]
1,2-Naphthoquinone [524-42-5]
1,4-Naphthoquinone [130-15-4]
1,2-Naphthoquinone 2-Semicarbazone* [31853-38-0]
Naphthoresorcinol [132-86-5]
2-Naphthoxyacetic Acid [120-23-0]
1-Naphthylamine [134-32-7]
2-Naphthylamine [91-59-8]
1-Naphthylamine-2,7-disulfonic Acid [486-54-4]
1-Naphthylamine-4,6-disulfonic Acid [85-74-5]
1-Naphthylamine-4,7-disulfonic Acid [85-75-6]
1-Naphthylamine-4-sulfonic Acid [84-86-6]
1-Naphthylamine-5-sulfonic Acid [84-89-9]
1-Naphthylamine-8-sulfonic Acid [82-75-7]
2-Naphthylamine-1-sulfonic Acid [81-16-3]
2-Naphthylamine-5-sulfonic Acid [81-05-0]
2-Naphthyl Benzoate [93-44-7]
N-(1-Naphthyl)ethylenediamine [551-09-7]
1-Naphthylisocyanate [86-84-0]

1-Naphthylisothiocyanate [551-06-4]
2-Naphthyl Lactate [93-43-6]
2-(2-Naphthyloxy)ethanol [93-20-9]
1-Naphthyl Salicylate [550-97-0]
2-Naphthyl Salicylate [613-78-5]
Napropamide [15299-99-7]
Naproxen [22204-53-1]
Naproxen Piperazine Salt* [70981-66-7]
Naproxen Sodium Salt* [26159-34-2]
Naptalam [132-66-1]
Narasin [55134-13-9]
Narbomycin [6036-25-5]
Narceine [131-28-2]
Narcobarbital [125-55-3]
Narcobarbital Sodium Salt* [3329-16-6]
Narcotoline [521-40-4]
Naringenin [480-41-1]
Naringin [10236-47-2]
Nartograstim* [134088-74-7]
Natamycin [7681-93-8]
Nauheim Salts (Artificial) [8007-50-9]
Nealbarbital [561-83-1]
Neamine [3947-65-7]
Nebivolol [99200-09-6]
Nebracetam [97205-34-0] (unspecified)
Nebracetam [116041-13-5] (±)
Nebracetam Fumarate* [97205-35-1]
Nebularine [550-33-4]
Neburon [555-37-3]
Nedocromil [69049-73-6]
Nedocromil Disodium Salt* [69049-74-7]
Nefazodone [83366-66-9]
Nefazodone Hydrochloride* [82752-99-6]
Nefiracetam [77191-36-7]
Nefopam [13669-70-0]
Nefopam Hydrochloride* [23327-57-3]
Negamycin [33404-78-3]
Negatol [9011-02-3]
Nemadectin [102130-84-7]
Nemonapride [75272-39-8]
Neoarsphenamine [457-60-3]
Neocembrene [31570-39-5]
Neo-cupferron [1013-20-3]
Neocuproine [484-11-7]
Neodymium [7440-00-8]
Neoergosterol [516-98-3]
Neohesperidin Dihydrochalcone [20702-77-6]
Neomethymycin [497-73-4]
Neomycin [1404-04-2]
Neomycin B* [119-04-0]
Neomycin C* [66-86-4]
Neomycin Neomycin B Hydrochloride* [25389-99-5]
Neomycin Neomycin B Sulfate* [1405-10-3]
Neomycin Undecylenate [1406-04-8]
Neon [7440-01-9]
Neopentane [463-82-1]
Neopentyl Alcohol [75-84-3]
Neopentyl Glycol [126-30-7]
Neophyl Chloride [515-40-2]
Neopine [467-14-1]
Neopterin [670-65-5]
Neopterin D-erythro-Form* [2009-64-5]
Neoquassin [76-77-7]
Neostigmine [59-99-4]
Neostigmine Bromide* [114-80-7]
Neostigmine Methyl Sulfate* [51-60-5]
Neotetrazolium Chloride [298-95-3]
Neovitamin A [36525-21-0]
Nepetalactone [490-10-8]

Nepetalactone cis-trans-Form* [21651-62-7]
Neptunium [7439-99-8]
Nequinate [13997-19-8]
Neriifolin [466-07-9]
Nerol [106-25-2]
Nerolidol [7212-44-4]
Nerolidol cis-Form* [3790-78-1]
Nerolidol trans-Form* [40716-66-3]
Nerve Growth Factor [9061-61-4]
Netilmicin [56391-56-1]
Netilmicin Sulfate* [56391-57-2]
Netobimin [88255-01-0]
Netropsin [1438-30-8]
Neuraminic Acid [114-04-5]
Neurine [463-88-7]
Neuropeptide Y [82785-45-3]
Neurotensin [39379-15-2]
Neurotrophin-3* [130939-66-1]
Neutral Red [553-24-2]
Nevirapine [129618-40-2]
Nialamide [51-12-7]
Niaprazine [27367-90-4]
Nicametate [3099-52-3]
Nicametate Hydrochloride* [56676-64-3]
Nicarbazin [330-95-0]
Nicardipine [55985-32-5]
Nicardipine Hydrochloride* [54527-84-3]
Nicergoline [27848-84-6]
Niceritrol [5868-05-3]
Nickel [7440-02-0]
Nickel Acetate [373-02-4]
Nickel Acetylacetonate [3264-82-2]
Nickel Bromide [13462-88-9]
Nickel Carbonate Hydroxide [12607-70-4]
Nickel Carbonyl [13463-39-3]
Nickel Chloride [7718-54-9]
Nickel Cyanide [557-19-7]
Nickel Dimethylglyoxime [13478-93-8]
Nickel Fluoride [10028-18-9]
Nickel Formate [3349-06-2]
Nickel Hydroxide [12054-48-7]
Nickel Iodide [13462-90-3]
Nickel Monoxide [1313-99-1]
Nickel Nitrate [13138-45-9]
Nickel Oxalate [547-67-1]
Nickel Phosphate [10381-36-9]
Nickel Sesquioxide [1314-06-3]
Nickel Sulfate [7786-81-4]
Niclofolan [10331-57-4]
Niclosamide [50-65-7]
Niclosamide Ethanolamine Salt* [1420-04-8]
Nicoclonate [10571-59-2]
Nicofibrate [31980-29-7]
Nicofibrate Hydrochloride* [17413-51-3]
Nicofuranose [15351-13-0]
Nicomol [27959-26-8]
Nicomorphine [639-48-5]
Nicorandil [65141-46-0]
Nicotelline [494-04-2]
Nicotinamide [98-92-0]
Nicotinamide Ascorbate [1987-71-9]
Nicotine [54-11-5]
Nicotine Bitartrate* [65-31-6]
Nicotine Dihydrochloride* [6019-02-9]
Nicotine Hydrochloride* [21361-93-3]
Nicotine Polacrilex* [96055-45-7]
Nicotine Salicylate* [29790-52-1]
Nicotine Sulfate* [65-30-5]
Nicotinic Acid [59-67-6]
Nicotinic Acid Benzyl Ester [94-44-0]
Nicotinyl Alcohol [100-55-0]

* Denotes derivative of title compound

Nicotinyl Alcohol d-Tartrate* [6164-87-0]
Nidroxyzone [405-22-1]
Nidulin [10089-10-8]
Nifedipine [21829-25-4]
Nifenalol [7413-36-7]
Nifenalol Hydrochloride* [5704-60-9]
Nifenazone [2139-47-1]
Niflumic Acid [4394-00-7]
Niflumic Acid β-Morpholinoethyl Ester* [65847-85-0]
Nifuradene [555-84-0]
Nifuraldezone [3270-71-1]
Nifuratel [4936-47-4]
Nifurfoline [3363-58-4]
Nifuroquine [57474-29-0]
Nifuroxazide [965-52-6]
Nifuroxime [6236-05-1]
Nifurpirinol [13411-16-0]
Nifurprazine [1614-20-6]
Nifurprazine Hydrochloride* [50832-74-1]
Nifurtimox [23256-30-6]
Nifurtoinol [1088-92-2]
Nifurzide [39978-42-2]
Nigericin [28380-24-7]
Nihydrazone [67-28-7]
Nikethamide [59-26-7]
Nilutamide [63612-50-0]
Nilvadipine [75530-68-6]
Nimbin [5945-86-8]
Nimbiol [561-95-5]
Nimesulide [51803-78-2]
Nimetazepam [2011-67-8]
Nimidane [50435-25-1]
Nimodipine [66085-59-4]
Nimorazole [6506-37-2]
Nimustine [42471-28-3]
Nimustine Hydrochloride* [55661-38-6]
Ninhydrin [485-47-2]
Ninopterin [2179-16-0]
Niobium [7440-03-1]
Niobium Pentachloride [10026-12-7]
Niobium Pentafluoride [7783-68-8]
Niobium Pentoxide [1313-96-8]
Niobium Potassium Oxypentafluoride [17523-77-2]
Nioxime [492-99-9]
Nipecotic Acid [498-95-3]
Nipradilol [81486-22-8]
Niridazole [61-57-4]
Nisin [1414-45-5]
Nisoldipine [63675-72-9]
Nitarsone [98-72-6]
Nithiazide [139-94-6]
Nitracrine [4533-39-5]
Nitracrine Dihydrochloride Monohydrate* [55429-45-3]
Nitralin [4726-14-1]
Nitramide [7782-94-7]
Nitramine [479-45-8]
Nitranilic Acid [479-22-1]
Nitrapyrin [1929-82-4]
Nitrazepam [146-22-5]
Nitrefazole [21721-92-6]
Nitrendipine [39562-70-4]
Nitric Acid [7697-37-2]
Nitric Acid, Anhydrous [7697-37-2]
Nitric Oxide [10102-43-9]
Nitrilotriacetic Acid [139-13-9]
Nitrin [553-74-2]
m-Nitroacetanilide* [122-28-1]
o-Nitroacetanilide* [552-32-9]
p-Nitroacetanilide* [104-04-1]
Nitroakridin 3582 [6035-39-8]
Nitroakridin 3582 Dihydrochloride* [73972-50-6]
m-Nitroaniline [99-09-2]
o-Nitroaniline [88-74-4]

p-Nitroaniline [100-01-6]
5-Nitrobarbituric Acid [480-68-2]
Nitrobenzene [98-95-3]
4-Nitrobenzoyl Chloride [122-04-3]
p-Nitrobenzyl Cyanide [555-21-5]
o-Nitrobiphenyl [86-00-0]
p-Nitrobiphenyl [92-93-3]
2-Nitro-1,1-bis(p-chlorophenyl)propane [117-27-1]
m-Nitrocinnamic Acid [555-68-0]
Nitrocobalamin* [20623-13-6]
Nitrodan [962-02-7]
Nitroethane [79-24-3]
Nitrofen [1836-75-5]
Nitrofurantoin [67-20-9]
Nitrofurazone [59-87-0]
Nitrogen [7727-37-9]
Nitrogen Chloride [10025-85-1]
Nitrogen Dioxide [10102-44-0]
Nitrogen Fluoride [7783-54-2]
Nitrogen Pentoxide [10102-03-1]
Nitrogen Selenide [12033-88-4]
Nitroglycerin [55-63-0]
Nitroguanidine [556-88-7]
Nitrohydrochloric Acid [8007-56-5]
Nitromersol [133-58-4]
Nitromethane [75-52-5]
Nitromide [121-81-3]
Nitron [2218-94-2]
1-Nitronaphthalene [86-57-7]
1-Nitro-2-naphthol [550-60-7]
3-Nitropentane [551-88-2]
5-Nitro-o-phenetidine [136-79-8]
Nitrophenide [537-91-7]
m-Nitrophenol [554-84-7]
o-Nitrophenol [88-75-5]
p-Nitrophenol [100-02-7]
p-Nitrophenylacetic Acid [104-03-0]
4-Nitro-o-phenylenediamine [99-56-9]
(4-Nitrophenyl)hydrazine [100-16-3]
o-Nitrophenylpropiolic Acid [530-85-8]
(4-Nitrophenyl)urea [556-10-5]
1-Nitropropane [108-03-2]
2-Nitropropane [79-46-9]
5'-Nitro-2'-propoxyacetanilide [553-20-8]
5-Nitro-2-propoxyaniline [553-79-7]
5-Nitroquinaldic Acid [525-47-3]
3-Nitrosalicylic Acid [85-38-1]
5-Nitrosalicylic Acid [96-97-9]
Nitroscanate [19881-18-6]
3-Nitrosobenzamide [144189-66-2]
N-Nitrosodiethanolamine [1116-54-7]
N-Nitrosodiethylamine [55-18-5]
N-Nitrosodimethylamine [62-75-9]
p-Nitroso-N,N-dimethylaniline [138-89-6]
p-Nitrosodiphenylamine [156-10-5]
N-Nitrosomorpholine [59-89-2]
1-Nitroso-2-naphthol [131-91-9]
4-Nitrosophenol [104-91-6]
N-Nitrosopyrrolidine [930-55-2]
Nitroso-R Salt [525-05-3]
2-Nitro-4-sulfobenzoic Acid [552-23-8]
Nitrosyl Chloride [2696-92-6]
Nitrosyl Fluoride [7789-25-5]
Nitrosylsulfuric Acid [7782-78-7]
Nitrosyl Tetrafluoroborate [14635-75-7]
Nitrourea [556-89-8]
Nitrous Acid [7782-77-6]
Nitrous Oxide [10024-97-2]
Nitrovin [804-36-4]
Nitroxoline [4008-48-4]
Nitroxynil [1689-89-0]
Nitryl Chloride [13444-90-1]
Nitryl Fluoride [10022-50-1]
Nivalenol [23282-20-4]

Nizatidine [76963-41-2]
Nizofenone [54533-85-6]
Nizofenone Fumarate* [54533-86-7]
NMDA [6384-92-5]
NMN [1094-61-7]
Nobelium [10028-14-5]
Nocardamin [26605-16-3]
Nocardicin A* [39391-39-4]
Nocardicin B* [60134-71-6]
Nodakenin [495-31-8]
Noformicin [155-38-4]
Nogalamycin [1404-15-5]
Nomifensine [24526-64-5]
Nomifensine Maleate* [32795-47-4]
Nomilin [1063-77-0]
Nonactin [6833-84-7]
2-Nonenal [18829-56-6]
Nonoxynol [26027-38-3]
n-Nonyl Acetate [143-13-5]
n-Nonyl Alcohol [143-08-8]
Nonyl Phenol [25154-52-3]
(p-Nonylphenoxy)acetic Acid [3115-49-9]
Noprylsulfamide [576-97-6]
Noracymethadol [1477-39-0]
Noracymethadol Hydrochloride* [5633-25-0]
Norbolethone [1235-15-0]
Norbormide [991-42-4]
Norcarane [286-08-8]
Norcholanic Acid [511-18-2]
Norcodeine [467-15-2]
Nordazepam [1088-11-5]
Nordefrin [6539-57-7] (unspecified)
Nordefrin [74812-63-8] (R*,S*)-(±)-Form
Nordefrin (—)-Form* [829-74-3]
Nordefrin Hydrochloride* [61-96-1] (R*,S*)-(±)-Form
Nordefrin Hydrochloride* [138-61-4] (unspecified)
Nordihydroguaiaretic Acid [500-38-9] (unspecified)
Nordihydroguaiaretic Acid [27686-84-6] (meso-form)
Norea [18530-56-8]
Norepinephrine [51-41-2]
Norepinephrine d-Bitartrate* [69815-49-2]
Norepinephrine Hydrochloride* [329-56-6]
Norethandrolone [52-78-8]
Norethindrone [68-22-4]
Norethindrone Acetate* [51-98-9]
Norethynodrel [68-23-5]
Norethynodrel Mixture with Mestranol* [8015-30-3]
Norfenefrine [536-21-0]
Norfenefrine dl-Form Hydrochloride* [15308-34-6]
Norfloxacin [70458-96-7]
Norflurazon [27314-13-2]
Norgesterone [13563-60-5]
Norgesterone Mixture with Ethinyl Estradiol* [8064-75-3]
Norgestimate [35189-28-7]
Norgestimate Mixture with Ethinyl Estradiol* [79871-54-8]
Norgestrel [6533-00-2]
Norgestrel (+)-Form* [797-64-8]
Norgestrienone [848-21-5]
Norhyoscyamine [537-29-1]
Norleucine [327-57-1]
Norlevorphanol [1531-12-0]
Norlobelanine [6035-31-0]
Normetanephrine [97-31-4]
Normethadone [467-85-6]
Normethadone Hydrochloride* [847-84-7]
Normethandrone [514-61-4]

Normorphine [466-97-7]
Nornicotine [494-97-3]
Nornidulin [33403-37-1]
Norphenazone [89-25-8]
Norpipanone [561-48-8]
Norpipanone Hydrobromide*
  [6033-42-7]
Norpipanone Hydrochloride*
  [6033-41-6]
Norpseudoephedrine [36393-56-3]
Nortriptyline [72-69-5]
Nortriptyline Hydrochloride*
  [894-71-3]
Norvaline [6600-40-4]
Norvinisterone [6795-60-4]
Noscapine [128-62-1]
Noscapine dl-Form* [6035-40-1]
Noscapine Hydrochloride* [912-60-7]
Nosiheptide [56377-79-8]
Novembichin [1936-40-9]
Novobiocin [303-81-1]
Novobiocin Monosodium Salt*
  [1476-53-5]
Novoldiamine [140-80-7]
Novonal [512-48-1]
Noxiptilin [3362-45-6]
Noxiptilin Hydrochloride*
  [4985-15-3]
Noxythiolin [15599-39-0]
Nucleocidin [24751-69-7]
Nupharidine [468-89-3]
Nybomycin [30408-30-1]
Nylidrin [447-41-6]
Nylidrin Hydrochloride* [849-55-8]
Nylon 6 [25038-54-4]
Nylon 46 [50327-22-5]
Nystatin [1400-61-9]
Nystatin A$_1$* [34786-70-4]
Obidoxime Chloride [114-90-9]
Ochratoxin A* [303-47-9]
Ochratoxin B* [4825-86-9]
Ochratoxin C* [4865-85-4]
Ocimene [29714-87-2]
Ocimene trans-β-Form* [3779-61-1]
Octabenzone [1843-05-6]
Octacaine [13912-77-1]
Octacaine Hydrochloride*
  [59727-70-7]
Octacosanol [557-61-9]
Octafluorocyclobutane [115-25-3]
2,2,3,3,4,4,5,5-Octafluoro-1-pentanol
  [355-80-6]
Octamethylcyclotetrasiloxane
  [556-67-2]
Octamethyltrisiloxane [107-51-7]
Octamoxin [4684-87-1]
Octamoxin Sulfate* [3845-07-6]
Octamylamine [502-59-0]
Octane [111-65-9]
Octanohydroxamic Acid [7377-03-9]
1-Octanol [111-87-5]
2-Octanol [123-96-6]
Octaverine [549-68-8]
Octenidine [71251-02-0]
Octenidine Dihydrochloride*
  [70775-75-6]
Octenidine Disaccharin* [86767-75-1]
Octhilinone [26530-20-1]
Octodrine [543-82-8]
Octogen [2691-41-0]
Octopamine [104-14-3]
Octopamine D(−)-Form* [876-04-0]
Octopamine DL-Form Hydrochloride*
  [770-05-8]
Octotiamine [137-86-0]
Octoxynol [9002-93-1]
Octreotide [83150-76-9]
Octyl Acetate [103-09-3]
n-Octyl Bromide [111-83-1]
sec-Octyl Bromide [557-35-7]

sec-Octyl Iodide [557-36-8]
Octyl Methoxycinnamate [5466-77-3]
Ofloxacin [82419-36-1]
Ofloxacin S-(−)-Form* [100986-85-4]
Okadaic Acid [78111-17-8]
Olanzapine [132539-06-1]
Olaquindox [23696-28-8]
Old Yellow Enzyme [9001-68-7]
Oleandomycin [3922-90-5]
Oleandomycin Hydrochloride*
  [6696-47-5]
Oleandomycin Phosphate*
  [7060-74-4]
Oleandrin [465-16-7]
Oleanolic Acid [508-02-1]
Oleic Acid [112-80-1]
Oleic Acid Sodium Salt* [143-19-1]
Oleuropein [32619-42-4]
Oleyl Alcohol [143-28-2]
Olibanum [8050-07-5]
Oligomycin A* [579-13-5]
Oligomycin B* [11050-94-5]
Oligomycin C* [11052-72-5]
Oligomycin D* [1404-59-7]
Olivacine [484-49-1]
Olivil [2955-23-9]
Olivomycin A* [6988-58-5]
Olsalazine [15722-48-2]
Olsalazine Disodium Salt*
  [6054-98-4]
Omeprazole [73590-58-6]
Omoconazole [74512-12-2]
Omoconazole Nitrate* [83621-06-1]
Ondansetron [99614-02-5]
Ondansetron Hydrochloride
  Dihydrate* [99614-01-4]
Ontianil [35727-72-1]
Oosporein [475-54-7]
Opianic Acid [519-05-1]
Opiniazide [2779-55-7]
Opipramol [315-72-0]
Opipramol Dihydrochloride*
  [909-39-7]
Opromazine [969-99-3]
Orange I [523-44-4]
Orange II [633-96-5]
Orange B [15139-76-1]
Orazamide [2574-78-9]
Orcein [1400-62-0]
Orcinol [504-15-4]
Oripavine [467-04-9]
Orlistat [96829-58-2]
Ormosinine [14350-67-5]
Ornidazole [16773-42-5]
Ornipressin [3397-23-7]
Ornithine [70-26-8]
Ornithine DL-Form* [616-07-9]
Ornithine L-Form Aspartate*
  [3230-94-2]
Ornithine L-Form Dihydrochloride*
  [6211-16-1]
Ornithine DL-Form Monohydrochlo-
  ride* [1069-31-4]
Ornithine L-Form Monohydrochlo-
  ride* [3184-13-2]
Ornithine L-Form Monosulfate*
  [6153-43-1]
Ornoprostil [70667-26-4]
Orotic Acid [65-86-1]
Orotic Acid Choline Salt*
  [24381-49-5]
Orotic Acid Ethyl Ester* [1747-53-1]
Orotic Acid Methyl Ester*
  [6153-44-2]
Orotic Acid Monohydrate*
  [50887-69-9]
Orotidine [314-50-1]
Oroxylin A [480-11-5]
Orphenadrine [83-98-7]
Orphenadrine Citrate* [4682-36-4]

o-Orsellinic Acid [480-64-8]
Orthanilic Acid [88-21-1]
Orthocaine [536-25-4]
Orthoformic Acid [463-78-5]
Oryzacidin [1400-67-5]
Oryzalin [19044-88-3]
γ-Oryzanol [11042-64-1]
Oryzanol A* [21238-33-5]
Oryzanol C* [469-36-3]
Osajin [482-53-1]
Osalmid [526-18-1]
Osaterone [105149-04-0]
Osaterone Acetate Ester*
  [105149-00-6]
Osmaron B [8031-66-1]
Osmium [7440-04-2]
Osmium Hexafluoride [13768-38-2]
Osmium Tetrachloride [10026-01-4]
Osmium Tetroxide [20816-12-0]
Osteocalcin [75757-02-7] (human
  reduced)
Osthole [484-12-8]
Ostruthin [148-83-4]
Ostruthol [642-08-0]
Otobain [3738-01-0]
Ouabagenin [508-52-1]
Ouabain [630-60-4]
Ovex [80-33-1]
Oxabolone [4721-69-1]
Oxabolone 17-Cyclopentanepro-
  pionate* [1254-35-9]
Oxaceprol [33996-33-7]
Oxacillin [66-79-5]
Oxacillin Sodium Salt Monohydrate*
  [7240-38-2]
1,3,4-Oxadiazole [288-99-3]
Oxadiazon [19666-30-9]
Oxaflozane [26629-87-8]
Oxaflozane Hydrochloride*
  [26629-86-7]
Oxaflumazine [16498-21-8]
Oxaflumazine Disuccinate*
  [7450-97-7]
Oxalacetic Acid [328-42-7]
Oxalenediuramidoxime [580-52-9]
Oxalic Acid [144-62-7]
Oxaliplatin [61825-94-3]
Oxalomolybdic Acid [53450-33-2]
Oxalyl Chloride [79-37-8]
Oxamarin [15301-80-1]
Oxamarin Dihydrochloride*
  [6830-17-7]
Oxametacine [27035-30-9]
Oxamic Acid [471-47-6]
Oxamide [471-46-5]
Oxamniquine [21738-42-1]
Oxamyl [23135-22-0]
Oxanamide [126-93-2]
Oxandrolone [53-39-4]
Oxantel [36531-26-7]
Oxantel Pamoate* [68813-55-8]
Oxapropanium Iodide [541-66-2]
Oxaprozin [21256-18-8]
Oxatomide [60607-34-3]
Oxazepam [604-75-1]
Oxazidione [27591-42-0]
Oxazolam [27167-30-2]
Oxcarbazepine [28721-07-5]
Oxeladin [468-61-1]
Oxeladin Citrate* [52432-72-1]
Oxendolone [33765-68-3]
Oxenin [3230-75-9]
Oxethazaine [126-27-2]
Oxethazaine Hydrochloride*
  [13930-31-9]
Oxetorone [26020-55-3]
Oxetorone Fumarate* [34522-46-8]
Oxfendazole [53716-50-0]
Oxibendazole [20559-55-1]
Oxiconazole Nitrate [64211-46-7]

Denotes derivative of title compound

Oxidimethiin [55290-64-7]
Oxidronic Acid [15468-10-7]
Oxidronic Acid Sodium Salt*
[14255-61-9]
Oxidronic Acid 99mTc-complex*
[72945-61-0]
Oxiniacic Acid [2398-81-4]
Oxiniacic Acid Ethanolamine Salt*
[36296-31-8]
Oxiracetam [62613-82-5]
Oxitropium Bromide [30286-75-0]
Oxolamine [959-14-8]
Oxolamine Citrate* [1949-20-8]
Oxolamine Hydrochloride*
[1219-20-1]
Oxolinic Acid [14698-29-4]
Oxomemazine [3689-50-7]
Oxonic Acid [937-13-3]
Oxophenarsine Hydrochloride
[538-03-4]
Oxophenylarsine [637-03-6]
L-2-Oxo-4-thiazolidinecarboxylic Acid
[19771-63-2]
Oxotremorine [70-22-4]
Oxprenolol [6452-71-7]
Oxprenolol Hydrochloride*
[6452-73-9]
Oxyacanthine [548-40-3]
Oxybenzone [131-57-7]
Oxybutynin Chloride [1508-65-2]
Oxychlorosene [8031-14-9]
Oxychlorosene Sodium Salt*
[52906-84-0]
Oxycinchophen [485-89-2]
Oxyclozanide [2277-92-1]
Oxycodone [76-42-6]
Oxycodone Hydrochloride*
[124-90-3]
4,4'-Oxydi-2-butanol [821-33-0]
10,10'-Oxydiphenoxarsine [58-36-6]
Oxyfedrine [15687-41-9]
Oxyfedrine DL-Form Hydrochloride*
[16648-69-4]
Oxyfedrine L-Form Hydrochloride*
[16777-42-7]
Oxyfluorfen [42874-03-3]
Oxygen [7782-44-7]
Oxymesterone [145-12-0]
Oxymetazoline [1491-59-4]
Oxymetazoline Hydrochloride*
[2315-02-8]
Oxymetholone [434-07-1]
Oxymethurea [140-95-4]
Oxymorphone [76-41-5]
Oxymorphone Hydrochloride*
[357-07-3]
Oxypendyl [5585-93-3]
Oxypendyl Dihydrochloride*
[17297-82-4]
Oxypertine [153-87-7]
Oxyphenbutazone [129-20-4]
Oxyphenbutazone Monohydrate*
[7081-38-1]
Oxyphencyclimine [125-53-1]
Oxyphencyclimine Hydrochloride*
[125-52-0]
Oxyphenisatin Acetate [115-33-3]
Oxyphenonium Bromide [50-10-2]
Oxypinocamphone [10136-65-9]
Oxytetracycline [79-57-2]
Oxytetracycline Hydrochloride*
[2058-46-0]
Oxythiamine [582-36-5]
Oxythioquinox [2439-01-2]
Oxytocin [50-56-6]
Oxytocin Citrate* [7563-62-4]
Ozagrel [82571-53-7]
Ozagrel Hydrochloride* [78712-43-3]
Ozone [10028-15-6]
Paclitaxel [33069-62-4]

Paclobutrazol [76738-62-0]
Pactamycin [23668-11-3]
Palitantin [15265-28-8]
Palladium [7440-05-3]
Palladium Chloride [7647-10-1]
Palladium Diacetate [19807-27-3]
Palladium Nitrate [10102-05-3]
Palladium Oxide [1314-08-5]
Palmatine [3486-67-7]
Palmidrol [544-31-0]
Palmitic Acid [57-10-3]
Palustric Acid [1945-53-5]
Palytoxin [77734-91-9]
Pamabrom [606-04-2]
Pamaquine [491-92-9]
Pamidronic Acid [40391-99-9]
Pamidronic Acid Disodium Salt*
[57248-88-1]
Pamoic Acid [130-85-8]
Pancreatin [8049-47-6]
Pancrelipase [53608-75-6]
Pancuronium Bromide [15500-66-0]
Panipenem [87726-17-8]
Panipenem Mixture with Betami-
pron* [138240-65-0]
Pankrin [9001-72-3]
Pantetheine [496-65-1]
Pantethine [16816-67-4]
Pantolactone [599-04-2]
Pantoprazole [102625-70-7]
Pantoprazole Sodium Salt*
[138786-67-1]
Pantothenic Acid [79-83-4]
Pantothenic Acid Calcium Salt*
[137-08-6]
Pantothenic Acid Sodium Salt*
[867-81-2]
Papain [9001-73-4]
Papaveraldine [522-57-6]
Papaveretum [8002-76-4]
Papaverine [58-74-2]
Papaverine Hydrochloride* [61-25-6]
Papaverine Nitrite* [132-40-1]
Papuamine [112455-84-2]
Papuamine Dihydrochloride*
[112347-74-7]
Parabanic Acid [120-89-8]
Paraflutizide [1580-83-2]
Paraformaldehyde [30525-89-4]
Paraherquamide [77392-58-6]
Paraldehyde [123-63-7]
Paramethadione [115-67-3]
Paramethasone [53-33-8]
Paramethasone 21-Acetate*
[1597-82-6]
Paranyline [1729-61-9]
Paranyline Hydrochloride*
[5585-60-4]
Paraoxon [311-45-5]
Paraquat [4685-14-7]
Parasorbic Acid [10048-32-5]
Parathion [56-38-2]
Parathyroid Hormone [9002-64-6]
Parbendazole [14255-87-9]
Parethoxycaine [94-23-5]
Parethoxycaine Hydrochloride*
[136-46-9]
Pargyline [555-57-7]
Pargyline Hydrochloride* [306-07-0]
Paromomycin [7542-37-2]
Paromomycin Sulfate* [1263-89-4]
Parotin [1392-81-0]
Paroxetine [61869-08-7]
Paroxypropione [70-70-2]
Parsalmide [30653-83-9]
Parthenin [508-59-8]
Parthenolide [20554-84-1]
Partricin [11096-49-4]
Parvaquone [4042-30-2]
Patchouli Alcohol [5986-55-0]

Patulin [149-29-1]
Pavoninin-5 [94480-49-6]
Pazinaclone [103255-66-9] (unspec-
ified)
Pazinaclone [151764-87-3] (±)
Pazufloxacin [127045-41-4]
Pazufloxacin Methanesulfonate*
[163680-77-1]
PBB* [49582-09-4]
PBP* [51005-85-7]
Pebulate [1114-71-2]
Pecilocin [19504-77-9]
Pectin [9000-69-5]
Pectolinarigenin [520-12-7]
Pederin [27973-72-4]
Pefloxacin [70458-92-3]
Pefloxacin Methanesulfonate*
[70458-95-6]
Pefloxacin Methanesulfonate
Dihydrate* [149676-40-4]
Pegaspargase* [130167-69-0]
Pegorgotein* [155773-57-2]
Pelargonic Acid [112-05-0]
Pelargonidin [134-04-3]
Pelargonidin 3,5-Diglucoside*
[17334-58-6]
Pelargonidin 3-Glucoside*
[18466-51-8]
Pelletierine [4396-01-4]
Pelletierine Hydrochloride*
[5984-61-2]
Pellitorine [18836-52-7]
Pellotine [83-14-7]
α-Peltatin [568-53-6]
β-Peltatin [518-29-6]
Pemirolast [69372-19-6]
Pemirolast Potassium Salt*
[100299-08-9]
Pemoline [2152-34-3]
Pemoline Magnesium Hydroxide
Mixture* [18968-99-5]
Pempidine [79-55-0]
Pempidine Tartrate* [546-48-5]
Penamecillin [983-85-7]
Penbutolol [38363-40-5]
Penbutolol Sulfate* [38363-32-5]
Penciclovir [39809-25-1]
Pendimethalin [40487-42-1]
Penethamate Hydriodide [808-71-9]
Penfluridol [26864-56-2]
Penicillamine [52-67-5]
Penicillamine Cysteine Disulfide
[18840-45-4]
Penicillamine Disulfide [312-10-7]
Penicillamine Disulfide D-Form*
[20902-45-8]
Penicillamine Disulfide DL-Form*
[21174-80-1]
Penicillamine Disulfide L-Form*
[113626-33-8]
Penicillamine DL-Form* [52-66-4]
Penicillamine L-Form* [1113-41-3]
Penicillamine DL-Form Hydrochlo-
ride* [22572-05-0]
Penicillamine Hydrochloride*
[2219-30-9]
Penicillanic Acid [87-53-6]
Penicillic Acid [90-65-3]
Penicillinase [9001-74-5]
Penicillin G Benethamine [751-84-8]
Penicillin G Benzathine [1538-09-6]
Penicillin G Benzhydrylamine
[1538-11-0]
Penicillin G Calcium [973-53-5]
Penicillin G Hydrabamine [3344-16-9]
Penicillin G Potassium [113-98-4]
Penicillin G Procaine [6130-64-9]
Penicillin N [525-94-0]
Penicillin O [87-09-2]

Penicillin O Potassium Salt* [897-61-0]
Penicillin O Sodium Salt* [7177-54-0]
Penicillin S Potassium [525-87-1]
Penicillin V [87-08-1]
Penicillin V Benzathine [5928-84-7]
Penicillin V Hydrabamine [6591-72-6]
Penicillin V Potassium Salt* [132-98-9]
Penicilloyl Polylysine [53608-77-8]
Penimepicycline [4599-60-4]
Pentaborane(9) [19624-22-7]
Pentaborane(11) [18433-84-6]
Pentabromoacetone [79-49-2]
Pentacene [135-48-8]
Pentachloroethane [76-01-7]
Pentachlorophenol [87-86-5]
Pentacynium Bis(methylsulfate) [3810-83-1]
3-Pentadecylcatechol [492-89-7]
Pentaerythritol [115-77-5]
Pentaerythritol Chloral [78-12-6]
Pentaerythritol Dichlorohydrin [2209-86-1]
Pentaerythritol Tetraacetate [597-71-7]
Pentaerythritol Tetranitrate [78-11-5]
Pentagastrin [5534-95-2]
Pentagestrone [7001-56-1]
Pentagestrone Acetate* [1178-60-5]
Pentahomoserine [533-88-0]
Pentamethonium Bromide [541-20-8]
Pentamidine [100-33-4]
Pentamidine Dimethanesulfonate* [6823-79-6]
Pentamidine Isethionate* [140-64-7]
Pentane [109-66-0]
1,5-Pentanediol [111-29-5]
1-Pentanol [71-41-0]
2-Pentanol [6032-29-7]
3-Pentanol [584-02-1]
Pentapiperide [7009-54-3]
Pentapiperide Methosulfate* [7681-80-3]
Pentazocine [359-83-1]
Pentazocine Hydrochloride* [64024-15-3]
Pentazocine Lactate* [17146-95-1]
1-Pentene [109-67-1]
2-Pentene [109-68-2]
2-Pentenylpenicillin Sodium [525-86-0]
Pentetate Calcium Trisodium [12111-24-9]
Pentetic Acid [67-43-6]
Pentetreotide [138661-02-6]
Pentetreotide [111]In Chelate* [139096-04-1]
Penthienate Bromide [60-44-6]
Pentifylline [1028-33-7]
Pentigetide [62087-72-3]
Pentisomide [78833-03-1] (unspecified)
Pentisomide [96513-83-6] (±)
Pentobarbital [76-74-4]
Pentobarbital Calcium Salt* [7563-42-0]
Pentobarbital Sodium Salt* [57-33-0]
1-Pentol [105-29-3]
Pentolinium Tartrate [52-62-0]
Pentorex [434-43-5] (unspecified)
Pentorex [46029-44-1] (DL-Form)
Pentorex Hydrochloride* [5585-52-4]
DL-Pentorex Hydrogen D-Tartrate* [22232-55-9]
Pentosan Polysulfate [37300-21-3]
Pentosan Polysulfate Sodium Salt* [37319-17-8]
Pentostatin [53910-25-1]
Pentoxifylline [6493-05-6]

Pentoxyl [147-61-5]
Pentrinitrol [1607-17-6]
Pentryl [4481-55-4]
tert-Pentyl Alcohol [75-85-4]
Pentylenetetrazole [54-95-5]
p-tert-Pentylphenol [80-46-6]
Peonidin [134-01-0]
Peplomycin [68247-85-8]
Peplomycin Sulfate Salt* [70384-29-1]
Pepsin [9001-75-6]
Pepstatin [26305-03-3]
Peptide T [106362-32-7]
Peracetic Acid [79-21-0]
Perazine [84-97-9]
Perbenzoic Acid [93-59-4]
Perchloric Acid [7601-90-3]
Perchloryl Fluoride [7616-94-6]
Perezone [3600-95-1]
Perflubron [423-55-2]
Perfluidone [37924-13-3]
Perfluorodecalin* [306-94-5]
Perfluorotripropylamine* [338-83-0]
Performic Acid [107-32-4]
Perfosfamide [39800-16-3] (unspecified)
Perfosfamide [62435-42-1] (±)
Pergolide [66104-22-1]
Pergolide Mesylate* [66104-23-2]
Perhexiline [6621-47-2]
Perhexiline Maleate* [6724-53-4]
Pericyazine [2622-26-6]
Perilla Ketone [553-84-4]
Perillaldehyde [2111-75-3]
Perimethazine [13093-88-4]
Perindopril [82834-16-0]
Perindoprilat* [95153-31-4]
Perindopril tert-Butylamine* [107133-36-8]
Periodic Acid [10450-60-9]
Periodyl [53586-99-5]
Periplanone A* [112709-47-4]
Periplanone B* [70613-99-9]
(−)-Periplanone B* [61228-92-0]
Periplocin [13137-64-9]
Periplocymarin [32476-67-8]
Periplogenin [514-39-6]
Perisoxal [2055-44-9]
Perisoxal Citrate* [2139-25-5]
Perivine [2673-40-7]
Perlapine [1977-11-3]
Permethrin [52645-53-1]
Perphenazine [58-39-9]
Peruvoside [1182-87-2]
Perylene [198-55-0]
Petroleum Benzin [8030-30-6]
Petroselinic Acid [593-39-5]
Petunidin [1429-30-7]
Petunidin 3,5-Diglucoside* [25846-73-5]
Peucedanin [133-26-6]
Peyonine [19717-25-0]
Pfeiffer's Substance [8015-18-7]
Phalloidin [17466-45-4]
Phanquinone [84-12-8]
Phaseolin [13401-40-6]
Phasin [1392-87-6]
α-Phellandrene [99-83-2]
β-Phellandrene [555-10-2]
α-Phellandrene d-Form* [2243-33-6]
α-Phellandrene l-Form* [4221-98-1]
β-Phellandrene d-Form* [6153-16-8]
β-Phellandrene l-Form* [6153-17-9]
Phenacaine Hydrochloride [620-99-5]
Phenacaine Hydrochloride Monohydrate* [6153-19-1]
Phenacemide [63-98-9]
Phenacetin [62-44-2]
Phenacetolin [1340-26-7]
Phenactropinium Chloride [3784-89-2]

Phenacylamine [613-89-8]
Phenadoxone [467-84-5]
Phenadoxone Hydrochloride* [545-91-5]
Phenallymal [115-43-5]
Phenamacide Hydrochloride [31031-74-0]
Phenamet [3819-34-9]
Phenamidine [101-62-2]
Phenamidine Isethionate* [620-90-6]
Phenampromid(e) [129-83-9]
Phenanthrene [85-01-8]
Phenanthrenequinone [84-11-7]
o-Phenanthroline [66-71-7]
Phenarsazine Chloride [578-94-9]
Phenatine [139-68-4]
Phenazine [92-82-0]
Phenazocine [127-35-5]
Phenazocine (±)-Hydrobromide* [1239-04-9]
Phenazopyridine Hydrochloride [136-40-3]
Phenbutamide [3149-00-6]
Phencarbamide [3735-90-8]
Phencarbamide Hydrochloride* [58-13-9]
Phencarbamide 1,5-Naphthalenedisulfonate* [72017-58-4]
Phencyclidine [77-10-1]
Phencyclidine Hydrochloride* [956-90-1]
Phendimetrazine [634-03-7]
Phendimetrazine Bitartrate* [50-58-8]
Phenelzine [51-71-8]
Phenelzine Acid Sulfate* [156-51-4]
Phenesterine [3546-10-9]
Phenetharbital [357-67-5]
Phenethicillin Potassium [132-93-4]
Phenethyl Alcohol [60-12-8]
Phenethylamine [64-04-0]
o-Phenetidine [94-70-2]
p-Phenetidine [156-43-4]
Phenetole [103-73-1]
Pheneturide [90-49-3]
Phenformin [114-86-3]
Phenformin Hydrochloride* [834-28-6]
Phenglutarimide [1156-05-4]
Phenglutarimide Hydrochloride* [1674-96-0]
Phenicarbazide [103-03-7]
Phenicin [128-68-7]
Phenindamine [82-88-2]
Phenindamine Hydrogen Tartrate* [569-59-5]
Phenindione [83-12-5]
Pheniprazine [55-52-7]
Pheniprazine DL-Form* [52031-11-5]
Pheniprazine DL-Form Hydrochloride* [54779-57-6]
Pheniramine [86-21-5]
Pheniramine Maleate* [132-20-7]
Phenmedipham [13684-63-4]
Phenmetrazine [134-49-6]
Phenmetrazine Hydrochloride* [1707-14-8]
Phenobarbital [50-06-6]
Phenobarbital Sodium Salt* [57-30-7]
Phenobutiodil [554-24-5]
Phenocoll [103-97-9]
Phenocoll Hydrochloride* [539-10-6]
Phenocoll Salicylate* [140-47-6]
Phenoctide [78-05-7]
Phenol [108-95-2]
Phenol Ammonium Salt* [5973-17-1]
Phenoldisulfonic Acid [96-77-5]
Phenolphthalein [77-09-8]
Phenolphthalein Sodium [518-51-4]
Phenolphthalin [81-90-3]
Phenolphthalol [81-92-5]

* Denotes derivative of title compound

Phenol Sodium Salt* [139-02-6]
p-Phenolsulfonic Acid [98-67-9]
Phenolsulfonphthalein [143-74-8]
Phenoltetrachlorophthalein [639-44-1]
Phenoltetrachlorophthalein Disodium
　Salt* [128-71-2]
Phenomorphan [468-07-5]
Phenoperidine [562-26-5]
Phenoperidine Hydrochloride*
　[3627-49-4]
Phenopyrazone [3426-01-5]
Phenosafranin [81-93-6]
Phenosulfazole [515-54-8]
Phenothiazine [92-84-2]
Phenothrin [26002-80-2]
Phenoxazine [135-67-1]
Phenoxyacetic Acid [122-59-8]
Phenoxyacetyl Cellulose [68332-77-4]
Phenoxybenzamine [59-96-1]
Phenoxybenzamine Hydrochloride*
　[63-92-3]
2-Phenoxyethanol [122-99-6]
Phenoxypropazine [3818-37-9]
Phenoxypropazine Maleate*
　[3941-06-8]
Phenprobamate [673-31-4]
Phenprocoumon [435-97-2]
Phensuximide [86-34-0]
Phentermine [122-09-8]
Phentermine Hydrochloride*
　[1197-21-3]
Phentetiothalein Sodium [18265-54-8]
Phentolamine [50-60-2]
Phentolamine Hydrochloride*
　[73-05-2]
Phentolamine Methanesulfonate*
　[65-28-1]
Phentydrone [634-19-5]
Phenylacetaldehyde [122-78-1]
α-Phenylacetamide [103-81-1]
Phenyl Acetate [122-79-2]
Phenylacetic Acid [103-82-2]
Phenylacetone [103-79-7]
Phenyl Acetylsalicylate [134-55-4]
Phenylalanine [63-91-2]
Phenylalanine D-Form* [673-06-3]
Phenylalanine DL-Form* [150-30-1]
Phenyl Aminosalicylate [133-11-9]
N-Phenylanthranilic Acid [91-40-7]
2-Phenyl-1H-benzimidazole [716-79-0]
Phenyl Benzoate [93-99-2]
Phenyl Biguanide [102-02-3]
Phenylbutazone [50-33-9]
Phenylbutazone Sodium Salt*
　[129-18-0]
α-Phenylbutyramide [90-26-6]
Phenyl Carbonate [102-09-0]
2-Phenyl-6-chlorophenol [85-97-2]
4-Phenyl-2-chlorophenol [92-04-6]
α-Phenylcinnamic Acid [3368-16-9]
α-Phenylcinnamic Acid cis-Form*
　[91-47-4]
α-Phenylcinnamic Acid trans-Form*
　[91-48-5]
m-Phenylenediamine [108-45-2]
o-Phenylenediamine [95-54-5]
p-Phenylenediamine [106-50-3]
Phenylephrine Hydrochloride
　[61-76-7]
Phenylethanolamine [7568-93-6]
Phenylethanolamine Sulfate*
　[613-82-1]
Phenyl Ether [101-84-8]
5-(α-Phenylethyl)semioxamazide
　[93-95-8]
Phenylglyceryl Ether [538-43-2]
N-Phenylglycine [103-01-5]
α-Phenylglycine [69-91-0]
Phenylhydrazine [100-63-0]

Phenylhydrazine Hydrochloride*
　[59-88-1]
Phenylhydroxylamine [100-65-2]
Phenyl Isocyanate [103-71-9]
Phenyl Isothiocyanate [103-72-0]
Phenylmagnesium Chloride [100-59-4]
N-Phenylmaleimide [941-69-5]
Phenylmercuric Acetate [62-38-4]
Phenylmercuric Chloride [100-56-1]
Phenylmercuric Nitrate, Basic
　[8003-05-2]
Phenylmercury Borate [102-98-7]
Phenylmethylbarbituric Acid
　[76-94-8]
o-Phenylphenol [90-43-7]
p-Phenylphenol [92-69-3]
Phenyl Phthalate [84-62-8]
(±)-Phenylpropanolamine*
　[14838-15-4]
Phenylpropanolamine Hydrochloride
　[154-41-6]
Phenylpropylmethylamine [93-88-9]
1-Phenyl-3-pyrazolidinone [92-43-3]
Phenyl Salicylate [118-55-8]
4-Phenylsemicarbazide [537-47-3]
N-Phenylsulfanilic Acid [101-57-5]
Phenyl Sulfide [139-66-2]
Phenylthiourea [103-85-5]
Phenyltoloxamine [92-12-6]
Phenyltrimethylammonium Iodide
　[98-04-4]
Phenylurea [64-10-8]
Phenylurethan(e) [101-99-5]
Phenyramidol [553-69-5]
Phenyramidol Hydrochloride*
　[326-43-2]
Phenytoin [57-41-0]
Phenytoin Sodium Salt* [630-93-3]
Phethenylate Sodium [510-34-9]
Phillyrin [487-41-2]
Phloionic Acid [23843-52-9]
Phloionolic Acid [583-86-8]
Phloretin [60-82-2]
Phloridzin [60-81-1]
Phloroglucinol [108-73-6]
Phlorol [90-00-6]
Pholcodine [509-67-1]
Pholedrine [370-14-9]
Pholedrine Sulfate* [6114-26-7]
Phorate [298-02-2]
Phorbol [17673-25-5]
Phorone [504-20-1]
Phosalone [2310-17-0]
Phosfolan [947-02-4]
Phosgene [75-44-5]
Phosmet [732-11-6]
Phosphamidon [13171-21-6]
Phosphine [7803-51-2]
Phosphinothricin [51276-47-2]
Phosphocreatine [67-07-2]
Phosphocysteamine [5746-40-7]
Phosphocysteamine Sodium Salt*
　[3724-89-8]
Phosphomolybdic Acid [11104-88-4]
Phosphonium Iodide [12125-09-6]
Phosphoric Acid [7664-38-2]
Phosphoric Acid, Meta [37267-86-0]
Phosphorous Acid [13598-36-2]
Phosphorus [7723-14-0]
Phosphorus Hemitriselenide
　[1314-81-4]
Phosphorus Oxybromide [7789-59-5]
Phosphorus Oxychloride [10025-87-3]
Phosphorus Pentabromide [7789-69-7]
Phosphorus Pentachloride
　[10026-13-8]
Phosphorus Pentafluoride [7647-19-0]
Phosphorus Pentaselenide [1314-82-5]
Phosphorus Pentasulfide [1314-80-3]
Phosphorus Pentoxide [1314-56-3]

Phosphorus Sulfochloride [3982-91-0]
Phosphorus Tribromide [7789-60-8]
Phosphorus Trichloride [7719-12-2]
Phosphorus Trifluoride [7783-55-3]
Phosphorus Trioxide [1314-24-5]
Phosphorus Triselenide [1314-86-9]
Phosphorylcholine [107-73-3]
Phosphorylcholine Barium Salt*
　[6484-71-5]
Phosphorylcholine Calcium Salt*
　[4826-71-5]
Phosphorylcholine Magnesium Salt*
　[17032-39-2]
Phosphoserine [407-41-0]
Phosphotungstic Acid [12067-99-1]
Phosvitin [9008-96-2]
Phoxim [14816-18-3]
Phrenosin [586-02-7]
Phthalamide [88-96-0]
Phthalazine [253-52-1]
Phthalic Acid [88-99-3]
Phthalic Anhydride [85-44-9]
Phthalimide [85-41-6]
Phthalofyne [131-67-9]
Phthaloyl Chloride [88-95-9]
Phthalylsulfacetamide [131-69-1]
Phthalylsulfathiazole [85-73-4]
Phthiocol [483-55-6]
Phylloquinone [84-80-0]
Phylloquinone 2,3-Epoxide*
　[25486-55-9]
Physalaemin [2507-24-6]
Physodic Acid [84-24-2]
Physostigmine [57-47-6]
Physostigmine Salicylate* [57-64-7]
Physostigmine Sulfate* [64-47-1]
Physovenine [6091-05-0]
Phytic Acid [83-86-3]
Phytic Acid Calcium Magnesium
　Salt* [3615-82-5]
Phytic Acid Sodium Salt* [7205-52-9]
Phytochlorin [19660-77-6]
Phytofluene [540-05-6]
Phytol [150-86-7]
Phytonadiol* [572-96-3]
Phytonadiol Sodium Diphosphate*
　[5988-22-7]
Piberaline [39640-15-8]
Picadex [99-00-3]
Picein [530-14-3]
Picene [213-46-7]
Picilorex [62510-56-9]
Picilorex Hydrochloride*
　[56109-02-5]
Picloram [1918-02-1]
Picloxydine [5636-92-0]
Picloxydine Dihydrochloride*
　[19803-62-4]
α-Picoline [109-06-8]
β-Picoline [108-99-6]
γ-Picoline [108-89-4]
Picolinic Acid [98-98-6]
Picoperine [21755-66-8]
Picoperine Hydrochloride*
　[24699-40-9]
Picoperine Tripalmitate* [24656-22-2]
Picosulfate Sodium [10040-45-6]
Picotamide [32828-81-2]
Picotamide Monohydrate*
　[80530-63-8]
Picramic Acid [96-91-3]
Picric Acid [88-89-1]
Picrocrocin [138-55-6]
Picrolichenic Acid [466-34-2]
Picrolonic Acid [550-74-3]
Picromycin [19721-56-3]
Picropodophyllic Acid* [477-67-8]
Picropodophyllin [477-47-4]
Picrotin [21416-53-5]
Picrotoxin [124-87-8]

Picrotoxinin [17617-45-7]
Picryl Chloride [88-88-0]
Picumast [39577-19-0]
Picumast Dihydrochloride*
  [39577-20-3]
Pidotimod [121808-62-6]
Pifarnine [56208-01-6]
Piketoprofen [60576-13-8]
Piketoprofen Hydrochloride*
  [59512-37-7]
Pildralazine [64000-73-3]
Pildralazine Dihydrochloride*
  [56393-22-7]
Pilocarpine [92-13-7]
Pilocarpine Hydrochloride* [54-71-7]
Pilocarpine Nitrate* [148-72-1]
Pilocereine [2552-47-8]
Pilosine* [13640-28-3]
Pilsicainide [88069-67-4]
Pilsicainide Hydrochloride Hemi-
  hydrate* [88069-49-2]
Pimaric Acid [127-27-5]
Pimeclone [534-84-9]
Pimeclone Hydrochloride*
  [6966-09-2]
Pimefylline [10001-43-1]
Pimefylline Nicotinate* [10058-07-8]
Pimelic Acid [111-16-0]
Piminodine [13495-09-5]
Piminodine Ethanesulfonate*
  [7081-52-9]
Pimobendan [74150-27-9]
Pimobendan Hydrochloride*
  [77469-98-8]
Pimozide [2062-78-4]
Pimpinellin [131-12-4]
Pinacidil [85371-64-8]
Pinacol [76-09-5]
Pinacolone [75-97-8]
Pinaverium Bromide [53251-94-8]
Pinazepam [52463-83-9]
Pindolol [13523-86-9]
Pindone [83-26-1]
α-Pinene [80-56-8]
β-Pinene [127-91-3]
Pinguinain [37288-97-4]
Pinosylvin [22139-77-1]
Pioglitazone [111025-46-8]
Pioglitazone Hydrochloride*
  [112529-15-4]
Pipacycline [1110-80-1]
Pipamazine [84-04-8]
Pipamperone [1893-33-0]
Pipamperone Dihydrochloride*
  [2448-68-2]
Pipazethate [2167-85-3]
Pipazethate Hydrochloride*
  [6056-11-7]
Pipebuzone [27315-91-9]
Pipecolic Acid [535-75-1]
Pipecurium Bromide [52212-02-9]
Pipemidic Acid [51940-44-4]
Pipemidic Acid Trihydrate*
  [72571-82-5]
Pipenzolate Bromide [125-51-9]
Piperacetazine [3819-00-9]
Piperacillin [61477-96-1]
Piperacillin Sodium Salt*
  [59703-84-3]
Piperazine [110-85-0]
Piperazine Adipate [142-88-1]
Piperazine Citrate* [144-29-6]
2,5-Piperazinedione [106-57-0]
Piperazine Edetate Calcium*
  [12002-30-1]
Piperazine Tartrate* [133-36-8]
Piperic Acid [5285-18-7]
Piperic Acid (E,E)-Form* [136-72-1]
Piperic Acid (E,Z)-Form* [495-87-4]
Piperic Acid (Z,E)-Form* [495-88-5]

Piperic Acid (Z,Z)-Form* [495-89-6]
Piperidine [110-89-4]
Piperidine Phosphate* [767-21-5]
Piperidione [77-03-2]
Piperidolate [82-98-4]
Piperidolate Hydrochloride*
  [129-77-1]
Piperilate [4546-39-8]
Piperilate Hydrochloride* [4544-15-4]
Piperine [94-62-2]
Piperitone [89-81-6]
Piperocaine [136-82-3]
Piperocaine Hydrochloride*
  [533-28-8]
Piperonal [120-57-0]
Piperonyl Butoxide [51-03-6]
Piperonylic Acid [94-53-1]
Piperoxan [59-39-2]
Piperylone [2531-04-6]
PIPES [5625-37-6]
PIPES Disodium Salt* [76836-02-7]
PIPES Monosodium Salt*
  [10010-67-0]
Pipobroman [54-91-1]
Piposulfan [2608-24-4]
Pipotiazine [39860-99-6]
Pipotiazine Palmitic Ester*
  [37517-26-3]
Pipotiazine Undecylenic Ester*
  [22178-11-6]
Pipoxolan Hydrochloride
  [18174-58-8]
Pipradrol [467-60-7]
Pipradrol Hydrochloride* [71-78-3]
Piprozolin [17243-64-0]
Pipsyl Chloride [98-61-3]
Piracetam [7491-74-9]
Pirarubicin [72496-41-4]
Pirazolac [71002-09-0]
Pirbuterol [38677-81-5]
Pirbuterol Dihydrochloride*
  [38029-10-6]
Pirenoxine [1043-21-6]
Pirenoxine Sodium Salt* [51410-30-1]
Pirenzepine [28797-61-7]
Pirenzepine Dihydrochloride*
  [29868-97-1]
Piretanide [55837-27-9]
Piribedil [3605-01-4]
Piridocaine [87-21-8]
Pirifibrate [55285-45-5]
Pirimicarb [23103-98-2]
Pirimiphos-ethyl [23505-41-1]
Pirimiphos-methyl* [29232-93-7]
Piritramide [302-41-0]
Piritrexim [72732-56-0]
Piritrexim Isethionate* [79483-69-5]
Pirlimycin [79548-73-5]
Pirlimycin Hydrochloride*
  [78822-40-9]
Pirmenol [68252-19-7]
Piroctone [50650-76-5]
Piroctone Ethanolamine Salt (1:1)*
  [68890-66-4]
Piroheptine [16378-21-5]
Piroheptine Hydrochloride*
  [16378-22-6]
Piromen [9008-99-5]
Piromidic Acid [19562-30-2]
Piroxicam [36322-90-4]
Piroxicam Cinnamic Acid Ester*
  [87234-24-0]
Piroxicam Compd with β-Cyclo-
  dextrin* [121696-62-6]
Pirozadil [54110-25-7]
Pirprofen [31793-07-4]
Piscidic Acid [35388-57-9]
Pithecolobine [22368-82-7]
Pivalic Acid [75-98-9]
Pivalylbenzhydrazine [306-19-4]

Pivampicillin [33817-20-8]
Pivampicillin Hydrochloride*
  [26309-95-5]
Pivcefalexin [63836-75-9]
Pivcefalexin Hydrochloride*
  [27726-31-4]
PIXY321* [137463-76-4]
Pizotyline [15574-96-6]
Pizotyline Hydrochloride*
  [73391-87-4]
Pizotyline Malate* [5189-11-7]
Plafibride [63394-05-8]
Plantisul [1407-93-8]
Plasmin [9001-90-5]
Plasminogen [9001-91-6]
Plasmocid [551-01-9]
Plastoquinone 9* [4299-57-4]
Platelet Activating Factor
  [65154-06-5]
Platinic Chloride [16941-12-1]
Platinic Iodide [7790-46-7]
Platinic Oxide [1314-15-4]
Platinous Chloride [10025-65-7]
Platinous Iodide [7790-39-8]
Platinum [7440-06-4]
Platonin [3571-88-8]
Platyphylline [480-78-4]
Plaunotol [64218-02-6]
Plegatil [53597-29-8]
Pleuromutilin [125-65-5]
Pleurotin(e) [1404-23-5]
Plicamycin [18378-89-7]
Plumbagin [481-42-5]
Plumericin [77-16-7]
Plumieride [511-89-7]
Plutonium [7440-07-5]
Podocarpic Acid [5947-49-9]
Pododacric Acid [32630-75-4]
Podophyllinic Acid* [1853-37-8]
Podophyllinic Acid 2-Ethylhydrazide*
  [1508-45-8]
Podophyllinic Acid Hydrazide*
  [78178-41-3]
Podophyllotoxin [518-28-5]
Polaprezinc [107667-60-7]
Polar Yellow [6372-96-9]
Poldine Methylsulfate [545-80-2]
Polidexide [9064-92-0]
Polidocanol [9002-92-0]
Polonium [7440-08-6]
Polonium Dioxide [7446-06-2]
Polonium Tetrachloride [10026-02-5]
Poloxalene [9003-11-6]
Polybenzarsol [54531-52-1]
Polydextrose [68424-04-4]
Polyestradiol Phosphate [28014-46-2]
Polyethylene [9002-88-4]
Polyethylene Glycol [25322-68-3]
Polyferose [9009-29-4]
Polygodial [6754-20-7]
Polylysine [25104-18-1]
Polymyxin [1406-11-7]
Polymyxin B* [1404-26-8]
Polymyxin B₁* [4135-11-9]
Polymyxin B₂* [34503-87-2]
Polymyxin B Sulfate* [1405-20-5]
Polymyxin D₁* [10072-50-1]
Polymyxin D₂* [34167-45-8]
Polynoxylin [9011-05-6]
Polyoxin A* [19396-03-3]
Polyoxin B* [19396-06-6]
Polyoxins [11113-80-7]
Polypropylene [9003-07-0]
Polysorbate 80* [9005-65-6]
Polytetrafluoroethylene [9002-84-0]
Polythiazide [346-18-9]
Polyvinyl Alcohol [9002-89-5]
Polyvinyl Chloride [9002-86-2]
Ponasterone A [13408-56-5]
Ponceau 3R [3564-09-8]

Denotes derivative of title compound

Ponceau SX [4548-53-2]
Populin [99-17-2]
Poractant Alfa [129069-19-8]
Porfimer Sodium [87806-31-3]
Porfiromycin [801-52-5]
Porofor BSH [80-17-1]
Porphine [101-60-0]
Porphobilinogen [487-90-1]
Porphyrillic Acid [129-65-7]
Porphyropsin [9009-58-9]
Potassium [7440-09-7]
Potassium Acetate [127-08-2]
Potassium Aluminate [1302-63-2]
Potassium p-Aminobenzoate
    [138-84-1]
Potassium Arsenate [7784-41-0]
Potassium Arsenite [10124-50-2]
Potassium Bicarbonate [298-14-6]
Potassium Bifluoride [7789-29-9]
Potassium Binoxalate [127-95-7]
Potassium Biphthalate [877-24-7]
Potassium Bisulfate [7646-93-7]
Potassium Bisulfide [1310-61-8]
Potassium Bitartrate [868-14-4]
Potassium Borohydride [13762-51-1]
Potassium Borotartrate [12001-68-2]
Potassium Bromate [7758-01-2]
Potassium Bromide [7758-02-3]
Potassium Carbonate [584-08-7]
Potassium Chlorate [3811-04-9]
Potassium Chloride [7447-40-7]
Potassium Chromate(VI) [7789-00-6]
Potassium Citrate [866-84-2]
Potassium Citrate, Monobasic
    [866-83-1]
Potassium Cobaltous Selenate
    [28041-86-3]
Potassium Cyanate [590-28-3]
Potassium Cyanide [151-50-8]
Potassium Dichromate(VI)
    [7778-50-9]
Potassium Dicyanoaurate(I)
    [13967-50-5]
Potassium Ferricyanide [13746-66-2]
Potassium Ferrocyanide [13943-58-3]
Potassium Fluoride [7789-23-3]
Potassium Formate [590-29-4]
Potassium Gluconate [299-27-4]
Potassium Glycerophosphate
    [1319-69-3]
Potassium Guaiacolsulfonate
    [1321-14-8]
Potassium Hexachloroosmate(IV)
    [16871-60-6]
Potassium Hexachloroplatinate(IV)
    [16921-30-5]
Potassium Hexacyanocobaltate(III)
    [13963-58-1]
Potassium Hexafluoromanganate(IV)
    [16962-31-5]
Potassium Hexafluorosilicate
    [16871-90-2]
Potassium Hexafluorozirconate(IV)
    [16923-95-8]
Potassium Hexathiocyanatoplatinate-
    (IV) [17069-38-4]
Potassium Hydroxide [1310-58-3]
Potassium Hypophosphite [7782-87-8]
Potassium Iodate [7758-05-6]
Potassium Iodide [7681-11-0]
Potassium Manganate(VI)
    [10294-64-1]
Potassium Metabisulfite [16731-55-8]
Potassium Methyl Sulfate [562-54-9]
Potassium Molybdate(VI)
    [13446-49-6]
Potassium Nitrate [7757-79-1]
Potassium Nitrite [7758-09-0]
Potassium Nitroprusside [14709-57-0]
Potassium Oleate [143-18-0]

Potassium Osmate(VI) [19718-36-6]
Potassium Oxalate [583-52-8]
Potassium Percarbonate [589-97-9]
Potassium Perchlorate [7778-74-7]
Potassium Periodate [7790-21-8]
Potassium Permanganate [7722-64-7]
Potassium Persulfate [7727-21-1]
Potassium Phenolsulfonate
    [12167-15-6]
Potassium Phenoxide [100-67-4]
Potassium Phosphate, Dibasic
    [7758-11-4]
Potassium Phosphate, Monobasic
    [7778-77-0]
Potassium Phosphate, Tribasic
    [7778-53-2]
Potassium Phosphite [13492-26-7]
Potassium Pyroantimonate, Acid
    [10090-54-7]
Potassium Pyrophosphate [7320-34-5]
Potassium Pyrosulfate [7790-62-7]
Potassium Salicylate [578-36-9]
Potassium Selenate [7790-59-2]
Potassium Selenide [1312-74-9]
Potassium Silicate [1312-76-1]
Potassium Silver Cyanide [506-61-6]
Potassium Sodium Tartrate [304-59-6]
Potassium Sorbate [24634-61-5]
Potassium Stannate(IV) [12142-33-5]
Potassium Stannosulfate [27790-37-0]
Potassium Stearate [593-29-3]
Potassium Sulfate [7778-80-5]
Potassium Sulfide [1312-73-8]
Potassium Sulfite [10117-38-1]
Potassium Sulfobenzoate [53608-78-9]
Potassium Tartrate [921-53-9]
Potassium Tellurate(IV) [7790-58-1]
Potassium Tellurate(VI) [15571-91-2]
Potassium Tetraborate [1332-77-0]
Potassium Tetrabromoaurate(III)
    [13966-47-7]
Potassium Tetrachloroaurate(III)
    [13682-61-6]
Potassium Tetrachloroplatinate(II)
    [10025-99-7]
Potassium Tetracyanomercurate(II)
    [591-89-9]
Potassium Tetracyanonickelate(II)
    [14220-17-8]
Potassium Tetracyanoplatinate(II)
    [562-76-5]
Potassium Tetrafluoroborate
    [14075-53-7]
Potassium Tetraiodoaurate(III)
    [7791-29-9]
Potassium Tetraiodocadmate
    [584-10-1]
Potassium Tetraiodomercurate(II)
    [7783-33-7]
Potassium Tetroxalate [127-96-8]
Potassium Thioantimonate(V)
    [14693-02-8]
Potassium Thiocarbonate
    [26750-66-3]
Potassium Thiocyanate [333-20-0]
Potassium Thiosulfate [10294-66-3]
Potassium Titanyl Oxalate
    [14402-67-6]
Potassium Triiodide [7790-42-3]
Potassium Triiodomercurate(II) Solu-
    tion [22330-18-3]
Potassium Triiodozincate [7790-43-4]
Potassium Tungstate(VI) [7790-60-5]
Potassium Uranate(VI) [7790-63-8]
Potassium Uranyl Nitrate
    [18078-40-5]
Potassium Uranyl Sulfate
    [27709-53-1]
Potassium Xanthogenate [140-89-6]
Potassium Zinc Sulfate [13932-17-7]

Potassium Zirconium Sulfate
    [53608-79-0]
Povidone [9003-39-8]
Povidone-Iodine [25655-41-8]
PPACK [71142-71-7]
Practolol [6673-35-4]
Prajmaline [35080-11-6]
Prajmaline Hydrogen Tartrate*
    [2589-47-1]
Pralidoxime Chloride [51-15-0]
Pralidoxime Iodide* [94-63-3]
Pralidoxime Mesylate* [154-97-2]
Pramipexole [104632-26-0]
Pramipexole Dihydrochloride*
    [104632-25-9]
Pramiracetam [68497-62-1]
Pramiracetam Sulfate* [72869-16-0]
Pramiverin [14334-40-8]
Pramiverin Hydrochloride*
    [14334-41-9]
Pramoxine [140-65-8]
Pramoxine Hydrochloride* [637-58-1]
Pranlukast [103177-37-3]
Pranoprofen [52549-17-4]
Praseodymium [7440-10-0]
Prasterone [53-43-0]
Prasterone Sodium Sulfate*
    [1099-87-2]
Pratensein [2284-31-3]
Pravastatin Sodium [81131-70-6]
Prazepam [2955-38-6]
Praziquantel [55268-74-1]
Prazosin [19216-56-9]
Prazosin Hydrochloride* [19237-84-4]
Prednicarbate [73771-04-7]
Prednimustine [29069-24-7]
Prednisolone [50-24-8]
Prednisolone 21-Acetate* [52-21-1]
Prednisolone 21-tert-Butylacetate*
    [7681-14-3]
Prednisolone 21-Diethylaminoacetate
    [5626-34-6]
Prednisolone 21-Diethylaminoacetate
    Hydrochloride* [17140-01-1]
Prednisolone 21-Hydrogen Succinate*
    [2920-86-7]
Prednisolone Sodium Phosphate
    [125-02-0]
Prednisolone 21-Stearoylglycolate*
    [5060-55-9]
Prednisolone 21-Succinate Sodium
    Salt* [1715-33-9]
Prednisolone 21-m-Sulfobenzoate
    Sodium Salt* [630-67-1]
Prednisolone 21-Trimethylacetate*
    [1107-99-9]
Prednisone [53-03-2]
Prednisone 21-Acetate* [125-10-0]
Prednival [15180-00-4]
Prednylidene [599-33-7]
Prednylidene 21-Diethylaminoacetate
    Hydrochloride* [22887-42-9]
Pregnane [481-26-5]
Pregnanediol [80-92-2]
3,20-Pregnanedione [128-23-4]
Pregnan-3α-ol-20-one [128-20-1]
4-Pregnene-20,21-diol-3,11-dione
    [116-56-3]
4-Pregnene-11β,17α,20β,21-tetrol-3-
    one [116-58-5]
4-Pregnene-17α,20β,21-triol-3,11-
    dione [116-59-6]
4-Pregnene-17α,20β,21-triol-3-one
    [128-19-8]
Pregnenolone [145-13-1]
Pregnenolone Methyl Ether*
    [511-26-2]
Prelog-Djerassi Lactone [69056-12-8]
Prenalterol [57526-81-5]

Prenalterol Hydrochloride* [61260-05-7]
Prenoxdiazine Hydrochloride [982-43-4]
Prenylamine [390-64-7]
Prenylamine Lactate* [69-43-2]
Prephenic Acid [126-49-8]
Pretilachlor [51218-49-6]
Pridinol [511-45-5]
Pridinol Hydrochloride* [968-58-1]
Pridinol Methanesulfonate* [6856-31-1]
Prifinium Bromide [4630-95-9]
Prilocaine [721-50-6]
Prilocaine Hydrochloride* [1786-81-8]
Primaquine [90-34-6]
Primaquine Diphosphate* [63-45-6]
Primeverose [26531-85-1]
Primidone [125-33-7]
Primisulfuron-methyl [86209-51-0]
Primocarcin [3750-26-3]
Primulaverin [154-61-0]
Pristane [1921-70-6]
Probenecid [57-66-9]
Probucol [23288-49-5]
Procainamide Hydrochloride [614-39-1]
Procaine [59-46-1]
Procaine Butyrate* [136-55-0]
Procaine Dihydrate* [6192-89-8]
Procaine Hydrochloride* [51-05-8]
Procaine Nitrate* [6192-92-3]
Procarbazine [671-16-9]
Procarbazine Hydrochloride* [366-70-1]
Procaterol [72332-33-3]
Procaterol Hydrochloride Hemihydrate* [81262-93-3]
Procerin [552-96-5]
Prochloraz [67747-09-5]
Prochlorperazine [58-38-8]
Prochlorperazine Dimaleate* [84-02-6]
Prochlorperazine Dimethanesulfonate* [51888-09-6]
Procodazole [23249-97-0]
Procyclidine [77-37-2]
Procyclidine Hydrochloride* [1508-76-5]
Procymate [13931-64-1]
Procymidone [32809-16-8]
Prodiamine [29091-21-2]
Prodigiosin [82-89-3]
Prodigiosin Hydrochloride* [112373-40-7]
Prodipine [31314-38-2]
Prodipine Hydrochloride* [31314-39-3]
Prodlure [50767-79-8]
Pro-Drone® [53905-38-7]
Proflavine [92-62-6]
Proflavine Dihydrochloride* [531-73-7]
Proflavine Sulfate* [553-30-0]
Profluralin [26399-36-0]
Progabide [62666-20-0]
Progesterone [57-83-0]
Proglumetacin [57132-53-3]
Proglumetacin Dimaleate* [59209-40-4]
Proglumide [6620-60-6]
Proheptazine [77-14-5]
Proinsulin [9035-68-1]
Prolactin [9002-62-4]
Proline [147-85-3]
Proline DL-Form* [609-36-9]
Prolintane [493-92-5]
Prolintane Hydrochloride* [1211-28-5]

Prolonium Iodide [123-47-7]
Promazine [58-40-2]
Promazine Hydrochloride* [53-60-1]
Promecarb [2631-37-0]
Promedol [64-39-1]
Promegestone [34184-77-5]
Promethazine [60-87-7]
Promethazine Compd with 8-Chlorotheophylline* [17693-51-5]
Promethazine Hydrochloride* [58-33-3]
Promethium [7440-12-2]
Prometon [1610-18-0]
Prometryn [7287-19-6]
Pronethalol [54-80-8]
Pronethalol Hydrochloride* [51-02-5]
Pro-opiomelanocortin [66796-54-1]
Propacetamol [66532-85-2]
Propacetamol Hydrochloride* [66532-86-3]
Propachlor [1918-16-7]
Propafenone [54063-53-5]
Propafenone Hydrochloride* [34183-22-7]
Propagermanium [12758-40-6]
Propallylonal [545-93-7]
Propamidine [104-32-5]
Propamidine Isethionate* [140-63-6]
Propane [74-98-6]
1-Propanearsonic Acid [107-34-6]
1,3-Propanedithiol [109-80-8]
Propanethial S-Oxide [32157-29-2]
Propanidid [1421-14-3]
Propanil [709-98-8]
Propanocaine [493-76-5]
Propantheline Bromide [50-34-0]
Propaquizafop [111479-05-1]
Proparacaine [499-67-2]
Proparacaine Hydrochloride* [5875-06-9]
Propargite [2312-35-8]
Propargyl Alcohol [107-19-7]
Propargyl Chloride [624-65-7]
Propatyl Nitrate [2921-92-8]
Propazine [139-40-2]
Propentofylline [55242-55-2]
Propenzolate [4354-45-4]
Propenzolate Hydrochloride* [1420-03-7]
Properdin [11016-39-0]
Propetamphos [31218-83-4]
Propham [122-42-9]
Propicillin [551-27-9]
Propicillin Potassium Salt* [1245-44-9]
Propiconazole [60207-90-1]
Propineb [12071-83-9]
β-Propiolactone [57-57-8]
Propiolic Acid [471-25-0]
Propiomazine [362-29-8]
Propiomazine Hydrochloride* [1240-15-9]
Propionaldehyde [123-38-6]
Propionamide [79-05-0]
Propionic Acid [79-09-4]
Propionic Anhydride [123-62-6]
Propionitrile [107-12-0]
Propionyl Chloride [79-03-8]
Propionylpromazine [3568-24-9]
Propiophenone [93-55-0]
Propipocaine [3670-68-6]
Propipocaine Hydrochloride* [1155-49-3]
Propiram [15686-91-6]
Propiram Fumarate* [13717-04-9]
Propiverine [60569-19-9]
Propiverine Hydrochloride* [54556-98-8]
Propizepine [10321-12-7]

Propizepine Hydrochloride* [14559-79-6]
Propofol [2078-54-8]
Propoxur [114-26-1]
Propoxycaine Hydrochloride [550-83-4]
Propoxyphene [469-62-5]
Propoxyphene Hydrochloride* [1639-60-7]
Propoxyphene Napsylate Monohydrate* [26570-10-5]
Propranolol [525-66-6]
Propranolol Hydrochloride* [318-98-9]
Propyl Acetate [109-60-4]
n-Propyl Alcohol [71-23-8]
Propylamine [107-10-8]
n-Propylbenzene [103-65-1]
Propyl Bromide [106-94-5]
Propyl Butyrate [105-66-8]
Propyl Chloride [540-54-5]
Propyl Chlorocarbonate [109-61-5]
Propylene [115-07-1]
Propylene Chlorohydrin [78-89-7]
sec-Propylene Chlorohydrin [127-00-4]
Propylenediamine [78-90-0]
Propylene Dibromide [78-75-1]
Propylene Dichloride [78-87-5]
Propylene Glycol [57-55-6]
Propylene Oxide [75-56-9]
Propyl Ether [111-43-3]
Propyl Formate [110-74-7]
Propyl Gallate [121-79-9]
Propylhexedrine [101-40-6]
Propylhexedrine d-Form* [6556-29-2]
Propylhexedrine dl-Form* [3595-11-7]
Propylhexedrine l-Form* [6192-97-8]
Propylhexedrine l-Form Ethylphenylbarbiturate* [4388-82-3]
Propylhexedrine d-Form Hydrochloride* [6192-96-7]
Propylhexedrine dl-Form Hydrochloride* [6192-95-6]
Propylhexedrine l-Form Hydrochloride* [6192-98-9]
Propylidene Chloride [78-99-9]
Propyl Iodide [107-08-4]
Propyliodone [587-61-1]
n-Propyl Nitrate [627-13-4]
n-Propyl Nitrite [543-67-9]
Propylparaben [94-13-3]
Propyl Propionate [106-36-5]
Propyl Sulfide [111-47-7]
Propylthiouracil [51-52-5]
Propylure [10297-61-7]
Propyphenazone [479-92-5]
Propyromazine [145-54-0]
Propyzamide [23950-58-5]
Proquazone [22760-18-5]
Proscillaridin [466-06-8]
Proscillaridin-4-methyl Ether* [33396-37-1]
Prostacyclin [35121-78-9]
Prostacyclin Sodium Salt* [61849-14-7]
Prostaglandin E$_1$ [745-65-3]
Prostaglandin E$_2$ [363-24-6]
Prostaglandin F$_{2\alpha}$ [551-11-1]
Prostaglandin F$_{2\alpha}$ Tromethamine Salt* [38562-01-5]
Prostalene [54120-61-5]
Prosulfuron [94125-34-5]
Prosultiamine [59-58-5]
Prosultiamine Hydrochloride* [973-99-9]
Protactinium [7440-13-3]
Protamine Zinc Insulin* [9004-17-5]
Protein C [60202-16-6]
Protheobromine [50-39-5]
Prothipendyl [303-69-5]

* Denotes derivative of title compound

Prothipendyl Hydrochloride* [1225-65-6]
Prothrombin [9001-26-7]
Protiofate [58416-00-5]
Protionamide [14222-60-7]
Protizinic Acid [54323-85-2]
Protoanemonin [108-28-1]
Protocatechualdehyde [139-85-5]
Protocatechuic Acid [99-50-3]
Protokosin [10091-70-0]
Protokylol [136-70-9]
Protokylol Hydrochloride* [136-69-6]
Protopine [130-86-9]
Protoporphyrin IX [553-12-8]
Protoporphyrin IX Dimethyl Ester* [5522-66-7]
Protoporphyrin IX Disodium Salt* [50865-01-5]
Protostephanine [549-28-0]
Protoveratrine A* [143-57-7]
Protoveratrine B* [124-97-0]
Protoverine [76-45-9]
Protriptyline [438-60-8]
Protriptyline Hydrochloride* [1225-55-4]
Pro-Urokinase [82657-92-9]
Proxazole [5696-09-3]
Proxazole Citrate* [132-35-4]
Proxibarbal [2537-29-3]
Proxyphylline [603-00-9]
Prozapine [3426-08-2]
Prunetin [552-59-0]
Pseudoaconitine [127-29-7]
Pseudobaptigenin [90-29-9]
Pseudococaine [478-73-9]
Pseudococaine Hydrochloride* [6363-57-1]
Pseudococaine n-Propyl Ester Analog* [55608-72-5]
Pseudococaine Tartrate* [1176-03-0]
Pseudocodeine [466-96-6]
Pseudoconhydrine [140-55-6]
Pseudocumene [95-63-6]
d-Pseudoephedrine* [90-82-4]
d-Pseudoephedrine Hydrochloride* [345-78-8]
d-Pseudoephedrine Sulfate* [7460-12-0]
Pseudohecogenin* [11005-20-2]
Pseudoionone [141-10-6]
Pseudojervine [36069-05-3]
Pseudomonic Acid C* [71980-98-8]
Pseudomonic Acid D* [85248-93-7]
Pseudomorphine [125-24-6]
Pseudopederin* [10352-73-5]
Pseudopelletierine [552-70-5]
Pseudotropine [135-97-7]
Pseudoyohimbine [84-37-7]
Psicofuranine [1874-54-0]
D-Psicose [551-68-8]
Psilocin [520-53-6]
Psilocybin [520-52-5]
PSK [66455-27-4]
Psoralen [66-97-7]
Psoralidin [18642-23-4]
Psychotrine [7633-29-6]
Psyllium Seed Gum* [8063-16-9]
Pteridine [91-18-9]
Pterocarpin [524-97-0]
Pteroic Acid [119-24-4]
Pteropterin [89-38-3]
Pteroylhexaglutamylglutamic Acid [6484-74-8]
Puberulic Acid [99-23-0]
Puberulonic Acid [82-83-7]
Pukateine [81-67-4]
Pulegone [89-82-7]
Purified Protein Derivative* [92129-86-7]
1H-Purine [120-73-0]

Puromycin [53-79-2]
Purothionin [9009-72-7]
Purpurin [81-54-9]
Purpurogallin [569-77-7]
Putrescine [110-60-1]
Pygeum Africanum Extract [85865-74-3]
Pymetrozine [123312-89-0]
Pyocyanine [85-66-5]
Pyracarbolid [24691-76-7]
Pyrantel [15686-83-6]
Pyrantel Pamoate* [22204-24-6]
Pyrantel Tartrate* [33401-94-4]
Pyrazinamide [98-96-4]
Pyrazine [290-37-9]
2,3-Pyrazinedicarboxylic Acid [89-01-0]
Pyrazinoic Acid [98-97-5]
Pyrazole [288-13-1]
2-Pyrazoline [109-98-8]
Pyrazophos [13457-18-6]
Pyrene [129-00-0]
Pyrethrin I* [121-21-1]
Pyrethrin II* [121-29-9]
Pyrethrosin [28272-18-6]
Pyridate [55512-33-9]
Pyridazine [289-80-5]
Pyridine [110-86-1]
3-Pyridineacetic Acid [501-81-5]
Pyridine 1-Oxide [694-59-7]
Pyridinium Bromide Perbromide [39416-48-3]
Pyridinium Chlorochromate [26299-14-9]
Pyridinol Carbamate [1882-26-4]
Pyridofylline [53403-97-7]
Pyridomycin [18791-21-4]
Pyridostigmine Bromide [101-26-8]
Pyridoxal [66-72-8]
Pyridoxal 5-Phosphate [54-47-7]
Pyridoxal 5-Phosphate Calcium Salt* [29956-24-9]
Pyridoxal 5-Phosphate O-Methyl-oxime* [20905-71-9]
Pyridoxal 5-Phosphate Oxime* [634-25-3]
Pyridoxamine Dihydrochloride [524-36-7]
4-Pyridoxic Acid [82-82-6]
Pyridoxine Hydrochloride [58-56-0]
Pyrifenox [88283-41-4]
Pyrilamine [91-84-9]
Pyrilamine Maleate* [59-33-6]
Pyrimethamine [58-14-0]
Pyrimethamine Combination with Sulfadoxine* [37338-39-9]
Pyrimidine [289-95-2]
Pyriminil [53558-25-1]
Pyrimithate [5221-49-8]
Pyrinoline [1740-22-3]
Pyriproxyfen [95737-68-1]
Pyrisuccideanol [33605-94-6]
Pyrisuccideanol Dimaleate* [53659-00-0]
Pyrithiamine [534-64-5]
Pyrithiobac [123342-93-8]
Pyrithiobac Sodium Salt* [123343-16-8]
Pyrithione [1121-30-8]
Pyrithione Zinc Derivative* [13463-41-7]
Pyrithyldione [77-04-3]
Pyritinol [1098-97-1]
Pyrocalciferol [128-27-8]
Pyrocarbonic Acid Diethyl Ester [1609-47-8]
Pyrocatechol [120-80-9]
Pyrogallol [87-66-1]
Pyrogallol Monoacetate* [1330-51-4]
Pyrogallol Triacetate* [525-52-0]

L-Pyroglutamic Acid [98-79-3]
Pyrolan [87-47-8]
Pyroligneous Acid [8030-97-5]
Pyromellitic Acid [89-05-4]
Pyronine B [2150-48-3]
Pyronine Y [92-32-0]
Pyrophosphoric Acid [2466-09-3]
Pyrosulfuric Acid [7783-05-3]
Pyrosulfuryl Chloride [7791-27-7]
Pyrovalerone [3563-49-3]
Pyrovalerone Hydrochloride* [1147-62-2]
Pyroxylin [9004-70-0]
Pyrrobutamine [91-82-7]
Pyrrobutamine Diphosphate* [135-31-9]
Pyrrocaine [2210-77-7]
Pyrrocaine Hydrochloride* [2210-64-2]
1H-Pyrrole [109-97-7]
Pyrrolidine [123-75-1]
2-Pyrrolidone [616-45-5]
3-Pyrroline [109-96-6]
Pyrrolnitrin [1018-71-9]
Pyruvaldehyde [78-98-8]
Pyruvate Decarboxylase [9001-04-1]
Pyruvic Acid [127-17-3]
Pyrvinium Chloride* [548-84-5]
Pyrvinium Pamoate [3546-41-6]
Q-Enzyme® [9001-97-2]
Quassin [76-78-8]
Quatrimycin [79-85-6]
Quazepam [36735-22-5]
Quebrachamine [4850-21-9]
Queen Substance [334-20-3]
Quercetagetin [90-18-6]
Quercetin [117-39-5]
Quercetin Dihydrate* [6151-25-3]
Quercetin Pentabenzyl Ether* [13157-90-9]
Quercimeritrin [491-50-9]
d-Quercitol [488-73-3]
Quercitrin [522-12-3]
Quillaic Acid [631-01-6]
Quinacillin [1596-63-0]
Quinacillin Disodium Salt* [985-32-0]
Quinacrine [83-89-6]
Quinacrine Dihydrochloride Dihydrate* [6151-30-0]
Quinacrine Methanesulfonate Monohydrate* [6598-46-5]
Quinagolide [87056-78-8]
Quinagolide Hydrochloride* [94424-50-7]
Quinaldic Acid [93-10-7]
Quinaldine [91-63-4]
Quinaldine Blue [2768-90-3]
Quinaldine Red [117-92-0]
Quinaldine Sulfate* [655-76-5]
Quinalizarin [81-61-8]
Quinamine [464-85-7]
Quinapril [85441-61-8]
Quinaprilat* [82768-85-2]
Quinapril Hydrochloride* [82586-55-8]
Quinapyramine [20493-41-8]
Quinazoline [253-82-7]
Quinbolone [2487-63-0]
Quinestradiol [1169-79-5]
Quinestrol [152-43-2]
Quinethazone [73-49-4]
Quinfamide [62265-68-3]
Quinhydrone [106-34-3]
Quinic Acid [77-95-2]
Quinidine [56-54-2]
Quinidine Gluconate* [7054-25-3]
Quinidine Hydrogen Sulfate Tetrahydrate* [6151-39-9]
Quinidine Polygalacturonate* [7681-28-9]

Denotes derivative of title compound

Quinidine Sulfate* [50-54-4]
Quinidine Sulfate Dihydrate*
  [6591-63-5]
Quinine [130-95-0]
Quinine Bisulfate* [549-56-4]
Quinine Bisulfate Heptahydrate*
  [6183-68-2]
Quinine Dihydrochloride* [60-93-5]
Quinine Hydrochloride* [130-89-2]
Quinine Hydrochloride Dihydrate*
  [6119-47-7]
Quinine Iodosulfate [7631-46-1]
Quinine Oleate [10486-11-0]
Quinine Sulfate* [804-63-7]
Quinine Sulfate Dihydrate*
  [6119-70-6]
Quininic Acid [86-68-0]
Quininone [84-31-1]
Quinizarin [81-64-1]
Quinizarin Green SS [128-80-3]
Quinocide [525-61-1]
Quinoline [91-22-5]
8-Quinolineboronic Acid [86-58-8]
8-Quinolinecarboxylic Acid [86-59-9]
Quinoline Yellow [8004-92-0]
Quinoline Yellow Spirit Soluble
  [8003-22-3]
Quinolinic Acid [89-00-9]
Quinone [106-51-4]
Quinovic Acid [465-74-7]
Quinovin [53516-73-7]
Quinovose [7658-08-4]
Quinoxaline [91-19-0]
Quintozene [82-68-8]
Quinuclidine [100-76-5]
3-Quinuclidinol [1619-34-7]
3-Quinuclidinol dl-Form* [3684-26-2]
3-Quinuclidinol l-Form* [25333-42-0]
3-Quinuclidinol dl-Form Acetate
  Hydrochloride* [6109-70-2]
3-Quinuclidinol dl-Form Benzoate
  (Ester)* [66-93-3]
3-Quinuclidinol dl-Form Benzoate
  Hydrochloride* [25333-48-6]
3-Quinuclidinol dl-Form
  Hydrochloride* [25333-43-1]
Quinupramine [31721-17-2]
Quisqualic Acid [52809-07-1]
Quizalofop-Ethyl [76578-14-8]
R-11 [126-15-8]
Rabeprazole [117976-89-3]
Rabeprazole Sodium Salt*
  [117976-90-6]
Racefemine [22232-57-1]
Racemethorphan [510-53-2]
Racemethorphan Hydrobromide*
  [6031-86-3]
Ractopamine [97825-25-7]
Ractopamine R,R-Form Hydrochlo-
  ride* [74432-68-1]
Ractopamine Hydrochloride*
  [90274-24-1]
Radicinin [10088-95-6]
Radium [7440-14-4]
Radium Bromide* [10031-23-9]
Radium Chloride* [10025-66-8]
Radon [10043-92-2]
Raffinose [512-69-6]
Rafoxanide [22662-39-1]
Raloxifene [84449-90-1]
Raloxifene Hydrochloride*
  [82640-04-8]
Ramifenazone [3615-24-5]
Ramifenazone Mixture with
  Phenylbutazone* [76120-72-4]
Ramipril [87333-19-5]
Ramiprilat* [87269-97-4]
Raney Nickel [7440-02-0]
Ranimustine [58994-96-0]
Ranitidine [66357-35-5]

Ranitidine Bismuth Citrate*
  [128345-62-0]
Ranitidine Hydrochloride*
  [66357-59-3]
Ranolazine [95635-55-5]
Ranolazine Dihydrochloride*
  [95635-56-6]
Rapamycin [53123-88-9]
Raubasine [483-04-5]
Raunescine [117-73-7]
Raunescine Monoacetate
  Monohydrate* [6105-86-8]
Raunescine Monohydrate*
  [6105-85-7]
Raunescine Nitrate* [6159-98-4]
Razoxane [21416-87-5]
Razoxane (+)-Form* [24584-09-6]
Rebamipide [90098-04-7]
Reductic Acid [80-72-8]
Regramostim* [127757-91-9]
Reinecke Salt [13573-16-5]
Relaxin [9002-69-1]
Remifentanil [132875-61-7]
Remifentanil Hydrochloride*
  [132539-07-2]
Remoxipride [80125-14-0]
Remoxipride Hydrochloride*
  [73220-03-8]
Remoxipride Hydrochloride Mono-
  hydrate* [117591-79-4]
Renin [9015-94-5]
Rennin [9001-98-3]
Repaglinide [135062-02-1]
Repirinast [73080-51-0]
Reposal [3625-25-0]
Reproterol [54063-54-6]
Reproterol Hydrochloride*
  [13055-82-8]
Resacetophenone [89-84-9]
Resazurin [550-82-3]
Rescimetol [73573-42-9]
Rescinnamine [24815-24-5]
Reserpic Acid [83-60-3]
Reserpilic Acid Dimethylaminoethyl
  Ester Dihydrochloride* [3735-84-0]
Reserpiline [131-02-2]
Reserpine [50-55-5]
Resibufogenin [465-39-4]
Resin Ipomea [9000-34-4]
Resin Jalap [9000-35-5]
Resin Scammony [9000-58-2]
Resistomycin [20004-62-0]
Resodec [9012-13-9]
Resorantel [20788-07-2]
Resorcinol [108-46-3]
Resorcinol Monoacetate* [102-29-4]
β-Resorcylaldehyde [95-01-2]
β-Resorcylic Acid [89-86-1]
Resveratrol [501-36-0]
cis-Resveratrol* [61434-67-1]
Resveratrol 3-β-Mono-D-glucoside*
  [27208-80-6]
Retamine [2122-29-4]
Retene [483-65-8]
Reticuline [485-19-8]
Retinal [116-31-4]
Retine [51848-43-2]
Retinoic Acid [302-79-4]
Retinoic Acid 13-cis-Form*
  [4759-48-2]
Retronecine [480-85-3]
Retrorsine [480-54-6]
Rhamnetin [90-19-7]
Rhamnose [3615-41-6]
Rhapontin [155-58-8]
Rheadine [2718-25-4]
Rhein [478-43-3]
Rhenium [7440-15-5]
Rhenium Heptoxide [1314-68-7]
Rhenium Hexafluoride [10049-17-9]

Rhenium Oxychloride [7791-09-5]
Rhenium Trioxide [1314-28-9]
Rhizopterin [119-20-0]
Rhodamine B [81-88-9]
Rhodanilic Acid [1332-53-2]
Rhodanine [141-84-4]
Rhodinol [6812-78-8]
Rhodium [7440-16-6]
Rhodium Carbonyl Chloride
  [14523-22-9]
Rhodium Chloride [10049-07-7]
Rhododendrin [497-78-9]
Rhodomycin A* [23666-50-4]
Rhodomycin B* [1404-52-0]
Rhodopin [105-92-0]
Rhodopsin [9009-81-8]
Rhodoquinone [5591-74-2]
Rhodoviolascin [34255-08-8]
Rhodoxanthin [116-30-3]
Rhynchophylline [76-66-4]
Ribavirin [36791-04-5]
α-Ribazole [132-13-8]
Riboflavin [83-88-5]
Riboflavin Monophosphate [146-17-8]
Riboflavin Monophosphate Mono-
  sodium Salt* [130-40-5]
Riboflavin 2′,3′,4′,5′-Tetrabutyrate*
  [752-56-7]
Ribonuclease [9001-99-4]
D-Ribose [50-69-1]
D-Ribose-5-phosphoric Acid
  [4300-28-1]
Ribostamycin [25546-65-0]
Ribostamycin Sulfate* [53797-35-6]
D-Ribulose [488-84-6]
Ricin [9009-86-3]
Ricinine [524-40-3]
Ricinoleic Acid [141-22-0]
Ricinoleic Acid Sodium Salt*
  [5323-95-5]
Ridogrel [110140-89-1]
Rifabutin [72559-06-9]
Rifamide [2750-76-7]
Rifampin [13292-46-1]
Rifamycin B* [13929-35-6]
Rifamycin O* [14487-05-9]
Rifamycin S* [13759-79-2]
Rifamycin SV [6998-60-3]
Rifamycin SV Sodium Salt*
  [14897-39-3]
Rifapentine [61379-65-5]
Rifaximin [80621-81-4]
Rilmazafone [99593-25-6]
Rilmenidine [54187-04-1]
Riluzole [1744-22-5]
Rimantadine [13392-28-4]
Rimantadine Hydrochloride*
  [1501-84-4]
Rimazolium Metilsulfate [28610-84-6]
Rimexolone [49697-38-3]
Rimiterol [32953-89-2]
Rimiterol Hydrobromide*
  [31842-61-2]
Rimocidin [1393-12-0]
Rioprostil [77287-05-9]
Risedronic Acid [105462-24-6]
Risedronic Acid Monosodium Salt*
  [115436-72-1]
Risperidone [106266-06-2]
Ristocetin [1404-55-3]
Ritanserin [87051-43-2]
Ritanserin L-Tartrate* [93076-39-2]
Ritipenem [84845-57-8]
Ritipenem Acetoxymethyl Ester*
  [87238-52-6]
Ritipenem Sodium Salt* [84845-58-9]
Ritodrine [26652-09-5]
Ritodrine Hydrochloride*
  [23239-51-2]
Ritonavir [155213-67-5]

Denotes derivative of title compound

Robenidine [25875-51-8]
Robenidine Hydrochloride* [25875-50-7]
Robinin [301-19-9]
Roccellic Acid [29838-46-8]
Rociverine [53716-44-2]
Rocuronium [143558-00-3]
Rocuronium Bromide* [119302-91-9]
Rokitamycin [74014-51-0]
Rolicyprine [2829-19-8]
Rolipram [61413-54-5]
Rolitetracycline [751-97-3]
Rolitetracycline Compd with Chloramphenicol Succinate* [4154-10-3]
Rolitetracycline Nitrate Sesquihydrate* [26657-13-6]
Roman Chamomile Oil* [8015-92-7]
Romurtide [78113-36-7]
Ronidazole [7681-76-7]
Ronifibrate [42597-57-9]
Ronnel [299-84-3]
Ropinirole [91374-21-9]
Ropinirole Hydrochloride* [91374-20-8]
Ropivacaine [84057-95-4]
Ropivacaine Monohydrochloride* [98717-15-8]
Ropivacaine Monohydrochloride Monohydrate* [132112-35-7]
Roquinimex [84088-42-6]
Rosaprostol [56695-65-9]
Rosaramicin [35834-26-5]
Rose Bengal [632-68-8]
Rose Bengal Disodium Salt* [632-69-9]
Rose Bengal ^{131}I-Labeled Disodium Salt* [50291-21-9]
Rosoxacin [40034-42-2]
Rotenone [83-79-4]
Rotraxate [92071-51-7]
Rotraxate Hydrochloride* [82085-94-7]
Rottlerin [82-08-6]
Roxarsone [121-19-7]
Roxatidine Acetate [78628-28-1]
Roxatidine Acetate Hydrochloride* [93793-83-0]
Roxindole [112192-04-8]
Roxindole Hydrochloride* [108050-82-4]
Roxindole Methanesulfonate* [119742-13-1]
Roxithromycin [80214-83-1]
Rubeanic Acid [79-40-3]
Ruberythric Acid [152-84-1]
Rubiadin [117-02-2]
Rubidium [7440-17-7]
Rubidium Bromide [7789-39-1]
Rubidium Chloride [7791-11-9]
Rubidium Hydroxide [1310-82-3]
Rubidium Iodide [7790-29-6]
Rubijervine [79-58-3]
Rubixanthin [3763-55-1]
Rufigallol [82-12-2]
Rufloxacin [101363-10-4]
Rufloxacin Hydrochloride* [106017-08-7]
Rugulovasine A* [26909-33-1]
Rutecarpine [84-26-4]
Ruthenium [7440-18-8]
Ruthenium Red [1307-52-4]
Ruthenium Tetroxide [20427-56-9]
Ruthenium Trichloride [10049-08-8]
Rutin [153-18-4]
Rutinose [90-74-4]
Ryania [8047-13-0]
Ryanodine [15662-33-6]
Sabadilla [8051-02-3]
Sabadine [124-80-1]

Sabeluzole [104383-17-7]
Saccharin [81-07-2]
L-Saccharopine [997-68-2]
Safranal [116-26-7]
Safrole [94-59-7]
SAICAR [3031-95-6]
Sakuranetin [2957-21-3]
Salacetamide [487-48-9]
Salatrim [64706-27-0]
Salazosulfadimidine [2315-08-4]
Salen [94-93-9] (unspecified)
Salen [129409-01-4] (E,E)
Salicin [138-52-3]
Salicyl Alcohol [90-01-7]
Salicylaldehyde [90-02-8]
Salicylaldoxime [94-67-7]
Salicylamide [65-45-2]
Salicylamide O-Acetic Acid [25395-22-6]
Salicylamide O-Acetic Acid Sodium Salt* [3785-32-8]
Salicylanilide [87-17-2]
Salicylhydroxamic Acid [89-73-6]
Salicylic Acid [69-72-7]
4-Salicyloylmorpholine [3202-84-4]
Salicylsulfuric Acid [89-45-2]
Salicylsulfuric Acid Monosodium Salt* [6155-64-2]
Salinazid [495-84-1]
Salinomycin [53003-10-4]
Salinomycin Sodium Salt* [55721-31-8]
Salmeterol [89365-50-4]
Salmeterol 1-Hydroxy-2-naphthoate* [94749-08-3]
Salmine Sulfate [53597-25-4]
Salsalate [552-94-3]
Salsoline [101467-40-7]
Salsoline (−)-Form* [89-31-6]
Salsoline Hydrochloride* [51424-33-0]
Salutaridine [1936-18-1]
Salverine [6376-26-7]
Samandarine [467-51-6]
Samarium [7440-19-9]
Sancycline [808-26-4]
Sanguinarine [2447-54-3]
α-Santalol [115-71-9]
β-Santalol [77-42-9]
Santonic Acid [510-35-0]
α-Santonin [481-06-1]
Saperconazole [110588-57-3]
Saponarin [20310-89-8]
Sapphire [1317-82-4]
Sapropterin [62989-33-7]
Sapropterin Dihydrochloride* [69056-38-8]
Saquinavir [127779-20-8]
Saquinavir Mesylate* [149845-06-7]
Sarafloxacin [98105-99-8]
Sarafloxacin Monohydrochloride* [91296-87-6]
Saralasin [34273-10-4]
Saralasin Hydrated Acetate* [39698-78-7]
Sarcosine [107-97-1]
Sargramostim* [123774-72-1]
Sarin [107-44-8]
Sarkomycin [11031-48-4]
Sarkomycin A* [489-21-4]
Sarmentogenin [76-28-8]
Sarmentose [13484-14-5]
Sarpagine [482-68-8]
Sarsasapogenin [126-19-2]
Saruplase* [99149-95-8]
Sarverogenin [22146-03-8]
Satumomab [144058-40-2]
Saxitoxin [35554-08-6]
Scabiolide [20055-57-6]
Scandium [7440-20-2]

Scarlet Red [85-83-6]
Schradan [152-16-9]
Schweizer's Reagent [17500-49-1]
Scillabiose [40525-07-3]
Scillaren [11003-70-6]
Scillarenin [465-22-5]
Scilliroside [507-60-8]
Scoparin [301-16-6]
Scoparone [120-08-1]
Scopolamine [51-34-3]
Scopolamine Hydrobromide Trihydrate* [114-49-8]
Scopolamine N-Oxide [97-75-6]
Scopoletin [92-61-5]
Scopolin [531-44-2]
Scopoline [487-27-4]
Scotophobin [33579-45-2]
Scutellarein [529-53-3]
Sebacic Acid [111-20-6]
Sebacil [96-01-5]
Sebacoin [96-00-4]
Secalonic Acid A* [35287-72-0]
Secnidazole [3366-95-8]
Secobarbital Sodium [309-43-3]
Secretin [1393-25-5]
Securinine [5610-40-2]
Sedecamycin [23477-98-7]
Seidlitz Mixture [8014-63-9]
Selegiline [14611-51-9]
Selegiline Hydrochloride* [14611-52-0]
Selenic Acid [7783-08-6]
Selenious Acid [7783-00-8]
Selenium [7782-49-2]
Selenium Bromide [7789-52-8]
Selenium Chloride [10025-68-0]
Selenium Disulfide* [7488-56-4]
Selenium Hexafluoride [7783-79-1]
Selenium Oxide [7446-08-4]
Selenium Oxybromide [7789-51-7]
Selenium Oxychloride [7791-23-3]
Selenium Oxyfluoride [7783-43-9]
Selenium Tetrabromide [7789-65-3]
Selenium Tetrachloride [10026-03-6]
Selenium Tetrafluoride [13465-66-2]
Selenocysteine [10236-58-5]
Selenomethionine [1464-42-2]
Selenomethionine ^{75}Se-Injection* [1187-56-0]
Sematilide [101526-83-4]
Sematilide Monohydrochloride* [101526-62-9]
Semduramicin [113378-31-7]
Semduramicin Sodium Salt* [119068-77-8]
Semicarbazide Hydrochloride [563-41-7]
Semioxamazide [515-96-8]
Semotiadil [116476-13-2]
Semotiadil Fumarate* [116476-14-3]
Sempervirine [6882-99-1]
Semtex* [52441-47-1]
Senecialdehyde [107-86-8]
Senecic Acid [13588-16-4]
Senecionine [130-01-8]
Seneciphylline [480-81-9]
Senegenin [2469-34-3]
Sennoside A* [81-27-6]
Sennoside B* [128-57-4]
Seratrodast [112665-43-7]
Serine [56-45-1]
Serine D-Form* [312-84-5]
Serine DL-Form* [302-84-1]
Sermorelin [86168-78-7]
Seroquel [111974-72-2]
Serotonin [50-67-9]
Serpentine (Alkaloid) [18786-24-8]
Sertaconazole [99592-32-2]
Sertaconazole Nitrate* [99592-39-9]
Sertindole [106516-24-9]

Sertraline [79617-96-2]
Sertraline Hydrochloride*
  [79559-97-0]
Serum Albumin [9048-46-8]
Sesamex [51-14-9]
Sesamin [607-80-7]
Sesamolin [526-07-8]
Sesin [94-83-7]
Setastine [64294-95-7]
Setastine Hydrochloride*
  [59767-13-4]
Sethoxydim [74051-80-2]
Sevoflurane [28523-86-6]
Shark Liver Oil [68990-63-6]
Shellac [9000-59-3]
Shellolic Acid [4448-95-7]
Shikimic Acid [138-59-0]
Shionone [10376-48-4]
Showdomycin [16755-07-0]
Sibutramine [106650-56-0]
Sibutramine Hydrochloride Mono-
  hydrate* [125494-59-9]
Siccanin [22733-60-4]
Siduron [1982-49-6]
Sikkimotoxin [18651-67-7]
Silane [7803-62-5]
Silicic Acid [1343-98-2]
Silicon [7440-21-3]
Silicon Carbide [409-21-2]
Silicon Dioxide [7631-86-9]
Silicon Disulfide [13759-10-9]
Silicon Monoxide [10097-28-6]
Silicon Nitride [12033-89-5]
Silicon Tetraacetate [562-90-3]
Silicon Tetrabromide [7789-66-4]
Silicon Tetrachloride [10026-04-7]
Silicon Tetrafluoride [7783-61-1]
Silicotungstic Acid [12520-88-6]
Silver [7440-22-4]
Silver Acetate [563-63-3]
Silver Bromide [7785-23-1]
Silver Carbonate [534-16-7]
Silver Chlorate [7783-92-8]
Silver Chloride [7783-90-6]
Silver Chromate(VI) [7784-01-2]
Silver Citrate [126-45-4]
Silver Cyanide [506-64-9]
Silver Difluoride [7783-95-1]
Silver Fluoride [7775-41-9]
Silver Iodate [7783-97-3]
Silver Iodide [7783-96-2]
Silver Lactate [128-00-7]
Silver Nitrate [7761-88-8]
Silver Nitrate, Toughened [8007-31-6]
Silver Nitrite [7783-99-5]
Silver Oxalate [533-51-7]
Silver Oxide [20667-12-3]
Silver(II) Oxide [1301-96-8]
Silver Perchlorate [7783-93-9]
Silver Permanganate [7783-98-4]
Silver Phosphate [7784-09-0]
Silver Picrate [146-84-9]
Silver Protein [9015-51-4]
Silver Selenate [7784-07-8]
Silver Selenide [1302-09-6]
Silver Selenite [7784-05-6]
Silver Subfluoride [1302-01-8]
Silver Sulfate [10294-26-5]
Silver Sulfide [21548-73-2]
Silver Tetraiodomercurate(II)
  [7784-03-4]
Silvex [93-72-1]
Silybin* [22888-70-6]
Silymarin [65666-07-1]
Simazine [122-34-9]
Simethicone* [8050-81-5]
Simetride [154-82-5]
Simetryn [1014-70-6]
Simfibrate [14929-11-4]
Simvastatin [79902-63-9]

Sinalbin [20196-67-2]
Sinapine [18696-26-9]
Sincalide [25126-32-3]
Sinefungin [58944-73-3]
Sinigrin [3952-98-5]
Sinomenine [115-53-7]
Sintropium Bromide [79467-19-9]
Sisomicin [32385-11-8]
Sisomicin Sulfate* [53179-09-2]
α_1-Sitosterol [474-40-8]
β-Sitosterol [83-46-5]
γ-Sitosterol [83-47-6]
Sizofiran [9050-67-3]
Skatole [83-34-1]
Skimmianine [83-95-4]
Skimmin [93-39-0]
Smilagenin [126-18-1]
Sobrerol [498-71-5]
Sobuzoxane [98631-95-9]
Soda Lime [8006-28-8]
Sodium [7440-23-5]
Sodium Acetate [127-09-3]
Sodiumacetic Acid Sodium Salt
  [534-12-3]
Sodium Acid Pyrophosphate
  [7758-16-9]
Sodium Alizarinesulfonate [130-22-3]
Sodium Aluminate [1302-42-7]
Sodium Amalgam [11110-52-4]
Sodium Amide [7782-92-5]
Sodium Amylosulfate [9010-01-9]
Sodium Antimonylgluconate*
  [12550-17-3]
Sodium Arsanilate [127-85-5]
Sodium Arsenate, Dibasic [7778-43-0]
Sodium Arsenite [7784-46-5]
Sodium Arsphenamine [1936-28-3]
Sodium Ascorbate [134-03-2]
Sodium Azide [26628-22-8]
Sodium Benzoate [532-32-1]
Sodium Bicarbonate [144-55-8]
Sodium Bifluoride [1333-83-1]
Sodium Bismuthate(V) [12232-99-4]
Sodium Bisulfate [7681-38-1]
Sodium Bisulfide [16721-80-5]
Sodium Bisulfite [7631-90-5]
Sodium Bitartrate [526-94-3]
Sodium Borate [1330-43-4]
Sodium Borohydride [16940-66-2]
Sodium Bromate [7789-38-0]
Sodium Bromide [7647-15-6]
Sodium Cacodylate [124-65-2]
Sodium Carbonate [497-19-8]
Sodium Cellulose Phosphate
  [9038-41-9]
Sodium Chlorate [7775-09-9]
Sodium Chloride [7647-14-5]
Sodium Chlorite [7758-19-2]
Sodium 6-Chloro-5-nitrotoluene-3-
  sulfonate [535-48-3]
Sodium Chromate(VI) [7775-11-3]
Sodium Chromate(VI) ^{51}Cr-labeled
  Compound* [10039-53-9]
Sodium Citrate [68-04-2]
Sodium Citrate, Acid [144-33-2]
Sodium Cobaltinitrite [13600-98-1]
Sodium Cyanate [917-61-3]
Sodium Cyanide [143-33-9]
Sodium Cyanoborohydride
  [25895-60-7]
Sodium Diacetate [126-96-5]
Sodium Dibunate [14992-59-7]
Sodium Dichromate(VI) [10588-01-9]
Sodium Dicyanoaurate(I) [15280-09-8]
Sodium Dithionate [7631-94-9]
Sodium Dodecylbenzenesulfonate
  [25155-30-0]
Sodium Ethoxide [141-52-6]
Sodium Ethyl Sulfate [546-74-7]
Sodium Ferrocyanide [13601-19-9]

Sodium Fluoborate [13755-29-8]
Sodium Fluoride [7681-49-4]
Sodium Folate [6484-89-5]
Sodium Formaldehydesulfoxylate
  [149-44-0]
Sodium Formate [141-53-7]
Sodium Gluconate [527-07-1]
Sodium Glycerophosphate [1334-74-3]
Sodium Hexachloroplatinate(IV)
  [16923-58-3]
Sodium Hexafluorosilicate
  [16893-85-9]
Sodium Hydride [7646-69-7]
Sodium Hydrosulfite [7775-14-6]
Sodium Hydroxide [1310-73-2]
Sodium Hypochlorite [7681-52-9]
Sodium Hypophosphite [7681-53-0]
Sodium Iodate [7681-55-2]
Sodium Iodide [7681-82-5]
Sodium Iodide, Radioactive
  [7790-26-3]
Sodium Iodomethamate [519-26-6]
Sodium Isopropyl Xanthate [140-93-2]
Sodium Lactate [72-17-3]
Sodium Lauryl Sulfate [151-21-3]
Sodium-Lead Alloy [12740-44-2]
Sodium Metabisulfite [7681-57-4]
Sodium Metaborate [7775-19-1]
Sodium Metaperiodate [7790-28-5]
Sodium Metasilicate [6834-92-0]
Sodium Methoxide [124-41-4]
Sodium Molybdate(VI) [7631-95-0]
Sodium Molybdate(VI) Dihydrate*
  [10102-40-6]
Sodium β-Naphthoquinone-4-sulfon-
  ate [521-24-4]
Sodium Nitrate [7631-99-4]
Sodium Nitrite [7632-00-0]
Sodium Nitroprusside [14402-89-2]
Sodium Nitroprusside Dihydrate*
  [13755-38-9]
Sodium Oxalate [62-76-0]
Sodium Oxide [1313-59-3]
Sodium Perborate [7632-04-4]
Sodium Perchlorate [7601-89-0]
Sodium Permanganate [10101-50-5]
Sodium Peroxide [1313-60-6]
Sodium Persulfate [7775-27-1]
Sodium Pertechnetate ^{99m}Tc
  [23288-60-0]
Sodium Phenolsulfonate [1300-51-2]
Sodium Phenoxide [139-02-6]
Sodium Phosphate, Dibasic
  [7558-79-4]
Sodium Phosphate, Monobasic
  [7558-80-7]
Sodium Phosphate, Radioactive
  [8027-28-9]
Sodium Phosphate, Tribasic
  [7601-54-9]
Sodium Phosphite [13708-85-5]
Sodium Phosphomolybdate
  [1313-30-0]
Sodium Phosphotungstate
  [51312-42-6]
Sodium Plumbate(IV) [12201-47-7]
Sodium Polyanetholesulfonate
  [52993-95-0]
Sodium Polymetaphosphate
  [50813-16-6]
Sodium Polystyrene Sulfonate
  [9003-59-2]
Sodium Propionate [137-40-6]
Sodium Pyroantimonate [10049-22-6]
Sodium Rhodizonate [523-21-7]
Sodium Salicylate [54-21-7]
Sodium Selenate [13410-01-0]
Sodium Selenide [1313-85-5]
Sodium Selenite [10102-18-8]
Sodium Sesquicarbonate [533-96-0]

Denotes derivative of title compound

Sodium Silicate [1344-09-8]
Sodium Silicate Solution [1344-09-8]
Sodium Stannate(IV) [12058-66-1]
Sodium Stearate [822-16-2]
Sodium Stibogluconate* [16037-91-5]
Sodium Succinate [150-90-3]
Sodium Sulfate [7757-82-6]
Sodium Sulfide [1313-82-2]
Sodium Sulfite [7757-83-7]
Sodium β-Sulfopropionitrile [513-15-5]
Sodium Tartrate [868-18-8]
Sodium Tellurate(IV) [10102-20-2]
Sodium Tellurate(VI) [10101-83-4]
Sodium Tetrachloroaluminate [7784-16-9]
Sodium Tetrachloroaurate(III) [15189-51-2]
Sodium Tetradecyl Sulfate [139-88-8]
Sodium Tetraphenylborate [143-66-8]
Sodium Thioantimonate(V) [13776-84-6]
Sodium Thiocarbonate [534-18-9]
Sodium Thioglycolate [367-51-1]
Sodium Thiophosphate [10101-88-9]
Sodium Thiosulfate [7772-98-7]
Sodium Trimetaphosphate [7785-84-4]
Sodium Tripolyphosphate [7758-29-4]
Sodium Tungstate(VI) [13472-45-2]
Sodium Uranate(VI) [13721-34-1]
Sodium Vanadate(V) [13718-26-8]
Sofalcone [64506-49-6]
Solan [2307-68-8]
Solanidine [80-78-4]
Solanine [20562-02-1]
Solanocapsine [639-86-1]
Solanone [1937-54-8]
Solasodine [126-17-0]
Solasonine [19121-58-5]
Solasulfone [133-65-3]
Soman [96-64-0]
Somatoliberin [83930-13-6]
Somatostatin [38916-34-6]
Somatotropin [9002-72-6]
Songorine [509-24-0]
Sonicated Human Serum Albumin* [150103-82-5]
Sophorabioside [2945-88-2]
Sophoricoside [152-95-4]
Sophorose [534-46-3]
Sorbic Acid [110-44-1]
Sorbic Alcohol [111-28-4]
Sorbinil [68367-52-2]
Sorbitan Laurate* [1338-39-2]
Sorbitan Oleate* [1338-43-8]
Sorbitan Stearate* [1338-41-6]
Sorbitol [50-70-4]
Sorbose [87-79-6]
Sorivudine [77181-69-2]
Sotalol [3930-20-9]
Sotalol Hydrochloride* [959-24-0]
Soterenol [13642-52-9]
Soterenol Monohydrochloride* [14816-67-2]
Sozoiodolic Acid [554-71-2]
Sozoiodolic Acid Mercury Deriv* [515-43-5]
Sozoiodolic Acid Sodium Salt Dihydrate* [6160-08-3]
Sozoiodolic Acid Trihydrate* [6160-10-7]
Sozoiodolic Acid Zinc Salt Hexahydrate* [6160-09-4]
Sparassol [520-43-4]
Sparfloxacin [110871-86-8]
Sparsiflorine [2128-61-2]
Sparsomycin [1404-64-4]
Sparteine [90-39-1]
Sparteine Sulfate Pentahydrate* [6160-12-9]

Spasmolytol [25333-96-4]
Spectinomycin [1695-77-8]
Spectinomycin Dihydrochloride Pentahydrate* [22189-32-8]
Spectinomycin Sulfate Tetrahydrate* [64058-48-6]
Spermidine [124-20-9]
Spermine [71-44-3]
Spherophysine [25978-54-5]
Sphingofungin C* [121025-46-5]
Sphingosine [123-78-4]
α-Spinasterol [481-18-5]
Spinulosin [85-23-4]
Spiperone [749-02-0]
Spiramycin [8025-81-8]
Spiramycin I* [24916-50-5]
Spiramycin II* [24916-51-6]
Spiramycin III* [24916-52-7]
Spirapril [83647-97-6]
Spiraprilat* [83602-05-5]
Spirapril Hydrochloride* [94841-17-5]
Spirit of Ammonia, Aromatic [8013-59-0]
Spirit of Camphor [76-22-2]
Spirit of Ether [8013-43-2]
Spirit of Ether Compound [8013-44-3]
Spirit of Ethyl Nitrite [109-95-5]
Spirit of Formic Acid [64-18-6]
Spirit of Glyceryl Trinitrate [55-63-0]
Spirogermanium [41992-23-8]
Spirogermanium Dihydrochloride* [41992-22-7]
Spironolactone [52-01-7]
Spizofurone [72492-12-7]
Splenin [1416-60-0]
Sporidesmin A* [1456-55-9]
Sporidesmolide I* [2900-38-1]
Sporidesmolide II* [3200-75-7]
Sporidesmolide III* [1803-67-4]
Sporidesmolide IV* [10252-34-3]
Squalamine [148717-90-2]
Squalane [111-01-3]
Squalene [111-02-4]
Stachydrine [471-87-4]
Stallimycin [636-47-5]
Stallimycin Hydrochloride* [6576-51-8]
Stannic Bromide [7789-67-5]
Stannic Chloride [7646-78-8]
Stannic Chromate(VI) [38455-77-5]
Stannic Fluoride [7783-62-2]
Stannic Iodide [7790-47-8]
Stannic Oxide [18282-10-5]
Stannic Selenide [20770-09-6]
Stannic Selenite [7446-25-5]
Stannic Sulfide [1315-01-1]
Stannous Acetate [638-39-1]
Stannous Bromide [10031-24-0]
Stannous Chloride [7772-99-8]
Stannous Fluoride [7783-47-3]
Stannous Hexafluorozirconate(IV) [12419-43-1]
Stannous Iodide [10294-70-9]
Stannous Oxalate [814-94-8]
Stannous Oxide [21651-19-4]
Stannous Pyrophosphate [15578-26-4]
Stannous Selenide [1315-06-6]
Stannous Sulfate [7488-55-3]
Stannous Sulfide [1314-95-0]
Stannous Tartrate [815-85-0]
Stanolone [521-18-6]
Stanozolol [302-96-5] (1'H form)
Stanozolol [10418-03-8] (2'H form)
Starch, Soluble [9005-84-9]
Statine [49642-07-1]
Statolon [11006-77-2]
Stavudine [3056-17-5]
Stearic Acid [57-11-4]
Stearyl Alcohol [112-92-5]

Stellar [151031-37-7]
Stenbolone [5197-58-0]
Stenbolone Acetate* [1242-56-4]
Stepronin [72324-18-6]
Stepronin Lysine Salt* [113790-28-6]
Stepronin Sodium Salt* [78126-10-0]
Steviol* [471-80-7]
Stevioside [57817-89-7]
Stibine [7803-52-3]
Stibocaptate [27279-76-1]
Stibophen [15489-16-4]
Stigmastanol [83-45-4]
Stigmasterol [83-48-7]
Stilbamidine [122-06-5]
Stilbamidine Isethionate* [140-59-0]
Stilbazium Iodide [3784-99-4]
Stilbene [588-59-0]
Storax [8023-62-9]
Strepogenin [11000-03-6]
Streptidine [85-17-6]
Streptobiosamine [126-05-6]
Streptodornase [37340-82-2]
Streptodornase Mixture with Streptokinase* [8048-16-6]
Streptokinase [9002-01-1]
Streptolydigin [7229-50-7]
Streptomycin [57-92-1]
Streptomycin B [128-45-0]
Streptomycin Sesquisulfate* [3810-74-0]
Streptonicozid [5667-71-0]
Streptonigrin [3930-19-6]
L-Streptose [13008-73-6]
Streptovaricin [1404-74-6]
Streptovaricin A* [23344-16-3]
Streptovaricin B* [11031-82-6]
Streptovaricin C* [23344-17-4]
Streptovaricin D* [32164-26-4]
Streptovaricin E* [35413-63-9]
Streptovaricin F* [35512-37-9]
Streptovaricin G* [11031-85-9]
Streptovaricin J* [52275-61-3]
Streptovirudin [56833-74-0]
Streptozocin [18883-66-4]
Strigol [11017-56-4]
Strobane [8001-50-1]
Strontium [7440-24-6]
Strontium Acetate [543-94-2]
Strontium Bromate [14519-18-7]
Strontium Bromide [10476-81-0]
Strontium Carbonate [1633-05-2]
Strontium Chlorate [7791-10-8]
Strontium Chloride [10476-85-4]
Strontium Chromate(VI) [7789-06-2]
Strontium Fluoride [7783-48-4]
Strontium Formate [592-89-2]
Strontium Hydroxide [18480-07-4]
Strontium Iodide [10476-86-5]
Strontium Lactate [29870-99-3]
Strontium Nitrate [10042-76-9]
Strontium Oxalate [814-95-9]
Strontium Oxide [1314-11-0]
Strontium Peroxide [1314-18-7]
Strontium Phosphate, Tribasic [7446-28-8]
Strontium Selenate [7446-21-1]
Strontium Sulfate [7759-02-6]
Strontium Sulfide [1314-96-1]
Strophanthidin [66-28-4]
Strophanthin [11005-63-3]
Strophanthobiose [7724-04-1]
Struvite [15490-91-2]
Strychnine [57-24-9]
Strychnine Nitrate [66-32-0]
Strychnine N⁶-Oxide [7248-28-4]
Strychnine Phosphate [509-42-2]
Strychnine Sulfate [60-41-3]
Stylopine [90-39-1]
Stylopine [7461-02-1]
Styphnic Acid [82-71-3]
Styramate [94-35-9]

Styrene [100-42-5]
Styrene Glycol [93-56-1]
Subathizone [121-55-1]
Suberic Acid [505-48-6]
Substance P [11035-08-8]
Subtilin [1393-38-0]
Succimer [304-55-2]
Succinamide [110-14-5]
Succinanil [83-25-0]
Succinanilic Acid [102-14-7]
Succinic Acid [110-15-6]
Succinic Anhydride [108-30-5]
Succinimide [123-56-8]
Succinonitrile [110-61-2]
Succinyl Chloride [543-20-4]
Succinylcholine Bromide [55-94-7]
Succinylcholine Chloride [71-27-2]
Succinylcholine Iodide [541-19-5]
Succinylsulfathiazole [116-43-8]
Succisulfone [5934-14-5]
Succisulfone 2,2′-Iminodiethanol
  Salt* [547-36-4]
Suclofenide [30279-49-3]
Sucralfate [54182-58-0]
Sucralose [56038-13-2]
Sucrose [57-50-1]
Sucrose Octaacetate [126-14-7]
Sudan III [85-86-9]
Sufentanil [56030-54-7]
Sufentanil Citrate* [60561-17-3]
Sulbactam [68373-14-8]
Sulbactam Mixture of Sodium Salt
  with Ampicillin Sodium*
  [117060-71-6]
Sulbactam Sodium Salt* [69388-84-7]
Sulbenicillin [41744-40-5]
Sulbenox [58095-31-1]
Sulbentine [350-12-9]
Sulconazole [61318-90-9]
Sulconazole Mononitrate*
  [61318-91-0]
Sulcotrione [99105-77-8]
Sulfabenz [127-77-5]
Sulfabenzamide [127-71-9]
Sulfabromomethazine [116-45-0]
Sulfacetamide [144-80-9]
Sulfacetamide Sodium Salt Mono-
  hydrate* [6209-17-2]
Sulfachlorpyridazine [80-32-0]
Sulfachlorpyridazine Sodium Salt*
  [23282-55-5]
Sulfachrysoidine [485-41-6]
Sulfacytine [17784-12-2]
Sulfadiazine [68-35-9]
Sulfadiazine Silver Salt* [22199-08-2]
Sulfadicramide [115-68-4]
Sulfadimethoxine [122-11-2]
Sulfadimethoxine Mixture with
  Trimethoprim* [39469-82-4]
Sulfadimethoxine Sodium Salt*
  [1037-50-9]
Sulfadimethoxine Sodium Salt Mix-
  ture with Trimethoprim*
  [131643-86-2]
Sulfadoxine [2447-57-6]
Sulfadoxine Mixture with Trimetho-
  prim* [39295-60-8]
Sulfaethidole [94-19-9]
Sulfaguanidine [57-67-0]
Sulfaguanole [27031-08-9]
Sulfalenc [152-47-6]
Sulfalene Mixture with Trimetho-
  prim* [50933-06-7]
Sulfallate [95-06-7]
Sulfaloxic Acid [14376-16-0]
Sulfaloxic Acid Calcium Salt*
  [59672-20-7]
Sulfamerazine [127-79-7]
Sulfamerazine Monosodium Salt*
  [127-58-2]

Sulfameter [651-06-9]
Sulfamethazine [57-68-1]
Sulfamethazine Sodium Salt*
  [1981-58-4]
Sulfamethizole [144-82-1]
Sulfamethomidine [3772-76-7]
Sulfamethoxazole [723-46-6]
Sulfamethoxazole Mixture with
  Trimethoprim* [8064-90-2]
Sulfamethoxypyridazine [80-35-3]
Sulfamethoxypyridazine Sodium Salt*
  [2577-32-4]
Sulfamethylthiazole [515-59-3]
Sulfamethylthiazole Sodium Salt*
  [6101-28-6]
Sulfametrole [32909-92-5]
Sulfametrole Mixture with Tri-
  methoprim* [63749-94-0]
Sulfamic Acid [5329-14-6]
Sulfamide [7803-58-9]
Sulfamidochrysoidine [103-12-8]
Sulfamidochrysoidine Hydrochloride*
  [33445-35-1]
Sulfamoxole [729-99-7]
Sulfamoxole Mixture with Trimetho-
  prim* [57197-43-0]
Sulfanilamide [63-74-1]
4-Sulfanilamidosalicylic Acid
  [6202-21-7]
Sulfanilic Acid [121-57-3]
Sulfanilic Acid Monohydrate*
  [6101-32-2]
Sulfanilic Acid Sodium Salt Di-
  hydrate* [6106-22-5]
Sulfanilic Acid Zinc Salt Tetrahy-
  drate* [31884-76-1]
2-p-Sulfanilylanilinoethanol [80-02-4]
p-Sulfanilylbenzylamine [4393-19-5]
Sulfanilyl Fluoride [98-62-4]
N⁴-Sulfanilylsulfanilamide [547-52-4]
Sulfanilylurea [547-44-4]
N-Sulfanily-3,4-xylamide [120-34-3]
Sulfanitran [122-16-7]
Sulfaperine [599-88-2]
Sulfaphenazole [526-08-9]
Sulfaproxyline [116-42-7]
Sulfapyrazine [116-44-9]
Sulfapyridine [144-83-2]
Sulfapyridine Sodium Salt Mono-
  hydrate* [127-57-1]
Sulfaquinoxaline [59-40-5]
Sulfarside [1134-98-1]
Sulfarsphenamine [618-82-6]
Sulfasalazine [599-79-1]
Sulfasomizole [632-00-8]
Sulfasymazine [1984-94-7]
Sulfathiazole [72-14-0]
Sulfathiourea [515-49-1]
Sulfathiourea Sodium Salt*
  [6101-34-4]
Sulfatolamide [1161-88-2]
Sulfazamet [852-19-7]
Sulfazecin [77912-79-9]
Sulfinalol [66264-77-5]
Sulfinalol Hydrochloride*
  [63251-39-8]
Sulfinpyrazone [57-96-5]
4,4′-Sulfinyldianiline [119-59-5]
Sulfiram [95-05-6]
Sulfisomidine [515-64-0]
Sulfisoxazole [127-69-5]
Sulfisoxazole Diethanolamine Salt*
  [4299-60-9]
Sulfitocobalamin* [15671-27-9]
Sulfoacetic Acid [123-43-3]
Sulfobromophthalein Sodium
  [71-67-0]
Sulfolane [126-33-0]
3-Sulfolene [77-79-2]
Sulfometuron Methyl [74222-97-2]

Sulfonethylmethane [76-20-0]
Sulfoniazide [3691-81-4]
Sulfonmethane [115-24-2]
Sulfonyldiacetic Acid [123-45-5]
Sulforaphane [4478-93-7] (unspecified)
Sulforaphane [142825-10-3]
Sulforaphen [592-95-0]
Sulforidazine [14759-06-9]
Sulfosalicylic Acid [97-05-2]
Sulfotep [3689-24-5]
Sulfoxide [120-62-7]
Sulfoxone Sodium [144-75-2]
Sulfur [7704-34-9]
Sulfur Chloride [10025-67-9]
Sulfur Dioxide [7446-09-5]
Sulfuretin [120-05-8]
Sulfur Hexafluoride [2551-62-4]
Sulfuric Acid [7664-93-9]
Sulfuric Acid, Fuming* [8014-95-7]
Sulfur Iodide [1312-15-8]
Sulfurous Acid [7782-99-2]
Sulfur, Pharmaceutical [7704-34-9]
Sulfur Tetrafluoride [7783-60-0]
Sulfur Trioxide [7446-11-9]
Sulfuryl Chloride [7791-25-5]
Sulfuryl Fluoride [2699-79-8]
Sulindac [38194-50-2]
Sulisatin [54935-03-4]
Sulisatin Disodium Salt* [54935-04-5]
Sulisobenzone [4065-45-6]
Sulmarin [29334-07-4]
Sulmazole [73384-60-8]
Suloctidil [54063-56-8]
Sulphan Blue [129-17-9]
Sulphan Blue 2,5-Disulfophenyl
  Isomer* [68238-36-8]
Sulphenone [80-07-2]
Sulpiride [15676-16-1]
Sulprofos [35400-43-2]
Sulprostone [60325-46-4]
Sultamicillin [76497-13-7]
Sultamicillin Tosylate* [83105-70-8]
Sulthiame [61-56-3]
Sultopride [53583-79-2]
Sultopride Hydrochloride*
  [23694-17-9]
Sultosilic Acid [57775-26-5]
Sultosilic Acid Piperazine Salt*
  [57775-27-6]
Sultroponium [15130-91-3]
Sumatriptan [103628-46-2]
Sumatriptan Succinate* [103628-48-4]
Sumatrol [82-10-0]
Sunset Yellow FCF [2783-94-0]
Superoxide Dismutase [9054-89-1]
Suplatast Tosylate [94055-76-2]
Suprasterol II [562-71-0]
Suprofen [40828-46-4]
Suramin Sodium [129-46-4]
Suriclone [53813-83-5]
Surinamine [537-49-5]
Suxethonium Bromide [111-00-2]
Suxibuzone [27470-51-5]
Swertiamarin [17388-39-5]
Sydnones [50927-09-8]
Symclosene [87-90-1]
Synephrine [94-07-5]
Synephrine Hydrochloride*
  [5985-28-4]
Synephrine Tartaric Acid Monoester*
  [6414-49-9]
Synephrine Tartrate* [16589-24-5]
Synhexyl [117-51-1]
Syringaldehyde [134-96-3]
Syringin [118-34-3]
Syrosingopine [84-36-6]
2,4,5-T [93-76-5]
Tabernanthine [83-94-3]
Tabun [77-81-6]
Tacalcitol [57333-96-7]

* Denotes derivative of title compound

Tachysterol [115-61-7]
Tacrine [321-64-2]
Tacrine Hydrochloride* [1684-40-8]
Tacrolimus [104987-11-3]
D-Tagatose [87-81-0]
Taka-Diastase [9001-19-8]
Talampicillin [47747-56-8]
Talampicillin Hydrochloride*
  [39878-70-1]
Talastine [16188-61-7]
Talastine Hydrochloride*
  [16188-76-4]
Talbutal [115-44-6]
Talc [14807-96-6]
Talinolol [57460-41-0]
Talipexole [101626-70-4]
Talipexole Dihydrochloride*
  [36085-73-1]
Tallysomycin [67995-68-0]
Tallysomycin A* [65057-90-1]
Tallysomycin B* [65057-91-2]
Talniflumate [66898-62-2]
Tamoxifen [10540-29-1]
Tamoxifen Citrate* [54965-24-1]
Tamsulosin [106133-20-4]
Tamsulosin (±)-Form Hydrochloride*
  [80223-99-0]
Tamsulosin (R)-Form Hydrochloride*
  [106463-17-6]
Tamsulosin (S)-Form Hydrochloride*
  [106463-19-8]
Tanacetin [1401-54-3]
Tandospirone [87760-53-0]
Tandospirone Citrate* [112457-95-1]
Tanghinigenin [6875-16-7]
Tanghinigenin 3-[(2-O-Acetyl-6-
  deoxy-3-O-methyl-α-L-gluco-
  pyranosyl)oxy]* [25390-16-3]
Tannoform [9010-29-1]
Tantalum [7440-25-7]
Tantalum Pentachloride [7721-01-9]
Tantalum Pentafluoride [7783-71-3]
Tantalum Pentoxide [1314-61-0]
Taprostene [108945-35-3]
Taprostene Sodium Salt* [87440-45-7]
Taraxasterol [1059-14-9]
Taraxein [9010-30-4]
Taraxerol [127-22-0]
D-Tartaric Acid [147-71-7]
DL-Tartaric Acid [133-37-9]
L-Tartaric Acid [87-69-4]
meso-Tartaric Acid [147-73-9]
Tartrazine [1934-21-0]
Tartronic Acid [80-69-3]
Taurine [107-35-7]
Taurocholic Acid [81-24-3]
Taurolidine [19388-87-5]
Taxicatin [90-71-1]
Taxicin-I* [5308-89-4]
Taxicin-II* [5308-90-7]
Taxine [12607-93-1]
Taxine A* [1361-49-5]
Taxodione [19026-31-4]
Tazanolast [82989-25-1]
Tazarotene [118292-40-3]
Tazettine [507-79-9]
Tazobactam [89786-04-9]
Tazobactam Sodium Salt*
  [89785-84-2]
TCDD [1746-01-6]
Tebuconazole [80443-41-0] (un-
  specified)
Tebuconazole [107534-96-3] (±)
Tebufenozide [112410-23-8]
Tebuthiuron [34014-18-1]
Technetium [7440-26-8]
Technetium ⁹⁹ᵐTc Bicisate
  [121281-41-2]
Technetium ⁹⁹ᵐTc Mertiatide
  [125224-05-7]

Technetium ⁹⁹ᵐTc Mertiatide Free
  Acid [104348-91-6]
Technetium ⁹⁹ᵐTc Sestamibi
  [109581-73-9]
Technetium ⁹⁹ᵐTc Teboroxime
  [104716-22-5]
Teclothiazide [4267-05-4]
Teclothiazide Potassium Salt*
  [5306-80-9]
Teclozan [5560-78-1]
Tecomanine [6878-83-7]
Tectorigenin [548-77-6]
Teflurane [124-72-1]
Tefluthrin [79538-32-2]
Tegafur [17902-23-7]
Teicoplanin [61036-64-4]
Telenzepine [80880-90-6]
Teleocidin B₄* [11032-05-6]
Teleocidins [78474-55-2]
Telluric(VI) Acid [7803-68-1]
Tellurium [13494-80-9]
Tellurium Dibromide [7789-54-0]
Tellurium Dichloride [10025-71-5]
Tellurium Dioxide [7446-07-3]
Tellurium Hexafluoride [7783-80-4]
Tellurium Tetrabromide [10031-27-3]
Tellurium Tetrachloride [10026-07-0]
Tellurium Tetraiodide [7790-48-9]
Tellurous Acid [10049-23-7]
Telomycin [19246-24-3]
Temafloxacin [108319-06-8]
Temafloxacin Hydrochloride*
  [105784-61-0]
Temazepam [846-50-4]
Temephos [3383-96-8]
Temocapril [111902-57-9]
Temocaprilat* [110221-53-9]
Temocapril Hydrochloride*
  [110221-44-8]
Temocillin [66148-78-5]
Temocillin Disodium Salt*
  [61545-06-0]
Temozolomide [85622-93-1]
Tenidap [120210-48-2]
Tenidap Sodium Salt* [119784-94-0]
Teniposide [29767-20-2]
Tenonitrozole [3810-35-3]
Tenoxicam [59804-37-4]
Tenuazonic Acid [610-88-8]
Tephrosin [76-80-2]
Teprenone [6809-52-5]
Terazosin [63590-64-7]
Terazosin Hydrochloride Dihydrate*
  [70024-40-7]
Terbacil [5902-51-2]
Terbinafine [91161-71-6]
Terbium [7440-27-9]
Terbufos [13071-79-9]
Terbutaline [23031-25-6]
Terbutaline Sulfate* [23031-32-5]
Terconazole [67915-31-5]
Terebene [8014-10-6]
Terebic Acid [79-91-4]
Terephthalic Acid [100-21-0]
Terfenadine [50679-08-8]
Terguride [37686-84-3]
Terguride Hydrogen Maleate*
  [37686-85-4]
Teriparatide Acetate [99294-94-7]
Terlipressin [14636-12-5]
Terodiline [15793-40-5]
Terodiline Hydrochloride*
  [7082-21-5]
Terofenamate [29098-15-5]
Terpenylic Acid [26754-48-3]
Terpin [80-53-5]
α-Terpineol [98-55-5]
Terpin trans-Form* [565-50-4]
Terpin cis-Form Hydrate* [2451-01-6]
Terreic Acid [121-40-4]

Tertatolol [34784-64-0]
Tertatolol Hydrochloride*
  [33580-30-2]
α-Terthienyl [1081-34-1]
Testolactone [968-93-4]
Testosterone [58-22-0]
Testosterone Acetate* [1045-69-8]
Testosterone 17β-Cyclopentanepro-
  pionate* [58-20-8]
Testosterone Enanthate* [315-37-7]
Testosterone Isobutyrate* [1169-49-9]
Testosterone Propionate* [57-85-2]
Testosterone Undecanoate*
  [5949-44-0]
Tetraamminecopper Sulfate
  [14283-05-7]
Tetrabarbital [76-23-3]
Tetrabenazine [58-46-8]
Tetrabenazine Hydrochloride*
  [2105-47-7]
Tetrabenazine Methanesulfonate*
  [804-53-5]
Tetraborane(10) [18283-93-7]
3,4,5,6-Tetrabromo-o-cresol [576-55-6]
sym-Tetrabromoethane [79-27-6]
3′,3″,5′,5″-Tetrabromophenolphtha-
  lein [76-62-0]
3′,3″,5′,5″-Tetrabromophenolphtha-
  lein Disodium Salt* [860-07-1]
Tetracaine Hydrochloride [136-47-0]
Tetrachloroethane [79-34-5]
Tetrachloroethylene [127-18-4]
3,3′,4′,5-Tetrachlorosalicylanilide
  [1154-59-2]
Tetrachlorvinphos [22248-79-9]
Tetracosamethylhendecasiloxane
  [107-53-9]
Tetracyanoethylene [670-54-2]
Tetracycline [60-54-8]
Tetracycline Hydrochloride*
  [64-75-5]
Tetracycline Phosphate Complex*
  [1336-20-5]
Tetradecamethylhexasiloxane
  [107-52-8]
Tetradifon [116-29-0]
Tetraethylammonium Bromide
  [71-91-0]
Tetraethylammonium Chloride
  [56-34-8]
Tetraethylammonium Hydroxide
  [77-98-5]
Tetraethyllead [78-00-2]
N,N,N′,N′-Tetraethylphthalamide
  [83-81-8]
Tetraethyl Pyrophosphate [107-49-3]
2,2,3,3-Tetrafluoro-1-propanol
  [76-37-9]
Tetraglycine Hydroperiodide
  [7097-60-1]
Tetraglyme [143-24-8]
Tetrahydrocannabinols (−)-Δ¹-3,4-
  trans-Form* [1972-08-3]
Tetrahydrocannabinols (−)-Δ⁶-3,4-
  trans-Form* [5957-75-5]
Tetrahydrocortisone [53-05-4]
Tetrahydrofuran [109-99-9]
2,5-Tetrahydrofurandimethanol
  [104-80-3]
Tetrahydrofurfuryl Alcohol [97-99-4]
Tetrahydropalmatine [10097-84-4]
Tetrahydropapaveroline [4747-99-3]
Tetrahydropyran [142-68-7]
Tetrahydrothiophene [110-01-0]
Tetrahydrozoline [84-22-0]
Tetrahydrozoline Hydrochloride*
  [522-48-5]
Tetraiodoethylene [513-92-8]
Tetralin [119-64-2]
Tetralol [530-91-6]

Tetramethrin [7696-12-0]
Tetramethylammonium Hydroxide [75-59-2]
Tetramethylammonium Iodide [75-58-1]
Tetramethyldiaminobutane [111-51-3]
Tetramethylenedisulfotetramine [80-12-6]
Tetramethyl-p-phenylenediamine [100-22-1]
Tetramethylurea [632-22-4]
Tetramisole* [5036-02-2]
Tetramisole Hydrochloride* [5086-74-8]
Tetrandrine [518-34-3]
Tetrandrine l-Form* [1263-79-2]
Tetranectin [109489-77-2]
Tetranitromethane [509-14-8]
Tetrantoin [52094-70-9]
Tetraphenylarsonium Bromide [507-27-7]
Tetraphenylarsonium Chloride [507-28-8]
Tetraphosphorus Trisulfide [1314-85-8]
Tetrasilane [7783-29-1]
Tetrasodium Pyrophosphate [7722-88-5]
Tetrasulfur Tetranitride [28950-34-7]
Tetrazepam [10379-14-3]
Tetrazolium Blue [1871-22-3]
Tetrodotoxin [4368-28-9]
Tetrofosmin [127502-06-1]
Tetrofosmin 99mTc-Complex* [127455-27-0]
Tetronasin [75139-06-9]
Tetroquinone [319-89-1]
Tetroxoprim [53808-87-0]
Tetroxoprim Mixture with Sulfadiazine* [73173-12-3]
Tevenel [4302-95-8]
TFM [88-30-2]
Thalicarpine [5373-42-2]
Thalidomide [50-35-1]
Thallium [7440-28-0]
Thallium Acetate [563-68-8]
Thallium Bromide [7789-40-4]
Thallium Carbonate [6533-73-9]
Thallium Chloride [7791-12-0]
Thallium Cyanide [13453-34-4]
Thallium Fluoride [7789-27-7]
Thallium Hydroxide [12026-06-1]
Thallium Iodide [7790-30-9]
Thallium Nitrate [10102-45-1]
Thallium Oxide [1314-12-1]
Thallium Selenate [7446-22-2]
Thallium Selenide [15572-25-5]
Thallium Sesquioxide [1314-32-5]
Thallium Sulfate [7446-18-6]
Thallium Sulfide [1314-97-2]
Thallium Trifluoride [7783-57-5]
Thaumatin [53850-34-3]
Theaflavine [4670-05-7]
Thearubigins [12698-96-3]
Thebaine [115-37-7]
Thebainone [467-98-1]
Thenaldine [86-12-4]
Thenium Closylate [4304-40-9]
3-Thenoic Acid [88-13-1]
Thenyldiamine [91-79-2]
Theobromine [83-67-0]
1-Theobromineacetic Acid [5614-56-2]
1-Theobromineacetic Acid Sodium Salt* [32245-40-2]
Theobromine Calcium Salicylate* [8065-51-8]
Theobromine Sodium Acetate* [8002-88-8]
Theobromine Sodium Salicylate* [8048-31-5]

Theofibrate [54504-70-0]
Theophylline [58-55-9]
Theophylline Ethanolamine* [573-41-1]
Theophylline Isopropanolamine* [5600-19-1]
Theophylline Sodium Acetate* [8002-89-9]
Theophylline Sodium Glycinate* [8000-10-0]
Thermolysin [9073-78-3]
Thermorubin [11006-83-0]
Thevetin A [37933-66-7]
Thevetose [5675-98-9]
Thiabendazole [148-79-8]
Thiabendazole Hypophosphite* [28558-32-9]
Thiacetazone [104-06-3]
Thialbarbital [467-36-7]
Thiambutene [86-14-6]
Thiamine [59-43-8]
Thiamine Diphosphate [154-87-0]
Thiamine Disulfide [67-16-3]
Thiamine Disulfide O,O-Diisobutyrate* [3286-46-2]
Thiamine Disulfide Hydrochloride* [18642-10-9]
Thiamine Disulfide Phosphate* [992-46-1]
Thiamine Hydrochloride* [67-03-8]
Thiamine Mononitrate* [532-43-4]
Thiamine Monophosphate* [532-40-1]
Thiamine Triphosphate [3475-65-8]
Thiamiprine [5581-52-2]
Thiamorpholine [123-90-0]
Thiamphenicol [15318-45-3]
Thiamphenicol DL-Form* [847-25-6]
Thiamylal [77-27-0]
Thiamylal Sodium Salt* [337-47-3]
Thianaphthene [95-15-8]
Thiarubrin(e) A* [63543-09-9]
Thiarubrin(e)B* [71539-72-5]
Thiazesim [5845-26-1]
Thiazesim Hydrochloride* [3122-01-8]
Thiazinamium Methylsulfate [58-34-4]
Thiazole [288-47-1]
Thiazolinobutazone [54749-86-9]
Thiazolsulfone [473-30-3]
Thiazol Yellow G [1829-00-1]
Thibenzazoline [6028-35-9]
Thidiazuron [51707-55-2]
Thienamycin [59995-64-1]
Thiethylperazine [1420-55-9]
Thiethylperazine Dimalate* [52239-63-1]
Thiethylperazine Dimaleate* [1179-69-7]
Thimerfonate Sodium [5964-24-9]
Thimerosal [54-64-8]
Thioacetaldehyde [2765-04-0]
Thioacetamide [62-55-5]
Thioacetic Acid [507-09-5]
Thiobarbital [77-32-7]
Thiobenzyl Alcohol [100-53-8]
Thiobutabarbital [2095-57-0]
Thiobutabarbital Sodium Salt* [947-08-0]
Thiocarbamizine [91-71-4]
Thiocarbarsone [120-02-5]
Thiocolchicine [2730-71-4]
Thiocolchicine 2-Glucoside Analog* [602-41-5]
Thiocresol [26445-03-4]
Thioctic Acid [62-46-4]
Thioctic Acid d-Form* [1200-22-2]
Thioctic Acid dl-Form* [1077-28-7]
Thioctic Acid l-Form* [1077-27-6]

Thioctic Acid Sodium Salt* [2319-84-8]
Thiocyanate Sodium [540-72-7]
Thiocyanic Acid [463-56-9]
Thiodicarb [59669-26-0]
2,2'-Thiodiethanol [111-48-8]
Thiodiglycolic Acid [123-93-3]
3,3'-Thiodipropionic Acid [111-17-1]
Thioformamide [115-08-2]
5-Thio-D-glucose [20408-97-3]
Thioglycerol [96-27-5]
Thioglycolic Acid [68-11-1]
Thioguanine [154-42-7]
Thioguanosine [85-31-4]
Thiolactic Acid [79-42-5]
2-Thiolhistidine [13552-61-9]
Thiolutin [87-11-6]
Thiomalic Acid [70-49-5]
Thionalide [93-42-5]
Thionaphthene-2-carboxylic Acid [6314-28-9]
Thionazin [297-97-2]
Thionine [135-59-1]
Thionyl Bromide [507-16-4]
Thionyl Chloride [7719-09-7]
Thionyl Fluoride [7783-42-8]
Thiopental Sodium [71-73-8]
Thiopeptin [12609-84-6]
Thiophanate [23564-06-9]
Thiophanate-methyl* [23564-05-8]
Thiophene [110-02-1]
2-Thiophenecarboxylic Acid [527-72-0]
Thiophenol [108-98-5]
Thiopropazate [84-06-0]
Thiopropazate Dihydrochloride* [146-28-1]
Thiopropazate Dimaleate* [104999-18-0]
Thioproperazine [316-81-4]
Thioproperazine Dimethanesulfonate* [2347-80-0]
Thioquinox [93-75-4]
Thioredoxin [52500-60-4]
Thioridazine [50-52-2]
Thioridazine Hydrochloride* [130-61-0]
Thiosalicylic Acid [147-93-3]
Thiosemicarbazide [79-19-6]
Thiosinamine [109-57-9]
1-Thiosorbitol [24531-57-5]
Thiostrepton [1393-48-2]
Thiothixene [5591-45-7]
2-Thiouracil [141-90-2]
Thiourea [62-56-6]
Thioxanthene [261-31-4]
Thioxanthone [492-22-8]
THIP [64603-91-4]
Thiphenamil [82-99-5]
Thiphenamil Hydrochloride* [548-68-5]
Thiram [137-26-8]
Thomas Phosphate [1306-01-0]
Thonzonium Bromide [553-08-2]
Thonzylamine Hydrochloride [63-56-9]
Thorium [7440-29-1]
Thorium Chloride [10026-08-1]
Thorium Iodide [7790-49-0]
Thorium Nitrate [13823-29-5]
Thorium Oxide [1314-20-1]
Thorium Sulfate [10381-37-0]
Thorium Tetracyanoplatinate(II) [14481-33-5]
Thozalinone [655-05-0]
Threonine [72-19-5]
D-Threose [95-43-2]
L-Threose [95-44-3]
Thrombin [9002-04-4]
Thromboplastin [9035-58-9]

Denotes derivative of title compound

Thromboxane A₂* [57576-52-0]
Thromboxane B₂* [54397-85-2]
Thujic Acid [499-89-8]
Thujopsene [470-40-6]
Thulium [7440-30-4]
Thurfyl Nicotinate [70-19-9]
Thymidine [50-89-5]
Thymine [65-71-4]
Thymol [89-83-8]
Thymol Acetate* [528-79-0]
Thymol Blue [76-61-9]
Thymol Carbonate* [552-93-2]
Thymolphthalein [125-20-2]
Thymomodulin [90803-92-2]
Thymopentin [69558-55-0]
Thymostimulin [117149-90-3]
o-Thymotic Acid [548-51-6]
Thymyl N-Isoamylcarbamate
   [578-20-1]
Thyropropic Acid [51-26-3]
Thyroprotein [9005-97-4]
Thyroxine [7488-70-2]
D-Thyroxine* [51-49-0]
DL-Thyroxine* [300-30-1]
L-Thyroxine* [51-48-9]
D-Thyroxine Sodium Salt* [137-53-1]
Tiadenol [6964-20-1]
Tiagabine [115103-54-3]
Tiagabine Hydrochloride*
   [145821-59-6]
Tiamenidine [31428-61-2]
Tiamenidine Hydrochloride*
   [51274-83-0]
Tiamulin [55297-95-5]
Tiamulin Fumarate* [55297-96-6]
Tianeptine [66981-73-5]
Tianeptine Sodium Salt* [30123-17-2]
Tiapride [51012-32-9]
Tiapride Hydrochloride* [51012-33-0]
Tiaprofenic Acid [33005-95-7]
Tiaprost [71116-82-0]
Tiaramide [32527-55-2]
Tiaramide Hydrochloride*
   [35941-71-0]
Tibezonium Iodide [54663-47-7]
Tibolone [5630-53-5]
Ticarbodine [31932-09-9]
Ticarcillin [34787-01-4]
Ticarcillin Disodium Salt* [4697-14-7]
Ticlopidine [55142-85-3]
Ticlopidine Hydrochloride*
   [53885-35-1]
Ticrynafen [40180-04-9]
Tiemonium Iodide [144-12-7]
Tigemonam [102507-71-1]
Tigemonam Dicholine Salt*
   [102916-21-2]
Tiglic Acid [80-59-1]
Tigloidine [495-83-0]
Tigogenin [77-60-1]
Tigonin [1329-83-5]
Tiliacorine [27073-72-9]
Tilidine [20380-58-9]
Tilidine cis-(±)-Form* [112244-09-4]
Tilidine trans-(±)-Hydrochloride*
   [27107-79-5]
Tilisolol [85136-71-6]
Tilisolol Hydrochloride* [62774-96-3]
Tilmicosin [108050-54-0]
Tilorone [27591-97-5]
Tilorone Dihydrochloride*
   [27591-69-1]
Tiludronic Acid [89987-06-4]
Timepidium Bromide [35035-05-3]
Timiperone [57648-21-2]
Timolol [26839-75-8]
Timolol Hydrogen Maleate Salt*
   [26921-17-5]
Timonacic [444-27-9]
Tin [7440-31-5]

Tinidazole [19387-91-8]
Tinofedrine [66788-41-8]
Tinofedrine Hydrochloride*
   [50776-39-1]
Tinoridine [24237-54-5]
Tinoridine Hydrochloride*
   [25913-34-2]
Tiocarlide [910-86-1]
Tioclomarol [22619-35-8]
Tioconazole [65899-73-2]
Tiomesterone [2205-73-4]
Tiopronin [1953-02-2]
Tiotropium Bromide [136310-93-5]
Tioxidazole [61570-90-9]
Tioxolone [4991-65-5]
Tioxolone Combination with Hydro-
   cortisone* [8064-73-1]
Tipepidine [5169-78-8]
Tiquizium Bromide [71731-58-3]
Tiratricol [51-24-1]
Tirilazad [110101-66-1]
Tirilazad Monomethanesulfonate
   Monohydrate* [111793-42-1]
Tirofiban [144494-65-5]
Tirofiban Hydrochloride*
   [142373-60-2]
Tirofiban Hydrochloride Mono-
   hydrate* [150915-40-5]
Tiron [149-45-1]
Tiropramide [55837-29-1]
Tiropramide Hydrochloride*
   [53567-47-8]
Tissue Plasminogen Activator
   (Recombinant Human)*
   [105857-23-6]
Titanic(IV) Acid [20338-08-3]
Titanium [7440-32-6]
Titanium Dichloride [10049-06-6]
Titanium Dioxide [13463-67-7]
Titanium Hydride [7704-98-5]
Titanium Isopropylate [546-68-9]
Titanium Sesquisulfate [10343-61-0]
Titanium Sulfate [13825-74-6]
Titanium Tetrabromide [7789-68-6]
Titanium Tetrachloride [7550-45-0]
Titanium Tetrafluoride [7783-63-3]
Titanium Trichloride [7705-07-9]
Titanocene Dichloride [1271-19-8]
Tixocortol [61951-99-3]
Tixocortol 21-Pivalate* [55560-96-8]
Tizanidine [51322-75-9]
TMD [60761-10-6]
TNAZ [97645-24-4]
Tobramycin [32986-56-4]
Tobramycin Sulfate* [79645-27-5]
Tocainide [41708-72-9]
Tocainide (±)-Form Hydrochloride*
   [71395-14-7]
Tocainide R-(−)-Form Hydrochlo-
   ride* [53984-74-0]
Tocainide S-(+)-Form Hydrochloride*
   [53984-76-2]
Tocamphyl [5634-42-4]
Tocol [119-98-2]
β-Tocopherol [148-03-8]
γ-Tocopherol [7616-22-0]
δ-Tocopherol [119-13-1]
ε-Tocopherol [490-23-3]
ζ₁-Tocopherol [1721-51-3]
ζ₂-Tocopherol [493-35-6]
η-Tocopherol [91-86-1]
Tocoretinate [40516-48-1]
Todralazine [14679-73-3]
Todralazine Hydrochloride*
   [3778-76-5]
Tofenacin [15301-93-6]
Tofenacin Hydrochloride*
   [10488-36-5]
Tofisopam [22345-47-7]
Tolan [501-65-5]

Tolazamide [1156-19-0]
Tolazoline [59-98-3]
Tolazoline Hydrochloride* [59-97-2]
Tolbutamide [64-77-7]
Tolbutamide Sodium Salt* [473-41-6]
Tolciclate [50838-36-3]
Tolcyclamide [664-95-9]
Toldimfos Sodium [575-75-7]
Toldimfos Sodium Trihydrate*
   [5787-63-3]
Tolfenamic Acid [13710-19-5]
o-Tolidine [119-93-7]
Tolindate [27877-51-6]
Toliprolol [2933-94-0]
Toliprolol Hydrochloride* [306-11-6]
Tolmetin [26171-23-3]
Tolmetin Sodium Salt Dihydrate*
   [64490-92-2]
Tolnaftate [2398-96-1]
Tolonidine [4201-22-3]
Tolonidine Nitrate* [57524-15-9]
Tolonium Chloride [92-31-9]
Toloxatone [29218-27-7]
Tolperisone [728-88-1]
Tolperisone Hydrochloride*
   [3644-61-9]
Tolpropamine [5632-44-0]
Tolrestat [82964-04-3]
Toltrazuril [69004-03-1]
o-Tolualdehyde [529-20-4]
o-Toluamide [527-85-5]
Toluene [108-88-3]
Toluene 2,4-Diisocyanate [584-84-9]
Toluene-3,4-dithiol [496-74-2]
p-Toluenesulfinic Acid [536-57-2]
p-Toluenesulfonic Acid [104-15-4]
p-Toluenesulfonyl Chloride [98-59-9]
m-Toluidine* [108-44-1]
o-Toluidine* [95-53-4]
p-Toluidine* [106-49-0]
o-Tolunitrile [529-19-1]
p-Tolunitrile [104-85-8]
2-(p-Toluyl)benzoic Acid [85-55-2]
Toluylene Blue [97-26-7]
Tolycaine [3686-58-6]
Tolycaine Hydrochloride* [7210-92-6]
p-Tolylsulfonylmethylnitrosamide
   [80-11-5]
Tomatidine [77-59-8]
Tomatine [17406-45-0]
Tomudex [112887-68-0]
Tonin [53414-68-9]
Topiramate [97240-79-4]
Topotecan [123948-87-8]
Topotecan Monohydrochloride*
   [119413-54-6]
Toremifene [89778-26-7]
Toremifene Citrate* [89778-27-8]
Torsemide [56211-40-6]
Torularhodin [514-92-1]
Tosufloxacin [100490-36-6] (unspeci-
   fied)
Tosufloxacin [120382-07-2] (±)
Tosufloxacin Hydrochloride*
   [104051-69-6]
Toxaphene [8001-35-2]
Toxiferine I [6888-23-9]
Toxoflavin [84-82-2]
Toxohormone [9014-44-2]
Toxopyrimidine [73-67-6]
Toyocamycin [606-58-6]
Tralkoxydim [87820-88-0]
Tralomethrin [66841-25-6]
Tramadol [27203-92-5]
Tramadol Hydrochloride*
   [22204-88-2]
Tramazoline [1082-57-1]
Tramazoline Hydrochloride
   Monohydrate* [3715-90-0]
Trandolapril [87679-37-6]

Denotes derivative of title compound

Trandolaprilat* [87679-71-8]
Tranexamic Acid [1197-18-8]
Tranexamic Acid cis-Form*
  [1197-17-7]
Tranilast [53902-12-8]
Tranylcypromine [155-09-9]
Tranylcypromine Sulfate*
  [13492-01-8]
Trapidil [15421-84-8]
Traumatic Acid [6402-36-4]
Traxanox [58712-69-9]
Trazodone [19794-93-5]
Trazodone Hydrochloride*
  [25332-39-2]
Trehalose [99-20-7]
Tremetone [4976-25-4]
Tremorine [51-73-0]
Trenbolone [10161-33-8]
Trenbolone Acetate* [10161-34-9]
Trenbolone Cyclohexylmethyl-
  carbonate* [23454-33-3]
Trengestone [5192-84-7]
Trepibutone [41826-92-0]
Tretoquinol [21650-42-0]
Tretoquinol dl-Form Hydrochloride*
  [18559-63-2]
Tretoquinol l-Form Hydrochloride*
  [18559-59-6]
TRH [24305-27-9]
TRH Tartrate* [54974-54-8]
Triacetin [102-76-1]
1-Triacontanol [593-50-0]
Triadimefon [43121-43-3]
Triadimenol [55219-65-3]
Triafur [712-68-5]
Triallate [2303-17-5]
Triamcinolone [124-94-7]
Triamcinolone Acetonide [76-25-5]
Triamcinolone Acetonide 21-Disodium
  Phosphate* [1997-15-5]
Triamcinolone Benetonide
  [31002-79-6]
Triamcinolone 16,21-Diacetate*
  [67-78-7]
Triamcinolone Hexacetonide
  [5611-51-8]
Triamterene [396-01-0]
Triasulfuron [82097-50-5]
Triaziquone [68-76-8]
Triazolam [28911-01-5]
1H-1,2,4-Triazole [288-88-0]
Triazophos [24017-47-8]
Triazoxide [72459-58-6]
Tribenoside [10310-32-4]
Tribromoacetic Acid [75-96-7]
2,4,6-Tribromoaniline [147-82-0]
Tribromo-tert-butyl Alcohol [76-08-4]
2,4,6-Tribromo-m-cresol [4619-74-3]
Tribromoethanol [75-80-9]
2,4,6-Tribromophenol [118-79-6]
1,2,3-Tribromopropane [96-11-7]
Tribromosilane [7789-57-3]
Tribromsalan [87-10-5]
Tributylamine [102-82-9]
Tributyl Phosphate [126-73-8]
Tributyrin [60-01-5]
Tricaine [886-86-2]
Tricarballylic Acid [99-14-9]
Trichlorfon [52-68-6]
Trichlormethiazide [133-67-5]
Trichloroacetaldehyde [75-87-6]
Trichloroacetic Acid [76-03-9]
Trichloroacetic Acid Ethyl Ester*
  [515-84-4]
Trichloroacetic Acid Sodium Salt*
  [650-51-1]
Trichloroacetonitrile [545-06-2]
2,4,6-Trichloroanisole [87-40-1]
1,2,3-Trichlorobenzene [87-61-6]
1,2,4-Trichlorobenzene [120-82-1]

1,3,5-Trichlorobenzene [108-70-3]
α,α,β-Trichloro-n-butyraldehyde
  [76-36-8]
2,3,6-Trichloro-p-cresol [551-77-9]
2,4,6-Trichloro-m-cresol [551-76-8]
4,5,6-Trichloro-o-cresol [551-78-0]
1,1,1-Trichloroethane [71-55-6]
1,1,2-Trichloroethane [79-00-5]
2,2,2-Trichloroethanol [115-20-8]
Trichloroethylene [79-01-6]
Trichlorofluoromethane [75-69-4]
3,4,6-Trichloro-2-nitrophenol
  [82-62-2]
2,4,5-Trichlorophenol [95-95-4]
2,4,6-Trichlorophenol [88-06-2]
1,1,1-Trichloro-2-propanol [76-00-6]
3',4',5-Trichlorosalicylanilide
  [642-84-2]
Trichlorosilane [10025-78-2]
2,2',2''-Trichlorotriethylamine
  [555-77-1]
2,2',2''-Trichlorotriethylamine
  Hydrochloride* [817-09-4]
Trichlorourethan [107-69-7]
Trichodermin [4682-50-2]
Trichosanthin [60318-52-7]
Trichostatin A* [58880-19-6]
Trichostatin B* [58895-00-4]
Trichostatin C* [68676-88-0]
Trichothecin [6379-69-7]
Tricine [5704-04-1]
Triclabendazole [68786-66-3]
Triclobisonium Chloride [79-90-3]
Triclocarban [101-20-2]
Triclofenol Piperazine [5714-82-9]
Triclofos [306-52-5]
Triclofos Monosodium Salt*
  [7246-20-0]
Triclopyr [55335-06-3]
Triclosan [3380-34-5]
Tricromyl [85-90-5]
Tridecylbenzene [123-02-4]
Tridemorph [24602-86-6]
Tridihexethyl Iodide [125-99-5]
Tridiphane [58138-08-2]
Trientine [112-24-3]
Trientine Dihydrochloride*
  [38260-01-4]
Trietazine [1912-26-1]
Triethanolamine [102-71-6]
Triethylamine [121-44-8]
Triethylammonium Formate
  [585-29-5]
Triethylenediamine [280-57-9]
Triethylene Glycol [112-27-6]
Triethylenemelamine [51-18-3]
Triethylenephosphoramide [545-55-1]
Triethylenethiophosphoramide
  [52-24-4]
Triethyl Phosphate [78-40-0]
Triethyl Phosphine [554-70-1]
Trifenmorph [1420-06-0]
Triflumuron [64628-44-0]
Trifluomeprazine [2622-37-9]
Trifluoperazine [117-89-5]
Trifluoperazine Dihydrochloride*
  [440-17-5]
Trifluoracetic Acid [76-05-1]
Trifluperidol [749-13-3]
Trifluperidol Hydrochloride*
  [2062-77-3]
Triflupromazine [146-54-3]
Triflupromazine Hydrochloride*
  [1098-60-8]
Trifluralin [1582-09-8]
Trifluridine [70-00-8]
Triflusal [322-79-2]
Triforine [26644-46-2]
Triglyme [112-49-2]
Trigonellamide Chloride [1005-24-9]

Trigonelline [535-83-1]
Trihexyphenidyl Hydrochloride
  [52-49-3]
Trihydrazine Dihydriodide
  [10034-92-1]
Trilobine [6138-73-4]
Trilostane [13647-35-3]
Trimazosin [35795-16-5]
Trimazosin Hydrochloride Mono-
  hydrate* [53746-46-6]
Trimebutine [39133-31-8]
Trimebutine Maleate* [34140-59-5]
Trimecaine [616-68-2]
Trimedlure [12002-53-8]
Trimellitic Acid [528-44-9]
Trimellitic Anhydride [552-30-7]
Trimeprazine [84-96-8]
Trimeprazine Tartrate* [4330-99-8]
Trimetazidine [5011-34-7]
Trimetazidine Dihydrochloride*
  [13171-25-0]
Trimethadione [127-48-0]
Trimethaphan Camsylate [68-91-7]
Trimethidinium Methosulfate
  [14149-43-0]
Trimethobenzamide [138-56-7]
Trimethobenzamide Hydrochloride*
  [554-92-7]
Trimethoprim [738-70-5]
Trimethylamine [75-50-3]
Trimethyl Borate [121-43-7]
N,N-1-Trimethyl-3,3-diphenylpropyl-
  amine [13957-55-6]
Trimethylene Bromide [109-64-8]
Trimethylene Glycol [504-63-2]
Trimethylene Oxide [503-30-0]
Trimethyl Isopropyl Butanamide
  [51115-67-4]
2,4,6-Trimethylpyridine [108-75-8]
Trimethylsilyl Triflate [27607-77-8]
Trimetozine [635-41-6]
Trimetrexate [52128-35-5]
Trimetrexate D-Glucuronate*
  [82952-64-5]
Trimetrexate Monoacetate Mono-
  hydrate* [117381-09-6]
Trimipramine [739-71-9]
Trimipramine Maleate* [521-78-8]
Trimoprostil [69900-72-7]
Trimyristin [555-45-3]
Trinexapac-ethyl [95266-40-3]
sym-Trinitrobenzene [99-35-4]
2,4,6-Trinitrobenzoic Acid [129-66-8]
2,4,7-Trinitrofluorenone [129-79-3]
Trinitromethane [517-25-9]
2,4,6-Trinitrotoluene [118-96-7]
Triolein [122-32-7]
s-Trioxane [110-88-3]
Trioxsalen [3902-71-4]
Tripalmitin [555-44-2]
Tripamide [73803-48-2]
Triparanol [78-41-1]
Tripelennamine [91-81-6]
Tripelennamine Citrate* [6138-56-3]
Tripelennamine Hydrochloride*
  [154-69-8]
Triphenylcarbinol [76-84-6]
Triphenylene [217-59-4]
Triphenylmethane [519-73-3]
Triphenyl Phosphate [115-86-6]
Triphenylphosphine [603-35-0]
Triphenyltetrazolium Chloride
  [298-96-4]
Triphenyltin Hydroxide [76-87-9]
Tripoli [1317-95-9]
Triprolidine [486-12-4]
Triprolidine Hydrochloride Mono-
  hydrate* [6138-79-0]
Triptorelin [57773-63-4]
Triptorelin Acetate* [140194-24-7]

* Denotes derivative of title compound

Denotes derivative of title compound

Vermiculite [1318-00-9]
Vernamycin B [9040-14-6]
Vernamycin B$_\beta$* [57206-54-9]
Vernamycin B$_\gamma$* [28979-74-0]
Vernamycin B$_\delta$* [29139-17-1]
Vernolate [1929-77-7]
Vernolepin [18542-37-5]
Vernolepin (±)-Form* [59598-29-7]
Vernolic Acid (−)-Form*
[32381-42-3]
Vernolic Acid (+)-Form* [503-07-1]
Verrucarin A* [3148-09-2]
Verrucarin B* [2290-11-1]
Verrucarin J* [4643-58-7]
Verrucarin K* [63739-93-5]
Versalide [88-29-9]
Versen-Ol [139-89-9]
Verticillin A* [32164-16-2]
Verticillin B* [52212-86-9]
Verticine [23496-41-5]
Vesnarinone [81840-15-5]
α-Vetivone* [15764-04-2]
β-Vetivone* [18444-79-6]
Vetrabutine [3735-45-3]
Vetrabutine Hydrochloride*
[5974-09-4]
Vicianin [155-57-7]
Vicianose [14116-69-9]
Vicine [152-93-2]
Vidarabine [5536-17-4]
Vigabatrin [60643-86-9]
Villikinin [9056-49-9]
Viloxazine [46817-91-8]
Viloxazine Hydrochloride*
[35604-67-2]
Viminol [21363-18-8]
Viminol p-Hydroxybenzoate*
[23784-10-3]
Vinbarbital Sodium [125-44-0]
Vinblastine [865-21-4]
Vinblastine Sulfate* [143-67-9]
Vincamine [1617-90-9]
Vincamine Hydrochloride*
[10592-03-7]
Vincetoxin [1401-78-1]
Vinclozolin [50471-44-8]
Vinconate [70704-03-9]
Vinconate Monohydrochloride*
[119600-43-0]
Vincristine [57-22-7]
Vincristine Sulfate* [2068-78-2]
Vindesine [53643-48-4]
Vindesine Sulfate Salt* [59917-39-4]
Vindoline [2182-14-1]
Vinorelbine [71486-22-1]
Vinpocetine [42971-09-5]
Vintiamol [26242-33-1]
Vinyl Acetate [108-05-4]
Vinylbital [2430-49-1]
Vinyl Chloride [75-01-4]
Vinyl Ether [109-93-3]
Vinylidene Chloride [75-35-4]
Violacein [548-54-9]
Violaxanthin [126-29-4]
Violuric Acid [87-39-8]
Viomycin [32988-50-4]
Viomycin Pantothenate Sulfate*
[1401-79-2]
Viomycin Sulfate* [37883-00-4]
VIP [37221-79-7]
Viquidil [84-55-9]
Viquidil Hydrochloride* [52211-63-9]
Virginiamycin M$_1$* [21411-53-0]
Virginiamycin S$_1$* [23152-29-6]
Viridicatin [129-24-8]
Viridin [3306-52-3]
Viscosin [27127-62-4]
Visnadine [477-32-7]
Visnagin [82-57-5]
Vital Red [574-65-2]

Vitamin A [68-26-8]
Vitamin A Acetate* [127-47-9]
Vitamin A Palmitate* [79-81-2]
Vitamin A$_2$ [79-80-1]
Vitamin B$_{12}$ [68-19-9]
Vitamin B$_{12r}$ [14463-33-3]
Vitamin B$_{12s}$* [18534-66-2]
Vitamin B$_{12}$-^{57}Co* [13115-03-2]
Vitamin B$_{12}$-^{58}Co* [18195-32-9]
Vitamin B$_{12}$-^{60}Co* [13422-53-2]
Vitamin D$_1$ [520-91-2]
Vitamin D$_2$ [50-14-6]
Vitamin D$_3$ [67-97-0]
Vitamin D$_4$ [511-28-4]
Vitamin E [59-02-9]
Vitamin E Acetate d-Form* [58-95-7]
Vitamin E Acetate dl-Form*
[52225-20-4]
Vitamin E Nicotinate* [43119-47-7]
Vitamin E Succinate* [4345-03-3]
Vitamin K$_5$ [83-70-5]
Vitamin K$_5$ N-Acetyl Analog*
[523-68-2]
Vitamin K$_5$ Hydrochloride* [130-24-5]
Vitamin T [1407-73-4]
Vitamin U [1115-84-0]
Vitamin U Bromide Analog*
[33515-32-1]
Vitride [22722-98-1]
Voacamine [3371-85-5]
Vomicine [125-15-5]
Vomitoxin [51481-10-8]
VX [50782-69-9]
Warburganal [62994-47-2]
Warfarin [81-81-2]
Warfarin Sodium Salt* [129-06-6]
Water [7732-18-5]
Wieland-Gumlich Aldehyde [466-85-3]
Wildfire Toxin [32190-57-1]
Withaferin A [5119-48-2]
Woodward's Reagent K [4156-16-5]
Worenine [38763-29-0]
Wortmannin [19545-26-7]
Xamoterol [81801-12-9]
Xamoterol Hemifumarate*
[73210-73-8]
Xanomeline [131986-45-3]
Xanomeline Oxalate* [141064-23-5]
Xanomeline Tartrate* [152854-19-8]
Xanthan Gum [11138-66-2]
Xanthatin [26791-73-1]
Xanthine [69-89-6]
Xanthinol Niacinate [437-74-1]
Xanthocillin [11042-38-9]
Xanthocillin X* [580-74-5]
Xanthocillin Y$_1$* [38965-69-4]
Xanthocillin Y$_2$* [38965-70-7]
Xanthone [90-47-1]
Xanthophyll [127-40-2]
Xanthopterin [119-44-8]
Xanthosine [146-80-5]
Xanthoxyletin [84-99-1]
Xanthoxylin [90-24-4]
Xanthurenic Acid [59-00-7]
Xanthyletin [553-19-5]
Xenazoic Acid [1174-11-4]
Xenbucin [959-10-4]
Xenbucin trans-4-Phenylcyclohexyl-
amine Salt (1:1)* [19992-80-4]
Xenon [7440-63-3]
Xenon-133* [14932-42-4]
Xenon Difluoride* [13709-36-9]
Xenon Hexafluoride* [13693-09-9]
Xenon Hydrate* [60212-94-4]
Xenon Tetrafluoride* [13709-61-0]
Xenon Trioxide* [13776-58-4]
p-Xenylcarbimide* [92-95-5]
Xenytropium Bromide [511-55-7]
Xibenolol [81584-06-7]

Xibenolol Hydrochloride*
[59708-57-5]
Xibornol [13741-18-9]
Ximoprofen [56187-89-4]
Xipamide [14293-44-8]
Xylazine [7361-61-7]
Xylazine Hydrochloride* [23076-35-9]
Xylene [1330-20-7]
Xylenol [1300-71-6]
Xylenol Blue [125-31-5]
Xylidine [1300-73-8]
Xylitol [87-99-0]
Xylometazoline [526-36-3]
Xylometazoline Hydrochloride*
[1218-35-5]
Xylose [58-86-6]
Xylulose [5962-29-8]
1-Xylylazo-2-naphthol [85-82-5]
m-Xylyl Bromide* [620-13-3]
o-Xylyl Bromide* [89-92-9]
p-Xylyl Bromide* [104-81-4]
m-Xylyl Chloride* [620-19-9]
o-Xylyl Chloride* [552-45-4]
p-Xylyl Chloride* [104-82-5]
Yangonin [500-62-9]
Yellow AB [85-84-7]
Yellow OB [131-79-3]
Yellow Phenolphthalein [8053-05-2]
Yig [12063-56-8]
Ylangene [14912-44-8]
Yohimbine [146-48-5]
allo-Yohimbine [522-94-1]
α-Yohimbine [131-03-3]
Yohimbine Hydrochloride* [65-19-0]
Ytterbium [7440-64-4]
Yttrium [7440-65-5]
Zafirlukast [107753-78-6]
Zalcitabine [7481-89-2]
Zaldaride [109826-26-8]
Zaldaride Maleate* [109826-27-9]
Zaltoprofen [74711-43-6] (unspecified)
Zaltoprofen [89482-00-8] (±)
Zaltoprofen (S)-Form* [89482-01-9]
Zatebradine [85175-67-3]
Zatebradine Hydrochloride*
[91940-87-3]
Zearalenone [17924-92-4]
Zeatin [1637-39-4]
Zeaxanthin [144-68-3]
Zeranol [55331-29-8]
Zidovudine [30516-87-1]
Zileuton [111406-87-2]
Zimeldine [56775-88-3]
Zimeldine Dihydrochloride Mono-
hydrate* [61129-30-4]
Zinc [7440-66-6]
Zinc Acetate [557-34-6]
Zinc Bromide [7699-45-8]
Zinc Caprylate [557-09-5]
Zinc Carbonate [3486-35-9]
Zinc Chloride [7646-85-7]
Zinc Chromate(VI) Hydroxide
[13530-65-9]
Zinc Citrate [546-46-3]
Zinc Cyanide [557-21-1]
Zinc Fluoride [7783-49-5]
Zinc Formate [557-41-5]
Zinc Hexafluorosilicate [16871-71-9]
Zinc Insulin* [8049-62-5]
Zinc Iodate [7790-37-6]
Zinc Iodide [10139-47-6]
Zinc Iodide-Starch [9010-71-3]
Zinc Lactate [16039-53-5]
Zinc Meta-arsenite [10326-24-6]
Zinc Nitrate [7779-88-6]
Zinc Nitride [1313-49-1]
Zinc Nitrite [10102-02-0]
Zinc Oleate [557-07-3]
Zinc Ortho-arsenate [13464-44-3]
Zinc Oxalate [547-68-2]

Denotes derivative of title compound

Zinc Oxide [1314-13-2]
Zinc Perchlorate [13637-61-1]
Zinc Permanganate [23414-72-4]
Zinc Peroxide [1314-22-3]
Zinc p-Phenolsulfonate [127-82-2]
Zinc Phosphate [7779-90-0]
Zinc Phosphide [1314-84-7]
Zinc Propionate [557-28-8]
Zinc Pyrophosphate [7446-26-6]
Zinc Salicylate [16283-36-6]
Zinc Selenate [13597-54-1]
Zinc Selenide [1315-09-9]
Zinc Silicate [13597-65-4]
Zinc Stearate [557-05-1]
Zinc Sulfate [7733-02-0]
Zinc Sulfate Heptahydrate*
 [7446-20-0]
Zinc Sulfate Monohydrate*
 [7446-19-7]
Zinc Sulfide [1314-98-3]
Zinc Tannate [8011-65-2]
Zinc Tartrate [551-64-4]

Zinc Telluride [1315-11-3]
Zinc Thiocyanate [557-42-6]
Zinc Valerate [556-38-7]
Zineb [12122-67-7]
Zingerone [122-48-5]
Zinostatin [9014-02-2]
Zipeprol [34758-83-3]
Zipeprol Dihydrochloride*
 [34758-84-4]
Ziprasidone [146939-27-7]
Ziprasidone Hydrochloride
 Monohydrate* [138982-67-9]
Ziram [137-30-4]
Zirconium [7440-67-7]
Zirconium Chloride [10026-11-6]
Zirconium Fluoride [7783-64-4]
Zirconium Hydride [7704-99-6]
Zirconium Hydroxide [14475-63-9]
Zirconium Iodide [13986-26-0]
Zirconium Nitrate [13746-89-9]
Zirconium Oxide [1314-23-4]
Zirconium Silicate [10101-52-7]

Zirconium Sulfate [14644-61-2]
Zirconyl Acetate [5153-24-2]
Zirconyl Chloride [7699-43-6]
Zoapatanol [71117-51-6]
Zolamine [553-13-9]
Zolamine Hydrochloride* [1155-03-9]
Zolimidine [1222-57-7]
Zolpidem [82626-48-0]
Zolpidem L-(+)-Hemitartrate*
 [99294-93-6]
Zomepirac [33369-31-2]
Zomepirac Sodium Salt Dihydrate*
 [64092-49-5]
Zonisamide [68291-97-4]
Zopiclone [43200-80-2]
Zopolrestat [110703-94-1]
Zorubicin [54083-22-6]
Zorubicin Hydrochloride*
 [36508-71-1]
Zotepine [26615-21-4]
Zoxazolamine [61-80-3]
Zygadenine [545-45-9]
Zymosan [9010-72-4]

[50-00-0] Formaldehyde, *4262*
[50-02-2] Dexamethasone, *2986*
[50-03-3] Hydrocortisone 21-Acetate*, *4828*
[50-04-4] Cortisone 21-Acetate*, *2602*
[50-06-6] Phenobarbital, *7386*
[50-07-7] Mitomycin C*, *6301*
[50-09-9] Hexobarbital Sodium Salt*, *4742*
[50-10-2] Oxyphenonium Bromide, *7109*
[50-11-3] Metharbital, *6029*
[50-12-4] Mephenytoin, *5898*
[50-13-5] Meperidine Hydrochloride*, *5894*
[50-14-6] Vitamin D$_2$, *10156*
[50-18-0] Cyclophosphamide (anhydrous form), *2816*
[50-19-1] Hydroxyphenamate, *4884*
[50-22-6] Corticosterone, *2601*
[50-23-7] Hydrocortisone, *4828*
[50-24-8] Prednisolone, *7901*
[50-27-1] Estriol, *3750*
[50-28-2] Estradiol, *3746*
[50-29-3] DDT, *2898*
[50-32-8] Benzo[a]pyrene, *1134*
[50-33-9] Phenylbutazone, *7431*
[50-34-0] Propantheline Bromide, *7989*
[50-35-1] Thalidomide, *9390*
[50-36-2] Cocaine, *2517*
[50-37-3] Lysergide, *5665*
[50-39-5] Protheobromine, *8072*
[50-41-9] Clomiphene Citrate*, *2446*
[50-42-0] Adiphenine Hydrochloride, *160*
[50-44-2] 6-Mercaptopurine, *5919*
[50-47-5] Desipramine, *2966*
[50-48-6] Amitriptyline, *511*
[50-49-7] Imipramine, *4955*
[50-50-0] Estradiol Benzoate*, *3746*
[50-52-2] Thioridazine, *9497*
[50-53-3] Chlorpromazine, *2238*
[50-54-4] Quinidine Sulfate*, *8244*
[50-55-5] Reserpine, *8314*
[50-56-6] Oxytocin, *7114*
[50-57-7] Lypressin, *5661*
[50-58-8] Phendimetrazine Bitartrate*, *7365*
[50-59-9] Cephaloridine, *2025*
[50-60-2] Phentolamine, *7417*
[50-62-4] Hexobendine Dihydrochloride*, *4743*
[50-63-5] Chloroquine Diphosphate*, *2215*
[50-65-7] Niclosamide, *6602*
[50-67-9] Serotonin, *8607*
[50-69-1] D-Ribose, *8371*
[50-70-4] Sorbitol, *8873*
[50-71-5] Alloxan, *290*
[50-76-0] Dactinomycin, *2867*
[50-78-2] Aspirin, *886*
[50-81-7] Ascorbic Acid, *867*
[50-85-1] m-Cresotic Acid, *2648*
[50-89-5] Thymidine, *9538*
[50-91-9] Floxuridine, *4148*
[50-98-6] Ephedrine l-Form Hydrochloride*, *3645*
[50-99-7] Glucose, *4467*

[51-02-5] Pronethalol Hydrochloride*, *7974*
[51-03-6] Piperonyl Butoxide, *7629*
[51-05-8] Procaine Hydrochloride*, *7937*
[51-12-7] Nialamide, *6575*
[51-14-9] Sesamex, *8615*
[51-15-0] Pralidoxime Chloride, *7884*
[51-17-2] Benzimidazole, *1111*
[51-18-3] Triethylenemelamine, *9803*
[51-20-7] 5-Bromouracil, *1464*
[51-21-8] Fluorouracil, *4219*
[51-24-1] Tiratricol, *9603*
[51-26-3] Thyropropic Acid, *9553*
[51-28-5] 2,4-Dinitrophenol, *3333*
[51-34-3] Scopolamine, *8550*
[51-35-4] Hydroxyproline, *4887*
[51-41-2] Norepinephrine, *6788*
[51-42-3] Epinephrine Bitartrate*, *3656*
[51-43-4] Epinephrine, *3656*
[51-45-6] Histamine, *4756*
[51-48-9] L-Thyroxine*, *9555*
[51-49-0] D-Thyroxine*, *9555*
[51-52-5] Propylthiouracil, *8054*
[51-55-8] Atropine, *907*
[51-56-9] Homatropine DL-Hydrobromide*, *4766*
[51-57-0] Methamphetamine Hydrochloride*, *6015*
[51-60-5] Neostigmine Methyl Sulfate*, *6553*
[51-61-6] Dopamine, *3479*
[51-63-8] Dextroamphetamine Sulfate, *2996*
[51-66-1] p-Acetanisidine, *48*
[51-67-2] Tyramine, *9966*
[51-68-3] Meclofenoxate, *5820*
[51-71-8] Phenelzine, *7366*
[51-73-0] Tremorine, *9715*
[51-74-1] Histamine Phosphate*, *4756*
[51-75-2] Mechlorethamine, *5815*
[51-77-4] Gefarnate, *4385*
[51-79-6] Urethan, *10013*
[51-83-2] Carbachol, *1823*
[51-85-4] Cystamine, *2846*
[51-98-9] Norethindrone Acetate*, *6790*
[52-01-7] Spironolactone, *8917*
[52-21-1] Prednisolone 21-Acetate*, *7901*
[52-24-4] Triethylenethiophosphoramide, *9805*
[52-26-6] Morphine Hydrochloride, *6360*
[52-28-8] Codeine Phosphate, *2528*
[52-31-3] Cyclobarbital, *2780*
[52-39-1] Aldosterone, *226*
[52-43-7] Allobarbital, *261*
[52-46-0] Apholate, *770*
[52-49-3] Trihexyphenidyl Hydrochloride, *9823*
[52-51-7] Bronopol, *1470*
[52-52-8] Cycloleucine, *2803*
[52-53-9] Verapamil, *10083*
[52-62-0] Pentolinium Tartrate, *7274*
[52-66-4] Penicillamine DL-Form*, *7214*
[52-67-5] Penicillamine, *7214*

[52-68-6] Trichlorfon, *9753*
[52-76-6] Lynestrenol, *5659*
[52-78-8] Norethandrolone, *6789*
[52-85-7] Famphur, *3973*
[52-86-8] Haloperidol, *4629*
[52-88-0] Atropine Methylnitrate*, *907*
[52-89-1] Cysteine Hydrochloride*, *2850*
[52-90-4] Cysteine, *2850*
[53-03-2] Prednisone, *7904*
[53-05-4] Tetrahydrocortisone, *9350*
[53-06-5] Cortisone, *2602*
[53-10-1] Hydroxydione Sodium, *4868*
[53-16-7] Estrone, *3751*
[53-18-9] Bietaserpine, *1254*
[53-19-0] Mitotane, *6302*
[53-31-6] Medibazine, *5831*
[53-33-8] Paramethasone, *7162*
[53-34-9] Fluprednisolone, *4229*
[53-36-1] Methylprednisolone 21-Acetate*, *6189*
[53-39-4] Oxandrolone, *7054*
[53-41-8] Androsterone, *680*
[53-43-0] Prasterone, *7892*
[53-46-3] Methantheline Bromide, *6025*
[53-57-6] NADPH*, *6433*
[53-59-8] NADP, *6433*
[53-60-1] Promazine Hydrochloride*, *7966*
[53-70-3] 1,2:5,6-Dibenzanthracene, *3054*
[53-73-6] Angiotensin Amide*, *689*
[53-79-2] Puromycin, *8130*
[53-84-9] NAD, *6429*
[53-86-1] Indomethacin, *4998*
[53-89-4] Benzpiperylon, *1152*
[53-96-3] N-2-Fluorenylacetamide, *4192*
[54-03-5] Hexobendine, *4743*
[54-04-6] Mescaline, *5965*
[54-05-7] Chloroquine, *2215*
[54-06-8] Adrenochrome, *171*
[54-11-5] Nicotine, *6611*
[54-21-7] Sodium Salicylate, *8819*
[54-25-1] 6-Azauridine, *937*
[54-30-8] Camylofine, *1784*
[54-31-9] Furosemide, *4331*
[54-32-0] Moxisylyte, *6374*
[54-36-4] Metyrapone, *6246*
[54-42-2] Idoxuridine, *4934*
[54-47-7] Pyridoxal 5-Phosphate, *8163*
[54-49-9] Metaraminol, *5993*
[54-62-6] Aminopterin, *493*
[54-64-8] Thimerosal, *9451*
[54-71-7] Pilocarpine Hydrochloride*, *7578*
[54-80-8] Pronethalol, *7974*
[54-85-3] Isoniazid, *5203*
[54-91-1] Pipobroman, *7634*
[54-92-2] Iproniazid, *5094*
[54-95-5] Pentylenetetrazole, *7283*
[54-96-6] 3,4-Diaminopyridine, *3029*
[55-03-8] Levothyroxine Sodium, *5497*
[55-06-1] Liothyronine Sodium Salt*, *5535*
[55-10-7] Vanilmandelic Acid, *10070*
[55-18-5] N-Nitrosodiethylamine, *6734*

Denotes derivative of title compound

[55-21-0] Benzamide, *1087*
[55-22-1] Isonicotinic Acid, *5204*
[55-31-2] Epinephrine Hydrochloride*, *3656*
[55-38-9] Fenthion, *4044*
[55-43-6] N-(2-Chloroethyl)dibenzylamine Hydrochloride, *2190*
[55-52-7] Pheniprazine, *7382*
[55-55-0] p-Methylaminophenol Sulfate, *6097*
[55-56-1] Chlorhexidine, *2140*
[55-63-0] Nitroglycerin, *6704*
[55-63-0] Spirit of Glyceryl Trinitrate, *8913*
[55-65-2] Guanethidine, *4589*
[55-73-2] Bethanidine, *1233*
[55-86-7] Mechlorethamine Hydrochloride*, *5815*
[55-91-4] Isoflurophate, *5192*
[55-94-7] Succinylcholine Bromide, *9043*
[55-97-0] Hexamethonium Bromide*, *4724*
[55-98-1] Busulfan, *1529*
[56-03-1] Biguanide, *1261*
[56-04-2] Methylthiouracil, *6210*
[56-06-4] 2,4-Diamino-6-hydroxypyrimidine, *3024*
[56-10-0] AET, *176*
[56-12-2] γ-Aminobutyric Acid, *450*
[56-23-5] Carbon Tetrachloride, *1864*
[56-25-7] Cantharidin, *1797*
[56-29-1] Hexobarbital, *4742*
[56-34-8] Tetraethylammonium Chloride, *9341*
[56-38-2] Parathion, *7167*
[56-40-6] Glycine, *4500*
[56-41-7] Alanine, *205*
[56-45-1] Serine, *8604*
[56-47-3] Deoxycorticosterone Acetate, *2948*
[56-49-5] 3-Methylcholanthrene, *6124*
[56-53-1] Diethylstilbestrol, *3177*
[56-54-2] Quinidine, *8244*
[56-55-3] 1,2-Benzanthracene, *1089*
[56-59-7] Felypressin, *3992*
[56-65-5] Adenosine Triphosphate, *154*
[56-69-9] 5-Hydroxytryptophan, *4895*
[56-72-4] Coumaphos, *2626*
[56-73-5] Glucose-6-phosphate, *4469*
[56-75-7] Chloramphenicol, *2120*
[56-81-5] Glycerol, *4493*
[56-84-8] Aspartic Acid, *875*
[56-85-9] Glutamine, *4479*
[56-86-0] Glutamic Acid, *4477*
[56-87-1] Lysine, *5667*
[56-88-2] L-Cystathionine, *2847*
[56-89-3] Cystine, *2851*
[56-94-0] Demecarium Bromide, *2936*
[56-95-1] Chlorhexidine Diacetate*, *2140*
[57-00-1] Creatine, *2637*
[57-03-4] Glycerophosphoric Acid, *4495*
[57-06-7] Allyl Isothiocyanate, *303*
[57-08-9] ε-Acetamidocaproic Acid, *43*
[57-09-0] Cetrimonium Bromide, *2068*
[57-10-3] Palmitic Acid, *7128*
[57-11-4] Stearic Acid, *8959*
[57-13-6] Urea, *10005*
[57-14-7] 1,1-Dimethylhydrazine, *3296*
[57-15-8] Chlorobutanol, *2180*
[57-22-7] Vincristine, *10124*
[57-24-9] Strychnine, *9020*
[57-27-2] Morphine, *6359*
[57-29-4] Nalorphine Hydrochloride*, *6448*
[57-30-7] Phenobarbital Sodium Salt*, *7386*

[57-33-0] Pentobarbital Sodium Salt*, *7272*
[57-37-4] Benactyzine Hydrochloride*, *1055*
[57-39-6] Metepa, *5998*
[57-41-0] Phenytoin, *7475*
[57-42-1] Meperidine, *5894*
[57-43-2] Amobarbital, *607*
[57-44-3] Barbital, *989*
[57-47-6] Physostigmine, *7540*
[57-48-7] Fructose, *4295*
[57-50-1] Sucrose, *9051*
[57-53-4] Meprobamate, *5908*
[57-55-6] Propylene Glycol, *8040*
[57-57-8] β-Propiolactone, *8005*
[57-62-5] Chlortetracycline, *2245*
[57-63-6] Ethinyl Estradiol, *3780*
[57-64-7] Physostigmine Salicylate*, *7540*
[57-66-9] Probenecid, *7934*
[57-67-0] Sulfaguanidine, *9076*
[57-68-1] Sulfamethazine, *9083*
[57-74-9] Chlordan(e), *2129*
[57-83-0] Progesterone, *7956*
[57-85-2] Testosterone Propionate*, *9322*
[57-87-4] Ergosterol, *3701*
[57-88-5] Cholesterol, *2256*
[57-91-0] α-Estradiol, *3747*
[57-92-1] Streptomycin, *8983*
[57-94-3] Tubocurarine Chloride, *9939*
[57-96-5] Sulfinpyrazone, *9121*
[57-97-6] 9,10-Dimethyl-1,2-benzanthracene, *3285*
[58-00-4] Apomorphine, *787*
[58-05-9] Folinic Acid, *4254*
[58-08-2] Caffeine, *1674*
[58-13-9] Phencarbamide Hydrochloride*, *7363*
[58-14-0] Pyrimethamine, *8169*
[58-15-1] Aminopyrine, *495*
[58-18-4] 17-Methyltestosterone, *6206*
[58-19-5] Dromostanolone, *3504*
[58-20-8] Testosterone 17β-Cyclopentanepropionate*, *9322*
[58-22-0] Testosterone, *9322*
[58-25-3] Chlordiazepoxide, *2132*
[58-27-5] Menadione, *5874*
[58-28-6] Desipramine Hydrochloride*, *2966*
[58-32-2] Dipyridamole, *3410*
[58-33-3] Promethazine Hydrochloride*, *7970*
[58-34-4] Thiazinamium Methylsulfate, *9441*
[58-36-6] 10,10'-Oxydiphenoxarsine, *7095*
[58-37-7] Aminopromazine, *487*
[58-38-8] Prochlorperazine, *7942*
[58-39-9] Perphenazine, *7323*
[58-40-2] Promazine, *7966*
[58-46-8] Tetrabenazine, *9325*
[58-54-8] Ethacrynic Acid, *3761*
[58-55-9] Theophylline, *9421*
[58-56-0] Pyridoxine Hydrochloride, *8166*
[58-61-7] Adenosine, *152*
[58-63-9] Inosine, *5005*
[58-64-0] Adenosine Diphosphate, *153*
[58-68-4] NADH*, *6429*
[58-71-9] Cephalothin Sodium Salt*, *2028*
[58-73-1] Diphenhydramine, *3367*
[58-74-2] Papaverine, *7151*
[58-82-2] Bradykinin, *1386*
[58-85-5] Biotin, *1272*
[58-86-6] Xylose, *10220*
[58-89-9] Lindane, *5526*
[58-93-5] Hydrochlorothiazide, *4822*
[58-94-6] Chlorothiazide, *2221*

[58-95-7] Vitamin E Acetate d-Form*, *10160*
[58-96-8] Uridine, *10016*
[58-97-9] 5'-Uridylic Acid, *10020*
[58-98-0] Uridine 5'-Diphosphate, *10017*
[59-00-7] Xanthurenic Acid, *10202*
[59-01-8] Kanamycin A*, *5293*
[59-02-9] Vitamin E, *10159*
[59-05-2] Methotrexate, *6065*
[59-06-3] Ethopabate, *3791*
[59-23-4] D-Galactose, *4353*
[59-26-7] Nikethamide, *6635*
[59-30-3] Folic Acid, *4253*
[59-31-4] Carbostyril, *1873*
[59-32-5] Chloropyramine, *2214*
[59-33-6] Pyrilamine Maleate*, *8168*
[59-39-2] Piperoxan, *7631*
[59-40-5] Sulfaquinoxaline, *9109*
[59-43-8] Thiamine, *9430*
[59-46-1] Procaine, *7937*
[59-47-2] Mephenesin, *5895*
[59-50-7] 4-Chloro-m-cresol, *2184*
[59-51-8] Methionine DL-Form*, *6053*
[59-52-9] Dimercaprol, *3255*
[59-56-3] α-Glucose-1-phosphate, *4470*
[59-58-5] Prosultiamine, *8068*
[59-63-2] Isocarboxazid, *5172*
[59-66-5] Acetazolamide, *50*
[59-67-6] Nicotinic Acid, *6612*
[59-87-0] Nitrofurazone, *6697*
[59-88-1] Phenylhydrazine Hydrochloride*, *7447*
[59-89-2] N-Nitrosomorpholine, *6738*
[59-92-7] Levodopa, *5490*
[59-96-1] Phenoxybenzamine, *7409*
[59-97-2] Tolazoline Hydrochloride*, *9645*
[59-98-3] Tolazoline, *9645*
[59-99-4] Neostigmine, *6553*
[60-00-4] Edetic Acid, *3559*
[60-01-5] Tributyrin, *9750*
[60-02-6] Guanethidine Sulfate*, *4589*
[60-09-3] p-Aminoazobenzene, *439*
[60-10-6] Dithizone, *3442*
[60-11-7] p-Dimethylaminoazobenzene, *3279*
[60-12-8] Phenethyl Alcohol, *7370*
[60-13-9] Amphetamine Sulfate*, *623*
[60-18-4] Tyrosine, *9970*
[60-19-5] Tyramine Hydrochloride*, *9966*
[60-23-1] Cysteamine, *2848*
[60-24-2] 2-Mercaptoethanol, *5917*
[60-26-4] Hexamethonium, *4724*
[60-27-5] Creatinine, *2638*
[60-29-7] Ethyl Ether, *3852*
[60-31-1] Acetylcholine Chloride, *88*
[60-32-2] ε-Aminocaproic Acid, *451*
[60-33-3] Linoleic Acid, *5529*
[60-34-4] Methylhydrazine, *6160*
[60-35-5] Acetamide, *41*
[60-39-9] Mazipredone Hydrochloride*, *5802*
[60-40-2] Mecamylamine, *5814*
[60-41-3] Strychnine Sulfate, *9024*
[60-44-6] Penthienate Bromide, *7268*
[60-46-8] Aminopentamide, *479*
[60-51-5] Dimethoate, *3269*
[60-54-8] Tetracycline, *9337*
[60-56-0] Methimazole, *6049*
[60-57-1] Dieldrin, *3152*
[60-70-8] Veratramine, *10085*
[60-79-7] Ergonovine, *3694*
[60-80-0] Antipyrine, *757*
[60-81-1] Phloridzin, *7481*
[60-82-2] Phloretin, *7480*
[60-87-7] Promethazine, *7970*
[60-89-9] Mepazine, *5892*

Denotes derivative of title compound

[70-70-2] Paroxypropione, 7176
[70-78-0] 3-Iodotyrosine, 5065
[71-00-1] Histidine, 4758
[71-23-8] n-Propyl Alcohol, 8027
[71-27-2] Succinylcholine Chloride, 9044
[71-30-7] Cytosine, 2864
[71-36-3] n-Butyl Alcohol, 1575
[71-41-0] 1-Pentanol, 7257
[71-43-2] Benzene, 1094
[71-44-3] Spermine, 8894
[71-48-7] Cobaltous Acetate, 2494
[71-55-6] 1,1,1-Trichloroethane, 9766
[71-58-9] Medroxyprogesterone 17-Acetate*, 5838
[71-62-5] Veratridine, 10087
[71-63-6] Digitoxin, 3206
[71-67-0] Sulfobromophthalein Sodium, 9127
[71-68-1] Hydromorphone Hydrochloride*, 4847
[71-73-8] Thiopental Sodium, 9487
[71-78-3] Pipradrol Hydrochloride*, 7638
[71-81-8] Isopropamide Iodide, 5221
[71-82-9] Levallorphan Tartrate*, 5485
[71-91-0] Tetraethylammonium Bromide, 9340
[72-14-0] Sulfathiazole, 9115
[72-17-3] Sodium Lactate, 8781
[72-18-4] Valine, 10046
[72-19-5] Threonine, 9522
[72-20-8] Endrin, 3618
[72-23-1] 11-Dehydrocorticosterone, 2923
[72-33-3] Mestranol, 5976
[72-43-5] Methoxychlor, 6070
[72-44-6] Methaqualone, 6028
[72-48-0] Alizarin, 247
[72-54-8] 1,1-Dichloro-2,2-bis(p-chlorophenyl)ethane, 3111
[72-57-1] Trypan Blue, 9923
[72-63-9] Methandrostenolone, 6018
[72-69-5] Nortriptyline, 6812
[72-80-0] Chlorquinaldol, 2243
[73-03-0] Cordycepin, 2593
[73-05-2] Phentolamine Hydrochloride*, 7417
[73-09-6] Etozolin, 3934
[73-22-3] Tryptophan, 9929
[73-24-5] Adenine, 150
[73-31-4] Melatonin, 5857
[73-32-5] Isoleucine, 5196
[73-40-5] Guanine, 4593
[73-48-3] Bendroflumethiazide, 1064
[73-49-4] Quinethazone, 8240
[73-67-6] Toxopyrimidine, 9697
[74-11-3] p-Chlorobenzoic Acid, 2177
[74-31-7] N,N'-Diphenyl-p-phenylenediamine, 3388
[74-39-5] Magneson, 5735
[74-55-5] Ethambutol, 3764
[74-61-3] 2,3-Dimercapto-1-propanesulfonic Acid, 3256
[74-79-3] Arginine, 817
[74-82-8] Methane, 6019
[74-83-9] Methyl Bromide, 6108
[74-84-0] Ethane, 3767
[74-85-1] Ethylene, 3837
[74-86-2] Acetylene, 91
[74-87-3] Methyl Chloride, 6121
[74-88-4] Methyl Iodide, 6161
[74-89-5] Methylamine, 6095
[74-90-8] Hydrogen Cyanide, 4836
[74-93-1] Methanethiol, 6023
[74-95-3] Methylene Bromide, 6138
[74-96-4] Ethyl Bromide, 3819
[74-98-6] Propane, 7982
[75-00-3] Ethyl Chloride, 3829

[75-01-4] Vinyl Chloride, 10132
[75-03-6] Ethyl Iodide, 3859
[75-04-7] Ethylamine, 3808
[75-05-8] Acetonitrile, 68
[75-07-0] Acetaldehyde, 37
[75-08-1] Ethanethiol, 3771
[75-09-2] Methylene Chloride, 6140
[75-11-6] Methylene Iodide, 6143
[75-12-7] Formamide, 4264
[75-13-8] Isocyanic Acid, 5181
[75-15-0] Carbon Disulfide, 1859
[75-18-3] Methyl Sulfide, 6204
[75-19-4] Cyclopropane, 2817
[75-20-7] Calcium Carbide, 1696
[75-21-8] Ethylene Oxide, 3848
[75-25-2] Bromoform, 1441
[75-26-3] Isopropyl Bromide, 5229
[75-27-4] Bromodichloromethane, 1438
[75-29-6] Isopropyl Chloride, 5230
[75-30-9] Isopropyl Iodide, 5233
[75-31-0] Isopropylamine, 5228
[75-34-3] Ethylidene Chloride, 3856
[75-35-4] Vinylidene Chloride, 10134
[75-36-5] Acetyl Chloride, 86
[75-39-8] Acetaldehyde Ammonia, 38
[75-44-5] Phosgene, 7491
[75-46-7] Fluoroform, 4211
[75-47-8] Iodoform, 5054
[75-50-3] Trimethylamine, 9841
[75-52-5] Nitromethane, 6708
[75-56-9] Propylene Oxide, 8041
[75-58-1] Tetramethylammonium Iodide, 9364
[75-59-2] Tetramethylammonium Hydroxide, 9363
[75-60-5] Cacodylic Acid, 1640
[75-64-9] tert-Butylamine, 1580
[75-65-0] tert-Butyl Alcohol, 1577
[75-66-1] tert-Butyl Mercaptan, 1613
[75-69-4] Trichlorofluoromethane, 9770
[75-71-8] Dichlorodifluoromethane, 3114
[75-73-0] Carbon Tetrafluoride, 1865
[75-75-2] Methanesulfonic Acid, 6021
[75-80-9] Tribromoethanol, 9743
[75-84-3] Neopentyl Alcohol, 6546
[75-85-4] tert-Pentyl Alcohol, 7282
[75-86-5] Acetone Cyanohydrin, 65
[75-87-6] Trichloroacetaldehyde, 9755
[75-91-2] tert-Butyl Hydroperoxide, 1604
[75-93-4] Methyl Sulfate, 6203
[75-96-7] Tribromoacetic Acid, 9739
[75-97-8] Pinacolone, 7594
[75-98-9] Pivalic Acid, 7667
[75-99-0] Dalapon, 2869
[76-00-6] 1,1,1-Trichloro-2-propanol, 9774
[76-01-7] Pentachloroethane, 7241
[76-03-9] Trichloroacetic Acid, 9756
[76-05-1] Trifluoroacetic Acid, 9812
[76-06-2] Chloropicrin, 2208
[76-08-4] Tribromo-tert-butyl Alcohol, 9741
[76-09-5] Pinacol, 7593
[76-14-2] Cryofluorane, 2672
[76-20-0] Sulfonethylmethane, 9131
[76-22-2] Camphor, 1779
[76-22-2] Spirit of Camphor, 8907
[76-23-3] Tetrabarbital, 9324
[76-24-4] Alloxantin, 291
[76-25-5] Triamcinolone Acetonide, 9728
[76-28-8] Sarmentogenin, 8522
[76-29-9] 3-Bromo-d-camphor, 1435
[76-30-2] Dihydroxytartaric Acid, 3231

[76-36-8] α,α,β-Trichloro-n-butyraldehyde, 9762
[76-37-9] 2,2,3,3-Tetrafluoro-1-propanol, 9346
[76-38-0] Methoxyflurane, 6072
[76-41-5] Oxymorphone, 7103
[76-42-6] Oxycodone, 7093
[76-43-7] Fluoxymesterone, 4223
[76-44-8] Heptachlor, 4690
[76-45-9] Protoverine, 8087
[76-47-1] Hydrocortamate, 4827
[76-49-3] Bornyl Acetate, 1367
[76-53-9] Veatchine, 10074
[76-57-3] Codeine, 2525
[76-58-4] Ethylmorphine, 3876
[76-59-5] Bromthymol Blue, 1469
[76-60-8] Bromcresol Green, 1407
[76-61-9] Thymol Blue, 9541
[76-62-0] 3',3'',5',5''-Tetrabromophenolphthalein, 9329
[76-65-3] Amolanone, 610
[76-66-4] Rhynchophylline, 8364
[76-68-6] Cyclopentobarbital, 2814
[76-74-4] Pentobarbital, 7272
[76-77-7] Neoquassin, 6552
[76-78-8] Quassin, 8209
[76-80-2] Tephrosin, 9295
[76-84-6] Triphenylcarbinol, 9869
[76-87-9] Triphenyltin Hydroxide, 9875
[76-90-4] Mepenzolate Bromide, 5893
[76-93-7] Benzilic Acid, 1109
[76-94-8] Phenylmethylbarbituric Acid, 7457
[77-02-1] Aprobarbital, 794
[77-03-2] Piperidione, 7622
[77-04-3] Pyrithyldione, 8179
[77-06-5] Gibberellic Acid, 4426
[77-07-6] Levorphanol, 5496
[77-09-8] Phenolphthalein, 7392
[77-10-1] Phencyclidine, 7364
[77-14-5] Proheptazine, 7959
[77-15-6] Ethoheptazine, 3789
[77-16-7] Plumericin, 7698
[77-20-3] Alphaprodine, 317
[77-21-4] Glutethimide, 4485
[77-22-5] Caramiphen, 1820
[77-23-6] Carbetapentane, 1840
[77-25-8] Ethyl Diethylmalonate, 3835
[77-26-9] Butalbital, 1536
[77-27-0] Thiamylal, 9437
[77-28-1] Butethal, 1550
[77-32-7] Thiobarbital, 9455
[77-36-1] Chlorthalidone, 2246
[77-37-2] Procyclidine, 7944
[77-38-3] Chlorphenoxamine, 2234
[77-40-7] Bisphenol B, 1339
[77-41-8] Methsuximide, 6085
[77-42-9] β-Santalol, 8506
[77-46-3] Acedapsone, 20
[77-51-0] Isoaminile, 5124
[77-52-1] Ursolic Acid, 10027
[77-53-2] Cedrol, 1961
[77-58-7] Dibutyltin Dilaurate, 3089
[77-59-8] Tomatidine, 9682
[77-60-1] Tigogenin, 9575
[77-65-6] Carbromal, 1879
[77-66-7] Acecarbromal, 18
[77-67-8] Ethosuximide, 3794
[77-75-8] Meparfynol, 5890
[77-78-1] Dimethyl Sulfate, 3305
[77-79-2] 3-Sulfolene, 9129
[77-81-6] Tabun, 9196
[77-86-1] Tromethamine, 9902
[77-91-8] Choline Dihydrogen Citrate, 2263
[77-92-9] Citric Acid, 2387
[77-94-1] Butyl Citrate, 1599
[77-95-2] Quinic Acid, 8243

[77-98-5] Tetraethylammonium Hydroxide, 9342
[78-00-2] Tetraethyllead, 9343
[78-05-7] Phenoctide, 7389
[78-10-4] Ethyl Silicate, 3895
[78-11-5] Pentaerythritol Tetranitrate, 7249
[78-12-6] Pentaerythritol Chloral, 7246
[78-28-4] Emylcamate, 3604
[78-30-8] Tri-o-tolyl Phosphate, 9893
[78-34-2] Dioxathion, 3355
[78-40-0] Triethyl Phosphate, 9806
[78-41-1] Triparanol, 9867
[78-44-4] Carisoprodol, 1889
[78-53-5] Amiton, 509
[78-57-9] Menazon, 5877
[78-59-1] Isophorone, 5213
[78-67-1] 2,2′-Azobisisobutyronitrile, 949
[78-70-6] Linalool, 5520
[78-75-1] Propylene Dibromide, 8038
[78-76-2] sec-Butyl Bromide, 1589
[78-77-3] Isobutyl Bromide, 5150
[78-79-5] Isoprene, 5218
[78-81-9] Isobutylamine, 5147
[78-82-0] Isobutyronitrile, 5171
[78-83-1] Isobutyl Alcohol, 5146
[78-84-2] Isobutyraldehyde, 5169
[78-86-4] sec-Butyl Chloride, 1596
[78-87-5] Propylene Dichloride, 8039
[78-89-7] Propylene Chlorohydrin, 8035
[78-90-0] Propylenediamine, 8037
[78-91-1] 2-Aminopropanol, 488
[78-92-2] sec-Butyl Alcohol, 1576
[78-93-3] Methyl Ethyl Ketone, 6149
[78-94-4] Methyl Vinyl Ketone, 6214
[78-95-5] Chloroacetone, 2164
[78-98-8] Pyruvaldehyde, 8203
[78-99-9] Propylidene Chloride, 8046
[79-00-5] 1,1,2-Trichloroethane, 9767
[79-01-6] Trichloroethylene, 9769
[79-03-8] Propionyl Chloride, 8013
[79-04-9] Chloracetyl Chloride, 2107
[79-05-0] Propionamide, 8009
[79-06-1] Acrylamide, 131
[79-07-2] Chloroacetamide, 2160
[79-08-3] Bromoacetic Acid, 1421
[79-09-4] Propionic Acid, 8010
[79-10-7] Acrylic Acid, 132
[79-11-8] Chloroacetic Acid, 2162
[79-14-1] Glycolic Acid, 4508
[79-15-2] N-Bromoacetamide, 1419
[79-17-4] Aminoguanidine, 461
[79-19-6] Thiosemicarbazide, 9499
[79-20-9] Methyl Acetate, 6089
[79-21-0] Peracetic Acid, 7293
[79-22-1] Methyl Chlorocarbonate, 6123
[79-24-3] Nitroethane, 6694
[79-27-6] sym-Tetrabromoethane, 9328
[79-31-2] Isobutyric Acid, 5170
[79-33-4] L-Lactic Acid, 5350
[79-34-5] Tetrachloroethane, 9331
[79-36-7] 2,2-Dichloroacetyl Chloride, 3103
[79-37-8] Oxalyl Chloride, 7046
[79-40-3] Rubeanic Acid, 8436
[79-41-4] Methacrylic Acid, 6005
[79-42-5] Thiolactic Acid, 9476
[79-43-6] Dichloroacetic Acid, 3100
[79-46-9] 2-Nitropropane, 6725
[79-49-2] Pentabromoacetone, 7239
[79-54-9] Levopimaric Acid, 5494
[79-55-0] Pempidine, 7207
[79-57-2] Oxytetracycline, 7111
[79-58-3] Rubijervine, 8444
[79-61-8] Dichlorisone 21-Acetate*, 3099

[79-63-0] Lanosterol, 5372
[79-64-1] Dimethisterone, 3268
[79-68-5] γ-Irone, 5110
[79-69-6] α-Irone, 5108
[79-70-9] β-Irone, 5109
[79-74-3] 2,5-Di-tert-pentylhydroquinone, 3359
[79-80-1] Vitamin A₂, 10151
[79-81-2] Vitamin A Palmitate*, 10150
[79-83-4] Pantothenic Acid, 7147
[79-85-6] Quatrimycin, 8210
[79-90-3] Triclobisonium Chloride, 9785
[79-91-4] Terebic Acid, 9305
[79-92-5] Camphene, 1777
[80-00-2] Sulphenone, 9162
[80-02-4] 2-p-Sulfanilylanilinoethanol, 9097
[80-03-5] Acediasulfone, 21
[80-05-7] Bisphenol A, 1338
[80-06-8] Chlorfenethol, 2136
[80-08-0] Dapsone, 2885
[80-11-5] p-Tolylsulfonylmethylnitrosamide, 9681
[80-12-6] Tetramethylenedisulfotetramine, 9366
[80-13-7] Halazone, 4620
[80-17-1] Porofor BSH, 7757
[80-32-0] Sulfachlorpyridazine, 9068
[80-33-1] Ovex, 7033
[80-35-3] Sulfamethoxypyridazine, 9087
[80-40-0] Ethyl p-Toluenesulfonate, 3903
[80-46-6] p-tert-Pentylphenol, 7284
[80-49-9] Homatropine DL-Methylbromide*, 4766
[80-50-2] Anisotropine Methylbromide, 709
[80-53-5] Terpin, 9314
[80-56-8] α-Pinene, 7599
[80-58-0] α-Bromobutyric Acid, 1434
[80-59-1] Tiglic Acid, 9573
[80-60-4] α-Aminobutyric Acid, 448
[80-62-6] Methacrylic Acid Methyl Ester*, 6005
[80-69-3] Tartronic Acid, 9240
[80-72-8] Reductic Acid, 8297
[80-73-9] 1,3-Dimethyl-2-imidazolidinone, 3298
[80-74-0] Acetyl Sulfisoxazole*, 9125
[80-77-3] Chlormezanone, 2155
[80-78-4] Solanidine, 8852
[80-92-2] Pregnanediol, 7908
[80-97-7] Cholestanol, 2255
[81-04-9] Armstrong's Acid, 824
[81-05-0] 2-Naphthylamine-5-sulfonic Acid, 6494
[81-07-2] Saccharin, 8463
[81-11-8] Amsonic Acid, 636
[81-13-0] Dexpanthenol, 2988
[81-16-3] 2-Naphthylamine-1-sulfonic Acid, 6493
[81-23-2] Dehydrocholic Acid, 2922
[81-24-3] Taurocholic Acid, 9242
[81-25-4] Cholic Acid, 2258
[81-27-6] Sennoside A*, 8600
[81-38-9] Laureline, 5393
[81-54-9] Purpurin, 8132
[81-61-8] Quinalizarin, 8231
[81-64-1] Quinizarin, 8250
[81-67-4] Pukateine, 8123
[81-77-6] Indanthrene, 4968
[81-81-2] Warfarin, 10174
[81-82-3] Coumachlor, 2623
[81-88-9] Rhodamine B, 8349
[81-90-3] Phenolphthalin, 7394
[81-92-5] Phenolphthalol, 7395
[81-93-6] Phenosafranin, 7402
[82-02-0] Khellin, 5321

[82-05-3] Benzanthrone, 1090
[82-08-6] Rottlerin, 8429
[82-10-0] Sumatrol, 9173
[82-12-2] Rufigallol, 8447
[82-22-4] Anthrimide, 729
[82-24-6] 1-Aminoanthraquinone-2-carboxylic Acid, 438
[82-40-6] Lunacrine, 5631
[82-45-1] 1-Aminoanthraquinone, 437
[82-54-2] Cotarnine, 2618
[82-57-5] Visnagin, 10148
[82-58-6] Lysergic Acid, 5664
[82-62-2] 3,4,6-Trichloro-2-nitrophenol, 9771
[82-66-6] Diphenadione, 3363
[82-68-8] Quintozene, 8264
[82-71-3] Styphnic Acid, 9026
[82-75-7] 1-Naphthylamine-8-sulfonic Acid, 6492
[82-76-8] 1-Anilino-8-naphthalenesulfonate, 699
[82-82-6] 4-Pyridoxic Acid, 8165
[82-83-7] Puberulonic Acid, 8122
[82-88-2] Phenindamine, 7380
[82-89-3] Prodigiosin, 7948
[82-92-8] Cyclizine, 2779
[82-93-9] Chlorcyclizine, 2128
[82-95-1] Buclizine, 1484
[82-98-4] Piperidolate, 7623
[82-99-5] Thiphenamil, 9509
[83-07-8] Ampyrone, 633
[83-12-5] Phenindione, 7381
[83-14-7] Pellotine, 7202
[83-25-0] Succinanil, 9036
[83-26-1] Pindone, 7598
[83-28-3] 2-Isovalerylindane-1,3-dione, 5254
[83-32-9] Acenaphthene, 28
[83-34-1] Skatole, 8700
[83-40-9] o-Cresotic Acid, 2649
[83-43-2] Methylprednisolone, 6189
[83-44-3] Deoxycholic Acid, 2946
[83-45-4] Stigmastanol, 8968
[83-46-5] β-Sitosterol, 8697
[83-47-6] γ-Sitosterol, 8698
[83-48-7] Stigmasterol, 8969
[83-49-8] Hyodeoxycholic Acid, 4906
[83-54-5] Echinopsine, 3546
[83-58-9] Lunacridine, 5630
[83-60-3] Reserpic Acid, 8312
[83-63-6] Diacetazotol, 3004
[83-67-0] Theobromine, 9418
[83-70-5] Vitamin K₅, 10162
[83-72-7] Lawsone, 5406
[83-73-8] Iodoquinol, 5063
[83-74-9] Ibogaine, 4920
[83-79-4] Rotenone, 8427
[83-81-8] N,N,N′,N′-Tetraethylphthalamide, 9344
[83-85-2] Fuscin, 4338
[83-86-3] Phytic Acid, 7542
[83-88-5] Riboflavin, 8367
[83-89-6] Quinacrine, 8225
[83-94-3] Tabernanthine, 9195
[83-95-4] Skimmianine, 8701
[83-98-7] Orphenadrine, 7007
[84-01-5] Chlorproethazine, 2236
[84-02-6] Prochlorperazine Dimaleate*, 7942
[84-04-8] Pipamazine, 7607
[84-06-0] Thiopropazate, 9493
[84-11-7] Phenanthrenequinone, 7355
[84-12-8] Phanquinone, 7337
[84-16-2] Hexestrol, 4738
[84-17-3] Dienestrol, 3153
[84-19-5] Dienestrol Diacetate*, 3153
[84-21-9] 3′-Adenylic Acid, 156
[84-22-0] Tetrahydrozoline, 9358
[84-24-2] Physodic Acid, 7538
[84-26-4] Rutecarpine, 8451

* Denotes derivative of title compound

Denotes derivative of title compound

[91-60-1] 2-Naphthalenethiol, 6466
[91-63-4] Quinaldine, 8228
[91-64-5] Coumarin, 2630
[91-66-7] Diethylaniline, 3162
[91-71-4] Thiocarbamizine, 9458
[91-75-8] Antazoline, 718
[91-76-9] Benzoguanamine, 1120
[91-79-2] Thenyldiamine, 9416
[91-80-5] Methapyrilene, 6027
[91-81-6] Tripelennamine, 9868
[91-82-7] Pyrrobutamine, 8196
[91-84-9] Pyrilamine, 8168
[91-86-1] η-Tocopherol, 9638
[91-94-1] 3,3'-Dichlorobenzidine, 3109
[92-04-6] 4-Phenyl-2-chlorophenol, 7435
[92-12-6] Phenyltoloxamine, 7469
[92-13-7] Pilocarpine, 7578
[92-24-0] Naphthacene, 6456
[92-28-4] Amido-R-Acid, 423
[92-31-9] Tolonium Chloride, 9658
[92-32-0] Pyronine Y, 8190
[92-41-1] 2,7-Naphthalenedisulfonic Acid, 6462
[92-43-3] 1-Phenyl-3-pyrazolidinone, 7463
[92-52-4] Diphenyl, 3372
[92-59-1] Ethylbenzylaniline, 3817
[92-61-5] Scopoletin, 8552
[92-62-6] Proflavine, 7953
[92-67-1] p-Biphenylamine, 1276
[92-69-3] p-Phenylphenol, 7459
[92-70-6] 3-Hydroxy-2-naphthoic Acid, 4882
[92-82-0] Phenazine, 7359
[92-84-2] Phenothiazine, 7404
[92-87-5] Benzidine, 1106
[92-93-3] p-Nitrobiphenyl, 6690
[92-95-5] p-Xenylcarbimide, 10207
[93-01-6] 2-Naphthol-6-sulfonic Acid, 6479
[93-04-9] 2-Methoxynaphthalene, 6076
[93-07-2] Veratric Acid, 10086
[93-09-4] 2-Naphthoic Acid, 6469
[93-10-7] Quinaldic Acid, 8227
[93-14-1] Guaifenesin, 4582
[93-18-5] 2-Ethoxynaphthalene, 3799
[93-20-9] 2-(2-Naphthyloxy)ethanol, 6500
[93-30-1] Methoxyphenamine, 6077
[93-35-6] Umbelliferone, 9981
[93-39-0] Skimmin, 8702
[93-42-5] Thionalide, 9480
[93-43-6] 2-Naphthyl Lactate, 6499
[93-44-7] 2-Naphthyl Benzoate, 6495
[93-47-0] Verazide, 10091
[93-51-6] Creosol, 2640
[93-54-9] α-Ethylbenzyl Alcohol, 3816
[93-55-0] Propiophenone, 8015
[93-56-1] Styrene Glycol, 9029
[93-58-3] Methyl Benzoate, 6104
[93-59-4] Perbenzoic Acid, 7295
[93-60-7] Methyl Nicotinate, 6174
[93-71-0] Allidochlor, 259
[93-72-1] Silvex, 8679
[93-75-4] Thioquinox, 9495
[93-76-5] 2,4,5-T, 9194
[93-88-9] Phenylpropylmethylamine, 7462
[93-89-0] Ethyl Benzoate, 3814
[93-95-8] 5-(α-Phenylethyl)semioxam-azide, 7443
[93-97-0] Benzoic Anhydride, 1123
[93-98-1] Benzanilide, 1088
[93-99-2] Phenyl Benzoate, 7429
[94-02-0] Ethyl Benzoylacetate, 3815
[94-07-5] Synephrine, 9189
[94-09-7] Benzocaine, 1116
[94-10-0] Ethoxazene, 3796
[94-13-3] Propylparaben, 8051

[94-14-4] Isobutyl p-Aminobenzoate, 5148
[94-15-5] Dimethocaine, 3270
[94-16-6] p-Aminohippuric Acid Sodium Salt*, 462
[94-19-9] Sulfaethidole, 9075
[94-20-2] Chlorpropamide, 2239
[94-23-5] Parethoxycaine, 7171
[94-25-7] Butamben, 1538
[94-26-8] Butylparaben, 1619
[94-35-9] Styramate, 9027
[94-36-0] Benzoyl Peroxide, 1149
[94-41-7] Chalcone, 2078
[94-44-0] Nicotinic Acid Benzyl Ester, 6613
[94-46-2] Isoamyl Benzoate, 5127
[94-53-1] Piperonylic Acid, 7630
[94-59-7] Safrole, 8468
[94-62-2] Piperine, 7625
[94-63-3] Pralidoxime Iodide*, 7884
[94-67-7] Salicylaldoxime, 8479
[94-70-2] o-Phenetidine, 7372
[94-74-6] MCPA, 5803
[94-75-7] 2,4-D, 2865
[94-82-6] 2,4-DB, 2893
[94-83-7] Sesin, 8618
[94-93-9] Salen (unspecified), 8474
[94-96-2] Ethohexadiol, 3790
[95-01-2] β-Resorcylaldehyde, 8324
[95-04-5] Ectylurea, 3552
[95-05-6] Sulfiram, 9123
[95-06-7] Sulfallate, 9079
[95-13-6] Indene, 4973
[95-14-7] 1H-Benzotriazole, 1140
[95-15-8] Thianaphthene, 9438
[95-16-9] Benzothiazole, 1139
[95-25-0] Chlorzoxazone, 2249
[95-38-5] Amine 220, 428
[95-43-2] D-Threose, 9523
[95-44-3] L-Threose, 9524
[95-45-4] Dimethylglyoxime, 3295
[95-48-7] o-Cresol*, 2645
[95-49-8] o-Chlorotoluene*, 2224
[95-50-1] o-Dichlorobenzene, 3106
[95-51-2] o-Chloroaniline*, 2169
[95-52-3] o-Fluorotoluene*, 4218
[95-53-4] o-Toluidine*, 9674
[95-54-5] o-Phenylenediamine, 7438
[95-55-6] o-Aminophenol, 481
[95-56-7] o-Bromophenol*, 1450
[95-63-6] Pseudocumene, 8100
[95-76-1] 3,4-Dichloroaniline, 3104
[95-86-3] 2,4-Diaminophenol, 3026
[95-92-1] Diethyl Oxalate, 3174
[95-93-2] Durene, 3520
[95-95-4] 2,4,5-Trichlorophenol, 9772
[96-00-4] Sebacoin, 8560
[96-01-5] Sebacil, 8559
[96-11-7] 1,2,3-Tribromopropane, 9745
[96-12-8] Dibromochloropropane, 3068
[96-20-8] 2-Amino-1-butanol, 447
[96-22-0] Diethyl Ketone, 3170
[96-23-1] 1,3-Dichloro-2-propanol, 3124
[96-24-2] α-Chlorohydrin, 2197
[96-26-4] Dihydroxyacetone, 3225
[96-27-5] Thioglycerol, 9471
[96-33-3] Methyl Acrylate, 6092
[96-34-4] Methyl Chloroacetate, 6122
[96-41-3] Cyclopentanol, 2810
[96-45-7] Ethylene Thiourea, 3849
[96-48-0] Butyrolactone, 1632
[96-50-4] 2-Aminothiazole, 501
[96-62-8] Dinsed, 3343
[96-64-0] Soman, 8860
[96-77-5] Phenoldisulfonic Acid, 7391
[96-82-2] Lactobionic Acid, 5354
[96-83-3] Iopanoic Acid, 5072

[96-84-4] Iophenoxic Acid, 5075
[96-88-8] Mepivacaine, 5905
[96-91-3] Picramic Acid, 7561
[96-97-9] 5-Nitrosalicylic Acid, 6730
[97-00-7] 1-Chloro-2,4-dinitrobenzene, 2187
[97-02-9] 2,4-Dinitroaniline, 3322
[97-05-2] Sulfosalicylic Acid, 9138
[97-11-0] Cyclethrin, 2774
[97-16-5] Genite, 4397
[97-17-6] Dichlofenthion, 3094
[97-18-7] Bithionol, 1343
[97-23-4] Dichlorophen, 3120
[97-24-5] Fenticlor, 4046
[97-25-6] Urochloralic Acid, 10023
[97-26-7] Toluylene Blue, 9678
[97-27-8] Chlorbetamide, 2126
[97-30-3] α-Methylglucoside, 6156
[97-31-4] Normetanephrine, 6803
[97-44-9] Acetarsone, 49
[97-53-0] Eugenol, 3944
[97-54-1] Isoeugenol, 5186
[97-56-3] o-Aminoazotoluene, 440
[97-59-6] Allantoin, 255
[97-62-1] Ethyl Isobutyrate, 3860
[97-64-3] Ethyl Lactate, 3863
[97-65-4] Itaconic Acid, 5261
[97-67-6] Malic Acid L-(−)-Form*, 5747
[97-75-6] Scopolamine N-Oxide, 8551
[97-77-8] Disulfiram, 3428
[97-85-8] Isobutyl Isobutyrate, 5159
[97-99-4] Tetrahydrofurfuryl Alcohol, 9353
[98-00-0] Furfuryl Alcohol, 4325
[98-01-1] Furfural, 4324
[98-04-4] Phenyltrimethylammonium Iodide, 7471
[98-05-5] Benzenearsonic Acid, 1095
[98-06-6] tert-Butylbenzene, 1586
[98-07-7] Benzotrichloride, 1141
[98-08-8] Benzotrifluoride, 1142
[98-09-9] Benzenesulfonyl Chloride, 1100
[98-11-3] Benzenesulfonic Acid, 1098
[98-50-0] Arsanilic Acid, 830
[98-54-4] p-tert-Butylphenol, 1620
[98-55-5] α-Terpineol, 9316
[98-58-8] p-Bromobenzenesulfonyl Chloride, 1428
[98-59-9] p-Toluenesulfonyl Chloride, 9672
[98-61-3] Pipsyl Chloride, 7640
[98-62-4] Sulfanilyl Fluoride, 9099
[98-66-8] p-Chlorobenzenesulfonic Acid, 2173
[98-67-9] p-Phenolsulfonic Acid, 7396
[98-71-5] 4-Hydrazinobenzenesulfonic Acid, 4813
[98-72-6] Nitarsone, 6659
[98-79-3] L-Pyroglutamic Acid, 8185
[98-80-6] Benzeneboronic Acid, 1096
[98-82-8] Cumene, 2683
[98-84-0] α-Methylbenzylamine, 6106
[98-86-2] Acetophenone, 71
[98-87-3] Benzal Chloride, 1084
[98-88-4] Benzoyl Chloride, 1145
[98-89-5] Cyclohexanecarboxylic Acid, 2793
[98-92-0] Nicotinamide, 6574
[98-95-3] Nitrobenzene, 6685
[98-96-4] Pyrazinamide, 8140
[98-97-5] Pyrazinoic Acid, 8143
[98-98-6] Picolinic Acid, 7557
[99-00-3] Picadex, 7548
[99-05-8] m-Aminobenzoic Acid, 441
[99-09-2] m-Nitroaniline, 6679
[99-11-6] Citrazinic Acid, 2386
[99-14-9] Tricarballylic Acid, 9752
[99-17-2] Populin, 7753

Denotes derivative of title compound

[99-20-7] Trehalose, 9713
[99-23-0] Puberulic Acid, 8121
[99-24-1] Gallic Acid Methyl Ester*, 4363
[99-32-1] Chelidonic Acid, 2094
[99-33-2] 3,5-Dinitrobenzoyl Chloride, 3329
[99-34-3] 3,5-Dinitrobenzoic Acid, 3328
[99-35-4] sym-Trinitrobenzene, 9856
[99-43-4] Benoxinate, 1076
[99-45-6] Adrenalone, 170
[99-49-0] Carvone, 1925
[99-50-3] Protocatechuic Acid, 8080
[99-56-9] 4-Nitro-o-phenylenediamine, 6720
[99-66-1] Valproic Acid, 10049
[99-73-0] p-Bromophenacyl Bromide, 1449
[99-76-3] Methylparaben, 6182
[99-79-6] Iophendylate, 5074
[99-83-2] α-Phellandrene, 7340
[99-88-7] Cumidine, 2686
[99-90-1] p-Bromoacetophenone, 1423
[99-91-2] p-Chloroacetophenone, 2165
[99-96-7] p-Hydroxybenzoic Acid, 4857
[99-98-9] Dimethyl-p-phenylenedi-amine, 3303
[100-01-6] p-Nitroaniline, 6681
[100-02-7] p-Nitrophenol, 6718
[100-07-2] p-Anisoyl Chloride, 711
[100-09-4] p-Anisic Acid, 704
[100-10-7] p-Dimethylaminobenzalde-hyde, 3280
[100-16-3] (4-Nitrophenyl)hydrazine, 6721
[100-21-0] Terephthalic Acid, 9306
[100-22-1] Tetramethyl-p-phenylene-diamine, 9367
[100-26-5] Isocinchomeronic Acid, 5175
[100-33-4] Pentamidine, 7254
[100-37-8] 2-Diethylaminoethanol, 3161
[100-39-0] Benzyl Bromide, 1163
[100-41-4] Ethylbenzene, 3812
[100-42-5] Styrene, 9028
[100-44-7] Benzyl Chloride, 1164
[100-46-9] Benzylamine, 1160
[100-47-0] Benzonitrile, 1128
[100-49-2] Cyclohexylcarbinol, 2800
[100-51-6] Benzyl Alcohol, 1159
[100-52-7] Benzaldehyde, 1085
[100-53-8] Thiobenzyl Alcohol, 9456
[100-55-0] Nicotinyl Alcohol, 6614
[100-56-1] Phenylmercuric Chloride, 7454
[100-59-4] Phenylmagnesium Chlo-ride, 7451
[100-61-8] Methylaniline, 6098
[100-63-0] Phenylhydrazine, 7447
[100-65-2] Phenylhydroxylamine, 7448
[100-66-3] Anisole, 707
[100-67-4] Potassium Phenoxide, 7827
[100-76-5] Quinuclidine, 8265
[100-79-8] Isopropylidene Glycerol, 5232
[100-88-9] Cyclamic Acid, 2770
[100-91-4] Eucatropine, 3943
[100-92-5] Mephentermine, 5897
[100-95-8] Dodecarbonium Chloride, 3466
[100-97-0] Methenamine, 6036
[101-05-3] Anilazine, 694
[101-14-4] 4,4'-Methylenebis[2-chloro-aniline], 6136
[101-20-2] Triclocarban, 9786
[101-21-3] Chlorpropham, 2240

[101-26-8] Pyridostigmine Bromide, 8161
[101-27-9] Barban, 986
[101-31-5] Hyoscyamine, 4907
[101-38-2] Gibbs Reagent, 4428
[101-40-6] Propylhexedrine, 8045
[101-42-8] Fenuron, 4050
[101-53-1] p-Benzylphenol, 1180
[101-57-5] N-Phenylsulfanilic Acid, 7466
[101-60-0] Porphine, 7758
[101-61-1] Michler's Base, 6264
[101-62-2] Phenamidine, 7352
[101-64-4] N-(p-Methoxyphenyl)-p-phenylenediamine, 6079
[101-71-3] Diphenane, 3365
[101-77-9] p,p'-Diaminodiphenylmeth-ane, 3023
[101-81-5] Diphenylmethane, 3387
[101-83-7] Dicyclohexylamine, 3145
[101-84-8] Phenyl Ether, 7442
[101-91-7] 4'-Hydroxybutyranilide, 4859
[101-97-3] Ethyl Phenylacetate, 3885
[101-99-5] Phenylurethan(e), 7473
[102-01-2] Acetoacetanilide, 55
[102-02-3] Phenyl Biguanide, 7430
[102-06-7] 1,3-Diphenylguanidine, 3383
[102-07-8] Carbanilide, 1829
[102-08-9] sym-Diphenylthiourea, 3393
[102-09-0] Phenyl Carbonate, 7433
[102-14-7] Succinanilic Acid, 9037
[102-29-4] Resorcinol Monoacetate*, 8323
[102-45-4] Cyclopentamine, 2808
[102-54-5] Ferrocene, 4084
[102-60-3] Entprol, 3635
[102-71-6] Triethanolamine, 9798
[102-76-1] Triacetin, 9721
[102-82-9] Tributylamine, 9748
[102-92-1] Cinnamoyl Chloride, 2360
[102-98-7] Phenylmercury Borate, 7456
[103-01-5] N-Phenylglycine, 7445
[103-03-7] Phenicarbazide, 7378
[103-09-3] Octyl Acetate, 6860
[103-12-8] Sulfamidochrysoidine, 9092
[103-16-2] Monobenzone, 6331
[103-17-3] Chlorbenside, 2124
[103-29-7] Bibenzyl, 1245
[103-32-2] Benzylaniline, 1161
[103-33-3] Azobenzene, 948
[103-34-4] 4,4'-Dithiodimorpholine, 3437
[103-41-3] Benzyl Cinnamate, 1165
[103-47-9] CHES, 2097
[103-49-1] Dibenzylamine, 3058
[103-50-4] Benzyl Ether, 1167
[103-65-1] n-Propylbenzene, 8029
[103-69-5] Ethylaniline, 3811
[103-70-8] Formanilide, 4265
[103-71-9] Phenyl Isocyanate, 7449
[103-72-0] Phenyl Isothiocyanate, 7450
[103-73-1] Phenetole, 7374
[103-79-7] Phenylacetone, 7423
[103-81-1] α-Phenylacetamide, 7420
[103-82-2] Phenylacetic Acid, 7422
[103-84-4] Acetanilide, 47
[103-85-5] Phenylthiourea, 7468
[103-88-8] p-Bromoacetanilide, 1420
[103-89-9] p-Acetotoluide*, 74
[103-90-2] Acetaminophen, 45
[103-97-9] Phenocoll, 7388
[104-03-0] p-Nitrophenylacetic Acid, 6719
[104-04-1] p-Nitroacetanilide*, 6677
[104-06-3] Thiacetazone, 9427

[104-14-3] Octopamine, 6856
[104-15-4] p-Toluenesulfonic Acid, 9671
[104-22-3] Benzylsulfamide, 1182
[104-28-9] Cinoxate, 2370
[104-29-0] Chlorphenesin, 2230
[104-31-4] Benzonatate, 1127
[104-32-5] Propamidine, 7981
[104-46-1] Anethole, 682
[104-51-8] n-Butylbenzene, 1584
[104-54-1] Cinnamyl Alcohol, 2362
[104-55-2] Cinnamaldehyde, 2356
[104-57-4] Benzyl Formate, 1169
[104-76-7] 2-Ethyl-1-hexanol, 3854
[104-80-3] 2,5-Tetrahydrofuran-dimethanol, 9352
[104-81-4] p-Xylyl Bromide*, 10223
[104-82-5] p-Xylyl Chloride*, 10224
[104-85-8] p-Tolunitrile, 9676
[104-90-5] 5-Ethyl-2-picoline, 3889
[104-91-6] 4-Nitrosophenol, 6740
[105-13-5] Anise Alcohol, 703
[105-20-4] Betazole, 1230
[105-29-3] 1-Pentol, 7273
[105-37-3] Ethyl Propionate, 3891
[105-39-5] Ethyl Chloroacetate, 3830
[105-41-9] Methylhexaneamine, 6159
[105-45-3] Methyl Acetoacetate, 6090
[105-46-4] sec-Butyl Acetate, 1571
[105-53-3] Ethyl Malonate, 3869
[105-54-4] Ethyl Butyrate, 3822
[105-56-6] Ethyl Cyanoacetate, 3833
[105-57-7] Acetal, 36
[105-58-8] Ethyl Carbonate, 3827
[105-60-2] Caprolactam, 1805
[105-61-3] Maleuric Acid, 5746
[105-66-8] Propyl Butyrate, 8031
[105-92-0] Rhodopin, 8358
[106-02-5] Exaltolide, 3956
[106-22-9] β-Citronellol, 2391
[106-23-0] Citronellal, 2390
[106-24-1] Geraniol, 4411
[106-25-2] Nerol, 6560
[106-27-4] Isoamyl Butyrate, 5129
[106-30-9] Ethyl Oenanthate, 3881
[106-31-0] Butyric Anhydride, 1630
[106-32-1] Ethyl Caprylate, 3825
[106-33-2] Ethyl Laurate, 3864
[106-34-3] Quinhydrone, 8242
[106-36-5] Propyl Propionate, 8052
[106-37-6] p-Dibromobenzene, 3066
[106-40-1] p-Bromoaniline, 1425
[106-41-2] p-Bromophenol*, 1450
[106-43-4] p-Chlorotoluene*, 2224
[106-44-5] p-Cresol*, 2645
[106-46-7] p-Dichlorobenzene, 3107
[106-47-8] p-Chloroaniline*, 2169
[106-49-0] p-Toluidine*, 9674
[106-50-3] p-Phenylenediamine, 7439
[106-51-4] Quinone, 8259
[106-57-0] 2,5-Piperazinedione, 7619
[106-60-5] δ-Aminolevulinic Acid, 466
[106-68-3] Ethyl Amyl Ketone, 3810
[106-89-8] Epichlorohydrin, 3648
[106-93-4] Ethylene Dibromide, 3842
[106-94-5] Propyl Bromide, 8030
[106-95-6] Allyl Bromide, 296
[106-97-8] Butane, 1541
[106-98-9] 1-Butene, 1548
[106-99-0] 1,3-Butadiene, 1534
[107-01-7] 2-Butene, 1549
[107-02-8] Acrolein, 130
[107-05-1] Allyl Chloride, 297
[107-06-2] Ethylene Dichloride, 3843
[107-07-3] Ethylene Chlorohydrin, 3839
[107-08-4] Propyl Iodide, 8047
[107-10-8] Propylamine, 8028
[107-11-9] Allylamine, 295
[107-12-0] Propionitrile, 8012

Denotes derivative of title compound

[107-13-1] Acrylonitrile, *133*
[107-15-3] Ethylenediamine, *3841*
[107-18-6] Allyl Alcohol, *294*
[107-19-7] Propargyl Alcohol, *7993*
[107-20-0] Chloroacetaldehyde, *2159*
[107-21-1] Ethylene Glycol, *3844*
[107-22-2] Glyoxal, *4519*
[107-27-7] Ethylmercuric Chloride, *3871*
[107-29-9] Acetaldoxime, *40*
[107-30-2] Chloromethyl Methyl Ether, *2198*
[107-31-3] Methyl Formate, *6153*
[107-32-4] Performic Acid, *7302*
[107-34-6] 1-Propanearsonic Acid, *7983*
[107-35-7] Taurine, *9241*
[107-36-8] Isethionic Acid, *5122*
[107-38-0] Arsonoacetic Acid, *850*
[107-41-5] Hexylene Glycol, *4748*
[107-43-7] Betaine, *1225*
[107-44-8] Sarin, *8520*
[107-49-3] Tetraethyl Pyrophosphate, *9345*
[107-51-7] Octamethyltrisiloxane, *6844*
[107-52-8] Tetradecamethylhexasiloxane, *9338*
[107-53-9] Tetracosamethylhendecasiloxane, *9335*
[107-59-5] *tert*-Butyl Chloroacetate, *1598*
[107-68-6] *N*-Methyltaurine, *6205*
[107-69-7] Trichlorourethan, *9778*
[107-73-3] Phosphorylcholine, *7519*
[107-82-4] Isoamyl Bromide, *5128*
[107-84-6] Isoamyl Chloride, *5130*
[107-85-7] Isoamylamine, *5126*
[107-86-8] Senecialdehyde, *8592*
[107-87-9] Methyl Propyl Ketone, *6193*
[107-88-0] 1,3-Butylene Glycol, *1601*
[107-89-1] Aldol, *225*
[107-91-5] Cyanoacetamide, *2758*
[107-92-6] Butyric Acid, *1629*
[107-93-7] Crotonic Acid, *2664*
[107-94-8] β-Chloropropionic Acid, *2211*
[107-95-9] β-Alanine, *206*
[107-97-1] Sarcosine, *8519*
[108-01-0] Deanol, *2900*
[108-03-2] 1-Nitropropane, *6724*
[108-05-4] Vinyl Acetate, *10130*
[108-10-1] Isopropylacetone, *5226*
[108-12-3] Isovaleryl Chloride, *5252*
[108-18-9] Diisopropylamine, *3240*
[108-19-0] Biuret, *1346*
[108-20-3] Isopropyl Ether, *5231*
[108-21-4] Isopropyl Acetate, *5224*
[108-22-5] Isopropenyl Acetate, *5222*
[108-24-7] Acetic Anhydride, *53*
[108-28-1] Protoanemonin, *8078*
[108-30-5] Succinic Anhydride, *9039*
[108-31-6] Maleic Anhydride, *5744*
[108-39-4] *m*-Cresol*, *2645*
[108-41-8] *m*-Chlorotoluene*, *2224*
[108-42-9] *m*-Chloroaniline*, *2169*
[108-44-1] *m*-Toluidine*, *9674*
[108-45-2] *m*-Phenylenediamine, *7437*
[108-46-3] Resorcinol, *8323*
[108-48-5] 2,6-Lutidine, *5643*
[108-56-5] Ethyl Oxalacetate, *3882*
[108-59-8] Methyl Malonate, *6169*
[108-64-5] Ethyl Isovalerate, *3862*
[108-67-8] Mesitylene, *5967*
[108-70-3] 1,3,5-Trichlorobenzene, *9761*
[108-73-6] Phloroglucinol, *7482*
[108-75-8] 2,4,6-Trimethylpyridine, *9848*

[108-78-1] Melamine, *5853*
[108-80-5] Cyanuric Acid, *2767*
[108-85-0] Cyclohexyl Bromide, *2799*
[108-86-1] Bromobenzene, *1427*
[108-88-3] Toluene, *9667*
[108-89-4] γ-Picoline, *7556*
[108-90-7] Chlorobenzene, *2172*
[108-91-8] Cyclohexylamine, *2798*
[108-93-0] Cyclohexanol, *2794*
[108-94-1] Cyclohexanone, *2795*
[108-95-2] Phenol, *7390*
[108-98-5] Thiophenol, *9492*
[108-99-6] β-Picoline, *7555*
[109-06-8] α-Picoline, *7554*
[109-21-7] *n*-Butyl *n*-Butyrate, *1591*
[109-52-4] Valeric Acid, Normal, *10042*
[109-57-9] Thiosinamine, *9500*
[109-60-4] Propyl Acetate, *8026*
[109-61-5] Propyl Chlorocarbonate, *8033*
[109-63-7] Boron Trifluoride Etherate, *1380*
[109-64-8] Trimethylene Bromide, *9844*
[109-65-9] *n*-Butyl Bromide, *1588*
[109-66-0] Pentane, *7255*
[109-67-1] 1-Pentene, *7262*
[109-68-2] 2-Pentene, *7263*
[109-69-3] *n*-Butyl Chloride, *1595*
[109-73-9] *n*-Butylamine, *1578*
[109-74-0] Butyronitrile, *1633*
[109-75-1] Allyl Cyanide, *298*
[109-77-3] Malononitrile, *5750*
[109-78-4] Ethylene Cyanohydrin, *3840*
[109-79-5] *n*-Butyl Mercaptan, *1611*
[109-80-8] 1,3-Propanedithiol, *7984*
[109-82-0] Methyleneaminoacetonitrile, *6134*
[109-83-1] 2-Methylaminoethanol, *6096*
[109-84-2] 2-Hydrazinoethanol, *4814*
[109-86-4] Methyl Cellosolve, *6118*
[109-87-5] Methylal, *6093*
[109-89-7] Diethylamine, *3160*
[109-93-3] Vinyl Ether, *10133*
[109-94-4] Ethyl Formate, *3853*
[109-95-5] Ethyl Nitrite, *3877*
[109-95-5] Spirit of Ethyl Nitrite, *8911*
[109-96-6] 3-Pyrroline, *8201*
[109-97-7] 1*H*-Pyrrole, *8198*
[109-98-8] 2-Pyrazoline, *8145*
[109-99-9] Tetrahydrofuran, *9351*
[110-00-9] Furan, *4316*
[110-01-0] Tetrahydrothiophene, *9357*
[110-02-1] Thiophene, *9490*
[110-04-3] 1,2-Ethanedisulfonic Acid, *3769*
[110-05-4] DTBP, *3515*
[110-13-4] Acetonylacetone, *69*
[110-14-5] Succinamide, *9035*
[110-15-6] Succinic Acid, *9038*
[110-16-7] Maleic Acid, *5743*
[110-17-8] Fumaric Acid, *4309*
[110-19-0] Isobutyl Acetate, *5145*
[110-27-0] Isopropyl Myristate, *5234*
[110-38-3] Ethyl Caprate, *3823*
[110-43-0] 2-Heptanone, *4698*
[110-44-1] Sorbic Acid, *8869*
[110-45-2] Isoamyl Formate, *5133*
[110-46-3] Isoamyl Nitrite, *5137*
[110-49-6] Methyl Cellosolve Acetate, *6119*
[110-53-2] *n*-Amyl Bromide, *643*
[110-54-3] *n*-Hexane, *4729*
[110-58-7] *n*-Amylamine, *639*
[110-59-8] Valeronitrile, *10043*
[110-60-1] Putrescine, *8134*
[110-61-2] Succinonitrile, *9041*

[110-62-3] *n*-Valeraldehyde, *10040*
[110-66-7] *n*-Amyl Mercaptan, *653*
[110-71-4] 1,2-Dimethoxyethane, *3274*
[110-74-7] Propyl Formate, *8043*
[110-75-8] 2-Chloroethyl Vinyl Ether, *2191*
[110-77-0] 2-(Ethylthio)ethanol, *3902*
[110-80-5] 2-Ethoxyethanol, *3797*
[110-82-7] Cyclohexane, *2792*
[110-83-8] Cyclohexene, *2796*
[110-85-0] Piperazine, *7617*
[110-86-1] Pyridine, *8153*
[110-88-3] *s*-Trioxane, *9863*
[110-89-4] Piperidine, *7621*
[110-91-8] Morpholine, *6362*
[110-94-1] Glutaric Acid, *4481*
[111-00-2] Suxethonium Bromide, *9184*
[111-01-3] Squalane, *8923*
[111-02-4] Squalene, *8924*
[111-13-7] Hexyl Methyl Ketone, *4749*
[111-14-8] Heptanoic Acid, *4695*
[111-15-9] 2-Ethoxyethyl Acetate, *3798*
[111-16-0] Pimelic Acid, *7585*
[111-17-1] 3,3'-Thiodipropionic Acid, *9468*
[111-20-6] Sebacic Acid, *8558*
[111-27-3] 1-Hexanol, *4732*
[111-28-4] Sorbic Alcohol, *8870*
[111-29-5] 1,5-Pentanediol, *7256*
[111-30-8] Glutaraldehyde, *4480*
[111-36-4] Butyl Isocyanate, *1609*
[111-42-2] Diethanolamine, *3156*
[111-43-3] Propyl Ether, *8042*
[111-44-4] *sym*-Dichloroethyl Ether, *3116*
[111-46-6] Diethylene Glycol, *3168*
[111-47-7] Propyl Sulfide, *8053*
[111-48-8] 2,2'-Thiodiethanol, *9466*
[111-51-3] Tetramethyldiaminobutane, *9365*
[111-55-7] Ethylene Glycol Diacetate, *3845*
[111-65-9] Octane, *6847*
[111-66-0] Caprylene, *1807*
[111-70-6] 1-Heptanol, *4696*
[111-71-7] Heptanal, *4692*
[111-76-2] Butyl Cellosolve, *1594*
[111-77-3] Methyl Carbitol, *6116*
[111-83-1] *n*-Octyl Bromide, *6861*
[111-87-5] 1-Octanol, *6849*
[111-90-0] Carbitol, *1847*
[111-92-2] *n*-Dibutylamine, *3083*
[111-96-6] Diglyme, *3208*
[112-05-0] Pelargonic Acid, *7198*
[112-12-9] Methyl Nonyl Ketone, *6178*
[112-24-3] Trientine, *9796*
[112-27-6] Triethylene Glycol, *9802*
[112-30-1] *n*-Decyl Alcohol, *2911*
[112-34-5] Butyl Carbitol, *1592*
[112-36-7] Diethyl Carbitol, *3167*
[112-38-9] Undecylenic Acid, *9983*
[112-47-0] Decamethylene Glycol, *2906*
[112-49-2] Triglyme, *9820*
[112-53-8] 1-Dodecanol, *3464*
[112-56-1] Lethane 384, *5472*
[112-63-0] Methyl Linoleate, *6168*
[112-72-1] Myristyl Alcohol, *6418*
[112-79-8] Elaidic Acid, *3574*
[112-80-1] Oleic Acid, *6965*
[112-85-5] Behenic Acid, *1051*
[112-86-7] Erucic Acid, *3713*
[112-92-5] Stearyl Alcohol, *8960*
[113-00-8] Guanidine, *4591*
[113-15-5] Ergotamine, *3703*
[113-18-8] Ethchlorvynol, *3774*

Denotes derivative of title compound

[113-22-4] Estriol 16,17-Bis(sodium Hemisuccinate)*, 3750
[113-38-2] Estradiol Dipropionate*, 3746
[113-42-8] Methylergonovine, 6147
[113-45-1] Methylphenidate, 6186
[113-52-0] Imipramine Hydrochloride*, 4955
[113-53-1] Dothiepin, 3485
[113-59-7] Chlorprothixene, 2241
[113-73-5] Gramicidin S, 4552
[113-78-0] Deaminooxytocin, 2899
[113-92-8] Chlorpheniramine Maleate*, 2232
[113-98-4] Penicillin G Potassium, 7225
[114-03-4] 5-Hydroxytryptophan DL-Form*, 4895
[114-04-5] Neuraminic Acid, 6566
[114-07-8] Erythromycin, 3720
[114-21-6] Mandelic Acid Sodium Salt*, 5757
[114-25-0] Biliverdine, 1264
[114-26-1] Propoxur, 8022
[114-42-1] Flavaspidic Acid, 4130
[114-43-2] Desaspidin BB, 2962
[114-49-8] Scopolamine Hydrobromide Trihydrate*, 8550
[114-80-7] Neostigmine Bromide*, 6553
[114-85-2] Bethanidine Sulfate*, 1233
[114-86-3] Phenformin, 7376
[114-90-9] Obidoxime Chloride, 6835
[114-91-0] Metyridine, 6247
[115-02-6] Azaserine, 932
[115-07-1] Propylene, 8034
[115-08-2] Thioformamide, 9469
[115-10-6] Methyl Ether, 6148
[115-11-7] Isobutylene, 5155
[115-17-3] Bromal, 1404
[115-19-5] 2-Methyl-3-butyn-2-ol, 6113
[115-20-8] 2,2,2-Trichloroethanol, 9768
[115-24-2] Sulfonmethane, 9133
[115-25-3] Octafluorocyclobutane, 6841
[115-26-4] Dimefox, 3250
[115-27-5] Chlorendic Anhydride, 2134
[115-29-7] Endosulfan, 3614
[115-31-1] Isobornyl Thiocyanoacetate, 5144
[115-32-2] Dicofol, 3136
[115-33-3] Oxyphenisatin Acetate, 7108
[115-37-7] Thebaine, 9411
[115-38-8] Mephobarbital, 5899
[115-39-9] Bromphenol Blue, 1468
[115-40-2] Bromcresol Purple, 1408
[115-43-5] Phenallymal, 7349
[115-44-6] Talbutal, 9206
[115-46-8] Azacyclonol, 925
[115-51-5] Ambutonium Bromide, 406
[115-53-7] Sinomenine, 8692
[115-56-0] Methallatal, 6011
[115-61-7] Tachysterol, 9198
[115-63-9] Hexocyclium Methyl Sulfate, 4744
[115-67-3] Paramethadione, 7161
[115-68-4] Sulfadicramide, 9072
[115-69-5] 2-Amino-2-methyl-1,3-propanediol, 468
[115-70-8] 2-Amino-2-ethyl-1,3-propanediol, 459
[115-71-9] α-Santalol, 8505
[115-77-5] Pentaerythritol, 7245
[115-79-7] Ambenonium Chloride, 396
[115-86-6] Triphenyl Phosphate, 9872
[115-90-2] Fensulfothion, 4042

[115-93-5] Cythioate, 2854
[115-95-7] Linalyl Acetate, 5521
[116-04-1] β-Humulene*, 4789
[116-06-3] Aldicarb, 223
[116-25-6] 1-(Hydroxymethyl)-5,5-dimethylhydantoin, 4877
[116-26-7] Safranal, 8467
[116-29-0] Tetradifon, 9339
[116-30-3] Rhodoxanthin, 8362
[116-31-4] Retinal, 8331
[116-38-1] Edrophonium Chloride, 3562
[116-42-7] Sulfaproxyline, 9106
[116-43-8] Succinylsulfathiazole, 9046
[116-44-9] Sulfapyrazine, 9107
[116-45-0] Sulfabromomethazine, 9066
[116-49-4] Glycobiarsol, 4503
[116-56-3] 4-Pregnene-20,21-diol-3,11-dione, 7911
[116-58-5] 4-Pregnene-11β,17α,20β,21-tetrol-3-one, 7912
[116-59-6] 4-Pregnene-17α,20β,21-triol-3,11-dione, 7913
[116-63-2] 1-Amino-2-naphthol-4-sulfonic Acid, 474
[117-02-2] Rubiadin, 8438
[117-10-2] Danthron, 2878
[117-12-4] Anthrarufin, 728
[117-26-0] Bulan, 1505
[117-27-1] 2-Nitro-1,1-bis(p-chlorophenyl)propane, 6691
[117-30-6] Dipiproverine, 3399
[117-34-0] Diphenylacetic Acid, 3374
[117-37-3] Anisindione, 706
[117-39-5] Quercetin, 8216
[117-51-1] Synhexyl, 9190
[117-52-2] Coumafuryl, 2621
[117-56-6] 1-Naphthol-4,8-disulfonic Acid, 6473
[117-68-0] 3'-Guanylic Acid, 4599
[117-73-7] Raunescine, 8292
[117-80-6] Dichlone, 3096
[117-81-7] Bis(2-ethylhexyl) Phthalate, 1291
[117-89-5] Trifluoperazine, 9811
[117-92-0] Quinaldine Red, 8230
[118-00-3] Guanosine, 4596
[118-08-1] Hydrastine, 4806
[118-10-5] Cinchonine, 2346
[118-23-0] Bromodiphenhydramine, 1439
[118-32-1] 2-Naphthol-6,8-disulfonic Acid, 6475
[118-34-3] Syringin, 9192
[118-42-3] Hydroxychloroquine, 4863
[118-52-5] 1,3-Dichloro-5,5-dimethylhydantoin, 3115
[118-55-8] Phenyl Salicylate, 7464
[118-56-9] Homosalate, 4776
[118-57-0] Acetaminosalol, 46
[118-58-1] Benzyl Salicylate, 1181
[118-61-6] Ethyl Salicylate, 3894
[118-65-0] Isocaryophyllene*, 1926
[118-68-3] Etryptamine Acetate*, 3937
[118-71-8] Maltol, 5752
[118-74-1] Hexachlorobenzene, 4714
[118-75-2] Chloranil, 2121
[118-78-5] Uramil, 9987
[118-79-6] 2,4,6-Tribromophenol, 9744
[118-91-2] o-Chlorobenzoic Acid, 2176
[118-92-3] o-Aminobenzoic Acid, 442
[118-96-7] 2,4,6-Trinitrotoluene, 9860
[119-04-0] Neomycin B*, 6542
[119-13-1] δ-Tocopherol, 9634
[119-20-0] Rhizopterin, 8348
[119-24-4] Pteroic Acid, 8118
[119-26-6] 2,4-Dinitrophenylhydrazine, 3336

[119-28-8] 1,7-Cleve's Acid, 2410
[119-29-9] Ambucaine, 402
[119-36-8] Methyl Salicylate, 6200
[119-38-0] Isolan, 5195
[119-41-5] Efloxate, 3565
[119-44-8] Xanthopterin, 10198
[119-48-2] Dimorpholamine, 3317
[119-53-9] Benzoin, 1124
[119-59-5] 4,4'-Sulfinyldianiline, 9122
[119-61-9] Benzophenone, 1129
[119-64-2] Tetralin, 9360
[119-65-3] Isoquinoline, 5241
[119-79-9] 1,6-Cleve's Acid, 2409
[119-90-4] Dianisidine, 3036
[119-93-7] o-Tolidine, 9651
[119-96-0] Arsthinol, 852
[119-98-2] Tocol, 9631
[120-02-5] Thiocarbarsone, 9459
[120-05-8] Sulfuretin, 9145
[120-08-1] Scoparone, 8548
[120-12-7] Anthracene, 721
[120-14-9] Veratraldehyde, 10084
[120-18-3] 2-Naphthalenesulfonic Acid, 6464
[120-23-0] 2-Naphthoxyacetic Acid, 6484
[120-29-6] Tropine, 9912
[120-32-1] Clorophene, 2469
[120-34-3] N-Sulfanilyl-3,4-xylamide, 9102
[120-36-5] Dichlorprop, 3128
[120-46-7] Dibenzoylmethane, 3056
[120-47-8] Ethylparaben, 3883
[120-51-4] Benzyl Benzoate, 1162
[120-57-0] Piperonal, 7628
[120-58-1] Isosafrole, 5243
[120-62-7] Sulfoxide, 9140
[120-66-1] o-Acetotoluide*, 74
[120-72-9] Indole, 4993
[120-73-0] 1H-Purine, 8129
[120-78-5] 2,2'-Dithiobis[benzothiazole], 3435
[120-80-9] Pyrocatechol, 8183
[120-82-1] 1,2,4-Trichlorobenzene, 9760
[120-83-2] 2,4-Dichlorophenol, 3122
[120-87-6] 9,10-Dihydroxystearic Acid, 3230
[120-89-8] Parabanic Acid, 7154
[120-92-3] Cyclopentanone, 2811
[120-93-4] 2-Imidazolidinone, 4950
[120-97-8] Dichlorphenamide, 3127
[121-19-7] Roxarsone, 8430
[121-20-0] Cinerin II*, 2351
[121-21-1] Pyrethrin I*, 8148
[121-25-5] Amprolium, 631
[121-29-9] Pyrethrin II*, 8148
[121-30-2] Chloraminophenamide, 2119
[121-32-4] Ethyl Vanillin, 3904
[121-33-5] Vanillin, 10069
[121-34-6] Vanillic Acid, 10068
[121-40-4] Terreic Acid, 9317
[121-43-7] Trimethyl Borate, 9842
[121-44-8] Triethylamine, 9799
[121-47-1] Metanilic Acid, 5988
[121-54-0] Benzethonium Chloride, 1103
[121-55-1] Subathizone, 9030
[121-57-3] Sulfanilic Acid, 9096
[121-59-5] Carbarsone, 1830
[121-60-8] N-Acetylsulfanilyl Chloride, 106
[121-61-9] N⁴-Acetylsulfanilamide, 104
[121-62-0] N-Acetylsulfanilic Acid, 105
[121-66-4] 2-Amino-5-nitrothiazole, 477
[121-69-7] N,N-Dimethylaniline, 3284
[121-75-5] Malathion, 5740

[121-79-9] Propyl Gallate, 8044
[121-81-3] Nitromide, 6709
[121-82-4] Cyclonite, 2805
[121-91-5] Isophthalic Acid, 5214
[122-03-2] Cuminaldehyde, 2687
[122-04-3] 4-Nitrobenzoyl Chloride, 6687
[122-06-5] Stilbamidine, 8970
[122-09-8] Phentermine, 7415
[122-10-1] Bomyl, 1360
[122-11-2] Sulfadimethoxine, 9073
[122-14-5] Fenitrothion, 4017
[122-15-6] Dimetan, 3259
[122-16-7] Sulfanitran, 9103
[122-18-9] Cetalkonium Chloride, 2059
[122-25-8] 5,5'-Methylenedisalicylic Acid, 6142
[122-28-1] m-Nitroacetanilide*, 6677
[122-32-7] Triolein, 9861
[122-34-9] Simazine, 8681
[122-39-4] Diphenylamine, 3375
[122-42-9] Propham, 8001
[122-46-3] m-Cresyl Acetate, 2651
[122-48-5] Zingerone, 10301
[122-57-6] Benzylideneacetone, 1172
[122-59-8] Phenoxyacetic Acid, 7407
[122-62-3] Bis(2-ethylhexyl) Sebacate, 1292
[122-69-0] Cinnamyl Cinnamate, 2364
[122-78-1] Phenylacetaldehyde, 7419
[122-79-2] Phenyl Acetate, 7421
[122-80-5] p-Aminoacetanilide, 431
[122-87-2] N-(4-Hydroxyphenyl)glycine, 4885
[122-99-6] 2-Phenoxyethanol, 7410
[123-02-4] Tridecylbenzene, 9792
[123-03-5] Cetylpyridinium Chloride, 2074
[123-08-0] p-Hydroxybenzaldehyde, 4856
[123-11-5] p-Anisaldehyde, 701
[123-19-3] Dipropyl Ketone, 3408
[123-29-5] Ethyl Pelargonate, 3884
[123-30-8] p-Aminophenol, 482
[123-31-9] Hydroquinone, 4853
[123-33-1] Maleic Hydrazide, 5745
[123-35-3] β-Myrcene, 6413
[123-38-6] Propionaldehyde, 8008
[123-39-7] N-Methylformamide, 6152
[123-42-2] Diacetone Alcohol, 3008
[123-43-3] Sulfoacetic Acid, 9126
[123-45-5] Sulfonyldiacetic Acid, 9134
[123-47-7] Prolonium Iodide, 7965
[123-51-3] Isopentyl Alcohol, 5212
[123-54-6] Acetylacetone, 82
[123-56-8] Succinimide, 9040
[123-62-6] Propionic Anhydride, 8011
[123-63-7] Paraldehyde, 7160
[123-66-0] Ethyl Caproate, 3824
[123-72-8] Butyraldehyde, 1627
[123-75-1] Pyrrolidine, 8199
[123-76-2] Levulinic Acid, 5498
[123-77-3] Azodicarbonamide, 950
[123-78-4] Sphingosine, 8899
[123-82-0] Tuaminoheptane, 9934
[123-84-2] 1-[(2-Aminoethyl)amino]-2-propanol, 458
[123-86-4] n-Butyl Acetate, 1570
[123-90-0] Thiamorpholine, 9435
[123-91-1] Dioxane, 3353
[123-92-2] Isoamyl Acetate, 5125
[123-93-3] Thiodiglycolic Acid, 9467
[123-95-5] Butyl Stearate, 1625
[123-96-2] 2-Octanol, 6850
[123-99-9] Azelaic Acid, 938
[124-02-7] Diallylamine, 3015
[124-03-8] Cetyldimethylethylammonium Bromide, 2071
[124-04-9] Adipic Acid, 161

[124-07-2] Caprylic Acid, 1808
[124-09-4] 1,6-Hexanediamine, 4730
[124-13-0] Caprylic Aldehyde, 1809
[124-20-9] Spermidine, 8893
[124-23-2] Cetrimonium Stearate, 2069
[124-28-7] Dymanthine, 3525
[124-38-9] Carbon Dioxide, 1857
[124-40-3] Dimethylamine, 3278
[124-41-4] Sodium Methoxide, 8789
[124-42-5] Acetamidine Hydrochloride, 42
[124-43-6] Urea Hydrogen Peroxide, 10007
[124-48-1] Chlorodibromomethane, 2186
[124-58-3] Methanearsonic Acid, 6020
[124-63-0] Methanesulfonyl Chloride, 6022
[124-65-2] Sodium Cacodylate, 8738
[124-68-5] 2-Amino-2-methyl-1-propanol, 469
[124-72-1] Teflurane, 9265
[124-76-5] Isoborneol, 5143
[124-80-1] Sabadine, 8461
[124-82-3] Mercamphamide Sodium Salt*, 5915
[124-87-8] Picrotoxin, 7570
[124-90-3] Oxycodone Hydrochloride*, 7093
[124-92-5] Metopon Hydrochloride*, 6234
[124-94-7] Triamcinolone, 9727
[124-97-0] Protoveratrine B*, 8086
[124-98-1] Cevine, 2076
[125-02-0] Prednisolone Sodium Phosphate, 7903
[125-03-1] Hydrocortamate Hydrochloride*, 4827
[125-04-2] Hydrocortisone 21-Sodium Succinate*, 4828
[125-10-0] Prednisone 21-Acetate*, 7904
[125-15-5] Vomicine, 10170
[125-20-2] Thymolphthalein, 9542
[125-24-6] Pseudomorphine, 8104
[125-25-7] Codeine Hydrobromide*, 2525
[125-27-9] Codeine Methyl Bromide, 2526
[125-28-0] Dihydrocodeine, 3214
[125-29-1] Hydrocodone, 4826
[125-31-5] Xylenol Blue, 10216
[125-33-7] Primidone, 7927
[125-44-0] Vinbarbital Sodium, 10118
[125-46-2] Usnic Acid, 10031
[125-51-9] Pipenzolate Bromide, 7614
[125-52-0] Oxyphencyclimine Hydrochloride*, 7107
[125-53-1] Oxyphencyclimine, 7107
[125-55-3] Narcobarbital, 6509
[125-56-4] Methadone Hydrochloride dl-Form*, 6008
[125-60-0] Fenpiverinium Bromide, 4032
[125-64-4] Methyprylon, 6216
[125-65-5] Pleuromutilin, 7694
[125-68-8] Levomethorphan Hydrobromide*, 8274
[125-69-9] Dextromethorphan Hydrobromide*, 8274
[125-73-5] Levorphanol d-Form*, 5496
[125-84-8] Aminoglutethimide, 460
[125-85-9] Caramiphen Hydrochloride*, 1820
[125-86-0] Caramiphen Ethanedisulfonate*, 1820
[125-88-2] Aprobarbital Sodium Salt*, 794

[125-99-5] Tridihexethyl Iodide, 9794
[126-00-1] Diphenolic Acid, 3370
[126-02-3] Cycrimine Hydrochloride, 2825
[126-05-6] Streptobiosamine, 8978
[126-07-8] Griseofulvin, 4571
[126-11-4] Tris(hydroxymethyl)-nitromethane, 9883
[126-12-5] Anileridine Dihydrochloride*, 695
[126-14-7] Sucrose Octaacetate, 9052
[126-15-8] R-11, 8271
[126-17-0] Solasodine, 8857
[126-18-1] Smilagenin, 8705
[126-19-2] Sarsasapogenin, 8526
[126-22-7] Butonate*, 9753
[126-27-2] Oxethazaine, 7067
[126-29-4] Violaxanthin, 10136
[126-30-7] Neopentyl Glycol, 6547
[126-31-8] Methiodal Sodium, 6051
[126-33-0] Sulfolane, 9128
[126-45-4] Silver Citrate, 8654
[126-49-8] Prephenic Acid, 7920
[126-52-3] Ethinamate, 3779
[126-72-7] Tris-BP, 9881
[126-73-8] Tributyl Phosphate, 9749
[126-81-8] 5,5-Dimethyl-1,3-cyclohexanedione, 3291
[126-93-2] Oxanamide, 7053
[126-96-5] Sodium Diacetate, 8752
[126-98-7] Methacrylonitrile, 6006
[127-00-4] sec-Propylene Chlorohydrin, 8036
[127-06-0] Acetoxime, 75
[127-07-1] Hydroxyurea, 4896
[127-08-2] Potassium Acetate, 7764
[127-09-3] Sodium Acetate, 8711
[127-17-3] Pyruvic Acid, 8205
[127-18-4] Tetrachloroethylene, 9332
[127-19-5] N,N-Dimethylacetamide, 3277
[127-22-0] Taraxerol, 9231
[127-25-3] Methyl Abietate, 6087
[127-27-5] Pimaric Acid, 7582
[127-29-7] Pseudoaconitine, 8095
[127-31-1] Fludrocortisone, 4166
[127-33-3] Demeclocycline, 2937
[127-35-5] Phenazocine, 7360
[127-39-9] Diisobutyl Sodium Sulfosuccinate, 3238
[127-40-2] Xanthophyll, 10197
[127-47-9] Vitamin A Acetate*, 10150
[127-48-0] Trimethadione, 9836
[127-50-4] Mercamphamide, 5915
[127-52-6] Chloramine-B, 2117
[127-57-1] Sulfapyridine Sodium Salt Monohydrate*, 9108
[127-58-2] Sulfamerazine Monosodium Salt*, 9081
[127-60-6] Acediasulfone Sodium Salt*, 21
[127-63-9] Diphenyl Sulfone, 3392
[127-65-1] Chloramine-T, 2118
[127-69-5] Sulfisoxazole, 9125
[127-71-9] Sulfabenzamide, 9065
[127-77-5] Sulfabenz, 9064
[127-79-7] Sulfamerazine, 9081
[127-82-2] Zinc p-Phenolsulfonate, 10283
[127-83-3] Calcium Phenolsulfonate, 1737
[127-85-5] Sodium Arsanilate, 8719
[127-91-3] β-Pinene, 7600
[127-95-7] Potassium Binoxalate, 7772
[127-96-8] Potassium Tetroxalate, 7863
[128-00-7] Silver Lactate, 8660
[128-08-5] N-Bromosuccinimide, 1462
[128-09-6] N-Chlorosuccinimide, 2217
[128-12-1] Acetosulfone Sodium, 73

Denotes derivative of title compound

[128-13-2] Ursodiol, *10026*
[128-19-8] 4-Pregnene-17α,20β,21-triol-3-one, *7914*
[128-20-1] Pregnan-3α-ol-20-one, *7910*
[128-23-4] 3,20-Pregnanedione, *7909*
[128-27-8] Pyrocalciferol, *8181*
[128-37-0] Butylated Hydroxytoluene, *1583*
[128-45-0] Streptomycin B, *8984*
[128-46-1] Dihydrostreptomycin, *3222*
[128-49-4] Docusate Calcium, *3459*
[128-53-0] N-Ethylmaleimide, *3868*
[128-57-4] Sennoside B*, *8600*
[128-62-1] Noscapine, *6815*
[128-68-7] Phenicin, *7379*
[128-71-2] Phenoltetrachlorophthalein Disodium Salt*, *7398*
[128-76-7] Laurotetanine, *5400*
[128-80-3] Quinizarin Green SS, *8251*
[128-87-0] Bis(4-amino-1-anthraquinonyl)amine, *1283*
[129-00-0] Pyrene, *8147*
[129-03-3] Cyproheptadine, *2842*
[129-06-6] Warfarin Sodium Salt*, *10174*
[129-16-8] Merbromin, *5914*
[129-17-9] Sulphan Blue, *9161*
[129-18-0] Phenylbutazone Sodium Salt*, *7431*
[129-20-4] Oxyphenbutazone, *7106*
[129-24-8] Viridicatin, *10143*
[129-46-4] Suramin Sodium, *9181*
[129-49-7] Methysergide Hydrogen Maleate*, *6217*
[129-51-1] Ergonovine Maleate*, *3694*
[129-63-5] Acetrizoate Sodium, *79*
[129-64-6] Carbic Anhydride, *1842*
[129-65-7] Porphyrillic Acid, *7760*
[129-66-8] 2,4,6-Trinitrobenzoic Acid, *9857*
[129-74-8] Buclizine Dihydrochloride*, *1484*
[129-77-1] Piperidolate Hydrochloride*, *7623*
[129-79-3] 2,4,7-Trinitrofluorenone, *9858*
[129-83-9] Phenampromid(e), *7353*
[130-01-8] Senecionine, *8595*
[130-15-4] 1,4-Naphthoquinone, *6482*
[130-16-5] Cloxyquin, *2483*
[130-22-3] Sodium Alizarinesulfonate, *8714*
[130-24-5] Vitamin K₅ Hydrochloride*, *10162*
[130-26-7] Iodochlorhydroxyquin, *5052*
[130-37-0] Menadione Sodium Bisulfite*, *5874*
[130-40-5] Riboflavin Monophosphate Monosodium Salt*, *8368*
[130-61-0] Thioridazine Hydrochloride*, *9497*
[130-73-4] Methestrol, *6045*
[130-79-0] Dimestrol, *3257*
[130-80-3] Diethylstilbestrol Dipropionate*, *3177*
[130-85-8] Pamoic Acid, *7136*
[130-86-9] Protopine, *8083*
[130-89-2] Quinine Hydrochloride*, *8245*
[130-95-0] Quinine, *8245*
[131-01-1] Deserpidine, *2964*
[131-02-2] Reserpiline, *8313*
[131-03-3] α-Yohimbine, *10238*
[131-11-3] Dimethyl Phthalate, *3304*
[131-12-4] Pimpinellin, *7591*
[131-13-5] Menadiol Sodium Diphosphate*, *5873*
[131-28-2] Narceine, *6508*

[131-49-7] Meglumine Diatrizoate, *5851*
[131-53-3] Dioxybenzone, *3357*
[131-54-4] Benzophenone-6, *1130*
[131-56-6] Benzoresorcinol, *1138*
[131-57-7] Oxybenzone, *7088*
[131-67-9] Phthalofyne, *7530*
[131-69-1] Phthalylsulfacetamide, *7532*
[131-73-7] Dipicrylamine, *3396*
[131-74-8] Ammonium Picrate, *581*
[131-79-3] Yellow OB, *10229*
[131-89-5] 2-Cyclohexyl-4,6-dinitrophenol, *2802*
[131-91-9] 1-Nitroso-2-naphthol, *6739*
[131-99-7] Inosinic Acid, *5007*
[132-13-8] α-Ribazole, *8366*
[132-17-2] Benztropine Mesylate, *1156*
[132-18-3] Diphenylpyraline Hydrochloride*, *3390*
[132-20-7] Pheniramine Maleate*, *7383*
[132-21-8] Brompheniramine d-Form*, *1467*
[132-22-9] Chlorpheniramine, *2232*
[132-35-4] Proxazole Citrate*, *8090*
[132-40-1] Papaverine Nitrite*, *7151*
[132-49-0] Magnesium Acetylsalicylate, *5689*
[132-57-0] Croceic Acid, *2656*
[132-58-1] Cinchophen Hydrochloride*, *2347*
[132-60-5] Cinchophen, *2347*
[132-66-1] Naptalam, *6505*
[132-69-4] Benzydamine Hydrochloride*, *1157*
[132-86-5] Naphthoresorcinol, *6483*
[132-89-8] Chlorthenoxazin(e), *2247*
[132-92-3] Methicillin Sodium, *6047*
[132-93-4] Phenethicillin Potassium, *7369*
[132-98-9] Penicillin V Potassium Salt*, *7230*
[133-04-0] Asarinin, *860*
[133-06-2] Captan, *1815*
[133-07-3] Folpet, *4255*
[133-09-5] p-Aminosalicylic Acid Potassium Salt*, *498*
[133-11-9] Phenyl Aminosalicylate, *7426*
[133-15-3] p-Aminosalicylic Acid Calcium Salt*, *498*
[133-17-5] o-Iodohippurate Sodium, *5055*
[133-26-6] Peucedanin, *7332*
[133-32-4] Indolebutyric Acid, *4995*
[133-36-8] Piperazine Tartrate*, *7617*
[133-37-9] DL-Tartaric Acid, *9236*
[133-51-7] N-Methylglucamine Antimonate*, *6154*
[133-53-9] Dichloroxylenol, *3126*
[133-58-4] Nitromersol, *6707*
[133-65-3] Solasulfone, *8859*
[133-67-5] Trichlormethiazide, *9754*
[133-89-1] Uridine Diphosphate Glucose, *10018*
[133-90-4] Chloramben, *2115*
[133-91-5] 3,5-Diiodosalicylic Acid, *3234*
[134-01-0] Peonidin, *7285*
[134-03-2] Sodium Ascorbate, *8723*
[134-04-3] Pelargonidin, *7199*
[134-20-3] Methyl Anthranilate, *6099*
[134-31-6] 8-Hydroxyquinoline Sulfate*, *4890*
[134-32-7] 1-Naphthylamine, *6485*
[134-36-1] Erythromycin Propionate, *3725*
[134-37-2] Amphenidone, *621*
[134-49-6] Phenmetrazine, *7385*

[134-50-9] Aminacrine Hydrochloride*, *427*
[134-53-2] Amprotropine Phosphate, *632*
[134-55-4] Phenyl Acetylsalicylate, *7424*
[134-58-7] 8-Azaguanine, *928*
[134-62-3] Deet, *2912*
[134-63-4] Lobeline Hydrochloride*, *5580*
[134-64-5] Lobeline Sulfate*, *5580*
[134-65-6] Lobeline (±)-Form*, *5580*
[134-71-4] Ephedrine dl-Form Hydrochloride*, *3645*
[134-72-5] Ephedrine l-Form Sulfate*, *3645*
[134-80-5] Diethylpropion Hydrochloride*, *3175*
[134-81-6] Benzil, *1107*
[134-95-2] Mandelic Acid Calcium Salt*, *5757*
[134-96-3] Syringaldehyde, *9191*
[135-07-9] Methyclothiazide, *6086*
[135-09-1] Hydroflumethiazide, *4830*
[135-19-3] 2-Naphthol, *6471*
[135-20-6] Cupferron, *2688*
[135-23-9] Methapyrilene Hydrochloride*, *6027*
[135-31-9] Pyrrobutamine Diphosphate*, *8196*
[135-43-3] Lauroguadine, *5398*
[135-44-4] Leucinocaine Mesylate, *5476*
[135-48-8] Pentacene, *7240*
[135-58-0] Mesulfen, *5977*
[135-59-1] Thionine, *9483*
[135-67-1] Phenoxazine, *7406*
[135-68-2] 4-Amino-4'-chlorodiphenyl, *454*
[135-97-7] Pseudotropine, *8106*
[135-98-8] sec-Butylbenzene, *1585*
[136-25-4] Erbon, *3681*
[136-35-6] Diazoaminobenzene, *3046*
[136-40-3] Phenazopyridine Hydrochloride, *7361*
[136-44-7] Glyceryl p-Aminobenzoate, *4496*
[136-46-9] Parethoxycaine Hydrochloride*, *7171*
[136-47-0] Tetracaine Hydrochloride, *9330*
[136-55-0] Procaine Butyrate*, *7937*
[136-60-7] n-Butyl Benzoate, *1587*
[136-69-6] Protokylol Hydrochloride*, *8082*
[136-70-9] Protokylol, *8082*
[136-72-1] Piperic Acid (E,E)-Form*, *7620*
[136-77-6] 4-Hexylresorcinol, *4750*
[136-78-7] Disul-sodium, *3430*
[136-79-8] 5-Nitro-o-phenetidine, *6714*
[136-82-3] Piperocaine, *7627*
[136-91-4] Calcium 2-Ethylbutanoate, *1707*
[136-95-8] 2-Aminobenzothiazole, *444*
[136-96-9] Diamthazole Dihydrochloride, *3033*
[137-00-8] 4-Methyl-5-thiazoleethanol, *6208*
[137-05-3] Mecrylate, *5827*
[137-08-6] Pantothenic Acid Calcium Salt*, *7147*
[137-16-6] Gardol®, *4379*
[137-26-8] Thiram, *9510*
[137-30-4] Ziram, *10305*
[137-32-6] 2-Methyl-1-butanol, *6109*
[137-40-6] Sodium Propionate, *8816*
[137-42-8] Metham Sodium, *6016*
[137-53-1] D-Thyroxine Sodium Salt*, *9555*

Denotes derivative of title compound

[137-58-6] Lidocaine, 5505
[137-76-8] Cetotiamine, 2064
[137-86-0] Octotiamine, 6857
[138-14-7] Deferoxamine Methanesulfonate*, 2914
[138-15-8] Glutamic Acid Hydrochloride*, 4477
[138-32-9] Cetrimonium Tosylate*, 2068
[138-37-4] Mafenide Hydrochloride*, 5683
[138-39-6] Mafenide, 5683
[138-41-0] Carzenide, 1927
[138-52-3] Salicin, 8476
[138-53-4] Mandelonitrile Glucoside, 5760
[138-55-6] Picrocrocin, 7563
[138-56-7] Trimethobenzamide, 9839
[138-58-9] Bismuth Iodosubgallate, 1310
[138-59-0] Shikimic Acid, 8625
[138-61-4] Nordefrin Hydrochloride* (unspecified), 6785
[138-84-1] Potassium p-Aminobenzoate, 7766
[138-86-3] Limonene, 5518
[138-89-6] p-Nitroso-N,N-dimethylaniline, 6736
[138-92-1] Betazole Dihydrochloride*, 1230
[139-02-6] Phenol Sodium Salt*, 7390
[139-02-6] Sodium Phenoxide, 8804
[139-06-0] Calcium Cyclamate, 1705
[139-10-6] Amphetamine Phosphate*, 623
[139-13-9] Nitrilotriacetic Acid, 6675
[139-33-3] Edetate Disodium, 3556
[139-40-2] Propazine, 7996
[139-42-4] Cerous Oxalate, 2045
[139-62-8] Cyclomethycaine, 2804
[139-66-2] Phenyl Sulfide, 7467
[139-68-4] Phenatine, 7358
[139-85-5] Protocatechualdehyde, 8079
[139-88-8] Sodium Tetradecyl Sulfate, 8838
[139-89-9] Versen-Ol, 10102
[139-91-3] Furaltadone, 4315
[139-93-5] Arsphenamine, 851
[139-94-6] Nithiazide, 6660
[140-11-4] Benzyl Acetate, 1158
[140-22-7] sym-Diphenylcarbazide, 3379
[140-28-3] Benzathine, 1092
[140-29-4] Benzyl Cyanide, 1166
[140-40-9] Aminitrozole, 430
[140-47-6] Phenocoll Salicylate*, 7388
[140-55-6] Pseudoconhydrine, 8099
[140-57-8] Aramite, 805
[140-59-0] Stilbamidine Isethionate*, 8970
[140-63-6] Propamidine Isethionate*, 7981
[140-64-7] Pentamidine Isethionate*, 7254
[140-65-8] Pramoxine, 7888
[140-67-0] Estragole, 3748
[140-80-7] Novoldiamine, 6819
[140-87-4] Cyacetacide, 2751
[140-88-5] Ethyl Acrylate, 3805
[140-89-6] Potassium Xanthogenate, 7876
[140-93-2] Sodium Isopropyl Xanthate, 8780
[140-95-4] Oxymethurea, 7102
[140-99-8] Calcium Succinate, 1752
[141-00-4] Cadmium Succinate, 1664
[141-01-5] Ferrous Fumarate, 4094
[141-03-7] Dibutyl Succinate, 3087
[141-05-9] Diethyl Maleate, 3172

[141-10-6] Pseudoionone, 8101
[141-20-8] Diethylene Glycol Monolaurate, 3169
[141-22-0] Ricinoleic Acid, 8378
[141-26-4] α-Citronellal*, 2390
[141-32-2] n-Butyl Acrylate, 1574
[141-43-5] Ethanolamine, 3772
[141-52-6] Sodium Ethoxide, 8758
[141-53-7] Sodium Formate, 8765
[141-62-8] Decamethyltetrasiloxane, 2907
[141-63-9] Dodecamethylpentasiloxane, 3463
[141-66-2] Dicrotophos, 3137
[141-75-3] n-Butyryl Chloride, 1634
[141-78-6] Ethyl Acetate, 3803
[141-79-7] Mesityl Oxide, 5968
[141-82-2] Malonic Acid, 5749
[141-84-4] Rhodanine, 8351
[141-90-2] 2-Thiouracil, 9504
[141-94-6] Hexetidine, 4741
[141-97-9] Ethyl Acetoacetate, 3804
[142-03-0] Aluminum Diacetate, 349
[142-10-9] Glyceraldehyde 3-Phosphate, 4491
[142-17-6] Calcium Oleate, 1731
[142-47-2] Monosodium Glutamate, 6337
[142-59-6] Nabam, 6426
[142-61-0] Caproyl Chloride, 1806
[142-62-1] n-Caproic Acid, 1803
[142-68-7] Tetrahydropyran, 9356
[142-71-2] Cupric Acetate, 2690
[142-72-3] Magnesium Acetate, 5688
[142-73-4] Iminodiacetic Acid, 4952
[142-82-5] n-Heptane, 4694
[142-84-7] n-Dipropylamine, 3406
[142-88-1] Piperazine Adipate, 7618
[142-96-1] n-Butyl Ether, 1603
[143-07-7] Lauric Acid, 5396
[143-08-8] n-Nonyl Alcohol, 6774
[143-13-5] n-Nonyl Acetate, 6773
[143-15-7] Lauryl Bromide, 5402
[143-18-0] Potassium Oleate, 7818
[143-19-1] Oleic Acid Sodium Salt*, 6965
[143-24-8] Tetraglyme, 9348
[143-28-2] Oleyl Alcohol, 6968
[143-33-9] Sodium Cyanide, 8750
[143-50-0] Chlordecone, 2131
[143-52-2] Metopon, 6234
[143-57-7] Protoveratrine A*, 8086
[143-62-4] Digitoxigenin, 3205
[143-66-8] Sodium Tetraphenylborate, 8839
[143-67-9] Vinblastine Sulfate*, 10119
[143-74-8] Phenolsulfonphthalein, 7397
[143-76-0] Cyclobarbital Calcium Salt*, 2780
[143-81-7] Butabarbital Sodium, 1530
[144-00-3] Hexethal Sodium, 4740
[144-02-5] Barbital Sodium Salt*, 989
[144-12-7] Tiemonium Iodide, 9571
[144-14-9] Anileridine, 695
[144-23-0] Magnesium Citrate, Dibasic, 5700
[144-29-6] Piperazine Citrate*, 7617
[144-33-2] Sodium Citrate, Acid, 8747
[144-49-0] Fluoroacetic Acid, 4205
[144-55-8] Sodium Bicarbonate, 8726
[144-62-7] Oxalic Acid, 7043
[144-68-3] Zeaxanthin, 10248
[144-75-2] Sulfoxone Sodium, 9141
[144-80-9] Sulfacetamide, 9067
[144-82-1] Sulfamethizole, 9084
[144-83-2] Sulfapyridine, 9108
[144-87-6] Glycarsamide, 4489
[145-12-0] Oxymesterone, 7099
[145-13-1] Pregnenolone, 7915

[145-41-5] Dehydrocholic Acid Sodium Salt*, 2922
[145-54-0] Propyromazine, 8057
[145-73-3] Endothall, 3615
[145-94-8] Chlorindanol, 2144
[146-14-5] Flavine-Adenine Dinucleotide, 4131
[146-17-8] Riboflavin Monophosphate, 8368
[146-22-5] Nitrazepam, 6667
[146-28-1] Thiopropazate Dihydrochloride*, 9493
[146-37-2] Laurolinium Acetate, 5399
[146-48-5] Yohimbine, 10236
[146-54-3] Triflupromazine, 9814
[146-56-5] Fluphenazine Dihydrochloride*, 4226
[146-80-5] Xanthosine, 10199
[146-84-9] Silver Picrate, 8670
[146-90-7] Cinnabarine, 2355
[147-14-8] Copper Phthalocyanine, 2584
[147-20-6] Diphenylpyraline, 3390
[147-24-0] Diphenhydramine Hydrochloride*, 3367
[147-27-3] Dimoxyline, 3318
[147-61-5] Pentoxyl, 7279
[147-71-7] D-Tartaric Acid, 9235
[147-73-9] meso-Tartaric Acid, 9238
[147-81-9] Arabinose, 799
[147-82-0] 2,4,6-Tribromoaniline, 9740
[147-85-3] Proline, 7963
[147-90-0] Morpholine Salicylate*, 6362
[147-93-3] Thiosalicylic Acid, 9498
[147-94-4] Cytarabine, 2853
[148-01-6] Dinitolmide, 3321
[148-03-8] β-Tocopherol, 9632
[148-18-5] Ditiocarb Sodium, 3443
[148-24-3] 8-Hydroxyquinoline, 4890
[148-25-4] Chromotropic Acid, 2298
[148-56-1] Flumethiazide, 4174
[148-64-1] Chlorothen Citrate*, 2220
[148-65-2] Chlorothen, 2220
[148-72-1] Pilocarpine Nitrate*, 7578
[148-75-4] 2-Naphthol-3,6-disulfonic Acid, 6474
[148-79-8] Thiabendazole, 9426
[148-82-3] Melphalan, 5871
[148-83-4] Ostruthin, 7027
[149-15-5] Butacaine Sulfate*, 1531
[149-16-6] Butacaine, 1531
[149-29-1] Patulin, 7185
[149-30-4] 2-Mercaptobenzothiazole, 5916
[149-32-6] Erythritol, 3714
[149-44-0] Sodium Formaldehydesulfoxylate, 8764
[149-45-1] Tiron, 9606
[149-64-4] N-Butylscopolammonium Bromide, 1624
[149-90-6] Acetylleucine Monoethanolamine, 96
[149-91-7] Gallic Acid, 4363
[150-13-0] p-Aminobenzoic Acid, 443
[150-25-4] Bicine, 1248
[150-30-1] Phenylalanine DL-Form*, 7425
[150-38-9] Edetate Trisodium, 3558
[150-59-4] Alverine, 386
[150-60-7] Dibenzyl Disulfide, 3060
[150-61-8] 1,2-Dianilinoethane, 3035
[150-68-5] Monuron, 6342
[150-69-6] Dulcin, 3517
[150-86-7] Phytol, 7545
[150-90-3] Sodium Succinate, 8828
[150-97-0] Mevalonic Acid, 6250
[151-06-4] Chlorphentermine Hydrochloride*, 2235
[151-07-5] Chloroarsenol, 2170

Denotes derivative of title compound

Denotes derivative of title compound REG-73

[313-67-7] Aristolochic Acid, *822*
[314-13-6] Evan's Blue, *3952*
[314-35-2] Etamiphyllin, *3754*
[314-40-9] Bromacil, *1402*
[314-42-1] Isocil, *5174*
[314-50-1] Orotidine, *7005*
[315-18-4] Mexacarbate, *6253*
[315-22-0] Monocrotaline, *6333*
[315-30-0] Allopurinol, *287*
[315-37-7] Testosterone Enanthate*, *9322*
[315-72-0] Opipramol, *6985*
[315-80-0] Dibenzepin Hydrochloride*, *3055*
[316-42-7] Emetine Dihydrochloride*, *3600*
[316-81-4] Thioproperazine, *9494*
[317-34-0] Aminophylline, *485*
[317-52-2] Hexafluorenium Bromide, *4721*
[318-23-0] Imolamine, *4959*
[318-98-9] Propranolol Hydrochloride*, *8025*
[319-89-1] Tetroquinone, *9385*
[320-67-2] Azacitidine, *923*
[321-54-0] Coroxon, *2600*
[321-55-1] Haloxon, *4636*
[321-64-2] Tacrine, *9199*
[322-35-0] Benserazide, *1079*
[322-79-2] Triflusal, *9817*
[326-43-2] Phenyramidol Hydrochloride*, *7474*
[327-57-1] Norleucine, *6800*
[327-97-9] Chlorogenic Acid, *2194*
[328-42-7] Oxalacetic Acid, *7041*
[328-50-7] α-Ketoglutaric Acid, *5314*
[329-56-6] Norepinephrine Hydrochloride*, *6788*
[329-63-5] Epinephrine *dl*-Form Hydrochloride*, *3656*
[329-65-7] Epinephrine *dl*-Form*, *3656*
[329-71-5] 2,5-Dinitrophenol, *3334*
[330-54-1] Diuron, *3447*
[330-55-2] Linuron, *5534*
[330-95-0] Nicarbazin, *6578*
[331-39-5] Caffeic Acid, *1673*
[333-20-0] Potassium Thiocyanate, *7866*
[333-36-8] Flurothyl, *4236*
[333-41-5] Diazinon, *3043*
[334-20-3] Queen Substance, *8214*
[334-48-5] *n*-Capric Acid, *1802*
[334-88-3] Diazomethane, *3048*
[337-47-3] Thiamylal Sodium Salt*, *9437*
[338-69-2] Alanine D-Form*, *205*
[338-83-0] Perfluorotripropylamine*, *4221*
[338-95-4] Isoflupredone, *5190*
[338-98-7] Isoflupredone 21-Acetate*, *5190*
[339-43-5] Carbutamide, *1881*
[339-44-6] Glymidine, *4517*
[340-56-7] Methaqualone Hydrochloride*, *6028*
[340-57-8] Mecloqualone, *5821*
[341-00-4] Etifelmin, *3909*
[341-70-8] Diethazine Hydrochloride*, *3157*
[341-88-8] HQNO, *4787*
[342-10-9] Kallidin, *5290*
[343-65-7] Kynurenine, *5340*
[345-78-8] *d*-Pseudoephedrine Hydrochloride*, *3645*
[346-18-9] Polythiazide, *7744*
[348-67-4] Methionine D-Form*, *6053*
[349-46-2] Cystine D-Form*, *2851*
[350-03-8] Methyl Pyridyl Ketone, *6195*

[350-12-9] Sulbentine, *9061*
[352-21-6] 4-Amino-3-hydroxybutyric Acid, *464*
[352-32-9] *p*-Fluorotoluene*, *4218*
[352-70-5] *m*-Fluorotoluene*, *4218*
[352-93-2] Ethyl Sulfide, *3898*
[352-97-6] Glycocyamine, *4505*
[353-50-4] Carbonyl Fluoride, *1868*
[355-80-6] 2,2,3,3,4,4,5,5-Octafluoro-1-pentanol, *6842*
[356-12-7] Fluocinonide, *4186*
[357-07-3] Oxymorphone Hydrochloride*, *7103*
[357-08-4] Naloxone Hydrochloride* (Dihydrate), *6449*
[357-56-2] Dextromoramide, *2997*
[357-57-3] Brucine, *1476*
[357-67-5] Phenetharbital, *7368*
[357-70-0] Galanthamine, *4357*
[358-52-1] Hexapropymate, *4733*
[359-83-1] Pentazocine, *7261*
[360-66-7] Androisoxazole, *674*
[360-68-9] Coprosterol, *2588*
[360-70-3] Nandrolone Decanoate*, *6452*
[361-37-5] Methysergide, *6217*
[362-29-8] Propiomazine, *8007*
[362-74-3] Bucladesine, *1483*
[363-24-6] Prostaglandin E₂, *8064*
[364-62-5] Metoclopramide, *6226*
[364-98-7] Diazoxide, *3051*
[365-26-4] *p*-Hydroxyephedrine, *4869*
[366-18-7] α,α'-Dipyridyl, *3411*
[366-70-1] Procarbazine Hydrochloride*, *7938*
[367-25-9] 2,4-Difluoroaniline, *3191*
[367-47-5] Glyceraldehyde, *4490*
[367-51-1] Sodium Thioglycolate, *8842*
[369-77-7] Cloflucarban, *2438*
[370-14-9] Pholedrine, *7485*
[371-40-4] *p*-Fluoroaniline, *4207*
[371-86-8] Mipafox, *6292*
[372-09-8] Cyanoacetic Acid, *2759*
[372-66-7] Heptaminol, *4691*
[372-75-8] Citrulline, *2392*
[373-02-4] Nickel Acetate, *6583*
[378-44-9] Betamethasone, *1226*
[379-79-3] Ergotamine Tartrate*, *3703*
[382-45-6] Adrenosterone, *174*
[382-67-2] Desoximetasone, *2975*
[386-17-4] Iodophthalein Sodium Free Acid*, *5059*
[388-51-2] Metofenazate, *6228*
[389-08-2] Nalidixic Acid, *6446*
[390-28-3] Methoxamine, *6067*
[390-64-7] Prenylamine, *7919*
[392-56-3] Hexafluorobenzene, *4722*
[395-28-8] Isoxsuprine, *5259*
[396-01-0] Triamterene, *9731*
[398-23-2] 4,4'-Difluorodiphenyl, *3193*
[402-61-9] 5-Methylpyrazole-3-carboxylic Acid, *6194*
[404-82-0] Fenfluramine Hydrochloride*, *4015*
[404-86-4] Capsaicin, *1811*
[405-22-1] Nidroxyzone, *6615*
[405-50-5] *p*-Fluorophenylacetic Acid, *4215*
[406-76-8] Carnitine DL-Form*, *1898*
[406-90-6] Fluroxene, *4237*
[407-41-0] Phosphoserine, *7520*
[409-21-2] Silicon Carbide, *8636*
[420-04-2] Cyanamide, *2754*
[420-05-3] Cyanic Acid, *2756*
[421-20-5] Methyl Fluorosulfonate, *6151*
[423-55-2] Perflubron, *7299*
[426-13-1] Fluorometholone, *4213*
[427-00-9] Desomorphine, *2972*
[427-01-0] Geissospermine, *4387*

[427-28-1] Gentrogenin, *4407*
[427-51-0] Cyproterone Acetate*, *2844*
[427-63-4] Erdin (+)-Form Methyl Ester*, *3682*
[428-07-9] Levomepate, *5491*
[431-03-8] Diacetyl, *3010*
[432-60-0] Allylestrenol, *299*
[432-68-8] Echinenone, *3543*
[434-03-7] Ethisterone, *3787*
[434-05-9] Methenolone 17-Acetate*, *6044*
[434-07-1] Oxymetholone, *7101*
[434-13-9] Lithocholic Acid, *5571*
[434-16-2] (3β)-7-Dehydrocholesterol, *2921*
[434-22-0] Nandrolone, *6452*
[434-43-5] Pentorex (unspecified), *7275*
[435-97-2] Phenprocoumon, *7413*
[436-05-5] Curine, *2745*
[437-38-7] Fentanyl, *4043*
[437-50-3] Gentisin, *4405*
[437-74-1] Xanthinol Niacinate, *10194*
[438-08-4] Etiocholanic Acid, *3913*
[438-22-2] Androstane, *675*
[438-23-3] Etiocholane, *3912*
[438-41-5] Chlordiazepoxide Hydrochloride*, *2132*
[438-60-8] Protriptyline, *8088*
[439-14-5] Diazepam, *3042*
[439-89-4] Gentianine, *4400*
[440-17-5] Trifluoperazine Dihydrochloride*, *9811*
[440-58-4] Iodamide, *5031*
[441-38-3] Benzoin Oxime, *1125*
[441-61-2] Ethylmethylthiambutene, *3875*
[442-16-0] Ethacridine, *3760*
[442-33-1] Cuspareine, *2749*
[442-51-3] Harmine, *4647*
[442-52-4] Clemizole, *2406*
[443-48-1] Metronidazole, *6242*
[444-27-9] Timonacic, *9586*
[445-30-7] Homarine, *4765*
[446-71-9] Homoeriodictyol, *4773*
[446-72-0] Genistein, *4395*
[446-86-6] Azathioprine, *935*
[446-95-7] Genisteine-Alkaloid, *4396*
[447-41-6] Nylidrin, *6830*
[451-13-8] Homogentisic Acid, *4775*
[451-71-8] Glyhexamide, *4516*
[452-35-7] Ethoxzolamide, *3801*
[454-14-8] Cuscohygrine, *2748*
[456-22-4] *p*-Fluorobenzoic Acid, *4209*
[456-59-7] Cyclandelate, *2771*
[457-60-3] Neoarsphenamine, *6534*
[457-87-4] *N*-Ethylamphetamine, *3809*
[458-24-2] Fenfluramine, *4015*
[458-35-5] Coniferyl Alcohol, *2568*
[458-37-7] Curcumin, *2744*
[458-88-8] Coniine, *2569*
[459-67-6] Hydnocarpic Acid, *4797*
[459-86-9] Mitoguazone, *6299*
[460-19-5] Cyanogen, *2761*
[461-05-2] Carnitine DL-Form Hydrochloride*, *1898*
[461-06-3] Carnitine, *1898*
[461-58-5] Dicyanodiamide, *3142*
[461-72-3] Hydantoin, *4796*
[461-78-9] Chlorphentermine, *2235*
[461-98-3] Kyanmethin, *5338*
[462-02-2] Cyamelide, *2752*
[462-06-6] Fluorobenzene, *4208*
[462-08-8] β-Aminopyridine*, *494*
[462-10-2] Homocystine, *4772*
[462-94-2] Cadaverine, *1645*
[463-40-1] Linolenic Acid, *5530*
[463-51-4] Ketene, *5309*
[463-56-9] Thiocyanic Acid, *9464*

Denotes derivative of title compound

[463-72-9] Carbamyl Chloride, 1827
[463-78-5] Orthoformic Acid, 7012
[463-82-1] Neopentane, 6545
[463-88-7] Neurine, 6567
[464-41-5] Bornyl Chloride, 1369
[464-72-2] Benzopinacol, 1131
[464-73-3] Isatide, 5115
[464-81-3] Bufotoxin, 1503
[464-85-7] Quinamine, 8232
[464-86-8] Conquinamine, 2572
[465-11-2] Gamabufotalin, 4370
[465-16-7] Oleandrin, 6963
[465-19-0] Bufogenin B, 1499
[465-21-4] Bufalin, 1493
[465-22-5] Scillarenin, 8544
[465-28-1] Carotol, 1905
[465-39-4] Resibufogenin, 8315
[465-42-9] Capsanthin, 1812
[465-61-2] Caldariomycin, 1761
[465-65-6] Naloxone, 6449
[465-74-7] Quinovic Acid, 8260
[465-92-9] Marrubiin, 5794
[465-99-6] Hederagenin, 4654
[466-06-8] Proscillaridin, 8060
[466-07-9] Neriifolin, 6559
[466-09-1] Uzarigenin*, 10035
[466-34-2] Picrolichenic Acid, 7564
[466-40-0] Isomethadone, 5201
[466-43-3] Atisine, 895
[466-49-9] Aspidospermine, 885
[466-61-5] Lycopodine, 5652
[466-80-8] α-Erythroidine, 3718
[466-81-9] β-Erythroidine, 3719
[466-85-3] Wieland-Gumlich Aldehyde, 10181
[466-90-0] Dihydrocodeinone Enol Acetate, 3215
[466-96-6] Pseudocodeine, 8098
[466-97-7] Normorphine, 6806
[466-99-9] Hydromorphone, 4847
[467-04-9] Oripavine, 6996
[467-14-1] Neopine, 6549
[467-15-2] Norcodeine, 6783
[467-18-5] Myrophine, 6420
[467-22-1] Carbiphene Hydrochloride*, 1846
[467-36-7] Thialbarbital, 9428
[467-43-6] Methitural, 6058
[467-51-6] Samandarine, 8497
[467-53-8] Diginin, 3199
[467-55-0] Hecogenin, 4652
[467-60-7] Pipradrol, 7638
[467-83-4] Dipipanone, 3398
[467-84-5] Phenadoxone, 7348
[467-86-7] Dioxaphetyl Butyrate, 3354
[467-98-1] Thebainone, 9412
[468-07-5] Phenomorphan, 7399
[468-10-0] Morphinan, 6358
[468-28-0] Lupulon, 5640
[468-45-1] Isorubijervine, 5242
[468-56-4] Hydroxypethidine, 4883
[468-61-1] Oxeladin, 7064
[468-65-5] Buthalital*, 1553
[468-76-8] Cassaine, 1938
[468-89-3] Nupharidine, 6826
[469-21-6] Doxylamine, 3497
[469-32-9] Hamamelitannin, 4639
[469-36-3] Oryzanol C*, 7016
[469-59-0] Jervine, 5277
[469-62-5] Propoxyphene, 8024
[469-79-4] Ketobemidone, 5312
[469-81-8] Morpheridine, 6357
[469-83-0] Cafestol, 1672
[470-40-6] Thujopsene, 9534
[470-82-6] Eucalyptol, 3540
[470-90-6] Chlorfenvinphos, 2137
[471-09-0] Felinine, 3990
[471-25-0] Propiolic Acid, 8006
[471-29-4] Methylguanidine, 6158

[471-34-1] Calcium Carbonate, 1697
[471-35-2] Cacodyl, 1639
[471-46-5] Oxamide, 7050
[471-47-6] Oxamic Acid, 7049
[471-53-4] Enoxolone, 3628
[471-71-6] Cassamine, 1939
[471-80-7] Steviol*, 8964
[471-87-4] Stachydrine, 8926
[471-95-4] Bufotalin, 1501
[472-61-7] Astaxanthin, 890
[472-70-8] Cryptoxanthin, 2676
[472-87-7] 3-Dehydroretinal, 2926
[472-92-4] δ-Carotene, 1904
[472-93-5] γ-Carotene, 1903
[473-06-3] Chrysanthenone, 2311
[473-30-3] Thiazolsulfone, 9444
[473-34-7] Dichloramine T, 3098
[473-41-6] Tolbutamide Sodium Salt*, 9646
[473-81-4] Glyceric Acid, 4492
[473-90-5] Mesoxalic Acid, 5971
[473-98-3] Betulin, 1237
[474-00-0] Eburnamonine, 3537
[474-07-7] Brazilin, 1392
[474-25-9] Chenodiol, 2096
[474-40-8] α1-Sitosterol, 8696
[474-45-3] Funtumine, 4314
[474-62-4] Campesterol, 1776
[474-69-1] Lumisterol, 5629
[474-70-4] Isopyrocalciferol, 5238
[474-73-7] 24-Hydroxycholesterol, 4865
[474-77-1] Epicholesterol, 3650
[474-86-2] Equilin, 3676
[475-20-7] Longifolene, 5597
[475-25-2] Hematein, 4667
[475-26-3] DFDT, 3000
[475-31-0] Glycocholic Acid, 4504
[475-54-7] Oosporein, 6982
[475-67-2] Isocorydine, 5178
[475-81-0] Glaucine, 4445
[476-28-8] Lycorine, 5656
[476-32-4] Chelidonine (+)-Form*, 2095
[476-33-5] Homochelidonine, 4769
[476-45-9] Javanicin, 5275
[476-66-4] Ellagic Acid, 3588
[476-69-7] Corydine, 2612
[476-70-0] Boldine, 1355
[476-71-1] Domesticine, 3472
[477-27-0] Colchiceine, 2535
[477-30-5] Demecolcine, 2938
[477-32-7] Visnadine, 10147
[477-47-4] Picropodophyllin, 7567
[477-58-7] Chondrocurine, 2268
[477-60-1] Bebeerine, 1042
[477-62-3] Isochondrodendrine, 5173
[477-67-8] Picropodophyllic Acid*, 7703
[477-75-8] Triptycene, 9879
[477-89-4] Casimiroin, 1936
[477-90-7] Bergenin, 1197
[477-93-0] Dimethoxanate, 3272
[478-05-7] Gladiolic Acid, 4442
[478-10-4] Elliptone, 3591
[478-43-3] Rhein, 8342
[478-53-5] Morphothebaine, 6364
[478-60-4] Citromycetin, 2389
[478-61-5] Berbamine, 1191
[478-73-9] Pseudococaine, 8097
[478-84-2] Bromolysergide, 1445
[478-85-3] Apo-β-erythroidine, 785
[478-94-4] Lysergamide, 5663
[478-95-5] Isolysergic Acid, 5197
[479-00-5] Ergometrinine, 3693
[479-13-0] Coumestrol, 2632
[479-18-5] Dyphylline, 3529
[479-20-9] Atranorin, 901
[479-21-0] Cotoin, 2620
[479-22-1] Nitranilic Acid, 6665

[479-23-2] Cholanthrene, 2252
[479-27-6] 1,8-Naphthalenediamine, 6459
[479-36-7] Daphnoline, 2883
[479-45-8] Nitramine, 6664
[479-68-5] Broparoestrol, 1471
[479-81-2] Bietamiverine, 1252
[479-92-5] Propyphenazone, 8056
[479-98-1] Aucubin, 910
[480-11-5] Oroxylin A, 7006
[480-15-9] Datiscetin, 2887
[480-16-0] Morin, 6352
[480-17-1] Leucocyanidin, 5477
[480-30-8] Dichloralphenazone, 3097
[480-36-4] Linarin, 5523
[480-40-0] Chrysin, 2316
[480-41-1] Naringenin, 6511
[480-44-4] Acacetin, 10
[480-49-9] Filipin III*, 4123
[480-54-6] Retrorsine, 8335
[480-64-8] o-Orsellinic Acid, 7009
[480-68-2] 5-Nitrobarbituric Acid, 6683
[480-78-4] Platyphylline, 7691
[480-81-9] Seneciphylline, 8596
[480-85-3] Retronecine, 8334
[481-05-0] Artemisin, 855
[481-06-1] α-Santonin, 8509
[481-17-4] Chondrillasterol, 2267
[481-18-5] α-Spinasterol, 8901
[481-20-9] Coprostane, 2587
[481-21-0] (5α)-Cholestane, 2254
[481-26-5] Pregnane, 7907
[481-29-8] Epiandrosterone, 3646
[481-37-8] Ecgonine, 3541
[481-39-0] Juglone, 5282
[481-42-5] Plumbagin, 7697
[481-49-2] Cepharanthine, 2031
[481-72-1] Aloe-Emodin, 313
[481-74-3] Chrysophanic Acid, 2318
[481-85-6] Menadiol, 5873
[481-97-0] Estrone Sulfate*, 3751
[482-05-3] Diphenic Acid, 3368
[482-15-5] Isothipendyl, 5248
[482-20-2] Tylophorine, 9962
[482-28-0] Cinchonamine, 2344
[482-44-0] Imperatorin, 4960
[482-53-1] Osajin, 7017
[482-66-6] 1,2-Cyclopentenophenanthrene, 2812
[482-68-8] Sarpagine, 8524
[482-70-2] Chimaphilin, 2100
[482-74-6] Cryptopine, 2675
[482-89-3] Indigo, 4977
[482-91-7] Aricine, 821
[483-04-5] Raubasine, 8291
[483-10-3] Corynanthine, 2614
[483-17-0] Cephaeline, 2020
[483-18-1] Emetine, 3600
[483-19-2] Emetamine, 3599
[483-49-8] Berbine, 1195
[483-54-5] Aurantioglioctadin, 912
[483-55-6] Phthiocol, 7534
[483-57-8] Fervenulin, 4111
[483-63-6] Crotamiton, 2661
[483-64-7] Damascenine, 2873
[483-65-8] Retene, 8328
[483-78-3] Cadalene, 1644
[484-11-7] Neocuproine, 6537
[484-12-8] Osthole, 7026
[484-20-8] Bergapten(e), 1196
[484-23-1] Dihydralazine, 3212
[484-29-7] Dictamnine, 3139
[484-31-1] Apiole (Dill), 776
[484-44-9] Olivacine, 6971
[484-67-3] Drosophilin A, 3509
[484-89-9] Fumigatin, 4310
[484-93-5] Ecgonidine, 3540
[485-19-8] Reticuline, 8330
[485-31-4] Binapacryl, 1265

Denotes derivative of title compound

[506-96-7] Acetyl Bromide, 83
[507-02-8] Acetyl Iodide, 95
[507-09-5] Thioacetic Acid, 9454
[507-16-4] Thionyl Bromide, 9484
[507-19-7] tert-Butyl Bromide, 1590
[507-20-0] tert-Butyl Chloride, 1597
[507-25-5] Carbon Tetraiodide, 1866
[507-27-7] Tetraphenylarsonium Bromide, 9373
[507-28-8] Tetraphenylarsonium Chloride, 9374
[507-32-4] Ethanearsonic Acid, 3768
[507-36-8] tert-Amyl Bromide, 644
[507-40-4] tert-Butyl Hypochlorite, 1605
[507-42-6] Bromal Hydrate, 1405
[507-45-9] Amylene Dichloride, 650
[507-47-1] Chloral Ammonia, 2110
[507-60-8] Scilliroside, 8545
[507-61-9] Azafrin, 927
[507-70-0] Borneol, 1366
[507-79-9] Tazettine, 9250
[508-02-1] Oleanolic Acid, 6964
[508-29-2] 3-Chloro-d-camphor, 2183
[508-44-1] Anemonin, 681
[508-52-1] Ouabagenin, 7030
[508-54-3] Hydroxycodeinone, 4867
[508-59-8] Parthenin, 7179
[508-65-6] Germine, 4420
[508-75-8] Convallatoxin, 2575
[508-77-0] Cymarin, 2830
[509-00-2] Cortisone 21-Cyclopentanepropionate*, 2602
[509-14-8] Tetranitromethane, 9371
[509-15-9] Gelsemine, 4390
[509-17-1] Ajacine, 192
[509-18-2] Delsoline, 2933
[509-20-6] Aconine, 117
[509-24-0] Songorine, 8865
[509-36-4] β-Colubrine*, 2556
[509-40-0] Diaboline, 3002
[509-42-2] Strychnine Phosphate, 9023
[509-44-4] α-Colubrine*, 2556
[509-60-4] Dihydromorphine, 3220
[509-67-1] Pholcodine, 7484
[509-71-7] Diacetyldihydromorphine, 3011
[509-74-0] Methadyl Acetate, 6009
[509-77-3] Anilinephthalein, 698
[509-78-4] Dimenoxadol, 3253
[509-86-4] Heptabarbital, 4689
[509-93-3] Ambrosin, 400
[510-15-6] Chlorobenzilate, 2174
[510-20-3] Diethylmalonic Acid, 3173
[510-25-8] Belladonnine, 1053
[510-34-9] Phethenylate Sodium, 7476
[510-35-0] Santonic Acid, 8508
[510-53-2] Racemethorphan, 8274
[510-90-7] Buthalital Sodium, 1553
[511-00-2] Erythrophlamine, 3727
[511-07-9] Ergocristinine, 3688
[511-08-0] Ergocristine, 3687
[511-09-1] α-Ergocryptine*, 3689
[511-10-4] Ergocryptine, 3690
[511-12-6] Dihydroergotamine, 3217
[511-13-7] Chlophedianol Hydrochloride*, 2106
[511-18-2] Norcholanic Acid, 6782
[511-20-6] Ergostane, 3696
[511-21-7] Coproergostane, 2585
[511-26-2] Pregnenolone Methyl Ether*, 7915
[511-28-4] Vitamin D₄, 10158
[511-34-2] Digitogenin, 3203
[511-45-5] Pridinol, 7922
[511-55-7] Xenytropium Bromide, 10208
[511-70-6] Diethylbromoacetamide, 3164
[511-89-7] Plumieride, 7699

[511-96-6] Gitogenin, 4438
[512-04-9] Diosgenin, 3348
[512-12-9] 5,5-Dipropyl-2,4-oxazolidinedione, 3409
[512-15-2] Cyclopentolate, 2815
[512-16-3] Cyclobutyrol, 2784
[512-29-8] Flavoxanthin, 4134
[512-35-6] Benzenesulfonic Anhydride, 1099
[512-39-0] Hepaxanthin, 4686
[512-48-1] Novonal, 6820
[512-64-1] Echinomycin, 3545
[512-69-6] Raffinose, 8279
[512-85-6] Ascaridole, 864
[513-10-0] Echothiophate Iodide, 3549
[513-12-2] 2-(Ethylsulfonyl)ethanol, 3899
[513-15-5] Sodium β-Sulfopropionitrile, 8832
[513-29-1] Glycine Sulfate, 4501
[513-31-5] 2,3-Dibromopropene, 3074
[513-35-9] Amylene, 649
[513-36-0] Isobutyl Chloride, 5153
[513-37-1] 1-Chloro-2-methyl-1-propene, 2199
[513-38-2] Isobutyl Iodide, 5158
[513-44-0] Isobutyl Mercaptan, 5161
[513-48-4] sec-Butyl Iodide, 1608
[513-53-1] sec-Butyl Mercaptan, 1612
[513-74-6] Ammonium Dithiocarbamate, 545
[513-77-9] Barium Carbonate, 996
[513-78-0] Cadmium Carbonate, 1652
[513-79-1] Cobaltous Carbonate, 2497
[513-80-4] Urea Nitrate, 10008
[513-81-5] 2,3-Dimethyl-1,3-butadiene, 3288
[513-85-9] 2,3-Butylene Glycol, 1602
[513-86-0] Acetoin, 61
[513-88-2] 1,1-Dichloroacetone, 3101
[513-92-8] Tetraiodoethylene, 9359
[513-96-2] Chlorocyanohydrin, 2185
[514-10-3] Abietic Acid, 3
[514-12-5] α,α'-Dibromo-d-camphor, 3067
[514-17-0] Androstane-3β,11β-diol-17-one, 676
[514-36-3] Fludrocortisone 21-Acetate*, 4166
[514-39-6] Periplogenin, 7317
[514-61-4] Normethandrone, 6805
[514-65-8] Biperiden, 1274
[514-73-8] Dithiazanine Iodide, 3434
[514-76-1] Astacin, 888
[514-78-3] Canthaxanthin, 1798
[514-92-1] Torularhodin, 9691
[515-24-2] Turicine, 9952
[515-25-3] Betonicine, 1234
[515-30-0] Atrolactic Acid, 905
[515-40-2] Neophyl Chloride, 6548
[515-43-5] Sozoiodolic Acid Mercury Deriv*, 8881
[515-49-1] Sulfathiourea, 9116
[515-54-8] Phenosulfazole, 7403
[515-59-3] Sulfamethylthiazole, 9088
[515-64-0] Sulfisomidine, 9124
[515-69-5] α-Bisabolol, 1281
[515-72-0] Barium Benzenesulfonate, 993
[515-76-4] Diphetarsone, 3394
[515-82-2] Chloral Formamide, 2112
[515-83-3] Chloral Alcoholate, 2109
[515-84-4] Trichloroacetic Acid Ethyl Ester*, 9756
[515-94-6] 2,3-Diaminopropionic Acid, 3027
[515-96-8] Semioxamazide, 8589
[516-02-9] Barium Oxalate, 1014
[516-03-0] Ferrous Oxalate, 4099
[516-12-1] N-Iodosuccinimide, 5064

[516-15-4] 11-Ketoprogesterone, 5317
[516-16-5] Allopregnane-3β,11β,21-triol-20-one, 281
[516-18-7] Asaprol, 859
[516-21-2] Cycloguanil, 2790
[516-38-1] Cortol, 2604
[516-39-2] Allopregnane-3β,11β,17α,-20β,21-pentol, 274
[516-41-6] Hydrallostane, 4801
[516-42-7] Cortolone, 2605
[516-45-0] Allopregnane-3β,17α,21-triol-11,20-dione, 280
[516-47-2] Allopregnane-3β,17α,21-triol-20-one, 282
[516-51-8] Uranediol, 9989
[516-53-0] Allopregnane-3β,20β-diol, 270
[516-54-1] Allopregnan-3α-ol-20-one, 283
[516-55-2] Allopregnan-3β-ol-20-one, 284
[516-58-5] Allopregnan-20β-ol-3-one, 286
[516-59-6] Allopregnan-20α-ol-3-one, 285
[516-78-9] γ-Ergostenol, 3700
[516-85-8] Dehydroergosterol, 2925
[516-95-0] Epicholestanol, 3649
[516-98-3] Neoergosterol, 6539
[517-04-4] Isoestradiol, 5183
[517-06-6] 8-Isoestrone, 5184
[517-09-9] Equilenin, 3675
[517-10-2] Allocholesterol, 262
[517-16-8] N-(Ethylmercuri)-p-toluenesulfonanilide, 3872
[517-18-0] Methallenestril, 6012
[517-21-5] Glyoxal-Sodium Bisulfite, 4520
[517-23-7] α-Acetylbutyrolactone, 84
[517-25-9] Trinitromethane, 9859
[517-28-2] Hematoxylin, 4670
[517-56-6] Corytuberine, 2616
[517-60-2] Mellitic Acid, 5868
[517-66-8] Dicentrine, 3092
[517-69-1] Actinodaphnine, 140
[517-82-8] Echinochrome A, 3544
[517-88-4] Alkannin, 253
[517-92-0] Chrysamminic Acid, 2308
[517-97-5] Tsuduranine, 9932
[518-05-8] Naphthalic Acid, 6467
[518-17-2] Evodiamine, 3954
[518-20-7] Cyclocumarol, 2786
[518-28-5] Podophyllotoxin, 7704
[518-29-6] β-Peltatin, 7204
[518-34-3] Tetrandrine, 9369
[518-44-5] Fluorescin, 4196
[518-47-8] Fluorescein Disodium Salt*, 4194
[518-51-4] Phenolphthalein Sodium, 7393
[518-60-5] Lunine, 5633
[518-61-6] Ahistan, 189
[518-69-4] Corydaline, 2610
[518-75-2] Citrinin, 2388
[518-77-4] Corybulbine, 2606
[518-82-1] Emodin, 3602
[519-02-8] Matrine, 5799
[519-04-0] Anhalonine, 693
[519-05-1] Opianic Acid, 6983
[519-09-5] Benzoylecgonine, 1146
[519-21-1] Mycaminose, 6398
[519-23-3] Ellipticine, 3589
[519-26-6] Sodium Iodomethamate, 8779
[519-30-2] Dimethazan, 3261
[519-34-6] Maclurin, 5676
[519-37-9] Etofylline, 3924
[519-40-4] Aspidinol, 882
[519-44-8] 2,4-Dinitroresorcinol, 3337
[519-56-2] Morphenol, 6356

Denotes derivative of title compound

[519-65-3] Dioxypyramidon, *3358*
[519-72-2] *N,N*-Diethylbenzhydryl-amine, *3163*
[519-73-3] Triphenylmethane, *9871*
[519-87-9] Diphenylacetamide, *3373*
[519-88-0] Ambucetamide, *403*
[519-95-9] Florantyrone, *4143*
[520-07-0] Antipyrine Salicylate, *758*
[520-09-2] Dopan, *3480*
[520-12-7] Pectolinarigenin, *7195*
[520-18-3] Kaempferol, *5288*
[520-26-3] Hesperidin, *4705*
[520-27-4] Diosmin, *3350*
[520-33-2] Hesperetin, *4704*
[520-34-3] Diosmetin, *3349*
[520-36-5] Apigenin, *773*
[520-43-4] Sparassol, *8883*
[520-45-6] Dehydroacetic Acid, *2919*
[520-52-5] Psilocybin, *8111*
[520-53-6] Psilocin, *8110*
[520-77-4] Ethadione, *3762*
[520-85-4] Medroxyprogesterone, *5838*
[520-86-5] Allopregnane-3β,17α,20β-triol, *279*
[520-91-2] Vitamin D₁, *10155*
[521-04-0] 7-Dehydrositosterol, *2927*
[521-10-8] Methandriol, *6017*
[521-11-9] Mestanolone, *5973*
[521-12-0] Dromostanolone Propionate*, *3504*
[521-17-5] Androstenediol, *677*
[521-18-6] Stanolone, *8950*
[521-24-4] Sodium β-Naphtho-quinone-4-sulfonate, *8791*
[521-31-3] Luminol, *5628*
[521-35-7] Cannabinol, *1793*
[521-40-4] Narcotoline, *6510*
[521-62-0] Frangulin A*, *4288*
[521-67-5] Cinnamoylcocaine, *2361*
[521-74-4] Broxyquinoline, *1474*
[521-78-8] Trimipramine Maleate*, *9852*
[521-85-7] Corycavamine, *2607*
[521-88-0] Karanjin, *5295*
[521-93-7] Corycavidine, *2608*
[522-00-9] Ethopropazine, *3793*
[522-12-3] Quercitrin, *8219*
[522-17-8] Deguelin, *2918*
[522-18-9] Chlorbenzoxamine, *2125*
[522-23-6] Metofenazate Difumarate*, *6228*
[522-24-7] Fenethazine, *4013*
[522-27-0] α-Furildioxime, *4328*
[522-40-7] Fosfestrol, *4279*
[522-48-5] Tetrahydrozoline Hydro-chloride*, *9358*
[522-51-0] Dequalinium Chloride, *2959*
[522-57-6] Papaveraldine, *7149*
[522-60-1] Ethylhydrocupreine, *3855*
[522-66-7] Hydroquinine, *4852*
[522-70-3] Antimycin A₃, *756*
[522-94-1] *allo*-Yohimbine, *10237*
[522-97-4] Canadine, *1785*
[523-21-7] Sodium Rhodizonate, *8818*
[523-27-3] 9,10-Dibromoanthracene, *3065*
[523-44-4] Orange I, *6989*
[523-54-6] Etymemazine, *3938*
[523-64-8] Flindersine, *4139*
[523-68-2] Vitamin K₅ *N*-Acetyl Analog*, *10162*
[523-80-8] Apiole (Parsley), *777*
[523-87-5] Dimenhydrinate, *3252*
[523-88-6] Dibromsalicil, *3080*
[524-15-2] Fagarine, *3970*
[524-17-4] Dauricine, *2891*
[524-20-9] Armepavine, *823*
[524-30-1] Fraxin, *4290*

[524-34-5] Cinchophen Allyl Ester*, *2347*
[524-36-7] Pyridoxamine Dihydro-chloride, *8164*
[524-40-3] Ricinine, *8377*
[524-42-5] 1,2-Naphthoquinone, *6481*
[524-46-9] Adlumine, *164*
[524-63-0] Cupreine, *2689*
[524-80-1] 9-Carbazoleacetic Acid, *1835*
[524-81-2] Mebhydroline, *5809*
[524-83-4] Ethybenztropine, *3802*
[524-84-5] Dimethylthiambutene, *3309*
[524-96-9] Benzylimidobis(*p*-methoxy-phenyl)methane, *1174*
[524-97-0] Pterocarpin, *8117*
[524-99-2] Medrylamine, *5839*
[525-05-3] Nitroso-R Salt, *6742*
[525-06-4] Diphenylketene, *3385*
[525-37-1] 1,6-Naphthalenedisulfonic Acid, *6460*
[525-47-3] 5-Nitroquinaldic Acid, *6728*
[525-52-0] Pyrogallol Triacetate*, *8184*
[525-57-5] Harmalol, *4645*
[525-61-1] Quinocide, *8252*
[525-64-4] 9*H*-Fluorene-2,7-diamine, *4191*
[525-66-6] Propranolol, *8025*
[525-68-8] Galipine, *4359*
[525-74-6] *N*-Methylmyosmine, *6173*
[525-79-1] Kinetin, *5324*
[525-82-6] Flavone, *4132*
[525-86-0] 2-Pentenylpenicillin Sodium, *7264*
[525-87-1] Penicillin S Potassium, *7229*
[525-94-0] Penicillin N, *7227*
[526-02-3] Actiphenol, *145*
[526-07-8] Sesamolin, *8617*
[526-08-9] Sulfaphenazole, *9105*
[526-13-6] Meteloidine, *5997*
[526-18-1] Osalmid, *7018*
[526-30-7] Tryptazan, *9928*
[526-31-8] Abrine, *6*
[526-35-2] Aloxidone, *315*
[526-36-3] Xylometazoline, *10219*
[526-55-6] Tryptophol, *9930*
[526-78-3] 2,3-Dibromosuccinic Acid, *3078*
[526-84-1] Dihydroxymaleic Acid, *3229*
[526-94-3] Sodium Bitartrate, *8732*
[526-95-4] Gluconic Acid, *4464*
[526-97-6] L-Gulonic Acid, *4605*
[526-98-7] 2-Keto-L-gulonic Acid, *5315*
[526-99-8] Galactaric Acid, *4349*
[527-07-1] Sodium Gluconate, *8766*
[527-09-3] Cupric Gluconate, *2706*
[527-17-3] Duroquinone, *3522*
[527-18-4] Durohydroquinone, *3521*
[527-38-8] Mycosamine, *6409*
[527-52-6] Digitoxose, *3207*
[527-53-7] Isodurene, *5182*
[527-62-8] 2-Amino-4,6-dichloro-phenol, *457*
[527-72-0] 2-Thiophenecarboxylic Acid, *9491*
[527-73-1] Azomycin, *952*
[527-76-4] 4-Amino-2-sulfobenzoic Acid, *500*
[527-85-5] *o*-Toluamide, *9666*
[527-89-9] Dithiosalicylic Acid, *3440*
[527-93-5] Diploicin, *3401*
[528-10-9] 5,5'-Dinitrosalicil, *3338*
[528-21-2] Gallacetophenone, *4360*
[528-43-8] Magnolol, *5738*
[528-44-9] Trimellitic Acid, *9832*

[528-45-0] 3,4-Dinitrobenzoic Acid, *3327*
[528-48-3] Fisetin, *4129*
[528-50-7] Cellobiose, *2008*
[528-53-0] Delphinidin, *2931*
[528-58-5] Cyanidin Chloride, *2757*
[528-71-2] Hemipyocyanine, *4679*
[528-75-6] 2,4-Dinitrobenzaldehyde, *3324*
[528-76-7] 2,4-Dinitrobenzenesulfenyl Chloride, *3326*
[528-79-0] Thymol Acetate*, *9540*
[528-81-4] 6-Deoxy-L-ascorbic Acid, *2945*
[528-88-1] D-Glucoascorbic Acid, *4459*
[528-92-7] Apronalide, *795*
[528-94-9] Ammonium Salicylate, *584*
[529-05-5] Chamazulene, *2081*
[529-17-9] Tropane, *9907*
[529-19-1] *o*-Tolunitrile, *9675*
[529-20-4] *o*-Tolualdehyde, *9665*
[529-21-5] 3-Ethyl-4-picoline, *3887*
[529-28-2] *o*-Iodoanisole, *5049*
[529-36-2] 1-Naphthalenethiol, *6465*
[529-38-4] Cocaethylene, *2516*
[529-44-2] Myricetin, *6414*
[529-53-3] Scutellarein, *8556*
[529-64-6] Tropic Acid, *9910*
[529-86-2] Anthranol, *725*
[529-92-0] Cusparine, *2750*
[530-08-5] Isoetharine, *5185*
[530-14-3] Picein, *7549*
[530-31-4] Ammonium Mandelate, *563*
[530-34-7] Isoladol, *5194*
[530-40-5] Isonicotinic Acid Diethyl-amide, *5205*
[530-43-8] Chloramphenicol Palmi-tate*, *2120*
[530-44-9] *p*-Dimethylaminobenzo-phenone, *3283*
[530-45-0] 1,1-Di-*p*-tolylethane, *3444*
[530-48-3] 1,1-Diphenylethene, *3381*
[530-50-7] 1,1-Diphenylhydrazine, *3384*
[530-55-2] 2,6-Dimethoxyquinone, *3275*
[530-62-1] *N,N'*-Carbonyldiimidazole, *1867*
[530-75-6] Acetylsalicylsalicylic Acid, *102*
[530-78-9] Flufenamic Acid, *4167*
[530-85-8] *o*-Nitrophenylpropiolic Acid, *6722*
[530-91-6] Tetralol, *9361*
[530-97-2] Heptoxime, *4700*
[531-02-2] Diresorcinol, *3417*
[531-06-6] Methafurylene, *6010*
[531-18-0] Hexamethylolmelamine, *4727*
[531-29-3] Coniferin, *2567*
[531-37-3] Guaiacol Benzoate, *4576*
[531-39-5] Guaiacol Valerate, *4579*
[531-44-2] Scopolin, *8553*
[531-53-3] Azure A, *957*
[531-55-5] Azure B, *958*
[531-57-7] Azure C, *959*
[531-58-8] Cichoriin, *2322*
[531-72-6] Arsenamide, *831*
[531-73-7] Proflavine Dihydrochlo-ride*, *7953*
[531-75-9] Esculin, *3739*
[531-81-7] Coumarin-3-carboxylic Acid, *2631*
[531-84-0] Hexadecyl 3-Hydroxy-2-naphthoate, *4718*
[531-91-9] *N,N'*-Diphenylbenzidine, *3377*
[531-95-3] Equol, *3677*

* Denotes derivative of title compound

[531-99-7] 5-Acetyl-2-methoxybenz-aldehyde, 98
[532-03-6] Methocarbamol, 6060
[532-05-8] 1,3-Di-6-quinolylurea, 3416
[532-11-6] Anethole Trithione, 683
[532-18-3] β,β'-Dinaphthylamine, 3319
[532-27-4] ω-Chloroacetophenone, 2166
[532-28-5] Mandelonitrile, 5759
[532-32-1] Sodium Benzoate, 8725
[532-34-3] Butopyronoxyl, 1564
[532-40-1] Thiamine Monophosphate*, 9431
[532-43-4] Thiamine Mononitrate*, 9430
[532-48-9] Euparin, 3946
[532-49-0] Dibutoline Sulfate, 3082
[532-52-5] Cyclexedrine, 2776
[532-54-7] Isonitrosoacetophenone, 5208
[532-55-8] Benzoyl Isothiocyanate, 1147
[532-59-2] Amylocaine Hydrochloride, 656
[532-65-0] ar-Turmerone, 9956
[532-76-3] Hexylcaine Hydrochloride, 4746
[532-82-1] Chrysoidine, 2317
[533-06-2] Mephenesin Carbamate*, 5895
[533-18-6] o-Cresyl Acetate, 2652
[533-22-2] Hydroxystilbamidine Isethionate*, 4892
[533-28-8] Piperocaine Hydrochloride*, 7627
[533-30-2] 6-Aminobenzothiazole, 445
[533-32-4] Isovaleryl Diethylamide, 5253
[533-45-9] Clomethiazole, 2444
[533-48-2] Desthiobiotin, 2978
[533-49-3] L-Erythrose, 3732
[533-50-6] L-Erythrulose, 3735
[533-51-7] Silver Oxalate, 8664
[533-58-4] o-Iodophenol, 5056
[533-62-0] Hydroxyglutamic Acid, 4871
[533-67-5] D-2-Deoxyribose, 2955
[533-70-0] 2,4-Diiodoaniline, 3232
[533-73-3] 1,2,4-Benzenetriol, 1101
[533-74-4] Dazomet, 2892
[533-76-6] Ketipic Acid, 5311
[533-77-7] L-Cystine S,S'-dioxide*, 2852
[533-87-9] Aleuritic Acid, 230
[533-88-0] Pentahomoserine, 7252
[533-96-0] Sodium Sesquicarbonate, 8823
[533-98-2] α-Butylene Dibromide, 1600
[534-00-9] d-Amyl Bromide, 642
[534-07-6] 1,3-Dichloroacetone, 3102
[534-08-7] Iothion, 5081
[534-12-3] Sodiumacetic Acid Sodium Salt, 8712
[534-13-4] N,N'-Dimethylthiourea, 3310
[534-15-6] Dimethylacetal, 3276
[534-16-7] Silver Carbonate, 8650
[534-17-8] Cesium Carbonate, 2053
[534-18-9] Sodium Thiocarbonate, 8841
[534-26-9] Lysidine, 5666
[534-30-5] Hercynine, 4702
[534-33-8] Acetarsone Diethylamine Salt*, 49
[534-46-3] Sophorose, 8868
[534-51-0] 3,5-Diiodothyronine, 3235
[534-52-1] Dinitrocresol, 3331
[534-59-8] n-Butylmalonic Acid, 1610
[534-64-5] Pyrithiamine, 8176

[534-82-7] 3-Methoxy-4-hydroxy-phenylglycol, 6074
[534-84-9] Pimeclone, 7583
[534-87-2] Etilefrin dl-Form Hydro-chloride*, 3911
[535-11-5] Ethyl α-Bromopropionate, 3820
[535-13-7] Ethyl α-Chloropropionate, 3832
[535-37-5] Gold Tricyanide, 4543
[535-46-6] Benzenestibonic Acid, 1097
[535-48-8] Sodium 6-Chloro-5-nitro-toluene-3-sulfonate, 8744
[535-55-7] Mercuric Sodium p-Phenol-sulfonate, 5939
[535-65-9] Glybuthiazol(e), 4487
[535-75-1] Pipecolic Acid, 7611
[535-80-8] m-Chlorobenzoic Acid, 2175
[535-83-1] Trigonelline, 9822
[535-86-4] Homocamfin, 4768
[535-87-5] 3,5-Diaminobenzoic Acid, 3020
[535-89-7] Crimidine, 2655
[536-08-3] Digallic Acid, 3195
[536-17-4] p-Dimethylaminobenzal-rhodanine, 3281
[536-21-0] Norfenefrine, 6792
[536-24-3] Ethylnorepinephrine, 3880
[536-25-4] Orthocaine, 7011
[536-29-8] Dichlorophenarsine Hydro-chloride, 3121
[536-33-4] Ethionamide, 3783
[536-37-8] 2-Hexyldecanoic Acid Sodium Salt*, 4747
[536-38-9] p-Chlorophenacyl Bromide, 2205
[536-43-6] Dyclonine Hydrochloride*, 3523
[536-50-5] p,α-Dimethylbenzyl Alcohol, 3287
[536-57-2] p-Toluenesulfinic Acid, 9670
[536-60-7] Cumic Alcohol, 2685
[536-66-3] Cumic Acid, 2684
[536-69-6] Fusaric Acid, 4336
[536-74-3] Ethynylbenzene, 3906
[536-75-4] 4-Ethylpyridine, 3893
[536-78-7] 3-Ethylpyridine, 3892
[536-88-9] 4-Ethyl-2-picoline, 3888
[536-95-8] p-(Benzylsulfonamido)-benzoic Acid, 1184
[537-01-9] Cerous Carbonate, 2040
[537-12-2] Diperodon Hydrochloride, 3360
[537-17-7] Amanozine, 388
[537-20-2] Benzoylpas Sodium Salt*, 1148
[537-21-3] Chlorproguanil, 2237
[537-26-8] Tropacocaine, 9904
[537-29-1] Norhyoscyamine, 6799
[537-45-1] 2,6-Dibromoquinone-4-chlorimide, 3075
[537-46-2] Methamphetamine, 6015
[537-47-3] 4-Phenylsemicarbazide, 7465
[537-49-5] Surinamine, 9183
[537-61-1] Cyclomethycaine Hydro-chloride*, 2804
[537-64-4] p-Ditolylmercury, 3446
[537-65-5] 4,4'-Diaminodiphenyl-amine, 3022
[537-91-7] Nitrophenide, 6715
[537-92-8] m-Acetotoluide*, 74
[538-03-4] Oxophenarsine Hydro-chloride, 7082
[538-08-9] Diallylcyanamide, 3016
[538-17-0] Aluminum Thiocyanate, 384

[538-28-3] S-Benzylthiuronium Chloride, 1185
[538-32-9] Benzylurea, 1186
[538-37-4] Dibenzyl Chlorophos-phonate, 3059
[538-39-6] 1,2-Di-p-tolylethane, 3445
[538-41-0] p-Diaminoazobenzene, 3019
[538-43-2] Phenylglyceryl Ether, 7444
[538-45-4] 3,3-Dithiodipyridine Dihy-drochloride, 3438
[538-51-2] Benzylideneaniline, 1173
[538-58-9] Dibenzalacetone, 3053
[538-62-5] Diphenylcarbazone, 3380
[538-64-7] Benzyl Fumarate, 1170
[538-68-1] Amylbenzene, 641
[538-71-6] Domiphen Bromide, 3474
[538-74-9] Benzyl Sulfide, 1183
[538-75-0] Dicyclohexylcarbodiimide, 3146
[538-86-3] Benzyl Methyl Ether, 1175
[538-90-9] β-Coniceine, 2565
[538-93-2] Isobutylbenzene, 5149
[539-00-4] Isopelletierine*, 7200
[539-08-2] p-Lactophenetide, 5355
[539-10-6] Phenocoll Hydrochloride*, 7388
[539-12-8] p-Anol, 715
[539-15-1] Hordenine, 4783
[539-21-9] Ambazone, 395
[539-30-0] Benzyl Ethyl Ether, 1168
[539-35-5] Mycobacidin, 6403
[539-47-9] 2-Furanacrylic Acid, 4317
[539-54-8] Antimony Sodium Thiogly-collate, 744
[539-64-0] N,N-Dimethylglycine Hy-drazide Hydrochloride, 3294
[539-68-4] Dihydroxyaluminum Sodium Carbonate, 3228
[539-86-6] Allicin, 258
[539-88-8] Ethyl Levulinate, 3865
[539-89-9] Isobutyl Urethane, 5168
[539-90-2] Isobutyl n-Butyrate, 5151
[540-05-6] Phytofluene, 7544
[540-07-8] n-Amyl Caproate, 646
[540-10-3] Cetyl Palmitate, 2073
[540-16-9] Cupric Butyrate, 2696
[540-18-1] n-Amyl Butyrate, 645
[540-36-3] p-Difluorobenzene, 3192
[540-37-4] p-Iodoaniline, 5048
[540-38-5] p-Iodophenol, 5057
[540-42-1] Isobutyl Propionate, 5164
[540-47-6] Cyclopropyl Methyl Ether, 2818
[540-49-8] Acetylene Dibromide, 92
[540-51-2] Ethylene Bromohydrin, 3838
[540-54-5] Propyl Chloride, 8032
[540-59-0] Acetylene Dichloride, 93
[540-61-4] Aminoacetonitrile, 432
[540-63-6] 1,2-Ethanedithiol, 3770
[540-67-0] Ethyl Methyl Ether, 3874
[540-69-2] Ammonium Formate, 554
[540-72-7] Thiocyanate Sodium, 9463
[540-73-8] 1,2-Dimethylhydrazine, 3297
[540-80-7] tert-Butyl Nitrite, 1618
[540-82-9] Ethyl Sulfate, 3897
[540-84-1] Isooctane, 5210
[540-88-5] tert-Butyl Acetate, 1572
[540-92-1] Acetone Sodium Bisulfite, 67
[540-97-6] Dodecamethylcyclohexa-siloxane, 3462
[541-02-6] Decamethylcyclopenta-siloxane, 2905
[541-04-8] sym-Diphenylpyrophos-phorodiamidic Acid, 3391
[541-07-1] Mevaldic Acid, 6249
[541-09-3] Uranyl Acetate, 9998
[541-14-0] Carnitine D-Form*, 1898

[541-15-1] Carnitine L-Form*, *1898*
[541-16-2] Di-*tert*-butyl Malonate, *3085*
[541-19-5] Succinylcholine Iodide, *9045*
[541-20-8] Pentamethonium Bromide, *7253*
[541-22-0] Decamethonium Bromide, *2904*
[541-25-3] Dichloro(2-chlorovinyl)arsine, *3112*
[541-28-6] Isoamyl Iodide, *5134*
[541-33-3] Butylidene Chloride, *1606*
[541-35-5] *n*-Butyramide, *1628*
[541-41-3] Ethyl Chloroformate, *3831*
[541-42-4] Isopropyl Nitrite, *5235*
[541-43-5] Barium Formate, *1004*
[541-46-8] Isovaleramide, *5250*
[541-48-0] β-Aminobutyric Acid, *449*
[541-50-4] Acetoacetic Acid, *56*
[541-53-7] 2,4-Dithiobiuret, *3436*
[541-64-0] Furtrethonium Iodide*, *4334*
[541-66-2] Oxapropanium Iodide, *7056*
[541-73-1] *m*-Dichlorobenzene, *3105*
[541-79-7] Carbocloral, *1849*
[541-88-8] Chloroacetic Anhydride, *2163*
[541-91-3] Muscone, *6392*
[542-05-2] Acetonedicarboxylic Acid, *66*
[542-08-5] Isopropyl Acetoacetate, *5225*
[542-10-9] Ethylidene Diacetate, *3857*
[542-18-7] Cyclohexyl Chloride, *2801*
[542-32-5] α-Aminoadipic Acid, *436*
[542-37-0] *tert*-Amyl Isovalerate, *652*
[542-42-7] Calcium Palmitate, *1734*
[542-46-1] Civetone, *2396*
[542-52-9] *n*-Butyl Carbonate, *1593*
[542-54-1] Isoamyl Cyanide, *5131*
[542-55-2] Isobutyl Formate, *5157*
[542-56-3] Isobutyl Nitrite, *5163*
[542-58-5] β-Chloroethyl Acetate, *2189*
[542-59-6] Ethylene Glycol Monoacetate, *3847*
[542-62-1] Barium Cyanide, *1000*
[542-69-8] *n*-Butyl Iodide, *1607*
[542-75-6] 1,3-Dichloropropene, *3125*
[542-76-7] β-Chloropropionitrile, *2212*
[542-83-6] Cadmium Cyanide, *1654*
[542-84-7] Cobaltous Cyanide, *2500*
[542-85-8] Ethyl Isothiocyanate, *3861*
[542-88-1] *sym*-Dichloromethyl Ether, *3119*
[542-91-6] Diethylsilane, *3176*
[542-92-7] Cyclopentadiene, *2807*
[543-15-7] Heptaminol Hydrochloride*, *4691*
[543-20-4] Succinyl Chloride, *9042*
[543-21-5] Cellocidin, *2009*
[543-24-8] Aceturic Acid, *81*
[543-27-1] Isobutyl Chlorocarbonate, *5154*
[543-28-2] Isobutyl Carbamate, *5152*
[543-29-3] Isobutyl Nitrate, *5162*
[543-38-4] Canavanine, *1786*
[543-49-7] 2-Heptanol, *4697*
[543-59-9] Amyl Chloride, *647*
[543-63-5] *n*-Butylmercuric Chloride, *1614*
[543-67-9] *n*-Propyl Nitrite, *8050*
[543-80-6] Barium Acetate, *992*
[543-81-7] Beryllium Acetate, *1202*
[543-82-8] Octodrine, *6854*
[543-83-9] Galegine, *4358*
[543-87-3] Isoamyl Nitrate, *5136*
[543-90-8] Cadmium Acetate, *1650*

[543-94-2] Strontium Acetate, *8995*
[544-00-3] Diisoamylamine, *3237*
[544-01-4] Isoamyl Ether, *5132*
[544-10-5] 1-Chlorohexane, *2196*
[544-12-7] 3-Hexen-1-ol, *4737*
[544-13-8] Glutaronitrile, *4482*
[544-16-1] *n*-Butyl Nitrite, *1617*
[544-17-2] Calcium Formate, *1711*
[544-18-3] Cobaltous Formate, *2502*
[544-19-4] Cupric Formate, *2705*
[544-27-4] Arsenoacetic Acid, *848*
[544-31-0] Palmidrol, *7127*
[544-35-4] Ethyl Linoleate, *3866*
[544-40-1] *n*-Butyl Sulfide, *1626*
[544-44-5] Agrocybin, *188*
[544-47-8] *p*-Chlorobenzylpseudothiuronium Chloride, *2179*
[544-51-4] Mycomycin, *6407*
[544-60-5] Ammonium Oleate, *569*
[544-62-7] Batyl Alcohol, *1039*
[544-63-8] Myristic Acid, *6416*
[544-92-3] Cuprous Cyanide, *2731*
[544-97-8] Dimethylzinc, *3312*
[545-06-2] Trichloroacetonitrile, *9757*
[545-26-6] Gitoxigenin, *4440*
[545-45-9] Zygadenine, *10329*
[545-47-1] Lupeol, *5636*
[545-55-1] Triethylenephosphoramide, *9804*
[545-56-2] Delcosine*, *2933*
[545-61-9] Ajaconine, *193*
[545-80-2] Poldine Methylsulfate, *7714*
[545-90-4] Dimepheptanol, *3254*
[545-91-5] Phenadoxone Hydrochloride*, *7348*
[545-93-7] Propallylonal, *7980*
[546-06-5] Conessine, *2561*
[546-43-0] Alantolactone, *208*
[546-46-3] Zinc Citrate, *10263*
[546-48-5] Pempidine Tartrate*, *7207*
[546-62-3] Verbascose, *10092*
[546-67-8] Lead Tetraacetate, *5447*
[546-68-9] Titanium Isopropylate, *9614*
[546-74-7] Sodium Ethyl Sulfate, *8759*
[546-88-3] Acetohydroxamic Acid, *60*
[546-89-4] Lithium Acetate, *5544*
[546-97-4] Columbin, *2558*
[547-25-1] Turanose, *9951*
[547-36-4] Succisulfone 2,2′-Iminodiethanol Salt*, *9047*
[547-44-4] Sulfanilylurea, *9101*
[547-52-4] *N*⁴-Sulfanilylsulfanilamide, *9100*
[547-53-5] 4′-(Methylsulfamoyl)sulfanilanilide, *6202*
[547-56-8] Cupric *p*-Phenolsulfonate, *2715*
[547-57-9] Tropaeolin O, *9905*
[547-58-0] Methyl Orange, *6180*
[547-63-7] Methyl Isobutyrate, *6162*
[547-64-8] Methyl Lactate, *6167*
[547-65-9] α-Methylene Butyrolactone, *6139*
[547-66-0] Magnesium Oxalate, *5712*
[547-67-1] Nickel Oxalate, *6597*
[547-68-2] Zinc Oxalate, *10278*
[547-77-3] Allotetrahydrocortisone, *289*
[547-81-9] 16-Epiestriol, *3652*
[547-91-1] 8-Hydroxy-7-iodo-5-quinolinesulfonic Acid, *4872*
[548-00-5] Ethyl Biscoumacetate, *3818*
[548-04-9] Hypericin, *4911*
[548-24-3] Eosine I Bluish, *3638*
[548-37-8] Verbenalin, *10093*
[548-40-3] Oxyacanthine, *7087*
[548-42-5] Agroclavine, *187*
[548-43-6] Elymoclavine, *3593*

[548-51-6] *o*-Thymotic Acid, *9549*
[548-54-9] Violacein, *10135*
[548-57-2] Lucanthone Hydrochloride, *5620*
[548-62-9] Gentian Violet, *4401*
[548-66-3] Drofenine Hydrochloride*, *3501*
[548-68-5] Thiphenamil Hydrochloride*, *9509*
[548-73-2] Droperidol, *3505*
[548-76-5] Irigenin, *5103*
[548-77-6] Tectorigenin, *9264*
[548-80-1] Chromotrope 2B, *2297*
[548-83-4] Galangin, *4356*
[548-84-5] Pyrvinium Chloride*, *8206*
[548-98-1] Cholane, *2250*
[549-07-5] Insularine, *5010*
[549-18-8] Amitriptyline Hydrochloride*, *511*
[549-28-0] Protostephanine, *8085*
[549-38-2] 3′-Methyl-1,2-cyclopentenophenanthrene, *6127*
[549-56-4] Quinine Bisulfate*, *8245*
[549-68-8] Octaverine, *6851*
[550-01-6] Metabutoxycaine Hydrochloride*, *5978*
[550-10-7] Hydrocotarnine, *4829*
[550-24-3] Embelin, *3595*
[550-28-7] Amisometradine, *507*
[550-33-4] Nebularine, *6521*
[550-49-2] Adlumidine, *163*
[550-54-9] Epicinchonidine*, *2345*
[550-60-7] 1-Nitro-2-naphthol, *6712*
[550-73-2] 3-Methyl-6,7-methylenedioxy-1-piperonylisoquinoline, *6171*
[550-74-3] Picrolonic Acid, *7565*
[550-82-3] Resazurin, *8309*
[550-83-4] Propoxycaine Hydrochloride, *8023*
[550-90-3] Lupanine *d*-Form*, *5635*
[550-97-0] 1-Naphthyl Salicylate, *6501*
[550-99-2] Naphazoline Hydrochloride*, *6455*
[551-01-9] Plasmocid, *7680*
[551-06-4] 1-Naphthylisothiocyanate, *6498*
[551-09-7] *N*-(1-Naphthyl)ethylenediamine, *6496*
[551-11-1] Prostaglandin F₂ₐ, *8065*
[551-16-6] 6-Aminopenicillanic Acid, *478*
[551-27-9] Propicillin, *8002*
[551-48-4] Guanochlor Sulfate*, *4595*
[551-64-4] Zinc Tartrate, *10296*
[551-68-8] D-Psicose, *8109*
[551-74-6] Mannomustine, *5790*
[551-76-8] 2,4,6-Trichloro-*m*-cresol, *9764*
[551-77-9] 2,3,6-Trichloro-*p*-cresol, *9763*
[551-78-0] 4,5,6-Trichloro-*o*-cresol, *9765*
[551-88-2] 3-Nitropentane, *6713*
[551-92-8] Dimetridazole, *3315*
[552-21-6] Methylenedigallic Acid, *6141*
[552-23-8] 2-Nitro-4-sulfobenzoic Acid, *6743*
[552-25-0] Diampromide, *3032*
[552-30-7] Trimellitic Anhydride, *9833*
[552-32-9] *o*-Nitroacetanilide*, *6677*
[552-33-0] Amidomycin, *422*
[552-45-4] *o*-Xylyl Chloride*, *10224*
[552-58-9] Eriodictyol, *3710*
[552-59-0] Prunetin, *8094*
[552-70-5] Pseudopelletierine, *8105*
[552-72-7] Lobelanidine, *5577*
[552-79-4] *N*-Methylephedrine, *6145*
[552-82-9] Methyldiphenylamine, *6131*
[552-93-2] Thymol Carbonate*, *9540*

Denotes derivative of title compound

[552-94-3] Salsalate, *8491*
[552-96-5] Procerin, *7940*
[552-98-7] Lithium Acetylsalicylate, *5545*
[553-03-7] Hydrocarbostyril, *4820*
[553-06-0] N-(1,2-Diphenylethyl)-nicotinamide, *3382*
[553-08-2] Thonzonium Bromide, *9512*
[553-12-8] Protoporphyrin IX, *8084*
[553-13-9] Zolamine, *10319*
[553-17-3] Guaiacol Carbonate, *4577*
[553-19-5] Xanthyletin, *10203*
[553-20-8] 5'-Nitro-2'-propoxyacet-anilide, *6726*
[553-24-2] Neutral Red, *6571*
[553-26-4] γ,γ'-Dipyridyl, *3412*
[553-27-5] Aniline Mustard, *697*
[553-30-0] Proflavine Sulfate*, *7953*
[553-39-9] Allenolic Acid, *256*
[553-54-8] Lithium Benzoate, *5547*
[553-58-2] Butethamine *meta*-Isomer Hydrochloride*, *1552*
[553-68-4] Butethamine Hydrochloride*, *1552*
[553-69-5] Phenyramidol, *7474*
[553-70-8] Magnesium Benzoate, *5691*
[553-74-2] Nitrin, *6676*
[553-75-3] Methylconiine, *6126*
[553-79-7] 5-Nitro-2-propoxyaniline, *6727*
[553-84-4] Perilla Ketone, *7307*
[553-90-2] Methyl Oxalate, *6181*
[553-91-3] Lithium Oxalate, *5563*
[554-01-8] 5-Methylcytosine, *6128*
[554-10-9] Glyceryl Iodide, *4497*
[554-12-1] Methyl Propionate, *6191*
[554-13-2] Lithium Carbonate, *5552*
[554-15-4] N-Methylpyrroline, *6198*
[554-18-7] Glucosulfone Sodium, *4471*
[554-24-5] Phenobutiodil, *7387*
[554-35-8] Linamarin, *5522*
[554-57-4] Methazolamide, *6031*
[554-70-1] Triethyl Phosphine, *9807*
[554-71-2] Sozoiodolic Acid, *8881*
[554-72-3] Tryparsamide, *9925*
[554-73-4] Tropaeolin OO, *9906*
[554-84-7] m-Nitrophenol, *6716*
[554-91-6] Gentiobiose, *4402*
[554-92-7] Trimethobenzamide Hydrochloride*, *9839*
[554-99-4] N-Methylepinephrine, *6146*
[555-10-2] β-Phellandrene, *7341*
[555-21-5] p-Nitrobenzyl Cyanide, *6688*
[555-30-6] Methyldopa, *6132*
[555-31-7] Aluminum Isopropoxide, *359*
[555-32-8] Aluminum Benzoate, *337*
[555-35-1] Aluminum Palmitate, *370*
[555-37-3] Neburon, *6523*
[555-43-1] Tristearin, *9885*
[555-44-2] Tripalmitin, *9865*
[555-45-3] Trimyristin, *9854*
[555-54-4] Diphenylmagnesium, *3386*
[555-57-7] Pargyline, *7172*
[555-59-9] Maleanilic Acid, *5742*
[555-68-0] m-Nitrocinnamic Acid, *6692*
[555-75-9] Aluminum Ethoxide, *350*
[555-76-0] Ferric Formate, *4066*
[555-77-1] 2,2',2''-Trichlorotriethyl-amine, *9777*
[555-84-0] Nifuradene, *6621*
[555-89-5] Bis(p-chlorophenoxy)meth-ane, *1287*
[555-92-0] Coniine Hydrochloride*, *2569*
[556-10-5] (4-Nitrophenyl)urea, *6723*
[556-22-9] Glyodin, *4518*

[556-24-1] Methyl Isovalerate, *6166*
[556-27-4] Alliin, *260*
[556-38-7] Zinc Valerate, *10299*
[556-50-3] N-Glycylglycine, *4513*
[556-52-5] Glycidol, *4499*
[556-56-9] Allyl Iodide, *302*
[556-61-6] Methyl Isothiocyanate, *6165*
[556-63-8] Lithium Formate, *5558*
[556-64-9] Methyl Thiocyanate, *6209*
[556-67-2] Octamethylcyclotetrasilox-ane, *6843*
[556-88-7] Nitroguanidine, *6705*
[556-89-8] Nitrourea, *6749*
[556-91-2] Aluminum *tert*-Butoxide, *343*
[556-99-0] Triuret, *9895*
[557-04-0] Magnesium Stearate, *5730*
[557-05-1] Zinc Stearate, *10292*
[557-07-3] Zinc Oleate, *10276*
[557-09-5] Zinc Caprylate, *10259*
[557-11-9] Allylurea, *306*
[557-17-5] Methyl Propyl Ether, *6192*
[557-18-6] Diethylmagnesium, *3171*
[557-19-7] Nickel Cyanide, *6589*
[557-20-0] Diethylzinc, *3179*
[557-21-1] Zinc Cyanide, *10264*
[557-24-4] Maleamic Acid, *5741*
[557-25-5] Monobutyrin, *6332*
[557-28-8] Zinc Propionate, *10286*
[557-31-3] Allyl Ethyl Ether, *301*
[557-34-6] Zinc Acetate, *10256*
[557-35-7] *sec*-Octyl Bromide, *6862*
[557-36-8] *sec*-Octyl Iodide, *6863*
[557-39-1] Magnesium Formate, *5702*
[557-40-4] Allyl Ether, *300*
[557-41-5] Zinc Formate, *10266*
[557-42-6] Zinc Thiocyanate, *10298*
[557-59-5] Lignoceric Acid, *5511*
[557-61-9] Octacosanol, *6840*
[557-66-4] Ethylamine Hydrochlo-ride*, *3808*
[559-48-8] Kopsine, *5332*
[559-49-9] Annotinine, *714*
[559-57-9] Diginatigenin, *3197*
[559-70-6] β-Amyrin, *660*
[559-74-0] Friedelin, *4294*
[560-88-3] Bornyl Salicylate, *1371*
[561-07-9] Delphinine, *2932*
[561-10-4] Isomethadone *l*-Form*, *5201*
[561-20-6] Cacotheline, *1641*
[561-22-8] Holarrhenine, *4762*
[561-25-1] Dihydrothebaine, *3224*
[561-27-3] Diacetylmorphine, *3012*
[561-48-8] Norpipanone, *6810*
[561-51-3] Garryine, *4381*
[561-77-3] Dihexyverine, *3211*
[561-78-4] Alphaprodine *dl*-Form Hydrochloride*, *317*
[561-79-5] Metcaraphen, *5996*
[561-83-1] Nealbarbital, *6516*
[561-86-4] Brallobarbital, *1387*
[561-94-4] Ergosine, *3695*
[561-95-5] Nimbiol, *6639*
[562-09-4] Chlorphenoxamine Hydro-chloride*, *2234*
[562-10-7] Doxylamine Succinate*, *3497*
[562-26-5] Phenoperidine, *7400*
[562-34-5] Chlorogenin, *2195*
[562-54-9] Potassium Methyl Sulfate, *7813*
[562-71-0] Suprasterol II, *9179*
[562-76-5] Potassium Tetracyano-platinate(II), *7858*
[562-81-2] Barium Platinous Cyanide, *1020*
[562-90-3] Silicon Tetraacetate, *8642*
[563-03-1] Guaiacol Phosphate, *4578*

[563-12-2] Ethion, *3782*
[563-41-7] Semicarbazide Hydro-chloride, *8588*
[563-47-3] 3-Chloro-2-methyl-1-prop-ene, *2200*
[563-52-0] 3-Chloro-1-butene, *2182*
[563-63-3] Silver Acetate, *8648*
[563-68-8] Thallium Acetate, *9392*
[563-72-4] Calcium Oxalate, *1732*
[563-80-4] Methyl Isopropyl Ketone, *6164*
[564-00-1] Erythritol Anhydride, *3715*
[564-25-0] Doxycycline (anhydrous), *3496*
[564-36-3] Ergocornine, *3685*
[564-37-4] Ergocorninine, *3686*
[565-50-4] Terpin *trans*-Form*, *9314*
[565-63-9] Angelic Acid, *685*
[565-74-2] α-Bromoisovaleric Acid, *1443*
[566-02-9] Allopregnane-3β,21-diol-11,20-dione, *271*
[566-41-6] Allopregnane-3β,17α,20β,-21-tetrol, *275*
[566-48-3] Formestane, *4267*
[566-56-3] Allopregnane-3β,20α-diol, *269*
[566-57-4] Allopregnane-3α,20β-diol, *268*
[566-58-5] Allopregnane-3α,20α-diol, *267*
[566-65-4] 3,20-Allopregnanedione, *273*
[566-78-9] 21-Acetoxypregnenolone, *77*
[567-18-0] 1-Naphthol-2-sulfonic Acid, *6477*
[568-02-5] Alizarine Blue, *249*
[568-21-8] Isothebaine, *5247*
[568-53-6] α-Peltatin, *7203*
[568-93-4] Alizarine Orange, *250*
[569-31-3] Meconin, *5825*
[569-57-3] Chlorotrianisene, *2225*
[569-58-4] Aluminon, *330*
[569-59-5] Phenindamine Hydrogen Tartrate*, *7380*
[569-64-2] Malachite Green, *5739*
[569-65-3] Meclizine, *5817*
[569-77-7] Purpurogallin, *8133*
[569-84-6] Antipyrine Acetylsalic-ylate*, *757*
[570-50-3] Allopregnane-3β,17α,20α-triol, *278*
[570-54-7] Allopregnane-3β,17α-diol-20-one, *272*
[572-09-8] Acetobromglucose, *57*
[572-48-5] Coumithoate, *2635*
[572-59-8] Epiquinidine, *3657*
[572-60-1] Epiquinine, *3658*
[572-76-9] Nandinine, *6451*
[572-96-3] Phytonadiol*, *7536*
[573-01-3] Menadoxime, *5875*
[573-20-6] Menadiol Diacetate*, *5873*
[573-35-3] Inositol Monophosphate*, *5008*
[573-41-1] Theophylline Ethanol-amine*, *9421*
[573-56-8] 2,6-Dinitrophenol, *3335*
[573-58-0] Congo Red, *2562*
[574-12-9] Isoflavone, *5189*
[574-64-1] Trypan Red, *9924*
[574-65-2] Vital Red, *10149*
[574-84-2] 3,7-Dinitro-5-oxopheno-thiazine, *3332*
[574-84-5] Fraxetin, *4289*
[574-95-8] Aureothricin, *914*
[575-19-9] 6,7-Benzomorphan, *1126*
[575-47-3] Amylpenicillin Sodium, *658*
[575-74-6] Buclosamide, *1485*
[575-75-7] Toldimfos Sodium, *9649*

* Denotes derivative of title compound

[576-11-4] Medmain, 5835
[576-19-2] Biocytin, 1268
[576-55-6] 3,4,5,6-Tetrabromo-o-cresol, 9327
[576-97-6] Noprylsulfamide, 6777
[577-11-7] Docusate Sodium, 3460
[577-27-5] Ledol, 5454
[577-33-3] Anthrarobin, 727
[577-37-7] Aphylline, 771
[577-91-3] Iodoalphionic Acid, 5047
[578-20-1] Thymyl N-Isoamylcarbamate, 9550
[578-36-9] Potassium Salicylate, 7835
[578-74-5] Apigetrin, 774
[578-94-9] Phenarsazine Chloride, 7357
[579-04-4] Cymarose, 2831
[579-10-2] N-Methylacetanilide, 6088
[579-13-5] Oligomycin A*, 6970
[579-21-5] Lobelanine, 5578
[579-23-7] Cyclovalone, 2823
[579-38-4] Diloxanide, 3246
[579-56-6] Isoxsuprine Hydrochloride*, 5259
[579-92-0] Diphenylamine-2,2'-dicarboxylic Acid, 3376
[580-02-9] Methyl Acetylsalicylate, 6091
[580-13-2] 2-Bromonaphthalene, 1448
[580-52-9] Oxalenediuramidoxime, 7042
[580-74-5] Xanthocillin X*, 10195
[581-49-7] Anatabine, 668
[581-75-9] 2,6-Naphthalenedisulfonic Acid, 6461
[581-88-4] Debrisoquin Sulfate*, 2901
[582-36-5] Oxythiamine, 7112
[582-52-5] Diacetoneglucose, 3009
[582-60-5] 5,6-Dimethylbenzimidazole, 3286
[583-03-9] Fenipentol, 4016
[583-15-3] Mercuric Benzoate, 5924
[583-39-1] 2-Benzimidazolethiol, 1112
[583-50-6] D-Erythrose, 3731
[583-52-8] Potassium Oxalate, 7820
[583-86-8] Phloionolic Acid, 7479
[583-91-5] Methionine Hydroxy Analog, 6054
[584-02-1] 3-Pentanol, 7259
[584-08-7] Potassium Carbonate, 7781
[584-10-1] Potassium Tetraiodocadmate, 7861
[584-18-9] Acetomeroctol, 63
[584-28-1] Aspidin, 881
[584-42-9] Metachrome Yellow, 5979
[584-43-0] Mercuric Succinimide, 5942
[584-79-2] Allethrin I*, 257
[584-84-9] Toluene 2,4-Diisocyanate, 9668
[584-85-0] Anserine, 717
[585-18-2] D-Erythrose 4-Phosphate, 3733
[585-21-7] Glutamine DL-Form*, 4479
[585-29-5] Triethylammonium Formate, 9800
[585-48-8] 2,6-Di-tert-butylpyridine, 3086
[585-86-4] Lactitol, 5352
[585-99-9] Melibiose, 5861
[586-02-7] Phrenosin, 7524
[586-06-1] Metaproterenol, 5992
[586-60-7] Dyclonine, 3523
[586-76-5] p-Bromobenzoic Acid, 1429
[586-84-5] 2-(Methoxymethyl)-5-nitrofuran, 6075
[587-23-5] Methenamine Mandelate, 6040
[587-33-7] m-Tyrosine, 9971
[587-46-2] Benzpyrinium Bromide, 1153

[587-61-1] Propyliodone, 8048
[587-63-3] Dihydrokawain*, 5299
[587-90-6] 4,4'-Dinitrocarbanilide, 3330
[587-98-4] Metanil Yellow, 5989
[588-36-3] Methioprim, 6055
[588-42-1] Trolnitrate Phosphate, 9900
[588-59-0] Stilbene, 8972
[589-15-1] p-Bromobenzyl Bromide, 1430
[589-17-3] p-Bromobenzyl Chloride, 1431
[589-21-9] p-Bromophenylhydrazine, 1451
[589-44-6] 3-Amino-4-hydroxybutyric Acid, 463
[589-57-1] Ethyl Phosphorochloridite, 3886
[589-59-3] Isobutyl Isovalerate, 5160
[589-97-9] Potassium Percarbonate, 7821
[590-01-2] n-Butyl Propionate, 1623
[590-28-3] Potassium Cyanate, 7788
[590-29-4] Potassium Formate, 7795
[590-46-5] Betaine Hydrochloride*, 1225
[590-47-6] Betaine Monohydrate*, 1225
[590-63-6] Bethanechol Chloride, 1232
[590-86-3] Isovaleraldehyde, 5249
[590-92-1] β-Bromopropionic Acid, 1455
[591-01-5] Dicyanodiamidine Sulfate, 3143
[591-09-3] Acetyl Nitrate, 99
[591-20-8] m-Bromophenol*, 1450
[591-27-5] m-Aminophenol, 480
[591-50-4] Iodobenzene, 5050
[591-64-0] Calcium Levulinate, 1724
[591-78-6] Methyl Butyl Ketone, 6112
[591-81-1] γ-Hydroxybutyrate, 4860
[591-84-4] Isobutyl Thiocyanate, 5167
[591-87-7] Allyl Acetate, 293
[591-89-9] Potassium Tetracyanomercurate(II), 7856
[591-97-9] 1-Chloro-2-butene, 2181
[592-01-8] Calcium Cyanide, 1704
[592-04-1] Mercuric Cyanide, 5928
[592-65-4] Isobutyl Sulfide, 5166
[592-85-8] Mercuric Thiocyanate, 5946
[592-87-0] Lead Thiocyanate, 5450
[592-88-1] Allyl Sulfide, 305
[592-89-2] Strontium Formate, 9003
[592-95-0] Sulforaphen, 9136
[593-26-0] Ammonium Palmitate, 572
[593-29-3] Potassium Stearate, 7844
[593-39-5] Petroselinic Acid, 7330
[593-50-0] 1-Triacontanol, 9722
[593-53-3] Fluoromethane, 4212
[593-74-8] Dimethylmercury, 3299
[594-08-1] Trithiocarbonic Acid, 9888
[595-05-1] Calycanthine, 1772
[595-33-5] Megestrol Acetate, 5849
[595-39-1] Isovaline, 5255
[595-40-4] Isovaline L-Form*, 5255
[595-77-7] Algestone, 238
[596-01-0] α-Naphtholphthalein, 6476
[596-03-2] 4',5'-Dibromofluorescein, 3070
[596-19-0] Morphine Mucate*, 6359
[596-27-0] o-Cresolphthalein, 2646
[596-51-0] Glycopyrrolate, 4511
[596-56-5] Isatropic Acid, 5117
[597-12-6] Melezitose, 5860
[597-71-7] Pentaerythritol Tetraacetate, 7248
[598-31-2] Bromoacetone, 1422
[598-54-9] Cuprous Acetate, 2728

[598-55-0] Methyl Carbamate, 6115
[598-58-3] Methyl Nitrate, 6175
[598-62-9] Manganese Carbonate, 5766
[598-75-4] 3-Methyl-2-butanol, 6110
[598-82-3] DL-Lactic Acid, 5349
[599-04-2] Pantolactone, 7145
[599-33-7] Prednylidene, 7906
[599-79-1] Sulfasalazine, 9112
[599-88-2] Sulfaperine, 9104
[602-41-5] Thiocolchicine 2-Glucoside Analog*, 9460
[602-64-2] Anthragallol, 722
[602-81-3] Apomorphine Methylbromide*, 787
[602-92-6] Dibromogallic Acid, 3071
[603-00-9] Proxyphylline, 8092
[603-35-0] Triphenylphosphine, 9873
[603-45-2] Aurin, 915
[603-50-9] Bisacodyl, 1282
[603-55-4] DFDD, 2999
[603-57-6] Aminopyrine Salicylate*, 495
[603-59-8] Antipyrine Salicylacetate*, 757
[603-64-5] Antipyrine Mandelate*, 757
[604-51-3] Deptropine, 2958
[604-75-1] Oxazepam, 7059
[605-50-5] Isoamyl Phthalate, 5138
[605-65-2] Dansyl Chloride, 2877
[606-04-2] Pamabrom, 7133
[606-17-7] Iodipamide, 5043
[606-21-3] 2-Chloro-1,3-dinitrobenzene, 2188
[606-22-4] 2,6-Dinitroaniline, 3323
[606-58-6] Toyocamycin, 9698
[606-90-6] Diphenylpyraline 8-Chlorotheophyllinate*, 3390
[607-80-7] Sesamin, 8616
[607-91-0] Myristicin, 6417
[608-07-1] 5-Methoxytryptamine, 6083
[608-66-2] Galactitol, 4350
[608-89-9] Ethyl Tartrate, Acid, 3901
[609-36-9] Proline DL-Form*, 7963
[609-78-9] Cycloguanil Pamoate*, 2790
[610-60-6] Methyl Benzoylsalicylate, 6105
[610-88-8] Tenuazonic Acid, 9294
[611-75-6] Bromhexine Hydrochloride*, 1412
[612-83-9] 3,3'-Dichlorobenzidine Dihydrochloride*, 3109
[613-78-5] 2-Naphthyl Salicylate, 6502
[613-82-1] Phenylethanolamine Sulfate*, 7441
[613-89-8] Phenacylamine, 7347
[614-39-1] Procainamide Hydrochloride, 7936
[614-87-9] Dibromopropamidine Isethionate*, 3073
[615-05-4] 2,4-Diaminoanisole, 3018
[616-05-7] α-Bromo-n-caproic Acid, 1436
[616-07-9] Ornithine DL-Form*, 7002
[616-38-6] Methyl Carbonate, 6117
[616-45-2] 2-Pyrrolidone, 8200
[616-68-2] Trimecaine, 9830
[616-91-1] Acetylcysteine, 89
[617-12-9] Chorismic Acid, 2274
[617-19-6] 2,6-Diamino-2'-butyloxy-3,5'-azopyridine, 3021
[617-48-1] Malic Acid DL-Form*, 5747
[617-52-7] Lactic Acid Lactate, 5351
[617-65-2] Glutamic Acid DL-Form*, 4477
[618-22-4] Arsacetin, 829
[618-27-9] Hydroxyproline cis-Form*, 4887

Denotes derivative of title compound

[618-65-5] Helicin, 4658
[618-67-7] Melilotoside, 5863
[618-82-6] Sulfarsphenamine, 9111
[619-84-1] 4-(Dimethylamino)benzoic Acid, 3282
[620-13-3] m-Xylyl Bromide*, 10223
[620-19-9] m-Xylyl Chloride*, 10224
[620-30-4] Metyrosine DL-Form*, 6248
[620-34-8] Methenamine Salicylate, 6041
[620-45-1] 2,6-Dichloroindophenol Sodium, 3118
[620-90-6] Phenamidine Isethionate*, 7352
[620-99-5] Phenacaine Hydrochloride, 7342
[621-72-7] Bendazol, 1062
[621-82-9] Cinnamic Acid, 2358
[622-91-3] Metoquinone, 6236
[623-32-5] Chaulmoogric Acid Ethyl Ester*, 2089
[623-42-7] Methyl Butyrate, 6114
[623-73-4] Diazoacetic Ester, 3045
[624-04-4] Glycol Dilaurate, 4507
[624-65-7] Propargyl Chloride, 7994
[624-83-9] Methyl Isocyanate, 6163
[625-04-7] Diacetonamine, 3006
[625-51-4] Formicin, 4269
[626-48-2] 6-Methyluracil, 6213
[627-13-4] n-Propyl Nitrate, 8049
[628-52-4] Chromous Acetate, 2299
[628-55-7] Isobutyl Ether, 5156
[628-96-6] Ethylene Glycol Dinitrate, 3846
[629-11-8] Hexamethylene Glycol, 4726
[630-08-0] Carbon Monoxide, 1861
[630-56-8] 17α-Hydroxyprogesterone Caproate*, 4886
[630-60-4] Ouabain, 7031
[630-64-8] Helveticoside, 4665
[630-67-1] Prednisolone 21-m-Sulfobenzoate Sodium Salt*, 7901
[630-93-3] Phenytoin Sodium Salt*, 7475
[631-01-6] Quillaic Acid, 8221
[631-60-7] Mercurous Acetate, 5948
[631-61-8] Ammonium Acetate, 520
[631-69-6] β-Boswellic Acid, 1382
[632-00-8] Sulfasomizole, 9113
[632-22-4] Tetramethylurea, 9368
[632-31-5] β-Ergostenol, 3699
[632-32-6] α-Ergostenol, 3698
[632-68-8] Rose Bengal, 8421
[632-69-9] Rose Bengal Disodium Salt*, 8421
[632-99-5] Magenta I, 5686
[633-03-4] Brilliant Green, 1398
[633-47-6] Cropropamide, 2659
[633-59-0] Clopenthixol Dihydrochloride*, 2455
[633-96-5] Orange II, 6990
[634-03-7] Phendimetrazine, 7365
[634-19-5] Phentydrone, 7418
[634-25-3] Pyridoxal 5-Phosphate Oxime*, 8163
[634-62-8] Hydrazine Tartrate, 4812
[635-41-6] Trimetozine, 9850
[635-65-4] Bilirubin, 1263
[636-46-4] 4-Hydroxyisophthalic Acid, 4873
[636-47-5] Stallimycin, 8927
[636-54-4] Clopamide, 2454
[636-61-3] Malic Acid D-(+)-Form*, 5747
[636-65-7] L-Isoglutamine, 5193
[637-03-6] Oxophenylarsine, 7083
[637-07-0] Clofibrate, 2436
[637-12-7] Aluminum Stearate, 379

[637-32-1] Chlorguanide Hydrochloride*, 2138
[637-49-0] Coniine Hydrobromide*, 2569
[637-58-1] Pramoxine Hydrochloride*, 7888
[637-92-3] Ethyl tert-Butyl Ether, 3821
[638-23-3] Carbocysteine, 1850
[638-38-0] Manganese Acetate, 5763
[638-39-1] Stannous Acetate, 8937
[638-94-8] Desonide, 2973
[638-95-9] α-Amyrin, 659
[639-14-5] Gypsogenin, 4614
[639-43-0] Akuammicine, 199
[639-44-1] Phenoltetrachlorophthalein, 7398
[639-48-5] Nicomorphine, 6607
[639-81-6] Ergotaminine, 3704
[639-86-1] Solanocapsine, 8854
[639-97-4] Apiose, 778
[640-19-7] Fluoroacetamide, 4204
[640-67-5] Manganese Oxalate, 5775
[641-36-1] Apocodeine, 781
[641-85-0] Allopregnane, 266
[642-08-0] Ostruthol, 7028
[642-15-9] Antimycin A₁, 755
[642-18-2] Alstonine, 324
[642-44-4] Aminometradine, 471
[642-58-0] Ethylbenzhydramine, 3813
[642-72-8] Benzydamine, 1157
[642-75-1] Adrenolutin, 173
[642-83-1] Aceglatone, 24
[642-84-2] 3',4',5-Trichlorosalicylanilide, 9775
[643-13-0] Fructose-6-phosphate, 4298
[643-22-1] Erythromycin Stearate, 3726
[643-60-7] Anhalamine, 691
[643-84-5] Malvidin Chloride, 5754
[644-31-5] Acetozone, 78
[644-62-2] Meclofenamic Acid, 5819
[644-64-4] Dimetilan, 3313
[645-05-6] Altretamine, 328
[645-17-0] Linoleic Acid Aluminum Salt*, 5529
[645-43-2] Guanethidine Monosulfate*, 4589
[645-74-9] Myoral, 6411
[645-99-8] Mercuric Stearate, 5940
[646-13-9] Isobutyl Stearate, 5165
[650-51-1] Trichloroacetic Acid Sodium Salt*, 9756
[650-69-1] β-Citraurin, 2385
[651-06-9] Sulfameter, 9082
[651-43-4] Allopregnane-3β,11β,17α,-21-tetrol-20-one, 277
[651-55-8] Dihydroequilin 17α-Form*, 3216
[652-37-9] Acefylline, 22
[652-67-5] Isosorbide, 5244
[653-03-2] Butaperazine, 1543
[655-05-0] Thozalinone, 9521
[655-35-6] Chromonar Hydrochloride*, 2296
[655-76-5] Quinaldine Sulfate*, 8228
[655-86-7] 2,3-Diaminophenazine, 3025
[657-24-9] Metformin, 6001
[657-27-2] L-Lysine Monohydrochloride*, 5667
[659-40-5] Hexamidine Isethionate*, 4728
[659-70-1] Isoamyl Isovalerate, 5135
[660-27-5] Diisopropylamine Dichloroacetate, 3241
[660-60-6] Cupric Stearate, 2721
[661-19-8] 1-Docosanol*, 8135
[664-95-9] Tolcyclamide, 9648
[665-66-7] Amantadine Hydrochloride*, 389

[666-99-9] Agaricic Acid, 184
[670-54-2] Tetracyanoethylene, 9336
[670-65-5] Neopterin, 6551
[671-16-9] Procarbazine, 7938
[671-88-5] Disulfamide, 3427
[671-95-4] Clofenamide, 2434
[672-15-1] Homoserine, 4777
[672-87-7] Metyrosine, 6248
[673-06-3] Phenylalanine D-Form*, 7425
[673-31-4] Phenprobamate, 7412
[679-84-5] Halopropane, 4632
[680-31-9] HMPA, 4761
[685-73-4] D-Galacturonic Acid, 4354
[688-37-9] Aluminum Oleate, 367
[689-13-4] Hadacidin, 4617
[692-13-7] Buformin, 1500
[693-30-1] Hemisulfur Mustard, 4680
[693-65-2] n-Amyl Ether, 651
[693-72-1] Vaccenic Acid, 10036
[694-59-7] Pyridine 1-Oxide, 8155
[695-34-1] 2-Amino-4-picoline, 486
[695-53-4] Dimethadione, 3260
[702-54-5] Diethadione, 3155
[703-95-7] 5-Fluoroorotic Acid, 4214
[705-16-8] Chrysanthemic Acid dl-trans-Form*, 2310
[709-55-7] Etilefrin, 3911
[709-98-8] Propanil, 7987
[712-68-5] Triafur, 9725
[716-79-0] 2-Phenyl-1H-benzimidazole, 7428
[721-50-6] Prilocaine, 7924
[723-46-6] Sulfamethoxazole, 9086
[723-61-5] Helminthosporal, 4662
[728-88-1] Tolperisone, 9660
[729-99-7] Sulfamoxole, 9093
[730-68-7] Methitural Sodium Salt*, 6058
[732-11-6] Phosmet, 7492
[737-31-5] Diatrizoate Sodium, 3040
[738-70-5] Trimethoprim, 9840
[738-99-8] Dipin, 3397
[739-71-9] Trimipramine, 9852
[742-20-1] Cyclopenthiazide, 2813
[745-65-3] Prostaglandin E₁, 8063
[747-36-4] Hydroxychloroquine Sulfate*, 4863
[749-02-0] Spiperone, 8903
[749-13-3] Trifluperidol, 9813
[751-84-8] Penicillin G Benethamine, 7220
[751-94-0] Fusidic Acid Sodium Salt*, 4340
[751-97-3] Rolitetracycline, 8411
[752-56-7] Riboflavin 2',3',4',5'-Tetrabutyrate*, 8367
[752-61-4] Digitalin, 3200
[759-73-9] N-Ethyl-N-nitrosourea, 3879
[759-94-4] EPTC, 3674
[764-05-6] Cyanogen Azide, 2762
[767-21-5] Piperidine Phosphate*, 7621
[768-94-5] Amantadine, 389
[770-05-8] Octopamine DL-Form Hydrochloride*, 6856
[773-76-2] Chloroxine, 2227
[777-11-7] Haloprogin, 4631
[786-19-6] Carbophenothion, 1869
[790-69-2] Loflucarban, 5589
[791-35-5] Chlophedianol, 2106
[795-13-1] N²-Formylsulfisomidine, 4275
[795-38-0] Dihydroisocodeine, 3219
[797-63-7] Levonorgestrel*, 6797
[797-64-8] Norgestrel (+)-Form*, 6797
[800-22-6] Chloracizine, 2108
[800-24-8] 2,5-Bis(1-aziridinyl)-3,6-

Denotes derivative of title compound

bis(2-methoxyethoxy)-1,4-benzo-
quinone, 1285
[801-52-5] Porfiromycin, 7756
[804-10-4] Chromonar, 2296
[804-30-8] Fursultiamine, 4333
[804-36-4] Nitrovin, 6752
[804-53-5] Tetrabenazine Methanesulfonate*, 9325
[804-63-7] Quinine Sulfate*, 8245
[805-84-5] 17-Hydroxy-16-methylene-
Δ⁶-progesterone 17-Acetate*, 4878
[807-31-8] Aceperone, 30
[808-26-4] Sancycline, 8500
[808-71-9] Penethamate Hydriodide,
7212
[811-54-1] Lead Formate, 5426
[813-94-5] Calcium Citrate, 1701
[814-71-1] Calcium Thioglycollate,
1758
[814-80-2] Calcium Lactate, 1723
[814-87-9] Aluminum Oxalate, 368
[814-89-1] Cobaltous Oxalate, 2506
[814-90-4] Chromous Oxalate, 2304
[814-91-5] Cupric Oxalate, 2712
[814-93-7] Lead Oxalate, 5436
[814-94-8] Stannous Oxalate, 8943
[814-95-9] Strontium Oxalate, 9008
[815-78-1] Aluminum Tartrate, 383
[815-82-7] Cupric Tartrate, 2725
[815-85-0] Stannous Tartrate, 8949
[817-09-4] 2,2',2''-Trichlorotriethyl-
amine Hydrochloride*, 9777
[819-17-0] Calcium Methionate, 1727
[819-73-8] Lead Butyrate, 5419
[821-33-0] 4,4'-Oxydi-2-butanol, 7094
[822-16-2] Sodium Stearate, 8827
[826-39-1] Mecamylamine Hydrochloride*, 5814
[828-00-2] Dimethoxane, 3273
[829-74-3] Nordefrin (−)-Form*,
6785
[830-89-7] Albutoin, 218
[834-12-8] Ametryn, 411
[834-28-6] Phenformin Hydrochloride*, 7376
[835-31-4] Naphazoline, 6455
[837-27-4] Acefylline Sodium Salt*, 22
[840-50-6] MADU, 5681
[841-73-6] Bucolome, 1487
[846-48-0] Boldenone, 1354
[846-49-1] Lorazepam, 5609
[846-50-4] Temazepam, 9285
[846-70-8] Naphthol Yellow S, 6480
[847-25-6] Thiamphenicol DL-Form*,
9436
[847-84-7] Normethadone Hydrochloride*, 6804
[848-21-5] Norgestrienone, 6798
[848-53-3] Homochlorcyclizine, 4770
[848-75-9] Lormetazepam, 5611
[849-55-8] Nylidrin Hydrochloride*,
6830
[850-52-2] Altrenogest, 327
[852-19-7] Sulfazamet, 9118
[852-42-6] Guaiapate, 4580
[853-34-9] Kebuzone, 5300
[855-19-6] Clostebol Acetate*, 2475
[855-96-9] Eupatorin, 3947
[856-87-1] Dipipanone Hydrochloride*, 3398
[860-07-1] 3',3'',5',5''-Tetrabromo-
phenolphthalein Disodium Salt*,
9329
[860-22-0] Indigo Carmine, 4978
[863-61-6] Menaquinone 4*, 5876
[865-04-3] Deserpidine 10-Methoxy
Deriv*, 2964
[865-21-4] Vinblastine, 10119
[865-28-1] Tyrocidine B*, 9967
[865-44-1] Iodine Trichloride, 5041

[866-82-0] Cupric Citrate, 2702
[866-83-1] Potassium Citrate, Monobasic, 7786
[866-84-2] Potassium Citrate, 7785
[867-81-2] Pantothenic Acid Sodium
Salt*, 7147
[868-14-4] Potassium Bitartrate, 7776
[868-16-6] Lithium Bitartrate, 5548
[868-18-8] Sodium Tartrate, 8833
[868-54-2] 2-Amino-1,1,3-tricyano-
propene, 502
[870-72-4] Formaldehyde Sodium
Bisulfite, 4263
[872-50-4] 1-Methylpyrrolidone, 6197
[875-83-2] 2-Naphthol Sodium Salt*,
6471
[876-04-0] Octopamine D(−)-Form*,
6856
[877-24-7] Potassium Biphthalate,
7773
[882-09-7] Clofibric Acid, 2437
[886-38-4] Diphencyprone, 3366
[886-65-7] 1,4-Diphenyl-1,3-butadi-
ene, 3378
[886-74-8] Chlorphenesin Carbamate,
2231
[886-86-2] Tricaine, 9751
[887-08-1] Daucol, 2889
[894-71-3] Nortriptyline Hydrochloride*, 6812
[897-15-4] Dothiepin Hydrochloride*,
3485
[897-61-0] Penicillin O Potassium
Salt*, 7228
[900-77-6] Drocarbil*, 815
[901-93-9] Estrone Acetate*, 3751
[904-04-1] Captodiamine Hydrochloride*, 1816
[908-54-3] Diminazene Aceturate,
3316
[909-39-7] Opipramol Dihydrochloride*, 6985
[910-86-1] Tiocarlide, 9593
[911-45-5] Clomiphene, 2446
[911-65-9] Etonitazene, 3929
[912-57-2] Nandrolone Cyclohexanepropionate*, 6452
[912-60-7] Noscapine Hydrochloride*,
6815
[914-00-1] Methacycline, 6007
[915-30-0] Diphenoxylate, 3371
[915-67-3] Amaranth (Dye), 391
[917-61-3] Sodium Cyanate, 8749
[917-69-1] Cobaltic Acetate, 2489
[918-04-7] Acetaldehyde Sodium
Bisulfite, 39
[919-16-4] Lithium Citrate, 5555
[921-53-9] Potassium Tartrate, 7849
[922-55-4] Lanthionine, 5375
[922-80-5] Diamyl Sodium Sulfosuccinate, 3034
[923-06-8] Bromosuccinic Acid, 1461
[923-32-0] Cystine DL-Form*, 2851
[926-26-1] Di-tert-butyl Succinate,
3088
[926-93-2] Methallibure, 6013
[930-55-2] N-Nitrosopyrrolidine, 6741
[932-53-6] 6-Azathymine, 936
[937-13-3] Oxonic Acid, 7081
[938-73-8] Ethenzamide, 3776
[941-69-5] N-Phenylmaleimide, 7452
[944-22-9] Fonofos, 4261
[947-02-4] Phosfolan, 7490
[947-08-0] Thiobutabarbital Sodium
Salt*, 9457
[950-10-7] Mephosfolan, 5900
[950-37-8] Methidathion, 6048
[951-78-0] Deoxyuridine, 2957
[952-54-5] Morphazinamide, 6355

[956-03-6] Meprylcaine Hydrochloride*, 5909
[956-90-1] Phencyclidine Hydrochloride*, 7364
[957-51-7] Diphenamid, 3364
[957-56-2] Fluindione, 4168
[957-68-6] 7-Aminocephalosporanic
Acid, 453
[959-10-4] Xenbucin, 10205
[959-14-8] Oxolamine, 7078
[959-24-0] Sotalol Hydrochloride*,
8876
[960-05-4] Carbubarb, 1880
[962-02-7] Nitrodan, 6693
[964-52-3] Moxisylyte Hydrochloride*, 6374
[965-52-6] Nifuroxazide, 6626
[965-90-2] Ethylestrenol, 3851
[965-93-5] Methyltrienolone, 6211
[968-58-1] Pridinol Hydrochloride*,
7922
[968-81-0] Acetohexamide, 59
[968-93-4] Testolactone, 9321
[969-99-3] Opromazine, 6987
[972-02-1] Diphenidol, 3369
[973-21-7] Dinobuton, 3339
[973-53-5] Penicillin G Calcium, 7223
[973-99-9] Prosultiamine Hydrochloride*, 8068
[975-64-4] Estrone Propionate*, 3751
[976-71-6] Canrenone, 1795
[977-79-7] Medrogestone, 5836
[979-32-8] Estradiol 17-Valerate*,
3746
[980-71-2] Brompheniramine Maleate*, 1467
[982-24-1] Clopenthixol, 2455
[982-43-4] Prenoxdiazine Hydrochloride, 7918
[982-57-0] Chloramphenicol Monosuccinate Sodium Salt*, 2120
[983-85-7] Penamecillin, 7208
[985-13-7] Ethaverine Hydrochloride*, 3773
[985-16-0] Nafcillin Sodium, 6438
[985-32-0] Quinacillin Disodium Salt*,
8224
[987-18-8] Flumedroxone Acetate,
4171
[987-24-6] Betamethasone 21-
Acetate*, 1226
[987-78-0] Citicoline, 2380
[989-51-5] EGCG, 3549
[990-73-8] Fentanyl Citrate*, 4043
[991-42-4] Norbormide, 6780
[992-21-2] Lymecycline, 5658
[992-46-1] Thiamine Disulfide
Phosphate*, 9432
[992-59-6] Benzopurpurine 4B, 1132
[994-05-8] tert-Amyl Methyl Ether,
654
[997-55-7] N-Acetyl-L-aspartic Acid*,
875
[997-68-2] L-Saccharopine, 8464
[999-81-5] Chlormequat Chloride,
2153
[999-97-3] Hexamethyldisilazane, 4725
[1002-89-7] Ammonium Stearate, 588
[1003-78-7] 2,4-Dimethylsulfolane,
3306
[1005-24-9] Trigonellamide Chloride,
9821
[1008-65-7] Fenadiazole, 3994
[1013-20-3] Neo-cupferron, 6536
[1014-70-6] Simetryn, 8683
[1018-71-9] Pyrrolnitrin, 8202
[1028-33-7] Pentifylline, 7269
[1037-50-9] Sulfadimethoxine Sodium
Salt*, 9073

Denotes derivative of title compound

[1041-90-3] Nalorphine Hydrobromide*, 6448
[1043-21-6] Pirenoxine, 7645
[1045-69-8] Testosterone Acetate*, 9322
[1050-48-2] Benzilonium Bromide, 1110
[1050-79-9] Moperone, 6343
[1055-55-6] Bunamidine Hydrochloride*, 1510
[1059-14-9] Taraxasterol, 9229
[1063-55-4] Butaperazine Dimaleate*, 1543
[1063-77-0] Nomilin, 6769
[1066-17-7] Colistin, 2542
[1066-30-4] Chromic Acetate, 2275
[1066-33-7] Ammonium Bicarbonate, 522
[1069-31-4] Ornithine DL-Form Monohydrochloride*, 7002
[1069-54-1] Bis(1,2-dimethylpropyl)-borane, 1290
[1069-55-2] Bucrylate, 1488
[1069-66-5] Valproic Acid Sodium Salt (1:1)*, 10049
[1070-11-7] Ethambutol Dihydrochloride*, 3764
[1070-83-3] tert-Butylacetic Acid, 1573
[1071-83-6] Glyphosate, 4522
[1072-35-1] Lead Stearate, 5442
[1077-27-6] Thioctic Acid l-Form*, 9462
[1077-28-7] Thioctic Acid dl-Form*, 9462
[1078-19-9] 6-Methoxy-α-tetralone, 6081
[1078-21-3] 4-Amino-3-phenylbutyric Acid, 484
[1081-34-1] α-Terthienyl, 9319
[1082-57-1] Tramazoline, 9702
[1083-57-4] Bucetin, 1479
[1084-65-7] Meticrane, 6220
[1085-98-9] Dichlofluanid, 3095
[1086-80-2] Lumichrome, 5626
[1088-11-5] Nordazepam, 6784
[1088-56-8] Lumiflavine, 5627
[1088-92-2] Nifurtoinol, 6631
[1091-94-7] Estrone Methyl Ether*, 3751
[1093-58-9] Clostebol, 2475
[1094-08-2] Ethopropazine Hydrochloride*, 3793
[1094-61-7] NMN, 6761
[1095-90-5] Methadone Hydrochloride, 6008
[1098-60-8] Triflupromazine Hydrochloride*, 9814
[1098-97-1] Pyritinol, 8180
[1099-87-2] Prasterone Sodium Sulfate*, 7892
[1107-99-9] Prednisolone 21-Trimethylacetate*, 7901
[1108-68-5] Cinobufotalin, 2367
[1110-40-3] Cortivazol, 2603
[1110-80-1] Pipacycline, 7606
[1111-39-3] Acetyldigitoxins α-Form*, 90
[1111-44-0] Bietaserpine Bitartrate*, 1254
[1111-67-7] Cuprous Thiocyanate, 2739
[1111-71-3] Beryllium Formate, 1210
[1111-78-0] Ammonium Carbamate, 533
[1113-38-8] Ammonium Oxalate, 571
[1113-41-3] Penicillamine L-Form*, 7214
[1114-34-7] D-Lyxose, 5673
[1114-41-6] Muramic Acid, 6384
[1114-71-2] Pebulate, 7192

[1115-70-4] Metformin Hydrochloride*, 6001
[1115-84-0] Vitamin U, 10165
[1116-54-7] N-Nitrosodiethanolamine, 6733
[1118-68-9] N,N-Dimethylglycine, 3293
[1119-33-1] L-Glutamic Acid 5-Ethyl Ester, 4478
[1119-34-2] Arginine Hydrochloride*, 817
[1119-97-7] Myristyltrimethylammonium Bromide, 6419
[1120-44-1] Cupric Oleate, 2711
[1120-46-3] Lead Oleate, 5435
[1121-30-8] Pyrithione, 8178
[1130-23-0] Cyclobutyrol Sodium Salt*, 2784
[1131-64-2] Debrisoquin, 2901
[1132-61-2] MOPS, 6346
[1134-47-0] Baclofen, 967
[1134-98-1] Sulfarside, 9110
[1135-24-6] Ferulic Acid, 4110
[1135-40-6] CAPS, 1810
[1142-70-7] Butallylonal, 1537
[1143-38-0] Anthralin, 723
[1146-04-9] Illudin M*, 4939
[1146-21-0] 5-Furfuryl-5-isopropylbarbituric Acid, 4326
[1146-95-8] Etifelmin Hydrochloride*, 3909
[1146-98-1] Bromindione, 1414
[1146-99-2] Clorindione, 2468
[1147-62-2] Pyrovalerone Hydrochloride*, 8194
[1149-99-1] Illudin S*, 4939
[1151-11-7] Ipodate Calcium Salt*, 5087
[1151-15-1] o-(p-Anisoyl)benzoic Acid, 710
[1153-51-1] (3α,5α)-Androst-16-en-3-ol, 679
[1154-59-2] 3,3',4',5-Tetrachlorosalicylanilide, 9333
[1155-03-9] Zolamine Hydrochloride*, 10319
[1155-49-3] Propipocaine Hydrochloride*, 8016
[1156-05-4] Phenglutarimide, 7377
[1156-19-0] Tolazamide, 9644
[1159-93-9] Clobenzepam, 2420
[1161-88-2] Sulfatolamide, 9117
[1162-15-9] Dioxadrol d-Form Hydrochloride*, 3352
[1162-65-8] Aflatoxin B₁*, 178
[1163-36-6] Clemizole Hydrochloride*, 2406
[1163-37-7] Moxaverine Hydrochloride*, 6371
[1164-38-1] Lachesine, 5343
[1165-48-6] Dimefline, 3249
[1169-49-9] Testosterone Isobutyrate*, 9322
[1169-79-5] Quinestradiol, 8238
[1172-02-7] 9-Dicyanomethylene-2,4,-7-trinitrofluorene, 3144
[1172-18-5] Flurazepam Dihydrochloride*, 4233
[1172-63-0] Jasmolin II*, 5272
[1174-11-4] Xenazoic Acid, 10204
[1176-03-0] Pseudococaine Tartrate*, 8097
[1177-30-6] Fenbenicillin Potassium Salt*, 4001
[1177-87-3] Dexamethasone 21-Acetate*, 2986
[1178-28-5] Metoserpate, 6237
[1178-29-6] Metoserpate Hydrochloride*, 6237

[1178-60-5] Pentagestrone Acetate*, 7251
[1179-69-7] Thiethylperazine Dimaleate*, 9449
[1180-71-8] Limonin, 5519
[1181-54-0] Clomocycline, 2448
[1182-87-2] Peruvoside, 7324
[1183-76-2] Daphnandrine, 2880
[1185-57-5] Ammonium Ferric Citrate, 547
[1187-56-0] Selenomethionine ⁷⁵Se-Injection*, 8585
[1190-53-0] Buformin Hydrochloride*, 1500
[1191-80-6] Mercuric Oleate, 5934
[1194-65-6] Dichlobenil, 3093
[1195-16-0] Citiolone, 2381
[1197-17-7] Tranexamic Acid cis-Form*, 9704
[1197-18-8] Tranexamic Acid, 9704
[1197-21-3] Phentermine Hydrochloride*, 7415
[1197-55-3] p-Aminophenylacetic Acid, 483
[1200-22-2] Thioctic Acid d-Form*, 9462
[1201-26-9] 3-Indolylacetone, 4997
[1209-98-9] Fencamfamine, 4006
[1210-56-6] Adrenoglomerulotropin, 172
[1211-28-5] Prolintane Hydrochloride*, 7964
[1212-48-2] Bendazol Hydrochloride*, 1062
[1213-06-5] Ethebenecid, 3775
[1215-83-4] Clobutinol Hydrochloride*, 2425
[1218-35-5] Xylometazoline Hydrochloride*, 10219
[1219-20-1] Oxolamine Hydrochloride*, 7078
[1219-77-8] Ujothion, 9978
[1221-56-3] Ipodate Sodium Salt*, 5087
[1222-57-7] Zolimidine, 10320
[1225-20-3] Iothalamic Acid Sodium Salt*, 5080
[1225-55-4] Protriptyline Hydrochloride*, 8088
[1225-60-1] Isothipendyl Hydrochloride*, 5248
[1225-65-6] Prothipendyl Hydrochloride*, 8073
[1227-61-8] Mefexamide, 5844
[1228-19-9] Glypinamide, 4524
[1229-29-4] Doxepin Hydrochloride*, 3492
[1229-35-2] Methdilazine Hydrochloride*, 6034
[1231-93-2] Ethynodiol, 3905
[1235-15-0] Norbolethone, 6779
[1235-82-1] Biperiden Hydrochloride*, 1274
[1239-04-9] Phenazocine (±)-Hydrobromide*, 7360
[1239-29-8] Furazabol, 4319
[1239-45-8] Homidium Bromide*, 4767
[1240-15-9] Propiomazine Hydrochloride*, 8007
[1242-56-4] Stenbolone Acetate*, 8962
[1245-44-9] Propicillin Potassium Salt*, 8002
[1247-42-3] Meprednisone, 5907
[1247-71-8] Colpormon, 2554
[1249-84-9] Azacosterol Dihydrochloride*, 924
[1253-28-7] Gestonorone Caproate, 4422

Denotes derivative of title compound

[1254-35-9] Oxabolone 17-Cyclopentanepropionate*, 7034
[1255-35-2] Fluprednidene Acetate, 4228
[1256-01-5] Homofenazine Dihydrochloride*, 4774
[1260-17-9] Carminic Acid, 1891
[1263-79-2] Tetrandrine l-Form*, 9369
[1263-89-4] Paromomycin Sulfate*, 7173
[1264-72-8] Colistin Sulfate*, 2542
[1271-19-8] Titanocene Dichloride, 9621
[1299-86-1] Aluminum Carbide, 345
[1300-32-9] Furoyl Chloride, 4332
[1300-51-2] Sodium Phenolsulfonate, 8803
[1300-71-6] Xylenol, 10215
[1300-73-8] Xylidine, 10217
[1300-81-8] Aluminum β-Naphtholdisulfonate, 363
[1301-16-2] Calcium Iodostearate, 1722
[1301-42-4] Euprocin, 3950
[1301-96-8] Silver(II) Oxide, 8666
[1302-01-8] Silver Subfluoride, 8675
[1302-09-6] Silver Selenide, 8673
[1302-42-7] Sodium Aluminate, 8715
[1302-63-2] Potassium Aluminate, 7765
[1302-78-9] Bentonite, 1082
[1302-81-4] Aluminum Sulfide, 382
[1302-82-5] Aluminum Selenide, 376
[1302-83-6] Lazurite, 5408
[1303-00-0] Gallium Arsenide, 4365
[1303-11-3] Indium Arsenide, 4982
[1303-28-2] Arsenic Pentoxide, 839
[1303-32-8] Arsenic Disulfide, 834
[1303-33-9] Arsenic Trisulfide, 846
[1303-34-0] Arsenic Pentasulfide, 838
[1303-35-1] Arsenic Hemiselenide, 835
[1303-36-2] Arsenic Triselenide, 845
[1303-37-3] Arsenic Pentaselenide, 837
[1303-52-2] Gold Trihydroxide, 4544
[1303-57-7] Gold Monoxide, 4532
[1303-58-8] Gold Trioxide, 4545
[1303-60-2] Gold Monosulfide, 4531
[1303-61-3] Gold Trisulfide, 4546
[1303-62-4] Gold Selenide, 4535
[1303-86-2] Boric Anhydride, 1365
[1303-94-2] Lithium Borate, 5549
[1304-28-5] Barium Oxide, 1015
[1304-29-6] Barium Peroxide, 1018
[1304-39-8] Barium Selenide, 1021
[1304-40-1] Barium Silicide, 1022
[1304-54-7] Beryllium Nitride, 1215
[1304-56-9] Beryllium Oxide, 1216
[1304-76-3] Bismuth Oxide, 1314
[1304-82-1] Bismuth Telluride, 1331
[1304-85-4] Bismuth Subnitrate, 1326
[1305-62-0] Calcium Hydroxide, 1716
[1305-78-8] Calcium Oxide, 1733
[1305-79-9] Calcium Peroxide, 1736
[1305-84-6] Calcium Selenide, 1748
[1305-99-3] Calcium Phosphide, 1742
[1306-01-0] Thomas Phosphate, 9511
[1306-06-5] Durapatite, 3519
[1306-19-0] Cadmium Oxide, 1659
[1306-23-6] Cadmium Sulfide, 1666
[1306-24-7] Cadmium Selenide, 1663
[1306-25-8] Cadmium Telluride, 1667
[1306-38-3] Ceric Oxide, 2035
[1307-52-4] Ruthenium Red, 8453
[1307-96-6] Cobaltous Oxide, 2507
[1308-06-1] Cobaltic-Cobaltous Oxide, 2490
[1308-14-1] Chromic Hydroxide, 2281
[1308-38-9] Chromic Oxide, 2283
[1308-56-1] Chalcopyrite, 2079
[1309-37-1] Ferric Oxide, 4072

[1309-42-8] Magnesium Hydroxide, 5706
[1309-48-4] Magnesium Oxide, 5713
[1309-60-0] Lead Dioxide, 5424
[1309-64-4] Antimony Trioxide, 752
[1310-32-3] Ferrous Selenide, 4103
[1310-43-6] Ferrous Phosphide, 4102
[1310-45-8] Ferric Subsulfate Solution, 4078
[1310-52-7] Magnesium Germanide, 5703
[1310-53-8] Germanium Dioxide, 4417
[1310-58-3] Potassium Hydroxide, 7806
[1310-61-8] Potassium Bisulfide, 7775
[1310-65-2] Lithium Hydroxide, 5560
[1310-73-2] Sodium Hydroxide, 8772
[1310-82-3] Rubidium Hydroxide, 8442
[1311-11-1] Lead Hydroxide, 5428
[1311-90-6] Ammonium Phosphotungstate, 580
[1312-03-4] Mercuric Subsulfate, 5941
[1312-15-8] Sulfur Iodide, 9148
[1312-41-0] Indium Antimonide, 4981
[1312-42-1] Indium Selenide, 4986
[1312-43-2] Indium Oxide, 4984
[1312-45-4] Indium Telluride, 4988
[1312-46-5] Iridium Sesquioxide, 5100
[1312-73-8] Potassium Sulfide, 7846
[1312-74-9] Potassium Selenide, 7837
[1312-76-1] Potassium Silicate, 7838
[1313-04-8] Magnesium Selenide, 5725
[1313-08-2] Magnesium Stannide, 5729
[1313-13-9] Manganese Dioxide, 5770
[1313-22-0] Manganese Selenide, 5779
[1313-27-5] Molybdenum Trioxide, 6321
[1313-29-7] Molybdenum Sesquioxide, 6320
[1313-30-0] Sodium Phosphomolybdate, 8810
[1313-49-1] Zinc Nitride, 10274
[1313-59-3] Sodium Oxide, 8796
[1313-60-6] Sodium Peroxide, 8800
[1313-82-2] Sodium Sulfide, 8830
[1313-85-5] Sodium Selenide, 8821
[1313-96-8] Niobium Pentoxide, 6651
[1313-99-1] Nickel Monoxide, 6595
[1314-06-3] Nickel Sesquioxide, 6599
[1314-08-5] Palladium Oxide, 7125
[1314-11-0] Strontium Oxide, 9009
[1314-12-1] Thallium Oxide, 9401
[1314-13-2] Zinc Oxide, 10279
[1314-15-4] Platinic Oxide, 7686
[1314-18-7] Strontium Peroxide, 9010
[1314-20-1] Thorium Oxide, 9518
[1314-22-3] Zinc Peroxide, 10282
[1314-23-4] Zirconium Oxide, 10313
[1314-24-5] Phosphorus Trioxide, 7517
[1314-27-8] Lead Sesquioxide, 5440
[1314-28-9] Rhenium Trioxide, 8347
[1314-32-5] Thallium Sesquioxide, 9404
[1314-34-7] Vanadium Trioxide, 10059
[1314-35-8] Tungsten Trioxide, 9947
[1314-41-6] Lead Tetroxide, 5449
[1314-56-3] Phosphorus Pentoxide, 7512
[1314-60-9] Antimony Pentoxide, 739
[1314-61-0] Tantalum Pentoxide, 9226
[1314-62-1] Vanadium Pentoxide, 10056
[1314-64-3] Uranyl Sulfate, 10002
[1314-68-7] Rhenium Heptoxide, 8344
[1314-80-3] Phosphorus Pentasulfide, 7511

[1314-81-4] Phosphorus Hemitriselenide, 7504
[1314-82-5] Phosphorus Pentaselenide, 7510
[1314-84-7] Zinc Phosphide, 10285
[1314-85-8] Tetraphosphorus Trisulfide, 9375
[1314-86-9] Phosphorus Triselenide, 7518
[1314-87-0] Lead Sulfide, 5445
[1314-91-6] Lead Telluride, 5446
[1314-95-0] Stannous Sulfide, 8948
[1314-96-1] Strontium Sulfide, 9014
[1314-97-2] Thallium Sulfide, 9406
[1314-98-3] Zinc Sulfide, 10294
[1315-01-1] Stannic Sulfide, 8936
[1315-03-3] Vanadium Trisulfide, 10061
[1315-04-4] Antimony Pentasulfide, 738
[1315-05-5] Antimony Triselenide, 753
[1315-06-6] Stannous Selenide, 8946
[1315-09-9] Zinc Selenide, 10290
[1315-11-3] Zinc Telluride, 10297
[1317-26-6] Magaldrate, 5685
[1317-33-5] Molybdenum Disulfide, 6318
[1317-34-6] Manganese Sesquioxide, 5780
[1317-35-7] Manganese Oxide, 5776
[1317-36-8] Lead Monoxide, 5433
[1317-37-9] Ferrous Sulfide, 4106
[1317-38-0] Cupric Oxide, 2713
[1317-39-1] Cuprous Oxide, 2734
[1317-40-4] Cupric Sulfide, 2724
[1317-41-5] Cupric Selenide, 2719
[1317-42-6] Cobaltous Sulfide, 2511
[1317-54-0] Ferrite, 4082
[1317-61-9] Ferrosoferric Oxide, 4087
[1317-82-4] Sapphire, 8514
[1317-95-9] Tripoli, 9876
[1317-97-1] Ultramarine, 9980
[1318-00-9] Vermiculite, 10095
[1318-93-0] Montmorillonite, 6341
[1319-69-3] Potassium Glycerophosphate, 7797
[1319-77-3] Cresol, 2645
[1319-91-1] Calcium Iodobehenate, 1721
[1321-14-8] Potassium Guaiacolsulfonate, 7798
[1327-39-5] Calcium Aluminosilicate, 1685
[1327-41-9] Aluminum Hydroxychloride, 356
[1327-53-3] Arsenic Trioxide, 844
[1329-77-7] Monardein Chloride, 6327
[1329-83-5] Tigonin, 9576
[1330-20-7] Xylene, 10214
[1330-43-4] Sodium Borate, 8733
[1330-51-4] Pyrogallol Monoacetate*, 8184
[1330-78-5] Tritolyl Phosphate, 9892
[1332-12-3] Gold, Explosive, 4527
[1332-14-5] Cupric Sulfate, Basic, 2723
[1332-52-1] Beryllium Acetate, Basic, 1203
[1332-53-2] Rhodanilic Acid, 8350
[1332-77-0] Potassium Tetraborate, 7852
[1333-38-6] Angelica Lactone, 686
[1333-74-0] Hydrogen, 4833
[1333-82-0] Chromium Trioxide, 2293
[1333-83-1] Sodium Bifluoride, 8727
[1334-74-3] Sodium Glycerophosphate, 8767
[1335-31-5] Mercuric Oxycyanide, 5937
[1335-32-6] Lead Subacetate, 5443

Denotes derivative of title compound

[1336-20-5] Tetracycline Phosphate Complex*, 9337
[1336-29-4] Bisacodyl Complex with Tannin*, 1282
[1336-80-7] Ferrocholinate, 4085
[1337-59-3] Undecoylium Chloride, 9982
[1338-16-5] Iron Sorbitex, 5112
[1338-39-2] Sorbitan Laurate*, 8872
[1338-41-6] Sorbitan Stearate*, 8872
[1338-43-8] Sorbitan Oleate*, 8872
[1338-54-1] Undecoylium Chloride Complex with Iodine*, 9982
[1338-98-3] Ethylstibamine, 3896
[1339-92-0] Basic Aluminum Carbonate Gel, 1033
[1340-26-7] Phenacetolin, 7345
[1340-35-8] Urea Stibamine, 10010
[1341-49-7] Ammonium Bifluoride, 523
[1343-98-2] Silicic Acid, 8634
[1344-09-8] Sodium Silicate, 8824
[1344-09-8] Sodium Silicate Solution, 8825
[1344-28-1] Aluminum Oxide, 369
[1344-36-1] Basic Lead Carbonate, 1034
[1344-48-5] Mercuric Sulfide, Black, 5944
[1344-48-5] Mercuric Sulfide, Red, 5945
[1344-57-6] Uranium Dioxide, 9991
[1344-58-7] Uranium Trioxide, 9997
[1344-95-2] Calcium Silicate, 1749
[1345-04-6] Antimony Trisulfide, 754
[1345-05-7] Lithopone, 5572
[1345-07-9] Bismuth Sulfide, 1329
[1345-24-0] Gold Stannate, 4538
[1345-25-1] Ferrous Oxide, 4100
[1361-49-5] Taxine A*, 9246
[1362-42-1] Absinthin, 8
[1390-93-8] Citrullol, 2393
[1391-14-6] Coptine, 2589
[1391-82-8] Grisein, 4570
[1392-21-8] Leucomycins, 5480
[1392-27-4] Litmocidin, 5573
[1392-46-7] Micrococcin P, 6268
[1392-81-0] Parotin, 7174
[1392-87-6] Phasin, 7339
[1393-12-0] Rimocidin, 8394
[1393-25-5] Secretin, 8564
[1393-38-0] Subtilin, 9033
[1393-48-2] Thiostrepton, 9502
[1393-62-0] Abrin, 5
[1393-63-1] Annatto, 713
[1393-64-2] Araroba, 807
[1393-68-6] Bottromycin, 1383
[1393-87-9] Fusafungine, 4335
[1394-02-1] Hachimycin, 4616
[1395-18-2] Azolitmin, 951
[1397-74-6] Acetyltannic Acid, 107
[1397-77-9] Actinorhodine, 144
[1397-84-8] Alazopeptin, 209
[1397-89-3] Amphotericin B, 627
[1398-17-0] Bakankosin, 972
[1398-61-4] Chitin, 2105
[1398-78-3] Colocynthin, 2551
[1399-57-1] Graphitic Acid, 4562
[1399-64-0] Gymnemic Acid, 4613
[1400-61-9] Nystatin, 6834
[1400-62-0] Orcein, 6994
[1400-67-5] Oryzacidin, 7014
[1401-54-3] Tanacetin, 9218
[1401-69-0] Tylosin, 9963
[1401-78-1] Vincetoxin, 10121
[1401-79-2] Viomycin Pantothenate Sulfate*, 10139
[1401-98-5] Condurangin, 2560
[1402-10-4] Lichenin, 5503
[1402-37-5] Actinomycetin, 141

[1402-44-4] Actinomycin F₁, 142
[1402-82-0] Amphomycin, 625
[1403-17-4] Candicidin, 1789
[1403-28-7] Carzinophilin, 1928
[1403-66-3] Gentamicin, 4398
[1403-71-0] Hamycin, 4641
[1404-00-8] Mitomycins, 6301
[1404-04-2] Neomycin, 6542
[1404-15-5] Nogalamycin, 6767
[1404-23-5] Pleurotin(e), 7695
[1404-26-8] Polymyxin B*, 7734
[1404-52-0] Rhodomycin B*, 8357
[1404-55-3] Ristocetin, 8398
[1404-59-7] Oligomycin D*, 6970
[1404-64-4] Sparsomycin, 8886
[1404-74-6] Streptovaricin, 8989
[1404-88-2] Tyrothricin, 9972
[1404-90-6] Vancomycin, 10066
[1404-93-9] Vancomycin Monohydrochloride*, 10066
[1405-10-3] Neomycin Neomycin B Sulfate*, 6542
[1405-20-5] Polymyxin B Sulfate*, 7734
[1405-36-3] Capreomycin Disulfate*, 1801
[1405-41-0] Gentamicin C Complex Sulfate*, 4398
[1405-76-1] Gitalin, 1437
[1405-86-3] Glycyrrhizic Acid, 4515
[1405-87-4] Bacitracin, 965
[1405-88-5] Bacitracin Methylenedisalicylic Acid, 966
[1405-90-9] Candidin, 1790
[1406-04-8] Neomycin Undecylenate, 6543
[1406-11-7] Polymyxin, 7734
[1406-65-1] Chlorophyll, 2207
[1407-04-1] Lututrin, 5644
[1407-47-2] Angiotensin, 689
[1407-73-4] Vitamin T, 10164
[1407-85-8] Amphetamine d-Form Tannate*, 623
[1407-93-8] Plantisul, 7675
[1414-39-7] Albomycin, 214
[1414-45-5] Nisin, 6657
[1415-94-7] Iodopsin, 5060
[1416-60-0] Splenin, 8919
[1420-03-7] Propenzolate Hydrochloride*, 7998
[1420-04-8] Niclosamide Ethanolamine Salt*, 6602
[1420-06-0] Trifenmorph, 9808
[1420-53-7] Codeine Sulfate, 2529
[1420-55-9] Thiethylperazine, 9449
[1421-14-3] Propanidid, 7986
[1421-68-7] Amidephrine Monomethanesulfonate*, 418
[1422-07-7] Codeine Hydrochloride*, 2525
[1424-00-6] Mesterolone, 5974
[1424-27-7] Acetazolamide Sodium Salt*, 50
[1429-30-7] Petunidin, 7331
[1435-55-8] Hydroquinidine, 4851
[1438-30-8] Netropsin, 6565
[1438-57-9] Grindelic Acid, 4569
[1443-44-3] α-Methylfentanyl Hydrochloride*, 6150
[1448-23-3] Glaucarubin, 4444
[1456-55-9] Sporidesmin A*, 8920
[1463-28-1] Guanacline, 4586
[1464-42-2] Selenomethionine, 8585
[1473-73-0] Morphazinamide Hydrochloride*, 6355
[1476-53-5] Novobiocin Monosodium Salt*, 6818
[1476-98-8] Hydroquinidine Hydrochloride*, 4851

[1477-10-7] Bietamiverine Hydrochloride*, 1252
[1477-19-6] Benzarone, 1091
[1477-39-0] Noracymethadol, 6778
[1477-40-3] Levomethadyl Acetate, 5492
[1477-57-2] Fertilysin, 4108
[1480-19-9] Fluanisone, 4150
[1481-70-5] Tyrocidine A*, 9967
[1483-07-4] Albizziin, 212
[1491-41-4] Naftalofos, 6441
[1491-59-4] Oxymetazoline, 7100
[1492-02-0] Glybuzole, 4488
[1493-02-3] Formyl Fluoride, 4274
[1501-84-4] Rimantadine Hydrochloride*, 8390
[1503-53-3] 5-Bromosalicylic Acid Acetate, 1459
[1506-12-3] Butidrine Hydrochloride, 1558
[1508-45-8] Podophyllinic Acid 2-Ethylhydrazide*, 7703
[1508-65-2] Oxybutynin Chloride, 7089
[1508-75-4] Tropicamide, 9911
[1508-76-5] Procyclidine Hydrochloride*, 7944
[1511-16-6] Haloperidol Hydrochloride*, 4629
[1518-86-1] Hydroxyamphetamine, 4855
[1524-88-5] Flurandrenolide, 4232
[1531-12-0] Norlevorphanol, 6801
[1538-09-6] Penicillin G Benzathine, 7221
[1538-11-0] Penicillin G Benzhydrylamine, 7222
[1553-60-2] Ibufenac, 4924
[1555-53-9] Magnesium Oleate, 5711
[1562-13-6] 3-O-Lauroylpyridoxol Diacetate, 5401
[1562-71-6] Guanacline Monosulfate*, 4586
[1562-85-2] Gallocyanine, 4368
[1562-90-9] Celestin Blue, 2006
[1563-66-2] Carbofuran, 1851
[1580-83-2] Paraflutizide, 7157
[1582-09-8] Trifluralin, 9815
[1590-87-0] Disilane, 3419
[1592-23-0] Calcium Stearate, 1750
[1593-77-7] Dodemorph, 3467
[1596-63-0] Quinacillin, 8224
[1596-70-9] Levopropoxyphene Hydrochloride*, 5495
[1596-84-5] Daminozide, 2874
[1597-82-6] Paramethasone 21-Acetate*, 7162
[1600-27-7] Mercuric Acetate, 5922
[1603-91-4] 2-Amino-4-methylthiazole, 470
[1604-01-9] γ-Coniceine, 2566
[1605-89-6] Bolasterone, 1353
[1607-17-6] Pentrinitrol, 7280
[1609-47-8] Pyrocarbonic Acid Diethyl Ester, 8182
[1610-18-0] Prometon, 7972
[1613-17-8] Dymanthine Hydrochloride*, 3525
[1614-20-6] Nifurprazine, 6629
[1617-90-9] Vincamine, 10120
[1618-26-4] Bis[methylthio]methane, 1296
[1619-29-0] Helminthosporol, 4663
[1619-34-7] 3-Quinuclidinol, 8266
[1620-21-9] Chlorcyclizine Hydrochloride*, 2128
[1622-61-3] Clonazepam, 2449
[1622-62-4] Flunitrazepam, 4181
[1622-79-3] Melperone Hydrochloride*, 5870

Denotes derivative of title compound

[1633-05-2] Strontium Carbonate, 8998
[1634-04-4] Methyl tert-Butyl Ether, 6111
[1637-39-4] Zeatin, 10247
[1639-43-6] Androstenediol 3-Acetate*, 677
[1639-60-7] Propoxyphene Hydrochloride*, 8024
[1641-17-4] Mexenone, 6255
[1642-54-2] Diethylcarbamazine Citrate (Hydrogen Citrate)*, 3165
[1649-18-9] Azaperone, 931
[1661-29-6] Meturedepa, 6245
[1665-48-1] Metaxalone, 5994
[1668-19-5] Doxepin, 3492
[1672-46-4] Digoxigenin, 3209
[1673-06-9] Amphotalide, 626
[1674-96-0] Phenglutarimide Hydrochloride*, 7377
[1679-76-1] Drofenine, 3501
[1684-40-8] Tacrine Hydrochloride*, 9199
[1684-42-0] Acranil, 123
[1689-84-5] Bromoxynil, 1465
[1689-89-0] Nitroxynil, 6754
[1690-72-8] Methyl N-Methylnipecotate, 6172
[1695-77-8] Spectinomycin, 8890
[1702-17-6] Clopyralid, 2462
[1702-77-8] Fusarubin, 4337
[1707-14-8] Phenmetrazine Hydrochloride*, 7385
[1715-33-9] Prednisolone 21-Succinate Sodium Salt*, 7901
[1721-51-3] ζ₁-Tocopherol, 9636
[1722-62-9] Mepivacaine Hydrochloride*, 5905
[1729-61-9] Paranyline, 7163
[1733-12-6] Cresol Red, 2647
[1740-22-3] Pyrinoline, 8173
[1744-22-5] Riluzole, 8389
[1746-01-6] TCDD, 9252
[1747-53-1] Orotic Acid Ethyl Ester*, 7004
[1762-95-4] Ammonium Thiocyanate, 597
[1764-85-8] Epithiazide, 3661
[1777-82-8] Dichlorobenzyl Alcohol, 3110
[1783-96-6] Aspartic Acid D-Form*, 875
[1786-81-8] Prilocaine Hydrochloride*, 7924
[1787-61-7] Eriochrome Black T, 3709
[1794-74-7] Cethexonium Bromide, 2061
[1798-50-1] Azacyclonol Hydrochloride*, 925
[1803-67-4] Sporidesmolide III*, 8921
[1805-85-2] Hasubanonine, 4649
[1805-86-3] Metaphanine, 5990
[1808-12-4] Bromodiphenhydramine Hydrochloride*, 1439
[1812-30-2] Bromazepam, 1406
[1821-16-5] Ethylidene Dicoumarol, 3858
[1824-50-6] Benzylhydrochlorothiazide, 1171
[1824-58-4] Ethiazide, 3778
[1829-00-1] Thiazol Yellow G, 9445
[1830-32-6] Azintamide, 945
[1836-75-5] Nitrofen, 6695
[1836-77-7] Chlornitrofen, 2158
[1841-19-6] Fluspirilene, 4241
[1843-05-6] Octabenzone, 6838
[1853-37-8] Podophyllinic Acid*, 7703
[1859-87-6] Echinuline, 3547
[1861-21-8] Enallylpropymal, 3607
[1861-32-1] DCPA, 2896

[1861-40-1] Benfluralin, 1067
[1863-63-4] Ammonium Benzoate, 521
[1867-58-9] Clomethiazole Ethanedisulfonate*, 2444
[1867-66-9] Ketamine Hydrochloride*, 5306
[1871-22-3] Tetrazolium Blue, 9380
[1873-77-4] Tris(trimethylsilyl)silane, 9886
[1874-54-0] Psicofuranine, 8108
[1882-26-4] Pyridinol Carbamate, 8158
[1891-29-8] Lactucin, 5359
[1892-56-4] Athamantin, 893
[1892-80-4] Fenethylline Hydrochloride*, 4014
[1893-33-0] Pipamperone, 7608
[1897-45-6] Chlorothalonil, 2219
[1898-66-4] 1,1-Diphenyl-2-picrylhydrazyl (Free Radical), 3389
[1904-98-9] 2,6-Diaminopurine, 3028
[1910-68-5] Methisazone, 6057
[1912-24-9] Atrazine, 902
[1912-26-1] Trietazine, 9797
[1918-00-9] Dicamba, 3090
[1918-02-1] Picloram, 7552
[1918-08-7] Dipropalin, 3404
[1918-16-7] Propachlor, 7977
[1921-70-6] Pristane, 7932
[1923-76-8] Bunamiodyl Sodium, 1511
[1926-48-3] Fenbenicillin, 4001
[1926-49-4] Clometocillin, 2445
[1929-77-7] Vernolate, 10097
[1929-82-4] Nitrapyrin, 6666
[1934-21-0] Tartrazine, 9239
[1936-18-1] Salutaridine, 8493
[1936-28-3] Sodium Arsphenamine, 8722
[1936-40-9] Novembichin, 6817
[1937-19-5] Aminoguanidine Hydrochloride*, 461
[1937-54-8] Solanone, 8855
[1944-10-1] Fenoterol Hydrochloride*, 4022
[1944-12-3] Fenoterol Hydrobromide*, 4022
[1945-53-5] Palustric Acid, 7131
[1949-20-8] Oxolamine Citrate*, 7078
[1949-45-7] Metrizoic Acid, 6241
[1950-31-8] Metcaraphen Hydrochloride*, 5996
[1950-39-6] Deferoxamine Hydrochloride*, 2914
[1951-25-3] Amiodarone, 504
[1953-02-2] Tiopronin, 9597
[1953-04-4] Galanthamine Hydrobromide*, 4357
[1954-28-5] Etoglucid, 3926
[1954-79-6] Mecloralurea, 5822
[1962-14-7] Acacic Acid, 12
[1972-08-3] Tetrahydrocannabinols (−)-Δ¹-3,4-trans-Form*, 9349
[1976-28-9] Aluminum Nicotinate, 364
[1977-10-2] Loxapine, 5617
[1977-11-3] Perlapine, 7320
[1980-45-6] Benzodepa, 1118
[1981-58-4] Sulfamethazine Sodium Salt*, 9083
[1982-36-1] Homochlorcyclizine Dihydrochloride*, 4770
[1982-37-2] Methdilazine, 6034
[1982-49-6] Siduron, 8631
[1983-72-8] Medicagol, 5832
[1984-15-2] Medronic Acid, 5837
[1984-94-7] Sulfasymazine, 9114
[1986-53-4] Bolandiol Dipropionate*, 1352
[1986-70-5] Calotropin, 1769

[1987-71-9] Nicotinamide Ascorbate, 6610
[1990-29-0] D-Altrose, 329
[1997-15-5] Triamcinolone Acetonide 21-Disodium Phosphate*, 9728
[2001-81-2] Diponium Bromide, 3402
[2001-95-8] Valinomycin, 10047
[2002-29-1] Flumethasone 21-Pivalate*, 4173
[2002-44-0] MIF-I*, 5855
[2008-41-5] Butylate, 1581
[2009-64-5] Neopterin D-erythro-Form*, 6551
[2011-67-8] Nimetazepam, 6641
[2013-58-3] Meclocycline, 5818
[2016-36-6] Choline Salicylate, 2265
[2016-63-9] Bamifylline, 982
[2019-25-2] Chlorazanil Hydrochloride*, 2123
[2019-68-3] Atrolactamide, 904
[2022-85-7] Flucytosine, 4161
[2030-53-7] Aporeine, 789
[2030-63-9] Clofazimine, 2433
[2032-59-9] Aminocarb, 452
[2032-65-7] Methiocarb, 6050
[2033-24-1] Meldrum's Acid, 5858
[2037-48-1] 2-Deoxystreptamine, 2956
[2037-95-8] Carsalam, 1915
[2043-38-1] Buthiazide, 1554
[2052-01-9] α-Bromoisobutyric Acid, 1442
[2055-44-9] Perisoxal, 7318
[2058-46-0] Oxytetracycline Hydrochloride*, 7111
[2058-52-8] Clothiapine, 2476
[2058-58-4] Asparagine D-Form*, 872
[2061-86-1] Methandriol Diacetate*, 6017
[2062-77-3] Trifluperidol Hydrochloride*, 9813
[2062-78-4] Pimozide, 7589
[2062-84-2] Benperidol, 1077
[2065-00-1] Filicinic Acid, 4122
[2066-89-9] Isoniazid 4-Aminosalicylate*, 5203
[2068-78-2] Vincristine Sulfate*, 10124
[2078-54-8] Propofol, 8020
[2079-00-7] Blasticidin S, 1350
[2086-83-1] Berberine, 1192
[2086-96-6] Canadine d-Form*, 1785
[2090-54-2] N-Hydroxyethylpromethazine Chloride, 4870
[2090-89-3] Butethamine, 1552
[2092-16-2] Calcium Thiocyanate, 1757
[2092-17-3] Barium Thiocyanate, 1027
[2095-57-0] Thiobutabarbital, 9457
[2096-42-6] Gougerotin, 4550
[2098-66-0] Cyproterone, 2844
[2099-26-5] Androstenediol Diacetate*, 677
[2103-64-2] Gallein, 4362
[2104-64-5] EPN, 3663
[2104-96-3] Bromophos, 1453
[2105-47-7] Tetrabenazine Hydrochloride*, 9325
[2109-73-1] Butacetin, 1532
[2111-75-3] Perillaldehyde, 7308
[2119-75-7] Fluperolone Acetate, 4225
[2122-29-4] Retamine, 8327
[2124-57-4] Menaquinone 7*, 5876
[2127-01-7] Clorexolone, 2466
[2128-61-2] Sparsiflorine, 8885
[2135-17-3] Flumethasone, 4173
[2139-25-5] Perisoxal Citrate*, 7318
[2139-47-1] Nifenazone, 6619
[2140-46-7] 25-Hydroxycholesterol, 4866
[2141-09-5] Magnoflorine, 5736
[2150-48-3] Pyronine B, 8189

Denotes derivative of title compound

[2152-34-3] Pemoline, 7206
[2152-44-5] Betamethasone 17-Valerate*, 1226
[2152-56-9] Arabitol, 800
[2152-76-3] Idose, 4933
[2154-02-1] Metofoline, 6229
[2156-27-6] Benproperine, 1078
[2164-08-1] Lenacil, 5459
[2164-09-2] Dicryl, 3138
[2164-17-2] Fluometuron, 4189
[2165-19-7] Guanoxan, 4598
[2167-85-3] Pipazethate, 7609
[2169-64-4] 6-Azauridine 2′,3′,5′-Triacetate*, 937
[2169-75-7] Deptropine Citrate*, 2958
[2179-16-0] Ninopterin, 6647
[2179-37-5] Bencyclane, 1060
[2180-92-9] Bupivacaine, 1520
[2181-04-6] Canrenone Free Acid Potassium Salt*, 1795
[2182-14-1] Vindoline, 10126
[2188-67-2] Naepaine, 6435
[2192-20-3] Hydroxyzine Dihydrochloride*, 4897
[2205-73-4] Tiomesterone, 9596
[2207-50-3] Aminorex, 497
[2209-86-1] Pentaerythritol Dichlorohydrin, 7247
[2210-63-1] Mofebutazone, 6312
[2210-64-2] Pyrrocaine Hydrochloride*, 8197
[2210-77-7] Pyrrocaine, 8197
[2217-44-9] Iodophthalein Sodium, 5059
[2218-68-0] Chloral Betaine, 2111
[2218-94-2] Nitron, 6710
[2219-30-9] Penicillamine Hydrochloride*, 7214
[2221-95-6] Fichtelite, 4120
[2227-17-0] Dienochlor, 3154
[2235-90-7] Etryptamine, 3937
[2240-14-4] Fencamfamine Hydrochloride*, 4006
[2241-90-9] Cyclobuxine D, 2785
[2243-33-6] α-Phellandrene d-Form*, 7340
[2243-76-7] Alizarine Yellow R, 251
[2244-21-5] Troclosene Potassium, 9896
[2255-39-2] Muscazone, 6390
[2259-14-5] Chrysanthemic Acid l-trans-Form*, 2310
[2259-96-3] Cyclothiazide, 2822
[2260-13-1] Methyl 4-Pyridyl Ketone Thiosemicarbazone Hydrochloride*, 6196
[2272-11-9] Ethanolamine Oleate*, 3772
[2276-90-6] Iothalamic Acid, 5080
[2277-92-1] Oxyclozanide, 7092
[2284-31-3] Pratensein, 7893
[2290-11-1] Verrucarin B*, 10100
[2292-79-7] Congressane, 2563
[2295-58-1] Flopropione, 4142
[2297-30-5] Androstenediol Dipropionate*, 677
[2303-16-4] Diallate, 3014
[2303-17-5] Triallate, 9726
[2306-22-1] Citramalic Acid, 2384
[2307-68-8] Solan, 8851
[2310-17-0] Phosalone, 7489
[2312-35-8] Propargite, 7992
[2313-87-3] Ethoxazene Hydrochloride*, 3796
[2315-02-8] Oxymetazoline Hydrochloride*, 7100
[2315-08-4] Salazosulfadimidine, 8473
[2316-64-5] Bromosaligenin, 1460
[2319-84-8] Thioctic Acid Sodium Salt*, 9462

[2319-97-3] [2.2]Metacyclophane, 5981
[2320-96-9] 4′,5′-Dichlorofluorescein, 3117
[2321-07-5] Fluorescein, 4194
[2323-36-6] (±)-Deprenyl*, 8569
[2326-89-8] Drimenin, 3500
[2338-05-8] Ferric Citrate, 4063
[2338-37-6] Levopropoxyphene, 5495
[2347-80-0] Thioproperazine Dimethanesulfonate*, 9494
[2350-32-5] Butoxycaine Hydrochloride*, 1566
[2353-45-9] Fast Green FCF, 3979
[2363-58-8] Epitiostanol, 3662
[2375-03-3] Methylprednisolone 21-Succinate Sodium Salt*, 6189
[2382-48-1] Ubichromenol 50*, 9974
[2385-81-1] Furethidine, 4323
[2385-85-5] Mirex, 6294
[2390-22-9] Hetolin, 4710
[2390-99-0] Chanoclavine, 2083
[2391-03-9] Brompheniramine d-Form Maleate*, 1467
[2392-39-4] Dexamethasone 21-Phosphate Disodium Salt*, 2986
[2398-81-4] Oxiniacic Acid, 7075
[2398-96-1] Tolnaftate, 9656
[2403-84-1] 2-Amino-4-picoline Hydrochloride*, 486
[2404-18-4] Dipiproverine Dihydrochloride*, 3399
[2410-93-7] Methopterin, 6064
[2413-38-9] Flupentixol Dihydrochloride*, 4224
[2425-06-1] Captafol, 1814
[2429-71-2] Benzo Azurine G, 1115
[2430-27-5] Valpromide, 10050
[2430-49-1] Vinylbital, 10131
[2435-76-9] 5-Diazouracil, 3050
[2438-32-6] Chlorpheniramine d-Form Maleate*, 2232
[2438-72-4] Bufexamac, 1497
[2438-80-4] L-Fucose, 4302
[2439-01-2] Oxythioquinox, 7113
[2439-10-3] Dodine, 3468
[2439-99-8] Glyphosine, 4523
[2440-22-4] Drometrizole, 3503
[2447-54-3] Sanguinarine, 8504
[2447-57-6] Sulfadoxine, 9074
[2448-68-2] Pipamperone Dihydrochloride*, 7608
[2451-01-6] Terpin cis-Form Hydrate*, 9314
[2454-11-7] Formebolone, 4266
[2463-84-5] Dicapthon, 3091
[2466-09-3] Pyrophosphoric Acid, 8191
[2468-21-5] Catharanthine, 1952
[2469-34-3] Senegenin, 8598
[2470-73-7] Dixyrazine, 3450
[2482-00-0] Agmatine Sulfate*, 186
[2487-63-0] Quinbolone, 8236
[2490-97-3] Aceglutamide, 25
[2493-02-9] p-Bromophenyl Isocyanate, 1452
[2497-78-1] Iopanoic Acid Sodium Salt*, 5072
[2507-24-6] Physalaemin, 7537
[2507-91-7] Kethoxal Bis(thiosemicarbazone)*, 5310
[2508-72-7] Antazoline Hydrochloride*, 718
[2508-79-4] Methyldopa Ethyl Ester Hydrochloride*, 6132
[2517-04-6] 2-Azetidinecarboxylic Acid, 940
[2520-21-0] Celesticetin, 2005
[2529-45-5] Flurogestone Acetate, 4235

[2531-04-6] Piperylone, 7632
[2533-89-3] Karsil, 5297
[2537-29-3] Proxibarbal, 8091
[2540-82-1] Formothion, 4273
[2550-75-6] Chlorbicyclen, 2127
[2551-62-4] Sulfur Hexafluoride, 9146
[2552-47-8] Pilocereine, 7580
[2552-55-8] Ibotenic Acid, 4922
[2557-49-5] Diflorasone, 3186
[2571-22-4] Tutin, 9958
[2571-86-0] Coriamyrtin, 2594
[2574-78-9] Orazamide, 6993
[2576-11-6] Methcathinone Hydrochloride* (unspecified), 6033
[2576-92-3] Isoetharine Hydrochloride*, 5185
[2577-32-4] Sulfamethoxypyridazine Sodium Salt*, 9087
[2580-88-3] Eburnamonine (±)-Form*, 3537
[2581-07-9] Isoapo-β-erythroidine, 5140
[2589-47-1] Prajmaline Hydrogen Tartrate*, 7883
[2591-17-5] Firefly Luciferin, 4128
[2595-97-3] D-Allose, 288
[2606-93-1] Blancophor R, 1349
[2607-06-9] Diflucortolone, 3189
[2608-24-4] Piposulfan, 7635
[2609-46-3] Amiloride, 426
[2612-33-1] Clonitrate, 2452
[2618-25-9] Ioglycamic Acid, 5067
[2618-26-0] Iodipamide Disodium Salt*, 5043
[2622-26-6] Pericyazine, 7306
[2622-30-2] Carphenazine, 1910
[2622-37-9] Trifluomeprazine, 9810
[2624-43-3] Cyclofenil, 2789
[2624-44-4] Ethamsylate, 3766
[2627-69-2] Acadesine, 13
[2631-37-0] Promecarb, 7967
[2636-26-2] Ciafos, 2321
[2638-46-2] Cortolone 3,20,21-Triacetate*, 2605
[2642-98-0] 6-Chrysenamine, 2313
[2652-13-3] Eicosamethylnonasiloxane, 3571
[2667-89-2] Bisbentiamine, 1286
[2668-66-8] Medrysone, 5840
[2673-40-7] Perivine, 7319
[2686-47-7] Amphenone B, 622
[2688-77-9] Laudanosine, 5391
[2691-41-0] Octogen, 6855
[2691-45-4] Hexestrol Bis(β-diethylaminoethyl Ether), 4739
[2691-46-5] Bietamiverine Dihydrochloride*, 1252
[2696-92-6] Nitrosyl Chloride, 6744
[2698-41-1] o-Chlorobenzylidenemalononitrile, 2178
[2699-79-8] Sulfuryl Fluoride, 9154
[2709-56-0] Flupentixol, 4224
[2718-25-4] Rheadine, 8341
[2726-03-6] Clobenzepam Hydrochloride*, 2420
[2730-71-4] Thiocolchicine, 9460
[2740-04-7] Dimefline Hydrochloride*, 3249
[2740-52-5] Anagestone, 664
[2746-81-8] Fluphenazine Enanthate*, 4226
[2750-76-7] Rifamide, 8381
[2751-09-9] Troleandomycin, 9899
[2751-68-0] Acetophenazine, 70
[2752-65-0] Gambogic Acid, 4373
[2753-45-9] Mebeverine Hydrochloride*, 5808
[2757-90-6] Agaritine, 185
[2763-96-4] Muscimol, 6391
[2765-04-0] Thioacetaldehyde, 9452

Denotes derivative of title compound

[2768-90-3] Quinaldine Blue, 8229
[2773-92-4] Dimethisoquin Hydrochloride*, 3267
[2779-55-7] Opiniazide, 6984
[2783-94-0] Sunset Yellow FCF, 9176
[2798-20-1] Gardenin B*, 4377
[2809-21-4] Etidronic Acid, 3908
[2825-00-5] Aureothin, 913
[2825-60-7] Formocortal, 4270
[2829-19-8] Rolicyprine, 8409
[2834-90-4] 4-Amino-1-naphthol, 473
[2869-83-2] Janus Green B, 5269
[2873-97-4] Diacetone Acrylamide, 3007
[2898-11-5] Medazepam Hydrochloride*, 5829
[2898-12-6] Medazepam, 5829
[2900-38-1] Sporidesmolide I*, 8921
[2919-66-6] Melengestrol Acetate*, 5859
[2920-86-7] Prednisolone 21-Hydrogen Succinate*, 7901
[2921-88-2] Chlorpyrifos, 2242
[2921-92-8] Propatyl Nitrate, 7995
[2924-67-6] Fluoresone, 4197
[2933-94-0] Toliprolol, 9653
[2935-23-1] Chrysanthemic Acid dl-cis-Form*, 2310
[2945-88-2] Sophorabioside, 8866
[2955-23-9] Olivil, 6974
[2955-38-6] Prazepam, 7895
[2957-21-3] Sakuranetin, 8470
[2970-95-8] Macromerine, 5677
[2971-90-6] Clopidol, 2458
[2975-34-0] Carphenazine Dimaleate*, 1910
[2998-57-4] Estramustine, 3749
[3011-89-0] Aklomide, 198
[3012-65-5] Ammonium Citrate, Dibasic, 540
[3017-60-5] Cobaltous Thiocyanate, 2512
[3031-94-5] AICAR, 190
[3031-95-6] SAICAR, 8469
[3051-09-0] Murexide, 6386
[3056-17-5] Stavudine, 8958
[3059-97-0] Isovaline D-Form*, 5255
[3060-89-7] Metobromuron, 6224
[3092-17-9] Midodrine Hydrochloride*, 6272
[3093-35-4] Halcinonide, 4621
[3093-71-8] Homofenazine Difumarate*, 4774
[3094-09-5] Doxifluridine, 3493
[3095-65-6] Ammonium Bitartrate, 529
[3099-52-3] Nicametate, 6577
[3101-51-7] Ergoflavin, 3691
[3102-00-9] Febuprol, 3985
[3105-97-3] Hycanthone, 4795
[3115-05-7] Iobenzamic Acid, 5028
[3115-21-7] Methyl 4-Pyridyl Ketone Thiosemicarbazone, 6196
[3115-49-9] (p-Nonylphenoxy)acetic Acid, 6776
[3116-76-5] Dicloxacillin, 3134
[3122-01-8] Thiazesim Hydrochloride*, 9440
[3137-73-3] Anagestone Acetate*, 664
[3144-16-9] d-Camphorsulfonic Acid, 1781
[3147-14-6] Calmagite, 1766
[3147-55-5] 3,5-Dibromosalicylic Acid, 3077
[3148-09-2] Verrucarin A*, 10100
[3149-00-6] Phenbutamide, 7362
[3150-28-5] D-Chalcose, 2080
[3155-48-4] Kawain (±)-Form*, 5299
[3155-57-5] Methysticin (±)-Dihydromethysticin*, 6218

[3160-91-6] Moroxydine Hydrochloride*, 6354
[3164-34-9] Calcium Tartrate, 1756
[3166-62-9] Benactyzine Methobromide*, 1055
[3167-49-5] 6-Aminonicotinic Acid, 476
[3175-90-4] Lycofawcine, 5648
[3176-03-2] Drotebanol, 3511
[3184-13-2] Ornithine L-Form Monohydrochloride*, 7002
[3198-07-0] Ethylnorepinephrine Hydrochloride*, 3880
[3200-06-4] Nafronyl Acid Oxalate*, 6440
[3200-75-7] Sporidesmolide II*, 8921
[3202-55-9] Benapryzine Hydrochloride*, 1057
[3202-84-4] 4-Salicyloylmorpholine, 8485
[3207-50-9] Linoleic Acid Cyclohexylamide*, 5529
[3215-70-1] Hexoprenaline, 4745
[3230-75-9] Oxenin, 7066
[3230-94-2] Ornithine L-Form Aspartate*, 7002
[3232-84-6] Urazole, 10004
[3239-44-9] Dexfenfluramine*, 4015
[3239-45-0] Dexfenfluramine Hydrochloride*, 4015
[3244-88-0] Acid Fuchsin, 109
[3251-23-8] Cupric Nitrate, 2710
[3252-29-7] Tyrocidine C*, 9967
[3253-60-9] Laudexium Methyl Sulfate, 5392
[3254-66-8] Diisopropyl Paraoxon, 3242
[3254-89-5] Diphenidol Hydrochloride*, 3369
[3254-93-1] Doxenitoin, 3491
[3264-82-2] Nickel Acetylacetonate, 6584
[3270-71-1] Nifuraldezone, 6622
[3279-54-7] Disodium Phenyl Phosphate, 3421
[3286-46-2] Thiamine Disulfide O,O-Diisobutyrate*, 9432
[3301-68-6] Irisolone, 5105
[3306-52-3] Viridin, 10144
[3317-61-1] DMPO, 3454
[3329-14-4] Fenpiprane Hydrochloride*, 4031
[3329-16-6] Narcobarbital Sodium Salt*, 6509
[3329-91-7] Dioscorine, 3347
[3342-61-8] Deanol Aceglumate*, 2900
[3344-16-9] Penicillin G Hydrabamine, 7224
[3344-18-1] Magnesium Citrate, 5699
[3347-22-6] Dithianone, 3433
[3349-06-2] Nickel Formate, 6592
[3351-86-8] Fucoxanthin, 4304
[3354-67-4] Amidephrine, 418
[3362-45-6] Noxiptilin, 6821
[3363-58-4] Nifurfoline, 6624
[3366-95-8] Secnidazole, 8562
[3368-16-9] α-Phenylcinnamic Acid, 7436
[3371-85-5] Voacamine, 10168
[3376-83-8] Biotin l-Sulfoxide, 1273
[3380-34-5] Triclosan, 9790
[3383-96-8] Temephos, 9286
[3385-03-3] Flunisolide, 4180
[3397-23-7] Ornipressin, 7001
[3403-42-7] Methoxypromazine Maleate*, 6080
[3405-51-4] Erdin (+)-Form*, 3682
[3413-58-9] Ethylhydrocupreine Hydrochloride*, 3855

[3416-24-8] Glucosamine, 4466
[3416-26-0] Lidoflazine, 5507
[3420-59-5] Isomaltol, 5198
[3426-01-5] Phenopyrazone, 7401
[3426-08-2] Prozapine, 8093
[3440-28-6] Betamipron, 1227
[3447-95-8] Benfurodil Hemisuccinate, 1070
[3458-22-8] Improsulfan Hydrochloride*, 4962
[3458-28-4] D-Mannose, 5791
[3459-06-1] Cyclopentamine Hydrochloride*, 2808
[3459-20-9] Glymidine Sodium Salt*, 4517
[3459-96-9] Amicarbalide, 414
[3460-67-1] Furonazide, 4330
[3463-92-1] Carpaine, 1908
[3475-65-8] Thiamine Triphosphate, 9433
[3478-44-2] Exiproben Sodium Salt*, 3959
[3483-12-3] 1,4-Dithiothreitol, 3441
[3484-46-6] Hyenanchin, 4898
[3485-14-1] Cyclacillin, 2769
[3485-62-9] Clidinium Bromide, 2412
[3486-35-9] Zinc Carbonate, 10260
[3486-66-6] Coptisine, 2591
[3486-67-7] Palmatine, 7126
[3486-86-0] Butallylonal Sodium Salt*, 1537
[3505-38-2] Carbinoxamine Maleate*, 1845
[3511-16-8] Hetacillin, 4706
[3512-87-6] Akuammine, 200
[3521-62-8] Erythromycin Estolate, 3722
[3521-84-4] Iodipamide Bis[N-methylglucamine] Salt*, 5043
[3540-95-2] Fenpiprane, 4031
[3544-35-2] Iproclozide, 5092
[3546-03-0] Cyamemazine, 2753
[3546-10-9] Phenesterine, 7367
[3546-21-2] Homidium, 4767
[3546-41-6] Pyrvinium Pamoate, 8206
[3552-16-7] Enteromycin, 3634
[3562-84-3] Benzbromarone, 1093
[3562-99-0] Menbutone, 5878
[3563-27-7] Dihydroequilin, 3216
[3563-49-3] Pyrovalerone, 8194
[3563-58-4] Chlorhexadol, 2139
[3563-84-6] Dihydrostreptomycin Pantothenate*, 3222
[3564-09-8] Ponceau 3R, 7749
[3565-72-8] Embramine, 3596
[3567-38-2] Carfimate, 1886
[3567-76-8] Aminochlorthenoxazin, 455
[3568-24-9] Propionylpromazine, 8014
[3569-99-1] N-(Hydroxymethyl)nicotinamide, 4880
[3571-53-7] Estradiol Undecylate*, 3746
[3571-88-8] Platonin, 7690
[3572-43-8] Bromhexine, 1412
[3572-44-9] Anot, 716
[3572-52-9] Biphenamine, 1275
[3572-60-9] Amidinomycin, 419
[3572-74-5] Moxastine, 6370
[3572-80-3] Cyclazocine, 2773
[3573-82-8] Hygrophylline, 4902
[3575-80-2] Melperone, 5870
[3576-64-5] Chlorphenoxamide, 2233
[3577-01-3] Cephaloglycin, 2023
[3583-64-0] Bumadizon, 1507
[3589-73-9] 10-Methoxyharmalan, 6073
[3590-05-4] Acetyl Sulfamethoxypyrazine, 103
[3590-16-7] Feclemine, 3986

Denotes derivative of title compound

[3593-85-9] Methandriol Dipropionate*, 6017
[3595-11-7] Propylhexedrine dl-Form*, 8045
[3598-37-6] Acepromazine Maleate*, 32
[3599-32-4] Indocyanine Green, 4992
[3600-95-1] Perezone, 7298
[3604-87-3] Ecdysone*, 3539
[3605-01-4] Piribedil, 7648
[3607-18-9] Doxepin trans-Form Hydrochloride*, 3492
[3611-72-1] Clobenfurol, 2418
[3614-30-0] Emepronium Bromide, 3598
[3614-47-9] Hydracarbazine, 4798
[3614-69-5] Dimethindene Maleate*, 3265
[3615-24-5] Ramifenazone, 8282
[3615-37-0] D-Fucose, 4301
[3615-41-6] Rhamnose, 8338
[3615-44-9] D-manno-Heptulose, 4701
[3615-82-5] Phytic Acid Calcium Magnesium Salt*, 7542
[3616-78-2] Fenfluramine l-Form Hydrochloride*, 4015
[3621-36-1] Columbamine, 2557
[3621-38-3] Jatrorrhizine, 5274
[3624-87-1] Metabutoxycaine, 5978
[3624-96-2] Bialamicol Dihydrochloride*, 1242
[3625-06-7] Mebeverine, 5808
[3625-25-0] Reposal, 8306
[3625-68-1] Diatretyne II, 3039
[3627-49-4] Phenoperidine Hydrochloride*, 7400
[3635-74-3] Deanol Acetamidobenzoate*, 2900
[3639-12-1] Butethamate Citrate*, 1551
[3644-61-9] Tolperisone Hydrochloride*, 9660
[3650-17-7] Hirsutic Acid C, 4753
[3653-48-3] MCPA Sodium Salt*, 5803
[3666-69-1] Dioxadrol Hydrochloride*, 3352
[3670-68-6] Propipocaine, 8016
[3679-64-9] Bromosalicylchloranilide, 1457
[3682-35-7] 2,4,6-Tripyridyl-s-triazine, 9880
[3684-26-2] 3-Quinuclidinol dl-Form*, 8266
[3685-84-5] Meclofenoxate Hydrochloride*, 5820
[3686-58-6] Tolycaine, 9679
[3688-62-8] Aminopromazine Fumarate*, 487
[3688-65-1] Codeine N-Oxide, 2527
[3689-24-5] Sulfotep, 9139
[3689-50-7] Oxomemazine, 7080
[3689-76-7] Chlormidazole, 2156
[3690-04-8] Aminopropylon, 492
[3690-12-8] Bacimethrin, 964
[3691-21-2] Buzepide, 1635
[3691-35-8] Chlorophacinone, 2204
[3691-74-5] Glyconiazide, 4510
[3691-81-4] Sulfoniazide, 9132
[3693-39-8] Flucoronide, 4157
[3697-42-5] Chlorhexidine Dihydrochloride*, 2140
[3703-76-2] Cloperastine, 2456
[3703-79-5] Bamethan, 981
[3704-09-4] Mibolerone, 6262
[3715-90-0] Tramazoline Hydrochloride Monohydrate*, 9702
[3717-88-2] Flavoxate Hydrochloride*, 4135

[3724-89-8] Phosphocysteamine Sodium Salt*, 7497
[3731-59-7] Moroxydine, 6354
[3733-81-1] Defosfamide, 2917
[3734-24-5] Betasine, 1228
[3734-33-6] Denatonium Benzoate, 2942
[3734-52-9] Metazocine, 5995
[3735-45-3] Vetrabutine, 10107
[3735-84-0] Reserpilic Acid Dimethylaminoethyl Ester Dihydrochloride*, 8313
[3735-90-8] Phencarbamide, 7363
[3736-08-1] Fenethylline, 4014
[3736-36-5] Tropine Benzylate, 9913
[3736-81-0] Diloxanide Furoate*, 3246
[3737-09-5] Disopyramide, 3424
[3737-33-5] Etymemazine Hydrochloride*, 3938
[3738-01-0] Otobain, 7029
[3748-77-4] Bunamidine, 1510
[3749-97-1] Laureline (±)-Form*, 5393
[3750-26-3] Primocarcin, 7929
[3754-19-6] Ambuside, 405
[3759-07-7] Dimetacrine Tartrate*, 3258
[3763-55-1] Rubixanthin, 8445
[3772-43-8] Butoxycaine, 1566
[3772-76-7] Sulfamethomidine, 9085
[3776-92-9] Furfurylmethylamphetamine d-Form Cyclohexylsulfamate*, 4327
[3776-93-0] Furfurylmethylamphetamine d-Form*, 4327
[3778-73-2] Ifosfamide, 4937
[3778-76-5] Todralazine Hydrochloride*, 9640
[3779-59-7] Cimigenol, 2340
[3779-61-1] Ocimene trans-β-Form*, 6837
[3784-89-2] Phenactropinium Chloride, 7346
[3784-92-7] 5-Methyl-5-(3-phenanthryl)hydantoin, 6184
[3784-99-4] Stilbazium Iodide, 8971
[3785-21-5] Butanilicaine, 1542
[3785-32-8] Salicylamide O-Acetic Acid Sodium Salt*, 8481
[3790-78-1] Nerolidol cis-Form*, 6561
[3801-06-7] Fluorometholone 17-Acetate*, 4213
[3804-89-5] Isoniazid Methanesulfonate Sodium*, 5203
[3810-35-3] Tenonitrozole, 9292
[3810-74-0] Streptomycin Sesquisulfate*, 8983
[3810-80-8] Diphenoxylate Hydrochloride*, 3371
[3810-83-1] Pentacynium Bis(methylsulfate), 7243
[3811-04-9] Potassium Chlorate, 7782
[3811-20-9] Cuminaldehyde Thiosemicarbazone*, 2687
[3811-25-4] Clorprenaline, 2470
[3811-56-1] Aminoquinuride, 496
[3811-75-4] Hexamidine, 4728
[3818-37-9] Phenoxypropazine, 7411
[3818-50-6] Bephenium Hydroxynaphthoate*, 1187
[3818-62-0] Betoxycaine, 1235
[3819-00-9] Piperacetazine, 7615
[3819-34-9] Phenamet, 7351
[3820-67-5] Glafenine, 4443
[3833-99-6] Homofenazine, 4774
[3835-52-7] O-Cinnamoyltaxicin-II triacetate*, 9245
[3844-45-9] Brilliant Blue FCF, 1397
[3845-07-6] Octamoxin Sulfate*, 6845

[3847-29-8] Erythromycin Lactobionate, 3724
[3856-25-5] Copaene, 2579
[3858-89-7] Chloroprocaine Hydrochloride, 2210
[3861-73-2] Anazolene Sodium, 669
[3861-76-5] Clonitazene, 2451
[3871-82-7] Moperone Hydrochloride*, 6343
[3900-31-0] Fludiazepam, 4164
[3902-71-4] Trioxsalen, 9864
[3922-90-5] Oleandomycin, 6962
[3930-19-6] Streptonigrin, 8986
[3930-20-9] Sotalol, 8876
[3941-06-8] Phenoxypropazine Maleate*, 7411
[3947-65-7] Neamine, 6517
[3952-98-5] Sinigrin, 8691
[3963-95-9] Methacycline Hydrochloride*, 6007
[3964-81-6] Azatadine, 934
[3978-86-7] Azatadine Dimaleate*, 934
[3982-91-0] Phosphorus Sulfochloride, 7513
[4008-48-4] Nitroxoline, 6753
[4042-30-2] Parvaquone, 7182
[4044-65-9] Bitoscanate, 1345
[4055-39-4] Mitomycin A*, 6301
[4055-40-7] Mitomycin B*, 6301
[4065-45-6] Sulisobenzone, 9157
[4075-81-4] Calcium Propionate, 1745
[4076-02-2] 2,3-Dimercapto-1-propanesulfonic Acid Sodium Salt*, 3256
[4080-31-3] N-(3-Chloroallyl)hexaminium Chloride, 2168
[4091-75-2] Clomestrone, 2442
[4093-35-0] Bromopride, 1454
[4097-22-7] Dideoxyadenosine, 3151
[4098-40-2] Mitragynine, 6304
[4112-89-4] Guaiacol Phenylacetate*, 4575
[4119-52-2] Ferric Thiocyanate, 4081
[4135-11-9] Polymyxin B₁*, 7734
[4147-51-7] Dipropetryn, 3405
[4154-10-3] Rolitetracycline Compd with Chloramphenicol Succinate*, 8411
[4154-53-4] d-Camphocarboxylic Acid Basic Bismuth Salt*, 1778
[4156-16-5] Woodward's Reagent K, 10186
[4163-15-9] Cyclorphan, 2819
[4170-30-3] Crotonaldehyde, 2663
[4171-13-5] Valnoctamide, 10048
[4201-22-3] Tolonidine, 9657
[4201-78-9] Choline Dehydrocholate, 2262
[4205-23-6] D-Gulose, 4606
[4205-90-7] Clonidine, 2450
[4205-91-8] Clonidine Hydrochloride*, 2450
[4214-72-6] Isaxonine, 5118
[4221-98-1] α-Phellandrene l-Form*, 7340
[4232-99-9] 3,5-Dibromo-4-hydroxybenzenesulfonic Acid, 3072
[4258-85-9] Clocortolone 21-Acetate*, 2430
[4267-05-4] Teclothiazide, 9261
[4268-36-4] Tybamate, 9960
[4271-30-1] N-(p-Aminobenzoyl)-glutamic Acid, 446
[4282-07-9] Chonemorphine, 2272
[4282-16-0] N,N'-Dimethylcyclobuxine D*, 2785
[4291-63-8] Cladribine, 2397
[4299-57-4] Plastoquinone 9*, 7681

[4299-60-9] Sulfisoxazole Diethanolamine Salt*, 9125
[4300-28-1] D-Ribose-5-phosphoric Acid, 8372
[4302-95-8] Tevenel, 9387
[4304-40-9] Thenium Closylate, 9414
[4317-14-0] Amitriptylinoxide, 512
[4320-30-3] Arginine Glutamate, 818
[4323-43-7] Hexoprenaline Dihydrochloride*, 4745
[4330-99-8] Trimeprazine Tartrate*, 9834
[4342-03-4] Dacarbazine, 2866
[4345-03-3] Vitamin E Succinate*, 10159
[4350-07-6] 5-Hydroxytryptophan D-Form*, 4895
[4350-09-8] 5-Hydroxytryptophan L-Form*, 4895
[4354-45-4] Propenzolate, 7998
[4356-43-8] Lupanine dl-Form*, 5635
[4360-12-7] Ajmaline, 194
[4366-18-1] Coumetarol, 2633
[4368-28-9] Tetrodotoxin, 9382
[4373-40-4] Aurantiogliocladin Dihydro Deriv*, 912
[4378-36-3] Fenbutrazate, 4005
[4386-35-0] Meralein Sodium, 5912
[4388-82-3] Propylhexedrine l-Form Ethylphenylbarbiturate*, 8045
[4393-19-5] p-Sulfanilylbenzylamine, 9098
[4394-00-7] Niflumic Acid, 6620
[4396-01-4] Pelletierine, 7200
[4403-90-1] Alizarin Cyanine Green F, 248
[4406-22-8] Cyprenorphine, 2838
[4412-09-3] Mucochloric Anhydride, 6380
[4419-39-0] Beclomethasone, 1047
[4438-22-6] Atropine N-Oxide, 908
[4439-84-3] HON, 4779
[4448-95-7] Shellolic Acid, 8624
[4461-30-7] Chloroacetyl Isocyanate, 2167
[4462-95-7] Truxillic Acid, 9922
[4466-14-2] Jasmolin I*, 5272
[4478-93-7] Sulforaphane (unspecified), 9135
[4481-08-7] Digitalose, 3202
[4481-55-4] Pentryl, 7281
[4482-83-1] BBB*, 4124
[4491-19-4] Indaconitine, 4964
[4493-23-6] Dodecahedrane, 3461
[4493-37-2] Chromous Formate, 2303
[4498-32-2] Dibenzepin, 3055
[4499-40-5] Choline Theophyllinate, 2266
[4516-93-2] Aurantiogliocladin Quinhydrone Deriv*, 912
[4533-39-5] Nitracrine, 6661
[4533-89-5] Flunisolide 21-Acetate*, 4180
[4544-15-4] Piperilate Hydrochloride*, 7624
[4546-39-8] Piperilate, 7624
[4548-53-2] Ponceau SX, 7750
[4551-59-1] Fenalamide, 3995
[4562-36-1] Gitoxin, 4441
[4564-87-8] Carbomycin A*, 1854
[4573-78-8] Hamamelose, 4640
[4574-60-1] Atropine N-Oxide Hydrochloride*, 908
[4589-33-7] Bostrycoidin, 1381
[4599-60-4] Penimepicycline, 7235
[4602-84-0] Farnesol, 3978
[4611-02-3] Chlorproethazine Hydrochloride*, 2236
[4618-18-2] Lactulose, 5360

[4618-41-1] 1-Butyl-3-metanilylurea, 1615
[4619-74-3] 2,4,6-Tribromo-m-cresol, 9742
[4626-00-0] Kitol, 5328
[4630-95-9] Prifinium Bromide, 7923
[4638-92-0] Chrysanthemic Acid d-trans-Form*, 2310
[4643-58-7] Verrucarin J*, 10100
[4655-30-5] Hemi-Dewar Biphenyl, 4677
[4663-83-6] Buramate, 1525
[4670-05-7] Theaflavine, 9409
[4671-03-8] Hexazole, 4735
[4673-18-1] Buphanitine, 1518
[4678-44-8] Grayanotoxin II*, 4566
[4678-45-9] Grayanotoxin III*, 4566
[4680-78-8] Guinea Green B, 4603
[4682-36-4] Orphenadrine Citrate*, 7007
[4682-48-8] Cytolipin H, 2863
[4682-50-2] Trichodermin, 9779
[4684-87-1] Octamoxin, 6845
[4685-14-7] Paraquat, 7165
[4695-62-9] d-Fenchone, 4008
[4696-76-8] Kanamycin B*, 5293
[4697-14-7] Ticarcillin Disodium Salt*, 9568
[4697-36-3] Carbenicillin, 1838
[4719-75-9] Fosfestrol Tetrasodium Salt*, 4279
[4720-09-6] Grayanotoxin I*, 4566
[4721-69-1] Oxabolone, 7034
[4726-14-1] Nitralin, 6662
[4747-99-3] Tetrahydropapaveroline, 9355
[4757-55-5] Dimetacrine, 3258
[4759-48-2] Retinoic Acid 13-cis-Form*, 8333
[4764-17-4] MDA, 5804
[4767-03-7] Dimethylolpropionic Acid, 3302
[4776-06-1] Fluorosalan, 4216
[4800-94-6] Carbenicillin Disodium Salt*, 1838
[4803-27-4] Anthramycin, 724
[4825-86-9] Ochratoxin B*, 6836
[4826-71-5] Phosphorylcholine Calcium Salt*, 7519
[4828-27-7] Clocortolone, 2430
[4846-91-7] Fenoxazoline, 4025
[4850-21-9] Quebrachamine, 8212
[4865-85-4] Ochratoxin C*, 6836
[4876-45-3] Camphotamide, 1782
[4880-88-0] Eburnamonine (−)-Form*, 3537
[4914-30-1] Dehydroemetine, 2924
[4936-47-4] Nifuratel, 6623
[4940-39-0] Chromocarb, 2294
[4945-47-5] Bamipine, 983
[4955-90-2] Gentisic Acid Sodium Salt*, 4404
[4956-37-0] Estradiol Enanthate*, 3746
[4968-09-6] Algestone Cyclic Acetal with Acetone*, 228
[4968-29-0] Guaiazulene Trinitrobenzene Deriv*, 4581
[4969-02-2] Methixene, 6059
[4976-25-4] Tremetone, 9714
[4985-15-3] Noxiptilin Hydrochloride*, 6821
[4991-65-5] Tioxolone, 9600
[5001-32-1] Guanochlor, 4595
[5001-33-2] Metanephrine, 5987
[5002-47-1] Fluphenazine Decanoate*, 4226
[5003-47-4] Betoxycaine Monohydrochloride*, 1235
[5003-48-5] Benorylate, 1074

[5008-52-6] Napelline, 6454
[5011-34-7] Trimetazidine, 9835
[5015-36-1] Methylprednisolone 21-Phosphate Disodium Salt*, 6189
[5036-02-2] Tetramisole*, 5486
[5051-62-7] Guanabenz, 4585
[5053-06-5] Fenspiride, 4041
[5053-08-7] Fenspiride Hydrochloride*, 4041
[5060-55-9] Prednisolone 21-Stearoylglycolate*, 7901
[5061-22-3] Nafiverine, 6439
[5072-26-4] Buthionine Sulfoximine, 1556
[5080-50-2] Acetylcarnitine L-Form Hydrochloride*, 85
[5086-74-8] Tetramisole Hydrochloride*, 5486
[5089-89-4] 7-Morpholinomethyltheophylline, 6363
[5090-37-9] Metizoline Hydrochloride*, 6223
[5096-57-1] Canadine l-Form*, 1785
[5103-42-4] Hydrindantin, 4816
[5104-49-4] Flurbiprofen, 4234
[5108-94-1] Mytatrienediol, 6424
[5118-17-2] Furazolium Chloride, 4321
[5118-29-6] Melitracen, 5866
[5119-48-2] Withaferin A, 10184
[5133-19-7] Aloin, 314
[5141-20-8] Light Green SF Yellowish, 5508
[5142-76-7] Bismuth Subacetate, 1323
[5153-24-2] Zirconyl Acetate, 10316
[5169-78-8] Tipepidine, 9601
[5175-83-7] Bismuth Tribromophenate, 1333
[5189-11-7] Pizotyline Malate*, 7671
[5192-84-7] Trengestone, 9717
[5197-58-0] Stenbolone, 8962
[5205-82-3] Bevonium Methyl Sulfate, 1239
[5221-49-8] Pyrimithate, 8172
[5221-53-4] Dimethirimol, 3266
[5231-60-7] Demethoxyvindoline*, 10126
[5234-68-4] Carboxin, 1874
[5250-39-5] Floxacillin, 4147
[5251-34-3] Cloprednol, 2460
[5285-18-7] Piperic Acid, 7620
[5289-74-7] 20-Hydroxyecdysone*, 3539
[5306-80-9] Teclothiazide Potassium Salt*, 9261
[5308-89-4] Taxicin-I*, 9245
[5308-90-7] Taxicin-II*, 9245
[5309-42-2] Monobutyrin (R)-Form*, 6332
[5310-55-4] Clomacran, 2441
[5318-76-3] Imidocarb Dihydrochloride*, 4951
[5321-32-4] Hetacillin Potassium Salt*, 4706
[5323-95-5] Ricinoleic Acid Sodium Salt*, 8378
[5329-14-6] Sulfamic Acid, 9090
[5341-61-7] Hydrazine Dihydrochloride*, 4809
[5355-16-8] Diaveridine, 3041
[5355-48-6] β-Acetyldigoxin*, 3210
[5370-01-4] Mexiletine Hydrochloride*, 6257
[5373-42-2] Thalicarpine, 9389
[5376-03-4] Tropylium Bromide, 9916
[5377-20-8] Metomidate, 6232
[5392-40-5] Citral, 2383
[5394-83-2] Camphoric Acid, 1780
[5408-52-6] L-Lysine L-Glutamate, 5669

Denotes derivative of title compound

[5411-22-3] Benzphetamine Hydrochloride*, 1151
[5421-04-5] Mandelic Acid Isoamyl Ester, 5758
[5429-41-4] Methafurylene Fumarate*, 6010
[5466-22-8] Aminoacetonitrile Sulfate (2:1)*, 432
[5466-77-3] Octyl Methoxycinnamate, 6864
[5470-11-1] Hydroxylamine Hydrochloride*, 4874
[5486-03-3] Buquinolate, 1524
[5486-77-1] Alloclamide, 263
[5490-27-7] Dihydrostreptomycin Sesquisulfate*, 3222
[5508-58-7] Andrographolide, 673
[5511-98-8] α-Acetyldigoxin*, 3210
[5522-66-7] Protoporphyrin IX Dimethyl Ester*, 8084
[5534-09-8] Beclomethasone Dipropionate*, 1047
[5534-95-2] Pentagastrin, 7250
[5536-17-4] Vidarabine, 10113
[5545-89-1] Chimonanthine, 2101
[5560-59-8] Alverine Citrate*, 386
[5560-62-3] Biphenamine Hydrochloride*, 1275
[5560-69-0] Ethyl Dibunate, 3834
[5560-72-5] Iprindole, 5091
[5560-78-1] Teclozan, 9262
[5576-62-5] Chlorbenzoxamine Dihydrochloride*, 2125
[5579-13-5] Capuride, 1818
[5579-16-8] Epinephryl Borate*, 3656
[5579-81-7] Aldioxa, 224
[5579-84-0] Betahistine Dihydrochloride*, 1224
[5579-92-0] Iopydol, 5078
[5579-93-1] Iopydone, 5079
[5579-94-2] Merisoprol Hg 197, 5959
[5581-35-1] Amphecloral, 620
[5581-52-2] Thiamiprine, 9434
[5585-52-4] Pentorex Hydrochloride*, 7275
[5585-60-4] Paranyline Hydrochloride*, 7163
[5585-64-8] Amotriphene, 615
[5585-73-9] Butriptyline Hydrochloride*, 1568
[5585-93-3] Oxypendyl, 7104
[5586-87-8] Mefenorex Hydrochloride*, 5843
[5587-89-3] Ipodate, 5087
[5588-10-3] Methoxyphenamine Hydrochloride*, 6077
[5588-16-9] Althiazide, 326
[5588-20-5] Chlordantoin, 2130
[5588-22-7] Clorprenaline Hydrochloride Monohydrate*, 2470
[5588-25-0] Dihexyverine Hydrochloride*, 3211
[5588-33-0] Mesoridazine, 5970
[5591-22-0] Becanthone Hydrochloride*, 1044
[5591-29-7] Etafedrine Hydrochloride*, 3752
[5591-45-7] Thiothixene, 9503
[5591-74-2] Rhodoquinone, 8360
[5593-20-4] Betamethasone 17,21-Dipropionate*, 1226
[5598-13-0] Chlorpyrifos O,O-Dimethyl Analog*, 2242
[5598-52-7] Fospirate, 4283
[5600-19-1] Theophylline Isopropanolamine*, 9421
[5610-40-2] Securinine, 8565
[5611-51-8] Triamcinolone Hexacetonide, 9730

[5614-56-2] 1-Theobromineacetic Acid, 9419
[5625-37-6] PIPES, 7633
[5626-34-6] Prednisolone 21-Diethylaminoacetate, 7902
[5627-46-3] Clobenztropine, 2422
[5630-53-5] Tibolone, 9566
[5631-00-5] Benzoylpas Calcium Salt Pentahydrate*, 1148
[5632-44-0] Tolpropamine, 9662
[5633-14-7] Benzetimide Hydrochloride*, 1104
[5633-16-9] Leiopyrrole, 5457
[5633-18-1] Melengestrol, 5859
[5633-25-0] Noracymethadol Hydrochloride*, 6778
[5634-34-4] Ambuphylline, 404
[5634-38-8] Guaithylline*, 4582
[5634-39-9] Iodinated Glycerol, 5033
[5634-42-4] Tocamphyl, 9630
[5636-83-9] Dimethindene, 3265
[5636-92-0] Picloxydine, 7553
[5638-76-6] Betahistine, 1224
[5639-34-9] 1-Amino-2-naphthol-6-sulfonic Acid, 475
[5650-44-2] Methcathinone (unspecified), 6033
[5667-71-0] Streptonicozid, 8985
[5668-06-4] Mccloxamine, 5823
[5675-98-9] Thevetose, 9425
[5695-98-7] Cotinine Fumarate*, 2619
[5696-06-0] Methetoin, 6046
[5696-09-3] Proxazole, 8090
[5697-56-3] Carbenoxolone, 1839
[5704-04-1] Tricine, 9783
[5704-60-9] Nifenalol Hydrochloride*, 6618
[5707-69-7] Drazoxolon, 3499
[5714-00-1] Acetophenazine Dimaleate*, 70
[5714-04-5] Guanoxan Sulfate*, 4598
[5714-73-8] Methenamine Hippurate, 6039
[5714-82-9] Triclofenol Piperazine, 9787
[5714-90-9] Levopropoxyphene 2-Naphthalenesulfonate*, 5495
[5716-20-1] Bamethan Sulfate*, 981
[5728-52-9] Felbinac, 3989
[5735-19-3] Galactoflavin, 4351
[5741-22-0] Moprolol, 6345
[5743-27-1] Calcium Ascorbate, 1688
[5743-29-3] Calcium 3-Aurothio-2-propanol-1-sulfonate, 1689
[5743-34-0] Calcium Borogluconate, 1692
[5746-40-7] Phosphocysteamine, 7497
[5749-67-7] Calcium Acetylsalicylate Complex with Urea*, 1684
[5779-39-5] Desosamine, 2974
[5779-54-4] Cyclarbamate, 2772
[5786-21-0] Clozapine, 2484
[5787-63-3] Toldimfos Sodium Trihydrate*, 9649
[5793-84-0] Calcium Phenoxide, 1738
[5793-88-4] Calcium D-Saccharate, 1747
[5794-16-1] Antipyrine Methylethylglycolate*, 757
[5794-88-7] 5-Bromoanthranilic Acid, 1426
[5796-17-8] Dopa D-Form*, 3478
[5796-30-5] Ecgonine l-Form Monohydrate*, 3541
[5796-31-6] Ecgonine Hydrochloride*, 3541
[5798-41-4] Bismuth Potassium Tartrate, 1318
[5798-43-6] Bismuth Sodium Triglycollamate, 1322

[5798-47-0] Bismuth Valerate, Basic, 1334
[5798-49-2] Bis(1-naphthylmethyl)-amine, 1335
[5798-78-7] p-Bromobenzyl Chloroformate, 1432
[5798-79-8] α-Bromobenzyl Cyanide, 1433
[5798-88-9] β-Bromoisovaleric Acid, 1444
[5798-94-7] 5-Bromosalicylhydroxamic Acid, 1458
[5807-87-4] Diphenylpyraline Hydrobromide*, 3390
[5835-26-7] Isopimaric Acid, 5217
[5843-53-8] Clobenzorex Hydrochloride*, 2421
[5845-26-1] Thiazesim, 9440
[5853-02-1] Casimiroedine, 1935
[5854-93-3] L-Alanosine, 207
[5868-05-3] Niceritrol, 6581
[5868-06-4] Fentonium Bromide, 4048
[5870-25-7] Ambucaine Hydrochloride*, 402
[5870-29-1] Cyclopentolate Hydrochloride*, 2815
[5874-97-5] Metaproterenol Sulfate*, 5992
[5875-06-9] Proparacaine Hydrochloride*, 7991
[5892-10-1] Bismuth Subcarbonate, 1324
[5892-41-1] Camylofine Dihydrochloride*, 1784
[5897-76-7] α-(α-Aminopropyl)benzyl Alcohol, 491
[5902-51-2] Terbacil, 9298
[5905-34-0] Acid Violet 7B, 110
[5905-52-2] Ferrous Lactate, 4098
[5907-38-0] Dipyrone, 3414
[5908-63-4] Baptigenin, 984
[5908-89-4] Belladonnine Bisulfate Monohydrate*, 1053
[5909-04-6] Chrysoidine Citrate*, 2317
[5912-86-7] Isoeugenol cis-Form*, 5186
[5913-71-3] Codeine Acetate*, 2525
[5913-82-6] Conessine Dihydrobromide*, 2561
[5928-84-7] Penicillin V Benzathine, 7231
[5932-68-3] Isoeugenol trans-Form*, 5186
[5934-14-5] Succisulfone, 9047
[5936-28-7] Hydrastine Hydrochloride*, 4806
[5937-72-4] Androstenediol 17-Acetate*, 677
[5942-95-0] Carpipramine, 1911
[5945-50-6] Monotropein, 6338
[5945-86-8] Nimbin, 6638
[5947-49-9] Podocarpic Acid, 7701
[5949-18-8] Cinchophen Sodium Salt*, 2347
[5949-44-0] Testosterone Undecanoate*, 9322
[5953-63-9] Androstenediol 3-Acetate-17-benzoate*, 677
[5957-75-5] Tetrahydrocannabinols (−)-Δ6-3,4-trans-Form*, 9349
[5959-52-4] 3-Amino-2-naphthoic Acid, 472
[5962-29-8] Xylulose, 10221
[5964-24-9] Thimerfonate Sodium, 9450
[5964-62-5] Diathymosulfone, 3037
[5965-09-3] Fosfestrol Disodium Salt*, 4279

[5965-33-3] Antimony Potassium Oxalate, *740*
[5965-40-2] Allocupreide Sodium, *265*
[5965-49-1] Ketobemidone Hydrochloride*, *5312*
[5966-41-6] Diisopromine, *3239*
[5967-52-2] Metaraminol Hydrochloride*, *5993*
[5967-53-3] Metaraminol Oxalate Dihydrate*, *5993*
[5967-73-7] Methadone Hydrochloride *l*-Form*, *6008*
[5970-32-1] Mercuric Salicylate, *5938*
[5972-71-4] Ammonium Bimalate, *524*
[5972-72-5] Ammonium Binoxalate, *525*
[5972-76-9] Ammonium Caprylate, *532*
[5973-17-1] Phenol Ammonium Salt*, *7390*
[5974-09-4] Vetrabutine Hydrochloride*, *10107*
[5977-10-6] Fencibutirol, *4009*
[5978-58-5] Guaiacol Phosphate Sodium Salt*, *4578*
[5978-87-0] D-Araboflavin, *801*
[5980-01-8] Lochneridine, *5583*
[5980-31-4] Hexedine, *4736*
[5982-52-5] Ethylbenzhydramine Methyl Iodide*, *3813*
[5982-55-8] Ethanearsonic Acid Disodium Salt*, *3768*
[5984-61-2] Pelletierine Hydrochloride*, *7200*
[5985-28-4] Synephrine Hydrochloride*, *9189*
[5985-35-3] Levorphanol *dl*-Form Hydrobromide*, *5496*
[5985-38-6] Levorphanol Tartrate Dihydrate*, *5496*
[5986-55-0] Patchouli Alcohol, *7184*
[5987-82-6] Benoxinate Hydrochloride*, *1076*
[5988-22-7] Phytonadiol Sodium Diphosphate*, *7536*
[5988-51-2] Deanol Bitartrate*, *2900*
[5989-43-5] Matricarin, *5798*
[5991-71-9] Clorazepic Acid Monopotassium Salt*, *2465*
[5995-96-0] Eseridine Salicylate*, *3740*
[5996-14-5] α-Glucose-1-phosphate Dipotassium Salt Dihydrate*, *4470*
[6000-74-4] Hydrocortisone 21-Phosphate Disodium Salt*, *4828*
[6001-51-0] Bis(*p*-dimethylaminobenzylidene)benzidine, *1289*
[6001-74-7] Clomethiazole Hydrochloride*, *2444*
[6001-97-4] Bis(1-methylamyl) Sodium Sulfosuccinate, *1295*
[6003-09-4] Aleuritic Acid Ethyl Ester*, *230*
[6003-10-7] Aleuritic Acid Hydrazide*, *230*
[6004-24-6] Cetylpyridinium Chloride Monohydrate*, *2074*
[6006-09-3] Gravitol(e), *4565*
[6009-67-2] Amolanone Hydrochloride*, *610*
[6010-09-9] Ferrous Thiocyanate, *4107*
[6011-14-9] Aminoacetonitrile Hydrochloride*, *432*
[6018-19-5] *p*-Aminosalicylic Acid Sodium Salt Dihydrate*, *498*
[6018-40-2] Corypalmine, *2615*
[6018-53-7] 6-Desoxy-D-glucosamine, *2976*
[6018-84-4] *p*-Aminobenzoic Acid Diethylamine Salt*, *443*

[6019-02-9] Nicotine Dihydrochloride*, *6611*
[6020-39-9] *meso*-Cystine*, *2851*
[6020-73-1] Codeine Salicylate*, *2525*
[6027-13-0] Homocysteine, *4771*
[6027-89-0] L-Gulose, *4607*
[6028-35-9] Thibenzazoline, *9446*
[6031-86-3] Racemethorphan Hydrobromide*, *8274*
[6032-06-0] Cyproheptadine Hydrochloride Monohydrate*, *2842*
[6032-14-0] *N*-Methylepinephrine DL-Form*, *6146*
[6032-29-7] 2-Pentanol, *7258*
[6032-32-2] Methylarbutin, *6101*
[6032-92-4] Mycarose, *6399*
[6033-08-5] Hematoporphyrin Hydrochloride Monohydrate*, *4669*
[6033-41-6] Norpipanone Hydrochloride*, *6810*
[6033-42-7] Norpipanone Hydrobromide*, *6810*
[6033-69-8] Deserpidine Hydrochloride*, *2964*
[6035-31-0] Norlobelanine, *6802*
[6035-39-8] Nitroakridin 3582, *6678*
[6035-40-1] Noscapine *dl*-Form*, *6815*
[6035-45-6] Folinic Acid Calcium Salt Pentahydrate*, *4254*
[6035-50-3] DL-Fructose, *4296*
[6036-25-5] Narbomycin, *6507*
[6043-86-3] Methiotriazamine, *6056*
[6054-98-4] Olsalazine Disodium Salt*, *6976*
[6055-19-2] Cyclophosphamide, *2816*
[6055-73-8] Adrenochrome Oxime Sesquihydrate*, *171*
[6056-11-7] Pipazethate Hydrochloride*, *7609*
[6056-70-8] α-Glucose-1-phosphate Barium Salt Trihydrate*, *4470*
[6059-26-3] Isoniazid Methanesulfonate Calcium*, *5203*
[6064-83-1] Fosfosal, *4281*
[6078-26-8] Bikhaconitine, *1262*
[6088-51-3] DDD (Analytical), *2897*
[6091-05-0] Physovenine, *7541*
[6092-18-8] Cycothiamin(e), *2824*
[6101-28-6] Sulfamethylthiazole Sodium Salt*, *9088*
[6101-32-2] Sulfanilic Acid Monohydrate*, *9096*
[6101-34-4] Sulfathiourea Sodium Salt*, *9116*
[6105-85-7] Raunescine Monohydrate*, *8292*
[6105-86-8] Raunescine Monoacetate Monohydrate*, *8292*
[6106-22-5] Sulfanilic Acid Sodium Salt Dihydrate*, *9096*
[6108-05-0] Lidocaine Hydrochloride Monohydrate*, *5505*
[6109-70-2] 3-Quinuclidinol *dl*-Form Acetate Hydrochloride*, *8266*
[6113-04-8] Adiphenine Methyl Bromide*, *160*
[6113-17-3] Histapyrrodine Hydrochloride*, *4757*
[6114-26-7] Pholedrine Sulfate*, *7485*
[6117-91-5] Crotyl Alcohol, *2668*
[6119-47-7] Quinine Hydrochloride Dihydrate*, *8245*
[6119-70-6] Quinine Sulfate Dihydrate*, *8245*
[6130-64-9] Penicillin G Procaine, *7226*
[6138-47-2] Liothyronine Hydrochloride*, *5535*

[6138-56-3] Tripelennamine Citrate*, *9868*
[6138-73-4] Trilobine, *9826*
[6138-79-0] Triprolidine Hydrochloride Monohydrate*, *9877*
[6147-11-1] Mangostin, *5785*
[6147-37-1] Menadione Sodium Bisulfite* (Trihydrate), *5874*
[6150-97-6] Dipyrone Magnesium Salt*, *3414*
[6151-25-3] Quercetin Dihydrate*, *8216*
[6151-30-0] Quinacrine Dihydrochloride Dihydrate*, *8225*
[6151-39-9] Quinidine Hydrogen Sulfate Tetrahydrate*, *8244*
[6153-16-8] β-Phellandrene *d*-Form*, *7341*
[6153-17-9] β-Phellandrene *l*-Form*, *7341*
[6153-19-1] Phenacaine Hydrochloride Monohydrate*, *7342*
[6153-33-9] Mebhydroline 1,5-Naphthalenedisulfonate Salt*, *5809*
[6153-43-1] Ornithine L-Form Monosulfate*, *7002*
[6153-44-2] Orotic Acid Methyl Ester*, *7004*
[6155-64-2] Salicylsulfuric Acid Monosodium Salt*, *8486*
[6159-55-3] Vasicine, *10072*
[6159-98-4] Raunescine Nitrate*, *8292*
[6160-08-3] Sozoiodolic Acid Sodium Salt Dihydrate*, *8881*
[6160-09-4] Sozoiodolic Acid Zinc Salt Hexahydrate*, *8881*
[6160-10-7] Sozoiodolic Acid Trihydrate*, *8881*
[6160-12-9] Sparteine Sulfate Pentahydrate*, *8887*
[6163-66-2] Di-*tert*-butyl Ether, *3084*
[6164-87-0] Nicotinyl Alcohol *d*-Tartrate*, *6614*
[6164-98-3] Chlordimeform, *2133*
[6168-76-9] Crotethamide, *2662*
[6168-86-1] Isometheptene Hydrochloride*, *5202*
[6170-29-2] Aminopyrine Hydrochloride*, *495*
[6170-42-9] Chloropyramine Hydrochloride*, *2214*
[6183-68-2] Quinine Bisulfate Heptahydrate*, *8245*
[6190-36-9] Cotarnine Phthalate*, *2618*
[6190-39-2] Dihydroergotamine Methanesulfonate*, *3217*
[6190-43-8] Methenamine Anhydromethylenecitrate, *6038*
[6190-60-9] Mephentermine Sulfate Dihydrate*, *5897*
[6192-89-8] Procaine Dihydrate*, *7937*
[6192-92-3] Procaine Nitrate*, *7937*
[6192-95-6] Propylhexedrine *dl*-Form Hydrochloride*, *8045*
[6192-96-7] Propylhexedrine *d*-Form Hydrochloride*, *8045*
[6192-97-8] Propylhexedrine *l*-Form*, *8045*
[6192-98-9] Propylhexedrine *l*-Form Hydrochloride*, *8045*
[6199-67-3] Cucurbitacin B*, *2681*
[6202-21-7] 4-Sulfanilamidosalicylic Acid, *9095*
[6202-23-9] Cyclobenzaprine Hydrochloride*, *2782*
[6209-17-2] Sulfacetamide Sodium Salt Monohydrate*, *9067*
[6209-37-6] Methylergonovine Tartrate*, *6147*

Denotes derivative of title compound

[7224-08-0] Imiclopazine, *4945*
[7225-61-8] Metrizoic Acid Sodium Salt*, *6241*
[7229-50-7] Streptolydigin, *8982*
[7234-48-2] ε-Acetamidocaproic Acid Sodium Salt*, *43*
[7235-40-7] β-Carotene, *1902*
[7236-47-7] β-Benzalbutyramide, *1083*
[7237-81-2] Hepronicate, *4688*
[7240-38-2] Oxacillin Sodium Salt Monohydrate*, *7036*
[7242-04-8] Gitoxin Pentaacetate*, *4441*
[7246-07-3] Actinoquinol Sodium Salt*, *143*
[7246-20-0] Triclofos Monosodium Salt*, *9788*
[7246-21-1] Tyropanoate Sodium, *9968*
[7247-57-6] Heteronium Bromide, *4709*
[7248-28-4] Strychnine *N⁶*-Oxide, *9022*
[7257-29-6] *C*-Calebassine, *1762*
[7261-97-4] Dantrolene, *2879*
[7262-75-1] Lefetamine, *5455*
[7270-12-4] 8-Hydroxy-7-iodo-5-quinolinesulfonic Acid Compd with Chloroquine (2:1)*, *4872*
[7280-37-7] Estrone Sulfate Piperazine Salt*, *3751*
[7287-19-6] Prometryn, *7973*
[7295-85-4] Catechin *dl*-Form*, *1950*
[7297-25-8] Erythrityl Tetranitrate, *3716*
[7309-58-2] Helenynolic Acid, *4657*
[7320-34-5] Potassium Pyrophosphate, *7833*
[7327-87-9] Dihydralazine Sulfate*, *3212*
[7361-61-7] Xylazine, *10213*
[7365-45-9] HEPES, *4687*
[7377-03-9] Octanohydroxamic Acid, *6848*
[7400-08-0] *p*-Coumaric Acid, *2628*
[7413-34-5] Methotrexate Disodium Salt*, *6065*
[7413-36-7] Nifenalol, *6618*
[7414-83-7] Etidronic Acid Disodium Salt*, *3908*
[7414-95-1] Imiclopazine Dihydrochloride*, *4945*
[7416-34-4] Molindone, *6315*
[7421-40-1] Carbenoxolone Disodium Salt*, *1839*
[7429-90-5] Aluminum, *331*
[7429-91-6] Dysprosium, *3531*
[7429-92-7] Einsteinium, *3573*
[7432-25-9] Etaqualone, *3757*
[7439-88-5] Iridium, *5098*
[7439-89-6] Iron, *5106*
[7439-90-9] Krypton, *5335*
[7439-91-0] Lanthanum, *5374*
[7439-92-1] Lead, *5410*
[7439-93-2] Lithium, *5543*
[7439-94-3] Lutetium, *5642*
[7439-95-4] Magnesium, *5687*
[7439-96-5] Manganese, *5762*
[7439-97-6] Mercury, *5957*
[7439-98-7] Molybdenum, *6317*
[7439-99-8] Neptunium, *6557*
[7440-00-8] Neodymium, *6538*
[7440-01-9] Neon, *6544*
[7440-02-0] Nickel, *6582*
[7440-02-0] Raney Nickel, *8284*
[7440-03-1] Niobium, *6648*
[7440-04-2] Osmium, *7021*
[7440-05-3] Palladium, *7121*
[7440-06-4] Platinum, *7689*
[7440-07-5] Plutonium, *7700*
[7440-08-6] Polonium, *7718*

[7440-09-7] Potassium, *7763*
[7440-10-0] Praseodymium, *7891*
[7440-11-1] Mendelevium, *5879*
[7440-12-2] Promethium, *7971*
[7440-13-3] Protactinium, *8069*
[7440-14-4] Radium, *8277*
[7440-15-5] Rhenium, *8343*
[7440-16-6] Rhodium, *8353*
[7440-17-7] Rubidium, *8439*
[7440-18-8] Ruthenium, *8452*
[7440-19-9] Samarium, *8498*
[7440-20-2] Scandium, *8537*
[7440-21-3] Silicon, *8635*
[7440-22-4] Silver, *8647*
[7440-23-5] Sodium, *8710*
[7440-24-6] Strontium, *8994*
[7440-25-7] Tantalum, *9223*
[7440-26-8] Technetium, *9256*
[7440-27-9] Terbium, *9300*
[7440-28-0] Thallium, *9391*
[7440-29-1] Thorium, *9514*
[7440-30-4] Thulium, *9535*
[7440-31-5] Tin, *9587*
[7440-32-6] Titanium, *9610*
[7440-33-7] Tungsten, *9945*
[7440-34-8] Actinium, *138*
[7440-35-9] Americium, *410*
[7440-36-0] Antimony, *733*
[7440-37-1] Argon, *820*
[7440-38-2] Arsenic, *832*
[7440-39-3] Barium, *991*
[7440-40-6] Berkelium, *1198*
[7440-41-7] Beryllium, *1201*
[7440-42-8] Boron, *1373*
[7440-43-9] Cadmium, *1649*
[7440-44-0] Carbon, *1855*
[7440-45-1] Cerium, *2037*
[7440-46-2] Cesium, *2051*
[7440-47-3] Chromium, *2288*
[7440-48-4] Cobalt, *2488*
[7440-50-8] Copper, *2583*
[7440-51-9] Curium, *2746*
[7440-52-0] Erbium, *3679*
[7440-53-1] Europium, *3951*
[7440-54-2] Gadolinium, *4346*
[7440-55-3] Gallium, *4364*
[7440-56-4] Germanium, *4415*
[7440-57-5] Gold, *4526*
[7440-58-6] Hafnium, *4618*
[7440-59-7] Helium, *4660*
[7440-60-0] Holmium, *4763*
[7440-61-1] Uranium, *9990*
[7440-62-2] Vanadium, *10053*
[7440-63-3] Xenon, *10206*
[7440-64-4] Ytterbium, *10239*
[7440-65-5] Yttrium, *10240*
[7440-66-6] Zinc, *10255*
[7440-67-7] Zirconium, *10306*
[7440-68-8] Astatine, *889*
[7440-69-9] Bismuth, *1297*
[7440-70-2] Calcium, *1682*
[7440-71-3] Californium, *1764*
[7440-72-4] Fermium, *4055*
[7440-73-5] Francium, *4286*
[7440-74-6] Indium, *4980*
[7446-06-2] Polonium Dioxide, *7719*
[7446-07-3] Tellurium Dioxide, *9276*
[7446-08-4] Selenium Oxide, *8576*
[7446-09-5] Sulfur Dioxide, *9144*
[7446-11-9] Sulfur Trioxide, *9152*
[7446-14-2] Lead Sulfate, *5444*
[7446-15-3] Lead Selenate, *5438*
[7446-18-6] Thallium Sulfate, *9405*
[7446-19-7] Zinc Sulfate Monohydrate*, *10293*
[7446-20-0] Zinc Sulfate Heptahydrate*, *10293*
[7446-21-1] Strontium Selenate, *9012*
[7446-22-2] Thallium Selenate, *9402*
[7446-25-5] Stannic Selenite, *8935*

[7446-26-6] Zinc Pyrophosphate, *10287*
[7446-27-7] Lead Phosphate, *5437*
[7446-28-8] Strontium Phosphate, Tribasic, *9011*
[7446-32-4] Antimony Sulfate, *745*
[7446-70-0] Aluminum Chloride, *348*
[7447-39-4] Cupric Chloride, *2699*
[7447-40-7] Potassium Chloride, *7783*
[7447-41-8] Lithium Chloride, *5553*
[7449-03-8] Erythropterin, *3730*
[7450-97-7] Oxaflumazine Disuccinate*, *7040*
[7455-39-2] Fonazine Methanesulfonate*, *4260*
[7456-24-8] Fonazine, *4260*
[7460-12-0] *d*-Pseudoephedrine Sulfate*, *3645*
[7461-02-1] Stylopine, *9025*
[7481-89-2] Zalcitabine, *10242*
[7487-81-2] Isomethadone *l*-Form Hydrochloride*, *5201*
[7487-88-9] Magnesium Sulfate, *5731*
[7487-94-7] Mercuric Chloride, *5926*
[7488-51-9] Lead Selenite, *5439*
[7488-55-3] Stannous Sulfate, *8947*
[7488-56-4] Selenium Disulfide*, *8580*
[7488-70-2] Thyroxine, *9555*
[7488-99-5] α-Carotene, *1901*
[7491-09-0] Docusate Potassium*, *3460*
[7491-74-9] Piracetam, *7641*
[7512-17-6] *N*-Acetylglucosamine*, *4466*
[7527-91-5] Acrisorcin, *128*
[7527-94-8] Alkofanone, *254*
[7535-00-4] D-Galactosamine, *4352*
[7542-37-2] Paromomycin, *7173*
[7550-35-8] Lithium Bromide, *5551*
[7550-45-0] Titanium Tetrachloride, *9618*
[7553-56-2] Iodine, *5034*
[7553-56-2] Iodine Colloidal, *5035*
[7554-65-6] Fomepizole, *4257*
[7558-79-4] Sodium Phosphate, Dibasic, *8805*
[7558-80-7] Sodium Phosphate, Monobasic, *8806*
[7563-42-0] Pentobarbital Calcium Salt*, *7272*
[7563-62-4] Oxytocin Citrate*, *7114*
[7568-93-6] Phenylethanolamine, *7441*
[7580-67-8] Lithium Hydride, *5559*
[7601-54-9] Sodium Phosphate, Tribasic, *8808*
[7601-55-0] Metocurine Iodide, *6227*
[7601-89-0] Sodium Perchlorate, *8798*
[7601-90-3] Perchloric Acid, *7296*
[7611-43-0] Lycomarasmine, *5649*
[7616-22-0] γ-Tocopherol, *9633*
[7616-94-6] Perchloryl Fluoride, *7297*
[7618-86-2] Furtrethonium, *4334*
[7631-46-1] Quinine Iodosulfate, *8246*
[7631-86-9] Silicon Dioxide, *8637*
[7631-90-5] Sodium Bisulfite, *8731*
[7631-94-9] Sodium Dithionate, *8756*
[7631-95-0] Sodium Molybdate(VI), *8790*
[7631-99-4] Sodium Nitrate, *8792*
[7632-00-0] Sodium Nitrite, *8793*
[7632-04-4] Sodium Perborate, *8797*
[7633-29-6] Psychotrine, *8115*
[7637-07-2] Boron Trifluoride, *1379*
[7642-64-0] Nandrolone Furylpropionate*, *6452*
[7646-69-7] Sodium Hydride, *8770*
[7646-78-8] Stannic Chloride, *8929*
[7646-79-9] Cobaltous Chloride, *2498*
[7646-85-7] Zinc Chloride, *10261*
[7646-93-7] Potassium Bisulfate, *7774*

Denotes derivative of title compound

[7647-01-0] Hydrochloric Acid, *4821*
[7647-01-0] Hydrogen Chloride, *4835*
[7647-10-1] Palladium Chloride, *7122*
[7647-14-5] Sodium Chloride, *8742*
[7647-15-6] Sodium Bromide, *8737*
[7647-17-8] Cesium Chloride, *2054*
[7647-18-9] Antimony Pentachloride, *736*
[7647-19-0] Phosphorus Pentafluoride, *7509*
[7654-03-7] Benmoxine, *1072*
[7658-08-4] Quinovose, *8262*
[7664-38-2] Phosphoric Acid, *7500*
[7664-39-3] Hydrofluoric Acid, *4831*
[7664-39-3] Hydrogen Fluoride, *4837*
[7664-41-7] Ammonia, *517*
[7664-93-9] Sulfuric Acid, *9147*
[7665-99-8] Cyclic GMP, *2778*
[7681-11-0] Potassium Iodide, *7809*
[7681-14-3] Prednisolone 2l-*tert*-Butyl-acetate*, *7901*
[7681-28-9] Quinidine Polygalact-uronate*, *8244*
[7681-38-1] Sodium Bisulfate, *8729*
[7681-49-4] Sodium Fluoride, *8762*
[7681-52-9] Sodium Hypochlorite, *8773*
[7681-53-0] Sodium Hypophosphite, *8775*
[7681-55-2] Sodium Iodate, *8776*
[7681-57-4] Sodium Metabisulfite, *8784*
[7681-65-4] Cuprous Iodide, *2732*
[7681-76-7] Ronidazole, *8413*
[7681-79-0] Etafedrine, *3752*
[7681-80-3] Pentapiperide Methosul-fate*, *7260*
[7681-82-5] Sodium Iodide, *8777*
[7681-93-8] Natamycin, *6513*
[7683-59-2] Isoproterenol, *5236*
[7689-03-4] Camptothecin, *1783*
[7696-12-0] Tetramethrin, *9362*
[7697-37-2] Nitric Acid, *6671*
[7697-37-2] Nitric Acid, Anhydrous, *6672*
[7699-43-6] Zirconyl Chloride, *10317*
[7699-45-8] Zinc Bromide, *10258*
[7700-17-6] Crotoxyphos, *2667*
[7704-34-9] Sulfur, *9142*
[7704-34-9] Sulfur, Pharmaceutical, *9150*
[7704-98-5] Titanium Hydride, *9613*
[7704-99-6] Zirconium Hydride, *10309*
[7705-07-9] Titanium Trichloride, *9620*
[7705-08-0] Ferric Chloride, *4061*
[7706-67-4] Dimecrotic Acid, *3248*
[7712-50-7] Myrtecaine, *6422*
[7716-60-1] Etisazol, *3918*
[7718-54-9] Nickel Chloride, *6588*
[7719-09-7] Thionyl Chloride, *9485*
[7719-12-2] Phosphorus Trichloride, *7515*
[7720-78-7] Ferrous Sulfate, *4105*
[7721-01-9] Tantalum Pentachloride, *9224*
[7722-64-7] Potassium Permanganate, *7824*
[7722-76-1] Ammonium Phosphate, Monobasic, *577*
[7722-84-1] Hydrogen Peroxide, *4839*
[7722-86-3] Caro's Acid, *1900*
[7722-88-5] Tetrasodium Pyrophos-phate, *9377*
[7723-14-0] Phosphorus, *7503*
[7724-04-1] Strophanthobiose, *9017*
[7726-95-6] Bromine, *1415*
[7727-15-3] Aluminum Bromide, *342*
[7727-18-6] Vanadyl Trichloride, *10064*

[7727-21-1] Potassium Persulfate, *7825*
[7727-37-9] Nitrogen, *6698*
[7727-43-7] Barium Sulfate, *1023*
[7727-54-0] Ammonium Peroxy-disulfate, *575*
[7732-18-5] Water, *10175*
[7733-02-0] Zinc Sulfate, *10293*
[7757-79-1] Potassium Nitrate, *7815*
[7757-82-6] Sodium Sulfate, *8829*
[7757-83-7] Sodium Sulfite, *8831*
[7757-86-0] Magnesium Phosphate, Dibasic, *5718*
[7757-87-1] Magnesium Phosphate, Tribasic, *5720*
[7757-88-2] Magnesium Sulfite, *5732*
[7757-93-9] Calcium Phosphate, Dibasic, *1739*
[7758-01-2] Potassium Bromate, *7779*
[7758-02-3] Potassium Bromide, *7780*
[7758-05-6] Potassium Iodate, *7808*
[7758-09-0] Potassium Nitrite, *7816*
[7758-11-4] Potassium Phosphate, Dibasic, *7828*
[7758-16-9] Sodium Acid Pyrophos-phate, *8713*
[7758-19-2] Sodium Chlorite, *8743*
[7758-23-8] Calcium Phosphate, Monobasic, *1740*
[7758-29-4] Sodium Tripolyphosphate, *8846*
[7758-87-4] Calcium Phosphate, Tribasic, *1741*
[7758-88-5] Cerous Fluoride, *2042*
[7758-89-6] Cuprous Chloride, *2730*
[7758-94-3] Ferrous Chloride, *4091*
[7758-95-4] Lead Chloride, *5421*
[7758-97-6] Lead Chromate(VI), *5422*
[7758-98-7] Cupric Sulfate, *2722*
[7759-00-4] Manganese Silicate, *5781*
[7759-02-6] Strontium Sulfate, *9013*
[7761-88-8] Silver Nitrate, *8661*
[7772-98-7] Sodium Thiosulfate, *8844*
[7772-99-8] Stannous Chloride, *8939*
[7773-01-5] Manganese Chloride, *5768*
[7773-06-0] Ammonium Sulfamate, *589*
[7774-29-0] Mercuric Iodide, Red, *5932*
[7775-09-9] Sodium Chlorate, *8741*
[7775-11-3] Sodium Chromate(VI), *8745*
[7775-14-6] Sodium Hydrosulfite, *8771*
[7775-19-1] Sodium Metaborate, *8785*
[7775-27-1] Sodium Persulfate, *8801*
[7775-41-9] Silver Fluoride, *8657*
[7778-18-9] Calcium Sulfate, *1753*
[7778-39-4] Arsenic Acid, *833*
[7778-43-0] Sodium Arsenate, Dibasic, *8720*
[7778-44-1] Calcium Arsenate, *1686*
[7778-50-9] Potassium Dichromate-(VI), *7790*
[7778-53-2] Potassium Phosphate, Tribasic, *7830*
[7778-54-3] Calcium Hypochlorite, *1717*
[7778-54-3] Chlorinated Lime, *2143*
[7778-74-7] Potassium Perchlorate, *7822*
[7778-77-0] Potassium Phosphate, Monobasic, *7829*
[7778-80-5] Potassium Sulfate, *7845*
[7779-88-6] Zinc Nitrate, *10273*
[7779-90-0] Zinc Phosphate, *10284*
[7782-39-0] Deuterium, *2983*
[7782-40-3] Diamond, *3030*
[7782-41-4] Fluorine, *4199*
[7782-42-5] Graphite, *4560*

[7782-44-7] Oxygen, *7098*
[7782-49-2] Selenium, *8572*
[7782-50-5] Chlorine, *2145*
[7782-63-0] Ferrous Sulfate Hepta-hydrate*, *4105*
[7782-64-1] Manganese Difluoride, *5769*
[7782-65-2] Germane, *4414*
[7782-68-5] Iodic Acid, *5032*
[7782-76-5] Manganese Phosphate, Dibasic, *5777*
[7782-77-6] Nitrous Acid, *6750*
[7782-78-7] Nitrosylsulfuric Acid, *6746*
[7782-79-8] Hydrazoic Acid, *4815*
[7782-87-8] Potassium Hypophos-phite, *7807*
[7782-89-0] Lithium Amide, *5546*
[7782-91-4] Molybdic(VI) Acid, *6322*
[7782-92-5] Sodium Amide, *8717*
[7782-94-7] Nitramide, *6663*
[7782-95-8] Disodium Dihydrogen Hypophosphate, *3420*
[7782-99-2] Sulfurous Acid, *9149*
[7783-00-8] Selenious Acid, *8571*
[7783-03-1] Tungstic(VI) Acid, *9948*
[7783-05-3] Pyrosulfuric Acid, *8192*
[7783-06-4] Hydrogen Sulfide, *4843*
[7783-07-5] Hydrogen Selenide, *4842*
[7783-08-6] Selenic Acid, *8570*
[7783-09-7] Hydrogen Telluride, *4844*
[7783-18-8] Ammonium Thiosulfate, *598*
[7783-19-9] Ammonium Selenite, *586*
[7783-20-2] Ammonium Sulfate, *590*
[7783-21-3] Ammonium Selenate, *585*
[7783-22-4] Ammonium Uranate(VI), *601*
[7783-26-8] Trisilane, *9884*
[7783-28-0] Ammonium Phosphate, Dibasic, *576*
[7783-29-1] Tetrasilane, *9376*
[7783-32-6] Mercuric Iodate, *5931*
[7783-33-7] Potassium Tetraiodomer-curate(II), *7862*
[7783-35-9] Mercuric Sulfate, *5943*
[7783-36-0] Mercurous Sulfate, *5956*
[7783-39-3] Mercuric Fluoride, *5930*
[7783-40-6] Magnesium Fluoride, *5701*
[7783-41-7] Fluorine Monoxide, *4201*
[7783-42-8] Thionyl Fluoride, *9486*
[7783-43-9] Selenium Oxyfluoride, *8579*
[7783-44-0] Fluorine Dioxide, *4200*
[7783-46-2] Lead Fluoride, *5425*
[7783-47-3] Stannous Fluoride, *8940*
[7783-48-4] Strontium Fluoride, *9002*
[7783-49-5] Zinc Fluoride, *10265*
[7783-50-0] Ferric Fluoride, *4065*
[7783-51-9] Gallium Trifluoride, *4367*
[7783-52-0] Indium Trifluoride, *4990*
[7783-53-1] Manganese Trifluoride, *5784*
[7783-54-2] Nitrogen Fluoride, *6701*
[7783-55-3] Phosphorus Trifluoride, *7516*
[7783-56-4] Antimony Trifluoride, *750*
[7783-57-5] Thallium Trifluoride, *9407*
[7783-58-6] Germanium Tetrafluoride, *4419*
[7783-59-7] Lead Tetrafluoride, *5448*
[7783-60-0] Sulfur Tetrafluoride, *9151*
[7783-61-1] Silicon Tetrafluoride, *8645*
[7783-62-2] Stannic Fluoride, *8931*
[7783-63-3] Titanium Tetrafluoride, *9619*
[7783-64-4] Zirconium Fluoride, *10308*
[7783-66-6] Iodine Pentafluoride, *5039*

[7783-68-8] Niobium Pentafluoride, 6650
[7783-70-2] Antimony Pentafluoride, 737
[7783-71-3] Tantalum Pentafluoride, 9225
[7783-72-4] Vanadium Pentafluoride, 10055
[7783-75-7] Iridium Hexafluoride, 5099
[7783-77-9] Molybdenum Hexafluoride, 6319
[7783-79-1] Selenium Hexafluoride, 8575
[7783-80-4] Tellurium Hexafluoride, 9277
[7783-81-5] Uranium Hexafluoride, 9992
[7783-82-6] Tungsten Hexafluoride, 9946
[7783-84-8] Ferric Hypophosphite, 4069
[7783-86-0] Ferrous Iodide, 4097
[7783-90-6] Silver Chloride, 8652
[7783-92-8] Silver Chlorate, 8651
[7783-93-9] Silver Perchlorate, 8667
[7783-95-1] Silver Difluoride, 8656
[7783-96-2] Silver Iodide, 8659
[7783-97-3] Silver Iodate, 8658
[7783-98-4] Silver Permanganate, 8668
[7783-99-5] Silver Nitrite, 8663
[7784-01-2] Silver Chromate(VI), 8653
[7784-03-4] Silver Tetraiodomercurate(II), 8678
[7784-05-6] Silver Selenite, 8674
[7784-07-8] Silver Selenate, 8672
[7784-09-0] Silver Phosphate, 8669
[7784-13-6] Aluminum Chloride Hexahydrate*, 348
[7784-14-7] Ammonium Tetrachloroaluminate, 595
[7784-15-8] Aluminum Chlorate Nonahydrate*, 347
[7784-16-9] Sodium Tetrachloroaluminate, 8836
[7784-18-1] Aluminum Fluoride, 351
[7784-19-2] Ammonium Hexafluoroaluminate, 555
[7784-21-6] Aluminum Hydride, 354
[7784-22-7] Aluminum Hypophosphite, 357
[7784-23-8] Aluminum Iodide, 358
[7784-25-0] Aluminum Ammonium Sulfate, 335
[7784-30-7] Aluminum Phosphate, 371
[7784-33-0] Arsenic Tribromide, 840
[7784-34-1] Arsenic Trichloride, 841
[7784-35-2] Arsenic Trifluoride, 842
[7784-36-3] Arsenic Pentafluoride, 836
[7784-37-4] Mercuric Arsenate, 5923
[7784-41-0] Potassium Arsenate, 7767
[7784-42-1] Arsine, 849
[7784-45-4] Arsenic Triiodide, 843
[7784-46-5] Sodium Arsenite, 8721
[7785-23-1] Silver Bromide, 8649
[7785-24-2] Cobaltous Arsenate, 2495
[7785-84-4] Sodium Trimetaphosphate, 8845
[7785-87-7] Manganese Sulfate, 5782
[7786-30-3] Magnesium Chloride, 5698
[7786-34-7] Mevinphos, 6252
[7786-81-4] Nickel Sulfate, 6600
[7787-32-8] Barium Fluoride, 1003
[7787-35-1] Barium Manganate(VI), 1010
[7787-36-2] Barium Permanganate, 1017
[7787-39-5] Barium Sulfite, 1026
[7787-46-4] Beryllium Bromide, 1206

[7787-47-5] Beryllium Chloride, 1208
[7787-49-7] Beryllium Fluoride, 1209
[7787-50-0] Beryllium Potassium Fluoride, 1218
[7787-52-2] Beryllium Hydride, 1211
[7787-53-3] Beryllium Iodide, 1213
[7787-57-7] Bismuth Bromide Oxide, 1300
[7787-58-8] Bismuth Bromide, 1299
[7787-59-9] Bismuth Chloride Oxide, 1303
[7787-60-2] Bismuth Chloride, 1302
[7787-61-3] Bismuth Fluoride, 1305
[7787-62-4] Bismuth Pentafluoride, 1315
[7787-63-5] Bismuth Iodide Oxide, 1309
[7787-64-6] Bismuth Iodide, 1308
[7787-68-0] Bismuth Sulfate, 1328
[7787-69-1] Cesium Bromide, 2052
[7787-70-4] Cuprous Bromide, 2729
[7787-71-5] Bromine Trifluoride, 1417
[7788-96-7] Chromyl Fluoride, 2307
[7788-97-8] Chromic Fluoride, 2279
[7788-98-9] Ammonium Chromate(VI), 538
[7789-00-6] Potassium Chromate(VI), 7784
[7789-04-0] Chromic Phosphate, 2284
[7789-06-2] Strontium Chromate(VI), 9001
[7789-08-4] Ammonium Ferric Chromate, 546
[7789-09-5] Ammonium Dichromate(VI), 544
[7789-10-8] Mercuric Dichromate(VI), 5929
[7789-17-5] Cesium Iodide, 2056
[7789-18-6] Cesium Nitrate, 2057
[7789-19-7] Cupric Fluoride, 2704
[7789-20-0] Deuterium Oxide, 2984
[7789-21-1] Fluorosulfonic Acid, 4217
[7789-23-3] Potassium Fluoride, 7794
[7789-24-4] Lithium Fluoride, 5557
[7789-25-5] Nitrosyl Fluoride, 6745
[7789-26-6] Fluorine Nitrate, 4202
[7789-27-7] Thallium Fluoride, 9397
[7789-28-8] Ferrous Fluoride, 4093
[7789-29-9] Potassium Bifluoride, 7771
[7789-30-2] Bromine Pentafluoride, 1416
[7789-31-3] Bromic Acid, 1413
[7789-33-5] Iodine Monobromide, 5037
[7789-36-8] Magnesium Bromate, 5694
[7789-38-0] Sodium Bromate, 8736
[7789-39-1] Rubidium Bromide, 8440
[7789-40-4] Thallium Bromide, 9393
[7789-41-5] Calcium Bromide, 1693
[7789-42-6] Cadmium Bromide, 1651
[7789-43-7] Cobaltous Bromide, 2496
[7789-45-9] Cupric Bromide, 2695
[7789-46-0] Ferrous Bromide, 4088
[7789-47-1] Mercuric Bromide, 5925
[7789-48-2] Magnesium Bromide, 5695
[7789-51-7] Selenium Oxybromide, 8577
[7789-52-8] Selenium Bromide, 8573
[7789-54-0] Tellurium Dibromide, 9274
[7789-57-3] Tribromosilane, 9746
[7789-59-5] Phosphorus Oxybromide, 7505
[7789-60-8] Phosphorus Tribromide, 7514
[7789-61-9] Antimony Tribromide, 747

[7789-65-3] Selenium Tetrabromide, 8581
[7789-66-4] Silicon Tetrabromide, 8643
[7789-67-5] Stannic Bromide, 8928
[7789-68-6] Titanium Tetrabromide, 9617
[7789-69-7] Phosphorus Pentabromide, 7507
[7789-74-4] Calcium Fluorophosphate, 1710
[7789-75-5] Calcium Fluoride, 1709
[7789-78-8] Calcium Hydride, 1715
[7789-79-9] Calcium Hypophosphite, 1718
[7789-80-2] Calcium Iodate, 1719
[7789-82-4] Calcium Molybdate(VI), 1728
[7790-21-8] Potassium Periodate, 7823
[7790-26-3] Sodium Iodide, Radioactive, 8778
[7790-28-5] Sodium Metaperiodate, 8786
[7790-29-6] Rubidium Iodide, 8443
[7790-30-9] Thallium Iodide, 9399
[7790-33-2] Manganese Iodide, 5772
[7790-37-6] Zinc Iodate, 10268
[7790-39-8] Platinous Iodide, 7688
[7790-42-3] Potassium Triiodide, 7869
[7790-43-4] Potassium Triiodozincate, 7871
[7790-44-5] Antimony Triiodide, 751
[7790-46-7] Platinic Iodide, 7685
[7790-47-8] Stannic Iodide, 8932
[7790-48-9] Tellurium Tetraiodide, 9280
[7790-49-0] Thorium Iodide, 9516
[7790-58-1] Potassium Tellurate(IV), 7850
[7790-59-2] Potassium Selenate, 7836
[7790-60-5] Potassium Tungstate(VI), 7872
[7790-62-7] Potassium Pyrosulfate, 7834
[7790-63-8] Potassium Uranate(VI), 7873
[7790-69-4] Lithium Nitrate, 5562
[7790-75-2] Calcium Tungstate(VI), 1760
[7790-79-6] Cadmium Fluoride, 1655
[7790-80-9] Cadmium Iodide, 1657
[7790-85-4] Cadmium Tungstate(VI), 1668
[7790-86-5] Cerous Chloride, 2041
[7790-87-6] Cerous Iodide, 2043
[7790-89-8] Chlorine Monofluoride, 2148
[7790-91-2] Chlorine Trifluoride, 2150
[7790-92-3] Hypochlorous Acid, 4912
[7790-93-4] Chloric Acid, 2141
[7790-94-5] Chlorosulfonic Acid, 2218
[7790-95-6] Chloroselenic Acid, 2216
[7790-98-9] Ammonium Perchlorate, 574
[7790-99-0] Iodine Monochloride, 5038
[7791-03-9] Lithium Perchlorate, 5565
[7791-08-4] Antimony Chloride Oxide, 734
[7791-09-5] Rhenium Oxychloride, 8346
[7791-10-8] Strontium Chlorate, 8999
[7791-11-9] Rubidium Chloride, 8441
[7791-12-0] Thallium Chloride, 9395
[7791-16-4] Antimony Dichlorotrifluoride, 735
[7791-21-1] Chlorine Monoxide, 2149
[7791-23-3] Selenium Oxychloride, 8578
[7791-25-5] Sulfuryl Chloride, 9153

Denotes derivative of title compound

[7791-26-6] Uranyl Chloride, *9999*
[7791-27-7] Pyrosulfuryl Chloride, *8193*
[7791-29-9] Potassium Tetraiodoaurate(III), *7860*
[7798-23-4] Cupric Phosphate, *2716*
[7803-49-8] Hydroxylamine, *4874*
[7803-51-2] Phosphine, *7494*
[7803-52-3] Stibine, *8965*
[7803-54-5] Magnesium Amide, *5690*
[7803-55-6] Ammonium Vanadate(V), *605*
[7803-57-8] Hydrazine Hydrate, *4810*
[7803-58-9] Sulfamide, *9091*
[7803-60-3] Hypophosphoric Acid, *4915*
[7803-62-5] Silane, *8633*
[7803-63-6] Ammonium Bisulfate, *526*
[7803-65-8] Ammonium Hypophosphite, *559*
[7803-68-1] Telluric(VI) Acid, *9272*
[8000-10-0] Theophylline Sodium Glycinate*, *9421*
[8000-61-1] Aluminum Subacetate Solution, *380*
[8001-08-9] Acerin, *33*
[8001-10-3] Ferrous Carbonate Saccharated, *4090*
[8001-11-4] Ferric Albuminate, *4058*
[8001-27-2] Hirudin, *4754*
[8001-35-2] Toxaphene, *9693*
[8001-39-6] Japan Wax, *5270*
[8001-40-9] Iodized Oil, *5045*
[8001-50-1] Strobane, *8993*
[8001-58-9] Creosote, Coal Tar, *2641*
[8001-60-3] Coparaffinate, *2582*
[8001-61-4] Copaiba, *2580*
[8001-75-0] Ceresin, *2033*
[8001-83-0] Creosote, Wood, Calcium deriv.*, *2642*
[8001-95-4] Alseroxylon*, *8293*
[8002-36-6] Ustilagic Acid, *10032*
[8002-43-5] Lecithin, *5452*
[8002-53-7] Montan Wax, *6339*
[8002-55-9] Myrtol, *6423*
[8002-66-2] German Chamomile Oil*, *2082*
[8002-76-4] Papaveretum, *7150*
[8002-88-8] Theobromine Sodium Acetate*, *9418*
[8002-89-9] Theophylline Sodium Acetate*, *9421*
[8003-05-2] Phenylmercuric Nitrate, Basic, *7455*
[8003-22-3] Quinoline Yellow Spirit Soluble, *8257*
[8004-38-4] Fenbutrazate Hydrochloride Mixture with Phenmetrazine Theoclate*, *4005*
[8004-92-0] Quinoline Yellow, *8256*
[8005-78-5] Bismark Brown R, *1293*
[8006-08-4] Ergotinine, *3706*
[8006-13-1] Aluminum Acetate Solution, *332*
[8006-25-5] Ergotoxine, *3707*
[8006-27-7] Ferric and Ammonium Acetate Solution, *4059*
[8006-28-8] Soda Lime, *8709*
[8006-35-7] Diamond Ink, *3031*
[8006-37-9] Turks Island Salt, *9954*
[8006-44-8] Candelilla Wax, *1787*
[8006-91-5] Dextri-Maltose, *2994*
[8007-14-5] Argol, *819*
[8007-31-6] Silver Nitrate, Toughened, *8662*
[8007-49-6] Carlsbad Salt Artificial, *1890*
[8007-50-9] Nauheim Salts (Artificial), *6515*

[8007-56-5] Nitrohydrochloric Acid, *6706*
[8008-20-6] Kerosene, *5305*
[8008-53-5] Ethiodized Oil, *3781*
[8011-44-7] Aluminum Boroformate, *340*
[8011-61-8] Tyrocidine, *9967*
[8011-62-9] Barium Sulfide, Black, *1025*
[8011-65-2] Zinc Tannate, *10295*
[8011-82-3] Calomelol, *1768*
[8012-34-8] Mercamphamide Mixture of Sodium Salt with Theophylline*, *5915*
[8012-54-2] Donovan's Solution, *3477*
[8012-89-3] Beeswax, *1049*
[8013-17-0] Invert Sugar, *5026*
[8013-43-2] Spirit of Ether, *8909*
[8013-44-3] Spirit of Ether Compound, *8910*
[8013-59-0] Spirit of Ammonia, Aromatic, *8906*
[8013-61-4] Ammonium Acetate Solution*, *520*
[8013-88-5] Calcium Cyanamide Citrated, *1703*
[8013-96-5] Dymixal, *3526*
[8014-10-6] Terebene, *9304*
[8014-63-9] Seidlitz Mixture, *8567*
[8014-95-7] Sulfuric Acid, Fuming*, *9147*
[8015-14-3] Lynestrenol Mixture with Mestranol*, *5659*
[8015-18-7] Pfeiffer's Substance, *7334*
[8015-19-8] Dimethisterone Mixture with Ethinyl Estradiol*, *3268*
[8015-30-3] Norethynodrel Mixture with Mestranol*, *6791*
[8015-51-8] Cropropamide Combination with Crotethamide*, *2659*
[8015-86-9] Carnauba Wax, *1895*
[8015-92-7] Roman Chamomile Oil*, *2082*
[8018-01-7] Mancozeb, *5756*
[8018-15-3] Mercumallylic Acid Sodium Salt Compd with Theophylline*, *5921*
[8021-39-4] Creosote, Wood, *2642*
[8022-00-2] Methyl Demeton, *6129*
[8023-62-9] Storax, *8974*
[8025-81-8] Spiramycin, *8904*
[8027-28-9] Sodium Phosphate, Radioactive, *8807*
[8028-98-6] Acetulan, *80*
[8029-68-3] Ichthammol, *4929*
[8030-30-6] Petroleum Benzin, *7329*
[8030-35-1] Ferrous Carbonate Mass, *4089*
[8030-97-5] Pyroligneous Acid, *8187*
[8031-14-9] Oxychlorosene, *7090*
[8031-21-8] Napalm, *6453*
[8031-66-1] Osmaron B, *7020*
[8032-32-4] Ligroin, *5513*
[8038-65-1] Ambergris, *398*
[8047-13-0] Ryania, *8458*
[8047-26-5] Uglow Black Silver, *9976*
[8047-67-4] Ferric Oxide, Saccharated, *4073*
[8048-16-6] Streptodornase Mixture with Streptokinase*, *8979*
[8048-31-5] Theobromine Sodium Salicylate*, *9418*
[8048-52-0] Acriflavine, *125*
[8049-11-4] Devarda's Metal, *2985*
[8049-20-5] Misch Metal, *6296*
[8049-21-6] Hydrone, *4848*
[8049-47-6] Pancreatin, *7137*
[8049-62-5] Zinc Insulin*, *5011*
[8049-99-8] Milorganite, *6282*
[8050-07-5] Olibanum, *6969*

[8050-81-5] Simethicone*, *3264*
[8050-88-2] Celluloid, *2011*
[8051-02-3] Sabadilla, *8460*
[8052-16-2] Cactinomycin, *1642*
[8052-27-5] Kino, *5325*
[8053-05-2] Yellow Phenolphthalein, *10230*
[8063-06-7] Curare, *2741*
[8063-07-8] Kanamycin, *5293*
[8063-16-9] Psyllium Seed Gum*, *7674*
[8063-24-9] Acriflavine Dihydrochloride*, *125*
[8064-12-8] Meglumine Diatrizoate Mixture with Sodium Diatrizoate*, *5851*
[8064-51-5] Megestrol Acetate Mixture with Mestranol*, *5849*
[8064-66-2] Megestrol Acetate Mixture with Ethinyl Estradiol*, *5849*
[8064-73-1] Tioxolone Combination with Hydrocortisone*, *9600*
[8064-75-3] Norgesterone Mixture with Ethinyl Estradiol*, *6795*
[8064-76-4] Lynestrenol Mixture with Ethinyl Estradiol*, *5659*
[8064-90-2] Sulfamethoxazole Mixture with Trimethoprim*, *9086*
[8065-36-9] Bufencarb, *1494*
[8065-48-3] Demeton, *2940*
[8065-51-8] Theobromine Calcium Salicylate*, *9418*
[8065-91-6] Chlormadinone Acetate Mixture with Mestranol*, *2152*
[8067-24-1] Ergoloid Mesylates, *3692*
[8067-69-4] Halquinol, *4637*
[8067-82-1] Alfaxalone Mixture with Alfadolone Acetate*, *235*
[8068-28-8] Colistin Sodium Methanesulfonate*, *2542*
[8069-64-5] Meralluride, *5913*
[8075-95-4] Clofibrate Mixture with Androsterone*, *2436*
[8075-98-7] Ascorbigen, *868*
[9000-01-5] Acacia, *11*
[9000-02-6] Amber, *397*
[9000-03-7] Ammoniacum, *518*
[9000-04-8] Asafetida, *858*
[9000-05-9] Gum Benzoin, *4608*
[9000-07-1] Carrageenan, *1914*
[9000-14-0] Copal, *2581*
[9000-16-2] Damar, *2872*
[9000-20-8] Eucalyptus Gum, *3942*
[9000-21-9] Furcellaran, *4322*
[9000-25-3] Gamboge, *4372*
[9000-28-6] Ghatti Gum, *4424*
[9000-29-7] Guaiac, *4573*
[9000-30-0] Guaran, *4601*
[9000-30-0] Guar Gum, *4602*
[9000-32-2] Gutta-Percha, *4611*
[9000-34-4] Resin Ipomea, *8317*
[9000-35-5] Resin Jalap, *8318*
[9000-36-6] Karaya Gum, *5296*
[9000-45-7] Myrrh, *6421*
[9000-47-9] Mesquite Gum, *5972*
[9000-58-2] Resin Scammony, *8319*
[9000-59-3] Shellac, *8623*
[9000-65-1] Gum Tragacanth, *4609*
[9000-69-5] Pectin, *7194*
[9000-70-8] Gelatin, *4388*
[9000-71-9] Casein, *1934*
[9000-85-5] α-Amylase (Bacterial)*, *640*
[9000-88-8] D-Amino Acid Oxidase, *434*
[9000-89-9] L-Amino Acid Oxidase, *435*
[9000-90-2] α-Amylase (Porcine)*, *640*
[9000-91-3] β-Amylase (Sweet Potato)*, *640*

* Denotes derivative of title compound

[9000-92-4] Amylase, *640*
[9000-95-7] Apyrase, *797*
[9001-00-7] Bromelain, *1409*
[9001-01-8] Kallikrein, *5291*
[9001-03-0] Carbonic Anhydrase, *1860*
[9001-04-1] Pyruvate Decarboxylase, *8204*
[9001-05-2] Catalase, *1948*
[9001-08-5] Choline Esterase, *2264*
[9001-09-6] Chymopapain, *2319*
[9001-11-0] Clarase, *2399*
[9001-12-1] Collagenase, *2544*
[9001-19-8] Taka-Diastase, *9203*
[9001-24-5] Factor V, *3961*
[9001-25-6] Factor VII, *3962*
[9001-26-7] Prothrombin, *8074*
[9001-27-8] Factor VIII, *3963*
[9001-28-9] Factor IX, *3964*
[9001-29-0] Factor X, *3965*
[9001-30-3] Factor XII, *3967*
[9001-31-4] Fibrin, *4115*
[9001-32-5] Fibrinogen, *4116*
[9001-33-6] Ficin, *4121*
[9001-37-0] Glucose Oxidase, *4468*
[9001-45-0] β-Glucuronidase, *4475*
[9001-53-0] Histaminase, *4755*
[9001-54-1] Hyaluronidases, *4794*
[9001-57-4] Invertase, *5025*
[9001-60-9] Lactate Dehydrogenase, *5347*
[9001-62-1] Lipase, *5536*
[9001-63-2] Lysozyme, *5671*
[9001-66-5] Monoamine Oxidase, *6330*
[9001-68-7] Old Yellow Enzyme, *6961*
[9001-72-3] Pankrin, *7142*
[9001-73-4] Papain, *7148*
[9001-74-5] Penicillinase, *7219*
[9001-75-6] Pepsin, *7289*
[9001-90-5] Plasmin, *7678*
[9001-91-6] Plasminogen, *7679*
[9001-97-2] Q-Enzyme®, *8207*
[9001-98-3] Rennin, *8303*
[9001-99-4] Ribonuclease, *8369*
[9002-01-1] Streptokinase, *8981*
[9002-04-4] Thrombin, *9525*
[9002-07-7] Trypsin, *9926*
[9002-10-2] Tyrosinase, *9969*
[9002-12-4] Uricase, *10015*
[9002-13-5] Urease, *10009*
[9002-18-0] Agar, *182*
[9002-60-2] ACTH, *136*
[9002-61-3] Human Chorionic Gonadotropin*, *2273*
[9002-62-4] Prolactin, *7961*
[9002-64-6] Parathyroid Hormone, *7168*
[9002-67-9] LH, *5499*
[9002-68-0] FSH, *4299*
[9002-69-1] Relaxin, *8299*
[9002-70-4] Equine Chorionic Gonadotropin*, *2273*
[9002-71-5] TSH, *9931*
[9002-72-6] Somatotropin, *8864*
[9002-84-0] Polytetrafluoroethylene, *7743*
[9002-86-2] Polyvinyl Chloride, *7746*
[9002-88-4] Polyethylene, *7728*
[9002-89-5] Polyvinyl Alcohol, *7745*
[9002-91-9] Metaldehyde, *5983*
[9002-92-0] Polidocanol, *7717*
[9002-93-1] Octoxynol, *6858*
[9003-07-0] Polypropylene, *7741*
[9003-11-6] Poloxalene, *7721*
[9003-24-1] Darvan, *2886*
[9003-39-8] Povidone, *7879*
[9003-59-2] Sodium Polystyrene Sulfonate, *8815*
[9003-97-8] Calcium Polycarbophil, *1744*
[9003-98-9] Deoxyribonuclease I, *2953*

[9004-02-8] Lipoprotein Lipase, *5537*
[9004-06-2] Elastase, *3576*
[9004-07-3] α-Chymotrypsin*, *2320*
[9004-10-8] Insulin, *5011*
[9004-17-5] Protamine Zinc Insulin*, *5011*
[9004-32-4] Carboxymethylcellulose Sodium, *1877*
[9004-34-6] Cellulose, *2012*
[9004-53-9] Dextrin, *2995*
[9004-54-0] Dextran, *2989*
[9004-57-3] Ethyl Cellulose, *3828*
[9004-58-4] Cellulose Ethyl Hydroxyethyl Ether, *2014*
[9004-61-9] Hyaluronic Acid, *4793*
[9004-62-0] Hetastarch, *4707*
[9004-64-2] Hydroxypropyl Cellulose, *4888*
[9004-65-3] Hydroxypropyl Methylcellulose, *4889*
[9004-67-5] Methylcellulose, *6120*
[9004-70-0] Pyroxylin, *8195*
[9005-32-7] Alginic Acid, *241*
[9005-35-0] Alginic Acid Calcium Salt*, *241*
[9005-36-1] Alginic Acid Potassium Salt*, *241*
[9005-38-3] Algin, *240*
[9005-48-5] Heparin Potassium Salt*, *4685*
[9005-49-6] Heparin, *4685*
[9005-53-2] Lignin, *5510*
[9005-65-6] Polysorbate 80*, *7742*
[9005-79-2] Glycogen, *4506*
[9005-80-5] Inulin, *5024*
[9005-84-9] Starch, Soluble, *8955*
[9005-97-4] Thyroprotein, *9554*
[9006-58-0] Lysalbinic Acid, *5662*
[9006-65-9] Dimethicone, *3264*
[9007-03-8] Bifidus Factor, *1258*
[9007-12-9] Calcitonin, *1680*
[9007-20-9] Carboxypolymethylene, *1878*
[9007-28-7] Chondroitin Sulfate, *2270*
[9007-31-2] Clupeine, *2485*
[9007-40-3] Crotoxin, *2666*
[9007-41-4] CRP, *2670*
[9007-43-6] Cytochrome c, *2860*
[9007-57-2] Edestin, *3554*
[9007-67-4] Enterogastrone, *3632*
[9007-73-2] Ferritin, *4083*
[9007-76-5] Fibroin, *4118*
[9007-90-3] Gliadin, *4446*
[9007-92-5] Glucagon, *4455*
[9007-93-6] Glycinin, *4502*
[9008-08-6] Hypalon, *4909*
[9008-19-9] Krebiozen, *5334*
[9008-22-4] Laminaran, *5363*
[9008-96-2] Phosvitin, *7522*
[9008-99-5] Piromen, *7659*
[9009-29-4] Polyferose, *7731*
[9009-58-9] Porphyropsin, *7761*
[9009-72-7] Purothionin, *8131*
[9009-81-8] Rhodopsin, *8359*
[9009-86-3] Ricin, *8376*
[9010-01-9] Sodium Amylosulfate, *8718*
[9010-29-1] Tannoform, *9222*
[9010-30-4] Taraxein, *9230*
[9010-71-3] Zinc Iodide-Starch, *10270*
[9010-72-4] Zymosan, *10330*
[9011-02-3] Negatol, *6531*
[9011-05-6] Polynoxylin, *7735*
[9011-05-6] Uramite, *9988*
[9011-18-1] Dextran Sulfate Sodium, *2993*
[9011-93-2] Lysostaphin, *5670*
[9011-97-6] Cholecystokinin, *2253*
[9012-13-9] Resodec, *8321*
[9013-42-7] Ebimar, *3535*

[9013-55-2] Factor XI, *3966*
[9013-56-3] Factor XIII, *3968*
[9013-83-6] Insulinase, *5012*
[9014-02-2] Zinostatin, *10302*
[9014-44-2] Toxohormone, *9696*
[9014-67-9] Aloxiprin, *316*
[9015-13-8] ECTEOLA-Cellulose, *3551*
[9015-51-4] Silver Protein, *8671*
[9015-68-3] L-Asparaginase, *871*
[9015-71-8] CRF, *2653*
[9015-73-0] Detaxtran, *2980*
[9015-94-5] Renin, *8302*
[9025-41-6] Keratinase, *5303*
[9025-70-1] Dextranase, *2990*
[9031-37-2] Ceruloplasmin, *2049*
[9034-40-6] LH-RH, *5500*
[9034-50-8] Vasopressin, *10073*
[9035-55-6] Lipotropic Hormone, *5539*
[9035-58-9] Thromboplastin, *9527*
[9035-68-1] Proinsulin, *7960*
[9037-22-3] Amioca, *503*
[9038-41-9] Sodium Cellulose Phosphate, *8740*
[9039-53-6] Urokinase, *10024*
[9039-61-6] Batroxobin, *1038*
[9040-14-6] Vernamycin B, *10096*
[9041-08-1] Heparin Sodium Salt*, *4685*
[9041-92-3] α₁-Antitrypsin, *760*
[9041-93-4] Bleomycins Sulfate*, *1351*
[9044-70-6] Coherin, *2534*
[9046-56-4] Ancrod, *671*
[9048-46-8] Serum Albumin, *8613*
[9050-67-3] Sizofiran, *8699*
[9054-89-1] Superoxide Dismutase, *9177*
[9056-49-9] Villikinin, *10115*
[9061-61-4] Nerve Growth Factor, *6562*
[9063-57-4] Tuftsin, *9941*
[9064-91-9] Detaxtran Hydrochloride*, *2980*
[9064-92-0] Polidexide, *7716*
[9066-59-5] Lysozyme Hydrochloride*, *5671*
[9067-32-7] Hyaluronic Acid Sodium Salt*, *4793*
[9073-78-3] Thermolysin, *9422*
[9079-25-8] Amberlite, *399*
[9083-38-9] Melanostatin, *5855*
[9087-70-1] Aprotinin, *796*
[10001-43-1] Pimefylline, *7584*
[10004-44-1] Hymexazol, *4905*
[10010-67-0] PIPES Monosodium Salt*, *7633*
[10017-44-4] Carnitine D-Form Hydrochloride*, *1898*
[10018-19-6] Cotarnine Chloride*, *2618*
[10022-31-8] Barium Nitrate, *1012*
[10022-50-1] Nitryl Fluoride, *6756*
[10023-07-1] Frenolicin, *4292*
[10024-97-2] Nitrous Oxide, *6751*
[10025-65-7] Platinous Chloride, *7687*
[10025-66-8] Radium Chloride*, *8277*
[10025-67-9] Sulfur Chloride, *9143*
[10025-68-0] Selenium Chloride, *8574*
[10025-71-5] Tellurium Dichloride, *9275*
[10025-73-7] Chromic Chloride, *2278*
[10025-78-2] Trichlorosilane, *9776*
[10025-82-8] Indium Trichloride, *4989*
[10025-83-9] Iridium Trichloride, *5101*
[10025-85-1] Nitrogen Chloride, *6699*
[10025-87-3] Phosphorus Oxychloride, *7506*
[10025-91-9] Antimony Trichloride *see* 739 and 740

Denotes derivative of title compound

[10025-93-1] Uranium Trichloride, 9996
[10025-99-7] Potassium Tetrachloro-platinate(II), 7855
[10026-01-4] Osmium Tetrachloride, 7023
[10026-02-5] Polonium Tetrachloride, 7720
[10026-03-6] Selenium Tetrachloride, 8582
[10026-04-7] Silicon Tetrachloride, 8644
[10026-07-0] Tellurium Tetrachloride, 9279
[10026-08-1] Thorium Chloride, 9515
[10026-10-5] Uranium Tetrachloride, 9994
[10026-11-6] Zirconium Chloride, 10307
[10026-12-7] Niobium Pentachloride, 6649
[10026-13-8] Phosphorus Pentachlo-ride, 7508
[10026-17-2] Cobaltous Fluoride, 2501
[10026-18-3] Cobaltic Fluoride, 2491
[10028-14-5] Nobelium, 6762
[10028-15-6] Ozone, 7116
[10028-17-8] Tritium, 9891
[10028-18-9] Nickel Fluoride, 6591
[10028-22-5] Ferric Sulfate, 4079
[10028-26-9] Magnesium Bisulfate, 5692
[10030-90-7] Ferrous Succinate, 4104
[10031-13-7] Lead Arsenite, 5414
[10031-18-2] Mercurous Bromide, 5949
[10031-22-8] Lead Bromide, 5418
[10031-23-9] Radium Bromide*, 8277
[10031-24-0] Stannous Bromide, 8938
[10031-25-1] Chromic Bromide, 2276
[10031-26-2] Ferric Bromide, 4060
[10031-27-3] Tellurium Tetrabromide, 9278
[10034-81-8] Magnesium Perchlorate, 5715
[10034-85-2] Hydriodic Acid, 4817
[10034-85-2] Hydrogen Iodide, 4838
[10034-92-1] Trihydrazine Dihydrio-dide, 9824
[10034-93-2] Hydrazine Sulfate, 4811
[10034-94-3] Magnesium Orthosili-cate*, 5727
[10034-99-8] Magnesium Sulfate Heptahydrate*, 5731
[10035-10-6] Hydrobromic Acid, 4819
[10035-10-6] Hydrogen Bromide, 4834
[10038-98-9] Germanium Tetra-chloride, 4418
[10039-31-3] Beryllium Selenate, 1220
[10039-53-9] Sodium Chromate(VI) ^{51}Cr-labeled Compound*, 8745
[10039-54-0] Hydroxylamine Sulfate*, 4874
[10040-45-6] Picosulfate Sodium, 7559
[10042-76-9] Strontium Nitrate, 9007
[10043-01-3] Aluminum Sulfate, 381
[10043-11-5] Boron Nitride, 1376
[10043-35-3] Boric Acid, 1364
[10043-49-9] Gold, Radioactive, Colloidal, 4533
[10043-52-4] Calcium Chloride, 1699
[10043-67-1] Aluminum Potassium Sulfate, 373
[10043-84-2] Manganese Hypophos-phite, 5771
[10043-92-2] Radon, 8278
[10045-86-0] Ferric Phosphate, 4074
[10045-87-1] Ferric Sodium Pyrophos-phate, 4077

[10045-89-3] Ammonium Ferrous Sulfate, 552
[10045-94-0] Mercuric Nitrate, 5933
[10048-32-5] Parasorbic Acid, 7166
[10048-98-3] Barium Phosphate, Dibasic, 1019
[10048-99-4] Barium Mercuric Iodide, 1011
[10049-01-1] Bismuth Phosphate, 1316
[10049-03-3] Fluorine Perchlorate, 4203
[10049-04-4] Chlorine Dioxide, 2146
[10049-05-5] Chromous Chloride, 2301
[10049-06-6] Titanium Dichloride, 9611
[10049-07-7] Rhodium Chloride, 8355
[10049-08-8] Ruthenium Trichloride, 8455
[10049-10-2] Chromous Fluoride, 2302
[10049-11-3] Chromium Tetrafluoride, 2292
[10049-12-4] Vanadium Trifluoride, 10058
[10049-14-6] Uranium Tetrafluoride, 9995
[10049-16-8] Vanadium Tetrafluoride, 10057
[10049-17-9] Rhenium Hexafluoride, 8345
[10049-22-6] Sodium Pyroantimonate, 8817
[10049-23-7] Tellurous Acid, 9281
[10049-25-9] Chromous Bromide, 2300
[10058-07-8] Pimefylline Nicotinate*, 7584
[10058-44-3] Ferric Pyrophosphate, 4075
[10060-10-3] Ceric Fluoride, 2034
[10060-11-4] Germanium Dichloride, 4416
[10072-50-1] Polymyxin D$_1$*, 7734
[10075-36-2] Isoaminile Cyclamate*, 5124
[10085-81-1] Benzoctamine Hydro-chloride*, 1117
[10087-54-4] 17-Hydroxy-16-methyl-ene-Δ^6-progesterone, 4878
[10088-68-3] Erdin, 3682
[10088-95-6] Radicinin, 8276
[10089-10-8] Nidulin, 6616
[10090-54-7] Potassium Pyroanti-monate, Acid, 7832
[10091-70-0] Protokosin, 8081
[10097-28-6] Silicon Monoxide, 8640
[10097-84-4] Tetrahydropalmatine, 9354
[10099-74-8] Lead Nitrate, 5434
[10099-79-3] Lead Vanadate(V), 5451
[10101-50-5] Sodium Permanganate, 8799
[10101-52-7] Zirconium Silicate, 10314
[10101-53-8] Chromic Sulfate, 2287
[10101-63-0] Lead Iodide, 5430
[10101-83-4] Sodium Tellurate(VI), 8835
[10101-88-9] Sodium Thiophosphate, 8843
[10101-94-7] Lead Sodium Thio-sulfate, 5441
[10102-02-0] Zinc Nitrite, 10275
[10102-03-1] Nitrogen Pentoxide, 6702
[10102-05-3] Palladium Nitrate, 7124
[10102-18-8] Sodium Selenite, 8822
[10102-20-2] Sodium Tellurate(IV), 8834
[10102-24-6] Lithium Silicate, 5568

[10102-40-6] Sodium Molybdate(VI) Dihydrate*, 8790
[10102-43-9] Nitric Oxide, 6674
[10102-44-0] Nitrogen Dioxide, 6700
[10102-45-1] Thallium Nitrate, 9400
[10102-48-4] Lead Arsenate, 5413
[10102-68-8] Calcium Iodide, 1720
[10102-71-3] Aluminum Sodium Sulfate, 378
[10108-64-2] Cadmium Chloride, 1653
[10108-73-3] Cerous Nitrate, 2044
[10112-91-1] Mercurous Chloride, 5951
[10114-58-6] Bismark Brown Y, 1294
[10116-22-0] Demegestone, 2939
[10117-38-1] Potassium Sulfite, 7847
[10118-56-6] Cascarillin, 1933
[10118-76-0] Calcium Permanganate, 1735
[10118-90-8] Minocycline, 6289
[10124-36-4] Cadmium Sulfate, 1665
[10124-37-5] Calcium Nitrate, 1729
[10124-41-1] Calcium Thiosulfate, 1759
[10124-43-3] Cobaltous Sulfate, 2510
[10124-48-8] Mercuric Chloride, Ammoniated, 5927
[10124-50-2] Potassium Arsenite, 7768
[10124-53-5] Magnesium Thiosulfate, 5734
[10128-36-6] Etilefrin dl-Form*, 3911
[10129-92-7] Eledoisin Trifluoroace-tate*, 3579
[10136-65-9] Oxypinocamphone, 7110
[10137-74-3] Calcium Chlorate, 1698
[10138-04-2] Ammonium Ferric Sulfate, 549
[10139-06-7] Linatine, 5524
[10139-47-6] Zinc Iodide, 10269
[10140-70-2] Curvularin, 2747
[10141-00-1] Chromic Potassium Sulfate, 2286
[10141-05-6] Cobaltous Nitrate, 2505
[10161-33-8] Trenbolone, 9716
[10161-34-9] Trenbolone Acetate*, 9716
[10170-69-1] Manganese Carbonyl, 5767
[10190-55-3] Lead Molybdate(VI), 5432
[10192-30-0] Ammonium Bisulfite, 528
[10196-04-0] Ammonium Sulfite, 594
[10199-21-0] Guanidinium Aluminum Sulfate Hexahydrate, 4592
[10210-64-7] Beryllium Acetylaceto-nate, 1204
[10210-68-1] Dicobalt Octacarbonyl, 3135
[10212-25-6] Ancitabine Hydrochlo-ride*, 670
[10213-09-9] Vanadyl Dichloride, 10062
[10214-39-8] Lead Borate, 5416
[10214-40-1] Cupric Selenite, 2720
[10233-88-2] Gold Sodium Thio-sulfate, 4537
[10236-47-2] Naringin, 6512
[10236-58-5] Selenocysteine, 8584
[10238-21-8] Glyburide, 4486
[10246-75-0] Hydroxyzine Pamoate*, 4897
[10252-34-3] Sporidesmolide IV*, 8921
[10257-55-3] Calcium Sulfite, 1755
[10262-69-8] Maprotiline, 5792
[10265-92-6] Methamidophos, 6014
[10290-12-7] Cupric Arsenite, 2693
[10294-26-5] Silver Sulfate, 8676
[10294-28-7] Gold Tribromide, 4539

Denotes derivative of title compound

[10294-29-8] Gold Monochloride, 4528
[10294-31-2] Gold Monoiodide, 4530
[10294-32-3] Gold Selenate, 4534
[10294-33-4] Boron Tribromide, 1377
[10294-34-5] Boron Trichloride, 1378
[10294-40-3] Barium Chromate(VI), 999
[10294-44-7] Mercurous Chlorate, 5950
[10294-47-0] Lead Chlorate, 5420
[10294-48-1] Chlorine Heptoxide, 2147
[10294-52-7] Ferric Chromate(VI), 4062
[10294-54-9] Cesium Sulfate, 2058
[10294-58-3] Lead Hypophosphite, 5429
[10294-64-1] Potassium Manganate-(VI), 7810
[10294-66-3] Potassium Thiosulfate, 7867
[10294-70-9] Stannous Iodide, 8942
[10297-61-7] Propylure, 8055
[10310-32-4] Tribenoside, 9738
[10311-84-9] Dialifor, 3013
[10318-26-0] Mitolactol, 6300
[10321-12-7] Propizepine, 8019
[10325-94-7] Cadmium Nitrate, 1658
[10326-21-3] Magnesium Chlorate, 5697
[10326-24-6] Zinc Meta-arsenite, 10272
[10326-41-7] D-Lactic Acid, 5348
[10331-57-4] Niclofolan, 6601
[10343-61-0] Titanium Sesquisulfate, 9615
[10347-81-6] Maprotiline Hydrochloride*, 5792
[10352-73-5] Pseudopederin*, 7196
[10361-03-2] Maddrell's Salt, 5680
[10361-29-2] Ammonium Carbonate, 534
[10361-37-2] Barium Chloride, 998
[10361-43-0] Bismuth Hydroxide, 1306
[10361-44-1] Bismuth Nitrate, 1311
[10373-81-6] 3-Hydroxycamphor, 4862
[10376-48-4] Shionone, 8626
[10377-48-7] Lithium Sulfate, 5569
[10377-51-2] Lithium Iodide, 5561
[10377-58-9] Magnesium Iodide, 5707
[10377-60-3] Magnesium Nitrate, 5710
[10377-62-5] Magnesium Permanganate, 5716
[10377-66-9] Manganese Nitrate, 5773
[10379-14-3] Tetrazepam, 9379
[10380-31-1] Barium Uranium Oxide, 1030
[10381-36-9] Nickel Phosphate, 6598
[10381-37-0] Thorium Sulfate, 9519
[10389-72-7] Clortermine Hydrochloride*, 2472
[10389-73-8] Clortermine, 2472
[10397-75-8] Iocarmic Acid, 5029
[10402-53-6] Eprazinone Dihydrochloride*, 3666
[10402-90-1] Eprazinone, 3666
[10405-02-4] Trospium Chloride, 9918
[10415-75-5] Mercurous Nitrate, 5954
[10417-94-4] (all-Z)-5,8,11,14,17-Eicosapentaenoic Acid, 3572
[10418-03-8] Stanozolol (2'H form), 8951
[10421-48-4] Ferric Nitrate, 4071
[10450-55-2] Ferric Acetate, Basic, 4057
[10450-60-9] Periodic Acid, 7312

[10453-89-1] Chrysanthemic Acid, 2310
[10457-66-6] Geranylhydroquinone, 4413
[10457-90-6] Bromperidol, 1466
[10476-81-0] Strontium Bromide, 8997
[10476-85-4] Strontium Chloride, 9000
[10476-86-5] Strontium Iodide, 9005
[10486-11-0] Quinine Oleate, 8247
[10488-36-5] Tofenacin Hydrochloride*, 9641
[10534-89-1] Hexaaminecobalt Trichloride, 4711
[10539-19-2] Moxaverine, 6371
[10540-29-1] Tamoxifen, 9216
[10553-31-8] Barium Bromide, 995
[10563-70-9] Melitracen Hydrochloride*, 5866
[10567-69-8] Barium Iodate, 1008
[10571-59-2] Nicoclonate, 6603
[10580-02-6] Ammonium Titanium Oxalate, 599
[10588-01-9] Sodium Dichromate(VI), 8754
[10592-03-7] Vincamine Hydrochloride*, 10120
[10596-23-3] Clodronic Acid, 2432
[10597-89-4] N-Acetylmuramic Acid*, 6384
[10605-21-7] Carbendazim, 1836
[11000-03-6] Strepogenin, 8976
[11003-38-6] Capreomycin, 1801
[11003-70-6] Scillaren, 8543
[11005-20-2] Pseudohecogenin*, 4652
[11005-63-3] Strophanthin, 9016
[11006-77-2] Statolon, 8957
[11006-83-0] Thermorubin, 9423
[11006-87-4] Essential Fatty Acids, 3743
[11015-37-5] Bambermycins, 979
[11016-39-0] Properdin, 7999
[11017-56-4] Strigol, 8992
[11024-24-1] Digitonin, 3204
[11028-71-0] Concanavalin A, 2559
[11030-31-2] Lanatoside D*, 5368
[11030-71-0] Amanitin, 387
[11031-48-4] Sarkomycin, 8521
[11031-82-6] Streptovaricin B*, 8989
[11031-85-9] Streptovaricin G*, 8989
[11032-05-6] Teleocidin B₄*, 9271
[11034-45-0] O-Cinnamoyltaxicin-I*, 9245
[11035-08-8] Substance P, 9032
[11041-12-6] Cholestyramine Resin, 2257
[11042-38-9] Xanthocillin, 10195
[11042-64-1] γ-Oryzanol, 7016
[11050-94-5] Oligomycin B*, 6970
[11051-71-1] Avilamycin, 921
[11052-72-5] Oligomycin C*, 6970
[11061-68-0] Insulin (Human)*, 5011
[11061-96-4] Cardiotoxin, 1884
[11070-73-8] Insulin (Bovine)*, 5011
[11072-93-8] β-Escin*, 3737
[11075-36-8] Tuberactinomycin, 9935
[11076-19-0] Bongkrekic Acid, 1361
[11078-21-0] Filipin, 4123
[11085-36-2] HCS, 4650
[11096-26-7] Erythropoietin, 3729
[11096-49-4] Partricin, 7181
[11104-88-4] Phosphomolybdic Acid, 7498
[11110-52-4] Sodium Amalgam, 8716
[11113-63-6] Graphite Fluoride, 4561
[11113-80-7] Polyoxins, 7736
[11115-82-5] Enduracidin, 3619
[11120-15-3] Dermostatin, 2960
[11120-25-5] Ammonium Tungstate-(VI), 600
[11121-16-7] Aluminum Borate, 339

[11121-32-7] Mepartricin, 5891
[11125-96-5] Burgundy Mixture, 1526
[11138-66-2] Xanthan Gum, 10191
[11141-17-6] Azadirachtin, 926
[12001-65-9] Methenamine Tetraiodine, 6043
[12001-68-2] Potassium Borotartrate, 7778
[12001-72-8] Mafenide Propionate*, 5683
[12002-03-8] Cupric Acetoarsenite, 2692
[12002-30-1] Piperazine Edetate Calcium*, 7617
[12002-43-6] Uintahite, 9977
[12002-53-8] Trimedlure, 9831
[12007-56-6] Calcium Borate, 1691
[12007-58-8] Ammonium Borate, 530
[12016-80-7] Cobaltic Oxide Monohydrate, 2492
[12018-01-8] Chromium Dioxide, 2290
[12018-10-9] Cupric Chromite, 2701
[12026-06-1] Thallium Hydroxide, 9398
[12026-66-3] Ammonium Phosphomolybdate, 579
[12027-06-4] Ammonium Iodide, 560
[12027-67-7] Ammonium Molybdate-(VI), 565
[12029-98-0] Iodine Pentoxide, 5040
[12033-88-4] Nitrogen Selenide, 6703
[12033-89-5] Silicon Nitride, 8641
[12040-44-7] Alcian Blue, 219
[12041-72-4] Formosulfathiazole*, 9115
[12047-27-7] Barium Titanate(IV), 1029
[12048-50-9] Bismuth Tetroxide, 1332
[12054-48-7] Nickel Hydroxide, 6593
[12057-24-8] Lithium Oxide, 5564
[12058-66-1] Sodium Stannate(IV), 8826
[12062-24-7] Cupric Hexafluorosilicate, 2708
[12063-56-8] Yig, 10231
[12063-98-8] Gallium Phosphide, 4366
[12067-99-1] Phosphotungstic Acid, 7521
[12068-69-8] Bismuth Selenide, 1319
[12069-32-8] Boron Carbide, 1374
[12069-69-1] Cupric Carbonate, Basic, 2697
[12071-83-9] Propineb, 8004
[12108-13-3] MMT, 6307
[12111-24-9] Pentetate Calcium Trisodium, 7265
[12122-67-7] Zineb, 10300
[12124-97-9] Ammonium Bromide, 531
[12124-99-1] Ammonium Bisulfide, 527
[12125-01-8] Ammonium Fluoride, 553
[12125-02-9] Ammonium Chloride, 537
[12125-08-5] Ammonium Osmium Chloride, 570
[12125-09-6] Phosphonium Iodide, 7499
[12135-76-1] Ammonium Sulfide, 591
[12141-46-7] Aluminum Silicate, 377
[12142-33-5] Potassium Stannate(IV), 7842
[12167-15-6] Potassium Phenolsulfonate, 7826
[12173-47-6] Hectorite, 4653
[12192-57-3] Aurothioglucose, 917
[12201-47-7] Sodium Plumbate(IV), 8812
[12228-91-0] Manganese Borate, 5764

Denotes derivative of title compound

[12232-99-4] Sodium Bismuthate(V), 8728
[12244-57-4] Gold Sodium Thiomalate, 4536
[12261-97-1] 2-Amino-4-picoline Camphorsulfonate*, 486
[12284-76-3] Bismuth Aluminate, 1298
[12286-76-9] Fructose Ferric Form*, 4295
[12304-65-3] Hydrotalcite*, 5696
[12321-44-7] Calcitonin, porcine*, 1680
[12419-43-1] Stannous Hexafluorozirconate(IV), 8941
[12427-38-2] Maneb, 5761
[12505-77-0] Boron Monoxide, 1375
[12511-31-8] Aluminum Magnesium Silicate, 362
[12520-88-6] Silicotungstic Acid, 8646
[12550-17-3] Sodium Antimonylgluconate*, 742
[12584-58-6] Insulin (Porcine)*, 5011
[12584-83-7] Cobrotoxin, 2514
[12607-70-4] Nickel Carbonate Hydroxide, 6586
[12607-92-0] Aceglutamide Aluminum Complex*, 25
[12607-93-1] Taxine, 9246
[12609-84-6] Thiopeptin, 9488
[12616-24-9] Ammonium Zirconyl Carbonate, 606
[12629-01-5] Human Growth Hormone*, 8864
[12650-69-0] Mupirocin, 6383
[12698-96-3] Thearubigins, 9410
[12704-90-4] Aurodox, 916
[12740-44-2] Sodium-Lead Alloy, 8783
[12751-04-1] Creolin, 2639
[12758-40-6] Propagermanium, 7979
[12771-68-5] Ancymidol, 672
[12772-35-9] Butirosin, 1559
[12772-57-5] Monorden, 6336
[13007-85-7] Glucoheptonic Acid Sodium Salt*, 4463
[13007-90-4] Bis(triphenylphosphine)-dicarbonylnickel, 1341
[13007-92-6] Chromium Carbonyl, 2289
[13007-93-7] Cuproxoline, 2740
[13008-73-6] L-Streptose, 8987
[13009-99-9] Mafenide Acetate*, 5683
[13010-47-4] Lomustine, 5594
[13011-54-6] Ammonium Sodium Phosphate, 587
[13040-98-7] Guamecycline Dihydrochloride*, 4584
[13042-18-7] Fendiline, 4011
[13051-01-9] Carbazochrome Salicylate, 1832
[13055-82-8] Reproterol Hydrochloride*, 8307
[13058-67-8] Lucensomycin, 5621
[13067-93-1] Cyanofenphos, 2760
[13071-79-9] Terbufos, 9301
[13073-35-3] Ethionine, 3784
[13085-08-0] Mazipredone, 5802
[13087-53-1] Iothalamic Acid N-Methylglucamine Salt*, 5080
[13092-66-5] Magnesium Phosphate, Monobasic, 5719
[13093-88-4] Perimethazine, 7309
[13100-82-8] Cysteic Acid, 2849
[13103-34-9] Boldenone 10-Undecenoate*, 1354
[13115-03-2] Vitamin B_{12}-^{57}Co*, 10153
[13121-70-5] Cyhexatin, 2829
[13123-37-0] Lyxoflavine, 5672
[13137-64-9] Periplocin, 7315

[13138-45-9] Nickel Nitrate, 6596
[13157-90-9] Quercetin Pentabenzyl Ether*, 8216
[13171-21-6] Phosphamidon, 7493
[13171-25-0] Trimetazidine Dihydrochloride*, 9835
[13190-88-0] Leucoglycodrin, 5479
[13194-48-4] Ethoprop, 3792
[13246-02-1] Febarbamate, 3983
[13254-33-6] Carpronium Chloride, 1913
[13289-18-4] Hellebrin, 4661
[13291-74-2] Gentamicin A*, 4398
[13292-46-1] Rifampin, 8382
[13311-84-7] Flutamide, 4242
[13327-32-7] Beryllium Hydroxide, 1212
[13347-42-7] Dowicide 9, 3486
[13356-08-6] Fenbutatin Oxide, 4004
[13364-32-4] Clobenzorex, 2421
[13392-18-2] Fenoterol, 4022
[13392-28-4] Rimantadine, 8390
[13401-40-6] Phaseolin, 7338
[13402-08-9] Acetylpheneturide, 101
[13405-60-2] β-Glucogallin, 4462
[13408-56-5] Ponasterone A, 7748
[13410-01-0] Sodium Selenate, 8820
[13411-16-0] Nifurpirinol, 6628
[13412-64-1] Dicloxacillin Sodium Salt Monohydrate*, 3134
[13422-51-0] Hydroxocobalamin, 4854
[13422-52-1] Aquacobalamin*, 4854
[13422-53-2] Vitamin B_{12}-^{60}Co*, 10153
[13422-55-4] Methylcobalamin, 6125
[13424-46-9] Lead Azide, 5415
[13425-22-4] Aminorex Fumarate*, 497
[13425-39-3] Etofylline Nicotinate, 3925
[13425-98-4] Improsulfan, 4962
[13444-24-1] 1-Ethyl-3-piperidinol, 3890
[13444-90-1] Nitryl Chloride, 6755
[13445-60-8] Furfurylmethylamphetamine, 4327
[13445-63-1] Itramin Tosylate, 5263
[13446-03-2] Manganese Bromide, 5765
[13446-24-7] Magnesium Pyrophosphate, 5722
[13446-49-6] Potassium Molybdate-(VI), 7814
[13447-95-5] Isoniazid Methanesulfonate*, 5203
[13452-36-3] O-Cinnamoyltaxicin-I triacetate*, 9245
[13453-07-1] Gold Trichloride, 4541
[13453-34-4] Thallium Cyanide, 9396
[13454-94-9] Cerous Sulfate, 2046
[13455-25-9] Cobaltous Chromate-(III), 2499
[13455-36-2] Cobaltous Phosphate, 2508
[13457-18-6] Pyrazophos, 8146
[13462-88-9] Nickel Bromide, 6585
[13462-90-3] Nickel Iodide, 6594
[13463-39-3] Nickel Carbonyl, 6587
[13463-40-6] Iron Pentacarbonyl, 5111
[13463-41-7] Pyrithione Zinc Derivative*, 8178
[13463-67-7] Titanium Dioxide, 9612
[13464-44-3] Zinc Ortho-arsenate, 10277
[13464-80-7] Dihydrazine Sulfate, 3213
[13464-82-9] Indium Sulfate, 4987
[13465-66-2] Selenium Tetrafluoride, 8583
[13465-94-6] Barium Nitrite, 1013
[13465-95-7] Barium Perchlorate, 1016

[13466-78-9] 3-Carene, 1885
[13471-78-8] Beclotiamine, 1048
[13472-45-2] Sodium Tungstate(VI), 8847
[13473-90-0] Aluminum Nitrate, 365
[13477-00-4] Barium Chlorate, 997
[13478-49-4] Erbium Sulfate, 3680
[13478-93-8] Nickel Dimethylglyoxime, 6590
[13479-54-4] Cupric Glycinate, 2707
[13484-14-5] Sarmentose, 8523
[13492-01-8] Tranylcypromine Sulfate*, 9708
[13492-26-7] Potassium Phosphite, 7831
[13494-80-9] Tellurium, 9273
[13494-90-1] Gallium Nitrate*, 4364
[13495-04-0] Heptanal Sodium Bisulfite, 4693
[13495-09-5] Piminodine, 7587
[13510-49-1] Beryllium Sulfate, 1222
[13510-89-9] Lead Antimonate(V), 5412
[13523-86-9] Pindolol, 7597
[13530-57-9] Aluminum Rubidium Sulfate, 374
[13530-65-9] Zinc Chromate(VI) Hydroxide, 10262
[13539-59-8] Apazone, 768
[13548-38-4] Chromic Nitrate, 2282
[13548-42-0] Cupric Chromate(VI), 2700
[13548-43-1] Ammonium Chromic Sulfate, 539
[13552-61-9] 2-Thiolhistidine, 9477
[13553-79-2] Rifamycin S*, 8383
[13560-89-9] Dechlorane Plus, 2908
[13563-60-5] Norgesterone, 6795
[13573-16-5] Reinecke Salt, 8298
[13587-35-4] Cupric Tungstate(VI), 2726
[13588-16-4] Senecic Acid, 8593
[13590-82-4] Ceric Sulfate, 2036
[13596-22-0] Cobaltous Potassium Sulfate, 2509
[13596-46-8] Ammonium Cobaltous Sulfate, 542
[13597-54-1] Zinc Selenate, 10289
[13597-65-4] Zinc Silicate, 10291
[13597-95-0] Beryllium Perchlorate, 1217
[13597-99-4] Beryllium Nitrate, 1214
[13598-36-2] Phosphorous Acid, 7502
[13600-89-0] Erdmann's Salt, 3683
[13600-98-1] Sodium Cobaltinitrite, 8748
[13601-13-3] Cupric Ferrocyanide, 2703
[13601-19-9] Sodium Ferrocyanide, 8760
[13609-67-1] Hydrocortisone 17-Butyrate*, 4828
[13614-98-7] Minocycline Hydrochloride*, 6289
[13636-18-5] Fendiline Hydrochloride*, 4011
[13637-61-1] Zinc Perchlorate, 10280
[13640-28-3] Pilosine*, 5216
[13642-52-9] Soterenol, 8877
[13647-35-3] Trilostane, 9827
[13655-52-2] Alprenolol, 321
[13665-88-8] Mopidamol, 6344
[13669-70-0] Nefopam, 6529
[13674-87-8] Fyrol FR-2, 4342
[13675-18-8] Diboron Tetrahydroxide, 3064
[13682-61-6] Potassium Tetrachloroaurate(III), 7854
[13682-73-0] Cuprous Potassium Cyanide, 2735

Denotes derivative of title compound

[13682-92-3] Dihydroxyaluminum Aminoacetate, 3227
[13684-63-4] Phenmedipham, 7384
[13693-09-9] Xenon Hexafluoride*, 10206
[13698-49-2] Delmadinone Acetate, 2930
[13701-67-2] Diboron Tetrachloride, 3063
[13701-70-7] Vanadium Trisulfate, 10060
[13703-82-7] Magnesium Borate, 5693
[13707-88-5] Alprenolol Hydrochloride*, 321
[13708-85-5] Sodium Phosphite, 8809
[13709-36-9] Xenon Difluoride*, 10206
[13709-61-0] Xenon Tetrafluoride*, 10206
[13710-19-5] Tolfenamic Acid, 9650
[13717-04-9] Propiram Fumarate*, 8017
[13718-26-8] Sodium Vanadate(V), 8849
[13718-50-8] Barium Iodide, 1009
[13721-34-1] Sodium Uranate(VI), 8848
[13739-02-1] Diacerein, 3003
[13741-18-9] Xibornol, 10210
[13746-66-2] Potassium Ferricyanide, 7792
[13746-89-9] Zirconium Nitrate, 10312
[13755-29-8] Sodium Fluoborate, 8761
[13755-38-9] Sodium Nitroprusside Dihydrate*, 8794
[13759-10-9] Silicon Disulfide, 8638
[13762-51-1] Potassium Borohydride, 7777
[13765-19-0] Calcium Chromate(VI), 1700
[13768-38-2] Osmium Hexafluoride, 7022
[13770-18-8] Cupric Perchlorate, 2714
[13773-81-4] Krypton Difluoride*, 5335
[13776-58-4] Xenon Trioxide*, 10206
[13776-74-4] Magnesium Metasilicate*, 5727
[13776-84-6] Sodium Thioantimonate-(V), 8840
[13780-03-5] Calcium Bisulfite, Solution, 1690
[13780-06-8] Calcium Nitrite, 1730
[13782-01-9] Cobaltic Potassium Nitrite, 2493
[13804-51-8] Juvenile Hormone I*, 5287
[13814-62-5] Cadmium Selenate, 1662
[13820-41-2] Ammonium Platinous Chloride, 583
[13821-06-2] Barium Ferrocyanide, 1002
[13823-29-5] Thorium Nitrate, 9517
[13825-74-6] Titanium Sulfate, 9616
[13825-86-0] Chromous Sulfate, 2305
[13838-08-9] Azidamfenicol, 941
[13838-16-9] Enflurane, 3621
[13843-81-7] Lithium Dichromate(VI), 5556
[13845-17-5] Barium Dithionate, 1001
[13862-07-2] Diphemethoxidine, 3362
[13867-82-8] Cloxotestosterone Acetate*, 2482
[13870-90-1] Cobamamide, 2513
[13871-27-7] Beryllium Sodium Fluoride, 1221
[13876-85-2] Cuprous Mercuric Iodide, 2733
[13898-58-3] Benzoylpas, 1148
[13912-77-1] Octacaine, 6839

[13925-12-7] Myxin, 6425
[13929-35-6] Rifamycin B*, 8383
[13930-31-9] Oxethazaine Hydrochloride*, 7067
[13931-64-1] Procymate, 7945
[13932-17-7] Potassium Zinc Sulfate, 7877
[13943-58-3] Potassium Ferrocyanide, 7793
[13946-02-6] Metron S, 6243
[13952-84-6] sec-Butylamine, 1579
[13956-29-1] Cannabidiol, 1792
[13957-33-0] Leptodactyline, 5467
[13957-36-3] Hydramitrazine, 4803
[13957-55-6] N,N-1-Trimethyl-3,3-diphenylpropylamine, 9843
[13963-58-1] Potassium Hexacyano-cobaltate(III), 7801
[13966-47-7] Potassium Tetrabromo-aurate(III), 7853
[13967-25-4] Mercurous Fluoride, 5952
[13967-50-5] Potassium Dicyano-aurate(I), 7791
[13967-90-3] Barium Bromate, 994
[13977-28-1] Embramine Hydrochloride*, 3596
[13982-53-1] Cuprous Sulfite, 2738
[13986-26-0] Zirconium Iodide, 10311
[13993-65-2] Metiazinic Acid, 6219
[13997-19-8] Nequinate, 6558
[14007-64-8] Butethamate, 1551
[14008-44-7] Metopimazine, 6233
[14008-48-1] Bisoxatin Acetate, 1337
[14028-44-5] Amoxapine, 616
[14038-43-8] Ferric Ferrocyanide, 4064
[14051-33-3] Benzetimide, 1104
[14062-23-8] Felbinac Ethyl Ester*, 3989
[14066-79-6] Chloroprednisone 21-Acetate*, 2209
[14073-97-3] l-Menthone, 5883
[14075-53-7] Potassium Tetrafluoro-borate, 7859
[14089-52-2] MDE (unspecified), 5805
[14101-04-3] Frangulin B*, 4288
[14116-69-9] Vicianose, 10111
[14148-99-3] Lefetamine Hydrochloride*, 5455
[14149-43-0] Trimethidinium Metho-sulfate, 9838
[14176-10-4] Cetiedil, 2062
[14188-82-0] Hexestrol Diphosphate*, 4738
[14217-01-7] Chromic Potassium Oxalate, 2285
[14220-17-8] Potassium Tetracyano-nickelate(II), 7857
[14221-47-7] Ammonium Ferric Oxalate, 548
[14221-48-8] Ammonium Ferricya-nide, 550
[14222-60-7] Protionamide, 8076
[14222-61-8] 3-(o-Methoxyphenyl)-2-phenylacrylic Acid Magnesium Salt*, 6078
[14225-07-1] Leucodrin, 5478
[14235-86-0] Hydrargaphen, 4805
[14252-80-3] Bupivacaine Hydrochloride Monohydrate*, 1520
[14255-61-9] Oxidronic Acid Sodium Salt*, 7074
[14255-87-9] Parbendazole, 7169
[14259-45-1] Asperuloside, 879
[14261-75-7] Cloforex, 2440
[14277-97-5] Domoic Acid, 3475
[14283-05-7] Tetraamminecopper Sulfate, 9323

[14284-36-7] Aluminum Cesium Sulfate, 346
[14286-84-1] Bencyclane Fumarate*, 1060
[14293-44-8] Xipamide, 10212
[14297-87-1] Benzylmorphine, 1176
[14307-33-6] Calcium Dichromate-(VI), 1706
[14307-35-8] Lithium Chromate(VI), 5554
[14317-18-1] Ioglycamic Acid Meglumine Salt*, 5067
[14334-40-8] Pramiverin, 7887
[14334-41-9] Pramiverin Hydrochloride*, 7887
[14350-67-5] Ormosinine, 6999
[14357-78-9] Diprenorphine, 3403
[14358-43-1] Dehydroemetine (±)-Form Dihydrochloride*, 2924
[14376-16-0] Sulfaloxic Acid, 9080
[14402-67-6] Potassium Titanyl Oxa-late, 7868
[14402-70-1] Ammonium Nitroferricy-anide, 568
[14402-73-4] Lithium Tetracyano-platinate(II), 5570
[14402-75-6] Cadmium Potassium Cy-anide, 1660
[14402-89-2] Sodium Nitroprusside, 8794
[14417-88-0] Melinamide, 5864
[14435-78-0] Leiopyrrole Hydrochloride*, 5457
[14437-41-3] Clioxanide, 2416
[14451-99-1] Menadione Dimethyl-pyrimidinol Bisulfite*, 5874
[14452-57-4] Magnesium Peroxide, 5717
[14457-87-5] Cerous Bromide, 2039
[14459-29-1] Hematoporphyrin, 4669
[14463-33-3] Vitamin B_{12r}, 10154
[14475-63-9] Zirconium Hydroxide, 10310
[14481-29-9] Ammonium Ferrocy-anide, 551
[14481-33-5] Thorium Tetracyano-platinate(II), 9520
[14484-47-0] Deflazacort, 2916
[14484-64-1] Ferbam, 4054
[14487-05-9] Rifamycin O*, 8383
[14504-73-5] Tritoqualine, 9894
[14519-18-7] Strontium Bromate, 8996
[14521-96-1] Etorphine, 3932
[14523-22-9] Rhodium Carbonyl Chloride, 8354
[14556-46-8] Bupranolol, 1521
[14559-79-6] Propizepine Hydrochloride*, 8019
[14590-13-7] Ammonium Cobaltous Phosphate, 541
[14611-51-9] Selegiline, 8569
[14611-52-0] Selegiline Hydrochloride*, 8569
[14611-84-8] Furfurylmethylamphet-amine dl-Form Cyclohexylsulfa-mate*, 4327
[14613-30-0] Clofibric Acid Mag-nesium Salt*, 2437
[14635-75-7] Nitrosyl Tetrafluoro-borate, 6747
[14635-87-1] Magnesium Perborate, 5714
[14636-12-5] Terlipressin, 9310
[14639-25-9] Chromium Picolinate, 2291
[14639-94-2] Ammonium Hexafluoro-gallate, 556
[14639-97-5] Ammonium Tetrachloro-zincate, 596

Denotes derivative of title compound

[14639-98-6] Ammonium Pentachlorozincate, 573
[14644-61-2] Zirconium Sulfate, 10315
[14679-73-3] Todralazine, 9640
[14693-02-8] Potassium Thioantimonate(V), 7864
[14694-95-2] Chlorotris(triphenylphosphine)rhodium, 2226
[14698-29-4] Oxolinic Acid, 7079
[14709-57-0] Potassium Nitroprusside, 7817
[14721-21-2] Cupric Chlorate, 2698
[14759-06-9] Sulforidazine, 9137
[14769-73-4] Levamisole, 5486
[14807-96-6] Talc, 9207
[14816-18-3] Phoxim, 7523
[14816-67-2] Soterenol Monohydrochloride*, 8877
[14817-09-5] Decimemide, 2909
[14817-79-9] Furfurylmethylamphetamine dl-Form*, 4327
[14838-15-4] (±)-Phenylpropanolamine*, 7461
[14855-76-6] Methyl Green, 6157
[14860-49-2] Clobutinol, 2425
[14871-79-5] Barium Hypophosphite, 1007
[14874-24-9] Tris(ethylenediamine)cadmium Dihydroxide, 9882
[14875-96-8] Heme, 4672
[14882-18-9] Bismuth Subsalicylate, 1327
[14885-29-1] Ipronidazole, 5095
[14897-39-3] Rifamycin SV Sodium Salt*, 8384
[14901-08-7] Cycasin, 2768
[14912-44-8] Ylangene, 10232
[14918-35-5] Destomycin A, 2979
[14919-77-8] Benserazide Hydrochloride*, 1079
[14929-11-4] Simfibrate, 8684
[14930-96-2] Cytochalasin B*, 2859
[14932-42-4] Xenon-133*, 10206
[14940-41-1] Ferrous Phosphate, 4101
[14959-86-5] Looplure, 5600
[14976-57-9] Clemastine Hydrogen Fumarate*, 2405
[14977-61-8] Chromyl Chloride, 2306
[14984-68-0] Cloperastine Hydrochloride*, 2456
[14986-91-5] Magnesium Selenate, 5724
[14987-04-3] Magnesium Trisilicate*, 5727
[14992-59-7] Sodium Dibunate, 8753
[14992-62-2] Acetylcarnitine, 85
[15096-52-3] Cryolite, 2673
[15123-69-0] Cupric Selenate, 2718
[15130-91-3] Sultroponium, 9170
[15139-76-1] Orange B, 6991
[15148-80-8] Bupranolol Hydrochloride*, 1521
[15176-29-1] Edoxudine, 3561
[15180-00-4] Prednival, 7905
[15180-03-7] Alcuronium Dichloride*, 222
[15189-51-2] Sodium Tetrachloroaurate(III), 8837
[15238-00-3] Cobaltous Iodide, 2504
[15263-53-3] Cartap, 1916
[15265-28-8] Palitantin, 7120
[15280-09-8] Sodium Dicyanoaurate(I), 8755
[15291-75-5] Ginkgolide A*, 4433
[15291-76-6] Ginkgolide C*, 4433
[15291-77-7] Ginkgolide B*, 4433
[15299-99-7] Napropamide, 6503
[15301-40-3] Actinoquinol, 143
[15301-48-1] Bezitramide, 1241
[15301-52-7] Cyclexanone, 2775

[15301-69-6] Flavoxate, 4135
[15301-80-1] Oxamarin, 7047
[15301-93-6] Tofenacin, 9641
[15302-16-6] Fenozolone, 4028
[15304-14-0] Aluminum Hexaurea Sulfate Triiodide, 353
[15307-79-6] Diclofenac Sodium, 3132
[15308-34-6] Norfenefrine dl-Form Hydrochloride*, 6792
[15318-45-3] Thiamphenicol, 9436
[15320-30-6] Magnesium Sulfate Trihydrate*, 5731
[15351-04-9] Becanthone, 1044
[15351-05-0] Buzepide Methiodide*, 1635
[15351-09-4] Metamfepramone, 5984
[15351-13-0] Nicofuranose, 6605
[15358-48-2] Hydroxylupanine, 4875
[15375-94-7] Merphyrin, 5961
[15385-57-6] Mercurous Iodide, 5953
[15421-84-8] Trapidil, 9709
[15468-10-7] Oxidronic Acid, 7074
[15477-33-5] Aluminum Chlorate, 347
[15489-16-4] Stibophen, 8967
[15489-90-4] Hematin, 4668
[15490-91-2] Struvite, 9019
[15500-66-0] Pancuronium Bromide, 7139
[15518-82-8] Metescufylline, 6000
[15537-71-0] N-Acetylpenicillamine, 100
[15537-76-5] Chlorproguanil Hydrochloride*, 2237
[15571-91-2] Potassium Tellurate(VI), 7851
[15572-25-5] Thallium Selenide, 9403
[15574-96-6] Pizotyline, 7671
[15578-26-4] Stannous Pyrophosphate, 8945
[15585-70-3] Bibenzonium Bromide, 1244
[15593-51-8] Lithium Selenite, 5567
[15593-52-9] Lithium Selenate, 5566
[15593-61-0] Magnesium Selenite, 5726
[15599-36-7] Halethazole, 4622
[15599-39-0] Noxythiolin, 6822
[15599-51-6] Apicycline, 772
[15610-76-1] Ammonium Cupric Chloride, 543
[15622-65-8] Molindone Hydrochloride*, 6315
[15662-36-6] Ryanodine, 8459
[15663-27-1] Cisplatin, 2378
[15671-27-9] Sulfitocobalamin*, 4854
[15676-16-1] Sulpiride, 9163
[15686-51-8] Clemastine, 2405
[15686-61-0] Fenproporex, 4036
[15686-71-2] Cephalexin, 2021
[15686-83-6] Pyrantel, 8139
[15686-91-6] Propiram, 8017
[15687-16-8] Carbiphene, 1846
[15687-18-0] Fenpentadiol, 4029
[15687-22-6] Folescutol, 4252
[15687-27-1] Ibuprofen, 4925
[15687-41-9] Oxyfedrine, 7096
[15699-18-0] Ammonium Nickel Sulfate, 566
[15708-41-5] Ferric Sodium Edetate, 4076
[15722-48-2] Olsalazine, 6976
[15764-04-2] α-Vetivone*, 10106
[15793-40-5] Terodiline, 9311
[15823-89-9] Imolamine Hydrochloride*, 4959
[15825-70-4] Mannitol Hexanitrate, 5789
[15826-37-6] Cromolyn Disodium Salt*, 2658
[15829-53-5] Mercurous Oxide, 5955

[15845-98-4] Iothalamic Acid [131]I Labeled Sodium Salt*, 5080
[15876-67-2] Distigmine Bromide, 3426
[15879-93-3] α-Chloralose, 2114
[15972-60-8] Alachlor, 203
[16009-13-5] Hemin, 4678
[16034-77-8] Iocetamic Acid, 5030
[16037-91-5] Sodium Stibogluconate*, 742
[16039-53-5] Zinc Lactate, 10271
[16048-96-7] Cupric Salicylate, 2717
[16051-77-7] Isosorbide-5-mononitrate*, 5245
[16110-51-3] Cromolyn, 2658
[16188-61-7] Talastine, 9205
[16188-76-4] Talastine Hydrochloride*, 9205
[16203-97-7] Anthralin Triacetate*, 723
[16283-36-6] Zinc Salicylate, 10288
[16286-69-4] Cetiedil Citrate*, 2062
[16298-90-1] Jesaconitine, 5278
[16320-04-0] Gestrinone, 4423
[16357-59-8] EEDQ, 3563
[16377-00-7] Indospicine, 5001
[16378-21-5] Piroheptine, 7658
[16378-22-6] Piroheptine Hydrochloride*, 7658
[16423-68-0] Erythrosine, 3734
[16478-59-4] Hemicholinium, 4676
[16498-21-8] Oxaflumazine, 7040
[16545-11-2] Guamecycline, 4584
[16550-22-4] Cyprenorphine Hydrochloride*, 2838
[16589-24-5] Synephrine Tartrate*, 9189
[16590-41-3] Naltrexone, 6450
[16595-80-5] Levamisole Hydrochloride*, 5486
[16604-45-8] Medifoxamine Fumarate*, 5834
[16625-20-0] Haplophytine, 4642
[16643-66-6] Menbutone Magnesium Salt*, 5878
[16648-69-4] Oxyfedrine DL-Form Hydrochloride*, 7096
[16649-79-9] Calcium N-Carbamoylaspartate, 1695
[16649-82-4] Calcium N-Carbamoylaspartate D-Form*, 1695
[16662-46-7] Gallopamil Hydrochloride*, 4369
[16662-47-8] Gallopamil, 4369
[16672-87-0] Ethephon, 3777
[16676-29-2] Naltrexone Hydrochloride*, 6450
[16679-58-6] Desmopressin, 2969
[16683-27-5] Convallamarogenin, 2573
[16721-80-5] Sodium Bisulfide, 8730
[16731-55-8] Potassium Metabisulfite, 7811
[16752-77-5] Methomyl, 6062
[16755-07-0] Showdomycin, 8627
[16773-42-5] Ornidazole, 7000
[16774-21-3] Ammonium Ceric Nitrate, 535
[16777-42-7] Oxyfedrine L-Form Hydrochloride*, 7096
[16790-49-1] Butazolamide, 1545
[16816-67-4] Pantethine, 7144
[16830-15-2] Asiaticoside, 869
[16846-24-5] Josamycin, 5280
[16853-85-3] Aluminum Lithium Hydride, 361
[16871-60-6] Potassium Hexachloroosmate(IV), 7799
[16871-71-9] Zinc Hexafluorosilicate, 10267

[16871-90-2] Potassium Hexafluoro-silicate, 7803
[16872-11-0] Fluoboric Acid, 4184
[16893-85-9] Sodium Hexafluoro-silicate, 8769
[16903-35-8] Gold Trichloride, Acid, 4542
[16919-19-0] Ammonium Hexafluoro-silicate, 558
[16919-58-7] Ammonium Platinic Chloride, 582
[16921-30-5] Potassium Hexachloro-platinate(IV), 7800
[16921-96-3] Iodine Heptafluoride, 5036
[16923-58-3] Sodium Hexachloro-platinate(IV), 8768
[16923-95-8] Potassium Hexafluoro-zirconate(IV), 7804
[16925-39-6] Calcium Hexafluoro-silicate, 1714
[16925-51-2] Aurothioglycanide, 918
[16940-66-2] Sodium Borohydride, 8735
[16941-10-9] Aluminum Calcium Hy-dride, 344
[16941-11-0] Ammonium Hexafluoro-phosphate, 557
[16941-12-1] Platinic Chloride, 7684
[16949-15-8] Lithium Borohydride, 5550
[16949-65-8] Magnesium Hexafluoro-silicate, 5704
[16960-16-0] Cosyntropin, 2617
[16961-83-4] Fluosilicic Acid, 4220
[16962-07-5] Aluminum Borohydride, 341
[16962-31-5] Potassium Hexafluoro-manganate(IV), 7802
[16980-89-5] Bucladesine Sodium Salt*, 1483
[17021-26-0] Calusterone, 1771
[17032-39-2] Phosphorylcholine Mag-nesium Salt*, 7519
[17069-38-4] Potassium Hexathiocy-anatoplatinate(IV), 7805
[17083-68-0] Gold Tribromide, Acid, 4540
[17086-28-1] Doxycycline (monohy-drate), 3496
[17090-79-8] Monensin, 6329
[17099-70-6] Aluminum Hexafluoro-silicate, 352
[17109-49-8] Edifenphos, 3560
[17125-80-3] Barium Hexafluoro-silicate, 1005
[17140-01-1] Prednisolone 21-Diethyl-aminoacetate Hydrochloride*, 7902
[17140-68-0] Etamiphyllin Hydrochlo-ride*, 3754
[17146-95-1] Pentazocine Lactate*, 7261
[17155-31-6] Veralkamine, 10082
[17162-36-6] Arecoline p-Stibonoben-zoic Acid*, 815
[17169-60-7] Ferroglycine Sulfate, 4086
[17176-77-1] Dibenzyl Phosphite, 3061
[17185-29-4] Tristriphenylphosphine Rhodium Carbonyl Hydride, 9887
[17194-00-2] Barium Hydroxide, 1006
[17226-75-4] Khellol Glucoside, 5322
[17230-88-5] Danazol, 2875
[17243-38-8] Azidocillin, 942
[17243-39-9] Benzoctamine, 1117
[17243-57-1] Mefenorex, 5843
[17243-64-0] Piprozolin, 7639
[17297-82-4] Oxypendyl Dihydro-chloride*, 7104

[17321-77-6] Clomipramine Hydro-chloride*, 2447
[17334-58-6] Pelargonidin 3,5-Diglu-coside*, 7199
[17365-01-4] Etiroxate, 3917
[17372-87-1] Eosine Yellowish—(YS), 3639
[17375-41-6] Ferrous Sulfate Mono-hydrate*, 4105
[17388-39-5] Swertiamarin, 9186
[17397-89-6] Cerulenin, 2047
[17406-45-0] Tomatine, 9683
[17413-51-3] Nicofibrate Hydrochlo-ride*, 6604
[17440-83-4] Amiloride Hydrochlo-ride Dihydrate*, 426
[17440-85-6] Beryllium Borohydride, 1205
[17440-90-3] Hydrogen Tetracarbon-ylferrate(II), 4845
[17466-45-4] Phalloidin, 7336
[17500-49-1] Schweizer's Reagent, 8541
[17523-77-2] Niobium Potassium Oxy-pentafluoride, 6652
[17560-51-9] Metolazone, 6231
[17575-20-1] Lanatoside A*, 5368
[17575-21-2] Lanatoside B*, 5368
[17575-22-3] Lanatoside C*, 5368
[17590-01-1] Amphetaminil, 624
[17598-65-1] Deslanoside, 2967
[17605-67-3] Fucosterol, 4303
[17617-23-1] Flurazepam, 4233
[17617-45-7] Picrotoxinin, 7571
[17627-77-9] Anhalonidine, 692
[17627-78-0] Lophophorine, 5602
[17650-86-1] Amicetin, 415
[17650-98-5] Ceruletide, 2048
[17651-61-5] Adonitoxin, 167
[17673-25-5] Phorbol, 7487
[17692-22-7] Metizoline, 6223
[17692-31-8] Dropropizine, 3507
[17692-34-1] Etodroxizine, 3921
[17692-39-6] Fomocaine, 4259
[17692-51-2] Metergoline, 5999
[17692-71-6] Vanitiolide, 10071
[17693-51-5] Promethazine Compd with 8-Chlorotheophylline*, 7970
[17702-41-9] Decaborane(14), 2902
[17737-65-4] Clonixin, 2453
[17754-44-8] Atractyloside, 899
[17784-12-2] Sulfacytine, 9070
[17804-35-2] Benomyl, 1073
[17824-89-4] Butaverine Hydro-chloride*, 1544
[17854-59-0] Mepixanox, 5906
[17879-96-8] Iopanoic Acid d-Form*, 5072
[17879-97-9] Iopanoic Acid l-Form*, 5072
[17902-23-7] Tegafur, 9267
[17924-92-4] Zearalenone, 10246
[17946-87-1] Desacetoxymatricarin*, 5798
[17951-19-8] Justicidin B*, 5286
[17969-20-9] Fenclozic Acid, 4010
[18016-80-3] Lisuride, 5541
[18046-21-4] Fentiazac, 4045
[18053-31-1] Fominoben, 4258
[18059-10-4] Imperialine, 4961
[18059-99-9] Demexiptiline Hydro-chloride*, 2941
[18077-77-5] Ammonium Uranium Carbonate, 602
[18078-40-5] Potassium Uranyl Nitrate, 7874
[18109-80-3] Butamirate, 1539
[18109-81-4] Butamirate Citrate*, 1539

[18174-58-8] Pipoxolan Hydrochlo-ride, 7637
[18181-70-9] Iodofenphos, 5053
[18181-80-1] Bromopropylate, 1456
[18195-32-9] Vitamin B_{12}-^{58}Co*, 10153
[18265-54-8] Phentetiothalein Sodium, 7416
[18282-10-5] Stannic Oxide, 8933
[18283-93-7] Tetraborane(10), 9326
[18288-22-7] Bismuthine, 1307
[18305-29-8] Fenproporex Hydrochlo-ride*, 4036
[18309-32-5] d-Verbenone, 10094
[18323-44-9] Clindamycin, 2414
[18378-89-7] Plicamycin, 7696
[18397-07-4] Geissoschizoline, 4386
[18423-69-3] Cubebin, 2680
[18433-40-4] Ammonium Uranium Fluoride, 603
[18433-48-2] Uranyl Phosphate, 10001
[18433-84-6] Pentaborane(11), 7238
[18444-66-1] Cucurbitacin E*, 2681
[18444-79-6] β-Vetivone*, 10106
[18454-12-1] Lead Chromate(VI) Oxide, 5423
[18456-04-7] Cephalonic Acid, 2024
[18464-39-6] Caroxazone, 1907
[18466-51-8] Pelargonidin 3-Gluco-side*, 7199
[18467-77-1] Dikegulac, 3243
[18470-94-5] Nandrolone Cyclohexanecarboxylate*, 6452
[18471-20-0] Ditazol, 3432
[18472-51-0] Chlorhexidine D-Diglu-conate*, 2140
[18480-07-4] Strontium Hydroxide, 9004
[18493-30-6] Metochalcone, 6225
[18499-92-8] Kermesic Acid, 5304
[18507-89-6] Decoquinate, 2910
[18524-67-9] Mycobacillin, 6404
[18524-94-2] Loganin, 5590
[18530-30-8] d-Camphocarboxylic Acid, 1778
[18530-56-8] Norea, 6787
[18534-66-2] Vitamin B_{12s}*, 10154
[18542-37-5] Vernolepin, 10098
[18559-59-6] Tretoquinol l-Form Hydrochloride*, 9719
[18559-63-2] Tretoquinol dl-Form Hydrochloride*, 9719
[18559-94-9] Albuterol, 217
[18598-63-5] Mecysteine Hydrochlo-ride, 5828
[18624-44-7] Ferrous Hydroxide, 4096
[18642-10-9] Thiamine Disulfide Hydrochloride*, 9432
[18642-23-4] Psoralidin, 8114
[18651-67-5] Sikkimotoxin, 8632
[18656-21-8] Iodamide N-Methyl-D-glucamine Salt*, 5031
[18679-90-8] Hopantenic Acid, 4781
[18683-91-5] Ambroxol, 401
[18691-97-9] Methabenzthiazuron, 6002
[18694-40-1] Epirizole, 3659
[18696-26-9] Sinapine, 8688
[18699-02-0] Actarit, 135
[18700-78-2] Cervicarcin, 2050
[18786-24-8] Serpentine (Alkaloid), 8609
[18791-21-4] Pyridomycin, 8160
[18794-84-8] β-Farnesene, 3977
[18820-29-6] Manganese Sulfide, 5783
[18829-56-6] 2-Nonenal, 6771
[18829-70-4] Catechin l-Form*, 1950
[18836-52-7] Pellitorine, 7201
[18837-96-2] Bucladesine Barium Salt*, 1483
[18839-90-2] Mestilbol, 5975

Denotes derivative of title compound

[18840-45-4] Penicillamine Cysteine Disulfide, 7215
[18840-47-6] Gepefrine, 4409
[18883-66-4] Streptozocin, 8991
[18904-54-6] Corynantheine, 2613
[18917-82-3] Lead Lactate, 5431
[18917-89-0] Magnesium Salicylate, 5723
[18917-91-4] Aluminum Lactate, 360
[18917-93-6] Magnesium Lactate, 5708
[18937-33-2] Magnesium Mandelate, 5709
[18968-99-5] Pemoline Magnesium Hydroxide Mixture*, 7206
[19010-79-8] Cadmium Salicylate, 1661
[19014-05-2] Halimide®, 4624
[19026-31-4] Taxodione, 9247
[19044-88-3] Oryzalin, 7015
[19057-60-4] Dioscin, 3345
[19121-58-5] Solasonine, 8858
[19130-96-2] 1-Deoxynojirimycin, 2952
[19172-47-5] Lawesson's Reagent, 5404
[19216-56-9] Prazosin, 7897
[19237-84-4] Prazosin Hydrochloride*, 7897
[19246-24-3] Telomycin, 9283
[19286-37-4] Convicine, 2576
[19287-45-7] Diborane(6), 3062
[19295-31-9] Urothion, 10025
[19356-17-3] Calcifediol, 1677
[19379-90-9] Benzoxonium Chloride, 1144
[19387-91-8] Tinidazole, 9588
[19388-87-5] Taurolidine, 9243
[19395-58-5] Moquizone, 6347
[19395-78-9] Moquizone Hydrochloride*, 6347
[19396-03-3] Polyoxin A*, 7736
[19396-06-6] Polyoxin B*, 7736
[19428-14-9] Benproperine Trihydrogen Phosphate*, 1078
[19457-37-5] Batrachotoxinin A*, 1037
[19485-08-6] Cyproquinate, 2843
[19504-77-9] Pecilocin, 7193
[19525-15-6] Uranium Peroxide, 9993
[19545-26-7] Wortmannin, 10188
[19554-22-4] Cytohemin, 2862
[19562-30-2] Piromidic Acid, 7660
[19624-22-7] Pentaborane(9), 7237
[19625-10-6] Lactobacillic Acid, 5353
[19660-77-6] Phytochlorin, 7543
[19666-30-9] Oxadiazon, 7038
[19700-21-1] Geosmin, 4408
[19717-25-0] Peyonine, 7333
[19718-36-6] Potassium Osmate(VI), 7819
[19721-56-3] Picromycin, 7566
[19728-88-2] Metitepine Maleate*, 6222
[19767-45-4] Mesna, 5969
[19771-63-2] L-2-Oxo-4-thiazolidinecarboxylic Acid, 7084
[19774-82-4] Amiodarone Hydrochloride*, 504
[19793-20-5] Bolandiol, 1352
[19794-93-5] Trazodone, 9712
[19803-62-4] Picloxydine Dihydrochloride*, 7553
[19804-27-4] p-Methyldiphenhydramine, 6130
[19807-27-3] Palladium Diacetate, 7123
[19809-78-0] Uliginosin A*, 9979
[19809-79-1] Uliginosin B*, 9979
[19875-60-6] Lisuride Maleate*, 5541

[19879-03-9] Granaticin B*, 4555
[19879-06-2] Granaticin, 4555
[19881-18-6] Nitroscanate, 6731
[19888-56-3] Fluazacort, 4151
[19891-74-8] Lycoxanthin, 5657
[19891-75-9] Lycophyll, 5651
[19973-76-3] Estrone dl-Form*, 3751
[19982-08-2] Memantine, 5872
[19992-80-4] Xenbucin trans-4-Phenylcyclohexylamine Salt (1:1)*, 10205
[20004-62-0] Resistomycin, 8320
[20055-57-6] Scabiolide, 8535
[20123-80-2] Dobesilate Calcium, 3455
[20153-98-4] Dilazep Dihydrochloride*, 3244
[20168-99-4] Cinmetacin, 2353
[20170-20-1] Difenamizole, 3181
[20187-55-7] Bendazac, 1061
[20196-67-2] Sinalbin, 8687
[20229-30-5] Metitepine, 6222
[20231-81-6] Uzarin, 10035
[20236-82-2] Dihydrocodeinone Enol Acetate Hydrochloride*, 3215
[20246-33-7] D-Gulonic Acid, 4604
[20267-87-2] Chelidonine (±)-Form*, 2095
[20283-48-1] Chalcomycin, 2077
[20284-40-6] Murexine, 6387
[20287-37-0] Fenquizone, 4039
[20310-89-8] Saponarin, 8512
[20311-78-8] Enanthotoxin, 3608
[20315-46-2] β-Ergocryptine*, 3689
[20316-18-1] Lycodine, 5647
[20326-12-9] Mepiprazole, 5902
[20338-08-3] Titanic(IV) Acid, 9609
[20344-15-4] Mepiprazole Dihydrochloride*, 5902
[20344-49-4] Ferric Hydroxide, 4068
[20350-15-6] Brefeldin A, 1393
[20354-26-1] Methazole, 6032
[20380-58-9] Tilidine, 9578
[20405-64-5] Cuprous Selenide, 2736
[20408-97-3] 5-Thio-D-glucose, 9470
[20423-87-4] Hydramitrazine d-Tartrate*, 4803
[20427-56-9] Ruthenium Tetroxide, 8454
[20427-59-2] Cupric Hydroxide, 2709
[20432-64-8] Iprindole Hydrochloride*, 5091
[20438-98-6] Imipramine N-Oxide Hydrochloride*, 4956
[20449-79-0] Melittin, 5867
[20480-93-7] Methenamine Sulfosalicylate, 6042
[20493-41-8] Quinapyramine, 8234
[20537-88-6] Amifostine, 424
[20548-54-3] Calcium Sulfide, 1754
[20554-84-1] Parthenolide, 7180
[20559-55-1] Oxibendazole, 7070
[20562-02-1] Solanine, 8853
[20574-50-9] Morantel, 6348
[20594-83-6] Nalbuphine, 6444
[20623-13-6] Nitrocobalamin*, 4854
[20644-87-5] Vanadium Carbonyl, 10054
[20667-12-3] Silver Oxide, 8665
[20684-06-4] Bamifylline Hydrochloride*, 982
[20702-77-6] Neohesperidin Dihydrochalcone, 6540
[20725-03-5] Fustin, 4341
[20770-09-6] Stannic Selenide, 8934
[20788-07-2] Resorantel, 8322
[20816-12-0] Osmium Tetroxide, 7024
[20830-75-5] Digoxin, 3210
[20830-81-3] Daunorubicin, 2890
[20831-76-9] Gentiopicrin, 4403

[20859-73-8] Aluminum Phosphide, 372
[20867-01-0] Fulvoplumierin, 4307
[20902-45-8] Penicillamine Disulfide D-Form*, 7216
[20905-71-9] Pyridoxal 5-Phosphate O-Methyloxime*, 8163
[20972-43-4] Azoxybenzene trans-Form*, 954
[21018-84-8] Amarogentin, 393
[21040-59-5] Codamine, 2524
[21041-93-0] Cobaltous Hydroxide, 2503
[21041-95-2] Cadmium Hydroxide, 1656
[21056-98-4] Calcium Phosphite, 1743
[21085-60-9] Calcium Mesoxalate, 1726
[21087-64-9] Metribuzin, 6239
[21109-95-5] Barium Sulfide, 1024
[21133-52-8] Lycoramine, 5655
[21150-22-1] β-Amanitin*, 387
[21174-80-1] Penicillamine Disulfide DL-Form*, 7216
[21187-73-5] Gardenin A*, 4377
[21187-98-4] Gliclazide, 4448
[21200-24-8] Indolmycin, 4996
[21215-62-3] Calcitonin, human synthetic*, 1680
[21238-30-2] Carbomycin B*, 1854
[21238-33-5] Oryzanol A*, 7016
[21245-01-2] 4-(Dimethylamino)benzoic Acid 3-Methylbutyl Ester*, 3282
[21245-02-3] 4-(Dimethylamino)benzoic Acid 2-Ethylhexyl Ester*, 3282
[21256-18-8] Oxaprozin, 7057
[21259-20-1] T 2 Toxin, 9933
[21259-76-7] Mercaptomerin Sodium, 5918
[21293-20-9] Isoquassin, 5239
[21293-29-8] Abscisic Acid, 7
[21302-79-4] Ceanothic Acid, 1959
[21312-10-7] N⁴-Acetylsulfamethoxazole*, 9086
[21351-79-1] Cesium Hydroxide, 2055
[21361-93-3] Nicotine Hydrochloride*, 6611
[21362-69-6] Mepitiostane, 5904
[21363-18-8] Viminol, 10117
[21370-21-8] Fenoxazoline Hydrochloride*, 4025
[21411-53-0] Virginiamycin M₁*, 10142
[21416-53-5] Picrotin, 7569
[21416-87-5] Razoxane, 8295
[21434-91-3] Capobenic Acid, 1799
[21466-07-9] Bromofenofos, 1440
[21498-08-8] Lofexidine Hydrochloride*, 5588
[21535-47-7] Mianserin Hydrochloride*, 6260
[21548-32-3] Fosthietan, 4284
[21548-73-2] Silver Sulfide, 8677
[21590-92-1] Etomidoline, 3928
[21609-90-5] Leptophos, 5468
[21637-25-2] Isoquercitrin, 5240
[21645-51-2] Aluminum Hydroxide, 355
[21649-57-0] Carbenicillin Phenyl Sodium*, 1838
[21650-42-0] Tretoquinol, 9719
[21651-19-4] Stannous Oxide, 8944
[21651-62-7] Nepetalactone cis-trans-Form*, 6556
[21679-14-1] Fludarabine, 4162
[21715-46-8] Etifoxine, 3910
[21721-92-6] Nitrefazole, 6668
[21725-46-2] Cyanazine, 2755
[21730-16-5] Metapramine, 5991
[21738-42-1] Oxamniquine, 7051

[21755-66-8] Picoperine, 7558
[21829-25-4] Nifedipine, 6617
[21888-96-0] Dexetimide Hydrochloride*, 2987
[21888-98-2] Dexetimide, 2987
[21898-19-1] Clenbuterol Hydrochloride*, 2407
[21908-53-2] Mercuric Oxide, Red, 5935
[21908-53-2] Mercuric Oxide, Yellow, 5936
[21995-38-0] Ammonium Cerous Sulfate, 536
[22006-84-4] Denopterin, 2944
[22059-60-5] Disopyramide Phosphate*, 3424
[22071-15-4] Ketoprofen, 5316
[22089-22-1] Trofosfamide, 9897
[22103-14-6] Bufeniode, 1495
[22131-35-7] Butalamine, 1535
[22131-79-9] Alclofenac, 220
[22131-80-2] Alclofenac Monoethanolamine Complex*, 220
[22139-77-1] Pinosylvin, 7604
[22146-03-8] Sarverogenin, 8527
[22150-76-1] Biopterin, 1270
[22151-68-4] Methohexital Sodium, 6061
[22154-43-4] Meglumine Acetrizoate, 5850
[22178-11-6] Pipotiazine Undecylenic Ester*, 7636
[22189-32-8] Spectinomycin Dihydrochloride Pentahydrate*, 8890
[22195-34-2] Guanadrel Sulfate*, 4587
[22199-08-2] Sulfadiazine Silver Salt*, 9071
[22199-46-8] Clomacran Phosphate*, 2441
[22202-75-1] Cephaloglycin Dihydrate*, 2023
[22204-24-6] Pyrantel Pamoate*, 8139
[22204-29-1] Cetoxime Hydrochloride*, 2065
[22204-53-1] Naproxen, 6504
[22204-88-2] Tramadol Hydrochloride*, 9701
[22205-45-4] Cuprous Sulfide, 2737
[22224-92-6] Fenamiphos, 3997
[22232-54-8] Carbimazole, 1844
[22232-55-9] DL-Pentorex Hydrogen D-Tartrate*, 7275
[22232-57-1] Racefemine, 8273
[22232-71-9] Mazindol, 5801
[22248-79-9] Tetrachlorvinphos, 9334
[22254-24-6] Ipratropium Bromide, 5089
[22260-51-1] Bromocriptine Methanesulfonate*, 1437
[22263-79-2] Antheridiol, 719
[22298-29-9] Betamethasone 17-Benzoate*, 1226
[22304-30-9] Apazone Dihydrate*, 768
[22316-47-8] Clobazam, 2417
[22330-18-3] Potassium Triiodomercurate(II) Solution, 7870
[22345-47-7] Tofisopam, 9642
[22368-82-7] Pithecolobine, 7665
[22398-80-7] Indium Phosphide, 4985
[22457-89-2] Benfotiamine, 1068
[22465-48-1] Acetatocobalamin*, 4854
[22487-42-9] Benapryzine, 1057
[22494-42-4] Diflunisal, 3190
[22494-47-9] Clobuzarit, 2426
[22537-19-5] Lawrencium, 5405
[22560-50-5] Clodronic Acid Disodium Salt*, 2432
[22572-05-0] Penicillamine DL-Form Hydrochloride*, 7214

[22573-93-9] Alexidine, 231
[22619-35-8] Tioclomarol, 9594
[22647-32-1] Azidocillin Potassium Salt*, 942
[22650-86-8] Bismuth Subgallate, 1325
[22661-76-3] Amoproxan, 611
[22661-96-7] Amoproxan Hydrochloride*, 611
[22662-39-1] Rafoxanide, 8280
[22664-55-7] Metipranolol, 6221
[22668-01-5] Etanidazole, 3756
[22672-74-8] Isocorybulbine, 5177
[22722-98-1] Vitride, 10166
[22733-60-4] Siccanin, 8630
[22760-18-5] Proquazone, 8059
[22775-12-8] Dimetofrine Hydrochloride*, 3314
[22775-52-6] Mycelianamide, 6400
[22781-23-3] Bendiocarb, 1063
[22831-39-6] Magnesium Silicide, 5728
[22832-87-7] Miconazole Nitrate*, 6266
[22839-47-0] Aspartame, 874
[22862-76-6] Anisomycin, 708
[22887-42-9] Prednylidene 21-Diethylaminoacetate Hydrochloride*, 7906
[22888-70-6] Silybin, 8680
[22916-47-8] Miconazole, 6266
[22950-29-4] Dimetofrine, 3314
[22963-93-5] Juvenile Hormone III*, 5287
[22992-10-5] Aluminum Zinc Sulfate, 385
[22994-85-0] Benznidazole, 1114
[23031-25-6] Terbutaline, 9302
[23031-32-5] Terbutaline Sulfate*, 9302
[23047-25-8] Lofepramine, 5587
[23049-93-6] Enfenamic Acid, 3620
[23067-13-2] Erythromycin Glucoheptonate, 3723
[23076-35-9] Xylazine Hydrochloride*, 10213
[23089-26-1] α-Bisabolol (−)-Form*, 1281
[23092-17-3] Halazepam, 4619
[23093-74-5] Bunitrolol Hydrochloride*, 1514
[23103-98-2] Pirimicarb, 7651
[23109-05-9] α-Amanitin*, 387
[23110-15-8] Fumagillin, 4308
[23135-22-0] Oxamyl, 7052
[23142-01-0] Carbetapentane Citrate*, 1840
[23152-29-6] Virginiamycin S_1*, 10142
[23155-02-4] Fosfomycin, 4280
[23178-88-3] α-Bisabolol (+)-Form*, 1281
[23184-66-9] Butachlor, 1533
[23192-42-9] Hexalure, 4723
[23210-56-2] Ifenprodil, 4936
[23214-92-8] Doxorubicin, 3495
[23214-96-2] Alcuronium, 222
[23239-37-4] Etoxadrol (+)-Hydrochloride*, 3933
[23239-41-0] Cephacetrile Sodium, 2019
[23239-51-2] Ritodrine Hydrochloride*, 8401
[23249-97-0] Procodazole, 7943
[23250-73-9] Manganese Oleate, 5774
[23255-54-1] Dihydro-β-erythroidine, 3218
[23255-59-6] Lunularic Acid, 5634
[23255-93-8] Hycanthone Mesylate*, 4795
[23256-30-6] Nifurtimox, 6630
[23256-40-8] Guanoxabenz Hydrochloride*, 4597

[23256-50-0] Guanabenz Monoacetate*, 4585
[23257-56-9] Levophacetoperane Hydrochloride*, 5493
[23271-63-8] Amicibone, 416
[23277-43-2] Nalbuphine Hydrochloride*, 6444
[23282-20-4] Nivalenol, 6757
[23282-55-5] Sulfachlorpyridazine Sodium Salt*, 9068
[23288-49-5] Probucol, 7935
[23288-60-0] Sodium Pertechnetate ^{99m}Tc, 8802
[23315-05-1] Elaiomycin, 3575
[23325-78-2] Cephalexin Monohydrate*, 2021
[23327-57-3] Nefopam Hydrochloride*, 6529
[23344-16-3] Streptovaricin A*, 8989
[23344-17-4] Streptovaricin C*, 8989
[23383-11-1] Ferrous Citrate, 4092
[23413-80-1] Aluminum Bis(acetylsalicylate), 338
[23414-72-4] Zinc Permanganate, 10281
[23454-33-3] Trenbolone Cyclohexylmethylcarbonate*, 9716
[23465-76-1] Caroverine, 1906
[23477-98-7] Sedecamycin, 8566
[23496-41-5] Verticine, 10104
[23505-41-1] Pirimiphos-ethyl, 7652
[23509-16-2] Batrachotoxin, 1037
[23509-17-3] Homobatrachotoxin*, 1037
[23513-14-6] [6]-Gingerol, 4431
[23541-50-6] Daunorubicin Hydrochloride*, 2890
[23560-59-0] Heptenophos, 4699
[23564-05-8] Thiophanate-methyl*, 9489
[23564-06-9] Thiophanate, 9489
[23593-75-1] Clotrimazole, 2478
[23602-78-0] Benfluorex, 1066
[23605-05-2] α-Antiarin, 731
[23642-66-2] Benfluorex Hydrochloride*, 1066
[23651-95-8] Droxidopa, 3513
[23666-50-4] Rhodomycin A*, 8357
[23668-11-3] Pactamycin, 7119
[23672-07-3] Levosulpiride*, 9163
[23674-86-4] Difluprednate, 3194
[23694-17-9] Sultopride Hydrochloride*, 9168
[23694-81-7] Mepindolol, 5901
[23696-28-8] Olaquindox, 6960
[23736-58-5] Cloxacillin Benzathine Salt*, 2480
[23777-80-2] Hexaborane(10), 4712
[23779-99-9] Floctafenine, 4140
[23784-10-3] Viminol p-Hydroxybenzoate*, 10117
[23828-92-4] Ambroxol Hydrochloride*, 401
[23843-52-9] Phloionic Acid, 7478
[23873-81-6] Benzil Dioxime, 1108
[23887-31-2] Clorazepic Acid, 2465
[23887-41-4] Cinepazet, 2349
[23887-46-9] Cinepazide, 2350
[23910-07-8] Mebiquine, 5810
[23915-80-2] Chromocarb Diethylamine*, 2294
[23918-98-1] Eritadenine, 3712
[23930-19-0] Alfaxalone, 235
[23930-37-2] Alfadolone Acetate, 233
[23947-60-6] Ethirimol, 3786
[23950-58-5] Propyzamide, 8058
[23964-57-0] Carticaine Hydrochloride*, 1920
[23964-58-1] Carticaine, 1920
[24017-47-8] Triazophos, 9736

Denotes derivative of title compound

[24047-25-4] Guanoxabenz, 4597
[24159-07-7] Febrifugine, 3984
[24166-13-0] Cloxazolam, 2481
[24168-96-5] Isoconazole Nitrate*, 5176
[24169-02-6] Econazole Nitrate* (unspecified), 3550
[24219-97-4] Mianserin, 6260
[24237-54-5] Tinoridine, 9590
[24243-97-8] Tymazoline, 9965
[24279-91-2] Carboquone, 1872
[24280-93-1] Mycophenolic Acid, 6408
[24304-00-5] Aluminum Nitride, 366
[24305-27-9] TRH, 9720
[24307-26-4] Mepiquat Chloride, 5903
[24321-47-9] Uscharidin, 10030
[24345-16-2] Apamin, 767
[24356-60-3] Cephapirin Sodium, 2030
[24356-94-3] Algestone Acetophenide, 239
[24358-65-4] Diisopromine Hydrochloride*, 3239
[24381-49-5] Orotic Acid Choline Salt*, 7004
[24381-60-0] Chromic Phosphate Radioactive Form*, 2284
[24390-14-5] Doxycycline Hyclate*, 3496
[24394-09-0] Cnicin, 2486
[24397-89-5] Actinobolin, 139
[24526-64-5] Nomifensine, 6768
[24527-27-3] Clospirazine, 2474
[24531-57-5] 1-Thiosorbitol, 9501
[24558-01-8] Levophacetoperane, 5493
[24582-91-0] Elenolide, 3586
[24584-09-6] Razoxane (+)-Form*, 8295
[24600-36-0] Fominoben Hydrochloride*, 4258
[24602-86-6] Tridemorph, 9793
[24622-52-4] Amixetrine Hydrochloride*, 514
[24622-72-8] Amixetrine, 514
[24634-61-5] Potassium Sorbate, 7841
[24656-22-2] Picoperine Tripalmitate*, 7558
[24691-76-7] Pyracarbolid, 8138
[24697-74-3] Leonurine, 5463
[24699-40-9] Picoperine Hydrochloride*, 7558
[24701-51-7] Demexiptiline, 2941
[24724-90-1] Fucosamine, 4300
[24730-10-7] Dihydroergocristine Methanesulfonate, 3692
[24751-69-7] Nucleocidin, 6824
[24815-24-5] Rescinnamine, 8311
[24815-25-6] Ethybenztropine Hydrobromide*, 3802
[24818-79-9] Clofibric Acid Basic Aluminum Salt*, 2437
[24828-59-9] Chaulmoogric Acid Methyl Ester*, 2089
[24868-20-0] Dantrolene Sodium Salt Hemiheptahydrate*, 2879
[24870-04-0] Giractide, 4435
[24880-43-1] Mesembrine, 5966
[24915-04-6] β-Isospartcine, 5246
[24916-50-5] Spiramycin I*, 8904
[24916-51-6] Spiramycin II*, 8904
[24916-52-7] Spiramycin III*, 8904
[24967-93-9] Chondroitin 4-Sulfate*, 2270
[24967-94-0] Dermatan Sulfate*, 2270
[25001-57-4] Justicidin A*, 5286
[25013-16-5] Butylated Hydroxyanisole, 1582
[25038-54-4] Nylon 6, 6832
[25046-79-1] Glisoxepid, 4453
[25053-27-4] Lyapolate Sodium, 5645

[25057-89-0] Bentazon, 1080
[25102-12-9] Edetic Acid Dipotassium Salt Dihydrate*, 3559
[25104-18-1] Polylysine, 7733
[25117-79-1] Isomethadol, 5200
[25122-41-2] Clobetasol, 2423
[25122-46-7] Clobetasol 17-Propionate*, 2423
[25122-57-0] Clobetasone 17-Butyrate*, 2424
[25126-32-3] Sincalide, 8689
[25127-31-5] Doxepin cis-Form Hydrochloride*, 3492
[25152-52-7] Aluminum Antimonide, 336
[25154-52-3] Nonyl Phenol, 6775
[25155-15-1] Cymene, 2832
[25155-18-4] Methylbenzethonium Chloride, 6103
[25155-30-0] Sodium Dodecylbenzenesulfonate, 8757
[25161-91-5] Loline, 5591
[25229-42-9] Cicrotoic Acid, 2327
[25301-02-4] Tyloxapol, 9964
[25311-71-1] Isofenphos, 5187
[25312-65-6] Cholanic Acid, 2251
[25316-40-9] Doxorubicin Hydrochloride*, 3495
[25322-46-7] Chondroitin 6-Sulfate*, 2270
[25322-68-3] Polyethylene Glycol, 7729
[25332-39-2] Trazodone Hydrochloride*, 9712
[25333-25-9] 3-(o-Methoxyphenyl)-2-phenylacrylic Acid, 6078
[25333-42-0] 3-Quinuclidinol l-Form*, 8266
[25333-43-1] 3-Quinuclidinol dl-Form Hydrochloride*, 8266
[25333-48-6] 3-Quinuclidinol dl-Form Benzoate Hydrochloride*, 8266
[25333-96-4] Spasmolytol, 8888
[25354-97-6] 2-Hexyldecanoic Acid, 4747
[25384-17-2] Allylprodine, 304
[25384-21-8] 3-Methylarsacetin, 6102
[25389-94-0] Kanamycin A Sulfate*, 5293
[25389-99-5] Neomycin Neomycin B Hydrochloride*, 6542
[25390-16-3] Tanghinigenin 3-[(2-O-Acetyl-6-deoxy-3-O-methyl-α-L-glucopyranosyl)oxy]*, 9220
[25394-57-4] Affinin, 177
[25394-75-6] Alstonidine, 323
[25394-78-9] Cetoxime, 2065
[25395-22-6] Salicylamide O-Acetic Acid, 8481
[25395-31-7] Diacetin, 3005
[25402-06-6] Cinerin I*, 2351
[25428-43-7] α-Bisabolol (±)-form, 1281
[25451-15-4] Felbamate, 3988
[25486-55-9] Phylloquinone 2,3-Epoxide*, 7536
[25523-97-1] Chlorpheniramine d-Form*, 2232
[25546-65-0] Ribostamycin, 8373
[25573-43-7] Eseridine, 3740
[25614-03-3] Bromocriptine, 1437
[25635-88-5] AIR, 191
[25655-41-8] Povidone-Iodine, 7880
[25663-23-4] A-Denopterin, 151
[25672-33-7] Metformin p-Chlorophenoxyacetate (Salt)*, 6001
[25717-80-0] Molsidomine, 6316
[25803-14-9] Clometacin, 2443
[25808-74-6] Lead Hexafluorosilicate, 5427

[25812-30-0] Gemfibrozil, 4394
[25827-76-3] Iomeglamic Acid, 5069
[25846-73-5] Petunidin 3,5-Diglucoside*, 7331
[25867-19-0] Lycoctonine N-Succinylanthranilic Acid Ester*, 5646
[25875-50-7] Robenidine Hydrochloride*, 8403
[25875-51-8] Robenidine, 8403
[25876-10-2] Gentamicin C₁*, 4398
[25876-11-3] Gentamicin C₂*, 4398
[25895-60-7] Sodium Cyanoborohydride, 8751
[25905-77-5] Minaprine, 6287
[25913-34-2] Tinoridine Hydrochloride*, 9590
[25953-17-7] Minaprine Dihydrochloride*, 6287
[25953-19-9] Cefazolin, 1967
[25956-17-6] Allura Red AC, 292
[25967-29-7] Flutoprazepam, 4245
[25978-54-5] Spherophysine, 8896
[25999-31-9] Lasalocid A, 5384
[26000-17-9] Lycoctonine, 5646
[26002-80-2] Phenothrin, 7405
[26020-55-3] Oxetorone, 7068
[26027-38-3] Nonoxynol, 6772
[26086-49-7] Deoxydihydrostreptomycin, 2949
[26097-80-3] Cambendazole, 1774
[26098-04-4] Gentamicin C₁ₐ*, 4398
[26155-31-7] Morantel Tartrate*, 6348
[26159-34-2] Naproxen Sodium Salt*, 6504
[26171-23-3] Tolmetin, 9655
[26209-07-4] Ciclosidomine Hydrochloride*, 2326
[26225-79-6] Ethofumesate, 3788
[26241-81-6] Coumingine, 2634
[26242-33-1] Vintiamol, 10129
[26281-69-6] Exiproben, 3959
[26296-41-3] Cassaidine, 1937
[26299-14-9] Pyridinium Chlorochromate, 8157
[26305-03-3] Pepstatin, 7290
[26309-95-5] Pivampicillin Hydrochloride*, 7669
[26328-04-1] Cinepazide Maleate*, 2350
[26328-53-0] Amoscanate, 613
[26382-17-2] Chlorbicyclen trans-Isomer*, 2127
[26399-36-0] Profluralin, 7954
[26445-03-4] Thiocresol, 9461
[26446-35-5] Monacetin, 6235
[26472-41-3] Humulon, 4790
[26490-31-3] Nandrolone Dodecanoate*, 6452
[26530-20-1] Octhilinone, 6853
[26531-85-1] Primeverose, 7926
[26532-22-9] Grandisol, 4556
[26544-34-3] Apiin, 775
[26570-10-5] Propoxyphene Napsylate Monohydrate*, 8024
[26591-04-8] COP, 2577
[26594-41-2] Isomethadone d-Form*, 5201
[26598-44-7] Ethybenztropine Hydrochloride*, 3802
[26605-16-3] Nocardamin, 6763
[26605-69-6] Carindacillin Sodium Salt*, 1888
[26608-34-4] Etioporphyrin, 3915
[26615-21-4] Zotepine, 10327
[26628-22-8] Sodium Azide, 8724
[26629-86-7] Oxaflozane Hydrochloride*, 7039
[26629-87-8] Oxaflozane, 7039
[26644-46-2] Triforine, 9819
[26652-09-5] Ritodrine, 8401

Denotes derivative of title compound

[26657-13-6] Rolitetracycline Nitrate Sesquihydrate*, 8411
[26658-42-4] Colestipol, 2538
[26675-46-7] Isoflurane, 5191
[26750-66-3] Potassium Thiocarbonate, 7865
[26750-81-2] Alibendol, 243
[26754-48-3] Terpenylic Acid, 9313
[26774-90-3] Epicillin, 3651
[26786-32-3] Lofepramine Hydrochloride*, 5587
[26787-78-0] Amoxicillin, 617
[26791-73-1] Xanthatin, 10192
[26807-65-8] Indapamide, 4969
[26839-75-8] Timolol, 9585
[26844-12-2] Indoramin, 5000
[26864-56-2] Penfluridol, 7213
[26871-30-7] Herqueinone, 4703
[26909-33-1] Rugulovasine A*, 8449
[26921-17-5] Timolol Hydrogen Maleate Salt*, 9585
[26944-48-9] Glibornuride, 4447
[26952-21-6] Isooctyl Alcohol, 5211
[26973-24-0] Ceftezole, 1997
[27007-85-8] Clospirazine Hydrochloride*, 2474
[27013-91-8] α-Hederin, 4655
[27025-41-8] Glutathione Disulfide*, 4483
[27031-08-9] Sulfaguanole, 9077
[27035-30-9] Oxametacine, 7048
[27058-84-0] Moprolol Hydrochloride*, 6345
[27060-91-9] Flutazolam, 4243
[27073-72-9] Tiliacorine, 9577
[27107-79-5] Tilidine trans-(±)-Hydrochloride*, 9578
[27115-36-2] Chromic Formate, 2280
[27127-62-4] Viscosin, 10146
[27164-46-1] Cefazolin Sodium Salt*, 1967
[27167-30-2] Oxazolam, 7061
[27200-12-0] Ampelopsin, 618
[27203-92-5] Tramadol, 9701
[27208-80-6] Resveratrol 3-β-Mono-D-glucoside*, 8326
[27214-00-2] Calcium Glycerophosphate, 1713
[27220-47-9] Econazole, 3550
[27223-35-4] Ketazolam, 5308
[27267-69-2] Collinomycin, 2545
[27276-25-1] Capobenic Acid Sodium Salt*, 1799
[27279-76-1] Stibocaptate, 8966
[27314-13-2] Norflurazon, 6794
[27315-91-9] Pipebuzone, 7610
[27367-90-4] Niaprazine, 6576
[27469-53-0] Almitrine, 309
[27470-51-5] Suxibuzone, 9185
[27471-60-9] Fenoxedil Hydrochloride*, 4026
[27511-99-5] Eterobarb, 3758
[27519-02-4] Muscalure, 6388
[27523-40-6] Isoconazole, 5176
[27588-43-8] Eprozinol Dihydrochloride*, 3670
[27589-33-9] Azosemide, 953
[27591-42-0] Oxazidione, 7060
[27591-69-1] Tilorone Dihydrochloride*, 9581
[27591-97-5] Tilorone, 9581
[27607-77-8] Trimethylsilyl Triflate, 9849
[27686-84-6] Nordihydroguaiaretic Acid (meso-form), 6786
[27709-53-1] Potassium Uranyl Sulfate, 7875
[27724-96-5] Cetraxate Hydrochloride*, 2067

[27726-31-4] Pivcefalexin Hydrochloride*, 7670
[27762-78-3] Kethoxal, 5310
[27774-13-6] Vanadyl Sulfate, 10063
[27780-11-6] Chrysanthemaxanthin, 2309
[27790-37-0] Potassium Stannosulfate, 7843
[27792-36-5] Etiocobalamin, 3914
[27833-64-3] Loxapine Succinate*, 5617
[27848-84-6] Nicergoline, 6580
[27876-94-4] Crocetin, 2657
[27877-51-6] Tolindate, 9652
[27885-92-3] Imidocarb, 4951
[27912-14-7] Levobunolol Hydrochloride*, 5488
[27959-26-8] Nicomol, 6606
[27973-72-4] Pederin, 7196
[28008-04-0] Acetiamine Hydrochloride*, 51
[28014-46-2] Polyestradiol Phosphate, 7727
[28041-84-1] Magnesium Potassium Selenate, 5721
[28041-86-3] Potassium Cobaltous Selenate, 7787
[28189-85-7] Etoxadrol, 3933
[28272-18-6] Pyrethrosin, 8149
[28289-54-5] MPTP, 6376
[28300-74-5] Antimony Potassium Tartrate, 741
[28319-77-9] Choline Alfoscerate, 2260
[28380-24-7] Nigericin, 6633
[28384-96-5] Enterobactin, 3631
[28393-42-2] Cephalosporin P₁, 2027
[28395-03-1] Bumetanide, 1508
[28416-66-2] Isoaminile Citrate*, 5124
[28434-01-7] Bioresmethrin, 1271
[28521-94-0] Methcathinone (±), 6033
[28523-86-6] Sevoflurane, 8621
[28558-32-9] Thiabendazole Hypophosphite*, 9426
[28591-01-7] F-Gitonin, 4439
[28610-84-6] Rimazolium Metilsulfate, 8391
[28657-80-9] Cinoxacin, 2369
[28704-27-0] COP 1, 2578
[28721-07-5] Oxcarbazepine, 7063
[28728-55-4] Hexadimethrine Bromide, 4719
[28772-56-7] Bromadiolone, 1403
[28782-19-6] Flavoxate Succinate*, 4135
[28782-42-5] Difenoxin, 3183
[28797-61-7] Pirenzepine, 7646
[28860-95-9] Carbidopa, 1843
[28905-10-4] Glucosamine β-Form*, 4466
[28905-11-5] Glucosamine α-Form*, 4466
[28911-01-5] Triazolam, 9734
[28947-50-4] Fencamine, 4007
[28950-34-7] Tetrasulfur Tetranitride, 9378
[28968-07-2] Clidanac, 2411
[28979-74-0] Vernamycin B_γ*, 10096
[28981-97-7] Alprazolam, 320
[28983-56-4] Methyl Blue, 6107
[28994-41-4] o-Benzylphenol, 1179
[29025-14-7] Butropium Bromide, 1569
[29031-19-4] Glucosamine Sulfate Salt*, 4466
[29069-24-7] Prednimustine, 7900
[29074-38-2] Canadine dl-Form*, 1785
[29091-21-2] Prodiamine, 7947
[29094-61-9] Glipizide, 4451
[29098-15-5] Terofenamate, 9312

[29106-32-9] Chaulmoogric Acid, 2089
[29110-47-2] Guanfacine, 4590
[29110-48-3] Guanfacine Hydrochloride*, 4590
[29119-03-7] Frequentin, 4293
[29122-68-7] Atenolol, 892
[29139-17-1] Vernamycin B_δ*, 10096
[29177-84-2] Ethyl Loflazepate, 3867
[29202-00-4] Gardenin D*, 4377
[29216-28-2] Mequitazine, 5911
[29218-27-7] Toloxatone, 9659
[29232-93-7] Pirimiphos-methyl*, 7652
[29334-07-4] Sulmarin, 9158
[29342-05-0] Ciclopirox, 2325
[29365-11-5] Giractide Hexaacetate Salt*, 4435
[29393-20-2] Bacilysin, 963
[29400-42-8] Helvolic Acid, 4666
[29550-05-8] Gardenin C*, 4377
[29550-07-0] Gardenin E*, 4377
[29560-58-5] Moricizine Hydrochloride*, 6351
[29608-49-9] Almitrine Dimethanesulfonate*, 309
[29689-14-3] Chromic Carbonate, 2277
[29714-87-2] Ocimene, 6837
[29728-34-5] Acetoxolone Aluminium Salt*, 76
[29767-20-2] Teniposide, 9291
[29790-52-1] Nicotine Salicylate*, 6611
[29804-22-6] Disparlure (2S-cis), 3425
[29838-46-8] Roccellic Acid, 8405
[29868-97-1] Pirenzepine Dihydrochloride*, 7646
[29870-99-3] Strontium Lactate, 9006
[29883-15-6] Amygdalin, 638
[29899-95-4] Clobenoside, 2419
[29908-03-0] S-Adenosylmethionine, 155
[29913-86-8] Amarolide, 394
[29952-87-2] Clofibric Acid Pyridoxine Salt*, 2437
[29956-24-9] Pyridoxal 5-Phosphate Calcium Salt*, 8163
[29975-16-4] Estazolam, 3744
[30042-37-6] Lankamycin, 5369
[30073-40-6] Bucumolol Hydrochloride* (unspecified), 1489
[30116-80-4] Floredil Hydrochloride*, 4144
[30123-17-2] Tianeptine Sodium Salt*, 9560
[30223-48-4] Fluacizine, 4149
[30272-08-3] Amineptine Hydrochloride*, 429
[30279-49-3] Suclofenide, 9048
[30286-75-0] Oxitropium Bromide, 7077
[30299-08-2] Clinofibrate, 2415
[30392-40-6] Bitolterol, 1344
[30392-41-7] Bitolterol Methanesulfonate*, 1344
[30408-30-1] Nybomycin, 6829
[30452-69-8] L-Cystine S,S-dioxide*, 2852
[30484-77-6] Flunarizine Dihydrochloride*, 4179
[30516-87-1] Zidovudine, 10252
[30525-89-4] Paraformaldehyde, 7158
[30544-47-9] Etofenamate, 3922
[30544-61-7] Clanobutin, 2398
[30560-19-1] Acephate, 31
[30578-37-1] Amezinium Methyl Sulfate, 412
[30653-83-9] Parsalmide, 7177
[30685-43-9] β-Methyldigoxin*, 3210

* Denotes derivative of title compound

[30748-29-9] Feprazone, *4053*
[30905-63-6] Calcium *N*-Carbamoyl-aspartate DL-Form*, *1695*
[30909-51-4] Flupentixol Decanoate*, *4224*
[30924-31-3] Cafaminol, *1671*
[30964-13-7] Cynarin(e), *2835*
[31002-79-6] Triamcinolone Benetonide, *9729*
[31031-74-0] Phenamacide Hydrochloride, *7350*
[31036-80-3] Lofexidine, *5588*
[31112-62-6] Metrizamide, *6240*
[31218-83-4] Propetamphos, *8000*
[31282-04-9] Hygromycin B, *4901*
[31314-38-2] Prodipine, *7949*
[31314-39-3] Prodipine Hydrochloride*, *7949*
[31329-57-4] Nafronyl, *6440*
[31342-36-6] Chloramphenicol Pantothenate Calcium Complex (4:1)*, *2120*
[31362-50-2] Bombesin, *1359*
[31418-71-0] Coprogen, *2586*
[31428-61-2] Tiamenidine, *9558*
[31430-15-6] Flubendazole, *4154*
[31431-39-7] Mebendazole, *5807*
[31431-43-3] Cyclobendazole, *2781*
[31566-31-1] Glyceryl Monostearate, *4498*
[31570-39-5] Neocembrene, *6535*
[31586-77-3] Bismuth Sodium Tartrate, *1321*
[31637-97-5] Etofibrate, *3923*
[31677-93-7] Bupropion Hydrochloride*, *1523*
[31698-14-3] Ancitabine, *670*
[31721-17-2] Quinupramine, *8267*
[31774-33-1] Abikoviromycin, *4*
[31793-07-4] Pirprofen, *7663*
[31828-50-9] Cephradine Dihydrate*, *2032*
[31828-71-4] Mexiletine, *6257*
[31842-01-0] Indoprofen, *4999*
[31842-61-2] Rimiterol Hydrobromide*, *8393*
[31848-01-8] Morclofone, *6350*
[31853-38-0] 1,2-Naphthoquinone 2-Semicarbazone*, *6481*
[31868-18-5] Mexazolam, *6254*
[31879-05-7] Fenoprofen, *4021*
[31883-05-3] Moricizine, *6351*
[31884-76-1] Sulfanilic Acid Zinc Salt Tetrahydrate*, *9096*
[31884-77-2] Meclizine Dihydrochloride Monohydrate*, *5817*
[31932-09-9] Ticarbodine, *9567*
[31944-97-5] Chondrofoline, *2269*
[31980-29-7] Nicofibrate, *6604*
[32059-15-7] Guanazodine, *4588*
[32059-24-8] Homochlorcyclizine Hydrochloride*, *4770*
[32157-29-2] Propanethial *S*-Oxide, *7985*
[32164-16-2] Verticillin A*, *10103*
[32164-26-4] Streptovaricin D*, *8989*
[32190-57-1] Wildfire Toxin, *10183*
[32222-06-3] Calcitriol, *1681*
[32245-40-2] 1-Theobromineacetic Acid Sodium Salt*, *9419*
[32266-10-7] Hexoprenaline Sulfate*, *4745*
[32266-82-3] Butoctamide Hydrogen Succinate Calcium Salt*, *1562*
[32267-39-3] Divicine, *3448*
[32359-34-5] Medifoxamine, *5834*
[32381-42-3] Vernolic Acid (−)-Form*, *10099*
[32383-76-9] Medicarpin, *5833*
[32385-11-8] Sisomicin, *8695*

[32421-46-8] Bunaftine, *1509*
[32449-92-6] D-Glucuronolactone, *4476*
[32467-88-2] ACV, *147*
[32476-67-8] Periplocymarin, *7316*
[32527-55-2] Tiaramide, *9564*
[32619-42-4] Oleuropein, *6967*
[32630-75-4] Pododacric Acid, *7702*
[32665-36-4] Eprozinol, *3670*
[32672-69-8] Mesoridazine Benzenesulfonate*, *5970*
[32728-78-2] Heliosupine, *4659*
[32780-64-6] Labetalol Hydrochloride*, *5341*
[32784-82-0] Improsulfan Tosylate*, *4962*
[32795-44-1] Acecainide, *17*
[32795-47-4] Nomifensine Maleate*, *6768*
[32808-51-8] Bucloxic Acid, *1486*
[32808-53-0] Bucloxic Acid Calcium Salt*, *1486*
[32809-16-8] Procymidone, *7946*
[32828-81-2] Picotamide, *7560*
[32829-58-6] Gigantine, *4429*
[32838-26-9] Butoctamide, *1562*
[32838-28-1] Butoctamide Hydrogen Succinate*, *1562*
[32854-75-4] Lappaconitine, *5378*
[32886-97-8] Amdinocillin Pivoxil, *409*
[32887-01-7] Amdinocillin, *408*
[32887-03-9] Amdinocillin Pivoxil Hydrochloride*, *409*
[32909-92-5] Sulfametrole, *9089*
[32953-89-2] Rimiterol, *8393*
[32986-56-4] Tobramycin, *9628*
[32988-50-4] Viomycin, *10139*
[33005-95-7] Tiaprofenic Acid, *9562*
[33032-12-1] Methapyrilene Fumarate*, *6027*
[33069-62-4] Paclitaxel, *7117*
[33089-61-1] Amitraz, *510*
[33103-21-8] Tuberactinomycin A*, *9935*
[33103-22-9] Enviomycin, *3636*
[33125-97-2] Etomidate, *3927*
[33137-73-4] Tuberactinomycin O*, *9935*
[33159-27-2] Ecabet, *3538*
[33178-86-8] Alinidine, *244*
[33237-74-0] Aprindine Hydrochloride*, *793*
[33245-39-5] Fluchloralin, *4156*
[33286-22-5] Diltiazem *d-cis* Form Hydrochloride*, *3247*
[33342-05-1] Gliquidone, *4452*
[33368-20-6] Enduracidin A Hydrochloride*, *3619*
[33369-31-2] Zomepirac, *10322*
[33386-08-2] Buspirone Hydrochloride*, *1528*
[33396-37-1] Proscillaridin-4-methyl Ether*, *8060*
[33401-94-4] Pyrantel Tartrate*, *8139*
[33402-03-8] Metaraminol Bitartrate (Hydrogen L-Tartrate)*, *5993*
[33403-37-1] Nornidulin, *6808*
[33404-78-3] Negamycin, *6530*
[33419-42-0] Etoposide, *3931*
[33445-15-7] Ammonium Mercuric Chloride, *564*
[33445-35-1] Sulfamidochrysoidine Hydrochloride*, *9092*
[33490-33-4] Capreomycin IB*, *1801*
[33515-32-1] Vitamin U Bromide Analog*, *10165*
[33538-71-5] Venturicidin A*, *10080*
[33538-72-6] Venturicidin B*, *10080*

[33564-30-6] Cefoxitin Sodium Salt*, *1986*
[33564-31-7] Diflorasone Diacetate*, *3186*
[33579-45-2] Scotophobin, *8555*
[33580-30-2] Tertatolol Hydrochloride*, *9318*
[33605-67-3] Cargutocin, *1887*
[33605-94-6] Pyrisuccideanol, *8175*
[33629-47-9] Butralin, *1567*
[33659-28-8] Calcium Bromolactobionate, *1694*
[33665-90-6] Acesulfame, *35*
[33671-46-4] Clotiazepam, *2477*
[33765-68-3] Oxendolone, *7065*
[33797-51-2] Dimethyl(methylene)ammonium Iodide, *3300*
[33817-20-8] Pivampicillin, *7669*
[33818-15-4] Citicoline Sodium Salt*, *2380*
[33820-53-0] Isopropalin, *5220*
[33996-33-7] Oxaceprol, *7035*
[34014-18-1] Tebuthiuron, *9255*
[34018-28-5] Lead Bromate, *5417*
[34029-94-2] Laurepukine, *5395*
[34031-32-8] Auranofin, *911*
[34097-16-0] Clocortolone 21-Pivalate*, *2430*
[34118-92-8] Acecainide Hydrochloride*, *17*
[34123-59-6] Isoproturon, *5237*
[34135-85-8] Methyl Allyl Trisulfide, *6094*
[34140-59-5] Trimebutine Maleate*, *9829*
[34150-62-4] Ferriclate Calcium Sodium, *4070*
[34161-23-4] Fipexide Hydrochloride*, *4126*
[34161-24-5] Fipexide, *4126*
[34167-45-8] Polymyxin D_2*, *7734*
[34183-22-7] Propafenone Hydrochloride*, *7978*
[34184-77-5] Promegestone, *7969*
[34195-34-1] Hydrocodone Bitartrate Hemipentahydrate*, *4826*
[34214-51-2] Floxacillin Sodium Monohydrate*, *4147*
[34218-61-6] Juvenile Hormone II*, *5287*
[34255-08-8] Rhodoviolascin, *8361*
[34256-82-1] Acetochlor, *58*
[34273-10-4] Saralasin, *8518*
[34291-02-6] Butirosin A*, *1559*
[34291-03-7] Butirosin B*, *1559*
[34301-55-8] Isometamidium Chloride, *5199*
[34312-10-2] Citrulline Hydrochloride*, *2392*
[34316-15-9] Chelerythrine, *2093*
[34327-18-9] Chloramphenicol Monosuccinate Arginine Salt*, *2120*
[34368-04-2] Dobutamine, *3456*
[34381-68-5] Acebutolol Hydrochloride*, *16*
[34433-31-3] Asoxime Chloride, *870*
[34444-01-4] Cefamandole, *1964*
[34461-22-8] Metformin Embonate*, *6001*
[34493-98-6] Dibekacin, *3052*
[34503-87-2] Polymyxin B_2*, *7734*
[34521-09-0] Antimony Sodium Tartrate, *743*
[34522-46-8] Oxetorone Fumarate*, *7068*
[34524-20-4] Boromycin, *1372*
[34535-83-6] Fenalcomine Hydrochloride*, *3996*
[34552-83-5] Loperamide Hydrochloride*, *5601*

Denotes derivative of title compound

[34552-84-6] Isoxicam, 5258
[34580-13-7] Ketotifen, 5319
[34580-14-8] Ketotifen Fumarate*, 5319
[34590-94-8] Dipropylene Glycol Monomethyl Ether, 3407
[34616-39-2] Fenalcomine, 3996
[34633-34-6] Bifluranol, 1259
[34661-75-1] Urapidil, 10003
[34675-84-8] Cetraxate, 2067
[34707-92-1] Chlorothricin, 2222
[34758-83-3] Zipeprol, 10303
[34758-84-4] Zipeprol Dihydrochloride*, 10303
[34765-96-3] Alsactide, 322
[34765-98-5] Enduracidin B Hydrochloride*, 3619
[34784-64-0] Tertatolol, 9318
[34786-70-4] Nystatin A_1*, 6834
[34787-01-4] Ticarcillin, 9568
[34816-55-2] Moxestrol, 6372
[34866-46-1] Carbuterol Hydrochloride*, 1882
[34866-47-2] Carbuterol, 1882
[34911-55-2] Bupropion, 1523
[34915-68-9] Bunitrolol, 1514
[34919-98-7] Cetamolol, 2060
[34994-11-1] Hypusine, 4918
[35035-05-3] Timepidium Bromide, 9583
[35080-11-6] Prajmaline, 7883
[35108-88-4] Bufetolol Hydrochloride*, 1496
[35112-53-9] Barium Thiosulfate, 1028
[35121-78-9] Prostacyclin, 8061
[35189-28-7] Norgestimate, 6796
[35212-22-7] Ipriflavone, 5090
[35274-05-6] Cetyl Lactate, 2072
[35287-72-0] Secalonic Acid A*, 8561
[35334-12-4] Azidocillin Sodium Salt*, 942
[35354-74-6] Honokiol, 4780
[35367-38-5] Diflubenzuron, 3188
[35388-57-9] Piscidic Acid, 7664
[35400-43-2] Sulprofos, 9164
[35405-51-7] Calcium Pyrophosphate, 1746
[35413-63-9] Streptovaricin E*, 8989
[35455-20-0] 3-Fluoro-D-alanine, 4206
[35457-80-8] Midecamycin A_1*, 6271
[35512-37-9] Streptovaricin F*, 8989
[35531-88-5] Carindacillin, 1888
[35543-24-9] Buflomedil Hydrochloride*, 1498
[35554-08-6] Saxitoxin, 8533
[35554-44-0] Enilconazole, 3622
[35595-03-0] Centaurein, 2015
[35604-67-2] Viloxazine Hydrochloride*, 10116
[35607-66-0] Cefoxitin, 1986
[35619-65-9] Trithiozine, 9889
[35691-65-7] 1,2-Dibromo-2,4-dicyanobutane, 3069
[35700-21-1] Carboprost Methyl Ester*, 1871
[35700-23-3] Carboprost, 1871
[35703-32-3] Cinametic Acid, 2341
[35727-72-1] Ontianil, 6981
[35795-16-5] Trimazosin, 9828
[35834-26-5] Rosaramicin, 8420
[35846-53-8] Maytansine, 5800
[35898-87-4] Dilazep, 3244
[35941-65-2] Butriptyline, 1568
[35941-71-0] Tiaramide Hydrochloride*, 9564
[35991-93-6] Nadoxolol Hydrochloride*, 6432
[36002-19-4] Folescutol Hydrochloride*, 4252
[36025-69-1] Midecamycin A_3*, 6271

[36069-05-3] Pseudojervine, 8102
[36085-73-1] Talipexole Dihydrochloride*, 9209
[36104-64-0] Micranthine, 6267
[36104-80-0] Camazepam, 1773
[36141-82-9] Diamfenetide, 3017
[36150-73-9] Erythrophleine, 3728
[36167-63-2] Halofantrine Hydrochloride*, 4626
[36222-39-6] Gallopamil l-Form Hydrochloride*, 4369
[36296-31-8] Oxiniacic Acid Ethanolamine Salt*, 7075
[36304-84-4] Dimemorfan Phosphate*, 3251
[36309-01-0] Dimemorfan, 3251
[36322-90-4] Piroxicam, 7661
[36330-85-5] Fenbufen, 4003
[36338-96-2] Carthamin, 1918
[36364-49-5] Imidazole Salicylate, 4949
[36393-56-3] Norpseudoephedrine, 6811
[36478-76-9] Uranyl Nitrate, 10000
[36505-84-7] Buspirone, 1528
[36508-71-1] Zorubicin Hydrochloride*, 10326
[36525-21-0] Neovitamin A, 6555
[36531-26-7] Oxantel, 7055
[36556-75-9] Bucumolol Hydrochloride* (±)-form, 1489
[36590-19-9] Amocarzine, 608
[36592-77-5] Metipranolol Hydrochloride*, 6221
[36637-18-0] Etidocaine, 3907
[36637-19-1] Etidocaine Hydrochloride*, 3907
[36647-02-6] Cotarnine Hydrochloride*, 2618
[36653-82-4] Cetyl Alcohol, 2070
[36703-88-5] Inosine Pranobex, 5006
[36734-19-7] Iprodione, 5093
[36735-22-5] Quazepam, 8211
[36791-04-5] Ribavirin, 8365
[36798-79-5] Budralazine, 1492
[36894-69-6] Labetalol, 5341
[36895-62-2] Methenamine Allyl Iodide, 6037
[37025-55-1] Carbetocin, 1841
[37065-29-5] Miloxacin, 6283
[37091-65-9] Azlocillin Sodium Salt*, 947
[37091-66-0] Azlocillin, 947
[37106-97-1] Bentiromide, 1081
[37115-32-5] Adinazolam, 159
[37134-80-8] Dopastin, 3481
[37148-27-9] Clenbuterol, 2407
[37196-57-9] Crotamine, 2660
[37203-40-0] Colostrokinin, 2553
[37221-79-7] VIP, 10140
[37248-47-8] Validamycin A*, 10045
[37267-86-0] Phosphoric Acid, Meta, 7501
[37270-69-2] Benserazide Combination with Levodopa*, 1079
[37270-89-6] Heparin Calcium Salt*, 4685
[37280-35-6] Capreomycin IA*, 1801
[37288-97-4] Pinguinain, 7603
[37291-07-9] Crilanomer, 2654
[37296-80-3] Colestipol Hydrochloride*, 2538
[37300-21-3] Pentosan Polysulfate, 7276
[37301-55-6] Chlormadinone Acetate Mixture with Ethinyl Estradiol*, 2152
[37319-17-8] Pentosan Polysulfate Sodium Salt*, 7276
[37321-09-8] Apramycin, 792

[37332-99-3] Avoparcin, 922
[37338-39-9] Pyrimethamine Combination with Sulfadoxine*, 8169
[37339-90-5] Lentinan, 5462
[37340-82-2] Streptodornase, 8979
[37350-58-6] Metoprolol, 6235
[37517-26-3] Pipotiazine Palmitic Ester*, 7636
[37517-28-5] Amikacin, 425
[37517-30-9] Acebutolol, 16
[37561-27-6] Fenoverine, 4023
[37577-24-5] Fenfluramine l-Form*, 4015
[37640-71-4] Aprindine, 793
[37661-08-8] Bacampicillin Hydrochloride*, 961
[37686-84-3] Terguride, 9308
[37686-85-4] Terguride Hydrogen Maleate*, 9308
[37693-01-9] Clofoctol, 2439
[37758-47-7] Ganglioside G_{m1}*, 4376
[37878-19-6] Detoxin D_1*, 2982
[37883-00-4] Viomycin Sulfate*, 10139
[37893-02-0] Flubenzimine, 4155
[37895-35-5] Albofungin, 213
[37924-13-3] Perfluidone, 7300
[37933-66-7] Thevetin A, 9424
[38029-10-6] Pirbuterol Dihydrochloride*, 7644
[38129-37-2] Bicozamycin, 1249
[38142-58-4] ar-Turmerone (±)-Form*, 9956
[38176-09-9] Gallopamil d-Form Hydrochloride*, 4369
[38183-12-9] Fluorescamine, 4193
[38194-50-2] Sulindac, 9155
[38234-21-8] Fertirelin, 4109
[38260-01-4] Trientine Dihydrochloride*, 9796
[38260-54-7] Etrimfos, 3936
[38304-91-5] Minoxidil, 6290
[38363-32-5] Penbutolol Sulfate*, 7209
[38363-40-5] Penbutolol, 7209
[38455-77-5] Stannic Chromate(VI), 8930
[38562-01-5] Prostaglandin $F_{2\alpha}$ Tromethamine Salt*, 8065
[38577-97-8] 4',5'-Diiodofluorescein, 3233
[38677-81-5] Pirbuterol, 7644
[38677-85-9] Flunixin, 4182
[38763-29-0] Worenine, 10187
[38821-52-2] Indoramin Hydrochloride*, 5000
[38821-53-3] Cephradine, 2032
[38916-34-6] Somatostatin, 8863
[38932-40-0] Cephalexin Sodium Salt*, 2021
[38957-41-4] Emorfazone, 3603
[38965-69-4] Xanthocillin Y_1*, 10195
[38965-70-7] Xanthocillin Y_2*, 10195
[38966-21-1] Aphidicolin, 769
[39133-31-8] Trimebutine, 9829
[39148-24-8] Fosetyl Al, 4278
[39156-41-7] 2,4-Diaminoanisole Sulfate*, 3018
[39290-85-2] Cupric Borate, 2694
[39295-60-8] Sulfadoxine Mixture with Trimethoprim*, 9074
[39300-45-3] Dinocap, 3340
[39379-15-2] Neurotensin, 6569
[39391-39-4] Nocardicin A*, 6764
[39409-82-0] Magnesium Carbonate Hydroxide, 5696
[39416-48-3] Pyridinium Bromide Perbromide, 8156
[39455-18-0] Chondroitin 4-Sulfate Disodium Salt*, 2270

Denotes derivative of title compound

[39469-82-4] Sulfadimethoxine Mixture with Trimethoprim*, 9073
[39492-01-8] Gabexate, 4344
[39515-40-7] Cyphenothrin, 2837
[39515-41-8] Fenpropathrin, 4033
[39543-79-8] Befunolol Hydrochloride*, 1050
[39552-01-7] Befunolol, 1050
[39562-70-4] Nitrendipine, 6669
[39563-28-5] Cloranolol, 2464
[39577-19-0] Picumast, 7573
[39577-20-3] Picumast Dihydrochloride*, 7573
[39589-98-5] Dimethyl Carbate, 3290
[39640-15-8] Piberaline, 7547
[39664-27-2] Ethylamine Oleate*, 3808
[39698-78-7] Saralasin Hydrated Acetate*, 8518
[39711-79-0] Ethyl Menthane Carboxamide, 3870
[39715-02-1] Endralazine, 3617
[39718-89-3] Alminoprofen, 308
[39733-35-2] Ammonium Magnesium Chloride, 562
[39800-16-3] Perfosfamide (unspecified), 7303
[39809-25-1] Penciclovir, 7210
[39831-55-5] Amikacin Sulfate*, 425
[39860-99-6] Pipotiazine, 7636
[39878-70-1] Talampicillin Hydrochloride*, 9204
[39978-42-2] Nifurzide, 6632
[40034-42-2] Rosoxacin, 8426
[40054-69-1] Etizolam, 3919
[40164-67-8] Amidochlor, 420
[40180-04-9] Ticrynafen, 9570
[40391-99-9] Pamidronic Acid, 7135
[40487-42-1] Pendimethalin, 7211
[40507-23-1] Fluproquazone, 4230
[40507-78-6] Indanazoline, 4967
[40507-80-0] Indanazoline Hydrochloride*, 4967
[40516-48-1] Tocoretinate, 9639
[40525-07-3] Scillabiose, 8542
[40580-59-4] Guanadrel, 4587
[40596-69-8] Methoprene (unspecified), 6063
[40665-92-7] Cloprostenol, 2461
[40666-16-8] Fluprostenol, 4231
[40716-66-3] Nerolidol trans-Form*, 6561
[40762-15-0] Doxefazepam, 3490
[40819-93-0] Lorajmine Hydrochloride*, 5607
[40828-46-4] Suprofen, 9180
[41096-46-2] Hydroprene (unspecified), 4850
[41100-52-1] Memantine Hydrochloride*, 5872
[41136-22-5] Ceftezole Sodium Salt*, 1997
[41183-64-6] Gallium Citrate*, 4364
[41205-06-5] Methoprene (±), 6063
[41205-09-8] Hydroprene (±), 4850
[41294-56-8] 1α-Hydroxycholecalciferol, 4864
[41340-25-4] Etodolac, 3920
[41342-53-4] Erythromycin Ethylsuccinate*, 3720
[41354-29-4] Cyproheptadine Hydrochloride Sesquihydrate*, 2842
[41372-20-7] Apomorphine Hydrochloride*, 787
[41394-05-2] Metamitron, 5985
[41451-75-6] Bruceantin, 1475
[41473-08-9] Iopronic Acid, 5077
[41483-43-6] Bupirimate, 1519
[41544-24-5] Tromantadine Hydrochloride*, 9901

[41570-61-0] Tulobuterol, 9942
[41575-94-4] Carboplatin, 1870
[41621-49-2] Ciclopirox Ethanolamine Salt (1:1)*, 2325
[41663-50-1] Isobutol*, 3764
[41708-72-9] Tocainide, 9629
[41744-40-5] Sulbenicillin, 9059
[41767-29-7] Fluocortin Butyl, 4187
[41826-92-0] Trepibutone, 9718
[41859-67-0] Bezafibrate, 1240
[41944-01-8] Bismuth Potassium Iodide, 1317
[41992-22-7] Spirogermanium Dihydrochloride*, 8916
[41992-23-8] Spirogermanium, 8916
[42116-76-7] Carnidazole, 1897
[42200-33-9] Nadolol, 6431
[42399-41-7] Diltiazem, 3247
[42408-82-2] Butorphanol, 1565
[42461-84-7] Flunixin Meglumine Salt*, 4182
[42471-28-3] Nimustine, 6645
[42509-80-8] Isazofos, 5119
[42540-40-9] Cefamandole Nafate*, 1964
[42542-10-9] MDMA (unspecified), 5806
[42553-65-1] Crocetin Di-gentiobiose Ester*, 2657
[42576-02-3] Bifenox, 1256
[42588-37-4] Kinoprene, 5326
[42597-57-9] Ronifibrate, 8414
[42739-38-8] Ammonium Valerate, 604
[42779-82-8] Clopirac, 2459
[42794-76-3] Midodrine, 6272
[42835-25-6] Flumequine, 4172
[42852-95-9] N-Methyl-α-L-glucosamine, 6155
[42864-78-8] Bevantolol Hydrochloride*, 1238
[42874-03-3] Oxyfluorfen, 7097
[42924-53-8] Nabumetone, 6428
[42971-09-5] Vinpocetine, 10128
[43033-72-3] Levomethadyl Acetate Hydrochloride*, 5492
[43119-47-7] Vitamin E Nicotinate*, 10159
[43121-43-3] Triadimefon, 9723
[43200-80-2] Zopiclone, 10324
[43210-67-9] Fenbendazole, 4000
[43222-48-6] Difenzoquat Methyl Sulfate*, 3185
[46029-44-1] Pentorex (DL-Form), 7275
[46464-11-3] Meobentine, 5889
[46817-91-8] Viloxazine, 10116
[47000-92-0] Fluoridamid, 4198
[47141-42-4] Levobunolol, 5488
[47562-08-3] Lorajmine, 5607
[47739-98-0] Clocapramine, 2427
[47747-56-8] Talampicillin, 9204
[47931-85-1] Calcitonin, salmon synthetic*, 1680
[49562-28-9] Fenofibrate, 4019
[49564-56-9] Fazadinium Bromide, 3981
[49582-09-4] PBB*, 4124
[49642-07-1] Statine, 8956
[49656-78-2] Methcathinone Hydrochloride* (±), 6033
[49697-38-3] Rimexolone, 8392
[49745-95-1] Dobutamine Hydrochloride*, 3456
[49746-00-1] Carbinoxamine l-Form d-Tartrate*, 1845
[49866-87-7] Difenzoquat, 3185
[50264-69-2] Lonidamine, 5598
[50270-33-2] Isofezolac, 5188
[50276-98-7] Erythrocentaurin, 3717

[50291-21-9] Rose Bengal 131I-Labeled Disodium Salt*, 8421
[50321-35-2] Morazone Hydrochloride*, 6349
[50327-22-5] Nylon 46, 6833
[50435-25-1] Nimidane, 6642
[50471-44-8] Vinclozolin, 10122
[50594-66-6] Acifluorfen, 111
[50629-82-8] Halometasone, 4628
[50650-76-5] Piroctone, 7657
[50679-07-7] Cinepazet Maleate*, 2349
[50679-08-8] Terfenadine, 9307
[50700-49-7] Acetimidoquinone, 54
[50700-72-6] Vecuronium Bromide, 10075
[50767-79-8] Prodlure, 7950
[50776-39-1] Tinofedrine Hydrochloride*, 9589
[50782-69-9] VX, 10172
[50813-16-6] Sodium Polymetaphosphate, 8814
[50832-74-1] Nifurprazine Hydrochloride*, 6629
[50838-36-3] Tolciclate, 9647
[50847-11-5] Ibudilast, 4923
[50865-01-5] Protoporphyrin IX Disodium Salt*, 8084
[50887-69-9] Orotic Acid Monohydrate*, 7004
[50924-49-7] Mizoribine, 6306
[50927-09-8] Sydnones, 9187
[50933-06-7] Sulfalene Mixture with Trimethoprim*, 9078
[50933-33-0] Gossyplure, 4548
[50935-04-1] Carubicin, 1921
[50935-71-2] Mocimycin, 6308
[50972-17-3] Bacampicillin, 961
[51005-85-7] PBP*, 4124
[51012-32-9] Tiapride, 9561
[51012-33-0] Tiapride Hydrochloride*, 9561
[51016-68-3] Josamycin Propionate*, 5280
[51022-69-6] Amcinonide, 407
[51022-70-9] Albuterol Sulfate*, 217
[51022-71-0] Nabilone, 6427
[51022-98-1] Butirosin Sulfate Dihydrate*, 1559
[51023-56-4] Centchroman Hydrochloride*, 2018
[51025-85-5] Arbekacin, 809
[51037-30-0] Acipimox, 113
[51115-67-4] Trimethyl Isopropyl Butanamide, 9847
[51218-45-2] Metolachlor, 6230
[51218-49-6] Pretilachlor, 7921
[51235-04-2] Hexazinone, 4734
[51264-14-3] Amsacrine, 635
[51274-83-0] Tiamenidine Hydrochloride*, 9558
[51276-47-2] Phosphinothricin, 7495
[51308-54-4] Buthiobate, 1555
[51312-42-6] Sodium Phosphotungstate, 8811
[51322-75-9] Tizanidine, 9624
[51333-22-3] Budesonide, 1490
[51338-27-3] Diclofop-Methyl, 3133
[51395-42-7] Butedronic Acid, 1546
[51410-30-1] Pirenoxine Sodium Salt*, 7645
[51424-33-0] Salsoline Hydrochloride*, 8492
[51460-26-5] Carbazochrome Sodium Sulfonate, 1833
[51481-10-8] Vomitoxin, 10171
[51481-60-8] Naloxone Hydrochloride* (Dihydrate), 6449
[51481-61-9] Cimetidine, 2337
[51481-65-3] Mezlocillin, 6259

Denotes derivative of title compound

[51484-40-3] Difenpiramide, 3184
[51503-61-8] Ammonium Phosphite, 578
[51570-36-6] Milbemycins, 6278
[51579-82-9] Amfenac, 413
[51596-10-2] Milbemycin A$_3$*, 6276
[51596-11-3] Milbemycin A$_4$*, 6276
[51596-16-8] Milbemycin β$_1$*, 6278
[51598-60-8] Cimetropium Bromide, 2338
[51627-14-6] Cefatrizine, 1965
[51630-58-1] Fenvalerate, 4051
[51707-55-2] Thidiazuron, 9447
[51762-05-1] Cefroxadine, 1993
[51773-92-3] Mefloquine Hydrochloride, 5845
[51781-06-7] Carteolol, 1917
[51781-21-6] Carteolol Hydrochloride*, 1917
[51803-78-2] Nimesulide, 6640
[51848-43-2] Retine, 8332
[51873-93-9] Azulene Sodium Sulfonate*, 956
[51888-09-6] Prochlorperazine Dimethanesulfonate*, 7942
[51940-44-4] Pipemidic Acid, 7613
[51952-41-1] LH-RH Hydrochloride*, 5500
[52003-58-4] Ammonium Lactate, 561
[52031-11-5] Pheniprazine DL-Form*, 7382
[52080-56-5] Endobenzyline Bromide, 3612
[52080-57-6] Chloroprednisone, 2209
[52093-21-7] Micronomicin, 6269
[52094-70-9] Tetrantoin, 9372
[52109-93-0] Cyclodrine, 2788
[52128-35-5] Trimetrexate, 9851
[52152-93-9] Cefsulodin Sodium Salt*, 1994
[52205-73-9] Estramustine 17-(Dihydrogenphosphate) Disodium Salt*, 3749
[52207-99-5] Gossyplure (Z,Z)-Form*, 4548
[52211-63-9] Viquidil Hydrochloride*, 10141
[52212-02-9] Pipecurium Bromide, 7612
[52212-86-9] Verticillin B*, 10103
[52214-84-3] Ciprofibrate, 2373
[52225-20-4] Vitamin E Acetate dl-Form*, 10160
[52239-63-1] Thiethylperazine Dimalate*, 9449
[52246-40-9] Fenquizone Monopotassium Salt*, 4039
[52260-69-2] Dicyanine, 3141
[52275-61-3] Streptovaricin J*, 8989
[52279-57-9] Nandrolone p-Hexyloxyphenylpropionate*, 6452
[52306-35-1] Lunacridine (±)-Form*, 5630
[52315-07-8] Cypermethrin, 2836
[52365-63-6] Dipivefrin, 3400
[52423-56-0] Bromopride Hydrochloride*, 1454
[52432-72-1] Oxeladin Citrate*, 7064
[52441-47-1] Semtex*, 2805
[52443-21-7] Glucametacin, 4456
[52463-83-9] Pinazepam, 7596
[52468-60-7] Flunarizine, 4179
[52479-85-3] Exifone, 3958
[52485-79-7] Buprenorphine, 1522
[52486-78-9] Mercurophen, 5947
[52500-59-1] Mexicain, 6256
[52500-60-4] Thioredoxin, 9496
[52500-61-5] β-Amino-α-methylphenethyl Alcohol, 467

[52503-64-7] Cupric Acetate, Basic, 2691
[52549-17-4] Pranoprofen, 7890
[52589-12-5] Diginatin, 3198
[52645-53-1] Permethrin, 7321
[52665-69-7] Calcimycin, 1678
[52705-43-8] α-Hydroxybenzylphosphinic Acid, 4858
[52712-76-2] Bunazosin Hydrochloride*, 1512
[52731-38-1] Glucofrangulin, 4460
[52740-16-6] Calcium Arsenite, 1687
[52775-76-5] Methylenomycin A*, 6144
[52775-77-6] Methylenomycin B*, 6144
[52794-97-5] Carubicin Hydrochloride*, 1921
[52809-07-1] Quisqualic Acid, 8268
[52906-84-0] Oxychlorosene Sodium Salt*, 7090
[52906-92-0] Motilin, 6366
[52918-63-5] Deltamethrin, 2934
[52942-31-1] Etoperidone, 3930
[52951-36-7] Bisdequalinium Chloride, 1288
[52951-37-8] Bismuth Ethyl Camphorate, 1304
[52951-38-9] Bismuth Oleate, 1312
[52964-40-6] d-Bornyl α-Bromoisovalerate, 1368
[52993-95-0] Sodium Polyanetholesulfonate, 8813
[53003-10-4] Salinomycin, 8488
[53022-14-3] d-Bornyl Isovalerate, 1370
[53042-79-8] Gossyplure (Z,E)-Form*, 4548
[53043-14-4] 6-n-Amyl-m-cresol, 648
[53078-86-7] Arogenic Acid, 826
[53086-13-8] Indoprofen (+)-Form*, 4999
[53086-14-9] Indoprofen (−)-Form*, 4999
[53092-86-7] Lindlar Catalyst, 5527
[53092-91-4] Chrysergonic Acid, 2315
[53123-88-9] Rapamycin, 8288
[53152-21-9] Buprenorphine Hydrochloride*, 1522
[53164-05-9] Acemetacin, 27
[53179-09-2] Sisomicin Sulfate*, 8695
[53179-11-6] Loperamide, 5601
[53216-90-3] Griseoviridin, 4572
[53230-06-1] Dihydroxyaluminum Acetylsalicylate, 3226
[53251-94-8] Pinaverium Bromide, 7595
[53260-52-9] Heparamine, 4684
[53260-54-1] Fungisterol, 4313
[53267-01-9] Cifenline, 2330
[53274-53-6] N^4-β-D-Glucosylsulfanilamide, 4472
[53318-35-7] Diatretyne I, 3038
[53318-36-8] α-Glucogallin, 4461
[53370-43-7] Methargen, 6030
[53370-44-8] Menadiol Dibutyrate*, 5873
[53370-45-9] Menthyl Borate, 5885
[53370-48-2] p-(2-Methylpropenyl)-phenol Acetate, 6190
[53370-90-4] Exalamide, 3955
[53403-97-7] Pyridofylline, 8159
[53414-68-9] Tonin, 9685
[53447-14-6] Isocorypalmine, 5179
[53449-58-4] Ciclonicate, 2324
[53450-33-2] Oxalomolybdic Acid, 7045
[53516-73-7] Quinovin, 8261
[53555-68-3] 4,4'-Iminodicyclohexanecarboxylic Acid, 4953

[53558-25-1] Pyriminil, 8171
[53567-47-8] Tiropramide Hydrochloride*, 9607
[53583-79-2] Sultopride, 9168
[53586-99-5] Periodyl, 7313
[53597-25-4] Salmine Sulfate, 8490
[53597-27-6] Fendosal, 4012
[53597-29-8] Plegatil, 7693
[53608-75-6] Pancrelipase, 7138
[53608-77-8] Penicilloyl Polylysine, 7233
[53608-78-9] Potassium Sulfobenzoate, 7848
[53608-79-0] Potassium Zirconium Sulfate, 7878
[53608-96-1] Cloxotestosterone, 2482
[53643-48-4] Vindesine, 10125
[53643-53-1] Tuberin, 9938
[53648-05-8] Ibuproxam, 4926
[53648-55-8] Dezocine, 2998
[53659-00-0] Pyrisuccideanol Dimaleate*, 8175
[53678-77-6] Muramyl Dipeptide, 6385
[53684-48-3] Beryllium Potassium Sulfate, 1219
[53684-49-4] Bufetolol, 1496
[53684-61-0] Calcium Ferrous Citrate, 1708
[53714-56-0] Leuprolide, 5484
[53716-44-2] Rociverine, 8406
[53716-49-7] Carprofen, 1912
[53716-50-0] Oxfendazole, 7069
[53731-35-4] Manganese Pyrophosphate, 5778
[53731-36-5] Floredil, 4144
[53734-79-5] Metralindole Hydrochloride*, 6238
[53746-45-5] Fenoprofen Calcium Salt Dihydrate*, 4021
[53746-46-6] Trimazosin Hydrochloride Monohydrate*, 9828
[53762-90-6] Gratioside, 4564
[53762-92-8] Valacidin, 10038
[53772-83-1] Clopenthixol cis(Z)-Form*, 2455
[53778-50-0] Bismuth Sodium Iodide, 1320
[53780-34-0] Mefluidide, 5846
[53783-83-8] Tromantadine, 9901
[53797-35-6] Ribostamycin Sulfate*, 8373
[53808-87-0] Tetroxoprim, 9386
[53808-88-1] Lonazolac, 5596
[53813-83-5] Suriclone, 9182
[53834-53-0] Tropenzile, 9908
[53850-34-3] Thaumatin, 9408
[53850-35-4] Element 105, 3581
[53850-36-5] Element 104, 3580
[53861-57-7] γ-Carboxyglutamic Acid, 1876
[53882-12-5] Lodoxamide, 5585
[53882-13-6] Lodoxamide Diethyl Ester*, 5585
[53885-35-1] Ticlopidine Hydrochloride*, 9569
[53897-25-9] Bismuth Butylthiolaurate, 1301
[53902-12-8] Tranilast, 9705
[53905-38-7] Pro-Drone®, 7951
[53910-25-1] Pentostatin, 7277
[53943-88-7] Letosteine, 5473
[53984-74-0] Tocainide R-(−)-Form Hydrochloride*, 9629
[53984-76-2] Tocainide S-(+)-Form Hydrochloride*, 9629
[54024-22-5] Desogestrel, 2971
[54037-14-8] Element 107, 3583
[54037-57-9] Element 108, 3584
[54038-01-6] Element 109, 3585

Denotes derivative of title compound

[54038-81-2] Element 106, *3582*
[54063-32-0] Clobetasone, *2424*
[54063-40-0] Fenoxedil, *4026*
[54063-51-3] Nadoxolol, *6432*
[54063-53-5] Propafenone, *7978*
[54063-54-6] Reproterol, *8307*
[54063-56-8] Suloctidil, *9160*
[54083-22-6] Zorubicin, *10326*
[54110-25-7] Pirozadil, *7662*
[54120-61-5] Prostalene, *8066*
[54125-02-9] 1-Methoxy-3-(trimethyl-silyloxy)-1,3-butadiene (*E*)-form, *6082*
[54143-55-4] Flecainide (unspecified), *4136*
[54143-56-5] Flecainide Mono-acetate*, *4136*
[54143-57-6] Metoclopramide Mono-hydrochloride Monohydrate*, *6226*
[54182-58-0] Sucralfate, *9049*
[54187-04-1] Rilmenidine, *8388*
[54188-38-4] Metralindole, *6238*
[54239-37-1] Cimaterol, *2336*
[54247-25-5] Cloranolol Hydrochloride*, *2464*
[54283-65-7] Dimecrotic Acid Magnesium Salt*, *3248*
[54323-85-2] Protizinic Acid, *8077*
[54340-58-8] Meptazinol, *5910*
[54340-62-4] Bufuralol, *1504*
[54350-48-0] Etretinate, *3935*
[54397-85-2] Thromboxane B₂*, *9530*
[54400-59-8] Butamisole, *1540*
[54400-62-3] Butamisole Hydrochloride*, *1540*
[54479-70-8] Heparin Magnesium Salt*, *4685*
[54504-70-0] Theofibrate, *9420*
[54527-84-3] Nicardipine Hydrochloride*, *6579*
[54530-86-8] Clometocillin Sodium Salt*, *2445*
[54531-52-1] Polybenzarsol, *7723*
[54533-85-6] Nizofenone, *6759*
[54533-86-7] Nizofenone Fumarate*, *6759*
[54556-98-8] Propiverine Hydrochloride*, *8018*
[54605-45-7] Iocarmic Acid Di-*N*-methylglucamine Salt*, *5029*
[54663-47-7] Tibezonium Iodide, *9565*
[54739-18-3] Fluvoxamine, *4251*
[54749-86-9] Thiazolinobutazone, *9443*
[54749-90-5] Chlorozotocin, *2229*
[54779-57-6] Pheniprazine DL-Form Hydrochloride*, *7382*
[54783-95-8] Fortimicin B*, *4276*
[54910-51-9] Disparlure (2*S*-cis), *3425*
[54910-89-3] Fluoxetine, *4222*
[54935-03-4] Sulisatin, *9156*
[54935-04-5] Sulisatin Disodium Salt*, *9156*
[54940-97-5] Citrulline Malate (Salt)*, *2392*
[54965-21-8] Albendazole, *211*
[54965-24-1] Tamoxifen Citrate*, *9216*
[54974-54-8] TRH Tartrate*, *9720*
[55028-70-1] Arbaprostil, *808*
[55028-71-2] Fluprostenol Sodium Salt*, *4231*
[55028-72-3] Cloprostenol Sodium Salt*, *2461*
[55077-30-0] Aclatonium Napadisilate, *116*
[55079-83-9] Acitretin, *114*
[55096-26-9] Nalmefene, *6447*
[55134-13-9] Narasin, *6506*
[55142-85-3] Ticlopidine, *9569*
[55179-31-2] Bitertanol, *1342*

[55219-65-3] Triadimenol, *9724*
[55242-55-2] Propentofylline, *7997*
[55268-74-1] Praziquantel, *7896*
[55268-75-2] Cefuroxime, *2002*
[55283-68-6] Ethalfluralin, *3763*
[55285-45-5] Pirifibrate, *7650*
[55290-64-7] Oxidimethiin, *7072*
[55294-15-0] Muzolimine, *6397*
[55297-95-5] Tiamulin, *9559*
[55297-96-6] Tiamulin Fumarate*, *9559*
[55300-29-3] Antrafenine, *763*
[55327-22-5] Etiroxate Hydrochloride*, *3917*
[55331-29-8] Zeranol, *10251*
[55335-06-3] Triclopyr, *9789*
[55429-45-3] Nitracrine Dihydrochloride Monohydrate*, *6661*
[55453-87-7] Isoxepac, *5257*
[55482-89-8] Aspirin Guaiacol Ester*, *886*
[55512-33-9] Pyridate, *8151*
[55560-96-8] Tixocortol 21-Pivalate*, *9623*
[55589-62-3] Acesulfame Potassium Salt*, *35*
[55608-72-5] Pseudococaine *n*-Propyl Ester Analog*, *8097*
[55661-38-6] Nimustine Hydrochloride*, *6645*
[55721-31-8] Salinomycin Sodium Salt*, *8488*
[55726-47-1] Enocitabine, *3624*
[55769-64-7] Butobendine Dihydrochloride*, *1560*
[55769-65-8] Butobendine, *1560*
[55774-33-9] Azathioprine Sodium Salt*, *935*
[55779-06-1] Fortimicin A*, *4276*
[55779-18-5] Arprinocid, *828*
[55837-14-4] Dutavcrine, *1544*
[55837-18-8] Butibufen, *1557*
[55837-20-2] Halofuginone, *4627*
[55837-25-7] Buflomedil, *1498*
[55837-27-9] Piretanide, *7647*
[55837-29-1] Tiropramide, *9607*
[55837-30-4] Clonixin Lysine Salt*, *2453*
[55881-07-7] Miokamycin, *6291*
[55893-12-4] Gephyrotoxin, *4410*
[55905-53-8] Clebopride, *2404*
[55937-99-0] Beclobrate, *1046*
[55965-63-4] *O*-(β-Hydroxyethyl)-rutinosides (mixture)*, *9920*
[55985-32-5] Nicardipine, *6579*
[56030-54-7] Sufentanil, *9056*
[56038-13-2] Sucralose, *9050*
[56073-10-0] Brodifacoum, *1400*
[56087-11-7] Dextranomer, *2992*
[56109-02-5] Picilorex Hydrochloride*, *7551*
[56180-94-0] Acarbose, *15*
[56187-47-4] Cefazedone, *1966*
[56187-89-4] Ximoprofen, *10211*
[56208-01-6] Pifarnine, *7575*
[56211-40-6] Torsemide, *9690*
[56222-04-9] Femoxetine Hydrochloride*, *3993*
[56238-63-2] Cefuroxime Sodium Salt*, *2002*
[56281-36-8] Motretinide, *6367*
[56281-37-9] 2C-B Hydrochloride*, *1958*
[56283-74-0] Laidlomycin, *5361*
[56287-74-2] Afloqualone, *181*
[56335-21-8] Etodroxizine Dimaleate*, *3921*
[56341-08-3] Mabuterol, *5674*
[56377-79-8] Nosiheptide, *6816*

[56390-09-1] Epirubicin Hydrochloride*, *3660*
[56391-56-1] Netilmicin, *6563*
[56391-57-2] Netilmicin Sulfate*, *6563*
[56392-17-7] Metoprolol Tartrate*, *6235*
[56393-22-7] Pildralazine Dihydrochloride*, *7577*
[56396-94-2] Mepindolol Sulfate Salt*, *5901*
[56420-45-2] Epirubicin, *3660*
[56425-91-3] Flurprimidol, *4239*
[56430-99-0] Flumecinol, *4170*
[56518-41-3] Brodimoprim, *1401*
[56592-32-6] Efrotomycin, *3567*
[56676-64-3] Nicametate Hydrochloride*, *6577*
[56695-65-9] Rosaprostol, *8419*
[56775-88-3] Zimeldine, *10254*
[56776-01-3] Tulobuterol Hydrochloride*, *9942*
[56776-32-0] Etifoxine Hydrochloride*, *3910*
[56796-20-4] Cefmetazole, *1977*
[56796-39-5] Cefmetazole Sodium Salt*, *1977*
[56824-20-5] Amiprilose, *506*
[56833-74-0] Streptovirudin, *8990*
[56839-43-1] Eperisone Hydrochloride*, *3642*
[56974-46-0] Butalamine Hydrochloride*, *1535*
[56974-61-9] Gabexate Monomethanesulfonate*, *4344*
[56980-93-9] Celiprolol, *2007*
[56995-20-1] Flupirtine, *4227*
[57021-61-1] Isonixin, *5209*
[57041-67-5] Desflurane, *2965*
[57109-90-7] Clorazepic Acid Dipotassium Salt*, *2465*
[57132-53-3] Proglumetacin, *7957*
[57149-07-2] Naftopidil (unspecified), *6443*
[57149-08-3] Naftopidil Dihydrochloride*, *6443*
[57197-43-0] Sulfamoxole Mixture with Trimethoprim*, *9093*
[57206-54-9] Vernamycin Bβ*, *10096*
[57248-88-1] Pamidronic Acid Disodium Salt*, *7135*
[57249-13-5] Heptaminol 5'-Adenylate*, *4691*
[57308-51-7] Carbidopa Combination with Levodopa*, *1843*
[57333-96-7] Tacalcitol, *9197*
[57381-26-7] Irsogladine, *5113*
[57432-61-8] Methylergonovine Maleate*, *6147*
[57460-41-0] Talinolol, *9208*
[57469-78-0] Ketoprofen Lysine Salt*, *5316*
[57470-78-7] Celiprolol Hydrochloride*, *2007*
[57474-29-0] Nifuroquine, *6625*
[57475-17-9] Brovincamine, *1473*
[57524-15-9] Tolonidine Nitrate*, *9657*
[57524-89-7] Hydrocortisone 17-Valerate*, *4828*
[57526-81-5] Prenalterol, *7917*
[57574-09-1] Amineptine, *429*
[57576-44-0] Aclacinomycin A*, *115*
[57576-52-0] Thromboxane A₂*, *9530*
[57596-79-9] Aclacinomycin B*, *115*
[57644-54-9] Colloidal Bismuth Subcitrate, *2549*
[57645-91-7] Clebopride Malate*, *2404*
[57647-35-5] Binedaline Hydrochloride*, *1266*

* Denotes derivative of title compound

[57648-21-2] Timiperone, 9584
[57653-27-7] Droprenilamine, 3506
[57704-10-6] Bufuralol (−)-Hydrochloride*, 1504
[57704-11-7] Bufuralol (+)-Hydrochloride*, 1504
[57760-36-8] Alborixin, 215
[57773-63-4] Triptorelin, 9878
[57773-65-6] Deslorelin, 2968
[57775-22-1] Etoperidone Hydrochloride*, 3930
[57775-26-5] Sultosilic Acid, 9169
[57775-27-6] Sultosilic Acid Piperazine Salt*, 9169
[57775-29-8] Carazolol, 1822
[57781-14-3] Halopredone Acetate, 4630
[57801-81-7] Brotizolam, 1472
[57808-65-8] Closantel, 2473
[57808-66-9] Domperidone, 3476
[57817-89-7] Stevioside, 8964
[57837-19-1] Metalaxyl, 5982
[57852-57-0] Idarubicin Hydrochloride*, 4931
[57938-82-6] Adinazolam Monomethanesulfonate*, 159
[57982-77-1] Buserelin, 1527
[57982-78-2] Budipine, 1491
[57998-68-2] Diaziquone, 3044
[58001-44-8] Clavulanic Acid, 2402
[58045-23-1] Clopenthixol cis(Z)-Form Dihydrochloride*, 2455
[58066-85-6] Miltefosine, 6285
[58095-31-1] Sulbenox, 9060
[58138-08-2] Tridiphane, 9795
[58152-03-7] Isepamicin, 5121
[58158-77-3] Amantanium Bromide, 390
[58186-27-9] Idebenone, 4932
[58207-19-5] Clindamycin Hydrochloride Monohydrate*, 2414
[58306-30-2] Febantel, 3982
[58337-35-2] Elliptinium Acetate, 3590
[58409-52-2] Albaspidin, 210
[58409-59-9] Bucumolol (unspecified), 1489
[58416-00-5] Protiofate, 8075
[58473-74-8] Cinromide, 2371
[58503-79-0] Meobentine Sulfate*, 5889
[58551-69-2] Carboprost Tromethamine Salt*, 1871
[58579-51-4] Anagrelide Hydrochloride*, 665
[58580-55-5] Dibekacin Sulfate*, 3052
[58581-89-8] Azelastine, 939
[58670-63-6] Dinosterol, 3342
[58712-69-9] Traxanox, 9711
[58785-63-0] Lonomycins Lonomycin A*, 5599
[58786-99-5] Butorphanol Tartrate*, 1565
[58795-03-2] Apalcillin Sodium Salt*, 766
[58798-97-3] Berninamycin, 1200
[58845-80-0] Lonomycins Lonomycin A Sodium Salt*, 5599
[58880-19-6] Trichostatin A*, 9781
[58895-00-4] Trichostatin B*, 9781
[58934-46-6] Lorcainide Hydrochloride*, 5610
[58944-73-3] Sinefungin, 8690
[58957-92-9] Idarubicin, 4931
[58970-76-6] Ubenimex, 9973
[58994-96-0] Ranimustine, 8285
[59005-68-4] Enallylpropymal Sodium Salt*, 3607
[59017-64-0] Ioxaglic Acid, 5084
[59122-46-2] Misoprostol, 6297

[59128-97-1] Haloxazolam, 4635
[59160-29-1] Lidofenin, 5506
[59170-23-9] Bevantolol, 1238
[59182-63-7] Droprenilamine Hydrochloride*, 3506
[59198-70-8] Diflucortolone 21-Valerate*, 3189
[59209-40-4] Proglumetacin Dimaleate*, 7957
[59227-89-3] Laurocapram, 5397
[59263-76-2] Meptazinol Hydrochloride*, 5910
[59277-89-3] Acyclovir, 148
[59333-67-4] Fluoxetine Hydrochloride*, 4222
[59338-87-3] Alizapride Hydrochloride*, 246
[59338-93-1] Alizapride, 246
[59414-23-2] 1-Methoxy-3-(trimethylsilyloxy)-1,3-butadiene (undefined), 6082
[59467-70-8] Midazolam, 6270
[59467-94-6] Midazolam Maleate*, 6270
[59467-96-8] Midazolam Hydrochloride*, 6270
[59512-37-7] Piketoprofen Hydrochloride*, 7576
[59547-58-9] Etamiphyllin Heparinate*, 3754
[59598-29-7] Vernolepin (±)-Form*, 10098
[59619-81-7] Etiproston, 3916
[59625-89-7] Gluconic Acid Magnesium Salt Dihydrate*, 4464
[59652-29-8] Bufuralol Hydrochloride*, 1504
[59669-26-0] Thiodicarb, 9465
[59672-20-7] Sulfaloxic Acid Calcium Salt*, 9080
[59703-84-3] Piperacillin Sodium Salt*, 7616
[59708-57-5] Xibenolol Hydrochloride*, 10209
[59721-28-7] Camostat, 1775
[59721-29-8] Camostat Monomethanesulfonate*, 1775
[59727-70-7] Octacaine Hydrochloride*, 6839
[59729-31-6] Lorcainide, 5610
[59729-32-7] Citalopram Hydrobromide*, 2379
[59729-33-8] Citalopram, 2379
[59767-13-4] Setastine Hydrochloride*, 8619
[59787-61-0] Cyclosporin C*, 2821
[59798-30-0] Mezlocillin Sodium Salt Monohydrate*, 6259
[59803-98-4] Brimonidine, 1399
[59804-37-4] Tenoxicam, 9293
[59859-58-4] Femoxetine, 3993
[59865-13-3] Cyclosporin A*, 2821
[59865-23-5] MeBmt, 5811
[59917-39-4] Vindesine Sulfate Salt*, 10125
[59937-28-9] Malotilate, 5751
[59995-64-1] Thienamycin, 9448
[60134-71-6] Nocardicin B*, 6764
[60142-96-3] Gabapentin, 4343
[60166-93-0] Iopamidol, 5071
[60168-88-9] Fenarimol, 3999
[60200-06-8] Clorsulon, 2471
[60202-16-6] Protein C, 8070
[60207-90-1] Propiconazole, 8003
[60212-94-4] Xenon Hydrate*, 10206
[60282-87-3] Gestodene, 4421
[60296-01-7] Methyl N-Methylnipecotate Methobromide*, 6172
[60318-52-7] Trichosanthin, 9780
[60325-46-4] Sulprostone, 9165

[60414-06-4] Amiprilose Hydrochloride*, 506
[60450-21-7] Morindin, 6353
[60529-33-1] Frangulin, 4288
[60539-06-2] Diphenhydramine Compd with p-Sulfonamide Benzoate*, 3367
[60561-17-3] Sufentanil Citrate*, 9056
[60569-19-9] Propiverine, 8018
[60576-13-8] Piketoprofen, 7576
[60607-34-3] Oxatomide, 7058
[60607-68-3] Indenolol, 4974
[60616-74-2] Magnesium Hydride, 5705
[60628-96-8] Bifonazole, 1260
[60643-86-9] Vigabatrin, 10114
[60662-16-0] Binedaline, 1266
[60668-24-8] Alafosfalin, 204
[60719-84-8] Amrinone, 634
[60731-46-6] Elcatonin, 3578
[60761-10-6] TMD, 9625
[60763-48-6] Gepefrine Tartrate*, 4409
[60789-62-0] Clocapramine Dihydrochloride Monohydrate*, 2427
[60925-61-3] Ceforanide, 1982
[61019-78-1] Fenapanil, 3998
[61036-64-4] Teicoplanin, 9269
[61129-30-4] Zimeldine Dihydrochloride Monohydrate*, 10254
[61197-73-7] Loprazolam, 5604
[61228-92-0] (−)-Periplanone B*, 7314
[61230-25-9] Aplasmomycin, 779
[61260-05-7] Prenalterol Hydrochloride*, 7917
[61270-58-4] Cefonicid, 1980
[61270-78-8] Cefonicid Disodium Salt*, 1980
[61318-90-9] Sulconazole, 9062
[61318-91-0] Sulconazole Mononitrate*, 9062
[61336-41-2] Ceforanide Sodium Salt*, 1982
[61336-70-7] Amoxicillin Trihydrate*, 617
[61337-67-5] Mirtazepine, 6295
[61379-65-5] Rifapentine, 8385
[61380-40-3] Lofentanil, 5586
[61380-41-4] Lofentanil Oxalate*, 5586
[61413-54-5] Rolipram, 8410
[61422-45-5] Carmofur, 1892
[61434-67-1] cis-Resveratrol*, 8326
[61477-96-1] Piperacillin, 7616
[61512-20-7] Cord Factor, 2592
[61545-06-0] Temocillin Disodium Salt*, 9288
[61570-90-9] Tioxidazole, 9599
[61618-27-7] Amfenac Sodium Salt Monohydrate*, 413
[61622-34-2] Cefotiam, 1985
[61676-87-7] Cymiazole, 2833
[61718-82-9] Fluvoxamine Maleate*, 4251
[61732-85-2] Bamipine Dihydrochloride*, 983
[61802-93-5] Metaclazepam Hydrochloride*, 5980
[61825-94-3] Oxaliplatin, 7044
[61849-14-7] Prostacyclin Sodium Salt*, 8061
[61869-07-6] Domiodol, 3473
[61869-08-7] Paroxetine, 7175
[61891-34-7] Flufenamic Acid Aluminum Salt*, 4167
[61951-99-3] Tixocortol, 9623
[62013-04-1] Dirithromycin, 3418
[62031-54-3] Fibroblast Growth Factor, 4117

Denotes derivative of title compound

[62087-72-3] Pentigetide, 7270
[62232-46-6] Bifemelane Hydrochloride*, 1255
[62265-68-3] Quinfamide, 8241
[62288-83-9] Desmopressin Monoacetate*, 2969
[62435-42-1] Perfosfamide (±), 7303
[62510-56-9] Picilorex, 7551
[62534-68-3] Mepartricin A*, 5891
[62534-69-4] Mepartricin B*, 5891
[62568-57-4] DSIP, 3514
[62571-86-2] Captopril, 1817
[62587-73-9] Cefsulodin, 1994
[62610-77-9] Methacrifos, 6004
[62613-82-5] Oxiracetam, 7076
[62640-05-5] Ajugarin-I*, 197
[62640-06-6] Ajugarin-II*, 197
[62640-07-7] Ajugarin-III*, 197
[62658-63-3] Bopindolol, 1362
[62658-64-4] Bopindolol Maleate*, 1362
[62666-20-0] Progabide, 7955
[62774-96-3] Tilisolol Hydrochloride*, 9579
[62893-19-0] Cefoperazone, 1981
[62893-20-3] Cefoperazone Sodium Salt*, 1981
[62952-06-1] Lysine Acetylsalicylate, 5668
[62973-76-6] Azanidazole, 930
[62989-33-7] Sapropterin, 8515
[62992-61-4] Etersalate, 3759
[62994-47-2] Warburganal, 10173
[63075-47-8] Fepradinol, 4052
[63147-28-4] DBHBT, 2894
[63251-39-8] Sulfinalol Hydrochloride*, 9120
[63269-31-8] Ciramadol, 2375
[63323-46-6] Ciramadol Hydrochloride*, 2375
[63329-53-3] Lobenzarit, 5581
[63333-35-7] Bromethalin, 1410
[63358-49-6] Aspoxicillin, 887
[63394-05-8] Plafibride, 7673
[63469-19-2] Apalcillin, 766
[63516-07-4] Flutropium Bromide, 4248
[63521-15-3] Cefazedone Sodium Salt*, 1966
[63527-52-6] Cefotaxime, 1983
[63543-09-9] Thiarubrin(e) A*, 9439
[63547-13-7] Adrafinil, 168
[63585-09-1] Foscarnet Sodium, 4277
[63590-64-7] Terazosin, 9297
[63610-08-2] Indobufen, 4991
[63610-09-3] Lodoxamide Tromethamine Salt*, 5585
[63612-50-0] Nilutamide, 6636
[63659-18-7] Betaxolol, 1229
[63659-19-8] Betaxolol Hydrochloride*, 1229
[63661-61-0] Budipine Hydrochloride*, 1491
[63675-72-9] Nisoldipine, 6658
[63717-27-1] Amifostine Monohydrate*, 424
[63739-93-5] Verrucarin K*, 10100
[63749-94-0] Sulfametrole Mixture with Trimethoprim*, 9089
[63758-79-2] Indalpine, 4965
[63775-95-1] Cyclosporin B*, 2821
[63775-96-2] Cyclosporin D*, 2821
[63814-06-2] Isomethadone d-Form Hydrochloride*, 5201
[63836-75-9] Pivcefalexin, 7670
[63918-50-3] Fencamine Hydrochloride*, 4007
[63968-64-9] Artemisinin, 856
[64000-73-3] Pildralazine, 7577

[64011-61-6] Cyclexedrine Hydrochloride*, 2776
[64019-93-8] Dipivefrin Hydrochloride*, 3400
[64024-15-3] Pentazocine Hydrochloride*, 7261
[64057-70-1] MDMA Hydrochloride* (unspecified), 5806
[64058-48-6] Spectinomycin Sulfate Tetrahydrate*, 8890
[64092-49-5] Zomepirac Sodium Salt Dihydrate*, 10322
[64204-55-3] Esaprazole, 3736
[64211-46-7] Oxiconazole Nitrate, 7071
[64217-62-5] Cefatrizine Compd with Propylene Glycol*, 1965
[64218-02-6] Plaunotol, 7692
[64228-81-5] Atracurium Besylate, 900
[64238-92-2] Benproperine Pamoate*, 1078
[64241-34-5] Cadralazine, 1669
[64294-95-7] Setastine, 8619
[64318-79-2] Gemeprost, 4393
[64475-85-0] Mineral Spirits, 6288
[64485-93-4] Cefotaxime Sodium Salt (syn-Isomer)*, 1983
[64490-92-2] Tolmetin Sodium Salt Dihydrate*, 9655
[64506-49-6] Sofalcone, 8850
[64529-56-2] Ethiozin, 3785
[64552-17-6] Butofilolol, 1563
[64603-91-4] THIP, 9508
[64628-44-0] Triflumuron, 9809
[64706-27-0] Salatrim, 8472
[64706-54-3] Bepridil, 1188
[64726-91-6] Japonine, 5271
[64808-48-6] Lobenzarit Disodium Salt*, 5581
[64840-90-0] Eperisone, 3642
[64872-76-0] Butoconazole, 1561
[64872-77-1] Butoconazole Nitrate*, 1561
[64902-72-3] Chlorsulfuron, 2244
[64924-67-0] Halofuginone Hydrobromide*, 4627
[64952-97-2] Moxalactam, 6369
[64953-12-4] Moxalactam Disodium Salt*, 6369
[65002-17-7] Bucillamine, 1481
[65009-35-0] Lidamidine Hydrochloride*, 5504
[65035-34-9] Lineatin, 5528
[65043-22-3] Indeloxazine Hydrochloride, 4972
[65052-63-3] Cefetamet, 1974
[65057-90-1] Tallysomycin A*, 9213
[65057-91-2] Tallysomycin B*, 9213
[65085-01-0] Cefmenoxime, 1976
[65085-02-1] Cefmenoxime Sodium Salt (syn-Isomer)*, 1976
[65141-46-0] Nicorandil, 6608
[65154-06-5] Platelet Activating Factor, 7682
[65189-78-8] Tropesin (±), 9909
[65243-33-6] Cefetamet Pivaloyloxymethyl Ester*, 1974
[65271-80-9] Mitoxantrone, 6303
[65277-42-1] Ketoconazole, 5313
[65322-72-7] Endralazine Methanesulfonate*, 3617
[65472-88-0] Naftifine, 6442
[65473-14-5] Naftifine Hydrochloride*, 6442
[65474-79-5] 3-Hydroxymethyl-β-carboline*, 3826
[65606-21-5] Isaxonine Phosphate*, 5118
[65646-68-6] Fenretinide, 4040
[65666-07-1] Silymarin, 8680

[65717-97-7] Disofenin, 3422
[65733-18-8] Hydroprene (S)-Form*, 4850
[65807-02-5] Goserelin, 4547
[65828-79-7] Tropesin (unspecified), 9909
[65847-85-0] Niflumic Acid β-Morpholinoethyl Ester*, 6620
[65899-73-2] Tioconazole, 9595
[65981-49-9] Laureline (S)-Form*, 5393
[66002-66-2] Fertirelin Monoacetate*, 4109
[66085-59-4] Nimodipine, 6643
[66104-22-1] Pergolide, 7304
[66104-23-2] Pergolide Mesylate*, 7304
[66108-95-0] Iohexol, 5068
[66142-81-2] 2C-B, 1958
[66148-78-5] Temocillin, 9288
[66195-31-1] Ibopamine, 4921
[66215-27-8] Cyromazine, 2845
[66230-04-4] Fenvalerate (S,S)-Isomer*, 4051
[66264-77-5] Sulfinalol, 9120
[66309-69-1] Cefotiam Dihydrochloride*, 1985
[66357-35-5] Ranitidine, 8286
[66357-59-3] Ranitidine Hydrochloride*, 8286
[66376-36-1] Alendronic Acid, 228
[66455-27-4] PSK, 8112
[66532-85-2] Propacetamol, 7976
[66532-86-3] Propacetamol Hydrochloride*, 7976
[66535-86-2] Lotrifen, 5615
[66564-14-5] Cinitapride, 2352
[66564-16-7] Ciclosidomine, 2326
[66575-29-9] Colforsin, 2539
[66592-87-8] Cefadroxil, 1963
[66644-81-3] Veralipride, 10081
[66685-79-8] Befunolol (R)-(+)-Form Hydrochloride*, 1050
[66711-21-5] Apraclonidine, 791
[66717-59-7] Befunolol (S)-(−)-Form Hydrochloride*, 1050
[66722-44-9] Bisoprolol, 1336
[66734-13-2] Alclometasone 17,21-Dipropionate*, 221
[66778-36-7] Encainide, 3609
[66788-41-8] Tinofedrine, 9589
[66794-74-9] Encainide Hydrochloride*, 3609
[66795-86-6] α-Escin*, 3737
[66796-54-1] Pro-opiomelanocortin, 7975
[66813-51-2] Alexitol Sodium, 232
[66827-12-1] Almagate, 307
[66841-25-6] Tralomethrin, 9700
[66852-54-8] Halobetasol Propionate, 4625
[66871-56-5] Lidamidine, 5504
[66898-62-2] Talniflumate, 9214
[66934-18-7] Flunoxaprofen, 4183
[66981-73-5] Tianeptine, 9560
[67037-37-0] Eflornithine, 3564
[67227-56-9] Fenoldopam, 4020
[67227-57-0] Fenoldopam Monomethanesulfonate*, 4020
[67306-00-7] Fenpropidin, 4034
[67306-03-0] Fenpropimorph, 4035
[67330-25-0] Flufenamic Acid Butyl Ester*, 4167
[67392-87-4] Drospirenone, 3510
[67434-14-4] Benoxaprofen, 1075
[67452-97-5] Alclometasone, 221
[67485-29-4] Hydramethylnon, 4802
[67527-71-3] Mildiomycin, 6279
[67564-91-4] Fenpropimorph cis-Form*, 4035

[67747-09-5] Prochloraz, 7941
[67769-47-5] Lucifer Yellow CH, 5623
[67814-76-0] Isepamicin Sulfate*, 5121
[67915-31-5] Terconazole, 9303
[67995-68-0] Tallysomycin, 9213
[68085-85-8] Cyhalothrin, 2827
[68206-94-0] Cloricromen, 2467
[68238-36-8] Sulphan Blue 2,5-Disulfophenyl Isomer*, 9161
[68247-85-8] Peplomycin, 7286
[68252-19-7] Pirmenol, 7656
[68291-97-4] Zonisamide, 10323
[68302-57-8] Amlexanox, 515
[68332-77-4] Phenoxyacetyl Cellulose, 7408
[68359-37-5] Cyfluthrin, 2826
[68367-52-2] Sorbinil, 8871
[68373-14-8] Sulbactam, 9058
[68377-91-3] Arotinolol Hydrochloride*, 827
[68377-92-4] Arotinolol, 827
[68401-81-0] Ceftizoxime, 2000
[68401-82-1] Ceftizoxime Sodium Salt*, 2000
[68424-04-8] Polydextrose, 7726
[68475-42-3] Anagrelide, 665
[68497-62-1] Pramiracetam, 7886
[68576-86-3] Enciprazine, 3610
[68576-88-5] Enciprazine Dihydrochloride*, 3610
[68630-75-1] Buserelin Monoacetate*, 1527
[68676-88-0] Trichostatin C*, 9781
[68693-11-8] Modafinil, 6311
[68767-14-6] Loxoprofen, 5619
[68786-66-3] Triclabendazole, 9784
[68797-31-9] Econazole Nitrate* (±), 3550
[68813-55-8] Oxantel Pamoate*, 7055
[68844-77-9] Astemizole, 891
[68890-66-4] Piroctone Ethanolamine Salt (1:1)*, 7657
[68990-63-6] Shark Liver Oil, 8622
[69004-03-1] Toltrazuril, 9664
[69047-39-8] Binifibrate, 1267
[69049-73-6] Nedocromil, 6524
[69049-74-7] Nedocromil Disodium Salt*, 6524
[69056-12-8] Prelog-Djerassi Lactone, 7916
[69056-38-8] Sapropterin Dihydrochloride*, 8515
[69123-98-4] Fialuridine, 4114
[69191-18-0] Mefloquine Hydrochloride Combination with Sulfadoxine and Pyrimethamine*, 5845
[69198-10-3] Metronidazole Hydrochloride*, 6242
[69365-73-7] Bumadizon Calcium Salt Hemihydrate*, 1507
[69372-19-6] Pemirolast, 7205
[69377-81-7] Fluroxypyr, 4238
[69381-94-8] Fenprostalene, 4037
[69388-84-7] Sulbactam Sodium Salt*, 9058
[69409-94-5] Fluvalinate, 4249
[69558-55-0] Thymopentin, 9544
[69610-10-2] MDMA (±), 5806
[69655-05-6] Didanosine, 3148
[69712-56-7] Cefotetan, 1984
[69739-16-8] Cefodizime, 1979
[69756-53-2] Halofantrine, 4626
[69770-45-2] Flumethrin, 4175
[69806-50-4] Fluazifop-butyl, 4152
[69815-49-2] Norepinephrine d-Bitartrate*, 6788
[69900-72-7] Trimoprostil, 9853
[69975-86-6] Doxofylline, 3494
[70020-71-2] ε-Acetamidocaproic Acid Zinc Salt*, 43

[70024-40-7] Terazosin Hydrochloride Dihydrate*, 9297
[70059-30-2] Cimetidine Hydrochloride*, 2337
[70111-54-5] Loprazolam Methanesulfonate*, 5604
[70124-77-5] Flucythrinate, 4160
[70280-88-5] Difemerine Hydrochloride*, 3180
[70288-86-7] Ivermectin, 5264
[70356-03-5] Cefaclor, 1962
[70356-09-1] Butyl Methoxydibenzoylmethane, 1616
[70359-46-5] Brimonidine D-Tartrate*, 1399
[70374-39-9] Lornoxicam, 5612
[70384-29-1] Peplomycin Sulfate Salt*, 7286
[70458-92-3] Pefloxacin, 7197
[70458-95-6] Pefloxacin Methanesulfonate*, 7197
[70458-96-7] Norfloxacin, 6793
[70476-82-3] Mitoxantrone Dihydrochloride*, 6303
[70608-72-9] 5-HETE, 4708
[70613-99-9] Periplanone B*, 7314
[70667-26-4] Ornoprostil, 7003
[70704-03-9] Vinconate, 10123
[70775-75-6] Octenidine Dihydrochloride*, 6852
[70797-11-4] Cefpiramide, 1989
[70879-28-6] Alfentanil Hydrochloride Monohydrate*, 236
[70981-66-7] Naproxen Piperazine Salt*, 6504
[71002-09-0] Pirazolac, 7643
[71010-52-1] Gellan Gum, 4389
[71031-15-7] Cathinone, 1954
[71116-82-0] Tiaprost, 9563
[71117-51-6] Zoapatanol, 10318
[71125-38-7] Meloxicam, 5869
[71138-35-7] Desogestrel Mixture with Ethinyl Estradiol*, 2971
[71142-71-7] PPACK, 7881
[71160-24-2] Leukotriene B₄*, 5482
[71195-58-9] Alfentanil, 236
[71247-25-1] Ceruletide Diethylamine Salt*, 2048
[71251-02-0] Octenidine, 6852
[71320-77-9] Moclobemide, 6309
[71395-14-7] Tocainide (±)-Form Hydrochloride*, 9629
[71439-68-4] Bisantrene Dihydrochloride*, 1284
[71486-22-1] Vinorelbine, 10127
[71539-72-5] Thiarubrin(e)B*, 9439
[71626-11-4] Benalaxyl, 1056
[71628-96-1] Menogaril, 5881
[71675-85-9] Amisulpride, 508
[71731-58-3] Tiquizium Bromide, 9602
[71751-41-2] Abamectin, 1
[71771-90-9] Denopamine, 2943
[71789-16-7] Clopamide Mixture with Pindolol*, 2454
[71963-77-4] Artemether, 854
[71980-98-8] Pseudomonic Acid C*, 8103
[71990-71-1] Etofylline Combination with Theophylline*, 3924
[72017-58-4] Phencarbamide 1,5-Naphthalenedisulfonate*, 7363
[72025-60-6] Leukotriene C₄*, 5482
[72058-36-7] Lochnericine, 5582
[72059-45-1] Leukotriene A₄*, 5482
[72301-79-2] Enviroxime, 3637
[72324-18-6] Stepronin, 8963
[72332-33-3] Procaterol, 7939
[72420-38-3] Acifran, 112
[72432-03-2] Miglitol, 6274
[72432-10-1] Aniracetam, 700

[72459-58-6] Triazoxide, 9737
[72479-26-6] Fenticonazole, 4047
[72490-01-8] Fenoxycarb, 4027
[72492-12-7] Spizofurone, 8918
[72496-41-4] Pirarubicin, 7642
[72509-76-3] Felodipine, 3991
[72522-13-5] Eptazocine, 3673
[72558-82-8] Ceftazidime, 1995
[72559-06-9] Rifabutin, 8380
[72571-82-5] Pipemidic Acid Trihydrate*, 7613
[72732-56-0] Piritrexim, 7654
[72822-12-9] Dapiprazole, 2884
[72822-13-0] Dapiprazole Monohydrochloride*, 2884
[72869-16-0] Pramiracetam Sulfate*, 7886
[72945-61-0] Oxidronic Acid ⁹⁹ᵐTc-complex*, 7074
[72956-09-3] Carvedilol, 1924
[72962-43-7] Brassinolide, 1390
[73080-51-0] Repirinast, 8305
[73121-56-9] Enprostil, 3629
[73151-29-8] Fenticonazole Mononitrate*, 4047
[73173-12-3] Tetroxoprim Mixture with Sulfadiazine*, 9386
[73210-73-8] Xamoterol Hemifumarate*, 10189
[73217-88-6] Apraclonidine Dihydrochloride*, 791
[73218-79-8] Apraclonidine Hydrochloride*, 791
[73220-03-8] Remoxipride Hydrochloride*, 8301
[73231-34-2] Florfenicol, 4145
[73334-07-3] Iopromide, 5076
[73384-59-5] Ceftriaxone, 2001
[73384-60-8] Sulmazole, 9159
[73391-87-4] Pizotyline Hydrochloride*, 7671
[73523-00-9] Luprostiol, 5638
[73573-42-9] Rescimetol, 8310
[73573-87-2] Formoterol, 4272
[73573-88-3] Mevastatin, 6251
[73590-58-6] Omeprazole, 6977
[73681-12-6] Indecainide Hydrochloride*, 4971
[73771-04-7] Prednicarbate, 7899
[73803-48-2] Tripamide, 9866
[73816-42-9] Meclocycline 5-Sulfosalicylate*, 5818
[73836-78-9] Leukotriene D₄*, 5482
[73963-72-1] Cilostazol, 2335
[73972-50-6] Nitroakridin 3582 Dihydrochloride*, 6678
[74011-58-8] Enoxacin, 3625
[74014-51-0] Rokitamycin, 8408
[74050-97-8] Haloperidol Decanoate*, 4629
[74050-98-9] Ketanserin, 5307
[74051-80-2] Sethoxydim, 8620
[74103-06-3] Ketorolac, 5318
[74103-07-4] Ketorolac (±)-Form Tromethamine Salt*, 5318
[74115-24-5] Clofentezine, 2435
[74150-27-9] Pimobendan, 7588
[74176-31-1] Alfaprostol, 234
[74191-85-8] Doxazosin, 3489
[74214-62-3] Ethyl β-Carboline-3-carboxylate, 3826
[74222-97-2] Sulfometuron Methyl, 9130
[74223-64-6] Metsulfuron Methyl, 6244
[74252-25-8] Indomethacin Sodium Salt Trihydrate*, 4998
[74258-86-9] Alacepril, 202
[74298-63-8] Chlormidazole Hydrochloride*, 2156

Denotes derivative of title compound

[74341-78-9] MDE Hydrochloride*, 5805

[74356-00-6] Cefotetan Disodium Salt*, 1984

[74381-53-6] Leuprolide Acetate*, 5484

[74397-12-9] Limaprost, 5514

[74431-23-5] Imipenem, 4954

[74432-68-1] Ractopamine R,R-Form Hydrochloride*, 8275

[74436-00-3] Cyclosporin G*, 2821

[74449-59-5] 5-Methyl-5-(3-phenanthryl)hydantoin Sodium Salt*, 6184

[74512-12-2] Omoconazole, 6978

[74517-78-5] Indecainide, 4971

[74545-79-2] Aloeresin A*, 312

[74591-03-0] Endiandric Acid A*, 3611

[74639-40-0] Docarpamine, 3457

[74697-28-2] Cloricromen Hydrochloride*, 2467

[74711-43-6] Zaltoprofen (unspecified), 10244

[74717-53-6] Detoxin C₁*, 2982

[74738-17-3] Fenpiclonil, 4030

[74764-40-2] Bepridil Hydrochloride Monohydrate*, 1188

[74812-63-8] Nordefrin (R*,S*)-(±)-Form, 6785

[74849-93-7] Cefpiramide Sodium Salt*, 1989

[74863-84-6] Argatroban, 816

[74913-18-1] Dynorphin, 3528

[75011-65-3] Ibopamine Hydrochloride*, 4921

[75139-06-9] Tetronasin, 9384

[75199-98-3] Buclosamide Combination with Salicylic Acid*, 1485

[75216-20-5] Fluosol DA, 4221

[75272-39-8] Nemonapride, 6533

[75330-75-5] Lovastatin, 5616

[75438-57-2] Moxonidine, 6375

[75438-58-3] Moxonidine Hydrochloride*, 6375

[75507-68-5] Flupirtine Maleate*, 4227

[75530-68-6] Nilvadipine, 6637

[75558-90-6] Amperozide, 619

[75607-67-9] Fludarabine 5'-Monophosphate*, 4162

[75621-03-3] CHAPS, 2085

[75659-07-3] Dilevalol, 3245

[75659-08-4] Dilevalol Hydrochloride*, 3245

[75695-93-1] Isradipine, 5260

[75696-02-5] Cinolazepam, 2368

[75715-89-8] Leukotriene E₄*, 5482

[75736-33-3] Diclobutrazol, 3131

[75738-58-8] Cefmenoxime Hydrochloride (syn-Isomer)*, 1976

[75757-02-7] Osteocalcin (human reduced), 7025

[75821-71-5] Lonazolac Calcium Salt*, 5596

[75847-73-3] Enalapril, 3605

[75887-54-6] Arteether, 853

[75898-26-9] β-Carboxyaspartic Acid, 1875

[75917-92-9] Iofetamine ¹²³I, 5066

[75975-70-1] Cephradine Monohydrate*, 2032

[76025-73-5] Carpetimycin A*, 1909

[76060-33-8] Endiandric Acid B*, 3611

[76060-34-9] Endiandric Acid C*, 3611

[76094-36-5] Carpetimycin B*, 1909

[76095-16-4] Enalapril Maleate*, 3605

[76120-72-4] Ramifenazone Mixture with Phenylbutazone*, 8282

[76123-46-1] Calcium Magnesium Acetate, 1725

[76343-93-6] Latrunculin A*, 5388

[76343-94-7] Latrunculin B*, 5388

[76420-72-9] Enalaprilat, 3606

[76497-13-7] Sultamicillin, 9166

[76543-88-9] Interferon Alfa-2a*, 5016

[76568-02-0] Flosequinan, 4146

[76578-14-8] Quizalofop-Ethyl, 8269

[76584-70-8] Valproic Acid Sodium Salt (2:1)*, 10049

[76610-84-9] Cefbuperazone, 1968

[76631-46-4] Detomidine, 2981

[76648-01-6] Cefbuperazone Sodium Salt*, 1968

[76674-21-0] Flutriafol, 4246

[76712-82-8] Histrelin, 4760

[76738-62-0] Paclobutrazol, 7118

[76738-75-5] α-Bisabolol (+)-Epi-form*, 1281

[76820-74-1] Ioxaglate Meglumine mixture with Ioxaglate Sodium*, 5084

[76824-35-6] Famotidine, 3972

[76836-02-7] PIPES Disodium Salt*, 7633

[76932-56-4] Nafarelin, 6437

[76963-41-2] Nizatidine, 6758

[77086-19-2] Dizocilpine (−)-Form*, 3451

[77086-21-6] Dizocilpine, 3451

[77086-22-7] Dizocilpine Maleate*, 3451

[77103-91-4] Acifran (+)-Form*, 112

[77103-92-5] Acifran (−)-Form*, 112

[77107-46-1] Calmodulin, 1767

[77164-20-6] Moprolol l-Form*, 6345

[77174-66-4] Cloconazole Monohydrochloride*, 2429

[77175-51-0] Cloconazole, 2429

[77181-69-2] Sorivudine, 8875

[77191-36-7] Nefiracetam, 6528

[77287-05-9] Rioprostil, 8395

[77327-04-9] Didemnin A*, 3150

[77327-05-0] Didemnin B*, 3150

[77327-06-1] Didemnin C*, 3150

[77337-73-6] Acamprosate Calcium, 14

[77392-58-6] Paraherquamide, 7159

[77469-98-8] Pimobendan Hydrochloride*, 7588

[77590-95-5] Cetamolol Hydrochloride*, 2060

[77671-31-9] Enoximone, 3627

[77734-91-9] Palytoxin, 7132

[77855-81-3] Milbemycin D*, 6278

[77883-43-3] Doxazosin Monomethanesulfonate*, 3489

[77912-79-9] Sulfazecin, 9119

[78110-38-0] Aztreonam, 955

[78111-17-8] Okadaic Acid, 6958

[78113-36-7] Romurtide, 8412

[78126-10-0] Stepronin Sodium Salt*, 8963

[78148-59-1] α-Bisabolol (−)-Epi-form*, 1281

[78178-41-3] Podophyllinic Acid Hydrazide*, 7703

[78186-34-2] Bisantrene, 1284

[78266-06-5] Mebrofenin, 5812

[78415-72-2] Milrinone, 6284

[78439-06-2] Ceftazidime Pentahydrate*, 1995

[78473-71-9] Enterolactone, 3633

[78474-55-2] Teleocidins, 9271

[78499-27-1] Bermoprofen, 1199

[78587-05-0] Hexythiazox, 4751

[78613-35-1] Amorolfine, 612

[78613-38-4] Amorolfine Hydrochloride*, 612

[78628-28-1] Roxatidine Acetate, 8431

[78654-44-1] Amicoumacin A, 417

[78697-56-0] Lophotoxin, 5603

[78712-43-3] Ozagrel Hydrochloride*, 7115

[78718-25-9] Benexate Hydrochloride, 1065

[78755-81-4] Flumazenil, 4169

[78822-40-9] Pirlimycin Hydrochloride*, 7655

[78833-03-1] Pentisomide (unspecified), 7271

[78853-39-1] Cyclodrine Hydrochloride*, 2788

[78919-13-8] Iloprost, 4940

[78964-85-9] Fosfomycin Tromethanol*, 4280

[78967-07-4] Mofezolac, 6314

[78994-24-8] Centchroman, 2018

[79094-20-5] Daltroban, 2871

[79307-93-0] Azelastine Monohydrochloride*, 939

[79350-37-1] Cefixime, 1975

[79467-19-9] Sintropium Bromide, 8693

[79483-69-5] Piritrexim Isethionate*, 7654

[79516-68-0] Levocabastine, 5489

[79538-32-2] Tefluthrin, 9266

[79547-78-7] Levocabastine Hydrochloride*, 5489

[79548-73-5] Pirlimycin, 7655

[79559-97-0] Sertraline Hydrochloride*, 8612

[79617-96-2] Sertraline, 8612

[79622-59-6] Fluazinam, 4153

[79645-27-5] Tobramycin Sulfate*, 9628

[79660-72-3] Fleroxacin, 4137

[79704-88-4] α-Methylfentanyl, 6150

[79770-24-4] Iotrolan, 5082

[79794-75-5] Loratadine, 5608

[79831-76-8] Castanospermine, 1944

[79871-54-8] Norgestimate Mixture with Ethinyl Estradiol*, 6796

[79902-63-9] Simvastatin, 8686

[80012-43-7] Epinastine, 3655

[80125-14-0] Remoxipride, 8301

[80214-83-1] Roxithromycin, 8433

[80223-99-0] Tamsulosin (±)-Form Hydrochloride*, 9217

[80370-57-6] Ceftiofur, 1999

[80382-23-6] Loxoprofen Sodium Salt*, 5619

[80387-96-8] Difemerine, 3180

[80433-71-2] Folinic Acid l-Form Calcium Salt*, 4254

[80443-41-0] Tebuconazole (unspecified), 9253

[80449-31-6] Urinastatin, 10021

[80455-68-1] Fredericamycin A, 4291

[80471-63-2] Epostane, 3664

[80474-14-2] Fluticasone Propionate, 4244

[80529-93-7] Gadopentetic Acid, 4347

[80530-63-8] Picotamide Monohydrate*, 7560

[80576-83-6] Edatrexate, 3553

[80621-81-4] Rifaximin, 8386

[80663-95-2] Iobenguane, 5027

[80734-02-7] Lenampicillin Hydrochloride*, 5460

[80750-21-6] Bucumolol (±)-form, 1489

[80755-51-7] Bunazosin, 1512

[80863-62-3] Alitame, 245

[80880-90-6] Telenzepine, 9270

[81025-03-8] Lactitol Dihydrate*, 5352

[81025-04-9] Lactitol Monohydrate*, 5352
[81098-60-4] Cisapride, 2377
[81103-11-9] Clarithromycin, 2400
[81110-73-8] Acetorphan, 72
[81129-83-1] Cilastatin Sodium Salt*, 2331
[81131-70-6] Pravastatin Sodium, 7894
[81147-92-4] Esmolol, 3741
[81161-17-3] Esmolol Hydrochloride*, 3741
[81262-93-3] Procaterol Hydrochloride Hemihydrate*, 7939
[81334-34-1] Imazapyr, 4942
[81335-37-7] Imazaquin, 4943
[81335-77-5] Imazethapyr, 4944
[81403-68-1] Alfuzosin Hydrochloride*, 237
[81403-80-7] Alfuzosin, 237
[81405-85-8] Imazamethabenz, 4941
[81406-37-3] Fluroxypyr 1-Methylheptyl Ester*, 4238
[81409-90-7] Cabergoline, 1637
[81486-22-8] Nipradilol, 6655
[81510-83-0] Imazapyr Isopropylamine Salt*, 4942
[81525-10-2] Nafamostat, 6436
[81584-06-7] Xibenolol, 10209
[81627-83-0] Macrophage Colony-Stimulating Factor, 5678
[81669-57-0] Anistreplase, 712
[81732-46-9] Bambuterol Hydrochloride*, 980
[81732-65-2] Bambuterol, 980
[81777-89-1] Dimethazone, 3262
[81789-85-7] Indenolol Hydrochloride*, 4974
[81801-12-9] Xamoterol, 10189
[81840-15-5] Vesnarinone, 10105
[81982-32-3] Alpropride, 319
[82009-34-5] Cilastatin, 2331
[82030-87-3] Methionyl Human Growth Hormone*, 8864
[82034-46-6] Loteprednol Etabonate, 5614
[82085-94-7] Rotraxate Hydrochloride*, 8428
[82097-50-5] Triasulfuron, 9732
[82110-72-3] Fenoxaprop-ethyl, 4024
[82115-62-6] Interferon-γ, 5018
[82159-09-9] Epalrestat, 3640
[82219-78-1] Cefuzonam, 2003
[82225-47-6] Ajugarin-IV*, 197
[82231-14-9] Ajugarin-V*, 197
[82310-93-8] Hypusine Dihydrochloride*, 4918
[82410-32-0] Ganciclovir, 4374
[82413-20-5] Droloxifene, 3502
[82419-36-1] Ofloxacin, 6865
[82473-24-3] CHAPS Hydroxy Analog*, 2085
[82547-58-8] Cefteram, 1996
[82547-81-7] Cefteram Pivaloyloxymethyl Ester*, 1996
[82558-50-7] Isoxaben, 5256
[82560-54-1] Benfuracarb, 1069
[82571-53-7] Ozagrel, 7115
[82586-55-8] Quinapril Hydrochloride*, 8233
[82626-01-5] Alpidem, 318
[82626-48-0] Zolpidem, 10321
[82640-04-8] Raloxifene Hydrochloride*, 8281
[82657-04-3] Bifenthrin, 1257
[82657-92-9] Pro-Urokinase, 8089
[82747-56-6] Cicletanine Hydrochloride*, 2323
[82752-99-6] Nefazodone Hydrochloride*, 6527

[82768-85-2] Quinaprilat*, 8233
[82785-45-3] Neuropeptide Y, 6568
[82801-81-8] MDE (±), 5805
[82834-16-0] Perindopril, 7311
[82857-38-3] Bopindolol Malonate*, 1362
[82952-64-5] Trimetrexate D-Glucuronate*, 9851
[82956-11-4] Nafamostat Dimethanesulfonate*, 6436
[82964-04-3] Tolrestat, 9663
[82989-25-1] Tazanolast, 9248
[83105-70-8] Sultamicillin Tosylate*, 9166
[83150-76-9] Octreotide, 6859
[83314-01-6] Bryostatin 1*, 1478
[83366-66-9] Nefazodone, 6527
[83435-66-9] Delapril, 2928
[83435-67-0] Delapril Hydrochloride*, 2928
[83602-05-5] Spiraprilat*, 8905
[83621-06-1] Omoconazole Nitrate*, 6978
[83647-97-6] Spirapril, 8905
[83657-18-5] Diniconazole (−)-Form*, 3320
[83657-19-6] Diniconazole (+)-Form*, 3320
[83657-24-3] Diniconazole, 3320
[83712-60-1] Defibrotide, 2917
[83799-24-0] Fexofenadine, 4113
[83846-83-7] Ketanserin Tartrate*, 5307
[83869-56-1] Granulocyte-Macrophage Colony-Stimulating Factor, 4559
[83881-51-0] Cetirizine, 2063
[83881-52-1] Cetirizine Dihydrochloride*, 2063
[83905-01-5] Azithromycin, 946
[83915-83-7] Lisinopril, 5540
[83919-23-7] Mometasone Furoate, 6324
[83930-13-6] Somatoliberin, 8861
[84031-17-4] Metaclazepam, 5980
[84057-84-1] Lamotrigine, 5367
[84057-95-4] Ropivacaine, 8417
[84088-42-6] Roquinimex, 8418
[84294-96-2] Enoxacin Sesquihydrate*, 3625
[84297-59-6] Cabenegrin A-I*, 1636
[84297-60-9] Cabenegrin A-II*, 1636
[84305-41-9] Cefminox, 1978
[84371-65-3] Mifepristone, 6273
[84449-90-1] Raloxifene, 8281
[84504-69-8] Irsogladine Maleate*, 5113
[84611-23-4] Erdosteine, 3684
[84625-61-6] Itraconazole, 5262
[84799-02-0] Laidlomycin Propionate Potassium*, 5361
[84845-57-8] Ritipenem, 8400
[84845-58-9] Ritipenem Sodium Salt*, 8400
[84878-61-5] Maduramicin, 5682
[84880-03-5] Cefpimizole, 1988
[84937-45-1] Gusperimus (−)-Form Trihydrochloride*, 4610
[84957-29-9] Cefpirome, 1990
[84964-12-5] Brovincamine Hydrogen Fumarate*, 1473
[85068-76-4] Iofetamine 123I Hydrochloride*, 5066
[85136-71-6] Tilisolol, 9579
[85175-67-3] Zatebradine, 10245
[85248-93-7] Pseudomonic Acid D*, 8103
[85287-61-2] Cefpimizole Monosodium Salt*, 1988
[85320-68-9] Amosulalol, 614

[85329-89-1] Cabergoline Diphosphate*, 1637
[85371-64-8] Pinacidil, 7592
[85441-61-8] Quinapril, 8233
[85468-01-5] Gusperimus Trihydrochloride*, 4610
[85509-19-9] Flusilazole, 4240
[85622-93-1] Temozolomide, 9289
[85637-73-6] Atrial Natriuretic Factor, 903
[85649-12-3] Cyfluthrin (1R,3R,αR)-Form*, 2826
[85649-15-6] Cyfluthrin (1R,3R,αS)-Form*, 2826
[85649-19-0] Cyfluthrin (1R,3S,αS)-Form*, 2826
[85721-33-1] Ciprofloxacin, 2374
[85856-54-8] Moveltipril, 6368
[85865-74-3] Pygeum Africanum Extract, 8135
[85898-30-2] Interleukin-2, 5020
[85921-53-5] Moveltipril Calcium Salt*, 6368
[85960-17-4] Imipenem Combination with Cilastatin Sodium*, 4954
[86050-77-3] Gadopentetic Acid Dimeglumine Salt*, 4347
[86168-78-7] Sermorelin, 8605
[86197-47-9] Dopexamine, 3482
[86209-51-0] Primisulfuron-methyl, 7928
[86220-42-0] Nafarelin Acetate Hydrate*, 6437
[86227-47-6] (all-Z)-5,8,11,14,17-Eicosapentaenoic Acid Ethyl Ester*, 3572
[86273-18-9] Lenampicillin, 5460
[86329-79-5] Cefodizime Disodium Salt*, 1979
[86347-14-0] Medetomidine, 5830
[86347-15-1] Medetomidine Hydrochloride*, 5830
[86386-73-4] Fluconazole, 4158
[86393-32-0] Ciprofloxacin Monohydrochloride Monohydrate*, 2374
[86401-95-8] Methylprednisolone Aceponate*, 6189
[86408-72-2] Ecabet Sodium Salt*, 3538
[86479-06-3] Hexaflumuron, 4720
[86484-91-5] Dopexamine Dihydrochloride*, 3482
[86541-74-4] Benazepril Hydrochloride*, 1058
[86541-75-5] Benazepril, 1058
[86541-78-8] Benazeprilat*, 1058
[86767-75-1] Octenidine Disaccharin*, 6852
[86780-90-7] Aranidipine, 806
[86832-68-0] Carumonam Disodium Salt*, 1922
[86880-51-5] Epanolol, 3641
[87041-58-5] Hementin, 4673
[87051-43-2] Ritanserin, 8399
[87056-78-8] Quinagolide, 8226
[87130-20-9] Diethofencarb, 3158
[87233-61-2] Emedastine, 3597
[87233-62-3] Emedastine Difumarate*, 3597
[87234-24-0] Piroxicam Cinnamic Acid Ester*, 7661
[87238-52-6] Ritipenem Acetoxymethyl Ester*, 8400
[87239-81-4] Cefpodoxime Proxetil, 1991
[87269-97-4] Ramiprilat*, 8123
[87333-19-5] Ramipril, 8283
[87344-06-7] Amtolmetin Guacil, 637
[87440-45-7] Taprostene Sodium Salt*, 9227

Denotes derivative of title compound

[87546-18-7] Flumiclorac-pentyl*, 4178
[87547-04-4] Flumiclorac, 4178
[87638-04-8] Carumonam, 1922
[87674-68-8] Dimethenamid, 3263
[87679-37-6] Trandolapril, 9703
[87679-71-8] Trandolaprilat*, 9703
[87726-17-8] Panipenem, 7141
[87760-53-0] Tandospirone, 9219
[87771-40-2] Ioversol, 5083
[87806-31-3] Porfimer Sodium, 7755
[87818-31-3] Cinmethylin, 2354
[87820-88-0] Tralkoxydim, 9699
[87848-99-5] Acrivastine, 129
[88040-23-7] Cefepime, 1973
[88069-49-2] Pilsicainide Hydrochloride Hemihydrate*, 7581
[88069-67-4] Pilsicainide, 7581
[88150-42-9] Amlodipine, 516
[88150-47-4] Amlodipine (±)-Form Maleate*, 516
[88200-01-5] Chelidonine (−)-Form*, 2095
[88255-01-0] Netobimin, 6564
[88283-41-4] Pyrifenox, 8167
[88426-33-9] Buparvaquone, 1516
[88430-50-6] Beraprost, 1190
[88475-69-8] Beraprost Sodium Salt*, 1190
[88495-63-0] Artesunate, 857
[88606-96-6] Butofilolol Maleate*, 1563
[88669-04-9] Trospectomycin, 9917
[88671-89-0] Myclobutanil, 6402
[88851-61-0] Trospectomycin Sulfate Pentahydrate*, 9917
[88889-14-9] Fosinopril Sodium Salt*, 4282
[89226-50-6] Manidipine, 5786
[89226-75-5] Manidipine Dihydrochloride*, 5786
[89365-50-4] Salmeterol, 8489
[89371-37-9] Imidapril, 4947
[89371-44-8] Imidaprilat*, 4947
[89396-94-1] Imidapril Monohydrochloride*, 4947
[89419-40-9] Mosapramine, 6365
[89482-00-8] Zaltoprofen (±), 10244
[89482-01-9] Zaltoprofen (S)-Form*, 10244
[89565-68-4] Tropisetron, 9914
[89667-40-3] Isbogrel, 5120
[89672-11-7] Cioteronel, 2372
[89778-26-7] Toremifene, 9688
[89778-27-8] Toremifene Citrate*, 9688
[89785-84-2] Tazobactam Sodium Salt*, 9251
[89786-04-9] Tazobactam, 9251
[89796-99-6] Aceclofenac, 19
[89797-00-2] Iopentol, 5073
[89899-81-0] HPA-23, 4786
[89943-82-8] Cicletanine, 2323
[89987-06-4] Tiludronic Acid, 9582
[90098-04-7] Rebamipide, 8296
[90101-16-9] Droxicam, 3512
[90274-24-1] Ractopamine Hydrochloride*, 8275
[90293-01-9] Bifemelane, 1255
[90357-06-5] Bicalutamide, 1247
[90729-43-4] Ebastine, 3534
[90803-92-2] Thymomodulin, 9543
[90878-85-6] Inaperisone Hydrochloride* (unspecified), 4963
[90881-70-2] D-Galactose Microparticulate Form*, 4353
[90982-32-4] Chlorimuron Ethyl, 2142
[91161-71-6] Terbinafine, 9299
[91296-86-5] Difloxacin Monohydrochloride*, 3187

[91296-87-6] Sarafloxacin Monohydrochloride*, 8517
[91374-20-8] Ropinirole Hydrochloride*, 8416
[91374-21-9] Ropinirole, 8416
[91423-46-0] [S-(all-E)]-3-(1,3,5,7,9-Dodecapentaenyloxy)-1,2-propanediol, 3465
[91431-42-4] Lonapalene, 5595
[91574-91-3] Benexate Hydrochloride Compd with β-Cyclodextrin (1:1)*, 1065
[91599-74-5] Benidipine (±)-α-Form Hydrochloride*, 1071
[91618-36-9] Ibafloxacin, 4919
[91714-93-1] Bromfenac Monosodium Salt*, 1411
[91714-94-2] Bromfenac, 1411
[91832-40-5] Cefdinir, 1971
[91940-87-3] Zatebradine Hydrochloride*, 10245
[92071-51-7] Rotraxate, 8428
[92077-78-6] Cilazapril, 2332
[92118-27-9] Fotemustine, 4285
[92129-86-7] Purified Protein Derivative*, 9937
[92279-84-0] MDMA Hydrochloride* (±), 5806
[92285-01-3] Ajoene, 195
[92339-11-2] Iodixanol, 5044
[92589-98-5] Ipsapirone Hydrochloride*, 5096
[92623-85-3] Milnacipran, 6281
[92665-29-7] Cefprozil (anhydrous), 1992
[92676-86-3] Cefprozil E-Form*, 1992
[92823-03-5] Flomoxef Sodium Salt*, 4141
[93076-39-2] Ritanserin L-Tartrate*, 8399
[93106-60-6] Enrofloxacin, 3630
[93227-64-6] Betaine Sodium Aspartate*, 1225
[93384-43-1] Botulin A*, 1384
[93413-04-8] Asperlicin, 878
[93413-05-9] Asperlicin E*, 878
[93413-06-0] Asperlicin C*, 878
[93413-07-1] Asperlicin D*, 878
[93413-08-2] Asperlicin B*, 878
[93413-69-5] Venlafaxine, 10079
[93479-97-1] Glimepiride, 4449
[93633-92-2] Amosulalol Monohydrochloride*, 614
[93793-83-0] Roxatidine Acetate Hydrochloride*, 8431
[93957-54-1] Fluvastatin, 4250
[93957-55-2] Fluvastatin Sodium Salt*, 4250
[94050-52-9] Flucycloxuron (E-isomer), 4159
[94050-53-0] Flucycloxuron (Z-isomer), 4159
[94055-76-2] Suplatast Tosylate, 9178
[94125-34-5] Prosulfuron, 8067
[94134-60-8] Eburnamonine (−)-Form Phosphate*, 3537
[94361-06-5] Cyproconazole, 2840
[94424-50-7] Quinagolide Hydrochloride*, 8226
[94442-12-3] Aminopyrine Bicamphorate*, 495
[94470-67-4] Cromakalim*, 5487
[94480-49-6] Pavoninin-5, 7186
[94535-50-9] Levcromakalim, 5487
[94749-08-3] Salmeterol 1-Hydroxy-2-naphthoate*, 8489
[94841-17-5] Spirapril Hydrochloride*, 8905
[95058-81-4] Gemcitabine, 4392
[95153-31-4] Perindoprilat*, 7311

[95233-18-4] Atovaquone, 898
[95266-40-3] Trinexapac-ethyl, 9855
[95399-71-6] Fosinopril Diacid*, 4282
[95465-99-9] Cadusafos, 1670
[95635-55-5] Ranolazine, 8287
[95635-56-6] Ranolazine Dihydrochloride*, 8287
[95656-48-2] Mabuterol dl-Form Hydrochloride*, 5674
[95656-54-5] Mabuterol d-Form Hydrochloride*, 5674
[95656-55-6] Mabuterol l-Form Hydrochloride*, 5674
[95723-59-4] Lunacrine dl-Form*, 5631
[95737-68-1] Pyriproxyfen, 8174
[95751-30-7] Charybdotoxin, 2087
[95789-30-3] Cefotiam Hexetil Hydrochloride*, 1985
[95847-70-4] Ipsapirone, 5096
[96020-91-6] Eflornithine Hydrochloride Monohydrate*, 3564
[96036-03-2] Meropenem (anhydrous), 5960
[96055-45-7] Nicotine Polacrilex*, 6611
[96125-53-0] Clentiazem, 2408
[96128-89-1] Erythromycin Acistrate, 3721
[96128-92-6] Clentiazem Maleate*, 2408
[96187-53-0] Brequinar, 1394
[96201-88-6] Brequinar Sodium Salt*, 1394
[96513-83-6] Pentisomide (±), 7271
[96588-03-3] Medibazine Dihydrochloride*, 5831
[96623-56-2] Cinitapride Tartrate*, 2352
[96829-58-2] Orlistat, 6997
[96946-42-8] Atracurium Besylate 1R-cis,1R′ cis Form*, 900
[97205-34-0] Nebracetam (unspecified), 6520
[97205-35-1] Nebracetam Fumarate*, 6520
[97240-79-4] Topiramate, 9686
[97322-87-7] Troglitazone, 9898
[97519-39-6] Ceftibuten, 1998
[97621-73-3] Medullipin I, 5841
[97645-24-4] TNAZ, 9626
[97682-44-5] Irinotecan, 5104
[97689-87-7] Tunichrome B-1, 9950
[97752-20-0] Droloxifene Citrate*, 3502
[97772-98-0] Butedronic Acid Tetrasodium Salt*, 1546
[97825-25-7] Ractopamine, 8275
[97886-45-8] Dithiopyr, 3439
[97987-88-7] Vamicamide (unspecified), 10052
[98043-60-8] Mosapramine Dihydrochloride*, 6365
[98048-97-6] Fosinopril, 4282
[98059-18-8] Cys-Tyr-Cys-interferon γ*, 5018
[98059-61-1] Interferon Gamma-1b*, 5018
[98079-51-7] Lomefloxacin, 5592
[98079-52-8] Lomefloxacin Monohydrochloride*, 5592
[98105-99-8] Sarafloxacin, 8517
[98106-17-3] Difloxacin, 3187
[98123-83-2] Epsiprantel, 3671
[98205-89-1] Flesinoxan Hydrochloride*, 4138
[98206-09-8] Eltoprazine Hydrochloride*, 3592
[98206-10-1] Flesinoxan, 4138
[98224-03-4] Eltoprazine, 3592

Denotes derivative of title compound

[98319-26-7] Finasteride, *4125*
[98323-83-2] Carmoxirole, *1893*
[98418-47-4] Metoprolol Succinate*, *6235*
[98631-95-9] Sobuzoxane, *8708*
[98717-15-8] Ropivacaine Monohydrochloride*, *8417*
[98753-19-6] Cefpirome Sulfate*, *1990*
[98967-40-9] Flumetsulam, *4177*
[99011-02-6] Imiquimod, *4957*
[99105-77-8] Sulcotrione, *9063*
[99149-95-8] Saruplase*, *8089*
[99200-09-6] Nebivolol, *6519*
[99210-65-8] Interferon Alfa-2b*, *5016*
[99283-10-0] Molgramostim*, *4559*
[99291-24-4] Dropropizine S-Form*, *3507*
[99294-93-6] Zolpidem L-(+)-Hemitartrate*, *10321*
[99294-94-7] Teriparatide Acetate, *9309*
[99300-78-4] Venlafaxine Hydrochloride*, *10079*
[99323-21-4] Inaperisone (unspecified), *4963*
[99464-64-9] Ampiroxicam, *629*
[99495-87-1] Flecainide (±), *4136*
[99592-32-2] Sertaconazole, *8610*
[99592-39-9] Sertaconazole Nitrate*, *8610*
[99593-25-6] Rilmazafone, *8387*
[99607-70-2] Cloquintocet-Mexyl, *2463*
[99614-01-4] Ondansetron Hydrochloride Dihydrate*, *6979*
[99614-02-5] Ondansetron, *6979*
[99665-00-6] Flomoxef, *4141*
[99685-96-8] Buckminsterfullerene, *1482*
[99732-49-7] Exosurf, *3960*
[99751-63-0] AL 721, *201*
[99777-81-8] Troxipide, *9921*
[100286-90-6] Irinotecan Hydrochloride*, *5104*
[100286-97-3] Milrinone Lactate*, *6284*
[100299-08-9] Pemirolast Potassium Salt*, *7205*
[100427-26-7] Lercanidipine, *5469*
[100490-36-6] Tosufloxacin (unspecified), *9692*
[100678-32-8] Cifenline Succinate*, *2330*
[100981-43-9] Ebrotidine, *3536*
[100986-85-4] Ofloxacin S-(−)-Form*, *6865*
[101007-06-1] Acrinathrin, *127*
[101152-94-7] Milnacipran Hydrochloride*, *6281*
[101246-68-8] Eptastigmine, *3672*
[101363-10-4] Rufloxacin, *8448*
[101467-40-7] Salsoline, *8492*
[101477-54-7] Lomerizine Dihydrochloride*, *5593*
[101477-55-8] Lomerizine, *5593*
[101526-62-9] Sematilide Monohydrochloride*, *8586*
[101526-83-4] Sematilide, *8586*
[101530-10-3] Lanoconazole, *5370*
[101626-70-4] Talipexole, *9209*
[101827-46-7] Butenafine Hydrochloride*, *1547*
[101828-21-1] Butenafine, *1547*
[101831-36-1] Clazuril, *2403*
[101831-37-2] Diclazuril, *3130*
[101917-66-2] Imazethapyr Ammonium Salt*, *4944*
[102106-21-8] Cilnidipine (unspecified), *2334*
[102130-84-7] Nemadectin, *6532*

[102507-71-1] Tigemonam, *9572*
[102518-79-6] Huperzine A, *4791*
[102625-70-7] Pantoprazole, *7146*
[102676-31-3] Fadrozole Hydrochloride* (unspecified), *3969*
[102676-47-1] Fadrozole, *3969*
[102676-96-0] Fadrozole Hydrochloride* (±), *3969*
[102916-21-2] Tigemonam Dicholine Salt*, *9572*
[103055-07-8] Lufenuron, *5624*
[103177-37-3] Pranlukast, *7889*
[103220-14-0] Defensins, *2913*
[103255-66-9] Pazinaclone (unspecified), *7187*
[103577-45-3] Lansoprazole, *5373*
[103628-46-2] Sumatriptan, *9172*
[103628-48-4] Sumatriptan Succinate*, *9172*
[103745-39-7] Fasudil, *3980*
[103775-75-3] Miboplatin, *6263*
[103810-45-3] Bidisomide (unspecified), *1251*
[103878-83-7] Lazabemide Monohydrochloride*, *5407*
[103878-84-8] Lazabemide, *5407*
[103890-78-4] Lacidipine, *5344*
[104010-37-9] Ceftiofur Monosodium Salt*, *1999*
[104051-69-6] Tosufloxacin Hydrochloride*, *9692*
[104054-27-5] Atipamezole, *894*
[104075-48-1] Atipamezole Hydrochloride*, *894*
[104118-48-1] Adipsin, *162*
[104145-95-1] Cefditoren, *1972*
[104227-87-4] Famciclovir, *3971*
[104317-84-2] Gusperimus, *4610*
[104344-23-2] Bisoprolol Hemifumarate*, *1336*
[104348-91-6] Technetium ⁹⁹ᵐTc Mertiatide Free Acid, *9258*
[104376-79-6] Ceftriaxone Disodium Salt Hemiheptahydrate*, *2001*
[104383-17-7] Sabeluzole, *8462*
[104632-25-9] Pramipexole Dihydrochloride*, *7885*
[104632-26-0] Pramipexole, *7885*
[104675-29-8] Velnacrine, *10077*
[104713-75-9] Barnidipine, *1031*
[104716-22-5] Technetium ⁹⁹ᵐTc Teboroxime, *9260*
[104757-53-1] Barnidipine Hydrochloride*, *1031*
[104987-11-3] Tacrolimus, *9200*
[104999-18-0] Thiopropazate Dimaleate*, *9493*
[105149-00-6] Osaterone Acetate Ester*, *7019*
[105149-04-0] Osaterone, *7019*
[105219-56-5] Apafant, *765*
[105239-91-6] Cefclidin, *1970*
[105462-24-6] Risedronic Acid, *8396*
[105512-06-9] Clodinafop-Propargyl, *2431*
[105613-48-7] Exametazime, *3957*
[105628-07-7] Fasudil Hydrochloride*, *3980*
[105784-61-0] Temafloxacin Hydrochloride*, *9284*
[105826-92-4] Tropisetron Monohydrochloride*, *9914*
[105827-78-9] Imidacloprid, *4946*
[105851-17-0] Fludeoxyglucose F¹⁸, *4163*
[105857-23-6] Tissue Plasminogen Activator (Recombinant Human)*, *9608*
[105879-42-3] Cephalexin Monohydrochloride Monohydrate*, *2021*

[105889-45-0] Cefcapene Pivoxil, *1969*
[105956-97-6] Clinafloxacin, *2413*
[105956-99-8] Clinafloxacin Hydrochloride*, *2413*
[105979-17-7] Benidipine, *1071*
[106017-08-7] Rufloxacin Hydrochloride*, *8448*
[106096-92-8] Acidic Fibroblast Growth Factor*, *4117*
[106096-93-9] Basic Fibroblast Growth Factor*, *4117*
[106133-20-4] Tamsulosin, *9217*
[106266-06-2] Risperidone, *8397*
[106325-08-0] Epoxiconazole (undefined), *3665*
[106362-32-7] Peptide T, *7291*
[106463-17-6] Tamsulosin (R)-Form Hydrochloride*, *9217*
[106463-19-8] Tamsulosin (S)-Form Hydrochloride*, *9217*
[106516-24-9] Sertindole, *8611*
[106650-56-0] Sibutramine, *8629*
[106730-54-5] Loprinone, *5605*
[106819-53-8] Doxacurium Chloride, *3487*
[106861-44-3] Mivacurium Chloride, *6305*
[107007-99-8] Granisetron Hydrochloride*, *4557*
[107097-80-3] Loxiglumide, *5618*
[107133-36-8] Perindopril tert-Butylamine*, *7311*
[107534-96-3] Tebuconazole (±), *9253*
[107667-60-7] Polaprezinc, *7712*
[107753-78-6] Zafirlukast, *10241*
[107793-72-6] Ioxilan, *5085*
[107910-75-8] Ganciclovir Sodium Salt*, *4374*
[108050-54-0] Tilmicosin, *9580*
[108050-82-4] Roxindole Hydrochloride*, *8432*
[108319-06-8] Temafloxacin, *9284*
[108424-49-3] Indium Gallium Aluminium Phosphide, *4983*
[108730-13-8] Indium Gallium Aluminium Phosphide, *4983*
[108778-82-1] Beractant, *1189*
[108929-04-0] Epinastine Hydrochloride*, *3655*
[108945-35-3] Taprostene, *9227*
[109489-77-2] Tetranectin, *9370*
[109581-73-9] Technetium ⁹⁹ᵐTc Sestamibi, *9259*
[109784-48-7] Doisynolic Acid, *3470*
[109826-26-8] Zaldaride, *10243*
[109826-27-9] Zaldaride Maleate*, *10243*
[109852-02-0] Gestodene Mixture with Ethinyl Estradiol*, *4421*
[109889-09-0] Granisetron, *4557*
[110075-07-5] Captopril Combination with Hydrochlorothiazide*, *1817*
[110101-66-1] Tirilazad, *9604*
[110140-89-1] Ridogrel, *8379*
[110221-44-8] Temocapril Hydrochloride*, *9287*
[110221-53-9] Temocaprilat*, *9287*
[110488-70-5] Dimethomorph, *3271*
[110588-57-3] Saperconazole, *8510*
[110690-43-2] Emitefur, *3601*
[110703-94-1] Zopolrestat, *10325*
[110871-86-8] Sparfloxacin, *8884*
[110942-02-4] Interleukin-2 (Recombinant Human)*, *5020*
[111011-53-1] Efonidipine Hydrochloride*, *3566*
[111011-63-3] Efonidipine, *3566*
[111011-76-8] Efonidipine Hydrochloride Ethanol*, *3566*
[111025-46-8] Pioglitazone, *7605*

Denotes derivative of title compound

[111223-26-8] Ceronapril, 2038
[111406-87-2] Zileuton, 10253
[111479-05-1] Propaquizafop, 7990
[111793-42-1] Tirilazad Monometh-anesulfonate Monohydrate*, 9604
[111841-85-1] Abecarnil, 2
[111902-57-9] Temocapril, 9287
[111974-72-2] Seroquel, 8606
[112192-04-8] Roxindole, 8432
[112244-09-4] Tilidine cis-(±)-Form*, 9578
[112347-74-7] Papuamine Dihydro-chloride*, 7153
[112373-40-7] Prodigiosin Hydro-chloride*, 7948
[112398-08-0] Danofloxacin, 2876
[112410-23-8] Tebufenozide, 9254
[112455-84-2] Papuamine, 7153
[112457-95-1] Tandospirone Citrate*, 9219
[112529-15-4] Pioglitazone Hydro-chloride*, 7605
[112573-73-6] Ecadotril*, 72
[112665-43-7] Seratrodast, 8603
[112709-47-4] Periplanone A*, 7314
[112809-51-5] Letrozole, 5474
[112887-68-0] Tomudex, 9684
[112901-68-5] Amifostine Trihydrate*, 424
[112965-21-6] Calcipotriene, 1679
[113036-88-7] Flucycloxuron, 4159
[113185-69-6] Adrenochrome Thiosemicarbazone*, 171
[113359-04-9] Cefozopran, 1987
[113378-31-7] Semduramicin, 8587
[113427-24-0] Epoetin Alfa*, 3729
[113482-87-4] Moprolol l-Form Hydrochloride*, 6345
[113507-06-5] Moxidectin, 6373
[113626-33-8] Penicillamine Disulfide L-Form*, 7216
[113665-84-2] Clopidogrel, 2457
[113775-47-6] Dexmedetomidine*, 5830
[113790-28-6] Stepronin Lysine Salt*, 8963
[113852-37-2] Cidofovir, 2329
[114369-43-6] Fenbuconazole, 4002
[114432-13-2] Fantofarone, 3975
[114798-26-4] Losartan, 5613
[114977-28-5] Docetaxel, 3458
[115092-85-8] Carmoxirole Hydro-chloride*, 1893
[115097-93-3] Feclemine Dihydro-chloride*, 3986
[115103-54-3] Tiagabine, 9557
[115256-11-6] Dofetilide, 3469
[115436-72-1] Risedronic Acid Mono-sodium Salt*, 8396
[115956-12-2] Dolasetron, 3471
[115956-13-3] Dolasetron Methanesul-fonate*, 3471
[116041-13-5] Nebracetam (±), 6520
[116078-65-0] Bidisomide (±), 1251
[116355-83-0] Fumonisin B₁, 4311
[116476-13-2] Semotiadil, 8590
[116476-14-3] Semotiadil Fumarate*, 8590
[116539-59-4] Duloxetine, 3518
[116644-53-2] Mibefradil, 6261
[116666-63-8] Mibefradil Dihydro-chloride*, 6261
[116836-09-0] Isomethadone dl-Form*, 5201
[116876-37-0] Clavulanic Acid Combi-nation of Potassium Salt with Ticar-cillin Disodium*, 2402
[117060-71-6] Sulbactam Mixture of

Sodium Salt with Ampicillin Sodium*, 9058
[117149-90-3] Thymostimulin, 9548
[117276-75-2] 4-221-Colony-stimu-lating Factor 1 (Human Clone p3ACSF-69 Protein Moiety Re-duced)*, 5678
[117381-09-6] Trimetrexate Mono-acetate Monohydrate*, 9851
[117399-94-7] Endothelin, 3616
[117467-28-4] Cefditoren Pivaloyloxy-methyl Ester*, 1972
[117591-79-4] Remoxipride Hydro-chloride Monohydrate*, 8301
[117704-25-3] Doramectin, 3483
[117976-89-3] Rabeprazole, 8272
[117976-90-6] Rabeprazole Sodium Salt*, 8272
[118081-34-8] Ceftibuten Dihydrate*, 1998
[118230-97-0] Inaperisone Hydrochlo-ride* (±), 4963
[118292-40-3] Tazarotene, 9249
[118909-22-1] Velnacrine Maleate*, 10077
[119006-77-8] Flutrimazole, 4247
[119068-77-8] Semduramicin Sodium Salt*, 8587
[119119-85-6] α-Glucose-1-phosphate Calcium Salt*, 4470
[119169-78-7] Episteride, 3668
[119257-34-0] Besipirdine, 1223
[119257-40-8] Besipirdine Maleate*, 1223
[119302-91-9] Rocuronium Bromide*, 8407
[119386-96-8] Mofegiline, 6313
[119413-54-6] Topotecan Monohydro-chloride*, 9687
[119413-55-7] Elgodipine (±), 3587
[119446-68-3] Difenoconazole, 3182
[119478-55-6] Danofloxacin Mono-methanesulfonate*, 2876
[119478-56-7] Meropenem (trihy-drate), 5960
[119506-01-3] Deserpidine Oxalate*, 2964
[119600-43-0] Vinconate Monohydro-chloride*, 10123
[119742-13-1] Roxindole Methane-sulfonate*, 8432
[119784-94-0] Tenidap Sodium Salt*, 9290
[119914-60-2] Grepafloxacin (un-specified), 4567
[120011-70-3] E-2020, 3533
[120066-54-8] Gadoteridol, 4348
[120068-37-3] Fipronil, 4127
[120210-48-2] Tenidap, 9290
[120279-96-1] Dorzolamide, 3484
[120373-24-2] Unoprostone Isopropyl Ester*, 9984
[120373-36-6] Unoprostone, 9984
[120382-07-2] Tosufloxacin (±), 9692
[120410-24-4] Biapenem, 1243
[120511-73-1] Anastrozole, 667
[120635-25-8] Mofegiline Hydrochlo-ride*, 6313
[120638-55-3] Bromfenac Mono-sodium Salt Sesquihydrate*, 1411
[120993-53-5] Desirudin, 4754
[121025-46-5] Sphingofungin C*, 8897
[121123-17-9] Cefprozil (monohy-drate), 1992
[121124-29-6] Emamectin (major component), 3594
[121181-53-1] Filgrastim*, 4558
[121268-17-5] Alendronic Acid Mono-sodium Salt Trihydrate*, 228

[121281-41-2] Technetium ⁹⁹ᵐTc Bicisate, 9257
[121412-77-9] Cefprozil Z-Form*, 1992
[121424-52-0] Emamectin (minor component), 3594
[121488-95-7] Elgodipine (unspec-ified), 3587
[121489-04-1] Elgodipine Hydrochlo-ride*, 3587
[121543-68-8] Lunasine (−)-Form Perchlorate*, 5632
[121547-04-4] Mirimostim*, 5678
[121552-61-2] Cyprodinil, 2841
[121652-76-4] Eptastigmine Tartrate*, 3672
[121696-62-6] Piroxicam Compd with β-Cyclodextrin*, 7661
[121808-62-6] Pidotimod, 7574
[121961-22-6] Loracarbef, 5606
[122111-03-9] Gemcitabine Hydro-chloride*, 4392
[122312-54-3] Epoetin Beta*, 3729
[122392-03-4] Medullipin II*, 10251
[122647-31-8] Ibutilide, 4927
[122647-32-9] Ibutilide Fumarate*, 4927
[123040-16-4] Azasetron Hydrochlo-ride*, 933
[123040-69-7] Azasetron, 933
[123171-59-5] Cefepime Hydro-chloride Monohydrate*, 1973
[123312-89-0] Pymetrozine, 8136
[123342-93-8] Pyrithiobac, 8177
[123343-16-8] Pyrithiobac Sodium Salt*, 8177
[123618-00-8] Fedotozine, 3987
[123774-72-1] Sargramostim*, 4559
[123948-87-8] Topotecan, 9687
[123997-26-2] Eprinomectin, 3667
[124750-99-8] Losartan Mono-potassium Salt*, 5613
[124832-26-4] Valacyclovir, 10039
[124832-27-5] Valacyclovir Hydro-chloride*, 10039
[124858-35-1] Nadifloxacin, 6430
[125224-05-7] Technetium ⁹⁹ᵐTc Mer-tiatide, 9258
[125251-66-3] Arbutamine Hydrochloride*, 811
[125494-59-9] Sibutramine Hydro-chloride Monohydrate*, 8629
[125676-39-3] Inaperisone (±), 4963
[125995-03-1] Atorvastatin Lactone*, 897
[127045-41-4] Pazufloxacin, 7188
[127455-27-0] Tetrofosmin ⁹⁹ᵐTc-Complex*, 9383
[127502-06-1] Tetrofosmin, 9383
[127757-91-9] Regramostim*, 4559
[127779-20-8] Saquinavir, 8516
[127786-29-2] Epinastine Hydrobro-mide*, 3655
[127829-86-1] Kistrin, 5327
[128345-62-0] Ranitidine Bismuth Citrate*, 8286
[128470-16-6] Arbutamine, 811
[129069-19-8] Poractant Alfa, 7754
[129409-01-4] Salen (E,E), 8474
[129496-10-2] Milbemycin Oxime, 6277
[129618-40-2] Nevirapine, 6573
[130167-69-0] Pegaspargase*, 871
[130209-82-4] Latanoprost, 5387
[130370-60-4] Batimastat, 1036
[130693-82-2] Dorzolamide Hydro-chloride*, 3484
[130939-66-1] Neurotrophin-3*, 6570
[130953-69-4] Besipirdine Hydro-chloride*, 1223

* Denotes derivative of title compound

Denotes derivative of title compound

THERAPEUTIC CATEGORY AND BIOLOGICAL ACTIVITY INDEX

THERAPEUTIC CATEGORY AND BIOLOGICAL ACTIVITY INDEX

This index is designed to serve as an additional entry point to the monographs of this edition for which a therapeutic category (THERAP CAT) has been given. It contains listings for discrete chemical entities; plants and plant portions, *e.g.* seeds or leaves, although used medicinally, have generally been excluded. An attempt has been made to standardize the nomenclature used for index terms. Therefore, the heading under which a compound is listed may differ slightly from the therapeutic category found within the monograph.

Biological activities indicating mechanism of action have also been included as index headings. Whenever possible, cross references to appropriate therapeutic categories have been given for these listings. Cross references have also been provided for synonyms of preferred terms or, in some cases, for closely related entries. In addition, many therapeutic category headings have been subclassified according to structural features.

Monographs are listed alphabetically by title under each heading. Selected derivatives and isomers of title compounds have been listed by generic or trivial name and are referenced to the appropriate monograph by number.

Therapeutic categories have been assigned to reflect the major indications listed in manufacturer's product information, reported in the clinical literature, or published by the USAN Council. Inclusion of a drug in this index does not imply efficacy or endorsement, nor does it imply that the compound is *presently* being marketed. Omission of a compound from a particular category does not imply that the substance is not in use or under investigation for that indication. Additional information on each entry can be found in the cited monograph. Reference to primary sources is encouraged.

ABORTIFACIENT/INTERCEPTIVE

Epostane, *3664*
Gemeprost, *4393*
Mifepristone, *6273*
Prostaglandin E₂, *8064*
Prostaglandin F₂ₐ, *8065*
Sulprostone, *9165*
Trichosanthin, *9780*

ACE-INHIBITOR *see also Antihypertensive*

Alacepril, *202*
Benazepril, *1058*
Captopril, *1817*
Ceronapril, *2038*
Cilazapril, *2332*
Delapril, *2928*
Enalapril, *3605*
Enalaprilat, *3606*
Fosinopril, *4282*
Imidapril, *4947*
Lisinopril, *5540*
Moveltopril, *6368*
Perindopril, *7311*
Quinapril, *8233*
Ramipril, *8283*
Spirapril, *8905*
Temocapril, *9287*
Trandolapril, *9703*

α-ADRENERGIC AGONIST *see also Antihypotensive; Antihypertensive; Decongestant; Mydriatic*

Adrafinil, *168*
Adrenolone, *170*
Amidephrine, *418*
Apraclonidine, *791*
Budralazine, *1492*
Clonidine, *2450*
Cyclopentamine, *2808*
Detomidine, *2981*
Dimetofrine, *3314*
Dipivefrin, *3400*
Ephedrine, *3645*
Epinephrine, *3656*
Fenoxazoline, *4025*
Guanabenz, *4585*
Guanfacine, *4590*
Hydroxyamphetamine, *4855*
Ibopamine, *4921*
Indanazoline, *4967*
Isometheptene, *5202*
Mephentermine, *5897*
Metaraminol, *5993*
Methoxamine, *6067*
Methylhexaneamine, *6159*
Metizoline, *6223*
Midodrine, *6272*
Modafinil, *6311*
Moxonidine, *6375*
Naphazoline, *6455*
Norepinephrine, *6788*
Norfenefrine, *6792*
Octodrine, *6854*
Octopamine, *6856*
Oxymetazoline, *7100*
Phenylephrine Hydrochloride, *7440*
Phenylpropanolamine Hydrochloride, *7461*
Phenylpropylmethylamine, *7462*
Pholedrine, *7485*
Propylhexedrine, *8045*
Pseudoephedrine *see 3645*
Rilmenidine, *8388*

Synephrine, *9189*
Talipexole, *9209*
Tetrahydrozoline, *9358*
Tiamenidine, *9558*
Tramazoline, *9702*
Tuaminoheptane, *9934*
Tymazoline, *9965*
Tyramine, *9966*
Xylometazoline, *10219*

β-ADRENERGIC AGONIST *see also Bronchodilator; Cardiotonic; Tocolytic*

Albuterol, *217*
Bambuterol, *980*
Bitolterol, *1344*
Carbuterol, *1882*
Clenbuterol, *2407*
Clorprenaline, *2470*
Denopamine, *2943*
Dioxethedrine, *3356*
Dopexamine, *3482*
Ephedrine, *3645*
Epinephrine, *3656*
Etafedrine, *3752*
Ethylnorepinephrine, *3880*
Fenoterol, *4022*
Formoterol, *4272*
Hexoprenaline, *4745*
Ibopamine, *4921*
Isoetharine, *5185*
Isoproterenol, *5236*
Mabuterol, *5674*
Metaproterenol, *5992*
Methoxyphenamine, *6077*
Oxyfedrine, *7096*
Pirbuterol, *7644*
Prenalterol, *7917*
Procaterol, *7939*
Protokylol, *8082*
Reproterol, *8307*
Rimiterol, *8393*
Ritodrine, *8401*
Salmeterol, *8489*
Soterenol, *8877*
Terbutaline, *9302*
Tretoquinol, *9719*
Tulobuterol, *9942*
Xamoterol, *10189*

α-ADRENERGIC BLOCKER *see also Antihypertensive*

Amosulalol, *614*
Arotinolol, *827*
Dapiprazole, *2884*
Doxazosin, *3489*
Ergoloid Mesylates, *3692*
Fenspiride, *4041*
Indoramin, *5000*
Labetalol, *5341*
Naftopidil, *6443*
Nicergoline, *6560*
Prazosin, *7897*
Tamsulosin, *9217*
Terazosin, *9297*
Tolazoline, *9645*
Trimazosin, *9828*
Yohimbine, *10236*

β-ADRENERGIC BLOCKER *see also Antianginal; Antiarrhythmic; Antiglaucoma; Antihypertensive*

Acebutolol, *16*
Alprenolol, *321*
Amosulalol, *614*

Arotinolol, *827*
Atenolol, *892*
Befunolol, *1050*
Betaxolol, *1229*
Bevantolol, *1238*
Bisoprolol, *1336*
Bopindolol, *1362*
Bucumolol, *1489*
Bufetolol, *1496*
Bufuralol, *1504*
Bunitrolol, *1514*
Bupranolol, *1521*
Butidrine Hydrochloride, *1558*
Butofilolol, *1563*
Carazolol, *1822*
Carteolol, *1917*
Carvedilol, *1924*
Celiprolol, *2007*
Cetamolol, *2060*
Cloranolol, *2464*
Dilevalol, *3245*
Epanolol, *3641*
Esmolol, *3741*
Indenolol, *4974*
Labetalol, *5341*
Levobunolol, *5488*
Mepindolol, *5901*
Metipranolol, *6221*
Metoprolol, *6235*
Moprolol, *6345*
Nadolol, *6431*
Nadoxolol, *6432*
Nebivalol, *6519*
Nifenalol, *6618*
Nipradilol, *6655*
Oxprenolol, *7086*
Penbutolol, *7209*
Pindolol, *7597*
Practolol, *7882*
Pronethalol, *7974*
Propranolol, *8025*
Sotalol, *8876*
Sulfinalol, *9120*
Talinolol, *9208*
Tertatolol, *9318*
Tilisolol, *9579*
Timolol, *9585*
Toliprolol, *9653*
Xibenolol, *10209*

ADRENOCORTICAL STEROID *see Glucocorticoid; Mineralocorticoid*

ADRENOCORTICAL SUPPRESSANT

Trilostane, *9827*

ADRENOCORTICOTROPIC HORMONE

ACTH, *136*
Cosyntropin, *2617*
Giractide, *4435*

ALCOHOL DETERRENT

Calcium Cyanamide Citrated, *1703*
Disulfiram, *3428*
Nitrefazole, *6668*

ALDOSE REDUCTASE INHIBITOR

Epalrestat, *3640*
Sorbinil, *8871*
Tolrestat, *9663*
Zopolrestat, *10325*

ALDOSTERONE ANTAGONIST *see*
also **Diuretic**

Canrenone, *1795*
Spironolactone, *8917*

5-ALPHA REDUCTASE INHIBITOR
see 5α-Reductase Inhibitor

ANABOLIC

Androisoxazole, *674*
Androstenediol, *677*
Bolandiol, *1352*
Bolasterone, *1353*
Clostebol, *2475*
Ethylestrenol, *3851*
Formebolone, *4266*
Methandriol, *6017*
Methenolone, *6044*
Methyltrienolone, *6211*
Nandrolone, *6452*
Norbolethone, *6779*
Oxabolone, *7034*
Oxymesterone, *7099*
Pizotyline, *7671*
Quinbolone, *8236*
Stenbolone, *8962*
Trenbolone, *9716*

ANALEPTIC *see CNS Stimulant*

ANALGESIC (DENTAL)

Chlorobutanol, *2180*
Clove *see 2479 and 6893*
Eugenol, *3944*

ANALGESIC (NARCOTIC)

Alfentanil, *236*
Allylprodine, *304*
Alphaprodine, *317*
Anileridine, *695*
Benzylmorphine, *1176*
Bezitramide, *1241*
Buprenorphine, *1522*
Butorphanol, *1565*
Clonitazene, *2451*
Codeine, *2525*
Codeine Methyl Bromide, *2526*
Codeine Phosphate, *2528*
Codeine Sulfate, *2529*
Desomorphine, *2972*
Dextromoramide, *2997*
Dezocine, *2998*
Diampromide, *3032*
Dihydrocodeine, *3214*
Dihydrocodeinone Enol Acetate,
    3215
Dihydromorphine, *3220*
Dimenoxadol, *3253*
Dimepheptanol, *3254*
Dimethylthiambutene, *3309*
Dioxaphetyl Butyrate, *3354*
Dipipanone, *3398*
Eptazocine, *3673*
Ethoheptazine, *3789*
Ethylmethylthiambutene, *3875*
Ethylmorphine, *3876*
Etonitazene, *3929*
Fentanyl, *4043*
Hydrocodone, *4826*
Hydromorphone, *4847*
Hydroxypethidine, *4883*
Isomethadone, *5201*
Ketobemidone, *5312*

Levorphanol, *5496*
Lofentanil, *5586*
Meperidine, *5894*
Meptazinol, *5910*
Metazocine, *5995*
Methadone Hydrochloride,
    6008
Metopon, *6234*
Morphine, *6359*
Morphine Hydrochloride, *6360*
Morphine Sulfate, *6361*
Myrophine, *6420*
Nalbuphine, *6444*
Narceine, *6508*
Nicomorphine, *6607*
Norlevorphanol, *6801*
Normethadone, *6804*
Normorphine, *6806*
Norpipanone, *6810*
Opium, *6986*
Oxycodone, *7093*
Oxymorphone, *7103*
Papaveretum, *7150*
Pentazocine, *7261*
Phenadoxone, *7348*
Phenazocine, *7360*
Phenoperidine, *7400*
Piminodine, *7587*
Piritramide, *7653*
Proheptazine, *7959*
Promedol, *7968*
Propiram, *8017*
Propoxyphene, *8024*
Remifentanil, *8300*
Sufentanil, *9056*
Tilidine, *9578*

ANALGESIC (NON-NARCOTIC)

Aceclofenac, *19*
Acetaminophen, *45*
Acetaminosalol, *46*
Acetanilide, *47*
Acetylsalicylsalicylic Acid, *102*
Alclofenac, *220*
Alminoprofen, *308*
Aloxiprin, *316*
Aluminum Bis(acetylsalicylate),
    338
Aminochlorthenoxazin, *455*
2-Amino-4-picoline, *486*
Aminopropylon, *492*
Aminopyrine, *495*
Ammonium Salicylate, *584*
Amtolmetin Guacil, *637*
Antipyrine, *757*
Antipyrine Salicylate, *758*
Antrafenine, *763*
Apazone, *768*
Aspirin, *886*
Benorylate, *1074*
Benoxaprofen, *1075*
Benzpiperylon, *1152*
Benzydamine, *1157*
Bermoprofen, *1199*
Bromfenac, *1411*
p-Bromoacetanilide, *1420*
5-Bromosalicylic Acid Acetate,
    1459
Bucetin, *1479*
Bufexamac, *1497*
Bumadizon, *1507*
Butacetin, *1532*
Calcium Acetylsalicylate, *1684*
Carbamazepine, *1826*
Carbiphene, *1846*
Carsalam, *1915*
Chlorthenoxazin(e), *2247*
Choline Salicylate, *2265*
Cinchophen, *2347*

Ciramadol, *2375*
Clometacin, *2443*
Clonixin, *2453*
Cropropamide, *2659*
Crotethamide, *2662*
Dexoxadrol *see 3352*
Difenamizole, *3181*
Diflunisal, *3190*
Dihydroxyaluminum Acetylsalicy-
    late, *3226*
Dipyrocetyl, *3413*
Dipyrone, *3414*
Emorfazone, *3603*
Enfenamic Acid, *3620*
Epirizole, *3659*
Etersalate, *3759*
Ethenzamide, *3776*
Ethoxazene, *3796*
Etodolac, *3920*
Felbinac, *3989*
Fenoprofen, *4021*
Floctafenine, *4140*
Flufenamic Acid, *4167*
Fluoresone, *4197*
Flupirtine, *4227*
Fluproquazone, *4230*
Flurbiprofen, *4234*
Fosfosal, *4281*
Gentisic Acid, *4404*
Glafenine, *4443*
Ibufenac, *4924*
Imidazole Salicylate, *4949*
Indomethacin, *4998*
Indoprofen, *4990*
Isofezolac, *5188*
Isoladol, *5194*
Isonixin, *5209*
Ketoprofen, *5316*
Ketorolac, *5318*
p-Lactophenetide, *5355*
Lefetamine, *5348*
Lornoxicam, *5612*
Loxoprofen, *5619*
Lysine Acetylsalicylate, *5668*
Magnesium Acetylsalicylate,
    5689
Methotrimeprazine, *6066*
Metofoline, *6229*
Mofezolac, *6314*
Morazone, *6349*
Morpholine Salicylate *see 6362*
Naproxen, *6504*
Nefopam, *6529*
Nifenazone, *6619*
5'-Nitro-2'-propoxyacetanilide,
    6726
Parsalmide, *7177*
Perisoxal, *7318*
Phenacetin, *7344*
Phenazopyridine Hydrochloride,
    7361
Phenocoll, *7388*
Phenopyrazone, *7401*
Phenyl Acetylsalicylate, *7424*
Phenyl Salicylate, *7464*
Phenyramidol, *7474*
Pipebuzone, *7610*
Piperylone, *7632*
Propacetamol, *7976*
Propyphenazone, *8056*
Ramifenazone, *8282*
Rimazolium Metilsulfate, *8391*
Salacetamide, *8471*
Salicin, *8476*
Salicylamide, *8480*
Salicylamide O-Acetic Acid,
    8481
Salicylsulfuric Acid, *8486*
Salsalate, *8491*
Salverine, *8494*
Simetride, *8682*
Sodium Salicylate, *8819*

Suprofen, *9180*
Talniflumate, *9214*
Tenoxicam, *9293*
Terofenamate, *9312*
Tetrandrine, *9369*
Tinoridine, *9589*
Tolfenamic Acid, *9650*
Tramadol, *9701*
Tropesin, *9909*
Viminol, *10117*
Xenbucin, *10205*
Zomepirac, *10322*

ANDROGEN

Boldenone, *1354*
Cloxotestosterone, *2482*
Fluoxymesterone, *4223*
Mestanolone, *5973*
Mesterolone, *5974*
Methandrostenolone, *6018*
17-Methyltestosterone, *6206*
17α-Methyltestosterone 3-Cyclo-
  pentyl Enol Ether, *6207*
Norethandrolone, *6789*
Normethandrone, *6805*
Oxandrolone, *7054*
Oxymesterone, *7099*
Oxymetholone, *7101*
Prasterone, *7892*
Stanolone, *8950*
Stanozolol, *8951*
Testosterone, *9322*
Tiomesterone, *9596*

ANESTHETIC (INHALATION)

Cyclopropane, *2817*
Desflurane, *2965*
Enflurane, *3621*
Ethylene, *3837*
Ethyl Ether, *3852*
Fluroxene, *4237*
Halopropane, *4632*
Halothane, *4634*
Isoflurane, *5191*
Methoxyflurane, *6072*
Methyl Propyl Ether, *6192*
Nitrous Oxide, *6751*
Sevoflurane, *8621*
Teflurane, *9265*
Trichloroethylene, *9769*
Vinyl Ether, *10133*
Xenon, *10206*

ANESTHETIC (INTRAVENOUS)

Acetamidoeugenol, *44*
Alfadolone Acetate, *233*
Alfaxalone, *235*
Buthalital Sodium, *1553*
Etoxadrol, *3933*
Hexobarbital, *4742*
Hydroxydione Sodium, *4868*
Ketamine, *5306*
Methohexital Sodium, *6061*
Midazolam, *6270*
Phencyclidine, *7364*
Propanidid, *7986*
Propofol, *8020*
Sodium Oxybate *see 4860*
Thialbarbital, *9428*
Thiamylal, *9437*
Thiobutabarbital, *9257*
Thiopental Sodium, *9487*

ANESTHETIC (LOCAL)

Ambucaine, *402*
Amolanone, *610*
Amylocaine Hydrochloride, *656*
Benoxinate, *1076*
Benzocaine, *1116*
Betoxycaine, *1235*
Biphenamine, *1275*
Bupivacaine, *1520*
Butacaine, *1531*
Butamben, *1538*
Butanilicaine, *1542*
Butethamine, *1552*
Butoxycaine, *1566*
Carticaine, *1920*
Chloroprocaine Hydrochloride, *2210*
Cocaethylene, *2516*
Cocaine, *2517*
Cyclomethycaine, *2804*
Dibucaine Hydrochloride, *3081*
Dimethisoquin, *3267*
Dimethocaine, *3270*
Diperodon Hydrochloride, *3360*
Dyclonine, *3523*
Ecgonidine, *3540*
Ecgonine, *3541*
Ethyl Chloride, *3829*
Etidocaine, *3907*
β-Eucaine, *3939*
Euprocin, *3950*
Fenalcomine, *3996*
Fomocaine, *4259*
Hexylcaine Hydrochloride, *4746*
Hydroxytetracaine, *4894*
Isobutyl *p*-Aminobenzoate, *5148*
Leucinocaine Mesylate, *5476*
Levoxadrol *see 3352*
Lidocaine, *5505*
Mepivacaine, *5905*
Meprylcaine, *5909*
Metabutoxycaine, *5978*
Methyl Chloride, *6121*
Myrtecaine, *6422*
Naepaine, *6435*
Octacaine, *6839*
Orthocaine, *7011*
Oxethazaine, *7067*
Parethoxycaine, *7171*
Phenacaine Hydrochloride, *7342*
Phenol, *7390*
Piperocaine, *7627*
Piridocaine, *7649*
Polidocanol, *7717*
Pramoxine, *7888*
Prilocaine, *7924*
Procaine, *7937*
Propanocaine, *7988*
Proparacaine, *7991*
Propipocaine, *8016*
Propoxycaine Hydrochloride, *8023*
Pseudococaine, *8097*
Pyrrocaine, *8197*
Ropivacaine, *8417*
Salicyl Alcohol, *8477*
Tetracaine Hydrochloride, *9330*
Tolycaine, *9679*
Trimecaine, *9830*
Zolamine, *10319*

ANGIOTENSIN CONVERTING ENZYME INHIBITOR *see ACE-Inhibitor*

ANGIOTENSIN II RECEPTOR ANTAGONIST *see also Antihypertensive*

Candesartan, *1788*
Eprosartan, *3669*

Irbesartan, *5097*
Losartan, *5613*
Valsartan, *10051*

ANOREXIC

Aminorex, *497*
Amphecloral, *620*
Amphetamine, *623*
Benzphetamine, *1151*
Chlorphentermine, *2235*
Clobenzorex, *2421*
Cloforex, *2440*
Clortermine, *2472*
Cyclexedrine, *2776*
Dextroamphetamine Sulfate, *2996*
Diethylpropion, *3175*
Diphemethoxidine, *3362*
N-Ethylamphetamine, *3809*
Fenbutrazate, *4005*
Fenfluramine, *4015*
Fenproporex, *4036*
Furfurylmethylamphetamine, *4327*
Levophacetoperane, *5493*
Mazindol, *5801*
Mefenorex, *5843*
Metamfepramone, *5984*
Methamphetamine, *6015*
Norpseudoephedrine, *6811*
Pentorex, *7275*
Phendimetrazine, *7365*
Phenmetrazine, *7385*
Phentermine, *7415*
Phenylpropanolamine Hydrochlo-
  ride, *7461*
Picilorex, *7551*
Sibutramine, *8629*

ANTACID

Alexitol Sodium, *232*
Almagate, *307*
Aluminum Hydroxide, *355*
Aluminum Magnesium Silicate,
  362
Aluminum Phosphate, *371*
Azulene, *956*
Basic Aluminum Carbonate Gel,
  1033
Bismuth Aluminate, *1298*
Bismuth Phosphate, *1316*
Bismuth Subgallate, *1325*
Bismuth Subnitrate, *1326*
Calcium Carbonate, *1697*
Dihydroxyaluminum Aminoacetate,
  3227
Dihydroxyaluminum Sodium
  Carbonate, *3228*
Ebimar, *3535*
Magaldrate, *5685*
Magnesium Carbonate Hydroxide,
  5696
Magnesium Hydroxide, *5706*
Magnesium Oxide, *5713*
Magnesium Peroxide, *5717*
Magnesium Phosphate, Tribasic,
  5720
Magnesium Silicates, *5727*
Potassium Citrate, *7785*
Sodium Bicarbonate, *8726*

ANTHELMINTIC (CESTODES)

Arecoline, *815*
Aspidin, *881*
Aspidinol, *882*
Dichlorophen(e), *3120*
Embelin, *3595*
Kosin, *5333*

ANTHELMINTIC (CESTODES)
(continued)

Naphthalene, *6457*
Niclosamide, *6600*
Pelletierine, *7200*
Quinacrine, *8225*

ANTHELMINTIC (NEMATODES)

Alantolactone, *208*
Amocarzine, *608*
Amoscanate, *613*
Ascaridole, *864*
Bephenium, *1187*
Bitoscanate, *1345*
Carbon Tetrachloride, *1864*
Carvacrol, *1923*
Cyclobendazole, *2781*
Diethylcarbamazine, *3165*
Diphenane, *3365*
Dithiazanine Iodide, *3434*
Dymanthine, *3525*
Gentian Violet, *4401*
4-Hexylresorcinol, *4750*
Ivermectin, *5264*
Kainic Acid, *5289*
Levamisole, *5486*
Mebendazole, *5807*
2-Naphthol, *6471*
Oxantel, *7055*
Papain, *7148*
Piperazine, *7617*
Pyrantel, *8139*
Pyrvinium Pamoate, *8206*
α-Santonin, *8509*
Stilbazium Iodide, *8971*
Suramin Sodium, *9181*
Tetrachloroethylene, *9332*
Thiabendazole, *9426*
Thymol, *9540*
Thymyl *N*-Isoamylcarbamate, *9550*
Triclofenol Piperazine, *9787*
Urea Stibamine, *10010*

ANTHELMINTIC (SCHISTOSOMA)

Amoscanate, *613*
Amphotalide, *626*
Antimony Potassium Tartrate, *741*
Antimony Sodium Tartrate, *743*
Antimony Sodium Thioglycollate,
    744
Antimony Thioglycollamide, *746*
Becanthone, *1044*
Hycanthone, *4795*
Lucanthone Hydrochloride, *5620*
Niridazole, *6656*
Oxamniquine, *7051*
Praziquantel, *7896*
Stibocaptate, *8966*
Stibophen, *8967*
Sodium Antimonylgluconate *see*
    742
Urea Stibamine, *10010*

ANTHELMINTIC (TREMATODES)

Anthiolimine, *720*
Tetrachloroethylene, *9332*

ANTIACNE *see also **Keratolytic***

Algestone Acetophenide, *239*
Azelaic Acid, *938*
Benzoyl Peroxide, *1149*
Cioteronel, *2372*
Cyproterone, *2844*
Motretinide, *6367*

Resorcinol, *8323*
Retinoic Acid, *8333*
Tazarotene, *9249*
Tetroquinone, *9385*
Tioxolone, *9600*

ANTIALLERGIC *see also **Antihista-minic; Decongestant; Glucocorticoid***

Amlexanox, *515*
Astemizole, *891*
Azelastine, *939*
Cromolyn, *2658*
Fenpiprane, *4031*
Ibudilast, *4923*
Lodoxamide, *5585*
Nedocromil, *6524*
Oxatomide, *7058*
Pemirolast, *7205*
Pentigetide, *7270*
Picumast, *7573*
Repirinast, *8305*
Suplatast Tosylate, *9178*
Tazanolast, *9248*
Tranilast, *9705*
Traxanox, *9711*

ANTIALLERGIC (HYPOSENSITIZA-TION THERAPY)

Histamine, *4756*
Poison Ivy Extract *see 7709*
Poison Oak Extract *see 7710*
Poison Sumac Extract *see 7711*
Urushiol, *10028*

ANTIALLERGIC (STEROIDAL, NASAL) *see also **Glucocorticoid***

Beclomethasone, *1047*
Dexamethasone, *2986*
Flunisolide, *4180*
Fluticasone Propionate, *4244*
Triamcinolone Acetonide, *9728*

ANTIALOPECIA AGENT

Cioteronel, *2372*
Diphencyprone, *3366*
Minoxidil, *6290*

ANTIAMEBIC

Arsthinol, *852*
Bialamicol, *1242*
Carbarsone, *1830*
Cephaeline, *2020*
Chlorbetamide, *2126*
Chloroquine, *2215*
Chlorphenoxamide, *2233*
Chlortetracycline, *2245*
Dehydroemetine, *2924*
Dibromopropamidine, *3073*
Diloxanide, *3246*
Diphetarsone, *3394*
Emetine, *3600*
Fumagillin, *4308*
Glaucarubin, *4444*
Glycobiarsol, *4503*
8-Hydroxy-7-iodo-5-quinoline-
    sulfonic Acid, *4872*
Iodochlorhydroxyquin, *5052*
Iodoquinol, *5063*
Paromomycin, *7173*
Phanquinone, *7337*
Polybenzarsol, *7723*
Propamidine, *7981*

Quinfamide, *8241*
Secnidazole, *8562*
Sulfarside, *9110*
Teclozan, *9262*
Tetracycline, *9337*
Thiocarbamizine, *9458*
Thiocarbarsone, *9459*
Tinidazole, *9588*

ANTIANDROGEN *see also **Antiacne; Anti-alopecia; Antineoplastic (Hormonal); Antiprostatic Hypertrophy***

Bicalutamide, *1247*
Bifluranol, *1259*
Cioteronel, *2372*
Cyproterone, *2844*
Delmadinone Acetate, *2930*
Flutamide, *4242*
Nilutamide, *6636*
Osaterone, *7019*
Oxendolone, *7065*

ANTIANGINAL *see also **Vasodilator (Coronary)***

Acebutolol, *16*
Alprenolol, *321*
Amiodarone, *504*
Amlodipine, *516*
Arotinolol, *827*
Atenolol, *892*
Barnidipine, *1031*
Bepridil, *1188*
Bevantolol, *1238*
Bucumolol, *1489*
Bufetolol, *1496*
Bufuralol, *1504*
Bunitrolol, *1514*
Bupranolol, *1521*
Carazolol, *1822*
Carteolol, *1917*
Celiprolol, *2007*
Cinepazet Maleate, *2349*
Diltiazem, *3247*
Elgodipine, *3587*
Epanolol, *3641*
Felodipine, *3991*
Gallopamil, *4369*
Imolamine, *4959*
Indenolol, *4974*
Isosorbide Dinitrate, *5245*
Isradipine, *5260*
Limaprost, *5514*
Mepindolol, *5901*
Metoprolol, *6235*
Molsidomine, *6316*
Nadolol, *6431*
Nicardipine, *6579*
Nicorandil, *6608*
Nifedipine, *6617*
Nifenalol, *6618*
Nilvadipine, *6637*
Nipradilol, *6655*
Nisoldipine, *6658*
Nitroglycerin, *6704*
Oxprenolol, *7086*
Oxyfedrine, *7096*
Ozagrel, *7115*
Penbutolol, *7209*
Pentaerythritol Tetranitrate, *7249*
Pindolol, *7597*
Pronethalol, *7974*
Propranolol, *8025*
Ranolazine, *8287*
Semotiadil, *8590*
Sotalol, *8876*
Terodiline, *9311*
Timolol, *9585*

Toliprolol, *9653*
Trolnitrate Phosphate, *9900*
Verapamil, *10083*
Zatebradine, *10245*

ANTIARRHYTHMIC

Acebutolol, *16*
Acecainide, *17*
Adenosine, *152*
Ajmaline, *194*
Alprenolol, *321*
Amiodarone, *504*
Amoproxan, *611*
Aprindine, *793*
Arotinolol, *827*
Atenolol, *892*
Azimilide, *943*
Bevantolol, *1238*
Bidisomide, *1251*
Bretylium Tosylate, *1395*
Bucumolol, *1489*
Bufetolol, *1496*
Bunaftine, *1509*
Bunitrolol, *1514*
Bupranolol, *1521*
Butidrine Hydrochloride, *1558*
Butobendine, *1560*
Capobenic Acid, *1799*
Carazolol, *1822*
Carteolol, *1917*
Cifenline, *2330*
Cloranolol, *2464*
Disopyramide, *3424*
Dofetilide, *3469*
Encainide, *3609*
Esmolol, *3741*
Flecainide, *4136*
Hydroquinidine, *4851*
Ibutilide, *4927*
Indecainide, *4971*
Indenolol, *4974*
Ipratropium Bromide, *5089*
Lidocaine, *5505*
Lorajmine, *5607*
Lorcainide, *5610*
Meobentine, *5889*
Mexiletine, *6257*
Moricizine, *6351*
Nadoxolol, *6432*
Nifenalol, *6618*
Oxprenolol, *7086*
Penbutolol, *7209*
Pentisomide, *7271*
Pilsicainide, *7581*
Pindolol, *7597*
Pirmenol, *7656*
Practolol, *7882*
Prajmaline, *7883*
Procainamide Hydrochloride, *7936*
Pronethalol, *7974*
Propafenone, *7978*
Propranolol, *8025*
Pyrinoline, *8173*
Quinidine, *8244*
Sematilide, *8586*
Sotalol, *8876*
Talinolol, *9208*
Tilisolol, *9579*
Timolol, *9585*
Tocainide, *9629*
Verapamil, *10083*
Viquidil, *10141*
Xibenolol, *10209*

ANTIARTERIOSCLEROTIC *see also Antihyperlipoproteinemic*

Pyridinol Carbamate, *8158*

ANTIARTHRITIC/ANTIRHEUMATIC *see also Anti-inflammatory (Nonsteroidal); Glucocorticoid*

Actarit, *135*
Allocupreide Sodium, *265*
Auranofin, *911*
Aurothioglucose, *917*
Aurothioglycanide, *918*
Azathioprine, *935*
Bucillamine, *1481*
Calcium 3-Aurothio-2-propanol-1-sulfonate, *1689*
Chloroquine, *2215*
Clobuzarit, *2426*
Cuproxoline, *2740*
Diacerein, *2996*
Glucosamine, *4466*
Gold Sodium Thiomalate, *4536*
Gold Sodium Thiosulfate, *4537*
Hydroxychloroquine, *4836*
Kebuzone, *5300*
Lobenzarit, *5581*
Melittin, *5867*
Methotrexate, *6065*
Myoral, *6411*
Penicillamine, *7214*

ANTIASTHMATIC (NONBRONCHO-DILATOR) *see also Bronchodilator; Glucocorticoid*

Amlexanox, *515*
Azelastine, *939*
Cromolyn, *2658*
Ibudilast, *4923*
Ketotifen, *5319*
Montelukast, *6340*
Nedocromil, *6524*
Oxatomide, *7058*
Pranlukast, *7889*
Seratrodast, *8603*
Suplatast Tosylate, *9178*
Tiaramide, *9564*
Traxanox, *9711*
Zafirlukast, *10241*
Zileuton, *10253*

ANTIASTHMATIC (STEROIDAL, IN-HALANT) *see also Glucocorticoid*

Beclomethasone, *1047*
Budesonide, *1490*
Dexamethasone, *2986*
Flunisolide, *4180*
Triamcinolone Acetonide, *9728*

ANTIBACTERIAL (ANTIBIOTICS)

Aminoglycosides
Amikacin, *425*
Apramycin, *792*
Arbekacin, *809*
Bambermycins, *979*
Butirosin, *1559*
Dibekacin, *3052*
Dihydrostreptomycin, *3222*
Fortimicin(s), *4276*
Gentamicin, *4398*
Isepamicin, *5121*
Kanamycin, *5298*
Micronomicin, *6269*
Neomycin, *6542*
Neomycin Undecylenate, *6543*
Netilmicin, *6563*
Paromomycin, *7173*
Ribostamycin, *8373*
Sisomicin, *8695*
Spectinomycin, *8890*

Streptomycin, *8983*
Tobramycin, *9328*
Trospectomycin, *9917*

Amphenicols
Azidamfenicol, *941*
Chloramphenicol, *2120*
Florfenicol, *4145*
Thiamphenicol, *9436*

Ansamycins
Rifamide, *8381*
Rifampin, *8382*
Rifamycin SV, *8384*
Rifapentine, *8385*
Rifaximin, *8386*

β-Lactams

Carbacephems
Loracarbef, *5606*

Carbapenems
Biapenem, *1243*
Imipenem, *4954*
Meropenem, *5960*
Panipenem, *7141*

Cephalosporins
Cefaclor, *1962*
Cefadroxil, *1963*
Cefamandole, *1964*
Cefatrizine, *1965*
Cefazedone, *1966*
Cefazolin, *1967*
Cefcapene Pivoxil, *1969*
Cefclidin, *1970*
Cefdinir, *1971*
Cefditoren, *1972*
Cefepime, *1973*
Cefetamet, *1974*
Cefixime, *1975*
Cefmenoxime, *1976*
Cefodizime, *1979*
Cefonicid, *1980*
Cefoperazone, *1981*
Ceforanide, *1982*
Cefotaxime, *1983*
Cefotiam, *1985*
Cefozopran, *1987*
Cefpimizole, *1988*
Cefpiramide, *1989*
Cefpirome, *1990*
Cefpodoxime Proxetil, *1991*
Cefprozil, *1992*
Cefroxadine, *1993*
Cefsulodin, *1994*
Ceftazidime, *1995*
Cefteram, *1996*
Ceftezole, *1997*
Ceftibuten, *1998*
Ceftizoxime, *2000*
Ceftriaxone, *2001*
Cefuroxime, *2002*
Cefuzonam, *2003*
Cephacetrile Sodium, *2019*
Cephalexin, *2021*
Cephaloglycin, *2023*
Cephaloridine, *2025*
Cephalosporin C, *2026*
Cephalothin, *2028*
Cephapirin Sodium, *2030*
Cephradine, *2032*
Pivcefalexin, *7670*

Cephamycins
Cefbuperazone, *1968*
Cefmetazole, *1977*
Cefminox, *1978*

ANTIBACTERIAL (ANTIBIOTICS)

β-Lactams

Cephamycins (continued)
Cefotetan, *1984*
Cefoxitin, *1986*

Monobactams
Aztreonam, *955*
Carumonam, *1922*
Tigemonam, *9572*

Oxacephems
Flomoxef, *4141*
Moxalactam, *6369*

Penicillins
Amdinocillin, *408*
Amdinocillin Pivoxil, *409*
Amoxicillin, *617*
Ampicillin, *628*
Apalcillin, *766*
Aspoxicillin, *887*
Azidocillin, *942*
Azlocillin, *947*
Bacampicillin, *961*
Benzylpenicillinic Acid, *1177*
Benzylpenicillin Sodium, *1178*
Carbenicillin, *1838*
Carindacillin, *1888*
Clometocillin, *2445*
Cloxacillin, *2480*
Cyclacillin, *2769*
Dicloxacillin, *3134*
Epicillin, *3651*
Fenbenicillin, *4001*
Floxacillin, *4147*
Hetacillin, *4706*
Lenampicillin, *5460*
Metampicillin, *5986*
Methicillin Sodium, *6047*
Mezlocillin, *6259*
Nafcillin Sodium, *6438*
Oxacillin, *7036*
Penamecillin, *7208*
Penethamate Hydriodide, *7212*
Penicillin G Benethamine, *7220*
Penicillin G Benzathine, *7221*
Penicillin G Benzhydrylamine, *7222*
Penicillin G Calcium, *7223*
Penicillin G Hydrabamine, *7224*
Penicillin G Potassium, *7225*
Penicillin G Procaine, *7226*
Penicillin N, *7227*
Penicillin O, *7228*
Penicillin V, *7230*
Penicillin V Benzathine, *7231*
Penicillin V Hydrabamine, *7232*
Penimepicycline, *7235*
Phenethicillin Potassium, *7369*
Piperacillin, *7616*
Pivampicillin, *7669*
Propicillin, *8002*
Quinacillin, *8224*
Sulbenicillin, *9059*
Sultamicillin, *9166*
Talampicillin, *9204*
Temocillin, *9288*
Ticarcillin, *9568*

Other
Ritipenem, *8400*

Lincosamides
Clindamycin, *2414*
Lincomycin, *5525*

Macrolides
Azithromycin, *946*
Carbomycin, *1854*
Clarithromycin, *2400*
Dirithromycin, *3418*
Erythromycin, *3720*
Erythromycin Acistrate, *3721*
Erythromycin Estolate, *3722*
Erythromycin Glucoheptonate, *3723*
Erythromycin Lactobionate, *3724*
Erythromycin Propionate, *3725*
Erythromycin Stearate, *3726*
Josamycin, *5280*
Leucomycins, *5480*
Midecamycins, *6271*
Miokamycin, *6291*
Oleandomycin, *6962*
Primycin, *7931*
Rokitamycin, *8408*
Rosaramicin, *8420*
Roxithromycin, *8433*
Spiramycin, *8904*
Troleandomycin, *9899*

Polypeptides
Amphomycin, *625*
Bacitracin, *965*
Capreomycin, *1801*
Colistin, *2542*
Enduracidin, *3619*
Enviomycin, *3636*
Fusafungine, *4335*
Gramicidin S, *4552*
Gramicidin(s), *4553*
Mikamycin, *6275*
Polymyxin, *7734*
Pristinamycin, *7933*
Ristocetin, *8398*
Teicoplanin, *9269*
Thiostrepton, *9502*
Tuberactinomycin, *9935*
Tyrocidine, *9967*
Tyrothricin, *9972*
Vancomycin, *10066*
Viomycin, *10139*
Virginiamycin, *10142*
Zinc Bacitracin, *10257*

Tetracyclines
Apicycline, *772*
Chlortetracycline, *2245*
Clomocycline, *2448*
Demeclocycline, *2937*
Doxycycline, *3496*
Guamecycline, *4584*
Lymecycline, *5658*
Meclocycline, *5818*
Methacycline, *6007*
Minocycline, *6289*
Oxytetracycline, *7111*
Penimepicycline, *7235*
Pipacycline, *7606*
Rolitetracycline, *8411*
Sancycline, *8500*
Tetracycline, *9337*

Others
Cycloserine, *2820*
Mupirocin, *6383*
Tuberin, *9938*

ANTIBACTERIAL (SYNTHETIC)

2,4-Diaminopyrimidines
Brodimoprim, *1401*
Tetroxoprim, *9386*
Trimethoprim, *9840*

Nitrofurans
Furaltadone, *4315*
Furazolium Chloride, *4321*
Nifuradene, *6621*
Nifuratel, *6623*
Nifurfoline, *6624*
Nifurpirinol, *6628*
Nifurprazine, *6629*
Nifurtoinol, *6631*
Nitrofurantoin, *6696*

Quinolones and Analogs
Cinoxacin, *2369*
Ciprofloxacin, *2374*
Clinafloxacin, *2413*
Difloxacin, *3187*
Enoxacin, *3625*
Fleroxacin, *4137*
Flumequine, *4172*
Grepafloxacin, *4567*
Lomefloxacin, *5592*
Miloxacin, *6283*
Nadifloxacin, *6430*
Nalidixic Acid, *6446*
Norfloxacin, *6793*
Ofloxacin, *6865*
Oxolinic Acid, *7079*
Pazufloxacin, *7188*
Pefloxacin, *7197*
Pipemidic Acid, *7613*
Piromidic Acid, *7660*
Rosoxacin, *8426*
Rufloxacin, *8448*
Sparfloxacin, *8884*
Temafloxacin, *9284*
Tosufloxacin, *9692*
Trovafloxacin, *9919*

Sulfonamides
Acetyl Sulfamethoxypyrazine, *103*
Benzylsulfamide, *1182*
Chloramine-B, *2117*
Chloramine-T, *2118*
Dichloramine T, *3098*
N^2-Formylsulfisomidine, *4275*
N^4-β-D-Glucosylsulfanilamide, *4472*
Mafenide, *5683*
4'-(Methylsulfamoyl)sulfanilanilide, *6202*
Noprylsulfamide, *6777*
Phthalylsulfacetamide, *7532*
Phthalylsulfathiazole, *7533*
Salazosulfadimidine, *8473*
Succinylsulfathiazole, *9046*
Sulfabenzamide, *9065*
Sulfacetamide, *9067*
Sulfachlorpyridazine, *9068*
Sulfachrysoidine, *9069*
Sulfacytine, *9070*
Sulfadiazine, *9071*
Sulfadicramide, *9072*
Sulfadimethoxine, *9073*
Sulfadoxine, *9074*
Sulfaethidole, *9075*
Sulfaguanidine, *9076*
Sulfaguanol, *9077*
Sulfalene, *9078*
Sulfaloxic Acid, *9080*
Sulfamerazine, *9081*
Sulfameter, *9082*
Sulfamethazine, *9083*
Sulfamethizole, *9084*
Sulfamethomidine, *9085*
Sulfamethoxazole, *9086*
Sulfamethoxypyridazine, *9087*
Sulfametrole, *9089*
Sulfamidochrysoidine, *9092*
Sulfamoxole, *9093*
Sulfanilamide, *9094*
4-Sulfanilamidosalicylic Acid, *9095*
N^4-Sulfanilylsulfanilamide, *9100*
Sulfanilylurea, *9101*

N-Sulfanilyl-3,4-xylamide, *9102*
Sulfanitran, *9103*
Sulfaperine, *9104*
Sulfaphenazole, *9105*
Sulfaproxyline, *9106*
Sulfapyrazine, *9107*
Sulfapyridine, *9108*
Sulfasomizole, *9113*
Sulfasymazine, *9114*
Sulfathiazole, *9115*
Sulfathiourea, *9116*
Sulfatolamide, *9117*
Sulfisomidine, *9124*
Sulfisoxazole, *9125*

Sulfones
Acedapsone, *20*
Acediasulfone, *21*
Acetosulfone Sodium, *73*
Dapsone, *2885*
Diathymosulfone, *3037*
Glucosulfone Sodium, *4471*
Solasulfone, *8859*
Succisulfone, *9047*
Sulfanilic Acid, *9096*
p-Sulfanilylbenzylamine, *9098*
Sulfoxone Sodium, *9141*
Thiazolsulfone, *9444*

Others
Clofoctol, *2439*
Hexedine, *4735*
Methenamine, *6036*
Methenamine Anhydromethylene-
citrate, *6038*
Methenamine Hippurate, *6039*
Methenamine Mandelate, *6040*
Methenamine Sulfosalicylate,
6042
Nitroxoline, *6734*
Taurolidine, *9243*
Xibornol, *10210*

ANTIBACTERIAL (LEPROSTATIC)

Acedapsone, *20*
Acetosulfone Sodium, *73*
Clofazimine, *2433*
Dapsone, *2885*
Diathymosulfone, *3037*
Glucosulfone Sodium, *4471*
Hydnocarpic Acid, *4797*
Solasulfone, *8859*
Succisulfone, *9047*
Sulfoxone Sodium, *9141*

ANTIBACTERIAL (RICKETTSIA) see *Antirickettsial*

ANTIBACTERIAL (TUBERCULO-STATIC)

p-Aminosalicylic Acid, *498*
p-Aminosalicylic Acid Hydrazide,
499
Benzoylpas, *1148*
5-Bromosalicylhydroxamic Acid,
1458
Capreomycin, *1801*
Clofazimine, *2433*
Cyacetacide, *2751*
Cycloserine, *2820*
Dihydrostreptomycin, *3222*
Enviomycin, *3636*
Ethambutol, *3764*
Ethionamide, *3783*

Furonazide, *4330*
Glyconiazide, *4510*
Isoniazid, *5203*
Morphazinamide, *6355*
Opiniazide, *6984*
Phenyl Aminosalicylate, *7426*
Protionamide, *8076*
Pyrazinamide, *8140*
Rifabutin, *8380*
Rifampin, *8382*
Rifapentine, *8385*
Salinazid, *8487*
Streptomycin, *8983*
Streptonicozid, *8985*
Subathizone, *9030*
Sulfoniazide, *9132*
Thiacetazone, *9427*
Tiocarlide, *9593*
Tuberactinomycin, *9935*
Tubercidin, *9936*
Tuberin, *9938*
Verazide, *10091*
Viomycin, *10139*

ANTIBACTERIAL ADJUNCTS

β-Lactamase Inhibitors
Clavulanic Acid, *2402*
Sulbactam, *9058*
Sultamicillin, *9166*
Tazobactam, *9251*

Renal Dipeptidase Inhibitors
Cilastatin, *2331*

Renal Protectant
Betamipron, *1227*

ANTIBIOTIC see *Antibacterial (Antibiotic); Antifungal (Antibiotic); Antineoplastic (Antibiotic)*

ANTICANCER see *Antineoplastic*

ANTICHOLELITHOGENIC see *Cholelitholytic*

ANTICHOLESTEREMIC see *Antihyperlipoproteinemic*

ANTICHOLINERGIC see also *Antispasmodic; Bronchodilator*

Adiphenine Hydrochloride, *160*
Alverine, *386*
Ambutonium Bromide, *406*
Aminopentamide, *479*
Amixetrine, *514*
Amprotropine Phosphate, *632*
Anisotropine Methylbromide, *709*
Apoatropine, *780*
Atropine, *907*
Atropine N-Oxide, *908*
Benactyzine, *1050*
Benapryzine, *1057*
Benzetimide, *1104*
Benzilonium Bromide, *1110*
Benztropine Mesylate, *1156*
Bevonium Methyl Sulfate, *1239*
Biperiden, *1274*
Butropium Bromide, *1569*
N-Butylscopolammonium Bromide,
1324
Buzepide, *1635*
Camylofine, *1784*

Caramiphen, *1820*
Chlorbenzoxamine, *2125*
Chlorphenoxamine, *2234*
Cimetropium Bromide, *2338*
Clidinium Bromide, *2412*
Cyclodrine, *2788*
Cyclonium Iodide, *2806*
Cycrimine Hydrochloride, *2825*
Deptropine, *2958*
Dexetimide, *2987*
Dibutoline Sulfate, *3082*
Dicyclomine Hydrochloride, *3147*
Diethazine, *3157*
Difemerine, *3180*
Dihexyverine, *3211*
Diphemanil Methylsulfate, *3361*
N-(1,2-Diphenylethyl)nicotinamide,
3382
Dipiproverine, *3399*
Diponium Bromide, *3402*
Emepronium Bromide, *3598*
Endobenzyline Bromide, *3612*
Ethopropazine, *3793*
Ethybenztropine, *3802*
Ethylbenzhydramine, *3813*
Etomidoline, *3928*
Eucatropine, *3943*
Fenpiverinium Bromide, *4032*
Fentonium Bromide, *4048*
Flutropium Bromide, *4248*
Glycopyrrolate, *4511*
Heteronium Bromide, *4709*
Hexocyclium Methyl Sulfate, *4744*
Homatropine, *4766*
Hyoscyamine, *4907*
Ipratropium Bromide, *5089*
Isopropamide Iodide, *5221*
Levomepate, *5491*
Mecloxamine, *5823*
Mepenzolate Bromide, *5893*
Metcaraphen, *5996*
Methantheline Bromide, *6025*
Methixene, *6059*
Methscopolamine Bromide, *6084*
Octamylamine, *6846*
Oxybutynin Chloride, *7089*
Oxyphencyclimine, *7107*
Oxyphenonium Bromide, *7109*
Pentapiperide, *7260*
Penthienate Bromide, *7268*
Phencarbamide, *7363*
Phenglutarimide, *7377*
Pipenzolate Bromide, *7614*
Piperidolate, *7623*
Piperilate, *7624*
Poldine Methylsulfate, *7714*
Pridinol, *7922*
Prifinium Bromide, *7923*
Procyclidine, *7944*
Propantheline Bromide, *7989*
Propenzolate, *7998*
Propiverine, *8018*
Propyromazine, *8057*
Scopolamine, *8550*
Scopolamine N-Oxide, *8551*
Stramonium, *8975*
Sultroponium, *9170*
Thiphenamil, *9509*
Tiemonium Iodide, *9571*
Timepidium Bromide, *9583*
Tiquizium Bromide, *9602*
Tridihexethyl Iodide, *9794*
Trihexyphenidyl Hydrochloride,
9823
Trimebutine, *9829*
Tropacine, *9903*
Tropenzile, *9908*
Tropicamide, *9911*
Trospium Chloride, *9918*
Valethamate Bromide, *10044*
Vamicamide, *10052*
Xenytropium Bromide, *10208*

ANTICOAGULANT see also Antithrombotic; Thrombolytic

Acenocoumarol, 29
Ancrod, 671
Anisindione, 706
Bromindione, 1414
Clorindione, 2468
Coumetarol, 2633
Cyclocumarol, 2786
Dextran Sulfate Sodium, 2993
Dicumarol, 3140
Diphenadione, 3363
Ethyl Biscoumacetate, 3818
Ethylidene Dicoumarol, 3858
Fluindione, 4168
Heparin, 4685
Hirudin, 4754
Lyapolate Sodium, 5645
Oxazidione, 7060
Pentosan Polysulfate, 7276
Phenindione, 7381
Phenprocoumon, 7413
Phosvitin, 7522
Picotamide, 7560
Tioclomarol, 9594
Warfarin, 10174

ANTICONVULSANT

Acetylpheneturide, 101
Albutoin, 218
Aloxidone, 315
Aminoglutethimide, 460
4-Amino-3-hydroxybutyric Acid, 464
Atrolactamide, 904
Beclamide, 1045
Buramate, 1525
Calcium Bromide, 1693
Carbamazepine, 1826
Cinromide, 2371
Clomethiazole, 2444
Clonazepam, 2449
Decimemide, 2909
Diethadione, 3155
Dimethadione, 3260
Doxenitoin, 3491
Eterobarb, 3758
Ethadione, 3762
Ethosuximide, 3794
Ethotoin, 3795
Felbamate, 3988
Fluoresone, 4197
Gabapentin, 4343
5-Hydroxytryptophan, 4895
Lamotrigine, 5367
Magnesium Bromide, 5695
Magnesium Sulfate, 5731
Mephenytoin, 5898
Mephobarbital, 5899
Metharbital, 6029
Methetoin, 6046
Methsuximide, 6085
5-Methyl-5-(3-phenanthryl)-hydantoin, 6184
3-Methyl-5-phenylhydantoin, 6187
Narcobarbital, 6509
Nimetazepam, 6642
Nitrazepam, 6667
Oxcarbazepine, 7063
Paramethadione, 7150
Phenacemide, 7343
Phenetharbital, 7368
Pheneturide, 7375
Phenobarbital, 7386
Phensuximide, 7414
Phenylmethylbarbituric Acid, 7457
Phenytoin, 7475
Phethenylate Sodium, 7476

Potassium Bromide, 7780
Primidone, 7927
Progabide, 7955
Sodium Bromide, 8737
Solanum, 8856
Strontium Bromide, 8997
Suclofenide, 9048
Sulthiame, 9167
Tetrantoin, 9372
Tiagabine, 9557
Topiramate, 9686
Trimethadione, 9836
Valproic Acid, 10049
Valpromide, 10050
Vigabatrin, 10114
Zonisamide, 10323

ANTIDEPRESSANT see also Antimanic

Bicyclics
Binedaline, 1266
Caroxazone, 1907
Citalopram, 2379
Dimethazan, 3261
Fencamine, 4007
Indalpine, 4965
Indeloxazine Hydrochloride, 4972
Nefopam, 6529
Nomifensine, 6768
Oxitriptan see 4895
Oxypertine, 7105
Paroxetine, 7175
Sertraline, 8612
Thiazesim, 9440
Trazodone, 9712

Hydrazides/Hydrazines
Benmoxine, 1072
Iproclozide, 5092
Iproniazid, 5094
Isocarboxazid, 5172
Nialamide, 6575
Octamoxin, 6845
Phenelzine, 7366

Pyrrolidones
Cotinine, 2619
Rolicyprine, 8409
Rolipram, 8410

Tetracyclics
Maprotiline, 5792
Metralindole, 6238
Mianserin, 6260
Mirtazepine, 6295

Tricyclics
Adinazolam, 159
Amitriptyline, 511
Amitriptylinoxide, 512
Amoxapine, 616
Butriptyline, 1568
Clomipramine, 2447
Demexiptiline, 2941
Desipramine, 2966
Dibenzepin, 3055
Dimetacrine, 3258
Dothiepin, 3485
Doxepin, 3492
Fluacizine, 4149
Imipramine, 4955
Imipramine N-Oxide, 4956
Iprindole, 5091
Lofepramine, 5587
Melitracen, 5866
Metapramine, 5991
Nortriptyline, 6812
Noxiptilin, 6821

Opipramol, 6985
Pizotyline, 7671
Propizepine, 8019
Protriptyline, 8088
Quinupramine, 8267
Tianeptine, 9560
Trimipramine, 9852

Others
Adrafinil, 168
Benactyzine, 1050
Bupropion, 1523
Butacetin, 1532
Dioxadrol, 3352
Duloxetine, 3518
Etoperidone, 3930
Febarbamate, 3983
Femoxetine, 3993
Fenpentadiol, 4029
Fluoxetine, 4222
Fluvoxamine, 4251
Hematoporphyrin, 4669
Hypericin, 4911
Levophacetoperane, 5493
Medifoxamine, 5834
Milnacipran, 6281
Minaprine, 6287
Moclobemide, 6309
Nefazodone, 6527
Oxaflozane, 7039
Piberaline, 7547
Prolintane, 7964
Pyrisuccideanol, 8175
Ritanserin, 8399
Roxindole, 8432
Rubidium Chloride, 8441
Sulpiride, 9163
Tandospirone, 9219
Thozalinone, 9521
Tofenacin, 9641
Toloxatone, 9659
Tranylcypromine, 9708
L-Tryptophan, 9929
Venlafaxine, 10079
Viloxazine, 10116
Zimeldine, 10254

ANTIDIABETIC

Biguanides
Buformin, 1500
Metformin, 6001
Phenformin, 7376

Hormones
Insulin, 5011

Sulfonylurea Derivatives
Acetohexamide, 59
1-Butyl-3-metanilylurea, 1615
Carbutamide, 1881
Chlorpropamide, 2239
Glibornuride, 4447
Gliclazide, 4448
Glimepiride, 4449
Glipizide, 4451
Gliquidone, 4451
Glisoxepid, 4453
Glyburide, 4486
Glybuthiazol(e), 4487
Glybuzole, 4488
Glyhexamide, 4516
Glymidine, 4517
Glypinamide, 4524
Phenbutamide, 7362
Tolazamide, 9644
Tolbutamide, 9646
Tolcyclamide, 9648

Thiazolidinediones

Pioglitazone, *7605*
Troglitazone, *9898*

Others

Acarbose, *15*
Calcium Mesoxalate, *1726*
Miglitol, *6274*
Repaglinide, *8305*

ANTIDIARRHEAL

Acetorphan, *72*
Acetyltannic Acid, *107*
Alkofanone, *254*
Aluminum Salicylates, Basic, *375*
Catechin, *1950*
Difenoxin, *3183*
Diphenoxylate, *3371*
Lidamidine, *5504*
Loperamide, *5601*
Mebiquine, *5810*
Trillium, *9825*
Uzarin, *10035*
Zaldaride, *10243*

ANTIDIURETIC

Desmopressin, *2969*
Felypressin, *3992*
Lypressin, *5661*
Ornipressin, *7001*
Oxycinchophen, *7091*
Pituitary, Posterior, *7666*
Terlipressin, *9310*
Vasopressin, *10073*

ANTIDOTE (ACETAMINOPHEN POISONING)

Acetylcysteine, *89*
Cysteamine, *2848*
Methionine, *6053*

ANTIDOTE (CURARE)

Edrophonium Chloride, *3562*
Neostigmine, *6553*
Tacrine, *9199*

ANTIDOTE (CYANIDE)

p-Aminopropiophenone, *490*
Methylene Blue, *6137*
Potassium Nitrite, *7816*
Sodium Nitrite, *8793*
Sodium Thiosulfate, *8844*

ANTIDOTE (FOLIC ACID ANTAGO-NISTS)

Folinic Acid, *4254*

ANTIDOTE (HEAVY METAL POI-SONING)

Albumen, *216*
Deferoxamine, *2914*
Dimercaprol, *3255*
2,3-Dimercapto-1-propanesulfonic Acid, *3256*
Ditiocarb Sodium, *3443*
Succimer, *9034*
Tiopronin, *9597*

ANTIDOTE (METHANOL AND ETH-YLENE GLYCOL POISONING)

Fomepizole, *4257*

ANTIDOTE (ORGANOPHOSPHATE POISONING)

Asoxime Chloride, *870*

ANTIDYSKINETIC *see also Antiparkinsonian*

Amantadine, *389*
Clonidine, *2450*
Haloperidol, *4629*
Pimozide, *7589*
Tetrabenazine, *9325*
Tiapride, *9561*

ANTIECZEMATIC

Evening Primrose Oil, *3953*
γ-Linolenic Acid, *5531*

ANTIEMETIC

Acetylleucine Monoethanolamine, *96*
Alizapride, *246*
Azasetron, *933*
Benzquinamide, *1154*
Bietanautine, *1253*
Bromopride, *1454*
Buclizine, *1484*
Chlorpromazine, *2238*
Clebopride, *2404*
Cyclizine, *2779*
Dimenhydrinate, *3252*
Diphenidol, *3369*
Dolasetron, *3471*
Domperidone, *3476*
Granisetron, *4557*
Meclizine, *5817*
Methallatal, *6011*
Metoclopramide, *6226*
Metopimazine, *6233*
Nabilone, *6427*
Ondansetron, *6979*
Oxypendyl, *7104*
Pipamazine, *7607*
Prochlorperazine, *7942*
Scopolamine, *8550*
Sulpiride, *9163*
Tetrahydrocannabinols, *9349*
Thiethylperazine, *9449*
Thioproperazine, *9494*
Trimethobenzamide, *9839*
Tropisetron, *9914*

ANTIEPILEPTIC *see Anticonvulsant*

ANTIESTROGEN

Centchroman, *2018*
Delmadinone Acetate, *2930*
Tamoxifen, *9216*
Toremifene, *9688*

ANTIFIBROTIC

Potassium *p*-Aminobenzoate, *7766*

ANTIFLATULENT

Dimethicone, *3264*

ANTIFUNGAL (ANTIBIOTICS)

Polyenes

Amphotericin B, *627*
Candicidin, *1789*
Dermostatin, *2960*
Filipin, *4123*
Fungichromin, *4312*
Hachimycin, *4616*
Hamycin, *4641*
Lucensomycin, *5621*
Mepartricin, *5891*
Natamycin, *6513*
Nystatin, *6834*
Pecilocin, *7193*
Perimycin, *7310*

Others

Azaserine, *932*
Griseofulvin, *4571*
Oligomycins, *6970*
Neomycin Undecylenate, *6543*
Pyrrolnitrin, *8202*
Siccanin, *8630*
Tubercidin, *9936*
Viridin, *10174*

ANTIFUNGAL (SYNTHETIC)

Allylamines

Butenafine, *1547*
Naftifine, *6442*
Terbinafine, *9299*

Imidazoles

Bifonazole, *1260*
Butoconazole, *1561*
Chlordantoin, *2130*
Chlormidazole, *2156*
Cloconazole, *2429*
Clotrimazole, *2478*
Econazole, *3550*
Enilconazole, *3622*
Fenticonazole, *4047*
Flutrimazole, *4247*
Isoconazole, *5176*
Ketoconazole, *5313*
Lanoconazole, *5370*
Miconazole, *6266*
Omoconazole, *6978*
Oxiconazole Nitrate, *7071*
Sertaconazole, *8610*
Sulconazole, *9062*
Tioconazole, *9595*

Thiocarbamates

Tolciclate, *9647*
Tolindate, *9652*
Tolnaftate, *9656*

Triazoles

Fluconazole, *4158*
Itraconazole, *5262*
Saperconazole, *8510*
Terconazole, *9303*

Others

Acrisorcin, *128*
Amorolfine, *612*
Biphenamine, *1275*
Bromosalicylchloranilide, *1457*
Buclosamide, *1485*
Calcium Propionate, *1745*
Chlorphenesin, *2230*
Ciclopirox, *2325*
Cloxyquin, *2483*
Coparaffinate, *2582*
Diamthazole Dihydrochloride, *3033*
Exalamide, *3955*
Flucytosine, *4161*
Halethazole, *4622*

ANTIFUNGAL (SYNTHETIC)

Others (continued)
Hexetidine, 4741
Loflucarban, 5589
Nifuratel, 6623
Potassium Iodide, 7809
Propionic Acid, 8010
Pyrithione, 8178
Salicylanilide, 8482
Sodium Propionate, 8816
Sulbentine, 9061
Tenonitrozole, 9292
Triacetin, 9721
Ujothion, 9978
Undecylenic Acid, 9983
Zinc Propionate, 10286

ANTIGLAUCOMA

Acetazolamide, 50
Befunolol, 1050
Betaxolol, 1229
Brimonidine, 1399
Bupranolol, 1521
Carteolol, 1917
Dapiprazole, 2884
Dichlorphenamide, 3127
Dipivefrin, 3400
Dorzolamide, 3484
Epinephrine, 3656
Latanoprost, 5387
Levobunolol, 5488
Methazolamide, 6031
Metipranolol, 6221
Pilocarpine, 7578
Pindolol, 7597
Timolol, 9585
Unoprostone, 9984

ANTIGONADOTROPIN

Danazol, 2875
Gestrinone, 4423
Paroxypropione, 7176

ANTIGOUT

Allopurinol, 287
Carprofen, 1912
Colchicine, 2536
Probenecid, 7934
Sulfinpyrazone, 9121

ANTIHEMORRHAGIC see Hemostatic

ANTIHISTAMINIC see also Antiallergic

Alkylamine Derivatives
Acrivastine, 129
Bamipine, 983
Brompheniramine, 1467
Chlorpheniramine, 2232
Dimethindene, 3265
Metron S, 6243
Pheniramine, 7383
Pyrrobutamine, 8196
Thenaldine, 9413
Tolpropamine, 9662
Triprolidine, 9877

Aminoalkylethers
Bietanautine, 1253
Bromodiphenhydramine, 1439
Carbinoxamine, 1845
Clemastine, 2405
Diphenhydramine, 3367

Diphenylpyraline, 3390
Doxylamine, 3497
Embramine, 3596
Medrylamine, 5839
Moxastine, 6370
p-Methyldiphenhydramine, 6130
Orphenadrine, 7007
Phenyltoloxamine, 7469
Setastine, 8619

Ethylenediamine Derivatives
Alloclamide, 263
Chloropyramine, 2214
Chlorothen, 2220
Histapyrrodine, 4757
Methafurylene, 6010
Methaphenilene, 6026
Methapyrilene, 6027
Pyrilamine, 8168
Talastine, 9205
Thenyldiamine, 9416
Thonzylamine Hydrochloride, 9513
Tripelennamine, 9868
Zolamine, 10319

Piperazines
Cetirizine, 2063
Chlorcyclizine, 2128
Cinnarizine, 2365
Clocinizine, 2428
Hydroxyzine, 4897

Tricyclics

Phenothiazines
Ahistan, 189
Etymemazine, 3938
Fenethazine, 4013
N-Hydroxyethylpromethazine
  Chloride, 4870
Isopromethazine, 5219
Mequitazine, 5911
Promethazine, 7970
Thiazinamium Methyl Sulfate,
  9441

Other Tricyclics
Azatadine, 934
Clobenzepam, 2420
Cyproheptadine, 2842
Deptropine, 2958
Isothipendyl, 5248
Loratadine, 5608

Others
Antazoline, 718
Astemizole, 891
Azelastine, 939
Cetoxime, 2065
Clemizole, 2406
Clobenztropine, 2422
Ebastine, 3534
Emedastine, 3597
Epinastine, 3655
Fexofenadine, 4113
Levocabastine, 5489
Mebhydroline, 5809
Phenindamine, 7380
Terfenadine, 9307
Tritoqualine, 9894

ANTIHYPERCHOLESTEROLEMIC
see Antihyperlipoproteinemic

ANTIHYPERLIPIDEMIC see Antihy-
perlipoproteinemic

ANTIHYPERLIPOPROTEINEMIC

Bile Acid Sequesterants
Cholestyramine Resin, 2257
Colestipol, 2538
Polidexide, 7716

Fibrates
Beclobrate, 1046
Bezafibrate, 1240
Binifibrate, 1267
Ciprofibrate, 2373
Clinofibrate, 2415
Clofibrate, 2436
Clofibric Acid, 2437
Etofibrate, 3923
Fenofibrate, 4019
Gemfibrozil, 4394
Nicofibrate, 6604
Pirifibrate, 7650
Ronifibrate, 8414
Simfibrate, 8684
Theofibrate, 9420

HMG CoA Reductase Inhibitors
Atorvastatin, 897
Fluvastatin, 4250
Lovastatin, 5616
Pravastatin Sodium, 7894
Simvastatin, 8686

Nicotinic Acid Derivatives
Acipimox, 113
Aluminum Nicotinate, 364
Niceritrol, 6581
Nicoclonate, 6603
Nicomol, 6606
Oxiniacic Acid, 7075

Thyroid Hormones/Analogs
Etiroxate, 3917
Thyropropic Acid, 9553
Thyroxine, 9555

Others
Acifran, 112
Azacosterol, 924
Benfluorex, 1066
β-Benzalbutyramide, 1083
Carnitine, 1898
Chondroitin Sulfate, 2270
Clomestrone, 2442
Detaxtran, 2980
Dextran Sulfate Sodium, 2993
5,8,11,14,17-Eicosapentaenoic
  Acid, 3572
Eritadenine, 3712
Furazabol, 4319
Meglutol, 5852
Melinamide, 5864
Mytatrienediol, 6424
Ornithine, 7002
γ-Oryzanol, 7016
Pantethine, 7144
Pentaerythritol Tetraacetate, 7248
α-Phenylbutyramide, 7432
Pirozadil, 7662
Probucol, 7935
β-Sitosterol, 8697
Sultosilic Acid, Piperazine Salt,
  9169
Tiadenol, 9556
Triparanol, 9867
Xenbucin, 10205

ANTIHYPERPHOSPHATEMIC

Aluminum Hydroxide, 355
Aluminum Hydroxychloride, 356

ANTIHYPERTENSIVE *see also Diuretic*

Arylethanolamine Derivatives
Amosulalol, *614*
Bufuralol, *1504*
Dilevalol, *3245*
Labetalol, *5341*
Pronethalol, *7974*
Sotalol, *8876*
Sulfinalol, *9120*

Aryloxypropanolamine Derivatives
Acebutolol, *16*
Alprenolol, *321*
Arotinolol, *827*
Atenolol, *892*
Betaxolol, *1229*
Bevantolol, *1238*
Bisoprolol, *1336*
Bopindolol, *1362*
Bunitrolol, *1514*
Bupranolol, *1521*
Butofilolol, *1563*
Carazolol, *1822*
Carteolol, *1917*
Carvedilol, *1924*
Celiprolol, *2007*
Cetamolol, *2060*
Epanolol, *3641*
Indenolol, *4974*
Mepindolol, *5901*
Metipranolol, *6221*
Metoprolol, *6235*
Moprolol, *6345*
Nadolol, *6431*
Nebivalol, *6519*
Nipradilol, *6655*
Oxprenolol, *7086*
Penbutolol, *7209*
Pindolol, *7597*
Propranolol, *8025*
Talinolol, *9208*
Tertatolol, *9318*
Tilisolol, *9579*
Timolol, *9585*
Toliprolol, *9653*

Benzothiadiazine Derivatives *(Thiazides)*
Althiazide, *326*
Bendroflumethiazide, *1064*
Benzthiazide, *1155*
Benzylhydrochlorothiazide, *1171*
Buthiazide, *1554*
Chlorothiazide, *2221*
Chlorthalidone, *2246*
Cyclopenthiazide, *2813*
Cyclothiazide, *2822*
Diazoxide, *3051*
Epithiazide, *3661*
Ethiazide, *3778*
Fenquizone, *4039*
Hydrochlorothiazide, *4822*
Hydroflumethiazide, *4830*
Indapamide, *4969*
Methyclothiazide, *6086*
Meticrane, *6220*
Metolazone, *6231*
Paraflutizide, *7157*
Polythiazide, *7744*
Quinethazone, *8240*
Teclothiazide, *9261*
Trichlormethiazide, *9754*

Biphenyltetrazole Derivatives
Candesartan, *1788*
Irbesartan, *5097*
Losartan, *5613*
Valsartan, *10051*

N-Carboxyalkyl(peptide/lactam) Derivatives
Alacepril, *202*
Benazepril, *1058*
Captopril, *1817*
Ceronapril, *2038*
Cilazapril, *2332*
Delapril, *2928*
Enalapril, *3605*
Enalaprilat, *3606*
Fosinopril, *4282*
Imidapril, *4947*
Lisinopril, *5540*
Moveltipril, *6368*
Perindopril, *7311*
Quinapril, *8233*
Ramipril, *8283*
Spirapril, *8905*
Temocapril, *9287*
Trandolapril, *9703*

Dihydropyridine Derivatives
Amlodipine, *516*
Aranidipine, *806*
Barnidipine, *1031*
Benidipine, *1071*
Cilnidipine, *2334*
Efonidipine, *3566*
Felodipine, *3991*
Isradipine, *5260*
Lacidipine, *5344*
Lercanidipine, *5469*
Manidipine, *5786*
Nicardipine, *6579*
Nifedipine, *6617*
Nilvadipine, *6637*
Nisoldipine, *6658*
Nitrendipine, *6669*

Guanidine Derivatives
Bethanidine, *1233*
Debrisoquin, *2901*
Guanabenz, *4585*
Guanacline, *4586*
Guanadrel, *4587*
Guanazodine, *4588*
Guanethidine, *4589*
Guanfacine, *4590*
Guanochlor, *4595*
Guanoxabenz, *4597*
Guanoxan, *4598*

Hydrazines/Phthalazines
Budralazine, *1492*
Cadralazine, *1669*
Dihydralazine, *3212*
Endralazine, *3617*
Hydracarbazine, *4798*
Hydralazine, *4800*
Pheniprazine, *7382*
Pildralazine, *7577*
Todralazine, *9640*

Imidazole Derivatives
Clonidine, *2450*
Lofexidine, *5588*
Moxonidine, *6375*
Phentolamine, *7417*
Tiamenidine, *9558*
Tolonidine, *9657*

Imidazoleacrylic Acid Derivatives
Eprosartan, *3669*

Quaternary Ammonium Compounds
Azamethonium Bromide, *929*
Chlorisondamine Chloride, *2151*
Hexamethonium, *4724*
Pentacynium Bis(methyl sulfate), *7243*

Pentamethonium Bromide, *7253*
Pentolinium Tartrate, *7274*
Phenactropinium Chloride, *7346*
Trimethidinium Methosulfate, *9838*

Quinazoline Derivatives
Alfuzosin, *237*
Bunazosin, *1512*
Doxazosin, *3489*
Prazosin, *7897*
Terazosin, *9297*
Trimazosin, *9828*

Reserpine Derivatives
Bietaserpine, *1254*
Deserpidine, *2964*
Rescinnamine, *8311*
Reserpine, *8314*
Syrosingopine, *9193*

Sulfonamide Derivatives
Ambuside, *405*
Clopamide, *2454*
Furosemide, *4331*
Quinethazone, *8240*
Tripamide, *9866*
Xipamide, *10212*

Others
Ajmaline, *194*
γ-Aminobutyric Acid, *450*
Bufeniode, *1495*
Carmoxirole, *1893*
Chlorthalidone, *2246*
Cicletanine, *2323*
Ciclosidomine, *2326*
Clentiazem, *2408*
Cryptenamine Tannates, *2674*
Fantofarone, *3975*
Fenoldopam, *4020*
Flosequinan, *4146*
Indoramin, *5000*
Ketanserin, *5307*
Levcromakalim, *5487*
Mebutamate, *5813*
Mecamylamine, *5814*
Methyldopa, *6132*
Methyl 4-Pyridyl Ketone
  Thiosemicarbazone, *6196*
Metolazone, *6231*
Mibefradil, *6261*
Minoxidil, *6290*
Muzolimine, *6397*
Naftopidil, *6443*
Pargyline, *7172*
Pempidine, *7207*
Pinacidil, *7592*
Piperoxan, *7631*
Protoveratrines, *8086*
Raubasine, *8291*
Rescimetol, *8310*
Rilmenidene, *8388*
Saralasin, *8518*
Semotiadil, *8590*
Sodium Nitroprusside, *8794*
Ticrynafen, *9570*
Trimethaphan Camsylate, *9837*
Tyrosinase, *9969*
Urapidil, *10003*

ANTIHYPERTHYROID

2-Amino-4-methylthiazole, *470*
2-Aminothiazole, *501*
Carbimazole, *1844*
3,5-Dibromo-L-tyrosine, *3079*
3,5-Diiodotyrosine, *3236*
Iodine, *5034*
Methimazole, *6049*
Methylthiouracil, *6210*

ANTIHYPERTHYROID (*continued*)

Propylthiouracil, *8054*
Sodium Perchlorate, *8798*
Thibenzazoline, *9446*
Thiobarbital, *9455*
2-Thiouracil, *9504*

ANTIHYPOTENSIVE

Amezinium Methyl Sulfate, *412*
Angiotensin Amide *see 689*
Dimetofrine, *3314*
Dopamine, *3479*
Etifelmin, *3909*
Etilefrin, *3911*
Gepefrine, *4409*
Metaraminol, *5993*
Methoxamine, *6067*
Midodrine, *6272*
Norepinephrine, *6788*
Pholedrine, *7485*
Synephrine, *9189*

ANTIHYPOTHYROID

Levothyroxine Sodium, *5497*
Liothyronine, *5535*
Thyroid, *9551*
Thyroidin, *9552*
Thyroxine, *9555*
Tiratricol, *9603*
TSH, *9931*

ANTI-INFECTIVE *see Antiseptic/Disinfectant*

ANTI-INFLAMMATORY (GASTRO-INTESTINAL)

Mesalamine, *5964*
Olsalazine, *6976*
Sulfasalazine, *9112*

ANTI-INFLAMMATORY (NONSTE-ROIDAL) *see also Antiarthritic/Antirheumatic*

Aminoarylcarboxylic Acid Derivatives
Enfenamic Acid, *3620*
Etofenamate, *3922*
Flufenamic Acid, *4167*
Isonixin, *5209*
Meclofenamic Acid, *5819*
Mefenamic Acid, *5842*
Niflumic Acid, *6620*
Talniflumate, *9214*
Terofenamate, *9312*
Tolfenamic Acid, *9650*

Arylacetic Acid Derivatives
Aceclofenac, *19*
Acemetacin, *27*
Alclofenac, *220*
Amfenac, *413*
Amtolmetin Guacil, *637*
Bromfenac, *1411*
Bufexamac, *1497*
Cinmetacin, *2353*
Clopirac, *2459*
Diclofenac Sodium, *3132*
Etodolac, *3920*
Felbinac, *3989*
Fenclozic Acid, *4010*
Fentiazac, *4045*
Glucametacin, *4456*
Ibufenac, *4924*
Indomethacin, *4998*
Isofezolac, *5188*

Isoxepac, *5257*
Lonazolac, *5596*
Metiazinic Acid, *6219*
Mofezolac, *6314*
Oxametacine, *7048*
Pirazolac, *7643*
Proglumetacin, *7957*
Sulindac, *9155*
Tiaramide, *9564*
Tolmetin, *9655*
Tropesin, *9909*
Zomepirac, *10322*

Arylbutyric Acid Derivatives
Bumadizon, *1507*
Butibufen, *1557*
Fenbufen, *4003*
Xenbucin, *10205*

Arylcarboxylic Acids
Clidanac, *2411*
Ketorolac, *5318*
Tinoridine, *9590*

Arylpropionic Acid Derivatives
Alminoprofen, *308*
Benoxaprofen, *1075*
Bermoprofen, *1199*
Bucloxic Acid, *1486*
Carprofen, *1912*
Fenoprofen, *4021*
Flunoxaprofen, *4183*
Flurbiprofen, *4234*
Ibuprofen, *4925*
Ibuproxam, *4926*
Indoprofen, *4999*
Ketoprofen, *5316*
Loxoprofen, *5619*
Naproxen, *6504*
Oxaprozin, *7057*
Piketoprofen, *7576*
Pirprofen, *7663*
Pranoprofen, *7890*
Protizinic Acid, *8077*
Suprofen, *9180*
Tiaprofenic Acid, *9562*
Ximoprofen, *10211*
Zaltoprofen, *10244*

Pyrazoles
Difenamizole, *3181*
Epirizole, *3659*

Pyrazolones
Apazone, *768*
Benzpiperylon, *1152*
Feprazone, *4053*
Mofebutazone, *6312*
Morazone, *6349*
Oxyphenbutazone, *7106*
Phenylbutazone, *7431*
Pipebuzone, *7610*
Propyphenazone, *8056*
Ramifenazone, *8282*
Suxibuzone, *9185*
Thiazolinobutazone, *9443*

Salicylic Acid Derivatives
Acetaminosalol, *46*
Aspirin, *886*
Benorylate, *1074*
Bromosaligenin, *1460*
Calcium Acetylsalicylate, *1684*
Diflunisal, *3190*
Etersalate, *3759*
Fendosal, *4012*
Gentisic Acid, *4404*

Glycol Salicylate, *4509*
Imidazole Salicylate, *4949*
Lysine Acetylsalicylate, *5668*
Mesalamine, *5964*
Morpholine Salicylate *see 6362*
1-Naphthyl Salicylate, *6501*
Olsalazine, *6976*
Parsalmide, *7177*
Phenyl Acetylsalicylate, *7424*
Phenyl Salicylate, *7464*
Salacetamide, *8471*
Salicylamide *O*-Acetic Acid, *8481*
Salicylsulfuric Acid, *8486*
Salsalate, *8491*
Sulfasalazine, *9112*

Thiazinecarboxamides
Ampiroxicam, *629*
Droxicam, *3512*
Isoxicam, *5258*
Lornoxicam, *5612*
Piroxicam, *7661*
Tenoxicam, *9293*

Others
ε-Acetamidocaproic Acid, *43*
S-Adenosylmethionine, *155*
3-Amino-4-hydroxybutyric Acid, *463*
Amixetrine, *514*
Bendazac, *1061*
Benzydamine, *1157*
α-Bisabolol, *1281*
Bucolome, *1487*
Difenpiramide, *3184*
Ditazol, *3432*
Emorfazone, *3603*
Fepradinol, *4052*
Guaiazulene, *4581*
Nabumetone, *6428*
Nimesulide, *6640*
Oxaceprol, *7035*
Paranyline, *7163*
Perisoxal, *7318*
Proquazone, *8059*
Superoxide Dismutase, *9177*
Tenidap, *9290*
Zileuton, *10253*

ANTI-INFLAMMATORY (STE-ROIDAL) *see Glucocorticoid*

ANTILEPROTIC *see Antibacterial (Leprostatic)*

ANTILEUKEMIC *see Antineoplastic*

ANTILIPEMIC *see Antihyperlipoproteinemic*

ANTILIPIDEMIC *see Antihyperlipoproteinemic*

ANTIMALARIAL

Acedapsone, *24*
Amodiaquin, *609*
Arteether, *853*
Artemether, *854*
Artemisinin, *856*
Artesunate, *857*
Atovaquone, *898*
Bebeerine, *1042*
Berberine, *1192*
Chirata, *2104*

Chlorguanide, *2138*
Chloroquine, *2215*
Chlorproguanil, *2237*
Cinchona, *2343*
Cinchonidine, *2345*
Cinchonine, *2346*
Cycloguanil, *2790*
Gentiopicrin, *4403*
Halofantrine, *4626*
Hydroxychloroquine, *4863*
Mefloquine Hydrochloride, *5845*
3-Methylarsacetin, *6102*
Pamaquine, *7134*
Plasmocid, *7680*
Primaquine, *7925*
Pyrimethamine, *8169*
Quinacrine, *8225*
Quinidine, *8244*
Quinine, *8245*
Quinocide, *8252*
Quinoline, *8253*
Sodium Arsenate, Dibasic, *8720*

ANTIMANIC

Lithium Acetate, *5544*
Lithium Carbonate, *5552*
Lithium Chloride, *5553*
Lithium Citrate, *5555*
Lithium Sulfate, *5569*

ANTIMETHEMOGLOBINEMIC

Methylene Blue, *6137*

ANTIMIGRAINE

Alpiropride, *319*
Dihydroergotamine, *3217*
Dolasetron, *3471*
Ergocornine, *3685*
Ergocorninine, *3686*
Ergocryptine, *3689*
Ergot, *3702*
Ergotamine, *3703*
Flumedroxone acetate, *4171*
Fonazine, *4260*
Lisuride, *5541*
Lomerizine, *5593*
Methysergid(e), *6217*
Oxetorone, *7068*
Pizotyline, *7671*
Sumatriptan, *9172*

ANTIMYCOTIC *see Antifungal*

ANTINAUSEANT *see Antiemetic*

ANTINEOPLASTIC

Alkaloids
Docetaxel, *3458*
Etoposide, *3931*
Irinotecan, *5104*
Paclitaxel, *7117*
Teniposide, *9291*
Topotecan, *9687*
Vinblastine, *10119*
Vincristine, *10124*
Vindesine, *10125*

Alkylating Agents

Alkyl Sulfonates
Busulfan, *1529*
Improsulfan, *4962*
Piposulfan, *7635*

Aziridines
Benzodepa, *1118*
Carboquone, *1872*
Meturedepa, *6245*
Uredepa, *10011*

Ethylenimines and Methylmelamines
Altretamine, *328*
Triethylenemelamine, *9803*
Triethylenephosphoramide, *9804*
Triethylenethiophosphoramide, *9805*

Nitrogen Mustards
Chlorambucil, *2116*
Chlornaphazine, *2157*
Cyclophosphamide, *2816*
Estramustine, *3749*
Ifosfamide, *4937*
Mechlorethamine, *5815*
Mechlorethamine Oxide Hydrochloride, *5816*
Melphalan, *5871*
Novembichin, *6817*
Perfosfamide, *7303*
Phenesterine, *7367*
Prednimustine, *7900*
Trofosfamide, *9897*
Uracil Mustard, *9986*

Nitrosoureas
Carmustine, *1894*
Chlorozotocin, *2229*
Fotemustine, *4285*
Lomustine, *5594*
Nimustine, *6645*
Ranimustine, *8285*

Others
Dacarbazine, *2866*
Mannomustine, *5790*
Mitobronitol, *6298*
Mitolactol, *6300*
Pipobroman, *7634*
Temozolomide, *9289*

Antibiotics and Analogs
Aclacinomycins, *115*
Actinomycin F_1, *142*
Anthramycin, *724*
Azaserine, *932*
Bleomycins, *1351*
Cactinomycin, *1642*
Carubicin, *1921*
Carzinophilin, *1928*
Chromomycins, *2295*
Dactinomycin, *2867*
Daunorubicin, *2890*
6-Diazo-5-oxo-L-norleucine, *3049*
Doxorubicin, *3495*
Epirubicin, *3660*
Idarubicin, *4931*
Menogaril, *5881*
Mitomycins, *6301*
Mycophenolic Acid, *6408*
Nogalamycin, *6767*
Olivomycins, *6975*
Peplomycin, *7286*
Pirarubicin, *7642*
Plicamycin, *7696*
Porfiromycin, *7756*
Puromycin, *8130*
Streptonigrin, *8986*
Streptozocin, *8991*
Tubercidin, *9936*
Zinostatin, *10302*
Zorubicin, *10326*

Antimetabolites

Folic Acid Analogs
Denopterin, *2944*
Edatrexate, *3553*
Methotrexate, *6065*
Piritrexim, *7654*
Pteropterin, *8119*
Tomudex®, *9684*
Trimetrexate, *9851*

Purine Analogs
Cladribine, *2397*
Fludarabine, *4162*
6-Mercaptopurine, *5919*
Thiamiprine, *9434*
Thioguanine, *9473*

Pyrimidine Analogs
Ancitabine, *670*
Azacitidine, *923*
6-Azauridine, *937*
Carmofur, *1892*
Cytarabine, *2853*
Doxifluridine, *3493*
Emitefur, *3601*
Enocitabine, *3624*
Floxuridine, *4148*
Fluorouracil, *4219*
Gemcitabine, *4392*
Tegafur, *9267*

Enzymes
L-Asparaginase, *871*

Immunomodulators
Interferon-α, *5016*
Interferon β, *5017*
Interferon-γ, *5018*
Interleukin-2, *5020*
Lentinan, *5462*
Propagermanium, *7979*
PSK, *8112*
Roquinimex, *8418*
Sizofiran, *8699*
Ubenimex, *9973*

Platinum Complexes
Carboplatin, *1870*
Cisplatin, *2378*
Miboplatin, *6263*
Oxaliplatin, *7044*

Others
Aceglatone, *24*
Amsacrine, *635*
Bisantrene, *1284*
Defosfamide, *2917*
Demecolcine, *2938*
Diaziquone, *3044*
Eflornithine, *3564*
Elliptinium Acetate, *3590*
Etoglucid, *3926*
Fenretinide, *4040*
Gallium Nitrate *see 4364*
Hydroxyurea, *4896*
Lonidamine, *5598*
Miltefosine, *6285*
Mitoguazone, *6299*
Mitoxantrone, *6303*
Mopidamol, *6344*
Nitracrine, *6661*
Pentostatin, *7277*
Phenamet, *7351*
Podophyllinic Acid 2-Ethylhydrazide, *see 7703*
Procarbazine, *7938*
Razoxane, *8295*
Sobuzoxane, *8708*
Spirogermanium, *8916*

ANTINEOPLASTIC

Others (*continued*)
Tenuazonic Acid, *9294*
Triaziquone, *9733*
2,2',2''-Trichlorotriethylamine, *9777*
Urethan, *10013*

ANTINEOPLASTIC (HORMONAL)

Androgens
Calusterone, *1771*
Dromostanolone, *3504*
Epitiostanol, *3662*
Mepitiostane, *5904*
Testolactone, *9321*

Antiadrenals
Aminoglutethimide, *460*
Mitotane, *6302*
Trilostane, *9827*

Antiandrogens
Bicalutamide, *1247*
Flutamide, *4242*
Nilutamide, *6636*

Antiestrogens
Droloxifene, *3502*
Tamoxifen, *9216*
Toremifene, *9688*

Aromatase Inhibitors
Aminoglutethimide, *460*
Anastrozole, *667*
Fadrozole, *3969*
Formestane, *4267*
Letrozole, *5474*

Estrogens
Fosfestrol, *4279*
Hexestrol, *4738*
Polyestradiol Phosphate, *7727*

LHRH Analogs
Buserelin, *1527*
Goserelin, *4547*
Leuprolide, *5484*
Triptorelin, *9878*

Progestogens
Chlormadinone Acetate, *2152*
Medroxyprogesterone, *5838*
Megestrol Acetate, *5849*
Melengestrol, *5859*

ANTINEOPLASTIC (PHOTO-SENSITIZER)

Porfimer Sodium, *7755*

ANTINEOPLASTIC (RADIATION SOURCE)

Americium, *410*
Chromic Phosphate, Radioactive *see 2284*
Cobalt, *2488*
[131]I-Ethiodized Oil *see 3781*
Gold, Radioactive, Colloidal, *4533*
Iobenguane, *5027*
Radium, *8277*
Radon, *8278*
Sodium Iodide, Radioactive, *8778*
Sodium Phosphate, Radioactive, *8807*

ANTINEOPLASTIC ADJUNCT

Antimetastatic Agent
Batimastat, *1036*

Folic Acid Replenisher
Folinic Acid, *4254*

Radioprotective
Amifostine, *424*

Radiosensitizer
Etanidazole, *3756*

Uroprotective
Mesna, *5969*

ANTINEUTROPENIC

Granulocyte Colony Stimulating Factor, *4558*
Granulocyte-Macrophage Colony Stimulating Factor, *4559*

ANTIOSTEOPOROTIC *see also Bone Resorption Inhibitor; Calcium Regulator*

Alendronic Acid, *228*
Clodronic Acid, *2432*
Etidronic Acid, *3908*
Pamidronic Acid, *7135*
Raloxifene, *8281*
Risedronic Acid, *8396*
Sodium Fluoride, *8762*
Teriparatide Acetate, *9309*
Tiludronic Acid, *9582*

ANTIPAGETIC *see also Bone Resorption Inhibitor; Calcium Regulator*

Alendronic Acid, *228*
Elcatonin, *3578*
Tiludronic Acid, *9582*

ANTIPARKINSONIAN

Amantadine, *389*
Benserazide, *1079*
Bietanautine, *1253*
Biperiden, *1274*
Bromocriptine, *1437*
Budipine, *1491*
Carbidopa, *1843*
Dexetimide, *2987*
Diethazine, *3157*
Droxidopa, *3513*
Ethopropazine, *3793*
Ethylbenzhydramine, *3813*
Lazabemide, *5407*
Levodopa, *5490*
Mofegiline, *6313*
Pergolide, *7304*
Piroheptine, *7658*
Pramipexole, *7885*
Pridinol, *7922*
Prodipine, *7949*
Ropinirole, *8416*
Selegiline, *8569*
Talipexole, *9209*
Terguride, *9308*
Trihexyphenidyl Hydrochloride, *9823*

ANTIPERISTALTIC *see Antidiarrheal*

ANTIPHEOCHROMOCYTOMA

Metyrosine, *6248*
Phenoxybenzamine, *7409*
Phentolamine, *7417*

ANTIPNEUMOCYSTIS

Atovaquone, *898*
Eflornithine, *3564*
Pentamidine, *7254*
Sulfamethoxazole, *9086*

ANTIPROSTATIC HYPERTROPHY

Epristeride, *3668*
Finasteride, *4125*
Gestonorone Caproate, *4422*
Mepartricin, *5891*
Osaterone, *7019*
Oxendolone, *7065*
Tamsulosin, *9217*
Terazosin, *9297*

ANTIPROTOZOAL (AMEBA) *see Antiamebic*

ANTIPROTOZOAL (GIARDIA)

Acranil, *123*
Tinidazole, *9588*

ANTIPROTOZOAL (HISTOMONAS)

Ipronidazole, *5095*

ANTIPROTOZOAL (LEISHMANIA)

Ethylstibamine, *3896*
Hydroxystilbamidine, *4892*
N-Methylglucamine, *6154*
Pentamidine, *7254*
Stilbamidine, *8970*
Sodium Stibogluconate *see 742*
Urea Stibamine, *10010*

ANTIPROTOZOAL (MALARIA) *see Antimalarial*

ANTIPROTOZOAL (PNEUMOCYSTIS) *see Antipneumocystis*

ANTIPROTOZOAL (TOXOPLASMA)

Atovaquone, *898*
Pyrimethamine, *8169*

ANTIPROTOZOAL (TRICHOMONAS)

Acetarsone, *49*
Aminitrozole, *430*
Anisomycin, *708*
Azanidazole, *930*
Furazolidone, *4320*
Hachimycin, *4616*
Lauroguadine, *5398*
Mepartricin, *5891*
Metronidazole, *6242*
Nifuratel, *6623*
Nifuroxime, *6627*
Nimorazole, *6644*

Secnidazole, 8562
Silver Picrate, 8670
Tenonitrozole, 9292
Tinidazole, 9588

ANTIPROTOZOAL (TRYPANO-SOMA)

Benznidazole, 1114
Eflornithine, 3564
Melarsoprol, 5856
Nifurtimox, 6630
Oxophenarsine Hydrochloride, 7082
Pentamidine, 7254
Propamidine, 7981
Puromycin, 8130
Quinapyramine, 8234
Stilbamidine, 8970
Suramin Sodium, 9181
Trypan Red, 9924
Tryparsamide, 9925

ANTIPRURITIC

Camphor, 1779
Cyproheptadine, 2842
Dichlorisone, 3099
Glycine, 4500
Halometasone, 4628
3-Hydroxycamphor, 4862
Menthol, 5882
Mesulfen, 5977
Methdilazine, 6034
Phenol, 7390
Polidocanol, 7717
Spirit of Camphor, 8907
Thenaldine, 9413
Tolpropamine, 9662
Trimeprazine, 9834

ANTIPSORIATIC

Acitretin, 114
Ammonium Salicylate, 584
Anthralin, 723
6-Azauridine, 937
Bergapten(e), 1196
Calcipotriene, 1679
Chrysarobin, 2312
Etretinate, 3935
Lonapalene, 5595
Pyrogallol, 8184
Tacalcitol, 9197
Tazarotene, 9249

ANTIPSYCHOTIC

Benzamides
Alizapride, 246
Amisulpride, 508
Nemonapride, 6533
Remoxipride, 8301
Sulpiride, 9163
Sultopride, 9168

Benzisoxazoles
Risperidone, 8397

Butyrophenones
Benperidol, 1077
Bromperidol, 1466
Droperidol, 3505
Fluanisone, 4150
Haloperidol, 4629
Melperone, 5870
Moperone, 6343
Pipamperone, 7608

Spiperone, 8903
Timiperone, 9584
Trifluperidol, 9813

Phenothiazines
Acetophenazine, 70
Butaperazine, 1543
Carphenazine, 1910
Chlorproethazine, 2236
Chlorpromazine, 2238
Clospirazine, 2474
Cyamemazine, 2753
Dixyrazine, 3450
Fluphenazine, 4226
Imiclopazine, 4945
Mepazine, 5892
Mesoridazine, 5970
Methoxypromazine, 6080
Metofenazate, 6228
Oxaflumazine, 7040
Perazine, 7294
Pericyazine, 7306
Perimethazine, 7309
Perphenazine, 7323
Piperacetazine, 7615
Pipotiazine, 7635
Prochlorperazine, 7942
Promazine, 7966
Sulforidazine, 9137
Thiopropazate, 9493
Thioproperazine, 9494
Thioridazine, 9497
Trifluoperazine, 9811
Triflupromazine, 9814

Thioxanthenes
Chlorprothixene, 2241
Clopenthixol, 2455
Flupentixol, 4224
Thiothixene, 9503

Other Tricyclics
Benzquinamide, 1154
Carpipramine, 1911
Clocapramine, 2427
Clomacran, 2441
Clothiapine, 2476
Clozapine, 2484
Mosapramine, 6365
Olanzapine, 6959
Opipramol, 6985
Prothipendyl, 8073
Seroquel®, 8606
Tetrabenazine, 9325
Zotepine, 10327

Others
Buramate, 1525
Fluspirilene, 4241
Molindone, 6315
Penfluridol, 7213
Pimozide, 7589
Ziprasidone, 10304

ANTIPYRETIC

Acetaminophen, 45
Acetaminosalol, 46
Acetanilide, 47
Alclofenac, 220
Aluminum Bis(acetylsalicylate), 338
Aminochlorthenoxazin, 455
Aminopyrine, 495
Aspirin, 886
Benorylate, 1074
Benzydamine, 1157
Berberine, 1192
Bermoprofen, 1199
p-Bromoacetanilide, 1420

Bufexamac, 1497
Bumadizon, 1507
Calcium Acetylsalicylate, 1684
Chlorthenoxazin(e), 2247
Choline Salicylate, 2265
Clidanac, 2911
Dihydroxyaluminum Acetyl-salicylate, 3226
Dipyrocetyl, 3413
Dipyrone, 3414
Epirizole, 3659
Etersalate, 3759
Imidazole Salicylate, 4949
Indomethacin, 4998
Isofezolac, 5188
p-Lactophenetide, 5355
Lysine Acetylsalicylate, 5668
Magnesium Acetylsalicylate, 5689
Meclofenamic Acid, 5819
Morazone, 6349
Morpholine Salicylate see 6362
Naproxen, 6504
Nifenazone, 6619
5'-Nitro-2'-propoxyacetanilide, 6726
Phenacetin, 7344
Phenicarbazide, 7378
Phenocoll, 7388
Phenopyrazone, 7401
Phenyl Acetylsalicylate, 7424
Phenyl Salicylate, 7464
Pipebuzone, 7610
Propacetamol, 7976
Propyphenazone, 8056
Ramifenazone, 8282
Salacetamide, 8471
Salicylamide O-Acetic Acid, 8481
Sodium Salicylate, 8819
Tetrandrine, 9369
Tinoridine, 9590

ANTIRHEUMATIC see *Antiarthritic/ Antirheumatic*

ANTIRICKETTSIAL

p-Aminobenzoic Acid, 443
Chloramphenicol, 2120
Tetracycline, 9337

ANTISEBORRHEIC

Chloroxine, 2227
3-O-Lauroylpyridoxol Diacetate, 5401
Piroctone, 7657
Pyrithione, 8178
Resorcinol, 8323
Selenium Sulfides, 8580
Tioxolone, 9600

ANTISEPTIC/DISINFECTANT

Alcohols
Dichlorobenzyl Alcohol, 3110
Ethyl alcohol, 3806
Isopropyl Alcohol, 5227

Aldehydes
Formaldehyde Solution see 4262
Glutaraldehyde, 4480

Dyes
Acriflavine, 125
Aminacrine, 427
Brilliant Green, 1398
Ethacridine, 3760

ANTISEPTIC/DISINFECTANT

Dyes (*continued*)
Gentian Violet, *4401*
Magenta I, *5686*
Methyl Blue, *6107*

Guanidines
Alexidine, *231*
Ambazone, *395*
Chlorhexidine, *2140*
Picloxydine, *7553*

**Halogens/Halogen Containing
Compounds**
Bismuth Iodide Oxide, *1309*
Bismuth Iodosubgallate, *1310*
Bismuth Tribromophenate, *1333*
Bornyl Chloride, *1369*
Calcium Iodate, *1719*
Chlorinated Lime, *2143*
Cloflucarban, *2438*
Fluorosalan, *4216*
Iodic Acid, *5032*
Iodine, *5034*
Iodine Monochloride, *5038*
Iodine Trichloride, *5041*
Iodoform, *5054*
Methenamine Tetraiodine, *6043*
Oxychlorosene, *7090*
Povidone-Iodine, *7880*
Sodium Hypochlorite *see 8773 and
8774*
Sodium Iodate, *8776*
Symclosene, *9188*
Triclocarban, *9786*
Triclosan, *9790*
Troclosene Potassium, *9896*

Mercurial compounds
Hydrargaphen, *4805*
Meralein Sodium, *5912*
Merbromin, *5914*
Mercuric Chloride, *5926*
Mercuric Chloride, Ammoniated,
5927
Mercuric Sodium *p*-Phenol-
sulfonate, *5939*
Mercuric Succinimide, *5942*
Mercuric Sulfide, Red, *5945*
Mercurophen, *5947*
Mercurous Acetate, *5948*
Mercurous Chloride, *5951*
Mercurous Iodide, *5953*
Nitromersol, *6707*
Potassium Tetraiodo-
mercurate(II), *7862*
Potassium Triiodomercurate(II),
Solution, *7870*
Thimerfonate Sodium, *9450*
Thimerosal, *9451*

Nitrofurans
Furazolidone, *4320*
2-(Methoxymethyl)-5-nitrofuran,
6075
Nidroxyzone, *6615*
Nifuroxime, *6627*
Nifurzide, *6632*
Nitrofurazone, *6697*

Peroxides/Permanganates
Calcium Peroxide, *1737*
Hydrogen Peroxide, *4839*
Hydrogen Peroxide Solution 3%,
4840
Magnesium Peroxide, *5717*
Potassium Permanganate, *7824*
Strontium Peroxide, *9010*
Urea Hydrogen Peroxide, *10007*
Zinc Permanganate, *10281*

Phenols
Acetomeroctol, *63*
Bithionol, *1343*
Cadmium Salicylate, *1661*
Carvacrol, *1923*
Chloroxylenol, *2228*
Clorophene, *2469*
Creosote, *2642*
Cresol, *2645*
Fenticlor, *4046*
Hexachlorophene, *4716*
1-Naphthyl Salicylate, *6501*
2-Naphthyl Salicylate, *6502*
2,4,6-Tribromo-*m*-cresol, *9742*
3′,4′,5-Trichlorosalicylanilide,
9775

Quaternary Ammonium Compounds
Amantanium Bromide, *390*
Benzethonium Chloride, *1103*
Benzoxonium Chloride, *1144*
Bisdequalinium Chloride, *1288*
Cetalkonium Chloride, *2059*
Cethexonium Bromide, *2061*
Cetylpyridinium Chloride, *2074*
Dequalinium Chloride, *2959*
Dodecarbonium Chloride, *3466*
Domiphen Bromide, *3474*
Halimide, *4624*
Laurolinium Acetate, *5399*
Methylbenzethonium Chloride,
6103
Phenoctide, *7389*
Tibezonium Iodide, *9565*
Triclobisonium Chloride, *9785*

Quinolines
Aminoquinuride, *496*
Benzoxiquine, *1143*
Broxyquinoline, *1474*
Chloroxine, *2227*
Chlorquinaldol, *2243*
Cloxyquin, *2483*
Ethylhydrocupreine, *3855*
Euprocin, *3950*
Halquinol, *4637*
Hydrastine, *4806*
8-Hydroxyquinoline, *4890*
Iodochlorhydroxyquin, *5052*

Silver compounds
Silver Bromide, *8649*
Silver Fluoride, *8657*
Silver Lactate, *8660*
Silver Nitrate, *8661*
Silver Protein, *8671*

Others
Aluminum Acetate Solution, *332*
Aluminum Subacetate Solution,
380
Aluminum Sulfate, *381*
3-Amino-4-hydroxybutyric Acid,
463
Boric Acid, *1364*
Chlorhexidine, *2140*
Chloroazodin, *2171*
m-Cresyl Acetate, *2651*
Cupric Sulfate, *2722*
Dibromopropamidine, *3073*
Ichthammol, *4929*
Negatol, *6531*
Noxythiolin, *6822*
Octenidine, *6852*
Ornidazole, *7000*
β-Propiolactone, *8005*
α-Terpineol, *9316*

ANTISPASMODIC *see also Anticholin-
ergic*

Alibendol, *243*
Ambucetamide, *403*
Aminopromazine, *487*
Apoatropine, *780*
Bevonium Methyl Sulfate, *1239*
Bietamiverine, *1252*
Butaverine, *1544*
Butropium Bromide, *1569*
N-Butylscopolammonium Bromide,
1624
Caroverine, *1906*
Cimetropium Bromide, *2338*
Cinnamedrine, *2357*
Clebopride, *2404*
Cyclonium Iodide, *2806*
Difemerine, *3180*
Diisopromine, *3239*
Dioxaphetyl Butyrate, *3354*
Diponium Bromide, *3402*
Drofenine, *3501*
Emepronium Bromide, *3598*
Ethaverine, *3773*
Etomidoline, *3928*
Feclemine, *3986*
Fenalamide, *3995*
Fenoverine, *4023*
Fenpiprane, *4031*
Fenpiverinium Bromide, *4032*
Fentonium Bromide, *4048*
Flavoxate, *4135*
Flopropione, *4142*
Gluconic Acid, *4464*
Hydramitrazine, *4803*
Hymecromone, *4903*
Leiopyrrole, *5457*
Mebeverine, *5808*
Moxaverine, *6371*
Nafiverine, *6439*
Octamylamine, *6846*
Octaverine, *6851*
Pentapiperide, *7260*
Phenamacide Hydrochloride,
7350
Phloroglucinol, *7482*
Pinaverium Bromide, *7595*
Piperilate, *7624*
Pipoxolan Hydrochloride, *7637*
Pramiverin, *7887*
Prifinium Bromide, *7923*
Propyromazine, *8057*
Proxazole, *8090*
Prozapine, *8093*
Racefemine, *8273*
Rociverine, *8406*
Sintropium Bromide, *8693*
Spasmolytol, *8888*
Sultroponium, *9170*
Tiemonium Iodide, *9571*
Tigloidine, *9574*
Tiquizium Bromide, *9602*
Tiropramide, *9607*
Trepibutone, *9718*
Tricromyl, *9791*
Trifolium, *9818*
Trimebutine, *9829*
N,N-1-Trimethyl-3,3-diphenyl-
propylamine, *9842*
Tropenzile, *9908*
Trospium Chloride, *9918*
Xenytropium Bromide, *10208*

ANTISYPHILITIC *see also Antibacterial*

Acetarsone, *49*
Arsacetin, *829*
Arsphenamine, *851*
Bismuth Butylthiolaurate, *1301*
Bismuth Chloride Oxide, *1303*

Bismuth Ethyl Camphorate, *1304*
Bismuth Potassium Tartrate, *1318*
Bismuth Sodium Iodide, *1320*
Bismuth Sodium Tartrate, *1321*
d-Camphocarboxylic Acid, *1778*
Dichlorophenarsine Hydrochloride, *3121*
Ethanearsonic Acid, *3768*
Mercuric Benzoate, *5924*
Mercurous Chloride, *5951*
Sodium Arsanilate, *8719*
Sodium Arsphenamine, *8722*
Sozoiodole-mercury *see 8881*
Sulfarsphenamine, *9111*

ANTITHROMBOCYTHEMIC

Anagrelide, *665*

ANTITHROMBOTIC *see also Anticoagulant; Thrombolytic*

Argatroban, *816*
Cilostazol, *2335*
Clopidogrel, *2457*
Cloricromen, *2467*
Dalteparin, *2870*
Daltroban, *2871*
Defibrotide, *2915*
Enoxaparin, *3626*
Indobufen, *4991*
Iloprost, *4940*
Integrelin, *5013*
Isbogrel, *5120*
Lamifiban, *5362*
Lamoparan, *5366*
Nadroparin, *6434*
Ozagrel, *7115*
Picotamide, *7560*
Plafibride, *7673*
Reviparin Sodium, *8336*
Ridogrel, *8379*
Sulfinpyrazone, *9121*
Taprostene, *9227*
Ticlopidine, *9569*
Tinzaparin, *9592*
Tirofiban, *9605*
Triflusal, *9817*

ANTITUBERCULAR *see Antibacterial (Tuberculostatic)*

ANTITUMOR *see Antineoplastic*

ANTITUSSIVE

Alloclamide, *263*
Amicibone, *416*
Benproperine, *1078*
Benzonatate, *1127*
Bibenzonium Bromide, *1244*
Bromoform, *1441*
Butamirate, *1539*
Butethamate, *1551*
Caramiphen Ethanedisulfonate *see 1820*
Carbetapentane, *1840*
Chlophedianol, *2106*
Clobutinol, *2425*
Cloperastine, *2456*
Codeine, *2525*
Codeine Methyl Bromide, *2526*
Codeine N-Oxide, *2527*
Codeine Phosphate, *2528*
Codeine Sulfate, *2529*
Cyclexanone, *2775*
Dextromethorphan *see 8274*

Dihydrocodeine, *3214*
Dihydrocodeinone Enol Acetate, *3215*
Dimemorfan, *3251*
Dimethoxanate, *3272*
Dropropizine, *3507*
Drotebanol, *3511*
Eprazinone, *3666*
Ethyl Dibunate, *3834*
Ethylmorphine, *3876*
Fominoben, *4258*
Guaiapate, *4580*
Hydrocodone, *4826*
Isoaminile, *5124*
Levopropoxyphene, *5495*
Morclofone, *6350*
Narceine, *6508*
Normethadone, *6804*
Noscapine, *6815*
Oxeladin, *7064*
Oxolamine, *7078*
Pholcodine, *7484*
Picoperine, *7558*
Pipazethate, *7609*
Piperidione, *7622*
Prenoxdiazine Hydrochloride, *7918*
Racemethorphan, *8274*
Sodium Dibunate, *8753*
Tipepidine, *9601*
Zipeprol, *10303*

ANTIULCERATIVE *see also Antacid*

Aceglutamide Aluminum Complex *see 25*
ε-Acetamidocaproic Acid Zinc Salt *see 43*
Acetoxolone, *76*
Aldioxa, *224*
Arbaprostil, *808*
Benexate Hydrochloride, *1065*
Carbenoxolone, *1839*
Cetraxate, *2067*
Cimetidine, *2337*
Colloidal Bismuth Subcitrate, *2549*
Ebrotidine, *3536*
Ecabet, *3538*
Enprostil, *3629*
Esaprazole, *3736*
Famotidine, *3972*
Gefarnate, *4385*
Guaiazulene, *4581*
Irsogladine, *5113*
Lansoprazole, *5373*
Misoprostol, *6297*
Nizatidine, *6758*
Omeprazole, *6977*
Ornoprostil, *7003*
γ-Oryzanol, *7016*
Pantoprazole, *7146*
Pifarnine, *7575*
Pirenzepine, *7646*
Plaunotol, *7692*
Polaprezinc, *7712*
Rabeprazole, *8272*
Ranitidine, *8286*
Rebamipide, *8296*
Rioprostil, *8395*
Rosaprostol, *8419*
Rotraxate, *8428*
Roxatidine Acetate, *8431*
Sofalcone, *8850*
Spizofurone, *8918*
Sucralfate, *9049*
Telenzepine, *9270*
Teprenone, *9296*
Trimoprostil, *9853*
Trithiozine, *9889*
Troxipide, *9921*
Zolimidine, *10320*

ANTIUROLITHIC

Acetohydroxamic Acid, *60*
Allopurinol, *287*
Potassium Citrate, *7785*
Succinimide, *9040*

ANTIVENIN

Lyovac, Antivenin, *5660*

ANTIVERTIGO *see Antiemetic*

ANTIVIRAL

Purines/Pyrimidinones
Acyclovir, *148*
Cidofovir, *2329*
Cytarabine, *2853*
Dideoxyadenosine, *3151*
Didanosine, *3148*
Edoxudine, *3561*
Famciclovir, *3971*
Floxuridine, *4148*
Ganciclovir, *4374*
Idoxuridine, *4934*
Inosine Pranobex, *5006*
Lamivudine, *5365*
MADU, *5681*
Penciclovir, *7210*
Sorivudine, *8875*
Stavudine, *8958*
Trifluridine, *9816*
Valacyclovir, *10039*
Vidarabine, *10113*
Zalcitabine, *10242*
Zidovudine, *10252*

Others
Acemannan, *26*
Acetylleucine Monoethanolamine, *96*
Amantadine, *389*
Amidinomycin, *419*
Delavirdine, *2929*
Foscarnet Sodium, *4277*
Indinavir, *4979*
Interferon-α, *5016*
Interferon-β, *5017*
Interferon-γ, *5018*
Kethoxal, *5310*
Lysozyme, *5671*
Methisazone, *6057*
Moroxydine, *6354*
Nevirapine, *6573*
Podophyllotoxin, *7704*
Ribavirin, *8365*
Rimantadine, *8390*
Ritonavir, *8402*
Saquinavir, *8516*
Stallimycin, *8927*
Statolon, *8957*
Tromantadine, *9901*
Xenazoic Acid, *10204*

ANXIOLYTIC

Arylpiperazines
Buspirone, *1528*
Enciprazine, *3610*
Flesinoxan, *4138*
Ipsapirone, *5096*
Lesopitron, *5470*
Tandospirone, *9219*

Benzodiazepine Derivatives
Alprazolam, *320*
Bromazepam, *1406*
Camazepam, *1773*

ANXIOLYTIC

Benzodiazepine Derivatives (*continued*)
Chlordiazepoxide, 2132
Clobazam, 2417
Clorazepate, 2465
Clotiazepam, 2477
Cloxazolam, 2481
Diazepam, 3042
Ethyl Loflazepate, 3867
Etizolam, 3919
Fludiazepam, 4164
Flutazolam, 4243
Flutoprazepam, 4245
Halazepam, 4619
Ketazolam, 5308
Lorazepam, 5609
Loxapine, 5617
Medazepam, 5829
Metaclazepam, 5980
Mexazolam, 6254
Nordazepam, 6784
Oxazepam, 7059
Oxazolam, 7061
Pinazepam, 7596
Prazepam, 7895
Tofisopam, 9642

Carbamates
Cyclarbamate, 2772
Emylcamate, 3604
Hydroxyphenamate, 4884
Meprobamate, 5908
Phenprobamate, 7412
Tybamate, 9960

Others
Abecarnil, 2
Alpidem, 318
Benzoctamine, 1117
Captodiamine, 1816
Chlormezanone, 2155
Etifoxine, 3910
Fluoresone, 4197
Glutamic Acid, 4477
Hydroxyzine, 4897
Mecloralurea, 5822
Mephenoxalone, 5896
Oxanamide, 7053
Pazinaclone, 7187
Suriclone, 9182

AROMATASE INHIBITORS *see also Antineoplastic (Hormonal)*

Aminoglutethimide, 460
Anastrozole, 667
Fadrozole, 3969
Formestane, 4267
Letrozole, 5474

ASTRINGENT

Alkannin, 253
Aluminum Acetate Solution, 332
Aluminum Acetotartrate, 333
Aluminum Ammonium Sulfate, 335
Aluminum Chlorate, 347
Aluminum Chloride, 348
Aluminum Hydroxychloride, 356
Aluminum β-Naphtholdisulfonate, 363
Aluminum Potassium Sulfate, 373
Aluminum Sodium Sulfate, 378
Aluminum Subacetate Solution, 380
Ammonium Ferric Sulfate, 549
Baicalein, 971
Bismuth Oxide, 1314
Bismuth Subgallate, 1325

Bismuth Tannate, 1330
Boric Acid, 1364
Calcium Hydroxide, 1716
Cupric Citrate, 2702
Dichloroacetic Acid, 3100
Ferric Chloride, 4061
Formic Acid, 4268
Gallic Acid, 4363
Iodic Acid, 5032
Lead Acetate, 5411
Methionic Acid, 6052
Silver Bromide, 8649
Silver Lactate, 8660
Sodium Formate, 8765
Tannic Acid, 9221
Tannoform, 9222
Zinc Acetate, 10256
Zinc Carbonate, 10260
Zinc Chloride, 10261
Zinc Iodide, 10269
Zinc Oxide, 10279
Zinc Permanganate, 10281
Zinc Peroxide, 10282
Zinc *p*-Phenolsulfonate, 10283
Zinc Salicylate, 10288
Zinc Sulfate, 10293
Zinc Tannate, 10295

BENZODIAZEPINE ANTAGONIST

Flumazenil, 4169

BETA-BLOCKER *see β-Adrenergic Blocker*

BONE RESORPTION INHIBITOR *see also Calcium Regulator*

Alendronic Acid, 228
Clodronic Acid, 2432
Etidronic Acid, 3908
Pamidronic Acid, 7127
Risedronic Acid, 8396
Tiludronic Acid, 9582

BRADYCARDIC AGENT *see also Antianginal*

Zatebradine, 10245

BRADYKININ ANTAGONIST

Bradycor, 1385

BRONCHODILATOR *see also Antiasthmatic; Glucocorticoid*

Ephedrine Derivatives
Albuterol, 217
Bambuterol, 980
Bitolterol, 1344
Carbuterol, 1882
Clenbuterol, 2407
Clorprenaline, 2470
Dioxethedrine, 3356
Ephedrine, 3645
Epinephrine, 3656
Eprozinol, 3670
Etafedrine, 3752
Ethylnorepinephrine, 3880
Fenoterol, 4022
Formoterol, 4272
Hexoprenaline, 4745
Isoetharine, 5185
Isoproterenol, 5236
Mabuterol, 5674
Metaproterenol, 5992
N-Methylephedrine, 6145

Pirbuterol, 7644
Procaterol, 7939
Protokylol, 8082
Reproterol, 8307
Rimiterol, 8393
Salmeterol, 8489
Soterenol, 8877
Terbutaline, 9302
Tulobuterol, 9942

Quaternary Ammonium Compounds
Bevonium Methyl Sulfate, 1239
Flutropium Bromide, 4248
Ipratropium Bromide, 5089
Oxitropium Bromide, 7077
Tiotropium Bromide, 9598

Xanthine Derivatives
Acefylline, 22
Acefylline Piperazine, 23
Ambuphylline, 404
Aminophylline, 485
Bamifylline, 982
Choline Theophyllinate, 2266
Doxofylline, 3494
Dyphylline, 3529
Etamiphyllin, 3754
Etofylline, 3924
Guaithylline *see* 4582
Proxyphylline, 8092
Theobromine, 9418
1-Theobromineacetic Acid, 9419
Theophylline, 9421

Others
Fenspiride, 4041
Medibazine, 5831
Methoxyphenamine, 6077
Tretoquinol, 9719

CALCIUM CHANNEL BLOCKER *see also Antianginal; Antihypertensive; Vasodilator (Coronary)*

Arylalkylamines
Bepridil, 1188
Clentiazem, 2408
Diltiazem, 3247
Fendiline, 4011
Gallopamil, 4369
Mibefradil, 6261
Prenylamine, 7919
Semotiadil, 8590
Terodiline, 9311
Verapamil, 10083

Dihydropyridine Derivatives
Amlodipine, 516
Aranidipine, 806
Barnidipine, 1031
Benidipine, 1071
Cilnidipine, 2334
Efonidipine, 3566
Elgodipine, 3587
Felodipine, 3991
Isradipine, 5260
Lacidipine, 5344
Lercanidipine, 5469
Manidipine, 5786
Nicardipine, 6579
Nifedipine, 6637
Nilvadipine, 6637
Nimodipine, 6643
Nisoldipine, 6658
Nitrendipine, 6669

Piperazine Derivatives
Cinnarizine, 2365
Flunarizine, 4179
Lidoflazine, 5507
Lomerizine, 5593

Others

Bencyclane, *1060*
Etafenone, *3753*
Fantofarone, *3975*
Perhexiline, *7305*

CALCIUM REGULATOR

Calcifediol, *1677*
Calcitonin, *1680*
Calcitriol, *1681*
Dihydrotachysterol, *3223*
Elcatonin, *3578*
Ipriflavone, *5090*
Parathyroid Hormone, *7168*
Teriparatide Acetate, *9309*

CALCIUM SUPPLEMENT *see Replenishers/Supplements*

CANCER CHEMOTHERAPY *see Antineoplastic*

CAPILLARY PROTECTANT *see Vasoprotectant*

CARBONIC ANHYDRASE INHIBITOR *see also Antiglaucoma; Diuretic*

Acetazolamide, *50*
Butazolamide, *1545*
Dichlorphenamide, *3127*
Dorzolamide, *3481*
Ethoxzolamide, *3801*
Flumethiazide, *4174*
Methazolamide, *6031*

CARDIAC DEPRESSANT (ANTIARRHYTHMIC) *see Antiarrhythmic*

CARDIOTONIC

Acefylline, *22*
Acetyldigitoxins, *90*
2-Amino-4-picoline, *486*
Amrinone, *634*
Benfurodil Hemisuccinate, *1070*
Bucladesine, *1782*
Camphotamide, *1782*
Convallatoxin, *2575*
Cymarin, *2830*
Denopamine, *2943*
Deslanoside, *2967*
Digitalin, *3200*
Digitalis, *3201*
Digitoxin, *3206*
Digoxin, *3210*
Dobutamine, *3456*
Docarpamine, *3457*
Dopamine, *3479*
Dopexamine, *3482*
Enoximone, *3627*
Erythrophleine, *3728*
Fenalcomine, *3996*
Gitalin, *4437*
Gitoxin, *4441*
Glycocyamine, *4505*
Heptaminol, *4691*
Hydrastinine, *4807*
Ibopamine, *4921*
Lanatosides, *5368*
Loprinone, *5605*
Milrinone, *6284*
Neriifolin, *6559*
Oleandrin, *6963*

Ouabain, *7031*
Oxyfedrine, *7096*
Pimobendan, *7588*
Prenalterol, *7917*
Proscillaridin, *8060*
Resibufogenin, *8315*
Scillaren, *8543*
Scillarenin, *8544*
Strophanthin, *9016*
Sulmazole, *9159*
Theobromine, *9418*
Vesnarinone, *10105*
Xamoterol, *10189*

CATHARTIC *see Laxative/Cathartic*

CATION-EXCHANGE RESIN *see Ion-exchange Resin*

CCK ANTAGONIST

Loxiglumide, *5618*
Proglumide, *7958*

CENTRAL STIMULANT *see CNS Stimulant*

CEREBRAL VASODILATOR *see Vasodilator (Cerebral)*

CHELATING AGENT

Deferoxamine, *2914*
Ditiocarb Sodium, *3443*
Edetate Calcium Disodium, *3555*
Edetate Disodium, *3556*
Edetate Sodium, *3557*
Edetate Trisodium, *3558*
Penicillamine, *7214*
Pentetate Calcium Trisodium, *7265*
Pentetic Acid, *7266*
Succimer, *9034*
Trientine, *9796*

CHOLECYSTOKININ ANTAGONIST *see CCK Antagonist*

CHOLELITHOLYTIC AGENT

Chenodiol, *2096*
Methyl *tert*-Butyl Ether, *6111*
Monoctanoin, *6335*
Ursodiol, *10026*

CHOLERETIC

Alibendol, *243*
Anethole Trithion, *683*
Azintamide, *945*
Cholic Acid, *2258*
Cicrotoic Acid, *2327*
Clanobutin, *2398*
Cyclobutyrol, *2784*
Cyclovalone, *2823*
Cynarin(e), *2835*
Dehydrocholic Acid, *2922*
Deoxycholic Acid, *2946*
Dimecrotic Acid, *3248*
α-Ethylbenzyl Alcohol, *3816*
Exiproben, *3959*
Febuprol, *3985*
Fencibutirol, *4009*
Fenipentol, *4016*

Florantyrone, *4143*
Hymecromone, *4903*
Menbutone, *5878*
3-(o-Methoxyphenyl)-2-phenyl-acrylic Acid, *5880*
Metochalcone, *6225*
Moquizone, *6347*
Osalmid, *7018*
Ox Bile Extract, *7062*
4,4'-Oxydi-2-butanol, *7094*
Piprozolin, *7639*
4-Salicyloylmorpholine, *8485*
Sincalide, *8689*
Taurocholic Acid, *9242*
Tocamphyl, *9630*
Trepibutone, *9718*
Vanitiolide, *10071*

CHOLINERGIC

Aceclidine *see 8266*
Acetylcholine Bromide, *87*
Acetylcholine Chloride, *88*
Aclatonium Napadisilate, *116*
Benzpyrinium Bromide, *1153*
Bethanechol Chloride, *1232*
Carbachol, *1823*
Carpronium Chloride, *1913*
Demecarium Bromide, *2936*
Dexpanthenol, *2988*
Diisopropyl Paraoxon, *3242*
Echothiophate Iodide, *3549*
Edrophonium Chloride, *3562*
Eptastigmine, *3672*
Eseridine, *3740*
Furtrethonium, *4334*
Isoflurophate, *5192*
Methacholine Chloride, *6003*
Muscarine, *6389*
Neostigmine, *6553*
Oxapropanium Iodide, *7056*
Physostigmine, *7540*
Pyridostigmine Bromide, *8161*
Xanomeline, *10190*

CHOLINESTERASE INHIBITOR

Ambenonium Chloride, *396*
Distigmine Bromide, *3426*
Eptastigmine, *3672*
Galanthamine, *4357*

CHOLINESTERASE REACTIVATOR

Asoxime Chloride, *870*
Obidoxime Chloride, *6835*
Pralidoxime Chloride, *7884*

CNS STIMULANT

Amineptine, *429*
Amphetamine, *623*
Amphetaminil, *624*
Bemegride, *1054*
Benzphetamine, *1151*
Brucine, *1476*
Caffeine, *1674*
Chlorphentermine, *2235*
Clortermine, *2472*
Coca, *2515*
Deanol, *2900*
Demanyl Phosphate, *2935*
Dexoxadrol *see 3352*
Dextroamphetamine Sulfate, *2996*
Diethylpropion, *3175*
N-Ethylamphetamine, *3809*
Ethamivan, *3765*
Etifelmin, *3919*
Etryptamine, *3937*
Fencamfamine, *4006*

CNS STIMULANT (*continued*)

Fenethylline, *4014*
Fenozolone, *4028*
Flurothyl, *4236*
Hexacyclonate Sodium, *4717*
Homocamfin, *4768*
Mazindol, *5801*
Mefexamide, *5844*
Methamphetamine, *6015*
Methylphenidate, *6186*
Modafinil, *6311*
Nikethamide, *6635*
Pemoline, *7206*
Pentylenetetrazole, *7283*
Phendimetrazine, *7365*
Phenmetrazine, *7385*
Phentermine, *7415*
Picrotoxin, *7570*
Pipradrol, *7638*
Prolintane, *7964*
Pyrovalerone, *8194*

COGNITION ACTIVATOR *see
Nootropic*

CONTRACEPTIVE (INJECTABLE)

Medroxyprogesterone, *5838*
Norethindrone, *6790*

CONTRACEPTIVE (ORAL)

Desogestrel, *2971*
Ethinyl Estradiol, *3780*
Ethynodiol, *3905*
Gestodene, *4421*
Lynestrenol, *5659*
Mestranol, *5976*
Norethindrone, *6790*
Norethynodrel, *6791*
Norgestimate, *6796*
Norgestrel, *6797*

**CONTROL OF INTRAOCULAR
PRESSURE** *see also* **Antiglaucoma**

Apraclonidine, *791*

CONVERTING ENZYME INHIBITOR
see ACE-Inhibitor

CORONARY VASODILATOR *see
Vasodilator (Coronary)*

CYTOPROTECTANT (GASTRIC) *see
also* **Antiulcerative**

Aceglutamide Aluminum Complex
see 25
Acetoxolone, *76*
Benexate Hydrochloride, *4065*
Carbenoxolone, *1839*
Cetraxate, *2067*
Guaiazulene, *4581*
Irsogladine, *5113*
Plaunotol, *7692*
Polaprezinc, *7712*
Rebamipide, *8296*
Sofalcone, *8850*
Spizofurone, *8918*
Sucralfate, *9049*
Teprenone, *9296*
Troxipide, *9921*
Zolimidine, *10320*

DEBRIDING AGENT

Collagenase, *2544*
Deoxyribonuclease I, *2953*
Papain, *7148*

DECONGESTANT

Amidephrine, *418*
Cafaminol, *1671*
Cyclopentamine, *2808*
Ephedrine, *3645*
Epinephrine, *3656*
Fenoxazoline, *4025*
Indanazoline, *4967*
Metizoline, *6223*
Naphazoline, *6455*
Nordefrin Hydrochloride, *6785*
Octodrine, *6854*
Oxymetazoline, *7100*
Phenylephrine Hydrochloride, *7440*
Phenylpropanolamine Hydro-
chloride, *7461*
Phenylpropylmethylamine, *7462*
Propylhexedrine, *8045*
Pseudoephedrine *see 3641*
Tetrahydrozoline, *9358*
Tramazoline, *9702*
Tuaminoheptane, *9934*
Tymazoline, *9965*
Xylometazoline, *10219*

DEPIGMENTOR

Hydroquinine, *4852*
Hydroquinone, *4853*
Monobenzone, *6331*

**DERMATITIS HERPETIFORMIS
SUPPRESSANT**

Dapsone, *2885*
Sulfapyridine, *9108*

DIAGNOSTIC AID

Alsactide, *322*
Americium, *410*
p-Aminohippuric Acid, *462*
Anazolene Sodium, *669*
Arbutamine, *811*
Arginine, *817*
Bentiromide, *1081*
Betazole, *1230*
Ceruletide, *2048*
Colfosceril Palmitate, *2540*
Congo Red, *2562*
Dexamethasone, *2986*
Edrophonium Chloride, *3562*
Evan's Blue, *3952*
Fluorescein, *4194*
Galactose, *4353*
Glycerol, *4493*
Histamine, *4756*
Indocyanine Green, *4992*
Inulin, *5024*
Iodinated Serum Albumin *see 8613*
Isosulphan Blue *see 9161*
Mannitol, *5788*
Merisoprol Hg 197, *5959*
Methacholine Chloride, *6003*
Metyrapone, *6246*
Oleic Acid, *6965*
Penicilloyl Polylysine, *7233*
3-Pentadecylcatechol, *7244*
Pentagastrin, *7250*
Phenolsulfonphthalein, *7397*
Phenoltetrachlorophthalein, *7398*
Phentolamine, *7417*

Piperoxan, *7631*
Porfimer Sodium, *7755*
Rose Bengal, *8421*
Saralasin, *8518*
Sodium Benzoate, *8725*
Sodium Chromate(VI), Radio-
active, *see 8745*
Sodium Iodide, Radioactive, *8778*
Sulfobromophthalein Sodium, *9127*
Teriparatide Acetate, *9309*
Tolonium Chloride, *9658*
TSH, *9931*
Tuberculin, *9937*
Tubocurarine Chloride, *9939*
Vitamin B$_{12}$, Radioactive, *10153*
Xylose, *10220*

**DIAGNOSTIC AID (CONTRAST
AGENT)**

Gadodiamide, *4345*
Gadopentetic Acid, *4347*
Gadoteridol, *4348*
Perflubron, *7299*

**DIAGNOSTIC AID (RADIOACTIVE
IMAGING AGENT)**

Butedronic Acid Complex with
^{99m}Tc *see 1546*
Disofenin Complex with
^{99m}Tc *see 3422*
Exametazime Complex with
^{99m}Tc *see 3957*
Fludeoxyglucose F^{18}, *4163*
Iobenguane, *5027*
Iofetamine ^{123}I, *5066*
Lidofenin Complex with ^{99m}Tc *see
5506*
Medronic Acid Complex with
^{99m}Tc *see 5837*
Oxidronic Acid Complex with
^{99m}Tc *see 7074*
Pamidronic Acid Complex with
^{99m}Tc *see 7135*
Pentetreotide Chelate with ^{111}In *see
7267*
Satumomab Pendetide ^{111}In *see
8530*
Sodium Pertechnetate ^{99m}Tc, *8802*
Sodium Phosphate, Radioactive,
8807
Stannous Pyrophosphate Complex
with ^{99m}Tc *see 8945*
Succimer Complex with ^{99m}Tc *see
9034*
Technetium, *9256*
Technetium ^{99m}Tc Bicisate, *9257*
Technetium ^{99m}Tc Mertiatide, *9258*
Technetium ^{99m}Tc Sestamibi, *9259*
Technetium ^{99m}Tc Teboroxime,
9260
Tetrofosmin Complex with ^{99m}Tc
see 9383
^{133}Xenon *see 10206*

**DIAGNOSTIC AID (RADIOPAQUE
MEDIUM)**

Acetrizoate Sodium, *79*
Barium Sulfate, *1023*
Bunamiodyl Sodium, *1511*
Diatrizoate Sodium, *3040*
Ethiodized Oil, *3781*
Iobenzamic Acid, *5028*
Iocarmic Acid, *5029*
Iocetamic Acid, *5030*
Iodipamide, *5043*
Iodixanol, *5044*

Iodized Oil, *5045*
Iodoalphionic Acid, *5047*
Iodophthalein Sodium, *5059*
Iodopyracet, *5061*
Ioglycamic Acid, *5067*
Iohexol, *5068*
Iomeglamic Acid, *5069*
Iopamidol, *5071*
Iopanoic Acid, *5072*
Iopentol, *5073*
Iophendylate, *5074*
Iophenoxic Acid, *5075*
Iopromide, *5076*
Iopronic Acid, *5077*
Iopydol, *5078*
Iopydone, *5079*
Iothalamic Acid, *5080*
Iotrolan, *5082*
Ioversol, *5083*
Ioxaglic Acid, *5084*
Ioxilan, *5085*
Ipodate, *5087*
Meglumine Acetrizoate, *5850*
Meglumine Diatrizoate, *5851*
Methiodal Sodium, *6051*
Metrizamide, *6240*
Metrizoic Acid, *6241*
Phenobutiodil, *7387*
Phentetiothalein Sodium, *7416*
Propyliodone, *8048*
Sodium Iodomethamate, *8779*
Sozoiodolic Acid, *8881*
3',3'',5',5''-Tetrabromophenol-
    phthalein, *9329*
Thorium Oxide, *9518*
Tyropanoate Sodium, *9968*

DIAGNOSTIC AID (ULTRASOUND CONTRAST AGENT)

Serum Albumin, *8613*

DIGESTIVE AID

Amylase, *640*
Lipase, *5536*
Pancreatin, *7137*
Pancrelipase, *7138*
Papain, *7148*
Pepsin, *7289*
Rennin, *8303*

DISINFECTANT *see Antiseptic/Disinfectant*

DIURETIC

Benzothiadiazine Derivatives *(Thiazides)*
Althiazide, *326*
Bendroflumethiazide, *1064*
Benzthiazide, *1155*
Benzylhydrochlorothiazide, *1171*
Buthiazide, *1554*
Chlorothiazide, *2221*
Chlorthalidone, *2246*
Cyclopenthiazide, *2813*
Cyclothiazide, *2822*
Epithiazide, *3661*
Ethiazide, *3778*
Fenquizone, *4039*
Indapamide, *4969*
Hydrochlorothiazide, *4822*
Hydroflumethiazide, *4830*
Methyclothiazide, *6086*
Meticrane, *6220*
Metolazone, *6231*
Paraflutizide, *7157*
Polythiazide, *7744*

Quinethazone, *8240*
Teclothiazide, *9261*
Trichlormethiazide, *9754*

Organomercurials
Chlormerodrin, *2154*
Meralluride, *5913*
Mercamphamide, *5915*
Mercaptomerin Sodium, *5918*
Mercumatilin Sodium *see 5921*
Mercurous Chloride, *5951*
Mersalyl, *5962*

Purines
Acefylline, *22*
7-Morpholinomethyltheophylline,
    6363
Pamabrom, *7133*
Protheobromine, *8072*
Theobromine, *9418*

Steroids
Canrenone, *1795*
Oleandrin, *6963*
Spironolactone, *8917*

Sulfonamide Derivatives
Acetazolamide, *50*
Ambuside, *405*
Azosemide, *953*
Bumetanide, *1508*
Butazolamide, *1545*
Chloraminophenamide, *2119*
Clofenamide, *2434*
Clopamide, *2454*
Clorexolone, *2466*
Disulfamide, *3427*
Ethoxzolamide, *3801*
Furosemide, *4331*
Mefruside, *5847*
Methazolamide, *6031*
Piretanide, *7647*
Torsemide, *9690*
Tripamide, *9866*
Xipamide, *10212*

Uracils
Aminometradine, *471*
Amisometradine, *507*

Others
Amanozine, *388*
Amiloride, *426*
Arbutin, *812*
Chlorazanil, *2123*
Ethacrynic Acid, *3761*
Etozolin, *3934*
Hydracarbazine, *4798*
Isosorbide, *5244*
Mannitol, *5788*
Metochalcone, *6225*
Muzolimine, *6397*
Perhexiline, *7305*
Ticrynafen, *9570*
Triamterene, *9731*
Urea, *10005*

DOPAMINE RECEPTOR AGONIST
see also Antihypertensive; Antimigraine; Antiparkinsonian; Prolactin Inhibitor

Bromocriptine, *1437*
Cabergoline, *1637*
Carmoxirole, *1893*
Dopexamine, *3482*
Fenoldopam, *4020*
Ibopamine, *4921*
Lisuride, *5541*
Pergolide, *7304*

Pramipexole, *7885*
Quinagolide, *8226*
Ropinirole, *8416*
Roxindole, *8432*
Talipexole, *9209*

DOPAMINE RECEPTOR ANTAGONIST *see also Antiemetic; Antipsychotic*

Amisulpride, *508*
Clebopride, *2404*
Domperidone, *3476*
Metoclopramide, *6226*
Mosapramine, *6365*
Nemonapride, *6533*
Remoxipride, *8301*
Risperidone, *8397*
Sulpiride, *9163*
Sultopride, *9168*
Ziprasidone, *10304*

ECTOPARASITICIDE

Amitraz, *510*
Benzyl Benzoate, *1162*
Carbaryl, *1831*
Crotamiton, *2661*
DDT, *2898*
Dixanthogen, *3449*
Lime Sulfurated Solution, *5516*
Lindane, *5526*
Malathion, *5740*
Mercuric Oleate, *5934*
Mesulfen, *5977*
Sulfiram, *9123*
Sulfur, Pharmaceutical, *9150*

ELECTROLYTE REPLENISHER *see Replenishers/Supplements*

EMETIC

Apocodeine, *781*
Apomorphine, *787*
Cephaeline, *2020*
Ipecac, *5086*
Sodium Chloride, *8742*
Zinc Acetate, *10256*

ENKEPHALINASE INHIBITOR

Acetorphan, *72*

ENZYME

Digestive
Amylase, *640*
Lipase, *5536*
Pancrelipase, *7138*
Pepsin, *7289*
Rennin, *8303*

Mucolytic
Lysozyme, *5671*

Penicillin inactivating
Penicillinase, *7219*

Proteolytic
Collagenase, *2544*
Chymopapain, *2319*
Chymotrypsins, *2320*
Papain, *7148*
Trypsin, *9926*

ENZYME COFACTOR

Acetiamine, 51
Benfotiamine, 1068
Bisbentiamine, 1286
Cetotiamine, 2064
Cycothiamin(e), 2824
Dexpanthenol, 2988
Fursultiamine, 4333
Methylol Riboflavin, 6179
Niacinamide, 6574
Nicotinamide Ascorbate, 6610
Nicotinic Acid, 6612
Octotiamine, 6857
Pantothenic Acid, 7147
Prosultiamine, 8068
Pyridoxal 5-Phosphate, 8163
Pyridoxine Hydrochloride, 8166
Riboflavin, 8367
Riboflavine Monophosphate, 8368
Sapropterin, 8515
Thiamine, 9430
Thiamine Diphosphate, 9431
Thiamine Disulfide, 9432
Vintiamol, 10129

ENZYME INDUCER (HEPATIC)

Flumecinol, 4170

ESTROGEN

Nonsteroidal
Benzestrol, 1102
Broparoestrol, 1471
Chlorotrianisene, 2225
Dienestrol, 3153
Diethylstilbestrol, 3177
Dimestrol, 3257
Fosfestrol, 4279
Hexestrol, 4738
Methallenestril, 6012
Methestrol, 6045

Steroidal
Colpormon, 2554
Conjugated Estrogenic Hormones, 2571
Equilenin, 3675
Equilin, 3676
Estradiol, 3746
Estriol, 3750
Estrone, 3751
Ethinyl Estradiol, 3780
Mestranol, 5976
Moxestrol, 6372
Mytatrienediol, 6424
Quinestradiol, 8238
Quinestrol, 8239

ESTROGEN ANTAGONIST see Antiestrogen

EXPECTORANT

Ambroxol, 401
Ammonium Bicarbonate, 522
Ammonium Carbonate, 534
Bromhexine, 1412
Calcium Iodide, 1720
Carbocysteine, 1850
Guaiacol, 4575
Guaiacol Benzoate, 4576
Guaiacol Carbonate, 4577
Guaiacol Phosphate, 4578
Guaifenesin, 4582
Guaithylline see 4582
Hydriodic Acid, 4817

Iodinated Glycerol, 5033
Potassium Guaiacolsulfonate, 7798
Potassium Iodide, 7809
Sodium Citrate, 8746
Sodium Iodide, 8777
Storax, 8974
Terebene, 9304
Terpin, 9314
Trifolium, 9818

FIBRINOGEN RECEPTOR ANTAGONIST see also Antithrombotic

Lamifiban, 5362
Tirofiban, 9605

GASTRIC AND PANCREATIC SECRETION STIMULANT

Carnitine, 1898
Ceruletide, 2048
Secretin, 8564
Sincalide, 8689

GASTRIC PROTON PUMP INHIBITOR see also Antiulcerative

Lansoprazole, 5373
Omeprazole, 6977
Pantoprazole, 7146
Rabeprazole, 8272

GASTRIC SECRETION INHIBITOR

Enterogastrone, 3632
Octreotide, 6859
Telenzepine, 9270

GASTROPROKINETIC

Cinitapride, 2352
Cisapride, 2377
Fedotozine, 3987
Loxiglumide, 5618

GLUCOCORTICOID

21-Acetoxypregnenolone, 77
Alclometasone, 221
Algestone, 238
Amcinonide, 407
Beclomethasone, 1047
Betamethasone, 1226
Budesonide, 1490
Chloroprednisone, 2209
Clobetasol, 2423
Clobetasone, 2424
Clocortolone, 2430
Cloprednol, 2460
Corticosterone, 2600
Cortisone, 2602
Cortivazol, 2603
Deflazacort, 2916
Desonide, 2973
Desoximetasone, 2975
Dexamethasone, 2986
Diflorasone, 3186
Diflucortolone, 3189
Difluprednate, 3194
Enoxolone, 3628
Fluazacort, 4151
Flucloronide, 4157
Flumethasone, 4173
Flunisolide, 4180
Fluocinolone Acetonide, 4185

Fluocinonide, 4186
Fluocortin Butyl, 4187
Fluocortolone, 4188
Fluorometholone, 4213
Fluperolone Acetate, 4225
Fluprednidene Acetate, 4228
Fluprednisolone, 4229
Flurandrenolide, 4232
Fluticasone Propionate, 4244
Formocortal, 4270
Halcinonide, 4621
Halobetasol Propionate, 4625
Halometasone, 4628
Halopredone Acetate, 4630
Hydrocortamate, 4827
Hydrocortisone, 4828
Loteprednol Etabonate, 5614
Mazipredone, 5802
Medrysone, 5840
Meprednisone, 5907
Methylprednisolone, 6189
Mometasone Furoate, 6324
Paramethasone, 7162
Prednicarbate, 7899
Prednisolone, 7901
Prednisolone 25-Diethylamino-
   acetate, 7902
Prednisolone Sodium Phosphate, 7903
Prednisone, 7904
Prednival, 7905
Prednylidene, 7906
Rimexolone, 8392
Tixocortol, 9623
Triamcinolone, 9727
Triamcinolone Acetonide, 9728
Triamcinolone Benetonide, 9729
Triamcinolone Hexacetonide, 9730

α-GLUCOSIDASE INHIBITOR see also Antidiabetic

Acarbose, 15
Miglitol, 6274

GONAD-STIMULATING PRINCIPLE

Buserelin, 1527
Chorionic Gonadotropin, 2273
Clomiphene, 2446
Cyclofenil, 2789
Epimestrol, 3654
FSH, 4299
LH, 5499
LH-RH, 5500

GOUT SUPPRESSANT see Antigout

GROWTH HORMONE INHIBITOR

Octreotide, 6859
Somatostatin, 8863

GROWTH HORMONE RELEASING FACTOR

Sermorelin, 8605

GROWTH STIMULANT

Somatotropin, 8864

HEMATINIC

Ammonium Ferric Citrate, 547
Calcium Ferrous Citrate, 1708
Cobaltous Chloride, 2498

Dextran Iron Complex, *2991*
Ferric Albuminate, *4058*
Ferric and Ammonium Acetate
  Solution, *4059*
Ferric Citrate, *4063*
Ferric Fructose, *see 4295*
Ferriclate Calcium Sodium, *4070*
Ferric Oxide, Saccharated, *4073*
Ferric Pyrophosphate, *4075*
Ferric Sodium Edetate, *4076*
Ferritin, *4083*
Ferrocholinate, *4085*
Ferroglycine Sulfate, *4086*
Ferrous Carbonate Mass, *4089*
Ferrous Carbonate Saccharated,
  4090
Ferrous Citrate, *4092*
Ferrous Fumarate, *4094*
Ferrous Gluconate, *4095*
Ferrous Lactate, *4098*
Ferrous Succinate, *4104*
Ferrous Sulfate, *4105*
Iron Sorbitex, *5112*
Liver Extract, *5575*
Peptonized Iron, *7292*
Polyferose, *7731*

HEMATOPOIETIC

Erythropoietin, *3729*
Granulocyte Colony Stimulating
  Factor, *4558*
Granulocyte-Macrophage Colony
  Stimulating Factor, *4559*

HEMOLYTIC

Phenylhydrazine, *7447*

HEMOSTATIC

Adrenalone, *170*
Adrenochrome, *171*
Algin, *240*
Alginic Acid, *241*
ε-Aminocaproic Acid, *451*
Aminochromes, *456*
Batroxobin, *1038*
Carbazochrome Salicylate, *1832*
Carbazochrome Sodium Sulfonate,
  1833
Cephalins, *2022*
Cotarnine, *2618*
Ellagic Acid, *3588*
Ethamsylate, *3766*
Factor VIII, *3963*
Factor IX, *3964*
Factor XIII, *3968*
Fibrinogen, *4116*
1,2-Naphthoquinone, *6481*
1-Naphthylamine-4-sulfonic Acid,
  6490
Oxamarin, *7047*
Oxidized Cellulose, *7073*
Styptic Collodion *see 2547*
Sulmarin, *9158*
Thrombin, *9525*
Thromboplastin, *9527*
Tolonium Chloride, *9658*
Tranexamic Acid, *9704*
Vasopressin, *10073*

HEPARIN ANTAGONIST

Hexadimethrine Bromide, *4719*

HEPATOPROTECTANT

S-Adenosylmethionine, *155*
Betaine, *1225*
Catechin, *1950*
Citiolone, *2381*
Malotilate, *5751*
Methionine, *6053*
Orazamide, *6993*
Phosphorylcholine, *7519*
Protoporphyrin IX, *8084*
Silymarin-Group, *8680*
Thioctic Acid, *9462*
Timonacic, *9586*
Tiopronin, *9597*

HISTAMINE H₁-RECEPTOR ANTAG-ONIST *see Antihistaminic*

HISTAMINE H₂-RECEPTOR ANTAG-ONIST *see also Antiulcerative*

Cimetidine, *2337*
Ebrotidine, *3536*
Famotidine, *3972*
Nizatidine, *6758*
Ranitidine, *8286*
Roxatidine Acetate, *8431*

HIV PROTEINASE INHIBITOR *see also Antiviral*

Saquinavir, *8516*

HMG CoA REDUCTASE INHIBITOR *see also Antihyperlipoproteinemic*

Atorvastatin, *897*
Fluvastatin, *4250*
Lovastatin, *5616*
Pravastatin Sodium, *7894*
Simvastatin, *8686*

HYPNOTIC *see Sedative/Hypnotic*

HYPOCHOLESTEREMIC *see Antihyperlipoproteinemic*

HYPOLIPIDEMIC *see Antihyperlipo-proteinemic*

HYPOTENSIVE *see Antihypertensive*

IMMUNOMODULATOR

Acemannan, *26*
Amiprilose, *506*
Bucillamine, *1481*
Ditiocarb Sodium, *3443*
Imiquimod, *4957*
Inosine Pranobex, *5006*
Interferon-β, *5017*
Interferon-γ, *5018*
Interleukin-2, *5020*
Lentinan, *5462*
Levamisole, *5486*
Macrophage Colony Stimulating
  Factor, *5678*
Pidotimod, *7574*
Romurtide, *8412*
Platonin, *7690*
Procodazole, *7943*
Propagermanium, *7979*

Thymomodulin, *9543*
Thymopentin, *9544*
Ubenimex, *9973*

IMMUNOSUPPRESSANT

Azathioprine, *935*
Brequinar, *1394*
Cyclosporins, *2821*
Gusperimus, *4610*
6-Mercaptopurine, *5919*
Mizoribine, *6306*
Rapamycin, *8288*

INOTROPIC AGENT *see Cardiotonic*

INSULIN SENSITIZER *see also Anti-diabetic*

Pioglitazone, *7605*
Troglitazone, *9898*

ION EXCHANGE RESIN

Carbacrylic Resins, *1824*
Cholestyramine Resin, *2257*
Colestipol, *2538*
Polidexide, *7716*
Resodec, *8321*
Sodium Polystyrene Sulfonate,
  8815

KERATOLYTIC *see also Antiacne*

Benzoyl Peroxide, *1149*
Dichloroacetic Acid, *3100*
Resorcinol, *8323*
Retinoic Acid, *8333*
Salicylic Acid, *8484*
Tetroquinone, *9385*

LACTATION STIMULATING HOR-MONE

Prolactin, *7961*

LAXATIVE/CATHARTIC

Agar, *182*
Aloe-emodin, *313*
Aloin, *314*
Bisacodyl, *1282*
Bisoxatin Acetate, *1337*
Calcium Polycarbophil, *1744*
Calomelol, *1768*
Carlsbad Salt Artificial, *1890*
Casanthranol, *1929*
Castor Oil, *1946*
Cellulose Ethyl Hydroxyethyl
  Ether, *2014*
Colocynthin, *2551*
Danthron, *2878*
Docusate Calcium, *3459*
Docusate Sodium, *3460*
Emodin, *3602*
Frangulin, *4288*
Glucofrangulin, *4460*
Lactitol, *5352*
Lactulose, *5360*
Magnesium Carbonate Hydroxide,
  5696
Magnesium Chloride, *5698*
Magnesium Citrate, *5699*
Magnesium Hydroxide, *5706*
Magnesium Lactate, *5708*
Magnesium Phosphate, Dibasic,
  5718

LAXATIVE/CATHARTIC (continued)

Magnesium Sulfate, 5731
Mercurous Chloride, 5951
Mercury Mass, 5958
Oxyphenisatin Acetate, 7108
Petrolatum, Liquid, 7327
Phenolphthalein, 7392
Phenolphthalol, 7395
Phenoltetrachlorophthalein, 7398
Picosulfate Sodium, 7559
Poloxamers, 7722
Potassium Bisulfate, 7774
Potassium Bitartrate, 7776
Potassium Phosphate, Dibasic, 7828
Potassium Sodium Tartrate, 7840
Potassium Sulfate, 7845
Potassium Sulfite, 7847
Potassium Tartrate, 7849
Seidlitz Mixture, 8567
Senna, 8599
Sennosides, 8600
Sodium Phosphate, Dibasic, 8805
Sodium Succinate, 8828
Sodium Sulfate, 8829
Sodium Tartrate, 8833
Sorbitol, 8873
Sulisatin, 9156
Yellow Phenolphthalein, 10230

LEUKOTRIENE ANTAGONIST see also Antiasthmatic

Ibudilast, 4923
Montelukast, 6340
Pranlukast, 7889
Zafirlukast, 10241

LH-RH AGONIST see also Antineoplastic; Gonad-Stimulating Principle

Buserelin, 1527
Deslorelin, 2968
Goserelin, 4547
Histrelin, 4760
Leuprolide, 5484
Nafarelin, 6437
Triptorelin, 9878

LIPOTROPIC

N-Acetylmethionine, 97
Choline Chloride, 2261
Choline Dehydrocholate, 2262
Choline Dihydrogen Citrate, 2263
Inositol, 5008
Lecithin, 5452
Methionine, 6053

5-LIPOXYGENASE INHIBITOR see also Antiallergic; Antiasthmatic; Anti-inflammatory; Antipsoriatic

Amlexanox, 515
Lonapalene, 5595
Tenidap, 9290
Zileuton, 10253

LOCAL ANESTHETIC see Anesthetic (Local)

LUPUS ERYTHEMATOSUS SUPPRESSANT

Bismuth Sodium Triglycollamate, 1322
Bismuth Subsalicylate, 1327
Chloroquine, 2215
Hydroxychloroquine, 4863

MAJOR TRANQUILIZER see Antipsychotic

MATRIX METALLOPROTEINASE INHIBITOR

Batimastat, 1036

MINERALOCORTICOID

Aldosterone, 226
Deoxycorticosterone, 2947
Deoxycorticosterone Acetate, 2948
Fludrocortisone, 4166

MINOR TRANQUILIZER see Anxiolytic

MIOTIC

Carbachol, 1823
Neostigmine, 6553
Physostigmine, 7540
Pilocarpine, 7578

MONOAMINE OXIDASE INHIBITOR see also Antidepressant; Antihypertensive; Antiparkinsonian

Iproclozide, 5092
Iproniazid, 5094
Isocarboxazid, 5172
Lazabemide, 5407
Mofegiline, 6313
Moclobemide, 6309
Octamoxin, 6845
Pargyline, 7172
Phenelzine, 7366
Phenoxypropazine, 7411
Pivalylbenzhydrazine, 7668
Prodipine, 7949
Selegiline, 8569
Toloxatone, 9659
Tranylcypromine, 9708

MUCOLYTIC

Acetylcysteine, 89
Bromhexine, 1412
Carbocysteine, 1850
Domiodol, 3473
Erdosteine, 3684
Letosteine, 5473
Lysozyme, 5671
Mecysteine Hydrochloride, 5828
Mesna, 5969
Sobrerol, 8710
Stepronin, 8963
Tiopronin, 9597
Tyloxapol, 9964

MUSCLE RELAXANT (SKELETAL)

Afloqualone, 181
Alcuronium, 222
Atracurium Besylate, 900
Baclofen, 967
Benzoctamine, 1117
Benzoquinonium Chloride, 1137
C-Calebassine, 1762
Carisoprodol, 1889
Chlormezanone, 2155
Chlorphenesin Carbamate, 2231
Chlorproethazine, 2236
Chlorzoxazone, 2249
Curare, 2741

Cyclarbamate, 2772
Cyclobenzaprine, 2782
Dantrolene, 2879
Decamethonium Bromide, 2904
Diazepam, 3042
Doxacurium Chloride, 3487
Eperisone, 3642
Fazadinium Bromide, 3981
Flumetramide, 4176
Gallamine Triethiodide, 4361
Hexacarbacholine Bromide, 4713
Hexafluorenium Bromide, 4721
Idrocilamide, 4935
Inaperisone, 4963
Laudexium Methyl Sulfate, 5392
Leptodactyline, 5467
Memantine, 5872
Mephenesin, 5895
Mephenoxalone, 5896
Metaxalone, 5994
Methocarbamol, 6060
Metocurine Iodide, 6227
Mivacurium Chloride, 6305
Nimetazepam, 6641
Orphenadrine, 7007
Pancuronium Bromide, 7139
Phenprobamate, 7412
Phenyramidol, 7474
Pipecurium Bromide, 7612
Quinine, 8245
Rocuronium, 8407
Styramate, 9027
Succinylcholine Bromide, 9043
Succinylcholine Chloride, 9044
Succinylcholine Iodide, 9045
Suxethonium Bromide, 9184
Tetrazepam, 9479
Thiocolchicoside see 9460
Tizanidine, 9624
Tolperisone, 9660
Tubocurarine Chloride, 9939
Vecuronium Bromide, 10075
Zoxazolamine, 10328

MUSCLE RELAXANT (SMOOTH) see Anticholinergic; Antispasmodic; Bronchodilator; Vasodilator

MYDRIATIC

Atropine, 907
Cyclopentolate, 2815
Epinephrine, 3656
Hydroxyamphetamine, 4855
Phenylephrine Hydrochloride, 7440
Yohimbine, 10236

NARCOTIC ANALGESIC see Analgesic (Narcotic)

NARCOTIC ANTAGONIST

Amiphenazole, 505
Cyclazocine, 2773
Levallorphan, 5485
Nalmefene, 6447
Nalorphine, 6448
Naloxone, 6449
Naltrexone, 6450

NASAL DECONGESTANT see Decongestant

NEUROLEPTIC see Antipsychotic

NEUROMUSCULAR BLOCKING AGENT see **Muscle Relaxant (Skeletal)**

NEUROPROTECTIVE

Riluzole, *8389*

NOOTROPIC

Aceglutamide, *25*
Acetylcarnitine, *85*
Aniracetam, *700*
Besipirdine, *1223*
Bifemelane, *1255*
Choline Alfoscerate, *2260*
E-2020, *3533*
Exifone, *3958*
Fipexide, *4126*
Idebenone, *4932*
Indeloxazine Hydrochloride, *4972*
Nebracetam, *6520*
Nefiracetam, *6528*
Nizofenone, *6759*
Oxiracetam, *7076*
Piracetam, *7641*
Pramiracetam, *7886*
Propentofylline, *7997*
Pyritinol, *8180*
Sabeluzole, *8462*
Tacrine, *9199*
Velnacrine, *10077*
Vinconate, *10123*
Xanomeline, *10190*

NSAID see **Anti-inflammatory (Nonsteroidal)**

OPIOID ANALGESIC see **Analgesic (Narcotic)**

ORAL CONTRACEPTIVE see **Contraceptive (Oral)**

OVARIAN HORMONE

Relaxin, *8299*

OXYTOCIC

Carboprost, *1871*
Cargutocin, *1887*
Deaminooxytocin, *2899*
Ergonovine, *3694*
Gemeprost, *4393*
Methylergonovine, *6147*
Oxytocin, *7114*
Pituitary, Posterior, *7666*
Prostaglandin E$_2$, *8064*
Prostaglandin F$_{2a}$, *8065*
Sparteine, *8887*

PARASYMPATHOMIMETIC see **Cholinergic**

PEDICULICIDE see **Ectoparasiticide**

PEPSIN INHIBITOR

Sodium Amylosulfate, *8718*

PERIPHERAL VASODILATOR see **Vasodilator (Peripheral)**

PERISTALTIC STIMULANT see **Gastroprokinetic**

PIGMENTATION AGENT

Methoxsalen, *6068*
Trioxsalen, *9864*

PLASMA VOLUME EXPANDER

Serum Albumin, *8613*
Dextran, *2989*
Hetastarch, *4707*

POTASSIUM CHANNEL ACTIVATOR/OPENER see also **Antihypertensive; Antianginal**

Levcromakalim, *5487*
Nicorandil, *6608*
Pinacidil, *7592*

PRESSOR AGENT see **Antihypotensive**

PROGESTOGEN

Allylestrenol, *299*
Anagestone, *664*
Chlormadinone Acetate, *2152*
Delmadinone Acetate, *2930*
Demegestone, *2939*
Desogestrel, *2971*
Dimethisterone, *3268*
Drospirenone, *3510*
Dydrogesterone, *3524*
Ethisterone, *3787*
Ethynodiol, *3905*
Flurogestone Acetate, *4235*
Gestodene, *4421*
Gestonorone Caproate, *4422*
17-Hydroxy-16-methylene-Δ^6-progesterone, *4878*
17α-Hydroxyprogesterone, *4886*
Lynestrenol, *5659*
Medrogestone, *5836*
Medroxyprogesterone, *5838*
Megestrol Acetate, *5849*
Melengestrol, *5859*
Norethindrone, *6790*
Norethynodrel, *6791*
Norgesterone, *6795*
Norgestimate, *6796*
Norgestrel, *6797*
Norgestrienone, *6798*
Norvinisterone, *6814*
Pentagestrone, *7251*
Progesterone, *7956*
Promegestone, *7969*
Trengestone, *9717*

PROLACTIN INHIBITOR

Bromocriptine, *1437*
Cabergoline, *1637*
Lisuride, *5541*
Metergoline, *5999*
Quinagolide, *8226*
Terguride, *9308*

PROSTAGLANDIN/PROSTAGLANDIN ANALOG see also **Abortifacient; Antiglaucoma; Antiulcerative; Oxytocic**

Arbaprostil, *808*
Beraprost, *1190*

Carboprost, *1871*
Enprostil, *3629*
Gemeprost, *4393*
Latanoprost, *5387*
Limaprost, *5514*
Misoprostol, *6297*
Ornoprostil, *7003*
Prostacyclin, *8061*
Prostaglandin E$_1$, *8063*
Prostaglandin E$_2$, *8064*
Prostaglandin F$_{2a}$, *8065*
Rioprostil, *8395*
Rosaprostol, *8419*
Sulprostone, *9165*
Trimoprostil, *9853*
Unoprostone, *9984*

PROTEASE INHIBITOR see also **Retroviral Protease Inhibitor**

Aprotinin, *796*
Camostat, *1775*
Gabexate, *4344*
Nafamostat, *6436*
Urinastatin, *10021*

PROTON PUMP INHIBITOR see **Gastric Proton Pump Inhibitor**

PULMONARY SURFACTANT

Beractant, *1189*
Colfosceril Palmitate, *2540*
Exosurf®, *3960*
Infasurf, *5002*
Poractant Alfa, *7754*
Pumactant, *8126*

5α-REDUCTASE INHIBITOR see also **Antiprostatic Hypertrophy**

Epristeride, *3668*
Finasteride, *4125*

REPLENISHERS/SUPPLEMENTS

Calcium
Calcium Carbonate, *1697*
Calcium Gluconate, *1712*
Calcium Hypophosphite, *1718*
Calcium Lactate, *1723*
Calcium Levulinate, *1724*
Calcium Phosphate, Dibasic, *1739*
Calcium Phosphate, Tribasic, *1741*

Electrolyte
Calcium Chloride, *1699*
Potassium Chloride, *7783*
Potassium Gluconate, *7796*
Sodium Chloride, *8742*
Sodium Lactate, *8781*

Fluid and Nutrient
Protein Hydrolysates, *8701*

Iodine
Betasine, *1228*
Calcium Iodostearate, *1722*
Methenamine Tetraiodine, *6043*
Periodyl, *7313*
Potassium Iodide, *7809*
Prolonium Iodide, *7965*
Rubidium Iodide, *8443*
Sodium Iodide, *8777*
Strontium Iodide, *9005*

Magnesium
Magnesium Gluconate see *4464*

REPLENISHERS/SUPPLEMENTS
(*continued*)

Phosphorus
Calcium Glycerophosphate, *1713*
Durapatite, *3519*

Potassium
Potassium Bicarbonate, *7770*

Zinc
Zinc Sulfate, *10293*

RESPIRATORY STIMULANT
Almitrine, *309*
Bemegride, *1054*
Carbon Dioxide, *1857*
Cropropamide, *2659*
Crotethamide, *2662*
Dimefline, *3249*
Dimorpholamine, *3317*
Doxapram, *3488*
Ethamivan, *3765*
Fominoben, *4258*
Lobeline, *5580*
Mepixanox, *5906*
Nikethamide, *6635*
Picrotoxin, *7570*
Pimeclone, *7583*
Pyridofylline, *8159*
Sodium Succinate, *8828*
Tacrine, *9199*

RETROVIRAL PROTEASE INHIBITOR *see also Antiviral*
Indinavir, *4979*
Ritonavir, *8402*

REVERSE TRANSCRIPTASE INHIBITOR *see also Antiviral*
Delavirdine, *2929*
Didanosine, *3148*
Dideoxyadenosine, *3151*
Foscarnet Sodium, *4277*
Lamivudine, *5365*
Nevirapine, *6573*
Stavudine, *8958*
Suramin Sodium, *9181*
Zalcitabine, *10242*
Zidovudine, *10252*

SCABICIDE *see Ectoparasiticide*

SCLEROSING AGENT
Ethanolamine, *3772*
Ethylamine, *3808*
2-Hexyldecanoic Acid, *4747*
Polidocanol, *7717*
Sodium Ricinoleate *see 8378*
Sodium Tetradecyl Sulfate, *8838*
Tribenoside, *9738*

SEDATIVE/HYPNOTIC *see also Anxiolytic*

Acylic Ureides
Acecarbromal, *18*
Apronalide, *795*
Bromisovalum, *1418*
Capuride, *1818*
Carbromal, *1879*
Ectylurea, *3552*

Alcohols
Chlorhexadol, *2139*
Ethchlorvynol, *3774*
Meparfynol, *5890*
4-Methyl-5-thiazoleethanol, *6208*
tert-Pentyl Alcohol, *7282*
2,2,2-Trichloroethanol, *9768*

Amides
Butoctamide, *1562*
Diethylbromoacetamide, *3164*
Isovaleryl Diethylamide, *5253*
Niaprazine, *6576*
Trimetozine, *9850*
Zolpidem, *10321*
Zopiclone, *10395*

Barbituric Acid Derivatives
Allobarbital, *261*
Amobarbital, *607*
Aprobarbital, *794*
Barbital, *989*
Brallobarbital, *1387*
Butabarbital Sodium, *1530*
Butalbital, *1536*
Butallylonal, *1537*
Butethal, *1550*
Carbubarb, *1880*
Cyclobarbital, *2780*
Cyclopentobarbital, *2814*
Enallylpropymal, *3607*
5-Furfuryl-5-isopropylbarbituric Acid, *4326*
Heptabarbital, *4689*
Hexethal Sodium, *4740*
Hexobarbital, *4742*
Mephobarbital, *5899*
Methitural, *6058*
Narcobarbital, *6509*
Nealbarbital, *6516*
Pentobarbital, *7272*
Phenallymal, *7349*
Phenobarbital, *7386*
Phenylmethylbarbituric Acid, *7457*
Propallylonal, *7980*
Proxibarbal, *8091*
Reposal, *8306*
Secobarbital Sodium, *8563*
Talbutal, *9206*
Tetrabarbital, *9324*
Vinbarbital Sodium, *10118*
Vinylbital, *10131*

Benzodiazepine Derivatives
Brotizolam, *1472*
Cinolazepam, *2368*
Doxefazepam, *3490*
Estazolam, *3744*
Flunitrazepam, *4181*
Flurazepam, *4233*
Haloxazolam, *4635*
Loprazolam, *5604*
Lormetazepam, *5611*
Nitrazepam, *6667*
Quazepam, *8211*
Temazepam, *9285*
Triazolam, *9734*

Bromides
Ammonium Bromide, *531*
Calcium Bromide, *1693*
Calcium Bromolactobionate, *1694*
Lithium Bromide, *5551*
Magnesium Bromide, *5695*
Potassium Bromide, *7780*
Sodium Bromide, *8734*

Carbamates
Carfimate, *1886*
Ethinamate, *3779*
Hexapropymate, *4733*
Novonal, *6820*
Trichlorourethan, *9778*

Chloral Derivatives
Carbocloral, *1849*
Chloral Betaine, *2111*
Chloral Formamide, *2112*
Chloral Hydrate, *2113*
Dichloralphenazone, *3097*
Pentaerythritol Chloral, *7246*
Triclofos, *9788*

Quinazolone Derivatives
Etaqualone, *3757*
Mecloqualone, *5821*
Methaqualone, *6028*

Piperidinediones
Glutethimide, *4485*
Methyprylon, *6216*
Piperidione, *7622*
Pyrithyldione, *8179*
Thalidomide, *9390*

Others
Acetal, *36*
Acetophenone, *71*
Aldol, *225*
Ammonium Valerate, *604*
Amphenidone, *621*
d-Bornyl α-Bromoisovalerate, *1368*
d-Bornyl Isovalerate, *1370*
Bromoform, *1441*
Calcium 2-Ethylbutanoate, *1707*
α-Chloralose, *2114*
Clomethiazole, *2444*
Cypripedium, *2839*
Doxylamine, *3497*
Etodroxizine, *3921*
Etomidate, *3927*
Fenadiazole, *3994*
Homofenazine, *4774*
Hydrobromic Acid, *4819*
Mecloxamine, *5823*
Menthyl Valerate, *5887*
Opium, *6986*
Paraldehyde, *7160*
Perlapine, *7320*
Propiomazine, *8007*
Rilmazafone, *8387*
Sulfonethylmethane, *9131*
Sulfonmethane, *9133*

SERENIC
Eltoprazine, *3592*

SEROTONIN NORADRENALINE REUPTAKE INHIBITOR *see also Antidepressant*
Duloxetine, *3518*
Venlafaxine, *10079*

SEROTONIN RECEPTOR AGONIST *see also Anxiolytic; Antidepressant; Antimigraine; Serenic*
Buspirone, *1598*
Dihydroergotamine, *3217*
Eltoprazine, *3592*
Ergotamine, *3703*
Flesinoxan, *4138*
Ipsapirone, *5096*

Lesopitron, *5470*
Methysergid(e), *6217*
Sumatriptan, *9172*
Tandospirone, *9219*

**SEROTONIN RECEPTOR ANTAGO-
NIST** *see also* **Antidepressant; Antiemetic; Antihypertensive; Antimigraine**

Azasetron, *933*
Dolasetron, *3471*
Granisetron, *4557*
Ketanserin, *5307*
Methysergid(e), *6217*
Nefazodone, *6527*
Ondansetron, *6979*
Oxetorone, *7068*
Ritanserin, *8399*
Tropisetron, *9914*

SEROTONIN UPTAKE INHIBITOR
see also **Anorexic; Antidepressant**

Femoxetine, *3993*
Fluoxetine, *4222*
Fluvoxamine, *4251*
Indalpine, *4965*
Indeloxazine Hydrochloride, *4972*
Milnacipran, *6281*
Paroxetine, *7175*
Sertraline, *8612*
Sibutramine, *8629*
Zimeldine, *10254*

SKELETAL MUSCLE RELAXANT *see*
Muscle Relaxant (Skeletal)

SOMATOSTATIN ANALOG

Octreotide, *6859*

SPASMOLYTIC *see* **Antispasmodic**

STOOL SOFTENER *see* **Laxative/
Cathartic**

SUCCINYLCHOLINE SYNERGIST

Hexafluorenium Bromide, *4721*

SYMPATHOMIMETIC *see* **α-Adrenergic Agonist; β-Adrenergic Agonist**

THROMBOLYTIC *see also* **Anticoagulant; Antithrombotic**

Anistreplase, *712*
Plasmin, *7678*
Pro-Urokinase, *8089*
Streptokinase, *8981*
Tissue Plasminogen Activator,
9608
Urokinase, *10024*

**THROMBOXANE A$_2$-RECEPTOR AN-
TAGONIST** *see also* **Antithrombotic;
Antiasthmatic**

Daltroban, *2871*
Ridogrel, *8379*
Seratrodast, *8603*

**THROMBOXANE A$_2$-SYTHETASE IN-
HIBITOR** *see also* **Antithrombotic**

Isbogrel, *5120*
Ozagrel, *7115*
Ridogrel, *8379*

THYROID HORMONE *see also* **Antihypothyroid**

Liothyronine, *5535*
Thyroid, *9551*
Thyroidin, *9552*
Thyroxine, *9555*

THYROID INHIBITOR *see* **Antihyperthyroid**

THYROTROPIC HORMONE

TRH, *9720*
TSH, *9931*

TOCOLYTIC

Albuterol, *217*
Fenoterol, *4022*
Hexoprenaline, *4745*
Ritodrine, *8401*
Terbutaline, *9302*

TOPICAL PROTECTANT

Allantoin, *255*
Balsam Peru, *976*
Balsam Traumatic, *978*
Bismuth Phosphate, *1316*
Bismuth Subcarbonate, *1324*
Bismuth Subgallate, *1325*
Bismuth Tannate, *1330*
Calamine, *1675*
Collodion, *2547*
Esculin, *3739*
Gum Benzoin, *4608*
Hydroxypropyl Cellulose, *4888*
Pyroxylin, *8195*
Shark Liver Oil, *8622*
Storax, *8974*
Titanium Dioxide, *9612*
Zinc Oxide, *10279*

TOPOISOMERASE I INHIBITOR *see
also* **Antineoplastic**

Irinotecan, *5104*
Topotecan, *9687*

TOPOISOMERASE II INHIBITOR *see
also* **Antineoplastic**

Daunorubicin, *2890*
Doxorubicin, *3495*
Etoposide, *3931*
Sobuzoxane, *8708*
Teniposide, *9291*

TRANQUILIZER *see* **Antipsychotic;
Anxiolytic**

ULTRAVIOLET SCREEN

Actinoquinol, *143*
p-Aminobenzoic Acid, *443*
Butyl Methoxydibenzoylmethane,
1616

β-Carotene, *1902*
Cinoxate, *2370*
4-(Dimethylamino)benzoic Acid,
3282
Dioxybenzone, *3357*
Drometrizole, *3503*
Lawsone, *5406*
Mexenone, *6255*
Octabenzone, *6838*
Octyl Methoxycinnamate, *6864*
Oxybenzone, *7088*
Sulisobenzone, *9157*

URICOSURIC *see also* **Antigout**

Benzbromarone, *1093*
Ethebenecid, *3775*
Orotic Acid, *7004*
Oxycinchophen, *7091*
Probenecid, *7934*
Sulfinpyrazone, *9121*
Ticrynafen, *9570*
Zoxazolamine, *10328*

VASODILATOR (CEREBRAL)

Bencyclane, *1060*
Cinnarizine, *2365*
Citicoline, *2380*
Cyclandelate, *2771*
Ciclonicate, *2324*
Diisopropylamine Dichloroacctate,
3241
Eburnamonine, *3537*
Fasudil, *3980*
Fenoxedil, *4026*
Flunarizine, *4179*
Ibudilast, *4923*
Ifenprodil, *4936*
Lomerizine, *5593*
Nafronyl, *6440*
Nicametate, *6577*
Nicergoline, *6580*
Nimodipine, *6643*
Papaverine, *7151*
Pentifylline, *7269*
Tinofedrine, *9589*
Vincamine, *10120*
Vinpocetine, *10128*
Viquidil, *10141*

VASODILATOR (CORONARY)

Amotriphene, *615*
Bendazol, *1062*
Benfurodil Hemisuccinate, *1070*
Benziodarone, *1113*
Chloracizine, *2108*
Chromonar, *2296*
Clobenfurol, *2418*
Clonitrate, *2452*
Cloricromen, *2467*
Dilazep, *3244*
Dipyridamole, *3410*
Droprenilamine, *3506*
Efloxate, *3565*
Erythrityl Tetranitrate, *3716*
Etafenone, *3753*
Fendiline, *4011*
Floredil, *4144*
Ganglefene, *4375*
Heart Muscle Extract, *4651*
Hexestrol Bis(β-diethylaminoethyl
ether), *4739*
Hexobendine, *4743*
Itramin Tosylate, *5263*
Khellin, *5321*
Lidoflazine, *5507*
Mannitol Hexanitrate, *5789*

VASODILATOR (CORONARY) (continued)

Medibazine, 5831
Nitroglycerin, 6704
Pentaerythritol Tetranitrate, 7249
Pentrinitrol, 7280
Perhexiline, 7305
Pimefylline, 7584
Prenylamine, 7919
Propatyl Nitrate, 7995
Pyridofylline, 8159
Trapidil, 9707
Tricromyl, 9791
Trimetazidine, 9835
Trolnitrate Phosphate, 9900
Visnadine, 10147

VASODILATOR (PERIPHERAL)

Aluminum Nicotinate, 364
Bamethan, 981
Bencyclane, 1060
Betahistine, 1224
Bradykinin, 1386
Brovincamine, 1473
Bufeniode, 1495
Buflomedil, 1498
Butalamine, 1535
Cetiedil, 2062
Ciclonicate, 2324
Cinepazide, 2350
Cinnarizine, 2365
Cyclandelate, 2771
Diisopropylamine Dichloroacetate, 3241
Eledoisin, 3579
Fenoxedil, 4026
Flunarizine, 4179
Hepronicate, 4688
Ifenprodil, 4936
Iloprost, 4940
Inositol Niacinate, 5009
Isoxsuprine, 5259
Kallidin, 5290
Kallikrein, 5291
Moxisylyte, 6374
Nafronyl, 6440
Nicametate, 6577
Nicergoline, 6580
Nicofuranose, 6605

Nicotinyl Alcohol, 6614
Nylidrin, 6830
Pentifylline, 7269
Pentoxifylline, 7278
Piribedil, 7648
Prostaglandin E_1, 8063
Suloctidil, 9160
Tolazoline, 9645
Xanthinol Niacinate, 10194

VASOPRESSOR see Antihypotensive

VASOPROTECTANT

Benzarone, 1091
Bioflavonoids, 1269
Chromocarb, 2294
Clobenoside, 2419
Diosmin, 3350
Dobesilate Calcium, 3455
Escin, 3737
Folescutol, 4252
Leucocyanidin, 5477
Metescufylline, 6000
Quercetin, 8216
Rutin, 8456
Troxerutin, 9920

VITAMIN/VITAMIN SOURCE see also Enzyme Cofactor

Antirachitic
Calcipotriene, 1679
Calcitriol, 1681
Ergosterol, 3701
1α-Hydroxycholecalciferol, 4864
Vitamin D_2, 10156
Vitamin D_3, 10157

Antiscorbutic
Ascorbic Acid, 867
Calcium Ascorbate, 1688
Nicotinamide Ascorbate, 6610
Sodium Ascorbate, 8723

Antixerophthalmic
α-Carotene, 1901
β-Carotene, 1902
δ-Carotene, 1903
Vitamin A, 10150

Hematopoietic
Cobamamide, 2513
Folic Acid, 4253
Hydroxocobalamin, 4854
Methylcobalamin, 6125
Sodium Folate, 8763
Vitamin B_{12}, 10152

Prothrombogenic
Menadiol, 5873
Menadione, 5874
Menadoxime, 5875
Menaquinones, 5876
Phylloquinone, 7536
Vitamin K_5, 10162

Others
Inositol, 5008
β-Tocopherol, 9632
γ-Tocopherol, 9633
δ-Tocopherol, 9634
Vitamin E, 10159
Vitamin E Acetate, 10160
Vitamin U, 10165

VULNERARY

Acetylcysteine, 89
Allantoin, 255
Asiaticoside, 869
Cadexomer Iodine, 1649
Chitin, 2105
Dextranomer, 2992
Oxaceprol, 7035
Tocoretinate, 9639

WILSON'S DISEASE TREATMENT

Penicillamine, 7214
Trientine, 9796

XANTHINE OXIDASE INHIBITOR see also Antigout

Allopurinol, 287

FORMULA INDEX

FORMULA INDEX

A

AgBr
Silver Bromide, 8649
AgCl
Silver Chloride, 8652
AgClO₃
Silver Chlorate, 8651
AgClO₄
Silver Perchlorate, 8667
AgF
Silver Fluoride, 8657
AgF₂
Silver Difluoride, 8656
AgI
Silver Iodide, 8659
AgIO₃
Silver Iodate, 8658
AgMnO₄
Silver Permanganate, 8668
AgNO₂
Silver Nitrite, 8663
AgNO₃
Silver Nitrate, 8661
AgO
Silver(II) Oxide, 8666
Ag₂CrO₄
Silver Chromate(VI), 8653
Ag₂F
Silver Subfluoride, 8675
Ag₂HgI₄
Silver Tetraiodomercurate(II), 8678
Ag₂O
Silver Oxide, 8665
Ag₂O₃Se
Silver Selenite, 8674
Ag₂O₄S
Silver Sulfate, 8676
Ag₂O₄Se
Silver Selenate, 8672
Ag₂S
Silver Sulfide, 8677
Ag₂Se
Silver Selenide, 8673
Ag₃O₄P
Silver Phosphate, 8669
AlB₃H₁₂
Aluminum Borohydride, 341
AlBr₃
Aluminum Bromide, 342
AlCl₃
Aluminum Chloride, 348
AlCl₃O₉
Aluminum Chlorate, 347
AlCl₄H₄N
Ammonium Tetrachloroaluminate, 595
AlCl₄Na
Sodium Tetrachloroaluminate, 8836
AlCsO₈S₂
Aluminum Cesium Sulfate, 346
AlF₃
Aluminum Fluoride, 351
AlF₆H₁₂N₃
Ammonium Hexafluoroaluminate, 555

AlF₆Na₃
Cryolite, 2673
AlH₃
Aluminum Hydride, 354
AlH₃O₃
Aluminum Hydroxide, 355
AlH₄Li
Aluminum Lithium Hydride, 361
AlH₄NO₈S₂
Aluminum Ammonium Sulfate, 335
AlH₆O₆P₃
Aluminum Hypophosphite, 357
AlI₃
Aluminum Iodide, 358
AlKO₈S₂
Aluminum Potassium Sulfate, 373
AlN
Aluminum Nitride, 366
AlN₃O₉
Aluminum Nitrate, 365
AlNaO₂
Sodium Aluminate, 8715
AlNaO₈S₂
Aluminum Sodium Sulfate, 378
AlO₄P
Aluminum Phosphate, 371
AlO₈RbS₂
Aluminum Rubidium Sulfate, 374
AlP
Aluminum Phosphide, 372
AlSb
Aluminum Antimonide, 336
Al₂CaH₈
Aluminum Calcium Hydride, 344
Al₂F₁₈Si₃
Aluminum Hexafluorosilicate, 352
Al₂K₂O₄
Potassium Aluminate, 7765
Al₂MgO₈Si₂
Aluminum Magnesium Silicate, 362
Al₂O₃
Aluminum Oxide, 369
Al₂O₅Si
Aluminum Silicate, 377
Al₂O₁₂S₃
Aluminum Sulfate, 381
Al₂O₁₆S₄Zn
Aluminum Zinc Sulfate, 385
Al₂S₃
Aluminum Sulfide, 382
Al₂Se₃
Aluminum Selenide, 376
Al₆Bi₂O₁₂
Bismuth Aluminate, 1298
AsBr₃
Arsenic Tribromide, 840
AsCl₃
Arsenic Trichloride, 841
AsF₃
Arsenic Trifluoride, 842
AsF₅
Arsenic Pentafluoride, 836
AsGa
Gallium Arsenide, 4365
AsHHgO₄
Mercuric Arsenate, 5923
AsHNa₂O₄
Sodium Arsenate, Dibasic, 8720

AsH₂KO₄
Potassium Arsenate, 7767
AsH₃
Arsine, 849
AsH₃O₄
Arsenic Acid, 833
AsI₃
Arsenic Triiodide, 843
AsIn
Indium Arsenide, 4982
As₂Ca₃O₈
Calcium Arsenate, 1686
As₂Co₃O₈
Cobaltous Arsenate, 2495
As₂O₃
Arsenic Trioxide, 844
As₂O₄Zn
Zinc Meta-arsenite, 10272
As₂O₅
Arsenic Pentoxide, 839
As₂O₈Zn₃
Zinc Ortho-arsenate, 10277
As₂S₃
Arsenic Trisulfide, 846
As₂S₅
Arsenic Pentasulfide, 838
As₂Se
Arsenic Hemiselenide, 835
As₂Se₃
Arsenic Triselenide, 845
As₂Se₅
Arsenic Pentaselenide, 837
As₄S₄
Arsenic Disulfide, 834
AuBr₃
Gold Tribromide, 4539
AuBr₄H
Gold Tribromide, Acid, 4540
AuBr₄K
Potassium Tetrabromoaurate(III), 7853
AuCl
Gold Monochloride, 4528
AuCl₃
Gold Trichloride, 4541
AuCl₄H
Gold Trichloride, Acid, 4542
AuCl₄K
Potassium Tetrachloroaurate(III), 7854
AuCl₄Na
Sodium Tetrachloroaurate(III), 8837
AuH₃O₃
Gold Trihydroxide, 4544
AuI
Gold Monoiodide, 4530
AuI₄K
Potassium Tetraiodoaurate(III), 7860
AuNa₃O₆S₄
Gold Sodium Thiosulfate, 4537
Au₂O₃
Gold Trioxide, 4545
Au₂O₁₂Se₃
Gold Selenate, 4534
Au₂S
Gold Monosulfide, 4531
Au₂S₃
Gold Trisulfide, 4546

Denotes derivative of title compound

Br₂Cu
Cupric Bromide, 2695
Br₂Fe
Ferrous Bromide, 4088
Br₂Hg
Mercuric Bromide, 5925
Br₂Hg₂
Mercurous Bromide, 5949
Br₂Mg
Magnesium Bromide, 5695
Br₂MgO₆
Magnesium Bromate, 5694
Br₂Mn
Manganese Bromide, 5765
Br₂Ni
Nickel Bromide, 6585
Br₂OS
Thionyl Bromide, 9484
Br₂OSe
Selenium Oxybromide, 8577
Br₂O₆Pb
Lead Bromate, 5417
Br₂O₆Sr
Strontium Bromate, 8996
Br₂Pb
Lead Bromide, 5418
Br₂Se₂
Selenium Bromide, 8573
Br₂Sn
Stannous Bromide, 8938
Br₂Sr
Strontium Bromide, 8997
Br₂Te
Tellurium Dibromide, 9274
Br₂Zn
Zinc Bromide, 10258
Br₃Ce
Cerous Bromide, 2039
Br₃Cr
Chromic Bromide, 2276
Br₃Fe
Ferric Bromide, 4060
Br₃HSi
Tribromosilane, 9746
Br₃OP
Phosphorus Oxybromide, 7505
Br₃P
Phosphorus Tribromide, 7514
Br₃Sb
Antimony Tribromide, 747
Br₄Se
Selenium Tetrabromide, 8581
Br₄Si
Silicon Tetrabromide, 8643
Br₄Sn
Stannic Bromide, 8928
Br₄Te
Tellurium Tetrabromide, 9278
Br₄Ti
Titanium Tetrabromide, 9617
Br₅P
Phosphorus Pentabromide, 7507

C

CAgN
Silver Cyanide, 8655
CAg₂O₃
Silver Carbonate, 8650
CAuN
Gold Monocyanide, 4529
CB₄
Boron Carbide, 1374
CBaO₃
Barium Carbonate, 996
CBe₂
Beryllium Carbide, 1207

CBi₂O₅
Bismuth Subcarbonate, 1324
CBrN
Cyanogen Bromide, 2763
CCaN₂
Calcium Cyanamide, 1702
CCaO₃
Calcium Carbonate, 1697
CCdO₃
Cadmium Carbonate, 1652
CClN
Cyanogen Chloride, 2764
CCl₂F₂
Dichlorodifluoromethane, 3114
CCl₂O
Phosgene, 7491
CCl₃F
Trichlorofluoromethane, 9770
CCl₃NO₂
Chloropicrin, 2208
CCl₄
Carbon Tetrachloride, 1864
CCoO₃
Cobaltous Carbonate, 2497
CCs₂O₃
Cesium Carbonate, 2053
CCuN
Cuprous Cyanide, 2731
CCuNS
Cuprous Thiocyanate, 2739
CF₂O
Carbonyl Fluoride, 1868
CF₄
Carbon Tetrafluoride, 1865
CHBrCl₂
Bromodichloromethane, 1438
CHBr₂Cl
Chlorodibromomethane, 2186
CHBr₃
Bromoform, 1441
CHCl₃
Chloroform, 2193
CHFO
Formyl Fluoride, 4274
CHF₃
Fluoroform, 4211
CHI₃
Iodoform, 5054
CHKO₂
Potassium Formate, 7795
CHKO₃
Potassium Bicarbonate, 7770
CHLiO₂
Lithium Formate, 5558
CHN
Hydrogen Cyanide, 4836
CHNO
Cyanic Acid, 2756
Isocyanic Acid, 5181
CHNS
Thiocyanic Acid, 9464
CHN₃O₆
Trinitromethane, 9859
CHNaO₂
Sodium Formate, 8765
CHNaO₃
Sodium Bicarbonate, 8726
CH₂AlNaO₅
Dihydroxyaluminum Sodium Carbonate, 3228
CH₂Br₂
Methylene Bromide, 6138
CH₂CaO₆S₂
Calcium Methionate, 1727
CH₂ClNO
Carbamyl Chloride, 1827
CH₂Cl₂
Methylene Chloride, 6140
CH₂Cl₂Na₂O₆P₂
Clodronic Acid*, 2432

CH₂Cu₂O₅
Cupric Carbonate, Basic, 2697
CH₂INaO₃S
Methiodal Sodium, 6051
CH₂I₂
Methylene Iodide, 6143
CH₂N₂
Cyanamide, 2754
Diazomethane, 3048
CH₂O
Formaldehyde, 4262
CH₂O₂
Formic Acid, 4268
CH₂O₃
Performic Acid, 7302
CH₂S₃
Trithiocarbonic Acid, 9888
CH₃AsNa₂O₃
Methanearsonic Acid*, 6020
CH₃BNNa
Sodium Cyanoborohydride, 8751
CH₃Br
Methyl Bromide, 6108
CH₃Cl
Methyl Chloride, 6121
CH₃ClO₂S
Methanesulfonyl Chloride, 6022
CH₃F
Fluoromethane, 4212
CH₃FO₃S
Methyl Fluorosulfonate, 6151
CH₃I
Methyl Iodide, 6161
CH₃KO₄S
Potassium Methyl Sulfate, 7813
CH₃NO
Formamide, 4264
CH₃NO₂
Nitromethane, 6708
CH₃NO₃
Methyl Nitrate, 6175
CH₃NS
Thioformamide, 9469
CH₃N₃O₃
Nitrourea, 6749
CH₃NaO
Sodium Methoxide, 8789
CH₃NaO₃S
Sodium Formaldehydesulfoxylate, 8764
CH₃NaO₄S
Formaldehyde Sodium Bisulfite, 4263
CH₄
Methane, 6019
CH₄AsNaO₃
Methanearsonic Acid*, 6020
CH₄Cl₂O₆P₂
Clodronic Acid, 2432
CH₄N₂O
Urea, 10005
CH₄N₂O₂
Hydroxyurea, 4896
CH₄N₂S
Ammonium Thiocyanate, 597
Thiourea, 9505
CH₄N₄O₂
Nitroguanidine, 6705
CH₄Ni₃O₇
Nickel Carbonate Hydroxide, 6586
CH₄O
Methanol, 6024
CH₄O₃S
Methanesulfonic Acid, 6021
CH₄O₄S
Methyl Sulfate, 6203
CH₄O₆S₂
Methionic Acid, 6052
CH₄S
Methanethiol, 6023

CH₅AsO₃
Methanearsonic Acid, 6020
CH₅ClN₂O
Urea Hydrochloride, 10006
CH₅N
Methylamine, 6095
CH₅NO
Methoxyamine, 6069
CH₅NO₂
Ammonium Formate, 554
CH₅NO₃
Ammonium Bicarbonate, 522
CH₅N₃
Guanidine, 4591
CH₅N₃O₄
Urea Nitrate, 10008
CH₅N₃S
Thiosemicarbazide, 9499
CH₆ClN₃O
Semicarbazide Hydrochloride, 8588
CH₆N₂
Methylhydrazine, 6160
CH₆N₂O₂
Ammonium Carbamate, 533
CH₆N₂O₃
Urea Hydrogen Peroxide, 10007
CH₆N₂S₂
Ammonium Dithiocarbamate, 545
CH₆N₄
Aminoguanidine, 461
CH₆N₄O
Carbohydrazide, 1852
CH₆O₆P₂
Medronic Acid, 5837
CH₆O₇P₂
Oxidronic Acid, 7074
CH₁₈AlN₃O₁₄S₂
Guanidinium Aluminum Sulfate
Hexahydrate, 4592
CIN
Cyanogen Iodide, 2765
CI₄
Carbon Tetraiodide, 1866
CKN
Potassium Cyanide, 7789
CKNO
Potassium Cyanate, 7788
CKNS
Potassium Thiocyanate, 7866
CK₂O₃
Potassium Carbonate, 7781
CK₂S₃
Potassium Thiocarbonate, 7865
CLi₂O₃
Lithium Carbonate, 5552
CMnO₃
Manganese Carbonate, 5766
CNNa
Sodium Cyanide, 8750
CNNaO
Sodium Cyanate, 8749
CNNaS
Thiocyanate Sodium, 9463
CNTl
Thallium Cyanide, 9396
CN₄
Cyanogen Azide, 2762
CN₄O₈
Tetranitromethane, 9371
CNa₂O₃
Sodium Carbonate, 8739
CNa₂S₃
Sodium Thiocarbonate, 8841
CNa₃O₅P
Foscarnet Sodium, 4277
CO
Carbon Monoxide, 1861
CO₂
Carbon Dioxide, 1857

CO₃Sr
Strontium Carbonate, 8998
CO₃Tl₂
Thallium Carbonate, 9394
CO₃Zn
Zinc Carbonate, 10260
CS₂
Carbon Disulfide, 1859
CSe₂
Carbon Diselenide, 1858
CSi
Silicon Carbide, 8636
C₂AgKN₂
Potassium Silver Cyanide, 7839
C₂Ag₂O₄
Silver Oxalate, 8664
C₂AuKN₂
Potassium Dicyanoaurate(I), 7791
C₂AuN₂Na
Sodium Dicyanoaurate(I), 8755
C₂BaN₂
Barium Cyanide, 1000
C₂BaN₂S₂
Barium Thiocyanate, 1027
C₂BaO₄
Barium Oxalate, 1014
C₂Ca
Calcium Carbide, 1696
C₂CaN₂
Calcium Cyanide, 1704
C₂CaN₂S₂
Calcium Thiocyanate, 1757
C₂CaO₄
Calcium Oxalate, 1732
C₂CdN₂
Cadmium Cyanide, 1654
C₂Cl₂F₄
Cryofluorane, 2672
C₂Cl₂O₂
Oxalyl Chloride, 7046
C₂Cl₃N
Trichloroacetonitrile, 9757
C₂Cl₃NaO₂
Trichloroacetic Acid*, 9756
C₂Cl₄
Tetrachloroethylene, 9332
C₂Cl₄O₂
Diphosgene, 3395
C₂Cl₆
Hexachloroethane, 4715
C₂CoN₂
Cobaltous Cyanide, 2500
C₂CoN₂S₂
Cobaltous Thiocyanate, 2512
C₂CoO₄
Cobaltous Oxalate, 2506
C₂CrO₄
Chromous Oxalate, 2304
C₂CuKN₂
Cuprous Potassium Cyanide, 2735
C₂CuO₄
Cupric Oxalate, 2712
C₂FeN₂S₂
Ferrous Thiocyanate, 4107
C₂FeO₄
Ferrous Oxalate, 4099
C₂HBrClF₃
Halothane, 4634
C₂HBrF₄
Teflurane, 9265
C₂HBr₃O
Bromal, 1404
C₂HBr₃O₂
Tribromoacetic Acid, 9739
C₂HCl₃
Trichloroethylene, 9769
C₂HCl₃O
2,2-Dichloroacetyl Chloride, 3103
Trichloroacetaldehyde, 9755

C₂HCl₃O₂
Trichloroacetic Acid, 9756
C₂HCl₅
Pentachloroethane, 7241
C₂HF₃O₂
Trifluoracetic Acid, 9812
C₂HKO₄
Potassium Binoxalate, 7772
C₂HNa₃O₆
Sodium Sesquicarbonate, 8823
C₂H₂
Acetylene, 91
C₂H₂AsCl₃
Dichloro(2-chlorovinyl)arsine, 3112
C₂H₂BaO₄
Barium Formate, 1004
C₂H₂BeO₄
Beryllium Formate, 1210
C₂H₂Br₂
Acetylene Dibromide, 92
C₂H₂Br₄
sym-Tetrabromoethane, 9328
C₂H₂CaO₂S
Calcium Thioglycollate, 1758
C₂H₂CaO₄
Calcium Formate, 1711
C₂H₂ClNaO₂
Chloroacetic Acid*, 2162
C₂H₂Cl₂
Acetylene Dichloride, 93
Vinylidene Chloride, 10134
C₂H₂Cl₂O
Chloracetyl Chloride, 2107
C₂H₂Cl₂O₂
Dichloroacetic Acid, 3100
C₂H₂Cl₄
Tetrachloroethane, 9331
C₂H₂CoO₄
Cobaltous Formate, 2502
C₂H₂CrO₄
Chromous Formate, 2303
C₂H₂CuO₄
Cupric Formate, 2705
C₂H₂FNaO₂
Fluoroacetic Acid*, 4205
C₂H₂MgO₄
Magnesium Formate, 5702
C₂H₂MoO₇
Oxalomolybdic Acid, 7045
C₂H₂N₂O
1,3,4-Oxadiazole, 7037
C₂H₂Na₂O₂
Sodiumacetic Acid Sodium Salt,
8712
C₂H₂NiO₄
Nickel Formate, 6592
C₂H₂O
Ketene, 5309
C₂H₂O₂
Glyoxal, 4519
C₂H₂O₃
Glyoxylic Acid, 4521
C₂H₂O₄
Oxalic Acid, 7043
C₂H₂O₄Pb
Lead Formate, 5426
C₂H₂O₄Sr
Strontium Formate, 9003
C₂H₂O₄Zn
Zinc Formate, 10266
C₂H₃AgO₂
Silver Acetate, 8648
C₂H₃BiO₃
Bismuth Subacetate, 1323
C₂H₃BrO
Acetyl Bromide, 83
C₂H₃BrO₂
Bromoacetic Acid, 1421
C₂H₃Br₃O
Tribromoethanol, 9743

Denotes derivative of title compound

Denotes derivative of title compound

C₂H₈O₇P₂
Etidronic Acid, 3908

$C_2H_8O_7P_2$
Etidronic Acid, 3908

$C_2H_{14}Al_2Mg_6O_{20}.4H_2O$
Almagate, 307

C_2HgN_2
Mercuric Cyanide, 5928

$C_2HgN_2S_2$
Mercuric Thiocyanate, 5946

$C_2Hg_2N_2O$
Mercuric Oxycyanide, 5937

C_2I_4
Tetraiodoethylene, 9359

$C_2K_2O_4$
Potassium Oxalate, 7820

$C_2K_2O_6$
Potassium Percarbonate, 7821

$C_2Li_2O_4$
Lithium Oxalate, 5563

$C_2MgN_2S_2$
Magnesium Thiocyanate, 5733

C_2MgO_4
Magnesium Oxalate, 5712

C_2MnO_4
Manganese Oxalate, 5775

C_2N_2
Cyanogen, 2761

C_2N_2Ni
Nickel Cyanide, 6589

$C_2N_2PbS_2$
Lead Thiocyanate, 5450

$C_2N_2S_2Zn$
Zinc Thiocyanate, 10298

C_2N_2Zn
Zinc Cyanide, 10264

$C_2Na_2O_4$
Sodium Oxalate, 8795

C_2NiO_4
Nickel Oxalate, 6597

C_2O_4Pb
Lead Oxalate, 5436

C_2O_4Sn
Stannous Oxalate, 8943

C_2O_4Sr
Strontium Oxalate, 9008

C_2O_4Zn
Zinc Oxalate, 10278

$C_3AlN_3S_3$
Aluminum Thiocyanate, 384

C_3Al_4
Aluminum Carbide, 345

C_3AuN_3
Gold Tricyanide, 4543

C_3CaO_5
Calcium Mesoxalate, 1726

$C_3Ce_2O_9$
Cerous Carbonate, 2040

$C_3Cl_2KN_3O_3$
Troclosene Potassium, 9896

$C_3Cl_3N_3O_3$
Symclosene, 9188

$C_3FeN_3S_3$
Ferric Thiocyanate, 4081

C_3HBr_5O
Pentabromoacetone, 7239

$C_3H_2ClF_5O$
Enflurane, 3621
Isoflurane, 5191

$C_3H_2ClNO_2$
Chloroacetyl Isocyanate, 2167

$C_3H_2Cl_3NO$
Chlorocyanohydrin, 2185

$C_3H_2F_6O$
Desflurane, 2965

$C_3H_2N_2$
Malononitrile, 5750

$C_3H_2N_2O_3$
Parabanic Acid, 7154

$C_3H_2O_2$
Propiolic Acid, 8006

$C_3H_2O_5$
Mesoxalic Acid, 5971

$C_3H_3BrF_4$
Halopropane, 4632

C_3H_3Cl
Propargyl Chloride, 7994

$C_3H_3Cl_2NaO_2$
Dalapon*, 2869

$C_3H_3CrO_6$
Chromic Formate, 2280

$C_3H_3FeO_6$
Ferric Formate, 4066

C_3H_3N
Acrylonitrile, 133

$C_3H_3NOS_2$
Rhodanine, 8351

$C_3H_3NO_2$
Cyanoacetic Acid, 2759

C_3H_3NS
Thiazole, 9442

$C_3H_3N_3O_2$
Azomycin, 952

$C_3H_3N_3O_2S$
2-Amino-5-nitrothiazole, 477

$C_3H_3N_3O_3$
Cyamelide, 2752
Cyanuric Acid, 2767

$C_3H_4Br_2$
2,3-Dibromopropene, 3074

C_3H_4ClN
β-Chloropropionitrile, 2212

$C_3H_4Cl_2$
1,3-Dichloropropene, 3125

$C_3H_4Cl_2F_2O$
Methoxyflurane, 6072

$C_3H_4Cl_2O$
1,1-Dichloroacetone, 3101
1,3-Dichloroacetone, 3102

$C_3H_4Cl_2O_2$
Dalapon, 2869

$C_3H_4Cl_3NO_2$
Chloral Formamide, 2112
Trichlorourethan, 9778

$C_3H_4F_4O$
2,2,3,3-Tetrafluoro-1-propanol, 9346

$C_3H_4NNaO_3S$
Sodium β-Sulfopropionitrile, 8832

$C_3H_4N_2$
Imidazole, 4948
Methyleneaminoacetonitrile, 6134
Pyrazole, 8144

$C_3H_4N_2O$
Cyanoacetamide, 2758

$C_3H_4N_2O_2$
Hydantoin, 4796

$C_3H_4N_2S$
2-Aminothiazole, 501

$C_3H_4N_4O_6$
TNAZ, 9626

C_3H_4O
Acrolein, 130
Propargyl Alcohol, 7993

$C_3H_4O_2$
Acrylic Acid, 132
β-Propiolactone, 8005
Pyruvaldehyde, 8203

$C_3H_4O_3$
Pyruvic Acid, 8205

$C_3H_4O_4$
Malonic Acid, 5749

$C_3H_4O_5$
Tartronic Acid, 9240

$C_3H_5AgO_3$
Silver Lactate, 8660

C_3H_5Br
Allyl Bromide, 296

C_3H_5BrO
Bromoacetone, 1422

$C_3H_5BrO_2$
β-Bromopropionic Acid, 1455

$C_3H_5Br_2Cl$
Dibromochloropropane, 3068

$C_3H_5Br_3$
1,2,3-Tribromopropane, 9745

C_3H_5Cl
Allyl Chloride, 297

$C_3H_5ClN_2O_6$
Clonitrate, 2452

C_3H_5ClO
Chloroacetone, 2164
Epichlorohydrin, 3648
Propionyl Chloride, 8013

$C_3H_5ClO_2$
β-Chloropropionic Acid, 2211
Ethyl Chloroformate, 3831
Methyl Chloroacetate, 6122

$C_3H_5Cl_3O$
1,1,1-Trichloro-2-propanol, 9774

C_3H_5I
Allyl Iodide, 302

$C_3H_5KOS_2$
Potassium Xanthogenate, 7876

C_3H_5N
Propionitrile, 8012

C_3H_5NO
Acrylamide, 131
Ethylene Cyanohydrin, 3840

$C_3H_5NO_2$
Isonitrosoacetone, 5207

$C_3H_5NO_4$
Hadacidin, 4617

C_3H_5NS
Ethyl Isothiocyanate, 3861
Thiocyanic Acid*, 9464

$C_3H_5N_3O$
Cyacetacide, 2751

$C_3H_5N_3O_9$
Nitroglycerin, 6704

$C_3H_5NaO_2$
Sodium Propionate, 8816

$C_3H_5NaO_3$
Sodium Lactate, 8781

C_3H_6
Cyclopropane, 2817
Propylene, 8034

$C_3H_6BrNO_4$
Bronopol, 1470

$C_3H_6Br_2$
Propylene Dibromide, 8038
Trimethylene Bromide, 9844

$C_3H_6Cl_2$
Propylene Dichloride, 8039
Propylidene Chloride, 8046

$C_3H_6Cl_2O$
1,3-Dichloro-2-propanol, 3124

$C_3H_6FNO_2$
3-Fluoro-D-alanine, 4206

$C_3H_6I_2O$
Iothion, 5081

$C_3H_6N_2$
3-Aminopropionitrile, 489
2-Pyrazoline, 8145

$C_3H_6N_2O$
2-Imidazolidinone, 4950

$C_3H_6N_2O_2$
Cycloserine, 2820

$C_3H_6N_2S$
Ethylene Thiourea, 3849

$C_3H_6N_4O_3$
Triuret, 9895

$C_3H_6N_6$
Melamine, 5853

$C_3H_6N_6O_6$
Cyclonite, 2805

C_3H_6O
Acetone, 64
Allyl Alcohol, 294
Propionaldehyde, 8008
Propylene Oxide, 8041
Trimethylene Oxide, 9846

Denotes derivative of title compound

C₃H₆OS
Propanethial S-Oxide, 7985

C₃H₆O₂
Ethyl Formate, 3853
Glycidol, 4499
Methyl Acetate, 6089
Propionic Acid, 8010

C₃H₆O₂S
Thiolactic Acid, 9476

C₃H₆O₃
Dihydroxyacetone, 3225
Glyceraldehyde, 4490
Hydracrylic Acid, 4799
D-Lactic Acid, 5348
DL-Lactic Acid, 5349
L-Lactic Acid, 5350
Methyl Carbonate, 6117
s-Trioxane, 9863

C₃H₆O₄
Glyceric Acid, 4492

C₃H₇Br
Isopropyl Bromide, 5229
Propyl Bromide, 8030

C₃H₇CaO₆P
Calcium Glycerophosphate, 1713

C₃H₇Cl
Isopropyl Chloride, 5230
Propyl Chloride, 8032

C₃H₇ClO
Propylene Chlorohydrin, 8035
sec-Propylene Chlorohydrin, 8036

C₃H₇ClO₂
α-Chlorohydrin, 2197

C₃H₇I
Isopropyl Iodide, 5233
Propyl Iodide, 8047

C₃H₇IO₂
Glyceryl Iodide, 4497

C₃H₇K₂O₆P
Potassium Glycerophosphate, 7797

C₃H₇N
Allylamine, 295

C₃H₇NO
Acetoxime, 75
N,N-Dimethylformamide, 3292
Propionamide, 8009

C₃H₇NO₂
Alanine, 205
β-Alanine, 206
Formicin, 4269
Isopropyl Nitrite, 5235
1-Nitropropane, 6724
2-Nitropropane, 6725
n-Propyl Nitrite, 8050
Sarcosine, 8519
Urethan, 10013

C₃H₇NO₂S
Cysteine, 2850

C₃H₇NO₂Se
Selenocysteine, 8584

C₃H₇NO₃
n-Propyl Nitrate, 8049
Serine, 8604

C₃H₇NO₅S
Cysteic Acid, 2849

C₃H₇N₃O₂
N-Ethyl-N-nitrosourea, 3879
Glycocyamine, 4505

C₃H₇N₃O₄
L-Alanosine, 207

C₃H₇NaO₃S₃
2,3-Dimercapto-1-propanesulfonic
Acid*, 3256

C₃H₇NaO₄S
Acetone Sodium Bisulfite, 67

C₃H₇Na₂O₆P
Sodium Glycerophosphate, 8767

C₃H₇O₄P
Fosfomycin, 4280

C₃H₇O₆P
Glyceraldehyde 3-Phosphate, 4491

C₃H₈
Propane, 7982

C₃H₈HgO₂
Merisoprol Hg 197, 5959

C₃H₈IN
Dimethyl(methylene)ammonium
Iodide, 3300

C₃H₈NO₅P
Glyphosate, 4522

C₃H₈NO₆P
Phosphoserine, 7520

C₃H₈N₂OS
Noxythiolin, 6822

C₃H₈N₂O₂
2,3-Diaminopropionic Acid, 3027

C₃H₈N₂O₃
Oxymethurea, 7102

C₃H₈N₂S
N,N'-Dimethylthiourea, 3310

C₃H₈O
Ethyl Methyl Ether, 3874
Isopropyl Alcohol, 5227
n-Propyl Alcohol, 8027

C₃H₈OS₂
Dimercaprol, 3255

C₃H₈O₂
Methylal, 6093
Methyl Cellosolve, 6118
Propylene Glycol, 8040
Trimethylene Glycol, 9845

C₃H₈O₂S
Thioglycerol, 9471

C₃H₈O₃
Glycerol, 4493

C₃H₈O₃S
Ethyl Methanesulfonate, 3873

C₃H₈O₃S₃
2,3-Dimercapto-1-propanesulfonic
Acid, 3256

C₃H₈S₂
Bis[methylthio]methane, 1296
1,3-Propanedithiol, 7984

C₃H₉AsO₃
1-Propanearsonic Acid, 7983

C₃H₉BO₃
Trimethyl Borate, 9842

C₃H₉N
Isopropylamine, 5228
Propylamine, 8028
Trimethylamine, 9841

C₃H₉NNa₂O₇P₂
Pamidronic Acid*, 7135

C₃H₉NO
2-Aminopropanol, 488
2-Methylaminoethanol, 6096
Trimethylamine*, 9841

C₃H₉NO₃
Ammonium Lactate, 561

C₃H₉NO₃S
N-Methyltaurine, 6205

C₃H₉O₆P
Glycerophosphoric Acid, 4495

C₃H₁₀N₂
Propylenediamine, 8037

C₃H₁₁Br₂N₃S
AET, 176

C₃H₁₁NO₇P₂
Pamidronic Acid, 7135

C₃H₁₃N₃O₁₀Zr
Ammonium Zirconyl Carbonate,
606

C₃H₁₆N₄O₁₁U
Ammonium Uranium Carbonate,
602

C₃La₂O₉
Lanthanum*, 5374

C₃N₄
Carbon Nitride, 1862

C₃O₂
Carbon Suboxide, 1863

C₄BaN₄Pt
Barium Platinous Cyanide, 1020

C₄CdK₂N₄
Cadmium Potassium Cyanide, 1660

C₄Cl₂O₄Rh₂
Rhodium Carbonyl Chloride, 8354

C₄F₈
Octafluorocyclobutane, 6841

C₄HCoO₄
Dicobalt Octacarbonyl*, 3135

C₄HI₄N
Iodopyrrole, 5062

C₄H₂Cl₂O₂
Mucochloric Acid, 6379

C₄H₂FeO₄
Ferrous Fumarate, 4094
Hydrogen Tetracarbonylferrate(II),
4845

C₄H₂N₂O₄
Alloxan, 290

C₄H₂N₄O₂
5-Diazouracil, 3050

C₄H₂O₃
Maleic Anhydride, 5744

C₄H₃BrN₂O₂
5-Bromouracil, 1464

C₄H₃FN₂O₂
Fluorouracil, 4219

C₄H₃F₃O
Sevoflurane, 8621

C₄H₃KO₈
Potassium Tetroxalate, 7863

C₄H₃N₃O₄
Oxonic Acid, 7081
Violuric Acid, 10138

C₄H₃N₃O₅
5-Nitrobarbituric Acid, 6683

C₄H₄As₂Na₂O₄
Arsenoacetic Acid*, 848

C₄H₄Au₂CaO₄S₂
Myoral, 6411

C₄H₄BrNO₂
N-Bromosuccinimide, 1462

C₄H₄Br₂O₄
2,3-Dibromosuccinic Acid, 3078

C₄H₄CaO₄
Calcium Succinate, 1752

C₄H₄CaO₆
Calcium Tartrate, 1756

C₄H₄CdO₄
Cadmium Succinate, 1664

C₄H₄ClNO₂
N-Chlorosuccinimide, 2217

C₄H₄Cl₂O₂
Succinyl Chloride, 9042

C₄H₄Cl₂O₃
Chloroacetic Anhydride, 2163

C₄H₄CuO₆
Cupric Tartrate, 2725

C₄H₄FN₃O
Flucytosine, 4161

C₄H₄F₆O
Flurothyl, 4236

C₄H₄FeO₄
Ferrous Succinate, 4104

C₄H₄INO₂
N-Iodosuccinimide, 5064

C₄H₄KNO₄S
Acesulfame*, 35

C₄H₄KNaO₆
Potassium Sodium Tartrate, 7840

C₄H₄K₂O₃.3H₂O
Succinic Acid*, 9038

C₄H₄K₂O₆
Potassium Tartrate, 7849

C₄H₄N₂
Pyrazine, 8141
Pyridazine, 8152

Denotes derivative of title compound

C₄H₈Cl₂
Butylidene Chloride, 1606

C₄H₈Cl₂O
sym-Dichloroethyl Ether, 3116

C₄H₈Cl₂S
Mustard Gas, 6395

C₄H₈Cl₃O₄P
Trichlorfon, 9753

C₄H₈CuN₂O₄
Cupric Glycinate, 2707

C₄H₈N₂
Lysidine, 5666

C₄H₈N₂O
Allylurea, 306
N-Nitrosopyrrolidine, 6741

C₄H₈N₂O₂
Dimethylglyoxime, 3295
N-Nitrosomorpholine, 6738
Succinamide, 9035

C₄H₈N₂O₃
Asparagine, 872
N-Glycylglycine, 4513

C₄H₈N₂O₉Ti
Ammonium Titanium Oxalate, 599

C₄H₈N₂S
Thiosinamine, 9500

C₄H₈N₃Na₂O₅P
Phosphocreatine*, 7496

C₄H₈N₄O₄S₂
Tetramethylenedisulfotetramine, 9366

C₄H₈N₆O₄
Oxalenediuramidoxime, 7042

C₄H₈N₈O₈
Octogen, 6855

C₄H₈O
Butyraldehyde, 1627
Crotyl Alcohol, 2668
Cyclopropyl Methyl Ether, 2818
Isobutyraldehyde, 5169
Methyl Ethyl Ketone, 6149
Tetrahydrofuran, 9351

C₄H₈O₂
Acetoin, 61
Aldol, 225
Butyric Acid, 1629
Dioxane, 3353
Ethyl Acetate, 3803
Isobutyric Acid, 5170
Methyl Propionate, 6191
Propyl Formate, 8043

C₄H₈O₂S
Sulfolane, 9128

C₄H₈O₃
Ethylene Glycol Monoacetate, 3847
Glycerol Formal, 4494
γ-Hydroxybutyrate, 4860
β-Hydroxybutyric Acid, 4861
Methyl Lactate, 6167

C₄H₈O₄
D-Erythrose, 3731
L-Erythrose, 3732
L-Erythrulose, 3735
D-Threose, 9523
L-Threose, 9524

C₄H₈S
Tetrahydrothiophene, 9357

C₄H₈S₃
Methyl Allyl Trisulfide, 6094

C₄H₉Br
n-Butyl Bromide, 1588
sec-Butyl Bromide, 1589
tert-Butyl Bromide, 1590
Isobutyl Bromide, 5150

C₄H₉Cl
n-Butyl Chloride, 1595
sec-Butyl Chloride, 1596
tert-Butyl Chloride, 1597
Isobutyl Chloride, 5153

C₄H₉ClHg
n-Butylmercuric Chloride, 1614

C₄H₉ClO
tert-Butyl Hypochlorite, 1605

C₄H₉ClOS
Hemisulfur Mustard, 4680

C₄H₉F₃O₃SSi
Trimethylsilyl Triflate, 9849

C₄H₉I
n-Butyl Iodide, 1607
sec-Butyl Iodide, 1608
Isobutyl Iodide, 5158

C₄H₉N
Pyrrolidine, 8199

C₄H₉NO
n-Butyramide, 1628
N,N-Dimethylacetamide, 3277
Morpholine, 6362

C₄H₉NO₂
α-Aminobutyric Acid, 448
β-Aminobutyric Acid, 449
γ-Aminobutyric Acid, 450
α-Aminoisobutyric Acid, 465
n-Butyl Nitrite, 1617
tert-Butyl Nitrite, 1618
N,N-Dimethylglycine, 3293
Isobutyl Nitrite, 5163

C₄H₉NO₂S
Homocysteine, 4771

C₄H₉NO₃
3-Amino-4-hydroxybutyric Acid, 463
4-Amino-3-hydroxybutyric Acid, 464
Homoserine, 4777
Isobutyl Nitrate, 5162
Threonine, 9522

C₄H₉NO₅
Ammonium Bimalate, 524
Tris(hydroxymethyl)nitromethane, 9883

C₄H₉NO₆
Ammonium Bitartrate, 529

C₄H₉NS
Thiamorpholine, 9435

C₄H₉N₃O₂
Creatine, 2637

C₄H₉N₃O₃
Albizziin, 212

C₄H₉O₇P
D-Erythrose 4-Phosphate, 3733

C₄H₁₀
Butane, 1541

C₄H₁₀BF₃O
Boron Trifluoride Etherate, 1380

C₄H₁₀BaO₈S₂
Ethyl Sulfate*, 3897

C₄H₁₀CaO₈S₂
Ethyl Sulfate*, 3897

C₄H₁₀ClNO₂S
Mecysteine Hydrochloride, 5828

C₄H₁₀ClO₂P
Ethyl Phosphorochloridite, 3886

C₄H₁₀CrN₇S₄
Reinecke Salt, 8298

C₄H₁₀FO₂P
Sarin, 8520

C₄H₁₀Mg
Diethylmagnesium, 3171

C₄H₁₀NO₃PS
Acephate, 31

C₄H₁₀N₂
Piperazine, 7617

C₄H₁₀N₂O
N-Nitrosodiethylamine, 6734

C₄H₁₀N₂O₃
N-Nitrosodiethanolamine, 6733

C₄H₁₀N₂O₆
Hydrazine Tartrate, 4812

C₄H₁₀N₃O₅P
Phosphocreatine, 7496

C₄H₁₀O
n-Butyl Alcohol, 1575
sec-Butyl Alcohol, 1576
tert-Butyl Alcohol, 1577
Ethyl Ether, 3852
Isobutyl Alcohol, 5146
Methyl Propyl Ether, 6192

C₄H₁₀OS
2-(Ethylthio)ethanol, 3902

C₄H₁₀O₂
1,3-Butylene Glycol, 1601
2,3-Butylene Glycol, 1602
tert-Butyl Hydroperoxide, 1604
1,2-Dimethoxyethane, 3274
Dimethylacetal, 3276
2-Ethoxyethanol, 3797

C₄H₁₀O₂S
2,2'-Thiodiethanol, 9466

C₄H₁₀O₂S₂
1,4-Dithiothreitol, 3441

C₄H₁₀O₃
Diethylene Glycol, 3168
Orthoformic Acid*, 7012

C₄H₁₀O₃S
2-(Ethylsulfonyl)ethanol, 3899

C₄H₁₀O₄
Erythritol, 3714

C₄H₁₀O₄S
Diethyl Sulfate, 3178

C₄H₁₀O₈Pb₃
Lead Subacetate, 5443

C₄H₁₀S
n-Butyl Mercaptan, 1611
sec-Butyl Mercaptan, 1612
tert-Butyl Mercaptan, 1613
Ethyl Sulfide, 3898
Isobutyl Mercaptan, 5161

C₄H₁₀Zn
Diethylzinc, 3179

C₄H₁₁N
n-Butylamine, 1578
sec-Butylamine, 1579
tert-Butylamine, 1580
Diethylamine, 3160
Isobutylamine, 5147

C₄H₁₁NO
2-Amino-1-butanol, 447
2-Amino-2-methyl-1-propanol, 469
Deanol, 2900

C₄H₁₁NO₂
2-Amino-2-methyl-1,3-propanediol, 468
Diethanolamine, 3156

C₄H₁₁NO₃
Tromethamine, 9902

C₄H₁₁NO₈P₂
Glyphosine, 4523

C₄H₁₁N₅
Metformin, 6001

C₄H₁₂As₂
Cacodyl, 1639

C₄H₁₂As₂HgO₄
Cacodylic Acid*, 1640

C₄H₁₂ClN₃O
N,N-Dimethylglycine Hydrazide Hydrochloride, 3294

C₄H₁₂FN₂OP
Dimefox, 3250

C₄H₁₂IN
Tetramethylammonium Iodide, 9364

C₄H₁₂NNaO₇P₂.3H₂O
Alendronic Acid*, 228

C₄H₁₂NO₄P
Demanyl Phosphate, 2935

C₄H₁₂N₂
Putrescine, 8134

Denotes derivative of title compound

Denotes derivative of title compound

Denotes derivative of title compound

Ethadione, 3762
Paramethadione, 7161
C₇H₁₁NO₄
Oxaceprol, 7035
C₇H₁₁NO₇P₂
Risedronic Acid, 8396
C₇H₁₁N₃
Isaxonine, 5118
C₇H₁₁N₃O₂
Ipronidazole, 5095
C₇H₁₁N₃O₃
Secnidazole, 8562
C₇H₁₂
Norcarane, 6781
C₇H₁₂ClN₅
Simazine, 8681
C₇H₁₂N₂O₂
Ectylurea, 3552
Heptoxime, 4700
C₇H₁₂N₂O₄
Aceglutamide, 25
C₇H₁₂O
Cycloheptanone, 2791
C₇H₁₂O₂
n-Butyl Acrylate, 1574
Cyclohexanecarboxylic Acid, 2793
C₇H₁₂O₃
Ethyl Levulinate, 3865
Isopropyl Acetoacetate, 5225
C₇H₁₂O₄
n-Butylmalonic Acid, 1610
Diethylmalonic Acid, 3173
Ethyl Malonate, 3869
Pimelic Acid, 7585
C₇H₁₂O₅
Diacetin, 3005
C₇H₁₂O₆
Quinic Acid, 8243
C₇H₁₃BrN₂O₂
Carbromal, 1879
C₇H₁₃N
Quinuclidine, 8265
C₇H₁₃NO
Cycloheptanone*, 2791
3-Quinuclidinol, 8266
C₇H₁₃NO₂
Stachydrine, 8926
C₇H₁₃NO₃
Betonicine, 1234
Turicine, 9952
C₇H₁₃NO₃S
N-Acetylmethionine, 97
N-Acetylpenicillamine, 100
C₇H₁₃NO₃S₂
Bucillamine, 1481
C₇H₁₃NO₄
L-Glutamic Acid 5-Ethyl Ester, 4478
C₇H₁₃N₃O₃S
Oxamyl, 7052
C₇H₁₃NaO₈
Glucoheptonic Acid*, 4463
C₇H₁₃O₅PS
Methacrifos, 6004
C₇H₁₃O₆P
Mevinphos, 6252
C₇H₁₄AsClO₃
Chloroarsenol, 2170
C₇H₁₄Cl₃NO₄
Chloral Betaine, 2111
C₇H₁₄NO₃PS₂
Phosfolan, 7490
C₇H₁₄NO₅P
Monocrotophos, 6334
C₇H₁₄N₂O₂S
Aldicarb, 223
C₇H₁₄N₂O₄S
L-Cystathionine, 2847
C₇H₁₄N₂O₄S₂
Djenkolic Acid, 3452

C₇H₁₄N₃O₃P
Uredepa, 10011
C₇H₁₄N₄S₂
Methallibure, 6013
C₇H₁₄O
Cyclohexylcarbinol, 2800
Dipropyl Ketone, 3408
Heptanal, 4692
2-Heptanone, 4698
C₇H₁₄O₂
n-Butyl Propionate, 1623
Ethyl Isovalerate, 3862
Heptanoic Acid, 4695
Isoamyl Acetate, 5125
Isobutyl Propionate, 5164
Propyl Butyrate, 8031
Valeric Acid, Normal*, 10042
C₇H₁₄O₄
D-Chalcose, 2080
Cymarose, 2831
Monobutyrin, 6332
Mycarose, 6399
Sarmentose, 8523
C₇H₁₄O₅
Digitalose, 3202
Thevetose, 9425
C₇H₁₄O₆
α-Methylglucoside, 6156
C₇H₁₄O₇
D-manno-Heptulose, 4701
C₇H₁₄O₈
Glucoheptonic Acid, 4463
C₇H₁₅Cl₂N₂O₂P
Ifosfamide, 4937
C₇H₁₅Cl₂N₂O₂P.H₂O
Cyclophosphamide, 2816
C₇H₁₅Cl₂N₂O₄P
Perfosfamide, 7303
C₇H₁₅Cl₄N
Novembichin, 6817
C₇H₁₅NO
1-Ethyl-3-piperidinol, 3890
C₇H₁₅NO₂
Emylcamate, 3604
Isobutyl Urethane, 5168
C₇H₁₅NO₃
Carnitine, 1898
C₇H₁₅NO₄S
MOPS, 6346
C₇H₁₅NO₅
N-Methyl-α-L-glucosamine, 6155
C₇H₁₅N₃O₂
Indospicine, 5001
C₇H₁₅N₃O₂S₂
Cartap, 1916
C₇H₁₅NaO₄S
Heptanal Sodium Bisulfite, 4693
C₇H₁₆
n-Heptane, 4694
C₇H₁₆BrNO₂
Acetylcholine Bromide, 87
C₇H₁₆ClN
Mepiquat Chloride, 5903
C₇H₁₆ClNO₂
Acetylcholine Chloride, 88
C₇H₁₆ClN₃O₂S₂
Cartap*, 1916
C₇H₁₆FO₂P
Soman, 8860
C₇H₁₆INO₂
Oxapropanium Iodide, 7056
C₇H₁₆N₄O₄S₂
Taurolidine, 9243
C₇H₁₆O
1-Heptanol, 4696
2-Heptanol, 4697
C₇H₁₆O₃
Dipropylene Glycol Monomethyl Ether, 3407
Orthoformic Acid*, 7012

C₇H₁₆O₄S₂
Sulfonmethane, 9133
C₇H₁₇AsClNO₃
Chloroarsenol*, 2170
C₇H₁₇ClN₂O₂
Bethanechol Chloride, 1232
C₇H₁₇N
Methylhexaneamine, 6159
Tuaminoheptane, 9934
C₇H₁₇NO₂
Triethylammonium Formate, 9800
C₇H₁₇NO₅
N-Methylglucamine, 6154
C₇H₁₇O₂PS₃
Phorate, 7486
C₇H₁₈NO₇P
Fosfomycin*, 4280
C₇H₁₉N₃
Spermidine, 8893
C₈BrF₁₇
Perflubron, 7299
C₈Cl₂N₂O₂
2,3-Dichloro-5,6-dicyanobenzoquinone, 3113
C₈Cl₄N₂
Chlorothalonil, 2219
C₈Co₂O₈
Dicobalt Octacarbonyl, 3135
C₈H₂Cl₄O₅
Mucochloric Anhydride, 6380
C₈H₃I₂NNa₂O₅
Sodium Iodomethamate, 8779
C₈H₃NO₂
Diatretyne II, 3039
C₈H₄Cl₂O₂
Phthaloyl Chloride, 7531
C₈H₄Cl₆
1,4-Bis(trichloromethyl)benzene, 1340
C₈H₄K₂O₁₂Sb₂.3H₂O
Antimony Potassium Tartrate, 741
C₈H₄N₂S₂
Bitoscanate, 1345
C₈H₄O₃
Phthalic Anhydride, 7528
C₈H₅Cl₃O₂
Chlorfenac, 2135
C₈H₅Cl₃O₃
2,4,5-T, 9194
C₈H₅F₃N₂OS
Riluzole, 8389
C₈H₅KO₄
Potassium Biphthalate, 7773
C₈H₅NOS
Benzoyl Isothiocyanate, 1147
C₈H₅NO₂
Agrocybin, 188
Isatin, 5116
Phthalimide, 7529
C₈H₅NO₃
Carsalam, 1915
Diatretyne I, 3038
C₈H₅N₃O₃S₂
Tenonitrozole, 9292
C₈H₆
Ethynylbenzene, 3906
C₈H₆BrClO
p-Chlorophenacyl Bromide, 2205
C₈H₆BrClO₂
p-Bromobenzyl Chloroformate, 1432
C₈H₆BrN
α-Bromobenzyl Cyanide, 1433
C₈H₆Br₂O
p-Bromophenacyl Bromide, 1449
C₈H₆Cl₂O₃
2,4-D, 2865
Dicamba, 3090
C₈H₆F₃N₃O₄S₂
Flumethiazide, 4174

C₉H₈ClNaO₃

$C_9H_8ClNaO_3$
  MCPA*, 5803

$C_9H_8Cl_2O_3$
  Dichlorprop, 3128

$C_9H_8Cl_3NO_2S$
  Captan, 1815

$C_9H_8I_2O_3$
  3,5-Diiodosalicylic Acid*, 3234

$C_9H_8NNaO_4$
  Salicylamide O-Acetic Acid*, 8481

$C_9H_8N_2O_2$
  Pemoline, 7206

$C_9H_8N_2O_3S_2$
  Phenosulfazole, 7403

$C_9H_8N_3NaO_2S_2.1\frac{1}{2}H_2O$
  Sulfathiazole*, 9115

$C_9H_8N_4OS$
  Thidiazuron, 9447

$C_9H_8N_4O_6$
  Nifurtoinol, 6631

$C_9H_8N_8O_2S$
  Thiamiprine, 9434

C_9H_8O
  1,2-Benzopyran, 1133
  Cinnamaldehyde, 2356

$C_9H_8O_2$
  Atropic Acid, 906
  Cinnamic Acid, 2358

$C_9H_8O_3$
  Carbic Anhydride, 1842
  p-Coumaric Acid, 2628

$C_9H_8O_4$
  Acetozone, 78
  Aspirin, 886
  Caffeic Acid, 1673

$C_9H_9AlO_6$
  Dihydroxyaluminum Acetylsalicy-
    late, 3226

$C_9H_9Br_2NO_3$
  3,5-Dibromo-L-tyrosine, 3079

C_9H_9ClO
  Chlorindanol, 2144

$C_9H_9ClO_3$
  MCPA, 5803

$C_9H_9Cl_2NO$
  Propanil, 7987

$C_9H_9Cl_2NO_2$
  Diloxanide, 3246

$C_9H_9Cl_2N_3$
  Clonidine, 2450

$C_9H_9Cl_2N_3O$
  Guanfacine, 4590

$C_9H_9HgNaO_2S$
  Thimerosal, 9451

$C_9H_9I_2NO_3$
  Betasine, 1228
  3,5-Diiodotyrosine, 3236

C_9H_9N
  Skatole, 8700

C_9H_9NO
  Hydrocarbostyril, 4820

$C_9H_9NO_3$
  Adrenochrome, 171
  Adrenolutin, 173
  Hippuric Acid, 4752
  Salacetamide, 8471

$C_9H_9NO_4$
  Ethyl Nitrobenzoate, 3878
  Salicylamide O-Acetic Acid, 8481

$C_9H_9N_2NaO_3$
  p-Aminohippuric Acid*, 462

$C_9H_9N_3O_2$
  Carbendazim, 1836

$C_9H_9N_3O_2S_2$
  Sulfathiazole, 9115
  Thiazolsulfone, 9444

$C_9H_9N_3S$
  Amiphenazole, 505

$C_9H_9N_4NaO_4$
  Acefylline*, 22
  1-Theobromineacetic Acid*, 9419

$C_9H_9N_5$
  Amanozine, 388
  Benzoguanamine, 1120

C_9H_{10}
  Indan, 4966

$C_9H_{10}ClN_5O_2$
  Imidacloprid, 4946

$C_9H_{10}Cl_2N_2O$
  Diuron, 3447

$C_9H_{10}Cl_2N_2O_2$
  Linuron, 5534

$C_9H_{10}Cl_2N_4$
  Apraclonidine, 791

$C_9H_{10}FIN_2O_5$
  Fialuridine, 4114

$C_9H_{10}INO_3$
  3-Iodotyrosine, 5065

$C_9H_{10}NO_3PS$
  Ciafos, 2321

$C_9H_{10}N_2$
  5,6-Dimethylbenzimidazole, 3286

$C_9H_{10}N_2O$
  Aminorex, 497
  1-Phenyl-3-pyrazolidinone, 7463

$C_9H_{10}N_2O_2$
  Phenacemide, 7343

$C_9H_{10}N_2O_2S$
  Sulbenox, 9060
  Thibenzazoline, 9446

$C_9H_{10}N_2O_2S_2$
  Aureothricin, 914

$C_9H_{10}N_2O_3$
  p-Aminohippuric Acid, 462

$C_9H_{10}N_2O_3S_2$
  Ethoxzolamide, 3801

$C_9H_{10}N_2S$
  Etisazol, 3918

$C_9H_{10}N_4O_2S_2$
  Sulfamethizole, 9084

$C_9H_{10}N_4O_3S_2$
  Sulfametrole, 9089

$C_9H_{10}N_4O_4$
  Acefylline, 22
  1-Theobromineacetic Acid, 9419

$C_9H_{10}N_6O_5$
  Orazamide, 6993

$C_9H_{10}O$
  p-Anol, 715
  Chavicol, 2091
  Cinnamyl Alcohol, 2362
  Phenylacetone, 7423
  Propiophenone, 8015

$C_9H_{10}O_2$
  Benzyl Acetate, 1158
  m-Cresyl Acetate, 2651
  o-Cresyl Acetate, 2652
  Ethyl Benzoate, 3814
  Hydrocinnamic Acid, 4825
  Paroxypropione, 7176

$C_9H_{10}O_3$
  Apocynin, 782
  Atrolactic Acid, 905
  Ethylparaben, 3883
  Ethyl Salicylate, 3894
  Ethyl Vanillin, 3904
  Tropic Acid, 9910
  Veratraldehyde, 10084

$C_9H_{10}O_4$
  Flopropione, 4142
  Glycol Salicylate, 4509
  Homovanillic Acid, 4778
  Methylenomycins*, 6144
  Syringaldehyde, 9191
  Veratric Acid, 10086

$C_9H_{10}O_5$
  Vanilmandelic Acid, 10070

$C_9H_{11}BrN_2O_2$
  Metobromuron, 6224

$C_9H_{11}ClN_2O$
  Monuron, 6342

$C_9H_{11}ClO_3$
  Chlorphenesin, 2230

$C_9H_{11}Cl_2FN_2O_2S_2$
  Dichlofluanid, 3095

$C_9H_{11}ClN_3O_4S_2$
  Methyclothiazide, 6086

$C_9H_{11}Cl_3NO_3PS$
  Chlorpyrifos, 2242

$C_9H_{11}FN_2O_5$
  Doxifluridine, 3493
  Floxuridine, 4148

$C_9H_{11}F_2N_3O_4$
  Gemcitabine, 4392

$C_9H_{11}IN_2O_5$
  Idoxuridine, 4934

$C_9H_{11}N$
  Tranylcypromine, 9708

$C_9H_{11}NO$
  Acetotoluide, 74
  p-Aminopropiophenone, 490
  Cathinone, 1954
  p-Dimethylaminobenzaldehyde,
    3280
  N-Methylacetanilide, 6088

$C_9H_{11}NO_2$
  p-Acetanisidine, 48
  Atrolactamide, 904
  Benzocaine, 1116
  4-(Dimethylamino)benzoic Acid,
    3282
  Ethenzamide, 3776
  Phenylalanine, 7425
  Phenylurethan(e), 7473

$C_9H_{11}NO_3$
  Adrenalone, 170
  Styramate, 9027
  Tyrosine, 9970
  m-Tyrosine, 9971

$C_9H_{11}NO_4$
  Dopa, 3478
  Levodopa, 5490

$C_9H_{11}NO_5$
  Droxidopa, 3513

$C_9H_{11}NO_6$
  Showdomycin, 8627

$C_9H_{11}N_3O_4$
  Ancitabine, 670

$C_9H_{11}N_5O_3$
  Biopterin, 1270

$C_9H_{11}N_5O_4$
  Eritadenine, 3712
  Ichthyopterin, 4930
  Neopterin, 6551

C_9H_{12}
  Cumene, 2683
  Mesitylene, 5967
  n-Propylbenzene, 8029
  Pseudocumene, 8100

$C_9H_{12}AsNO_4$
  3-Methylarsacetin, 6102

$C_9H_{12}BNO_4$
  Epinephrine*, 3656

$C_9H_{12}ClN_3O_4S_2$
  Ethiazide, 3778

$C_9H_{12}ClN_5O$
  Moxonidine, 6375

$C_9H_{12}ClO_4P$
  Heptenophos, 4699

$C_9H_{12}Cl_2N_4O$
  Guanochlor, 4595

$C_9H_{12}NO_5PS$
  Fenitrothion, 4017

$C_9H_{12}N_2$
  Nornicotine, 6807

$C_9H_{12}N_2Na_3O_{15}P_3.2H_2O$
  Uridine 5′-Triphosphate*, 10019

$C_9H_{12}N_2O$
  Fenuron, 4050

C$_9$H$_{12}$N$_2$O$_2$
Dulcin, 3517

C$_9$H$_{12}$N$_2$O$_3$
5-Nitro-2-propoxyaniline, 6727

C$_9$H$_{12}$N$_2$O$_4$S
Pidotimod, 7574

C$_9$H$_{12}$N$_2$O$_5$
Deoxyuridine, 2957

C$_9$H$_{12}$N$_2$O$_6$
Uridine, 10016

C$_9$H$_{12}$N$_2$S
Protionamide, 8076

C$_9$H$_{12}$N$_4$O$_3$
Etofylline, 3924

C$_9$H$_{12}$N$_5$NaO$_4$
Ganciclovir*, 4374

C$_9$H$_{12}$N$_6$
Triethylenemelamine, 9803

C$_9$H$_{12}$O
Benzyl Ethyl Ether, 1168
p,α-Dimethylbenzyl Alcohol, 3287
α-Ethylbenzyl Alcohol, 3816

C$_9$H$_{12}$O$_3$
Phenylglyceryl Ether, 7444

C$_9$H$_{12}$O$_3$S
Ethyl p-Toluenesulfonate, 3903

C$_9$H$_{12}$O$_4$
3-Methoxy-4-hydroxyphenylglycol, 6074

C$_9$H$_{13}$BO$_3$
Methoxyethylbenzeneboronic Acid, 6071

C$_9$H$_{13}$BrN$_2$O$_2$
Bromacil, 1402
Pyridostigmine Bromide, 8161

C$_9$H$_{13}$ClN$_2$O$_2$
Terbacil, 9298

C$_9$H$_{13}$ClN$_6$
Cyanazine, 2755

C$_9$H$_{13}$ClN$_6$O$_2$
Nimustine, 6645

C$_9$H$_{13}$Cl$_2$N$_3$O$_2$
Dopan, 3480

C$_9$H$_{13}$N
Amphetamine, 623
Cumidine, 2686

C$_9$H$_{13}$NNaO$_2$P
Toldimfos Sodium, 9649

C$_9$H$_{13}$NO
β-Amino-α-methylphenethyl Alcohol, 467
Gepefrine, 4409
Halostachine, 4633
Hydroxyamphetamine, 4855
Norpseudoephedrine, 6811

C$_9$H$_{13}$NO$_2$
Deoxyepinephrine, 2950
Ecgonidine, 3540
Ethinamate, 3779
Metaraminol, 5993
Pyrithyldione, 8179
Synephrine, 9189

C$_9$H$_{13}$NO$_3$
Epinephrine, 3656
Nordefrin, 6785
Normetanephrine, 6803

C$_9$H$_{13}$N$_2$O$_9$P
5'-Uridylic Acid, 10020

C$_9$H$_{13}$N$_3$O
Iproniazid, 5094

C$_9$H$_{13}$N$_3$O$_2$
Aminometradine, 471
Amisometradine, 507

C$_9$H$_{13}$N$_3$O$_3$
Zalcitabine, 10242

C$_9$H$_{13}$N$_3$O$_5$
Cytarabine, 2853
Cytidine, 2855

C$_9$H$_{13}$N$_3$O$_6$
Mizoribine, 6306

C$_9$H$_{13}$N$_5$O$_4$
Ganciclovir, 4374

C$_9$H$_{14}$ClNO
Phenylpropanolamine Hydrochloride, 7461

C$_9$H$_{14}$ClNO$_2$
Phenylephrine Hydrochloride, 7440

C$_9$H$_{14}$IN
Phenyltrimethylammonium Iodide, 7471

C$_9$H$_{14}$N$_2$
Pheniprazine, 7382

C$_9$H$_{14}$N$_2$O
Phenoxypropazine, 7411

C$_9$H$_{14}$N$_2$O$_3$
Metharbital, 6029

C$_9$H$_{14}$N$_2$O$_4$S
Mafenide*, 5683

C$_9$H$_{14}$N$_2$O$_6$S
Itramin Tosylate, 5263

C$_9$H$_{14}$N$_4$O$_{12}$P$_2$
Uridine 5'-Diphosphate, 10017

C$_9$H$_{14}$N$_3$O$_8$P
2'-Cytidylic Acid, 2856
3'-Cytidylic Acid, 2857

C$_9$H$_{14}$N$_4$O$_3$
Carnosine, 1899
Nimorazole, 6644

C$_9$H$_{14}$N$_4$O$_4$
Molsidomine, 6316

C$_9$H$_{14}$N$_4$O$_5$
Acadesine, 13

C$_9$H$_{14}$O
Isophorone, 5213
Phorone, 7488

C$_9$H$_{14}$OS$_3$
Ajoene, 195

C$_9$H$_{14}$O$_6$
Triacetin, 9721

C$_9$H$_{15}$AlO$_9$
Aluminum Lactate, 360

C$_9$H$_{15}$BrN$_2$O$_3$
Acecarbromal, 18

C$_9$H$_{15}$Br$_6$O$_4$P
Tris-BP, 9881

C$_9$H$_{15}$Cl$_6$O$_4$P
Fyrol FR-2, 4342

C$_9$H$_{15}$NO
Pseudopelletierine, 8105

C$_9$H$_{15}$NO$_2$
Diacetone Acrylamide, 3007
Piperidione, 7622
3-Quinuclidinol*, 8266

C$_9$H$_{15}$NO$_3$
5,5-Dipropyl-2,4-oxazolidinedione, 3409
Ecgonine, 3541

C$_9$H$_{15}$NO$_3$S
Captopril, 1817
Mycobacidin, 6403

C$_9$H$_{15}$N$_2$O$_{15}$P$_3$
Uridine 5'-Triphosphate, 10019

C$_9$H$_{15}$N$_3$O$_2$
Hercynine, 4702

C$_9$H$_{15}$N$_3$O$_2$S
Ergothioneine, 3705

C$_9$H$_{15}$N$_3$O$_7$
Lycomarasmine, 5649

C$_9$H$_{15}$N$_4$O$_8$P
AICAR, 190

C$_9$H$_{15}$N$_5$O
Minoxidil, 6290

C$_9$H$_{15}$N$_5$O$_3$
Sapropterin, 8515
Theophylline*, 9421

C$_9$H$_{15}$NaO$_3$
Hexacyclonate Sodium, 4717

C$_9$H$_{15}$O$_8$P
Bomyl, 1360

C$_9$H$_{16}$ClN$_3$O$_2$
Lomustine, 5594

C$_9$H$_{16}$ClN$_3$O$_7$
Chlorozotocin, 2229

C$_9$H$_{16}$ClN$_5$
Propazine, 7996
Trietazine, 9797

C$_9$H$_{16}$Cl$_2$N$_4$
N-(3-Chloroallyl)hexaminium Chloride, 2168

C$_9$H$_{16}$NNaO$_5$
Pantothenic Acid*, 7147

C$_9$H$_{16}$N$_2$O$_2$
Apronalide, 795

C$_9$H$_{16}$N$_4$OS
Ethiozin, 3785
Tebuthiuron, 9255

C$_9$H$_{16}$O
2-Nonenal, 6771

C$_9$H$_{16}$O$_4$
Azelaic Acid, 938

C$_9$H$_{17}$ClN$_3$O$_3$PS
Isazofos, 5119

C$_9$H$_{17}$IN$_4$
Methenamine Allyl Iodide, 6037

C$_9$H$_{17}$NO
Novonal, 6820
Pelletierine*, 7200

C$_9$H$_{17}$NO$_2$
Gabapentin, 4343

C$_9$H$_{17}$NO$_2$S
Lethane 384, 5472

C$_9$H$_{17}$NO$_4$
Acetylcarnitine, 85

C$_9$H$_{17}$NO$_5$
Pantothenic Acid, 7147

C$_9$H$_{17}$NO$_7$
Muramic Acid, 6384

C$_9$H$_{17}$NO$_8$
Neuraminic Acid, 6566

C$_9$H$_{17}$N$_2$NaO$_6$
Betaine*, 1225

C$_9$H$_{17}$N$_3$O$_3$
Dopastin, 3481

C$_9$H$_{17}$N$_5$S
Ametryn, 411

C$_9$H$_{18}$BrNO$_2$
Methyl N-Methylnipecotate*, 6172

C$_9$H$_{18}$Cl$_3$N$_2$O$_2$P
Trofosfamide, 9897

C$_9$H$_{18}$FeN$_3$S$_6$
Ferbam, 4054

C$_9$H$_{18}$N$_2$O$_2$
Capuride, 1818

C$_9$H$_{18}$N$_2$O$_4$
Meprobamate, 5908

C$_9$H$_{18}$N$_3$OP
Metepa, 5998

C$_9$H$_{18}$N$_4$
Guanacline, 4586

C$_9$H$_{18}$N$_4$O
Amidinomycin, 419

C$_9$H$_{18}$N$_6$
Altretamine, 328

C$_9$H$_{18}$N$_6$O$_6$
Hexamethylolmelamine, 4727

C$_9$H$_{18}$O$_2$
n-Amyl Butyrate, 645
Ethyl Oenanthate, 3881
Isoamyl Butyrate, 5129
Isobutyl Isovalerate, 5160
Pelargonic Acid, 7198

C$_9$H$_{18}$O$_3$
n-Butyl Carbonate, 1593

C$_9$H$_{19}$ClN$_3$O$_5$P
Fotemustine, 4285

C$_9$H$_{19}$N
Cyclopentamine, 2808
Isometheptene, 5202
Methylconiine, 6126

Denotes derivative of title compound

C₉H₁₉NO
Isovaleryl Diethylamide, 5253

C₉H₁₉NOS
EPTC, 3674

C₉H₁₉NO₃S
CAPS, 1810

C₉H₁₉NO₄
Dexpanthenol, 2988

C₉H₂₀Cl₃N₂O₃P
Defosfamide, 2917

C₉H₂₀NO₂⁺
Muscarine, 6389

C₉H₂₀N₄
Guanazodine, 4588

C₉H₂₀N₄O₄
Negamycin, 6530

C₉H₂₀O
n-Nonyl Alcohol, 6774

C₉H₂₁AlO₃
Aluminum Isopropoxide, 359

C₉H₂₁O₂PS₃
Terbufos, 9301

C₉H₂₂N₂
Novoldiamine, 6819

C₉H₂₂O₄P₂S₄
Ethion, 3782

C₉H₂₃INO₃PS
Echothiophate Iodide, 3549

C₉H₂₄I₂N₂O
Prolonium Iodide, 7965

C₉H₂₈Si₄
Tris(trimethylsilyl)silane, 9886

C₁₀Cl₁₀
Dienochlor, 3154

C₁₀Cl₁₀O
Chlordecone, 2131

C₁₀Cl₁₂
Mirex, 6294

C₁₀F₁₈
Fluosol DA*, 4221

C₁₀H₄Cl₂O₂
Dichlone, 3096

C₁₀H₄N₂Na₂O₈S
Naphthol Yellow S, 6480

C₁₀H₅ClN₂
o-Chlorobenzylidenemalononitrile, 2178

C₁₀H₅Cl₇
Heptachlor, 4690

C₁₀H₅NNa₂O₈S₂
Nitroso-R Salt, 6742

C₁₀H₅NaO₅S
Sodium β-Naphthoquinone-4-sulfonate, 8791

C₁₀H₆ClN₅O
Triazoxide, 9737

C₁₀H₆Cl₂N₂O₂
Pyrrolnitrin, 8202

C₁₀H₆Cl₄O₄
DCPA, 2896

C₁₀H₆Cl₈
Chlordan(e), 2129

C₁₀H₆N₂OS₂
Oxythioquinox, 7113

C₁₀H₆N₂O₄
5-Nitroquinaldic Acid, 6728

C₁₀H₆O₂
1,2-Naphthoquinone, 6481
1,4-Naphthoquinone, 6482

C₁₀H₆O₃
Juglone, 5282
Lawsone, 5406

C₁₀H₆O₄
Chromocarb, 2294
Coumarin-3-carboxylic Acid, 2631

C₁₀H₆O₈
Pyromellitic Acid, 8188

C₁₀H₇Br
1-Bromonaphthalene, 1447
2-Bromonaphthalene, 1448

C₁₀H₇Cl
1-Chloronaphthalene, 2201
2-Chloronaphthalene, 2202

C₁₀H₇Cl₂NO
Chlorquinaldol, 2243

C₁₀H₇Cl₂N₃O
Anagrelide, 665

C₁₀H₇Cl₅O
Tridiphane, 9795

C₁₀H₇F₃O₄
Triflusal, 9817

C₁₀H₇KN₆O
Pemirolast*, 7205

C₁₀H₇NO₂
1-Nitronaphthalene, 6711
1-Nitroso-2-naphthol, 6739
N-Phenylmaleimide, 7452
Quinaldic Acid, 8227
8-Quinolinecarboxylic Acid, 8255

C₁₀H₇NO₃
Kynurenic Acid, 5339
1-Nitro-2-naphthol, 6712

C₁₀H₇NO₄
Xanthurenic Acid, 10202

C₁₀H₇N₃S
Thiabendazole, 9426

C₁₀H₇NaO
2-Naphthol*, 6471

C₁₀H₇NaO₄S
2-Naphthol-6-sulfonic Acid*, 6479

C₁₀H₈
Azulene, 956
Naphthalene, 6457

C₁₀H₈ClN₃O₂
Drazoxolon, 3499

C₁₀H₈ClN₄NaO₂S
Sulfachlorpyridazine*, 9068

C₁₀H₈NNaO₃S.4H₂O
1-Naphthylamine-4-sulfonic Acid*, 6490

C₁₀H₈NNaO₄S.2½H₂O
1-Amino-2-naphthol-6-sulfonic Acid*, 475

C₁₀H₈N₂
α,α'-Dipyridyl, 3411
γ,γ'-Dipyridyl, 3412

C₁₀H₈N₂O₂S₂
Pyrithione*, 8178

C₁₀H₈N₂O₂S₂Zn
Pyrithione*, 8178

C₁₀H₈N₂O₄
α-Furildioxime, 4328

C₁₀H₈N₄O₃
Nifurprazine, 6629

C₁₀H₈N₄O₃S₂
Nitrodan, 6693

C₁₀H₈N₄O₄
Nitrefazole, 6668

C₁₀H₈N₄O₅
Picrolonic Acid, 7565

C₁₀H₈N₆O
Pemirolast, 7205

C₁₀H₈O
1-Naphthol, 6470
2-Naphthol, 6471

C₁₀H₈OS₃
Anethole Trithione, 683

C₁₀H₈O₂
Naphthoresorcinol, 6483
Tricromyl, 9791

C₁₀H₈O₃
Erythrocentaurin, 3717
Hymecromone, 4903

C₁₀H₈O₃S
1-Naphthalenesulfonic Acid, 6463
2-Naphthalenesulfonic Acid, 6464

C₁₀H₈O₄
Anemonin, 681
Scopoletin, 8552

C₁₀H₈O₄S

Cassella's Acid, 1940
Croceic Acid, 2656
1-Naphthol-2-sulfonic Acid, 6477
1-Naphthol-4-sulfonic Acid, 6478
2-Naphthol-6-sulfonic Acid, 6479

C₁₀H₈O₅
Fraxetin, 4289

C₁₀H₈O₆S₂
Armstrong's Acid, 824
1,6-Naphthalenedisulfonic Acid, 6460
2,6-Naphthalenedisulfonic Acid, 6461
2,7-Naphthalenedisulfonic Acid, 6462

C₁₀H₈O₇S₂
1-Naphthol-4,8-disulfonic Acid, 6473
2-Naphthol-3,6-disulfonic Acid, 6474
2-Naphthol-6,8-disulfonic Acid, 6475

C₁₀H₈O₈S₂
Chromotropic Acid, 2298

C₁₀H₈O₁₀S₂
Sulmarin, 9158

C₁₀H₈S
1-Naphthalenethiol, 6465
2-Naphthalenethiol, 6466

C₁₀H₉ClN₄O₂S
Sulfachlorpyridazine, 9068

C₁₀H₉Cl₂NO
Dicryl, 3138

C₁₀H₉Cl₂NaO₃
2,4-DB*, 2893

C₁₀H₉Cl₄NO₂S
Captafol, 1814

C₁₀H₉Cl₄O₄P
Tetrachlorvinphos, 9334

C₁₀H₉I₃O₃
Phenobutiodil, 7387

C₁₀H₉N
Lepidine, 5464
1-Naphthylamine, 6485
2-Naphthylamine, 6486
Quinaldine, 8228

C₁₀H₉NO
4-Amino-1-naphthol, 473
Echinopsine, 3546

C₁₀H₉NO₂
Carfimate, 1886
Gentianine, 4400
Indoleacetic Acid, 4994
Succinanil, 9036

C₁₀H₉NO₃
Maleanilic Acid, 5742

C₁₀H₉NO₃S
Badische Acid, 969
Cassella's Acid F, 1941
1,6-Cleve's Acid, 2409
1,7-Cleve's Acid, 2410
1-Naphthylamine-4-sulfonic Acid, 6490
1-Naphthylamine-5-sulfonic Acid, 6491
1-Naphthylamine-8-sulfonic Acid, 6492
2-Naphthylamine-1-sulfonic Acid, 6493
2-Naphthylamine-5-sulfonic Acid, 6494

C₁₀H₉NO₄S
1-Amino-2-naphthol-4-sulfonic Acid, 474
1-Amino-2-naphthol-6-sulfonic Acid, 475

C₁₀H₉NO₆S₂
Amido-G-Acid, 421
Amido-R-Acid, 423
1-Naphthylamine-2,7-disulfonic

* *Denotes derivative of title compound*

C$_{10}$H$_{16}$N$_2$O$_4$S$_3$
Dorzolamide, 3484

C$_{10}$H$_{16}$N$_2$O$_8$
Edetic Acid, 3559

C$_{10}$H$_{16}$N$_4$O$_3$
Anserine, 717
Dimetilan, 3313

C$_{10}$H$_{16}$N$_4$O$_7$
Vicine, 10112

C$_{10}$H$_{16}$N$_5$O$_{13}$P$_3$
Adenosine Triphosphate, 154

C$_{10}$H$_{16}$N$_6$S
Cimetidine, 2337

C$_{10}$H$_{16}$O
Camphor, 1779
Citral, 2383
d-Fenchone, 4008
Homocamfin, 4768
Piperitone, 7626
Pulegone, 8124
Thujone, 9533

C$_{10}$H$_{16}$O$_2$
Ascaridole, 864
Chrysanthemic Acid, 2310
Cicrotoic Acid, 2327
Diosphenol, 3351
3-Hydroxycamphor, 4862
Iridomyrmecin, 5102
Lineatin, 5528
Oxypinocamphone, 7110
Sebacil, 8559

C$_{10}$H$_{16}$O$_3$
Queen Substance, 8214

C$_{10}$H$_{16}$O$_4$
Camphoric Acid, 1780
Prelog-Djerassi Lactone, 7916

C$_{10}$H$_{16}$O$_4$S
d-Camphorsulfonic Acid, 1781

C$_{10}$H$_{16}$O$_5$
Senecic Acid, 8593

C$_{10}$H$_{17}$BrMgN$_2$O$_8$.H$_2$O
Glutamic Acid*, 4477

C$_{10}$H$_{17}$Cl
Bornyl Chloride, 1369

C$_{10}$H$_{17}$Cl$_2$NOS
Diallate, 3014

C$_{10}$H$_{17}$N
Amantadine, 389

C$_{10}$H$_{17}$NO$_2$
Methyprylon, 6216

C$_{10}$H$_{17}$NO$_4$S$_2$
Letosteine, 5473

C$_{10}$H$_{17}$NO$_5$S
Ethamsylate, 3766

C$_{10}$H$_{17}$NO$_6$
Linamarin, 5522

C$_{10}$H$_{17}$N$_2$O$_4$PS
Etrimfos, 3936

C$_{10}$H$_{17}$N$_3$
Hexazole, 4735

C$_{10}$H$_{17}$N$_3$O$_2$
Isolan, 5195

C$_{10}$H$_{17}$N$_3$O$_5$
Linatine, 5524

C$_{10}$H$_{17}$N$_3$O$_6$S
Glutathione, 4483

C$_{10}$H$_{17}$N$_3$S
Pramipexole, 7885

C$_{10}$H$_{17}$N$_5$O$_3$
Theophylline*, 9421

C$_{10}$H$_{17}$N$_7$O$_4$$^{2+}$
Saxitoxin, 8533

C$_{10}$H$_{17}$NaO$_3$
Cyclobutyrol*, 2784

C$_{10}$H$_{18}$
Decalin, 2903

C$_{10}$H$_{18}$CaO$_6$S$_2$
Methionine Hydroxy Analog*, 6054

C$_{10}$H$_{18}$ClN$_3$O$_7$
Ranimustine, 8285

C$_{10}$H$_{18}$N$_2$O$_3$
Desthiobiotin, 2978

C$_{10}$H$_{18}$N$_2$O$_7$
α-Ketoglutaric Acid*, 5314

C$_{10}$H$_{18}$N$_4$O$_4$S$_3$
Thiodicarb, 9465

C$_{10}$H$_{18}$O
Borneol, 1366
Citronellal, 2390
Eucalyptol, 3940
Geraniol, 4411
Grandisol, 4556
Isoborneol, 5143
Linalool, 5520
l-Menthone, 5883
Nerol, 6560
α-Terpineol, 9316

C$_{10}$H$_{18}$O$_2$
Sebacoin, 8560
Sobrerol, 8707

C$_{10}$H$_{18}$O$_3$
Cyclobutyrol, 2784
Royal Jelly*, 8434

C$_{10}$H$_{18}$O$_4$
Adipic Acid*, 161
Sebacic Acid, 8558

C$_{10}$H$_{18}$O$_4$Zn
Zinc Valerate, 10299

C$_{10}$H$_{19}$ClNO$_5$P
Phosphamidon, 7493

C$_{10}$H$_{19}$NO
Lupinine, 5637

C$_{10}$H$_{19}$NO$_2$
Procymate, 7945

C$_{10}$H$_{19}$NO$_3$
MeBmt, 5811

C$_{10}$H$_{19}$NO$_5$
Hopantenic Acid, 4781

C$_{10}$H$_{19}$N$_3$O$_2$
Guanadrel, 4587

C$_{10}$H$_{19}$N$_5$O
Prometon, 7972

C$_{10}$H$_{19}$N$_5$S
Prometryn, 7973

C$_{10}$H$_{19}$O$_6$PS$_2$
Malathion, 5740

C$_{10}$H$_{20}$CaN$_2$O$_8$S$_2$
Acamprosate Calcium, 14

C$_{10}$H$_{20}$NO$_4$PS
Propetamphos, 8000

C$_{10}$H$_{20}$N$_2$O$_4$
Mebutamate, 5813
Piperazine Adipate, 7618

C$_{10}$H$_{20}$N$_2$O$_4$S$_2$
Penicillamine Disulfide, 7216

C$_{10}$H$_{20}$N$_2$S$_3$
Sulfiram, 9123

C$_{10}$H$_{20}$N$_2$S$_4$
Disulfiram, 3428

C$_{10}$H$_{20}$O
β-Citronellol, 2391
Menthol, 5882
Rhodinol, 8352

C$_{10}$H$_{20}$O$_2$
tert-Amyl Isovalerate, 652
n-Capric Acid, 1802
Ethyl Caprylate, 3825
Isoamyl Isovalerate, 5135
Octyl Acetate, 6860
Terpin, 9314

C$_{10}$H$_{21}$N
Cyclexedrine, 2776
Pempidine, 7207
Propylhexedrine, 8045

C$_{10}$H$_{21}$NO
Trimethyl Isopropyl Butanamide, 9847

C$_{10}$H$_{21}$NOS
Pebulate, 7192
Vernolate, 10097

C$_{10}$H$_{21}$N$_3$O
Diethylcarbamazine, 3165

C$_{10}$H$_{21}$N$_5$O$_6$
Aspartic Acid*, 875

C$_{10}$H$_{22}$N$_2$O$_4$
Acetylleucine Monoethanolamine, 96

C$_{10}$H$_{22}$N$_4$
Guanethidine, 4589
Spherophysine, 8896

C$_{10}$H$_{22}$O
n-Amyl Ether, 651
n-Decyl Alcohol, 2911
Isoamyl Ether, 5132

C$_{10}$H$_{22}$O$_2$
Decamethylene Glycol, 2906

C$_{10}$H$_{22}$O$_5$
Tetraglyme, 9348

C$_{10}$H$_{23}$B
Bis(1,2-dimethylpropyl)borane, 1290

C$_{10}$H$_{23}$N
Diisoamylamine, 3237

C$_{10}$H$_{23}$N$_3$O$_3$
Hypusine, 4918

C$_{10}$H$_{23}$O$_2$PS$_2$
Cadusafos, 1670

C$_{10}$H$_{24}$Cl$_4$N$_2$O$_4$
Mannomustine, 5790

C$_{10}$H$_{24}$NO$_3$PS
Amiton, 509

C$_{10}$H$_{24}$N$_2$O$_2$
Ethambutol, 3764

C$_{10}$H$_{26}$N$_4$
Spermine, 8894

C$_{10}$H$_{30}$O$_3$Si$_4$
Decamethyltetrasiloxane, 2907

C$_{10}$H$_{30}$O$_5$Si$_5$
Decamethylcyclopentasiloxane, 2905

C$_{10}$H$_{32}$N$_4$O$_8$P$_2$.6H$_2$O
Spermine*, 8894

C$_{10}$Mn$_2$O$_{10}$
Manganese Carbonyl, 5767

C$_{11}$H$_6$ClN$_3$O$_6$
Lodoxamide, 5585

C$_{11}$H$_6$Cl$_2$N$_2$
Fenpiclonil, 4030

C$_{11}$H$_6$O$_3$
Psoralen, 8113

C$_{11}$H$_7$NO
1-Naphthylisocyanate, 6497

C$_{11}$H$_7$NS
1-Naphthylisothiocyanate, 6498

C$_{11}$H$_8$ClNO$_2$
Cloxyquin*, 2483

C$_{11}$H$_8$ClNO$_2$S
Fenclozic Acid, 4010

C$_{11}$H$_8$I$_3$N$_2$NaO$_4$
Diatrizoate Sodium, 3040
Iothalamic Acid*, 5080

C$_{11}$H$_8$N$_2$
γ-Carboline, 1853

C$_{11}$H$_8$N$_2$O$_3$S$_2$
Firefly Luciferin, 4128

C$_{11}$H$_8$Na$_4$O$_8$P$_2$
Menadiol*, 5873

C$_{11}$H$_8$O$_2$
Menadione, 5874
1-Naphthoic Acid, 6468
2-Naphthoic Acid, 6469

C$_{11}$H$_8$O$_3$
1-Hydroxy-2-naphthoic Acid, 4881
3-Hydroxy-2-naphthoic Acid, 4882
Phthiocol, 7534
Plumbagin, 7697

C$_{11}$H$_8$O$_5$
Purpurogallin, 8133

C$_{11}$H$_9$Cl$_2$NO$_2$
Barban, 986

Denotes derivative of title compound

C₁₁H₉Cl₅O₃
Erbon, 3681
C₁₁H₉FN₂O₃
Sorbinil, 8871
C₁₁H₉I₃N₂O₄
Iothalamic Acid, 5080
C₁₁H₉NO₂
3-Amino-2-naphthoic Acid, 472
C₁₁H₉NO₃
Quininic Acid, 8248
C₁₁H₉N₃O₂
1,2-Naphthoquinone*, 6481
C₁₁H₉NaO₅S
Menadione*, 5874
C₁₁H₁₀BrN₅
Brimonidine, 1399
C₁₁H₁₀FNO₂S
Flosequinan, 4146
C₁₁H₁₀F₃NO₂
Flumetramide, 4176
C₁₁H₁₀NNaO₄S
Actinoquinol*, 143
C₁₁H₁₀N₂O
Amphenidone, 621
C₁₁H₁₀N₂O₃
Phenylmethylbarbituric Acid, 7457
C₁₁H₁₀N₂S
ANTU, 764
C₁₁H₁₀N₃NaO₂S.H₂O
Sulfapyridine*, 9108
C₁₁H₁₀N₄O₄
Carbadox, 1825
C₁₁H₁₀O
2-Methoxynaphthalene, 6076
C₁₁H₁₀O₂
Menadiol, 5873
C₁₁H₁₀O₄
Limettin, 5517
Scoparone, 8548
C₁₁H₁₀O₅
Gladiolic Acid, 4442
C₁₁H₁₀O₆
Dipyrocetyl, 3413
C₁₁H₁₁ClO₃
Alclofenac, 220
C₁₁H₁₁Cl₂N₃O
Muzolimine, 6397
C₁₁H₁₁Cl₄NO₂
Chlorbetamide, 2126
C₁₁H₁₁CuN₂NaO₂S
Allocupreide Sodium, 265
C₁₁H₁₁F₃N₂O₃
Flutamide, 4242
C₁₁H₁₁I₃O₃
Iophenoxic Acid, 5075
C₁₁H₁₁NO
3-Indolylacetone, 4997
Vitamin K₅, 10162
C₁₁H₁₁NO₂
Phensuximide, 7414
C₁₁H₁₁NO₄S
Actinoquinol, 143
Woodward's Reagent K, 10186
C₁₁H₁₁N₃O₂S
Sulfapyridine, 9108
C₁₁H₁₁N₄NaO₂S
Sulfamerazine*, 9081
C₁₁H₁₁N₄NaO₃S
Sulfamethoxypyridazine*, 9087
C₁₁H₁₁N₅O₃S₂
Urothion, 10025
C₁₁H₁₂AsNO₅S₂
Arsenamide, 831
C₁₁H₁₂BrNO
Cinromide, 2371
C₁₁H₁₂ClNO₃S
Chlormezanone, 2155
C₁₁H₁₂ClN₅
Phenazopyridine Hydrochloride, 7361

C₁₁H₁₂Cl₂N₂O
Lofexidine, 5588
C₁₁H₁₂Cl₂N₂O₅
Chloramphenicol, 2120
C₁₁H₁₂Cl₂O₃
2,4-D*, 2865
C₁₁H₁₂Cl₃N
Amphecloral, 620
C₁₁H₁₂I₃NO₂
Iopanoic Acid, 5072
C₁₁H₁₂NO₄PS₂
Phosmet, 7492
C₁₁H₁₂N₂O
Antipyrine, 757
Vasicine, 10072
C₁₁H₁₂N₂O.C₈H₈O₃
Antipyrine*, 757
C₁₁H₁₂N₂O₂
Ethotoin, 3795
Fenozolone, 4028
Thozalinone, 9521
Tryptophan, 9929
C₁₁H₁₂N₂O₂S
Zileuton, 10253
C₁₁H₁₂N₂O₃
5-Hydroxytryptophan, 4895
C₁₁H₁₂N₂S
Levamisole, 5486
C₁₁H₁₂N₄O₂
Todralazine, 9640
C₁₁H₁₂N₄O₂S
Sulfamerazine, 9081
Sulfaperine, 9104
C₁₁H₁₂N₄O₃S
Sulfalene, 9078
Sulfameter, 9082
Sulfamethoxypyridazine, 9087
C₁₁H₁₂O₂
Cinnamic Acid*, 2358
6-Methoxy-α-tetralone, 6081
C₁₁H₁₂O₃
Ethyl Benzoylacetate, 3815
Myristicin, 6417
C₁₁H₁₂O₅
Elenolide, 3586
C₁₁H₁₂O₇
Piscidic Acid, 7664
C₁₁H₁₃AsN₂O₅S₂
Thiocarbarsone, 9459
C₁₁H₁₃BrN₂O₆
Sorivudine, 8875
C₁₁H₁₃ClF₃N₃O₄S₃
Polythiazide, 7744
C₁₁H₁₃ClN₂
Epibatidine, 3647
C₁₁H₁₃ClO
Dowicide 9, 3486
C₁₁H₁₃F₂N
Mofegiline, 6313
C₁₁H₁₃F₃N₂O₃S
Mefluidide, 5846
C₁₁H₁₃N
Pargyline, 7172
C₁₁H₁₃NO
β-Benzalbutyramide, 1083
C₁₁H₁₃NO₂
Hydrohydrastinine, 4846
Idrocilamide, 4935
C₁₁H₁₃NO₃
Corydaldine, 2609
Hydrastinine, 4807
4-Salicyloylmorpholine, 8485
Thurfyl Nicotinate, 9536
Toloxatone, 9659
C₁₁H₁₃NO₄
Bendiocarb, 1063
Mephenoxalone, 5896
C₁₁H₁₃N₃O
Ampyrone, 633
C₁₁H₁₃N₃O₃S

Sulfamoxole, 9093
Sulfisoxazole, 9125
C₁₁H₁₃N₅O₅
Azidamfenicol, 941
C₁₁H₁₄AsNO₃S₂
Arsthinol, 852
C₁₁H₁₄BrN₂NaO₃
Butallylonal*, 1537
C₁₁H₁₄ClNO
Propachlor, 7977
C₁₁H₁₄ClNO₂
Buclosamide, 1485
C₁₁H₁₄ClN₃O₄S₃
Althiazide, 326
C₁₁H₁₄ClN₅
Cycloguanil, 2790
C₁₁H₁₄Cl₂N₂O₅S
Tevenel, 9387
C₁₁H₁₄N₂
Gramine, 4554
C₁₁H₁₄N₂O
Cytisine, 2858
5-Methoxytryptamine, 6083
C₁₁H₁₄N₂O₂
Levulinic Acid*, 5498
Pheneturide, 7375
C₁₁H₁₄N₂O₃S
Sulfadicramide, 9072
C₁₁H₁₄N₂O₄
Felbamate, 3988
5'-Nitro-2'-propoxyacetanilide, 6726
C₁₁H₁₄N₂S
Pyrantel, 8139
C₁₁H₁₄N₄O₂
Epirizole, 3659
C₁₁H₁₄N₄O₄
Doxofylline, 3494
Tubercidin, 9936
C₁₁H₁₄O₂
n-Butyl Benzoate, 1587
C₁₁H₁₄O₃
Butylparaben, 1619
o-Thymotic Acid, 9549
Zingerone, 10301
C₁₁H₁₄O₄
Dimethyl Carbate, 3290
C₁₁H₁₅BrN₂O₃
Butallylonal, 1537
Narcobarbital, 6509
C₁₁H₁₅ClN₂O₂
Iproclozide, 5092
C₁₁H₁₅Cl₂N₅
Chlorproguanil, 2237
C₁₁H₁₅NO
Metamfepramone, 5984
Phenmetrazine, 7385
C₁₁H₁₅NO₂
Butamben, 1538
Isobutyl p-Aminobenzoate, 5148
MDMA, 5806
Salsoline, 8492
C₁₁H₁₅NO₂S
Methiocarb, 6050
C₁₁H₁₅NO₃
Anhalamine, 691
Hydroxyphenamate, 4884
p-Lactophenetide, 5355
Propoxur, 8022
C₁₁H₁₅NO₄
Mephenesin*, 5895
C₁₁H₁₅NO₄S
Ethebenecid, 3775
C₁₁H₁₅NO₅
Methocarbamol, 6060
C₁₁H₁₅N₂NaO₂S
Buthalital Sodium, 1553
C₁₁H₁₅N₂NaO₃
Enallylpropymal*, 3607
Vinbarbital Sodium, 10118

$C_{11}H_{15}N_2O_8P$
NMN, 6761
$C_{11}H_{15}N_3O_4$
Pyridinol Carbamate, 8158
$C_{11}H_{15}N_5O_5$
Psicofuranine, 8108
$C_{11}H_{16}$
Amylbenzene, 641
$C_{11}H_{16}ClNO$
Clorprenaline, 2470
$C_{11}H_{16}ClN_3O_4S_2$
Buthiazide, 1554
$C_{11}H_{16}ClN_5$
Chlorguanide, 2138
$C_{11}H_{16}ClO_2PS_3$
Carbophenothion, 1869
$C_{11}H_{16}FN_3O_3$
Carmofur, 1892
$C_{11}H_{16}I_2N_2O_5$
Iodopyracet, 5061
$C_{11}H_{16}N_2O$
Tocainide, 9629
$C_{11}H_{16}N_2O_2$
Aminocarb, 452
Pilocarpine, 7578
$C_{11}H_{16}N_2O_3$
Butalbital, 1536
Enallylpropymal, 3607
Nifenalol, 6618
Talbutal, 9206
Vinylbital, 10131
$C_{11}H_{16}N_2O_3S$
Etozolin*, 3934
Phenbutamide, 7362
$C_{11}H_{16}N_2O_4S$
Thienamycin, 9448
$C_{11}H_{16}N_2O_5$
Edoxudine, 3561
$C_{11}H_{16}N_4O$
Lidamidine, 5504
$C_{11}H_{16}N_4O_4$
Pentostatin, 7277
Razoxane, 8295
$C_{11}H_{16}O$
Fenipentol, 4016
Jasmone, 5273
p-tert-Pentylphenol, 7284
$C_{11}H_{16}O_2$
Butylated Hydroxyanisole, 1582
$C_{11}H_{16}O_3$
d-Camphocarboxylic Acid, 1778
$C_{11}H_{17}Cl_3N_2O_2S$
Chlordantoin, 2130
$C_{11}H_{17}N$
N-Ethylamphetamine, 3809
Mephentermine, 5897
Pentorex, 7275
$C_{11}H_{17}NO$
Methoxyphenamine, 6077
N-Methylephedrine, 6145
Mexiletine, 6257
Tecomanine, 9263
$C_{11}H_{17}NO_3$
Dimetan, 3259
Dioxethedrine, 3356
Isoproterenol, 5236
Mescaline, 5965
Metaproterenol, 5992
Methoxamine, 6067
$C_{11}H_{17}NO_4$
Dimetofrine, 3314
$C_{11}H_{17}N_2NaO_2S$
Thiopental Sodium, 9487
$C_{11}H_{17}N_2NaO_3$
Amobarbital*, 607
Pentobarbital*, 7272
$C_{11}H_{17}N_3O$
Meobentine, 5889
$C_{11}H_{17}N_3O_3$
Emorfazone, 3603

$C_{11}H_{17}N_3O_3S$
1-Butyl-3-metanilylurea, 1615
Carbutamide, 1881
$C_{11}H_{17}N_3O_5$
Carbubarb, 1880
$C_{11}H_{17}N_3O_8$
Tetrodotoxin, 9382
$C_{11}H_{17}N_5O_2$
Dimethazan, 3261
$C_{11}H_{17}N_5O_3$
Cafaminol, 1671
$C_{11}H_{17}O_4PS_2$
Fensulfothion, 4042
$C_{11}H_{18}BrN_5O_3$
Pamabrom, 7133
$C_{11}H_{18}NO^+$
Leptodactyline, 5467
$C_{11}H_{18}N_2O_2$
p-Aminobenzoic Acid*, 443
$C_{11}H_{18}N_2O_3$
Amobarbital, 607
Pentobarbital, 7272
$C_{11}H_{18}N_2O_4Pt$
Miboplatin, 6263
$C_{11}H_{18}N_3O_2^+$
Murexine, 6387
$C_{11}H_{18}N_4O_2$
Pirimicarb, 7651
$C_{11}H_{18}N_4O_6$
Lysine*, 5667
$C_{11}H_{18}O_5$
Apiose*, 778
$C_{11}H_{19}NOS$
Octhilinone, 6853
$C_{11}H_{19}NO_3$
d-Camphocarboxylic Acid*, 1778
$C_{11}H_{19}NO_9$
Sialic Acids*, 8628
$C_{11}H_{19}N_3O$
Dimethirimol, 3266
Ethirimol, 3786
$C_{11}H_{19}N_3O_6$
Wildfire Toxin, 10183
$C_{11}H_{19}N_5O_3$
Ambuphylline, 404
$C_{11}H_{20}N_2O_6$
L-Saccharopine, 8464
$C_{11}H_{20}N_3O_3PS$
Pirimiphos-ethyl*, 7652
Pyrimithate, 8172
$C_{11}H_{20}O_2$
Undecylenic Acid, 9983
$C_{11}H_{20}O_4$
Di-tert-butyl Malonate, 3085
Ethyl Diethylmalonate, 3835
$C_{11}H_{20}O_{10}$
Primeverose, 7926
Vicianose, 10111
$C_{11}H_{21}N$
Mecamylamine, 5814
$C_{11}H_{21}NO_8$
Choline Dihydrogen Citrate, 2263
$C_{11}H_{21}N_5S$
Dipropetryn, 3405
$C_{11}H_{22}N_2O_4S$
Pantetheine, 7143
$C_{11}H_{22}N_3O_3P$
Meturedepa, 6245
$C_{11}H_{22}O$
Methyl Nonyl Ketone, 6178
$C_{11}H_{22}O_2$
n-Amyl Caproate, 646
Ethyl Pelargonate, 3884
n-Nonyl Acetate, 6773
$C_{11}H_{22}O_4$
Monoctanoin*, 6335
$C_{11}H_{23}NOS$
Butylate, 1581
$C_{11}H_{23}N_3O_6$
L-Lysine L-Glutamate, 5669

$C_{11}H_{23}N_5O_6$
Arginine Glutamate, 818
$C_{11}H_{23}N_7$
Hydramitrazine, 4803
$C_{11}H_{24}FeNO_{11}$
Ferrocholinate, 4085
$C_{11}H_{25}N$
Metron S, 6243
$C_{11}H_{26}NO_2PS$
VX, 10172
$C_{11}H_{28}Br_2N_2$
Pentamethonium Bromide, 7253
$C_{12}H_4Cl_2F_6N_4OS$
Fipronil, 4127
$C_{12}H_4Cl_4Na_2O_2S$
Bithionol*, 1343
$C_{12}H_4Cl_4O_2$
TCDD, 9252
$C_{12}H_4Na_5O_{16}S_4Sb.7H_2O$
Stibophen, 8967
$C_{12}H_5N_7O_{12}$
Dipicrylamine, 3396
$C_{12}H_6Cl_2NNaO_2$
2,6-Dichloroindophenol Sodium, 3118
$C_{12}H_6Cl_2N_2O_6$
Niclofolan, 6601
$C_{12}H_6Cl_3NO_3$
Chlornitrofen, 2158
$C_{12}H_6Cl_4O_2S$
Bithionol, 1343
Tetradifon, 9339
$C_{12}H_6Cl_4O_3S$
Bithionol*, 1343
$C_{12}H_6F_2N_2O_2$
Fludioxonil, 4165
$C_{12}H_6I_4O_8S_2Zn.6H_2O$
Sozoiodolic Acid*, 8881
$C_{12}H_6N_2O_2$
Phanquinone, 7337
$C_{12}H_6N_2O_5S$
Berninamycin*, 1200
$C_{12}H_6Na_6O_{12}S_6Sb_2$
Stibocaptate, 8966
$C_{12}H_6O_{12}$
Mellitic Acid, 5868
$C_{12}H_7Br_4O_5P$
Bromofenofos, 1440
$C_{12}H_7Cl_2NO_3$
Nitrofen, 6695
$C_{12}H_7Cl_3O_2$
Triclosan, 9790
$C_{12}H_7NO_4$
Resazurin, 8309
$C_{12}H_7N_3O_5S$
3,7-Dinitro-5-oxophenothiazine, 3332
$C_{12}H_8Cl_2O_2S$
Fenticlor, 4046
$C_{12}H_8Cl_2O_3S$
Genite, 4397
Ovex, 7033
$C_{12}H_8Cl_6$
Aldrin, 227
$C_{12}H_8Cl_6O$
Dieldrin, 3152
Endrin, 3618
$C_{12}H_8F_2$
4,4'-Difluorodiphenyl, 3193
$C_{12}H_8HgNa_2O_8S_2$
Mercuric Sodium p-Phenolsulfonate, 5939
$C_{12}H_8N_2$
o-Phenanthroline, 7356
Phenazine, 7359
$C_{12}H_8N_2O$
Hemipyocyanine, 4679
$C_{12}H_8N_2O_4$
Iodinin, 5042

* Denotes derivative of title compound

C₁₂H₈N₂O₄S₂
Nitrophenide, 6715
$C_{12}H_8N_4O_4S$
Nifurzide, 6632
$C_{12}H_8O_4$
Bergapten(e), 1196
Methoxsalen, 6068
Naphthalic Acid, 6467
$C_{12}H_8S_3$
α-Terthienyl, 9319
$C_{12}H_9AsClN$
Phenarsazine Chloride, 7357
$C_{12}H_9ClFN_5$
Arprinocid, 828
$C_{12}H_9ClF_3N_3O$
Norflurazon, 6794
$C_{12}H_9ClO$
2-Phenyl-6-chlorophenol, 7434
$C_{12}H_9ClO_2S$
Sulphenone, 9162
$C_{12}H_9Cl_2NO_3$
Vinclozolin, 10122
$C_{12}H_9F_2N_5O_2S$
Flumetsulam, 4177
$C_{12}H_9Li_6O_{12}S_3Sb$
Anthiolimine, 720
$C_{12}H_9N$
Carbazole, 1834
$C_{12}H_9NO$
Phenoxazine, 7406
$C_{12}H_9NO_2$
Dictamnine, 3139
o-Nitrobiphenyl, 6689
p-Nitrobiphenyl, 6690
$C_{12}H_9NO_6$
Miloxacin, 6283
$C_{12}H_9NS$
Phenothiazine, 7404
$C_{12}H_9N_2NaO_5S$
Tropaeolin O, 9905
$C_{12}H_9N_3O$
Milrinone, 6284
$C_{12}H_9N_3O_4$
Magneson, 5735
$C_{12}H_9N_3O_5$
Nifuroxazide, 6626
$C_{12}H_9NaO.4H_2O$
o-Phenylphenol*, 7458
$C_{12}H_{10}$
Acenaphthene, 28
Diphenyl, 3372
Hemi-Dewar Biphenyl, 4677
$C_{12}H_{10}As_2N_2Na_2O_2$
Sodium Arsphenamine, 8722
$C_{12}H_{10}BaO_6S_2$
Barium Benzenesulfonate, 993
$C_{12}H_{10}BaO_8S_2$
p-Phenolsulfonic Acid*, 7396
$C_{12}H_{10}Ba_3O_{14}.7H_2O$
Citric Acid*, 2387
$C_{12}H_{10}CaO_2$
Calcium Phenoxide, 1738
$C_{12}H_{10}CaO_8S_2$
Calcium Phenolsulfonate, 1737
$C_{12}H_{10}CaO_{10}S_2$
Dobesilate Calcium, 3455
$C_{12}H_{10}Ca_2FeO_{14}$
Calcium Ferrous Citrate, 1708
$C_{12}H_{10}Ca_3O_{14}$
Calcium Citrate, 1701
$C_{12}H_{10}ClN$
4-Amino-4′-chlorodiphenyl, 454
$C_{12}H_{10}ClN_3S$
Thionine, 9483
$C_{12}H_{10}Cl_2N_2$
2,2′-Dichlorobenzidine, 3108
3,3′-Dichlorobenzidine, 3109
$C_{12}H_{10}CuO_8S_2$
Cupric p-Phenolsulfonate, 2715

$C_{12}H_{10}F_3N_3O_4$
Nilutamide, 6636
$C_{12}H_{10}Mg$
Diphenylmagnesium, 3386
$C_{12}H_{10}Mg_3O_{14}$
Magnesium Citrate, 5699
$C_{12}H_{10}N_2$
Azobenzene, 948
Harman, 4646
$C_{12}H_{10}N_2O$
Azoxybenzene, 954
Ethyl β-Carboline-3-carboxylate*, 3826
p-Nitrosodiphenylamine, 6737
$C_{12}H_{10}N_2O_4$
Nifurpirinol, 6628
$C_{12}H_{10}N_2O_5$
Cinoxacin, 2369
$C_{12}H_{10}N_4$
2,3-Diaminophenazine, 3025
$C_{12}H_{10}N_4O_2$
Lumichrome, 5626
$C_{12}H_{10}O$
Phenyl Ether, 7442
o-Phenylphenol, 7458
$C_{12}H_{10}O_2$
Chimaphilin, 2100
1-Naphthaleneacetic Acid, 6458
$C_{12}H_{10}O_2S$
Diphenyl Sulfone, 3392
$C_{12}H_{10}O_3$
2-Naphthoxyacetic Acid, 6484
Spizofurone, 8918
$C_{12}H_{10}O_4$
Acifran, 112
Diresorcinol, 3417
Piperic Acid, 7620
Quinhydrone, 8242
$C_{12}H_{10}O_5S_2$
Benzenesulfonic Anhydride, 1099
$C_{12}H_{10}O_7$
Echinochrome A, 3544
$C_{12}H_{10}O_8S_2Zn$
Zinc p-Phenolsulfonate, 10283
$C_{12}H_{10}O_{14}Zn_3$
Zinc Citrate, 10263
$C_{12}H_{10}S$
Phenyl Sulfide, 7467
$C_{12}H_{11}ClN_2O_5S$
Furosemide, 4331
$C_{12}H_{11}ClN_6O_2S_2$
Azosemide, 953
$C_{12}H_{11}Cl_2NO$
Propyzamide, 8058
$C_{12}H_{11}Hg_2NO_4$
Phenylmercuric Nitrate, Basic, 7455
$C_{12}H_{11}I_3N_2O_4$
Iodamide, 5031
Metrizoic Acid, 6241
$C_{12}H_{11}N$
p-Biphenylamine, 1276
Diphenylamine, 3375
$C_{12}H_{11}NOS$
Thionalide, 9480
$C_{12}H_{11}NO_2$
4-Amino-1-naphthol*, 473
Carbaryl, 1831
$C_{12}H_{11}NO_3S$
N-Phenylsulfanilic Acid, 7466
$C_{12}H_{11}NO_4$
Casimiroin, 1936
$C_{12}H_{11}N_2NaO_3$
Phenobarbital*, 7386
$C_{12}H_{11}N_3$
p-Aminoazobenzene, 439
Diazoaminobenzene, 3046
$C_{12}H_{11}N_3O_2$
Furonazide, 4330
$C_{12}H_{11}N_7$
Triamterene, 9731

$C_{12}H_{12}Al_2O_{18}$
Aluminum Tartrate, 383
$C_{12}H_{12}BrN_4NaO_2S.H_2O$
Sulfabromomethazine*, 9066
$C_{12}H_{12}Br_2N_2$
Diquat Dibromide, 3415
$C_{12}H_{12}ClNO_2S$
Dansyl Chloride, 2877
$C_{12}H_{12}ClN_5O_4S$
Chlorsulfuron, 2244
$C_{12}H_{12}I_3N_2NaO_2$
Ipodate*, 5087
$C_{12}H_{12}N_2$
Benzidine, 1106
2,4′-Biphenyldiamine, 1277
1,1-Diphenylhydrazine, 3384
$C_{12}H_{12}N_2O$
Harmalol, 4645
$C_{12}H_{12}N_2OS$
4,4′-Sulfinyldianiline, 9122
$C_{12}H_{12}N_2OS_2$
p-Dimethylaminobenzalrhodanine, 3281
$C_{12}H_{12}N_2O_2$
Tetrantoin, 9372
$C_{12}H_{12}N_2O_2S$
Dapsone, 2885
Enoximone, 3627
Sulfabenz, 9064
$C_{12}H_{12}N_2O_3$
Nalidixic Acid, 6446
Phenobarbital, 7386
$C_{12}H_{12}N_4$
Chrysoidine*, 2317
p-Diaminoazobenzene, 3019
$C_{12}H_{12}N_4O_3$
Benzmidazole, 1114
$C_{12}H_{12}N_6Na_2O_{10}S_2$
Carumonam*, 1922
$C_{12}H_{12}O$
2-Ethoxynaphthalene, 3799
$C_{12}H_{12}O_2$
2-(2-Naphthyloxy)ethanol, 6500
$C_{12}H_{12}O_5$
Fraxetin*, 4289
Radicinin, 8276
$C_{12}H_{12}O_6$
Pyrogallol*, 8184
$C_{12}H_{13}BrN_4O_2S$
Sulfabromomethazine, 9066
$C_{12}H_{13}ClF_3N_3O_4$
Fluchloralin, 4156
$C_{12}H_{13}ClN_4$
Chrysoidine, 2317
Pyrimethamine, 8169
$C_{12}H_{13}Cl_2N_3$
Alinidine, 244
$C_{12}H_{13}I_3N_2O_2$
Ipodate, 5087
$C_{12}H_{13}I_3N_2O_3$
Iocetamic Acid, 5030
Iomeglamic Acid, 5069
$C_{12}H_{13}N$
N,N-Dimethyl-1-naphthylamine, 3301
$C_{12}H_{13}NO_2$
Indolebutyric Acid, 4995
Methsuximide, 6085
$C_{12}H_{13}NO_2S$
Carboxin, 1874
$C_{12}H_{13}NO_3$
Aniracetam, 700
$C_{12}H_{13}N_2NaO_3$
Cyclopentobarbital*, 2814
$C_{12}H_{13}N_2NaO_5$
2-Cyclohexyl-4,6-dinitrophenol*, 2802
$C_{12}H_{13}N_3$
4,4′-Diaminodiphenylamine, 3022
$C_{12}H_{13}N_3O_2$

Denotes derivative of title compound

Isocarboxazid, 5172
Triaziquone, 9733
$C_{12}H_{13}N_3O_4$
  Olaquindox, 6960
$C_{12}H_{13}N_3O_4S_2$
  N^4-Sulfanilylsulfanilamide, 9100
$C_{12}H_{13}N_3O_6$
  Glyconiazide, 4510
$C_{12}H_{13}N_4NaO_2S$
  Sulfamethazine*, 9083
$C_{12}H_{13}N_5Na_2O_9S_2$
  Tigemonam*, 9572
$C_{12}H_{13}N_5O_2S$
  Sulfamidochrysoidine, 9092
$C_{12}H_{13}N_5O_4$
  Toyocamycin, 9698
$C_{12}H_{14}As_2Cl_2N_2O_2$
  Arsphenamine, 851
$C_{12}H_{14}CaO_{12}$
  Calcium Ascorbate, 1688
$C_{12}H_{14}ClNO_2$
  Dimethazone, 3262
$C_{12}H_{14}ClNO_3$
  Cotarnine*, 2618
$C_{12}H_{14}Cl_2FNO_4S$
  Florfenicol, 4145
$C_{12}H_{14}Cl_2N_2$
  Paraquat*, 7165
$C_{12}H_{14}Cl_2O_3$
  2,4-D*, 2865
$C_{12}H_{14}Cl_3O_4P$
  Chlorfenvinphos, 2137
$C_{12}H_{14}N_2$
  Detomidine, 2981
  N-(1-Naphthyl)ethylenediamine, 6496
$C_{12}H_{14}N_2^{2+}$
  Paraquat, 7165
$C_{12}H_{14}N_2O_2$
  Abrine, 6
  Mephenytoin, 5898
  Methetoin, 6046
  Primidone, 7927
$C_{12}H_{14}N_2O_2S_2$
  Ujothion, 9978
$C_{12}H_{14}N_2O_3$
  Cyclopentobarbital, 2814
$C_{12}H_{14}N_2O_3S$
  Tioxidazole, 9599
$C_{12}H_{14}N_2O_4$
  5-Furfuryl-5-isopropylbarbituric Acid, 4326
$C_{12}H_{14}N_2O_5$
  N-(p-Aminobenzoyl)glutamic Acid, 446
  2-Cyclohexyl-4,6-dinitrophenol, 2802
$C_{12}H_{14}N_2O_5P_2$
  sym-Diphenylpyrophosphorodi-amidic Acid, 3391
$C_{12}H_{14}N_2O_6$
  Dinoseb*, 3341
$C_{12}H_{14}N_2O_7$
  Nicotinamide Ascorbate, 6610
$C_{12}H_{14}N_2S$
  Cymiazole, 2833
$C_{12}H_{14}N_4O_2S$
  Sulfamethazine, 9083
  Sulfisomidine, 9124
$C_{12}H_{14}N_4O_3S$
  Sulfacytine, 9070
  Sulfamethomidine, 9085
$C_{12}H_{14}N_4O_4S$
  Sulfadimethoxine, 9073
  Sulfadoxine, 9074
$C_{12}H_{14}N_4O_4S_2$
  Thiophanate*, 9489
$C_{12}H_{14}N_6O_{10}S_2$
  Carumonam, 1922
$C_{12}H_{14}O_2$

p-(2-Methylpropenyl)phenol Acetate, 6190
Precocenes*, 7898
$C_{12}H_{14}O_4$
  Apiole (Dill), 776
  Apiole (Parsley), 777
  Dimecrotic Acid, 3248
  Phthalic Acid*, 7527
$C_{12}H_{14}O_5$
  Cinametic Acid, 2341
$C_{12}H_{14}O_7$
  Piscidic Acid*, 7664
$C_{12}H_{15}AsN_6OS_2$
  Melarsoprol, 5856
$C_{12}H_{15}ClNO_4PS_2$
  Phosalone, 7489
$C_{12}H_{15}ClO_3$
  Clofibrate, 2436
$C_{12}H_{15}Cl_2NO$
  Karsil, 5297
$C_{12}H_{15}Cl_2NO_5S$
  Thiamphenicol, 9436
$C_{12}H_{15}N$
  6,7-Benzomorphan, 1126
  MPTP, 6376
$C_{12}H_{15}NO_3$
  Anhalonine, 693
  Carbofuran, 1851
  Hydrocotarnine, 4829
  Metaxalone, 5994
$C_{12}H_{15}NO_3S$
  Vanitiolide, 10071
$C_{12}H_{15}NO_4$
  Cotarnine, 2618
  Ethopabate, 3791
$C_{12}H_{15}N_2NaO_3$
  Hexobarbital*, 4742
$C_{12}H_{15}N_2O_3PS$
  Phoxim, 7523
$C_{12}H_{15}N_3$
  Indanazoline, 4967
$C_{12}H_{15}N_3O_2S$
  Albendazole, 211
$C_{12}H_{15}N_3O_2S_2$
  Glybuzole, 4488
$C_{12}H_{15}N_3O_3$
  Oxibendazole, 7070
$C_{12}H_{15}N_3O_5S$
  Amezinium Methyl Sulfate, 412
$C_{12}H_{15}N_5O_3S$
  Sulfaguanole, 9077
$C_{12}H_{15}N_5O_9S_2$
  Tigemonam, 9572
$C_{12}H_{16}ClNO_3$
  Meclofenoxate, 5820
$C_{12}H_{16}ClN_3O_2S$
  Oxythiamine, 7112
$C_{12}H_{16}Cl_2N_2O$
  Neburon, 6523
$C_{12}H_{16}Cl_2N_4S$
  Beclotiamine, 1048
$C_{12}H_{16}F_3N$
  Fenfluramine, 4015
$C_{12}H_{16}N_2$
  N,N-Dimethyltryptamine, 3311
  Etryptamine, 3937
  Fenproporex, 4036
$C_{12}H_{16}N_2O$
  Bufotenine, 1502
  Caulophylline, 1956
  Nebracetam, 6520
  Psilocin, 8110
$C_{12}H_{16}N_2O_2$
  Eltoprazine, 3592
$C_{12}H_{16}N_2O_3$
  Cyclobarbital, 2780
  Hexobarbital, 4742
$C_{12}H_{16}N_2S$
  Morantel, 6348
  Xylazine, 10213

$C_{12}H_{16}N_3O_3P$
  Benzodepa, 1118
$C_{12}H_{16}N_3O_3PS$
  Triazophos, 9736
$C_{12}H_{16}N_3O_3PS_2$
  Azinphos-methyl*, 944
$C_{12}H_{16}N_4O_2S_2$
  Glybuthiazol(e), 4487
$C_{12}H_{16}O_2$
  Ibufenac, 4924
  Isoamyl Benzoate, 5127
  Thymol*, 9540
$C_{12}H_{16}O_3$
  Asarones, 861
  Guaiacol Valerate, 4579
  Isoamyl Salicylate, 5139
$C_{12}H_{16}O_4$
  Aspidinol, 882
$C_{12}H_{16}O_4S_2$
  Malotilate, 5751
$C_{12}H_{16}O_6S$
  Protiofate, 8075
$C_{12}H_{16}O_7$
  Arbutin, 812
$C_{12}H_{17}ClN_4OS$
  Thiamine, 9430
$C_{12}H_{17}ClO_2$
  Fenpentadiol, 4029
$C_{12}H_{17}Cl_2NO_3$
  2,4-DB*, 2893
$C_{12}H_{17}NO$
  Deet, 2912
  Phendimetrazine, 7365
$C_{12}H_{17}NO_2$
  Butacetin, 1532
  Ciclopirox, 2325
  MDE, 5805
  Promecarb, 7967
$C_{12}H_{17}NO_3$
  Anhalonidine, 692
  Anisomycin*, 708
  Bucetin, 1479
  Bufexamac, 1497
  Cerulenin, 2047
  Ethamivan, 3765
  Rimiterol, 8393
$C_{12}H_{17}N_2NaO_2S$
  Thiamylal*, 9437
$C_{12}H_{17}N_2NaO_3$
  Secobarbital Sodium, 8563
$C_{12}H_{17}N_2O_4P$
  Psilocybin, 8111
$C_{12}H_{17}N_3O$
  Cimaterol, 2336
$C_{12}H_{17}N_3O_4$
  Agaritine, 185
$C_{12}H_{17}N_3O_4S.H_2O$
  Imipenem, 4954
$C_{12}H_{17}N_5O_3$
  7-Morpholinomethyltheophylline, 6363
$C_{12}H_{17}N_5O_4S$
  Thiamine*, 9430
$C_{12}H_{17}N_5S$
  Methiotriazamine, 6056
$C_{12}H_{17}NaO_7$
  Dikegulac*, 3243
$C_{12}H_{17}Na_3O_{17}Sb_2.9H_2O$
  Antimony Sodium Gluconate*, 742
$C_{12}H_{18}Be_4O_{13}$
  Beryllium Acetate, Basic, 1203
$C_{12}H_{18}ClN$
  Mefenorex, 5843
$C_{12}H_{18}ClNO$
  Tulobuterol, 9942
$C_{12}H_{18}ClNO_2S$
  Dimethenamid, 3263
$C_{12}H_{18}ClNO_4$
  Methyldopa*, 6132

Denotes derivative of title compound

C₁₂H₁₈ClN₄O₄PS
Thiamine Diphosphate*, 9431
C₁₂H₁₈Cl₂N₂O
Clenbuterol, 2407
C₁₂H₁₈¹²³IN
Iofetamine ¹²³I, 5066
C₁₂H₁₈NO₆P
Diisopropyl Paraoxon, 3242
C₁₂H₁₈N₂O
Isoproturon, 5237
Oxotremorine, 7085
Pivalylbenzhydrazine, 7668
C₁₂H₁₈N₂O₂
Mexacarbate, 6253
Nicametate, 6577
C₁₂H₁₈N₂O₂S
Thiamylal, 9437
C₁₂H₁₈N₂O₃
Nealbarbital, 6516
C₁₂H₁₈N₂O₃S
Tolbutamide, 9646
C₁₂H₁₈N₂O₄
Midodrine, 6272
C₁₂H₁₈N₂O₅
Bacilysin, 963
Hypoglycine B, 4914
C₁₂H₁₈N₂O₇
Bicozamycin, 1249
C₁₂H₁₈N₂O₇S
N⁴-β-D-Glucosylsulfanilamide, 4472
C₁₂H₁₈N₄O₆S
Oryzalin, 7015
C₁₂H₁₈O
6-n-Amyl-m-cresol, 648
Propofol, 8020
C₁₂H₁₈O₂
4-Hexylresorcinol, 4750
C₁₂H₁₈O₄
Butopyronoxyl, 1564
C₁₂H₁₈O₇
Dikegulac, 3243
C₁₂H₁₉BrN₂O₂
Neostigmine*, 6553
C₁₂H₁₉ClNO₃P
Crufomate, 2671
C₁₂H₁₉ClN₄O₇P₂S
Thiamine Diphosphate, 9431
C₁₂H₁₉Cl₃O₈
Sucralose, 9050
C₁₂H₁₉NO
Etafedrine, 3752
C₁₂H₁₉NO₂
Bamethan, 981
Fepradinol, 4052
C₁₂H₁₉NO₃
Macromerine, 5677
Prenalterol, 7917
Terbutaline, 9302
C₁₂H₁₉NO₄
Choline Salicylate, 2265
C₁₂H₁₉N₂NaO₂S₂
Methitural*, 6058
C₁₂H₁₉N₂NaO₃
Hexethal Sodium, 4740
C₁₂H₁₉N₂O₂⁺
Neostigmine, 6553
C₁₂H₁₉N₃O
Procarbazine, 7938
C₁₂H₁₉N₄O₁₀P₃S
Thiamine Triphosphate, 9433
C₁₂H₁₉O₂PS₃
Sulprofos, 9164
C₁₂H₂₀B₂CaO₁₆
Calcium Borogluconate, 1692
C₁₂H₂₀Cl₄N₂O₂
Fertilysin, 4108
C₁₂H₂₀N₂
Tremorine, 9715
C₁₂H₂₀N₂O₂
Aspergillic Acid, 876

C₁₂H₂₀N₂O₂S₂
Methitural, 6058
C₁₂H₂₀N₂O₃
Pirbuterol, 7644
Tetrabarbital, 9324
C₁₂H₂₀N₂O₃S
Sotalol, 8876
C₁₂H₂₀N₂O₄S
Soterenol, 8877
C₁₂H₂₀N₄O₂
Hexazinone, 4734
C₁₂H₂₀N₄O₉S
Sulfazecin, 9119
C₁₂H₂₀O₂
Bornyl Acetate, 1367
Linalyl Acetate, 5521
C₁₂H₂₀O₄
Traumatic Acid, 9710
C₁₂H₂₀O₆
Diacetoneglucose, 3009
C₁₂H₂₀O₇
Citric Acid*, 2387
C₁₂H₂₁AsN₂O₅
Acetarsone*, 49
C₁₂H₂₁ClO₂
Trimedlure, 9831
C₁₂H₂₁N
Memantine, 5872
Rimantadine, 8390
C₁₂H₂₁NO
Pimeclone, 7583
C₁₂H₂₁NO₈S
Topiramate, 9686
C₁₂H₂₁NO₁₁
Chondrosine, 2271
Hyalobiuronic Acid, 4792
C₁₂H₂₁N₂O₃PS
Diazinon, 3043
C₁₂H₂₁N₂O₅S₂⁹⁹ᵐTc
Technetium ⁹⁹ᵐTc Bicisate, 9257
C₁₂H₂₁N₅O₂S₂
Nizatidine, 6758
C₁₂H₂₁N₅O₃
Cadralazine, 1669
C₁₂H₂₁NaO₇S
Diisobutyl Sodium Sulfosuccinate, 3238
C₁₂H₂₂CaO₄
Calcium 2-Ethylbutanoate, 1707
C₁₂H₂₂CaO₁₄
Calcium Gluconate, 1712
C₁₂H₂₂Cl₂N₂O₃
Pirbuterol*, 7644
C₁₂H₂₂CuO₁₄
Cupric Gluconate, 2706
C₁₂H₂₂FeO₁₄
Ferrous Gluconate, 4095
C₁₂H₂₂MgO₁₄.2H₂O
Gluconic Acid*, 4464
C₁₂H₂₂N₂O₂
Crotethamide, 2662
C₁₂H₂₂N₂O₈S
Piposulfan, 7635
C₁₂H₂₂O
Geosmin, 4408
C₁₂H₂₂O₂
Menthyl Acetate, 5884
C₁₂H₂₂O₄
Dibutyl Succinate, 3087
Di-tert-butyl Succinate, 3088
C₁₂H₂₂O₆
Etoglucid, 3926
C₁₂H₂₂O₁₀
Rutinose, 8457
Scillabiose, 8542
C₁₂H₂₂O₁₁
Cellobiose, 2008
Gentiobiose, 4402
Lactose, 5356
Lactulose, 5360

Maltose, 5753
Melibiose, 5861
Sophorose, 8868
Sucrose, 9051
Trehalose, 9713
Turanose, 9951
C₁₂H₂₂O₁₂
Lactobionic Acid, 5354
C₁₂H₂₂O₁₄Zn
Gluconic Acid*, 4464
C₁₂H₂₃N
Dicyclohexylamine, 3145
C₁₂H₂₃N₃O
Esaprazole, 3736
C₁₂H₂₄CaN₂O₆S₂
Calcium Cyclamate, 1705
C₁₂H₂₄N₂O₄
Carisoprodol, 1889
C₁₂H₂₄N₆O₂P₂
Dipin, 3397
C₁₂H₂₄N₉P₃
Apholate, 770
C₁₂H₂₄O₂
Ethyl Caprate, 3823
Lauric Acid, 5396
C₁₂H₂₄O₆
Crown Ethers*, 2669
C₁₂H₂₄O₁₁
Lactitol, 5352
C₁₂H₂₅Br
Lauryl Bromide, 5402
C₁₂H₂₅NO₂
Butoctamide, 1562
C₁₂H₂₅NaO₄S
Sodium Lauryl Sulfate, 8782
C₁₂H₂₆N₄O₆
Neamine, 6317
C₁₂H₂₆O
1-Dodecanol, 3464
C₁₂H₂₆O₆P₂S₄
Dioxathion, 3355
C₁₂H₂₇AlO₃
Aluminum tert-Butoxide, 343
C₁₂H₂₇N
Tributylamine, 9748
C₁₂H₂₇O₄P
Tributyl Phosphate, 9749
C₁₂H₂₈O₄Ti
Titanium Isopropylate, 9614
C₁₂H₃₀Br₂N₂
Hexamethonium*, 4724
C₁₂H₃₀Cl₂N₂
Hexamethonium*, 4724
C₁₂H₃₀I₂N₂
Hexamethonium*, 4724
C₁₂H₃₀N₂²⁺
Hexamethonium, 4724
C₁₂H₃₆O₄Si₅
Dodecamethylpentasiloxane, 3463
C₁₂H₃₆O₆Si₆
Dodecamethylcyclohexasiloxane, 3462
C₁₂H₄₄CaFe₆Na₄O₃₆
Ferriclate Calcium Sodium, 4070
C₁₂H₅₄Al₁₆O₇₅S₈
Sucralfate, 9049
C₁₃H₄Cl₂F₆N₄O₄
Fluazinam, 4153
C₁₃H₅ClN₅NaO₂.5H₂O
Traxanox*, 9711
C₁₃H₅N₃O₇
2,4,7-Trinitrofluorenone, 9858
C₁₃H₆ClN₅O₂
Traxanox, 9711
C₁₃H₆Cl₅NO₃
Oxyclozanide, 7092
C₁₃H₆Cl₆O₂
Hexachlorophene, 4716
C₁₃H₇Cl₄NO₂

3,3′,4′,5-Tetrachlorosalicylanilide,
9333

$C_{13}H_8Br_3NO_2$
Tribromsalan, 9747

$C_{13}H_8Cl_2N_2O_4$
Niclosamide, 6602

$C_{13}H_8Cl_2O_4S$
Ticrynafen, 9570

$C_{13}H_8Cl_3NO_2$
3′,4′,5-Trichlorosalicylanilide, 9775

$C_{13}H_8F_2O_3$
Diflunisal, 3190

$C_{13}H_8N_2O_3S$
Nitroscanate, 6731

$C_{13}H_8N_3NaO_5$
Alizarine Yellow R*, 251
Metachrome Yellow, 5979

$C_{13}H_8OS$
Thioxanthone, 9507

$C_{13}H_8O_2$
Xanthone, 10196

$C_{13}H_8S_2$
Thiarubrines, 9439

$C_{13}H_9BrClNO_2$
Bromosalicylchloranilide, 1457

$C_{13}H_9Cl_2FN_2S$
Loflucarban, 5589

$C_{13}H_9Cl_3N_2O$
Triclocarban, 9786

$C_{13}H_9F_3N_2O_2$
Niflumic Acid, 6620

$C_{13}H_9Li_2N_5O_9S_2$
Lucifer Yellow CH, 5623

$C_{13}H_9N$
Acridine, 124
Benzo[f]quinoline, 1136

$C_{13}H_9NO$
p-Xenylcarbimide, 10207

$C_{13}H_9N_2NaO_2S$
Phethenylate Sodium, 7476

$C_{13}H_9N_3O_2S$
Amoscanate, 613

$C_{13}H_9N_3O_5$
Alizarine Yellow R, 251

$C_{13}H_{10}$
9H-Fluorene, 4190

$C_{13}H_{10}BrCl_2O_2PS$
Leptophos, 5468

$C_{13}H_{10}BrNO_3$
Resorantel, 8322

$C_{13}H_{10}ClN_3O_4S_2$
Lornoxicam, 5612

$C_{13}H_{10}Cl_2O_2$
Bis(p-chlorophenoxy)methane, 1287
Dichlorophen, 3120

$C_{13}H_{10}Cl_2S$
Chlorbenside, 2124

$C_{13}H_{10}N_2$
Aminacrine, 427
2-Phenyl-1H-benzimidazole, 7428

$C_{13}H_{10}N_2O$
Pyocyanine, 8137

$C_{13}H_{10}N_2O_4$
Myxin, 6425
Thalidomide, 9390

$C_{13}H_{10}N_3NaO_4S.3H_2O$
Sulfoniazide*, 9132

$C_{13}H_{10}N_4O_5$
4,4′-Dinitrocarbanilide, 3330

$C_{13}H_{10}O$
Benzophenone, 1129

$C_{13}H_{10}O_2$
Mycomycin, 6407
Phenyl Benzoate, 7429

$C_{13}H_{10}O_3$
Benzoresorcinol, 1138
Phenyl Carbonate, 7433
Phenyl Salicylate, 7464

$C_{13}H_{10}O_4$
Visnagin, 10148

$C_{13}H_{10}O_5$
Pimpinellin, 7591

$C_{13}H_{10}O_6$
Maclurin, 5676

$C_{13}H_{10}O_7$
Exifone, 3958

$C_{13}H_{10}S$
Thioxanthene, 9506

$C_{13}H_{11}ClN_2O_2$
Clonixin, 2453

$C_{13}H_{11}ClN_2O_4S$
Pyrithiobac, 8177

$C_{13}H_{11}ClO$
Clorophene, 2469

$C_{13}H_{11}Cl_2NO_2$
Procymidone, 7946

$C_{13}H_{11}Cl_2NaO_4$
Ethacrynic Acid*, 3761

$C_{13}H_{11}N$
Benzylideneaniline, 1173

$C_{13}H_{11}NO$
Benzanilide, 1088

$C_{13}H_{11}NO_2$
Nicotinic Acid Benzyl Ester, 6613
N-Phenylanthranilic Acid, 7427
Salicylanilide, 8482

$C_{13}H_{11}NO_3$
Fagarine, 3970
Osalmid, 7018
Phenyl Aminosalicylate, 7426

$C_{13}H_{11}NO_5$
Oxolinic Acid, 7079

$C_{13}H_{11}N_2NaO_2$
Ozagrel*, 7115

$C_{13}H_{11}N_3$
Proflavine, 7953

$C_{13}H_{11}N_3O$
Drometrizole, 3503

$C_{13}H_{11}N_3O_2$
Salinazid, 8487

$C_{13}H_{11}N_3O_4S$
Sulfoniazide, 9132

$C_{13}H_{11}N_3O_4S_2$
Tenoxicam, 9293

$C_{13}H_{11}N_8NaO_4S_3$
Ceftezole*, 1997

$C_{13}H_{12}$
Diphenylmethane, 3387

$C_{13}H_{12}ClKNO_2S$
Ontianil*, 6981

$C_{13}H_{12}ClNO_2S$
Ontianil, 6981

$C_{13}H_{12}ClN_3S$
Azure C, 959

$C_{13}H_{12}Cl_2N_2$
4,4′-Methylenebis[2-chloroaniline],
6136

$C_{13}H_{12}Cl_2O_4$
Ethacrynic Acid, 3761

$C_{13}H_{12}F_2N_6O$
Fluconazole, 4158

$C_{13}H_{12}N_2$
9H-Fluorene-2,7-diamine, 4191

$C_{13}H_{12}N_2O$
Carbanilide, 1829
Harmine, 4647

$C_{13}H_{12}N_2O_2$
Ozagrel, 7115

$C_{13}H_{12}N_2O_3$
Phenallymal, 7349

$C_{13}H_{12}N_2O_3S$
Sulfabenzamide, 9065

$C_{13}H_{12}N_2O_5S$
Nimesulide, 6640
4-Sulfanilamidosalicylic Acid, 9095

$C_{13}H_{12}N_2S$
sym-Diphenylthiourea, 3393

$C_{13}H_{12}N_3NaO_6S$
Cephacetrile Sodium, 2019

$C_{13}H_{12}N_4O$
Diphenylcarbazone, 3380

$C_{13}H_{12}N_4O_2$
Lumiflavine, 5627

$C_{13}H_{12}N_4O_3$
Pyriminil, 8171

$C_{13}H_{12}N_4S$
Dithizone, 3442

$C_{13}H_{12}N_5NaO_5S_2$
Ceftizoxime*, 2000

$C_{13}H_{12}N_8O_4S_3$
Ceftezole, 1997

$C_{13}H_{12}O$
Benzohydrol, 1121
o-Benzylphenol, 1179
p-Benzylphenol, 1180
Phentydrone, 7418

$C_{13}H_{12}O_2$
Monobenzone, 6331

$C_{13}H_{12}O_3$
Allenolic Acid, 256
Euparin, 3946
2-Naphthyl Lactate, 6499

$C_{13}H_{12}O_7S_2$
Sultosilic Acid, 9169

$C_{13}H_{13}As_2N_2NaO_4S$
Neoarsphenamine, 6534

$C_{13}H_{13}Cl_2N_3O_3$
Iprodione, 5093

$C_{13}H_{13}N$
Benzhydrylamine, 1105
Benzylaniline, 1161
Methyldiphenylamine, 6131

$C_{13}H_{13}NO_2$
Vitamin K_5*, 10162

$C_{13}H_{13}N_3$
1,3-Diphenylguanidine, 3383
Nitrin, 6676

$C_{13}H_{13}N_3O_3$
Cyclobendazole, 2781

$C_{13}H_{13}N_3O_5S_2$
Succinylsulfathiazole, 9046

$C_{13}H_{13}N_5O_4S$
Sulfachrysoidine, 9069

$C_{13}H_{13}N_5O_5S_2$
Ceftizoxime, 2000

$C_{13}H_{14}ClNO_2$
Pirprofen, 7663

$C_{13}H_{14}Cl_2O_3$
Ciprofibrate, 2373

$C_{13}H_{14}F_3N_3O_4$
Ethalfluralin, 3763

$C_{13}H_{14}N_2$
p,p′-Diaminodiphenylmethane, 3023
Tacrine, 9199

$C_{13}H_{14}N_2O$
Harmaline, 4644
10-Methoxyharmalan, 6073
N-(p-Methoxyphenyl)-p-phenylene-
diamine, 6079
Phenyramidol, 7474
Velnacrine, 10077

$C_{13}H_{14}N_2O_2$
Metomidate, 6232

$C_{13}H_{14}N_2O_2S$
Benzylsulfamide, 1182
p-Sulfanilylbenzylamine, 9098

$C_{13}H_{14}N_2O_3$
Mephobarbital, 5899

$C_{13}H_{14}N_2O_4$
Menadoxime, 5875

$C_{13}H_{14}N_2O_4S_2$
Gliotoxin, 4450

$C_{13}H_{14}N_2S$
Metizoline, 6223

$C_{13}H_{14}N_3NaO_4S$
Glymidine*, 4517

$C_{13}H_{14}N_4O$
sym-Diphenylcarbazide, 3379

Denotes derivative of title compound

Denotes derivative of title compound

C₁₃H₂₂N₂O
  Norea, 6787
C₁₃H₂₂N₂O₄
  Salicylamide *O*-Acetic Acid*, 8481
C₁₃H₂₂N₂O₆S
  Neostigmine*, 6553
C₁₃H₂₂N₄O₃S
  Ranitidine, 8286
C₁₃H₂₂O
  Solanone, 8855
C₁₃H₂₃NO₉
  Streptobiosamine, 8978
C₁₃H₂₄N₂O
  Cuscohygrine, 2748
C₁₃H₂₄N₂O₂
  Cropropamide, 2659
C₁₃H₂₄N₃O₃PS
  Pirimiphos-ethyl, 7652
C₁₃H₂₄N₄O₃
  Melanostatin*, 5855
C₁₃H₂₄N₄O₃S
  Bupirimate, 1519
  Timolol, 9585
C₁₃H₂₄O
  TMD, 9625
C₁₃H₂₄O₉
  Strophanthobiose, 9017
C₁₃H₂₄O₁₀
  Gaultherin*, 4384
C₁₃H₂₅NO
  Ethyl Menthane Carboxamide, 3870
C₁₃H₂₆N₂O₃
  Elaiomycin, 3575
C₁₃H₂₆N₂O₄
  Tybamate, 9960
C₁₃H₂₈N₄O₂
  Exametazime, 3957
C₁₃H₂₉N
  Octamylamine, 6846
C₁₃H₃₂I₂N₂O
  Plegatil, 7693
C₁₃H₃₃Br₂N₃
  Azamethonium Bromide, 929
C₁₄H₄N₂O₂S₂
  Dithianone, 3433
C₁₄H₄N₄O₁₂
  Chrysamminic Acid, 2308
C₁₄H₆ClF₃NNaO₅
  Acifluorfen*, 111
C₁₄H₆O₈
  Ellagic Acid, 3588
C₁₄H₇Br₃F₃N₃O₄
  Bromethalin, 1410
C₁₄H₇ClF₃NO₅
  Acifluorfen, 111
C₁₄H₇NO₆
  Alizarine Orange, 250
C₁₄H₇NaO₇S
  Sodium Alizarinesulfonate, 8714
C₁₄H₈Br₂
  9,10-Dibromoanthracene, 3065
C₁₄H₈Br₂F₃NO₂
  Fluorosalan, 4216
C₁₄H₈Br₂O₄
  Dibromsalicil, 3080
C₁₄H₈ClNNa₂O₄
  Lobenzarit*, 5581
C₁₄H₈ClN₂NaO₃S
  Tenidap*, 9290
C₁₄H₈Cl₂N₄
  Clofentezine, 2435
C₁₄H₈N₂Na₂O₆
  Olsalazine*, 6976
C₁₄H₈N₂O₆
  Nifuroquine, 6625
C₁₄H₈N₂O₈
  5,5′-Dinitrosalicil, 3338
C₁₄H₈N₂S₄
  2,2′-Dithiobis[benzothiazole], 3435
C₁₄H₈O₂

Anthraquinone, 726
Morphenol, 6356
Phenanthrenequinone, 7355
C₁₄H₈O₄
  Alizarin, 247
  Anthrarufin, 728
  Danthron, 2878
  Quinizarin, 8250
C₁₄H₈O₅
  Anthragallol, 722
  Purpurin, 8132
C₁₄H₈O₆
  Quinalizarin, 8231
C₁₄H₈O₈
  Rufigallol, 8447
C₁₄H₉ClF₂N₂O₂
  Diflubenzuron, 3188
C₁₄H₉ClN₂O₃S
  Tenidap, 9290
C₁₄H₉Cl₂F₃N₂O
  Cloflucarban, 2438
C₁₄H₉Cl₂NO₅
  Bifenox, 1256
C₁₄H₉Cl₃F₂
  DFDT, 3000
C₁₄H₉Cl₃N₂OS
  Triclabendazole, 9784
C₁₄H₉Cl₅
  DDT, 2898
C₁₄H₉Cl₅O
  Dicofol, 3136
C₁₄H₉I₃O₄
  Tiratricol, 9603
C₁₄H₉NO₂
  1-Aminoanthraquinone, 437
C₁₄H₉N₄NaO₅.4H₂O
  Dantrolene*, 2879
C₁₄H₁₀
  Anthracene, 721
  Phenanthrene, 7354
  Tolan, 9643
C₁₄H₁₀BaO₄.2H₂O
  Benzoic Acid*, 1122
C₁₄H₁₀BrN₃O
  Bromazepam, 1406
C₁₄H₁₀CaO₄.3H₂O
  Benzoic Acid*, 1122
C₁₄H₁₀CdO₆
  Cadmium Salicylate, 1661
C₁₄H₁₀ClNO₄
  Lobenzarit, 5581
C₁₄H₁₀ClN₃S₂
  Lanoconazole, 5370
C₁₄H₁₀Cl₂F₂
  DFDD, 2999
C₁₄H₁₀Cl₂NNaO₂
  Diclofenac Sodium, 3132
C₁₄H₁₀Cl₄
  1,1-Dichloro-2,2-bis(*p*-chlorophenyl)ethane, 3111
  Mitotane, 6302
C₁₄H₁₀CuO₄.2H₂O
  Benzoic Acid*, 1122
C₁₄H₁₀CuO₆
  Cupric Salicylate, 2717
C₁₄H₁₀F₃NO₂
  Flufenamic Acid, 4167
C₁₄H₁₀HgO₄
  Mercuric Benzoate, 5924
C₁₄H₁₀MgO₄
  Magnesium Benzoate, 5691
C₁₄H₁₀MgO₆
  Magnesium Salicylate, 5723
C₁₄H₁₀MnO₄.4H₂O
  Benzoic Acid*, 1122
C₁₄H₁₀NNaO₄
  Benzoylpas*, 1148
C₁₄H₁₀N₂O₅
  Cinnabarine, 2355

C₁₄H₁₀N₂O₆
  Olsalazine, 6976
C₁₄H₁₀N₄O
  Loprinone, 5605
C₁₄H₁₀N₄O₅
  Dantrolene, 2879
C₁₄H₁₀NiO₄.3H₂O
  Benzoic Acid*, 1122
C₁₄H₁₀O
  Anthranol, 725
  Anthrone, 730
  Diphenylketene, 3385
C₁₄H₁₀O₂
  Benzil, 1107
C₁₄H₁₀O₃
  Anthralin, 723
  Anthrarobin, 727
  Benzoic Anhydride, 1123
C₁₄H₁₀O₄
  Benzoyl Peroxide, 1149
  Diphenic Acid, 3368
C₁₄H₁₀O₄Pb.2H₂O
  Benzoic Acid*, 1122
C₁₄H₁₀O₅
  Gentisin, 4405
  Salsalate, 8491
C₁₄H₁₀O₆
  Phenicin, 7379
C₁₄H₁₀O₆U
  Benzoic Acid*, 1122
C₁₄H₁₀O₆Zn
  Zinc Salicylate, 10288
C₁₄H₁₀O₇
  Citromycetin, 2389
C₁₄H₁₀O₈
  Oosporein, 6982
C₁₄H₁₀O₉
  Digallic Acid, 3195
C₁₄H₁₁AlO₇.H₂O
  Aluminum Salicylates, Basic*, 375
C₁₄H₁₁ClN₂O₄S
  Chlorthalidone, 2246
C₁₄H₁₁Cl₂NO₂
  Meclofenamic Acid, 5819
C₁₄H₁₁F₃N₂O₂
  Flunixin, 4182
C₁₄H₁₁N
  Diphenylacetic Acid*, 3374
C₁₄H₁₁NO₂
  9-Carbazoleacetic Acid, 1835
C₁₄H₁₁NO₄
  Benzoylpas, 1148
  Diphenylamine-2,2′-dicarboxylic Acid, 3376
C₁₄H₁₁N₄NaO₂S
  Sulfaquinoxaline*, 9109
C₁₄H₁₂
  1,1-Diphenylethene, 3381
  Stilbene, 8972
C₁₄H₁₂CaN₂O₆
  p-Aminosalicylic Acid*, 498
C₁₄H₁₂ClKN₃O₃S
  Fenquizone*, 4039
C₁₄H₁₂ClNO₂
  Cicletanine, 2323
  Tolfenamic Acid, 9650
C₁₄H₁₂ClN₃O₃S
  Fenquizone, 4039
C₁₄H₁₂Cl₂N₂O
  Pyrifenox, 8167
C₁₄H₁₂Cl₂O
  Chlorfenethol, 2136
C₁₄H₁₂FNO₃
  Flumequine, 4172
C₁₄H₁₂F₃NO₄S₂
  Perfluidone, 7300
C₁₄H₁₂N₂
  Bendazol, 1062
  Neocuproine, 6537
C₁₄H₁₂N₂O₂

* *Denotes derivative of title compound*

Denotes derivative of title compound   **FI-35**

$C_{14}H_{18}CaN_3Na_3O_{10}$
Pentetate Calcium Trisodium, 7265

$C_{14}H_{18}ClKN_2O_4S_2$
Penicillin S Potassium, 7229

$C_{14}H_{18}ClN$
Picilorex, 7551

$C_{14}H_{18}ClNO_2$
Indeloxazine Hydrochloride, 4972

$C_{14}H_{18}ClN_3O_2$
Triadimenol, 9724

$C_{14}H_{18}ClN_3S$
Chlorothen, 2220

$C_{14}H_{18}F_3NO$
Oxaflozane, 7039

$C_{14}H_{18}N_2O$
Ibudilast, 4923
Propyphenazone, 8056

$C_{14}H_{18}N_2O_2$
Hypaphorine, 4910
Nefiracetam, 6528
Parsalmide, 7177

$C_{14}H_{18}N_2O_3$
Physovenine, 7541
Reposal, 8306

$C_{14}H_{18}N_2O_4$
α-Ribazole, 8366

$C_{14}H_{18}N_2O_5$
Aspartame, 874
Lidofenin, 5506

$C_{14}H_{18}N_2O_6S$
Carpetimycins*, 1909

$C_{14}H_{18}N_2O_7$
Dinobuton, 3339

$C_{14}H_{18}N_2O_9S_2$
Carpetimycins*, 1909

$C_{14}H_{18}N_4O_3$
Benomyl, 1073
Trimethoprim, 9840

$C_{14}H_{18}N_4O_4S_2$
Thiophanate, 9489

$C_{14}H_{18}N_6O$
2,6-Diamino-2'-butyloxy-3,5'-azopy-
ridine, 3021

$C_{14}H_{18}O_4$
Cinoxate, 2370

$C_{14}H_{18}O_7$
Picein, 7549

$C_{14}H_{18}O_8$
Glucovanillin, 4473

$C_{14}H_{19}BrO_9$
Acetobromglucose, 57

$C_{14}H_{19}ClN_4$
Amprolium, 631

$C_{14}H_{19}Cl_2NO_2$
Chlorambucil, 2116

$C_{14}H_{19}NO$
Ethoxyquin, 3800

$C_{14}H_{19}NO_2$
Levophacetoperane, 5493
Methylphenidate, 6186
Piperoxan, 7631

$C_{14}H_{19}NO_4$
Anisomycin, 708

$C_{14}H_{19}NO_4S$
Furtrethonium*, 4334
Trithiozine, 9889

$C_{14}H_{19}NO_5$
Trimetozine, 9850

$C_{14}H_{19}N_2NaO_4S$
2-Pentenylpenicillin Sodium, 7264

$C_{14}H_{19}N_3O$
Methafurylene, 6010
Oxolamine, 7078
Ramifenazone, 8282

$C_{14}H_{19}N_3S$
Methapyrilene, 6027
Thenyldiamine, 9416

$C_{14}H_{19}N_5O_4$
Famciclovir, 3971

$C_{14}H_{19}N_5O_4S_3$
Sulfatolamide, 9117

$C_{14}H_{19}O_6P$
Crotoxyphos, 2667

$C_{14}H_{20}$
Congressane, 2563

$C_{14}H_{20}Br_2N_2$
Bromhexine, 1412

$C_{14}H_{20}Br_2N_4O$
Pyrithiamine, 8176

$C_{14}H_{20}ClNO_2$
Acetochlor, 58
Alachlor, 203

$C_{14}H_{20}ClN_3O_3S$
Clopamide, 2454

$C_{14}H_{20}Cl_6N_2$
Chlorisondamine Chloride, 2151

$C_{14}H_{20}GdN_3O_{10}$
Gadopentetic Acid, 4347

$C_{14}H_{20}IN_3O_8$
Nitroxynil*, 6754

$C_{14}H_{20}N_2O$
Pyrrocaine, 8197
Siduron, 8631
Tymazoline, 9965

$C_{14}H_{20}N_2O_2$
Bunitrolol, 1514
Pindolol, 7597
Piridocaine, 7649

$C_{14}H_{20}N_2O_3$
Propacetamol, 7976

$C_{14}H_{20}N_2O_3S$
Tolcyclamide, 9648

$C_{14}H_{20}N_2O_6S$
p-Methylaminophenol Sulfate, 6097

$C_{14}H_{20}N_2O_8S_2$
Paraquat*, 7165

$C_{14}H_{20}N_3O_5PS$
Pyrazophos, 8146

$C_{14}H_{20}N_4O$
Imolamine, 4959

$C_{14}H_{20}N_4O_3$
Methenamine Mandelate, 6040

$C_{14}H_{20}N_4O_7S_2$
Netobimin, 6564

$C_{14}H_{20}O_2$
Butibufen, 1557

$C_{14}H_{20}O_4$
Frequentin, 4293

$C_{14}H_{20}O_8$
Taxicatin, 9244

$C_{14}H_{21}Br_2NO_2$
Spasmolytol, 8888

$C_{14}H_{21}NO_2$
4-(Dimethylamino)benzoic Acid*,
3282
Meprylcaine, 5909

$C_{14}H_{21}NO_4$
Diethofencarb, 3158

$C_{14}H_{21}NO_8$
6-Desoxy-D-glucosamine*, 2976

$C_{14}H_{21}N_2NaO_4S$
Amylpenicillin Sodium, 658

$C_{14}H_{21}N_3O_2S$
Sumatriptan, 9172

$C_{14}H_{21}N_3O_3$
Oxamniquine, 7051

$C_{14}H_{21}N_3O_3S$
Tolazamide, 9644

$C_{14}H_{21}N_3O_4$
Butralin, 1567

$C_{14}H_{21}N_3O_5$
Leonurine, 5463

$C_{14}H_{21}N_3O_6S$
Penicillin N, 7227

$C_{14}H_{21}N_5O_6.S.H_2O$
Serotonin*, 8607

$C_{14}H_{22}BrN_3O_2$
Bromopride, 1454

$C_{14}H_{22}ClNO$
Clobutinol, 2425

$C_{14}H_{22}ClNO_2$
Amylocaine Hydrochloride, 656
Bupranolol, 1521

$C_{14}H_{22}ClNO_3$
Mecoprop*, 5826

$C_{14}H_{22}ClN_3O_2$
Metoclopramide, 6226

$C_{14}H_{22}Cl_2N_2O_6S_4$
Clomethiazole*, 2444

$C_{14}H_{22}N_2O$
Lidocaine, 5505
Octacaine, 6839

$C_{14}H_{22}N_2O_2$
Naepaine, 6435

$C_{14}H_{22}N_2O_3$
Atenolol, 892
Bucolome, 1487
Practolol, 7882
Trimetazidine, 9835

$C_{14}H_{22}N_2O_3S$
Piprozolin, 7639

$C_{14}H_{22}N_2O_7S$
Rimazolium Metilsulfate, 8391

$C_{14}H_{22}O$
Irone, 5107

$C_{14}H_{22}O_4$
Palitantin, 7120

$C_{14}H_{23}NO$
Affinin, 177

$C_{14}H_{23}NO_2$
Piroctone, 7657

$C_{14}H_{23}NO_4$
4,4'-Iminodicyclohexanecarboxylic
Acid, 4953

$C_{14}H_{23}N_3OS$
Xanomeline, 10190

$C_{14}H_{23}N_3O_3S$
Sematilide, 8586

$C_{14}H_{23}N_3O_{10}$
Pentetic Acid, 7266

$C_{14}H_{24}CaN_4O_8$
Piperazine*, 7617

$C_{14}H_{24}N_2O_3$
Ciclopirox*, 2325

$C_{14}H_{24}N_2O_5$
Pirbuterol*, 7644

$C_{14}H_{24}N_2O_7$
Spectinomycin, 8890

$C_{14}H_{24}O_2$
Japonilure, 5271

$C_{14}H_{25}HgNO_5$
Mercamphamide, 5915

$C_{14}H_{25}NO$
Pellitorine, 7201

$C_{14}H_{25}N_3O_4S$
Alitame, 245

$C_{14}H_{25}N_3O_6S$
ACV, 147

$C_{14}H_{25}N_4NaO_{11}P_2$
Citicoline*, 2380

$C_{14}H_{25}NaO_7S$
Diamyl Sodium Sulfosuccinate, 3034

$C_{14}H_{26}CaO_{16}$
Glucoheptonic Acid*, 4463

$C_{14}H_{26}Cl_2N_2$
Bipiperidyl Mustard, 1278

$C_{14}H_{26}MgO_{16}$
Glucohcptonic Acid*, 4463

$C_{14}H_{26}N_4O_{11}P_2$
Citicoline, 2380

$C_{14}H_{26}O_2$
Looplure, 5600

$C_{14}H_{26}O_4$
Sebacic Acid*, 8558

$C_{14}H_{27}NO_6$
Amiprilose, 506

$C_{14}H_{27}N_3O_2$
Pramiracetam, 7886

Denotes derivative of title compound

$C_{14}H_{28}O_2$
Ethyl Laurate, 3864
Myristic Acid, 6416
$C_{14}H_{29}NaO_4S$
Sodium Tetradecyl Sulfate, 8838
$C_{14}H_{30}Br_2N_2O_4$
Succinylcholine Bromide, 9043
$C_{14}H_{30}Cl_2N_2O_4$
Succinylcholine Chloride, 9044
$C_{14}H_{30}I_2N_2O_4$
Succinylcholine Iodide, 9045
$C_{14}H_{30}O$
Myristyl Alcohol, 6418
$C_{14}H_{30}O_2$
Decamethylene Glycol*, 2906
$C_{14}H_{30}O_2S_2$
Tiadenol, 9556
$C_{14}H_{32}N_2O_4$
Entprol, 3635
$C_{14}H_{42}O_5Si_6$
Tetradecamethylhexasiloxane, 9338
$C_{15}H_8O_5$
Coumestrol, 2632
$C_{15}H_8O_6$
Rhein, 8342
$C_{15}H_9BrO_2$
Bromindione, 1414
$C_{15}H_9ClO_2$
Clorindione, 2468
$C_{15}H_9FO_2$
Fluindione, 4168
$C_{15}H_9NO_4$
1-Aminoanthraquinone-2-carboxylic
Acid, 438
$C_{15}H_{10}BrClN_4S$
Brotizolam, 1472
$C_{15}H_{10}ClF_3N_2O_3$
Triflumuron, 9809
$C_{15}H_{10}ClI_2NO_3$
Clioxanide, 2416
$C_{15}H_{10}ClN_2O_3$
Clonazepam, 2449
$C_{15}H_{10}Cl_2N_2O_2$
Lonidamine, 5598
Lorazepam, 5609
$C_{15}H_{10}I_4NNaO_4$
Levothyroxine Sodium, 5497
Thyroxine*, 9555
$C_{15}H_{10}O$
Diphencyprone, 3366
$C_{15}H_{10}O_2$
Flavone, 4132
Isoflavone, 5189
2-Methylanthraquinone, 6100
Phenindione, 7381
$C_{15}H_{10}O_4$
Chrysin, 2316
Chrysophanic Acid, 2318
Daidzein, 2868
Rubiadin, 8438
$C_{15}H_{10}O_5$
Aloe-Emodin, 313
Apigenin, 773
Baicalein, 971
Emodin, 3602
Galangin, 4356
Genistein, 4395
Sulfuretin, 9145
$C_{15}H_{10}O_6$
Baptigenin, 984
Datiscetin, 2887
Fisetin, 4129
Kaempferol, 5288
Luteolin, 5641
Scutellarein, 8556
$C_{15}H_{10}O_7$
Morin, 6352
Quercetin, 8216
$C_{15}H_{10}O_8$

Myricetin, 6414
Quercetagetin, 8215
$C_{15}H_{11}BrNNaO_3 \cdot 1\frac{1}{2}H_2O$
Bromfenac*, 1411
$C_{15}H_{11}ClF_3NO_4$
Oxyfluorfen, 7097
$C_{15}H_{11}ClN_2O$
Mecloqualone, 5821
Nordazepam, 6784
$C_{15}H_{11}ClN_2O_2$
Oxazepam, 7059
$C_{15}H_{11}ClO_2$
Clobenfurol, 2418
$C_{15}H_{11}ClO_5$
Pelargonidin, 7199
$C_{15}H_{11}ClO_6$
Cyanidin Chloride, 2757
$C_{15}H_{11}ClO_7$
Delphinidin, 2931
$C_{15}H_{11}I_3NNaO_4$
Liothyronine*, 5535
$C_{15}H_{11}I_3O_4$
Thyropropic Acid, 9553
$C_{15}H_{11}I_4NO_4$
Thyroxine, 9555
$C_{15}H_{11}NO_2$
Viridicatin, 10143
$C_{15}H_{11}NO_5$
Bostrycoidin, 1381
$C_{15}H_{11}N_2NaO_2$
Phenytoin*, 7475
$C_{15}H_{11}N_3$
Nicotelline, 6609
$C_{15}H_{11}N_3O_3$
Nitrazepam, 6667
$C_{15}H_{12}BrNO_3$
Bromfenac, 1411
$C_{15}H_{12}ClNO_2$
Carprofen, 1912
$C_{15}H_{12}Cl_2O_3$
Sesin, 8618
$C_{15}H_{12}F_4N_4O_7S$
Primisulfuron-methyl, 7928
$C_{15}H_{12}I_2O_3$
Iodoalphionic Acid, 5047
$C_{15}H_{12}I_3NO_4$
Liothyronine, 5535
$C_{15}H_{12}NNaO_3 \cdot H_2O$
Amfenac*, 413
$C_{15}H_{12}N_2O$
Carbamazepine, 1826
$C_{15}H_{12}N_2O_2$
Oxcarbazepine, 7063
Phenopyrazone, 7401
Phenytoin, 7475
$C_{15}H_{12}N_2O_3$
Hydrofuramide, 4832
$C_{15}H_{12}N_6O_4$
Rhizopterin, 8348
$C_{15}H_{12}O$
Chalcone, 2078
$C_{15}H_{12}O_2$
Dibenzoylmethane, 3056
α-Phenylcinnamic Acid, 7436
$C_{15}H_{12}O_3$
2-(p-Toluyl)benzoic Acid, 9677
$C_{15}H_{12}O_4$
o-(p-Anisoyl)benzoic Acid, 710
Methyl Benzoylsalicylate, 6105
Phenyl Acetylsalicylate, 7424
$C_{15}H_{12}O_5$
Naringenin, 6511
$C_{15}H_{12}O_6$
Eriodictyol, 3710
Fustin, 4341
5,5'-Methylenedisalicylic Acid, 6142
$C_{15}H_{12}O_8$
Ampelopsin, 618
$C_{15}H_{12}O_{10}$
Methylenedigallic Acid, 6141

$C_{15}H_{13}ClNNaO_3 \cdot 2H_2O$
Zomepirac*, 10322
$C_{15}H_{13}ClN_2$
Chlormidazole, 2156
$C_{15}H_{13}ClN_2O_5$
Gallocyanine, 4368
$C_{15}H_{13}Cl_2NO_2$
2-Nitro-1,1-bis(p-chlorophenyl)pro-
pane, 6691
$C_{15}H_{13}Cl_2N_5$
Robenidine, 8403
$C_{15}H_{13}FO_2$
Flurbiprofen, 4234
$C_{15}H_{13}I_2NO_4$
3,5-Diiodothyronine, 3235
$C_{15}H_{13}NO$
N-2-Fluorenylacetamide, 4192
$C_{15}H_{13}NO_2S$
Metiazinic Acid, 6219
$C_{15}H_{13}NO_3$
Amfenac, 413
Ketorolac, 5318
3'-Methylphthalanilic Acid, 6188
Pranoprofen, 7890
$C_{15}H_{13}NO_3S_2$
Epalrestat, 3640
$C_{15}H_{13}NO_4$
Acetaminosalol, 46
$C_{15}H_{13}N_3O_2S$
Fenbendazole, 4000
$C_{15}H_{13}N_3O_3S$
Oxfendazole, 7069
$C_{15}H_{13}N_3O_4S$
Piroxicam, 7661
$C_{15}H_{13}N_4NaO_2S \cdot H_2O$
Sulfaphenazole*, 9105
$C_{15}H_{14}ClNO_3$
Zomepirac, 10322
$C_{15}H_{14}ClN_3O_4S \cdot H_2O$
Cefaclor, 1962
$C_{15}H_{14}ClN_3O_4S_3$
Benzthiazide, 1155
$C_{15}H_{14}ClN_3O_6$
Lodoxamide*, 5585
$C_{15}H_{14}FNO_3$
Ibafloxacin, 4919
$C_{15}H_{14}FN_3O_3$
Flumazenil, 4169
$C_{15}H_{14}F_3N_3O_4S_2$
Bendroflumethiazide, 1064
$C_{15}H_{14}NNaO_3 \cdot 2H_2O$
Tolmetin*, 9655
$C_{15}H_{14}NO_2PS$
Cyanofenphos, 2760
$C_{15}H_{14}N_2Na_2O_6S_2$
Ticarcillin*, 9568
$C_{15}H_{14}N_2O$
Doxenitoin, 3491
$C_{15}H_{14}N_4O$
Nevirapine, 6573
$C_{15}H_{14}N_4O_2S$
Sulfaphenazole, 9105
$C_{15}H_{14}N_4O_6S_2$
Ceftibuten, 1998
$C_{15}H_{14}O$
Lactaroviolin, 5346
$C_{15}H_{14}O_3$
Equol, 3677
Fenoprofen, 4021
Guaiacol*, 4575
Lapachol, 5376
Mexenone, 6255
$C_{15}H_{14}O_4$
Lunularic Acid, 5634
Menadiol*, 5873
Menbutone, 5878
Peucedanin, 7332
Xanthoxyletin, 10200
Yangonin, 10226
$C_{15}H_{14}O_5$

Benzophenone-6, 1130
Guaiacol Carbonate, 4577
Methysticin, 6218
Phloretin, 7480
$C_{15}H_{14}O_6$
  Catechin, 1950
  Javanicin, 5275
  Plumericin, 7698
$C_{15}H_{14}O_7$
  Fusarubin, 4337
  Leucocyanidin, 5477
$C_{15}H_{15}ClN_2O_4S$
  Xipamide, 10212
$C_{15}H_{15}ClN_4O_6S$
  Chlorimuron Ethyl, 2142
$C_{15}H_{15}Cl_2N_2NaO_8$
  Chloramphenicol*, 2120
$C_{15}H_{15}F_3N_2O_2$
  Flurprimidol, 4239
$C_{15}H_{15}I_3NNaO_3$
  Bunamiodyl Sodium, 1511
$C_{15}H_{15}NO$
  p-Dimethylaminobenzophenone,
    3283
$C_{15}H_{15}NO_2$
  Apo-β-erythroidine, 785
  Enfenamic Acid, 3620
  Isoapo-β-erythroidine, 5140
  Mefenamic Acid, 5842
$C_{15}H_{15}NO_2S$
  Modafinil, 6311
$C_{15}H_{15}NO_3$
  Tolmetin, 9655
$C_{15}H_{15}NO_3S$
  Adrafinil, 168
$C_{15}H_{15}NO_6$
  Ascorbigen, 868
$C_{15}H_{15}N_3O$
  Ethacridine, 3760
$C_{15}H_{15}N_3O_2$
  Methyl Red, 6199
$C_{15}H_{15}N_3O_3$
  Verazide, 10091
$C_{15}H_{15}N_5O_4$
  Etofylline Nicotinate, 3925
$C_{15}H_{16}ClN_3S$
  Azure B, 958
  Tolonium Chloride, 9658
$C_{15}H_{16}Cl_3N_3O_2$
  Prochloraz, 7941
$C_{15}H_{16}F_3N_5O_4S$
  Prosulfuron, 8067
$C_{15}H_{16}F_5NO_2S_2$
  Dithiopyr, 3439
$C_{15}H_{16}N_2Na_2O_8S_3$
  Noprylsulfamide, 6777
$C_{15}H_{16}N_2O$
  Benmoxine, 1072
  Phenatine, 7358
$C_{15}H_{16}N_2O_2$
  Ancymidol, 672
$C_{15}H_{16}N_2O_3S$
  N-Sulfanilyl-3,4-xylamide, 9102
$C_{15}H_{16}N_2O_6S_2$
  Ticarcillin, 9568
$C_{15}H_{16}N_4O_5S$
  Sulfometuron Methyl, 9130
$C_{15}H_{16}N_5NaO_6S_2$
  Cefpodoxime Proxetil*, 1991
$C_{15}H_{16}N_6O$
  Amicarbalide, 414
$C_{15}H_{16}N_7NaO_5S_3$
  Cefmetazole*, 1977
$C_{15}H_{16}O_2$
  Bisphenol A, 1338
  Nabumetone, 6428
$C_{15}H_{16}O_3$
  Osthole, 7026
$C_{15}H_{16}O_5$
  Fuscin, 4338

Lactucin, 5359
Vernolepin, 10098
$C_{15}H_{16}O_6$
  Picrotoxinin, 7571
$C_{15}H_{16}O_8$
  Leucodrin, 5478
  Skimmin, 8702
$C_{15}H_{16}O_9$
  Cichoriin, 2322
  Daphnin, 2882
  Esculin, 3739
$C_{15}H_{17}BrN_2O_2$
  Benzpyrinium Bromide, 1153
$C_{15}H_{17}Br_2NO_2$
  Bromoxynil*, 1465
$C_{15}H_{17}ClN_4$
  Myclobutanil, 6402
  Neutral Red, 6571
$C_{15}H_{17}Cl_2N_3O$
  Diniconazole, 3320
$C_{15}H_{17}Cl_2N_3O_2$
  Propiconazole, 8003
$C_{15}H_{17}FN_4O_2$
  Flupirtine, 4227
$C_{15}H_{17}FN_4O_3$
  Enoxacin, 3625
$C_{15}H_{17}F_2N_6NaO_7S_2$
  Flomoxef*, 4141
$C_{15}H_{17}HgNO_2S$
  N-(Ethylmercuri)-p-toluenesulfon-
    anilide, 3872
$C_{15}H_{17}I_3NNaO_3$
  Tyropanoate Sodium, 9968
$C_{15}H_{17}N$
  Ethylbenzylaniline, 3817
$C_{15}H_{17}NO_4$
  Actiphenol, 145
$C_{15}H_{17}NS_2$
  Tipepidine, 9601
$C_{15}H_{17}N_3O$
  Cetoxime, 2065
  Metralindole, 6238
$C_{15}H_{17}N_5O_6S_2$
  Cefpodoxime Proxetil*, 1991
$C_{15}H_{17}N_7O_5S_3$
  Cefmetazole, 1977
$C_{15}H_{17}NaO_3$
  Loxoprofen*, 5619
$C_{15}H_{17}NaO_3S$
  Guaiazulene*, 4581
$C_{15}H_{18}$
  Cadalene, 1644
  Guaiazulene, 4581
$C_{15}H_{18}ClN_3O$
  Cyproconazole, 2840
$C_{15}H_{18}ClN_3O_3S$
  Tiaramide, 9564
$C_{15}H_{18}Cl_2N_2O_3$
  Oxadiazon, 7038
$C_{15}H_{18}Cl_6N_2O_5$
  Dichloralphenazone, 3097
$C_{15}H_{18}F_2N_6O_7S_2$
  Flomoxef, 4141
$C_{15}H_{18}I_3NO_5$
  Iopronic Acid, 5077
$C_{15}H_{18}N_2O$
  Huperzine A, 4791
$C_{15}H_{18}N_2O_6$
  Binapacryl, 1265
$C_{15}H_{18}N_4O_4S$
  Biapenem, 1243
$C_{15}H_{18}N_4O_5$
  Mitomycins*, 6301
$C_{15}H_{18}N_6O_2$
  Pimefylline, 7584
$C_{15}H_{18}O_2$
  Procerin, 7940
$C_{15}H_{18}O_3$
  Ambrosin, 400
  Loxoprofen, 5619

α-Santonin, 8509
Xanthatin, 10192
$C_{15}H_{18}O_4$
  Artemisin, 855
  Helenalin, 4656
  Parthenin, 7179
$C_{15}H_{18}O_5$
  Coriamyrtin, 2594
$C_{15}H_{18}O_6$
  Tutin, 9958
$C_{15}H_{18}O_7$
  Hyenanchin, 4898
  Picrotin, 7569
$C_{15}H_{18}O_8$
  Melilotoside, 5863
$C_{15}H_{19}BrN_2O_5$
  Mebrofenin, 5812
$C_{15}H_{19}ClN_4$
  Toluylene Blue, 9678
$C_{15}H_{19}Cl_2N_3O$
  Diclobutrazol, 3131
$C_{15}H_{19}F_3N_2S$
  Ticarbodine, 9567
$C_{15}H_{19}NO$
  Furfurylmethylamphetamine, 4327
  Pronethalol, 7974
$C_{15}H_{19}NO_2$
  Tropacocaine, 9904
$C_{15}H_{19}NO_3$
  Ximoprofen, 10211
$C_{15}H_{19}NS_2$
  Ethylmethylthiambutene, 3875
$C_{15}H_{19}N_3OS$
  Butamisole, 1540
$C_{15}H_{19}N_3O_3$
  Imazethapyr, 4944
$C_{15}H_{19}N_3O_5$
  Carboquone, 1872
$C_{15}H_{20}ClN_3O$
  Paclobutrazol, 7118
$C_{15}H_{20}N_2$
  Indalpine, 4965
$C_{15}H_{20}N_2O$
  Anagyrine, 666
$C_{15}H_{20}N_2O_2$
  Fenspiride, 4041
$C_{15}H_{20}N_2O_4S$
  Acetohexamide, 59
$C_{15}H_{20}N_2S$
  Methaphenilene, 6026
$C_{15}H_{20}N_6O_5$
  Alazopeptin, 209
$C_{15}H_{20}O$
  ar-Turmerone, 9956
$C_{15}H_{20}O_2$
  Alantolactone, 208
  Periplanones*, 7314
$C_{15}H_{20}O_3$
  Arborescin, 810
  Illudins*, 4939
  Parthenolide, 7180
  Perezone, 7298
  Periplanones*, 7314
$C_{15}H_{20}O_4$
  Abscisic Acid, 7
  Hirsutic Acid C, 4753
  Illudins*, 4939
  Santonic Acid, 8508
  Tanacetin, 9218
$C_{15}H_{20}O_6$
  Shellolic Acid, 8624
  Vomitoxin, 10171
$C_{15}H_{20}O_7$
  Nivalenol, 6757
$C_{15}H_{21}ClN_2O_2$
  Amidochlor, 420
$C_{15}H_{21}ClN_6$
  Lesopitron, 5470
$C_{15}H_{21}Cl_2FN_2O_3$
  Fluroxypyr*, 4238

Denotes derivative of title compound

$C_{16}H_{12}Cl_2N_2O_2$
Lormetazepam, 5611
$C_{16}H_{12}FNO_3$
Flunoxaprofen, 4183
$C_{16}H_{12}FN_3O_3$
Flubendazole, 4154
Flunitrazepam, 4181
$C_{16}H_{12}N_2O_4$
Isatide, 5115
$C_{16}H_{12}O_3$
Anisindione, 706
$C_{16}H_{12}O_4$
Chrysin*, 2316
Formononetin, 4271
Isoxepac, 5257
$C_{16}H_{12}O_5$
Acacetin, 10
Emodin*, 3602
Genistein*, 4395
Oroxylin A, 7006
Prunetin, 8094
$C_{16}H_{12}O_6$
Acetylsalicylsalicylic Acid, 102
Diosmetin, 3349
Hematein, 4667
Pratensein, 7893
Tectorigenin, 9264
$C_{16}H_{12}O_7$
Rhamnetin, 8337
$C_{16}H_{13}ClFNO_5$
Flumiclorac, 4178
$C_{16}H_{13}ClN_2O$
Diazepam, 3042
Mazindol, 5801
$C_{16}H_{13}ClN_2O_2$
Clobazam, 2417
Temazepam, 9285
$C_{16}H_{13}ClN_2O_4S$
Suclofenide, 9048
$C_{16}H_{13}ClO_6$
Peonidin, 7285
$C_{16}H_{13}ClO_7$
Petunidin, 7331
$C_{16}H_{13}Cl_2NO_4$
Aceclofenac, 19
Quinfamide, 8241
$C_{16}H_{13}Cl_3N_2OS$
Tioconazole, 9595
$C_{16}H_{13}F_2N_3O$
Flutriafol, 4246
$C_{16}H_{13}I_3N_2O_3$
Iobenzamic Acid, 5028
$C_{16}H_{13}NO_3S$
1-Anilino-8-naphthalenesulfonate,
699
$C_{16}H_{13}N_3$
Yellow AB, 10228
$C_{16}H_{13}N_3O_3$
Mebendazole, 5807
Nimetazepam, 6641
$C_{16}H_{14}$
1,4-Diphenyl-1,3-butadiene, 3378
$C_{16}H_{14}CaO_6$
Mandelic Acid*, 5757
$C_{16}H_{14}ClN_3O$
Chlordiazepoxide, 2132
$C_{16}H_{14}Cl_2O_3$
Chlorobenzilate, 2174
$C_{16}H_{14}Cl_2O_4$
Diclofop-Methyl, 3133
$C_{16}H_{14}FN_3$
Afloqualone, 181
$C_{16}H_{14}F_2N_3NaO_4S$
Pantoprazole*, 7146
$C_{16}H_{14}F_3NO_3S$
Tolrestat, 9663
$C_{16}H_{14}F_3N_3O_2S$
Lansoprazole, 5373
$C_{16}H_{14}MgO_6$
Magnesium Mandelate, 5709

$C_{16}H_{14}N_2O$
Glycosine, 4512
Methaqualone, 6028
$C_{16}H_{14}N_2O_3$
Bendazac, 1061
$C_{16}H_{14}N_2O_4$
Amlexanox, 515
Nybomycin, 6829
$C_{16}H_{14}N_2O_6S$
Phthalylsulfacetamide, 7532
$C_{16}H_{14}N_7NaO_5S_4.2H_2O$
Cefuzonam*, 2003
$C_{16}H_{14}O$
Dypnone, 3530
$C_{16}H_{14}O_2$
Benzyl Cinnamate, 1165
$C_{16}H_{14}O_3$
Fenbufen, 4003
Ketoprofen, 5316
3-(o-Methoxyphenyl)-2-phenylacryl-
ic Acid, 6078
$C_{16}H_{14}O_4$
Imperatorin, 4960
Medicarpin, 5833
$C_{16}H_{14}O_5$
Aspirin*, 886
Brazilin, 1392
Sakuranetin, 8470
$C_{16}H_{14}O_6$
Hematoxylin, 4670
Hesperetin, 4704
Homoeriodictyol, 4773
$C_{16}H_{15}ClN_2$
Medazepam, 5829
$C_{16}H_{15}ClN_2OS$
Clotiazepam, 2477
$C_{16}H_{15}ClO_6$
Lonapalene, 5595
$C_{16}H_{15}Cl_2NO_2$
Bulan, 1505
$C_{16}H_{15}Cl_3O_2$
Methoxychlor, 6070
$C_{16}H_{15}CrN_6S_4$
Rhodanilic Acid, 8350
$C_{16}H_{15}FN_2O_4$
Pazufloxacin, 7188
$C_{16}H_{15}F_2N_3O_4S$
Pantoprazole, 7146
$C_{16}H_{15}F_2N_3Si$
Flusilazole, 4240
$C_{16}H_{15}F_3O$
Flumecinol, 4170
$C_{16}H_{15}N$
Dizocilpine, 3451
$C_{16}H_{15}NO$
Cyheptamide, 2828
$C_{16}H_{15}NO_2$
Cinnamyl Anthranilate, 2363
$C_{16}H_{15}N_2NaO_6S_2$
Cephalothin*, 2028
$C_{16}H_{15}N_3$
Epinastine, 3655
$C_{16}H_{15}N_3O_5$
Opiniazide, 6984
$C_{16}H_{15}N_3O_7S$
Sulfaloxic Acid, 9080
$C_{16}H_{15}N_4NaO_8S$
Cefuroxime*, 2002
$C_{16}H_{15}N_5O_7S_2$
Cefixime, 1975
$C_{16}H_{15}N_7O_5S_4$
Cefuzonam, 2003
$C_{16}H_{16}$
[2.2]Metacyclophane, 5981
$C_{16}H_{16}ClNO_2$
Nicoclonate, 6603
$C_{16}H_{16}ClNO_2S$
Clopidogrel, 2457
$C_{16}H_{16}ClNO_3$

Fenoldopam, 4020
Nicofibrate, 6604
$C_{16}H_{16}ClNO_4S$
Daltroban, 2871
$C_{16}H_{16}ClN_3O_3S$
Indapamide, 4969
Metolazone, 6231
$C_{16}H_{16}ClN_3O_4.H_2O$
Loracarbef, 5606
$C_{16}H_{16}Cl_6N_2O_2$
Triclofenol Piperazine, 9787
$C_{16}H_{16}KN_5O_4S$
Azidocillin*, 942
$C_{16}H_{16}NO_6P$
Naftalofos, 6441
$C_{16}H_{16}N_2Na_2O_7S_2$
Sulbenicillin*, 9059
Temocillin*, 9288
$C_{16}H_{16}N_2OS$
Ahistan, 189
$C_{16}H_{16}N_2O_2$
Isolysergic Acid, 5197
Lysergic Acid, 5664
Rugulovasines, 8449
Salen, 8474
$C_{16}H_{16}N_2O_4$
Phenmedipham, 7384
$C_{16}H_{16}N_2O_4S$
Acedapsone, 20
$C_{16}H_{16}N_2O_5S$
Succisulfone, 9047
$C_{16}H_{16}N_2O_6S_2$
Cephalothin, 2028
$C_{16}H_{16}N_3NaO_4S$
Cephalexin*, 2021
$C_{16}H_{16}N_3NaO_7S_2$
Cefoxitin*, 1986
$C_{16}H_{16}N_4$
Stilbamidine, 8970
$C_{16}H_{16}N_4O$
Hydroxystilbamidine, 4892
$C_{16}H_{16}N_4O_2S$
Sulfazamet, 9118
$C_{16}H_{16}N_4O_8S$
Cefuroxime, 2002
$C_{16}H_{16}N_5NaO_4S$
Azidocillin*, 942
$C_{16}H_{16}N_5NaO_7S_2$
Cefotaxime*, 1983
$C_{16}H_{16}O_2$
Felbinac*, 3989
Xenbucin, 10205
$C_{16}H_{16}O_3$
Parvaquone, 7182
$C_{16}H_{16}O_5$
Alkannin, 253
$C_{16}H_{17}BrClN_3O_3$
Halofuginone, 4627
$C_{16}H_{17}BrN_2$
Zimeldine, 10254
$C_{16}H_{17}ClN_2O$
Tetrazepam, 9379
$C_{16}H_{17}KN_2O_4S$
Penicillin G Potassium, 7225
$C_{16}H_{17}KN_2O_5S$
Penicillin V*, 7230
$C_{16}H_{17}NO$
Diphenamid, 3364
$C_{16}H_{17}NO_3$
Normorphine, 6806
$C_{16}H_{17}NO_4$
Lunine, 5633
Lycorine, 5656
Tetrahydropapaveroline, 9355
$C_{16}H_{17}N_2NaO_4S$
Benzylpenicillin Sodium, 1178
$C_{16}H_{17}N_3$
Besipirdine, 1223
$C_{16}H_{17}N_3O$
Lysergamide, 5663

Denotes derivative of title compound

C$_{16}$H$_{23}$N$_3$O$_4$
Gabexate, 4344
C$_{16}$H$_{23}$NaO$_5$
Exiproben*, 3959
C$_{16}$H$_{24}$ClNO$_2$
Hexylcaine Hydrochloride, 4746
C$_{16}$H$_{24}$HgO$_3$
Acetomeroctol, 63
C$_{16}$H$_{24}$NO$_5$$^+$
Sinapine, 8688
C$_{16}$H$_{24}$N$_2$
Isoaminile, 5124
Xylometazoline, 10219
C$_{16}$H$_{24}$N$_2$O
Oxymetazoline, 7100
Ropinirole, 8416
C$_{16}$H$_{24}$N$_2$O$_2$
Molindone, 6315
N,N,N',N'-Tetraethylphthalamide, 9344
C$_{16}$H$_{24}$N$_2$O$_3$
Carteolol, 1917
C$_{16}$H$_{24}$N$_2$O$_4$
Ubenimex, 9973
C$_{16}$H$_{24}$N$_2$O$_4$S
2-Amino-4-picoline*, 486
C$_{16}$H$_{24}$N$_2$O$_6$
Pyrisuccideanol, 8175
C$_{16}$H$_{24}$N$_{10}$O$_4$
Aminophylline, 485
C$_{16}$H$_{24}$O$_4$
Brefeldin A, 1393
C$_{16}$H$_{24}$O$_5$
Exiproben, 3959
C$_{16}$H$_{24}$O$_7$
Rhododendrin, 8356
C$_{16}$H$_{25}$HgNNa$_2$O$_6$S
Mercaptomerin Sodium, 5918
C$_{16}$H$_{25}$NO
Lycopodine, 5652
C$_{16}$H$_{25}$NO$_2$
Butethamate, 1551
Cyclexanone, 2775
Gravitol(e), 4565
Thymyl N-Isoamylcarbamate, 9550
Tramadol, 9701
C$_{16}$H$_{25}$NO$_2$S
Tertatolol, 9318
C$_{16}$H$_{25}$NO$_3$
Moxisylyte, 6374
C$_{16}$H$_{25}$NO$_4$
Esmolol, 3741
Floredil, 4144
C$_{16}$H$_{25}$N$_2$NaO$_5$S
Cilastatin*, 2331
C$_{16}$H$_{25}$N$_3$O
Propiram, 8017
C$_{16}$H$_{25}$N$_3$O$_5$
Xamoterol, 10189
C$_{16}$H$_{25}$N$_7$O$_8$
Gougerotin, 4550
C$_{16}$H$_{26}$ClNO
Butidrine Hydrochloride, 1558
C$_{16}$H$_{26}$GdN$_5$O$_8$
Gadodiamide, 4345
C$_{16}$H$_{26}$NO$_4$S.C$_7$H$_7$O$_3$S
Suplatast Tosylate, 9178
C$_{16}$H$_{26}$N$_2$O$_2$
Dimethocaine, 3270
C$_{16}$H$_{26}$N$_2$O$_3$
Proparacaine, 7991
C$_{16}$H$_{26}$N$_2$O$_4$
Cetamolol, 2060
C$_{16}$H$_{26}$N$_2$O$_5$S
Cilastatin, 2331
C$_{16}$H$_{26}$O$_2$
2,5-Di-tert-pentylhydroquinone, 3359
C$_{16}$H$_{26}$O$_3$
Juvenile Hormones*, 5287

C$_{16}$H$_{26}$O$_5$
Artemether, 854
C$_{16}$H$_{26}$O$_7$
Picrocrocin, 7563
C$_{16}$H$_{27}$ClN$_2$O$_3$
Propoxycaine Hydrochloride, 8023
C$_{16}$H$_{27}$N$_3$O$_8$
Dinoseb*, 3341
C$_{16}$H$_{28}$N$_2$O$_2$
Tromantadine, 9901
C$_{16}$H$_{28}$N$_2$O$_6$Zn
ε-Acetamidocaproic Acid*, 43
C$_{16}$H$_{28}$N$_4$O$_4$S
Biocytin, 1268
C$_{16}$H$_{28}$O$_2$
Cioteronel, 2372
Hydnocarpic Acid, 4797
Prodlure, 7950
C$_{16}$H$_{29}$NO$_5$
Butoctamide*, 1562
C$_{16}$H$_{29}$NaO$_7$S
Bis(1-methylamyl) Sodium Sulfosuccinate, 1295
C$_{16}$H$_{30}$MgO$_4$
Valproic Acid*, 10049
C$_{16}$H$_{30}$N$_2$O$_3$
Piroctone*, 7657
C$_{16}$H$_{30}$O
Muscone, 6392
C$_{16}$H$_{30}$O$_4$Zn
Zinc Caprylate, 10259
C$_{16}$H$_{31}$NaO$_4$
Valproic Acid*, 10049
C$_{16}$H$_{32}$O$_2$
2-Hexyldecanoic Acid, 4747
Myristic Acid*, 6416
Palmitic Acid, 7128
C$_{16}$H$_{32}$O$_4$
Diethylene Glycol Monolaurate, 3169
C$_{16}$H$_{32}$O$_5$
Aleuritic Acid, 230
C$_{16}$H$_{33}$BiO$_4$S
Bismuth Butylthiolaurate, 1301
C$_{16}$H$_{34}$Br$_2$N$_2$O$_4$
Suxethonium Bromide, 9184
C$_{16}$H$_{34}$O
Cetyl Alcohol, 2070
C$_{16}$H$_{35}$NO$_2$
Ammonium Palmitate, 572
C$_{16}$H$_{38}$Br$_2$N$_2$
Decamethonium Bromide, 2904
C$_{16}$H$_{42}$I$_7$N$_8$O$_{16}$
Tetraglycine Hydroperiodide, 9347
C$_{17}$H$_8$Cl$_2$F$_8$N$_2$O$_3$
Lufenuron, 5624
C$_{17}$H$_9$Cl$_3$N$_4$O$_2$
Diclazuril, 3130
C$_{17}$H$_9$NO$_4$
Alizarine Blue, 249
C$_{17}$H$_{10}$Cl$_2$N$_4$O$_2$
Clazuril, 2403
C$_{17}$H$_{10}$F$_6$N$_4$S
Flubenzimine, 4155
C$_{17}$H$_{10}$O
Benzanthrone, 1090
C$_{17}$H$_{10}$O$_4$
Fluorescamine, 4193
C$_{17}$H$_{11}$ClF$_4$N$_2$S
Quazepam, 8211
C$_{17}$H$_{11}$NO$_7$
Aristolochic Acid, 822
C$_{17}$H$_{11}$N$_5$
Letrozole, 5474
C$_{17}$H$_{12}$Br$_2$O$_3$
Benzbromarone, 1093
C$_{17}$H$_{12}$ClFN$_2$O$_2$
Pirazolac, 7643
C$_{17}$H$_{12}$ClF$_3$N$_2$O
Halazepam, 4619

C$_{17}$H$_{12}$ClNO$_2$S
Fentiazac, 4045
C$_{17}$H$_{12}$Cl$_2$N$_2$O
Fenarimol, 3999
C$_{17}$H$_{12}$Cl$_2$N$_4$
Triazolam, 9734
C$_{17}$H$_{12}$Cl$_2$O$_7$
Erdin*, 3682
C$_{17}$H$_{12}$I$_2$O$_3$
Benziodarone, 1113
C$_{17}$H$_{12}$O$_2$
2-Naphthyl Benzoate, 6495
C$_{17}$H$_{12}$O$_3$
1-Naphthyl Salicylate, 6501
2-Naphthyl Salicylate, 6502
C$_{17}$H$_{12}$O$_6$
Aflatoxins B*, 178
Irisolone, 5105
C$_{17}$H$_{12}$O$_7$
Aflatoxins G*, 179
Aflatoxins M*, 180
C$_{17}$H$_{13}$ClFNO$_4$
Clodinafop-Propargyl, 2431
C$_{17}$H$_{13}$ClFN$_3$O
Epoxiconazole, 3665
C$_{17}$H$_{13}$ClN$_2$O$_2$
Lonazolac, 5596
C$_{17}$H$_{13}$ClN$_4$
Alprazolam, 320
C$_{17}$H$_{13}$N$_3$O$_5$S$_2$
Phthalylsulfathiazole, 7533
C$_{17}$H$_{14}$
1,2-Cyclopentenophenanthrene, 2812
C$_{17}$H$_{14}$BrFN$_2$O$_2$
Haloxazolam, 4635
C$_{17}$H$_{14}$ClFN$_2$O$_3$
Doxefazepam, 3490
C$_{17}$H$_{14}$ClF$_7$O$_2$
Tefluthrin, 9266
C$_{17}$H$_{14}$Cl$_2$N$_2$O$_2$
Cloxazolam, 2481
C$_{17}$H$_{14}$N$_2$
Ellipticine, 3589
Flavopereirine, 4133
Olivacine, 6971
C$_{17}$H$_{14}$N$_2$O$_3$
Rosoxacin, 8426
C$_{17}$H$_{14}$N$_2$O$_5$S
Calmagite, 1766
C$_{17}$H$_{14}$O
Dibenzalacetone, 3053
C$_{17}$H$_{14}$O$_3$
Benzarone, 1091
C$_{17}$H$_{14}$O$_3$S
Zaltoprofen, 10244
C$_{17}$H$_{14}$O$_5$
Coumafuryl, 2624
Pterocarpin, 8117
C$_{17}$H$_{14}$O$_6$
Aflatoxins B*, 178
Pectolinarigenin, 7195
C$_{17}$H$_{14}$O$_7$
Aflatoxins G*, 179
Aflatoxins M*, 180
C$_{17}$H$_{15}$ClN$_4$S
Etizolam, 3919
C$_{17}$H$_{15}$ClO$_7$
Malvidin Chloride, 5754
C$_{17}$H$_{15}$NO$_3$
Indoprofen, 4999
C$_{17}$H$_{15}$NO$_5$
Benorylate, 1074
C$_{17}$H$_{15}$N$_3$
Yellow OB, 10229
C$_{17}$H$_{15}$N$_7$Na$_2$O$_8$S$_4$
Cefotetan*, 1984
C$_{17}$H$_{16}$Br$_2$O$_3$
Bromopropylate, 1456

* Denotes derivative of title compound

$C_{17}H_{16}ClFN_2O_2$
Progabide, 7955
$C_{17}H_{16}ClN_3O$
Amoxapine, 616
$C_{17}H_{16}Cl_2N_2O_5$
Chlorphenoxamide, 2233
$C_{17}H_{16}N_2Na_2O_6S$
Carbenicillin*, 1838
$C_{17}H_{16}N_2O$
Etaqualone, 3757
$C_{17}H_{16}N_3NaO_6S_2$
Cephapirin Sodium, 2030
$C_{17}H_{16}N_4O_2$
Nifenazone, 6619
$C_{17}H_{16}O_3$
Eugenol*, 3944
$C_{17}H_{17}ClFN_3O_3$
Clinafloxacin, 2413
$C_{17}H_{17}ClF_6N_2O$
Mefloquine Hydrochloride, 5845
$C_{17}H_{17}ClN_2O$
Etifoxine, 3910
$C_{17}H_{17}ClN_6O_3$
Zopiclone, 10324
$C_{17}H_{17}ClO_3$
Clobuzarit, 2426
$C_{17}H_{17}ClO_6$
Griseofulvin, 4571
$C_{17}H_{17}Cl_2N$
Sertraline, 8612
$C_{17}H_{17}Cl_2NO_3$
Terofenamate, 9312
$C_{17}H_{17}Cl_2N_2NaO_5S$
Clometocillin*, 2445
$C_{17}H_{17}N$
Berbine, 1195
$C_{17}H_{17}NO_2$
Apomorphine, 787
$C_{17}H_{17}NO_3$
Sparsiflorine, 8885
$C_{17}H_{17}NO_3S$
Protizinic Acid, 8077
$C_{17}H_{17}N_2^+$
Difenzoquat, 3185
$C_{17}H_{17}N_3O_3$
Imazaquin, 4943
$C_{17}H_{17}N_4NaO_6$
Carbazochrome Salicylate, 1832
$C_{17}H_{17}N_7O_8S_4$
Cefotetan, 1984
$C_{17}H_{18}Br_2N_4O_2$
Dibromopropamidine, 3073
$C_{17}H_{18}ClNO_4$
Pirifibrate, 7650
$C_{17}H_{18}ClN_3O$
Clobenzepam, 2420
$C_{17}H_{18}ClN_3O_4$
Celestin Blue, 2006
$C_{17}H_{18}Cl_2N_2O_5S$
Clometocillin, 2445
$C_{17}H_{18}FN_3O_3$
Ciprofloxacin, 2374
$C_{17}H_{18}FN_3O_3S$
Rufloxacin, 8448
$C_{17}H_{18}F_2O_2$
Bifluranol, 1259
$C_{17}H_{18}F_3NO$
Fluoxetine, 4222
$C_{17}H_{18}F_3N_3O_3$
Fleroxacin, 4137
$C_{17}H_{18}N_2$
Amphetaminil, 624
$C_{17}H_{18}N_2O_5S$
Piretanide, 7647
$C_{17}H_{18}N_2O_6$
Nifedipine, 6617
$C_{17}H_{18}N_2O_6S$
Carbenicillin, 1838
$C_{17}H_{18}N_2S_2$
Sulbentine, 9061

$C_{17}H_{18}N_3NaO_4S$
Metampicillin*, 5986
$C_{17}H_{18}N_4O_3S$
Enviroxime, 3637
$C_{17}H_{18}O_3$
4-tert-Butylphenyl Salicylate, 1621
$C_{17}H_{18}O_4$
Diphenolic Acid, 3370
$C_{17}H_{19}ClN_2O$
Pyronine Y, 8190
$C_{17}H_{19}ClN_2OS$
Opromazine, 6987
$C_{17}H_{19}ClN_2S$
Chlorpromazine, 2238
$C_{17}H_{19}F_2N_3O_3$
Lomefloxacin, 5592
$C_{17}H_{19}KN_2O_5S$
Phenethicillin Potassium, 7369
$C_{17}H_{19}N$
Etifelmin, 3909
$C_{17}H_{19}NO$
Nefopam, 6529
$C_{17}H_{19}NO_3$
Chavicine, 2090
Coclaurine, 2521
Hydromorphone, 4847
Morphine, 6359
Norcodeine, 6783
Piperine, 7625
$C_{17}H_{19}NO_4$
Buphanamine, 1517
Fenoxycarb, 4027
Oxymorphone, 7103
$C_{17}H_{19}N_2NaO_6S$
Methicillin Sodium, 6047
$C_{17}H_{19}N_3$
Antazoline, 718
Mirtazepine, 6295
$C_{17}H_{19}N_3O$
Phentolamine, 7417
Piberaline, 7547
$C_{17}H_{19}N_3O_3S$
Omeprazole, 6977
$C_{17}H_{19}N_3O_4S$
Metampicillin, 5986
$C_{17}H_{19}N_5$
Anastrozole, 667
$C_{17}H_{19}N_5O_2$
Piritrexim, 7654
$C_{17}H_{20}BrNO$
Bromodiphenhydramine, 1439
$C_{17}H_{20}ClNO$
Chlophedianol, 2106
$C_{17}H_{20}ClNO_3$
Morphine Hydrochloride, 6360
$C_{17}H_{20}ClN_3O_3$
Azasetron, 933
$C_{17}H_{20}FN_3O_3$
Pefloxacin, 7197
$C_{17}H_{20}F_6N_2O_3$
Flecainide, 4136
$C_{17}H_{20}N_2O$
N,N'-Diethylcarbanilide, 3166
Michler's Ketone, 6265
$C_{17}H_{20}N_2O_2$
Tropicamide, 9911
Tropisetron, 9914
$C_{17}H_{20}N_2O_2S$
Tinoridine, 9590
$C_{17}H_{20}N_2O_3$
Nicotine*, 6611
$C_{17}H_{20}N_2O_5S$
Bumetanide, 1508
$C_{17}H_{20}N_2S$
Isopromethazine, 5219
Promazine, 7966
Promethazine, 7970
$C_{17}H_{20}N_4O$
Propizepine, 8019

$C_{17}H_{20}N_4O_2$
Propamidine, 7981
$C_{17}H_{20}N_4O_3$
Imazaquin*, 4943
$C_{17}H_{20}N_4O_6$
D-Araboflavin, 801
Lyxoflavine, 5672
Riboflavin, 8367
$C_{17}H_{20}N_4S$
Olanzapine, 6959
$C_{17}H_{20}O_5$
Matricarin, 5798
$C_{17}H_{20}O_6$
Mycophenolic Acid, 6408
$C_{17}H_{20}O_9$
Chlorogenic Acid*, 2194
$C_{17}H_{21}ClN_2O_2S$
Hexythiazox, 4751
$C_{17}H_{21}N$
Benzphetamine, 1151
N,N-Diethylbenzhydrylamine, 3163
$C_{17}H_{21}NO$
Diphenhydramine, 3367
Phenyltoloxamine, 7469
Tofenacin, 9641
$C_{17}H_{21}NO_2$
Apoatropine, 780
Desomorphine, 2972
Napropamide, 6503
$C_{17}H_{21}NO_3$
Dihydromorphine, 3220
Etodolac, 3920
Galanthamine, 4357
Ritodrine, 8401
$C_{17}H_{21}NO_4$
Cocaine, 2517
Fenoterol, 4022
Pseudococaine, 8097
Scopolamine, 8550
$C_{17}H_{21}NO_5$
Buphanitine, 1518
Scopolamine N-Oxide, 8551
$C_{17}H_{21}N_4O_9P$
Riboflavin Monophosphate, 8368
$C_{17}H_{21}O_5PS$
Coumithoate, 2635
$C_{17}H_{22}BrNOS_2$
Timepidium Bromide, 9583
$C_{17}H_{22}I_3N_3O_8$
Iopamidol, 5071
$C_{17}H_{22}NO_3^+$
Lunasine, 5632
$C_{17}H_{22}N_2$
Michler's Base, 6264
$C_{17}H_{22}N_2O$
Doxylamine, 3497
$C_{17}H_{22}N_2O_3$
Trichostatin(s)*, 9781
$C_{17}H_{22}N_2S$
Thenaldine, 9413
$C_{17}H_{22}N_4O$
Minaprine, 6287
$C_{17}H_{22}O_2$
Cicutoxin, 2328
Enanthotoxin, 3608
$C_{17}H_{22}O_3$
Bornyl Salicylate, 1371
Podocarpic Acid, 7701
$C_{17}H_{22}O_5$
Pyrethrosin, 8149
$C_{17}H_{23}NO$
Levorphanol, 5496
$C_{17}H_{23}NO_2$
Tilidine, 9578
$C_{17}H_{23}NO_3$
Atropine, 907
Hyoscyamine, 4907
Lycoramine, 5655
Mesembrine, 5966
Rotraxate, 8428

Denotes derivative of title compound

Denotes derivative of title compound

$C_{19}H_9HgI_2NaO_7S$
Meralein Sodium, 5912

$C_{19}H_{10}Br_4O_5S$
Bromphenol Blue, 1468

$C_{19}H_{11}Cl_2I_2NO_3$
Rafoxanide, 8280

$C_{19}H_{12}F_3N_3O_3S$
Zopolrestat, 10325

$C_{19}H_{12}O_6$
Dicumarol, 3140

$C_{19}H_{12}O_8$
Diacerein, 3003

$C_{19}H_{14}NO_4^+$
Coptisine, 2591

$C_{19}H_{14}N_4O$
1,3-Di-6-quinolylurea, 3416

$C_{19}H_{14}O_3$
Aurin, 915

$C_{19}H_{14}O_5S$
Phenolsulfonphthalein, 7397

$C_{19}H_{15}ClNNaO_4.3H_2O$
Indomethacin*, 4998

$C_{19}H_{15}ClN_2O_4$
Rebamipide, 8296

$C_{19}H_{15}ClN_4$
Triphenyltetrazolium Chloride, 9874

$C_{19}H_{15}ClO_4$
Coumachlor, 2623

$C_{19}H_{15}Cl_3O_5$
Nornidulin, 6808

$C_{19}H_{18}F_3N_4O_3$
Tosufloxacin, 9692

$C_{19}H_{15}KO_4$
Warfarin*, 10174

$C_{19}H_{15}NNa_2O_7$
Nedocromil*, 6524

$C_{19}H_{15}NO_2$
Cinchophen*, 2347

$C_{19}H_{15}NO_4$
3-Methyl-6,7-methylenedioxy-1-
piperonylisoquinoline, 6171

$C_{19}H_{15}NO_6$
Acenocoumarol, 29

$C_{19}H_{15}NaO_4$
Warfarin*, 10174

$C_{19}H_{16}$
Triphenylmethane, 9871

$C_{19}H_{16}ClFN_2O$
Flutoprazepam, 4245

$C_{19}H_{16}ClFN_3NaO_5S.H_2O$
Floxacillin*, 4147

$C_{19}H_{16}ClNO_4$
Clometacin, 2443
Indomethacin, 4998

$C_{19}H_{16}Cl_2N_3NaO_5S.H_2O$
Dicloxacillin*, 3134

$C_{19}H_{16}N_2$
Sempervirine, 8591

$C_{19}H_{16}N_2Na_2O_7S_2$
Ponceau 3R, 7749

$C_{19}H_{16}N_2O$
Difenpiramide, 3184

$C_{19}H_{16}N_5NaO_7S_3$
Ceftiofur*, 1999

$C_{19}H_{16}O$
Triphenylcarbinol, 9869

$C_{19}H_{16}O_4$
Warfarin, 10174

$C_{19}H_{16}O_5$
Efloxate, 3565

$C_{19}H_{17}ClFN_3O_5S$
Floxacillin, 4147

$C_{19}H_{17}ClN_2O$
Prazepam, 7895

$C_{19}H_{17}ClN_2O_4$
Glafenine, 4443
Oxametacine, 7048
Quizalofop-Ethyl, 8269

$C_{19}H_{17}ClN_3NaO_5S.H_2O$
Cloxacillin*, 2480

$C_{19}H_{17}ClN_4$
Fenbuconazole, 4002

$C_{19}H_{17}Cl_2N_3O_3$
Difenoconazole, 3182

$C_{19}H_{17}Cl_2N_3O_5S$
Dicloxacillin, 3134

$C_{19}H_{17}Cl_3N_2S$
Butoconazole, 1561

$C_{19}H_{17}NOS$
Tolnaftate, 9656

$C_{19}H_{17}NO_3$
Cusparine, 2750

$C_{19}H_{17}NO_4$
Stylopine, 9025

$C_{19}H_{17}NO_5$
Mofezolac, 6314

$C_{19}H_{17}NO_7$
Nedocromil, 6524

$C_{19}H_{17}N_3O$
Evodiamine, 3954

$C_{19}H_{17}N_3O_4S_2$
Cephaloridine, 2025

$C_{19}H_{17}N_5O_2$
Nafamostat, 6436

$C_{19}H_{17}N_5O_5S$
Salazosulfadimidine, 8473

$C_{19}H_{17}N_5O_7S_3$
Ceftiofur, 1999

$C_{19}H_{17}N_6NaO_5S_3.1\frac{1}{2}H_2O$
Cefditoren*, 1972

$C_{19}H_{17}N_6NaO_6S_2$
Cefamandole*, 1964

$C_{19}H_{17}N_9O_5S_2$
Cefozopran, 1987

$C_{19}H_{18}CaN_2O_9$
Calcium Acetylsalicylate*, 1684

$C_{19}H_{18}ClFN_2O_3$
Flutazolam, 4243

$C_{19}H_{18}ClN_3O_3$
Camazepam, 1773

$C_{19}H_{18}ClN_3O_5S$
Cloxacillin, 2480

$C_{19}H_{18}ClN_5$
Adinazolam, 159

$C_{19}H_{18}N_2O_3$
Kebuzone, 5300

$C_{19}H_{18}N_3NaO_5S.H_2O$
Oxacillin*, 7036

$C_{19}H_{18}N_4O_2$
Pimobendan, 7588

$C_{19}H_{18}N_6O_5S_3$
Cefditoren, 1972

$C_{19}H_{18}N_6O_6$
Nicarbazin, 6578

$C_{19}H_{18}N_7NaO_6$
Sodium Folate, 8763

$C_{19}H_{18}O_7$
Benfurodil Hemisuccinate, 1070
Gardenins*, 4377

$C_{19}H_{18}O_8$
Atranorin, 901
Gardenins*, 4377

$C_{19}H_{18}O_9$
Gardenins*, 4377

$C_{19}H_{19}N$
Phenindamine, 7380

$C_{19}H_{19}NOS$
Ketotifen, 5319

$C_{19}H_{19}NO_3$
Laureline, 5393

$C_{19}H_{19}NO_4$
Bulbocapnine, 1506
Domesticine, 3472
Nandinine, 6451

$C_{19}H_{19}NO_6$
Etersalate, 3759

$C_{19}H_{19}N_3O_5S$
Oxacillin, 7036

$C_{19}H_{19}N_3O_6$
Nilvadipine, 6637

$C_{19}H_{19}N_7O_6$
Folic Acid, 4253

$C_{19}H_{20}ClNO_4$
Bezafibrate, 1240

$C_{19}H_{20}ClNO_5$
Ronifibrate, 8414

$C_{19}H_{20}ClN_3$
Clemizole, 2406

$C_{19}H_{20}FNO_3$
Paroxetine, 7175

$C_{19}H_{20}FN_3O_3$
Danofloxacin, 2876

$C_{19}H_{20}F_3NO_2$
Benfluorex, 1066

$C_{19}H_{20}F_3NO_4$
Fluazifop-butyl, 4152

$C_{19}H_{20}F_3N_3O_3$
Niflumic Acid*, 6620

$C_{19}H_{20}N_2$
Mebhydroline, 5809

$C_{19}H_{20}N_2O$
Vellosimine, 10076

$C_{19}H_{20}N_2O_2$
Phenylbutazone, 7431

$C_{19}H_{20}N_2O_3$
Amphotalide, 626
Ditazol, 3432
Dolasetron, 3471
Oxyphenbutazone, 7106

$C_{19}H_{20}N_2O_3S$
Pioglitazone, 7605

$C_{19}H_{20}N_2O_7$
Aranidipine, 806

$C_{19}H_{20}N_6O$
Imidocarb, 4951

$C_{19}H_{20}N_8O_5$
Aminopterin, 493

$C_{19}H_{20}O_2$
Cyclofenil*, 2789

$C_{19}H_{20}O_9$
Cervicarcin, 2050

$C_{19}H_{20}O_{10}$
Khellol Glucoside, 5322

$C_{19}H_{21}ClN_2OS$
Chloracizine, 2108
Halethazole, 4622

$C_{19}H_{21}ClN_4O_5$
Theofibrate, 9420

$C_{19}H_{21}FN_2O_4$
Nadifloxacin, 6430

$C_{19}H_{21}F_3N_2S$
Trifluomeprazine, 9810

$C_{19}H_{21}KO_6$
Gibberellic Acid*, 4426

$C_{19}H_{21}N$
Nortriptyline, 6812
Protriptyline, 8088

$C_{19}H_{21}NO$
Doxepin, 3492

$C_{19}H_{21}NO_3$
Isothebaine, 5247
Nalorphine, 6448
Thebaine, 9411

$C_{19}H_{21}NO_4$
Boldine, 1355
Corytuberine, 2616
Laurotetanine, 5400
Naloxone, 6449
Salutaridine, 8493

$C_{19}H_{21}NS$
Dothiepin, 3485
Pizotyline, 7671

$C_{19}H_{21}N_3$
Perlapine, 7320

$C_{19}H_{21}N_3O$
Talastine, 9205
Zolpidem, 10321

$C_{19}H_{21}N_3O_5$
Isradipine, 5260

C₁₉H₂₉NO₂
Samandarine*, 8497

C₁₉H₂₉NO₃
Cyclodrine, 2788

C₁₉H₂₉NO₅
Dipivefrin, 3400

C₁₉H₂₉N₃O
Pamaquine, 7134

C₁₉H₃₀ClNO
Cycrimine Hydrochloride, 2825

C₁₉H₃₀N₂O₂
Bietamiverine, 1252

C₁₉H₃₀N₂O₃
Fenalamide, 3995

C₁₉H₃₀N₂O₅S
Moveltipril, 6368

C₁₉H₃₀N₈O₉
Mildiomycin, 6279

C₁₉H₃₀O
(3α,5α)-Androst-16-en-3-ol, 679

C₁₉H₃₀OS
Epitiostanol, 3662

C₁₉H₃₀O₂
Androstenediol, 677
Androsterone, 680
Epiandrosterone, 3646
Stanolone, 8950

C₁₉H₃₀O₃
Androstane-3β,11β-diol-17-one, 676
Oxandrolone, 7054

C₁₉H₃₀O₅
Idebenone, 4932
Piperonyl Butoxide, 7629

C₁₉H₃₁N
Fenpropidin, 4034

C₁₉H₃₁NO
Bencyclane, 1060

C₁₉H₃₁NO₂
Samandarine, 8497

C₁₉H₃₁NO₄
Decimemide, 2909

C₁₉H₃₁N₇O₄
Mopidamol, 6344

C₁₉H₃₂
Androstane, 675
Etiocholane, 3912
Tridecylbenzene, 9792

C₁₉H₃₂BrNO₂
Valethamate Bromide, 10044

C₁₉H₃₂N₂O₂
Camylofine, 1784

C₁₉H₃₂N₂O₄
Betoxycaine, 1235

C₁₉H₃₂N₂O₅
Perindopril, 7311

C₁₉H₃₂N₄O₁₁
Muramyl Dipeptide, 6385

C₁₉H₃₃N₃O
Pentisomide, 7271

C₁₉H₃₄
Fichtelite, 4120

C₁₉H₃₄N₈O₈
Streptothricins*, 8988

C₁₉H₃₄O₂
Methyl Linoleate, 6168

C₁₉H₃₄O₃
Methoprene, 6063

C₁₉H₃₅N
Perhexiline, 7305

C₁₉H₃₆BrNO₂
Sintropium Bromide, 8693

C₁₉H₃₆ClNO₂
Dicyclomine Hydrochloride, 3147

C₁₉H₃₆O₂
Lactobacillic Acid, 5353
Oleic Acid*, 6965

C₁₉H₃₇NO
Lactobacillic Acid*, 5353

C₁₉H₃₇N₅O₇
Sisomicin, 8695

C₁₉H₃₈O
Disparlure, 3425

C₁₉H₃₈O₂
Stearic Acid*, 8959

C₁₉H₃₈O₃
Cetyl Lactate, 2072

C₁₉H₃₉NO
Tridemorph, 9793

C₁₉H₃₉N₅O₇
Gentamicin*, 4398

C₁₉H₄₀
Pristane, 7932

C₁₉H₄₀O₃
Chimyl Alcohol, 2102

C₁₉H₄₂BrN
Cetrimonium Bromide, 2068

C₁₉H₄₂N₂O₈S₂
Trimethidinium Methosulfate, 9838

C₂₀H₂Cl₄I₄K₂O₅
Rose Bengal, 8421

C₂₀H₂Cl₄I₄Na₂O₅
Rose Bengal*, 8421

C₂₀H₆Br₂N₂Na₂O₉
Eosine I Bluish, 3638

C₂₀H₆Br₄Na₂O₅
Eosine Yellowish—(YS), 3639

C₂₀H₆I₄Na₂O₅
Erythrosine, 3734

C₂₀H₈Br₂HgNa₂O₆
Merbromin, 5914

C₂₀H₈Br₂Na₂O₅
4′,5′-Dibromofluorescein*, 3070

C₂₀H₈Br₄Na₂O₄
3′,3″,5′,5″-Tetrabromophenolphtha-
lein*, 9329

C₂₀H₈Br₄Na₂O₁₀S₂
Sulfobromophthalein Sodium, 9127

C₂₀H₈Br₄O₅
Eosine Yellowish—(YS)*, 3639

C₂₀H₈Cl₂Na₂O₅
4′,5′-Dichlorofluorescein*, 3117

C₂₀H₈Cl₄Na₂O₄
Phenoltetrachlorophthalein*, 7398

C₂₀H₈I₂Na₂O₅
4′,5′-Diiodofluorescein*, 3233

C₂₀H₈I₄Na₂O₄
Iodophthalein Sodium, 5059
Phentetiothalein Sodium, 7416

C₂₀H₁₀Br₂O₅
4′,5′-Dibromofluorescein, 3070

C₂₀H₁₀Br₄O₄
3′,3″,5′,5″-Tetrabromophenolphtha-
lein, 9329

C₂₀H₁₀Cl₂O₅
4′,5′-Dichlorofluorescein, 3117

C₂₀H₁₀Cl₄O₄
Phenoltetrachlorophthalein, 7398

C₂₀H₁₀I₂O₅
4′,5′-Diiodofluorescein, 3233

C₂₀H₁₀I₄O₄
Iodophthalein Sodium*, 5059

C₂₀H₁₀Na₂O₅
Fluorescein*, 4194

C₂₀H₁₀Na₂O₇
Gallein*, 4362

C₂₀H₁₁N₂Na₃O₁₀S₃
Amaranth (Dye), 391

C₂₀H₁₂
Benzo[a]pyrene, 1134
Benzo[e]pyrene, 1135
Perylene, 7325

C₂₀H₁₂N₃NaO₇S
Eriochrome Black T, 3709

C₂₀H₁₂Na₂O₄
Phenolphthalein Sodium, 7393

C₂₀H₁₂O₅
Fluorescein, 4194

C₂₀H₁₂O₇
Gallein, 4362

C₂₀H₁₃N₃O₃
Violacein, 10135

C₂₀H₁₄
Cholanthrene, 2252
Triptycene, 9879

C₂₀H₁₄CaO₈S₂
Asaprol, 859

C₂₀H₁₄ClNO₄
Sanguinarine*, 8504

C₂₀H₁₄I₆N₂O₆
Iodipamide, 5043

C₂₀H₁₄NO₄⁺
Sanguinarine, 8504

C₂₀H₁₄N₄
Porphine, 7758

C₂₀H₁₄O₂S₂
DDD (Analytical), 2897

C₂₀H₁₄O₃
Florantyrone, 4143

C₂₀H₁₄O₄
Phenolphthalein, 7392
Phenyl Phthalate, 7460

C₂₀H₁₄O₅
Fluorescin, 4196

C₂₀H₁₄O₆
Ethylidene Dicoumarol, 3858

C₂₀H₁₅Cl₃N₂OS
Sertaconazole, 8610

C₂₀H₁₅F₃N₄O₃
Trovafloxacin, 9919

C₂₀H₁₅N
β,β′-Dinaphthylamine, 3319

C₂₀H₁₆
9,10-Dimethyl-1,2-benzanthracene,
3285

C₂₀H₁₆NO₄⁺
Woreninc, 10187

C₂₀H₁₆N₂O₂
Anilinephthalein, 698

C₂₀H₁₆N₂O₄
Camptothecin, 1783

C₂₀H₁₆N₄
Nitron, 6710

C₂₀H₁₆O₄
Phenolphthalin, 7394

C₂₀H₁₆O₅
Psoralidin, 8114

C₂₀H₁₆O₆
Anthralin*, 723
Elliptone, 3591
Viridin, 10144

C₂₀H₁₇ClN₂O₃
Ketazolam, 5308

C₂₀H₁₇Cl₃N₂O₂
Omoconazole, 6978

C₂₀H₁₇Cl₃O₅
Nidulin, 6616

C₂₀H₁₇FO₃S
Sulindac, 9155

C₂₀H₁₇F₂N₃O₃
Sarafloxacin, 8517

C₂₀H₁₇F₃N₂O₄
Floctafenine, 4140

C₂₀H₁₇NO₆
Adlumidine, 163
Bicuculline, 1250

C₂₀H₁₇N₃Na₂O₉S₃
Acid Fuchsin, 109

C₂₀H₁₈ClNO₆
Ochratoxins*, 6836

C₂₀H₁₈NO₄⁺
Berberine, 1192

C₂₀H₁₈N₂O
N-(1,2-Diphenylethyl)nicotinamide,
3382

C₂₀H₁₈N₆Na₂O₇S₄
Cefodizime*, 1979

C₂₀H₁₈N₆Na₂O₉S
Moxalactam*, 6369

Denotes derivative of title compound

Denotes derivative of title compound

C₂₀H₂₅ClN₂O₅
Amlodipine, 516

C₂₀H₂₅ClO₄
Osaterone, 7019

C₂₀H₂₅FN₄O
Niaprazine, 6576

C₂₀H₂₅N
Fenpiprane, 4031
Prodipine, 7949

C₂₀H₂₅NO
Diphemethoxidine, 3362
Normethadone, 6804
Pridinol, 7922

C₂₀H₂₅NOS
Thiphenamil, 9509

C₂₀H₂₅NO₂
Cuspareine, 2749
Femoxetine, 3993
Fomocaine, 4259
Propanocaine, 7988

C₂₀H₂₅NO₂S₂
Tiagabine, 9557

C₂₀H₂₅NO₃
Benactyzine, 1055
Difemerine, 3180
Dimenoxadol, 3253

C₂₀H₂₅NO₄
Codamine, 2524
Laudanidine, 5389
Laudanine, 5390

C₂₀H₂₅N₂O⁺
Macusines*, 5679

C₂₀H₂₅N₃O
Lysergide, 5665
Prodigiosin, 7948

C₂₀H₂₅N₃O₂
Methylergonovine, 6147

C₂₀H₂₅N₃S
Perazine, 7294

C₂₀H₂₅N₅O₇S₂
Cefetamet*, 1974

C₂₀H₂₆ClNO₂
Adiphenine Hydrochloride, 160

C₂₀H₂₆ClNO₃
Lachesine, 5343

C₂₀H₂₆ClNO₅
Cloricromen, 2467

C₂₀H₂₆N₂
Dimetacrine, 3258
Trimipramine, 9852

C₂₀H₂₆N₂O
Ibogaine, 4920
Tabernanthine, 9195

C₂₀H₂₆N₂O₂
Ajmaline, 194
Epsiprantel, 3671
Hydroquinidine, 4851
Hydroquinine, 4852

C₂₀H₂₆N₂O₅S
Alacepril, 202

C₂₀H₂₆N₂O₅S₂
Spirapril*, 8905

C₂₀H₂₆N₂S
Etymemazine, 3938

C₂₀H₂₆N₄O
Lisuride, 5541

C₂₀H₂₆N₄OS
Oxypendyl, 7104

C₂₀H₂₆N₄O₂
Hexamidine, 4728

C₂₀H₂₆O
3-Dehydroretinal, 2926

C₂₀H₂₆O₂
Benzestrol, 1102
Methestrol, 6045
Norethindrone, 6790
Norethynodrel, 6791

C₂₀H₂₆O₃
Taxodione, 9247

C₂₀H₂₆O₅
Allethrins*, 257

C₂₀H₂₆O₇
Cnicin, 2486

C₂₀H₂₇BrN₂O
Ambutonium Bromide, 406

C₂₀H₂₇N
Alverine, 386
Terodiline, 9311

C₂₀H₂₇NO
Cyclorphan, 2819

C₂₀H₂₇NO₂
Fenalcomine, 3996
Vetrabutine, 10107

C₂₀H₂₇NO₃
Tralkoxydim, 9699
Trilostane, 9827

C₂₀H₂₇NO₄
Bevantolol, 1238

C₂₀H₂₇NO₄S
Sulfinalol, 9120

C₂₀H₂₇NO₅
Buquinolate, 1524
Chromonar, 2296

C₂₀H₂₇NO₁₁
Amygdalin, 638

C₂₀H₂₇N₃O₅
Cilazapril*, 2332

C₂₀H₂₇N₃O₆
Febarbamate, 3983
Imidapril, 4947

C₂₀H₂₇N₃O₇S
Succisulfone*, 9047

C₂₀H₂₇N₅O₂
Cilostazol, 2335

C₂₀H₂₇N₅O₃
Bamifylline, 982

C₂₀H₂₇N₅O₅S
Glisoxepid, 4453

C₂₀H₂₇N₇O₆
Diminazene Aceturate*, 3316

C₂₀H₂₈BrN
Emepronium Bromide, 3598

C₂₀H₂₈BrNO₃
Endobenzyline Bromide, 3612

C₂₀H₂₈Cl₄N₂O₄
Teclozan, 9262

C₂₀H₂₈I₃N₃O₉
Iopentol, 5073

C₂₀H₂₈N₂O₃
Oxyphencyclimine, 7107

C₂₀H₂₈N₂O₅
Enalapril, 3605
Remifentanil, 8300

C₂₀H₂₈N₂O₅S
Tamsulosin, 9217

C₂₀H₂₈N₂O₆
Cinepazet, 2349

C₂₀H₂₈N₄O
Terguride, 9308

C₂₀H₂₈N₄O₈S₂
Stilbamidine*, 8970

C₂₀H₂₈N₆O₂
Fencamine, 4007

C₂₀H₂₈O
Lynestrenol, 5659
Retinal, 8331
Vitamin A₂, 10151

C₂₀H₂₈O₂
Ethynodiol, 3905
Methandrostenolone, 6018
Norgesterone, 6795
Norvinisterone, 6814
Retinoic Acid, 8333

C₂₀H₂₈O₃
Cafestol, 1672
Cinerins*, 2351
Mytatrienediol, 6424

C₂₀H₂₈O₃S
Ethyl Dibunate, 3834

C₂₀H₂₈O₄
Marrubiin, 5794

C₂₀H₂₈O₅
Pododacric Acid, 7702

C₂₀H₂₈O₅S
Ecabet, 3538

C₂₀H₂₈O₆
Amarolide, 394
Phorbol, 7487

C₂₀H₂₈O₆S
Tiaprost, 9563

C₂₀H₂₈O₁₃
Primulaverin, 7930

C₂₀H₂₉FO₃
Fluoxymesterone, 4223

C₂₀H₂₉NO₃
Propenzolate, 7998

C₂₀H₂₉NO₅S
Latrunculins*, 5388

C₂₀H₂₉NO₆S
Sultroponium, 9170

C₂₀H₂₉N₃O₅
Propiram*, 8017

C₂₀H₂₉N₃O₇
Amicoumacin A, 417

C₂₀H₂₉N₅O₃
Urapidil, 10003

C₂₀H₂₉N₅O₃
Trimazosin, 9828

C₂₀H₃₀BrNO₃
Ipratropium Bromide, 5089

C₂₀H₃₀ClN₃O₂
Dibucaine Hydrochloride, 3081

C₂₀H₃₀N₂O₂
Dipiproverine, 3399
Furazabol, 4319

C₂₀H₃₀N₂O₃
Morpheridine, 6357

C₂₀H₃₀N₂O₅S
Benfuracarb, 1069

C₂₀H₃₀O
Neovitamin A, 6555
Vitamin A, 10150

C₂₀H₃₀O₂
Abietic Acid, 3
(all-Z)-5,8,11,14,17-Eicosapen-
taenoic Acid, 3572
Hepaxanthin, 4686
Isopimaric Acid, 5217
Levopimaric Acid, 5494
Methenolone, 6044
17-Methyltestosterone, 6206
Mibolerone, 6262
Norethandrolone, 6789
Oxendolone, 7065
Oxenin, 7066
Palustric Acid, 7131
Pimaric Acid, 7582
Stenbolone, 8962

C₂₀H₃₀O₃
Leukotrienes*, 5482
Oxymesterone, 7099

C₂₀H₃₀O₅
Andrographolide, 673
Taxicins*, 9245

C₂₀H₃₀O₆
Taxicins*, 9245

C₂₀H₃₁NO₂
Drofenine, 3501
Metcaraphen, 5996

C₂₀H₃₁NO₂S
Cetiedil, 2062

C₂₀H₃₁NO₃
Carbetapentane, 1840

C₂₀H₃₁NO₇
Heliosupine, 4659

C₂₀H₃₁NaO₅
Prostacyclin*, 8061

C₂₀H₃₂
Neocembrene, 6535

Denotes derivative of title compound

$C_{22}H_{14}Cl_2I_2N_2O_2$
Closantel, 2473

$C_{22}H_{16}Cl_2O_4S$
Tioclomarol, 9594

$C_{22}H_{16}F_2N_2$
Flutrimazole, 4247

$C_{22}H_{16}N_4Na_2O_9S_2$
Orange B, 6991

$C_{22}H_{16}N_4O$
Sudan III, 9054

$C_{22}H_{16}O_6$
Resistomycin, 8320

$C_{22}H_{16}O_8$
Ethyl Biscoumacetate, 3818

$C_{22}H_{17}ClN_2$
Clotrimazole, 2478

$C_{22}H_{18}Cl_2FNO_3$
Cyfluthrin, 2826

$C_{22}H_{18}N_2$
Bifonazole, 1260

$C_{22}H_{18}O_4$
o-Cresolphthalein, 2646

$C_{22}H_{18}O_7$
Justicidins*, 5286

$C_{22}H_{18}O_{11}$
EGCG, 3568

$C_{22}H_{19}Br$
Broparoestrol, 1471

$C_{22}H_{19}Br_2NO_3$
Deltamethrin, 2934

$C_{22}H_{19}Br_4NO_3$
Tralomethrin, 9700

$C_{22}H_{19}ClO_3$
Atovaquone, 898

$C_{22}H_{19}Cl_2NO_3$
Cypermethrin, 2836

$C_{22}H_{19}N$
Bis(1-naphthylmethyl)amine, 1335

$C_{22}H_{19}NO_4$
Bisacodyl, 1282

$C_{22}H_{19}N_4NaO_8S_2$
Cefsulodin*, 1994

$C_{22}H_{20}N_2O_2$
Piketoprofen, 7576

$C_{22}H_{20}N_4O_8S_2$
Cefsulodin, 1994

$C_{22}H_{20}O_{10}$
Granaticin, 4555

$C_{22}H_{20}O_{13}$
Carminic Acid, 1891

$C_{22}H_{21}ClN_2O_8$
Meclocycline, 5818

$C_{22}H_{21}KN_2O_5S$
Fenbenicillin*, 4001

$C_{22}H_{21}NO_2$
Benzylimidobis(p-methoxyphenyl)-
methane, 1174

$C_{22}H_{22}ClNO_6$
Ochratoxins*, 6836

$C_{22}H_{22}ClN_3O_5$
Propaquizafop, 7990

$C_{22}H_{22}ClN_5O_2S$
Apafant, 765

$C_{22}H_{22}FN_3O_2$
Droperidol, 3505

$C_{22}H_{22}FN_3O_3$
Ketanserin, 5307

$C_{22}H_{22}N_2O_5S$
Fenbenicillin, 4001

$C_{22}H_{22}N_2O_8$
Methacycline, 6007

$C_{22}H_{22}N_6O_5S_2$
Cefpirome, 1990

$C_{22}H_{22}N_6O_7S_2$
Ceftazidime, 1995

$C_{22}H_{22}N_8$
Bisantrene, 1284

$C_{22}H_{22}O_4$
Dienestrol*, 3153

$C_{22}H_{22}O_5$
Cyclovalone, 2823

$C_{22}H_{22}O_8$
β-Peltatin, 7204
Picropodophyllin, 7567
Podophyllotoxin, 7704

$C_{22}H_{22}O_9$
Formononetin*, 4271

$C_{22}H_{22}O_{10}$
Prunetin*, 8094

$C_{22}H_{22}O_{11}$
Scoparin, 8546
Tectorigenin*, 9264

$C_{22}H_{23}ClN_2O_2$
Loratadine, 5608

$C_{22}H_{23}ClN_2O_8$
Chlortetracycline, 2245

$C_{22}H_{23}ClN_6O$
Losartan, 5613

$C_{22}H_{23}F_4NO_2$
Trifluperidol, 9813

$C_{22}H_{23}NO_3$
Fenpropathrin, 4033

$C_{22}H_{23}NO_4$
Nequinate, 6558

$C_{22}H_{23}NO_6$
Aureothin, 913

$C_{22}H_{23}NO_7$
Noscapine, 6815

$C_{22}H_{23}N_3O_9$
Aluminon, 330

$C_{22}H_{24}ClN_3O$
Azelastine, 939

$C_{22}H_{24}ClN_3OS_2$
Clospirazine, 2474

$C_{22}H_{24}ClN_5O_2$
Domperidone, 3476

$C_{22}H_{24}FN_3OS$
Timiperone, 9584

$C_{22}H_{24}FN_3O_2$
Benperidol, 1077

$C_{22}H_{24}N_2O_2$
Acrivastine, 129

$C_{22}H_{24}N_2O_3$
Colubrines, 2556

$C_{22}H_{24}N_2O_4$
Alstonidine, 323
Kopsine, 5332
Vomicine, 10170

$C_{22}H_{24}N_2O_5$
Benazepril*, 1058

$C_{22}H_{24}N_2O_8$
Quatrimycin, 8210
Tetracycline, 9337

$C_{22}H_{24}N_2O_8.H_2O$
Doxycycline, 3496

$C_{22}H_{24}N_2O_9$
Oxytetracycline, 7111

$C_{22}H_{24}O_8$
Lophotoxin, 5603

$C_{22}H_{24}O_9$
Podophyllic Acids, 7703

$C_{22}H_{24}O_{10}$
Sakuranetin*, 8470

$C_{22}H_{25}ClN_2OS$
Clopenthixol, 2455

$C_{22}H_{25}ClN_2O_4S$
Clentiazem, 2408

$C_{22}H_{25}F_2NO_4$
Nebivolol, 6519

$C_{22}H_{25}N$
Piroheptine, 7658

$C_{22}H_{25}NO_2$
Lobelanine, 5578
Tropacine, 9903

$C_{22}H_{25}NO_3$
Tropine Benzylate, 9913

$C_{22}H_{25}NO_4$
Dimoxyline, 3318

$C_{22}H_{25}NO_5$
Corycavidine, 2608

$C_{22}H_{25}NO_5S$
Thiocolchicine, 9460

$C_{22}H_{25}NO_6$
Colchicine, 2536

$C_{22}H_{25}N_3O$
Benzpiperylon, 1152
Indoramin, 5000

$C_{22}H_{25}N_3O_4$
Vesnarinone, 10105

$C_{22}H_{25}N_3O_4S$
Moricizine, 6351

$C_{22}H_{25}N_7O_5$
Edatrexate, 3553

$C_{22}H_{26}BrNO_3$
Clidinium Bromide, 2412

$C_{22}H_{26}ClFO_4$
Clobetasone, 2424

$C_{22}H_{26}ClNO_3$
Pipoxolan Hydrochloride, 7637

$C_{22}H_{26}FNO_2$
Moperone, 6343

$C_{22}H_{26}FN_3O_2S$
Sabeluzole, 8462

$C_{22}H_{26}FN_3O_4$
Flesinoxan, 4138

$C_{22}H_{26}F_3N_3OS$
Fluphenazine, 4226

$C_{22}H_{26}N_2O_2$
Vinpocetine, 10128

$C_{22}H_{26}N_2O_3$
Corynantheine, 2613

$C_{22}H_{26}N_2O_4$
Akuammine, 200
Aricine, 821
Tofisopam, 9642

$C_{22}H_{26}N_2O_4S$
Diltiazem, 3247

$C_{22}H_{26}N_4$
Calycanthine, 1772
Chimonanthine, 2101

$C_{22}H_{26}N_4O_2S$
Thiazolinobutazone, 9443

$C_{22}H_{26}O_3$
Bioresmethrin, 1271

$C_{22}H_{26}O_4$
Seratrodast, 8603

$C_{22}H_{26}O_5$
Colpormon, 2554

$C_{22}H_{26}O_{12}$
Catalposide, 1949

$C_{22}H_{27}ClF_2O_5$
Halometasone, 4628

$C_{22}H_{27}ClN_2O$
Lorcainide, 5610

$C_{22}H_{27}ClN_2O_3$
Lorajmine, 5607

$C_{22}H_{27}ClO_3$
Cyproterone, 2844

$C_{22}H_{27}ClO_5$
Osaterone*, 7019

$C_{22}H_{27}NO$
Ethybenztropine, 3802
Phenazocine, 7360

$C_{22}H_{27}NO_2$
Amineptine, 429
Danazol, 2875
Lobeline, 5580

$C_{22}H_{27}NO_3$
Dioxaphetyl Butyrate, 3354

$C_{22}H_{27}NO_4$
Corydaline, 2610

$C_{22}H_{27}NO_5$
Racefemine*, 8273

$C_{22}H_{27}N_2O_3^+$
Macusines*, 5679
Melinonine A, 5865

$C_{22}H_{27}N_3O_2$
Caroverine, 1906

Denotes derivative of title compound

FI-55

Denotes derivative of title compound

$C_{22}H_{38}O_4$
Citrullol, 2393
$C_{22}H_{38}O_4Zn$
Undecylenic Acid*, 9983
$C_{22}H_{38}O_5$
Carboprost*, 1871
Misoprostol, 6297
Unoprostone, 9984
$C_{22}H_{40}BrNO$
Domiphen Bromide, 3474
$C_{22}H_{40}ClN$
Halimide®, 4624
$C_{22}H_{40}O_7$
Agaricic Acid, 184
$C_{22}H_{41}NO_7$
Sphingofungins*, 8897
$C_{22}H_{41}N_2O_{11}S_2$
Tigemonam*, 9572
$C_{22}H_{42}N_2O$
Amine 220, 428
$C_{22}H_{42}N_4O_8S_2$
Pantethine, 7144
$C_{22}H_{42}O_2$
Brassidic Acid, 1389
Erucic Acid, 3713
$C_{22}H_{43}N_5O_{12}$
Isepamicin, 5121
$C_{22}H_{43}N_5O_{13}$
Amikacin, 425
$C_{22}H_{44}N_2O_2$
Glyodin, 4518
$C_{22}H_{44}N_6O_{10}$
Arbekacin, 809
$C_{22}H_{44}O_2$
Behenic Acid, 1051
Butyl Stearate, 1625
Isobutyl Stearate, 5165
$C_{22}H_{45}N_3$
Hexedine, 4736
$C_{22}H_{46}BrN$
Ethyldimethyl-9-octadecenylam-
monium Bromide, 3836
$C_{22}H_{46}N_4O_2$
Pithecolobine, 7665
$C_{22}H_{46}O$
Pygeum Africanum Extract*, 8135
$C_{22}H_{48}ClN$
Didecyldimethylammonium
Chloride, 3149
$C_{23}H_{14}F_2NNaO_2$
Brequinar*, 1394
$C_{23}H_{14}Na_2O_{11}$
Cromolyn*, 2658
$C_{23}H_{15}ClO_3$
Chlorophacinone, 2204
$C_{23}H_{15}F_2NO_2$
Brequinar, 1394
$C_{23}H_{16}O_3$
Diphenadione, 3363
$C_{23}H_{16}O_6$
Pamoic Acid, 7136
$C_{23}H_{16}O_{11}$
Cromolyn, 2658
$C_{23}H_{18}ClN_4NaO_7S_2$
Polar Yellow, 7713
$C_{23}H_{18}N_2O_2$
Isofezolac, 5188
$C_{23}H_{19}ClF_3NO_3$
Cyhalothrin, 2827
$C_{23}H_{20}N_2O_3S$
Sulfinpyrazone, 9121
$C_{23}H_{20}N_2O_5$
Bentiromide, 1081
$C_{23}H_{21}ClN_6O_3$
Loprazolam, 5604
$C_{23}H_{21}ClO_3$
Chlorotrianisene, 2225
$C_{23}H_{21}NO_4$
Xenazoic Acid, 10204

$C_{23}H_{21}N_2NaO_6S$
Carbenicillin*, 1838
$C_{23}H_{22}ClF_3O_2$
Bifenthrin, 1257
$C_{23}H_{22}O_5S$
Xylenol Blue, 10216
$C_{23}H_{22}O_6$
Deguelin, 2918
Rotenone, 8427
$C_{23}H_{22}O_7$
Lactucin*, 5359
Sumatrol, 9173
Tephrosin, 9295
$C_{23}H_{23}IN_2S_2$
Dithiazanine Iodide, 3434
$C_{23}H_{23}NO$
Trifenmorph, 9808
$C_{23}H_{23}N_3O_5$
Topotecan, 9687
$C_{23}H_{24}N_2O_4S$
Eprosartan, 3669
$C_{23}H_{24}N_4O_9$
Nocardicin(s)*, 6764
$C_{23}H_{24}O_2$
Endiandric Acids*, 3611
$C_{23}H_{24}O_4$
Cyclofenil, 2789
$C_{23}H_{24}O_8$
Wortmannin, 10188
$C_{23}H_{25}ClN_2$
Malachite Green, 5739
$C_{23}H_{25}ClN_2O_9$
Clomocycline, 2448
$C_{23}H_{25}ClO_{12}.5\frac{1}{2}H_2O$
Malvidin Chloride*, 5754
$C_{23}H_{25}ClO_{12}.4H_2O$
Malvidin Chloride*, 5754
$C_{23}H_{25}F_3N_2OS$
Flupentixol, 4224
$C_{23}H_{25}N$
Fendiline, 4011
$C_{23}H_{25}N_5O_5$
Doxazosin, 3489
$C_{23}H_{26}Cl_2O_6$
Simfibrate, 8684
$C_{23}H_{26}FN_3O_2$
Spiperone, 8903
$C_{23}H_{26}N_2O$
Roxindole, 8432
$C_{23}H_{26}N_2O_2$
Benzetimide, 1104
Dexetimide, 2987
$C_{23}H_{26}N_2O_4$
Brucine, 1476
$C_{23}H_{26}N_2O_5$
Quinapril*, 8233
$C_{23}H_{26}N_4O_9S_2$
1,3-Di-6-quinolylurea*, 3416
$C_{23}H_{26}N_5NaO_7S$
Piperacillin*, 7616
$C_{23}H_{26}O_3$
Phenothrin, 7405
$C_{23}H_{26}O_8$
Sikkimotoxin, 8632
$C_{23}H_{27}ClO_4$
Delmadinone Acetate, 2930
$C_{23}H_{27}ClO_6$
Chloroprednisone*, 2209
$C_{23}H_{27}FN_4O_2$
Risperidone, 8397
$C_{23}H_{27}N$
Butenafine, 1547
$C_{23}H_{27}NO$
Deptropine, 2958
$C_{23}H_{27}NO_4$
Tropenzile, 9908
$C_{23}H_{27}NO_5$
Octaverine, 6851
Warfarin*, 10174

$C_{23}H_{27}NO_8$
Narceine, 6508
$C_{23}H_{27}N_3O_2$
Morazone, 6349
$C_{23}H_{27}N_3O_7$
Minocycline, 6289
$C_{23}H_{27}N_5O_7S$
Piperacillin, 7616
$C_{23}H_{28}ClN_3O$
Prenoxdiazine Hydrochloride, 7918
$C_{23}H_{28}ClN_3O_2S$
Thiopropazate, 9493
$C_{23}H_{28}ClN_3O_4$
Benexate Hydrochloride, 1065
$C_{23}H_{28}ClN_3O_5S$
Glyburide, 4486
$C_{23}H_{28}ClN_5O_3$
Azimilide, 943
$C_{23}H_{28}Cl_2O_5$
Dichlorisone*, 3099
$C_{23}H_{28}F_3N_3OS$
Homofenazine, 4774
$C_{23}H_{28}F_3NaO_6$
Fluprostenol*, 4231
$C_{23}H_{28}N_2O$
Leiopyrrole, 5457
$C_{23}H_{28}N_2O_3$
Bopindolol, 1362
$C_{23}H_{28}N_2O_5$
Reserpiline, 8313
$C_{23}H_{28}N_2O_5S_2$
Temocapril, 9287
$C_{23}H_{28}O_6$
Enprostil, 3629
Prednisone*, 7904
$C_{23}H_{28}O_{15}$
Purpurogallin*, 8133
$C_{23}H_{29}ClFN_3O_4$
Cisapride, 2377
$C_{23}H_{29}ClO_4$
Chlormadinone Acetate, 2152
$C_{23}H_{29}FO_6$
Isoflupredone*, 5190
$C_{23}H_{29}F_2N_3O$
Amperozide, 619
$C_{23}H_{29}F_3O_6$
Fluprostenol, 4231
$C_{23}H_{29}NO$
Norpipanone, 6810
$C_{23}H_{29}NO_2$
Phenadoxone, 7348
$C_{23}H_{29}NO_3$
Fenbutrazate, 4005
Phenoperidine, 7400
Propiverine, 8018
$C_{23}H_{29}NO_{12}$
Hygromycin, 4900
$C_{23}H_{29}N_3O$
Opipramol, 6985
$C_{23}H_{29}N_3O_2$
Etomidoline, 3928
Oxypertine, 7105
$C_{23}H_{29}N_3O_2S$
Acetophenazine, 70
$C_{23}H_{29}N_3O_2S_2$
Thiothixene, 9503
$C_{23}H_{29}N_5O_8S_2$
Cefcapene Pivoxil, 1969
$C_{23}H_{30}BrNO_3$
Piperilate*, 7624
Propantheline Bromide, 7989
$C_{23}H_{30}ClN_3O$
Quinacrine, 8225
$C_{23}H_{30}Cl_2NNa_2O_6P$
Estramustine*, 3749
$C_{23}H_{30}N_2O$
α-Methylfentanyl, 6150
$C_{23}H_{30}N_2O_2$
Piminodine, 7587
$C_{23}H_{30}N_2O_4$

Denotes derivative of title compound

FI-57

Mitragynine, 6304
Pholcodine, 7484

$C_{23}H_{30}O_3$
Etretinate, 3935
Melengestrol, 5859

$C_{23}H_{30}O_6$
Cortisone*, 2602
Fenprostalene, 4037
Prednisolone*, 7901

$C_{23}H_{30}O_7$
Sarverogenin, 8527

$C_{23}H_{31}ClN_2O_3$
Etodroxizine, 3921

$C_{23}H_{31}Cl_2NO_3$
Estramustine, 3749

$C_{23}H_{31}Cl_3O_4$
Cloxotestosterone*, 2482

$C_{23}H_{31}FO_5$
Flurogestone Acetate, 4235

$C_{23}H_{31}FO_6$
Fludrocortisone*, 4166

$C_{23}H_{31}NO_2$
Levomethadyl Acetate, 5492
Methadyl Acetate, 6009
Motretinide, 6367

$C_{23}H_{31}NO_3$
Norgestimate, 6796

$C_{23}H_{31}NO_7S$
Bevonium Methyl Sulfate, 1239
Sulprostone, 9165

$C_{23}H_{31}N_3O_4S_2$
Batimastat, 1036

$C_{23}H_{32}Cl_2NO_6P$
Estramustine*, 3749

$C_{23}H_{32}N_2O_2S$
Tiocarlide, 9593

$C_{23}H_{32}N_2O_3$
Zipeprol, 10303

$C_{23}H_{32}N_2O_5$
Ramipril, 8283

$C_{23}H_{32}N_2O_6$
Enciprazine, 3610

$C_{23}H_{32}N_2O_8$
Trichostatin(s)*, 9781

$C_{23}H_{32}N_6O_{14}$
Polyoxins*, 7736

$C_{23}H_{32}O_2$
Dimethisterone, 3268
Medrogestone, 5836

$C_{23}H_{32}O_3$
Estradiol*, 3746
Quinestradiol, 8238

$C_{23}H_{32}O_4$
Deoxycorticosterone Acetate, 2948
17α-Hydroxyprogesterone*, 4886

$C_{23}H_{32}O_5$
Tanghinigenin, 9220

$C_{23}H_{32}O_6$
Strophanthidin, 9015

$C_{23}H_{32}O_7$
Strophanthidin*, 9015

$C_{23}H_{33}IN_2O$
Isopropamide Iodide, 5221

$C_{23}H_{33}N_2O_2{}^+$
Prajmaline, 7883

$C_{23}H_{34}NO_5P$
Fosinopril*, 4282

$C_{23}H_{34}O_3$
Testosterone*, 9322

$C_{23}H_{34}O_4$
21-Acetoxypregnenolone, 77
Digitoxigenin, 3205
Uzarin*, 10035

$C_{23}H_{34}O_5$
Alfadolone Acetate, 233
Digoxigenin, 3209
Gitoxigenin, 4440
Mevastatin, 6251
Periplogenin, 7317
Sarmentogenin, 8522

$C_{23}H_{34}O_6$
Ajugarins*, 197
Diginatigenin, 3197

$C_{23}H_{34}O_8$
Ouabagenin, 7030

$C_{23}H_{35}NaO_7$
Pravastatin Sodium, 7894

$C_{23}H_{36}N_2O_2$
Finasteride, 4125

$C_{23}H_{36}N_4O_5S_3$
Octotiamine, 6857

$C_{23}H_{36}N_4O_{10}S_2$
Pentamidine*, 7254

$C_{23}H_{36}N_6O_5S$
Argatroban, 816

$C_{23}H_{36}O_3$
Dromostanolone*, 3504

$C_{23}H_{37}NO_5S$
Leukotrienes*, 5482

$C_{23}H_{37}NO_6$
Tocamphyl, 9630

$C_{23}H_{37}N_5O_6S$
Etamiphyllin*, 3754

$C_{23}H_{38}O$
Teprenone, 9296

$C_{23}H_{38}O_2$
Norcholanic Acid, 6782

$C_{23}H_{38}O_3$
Pregnanediol*, 7908

$C_{23}H_{38}O_4$
Trimoprostil, 9853

$C_{23}H_{38}O_5$
Gemeprost, 4393

$C_{23}H_{38}O_6$
Ornoprostil, 7003

$C_{23}H_{41}ClN_2O$
Dodecarbonium Chloride, 3466

$C_{23}H_{42}ClNO_2$
Benzoxonium Chloride, 1144

$C_{23}H_{42}N_2$
Chonemorphine, 2272

$C_{23}H_{42}N_2O_{12}$
Pentolinium Tartrate, 7274

$C_{23}H_{45}N_5O_{14}$
Paromomycin, 7173

$C_{23}H_{46}$
Muscalure, 6388

$C_{23}H_{46}N_6O_{13}$
Neomycin*, 6542

$C_{24}H_{16}As_2O_3$
10,10'-Oxydiphenoxarsine, 7095

$C_{24}H_{16}O_{12}$
Laccaic Acid*, 5342

$C_{24}H_{19}NO_5$
Oxyphenisatin Acetate, 7108

$C_{24}H_{19}NO_6$
Bisoxatin Acetate, 1337

$C_{24}H_{19}N_3O_5S$
Piroxicam*, 7661

$C_{24}H_{20}AsBr$
Tetraphenylarsonium Bromide, 9373

$C_{24}H_{20}AsCl$
Tetraphenylarsonium Chloride, 9374

$C_{24}H_{20}BNa$
Sodium Tetraphenylborate, 8839

$C_{24}H_{20}Cl_2N_2OS$
Fenticonazole, 4047

$C_{24}H_{20}I_6N_4O_8$
Iocarmic Acid, 5029

$C_{24}H_{20}N_2$
N,N'-Diphenylbenzidine, 3377

$C_{24}H_{20}N_4O$
Scarlet Red, 8538

$C_{24}H_{20}N_6O_3$
Candesartan, 1788

$C_{24}H_{21}I_6N_5O_8$
Ioxaglic Acid, 5084

$C_{24}H_{23}N_3O_6S$
Talampicillin, 9204

$C_{24}H_{24}CaI_6N_4O_4$
Ipodate*, 5087

$C_{24}H_{24}N_2O_4$
Abecarnil, 2

$C_{24}H_{24}N_2O_5$
Amtolmetin Guacil, 637

$C_{24}H_{25}FNNaO_4$
Fluvastatin*, 4250

$C_{24}H_{25}NO_3$
Benzylmorphine, 1176
Cyphenothrin, 2837

$C_{24}H_{25}NO_4$
Flavoxate, 4135

$C_{24}H_{26}BrN_3O_3$
Nicergoline, 6580

$C_{24}H_{26}ClFN_4O$
Sertindole, 8611

$C_{24}H_{26}FNO_4$
Fluvastatin, 4250

$C_{24}H_{26}MgO_8$
Dimecrotic Acid*, 3248

$C_{24}H_{26}N_2O_2$
Carmoxirole, 1893

$C_{24}H_{26}N_2O_4$
Carvedilol, 1924

$C_{24}H_{26}N_2O_6$
Suxibuzone, 9185

$C_{24}H_{26}N_8O_9$
Diopterin, 3344

$C_{24}H_{26}O_6$
Mangostin, 5785

$C_{24}H_{26}O_{13}$
Centaurein, 2015
Irigenin*, 5103

$C_{24}H_{27}N$
Prenylamine, 7919

$C_{24}H_{27}NO_4$
Tylocrebrine, 9961
Tylophorine, 9962

$C_{24}H_{27}NO_5S$
Troglitazone, 9898

$C_{24}H_{27}N_3O_4$
Fenpyroximate, 4038

$C_{24}H_{28}BiN_4Na_7O_{25}$
Bismuth Sodium Triglycollamate, 1322

$C_{24}H_{28}ClNO_4$
Phenactropinium Chloride, 7346

$C_{24}H_{28}ClN_5O_3$
Dimenhydrinate, 3252

$C_{24}H_{28}N_2O_3$
Naftopidil, 6443

$C_{24}H_{28}N_2O_5$
Benazepril, 1058

$C_{24}H_{28}N_4O_6$
Lamifiban, 5362

$C_{24}H_{28}O_4$
Diethylstilbestrol*, 3177

$C_{24}H_{29}BrFNO_3$
Flutropium Bromide, 4248

$C_{24}H_{29}ClO_4$
Cyproterone*, 2844

$C_{24}H_{29}Cl_2FO_5$
Flucloronide, 4157

$C_{24}H_{29}FN_2O_2$
Aceperone, 30

$C_{24}H_{29}FO_6$
Fluprednidene Acetate, 4228

$C_{24}H_{29}NO$
Phenomorphan, 7399

$C_{24}H_{29}NO_3·HCl$
E-2020, 3533

$C_{24}H_{29}NO_4$
Ethaverine, 3773

$C_{24}H_{29}N_5O_3$
Valsartan, 10051

$C_{24}H_{29}NaO_5$
Beraprost*, 1190
Taprostene*, 9227

Denotes derivative of title compound

C$_{24}$H$_{30}$CaN$_4$O$_6$
Cyclobarbital*, 2780
C$_{24}$H$_{30}$ClFO$_5$
Clocortolone*, 2430
C$_{24}$H$_{30}$FNa$_2$O$_9$P
Triamcinolone Acetonide*, 9728
C$_{24}$H$_{30}$F$_2$O$_6$
Fluocinolone Acetonide, 4185
C$_{24}$H$_{30}$N$_2$O$_2$
Doxapram, 3488
C$_{24}$H$_{30}$N$_2$O$_2$S
Piperacetazine, 7615
C$_{24}$H$_{30}$N$_2$O$_5$
Vindoline*, 10126
C$_{24}$H$_{30}$O$_3$
Drospirenone, 3510
C$_{24}$H$_{30}$O$_4$
17-Hydroxy-16-methylene-Δ^6-pro-
gesterone*, 4878
C$_{24}$H$_{30}$O$_5$
Beraprost, 1190
Taprostene, 9227
C$_{24}$H$_{30}$O$_7$
Athamantin, 893
C$_{24}$H$_{30}$O$_8$
Desaspidin BB, 2962
Flavaspidic Acid, 4130
C$_{24}$H$_{31}$ClO$_7$
Loteprednol Etabonate, 5614
C$_{24}$H$_{31}$FO$_6$
Betamethasone*, 1226
Dexamethasone*, 2986
Flunisolide, 4180
Fluperolone Acetate, 4225
Paramethasone*, 7162
Triamcinolone Acetonide, 9728
C$_{24}$H$_{31}$F$_3$O$_4$
Flumedroxone Acetate, 4171
C$_{24}$H$_{31}$NO
Dipipanone, 3398
C$_{24}$H$_{31}$N$_3$OS
Butaperazine, 1543
C$_{24}$H$_{31}$N$_3$O$_2$S
Carphenazine, 1910
C$_{24}$H$_{32}$ClFO$_5$
Halcinonide, 4621
C$_{24}$H$_{32}$N$_2$O$_2$
Eprazinone, 3666
C$_{24}$H$_{32}$N$_2$O$_5$
Metoserpate, 6237
C$_{24}$H$_{32}$N$_4$O$_5$
Terguride*, 9308
C$_{24}$H$_{32}$O$_2$
Quinbolone, 8236
C$_{24}$H$_{32}$O$_4$
Estradiol*, 3746
Ethynodiol*, 3905
Megestrol Acetate, 5849
Resibufogenin, 8315
Scillarenin, 8544
C$_{24}$H$_{32}$O$_4$S
Spironolactone, 8917
C$_{24}$H$_{32}$O$_6$
Desonide, 2973
Methylprednisolone*, 6189
C$_{24}$H$_{32}$O$_7$
Etiproston, 3916
C$_{24}$H$_{33}$FO$_6$
Flurandrenolide, 4232
C$_{24}$H$_{33}$N
Droprenilamine, 3506
C$_{24}$H$_{33}$NO$_3$
Nafronyl, 6440
C$_{24}$H$_{33}$N$_3$O$_2$S
Dixyrazine, 3450
C$_{24}$H$_{33}$N$_3$O$_3$S$_2$
Pipotiazine, 7636
C$_{24}$H$_{33}$N$_3$O$_4$
Ranolazine, 8287

C$_{24}$H$_{33}$NaO$_5$
Dehydrocholic Acid*, 2922
C$_{24}$H$_{33}$NaO$_7$S
Bufalin*, 1493
C$_{24}$H$_{34}$Br$_2$N$_2$O$_4$
Hemicholinium*, 4676
C$_{24}$H$_{34}$N$_2$Na$_2$O$_{18}$S$_3$
Glucosulfone Sodium, 4471
C$_{24}$H$_{34}$N$_2$O
Bepridil, 1188
C$_{24}$H$_{34}$N$_2$O$_2$
Euprocin, 3950
C$_{24}$H$_{34}$N$_2$O$_4$$^{2+}$
Hemicholinium, 4676
C$_{24}$H$_{34}$N$_2$O$_5$
Trandolapril, 9703
C$_{24}$H$_{34}$N$_4$O$_5$S
Glimepiride, 4449
C$_{24}$H$_{34}$N$_8$O$_4$S$_2$
Thiamine Disulfide, 9432
C$_{24}$H$_{34}$O$_3$
Rimexolone, 8392
C$_{24}$H$_{34}$O$_4$
Algestone*, 238
Bufalin, 1493
Medroxyprogesterone*, 5838
C$_{24}$H$_{34}$O$_4$S$_2$
Tiomesterone, 9596
C$_{24}$H$_{34}$O$_5$
Bufogenin B, 1499
Dehydrocholic Acid, 2922
Gamabufotalin, 4370
C$_{24}$H$_{34}$O$_7$
Ajugarins*, 197
C$_{24}$H$_{34}$O$_9$
T-2 Toxin, 9933
C$_{24}$H$_{35}$NO$_5$
Decoquinate, 2910
C$_{24}$H$_{36}$N$_2$O$_9$S
Celesticetin, 2005
C$_{24}$H$_{36}$N$_8$O$_{10}$P$_2$S$_2$
Thiamine Disulfide*, 9432
C$_{24}$H$_{36}$O$_3$
Anagestone*, 664
Nabilone, 6427
C$_{24}$H$_{36}$O$_4$
Bolandiol*, 1352
C$_{24}$H$_{36}$O$_5$
Lovastatin, 5616
C$_{24}$H$_{36}$O$_8$
Ajugarins*, 197
C$_{24}$H$_{37}$NO$_6$
3-O-Lauroylpyridoxol Diacetate,
5401
C$_{24}$H$_{38}$N$_2$O$_2$
Laurolinium Acetate, 5399
C$_{24}$H$_{38}$N$_4$O$_{10}$S$_2$
Hexamidine*, 4728
C$_{24}$H$_{38}$O$_4$
Bis(2-ethylhexyl) Phthalate, 1291
C$_{24}$H$_{38}$O$_5$
Alfaprostol, 234
C$_{24}$H$_{39}$NO$_4$
Cassaine, 1938
C$_{24}$H$_{39}$NO$_5$
Erythrophleine, 3728
C$_{24}$H$_{39}$NO$_9$S
Tiaprost*, 9563
C$_{24}$H$_{39}$NaO$_4$
Deoxycholic Acid*, 2946
C$_{24}$H$_{39}$NaO$_5$
Cholic Acid*, 2258
C$_{24}$H$_{40}$N$_2$
Conessine, 2561
C$_{24}$H$_{40}$N$_2$O
Holarrhenine, 4762
C$_{24}$H$_{40}$N$_8$O$_4$
Dipyridamole, 3410
C$_{24}$H$_{40}$O$_2$
Cholanic Acid, 2251

C$_{24}$H$_{40}$O$_3$
Lithocholic Acid, 5571
C$_{24}$H$_{40}$O$_4$
Chenodiol, 2096
Deoxycholic Acid, 2946
Hyodeoxycholic Acid, 4906
Ursodiol, 10026
C$_{24}$H$_{40}$O$_5$
Cholic Acid, 2258
C$_{24}$H$_{41}$NO$_4$
Cassaidine, 1937
C$_{24}$H$_{42}$
Cholane, 2250
C$_{24}$H$_{42}$Br$_2$Ca$_2$O$_{24}$
Calcium Bromolactobionate, 1694
C$_{24}$H$_{42}$CaO$_{24}$
Lactobionic Acid*, 5354
C$_{24}$H$_{42}$N$_2$
Feclemine, 3986
C$_{24}$H$_{43}$NO
Linoleic Acid*, 5529
C$_{24}$H$_{43}$NO$_9$
Pederin*, 7196
C$_{24}$H$_{44}$O$_6$
Sorbitan Esters*, 8872
C$_{24}$H$_{46}$O$_6$
Sorbitan Esters*, 8872
C$_{24}$H$_{48}$O$_2$
Lignoceric Acid, 5511
C$_{24}$H$_{50}$BrNO
Cethexonium Bromide, 2061
C$_{24}$H$_{72}$O$_{10}$Si$_{11}$
Tetracosamethylhendecasiloxane,
9335
C$_{25}$H$_{17}$NO$_{13}$
Laccaic Acid*, 5342
C$_{25}$H$_{18}$N$_4$O$_3$
Asperlicin*, 878
C$_{25}$H$_{18}$N$_4$O$_3$
Asperlicin*, 878
C$_{25}$H$_{19}$NO$_3$
Fendosal, 4012
C$_{25}$H$_{20}$ClF$_2$N$_3$O$_3$
Flucycloxuron, 4159
C$_{25}$H$_{22}$ClNO$_3$
Fenvalerate, 4051
C$_{25}$H$_{22}$N$_4$O$_8$
Streptonigrin, 8986
C$_{25}$H$_{22}$N$_5$NaO$_6$S
Apalcillin*, 766
C$_{25}$H$_{22}$O$_{10}$
Silymarin*, 8680
C$_{25}$H$_{23}$ClN$_2$O$_7$
Binifibrate, 1267
C$_{25}$H$_{23}$ClN$_4$O$_4$
Pazinaclone, 7187
C$_{25}$H$_{23}$N$_5$O$_6$S
Apalcillin, 766
C$_{25}$H$_{23}$N$_8$NaO$_7$S$_2$
Cefpiramide*, 1989
C$_{25}$H$_{24}$F$_6$N$_4$
Hydramethylnon, 4802
C$_{25}$H$_{24}$N$_8$O$_7$S$_2$
Cefpiramide, 1989
C$_{25}$H$_{24}$O$_5$
Osajin, 7017
C$_{25}$H$_{24}$O$_{12}$
Cynarin(e), 2835
C$_{25}$H$_{25}$ClN$_2$
Quinaldine Blue, 8229
C$_{25}$H$_{25}$NO$_6$
Oxetorone*, 7068
C$_{25}$H$_{26}$N$_2$O$_2$
Medibazine, 5831
C$_{25}$H$_{26}$N$_9$NaO$_8$S$_2$
Cefoperazone*, 1981
C$_{25}$H$_{26}$O$_2$
Diethylstilbestrol*, 3177
C$_{25}$H$_{26}$O$_{13}$
Ruberythric Acid, 8437

Denotes derivative of title compound

$C_{26}H_{30}Na_2O_9$
Estriol*, 3750
$C_{26}H_{30}O_8$
Limonin, 5519
Physodic Acid, 7538
$C_{26}H_{31}Cl_2N_5O_3$
Terconazole, 9303
$C_{26}H_{31}N_5O_2$
Physostigmine*, 7540
$C_{26}H_{32}ClFO_5$
Clobetasone*, 2424
$C_{26}H_{32}F_2O_7$
Diflorasone*, 3186
Fluocinonide, 4186
$C_{26}H_{32}F_3N_3O_2S$
Oxaflumazine, 7040
$C_{26}H_{32}MgN_6O_8S_2$
Dipyrone*, 3414
$C_{26}H_{32}N_2O_5$
Delapril, 2928
$C_{26}H_{32}N_2O_8$
Tritoqualine, 9894
$C_{26}H_{32}O_4$
Bixin*, 1347
$C_{26}H_{33}FO_7$
Flunisolide*, 4180
$C_{26}H_{33}NO_2$
Fenretinide, 4040
$C_{26}H_{33}NO_4$
Cyprenorphine, 2838
$C_{26}H_{33}NO_6$
Lacidipine, 5344
$C_{26}H_{33}NaO_8$
Methylprednisolone*, 6189
$C_{26}H_{34}O_4$
Methestrol*, 6045
Trenbolone*, 9716
$C_{26}H_{34}O_7$
Cinobufotalin, 2367
Fumagillin, 4308
$C_{26}H_{35}FO_5$
Fluocortin Butyl, 4187
$C_{26}H_{35}NO_4$
Diprenorphine, 3403
$C_{26}H_{36}N_2O_5$
Zatebradine, 10245
$C_{26}H_{36}N_2O_9$
Antimycin A_3, 756
Quinidine*, 8244
$C_{26}H_{36}O_3$
Estradiol*, 3746
$C_{26}H_{36}O_6$
Bufotalin, 1501
Prednisolone*, 7901
Prednival, 7905
$C_{26}H_{37}ClN_4O_6S_2 \cdot H_2O$
Penicillin O*, 7228
$C_{26}H_{37}N_5O_2$
Cabergoline, 1637
$C_{26}H_{38}N_2O_4$
Mazipredone, 5802
$C_{26}H_{38}O_3$
Pentagestrone, 7251
$C_{26}H_{38}O_4$
Gestonorone Caproate, 4422
Lupulon, 5640
Oxabolone*, 7034
$C_{26}H_{38}O_5S$
Tixocortol*, 9623
$C_{26}H_{38}O_6$
Hydrocortisone*, 4828
$C_{26}H_{40}O_3$
Testosterone*, 9322
$C_{26}H_{40}O_4$
Methandriol*, 6017
$C_{26}H_{40}O_5$
Latanoprost, 5387
$C_{26}H_{41}Br_2NO_4$
Pinaverium Bromide, 7595

$C_{26}H_{41}NO$
Melinamide, 5864
$C_{26}H_{42}O_9$
Pseudomonic Acids*, 8103
$C_{26}H_{43}NO_6$
Glycocholic Acid, 4504
$C_{26}H_{44}NNaO_7S$
Taurocholic Acid*, 9242
$C_{26}H_{44}O_2$
Tocol, 9631
$C_{26}H_{44}O_8$
Pseudomonic Acids*, 8103
$C_{26}H_{44}O_9$
Mupirocin, 6383
$C_{26}H_{45}NO_7S$
Taurocholic Acid, 9242
$C_{26}H_{49}NO_3S$
Cetrimonium Bromide*, 2068
$C_{26}H_{50}O_4$
Bis(2-ethylhexyl) Sebacate, 1292
Glycol Dilaurate, 4507
$C_{26}H_{54}NO_7P$
Platelet Activating Factor*, 7682
$C_{26}H_{56}N_{10}$
Alexidine, 231
$C_{27}H_{20}N_4O$
Pyrinoline, 8173
$C_{27}H_{20}O_{12}$
Collinomycin, 2545
$C_{27}H_{22}Cl_2N_4$
Clofazimine, 2433
$C_{27}H_{22}O_{18}$
Tannic Acid*, 9221
$C_{27}H_{23}N_5O_4$
Pranlukast, 7889
$C_{27}H_{24}AlN_3O_{15}S_3$
8-Hydroxyquinoline*, 4890
$C_{27}H_{24}N_2O_9$
Albofungin, 213
$C_{27}H_{25}F_2N_3OS$
Ritanserin, 8399
$C_{27}H_{28}Br_2O_5S$
Bromthymol Blue, 1469
$C_{27}H_{28}N_2O_7$
Cilnidipine, 2334
$C_{27}H_{29}IN_2$
Dicyanine, 3141
$C_{27}H_{29}NO_{10}$
Daunorubicin, 2890
Pirozadil, 7662
$C_{27}H_{29}NO_{11}$
Doxorubicin, 3495
Epirubicin, 3660
$C_{27}H_{29}N_3O_6$
Barnidipine, 1031
$C_{27}H_{30}Cl_2O_6$
Mometasone Furoate, 6324
$C_{27}H_{30}F_2N_2O_3$
Lomerizine, 5593
$C_{27}H_{30}N_4O$
Oxatomide, 7058
$C_{27}H_{30}O_5S$
Thymol Blue, 9541
$C_{27}H_{30}O_6$
Sofalcone, 8850
$C_{27}H_{30}O_{14}$
Glucofrangulin, 4460
Kaempferol*, 5288
Morindin, 6353
Sophorabioside, 8866
$C_{27}H_{30}O_{15}$
Saponarin, 8512
$C_{27}H_{30}O_{16}$
Rutin, 8456
$C_{27}H_{31}ClN_2O$
Chlorbenzoxamine, 2125
$C_{27}H_{31}ClO_{15}$
Cyanidin Chloride*, 2757
Pelargonidin*, 7199

$C_{27}H_{31}ClO_{16}$
Cyanidin Chloride*, 2757
$C_{27}H_{31}ClO_{17}$
Delphinidin*, 2931
$C_{27}H_{31}N_2NaO_6S_2$
Sulphan Blue, 9161
$C_{27}H_{32}ClNO_2$
Triparanol, 9867
$C_{27}H_{32}N_2O_4S_3$
Mesoridazine*, 5970
$C_{27}H_{32}N_4O_8$
Pyridomycin, 8160
$C_{27}H_{32}O_8$
Verrucarins*, 10100
$C_{27}H_{32}O_9$
Verrucarins*, 10100
$C_{27}H_{32}O_{14}$
Naringin, 6512
$C_{27}H_{33}NO_3$
Prenylamine*, 7919
$C_{27}H_{33}NO_{10}S$
Thiocolchicine*, 9460
$C_{27}H_{33}N_3O_6S$
Gliquidone, 4452
$C_{27}H_{33}N_3O_8$
Rolitetracycline, 8411
$C_{27}H_{33}N_9O_{15}P_2$
Flavine-Adenine Dinucleotide, 4131
$C_{27}H_{34}F_7O_7$
Difluprednate, 3194
$C_{27}H_{34}N_2O_4S$
Brilliant Green, 1398
$C_{27}H_{34}N_2O_7$
Lofentanil*, 5586
$C_{27}H_{34}N_4O$
Piritramide, 7653
$C_{27}H_{34}O_3$
Nandrolone*, 6452
$C_{27}H_{34}O_4$
Bixin*, 1347
$C_{27}H_{34}O_8$
Verrucarins*, 10100
$C_{27}H_{34}O_9$
Verrucarins*, 10100
$C_{27}H_{34}O_{11}$
Phillyrin, 7477
$C_{27}H_{35}BrClN_3$
Methyl Green, 6157
$C_{27}H_{35}NO_5$
Etorphine*, 3932
$C_{27}H_{35}NO_8$
Sedecamycin, 8566
$C_{27}H_{35}N_5O_7S$
Endorphins*, 3613
$C_{27}H_{36}ClFO_5$
Clocortolone*, 2430
$C_{27}H_{36}F_2O_5$
Diflucortolone*, 3189
$C_{27}H_{36}F_2O_6$
Flumethasone*, 4173
$C_{27}H_{36}N_2O_4$
Repaglinide, 8304
$C_{27}H_{36}O_8$
Prednicarbate, 7899
$C_{27}H_{37}FO_6$
Betamethasone*, 1226
$C_{27}H_{38}N_2O_4$
Verapamil, 10083
$C_{27}H_{38}O_3$
Norethindrone*, 6790
$C_{27}H_{38}O_4$
Azafrin, 927
$C_{27}H_{38}O_6$
Prednisolone*, 7901
$C_{27}H_{39}NO_2$
Veratramine, 10085
$C_{27}H_{39}NO_3$
Jervine, 5277
$C_{27}H_{39}NO_6$

* Denotes derivative of title compound

C$_{28}$H$_{39}$FO$_5$
 Fluocortolone*, 4188
C$_{28}$H$_{39}$FO$_6$
 Dexamethasone*, 2986
C$_{28}$H$_{40}$N$_2$O$_2$
 Bialamicol, 1242
C$_{28}$H$_{40}$N$_2$O$_5$
 Gallopamil, 4369
C$_{28}$H$_{40}$N$_2$O$_9$
 Antimycin A$_1$, 755
C$_{28}$H$_{40}$O$_4$
 Pentagestrone*, 7251
C$_{28}$H$_{40}$O$_7$
 Diginin, 3199
C$_{28}$H$_{41}$FNO$_6$
 Dexamethasone*, 2986
C$_{28}$H$_{41}$N$_3$O$_3$
 Oxethazaine, 7067
 Tiropramide, 9607
C$_{28}$H$_{41}$N$_3$O$_8$
 Detoxin Complex*, 2982
C$_{28}$H$_{42}$Cl$_4$N$_4$O$_2$
 Ambenonium Chloride, 396
C$_{28}$H$_{42}$N$_2$O$_5$
 Fenoxedil, 4026
C$_{28}$H$_{42}$O
 Dehydroergosterol, 2925
C$_{28}$H$_{42}$O$_2$
 ε-Tocopherol, 9635
C$_{28}$H$_{42}$O$_6$
 Cynanchogenin, 2834
C$_{28}$H$_{44}$ClNO$_2$
 Methylbenzethonium Chloride,
 6103
C$_{28}$H$_{44}$O
 Ergosterol, 3701
 Fungisterol, 4313
 Isopyrocalciferol, 5238
 Lumisterol, 5629
 Pyrocalciferol, 8181
 Suprasterol II, 9179
 Tachysterol, 9198
 Vitamin D$_2$, 10156
C$_{28}$H$_{44}$O$_3$
 Nandrolone*, 6452
C$_{28}$H$_{46}$O
 Dihydrotachysterol, 3223
 Ergosterol*, 3701
 Vitamin D$_4$, 10158
C$_{28}$H$_{47}$NO$_4$S
 Tiamulin, 9559
C$_{28}$H$_{47}$NO$_7$
 Narbomycin, 6507
C$_{28}$H$_{47}$NO$_8$
 Picromycin, 7566
C$_{28}$H$_{48}$O
 Campesterol, 1776
 α-Ergostenol, 3698
 β-Ergostenol, 3699
 γ-Ergostenol, 3700
C$_{28}$H$_{48}$O$_2$
 β-Tocopherol, 9632
 γ-Tocopherol, 9633
 ζ$_2$-Tocopherol, 9637
C$_{28}$H$_{48}$O$_6$
 Brassinolide, 1390
C$_{28}$H$_{50}$
 Coproergostane, 2585
 Ergostane, 3696
C$_{28}$H$_{50}$N$_2$O$_4$
 Carpaine, 1908
C$_{28}$H$_{50}$O
 Ergostanol, 3697
C$_{28}$H$_{54}$GdN$_5$O$_{20}$
 Gadopentetic Acid*, 4347
C$_{28}$H$_{58}$NO$_7$P
 Platelet Activating Factor*, 7682
C$_{28}$H$_{58}$O
 Octacosanol, 6840

C$_{29}$H$_{24}$N$_4$O$_8$
 Niceritrol, 6581
C$_{29}$H$_{24}$O$_{12}$
 Theaflavine, 9409
C$_{29}$H$_{25}$N$_3$O$_5$
 Nicomorphine, 6607
C$_{29}$H$_{27}$ClN$_2$O$_{14}$S
 Meclocycline*, 5818
C$_{29}$H$_{27}$NO$_4$S$_2$
 Tipepidine*, 9601
C$_{29}$H$_{30}$O$_{13}$
 Amarogentin, 393
C$_{29}$H$_{31}$F$_2$N$_3$O
 Fluspirilene, 4241
C$_{29}$H$_{31}$N$_3$O$_4$S
 Penicillin G Benzhydrylamine, 7222
C$_{29}$H$_{32}$N$_2$O$_6$S
 Semotiadil, 8590
C$_{29}$H$_{32}$O$_{13}$
 Etoposide, 3931
C$_{29}$H$_{33}$ClN$_2$O$_2$
 Loperamide, 5601
C$_{29}$H$_{33}$FN$_2$O$_6$
 Elgodipine, 3587
C$_{29}$H$_{33}$FO$_6$
 Betamethasone*, 1226
C$_{29}$H$_{33}$N$_9$O$_{12}$
 Pteropterin, 8119
C$_{29}$H$_{34}$O$_6$
 Tribenoside, 9738
C$_{29}$H$_{34}$O$_{15}$
 Pectolinarigenin*, 7195
C$_{29}$H$_{35}$ClO$_{17}$
 Malvidin Chloride*, 5754
C$_{29}$H$_{35}$NO$_2$
 Mifepristone, 6273
C$_{29}$H$_{36}$N$_2$O$_4$
 Emetamine, 3599
C$_{29}$H$_{36}$O$_4$
 Algestone Acetophenide, 239
C$_{29}$H$_{37}$NO$_5$
 Cytochalasins*, 2859
C$_{29}$H$_{37}$N$_3$O$_6$
 Calcimycin, 1678
C$_{29}$H$_{38}$ClFO$_8$
 Formocortal, 4270
C$_{29}$H$_{38}$FN$_3$O$_3$
 Mibefradil, 6261
C$_{29}$H$_{38}$F$_3$N$_3$O$_2$S
 Fluphenazine*, 4226
C$_{29}$H$_{38}$N$_2$O$_4$
 Dehydroemetine, 2924
C$_{29}$H$_{38}$N$_4$O$_6$S.H$_2$O
 Penicillin G Procaine, 7226
C$_{29}$H$_{38}$N$_4$O$_9$
 Pipacycline, 7606
C$_{29}$H$_{38}$N$_4$O$_{10}$
 Lymecycline, 5658
C$_{29}$H$_{38}$N$_8$O$_8$
 Guamecycline, 4584
C$_{29}$H$_{38}$O$_9$
 Uscharidin, 10030
C$_{29}$H$_{39}$N$_3$O$_2$
 Echinuline, 3547
C$_{29}$H$_{40}$N$_2$O$_4$
 Emetine, 3600
C$_{29}$H$_{40}$O$_6$
 Cortisone*, 2602
C$_{29}$H$_{40}$O$_9$
 Calotropin, 1769
C$_{29}$H$_{41}$NO$_4$
 Buprenorphine, 1522
C$_{29}$H$_{42}$N$_6$O$_9$
 Amicetin, 415
C$_{29}$H$_{42}$N$_6$O$_{19}$S$_5$
 S-Adenosylmethionine*, 155
C$_{29}$H$_{42}$O$_5$
 Antheridiol, 719
C$_{29}$H$_{42}$O$_9$
 Helveticoside, 4665

C$_{29}$H$_{42}$O$_{10}$
 Adonitoxin, 167
 Convallatoxin, 2575
C$_{29}$H$_{42}$O$_{11}$
 α-Antiarin, 731
C$_{29}$H$_{44}$O$_2$
 ζ$_1$-Tocopherol, 9636
C$_{29}$H$_{44}$O$_3$
 Estradiol*, 3746
C$_{29}$H$_{44}$O$_{12}$
 Ouabain, 7031
C$_{29}$H$_{48}$N$_3$O$_9$S$_2$
 Pentacynium Bis(methylsulfate),
 7243
C$_{29}$H$_{47}$NO$_6$
 Coumingine, 2634
C$_{29}$H$_{47}$NO$_8$
 Sabadine, 8461
C$_{29}$H$_{48}$NO$_6$
 Choline Dehydrocholate, 2262
C$_{29}$H$_{48}$O
 Chondrillasterol, 2267
 7-Dehydrositosterol, 2927
 Fucosterol, 4303
 α-Spinasterol, 8901
 Stigmasterol, 8969
C$_{29}$H$_{48}$O$_3$
 Testosterone*, 9322
C$_{29}$H$_{50}$O
 β-Sitosterol, 8697
 γ-Sitosterol, 8698
C$_{29}$H$_{50}$O$_2$
 Vitamin E, 10159
C$_{29}$H$_{52}$O
 Stigmastanol, 8968
C$_{29}$H$_{53}$NO$_5$
 Orlistat, 6997
C$_{30}$H$_{16}$O$_8$
 Hypericin, 4911
C$_{30}$H$_{18}$Al$_2$O$_{21}$S$_6$
 Aluminum β-Naphtholdisulfonate,
 363
C$_{30}$H$_{21}$FeN$_3$O$_{15}$$^{3-}$
 Enterobactin*, 3631
C$_{30}$H$_{21}$NO$_9$
 Fredericamycin A, 4291
C$_{30}$H$_{23}$BrO$_4$
 Bromadiolone, 1403
C$_{30}$H$_{24}$N$_4$O$_{10}$
 Nicofuranose, 6605
C$_{30}$H$_{26}$CaO$_6$.2H$_2$O
 Fenoprofen*, 4021
C$_{30}$H$_{26}$F$_6$N$_4$O$_2$
 Antrafenine, 763
C$_{30}$H$_{26}$MgO$_8$
 Menbutone*, 5878
C$_{30}$H$_{26}$O$_{14}$
 Ergoflavin, 3691
C$_{30}$H$_{27}$N$_3$O$_{15}$
 Enterobactin, 3631
C$_{30}$H$_{28}$N$_2$Na$_4$O$_{14}$S$_5$
 Solasulfone, 8859
C$_{30}$H$_{28}$N$_6$O$_6$S$_4$
 Verticillins*, 10103
C$_{30}$H$_{28}$N$_6$O$_7$S$_4$
 Verticillins*, 10103
C$_{30}$H$_{28}$N$_6$O$_7$S$_5$
 Verticillins*, 10103
C$_{30}$H$_{28}$O$_8$
 Rottlerin, 8429
C$_{30}$H$_{30}$N$_4$
 Bis(p-dimethylaminobenzylidene)-
 benzidine, 1289
C$_{30}$H$_{30}$O$_8$
 Gossypol, 4549
C$_{30}$H$_{31}$ClN$_6$
 Janus Green B, 5269
C$_{30}$H$_{32}$N$_2$O$_2$
 Diphenoxylate, 3371

C$_{30}$H$_{34}$BrNO$_3$
Xenytropium Bromide, 10208
C$_{30}$H$_{34}$O$_{13}$
Picrotoxin, 7570
C$_{30}$H$_{35}$F$_2$N$_3$O
Lidoflazine, 5507
C$_{30}$H$_{35}$NO$_3$
Centchroman, 2018
C$_{30}$H$_{36}$O$_9$
Nimbin, 6638
C$_{30}$H$_{37}$N$_5$O$_5$
Ergosine, 3695
C$_{30}$H$_{38}$N$_4$O$_{11}$
Apicycline, 772
C$_{30}$H$_{40}$Cl$_2$N$_4$
Dequalinium Chloride, 2959
C$_{30}$H$_{40}$O$_2$
β-Citraurin, 2385
C$_{30}$H$_{40}$O$_6$
Absinthin, 8
Anabsinthin, 662
C$_{30}$H$_{41}$FO$_7$
Triamcinolone Hexacetonide, 9730
C$_{30}$H$_{42}$N$_2$O$_{15}$S$_2$
Sinalbin, 8687
C$_{30}$H$_{42}$O$_8$
Proscillaridin, 8060
C$_{30}$H$_{44}$K$_2$O$_{16}$S$_2$
Atractyloside, 899
C$_{30}$H$_{44}$N$_2$O$_{10}$
Hexobendine, 4743
C$_{30}$H$_{44}$O$_3$
Boldenone*, 1354
C$_{30}$H$_{44}$O$_9$
Cymarin, 2830
Peruvoside, 7324
C$_{30}$H$_{45}$ClO$_6$
Senegenin, 8598
C$_{30}$H$_{45}$NNaO$_7$P
Fosinopril*, 4282
C$_{30}$H$_{46}$NO$_7$P
Fosinopril, 4282
C$_{30}$H$_{46}$N$_2$O$_{14}$S$_2$
Aclatonium Napadisilate, 116
C$_{30}$H$_{46}$O$_4$
Enoxolone, 3628
Gypsogenin, 4614
C$_{30}$H$_{46}$O$_5$
Ceanothic Acid, 1959
Quillaic Acid, 8221
Quinovic Acid, 8260
C$_{30}$H$_{46}$O$_8$
Neriifolin, 6559
Periplocymarin, 7316
C$_{30}$H$_{47}$N$_3$O$_9$S
Leukotrienes*, 5482
C$_{30}$H$_{48}$N$_2$O$_2$
Hexestrol Bis(β-diethylaminoethyl Ether), 4739
C$_{30}$H$_{48}$O$_3$
β-Boswellic Acid, 1382
Nandrolone*, 6452
Oleanolic Acid, 6964
Ursolic Acid, 10027
C$_{30}$H$_{48}$O$_4$
Gratiogenin, 4563
Hederagenin, 4654
C$_{30}$H$_{48}$O$_5$
Acacic Acid, 12
Cimigenol, 2340
C$_{30}$H$_{49}$N$_9$O$_9$
Thymopentin, 9544
C$_{30}$H$_{50}$
Squalene, 8924
C$_{30}$H$_{50}$O
α-Amyrin, 659
β-Amyrin, 660
Friedelin, 4294
Lanosterol, 5372
Lupeol, 5636

Shionone, 8626
α$_1$-Sitosterol, 8696
Taraxasterol, 9229
Taraxerol, 9231
C$_{30}$H$_{50}$O$_2$
Betulin, 1237
C$_{30}$H$_{51}$B
Longifolene*, 5597
C$_{30}$H$_{52}$O
Dinosterol, 3342
C$_{30}$H$_{52}$O$_{26}$
Verbascose, 10092
C$_{30}$H$_{53}$NO$_{11}$
Benzonatate, 1127
C$_{30}$H$_{57}$BO$_3$
Menthyl Borate, 5885
C$_{30}$H$_{60}$I$_3$N$_3$O$_3$
Gallamine Triethiodide, 4361
C$_{30}$H$_{62}$
Squalane, 8923
C$_{30}$H$_{62}$O
1-Triacontanol, 9722
C$_{30}$H$_{66}$N$_4$O$_8$S
Dibutoline Sulfate, 3082
C$_{31}$H$_{23}$BrO$_3$
Brodifacoum, 1400
C$_{31}$H$_{29}$N$_3$O$_4$
Asperlicin, 878
C$_{31}$H$_{29}$N$_5$O$_5$
Asperlicin*, 878
C$_{31}$H$_{32}$N$_4$O$_2$
Bezitramide, 1241
C$_{31}$H$_{33}$N$_3$O$_6$S
Zafirlukast, 10241
C$_{31}$H$_{34}$BrNO$_4$
Fentonium Bromide, 4048
C$_{31}$H$_{35}$N$_2$NaO$_{11}$
Novobiocin*, 6818
C$_{31}$H$_{35}$N$_3$O$_4$S
Penicillin G Benethamine, 7220
C$_{31}$H$_{36}$ClN$_3$O$_5$S
Metofenazate, 6228
C$_{31}$H$_{36}$IN$_3$
Stilbazium Iodide, 8971
C$_{31}$H$_{36}$N$_2$O$_8$
Raunescine, 8292
C$_{31}$H$_{36}$N$_2$O$_{11}$
Novobiocin, 6818
C$_{31}$H$_{38}$N$_2$O$_5$S
Fantofarone, 3975
C$_{31}$H$_{39}$N$_5$O$_5$
Ergocornine, 3685
Ergocorninine, 3686
C$_{31}$H$_{40}$O$_2$
Menaquinones*, 5876
C$_{31}$H$_{41}$ClFNO$_3$
Haloperidol*, 4629
C$_{31}$H$_{42}$N$_2$O$_6$
Batrachotoxin, 1037
C$_{31}$H$_{43}$NO$_7$
Milbemycin Oxime*, 6277
C$_{31}$H$_{44}$N$_2$O$_{10}$
Dilazep, 3244
C$_{31}$H$_{44}$O$_7$
Milbemectin*, 6276
C$_{31}$H$_{44}$O$_8$
Proscillaridin*, 8060
C$_{31}$H$_{46}$O$_2$
Phylloquinone, 7536
C$_{31}$H$_{46}$O$_3$
Phylloquinone*, 7536
C$_{31}$H$_{47}$NaO$_6$
Fusidic Acid*, 4340
C$_{31}$H$_{48}$Na$_2$O$_8$P$_2$
Phylloquinone*, 7536
C$_{31}$H$_{48}$O$_2$S$_2$
Probucol, 7935
C$_{31}$H$_{48}$O$_6$
Fusidic Acid, 4340

C$_{31}$H$_{51}$NO$_9$
Rosaramicin, 8420
C$_{31}$H$_{52}$O$_3$
Vitamin E Acetate, 10160
C$_{31}$H$_{55}$N$_3$O$_6$
Enocitabine, 3624
C$_{32}$H$_{16}$CuN$_8$
Copper Phthalocyanine, 2584
C$_{32}$H$_{19}$N$_6$Na$_5$O$_{15}$S$_5$
Trypan Red, 9924
C$_{32}$H$_{22}$N$_6$Na$_2$O$_6$S$_2$
Congo Red, 2562
C$_{32}$H$_{24}$O$_{12}$
Thermorubin*, 9423
C$_{32}$H$_{26}$MgO$_6$
3-(o-Methoxyphenyl)-2-phenylacryl-
ic Acid*, 6078
C$_{32}$H$_{26}$O$_{14}$
Actinorhodine, 144
C$_{32}$H$_{28}$CaN$_6$O$_{14}$S$_2$
Sulfaloxic Acid*, 9080
C$_{32}$H$_{30}$O$_{14}$
Chrysergonic Acid, 2315
Secalonic Acids, 8561
C$_{32}$H$_{32}$Ag$_2$N$_4$O$_4$S
Diathymosulfone*, 3037
C$_{32}$H$_{32}$N$_2$O$_{10}$
Cotarnine*, 2618
C$_{32}$H$_{32}$O$_{13}$S
Teniposide, 9291
C$_{32}$H$_{32}$O$_{14}$
Chartreusin, 2086
C$_{32}$H$_{34}$CaN$_4$O$_8$S$_2$
Penicillin G Calcium, 7223
C$_{32}$H$_{34}$CaN$_4$O$_{10}$S$_2$
Penicillin V*, 7230
C$_{32}$H$_{34}$N$_4$O$_4$S
Diathymosulfone, 3037
C$_{32}$H$_{36}$CaCl$_2$O$_6$
Bucloxic Acid*, 1486
C$_{32}$H$_{37}$NO$_{12}$
Pirarubicin, 7642
C$_{32}$H$_{38}$N$_2$O$_5$
Cortivazol, 2603
C$_{32}$H$_{38}$N$_2$O$_8$
Deserpidine, 2964
C$_{32}$H$_{38}$N$_4$
Etioporphyrin, 3915
C$_{32}$H$_{39}$NO$_2$
Ebastine, 3534
C$_{32}$H$_{39}$NO$_4$
Fexofenadine, 4113
C$_{32}$H$_{39}$NO$_{11}$S
Noscapine*, 6815
C$_{32}$H$_{40}$BrN$_5$O$_5$
Bromocriptine, 1437
C$_{32}$H$_{40}$N$_2$O$_5$S$_2$
Trimethaphan Camsylate, 9837
C$_{32}$H$_{41}$NO$_2$
Terfenadine, 9307
C$_{32}$H$_{41}$N$_5$O$_5$
Ergocryptine, 3689
Ergocryptinine, 3690
C$_{32}$H$_{44}$F$_3$N$_3$O$_2$S
Fluphenazine*, 4226
C$_{32}$H$_{44}$N$_2$O$_8$
Lappaconitine, 5378
C$_{32}$H$_{44}$N$_2$O$_9$
Streptolydigin, 8982
C$_{32}$H$_{44}$O$_8$
Cucurbitacins*, 2681
C$_{32}$H$_{44}$O$_{12}$
Scilliroside, 8545
C$_{32}$H$_{45}$NO$_7$
Milbemycin Oxime*, 6277
C$_{32}$H$_{46}$N$_8$O$_6$S$_2$
Thiamine Disulfide*, 9432
C$_{32}$H$_{46}$O$_8$
Cucurbitacins*, 2681

*Denotes derivative of title compound

Denotes derivative of title compound

$C_{35}H_{58}O_{11}$
Filipin*, 4123
$C_{35}H_{58}O_{12}$
Fungichromin, 4312
$C_{35}H_{59}Al_3N_{10}O_{24}$
Aceglutamide*, 25
$C_{35}H_{60}Br_2N_2O_4$
Pancuronium Bromide, 7139
$C_{35}H_{61}NO_{12}$
Oleandomycin, 6962
$C_{35}H_{62}Br_2N_2O_4$
Pipecurium Bromide, 7612
$C_{36}H_{36}N_2O_5$
Tiliacorine, 9577
Trilobine*, 9826
$C_{36}H_{37}ClO_{17}$
Monardein Chloride, 6327
$C_{36}H_{37}ClO_{18}$
Delphinidin*, 2931
$C_{36}H_{38}N_2O_6$
Bebeerine, 1042
Chondrocurine, 2268
Curine, 2745
Daphnandrine, 2880
Isochondrodendrine, 5173
$C_{36}H_{40}N_2O_6$
Magnoline, 5737
$C_{36}H_{41}N_3O_6$
Lercanidipine, 5469
$C_{36}H_{42}Br_2N_2$
Hexafluorenium Bromide, 4721
$C_{36}H_{44}N_2O_{10}S$
Codeine Sulfate, 2529
$C_{36}H_{44}O_{12}$
Filixic Acids*, 4124
$C_{36}H_{45}N_5O_8S$
Ergoloid Mesylates*, 3692
$C_{36}H_{47}N_5O_4$
Indinavir, 4979
$C_{36}H_{48}N_2O_{10}$
Lycoctonine*, 5646
$C_{36}H_{48}N_2O_{12}$
Rhodomycins*, 8357
$C_{36}H_{51}NO_{11}$
Bikhaconitine, 1262
Veratridine, 10087
$C_{36}H_{51}NO_{12}$
Pseudoaconitine, 8095
$C_{36}H_{52}O_8$
Nemadectin, 6532
$C_{36}H_{52}O_{13}$
Scillaren*, 8543
$C_{36}H_{52}O_{15}$
Hellebrin, 4661
$C_{36}H_{53}NO_{13}$
Lucensomycin, 5621
$C_{36}H_{55}NO_{12}$
Desatrine, 2963
$C_{36}H_{56}O_8$
Phorbol*, 7487
$C_{36}H_{56}O_9$
Quinovin*, 8261
$C_{36}H_{56}O_{10}$
Quinovin*, 8261
$C_{36}H_{56}O_{13}$
Periplocin, 7315
$C_{36}H_{56}O_{14}$
Digitalin, 3200
$C_{36}H_{57}BiO_{12}$
Bismuth Ethyl Camphorate, 1304
$C_{36}H_{60}O_2$
Vitamin A*, 10150
$C_{36}H_{60}O_{30}$
Cyclodextrins*, 2787
$C_{36}H_{61}NaO_{11}$
Monensin*, 6329
$C_{36}H_{62}N_4$
Octenidine, 6852
$C_{36}H_{62}O_{11}$
Monensin, 6329

$C_{36}H_{63}N_3O_9$
Enniatins*, 3623
$C_{36}H_{66}BaO_4$
Oleic Acid*, 6965
$C_{36}H_{66}CaO_4$
Calcium Oleate, 1731
$C_{36}H_{66}CuO_4$
Cupric Oleate, 2711
$C_{36}H_{66}HgO_4$
Mercuric Oleate, 5934
$C_{36}H_{66}MgO_4$
Magnesium Oleate, 5711
$C_{36}H_{66}N_6O_6{}^{99m}Tc$
Technetium ^{99m}Tc Sestamibi, 9259
$C_{36}H_{66}O_4Zn$
Zinc Oleate, 10276
$C_{36}H_{68}CaI_2O_4$
Calcium Iodostearate, 1722
$C_{36}H_{70}CaO_4$
Calcium Stearate, 1750
$C_{36}H_{70}CuO_4$
Cupric Stearate, 2721
$C_{36}H_{70}HgO_4$
Mercuric Stearate, 5940
$C_{36}H_{70}MgO_4$
Magnesium Stearate, 5730
$C_{36}H_{70}O_4Zn$
Zinc Stearate, 10292
$C_{36}H_{74}Cl_2N_2$
Triclobisonium Chloride, 9785
$C_{37}H_{27}N_3Na_2O_9S_3$
Methyl Blue, 6107
$C_{37}H_{34}N_2Na_2O_9S_3$
Brilliant Blue FCF, 1397
Light Green SF Yellowish, 5508
$C_{37}H_{34}N_2Na_2O_{10}S_3$
Fast Green FCF, 3979
$C_{37}H_{35}N_2NaO_6S_2$
Guinea Green B, 4603
$C_{37}H_{36}N_3NaO_6S_2$
Acid Violet 7B, 110
$C_{37}H_{38}N_2O_6$
Cepharanthine, 2031
$C_{37}H_{40}N_2O_6$
Berbamine, 1191
Chondrofoline, 2269
Oxyacanthine, 7087
$C_{37}H_{40}N_4O_7$
Haplophytine, 4642
$C_{37}H_{42}Cl_2N_2O_6$
Tubocurarine Chloride, 9939
$C_{37}H_{45}NO_{12}$
Rifamycins*, 8383
$C_{37}H_{45}N_3O_{11}$
Rifamycins*, 8383
$C_{37}H_{47}NO_{12}$
Rifamycin SV, 8384
$C_{37}H_{48}I_6N_6O_{18}$
Iotrolan, 5082
$C_{37}H_{48}N_6O_5S_2$
Ritonavir, 8402
$C_{37}H_{49}N_7O_9S$
Pentagastrin, 7250
$C_{37}H_{53}NO_8$
Moxidectin, 6373
$C_{37}H_{61}NO_9$
Pavoninin-5, 7186
$C_{37}H_{62}O_{12}$
Laidlomycin, 5361
$C_{37}H_{67}NO_{13}$
Erythromycin, 3720
$C_{37}H_{77}NO_2$
Cetrimonium Stearate, 2069
$C_{38}H_{28}Cl_2N_8$
Neotetrazolium Chloride, 6554
$C_{38}H_{30}NiO_2P_2$
Bis(triphenylphosphine)dicarbonyl-
nickel, 1341
$C_{38}H_{36}CaN_4O_4$
Phenylbutazone*, 7431

$C_{38}H_{40}N_2O_6$
Insularine, 5010
$C_{38}H_{42}CaN_4O_6 \cdot \frac{1}{2}H_2O$
Bumadizon*, 1507
$C_{38}H_{42}N_2O_6$
Berbamine*, 1191
Tetrandrine, 9369
$C_{38}H_{42}N_8O_6S_2$
Bisbentiamine, 1286
$C_{38}H_{44}N_2O_6$
Dauricine, 2891
$C_{38}H_{44}O_8$
Gambogic Acid, 4373
$C_{38}H_{50}N_6O_5$
Saquinavir, 8516
$C_{38}H_{51}NO_4$
Myrophine, 6420
$C_{38}H_{52}Br_2N_2O_4 \cdot 4H_2O$
Belladonnine*, 1053
$C_{38}H_{52}N_6O_2$
Tirilazad, 9604
$C_{38}H_{54}I_6N_6O_{18}$
Iocarmic Acid*, 5029
$C_{38}H_{54}O_{13}$
Colocynthin, 2551
$C_{38}H_{58}CaN_4O_{10}S_2$
Moveltipril*, 6368
$C_{38}H_{59}FO_6$
Dexamethasone*, 2986
$C_{38}H_{60}O_{18}$
Stevioside, 8964
$C_{38}H_{61}I_2N_3S_3$
Platonin, 7690
$C_{38}H_{69}NO_{13}$
Clarithromycin, 2400
$C_{38}H_{72}N_2O_{12}$
Azithromycin, 946
$C_{39}H_{46}Cl_2N_2O_6$
Tubocurarine Chloride*, 9939
$C_{39}H_{47}NO_{14}$
Rifamycins*, 8383
Streptovaricin*, 8989
$C_{39}H_{49}NO_{14}$
Rifamycins*, 8383
$C_{39}H_{49}NO_{16}$
Nogalamycin, 6767
$C_{39}H_{53}N_3O_9$
Bietaserpine, 1254
$C_{39}H_{53}N_9O_{15}S$
Amanitin*, 387
$C_{39}H_{54}N_{10}O_{14}S$
Amanitin*, 387
$C_{39}H_{59}Cl_2NO_2$
Phenesterine, 7367
$C_{39}H_{61}NO_{14}$
Rimocidin, 8394
$C_{40}H_{32}Cl_2N_8O_2$
Tetrazolium Blue, 9380
$C_{40}H_{44}N_4O^{2+}$
C-Curarine I, 2742
$C_{40}H_{46}N_4O_2{}^{2+}$
Toxiferine I, 9694
$C_{40}H_{48}I_2N_2O_6$
Metocurine Iodide, 6227
$C_{40}H_{48}N_4O_2{}^{2+}$
C-Calebassine, 1762
$C_{40}H_{48}N_4O_3$
Geissospermine, 4387
$C_{40}H_{48}O_4$
Astacin, 888
$C_{40}H_{49}NO_{14}$
Streptovaricin*, 8989
$C_{40}H_{50}O_2$
Rhodoxanthin, 8362
$C_{40}H_{51}NO_{13}$
Streptovaricin*, 8989
$C_{40}H_{51}NO_{14}$
Streptovaricin*, 8989
$C_{40}H_{51}NO_{15}$
Streptovaricin*, 8989

Denotes derivative of title compound

$C_{40}H_{52}O_2$
Canthaxanthin, 1798
Torularhodin, 9691

$C_{40}H_{52}O_4$
Astaxanthin, 890

$C_{40}H_{54}O$
Echinenone, 3543

$C_{40}H_{56}$
α-Carotene, 1901
β-Carotene, 1902
γ-Carotene, 1903
δ-Carotene, 1904
Lycopene, 5650

$C_{40}H_{56}O$
Cryptoxanthin, 2676
Lycoxanthin, 5657
Rubixanthin, 8445

$C_{40}H_{56}O_2$
Lycophyll, 5651
Xanthophyll, 10197
Zeaxanthin, 10248

$C_{40}H_{56}O_3$
Capsanthin, 1812
Chrysanthemaxanthin, 2309
Flavoxanthin, 4134

$C_{40}H_{56}O_4$
Violaxanthin, 10136

$C_{40}H_{58}Cl_2N_4$
Bisdequalinium Chloride, 1288

$C_{40}H_{58}O$
Rhodopin, 8358

$C_{40}H_{58}O_4$
γ-Oryzanol*, 7016

$C_{40}H_{60}BNaO_{14}$
Aplasmomycin, 779

$C_{40}H_{60}N_4O_{10}$
Bufotoxin, 1503

$C_{40}H_{60}O_2$
Kitol, 5328

$C_{40}H_{63}N_3O_4S_2$
Pipotiazine*, 7636

$C_{40}H_{64}O_{11}$
Dermostatin*, 2960

$C_{40}H_{64}O_{12}$
Nonactin, 6770

$C_{40}H_{66}N_6$
Ormosinine, 6999

$C_{40}H_{66}O_{10}$
Venturicidins*, 10080

$C_{40}H_{67}NO_{14}$
Leucomycins*, 5480

$C_{40}H_{68}$
Phytofluene, 7544

$C_{40}H_{68}N_4O_{12}$
Amidomycin, 422

$C_{40}H_{68}O_{11}$
Nigericin, 6633

$C_{40}H_{71}NO_{14}$
Erythromycin Propionate, 3725

$C_{40}H_{74}CaO_{14}S_2$
Docusate Calcium, 3459

$C_{40}H_{80}NO_8P$
Colfosceril Palmitate, 2540

$C_{41}H_{39}ClO_{21}$
Delphinidin*, 2931

$C_{41}H_{48}N_2O_8$
Thalicarpine, 9389

$C_{41}H_{52}N_6O_7$
Ergonovine*, 3694

$C_{41}H_{56}O_2$
Menaquinones*, 5876

$C_{41}H_{58}FeN_9O_{20}$
Ferrichromes*, 4067

$C_{41}H_{60}O_4$
γ-Oryzanol*, 7016

$C_{41}H_{63}NO_{14}$
Protoveratrines*, 8086

$C_{41}H_{63}NO_{15}$
Protoveratrines*, 8086

$C_{41}H_{64}O_8$
Prednisolone*, 7901

$C_{41}H_{64}O_{13}$
Digitoxin, 3206

$C_{41}H_{64}O_{14}$
Digoxin, 3210
Gitoxin, 4441

$C_{41}H_{64}O_{15}$
Diginatin, 3198

$C_{41}H_{65}NO_{15}$
Midecamycins*, 6271

$C_{41}H_{66}O_{11}$
Dermostatin*, 2960

$C_{41}H_{66}O_{12}$
α-Hederin, 4655

$C_{41}H_{67}NO_{11}$
Venturicidins*, 10080

$C_{41}H_{67}NO_{15}$
Midecamycins*, 6271
Troleandomycin, 9899

$C_{41}H_{76}N_2O_{15}$
Roxithromycin, 8433

$C_{42}H_{27}AlF_9N_3O_6$
Flufenamic Acid*, 4167

$C_{42}H_{30}N_6O_{12}$
Inositol Niacinate, 5009

$C_{42}H_{38}O_{20}$
Sennosides, 8600

$C_{42}H_{45}N_3O_7$
Pamaquine*, 7134

$C_{42}H_{46}N_4O_8S$
Strychnine Sulfate, 9024

$C_{42}H_{51}NO_{15}$
Aclacinomycins*, 115

$C_{42}H_{53}NO_{15}$
Aclacinomycins*, 115
Streptovaricin*, 8989

$C_{42}H_{53}NO_{16}$
Streptovaricin*, 8989

$C_{42}H_{58}O_6$
Fucoxanthin, 4304

$C_{42}H_{60}K_2O_{16}$
Glycyrrhizic Acid*, 4515

$C_{42}H_{60}O_2$
Rhodoviolascin, 8361

$C_{42}H_{62}N_8O_7S$
Bottromycin*, 1383

$C_{42}H_{62}O_{16}$
Glycyrrhizic Acid, 4515

$C_{42}H_{64}O_{19}$
Thevetin A, 9424

$C_{42}H_{65}N_{11}O_{12}$
Cargutocin, 1887

$C_{42}H_{65}N_{13}O_{10}$
Saralasin, 8518

$C_{42}H_{66}O_{14}$
Digoxin*, 3210

$C_{42}H_{67}NO_{15}$
Carbomycin*, 1854

$C_{42}H_{67}NO_{16}$
Carbomycin*, 1854

$C_{42}H_{68}O_{14}$
Gratioside, 4564

$C_{42}H_{69}NO_{15}$
Josamycin, 5280
Rokitamycin, 8408

$C_{42}H_{69}NaO_{11}$
Salinomycin*, 8488

$C_{42}H_{70}O_{11}$
Salinomycin, 8488

$C_{42}H_{70}O_{35}$
Cyclodextrins*, 2787

$C_{42}H_{72}O_{16}$
Lankamycin, 5369

$C_{42}H_{78}N_2O_{14}$
Dirithromycin, 3418

$C_{43}H_{42}O_{22}$
Carthamin, 1918

$C_{43}H_{47}N_2NaO_6S_2$
Indocyanine Green, 4992

$C_{43}H_{49}N_7O_{10}$
Virginiamycin*, 10142

$C_{43}H_{51}N_3O_{11}$
Rifaximin, 8386

$C_{43}H_{52}N_4O_5$
Voacamine, 10168

$C_{43}H_{53}NO_{14}$
Docetaxel, 3458

$C_{43}H_{55}N_5O_7$
Vindesine, 10125

$C_{43}H_{58}N_2O_{13}$
Rifamide, 8381

$C_{43}H_{58}N_4O_{12}$
Rifampin, 8382

$C_{43}H_{60}N_2O_{12}$
Mocimycin, 6308

$C_{43}H_{62}N_2O_{12}$
Mocimycin*, 6308

$C_{43}H_{65}N_{11}O_{12}S_2$
Deaminooxytocin, 2899

$C_{43}H_{66}N_{12}O_{12}S_2$
Oxytocin, 7114

$C_{43}H_{66}O_{14}$
Acetyldigitoxins, 90

$C_{43}H_{66}O_{15}$
Digoxin*, 3210

$C_{43}H_{72}O_{11}$
Narasin, 6506

$C_{43}H_{74}N_2O_{14}$
Spiramycin*, 8904

$C_{43}H_{75}NO_{16}$
Erythromycin*, 3720

$C_{43}H_{78}N_6O_{13}$
Romurtide, 8412

$C_{44}H_{43}ClN_2O_8$
Hydroxyzine*, 4897

$C_{44}H_{50}Cl_2N_4O_2$
Alcuronium*, 222

$C_{44}H_{50}N_4O_2^{2+}$
Alcuronium, 222

$C_{44}H_{62}N_2O_{12}$
Aurodox, 916

$C_{44}H_{62}N_8O_{11}$
Etamycin, 3755

$C_{44}H_{68}O_{13}$
Okadaic Acid, 6958

$C_{44}H_{69}NO_{12}$
Tacrolimus, 9200

$C_{44}H_{72}O_{11}$
Oligomycins*, 6970

$C_{44}H_{81}NO_{21}$
Erythromycin Glucoheptonate, 3723

$C_{44}H_{84}CaI_2O_4$
Calcium Iodobehenate, 1721

$C_{45}H_{44}Cl_2N_{10}O_6$
Cycloguanil*, 2790

$C_{45}H_{54}N_4O_8$
Vinorelbine, 10127

$C_{45}H_{54}N_8O_{10}$
Mikamycin*, 6275

$C_{45}H_{56}N_6O_{14}S$
Penimepicycline, 7235

$C_{45}H_{63}N_{13}O_{12}S_2$
Ornipressin, 7001

$C_{45}H_{65}N_3O_6$
Pilocereine, 7580

$C_{45}H_{69}N_{11}O_{12}S$
Carbetocin, 1841

$C_{45}H_{71}NO_{17}$
Miokamycin, 6291

$C_{45}H_{72}O_{12}$
Oligomycins*, 6970

$C_{45}H_{72}O_{16}$
Dioscin, 3345

$C_{45}H_{73}NO_{15}$
Solanine, 8853

$C_{45}H_{73}NO_{16}$
Josamycin*, 5280
Solasonine, 8858

$C_{57}H_{60}N_2O_8$
  Bephenium*, 1187
$C_{57}H_{82}O_{26}$
  Chromomycins*, 2295
$C_{57}H_{83}N_3O_{39}S$
  Piroxicam*, 7661
$C_{57}H_{89}N_7O_{15}$
  Didemnins*, 3150
$C_{57}H_{104}O_6$
  Petroselinic Acid*, 7330
  Triolein, 9861
$C_{57}H_{105}NO_{16}$
  Erythromycin Acistrate, 3721
$C_{57}H_{110}O_6$
  Tristearin, 9885
$C_{58}H_{73}N_{13}O_{21}S_2$
  Ceruletide, 2048
$C_{58}H_{80}Cl_2N_2O_{14}$
  Mivacurium Chloride, 6305
$C_{58}H_{84}N_2O_{19}\cdot2\frac{1}{2}H_2O$
  Partricin*, 7181
$C_{58}H_{84}N_{14}O_{16}S$
  Physalaemin, 7537
$C_{58}H_{84}O_{26}$
  Olivomycins*, 6975
$C_{58}H_{89}NO_3$
  Rhodoquinone, 8360
$C_{58}H_{91}N_{13}O_{20}$
  Amphomycin, 625
$C_{58}H_{105}N_{16}Na_5O_{28}S_5$
  Colistin*, 2542
$C_{59}H_{77}N_{13}O_{19}$
  Telomycin, 9283
$C_{59}H_{84}N_2O_{18}$
  Candicidin*, 1789
$C_{59}H_{84}N_{16}O_{12}$
  Leuprolide, 5484
$C_{59}H_{84}N_{18}O_{14}$
  Goserelin, 4547
$C_{59}H_{86}N_2O_{19}$
  Mepartricin*, 5891
$C_{59}H_{86}N_2O_{19}\cdot4H_2O$
  Partricin*, 7181
$C_{59}H_{88}N_2O_{17}$
  Perimycin*, 7310
$C_{59}H_{88}N_2O_{20}$
  Efrotomycin, 3567
$C_{59}H_{90}O_4$
  Ubiquinones*, 9974
C_{60}
  Buckminsterfullerene, 1482
$C_{60}H_{78}OSn_2$
  Fenbutatin Oxide, 4004
$C_{60}H_{86}N_{16}O_{13}$
  Buserelin, 1527
$C_{60}H_{88}N_2O_{19}$
  Mepartricin*, 5891
$C_{60}H_{92}N_{12}O_{10}$
  Gramicidin S, 4552
$C_{61}H_{88}Cl_2O_{32}$
  Avilamycin*, 921
$C_{61}H_{88}N_{18}O_{16}$
  Goserelin*, 4547
$C_{61}H_{88}N_{18}O_{21}S_2$
  Peplomycin, 7286
$C_{61}H_{90}Cl_2O_{32}$
  Avilamycin*, 921
$C_{61}H_{109}N_{11}O_{12}$
  Cyclosporins*, 2821
$C_{62}H_{80}CaCl_8N_{10}O_{30}$
  Chloramphenicol*, 2120
$C_{62}H_{86}N_{12}O_{16}$
  Dactinomycin, 2867
$C_{62}H_{88}CoN_{13}O_{14}P$
  Vitamin B_{12r}, 10154
$C_{62}H_{88}CoN_{14}O_{16}P$
  Hydroxocobalamin*, 4854
$C_{62}H_{89}CoN_{13}O_{15}P$
  Hydroxocobalamin, 4854

$C_{62}H_{89}CoN_{13}O_{17}PS$
  Hydroxocobalamin*, 4854
$C_{62}H_{97}N_{23}O_{26}$
  Scotophobin, 8555
$C_{62}H_{98}N_{20}O_{26}S_2$
  Tallysomycin*, 9213
$C_{62}H_{111}N_{11}O_{12}$
  Cyclosporins*, 2821
$C_{62}H_{111}N_{11}O_{13}$
  Cyclosporins*, 2821
$C_{63}H_{87}N_{13}O_{19}S_2$
  Pentetreotide, 7267
$C_{63}H_{88}CoN_{14}O_{14}P$
  Vitamin B_{12}, 10152
$C_{63}H_{88}N_{12}O_{16}$
  Cactinomycin*, 1642
$C_{63}H_{91}CoN_{13}O_{14}P$
  Methylcobalamin, 6125
$C_{63}H_{98}N_{18}O_{13}S$
  Substance P, 9032
$C_{63}H_{113}N_{11}O_{12}$
  Cyclosporins*, 2821
$C_{64}H_{82}N_{18}O_{13}$
  Triptorelin, 9878
$C_{64}H_{83}N_{17}O_{12}$
  Deslorelin, 2968
$C_{64}H_{90}N_{12}O_{16}$
  Cactinomycin*, 1642
$C_{64}H_{91}CoN_{13}O_{16}P$
  Hydroxocobalamin*, 4854
$C_{65}H_{82}N_2O_{18}S_2$
  Atracurium Besylate, 900
$C_{65}H_{85}N_{13}O_{30}$
  Mycobacillin, 6404
$C_{65}H_{98}ClN_3O_{39}$
  Benexate Hydrochloride*, 1065
$C_{66}H_{68}CaF_2N_4O_{10}$
  Atorvastatin*, 897
$C_{66}H_{75}Cl_2N_9O_{24}$
  Vancomycin, 10066
$C_{66}H_{83}N_{17}O_{13}$
  Nafarelin, 6437
$C_{66}H_{86}N_{18}O_{12}$
  Histrelin, 4760
$C_{66}H_{87}N_{13}O_{13}$
  Tyrocidine*, 9967
$C_{66}H_{103}N_{17}O_{16}S$
  Bacitracin*, 965
$C_{68}H_{84}{}^{111}InN_{13}O_{19}S_2$
  Pentetreotide*, 7267
$C_{68}H_{88}N_{14}O_{13}$
  Tyrocidine*, 9967
$C_{68}H_{110}N_{22}O_{27}S_2$
  Tallysomycin*, 9213
$C_{70}H_{89}N_{15}O_{13}$
  Tyrocidine*, 9967
$C_{71}H_{110}N_{24}O_{18}S$
  Bombesin, 1359
$C_{72}H_{85}N_{19}O_{18}S_5$
  Thiostrepton, 9502
$C_{72}H_{100}CoN_{18}O_{17}P$
  Cobamamide, 2513
$C_{72}H_{116}O_4$
  Xanthophyll*, 10197
  Zeaxanthin*, 10248
$C_{73}H_{131}N_3O_{31}$
  Gangliosides*, 4376
$C_{74}H_{100}N_6O_8S_2$
  Penicillin G Hydrabamine, 7224
$C_{74}H_{100}N_6O_{10}S_2$
  Penicillin V Hydrabamine, 7232
$C_{75}H_{70}N_6O_6$
  Pyrvinium Pamoate, 8206
$C_{75}H_{126}N_{24}O_{15}$
  Dynorphin*, 3528
$C_{76}H_{52}O_{46}$
  Tannic Acid*, 9221
$C_{76}H_{104}N_{18}O_{19}S_2$
  Somatostatin, 8863

$C_{79}H_{131}N_{31}O_{24}S_4$
  Apamin, 767
$C_{80}H_{104}I_6N_8O_{20}S_3$
  Quinine Iodosulfate, 8246
$C_{84}H_{133}N_{21}O_{26}\cdot6H_2O$
  Neurotensin*, 6569
$C_{88}H_{95}Cl_2N_9O_{33}$
  Teicoplanin*, 9269
$C_{88}H_{97}Cl_2N_9O_{33}$
  Teicoplanin*, 9269
$C_{89}H_{99}Cl_2N_9O_{33}$
  Teicoplanin*, 9269
$C_{89}H_{101}Cl_2N_9O_{36}$
  Avoparcin*, 922
$C_{89}H_{102}ClN_9O_{36}$
  Avoparcin*, 922
$C_{95}H_{110}N_8O_{44}$
  Ristocetin*, 8398
$C_{96}H_{141}AlO_{15}$
  Acetoxolone*, 76
$C_{99}H_{155}N_{29}O_{21}S$
  Alsactide, 322
$C_{100}H_{156}N_{34}O_{22}S$
  Giractide, 4435
$C_{128}H_{194}N_{40}O_{28}S_2$
  Bradycor*, 1385
$C_{129}H_{223}N_3O_{54}$
  Palytoxin, 7132
$C_{130}H_{220}N_{44}O_{41}$
  Secretin*, 8564
$C_{131}H_{229}N_{39}O_{31}$
  Melittin, 5867
$C_{136}H_{210}N_{40}O_{31}S$
  Cosyntropin, 2617
$C_{143}H_{230}N_{42}O_{37}S_7$
  Nisin, 6657
$C_{148}H_{244}N_{42}O_{47}$
  Elcatonin, 3578
$C_{149}H_{246}N_{44}O_{42}S$
  Sermorelin, 8605
$C_{176}H_{277}N_{57}O_{55}S_7$
  Charybdotoxin, 2087
$C_{198}H_{340}N_{68}O_{56}S_8$
  Purothionin*, 8131
$C_{202}H_{348}N_{68}O_{59}S_8$
  Purothionin*, 8131
$C_{203}H_{339}N_{67}O_{59}S_8$
  Purothionin*, 8131
$C_{284}H_{432}N_{84}O_{79}S_7$
  Aprotinin, 796
$C_{734}H_{1166}N_{204}O_{216}S_5$
  Interferon-γ*, 5018
$C_{860}H_{1353}N_{227}O_{255}S_9$
  Interferon-α*, 5016
$C_{860}H_{1353}N_{229}O_{255}S_9$
  Interferon-α*, 5016
$C_{938}H_{1465}N_{257}O_{278}S_6$
  Somatotropin*, 8864
$C_{938}H_{1469}N_{255}O_{275}S_7$
  Somatotropin*, 8864
$C_{976}H_{1533}N_{265}O_{286}S_8$
  Somatotropin*, 8864
$C_{977}H_{1527}N_{265}O_{287}S_7$
  Somatotropin*, 8864
$C_{978}H_{1537}N_{265}O_{286}S_9$
  Somatotropin*, 8864
$C_{979}H_{1527}N_{265}O_{287}S_8$
  Somatotropin*, 8864
$C_{987}H_{1550}N_{268}O_{291}S_9$
  Somatotropin*, 8864
$C_{990}H_{1529}N_{263}O_{299}S_7$
  Somatotropin*, 8864
$C_{995}H_{1537}N_{263}O_{301}S_8$
  Somatotropin*, 8864
$C_{1020}H_{1596}N_{274}O_{302}S_9$
  Somatotropin*, 8864
$CaCl_2$
  Calcium Chloride, 1699
$CaCl_2O_2$
  Calcium Hypochlorite, 1717

Denotes derivative of title compound

CaCl₂O₆
Calcium Chlorate, 1698
CaCrO₄
Calcium Chromate(VI), 1700
CaCr₂O₇
Calcium Dichromate(VI), 1706
CaFO₃P
Calcium Fluorophosphate, 1710
CaF₂
Calcium Fluoride, 1709
CaF₆Si
Calcium Hexafluorosilicate, 1714
CaHO₃P
Calcium Phosphite, 1743
CaHO₄P
Calcium Phosphate, Dibasic, 1739
CaH₂
Calcium Hydride, 1715
CaH₂O₂
Calcium Hydroxide, 1716
CaH₄O₄P₂
Calcium Hypophosphite, 1718
CaH₄O₈P₂
Calcium Phosphate, Monobasic, 1740
CaI₂
Calcium Iodide, 1720
CaI₂O₆
Calcium Iodate, 1719
CaMn₂O₈
Calcium Permanganate, 1735
CaMoO₄
Calcium Molybdate(VI), 1728
CaN₂O₄
Calcium Nitrite, 1730
CaN₂O₆
Calcium Nitrate, 1729
CaO
Calcium Oxide, 1733
CaO₂
Calcium Peroxide, 1736
CaO₃S
Calcium Sulfite, 1755
CaO₃S₂
Calcium Thiosulfate, 1759
CaO₄S
Calcium Sulfate, 1753
CaO₄W
Calcium Tungstate(VI), 1760
CaS
Calcium Sulfide, 1754
CaSe
Calcium Selenide, 1748
Ca₂O₇P₂
Calcium Pyrophosphate, 1746
Ca₃O₈P₂
Calcium Phosphate, Tribasic, 1741
Ca₃P₂
Calcium Phosphide, 1742
CdCl₂
Cadmium Chloride, 1653
CdF₂
Cadmium Fluoride, 1655
CdH₂O₂
Cadmium Hydroxide, 1656
CdI₂
Cadmium Iodide, 1657
CdI₄K₂
Potassium Tetraiodocadmate, 7861
CdN₂O₆
Cadmium Nitrate, 1658
CdO
Cadmium Oxide, 1659
CdO₄S
Cadmium Sulfate, 1665
CdO₄Se
Cadmium Selenate, 1662
CdO₄W
Cadmium Tungstate(VI), 1668

CdS
Cadmium Sulfide, 1666
CdSe
Cadmium Selenide, 1663
CdTe
Cadmium Telluride, 1667
CeCl₃
Cerous Chloride, 2041
CeF₃
Cerous Fluoride, 2042
CeF₄
Ceric Fluoride, 2034
CeH₄NO₈S₂
Ammonium Cerous Sulfate, 536
CeH₈N₈O₁₈
Ammonium Ceric Nitrate, 535
CeI₃
Cerous Iodide, 2043
CeN₃O₉
Cerous Nitrate, 2044
CeO₂
Ceric Oxide, 2035
CeO₈S₂
Ceric Sulfate, 2036
Ce₂O₁₂S₃
Cerous Sulfate, 2046
ClCs
Cesium Chloride, 2054
ClCu
Cuprous Chloride, 2730
ClF
Chlorine Monofluoride, 2148
ClFO₃
Perchloryl Fluoride, 7297
ClFO₄
Fluorine Perchlorate, 4203
ClF₃
Chlorine Trifluoride, 2150
ClH
Hydrogen Chloride, 4835
ClHO
Hypochlorous Acid, 4912
ClHO₃
Chloric Acid, 2141
ClHO₃S
Chlorosulfonic Acid, 2218
ClHO₃Se
Chloroselenic Acid, 2216
ClHO₄
Perchloric Acid, 7296
ClH₂HgN
Mercuric Chloride, Ammoniated, 5927
ClH₄N
Ammonium Chloride, 537
ClH₄NO₄
Ammonium Perchlorate, 574
ClI
Iodine Monochloride, 5038
ClK
Potassium Chloride, 7783
ClKO₃
Potassium Chlorate, 7782
ClKO₄
Potassium Perchlorate, 7822
ClLi
Lithium Chloride, 5553
ClLiO₄
Lithium Perchlorate, 5565
ClNO
Nitrosyl Chloride, 6744
ClNO₂
Nitryl Chloride, 6755
ClNa
Sodium Chloride, 8742
ClNaO
Sodium Hypochlorite, 8773
ClNaO₂
Sodium Chlorite, 8743

ClNaO₃
Sodium Chlorate, 8741
ClNaO₄
Sodium Perchlorate, 8798
ClOSb
Antimony Chloride Oxide, 734
ClO₂
Chlorine Dioxide, 2146
ClO₃Re
Rhenium Oxychloride, 8346
ClRb
Rubidium Chloride, 8441
ClTl
Thallium Chloride, 9395
Cl₂Co
Cobaltous Chloride, 2498
Cl₂Cr
Chromous Chloride, 2301
Cl₂CrO₂
Chromyl Chloride, 2306
Cl₂Cu
Cupric Chloride, 2699
Cl₂CuO₆
Cupric Chlorate, 2698
Cl₂CuO₈
Cupric Perchlorate, 2714
Cl₂F₃Sb
Antimony Dichlorotrifluoride, 735
Cl₂Fe
Ferrous Chloride, 4091
Cl₂Ge
Germanium Dichloride, 4416
Cl₂H₆N₂Pt
Cisplatin, 2378
Cl₂Hg
Mercuric Chloride, 5926
Cl₂Hg₂
Mercurous Chloride, 5951
Cl₂Hg₂O₆
Mercurous Chlorate, 5950
Cl₂Mg
Magnesium Chloride, 5698
Cl₂MgO₆
Magnesium Chlorate, 5697
Cl₂MgO₈
Magnesium Perchlorate, 5715
Cl₂Mn
Manganese Chloride, 5768
Cl₂Ni
Nickel Chloride, 6588
Cl₂O
Chlorine Monoxide, 2149
Cl₂OS
Thionyl Chloride, 9485
Cl₂OSe
Selenium Oxychloride, 8578
Cl₂OV
Vanadyl Dichloride, 10062
Cl₂OZr
Zirconyl Chloride, 10317
Cl₂O₂S
Sulfuryl Chloride, 9153
Cl₂O₂U
Uranyl Chloride, 9999
Cl₂O₅S₂
Pyrosulfuryl Chloride, 8193
Cl₂O₆Pb
Lead Chlorate, 5420
Cl₂O₆Sr
Strontium Chlorate, 8999
Cl₂O₇
Chlorine Heptoxide, 2147
Cl₂O₈Zn
Zinc Perchlorate, 10280
Cl₂Pb
Lead Chloride, 5421
Cl₂Pd
Palladium Chloride, 7122
Cl₂Pt
Platinous Chloride, 7687

Denotes derivative of title compound

Cl$_2$S$_2$
Sulfur Chloride, 9143
Cl$_2$Se$_2$
Selenium Chloride, 8574
Cl$_2$Sn
Stannous Chloride, 8939
Cl$_2$Sr
Strontium Chloride, 9000
Cl$_2$Te
Tellurium Dichloride, 9275
Cl$_2$Ti
Titanium Dichloride, 9611
Cl$_2$Zn
Zinc Chloride, 10261
Cl$_3$CoH$_{18}$N$_6$
Hexaaminecobalt Trichloride, 4711
Cl$_3$Cr
Chromic Chloride, 2278
Cl$_3$Fe
Ferric Chloride, 4061
Cl$_3$HSi
Trichlorosilane, 9776
Cl$_3$H$_4$MgN
Ammonium Magnesium Chloride,
562
Cl$_3$I
Iodine Trichloride, 5041
Cl$_3$In
Indium Trichloride, 4989
Cl$_3$Ir
Iridium Trichloride, 5101
Cl$_3$N
Nitrogen Chloride, 6699
Cl$_3$OP
Phosphorus Oxychloride, 7506
Cl$_3$OV
Vanadyl Trichloride, 10064
Cl$_3$P
Phosphorus Trichloride, 7515
Cl$_3$PS
Phosphorus Sulfochloride, 7513
Cl$_3$Rh
Rhodium Chloride, 8355
Cl$_3$Ru
Ruthenium Trichloride, 8455
Cl$_3$Sb
Antimony Trichloride, 748
Cl$_3$Ti
Titanium Trichloride, 9620
Cl$_3$U
Uranium Trichloride, 9996
Cl$_4$CuH$_8$N$_2$
Ammonium Cupric Chloride, 543
Cl$_4$Ge
Germanium Tetrachloride, 4418
Cl$_4$H$_8$HgN$_2$
Ammonium Mercuric Chloride, 564
Cl$_4$H$_8$N$_2$Pt
Ammonium Platinous Chloride, 583
Cl$_4$H$_8$N$_2$Zn
Ammonium Tetrachlorozincate, 596
Cl$_4$K$_2$Pt
Potassium Tetrachloroplatinate(II),
7855
Cl$_4$Os
Osmium Tetrachloride, 7023
Cl$_4$Po
Polonium Tetrachloride, 7720
Cl$_4$Se
Selenium Tetrachloride, 8582
Cl$_4$Si
Silicon Tetrachloride, 8644
Cl$_4$Sn
Stannic Chloride, 8929
Cl$_4$Te
Tellurium Tetrachloride, 9279
Cl$_4$Th
Thorium Chloride, 9515
Cl$_4$Ti
Titanium Tetrachloride, 9618

Cl$_4$U
Uranium Tetrachloride, 9994
Cl$_4$Zr
Zirconium Chloride, 10307
Cl$_5$H$_{12}$N$_3$Zn
Ammonium Pentachlorozincate, 573
Cl$_5$Nb
Niobium Pentachloride, 6649
Cl$_5$P
Phosphorus Pentachloride, 7508
Cl$_5$Sb
Antimony Pentachloride, 736
Cl$_5$Ta
Tantalum Pentachloride, 9224
Cl$_6$H$_2$Pt
Platinic Chloride, 7684
Cl$_6$H$_8$N$_2$Os
Ammonium Osmium Chloride, 570
Cl$_6$H$_8$N$_2$Pt
Ammonium Platinic Chloride, 582
Cl$_6$H$_{42}$N$_{14}$O$_2$Ru$_3$
Ruthenium Red, 8453
Cl$_6$K$_2$Os
Potassium Hexachloroosmate(IV),
7799
Cl$_6$K$_2$Pt
Potassium Hexachloroplatinate(IV),
7800
Cl$_6$Na$_2$Pt
Sodium Hexachloroplatinate(IV),
8768
CoCr$_2$O$_4$
Cobaltous Chromate(III), 2499
CoF$_2$
Cobaltous Fluoride, 2501
CoF$_3$
Cobaltic Fluoride, 2491
CoHO$_2$
Cobaltic Oxide Monohydrate, 2492
CoH$_2$O$_2$
Cobaltous Hydroxide, 2503
CoH$_4$NO$_4$P
Ammonium Cobaltous Phosphate,
541
CoH$_8$N$_2$O$_8$S$_2$
Ammonium Cobaltous Sulfate, 542
CoH$_{10}$N$_7$O$_8$
Erdmann's Salt, 3683
CoI$_2$
Cobaltous Iodide, 2504
CoK$_2$O$_8$S$_2$
Cobaltous Potassium Sulfate, 2509
CoK$_2$O$_8$Se$_2$
Potassium Cobaltous Selenate, 7787
CoK$_3$N$_6$O$_{12}$
Cobaltic Potassium Nitrite, 2493
CoN$_2$O$_6$
Cobaltous Nitrate, 2505
CoN$_6$Na$_3$O$_{12}$
Sodium Cobaltinitrite, 8748
CoO
Cobaltous Oxide, 2507
CoO$_4$S
Cobaltous Sulfate, 2510
CoS
Cobaltous Sulfide, 2511
Co$_3$O$_4$
Cobaltic-Cobaltous Oxide, 2490
Co$_3$O$_8$P$_2$
Cobaltous Phosphate, 2508
CrCuO$_4$
Cupric Chromate(VI), 2700
CrF$_2$
Chromous Fluoride, 2302
CrF$_2$O$_2$
Chromyl Fluoride, 2307
CrF$_3$
Chromic Fluoride, 2279
CrF$_4$
Chromium Tetrafluoride, 2292

CrH$_3$O$_3$
Chromic Hydroxide, 2281
CrH$_4$NO$_8$S$_2$
Ammonium Chromic Sulfate, 539
CrH$_8$N$_2$O$_4$
Ammonium Chromate(VI), 538
CrKO$_8$S$_2$
Chromic Potassium Sulfate, 2286
CrK$_2$O$_4$
Potassium Chromate(VI), 7784
CrLi$_2$O$_4$
Lithium Chromate(VI), 5554
CrN$_3$O$_9$
Chromic Nitrate, 2282
CrNa$_2$O$_4$
Sodium Chromate(VI), 8745
CrO$_2$
Chromium Dioxide, 2290
CrO$_3$
Chromium Trioxide, 2293
CrO$_4$P
Chromic Phosphate, 2284
CrO$_4$Pb
Lead Chromate(VI), 5422
CrO$_4$S
Chromous Sulfate, 2305
CrO$_4$Sr
Strontium Chromate(VI), 9001
CrO$_5$Pb$_2$
Lead Chromate(VI) Oxide, 5423
Cr$_2$CuO$_4$
Cupric Chromite, 2701
Cr$_2$FeH$_4$NO$_8$
Ammonium Ferric Chromate, 546
Cr$_2$H$_8$N$_2$O$_7$
Ammonium Dichromate(VI), 544
Cr$_2$HgO$_7$
Mercuric Dichromate(VI), 5929
Cr$_2$K$_2$O$_7$
Potassium Dichromate(VI), 7790
Cr$_2$Li$_2$O$_7$
Lithium Dichromate(VI), 5556
Cr$_2$Na$_2$O$_7$
Sodium Dichromate(VI), 8754
Cr$_2$O$_3$
Chromic Oxide, 2283
Cr$_2$O$_8$Sn
Stannic Chromate(VI), 8930
Cr$_2$O$_{12}$S$_3$
Chromic Sulfate, 2287
Cr$_3$Fe$_2$O$_{12}$
Ferric Chromate(VI), 4062
CsHO
Cesium Hydroxide, 2055
CsI
Cesium Iodide, 2056
CsNO$_3$
Cesium Nitrate, 2057
Cs$_2$O$_4$S
Cesium Sulfate, 2058
CuF$_2$
Cupric Fluoride, 2704
CuF$_6$Si
Cupric Hexafluorosilicate, 2708
CuFeS$_2$
Chalcopyrite, 2079
CuH$_2$O$_2$
Cupric Hydroxide, 2709
CuH$_{12}$N$_4$O$_4$S
Tetraamminecopper Sulfate, 9323
CuI
Cuprous Iodide, 2732
CuN$_2$O$_6$
Cupric Nitrate, 2710
CuO
Cupric Oxide, 2713
CuO$_3$Se
Cupric Selenite, 2720
CuO$_4$S
Cupric Sulfate, 2722

Denotes derivative of title compound

CuO₄Se
Cupric Selenate, 2718
CuO₄W
Cupric Tungstate(VI), 2726
CuS
Cupric Sulfide, 2724
CuSe
Cupric Selenide, 2719
Cu₂HgI₄
Cuprous Mercuric Iodide, 2733
Cu₂O
Cuprous Oxide, 2734
Cu₂O₃S
Cuprous Sulfite, 2738
Cu₂S
Cuprous Sulfide, 2737
Cu₂Se
Cuprous Selenide, 2736
Cu₃H₄O₈S
Cupric Sulfate, Basic*, 2723
Cu₃O₆S₂
Cuprous Sulfite*, 2738
Cu₃O₈P₂
Cupric Phosphate, 2716
Cu₄H₆O₁₀S
Cupric Sulfate, Basic*, 2723

D

D₂O
Deuterium Oxide, 2984
Dy₂O₃
Dysprosium*, 3531

E

Er₂O₃
Erbium*, 3679
Er₂O₁₂S₃
Erbium Sulfate, 3680
Eu₂O₃
Europium*, 3951

F

FH
Hydrogen Fluoride, 4837
FHO₃S
Fluorosulfonic Acid, 4217
FH₄N
Ammonium Fluoride, 553
FK
Potassium Fluoride, 7794
FLi
Lithium Fluoride, 5557
FNO
Nitrosyl Fluoride, 6745
FNO₂
Nitryl Fluoride, 6756
FNO₃
Fluorine Nitrate, 4202
FNa
Sodium Fluoride, 8762
FTl
Thallium Fluoride, 9397
F₂Fe
Ferrous Fluoride, 4093
F₂HK
Potassium Bifluoride, 7771

F₂HNa
Sodium Bifluoride, 8727
F₂H₅N
Ammonium Bifluoride, 523
F₂Hg
Mercuric Fluoride, 5930
F₂Hg₂
Mercurous Fluoride, 5952
F₂Kr
Krypton*, 5335
F₂Mg
Magnesium Fluoride, 5701
F₂Mn
Manganese Difluoride, 5769
F₂Ni
Nickel Fluoride, 6591
F₂O
Fluorine Monoxide, 4201
F₂OS
Thionyl Fluoride, 9486
F₂OSe
Selenium Oxyfluoride, 8579
F₂O₂
Fluorine Dioxide, 4200
F₂O₂S
Sulfuryl Fluoride, 9154
F₂Pb
Lead Fluoride, 5425
F₂Sn
Stannous Fluoride, 8940
F₂Sr
Strontium Fluoride, 9002
F₂Xe
Xenon*, 10206
F₂Zn
Zinc Fluoride, 10265
F₃Fe
Ferric Fluoride, 4065
F₃Ga
Gallium Trifluoride, 4367
F₃In
Indium Trifluoride, 4990
F₃Mn
Manganese Trifluoride, 5784
F₃N
Nitrogen Fluoride, 6701
F₃P
Phosphorus Trifluoride, 7516
F₃Sb
Antimony Trifluoride, 750
F₃Tl
Thallium Trifluoride, 9407
F₃V
Vanadium Trifluoride, 10058
F₄Ge
Germanium Tetrafluoride, 4419
F₄Pb
Lead Tetrafluoride, 5448
F₄S
Sulfur Tetrafluoride, 9151
F₄Se
Selenium Tetrafluoride, 8583
F₄Si
Silicon Tetrafluoride, 8645
F₄Sn
Stannic Fluoride, 8931
F₄Ti
Titanium Tetrafluoride, 9619
F₄U
Uranium Tetrafluoride, 9995
F₄V
Vanadium Tetrafluoride, 10057
F₄Xe
Xenon*, 10206
F₄Zr
Zirconium Fluoride, 10308
F₅H₁₂N₃O₂U
Ammonium Uranium Fluoride, 603
F₅I
Iodine Pentafluoride, 5039

F₅Nb
Niobium Pentafluoride, 6650
F₅P
Phosphorus Pentafluoride, 7509
F₅Sb
Antimony Pentafluoride, 737
F₅Ta
Tantalum Pentafluoride, 9225
F₅V
Vanadium Pentafluoride, 10055
F₆GaH₁₂N₃
Ammonium Hexafluorogallate, 556
F₆H₂Si
Fluosilicic Acid, 4220
F₆H₄NP
Ammonium Hexafluorophosphate, 557
F₆H₈N₂Si
Ammonium Hexafluorosilicate, 558
F₆Ir
Iridium Hexafluoride, 5099
F₆K₂Mn
Potassium Hexafluoromanganate-(IV), 7802
F₆K₂Si
Potassium Hexafluorosilicate, 7803
F₆K₂Zr
Potassium Hexafluorozirconate(IV), 7804
F₆MgSi
Magnesium Hexafluorosilicate, 5704
F₆Mo
Molybdenum Hexafluoride, 6319
F₆Na₂Si
Sodium Hexafluorosilicate, 8769
F₆Os
Osmium Hexafluoride, 7022
F₆PbSi
Lead Hexafluorosilicate, 5427
F₆Re
Rhenium Hexafluoride, 8345
F₆S
Sulfur Hexafluoride, 9146
F₆Se
Selenium Hexafluoride, 8575
F₆SiZn
Zinc Hexafluorosilicate, 10267
F₆SnZr
Stannous Hexafluorozirconate(IV), 8941
F₆Te
Tellurium Hexafluoride, 9277
F₆U
Uranium Hexafluoride, 9992
F₆W
Tungsten Hexafluoride, 9946
F₆Xe
Xenon*, 10206
F₇I
Iodine Heptafluoride, 5036
F₇K₂NbO
Niobium Potassium Oxypentafluoride, 6652
FeHO₂
Ferric Hydroxide, 4068
FeH₂O₂
Ferrous Hydroxide, 4096
FeH₄NO₈S₂
Ammonium Ferric Sulfate, 549
FeH₆O₆P₃
Ferric Hypophosphite, 4069
FeH₈N₂O₈S₂
Ammonium Ferrous Sulfate, 552
FeI₂
Ferrous Iodide, 4097
FeN₃O₉
Ferric Nitrate, 4071
FeO
Ferrous Oxide, 4100

Denotes derivative of title compound

FeO₄P
Ferric Phosphate, 4074
FeO₄S
Ferrous Sulfate, 4105
FeS
Ferrous Sulfide, 4106
FeSe
Ferrous Selenide, 4103
Fe₂O₃
Ferric Oxide, 4072
Fe₂O₁₂S₃
Ferric Sulfate, 4079
Fe₂P
Ferrous Phosphide, 4102
Fe₃O₄
Ferrosoferric Oxide, 4087
Fe₃O₈P₂
Ferrous Phosphate, 4101
Fe₄Na₈O₃₅P₁₀
Ferric Sodium Pyrophosphate, 4077
Fe₄O₂₁P₆
Ferric Pyrophosphate, 4075
Fe₅O₁₂Y₃
Yig, 10231

G

GaN₃O₉
Gallium*, 4364
GaP
Gallium Phosphide, 4366
Gd₂O₃
Gadolinium*, 4346
GeH₄
Germane, 4414
GeMg₂
Magnesium Germanide, 5703
GeO₂
Germanium Dioxide, 4417

H

HI
Hydrogen Iodide, 4838
HIO₃
Iodic Acid, 5032
HKO
Potassium Hydroxide, 7806
HKO₄S
Potassium Bisulfate, 7774
HKS
Potassium Bisulfide, 7775
HK₂O₃P
Potassium Phosphite, 7831
HK₂O₄P
Potassium Phosphate, Dibasic, 7828
HLi
Lithium Hydride, 5559
HLiO
Lithium Hydroxide, 5560
HMgO₄P
Magnesium Phosphate, Dibasic, 5718
HMnO₄P
Manganese Phosphate, Dibasic, 5777
HNO₂
Nitrous Acid, 6750
HNO₃
Nitric Acid, 6671
HNO₅S
Nitrosylsulfuric Acid, 6746
HN₃
Hydrazoic Acid, 4815

HNa
Sodium Hydride, 8770
HNaO
Sodium Hydroxide, 8772
HNaO₃S
Sodium Bisulfite, 8731
HNaO₄S
Sodium Bisulfate, 8729
HNaS
Sodium Bisulfide, 8730
HNa₂O₃P
Sodium Phosphite, 8809
HNa₂O₄P
Sodium Phosphate, Dibasic, 8805
HORb
Rubidium Hydroxide, 8442
HOTl
Thallium Hydroxide, 9398
HO₆PU
Uranyl Phosphate, 10001
H₂KO₂P
Potassium Hypophosphite, 7807
H₂KO₄P
Potassium Phosphate, Monobasic, 7829
H₂LiN
Lithium Amide, 5546
H₂Mg
Magnesium Hydride, 5705
H₂MgO₂
Magnesium Hydroxide, 5706
H₂MgO₈S₂
Magnesium Bisulfate, 5692
H₂MoO₄
Molybdic(VI) Acid, 6322
H₂NNa
Sodium Amide, 8717
H₂N₂O₂
Nitramide, 6663
H₂NaO₂P
Sodium Hypophosphite, 8775
H₂NaO₄P
Sodium Phosphate, Monobasic, 8806
H₂Na₂O₆P₂
Disodium Dihydrogen Hypophosphate, 3420
H₂Na₂O₇P₂
Sodium Acid Pyrophosphate, 8713
H₂NiO₂
Nickel Hydroxide, 6593
H₂O
Water, 10175
H₂O₂
Hydrogen Peroxide, 4839
H₂O₂Sr
Strontium Hydroxide, 9004
H₂O₃Se
Selenious Acid, 8571
H₂O₃Te
Tellurous Acid, 9281
H₂O₄Pb₃
Lead Hydroxide, 5428
H₂O₄S
Sulfuric Acid, 9147
H₂O₄Se
Selenic Acid, 8570
H₂O₄W
Tungstic(VI) Acid, 9948
H₂O₅S
Caro's Acid, 1900
H₂O₇S₂
Pyrosulfuric Acid, 8192
H₂S
Hydrogen Sulfide, 4843
H₂Se
Hydrogen Selenide, 4842
H₂Te
Hydrogen Telluride, 4844

H₂Ti
Titanium Hydride, 9613
H₃N
Ammonia, 517
H₃NO
Hydroxylamine, 4874
H₃NO₃S
Sulfamic Acid, 9090
H₃O₂P
Hypophosphorous Acid, 4916
H₃O₃P
Phosphorous Acid, 7502
H₃O₄P
Phosphoric Acid, 7500
H₃P
Phosphine, 7494
H₃Sb
Stibine, 8965
H₄IN
Ammonium Iodide, 560
H₄IP
Phosphonium Iodide, 7499
H₄MgN₂
Magnesium Amide, 5690
H₄MgO₈P₂
Magnesium Phosphate, Monobasic, 5719
H₄MnO₄P₂
Manganese Hypophosphite, 5771
H₄NO₃V
Ammonium Vanadate(V), 605
H₄N₂
Hydrazine, 4809
H₄N₂O₂S
Sulfamide, 9091
H₄N₂O₃
Ammonium Nitrate, 567
H₄O₄P₂Pb
Lead Hypophosphite, 5429
H₄O₄Zr
Zirconium Hydroxide, 10310
H₄O₆P₂
Hypophosphoric Acid, 4915
H₄O₇P₂
Pyrophosphoric Acid, 8191
H₄O₄₀SiW₁₂
Silicotungstic Acid, 8646
H₄Si
Silane, 8633
H₅IO₆
Periodic Acid, 7312
H₅NNaO₄P
Ammonium Sodium Phosphate, 587
H₅NO₃S
Ammonium Bisulfite, 528
H₅NO₄S
Ammonium Bisulfate, 526
H₅NS
Ammonium Bisulfide, 527
H₆MgNO₅P.6H₂O
Struvite, 9019
H₆NO₂P
Ammonium Hypophosphite, 559
H₆NO₄P
Ammonium Phosphate, Monobasic, 577
H₆N₂O
Hydrazine Hydrate, 4810
H₆N₂O₃S
Ammonium Sulfamate, 589
H₆N₂O₄S
Hydrazine Sulfate, 4811
H₆O₆Te
Telluric(VI) Acid, 9272
H₆Si₂
Disilane, 3419
H₈N₂NiO₈S₂
Ammonium Nickel Sulfate, 566
H₈N₂O₃S
Ammonium Sulfite, 594

MgO₂
Magnesium Peroxide, 5717
MgO₃S
Magnesium Sulfite, 5732
MgO₃S₂
Magnesium Thiosulfate, 5734
MgO₃Se
Magnesium Selenite, 5726
MgO₃Si
Magnesium Silicates*, 5727
MgO₄S
Magnesium Sulfate, 5731
MgO₄Se
Magnesium Selenate, 5724
MgSe
Magnesium Selenide, 5725
Mg₂O₄Si
Magnesium Silicates*, 5727
Mg₂O₇P₂
Magnesium Pyrophosphate, 5722
Mg₂O₈Si₃
Magnesium Silicates*, 5727
Mg₂Si
Magnesium Silicide, 5728
Mg₂Sn
Magnesium Stannide, 5729
Mg₃O₇Si₂
Magnesium Silicates*, 5727
Mg₃O₈P₂
Magnesium Phosphate, Tribasic, 5720
MnN₂O₆
Manganese Nitrate, 5773
MnNaO₄
Sodium Permanganate, 8799
MnO₂
Manganese Dioxide, 5770
MnO₄S
Manganese Sulfate, 5782
MnO₇P₂
Manganese Pyrophosphate, 5778
MnS
Manganese Sulfide, 5783
MnSe
Manganese Selenide, 5779
Mn₂O₃
Manganese Sesquioxide, 5780
Mn₂O₈Zn
Zinc Permanganate, 10281
Mn₃O₄
Manganese Oxide, 5776
MoNa₂O₄
Sodium Molybdate(VI), 8790
MoO₃
Molybdenum Trioxide, 6321
MoO₄Pb
Lead Molybdate(VI), 5432
MoS₂
Molybdenum Disulfide, 6318
Mo₂O₃
Molybdenum Sesquioxide, 6320
Mo₁₂Na₃O₄₀P
Sodium Phosphomolybdate, 8810

N

NNaO₂
Sodium Nitrite, 8793
NNaO₃
Sodium Nitrate, 8792
NO
Nitric Oxide, 6674
NO₂
Nitrogen Dioxide, 6700
NO₃Tl
Thallium Nitrate, 9400

N₂NiO₆
Nickel Nitrate, 6596
N₂O
Nitrous Oxide, 6751
N₂O₄Zn
Zinc Nitrite, 10275
N₂O₅
Nitrogen Pentoxide, 6702
N₂O₆Pb
Lead Nitrate, 5434
N₂O₆Pd
Palladium Nitrate, 7124
N₂O₆Sr
Strontium Nitrate, 9007
N₂O₆Zn
Zinc Nitrate, 10273
N₂O₈U
Uranyl Nitrate, 10000
N₂Zn₃
Zinc Nitride, 10274
N₃Na
Sodium Azide, 8724
N₄O₁₂Th
Thorium Nitrate, 9517
N₄O₁₂Zr
Zirconium Nitrate, 10312
N₄S₄
Tetrasulfur Tetranitride, 9378
N₄Se₄
Nitrogen Selenide, 6703
N₄Si₃
Silicon Nitride, 8641
N₆Pb
Lead Azide, 5415
NaO₃V
Sodium Vanadate(V), 8849
NaO₄⁹⁹ᵐTc
Sodium Pertechnetate ⁹⁹ᵐTc, 8802
Na₂O
Sodium Oxide, 8796
Na₂O₂
Sodium Peroxide, 8800
Na₂O₃Pb
Sodium Plumbate(IV), 8812
Na₂O₃S
Sodium Sulfite, 8831
Na₂O₃S₂
Sodium Thiosulfate, 8844
Na₂O₃Se
Sodium Selenite, 8822
Na₂O₃Si
Sodium Metasilicate, 8788
Na₂O₃Sn
Sodium Stannate(IV), 8826
Na₂O₃Te
Sodium Tellurate(IV), 8834
Na₂O₄S
Sodium Sulfate, 8829
Na₂O₄S₂
Sodium Hydrosulfite, 8771
Na₂O₄Se
Sodium Selenate, 8820
Na₂O₄Te
Sodium Tellurate(VI), 8835
Na₂O₄W
Sodium Tungstate(VI), 8847
Na₂O₅S₂
Sodium Metabisulfite, 8784
Na₂O₆S₂
Sodium Dithionate, 8756
Na₂O₇U₂
Sodium Uranate(VI), 8848
Na₂O₈S₂
Sodium Persulfate, 8801
Na₂S
Sodium Sulfide, 8830
Na₂Se
Sodium Selenide, 8821
Na₃O₃PS
Sodium Thiophosphate, 8843

Na₃O₄P
Sodium Phosphate, Tribasic, 8808
Na₃O₉P₃
Sodium Trimetaphosphate, 8845
Na₃S₄Sb
Sodium Thioantimonate(V), 8840
Na₄O₇P₂
Tetrasodium Pyrophosphate, 9377
Na₄O₉PbS₆
Lead Sodium Thiosulfate, 5441
Na₅O₁₀P₃
Sodium Tripolyphosphate, 8846
Nb₂O₅
Niobium Pentoxide, 6651
NiO
Nickel Monoxide, 6595
NiO₄S
Nickel Sulfate, 6600
Ni₂O₃
Nickel Sesquioxide, 6599
Ni₃O₈P₂
Nickel Phosphate, 6598

O

OPb
Lead Monoxide, 5433
OPd
Palladium Oxide, 7125
OSi
Silicon Monoxide, 8640
OSn
Stannous Oxide, 8944
OSr
Strontium Oxide, 9009
OTl₂
Thallium Oxide, 9401
OZn
Zinc Oxide, 10279
O₂Pb
Lead Dioxide, 5424
O₂Po
Polonium Dioxide, 7719
O₂Pt
Platinic Oxide, 7686
O₂S
Sulfur Dioxide, 9144
O₂Se
Selenium Oxide, 8576
O₂Si
Silicon Dioxide, 8637
O₂Sn
Stannic Oxide, 8933
O₂Sr
Strontium Peroxide, 9010
O₂Te
Tellurium Dioxide, 9276
O₂Th
Thorium Oxide, 9518
O₂Ti
Titanium Dioxide, 9612
O₂U
Uranium Dioxide, 9991
O₂Zn
Zinc Peroxide, 10282
O₂Zr
Zirconium Oxide, 10313
O₃P₂
Phosphorus Trioxide, 7517
O₃PbSe
Lead Selenite, 5439
O₃Pb₂
Lead Sesquioxide, 5440
O₃Re
Rhenium Trioxide, 8347
O₃S
Sulfur Trioxide, 9152

O_3Sb_2
Antimony Trioxide, 752
O_3Sc_2
Scandium*, 8537
O_3Tb_2
Terbium*, 9300
O_3Tl_2
Thallium Sesquioxide, 9404
O_3Tm_2
Thulium*, 9535
O_3U
Uranium Trioxide, 9997
O_3V_2
Vanadium Trioxide, 10059
O_3W
Tungsten Trioxide, 9947
O_3Xe
Xenon*, 10206
O_3Y_2
Yttrium*, 10240
O_3Yb_2
Ytterbium*, 10239
O_4Os
Osmium Tetroxide, 7024
O_4PbS
Lead Sulfate, 5444
O_4PbSe
Lead Selenate, 5438
O_4Pb_3
Lead Tetroxide, 5449
O_4Ru
Ruthenium Tetroxide, 8454
O_4SSn
Stannous Sulfate, 8947
O_4SSr
Strontium Sulfate, 9013
O_4STl_2
Thallium Sulfate, 9405
O_4SZn
Zinc Sulfate, 10293
O_4SeSr
Strontium Selenate, 9012
O_4SeTl_2
Thallium Selenate, 9402
O_4SeZn
Zinc Selenate, 10289
O_4SiZn_2
Zinc Silicate, 10291
O_4SiZr
Zirconium Silicate, 10314
O_4U
Uranium Peroxide, 9993

O_5P_2
Phosphorus Pentoxide, 7512
O_5STi
Titanium Sulfate, 9616
O_5SV
Vanadyl Sulfate, 10063
O_5Sb_2
Antimony Pentoxide, 739
O_5Ta_2
Tantalum Pentoxide, 9226
O_5V_2
Vanadium Pentoxide, 10056
O_6PbV_2
Lead Vanadate(V), 5451
O_6SU
Uranyl Sulfate, 10002
O_6Se_2Sn
Stannic Selenite, 8935
$O_7P_2Sn_2$
Stannous Pyrophosphate, 8945
$O_7P_2Zn_2$
Zinc Pyrophosphate, 10287
O_7Re_2
Rhenium Heptoxide, 8344
$O_8P_2Pb_3$
Lead Phosphate, 5437
$O_8P_2Sr_3$
Strontium Phosphate, Tribasic, 9011
$O_8P_2Zn_3$
Zinc Phosphate, 10284
O_8S_2Th
Thorium Sulfate, 9519
O_8S_2Zr
Zirconium Sulfate, 10315
$O_{12}S_3Sb_2$
Antimony Sulfate, 745
$O_{12}S_3Ti_2$
Titanium Sesquisulfate, 9615
$O_{12}S_3V_2$
Vanadium Trisulfate, 10060

P

P_2S_5
Phosphorus Pentasulfide, 7511
P_2Se_3
Phosphorus Hemitriselenide, 7504
P_2Se_5
Phosphorus Pentaselenide, 7510

P_2Zn_3
Zinc Phosphide, 10285
P_4S_3
Tetraphosphorus Trisulfide, 9375
P_4Se_3
Phosphorus Triselenide, 7518
PbS
Lead Sulfide, 5445
$PbTe$
Lead Telluride, 5446

S

SSn
Stannous Sulfide, 8948
SSr
Strontium Sulfide, 9014
STl_2
Thallium Sulfide, 9406
SZn
Zinc Sulfide, 10294
S_2Se
Selenium Sulfides*, 8580
S_2Si
Silicon Disulfide, 8638
S_2Sn
Stannic Sulfide, 8936
S_3Sb_2
Antimony Trisulfide, 754
S_3V_2
Vanadium Trisulfide, 10061
S_5Sb_2
Antimony Pentasulfide, 738
Sb_2Se_3
Antimony Triselenide, 753
$SeSn$
Stannous Selenide, 8946
$SeTl_2$
Thallium Selenide, 9403
$SeZn$
Zinc Selenide, 10290
Se_2Sn
Stannic Selenide, 8934

T

$TeZn$
Zinc Telluride, 10297

Denotes derivative of title compound

NAME INDEX

NAME INDEX

The Name Index refers the reader to monograph numbers with three types of citations:

Ethacrynic Acid, 3643 (format for titles)
Ethacrynate Sodium *see* 3643 (format for all synonyms)

Company name included in citation denotes trademark owner or its subsidiary, affiliate, licensee or distributor.

Edecrin [Merck & Co.] *see* 3643

1C50 *see* 3875
02-115 *see* 9208
2-5410-3A *see* 5044
3PAA *see* 8154
3TC *see* 5365
4A65 *see* 4951
4-AP *see* 3974
4-C-32 *see* 9569
6-12 [Union Carbide] *see* 3790
6-APA *see* 478
6-H.E.S. [Morishita] *see* 4707
8-AL *see* 6581
27-400 *see* 2821
51W89 *see* 900
53-32 C *see* 9569
67/20 CDRI *see* 2018
84L *see* 3165
96-H-60 *see* 4636
101-E *see* 6490
101-G *see* 6490
106-7 *see* 2820
191C49 *see* 9429
194-B *see* 10319
217-MI *see* 3549
256U *see* 10039
295C51 *see* 9877
338C48 *see* 3309
374-JL *see* 6815
378C48 *see* 3398
501-P *see* 4471
516-MD *see* 2365
566C80 *see* 898
611C55 *see* 9414
683-M *see* 7078
688-A *see* 7409
710-F *see* 7680
722 D *see* 7662
746-CE *see* 3666
00836 *see* 6026
864T *see* 4962
921 C *see* 9794
1080 [Monsanto] *see* 4205
1081 *see* 4204
1162-F *see* 9094
1314-Th *see* 3783
1358F *see* 2885
1399F *see* 20
1489-RB *see* 7613
1497-CB *see* 8014
1522-CB *see* 32
1589mRB *see* 7197
1589RB *see* 7197
1600 Antibiotic *see* 7173
1609-RB *see* 7642
1633 Labaz *see* 8753
1665-RB *see* 809
1703-18B *see* 4900
1766 *see* 2234
2249-F *see* 7056
2329 Labaz *see* 1113
2601-A *see* 2238
3123-L *see* 8130
4114 Th *see* 7387
5512-M *see* 718
6063 *see* 50
6315-S *see* 4141
7432-S *see* 1998
8053CB *see* 9436

8102 CB *see* 982
10040 *see* 2140
10275-S *see* 3662
10364-S *see* 5904
12494 Hoechst *see* 4032
13832-JL *see* 10211
29060-LE *see* 10119
47657 *see* 792
81723 hfu *see* 9559
450191-S *see* 8387
710674-S *see* 2429
A *see* 205
A-16 *see* 403
A-21 *see* 5236
A-41-304 *see* 2975
A-65 *see* 7653
A-66 *see* 7385
A-101 *see* 6784
A-118 *see* 9170
A-124 *see* 8888
A-145 *see* 7966
A-272 *see* 6970
A-300-I *see* 9781
A-350 *see* 455
A-363 *see* 452
A-585 *see* 9169
A-820 *see* 1567
A-3823A *see* 6329
A-4696 *see* 134
A-4942 *see* 4937
A-5610 *see* 939
A-8103 *see* 7634
A-9145 *see* 8690
A-19120 *see* 7172
A-20968 *see* 7635
A-23187 *see* 1678
A-29622 *see* 4277
A-33547 *see* 8301
A-35957 *see* 327
A-38414 *see* 671
A-43818 *see* 5484
A-46745 *see* 4423
A-56268 *see* 2400
A-56619 *see* 3187
A-56620 *see* 8517
A-60386X *see* 1189
A-60969 *see* 9692
A-61827 *see* 9692
A-62254 *see* 9284
A-63004 *see* 9284
A-64077 *see* 10253
A-64730 *see* 9692
A-65006 *see* 5373
A-70569 *see* 9557
A-84538 *see* 8402
α-0817185 *see* 8386
Aα *see* 4051
AA-149 *see* 9718
AA-673 *see* 515
AA-2414 *see* 8603
Aa-5648 *see* 9130
Aacifemine [Aaciphar] *see* 3750
AAF *see* 4192
AA-protein *see* 657
Aarane [Syntex] *see* 2658
AAT *see* 760
A1AT *see* 760
AAtrex [Geigy] *see* 902

AB-100 *see* 10011
AB-103 *see* 1118
AB-132 *see* 6245
AB-206 *see* 6283
ABA *see* 7
AB-A-663 *see* 2336
Abacin [Benedetti] *see* 9086
Abadol [Specia] *see* 501
Abalyn *see* 6087
Abamectin, 1
Abapresin [Polfa] *see* 4589
Abar [Velsicol] *see* 5468
Abasin [Bayer] *see* 18
Abate [Am. Cyanamid] *see* 9286
Abbocillin-DC [Abbott] *see* 7226
Abbocillin V [Abbott] *see* 7232
Abbocin [Abbott] *see* 7111
Abboflox [Abbott] *see* 4147
Abbokinase [Abbott] *see* 10024
Abbolexin [Abbott] *see* 7474
Abboticine [Abbott] *see* 3726
Abbott 16900 *see* 9265
Abbott 30400 *see* 7206
Abbott 35616 *see* 2465
Abbott 36581 *see* 1539
Abbott 38414 *see* 671
Abbott 43326 *see* 1917
Abbott 43818 *see* 5484
Abbott 44747 *see* 4276
Abbott 45975 *see* 9297
Abbott 46811 *see* 1994
Abbott 48999 *see* 1985
Abbott 50192 *see* 1976
Abbott 50711 *see* 10049
Abbott 56620 *see* 3187
Abbott 56620 *see* 8517
Abbott 64077 *see* 10253
Abbott 73001 *see* 8603
Abbott 84538 *see* 8402
Abboxapam [Abbott] *see* 7059
Abbsa [Sanko] *see* 9660
ABC-12/3 *see* 3494
ABDP *see* 228
Abecarnil, 2
Abenox [Abbott] *see* 3625
Abensanil *see* 45
Abequito [Am. Cyanamid] *see* 6642
Aberel [Cilag] *see* 8333
ABG-6215 *see* 4027
Abicoviromycin *see* 4
Δ6,8(14)-Abietadienoic Acid *see* 5494
Abietic Acid, 3
Abietic Anhydride *see* 8424
Abietin *see* 2567
Abikoviromycin, 4
Abil [Goldschmidt Chem.] *see* 8639
Abilit [Sumitomo] *see* 9163
Abimasten 100 [ABIC] *see* 6625
Abiocine [Lepetit] *see* 3222
Abitilguanide *see* 6354
Abitylguanide *see* 6354
Abminthic [Pfizer] *see* 3434
ABOB *see* 6354
Abomacetin [Mochida] *see* 3720
Abomasal Enzyme *see* 8303
Aboren [Promeco] *see* 6271
Abor K [Gellini] *see* 5874
Abovis [Toyama] *see* 116

Abramen [Mepha] *see* 401
Abrastol *see* 859
Abricycline *see* 9337
Abrifam [Abbott] *see* 8382
Abrin, 5
Abrine, 6
Abrodil [Schering] *see* 6051
Abrus Agglutinin *see* 5
Abscisic Acid, 7
Abscisin I *see* 7
Abscisin II *see* 7
Absentol [Nourypharma] *see* 9836
Absinthe *see* 9
Absinthiin *see* 8
Absinthin, 8
Absinthium, 9
Absolute Alcohol *see* 3806
Absonal V [Abbott] *see* 1144
Absorbable Cellulose *see* 7073
Abstem *see* 1703
Abstensil *see* 3428
Abstinyl [Pharmacia] *see* 3428
Absynthin *see* 8
ABT-538 *see* 8402
Abufène [Théraplix] *see* 206
Abuphenine *see* 1551
Aburamycin B *see* 2295
AC-17 *see* 1833
AC-223 *see* 5864
AC-293 *see* 4941
AC-528 *see* 3355
AC-601 *see* 1525
AC-1198 *see* 3260
AC-1370 *see* 1988
AC-1802 *see* 793
AC-3422 *see* 7167
AC-3810 *see* 982
AC-3911 *see* 7486
AC-4464 *see* 9690
AC-5223 *see* 3468
AC-38555 *see* 2153
AC-47031 *see* 7490
AC-47300 *see* 4017
AC-52160 *see* 9286
AC-64475 *see* 4284
AC-84633 *see* 6642
AC-84777 *see* 3185
AC-92100 *see* 9301
AC-92553 *see* 7211
AC-217300 *see* 4802
AC-222293 *see* 4941
AC-222705 *see* 4160
AC-252214 *see* 4943
AC-252925 *see* 4944
AC-263499 *see* 4944
AC-263780 *see* 2336
7-ACA *see* 453
Acabel [Grünenthal] *see* 1239
Acacetin, 10
Acacetin-β-rutinoside *see* 5523
Acacia, 11
Acacic Acid, 12
Acaciin *see* 10
Acadesine, 13
Acadione [Cassenne] *see* 9597
Acamol [Norgine] *see* 45
Acamprosate Calcium, 14
Acamylophenine *see* 1784
Acaprin *see* 3416
Acaraben [Ciba-Geigy] *see* 2174
Acarbose, 15
Acarflor [Sipcam] *see* 4751
Acarin [Makhteshim-Agan] *see* 3136
Acarol [Ciba-Geigy] *see* 1456
Accelerase [Organon] *see* 7138
Accelerator Globulin *see* 3961
Accelerin *see* 3961
Accelerine *see* 6736
Accent *see* 6337
Acclaim [Hoechst] *see* 4024

Accolate [ICI] *see* 10241
Accothion [Am. Cyanamid] *see* 4017
Accroibile [Théraplix] *see* 2327
Accupril [Parke-Davis] *see* 8233
Accuprin [Parke-Davis] *see* 8233
Accupro [Parke-Davis] *see* 8233
Accurbron [Merrell Dow] *see* 9421
Accuretic [Parke-Davis] *see* 8233
Accutane [Roche] *see* 8333
Acdeam [Grelan] *see* 5671
Acdrile *see* 5828
ACE *see* 169
Acebutolol, 16
Acecainide, 17
Acecarbromal, 18
Aceclidine *see* 8266
Aceclofenac, 19
Acecoline [Lematte et Boinot] *see* 88
Acecor [Squibb] *see* 4282
Acedapsone, 20
Acediamine Hydrochloride *see* 42
Acediasulfone, 21
Acedicon [Boehringer, Ing.] *see* 3215
Acedist [Chemiefarma] *see* 1440
Acediur [Guidotti] *see* 1817
Acef [Tiber] *see* 1967
Acefylline, 22
Acefylline Piperazine, 23
Aceglaton [Chugai] *see* 24
Aceglatone, 24
Aceglumide, 25
Aceglumide Aluminum *see* 25
Acemannan, 26
Acemetacin, 27
Acemethadone *see* 6009
Acemix [Proter] *see* 27
Acenalin [Janssen Kyowa] *see* 2377
Acenaphthene, 28
Acenit [Chemol] *see* 58
Acenocoumarin *see* 29
Acenocoumarol, 29
Acenterine *see* 886
Aceon [Ortho] *see* 7311
Aceperone, 30
Acephate, 31
Acephen *see* 5820
Acepifylline *see* 23
Aceplus [Squibb] *see* 1817
Acepress [Guidotti] *see* 1817
Acepreval *see* 7905
Acepril [Duncan, Flockart] *see* 1817
Acepromazine, 32
Acequide [Recordati] *see* 8233
Acequin [Recordati] *see* 8233
Acerbon [ICI] *see* 5540
Acerin, 33
Acerola, 34
Aceroxatidine *see* 8431
Acesistem [Sigma-Tau] *see* 3605
Acesulfame, 35
Acesulfame-K *see* 35
Acetabutone *see* 30
Acetadeps *see* 5371
Acetal, 36
Acetaldehyde, 37
Acetaldehyde Ammonia, 38
Acetaldehyde Diethyl Acetal *see* 36
Acetaldehyde Dimethyl Acetal *see* 3276
Acetaldehyde 2-(2-Ethoxyethoxy)ethyl 3,4-(Methylenedioxy)phenyl Acetal *see* 8615
Acetaldehyde Oxime *see* 40
Acetaldehyde Sodium Bisulfite, 39
Acetaldol *see* 225
Acetaldoxime, 40
Acetalgin [Streuli] *see* 45
Acetamide, 41
3-Acetamide-4*H*-1,3-benzoxazin-2-one *see* 1907

Acetamidine Hydrochloride, 42
3-Acetamido-5-(acetamidomethyl)-2,-4,6-triiodobenzoic Acid *see* 5031
Acetamidoacetic Acid *see* 81
p-Acetamidobenzaldehyde Thiosemicarbazone *see* 9427
p-Acetamidobenzenearsonic Acid *see* 829
4-Acetamidobenzenesulfonic Acid *see* 105
p-Acetamidobenzenesulfonyl Chloride *see* 106
ε-Acetamidocaproic Acid, 43
2-Acetamido-2-deoxy-3-*O*-(D-2-propionyl-L-alanyl-D-isoglutamine)-D-glucopyranose *see* 6385
(—)-(*S*)-2-Acetamido-*N*-(3,4-dihydroxyphenethyl)-4-(methylthio)-butyramide Bis(ethylcarbonate) (Ester) *see* 3457
5-Acetamido-2,4-dimethyltrifluoro-methanesulfonanilide *see* 5846
6-Acetamido-1,2-dithiolo[4,3-*b*]pyrrol-5(4*H*)-one *see* 4764
4-Acetamido-2-ethoxybenzoic Acid Methyl Ester *see* 3791
Acetamidoeugenol, 44
6-Acetamidohexanoic Acid *see* 43
3-Acetamido-4-hydroxybenzene-arsonic Acid *see* 49
3-Acetamido-4-hydroxydithiobenzene-arsonous Acid, Cyclic (Hydroxy-methyl)ethylene Ester *see* 852
3-Acetamido-4-hydroxydithiobenzene-arsonous Acid Cyclic 3-Hydroxy-propylene Ester *see* 852
3-Acetamido-4-hydroxyphenylarsonic Acid *see* 49
2-(3'-Acetamido-4'-hydroxyphenyl)-1,3-dithia-2-arsacyclopentane-4-methanol *see* 852
7-Acetamido-10-hydroxy-1,2,3-tri-methoxy-6,7-dihydrobenzo[*a*]hepta-len-9(5*H*)-one *see* 2535
2-Acetamido-4-mercaptobutyric Acid γ-Thiolactone *see* 2381
L-α-Acetamido-β-mercaptopropionic Acid *see* 89
N-Acetamidomethyl-2-chloro-2',6'-diethylacetanilide *see* 420
6-(Acetamido)-4-methyl-1,2-dithiolo-[4,3-*b*]pyrrol-5(4*H*)-one *see* 9478
4-Acetamido-2-methyl-1-naphthol *see* 10162
3-Acetamido-5-methylpyrrolin-4-one-[4,3-*d*]-1,2-dithiole *see* 9478
5-Acetamido-2-methyltrifluorometh-anesulfonanilide *see* 4198
4-Acetamido-1-naphthol *see* 473
2-Acetamido-5-nitrothiazole *see* 430
4-Acetamido-4-penten-3-one-1-carbox-amide *see* 7929
p-Acetamidophenol *see* 45
2-(4-Acetamidophenoxy)ethyl 2-Acetoxybenzoate *see* 3759
1-(4-Acetamidophenoxy)-3-isopropyl-amino-2-propanol *see* 7882
(*p*-Acetamidophenyl)acetic Acid *see* 135
4'-(Acetamido)phenyl-2-acetoxybenzo-ate *see* 1074
p-Acetamidophenyl Acetylsalicylate *see* 1074
4-Acetamidophenyl (Diethylamino)-acetate *see* 7976
p-Acetamidophenyl Salicylate *see* 46
1-Acetamido-2-pyrrolidinone *see* 7641
5-Acetamido-1,3,4-thiadiazole-2-sul-fonamide *see* 50

α-Acetamido-γ-thiobutyrolactone *see* 2381

3-Acetamido-2,4,6-triiodobenzoic Acid Methylglucamine Salt *see* 5850

3-Acetamido-2,4,6-triiodobenzoic Acid Sodium Salt *see* 79

2-[3-Acetamido-2,4,6-triiodo-5-(*N*-methylacetamido)benzamido]-2-deoxy-D-glucopyranose *see* 6240

2-[3-Acetamido-2,4,6-triiodo-5-(*N*-methylacetamido)benzamido]-2-deoxy-D-glucose *see* 6240

3-Acetamido-2,4,6-triiodo-5-(*N*-methylacetamido)benzoic Acid *see* 6241

5-Acetamido-2,4,6-triiodo-*N*-methyl-isophthalamic Acid *see* 5080

(±)-2-[[2-(3-Acetamido-2,4,6-triiodophenoxy)ethoxy]methyl]butyric Acid *see* 5077

4-Acetaminobenzenesulfon-4'-nitroanilide *see* 9103

p-Acetaminobenzenesulfonyl Chloride *see* 106

p-Acetaminobenzylidenethiosemicarbazone *see* 9427

Acetaminocaproic Acid *see* 43

Acetaminophen, 45

p-Acetaminophenol *see* 45

Acetaminosalol, 46

Acetamox [Tobishi-Santen] *see* 50

Acetanilide, 47

Acetanilide-*p*-sulfonyl Chloride *see* 106

Acetanilidsulfonic Acid *see* 105

p-Acetanisidide *see* 48

p-Acetanisidine, 48

Acetannin *see* 107

Acetanol [Rhône-Poulenc Rorer] *see* 16

Acetarsin *see* 49

Acetarsol *see* 49

Acetarsone, 49

Acetarsonic Acid Disodium Salt *see* 850

Acetasol *see* 757

Acetate C-9 *see* 6773

Acetate Replacing Factor *see* 9462

Acetatocobalamin *see* 4854

(Acetato-*O*)[2-hydroxy-5-(1,1,3,3-tetramethylbutyl)phenyl]mercury *see* 63

(Acetato)phenylmercury *see* 7453

(Acetato)trimetaarsenitodicopper *see* 2692

Acetazine *see* 32

Acetazolamide, 50

Acetexa [Lilly] *see* 6812

Acethropan [Hoechst] *see* 136

Acetiamine, 51

Acetic Acid Amide *see* 41

Acetic Acid Ammonium Salt *see* 520

Acetic Acid Anhydride with Nitric Acid *see* 99

Acetic Acid Basic Bismuth Salt *see* 1323

Acetic Acid Benzyl Ester *see* 1158

Acetic Acid Beryllium Salt *see* 1202

Acetic Acid Butyl Ester *see* 1570

Acetic Acid *sec*-Butyl Ester *see* 1571

Acetic Acid *tert*-Butyl Ester *see* 1572

Acetic Acid, Calcium Magnesium Salt *see* 1725

Acetic Acid Copper(1+) Salt *see* 2728

Acetic Acid *m*-Cresol Ester *see* 2651

Acetic Acid Dimethylamide *see* 3277

Acetic Acid 1,1-Dimethylethyl Ester *see* 1572

Acetic Acid Ethenyl Ester *see* 10130

Acetic Acid Ethyl Ester *see* 3803

Acetic Acid 2-Ethylhexyl Ester *see* 6860

Acetic Acid Glacial, 52

Acetic Acid 2-Methylphenyl Ester *see* 2652

Acetic Acid 3-Methylphenyl Ester *see* 2651

Acetic Acid 1-Methylpropyl Ester *see* 1571

Acetic Acid [(5-Nitro-2-furanyl)methylene]hydrazide *see* 6634

Acetic Acid 5-(Nitrofurfurylidene)hydrazide *see* 6634

Acetic Acid *n*-Nonyl Ester *see* 6773

Acetic Acid Oxime *see* 60

Acetic Acid Palladium Salt *see* 7123

Acetic Acid Phenyl Ester *see* 7421

Acetic Acid Phenylmethyl Ester *see* 1158

Acetic Acid α-Phenyl-2-piperidylmethyl Ester *see* 5493

Acetic Acid 2-Propenyl Ester *see* 293

Acetic Acid *n*-Propyl Ester *see* 8026

Acetic Acid Tetraanhydride with Silicic Acid *see* 8642

Acetic Acid *o*-Tolyl Ester *see* 2652

Acetic Acid Vinyl Ester *see* 10130

Acetic Aldehyde *see* 37

Acetic Anhydride, 53

Acetic Ether *see* 3803

Acetic Oxide *see* 53

Aceticyl *see* 886

Acetilarsano *see* 49

Acetimidoquinone, 54

(+)-(5*R*,6*S*)-3-[[(*S*)-1-Acetimidoyl-3-pyrrolidinyl]thio]-6-[(*R*)-1-hydroxy-ethyl]-7-oxo-1-azabicyclo[3.2.0]-hept-2-ene-2-carboxylic Acid *see* 7141

Acetin *see* 6325

Acétiodone *see* 79

Acetoacetanilide, 55

Acetoacetic Acid, 56

Acetoacetic Acid Ethyl Ester *see* 3804

Acetoacetic Anilide *see* 55

Acetoacetic Ester *see* 3804

Aceto-*m*-aminotoluene *see* 74

Acetobromamide *see* 1419

Acetobromglucose, 57

α-Acetobromoglucose *see* 57

Acetobuton *see* 30

α-Acetobutyrolactone *see* 84

Acetochlor, 58

Acetocinnamone *see* 1172

Acetohexamide, 59

Acetohydroxamic Acid, 60

Acetoin, 61

Acetol®, 62

Acetomenaphthone *see* 5873

Acetomeroctol, 63

Acetomethoxane *see* 3273

Acetomethylanilide *see* 6088

Acetomorphine *see* 3012

Acetone, 64

Acetone Bis(3,5-di-*tert*-butyl-4-hydroxyphenyl)mercaptole *see* 7935

Acetone-bromoform *see* 9741

Acetonecarboxylic Acid *see* 56

Acetone Chloroform *see* 2180

Acetone Cyanohydrin, 65

Acetonedicarboxylic Acid, 66

Acetone Glycerol *see* 5232

Acetone Oxime *see* 75

Acetone Sodium Bisulfite, 67

Acetone Sulfite *see* 67

Acetonitrile, 68

Acetonylacetone, 69

3-(α-Acetonylbenzyl)-4-hydroxy-coumarin *see* 10174

Acetonyl Chloride *see* 2164

3-(α-Acetonyl-*p*-chlorobenzyl)-4-hydroxycoumarin *see* 2623

3-Acetonyl-5,8-dihydroxy-2-(hydroxymethyl)-6-methoxy-1,4-naphthoquinone *see* 4337

3-Acetonyl-5,8-dihydroxy-6-methoxy-2-methyl-1,4-naphthoquinone *see* 5275

3-(α-Acetonylfurfuryl)-4-hydroxycoumarin *see* 2624

(±)-Acetonyl Methyl 1,4-Dihydro-2,6-dimethyl-4-(*o*-nitrophenyl)-3,5-pyridinedicarboxylate *see* 806

2-Acetonyl-1-methylpyrrolidine *see* 4899

3-(α-Acetonyl-*p*-nitrobenzyl)-4-hydroxycoumarin *see* 29

2-Acetonylpiperidine *see* 7200

Acetophen [Dorsey; Frosst] *see* 886

Acetophenazine, 70

p-Acetophenetide *see* 7344

p-Acetophenetidide *see* 7344

Acetophenetidin *see* 7344

Acetophenone, 71

Acetopromazine *see* 32

Acetopyrine *see* 757

Acetopyrrothine *see* 9478

Acetoquat CDAC *see* 2059

Acetorphan, 72

Acetorphine *see* 3932

Acetosal *see* 886

Acetosalic Acid *see* 886

Acetosalin *see* 886

Acetosulfam *see* 35

Acetosulfone Sodium, 73

Acetosulphone *see* 73

Acetotoluide, 74

m-Acetotoluide *see* 74

o-Acetotoluide *see* 74

p-Acetotoluide *see* 74

m-Acetotoluidide *see* 74

Acetovanillon *see* 782

Acetoxanon [Organon] *see* 77

Acetoxime, 75

Acetoxolone, 76

2-Acetoxy-4'-(acetamino)phenylbenzoate *see* 1074

9-[4-Acetoxy-3-(acetoxymethyl)but-1-yl]-2-aminopurine *see* 3971

5-Acetoxy-3-acetylthio-2-[(4-amino-2-methyl-5-pyrimidinyl)methyl-*N*-formylamino]-2-pentene *see* 51

2-Acetoxybenzoic Acid *see* 886

17α-Acetoxy-6-chloro-6,7-dehydro-progesterone *see* 2152

Acetoxy-4'-chloro-3,5-diiodobenzanilide *see* 2416

17α-Acetoxy-6-dehydro-16-methyleneprogesterone *see* 4878

2-Acetoxy-3-(*N*,*N*-diethylcarboxamido)-9,10-dimethoxy-1,2,3,4,6,7-hexahydro-11b*H*-benzopyridocoline *see* 1154

21-Acetoxy-6α,9α-difluoro-11β-hydroxy-16α,17α-isopropylidenedioxy-1,4-pregnadiene-3,20-dione *see* 4186

3-Acetoxy-6-dimethylamino-4,4-diphenylheptane *see* 6009

5-Acetoxy-2-dimethylamino-4,4-diphenylheptane *see* 6009

6-Acetoxy-2,4-dimethyl-*m*-dioxane *see* 3273

α-*dl*-3-Acetoxy-4,4-diphenyl-6-methylaminoheptane *see* 6778

cis-1-Acetoxy-7-dodecene *see* 5600

7β-Acetoxy-8,13-epoxy-1α,6β,9α-trihydroxylabd-14-en-11-one *see* 2539

17β-Acetoxyestra-4,9,11-trien-3-one *see* 9716

1-(2-Acetoxyethyl)-4-[3-(2-chloro-10-

Acetylcholine Bromide, 87
Acetylcholine Chloride, 88
Acetylcholinesterase see 2264
Acetyl-o-cresol see 2652
4-Acetyl-N-[(cyclohexylamino)carbon-
yl]benzenesulfonamide see 59
Acetylcysteine, 89
N-Acetyl-L-cysteine see 89
N-Acetyldeferoxamine see 2914
Acetyldemethyldihydrothebaine see
3215
N-Acetyldesferrioxamine B see 2914
cis-1-Acetyl-4-[4-[[2-(2,4-dichloro-
phenyl)-2-(1H-imidazol-1-ylmethyl)-
1,3-dioxolan-4-yl]methoxy]phenyl]-
piperazine see 5313
1-Acetyl-19,20-didehydro-17,18-
epoxycuran-17-ol see 3002
3^B-O-(4-O-Acetyl-2,6-dideoxy-3-C-
methyl-α-L-arabino-hexopyranosyl)-
7-methylolivomycin D see 2295
Acetyldigitoxins, 90
α-Acetyldigoxin see 3210
β-Acetyldigoxin see 3210
Acetyldihydrocodeinone see 3215
3-Acetyldihydro-2(3H)-furanone see
84
O^3-Acetyl-7,8-dihydro-7α-[1(R)-hydr-
oxy-1-methylbutyl]-O^6-methyl-6,14-
endo-ethenomorphine see 3932
3-Acetyl-1,5-dihydro-4-hydroxy-5-(1-
methylpropyl)-2H-pyrrol-2-one see
9294
5-[Acetyl(2,3-dihydroxypropyl)-
amino]-N,N'-bis(2,3-dihydroxyprop-
yl)-2,4,6-triiodo-1,3-benzenedicar-
boxamide see 5068
5-[Acetyl(2,3-dihydroxypropyl)-
amino]-N-(2,3-dihydroxypropyl)-N'-
(2-hydroxyethyl)-2,4,6-triiodo-1,3-
benzenedicarboxamide see 5085
O-Acetyl-6-dimethylamino-4,4-diphen-
yl-3-heptanol see 6009
2-Acetyl-10-(3-dimethylaminopropyl)-
phenothiazine see 32
3-Acetyl-10-(3-dimethylaminopropyl)-
phenothiazine see 32
N'-[3-Acetyl-4-[3-[(1,1-dimethylethyl)-
amino]-2-hydroxypropoxy]phenyl]-
N,N-diethylurea see 2007
Acetyldiphenylamine see 3373
Acetylene, 91
Acetylene Carbamide see 94
Acetylenecarboxylic Acid see 8006
Acetylene Dibromide, 92
Acetylenedicarboxamide see 2009
Acetylenedicarboxylic Acid Diamide
see 2009
Acetylene Dichloride, 93
Acetylenediureine see 94
Acetylene Tetrabromide see 9328
Acetylene Tetrachloride see 9331
Acetyleneurea, 94
Acetylenogen see 1696
2'-Acetylerythromycin Stearate see
3721
Acetyl Ethylene see 6214
Acetylformaldehyde see 8203
Acetylformic Acid see 8205
N^2-Acetyl-L-glutamine see 25
α-N-Acetyl-L-glutamine see 25
Acetyl Glyceryl Ether Phosphoryl-
choline see 7682
N-Acetylglycine see 81
N-Acetylglycine Compd with 4,4'-
(Diazoamino)dibenzamidine see
3316
N-Acetylglycine Compd with 4,4'-(1-
Triazene-1,3-diyl)bis(benzenecar-
boximidamide) (2:1) see 3316

Acetylglycocoll see 81
3-O-Acetyl-18β-glycyrrhetic Acid see
76
Acetylglycyrrhetinic Acid see 76
(1S,3S)-3-Acetyl-1,2,3,4,6,11-hexahy-
dro-3,5,10,12-tetrahydroxy-6,11-di-
oxo-1-naphthacenyl 3-Amino-2,3,6-
trideoxy-α-L-lyxo-hexopyranoside
see 1921
(1S,3S)-3-Acetyl-1,2,3,4,6,11-hexahy-
dro-3,5,12-trihydroxy-6,11-dioxo-1-
naphthacenyl-3-amino-2,3,6-tride-
oxy-α-L-lyxo-hexopyranoside see
4931
N-Acetylhomocysteinethiolactone see
2381
Acetyl Hydroperoxide see 7293
N'-[5-[[4-[[5-(Acetylhydroxyamino)-
pentyl]amino]-1,4-dioxobutyl]hydr-
oxyamino]pentyl]-N-(5-aminopent-
yl)-N-hydroxybutanediamide see
2914
N-Acetyl-4-hydroxy-m-arsanilic Acid
see 49
N-Acetyl-2-hydroxybenzamide see
8471
α-Acetyl-γ-hydroxybutyric Acid γ-
Lactone see 84
17β-Acetyl-17-hydroxyestr-4-ene-3-
one Hexanoate see 4422
2-Acetyl-10-[3-[4-(β-hydroxyethyl)-
piperazinyl]propyl]phenothiazine see
70
2-Acetyl-10-[3-[4-(β-hydroxyethyl)-
piperidino]propyl]phenothiazine see
7615
2-Acetyl-10-[3-[γ-(2-hydroxyethyl)-
piperidino]propyl]phenothiazine see
7615
5-Acetyl-6-hydroxy-2-isopropenylben-
zofuran see 3946
2-Acetyl-7-(2-hydroxy-3-isopropyl-
aminopropoxy)benzofuran see 1050
3'-Acetyl-4'-[2-hydroxy-3-(isopropylamino)propoxy]butyranilide see 16
N-Acetylhydroxylamine see 60
5-[Acetyl(2-hydroxy-3-methoxyprop-
yl)amino]-N,N'-bis(2,3-dihydroxy-
propyl)-2,4,6-triiodo-1,3-benzene-
dicarboxamide see 5073
3-O-Acetyl-7α-[1(R)-hydroxy-1-
methylbutyl]-6,14-endo-ethenotetra-
hydrooripavine see 3932
N-[3-Acetyl-4-[2-hydroxy-3-[(1-meth-
ylethyl)amino]propoxy]phenyl]-
butanamide see 16
2-Acetyl-5-hydroxy-3-oxo-4-hexenoic
Acid δ-Lactone see 2919
trans-1-Acetyl-4-hydroxy-L-proline see
7035
1-Acetyl-4-hydroxy-2-pyrrolidinecar-
boxylic Acid see 7035
N-Acetylimidoquinone see 54
Acetylin [Heyden] see 886
Acetyllactoylcholine 1,5-Naphthalene-
disulfonate see 116
N-Acetyl-DL-leucine Compd with 2-
Aminoethanol (1:1) see 96
Acetylleucine Monoethanolamine, 96
DL-Acetylleucine Monoethanolamine
Salt see 96
N-Acetyl-3-mercaptoalanine see 89
N-Acetyl-3-mercapto-D-valine see 100
Acetylmetacresol see 2651
Acetylmethadol see 6009
levo-α-Acetylmethadol see 5492
α-l-Acetylmethadol see 5492
N-Acetylmethionine, 97

N-(N-Acetyl-L-methionyl)-O,O-bis-
(ethoxycarbonyl)dopamine see 3457
N-(N-Acetyl-L-methionyl)-3,4-dieth-
oxycarboxyphenethylamine see 3457
1-Acetyl-17-methoxyaspidospermidine
see 885
3-Acetyl-6-methoxybenzaldehyde see
98
5-Acetyl-2-methoxybenzaldehyde, 98
N^1-Acetyl-N^1-(3-methoxypyrazinyl)-
sulfanilamide see 103
N-Acetyl-5-methoxytryptamine see
5857
3-[[[[3-(Acetylmethylamino)-2,4,6-
triiodo-5-[(methylamino)carbonyl]-
benzoyl]amino]acetyl]amino]-5-[[(2-
hydroxyethyl)amino]carbonyl]-2,4,-
6-triiodobenzoic Acid see 5084
N-Acetyl-3-methylarsanilic Acid see
6102
Acetyl Methyl Carbinol see 61
O-Acetyl-β-methylcholine Chloride
see 6003
Acetyl-β-methylcholine Chloride see
6003
(±)-α-[2-[Acetyl(1-methylethyl)-
amino]ethyl]-α-(2-chlorophenyl)-1-
piperidinebutanamide see 1251
N-Acetyl-N-[2-methyl-4-[(2-methyl-
phenyl)azo]phenyl]acetamide see
3004
3-Acetyl-6-methyl-2H-pyran-2,4(3H)-
dione see 2919
N-Acetylmuramic Acid see 6384
N-Acetylmuramide Glycanohydrolase
see 5671
N^2-[N-(N-Acetylmuramoyl) L-alanyl]-
D-α-glutamine see 6385
N^2-[N^2-[N-(N-Acetylmuramoyl)-L-alan-
yl]-D-α-glutaminyl]-N^6-(1-oxoocta-
decyl)-L-lysine see 8412
N^2-(N-Acetylmuramoyl-L-alanyl-D-isoglutaminyl)-N^6-stearoyl-L-lysine see
8412
N-Acetylmuramyl-L-alanyl-D-isoglut-
amine see 6385
N-Acetylmuramyl Hydrolase see 5671
N-Acetylneuraminic Acid see 8628
Acetyl Nitrate, 99
N-Acetyl-5-nitro-2-furaldehyde Hydrazide see 6634
1-Acetyl-2-(5-nitro-2-furfurylidene)-
hydrazide see 6634
N^4-Acetyl-N^1-(p-nitrophenyl)sulfanil-
amide see 9103
Acetyl Oxide see 53
[1R-(1α,4aβ,5β,6α,8α,8aα)]-4-[2-[8-
(Acetyloxy)-8a-[(acetyloxy)methyl]-
octahydro-5,6-dimethylspiro[naph-
thalene-1(2H),2'-oxiran]-5-yl]ethyl]-
2(5H)furanone see 197
[3β(S),5β,16β]-16-(Acetyloxy)-3-[[8-
[[4-[(aminoiminomethyl)amino]-1-
carboxybutyl]amino]-1,8-dioxooct-
yl]oxy]-14-hydroxybufa-20,22-
dienolide see 1503
5-(Acetyloxy)-2-amino-3,4,14-trihydr-
oxy-6-eicosenoic Acid see 8897
2-(Acetyloxy)benzoic Acid see 886
2-(Acetyloxy)benzoic Acid 2-[4-(Acetylamino)phenoxy]ethyl Ester see
3759
2-(Acetyloxy)benzoic Acid 4-(Acetylamino)phenyl Ester see 1074
2-(Acetyloxy)benzoic Acid Calcium
Salt see 1684
2-(Acetyloxy)benzoic Acid 2-Carboxy-
phenyl Ester see 102
2-(Acetyloxy)benzoic Acid Magnesium Salt see 5689

2-(Acetyloxy)-*N*,*N*,*N*-trimethylethan-
aminium Bromide *see* 87
2-(Acetyloxy)-*N*,*N*,*N*-trimethylethan-
aminium Chloride *see* 88
2-(Acetyloxy)-*N*,*N*,*N*-trimethyl-1-pro-
panaminium Chloride *see* 6003
N-Acetylpenicillamine, 100
Acetylpheneturide, 101
Acetylphenol *see* 7421
[2*S*-[1(*R**),2α(*R**),3α]]-*N*-Acetyl-L-
phenylalanine 1-[3-(Acetyloxy)-1-(2-
amino-3-methyl-1-oxobutyl)-2-
pyrrolidinyl]-2-carboxyethyl Ester
see 2982
1-Acetyl-3-(2-phenylbutyryl)urea *see*
101
Acetyl-*p*-phenylenediamine *see* 431
p-(Acetylphenyl)-β-D-glucopyranoside
see 7549
Acetylphenylsalicylate *see* 7424
1-[(*p*-Acetylphenyl)sulfonyl]-3-cyclo-
hexylurea *see* 59
Acetylphosphoramidothioic Acid *O*,*S*-
Dimethyl Ester *see* 31
*N*¹-Acetyl-*N*⁴-phthaloylsulfanilamide
see 7532
*N*¹-Acetyl-*N*⁴-phthalylsulfanilamide *see*
7532
N-Acetylprocainamide *see* 17
Acetylpromazine *see* 32
2-Acetylpropane *see* 6164
β-Acetylpropionic Acid *see* 5498
3-*O*-Acetyl-17-propylorvinol *see* 3932
3-Acetylpyridine *see* 6195
β-Acetylpyridine *see* 6195
Acetylresorcinol *see* 8323
Acetyl-SAL *see* 886
N-Acetylsalicylamide *see* 8471
Acetylsalicylate Aluminum *see* 338
Acetylsalicylic Acid *see* 886
Acetylsalicylic Acid Calcium Salt *see*
1684
Acetylsalicylic Acid Calcium Salt
Complex with Urea *see* 1684
Acetylsalicylic Acid Methyl Ester *see*
6091
Acetylsalicylsalicylic Acid, 102
Acetylsalol *see* 7424
5-Acetylspiro[benzofuran-2(3*H*),1'-
cyclopropan]-3-one *see* 8918
Acetyl Sulfamethoxypyrazine, 103
4'-(Acetylsulfamyl)phthalanilic Acid
see 7532
*N*¹-Acetylsulfanilamide *see* 9067
*N*⁴-Acetylsulfanilamide, 104
2-(*N*¹-Acetylsulfanilamido)-3-meth-
oxypyrazine *see* 103
N-Acetylsulfanilic Acid, 105
N-Acetylsulfanilyl Chloride, 106
*N*¹-Acetyl-6-(sulfanilylmetanilamido)-
sodium *see* 73
Acetyl Sulfisoxazole *see* 9125
Acetyltanghinigenin *see* 9220
Acetyltannic Acid, 107
6-(Acetylthio)-8-[[2-[[(4-amino-2-
methyl-5-pyrimidinyl)methyl]for-
mylamino]-1-(2-hydroxyethyl)-1-
propenyl]dithio]octanoic Acid
Methyl Ester *see* 6857
3-Acetylthio-4-[(4-amino-2-methyl-5-
pyrimidinyl)methyl-*N*-formyl-
amino]-3-pentenyl Acetate *see* 51
N-[(*R*,*S*)-3-Acetylthio-2-benzylpropa-
noyl]glycine Benzyl Ester *see* 72
S-(3-Acetylthio-7-carbomethoxyhept-
ylthio)thiamine *see* 6857
(7α,17α)-7-(Acetylthio)-17-hydroxy-3-
oxopregn-4-ene-21-carboxylic Acid
γ-Lactone *see* 8917
(±)-*N*-[2-[(Acetylthio)methyl]-1-oxo-3-

phenylpropyl]glycine Phenylmethyl
Ester *see* 72
(*S*)-*N*-[1-[3-(Acetylthio)-2-methyl-1-
oxopropyl]-L-prolyl]-L-phenylala-
nine *see* 202
1-(D-3-Acetylthio-2-methylpropanoyl)-
L-prolyl-L-phenylalanine *see* 202
Acetylthymol *see* 9540
N-Acetyl-*m*-toluidine *see* 74
Acetyl-4-trifluoromethylsalicylic Acid
see 9817
(*E*)-1-[6-[(3-Acetyl-2,4,6-trihydroxy-5-
methylphenyl)methyl]-5,7-dihydr-
oxy-2,2-dimethyl-2*H*-1-benzopyran-
8-yl]-3-phenyl-2-propen-1-one *see*
8429
N-Acetyl-*N*-(2,4,6-triiodo-3-amino-
phenyl)-β-aminoisobutyric Acid *see*
5030
1-Acetyl-1,5,5-trimethyl-2-phenyl-
semioxamazide *see* 3358
Acetylveratroylbikhaconine *see* 1262
N-Acetyl-Wieland-Gumlich Aldehyde
see 3002
Acexamic Acid *see* 43
Acezide [Duncan, Flockhart] *see* 1817
Ac-globulin *see* 3961
Achieve [Zeneca] *see* 9699
Achillea, 108
Achilleic Acid *see* 119
Achless [Tatsumi] *see* 4167
Achletin [Toyama] *see* 9754
Achro [Lederle] *see* 9337
Achromycin [Lederle] *see* 9337
Achromycin V [Lederle] *see* 9337
Acibilin [Exa] *see* 2337
Acicloftal [Bruschettini] *see* 148
1,2,4-Acid *see* 474
Acid Acriflavine *see* 125
Acid Ammonium Carbonate *see* 522
Acid Ammonium Fluoride *see* 523
Acid Ammonium Purpurate *see* 6386
Acid Ammonium Sulfate *see* 526
Acid Blue 74 *see* 4978
Acid Calcium Phosphate *see* 1740
Acid Fuchsin, 109
Acidic Fibroblast Growth Factor *see*
4117
Acid Magenta *see* 109
Acid Magnesium Citrate *see* 5700
Acid Magnesium Phosphate *see* 5719
Acid Magnesium Sulfate *see* 5692
Acid Methyl Sulfate *see* 6203
Acidogen *see* 4477
Acidogen Nitrate *see* 10008
Acidol [Winthrop] *see* 1225
Acidomycin *see* 6403
Acidoride [Abbott] *see* 4477
Acid Platinic Chloride *see* 7684
Acid Potassium Phthalate *see* 7773
Acid Potassium Tartrate *see* 7776
Acid Quinine Hydrochloride *see* 8245
Acid Roseine *see* 109
Acid Rubin *see* 109
Acid Sodium Phosphate *see* 8806
Acid Trypaflavine *see* 125
Acidulin [Lilly] *see* 4477
Acidum Fenclozicum *see* 4010
Acid Violet 7B, 110
Acid Yellow 3 *see* 8256
Acid Yellow D *see* 9906
Acid Yellow S *see* 6480
Acifluorfen, 111
Acifran, 112
Aciglumin *see* 4477
Aci-Jel [Ortho-Cilag] *see* 52
Aciloc [Orion] *see* 2337
Acimethin [Gry] *see* 6053
Acimetten [Kwizda] *see* 886
Acinil [GEA] *see* 2337

Acinitrazole *see* 430
Acino [IMA] *see* 4015
Acintol C *see* 9210
Acipen-V [Mycopharm] *see* 7230
Acipimox, 113
Acitretin, 114
ACL-59 *see* 9896
ACL-85 *see* 9188
Aclacin [Lundbeck] *see* 115
Aclacinomycin A *see* 115
Aclacinomycin B *see* 115
Aclacinomycins, 115
Aclacinon [Sanraku] *see* 115
Aclaplastin [Behringwerke] *see* 115
Aclarubicin *see* 115
Aclarubicin Hydrochloride *see* 115
Aclatonium Napadisilate, 116
Aclor *see* 4477
Aclosone [Schering] *see* 221
Aclovate [Glaxo] *see* 221
Aclovir [Lek] *see* 148
Acnegel [Stiefel] *see* 1149
Acnestrol [Dermik] *see* 1471
ACNU [Asta] *see* 6645
Acocantherin *see* 7031
Acodeen [Hommel] *see* 1539
Acolen *see* 2922
Acon [Endo] *see* 10150
Aconine, 117
Aconite, 118
Aconitic Acid, 119
Aconitine, 120
Aconitine, Amorphous, 121
Aconitum Ferox, 122
Acopyrine *see* 757
"Acorn Sugar" *see* 8218
Acortan [Ferring] *see* 136
Acorto *see* 136
Acosterina *see* 9867
Acozid [ACO] *see* 8487
ACP-332 *see* 6505
ACPC *see* 2803
ACPM-629 *see* 2115
ACPMD *see* 9582
Acquat CDAC *see* 2059
Acquinite *see* 2208
ACR-3453A *see* 8167
Acraconitine *see* 8095
Acraldehyde *see* 130
Acramine Yellow *see* 427
Acranil [Bayer], 123
Acrasin *see* 2777
Acrex [Murphy] *see* 3339
Acricid [Hoechst] *see* 1265
9-Acridinamine *see* 427
Acridine, 124
3,6-Acridinediamine *see* 7953
4'-(9-Acridinylamino)methanesulfon-
m-anisidide *see* 635
N-[4-(9-Acridinylamino)-3-methoxy-
phenyl]methanesulfonamide *see* 635
Acriflavine, 125
Acriflavine Hydrochloride *see* 125
Acrilan [Monsanto], 126
Acrinathrin, 127
Acrinol *see* 3760
Acrisorcin, 128
Acritet [Stauffer] *see* 133
Acrivastine, 129
Acrobat [Shell] *see* 3271
Acrol *see* 3026
Acrolactine *see* 3760
Acrolein, 130
Acronize [Am. Cyanamid] *see* 2245
α-Acrose *see* 4296
Acrosoxacin *see* 8426
Acrylaldehyde *see* 130
Acrylamide, 131
Acrylic Acid, 132
Acrylic Acid *n*-Butyl Ester *see* 1574

Acrylic Acid Ethyl Ester *see* 3805
Acrylic Acid Methyl Ester *see* 6092
Acrylic Aldehyde *see* 130
Acrylonitrile, 133
ACS *see* 759
Actaea *see* 2339
Actal [Winthrop] *see* 232
Actamer [Monsanto] *see* 1343
Actaplanins, 134
Actarit, 135
Actasal [Purdue Frederick] *see* 2265
Actase [Ortho] *see* 7678
Actedron *see* 623
Actellic [ICI] *see* 7652
Actellifog [ICI] *see* 7652
Actemin *see* 623
Acterol [Lappe] *see* 6644
ACTH, 136
Acthar [Armour Pharm.] *see* 136
Acthiazidum *see* 3778
ACTH-β-lipotropin Common Precursor *see* 7975
Actholain *see* 2617
Acthormon [Shionogi] *see* 4435
Acticolin [Farge] *see* 2380
Actidil [Wellcome] *see* 9877
Actidilon [Wellcome] *see* 9877
Actidione [Upjohn] *see* 2797
Actifed-C [Burroughs Wellcome] *see* 4582
Actigall [Ciba-Geigy] *see* 10026
Actigam [Daltan] *see* 4470
Actilin *see* 6542
Actilyse [Boehringer, Ing.] *see* 9608
Actimide [Tobishi] *see* 2513
Actimmune [Genentech] *see* 5018
Actin, 137
F-Actin *see* 137
G-Actin *see* 137
α-Actin *see* 137
β-Actin *see* 137
γ-Actin *see* 137
Actinamin [Daiichi] *see* 1913
Actinex [Block] *see* 6786
Actinium, 138
Actinium K *see* 4286
Actinium X *see* 8277
Actinobolin, 139
Actinodaphnine, 140
Actinomycetin, 141
Actinomycin A$_{IV}$ *see* 2867
Actinomycin C *see* 1642
Actinomycin C$_1$ *see* 2867
Actinomycin C$_2$ *see* 1642
Actinomycin C$_3$ *see* 1642
Actinomycin D *see* 2867
Actinomycin F$_1$, 142
Actinomycin I$_1$ *see* 2867
Actinomycin IV *see* 2867
Actinomycin KS4 *see* 142
Actinomycin-[thr-val-pro-sar-meval] *see* 2867
Actinomycin X$_1$ *see* 2867
Actinon *see* 8278
Actinoquinol, 143
Actinorhodine, 144
Actinospectacin *see* 8890
Actiphenol, 145
Actiplas [Dompé] *see* 9608
Actithiazic Acid *see* 6403
Activase [Genentech] *see* 9608
Activated Carbon *see* 1856
Activated Charcoal *see* 1856
Activated 7-Dehydrocholesterol *see* 10157
Activated Dimethicone *see* 3264
Activated Ergosterol *see* 10156
Active Amyl Alcohol *see* 6109
Active Lipid *see* 201
"Active Methionine" *see* 155

Activin *see* 6452
Activin A *see* 146
Activin AB *see* 146
Activins, 146
Activol *see* 482
Actol [Beecham] *see* 6620
Actomyosin *see* 137
Acton [Ferring] *see* 136
Actonar *see* 136
Actosin [Daiichi Seiyaku] *see* 1483
Actosolv [Behringwerke] *see* 10024
Actozine *see* 1055
Actriol [Organon] *see* 3652
Actybaryte [Labaz] *see* 1023
Actylamide [Rougier] *see* 8471
Acuitel [Parke-Davis] *see* 8233
Acular [Allergan] *see* 5318
Acupan [3M] *see* 6529
Acupaque [Sterling Winthrop] *see* 5044
Acutil-S [ISF] *see* 25
Acutran [Mallinckrodt] *see* 620
Acuvel [Burroughs-Wellcome] *see* 898
ACV, 147
Acycloguanosine *see* 148
Acyclovir, 148
Acylanid [Sandoz] *see* 90
Acylcolaminoformylmethylpyridinium Chloride *see* 9982
N-(Acylcolaminoformylmethyl)pyridinium Chloride *see* 5379
Acylpyrin [Spofa] *see* 886
AD$_6$ *see* 2467
AD-106 *see* 2327
AD-810 *see* 10323
AD-1590 *see* 1199
AD-4833 *see* 7605
Ada *see* 3227
ADAH *see* 1284
Adalat(e) [Bayer] *see* 6617
Adalin [Bayer] *see* 1879
"Adam" *see* 5806
1-Adamantanamine *see* 389
Adamantane, 149
2-(1'-Adamantanecarbonyloxy)ethyldimethyldecylammonium Bromide *see* 390
N-1-Adamantyl-N-[2-(dimethylamino)ethoxy]acetamide *see* 9901
Adamite *see* 10277
Adams' Catalyst *see* 7686
Adamsite *see* 7357
Adapin [Lotus] *see* 3492
Adapress [Lagap] *see* 6617
Adaptinol [Bayer] *see* 10197
Adazine [Upjohn] *see* 9814
ADCA *see* 1284
Adcortyl [Bristol-Myers Squibb] *see* 9728
ADD-03055 *see* 3988
Addisomnol [Synochem] *see* 1879
Adecut [Takeda] *see* 2928
Adelir [Nagase] *see* 9974
Ademetionine *see* 155
Ademide [Toyo Jozo] *see* 2513
Ademin(e) *see* 9731
Ademol [Squibb] *see* 4174
Adenaron [Kowa] *see* 1833
Adenex [ICN] *see* 867
Adenine, 150
Adenine Arabinoside *see* 10113
Adenine Riboside *see* 152
Adenocard [Fujisawa] *see* 152
Adenock [Shiraimatsu] *see* 287
Adenocor [Sanofi] *see* 152
Adenogen *see* 1832
Adenohypophyseal Growth Hormone *see* 8864
Adenohypophysial Luteotropin *see* 7961

A-Denopterin, 151
Adenoscan [Medco] *see* 152
Adenosine, 152
Adenosine Cyclic 3',5'-(Hydrogen Phosphate) *see* 2777
Adenosine 3',5'-Cyclic Monophosphate *see* 2777
Adenosine 3',5'-Cyclic Phosphate *see* 2777
Adenosine Diphosphate, 153
Adenosine 3'-Monophosphate *see* 156
Adenosine 3',5'-Monophosphate *see* 2777
Adenosine 5'-Monophosphate *see* 157
Adenosine-3'-monophosphoric Acid *see* 156
Adenosine-5'-monophosphoric Acid *see* 157
Adenosine Phosphate *see* 157
Adenosine 3',5'-Phosphate *see* 2777
Adenosine-3'-phosphoric Acid *see* 156
Adenosine-5'-phosphoric Acid *see* 157
Adenosine 5'-(Tetrahydrogen Triphosphate) *see* 154
Adenosine 5'-(Trihydrogen Diphosphate) *see* 153
Adenosine 5'-(Trihydrogen Diphosphate) 2'-(Dihydrogen Phosphate) P'→5'-Ester with 3-(Aminocarbonyl)-1-β-D-ribofuranosylpyridinium Inner Salt *see* 6433
Adenosine 5'-(Trihydrogen Diphosphate) P'→5'-Ester with 3-(Aminocarbonyl)-1-β-D-ribofuranosylpyridinium Inner Salt *see* 6429
Adenosine 5'-(Trihydrogen Diphosphate) 5'→5'-Ester with Riboflavine *see* 4131
Adenosine Triphosphate, 154
Adenosine 5'-Triphosphoric Acid *see* 154
Adenosyl-B$_{12}$ *see* 2513
S-Adenosylmethionine, 155
Adényl [Wyeth France] *see* 157
Adenylate Cyclase *see* 2777
3'-Adenylic Acid, 156
5'-Adenylic Acid, 157
Adenylic Acid b *see* 156
h-Adenylic Acid *see* 156
t-Adenylic Acid *see* 157
Adephos *see* 154
Adepril [Lepetit] *see* 511
Adeps *see* 5380
Adepsine Oil *see* 7327
Adergon *see* 4401
Adermine Hydrochloride *see* 8166
Adermykon *see* 2230
Aderoxal *see* 8163
Adesitrin [Erba] *see* 6704
Adetol *see* 154
Adetphos [Kowa] *see* 154
Adfeed [Lead Chem.] *see* 4234
Adhatoda, 158
Adhatodai *see* 158
AD/here *see* 5827
Adiab [Guidotti] *see* 4486
Adiaben *see* 2239
Adiazine [Theraplix] *see* 9071
Adicillin *see* 7227
Adifax [Servier] *see* 4015
Adigal [Beiersdorf] *see* 5368
Adinazolam, 159
Adinazolam Mesylate *see* 159
Adipan *see* 623
Adiparthrol [Riker] *see* 3809
Adipex-P [Gate] *see* 7415
Adiphenine Hydrochloride, 160
Adipic Acid, 161
Adipic Acid Di(3-carboxy-2,4,6-triiodoanilide) *see* 5043

Adipic Ketone see 2811
Adipiodone see 5043
Adipokinetic Hormone see 5539
Adipomin [Streuli] see 4015
Adiposetten [Reiss] see 6811
3,3'-(Adipoyldiimino)bis[2,4,6-triiodo-benzoic Acid] see 5043
5,5'-(Adipoyldiimino)bis[2,4,6-triiodo-N-methylisophthalamic Acid] see 5029
Adipsin, 162
Adipsin [Heilmittelwerke] see 6172
N,N'-Adipylbis(3-amino-2,4,6-triiodo-benzoic Acid) see 5043
Adiuretin SD [Spofa] see 2969
Adizem [Napp] see 3247
Adlone [Pharmascience] see 3958
Adlumidine, 163
Adlumine, 164
Admire [Nihon Bayer] see 4946
Admon [Esteve] see 6643
A-DNA see 2954
Adnexol [Protina] see 4929
Adobiol [Yoshitomi] see 1496
Adofen [Ferrer] see 4222
Adoisine see 10174
Adomal [Malesci] see 3190
AdoMet see 155
Adona [Tanabe Seiyaku] see 1833
Adonigen see 165
Adonilen see 165
Adonisid see 165
Adonis Vernalis, 165
Adonite see 166
Adonitol, 166
Adonitoxin, 167
Adonival see 165
D-Adonose see 8374
Adopon see 1784
Adovern [Roche] see 165
ADP see 153 and 7135
Adphen [Ferndale] see 7365
ADR-033 see 9866
ADR-529 see 8295
Adrafinil, 168
Adral [Lifepharma] see 6626
Adran [Takata] see 4925
Adrenal Cortical Extract, 169
Adrenalin [Parke, Davis] see 3656
Adrenaline 3656
Adrenaline Oxidase see 6330
Adrenalone, 170
Adrenochrome, 171
Adrenochrome Monosemicarbazone Sodium Salicylate Complex see 1832
Adrenochrome Semicarbazone Compd with Sodium Salicylate see 1832
Adrenocorticotrop(h)ic Hormone of the Pituitary Gland see 136
Adrenocorticotrop(h)in see 136
Adrenoglomerulotropin, 172
Adrenolutin, 173
Adrenomone [Armour Pharm.] see 136
Adrenone see 170
Adrenor see 6788
Adrenosem [Beecham] see 1832
Adrenosterone, 174
Adrenoxyl [Nordmark] see 171
Adreson see 2602
Adrestat-F [Organon] see 1832
Adrevil [Zyma] see 1535
Adriacin [Kyowa] see 3495
Adriamicina [Farmochim. Ital.] see 6007
Adriamycin [Farmitalia] see 3495
Adriamycinone see 3495
Adrianol [Boehringer, Ing.] see 7440
Adriblastina [Farmitalia] see 3495
Adronat [Neopharmed] see 228
Adroyd [Parke, Davis] see 7101

Adrucil [Adria] see 4219
Adulsa see 158
Adumbran [Thomae] see 7059
Adurix [Benzon] see 2454
Advantan [Schering] see 6189
Advil [Whitehall] see 4925
Advocin [Pfizer] see 2876
AE-17 see 9185
Aedurid [Robugen] see 3561
Aequamen [Promonta] see 1224
Aequorin, 175
Aerobec [3M] see 1047
Aerobid [Forest] see 4180
Aerobin [Farmasan] see 9421
Aerolate [Fleming] see 9421
Aerolin [Glaxo] see 217
Aerolone [Lilly] see 5236
Aeromatt see 1697
Aeropent [Fisons] see 7254
Aeroseb-Dex [Herbert] see 2986
Aeroseb-HC [Allergan] see 4828
Aerosol AY see 3034
Aerosol IB see 3238
Aerosol MA see 1295
Aerosol OT [Cyanamid] see 3460
Aerosporin [Burroughs Wellcome] see 7734
Aerotrol [Abbott] see 5236
Aerrane [Anaquest] see 5191
Aerugipen [Wülfing] see 9568
Aescin see 3737
Aescusan see 3737
AET, 176
Aether Oenanthicus see 3881
Aethoform see 1116
Aethone [Sopar] see 7012
Aethoxysklerol [Kreussler] see 7717
Aethylis Chloridum see 3829
Aetina see 3783
Aetiocholanic Acid see 3913
Aetoxisclerol [Dexo] see 7717
AF-438 see 7078
AF-634 see 8090
AF-983 see 1061
AF-1161 see 9712
AF-1191 see 3930
AF-1890 see 5598
AF-2139 see 2884
Afalon [Hoechst] see 5534
Afatin see 2996
Afaxin [Winthrop] see 10150
Affinin, 177
Affinitins see 5453
Affirm [Merck & Co.] see 1
Affirm [Merck & Co.] see 3594
Afibrin [Lederle] see 451
AFI-Ftalyl see 7533
AFI-phyllin see 3529
AFI-Tiazin see 7404
Aflatoxicol see 178
Aflatoxin B₁ see 178
Aflatoxin B₂ see 178
Aflatoxin G₁ see 179
Aflatoxin G₂ see 179
Aflatoxin M₁ see 180
Aflatoxin M₂ see 180
Aflatoxins B, 178
Aflatoxins G, 179
Aflatoxins M, 180
Aflix [Sandoz] see 4273
Afloben [Esseti] see 1157
Aflodac [Janus] see 9155
Afloqualone, 181
Aflorix [Gramon] see 6266
Afloxan [Rotta] see 7957
Afluon [Asta] see 5701
Afonilum [Minden] see 9421
Afos [Tiber] see 4280
African Saffron see 1919

Afrin [Schering] see 7100
Afrinol [Schering] see 3645
Afsillin [Squibb] see 7226
Aftate [Schering] see 9656
Afugan [Hoechst] see 8146
Afungil [EGYT] see 2243
Afwillite see 1749
Afzelechin see 1950
AG-629 see 8918
AG-1749 see 5373
Agar, 182
Agar-agar see 182
Agaric, 183
Agaric Acid see 184
Agaricic Acid, 184
Agaricin see 184
Agaricinic Acid see 184
Agarin see 6391
Agaritine, 185
Agaropectin see 182
Agarose see 182
Agate see 8637
Agbayun see 6293
Agedal [Bayer] see 6821
Agedoite see 872
AG-EE-623 ZW see 8304
Agene see 6699
Agent Orange see 9252
AGEPC see 7682
Ageratochromene see 7898
Agerite see 6331
Ageroplas [Serono] see 3432
Agglutinin see 5
Agglutinins see 5453
Aggrastat [Merck & Co.] see 9605
Agifutol S [Kyorin] see 4483
Agil [Ciba] see 7990
Agilene [Am. Agile] see 7728
Agiolan see 10150
Agit [Sanofi Winthrop] see 3217
Agkistrodon rhodostoma Venom Proteinase see 671
Agkistrodon Serine Proteinase see 671
Aglumin see 3766
Agmatine, 186
AGN-190168 see 9249
AGN-190342 see 1399
AGN-190342-LF see 1399
AGN-191622 see 1384
Agnin see 5371
Agnolin see 5371
Agofell [Janssen] see 3239
Agofollin [Leciva] see 3746
Agolanid [Sandoz] see 3210
Agon [Hoechst] see 3991
Agoniadin see 7699
Agontan [Knoll] see 3236
Agopton [Takeda] see 5373
Agotan see 2347
Agr-1240 see 6287
Agradil [Vita] see 10081
Agram [Inava] see 617
Agréal [Delagrange] see 10081
Agrelin [Roberts] see 665
Agribon [Roche] see 9073
Agricultural Limestone see 5515
Agrimek [Merck & Co.] see 1
AgriStrep [Merck & Co.] see 8983
Agritan see 2898
Agritol see 962
Agritox [M & B] see 5803
Agroclavine, 187
Agrocybin, 188
Agropyrum see 9890
Agrosan GN [ICI] see 7453
Agrothrin [Sumitomo] see 2836
Agroxone [ICI] see 5803
Agrypnal [Eggochemia] see 7386
Agstone see 5515
Ague Tree see 8528

AH-42 see 6027
AH-289 see 2128
AH-2250 see 1520
AH-3365 see 217
AH-5158A see 5341
AH-8165 see 3981
AH-19065 see 8286
AHA see 60
Ahanon [Rodleben] see 9205
AHB-DKB see 809
AHCTL see 2381
AHF see 3963
AHG see 3963
Ahistan, 189
AHMHA see 8956
AHP-200 see 7035
AHR-85 see 6060
AHR-233 see 5896
AHR-438 see 5994
AHR-504 see 4511
AHR-619 see 3488
AHR-857 see 9082
AHR-3018 see 768
AHR-3053 see 1850
AHR-3070-C see 6226
AHR-5850D see 413
AHR-10282 see 1411
AHR-10282B see 1411
A-hydroCort [Abbott] see 4828
AI-3-36206 see 7951
AI-27303 see 2060
AIBN see 949
Aicamin [Crinos] see 6993
AICA Orotate see 6993
AICAR, 190
AICA-riboside see 13
Aicorat [Mack, Illert.] see 6993
Aida [Paraphar] see 4853
Aiglonyl [Fumouze] see 9163
Aimax [Ayerst] see 6013
AIR, 191
Airbron [BDH] see 89
Airoform see 1310
Airogen see 1310
Airol [Roche] see 8333
Airtal [Prodes] see 19
Airum [Promeco] see 4022
Aisemide see 4331
Aizumycin see 1249
AJ-1590 see 1199
Ajacine, 192
Ajaconine, 193
Ajan [3M] see 6529
Ajicef [Ajinomoto] see 1988
Ajinomoto see 6337
Ajmalan-17,21-diol see 194
(17R,21α)-Ajmalan-17,21-diol 17-
(Chloroacetate) see 5607
Ajmalicine see 8291
Ajmaline, 194
Ajoene, 195
Ajowan Oil, 196
Ajugarin-I see 197
Ajugarin-II see 197
Ajugarin-III see 197
Ajugarin-IV see 197
Ajugarin-V see 197
Ajugarins, 197
AK-33X see 6307
Akar [Ciba-Geigy] see 2174
Akarpine [Akorn] see 7578
Akatinol [Merz] see 5872
Ak-Chlor [Akorn] see 2120
Ak-Con [Akorn] see 6455
Ak-Dex [Akorn] see 2986
Ak-Dilate [Akorn] see 7440
Akermanite see 1749
Ak-Fluor [Akorn] see 4194
Akineton [Knoll] see 1274
Akinophyl [Knoll] see 1274

Akistin [Nycomed] see 8481
Aklavinone see 115
Aklomide, 198
Aklomix [Salsbury] see 198
Aklonin [DDR] see 7350
Ak-Mycin [Akorn] see 3720
Ak-Nefrin [Akorn] see 7440
Akneroxid L [Lubapharm] see 1149
Aknin [Schwarzhaupt] see 3720
Aknoten [Krka] see 8333
Akotin [Glaxo] see 6612
Ak-Pentolate [Akorn] see 2815
Ak-Pred [Akorn] see 7903
Akrinol [Schering] see 128
Akrochem ETU-22 [Akron Chem.] see
3849
Ak-Sulf [Akorn] see 9067
Ak-Taine [Akorn] see 7991
Aktamin see 6788
Ak-Tate [Akorn] see 7901
Aktedron [Chinoin] see 623
Aktil [Lek] see 911
Aktiven see 2118
Aktivin [Kobayashi] see 9432
Akuammicine, 199
Akuammine, 200
Akwa Tears [Akorn] see 7745
AL-132 see 7362
AL-0361 see 4884
AL 721, 201
AL-1241 see 7943
Ala see 205
Alabandite see 5783
Alabaster see 1753
Alacepril, 202
Alachlor, 203
Ala-Cort [Del-Ray] see 4828
ALAcortril [Pfizer] see 4225
Ala²-cyclosporine see 2821
Alafosfalin, 204
[β-Ala¹,lys¹⁷]ACTH¹⁻¹⁷-4-amino-N-bu-
tylamide see 322
[β-Ala¹,lys¹⁷]corticotropin-(1-17)-hep-
tadecapeptide-4-amino-N-butyl-
amide see 322
Alamine see 3227
Alamon [Grelan] see 4897
Alamycin [Norbrook] see 7111
Alanap see 6505
Alanetorin [Takata; Nippon Kayaku]
see 224
Alanex [Makhteshim-Agan] see 203
Alanine, 205
L-Alanine see 205
L-α-Alanine see 205
β-Alanine, 206
1-β-Alanine-17-[N-(4-aminobutyl)-L-
lysinamide]-α¹⁻¹⁷-corticotropin see
322
1-β-Alanine-17-[L-2,6-diamino-N-(4-
aminobutyl)hexanamide]-α¹⁻¹⁷-cor-
ticotropin see 322
2-D-Alanine-4-[4-(dimethylamino)-N-
methyl-L-phenylalanine]virginia-
mycin S₁ see 10096
2-D-Alanine-4-[N-methyl-4-(methyl-
amino)-L-phenylalanine]virginia-
mycin S₁ see 10096
Alanine Nitrogen Mustard see 5871
L-Alanosine, 207
Alant Camphor see 208
Alantin see 5024
Alantolactone, 208
Alant Starch see 5024
1R-1-(L-Alanylamino)ethylphosphonic
Acid see 204
L-Alanyl-(6-diazo-5-oxo)-L-norleucyl-
(6-diazo-5-oxo)-L-norleucine see 209
[N-β-Alanyl-L-histidinato(2—)-N,Nᴺ,-
Oᵅ]zinc see 7712

β-Alanyl-L-histidinato Zinc see 7712
N-β-Alanyl-L-histidine see 1899
N-β-Alanyl-3-methyl-L-histidine see
717
[1R-(1α,2β,6α)]-N-L-Alanyl-3-(5-oxo-
7-oxabicyclo[4.1.0]hept-2-yl)-L-
alanine see 963
[D-Ala¹] Peptide T Amide see 7291
Alaphosphin see 204
Alapril [Sigma-Tau] see 5540
Alaprin see 316
Alapurin see 5371
Alar [Uniroyal] see 2874
Ala Tet [Del Ray] see 9337
Alathon [Du Pont] see 7728
Alazopeptin, 209
Albacar see 1697
Albalon [Allergan] see 6455
Albamycin [Upjohn] see 6818
Albasapogenin see 4614
Albasil C see 9100
Albaspidin, 210
Albaspidin-BB see 210
Albego [Boehringer, Ing.] see 1773
Albemap see 2996
Albendazole, 211
Albert-285 see 7997
Albexan see 9094
Albiotic [Upjohn] see 5525
Albipen [Veterinaria] see 628
Albistat [Ortho] see 6266
Albizziin, 212
Albocresil [Byk Liprandi] see 6531
Albofungin, 213
Alboline see 7327
Albomycetin see 7566
Albomycin, 214
Albon [Roche] see 9073
Albone see 4839
Alborixin, 215
Albothyl [Byk-Gulden] see 6531
Albucid [Ankerpharm] see 9067
Albumen, 216
Albuminar [Armour] see 8613
Albuminate see 8613
Albumisol [Merck & Co.] see 8613
Albumized Iron see 4058
Albumose Silver see 8671
Albumotope-¹³¹I [Squibb] see 8613
Albunex [Molecular Biosystems] see
8613
Albuspan [Parke, Davis] see 8613
Albuterol, 217
Albutest [Miles] see 1468
Albutoin, 218
Alcacyl [Wander] see 1684
Alcaine [Alcon] see 7991
Alcar [Sigma-Tau] see 85
Alchloquin see 5052
Alcian Blue, 219
Alcian Blue 8GX see 219
Alclofenac, 220
Alclometasone, 221
Alcobon [Roche] see 4161
Alcohol see 3806
95% Alcohol see 3806
Alcohol Ethoxylates see 7737
Alcohol, USP see 3806
Alcomicin [Alcon] see 4398
Alcopan-250 see 2988
Alcopar(a) [Burroughs Wellcome] see
1187
Alcopon see 1820
Alcuronium, 222
Aldace [Searle] see 8917
Aldactazine see 326
Aldactone [Boehringer, Mann.] see
8917
Aldanil see 8764
Aldazine [Alphapharm] see 9497

Aldebaranium *see* 10239
Aldecin [Schering] *see* 1047
"Aldehyde" *see* 37
Aldehyde Ammonia *see* 38
Aldehyde C-7 *see* 4692
Aldehyde-collidine *see* 3889
Aldehydine *see* 3889
Alderlin [ICI] *see* 7974
Aldesleukin *see* 5020
Aldesulfone Sodium *see* 9141
Aldicarb, 223
Aldifen *see* 3333
Aldioxa, 224
Aldipin [Helvepharm] *see* 6617
Aldocorten [Ciba] *see* 226
Aldocortin [Ciba] *see* 226
Aldol, 225
Aldomet [Merck & Co.] *see* 6132
Aldometil [Merck & Co.] *see* 6132
Aldomine [Merck & Co.] *see* 6132
Aldomycin *see* 6697
Aldopur [Hormosan] *see* 8917
Aldosterone, 226
Aldosterone-stimulating Hormone *see* 172
Aldoxime *see* 40
Aldrich Dye Mixture *see* 3526
Aldrin, 227
Aldrox [Wyeth] *see* 355
ALEC [Britannia] *see* 8126
Alecor [Andromaco] *see* 3670
Alegan S [Ind. Med.] *see* 2214
Alendronate Sodium *see* 228
Alendronic Acid, 228
Alendros [Gentili] *see* 228
Alentol *see* 623
Aleppo-galls *see* 6827
Alercrom [Osiris] *see* 2658
Alergicide *see* 2128
Alerion [Bouchara] *see* 2658
Alerlisin [Prodes] *see* 2063
Alesion [Boehringer, Ing.] *see* 3655
Aletris, 229
Aleudrin [Lewis] *see* 5236
Aleuritic Acid, 230
Alevaire [Winthrop] *see* 9964
Aleve [Procter-Syntex] *see* 6504
Aleviatin [Warner-Lambert] *see* 10323
Alexan [Pfizer] *see* 2853
Alexidine, 231
Alexitol Sodium, 232
Alfacalcidol *see* 4864
Alfacet [Galenika] *see* 1962
Alfacillin *see* 7369
Alfadat [Alfa] *see* 6617
Alfadil [Pfizer] *see* 3489
Alfadione *see* 235
Alfadolone Acetate, 233
Alfadryl [Leciva] *see* 6370
Alfa-interferon *see* 5016
Alfamox [Alfa] *see* 617
Alfaprostol, 234
Alfarol [Chugai] *see* 4864
Alfason [Yamanouchi] *see* 4828
Alfaspoven [Alfa] *see* 2021
Alfatésin [Glaxo] *see* 235
Alfatil [Lilly] *see* 1962
Alfatrofin *see* 136
Alfavet [Hoffmann-La Roche] *see* 234
Alfaxalone, 235
Alfenamin *see* 4167
Alfenta [Janssen] *see* 236
Alfentanil, 236
Alfentanil Hydrochloride *see* 236
Alferon [Interferon Sciences] *see* 5016
Alficetin [Argentia] *see* 2542
Alflorone [Merck & Co.] *see* 4166
Alfone *see* 254
Alfonic [Vista] *see* 7737
Alfospas [Rorer] *see* 9607

Alfoten [Synthelabo] *see* 237
Alfuzosin, 237
Alganex [Roche] *see* 9293
Algeldrate *see* 355
Algeril [Bayer] *see* 8017
Algestone, 238
Algestone Acetonide *see* 238
Algestone Acetophenide, 239
Algidon *see* 6008
Algil [Maggioni] *see* 5894
Algin, 240
Alginic Acid, 241
Alginic Acid Sodium Salt *see* 240
Alginodia [Beta] *see* 3414
Alginor [De Angeli] *see* 2338
Alglucerase, 242
Algobaz [Labaz] *see* 6125
Algocalmin *see* 3414
Algocetil [Francia] *see* 9155
Algocor [Ravizza] *see* 4369
Algolysin *see* 6008
Algylen *see* 9769
Alibendol, 243
Alidase [Searle] *see* 4794
Alidine *see* 695
Aliette [Rhône-Poulenc] *see* 4278
Alimemazine *see* 9834
Alimet [Monsanto] *see* 6054
Alimix [Cilag] *see* 2377
Alinam [Lucien] *see* 2155
Alinamin [Takeda] *see* 8068
Alinamin F [Takeda] *see* 4333
Alindapril *see* 2928
Alinidine, 244
Alisactide *see* 322
Alisobumal *see* 1536
Alitame, 245
Alival [Hoechst] *see* 6768
Alivin *see* 7212
Alizapride, 246
Alizarin, 247
Alizarin Blue R *see* 249
Alizarin Bordeaux B *see* 8231
Alizarin Cyanine Green F, 248
Alizarine Blue, 249
Alizarine Carmine *see* 8714
Alizarine Orange, 250
Alizarine S *see* 8714
Alizarine Yellow C *see* 4360
Alizarine Yellow GG *see* 5979
Alizarine Yellow R, 251
Alizarine Yellow RW *see* 251
β-2-Alizarin Primeveroside *see* 8437
Alizarin Violet *see* 4362
Alkacitron *see* 8747
Alkagel I.S. [Quim. Farm.] *see* 355
Alkalovert [Klein] *see* 7542
Alkamuls SML [Alkaril] *see* 8872
Alkamuls SMO [Alkaril] *see* 8872
Alkamuls SMS [Alkaril] *see* 8872
Alkanet, 252
Alkanet Extract *see* 253
Alkanna *see* 252
Alkanna Red *see* 253
Alkannin, 253
Alkannin Paper *see* 252
Alkathene [ICI] *see* 7728
Alkavervir *see* 10090
Alkeran [Burroughs Wellcome] *see* 5871
Alkiron *see* 6210
Alkofanone, 254
Alkron *see* 7167
1-*O*-Alkyl-2-acetyl-*sn*-glycero-3-phosphorylcholine *see* 7682
Alkylaluminum Halide *see* 334
Alkylaluminum Sesquihalide *see* 334
Allantoin, 255
(Allantoinato)dihydroxyaluminum *see* 224

Allantoxanic Acid *see* 7081
All-*trans*-cryptoxanthin *see* 2676
Alledryl *see* 3367
Allegan *see* 4489
Allegra [Hoechst Marion Roussel] *see* 4113
Allegron [Dista] *see* 6812
Allenolic Acid, 256
Alleoside A *see* 4665
Allercorb [Plessner] *see* 867
Allercur [Roerig] *see* 2406
Allergefon [Lafon] *see* 1845
Allergina [De Angeli] *see* 3367
Allergisan [Pharmacia] *see* 2232
Allergocrom [Ursapharm] *see* 2658
Allergodil [Asta] *see* 939
Allerkif [Edmond] *see* 5319
Alleron *see* 7167
Allerplus [Astra Simes] *see* 9307
Allethrin I *see* 257
Allethrin II *see* 257
Allethrins, 257
Allethrolone Ester of Chrysanthemumdicarboxylic Acid Monomethyl Ester *see* 257
Allethrolone Ester of Chrysanthemummonocarboxylic Acid *see* 257
All-*trans*-fucoxanthin *see* 4304
All-*trans*-N-4'-hydroxyretinanilide *see* 4040
Allicin, 258
Allidochlor, 259
Allie [Du Pont] *see* 6244
Alliin, 260
Allium *see* 4380
Allobarbital, 261
Allobarbitone *see* 261
Alloca [Upjohn] *see* 7003
Allocholesterol, 262
Alloclamide, 263
Allocor [Kwizda] *see* 5368
Allocryptopine, 264
Allocupreide Sodium, 265
Allodene *see* 623
Allodihydro F *see* 4801
Allodihydrohydrocortisone *see* 4801
Allofenyl *see* 7349
Alloferin [Roche] *see* 222
Allohydroxyproline *see* 4887
Allohydroxy-D-proline Betaine *see* 9952
Allohydroxy-L-proline Betaine *see* 9952
Alloid [Kyosei] *see* 240
L(+)-Alloisoleucine *see* 5196
2-D-Alloisoleucine Sporidesmolide I *see* 8921
Allomaleic Acid *see* 4309
Allomelanins *see* 5854
Allomethadione *see* 315
Allophycocyanins *see* 7535
Allopregnane, 266
Allopregnane-3α,20α-diol, 267
Allopregnane-3α,20β-diol, 268
Allopregnane-3β,20α-diol, 269
Allopregnane-3β,20β-diol, 270
Allopregnane-3β,21-diol-11,20-dione, 271
Allopregnane-3β,17α-diol-20-one, 272
3,20-Allopregnanedione, 273
Allopregnane-3β,11β,17α,20β,21-pentol, 274
Allopregnane-3β,17α,20β,21-tetrol, 275
Allopregnane-3α,11β,17α,21-tetrol-20-one, 276
Allopregnane-3β,11β,17α,21-tetrol-20-one, 277
Allopregnane-3β,17α,20α-triol, 278
Allopregnane-3β,17α,20β-triol, 279

Allopregnane-3α,17α,21-triol-11,20-dione see 289
Allopregnane-3β,17α,21-triol-11,20-dione, 280
Allopregnane-11β,17α,21-triol-3,20-dione see 4801
Allopregnane-3β,11β,21-triol-20-one, 281
Allopregnane-3β,17α,21-triol-20-one, 282
Allopregnan-3α-ol-20-one, 283
Allopregnan-3β-ol-20-one, 284
Allopregnan-20α-ol-3-one, 285
Allopregnan-20β-ol-3-one, 286
Allo-Puren [Klinge-Nattermann] see 287
Allopurinol, 287
Allopydin [Chugai] see 220
β-D-Allopyranose see 288
Allorphine see 6448
D-Allose, 288
Allose [Kyosei] see 240
Allotelluric Acid see 9272
3α-Allotetrahydrocortisol see 276
Allotetrahydrocortisone, 289
Allotropal [Heyl] see 5890
Alloxan, 290
Alloxan 5-Oxime see 10138
Alloxantin, 291
Alloyohimbine see 10237
Allozym [Sawai] see 287
All-trans-phytofluene see 7544
All-trans-retinol see 10150
4-(All-trans-retinoyl)aminophenol see 4040
Allspice see 7586
Alltox see 9693
D-Allulose see 8109
Allural [Rovi] see 287
Allura Red AC [Allied], 292
Alluval [Berlin-Chemie] see 1418
Allvoran [TAD] see 3132
Ally [Du Pont] see 6244
Allyl Acetate, 293
Allyl Alcohol, 294
Allylamine, 295
1-Allyl-6-amino-3-ethyl-2,4(1H,3H)-pyrimidinedione see 471
1-Allyl-6-amino-3-ethyluracil see 471
6-Allyl-2-amino-5,6,7,8-tetrahydro-4H-thiazolo[4,5-d]azepine see 9209
p-Allylanisole see 3748
Allylbarbital see 1536
Allyl Bromide, 296
5-Allyl-5-(2-bromoallyl)barbituric Acid see 1387
5-Allyl-5-sec-butylbarbituric Acid see 9206
3-Allyl-5-sec-butyl-2-thiohydantoin see 218
Allylcarbamide see 306
Allylcatechol Methylene Ether see 8468
Allyl Chloride, 297
N'-Allyl-4-chloro-6-[(3-hydroxy-2-butenylidene)amino]-m-benzene-disulfonamide see 405
Allyl Cinerins see 257
3-[1-(3-Allyl-S-cupropseudothio-ureido)]benzoic Acid Sodium Salt see 265
Allyl Cyanide, 298
5-Allyl-5-(2-cyclohexen-1-yl)-2-thiobarbituric Acid see 9428
5-Allyl-5-(2-cyclopenten-1-yl)barbituric Acid see 2814
S-Allyl-L-cysteine Sulfoxide see 260
2-(N-Allyl-2,6-dichloroanilino)-2-imidazoline see 244

2-[N-Allyl-N-(2,6-dichlorophenyl)-amino]-2-imidazoline see 244
17-Allyl-1,14-dihydroxy-12-[2-(4-hydroxy-3-methoxycyclohexyl)-1-methylvinyl]-23,25-dimethoxy-13,-19,21,27-tetramethyl-11,28-dioxa-4-azatricyclo[22.3.1.0^{4,9}]octacos-18-ene-2,3,10,16-tetraone see 9200
1-Allyl-1-(3α,17β-dihydroxy-2β-morpholino-5α-androstan-16β-yl)pyrrolidinium 17-Acetate see 8407
1-Allyl-2,3-dimethoxy-4,5-(methylene-dioxy)benzene see 776
1-Allyl-2,5-dimethoxy-3,4-methylene-dioxybenzene see 777
Allyldioxybenzene Methylene Ether see 8468
17-Allyl-4,5α-epoxy-3,14-dihydroxy-morphinan-6-one see 6449
1-[(6-Allylergoline-8β-yl)carbonyl]-1-[3-(dimethylamino)propyl]-3-ethyl-urea see 1637
Allylestrenol, 299
17-Allylestr-4-en-17β-ol see 299
Allyl Ether, 300
1-Allyl-3-ethyl-6-amino-1,2,3,4-tetra-hydro-2,4-pyrimidinedione see 471
Allyl Ethyl Ether, 301
Allyl Glucosinolate see 8691
Allylguaiacol see 3944
Allyl 1,5-Hexadienyltrisulfide see 195
17α-Allyl-17-hydroxyestra-4,9,11-trien-3-one see 327
5-Allyl-5-hydroxy-N-(2-hydroxyeth-yl)-m-anisamide see 243
l,N-Allyl-3-hydroxymorphinan see 5485
17α-Allyl-17-hydroxy-19-nor-4-andro-stene see 299
(—)-N-Allyl-14-hydroxynordihydro-morphinone see 6449
5-Allyl-5-(2-hydroxypropyl)barbituric Acid see 8091
5-Allyl-5-(β-hydroxypropyl)barbituric Acid see 8091
5-Allyl-5-(β-hydroxypropyl)malonyl-urea see 8091
m-[[(Allylimino)mercaptomethyl]-amino]benzoic Acid S-Copper Derivative Sodium Salt see 265
Allyl Iodide, 302
Allyl Iodide Hexamethylenetetramine see 6037
Allyl Iodide Hexamine see 6037
Allyliodourotropine see 6037
5-Allyl-5-isobutylbarbituric Acid see 1536
5-Allyl-5-isobutyl-2-thiobarbituric Acid Sodium Salt see 1553
3-Allyl-5-isobutyl-2-thiohydantoin see 218
Allylisopropylacetylurea see 795
5-Allyl-5-isopropylbarbituric Acid see 794
5-Allyl-5-isopropyl-1-methylbarbituric Acid see 3607
Allyl Isosulfocyanate see 303
Allyl Isothiocyanate, 303
m-[(N-Allyl-1-mercaptoformimidoyl)-amino]benzoic Acid Copper Derivative Sodium Salt see 265
Allylmercaptomethylpenicillin see 7228
Allylmercaptomethylpenicillinic Acid see 7228
5-Allyl-1-methoxy-2,3-(methylenedi-oxy)benzene see 6417
4-Allyl-2-methoxyphenol see 3944
2-(4-Allyl-2-methoxyphenoxy)-N,N-di-ethylacetamide see 44

2-(2-Allyl-6-methoxyphenoxy)triethyl-amine see 4565
5-Allyl-5-(1-methylbutyl)barbituric Acid Sodium Salt see 8563
5-Allyl-5-(1-methylbutyl)malonylurea Sodium Salt see 8563
5-Allyl-5-(1-methylbutyl)-2-thiobarbi-turic Acid see 9437
4-Allyl-1,2-methylenedioxybenzene see 8468
5-Allyl-1-methyl-5-(1-methyl-2-pentynyl)barbituric Acid Sodium Salt see 6061
3-Allyl-5-methyl-2,4-oxazolidinedione see 315
α-3-Allyl-1-methyl-4-phenyl-4-piperi-dinol Propionate see 304
α-3-Allyl-1-methyl-4-phenyl-4-pro-pionoxypiperidine see 304
5-Allyl-5-(1-methylpropyl)barbituric Acid see 9206
5-Allyl-5-(2-methylpropyl)barbituric Acid see 1536
5-Allyl-5-(2-methylpropyl)-2-thiobar-bituric Acid Sodium Salt see 1553
Allyl Methyl Trisulfide see 6094
Allyl Mustard Oil see 303
5-Allyl-5-neopentylbarbituric Acid see 6516
N-Allylnormorphine see 6448
N-Allylnoroxymorphone see 6449
2-(Allyloxy)-4-chloro-N-[2-(diethyl-amino)ethyl]benzamide see 263
[4-(Allyloxy)-3-chlorophenyl]acetic Acid see 220
(±)-1-[β-(Allyloxy)-2,4-dichlorophen-ethyl]imidazole see 3622
1-[o-(Allyloxy)phenyl]-3-(isopropyl-amino)-2-propanol see 7086
p-Allylphenol see 2091
1-(o-Allylphenoxy)-3-(isopropylami-no)-2-propanol see 321
5-Allyl-5-phenylbarbituric Acid see 7349
Allyl 2-Phenylcinchoninate see 2347
α-Allylpiperidine see 2565
Allylprodine, 304
m-Allylpyrocatechin Methylene Ether see 8468
(±)-N-[(1-Allyl-2-pyrrolidinyl)meth-yl]-4-amino-5-(methylsulfamoyl)-o-anisamide see 319
N-(1-Allyl-2-pyrrolidinylmethyl)-2-methoxy-4-amino-5-methylsulfam-oylbenzamide see 319
N-[(1-Allyl-2-pyrrolidinyl)methyl]-6-methoxy-1H-benzotriazole-5-car-boxamide see 246
N-[(1-Allyl-2-pyrrolidinyl)methyl]-5-sulfamoyl-o-veratramide see 10081
5-Allylsulfamoyl-2-chloro-4-(3-hydr-oxybut-2-enylideneamino)benzene-sulfonamide see 405
2-(Allylsulfamoyl)-5-chloro-4-sulfamo-yl-N-(3-hydroxy-2-butenylidene)-aniline see 405
2-Allylsulfamyl-4-sulfamyl-5-chloro-N-(3-hydroxy-2-butenylidene)aniline see 405
Allyl Sulfide, 305
3-((S)-Allylsulfinyl)-L-alanine see 260
Allylthiocarbamide see 9500
3-[(Allylthio)methyl]-6-chloro-3,4-dihydro-2H-1,2,4-benzothiadiazine-7-sulfonamide 1,1-Dioxide see 326
3-Allylthiomethyl-6-chloro-7-sulfamo-yl-3,4-dihydrobenzothiadiazine 1,1-Dioxide see 326
[(Allylthio)methyl] Penicillin see 7228
Allyl Thiourea see 9500

Name Index

Allyl Tribromide see 9745
Allylurea, 306
Allypropymal see 794
Allytrenbolone see 327
Almagate, 307
Alman [Kyosei] see 240
Almarl [Sumitomo] see 827
Almarytm [Synthelabo] see 4136
Almatol [Fujisawa] see 8917
Almax [Almirall] see 307
Almeta [Shionogi] see 221
Alminate [Bristol] see 3227
Alminoprofen, 308
Almitrine, 309
Almitrine Dimesylate see 309
Almocarpine [Ayerst] see 7578
Almodan [Rorer] see 617
Almond, Bitter, 310
Almond, Sweet, 311
Almora [Forest] see 4464
Almuth see 1298
Alnert [Mitsubishi; Eisai] see 1255
Alnide [Cusi] see 2815
Alnovin [Hoechst] see 4012
Alnulin see 9231
ALO-2145 see 791
Alodan [Gerot] see 5894
Alodan (obsolete) [Hoechst] see 2127
Aloe, 312
Aloe-Emodin, 313
Aloeresin A see 312
Aloginan [Tobishi] see 2405
Aloin, 314
Alomide [Alcon] see 5585
Aloperidin see 4629
Alopexil [Fabre] see 6290
Alopresin [Alonga] see 1817
Aloral [Lagap] see 287
Alositol [Tanabe] see 287
Alostil [Sanofi] see 6290
Alotec [Boehringer, Ing.] see 5992
Aloxidone, 315
Aloxiprin, 316
Aloxyn see 4890
Alpen [Lederle] see 628
Alpen-N [Lederle] see 628
Alpen (obsolete) see 7369
Alperidine see 304
Alpha$_1$-antitrypsin see 760
Alphacept [Beecham] see 234
Alphacillin [Merck & Co.] see 7669
alphaCobione [Merck & Co.] see 4854
Alpha D$_3$ [Teva] see 4864
Alphadione see 235
Alphadolone Acetate see 233
Alphadrol [Upjohn] see 4229
Alpha-hypophamine see 7114
Alpha-interferon see 5016
Alphakil [Rentokil] see 2114
Alphalin [Lilly] see 10150
Alphamide see 9098
Alphamin [SS Pharm.] see 2405
Alphamine [Centerchem] see 6272
Alpha-naphthol see 6470
AlphaNine [Alpha Ther.] see 3964
Alpha-pipradrol see 7638
Alphaprodine, 317
alphaRedisol [Merck & Co.] see 4854
Alpha-Ruvite [Savage] see 4854
Alphasol AY see 3034
Alphasol IB see 3238
Alphasol MA see 1295
Alphasone see 238
Alphasone Acetonide see 238
Alphasone Acetophenide see 239
Alphaxalone see 235

Alphenal see 7349
Alphenate see 7349
Alphol see 6501
Alpidem, 318
Alpiny [SS Pharm.] see 45
Alpiropride, 319
Alplax [Gador] see 320
Alprazolam, 320
Alprenolol, 321
Alpress LP [Pfizer] see 7897
Alprostadil see 8063
AL-protein see 657
Alredase [Am. Home] see 9663
Alrheumat [Bayer] see 5316
Alrheumun [Bayer] see 5316
Alsactide, 322
Alsadorm [Woelm] see 3497
Alsanate [Dainippon] see 4385
Alseroxylon see 8293
Alsol [Athenstaedt] see 333
Alstonia Cortex see 3431
Alstonidine, 323
Alstonine, 324
Alsystin [Bayer] see 9809
Altabactina see 4315
Altace [Hoechst-Roussel] see 8283
Altacite [Roussel-UCLAF] see 5696
Altafur [Norwich] see 4315
Altaite see 5446
Altan see 2878
Altapin [Marion] see 3485
Altat [Teikoku Hormone] see 8431
Alteplase see 9608
ALternaGEL [J & J-Merck] see 355
Altex [Cenci] see 8917
Althea, 325
Altheine see 872
Althesin [Glaxo] see 235
Althiazide, 326
Altiazem [Lusofarmaco] see 3247
Altilev [Squibb] see 6812
Altim [Roussel-UCLAF] see 2603
Altimina [Miquel] see 4007
Altimol [E. Merck] see 6668
Altinil [Squibb] see 9440
Altiopril see 6368
Altizide see 326
Alto [Kyosei] see 240
Alto [Sandoz] see 2840
Altodor [Delalande] see 3766
Altosid [Sandoz] see 6063
Altozar IGR [Zoecon] see 4850
Altracin [A. L. Labs] see 965
Altrenogest, 327
Altretamine, 328
Altrigen [Ethigen] see 201
D-Altropyranose see 329
D-Altrose, 329
Alubasine see 3227
Alucol [Wander] see 355
Al-U-Creme [MacAllister] see 355
Aluctyl see 360
Aludrin [Boehringer, Ing.] see 5236
Aludrine see 5236
Aludrox [Wyeth-Ayerst] see 355
Aludyal [Sandoz] see 355
Alufibrate see 2437
Alum see 373
Alum Flour see 373
Alumina see 369
Aluminium see 331
Aluminon, 330
Aluminum, 331
Aluminum Acetate Solution, 332
Aluminum Acetotartrate, 333
Aluminum Acetylsalicylate N.F. see 338
Aluminum Alkyls, 334
Aluminum Aminoacetate (Basic) see 3227

Aluminum Ammonium Chloride see 595
Aluminum Ammonium Sulfate, 335
Aluminum Antimonide, 336
Aluminum Aspirin see 338
Aluminum Benzoate, 337
Aluminum Bis(acetylsalicylate), 338
Aluminum Bismuth Oxide see 1298
Aluminum Borate, 339
Aluminum Boroformate, 340
Aluminum Boro-formicicum see 340
Aluminum Borohydride, 341
Aluminum Bromide, 342
Aluminum tert-Butoxide, 343
Aluminum Calcium Hydride, 344
Aluminum Calcium Silicate see 1685
Aluminum Carbide, 345
Aluminum Cesium Sulfate, 346
Aluminum Chlorate, 347
Aluminum Chloride, 348
Aluminum Chlorohydrate see 356
Aluminum Chlorohydroxide see 356
Aluminum Diacetate, 349
Aluminum Diacetylsalicylate see 338
Aluminum Diaspirin see 338
Aluminum Dihydroxy Allantoinate see 224
Aluminum Dihydroxyaminoacetate see 3227
Aluminum Ethoxide, 350
Aluminum Ethylate see 350
Aluminum Flufenamate see 4167
Aluminum Fluoride, 351
Aluminum Fluosilicate see 352
Aluminum Glycinate see 3227
Aluminum Hexacarbamide Sulfate Triiodide see 353
Aluminum Hexafluorosilicate, 352
Aluminum Hexaurea Sulfate Tri-iodide, 353
Aluminum Hydrate see 355
Aluminum Hydride, 354
Aluminum Hydroxide, 355
Aluminum Hydroxyacetate see 349
Aluminum Hydroxychloride, 356
Aluminum Hypophosphite, 357
Aluminum Iodide, 358
Aluminum Isopropoxide, 359
Aluminum Isopropylate see 359
Aluminum Lactate, 360
Aluminum Lithium Hydride, 361
Aluminum Magnesium Hydroxide (AlMg(OH)$_7$) Monohydrate see 5685
Aluminum Magnesium Silicate, 362
Aluminum β-Naphtholdisulfonate, 363
Aluminum Nicotinate, 364
Aluminum Nitrate, 365
Aluminum Nitride, 366
Aluminum Oleate, 367
Aluminum Orthophosphate see 371
Aluminum Oxalate, 368
Aluminum Oxide, 369
Aluminum Palmitate, 370
Aluminum Phosphate, 371
Aluminum Phosphide, 372
Aluminum Potassium Oxide see 7765
Aluminum Potassium Sulfate, 373
Aluminum Rubidium Sulfate, 374
Aluminum Salicylates, Basic, 375
Aluminum Selenide, 376
Aluminum Silicate, 377
Aluminum Silicofluoride see 352
Aluminum Sodium Carbonate Hexitol Complex see 232
Aluminum Sodium Carbonate Hydroxide see 3228
Aluminum Sodium Chloride see 8836
Aluminum Sodium Oxide see 8715
Aluminum Sodium Sulfate, 378

Aluminum Stearate, 379
Aluminum Subacetate *see* 349
Aluminum Subacetate Solution, 380
Aluminum Sulfate, 381
Aluminum Sulfide, 382
Aluminum Sulfocyanate *see* 384
Aluminum Tartrate, 383
Aluminum Tetrahydroborate *see* 341
Aluminum Thiocyanate, 384
Aluminum Trifluoride *see* 351
Aluminum Trihydrate *see* 355
Aluminum Tris(ethyl Phosphite) *see* 4278
Aluminum Tris(*O*-ethylphosphonate) *see* 4278
Aluminum Tristearate *see* 379
Aluminum Zinc Sulfate, 385
Alum Meal *see* 373
Alumnol [Winthrop] *see* 363
Alum Root *see* 4412
Alundum *see* 369
Alunitine *see* 364
Alunogenite *see* 381
Alunozal *see* 375
Alupent [Boehringer, Ing.] *see* 5992
Aluphos *see* 371
Alurate [Roche] *see* 794
Alurate Sodium [Roche] *see* 794
Alure *see* 353
Alurene *see* 2221
Alusa [Towa Yakuhin] *see* 224
"Alutiae" *see* 7689
Alutyl *see* 2347
Aluwets [Stiefel] *see* 348
Alven [Firma] *see* 9738
Alveograf *see* 3519
Alverine, 386
Alvo [Wyeth; Taisho] *see* 7057
Alvodine [Winthrop] *see* 7587
Alvonal MR [Gödecke] *see* 2830
Alyrane [Anaquest] *see* 3621
Alysine [Merrell] *see* 8819
Alzinox [Berlex] *see* 3227
AM-109 *see* 6058
AM-715 *see* 6793
AM-725 *see* 7197
AM-833 *see* 4137
AM-1091 *see* 2413
AMA-1080 *see* 1922
Amabevan *see* 1830
Amacetam *see* 7886
Amadil *see* 45
Amadou *see* 183
Amal *see* 607
Amanin *see* 387
Amanitin, 387
α-Amanitin *see* 387
β-Amanitin *see* 387
Amanozine, 388
Amantadine, 389
Amantanium Bromide, 390
Amantol [Rotta] *see* 390
Amaranth (Dye), 391
Amaranth (Plant), 392
Amarogentin, 393
Amarolide, 394
Amaromycin *see* 7566
Amarsan *see* 49
Amarylline *see* 5656
Amasulin [Takeda] *see* 1922
Amasust *see* 607
Amatine [Roberts] *see* 6272
Amaze [Mobay] *see* 5187
Amazolon [Sawai] *see* 389
Ambacamp [Upjohn] *see* 961
Ambamide *see* 5683
Ambaxin [Upjohn] *see* 961
Ambazone, 395
Amben *see* 443
Ambene [Merckle] *see* 7431

Ambenonium Chloride, 396
Amber, 397
Amber [Ciba-Geigy] *see* 9732
Amber Acid *see* 9038
Ambergris, 398
Amberlite [Rohm & Haas], 399
Ambien [Searle] *see* 10321
Ambilhar [Ciba] *see* 6656
Ambinon [Organon] *see* 2273
Ambisome [Vestar] *see* 627
Ambivalon [Nattermann] *see* 512
Amblosin [Hoechst] *see* 628
Amblygonite *see* 5543
Amboclorin *see* 2116
Ambodryl [Parke, Davis] *see* 1439
Ambox [Hoechst] *see* 1265
Ambracyn [Biomedis] *see* 9337
Ambramicina [Marion Merrell Dow] *see* 9337
Ambramycin [Galenika] *see* 9337
Ambra-Vena [Lepetit] *see* 7606
Ambril [Cascan] *see* 401
Ambrocef [Lusofarmaco] *see* 2030
Ambrosia *see* 6888
Ambrosin, 400
Ambroxol, 401
Ambucaine, 402
Ambucetamide, 403
Ambuphylline, 404
Ambush [ICI] *see* 7321
Ambuside, 405
Ambuterol *see* 5674
Ambutonium Bromide, 406
Ambutoxate *see* 402
Ambutyrosin *see* 1559
Ambylan [Lilly] *see* 792
AMCHA *see* 9704
Amchem 65-81B *see* 2115
Amchem 70-25 *see* 1567
Amchlor [Cooper] *see* 537
Amcide *see* 589
Amciderm [E. Merck] *see* 407
Amcill [Parke, Davis] *see* 628
Amcill-S [Parke, Davis] *see* 628
Amcinonide, 407
Amco [Am. Mfg.] *see* 7741
AMD *see* 6132
Amdinocillin, 408
Amdinocillin Pivoxil, 409
Amdro [Am. Cyanamid] *see* 4802
Amebacilin *see* 4308
Ameban *see* 1830
Amebarsin *see* 3394
Amebarsone [Lilly] *see* 1830
Amebil *see* 5052
Ame-Boots *see* 3246
Amechol [Moorsfields] *see* 6003
Amedel [Dainippon] *see* 7634
Ameisensäure (German) *see* 4268
Amekrin [Parke, Davis] *see* 635
Ameliaroside *see* 7549
Amen [Carnrick] *see* 5838
Amenide [Sterling Winthrop] *see* 8241
Amenox [Sanofi-Winthrop] *see* 8241
Amenyl *see* 2152
Amerfil [Am. Thermoplastic] *see* 7741
Americaine [Arnar-Stone] *see* 1116
American Cyanamid 4124 *see* 3091
American Cyanamid 12880 *see* 3269
American Cyanamid 18133 *see* 9482
American Cyanamid 38023 *see* 3973
American Elder *see* 8499
American Hellebore *see* 10090
American Horsemint *see* 6326
American Indian Hemp *see* 784
American Nightshade Root *see* 7546
American Penicillin *see* 1178
American Saffron *see* 1919
American Spikenard *see* 804
American Valerian *see* 2839

American Veratrum *see* 10090
American Wormseed *see* 6888
Americium, 410
Amerscan MDP [Amersham] *see* 5837
Ametazole *see* 1230
Amethocaine Hydrochloride *see* 9330
Amethone [Abbott] *see* 610
Amethopterin *see* 6065
A-Methopterin *see* 6065
Amethyst *see* 8637
Ametox *see* 8844
Ametrex [Makhteshim-Agan] *see* 411
Ametriodinic Acid *see* 5031
Ametryn, 411
Ametryne *see* 411
Ametycine [Sanofi Winthrop] *see* 6301
Amexine [CFPI] *see* 1567
Amezinium Methyl Sulfate, 412
Amfamox [AMRAD] *see* 3972
Amfebutamon(e) *see* 1523
Amfecloral *see* 620
Amfenac, 413
Amfepramone *see* 3175
d-Amfetasul [Pitman-Moore] *see* 2996
Amfipen [Brocades] *see* 628
Amfomycin *see* 625
Amianthus *see* 863
Amiben [Rhone-Poulenc] *see* 2115
Amibiarson *see* 1830
Amibufen [Harris] *see* 4925
Amicar [Lederle] *see* 451
Amicarbalide, 414
Amicardine *see* 5321
Amicetin, 415
Amicibone, 416
Amicos [Banyu] *see* 2404
Amicoumacin A, 417
Amidalgon *see* 3354
Amidate [Abbott] *see* 3927
Amidazine *see* 3783
Amidazophen *see* 495
Amidephrine, 418
Amidephrine Mesylate *see* 418
[[1-[*N*-(*p*-Amidinobenzoyl)-L-tyrosyl]-4-piperidinyl]oxy]acetic Acid *see* 5362
N-Amidino-3,5-diamino-6-chloropyr-azinamide *see* 426
N-Amidino-3,5-diamino-6-chloropyr-azinecarboxamide *see* 426
N-Amidino-2-(2,6-dichlorophenyl)acet-amide *see* 4590
N-(2'-Amidinoethyl)-3-aminocyclopen-tanecarboxamide *see* 419
2-[*N*-(2-Amidinoethyl)carbamoyl]-5-iminopyrrolidine *see* 6766
N'-(2-Amidinoethyl)-4-(2-guanidino-acetamido)-1,1'-dimethyl-*N*,4'-bi-[pyrrole-2-carboxamide] *see* 6565
N-(2-Amidinoethyl)-5-imino-2-pyrrol-idinecarboxamide *see* 6766
N-Amidinoglycine *see* 4505
Amidinoguanidine *see* 1261
1-Amidinohydrazono-4-thiosemicarb-azono-2,5-cyclohexadiene *see* 395
Amidinomycin, 419
6-Amidino-2-naphthyl-4-guanidino-benzoate *see* 6436
L-6-Amidinonorleucine *see* 5001
7-*m*-Amidinophenyldiazoamino-2-amino-10-ethyl-9-phenylphenan-thridinium Chloride *see* 5199
8-[3-(*m*-Amidinophenyl)-2-triazeno]-3-amino-5-ethyl-6-phenylphenan-thridinium Chloride *see* 5199
N-Amidinosarcosine *see* 2637
N[1]-Amidinosulfanilamide *see* 9076
2-Amidino-1,2,3,4-tetrahydroiso-quinoline *see* 2901
Amidochlor, 420

4-Aminobenzenesulfonamide see 9094
4-(4′-Aminobenzenesulfonamido)benzenesulfonamide see 9100
6-(4-Aminobenzenesulfonamido)-4,5-dimethoxypyrimidine see 9074
5-(p-Aminobenzenesulfonamido)-3,4-dimethylisooxazole see 9125
2-(p-Aminobenzenesulfonamido)-4,5-dimethyloxazole see 9093
2-(p-Aminobenzenesulfonamido)-5-ethylthiadiazole see 9075
4-(p-Aminobenzenesulfonamido)-2-hydroxybenzoic Acid see 9095
2-(p-Aminobenzenesulfonamido)-3-methoxypyrazine see 9078
2-(p-Aminobenzenesulfonamido)-5-methoxypyrimidine see 9082
5-p-Aminobenzenesulfonamido-3-methylisothiazole see 9113
5-(p-Aminobenzenesulfonamido)-3-methyl-1-phenylpyrazole see 9118
2-(p-Aminobenzenesulfonamido)-5-methylthiadiazole see 9084
2-(p-Aminobenzenesulfonamido)-4-methylthiazole see 9088
3-(p-Aminobenzenesulfonamido)-2-phenyl-5-methylpyrazole see 9118
3-(p-Aminobenzenesulfonamido)-2-phenylpyrazole see 9105
2-(p-Aminobenzenesulfonamido)pyridine see 9108
2-(p-Aminobenzenesulfonamido)thiazole see 9115
p-Aminobenzenesulfonanilide see 9064
2-Aminobenzenesulfonic Acid see 7010
3-Aminobenzenesulfonic Acid see 5988
4-Aminobenzenesulfonic Acid see 9096
p-Aminobenzenesulfonoacetamide see 9067
4-Aminobenzenesulfono-p-sulfamoylanilide see 9100
p-Aminobenzenesulfonyl-2-amino-4,5-dimethyloxazole see 9093
6-(p-Aminobenzenesulfonyl)amino-2,4-dimethylpyrimidine see 9124
4-(p-Aminobenzenesulfonyl)amino-2-methyl-6-methoxypyrimidine see 9085
N-(p-Aminobenzenesulfonyl)benzamide see 9065
N¹-(3-Aminobenzenesulfonyl)-N²-n-butylurea see 1615
N-(4-Aminobenzenesulfonyl)-N′-butylurea see 1881
p-Aminobenzenesulfonyl Fluoride see 9099
p-Aminobenzenesulfonylguanidine see 9076
p-Aminobenzoate of Diethylaminoneopentyl Alcohol see 3270
4-Amino-2-[4-(1,4-benzodioxan-2-carbonyl)piperazin-1-yl]-6,7-dimethoxyquinazoline see 3489
2-Aminobenzoic Acid see 442
3-Aminobenzoic Acid see 441
4-Aminobenzoic Acid see 443
m-Aminobenzoic Acid, 441
o-Aminobenzoic Acid, 442
p-Aminobenzoic Acid, 443
p-Aminobenzoic Acid 2-n-Amylaminoethyl Ester see 6435
4-Aminobenzoic Acid Butyl Ester see 1538
4-Aminobenzoic Acid 2-(Diethylamino)ethyl Ester see 7937
p-Aminobenzoic Acid β-Diethylaminoisohexyl Ester Methanesulfonate see 5476
p-Aminobenzoic Acid N,N-Diethyl-

leucinol Ester Methanesulfonate see 5476
4-Aminobenzoic Acid Ethyl Ester see 1116
3-Aminobenzoic Acid Ethyl Ester Methanesulfonate see 9751
2-Aminobenzoic Acid Methyl Ester see 6099
4-Aminobenzoic Acid 2-Methylpropyl Ester see 5148
p-Aminobenzoic Acid Monoglyceryl Ester see 4496
p-Aminobenzoic Acid 2-n-Pentylaminoethyl Ester see 6435
2-Aminobenzoic Acid 3-Phenyl-2-propenyl Ester see 2363
p-Aminobenzoic Acid Potassium Salt see 7766
2-Aminobenzothiazole, 444
6-Aminobenzothiazole, 445
3-(p-Aminobenzoyl)-1-di-n-butylaminopropane see 1531
N-(p-Aminobenzoyl)aminoacetic Acid see 462
2-Amino-3-benzoylbenzeneacetic Acid see 413
p-Aminobenzoyldibutylaminopropanol see 1531
p-Aminobenzoyldiethylaminoethanol see 7937
p-Aminobenzoyl-N-1-diethylamino-1-isobutylethanol Methanesulfonate see 5476
1-Aminobenzoyl-2,2-dimethyl-3-diethylaminopropanol see 3270
N-(4-Aminobenzoyl)-L-glutamic Acid see 446
N-(p-Aminobenzoyl)glutamic Acid, 446
N-(4-Aminobenzoyl)glycine see 462
(E)-2-Amino-6-benzoyl-1-(isopropylsulfonyl)benzimidazole Oxime see 3637
2-Amino-3-benzoylphenylacetic Acid see 413
D(—)-α-Aminobenzylpenicillin see 628
4-Amino-N-(1-benzyl-4-piperidyl)-5-chloro-o-anisamide see 2404
2-Amino-6-benzyl-4,5,6,7-tetrahydro-thieno[2,3-c]pyridine-3-carboxylic Acid Ethyl Ester see 9590
4-Aminobiphenyl see 1276
5-Amino-(3,4′-bipyridin)-6(1H)-one see 634
5-Amino-1,3-bis(2-ethylhexyl)hexahydro-5-methylpyrimidine see 4741
2-Amino-1,2-bis(p-methoxyphenyl)ethanol see 5194
2-Amino-5-bromobenzoic Acid see 1426
2-Amino-3-(4-bromobenzoyl)benzeneacetic Acid see 1411
4-Amino-5-bromo-N-[2-(diethylamino)ethyl]-o-anisamide see 1454
4-Amino-5-bromo-N-[2-(diethylamino)ethyl]-2-methoxybenzamide see 1454
4-Amino-N-(5-bromo-4,6-dimethyl-2-pyrimidinyl)benzenesulfonamide see 9066
1-Aminobutane see 1578
2-Aminobutane see 1579
(S)-Aminobutanedioic Acid see 875
2-Aminobutanoic Acid see 448
3-Aminobutanoic Acid see 449
4-Aminobutanoic Acid see 450
2-Amino-1-butanol, 447
Aminobutene see 3309
3-Amino-4-butoxybenzoic Acid 2-[2-

(Diethylamino)ethoxy]ethyl Ester see 1235
3-Amino-2-butoxybenzoic Acid 2-(Diethylamino)ethyl Ester see 5978
4-Amino-2-butoxybenzoic Acid 2-Diethylaminoethyl Ester see 402
4-Amino-3-butoxybenzoic Acid 2-(Diethylamino)ethyl Ester see 1076
2-Amino-n-butyl Alcohol see 447
4-Amino-N-[(butylamino)carbonyl]-benzenesulfonamide see 1881
4-Amino-α-[(tert-butylamino)methyl]-3-chloro-5-(trifluoromethyl)benzyl Alcohol see 5674
4-Amino-α-[(tert-butylamino)methyl]-3,5-dichlorobenzyl Alcohol see 2407
(3β,5α,7α)-3-[[3-((4-Aminobutyl)-amino)propyl]amino]-cholestane-7,-24-diol 24-Hydrogen Sulfate see 8922
N¹-[3-[(4-Aminobutyl)amino]propyl]-13-[[4-amino-4,6-dideoxy-α-L-talo-pyranosyl)oxy]-19-demethyl-12-hydroxybleomycinamide see 9213
3β-N-1-[N-[3-(4-Aminobutyl)]-1,3-diaminopropane]-7α,24ζ-dihydroxy-5α-cholestane 24-Sulfate see 8922
4-(Aminobutyl)guanidine see 186
3-(2-Aminobutyl)indole see 3937
N-(4-Aminobutyl)-N-(3-methyl-2-butenyl)guanidine see 8896
4-Amino-6-tert-butyl-3-(methylthio)-as-triazin-5(4H)-one see 6239
5-Amino-N-butyl-2-(2-propynyloxy)-benzamide see 7177
2-Amino-4-(S-butylsulfonimidoyl)-butanoic Acid see 1556
α-Aminobutyric Acid, 448
α-Amino-n-butyric Acid see 448
β-Aminobutyric Acid, 449
β-Amino-n-butyric Acid see 449
γ-Aminobutyric Acid, 450
γ-Amino-n-butyric Acid see 450
α-Aminocaproic Acid see 6800
ε-Aminocaproic Acid, 451
Aminocaproic Lactam see 1805
Aminocarb, 452
3-[(Aminocarbonyl)amino]-L-alanine see 212
[3-[(Aminocarbonyl)amino]-2-methoxypropyl]chloromercury see 2154
(Z)-4-[(Aminocarbonyl)amino]-4-oxo-2-butenoic Acid see 5746
2,2′-[[[4-[(Aminocarbonyl)amino]-phenyl]arsinidene]bis(thio)]bis-[acetic Acid] see 9459
2,2′-[[[4-[(Aminocarbonyl)amino]-phenyl]arsinidene]bis(thio)]bis-[benzoic Acid] see 9458
[4-[(Aminocarbonyl)amino]phenyl]arsonic Acid see 1830
[6R-[6α,7β(Z)]]-4-(Aminocarbonyl)-1-[[7-[[(5-amino-1,2,4-thiadiazol-3-yl)-(methoxyimino)acetyl]amino]-2-carboxy-8-oxo-5-thia-1-azabicyclo-[4.2.0]oct-2-en-3-yl]methyl]-1-azo-niabicyclo[2.2.2]octane Inner Salt see 1970
N-(Aminocarbonyl)-L-aspartic Acid Calcium Salt (1:1) see 1695
N-(Aminocarbonyl)benzeneacetamide see 7343
N-(Aminocarbonyl)-2-bromo-2-ethylbutanamide see 1879
N-(Aminocarbonyl)-2-bromo-3-methylbutanamide see 1418
[6R-[6α,7β(R*)]]-4-(Aminocarbonyl)-1-[[2-carboxy-8-oxo-7-[(phenylsulfoacetyl)amino]-5-thia-1-azabicyclo-

4-Amino-*N*-(5-ethyl-1,3,4-thiadiazol-2-yl)benzenesulfonamide *see* 9075
2-Amino-4-(ethylthio)butyric Acid *see* 3784
[5*R*-[5α,6α(*R**)]]-3-[(2-Aminoethyl)-thio]-6-(1-hydroxyethyl)-7-oxo-1-azabicyclo[3.2.0]hept-2-ene-2-carboxylic Acid *see* 9448
2-(2-Aminoethyl)-2-thiopseudourea Dihydrobromide *see* 176
3-Amino-α-ethyl-2,4,6-triiodobenzenepropanoic Acid *see* 5072
3-Amino-α-ethyl-2,4,6-triiodohydrocinnamic Acid *see* 5072
2-Amino-6-[(*p*-fluorobenzyl)amino]-3-pyridinecarbamic Acid Ethyl Ester *see* 4227
6-Amino-2-(fluoromethyl)-3-(2-methylphenyl)-4(3*H*)-quinazolinone *see* 181
6-Amino-2-fluoromethyl-3-(*o*-tolyl)-4(3*H*)-quinazolinone *see* 181
[2-Amino-6-[[(4-fluorophenyl)methyl]-amino]-3-pyridinyl]carbamic Acid Ethyl Ester *see* 4227
S-2-Amino-3-fluoropropanoic Acid *see* 4206
4-Amino-5-fluoro-2(1*H*)-pyrimidinone *see* 4161
4-Aminofolic Acid *see* 493
Aminoform *see* 6036
Aminoformamidine *see* 4591
N-[4-[[(2-Amino-5-formyl-1,4,5,6,7,8-hexahydro-4-oxo-6-pteridinyl)meth-yl]amino]benzoyl]-L-glutamic Acid *see* 4254
N-[*p*-[[(2-Amino-5-formyl-5,6,7,8-tetrahydro-4-hydroxy-6-pteridinyl)-methyl]amino]benzoyl]glutamic Acid *see* 4254
Amino-G-acid *see* 421
6-Amino-5-(β-D-glucopyranosyloxy)-2,4(1*H*,3*H*)-pyrimidinedione *see* 2576
2-Aminoglutaramic Acid *see* 4479
4-Amino-L-glutaramic Acid *see* 5193
α-Aminoglutaric Acid *see* 4477
α-Aminoglutaric Acid Lactam *see* 8185
Aminoglutethimide, 460
Aminoguanidine, 461
1-Amino-19-guanidino-11-hydroxy-4,-9,12-triazanonadecane-10,13-dione *see* 4610
2-Amino-4-(guanidinooxy)butyric Acid *see* 1786
2-Amino-5-guanidinovaleric Acid *see* 817
1-Amino-4-guanidobutane *see* 186
2-Aminoheptane *see* 9934
[3*R*-[3α,4α(*S**),4aβ,5β,6α]]-2-Amino-*N*-(3,4,4a,5,6,7-hexahydro-5,6,8-tri-hydroxy-3-methyl-1-oxo-1*H*-2-ben-zopyran-4-yl)propanamide *see* 139
2-Aminohexanedioic Acid *see* 436
2-Aminohexanoic Acid *see* 6800
6-Aminohexanoic Acid *see* 451
4-Amino-5-hexenoic Acid *see* 10114
p-Aminohippuric Acid, 462
6-Amino-2-[(hydrazinocarbonyl)-amino]-2,3-dihydro-1,3-dioxo-1*H*-benz[*de*]isoquinoline-5,8-disulfonic Acid Dilithium Salt *see* 5623
α-Aminohydrocinnamic Acid *see* 7425
o-Aminohydrocinnamic Acid Lactam *see* 4820
2-Amino-1-hydroxybenzene *see* 481
3-Amino-1-hydroxybenzene *see* 480
4-Amino-1-hydroxybenzene *see* 482

5-Amino-2-hydroxybenzene-1-carboxylic Acid *see* 5964
(*S*)-α-Amino-4-hydroxybenzenepropanoic Acid *see* 9970
4-Amino-2-hydroxybenzoic Acid *see* 498
5-Amino-2-hydroxybenzoic Acid *see* 5964
4-Amino-2-hydroxybenzoic Acid Hydrazide *see* 499
3-Amino-4-hydroxybenzoic Acid Methyl Ester *see* 7011
4-Amino-2-hydroxybenzoic Acid Phenyl Ester *see* 7426
α-Amino-*p*-hydroxybenzylpenicillin *see* 617
2-Amino-4-hydroxybutanoic Acid *see* 4777
3-Amino-4-hydroxybutanoic Acid *see* 463
4-Amino-3-hydroxybutanoic Acid *see* 464
[*R*-(*R**,*S**)]-2-Amino-3-hydroxybutanoic Acid *see* 9522
4-(2-Amino-1-hydroxybutyl)-1,2-benzenediol *see* 3880
*N*⁶-(4-Amino-2-hydroxybutyl)-2,6-diaminohexanoic Acid *see* 4918
(4-Amino-1-hydroxybutylidene)-bisphosphonic Acid *see* 228
(*R*)-*N*⁶-(4-Amino-2-hydroxybutyl)-L-lysine *see* 4918
2-Amino-3-hydroxybutyric Acid *see* 9522
2-Amino-4-hydroxybutyric Acid *see* 4777
3-Amino-4-hydroxybutyric Acid, 463
4-Amino-3-hydroxybutyric Acid, 464
α-Amino-β-hydroxybutyric Acid *see* 9522
α-Amino-γ-hydroxy-*n*-butyric Acid *see* 4777
γ-Amino-β-hydroxybutyric Acid *see* 464
γ-Amino-β-hydroxybutyric Acid Trimethylbetaine *see* 1898
1-*N*-[(*S*)-4-Amino-2-hydroxybutyryl]-dibekacin *see* 809
1-*N*-[(*S*)-4-Amino-2-hydroxybutyryl]-3′,4′-dideoxykanamycin B *see* 809
1-*N*-[L(—)-4-Amino-2-hydroxybutyr-yl]kanamycin A *see* 425
1-*N*-[(*S*)-4-Amino-2-(hydroxybutyryl)]-ribostamycin *see* 1559
(—)-(2*S*,3*R*)-2-Amino-3-hydroxy-3-(3,-4-dihydroxyphenyl)propionic Acid *see* 3513
2-Amino-4-hydroxy-6-(1,2-dihydroxy-propyl)pteridine *see* 1270
β-Amino-4-hydroxy-3,5-diiodobenzenepropanoic Acid *see* 1228
β-Amino-4-hydroxy-3,5-diiodohydrocinnamic Acid *see* 1228
4-Amino-4′-(β-hydroxyethylamino)-diphenyl Sulfone *see* 9097
(*R*)-4-(2-Amino-1-hydroxyethyl)-1,2-benzenediol *see* 6788
5-*O*-[2,3-*O*-[6-(1-Amino-2-hydroxyeth-yl)tetrahydro-3,4,5-trihydroxy-2*H*-pyran-2-ylidene]-β-D-talopyranos-yl]-2-deoxy-*N*³-methyl-D-strept-amine *see* 2979
α-Amino-β-hydroxyglutaric Acid *see* 4871
1-[(2*S*)-6-Amino-2-hydroxyhexanoyl]-L-proline, Hydrogen (4-Phenylbut-yl)phosphonate (Ester) *see* 2038
α-Amino-3-hydroxyhydrocinnamic Acid *see* 9971

α-Amino-*p*-hydroxyhydrocinnamic Acid *see* 9970
2-Amino-1,9-[[2-hydroxy-1-(hydroxy-methyl)ethoxy]methyl]-6*H*-purin-6-one *see* 4374
α-Amino-3-hydroxy-5-isoxazoleacetic Acid *see* 4922
Amino-(3-hydroxy-5-isoxazolyl)acetic Acid *see* 4922
2-Amino-5-hydroxylevulinic Acid *see* 4779
2-Amino-5-[1-hydroxy-2-[(1-methyl-ethyl)amino]ethyl]benzonitrile *see* 2336
[*S*-(*R**,*R**)]-4-Amino-3-hydroxy-6-methylheptanoic Acid *see* 8956
4-Amino-5-hydroxymethyl-2-meth-oxypyrimidine *see* 964
4-Amino-5-hydroxymethyl-2-methyl-pyrimidine *see* 9697
6-Amino-5-hydroxymethyl-2-methyl-pyrimidine *see* 9697
4-Amino-5-hydroxymethyl-2-methyl-thiopyrimidine *see* 6055
(2*R*-*cis*)-4-Amino-1-[2-(hydroxymeth-yl)-1,3-oxathiolan-5-yl]-2(1*H*)-pyri-midinone *see* 5365
2-Amino-9-hydroxymethyl-3-oxo-3*H*-phenoxazine-1-carboxylic Acid *see* 2355
[2*R*-[2α(*S**),5β,6α]]-4-[(2-Amino-3-hydroxy-2-methyl-1-oxopropyl)-amino]-*N*-[1-[5-[[4,6-dideoxy-4-(dimethylamino)-α-D-glucopyrano-syl]oxy]tetrahydro-6-methyl-2*H*-pyran-2-yl]-1,2-dihydro-2-oxo-4-pyrimidinyl]benzamide *see* 415
7-Amino-3-(hydroxymethyl)-8-oxo-5-thia-1-azabicyclo[4.2.0]oct-2-ene-2-carboxylic Acid Acetate Ester *see* 453
2-Amino-4-(hydroxymethylphosphin-yl)butanoic Acid *see* 7495
2-Amino-2-hydroxymethyl-1,3-pro-panediol *see* 9902
4-Amino-5-hydroxy-2,7-naphthalene-disulfonic Acid *see* 6472
4-Amino-3-hydroxy-1-naphthalene-sulfonic Acid *see* 474
5-Amino-6-hydroxy-2-naphthalene-sulfonic Acid *see* 475
L-2-Amino-3-(hydroxynitrosoamino)-propanoic Acid *see* 207
N-[2-Amino-4-(3-hydroxy-2-oxo-3-azetidinyl)-1-oxobutyl]-L-threonine *see* 10183
2-Amino-5-hydroxy-4-oxopentanoic Acid *see* 4779
[*S*-(*R**,*S**)]-*N*-(3-Amino-2-hydroxy-1-oxo-4-phenylbutyl)-L-leucine *see* 9973
α-Amino-3-hydroxy-4-oxo-1(4*H*)-pyri-dinepropanoic Acid *see* 6286
2-Amino-5-hydroxypentanoic Acid *see* 7252
(—)-6-[2-Amino-2-(*p*-hydroxyphenyl)-acetamido]-3,3-dimethyl-7-oxo-4-thia-1-azabicyclo[3.2.0]heptane-2-carboxylic Acid *see* 617
7-[D-(—)-α-Amino-α-(4-hydroxyphen-yl)acetamido]-3-methyl-3-cephem-4-carboxylic Acid Monohydrate *see* 1963
(6*R*,7*R*)-7-[(*R*)-2-Amino-2-(*p*-hydroxy-phenyl)acetamido]-8-oxo-3-propen-yl-5-thia-1-azabicyclo[4.2.0]oct-2-ene-2-carboxylic Acid Monohydrate *see* 1992
6-[D(—)-α-Amino-*p*-hydroxyphenyl-acetamido]penicillanic Acid *see* 617

6-[D(—)-α-Aminophenylacetamido]-
  penicillanic Acid (5-Methyl-2-oxo-
  1,3-dioxol-4-yl)methyl Ester *see*
  5460
6-[D(—)-α-Aminophenylacetamido]-
  penicillanic Acid Phthalide Ester *see*
  9204
6-[D-α-Aminophenylacetamido]peni-
  cillanic Acid Pivaloyloxymethyl
  Ester *see* 7669
6′-(2-Amino-2-phenylacetamido)-
  penicillanoyloxymethyl Penicillanate 1,1-Dioxide *see* 9166
p-Aminophenylacetic Acid, 483
α-Aminophenylacetic Acid *see* 7446
[6*R*-[6α,7β(*R**)]]-7-[(Aminophenylacetyl)amino]-3-chloro-8-oxo-1-aza-
  bicyclo[4.2.0]oct-2-ene-2-carboxylic
  Acid Monohydrate *see* 5606
[6*R*-[6α,7β(*R**)]]-7-[(Aminophenylacetyl)amino]-3-chloro-8-oxo-5-thia-
  1-azabicyclo[4.2.0]oct-2-ene-2-carb-
  oxylic Acid Monohydrate *see* 1962
[2*S*-[2α,5α,6β(*S**)]]-6-[(Aminophenylacetyl)amino]-3,3-dimethyl-7-oxo-4-
  thia-1-azabicyclo[3.2.0]heptane-2-
  carboxylic Acid *see* 628
[2*S*-[2α,5α,6β(*S**)]]-6-[(Aminophenylacetyl)amino]-3,3-dimethyl-7-oxo-4-
  thia-1-azabicyclo[3.2.0]heptane-2-
  carboxylic Acid 1,3-Dihydro-3-oxo-
  1-isobenzofuranyl Ester *see* 9204
[2*S*-[2α,5α,6β(*S**)]]-6-[(Aminophenylacetyl)amino]-3,3-dimethyl-7-oxo-4-
  thia-1-azabicyclo[3.2.0]heptane-2-
  carboxylic Acid (2,2-Dimethyl-1-
  oxopropoxy)methyl Ester *see* 7669
[2*S*-[2α(2*R**,5*S**),5α,6β(*S**)]]-6-
  [(Aminophenylacetyl)amino]-3,3-
  dimethyl-7-oxo-4-thia-1-azabicyclo-
  [3.2.0]heptane-2-carboxylic Acid
  [[(3,3-Dimethyl-7-oxo-4-thia-1-aza-
  bicyclo[3.2.0]hept-2-yl)carbonyl]-
  oxy]methyl Ester, *S*,*S*-Dioxide *see*
  9166
[2*S*-[2α,5α,6β(*S**)]]-6-[(Aminophenylacetyl)amino]-3,3-dimethyl-7-oxo-4-
  thia-1-azabicyclo[3.2.0]heptane-2-
  carboxylic Acid 1-[(Ethoxycarbonyl)oxy]ethyl Ester *see* 961
[2*S*-[2α,5α,6β(*S**)]]-6-[(Aminophenylacetyl)amino]-3,3-dimethyl-7-oxo-4-
  thia-1-azabicyclo[3.2.0]heptane-2-
  carboxylic Acid (5-Methyl-2-oxo-
  1,3-dioxol-4-yl)methyl Ester *see*
  5460
[6*R*-[6α,7β(*R**)]]-7-[(Aminophenylacetyl)amino]-3-methyl-8-oxo-5-
  thia-1-azabicyclo[4.2.0]oct-2-ene-2-
  carboxylic Acid *see* 2021
[6*R*-[6α,7β(*R**)]]-7-[(Aminophenylacetyl)amino]-3-methyl-8-oxo-5-
  thia-1-azabicyclo[4.2.0]oct-2-ene-2-
  carboxylic Acid (2,2-Dimethyl-1-
  oxopropoxy)methyl Ester *see* 7670
p-Aminophenyl-*p*-aminomethylphenyl
  Sulfone *see* 9098
4-Aminophenyl-2′-aminothiazolyl-5′-
  sulfone *see* 9444
(4-Aminophenyl)arsonic Acid *see* 830
(4-Aminophenyl)arsonic Acid Sodium
  Salt *see* 8719
N-(4-Aminophenyl)-1,4-benzenedi-
  amine *see* 3022
4-Amino-*N*-phenylbenzenesulfonamide
  see 9064
p-Aminophenyl(2-benzoyl-1-phenyl-
  ethyl)sulfone *see* 254
4-Amino-3-phenylbutanoic Acid *see*
  484

2-Amino-1-phenyl-1-butanol *see* 491
β-Amino-α-phenylbutyl Alcohol *see*
  491
4-Amino-3-phenylbutyric Acid, 484
1-Amino-2-phenylethane *see* 7371
1-(3-Aminophenyl)ethanone *see* 433
2-Amino-1-phenylethanone *see* 7347
p-Aminophenyl Ethyl Ether *see* 7373
2-(*p*-Aminophenyl)-2-ethylglutarimide
  see 460
N-β-(*p*-Aminophenyl)ethylnormeperi-
  dine *see* 695
N-[β-(*p*-Aminophenyl)ethyl]-4-phenyl-
  4-carbethoxypiperidine *see* 695
1-[2-(4-Aminophenyl)ethyl]-4-phenyl-
  4-piperidinecarboxylic Acid Ethyl
  Ester *see* 695
3-(4-Aminophenyl)-3-ethyl-2,6-piperi-
  dinedione *see* 460
p-Aminophenyl *p*-(2-Hydroxyethyl-
  amino)phenyl Sulfone *see* 9097
4-[(4-Aminophenyl)(4-imino-2,5-cyclo-
  hexadien-1-ylidene)methyl]-2-meth-
  ylbenzenamine Monohydrochloride
  see 5686
α⁴-(*p*-Aminophenyl)-α⁴-(4-imino-2,5-
  cyclohexadien-1-ylidene)-2,4-xyli-
  dine Hydrochloride *see* 5686
2-Amino-5-phenyl-2-oxazoline *see* 497
2-Amino-5-phenyl-4(5*H*)-oxazolone *see*
  7206
β-(4-Aminophenyl)propane *see* 2686
(*S*)-2-Amino-3-phenylpropanoic Acid
  see 7425
1-Amino-1-phenyl-2-propanol *see* 467
2-Amino-1-phenyl-1-propanol Hydrochloride *see* 7461
1-(4-Aminophenyl)-1-propanone *see*
  490
(*S*)-2-Amino-1-phenyl-1-propanone *see*
  1954
α-Amino-β-phenylpropionic Acid *see*
  7425
4-Amino-*N*-(1-phenyl-1*H*-pyrazol-5-
  yl)benzenesulfonamide *see* 9105
1-(3-Aminophenyl)-2(1*H*)-pyridinone
  see 621
1-(*m*-Aminophenyl)-2(1*H*)-pyridone
  see 621
p-Aminophenylsulfonamidobenzene
  see 9064
3-(*p*-Aminophenylsulfonamido)-6-
  chloropyridazine *see* 9068
5-(4-Aminophenylsulfonamido)-3,4-
  dimethylisoxazole *see* 9125
6-(*p*-Aminophenylsulfonamido)-2,4-
  dimethylpyrimidine *see* 9124
3-(*p*-Aminophenylsulfonamido)-5-
  methylisoxazole *see* 9086
N-[(4-Aminophenyl)sulfonyl]acet-
  amide *see* 9067
N-[(4-Aminophenyl)sulfonyl]benz-
  amide *see* 9065
4-[(4-Aminophenyl)sulfonyl]benzene-
  methanamine *see* 9098
N-[(3-Aminophenyl)sulfonyl]-*N*′-butyl-
  urea *see* 1615
N-[(-4-Aminophenyl)sulfonyl]-3,4-
  dimethylbenzamide *see* 9102
N-[(4-Aminophenyl)sulfonyl]-*N*-(3,4-
  dimethyl-5-isoxazolyl)acetamide *see*
  9125
1-(*p*-Aminophenylsulfonyl)-3-(4,5-
  dimethyl-2-oxazolyl)guanidine *see*
  9077
3-[(4-Aminophenyl)sulfonyl]-1,3-
  diphenyl-1-propanone *see* 254
N-[(4-Aminophenyl)sulfonyl]-*N*-(3-
  methoxypyrazinyl)acetamide *see* 103

N-[(4-Aminophenyl)sulfonyl]-3-meth-
  yl-2-butenamide *see* 9072
2-[[4-[(4-Aminophenyl)sulfonyl]phen-
  yl]amino]ethanol *see* 9097
4-[[4-[(4-Aminophenyl)sulfonyl]phen-
  yl]amino]-4-oxobutanoic Acid *see*
  9047
N-[4-[(4-Aminophenyl)sulfonyl]phen-
  yl]glycine *see* 21
5-[(4-Aminophenyl)sulfonyl]-2-thiazol-
  amine *see* 9444
5-Amino-1-(5′-phosphofuranoribosyl)-
  imidazole *see* 191
5-Amino-1-(5′-phosphofuranoribosyl)-
  4-imidazolecarboxamide *see* 190
5-Amino-1-(5-*O*-phosphono-β-D-ribo-
  furanosyl)-1*H*-imidazole-4-carbox-
  amide *see* 190
N-[[5-Amino-1-(5-*O*-phosphono-β-D-
  ribofuranosyl)-1*H*-imidazol-4-yl]-
  carbonyl]-L-aspartic Acid *see* 8469
N-[5-Amino-1-(5′-phosphoribofuranos-
  yl)-4-imidazolecarbonyl]aspartic
  Acid *see* 8469
3-Aminophthalic Hydrazide *see* 5628
o-Aminophthaloyl Hydrazide *see* 5628
o-Aminophthalylhydrazide *see* 5628
Aminophylline, 485
2-Amino-4-picoline, 486
α-Amino-γ-picoline *see* 486
(3α,5α)-3-Aminopregnan-20-one *see*
  4314
Aminopromazine, 487
1-Aminopropane *see* 8028
2-Aminopropane *see* 5228
1-Aminopropane-1,3-dicarboxylic
  Acid *see* 4477
3-Aminopropanenitrile *see* 489
3-Amino-1,1,3-propanetricarboxylic
  Acid *see* 1876
3-Aminopropanoic Acid *see* 206
(*S*)-2-Aminopropanoic Acid *see* 205
2-Aminopropanol, 488
2-Amino-1-propanol *see* 488
Aminopropanolpyrocatechol *see* 6785
2-Amino-1-propene-1,1,3-tricarboni-
  trile *see* 502
4-(2-Aminopropionamido)-3,4,4a,5,6,7-
  hexahydro-5,6,8-trihydroxy-3-meth-
  ylisocoumarin *see* 139
α-(2-Aminopropionamido)-5-oxo-7-
  oxabicyclo[4.1.0]heptane-2-propion-
  ic Acid, Stereoisomer *see* 963
3-Aminopropionic Acid *see* 206
L-α-Aminopropionic Acid *see* 205
β-Aminopropionic Acid *see* 206
3-Aminopropionitrile *see* 489
β-Aminopropionitrile *see* 489
2-Aminopropiono-2′,6′-xylide *see*
  9629
α-(β-Aminopropionylamino)-β-(1-
  methyl-5-imidazolyl)propionic Acid
  see 717
catena-(*S*)-[μ-[*N*α-(3-Aminopropionyl)-
  L-histidinato(2—)-*N*¹,*N*²,*O*:*N*ᵀ]zinc]
  see 7712
N-(3-Aminopropionyl)-L-histidinato
  Zinc *see* 7712
p-Aminopropiophenone, 490
α-Aminopropiophenone *see* 1954
3-Amino-4-propoxybenzoic Acid 2-
  (Diethylamino)ethyl Ester *see* 7991
4-Amino-2-propoxybenzoic Acid 2-
  Diethylaminoethyl Ester Hydrochloride *see* 8023
2-Aminopropyl Alcohol *see* 488
(±)-*N*-[[[4-[(3-Aminopropyl)amino]-
  butyl]carbamoyl]hydroxymethyl]-7-
  guanidinoheptanamide *see* 4610

tetrazol-5-yl]thio]methyl]-8-oxo-5-
thia-1-azabicyclo[4.2.0]oct-2-ene-2-
carboxylic Acid *see* 1982
Aminotriazole *see* 513
3-Amino-1*H*-1,2,4-triazole *see* 513
5-Amino-1*H*-*v*-triazolo[*d*]pyrimidin-7-
ol *see* 928
1-Amino-2,2,2-trichloroethanol *see*
2110
4-Amino-6-(trichloroethenyl)-1,3-
benzenedisulfonamide *see* 2471
4-Amino-3,5,6-trichloropicolinic Acid
see 7552
4-Amino-3,5,6-trichloro-2-pyridinecar-
boxylic Acid *see* 7552
4-Amino-6-(trichlorovinyl)-*m*-benz-
enedisulfonamide *see* 2471
2-Amino-1,1,3-tricyanopropene, 502
2-Amino-1,3,3-tricyano-2-propene *see*
502
1-Aminotricyclo[3.3.1.1^{3,7}]decane *see*
389
(8*S*-*cis*)-10-[(3-Amino-2,3,6-trideoxy-
α-L-*arabino*-hexopyranosyl)oxy]-7,-
8,9,10-tetrahydro-6,8,11-trihydroxy-
8-(hydroxyacetyl)-1-methoxy-5,12-
naphthacenedione *see* 3660
(8*S*-*cis*)-10-[(3-Amino-2,3,6-trideoxy-
α-L-*lyxo*-hexopyranosyl)oxy]-7,8,9,-
10-tetrahydro-6,8,11-trihydroxy-8-
(hydroxyacetyl)-1-methoxy-5,12-
naphthacenedione *see* 3495
(8*S*,10*S*)-10-[[3-Amino-2,3,6-trideoxy-
4-*O*-(tetrahydro-2*H*-pyran-2-yl)-α-L-
lyxo-hexopyranosyl]oxy]-8-glycol-
oyl-7,8,9,10-tetrahydro-6,8,11-tri-
hydroxy-1-methoxy-5,12-naphtha-
cenedione *see* 7642
[8*S*-[8α,10α(*S**)]]-10-[[3-Amino-2,3,6-
trideoxy-4-*O*-(tetrahydro-2*H*-pyran-
2-yl)-α-L-*lyxo*-hexopyranosyl]oxy]-
7,8,9,10-tetrahydro-6,8,11-trihydr-
oxy-8-(hydroxyacetyl)-1-methoxy-
5,12-naphthacenedione *see* 7642
7-Amino-4,5,6-triethoxy-3-(5,6,7,8-
tetrahydro-4-methoxy-6-methyl-1,3-
dioxolo[4,5-*g*]isoquinolin-5-yl)-
1(3*H*)-isobenzofuranone *see* 9894
7-Amino-4,5,6-triethoxy-3-(5,6,7,8-
tetrahydro-4-methoxy-6-methyl-1,3-
dioxolo[4,5-*g*]isoquinolin-5-yl)-
phthalide *see* 9894
2-Amino-6-(trifluoromethoxy)benzo-
thiazole *see* 8389
2-Amino-6-(1,2,3-trihydroxypropyl)-
4(3*H*)-pteridinone *see* 6551
2-Amino-4,6,7-trihydroxpteridine *see*
5481
2-Amino-4,6,7-trihydroxypyrimido[4,-
5-*b*]pyrazine *see* 5481
N-(3-Amino-2,4,6-triiodobenzoyl)-*N*-
phenyl-β-alanine *see* 5028
β-[*N*-(3-Amino-2,4,6-triiodobenz-
oyl)phenylamino]propionic Acid *see*
5028
3'-Amino-2',4',6'-triiodo-*N*-methyl-
glutaranilic Acid *see* 5069
3-(3-Amino-2,4,6-triiodophenyl)-2-
ethylpropanoic Acid *see* 5072
3-(3-Amino-2,4,6-triiodophenyl)-2-
ethylpropionic Acid *see* 5072
5-[(3-Amino-2,4,6-triiodophenyl)meth-
ylamino]-5-oxopentanoic Acid *see*
5069
4-(4-Amino-6,7,8-trimethoxy-2-quin-
azolinyl)-1-piperazinecarboxylic
Acid 2-Hydroxy-2-methylpropyl
Ester *see* 9828
Aminourea Hydrochloride *see* 8588

2-Amino-3-ureidopropionic Acid *see*
212
α-Amino-δ-ureidovaleric Acid *see*
2392
2-Aminovaleric Acid *see* 6813
α-Aminovaleric Acid *see* 6813
Aminoxafen *see* 497
Aminoxaphen *see* 497
Aminoxidin [Farmitalia] *see* 7173
Aminoxytriphene *see* 615
Aminoxytropine Tropate *see* 908
Amioca, 503
Amiodar [Midy] *see* 504
Amiodarone, 504
Amipan T *see* 3317
Amipaque [Winthrop] *see* 6240
Amipenix [Toyo Jozo] *see* 628
Amiphenazole, 505
Amipramidin *see* 426
Amipramizide *see* 426
Amipress [Dox-Al] *see* 5341
Amiprilose, 506
Amipylo *see* 492
Amisalin [Daiichi] *see* 7936
Amisepan [Nippon Shoji] *see* 1624
Amisometradine, 507
Amisulpride, 508
Amithiozone *see* 9427
Amitid [Squibb] *see* 511
Amiton, 509
Amitraz, 510
Amitril [Warner-Chilcott] *see* 511
Amitriptyline, 511
Amitriptyline *N*-Oxide *see* 512
Amitriptylinoxide, 512
Amitrole, 513
Amixetrine, 514
Amizil *see* 1055
Amizol [Rhone-Poulenc] *see* 513
Amlexanox, 515
Amlodipine, 516
Amlodipine Besylate *see* 516
Ammate [Du Pont] *see* 589
Ammicardine *see* 5321
Ammidin *see* 4960
Amminosidin *see* 7173
Ammipuran *see* 5321
Ammivin [Merrell] *see* 5321
Ammivisnagen *see* 5321
Ammo [FMC] *see* 2836
Ammoform *see* 6036
Ammoidin *see* 6068
Ammonia, 517
Ammoniacum, 518
Ammonia Soap *see* 569
Ammoniated Mercury *see* 5927
Ammonia Water, 519
Ammonia Water—10% *see* 519
Ammonium Acetate, 520
Ammonium Acetate Solution *see* 520
Ammonium Acid Malate *see* 524
Ammonium Acid Oxalate *see* 525
Ammonium Acid Tartrate *see* 529
Ammonium Alum *see* 335
Ammonium Aluminum Fluoride *see*
555
Ammonium Aminoformate *see* 533
Ammonium Antimony Tungsten
Oxide *see* 4786
Ammonium Benzoate, 521
Ammonium Biborate *see* 530
Ammonium Bicarbonate, 522
Ammonium Bichromate *see* 544
Ammonium Bifluoride, 523
Ammonium Bimalate, 524
Ammonium Binoxalate, 525
Ammonium Biphosphate *see* 577
Ammonium Bisulfate, 526
Ammonium Bisulfide, 527
Ammonium Bisulfide Soln *see* 593

Ammonium Bisulfite, 528
Ammonium Bitartrate, 529
Ammonium Bithiolicum *see* 4929
Ammonium Bituminosulfonate *see*
4929
Ammonium Borate, 530
Ammonium Bromide, 531
Ammonium Caprylate, 532
Ammonium Carbamate, 533
Ammonium Carbazoate *see* 581
Ammonium Carbolate *see* 7390
Ammonium Carbonate, 534
Ammonium Ceric Nitrate, 535
Ammonium Cerous Sulfate, 536
Ammonium Chloride, 537
Ammonium Chloroaluminate *see* 595
Ammonium Chlorocuprate *see* 543
Ammonium Chloroheptenearsonate *see*
2170
Ammonium Chloroosmate *see* 570
Ammonium Chloroplatinate *see* 582
Ammonium Chloroplatinite *see* 583
Ammonium Chromate(VI), 538
Ammonium Chromic Sulfate, 539
Ammonium Citrate, Dibasic, 540
Ammonium Cobaltous Phosphate, 541
Ammonium Cobaltous Sulfate, 542
Ammonium Cryolite *see* 555
Ammonium Cupric Chloride, 543
Ammonium Cupric Sulfate *see* 9323
Ammonium Diamminetetrakis(nitrito-
N)cobaltate(1−) *see* 3683
Ammonium Diamminetetrakis(thiocy-
anato-*N*)chromate(1−) *see* 8298
Ammonium Dichromate(VI), 544
Ammonium Dihydrogen Phosphate *see*
577
Ammonium Dioxopentafluoroura-
nate(VI) *see* 603
Ammonium Dioxotricarbonatoura-
nate(VI) *see* 602
Ammonium Disulfatocerate(III) *see*
536
Ammonium Disulfatochromate(III)
see 539
Ammonium Disulfatocobaltate(II) *see*
542
Ammonium Disulfatonickelate(II) *see*
566
Ammonium Dithiocarbamate, 545
Ammonium Diuranate *see* 601
Ammonium Embelate *see* 3595
Ammonium Ferric Chromate, 546
Ammonium Ferric Citrate, 547
Ammonium Ferric Oxalate, 548
Ammonium Ferric Sulfate, 549
Ammonium Ferricyanide, 550
Ammonium Ferrocyanide, 551
Ammonium Ferrous Sulfate, 552
Ammonium Fluoaluminate *see* 555
Ammonium Fluoride, 553
Ammonium Fluosilicate *see* 558
Ammonium Formate, 554
Ammonium Hexachloroosmate(IV) *see*
570
Ammonium Hexachloroplatinate(IV)
see 582
Ammonium Hexacyanoferrate (II) *see*
551
Ammonium Hexacyanoferrate(III) *see*
550
Ammonium Hexafluogallate *see* 556
Ammonium Hexafluophosphate *see*
557
Ammonium Hexafluoroaluminate, 555
Ammonium Hexafluorogallate, 556
Ammonium Hexafluorophosphate, 557
Ammonium Hexafluorosilicate, 558
Ammonium Hexanitratocerate(IV) *see*
535

Ammonium Hydrogen Carbonate *see* 522
Ammonium Hydrogen Fluoride *see* 523
Ammonium Hydrogen Oxalate *see* 525
Ammonium Hydrogen Sulfate *see* 526
Ammonium Hydrogen Sulfide *see* 527
Ammonium Hydrogen Tartrate *see* 529
Ammonium Hydrosulfide *see* 527
Ammonium Hydrosulfide Soln *see* 593
Ammonium Hydroxide *see* 519
Ammonium Hypophosphite, 559
Ammonium Hyposulfite *see* 598
Ammonium Ichthosulfonate *see* 4929
Ammonium Iodide, 560
Ammonium Lactate, 561
Ammonium Magnesium Chloride, 562
Ammonium Mandelate, 563
Ammonium Mercuric Chloride, 564
Ammonium Metaborate *see* 530
Ammonium Metavanadate *see* 605
Ammonium Molybdate(VI), 565
Ammonium Molybdophosphate *see* 579
Ammonium Muriate *see* 537
Ammonium α-Naphthylnitrosohydroxylamine *see* 6536
Ammonium Nickel Sulfate, 566
Ammonium Nitrate, 567
Ammonium Nitratocerate(IV) *see* 535
Ammonium Nitroferricyanide, 568
Ammonium Nitroprusside *see* 568
Ammonium Oleate, 569
Ammonium Osmium Chloride, 570
Ammonium Oxalate, 571
Ammonium Oxodioxalatotitanate(IV) *see* 599
Ammonium Palmitate, 572
Ammonium Paramolybdate *see* 565
Ammonium Paratungstate *see* 600
Ammonium Pentachlorozincate, 573
Ammonium Pentacyanonitrosylferrate(III) *see* 568
Ammonium Perchlorate, 574
Ammonium Peroxydisulfate, 575
Ammonium Persulfate *see* 575
Ammonium Phenate *see* 7390
Ammonium *p*-Phenolsulfonate *see* 7396
Ammonium Phosphate, Dibasic, 576
Ammonium Phosphate, Monobasic, 577
Ammonium Phosphite, 578
Ammonium Phosphomolybdate, 579
Ammonium Phosphorus Hexafluoride *see* 557
Ammonium Phosphotungstate, 580
Ammonium Phosphowolframate *see* 580
Ammonium Picrate, 581
Ammonium Picronitrate *see* 581
Ammonium Platinic Chloride, 582
Ammonium Platinochloride *see* 583
Ammonium Platinous Chloride, 583
Ammonium Polysulfide Soln *see* 592
Ammonium Purpurate *see* 6386
Ammonium Reineckate *see* 8298
Ammonium Rhodanide *see* 597
Ammonium Rhodanilate *see* 8350
Ammonium Salicylate, 584
Ammonium Salt of Aurintricarboxylic Acid *see* 330
Ammonium Selenate, 585
Ammonium Selenite, 586
Ammonium Silicofluoride *see* 558
Ammonium Sodium Phosphate, 587
Ammonium Stearate, 588
Ammonium Sulfamate, 589
Ammonium Sulfate, 590

Ammonium Sulfhydrate *see* 527
Ammonium Sulfhydrate Soln *see* 593
Ammonium Sulfide, 591
Ammonium Sulfide Solution, Red, 592
Ammonium Sulfide Solution, Yellow, 593
Ammonium Sulfite, 594
Ammonium Sulfobituminate *see* 4929
Ammonium Sulfocarbamate *see* 545
Ammonium Sulfocyanate *see* 597
Ammonium Sulfocyanide *see* 597
Ammonium Sulfoichthyolate *see* 4929
Ammonium Tetraborate *see* 530
Ammonium Tetrachloroaluminate, 595
Ammonium Tetrachlorocuprate(II) *see* 543
Ammonium Tetrachlorodiaquocuprate(II) *see* 543
Ammonium Tetrachloromercurate(II) *see* 564
Ammonium Tetrachloroplatinate(II) *see* 583
Ammonium Tetrachlorozincate, 596
Ammonium Tetrathiocyanodiammonochromate *see* 8298
Ammonium Thiocyanate, 597
Ammonium Thiosulfate, 598
Ammonium Titanium Oxalate, 599
Ammonium Tricarbonatozirconate *see* 606
Ammonium Trioxalatoferrate(III) *see* 548
Ammonium Tungstate(VI), 600
Ammonium 5-Tungsto-2-antimonate *see* 4786
Ammonium Tungstophosphate *see* 580
Ammonium Uranate(VI), 601
Ammonium Uranium Carbonate, 602
Ammonium Uranium Fluoride, 603
Ammonium Valerate, 604
Ammonium Valerianate *see* 604
Ammonium Vanadate(V), 605
Ammonium Zirconyl Carbonate, 606
Ammonyx DME [Onyx] *see* 2071
Ammonyx G *see* 2059
Ammonyx T *see* 2059
Ammophyllin *see* 485
Amnestrogen [Squibb] *see* 2571
Amoban [Rhône-Poulenc] *see* 10324
Amobarbital, 607
Amocaine *see* 610
Amocarzine, 608
Amocilline [Inpharzam] *see* 617
Amodex [Robert et Carrière] *see* 617
Amodiaquin, 609
Amoenol *see* 5052
Amoksiklav [Lek] *see* 2402
Amolanone, 610
Amolin [Takeda] *see* 617
Amopenixin [Hishiyama] *see* 617
Amoproxan, 611
Amoram [Eastern] *see* 617
Amorolfine, 612
Amorphan [Heumann] *see* 6811
Amosamine *see* 415
Amoscanate, 613
Amosene [Ferndale] *see* 5908
Amosulalol, 614
Amosyt [Abigo] *see* 3252
Amotril *see* 2436
Amotriphene, 615
Amovane [Rhône-Poulenc] *see* 10324
Amoxanox *see* 515
Amoxapine, 616
Amoxi [Beecham] *see* 617
Amoxicillin, 617
Amoxidal [De Angeli] *see* 617
Amoxidin [Lagap] *see* 617
Amoxil [Beecham] *see* 617

Amoxillat [Azupharma] *see* 617
Amoxipen [Lorenzini] *see* 617
Amoxi-Wolff [Wolff] *see* 617
Amoxycillin *see* 617
Amoxypen [Grünenthal] *see* 617
AMP *see* 157 and 469
3′,5′-AMP *see* 2777
AMP-3 *see* 9387
A-5MP *see* 157
AMPC *see* 617
AMPD *see* 468
Ampecyclal [Sarget] *see* 4691
Ampelopsin, 618
Ampeloptin *see* 618
Amperozide, 619
Amphecloral, 620
Amphedroxyn [Lilly] *see* 6015
Amphenidone, 621
Amphenone *see* 622
Amphenone B, 622
Amphetamine, 623
d-Amphetamine Sulfate *see* 2996
Amphetaminil, 624
Amphicol [McKesson] *see* 2120
Amphihydroxynaphthyl-β-propionic Acid *see* 256
Amphojel [Wyeth-Ayerst] *see* 355
Ampho-Moronal [Squibb-Heyden] *see* 627
Amphomycin, 625
Amphotalide, 626
Amphotericin B, 627
Amphothalide *see* 626
Amphozone [Squibb] *see* 627
Ampicillin, 628
Ampicillin (5-Methyl-2-oxo-1,3-dioxolen-4-yl)methyl Ester *see* 5460
Ampicillin 1-Oxo-1,3-dihydroisobenzofuran-3-yl Ester *see* 9204
Ampicillin Pivaloyloxymethyl Ester *see* 7669
Ampicin [Bristol] *see* 628
Ampilag [Dumex] *see* 628
Ampilar [Lagap] *see* 628
Ampin-penicillin *see* 7226
Ampipenin [Kabi] *see* 628
Ampiroxicam, 629
Ampi-Tablinen [Sanorania] *see* 628
Ampliactil *see* 2238
Amplicain *see* 6839
Amplictil *see* 2238
Amplidione [Dausse] *see* 7060
Ampligen, 630
Ampligram [Hermes] *see* 2025
Amplit [Daiichi] *see* 5587
Amplital [Farmitalia] *see* 628
Amplivix [Sigma-Tau] *see* 1113
Amprace [Amrad] *see* 3605
Amprol [Merck & Co.] *see* 631
Amprolium, 631
Amprol Plus [Merck & Co.] *see* 3791
Amprotropine Phosphate, 632
Ampyrone, 633
Amrinone, 634
AMS *see* 589
m-AMSA *see* 635
Amsacrine, 635
Amsebarb *see* 607
Amsidine [Parke, Davis] *see* 635
Amsidyl [Parke, Davis] *see* 635
Amsonic Acid, 636
Amsubit *see* 4929
Amsulosin *see* 9217
Amsustain *see* 2996
Amtolmetin Guacil, 637
Amudane *see* 4571
Amuno [Merck & Co.] *see* 4998
Amurex [Reid-Provident] *see* 6053
Amycor [Lipha] *see* 1260

Amygdalic Acid see 5757
Amygdalin, 638
Amygdalinic Acid see 5757
Amygdalose see 4402
Amygdaloside see 638
Amylacetic Ester see 5125
dl-sec-Amyl Alcohol see 7258
n-Amyl Alcohol see 7257
tert-Amyl Alcohol see 7282
Amyl Alcohol (Commercial) see 4339
n-Amylamine, 639
2-n-Aminoethyl p-Aminobenzoate see 6435
Amylase, 640
α-Amylase (Aspergillus oryzae) see 9203
γ-Amylases see 640
Amylbenzene, 641
n-Amylbenzene see 641
d-Amyl Bromide, 642
d-pri-act-Amyl Bromide see 642
n-Amyl Bromide, 643
tert-Amyl Bromide, 644
n-Amyl Butyrate, 645
n-Amyl Caproate, 646
Amylcarbinol see 4732
Amyl Chloride, 647
n-Amyl Chloride see 647
6-n-Amyl-m-cresol, 648
Amyleine Hydrochloride see 656
Amylene, 649
α-n-Amylene see 7262
β-n-Amylene see 7263
Amylene Dichloride, 650
tert-Amylene Dichloride see 650
Amylene Hydrate see 7282
n-Amyl Ether, 651
Amyl Ethyl Ketone see 3810
3-Amyl-1-hydroxy-6,6,9-trimethyl-6H-dibenzo[b,d]pyran see 1793
Amyl Isovalerate see 5135
tert-Amyl Isovalerate, 652
n-Amyl Mercaptan, 653
Amylmethylcarbinol see 4697
tert-Amyl Methyl Ether, 654
Amyl Nitrite, 655
Amylobarbitone see 607
Amylocaine Hydrochloride, 656
Amylodextrin see 8955
Amylogen see 8955
Amyloid, 657
Amyloid Substance see 657
Amylopectin see 503 and 8954
Amylose see 8954
Amyl Oxide see 651
Amylpenicillin Sodium, 658
p-tert-Amylphenol see 7284
4-[3-(4-tert-Amylphenyl)-2-methylpro-pyl]-2,6-dimethylmorpholine see 612
Amyl Phthalate see 5138
Amylsine see 6435
Amyl Thioalcohol see 653
Amylum see 8954
Amyl Valerate see 5135
α-Amyrenol see 659
β-Amyrenol see 660
α-Amyrin, 659
β-Amyrin, 660
Amytal [Lilly] see 607
Amytal Sodium [Lilly] see 607
AN 1 [Voight] see 624
AN-148 see 6008
AN-1022 see 6078
AN-1317 see 7309
AN-1324 see 4488
Anabactyl [Beecham] see 1838
Anabasi [Pierrel] see 2513
Anabasine, 661
Anabet [Squibb] see 6431
Anabiol [Searle] see 1352

Anabloc [Irbi] see 7474
Anabolex [Samil] see 8950
Anabolicum Vister [Parke, Davis] see 8236
Anabsinthin, 662
Anabsynthin see 662
Anacardic Acid, 663
Anacardone see 6635
Anacetin [Philips Roxane] see 2120
Anacobin [BDH] see 10152
Anacyclin [Ciba] see 5659
Anadonis Green see 2283
Anador [Logeais] see 6452
Anadrol [Syntex] see 7101
Anadur [Kabi Pharmacia] see 6452
Anafebrina see 495
Anaflex [Geistlich] see 7735
Anaflon [Sterling-Winthrop] see 45
Anafranil [Geigy] see 2447
Anagestone, 664
Anagregal [Gentili] see 9569
Anagrelide, 665
Anagyrine, 666
Anahaemin [Bardin] see 5575
Anahist [Warner-Lambert] see 9513
Anakinra see 5022
Analate [Winston] see 5723
Analeptin see 9189
Analetil see 9344
Analexin [Biotrading] see 7474
Analgesine [Elder] see 757
Analgin see 3414
Analud [Unifa] see 4053
Analux [Polfa] see 5820
Anamidol [Iwaki] see 7099
Anamycin [Chephasaar] see 3720
Ananase [Rorer] see 1409
Anandron [Roussel-UCLAF] see 6636
Ananxyl [Synthelabo] see 318
Anapolon [Syntex] see 7101
Anaprel [Eutherapie] see 8311
Anaprox [Syntex] see 6504
Anaptivan [Help] see 2002
Anarel [Miles] see 4587
Anarexol [Merck & Co.] see 2842
Anasterone see 7101
Anastil [Eberth] see 4575
Anastrozole, 667
Anatabine, 668
Anatase see 9612
Anatensol [Bristol-Myers Squibb] see 4226
Anatola [Parke, Davis] see 10150
Anatran [Tobishi] see 9754
Anatrofin [Syntex] see 8962
Anatropin [Ortho] see 664
Anaus [Molteni] see 9839
Anautine see 3252
Anauxite see 377
Anavar [Searle] see 7054
Anavenol [ICI] see 6500
Anayodin see 8777
Anazolene Sodium, 669
Ancef [SK & F] see 1967
Anceron [Essex] see 1047
Anchoic Acid see 938
Anchovyxanthin see 10248
Anchusa see 252
Anchusa Acid see 253
Anchusin see 253
Anchusin Paper see 252
Ancitabine, 670
Anco [Boots] see 4925
Ancobon [Roche] see 4161
Ancolan see 5817
Ancoron [Libbs] see 504
Ancortone [Merck & Co.] see 7904
Ancotil [Roche] see 4161
Ancrod, 671
Ancylol [Am. Cyanamid] see 3423

Ancymidol, 672
Ancytabine see 670
Ancyte [Abbott] see 7635
Andalin [Solvay Duphar] see 4159
Andalusite see 377
Andantol [Sumitomo] see 5248
Andanton [Lacer] see 5248
Andaxin [EGYT] see 5908
Andere [Toyama] see 1500
Andergin [ISOM] see 6266
Andiamine [Polfa] see 4743
Andion [Gea] see 1047
Andirine see 9183
Andolex [Riker] see 1157
Andozac [DuPont Pharma] see 4125
Andractim [Besins-Iscovesco] see 8950
Andrade Indicator see 109
Andramine see 3252
Andriol [Organon] see 9322
Andro [Forest] see 9322
Androcur [Schering AG] see 2844
Androdiol see 6017
Androfluorene see 4223
Androfluorone see 4223
Androfurazanol see 4319
Andrographolide, 673
Android [Brown] see 6206
Androisoxazole, 674
Andro LA [Forest] see 9322
Androlone [Orma] see 8950
Andromedotoxin see 4566
1,4-Androstadien-17β-ol-3-one see 1354
Androstalone [Roussel] see 5973
Androstanazole see 8951
Androstane, 675
(5α)-Androstane see 675
(5β)-Androstane see 3912
(5β,17β)-Androstane-17-carboxylic Acid see 3913
Androstane-3β,11β-diol-17-one, 676
Androstanolone see 8950
Androstan-3α-ol-17-one see 680
3β-Androstanol-17-one see 3646
Androstan-17β-ol-3-one see 8950
Androstenediol, 677
(3β,17β)-Androst-5-ene-3,17-diol see 677
Δ5-Androstene-3β,17β-diol see 677
Androst-4-ene-3,17-dione see 678
4-Androstene-3,17-dione, 678
Androst-4-ene-3,11,17-trione see 174
(3α,5α)-Androst-16-en-3-ol, 679
Δ16-Androsten-3-ol see 679
Δ4-Androsten-17β-ol-3-one see 9322
Δ5-Androsten-3β-ol-17-one see 7892
Androsterolo [Pierrel] see 4223
Androsterone, 680
cis-Androsterone see 680
Androtardyl [Schering AG] see 9322
Androteston-M see 6017
Androtex see 678
Androviron [Schering AG] see 5974
Anectine [Burroughs Wellcome] see 9044
Anekain [Pliva] see 1520
Anelmid [Lilly] see 3434
Anemolin [Toyama] see 617
Anemone Camphor see 681
Anemonin, 681
Anergan see 4013
Anergen see 4013
Anergex [Mulford] see 7710
Anertan [Boehringer, Mann.] see 9322
Anesthesin [Abbott] see 1116
Anesthetic Ether see 3852
Anesthone [Parke, Davis] see 1116
Anestil see 7937
Anethaine [Glaxo] see 9330
Anethole, 682

Ansiced [Abello] see 1528
Ansieten [Exa] see 5308
Ansilan [Lek] see 5829
Ansimar [ABC] see 3494
Ansmin [SS Pharm.] see 3369
Ansolysen Bitartrate [Wyeth] see 7274
Ansolysen Tartrate [Wyeth] see 7274
Anspor [SK & F] see 2032
Antabuse [Ayerst; Tosse] see 3428
Antacal [Errekappa] see 516
Antadix see 3428
Antadys [Théramex] see 4234
Antagonate [Miles] see 2232
Antagosan [Behringwerke] see 796
Antalka see 4477
Antallin see 3555
Antalvic [Houdé] see 8024
Antalzyme see 5671
Antamine [Grove] see 8168
Antapentan [Ayerst] see 7365
Antastan see 718
Antasten see 718
Antaxone [Simes] see 6450
Antazoline, 718
Antébor [Biologiques] see 9067
Antees [Oriental] see 1840
Antegan [Merck & Co.] see 2842
Antemin [Streuli] see 3252
Antemovis [Vister] see 8607
Antepan [Henning] see 9720
Antepar (tabl.) [Burroughs Wellcome]
    see 7617
Antepsin [Wyeth] see 9049
Anterior Pituitary Growth Hormone
    see 8864
Anterior Pituitary Luteotropin see
    7961
Anteron [Schering AG] see 2273
Antex [Leo Pharm] see 2273
Anthelcide EQ [Norden] see 7070
Anthelin [Jensen-Salsbery] see 815
Anthelone see 3569
Anthelone E see 3632
Anthelone U see 3569
Anthen [Byk-Gulden] see 7949
Antheridiol, 719
Anthesterin see 9229
Anthio see 4273
Anthiolimine, 720
Anthiomaline [Specia] see 720
Anthion see 7825
Anthiphen see 3120
Anthisan [M & B] see 8168
Anthorine see 895
Anthracene, 721
Anthracene Brown see 722
9,10-Anthracenedicarboxaldehyde
    Bis[(4,5-dihydro-1H-imidazol-2-yl)-
    hydrazone] see 1284
9,10-Anthracenedicarboxaldehyde
    Bis(2-imidazolin-2-ylhydrazone) see
    1284
9,10-Anthracenedione see 726
1,2,10-Anthracenetriol see 727
1,8,9-Anthracenetriol Triacetate see
    723
9-Anthracenol see 725
9(10H)-Anthracenone see 730
Anthra-Derm [Dermik] see 723
Anthragallic Acid see 722
Anthragallol, 722
Anthralin, 723
Anthramycin, 724
Anthranilic Acid see 442
Anthranilic Acid Cinnamyl Ester see
    2363
Anthranilic Acid 2-(2-Piperidyl)ethyl
    Ester see 4649
3-Anthraniloylalanine see 5340
Anthranol, 725

Anthraquinone, 726
9,10-Anthraquinone see 726
6,6'-(1,4-Anthraquinonylenediimino)-
    di-m-toluenesulfonic Acid Disodium
    Salt see 248
Anthrarobin, 727
Anthrarufin, 728
1,2,10-Anthratriol see 727
Anthrimide, 729
9-Anthrol see 725
Anthrone, 730
Anthropodesoxycholic Acid see 2096
Antiangor [I.S.O.M.] see 2296
α-Antiarin, 731
Antibason see 6210
Antibiocin [Nycomed] see 7230
3C Antibiotic see 6425
Antibiotic 67-694 see 8420
Antibiotic 1037 see 9698
Antibiotic 5879 see 1249
Antibiotic 6640 see 8695
Antibiotic 32232RP see 8690
Antibiotic A 246 see 4312
Antibiotic A-300 see 9781
Antibiotic A 5283 see 6513
Antibiotic A-23187 see 1678
Antibiotic A-28086 Factor A see 6506
Antibiotic AB 206 see 6283
Antibiotic Ab 651 see 1909
Antibiotic B-41 see 6278
Antibiotic B-41A1 see 6278
Antibiotic B-41A3 see 6276
Antibiotic B-41A4 see 6276
Antibiotic B-41D see 6278
Antibiotic C-19393-II₂ see 1909
Antibiotic C-19393-S₂ see 1909
Antibiotic DE-3936 see 5599
Antibiotic FI 1163 see 5621
Antibiotic G-6302 see 9119
Antibiotic K 178 see 6633
Antibiotic KA-6643-A see 1909
Antibiotic KA-6643-B see 1909
Antibiotic KW-1062 see 6269
Antibiotic LA-7017 see 7696
Antibiotic M139603 see 9384
Antibiotic MA 144A1 see 115
Antibiotic MA 144B1 see 115
Antibiotic MYC 8003 see 6308
Antibiotic No. 899 see 10142
Antibiotic S-541A see 6532
Antibiotic S 14750A see 215
Antibiotic SF-837A3 see 6271
Antibiotic T-2636A see 8566
Antibiotic WR 141 see 4555
Antibiotic X-464 see 6633
Antibiotic X-465A see 2086
Antibiotic X-537A see 5384
Antibiotic X-5108 see 916
Antibiotic X-14868A Ammonium Salt
    see 5682
Antibiotique EF 185 see 6542
Anticanitic Vitamin see 443
Anticarie see 4714
Antichlor see 8844
Antichoc Hipmag see 5734
Anti-chromotrichia Factor see 443
Antidiar [Armour] see 355
Antidipsin see 3775
Antidiuretic Hormone see 10073
Antidrasi see 3127
Antietanol see 3428
Antifebrin see 47
Antifoam A [Dow] see 3264
"Antifungin" see 5693
Antigal [Streuli] see 3043
Antigens, 732
Antigestil see 3177
Antigorite see 5727
Anti-gray-hair Factor see 443
Antihemophilic Factor A see 3963

Antihemophilic Factor B see 3964
Antihemophilic Globulin see 3963
Antihistal see 718
Antihypertensive Neutral Renomedul-
    lary Lipid see 5841
Antihypertensive Polar Renomedul-
    lary Lipid see 7682
Anti-hypo see 7821
Anti-infective Vitamin see 10150
Anti-inflammatory Hormone see 4828
Anti-JH see 7898
Antikrein [Teikoku Zoki] see 796
Antilirium [Forest] see 7540
Antilon [Lakeside] see 3758
Antimalarine see 7680
Antiminth [Roerig] see 8139
Antimonial Saffron see 738
Antimonic "Acid" see 739
Antimonic Oxide see 739
Antimonic Sulfide see 738
Antimonous Fluoride see 750
Antimonous Sulfate see 745
Antimonous Sulfide see 754
Antimony, 733
Antimony Chloride Fluoride see 735
Antimony Chloride Oxide, 734
Antimony Chloride Solution see 749
Antimony Dichlorofluoride see 735
Antimony Dichlorotrifluoride, 735
"Antimony Dimercaptosuccinate" see
    8966
Antimony Fluoride see 750
Antimony Glance see 754
Antimony Hydride see 8965
Antimony Oxychloride see 734
Antimony Pentachloride, 736
Antimony Pentafluoride, 737
Antimony Pentasulfide, 738
Antimony Pentoxide, 739
Antimony Potassium Oxalate, 740
Antimony Potassium Tartrate, 741
Antimony Pyrocatechol Sodium Di-
    sulfonate Heptahydrate see 8967
Antimony Red see 738
"Antimony Salt" see 740
Antimony Sodium Gluconate, 742
Antimony Sodium Oxide L(+)-Tar-
    trate see 743
Antimony Sodium Tartrate, 743
Antimony Sodium Thioacetate see 744
Antimony Sodium Thioglycollate, 744
Antimony Sulfate, 745
Antimony Sulfide see 754
Antimony Thioglycollamide, 746
Antimony Thioglycollic Acid Tri-
    amide see 746
Antimony Tribromide, 747
Antimony Trichloride, 748
Antimony Trichloride Solution, 749
Antimony Trifluoride, 750
Antimony Trifluorodichloride see 735
Antimony Triiodide, 751
Antimony Trioxide, 752
Antimony Triselenide, 753
Antimony Trisulfate see 745
Antimony Trisulfide, 754
Antimosan see 8967
Antimycin see 2388
Antimycin A₁, 755
Antimycin A₃, 756
Antin see 7469
Antinonnin see 3331
Antinosin see 5059
Antipar [Farmitalia] see 3157
Antiparkin see 3813
Antipellagra Vitamin see 6612
Antipernicin [Galenika] see 10152
Antipernicious Anemia Principle see
    10152
α₂-Antiplasmin see 7678

Antipyrine, 757
Antipyrine Amygdalate *see* 757
Antipyrine 2-Hydroxy-2-methyl-
  butyrate *see* 757
Antipyrine Salicylate, 758
N-(Antipyrinyl)-2-(dimethylamino)pro-
  pionamide *see* 492
(Antipyrinylmethylamino)methane-
  sulfonic Acid Sodium Salt *see* 3414
N-Antipyrinylnicotinamide *see* 6619
Antirad *see* 176
Antiradon *see* 176
Antireticular Cytotoxic Serum, 759
Antirex *see* 3562
Antirobe [Upjohn] *see* 2414
Antirrhinin *see* 2757
Antisacer [Wander] *see* 7475
Antiscorbutic Vitamin *see* 867
Antisedan [Farmos] *see* 894
Antisepsin *see* 1420
Antiserotonin *see* 5835
Antispasmin *see* 386
Antisterility Vitamin *see* 10159
Antistin [Ciba] *see* 718
Antistine [Ciba] *see* 718
Anti-Stress [Sintyal] *see* 5890
Antitanil *see* 3223
Anti-tetany Substance 10 *see* 3223
Antithermin *see* 5498
Antithrombin III *see* 4685
α_1-Antitrypsin, 760
Antituxil-Z [Ghimas] *see* 10303
Antiulcera Master [Coli] *see* 2125
Antivariz *see* 3772
Antivenin *(Crotalidae)* Polyvalent, 761
Antiverm *see* 7404
Antivert [Roerig] *see* 5817
Antivirin, 762
Antiweinsäure (German) *see* 9238
Antixerophthalmic Vitamin *see* 10150
Antlerite *see* 2723
Antodyne *see* 7444
Antoral [Recordati] *see* 9565
Antorphine *see* 6448
Antra [Astra] *see* 6977
Antracol [Bayer] *see* 8004
Antraderm [Brocades] *see* 723
Antrafenine, 763
Antrancine 8 [Jandekker] *see* 1583
Antrancine 12 [Jandekker] *see* 1582
Antrapurol *see* 2878
Antrenyl [Ciba] *see* 7109
Antril [Synergen] *see* 5022
Antrycide [ICI] *see* 8234
Antrypol [ICI] *see* 9181
ANTU, 764
Antuitrin S [Parke, Davis] *see* 2273
Antulcus *see* 7107
Anturan [Ciba-Geigy] *see* 9121
Anturane [Ciba-Geigy] *see* 9121
Anturano [Ciba-Geigy] *see* 9121
Anturat *see* 764
Anvene [Searle] *see* 6424
Anvitoff [Knoll] *see* 9704
Anxon [Beecham] *see* 5308
Anymol *see* 1281
Anzemet [Marion Merrell Dow] *see*
  3471
Anzief [Nippon Chemiphar] *see* 287
AO-12 *see* 6190
Aolan [Beiersdorf] *see* 3757
Aolept [Bayer] *see* 7306
A.O.P. *see* 77
Aoral *see* 10150
Aotal [Meram] *see* 14
AP2 *see* 8682
AP-14 *see* 3181
AP-43 *see* 610
AP-67 *see* 2247
AP-1288 *see* 1057

Apache [FMC] *see* 1670
Apacil *see* 498
Apacizin *see* 499
Apacizina [Farmitalia] *see* 499
Apafant, 765
Apalcillin, 766
Apamide [Miles] *see* 45
Apamin, 767
APAP [Rugby] *see* 45
Aparasin *see* 5526
Aparkane [ICN] *see* 9823
Aparkazin *see* 3157
Apas *see* 498
Apatef [ICI] *see* 1984
Apaurin *see* 3042
Apazone, 768
APC *see* 7344
Apegmone [Oberval] *see* 9594
Aperdor [Tokyo Tanabe] *see* 9640
Apertase *see* 4794
Apesan *see* 1889 .
Apetinil [Syntex] *see* 3809
Apex 462-5 *see* 9881
Apexol [Pfizer] *see* 10150
APF-1 *see* 9975
Aphamite *see* 7167
Aphanin *see* 3543
Aphidicolin, 769
Apholate, 770
Aphox [ICI] *see* 7651
Aphoxide *see* 9804
Aphrodine *see* 10236
Aphrodyne [Star] *see* 10236
Aphtiria [Debat] *see* 5526
Aphylline, 771
A1PI *see* 760
Apicycline, 772
Apigenin, 773
Apigenin-7-apiosylglucoside *see* 775
Apigenin-7-D-glucoside *see* 774
Apigenin-4′-methyl Ether *see* 10
Apigetrin, 774
Apihepar [Asta] *see* 8680
Apiin, 775
Apilak *see* 8434
Apinol *see* 5883
7-[(2-*O*-D-Apio-β-D-furanosyl-β-D-
  glucopyranosyl)oxy]-5-hydroxy-2-
  (4-hydroxyphenyl)-4*H*-1-benzo-
  pyran-4-one *see* 775
6-*O*-(D-Apiofuranosyl)-1,6,8-trihydr-
  oxy-3-methylanthraquinone *see* 4288
Apiol *see* 777
Apiole (Dill), 776
Apiole (Parsley), 777
Apioline *see* 777
Apiose, 778
D-Apiose *see* 778
Apioside *see* 775
Apiquel [McNeil] *see* 497
Apiracohl [Kyowa] *see* 9640
Apirazin *see* 2247
A.P.L. [Wyeth-Ayerst] *see* 2273
Aplace [Kyorin] *see* 9921
Aplactan [Eisai] *see* 2365
Aplakil *see* 7059
Aplasmomycin, 779
Aplexal [Taiyo] *see* 2365
Aplisol [Parke, Davis] *see* 9937
Apllobal [Fujisawa] *see* 321
Aplonidine *see* 791
APM *see* 874
APNPS *see* 9103
APO *see* 9804
Apoaequorin *see* 175
Apoatropine, 780
Apocard [Esteve] *see* 4136
Apocodeine, 781
Apocretin [Kyowa] *see* 3911
Apocynamarin *see* 9015

Apocynin, 782
Apocynum Androsaemifolium, 783
Apocynum Cannabinum, 784
Apodel [Meiji] *see* 5541
Apodol [Squibb] *see* 695
ApoE *see* 786
Apo-β-erythroidine, 785
Apoferritin *see* 4083
Apohemocyanin *see* 4681
Apolan [Stockli] *see* 2436
Apolipoprotein E, 786
Apollo [FBC] *see* 2435
Apomorphine, 787
Aponal [Boehringer, Mann.] *see* 3492
Aponeuron [Apogepha] *see* 624
Aponorin [Kodama] *see* 9754
Apopain, 788
Apophedrin *see* 7441
Aporeine, 789
Aporheine *see* 789
6αβ-Aporphine-10,11-diol *see* 787
Apostavit *see* 10150
Apo-Sulfatrim [Apotex] *see* 9086
Apoterin S [Seiko] *see* 8311
Apotomin [Kowa] *see* 2365
3α,16α-Apovincaminic Acid Ethyl
  Ester *see* 10128
A-Poxide [Abbott] *see* 2132
Appertex [Janssen] *see* 2403
Appetize [Science Food Technology],
  790
Apple Acid *see* 5747
Apple of Peru *see* 8975
Apple Oil *see* 5135
Apraclonidine, 791
Apralan [Lilly] *see* 792
Apramycin, 792
Apranax [Laroche Navarron] *see* 6504
Aprecon *see* 5124
Apresoline [Ciba] *see* 4800
Aprical [Rentschler] *see* 6617
Apride [Kodama] *see* 9640
Apridol *see* 5890
Aprindine, 793
Aprinox [Boots] *see* 1064
APRL *see* 7682
Aprobal [Fujisawa] *see* 321
Aprobarbital, 794
Aprobarbital Sodium *see* 794
Aprobit *see* 4870
Aprocarb *see* 8022
Apron [Ciba-Geigy] *see* 5982
Apronal *see* 795
Apronalide, 795
Aprotinin, 796
Aprozal *see* 794
APSAC *see* 712
Apsifen [APS] *see* 4925
Apsin VK [APS] *see* 7230
Apsolol [APS] *see* 8025
Aptine [Pharma-Stern] *see* 321
Aptol Duriles [Astra] *see* 321
Apulonga [Dorsch] *see* 287
Apurin [GEA] *see* 287
Apurol [Siegfried] *see* 287
Apurone [Riker] *see* 4172
APY-606 *see* 2474
Apyrase, 797
Apyron [Wülfing] *see* 5689
AQ-110 *see* 9719
AQP1 *see* 798
AQP2 *see* 798
AQP.At1 *see* 798
AQP-CD *see* 798
AQP-CHIP *see* 798
Aqua Ammonia *see* 519
Aquacare [Herbert] *see* 10005
Aquacide *see* 3415
Aquacillin [Riker] *see* 7226
Aquacobalamin *see* 4854

Aquacycline [Syntex] *see* 7111
Aquadrate [Eaton] *see* 10005
Aquafen *see* 4145
Aquafortis *see* 6671
Aqualin [Shell] *see* 130
Aquamarine *see* 1201
AquaMephyton [Merck & Co.] *see* 7536
Aquamollin *see* 3557
Aquamox [Lederle] *see* 8240
Aquamycetin [Winzer] *see* 2120
Aquamycin *see* 2009
Aquanil [Selena] *see* 9585
Aquaphor [Beiersdorf] *see* 10212
Aquaporin-CHIP *see* 798
Aquaporins, 798
Aquareduct [Azupharma] *see* 8917
Aqua Regia *see* 6706
Aquaretic [Azupharma] *see* 426
Aquarius [Merck & Co.] *see* 4822
Aquasept [Hough, Hoseason] *see* 9790
Aquasuspen *see* 7226
Aquatag [Tutag] *see* 1155
Aquatensen [Mallinckrodt] *see* 6086
Aquathol [ELF Atochem] *see* 3615
Aquathol K [ELF Atochem] *see* 3615
Aquazine [Ciba-Geigy] *see* 8681
Aquedux [Delagrange] *see* 2434
Aquex [Sandoz] *see* 2454
Aquirel *see* 2822
Aquocobalamin *see* 4854
Aquo-Trinitrosan [E. Merck] *see* 6704
AR-12008 *see* 9709
Ara-A *see* 10113
Arabinitol *see* 800
9-β-D-Arabinofuranosyladenine Monohydrate *see* 10113
(*E*)-1-β-D-Arabinofuranosyl-5-(2-bromoethenyl)-2,4(1*H*,3*H*)-pyrimidinedione *see* 8875
1-β-D-Arabinofuranosyl-(*E*)-5-(2-bromovinyl)uracil *see* 8875
1-β-D-Arabinofuranosylcytosine *see* 2853
N-(1-β-D-Arabinofuranosyl-1,2-dihydro-2-oxo-4-pyrimidinyl)docosanamide *see* 3624
9-β-D-Arabinofuranosyl-2-fluoroadenine *see* 4162
9-β-D-Arabinofuranosyl-2-fluoro-9*H*-purin-6-amine *see* 4162
9-β-D-Arabinofuranosyl-9*H*-purine-6-amine Monohydrate *see* 10113
6-*O*-α-L-Arabinopyranosyl-D-glucose *see* 10111
Arabinose, 799
L-Arabinose *see* 799
6-(α-L-Arabinosido)-D-glucose *see* 10111
Arabinosyladenine *see* 10113
Arabite *see* 800
Arabitin [Sankyo] *see* 2853
Arabitol, 800
D-Araboascorbic Acid *see* 5141
D-Araboflavin, 801
Ara-C *see* 2853
Arachic Acid *see* 802
Arachidic Acid, 802
Arachidonic Acid, 803
Arachis Oil *see* 7191
Aracytidine *see* 2853
Aracytine [Upjohn] *see* 2853
Aragonite *see* 1697
Aralen [Winthrop] *see* 2215
Aralia, 804
Aramine [Merck & Co.] *see* 5993
Aramite [Uniroyal], 805
Aranidipine, 806
Arantoick [Ohta] *see* 9660
"Arariba" *see* 807

Araroba, 807
Arasan [Du Pont] *see* 9510
Arasena-A [Mochida] *see* 10113
Arasine [Gensia] *see* 13
Arathane *see* 3340
Aratron *see* 805
Arbacet [Upjohn] *see* 808
Arbaprostil, 808
Arbekacin, 809
Arborescin, 810
Arborine *see* 4512
Arbor Vitae *see* 9531
Arbotect [Merck & Co.] *see* 9426
Arbuse *see* 10177
Arbutamine, 811
Arbutin, 812
Arbutoside *see* 812
Arbuz *see* 7148
Arcadine *see* 1055
Arcalion [Servier] *see* 9432
Arcanum Duplicatum *see* 7845
Arcasin [Engelhard KG] *see* 7230
Archil *see* 6994
Archin *see* 3602
Arcilla *see* 819
Arcomonol Tablets *see* 6312
Arcosolv DPM [Arco] *see* 3407
Arcton [ICI] *see* 2192
Arcton 11 [ICI] *see* 9770
Arcton 12 [ICI] *see* 3114
Arcton 114 [ICI] *see* 2672
Ardésyl [Beytout] *see* 10165
Ardex *see* 2996
Ardeytropin [Ardeypharm] *see* 9929
AR-DF 26 *see* 4452
Ardine [Antibioticos] *see* 617
Arduan [Organon] *see* 7612
Areca, 813
Arecaidine, 814
Arecaine *see* 814
Arecaline *see* 815
Arechin [Polfa] *see* 2215
Arecholine *see* 815
Arecoline, 815
Arecoline-acetarsol *see* 815
Arecoline-acetarsone *see* 815
Arecoline *p*-Stibonobenzoic Acid *see* 815
Aredia [Ciba] *see* 7135
Arelix [Cassella-Riedel] *see* 7647
Arelon [Hoechst] *see* 5237
A-Rest [Elanco] *see* 672
Aretit [Hoechst] *see* 3341
Arfonad [Roche] *see* 9837
Arg *see* 817
Argatroban, 816
Argentic Fluoride *see* 8656
Argentic Oxide *see* 8666
Argentite *see* 8677
Argentoproteinum *see* 8671
Argentous Chlorate *see* 8651
Argentous Fluoride *see* 8657
Argentous Oxide *see* 8665
Argentous Sulfide *see* 8677
Argentum Vitellinum *see* 8671
Argil *see* 819
Argilla *see* 5294
Argilla Vini *see* 819
Arginine, 817
L-Arginine *see* 817
Arginine L-Aspartate *see* 875
Arginine Glutamate, 818
5-L-Arginine-27-glycine-33-L-isoleucine-34-L-serine-42-glycine-purothionin A I (Reduced) *see* 8131
8-D-Arginine-1-(3-mercaptopropanoic Acid) Vasopressin *see* 2969
5-L-Arginine-6-L-threonine-18-L-serine-26-L-serine-27-L-threonine-42-

glycine-purothionin A I (Reduced) *see* 8131
Arginine Vasopressin *see* 10073
N-(*N*-(*N*-(*N*²-L-Arginyl-L-lysyl)-L-α-aspartyl)-L-valyl)-L-tyrosine *see* 9544
Argipidine *see* 816
Argipressin *see* 10073
Argobyl [Specia] *see* 4142
Argol, 819
Argon, 820
Argun [Merckle] *see* 4998
Argun L [Merckle] *see* 5596
Argyrol [J & J] *see* 8671
Arheol *see* 6940
Arial [Dompé] *see* 8489
Aribine *see* 4646
Aricine, 821
Aricyl [Winthrop] *see* 850
Arilin [Wolff] *see* 6242
Arimidex [Zeneca] *see* 667
Aristamid [Nordmark] *see* 9124
Aristocort [Lederle] *see* 9727
Aristolochic Acid, 822
Aristolochic Acid-I *see* 822
Aristolochine *see* 822
Aristololactams *see* 822
Aristosol [Lederle] *see* 9728
Aristospan [Fujisawa] *see* 9730
Arkin [Otsuka] *see* 10105
Arkitropin *see* 4766
ARL *see* 827
AR-L 115BS *see* 9159
Arlacel 20 [ICI] *see* 8872
Arlacel 60 [ICI] *see* 8872
Arlacel 80 [ICI] *see* 8872
Arlanto [Nichiiko] *see* 224
Arlef [Parke, Davis] *see* 4167
Arlidin [USV] *see* 6830
Arliflav *see* 1269
Arlitene [Chinoin] *see* 6374
Armazal *see* 2751
Armepavine, 823
Armoise *see* 9
Armol *see* 6097
Armophylline [Bellon] *see* 9421
Armstrong's Acid, 824
Armyl [Armour Pharm.] *see* 5658
Arnaudon's Green *see* 2284
Arnica, 825
Arnica Flowers *see* 825
Arnold's Base *see* 6264
Arnosulfan [Beocet] *see* 9073
Arnotta *see* 713
Arobon *see* 5584
Aroclor 1242 *see* 7725
Aroclor 1254 *see* 7725
Aroclor 1260 *see* 7725
Aroclor [Monsanto] *see* 7725
Aroft [Tanabe] *see* 181
Arofuto [Tanabe] *see* 181
Arogenate *see* 826
Arogenic Acid, 826
Aropax [Beecham] *see* 7175
Arotinolol, 827
Arovit [Roche] *see* 10150
Arpamyl [Jouveinal] *see* 10083
Arpicolin [Rosemont] *see* 7944
Arpimycin [R. P. Drugs] *see* 3720
Arpocox [Merck & Co.] *see* 828
Arprinocid, 828
Arquad 16 Stearate *see* 2069
Arquel [Parke, Davis] *see* 5819
Arresten [Nippon Shinyaku] *see* 6220
Arret [Janssen] *see* 5601
Arrhenal [Cleary] *see* 6020
Arrivo [FMC] *see* 2836
Arsacetin, 829
Arsacol [Zambon] *see* 10026
Arsambide *see* 1830

Arsamin see 8719
Arsaminol see 851
Arsanilic Acid, 830
Arsanilic Acid Sodium Salt see 8719
Arsanyl [Taiyo] see 4385
Arsaphen see 49
Arsaphenan see 49
Arseclor see 3121
Arsecodile see 8738
Arsemétine see 3600
Arsen (German) see 832
Arsenal [Am. Cyan.] see 4942
Arsenamide, 831
Arsenic, 832
Arsenic Acid, 833
Arsenic Acid Anhydride see 839
Arsenical Soln see 7769
Arsenic Chloride Solution see 847
Arsenic Disulfide, 834
Arsenic Hemiselenide, 835
Arsenic Pentafluoride, 836
Arsenic Pentaselenide, 837
Arsenic Pentasulfide, 838
Arsenic Pentoxide, 839
Arsenic Sesquioxide see 844
Arsenic Sulfide see 834
Arsenic Tribromide, 840
Arsenic Trichloride, 841
Arsenic Trifluoride, 842
Arsenic Trihydride see 849
Arsenic Triiodide, 843
Arsenic Trioxide, 844
Arsenic Triselenide, 845
Arsenic Trisulfide, 846
Arsenic Yellow see 846
Arsenik see 832
Arsenious Acid Copper(2+) Salt (1:1)
  see 2693
Arsenious Acid Solution, 847
Arsenious Selenide see 845
Arseno 39 see 7082
Arsenoacetic Acid, 848
4,4′-Arsenobis(2-aminophenol) Dihy-
  drochloride see 851
4,4′-Arsenobis[2-aminophenol] Dis-
  odium Salt see 8722
Arsenodiacetic Acid see 848
Arsenolite see 844
Arsenosan see 7082
Arsenosobenzene see 7083
Arsenous Acid see 844
Arsenous Acid Anhydride see 844
Arsenous Oxide see 844
Arsenous Selenide see 845
Arsenoxide see 7082
Arsenphenolamine Hydrochloride see
  851
Arsevan see 6534
Arsicodile see 8738
Arsine, 849
Arsinyl [Cleary] see 6020
Arsion see 2170
Arsobal [Specia] see 5856
Arsonic Acid Copper(2+) Salt (1:1) see
  2693
Arsonoacetic Acid, 850
p-Arsonophenylurea see 1830
Arsphenamine, 851
Arsphenamine Methylenesulfoxylic
  Acid Sodium Salt see 6534
Arsphenamine Sodium see 8722
Arsthinol, 852
Arsycodile see 8738
Artabotrine see 5178
Artam see 2347
Artane [Lederle] see 9823
Artate [Nippon Chemiphar] see 2365
Arteannuin see 856
Arteether, 853
Artegodan [Artesan] see 7151

Artemether, 854
Artemisin, 855
Artemisine see 856
Artemisinin, 856
Arteoptic [Novopharma] see 1917
Arterenol [Hoechst] see 6788
Arteriohom [Weisskopf] see 2437
Arteriovinca [Farma-Lepori] see
  10120
Arterium V [ICI] see 6604
Arterocoline see 88
Arterolo [Nuovo Cons.] see 2442
Artes [Sumitomo] see 5864
Artesunate, 857
Artesunic Acid see 857
Arteven [Boehringer, Ing. Italia] see
  4685
Artevil see 2436
Artex [Servier] see 9318
Artexal [Servier] see 9318
Arthaxan [SKB] see 6428
Arthrisin [Sandoz] see 886
Arthrocine [Merck & Co.] see 9155
Arthrodont [Veyron-Froment] see
  3628
Arthropan [Purdue Frederick] see
  2265
Arthrosan see 2347
Artificial Cognac Essence see 3881
Artificial Essential Oil of Almond see
  1085
Artificial Ethyl Oenanthate see 3881
Artificial Lanthanite see 5374
Artificial Oil of Ants see 4324
Artil [Hoechst] see 5257
Artisone Acetate [Wyeth] see 77
Artolon [Roter] see 5908
Artosin [Boehringer, Mann.] see 9646
Artracin [DDSA] see 4998
Artrene [Irbi] see 4925
Artribid [Merck & Co.] see 9155
Artrichin see 2215
Artril 300 [Farmasa] see 4925
Artrinovo [Llorns, Spain] see 4998
Artrizin [Leo Pharm] see 7431
Artrobione see 2265
Artrodar [Proter] see 3003
Artrolasi [Lenza] see 9177
Artromialgina see 3413
Artrosilene [Dompé] see 5316
Artroxicam [Coli] see 7661
Artume see 1752
ARTZ [Seikagaku] see 4793
Arumel [SS Pharm] see 4219
Arusa see 158
Arusal see 1889
Arvigol [Zyma] see 2419
Arvin [Toyford] see 671
Arvynol [Pfizer] see 3774
Arwin [Knoll] see 671
Arylam see 1831
Arythmol [Knoll] see 7978
Arzene [Am. Cyanamid] see 7083
AS see 657
AS-101 see 830
Asa see 1875
A.S.A. [Lilly] see 886
ASA-226 see 2123
5-ASA see 5964
Asabaine see 6025
Asacol [Tillotts] see 5964
Asacolitin [Röhm] see 5964
Asafetida, 858
Asafoetida see 858
Asamedol [Maruko] see 3753
Asant see 858
Asaprol, 859
Asarabacca Camphor see 861
Asarin see 861
Asarinin, 860

l-Asarinin see 860
α-Asarone see 861
β-Asarone see 861
Asarones, 861
Asarum, 862
Asarum Camphor see 861
Asatard [De Angeli] see 886
Asbestos, 863
ASC see 106
Ascabin see 1162
Ascabiol [May & Baker] see 1162
Ascal see 1684
Ascaridil [Cilag] see 5486
Ascaridole, 864
Ascarisin see 864
Ascarite [Baker] see 863
Ascaryl [Wander] see 4750
Ascensil see 486
Asceptichrome see 5914
Ascharite see 5693
Asclepias, 865
Asclepias Syriaca, 866
Ascomp [Nippon Chemiphar] see 224
Ascorbic Acid, 867
L-Ascorbic Acid see 867
Ascorbic Acid Calcium Salt see 1688
L-Ascorbic Acid Mixt. with 3-Pyri-
  dinecarboxamide see 6610
Ascorbic Acid Nicotinamide Complex
  see 6610
Ascorbic Acid Sodium Derivative see
  8723
Ascorbicin [Squibb] see 8723
Ascorbigen, 868
Ascorbigen A see 868
Ascorbigen B see 868
Ascorbin [Lakeside] see 8723
Ascorin see 867
Ascorteal see 867
Ascorvit see 867
Ascotoxin see 1393
AS-E 136 see 3418
Asebotoxin see 4566
Asellacrin [Serono] see 8864
Asendin [Lederle] see 616
Asendis [Lederle] see 616
Asepsin see 1420
Aseptamide [Merminod] see 4620
ASH see 172
Asiatic Acid see 869
Asiaticoside, 869
Asil see 4929
Asiprenol see 5236
Aska-Rid see 1659
Askensil see 486
ASL-279 see 3479
ASL-601 see 17
ASL-8052 see 3741
Asmalar see 5236
Asmatane Mist [3M] see 3656
Asmaten [Riker] see 8393
Asmaterolo [Lusofarmaco] see 8307
Asmaven [Berk] see 217
Asn see 872
L-Asnase see 871
Asoxime Chloride, 870
Asp see 875
D-Asp see 875
ASP-47 see 9139
Aspara see 875
Asparagic Acid see 875
L-Asparaginase, 871
Asparaginate Calcium see 875
Asparagine, 872
L-Asparagine, 872
L-β-Asparagine see 872
L-Asparagine Amidohydrolase see 871
1-Asparagine-5-valine-angiotensin II
  see 689
Asparaginic Acid see 875

Asparaginsäure (German) see 875
Asparagus, 873
Aspara K see 875
Asparamide see 872
Aspartame, 874
Aspartat [Debat] see 875
Aspartic Acid, 875
L-Aspartic Acid see 875
Aspartic Acid β-Amide see 872
3-(L-Aspartyl-D-alaninamido)-2,2,4,4-
  tetramethylthietane see 245
L-Aspartyl-D-alanine-N-(2,2,4,4-tetra-
  methylthietan-3-yl)amide see 245
N-L-α-Aspartyl-L-phenylalanine 1-
  Methyl Ester see 874
N²-[1-[N-(N-L-α-Aspartyl-L-seryl)-L-α-
  aspartyl]-L-prolyl]-L-arginine see
  7270
L-α-Aspartyl-N-(2,2,4,4-tetramethyl-3-
  thietanyl)-D-alaninamide see 245
Aspasan [Hoechst] see 4031
ASPC see 887
Aspenil [Chemil] see 617
Aspenon [Mitsui] see 793
Aspergillic Acid, 876
Aspergillin, 877
Aspergillus Diastase see 9203
Asperlicin, 878
Asperlicin B see 878
Asperlicin C see 878
Asperlicin D see 878
Asperlicin E see 878
Asperule Absolute see 10185
Asperuloside, 879
Asphalt, 880
Asphaltum see 880
Asphocalcium see 867
Aspiculamycin see 4550
Aspidin, 881
Aspidinol, 882
Aspidium, 883
Aspidol [Piam] see 5668
Aspidosperma, 884
Aspidospermine, 885
Aspirin, 886
Aspirin Aluminum see 338
Aspirin Lysine Salt see 5668
Aspisol [Bayer] see 5668
Aspogen [Eaton] see 3227
Aspoxicillin, 887
Aspro [Nicholas] see 886
Assaren [Permamed] see 3132
Assert [Am. Cyanamid] see 4941
Assiprenol see 5236
Assoral [Glaxo] see 8433
Assugrin see 2770
Assure [Du Pont] see 8269
Asta C 4898 see 3244
Astacene see 888
Astacin, 888
Astaril see 3896
Astat [Nihon Nohyaku] see 5370
Astatine, 889
Astaxanthin, 890
Asta Z-4942 see 4937
Astemisan [Zdravlje] see 891
Astemizole, 891
Astenile [Recordati] see 7892
Asteric [Cooper] see 886
Asterin see 2757
Asterol Dihydrochloride [Roche] see
  3033
Asteromycin see 4550
Asthmahaler [Menley & James] see
  3656
Asthmalitan [Kettelhack] see 5185
Asthmanefrin [Menley & James] see
  3656
Asthma Weed see 5579
Asthmolysin [Kade] see 3529

Astiban [Roche] see 8966
Astix CMPP [Agrotec] see 5826
Astmamasit [Showa Yakuhin] see 3529
Astomin [Yamanouchi] see 3251
Astonin H [E. Merck] see 4166
Astragalin see 5288
Astridine [Astra] see 5245
Astringen see 356
Astrobot [Arnolds] see 3129
Astrocar see 6635
Astroderm [Medix] see 3099
Astrolin see 757
Astromicin see 4276
Astrophyllin see 3529
Astryl see 4489
Asturidon [Cilag] see 1530
Astyn see 3552
Asuccin see 9038
Asucrol see 2239
[ASU¹,⁷]-E-CT see 3578
Asuntol [Bayer] see 2626
Asuro [Nippon Kayaku] see 2993
Asverin [Tanabe Seiyaku] see 9601
Asymmetrical Trimethylbenzene see
  8100
AT 10 see 3223
AT-7 see 4716
AT-17 see 3251
AT-101 see 5244
AT-327 see 9601
AT-877 see 3980
AT-2266 see 3625
AT-4140 see 8884
ATA see 513
Atabrine Hydrochloride [Sanofi Win-
  throp] see 8225
Atapren [Sumitomo] see 9640
Ataractan see 925
Atarax [Pfizer] see 4897
Atav see 10150
ATBAC see 1799
ATC see 9586
Ateben [Chem-Sintyal] see 6812
Ateculon [Nippon Chemiphar] see
  2436
AteHexal [Hexal] see 892
Atelor [Roche] see 3033
Atem [Chiesi] see 5089
Atemi 50 SL [Sandoz] see 2840
Atemorin see 5890
Atempol [Norgine] see 5890
Atenen [Tsuruhara] see 154
Atenezol [Tsuruhara] see 50
Atenol [CT] see 892
Atenolol, 892
Atenos [UCB] see 9942
Atensil see 7577
Atensin see 5895
Atensine [Berk] see 3042
Aterax see 4897
Ateriosan see 2436
Aterosan [Aanderson] see 8158
[1'R-(1'α,5'Aβ,7'β,8'aβ,9'aβ)]-2',3',-
  8'a,9'-tetrahydro-1'-hydroxy-1',4,4,-
  8',8',11'-hexamethylspiro[4H,8H-
  [1,4]dioxepino[2,3-g]indole-8,7'-
  (8'H)-[5H,6H-5a,9a](iminomethano)-
  [1H]cyclopent[f]indolizine]-9,10'-
  (10H)-dione see 7159
Atgard [Shell] see 3129
Athamantin, 893
Atheran see 2442
Atheroitin see 2270
Atherolip see 2437
Atherolipin [Schwarz] see 2437
Atherophylline [Merrell] see 8159
Atheropront [Mack, Illert.] see 2436
Athrombin-K [Purdue Frederick] see
  10174
Athymil [Organon] see 6260

Athyromazole see 1844
Atilen [Spofa] see 3042
Atipamezole, 894
Atipi see 154
Atiramin [Juste] see 934
Atirin [Intersint] see 1967
Atisine, 895
Ativan [Wyeth] see 5609
Atlacide [Chipman] see 8741
Atlansil [Roemmers] see 504
Atlas G-4829 [ICI] see 7717
Atmosgen [Maruko] see 9660
Atmosphere, 896
Atocin see 2347
Atock [Yamanouchi] see 4272
Atophan [Warner-Chilcott] see 2347
Atophanyl [Warner-Chilcott] see 2347
Atoquinol see 2347
Atorvastatin, 897
Atosil [Bayer] see 7970
Atovaquone, 898
Atover [Recordati] see 8158
Atoxyl see 8719
Atoxylic Acid see 830
ATP see 154
ATP-dependent Proteolytic Factor see
  9975
Atractyl see 5758
Atractylin (C₃₀ Glucoside) see 899
Atractyloside, 899
Atracurium Besylate, 900
Atral see 3416
Atranex [Makhteshim-Agan] see 902
Atranoric Acid see 901
Atranorin, 901
Atravet [Ayerst] see 32
Atraxin [Daiichi] see 5908
Atrazine, 902
Atrial Natriuretic Factor, 903
Atrican [Innotech] see 9292
Atrilon 5 [Winthrop] see 7995
Atrimycon [Bioindustria] see 8075
Atrinal [Roche] see 3243
Atriopeptigen see 903
Atriopeptin see 903
Atriphos see 154
Atrolactamide, 904
Atrolactic Acid, 905
Atromid [Ayerst] see 2436
Atromidin see 2436
Atromid-S [Ayerst] see 2436
Atropamine see 780
Atropic Acid, 906
Atropine, 907
Atropine Aminoxide see 908
Atropine Hyperduric see 907
Atropine N-Oxide, 908
Atropisol [CooperVision] see 907
Atropyltropeine see 780
Atroscine see 8550
Atrovent [Boehringer, Ing.] see 5089
Attacins, 909
Attar of Rose see 6937
Attentil [Ravizza] see 4126
Atumin [Merrell] see 3147
Aturbal see 7377
Aturban(e) [Ciba] see 7377
Aturgyl [Dausse] see 4025
Atussil [Squibb] see 1840
Atysmal [Schaefer] see 3794
Aubépine see 2636
Aucubin, 910
Aucuboside see 910
Audax [Napp] see 2265
Audes [Nippon Kayaku] see 2380
Augmentin [SKB] see 2402
Auligen see 3449
Aulin [Robin] see 6640
Aulinogen [Boehringer, Ing.] see 3449
Auranofin, 911

Aurantiin see 6512
Aurantiogliocladin, 912
Aurcoloid [Abbott] see 4533
Aurcoscan-198 [Abbott] see 4533
Aureine see 8595
Aurelic Acid see 7696
Aureociclina see 2245
Aureocina see 2245
Aureolic Acid see 7696
Aureomycin [Lederle] see 2245
Aureotan [Byk-Gulden] see 917
Aureothin, 913
Aureothricin, 914
Aureotope [Squibb] see 4533
Auric Bromide see 4539
Auric Chloride see 4541
Auric Cyanide see 4543
Auric Hydroxide see 4544
Auricidine see 4537
Auric Oxide see 4545
Auric Selenate see 4534
Auric Selenide see 4535
Auric Sulfide see 4546
Auriculin see 903
Aurin, 915
Aurinol see 4378
Auripigment see 846
Aurochlorohydric Acid see 4542
Aurocidin see 4537
Aurodox, 916
Aurolin see 4537
α-Auromercaptoacetanilide see 918
Auropex see 4537
Auropin see 4537
Aurorix [Roche] see 6309
Aurosan see 4537
Aurothioglucose, 917
Aurothioglycanide, 918
Aurothioglycolanilide see 918
Aurothioglycolic Acid Anilide see 918
Aurothion see 4537
Aurothiosulfate Natrium see 4537
Aurothiosulfate Sodium see 4537
Aurous Chloride see 4528
Aurous Cyanide see 4529
Aurous Iodide see 4530
Aurous Oxide see 4532
Aurous Stannate see 4538
Aurous Sulfide see 4531
Aurumine see 917
Aurum Paradoxum see 9273
Ausocef [Ausonia] see 7670
Australian Fever Bark see 3431
Australian Fever Tree see 3941
Austrapen [CSL] see 628
Austrawolf [Commonwealth] see 8293
Austrian Cinnabar see 5423
Austyn [Faulding] see 9421
Autan see 2912
Auteral [Boehringer, Mann.] see 7573
Autoprothrombin II see 3964
Autoprothrombin II-A see 8070
Autoworm [Coopers] see 7069
Autumn Crocus see 2537
Auxiloson [Thomae] see 2986
Auxisone [Boehringer, Ing.] see 2986
Auxit [Bristol-Myers Squibb] see 1412
AV see 762
AV-290 see 922
Ava-ava see 5298
Avacan [Asta] see 1784
Avadex see 3014
Avadex BW [Monsanto] see 9726
Avadyl see 1784
Avagal see 6025
Avan [Takeda] see 4932
Avantrin [Inverni] see 9709
Avantyl see 6812
Avatec [Roche] see 5384
Avazyme [Wampole] see 2320

Avenein see 4473
Avenge [Am. Cyanamid] see 3185
Avenin see 1934
Aventox [DowElanco] see 9797
Aventyl [Lilly] see 6812
Avermectin A$_{1a/b}$ see 919
Avermectin A$_{2a/b}$ see 919
Avermectin B$_1$ see 1
Avermectin B$_{1a/b}$ see 1
Avermectin B$_{2a/b}$ see 919
Avermectins, 919
Averon-1 [Alfar] see 2028
Avertin [Winthrop] see 9743
Aviax [Pfizer] see 8587
Avibon see 10150
Avicalm [Squibb] see 6237
Avicel see 2012
Avicol SL [Troponwerke] see 2440
Avid [Merck & Co.] see 1
Avidin, 920
Avigilen [Brenner-Efeka] see 7641
Avil [Albert-Roussel] see 7383
Avilamycin, 921
Avilamycin A see 921
Avilamycin C see 921
Avinar [Armour Pharm.] see 10011
Aviochina [Agrimont-Vetem] see 9109
Aviral [Medici Domus] see 5006
Avisco Rayon [FMC] see 8294
Avishot [Kanebo] see 6443
Avita see 10150
Avitol see 10150
Avlocardyl [Zeneca] see 8025
Avloclor [ICI] see 2215
Avloprocil see 7226
Avlosulfon [ICI] see 2885
AVM see 919
Avogadrite see 7859
Avolin see 3304
Avomec [Merck & Co.] see 1
Avomine [Rhone-Poulenc Rorer] see 7970
Avoparcin, 922
Avornin see 4288
Avosyl see 5895
Avotan [Am. Cyanamid] see 922
Avoxyl see 5895
AW-14'2333 see 7320
AW-105-843 see 6442
Award [Ciba-Geigy] see 4027
AX 250 [Durachemie] see 617
Axeen [Hommel] see 8091
Axepim [Bristol-Myers Squibb] see 1973
Axer Alfa [Alfa Wassermann] see 6504
Axerol see 10150
Axerophthal see 8331
Axerophthol see 10150
Axid [Lilly] see 6758
Axil [Boehringer, Mann.] see 7574
Axiomin [Promeco] see 3930
Axiquel [McNeil] see 10048
Axiten see 5813
Axlon [Albert-Roussel] see 4854
Axonyl [Parke-Davis] see 7641
Axoren [Glaxo] see 1528
Axsain [Euroderma] see 1811
Axungia Porci see 5380
Axuris see 4401
AY-5406-1 see 1055
AY-5710 see 5685
AY-6108 see 628
AY-6608 see 7250
AY-8682 see 2828
AY-21011 see 7882
AY-22989 see 8288
AY-24031 see 5500
AY-24236 see 3920
AY-25650 see 9878

AY-25712 see 112
AY-27773 see 9663
AY-62014 see 1568
AY-62021 see 2455
AY-62022 see 5836
AY-64043 see 8025
Ayeramate see 5908
Ayermicina [Ayerst] see 5480
Ayfactin see 7181
Ayfivin see 965
Aygestin [Wyeth-Ayerst] see 6790
AZ 8 [Millet Roux] see 4581
AZ 8 Beris [Weimer] see 4581
10-Azaanthracene see 124
5-Aza-10-arsenaanthracene Chloride see 7357
1-Azabicyclo[2.2.2]octane see 8265
1-Azabicyclo[2.2.2]octan-3-ol see 8266
(±)-N-1-Azabicyclo[2.2.2]oct-3-yl-6-chloro-3,4-dihydro-4-methyl-3-oxo-2H-1,4-benzoxazine-8-carboxamide see 933
5-(1-Azabicyclo[2.2.2]oct-3-yl)-10,11-dihydro-5H-dibenz[b,f]azepine see 8267
10-(1-Azabicyclo[2.2.2]oct-3-ylmethyl)-10H-phenothiazine see 5911
1-(3-Azabicyclo[3.3.0]oct-3-yl)-3-(p-tolylsulfonyl)urea see 4448
Azacitidine, 923
Azacort see 2916
Azacortid [Lepetit] see 4151
Azacosterol, 924
Azactam [Squibb] see 955
Azacyclonol, 925
2-(1'-Azacyclooctyl)ethylguanidine see 4589
1-Azacyclooct-2-ylmethylguanidine see 4588
Azacyclopropane see 3850
5-Azacytidine see 923
Azadirachtin, 926
9-Azafluorene see 1834
Azafrin, 927
8-Azaguanine, 928
Azalomycin M see 6633
Azalone [SM & P] see 718
Azamethone see 929
Azamethonium Bromide, 929
4-Aza-5-(N-methyl-4-piperidinyliden-ene)-10,11-dihydro-5H-dibenzo-[a,d]cycloheptene see 934
Azameton see 929
6-Azamianserin see 6295
Azamin 4B see 1132
Azamune [Penn] see 935
Azanidazole, 930
Azanin [Tanabe] see 935
Azantac [Glaxo] see 8286
5-Azaorotic Acid see 7081
Azapen [Pfizer] see 6047
Azaperone, 931
1-Azaphenothiazine-10-carboxylic Acid 2-(2-Piperidinoethoxy)ethyl Ester see 7609
Azapren [Robins] see 768
Azapropazone see 768
Azaribine see 937
Azaron [Chefaro] see 9868
Azaserine, 932
Azasetron, 933
5α,20β$_F$,22α$_F$,25β$_F$,27-azaspirostan-3β-ol see 9682
Δ^5-20β$_F$,22α$_F$,25α$_F$,27-azaspirosten-3β-ol see 8857
Azasterol see 924
Azatadine, 934
Azathioprine, 935
6-Azathymine, 936
6-Azauracil Riboside see 937

6-Azauridine, 937
Azelaic Acid, 938
Azelastine, 939
Azene [Endo] see 2465
Azepinamide see 4524
Azeptin [Eisai] see 939
2-Azetidinecarboxylic Acid, 940
Azidamfenicol, 941
Azidin [USSR] see 3316
D-(—)-threo-2-Azidoacetamido-1-p-
   nitrophenyl-1,3-propanediol see 941
Azidoamphenicol see 941
α-Azidobenzylpenicillin see 942
Azidocillin, 942
3'-Azido-3'-deoxythymidine see 10252
2-Azido-N-[2-hydroxy-1-(hydroxy-
   methyl)-2-(4-nitrophenyl)ethyl]-
   acetamide see 941
6-[D-α-Azidophenylacetamido]penicil-
   lanic Acid see 942
[2S-[2α,5α,6β(S*)]]-6-[(Azidophenyl-
   acetyl)amino]-3,3-dimethyl-7-oxo-4-
   thia-1-azabicyclo[3.2.0]heptane-2-
   carboxylic Acid see 942
Azidothymidine see 10252
Azimethylene see 3048
Azimidobenzene see 1140
Azimilide, 943
Azindole see 1111
Azinepurine see 8116
Azinphos-ethyl see 944
Azinphos-methyl, 944
Azintamide, 945
Aziridine see 3850
Aziridinylbenzoquinone see 3044
1-Aziridinylphosphonitrile Trimer see
   770
Azithromycin, 946
Azitrocin [Roerig] see 946
Azlin [Miles] see 947
Azlocillin, 947
Azmacort [Rhone-Poulenc Rorer] see
   9728
Azoangin [Ursapharm] see 2317
p-Azoaniline see 3019
Azobenzene, 948
Azobenzide see 948
Azobenzol see 948
4,4'-Azobisbenzenamine see 3019
1,1'-Azobiscarbamide see 950
Azobiscarbonamide see 950
Azobiscarboxamide see 950
α,α'-Azobis[chloroformamidine] see
   2171
1,1'-Azobisformamide see 950
3,3'-Azobis(6-hydroxybenzoic Acid)
   see 6976
2,2'-Azobisisobutyronitrile, 949
1,1'-Azobis[3-methyl-2-phenylimid-
   azo[1,2-a]pyridinium] Dibromide see
   3981
2,2'-Azobis[2-methylpropanenitrile]
   see 949
5,5'-Azobis(salicylic Acid) see 6976
Azochloramide [Pennwalt] see 2171
Azo Compd No. 4 see 9069
5-Azocytosine see 923
4,4'-Azodianiline see 3019
Azodicarbonamide, 950
Azodicarboxamide see 950
α,α'-Azodiisobutyronitrile see 949
Azodisal see 6976
Azodisal Sodium see 6976
Azodolen see 3004
Azodrin [Shell] see 6334
Azohel [Ursapharm] see 2317
Azoksodon see 7206
Azol see 482
Azole see 8198
Azolid [USV] see 7431

Azolitmin, 951
Azolmen [Menarini] see 1260
Azoman see 4735
Azomycin, 952
Azonam [Krka] see 955
Azone [Nelson] see 5397
Azoniaspiro(3α-benziloyloxynortro-
   pane-8,1'-pyrrolidine) Chloride see
   9918
Azoniaspiro(3α-diphenylglycoloyloxy-
   nortropan-8,1'-pyrrolidine) Chloride
   see 9918
Azophenylene see 7359
Azoran [Searle] see 935
Azosemide, 953
Azothioprine see 935
Azovan Blue see 3952
Azoxodone see 7206
Azoxybenzene, 954
Azoxybenzide see 954
AZQ see 3044
AZT see 10252
Aztec [Verex] see 10252
Azthreonam see 955
Aztreon [Squibb] see 955
Aztreonam, 955
Azubromaron [Azupharma] see 1093
Azucaps see 7376
Azudimidine [Pharmacia] see 8473
Azudoxat [Azupharma] see 3496
Azuglucon [Azupharma] see 4486
Azulene, 956
Azulfidine [Kabi Pharmacia] see 9112
Azulon [Rorer] see 4581
Azupentat [Azupharma] see 7278
AzUR see 937
Azure A, 957
Azure B, 958
Azure Blue II see 6135
Azure C, 959
Azure I see 6135
Azure II see 6135
Azure II Eosin see 6135
Azurene [Cilag-Chemie] see 1466
Azurite see 6135
Azusalen [Ohta] see 956
Azutranquil [Azupharma] see 7059
B-9 see 2874
B72.3 see 8530
B72.3-Glycyl-tyrosyl-N-ε-diethylene-
   triaminepentaacetic Acid-In-111 see
   8530
B-360 see 7176
B-436 see 7919
B-518 see 2816
B-577 see 3922
B-612 see 2917
B-663 see 2433
B-995 see 2874
B-1500 see 5847
B-2310 see 6355
B-2311 see 6355
B-4576 see 7954
B-11420 see 5077
B-15000 see 5071
B-66256 see 10003
B-68138 see 3997
BA see 1361
Ba-168 see 5588
Ba-253 see 7077
Ba-253-BR-L see 7077
Ba-598Br see 4248
Ba-679 see 9598
Ba-679 BR see 9598
Ba-1355 see 7363
Ba-2758 see 2951
Ba-5473 see 7109
Ba-7205 see 2421
BA-7602-06 see 9214
Ba-10370 see 9068

Ba-18189 see 3362
Ba-21401 see 9738
Ba-29038 see 1354
Ba-29837 see 2914
Ba-30803 see 1117
Ba-32644 see 6656
Ba-34276 see 5792
Ba-34647 see 967
Ba-39089 see 7086
Ba-180265 see 213
BAAM [Upjohn] see 510
Babbitt Metal, 960
Babesan see 3416
Babesin [Hoechst] see 3316
Babidium see 4767
Bab-O see 9188
Baburan see 3416
Bacacil [Pfizer] see 961
Bacampicillin, 961
Bacampicine [Upjohn] see 961
Bacancosin see 972
Bacarate [Tutag] see 7365
Baccidal [Kyorin] see 6793
Bacfeed [Fujisawa] see 1249
Bachelor's Buttons see 6828
Baciferm see 10257
Bacifurane [Meram] see 6626
Bacillin see 963
Bacillus Calmette-Guérin see 1041
Bacillus Thermoproteolyticus Neutral
   Proteinase see 9422
Bacillus thuringiensis, 962
Bacilysin, 963
Bacimethrin, 964
Bacimex [Alter] see 9166
Bacitracin, 965
Bacitracin A see 965
Bacitracin Methylenebis[2-hydroxy-
   benzoate] see 966
Bacitracin Methylenedisalicylate see
   966
Bacitracin Methylenedisalicylic Acid,
   966
Bacitracin Zinc Complex see 10257
Bacitracin Zinc Salt see 10257
Baclofen, 967
Baclon [Medica] see 967
Bacterial Vitamin H[1] see 443
Bacterio-opsin see 968
Bacteriopsins see 6988
Bacteriorhodopsin, 968
Bacteron [Ciba-Geigy] see 1249
Bacticlens [Smith & Nephew] see 2140
Bactidan [Recordati] see 3625
Bactocill [Beecham] see 7036
Bactoderm [SKB] see 6383
Bactol see 5052
Bactopen [Beecham] see 2480
Bactospeine see 962
Bactramin [Roche] see 9086
Bactramyl [Carrion] see 7660
Bactrim [Roche] see 9086
Bactroban [SKB] see 6383
Bactrovet [Pitman-Moore] see 9073
Bactylan see 498
Baddeleyite see 10313
Badil [Bayer] see 4401
Badional see 9116
Badische Acid, 969
Bad Nauheim Salts (Artificial) see
   6515
BAF see 5019
Bafhameritin-M [Hishiyama] see 5842
Bagasse, 970
Bagodryl see 3367
Bagolax see 6120
Bagren [Serono] see 1437
Bagrosin-Natrium [Cassella-Riedel]
   see 6184
Bahama White Wood see 1791

BAHS *see* 1562
Baicalein, 971
Bajaten [Volpino] *see* 4969
Bakankoside *see* 972
Bakankosin, 972
Bakash *see* 158
Baking Soda *see* 8726
Bakontal *see* 1023
Baktar [Shionogi] *see* 9086
Bakthane *see* 962
BAL [HW & D] *see* 3255
Balan [Lilly] *see* 1067
Balance [Yamanouchi] *see* 2132
Balanophorin *see* 660
Balarsen [Endo] *see* 852
Baldex [Bausch & Lomb] *see* 2986
Balfin [Lilly] *see* 1067
"Balm of Gilead" *see* 973
Balm of Gilead *see* 975
Balmox [SKB] *see* 6428
Balsam Canada, 973
Balsam Capivi *see* 2580
Balsam Copaiba *see* 2580
Balsam Gurjun, 974
Balsam Mecca, 975
Balsam of Fir *see* 973
Balsam of Gilead *see* 975
Balsam Peru, 976
Balsam Tolu, 977
Balsam Traumatic, 978
Balsam Tree *see* 5795
Balsán-Katél *see* 975
Baltic Amber *see* 397
Baluvet [Veterinaria] *see* 375
Bambec [Astra] *see* 980
Bambermycins, 979
Bamboo Curare *see* 2741
Bambuterol, 980
Bamethan, 981
Bamifylline, 982
Bamiphylline *see* 982
Bamipine, 983
Bamo 400 [Misemer] *see* 5908
Banabins *see* 7675
Banabin-Sintyal *see* 2155
Banambax [Rhône-Poulenc Rorer] *see* 7254
Banamine [Schering] *see* 4182
Banan [Sankyo] *see* 1991
Banana Oil *see* 5125
Bancaris [Burroughs Wellcome] *see* 9414
Bandol [Squibb] *see* 1846
Banewort *see* 1052
Banflex [Forest] *see* 7007
Banicol *see* 2059
Banisterine *see* 4647
Banistyl [M & B] *see* 4260
Banlene [Fisons] *see* 3090
Ban-Lep [MSD] *see* 3594
Banminth [Pfizer] *see* 8139
Bannal *see* 8547
Banner [Ciba-Geigy] *see* 8003
Banocide [Wellcome] *see* 3165
Bantenol [Janssen] *see* 5807
Bantex *see* 5916
Banthine Bromide [Searle] *see* 6025
Banthionine [Banyu] *see* 6053
Bantron *see* 5580
Bantu *see* 764
Banvel D [Velsicol] *see* 3090
Baptigenin, 984
Baptisia, 985
Baptitoxine *see* 2858
BAQD-10 *see* 2959
Barakshin *see* 4929
Baratol [Wyeth] *see* 5000
Barazan [Dieckmann] *see* 6793
Barbaloin *see* 314
Barbamate *see* 986

Barbamil *see* 607
Barbamyl *see* 607
Barban, 986
Barbane *see* 986
Barbasco, 987
Barberry Bark, 988
Barbexaclone *see* 8045
Barbiphenyl [Lääkefarmos] *see* 7386
Barbipil [Conal] *see* 7386
Barbital, 989
Barbital Sodium *see* 989
Barbitone *see* 989
Barbitone Sodium *see* 989
Barbituric Acid, 990
Barbonin [Knoll] *see* 3773
Barbosec [Rowell] *see* 8563
Barbourin *see* 5013
Bardac 2250/2280 [Lonza] *see* 3149
Bardana *see* 5377
Bareon [Hokuriku] *see* 5592
Baridol *see* 1023
Barite *see* 1023
Baritop [Kaigen] *see* 1023
Barium, 991
Barium Acetate, 992
Barium Benzenesulfonate, 993
Barium Benzoate *see* 1122
Barium Bromate, 994
Barium Bromide, 995
Barium Carbonate, 996
Barium Chlorate, 997
Barium Chloride, 998
Barium Chromate(VI), 999
Barium Citrate *see* 2387
Barium Cyanide, 1000
Barium Cyanoplatinate(II) *see* 1020
Barium Dioxide *see* 1018
Barium Dithionate, 1001
Barium Diuranate *see* 1030
Barium Ethyl Sulfate *see* 3897
Barium Ferrocyanide, 1002
Barium Fluoride, 1003
Barium Fluosilicate *see* 1005
Barium Formate, 1004
Barium Hexacyanoferrate(II) *see* 1002
Barium Hexafluorosilicate, 1005
Barium Hydrate *see* 1006
Barium Hydroxide, 1006
Barium Hypophosphite, 1007
Barium "Hyposulfate" *see* 1001
Barium Hyposulfite *see* 1028
Barium Iodate, 1008
Barium Iodide, 1009
Barium Lactate *see* 5349
Barium Manganate(VI), 1010
Barium Manganate(VII) *see* 1017
Barium Mercuric Iodide, 1011
Barium Metatitanate *see* 1029
Barium Methyl Sulfate *see* 6203
Barium Monoxide *see* 1015
Barium Nitrate, 1012
Barium Nitrite, 1013
Barium Oleate *see* 6965
Barium Oxalate, 1014
Barium Oxide, 1015
Barium Perchlorate, 1016
Barium Permanganate, 1017
Barium Peroxide, 1018
Barium *p*-Phenolsulfonate *see* 7396
Barium Phosphate, Dibasic, 1019
Barium Platinocyanide *see* 1020
Barium Platinous Cyanide, 1020
Barium Propionate *see* 8010
Barium Protoxide *see* 1015
Barium Rhodanide *see* 1027
Barium Selenide, 1021
Barium Silicide, 1022
Barium Silicofluoride *see* 1005
Barium Sulfate, 1023
Barium Sulfide, 1024

Barium Sulfide, Black, 1025
Barium Sulfite, 1026
Barium Sulfocyanate *see* 1027
Barium Sulfocyanide *see* 1027
Barium Superoxide *see* 1018
Barium Taurocholate *see* 9242
Barium Tetracyanoplatinate(II) *see* 1020
Barium Tetraiodomercurate(II) *see* 1011
Barium Thiocyanate, 1027
Barium Thiosulfate, 1028
Barium Titanate(IV), 1029
Barium Uranate(VI) *see* 1030
Barium Uranium Oxide, 1030
Barizin [Lek] *see* 6579
Barnetil [Delagrange] *see* 9168
Barnidipine, 1031
Barnotil [Vita] *see* 9168
Baron [Dow] *see* 3681
Baros [Horii] *see* 3264
Baros Camphor *see* 1366
Barosma Camphor *see* 3351
Barosmin *see* 3350
Barosperse [Mallinckrodt] *see* 1023
Barquinol *see* 5052
Barricade [Shell] *see* 2836
Barthrin, 1032
Baryta Yellow *see* 999
Barytes *see* 1023
BAS-083 *see* 5903
BAS-238F *see* 3467
BAS-352F *see* 10122
BAS-392H *see* 4156
BAS-480-F *see* 3665
BAS-3460 *see* 1836
BAS-9052 *see* 8620
BAS-42100F *see* 4035
BAS-67054 *see* 1836
BAS-85559X *see* 5903
Basagran [BASF] *see* 1080
Basalin [BASF] *see* 4156
Basaljel [Wyeth] *see* 1033
Basanite [BASF] *see* 3341
Basecil *see* 6210
Basedol *see* 501
Base L *see* 5648
Basergin *see* 3694
Basethyrin *see* 6210
Basfapon B [BASF] *see* 2869
Basforin [BASF] *see* 9819
Basham's Mixture *see* 4059
Basic Aluminum Acetate *see* 349
Basic Aluminum Carbonate Gel, 1033
Basic Aluminum Chloride *see* 356
Basic Aluminum Salicylates *see* 375
Basic Antimony Chloride *see* 734
Basic Bismuth Bromide *see* 1300
Basic Bismuth Chloride *see* 1303
Basic Bismuth Iodide *see* 1309
Basic Bismuth Valerate *see* 1334
Basic Chromic Sulfates *see* 2287
Basic Cupric Chromates *see* 2700
Basic Dextran *see* 2980
Basic Ferric Sulfate Soln *see* 4078
Basic Fibroblast Growth Factor *see* 4117
Basic Lead Carbonate, 1034
Basic Lead Chromate *see* 5423
Basic Lead Hydroxide *see* 5428
Basic Mercuric Sulfate *see* 5941
Basi-Cop *see* 2723
Basic Zirconium Chloride *see* 10317
Basodexan [Rohm] *see* 10005
Basofortina [Sandoz] *see* 6147
Basolan *see* 6049
Bassado [Poli] *see* 3496
Basswood, 1035
Basta [Hoechst] *see* 7495
Bastard Saffron *see* 1919

Bastel [Rhone-Poulenc Rorer] see 3534
Bastiverit [Bastian-Werk] see 4486
Basudin [Geigy] see 3043
BAT see 9913
Batidrol [Mtadier] see 723
Batimastat, 1036
Batrachotoxin, 1037
Batrachotoxinin A see 1037
Batrachotoxinin A 20-(2,4-Dimethyl-1H-pyrrole-3-carboxylate) see 1037
Batrafen [Hoechst] see 2325
Batroxobin, 1038
Batyl Alcohol, 1039
Baum's Acid see 6477
Bauxite see 369
Bavenite see 1685
Bavistin [BASF] see 1836
Bax [McKesson] see 3367
BAX-1400Z see 3260
BAX-1526 see 2319
BAX-2739Z see 982
Baxacor [Mack, Illert.] see 3753
Baxan [Bristol-Myers] see 1963
Baxo [Toyama] see 7661
Bay 1470 see 10213
Bay 1521 see 6821
Bay 2502 see 6630
Bay 4503 see 8017
Bay 5621 see 7523
Bay 9002 see 6441
Bay 9010 see 8022
Bay 19639 see 3429
Bay 25141 see 4042
Bay 34727 see 2321
Bay 39007 see 8022
Bay 44646 see 452
Bay 68138 see 3997
Bay 70143 see 1851
Bay 77488 see 7523
Bay 92114 see 5187
Bay 94337 see 6239
Bay a 1040 see 6617
Bay b 4343 b see 1911
Bay b 5097 see 2478
Bay e 5009 see 6669
Bay e 6905 see 947
Bay e 9736 see 6643
Bay f 1353 see 6259
Bay g 2821 see 6397
Bay g 5421 see 15
Bay h 4502 see 1260
Bay h 5757 see 3982
Bay k 5552 see 6658
Bay m 1099 see 6274
Bay n 5595 see 2952
Bay o 6893 see 8395
Bay o 9867 see 2374
Bay q 3939 see 2374
Bay q 7821 see 5096
Bay Va 1470 see 10213
Bay Va 5387 see 3918
Bay Va 9391 see 6960
Bay Vh 5757 see 3982
Bay Vi 9142 see 9664
Bay Vp 2674 see 3630
BAY DRW 1139 see 5985
BAY FCR 1272 see 2826
BAY HWG 1608 see 9253
BAY KWG 0519 see 9724
BAY KWG 0599 see 1342
BAY MEB 6447 see 9723
BAY NTN 9306 see 9164
BAY NTN 33893 see 4946
BAY SAS 9244 see 9737
BAY SIR 8514 see 9809
BAY SMY 1500 see 3785
Bayberry Bark, 1040
Baycain [Bayer] see 9679
Baycaron [Bayer] see 5847

Baycid [Bayer] see 4044
Baycidal [Bayer] see 9809
Baycillin [Bayer] see 8002
Baycip [Bayer] see 2374
Baycipen [Bayer] see 6259
Baycor [Bayer] see 1342
Baycovin see 8182
Baycox [Bayer] see 9664
Bayer 21/116 see 6129
Bayer 21/199 see 2626
Bayer 205 see 9181
Bayer 693 see 3896
Bayer 1219 see 9834
Bayer 1362 see 1543
Bayer 1420 see 7986
Bayer 2353 see 6602
Bayer 3231 see 9733
Bayer 5312 see 3783
Bayer 5630 see 6242
Bayer 5633 see 6002
Bayer 8169 see 2940
Bayer 9015 see 6601
Bayer 9051 see 5486
Bayer 16259 see 944
Bayer 17147 see 944
Bayer 25820 see 6441
Bayer 29493 see 4044
Bayer 36205 see 7113
Bayer 37344 see 6050
Bayer 41831 see 4017
Bayer 46131 see 8004
Bayer 47531 see 3095
Bayer 68138 see 3997
Bayer 71628 see 6014
Bayer 74283 see 6002
Bayer 78418 see 3560
Bayer A 128 see 796
Bayer A 173 see 7942
Bayer E 39 Soluble see 1285
Bayer E 393 see 9139
Bayer L 1359 see 9753
Bayer S 5660 see 4017
Bayerite see 369
Bayer's Acid see 2656
Bayfidan [Bayer] see 9724
Baygnostil see 7387
Baygon [Bayer] see 8022
Bayleton [Bayer] see 9723
Bayluscid [Bayer] see 6602
Baymicin [Bayer] see 8695
Baymix [Chemagro] see 2626
Baymycard [Bayer] see 6658
Bayo-n-ox [Bayer] see 6960
Bayotensin [Bayer] see 6669
Baypen [Bayer] see 6259
Baypress [Bayer] see 6669
Bayrena [Bayer] see 9082
Bayrogel [Bayropharm] see 3922
Baytan [Bayer] see 9724
Baytex [Parke, Davis] see 4044
Baythion [Bayer] see 7523
Baythroid [Bayer] see 2826
Bayticol [Bayer] see 4175
Baytinal [Bayer] see 1553
Baytril [Mobay] see 3630
Bayvarol [Provet] see 4175
BB-94 see 1036
BBB see 4124
B.B.C. see 1433
BB-K8 [Bristol] see 425
BB Powder see 7841
BBS-1067 see 1992
BC-16 see 4713
BC-48 see 2936
BC-51 see 3426
BC-105 see 7671
BC-681 see 3563
BC-757 see 7694
levo-BC-2627 see 1565
Bc Conjugate see 8120

B-cell Activating Factor see 5019
BCG, 1041
(—)-BCH-189 see 5365
BCM see 1836 and 5790
BCME see 3119
BCNU see 1894
BCP see 1487
BD-40A see 4272
BDCM see 1438
BDF-5895 see 6375
BDH-312 see 5895
BDMPEA see 1958
B-DNA see 2954
BDNF see 6570
B.D.P.E. see 1471
Beacillin see 7221
Beacon [Ciba-Geigy] see 7928
Beamette [Dawbarn] see 7741
Beaprine see 1915
Bearberry see 10034
Bearberry Bark see 1931
Bear's Weed see 3711
Bearwood see 1931
Bebate see 1226
Bebeerine, 1042
d-Bebeerine see 1042
l-Bebeerine see 2745
Bebeeru Bark, 1043
Beben [Warner-Lambert] see 1226
Becantal see 8753
Becantex see 8753
Becanthone, 1044
Becantone see 1044
Becaptan [Labaz] see 2848
Becenun [Bristol-Myers] see 1894
Bécilan see 8166
Beclacin [Kaigai] see 1047
Beclamide, 1045
Beclipur [Siegfried] see 1046
Beclobrate, 1046
Becloforte [Allen & Hanburys] see 1047
Beclomet [Orion] see 1047
Beclomethasone, 1047
Beclorhinol [Desitin] see 1047
Beclotiamine, 1048
Becloval [Valeas] see 1047
Beclovent [Glaxo] see 1047
Becodisks [Allen & Hanburys] see 1047
Beconase [Allen & Hanburys] see 1047
Beconasol [Glaxo] see 1047
Becort [Rachelle] see 1226
Becotide [Allen & Hanburys] see 1047
Bedermin [Damor] see 1226
Bedodeka [Teva] see 10152
Bedoz [Nadeau] see 10152
Bedranol [Lagap] see 8025
Bedriol [Andromaco] see 1260
Bee Bread see 8021
Beechwood Creosote see 2642
Beeswax, 1049
Beet Sugar see 9051
Befaprost see 1190
Befeniol see 1187
Befizal [Oberval] see 1240
Beflavin [Roche] see 8367
Befunolol, 1050
Behenic Acid, 1051
Behenic Alcohol see 8135
N(4)-Behenoyl-1-β-D-arabinofuranosylcytosine see 3624
Behenoylcytosine Arabinoside see 3624
Behenyl Alcohol see 8135
Behepan [Kabi] see 10152
Behyd [Kyorin] see 1171
Bei-1293 see 10212
Bekadid see 4826

Acid Methyl 1-Methylethyl Ester
see 5260
3-Benzofuro[3,2-*c*][1]benzoxepin-
6(12*H*)-ylidene-*N,N*-dimethyl-1-pro-
panamine *see* 7068
Benzofurodil *see* 1070
2-Benzofuryl-*p*-chlorophenylcarbinol
see 2418
Benzoglyoxaline *see* 1111
Benzoguanamine, 1120
Benzo-Gynoestryl [Roussel] *see* 3746
Benzohydrol, 1121
Benzoic Acid, 1122
(2*S-cis*)-Benzoic Acid [1-[4-[(3-Amino-
2,3,6-trideoxy-α-L-*lyxo*-hexopyrano-
syl)oxy]-1,2,3,4,6,11-hexahydro-2,-
5,12-trihydroxy-7-methoxy-6,11-
dioxo-2-napthacenyl]ethylidene]hy-
drazide *see* 10326
Benzoic Acid, Ammonium Salt *see* 521
Benzoic Acid Anhydride *see* 1123
Benzoic Acid Benzyl Ester *see* 1162
Benzoic Acid Butyl Ester *see* 1587
Benzoic Acid 2,4-Dichlorophenoxy-
ethyl Ester *see* 8618
Benzoic Acid Ethyl Ester *see* 3814
Benzoic Acid Hydrazide 3-Hydrazone
with Daunorubicin *see* 10326
Benzoic Acid Methyl Ester *see* 6104
Benzoic Acid Phenyl Ester *see* 7429
Benzoic Acid 2-(1-Phenylethyl)hydra-
zide *see* 1072
Benzoic Acid Phenylmethyl Ester *see*
1162
Benzoic Acid *p*-Sulfamide *see* 1927
Benzoic Aldehyde *see* 1085
Benzoic Anhydride, 1123
Benzoic Sulfimide *see* 8463
Benzoin, 1124
Benzoin Oxime, 1125
Benzol *see* 1094
Benzolin *see* 3081
Benzoline *see* 5513
Benzometan [Castejon] *see* 1152
6,7-Benzomorphan, 1126
Benzo[*b*]naphthacene *see* 7240
Benzonaphthol *see* 6495
Benzonatate, 1127
Benzonitrile, 1128
Benzononatine *see* 1127
Benzoparadiazine *see* 8263
Benzo[*c*]phenanthrene *see* 1150
Benzo[*def*]phenanthrene *see* 8147
Benzophenone, 1129
Benzophenone-1 *see* 1138
Benzophenone-3 *see* 7088
Benzophenone-4 *see* 9157
Benzophenone-6, 1130
Benzophenone-8 *see* 3357
Benzophenone-10 *see* 6255
Benzophenone-12 *see* 6838
Benzopinacol, 1131
Benzopurpurine 4B, 1132
1,2-Benzopyran, 1133
2*H*-1-Benzopyran *see* 1133
α-5:6-Benzopyran *see* 1133
2*H*-1-Benzopyran-2-one *see* 2630
2-(5*H*-[1]Benzopyrano[2,3-*b*]pyridin-
7-yl)propionic Acid *see* 7890
Benzo[*a*]pyrazine *see* 8263
Benzopyrazole *see* 4970
1,2-Benzopyrene *see* 1134 and 1135
Benzo[*a*]pyrene, 1134
Benzo[*e*]pyrene, 1135
Benzo[*a*]pyrene-7,8-diol 9,10-Epoxide
see 1134
Benzo[*c*]pyridazine *see* 2366
Benzo[*d*]pyridazine *see* 7526
Benzo[*b*]pyridine *see* 8253
Benzo[*c*]pyridine *see* 5241

5,6-Benzopyrimidine *see* 8235
Benzo[*a*]pyrimidine *see* 8235
1,2-Benzopyrone *see* 2630
Benzo-γ-pyronecarboxylic Acid *see*
2294
2,3-Benzopyrrole *see* 4993
Benzoquin *see* 6331
5,6-Benzoquinoline *see* 1136
Benzo[*f*]quinoline, 1136
1,4-Benzoquinone *see* 8259
p-Benzoquinone Amidinohydrazone
Thiosemicarbazone *see* 395
Benzoquinone Guanylhydrazone Thiosemicarbazone *see* 395
Benzoquinonium Chloride, 1137
[2,5-*p*-Benzoquinonylenebis(iminotrimethylene)]bis[benzyldiethylammo-
nium Chloride] *see* 1137
Benzoresorcinol, 1138
Benzosalin *see* 6105
Benzosol *see* 4576
Benzosulfimide *see* 8463
Benzotetrazine *see* 8116
Benzothiazide *see* 1155
2-Benzothiazolamine *see* 444
6-Benzothiazolamine *see* 445
Benzothiazole, 1139
2-Benzothiazolethiol *see* 5916
2(3*H*)-Benzothiazolethione *see* 5916
1-(2-Benzothiazolyl)-1,3-dimethylurea
see 6002
N-2-Benzothiazolyl-*N,N'*-dimethylurea
see 6002
(±)-4-(2-Benzothiazolylmethylamino)-
α-[(4-fluorophenoxy)methyl]-1-
piperidineethanol *see* 8462
Benzothiazyl Disulfide *see* 3435
N-(1-Benzo[*b*]thien-2-ylethyl)-*N*-hy-
droxyurea *see* 10253
Benzothiofuran *see* 9438
Benzo[*b*]thiophene *see* 9438
2-Benzothiophenecarboxylic Acid *see*
9481
Benzo[*b*]thiophene-2-carboxylic Acid
see 9481
1,2,3-Benzotriazole *see* 1140
1*H*-Benzotriazole, 1140
2-(2*H*-Benzotriazol-2-yl)-*p*-cresol *see*
3503
2-(2*H*-Benzotriazol-2-yl)-4-methyl-
phenol *see* 3503
Benzotrichloride, 1141
Benzotrifluoride, 1142
4-(2,1,3-Benzoxadiazol-4-yl)-2,6-di-
methyl-1,4-dihydro-3-isopropyloxy-
carbonylpyridine-5-carboxylic Acid
Methyl Ester *see* 5260
4,4'-(3*H*-2,1-Benzoxathiol-3-ylidene)-
bis[2-bromo-3-methyl-6-(1-methyl-
ethyl)phenol] *S,S*-Dioxide *see* 1469
4,4'-(3*H*-2,1-Benzoxathiol-3-ylidene)-
bis[2-bromo-6-methylphenol] *S,S*-
Dioxide *see* 1408
4,4'-(3*H*-2,1-Benzoxathiol-3-ylidene)-
bis[2,6-dibromo-3-methylphenol]
S,S-Dioxide *see* 1407
4,4'-(3*H*-2,1-Benzoxathiol-3-ylidene)-
bis[2,6-dibromophenol] *S,S*-Dioxide
see 1468
4,4'-(3*H*-2,1-Benzoxathiol-3-ylidene)-
bis[2,5-dimethylphenol] *S,S*-Dioxide
see 10216
4,4'-(3*H*-2,1-Benzoxathiol-3-ylidene)-
bis[5-methyl-2-(1-methylethyl)phen-
ol] *S,S*-Dioxide *see* 9541
4,4'-(3*H*-2,1-Benzoxathiol-3-ylidene)-
bis(2-methylphenol) *S,S*-Dioxide *see*
2647
4,4'-(3*H*-2,1-Benzoxathiol-3-ylidene)-
bisphenol *S,S*-Dioxide *see* 7397

2*H*-1,3-Benzoxazine-2,4(3*H*)-dione *see*
1915
Benzoxine [Sanwa] *see* 1233
Benzoxiquine, 1143
Benzoxonium Chloride, 1144
Benzoxyline *see* 1143
3-Benzoxy-1-(2-methylpiperidino)-
propane *see* 7627
ω-Benzoylacetophenone *see* 3056
Benzoyl Acetyl Peroxide *see* 78
N-Benzoyl-β-alanine *see* 1227
Benzoylamide *see* 1087
Benzoylaminoacetic Acid *see* 4752
α-(Benzoylamino)-4-[2-(diethylamino)ethoxy]-*N,N*-dipropylbenzene-
propanamide *see* 9607
(±)-4-(Benzoylamino)-5-(dipropylamino)-5-oxopentanoic Acid *see* 7958
[2a*R*-[2aα,4β,4aβ,6β,9α(α*R**,β*S**),-
11α,12α,12aα,12bα]]-β-(Benzoylamino)-α-hydroxybenzenepropanoic
Acid 6,12b-Bis(acetyloxy)-12-(benzoyloxy)-2a,3,4,4a,5,6,9,10,11,12,-
12a,12b-dodecahydro-4,11-dihydr-
oxy-4a,8,13,13-tetramethyl-5-oxo-
7,11-methano-1*H*-cyclodeca[3,4]-
benz[1,2-*b*]oxet-9-yl Ester *see* 7117
4-(Benzoylamino)-2-hydroxybenzoic
Acid *see* 1148
(*S*)-4-[[2-(Benzoylamino)-3-(4-hydr-
oxyphenyl)-1-oxopropyl]amino]-
benzoic Acid *see* 1081
(11β,16α)-21-[3-(Benzoylamino)-2-
methyl-1-oxopropoxy]-9-fluoro-11-
hydroxy-16,17-[(1-methylethylid-
ene)bis(oxy)]pregna-1,4-diene-3,20-
dione *see* 9729
3-(Benzoylamino)propionic Acid *see*
1227
N-Benzoyl-*p*-aminosalicylic Acid *see*
1148
N-Benzoylaniline *see* 1088
Benzoylbenzene *see* 1129
5-Benzoyl-2-benzimidazolecarbamic
Acid Methyl Ester *see* 5807
(5-Benzoyl-1*H*-benzimidazol-2-yl)car-
bamic Acid Methyl Ester *see* 5807
Benzoyl Chloride, 1145
(±)-5-Benzoyl-2,3-dihydro-1*H*-pyrrol-
izine-1-carboxylic Acid *see* 5318
5-Benzoyl-1,2-dihydro-3*H*-pyrrolo[1,-
2-a]pyrrole-1-carboxylic Acid *see*
5318
Benzoyl-α-(dimethylamino)ethane *see*
5984
N-Benzoyl-*N',N'*-di-*n*-propyl-DL-iso-
glutamine *see* 7958
Benzoylecgonine, 1146
O-Benzoyl-*l*-ecgonine Ethyl Ester *see*
2516
O-Benzoyleugenol *see* 3944
Benzoylformaldoxime *see* 5208
6-Benzoyl-D-glucose *see* 10037
N-Benzoylglycine *see* 4752
Benzoylguaiacol *see* 4576
m-Benzoylhydratropic Acid *see* 5316
6-Benzoyl-3-hydrazino-5,6,7,8-tetra-
hydropyrido[4,3-*c*]pyridazine *see*
3617
Benzoyl Hydroperoxide *see* 7295
3-Benzoyl-4-hydroxy-6-methoxybenz-
enesulfonic Acid *see* 9157
5-Benzoyl-4-hydroxy-2-methoxybenz-
enesulfonic Acid *see* 9157
Benzoyl Isothiocyanate, 1147
2-Benzoyl-5-methoxy-1-phenol-4-
sulfonic Acid *see* 9157
3-Benzoyl-α-methylbenzeneacetic
Acid *see* 5316

Benzylhydrochlorothiazide, 1171
Benzylhydroflumethiazide see 1064
Benzyl Hydroquinone see 6331
4-Benzyl-α-(p-hydroxyphenyl)-β-methyl-1-piperidineethanol see 4936
1-Benzyl-3-hydroxypyridinium Bromide Dimethylcarbamate see 1153
Benzylideneacetone, 1172
Benzylideneacetophenone see 2078
Benzylideneaniline, 1173
β-Benzylidenebutyramide see 1083
Benzylidene Chloride see 1084
2-Benzyl-2-imidazoline see 9645
2-Benzyl-4,5-imidazoline see 9645
Benzylimidobis(p-methoxyphenyl)-methane, 1174
2-Benzyl-2-iminazoline see 9645
1-Benzyl-1H-indazol-3-yl 3-(Dimethylamino)propyl Ether see 1157
[(1-Benzyl-1H-indazol-3-yl)oxy]acetic Acid see 1061
N-Benzyl-β-(isonicotinoylhydrazino)-propionamide see 6575
α-Benzylisopropylamine see 7415
Benzyl Mercaptan see 9456
2-Benzyl-1-[4-(methylamino)butoxy]-benzene see 1255
Benzyl Methyl Ether, 1175
1-Benzyl-2-(5-methyl-3-isoxazolylcarbonyl)hydrazine see 5172
Benzyl Methyl Ketone see 7423
4-Benzyl-1-(1-methyl-4-piperidyl)-3-phenyl-3-pyrazolin-5-one see 1152
N-Benzyl-N-methyl-2-propynylamine see 7172
cis-N-(1-Benzyl-2-methylpyrrolidin-3-yl)-5-chloro-2-methoxy-4-(methylamino)benzamide see 6533
(±)-cis-N-(1-Benzyl-2-methyl-3-pyrrolidinyl)-5-chloro-4-(methylamino)-o-anisamide see 6533
2-Benzyl-1-methylquinazol-4-one see 4512
Benzylmorphine, 1176
3-Benzylmorphine 6-Myristate see 6420
Benzylmorphine Myristic Acid Ester see 6420
Benzylmorphinyl Myristate see 6420
Benzyl Nicotinate see 6613
N-Benzyl-2-nitroimidazole-1-acetamide see 1114
7-(Benzyloxy)-6-n-butyl-1,4-dihydro-4-oxo-3-quinolinecarboxylic Acid Methyl Ester see 6558
7-(Benzyloxy)-6-n-butyl-4-hydroxy-3-quinolinecarboxylic Acid Methyl Ester see 6558
7-(Benzyloxy)-6-n-butyl-3-methoxycarbonyl-4-quinolone see 6558
(2′-Benzyloxycarbonyl)phenyl trans-4-(Guanidinomethyl)cyclohexanecarboxylate Hydrochloride see 1065
3-Benzyloxy-6-myristoyloxy-N-methyl-4,5-epoxy-7-morphinene see 6420
3-Benzyloxy-6-myristoyloxymorphine see 6420
p-(Benzyloxy)phenol see 6331
Benzylpenicillin Acetoxymethyl Ester see 7208
Benzylpenicillin Benzhydrylamine see 7222
Benzylpenicillin Calcium see 7223
Benzylpenicillin β-Diethylaminoethyl Ester Hydriodide see 7212
Benzylpenicillinic Acid, 1177
Benzylpenicillinic Acid Benzhydrylamine Salt see 7222
Benzylpenicillinic Acid N-Benzyl-β-phenylethylamine Salt see 7220

Benzylpenicillinic Acid N,N′-Bis(dehydroabietyl)ethylenediamine Double Salt see 7224
Benzylpenicillinic Acid Calcium Salt see 7223
Benzylpenicillinic Acid Potassium Salt see 7225
Benzylpenicillinic Acid Sodium Salt see 1178
Benzylpenicillin Potassium see 7225
Benzylpenicillin Procaine see 7226
Benzylpenicillin Sodium, 1178
Benzylpenicilloyl Polylysine see 7233
o-Benzylphenol, 1179
p-Benzylphenol, 1180
4-(o-Benzylphenoxy)-N-methylbutylamine see 1255
1-[2-(2-Benzylphenoxy)-1-methylethyl]piperidine see 1078
Benzylphenylamine see 1161
α-(N-Benzyl-N-phenylamino)acetamidoxime see 2065
p-Benzylphenyl Carbamate see 3365
2-Benzylphenyl β-Dimethylaminoethyl Ether see 7469
N-Benzyl-2-phenylethylamine Salt of Benzylpenicillin see 7220
N-Benzyl-N-phenylpyrrolidinoethylamine see 4757
p-Benzylphenylurethan see 3365
Benzyl Phosphite see 3061
Benzyl Phosphorochloridate see 3059
1-Benzyl-4-picolinoylpiperazine see 7547
2-(4-Benzylpiperidino)-1-(4-hydroxyphenyl)-1-propanol see 4936
N-(1′-Benzyl-4′-piperidyl)-2-methoxy-4-amino-5-chlorobenzamide see 2404
(±)-(R*)-3-[(R*)-1-Benzyl-3-piperidyl] Methyl 1,4-Dihydro-2,6-dimethyl-4-(m-nitrophenyl)-3,5-pyridinedicarboxylate see 1071
2-(1-Benzyl-4-piperidyl)-2-phenylglutarimide see 1104
(S)-(+)-2-(1-Benzyl-4-piperidyl)-2-phenylglutarimide see 2987
(+)-3-(1-Benzyl-4-piperidyl)-3-phenyl-piperidine-2,6-dione see 2987
1-Benzyl-2-pivalylhydrazine see 7668
N-Benzyl-N-pyrrolidinoethylaniline see 4757
(+)-(3′S,4S)-1-Benzyl-3-pyrrolidinyl Methyl 1,4-Dihydro-2,6-dimethyl-4-(m-nitrophenyl)-3,5-pyridinedicarboxylate see 1031
(3S)-1-Benzyl-3-pyrrolidinyl Methyl (4S)-2,6-Dimethyl-4-(m-nitrophenyl)-1,4-dihydropyridine-3,5-dicarboxylate see 1031
Benzyl-Rodiuran [Boehringer, Ing.] see 1064
Benzyl Salicylate, 1181
Benzyl Salicylate trans-4-(Guanidinomethyl)cyclohexanecarboxylate Hydrochloride see 1065
Benzylsulfamide, 1182
N⁴-Benzylsulfanilamide see 1182
Benzyl Sulfide, 1183
p-(Benzylsulfonamido)benzoic Acid, 1184
5-Benzyl-1,2,3,4-tetrahydro-2-methyl-γ-carboline see 5809
5-Benzyl-2,3,4,5-tetrahydro-2-methyl-1H-pyrido[4,3-b]indole see 5809
3-[(Benzylthio)methyl]-6-chloro-2H-1,2,4-benzothiadiazine-7-sulfonamide 1,1-Dioxide see 1155
3-[(Benzylthio)methyl]-6-chloro-7-sulfamoyl-2H-benzo-1,2,4-thiadiazine 1,1-Dioxide see 1155

S-Benzylthiuronium Chloride, 1185
3-Benzyl-6-trifluoromethyl-3,4-dihydro-7-sulfamoyl-2H-1,2,4-benzothiadiazine 1,1-Dioxide see 1064
1-Benzyl-2-trimethylacetylhydrazine see 7668
Benzylurea, 1186
Benzyl Viologen see 10137
Benzyrin [Yoshitomi] see 1157
Benzytol see 2228
Beocid-Puroptal [Metochem] see 9067
Bepadin [Carter Wallace] see 1188
Bepanthen [Roche] see 2988
Bephenium, 1187
Bephenium Embonate see 1187
Beprane [Biotherapie] see 8025
Bepridil, 1188
Béprochine see 7134
Béprocin see 1286
Berachin [Tokyo Tanabe] see 9942
Beractant, 1189
Béradia see 4413
Beraprost, 1190
Beraunite see 4074
Berbamine see 1191
Berberine, 1192
Berberine Bisulfate. see 1192
Berberis, 1193
Berberis Aristata, 1194
Berberis Bark see 988
Berbine, 1195
Berculon A [ICI] see 9427
Berenil [Hoechst] see 3316
Beret [Ciba] see 4030
Bergacef [Berganon] see 1964
Bergamol [Synfleur] see 5521
Bergaptan see 1196
Bergapten(e), 1196
Bergenin, 1197
Bergenit see 1197
Berkatens [Berk] see 10083
Berkelium, 1198
Berkfurin [Berk] see 6696
Berkmycen [Berk] see 7111
Berkolol [Berk] see 8025
Berkomine [Berk] see 4955
Berkozide [Berk] see 1064
Berlin Blue see 4064
Bermoprofen, 1199
Berninamycin, 1200
Berninamycin A see 1200
Berninamycin B see 1200
Bernstein see 397
Bernsteinsäure (German) see 9038
Bernsteinsäureanhydrid (German) see 9039
Berocillin [Boehringer, Ing.] see 7669
Berofor Alpha 2 [Boehringer, Ing.] see 5016
Berolase [Roche] see 9431
Beromycin [Boehringer, Ing.] see 7230
Beronald [Kowa] see 4331
Berotec [Boehringer, Ing.] see 4022
Berry Wax see 1040
Bersen see 403
Bertholite see 2145
Berubi [Redel] see 10152
Berubigen [Upjohn] see 10152
Beryl see 1201
Beryllia see 1216
Beryllium, 1201
Beryllium Acetate, 1202
Beryllium Acetate, Basic, 1203
Beryllium Acetylacetonate, 1204
Beryllium Borohydride, 1205
Beryllium Bromide, 1206
Beryllium Carbide, 1207
Beryllium Chloride, 1208
Beryllium Fluoride, 1209
Beryllium Formate, 1210

Bietanautine, 1253
Bietaserpine, 1254
Biethylene see 1534
Bifemelane, 1255
Bifenox, 1256
Bifenthrin, 1257
Bifex [Provet] see 8022
Bifidus Factor, 1258
Bifiteral [Philips Roxane] see 5360
Bifluranol, 1259
Bifonazole, 1260
Biformyl see 4519
Biforon [Meiji] see 1500
Bigitalin see 4441
Bigunal [Nikken] see 1500
Bihypnal see 3097
[Δ²,²'-Biindoline]-3,3'-dione see 4977
Bijosal see 3234
Bi-Ketolan see 7062
Bikhaconitine, 1262
Bikhroot see 122
Biklin [Bristol; Grünenthal] see 425
Bikunin Trypsin Inhibitor see 10021
Bilagol [Janssen] see 3239
Bilamid(e) [Cilag-Chemie] see 4880
Bilcolic [Unifa] see 4903
Bilcrine [Bellon] see 6078
Bildux [Perrier] see 10071
Bilein see 7062
Biletan [Gador] see 9462
Bilevon see 4716
Bilevon-M see 6601
Bilibyk [Byk-Gulden] see 5028
Bilicholan see 7062
Bilidren see 2922
Biligen see 4009
Biligrafin [Schering AG] see 5043
Biligram [Schering AG] see 5067
Bilijodon-Natrium see 5072
Bilimiro [Bracco] see 5077
Bilimiron [Bracco] see 5077
Bilimix [Gero] see 2784
Bilineurine see 2259
Biliognost see 5047
Biliphorine [Couvreur] see 9630
Bilirubin, 1263
Bilirubin IXα see 1263
Biliselectan see 5047
Biliton H [Berlin-Chemie] see 4903
Bilitrast see 5059
Biliverdine, 1264
Bilopac see 9968
Bilopaque [Winthrop] see 9968
Biloptin [Schering AG] see 5087
Bilordyl [Fisons] see 9421
Bilostat [Pennwalt] see 2922
Biloxazol see 1342
Biltricide [Bayer] see 7896
Bilutac [SK&F] see 211
Δ²,²'-Bimalononitrile see 9336
Bimaran [Roux-Ocefa] see 9712
Bimethadol see 3254
Bimethyl see 3767
2,2'-Bimorphine see 8104
Binapacryl, 1265
β,β-Binaphthyleneethene see 7550
Binazin [Polfa] see 9640
Bindan see 7381
Bindazac see 1061
Binedaline, 1266
Binifibrate, 1267
Biniwas [Wassermann] see 1267
Binodaline see 1266
Binodrenal see 6788
Binotal [Bayer] see 628
Binova [Gentili] see 8068
Binovum [Ortho-Cilag] see 6790
Biobamat [Biochemie] see 5908
Biobenzyfuroline [Sumitomo] see 1271

Biocefalin [Benvegna] see 8180
Biochanin A see 4395
Biochanin B see 4271
Biociclin [Del Saz & Filippini] see 2002
Biocidan [Clin-Comar-Byla] see 2061
Biocin [Ibirn] see 4280
Biocolina see 2261
Biocytin, 1268
Biodermatin [LaFare] see 1272
Biofanal [Pfleger] see 6834
Biofenac [Prodes] see 19
Bioflavonoids, 1269
Biofos [Von Boch] see 4280
Biofurex [Lenza] see 2002
Biofusal see 4335
Biogastrone [Winthrop] see 1839
Biohulin [Biobras] see 5011
BIOLF-62 see 4374
Biomag [Pulitzer] see 2337
Biomioran see 2249
Biomitsin see 2245
Bio-Mycin [Philips Roxane] see 7111
Bionicard [Bioindustria] see 6579
Biopal CVL-10 see 5058
Bioperazone [Von Boch] see 1981
Bioperidolo [Firma] see 4629
Biophylline [Delandale] see 9421
Bioplex [Thames] see 1839
Bioplex [Torre] see 8071
Biopten [Maruho] see 8515
Biopterin, 1270
Bioral [Winthrop] see 1839
Bioresmethrin, 1271
Biorphen see 7007
Bios I see 5008
Bios II see 1272
Bioscleran [Pfleger] see 2436
Biosinon [Farmakhim] see 3766
Bio Soft EA [Stepan] see 7737
Biosol [Biokema] see 6542
Biosone [Berk] see 3628
Biosterol see 10150
Biosynthetic Human Insulin see 5011
Biotamin [Sankyo] see 1068
Bioterciclin see 2937
Biotertussin [Lisapharma] see 2425
Bio-Tetra [General Pharm.] see 9337
Biotexin see 6818
Biothion [Am. Cyanamid] see 9286
Biotin, 1272
Biotin Complex of Yeast see 1268
Biotinin [Sankyo] see 9432
Biotin l-Sulfoxide, 1273
ε-N-Biotinyl-L-lysine see 1268
Biotiren [Benvegna] see 3079
Biotirmone [Sanofi] see 9555
Biotussal see 416
Bioxima [Ital. Suisse] see 2002
Bioxiran see 3715
(R*,S*)-2,2'-Bioxirane see 3715
Biozolene [Bioindustria] see 4158
Biperiden, 1274
Biphenal see 4883
Biphenamine, 1275
Biphenate see 1257
Biphenthrin see 1257
Biphentrin see 1257
1,1'-Biphenyl see 3372
(1,1'-Biphenyl)-4-acetic Acid see 3989
4-Biphenylacetic Acid see 3989
[1,1'-Biphenyl]-4-amine see 1276
p-Biphenylamine see 1276
[1,1'-Biphenyl]-4,4'-diamine see 1106
2,4'-Biphenyldiamine, 1277
[1,1'-Biphenyl]-2,2'-dicarboxylic Acid see 3368
3,3'-[[1,1'Biphenyl]-4,4'-diylbis(azo)]-bis[4-amino-1-naphthalenesulfonic Acid] Disodium Salt see 2562

3,3'-[1,1'-Biphenyl]-4,4'-diylbis[2,5-di-phenyl-2H-tetrazolium] Dichloride see 6554
2,2'-[1,1'-Biphenyl]-4,4'-diylbis[2-hydroxy-4,4-dimethylmorpholinium] see 4676
3,3'-[4,4'-Biphenylene]bis[2,5-diphen-yl-2H-tetrazolium] Dichloride see 6554
o-Biphenylenemethane see 4190
p-Biphenylmethyl-(dl-tropyl-α-tropin-ium)bromide see 10208
(1,1'-Biphenyl)-2-ol see 7458
[1,1'-Biphenyl]-4-ol see 7459
2-Biphenylol see 7458
1-(3,3-Biphenylpropyl)piperidine see 4031
[1,1'-Biphenyl]-3,3',5,5'-tetrol see 3417
2-(4-Biphenylyl)butyric Acid see 10205
α-(4-Biphenylyl)butyric Acid see 10205
3-(4-Biphenylylcarbonyl)propionic Acid see 4003
2,2'-(4,4'-Biphenylylene)bis[2-hydr-oxy-4,4-dimethylmorpholinium] see 4676
4-[(2-[1,1'-Biphenyl]-4-yl-1-ethoxy-2-oxoethyl)amino]benzoic Acid see 10204
4-Biphenylylethylacetic Acid see 10205
endo-(±)-8-([1,1'-Biphenyl]-4-ylmeth-yl)-3-(3-hydroxy-2-phenyl-1-oxopro-poxy)-8-methyl-8-azoniabicyclo-[3.2.1]octane Bromide see 10208
4-(4-Biphenylyl)-4-oxobutyric Acid see 4003
[3(S)-endo]-8-(2-[1,1'-Biphenyl]-4-yl-2-oxoethyl)-3-(3-hydroxy-1-oxo-2-phenylpropoxy)-8-methyl-8-azonia-bicyclo[3.2.1]octane Bromide see 4048
β-[(1,1'-Biphenyl)-4-yloxy]-α-(1,1-di-methylethyl)-1H-1,2,4-triazol-1-ethanol see 1342
1-(Biphenyl-4-yloxy)-3,3-dimethyl-1-(1H-1,2,4-triazol-1-yl)butan-2-ol see 1342
1-([1,1'-Biphenyl]-4-ylphenylmethyl)-1H-imidazole see 1260
[1,4'-Bipiperidine]-1'-carboxylic Acid (S)-4,11-Diethyl-3,4,12,14-tetrahy-dro-4-hydroxy-3,14-dioxo-1H-pyra-no[3',4':6,7]indolizino[1,2-b]quino-lin-9-yl Ester see 5104
Bipiperidyl Mustard, 1278
Δ²,²'-Bipseudoindoxyl see 4977
2,2'-Bipyridine see 3411
4,4'-Bipyridine see 3412
Biquinate [Rhone-Poulenc Rorer] see 8245
Birch Tar Oil, Empyreumatic, 1279
Birch Tar Oil, Rectified, 1280
Bird Manure see 4594
Bird's Eye see 165
BI-RG-587 see 6573
Birlane [Shell] see 2137
Birodan [Lundbeck] see 9553
Biroxin [Roche] see 9386
Birthroot see 9825
Birthwort see 8608
Birutan [E. Merck] see 8456
(+)-α-Bisabolol see 1281
[S-(R*,R*)]-α-Bisabolol see 1281
α-Bisabolol, 1281
β,β'-Bis(4-acetamidophenyloxy)ethyl Ether see 3017
Bis(4-acetamidophenyl)sulfone see 20
Bis(acetato)palladium see 7123

5-[Bis(2-chloroethyl)amino]-2,4-(1H,-3H)-pyrimidinedione *see* 9986

cis-(±)-2-[Bis(2-chloroethyl)amino]-tetrahydro-2H-1,3,2-oxazaphosphorin-4-yl Hydroperoxide P-oxide *see* 7303

2-[Bis(2-chloroethyl)amino]tetrahydro-2H-1,3,2-oxazaphosphorine 2-Oxide Monohydrate *see* 2816

5-[Bis(2-chloroethyl)amino]uracil *see* 9986

N,N-Bis(2-chloroethyl)aniline *see* 697

N,N-Bis(2-chloroethyl)benzenamine *see* 697

1,1'-Bis(2-chloroethyl)-4,4'-bipiperidine *see* 1278

O,O-Bis(β-chloroethyl) O-(3-Chloro-4-methyl-7-coumarinyl) Phosphate *see* 4636

Bis(2-chloroethyl) Ether *see* 3116

N,N-Bis(2-chloroethyl)-N'-(3-hydroxypropyl)phosphorodiamidic Acid 2-Chloroethyl Ester *see* 2917

N,N-Bis(2-chloroethyl)-2-naphthylamine *see* 2157

N,N'-Bis(2-chloroethyl)-N-nitrosourea *see* 1894

Bis(2-chloroethyl)phosphamide Cyclic Propanolamide Ester Monohydrate *see* 2816

Bis(2-chloroethyl)phosphoramide Cyclic Propanolamide Ester Monohydrate *see* 2816

N,N-Bis(β-chloroethyl)-N',O-propylenephosphoric Acid Ester Diamide Monohydrate *see* 2816

Bis(2-chloroethyl)sulfide *see* 6395

Bis(β-chloroethyl)sulfide *see* 6395

N,3-Bis(2-chloroethyl)tetrahydro-2H-1,3,2-oxazaphosphorin-2-amine 2-Oxide *see* 4937

N,N-Bis(2-chloroethyl)tetrahydro-2H-1,3,2-oxazaphosphorin-2-amine 2-Oxide Monohydrate *see* 2816

N,N-Bis(β-chloroethyl)-N',O-trimethylenephosphoric Acid Ester Diamide Monohydrate *see* 2816

Bis[5-chloro-2-hydroxyphenyl]methane *see* 3120

Bis(chloromethyl) Ether *see* 3119

5,6-Bis(chloromethyl)-1,2,3,4,7,7-hexachlorobicyclo[2.2.1]hept-2-ene *see* 2127

5,6-Bis(chloromethyl)-1,2,3,4,7,7-hexachloro-2-norbornene *see* 2127

Bis(chloromethyl) Ketone *see* 3102

2,2-Bis(chloromethyl)-1,3-propanediol *see* 7247

Bis(p-chlorophenoxy)methane, 1287

1,4-Bis(N^d-p-chlorophenylamidinoamidinyl)piperazine *see* 7553

N,N'''-Bis[(p-chlorophenyl)amidino]-1,4-piperazinedicarboxamidine *see* 7553

N,N'''-Bis[[(4-chlorophenyl)amino]iminomethyl]-1,4-piperazine Dicarboximidamide *see* 7553

1,6-Bis[N'-(p-chlorophenyl)-N^5-biguanido]hexane *see* 2140

2,2-Bis(2-chlorophenyl-4-chlorophenyl)-1,1-dichloroethane *see* 6302

1,6-Bis(N^5-p-chlorophenyl-N'-diguanido)hexane *see* 2140

N,5-Bis(4-chlorophenyl)-3,5-dihydro-3-[(1-methylethyl)imino]-2-phenazinamine *see* 2433

N,N'''-Bis(4-chlorophenyl)-3,12-diimino-2,4,11,13-tetraazatetradecanediimidamide *see* 2140

1,1-Bis(p-chlorophenyl)ethanol *see* 2136

1,1-Bis(p-chlorophenyl)methyl Carbinol *see* 2136

Bis[(4-chlorophenyl)methylene]carbonimidic Dihydrazide *see* 8403

1,1-Bis(p-chlorophenyl)-2-nitrobutane *see* 1505

1,1-Bis(p-chlorophenyl)-2-nitropropane *see* 6691

3,6-Bis(2-chlorophenyl)-1,2,4,5-tetrazine *see* 2435

α,α-Bis(p-chlorophenyl)-β,β,β-trichlorethane *see* 2898

1,1-Bis(p-chlorophenyl)-2,2,2-trichloroethanol *see* 3136

Bisclofentezin *see* 2435

Biscolan *see* 2262

Biscosal [Ohta] *see* 4186

1-[Bis(4-cyanophenyl)methyl]-1,2,4-triazole *see* 5474

(TPS-7-1-232'4'54)-[Bis[(1,2-cyclohexanedione Dioximato)(1—)-O] [(1,2-Cyclohexanedione Dioximato)-(2—)-O]methylborato(2—)-N,N',N'',-N''',N'''',N''''']chlorotechnetium-99mTc *see* 9260

Bis(cyclohexyl)carboxylic Acid Diethylaminoethyl Ester Hydrochloride *see* 3147

Biscyclopentadienyliron *see* 4084

Biscyclopentadienyltitanium(IV) Dichloride *see* 9621

6,7-Bis(cyclopropylmethoxy)-4-hydroxy-3-quinolinecarboxylic Acid Ethyl Ester *see* 2843

Bis-DEAE-fluorenone *see* 9581

N,N'-Bis(dehydroabietyl)ethylenediamine Bis(phenoxymethylpenicillin) *see* 7232

N,N'-Bis(dehydroabietyl)ethylenediamine Dipenicillin G *see* 7224

1,6-Bisdehydro-6-chloro-17α-acetoxyprogesterone *see* 2930

Bisdequalinium Chloride, 1288

α,α-Bis(3,5-dibromo-4-hydroxyphenyl)-α-hydroxy-o-toluenesulfonic Acid γ-Sultone *see* 1468

3,3-Bis(3,5-dibromo-4-hydroxyphenyl)-1(3H)-isobenzofuranone *see* 9329

α,α-Bis(3,5-dibromo-4-hydroxy-o-tolyl)-α-hydroxytoluenesulfonic Acid, γ-Sultone *see* 1407

N,N'-Bis(dichloroacetyl)-N,N'-bis(2-ethoxyethyl)-1,4-bis(aminomethyl)-benzene *see* 9262

N,N'-Bis(dichloroacetyl)-1,8-octamethylenediamine *see* 4108

Bis[3,4-dichloro-2(5)-furanonyl] Ether *see* 6380

2,6-Bis(diethanolamino)-4,8-dipiperidinopyrimido-[5,4-d]pyrimidine *see* 3410

2,6-Bis(diethanolamino)-4-piperidinopyrimido[5,4-d]pyrimidine *see* 6344

Bis[S-(diethoxyphosphinothioyl)mercapto]methane *see* 3782

4,4'-Bis(β-diethylaminoethoxy)-α,β-diethyldiphenylethane *see* 4739

2,7-Bis[2-(diethylamino)ethoxy]-9H-fluoren-9-one *see* 9581

6,7-Bis[2-(diethylamino)ethoxy]-4-methyl-2H-1-benzopyran-2-one *see* 7047

6,7-Bis[2-(diethylamino)ethoxy]-4-methylcoumarin *see* 7047

3,4-Bis[p-(β-diethylaminoethoxy)phenyl]hexane *see* 4739

N,N'-Bis[2-diethylaminoethyl]oxamide

Bis[2-chlorobenzyl Chloride] *see* 396

2,4-Bis(diethylamino)-6-hydrazino-s-triazine *see* 4803

3,3'-Bis[(diethylamino)methyl]-5,5'-di-2-propenyl-[1,1'-biphenyl]-4,4'-diol *see* 1242

1,3-Bis(diethylamino)-2-(α-phenyl-α-cyclohexylmethyl)propane *see* 3986

2,5-Bis(3-diethylaminopropylamino)-benzoquinone Bis(benzyl Chloride) *see* 1137

4,6-Bis(diethylamino)-1,3,5-triazin-2(1H)-one Hydrazone *see* 4803

3,6-Bis(diethylamino)xanthylium Chloride *see* 8189

Bis-O,O-diethylphosphoric Anhydride *see* 9345

Bis(diethylthiocarbamoyl) Disulfide *see* 3428

Bis(diethylthiocarbamoyl)sulfide *see* 9123

Bis(diethylthiocarbamyl) Disulfide *see* 3428

2-[[[[[4,6-Bis(difluoromethoxy)-2-pyrimidinyl]amino]carbonyl]amino]-sulfonyl]benzoic Acid Methyl Ester *see* 7928

2-[3-(4,6-Bis(difluoromethoxy)pyrimidin-2-yl)ureidosulfonyl]benzoic Acid Methyl Ester *see* 7928

N,N'-Bis[3-(4,5-dihydro-1H-imidazol-2-yl)phenyl]urea *see* 4951

[2S-[2α(2S*,3S*),3α]]-1,8-Bis(3,4-dihydro-3,5,7-trihydroxy-2H-1-benzopyran-2-yl)-3,4,6-trihydroxy-5H-benzocyclohepten-5-one *see* 9409

(T-4)-Bis[4,5-dihydroxy-1,3-benzenedisulfonato(4—)-O^4,O^5]-antimonate-(5—) Pentasodium Heptahydrate *see* 8967

2,3-Bis(3,4-dihydroxybenzyl)butane *see* 6786

2,2-Bisdihydroxymethyl-1,3-propanediol Tetranitrate *see* 7249

N,N'-Bis[2-(3,4-dihydroxyphenyl)-2-hydroxyethyl]hexamethylenediamine *see* 4745

[1R-(1α,3α,4α,5β)]-1,3-Bis[[3-(3,4-dihydroxyphenyl)-1-oxo-2-propenyl]oxy]-4,5-dihydroxycyclohexanecarboxylic Acid *see* 2835

N,N'-Bis(2,3-dihydroxypropyl)-5-[N-(2,3-dihydroxypropyl)acetamido]-2,4,6-triiodoisophthalamide *see* 5068

N,N'-Bis(2,3-dihydroxypropyl)-5-[(hydroxyacetyl)(2-hydroxyethyl)amino]-2,4,6-triiodo-1,3-benzenedicarboxamide *see* 5083

N,N'-Bis(2,3-dihydroxypropyl)-5-[N-(2-hydroxyethyl)glycolamido]-2,4,6-triiodoisophthalamide *see* 5083

N,N'-Bis(2,3-dihydroxypropyl)-5-[N-(2-hydroxy-3-methoxypropyl)acetamido]-2,4,6-triiodoisophthalamide *see* 5073

N,N'-Bis(2,3-dihydroxypropyl)-2,4,6-triiodo-5-(2-methoxyacetamido)-N-methylisophthalamide *see* 5076

N,N'-Bis(2,3-dihydroxypropyl)-2,4,6-triiodo-5-[(methoxyacetyl)amino]-N-methyl-1,3-benzenedicarboxamide *see* 5076

Bis(dimethylamido)fluorophosphate *see* 3250

Bis(dimethylamido)phosphoryl Fluoride *see* 3250

N,N'-Bis(p-dimethylaminobenzal)benzidine *see* 1289

4,4'-Bis(dimethylamino)benzophenone
see 6265
Bis(p-dimethylaminobenzylidene)benz-
idine, 1289
Bis(4-dimethylaminobenzylidene)-p,p'-
diaminodiphenyl see 1289
Bis[6-(dimethylamino)-2-[2-(2,5-di-
methyl-1-phenylpyrrol-3-yl)vinyl]-1-
methylquinolinium] 4,4'-Methylene-
bis(3-hydroxy-2-naphthoate) see
8206
Bis[2-dimethylaminoethyl]succinate
Bis[ethobromide] see 9184
Bis[2-dimethylaminoethyl]succinate
Bis[methobromide] see 9043
Bis[2-dimethylaminoethyl]succinate
Bis[methochloride] see 9044
Bis(β-dimethylaminoethyl)succinate
Bis(methyl Iodide) see 9045
Bisdimethylaminofluorophosphine
Oxide see 3250
[4S-(4α,4aα,5aα,12aα)]-4,7-Bis(dimethylamino)-1,4,4a,5,5a,6,11,12a-
octahydro-3,10,12,12a-tetrahy-
droxy-1,11-dioxo-2-naphthacenecar-
boxamide see 6289
3,7-Bis(dimethylamino)phenazathio-
nium Chloride see 6137
3,7-Bis(dimethylamino)phenothiazin-
5-ium Chloride see 6137
Bis(p-dimethylaminophenyl)methane
see 6264
Bis[4-(dimethylamino)phenyl]methan-
one see 6265
N,N'-Bis[[4-(dimethylamino)phenyl]-
methylene][1,1'-biphenyl]-4,4'-
diamine see 1289
N-[4-[Bis[4-(dimethylamino)phenyl]-
methylene]-2,5-cyclohexadien-1-
ylidene]-N-methylmethanaminium
Chloride see 4401
Bis[p-(dimethylamino)phenyl]phenyl-
methylium Chloride see 5739
3β,20α-Bis(dimethylamino)-5-pregn-
ene see 5336
10-[2,3-Bis(dimethylamino)propyl]-
phenothiazine see 487
3,6-Bis(dimethylamino)xanthylium
Chloride see 8190
[Bis(2,2-dimethyl-1-aziridinyl)phos-
phinyl]carbamic Acid Ethyl Ester
see 6245
(T-4)-Bis(dimethylcarbamodithioato-
S,S')zinc see 10305
1-[Bis(3',5'-N,N-dimethylcarbamoyl-
oxy)phenyl]-2-N-tert-butylamino-
ethanol see 980
Bis(dimethyldithiocarbamato)zinc see
10305
2,4-Bis(1,1-dimethylethyl)-5-methyl-
phenol see 2895
2,6-Bis(1,1-dimethylethyl)-4-methyl-
phenol see 1583
3,6-Bis(1,1-dimethylethyl)-1-naphtha-
lenesulfonic Acid Ethyl Ester see
3834
2,6-Bis(1,1-dimethylethyl)-1-naph-
thalenesulfonic Acid Sodium Salt
see 8753
Bis(1,1-Dimethylethyl) Peroxide see
3515
2,6-Bis(1,1-dimethylethyl)pyridine see
3086
Bis(dimethylglyoximato)nickel see
6590
2,5-Bis(1,1-dimethylpropyl)-1,4-benz-
enediol see 3359
Bis(1,2-dimethylpropyl)borane, 1290
2,5-Bis(1,1-dimethylpropyl)hydro-
quinone see 3359

Bis(dimethylthiocarbamoyl) Disulfide
see 9510
Bis(dimethylthiocarbamyl) Disulfide
see 9510
Bis(1,1-dioxoperhydro-1,2,4-thia-
diazin-4-yl)methane see 9243
(±)-1,2-Bis(3,5-dioxopiperazinyl)prop-
ane see 8295
1,2-Bis[2-(2,3-epoxypropoxy)ethoxy]-
ethane see 3926
O,S-Bis(ethoxycarbonyl)thiamine see
2064
1,2-Bis(3-ethoxycarbonyl-2-thio-
ureido)benzene see 9489
N,N'-Bis(ethoxycarbonyl)-N,N'-bis(dichloroacetyl)-1,4-xylylenediamine see
9262
6,9-Bis(2-ethoxyethyl)-3,12-dioxa-6,9-
diphosphatetradecane see 9383
N¹,N²-Bis(p-ethoxyphenyl)acetamidine
Hydrochloride see 7342
N,N'-Bis(4-ethoxyphenyl)ethanimid-
amide Monohydrochloride see 7342
2,4-Bis(ethylamino)-6-chloro-s-tria-
zine see 8681
2,4-Bis(ethylamino)-6-(methylthio)-
1,3,5-triazine see 8683
2,4-Bis(ethylamino)-6-(methylthio)-s-
triazine see 8683
Bis(ethylenimido)phosphorylurethane
see 10011
Bis[2-ethylhexyl]calcium Sulfosuccinate see 3459
N,N''-Bis(2-ethylhexyl)-3,12-diimino-
2,4,11,13-tetraazatetradecanedi-
imidamide see 231
2,6-Bis(2-ethylhexyl)hexahydro-7a-
methyl-1H-imidazo[1,5-c]imidazole
see 4736
1,3-Bis(2-ethylhexyl)hexahydro-5-
methyl-5-pyrimidinamine see 4741
1,3-Bis(β-ethylhexyl)-5-methyl-5-
aminohexahydropyrimidine see 4741
3,7-Bis(2-ethylhexyl)-5-methyl-1,3,7-
triazabicyclo[3.3.0]octane see 4736
Bis(2-ethylhexyl) Phthalate, 1291
Bis(2-ethylhexyl) Sebacate, 1292
Bis(2-ethylhexyl)sodium Sulfosuccinate see 3460
Bis[ethyl(2-hydroxyethyl)dimethyl-
ammonium] Sulfate Bis(dibutylcarbamate) see 3082
Bis[1-ethylquinoline-(2)]trimethinecy-
anine Chloride see 8229
2,2-Bis(ethylsulfonyl)butane see 9131
2,2-Bis(ethylsulfonyl)propane see 9133
N-[4-[Bis[4-(ethyl(4-sulfophenyl)-
amino]phenyl]methylene]-2,5-cyclo-
hexadien-1-ylidene]-N-methylmeth-
anaminium Inner Salt, Sodium Salt
see 110
Bisethylxanthogen see 3449
Bisexovis see 677
1-[4,4-Bis(4-fluorophenyl)butyl]-4-[4-
chloro-3-(trifluoromethyl)phenyl]-4-
piperidinol see 7213
1-[4,4-Bis(p-fluorophenyl)butyl]-4-(4-
chloro-α,α,α-trifluoro-m-tolyl)-4-
piperidinol see 7213
4-[4,4-Bis(4-fluorophenyl)butyl]-N-(2,-
6-dimethylphenyl)-1-piperazine-
acetamide see 5507
4-[4,4-Bis(4-fluorophenyl)butyl]-N-
ethyl-1-piperazinecarboxamide see
619
1-(4,4-Bis(4-fluorophenyl)butyl)-4-
hydroxy-4-(3-trifluoromethyl-4-
chlorophenyl)piperidine see 7213
8-[4,4-Bis(4-fluorophenyl)butyl]-1-

phenyl-1,3,8-triazaspiro[4.5]decan-
4-one see 4241
4-[4,4-Bis(p-fluorophenyl)butyl]-1-
piperazineaceto-2',6'-xylidide see
5507
1-[1-[4,4-Bis(4-fluorophenyl)butyl]-4-
piperidinyl]-1,3-dihydro-2H-benz-
imidazol-2-one see 7589
1-[1-[4,4-Bis(p-fluorophenyl)butyl]-4-
piperidyl]-2-benzimidazolinone see
7589
(E)-1-[Bis(p-fluorophenyl)-methyl]-4-
cinnamylpiperazine see 4179
6-[2-[4-[Bis(4-fluorophenyl)methyl-
ene]-1-piperidinyl]ethyl]-7-methyl-
5H-thiazolo[3,2-a]pyrimidin-5-one
see 8399
(E)-1-[Bis(4-fluorophenyl)methyl]-4-
(3-phenyl-2-propenyl)piperazine see
4179
6-[4-[Bis(4-fluorophenyl)methyl]-1-
piperazinyl]-N,N'-di-2-propenyl-1,3,-
5-triazine-2,4-diamine see 309
1-[[Bis(4-fluorophenyl)methylsilyl]-
methyl]-1H-1,2,4-triazole see 4240
Bis(4-fluorophenyl)methyl(1H-1,2,4-
triazol-1-ylmethyl)silane see 4240
1-[Bis(p-fluorophenyl)methyl]-4-(2,3,-
4-trimethoxybenzyl)piperazine see
5593
1-[Bis(4-fluorophenyl)methyl]-4-[(2,3,-
4-trimethoxyphenyl)methyl]pipera-
zine see 5593
1,4-Bis(1-formamido-2,2,2-trichloro-
ethyl)piperazine see 9819
2,2'-Bis[8-formyl-1,6,7-trihydroxy-5-
isopropyl-3-methylnaphthalene] see
4549
Bis(D-gluconato-O¹,O²)-iron see 4095
3,5-Bis(β-D-glucopyranosyloxy)-7-
hydroxy-2-(4-hydroxy-3-methoxy-
phenyl)-1-benzopyrylium Chloride
see 7285
3,5-Bis(β-D-glucopyranosyloxy)-7-
hydroxy-2-(4-hydroxyphenyl)-1-
benzopyrylium Chloride see 7199
(R*,R*)-5,5'-Bis(β-D-glucopyranosyloxy)-9,9',10,10'-tetrahydro-4,4'-di-
hydroxy-10,10'-dioxo[9,9'-bianthra-
cene]-2,2'-dicarboxylic Acid see
8600
(R*,S*)-5,5'-Bis(β-D-glucopyranosyloxy)-9,9',10,10'-tetrahydro-4,4'-di-
hydroxy-10,10'-dioxo[9,9'-bianthra-
cene]-2,2'-dicarboxylic Acid see
8600
3,5-Bis(glucosyloxy)-4',7-dihydroxy-
3',5'-dimethoxyflavylium Chloride
see 5754
3,5-Bis(glucosyloxy)-3',4',5',7-tetrahy-
droxyflavylium Chloride see 2931
Bis(glycinato)copper see 2707
Bisguadine [Sterwin] see 231
1,3-Bis[4-guanylphenyl]triazene Diaceturate see 3316
Bish see 122
Bis(hexachloro-
cyclopentadieno)cyclooctane see
2908
Bishma see 122
trans-(±)-2,3-Bis(3'-hydroxybenzyl)-γ-
butyrolactone see 3633
Bis(p-hydroxybenzylidene)ethylene
Isocyanide see 10195
Bis[3-hydroxybutyl] Ether see 7094
Bis[2-hydroxy-5-chlorophenyl]sulfide
see 4046
Bishydroxycoumarin (Rescinded) see
3140

Bistrimin see 7469
Bis(triphenylphosphine)dicarbonyl-
nickel, 1341
Bistrium Bromide [Squibb] see 4724
Bistrium Chloride [Squibb] see 4724
Bistyryl see 3378
Bitensil [UCB] see 3605
Biterol [Winthrop] see 1344
Bitertanol, 1342
Bithin see 1343
Bithiodine see 9601
Bithionol, 1343
Bithionolate Sodium see 1343
Bitin-S [Tanabe Seiyaku] see 1343
Bitolterol, 1344
Bitolterol Mesylate see 1344
Bitoscanate, 1345
Bitrex [Macfarlan Smith] see 2942
Bitriben see 1340
Bitrop [Boehringer, Ing.] see 5089
α-Bitter Acid see 4790
β-Bitter Acid see 5640
Bitter-almond-oil Camphor see 1124
Bitter Apple see 2550
Bitter Ash see 3945 and 8208
Bitter-bloom see 2017
Bitter Cucumber see 2550
Bitter Gourd see 2550
Bitter Herb see 2016
Bitter Orange Peel see 6992
Bitter Root see 783 and 4399
Bitter Salts see 5731
Bitter Stick see 2104
Bittersweet see 3516
Bitter Wintergreen see 2099
Bitter Wood see 8208
Bitulan [Adroka] see 4929
Bitumen see 880
Bituminol see 4929
Bitumol see 4929
Bituvitan [Hishiyama] see 8367
Biuret, 1346
Biuretamidine Sulfate see 3143
Bivinyl see 1534
Bivitasi [ISI] see 9431
Bixin, 1347
Bizolin [Boehringer, Ing.] see 7431
BL-5 see 2786
BL-139 see 479
BL-191 see 7278
BL-4162A see 665
BL-5583 see 9481
Black and White Bleaching Cream
[Plough] see 4853
Black Ash see 1025
Black Balsam see 976
Blackberry Bark see 8446
Black Cohosh see 2339
Black Copper Oxide see 2713
Black Hawk see 10109
Black Henbane see 4908
Black Indian Hemp see 784
Black Iron Oxide see 4087
Black Lead see 4560
Black Leaf 40 see 6611
Black Manganese Oxide see 5770
Black Nickel Oxide see 6599
Black Pepper see 7287
Black Phosphorus see 7503
Black Root see 5465
Black Sea Sterol see 3342
Black Snake Root see 2339
Blackstrap Molasses, 1348
Black-tang see 4305
Black Uranium Oxide see 9991
Bladafum [Bayer] see 9139
Bladan [Bayer] see 9345
Bladder Fucus see 4305
Bladderon [Nippon Shinyaku] see
4135

Bladder Pod see 5579
Bladder-wrack see 4305
Bladex [Shell] see 2755
Blanc de Perle see 1303
Blanc d'Espagne see 1303
Blanc Fixe see 1023
Blancol C see 1349
Blancophor R [GAF], 1349
Blankophor R see 1349
Blascorid [Guidotti] see 1078
Blasticidin S, 1350
Blasting Gelatin see 6704
Blasting Oil see 6704
Blastmycin see 756
Blattanex [Bayer] see 8022
Blätteralkohol see 4737
Blaud's Mass see 4089
Blausäure (German) see 4836
Blazer [Rohm & Haas] see 111
Blazing Star see 229 and 4664
Bleached Beeswax see 1049
Bleached Yellow Wax see 1049
Bleaching Powder see 2143
Bled [Poli] see 2324
Bleiweiss (German) see 1034
Bleminol [Desitin] see 287
Blenoxane [Bristol] see 1351
Bleo [Nippon Kayaku] see 1351
Bleomycin A₂ see 1351
Bleomycins, 1351
Bleph-10 [Allergan] see 9067
Blex [ICI] see 7652
Blistering Beetle see 1796
Blistering Collodion see 2548
Blistering Fly see 1796
Bloat Guard [Agrimin] see 7721
Bloc [Elanco] see 3999
Blocadren [Merck & Co.] see 9585
Blockade [Sandoz] see 7947
Blockain Hydrochloride [Sterling
Winthrop] see 8023
Blocklin L [Shionogi] see 7597
Blood-coagulation Factor II see 8074
Blood-coagulation Factor III see 9527
Blood-coagulation Factor IIa see 9525
Blood-coagulation Factor V see 3961
Blood-coagulation Factor VII see 3962
Blood-coagulation Factor VIII see
3963
Blood-coagulation Factor IX see 3964
Blood-coagulation Factor X see 3965
Blood-coagulation Factor XI see 3966
Blood-coagulation Factor XII see 3967
Blood-coagulation Factor XIII see
3968
Blood-coagulation Factor XIV see
8070
Bloodroot see 8503
Blood Sugar see 4467
Blotic [Sandoz] see 8000
Blow Gas see 7952
Blox [Biomed Foscama] see 5601
Bloxanth [Calmic] see 287
BL-P-1322 see 2030
BL-S578 see 1963
BL-S640 see 1965
BL-S786 see 1982
Blue Cohosh see 1957
Blue Gas see 10176
Blue-gum Tree see 3941
Blue Mass see 5958
Blue Pill see 5958
Bluestone see 2722
Blue Tetrazolium see 9380
Blue Verdigris see 2691
Blue Vitriol see 2722
Blusalt see 8742
Blutene Chloride [Abbott] see 9658
Bluton [Morishita] see 4925
BM-02015 see 9690

BM-13505 see 2871
BM-14190 see 1924
BM-15075 see 1240
BM-15100 see 7573
BM-51052 see 1822
BMC see 1836
BMS-181173 see 4610
BMS-186295 see 5097
BMY-13754 see 6527
BMY-26517 see 7205
BMY-26538-01 see 665
BMY-27857 see 8958
BMY-28100 see 1992
BMY-28100-03-800 see 1992
BMY-28142 see 1973
BMY-28167 see 1992
BMY-28488 see 1971
BMY-30056 see 4625
B-mycin see 1383
BN-1270 see 2323
BN-52020 see 4433
BN-52021 see 4433
BN-52022 see 4433
B-Nine [Uniroyal] see 2874
BO-714 see 2381
Bobierrite see 5720
Boc-β-Ala-Trp-Met-Asp-Phe(NH₂)
see 7250
B.O.E.A. see 3818
Boehmite see 369
Boettger's Paper see 252
BOF-A2 see 3601
Boforsin (obsolete) see 2539
Bog Bean see 5888
Bogomolets' Serum see 759
Bogoserum see 759
Bohrium see 3581 and 3583
BOL-148 see 1445
Bolandiol, 1352
Bolasterone, 1353
Boldea see 1356
Boldenone, 1354
Boldenone Undecylenate see 1354
Boldine, 1355
Boldine Dimethyl Ether see 4445
Boldo, 1356
Boldoa see 1356
Boldu see 1356
Boldus see 1356
Bole, Armenian, 1357
Bolecic Acid see 5114
Boleko Oil, 1358
Boletic Acid see 4309
Bolstar [Miles] see 9164
Bolus Alba see 5294
Bolus Armena see 1357
Bolus Rubra see 1357
Bolvidon [Organon] see 6260
Bombesin, 1359
Bomyl [Hopkins], 1360
Bonabol [Sawai] see 5842
Bonacid [Lilly] see 6617
Bonadorm see 3097
Bonaid [Norwich] see 1524
Bonalfa [Teijin] see 9197
Bonamid [Heilmittelwerke] see 934
Bonamine [Roerig] see 5817
Bonapicillin [Taiyo] see 628
Bonare see 7059
Bonasanit [Weimer] see 8166
Bondou see 8529
"Bone Ash" see 1741
Bonefos [Astra] see 2432
Bone Gla Protein see 7025
Boneset see 3948
Bongast see 1325
Bongkrekic Acid, 1361
Bonicor [Farmetrusca] see 8072
Bonidon [Mepha] see 4998
Bonifen [E. Merck] see 8180

Bonine [Pfizer] *see* 5817
Bonlam [Merck & Co.] *see* 1774
Bonoform *see* 9331
Bonomycin [Pfizer] *see* 8500
Bonopen [Belupo] *see* 5986
Bonpac [Hoei] *see* 4260
Bonpyrin [Takeda] *see* 3414
Bontril [Carnrick] *see* 2996
Bonyl [Erco] *see* 6504
Bonzi [ICI] *see* 7118
Bonzol [Tokyo Tanabe] *see* 2875
Bopindolol, 1362
Boracic Acid *see* 1364
Borated Cream of Tartar *see* 7778
Borax *see* 8733
Borax Glass *see* 8733
Borazane *see* 1363
Borazine, 1363
Borazole *see* 1363
Borazon *see* 1376
Bor-Cefazol [Proter] *see* 1967
Bordeaux Mixture *see* 2723
Borgal [Hoechst Italia] *see* 9074
Boric Acid, 1364
Boric Acid Trimethyl Ester *see* 9842
Boric Anhydride, 1365
Boric Oxide *see* 1365
Bor-Ind [Borromeo] *see* 4999
endo-2-Bornanol *see* 1366
exo-2-Bornanol *see* 5143
2-Bornanone *see* 1779
Borneo Camphor *see* 1366
Borneol, 1366
Borneol Acetate *see* 1367
Borneol α-Bromoisovalerate *see* 1368
Borneol Isovalerate *see* 1370
Borneol Salicylate *see* 1371
Bornite *sec* 2079
Bornyl Acetate, 1367
Bornyl Alcohol *see* 1366
d-Bornyl α-Bromoisovalerate, 1368
d-Bornyl 2-Bromo-3-methylbutyrate *see* 1368
Bornyl Chloride, 1369
d-Bornyl Isovalerate, 1370
Bornyl Salicylate, 1371
Bornyval *see* 1370
Borobutane *see* 9326
Boroethane *see* 3062
Borofax [Burroughs Wellcome] *see* 1364
Borofluoric Acid *see* 4184
Borohexane *see* 4712
Boromycin, 1372
Boron, 1373
Boron Carbide, 1374
Boron Chloride *see* 3063
Boron Fluoride Etherate *see* 1380
Boron Fluoride Ethyl Ether *see* 1380
Boron Monoxide, 1375
Boron Nitride, 1376
Boron Oxide *see* 1365
Boron Sesquioxide *see* 1365
Boron Tribromide, 1377
Boron Trichloride, 1378
Boron Trifluoride, 1379
Boron Trifluoride Etherate, 1380
Boron Trioxide *see* 1365
Bostrycoidin, 1381
β-Boswellic Acid, 1382
Bothrops atrox Serine Proteinase *see* 1038
Bothrops Venom Proteinase *see* 1038
Botogenin *see* 4407
BoTox [Allergan] *see* 1384
Botrilex [Bayer] *see* 8264
Botropase [Ravizza] *see* 1038
Bottromycic A₂ Acid Methyl Ester *see* 1383
Bottromycin, 1383

Bottromycin A₂ *see* 1383
Botulin A *see* 1384
Botulinium Toxin Type A *see* 1384
Botulin Toxins, 1384
Bouncing Bet *see* 8511
Bourbonal *see* 3904
Bovanide [Merck & Co.] *see* 8280
Bovatec *see* 5384
Bovicam [Merck & Co.] *see* 1774
Boviclox [Squibb] *see* 2480
Bovilene [Syntex] *see* 4037
Bovine Somatotropin *see* 8864
Bovizole [Merck & Co.] *see* 9426
Bo-Xan *see* 10197
Boxol [Rovi] *see* 2870
BP-662 *see* 4126
BPAA *see* 3989
BPAS *see* 1148
BPM *see* 1278
BPPS *see* 7992
BQ 22-708 *see* 3617
BQC Reagent *see* 3075
BR *see* 968
BR-18 *see* 7637
BR-700 *see* 4045
BRA *see* 8325
Brace [Ciba-Geigy] *see* 5119
Bradilan [Mundipharma] *see* 6605
Bradophen [Ciba-Geigy] *see* 1144
Bradosol Bromide [Ciba-Geigy] *see* 3474
Bradycor [Cortech], 1385
Bradykinin, 1386
Bradyl [Lafon] *see* 6432
Brain-derived Neurotrophic Factor *see* 6570
Brain Sugar *see* 4353
Brallobarbital, 1387
Branching Factor *see* 8207
Brandy, 1388
Brandy Mint *see* 7288
Branigen [Glaxo] *see* 85
Brantur [Dong-A] *see* 6287
Brasilin *see* 1392
Brassel [Schiapparelli-Searle] *see* 2380
Brassicol [Hoechst] *see* 8264
Brassidic Acid, 1389
Brassinolide, 1390
Bratenol [Elmu] *see* 7650
Braunite *see* 5762
Braunol [Braun Melsungen] *see* 7880
Braunosan H [Braun Melsungen] *see* 7880
Bravo [Diamond] *see* 2219
Brayera, 1391
Brazilin, 1392
Brazil Powder *see* 807
Brazil Wax *see* 1895
Brazil Wood *see* 7322
Bredinin [Toyo Jozo] *see* 6306
Bredon [Organon] *see* 7078
Brefeldin A, 1393
Brek [Irbi] *see* 5601
Brelomax [Abbott] *see* 9942
Bremax [Abbott] *see* 9942
Bremen Blue *see* 2697
Bremen Green *see* 2697
Bremfol *see* 6647
Bremil *see* 4822
Brenal [Tanabe Seiyaku] *see* 5820
Brendil [Daiichi] *see* 2350
Brentan [Janssen] *see* 6266
Brenzschleimsäure (German) *see* 4329
Brenztraubensäure (German) *see* 8205
Breokinase [Sanofi-Winthrop] *see* 10024
Breon *see* 7746
Brequinar, 1394
Brestan [Hoechst] *see* 9875
Brethaire [Ciba-Geigy] *see* 9302

Brethine [Ciba-Geigy] *see* 9302
Bretol [Fine Organics] *see* 2071
Bretylan [Burroughs Wellcome] *see* 1395
Bretylate [Burroughs Wellcome] *see* 1395
Bretylium Tosylate, 1395
Bretylol [DuPont Merck] *see* 1395
Brevetoxin A *see* 1396
Brevetoxin B *see* 1396
Brevetoxin C *see* 1396
Brevetoxins, 1396
Brevibloc [Ohmeda] *see* 3741
Brevicid [Arzneimittelwerk Dresden] *see* 10195
Brevicidin [Penick] *see* 9967
Brevicon [Syntex] *see* 6790
Brevidil E [M & B] *see* 9184
Brevidil M [M & B] *see* 9043
Brevifolin *see* 10201
Brevimytal Sodium [Lilly] *see* 6061
Brevinarcon [Heyden] *see* 9457
Brevinor [Syntex] *see* 6790
Brevital [Lilly] *see* 6061
Brevital Sodium [Lilly] *see* 6061
Brevium *see* 8069
Brexin [Chiesi] *see* 7661
Brezal [Sandoz] *see* 2260
Bricanyl [Pharma-stern] *see* 9302
Bricef [Bristol-Myers Squibb] *sec* 1965
Briem [Fabre] *see* 1058
Brietal Sodium [Lilly] *see* 6061
Brigade [FMC] *see* 1257
Brij [ICI] *see* 7737
Brij 58 [ICI] *see* 7737
Brilliant Blue FCF, 1397
Brilliant Congo R *see* 10149
Brilliant Cotton Blue *see* 6107
Brilliant Green, 1398
Brilliant Vital Red *see* 10149
Brimonidine, 1399
Brimstone *see* 9142
Brinaldix [Sandoz] *see* 2454
Briofil [Alfa] *see* 982
Bripadon *see* 4197
Briplatin [Bristol-Myers] *see* 2378
Brisfirina [Bristol] *see* 2030
Brispen *see* 3134
Bristab [Bristol] *see* 4830
Bristaciclina [Bristol] *see* 9337
Bristacin [Bristol] *see* 8411
Bristagen [Bristol] *see* 4398
Bristamin [Bristol] *see* 7469
Bristamox [Bristol] *see* 617
Bristamycin [Bristol] *see* 3726
Bristocef [Bristol] *see* 2030
Bristopen [Bristol] *see* 7036
Bristuric *see* 1064
Bristurin [Bristol] *see* 4830
Bristuron [Bristol] *see* 1064
Britacil [Wyeth] *see* 628
Britai [Bristol-Myers] *see* 2411
Britiazim [Thames] *see* 3247
British Anti-Lewisite *see* 3255
British Gum *see* 2995
Britlofex [Brittania] *see* 5588
BRL-284 *see* 8002
BRL-804 *see* 4706
BRL-1241 *see* 6047
BRL-1288 *see* 1057
BRL-1341 *see* 628
BRL-1400 *see* 7036
BRL-1621 *see* 2480
BRL-1702 *see* 3134
BRL-2039 *see* 4147
BRL-2064 *see* 1838
BRL-2288 *see* 9568
BRL-2333 *see* 617
BRL-2351 *see* 942
BRL-3475 *see* 1838

2-[p-Bromo-α-(2-dimethylaminoethyl)-benzyl]pyridine see 1467
α-Bromo-β-dimethylpropanoylurea see 1418
N¹-(5-Bromo-4,6-dimethyl-2-pyrimidinyl)sulfanilamide see 9066
5-Bromo-4,6-dimethyl-2-sulfanilamidopyrimidine see 9066
Bromodiphenhydramine, 1439
1-(2-Bromo-1,2-diphenylethenyl)-4-ethylbenzene see 1471
2-[1-(4-Bromodiphenyl)ethoxy]-N,N-dimethylethylamine see 3596
α-Bromo-α,β-diphenyl-β-(p-ethylphenyl)ethylene see 1471
1-Bromododecane see 5402
Bromoeosine see 3639
2-Bromoergocryptine see 1437
2-Bromo-α-ergokryptin see 1437
Bromoethane see 3819
2-Bromoethanol see 3838
β-Bromoethyl Alcohol see 3838
2-Bromo-2-ethylbutanamide see 3164
(α-Bromo-α-ethylbutyryl)carbamide see 1879
(α-Bromo-α-ethylbutyryl)urea see 1879
trans-3-Bromo-N-ethylcinnamamide see 2371
1-Bromo-2-(4-ethylphenyl)-1,2-diphenylethene see 1471
1-Bromo-2-(p-ethylphenyl)-1,2-diphenylethylene see 1471
(S)-3-Bromo-N-[(1-ethyl-2-pyrrolidinyl)methyl]-2,6-dimethoxybenzamide see 8301
Bromofenofos, 1440
Bromofluoresceic Acid see 3639
10-Bromo-11b-(2-fluorophenyl)-2,3,7,-11b-tetrahydrooxazolo[3,2-d][1,4]-benzodiazepin-6(5H)-one see 4635
Bromoform, 1441
1-Bromo 1,1,2,2,3,3,4,4,5,5,6,6,7,7,8,-8,8-heptadecafluorooctane see 7299
2-Bromohexanoic Acid see 1436
4-Bromo-α-hydroxybenzeneacetic Acid see 1446
5-Bromo-2-hydroxybenzenemethanol see 1460
5-Bromo-2-hydroxybenzyl Alcohol see 1460
(5'α)-2-Bromo-12'-hydroxy-2'-(1-methylethyl)-5'-(2-methylpropyl)-ergotaman-3',6',18-trione see 1437
α-(5-Bromo-4-hydroxy-m-tolyl)-α-(3-bromo-5-methyl-4-oxo-2,5-cyclohexadien-1-ylidene)-o-toluenesulfonic Acid see 1460
5-Bromo-6-(2-imidazolin-2-ylamino)-quinoxaline see 1399
2-Bromoisobutane see 1590
α-Bromoisobutyric Acid, 1442
1-Bromo-4-isocyanatobenzene see 1452
2-Bromoisopentane see 644
5-Bromo-3-isopropyl-6-methyluracil see 5174
α-Bromoisovaleric Acid, 1443
β-Bromoisovaleric Acid, 1444
(α-Bromoisovaleryl)urea see 1418
Bromol see 9744
Bromo-LSD see 1445
D-2-Bromolysergic Acid Diethylamide see 1445
Bromolysergide, 1445
p-Bromomandelic Acid, 1446
[[[(3-Bromomesityl)carbamoyl]methyl]imino]diacetic Acid see 5812
Bromomethane see 6108
(Bromomethyl)benzene see 1163

1-Bromo-3-methylbenzene see 1463
p-Bromo-α-methylbenzhydryl 2-Dimethylaminoethyl Ether see 3596
1-Bromo-2-methylbutane see 642
L-1-Bromo-2-methylbutane see 642
1-Bromo-3-methylbutane see 5128
2-Bromo-2-methylbutane see 644
2-Bromo-3-methylbutanoic Acid see 1443
3-Bromo-3-methylbutanoic Acid see 1444
2-Bromo-3-methylbutanoic Acid 1,7,7-Trimethylbicyclo[2.2.1]hept-2-yl Ester see 1368
2-Bromo-3-methylbutyric Acid Ester with d-Borneol see 1368
1-(Bromomethyl)-3-methylbenzene see 10223
5-Bromo-6-methyl-3-(1-methylethyl)-2,4(1H,3H)-pyrimidinedione see 5174
5-Bromo-6-methyl-3-(1-methylpropyl)-2,4(1H,3H)-pyrimidinedione see 1402
5-Bromo-6-methyl-3-(1-methylpropyl)-uracil see 1402
2-[(p-Bromo-α-methyl-α-phenylbenzyl)oxy]-N,N-dimethylethylamine see 3596
1-Bromo-2-methylpropane see 5150
2-Bromo-2-methylpropane see 1590
2-Bromo-2-methylpropanoic Acid see 1442
1-Bromonaphthalene, 1447
2-Bromonaphthalene, 1448
α-Bromonaphthalene see 1447
β-Bromonaphthalene see 1448
Bromone [Lipha] see 1403
2-Bromo-N-ethyl-N,N-dimethylbenzenemethanaminium Salt with 4-Methylbenzenesulfonic Acid (1:1) see 1395
8β-[(5-Bromonicotinoyloxy)methyl]-1,6-dimethyl-10α-methoxyergoline see 6580
2-Bromo-2-nitro-1,3-propanediol see 1470
β-Bromo-β-nitrotrimethyleneglycol see 1470
1-Bromooctane see 6861
2-Bromooctane see 6862
1-Bromopentane see 643
Bromoperfluorooctane see 7299
p-Bromophenacyl Bromide, 1449
p-Bromophenetole see 1450
Bromophenol, 1450
m-Bromophenol see 1450
o-Bromophenol see 1450
p-Bromophenol see 1450
Bromophenophos see 1440
N-(4-Bromophenyl)acetamide see 1420
α-Bromophenylacetonitrile see 1433
2-(p-Bromo-α-phenylbenzyloxy)-N,N-dimethylethylamine see 1439
4-Bromophenylcarbimide see 1452
p-Bromophenylcarbonimide see 1452
N-(4-Bromophenyl)-2,6-dihydroxybenzamide see 8322
γ-(4-Bromophenyl)-N,N-dimethyl-2-pyridinepropanamine see 1467
(Z)-3-(4-Bromophenyl)-N,N-dimethyl-3-(3-pyridinyl)-2-propen-1-amine see 10254
2-(4-Bromophenyl)-1,3-dioxohydrindene see 1414
1-(4-Bromophenyl)ethanone see 1423
2-Bromo-1-phenylethanone see 1424
(E)-3-(3-Bromophenyl)-N-ethyl-2-propenamide see 2371
p-Bromophenylglycolic Acid see 1446

p-Bromophenylhydrazine, 1451
3-[α-[p-(p-Bromophenyl)-β-hydroxyphenethyl]benzyl]-4-hydroxycoumarin see 1403
4-[4-(p-Bromophenyl)-4-hydroxypiperidino]-4'-fluorobutyrophenone see 1466
4-[4-(4-Bromophenyl)-4-hydroxy-1-piperidinyl]-1-(4-fluorophenyl)-1-butanone see 1466
2-(4-Bromophenyl)-1H-indene-1,3-(2H)-dione see 1414
p-Bromophenyl Isocyanate, 1452
3-(p-Bromophenyl)-1-methoxy-1-methylurea see 6224
N'-(4-Bromophenyl)-N-methoxy-N-methylurea see 6224
1-(p-Bromophenyl)-1-phenyl-1-(2-dimethylaminoethoxy)ethane see 3596
2-[1-(4-Bromophenyl)-1-phenylethoxy]-N,N-dimethylethanamine see 3596
[2-(1-p-Bromophenyl-1-phenylethoxy)ethyl]dimethylamine see 3596
2-[(4-Bromophenyl)phenylmethoxy]-N,N-dimethylethanamine see 1439
(Z)-3-(4'-Bromophenyl)-3-(3''-pyridyl)-dimethylallylamine see 10254
1-(p-Bromophenyl)-1-(2-pyridyl)-3-dimethylaminopropane see 1467
3-(p-Bromophenyl)-3-(2-pyridyl)-N,N-dimethylpropylamine see 1467
Bromophin see 787
Bromophos, 1453
Bromophos-ethyl see 1453
Bromophos-methyl see 1453
Bromopride, 1454
1-Bromopropane see 8030
2-Bromopropane see 5229
3-Bromopropanoic Acid see 1455
2-Bromopropanoic Acid Ethyl Ester see 3820
1-Bromo-2-propanone see 1422
3-Bromo-1-propene see 296
5-(2-Bromo-2-propenyl)-5-(1-methylethyl)-2,4,6(1H,3H,5H)pyrimidinetrione see 7980
5-(2-Bromo-2-propenyl)-1-methyl-5-(1-methylethyl)-2,4,6(1H,3H,5H)-pyrimidinetrione see 6509
5-(2-Bromo-2-propenyl)-5-(1-methylpropyl)-2,4,6(1H,3H,5H)pyrimidinetrione see 1537
5-(2-Bromo-2-propenyl)-5-(2-propenyl)-2,4,6(1H,3H,5H)-pyrimidinetrione see 1387
β-Bromopropionic Acid, 1455
Bromopropylate, 1456
3-Bromopropylene see 296
7-Bromo-5-(2-pyridyl)-3H-1,4-benzodiazepin-2(1H)-one see 1406
5-Bromo-2,4(1H,3H)-pyrimidinedione see 1464
1-Bromo-2,5-pyrrolidinedione see 1462
4'-Bromo-γ-resorcylanilide see 8322
Bromosalicylchloranilide, 1457
5-Bromosalicylhydroxamic Acid, 1458
5-Bromosalicylic Acid Acetate, 1459
N-5-Bromosalicyloyl-p-chloroaniline see 1457
Bromosaligenin, 1460
5-Bromosaligenin see 1460
Bromosuccinic Acid, 1461
N-Bromosuccinimide, 1462
5-Bromosulfamethazine see 9066
Bromosulfophthalein see 9127
2-Bromo-1,1,1,2-tetrafluoroethane see 9265
1-Bromo-2,2,3,3-tetrafluoropropane see 4632

3-Bromo-1,1,2,2-tetrafluoropropane see 4632
8-Bromotheophylline Compd with 2-Amino-2-methyl-1-propanol (1:1) see 7133
Bromotiren see 3079
Bromotoluene, 1463
3-Bromotoluene see 1463
m-Bromotoluene see 1463
o-Bromotoluene see 1463
p-Bromotoluene see 1463
α-Bromotoluene see 1163
ω-Bromotoluene see 1163
α-Bromo-α-tolunitrile see 1433
N-(3-Bromo-2,4,6-trimethylacetanilide)iminodiacetic Acid see 5812
3-Bromo-1,7,7-trimethylbicyclo-[2.2.1]heptan-2-one see 1435
N-[2-[(3-Bromo-2,4,6-trimethylphenyl)amino]-2-oxoethyl]-N-(carboxymethyl)glycine see 5812
5-Bromouracil, 1464
4-(6-Bromoveratryl)-4-[2-[2-(6,6-dimethyl-2-norpinyl)ethoxy]ethyl]-morpholinium Bromide see 7595
cis-11-Bromovincamine see 1473
5-Bromovinyl-araU see 8875
α-Bromo-m-xylene see 10223
ω-Bromo-m-xylene see 10223
Bromoxynil, 1465
Bromperidol, 1466
Brompheniramine, 1467
Bromphenol Blue, 1468
Bromphenphos see 1440
Bromsalizol [HW & D] see 1460
Bromsulphalein [HW & D] see 9127
Brom-Tetragnost see 9329
Bromthalein [E. Merck] see 9127
Bromthymol Blue, 1469
Bromural [Knoll] see 1418
Bromuvan see 1418
Bromvaletone see 1418
Bronalide [Syntex] see 4180
Bronalin [Byk-Liprandi] see 4745
Broncaspin [Bayer] see 886
Broncatar [Pulitzer] see 7078
Bronchocillin see 7212
Bronchodil [Asta] see 8307
Broncholin [Kaken] see 5674
Broncholysin [Spofa] see 89
Bronchopront [Mack, Illert.] see 401
Bronchoretard [Klinge] see 9421
Bronchoselectan [Schering AG] see 79
Bronchospasmin [Asta] see 8307
Broncocor [Irbi] see 7644
Broncon [Wakamoto] see 2470
Broncopen see 7212
Broncoplus [Sigma-Tau] see 8963
Broncovaleas [Valeas] see 217
Bronkaid Mist [Sterling Winthrop] see 3656
Bronkephrine [Sterling-Winthrop] see 3880
Bronkodyl [Sanofi-Winthrop] see 9421
Bronner's Acid see 1941
Bronopol, 1470
Bronosol [Green Cross] see 1470
Bronsecur [SK & F] see 1882
Brontine [Brocades-Stheeman] see 2958
Brontyl [Reckitt & Colman] see 8092
Brookite see 9612
Broom see 8547
Broparoestrol, 1471
Brosalamid see 1458
Brotizolam, 1472
Brotopon [Pfizer Taito] see 4629
Brovalol see 1368
Brovalurea see 1418
Brovavir see 8875

Brovel [Lepetit] see 3670
Brovincamine, 1473
"Brown Acetate of Lime" see 1683
Brown Mustard see 6394
Broxalax see 1282
Broxil [Beecham] see 7369
Broxykinolin see 1474
Broxynil see 1465
Broxyquinoline, 1474
BRS-640 see 1386
Bruceantin, 1475
Brucine, 1476
Brucite see 5706
Brufaneuxol see 495
Brufen [Boots] see 4925
Brufort [Lampugnani] see 4925
Bruisewort see 8511
Brulidine [M & B] see 3073
Brumetidina [Bruschettini] see 2337
Brumixol [Bruschettini] see 2325
Brunac [Bruschettini] see 89
Bruneomycin see 8986
Brunet Saxifrage see 7590
Brushite see 1739
Bruxicam [Bruschettini] see 7661
Bruzem [Bruschettini] see 3247
Bryamycin [Bristol] see 9502
Bryonia, 1477
Bryony see 1477
Bryostatins, 1478
BS-100-141 see 4590
BS-572 see 2771
BS-4231 see 4453
BS-5930 see 7007
BS-6987 see 2958
BS-7331 see 9641
BSF-2 see 5017
B.S.G. [Lemmon] see 1325
BSM-906M see 2772
BSO see 1556
BSP see 9127
BT see 962 and 9380
BT-621 see 9640
BTC see 1086
BTC-1010 [Onyx] see 3149
BTE see 9913
BTMP see 1068
BTPABA see 1081
BTS see 1343
BTS-18322 see 4234
BTS-27419 see 510
BTS-40542 see 7941
BTS-49465 see 4146
BTS-54524 see 8629
BTX see 1396
BTX-B see 1396
BTX-C see 1396
BU-2231 see 9213
BU-2231A see 9213
BU-2231B see 9213
Buban [Burroughs Wellcome] see 1510
Bubarbital Sodium [Philips Roxane] see 1530
Buburone [Towa Yakuhin] see 4925
Bucarban [Chinoin] see 1881
Buccalsone [Will-Pharma] see 4828
Buccastem [Reckitt & Colman] see 7942
Bucco see 1480
Bucetalon [Isei] see 1479
Bucetin, 1479
Buchu, 1480
Buchu Camphor see 3351
Buchu Resin see 3350
Bucillamine, 1481
Buck Bean see 5888
Buckminsterfullerene, 1482
Buckthorn Bark see 4287 and 8339
Bucku see 1480
Buckyball see 1482

Bucladesine, 1483
Buclamase [Rystan] see 640
Buclifen [Pfizer] see 1484
Buclina [Usafarma] see 1484
Buclizine, 1484
Buclosamide, 1485
Bucloxic Acid, 1486
Bucloxonic Acid see 1486
Bucolome, 1487
Bucrilate see 1488
Bucrylate, 1488
Buctril [Rhône-Poulenc] see 1465
Bucumarol [Sankyo] see 1489
Bucumolol, 1489
Buddleoflavonoloside see 5523
Budeson [Fujisawa] see 1490
Budesonide, 1490
Budipine, 1491
Budirol [Fher] see 8056
Budoform [Dolder] see 5052
Budralazine, 1492
Bueno [ISK Biosci] see 6020
Bufalin, 1493
Bufedil [Abbott] see 1498
Bufedon [Cosmopharma] see 6830
Bufemid [Lederle] see 4003
Bufencarb, 1494
Bufeniode, 1495
Bufetolol, 1496
Bufexamac, 1497
Buflan [Pierrel] see 1498
Buflocit [CT] see 1498
Buflomedil, 1498
Buflonat [Nattermann] see 1498
Bufogenin B, 1499
Bufon see 3177
Bufonamin [Kaken] see 1500
Bufopto Homatrocel [Softcon] see 4766
Bufor [Bofors] see 6581
Buformin, 1500
Bufotalin, 1501
Bufotalin 3-Suberoylarginine Ester see 1503
Bufotenine, 1502
Bufotoxin, 1503
Bufuralol, 1504
Bufylline (Rescinded USAN) see 404
Bugbane see 2339
Bugleweed see 5654
Bugwort see 2339
Buksamin see 464
Buku see 1480
Bulan [Comm. Solvents], 1505
Bulbocapnine, 1506
Bulbold [Sandoz-Wander] see 4493
Bulbonin [Sankyo] see 1500
Bulbus Scillae see 8925
Bull Nettle see 8856
Bumadizon, 1507
Bumaflex [Byk-Liprandi] see 1507
Bumetanide, 1508
Bumex [Roche] see 1508
Buminate [Hyland] see 8613
Bunafline, 1509
Bunaiod see 1511
Bunamidine, 1510
Bunamiodyl Sodium, 1511
Bunaphtide see 1509
Bunaphtine see 1509
Buna S [I. G. Farben] see 8534
Bunazosin, 1512
Bundlin B see 8566
α-Bungarotoxin see 1513
β-Bungarotoxin see 1513
Bungarotoxins, 1513
Buniodyl see 1511
Bunitrolol, 1514
l-Bunolol see 5488
Bunsenite see 6595

Bunt-cure *see* 4714
Bunte Salts, 1515
Bunt-no-more *see* 4714
Buparvaquone, 1516
Bupatol [Gedeon Richter] *see* 981
Buphanamine, 1517
Buphanitine, 1518
Buphedrin [Tatsumi] *see* 6830
Buphenine *see* 6830
Bupirimate, 1519
Bupivacaine, 1520
Bupranol *see* 1521
Bupranolol, 1521
Buprenex [Norwich-Eaton] *see* 1522
Buprenorphine, 1522
Bupropion, 1523
Buquinolate, 1524
Buramate, 1525
Burdock *see* 5377
Burgodin [Janssen] *see* 1241
Burgundy Mixture, 1526
Burinex [Leo Pharm.] *see* 1508
Burnt Alum *see* 373
Burnt Ammonium Alum *see* 335
Burnt Lime *see* 1733
Burnt Sugar *see* 1819
Burnt Sugar Coloring *see* 1819
Buronil [Ferrosan] *see* 5870
Burow's Solution *see* 332
Bursine *see* 2259
Burtonite 44 *see* 4322
Burtonite V-7-E *see* 4602
Buscapina *see* 1624
Buscol *see* 1624
Buscolysin [Pharmachim] *see* 1624
Buscopan [Boehringer, Ing.] *see* 1624
Buserelin, 1527
Busodium [Truxton] *see* 1530
Busotran [Galen] *see* 1530
Buspar [Mead Johnson] *see* 1528
Buspimen [Menarini] *see* 1528
Buspinol [Zdravlje] *see* 1528
Buspirone, 1528
Buspisal [Lesvi] *see* 1528
Busulfan, 1529
Busulphan *see* 1529
Butabar [Flar] *see* 1530
Butabarbital Sodium, 1530
Butabarbitone Sodium *see* 1530
Butabon *see* 1530
Butacaine, 1531
Butacetin, 1532
Butachlor, 1533
Butacide [Roussel-UCLAF] *see* 7629
Butacote [Ciba-Geigy] *see* 7431
1,3-Butadiene, 1534
α,γ-Butadiene *see* 1534
1,3-Butadiene-1,4-dicarboxylic Acid *see* 6381
Butadiene Diepoxide *see* 3715
Butadiene Dioxide *see* 3715
1,1′-(1,3-Butadiene-1,4-diyl)bisbenzene *see* 3378
Butadiene Sulfone *see* 9129
Butadion [Streuli] *see* 7431
Butadiona [Miquel-Otsuka] *see* 7431
Butak [Lemmon] *see* 1530
Buta-Kay [King] *see* 1530
Butalamine, 1535
Butalan [Lannett] *see* 1530
Butalbital, 1536
Butalex [Coopers] *see* 1516
Butalgin *see* 6008
Butaliret [Fatol] *see* 9302
Butallylonal, 1537
Butamben, 1538
Butamide *see* 1545
Butamirate, 1539
Butamisole, 1540
Butamiverine *see* 1544

Butamyrate *see* 1539
Butanal *see* 1627
Butanamide *see* 1628
1-Butanamine *see* 1578
2-Butanamine *see* 1579
Butane, 1541
n-Butane *see* 1541
Butanediamide *see* 9035
1,4-Butanediamine *see* 8134
1,4-Butanedicarboxylic Acid *see* 161
Butanedinitrile *see* 9041
Butanedioic Acid *see* 9038
Butanedioic Acid Bis(1,1-dimethylethyl) Ester *see* 3088
Butanedioic Acid Calcium Salt *see* 1752
Butanedioic Acid Dibutyl Ester *see* 3087
Butanedioic Acid 2-(Dimethylamino)-ethyl [5-Hydroxy-4-(hydroxymethyl)-6-methyl-3-pyridinyl]methyl Ester *see* 8175
Butanedioic Acid Mono[(4-butyl-3,5-dioxo-1,2-diphenyl-4-pyrazolidinyl)-methyl] Ester *see* 9185
[3*R*-(3α,5aβ,6β,8aβ,9α,10α,12β,-12a*R**)]-Butanedioic Acid Mono-(decahydro-3,6,9-trimethyl-3,12-epoxy-12*H*-pyrano[4,3-*j*]-1,2-benzodioxepin-10-yl) Ester *see* 857
Butanedioic Acid Mono[1-[5-(2,5-dihydro-5-oxo-3-furanyl)-3-methyl-2-benzofuranyl]ethyl] Ester *see* 1070
Butanedioic Acid Mono(2,2-dimethylhydrazide) *see* 2874
Butanedioic Acid Mono[3-[(2-ethylhexyl)amino]-1-methyl-3-oxopropyl] Ester *see* 1562
Butanedioic Anhydride *see* 9039
Butane-1,3-diol *see* 1601
1,3-Butanediol *see* 1601
2,3-Butanediol *see* 1602
1,4-Butanediol Dimethanesulfonate Esters *see* 1529
Butanedione *see* 3715
2,3-Butanedione *see* 3010
2,3-Butanedionedioxime *see* 3295
Butanedioyl Chloride *see* 9042
Butanenitrile *see* 1633
(*R**,*S**)-1,2,3,4-Butanetetrol *see* 3714
(*R**,*S**)-1,2,3,4-Butanetetroltetranitrate *see* 3716
1-Butanethiol *see* 1611
2-Butanethiol *see* 1612
Butanex [Makhteshim-Agan] *see* 1533
Butanilicaine, 1542
Butanimide *see* 9040
Butanoic Acid *see* 1629
Butanoic Acid Anhydride *see* 1630
1-Butanoic Acid-26-L-aspartic Acid-27-L-valine-29-L-alanine-1,7-dicarbacalcitonin (Salmon) *see* 3578
Butanoic Acid Butyl Ester *see* 1591
Butanoic Acid 2,3-Dihydroxypropyl Ester *see* 6332
Butanoic Acid Ethyl Ester *see* 3822
1-Butanoic Acid-7-glycine-1,6-dicarbaoxytocin *see* 1887
Butanoic Acid Methyl Ester *see* 6114
1-Butanoic Acid-2-(*O*-methyl-L-tyrosine)-1-carbaoxytocin *see* 1841
Butanoic Acid Pentyl Ester *see* 645
Butanoic Acid 1,2,3-Propanetriyl Ester *see* 9750
Butanoic Acid Propyl Ester *see* 8031
1-Butanol *see* 1575
2-Butanol *see* 1576
1,2-Butanolide *see* 1632
1,4-Butanolide *see* 1632
2,3-Butanolone *see* 61

2-Butanone *see* 6149
Butanotic [Medical Arts] *see* 1530
Butanoyl Chloride *see* 1634
L-α-Butanoyl Glycerol *see* 6332
Butaperazine, 1543
Butaphyllamine [Merrell] *see* 404
Butapirazol [Polfa] *see* 7431
Butased [Schlicksup] *see* 1530
Butatensin [Benvegna] *see* 5813
Butatran [Palmedico] *see* 1530
Butatron [Sanofi] *see* 7431
Butaverine, 1544
Butazem [Zemmer] *see* 1530
Butazolamide, 1545
Butazolidin [Ciba-Geigy] *see* 7431
Butedrin *see* 981
Butedronic Acid, 1546
Butelline *see* 1531
Butenafine, 1547
2-Butenal *see* 2663
1-Butene, 1548
2-Butene, 1549
(*E*)-2-Butenedioic Acid *see* 4309
(*Z*)-Butenedioic Acid *see* 5743
(*E*)-2-Butenedioic Acid Bis(phenylmethyl) Ester *see* 1170
(*Z*)-2-Butenedioic Acid, Diethyl Ester *see* 3172
cis-Butenedioic Anhydride *see* 5744
3-Butenenitrile *see* 298
cis-2-Butenoic Acid *see* 5180
(*E*)-2-Butenoic Acid *see* 2664
2-Buten-1-ol *see* 2668
3-Buten-2-one *see* 6214
Δ³-2-Butenone *see* 6214
β-Butenonitrile *see* 298
(4*R*)-4-[(*E*)-2-Butenyl]-4,*N*-dimethyl-L-threonine *see* 5811
3-(2-Butenylidene)-2-carboxy-α-(hydroxymethylene)-1,4-cyclopentadiene-1-acetic Acid δ-Lactone Methyl Ester *see* 4307
(*E,E*)-7-(2-Butenylidene)-1,7-dihydro-1-oxocyclopenta[*c*]pyran-4-carboxylic Acid Methyl Ester *see* 4307
3-(3-Buten-1-ynyl)-6-(1,3-pentadiynyl)-1,2-dithiin *see* 9439
Buterazine [Daiichi] *see* 1492
Butesin [Abbott] *see* 1538
Butethal, 1550
Butethamate, 1551
Butethamine, 1552
Butex [Phillips] *see* 1530
Butformin *see* 1500
Buthalital Sodium, 1553
Buthalitone Sodium *see* 1553
Buthiazide, 1554
Buthiobate, 1555
Buthionine Sulfoximine, 1556
Buthoid [Merrell] *see* 404
Butibufen, 1557
Buticaps [McNeil] *see* 1530
Butidrine Hydrochloride, 1558
Butilate *see* 1581
Butilopan [Juste] *see* 1557
Butirosin, 1559
Butirosin A *see* 1559
Butirosin B *see* 1559
Butisol Sodium [McNeil] *see* 1530
Butix [Fabre] *see* 5911
Butixirate *see* 10205
Butizide *see* 1554
Butobarbital *see* 1550
Butobarbitone *see* 1550
Butoben [Merck & Co.] *see* 1619
Butobendine, 1560
Butoconazole, 1561
Butoctamide, 1562
Butofilolol, 1563
Butoform *see* 1538

Butolan [Bayer] see 3365
Butolen see 3365
Butonate see 9753
Butopamine Hydrochloride see 8275
Butopyronoxyl, 1564
Butorphanol, 1565
Butox [Distrivet] see 2934
Butoxone. see 2893
4'-tert-Butoxyacetanilide see 1532
2-Butoxy-3-aminobenzoic Acid β-Diethylaminoethyl Ester see 5978
2-Butoxy-4-aminobenzoic Acid β-Diethylaminoethyl Ester see 402
3-Butoxy-4-aminobenzoic Acid 2-(Diethylamino)ethyl Ester see 1076
4-Butoxybenzoic Acid 2-(Diethylamino)ethyl Ester see 1566
8-(p-Butoxybenzyl)-3α-hydroxy-1αH,-5αH-tropanium Bromide (—)-Tropate see 1569
l-[1-(p-n-Butoxybenzyl)hyoscyaminium] Bromide see 1569
Butoxycaine, 1566
1-(3-Butoxy-2-carbamoyloxypropyl)-5-phenyl-5-ethylbarbituric Acid see 3983
N-[N-[N-[N-(N-tert-Butoxycarbonyl-β-alanyl)-L-tryptophanyl]-L-methionyl]-L-aspartyl]-L-phenylalaninamide see 7250
2-Butoxy-N-(2-diethylaminoethyl)-cinchoninamide Hydrochloride see 3081
2-Butoxy-N-[2-(diethylamino)ethyl]-4-quinolinecarboxamide Monohydrochloride see 3081
2-Butoxyethanol see 1594
2-(2-Butoxyethoxy)ethanol see 1592
α-[2-(2-Butoxyethoxy)ethoxy]-4,5-methylenedioxy-2-propyltoluene see 7629
5-[[2-(2-Butoxyethoxy)ethoxy]methyl]-6-propyl-1,3-benzodioxole see 7629
2-[2-(Butoxy)ethoxy]ethyl Ester of Thiocyanic Acid see 5472
4-Butoxy-N-hydroxybenzeneacetamide see 1497
1-(3-Butoxy-2-hydroxypropyl)-5-ethyl-5-phenylbarbituric Acid Carbamate see 3983
N-(Butoxymethyl)-2-chloro-2',6'-diethylacetanilide see 1533
N-(Butoxymethyl)-2-chloro-N-(2,6-diethylphenyl)acetamide see 1533
2-(4-Butoxyphenoxy)-N-(2,5-diethoxyphenyl)-N-[2-(diethylamino)ethyl]-acetamide see 4026
1-Butoxy-3-phenoxy-2-propanol see 3985
4-[3-(4-Butoxyphenoxy)propyl]morpholine see 7888
2-(p-Butoxyphenyl)acetohydroxamic Acid see 1497
2-(p-Butoxyphenyl)-N-[2-(diethylamino)ethyl]-2',5'-diethoxyacetanilide see 4026
[3(S)-endo]-8-[(4-Butoxyphenyl)methyl]-3-(3-hydroxy-1-oxo-2-phenylpropoxy)-8-methyl-8-azoniabicyclo-[3.2.1]octane Bromide see 1569
p-Butoxyphenyl γ-Morpholinopropyl Ether see 7888
4-Butoxyphenyl Piperidineethyl Ketone see 3523
1-(4-Butoxyphenyl)-3-(1-piperidinyl)-1-propanone see 3523
4-Butoxy-β-piperidinopropiophenone see 3523

4-n-Butoxy-β-(1-piperidyl)propiophenone see 3523
3-[(6-Butoxy-3-pyridinyl)azo]-2,6-pyridinediamine see 3021
2-Butoxy-2'-thiocyanodiethyl Ether see 5472
1-Butoxy-α-(2-thiocyanoethoxy)ethane see 5472
Butoz [Hamilton] see 7431
Butralin, 1567
Butrate [Kay] see 1530
Butriptylene see 1568
Butriptyline, 1568
Butropium Bromide, 1569
Butte [Scrip] see 1530
Buttercup Yellow see 10262
Butterfly Weed see 865
Butternut see 5281
Butter of Antimony see 748
Butter of Arsenic see 841
Butter of Zinc see 10261
Butter Yellow see 3279
β-BuTX see 1513
Butydrine Hydrochloride see 1558
n-Butyl Acetate, 1570
sec-Butyl Acetate, 1571
tert-Butyl Acetate, 1572
21-tert-Butylacetate-9α-fluoro-11β-hydroxy-16α,17α-(isopropylidenedioxy)pregna-1,4-diene-3,20-dione see 9730
tert-Butylacetic Acid, 1573
n-Butyl Acrylate, 1574
Butyl Alcohol see 1575
n-Butyl Alcohol, 1575
sec-Butyl Alcohol, 1576
tert-Butyl Alcohol, 1577
n-Butylamine, 1578
sec-Butylamine, 1579
tert-Butylamine, 1580
ω-n-Butylaminoacetic Acid 2-Methyl-6-chloroanilide see 1542
1-(Butylaminoacetylamino)-2-chloro-6-methylbenzene see 1542
N-(Butylaminoacetyl)-6-chloro-o-toluidine see 1542
Butyl Aminobenzoate see 1538
n-Butyl p-Aminobenzoate see 1538
4-(Butylamino)benzoic Acid 2-(Dimethylamino)ethyl Ester Monohydrochloride see 9330
p-Butylaminobenzoic Acid ω-O-Methylnonaethyleneglycol Ester see 1127
4-(Butylamino)benzoic Acid 3,6,9,12,-15,18,21,24,27-Nonaoxaoctacos-1-yl Ester see 1127
p-Butylaminobenzoyl-2-dimethylaminoethanol Hydrochloride see 9330
N-[(Butylamino)carbonyl]benzenesulfonamide see 7362
[1-[(Butylamino)carbonyl]-1H-benzimidazol-2-yl]carbamic Acid Methyl Ester see 1073
N-[(Butylamino)carbonyl]-4-methylbenzenesulfonamide see 9646
2-(Butylamino)-6'-chloro-o-acetotoluidide see 1542
1-tert-Butylamino-3-(2-chloro-5-methylphenoxy)-2-propanol see 1521
2-(Butylamino)-N-(2-chloro-6-methylphenyl)acetamide see 1542
(±)-2'-(tert-Butylamino)-3'-chloropropiophenone see 1523
1-(tert-Butylamino)-3-[(6-chloro-m-tolyl)oxy]-2-propanol see 1521
(S)-1-(tert-Butylamino)-3-(o-cyclopentylphenoxy)-2-propanol see 7209
1-(tert-Butylamino)-3-(2,5-dichlorophenoxy)-2-propanol see 2464

3-(tert-Butylamino)-3-(2',3'-dimethylphenoxy)-2-propanol see 10209
4-Butylamino-2-hydroxybenzoic Acid 2-Dimethylaminoethyl Ester see 4894
(—)-1-tert-Butylamino-2-hydroxy-3-(2'-cyclopentylphenoxy)propane see 7209
2-(2-tert-Butylamino-1-hydroxyethyl)-7-ethylbenzofuran see 1504
(±)-5-[2-(tert-Butylamino)-1-hydroxyethyl]-m-phenylene Bis(dimethylcarbamate) see 980
2-(tert-Butylamino)-1-(4-hydroxy-3-hydroxymethylphenyl)ethanol see 217
2-Butylamino-1-p-hydroxyphenylethanol see 981
o-[3-(tert-Butylamino)-2-hydroxypropoxy]benzonitrile see 1514
5-[3-(tert-Butylamino)-2-hydroxypropoxy]-3,4-dihydrocarbostyril see 1917
(—)-5-[3-(tert-Butylamino)-2-hydroxypropoxy]-3,4-dihydro-1(2H)-naphthalenone see 5488
(±)-2'-[3-(tert-Butylamino)-2-hydroxypropoxy]-5'-fluorobutyrophenone see 1563
(±)-8-[3-(t-Butylamino)-2-hydroxypropoxy]-5-methylcoumarin see 1489
4-[3-(tert-Butylamino)-2-hydroxypropoxy]-N-methylisocarbostyril see 9579
S-(—)-3-(3-tert-Butylamino-2-hydroxypropoxy)-4-morpholino-1,2,5-thiadiazole see 9585
2-[o-3-(t-Butylamino)-2-hydroxypropoxy]phenoxy]-N-methylacetamide see 2060
(±)-1-[p-[3-(tert-Butylamino)-2-hydroxypropoxy]phenyl]-3-cyclohexylurea see 9208
(2R,3S)-5-[3-(tert-Butylamino)-2-hydroxypropoxy]-1,2,3,4-tetrahydronaphthalene-2,3-diol see 6431
2-(3'-tert-Butylamino-2'-hydroxypropylthio)-4-(5'-carbamoyl-2'-thienyl)thiazole see 827
(±)-1-t-Butylamino-3-(o-N-methylcarbamoylmethoxyphenoxy)propan-2-ol see 2060
α-[(tert-Butylamino)methyl]-o-chlorobenzyl Alcohol see 9942
α-[(tert-Butylamino)methyl]-3,5-dihydroxybenzyl Alcohol see 9302
α-[(tert-Butylamino)methyl]-3,4-dihydroxybenzyl Alcohol 3,4-Di-p-toluate see 1344
α-[(tert-Butylamino)methyl]-7-ethyl-2-benzofuranmethanol see 1504
α-[(Butylamino)methyl]-4-hydroxybenzenemethanol see 981
α-[(Butylamino)methyl]-p-hydroxybenzyl Alcohol see 981
α-(t-Butylaminomethyl)-4-hydroxy-3-ureidobenzyl Alcohol see 1882
α¹-[(tert-Butylamino)methyl]-4-hydroxy-m-xylene-α,α'-diol see 217
(±)-1-(tert-Butylamino)methyl]-3-[(2-methylindol-4-yl)oxy]-2-propanol Benzoate (Ester) see 1362
α-[(sec-Butylamino)methyl]-5,6,7,8-tetrahydro-2-naphthalenemethanol Hydrochloride see 1558
3-(Butylamino)-4-phenoxy-5-sulfamoylbenzoic Acid see 1508
p-Butylaminosalicylic Acid 2-Dimethylaminoethyl Ester see 4894
1-(tert-Butylamino)-3-[(5,6,7,8-tetra-

Butylethylcarbamothioic Acid *S*-
Propyl Ester *see* 7192
N-Butyl-*N*-ethyl-2,6-dinitro-4-tri-
fluoromethylaniline *see* 1067
N-Butyl-*N*-ethyl-2,6-dinitro-4-(trifluoromethyl)benzenamine *see* 1067
tert-Butyl Ethyl Ether *see* 3821
5-Butyl-5-ethyl-2,4,6(1*H*,3*H*,5*H*)-pyri-
midinetrione *see* 1550
5-*sec*-Butyl-5-ethyl-2-thiobarbituric
Acid *see* 9457
N-Butyl-*N*-ethyl-α,α,α-trifluoro-2,6-
dinitro-*p*-toluidine *see* 1067
Butyl 6α-Fluoro-11β-hydroxy-16α-
methyl-3,20-dioxopregna-1,4-dien-
21-oate *see* 4187
S-(*n*-Butyl)homocysteine Sulfoximine
see 1556
tert-Butyl Hydroperoxide, 1604
2(3)-*tert*-Butyl-4-hydroxyanisole *see*
1582
n-Butyl *p*-Hydroxybenzoate *see* 1619
5-Butyl-5-(2-hydroxyethyl)barbituric
Acid Carbamate *see* 1880
6-*sec*-Butyl-1-hydroxy-3-isobutyl-
2(1*H*)-pyrazinone *see* 876
4-Butyl-4-(hydroxymethyl)-1,2-di-
phenyl-3,5-pyrazolidinedione Hydrogen Succinate (Ester) *see* 9185
[7(*S*)-(1α,2β,4β,5α,7β)]-9-Butyl-7-(3-
hydroxy-1-oxo-2-phenylpropoxy)-9-
methyl-3-oxa-9-azoniatricyclo[3.3.-
1.0²,⁴]nonane Bromide *see* 1624
4-Butyl-1-(4-hydroxyphenyl)-2-phen-
yl-3,5-pyrazolidinedione *see* 7106
4-Butyl-2-(*p*-hydroxyphenyl)-1-phen-
yl-3,5-pyrazolidinedione *see* 7106
t-Butyl Hypochlorite *see* 1605
tert-Butyl Hypochlorite, 1605
(11β,16α)-16,17-[Butylidenebis(oxy)]-
11,21-dihydroxypregna-1,4-diene-3,-
20-dione *see* 1490
Butylidene Chloride, 1606
p,*p*'-*sec*-Butylidenediphenol *see* 1339
6-*tert*-Butyl-3-(2-imidazolin-2-ylmeth-
yl)-2,4-dimethylphenol *see* 7100
N-Butylimidodicarbonimidic Diamide
see 1500
n-Butyl Iodide, 1607
sec-Butyl Iodide, 1608
6-*sec*-Butyl-3-isobutylpyrazinol 1-
Oxide *see* 876
Butyl Isocyanate, 1609
n-Butyl Isocyanate *see* 1609
2-[(3-Butyl-1-isoquinolinyl)oxy]-*N*,*N*-
dimethylethanamine *see* 3267
n-Butylmalonic Acid, 1610
Butylmalonic Acid Mono(1,2-diphen-
ylhydrazide) *see* 1507
n-Butyl Mercaptan, 1611
sec-Butyl Mercaptan, 1612
tert-Butyl Mercaptan, 1613
p-Butylmercaptobenzhydryl β-Dimethylaminoethyl Sulfide *see* 1816
n-Butylmercuric Chloride, 1614
Butyl Mesityl Oxide Oxalate *see* 1564
*N*¹-Butyl-*N*²-metanilylcarbamide *see*
1615
1-Butyl-3-metanilylurea, 1615
Butyl Methoxydibenzoylmethane,
1616
4-*tert*-Butyl-4'-methoxydibenzoyl-
methane *see* 1616
2-*tert*-Butyl-4-methoxyphenol *see* 1582
3-*tert*-Butyl-4-methoxyphenol *see* 1582
N-*tert*-Butyl-1-methyl-3,3-diphenyl-
propylamine *see* 9311
tert-Butyl Methyl Ether *see* 6111
tert-Butyl Methyl Ketone *see* 7594
4-Butyl-4-[(4-methyl-1-piperazinyl)-

methyl]-1,2-diphenyl-3,5-pyrazoli-
dinedione *see* 7610
2-*sec*-Butyl-2-methyl-1,3-propanediol
Dicarbamate *see* 5813
N-Butyl-2-methyl-2-propyl-1,3-pro-
panediol Dicarbamate *see* 9960
2-*sec*-Butyl-2-methyltrimethylenecar-
bamate *see* 5813
Butylmin [Nippon Kayaku] *see* 1624
n-Butyl Nitrite, 1617
tert-Butyl Nitrite, 1618
Butyl-Nor-Sympatol *see* 981
Butylon [Grelan] *see* 1479
2-[*p*-(Butyloxy)phenyl]acetohydrox-
amic Acid *see* 1497
Butylparaben, 1619
Butyl Parasept *see* 1619
Butylphen *see* 1620
p-*tert*-Butylphenol, 1620
2-(*p*-*tert*-Butylphenoxy)cyclohexyl
Propargyl Sulfite *see* 7992
2-(*p*-*tert*-Butylphenoxy)isopropyl 2-
Chloroethyl Sulfite *see* 805
4-Butyl-1-phenyl-3,5-dioxopyrazoli-
dine *see* 6312
1-(*p*-*tert*-Butylphenyl)-4-[4'-(α-
hydroxydiphenylmethyl)-1'-piperid-
yl]butanol *see* 9307
α-(*p*-*tert*-Butylphenyl)-4-(α-hydroxy-
α-phenylbenzyl)-1-piperidinebutanol
see 9307
α-Butyl-α-phenyl-1*H*-imidazole-1-pro-
panenitrile *see* 3998
(±)-*cis*-4-[3-(4-*tert*-Butylphenyl)-2-
methylpropyl]-2,6-dimethylmor-
pholine *see* 4035
(*RS*)-1-[3-(4-*tert*-Butylphenyl)-2-meth-
ylpropyl]piperidine *see* 4034
Butyl β-Phenyl-1-piperidinepropion-
ate *see* 1544
4-Butyl-1-phenyl-3,5-pyrazolidine-
dione *see* 6312
4-*tert*-Butylphenyl Salicylate, 1621
1-Butyl-3-(phenylsulfonyl)urea *see*
7362
n-Butyl Phthalate, 1622
5-Butylpicolinic Acid *see* 4336
dl-*N*-*n*-Butylpipecolic Acid 2,6-Xyli-
dide *see* 1520
dl-1-Butyl-2',6'-pipecoloxylidide *see*
1520
dl-1-*n*-Butylpiperidine-2-carboxylic
Acid 2,6-Dimethylanilide *see* 1520
Butyl β-Piperidinohydrocinnamate *see*
1544
n-Butyl β-(*N*-Piperidyl)-β-phenyl-
propionate *see* 1544
Butylpropanedioic Acid *see* 1610
Butylpropanedioic Acid Mono(1,2-di-
phenylhydrazide) *see* 1507
n-Butyl Propionate, 1623
5-Butyl-2-pyridinecarboxylic Acid *see*
4336
Butylscopolamine Bromide *see* 1624
N-Butylscopolammonium Bromide,
1624
2-*n*-Butyl-4-spirocyclopentane-1-[(2'-
(tetrazol-5-yl)biphenyl-4-yl)methyl]-
2-imidazolin-5-one *see* 5097
Butyl Stearate, 1625
1-Butyl-3-sulfanilylurea *see* 1881
n-Butyl Sulfide, 1626
2-*S*-(*n*-Butylsulfonylamino)-3[4-(piper-
idin-4-yl)butyloxyphenyl]propionic
Acid *see* 9605
N-(Butylsulfonyl)-*O*-[4-(4-piperidinyl)-
butyl]-L-tyrosine *see* 9605
N-(Butylsulfonyl)-4-[4-(4-piperidyl)-
butoxy]-L-phenylalanine *see* 9605
2-Butyl-3-[[2'-(1*H*-tetrazol-5-yl)[1,1'-

biphenyl]-4-yl]methyl]-1,3-diaza-
spiro[4.4]non-1-en-4-one *see* 5097
Butyl 3'-(1*H*-Tetrazol-5-yl)oxanilate
see 9248
2-Butyl-3-[*p*-(*o*-1*H*-tetrazol-5-ylphen-
yl)benzyl]-1,3-diazaspiro[4.4]non-1-
en-4-one *see* 5097
N-(5-*tert*-Butyl-1,3,4-thiadiazol-2-yl)-
benzenesulfonamide *see* 4488
1-(5-*tert*-Butyl-1,3,4-thiadiazol-2-yl)-
1,3-dimethylurea *see* 9255
*N*¹-(5-*tert*-Butyl-1,3,4-thiadiazol-2-yl)-
sulfanilamide *see* 4487
sec-Butyl Thioalcohol *see* 1612
Butylthiobutane *see* 1626
p-Butylthiodiphenylmethyl 2-Dimeth-
ylaminoethyl Sulfide *see* 1816
2-(Butylthio)dodecanoic Acid Bismuth
Basic Salt *see* 1301
2-[*p*-(Butylthio)-α-phenylbenzylthio]-
N,*N*-dimethylethylamine *see* 1816
2-[[[4-(Butylthio)phenyl]phenylmeth-
yl]thio]-*N*,*N*-dimethylethanamine *see*
1816
1-Butyl-3-(*p*-tolylsulfonyl)urea *see*
9646
N-*n*-Butyl-*N*'-tosylurea *see* 9646
1-*t*-Butyl-2-(1,2,4-triazol-1-yl)-2-(2',4'-
dichlorobenzyl)ethanol *see* 3131
Butyl 2-[4-(5-Trifluoromethyl-2-
pyridyloxy)phenoxy]propionate *see*
4152
1-Butyl-2-(2,6-xylylcarbamoyl)piper-
idine *see* 1520
2-Butynediamide *see* 2009
1,1'-(2-Butyne-1,4-diyl)bispyrrolidine
see 9715
Butynorate *see* 3089
Butyn Sulfate [Abbott] *see* 1531
1,1'-(2-Butynylene)dipyrrolidine *see*
9715
Butyrac [Union Carbide] *see* 2893
Butyraldehyde, 1627
n-Butyramide, 1628
3-Butyramido-α-ethyl-2,4,6-triiodo-
cinnamic Acid Sodium Salt *see* 1511
3-Butyramido-α-ethyl-2,4,6-triiodohy-
drocinnamic Acid Sodium Salt *see*
9968
5'-Butyramido-2'-(2-hydroxy-3-iso-
propylaminopropoxy)acetophenone
see 16
5-Butyramido-1,3,4-thiadiazole-2-
sulfonamide *see* 1545
Butyrchloral *see* 9762
Butyric Acid, 1629
n-Butyric Acid *see* 1629
1-Butyric Acid-6-(L-2-aminobutyric
Acid)-7-glycineoxytocin *see* 1887
Butyric Acid Butyl Ester *see* 1591
Butyric Acid Ethyl Ester *see* 3822
Butyric Acid Lead Salt *see* 5419
1-Butyric Acid-2-[3-(*p*-methoxyphen-
yl)-L-alanine]oxytocin *see* 1841
Butyric Acid Nitrile *see* 1633
Butyric Anhydride, 1630
γ-Butyrobetaine Chloride Methyl
Ester *see* 1913
Butyroin, 1631
Butyrolactone, 1632
γ-Butyrolactone *see* 1632
Butyrone *see* 3408
Butyronitrile, 1633
N-Butyroyl-*p*-aminophenol *see* 4859
3-(3-Butyrylamino-2,4,6-triiodophen-
yl)-2-ethylacrylic Acid Sodium Salt
see 1511
Butyryl Chloride *see* 1634
n-Butyryl Chloride, 1634
Butyrylcholinesterase *see* 2264

3'-[(5-Butyryl-2,4-dihydroxy-3,3-dimethyl-6-oxo-1,4-cyclohexadien-1-yl)methyl]-2',6'-dihydroxy-4'-methoxybutyrophenone *see* 2962
3'-[(5-Butyryl-2,4-dihydroxy-3,3-dimethyl-6-oxo-1,4-cyclohexadien-1-yl)methyl]-2',4'-dihydroxy-6'-methoxy-5'-methylbutyrophenone *see* 881
3'-[(5-Butyryl-2,4-dihydroxy-3,3-dimethyl-6-oxo-1,4-cyclohexadien-1-yl)methyl]-5'-methylphlorobutyrophenone *see* 4130
4-Butyryl-3,5-dihydroxy-1-methoxy-2-methylbenzene *see* 882
1-Butyrylglycerol *see* 6332
3-Butyryl-*sn*-glycerol *see* 6332
4-Butyryl-2-methylphloroglucinol 1-Methyl Ether *see* 882
2-Butyryl-10-[3-(4-methyl-1-piperazinyl)propyl]phenothiazine *see* 1543
Butyryl Oxide *see* 1630
Butyrylperazine *see* 1543
Butysal *see* 9324
Butysedal *see* 9324
Butyvinal *see* 10131
Bux [Chevron] *see* 1494
Buzepide, 1635
Buzon [Knoll] *see* 7431
BV-26-723 *see* 1473
BV-araU *see* 8875
BVAU *see* 8875
B.V.U. *see* 1418
BW-33A *see* 900
BW-33-T-57 *see* 6057
BW-57-322 *see* 935
BW-61-32 *see* 8971
BW-122U *see* 2371
BW-248U *see* 148
BW-256 *see* 10039
BW-256U87 *see* 10039
BW-270C *see* 129
BW-301U *see* 7654
BW-356-C-61 *see* 5310
BW-430C *see* 5367
BW-467C60 *see* 1233
BW-566C *see* 898
BW-566C-80 *see* 898
BW-720C *see* 1516
BW-759 *see* 4374
BW-759U *see* 4374
BW-825C *see* 129
BW-993C *see* 7182
BW-56158 *see* 287
BW-A509U *see* 10252
BW-A825C *see* 129
BW-A938U *see* 3487
BW-B759U *see* 4374
BW-B1090U *see* 6305
BX-341 *see* 1259
BY-1023 *see* 7146
Bydolax *see* 7108
BYK 1512 *see* 4745
Bykahepar [Byk-Gulden] *see* 2398
Bykomycin [Byk-Gulden] *see* 6542
Bylotensin [Yoshitomi] *see* 6669
BZ *see* 8266
BZ-55 *see* 1881
BZQ *see* 1154
C *see* 2850
C-3 *see* 1799
C-5 *see* 7253
C-6 *see* 4724
C-10 *see* 2904
C-63 *see* 3736
C-73 *see* 145
C-78 *see* 9942
C-83 *see* 8419
C-100 *see* 1053
C-238 *see* 7922

C-247 *see* 3356
C-254 *see* 922
C-283 *see* 6661
C-637 *see* 5480
C-709 *see* 3137
C-1065 *see* 3382
C-1414 *see* 6334
C-1656 *see* 2443
C-4675 *see* 7260
C-5473 *see* 7109
C-5581H *see* 7469
C-5720 *see* 1912
C-5968 *see* 4800
C-7337 *see* 7417
C-7819B *see* 6506
C-9333-Go *see* 613
C-9491 *see* 5053
C-19393-H$_2$ *see* 1909
C-19393-S$_2$ *see* 1909
C-48401-Ba *see* 4628
C9-related Protein *see* 7301
C.A. *see* 1433
Ca-AOTA *see* 14
Cabagin-U [Kowa] *see* 10165
(—)-Cabastine *see* 5489
Cabenegrin A-I *see* 1636
Cabenegrin A-II *see* 1636
Cabenegrins, 1636
Cabergoline, 1637
Cabermox [Caber] *see* 617
Cabeza De Negro *see* 10225
Cabimicina *see* 4616
Cabral [Kali-Chemie] *see* 7474
Cacao Butter *see* 9417
Cacao Shell, 1638
Cachectin *see* 9943
Cachou *see* 1951
Cacit [Norwich] *see* 1697
Caclate [Hokuriku] *see* 7016
Cacodyl, 1639
Cacodylic Acid, 1640
Cacotheline, 1641
Cacoxenite *see* 4074
CACP *see* 2378
Cactinomycin, 1642
Cactoid *see* 1643
Cactus Grandiflorus, 1643
Cadalene *see* 1648
Cadalin *see* 1644
Cadaverine, 1645
Caddy [Cleary] *see* 1653
Cadexomer Iodine, 1646
Cadherins, 1647
α-Cadinene *see* 1648
β-Cadinene *see* 1648
Cadinenes, 1648
Cadminate [Mallinckrodt] *see* 1664
Cadmium, 1649
Cadmium Acetate, 1650
Cadmium Bromide, 1651
Cadmium Carbonate, 1652
Cadmium Chloride, 1653
Cadmium Cyanide, 1654
Cadmium Dimethyl *see* 3289
Cadmium Fluoride, 1655
Cadmium Hydrate *see* 1656
Cadmium Hydroxide, 1656
Cadmium Iodide, 1657
Cadmium Nitrate, 1658
Cadmium Oxide, 1659
Cadmium Potassium Cyanide, 1660
Cadmium Potassium Iodide *see* 7861
Cadmium Salicylate, 1661
Cadmium Selenate, 1662
Cadmium Selenide, 1663
Cadmium Succinate, 1664
Cadmium Sulfate, 1665
Cadmium Sulfide, 1666
Cadmium Telluride, 1667
Cadmium Tungstate(VI), 1668

Cadmium Yellow *see* 1666
Cadoxen *see* 9882
Cadral [Ciba-Geigy] *see* 1669
Cadralazine, 1669
Cadraten [ISF] *see* 1669
Cadrilan [Geigy] *see* 1669
Caducid [Wassermann] *see* 4197
Cadusafos, 1670
Caerulein *see* 2048
Caeruloplasmin *see* 2049
Caesium *see* 2051
CAF *see* 4119
Cafaminol, 1671
Cafesterol *see* 1672
Cafestol, 1672
Caffearine *see* 9822
Caffeic Acid, 1673
Caffeic Acid 1-Carboxy-4,5-dihydroxy-1,3-cyclohexylene Ester *see* 2835
Caffeic Acid 3-Methyl Ether *see* 4110
Caffeine, 1674
"Caffeine Citrate" *see* 1674
3-Caffeoylquinic Acid *see* 2194
N-8-Caffeyl-*N'*-methyl-*N'*-(α-methylphenethyl)ethylenediamine *see* 4007
Cafide [Labaz] *see* 1563
Cafilon [Ravensberg] *see* 4005
Cafrolicycline *see* 8411
Cafron *see* 1055
Caid [Lipha] *see* 2204
Cajeputene *see* 5518
Cajeputi Oil *see* 6877
Cajeputol *see* 3940
Cajuput Oil *see* 6877
Cake Alum *see* 381
Calabar Bean *see* 7539
Calabash Curare *see* 2741
Calabron [Spofa] *see* 1694
Calamine, 1675
Calamus, 1676
Calan [Searle] *see* 10083
Cal-Aspirin [Winthrop] *see* 1684
Calavarite *see* 4526
Calcamine [Wander] *see* 3223
Calciben [Firma] *see* 1680
Calcibind [U. Tex./Mission] *see* 8740
Calcibromin *see* 1694
Calcibronat [Sandoz] *see* 1694
Calcicard [Riker] *see* 3247
Calcichew [Shire] *see* 1697
Calcic Liver of Sulfur *see* 1754
Calcidia [Lefrancq] *see* 1697
Calcifediol, 1677
Calciferol *see* 10156
Calciforte [Serolam] *see* 4463
Calcigenol Simple *see* 1741
Calcijex [Abbott] *see* 1681
Calcimar [Rhône-Poulenc Rorer] *see* 1680
Calcimux [Gador] *see* 3908
Calcimycin, 1678
Calcinaphthol *see* 859
Calcined Baryta *see* 1015
Calcined Magnesia *see* 5713
Calcinil [Sclavo] *see* 3578
Calciobrom *see* 1694
Calciofon *see* 1712
Calciopen [Kabi] *see* 7230
Calciparine [Sanofi Winthrop] *see* 4685
Calcipen [Farmabion] *see* 7230
Calcipotriene, 1679
Calcipotriol, 1679
Calciretard [Köhler] *see* 875
Calcisorb [Kettelhack] *see* 8740
Calcistin [Galenus] *see* 4757
Calcitar(e) [Rhône-Poulenc Rorer] *see* 1680
Calcite *see* 1697

Calcitetracemate Disodium [Laroche-Navarron] *see* 3555
Calcitonin, 1680
Calcitriol, 1681
Calcium, 1682
Calcium Acetate, 1683
Calcium Acetyl Homotaurinate *see* 14
Calcium Acetylsalicylate, 1684
Calcium Acetylsalicylate Carbamide *see* 1684
Calcium Aluminosilicate, 1685
Calcium Aluminum Hydride *see* 344
Calcium Arsenate, 1686
Calcium Arsenite, 1687
Calcium Ascorbate, 1688
Calcium Aspartate *see* 875
Calcium Aspirin *see* 1684
Calcium Aurothioglycolate *see* 6411
Calcium 3-Aurothio-2-propanol-1-sulfonate, 1689
Calcium Benzoate *see* 1122
Calcium Benzylpenicillinate *see* 7223
Calcium Bichromate *see* 1706
Calcium Biphosphate *see* 1740
Calcium Bisacetyl Homotaurine *see* 14
Calcium Bisulfite, Solution, 1690
Calcium Borate, 1691
Calcium Borogluconate, 1692
Calcium Bromide, 1693
Calcium Bromolactobionate, 1694
Calcium *N*-Carbamoylaspartate, 1695
Calcium Carbide, 1696
Calcium Carbimide *see* 1702
Calcium Carbolate *see* 1738
Calcium Carbonate, 1697
Calcium Chel 330 [Geigy] *see* 7265
Calcium Chlorate, 1698
Calcium Chloride, 1699
Calcium Chromate(VI), 1700
Calcium Chrome Yellow *see* 1700
Calcium Citrate, 1701
Calcium Creosotate *see* 2642
Calcium Cyanamide, 1702
Calcium Cyanamide Citrated, 1703
Calcium Cyanide, 1704
Calcium Cyclamate, 1705
Calcium Cyclohexanesulfamate *see* 1705
Calcium Cyclohexylsulfamate *see* 1705
Calcium-dependent Regulator Protein *see* 1767
Calcium Diborogluconate *see* 1692
Calcium Dichromate(VI), 1706
Calcium Dioctyl Sulfosuccinate *see* 3459
Calcium Dioxide *see* 1736
Calcium Diphosphate *see* 1746
Calcium Disodium Edetate *see* 3555
Calcium Disodium Ethylenediamine-tetraacetate *see* 3555
Calcium Disodium (Ethylenedinitrilo)-tetraacetate *see* 3555
Calcium Disodium Versenate [Riker] *see* 3555
Calcium Diuretin *see* 9418
Calcium Dobesilate *see* 3455
Calcium 2-Ethylbutanoate, 1707
Calcium Ethyl Sulfate *see* 3897
Calcium Ferrous Citrate, 1708
Calcium Fluoride, 1709
Calcium Fluorophosphate, 1710
Calcium Fluosilicate *see* 1714
Calcium Folinate *see* 4254
Calcium (6*S*)-Folinate *see* 4254
Calcium Formate, 1711
Calcium Galactogluconate Bromide *see* 1694
Calcium Glucoheptonate *see* 4463

Calcium Glucomonocarbonate *see* 4463
Calcium Gluconate, 1712
Calcium Glucosemonocarbonate *see* 4463
Calcium Glycerinophosphate *see* 1713
Calcium Glycerophosphate, 1713
Calcium Hexafluorosilicate, 1714
Calcium Homopantothenate *see* 4781
Calcium Hydrate *see* 1716
Calcium Hydride, 1715
Calcium Hydroxide, 1716
Calcium 2-Hydroxy-1-naphthalene-sulfonate *see* 859
Calcium Hypochlorite, 1717
Calcium Hypophosphite, 1718
Calcium Hyposulfite *see* 1759
Calcium Iodate, 1719
Calcium Iodide, 1720
Calcium Iodobehenate, 1721
Calcium Iodostearate, 1722
Calcium Ipodate *see* 5087
Calcium Ketomalonate *see* 1726
Calcium Lactate, 1723
Calcium Lactobionate *see* 5354
Calcium Levofolinate *see* 4254
Calcium Levulinate, 1724
Calcium Magnesium Acetate, 1725
Calcium Mandelate *see* 5757
Calcium [Mercaptoacetato(2—)-*O*,*S*]-aurate(1—) (1:2) *see* 6411
Calcium Mesoxalate, 1726
Calcium Methionate, 1727
Calcium Methyl Sulfate *see* 6203
Calcium Molybdate(VI), 1728
Calcium Monofluorophosphate *see* 1710
Calcium Monohydrogen Phosphate *see* 1739
Calcium β-Naphthol-α-monosulfonate *see* 859
Calcium 2-Naphthol-1-sulfonate *see* 859
Calcium Nitrate, 1729
Calcium Nitrite, 1730
Calcium Oleate, 1731
Calcium Orthophosphate Basic *see* 3519
Calcium Oxalate, 1732
Calcium Oxide, 1733
Calcium Oxomalonate *see* 1726
"Calcium Oxychloride" *see* 2143
Calcium Oxysulfide Solution *see* 5516
Calcium Palmitate, 1734
Calcium Pectolith *see* 1749
Calcium Penicillin G *see* 7223
Calcium Permanganate, 1735
Calcium Peroxide, 1736
Calcium Phenate *see* 1738
Calcium Phenolate, 1738
Calcium Phenolsulfonate, 1737
Calcium Phenoxide, 1738
Calcium Phenylate *see* 1738
Calcium Phosphate, Dibasic, 1739
Calcium Phosphate Hydroxide *see* 3519
Calcium Phosphate, Monobasic, 1740
Calcium Phosphate, Tetrabasic *see* 9511
Calcium Phosphate, Tribasic, 1741
Calcium Phosphide, 1742
Calcium Phosphite, 1743
Calcium Phosphoglycerate *see* 1713
Calcium Polycarbophil, 1744
Calcium Propionate, 1745
Calcium Pyroborate *see* 1691
Calcium Pyrophosphate, 1746
Calcium Rhodanate *see* 1757
Calcium D-Saccharate, 1747
Calcium Selenide, 1748

Calcium Silicate, 1749
Calcium Silicofluoride *see* 1714
Calcium Stearate, 1750
Calcium Stearyl-2 Lactylate, 1751
Calcium Stelate *see* 1751
Calcium Succinate, 1752
Calcium Sulfate, 1753
Calcium Sulfide, 1754
Calcium Sulfite, 1755
Calcium Sulfocarbolate *see* 1737
Calcium Sulfocyanate *see* 1757
Calcium Sulfomethylate *see* 6203
Calcium Sulfophenolate *see* 1737
Calcium Sulfovinate *see* 3897
"Calcium Superphosphate" *see* 1740
Calcium Tartrate, 1756
Calcium Tetraborate *see* 1691
Calcium Tetrahydroaluminate *see* 344
Calcium Thiocyanate, 1757
Calcium Thioglycollate, 1758
Calcium Thiosulfate, 1759
Calcium Trisodium Pentetate *see* 7265
Calcium Tungstate(VI), 1760
Calcium Ureidosuccinate *see* 1695
Calcort [Lepetit] *see* 2916
Calcreose [Pennwalt] *see* 2642
Caldariomycin, 1761
Calderol [Upjohn] *see* 1677
Caldine [Boehringer, Ing.] *see* 5344
Caldon [Hoechst] *see* 3341
C-Calebassine, 1762
Calendula, 1763
Calepsin [Orion] *see* 1826
Calglucon *see* 1712
Calgon *see* 8814
Calheptose *see* 4463
Calibène [Delalande] *see* 9185
Calico Yellow *see* 6352
Californit [Merckle] *see* 7106
Californium, 1764
Caliment [Apotex] *see* 7661
Calioben *see* 1721
Calisaya Bark *see* 2343
Calixarenes, 1765
Calixin [BASF] *see* 9793
Callicrein *see* 5291
Callidin I *see* 1386
Callistephin *see* 7199
Callusolve [Dermal] *see* 1086
Calmabel *see* 386
Calmagite, 1766
Calmalone [Cassenne] *see* 2772
Calmatel (Aerosol) [Almirall-Omega] *see* 7576
Calmatel (Cream) [Almirall-Omega] *see* 7576
Calmaxid [Lilly] *see* 6758
Calmday [Will-Pharma] *see* 6784
Calmeran *see* 925
Calmipan *see* 4582
Calmiren *see* 5908
Calmivet [Stricker] *see* 32
Calmodid *see* 4826
Calmodulin, 1767
Calmonal [Heyden] *see* 5817
Calmpose [Ranbaxy] *see* 3042
Calmus *see* 1676
Calnathal *see* 1840
Calnegyt [Toyo Jozo] *see* 4588
Calogreen [Mallinckrodt] *see* 5951
Calomel *see* 5951
Calomelol, 1768
Calomide [Yamanouchi] *see* 2513
Calorose *see* 5026
Calotropin, 1769
Calpanate [Consumers Vitamin] *see* 7147
Calphosan [Carlton] *see* 1713
Calpol [Calmic] *see* 45
Calscorbat [Biosedra] *see* 867

Calsekin [Mect] see 4260
Calslot [Takeda] see 5786
Calsmin [Upjohn] see 6667
Calsol see 3557
Calsyn [Rhône-Poulenc Rorer] see 1680
Calsynar [Rhône-Poulenc Rorer] see 1680
Calthor [Ayerst] see 2769
Calumba, 1770
Calurin [Dorsey] see 1684
Calusterone, 1771
Calvisken [Sandoz-Sankyo] see 7597
Calx see 1733
Calycanthine, 1772
Calystigine see 7126
CaM see 1767
CAM-AQ1 see 609
Camazepam, 1773
Cambendazole, 1774
Cambenzole [Merck & Co.] see 1774
Cambet [Merck & Co.] see 1774
Cambilene [Fisons] see 3090
Cambogia see 4372
Camcolit [Norgine] see 5552
Camdelate see 5757
Camganiba see 5755
Camilol [Maybrook] see 1281
Camite see 1433
Camleed [Tanabe Seiyaku] see 3629
Camoform [Parke, Davis] see 1242
Camolar [Parke, Davis] see 2790
Camomile see 2082
Camont [Klinge] see 6617
Camoquin [Parke, Davis] see 609
Camostat, 1775
Camostat Mesylate see 1775
cAMP see 2777
Campel [Farmitalia] see 2294
Campesterol, 1776
endo-2-Camphanol see 1366
exo-2-Camphanol see 5143
2-Camphanone see 1779
Camphechlor see 9693
Camphene, 1777
Camphetamide see 1782
Camphidonium see 9838
d-Camphocarboxylic Acid, 1778
Camphol see 1366
Camphophyline [Millot] see 3754
Camphor, 1779
d-3-Camphorcarboxylic Acid see 1778
Camphoric Acid, 1780
Camphoric Acid 1-(p,α-Dimethyl-benzyl) Ester, Compd with 2,2'-Imidodiethanol (1:1) see 9630
d-Camphoric Acid Ethyl Ester Bismuth Salt see 1304
Camphor Monobromated see 1435
Camphor Monochlorated see 2183
Camphor Solubilized see 1778
10-Camphorsulfonic Acid see 1781
d-Camphorsulfonic Acid, 1781
β-Camphorsulfonic Acid see 1781
Camphostene see 7110
Camphostyl see 1781
Camphosulfonyl-N-methylpyridine-β-diethylcarboxamide see 1782
Camphotamide, 1782
Camphramine see 1782
Camphydryl see 1778
Campolon [Winthrop; Bayer] see 5575
Camporit see 1717
Camposan see 3777
Campovit see 5575
Campoviton 6 see 8166
Campto [Yakult] see 5104
Camptothecin, 1783
Camsellite see 5693
Camsylate see 1781

Camylofine, 1784
Camyna [Nycomed] see 9600
Canada Snakeroot see 862
Canada Turpentine see 973
Canadian Hemp see 784
Canadine, 1785
Canadol see 5513
Cananga Oil see 10233
Canary Dextrin see 2995
Canavanine, 1786
Canchalagua see 2016
Candelilla Wax, 1787
Canderel [Searle] see 874
Candesartan, 1788
Candesartan Cilexetil see 1788
Candex [Miles] see 6834
Candicidin, 1789
Candicidin D see 1789
Candidin, 1790
Candidinin see 1790
Candidoin see 1790
Candimon [Ayerst] see 1789
Candio-Hermal [Hermal] see 6834
Candleberry Bark see 1040
Canella, 1791
Canescine see 2964
Canesten [Bayer] see 2478
Cane Sugar see 9051
Canex [Pitman-Moore] see 8427
Canferon [Takeda] see 5016
Canfodion [Gentili] see 8274
Canfoxll [Gentili] see 1778
Canifug [Wolff] see 2478
Cannabidiol, 1792
Cannabinol, 1793
Cannabis, 1794
Cannabiscetin see 6414
Cannoc [Takeda] see 3744
Cannogenin α-l.-Thevetoside see 7324
Canocenta [Byk Gulden] see 5615
Canogard [Shell] see 3129
Canopar [Cooper] see 9414
Canrenone, 1795
Cantabilin [Formenti] see 4903
Cantabiline see 4903
Cantan [Hoechst] see 867
Cantaxin [Winthrop] see 867
Cantharides, 1796
Cantharides Camphor see 1797
Cantharidin, 1797
Cantharis vesicatoria see 1796
Canthaxanthin, 1798
Cantil [Lakeside] see 5893
Canton's Phosphorus see 1754
Cantor [Sanofi Winthrop] see 6287
Cantralax [Ferrosan] see 1929
Cantrex see 5293
Cantricin [Corvi] see 4312
Cantril see 5893
Cantrodifene (obsolete) see 6731
Canutillo see 3644
Caocobre [Sandoz] see 2734
Caomet [Simes] see 9974
Caoutchouc see 8435
Caparlem see 5285
Caparol [Geigy] see 7973
Caparside see 831
Caparsolate [Abbott] see 831
Capastat [Lilly] see 1801
Capben [Schwarz] see 1799
Capen [Phoenix] see 9597
Caperase see 1948
Capisten [Kissei] see 5316
Capitol [Dermal] see 1086
Capitrol [Westwood] see 2227
Capitus [Berk] see 3794
Capla [Wallace Labs.] see 5813
Caplenal [Berk] see 287
Capmul 8210 [Stokely-Van Camp] see 6335

Capmul O [Capital City Prod.] see 8872
Capnoidine see 163
Capobenic Acid, 1799
Capoten [Squibb] see 1817
Capozide [Squibb] see 1817
Capquin [Ciba-Geigy] see 2215
Capramol [Choay] see 451
Caprenin, 1800
Capreomycin, 1801
Capreomycin IA see 1801
Capreomycin IB see 1801
Capreomycin IIA see 1801
Capreomycin IIB see 1801
n-Capric Acid, 1802
Caprin [Sinclair] see 886
Caprine see 6800
Caproaldehyde see 1804
Caprocid see 451
Caprocin [Lilly] see 1801
Caprodat see 1889
n-Caproic Acid, 1803
n-Caproic Acid n-Amyl Ester see 646
Caproic Aldehyde, 1804
Caprokol [Merck & Co.] see 4750
Caprolactam, 1805
ε-Caprolactam see 1805
Caprolan [AlliedSignal] see 6832
Caprosem [A. B. Leo] see 2482
Caproyl Chloride, 1806
Caprylaldehyde see 1809
Caprylene, 1807
Caprylic Acid, 1808
Caprylic Acid Ammonium Salt see 532
Caprylic Acid α-Monoglyceride see 6335
Caprylic Alcohol see 6849
Caprylic Aldehyde, 1809
Caprylohydroxamic Acid see 6848
CAPS, 1810
Capsaicin, 1811
Capsanthin, 1812
Capsebon [Pitman-Moore] see 1666
Capsicum, 1813
Captafol, 1814
Captagon [Asta] see 4014
Captan, 1815
Captax see 5916
Captea [Théraplix] see 1817
Captin [Krewel] see 45
Captodiam see 1816
Captodiamine, 1816
Captodramin see 1816
Captolane [Théraplix] see 1817
Captopril, 1817
Captoril [Sankyo] see 1817
Capture [FMC] see 1257
Capuride, 1818
Capval [Dreluso] see 6815
Carace [Morson] see 5540
Carachol see 2922
Caracurine VII see 10181
Caradrin [Asta] see 8060
Carafate [Marion Merrell Dow] see 9049
Caramel, 1819
Caramiphen, 1820
Caraway, 1821
Carazolol, 1822
Carbacefaclor see 5606
Carbachol, 1823
Carbacrylic Resins, 1824
Carbacylamine Resins see 1824
Carbadipimidine see 1911
Carbadox, 1825
Carbamaldehyde see 4264
Carbamamidine see 4591
3-Carbamate see 8400
Carbamate see 4054
Carbamate of 2-(Hydroxymethyl)-2-

Caricide [Am. Cyanamid] *see* 3165
Caridan *see* 7107
Caridolol [Sankyo Zoki] *see* 8025
Carinamide *see* 1184
Carindacillin, 1888
Carindapen [Pfizer] *see* 1888
Carisano [Roedler] *see* 4380
Carisoma [Wallace Labs.] *see* 1889
Carisoprodate *see* 1889
Carisoprodol, 1889
Carlsbad Salt Artificial, 1890
Carmazon [Nikken] *see* 8060
Carmethose [Ciba-Geigy] *see* 1877
Carmine *see* 1891
Carminic Acid, 1891
Carminomycin *see* 1921
Carminomycin I *see* 1921
Carmofur, 1892
Carmoxirole, 1893
Carmubris [Bristol] *see* 1894
Carmurit *see* 3796
Carmustine, 1894
Carnacid-Cor [Tad] *see* 4441
Carnallite *see* 7763
Carnauba Wax, 1895
Carnegine, 1896
Carnicor [Sigma-Tau] *see* 1898
Carnidazole, 1897
Carnigen [Albert Roussel] *see* 4869
Carnitene [Sigma-Tau] *see* 1898
Carnitine, 1898
Carnitine Acetyl Ester *see* 85
Carnitor [Sigma-Tau] *see* 1898
Carnosine, 1899
Carnotite *see* 10053
Carnum [Firma] *see* 1898
Carob Flour *see* 5584
Carofur [Boehringer, Mann.] *see* 6629
Carolina Pink *see* 8900
Caronamide *see* 1184
Carony Bark *see* 690
Carophyll Red [Roche] *see* 1798
Caro's Acid, 1900
Carotaben [Hermal] *see* 1902
Carotaben Plus *see* 1798
α-Carotene, 1901
β-Carotene, 1902
β,β-Carotene *see* 1902
γ-Carotene, 1903
δ-Carotene, 1904
ε,ψ-Carotene *see* 1904
ψ,ψ-Carotene *see* 5650
all trans-β-Carotene-3,3′-diol *see* 10248
β,β-Carotene-3,3′-diol *see* 10248
β,ε-Carotene-3,3′-diol *see* 10197
ψ,ψ-Carotene-16,16′-diol *see* 5651
β,β-Carotene-4,4′-dione *see* 1798
ψ,ψ-Carotene-16-ol *see* 5657
β,β-Carotene-3,3′,4,4′-tetrone *see* 888
(3R)-β,β-Caroten-3-ol *see* 2676
(3R)-β,ψ-Caroten-3-ol *see* 8445
β-Caroten-3-ol *see* 2676
β,β-Caroten-4-one *see* 3543
Carotol, 1905
Caroverine, 1906
Caroxazone, 1907
Carpaine, 1908
Carpene [Agrimont] *see* 3468
Carpetimycin A *see* 1909
Carpetimycin B *see* 1909
Carpetimycins, 1909
Carphenazine, 1910
Carphenol [Bayer] *see* 3365
Carpidine *see* 5216
Carpiline *see* 5216
Carpipramine, 1911
Carprofen, 1912
Carpronium Chloride, 1913
Carrageen *see* 1914

Carrageenan, 1914
Carrageenin *see* 1914
Carrbutabarb [Century] *see* 1530
Carrel-Dakin Soln *see* 8774
Carrier [Chiesi] *see* 1898
Carrisyn [Carrington] *see* 26
Carsalam, 1915
Cartap, 1916
Carteol [Chauvin] *see* 1917
Carteolol, 1917
Carthamic Acid *see* 1918
Carthamin, 1918
Carthamus, 1919
Carticaine, 1920
Cartonic [Yamanouchi] *see* 634
Cartric [Sanwa] *see* 8311
Cartrol [Otsuka] *see* 1917
Cartrophen [Univet] *see* 7276
Carubicin, 1921
Carubinose *see* 5791
Carudol [Fher] *see* 7431
Carumonam, 1922
Carvacrol, 1923
Carvacron [Taiyo] *see* 9754
Carvanil *see* 5245
Carvasin [Wyeth] *see* 5245
Carvedilol, 1924
Carvol *see* 1925
6,8-Carvomenthenediol *see* 8707
Carvone, 1925
Carwin [ICI] *see* 10189
Carylderm [Napp] *see* 1831
Caryolysine [Delagrange] *see* 5815
Caryophyllene, 1926
cis-Caryophyllene *see* 1926
trans-Caryophyllene *see* 1926
α-Caryophyllene *see* 4789
β-Caryophyllene *see* 1926
γ-Caryophyllene *see* 1926
Caryophyllic Acid *see* 3944
Caryophyllin *see* 6964
Caryophyllus *see* 2479
Carzenide, 1927
Carzinophilin, 1928
Carzinophilin A *see* 1928
Casakol [Upjohn] *see* 1929
Casamino Acids *see* 1934
Casanthranol, 1929
Casanthranol A *see* 1929
Casanthranol B *see* 1929
Casantin [Cassella-Riedel] *see* 3157
Casca Bark *see* 8529
Cascapride [Cascan] *see* 1454
Cascara Amara, 1930
Cascara Sagrada, 1931
Cascarilla, 1932
Cascarillin, 1933
Cascarin *see* 4288
Cascarosides *see* 1931
Caseanine *see* 9354
Casein, 1934
Cas-Evac *see* 1931
Cashoo *see* 1951
Casil *see* 349
Casimiroedine, 1935
Casimiroin, 1936
Casmalon *see* 2772
Casodex [ICI] *see* 1247
Casoron [Solvay Duphar] *see* 3093
Cassaidine, 1937
Cassaine, 1938
Cassamine, 1939
Cassella 4489 *see* 2296
Cassella's Acid, 1940
Cassella's Acid F, 1941
Cassel's Green *see* 1010
Cassia Fistula, 1942
Cassia Pods *see* 1942
Cassia Pulp *see* 1942
Cassic Acid *see* 8342

Cassiopeium *see* 5642
Cassiterite *see* 8933
Castanea, 1943
Castanospermine, 1944
Castle's Intrinsic Factor, 1945
Castor Oil, 1946
Castor Oil, Hydrogenated, 1947
Castorwax *see* 1947
Castrix *see* 2655
Cat *see* 6033
Cataclot [Ono] *see* 7115
Cataflam [Geigy] *see* 3132
Catalase, 1948
Catalin [Senju] *see* 7645
Catalpin *see* 1949
Catalposide, 1949
Catanil [DeAngeli] *see* 2239
Catapres [Boehringer, Ing.] *see* 2450
Catapresan [Boehringer, Ing.] *see* 2450
Catapyrin *see* 471
Catarase [Cooper] *see* 2320
Cataria *see* 1955
Catatrol [ICI] *see* 10116
Catavin C *see* 867
Catechin, 1950
Catechinic Acid *see* 1950
Catechol *see* 1950 and 8183
Catecholase *see* 9969
Catechu *see* 4371
Catechu Black, 1951
Catechuic Acid *see* 1950
Catenins *see* 1647
Catenulin *see* 7173
Catergen [Zyma] *see* 1950
Cateudyl *see* 6028
Catharanthine, 1952
Cathepsin C *see* 1953
Cathepsin D *see* 1953
Cathepsin G *see* 1953
Cathepsins, 1953
Cathine *see* 6811
Cathinone, 1954
Cathocin [Merck & Co.] *see* 6818
Cathomycin [Merck & Co.] *see* 6818
Catlep [Sumitomo] *see* 4998
Catmint *see* 1955
Catnep, 1955
Catnip *see* 1955
Catonin [Magis] *see* 1680
Catral *see* 7382
Catron [Lakeside] *see* 7382
Catroniazid *see* 7382
Cat's Hair *see* 3949
Cattlyst [Syntex] *see* 5361
Caudaline [Exa] *see* 9569
Caulophylline, 1956
Caulophyllum, 1957
Caulosapogenin *see* 4654
Caustic Alcohol *see* 8758
Caustic Barley *see* 8460
Caustic Baryta *see* 1006
Caustic Potash *see* 7806
Caustic Soda *see* 8772
Causyth [Inverni] *see* 8056
Cav-Ecol *see* 10178
Caverject [Upjohn] *see* 8063
Cavinton [Gedeon Richter] *see* 10128
Cavitands *see* 2401
Cavitol *see* 2401
Cavodil *see* 7382
Cavonyl *see* 2780
Cayenne Pepper *see* 1813
Cay Note *see* 3644
Caytine [Lakeside] *see* 8082
2C-B, 1958
CB-11 *see* 7348
CB-154 *see* 1437
CB-154 Mesylate *see* 1437
CB-302 *see* 4295
CB-304 *see* 937
CB-311 *see* 8864

CB-313 *see* 6302
CB-337 *see* 5852
CB-804 *see* 1486
CB-1048 *see* 2157
CB-1314 *see* 6225
CB-1348 *see* 2116
CB-1678 *see* 8007
CB-2041 *see* 1529
CB-3025 *see* 5871
CB-3026 *see* 5871
CB-3697 *see* 8273
CB-4261 *see* 9379
CB-4306 *see* 2465
CB-4311 *see* 2465
CB-8061 *see* 7276
CB-30038 *see* 6287
CBA-93626 *see* 2453
CBDCA *see* 1870
CBS *see* 2549
CC *see* 10157
CC-2481 *see* 2781
CCA *see* 5581
CCC *see* 1703
CCC [BASF] *see* 2153
β-CCE *see* 3826
CCI-15641 *see* 2002
CCI-18781 *see* 4244
CCK-179 *see* 3692
CCK-PZ *see* 2253
CCK *C*-Terminal Octapeptide *see* 8689
CCNU [Lundbeck] *see* 5594
CCP *see* 2427
CCRG-81045 *see* 9289
CCT *see* 2824
CD-68 *see* 2129
CD-3400 *see* 8310
CDA *see* 2071
2-CdA *see* 2397
CDAA *see* 259
CDC *see* 2096
CDDD-5604 *see* 5614
CDEC *see* 9079
CDM *see* 2133
CDO-MeB *see* 9260
CDP-choline *see* 2380
CDR *see* 1767
CE-3624 *see* 9570
Ceanothic Acid, 1959
Cebera [Bouchara] *see* 243
Cebesine [Chauvin] *see* 1076
Cebicure [Merck & Co.] *see* 867
Cebion [E. Merck] *see* 867
Cebitate [Merck & Co.] *see* 8723
Cebrogen [Walker] *see* 4479
Cebroton [Sancarlo] *see* 2380
Cebrum [SIFA] *see* 2132
Cebutid [Boots] *see* 4234
Cecenu [Medac] *see* 5594
Ceclor [Lilly] *see* 1962
Cecon [Abbott] *see* 867
Cecropins, 1960
CED-3 *see* 788
Cedad [Recordati] *see* 1055
Cedar Camphor *see* 1961
Cedar Leaf Oil *see* 6884
Cedax [Schering-Plough] *see* 1998
Cedilanid [Sandoz] *see* 5368
Cedilanid-D [Sandoz] *see* 2967
Cedocard [Tillotts] *see* 5245
Cedol [Tiber] *see* 1964
8βH-Cedran-8-ol *see* 1961
Cedrol, 1961
Cedro Oil *see* 6915
Cedulamin *see* 6040
Cedur [Boehringer, Mann.] *see* 1240
CeeNU [Bristol] *see* 5594
Ceepryn [Merrell] *see* 2074
Cefacidal [Allard] *see* 1967
Cefaclor, 1962
Cefadol [Nippon Shinyaku] *see* 3369

Cefa-Drops [Fort Dodge] *see* 1963
Cefadros [Proter] *see* 2021
Cefadroxil, 1963
Cefadyl [Bristol] *see* 2030
Cefa-Iskia [Iskia] *see* 2021
Cefa-Lak [Bristol] *see* 2030
Cefalen [Lenza] *see* 7670
Cefalex [Von Boch] *see* 7670
Cefaloridin *see* 2025
Cefalotin *see* 2028
Cefaloto [Lifepharma] *see* 2021
Cefam [Magis] *see* 1964
Cefamandole, 1964
Cefamedin [Fujisawa] *see* 1967
Cefamezin [Fujisawa] *see* 1967
Cefamox [Bristol] *see* 1963
Cefanex [Bristol-Myers] *see* 2021
Cefaperos [Allard] *see* 1965
Cefaseptin [Chassot] *see* 2021
Cefatrexyl [Bristol] *see* 2030
Cefatriaxone *see* 2001
Cefatrizine, 1965
Cefazedone, 1966
Cefazil [Giustini] *see* 1967
Cefazina [Chemil] *see* 1967
Cefazolin, 1967
Cefazone [Firma] *see* 1981
Cefbon-C₂F [Central Glass] *see* 4561
Cefbon-CF [Central Glass] *see* 4561
Cefbuperazone, 1968
Cefcapene Pivoxil, 1969
Cefclidin, 1970
Cefdinir, 1971
Cefditoren, 1972
Cefditoren Pivoxil *see* 1972
Cefepime, 1973
Cefepime Hydrochloride *see* 1973
Cefetamet, 1974
Cefetamet Pivoxil *see* 1974
Cefibacter [Rubio] *see* 2021
Cefiran [Pierrel] *see* 1964
Cefixime, 1975
Cefixoral [Menarini] *see* 1975
Cefizox [SmithKline] *see* 2000
Ceflorin [Glaxo] *see* 2025
Cefmax [TAP] *see* 1976
Cefmenoxime, 1976
Cefmetazole, 1977
Cefmetazon [Sankyo] *see* 1977
Cefminox, 1978
Cefobid [Pfizer] *see* 1981
Cefobine [Pfizer] *see* 1981
Cefobis [Pfizer] *see* 1981
Cefodie [SKB] *see* 1980
Cefodizime, 1979
Cefodox [Roussel] *see* 1991
Cefogram [Gibipharma] *see* 1981
Cefomonil [TAP] *see* 1994
Cefoneg [Tosi] *see* 1981
Cefonicid, 1980
Cefoperazone, 1981
Cefoprim [Esseti] *see* 2002
Ceforal [Farmoffer] *see* 1963
Ceforanide, 1982
Cefosint [Proter] *see* 1981
Cefotan [Yamanouchi] *see* 1984
Cefotax [Chugai] *see* 1983
Cefotaxime, 1983
Cefotetan, 1984
Cefotiam, 1985
Cefotiam Hexetil Hydrochloride *see* 1985
Cefotrizin [Firma] *see* 1965
Cefoxitin, 1986
Cefozopran, 1987
Cefpimizole, 1988
Cefpiramide, 1989
Cefpiran [Sumitomo] *see* 1989
Cefpirome, 1990

Cefpodoxime *see* 1991
Cefpodoxime Proxetil, 1991
Cefprozil, 1992
Cefracycline [Frosst] *see* 9337
Cefradex [Ausonia] *see* 2032
Cefradin *see* 2032
Cefrag [Magis] *see* 2032
Cefro [Sankyo] *see* 2032
Cefrom [Roussel-UCLAF] *see* 1990
Cefroxadine, 1993
Cefspan [Fujisawa] *see* 1975
Cefsulodin, 1994
Ceftazidime, 1995
Ceftenon [Biochemie] *see* 1984
Cefteram, 1996
Cefteram Pivoxil *see* 1996
Ceftetrame *see* 1996
Ceftezole, 1997
Ceftibuten, 1998
Ceftim [Bonomelli] *see* 1995
Ceftin [Glaxo] *see* 2002
Ceftiofur, 1999
Ceftix [Boehringer, Mann.] *see* 2000
Ceftizon [Bago] *see* 2000
Ceftizoxime, 2000
Ceftriaxone, 2001
Cefumax [Locatelli] *see* 2002
Cefurax [Glaxo] *see* 2002
Cefurex [Sarm] *see* 2002
Cefurin [Magis] *see* 2002
Cefuroxime, 2002
Cefuroxime Axetil *see* 2002
Cefuzonam, 2003
Cefyl [Takeda] *see* 1974
Cefzil [Bristol-Myers Squibb] *see* 1992
Cegiolan *see* 867
Ceglunat [DDR] *see* 5368
Ceglution [Ariston] *see* 5552
Cekufon [Cequisa] *see* 9753
Celadigal [Beiersdorf] *see* 5368
Celance [Lilly] *see* 7304
Celaskon [Spofa] *see* 867
Cela W-524 *see* 9819
Celbenin [Beecham] *see* 6047
Celectol [Rorer] *see* 2007
Celeport [Mitsubishi; Eisai] *see* 1255
Celery Seed, 2004
Celest [Ciba-Geigy] *see* 4165
Celestan [Essex] *see* 1226
Celestan-V [Essex] *see* 1226
Celestene [Schering] *see* 1226
Celesticetin, 2005
Celestin Blue, 2006
Celestine *see* 9013
Celestite *see* 9013
Celestoderm-V [Schering] *see* 1226
Celestone [Schering] *see* 1226
Celestovet [Biokema] *see* 1226
Celevac [WB Pharm] *see* 6120
Celex [Aristochimica] *see* 2032
Celexane [Rhône-Poulenc] *see* 9182
Celin *see* 867
Célio [Ciba] *see* 2463
Celiomycin *see* 10139
Celiprolol, 2007
Celiptium [Sanofi] *see* 3590
Celite *see* 5003
Cellative [Tobishi] *see* 5820
β-Cell Differentiation Factor *see* 5017
Cellidrin [Hennig] *see* 287
Cellobiose, 2008
β-Cellobiose *see* 2008
Cellocidin, 2009
Celloidin *see* 8195
Cellolax *see* 1877
Cellon *see* 9331
Cellophane, 2010
Cellose *see* 2008
Cellosolve *see* 3797
Cellosolve Acetate *see* 3798

Cellothyl [Warner-Chilcott] *see* 6120
Cellucon [Medo] *see* 6120
Celluflex 179 [Celanese] *see* 9892
Celluloid [Celanese], 2011
Cellulose, 2012
Cellulose Acetates, 2013
Cellulose-ECTEOLA *see* 3551
Cellulose Ethyl Ether *see* 3828
Cellulose Ethyl Hydroxyethyl Ether, 2014
Cellulose 2-Hydroxypropyl Ether *see* 4888
Cellulose 2-Hydroxypropyl Methyl Ether *see* 4889
Cellulose Methyl Ether *see* 6120
Cellulose Nitrate *see* 8195
Cellulose Phenoxyacetate *see* 7408
Cellulosic Acid *see* 7073
Cellumeth [Conal] *see* 6120
Celmidol [Tobishi] *see* 3369
Cel-O-Brandt *see* 1877
Celocurine [Pharmacia] *see* 9045
Celontin [Parke, Davis] *see* 6085
Celoslin [Hoechst] *see* 1997
Celospor [Ciba-Geigy] *see* 2019
Celphos [Excel] *see* 372
Celpillina *see* 6047
Celtect [Kyowa] *see* 7058
Celupan [Lacer] *see* 6450
Cemado [Francia] *see* 1964
Cemandil [SIT] *see* 1964
Cemastin [Ist. Vitamine] *see* 2028
Cembrene A *see* 6535
Cemix [Takeda] *see* 1976
Cenalene-M [Central Pharm.] *see* 7283
Cenazol *see* 7283
Cenetone *see* 867
Cenocort [Central Pharm.] *see* 9727
Cenolate [Abbott] *see* 8723
Cenomycin [Daiichi] *see* 1986
Censedal [M & B] *see* 6516
Censpar [Bristol-Myers] *see* 1528
Centaurein, 2015
Centaurin *see* 2486
Centaury, 2016
Centaury, American, 2017
Centchroman, 2018
trans-Centchroman *see* 2018
Centedrin [Gedeon Richter] *see* 6186
Centelase Dermatologico [Laroche Navarron] *see* 869
Centralgin [Amino] *see* 5894
Centralgol [Zyma] *see* 8091
Centralgyl *see* 8091
Centralite I *see* 3166
Centrallasite *see* 1749
Centraureo [Central] *see* 2245
Centrax [Parke, Davis] *see* 7895
Centrine [Bristol] *see* 479
Centrophenoxine *see* 5820
Centroton [Wander] *see* 8194
Centurina [Merck & Co.] *see* 7669
Centyl [Leo Pharm.] *see* 1064
Ceosunin [Kyowa] *see* 2048
CEPA *see* 3777
Cepacilina *see* 7221
Cepacol [Merrell] *see* 2074
Cepaloridin [Glaxo] *see* 2025
Cepalorin *see* 2025
Cepan [IBI] *see* 1984
Cepaverin [Globopharm] *see* 7151
Cepazine [Glaxo] *see* 2002
CEPH *see* 9640
CEPHA [GAF] *see* 3777
Cephacetrile Sodium, 2019
Cephaeline, 2020
Cephaeline Methyl Ether *see* 3600
Cephalexin, 2021
Cephalins, 2022
Cephaloglycin, 2023

Cephalonic Acid, 2024
Cephaloridine, 2025
Cephalosporin C, 2026
Cephalosporin C$_A$ *see* 2026
Cephalosporin C$_C$ *see* 2026
Cephalosporin N *see* 7227
Cephalosporin P$_1$, 2027
Cephalothin, 2028
Cephamycins, 2029
Cephapirin Sodium, 2030
Cepharanthine, 2031
Cephation [Meiji] *see* 2028
Cephoral [E. Merck] *see* 1975
Cephos [CT] *see* 1963
Cephradine, 2032
Cephrol *see* 2391
Cephulac [Merrell] *see* 5360
Ceporacin [Torii] *see* 2028
Ceporan [Glaxo] *see* 2025
Ceporex [Glaxo] *see* 2021
Ceporexin [Glaxo] *see* 2021
Ceporexine [Glaxo] *see* 2021
Ceporin [Glaxo] *see* 2025
Cepovenin [Hoechst; Glaxo] *see* 2028
Cepticol [Banyu] *see* 1965
Cequartyl *see* 1086
Cer [Nippon Glaxo] *see* 2025
Ceractin [Wyeth-Ayerst] *see* 10128
Ceramide-β-lactoside *see* 2863
Ceraphyl 28 *see* 2072
Cerargyrite *see* 8647
Cerberigenin *see* 3205
Cerberin *see* 6559
Cercobin [Nippon Soda] *see* 9489
Cercobin-M [Nippon Soda] *see* 9489
Cereb [Ohta] *see* 2380
Cerebid [Saron] *see* 7151
Cerebolan [Tobishi] *see* 2365
Cerebro [Sidus] *see* 9160
Cerebroforte [Azupharma] *see* 7641
Cerebronylsphingosylglucosidogalactoside *see* 2863
Cerebrose *see* 4353
Cerebrostenediol *see* 4865
Cerebrosterol *see* 4865
Cereclor *see* 7156
Ceredase [Genzyme] *see* 242
Ceregulart [Kaken] *see* 3042
Cerekinon [Tanabe] *see* 9829
Cereon *see* 867
Cerepar [Mepha] *see* 2365
Ceresan Dry [Bayer AG] *see* 7453
Ceresan M *see* 3872
Ceresan Slaked Lime [Bayer AG] *see* 7453
Ceresan Wet *see* 3871
Ceresin, 2033
Cerespan [USV] *see* 7151
Ceretec [Amersham] *see* 3957
Cerevon *see* 4104
Cerezyme [Genzyme] *see* 242
Cergem [Searle] *see* 4393
Cergodum [Duncan] *see* 6580
Cergona *see* 867
Ceria *see* 2035
Ceric Ammonium Nitrate *see* 535
Ceric Fluoride, 2034
Ceric Oxide, 2035
Ceric Sulfate, 2036
Cerin *see* 2033
Cerium, 2037
CERM-730 *see* 611
CERM-1766 *see* 7039
CERM-1978 *see* 1188
CERM-3024 *see* 10303
CERM-10137 *see* 9657
Cer-O-Cillin Sodium [Upjohn] *see* 7228
Cerocral [Funai] *see* 4936
Ceronapril, 2038

Cerone [Union Carbide] *see* 3777
Cerosin *see* 2033
Cerous Ammonium Sulfate *see* 536
Cerous Benzoate *see* 1122
Cerous Bromide, 2039
Cerous Carbonate, 2040
Cerous Chloride, 2041
Cerous Fluoride, 2042
Cerous Iodide, 2043
Cerous Nitrate, 2044
Cerous Oxalate, 2045
Cerous Sulfate, 2046
Certinal *see* 482
Certomycin [Essex] *see* 6563
Cerubidin [M & B] *see* 2890
Cérubidine [Rhône-Poulenc] *see* 2890
Cerucal [Arzneimittelwerk Dresden] *see* 6226
Cerulein *see* 2048
Cerulenin, 2047
Ceruletide, 2048
Ceruloplasmin, 2049
Ceruse *see* 1034
Cerussa *see* 1034
Cerussite *see* 5410
Cervagem(e) [M&B] *see* 4393
Cervicarcin, 2050
Cervilaxin *see* 8299
Cerviprost [Organon] *see* 8064
Cervoxan [Sorbio] *see* 3537
Cesamet [Lilly] *see* 6427
C′esar [Roussel-UCLAF] *see* 4751
Cescan-131 [Abbott] *see* 2054
Cescorbat *see* 867
Cesium, 2051
Cesium Alum *see* 346
Cesium Bromide, 2052
Cesium Carbonate, 2053
Cesium Chloride, 2054
Cesium Hydrate *see* 2055
Cesium Hydroxide, 2055
Cesium Iodide, 2056
Cesium Nitrate, 2057
Cesium Sulfate, 2058
Cesol [E. Merck] *see* 7896
Cesplon [Esteve] *see* 1817
Cesporan [Errekappa] *see* 2032
Cestex [Beecham] *see* 3671
Cestocide [Bayer] *see* 6602
Cetab *see* 2068
Cetaceum *see* 8892
Cetacillin *see* 8002
Cetacort [Galderma] *see* 4828
Cetadol [Rybar] *see* 45
Cetal [Parke, Davis] *see* 10120
Cetalkonium Chloride, 2059
Cetamid *see* 867
Cetamide [Alcon] *see* 9067
Cetamolol, 2060
Cetampin [Scarium] *see* 628
Cetapril [Dainippon] *see* 202
Cetats [Zeeland] *see* 2068
Cetavlon [Ayerst] *see* 2068
Ceteareth *see* 7737
Cetebe [Stroschein] *see* 867
Cetemican *see* 867
Ceteth *see* 7737
Ceteth 20 *see* 7737
Cethexonium Bromide, 2061
Cethylose [Ascher] *see* 1877
Cetiedil, 2062
Cetiprin [Kabi] *see* 3598
Cetirizine, 2063
Cetol *see* 2059
Cetosanol [Sanol] *see* 5368
Cetotiamine, 2064
Cetoxime, 2065
Cétraphylline [Schering Plough] *see* 9421
Cetraric Acid, 2066

Cetrarin *see* 2066
Cetraxate, 2067
Cetrexin [Leciva] *see* 5820
Cetrimide *see* 2068
Cetrimonium Bromide, 2068
Cetrimonium Stearate, 2069
Cetrimonium Tosylate *see* 2068
Cetsim [Saint-Germain] *see* 217
Cetyl Alcohol, 2070
Cetylamine *see* 2068
α-Cetylcitric Acid *see* 184
Cetyldimethylbenzylammonium Chloride *see* 2059
N-Cetyl-N,N-dimethyl-2-cyclohexanolammonium Bromide *see* 2061
Cetyldimethylethylammonium Bromide, 2071
Cetyldimethyl(2-hydroxycyclohexyl)-ammonium Bromide *see* 2061
Cetylic Acid *see* 7128
Cetyl Lactate, 2072
Cetyl[2-[(p-methoxybenzyl)-2-pyrimidinylamino]ethyl]dimethylammonium Bromide *see* 9512
Cetyl Palmitate, 2073
Cetylpyridinium Chloride, 2074
Cetyltrimethylammonium Bromide *see* 2068
Cetyltrimethylammonium Stearate *see* 2069
Cevadilla *see* 8460
Cevadine, 2075
Cevalin [Lilly] *see* 867
(3β,4α,16β)-Cevane-3,4,12,14,16,17,-20-heptol 3-Acetate *see* 8461
(3β,5α,6α)-Cevane-3,6,20-triol *see* 10104
Cevanol [ICI] *see* 1055
Cevatine *see* 867
Cevex [Merrell] *see* 867
Cevimin *see* 867
Cevine, 2076
Ce-Vi-Sol [Mead Johnson] *see* 867
Cevitamic Acid *see* 867
Cevitan *see* 867
Cevitex *see* 867
Cewin [Winthrop] *see* 867
Cex [Nippon Glaxo] *see* 2021
Ceylon Cinnamon *see* 2359
Ceylon Isinglass *see* 182
CEZ *see* 1967
CF *see* 4254
CFCs *see* 2192
C-Film [Arun] *see* 6772
CFPQ *see* 3630
CG *see* 2273
CG-113 *see* 7921
CG-201 *see* 1239
CG-315E *see* 9701
CG-635 *see* 3917
CG-1283 *see* 6294
CG-4203 *see* 9227
CG-4509 *see* 8089
CGA-10832 *see* 7954
CGA-12223 *see* 5119
CGA-18731 *see* 5237
CGA-20168 *see* 6004
CGA-23654 *see* 6731
CGA-24705 *see* 6230
CGA-26423 *see* 7921
CGA-45156 *see* 9465
CGA-48988 *see* 5982
CGA-50439 *see* 2833
CGA-64250 *see* 8003
CGA-72662 *see* 2845
CGA-89317 *see* 9784
CGA-131036 *see* 9732
CGA-136872 *see* 7928
CGA-142705 *see* 4030
CGA-152005 *see* 8067

CGA-163935 *see* 9855
CGA-169374 *see* 3182
CGA-173506 *see* 4165
CGA-179246 *see* 6277
CGA-184699 *see* 5624
CGA-184927 *see* 2431
CGA-185072 *see* 2463
CGA-215944 *see* 8136
CGA-219417 *see* 2841
CG-B3Q *see* 1997
cGMP *see* 2778
CGP *see* 4650
CGP-2175 *see* 6235
CGP-4540 *see* 613
CGP-6140 *see* 608
CGP-7174/E *see* 1994
CGP-7760B *see* 7917
CGP-8426 *see* 9048
CGP-9000 *see* 1993
CGP-14221/E *see* 1985
CGP-14458 *see* 4625
CGP-30694 *see* 3553
CGP-32349 *see* 4267
CGP-39393 *see* 4754
CGP-48933 *see* 10051
CGS-9343B *see* 10243
CGS-14824A *see* 1058
CGS-14831 *see* 1058
CGS-16949A *see* 3969
CGS-20267 *see* 5474
CH-800 *see* 4045
CH-846 *see* 3537
CH-3635 *see* 9790
Chaetomidin *see* 6982
Chalcanthite *see* 2722
Chalcedony *see* 8637
Chalcocite *see* 2737
Chalcomenite *see* 2720
Chalcomycin, 2077
Chalcone, 2078
Chalcopyrite, 2079
D-Chalcose, 2080
Chalkone *see* 2078
Chalkopyrite *see* 2079
Chamazulene, 2081
Chamber Crystals *see* 6746
Chameleon Mineral *see* 7824
Chamomile, 2082
Champaca Camphor *see* 4583
Champacol *see* 4583
Chanchalagua *see* 2016
Channel Black *see* 1856
Channel Forming Integral Protein *see* 798
Channing's Soln *see* 7870
Chanoclavin-I *see* 2083
Chanoclavine, 2083
Chanoclavine II *see* 2083
Chanodesethylapovincamine *see* 10123
Chaperonins, 2084
CHAPS, 2085
CHAPSO *see* 2085
Charas *see* 1794
Charcot-Neumann Crystals *see* 8894
Chartarin *see* 2086
Chartreusin, 2086
Charybdotoxin, 2087
Chat *see* 5320
Chaulmestrol *see* 2089
Chaulmoogra Oil, 2088
Chaulmoogric Acid, 2089
Chavicine, 2090
Chavicinic Acid *see* 7620
Chavicol, 2091
Chavicol Methyl Ether *see* 3748
Chebutan *see* 5300
Cheeseflower *see* 5748
Cheirolin, 2092
Chelafer *see* 4085
Chelaplex III *see* 3556

Chelen *see* 3829
Chelerythrine, 2093
ψ-Chelerythrine *see* 8504
Chelidonic Acid, 2094
Chelidonine, 2095
Chel-Iron [Kinney] *see* 4085
Chemcef [Chemil] *see* 1983
Chemestrogen *see* 1102
Chemet [McNeil] *see* 9034
Chem-Fish [Tifa] *see* 7629
Chemical 109 *see* 764
Chemical Mace *see* 2166
Chemicetina [Erba] *see* 2120
Chemifluor [Chemipharm] *see* 8762
Chemiofuran [Italfarmaco] *see* 6696
Chemipen [Squibb] *see* 7369
Chemodyn *see* 1182
Chemofuran *see* 6697
Chemotrim [Rosemont] *see* 9086
Chemox DN [Blue Spruce] *see* 3341
Chemox PE [Blue Spruce] *see* 3341
Chemox Selective [Blue Spruce] *see* 3341
Chem Rice [Tifa] *see* 7987
Chendol [Fisons] *see* 2096
Chenic Acid *see* 2096
Chenix [Rowell] *see* 2096
Chenocedon [Tillotts] *see* 2096
Chenocol [Caber] *see* 2096
Chenodeoxycholic Acid *see* 2096
Chenodex [Roussel-UCLAF] *see* 2096
Chenodiol, 2096
Chenofalk [Falk] *see* 2096
Chenopodiol [Pohl] *see* 6888
Chenoposan *see* 6888
Chenoposetten *see* 6888
Chenosäure [Falk] *see* 2096
Chenossil [Guiliani] *see* 2096
Chephalotin [Lilly] *see* 2028
Chepirol *see* 5300
Cheque [Upjohn] *see* 6262
CHES, 2097
Chess [Ciba] *see* 8136
Chessylite *see* 2697
Chestnut *see* 1943
Chetazolidin [Zeria] *see* 5300
Chetil *see* 5300
Chevreul's Salt. *see* 2738
Chibro-Amuno [Chibret] *see* 4998
Chibro Pilocarpine [Chibret] *see* 7578
Chibro-Proscar [Chibret] *see* 4125
Chibroxin(e) [Merck & Co.] *see* 6793
Chibroxol [Merck & Co.] *see* 6793
Chick Antidermatitis Factor *see* 7147
Chicle, 2098
Chilean Centaury *see* 2016
Chile Saltpeter *see* 8792
Chimaphila, 2099
Chimaphilin, 2100
Chimonanthine, 2101
Chimono [Lusofarmaco] *see* 5592
Chimyl Alcohol, 2102
China Bark *see* 8222
China Clay *see* 5294
China Green *see* 5739
China Oil *see* 976
"China White" *see* 3012 and 6150
China Wood Oil *see* 9944
Chinese Anise *see* 8953
Chinese Blistering Flies *see* 6410
Chinese Blue *see* 4064
Chinese Cantharides *see* 6410
Chinese Isinglass *see* 182
Chinese Pea *see* 8878
Chinese Red *see* 5945
Chinese Seasoning *see* 6337
Chinese Wax, 2103
Chinic Acid *see* 8243
Chinicine *see* 10141
Chinidin-Duriles [Astra] *see* 8244

6-Chloro-3,4-dihydro-3-trichloromethyl-2H-1,2,4-benzothiadiazine-7-sulfonamide 1,1-Dioxide see 9261

6-Chloro-3,4-dihydro-3-[[(2,2,2-trifluoroethyl)thio]methyl]-2H-1,2,4-benzothiadiazine-7-sulfonamide 1,1-Dioxide see 3661

6-Chloro-7,8-dihydroxy-1-p-hydroxyphenyl-2,3,4,5-tetrahydro-1H-3-benzazepine see 4020

(6α)-Chloro-17,21-dihydroxypregna-1,4-diene-3,11,20-trione see 2209

3-Chloro-1,2-dihydroxypropane see 2197

4'-Chloro-3,5-diiodosalicylanilide Acetate see 2416

7-Chloro-4,6-dimethoxycoumaran-3-one-2-spiro-1'-(2'-methoxy-6'-methylcyclohex-2'-en-4'-one) see 4571

4'-Chloro-3,5-dimethoxy-4-(2-morpholinoethoxy)benzophenone see 6350

6-Chloro-2,3-dimethoxy-1,4-naphthalenediol Diacetate see 5595

2-Chloro-6-[(4,6-dimethoxy-2-pyrimidinyl)thio]benzoic Acid see 8177

2-[p-Chloro-α-(2-dimethylaminoethoxy)benzyl]pyridine see 1845

2-Chloro-11-(2-dimethylaminoethoxy)-dibenzo[b,f]thiepine see 10327

2-Chloro-α-(2-dimethylaminoethyl)-benzhydrol see 2106

2-[p-Chloro-α-(2-dimethylaminoethyl)-benzyl]pyridine see 2232

7-Chloro-10-(2-dimethylaminoethyl)-dibenzo[b,e][1,4]diazepin-11(10H)-one see 2420

7-Chloro-10-[2-(dimethylamino)ethyl]-5,10-dihydro-11H-dibenzo[b,e][1,4]-diazepin-11-one see 2420

(+)-(2S,3S)-8-Chloro-5-[2-(dimethylamino)ethyl]-2,3-dihydro-3-hydroxy-2-(p-methoxyphenyl)-1,5-benzothiazepin-4(5H)-one Acetate (Ester) see 2408

7-Chloro-10-(β-dimethylaminoethyl)-10,11-dihydro-11-oxo-5H-dibenzo-[b,e][1,4]diazepine see 2420

2-Chloro-α-[2-(dimethylamino)ethyl]-α-phenylbenzenemethanol see 2106

4-Chloro-α-[2-(dimethylamino)-1-methylethyl]-α-methylbenzeneethanol see 2425

p-Chloro-α-[2-(dimethylamino)-1-methylethyl]-α-methylphenethyl Alcohol see 2425

8-Chloro-1-[(dimethylamino)methyl]-6-phenyl-4H-s-triazolo[4,3-a][1,4]-benzodiazepine see 159

2-Chloro-4-(dimethylamino)-6-methylpyrimidine see 2655

[4S-(4α,4aα,5aα,6β,12aα)]-7-Chloro-4-(dimethylamino)-1,4,4a,5,5a,6,11,-12a-octahydro-3,6,10,12,12a-pentahydroxy-1,11-dioxo-2-naphthacenecarboxamide see 2937

7-Chloro-4-(dimethylamino)-1,4,4a,5,-5a,6,11,12a-octahydro-3,6,10,12,-12a-pentahydroxy-N-(hydroxymethyl)-6-methyl-1,11-dioxo-2-naphthacenecarboxamide see 2448

[4S-(4α,4aα,5aα,6β,12aα)]-7-Chloro-4-dimethylamino-1,4,4a,5,5a,6,11,-12a-octahydro-3,6,10,12,12a-pentahydroxy-6-methyl-1,11-dioxo-2-naphthacenecarboxamide see 2245

[4S-(4α,4aα,5α,5aα,12aα)]-7-Chloro-4-(dimethylamino)-1,4,4a,5,5a,6,11,-12a-octahydro-3,5,10,12,12a-penta-

hydroxy-6-methylene-1,11-dioxo-2-naphthacenecarboxamide see 5818

2-Chloro-9-[3-(dimethylamino)propyl]-acridan see 2441

3-Chloro-5-[3-(dimethylamino)propyl]-10,11-dihydro-5H-dibenz[b,f]azepine see 2447

α-2-Chloro-10-(3-dimethylaminopropylidene)thiaxanthene see 2241

2-Chloro-9-(3'-dimethylaminopropylidene)thioxanthene see 2241

2-Chloro-10-(3-dimethylaminopropyl)-phenothiazine see 2238

3-Chloro-10-(3-dimethylaminopropyl)phenothiazine see 2238

2-Chloro-10-[3-(dimethylamino)propyl]phenothiazine 5-Oxide see 6987

2-Chloro-α,α-dimethylbenzeneethanamine see 2472

4-Chloro-α,α-dimethylbenzeneethanamine see 2235

Chlorodimethyl Ether see 2198

2-Chloro-α-[[(1,1-dimethylethyl)-amino]methyl]benzenemethanol see 9942

(2-Chloro-1,1-dimethylethyl)benzene see 6548

5-Chloro-3-(1,1-dimethylethyl)-6-methyl-2,4(1H,3H)-pyrimidinedione see 9298

4-Chloro-α,α-dimethylphenethylamine see 2235

o-Chloro-α,α-dimethylphenethylamine see 2472

(p-Chloro-α,α-dimethylphenethyl)carbamic Acid Ethyl Ester see 2440

4-Chloro-3,5-dimethylphenol see 2228

2-Chloro-N,N-dimethyl-10H-phenothiazine-10-propanamine see 2238

2-Chloro-N,N-dimethyl-10H-phenothiazine-10-propanamine 5-Oxide see 6987

8-Chloro-N,N-dimethyl-6-phenyl-4H-[1,2,4]triazolo[4,3-a][1,4]benzodiazepine-1-methanamine see 159

4-Chloro-N'-(2',6'-dimethyl-1'-piperidyl)-3-sulfamoylbenzamide see 2454

4-Chloro-2',6'-dimethyl-5-sulfamoylsalicylanilide see 10212

2-Chloro-N-(2,4-dimethyl-3-thienyl)-N-(2-methoxy-1-methylethyl)-acetamide see 3263

2-Chloro-N,N-dimethylthioxanthene-Δ9,γ-propylamine see 2241

2-Chloro-N,N-dimethyl-3-thioxanthen-9-ylidenepropylamine see 2241

1-Chloro-2,4-dinitrobenzene, 2187

1-Chloro-2,6-dinitrobenzene see 2188

2-Chloro-1,3-dinitrobenzene, 2188

4-Chloro-1,3-dinitrobenzene see 2187

6-Chloro-1,3-dinitrobenzene see 2187

3-Chloro-4-(2,5-dioxo-3-phenyl-1-pyrrolidinyl)benzenesulfonamide see 9048

4'-Chloro-2,6-dioxothiocyclohexanecarboxanilide see 6981

1-(o-Chloro-α,α-diphenylbenzyl)imidazole see 4770

(Z)-2-[4-(4-Chloro-1,2-diphenyl-1-butenyl)phenoxy]-N,N-dimethylethanamine see 9688

(Z)-4-Chloro-1,2-diphenyl-1-[4-[2-(N,-N-dimethylamino)ethoxy]phenyl]-1-butene see 9688

2-[4-(2-Chloro-1,2-diphenylethenyl)-phenoxy]-N,N-diethylethanamine see 2446

1-(p-Chlorodiphenylmethyl)-4-[2-(2-hydroxyethoxy)ethyl]piperazine see 4897

1-(4-Chlorodiphenylmethyl)-4-methyl-2,3,4,5,6,7-hexahydro-1,4-diazepine see 4770

4-Chlorodiphenyl Sulfone see 9162

2-[p-(2-Chloro-1,2-diphenylvinyl)phenoxy]triethylamine see 2446

2-Chloro-N,N-di-2-propenylacetamide see 259

5-Chloro-2,4-disulfamylaniline see 2119

5-Chloro-2,4-disulfamyltoluene see 3427

4-Chloro-N-1,3-dithietan-2-ylidene-2-methylbenzeneamine see 6642

Chlorodracylic Acid see 2177

5-Chloro-6-(7,8-epoxy-10-hydroxy-2-oxo-3,5-undecadienyl)-β-resorcylic Acid μ-Lactone see 6336

1-Chloro-2,3-epoxypropane see 3648

16α-Chloroestrone 3-Methyl Ether see 2442

2-Chloro-1-ethanal see 2159

Chloroethane see 3829

2-Chloroethanephosphonic Acid see 3777

Chloroethanoic Acid see 2162

2-Chloroethanol see 3839

2-Chloroethanol Acetate see 2189

Chloroethene Homopolymer see 7746

(2-Chloroethenyl)arsonous Dichloride see 3112

1,1',1''-(1-Chloro-1-ethenyl-2-ylidene)-tris[4-methoxybenzene] see 2225

(2-Chloroethoxy)ethene see 2191

3-(2-Chloroethoxy)-9α-fluoro-6-formyl-11β,21-dihydroxy-16α,17α-isopropylidenedioxypregna-3,5-dien-20-one see 4270

3-(2-Chloroethoxy)-9-fluoro-11β,16α,-17,21-tetrahydroxy-20-oxopregna-3,5-diene-6-carboxaldehyde, Cyclic 16,17-Acetal with Acetone, 21-Acetate see 4270

3-(2-Chloroethoxy)-6-formyl-9α-fluoropregna-3,5-diene-11β,16α,17,-21-tetrol-20-one 21-Acetate 16α,-17α-Acetonide see 4270

2-(2-Chloroethoxy)-N-[[(4-methoxy-6-methyl-1,3,5-triazin-2-yl)amino]-carbonyl]benzenesulfonamide see 9732

2-Chloro-N-(ethoxymethyl)-6'-ethyl-o-acetotoluidide see 58

2-Chloro-N-(ethoxymethyl)-N-(2-ethyl-6-methylphenyl)acetamide see 58

2-Chloro-1-(3-ethoxy-4-nitrophenoxy)-4-(trifluoromethyl)benzene see 7097

1-[2-(2-Chloroethoxy)phenylsulfonyl]-3-(4-methoxy-6-methyl-1,3,5-triazin-2-yl)urea see 9732

β-Chloroethyl Acetate, 2189

2-Chloroethyl Alcohol see 3839

2-Chloro-4-ethylamino-6-isopropylamine-s-triazine see 902

6-Chloro-2-(ethylamino)-4-methyl-4-phenyl-4H-3,1-benzoxazine see 3910

2-[[4-Chloro-6-(ethylamino)-1,3,5-triazin-2-yl]amino]-2-methylpropanenitrile see 2755

2-[[4-Chloro-6-(ethylamino)-s-triazin-2-yl]amino]-2-methylpropionitrile see 2755

2-Chloroethyl N,N-Bis(2-chloroethyl)-N'-(3-hydroxypropyl)phosphorodiamidate see 2917

β-Chloroethyl β-(p-tert-Butylphenoxy)-α-methylethyl Sulfite see 805

3-(2-Chloroethyl)-2-[(2-chloroethyl)-

amino]tetrahydro-2*H*-1,3,2-oxaza-
phosphorin-2-oxide *see* 4937
1-(2-Chloroethyl)-3-cyclohexyl-1-
nitrosourea *see* 5594
N-(2-Chloroethyl)-*N'*-cyclohexyl-*N*-
nitrosourea *see* 5594
N-(2-Chloroethyl)dibenzylamine
Hydrochloride, 2190
6-Chloro-3-ethyl-3,4-dihydro-2*H*-1,2,-
4-benzothiadiazine-7-sulfonamide
1,1-Dioxide *see* 3778
2-(2-Chloroethyl)-2,3-dihydro-4*H*-1,3-
benzoxazin-4-one *see* 2247
2-(β-Chloroethyl)-2,3-dihydro-4-keto-
(benzo-1,3-oxazine) *see* 2247
2-(β-Chloroethyl)-2,3-dihydro-4-oxo-
6-amino-1,3-benzoxazine *see* 455
2-(β-Chloroethyl)-2,3-dihydro-4-oxo-
(benzo-1,3-oxazine) *see* 2247
6-Chloro-3-ethyl-3,4-dihydro-7-sulf-
amoylbenzo-1,2,4-thiadiazine 1,1-
Dioxide *see* 3778
N-(2-Chloroethyl)-2,6-dinitro-*N*-prop-
yl-4-(trifluoromethyl)benzenamine
see 4156
Chloroethylene *see* 10132
Chloroethylene Polymer *see* 7746
7-Chloro-4-[4-[ethyl(2-hydroxyethyl)-
amino]-1-methylbutylamino]quin-
oline *see* 4863
7-Chloro-4-[4-(*N*-ethyl-*N*-β-hydroxy-
ethylamino)-1-methylbutylamino]-
quinoline *see* 4863
7-Chloro-4-[5-(*N*-ethyl-*N*-2-hydroxy-
ethylamino)-2-pentyl]aminoquino-
line *see* 4863
2-Chloroethyl 2-Hydroxyethyl Sulfide
see 4680
β-Chloroethyl β-Hydroxyethyl Thioether *see* 4680
Chloroethylmercury *see* 3871
2-Chloro-6'-ethyl-*N*-(2-methoxy-1-
methylethyl)acet-*o*-toluidide *see*
6230
2-Chloroethyl 1-Methyl-2-(*p-tert*-
butylphenoxy)ethyl Sulfite *see* 805
6-Chloro-*N*-ethyl-*N'*-(1-methylethyl)-
1,3,5-triazine-2,4-diamine *see* 902
α-Chloro-2'-ethyl-6'-methyl-*N*-(1-
methyl-2'-methoxyethyl)acetanilide
see 6230
N-(2-Chloroethyl)-*N*-(1-methyl-2-phen-
oxyethyl)benzenemethanamine *see*
7409
N-(2-Chloroethyl)-*N*-(1-methyl-2-phen-
oxyethyl)benzylamine *see* 7409
6-Chloro-methyl-4-methyl-4-phenyl-
4*H*-3,1-benzoxazin-2-amine *see* 3910
2-Chloro-*N*-(2-ethyl-6-methylphenyl)-
N-(2-methoxy-1-methylethyl)-
acetamide *see* 6230
5-(2-Chloroethyl)-4-methylthiazole *see*
2444
2-[[[(2-Chloroethyl)nitrosoamino]car-
bonyl]amino]-2-deoxy-D-glucose *see*
2229
[1-[[[(2-Chloroethyl)nitrosoamino]-
carbonyl]amino]ethyl]phosphonic
Acid Diethyl Ester *see* 4285
1-(2-Chloroethyl)-1-nitroso-3-(D-glu-
cos-2-yl)urea *see* 2229
2-[3-(2-Chloroethyl)-3-nitrosoureido]-
2-deoxy-D-glucopyranose *see* 2229
6-[3-(2-Chloroethyl)-3-nitrosoureido]-
6-deoxy-α-D-glucopyranoside *see*
8285
1-[*N*-(2-Chloroethyl)-*N*-nitrosoureido]-
ethylphosphonic Acid Diethyl Ester
see 4285

1-Chloro-3-ethyl-1-penten-4-yl-3-ol *see*
3774
5-Chloro-3-ethylpent-1-yn-4-en-3-ol
see 3774
Chloroethylphenamide *see* 1045
N-(2-Chloroethyl)-*N*-(phenylmethyl)-
benzenemethanamine Hydrochloride *see* 2190
(2-Chloroethyl)phosphonic Acid *see*
3777
7-Chloro-2-ethyl-6-sulfamoyl-1,2,3,4-
tetrahydro-4-quinazolinone *see* 8240
7-Chloro-2-ethyl-1,2,3,4-tetrahydro-4-
oxo-6-quinazolinesulfonamide *see*
8240
7-Chloro-2-ethyl-1,2,3,4-tetrahydro-4-
oxo-6-sulfamoylquinazoline *see* 8240
5-Chloroethylthiamine *see* 1048
2-(2-Chloroethylthio)ethanol *see* 4680
N-(2-Chloroethyl)-α,α,α-trifluoro-2,6-
dinitro-*N*-propyl-*p*-toluidine *see* 4156
(2-Chloroethyl)trimethylammonium
Chloride *see* 2153
2-Chloroethyl Vinyl Ether, 2191
2-Chloroflumethasone *see* 4628
9-(2-Chloro-6-fluorobenzyl)adenine *see*
828
Chlorofluorocarbons, 2192
(6α,11β,16α)-9-Chloro-6-fluoro-11,21-
dihydroxy-16-methylpregna-1,4-
diene-3,20-dione *see* 2430
(11β,16β)-21-Chloro-9-fluoro-11,17-
dihydroxy-16-methylpregna-1,4-
diene-3,20-dione *see* 2423
[2-Chloro-4-fluoro-5-(1,3,4,5,6,7-hexa-
hydro-1,3-dioxo-2*H*-isoindol-2-yl)-
phenoxy]acetic Acid *see* 4178
21-Chloro-9α-fluoro-11β-hydroxy-
16α,17α-isopropylidenedioxy-4-
pregnene-3,20-dione *see* 4621
(11β,16α)-21-Chloro-9-fluoro-11 hy-
droxy-16,17-[(1-methylethylidene)-
bis(oxy)]pregn-4-ene-3,20-dione *see*
4621
(16β)-21-Chloro-9-fluoro-17-hydroxy-
16-methylpregna-1,4-diene-3,11,20-
trione *see* 2424
9-Chloro-6α-fluoro-16α-methyl-1,4-
pregnadiene-11β,21-diol-3,20-dione
see 2430
7-Chloro-5-(2-fluorophenyl)-1,3-dihy-
dro-3-hydroxy-1-(2-hydroxyethyl)-
2*H*-1,4-benzodiazepin-2-one *see*
3490
7-Chloro-5-(2-fluorophenyl)-2,3-dihy-
dro-3-hydroxy-2-oxo-1*H*-1,4-benzo-
diazepine-1-propanenitrile *see* 2368
7-Chloro-5-(2-fluorophenyl)-1,3-dihy-
dro-1-methyl-2*H*-1,4-benzodiazepin-
2-one *see* 4164
7-Chloro-5-(2-fluorophenyl)-2,3-dihy-
dro-2-oxo-1*H*-1,4-benzodiazepine-3-
carboxylic Acid Ethyl Ester *see*
3867
7-Chloro-5-(2-fluorophenyl)-1,3-dihy-
dro-1-(2,2,2-trifluoroethyl)-2*H*-1,4-
benzodiazepine-2-thione *see* 8211
5-Chloro-1-(4-fluorophenyl)-3-[1-[2-(2-
imidazolidinon-1-yl)ethyl]-4-piperid-
yl]-1*H*-indole *see* 8611
1-[2-[4-[5-Chloro-1-(4-fluorophenyl)-
1*H*-indol-3-yl]-1-piperidinyl]ethyl]-
2-imidazolidinone *see* 8611
6-Chloro-3-[(4-Fluorophenyl)methyl]-
3,4-dihydro-2*H*-1,2,4-benzothiadia-
zine-7-sulfonamide 1,1-Dioxide *see*
7157
8-Chloro-6-(2-fluorophenyl)-1-methyl-
4*H*-imidazo[1,5-*a*][1,4]benzodiaze-
pine *see* 6270

6-[3-(2-Chloro-6-fluorophenyl)-5-
methyl-4-isoxazolecarboxamido]-
penicillanic Acid *see* 4147
[2*S*-(2α,5α,6β)]-6-[[[3-(2-Chloro-6-
fluorophenyl)-5-methyl-4-isoxazol-
yl]carbonyl]amino]-3,3-dimethyl-7-
oxo-4-thia-1-azabicyclo[3.2.0]hept-
ane-2-carboxylic Acid *see* 4147
3-(2-Chloro-6-fluorophenyl)-5-methyl-
4-isoxazolylpenicillin *see* 4147
9-[(2-Chloro-6-fluorophenyl)methyl]-
9*H*-purin-6-amine *see* 828
10-Chloro-11b-(2-fluorophenyl)-2,3,7,-
11b-tetrahydro-7-(2-hydroxyethyl)-
oxazolo[3,2-*d*][1,4]benzodiazepin-
6(5*H*)-one *see* 4243
(*R*)-2-[4-[(5-Chloro-3-fluoro-2-pyridin-
yl)oxy]phenoxy]propanoic Acid 2-
Propynyl Ester *see* 2431
2-Chloro-4-fluoro-5-[(3,4,5,6-tetrahy-
dro)phthalimido]phenoxyacetic Acid
see 4178
21-Chloro-9-fluoro-11β,16α,17-trihy-
droxypregn-4-ene-3,20-dione Cyclic
16,17-Acetal with Acetone *see* 4621
Chlorofolin *see* 2207
Chloroform, 2193
Chloroformamide *see* 1827
Chloroformic Acid Benzyl Ester *see*
1848
Chloroformic Acid *p*-Bromobenzyl
Ester *see* 1432
Chloroformic Acid Ethyl Ester *see*
3831
Chloroformic Acid Trichloromethyl
Ester *see* 3395
Chloroformyl Chloride *see* 7491
Chlorofos *see* 9753
4-Chloro-*N*-furfuryl-5-sulfamoylan-
thranilic Acid *see* 4331
4-Chloro-*N*-(2-furylmethyl)-5-sulfamo-
ylanthranilic Acid *see* 4331
Chlorogenic Acid, 2194
Chlorogenin, 2195
Chloroguanide *see* 2138
Chlorohemin *see* 4678
2-Chloro-1-heptene-1-arsonic Acid *see*
2170
(2-Chloro-1-heptenyl)arsonic Acid *see*
2170
4-Chloro-*N*-[[(hexahydro-1*H*-azepin-1-
yl)amino]carbonyl]benzenesulfon-
amide *see* 4524
(3α,5aα,10bα,11β,11aα)-9-Chloro-2,-
3,5a,6,10b,11-hexahydro-10b,11-
dihydroxy-7,8-dimethoxy-2,3,6-
trimethyl-3,11a-epidithio-11a*H*-py-
razino[1',2':1,5]pyrrolo[2,3-*b*]indole-
1,4-dione *see* 8920
4-Chloro-*N*-(*endo*-hexahydro-4,7-
methanoisoindolin-2-yl)-3-sulfamo-
ylbenzamide *see* 9866
10-Chloro-2,3,5,6,7,11b-hexahydro-2-
methyl-11b-phenylbenzo[6,7]-1,4-
diazepino[5,4-*b*]oxazol-6-one *see*
7061
1-Chlorohexane, 2196
α-Chlorohydrin, 2197
(17β)-4-Chloro-17-hydroxyandrost-4-
en-3-one *see* 2475
4-Chloro-2-hydroxybenzoic Acid *n*-
Butylamide *see* 1485
5-Chloro-2-hydroxybenzoxazole *see*
2249
4-Chloro-6-[(3-hydroxy-2-butenylid-
ene)amino]-*N¹*-2-propenyl-1,3-benz-
enedisulfonamide *see* 405
2-Chloro-2'-hydroxydiethyl Sulfide *see*
4680

2-Chloro-5-hydroxy-1,3-dimethylbenzene *see* 2228

5-Chloro-2-hydroxydiphenylmethane *see* 2469

5-Chloro-3-[4-(2-hydroxyethyl)-1-piperazinyl]carbonylmethyl-2-benzothiazolinone *see* 9564

2-Chloro-9-[3-[4-(2-hydroxyethyl)-1-piperazinyl]propylidene]thiaxanthene *see* 2455

2-Chloro-10-[3-[1-(2-hydroxyethyl)-4-piperazinyl]propyl]phenothiazine *see* 7323

5-Chloro-8-hydroxy-7-iodoquinoline *see* 5052

3-Chloro-7-hydroxy-4-methyl-2*H*-1-benzopyran-2-one Bis(chloroethyl) Phosphate *see* 4636

3-Chloro-7-hydroxy-4-methylcoumarin Bis(2-chloroethyl) Phosphate *see* 4636

3-Chloro-7-hydroxy-4-methylcoumarin Diethyl Phosphate *see* 2600

3-Chloro-7-hydroxy-4-methylcoumarin *O*-Ester with *O*,*O*-Diethyl Phosphorothioate *see* 2626

6-Chloro-17-hydroxy-1α,2α-methylenepregna-4,6-diene-3,20-dione *see* 2844

6-Chloro-4-hydroxy-2-methyl-*N*-2-pyridinyl-2*H*-thieno[2,3-*e*]-1,2-thiazine-3-carboxamide 1,1-Dioxide *see* 5612

6-Chloro-4-hydroxy-2-methyl-3-(2-pyridylcarbamoyl)-2*H*-thieno[2,3-*e*]-1,2-thiazine-1,1-dioxide *see* 5612

6-Chloro-17-hydroxy-2-oxapregna-4,6-diene-3,20-dione *see* 7019

2-Chloro-5-(1-hydroxy-3-oxo-1-isoindolinyl)benzenesulfonamide *see* 2246

2-Chloro-5-(3-hydroxy-1-oxoisoindolin-3-yl)benzenesulfonamide *see* 2246

7-Chloro-3-hydroxy-5-phenyl-1,3-dihydro-2*H*-1,4-benzodiazepin-2-one *see* 7059

(3-Chloro-4-hydroxyphenyl)hydroxymercury *see* 4876

6-Chloro-17-hydroxypregna-4,6-diene-3,20-dione Acetate *see* 2152

6-Chloro-17-hydroxypregna-1,4,6-triene-3,20-dione Acetate *see* 2930

1-(3-Chloro-2-hydroxypropyl)-2-methyl-5-nitroimidazole *see* 7000

5-Chloro-8-hydroxyquinoline *see* 2483

2-Chloro-5-hydroxytoluene *see* 2184

6-Chloro-3-hydroxytoluene *see* 2184

2-Chloro-5-hydroxy-*m*-xylene *see* 2228

5-Chloro-4-(2-imidazolin-2-ylamino)-2,1,3-benzothiadiazole *see* 9624

7-Chloro-3-(1*H*-imidazol-1-yl)-1,2,4-benzotriazine 1-Oxide *see* 9737

4-Chloro-2,2'-iminodibenzoic Acid *see* 5581

7-Chloro-4-indanol *see* 2144

Chloroiodoquin *see* 5052

5-Chloro-7-iodo-8-quinolinol *see* 5052

Chloro-IPC *see* 2240

2-Chloroisobutane *see* 1597

α-Chloroisobutylene *see* 2199

γ-Chloroisobutylene *see* 2200

2-Chloro-*N*-isopropylacetanilide *see* 7977

1-Chloroisopropyl Alcohol *see* 8036

o-Chloro-α-(isopropylaminomethyl)-benzyl Alcohol *see* 2470

6-Chloro-4-isopropyl-1-methyl-3-phenol *see* 2223

4'-Chloro-*N*-(1-isopropyl-4-piperidyl)-2-phenylacetanilide *see* 5610

O-(5-Chloro-1-isopropyl-1*H*-1,2,4-triazol-3-yl) *O*,*O*-Diethyl Phosphorothioate *see* 5119

1-Chloro-2-ketopropane *see* 2164

1-[3-(Chloromercuri)-2-methoxypropyl]urea *see* 2154

o-(Chloromercuri)phenol *see* 5920

Chloromethane *see* 6121

Chloromethapyrilene *see* 2220

1-[(6-Chloro-2-methoxy-9-acridinyl)-amino]-3-(diethylamino)-2-propanol Dihydrochloride *see* 123

*N*⁴-(6-Chloro-2-methoxy-9-acridinyl)-*N*¹,*N*¹-diethyl-1,4-pentanediamine *see* 8225

(16α)-16-Chloro-3-methoxyestra-1,3,5(10)-trien-17-one *see* 2442

Chloro-(2-methoxyethyl)mercury 3871

Chloromethoxymethane *see* 2198

cis-5-Chloro-2-methoxy-4-(methylamino)-*N*-[2-methyl-2-(phenylmethyl)-3-pyrrolidinyl]benzamide *see* 6533

3-Chloro-7-methoxy-9-(1-methyl-4-diethyl Aminobutylamino)acridine *see* 8225

4-Chloro-6-methoxy-2-methyl-5-(2-imidazolin-2-yl)aminopyrimidine *see* 6375

2-Chloro-*N*-[[(4-methoxy-6-methyl-1,3,5-triazin-2-yl)amino]carbonyl]-benzenesulfonamide *see* 2244

4-[*p*-Chloro-*N*-(*p*-methoxyphenyl)benzamido]butyric Acid *see* 2398

2-[[[[(4-Chloro-6-methoxy-2-pyrimidinyl)amino]carbonyl]amino]sulfonyl]benzoic Acid Ethyl Ester *see* 2142

7-Chloro-2-methylamino-5-phenyl-3*H*-1,4-benzodiazepine 4-Oxide *see* 2132

4-Chloro-5-(methylamino)-2-[3-(trifluoromethyl)phenyl]-3(2*H*)-pyridazinone *see* 6794

4-Chloro-5-(methylamino)-2-(α,α,α-trifluoro-*m*-tolyl)-3(2*H*)-pyridazinone *see* 6794

2-(3-Chloro-2-methylanilino)nicotinic Acid *see* 2453

(Chloromethyl)benzene *see* 1164

1-Chloro-2-methylbenzene *see* 2224

1-Chloro-3-methylbenzene *see* 2224

1-Chloro-4-methylbenzene *see* 2224

2-Chloro-4-methylbenzene-1,5-disulfonamide *see* 3427

4-Chloro-6-methyl-1,3-benzenedisulfonamide *see* 3427

N-Chloro-4-methylbenzenesulfonamide Sodium Salt *see* 2118

7-Chloro-3-methyl-2*H*-1,2,4-benzothiadiazine 1,1-Dioxide *see* 3051

1-Chloro-3-methylbutane *see* 5130

6-Chloro-α-methyl-9*H*-carbazole-2-acetic Acid *see* 1912

4(Or 5)-Chloro-2-methylcyclohexanecarboxylic Acid, 1,1-Dimethylethyl Ester *see* 9831

(2β,3β,4α,12α)-12-(Chloromethyl)-2,3-dihydroxy-27-norolean-13-ene-23,28-dioic Acid *see* 8598

7-Chloro-6-methylene-5-hydroxy-tetracycline *see* 5818

6-Chloro-1,2α-methylene-4,6-pregnadien-17α-ol-3,20-dione *see* 2844

Chloromethyl 17α-Ethoxycarbonyloxy-11β-hydroxyandrosta-1,4-diene-3-one-17β-carboxylate *see* 5614

2-Chloro-α-[[(1-methylethyl)amino]methyl]benzenemethanol *see* 2470

4-Chloro-α-(1-methylethyl)benzene-

acetic Acid Cyano(3-phenoxyphenyl)methyl Ester *see* 4051

2-Chloro-*N*-(1-methylethyl)-*N*-phenylacetamide *see* 7977

2-Chloro-4-methyl-3-(2'-imidazolin-2'-ylamino)thiophene *see* 9558

4-Chloro-*N*-(2-methyl-1-indolinyl)-3-sulfamoylbenzamide *see* 4969

1-(Chloromethyl)-3-methylbenzene *see* 10224

Chloromethyl Methyl Ether, 2198

4-Chloro-5-methyl-2-(1-methylethyl)-phenol *see* 2223

α-(Chloromethyl)-2-methyl-5-nitro-1*H*-imidazole-1-ethanol *see* 7000

3'-Chloro-α-[methyl[(morpholinocarbonyl)methyl]amino]-*o*-benzotoluidide *see* 4258

3'-Chloro-β-[*N*-methyl-*N*-[(morpholinocarbonyl)methyl]aminomethyl]-benzanilide *see* 4258

N-[3-Chloro-2-[[methyl-[2-(4-morpholinyl)-2-oxoethyl]amino]methyl]-phenyl]benzamide *see* 4258

Chloromethyloxirane *see* 3648

3-Chloro-10-[γ-[*N*'-[β-(1-methyl-2-oxoimidazolidin-3-yl)ethyl]-*N*-piperazinyl]propyl]phenothiazine *see* 4945

4-Chloro-3-methylphenol *see* 2184

(4-Chloro-2-methylphenoxy)acetic Acid *see* 5803

1-(6-Chloro-3-methylphenoxy)-3-*tert*-butylaminopropan-2-ol *see* 1521

1-(2-Chloro-5-methylphenoxy)-3-[(1,1-dimethylethyl)amino]-2-propanol *see* 1521

(±)-2-(4-Chloro-2-methylphenoxy)propanoic Acid *see* 5826

2-[(3-Chloro-2-methylphenyl)amino]-benzoic Acid *see* 9650

2-(2-Chloro-4-methylphenyl)amino-1,3-diazacyclopentene-2 *see* 9657

2-[(3-Chloro-2-methylphenyl)amino]-3-pyridinecarboxylic Acid *see* 2453

7-Chloro-*N*-methyl-5-phenyl-3*H*-1,4-benzodiazepin-2-amine 4-Oxide *see* 2132

7-Chloro-1-methyl-5-phenyl-1*H*-1,5-benzodiazepine-2,4(3*H*,5*H*)-dione *see* 2417

7-Chloro-1-methyl-5-phenyl-3*H*-1,4-benzodiazepin-2(1*H*)-one *see* 3042

2-[(*p*-Chloro-α-methyl-α-phenylbenzyl)oxy]-*N*,*N*-dimethylethylamine *see* 2234

2-[(*p*-Chloro-α-methyl-α-phenylbenzyl)oxy]-*N*,*N*-dimethylpropylamine *see* 5823

1-[2-(4-Chloro-α-methyl-α-phenylbenzyloxy)ethyl]azepam *see* 8619

1-[2-[(*p*-Chloro-α-methyl-α-phenylbenzyl)oxy]ethyl]hexahydro-1*H*-azepine *see* 8619

N-[2-[(4-Chloro-α-methyl-α-phenylbenzyl)oxy]ethyl]hexamethylenimine *see* 8619

(+)-2-[2-[(*p*-Chloro-α-methyl-α-phenylbenzyl)oxy]ethyl]-1-methylpyrrolidine *see* 2405

N-(2-Chloro-4-methylphenyl)-4,5-dihydro-1*H*-imidazol-2-amine *see* 9657

N'-(4-Chloro-2-methylphenyl)-*N*,*N*-dimethylmethanimidamide *see* 2133

N-(3-Chloro-4-methylphenyl)-2-methylpentanamide *see* 8851

1-Chloro-2-methyl-2-phenylpropane *see* 6548

8-Chloro-1-methyl-6-phenyl-4*H*-

N^l-*p*-Chlorophenyl-N^5-isopropyl-
diguanide *see* 2138
Chlorophenylmagnesium *see* 7451
Chlorophenylmercury *see* 7454
1-[2-[(4-Chlorophenyl)methoxy]-2-(2,-
4-dichlorophenyl)ethyl]-1*H*-imida-
zole *see* 3550
1-[1-[2-[(3-Chlorophenyl)methoxy]-
phenyl]ethenyl]-1*H*-imidazole *see*
2429
(±)-2-(2-Chlorophenyl)-2-(methylamino)cyclohexanone *see* 5306
1-(*o*-Chlorophenyl)-2-methyl-2-amino-
propane *see* 2472
1-(*p*-Chlorophenyl)-2-methyl-2-amino-
propane *see* 2235
2-(4-Chlorophenyl)-α-methyl-5-benz-
oxazoleacetic Acid *see* 1075
2-[(4-Chlorophenyl)methyl]-*N*,*N*-dieth-
yl-5-nitro-1*H*-benzimidazole-1-
ethanamine *see* 2451
(*R**,*R**)-(±)-β-[(4-Chlorophenyl)meth-
yl]-α-(1,1-dimethylethyl)-1*H*-1,2,4-
triazole-1-ethanol *see* 7118
2-[(2-Chlorophenyl)methyl]-4,4-di-
methyl-3-isoxazolidinone *see* 3262
N-[(4-Chlorophenyl)methyl]-*N*′,*N*′-
dimethyl-*N*-2-pyridinyl-1,2-ethane-
diamine *see* 2214
[(2-Chlorophenyl)methylene]propane-
dinitrile *see* 2178
N-(4-Chlorophenyl)-*N*′-(1-methyleth-
yl)imidodicarbonimidic Diamide *see*
2138
N-(4-Chlorophenyl)-*N*-[1-(1-methyleth-
yl)-4-piperidinyl]benzeneacetamide
see 5610
4-[(4-Chlorophenyl)methyl]-2-(hexa-
hydro-1-methyl-1*H*-azepin-4-yl)-
1(2*H*)-phthalazinone *see* 939
6-[3-(*o*-Chlorophenyl)-5-methyl-4-
isoxazolecarboxamido]penicillanic
Acid *see* 2480
[2*S*-(2α,5α,6β)]-6-[[[3-(2-Chlorophen-
yl)-5-methyl-4-isoxazolyl]carbonyl]-
amino]-3,3-dimethyl-7-oxo-4-thia-1-
azabicyclo[3.2.0]heptane-2-carbox-
ylic Acid *see* 2480
[3-(*o*-Chlorophenyl)-5-methyl-4-
isoxazolyl]penicillin *see* 2480
2-(4-Chlorophenyl)-3-methyl-4-meta-
thiazanone 1,1-Dioxide *see* 2155
(+)-*N*-[(4-Chlorophenyl)methyl]-α-
methylbenzeneethanamine *see* 2421
1-[(4-Chlorophenyl)methyl]-2-methyl-
1*H*-benzimidazole *see* 2156
2-(4-Chlorophenyl)-4-methyl-2,4-pen-
tanediol *see* 4029
4-Chloro-2-(phenylmethyl)phenol *see*
2469
(±)-2-[4-[(4-Chlorophenyl)methyl]-
phenoxy]-2-methylbutanoic Acid
Ethyl Ester *see* 1046
7-[3-[4-(4-Chlorophenyl)methyl]-1-
piperazinyl]propoxy]-3,4-dimethyl-
2*H*-1-benzopyran-2-one *see* 7573
3-Chloro-*N*-(phenylmethyl)propan-
amide *see* 1045
1-(*o*-Chlorophenyl)-2-methyl-2-propyl-
amine *see* 2472
1-(*p*-Chlorophenyl)-2-methyl-2-propyl-
amine *see* 2235
1-(3-Chlorophenyl)-4-[2-(5-methyl-1*H*-
pyrazol-3-yl)ethyl]piperazine *see*
5902
1-[(4-Chlorophenyl)methyl]-2-(1-pyr-
rolidinylmethyl)-1*H*-benzimidazole
see 2406
3-(2-Chlorophenyl)-2-methyl-4(3*H*)-
quinazolinone *see* 5821

5-[(2-Chlorophenyl)methyl]-4,5,6,7-
tetrahydrothieno[3,2-*c*]pyridine *see*
9569
3-[4-(2-Chlorophenyl)-9-methyl-6*H*-
thieno[3,2-*f*][1,2,4]triazolo[4,3-*a*][1,-
4]diazepin-2-yl]-1-(4-morpholinyl)-
1-propanone *see* 765
4-[3-[4-(2-Chlorophenyl)-9-methyl-6*H*-
thieno[3,2-*f*][1,2,4]triazolo[4,3-*a*][1,-
4]diazepin-2-yl]-1-oxopropyl]mor-
pholine *see* 765
1-[2-[[(4-Chlorophenyl)methyl]thio]-2-
(2,4-dichlorophenyl)ethyl]-1*H*-
imidazole *see* 9062
6-Chloro-3-[[(phenylmethyl)thio]meth-
yl]-2*H*-1,2,4-benzothiadiazine-7-
sulfonamide 1,1-Dioxide *see* 1155
1-(*p*-Chlorophenyl)-1-(nicotinoyloxy)-
2-methylpropane *see* 6603
3-[1-(4-Chlorophenyl)-3-oxobutyl]-4-
hydroxy-2*H*-1-benzopyran-2-one *see*
2623
2-(*p*-Chlorophenyl)perhydro-3-methyl-
1,3-thiazin-4-one 1,1-Dioxide *see*
2155
2-[(*p*-Chlorophenyl)phenylacetyl]-1,3-
indandione *see* 2204
2-(4-Chlorophenyl)phenylacetyl]-1*H*-
indene-1,3(2*H*)-dione *see* 2204
1-[4-(4-Chlorophenyl)-3-phenyl-2-
butenyl]pyrrolidine *see* 8196
1-*o*-Chlorophenyl-1-phenyl-3-dimeth-
ylamino-1-propanol *see* 2106
2-[1-(4-Chlorophenyl)-1-phenyleth-
oxy]-*N*,*N*-dimethylethanamine *see*
2234
2-[1-(4-Chlorophenyl)-1-phenyleth-
oxy]-*N*,*N*-dimethyl-1-propanamine
see 5823
1-[2-[1-(4-Chlorophenyl)-1-phenyleth-
oxy]ethyl]hexahydro-1*H*-azepine *see*
8619
[*R*-(*R**,*R**)]-2-[2-[1-(4-Chlorophenyl)-
1-phenylethoxy]ethyl]-1-methyl-
pyrrolidine *see* 2405
[1-(*p*-Chlorophenyl)-1-phenyl]ethyl (β-
Dimethylaminoethyl) Ether *see* 2234
1-[2-[(2-Chlorophenyl)phenylmeth-
oxy]ethyl]-4-[(2-methylphenyl)meth-
yl]piperazine *see* 2125
1-[2-[(4-Chlorophenyl)phenylmeth-
oxy]ethyl]piperidine *see* 2456
3-[(4-Chlorophenyl)phenylmethoxy]-8-
methyl-8-azabicyclo[3.2.1]octane *see*
2422
1-[(4-Chlorophenyl)phenylmethyl]-4-
[[4-(1,1-dimethylethyl)phenyl]meth-
yl]piperazine *see* 1484
1-[(4-Chlorophenyl)phenylmethyl]hex-
ahydro-4-methyl-1*H*-1,4-diazepine
see 4770
1-[(4-Chlorophenyl)phenylmethyl]-4-
[(3-methylphenyl)methyl]piperazine
see 5817
1-[(4-Chlorophenyl)phenylmethyl]-4-
methylpiperazine *see* 2128
1-[(4-Chlorophenyl)phenylmethyl]-4-
(3-phenyl-2-propenyl)piperazine *see*
2428
[2-[4-[(4-Chlorophenyl)phenylmethyl]-
1-piperazinyl]ethoxy]acetic Acid *see*
2063
2-[2-[4-[(4-Chlorophenyl)phenylmeth-
yl]-1-piperazinyl]ethoxy]ethanol *see*
4897
2-[2-[2-[4-[(4-Chlorophenyl)phenyl-
methyl]-1-piperazinyl]ethoxy]eth-
oxy]ethanol *see* 3921
3-(4-Chlorophenyl)-1-phenyl-1*H*-pyr-
azole-4-acetic Acid *see* 5596

1-*p*-Chlorophenyl-2-phenyl-4-pyrrolid-
yl-2-butene *see* 8196
p-Chlorophenyl Phenyl Sulfone *see*
9162
4-(4-Chlorophenyl)-2-phenyl-5-thia-
zoleacetic Acid *see* 4045
(*R*,*S*)-4-(4-Chlorophenyl)-2-phenyl-2-
[(1*H*-1,2,4-triazol-1-yl)methyl]buty-
ronitrile *see* 4002
3-[2-(*N*′-*m*-Chlorophenylpiperazino)-
ethyl]-5-methylpyrazole *see* 5902
2-[3-[4-(3-Chlorophenyl)-1-piperazin-
yl]propyl]-4,5-diethyl-2,4-dihydro-
3*H*-1,2,4-triazol-3-one *see* 3930
2-[3-[4-(3-Chlorophenyl)-1-piperazin-
yl]propyl]-5-ethyl-2,4-dihydro-4-(2-
phenoxyethyl)-3*H*-1,2,4-triazol-3-
one *see* 6527
2-[3-[4-(3-Chlorophenyl)-1-piperazin-
yl]propyl]-5-ethyl-4-(2-phenoxyeth-
yl)-2*H*-1,2,4-triazol-3(4*H*)-one *see*
6527
2-[3-[4-(3-Chlorophenyl)-1-piperazin-
yl]propyl]-1,2,4-triazolo[4,3-*a*]pyri-
din-3(2*H*)-one *see* 9712
O-(6-Chloro-3-phenyl-4-pyridazinyl)-
carbonothioic Acid *S*-Octyl Ester
see 8151
2-[(4-Chlorophenyl)-2-pyridinylmeth-
oxy]-*N*,*N*-dimethylethanamine *see*
1845
1-(*p*-Chlorophenyl)-1-(2-pyridyl)-3-
dimethylaminopropane *see* 2232
1-(*p*-Chlorophenyl)-1-(2-pyridyl)-3-
N,*N*-dimethylpropylamine *see* 2232
3-(*p*-Chlorophenyl)-3-(2-pyridyl)-*N*,*N*-
dimethylpropylamine *see* 2232
γ-(4-Chlorophenyl)-γ-(2-pyridyl)pro-
pyldimethylamine *see* 2232
2-[*p*-(β-Chloro-α-phenylstyryl)phen-
oxy]triethylamine *see* 2446
3-Chloro-4-(phenylsuccinimido)benz-
enesulfonamide *see* 9048
7-Chloro-2-phenyl-6-sulfamyl-1,2,3,4-
tetrahydro-4-quinazolinone *see* 4039
4-[2-[[(4-Chlorophenyl)sulfonyl]ami-
no]ethyl]benzeneacetic Acid *see*
2871
1-Chloro-4-(phenylsulfonyl)benzene
see 9162
1-(*p*-Chlorophenylsulfonyl)-3-(hexahy-
dro-1*H*-azepin-1-yl)urea *see* 4524
1-(*p*-Chlorophenylsulfonyl)-3-propyl-
urea *see* 2239
2-(4-Chlorophenyl)tetrahydro-3-meth-
yl-4*H*-1,3-thiazin-4-one 1,1-Dioxide
see 2155
2-(4-Chlorophenyl)-4-thiazoleacetic
Acid *see* 4010
2-(*p*-Chlorophenyl)thiazol-4-ylacetic
Acid *see* 4010
S-[[(*p*-Chlorophenyl)thio]methyl] *O*,*O*-
Diethyl Phosphorodithioate *see*
1869
[[(4-Chlorophenyl)thio]methylene]-
bisphosphonic Acid *see* 9582
N-(4-Chlorophenyl)-1,3,5-triazine-2,4-
diamine *see* 2123
2-(*p*-Chlorophenyl)-*s*-triazolo[5,1-a]-
isoquinoline *see* 5615
8-Chloro-6-phenyl-4*H*-[1,2,4]triazolo-
[4,3-*a*][1,4]benzodiazepine *see* 3744
8-Chloro-6-phenyl-4*H*-*s*-triazolo[4,3-
a][1,4]benzodiazepine *see* 3744
2-(4-Chlorophenyl)-(1,2,4)triazolo[5,1-
a]isoquinoline *see* 5615
2-(4-Chlorophenyl)-2-(1*H*-1,2,4-tria-
zol-1-ylmethyl)hexanenitrile *see*
6402

Choline Dichloride *see* 2153
Choline Dihydrogen Citrate, 2263
Choline Esterase, 2264
D-Choline Hydroxide 2,3-Dihydroxypropyl Hydrogen Phosphate Inner Salt *see* 2260
Choline 1,5-Naphthalenedisulfonate (2:1), Dilactate, Diacetate *see* 116
Choline Orotate *see* 7004
Choline Phosphate Chloride *see* 7519
Choline Phosphate Hexadecyl Ester, Hydroxide, Inner Salt *see* 6285
Choline Phosphoric Acid Ester (Chloride) *see* 7519
"Choline Plasmalogen" *see* 7677
Choline Salicylate, 2265
Choline Salicylic Acid Salt *see* 2265
Choline Salt *see* 4085
Choline Succinate Dichloride *see* 9044
Choline Theophyllinate, 2266
Cholinfall [Tanabe Seiyaku] *see* 6059
Cholinophylline *see* 2266
Cholit-Ursan [Fresenius] *see* 10026
Chologon *see* 2922
Cholografin [Squibb] *see* 5043
Cholonerton [Dolorgiet] *see* 4903
Cholospect *see* 5043
Choloxin [Boots] *see* 9555
N-Choloyltaurine *see* 9242
Cholspasmin [Lipha] *see* 4903
Cholumbrin *see* 5059
Cholybar [Warner-Lambert] *see* 2257
N-Cholylglycine *see* 4504
Cholyltaurine *see* 9242
d-Chondocurine *see* 2268
Chondodendrine *see* 1042
Chondrillasterol, 2267
Chondrocurine, 2268
Chondrofoline, 2269
Chondroitin Sulfate, 2270
Chondroitin 4-Sulfate *see* 2270
Chondroitin 6-Sulfate *see* 2270
Chondroitin Sulfate A *see* 2270
Chondroitin Sulfate B *see* 2270
Chondroitin Sulfate C *see* 2270
Chondroitin 4-Sulfate Disodium Salt *see* 2270
Chondroitinsulfuric Acid *see* 2270
Chondrosamine *see* 4352
Chondrosine, 2271
Chonemorphine, 2272
Chonsurid *see* 2270
Chop Nut *see* 7539
Chopper [Am. Cyan.] *see* 4942
Choragon [Ferring] *see* 2273
Chorex [Hyrex] *see* 2273
Choriogonadotropin *see* 2273
Choriogonin [Gedeon Richter] *see* 2273
Chorionic Gonadotropin, 2273
Chorionic Growth Hormone-prolactin *see* 4650
Chorismic Acid, 2274
Choron [Forest] *see* 2273
Chothyn [Tilden-Yates] *see* 2263
CHQ *see* 4637
Chrisanol *see* 1689
Christmas Factor *see* 3964
Chromargyre *see* 5914
Chromatophorotropic Hormone *see* 6377
Chrome Alum *see* 2286
Chrome Alum Ammonium *see* 539
Chrome Green *see* 2283
1,2-Chromene *see* 1133
3-Chromene *see* 1133
Chrome Ocher *see* 2283
Chrome Oxide Green *see* 2283
Chrome Red *see* 5423
Chrome Yellow *see* 5422

Chromia *see* 2283
Chromic Acetate, 2275
Chromic Acid *see* 2293
Chromic Acid Iron(3+) Salt (3:2) *see* 4062
Chromic Acid Lead(2+) Salt (1:2) *see* 5423
Chromic Ammonium Sulfate *see* 539
Chromic Anhydride *see* 2293
Chromic Bromide, 2276
Chromic Carbonate, 2277
Chromic Chloride, 2278
Chromic Fluoride, 2279
Chromic Formate, 2280
Chromic Hydroxide, 2281
Chromic Nitrate, 2282
Chromic Oxide, 2283
Chromic Oxide Gel *see* 2281
Chromic Oxide, Hydrous *see* 2281
Chromic Phosphate, 2284
Chromic Potassium Oxalate, 2285
Chromic Potassium Sulfate, 2286
Chromic Sulfate, 2287
Chromite *see* 2288
Chromitope Sodium [Squibb] *see* 8745
Chromium, 2288
Chromium Carbonyl, 2289
Chromium Dioxide, 2290
Chromium Dioxychloride *see* 2306
Chromium Hexacarbonyl *see* 2289
Chromium Hydroxide *see* 2281
Chromium Lead Oxide *see* 5423
Chromium Oxyfluoride *see* 2307
Chromium Picolinate, 2291
Chromium Sesquioxide *see* 2283
Chromium Tetrafluoride, 2292
Chromium Trioxide, 2293
Chromium(III) Trispicolinate *see* 2291
Chromocarb, 2294
574-Chromogen *see* 4686
Chromomycin A₃ *see* 2295
Chromomycinone *see* 7696
Chromomycins, 2295
Chromonar, 2296
2-Chromonecarboxylic Acid *see* 2294
Chromotrichia Factor *see* 443
Chromotrope 2B, 2297
Chromotropic Acid, 2298
Chromous Acetate, 2299
Chromous Bromide, 2300
Chromous Chloride, 2301
Chromous Fluoride, 2302
Chromous Formate, 2303
Chromous Oxalate, 2304
Chromous Sulfate, 2305
Chromyl Chloride, 2306
Chromyl Fluoride, 2307
Chronadalate [Bayer] *see* 6617
Chronexan [Sarget] *see* 10212
Chronogest [Intervet] *see* 4235
Chronogyn [Winthrop] *see* 2875
Chrono-Indocid [Merck & Co.] *see* 4998
Chronosyn [Veterinaria] *see* 2152
Chrysammic Acid *see* 2308
Chrysamminic Acid, 2308
Chrysanol *see* 1689
Chrysanthemaxanthin, 2309
Chrysanthemic Acid, 2310
Chrysanthemin *see* 2757
N-(Chrysanthemoxymethyl)-1-cyclohexene-1,2-dicarboximide *see* 9362
Chrysanthemumdicarboxylic Acid Monomethyl Ester Pyrethrolone Ester *see* 8148
Chrysanthemumic Acid *see* 2310
Chrysanthemummonocarboxylic Acid *see* 2310
Chrysanthemummonocarboxylic Acid 6-Chloropiperonyl Ester *see* 1032

Chrysanthemummonocarboxylic Acid Ester with 3-(2-Cyclopenten-1-yl)-2-methyl-4-oxo-2-cyclopenten-1-ol *see* 2774
Chrysanthemummonocarboxylic Acid Pyrethrolone Ester *see* 8148
Chrysanthenone, 2311
Chrysarobin, 2312
Chrysatropic Acid *see* 8552
Chrysazin *see* 2878
Chrysazin-3-carboxylic Acid *see* 8342
6-Chrysenamine, 2313
Chrysene, 2314
Chrysenex *see* 2313
6-Chrysenylamine *see* 2313
Chrysergonic Acid, 2315
Chrysidenon 1438 *see* 2316
Chrysin, 2316
Chrysoberyl *see* 1201
Chrysogen *see* 6456
Chrysoidine, 2317
Chrysoidine Orange *see* 2317
Chrysoidine Y *see* 2317
Chrysoine *see* 9905
Chrysomykine *see* 2245
Chrysophanein *see* 2318
Chrysophanic Acid, 2318
Chrysophaniin *see* 2318
Chrysophanol *see* 2318
Chrysotile *see* 863 and 5727
6-Chrysylamine *see* 2313
Chrytemin [Fujinaga] *see* 4955
ChTX *see* 2087
Chuanxinlian *see* 673
Chugaev's Reagent *see* 3295
CHX-100 *see* 6786
CHX-3101 *see* 8890
CHX-3311 *see* 2853
CHX-3673 *see* 515
Chymar [Armour Pharm.] *see* 2320
Chymetin [BDH] *see* 2320
Chymex [Adria] *see* 1081
Chymodiactin [Smith] *see* 2319
Chymolase *see* 2320
Chymopapain, 2319
Chymosin *see* 8303
Chymotase (tabl.) *see* 2320
Chymotrypsins, 2320
Chymozym [Teikoku Hormone] *see* 2320
CI-336 *see* 1849
CI-379 *see* 1110
CI-395 *see* 7364
CI-416 *see* 9787
CI-440 *see* 4167
CI-473 *see* 5842
CI-501 *see* 2790
CI-556 *see* 20
CI-581 *see* 5306
CI-583 *see* 5819
CI-588 *see* 3411
CI-633 *see* 2416
CI-673 *see* 10113
CI-719 *see* 4394
CI-775 *see* 1238
CI-825 *see* 7277
CI-845 *see* 7656
CI-879 *see* 7886
CI-880 *see* 635
CI-898 *see* 9851
CI-904 *see* 3044
CI-906 *see* 8233
CI-912 *see* 10323
CI-919 *see* 3625
CI-928 *see* 8233
CI-945 *see* 4343
CI-960 *see* 2413
CI-978 *see* 8884
CI-981 *see* 897
CI-991 *see* 9898

C.I. **10316** *see* 6480
C.I. **11000** *see* 439
C.I. **11020** *see* 3279
C.I. **11050** *see* 5269
C.I. **11160** *see* 440
C.I. **11270** *see* 2317
C.I. **11380** *see* 10228
C.I. **11390** *see* 10229
C.I. **12140** *see* 10222
C.I. **12156** *see* 2394
C.I. **13020** *see* 6199
C.I. **13025** *see* 6180
C.I. **13065** *see* 5989
C.I. **13080** *see* 9906
C.I. **13390** *see* 669
C.I. **14025** *see* 5979
C.I. **14030** *see* 251
C.I. **14130** *see* 6976
C.I. **14270** *see* 9905
C.I. **14600** *see* 6989
C.I. **14645** *see* 3709
C.I. **14700** *see* 7750
C.I. **15510** *see* 6990
C.I. **15985** *see* 9176
C.I. **16035** *see* 292
C.I. **16155** *see* 7749
C.I. **16185** *see* 391
C.I. **16575** *see* 2297
C.I. **18950** *see* 7713
C.I. **19140** *see* 9239
C.I. **19235** *see* 6991
C.I. **19540** *see* 9445
C.I. **21000** *see* 1294
C.I. **21010** *see* 1293
C.I. **21010:1** *see* 1293
C.I. **22120** *see* 2562
C.I. **22850** *see* 9924
C.I. **23500** *see* 1132
C.I. **23570** *see* 10149
C.I. **23850** *see* 9923
C.I. **23860** *see* 3952
C.I. **24140** *see* 1115
C.I. **26100** *see* 9054
C.I. **26105** *see* 8538
C.I. **37500** *see* 6471
C.I. **40600** *see* 1349
C.I. **40850** *see* 1798
C.I. **42000** *see* 5739
C.I. **42040** *see* 1398
C.I. **42045** *see* 9161
C.I. **42053** *see* 3979
C.I. **42085** *see* 4603
C.I. **42090** *see* 1397
C.I. **42095** *see* 5508
C.I. **42510** *see* 5686
C.I. **42555** *see* 4401
C.I. **42590** *see* 6157
C.I. **42685** *see* 109
C.I. **42745** *see* 110
C.I. **42780** *see* 6107
C.I. **43800** *see* 915
C.I. **45005** *see* 8190
C.I. **45010** *see* 8189
C.I. **45170** *see* 8349
C.I. **45350** *see* 4194
C.I. **45350:1** *see* 4194
C.I. **45365** *see* 3117
C.I. **45370:1** *see* 3070
C.I. **45380** *see* 3639
C.I. **45380:2** *see* 3639
C.I. **45400** *see* 3638
C.I. **45425** *see* 3233
C.I. **45425:1** *see* 3233
C.I. **45430** *see* 3734
C.I. **45440** *see* 8421
C.I. **45445** *see* 4362
C.I. **47000** *see* 8257
C.I. **47005** *see* 8256
C.I. **49410** *see* 9678
C.I. **50040** *see* 6571

C.I. **50200** *see* 7402
C.I. **51030** *see* 4368
C.I. **51050** *see* 2006
C.I. **51400** *see* 5345
C.I. **52000** *see* 9483
C.I. **52002** *see* 959
C.I. **52005** *see* 957
C.I. **52010** *see* 958
C.I. **52015** *see* 6137
C.I. **52040** *see* 9658
C.I. **57000** *see* 4360
C.I. **58000** *see* 247
C.I. **58005** *see* 8714
C.I. **58015** *see* 250
C.I. **58050** *see* 8250
C.I. **58205** *see* 8132
C.I. **58500** *see* 8231
C.I. **58600** *see* 8447
C.I. **61565** *see* 8251
C.I. **61570** *see* 248
C.I. **67410** *see* 249
C.I. **73000** *see* 4977
C.I. **73015** *see* 4978
C.I. **74160** *see* 2584
C.I. **74240** *see* 219
C.I. **75140** *see* 1918
C.I. **75280** *see* 1392
C.I. **75300** *see* 2744
C.I. **75410** *see* 8132
C.I. **75450** *see* 5342
C.I. **75460** *see* 5304
C.I. **75470** *see* 1891
C.I. **75500** *see* 5282
C.I. **75530** *see* 253
C.I. **75620** *see* 4129
C.I. **76050** *see* 3018
C.I. **76076** *see* 7439
C.I. **77007** *see* 9980
C.I. **77085** *see* 834
C.I. **77103** *see* 999
C.I. **77199** *see* 1666
C.I. **77223** *see* 1700
C.I. **77266** *see* 1856
C.I. **77288** *see* 2283
C.I. **77350** *see* 2495
C.I. **77357** *see* 2493
C.I. **77360** *see* 669
C.I. **77402** *see* 2734
C.I. **77482** *see* 4538
C.I. **77505** *see* 4062
C.I. **77510** *see* 4064
C.I. **77578** *see* 5449
C.I. **77597** *see* 1034
C.I. **77766** *see* 5945
C.I. **77800** *see* 8453
C.I. **77891** *see* 9612
C.I. **77947** *see* 10279
C.I. **Acid Blue 1** *see* 9161
C.I. **Acid Blue 9** *see* 1397
C.I. **Acid Blue 74** *see* 4978
C.I. **Acid Blue 92** *see* 669
C.I. **Acid Blue 93** *see* 6107
C.I. **Acid Green 3** *see* 4603
C.I. **Acid Green 5** *see* 5508
C.I. **Acid Green 25** *see* 248
C.I. **Acid Orange 5** *see* 9906
C.I. **Acid Orange 6** *see* 9905
C.I. **Acid Orange 7** *see* 6990
C.I. **Acid Orange 11** *see* 3070
C.I. **Acid Orange 20** *see* 6989
C.I. **Acid Orange 52** *see* 6180
C.I. **Acid Orange 137** *see* 6991
C.I. **Acid Red 2** *see* 6199
C.I. **Acid Red 27** *see* 391
C.I. **Acid Red 51** *see* 3734
C.I. **Acid Red 87** *see* 3639
C.I. **Acid Red 91** *see* 3638
C.I. **Acid Red 94** *see* 8421
C.I. **Acid Red 95** *see* 3233
C.I. **Acid Red 176** *see* 2297

C.I. **Acid Violet 19** *see* 109
C.I. **Acid Violet 25** *see* 110
C.I. **Acid Yellow 1** *see* 6480
C.I. **Acid Yellow 3** *see* 8256
C.I. **Acid Yellow 23** *see* 9239
C.I. **Acid Yellow 36** *see* 5989
C.I. **Acid Yellow 40** *see* 7713
C.I. **Acid Yellow 73** *see* 4194
C.I. **Azoic Coupling Component 1** *see* 6471
C.I. **Basic Blue 9** *see* 6137
C.I. **Basic Blue 17** *see* 9658
C.I. **Basic Brown 1** *see* 1294
C.I. **Basic Brown 4** *see* 1293
C.I. **Basic Green 1** *see* 1398
C.I. **Basic Green 4** *see* 5739
C.I. **Basic Orange 2** *see* 2317
C.I. **Basic Red 5** *see* 6571
C.I. **Basic Violet 3** *see* 4401
C.I. **Basic Violet 10** *see* 8349
C.I. **Basic Violet 14** *see* 5686
C.I. **Developer 1** *see* 6809
C.I. **Developer 5** *see* 6471
C.I. **Direct Blue 8** *see* 1115
C.I. **Direct Blue 14** *see* 9923
C.I. **Direct Blue 53** *see* 3952
C.I. **Direct Red 2** *see* 1132
C.I. **Direct Red 28** *see* 2562
C.I. **Direct Red 34** *see* 10149
C.I. **Direct Yellow 9** *see* 9445
C.I. **Fluorescent Brightener 30** *see* 1349
C.I. **Food Blue 1** *see* 4978
C.I. **Food Blue 2** *see* 1397
C.I. **Food Blue 3** *see* 9161
C.I. **Food Green 1** *see* 4603
C.I. **Food Green 3** *see* 3979
C.I. **Food Red 1** *see* 7750
C.I. **Food Red 6** *see* 7749
C.I. **Food Red 14** *see* 3734
C.I. **Food Red 17** *see* 292
C.I. **Food Yellow 3** *see* 9176
C.I. **Food Yellow 4** *see* 9239
C.I. **Food Yellow 8** *see* 9905
C.I. **Ingrain Blue 1** *see* 219
C.I. **Mordant Black 11** *see* 3709
C.I. **Mordant Blue 10** *see* 4368
C.I. **Mordant Blue 14** *see* 2006
C.I. **Mordant Orange 1** *see* 251
C.I. **Mordant Orange 14** *see* 250
C.I. **Mordant Red 3** *see* 8714
C.I. **Mordant Red 11** *see* 247
C.I. **Mordant Violet 26** *see* 8231
C.I. **Mordant Yellow 1** *see* 5979
C.I. **Mordant Yellow 5** *see* 6976
C.I. **Natural Brown 1** *see* 4129
C.I. **Natural Brown 7** *see* 5282
C.I. **Natural Red 3** *see* 5304
C.I. **Natural Red 4** *see* 1891
C.I. **Natural Red 8** *see* 8132
C.I. **Natural Red 16** *see* 8132
C.I. **Natural Red 20** *see* 253
C.I. **Natural Red 24** *see* 1392
C.I. **Natural Red 25** *see* 5342
C.I. **Natural Red 26** *see* 1918
C.I. **Natural Yellow 3** *see* 2744
C.I. **Natural Yellow 8** *see* 6352
C.I. **Natural Yellow 11** *see* 5676 and 6352
C.I. **Oxidation Base 12** *see* 3018
C.I. **Pigment Blue 15** *see* 2584
C.I. **Pigment Blue 27** *see* 4064
C.I. **Pigment Blue 29** *see* 9980
C.I. **Pigment Blue 66** *see* 4977
C.I. **Pigment Brown 9** *see* 2703
C.I. **Pigment Green 17** *see* 2283
C.I. **Pigment Red 83** *see* 247
C.I. **Pigment Red 105** *see* 5449
C.I. **Pigment Red 106** *see* 5945
C.I. **Pigment Red 109** *see* 4538

C.I. Pigment Violet 14 *see* 2508
C.I. Pigment White 1 *see* 1034
C.I. Pigment White 4 *see* 10279
C.I. Pigment White 5 *see* 5572
C.I. Pigment White 6 *see* 9612
C.I. Pigment Yellow 31 *see* 999
C.I. Pigment Yellow 33 *see* 1700
C.I. Pigment Yellow 36 *see* 10262
C.I. Pigment Yellow 39 *see* 834
C.I. Pigment Yellow 40 *see* 2493
C.I. Pigment Yellow 45 *see* 4062
C.I. Solvent Brown 12 *see* 1293
C.I. Solvent Green 3 *see* 8251
C.I. Solvent Orange 3 *see* 2317
C.I. Solvent Orange 7 *see* 10222
C.I. Solvent Orange 32 *see* 3117
C.I. Solvent Red 23 *see* 9054
C.I. Solvent Red 24 *see* 8538
C.I. Solvent Red 43 *see* 3639
C.I. Solvent Red 72 *see* 3070
C.I. Solvent Red 73 *see* 3233
C.I. Solvent Red 80 *see* 2394
C.I. Solvent Yellow 1 *see* 439
C.I. Solvent Yellow 2 *see* 3279
C.I. Solvent Yellow 3 *see* 440
C.I. Solvent Yellow 5 *see* 10228
C.I. Solvent Yellow 6 *see* 10229
C.I. Solvent Yellow 33 *see* 8257
C.I. Solvent Yellow 94 *see* 4194
C.I. Vat Blue 1 *see* 4977
Ciafos, 2321
Ciamin *see* 867
Cianatil *see* 2753
Cianurina *see* 5928
Ciatyl [Troponwerke] *see* 2455
Ciba 570 *see* 7493
Ciba 2059 *see* 4189
Ciba 3126 *see* 6224
Ciba 4311b *see* 6186
Ciba 5968 *see* 4800
Ciba 8514 *see* 2133
Ciba 9295 *see* 929
Ciba 10370 *see* 9068
Ciba 11925 *see* 7337
Ciba 32644-Ba *see* 6656
Ciba 36278-Ba *see* 2019
Cibacalcin [Ciba] *see* 1680
Cibacen [Ciba-Geigy] *see* 1058
Cibacène [Ciba-Geigy] *see* 1058
Cibacthen [Ciba] *see* 136
Cibalgina [Ciba] *see* 8056
Cibazol [Ciba] *see* 9115
Cibenol [Fujisawa] *see* 2330
Cibenzoline *see* 2330
Ciberon [Taisho] *see* 1845
Cibian [Yamanouchi] *see* 5016
Ciblor [Inava] *see* 2402
Cichorigenin *see* 3738
Cichoriin, 2322
Ciclacillin *see* 2769
Cicladol [Master] *see* 7661
Cicletanide *see* 2323
Cicletanine, 2323
Ciclobendazole *see* 2781
Ciclobiotic [Beta] *see* 6007
Ciclochem [Novag] *see* 2325
Ciclolysal *see* 5658
Ciclonicate, 2324
Ciclonium Iodide *see* 2806
Ciclopirox, 2325
Ciclopirox Olamine *see* 2325
Ciclosidomine, 2326
Ciclosporin *see* 2821
Cicloven [Agips] *see* 8158
Cicrotoic Acid, 2327
Cicutine *see* 2569
Cicutoxin, 2328
Cidal [CID] *see* 8480
Cidandopa [Cidán] *see* 5490
Cidex [Arbrook] *see* 4480

Cidifos [Neopharmed] *see* 2380
Cidine [Almirall] *see* 2352
Cidofovir, 2329
Cidomycin [Roussel] *see* 4398
Cidoxepin Hydrochloride *see* 3492
Cidrex *see* 4822
Cifenline, 2330
Ciflox [Bayer] *see* 2374
CIG *see* 4119
Cignolin [Bayer] *see* 723
Cigthranol [Bayer] *see* 723
Cilastatin, 2331
Cilastatin Sodium *see* 2331
Cilazapril, 2332
Cilazaprilat *see* 2332
Cilest [Cilag] *see* 6796
Ciliary Neurotrophic Factor, 2333
Cilicaine *see* 7226
Cilifor [C.E.P.A.] *see* 2025
Cillenta [Ayerst] *see* 7221
Cilleral [Bristol] *see* 628
Cilligen [Sigma] *see* 7233
Cillimycin [Wyeth] *see* 5525
Cilnidipine, 2334
Ciloprost *see* 4940
Cilostazol, 2335
Ciloxan [Alcon] *see* 2374
Cimal [AL] *see* 2337
Cimaterol, 2336
Cimectacarb *see* 9855
Cimedone [Specia] *see* 8859
Cimetag [Cehasol] *see* 2337
Cimetidine, 2337
Cimetropium Bromide, 2338
Cimetum [Sintyal] *see* 2337
Cimicifuga, 2339
Cimicifugol *see* 2340
Cimigenol, 2340
Cinametic Acid, 2341
Cinaperazine [Yakult] *see* 2365
Cinarine *see* 2835
Cinazyn [Italchimici] *see* 2365
Cincaine *see* 3081
Cinch [Du Pont] *see* 2354
Cinchamidine *see* 4823
Cinchocaine *see* 3081
Cinchol *see* 8697
Cincholepidine *see* 5464
Cincholic Acid β-D-Quinovoside *see* 8261
Cinchomeronic Acid, 2342
Cinchona, 2343
Cinchona Bark *see* 2343
Cinchonamine, 2344
(8α,9R)-Cinchonan-6′,9-diol *see* 2689
(8α,9R)-Cinchonan-9-ol *see* 2345
(8α,9S)-Cinchonan-9-ol *see* 2345
(9R)-Cinchonan-9-ol *see* 2346
(9S)-Cinchonan-9-ol *see* 2346
Cinchonicine *see* 2348
Cinchonidine, 2345
Cinchonine, 2346
Cinchophen, 2347
Cinchotine *see* 4824
Cinchotoxine, 2348
Cinchovatine *see* 821 and 2345
Cincofarm [Lepori] *see* 4895
Cinconal *see* 2347
Cinconifine *see* 4824
Cincosal *see* 2347
Cindomet [Chiesi] *see* 2353
Cinene *see* 5518
Cineole *see* 3940
Cinepazet, 2349
Cinepazic Acid Ethyl Ester *see* 2349
Cinepazide, 2350
Cinerin I *see* 2351
Cinerin II *see* 2351
Cinerins, 2351
Cinitapride, 2352

Cinmetacin, 2353
Cinmethylin, 2354
Cinnabar *see* 5945
Cinnabarine, 2355
Cinnacet [Schwarzhaupt] *see* 2365
Cinnageron [Streuli] *see* 2365
Cinnaloid [Pfizer] *see* 8311
Cinnamal *see* 2356
Cinnamaldehyde, 2356
Cinnamedrine, 2357
Cinnamein *see* 1165
Cinnamene *see* 9028
Cinnamic Acid, 2358
trans-Cinnamic Acid Benzyl Ester *see* 1165
Cinnamic Acid Cinnamyl Ester *see* 2364
Cinnamic Alcohol *see* 2362
Cinnamic Aldehyde *see* 2356
Cinnamic Anhydride *see* 2358
Cinnamin [Nippon Chemiphar] *see* 768
Cinnamol *see* 9028
Cinnamon, 2359
Cinnamonin *see* 6403
Cinnamon Wood *see* 8528
Cinnamoyl Chloride, 2360
Cinnamoylcocaine *see* 2361
Cinnamoylecgonine Methyl Ester *see* 2361
1-Cinnamoyl-5-methoxy-2-methylindole-3-acetic Acid *see* 2353
Cinnamoylmethylecgonine *see* 2361
1-Cinnamoyl-2-methyl-5-methoxy-3-indolylacetic Acid *see* 2353
Cinnamyl Alcohol, 2362
Cinnamyl *o*-Aminobenzoate *see* 2363
Cinnamyl Anthranilate, 2363
1-Cinnamyl-4-benzhydrylpiperazine *see* 2365
1-Cinnamyl-4-(4-chlorobenzhydryl)-piperazine *see* 2428
1-Cinnamyl-4-(*p*-chloro-α-phenylbenzyl)piperazine *see* 2428
Cinnamyl Cinnamate, 2364
Cinnamylcocaine *see* 2361
1-Cinnamyl-4-(di-*p*-fluorobenzhydryl)-piperazine *see* 4179
1-Cinnamyl-4-diphenylmethylpiperazine *see* 2365
1-*trans*-Cinnamyl-4-diphenylmethyl-piperazine *see* 2365
N-Cinnamylephedrine *see* 2357
(±)-(*E*)-Cinnamyl 2-Methoxyethyl 1,-4-Dihydro-2,6-dimethyl-4-(*m*-nitrophenyl)-3,5-pyridinedicarboxylate *see* 2334
α-[1-(*N*-Cinnamyl-*N*-methylamino)eth-yl]benzyl Alcohol *see* 2357
Cinnamyl Methyl Ketone *see* 1172
(*E*)-*N*-Cinnamyl-*N*-methyl-1-naphthalenemethylamine *see* 6442
Cinnarizine, 2365
Cinnipirine *see* 2365
Cinnoline, 2366
Cinnoxicam *see* 7661
Cinnyl Cinnamate *see* 2364
CINO-40 [Tutag] *see* 9727
Cinobac [Lilly] *see* 2369
Cinobufotalin, 2367
Cinolazepam, 2368
Cinolone [Pierrel] *see* 9727
Cinopal [Lederle] *see* 4003
Cinopenil [Hoechst] *see* 6047
Cinopol [Lederle] *see* 4003
Cinoxacin, 2369
Cinoxate, 2370
Cin-Quin [Rowell] *see* 8244
Cinromide, 2371
CiNU [Bristol] *see* 5594

Ciodrin [Shell] see 2667
Cioteronel, 2372
CIPC see 2240
Cipca see 867
Cipractin [Andromaco] see 2842
Cipralan [Roche] see 2330
Cipramil [Lundbeck] see 2379
Cipril [Fisons] see 2377
Ciprinol [Krka] see 2374
Cipro [Bayer] see 2374
Ciprobay [Bayer] see 2374
Ciprofibrate, 2373
Ciprofloxacin, 2374
Ciprol [Sterling] see 2373
Ciproxan [Bayer] see 2374
Ciproxin [Bayer] see 2374
Ciradol [Wyeth] see 2375
Ciramadol, 2375
Cirantin see 4705
Circair [Circle] see 3529
Circanol [3M] see 3692
Circleton [IBI] see 9160
Circolene [Inverni] see 8291
Circo-Maren [Krewel] see 6580
Circubid [W. F. Merchant] see 3773
Circulen "Kyorin" see 154
Circuletin see 5291
Circulin A see 2376
Circulin B see 2376
Circulins, 2376
Circupon [Troponwerke] see 3911
Cirotyl [Parke, Davis] see 7108
Cirpon see 5908
Cirrocolina see 2263
Cisapride, 2377
Cisatracurium Besylate see 900
Cismaplat [Mack, Illert.] see 2378
Cisordinol [Lundbeck] see 2455
Cisplatin, 2378
Cisplatyl [Bellon] see 2378
Cistobil [Bracco] see 5072
Citalopram, 2379
Citanest [Astra Pharm.] see 7924
Citation [Ciba-Geigy] see 2845
Citexal see 6028
Citicoline, 2380
Citifar [Lafare] see 2380
Citilat [CT] see 6617
Citiolase [Roussel-UCLAF] see 2381
Citiolone, 2381
Citireuma [CT] see 9155
Citizem [CT] see 3247
Citizeta [CT] see 10303
Citnatin see 8746
Citobaryum see 1023
Citodon see 4742
Citofur [Lusofarmaco] see 9267
Citogel [Zyma] see 9049
Citol see 482
Citopan see 4742
Citoplatino [Rhône-Poulenc] see 2378
Citosarin [Toyo Jozo] see 2769
Citoxid [Disprovent] see 6440
Citracholine see 2263
Citraconic Acid, 2382
Citral, 2383
Citramag [Bioglan] see 5699
Citramalic Acid, 2384
Citramin see 6038
Citrated Caffeine see 1674
Citrated Calcium Carbimide see 1703
Citraurin see 2385
β-Citraurin, 2385
Citrazinic Acid, 2386
Citresia see 5700
Citric Acid, 2387
Citric Acid Diammonium Salt see 540
Citric Acid Tributyl Ester see 1599
Citric Acid Trilithium Salt see 5555
Citrical [Shire] see 1697

Citridic Acid see 119
Citrinin, 2388
Citromycetin, 2389
Citronellal, 2390
α-Citronellal see 2390
Citronellol see 2391
1-Citronellol see 8352
α-Citronellol see 8352
β-Citronellol, 2391
Citronin A see 6480
Citrophen see 7373
Citropten see 5517
Citrosodine see 8746
Citrostadienol see 8696
Citrovorum Factor see 4254
Citrulline, 2392
Citrullol, 2393
Citrus Flavonoid Compounds see 1269
Citrus Red 2, 2394
Citrylideneacetone see 8101
Civet, 2395
Civetone, 2396
C-16 JH see 5287
C-17 JH see 5287
C-18 JH see 5287
CK-1752 see 8586
CK-1752A see 8586
CL-20 see 4731
CL-68 see 2430
CL-337 see 932
CL-636 see 9070
CL-1388R see 4587
CL-1848C see 3933
CL-11344 see 8151
CL-12625 see 6513
CL-13494 see 9087
CL-13900 see 8130
CL-14377 see 6065
CL-19823 see 9727
CL-26691 see 2854
CL-34433 see 9730
CL-34699 see 407
CL-36010 see 8240
CL-47031 see 7490
CL-62362 see 5617
CL-64475 see 4284
CL-67772 see 616
CL-71563 see 5617
CL-81588 see 922
CL-82204 see 4003
CL-84777 see 3185
CL-112302 see 1522
CL-118523 see 2003
CL-184116 see 7755
CL-186815 see 1243
CL-206214 see 1540
CL-206576 see 9060
CL-216942 see 1284
CL-222293 see 4941
CL-227193 see 7616
CL-232315 see 6303
CL-251931 see 2003
CL-252925 see 4942
CL-263499 see 4944
CL-263780 see 2336
CL-273703 see 5682
CL-284635 see 1975
CL-287088 see 6532
CL-291894 see 8864
CL-298741 see 9251
CL-301423 see 6373
CL-307579 see 9251
CL-67310456 see 7277
Clabber see 10235
Cladinose see 6399
Cladribine, 2397
Claforan [Hoechst] see 1983
Clamoxyl [Beecham] see 617
Clam Poison see 8533
Clanobutin, 2398

Clanzol [Orfi Farma] see 2404
Claradin [Nicholas] see 886
Claramid [Jouveinal] see 8433
Clarase [Miles], 2399
Claratal [Searle] see 45
Claresan see 2437
Clarified Butter see 4425
Claripex [Usafarma] see 2436
Clarin (formerly) [Pfizer] see 4685
Clarithromycin, 2400
Claritin [Schering] see 5608
Clarityn [Schering] see 5608
Clarvisan [Allergan] see 7645
Classic [DuPont] see 2142
Clast [Meiji Seika] see 2404
Clasteon [Gentili] see 2432
Clathrates, 2401
Clathromycin [Taisho] see 2400
Claudetite see 844
Clausthalite see 8572
Clavacin see 7185
Clavatin see 7185
Claversal [SmithKline] see 5964
Claviformin see 7185
Clavitol see 4565
Claviton see 9794
Clavulanic Acid, 2402
Clayton Yellow see 9445
Clazuril, 2403
CldAdo see 2397
Clearing Factor see 5537
Clearnal [Yoshitomi] see 9711
Clearteck [Penreco] see 7327
Clebopride, 2404
Cleboril [Almirall-Omega] see 2404
Clédial [Anphar-Rolland] see 5834
Clefamide see 2233
Cleiton [Sankyo Zoki] see 4828
Cleland's Reagent see 3441
Clemastine, 2405
Clemizole, 2406
Clenbuterol, 2407
Clenil-A [Chiesi] see 1047
Clentiazem, 2408
Cleocin [Upjohn] see 2414
Cleofil [Ciba] see 3362
Cleprid [Recordati] see 2404
Clera [Person & Covey] see 6455
Clérégil [Merck-Clevenot] see 2900
Cleridium [Marcofina] see 3410
1,6-Cleve's Acid, 2409
1,7-Cleve's Acid, 2410
Clexane [Rhône-Poulenc] see 9182
Clexon [Wellcome] see 7182
Cliacil [Hoechst] see 7230
Clidanac, 2411
Clidinium Bromide, 2412
Clift [Knoll] see 8060
Climaval [Sandoz] see 3746
Clinacox [Janssen] see 3130
Clinafarm [Janssen] see 3622
Clinafloxacin, 2413
Clinagel [Schering] see 2654
Clindamycin, 2414
Clinestrol see 3177
Clinicide [DeWitt] see 1831
Clinimycin [Glaxo] see 7111
Clinimycin (Rescinded) see 2414
Clinitar [Smith & Nephew] see 2487
Clinium [McNeil] see 5507
Clinoenstatite see 5727
Clinofibrate, 2415
Clinofug [Wolff] see 3496
Clinolamide see 5529
Clinoril [Merck & Co.] see 9155
Clinovir [Upjohn] see 5838
Clinozoisite see 1685
Cliocinizine see 2428
Clionasterol see 8698
Clioquinol see 5052

Clioxanide, 2416
CLIP see 7975
Clipper [ICI] see 7118
Cliquinol see 5052
Cliradon [Ciba] see 5312
Clistin [McNeil] see 1845
Clisundac [Lagap] see 9155
Clitizina see 4330
Clivarin(e) [Nordmark] see 8336
Clivoten [Lifepharma] see 5260
CLL see 5002
Cl₂MDP see 2432
Clobazam, 2417
Clobenfurol, 2418
Clobenoside, 2419
Clobenzepam, 2420
Clobenzorex, 2421
Clobenztropine, 2422
Clobesol [Glaxo] see 2423
Clobetasol, 2423
Clobetasone, 2424
Clobren-SF [Morishita] see 2436
Clobutinol, 2425
Clobuzarit, 2426
Clocapramine, 2427
Clocarpramine see 2427
Clocete [Toyama] see 5820
Clocinizine, 2428
Cloconazole, 2429
Clocortolone, 2430
Cloderm [Ortho] see 2430
Clodinafop-Propargyl, 2431
Clodronate Disodium see 2432
Clodronic Acid, 2432
Clofazimine, 2433
Clofedanol see 2106
Clofekton [Yoshitomi] see 2427
Clofenamide, 2434
Clofenotane see 2898
Clofenpyride see 6604
Clofentezine, 2435
Clofibrate, 2436
Clofibric Acid, 2437
Clofinit [Gentili] see 2436
Cloflucarban, 2438
Clofoctol, 2439
Cloforex, 2440
Clomacran, 2441
Clomag see 2437
Clomestrone, 2442
Clometacillin see 2445
Clometacin, 2443
Clometazin see 2443
Clomethacillin see 2445
Clomethiazole, 2444
Clometocillin, 2445
Clomid [Merrell] see 2446
Clomidazole see 2156
Clomifene see 2446
Clomiphene, 2446
Clomipramine, 2447
Clomivid [Draco] see 2446
Clomocycline, 2448
Clomphid see 2446
Clonazepam, 2449
Clonidine, 2450
Clonistada [Stadapharm] see 2450
Clonitazene, 2451
Clonitrate, 2452
Clonitrilide see 6602
Clonix [Medicamenta] see 2453
Clonixic Acid see 2453
Clonixin, 2453
Clonopin [Roche] see 2449
Clont [Bayer] see 6242
Clopamide, 2454
Clopane [Lilly] see 2808
Clopenthixol, 2455
α-Clopenthixol see 2455
Cloperastine, 2456

Clophen [Bayer] see 7725
Clopidogrel, 2457
Clopidol, 2458
Clopindol see 2458
Clopinerin [Nippon Shoji] see 2470
Clopirac, 2459
Clopiran [Continental Pharma] see 2459
Clopixol [Lundbeck] see 2455
Clopoxide see 2132
Cloprane [Luitpold] see 8414
Cloprednol, 2460
Clopromate [Purdue Frederick] see 6226
Cloprostenol, 2461
Clopyralid, 2462
Cloquinate see 4872
Cloquintocet-Mexyl, 2463
Cloradryn [Recordati] see 2460
Cloramfen [Sclavo] see 2120
Cloramin [Simes] see 5815
Cloranolol, 2464
Clorarsen see 3121
Clorazepate Dipotassium see 2465
Clorazepic Acid, 2465
Clorazolam see 9734
Clordelazin see 2238
Clorevan [Evans] see 2234
Clorexolone, 2466
Clorfentermina see 2235
Cloricromen, 2467
Clorindione, 2468
Clorociclin [Panthox & Burck] see 8411
Clorocyn [Laif] see 2120
Clorolifarina [St. Gall] see 2120
Cloronaftina see 2157
Clorophene, 2469
Cloropiril see 2232
Clorox see 8773
Clorpactin [Guardian] see 7090
Clorpactin WCS [Guardian] see 7090
Clorpactin XCB [Guardian] see 7090
Clorprenaline, 2470
Clorsulon, 2471
Clortermine, 2472
Clortetrin [Fargal] see 2937
Closantel, 2473
Closina see 2820
Clospipramine see 6365
Clospirazine, 2474
Clostebol, 2475
Clostilbegyt [EGYT] see 2446
Clostridiopeptidase A see 2544
Closylate see 2173
Clotam [Gea] see 9650
Clotbur see 5377
Clothiapine, 2476
Clotiamine see 1048
Clotiazepam, 2477
Clotride [Merck & Co.] see 2221
Clotrimazole, 2478
Clotrox see 9867
Clout [Scott] see 6020
Clove, 2479
Clove Oil see 6893
Cloxacillin, 2480
Cloxapen [Beecham] see 2480
Cloxazolam, 2481
Cloxiquine see 2483
Cloxotestosterone, 2482
Cloxypen [Allard] see 2480
Cloxyquin, 2483
Clozan [Roerig] see 2477
Clozapine, 2484
Clozaril [Sandoz] see 2484
Clozic [ICI] see 2426
CLSE see 5002
Cl-TMBP see 9582
Club-moss Spores see 5653

Clupeine, 2485
CLY-503 see 8684
Clysodrast [Barnes-Hind] see 1282
CM-6805 see 1563
CM-6912 see 3867
CM-7857 see 7271
CM-8282 see 6978
CM-9155 see 3194
CM-31916 see 1999
CMA see 1725
CMC [Stuart] see 1877
CME-151 see 3271
CME-74770 see 9819
C-Meton [SS Pharm] see 2232
CMME see 2198
2'-CMP see 2856
3'-CMP see 2857
c-Mpl Ligand see 9528
CMPP see 5826
CMT see 1964
CMU see 6342
CN see 2166
CN-11-2936 see 7947
CN-100 see 10244
CN-3123 see 9093
CN-15757 see 932
CNDP see 3601
Cnicin, 2486
CNP see 2158
CNTF see 2333
Co I see 6429
Co II see 6433
CO-61 see 7035
CO-1063 see 3736
CoA see 2531
Coal Tar, 2487
Coal Tar Creosote see 2641
Co-amilozide see 426
Co-amoxiclav see 2402
Coapt [Ethicon] see 5827
Coaxin [Tobishi] see 2028
Cobactin see 6405
Cobalamin see 10152
Cob(I)alamin see 10154
Cob(II)alamin see 10154
Cobalamins see 10152
Cobalaminsulfonic Acid see 4854
Cobalin [Finlay] see 10152
Cobalin-H [Link] see 4854
Cobalion [Houdé] see 2513
Cobalt, 2488
Cobaltamin S [Wakamoto] see 2513
Cobalt Carbonate Hydroxide. see 2497
Cobalt Carbonyl Hydride see 3135
Cobalt Chromite see 2499
Cobalt Cyanide see 2500
Cobalt Dibromide see 2496
Cobalt Dichloride see 2498
Cobalt Difluoride see 2501
Cobalt Diiodide see 2504
Cobalt Hydrocarbonyl see 3135
Cobalt Hydroxide Oxide see 2492
Cobaltic Acetate, 2489
Cobaltic-Cobaltous Oxide, 2490
Cobaltic Fluoride, 2491
Cobaltic Hydroxide see 2492
Cobaltic Oxide Monohydrate, 2492
Cobaltic Potassium Nitrite, 2493
Cobaltite see 2488
Cobalto-cobaltic Oxide see 2490
Cobalt Octacarbonyl see 3135
Cobaltosic Oxide see 2490
Cobaltous Acetate, 2494
Cobaltous Ammonium Phosphate see 541
Cobaltous Ammonium Sulfate see 542
Cobaltous Arsenate, 2495
Cobaltous Bromide, 2496
Cobaltous Carbonate, 2497
Cobaltous Carbonate Basic see 2497

Cobaltous Chloride, 2498
Cobaltous Chromate(III), 2499
Cobaltous Cyanide, 2500
Cobaltous Fluoride, 2501
Cobaltous Formate, 2502
Cobaltous Hydroxide, 2503
Cobaltous Iodide, 2504
Cobaltous Nitrate, 2505
Cobaltous Oxalate, 2506
Cobaltous Oxide, 2507
Cobaltous Phosphate, 2508
Cobaltous Potassium Sulfate, 2509
Cobaltous Rhodanide see 2512
Cobaltous Sulfate, 2510
Cobaltous Sulfide, 2511
Cobaltous Sulfocyanate see 2512
Cobaltous Thiocyanate, 2512
Cobalt Potassium Cyanide see 7801
Cobalt Spar see 2497
Cobalt Tetracarbonyl see 3135
Cobalt Trifluoride see 2491
Cobalt Yellow see 2493
Cobamamide, 2513
Cobamamidum see 2513
Cobamic Acid see 10152
Cobamide see 10152
Coban [Lilly] see 6329
Cobantril [Pfizer] see 8139
Cobanzyme [Bouchara] see 2513
Cobazymase [Bouchara] see 2513
Cobefrin [Winthrop] see 6785
Coben [Takeda] see 7558
Co-beneldopa see 1079
Coben P [Takeda] see 7558
Cobinamide see 10152
Cobinamide Acetate (Salt) Dihy-
drogen Phosphate (Ester) Inner Salt
3′-Ester with 5,6-Dimethyl-1-α-D-
ribofuranosyl-1H-benzimidazole see
4854
Cobinamide-Co(I) Dihydrogen Phos-
phate (Ester) 3′-Ester with 5,6-
Dimethyl-1-α-D-ribofuranosyl-1H-
benzimidazole see 10154
Cobinamide Co-methyl Deriv. Dihy-
drogen Phosphate (Ester) Inner Salt
3′-Ester with 5,6-Dimethyl-1-α-D-
ribofuranosyl-1H-benzimidazole see
6125
Cobinamide Cyanide Phosphate 3′-
Ester with 5,6-Dimethyl-1-α-D-ribo-
furanosylbenzimidazole Inner Salt
see 10152
Cobinamide Dicyanide see 3914
Cobinamide Hydroxide Dihydrogen
Phosphate (Ester) Inner Salt 3′-
Ester with 5,6-Dimethyl-1-α-D-
ribofuranosyl-1H-benzimidazole see
4854
Cobinamide Hydroxide Monohydrate
Dihydrogen Phosphate (Ester) Inner
Salt 3′-Ester with 5,6-Dimethyl-1-α-
D-ribofuranosyl-1H-benzimidazole
see 4854
Cobinamide Nitrite (Salt) Dihydrogen
Phosphate (Ester) Inner Salt 3′-
Ester with 5,6-Dimethyl-1-α-D-ribo-
furanosyl-1H-benzimidazole see
4854
Cobinamide Sulfite (1:1) (Salt) Dihy-
drogen Phosphate (Ester) Inner Salt
3′-Ester with 5,6-Dimethyl-1-α-D-
ribofuranosyl-1H-benzimidazole see
4854
Cobinic Acid see 10152
Cobiona [Esteve] see 7058
Cobrentin Methanesulfonate [Merck
& Co.] see 1156
Cobrotoxin, 2514
Cobutolin [Cox] see 217

Cobyric Acid see 10152
Cobyrinic Acid see 10152
Coca, 2515
Cocaethylene, 2516
Cocaic Acid see 9922
β-Cocaic Acid see 9922
Cocaine, 2517
l-Cocaine see 2517
β-Cocaine see 2517
Cocaine Muriate see 2517
Cocalose [Maruko] see 9431
Co-careldopa see 1843
Cocarboxylase see 9431
Cocarvit [CT] see 9431
Cocciden [Sankyo] see 1048
Coccoclase see 9108
Cocculin see 7570
Cocculus, 2518
Cocculus indicus see 2518
Cochineal, 2519
Cocillana, 2520
Coclanoline see 8330
Coclaurine, 2521
Cocoa, 2522
Cocoa Butter see 9417
Cocoa Shells see 1638
Coconut Oil, 2523
Coculine see 8692
Codamine, 2524
Codecarboxylase see 8163
Codehydrase II see 6433
Codehydrogenase see 6429
Codehydrogenase II see 6433
Codeigene see 2527
Codeine, 2525
β-Codeine see 6549
ψ-Codeine see 8098
Codeine Methyl Bromide, 2526
Codeine N-Oxide, 2527
Codeine Phosphate, 2528
Codeine Sulfate, 2529
Codelcortone [Merck & Co.] see 7901
Codelcortone-T.B.A. [Merck & Co.]
see 7901
Codelsol [Merck & Co.] see 7903
Co-(5′-deoxyadenosine-5′) Deriv of 3′-
Ester of Cobinamide Dihydrogen
Phosphate (Ester) with 5,6-Dimeth-
yl-1-α-D-ribofuranosyl-1H-benz-
imidazole, Inner Salt see 2513
Co-dergocrine Mesylate see 3692
Codethyline see 3876
Codhydrine see 3214
Codicept [Sanol] see 2525
Codinovo see 4826
Cod Liver Oil, 2530
Codons see 8370
Codroxomin [Chromalloy] see 4854
Codylin see 7484
Coelenteramide see 175
Coelentrazine see 175
Coenzile [A.T.I.] see 2513
Coenzyme I see 6429
Coenzyme II see 6433
Coenzyme A, 2531
Coenzyme B₁₂ see 2513
Coenzyme Q₁₀ see 9974
Coenzyme Q-199 see 9974
Coenzyme R see 1272
Coenzymes Q see 9974
Coeruleolactite see 371
Co-Ervonum [Glaxo] see 5849
Coffearine see 9822
Coffee Bean Oil see 2533
Coffee Beans see 2532
Coffee, Green, 2532
Coffee Oil, 2533
Coffeine see 1674
Cofosfolactamines see 4280
Co-Fram [Abbott] see 9093

Co-galactoisomerase see 10018
Cogentin [Merck & Co.] see 1156
Cogentinol [Pharma-Stern] see 1156
Cogmine [Forest] see 7540
Cognac Oil, Synthetic see 3881
Cognex [Parke-Davis] see 9199
Cogomycin see 4312
Coherin, 2534
Cola see 5331
Colace [Bristol] see 3460
Colalin see 2258
Colamine see 3772
Colaspase see 871
Colcemid [Ciba] see 2938
Colchamine see 2938
Colchiceine, 2535
Colchicine, 2536
Colchicine-binding Protein see 9940
Colchicum Corm, 2537
Coldan see 6455
Coldrin [Nippon Shinyaku] see 2106
Colecalciferol see 10157
Colectril (obsolete) [Merck & Co.] see
426
Coleflux [Finadiet] see 7639
Colemanite see 1691
Colenormol see 7094
Colepan [Panther-Osfa] see 2784
Colepax see 5072
Colepur [Draco] see 1474
Colerainite see 362
Colesterinex [Galenica] see 8158
Colestid [Upjohn] see 2538
Colestipol, 2538
Colestyramin see 2257
Coletrast see 5047
Coletyl see 1823
Colfarit [Bayer] see 886
Colforsin, 2539
Colfosceril Palmitate, 2540
Colicine see 2541
Colicins, 2541
Colicon [Resse] see 3264
Colic Root see 229, 3346 and 4355
Colidosan see 8378
Colifoam [Stafford-Miller] see 4828
Colifos see 7519
Colimune [Fisons] see 2658
Colimycin see 2542
Coliopan [Eisai] see 1569
Coliquifilm [Allergan] see 2180
Colisone [Frosst] see 7904
Colisticina see 2542
Colistimethate Sodium see 2542
Colistin, 2542
Colite [Nippon Chemiphar] see 2380
Collagen, 2543
Collagenase, 2544
Collastin see 6994
2,4,6-Collidine see 9848
sym-Collidine see 9848
α-Collidine see 3888
α,γ,α′-Collidine see 9848
β-Collidine see 3887
γ-Collidine see 9848
Collinomycin see 2545
Collinsonia, 2546
Colliron I.V. see 4073
Collodion, 2547
Collodion, Cantharidal, 2548
Collodion Cotton see 8195
Collodion Wool see 8195
Colloidal Bismuth Subcitrate, 2549
Colloidal Calomel see 1768
Colloidal Gold—¹⁹⁸Au see 4533
Colloxylin see 8195
Collubiazol [Roussel] see 9069
Collunosol see 9772
Collunovar see 6534
Colme [Lasa] see 1703

Colocynth, 2550
Colocynthin, 2551
Colofac [Duphar] see 5808
Colofoam [Stafford-Miller] see 4828
Cologel [Lilly] see 6120
Cologne Yellow see 5422
Colombo see 1770
Colonorm [Mundipharma] see 8600
Colony-stimulating Factor 1 see 5678
Colony-stimulating Factor-2 see 4559
Colony-Stimulating Factor 2 (Human
    Clone pCSF-1 Protein Moiety
    Reduced) Glycoform GMC 89-107
    see 4559
Colony-Stimulating Factor 2 (Human
    Clone pHG25 Protein Moiety
    Reduced) see 4559
4-221-Colony-stimulating Factor 1
    (Human Clone p3ACSF-69 Protein
    Moiety Reduced) see 5678
1-214-Colony-stimulating Factor 1
    (Human Clone p3ACSF-69 Protein
    Moiety Reduced) Homodimer see
    5678
Colony Stimulating Factors, 2552
Colophony see 8424
Colo-Pleon [Henning] see 9112
Colostrokinin, 2553
Colpermin [Tillotts] see 6933
Colpogyn [Angelini] see 3750
Colpogynon see 2554
Colpormon, 2554
Colpovis [SIT] see 8238
Colpro [Dagra] see 5836
Colprone [Ayerst] see 5836
Colrex Expectorant [Rowell] see 4582
Coltericin [Quimica Argentia] see
    5293
Coltirot see 9972
Coltramyl [Roussel] see 9460
Coltrax [Roussel] see 9460
Coltsfoot, 2555
α-Colubrine see 2556
β-Colubrine see 2556
Colubrines, 2556
Columbamine, 2557
Columbin, 2558
Columbite see 6648
Columbium see 6648
Columbium Pentachloride see 6649
Columbium Pentafluoride see 6650
Columbium Pentoxide see 6651
Columbium Potassium Oxypentafluo-
    ride see 6652
Colvasone [Norbrook] see 2986
Colybar [Warner-Lambert] see 2257
Coly-Mycin [Warner-Chilcott] see
    2542
Colyonal [Mochida] see 2993
Colza Oil see 8289
Combantrin [Pfizer] see 8139
Combat [Cyanamid] see 4802
Combec [Tokyo Tanabe] see 4167
Combot [Miles] see 9753
Comelian [Kowa] see 3244
Cometamine [Yamanouchi] see 2824
Comfolax [Searle] see 3460
Comite [Uniroyal] see 7992
Command [FMC] see 3262
Commetamin see 2824
Common Mallow see 5748
Common Salt see 8742
Common Sundew see 3508
Common Verdigris see 2691
Comox [Norton] see 9086
Compactin see 6251
Compazine [SK & F] see 7942
Compd 3-120 see 9109
Compd 20 see 5392
Compd 22/190 see 2248

Compd 43-663 see 4597
Compd 47-83 see 2779
Compd 47-282 see 2128
Compd 48/268 see 9043
Compd 88R see 805
Compd 118 see 227
Compd 180/442 see 3633
Compd 269 see 3618
Compd 338 see 2174
Compd 347 see 3621
Compd 469 see 5191
Compd 497 see 3152
Compd 545 see 5068
Compd 711 see 227
Compd 933F see 7631
Compd 1080 see 4205
Compd 4072 see 2137
Compd 08958 see 2825
Compd 57926 see 8690
Compd 64716 see 2369
Compd 72500 see 4239
Compd 79891 see 6506
Compd 83405 see 1964
Compd 83846 see 793
Compd 90459 see 1075
Compd 99170 see 793
Compd 99638 see 1962
Compd B see 2601
Compd F-2 see 10246
Compendium [Polifarma] see 1406
Compitox [M & B] see 5826
Compitox Plus [M & B] see 5826
Complamin [Italchimici] see 10194
Complexone see 3557
Compocillin [Abbott] see 7224
Compocillin-V [Abbott] see 7232
Compocillin VK [Abbott] see 7230
Component B$_{1a}$ see 5264
Component B$_{1b}$ see 5264
Compound 42 see 10174
Compound 81929 see 3456
Compound 112531 see 10125
Comprecin [Warner-Lambert] see
    3625
ConA see 2559
Conalbumin see 9706
Concanavalin A, 2559
Concemin see 867
"Concentrated Opium" see 7150
Conceplan [Grünenthal] see 6790
Conceptrol [Ortho] see 6772
Conceral [Takeda] see 4109
Conco NI [Continental Chem.] see
    6772
Conco NI-90 [Continental Chem.] see
    6772
Conco NIX-100 [Continental Chem.]
    see 6858
Concor [E. Merck] see 1336
Concordin [Merck & Co.] see 8088
Condol see 10156
Condor Vine see 2560
Condrosulf [IBSA] see 2270
Conducton [Klinge] see 1822
Condurangin, 2560
Condurangogenin A see 2560
Condyline [Brocades] see 7704
Condylox [Oclassen] see 7704
Cone Flower see 3542
Conessine, 2561
Conestron [Wyeth] see 2571
Confidor [Bayer] see 4946
Confirm [Rohm & Haas] see 9254
Conflictan [Riom] see 7039
Confortid [Dumex] see 4998
Congo Blue see 9923
Congocidine see 6565
Congo Red, 2562
Congressane, 2563
Conhydrine, 2564

β-Coniceine, 2565
γ-Coniceine, 2566
Conicine see 2569
Coniel [Kyowa] see 1071
Coniferin, 2567
Coniferyl Alcohol, 2568
Coniine, 2569
Conium Fruit, 2570
Conjugated Estrogenic Hormones,
    2571
Conjugol see 3751
Conjuncain [Mann] see 1076
Conludag [Syntex] see 6790
Conmel [Winthrop] see 3414
Connectin see 9622
Connettivina [Fidia] see 4793
Conoco C-50 see 8757
Conoco C-60 see 8757
Conoco SD 40 see 8757
Conoderm [C-Vet] see 6266
Conofite [Pitman-Moore] see 6266
Conotrane see 4805
Conova [Gold Cross] see 3905
Conovid E [Searle] see 6791
Conquinamine, 2572
Conquinine see 8244
Conray [Mallinckrodt] see 5080
Conray-400 [Mallinckrodt] see 5080
Conselt [Sana] see 2470
Consolan [Novo] see 6428
Constaphyl [Bristol] see 3134
Consult [Dow Elanco] see 4720
Consumptive's Weed see 3711
Contalax [Riker] see 1282
Contamex [Beecham-Wlfing] see 5308
Contax see 7108
Conteben [Bayer] see 9427
Contergan see 9390
Contomin see 2238
Contralac [Virbac] see 5999
Contralin see 3428
Contrapar [Burroughs Wellcome] see
    5310
Contrathion [Rhône-Poulenc Rorer]
    see 7884
Contratuss see 5495
Contrheuma Retard [Spitzner] see 886
Contristamine see 2234
Contrix "28" see 5080
Control-Om see 5896
Convacard [Madaus] see 2574
Convallamarin see 2573
Convallamarogenin, 2573
Convallan [Gödecke] see 2574
Convallaria, 2574
Convallaria Glycosides see 2574
Convallaton see 2575
Convallatoxigenin see 9015
Convallatoxin, 2575
Convallen see 2574
Convalyt see 2574
Convasid see 2574
Convenil [Hommel] see 1551
Convicine, 2576
Convulex [Byk-Gulden] see 10049
Coolspan [Hishiyama] see 9163
Coomassie Blue [Ayerst] see 669
Coomassie Blue Medicinal [Ayerst]
    see 669
Coomassie Blue RL [Ayerst] see 669
Coopex see 2600
CO-ORD [Baxter] see 218
COP, 2577
COP 1, 2578
Copaene, 2579
α-Copaene see 2579
Copaiba, 2580
Copal, 2581
Coparaffinate, 2582

Coparogin [Nippon Chemiphar] *see* 9267
Copaxone [Teva] *see* 2578
Copinal [Vinas] *see* 43
Copirene *see* 5300
Copolymer 1 *see* 2578
Copper, 2583
Copper Acetate Arsenite *see* 2692
Copperas *see* 4105
Copper Carbonate Hydroxide *see* 2697
Copper-chromium Oxide *see* 2701
Copper DOS *see* 2740
Copper Glance *see* 2737
Copper Hydrate *see* 2709
Copper Hydroxide Sulfate *see* 2723
Copper Phthalocyanine, 2584
Copper Pyrites *see* 2079
Copper-Sandoz [Sandoz] *see* 2734
Copper Sulfate Dibasic *see* 2723
Copper Sulfate Tribasic *see* 2723
Coppertone [Schering] *see* 4776
COP 1 (Polyamide) *see* 2578
Copra Oil *see* 2523
Copren [Fuso] *see* 4483
COP Resist *see* 2577
Coproergostane, 2585
Coprogen, 2586
Coprola [Dunster] *see* 3460
Coprostane, 2587
3β-Coprostanol *see* 2588
Coprostenol *see* 262
4:5-Coprosten-3-ol *see* 262
Coprosterol, 2588
Coptine, 2589
Coptis, 2590
Coptisine, 2591
Coracon *see* 6635
Coracten [SmithKline] *see* 6617
Coractiv N [Phyteia] *see* 6635
Corafurone *see* 5321
Corail [Bayer] *see* 9253
Co-ral [Chemagro] *see* 2626
Coralgil *see* 4739
Coralgina *see* 4739
Coralln *see* 915
Coralox *see* 2600
Coramedan [Medice] *see* 3206
Coramine [Ciba] *see* 6635
Corangin [Ciba-Geigy] *see* 5245
Coranormol *see* 7283
Corathiem [Ohta] *see* 2365
Corax [Empire] *see* 2132
Co-Rax [Prentiss] *see* 10174
Corazole *see* 7283
Corbadrine *see* 6785
Corbasil *see* 6785
Corbel [BASF] *see* 4035
Corchorgenin *see* 9015
Corchorin *see* 9015
Corchsularin *see* 9015
Cordalin [Asta] *see* 3924
Cordan [IBI] *see* 4011
Cordanum [Arzneimittelwerk Dresden] *see* 9208
Cordarex [Labaz] *see* 504
Cordarone [Labaz; Sigma-Tau] *see* 504
Cordarone X [Labaz] *see* 504
Cordes [Ichthyol] *see* 4828
Cordes Vas [Ichthyol] *see* 8333
Cord Factor, 2592
Cordiamin *see* 6635
Cordianine *see* 255
Cordicant [Mundipharma] *see* 6617
Cordilan [Andreu] *see* 6617
Cordilox [Pfizer] *see* 10083
Cordiomon *see* 4651
Cordioxil *see* 3210
Cordipatch [Théraplix] *see* 6704
Corditrine [Bellon] *see* 6704
Cordium [Riom] *see* 1188

Cordoval [Tempelhof] *see* 4441
Cordoxene [Laroche-Navarron] *see* 3996
Cordran *see* 4232
Cordycepic Acid *see* 5788
Cordycepin, 2593
Cordycepin-5'-triphosphate *see* 2593
9-Cordyceposidoadenine *see* 2593
Coredamin [Meiji] *see* 7919
Corediol [Meiji] *see* 6635
Corein 2R *see* 2006
Coreminal [Mitsui] *see* 4243
Corenalin [Kaken] *see* 2380
Co-Renitec [Merck & Co.] *see* 3605
Coretal [Polfa] *see* 7086
Corflazine [Daltan] *see* 5507
Corgard [Squibb] *see* 6431
Corglykon *see* 2575
Corhormon *see* 4651
Coriamyrtin, 2594
Coriander, 2595
Coriandrol *see* 5520
Coriantin [Richter] *see* 2273
Coriban [Burroughs Wellcome] *see* 3017
Coric [Merck & Co.] *see* 5540
Corid [Merck & Co.] *see* 631
Coridil [Delalande] *see* 3410
Cori Ester *see* 4470
Corilagin *see* 9221
Corindolan [Schering AG] *see* 5901
Coriovis [Solvay] *see* 2273
Coriphate [Tokyo Tanabe] *see* 4185
Coristin [San Carlo] *see* 3692
Coritat [Green Cross] *see* 6792
Cork, 2596
Corlan [Evans] *see* 4828
Corlin *see* 2602
Corliprol [Upjohn] *see* 2007
Corlopam [SKB] *see* 4020
Corlutina [Schering-Plough] *see* 7956
Corluvite [Vitarine] *see* 7956
Cormax [Rhône-Poulenc Rorer] *see* 3247
Cormed [Reiss] *see* 6635
Cormelian [Asta] *see* 3244
Cormid *see* 6635
Cornetite *see* 2716
Cornin *see* 10093
Cornocentin *see* 3694
Corn Oil, 2597
Cornox [Boots] *see* 5803
Cornox RK [Boots] *see* 3128
Corn Steep Liquor, 2598
Corn Steep Water *see* 2598
Corn Sugar *see* 4467
Curnus, 2599
Corodane *see* 2129
Corodenin *see* 143
Corodilan [Meiji] *see* 3753
Coronarin *see* 2273
Coronarine [NEGMA] *see* 3410
Coronin *see* 5321
Coro-Nitro [Boehringer, Mann.] *see* 6704
Corontin [Chinoin] *see* 7919
Corotrend [Siegfried] *see* 6617
Corotrope [Sterling-Winthrop] *see* 6284
Corovliss [Boehringer, Mann.] *see* 5245
Coroxon, 2600
Corozate *see* 10305
Corps De Pfeiffer *see* 7334
Cor-Puren [Klinge-Nattermann] *see* 3210
Corpus Luteum Hormone *see* 7956
Corrigast [Searle] *see* 7989
Corrigen *see* 6963
Corrin *see* 10152

Corrinoids *see* 10152
Corrole *see* 10152
Corrosive Mercury Chloride *see* 5926
Corrosive Sublimate *see* 5926
Corsair [Rhône-Poulenc] *see* 7321
Corsodyl [ICI] *see* 2140
Corson [Takeda] *see* 2986
Corstiline *see* 136
Cortadren [Schering] *see* 2602
Cortaid [Upjohn] *see* 4828
Cortalone [Halsey] *see* 7901
Cortancyl [Roussel] *see* 7904
Cortate [Schering] *see* 2948
Cort-Dome [Miles] *see* 4828
Cortef [Upjohn] *see* 4828
Cortelan [Glaxo] *see* 2602
Cortenema [Solvay] *see* 4828
Cortensor *see* 4691
Cortex Angosturae *see* 690
Cortex Cuspariae *see* 690
Cortexilar [Veterinaria] *see* 4173
Cortexolone *see* 2977
Cortexone *see* 2947
Cortexone Acetate *see* 2948
Cor-Theophyllin [Paramedical] *see* 3529
Corticoliberin(e) *see* 2653
Corticopan [Ist. Chim. Inter.] *see* 169
Cortico-Sol [Roussel] *see* 7901
Corticosteroid-binding Globulin *see* 4828
Corticosterone, 2601
Corticotrophin *see* 136
Corticotrophin Zinc Hydroxide Suspension *see* 136
Corticotropin *see* 136
α[1-24]-Corticotropin *see* 2617
β[1-24]-Corticotropin *see* 2617
Corticotropin-like Intermediate Lobe Peptide *see* 7975
Corticotropin-releasing Factor *see* 2653
Corticotropin-releasing Hormone *see* 2653
Cortidene [Syntex Latino] *see* 7162
Cortidin [Crinos] *see* 169
Cortidyn *see* 169
Cortifoam [Reed & Carnrick] *see* 4828
Cortilet *see* 4213
Cortine Naturelle [Laroche Navarron] *see* 169
Cortiphyson [Promonta] *see* 136
Cortiplastol [Medici] *see* 4185
Cortiron [Schering AG] *see* 2948
Cortisol *see* 4828
Cortisol 21-(*N,N*-Diethyl)glycinate *see* 4827
Cortisol 21-Ester with *N,N*-Diethyl-glycine *see* 4827
Cortisone, 2602
Δ[1]-Cortisone *see* 7904
Cortisone Acetate *see* 2602
Cortistab [Boots] *see* 2602
Cortisumman [Winzer] *see* 2986
Cortisural *see* 169
Cortisyl Artriona [Roussel-UCLAF] *see* 2602
Cortivazol, 2603
Cortivent [Leiras] *see* 1490
Cortocin-F [Eisai] *see* 4270
Cortogen [Schering] *see* 2602
Cortol, 2604
α-Cortol *see* 2604
β-Cortol *see* 2604
Cortolone, 2605
Cortone [Merck & Co.] *see* 2602
Cortril [Pfizer] *see* 4828
Cortrophin [Organon] *see* 136
Cortrophin-Z *see* 136
Cortrosinta [Organon] *see* 2617

Cortrosyn [Organon] see 2617
Corulon [Intervet] see 2273
Corundum see 369
Corvasal [Hoechst] see 6316
Corvasol [Rieswerke] see 7283
Corvasymton see 9189
Corvaton [Cassella-Riedel] see 6316
Corvitol see 6635
Corvotone see 6635
Corwin [ICI] see 10189
Corybulbine, 2606
Corycavamine, 2607
Corycavidine, 2608
Corycavine. see 2607
Corydaldine, 2609
Corydaline, 2610
d-Corydaline see 2610
Corydalis, 2611
Corydalis-G see 2606
Corydine, 2612
Corylophyline see 4468
Corynantheine, 2613
Corynanthidine see 10238
Corynanthine, 2614
Corynine see 10236
Corynomycolic Acids see 6406
Corypalmine, 2615
Corytuberine, 2616
Cosaldon [Albert-Roussel] see 7269
Cosaprin see 105
Cosbiol [Laserson & Sabetay] see 8923
Coscopin [BDH] see 6815
Coscotabs see 6815
Coslan see 5842
Cosmegen [Merck & Co.] see 2867
Cosmoline [E. F. Houghton] see 7326
Cosmopen see 7225
Cosmos see 6033
Cosmosin [Nippon Lederle] see 2003
Cospanon [Eisai] see 4142
Cossmetin see 774
Cossym [Egnaro] see 1840
Costilate [Green Cross] see 5678
Costimulator see 5020
Cosulid [Ciba] see 9068
Cosumix [Ciba-Geigy] see 9068
Cosuric [DDSA] see 287
Cosylan [Warner-Lambert] see 8274
Cosyntropin, 2617
Cotarnine, 2618
Cotarnine Chloride see 2618
Cotarninium Chloride see 2618
Cotazym [Organon] see 7138
Co-tetroxazine see 9386
Cothera Syrup [Ayerst] see 3272
Co-thromboplastin see 3962
Cotinazin [Pfizer] see 5203
Cotinine, 2619
Cotnion-methyl [Makhteshim-Agan] see 944
Cotofor [Ciba-Geigy] see 3405
Cotoin, 2620
Cotoran [Ciba] see 4189
Cotrane [Midy] see 3272
Co-trifamole see 9093
Co-trimazine see 9071
Co-trimoxazole see 9086
Cottonex [Makhteshim-Agan] see 4189
Cotton Red 4B see 1132
Cotton-root Bark, 2621
Cottonseed Oil, 2622
Cotunnite 5421
Couch Grass see 9890
Coughwort see 2555
Coumachlor, 2623
Coumadin [DuPont Pharma] see 10174
Coumafos see 2626
Coumafuryl, 2624

Coumalic Acid, 2625
Coumaphos, 2626
Coumaphos Oxygen Analog see 2600
Coumaran, 2627
p-Coumaric Acid, 2628
Coumarilic Acid, 2629
Coumarin, 2630
Coumarin-3-carboxylic Acid, 2631
cis-o-Coumarinic Acid Lactone see 2630
Coumarinic Anhydride see 2630
Coumarone see 1119
Coumarone-2-carboxylic Acid see 2629
Coumestrol, 2632
Coumetarol, 2633
Coumingine, 2634
Coumithoate, 2635
Counter [Am. Cyanamid] see 9301
Courlene [Courtaulds] see 7728
Courlene PY [Courtaulds] see 7741
Cousso see 1391
Covalan [Dausse] see 4252
Covatine [Bailly] see 1816
Covatix [Gödecke] see 1816
Covellite see 2724
Coverine [Astier] see 2323
Coversum [Itherapia] see 7311
Coversyl [Servier] see 7311
Co-waldenase see 10018
Cow Clover see 9818
Cowrie see 2581
Cow's Milk see 6280
Coxigon [Lilly] see 1075
Coxistat see 6697
Coxytrol [Squibb] see 2843
Coyden [Dow] see 2458
Cozaar [Du Pont] see 5613
Cozymase see 6429
Cozyme [Travenol] see 2988
CP-0127 see 1385
CP-172AP see 2459
CP-556S see 9160
CP-1001 see 2313
CP-1044-J3 see 1497
CP-4742 see 9079
CP-6343 see 259
CP-10423-16 see 8139
CP-10423-18 see 8139
CP-12009-18 see 6348
CP-12299-1 see 7897
CP-12574 see 9588
CP-14445 see 7055
CP-14445-16 see 7055
CP-15336 see 3014
CP-15464 see 1888
CP-15464-2 see 1888
CP-15639-2 see 1838
CP-16171 see 7661
CP-19106-1 see 9828
CP-23426 see 9726
CP-24314-1 see 7644
CP-31393 see 7977
CP-34089 see 9165
CP-41845 see 4523
CP-45634 see 8871
CP-45899 see 9058
CP-45899-2 see 9058
CP-49952 see 9166
CP-50144 see 203
CP-51974-1 see 8612
CP-52640-2 see 1981
CP-53619 see 1533
CP-54802 see 245
CP-62993 see 946
CP-65703 see 629
CP-66248 see 9290
CP-66248-2 see 9290
CP-73850 see 10325
CP-76136-27 see 2876

CP-88059 see 10304
CP-88059-1 see 10304
CP-99219 see 9919
CP-99219-27 see 9919
CPA see 2844
CPC-10997 see 2372
CPDC see 2378
CPIB see 2436
Cpiron see 4094
Cpns see 2084
CPP32 see 788
CPT-11 see 5104
C-Quens see 2152
CR-242 see 7958
CR-604 see 7957
CR-605 see 9607
CR-662 see 9601
CR-1505 see 5618
CR-1639 see 3340
Crab-E-Rad [Vineland] see 6020
Crag 974 see 2892
Crag Fruit Fungicide 341 [Union Carbide] see 4518
Crag Herbicide-1 [Union Carbide] see 3430
Crag Sesone [Union Carbide] see 3430
Cramp Bark see 10108
Crampol see 101
Cranberry Tree see 10108
Cranesbill see 4412
Crapinon [Sanzen] see 7623
Crasnitin [Bayer] see 871
Crataegus, 2636
Crataegus-Kreussler see 2636
Cravit [Daiichi] see 6865
Craviten [Polfa] see 1560
CRD-401 see 3247
C-Reactive Protein see 2670
Creamalin [Winthrop] see 355
Cream of Tartar see 7776
Creasote see 2642
Creatergyl [Midy] see 7496
Creatine, 2637
Creatine Phosphate see 7496
Creatinephosphoric Acid see 7496
Creatinine, 2638
Cremin [Yoshitomi] see 6365
Cremorin [Winthrop] see 355
Cremor Tartari see 7776
Creolin, 2639
Creon [Duphar] see 7137
Creosedin [Osiris] see 1406
Creosol, 2640
Creosote, Coal Tar, 2641
Creosote, Wood, 2642
Creosotic Acid, 2643
Crepasin [Hoei] see 7919
Cresatin [Merck & Co.] see 2651
Cresatin Metacresylacetate [Merck & Co.] see 2651
Cresatin-Sulzberger [Merck & Co.] see 2651
Crescefel [Ayerst] see 7062
Crescormon [Kabi] see 8864
Creslan [Am. Cyanamid], 2644
Cresol, 2645
m-Cresol see 2645
o-Cresol see 2645
p-Cresol see 2645
o-Cresol Acetate see 2652
m-Cresol Acetic Acid Ester see 2651
Cresolase see 9969
o-Cresolphthalein, 2646
Cresol Red, 2647
o-Cresolsulfonphthalein see 2647
2,3-Cresotic Acid, 2649
2,4-Cresotic Acid see 2648
2,5-Cresotic Acid see 2650
m-Cresotic Acid, 2648
o-Cresotic Acid, 2649

p-Cresotic Acid, 2650
γ-Cresotic Acid see 2648
m-Cresotinic Acid see 2648
o-Cresotinic Acid see 2649
p-Cresotinic Acid see 2650
Cresoxydiol see 5895
Cresoxypropanediol see 5895
Crestabolic see 6017
Crestmoreite see 1749
Crestomycin see 7173
m-Cresyl Acetate, 2651
o-Cresyl Acetate, 2652
o-Cresyl Glycerol Ether see 5895
o-Cresylic Acetate see 2652
Cresylic Acid see 2645
o-Cresylic Acid see 2645
Cresylol see 2645
CRF, 2653
CRH see 2653
Crilanomer, 2654
Crimidine, 2655
Crinovaryl see 3751
Crinovyl [Rhovyl] see 7746
Crinuryl [Assia] see 3761
Crisalbine see 4537
Criseociclina [Farmitalia] see 9337
Criseocil see 7235
Crisinor [Rubio] see 911
Crisofin [Allergan] see 911
Crispin [Kowa] see 9701
Cristal [Vifor] see 4493
Cristallovar see 3751
Cristalomicina see 5293
Cristaloxine see 4437
Cristapen see 7225
Cristapurat see 3206
Cristobalite see 8637
Crithidia Factor see 6551
Crixivan [Merck & Co.] see 4979
CRL-40028 see 168
CRL-40476 see 6311
Croceic Acid, 2656
Crocetin, 2657
trans-Crocetin see 2657
Crocin see 2657
Crocoite see 2288 and 5422
Croconazole see 2429
Crocus see 8466
Cromadrenal see 171
Cromakalim see 5487
(−)-Cromakalim see 5487
Cromaton see 5575
Cromocap [Fidia] see 2467
Cromoci see 2860
Cromoglycic Acid see 2658
Cromolyn, 2658
Cromolyn Sodium see 2658
Cromonalgina [Ceccarelli] see 9791
Cromosan see 5688
Cromosil see 171
Cromovet [Pharmacia] see 2658
Cronassial [Fidia] see 4376
Cronetal see 3428
Cronil [Farmigea] see 3552
Cronizat [Farmitalia] see 6758
Cronodione see 7381
Cronolone [Searle] see 4235
Cropotex [Bayer] see 4155
Cropropamide, 2659
Crotaline see 6333
Crotamine, 2660
Crotamitex [Troponwerke] see 2661
Crotamiton, 2661
Crotethamide, 2662
Crotonaldehyde, 2663
cis-Crotonaldehyde see 2663
Crotonchloral see 9762
Crotonic Acid, 2664
trans-Crotonic Acid see 2664
α-Crotonic Acid see 2664

Crotonic Aldehyde see 2663
Croton Oil, 2665
Crotonyl Alcohol see 2668
α-(N'-Crotonyl-N'-ethyl)amino-N,N-
  dimethylbutyramide see 2662
Crotonyl-N-ethyl-o-toluidine see 2661
α-(N'-Crotonyl-N'-propyl)amino-N,N-
  dimethylbutyramide see 2659
Crotothane [M & B] see 3340
Crotoxin, 2666
Crotoxyphos, 2667
Crottle see 2682
Crotyl Alcohol, 2668
Crotyl Chloride see 2181
Crovaril see 7106
Crovicina [Viti] see 8411
18-Crown-6 see 2669
Crown Compounds see 2669
Crown Ethers, 2669
Croysulfone see 2885
CRP, 2670
Crude Chrysarobin see 807
Crude Cream of Tartar see 819
Crude Oil see 7328
Crude Opium see 6986
Crude Potassium Bitartrate see 819
Crufomate, 2671
Crustecdysone see 3539
Crylène [Auclair] see 7260
Cryofluorane, 2672
Cryogenine see 7378
Cryolite, 2673
Cryptates see 2669
Cryptenamine Tannates, 2674
Cryptocavine see 2675
Cryptocillin [Hoechst] see 7036
Cryptohalite see 558
Cryptopine, 2675
Cryptoxanthin, 2676
Cryptoxanthol see 2676
Crystallinic Acid see 6818
Crystallins, 2677
Crystallized Verdigris see 2690
Crystallose [Tenneco] see 8463
Crystal Violet see 4401
Crystamine [Dunhall] see 10152
Crystapen see 7225
Crystex see 9142
Crysticillin [Squibb] see 7226
Crystodigin [Lilly] see 3206
Crystoids [Merck & Co.] see 4750
Crystoserpine [Dorsey] see 8314
Crytion see 4537
CS see 2178
CS-045 see 9898
CS-359 see 1489
CS-370 see 2481
CS-386 see 6254
CS-430 see 4635
CS-443 see 1227
CS-500 see 6251
CS-514 see 7894
CS-533 see 7141
CS-600 see 5619
CS-622 see 9287
CS-684 see 7692
CS-807 see 1991
CS-847 see 986
CS-976 see 7141
CS-1170 see 1977
CSA see 1915 and 2270
CSF-1 see 5678
CSF-2 see 4559
CSFα see 4559
CSF-β see 4558
CSFs see 2552
CSP see 4119
C-Stuff see 4810
CT see 1048
CT-848 see 7076

CT-1341 see 235
C.T.A.B. see 2068
CTR-6110 see 6693
CTR-6669 see 1836
CTX see 2087
CTZ see 1997
Cubane, 2678
Cube Alum see 373
Cubeb, 2679
Cubebin, 2680
Cubic Niter see 8792
Cuca see 2515
Cucoline see 8692
Cucurbitacin B see 2681
Cucurbitacin E see 2681
Cucurbitacins, 2681
Cudbear, 2682
Cuemid [Merck & Co.] see 2257
Cuivasil [IDC] see 5505
Cujec [Pitman-Moore] see 2740
Culpen [Beecham] see 4147
Cultar [ICI] see 7118
Culver's Root see 5465
Cu-lyt see 2730
Cumaldehyde see 2687
Cumaran see 2627
Cumarin see 2630
Cumarone see 1119
Cumarote C [Towa Yakuhin] see 4903
Cumelon [Teijin] see 8428
Cumene, 2683
3-p-Cumenyl-1,1-dimethylurea see
  5237
2-[(o-Cumenyloxy)methyl]-2-imidazo-
  line see 4025
Cumertilin Sodium [Endo] see 5921
Cumetharol see 2633
Cumethoxaethane see 2633
Cumic Acid, 2684
Cumic Alcohol, 2685
Cumidine, 2686
Cuminal see 2687
Cuminaldehyde, 2687
Cuminic Acid see 2684
Cuminol see 2685
Cuminyl Alcohol see 2685
Cumol see 2683
Cumopyran [Abbott] see 2786
Cumopyrin see 2786
Cumotocopherol see 9632
Cupferron, 2688
Cupralene [Specia] see 265
Cupralylnatrium see 265
Cupralylsodium see 265
Cuprammonium Sulfate see 9323
Cupreine, 2689
Cuprelon see 265
Cuprenil [Polfa] see 7214
Cupreol see 8697
Cupressin [Takeda] see 2928
Cuprex [Beecham] see 2711
Cupriaseptol see 2715
Cupric Acetate, 2690
Cupric Acetate, Basic, 2691
Cupric Acetoarsenite, 2692
Cupric Aminoacetate see 2707
Cupric Ammonium Chloride see 543
Cupric Arsenite, 2693
Cupric Benzoate see 1122
Cupric Bis[8-hydroxyquinoline Di(di-
  ethylammonium Sulfonate)] see
  2740
Cupric Borate, 2694
Cupric Bromide, 2695
Cupric Butyrate, 2696
Cupric Carbonate, Basic, 2697
Cupric Chlorate, 2698
Cupric Chloride, 2699
Cupric Chromate(III) see 2701
Cupric Chromate(VI), 2700

Cupric Chromite, 2701
Cupric Citrate, 2702
Cupric Ferrocyanide, 2703
Cupric Ferrous Sulfide *see* 2079
Cupric Fluoride, 2704
Cupric Fluosilicate *see* 2708
Cupric Formate, 2705
Cupric Gluconate, 2706
Cupric Glycinate, 2707
Cupric Hexacyanoferrate(II) *see* 2703
Cupric Hexafluorosilicate, 2708
Cupric Hydroxide, 2709
Cupricin *see* 2731
Cupric Lactate *see* 5349
Cupric Nitrate, 2710
Cupric Oleate, 2711
Cupric Oxalate, 2712
Cupric Oxide, 2713
Cupric Perchlorate, 2714
Cupric *p*-Phenolsulfonate, 2715
Cupric Phosphate, 2716
Cupric Salicylate, 2717
Cupric Selenate, 2718
Cupric Selenide, 2719
Cupric Selenite, 2720
Cupric Silicofluoride *see* 2708
Cupric Stearate, 2721
Cupric Subacetate *see* 2691
Cupric Subcarbonate *see* 2697
Cupric Subsulfate *see* 2723
Cupric Sulfate, 2722
Cupric Sulfate, Ammoniated *see* 9323
Cupric Sulfate, Basic, 2723
Cupric Sulfide, 2724
Cupric Sulfocarbolate *see* 2715
Cupric Tartrate, 2725
Cupric Tungstate(VI), 2726
Cupric Wolframate *see* 2726
Cuprid [Merck & Co.] *see* 9796
Cuprimine [Merck & Co.] *see* 7214
Cuprimyl [M&B] *see* 2740
Cuprimyxin *see* 6425
Cuprion *see* 265
Cuprite *see* 2734
Cuprobam, 2727
Cuprobamé *see* 2727
Cuprocitrol *see* 2702
Cupro-cupric Sulfate *see* 2738
Cupron [Ayerst] *see* 1125
Cuprothiosinamine *m*-Benzoate
  Sodium *see* 265
Cuprous Acetate, 2728
Cuprous Bromide, 2729
Cuprous Chloride, 2730
Cuprous Cyanide, 2731
Cuprous Dimethyldithiocarbamate
  Cuprous Chloride Complex *see* 2727
Cuprous Iodide, 2732
Cuprous Mercuric Iodide, 2733
Cuprous Oxide, 2734
Cuprous Potassium Cyanide, 2735
Cuprous Selenide, 2736
Cuprous Sulfide, 2737
Cuprous Sulfite, 2738
Cuprous Sulfocyanate *see* 2739
Cuprous Sulfocyanide *see* 2739
Cuprous Tetraiodomercurate(II) *see*
  2733
Cuprous Thiocyanate, 2739
Cuproxat [Agrolinz] *see* 2723
Cuproxoline, 2740
Curamil [Hoechst] *see* 8146
Curan-17-ol *see* 4386
Curantyl [Berlin-Chemie] *see* 3410
Curare, 2741
C-Curarine I, 2742
C-Curarine III, 2743
Curarin-HAF [Ethicon] *see* 9939
Curatin [Pfizer] *see* 3492
Curatrem [Merck & Co.] *see* 2471

Curcuma Longa *see* 9955
Curcumin, 2744
Curethyl *see* 5575
Curine, 2745
Curium, 2746
Curled Dock *see* 8450
Curling Factor *see* 4571
Curocef [Glaxo] *see* 2002
Curosajin [Mohan] *see* 4770
Curosurf [Chiesi] *see* 7754
Curoxim [Glaxo-Hoechst] *see* 2002
Curretab [Reid-Provident] *see* 5838
Curtacrat *see* 2636
Curvularin, 2747
Curythan *see* 5895
Cuscohygrine, 2748
Cus Cus Oil *see* 6953
Cuscutin *see* 1197
Cuskhygrine *see* 2748
Cuspareine, 2749
Cusparia Bark *see* 690
Cusparine, 2750
Cusso *see* 1391
Cutch *see* 1951
Cuteparine [Biosedra] *see* 4685
Cutisan [Boots] *see* 9786
Cutisterol [Farmitalia] *see* 4270
Cutivate [Glaxo] *see* 4244
Cutless [Elanco] *see* 4239
Cuttle-fish Bone *see* 8601
Cut-weed *see* 4305
Cuvalit [Schering AG] *see* 5541
Cuxacillin [TAD] *see* 617
Cuxanorm [TAD] *see* 892
CV-205-502 *see* 8226
CV-2619 *see* 4932
CV-3317 *see* 2928
CV-4093 *see* 5786
CV-4151 *see* 5120
CV-11974 *see* 1788
CV-58903 *see* 10204
C-Vimin [Astra] *see* 867
CVP *see* 2137
C.V.P. [USV] *see* 1269
CY-116 *see* 451
CY-216 *see* 6434
Cyacetacide, 2751
Cyamelide, 2752
Cyamemazine, 2753
Cyamepromazine *see* 2753
Cyamopsis Gum *see* 4602
Cyanacethydrazide *see* 2751
Cyanacethydrazine *see* 2751
Cyanacetylhydrazide *see* 2751
"Cyanamide" *see* 1702
Cyanamide, 2754
Cyanazide *see* 2751
Cyanazine, 2755
Cyanein *see* 1393
Cyanic Acid, 2756
Cyanic Acid Sodium Salt *see* 8749
(+)-Cyanidanol-3 *see* 1950
Cyanidanon-3-methyl Ether 1625 *see*
  4773
Cyanidanon 4'-Methyl Ether 1626 *see*
  4704
Cyanidenolon 1522 *see* 8216
Cyanidenolon-7-methyl Ether 1537 *see*
  8337
Cyanidenon 1470 *see* 5641
Cyanidenon-4'-methyl Ether 1479 *see*
  3349
Cyanidin Chloride, 2757
Cyanidol *see* 1950
Cyanin *see* 2757
Cyanite *see* 377
Cyanizide *see* 2751
Cyanoacetamide, 2758
Cyanoacetic Acid, 2759

Cyanoacetic Acid Ethyl Ester *see*
  3833
Cyanoacetic Acid Hydrazide *see* 2751
Cyanoacetic Ester *see* 3833
Cyanoacetohydrazide *see* 2751
Cyanoacetonitrile *see* 5750
2-Cyanoacrylic Acid Isobutyl Ester
  see 1488
2-Cyanoacrylic Acid Methyl Ester *see*
  5827
Cyanobenzene *see* 1128
1-Cyanobutane *see* 10043
Cyanocobalamin *see* 10152
Cyanocobalamin Zinc Tannate *see*
  10152
N-Cyanodiallylamine *see* 3016
4-Cyano-3-(2,3-dichlorophenyl)pyrrole
  see 4030
(—)-6-Cyano-3,4-dihydro-2,2-dimeth-
  yl-*trans*-4-(2-oxo-1-pyrrolidinyl)-
  2*H*-benzo[*b*]pyran-3-ol *see* 5487
2-Cyano-1,4-dihydro-6-methyl-4-(3-
  nitrophenyl)-3,5-pyridinedicarbox-
  ylic Acid 3-Methyl 5-(1-Methyl-
  ethyl) Ester *see* 6637
3-Cyano-2,6-dihydroxypyridine *see*
  3601
2-Cyano-10-(3-dimethylamino-2-meth-
  ylpropyl)phenothiazine *see* 2753
3-Cyano-5-dimethylamino-3-phenyl-2-
  methylhexane *see* 5124
α-Cyanodiphenylmethane *see* 3374
4-[2-[(5-Cyano-5,5-diphenylpentyl)-
  dimethylammonio]ethyl]-4-methyl-
  morpholinium Bis(methylsulfate) *see*
  7243
N¹-(5-Cyano-5,5-diphenylpentyl)-*N¹*,-
  N¹,*N²*-trimethylethylene-1-ammo-
  nium-2-morpholinium Bis(methylsulfate) *see* 7243
1'-(3-Cyano-3,3-diphenylpropyl)-[1,4'-
  bipiperidine]-4'-carboxamide *see*
  7653
1-(3-Cyano-3,3-diphenylpropyl)-4-(2-
  oxo-3-propionyl-1-benzimidazolin-
  yl)piperidine *see* 1241
1-(3-Cyano-3,3-diphenylpropyl)-4-
  phenylisonipecotic Acid *see* 3183
1-(3-Cyano-3,3-diphenylpropyl)-4-
  phenylisonipecotic Acid Ethyl Ester
  see 3371
1-(3-Cyano-3,3-diphenylpropyl)-4-
  phenyl-4-piperidinecarboxylic Acid
  see 3183
1-(3-Cyano-3,3-diphenylpropyl)-4-
  phenyl-4-piperidinecarboxylic Acid
  Ethyl Ester *see* 3371
1-[1-(3-Cyano-3,3-diphenylpropyl)-4-
  piperidinyl]-1,3-dihydro-3-(1-oxo-
  propyl)-2*H*-benzimidazol-2-one *see*
  1241
1-[1-(3-Cyano-3,3-diphenylpropyl)-4-
  piperidyl]-3-propionyl-2-benzimid-
  azolinone *see* 1241
2α-Cyano-4α,5α-epoxyandrostan-
  17β-ol-3-one *see* 9827
2-Cyanoethanesulfonic Acid Sodium
  Salt *see* 8832
Cyanoethydrazide *see* 2751
(±)-*N*-2-Cyanoethylamphetamine *see*
  4036
1-(2-Cyanoethyl)-7-chloro-3-hydroxy-
  5-(2'-fluorophenyl)-1,3-dihydro-2*H*-
  1,4-benzodiazepin-2-one *see* 2368
Cyanoethylene *see* 133
Cyanofenphos, 2760
(*R,S*)-α-Cyano-4-fluoro-3-phenoxy-
  benzyl-(1*R*,*S*)-*cis*,*trans*-3-(2,2-dichlo-
  rovinyl)-2,2-dimethylcyclopropane-
  carboxylate *see* 2826

Name Index

Cyclohexanehexyl Hexaphosphate *see* 7542

Cyclohexanemethanol *see* 2800

1,2,3,4,5-Cyclohexanepentol *see* 8218

Cyclohexanesulfamic Acid *see* 2770

Cyclohexanesulfamic Acid Calcium Salt *see* 1705

Cyclohexanol, 2794

1-Cyclohexanol-α-butyric Acid *see* 2784

Cyclohexanone, 2795

Cyclohexatriene *see* 1094

Cyclohexene, 2796

5-(2-Cyclohexen-1-yl)-5-allyl-2-thiobarbituric Acid *see* 9428

5-(2-Cyclohexen-1-yl)dihydro-5-(2-propenyl)-2-thioxo-4,6(1*H*,5*H*)-pyrimidinedione *see* 9428

5-Cyclohexenyl-3,5-dimethylbarbituric Acid *see* 4742

5-(1-Cyclohexen-1-yl)-1,5-dimethylbarbituric Acid *see* 4742

5-(1-Cyclohexen-1-yl)-1,5-dimethyl-2,4,6(1*H*,3*H*,5*H*)-pyrimidinetrione *see* 4742

5-(1-Cyclohexen-1-yl)-5-ethylbarbituric Acid *see* 2780

5-(1-Cyclohexen-1-yl)-5-ethyl-2,4,6-(1*H*,3*H*,5*H*)-pyrimidinetrione *see* 2780

Cycloheximide, 2797

Cyclohexitol *see* 5008

3-Cyclohexyl-1-(*p*-acetylphenylsulfonyl)urea *see* 59

Cyclohexylamine, 2798

N-[2-[4-[[[(Cyclohexylamino)carbonyl]amino]sulfonyl]phenyl]ethyl]-5-methylpyrazinecarboxamide *see* 4451

N-[(Cyclohexylamino)carbonyl]-2,3-dihydro-1*H*-indene-5-sulfonamide *see* 4516

N-[(Cyclohexylamino)carbonyl]-4-[2-(3,4-dihydro-7-methoxy-4,4-dimethyl-1,3-dioxo-2(1*H*)-isoquinolinyl)-ethyl]benzenesulfonamide *see* 4452

N-[(Cyclohexylamino)carbonyl]-4-methylbenzenesulfonamide *see* 9648

2-(Cyclohexylamino)ethanesulfonic Acid *see* 2097

3-(Cyclohexylamino)-1-propanesulfonic Acid *see* 1810

1-(Cyclohexylamino)-2-propanol Benzoate (Ester) Hydrochloride *see* 4746

25-Cyclohexylavermectin B₁ *see* 3483

α-Cyclohexylbenzeneacetic Acid 2-(Diethylamino)ethyl Ester *see* 3501

Cyclohexyl Bromide, 2799

3-Cyclohexyl-2-butenoic Acid *see* 2327

N-[(*N*-Cyclohexylcarbamoyl)methyl]-piperazine *see* 3736

Cyclohexylcarbinol, 2800

(—)-*N*-[(*S*)-[3-(*N*-Cyclohexylcarbonyl-ᴅ-alanyl)thio]-2-methylpropionyl]-ʟ-proline *see* 6368

[*R*-(*R**,*S**)]-1-[3-[[2-[(Cyclohexyl-carbonyl)amino]-1-oxopropyl]thio]-2-methyl-1-oxopropyl]-ʟ-proline *see* 6368

2-(Cyclohexylcarbonyl)-2,3,6,7,8,12b-hexahydropyrazino[2,1-*a*][2]benz-azepin-4(1*H*)-one *see* 3671

2-(Cyclohexylcarbonyl)-1,2,3,6,7,11b-hexahydro-4*H*-pyrazino[2,1-*a*]iso-quinolin-4-one *see* 7896

N-(Cyclohexylcarbonyl)-3-(4-morpholinyl)sydnone Imine *see* 2326

2-Cyclohexylcarbonyl-4-oxo-1,2,3,4,-

6,7,12b-octahydropyrazino[2,1-*a*]-[2]benzazepine *see* 3671

Cyclohexyl Chloride, 2801

4-(4-Cyclohexyl-3-chlorophenyl)-4-oxobutyric Acid *see* 1486

N-Cyclohexylcyclohexanamine *see* 3145

1-Cyclohexylcyclohexanecarboxylic Acid β-Piperidinoethyl Ester *see* 3211

25-Cyclohexyl-5-*O*-demethyl-25-de(1-methylpropyl)avermectin A₁ₐ *see* 3483

3-Cyclohexyl-6,7-dihydro-1*H*-cyclopentapyrimidine-2,4-(3*H*,5*H*)-dione *see* 5459

1-Cyclohexyl-3-[[*p*-[2-(3,4-dihydro-7-methoxy-4,4-dimethyl-1,3-dioxo-2(1*H*)-isoquinolyl)ethyl]phenyl]-sulfonyl]urea *see* 4452

3-Cyclohexyl-6-(dimethylamino)-1-methyl-1,3,5-triazine-2,4(1*H*,3*H*)-dione *see* 4734

(±)-*N*-Cyclohexyl-*N*′-[4-[3-[(1,1-di-methylethyl)amino]-2-hydroxyprop-oxy]phenyl]urea *see* 9208

2-Cyclohexyl-4,6-dinitrophenate *see* 2802

2-Cyclohexyl-4,6-dinitrophenol, 2802

4-Cyclohexyl-3-ethyl-4*H*-1,2,4-tri-azole *see* 4735

1-Cyclohexyl-3-(5-hydrindenylsulfon-yl)urea *see* 4516

α-Cyclohexyl-α-hydroxybenzeneace-tic Acid 4-(Diethylamino)-2-butynyl Ester Hydrochloride *see* 7089

α-Cyclohexyl-α-hydroxybenzeneace-tic Acid 1-Methyl-3-piperidinyl Ester *see* 7998

α-Cyclohexyl-α-hydroxybenzene-acetic Acid (1,4,5,6-Tetrahydro-1-methyl-2-pyrimidinyl)methyl Ester *see* 7107

(4*S*)-4-Cyclohexyl-1-[[[(*RS*)-1-hydr-oxy-2-methylpropoxy](4-phenylbut-yl)phosphinyl]acetyl]-ʟ-proline Pro-pionate (Ester) *see* 4282

6-Cyclohexyl-1-hydroxy-4-methyl-2(1*H*)-pyridinone *see* 2325

2-Cyclohexyl-3-hydroxy-1,4-naph-thalenedione *see* 7182

2-Cyclohexyl-3-hydroxy-1,4-naphtho-quinone *see* 7182

[1*R*-[1α(*Z*),2β(*S**),3α,5α]]-7-[2-(5-Cyclohexyl-3-hydroxy-1-pentynyl)-3,5-dihydroxycyclopentyl]-5-hepten-oic Acid Methyl Ester *see* 4282

4-(β-Cyclohexyl-β-hydroxy-β-phen-ethyl)-1,1-dimethylpiperazinium Methyl Sulfate *see* 4744

Cyclohexylhydroxyphenylacetic Acid Diethylmethylaminoethyl Ester Bromide *see* 7109

2-[(Cyclohexylhydroxyphenylacetyl)-oxy]-*N*,*N*-diethyl-*N*-methylethana-minium Bromide *see* 7109

4-(2-Cyclohexyl-2-hydroxy-2-phenyl-ethyl)-1,1-dimethylpiperazinium Methyl Sulfate (Salt) *see* 4744

N-(β-Cyclohexyl-β-hydroxy-β-phenyl-ethyl)-*N*′-methylpiperazine Dimeth-ylsulfate *see* 4744

N-(β-Cyclohexyl-β-hydroxy-β-phenyl-ethyl)-*N*′-methylpiperazine Metho-sulfate *see* 4744

1-(3-Cyclohexyl-3-hydroxy-3-phenyl-propyl)-1-methylpyrrolidinium Chloride *see* 7944

(3-Cyclohexyl-3-hydroxy-3-phenyl-

propyl)triethylammonium Iodide *see* 9794

[3a*R*-[2*Z*,3aα,4α(1*E*,3*S**)-5β,6aα]]-3-[[4-(3-Cyclohexyl-3-hydroxy-1-propenyl)hexahydro-5-hydroxy-2*H*-cyclopenta[*b*]furan-2-ylidene]meth-yl]benzoic Acid *see* 9227

α-[(2*Z*,3a*R*,4*R*,5*R*,6a*S*)-4-[(1*E*,3*S*)-3-Cyclohexyl-3-hydroxypropenyl]hex-ahydro-5-hydroxy-2*H*-cyclopenta-[*b*]furan-2-ylidene]-*m*-toluic Acid *see* 9227

2,2′-[Cyclohexylidenebis(4,1-phenyl-eneoxy)]bis[2-methylbutanoic Acid] *see* 2415

2,2′-(4,4′-Cyclohexylidenediphenoxy)-2,2′-dimethyldibutyric Acid *see* 2415

α-Cyclohexylidene-α-(*p*-hydroxyphen-yl)-*p*-cresol Diacetate *see* 2789

1-Cyclohexyl-3-(5-indanylsulfonyl)-urea *see* 4516

Cyclohexylisopropylmethylamine *see* 2776

N-Cyclohexyllinoleamide *see* 5529

β-Cyclohexyl-β-methacrylic Acid *see* 2327

(17β)-17-[(Cyclohexylmethoxy)-carbonyl]oxyestra-4,9,11-trien-3-one *see* 9716

N-Cyclohexyl-*N*-methyl-2-(2-amino-3,5-dibromo)benzylammonium *see* 1412

1-Cyclohexyl-2-methylaminopropane *see* 8045

l-1-Cyclohexyl-2-(methylamino)prop-ane Ethylphenylbarbiturate *see* 8045

N-(2-Cyclohexyl-1-methylethyl)-γ-phenylbenzenepropanamine *see* 3506

1-Cyclohexyl-1-methyl-2-methyl-aminoethane *see* 2776

[1[*S**(*R**)],2α,4β]-4-Cyclohexyl-1-[[[2-methyl-1-(1-oxopropoxy)prop-oxy](4-phenylbutyl)phosphinyl]acet-yl]-ʟ-proline *see* 4282

2-Cyclohexyl-*N*-methylpropylamine *see* 2776

1-Cyclohexyl-3-[[*p*-[2-(5-methylpyra-zinecarboxamido)ethyl]phenyl]-sulfonyl]urea *see* 4451

p-Cyclohexyloxybenzoic Acid Ester of *N*-(3-Hydroxypropyl)pipecoline *see* 2804

4-(Cyclohexyloxy)benzoic Acid 3-(2-Methyl-1-piperidinyl)propyl Ester *see* 2804

1-[(2-Cyclohexyl-2-phenyl-1,3-dioxo-lan-4-yl)methyl]-1-methylpiperidin-ium Iodide *see* 2806

2-(Cyclohexylphenylmethyl)-*N*,*N*,*N*′,-*N*′-tetraethyl-1,3-propanediamine *see* 3986

α-Cyclohexyl-α-phenyl-1-piperidine-propanol Hydrochloride *see* 9823

α-Cyclohexyl-α-phenyl-1-pyrrolidine-propanol *see* 7944

1-Cyclohexyl-1-phenyl-3-pyrrolidino-1-propanol *see* 7944

1-Cyclohexyl-1-phenyl-3-(1-pyrroli-dinyl)-1-propanol *see* 7944

N-Cyclohexyl-1-piperazineacetamide *see* 3736

1-Cyclohexylpropyl Carbamate *see* 7945

Cyclohexylsulfamic Acid *see* 2770

Cyclohexylsulfamic Acid Calcium Salt *see* 1705

N-Cyclohexyltaurine *see* 2097

3-Cyclohexyl-1,5,6,7-tetrahydro-2*H*-cyclopentapyrimidine-2,4(3*H*)-dione *see* 5459

N-(4-Cyclopropyl-6-methylpyrimidin-2-yl)aniline *see* 2841
N-Cyclopropylmethylscopolamine Bromide *see* 2338
N-(Cyclopropylmethyl)-α,α,α-trifluoro-2,6-dinitro-N-propyl-*p*-toluidine *see* 7954
(1α,3β,5Z,7E,22E,24S)-24-Cyclopropyl-9,10-secochola-5,7,10(19),22-tetraene-1,3,24-triol *see* 1679
21-Cyclopropyl-6,7,8,14-tetrahydro-7α-(1-hydroxy-1-methylethyl)-6,14-endo-ethanooripavine *see* 3403
N-Cyclopropyl-1,3,5-triazine-2,4,6-triamine *see* 2845
Cyclo-Prostin [Upjohn] *see* 8061
Cyclorphan, 2819
Cyclosa [Nourypharma] *see* 2971
Cyclosal *see* 4768
Cycloserine, 2820
Cyclospasmol [Brocades] *see* 2771
Cyclosporin A *see* 2821
Cyclosporin B *see* 2821
Cyclosporin C *see* 2821
Cyclosporin D *see* 2821
Cyclosporin G *see* 2821
Cyclosporine *see* 2821
Cyclosporins, 2821
Cyclostin [Farmitalia] *see* 2816
Cyclosulfyne *see* 7992
Cyclotetramethylene Tetranitramine *see* 6855
Cyclothiazide, 2822
Cyclotrimethylenetrinitramine *see* 2805
Cyclotriphosphazenes *see* 2401
Cyclovalone, 2823
Cyclo-Werrol [Werrol] *see* 4911
Cycocel [Am. Cyanamid] *see* 2153
Cycogan [Makhteshim-Agan] *see* 2153
Cycostat [Am. Cyanamid] *see* 8403
Cycothiamin(e), 2824
Cycotiamine *see* 2824
Cycrimine Hydrochloride, 2825
Cycrin [Ayerst] *see* 5838
Cydectin [Am. Cyanamid] *see* 6373
Cydonia Seed *see* 8237
Cydril [Tutag] *see* 623
Cyfen [Am. Cyanamid] *see* 4017
Cyfluthrin, 2826
Cyfoxylate *see* 2826
Cygon [Am. Cyanamid] *see* 3269
Cygro [Am. Cyanamid] *see* 5682
Cyhalothrin, 2827
Cyheptamide, 2828
Cyhexatin, 2829
Cyklodorm *see* 2780
Cyklokapron [Kabi] *see* 9704
Cyklosal *see* 2808
Cykobemin [ACO] *see* 10152
Cylan [Am. Cyanamid] *see* 7490
Cylert [Abbott] *see* 7206
Cymarigenin *see* 9015
Cymarin, 2830
Cymarose, 2831
Cymbi [Dolorgiet] *see* 628
Cymbush [ICI] *see* 2836
Cymene, 2832
m-Cymene *see* 2832
o-Cymene *see* 2832
p-Cymene *see* 2832
2-*p*-Cymenol *see* 1923
3-*p*-Cymenol *see* 9540
Cymerin [Tokyo Tanabe] *see* 8285
Cymevan [Syntex] *see* 4374
Cymevene [Syntex] *see* 4374
Cymiazole, 2833
Cymidon *see* 5312
m-Cym-5-yl Methylcarbamate *see* 7967

Cynanchin *see* 10121
Cynanchogenin, 2834
Cynarin(e), 2835
Cynaroside *see* 5641
Cynem [Am. Cyanamid] *see* 9482
Cynisin *see* 2486
Cynoff [FMC] *see* 2836
Cynoglossophine *see* 4659
Cynomel [SK & F] *see* 5535
Cynotoxin *see* 9015
Cynt [Beiersdorf] *see* 6375
Cyoctol [Chantal] *see* 2372
Cyolane [Am. Cyanamid] *see* 7490
Cyomel [SKB] *see* 5535
CYP *see* 2760
Cypercare [Virbac] *see* 2836
Cyperkill [Mitchell Cotts] *see* 2836
Cypermethrin, 2836
Cypersect [Mitchell Cotts] *see* 2836
Cyphenothrin, 2837
Cypip [Am. Cyanamid] *see* 3165
Cyprenorphine, 2838
Cypress Camphor *see* 1961
Cypripedium, 2839
Cyproconazole, 2840
Cyprodinil, 2841
Cyproheptadine, 2842
Cyprome Ether *see* 2818
Cypromin [Lakeside] *see* 8409
Cyproquinate, 2843
Cyproquinidate *see* 2843
Cyprostat [Schering AG] *see* 2844
Cyproterone, 2844
Cyproxyquine *see* 2843
Cyren A [Bayer] *see* 3177
Cyren B [Bayer] *see* 3177
Cyromazine, 2845
Cyrpon [Troponwerke] *see* 5908
Cys *see* 2850
Cyscholin [MECT] *see* 2380
Cystafos *see* 7497
Cystamin *see* 6036
Cystamine, 2846
Cystaphos *see* 7497
L-Cystathionine, 2847
Cysteamine, 2848
Cysteamine-S-phosphate *see* 7497
Cysteic Acid, 2849
Cysteine, 2850
L-Cysteine *see* 2850
L-Cysteine Methyl Ester Hydrochloride *see* 5828
L-Cysteine Thioacetal of Formaldehyde *see* 3452
N²-[N-(N-L-Cysteinyl-L-tyrosyl)-L-cysteinyl]-interferon γ (Human Lymphocyte Protein Moiety Reduced) *see* 5018
Cystine, 2851
L-Cystine *see* 2851
Cystine S,S-Dioxide *see* 2852
Cystine S,S'-Dioxide *see* 2852
Cystine S-Dioxides, 2852
Cystine Disulfoxide *see* 2852
sym-L-Cystine Disulfoxide *see* 2852
L-Cystine Thiosulfonate *see* 2852
Cystit [Squibb-Heyden] *see* 6696
Cysto-Conray [Mallinckrodt] *see* 5080
Cystogen *see* 6036
Cystografin [Squibb] *see* 5851
Cystokon [Mallinckrodt] *see* 79
Cystorelin [Abbott] *see* 5500
Cystosol [Baxter] *see* 8873
Cystospaz [Alcon] *see* 4907
Cystrin [Farmitalia] *see* 7089
Cystural *see* 3796
Cys-Tyr-Cys-interferon γ *see* 5018
CYT-099 *see* 8530
Cytacon [Goldshield] *see* 10152

Cytadren [Ciba-Geigy] *see* 460
Cytamen [Evans] *see* 10152
Cytarabine, 2853
Cytarbel [Bellon] *see* 2853
Cyten [Am. Cyanamid] *see* 4017
Cythioate, 2854
Cythion [Am. Cyanamid] *see* 5740
Cytidine, 2855
Cytidine Diphosphate Choline Ester *see* 2380
Cytidine-2'-monophosphate *see* 2856
Cytidine-3'-monophosphate *see* 2857
Cytidine-2'-phosphate *see* 2856
Cytidine-3'-phosphate *see* 2857
2'-Cytidinephosphoric Acid *see* 2856
3'-Cytidinephosphoric Acid *see* 2857
Cytidine 5'-(Trihydrogen Diphosphate) P'-[2-(trimethylammonio)ethyl] Ester Inner Salt *see* 2380
2'-Cytidylic Acid, 2856
3'-Cytidylic Acid, 2857
Cytidylic Acid *a* *see* 2856
Cytidylic Acid *b* *see* 2857
Cytisine, 2858
Cytiton [USSR] *see* 2858
Cytoadhesins *see* 5014
Cytobin [Norden] *see* 5535
Cytobion [E. Merck] *see* 10152
Cytochalasin B *see* 2859
Cytochalasins, 2859
Cytochrome c, 2860
Cytochrome c_1 *see* 2860
Cytochrome c_2 *see* 2860
Cytochrome c_3 *see* 2860
Cytochrome e *see* 2860
Cytochromes P_{450}, 2861
Cytofol [Lappe] *see* 4253
Cytohemin, 2862
Cytolipin H, 2863
Cytolysin *see* 7301
Cytomel [SKB] *see* 5535
Cytomine [Darby] *see* 5535
Cytonal [Berlin-Chemie] *see* 4279
Cytophosphane *see* 2816
Cytorest [Mochida] *see* 2860
Cytosar [Upjohn] *see* 2853
Cytosine, 2864
β-Cytosine Arabinoside *see* 2853
Cytosine Riboside *see* 2855
1-(1'-Cytosinyl)-4-[L-3'-amino-5'-(1''-N-methylguanidino)-valerylamino]-1,2,3,4-tetradeoxy-β-D-erythro-hex-2-enuronic Acid *see* 1350
Cytostatin *see* 4738
2'-Cytosylic Acid *see* 2856
3'-Cytosylic Acid *see* 2857
Cytotec [Searle] *see* 6297
Cytotect [Biotest] *see* 4958
Cytovene [Syntex] *see* 4374
Cytoxan [Mead Johnson] *see* 2816
Cytoxin-anticytotoxin Serum *see* 759
Cytozyme *see* 9527
Cytrol [Am. Cyanamid] *see* 513
Cytrolane [Am. Cyanamid] *see* 5900
CZ *see* 1928
D *see* 875
1,3-D *see* 3125
2,4-D, 2865
D & C Blue No. 6 *see* 4977
D & C Green No. 5 *see* 248
D & C Green No. 6 *see* 8251
D & C Orange No. 4 *see* 6990
D & C Orange No. 5 *see* 3070
D & C Orange No. 8 *see* 3117
D & C Orange No. 10 *see* 3233
D & C Orange No. 11 *see* 3233
D & C Red No. 17 *see* 9054
D & C Red No. 19 *see* 8349
D & C Red No. 21 *see* 3639
D & C Red No. 22 *see* 3639

Name Index

D & C Yellow No. 7 *see* 4194
D & C Yellow No. 8 *see* 4194
D & C Yellow No. 10 *see* 8256
D & C Yellow No. 11 *see* 8257
D4T *see* 8958
D-20 *see* 2790
D-32 *see* 10209
D-40 *see* 8779
D-40TA *see* 3744
D-41 *see* 9901
D-50 *see* 8140
D-65MT *see* 320
D-109 *see* 4746
D-138 *see* 6796
D-145 *see* 5872
D-201 *see* 5248
D-206 *see* 8073
D-254 *see* 7609
D-301 *see* 1144
D-365 *see* 10083
D-563 *see* 7096
D-600 *see* 4369
D-706 *see* 7104
D-860 *see* 9646
D-1959 *see* 8307
D-2083 *see* 2973
D-7093 *see* 5969
D-8955 *see* 9589
D-9998 *see* 4227
D-13112 *see* 3610
D-18506 *see* 6285
DA-398 *see* 3659
DA-688 *see* 4385
DA-708 *see* 9265
DA-759 *see* 6072
DA-2370 *see* 4053
DA-3177 *see* 2338
2,4-DAA *see* 3018
DAAO *see* 434
Dabco [Air Prods.] *see* 9801
Dabroson [Hoyer] *see* 287
DAC-2787 *see* 2219
Dacarbazine, 2866
Dacarel [Roemmers] *see* 6579
Dacatic [Läkefarmos] *see* 2866
Daconate [ISK Biosci] *see* 6020
Daconil 2787 [Diamond] *see* 2219
Dacoren [Nattermann] *see* 3692
Dacortilen [E. Merck] *see* 7906
Dacortin [E. Merck] *see* 7904
DACPM *see* 6136
Dacron [Du Pont] *see* 7730
Dacthal *see* 2896
Dactil [Lakeside] *see* 7623
Dactin *see* 3115
Dactinomycin, 2867
DADA *see* 3241
DADDS *see* 20
Dadex *see* 2996
DADPS *see* 2885
Dafalgan [UPSA] *see* 45
Daflon [Servier] *see* 3350
Dafnegin [Poli] *see* 2325
D:A-Friedo-18,19-secolup-19-en-3-one *see* 8626
Dagenan [Rhône-Poulenc Rorer] *see* 9108
Dagger [Am. Cyanamid] *see* 4941
Dagutan *see* 8463
Dahi *see* 10235
Dahlin *see* 5024
Dahl's Acid *see* 6494
Dahl's Acid II *see* 6488
Dahl's Acid III *see* 6489
Daidzein, 2868
Daidzin *see* 2868
Daipisate [Taisho] *see* 7624
Daital [Alter] *see* 3759
Daitec [Lederle] *see* 3989
Daivonex [Leo] *see* 1679

Daktar [Janssen] *see* 6266
Daktarin [Janssen] *see* 6266
Dalacin C [Upjohn] *see* 2414
Dalacine [Upjohn] *see* 2414
Dalacin T [Upjohn] *see* 2414
Dalalone [Forest] *see* 2986
Dalapon, 2869
Dal-E-Rad [Vineland] *see* 6020
Dalf (obsolete) [Bayer] *see* 6183
Dalgan [Wyeth] *see* 2998
Dalgen [Elmu] *see* 4052
Dalgol *see* 5890
Dalmadorm [Roche] *see* 4233
Dalmane [Roche] *see* 4233
Dalmate [Roche] *see* 4233
Dalmatian Insect Powder *see* 8150
Dalnate [USV] *see* 9652
Dalpac *see* 1583
Dalteparin, 2870
Daltroban, 2871
Dalyde [HW & D] *see* 3076
Dalysep [Syntex] *see* 9078
Dalzic [ICI] *see* 7882
Damar, 2872
Damascenine, 2873
Dambose *see* 5008
Dametin [E. Merck] *see* 2924
Damfin [Ciba-Geigy] *see* 6004
Damide [SIT] *see* 4969
Daminozide, 2874
Dammar *see* 2872
DAMP *see* 1282
D-AMP-3 *see* 9387
DAN-603 *see* 9156
DAN-2163 *see* 508
Danabol *see* 6018
Danantizol *see* 6049
Danazol, 2875
Dandelion *see* 9228
Daneral [Hoechst] *see* 7383
Danex [Makhtcshim-Agan] *see* 9753
Danilon [Esteve] *see* 9185
Danish Agar *see* 4322
Danishefsky's Diene *see* 6082
Danitol [Sumitomo] *see* 4033
Danka [Mediolanum] *see* 3507
Danocrine [Winthrop] *see* 2875
Danofloxacin, 2876
Danol [Winthrop] *see* 2875
Danoval [Krka] *see* 2875
Dansida [Upjohn] *see* 4925
Dansyl Chloride, 2877
Dantamacrin [Röhm] *see* 2879
Danten [McKesson] *see* 7475
Danthron, 2878
Dantrium [Bago] *see* 1060
Dantrium [Eaton] *see* 2879
Dantrix [SIT] *see* 2879
Dantrolene, 2879
Dantron *see* 2878
Danylen [Temmler] *see* 3175
Daonil [Hoechst] *see* 4486
3,4-DAP *see* 3029
Daphnandrine, 2880
Daphnetin, 2881
Daphnetin 7-β-D-Glucoside *see* 2882
Daphnin, 2882
Daphnoline, 2883
Dapiprazole, 2884
Dapocel *see* 3241
Dapotum [Squibb-Heyden] *see* 4226
Dapotum D [Squibb-Heyden] *see* 4226
Dapropterin *see* 8515
Dapsone, 2885
DAPT *see* 505
Daptazile *see* 505
Daptazole [Nicholas] *see* 505
Daquin [Riker] *see* 2123
DAR *see* 3003
Daramin *see* 8463

Darammon *see* 537
Daranide [Merck & Co.] *see* 3127
Daraprim [Burroughs Wellcome] *see* 8169
Darbid [SK & F] *see* 5221
D'Arcet Metal-fusible *see* 1297
Darcil [Wyeth] *see* 7369
Dardum [Lisapharma] *see* 1981
Darenthin [Burroughs Wellcome] *see* 1395
Daricol *see* 7107
Daricon [Pfizer] *see* 7107
Darlan *see* 2886
Darmol [Iromedica] *see* 7392
Darob [Knoll] *see* 8876
Darotol *see* 2207
Dartal [Searle] *see* 9493
Dartalan [Searle] *see* 9493
Daruma [Am. Cyanamid] *see* 4932
Darvan [Celanese], 2886
Darvilen [Schering] *see* 1984
Darvisul [Burroughs Wellcome] *see* 3041
Darvon [Lilly] *see* 8024
Darvon-N [Lilly] *see* 8024
Darvyl *see* 5667
Dasanit [Bayer] *see* 4042
Daserol *see* 5895
Daskil [Wander] *see* 6612
Dasovas [Erba] *see* 2915
Dasten [Duncan] *see* 7058
D.A.T. [Wander] *see* 51
Datanil [Wander] *see* 9593
DATC *see* 3014 and 9593
Datiscetin, 2887
Datril [Bristol-Myers] *see* 45
Datura, 2888
Daturine *see* 4907
Daucol, 2889
Daunoblastina [Farmitalia] *see* 2890
Daunomycin *see* 2890
Daunomycinone *see* 2890
Daunorubicin, 2890
Daunosamine *see* 2890 and 3495
Dauricine, 2891
Davainex *see* 3089
Davitamon C *see* 867
Davosin [Parke, Davis] *see* 9087
Davoxin [Lakeside] *see* 3210
DAV Ritter [Ritter] *see* 2969
Daxauten [Kettelhack] *see* 7919
Dayfen [Riker] *see* 3390
Daypro [Searle] *see* 7057
Dazomet, 2892
Dazzle *see* 8773
2,4-DB, 2893
DB-32 *see* 9100
DB87 *see* 6202
DBC *see* 2513
DBcAMP *see* 1483
DBCM *see* 2186
DBCP *see* 3068
DBD *see* 6300
DBED *see* 1092
DBED Penicillin *see* 7221
DBHBT, 2894
DBI [Geigy] *see* 7376
DBI-TD [Geigy] *see* 7376
DBM *see* 6298
DBMC, 2895
DBP *see* 1622
DBPCl *see* 3059
DBS *see* 3080
DC-826 *see* 1669
DC-2797 *see* 1483
DCA *see* 2948 and 3100
DCB *see* 3109
DCC *see* 3146
DCCI *see* 3146
DCCK [Rentschler] *see* 3692

DCEE *see* 3116
DCET *see* 2064
DCF *see* 7277
2'-dCF *see* 7277
DCH-21 *see* 3959
DClMDP *see* 2432
DCMO *see* 1874
DCMX *see* 3126
DCNU *see* 2229
d(COMOT) *see* 1841
3,6-DCP *see* 2462
DCPA, 2896
DCPM *see* 1287
D-D [Shell] *see* 3125
DD-3480 *see* 9584
DDA *see* 3151
ddAdo *see* 3151
DDAVP [Ferring] *see* 2969
DDC *see* 3443
ddCyd *see* 10242
o,p'-DDD *see* 6302
DDD (Analytical), 2897
ddI *see* 3148
ddIno *see* 3148
DDP *see* 2378
cis-DDP *see* 2378
DDQ *see* 1512 and 3113
DDS *see* 2885
DDT, 2898
p,p'-DDT *see* 2898
DDTC *see* 3443
DDVP *see* 3129
DE-019 *see* 1481
DE-473 *see* 4720
DE-498 *see* 4177
Deacetoxycefotaxime *see* 1974
Deacetylandromedotoxin *see* 4566
Deacetylanhydroandromedotoxin *see* 4566
Deacetylanisomycin *see* 708
10-Deacetylbaccatin III *see* 3458
1-Deacetyldiaboline *see* 10181
Deacetyllanatoside *C* *see* 2967
N-Deacetyl-*N*-methylcolchicine *see* 2938
Deacylated Gellan Gum *see* 4389
Deacylated PS-60 *see* 4389
Dead-burned Gypsum *see* 1753
Deadly Nightshade *see* 1052
Deadopa [DeAngeli] *see* 5490
DEAE-cellulose *see* 2012
DEAE-dextran *see* 2980
DEAE-Sephadex [Pharmacia] *see* 7716
Deal Pine *see* 10180
Deamino-2-*O*-methyltyrosine-1-carba-oxytocin *see* 1841
Deaminooxytocin, 2899
Deanase [Consolidated Chem.] *see* 2953
Deandros [Farmochim. Ital.] *see* 7892
Deaner [Riker] *see* 2900
Deanol, 2900
Deapasil *see* 498
Deapril-ST [Mead Johnson] *see* 3692
Death's Herb *see* 1052
7-Deazaadenosine *see* 9936
Deba *see* 989
Debecacin *see* 3052
Débékacyl [Bellon] *see* 3052
N-Debenzoyl-*N*-(*tert*-butoxycarbonyl)-10-deacetyltaxol *see* 3458
Debeone-DT [Armstrong] *see* 7376
Debetrol *see* 9555
Debinyl [Bellon] *see* 7376
Deblaston [Madaus] *see* 7613
Debricin [J & J] *see* 4121
Debridat [Jouveinal] *see* 9829
Debrisan [Pharmacia] *see* 2992
Debrisoquin, 2901

Debrisorb [Deutsche Pharmacia] *see* 2992
Debrox [Marion] *see* 10007
Debroxide [Alcon] *see* 1149
Dec *see* 2903
Dec [Wendt] *see* 3165
Decabid [Lilly] *see* 4971
Decaborane(14), 2902
Decaboron Tetradecahydride *see* 2902
Decacarbonyldimanganese *see* 5767
Decachlor *see* 3154
1,1',2,2',3,3',4,4',5,5'-Decachlorobi-2,-4-cyclopentadien-1-yl *see* 3154
1,1a,3,3a,4,5,5,5a,5b,6-Decachloro-octahydro-1,3,4-metheno-2*H*-cyclo-buta[*cd*]pentalen-2-one *see* 2131
Decacort [F.C.F.] *see* 2986
Decaderm [Merck & Co.] *see* 2986
Decadron [Merck & Co.] *see* 2986
Decadronal [Merck & Co.] *see* 2986
Decadron-LA [Merck & Co.] *see* 2986
Decadron TBA [Merck & Co.] *see* 2986
Deca-Durabol *see* 6452
Deca-Durabolin [Organon] *see* 6452
Decafar [Lafare] *see* 9974
Deca-Hybolin [Hyrex] *see* 6452
Decahydro-3,5,1,7-[1,2,3,4]butane-tetraylnaphthalene *see* 2563
[3a*S*-(3aα,5aβ,6β,9aα,9bβ)]-Decahy-dro-6,9a-dihydroxy-5a-methyl-3,9-bis(methylene)naphtho[1,2-*b*]furan-2(3*H*)-one *see* 9218
1,6,7,8,9,11a,12,13,14,14a-Decahydro-1,13-dihydroxy-6-methyl-4*H*-cyclo-pent[*f*]oxacyclotridecin-4-one *see* 1393
[1*R*-(1α,4aβ,4bα,10aα)]-1,2,3,4,4a,-4b,5,6,10,10a-Decahydro-1,4a-dimethyl-7-(1-methylethyl)-1-phen-anthrenecarboxylic Acid *see* 3
[1*R*-(1α,4aβ,4bα,10aα)]-1,2,3,4,4a,-4b,5,9,10,10a-Decahydro-1,4a-dimethyl-7-(1-methylethyl)-1-phen-anthrenecarboxylic Acid *see* 5494
[1*R*-(1α,4aβ,10aα)]-1,2,3,4,4a,5,6,9,-10,10a-Decahydro-1,4a-dimethyl-7-(1-methylethyl)-1-phenanthrenecar-boxylic Acid *see* 7131
3a,4,5,5,6a,7,7a,8,11a,11b,11c-Decahy-dro-5-hydroxy-2,10-dimethoxy-3,8,-11a,11c-tetramethylphenanthro[10,-1-*bc*]pyran-1,11-dione *see* 6552
[1a*R*-(1aα,2β,3aβ,3bα,5α,6aα,7a*S**)]-Decahydro-2-hydroxy-3a,5-dimeth-yl-3-methylenecyclopenta[4,5]penta-leno[1,6a-*b*]oxirene-5-carboxylic Acid *see* 4753
1,2,3,4,4a,4b,7,9,10,10a-Decahydro-2-hydroxy-2,4b-dimethyl-7-oxo-1-phenanthrenepropionic Acid δ-Lac-tone *see* 9321
[1*R*-[1α[*E*(*S**)],4aβ,5α,6α,8aα]]-3-[2-[Decahydro-6-hydroxy-5-(hydr-oxymethyl)-5,8a-dimethyl-2-methyl-ene-1-naphthalenyl]ethylidene]dihy-dro-4-hydroxy-2(3*H*)-furanone 673
[3*R*-(3α,5aβ,6β,8aβ,9α,10α,12β,-12a*R**)]-Decahydro-10-methoxy-3,6,9-trimethyl-3,12-epoxy-12*H*-pyrano[4,3-*j*]-1,2-benzodioxepin *see* 854
(2aα,4aβ,5β,6β,8aα,12bβ,12cβ,-12dβ)-(−)-2a,3,4,4a,5,6,7,8a,12b,-12c-Decahydro-6-methyl-2*H*-5,12d-ethanofuro[4',3',2':4,10]anthra[9,1-*bc*]oxepin-2,9,12-trione *see* 7695
Decahydronaphthalene *see* 2903
Decahydro-2-oxo-1,3-bis(phenylmeth-

yl)thieno[1',2':1,2]thieno[3,4-*d*]imid-azol-5-ium Salt with (1*S*)-7,7-Di-methyl-2-oxobicyclo[2.2.1]heptane-1-methanesulfonic Acid (1:1) *see* 9837
[3*S*-(3α,4aα,5α,6α,11β,12α,12aβ)]-1,3,4,4a,5,6,7,11,12,12a-Decahydro-3,5,6,11,12-pentahydroxy-9,12a,13,-13-tetramethyl-4-methylene-6,10-methanobenzocyclodecen-8(2*H*)-one *see* 9245
[1a*R*-(1aα,1bβ,4aβ,7aα,7bα,8α,9β,-9aα)]-1,1a,1b,4,4a,7a,7b,8,9,9a-Decahydro-4a,7b,9,9a-tetrahydroxy-3-(hydroxymethyl)-1,1,6,8-tetra-methyl-5*H*-cyclopropa[3,4]benz[1,2-*e*]azulen-5-one *see* 7487
[4a*S*-(4aα,6aα,11bα,13a*R**,13bα)]-1,-2,3,4,4a,5,6,6a,11b,13b-Decahydro-4,4,6a,9-tetramethyl-13*H*-benzo[*a*]-furo[2,3,4-*mn*]xanthen-11-ol *see* 8630
[1a*R*-(1aα,4α,4aβ,7α,7aβ,7bα)]-Deca-hydro-1,1,4,7-tetramethyl-1*H*-cyclo-prop[*e*]azulen-4-ol *see* 5454
3,4,5,6,7,8,9,10,11,12-Decahydro-7,-14,16-trihydroxy-3-methyl-1*H*-2-benzoxacyclotetradecin-1-one *see* 10251
[2*R*-(2α,4aβ,5aβ,6β,7β,8β,9α,9aα,-10aβ)]-Decahydro-4a,7,9-trihydr-oxy-2-methyl-6,8-bis(methylamino)-4*H*-pyrano[2,3-*b*][1,4]benzodioxin-4-one *see* 8890
[1*S*-(1α,3aβ,4α,8aβ)]-Decahydro-4,8,8-trimethyl-9-methylene-1,4-methanoazulene *see* 5597
Decahydro-1,1,4a-trimethyl-2-naph-thalenol *see* 9625
Decalin [DuPont], 2903
Decalix [Pharmed] *see* 2986
Decamethonium Bromide, 2904
Decamethrin *see* 2934
Decamethylcyclopentasiloxane, 2905
1,1'-Decamethylenebis[4-amino-quinaldinium Chloride] *see* 2959
Decamethylene-α-ω-bis[1-(3',4'-di-methoxybenzyl)-1,2,3,4-tetrahydro-6,7-dimethoxy-2-methylisoquinolin-ium Methosulfate] *see* 5392
Decamethylenebis[*m*-dimethylamino-phenyl *N*-Methylcarbamate] Di-methobromide *see* 2936
Decamethylenebis[*N*-methylcarbamic Acid *m*-Dimethylaminophenyl Ester] Bromomethylate *see* 2936
Decamethylenebis[1,2,3,4-tetrahydro-6,7-dimethoxy-1-(3,4-dimethoxy-benzyl)-2-methylisoquinolinium Methyl Sulfate] *see* 5392
2,2'-Decamethylenebis[1,2,3,4-tetra-hydro-6,7-dimethoxy-2-methyl-1-veratrylisoquinolinium Methyl Sulfate] *see* 5392
Decamethylenebis[trimethylammo-nium Bromide] *see* 2904
N¹,N¹'-Decamethylene-*N⁴,N⁴*-deca-methylenebis[4-aminoquinaldinium Chloride] *see* 1288
1,1'-Decamethylene-4,4'-(1,10-deca-methylenediimino)bis[quinaldinium Chloride] *see* 1288
2,2'-(Decamethylenedithio)diethanol *see* 9556
Decamethylene Glycol, 2906
Decamethyltetrasiloxane, 2907
Decamine *see* 2959
Decanedioic Acid *see* 8558
Decanedioic Acid Bis(2-ethylhexyl) Ester *see* 1292

1,10-Decanediol see 2906
1,1'-(1,10-Decanediyl)bis-[4-amino-2-methylquinolinium Chloride] see 2959
2,2'-(1,10-Decanediyl)bis[1-[(3,4-dimethoxyphenyl)methyl]-1,2,3,4-tetrahydro-6,7-dimethoxy-2-methylisoquinolium] Bis(methyl Sulfate) see 5392
3,3'-[1,10-Decanediylbis[(methylimino)carbonyloxy]]bis[N,N,N-trimethylbenzenaminium]dibromide see 2936
2,2'-[1,10-Decanediylbis(thio)]bisethanol see 9556
N,N'-(1,10-Decanediyldi-1(4H)-pyridinyl-4-ylidene)bis[1-octanamine] see 6852
Decanoic Acid see 1802
Decanoic Acid Ethyl Ester see 3823
1-Decanol see 2911
Decapeptyl [Ipsen] see 9878
Decaprednil [Nycomed] see 7901
Decapryn Succinate [Merrell] see 3497
Decaps see 10156
(3-Decarbamoyloxy)-3-hydroxyventuricidin A see 10080
Decarboxycysteine see 2848
Decarboxycystine see 2846
Decardil [Trenker] see 3210
Decaris [Janssen] see 5486
Decaserpyl see 2964
Decasone [Frosst] see 2986
Decaspir [Pulitzer] see 4041
Decaspiride see 4041
2,4,6,8-Decatetraenedioic Acid Mono-[4-(1,2-epoxy-1,5-dimethyl-4-hexenyl)-5-methoxy-1-oxaspiro[2.5]oct-6-yl] Ester see 4308
[3R-[3α,4α(2R*,3R*),5β,6β(all-E)]]-2,4,6,8-Decatetraenedioic Acid, Mono[5-methoxy-4-[2-methyl-3-(3-methyl-2-butenyl)oxiranyl]-1-oxaspiro[2.5]oct-6-yl] Ester see 4308
Decatylen [Mepha] see 2959
Deccotane see 1579
Deccox [Merrell] see 2910
1-Decene-1,10-dicarboxylic Acid see 9710
Decentan [E. Merck] see 7323
[R-(Z)]-5-(1-Decenyl)dihydro-2(3H)-furanone see 5271
(R,Z)-5-(Dec-1-enyl)oxacyclopentan-2-one see 5271
Dechlorane [Occidental] see 6294
Dechlorane Plus [Occidental], 2908
Decholin [Miles] see 2922
Decicain [Winthrop] see 9330
Decimemide, 2909
Decis [Roussel-UCLAF] see 2934
Declid see 9983
Declinax [Roche] see 2901
Declomycin [Lederle] see 2937
Decme [Spitzner] see 3692
Decoderm Sine Gentamycin [E. Merck] see 4228
Decolorized Phenolphthalein see 7394
Decolorizing Carbon see 1856
Decomoton [Aesculaap] see 1841
Decontractyl see 5895
Decoquinate, 2910
Decorenone [Lifepharma] see 9974
Decorpa [Norgine] see 4602
Decortancyl [Roussel] see 7904
Decortilen [E. Merck] see 7906
Decortilen Soluble [E. Merck] see 7906
Decortin [E. Merck] see 7904
Decortin H [E. Merck] see 7901
Decosteron [Kwizda] see 2948

Decrelip [Ferrer] see 4394
Decreten [Dumex] see 7597
Decril [Mediolanum] see 3692
Dectancyl [Roussel] see 2986
Dectomax [Pfizer] see 3483
Decumbin see 1393
n-Decyl Alcohol, 2911
N-Decyl-N,N-dimethyl-1-decanaminium Chloride see 3149
Decyl(2-hydroxyethyl)dimethylammonium Bromide 1-Adamantanecarboxylate see 390
(2S-cis)-2-Decyl-3-(5-methylhexyl)-oxirane see 3425
4-(Decyloxy)-3,5-dimethoxybenzamide see 2909
6-Decyloxy-7-ethoxy-4-hydroxy-3-quinolinecarboxylic Acid Ethyl Ester see 2910
DEDC see 3443
Dedevap [Bayer] see 3129
4^A-O-De(2,6-dideoxy-3-C-methyl-α-L-ribo-hexopyranosyl)-20-deoxo-20-(3,5-dimethyl-1-piperidinyl)tylosin see 9580
Dedrogyl [Roussel-UCLAF] see 1677
DeDTC see 3443
Dedyl see 3241
DeeO see 4468
9,10-Deepithio-9,10-didehydroacanthifolicin see 6958
12,13-Deepoxy-12,13-didehydroverrucarin A see 10100
Dee-Ron see 10156
Deer's Tongue see 5501
Deet, 2912
16-Deethyl-3-O-demethyl-16-methyl-3-O-(1-oxopropyl)monensin see 5361
Defekton [Yoshitomi] see 1911
Defencin [Bristol] see 5259
Defenidol see 3369
Defensins, 2913
Deferoxamine, 2914
Defibrase [Pharma-Zentrale] see 1038
Defibrinotide see 2915
Defibrotide, 2915
Défiltran [Gallier] see 50
Deflamene [Farmitalia] see 4270
Deflamon [SPA] see 6242
Deflan [Guidotti] see 2916
Deflazacort, 2916
Deflexol see 10328
Deflorin [Glaxo] see 2025
Defluina [Simes] see 3692
Defol [Drexel] see 8741
Defosfamide, 2917
Deganol see 2937
Degener's Indicator see 7345
Deglycobleomycin see 1351
Degranol see 5790
Dégryp [Houdé] see 263
Deguelia Root see 2961
Deguelin, 2918
Dehacodin see 3214
Dehistin [EGYT] see 9868
Dehychol [Rexall] see 2922
Dehydol [Henkel] see 7737
Dehydrated Alcohol see 3806
Dehydrite see 5715
Dehydroacetic Acid, 2919
trans-Dehydroandrosterone see 7892
Dehydroascorbic Acid, 2920
Dehydrobenzperidol [J & J] see 3505
Dehydrobilirubin see 1264
11-Dehydro C see 289
(3β)-7-Dehydrocholesterol, 2921
24-Dehydrocholesterol see 2970
Dehydrocholic Acid, 2922
Dehydrocholic Acid Salt of Choline see 2262

Dehydrocollinusin see 5286
11-Dehydrocorticosterone, 2923
Δ1-Dehydrocortisol see 7901
Δ1-Dehydrocortisone see 7904
Dehydrodichavicol see 5738
Dehydroemetine, 2924
2-Dehydroemetine see 2924
2,3-Dehydroemetine see 2924
Dehydroepiandrosterone see 7892
Dehydroergosterol, 2925
1-Dehydro-9α-fluorohydrocortisone see 5190
Δ1-Dehydrohydrocortisone see 7901
11-Dehydro-17-hydroxycorticosterone see 2602
6-Dehydro-17α-hydroxy-16-methyleneprogesterone see 4878
Dehydroisoandrosterone see 7892
2,3-Dehydroisoemetine see 2924
6-Dehydro-6-methyl-17α-acetoxyprogesterone see 5849
1-Dehydro-16α-methyl-9α-fluorohydrocortisone see 2986
1-Dehydro-6α-methylhydrocortisone see 6189
1-Dehydro-17α-methyltestosterone see 6018
Dehydronivalenol see 10171
Dehydroperillic Acid see 9532
6-Dehydro-retro-progesterone see 3524
3-Dehydroretinal, 2926
Dehydroretinol see 10151
7-Dehydrositosterol, 2927
Dehydro-6-sulfoabietic Acid see 3538
1-Dehydrotestololactone see 9321
Dehydrotestosterone see 1354
1-Dehydrotestosterone 17-Cyclopent-1'-enyl Ether see 8236
6-Dehydrotestosterone-17α-propionic Acid γ-Lactone see 1795
Deidrocolico Vita see 2922
Deiten [ABC] see 6669
Dejo see 3434
Dekacort [Farber-Ref] see 2986
Dekadin see 2959
DeKalin see 2903
Dekamin see 2959
Dekamin [Monico] see 8071
Dekelmin see 6247
Dekrysil see 3331
DEL-1267 see 6226
Delacillin [Sankyo] see 617
Delacurarine [Squibb] see 9939
Deladroxone (obsolete) [Squibb] see 239
Delalutin [Squibb] see 4886
Delan [Am. Cyanamid] see 3433
Delaprem [Savage] see 4745
Delapril, 2928
Delatestryl [Bristol-Myers Squibb] see 9322
Delavirdine, 2929
Delavirdine Mesylate see 2929
Delaxin [Ferndale] see 6060
Delcaine see 8097
Delcortin [Leo Pharm.] see 7904
Delcosine see 2933
Delecit [LPB] see 2260
Delestrec [Bristol-Myers Squibb] see 3746
Delestrogen [Bristol-Myers Squibb] see 3746
Delfen [Ortho] see 6772
Delgesic [Karlspharma] see 5668
Delimmun [Newport] see 5006
Delinal [Merrell] see 7998
Delipid [Coop. Farm.] see 9556
Delix [Hoechst; Casella-Riedel] see 8283
Delmadinone Acetate, 2930

1-(2-Deoxy-β-D-ribofuranosyl)-5-methyluracil *see* 9538
1-(2-Deoxy-α-D-ribofuranosyl)uracil *see* 2957
1-(2-Deoxy-β-D-ribofuranosyl)uracil *see* 2957
Deoxyribonuclease (Human Clone 18-1 Protein Moiety) *see* 2953
Deoxyribonuclease I, 2953
Deoxyribonuclease (Pancreatic) *see* 2953
Deoxyribonucleic Acid, 2954
D-2-Deoxyribose, 2955
1-[4-Deoxy-4-(sarcosyl-D-seryl)amino-β-D-glucopyranuronamide]cytosine *see* 4550
Deoxyspergualin *see* 4610
(±)-15-Deoxyspergualin *see* 4610
2-Deoxystreptamine, 2956
(—)-2'-Deoxy-3'-thiacytidine *see* 5365
3'-Deoxy-2'-thymidinene *see* 8958
2'-Deoxy-5-(trifluoromethyl)uridine *see* 9816
11-Deoxy-11α,16,16-trimethyl-PGE₂ *see* 9853
Deoxyuridine, 2957
2'-Deoxyuridine *see* 2957
Depakene [Abbott] *see* 10049
Depakin [Sigma-Tau] *see* 10049
Dépakine [Labaz] *see* 10049
Depakote [Abbott] *see* 10049
Dépamide [Labaz; Sigma-Tau] *see* 10050
Depamine [Berk] *see* 7214
DepAndro [Forest] *see* 9322
Deparal *see* 10157
Deparkin *see* 3157
Deparon [Aron] *see* 2941
Depas [Yoshitomi] *see* 3919
Depasan [Giulini] *see* 8887
DEPC *see* 8182
Depen [Wallace Labs.] *see* 7214
Depepsen [Searle] *see* 8718
Depersolone *see* 5802
Depgynogen [Forest] *see* 3746
d Epifrin [Allergan] *see* 3400
Depigman [Hermal] *see* 6331
Depil [Calmic] *see* 1758
Depinar [Armour Pharm.] *see* 10152
Depixol [Lundbeck] *see* 4224
Depleil [Promedica] *see* 9261
Depo-Cer-O-Cillin Chloroprocaine [Upjohn] *see* 7228
Depocillin [Brocades] *see* 7226
Depo-Clinovir [Upjohn] *see* 5838
Depocolin-S [Ohta] *see* 8084
Depocural *see* 2406
Depogamma [Cooper] *see* 4854
Depogen [Hyrex] *see* 3746
Depo-Medrate [Upjohn] *see* 6189
Depo-Medrol [Upjohn] *see* 6189
Depo-Medrone [Upjohn] *see* 6189
Deponit [Schwarz] *see* 6704
Depo-Provera [Upjohn] *see* 5838
Deposal [Benvegna] *see* 6362
Depostat [Schering AG] *see* 4422
Depotest [Hyrex] *see* 9322
Depo-Testosterone [Upjohn] *see* 9322
Depot-Glumorin *see* 5291
Depot-Novadral [Diwag] *see* 6792
Depotocin [Spofa] *see* 1841
Depot-Oestromenine [E. Merck] *see* 3257
Depot-Oestromon *see* 3257
Depovirin [Hoechst] *see* 9322
Depracer [Lepori] *see* 3930
Deprancol [Parke-Davis] *see* 8024
(—)-Deprenil *see* 8569
Deprenyl *see* 8569
L-Deprenyl *see* 8569

Déprényl [Schering] *see* 8569
Depressan [OPW] *see* 3212
Depresym *see* 3485
Depreton *see* 6196
Deprex [Mowatt & Moore] *see* 511
Depridol *see* 6008
Deprilept [Promonta] *see* 5792
Deprinol [Dumex] *see* 4955
Depsococaine *see* 8097
Deptropine, 2958
Dequadin Chloride [Allen & Hanburys] *see* 2959
Dequafungan [Kreussler] *see* 2959
Dequalinium Chloride, 2959
Dequavagyn [Kreussler] *see* 2959
Dequavet *see* 2959
Deracil *see* 9504
Deracyn [Upjohn] *see* 159
Deralbine [Andromaco] *see* 6266
Deralin [Alphapharm] *see* 8025
Derantel [Nippon Chemiphar] *see* 2021
De-Rat Concentrate [Rentokil] *see* 10156
Deratol [Brewer] *see* 10156
Derbac [International] *see* 1831
Derbac-M [International] *see* 5740
Dereuma [ICI] *see* 455
Derfon *see* 3175
Dergotamine [Abbott] *see* 3217
Dermabet [Taro] *see* 1226
Dermacid *see* 5916
Dermacort [Solvay] *see* 4828
Dermadex [Alconox] *see* 4716
Dermaflor [Brocchieri] *see* 3186
Dermafur [Eaton] *see* 4321
Dermagan *see* 3004
Dermairol [Roche] *see* 8333
Dermalar *see* 4185
Dermaplus [Ripari-Gero] *see* 4186
Dermastatin *see* 2960
Dermatan Sulfate *see* 2270
Dermathycin [Coopers] *see* 9931
Dermatol [Hoechst] *see* 1325
Dermaton [Burroughs Wellcome] *see* 2137
Dermatop [Cassella] *see* 7899
Dermevan *see* 5058
Dermocortal [Puropharma] *see* 4828
Dermofix [Ferrer] *see* 8610
Dermolate [Schering] *see* 4828
Dermonistat [Ortho] *see* 6266
Dermosol [Iwaki] *see* 1226
Dermostatin, 2960
Dermostatin A *see* 2960
Dermostatin B *see* 2960
Dermotricine [Bellon] *see* 9972
Dermoval *see* 2423
Dermovaleas [Valeas] *see* 1226
Dermovate [Glaxo] *see* 2423
Dermoxin [Glaxo] *see* 2423
Dermoxinale [Glaxo] *see* 2423
Deronil [Schering] *see* 2986
Derosal [Hoechst] *see* 1836
Derride *see* 3591
Derringer [Roussel-UCLAF] *see* 7629
Derris Root, 2961
Dervan [Caffaro] *see* 8741
DES *see* 3177
Desacchromin Dispersion *see* 7659
Desace [Dexo] *see* 2967
Desacetylbufotalin *see* 1499
Desacetyldigilanide C *see* 2967
Desacetylvinblastine Amide *see* 10125
Desaci [Simes] *see* 2967
Desaglybuzole *see* 4488
1-Desamino-8-D-arginine Vasopressin *see* 2969
1-Desamino-1-monocarba-[2-tyr-(OMe)]-OT *see* 1841

Desaminooxytocin *see* 2899
Desanden [Max Ritter] *see* 1149
Desaspidin *see* 2962
Desaspidin BB, 2962
Desatrine, 2963
Desatrine-3-(N,N-diethylaminoacetate) *see* 2963
Deschlorobiomycin *see* 9337
Desclidium [Rhone-Poulenc] *see* 10141
Desdemin [Vitacain] *see* 4331
Déselmine *see* 3434
Deseptyl *see* 9094
Deseril [Sandoz] *see* 6217
Désernil [Sandoz] *see* 6217
Deserol *see* 1439
Deseronil [SCA] *see* 2986
Deserpidine, 2964
Desfedrin *see* 6015
Desferal [Ciba] *see* 2914
Desferricoprogen *see* 2586
Desferrioxamine B *see* 2914
Desferrioxamine Mesylate *see* 2914
Desflurane, 2965
Desglucotransvaaline *see* 8060
Des-Gly¹⁰-NH₂-LH-RH-ethylamide *see* 4109
Desicol [Parke, Davis] *see* 7062
Desipramine, 2966
Desirudin *see* 4754
Desitriptilina *see* 6812
Desiver [USV] *see* 5575
Deslanoside, 2967
Deslorelin, 2968
Desmel [Ciba-Geigy] *see* 8003
Desmesterol *see* 2970
11-Desmethoxyreserpine *see* 2964
Desmethylamitriptyline *see* 6812
N-Desmethylcodeine *see* 6783
Desmethyldiazepam *see* 6784
α-Desmethyl DOB *see* 1958
Desmethyldopan *see* 9986
Desmethylemetine *see* 2020
Desmethylimipramine *see* 2966
Desmethylmethadone *see* 6804
Desmethylmorphine *see* 6806
Desmethylnarcotine *see* 6510
5-O-Desmethyltangeretin *see* 4377
Desmocollins *see* 1647
Desmofosfamide *see* 2917
Desmogleins *see* 1647
Desmopressin, 2969
Desmospray [Ferring] *see* 2969
Desmosterol, 2970
Desmycosin *see* 9963
2,4-DES-Na *see* 3430
Desogen [Organon] *see* 2971
Desogestrel, 2971
Desol [CT] *see* 10026
Desomedine [Chauvin] *see* 4728
Desomorphine, 2972
Desonide, 2973
Desopan [Mochida] *see* 9827
Desosamine, 2974
Desoximetasone, 2975
Desoxiribon *see* 2954
Desoxostrophanthidin *see* 7317
S-(5'-Desoxyadenosin-5'-yl)-L-methionine *see* 155
Desoxyalizarin *see* 727
2-Desoxy-D-altromethylose *see* 3207
Desoxycholic Acid *see* 2946
Desoxycorticosterone *see* 2947
Desoxycorticosterone Acetate *see* 2948
11-Desoxycortisone *see* 2977
Desoxycortone *see* 2947
Desoxycortone Acetate *see* 2948
14-Desoxy-14-[(2-diethylaminoethyl)-mercaptoacetoxy]mutilin *see* 9559

d-Desoxyephedrine see 6015
Desoxyepinephrine see 2950
Desoxyfed see 6015
21-Desoxy-9α-fluoro-6α-methylprednisolone see 4213
6-Desoxy-D-glucosamine, 2976
D-arabino-2-Desoxyhexose see 2951
11-Desoxy-17-hydroxycorticosterone, 2977
Desoxymethasone see 2975
6-Desoxy-6-methylenenaltrexone see 6447
21-Desoxy-6α-methyl-9α-fluoroprednisolone see 4213
Desoxymycin [Kaken] see 2949
Desoxyn [Abbott] see 6015
3-Desoxynorlutin see 5659
2-Desoxyphenobarbital see 7927
5-Desoxyquercetin see 4129
Desoxyribonucleic Acid see 2954
Desoxyribose see 2955
DESP see 4041
Despacilina [Squibb] see 7226
Desquaman [Hermal] see 8178
Dessin [Union Carbide] see 3339
Desthiobiotin, 2978
Destim [Central Pharm.] see 6015
Destolit [Sandoz S.A.R.L.] see 10026
Destomycin A, 2979
Destonate 20 [Meiji] see 2979
Destriol see 3750
Destrone see 3751
Destun [3M] see 7300
N-Desulfoheparin see 4684
63-Desulfohirudin (Hirudo medicinalis Isoform HV1) see 4754
Desuric [Sanofi] see 1093
Desyrel [Apothecon] see 9712
m-DETA see 2912
Detal see 3331
Detalup [Bayer] see 10156
Detamide see 2912
Detantol [Eisai] see 1512
Detaril [Isom] see 7638
Detaxtran, 2980
Detensiel [Merck-Clévenot] see 1336
Detergent Alkylate #5 see 9792
Dethylandiamine see 9416
Dethyrona [Boots] see 9555
Detia [Freyberg] see 372
Deticene [Bellon] see 2866
DET MS [Rentschler] see 3217
Detomidine, 2981
Detoxepa [Wyeth-Ayerst] see 9586
Detoxin C₁ see 2982
Detoxin C_{α1} see 2982
Detoxin Complex, 2982
Detoxin D₁ see 2982
Detoxinine see 2982
Detravis [Vis] see 2937
Detrovel [Geigy] see 44
Détryptoréline see 9878
Dettol [Reckitt & Colman] see 2228
Detulin [Woelm] see 10160
Detyroxin see 9555
Deumacard [Jenapharm] see 7283
Deursil [Giuliani] see 10026
Deuterium, 2983
Deuterium Oxide, 2984
Devarda's Alloy see 2985
Devarda's Metal, 2985
Devaricin see 4747
Devegan [Hoechst] see 49
Develin [Gödecke] see 8024
Developer Z see 6809
Deverol [Waldheim] see 8917
Devil's Apple see 8975
Devil's Dung see 858
Devincan [Gedeon Richter] see 10120
Devrinol [Stauffer] see 6503

Devryl [SK & F] see 2441
Dexabene [Merckle] see 2986
Dexacilina [Squibb] see 3651
Dexacillin [Squibb] see 3651
Dexacortal [Organon] see 2986
Dexacortin [Streuli] see 2986
Dexacortisyl [Roussel] see 2986
Dexafarma [Llano] see 2986
Dexal [Pulitzer] see 5316
Dexalone see 2996
Dexa-Mamallet [Showa] see 2986
Dexambutol [L'Arguenon] see 3764
Dexameth [USV] see 2986
Dexamethasone, 2986
Dexamethasone tert-Butylacetate see 2986
Dexamethasone 21-(Dihydrogen Phosphate) Disodium Salt see 2986
Dexamethasone 21-(4-Pyridinecarboxylate) see 2986
Dexamethasone Sodium Phosphate see 2986
Dexamisole see 5486
Dexamonozon [Medice] see 2986
Dexampex [Lemmon] see 2996
Dexamphetamine see 2996
Dexapos [Ursapharm] see 2986
Dexa-sine [Alcon-Thilo] see 2986
Dexasone [ICN] see 2986
Dexawin [Winthrop] see 9436
Dexbenzetimide see 2987
Dexbrompheniramine see 1467
Dexchlorpheniramine see 2232
Dexcyanidanol see 1950
Dexedrine Sulfate [SK & F] see 2996
Dexetimide, 2987
Dexfenfluramine see 4015
Dexibuprofen Lysine see 4925
Dexide [Delalande Isnardi] see 2980
Dexindoprofen see 4999
Dexinoral [Desitin] see 2986
Dexium [Delalande] see 3455
Dexivacaine see 5905
Dexmedetomidine see 5830
Dexnorgestrel Acetime see 6796
Dexnorgestrel (obsolete) see 6797
Dexol see 8797
Dexoxadrol Hydrochloride see 3352
Dexpanthenol, 2988
Dexrazoxane see 8295
Dexten see 2996
Dextran, 2989
Dextran 40 see 2989
Dextran 70 see 2989
Dextran 75 see 2989
Dextranase, 2990
Dextran 2-(Diethylamino)ethyl 2-[[2-(Diethylamino)ethyl]diethylammonio]ethyl Ether Chloride Hydrochloride Epichlorohydrin Crosslinked see 7716
Dextran 2-(Diethylamino)ethyl Ether see 2980
Dextran 2,3-Dihydroxypropyl 2-Hydroxy-1,3-propanediyl Ethers see 2992
Dextran Iron Complex, 2991
Dextranomer, 2992
Dextran Sulfate Sodium, 2993
Dextran Sulfuric Acid Ester Sodium Salt see 2993
Dextrarine see 2993
Dextraven [Fisons] see 2989
Dextri-Maltose [Mead Johnson], 2994
Dextrin, 2995
Dextroamphetamine Sulfate, 2996
Dextrobenzetimide see 2987
Dextrocaine see 8097
Dextrocamphoric Acid see 1780
Dextrofenfluramine see 4015

Dextroid see 9555
Dextromethorphan Hydrobromide see 8274
Dextromoramide, 2997
Dextronic Acid see 4464
Dextronorgestrel see 6797
Dextropimaric Acid see 7582
Dextropropoxyphene see 8024
Dextropur see 4467
Dextrorotatory Lactic Acid see 5350
Dextrorphan see 5496
Dextrose see 4467
Dextrosol see 4467
Dextrosulphenidol see 9436
Dextrotartaric Acid see 9237
Dextrothyroxine Sodium see 9555
Dexulate see 2993
Dezacor [Marion Merrell Dow] see 2916
Dezocine, 2998
Dezone [Reid-Powell] see 2986
DF 118 see 3214
dFdC see 4392
dFdCyd see 4392
DFDD, 2999
DFDT, 3000
DFMO see 3564
DFOM see 2914
DFP see 5192
5'-DFUR see 3493
5'-dFUrd see 3493
2-DG see 2951
DH-245 see 4319
DH-581 see 7935
DHA see 2919
DHAD see 6303
DHAQ see 6303
DHA-S see 7892
DHA-S [Dow] see 2919
DHBE see 7094
DHC [Napp] see 3214
1,25-DHCC see 1681
DH-codeine see 3214
D.H.E. 45 [Sandoz] see 3217
DHIC see 3219
DHPG see 4374
DHSM see 3222
Dhurrin, 3001
Dia-basan [Sagitta] see 4486
Diabechlor see 2239
Diaben [Chugai] see 9646
Diabenal see 2239
Diabeta [Hoechst] see 4486
Diabetoral [Boehringer, Mann.] see 2239
Diabetosan [Brocchieri] see 6001
Diabewas [Wassermann] see 9644
Diabex [Fisons] see 6001
Diabinese [Pfizer] see 2239
Diaboline, 3002
Diaboral [Farmitalia] see 9648
Diabrin [USV] see 1500
Diacepin see 3042
Diacerein, 3003
Diacerhein see 3003
3,5-Diacetamido-2,4,6-triiodobenzoic Acid Methylglucamine Salt see 5851
3,5-Diacetamido-2,4,6-triiodobenzoic Acid Sodium Salt see 3040
α,5-Diacetamido-2,4,6-triiodo-m-toluic Acid see 5031
Diacetatopalladium(II) see 7123
Diacetatozirconic Acid see 10316
Diacetazotol, 3004
Diacethiamine see 51
Diacetic Acid see 56
Diacetin, 3005
1,2-Diacetin see 3005
1,3-Diacetin see 3005
Diacetonamine, 3006

Diacetone Acrylamide, 3007
Diacetone Alcohol, 3008
Diacetoneglucose, 3009
Diacetone-2-ketogulonic Acid *see* 3243
Diacetone-2-oxo-L-gulonic Acid *see* 3243
Diacetotoluide *see* 3004
4,4'-Diacetoxybenzhydrylidenecyclohexane *see* 2789
2,3-Diacetoxybenzoic Acid *see* 3413
1,8-Diacetoxy-3-carboxyanthraquinone *see* 3003
Diacetoxydiphenylisatin *see* 7108
2-(4,4'-Diacetoxydiphenylmethyl)-pyridine *see* 1282
(4,4'-Diacetoxydiphenyl)(2-pyridyl)-methane *see* 1282
3α,17β-Diacetoxy-2β,16β-dipiperidino-5α-androstane Dimethobromide *see* 7139
3,16α-Diacetoxy-Δ^{1,3,5}-estratrien-17-one *see* 2554
1,1-Diacetoxyethane *see* 3857
3β,17β-Diacetoxy-17α-ethynyl-4-estrene *see* 3905
16α,21-Diacetoxy-9α-fluoro-11β,17α-dihydroxy-1,4-pregnadiene-3,20-dione *see* 9727
1,4-Diacetoxy-2-methylnaphthalene *see* 5873
Di(acetoxyphenyl)oxindole *see* 7108
3α,20α-Diacetoxypregnane *see* 7908
Diacetyl, 3010
Diacetylaminoazotoluene *see* 3004
4,4'-Diacetylaminodiphenyl Sulfone *see* 20
Diacetylcholine Dichloride *see* 9044
Diacetylcholine Diiodide *see* 9045
Diacetylcholine Iodide *see* 9045
Diacetyldapsone *see* 20
4,4'-Diacetyldiaminodiphenyl Sulfone *see* 20
N,N'-Diacetyl-4,4'-diaminodiphenyl Sulfone *see* 20
Diacetyldihydromorphine, 3011
2,6-Diacetyl-7,9-dihydroxy-8,9b-dimethyl-1,3(2H,9bH)-dibenzofurandione *see* 10031
Diacetyldihydroxydiphenylisatin *see* 7108
Diacetyldioxime *see* 3295
Diacetyldioxyphenylisatin *see* 7108
Diacetyldiphenolisatin *see* 7108
α,β-Diacetylethane *see* 69
2,5-Di-O-acetyl-D-glucaro-1,4:6,3-dilactone *see* 24
2,5-Di-O-acetyl-D-glucosaccharo-1,4:6,3-dilactone *see* 24
Diacetylhydroxyphenylisatin *see* 7108
Diacetylmethane *see* 82
9,3''-Diacetylmidecamycin *see* 6291
Diacetylmorphine, 3012
Diacetylpyrocatechol-3-carboxylic Acid *see* 3413
Diacetylrhein *see* 3003
Diacetyltannic Acid *see* 107
O,S-Diacetylthiamine *see* 51
N,N-Diacetyl-o-tolyazo-o-toluidine *see* 3004
Diacid [Daro] *see* 1879
α,γ-Diacipiperazine *see* 7619
Diacort [Sumitomo; Upjohn] *see* 3186
Diacta [Nicholas] *see* 9076
Di-Actane [Menarini] *see* 6440
Diactol *see* 10156
Diacylglycerol Lipase *see* 5537
Di-Ademil [Squibb] *see* 4830
Diadol [Tilden-Yates] *see* 261
Di-Adreson [Organon] *see* 7904
Di-Adreson-F [Organon] *see* 7901

Diadril [Pliva] *see* 5817
Diafen [Riker] *see* 3390
Diafusor [Fabre] *see* 6704
Diaginol [M & B] *see* 79
Diagnorenol *see* 6051
Dial [Ciba] *see* 261
Dialar [Lagap] *see* 3042
Dialicor [Kissei] *see* 3753
Dialifor, 3013
Dialifos *see* 3013
Diallate, 3014
Diallylamine, 3015
5,5-Diallylbarbituric Acid *see* 261
3',5-Diallyl-2,4'-biphenyldiol *see* 4780
5,5'-Diallyl-2,2'-biphenyldiol *see* 5738
6,6'-Diallyl-α,α'-bis(diethylamino)-4,4'-bi-o-cresol *see* 1242
5,5'-Diallyl-α,α'-bis(diethylamino)-m,m'-bitolyl-4,4'-diol *see* 1242
Diallylbis(nortoxiferine) *see* 222
N,N-Diallyl-2-chloroacetamide *see* 259
Diallylcyanamide, 3016
5,3'-Diallyl-2,4'-dihydroxydiphenyl *see* 4780
Diallyl Disulfide-oxide *see* 258
Diallyl Ether *see* 300
Diallylnortoxiferine *see* 222
N,N-Diallylnortoxiferinium *see* 222
N,N'-Diallylnortoxiferinium Dichloride *see* 222
Diallyl Sulfide *see* 305
Diallyltoxiferine *see* 222
Dialuramide *see* 9987
Diamantane *see* 2563
Diamantane (obsolete) *see* 149
Diamarin *see* 3252
Diamfenetide, 3017
Diamicron [Servier] *see* 4448
3,3'-Diamidinocarbanilide *see* 414
4,4'-Diamidinodiazoaminobenzene Diaceturate *see* 3316
4,4'-Diamidino-α,ω-diphenoxyhexane *see* 4728
4,4'-Diamidino-α,ω-diphenoxypentane *see* 7254
4,4'-Diamidino-α,ω-diphenoxypropane *see* 7981
4,4'-Diamidinodiphenyl Ether *see* 7352
N,N'-Di(m-amidinophenyl)urea *see* 414
4,4'-Diamidi nostilbene *see* 8970
N,N'-Diamidinostreptamine *see* 8977
Diamine Blue *see* 9923
2,6-Diamine-5-hydroxy-4(3H)-pyrimidinone *see* 3448
Diamine Oxidase *see* 4755
Diamine Penicillin *see* 7221
2,8-Diaminoacridine *see* 7953
3,6-Diaminoacridine *see* 7953
3,5-Diamino-N-(aminoiminomethyl)-6-chloropyrazinecarboxamide *see* 426
6,9-Diamino-1-(6-amino-9H-purin-9-yl)-1,5,6,7,8,9-hexadeoxy-D-glycero-α-L-talo-decofuranuronic Acid *see* 8690
3,5-Diamino-2-[[4-(aminosulfonyl)-phenyl]azo]benzoic Acid *see* 9069
2,4-Diaminoanisole, 3018
4,4'-Diamino-1,1'-anthrimide *see* 1283
4,4'-Diaminoazobenzene *see* 3019
p-Diaminoazobenzene, 3019
2,4-Diaminoazobenzene Hydrochloride *see* 2317
2,4-Diaminoazobenzene Hydrochloride Citrate *see* 2317
2,4-Diaminoazobenzene-4'-sulfonamide *see* 9092
m-Diaminobenzene *see* 7437
o-Diaminobenzene *see* 7438
p-Diaminobenzene *see* 7439
3,5-Diaminobenzoic Acid, 3020

2,4-Diamino-5-(4-bromo-3,5-dimethoxybenzyl)pyrimidine *see* 1401
2,6-Diamino-2'-butoxy-3,5'-azopyridine *see* 3021
2,6-Diamino-6'-butoxy-3,3'-azopyridine *see* 3021
2',6'-Diamino-2-butoxy-5,5'-azopyridine *see* 3021
2,6-Diamino-3-(6-butoxy-3-pyridylazo)pyridine *see* 3021
2,6-Diamino-2'-butyloxy-3,5'-azopyridine, 3021
2',6'-Diamino-2-butyloxy-5,5'-azopyridine *see* 3021
α,ε-Diaminocaproic Acid *see* 5667
2,4-Diamino-1-p-chlorophenyl-1,6-dihydro-6,6-dimethyl-1,3,5-triazine *see* 2790
4,6-Diamino-1-(p-chlorophenyl)-1,2-dihydro-2,2-dimethyl-s-triazine *see* 2790
2,4-Diamino-5-(p-chlorophenyl)-6-ethylpyrimidine *see* 8169
Di-(4-amino-3-chlorophenyl)methane *see* 6136
1-(3,5-Diamino-6-chloropyrazinecarboxyl)guanidine *see* 426
1-(3,5-Diamino-6-chloropyrazinoyl)-guanidine *see* 426
4,4'-Diamino-1,1'-dianthraquinonylamine *see* 1283
1,8-Diamino-3,6-diazaoctane *see* 9796
β,β'-Diamino-β,β'-dicarboxydiethyl Disulfide *see* 2851
4,4'-Diamino-3,3'-dichlorodiphenylmethane *see* 6136
2,4-Diamino-6-(2,5-dichlorophenyl)-s-triazine *see* 5113
3,5-Diamino-6-(2,3-dichlorophenyl)-1,2,4-triazine *see* 5367
O-2,6-Diamino-2,6-dideoxy-α-D-glucopyranosyl-(1→4)-O-[β-D-ribofuranosyl-(1→5)]-2-deoxy-D-streptamine *see* 8373
(S)-O-2,6-Diamino-2,6-dideoxy-α-D-glucopyranosyl-(1→4)-O-[β-D-xylofuranosyl-(1→5)]-N¹-(4-amino-2-hydroxy-1-oxobutyl)-2-deoxystreptamine *see* 1559
O-2,6-Diamino-2,6-dideoxy-β-L-idopyranosyl-(1→3)-O-β-D-ribofuranosyl-(1→5)-O-[2-amino-2-deoxy-α-D-glucopyranosyl-(1→4)]-2-deoxystreptamine *see* 7173
β,β'-Diaminodiethyl Disulfide *see* 2846
4,6-Diamino-1,2-dihydro-2,2-dimethyl-1-[p-(methylthio)phenyl]-s-triazine *see* 6056
2,6-Diamino-1,6-dihydro-4,5-pyrimidinedione *see* 3448
3,3'-Diamino-4,4'-dihydroxyarsenobenzene Dihydrochloride *see* 851
3,3'-Diamino-4,4'-dihydroxyarsenobenzenemethylenesulfoxylate Sodium *see* 6534
2,4-Diamino-5,6-dihydroxypyrimidine *see* 3448
2,4-Diamino-6-(2,5-dimethoxybenzyl)-5-methylpyrido[2,3-d]pyrimidine *see* 7654
2,4-Diamino-5-(3',4'-dimethoxybenzyl)pyrimidine *see* 3041
2,4-Diamino-5-[3,5-dimethoxy-4-(2-methoxyethoxy)benzyl]pyrimidine *see* 9386
4,4'-Diamino-3,3'-dimethylbiphenyl *see* 9651
1,6-Diamino-5-5-dimethyl-3,4-dithia-

Diastal [Bayropharm] see 1495
Diastase Vera see 7137
Diastatin see 6834
Diaster [Diamant] see 2603
Diasulfa [Crosara] see 9073
Diasulfon [Streuli] see 9646
Diatensec [Searle] see 8917
Diathymosulfone, 3037
Diatomaceous Earth see 5003
Diatox see 3037
Diatrast see 5061
Diatretyne I, 3038
Diatretyne II, 3039
Diatretyne Amide see 3038
Diatretyne Nitrile see 3039
Diatrin see 6026
Diatrin Base [Warner-Chilcott] see 6026
Diatrizoate Meglumine see 5851
Diatrizoate Methylglucamine see 5851
Diatrizoate Sodium, 3040
Diatropic Acid see 5117
Diaveridine, 3041
1,4-Diazabicyclo[2.2.2]octane see 9801
20,25-Diazacholesterol see 924
1,3-Diaza-2,4-cyclopentadiene see 4948
8,10-Diaza-6-hydroxy-5-methylene-1-(2-methyl-1,2,3-trihydroxypropyl)-2-oxabicyclo[4.2.2]decan-7,9-dione see 1249
Diazald [Aldrich] see 9681
Diazamine Golden Yellow T see 9445
1,2-Diazanaphthalene see 2366
3,6-Diazaoctane-1,8-diamine see 9796
Diazasterol see 924
(RR,SS)-4,8-Diaza-3,6,6,9-tetramethylundecane-2,10-dione Bisoxime see 3957
Diazemuls [Farmitalia] see 3042
Diazenedicarboxamide see 950
Diazepam, 3042
Diazil [Bracco] see 9083
1,2-Diazine see 8152
1,3-Diazine see 8170
1,4-Diazine see 8141
Diazinon, 3043
Diaziquone, 3044
2,5-Diaziridinyl-3,6-bis(ethoxycarbonylamino)-1,4-benzoquinone see 3044
Diazoacetic Ester, 3045
O-Diazoacetyl-L-serine see 932
Diazoaminobenzene, 3046
4,4'-(Diazoamino)dibenzamidine Diacetuate see 3316
2,2'-[(Diazoamino)di-p-phenylene]bis-[6-methyl-7-benzothiazolesulfonic Acid] Disodium Salt see 9445
p-Diazobenzenesulfonic Acid, 3047
5-Diazo-2,4-dioxopyrimidine see 3050
Diazol [Makhteshim-Agan] see 3043
1,2-Diazole see 8144
1,3-Diazole see 4948
Diazoline see 5809
Diazomethane, 3048
Diazon see 9141
6-Diazo-5-oxo-L-norleucine, 3049
5-Diazo-2,4(1H,3H)-pyrimidinedione see 3050
"Diazoresorcinol" see 8309
5-Diazouracil, 3050
Diazoxide, 3051
Diazyl [Abbott] see 9071
Dibactil [Roemmers] see 9093
Dibasic Sodium Phosphate see 8805
Dibasol see 1062
Dibasole see 1062
Dibein [Pharmacia] see 7376
Dibekacin, 3052

Dibenamine Hydrochloride [SK & F] see 2190
Dibencil see 7221
Dibencillin see 7221
Dibencozide see 2513
Dibenon [Dainippon] see 1199
Dibenyline [SK&F] see 7409
Dibenzalacetone, 3053
1,2:5,6-Dibenzanthracene, 3054
Dibenz[a,h]anthracene see 3054
Dibenz[de,kl]anthracene see 7325
5H-Dibenz[b,f]azepine-5-carboxamide see 1826
4-[3-(5H-Dibenz[b,f]azepin-5-yl)propyl]-1-piperazineethanol see 6985
Dibenzcozamide see 2513
Dibenzepin, 3055
Dibenzheptropine see 2958
1,2,3,4-Dibenznaphthalene see 9870
2,3,6,7-Dibenzoanthracene see 7240
4-[3-(5H-Dibenzo[b,f]azepin-5-yl)-propyl]-1-(2-hydroxyethyl)piperazine see 6985
Dibenzo-18-crown-6 see 2669
Dibenzo[a,d][1,4]cycloheptadiene-5-carboxamide see 2828
5H-Dibenzo[a,d]cyclohepten-5-one O-[2-(Methylamino)ethyl]oxime see 2941
3-(5H-Dibenzo[a,d]cyclohepten-5-ylidene)-N,N-dimethyl-1-propanamine see 2782
4-(5H-Dibenzo[a,d]cyclohepten-5-ylidene)-1-methylpiperidine see 2842
2,3,11,12-Dibenzo-1,4,7,10,13,16-hexaoxacyclooctadeca-2,11-diene see 2669
Dibenzoparadiazine see 7359
Dibenzopenthiophene see 9506
Dibenzo[a,i]phenanthrene see 7550
Dibenzo[b,h]phenanthrene see 3057
Dibenzopyrazine see 7359
Dibenzo[b,e]pyridine see 124
Dibenzo-γ-pyrone see 10196
Dibenzopyrrole see 1834
2-[2-(4-Dibenzo[b,f][1,4]thiazepin-11-yl-1-piperazinyl)ethoxy]ethanol (E)-2-Butenedioate (2:1) Salt see 8606
Dibenzothiazine see 7404
2,2'-Dibenzothiazyl Disulfide see 3435
3-Dibenzo[b,e]thiepin-11(6H)-ylidene-N,N-dimethyl-1-propanamine see 3485
Dibenzothiopyran see 9506
3-Dibenz[b,e]oxepin-11(6H)-ylidene-N,N-dimethyl-1-propanamine see 3492
Dibenzoxin see 6821
Dibenzoyl see 1107
Dibenzoylmethane, 3056
Dibenzoyl Peroxide see 1149
1,2,7,8-Dibenzphenanthrene see 7550
2,3:6,7-Dibenzphenanthrene, 3057
β,β'-Dibenzphenanthrene see 3057
Dibenzthiazyl Disulfide see 3435
Dibenzthione see 9061
Dibenzyl see 1245
Dibenzylamine, 3058
N,N-Dibenzylaminoethyl Chloride Hydrochloride see 2190
N,N-Dibenzyl-β-chloroethylamine Hydrochloride see 2190
Dibenzylchlorophosphate see 3059
Dibenzyl Chlorophosphonate, 3059
N,N'-Dibenzyl-γ,γ'-dipyridylium see 10137
Dibenzyl Disulfide, 3060
Dibenzyl Ether see 1167
N,N'-Dibenzylethylenediamine see 1092

N,N'-Dibenzylethylenediamine Bis-[benzylpenicillin] see 7221
Dibenzylethylenediamine Dipenicillin G see 7221
Dibenzyl Fumarate see 1170
Dibenzyl Hydrogen Phosphite see 3061
Dibenzylidene Acetone see 3053
Dibenzylin [SK&F] see 7409
d-3,4-(1',3'-Dibenzyl-2'-ketoimidazolido)-1,2-trimethylenethiophanium d-Camphorsulfonate see 9837
4,6-Dibenzyl-5-oxo-1-thia-4,6-diazatricyclo[6.3.0.0^{3.7}]undecanium (+)-β-Camphorsulfonate see 9837
Dibenzyl Phosphite, 3061
Dibenzylphosphoryl Chloride see 3059
Dibenzylsulfide see 1183
3,5-Dibenzyltetrahydro-2H-1,3,5-thiadiazine-2-thione see 9061
Dibenzyran [Röhm] see 7409
N,N'-Dibetaine-γ,γ'-dipyridylium see 10137
Dibetos [Kodama] see 1500
Diblocin [Astra] see 3489
Dibondrin [Montavit] see 3367
Diborane(6), 3062
Diboron Hexahydride see 3062
Diboron Tetrachloride, 3063
Diboron Tetrahydroxide, 3064
Dibotin [Winthrop] see 7376
Dibrom [Chevron] see 6445
Dibromated Camphor see 3067
2,4'-Dibromoacetophenone see 1449
p,α-Dibromoacetophenone see 1449
9,10-Dibromoanthracene, 3065
p-Dibromobenzene, 3066
4,4'-Dibromobenzilic Acid Isopropyl Ester see 1456
d-3,3-Dibromo-2-bornanone see 3067
3,5-Dibromo-N-(4-bromophenyl)-2-hydroxybenzamide see 9747
1,2-Dibromobutane see 1600
2,3-Dibromobutanedioic Acid see 3078
α,α'-Dibromo-d-camphor, 3067
[2,7-Dibromo-9-(o-carboxyphenyl)-6-hydroxy-3-oxo-3H-xanthen-4-yl]hydroxymercury Disodium Salt see 5914
2,6-Dibromo-N-chloro-p-benzoquinoneimine see 3075
2,6-Dibromo-4-(chloroimino)-2,5-cyclohexadien-1-one see 3075
Dibromochloromethane see 2186
Dibromochloropropane, 3068
1,2-Dibromo-3-chloropropane see 3068
2,6-Dibromo-N-chloroquinonimine see 3075
5,5'-Dibromo-o-cresolsulfonphthalein see 1408
2,6-Dibromo-4-cyanophenol see 1465
3,5-Dibromo-Nα-cyclohexyl-Nα-methyltoluene-α,2-diamine see 1412
2',2''-Dibromo-4',4''-diamidino-1,3-diphenoxypropane see 3073
1,2-Dibromo-2,4-dicyanobutane, 3069
1,6-Dibromo-1,6-dideoxydulcitol see 6300
1,6-Dibromo-1,6-dideoxygalactitol see 6300
1,6-Dibromo-1,6-dideoxy-D-mannitol see 6298
5,5'-Dibromo-2,2'-dihydroxybenzil see 3080
5,5'-Dibromo-2,2'-dihydroxybibenzoyl see 3080
4',5'-Dibromo-3',6'-dihydroxy-2',7'-dinitrospiro[isobenzofuran-1(3H),9'-[9H]xanthen]-3-one Disodium Salt see 3638

Dichlorisone, 3099
Dichlorman [Shell] *see* 3129
D-*d*-*threo*-2-Dichloroacetamido-1-(4-methylsulfonyl)phenyl-1,3-propanediol *see* 9436
D(—)-*threo*-2-Dichloroacetamido-1-*p*-nitrophenyl-1,3-propanediol *see* 2120
Dichloroacet-4-hydroxy-*N*-methylanilide *see* 3246
N-Dichloroacet-4-hydroxy-*N*-methylanilide *see* 3246
Dichloroacetic Acid, 3100
Dichloroacetic Acid Compd with *N*-(1-Methylethyl)-2-propanamine (1:1) *see* 3241
Dichloroacetic Acid Diisopropylammonium Salt *see* 3241
sym-Dichloroacetic Anhydride *see* 2163
1,1-Dichloroacetone, 3101
1,3-Dichloroacetone, 3102
sym-Dichloroacetone *see* 3102
uns-Dichloroacetone *see* 3101
α,α-Dichloroacetone *see* 3101
α,γ-Dichloroacetone *see* 3102
2,2-Dichloroacetyl Chloride, 3103
1-(Dichloroacetyl)-6-(2-furoyloxy)-1,2,3,4-tetrahydroquinoline *see* 8241
D-*threo*-*N*-Dichloroacetyl-1-*p*-nitrophenyl-2-amino-1,3-propanediol *see* 2120
S-2,3-Dichloroallyl Diisopropylthiocarbamate *see* 3014
2,4-Dichloro-6-aminophenol *see* 457
4,6-Dichloro-*o*-aminophenol *see* 457
4-[(Dichloroamino)sulfonyl]benzoic Acid *see* 4620
9α,11β-Dichloro Analog of Prednisolone *see* 3099
3,4-Dichloroanilide of 5-Chlorosalicylic Acid *see* 9775
3,4-Dichloroaniline, 3104
2-(2,6-Dichloroanilino)-1,3-diazacyclopentene-(2) *see* 2450
2-(2,6-Dichloroanilino)-2-imidazoline *see* 2450
[*o*-(2,6-Dichloroanilino)phenyl]acetic Acid Sodium Salt *see* 3132
3,6-Dichloro-*o*-anisic Acid *see* 3090
Dichloroazodicarbonamidine *see* 2171
(±)-4-(3,4-Dichlorobenzamido)-*N*-(3-methoxypropyl)-*N*-pentylglutaramic Acid *see* 5618
1,2-Dichlorobenzene *see* 3106
1,3-Dichlorobenzene *see* 3105
m-Dichlorobenzene, 3105
o-Dichlorobenzene, 3106
p-Dichlorobenzene, 3107
3,4-Dichlorobenzeneamine *see* 3104
4,5-Dichloro-1,3-benzenedisulfonamide *see* 3127
2,4-Dichlorobenzenemethanol *see* 3110
2,2'-Dichlorobenzidine, 3108
3,3'-Dichlorobenzidine, 3109
4,4'-Dichlorobenzilic Acid Ethyl Ester *see* 2174
2,6-Dichlorobenzonitrile *see* 3093
2,6-Dichloro-*p*-benzoquinone-4-chloroimine *see* 4428
2,6-Dichloro-1,4-benzoquinone-4-(4-hydroxyanil) Sodium *see* 3118
(±)-4-[(3,4-Dichlorobenzoyl)amino]-5-[(3-methoxypropyl)pentylamino]-5-oxopentanoic Acid *see* 5618
Dichlorobenzyl Alcohol, 3110
N-(2,4-Dichlorobenzyl)-*N*-(2-hydroxyethyl)dichloroacetamide *see* 2126
N-(2,6-Dichlorobenzylidene)-*N*'-amidinohydrazine *see* 4585

[(2,6-Dichlorobenzylidene)amino]-guanidine *see* 4585
1-[(2,6-Dichlorobenzylidene)amino]-3-hydroxyguanidine *see* 4597
1-(2,4-Dichlorobenzyl)-1*H*-indazole-3-carboxylic Acid *see* 5598
2,2'-Dichloro[1,1'-biphenyl]-4,4'-diamine *see* 3108
3,3-Dichloro-(1,1'-biphenyl)-4,4'-diamine *see* 3109
3,3'-Dichloro-4,4'-biphenyldiamine *see* 3109
1,1-Dichloro-2,2-bis(*p*-chlorophenyl)-ethane, 3111
Dichlorobis(η⁵-2,4-cyclopentadien-1-yl)titanium *see* 9621
1,1-Dichloro-2,2-bis(*p*-fluorophenyl)-ethane *see* 2999
Dichlorobromomethane *see* 1438
1,1-Dichlorobutane *see* 1606
2,4-Dichloro-6-(*o*-chloroanilino)-*s*-triazine *see* 694
(±)-1-[2,4-Dichloro-β-[(7-chlorobenzo[*b*]thien-3-yl)methoxy]phenethyl]imidazole *see* 8610
1-[2,4-Dichloro-β-[(*p*-chlorobenzyl)oxy]phenethyl]imidazole *see* 3550
(±)-1-[2,4-Dichloro-β-[(*p*-chlorobenzyl)thio]phenethyl]imidazole *see* 9062
2,6-Dichloro-4-(chloroimino)-2,5-cyclohexadien-1-one *see* 4428
2,4-Dichloro-α-(chloromethylene)-benzyl Alcohol Diethyl Phosphate *see* 2137
(Z)-1-[2,4-Dichloro-β-[2-(*p*-chlorophenoxy)ethoxy]-α-methylstyryl]-imidazole *see* 6978
1,1-Dichloro-2-(*o*-chlorophenyl)-2-(*p*-chlorophenyl)ethane *see* 6302
2,6-Dichloro-α-(4-chlorophenyl)-4-(4,5-dihydro-3,5-dioxo-1,2,4-triazin-2(3*H*)-yl)benzeneacetonitrile *see* 3130
4,6-Dichloro-*N*-(2-chlorophenyl)-1,3,5-triazin-2-amine *see* 694
1-[2,4-Dichloro-β-[(2-chloro-3-thenyl)oxy]phenethyl]imidazole *see* 9595
Dichloro(2-chlorovinyl)arsine, 3112
(1*S*-*trans*)-2,2-Dichloro-1,3-cyclopentanediol *see* 1761
2-[4-(2,2-Dichlorocyclopropyl)phenoxy]isobutyric Acid *see* 2373
2-[4-(2,2-Dichlorocyclopropyl)phenoxy]-2-methylpropanoic Acid *see* 2373
N,*N*''-Dichlorodiazenedicarboximidamide *see* 2171
1,3-Dichloro-α-[2-(dibutylamino)ethyl]-6-(trifluoromethyl)-9-phenanthrenemethanol *see* 4626
2,2-Dichloro-*N*-(2,4-dichlorobenzyl)-*N*-(2-hydroxyethyl)acetamide *see* 2126
1-[2,4-Dichloro-β-[(2,4-dichlorobenzyl)oxy]phenethyl]imidazole *see* 6266
1-[2,4-Dichloro-β-[(2,6-dichlorobenzyl)oxy]phenethyl]imidazole *see* 5176
3,5-Dichloro-6-[(3,5-dichloro-6-hydroxy-4-methoxy-*o*-tolyl)oxy]-4,2-cresotic Acid ε-Lactone *see* 3401
3,5-Dichloro-*N*-(3,4-dichlorophenyl)-2-hydroxybenzamide *see* 9333
2,2-Dichloro-*N*-[(2,4-dichlorophenyl)-methyl]-*N*-(2-hydroxyethyl)acetamide *see* 2126
2,3-Dichloro-5,6-dicyanobenzoquinone, 3113
Dichlorodi-π-cyclopentadienyltitanium *see* 9621
1,6-Dichloro-1,6-dideoxy-β-D-fructo-

furanosyl-4-chloro-4-deoxy-α-D-galactopyranoside *see* 9050
β,β'-Dichlorodiethylaniline *see* 697
2,2'-Dichlorodiethyl Sulfide *see* 6395
2,2-Dichloro-1,1-difluoroethyl Methyl Ether *see* 6072
Dichlorodifluoromethane, 3114
2,2-Dichloro-1,1-difluoro-1-methoxy-ethane *see* 6072
2,6-Dichloro-*N*¹-(4,5-dihydro-1*H*-imidazol-2-yl)-1,4-benzenediamine *see* 791
6,7-Dichloro-1,5-dihydroimidazo[2,1-*b*]quinazolin-2(3*H*)-one *see* 665
2,5-Dichloro-3,6-dihydroxy-*p*-benzoquinone *see* 2122
2,5-Dichloro-3,6-dihydroxy-2,5-cyclohexadiene-1,4-dione *see* 2122
5,5'-Dichloro-2,2'-dihydroxy-3,3'-dinitrobiphenyl *see* 6601
5,5'-Dichloro-2,2'-dihydroxydiphenylmethane *see* 3120
9,21-Dichloro-11β,17-dihydroxy-16α-methylpregna-1,4-diene-3,20-dione 17-(2-Furoate) *see* 6324
(11β)-9,11-Dichloro-17,21-dihydroxypregna-1,4-diene-3,20-dione *see* 3099
4',5'-Dichloro-3',6'-dihydroxyspiro-[isobenzofuran-1(3*H*),9'-[9*H*]xanthen]-3-one *see* 3117
1,1-Dichloro-*N*-[(dimethylamino)sulfonyl]-1-fluoro-*N*-phenylmethanesulfenamide *see* 3095
2',5-Dichloro-2-(3-dimethylcarbamoyl-5-glycylaminomethyl-1*H*-1,2,4-triazol-1-yl)benzophenone *see* 8387
1,3-Dichloro-5,5-dimethylhydantoin, 3115
1,3-Dichloro-5,5-dimethyl-2,4-imidazolidinedione *see* 3115
2,4-Dichloro-3,5-dimethylphenol *see* 3126
3,5-Dichloro-*N*-(1,1-dimethyl-2-propynyl)benzamide *see* 8058
3,5-Dichloro-2,6-dimethyl-4-pyridinol *see* 2458
4,4'-Dichloro-6,6'-dinitro-*o*,*o*'-biphenol *see* 6601
5,5'-Dichloro-3,3'-dinitro[1,1'-biphenyl]-2,2'-diol *see* 6601
Dichlorodioxochromium *see* 2306
4,5-Dichloro-3,6-dioxo-1,4-cyclohexadiene-1,2-dicarbonitrile *see* 3113
2,4'-Dichlorodiphenyldichloroethane *see* 6302
p,*p*'-Dichlorodiphenylmethyl Carbinol *see* 2136
Dichlorodiphenyltrichloroethane *see* 2898
4,5-Dichloro-1,3-disulfamoylbenzene *see* 3127
1,1-Dichloroethane *see* 3856
1,2-Dichloroethane *see* 3843
sym-Dichloroethane *see* 3843
Dichloroethanoyl Chloride *see* 3103
1,1-Dichloroethene *see* 10134
1,2-Dichloroethene *see* 93
3-(2,2-Dichloroethenyl)-2,2-dimethylcyclopropanecarboxylic Acid Cyano(4-fluoro-3-phenoxyphenyl)methyl Ester *see* 2826
3-(2,2-Dichloroethenyl)-2,2-dimethylcyclopropanecarboxylic Acid Cyano(3-phenoxyphenyl)methyl Ester *see* 2836
3-(2,2-Dichloroethenyl)-2,2-dimethylcyclopropanecarboxylic Acid (3-Phenoxyphenyl)methyl Ester *see* 7321
Dichloroethylaluminum *see* 334

3-(2,3-Dichlorophenyl)-4-cyanopyrrole *see* 4030

1-[2-(2,4-Dichlorophenyl)-2-[(2,4-dichlorophenyl)methoxy]ethyl]-1*H*-imidazole *see* 6266

1-[2-(2,4-Dichlorophenyl)-2-[(2,6-dichlorophenyl)methoxy]ethyl]-1*H*-imidazole *see* 5176

O-2,4-Dichlorophenyl *O,O*-Diethyl Phosphorothioate *see* 3094

1,6-Di(4′-chlorophenyldiguanido)hexane *see* 2140

4-(2,3-Dichlorophenyl)-1,4-dihydro-2,-6-dimethyl-3,5-pyridinedicarboxylic Acid Ethyl Methyl Ester *see* 3991

N-(2,6-Dichlorophenyl)-4,5-dihydro-1*H*-imidazol-2-amine *see* 2450

N-(2,6-Dichlorophenyl)-4,5-dihydro-*N*-2-propenyl-1*H*-imidazol-2-amine 244

3-(3,5-Dichlorophenyl)-1,5-dimethyl-3-azabicyclo[3.1.0]hexane-2,4-dione *see* 7946

N-(3,5-Dichlorophenyl)-1,2-dimethyl-1,2-cyclopropanedicarboximide *see* 7946

1-(2,4-Dichlorophenyl)-4,4-dimethyl-2-(1,2,4-triazol-1-yl)pentan-3-ol *see* 3131

(*E*)-1-(2,4-Dichlorophenyl)-4,4-dimethyl-2-(1,2,4-triazol-1-yl)-1-penten-3-ol *see* 3320

N′-(3,4-Dichlorophenyl)-*N,N*-dimethylurea *see* 3447

3-(3,5-Dichlorophenyl)-5-ethenyl-5-methyl-2,4-oxazolidinedione *see* 10122

N-(3,5-Dichlorophenyl)-*N*′-(4-fluorophenyl)thiourea *see* 5589

(*Z*)-1-(2,4-Dichlorophenyl)-2-(1*H*-imidazol-1-yl)ethanone *O*-[(2,4-Dichlorophenyl)methyl]oxime Mononitrate *see* 7071

1-(2,4-Dichlorophenyl)-2-(*N*-imidazolyl)ethyl-4-phenylthiobenzyl Ether *see* 4047

2-[(2,6-Dichlorophenyl)imino]-2-imidazoline *see* 2450

1-(3,4-Dichlorophenyl)-5-isopropylbiguanide *see* 2237

*N*¹-3,4-Dichlorophenyl-*N*⁵-isopropylbiguanide *see* 2237

*N*¹-3,4-Dichlorophenyl-*N*⁵-isopropyldiguanide *see* 2237

N-(3,4-Dichlorophenyl)methacrylamide *see* 3138

[2*S*-(2α,5α,6β)]-6-[[(3,4-Dichlorophenyl)methoxyacetyl]amino]-3,3-dimethyl-7-oxo-4-thia-1-azabicyclo-[3.2.0]heptane-2-carboxylic Acid *see* 2445

2,4-Dichlorophenyl 3-(Methoxycarbonyl)-4-nitrophenyl Ether *see* 1256

N′-(3,4-Dichlorophenyl)-*N*-methoxy-*N*-methylurea *see* 5534

3-(3,4-Dichlorophenyl)-1-methyl-1-*n*-butylurea *see* 6523

(*R**,*R**)-(±)-β-[(2,4-Dichlorophenyl)-methyl]-α-(1,1-dimethylethyl)-1*H*-1,2,4-triazole-1-ethanol *see* 3131

(*E*)-(±)-β-[(2,4-Dichlorophenyl)methylene]-α-(1,1-dimethylethyl)-1*H*-1,2,4-triazole-1-ethanol *see* 3320

2-[(2,6-Dichlorophenyl)methylene]hydrazinecarboximidamide *see* 4585

2-[(2,6-Dichlorophenyl)methylene]-*N*-hydroxyhydrazinecarboximidamide *see* 4597

3-(3,5-Dichlorophenyl)-*N*-(1-methylethyl)-2,4-dioxo-1-imidazolidinecarboxamide *see* 5093

N-(3,4-Dichlorophenyl)-*N*′-(1-methylethyl)imidodicarbonimidic Diamide *see* 2237

1-[(2,4-Dichlorophenyl)methyl]-1*H*-indazole-3-carboxylic Acid *see* 5598

O-(2,4-Dichlorophenyl) *O*-Methyl Isopropylphosphoramidothioate *see* 3453

6-[3-(2,6-Dichlorophenyl)-5-methyl-4-isoxazolecarboxamido]penicillanic Acid *see* 3134

[2*S*-(2α,5α,6β)]-6-[[[3-(2,6-Dichlorophenyl)-5-methyl-4-isoxazolyl]-carbonyl]amino]-3,3-dimethyl-7-oxo-4-thia-1-azabicyclo[3.2.0]hept-ane-2-carboxylic Acid *see* 3134

3-(2,6-Dichlorophenyl)-5-methyl-4-isoxazolylpenicillin *see* 3134

2-(3,4-Dichlorophenyl)-4-methyl-1,2,4-oxadiazolidine-3,5-dione *see* 6032

N-(3,4-Dichlorophenyl)-2-methylpentanamide *see* 5297

N-(3,4-Dichlorophenyl)-2-methyl-2-propenamide *see* 3138

2-[(2,4-Dichlorophenyl)methyl]-4-(1,1,3,3-tetramethylbutyl)phenol *see* 2439

3-(3,5-Dichlorophenyl)-5-methyl-5-vinyloxazolidine-2,4-dione *see* 10122

2,4-Dichlorophenyl-*p*-nitrophenyl Ether *see* 6695

1-[2-(2,4-Dichlorophenyl)-2-[[4-(phenylthio)phenyl]methoxy]ethyl]-1*H*-imidazole *see* 4047

N-(3,4-Dichlorophenyl)propanamide *see* 7987

1-[2-(2,4-Dichlorophenyl)-2-(2-propenyloxy)ethyl]-1*H*-imidazole *see* 3622

N-(3,4-Dichlorophenyl)propionamide *see* 7987

1-[[2-(2,4-Dichlorophenyl)-4-propyl-1,3-dioxolan-2-yl]methyl]-1*H*-1,2,4-triazole *see* 8003

1-(2,4-Dichlorophenyl)-2-(3-pyridinyl)ethanone *O*-Methyloxime *see* 8167

4-(2,3-Dichlorophenyl)-1*H*-pyrrole-3-carbonitrile *see* 4030

(1*S*-*cis*)-4-(3,4-Dichlorophenyl)-1,2,3,-4-tetrahydro-*N*-methyl-1-naphthalenamine *see* 8612

(1*S*,4*S*)-4-(3,4-Dichlorophenyl)-1,2,3,4-tetrahydro-*N*-methyl-1-naphthylamine *see* 8612

α-(2,4-Dichlorophenyl)-4-(1,1,3,3-tetramethylbutyl)-*o*-cresol *see* 2439

1-[2,4-Dichloro-β-[[*p*-(phenylthio)-benzyl]oxy]phenethyl]imidazole *see* 4047

2,4-Dichloro-4′-phenylthio-(*N*-imidazolylmethyl)dibenzyl Ether *see* 4047

6-(2,3-Dichlorophenyl)-1,2,4-triazine-3,5-diamine *see* 5367

6-(2,5-Dichlorophenyl)-1,3,5-triazine-2,4-diamine *see* 5113

cis-1-[4-[[2-(2,4-Dichlorophenyl)-2-(1*H*-1,2,4-triazol-1-ylmethyl)-1,3-dioxolan-4-yl]methoxy]phenyl]-4-(1-methylethyl)piperazine *see* 9303

4-[4-[4-[4-[[2-(2,4-Dichlorophenyl)-2-(1*H*-1,2,4-triazol-1-ylmethyl)-1,3-dioxolan-4-yl]methoxy]phenyl]-1-piperazinyl]phenyl]-2,4-dihydro-2-(1-methylpropyl)-3*H*-1,2,4-triazol-3-one *see* 5262

2-(3,5-Dichlorophenyl)-2-(2,2,2-trichloroethyl)oxirane *see* 9795

Di(*p*-chlorophenyl)trichloromethylcarbinol *see* 3136

Dichlorophos *see* 3129

3,6-Dichloropicolinic Acid *see* 2462

9α,11β-Dichloro-1,4-pregnadiene-17α,21-diol-3,20-dione *see* 3099

Dichloroprop *see* 3128

1,1-Dichloropropane *see* 8046

1,2-Dichloropropane *see* 8039

2,2-Dichloropropanoic Acid *see* 2869

2,2-Dichloropropanoic Acid 2-(2,4,5-Trichlorophenoxy)ethyl Ester *see* 3681

1,3-Dichloro-2-propanol, 3124

1,3-Dichloro-2-propanol Phosphate (3:1) *see* 4342

1,1-Dichloro-2-propanone *see* 3101

1,3-Dichloro-2-propanone *see* 3102

3′,4′-Dichloropropionanilide *see* 7987

α,α-Dichloropropionic Acid *see* 2869

2,2-Dichloropropionic Acid 2-(2,4,5-Trichlorophenoxy)ethyl Ester *see* 3681

1,3-Dichloropropylene, 3125

α,γ-Dichloropropylene *see* 3125

3,6-Dichloro-2-pyridinecarboxylic Acid *see* 2462

2′,4′-Dichloro-2-(3-pyridyl)acetophenone *O*-Methyloxime *see* 8167

2,4′-Dichloro-α-(pyrimidin-5-yl)-benzhydryl Alcohol *see* 3999

5,7-Dichloro-8-quinaldinol *see* 2243

5,7-Dichloro-8-quinolinol *see* 2227

5,7-Dichloro-8-quinolinol Mixt with 5-Chloro-8-quinolinol and 7-Chloro-8-quinolinol *see* 4637

2,6-Dichloroquinone Chloroimide *see* 4428

Dichlorosal *see* 4822

p-(Dichlorosulfamoyl)benzoic Acid *see* 4620

Dichlorotetraaquochromium Chloride Dihydrate *see* 2278

Dichlorotetracarbonyldirhodium *see* 8354

1,2-Dichloro-1,1,2,2-tetrafluoroethane *see* 2672

N-[[[3,5-Dichloro-4-(1,1,2,2-tetra-fluoroethoxy)phenyl]amino]carbonyl]-2,6-difluorobenzamide *see* 4720

1-[3,5-Dichloro-4-(1,1,2,2-tetrafluoroethoxy)phenyl]-(2,6-difluorobenzoyl)urea *see* 4720

6,7-Dichloro-1,2,3,5-tetrahydroimidazo[2,1-*b*]quinazolin-2-one *see* 665

3,5-Dichlorotetrahydro-2,4,6-trioxo-*s*-triazin-1(2*H*)-yl Potassium *see* 9896

[2,3-Dichloro-4-(2-thenoyl)phenoxy]-acetic Acid *see* 9570

[2,3-Dichloro-4-(2-thienylcarbonyl)-phenoxy]acetic Acid *see* 9570

[2,3-Dichloro-4-(2-thiophenecarbonyl)phenoxy]acetic Acid *see* 9570

α,α-Dichlorotoluene *see* 1084

N,N-Dichloro-*p*-toluenesulfonamide *see* 3098

N-(2,6-Dichloro-*m*-tolyl)anthranilic Acid *see* 5819

N-(2,6-Dichloro-*m*-tolyl)anthranilic Acid Ethoxymethyl Ester *see* 9312

1,3-Dichloro-1,3,5-triazine-2,4,6(1*H*,-3*H*,5*H*)-trione Potassium Salt *see* 9896

4,4′-Dichloro-α-(trichloromethyl)-benzhydrol *see* 3136

4,4′-Dichloro-3-(trifluoromethyl)carbanilide *see* 2438

1-(1,3-Dichloro-6-trifluoromethyl-9-

phenanthryl)-3-di(n-butyl)amino-
propanol see 4626
Dichlorovos see 3129
Dichloroxide see 2149
Dichloroxylenol, 3126
2,4-Dichloro-3,5-xylenol see 3126
Dichlorphenamide, 3127
Dichlorprop, 3128
Dichlor-Stapenor [Bayer] see 3134
Dichlorvos, 3129
Dichlotride [Merck & Co.] see 4822
β-Dichroine see 3984
Dichronic [Sana] see 3132
Dichystrolum see 3223
Dickite see 377
Diclazuril, 3130
Diclobenin [Chassot] see 3132
Diclobutrazol, 3131
Diclocil [Bristol] see 3134
Diclofenac Sodium, 3132
Diclofop-Methyl, 3133
Dicloguamine see 5113
Diclondazolic Acid see 5598
Diclo-Phlogont [Azupharma] see 3132
Diclo-Puren [Klinge-Nattermann] see
3132
Diclord [Wakamoto] see 3132
Dicloreum [Alfa] see 3132
Diclotride [Merck & Co.] see 4822
Dicloxacillin, 3134
Dico see 3214
Dicobalt Octacarbonyl, 3135
Dicodid [Knoll] see 4826
Dicoferin see 6626
Dicofol, 3136
Dicopac [Amersham] see 10153
Dicophane see 2898
Dicopper Tetraiodomercurate(2—) see
2733
Dicorantil [Silva Araujo-Roussel] see
3424
Dicoumarin see 3140
Dicoumarol see 3140
Dicoumoxyl [Labaz] see 2633
Dicroden see 4710
Dicromil [Organon] see 2971
Dicrotalic Acid see 5852
Dicrotophos, 3137
Dicryl, 3138
Dictamine see 3139
Dictamnine, 3139
Dictycide [Fort Dodge] see 2751
Dictyzide see 2751
Dicumacyl see 3818
Dicumarol, 3140
Dicumol see 3140
Dicumoxane [Boots] see 2633
Dicuprene [UCB] see 2740
Dicurone see 4476
Dicyan see 2761
Dicyanine, 3141
2,2'-Dicyano-2,2'-azopropane see 949
β,β-Dicyano-o-chlorostyrene see 2178
Dicyanodiamide, 3142
Dicyanodiamidine Sulfate, 3143
2,3-Dicyano-1,4-dithiaanthraquinone
see 3433
sym-Dicyanoethane see 9041
Dicyanomethane see 5750
9-Dicyanomethylene-2,4,7-trinitro-
fluorene, 3144
1,3-Dicyano-2,4,5,6-tetrachlorobenz-
ene see 2219
Dicyclidine see 7946
Dicyclohexano-18-crown-6 see 2669
Dicyclohexylacetic Acid 2-Piperidino-
ethyl Ester see 3211
Dicyclohexylamine, 3145
Dicyclohexylcarbodiimide, 3146
Dicyclohexyl-18-crown-6 see 2669

2-(2,2-Dicyclohexylethyl)piperidine
see 7305
1,1-Dicyclohexyl-2-(2-piperidyl)eth-
ane see 7305
Dicyclomine Hydrochloride, 3147
Dicyclopentadiene see 2807
Dicyclopentadienyliron see 4084
Dicyclopentylacetic Acid β-Diethylaminoethyl Ester Ethobromide see
3402
2-[(Dicyclopentylacetyl)oxy]-N,N,N-
triethylethanaminium Bromide see
3402
2-[N-(Dicyclopropylmethyl)amino]-
oxazoline see 8388
N-(Dicyclopropylmethyl)-4,5-dihydro-
2-oxazolamine see 8388
(2α,7α,8R,19α,22E)-7,19:8,19-Dicy-
clo-9,10-secoergosta-5(10),22-dien-
2-ol see 9179
Dicycloverin Hydrochloride see 3147
Dicynene [Travenol] see 3766
Dicynone [OM] see 3766
Dicysteine see 2851
Didakene see 9332
Didandin [Boots] see 3363
Didanosine, 3148
Didecyldimethylammonium Chloride,
3149
4',5'-Didehydro-4,5-retro-β,β-carot-
ene-3,3'-dione see 8362
3',4'-Didehydro-β,ψ-caroten-16'-oic
Acid see 9691
1',4-Didehydro-1-deoxy-1,4-dihydro-
5'-(2-methylpropyl)-1-oxorifamycin
XIV see 8380
(16R,19E)-19,20-Didehydro-16-[(10β,-
13β,21β)-23-deoxy-21,22-dihydro-
11-oxa-12,14-secostrychnidin-10-
yl]corynan-17-oic Acid Methyl
Ester see 4387
3',4'-Didehydro-4'-deoxy-C'-norvinca-
leukoblastine see 10127
2',3'-Didehydro-3'-deoxythymidine see
8958
(2'E)-2',3'-Didehydro-2'-deoxy-
verrucarin A see 10100
9,10-Didehydro-N,N-diethyl-6-methyl-
ergoline-8β-carboxamide see 5665
12,13-Didehydro-13,14-dihydro-α-ery-
throidine see 3719
11,12-Didehydro-7,10-dihydro-10-hy-
droxyretinol see 7066
(3β)-1,6-Didehydro-14,17-dihydro-3-
methoxy-16(15H)-oxaerythrinan-15-
one see 3218
(16E,20β)-16,17-Didehydro-9,17-
dimethoxycorynan-16-carboxylic
Acid Methyl Ester see 6304
(3β,19α,20α)-16,17-Didehydro-10,11-
dimethoxy-19-methyloxayohimban-
16-carboxylic Acid Methyl Ester see
8313
(E)-16,17-Didehydro-9,17-dimethoxy-
17,18-seco-20α-yohimban-16-carb-
oxylic Acid Methyl Ester see 6304
8,9-Didehydro-6,8-dimethylergoline
see 187
(5α,6α,7α,12β,19α)-2,3-Didehydro-6,-
7-epoxyaspidospermidine-3-carbox-
ylic Acid Methyl Ester see 5582
(17R)-19,20-Didehydro-17,18-epoxy-
curan-17-ol see 10181
(5α)-6,7-Didehydro-4,5-epoxy-3,6-
dimethoxy-17-methylmorphinan see
3224
7,8-Didehydro-4,5-epoxy-3-ethoxy-17-
methylmorphinan-6-ol see 3876
6',7'-Didehydro-5,6-epoxy-4',5,5',6,7,-

8-hexahydro-3,3',5'-trihydroxy-8-
oxo-α-carotene 3'-Acetate see 4304
5α-7,8-Didehydro-4,5α-epoxy-14-
hydroxy-3-methoxy-17-methylmor-
phinan-6-one see 4867
(5α,6α)-7,8-Didehydro-4,5-epoxy-3-
methoxy-17-methylmorphinan-6-ol
see 2525
(5α,6α)-8,14-Didehydro-4,5-epoxy-3-
methoxy-17-methylmorphinan-6-ol
see 6549
(5α,8β)-6,7-Didehydro-4,5-epoxy-3-
methoxy-17-methylmorphinan-8-ol
see 8098
(5α)-6,7-Didehydro-4,5-epoxy-3-meth-
oxy-17-methylmorphinan-6-ol Acetate (Ester) see 3215
(5α,6α)-7,8-Didehydro-4,5-epoxy-3-
methoxymorphinan-6-ol see 6783
(5α,6α)-7,8-Didehydro-4,5-epoxy-17-
methylmorphinan-3,6-diol see 6359
(5α,6α)-7,8-Didehydro-4,5-epoxy-17-
methylmorphinan-3,6-diol Diacetate
(Ester) see 3012
(5α,6α)-7,8-Didehydro-4,5-epoxy-17-
methylmorphinan-3,6-diol Di-3-pyr-
idinecarboxylate (Ester) see 6607
(5α,6α)-7,8-Didehydro-4,5-epoxy-17-
methylmorphinan-3,6-diol Sulfate
Salt (2:1) see 6361
(5α,6α)-7,8-Didehydro-4,5-epoxy-17-
methyl-3-[2-(4-morpholinyl)ethoxy]-
morphinan-6-ol see 7484
7,8-Didehydro-4,5-epoxy-17-methyl-3-
(phenylmethoxy)morphinan-6-ol see
1176
7,8-Didehydro-4,5-epoxy-17-methyl-3-
(phenylmethoxy)morphinan-6 ol
Tetradecanoate(ester) see 6420
(5α,6α)-7,8-Didehydro-4,5-epoxymor-
phinan-3,6-diol see 6806
(5α,6α)-7,8-Didehydro-4,5-epoxy-17-
(2-propenyl)morphinan-3,6-diol see
6448
19,20-Didehydro-16-(15-ethyl-3a,5,-
5a,7,8,13a-hexahydro-4H-4,6-etha-
no-1H,3H-[1,3]oxazino[3,4,5-lm]-
pyrrolo[2,3-d]carbazol-1-yl)-cory-
nan-17-oic Acid Methyl Ester see
4387
(20α)-2,16-Didehydro-20-hydroxy-
curan-17-oic Acid Methyl Ester see
5583
(9α,13α,14α)-7,8-Didehydro-4-hydr-
oxy-3,7-dimethoxy-17-methylmor-
phinan-6-one see 8692
[15(3aS,7R)]-3,4-Didehydro-19-hydr-
oxy-16,17-dimethoxy-1-methyl-15-
(2,3,5,6-tetrahydro-11-hydroxy-4-
methyl-1,13-dioxo-1H-3a,7-meth-
anopyrrolo[1,2-a][1,3]benzodiazo-
cin-7(4H)-yl)aspidospermidin-21-oic
Acid γ-Lactone see 4642
7,8-Didehydro-4-hydroxy-3-methoxy-
17-methylmorphinan-6-one see 9412
[8α(S)]-9,10-Didehydro-N-(2-hydroxy-
1-methylethyl)-6-methylergolinecar-
boxamide see 3693
[8β(S)]-9,10-Didehydro-N-(2-hydroxy-
1-methylethyl)-6-methylergoline-8-
carboxamide see 3694
[8β(S)]-9,10-Didehydro-N-[1-(hydroxymethyl)propyl]-1,6-dimethyl-
ergoline-8-carboxamide see 6217
[8β(S)]-9,10-Didehydro-N-[1-(hydroxymethyl)propyl]-6-methyler-
goline-8-carboxamide see 6147
13,19-Didehydro-12-hydroxysenecio-
nan-11,16-dione see 8596

3,3-Diethyl-5-methyl-2,4-piperidine-
dione *see* 6216
4,4'-(1,2-Diethyl-3-methyl-1,3-pro-
panediyl)bisphenol *see* 1102
5,5-Diethyl-1-methyl-2,4,6(1*H*,3*H*,-
5*H*)-pyrimidinetrione *see* 6029
O,*O*-Diethyl *O*-[*p*-(Methylsulfinyl)-
phenyl] Phosphorothioate *see* 4042
N,*N*'-Diethyl-6-(methylthio)-1,3,5-tri-
azine-2,4-diamine *see* 8683
N,*N*-Diethyl-5-methyl-[1,2,4]triazolo-
[1,5-*a*]pyrimidin-7-amine *see* 9709
N,*N*-Diethyl-*N*-methyl-2-[2-(trimethylammonio)ethoxy]-1-propanaminium
Diiodide *see* 7693
O,*O*-Diethyl *O*-(4-Methylumbellifer-
one) Phosphorothioate *see* 5975
N,*N*-Diethyl-*N*-methyl-2-[(9*H*-xan-
then-9-ylcarbonyl)oxy]ethanamini-
um Bromide *see* 6025
N,*N*-Diethyl-2-(1-naphthalenyloxy)-
propanamide *see* 6503
O,*O*-Diethyl *O*-Naphthaloximide
Phosphate *see* 6441
N,*N*-Diethyl-2-(1-naphthyloxy)pro-
pionamide *see* 6503
N,*N*-Diethylnicotinamide *see* 6635
Diethyl-*p*-nitrophenyl Monothiophos-
phate *see* 7167
Diethyl *p*-Nitrophenyl Phosphate *see*
7164
O,*O*-Diethyl *O*-*p*-Nitrophenyl Phos-
phorothioate *see* 7167
Diethylnitrosamine *see* 6734
(3α,4aα,10aβ)-(±)-*N*,*N*-Diethyl-*N*'-(1,-
2,3,4,4a,5,10,10a-octahydro-6-hydr-
oxy-1-propylbenzo[*g*]quinolin-3-yl)-
sulfamide *see* 8226
Diethylolamine *see* 3156
Diethylolglycine *see* 1248
Diethyl Oxalacetate *see* 3882
Diethyl Oxalate, 3174
Diethyl Oxide *see* 3852
Diethyl Oxydiformate *see* 8182
*N*¹,*N*¹-Diethyl-1,4-pentanediamine *see*
6819
2,2-Diethyl-4-pentenamide *see* 6820
N,*N*-Diethyl-10*H*-phenothiazine-10-
ethanamine *see* 3157
α,α-Diethylphenylacetic Acid 2-(2-
Diethylaminoethoxy)ethyl Ester *see*
7064
5,5-Diethyl-1-phenylbarbituric Acid
see 7368
N,*N*-Diethyl-α-phenylbenzenemethan-
amine *see* 3163
N,*N*-Diethyl-3-phenyl-1,2,4-oxadi-
azole-5-ethanamine *see* 7078
N,*N*-Diethyl-3-(1-phenylpropyl)-1,2,4-
oxadiazole-5-ethanamine *see* 8090
5,5-Diethyl-1-phenyl-2,4,6(1*H*,3*H*,5*H*)-
pyrimidinetrione *see* 7368
Diethylphosphorous Acid Chloride *see*
3886
Diethyl Phthalate *see* 7527
3,3-Diethyl-2,4-piperidinedione *see*
7622
Diethylpropanedioic Acid *see* 3173
Diethylpropanedioic Acid Diethyl
Ester *see* 3835
Diethylpropion, 3175
2,6-Diethyl-*N*-(2'-*n*-propoxyethyl)-
chloroacetanilide *see* 7921
O,*O*-Diethyl-*O*-(2-pyrazinyl) Phosphorothioate *see* 9482
N,*N*-Diethyl-3-pyridinecarboxamide
see 6635
3,3-Diethyl-2,4-(1*H*,3*H*)pyridinedione
see 8179

5,5-Diethyl-2,4,6(1*H*,3*H*,5*H*)-pyrimi-
dinetrione *see* 989
Diethyl Pyrocarbonate *see* 8182
N,*N*-Diethylrifomycin B Amide *see*
8381
Diethyl Sebacate *see* 8558
Diethylsilane, 3176
α,α'-Diethylstilbenediol *see* 3177
α,α'-Diethyl-4,4'-stilbenediol Diphos-
phoric Acid Ester *see* 4279
Diethylstilbestrol, 3177
Diethylstilbestrol Diphosphate *see*
4279
Diethylstilbestrol 4,4'-Diphosphoric
Ester *see* 4279
Diethylstilbestrol Monomethyl Ether
see 5975
Diethylstilbestryl Diphosphate *see*
4279
p-(Diethylsulfamoyl)benzoic Acid *see*
3775
4,6-Diethyl-2-sulfanilamido-1,3,5-
triazine *see* 9114
Diethyl Sulfate, 3178
Diethyl Sulfide *see* 3898
Diethyl Sulfide 2,2'-Dicarboxylic
Acid *see* 9468
Diethylsulfondimethylmethane *see*
9133
Diethylsulfonmethylethylmethane *see*
9131
Diethyl Tartrate *see* 3900
5,5-Diethyl-2,3,5,6-tetrahydro-4*H*-1,3-
oxazine-2,4-dione *see* 3155
1,8-Diethyl-1,3,4,9-tetrahydropyrano-
[3,4-*b*]indole-1-acetic Acid *see* 3920
N,*N*-Diethyl-*N*-[2-[4-(1,1,3,3-tetra-
methylbutyl)phenoxy]ethyl]benz-
enemethanaminium Chloride *see*
7389
N,*N*-Diethyl-*N*,*N*',*N*',*N*'-tetramethyl-
N,*N*'-(2-methyl-3-oxapentamethyl-
ene)bis[ammonium Iodide] *see* 7693
3,3'-Diethylthiadicarbocyanine Iodide
see 3434
Diethylthiambutene *see* 9429
5,5-Diethyl-2-thiobarbituric Acid *see*
9455
O,*O*-Diethylthiophosphoric Ester of
3,4-Tetramethyleneumbelliferone
see 2635
N,*N*-Diethyl-*m*-toluamide *see* 2912
*N*¹-(4,6-Diethyl-*s*-triazin-2-yl)sulfanil-
amide *see* 9114
O,*O*-Diethyl *O*-3,5,6-Trichloro-2-pyri-
dyl Phosphorothioate *see* 2242
1,1'-Diethyl-2,2'-trimethinequinocy-
anine Chloride *see* 8229
O,*O*-Diethyl *S*-(2-Trimethylammoni-
umethyl)phosphorothioate Iodide *see*
3549
5,5-Diethyl-2,4,6-trioxo-1-phenylhexa-
hydropyrimidine *see* 7368
9-(3,3-Diethylureido)-4,6,6a,7,8,9-
hexahydro-7-methylindolo[4,3-*f*,*g*]-
quinoline *see* 5541
N,*N*-Diethylvanillamide *see* 3765
Diethyl Xanthogenate *see* 3449
Diethylzinc, 3179
Dietrol [Elliott-Marion] *see* 7365
Difacil Hydrochloride *see* 160
2-Difarnesyl-3-methyl-1,4-naphthoqui-
none *see* 5876
Difaterol [Andreu] *see* 1240
Difemerine, 3180
Difenamizole, 3181
Difenax [Zambeletti] *see* 3184
Difenidolin [Taiyo] *see* 3369
Difenoconazole, 3182
Difenoxilic Acid *see* 3183

Difenoxin, 3183
Difenoxylic Acid *see* 3183
Difenpiramide, 3184
Difenzoquat, 3185
Diferuloylmethane *see* 2744
Difflam [Carnegie] *see* 1157
Diffu-K [Delagrange] *see* 7783
Diffumal [Malesci] *see* 9421
Diffusin [Ortho] *see* 4794
Diffusing Factor *see* 4794
Difhydan [Leo Pharm.] *see* 7475
Diflorasone, 3186
Difloxacin, 3187
Diflubenzuron, 3188
Diflucan [Pfizer] *see* 4158
Diflucortolone, 3189
Difludol [Edmond] *see* 3190
Diflunisal, 3190
N,*N*'-Di-9*H*-fluoren-9-yl-*N*,*N*,*N*',*N*'-
tetramethyl-1,6-hexanediaminium
Dibromide *see* 4721
2,4-Difluoroaniline, 3191
2,4-Difluorobenzenamine *see* 3191
p-Difluorobenzene, 3192
3-(2,2-Difluorobenzodioxol-4-yl)-4-
cyanopyrrole *see* 4165
4-(2,2-Difluoro-1,3-benzodioxol-4-yl)-
1*H*-pyrrole-3-carbonitrile *see* 4165
4,4'-Difluoro-1,1'-biphenyl *see* 3193
2,4-Difluoro-α,α-bis(1*H*-1,2,4-triazol-
1-ylmethyl)benzyl Alcohol *see* 4158
1,1-Difluoro-2,2-dichloroethyl Methyl
Ether *see* 6072
Difluorodichloromethane *see* 3114
erythro-3,3'-Difluoro-4,4'-dihydroxy-
α-ethyl-α'-methyldibenzyl *see* 1259
(6α,11β,16α)-6,9-Difluoro-11,21-
dihydroxy-16,17-[(1-methylethylid-
ene)bis(oxy)]-pregna-1,4-diene-3,20-
dione *see* 4185
(6α, 11β, 16α)-6,9-Difluoro-11,21-
dihydroxy-16-methylpregna-1,4-
diene-3,20-dione *see* 3189
Difluorodioxochromium *see* 2307
4,4'-Difluorodiphenyl, 3193
Difluorodiphenyldichloroethane *see*
2999
Difluorodiphenyltrichloroethane *see*
3000
6,8-Difluoro-1-(2-fluoroethyl)-1,4-di-
hydro-7-(4-methyl-1-piperazinyl)-4-
oxo-3-quinolinecarboxylic Acid *see*
4137
2',4'-Difluoro-4-hydroxy-[1,1'-biphen-
yl]-3-carboxylic Acid *see* 3190
2',4'-Difluoro-4-hydroxy-3-biphenyl-
carboxylic Acid *see* 3190
2',4'-Difluoro-4-hydroxy-[1',1-diphen-
yl]-3-carboxylic Acid *see* 3190
(6α,11β,16α,17α)-6,9-Difluoro-11-
hydroxy-16-methyl-3-oxo-17-(1-
oxopropoxy)androsta-1,4-diene-17-
carbothioic Acid *S*-(Fluoromethyl)
Ester *see* 4244
6α,9α-Difluoro-16α-hydroxypred-
nisolone 16,17-Acetonide *see* 4185
6α,9α-Difluoro-16α,17α-isopropyl-
idenedioxy-1,4-pregnadiene-3,20-
dione *see* 4185
5-(Difluoromethoxy)-2-[[(3,4-dimeth-
oxy-2-pyridinyl)methyl]sulfinyl]-
1*H*-benzimidazole *see* 7146
4-(Difluoromethoxy)-α-(1-methyleth-
yl)benzeneacetic Acid Cyano(3-
phenoxyphenyl)methyl Ester *see*
4160
2-(Difluoromethoxy)-1,1,1,2-tetra-
fluoroethane *see* 2965
6α,9α-Difluoro-16α-methyl-1-dehy-
drocorticosterone *see* 3189

2-(Difluoromethyl)-4-(2-methylprop-yl)-6-(trifluoromethyl)-3,5-pyridine-dicarbothioic Acid *S,S*-Dimethyl Ester *see* 3439
2-(Difluoromethyl)-DL-ornithine *see* 3564
α-Difluoromethylornithine *see* 3564
6α,9α-Difluoro-16α-methylpredniso-lone *see* 4173
6α,9α-Difluoro-16β-methylpredni-solone *see* 3186
6α,9α-Difluoro-16α-methyl-1,4-pregnadiene-11β,21-diol-3,20-dione *see* 3189
6α,9α-Difluoro-16β-methyl-Δ^{1,4}-pregnadiene-11β,17α,21-triol-3,20-dione *see* 3186
(±)-2-Difluoromethyl 1,2,2,2-Tetra-fluoroethyl Ether *see* 2965
7-β-Difluoromethylthioacetamido-7α-methoxy-3-[[1-(2-hydroxyethyl)-1*H*-tetrazol-5-yl]thiomethyl]-1-oxa-3-cephem-4-carboxylic Acid *see* 4141
(6*R-cis*)-7-[[[(Difluoromethyl)thio]-acetyl]amino]-3-[[[1-(2-hydroxyeth-yl)-1*H*-tetrazol-5-yl]thio]methyl]-7-methoxy-8-oxo-5-oxa-1-azabicyclo-[4.2.0]oct-2-ene-2-carboxylic Acid *see* 4141
2-(2,4-Difluorophenyl)-1,3-bis(1*H*-1,2,-4-triazol-1-yl)propan-2-ol *see* 4158
1-[4,4-Di(4-fluorophenyl)butyl]-4-[(2,-6-dimethylanilinocarbonyl)methyl]-piperazine *see* 5507
1-[4,4-Di-(4-fluorophenyl)butyl]-4-(2-oxo-1-benzimidazolinyl)piperidine *see* 7589
1-(2,4-Difluorophenyl)-6-fluoro-1,4-dihydro-7-(3-methyl-1-piperazinyl)-4-oxo-3-quinolinecarboxylic Acid *see* 9284
N-(2,6-Difluorophenyl)-5-methyl[1,2,-4]triazolo-[1,5-*a*]pyrimidine-2-sulfonamide *see* 4177
5-(2,4-Difluorophenyl)salicylic Acid *see* 3190
4-[4-[4-[4-[[2-(2,4-Difluorophenyl)-2-(1*H*-1,2,4-triazol-1-ylmethyl)-1,3-dioxolan-4-yl]methoxy]phenyl]-1-piperazinyl]phenyl]-2,4-dihydro-2-(1-methylpropyl)-3*H*-1,2,4-triazol-3-one *see* 8510
α-(2,4-Difluorophenyl)-α-(1*H*-1,2,4-triazol-1-ylmethyl)-1*H*-1,2,4-tria-zole-1-ethanol *see* 4158
6α,9α-Difluoroprednisolone-21-acetate-17-butyrate *see* 3194
6α,9α-Difluoro-11β,16α,17,21-tetra-hydroxypregna-1,4-diene-3,20-dione Cyclic 16,17-Acetal with Acetone *see* 4185
6α,9-Difluoro-11β,16α,17,21-tetrahy-droxypregna-1,4-diene-3,20-dione, Cyclic 16,17-Acetal with Acetone, 21-Acetate *see* 4186
(*RS*)-2,4'-Difluoro-α-(1*H*-1,2,4-triazol-1-ylmethyl)benzhydryl Alcohol *see* 4246
(6α,11β,16α)-6,9-Difluoro-11,17,21-trihydroxy-16-methylpregna-1,4-diene-3,20-dione *see* 4173
(6α,11β,16β)-6,9-Difluoro-11,17,21-trihydroxy-16-methylpregna-1,4-diene-3,20-dione *see* 3186
6α,9-Difluoro-11β,17,21-trihydroxy-pregna-1,4-diene-3,20-dione 21-Acetate 17-Butyrate *see* 3194
Difluprednate, 3194
Diflupyl *see* 5192
Diflurex [Anphar-Rolland] *see* 9570

Difluron *see* 3188
Difolatan *see* 1814
Diformyl *see* 4519
2,3-Diformyl-6-methoxy-5-methylben-zoic Acid *see* 4442
1,3-Diformylpropane *see* 4480
Difosfocin [Magis] *see* 2380
Difosfonal [Benedetti] *see* 2432
Difulal [Yamanouchi] *see* 3186
Di-2-furanylethanedione Dioxime *see* 4328
N,N'-Difurfurylidene-2-furanmethane-diamine *see* 4832
Digacin [Beiersdorf] *see* 3210
Digallic Acid, 3195
"Digallic Acid" *see* 9221
m-Digallic Acid *see* 3195
Digalogenin, 3196
Digenic Acid *see* 5289
Digenin *see* 5289
Digerent [Polifarma] *see* 9829
Digicor [Hennig] *see* 3206
Digifortis [Parke, Davis] *see* 3201
Digilanide A *see* 5368
Digilanide B *see* 5368
Digilanide C *see* 5368
Digilanide D *see* 5368
Digilong [Boehringer, Mann.] *see* 3206
Digimed [Trommsdorf] *see* 3206
Digimerck [E. Merck] *see* 3206
Digin *see* 4438
Diginatigenin, 3197
Diginatin, 3198
Diginigenin *see* 3199
Diginin, 3199
Diginorgin *see* 3200
Diginose *see* 3199
3β-(Diginosyloxy)-12α,20α-epoxy-14β,17α-pregn-5-ene-11,15-dione *see* 3199
Digipural [Schaper & Brümmer] *see* 3206
Digisidin [Winthrop] *see* 3206
Digitalin, 3200
Digitalin, Crystalline *see* 3206
Digitaline Nativelle [Nativelle; Savage] *see* 3206
Digitalinum True *see* 3200
Digitalinum Verum *see* 3200
Digitalis, 3201
Digitalose, 3202
Digitin *see* 3204
Digitoflavone *see* 5641
Digitogenin, 3203
Digitonin, 3204
Digitora [Upjohn] *see* 3201
Digitoxigenin, 3205
Digitoxin, 3206
Digitoxose, 3207
Diglycine *see* 4952
Diglycolic Acid Bis(2,4,6-triiodo-3-carboxanilide) *see* 5067
Diglycol Laurate *see* 3169
3,3'-(Diglycoloyldiimino)bis(2,4,6-triiodobenzoic Acid) *see* 5067
Diglycolyldiamide *see* 7619
Diglyme, 3208
Dignonitrat [Dignos] *see* 5245
Dignover [Dignos] *see* 10083
Digoxigenin, 3209
Digoxin, 3210
Diguanide *see* 1261
2,4-Diguanidino-1-dodecyloxybenzene *see* 5398
2,4-Diguanidino-1-lauryloxybenzene *see* 5398
2,4-Diguanidinophenyl Dodecyl Ether *see* 5398
2,4-Diguanidinophenyl Lauryl Ether *see* 5398

1,3-Diguanido-2,4,5,6-cyclohexane-tetrol *see* 8977
Diguanyl *see* 2138
p,p'-Diguanyldiazoaminobenzene Di-aceturate *see* 3316
Dihexiverine *see* 3211
Dihexyl Sodium Sulfosuccinate *see* 1295
Dihexyverine, 3211
Dihycon [CMC] *see* 7475
Di-Hydan [Carrion] *see* 7475
Dihydergot [Sandoz] *see* 3217
1,4-Dihydracryloylpiperazine Dimeth-anesulfonate *see* 7635
Dihydral [Duphar] *see* 3223
Dihydral [SCS] *see* 3367
Dihydralazine, 3212
Dihydralazine Mesylate *see* 3212
Dihydrazine Sulfate, 3213
1,4-Dihydrazinophthalazine *see* 3212
Dihydrex [Astra] *see* 1155
o-Dihydric Phenolase *see* 9969
Dihydrin, 3214
Dihydrite *see* 2716
22,23-Dihydro C-076B₁ *see* 5264
22,23-Dihydro C-076B$_{1a}$ *see* 5264
22,23-Dihydroabamectin *see* 5264
1,2-Dihydroacenaphthylene *see* 28
N,N'-Dihydro-1,2,1',2'-anthraquinon-azine *see* 4968
6,15-Dihydro-5,9,14,18-anthrazine-tetrone *see* 4968
Dihydroarecoline *see* 6172
Dihydroartemisinin *see* 856
Dihydroartemisinin Ethyl Ether *see* 853
Dihydroartemisinin Methyl Ether *see* 854
22,23-Dihydroavermectin B₁ *see* 5264
22,23-Dihydroavermectin B$_{1a}$ *see* 5264
22,23-Dihydroavermectin B$_{1b}$ *see* 5264
Dihydrobaikiaine *see* 7611
1,2-Dihydrobenz[*j*]aceanthrylene *see* 2252
9,10-Dihydro-9,10-*o*-benzenoanthra-cene *see* 9879
1,3-Dihydro-2*H*-benzimidazole-2-thi-one *see* 1112
cis-(+)-7,11b-Dihydrobenz[*b*]indeno-[1,2-*d*]pyran-3,4,6a,9,10(6*H*)-pentol *see* 4670
(6a*S-cis*)-7,11b-Dihydrobenz[*b*]indeno-[1,2-*d*]pyran-3,6a,9,10(6*H*)-tetrol *see* 1392
4-(9,10-Dihydro-4*H*-benzo[4,5]cyclo-hepta[1,2-*b*]thien-4-ylidene)-1-meth-ylpiperidine *see* 7671
[(2,3-Dihydro-1,4-benzodioxin-2-yl)-methyl]guanidine *see* 4598
1-[(2,3-Dihydro-1,4-benzodioxin-2-yl)-methyl]piperidine *see* 7631
1-(2,3-Dihydro-1,4-benzodioxin-5-yl)-piperazine *see* 3592
2,3-Dihydrobenzofuran *see* 2627
(±)-1-[(3,4-Dihydro-2*H*-1-benzothio-pyran-8-yl)oxy]-3-[(1,1-dimethyleth-yl)amino]-2-propanol *see* 9318
6,7-Dihydrobis[1,3]benzodioxolo[5,6-*a*:4',5'-*g*]quinolizinium *see* 2591
1,3-Dihydro-1,3-bis(hydroxymethyl)-2*H*-benzimidazole-2-thione *see* 9446
trans-Dihydro-3,4-bis[(3-hydroxy-phenyl)methyl]-2(3*H*)-furanone *see* 3633
Dihydrobis(2-methoxyethanolato-*O,-O'*)aluminate(1—) Sodium *see* 10166
3,4-Dihydrocarbostyril *see* 4820
2,3-Dihydro-5-carboxanilido-6-meth-yl-1,4-oxathiin *see* 1874
1,3-Dihydro-3-(4-chlorophenyl)-7-

methyl)-3-pyrrolidinyl Ester *see*
1031
(±)-10,11-Dihydro-α,8-dimethyl-11-
oxodibenz[*b,f*]oxepin-2-acetic Acid
see 1199
3,7-Dihydro-3,7-dimethyl-1-(5-oxo-
hexyl)-1*H*-purine-2,6-dione *see* 7278
N-(2,3-Dihydro-1,5-dimethyl-3-oxo-2-
phenyl-1*H*-pyrazol-4-yl)-2-(dimethylamino)propanamide *see* 492
[(2,3-Dihydro-1,5-dimethyl-3-oxo-2-
phenyl-1*H*-pyrazol-4-yl)methylami-
no]methanesulfonic Acid Sodium
Salt Monohydrate *see* 3414
N-(2,3-Dihydro-1,5-dimethyl-3-oxo-2-
phenyl-1*H*-pyrazol-4-yl)-3-pyridine-
carboxamide *see* 6619
3,4-Dihydro-2,2-dimethyl-4-oxo-2*H*-
pyran-6-carboxylic Acid Butyl
Ester *see* 1564
1,2-Dihydro-1,5-dimethyl-2-phenyl-
3*H*-pyrazol-3-one *see* 757
1,2-Dihydro-1,5-dimethyl-2-phenyl-
3*H*-pyrazol-3-one Compd with 2,2,2-
Trichloro-1,1-ethanediol(1:2) *see*
3097
3,7-Dihydro-1,3-dimethyl-1*H*-purine-
2,6-dione *see* 9421
3,7-Dihydro-3,7-dimethyl-1*H*-purine-
2,6-dione *see* 9418
3,7-Dihydro-1,3-dimethyl-1*H*-purine-
2,6-dione Compd with 2-Amino-2-
methyl-1-propanol (1:1) *see* 404
3,7-Dihydro-1,3-dimethyl-1*H*-purine-
2,6-dione Compd with 1,2-Ethane-
diamine (2:1) *see* 485
2,6-Dihydro-2,2-dimethyl-5*H*-pyrano-
[3,2-*c*]quinolin-5-one *see* 4139
3,7-Dihydro-1,3-dimethyl-7-[2-[(3-
pyridinylmethyl)amino]ethyl]-1*H*-
purine-2,6-dione *see* 7584
3,4-Dihydro-2,2-dimethyl-2*H*-pyrrole
1-Oxide *see* 3454
2-[7-[1,3-Dihydro-1,1-dimethyl-3-(4-
sulfobutyl)-2*H*-benz[*e*]indol-2-ylid
ene]-1,3,5-heptatrienyl]-1,1-dimeth-
yl-3-(4-sulfobutyl)-1*H*-benz[*e*]indoli-
um Inner Salt, Sodium Salt *see* 4992
3,7-Dihydro-1,3-dimethyl-7-[2-(sulfooxy)ethyl]-1*H*-purine-2,6-dione
Compd with 5-Hydroxy-6-methyl-
3,4-pyridinedimethanol (1:1) *see*
8159
[2*R*-[2*R**(4*R**,8*R**)]]-3,4-Dihydro-2,8-
dimethyl-2-(4,8,12-trimethyltridec-
yl)-2*H*-1-benzopyran-6-ol *see* 9634
3,4-Dihydro-2,7-dimethyl-2-(4,8,12-
trimethyltridecyl)-2*H*-1-benzopyran-
6-ol *see* 9638
2,2-[(9,10-Dihydro-9,10-dioxo-1,4-
anthracenediyl)diimino]bis(5-meth-
ylbenzenesulfonic Acid) Disodium
Salt *see* 248
2-(2,3-Dihydro-1,3-dioxo-1*H*-indene-2-
yl)-6-quinolinesulfonic Acid *see*
8256
2-(2,3-Dihydro-1,3-dioxo-1*H*-indene-2-
yl)-8-quinolinesulfonic Acid *see*
8256
1,3-Dihydro-1,3-dioxo-5-isobenzofur-
ancarboxylic Acid *see* 9833
3,4-Dihydro-3,4-dioxo-1-naphthalene-
sulfonic Acid Sodium Salt *see* 8791
5,10-Dihydro-5,10-dioxonaphtho[2,3-
b]-1,4-dithiin-2,3-dicarbonitrile *see*
3433
6,7-Dihydrodipyrido[1,2-*a*:2′,1′-*c*]pyr-
azinediium Dibromi *see* 3415
1,2-Dihydro-α-elaterin. *see* 2681
Dihydroequilin, 3216

22,23-Dihydroergocalciferol *see* 10158
Dihydroergocristine Methanesulfonate *see* 3692
22,23-Dihydroergosterol *see* 3701
Dihydroergotamine, 3217
Dihydroergotamine Mesylate *see* 3217
Dihydroergotoxine Monomethanesulfonate (Salt) *see* 3692
Dihydro-β-erythroidine, 3218
1,3-Dihydro-1-ethoxy-6-methylfuro-
[3,4-*c*]pyridin-7-ol Hydrochloride
see 8162
1,2-Dihydro-6-ethoxy-2,2,4-trimethyl-
quinoline *see* 3800
5,8-Dihydro-8-ethyl-5-oxo-2-pyrroli-
dinopyrido[2,3-*d*]pyrimidine-6-car-
boxylic Acid *see* 7660
Dihydrofisetin *see* 4341
Dihydroflumethiazide *see* 4830
Dihydrofollicular Hormone *see* 3746
Dihydrofolliculin *see* 3746
α-Dihydrofucosterol *see* 8697
Dihydro-2,5-furandione *see* 9039
Dihydro-2(3*H*)-furanone *see* 1632
1,2-Dihydrogalanthamine *see* 5655
3,4-Dihydrogalanthamine *see* 5655
(Dihydrogen Borato)phenylmercury
see 7456
5-(Dihydrogen Phosphate)-D-ribose *see*
8372
[Dihydrogen 3,7,12,17-Tetramethyl-
8,13-divinyl-2,18-porphinedipropio-
nato(2—)]hydroxyiron *see* 4668
[Dihydrogen 3,7,12,17-Tetramethyl-
8,13-divinyl-2,18-porphinedipropio-
nato(2—)]iron *see* 4672
Dihydrogen Thiocarbonate *see* 9888
Dihydrogeraniol *see* 2391
3,4-Dihydroharmine *see* 4644
Dihydroheroin *see* 3011
(*S*)-4,5-Dihydro-2-(6-hydroxy-2-benzo-
thiazolyl)-4-thiazolecarboxylic Acid
see 4128
1,2-Dihydro-1-hydroxy-ψ,ψ-carotene
see 8358
Dihydrohydroxycodeinone *see* 7093
(7a*R*-*cis*)-13,13a-Dihydro-7a-hydroxy-
9,10-dimethoxy-3,3-dimethyl-3*H*-
bis[1]benzopyrano[3,4-*b*:6′,5′-*e*]py-
ran-7(7a*H*)-one *see* 9295
6-[2-(4,9-Dihydro-8-hydroxy-5,7-
dimethoxy-4,9-dioxonaphtho[2,3-*b*]-
furan-2-yl)ethyl]-7,8-dihydroxy-1-
oxo-1*H*-2-benzopyran-3-carboxylic
Acid Methyl Ester *see* 2545
Dihydro-3-hydroxy-4*H*-dimethyl-
2(3*H*)-furanone *see* 7145
1,2-Dihydro-4-hydroxy-*N*,1-dimethyl-
2-oxo-*N*-phenyl-3-quinolinecarbox-
amide *see* 8418
(3*S*-*trans*)-3,4-Dihydro-4-hydroxy-2,2-
dimethyl-4-(2-oxo-1-pyrrolidinyl)-
2*H*-1-benzopyran-6-carbonitrile *see*
5487
1,2-Dihydro-4-hydroxy-*N*,1-dimethyl-
2-oxo-3-quinolinecarboxanilide *see*
8418
(3α,14β,16α)-14,15-Dihydro-14-hydr-
oxyeburnamenine-14-carboxylic
Acid Methyl Ester *see* 10120
3,7-Dihydro-7-(2-hydroxyethyl)-1,3-
dimethyl-1*H*-purine-2,6-dione *see*
3924
3,7-Dihydro-8-[(2-hydroxyethyl)meth-
ylamino]-1,3,7-trimethyl-1*H*-purine-
2,6-dione *see* 1671
2,3-Dihydro-5-hydroxy-2-(4-hydr-
oxyphenyl)-7-methoxy-4*H*-1-benzo-
pyran-4-one *see* 8470
2,3-Dihydro-3-hydroxy-2-imino-6-(1-

piperidinyl)-4-pyrimidinamine *see*
6290
(*E*)-3-[1,2-Dihydro-5-(5-hydroxy-1*H*-
indol-3-yl)-2-oxo-3*H*-pyrrol-3-ylid-
ene]-1,3-dihydro-2*H*-indol-2-one *see*
10135
3,4-Dihydro-8-(2-hydroxy-3-isopropyl-
amino)propoxy-3-nitroxy-2*H*-1-
benzopyran *see* 6655
1,2-Dihydro-1-hydroxylycopene *see*
8358
(*E*)-6-(1,3-Dihydro-4-hydroxy-6-meth-
oxy-7-methyl-3-oxo-5-isobenzo-
furanyl)-4-methyl-4-hexenoic Acid
see 6408
[2*R*-[2α,3β,6(2*R**,3*R**)]]-2-[2,3-Dihy-
dro-3-(4-hydroxy-3-methoxyphen-
yl)-2-(hydroxymethyl)-1,4-benzo-
dioxin-6-yl]-2,3-dihydro-3,5,7-
trihydroxy-4*H*-1-benzopyran-4-one
see 8680
(*R*)-*N*-[2-[4-[2,3-Dihydro-2-(hydroxymethyl)-1,4-benzodioxin-5-yl]-1-
piperazinyl]ethyl]-4-fluorobenza-
mide *see* 4138
[6a*R*-[4(*E*),6aα,12aα]]-6a,12a-Dihy-
dro-4-(4-hydroxy-3-methyl-2-buten-
yl)-6*H*-[1,3]dioxolo[5,6]benzofuro-
[3,2-*c*][1]benzopyran *see* 1636
7,8-Dihydro-7α-[1(*R*)-hydroxy-1-
methylbutyl]-*O*⁶-methyl-6,14-*endo*-
ethenomorphine *see* 3932
6a,12a-Dihydro-3-hydroxy-β-methyl-
6*H*-[1,3]dioxolo[5,6]benzofuro[3,2-
c][1]benzopyran-2-butanol *see* 1636
4,5α-Dihydro-2-hydroxymethylene-
17α-methyltestosterone *see* 7101
3,4-Dihydro-8-[2-hydroxy-3-[(1-
methylethyl)amino]propoxy]-2*H*-1-
benzopyran-3-ol 3-Nitrate *see* 6655
2,3-Dihydro-3-hydroxy-1-methyl-1*H*-
indole-5,6-dione *see* 171
2,3-Dihydro-3α-hydroxy-8-methylnor-
tropidine *see* 9912
(*R*)-*N*-[(3,4-Dihydro-8-hydroxy-3-
methyl-1-oxo-1*H*-2-benzopyran-7-
yl)carbonyl]-L-phenylalanine *see*
6836
9,10-Dihydro-12′-hydroxy-2′-methyl-
5′-(phenylmethyl)ergotaman-3′,6′,-
18-trione *see* 3217
[2*S*-[2α,3β,7(*E*)]]-2,3-Dihydro-3-hydr-
oxy-2-methyl-7-(1-propenyl)-4*H*,5*H*-
pyrano[4,3-*b*]pyran-4,5-dione *see*
8276
Dihydro-14-hydroxy-4-*O*-methyl-6β-
thebainol *see* 3511
Dihydrohydroxymorphinone *see* 7103
Dihydro-14-hydroxymorphinone *see*
7103
1,2-Dihydro-6-hydroxy-2-oxo-4-pyri-
dinecarboxylic Acid *see* 2386
3,4-Dihydro-3-(4-hydroxyphenyl)-2*H*-
1-benzopyran-7-ol *see* 3677
[3*S*-[3α(*S**),4β]]-Dihydro-3-(hydroxyphenylmethyl)-4-[(1-methyl-1*H*-
imidazol-5-yl)methyl]-2(3*H*)-
furanone *see* 5216
3,7-Dihydro-1-(2-hydroxypropyl)-3,7-
dimethyl-1*H*-purine-2,6-dione *see*
8072
3,7-Dihydro-7-(2-hydroxypropyl)-1,3-
dimethyl-1*H*-purine-2,6-dione *see*
8092
(1α)-1,2-Dihydro-12-hydroxysenecio-
nan-11,16-dione *see* 7691
5-[[4-[(3,4-Dihydro-6-hydroxy-2,5,7,8-
tetramethyl-2*H*-1-benzopyran-2-yl)-
methoxy]phenyl]methyl]-2,4-thiazo-
lidinedione *see* 9898

Name Index

(3α,5β,7α)-3,7-Dihydroxycholan-24-oic Acid *see* 2096
(3α,5β,7β)-3,7-Dihydroxycholan-24-oic Acid *see* 10026
(3α,5β,12α)-3,12-Dihydroxy-5-cholan-24-oic Acid *see* 2946
1α,24(*R*)-Dihydroxycholecalciferol *see* 9197
1α,25-Dihydroxycholecalciferol *see* 1681
21,25-Dihydroxycholecalciferol *see* 1681
3,4-Dihydroxycinnamic Acid *see* 1673
3,4-Dihydroxycinnamic Acid 1-Carboxy-4,5-dihydroxy-1,3-cyclohexylene Ester *see* 2835
3-(3,4-Dihydroxycinnamoyl)quinic Acid *see* 2194
6,7-Dihydroxycoumarin *see* 3738
7,8-Dihydroxycoumarin *see* 2881
6,7-Dihydroxycoumarin 6-Glucoside *see* 3739
6,7-Dihydroxycoumarin-7-glucoside *see* 2322
7,8-Dihydroxycoumarin 7-β-D-Glucoside *see* 2882
7′,6-Dihydroxycoumarino(3′,4′,3,2)-coumarone *see* 2632
1,2-Dihydroxy-1,2-cyclobutanediacrylic Acid Di-γ-lactone *see* 681
5,6-Dihydroxy-5-cyclohexene-1,2,3,4-tetrone Disodium Salt *see* 8818
2,3-Dihydroxy-2-cyclopenten-1-one *see* 8297
2,2′-Dihydroxy-5,5′-diallylbiphenyl *see* 5738
3,3′-Dihydroxydibutyl Ether *see* 7094
1,3-Dihydroxy-2,2-dichlorocyclopentane *see* 1761
2,2′-Dihydroxy-5,5′-dichlorodiphenylmethane *see* 3120
2,2′-Dihydroxy-5,5′-dichlorodiphenyl Sulfide *see* 4046
2,2′-Dihydroxydiethylamine *see* 3156
4,4′-Dihydroxy-α,β-diethyldiphenylethane *see* 4738
4,4′-Dihydroxy-α,β-diethylstilbene *see* 3177
Dihydroxy[(4,5-dihydro-2-hydroxy-5-oxo-1*H*-imidazol-4-yl)ureato]-aluminum *see* 224
3,3′-Dihydroxy-3,3′-dihydroisoindigo *see* 5115
6,7-Dihydroxy-1-(3,4-dihydroxybenzyl)-1,2,3,4-tetrahydroisoquinoline *see* 9355
(3′,6′-Dihydroxy-2′,7′-diiodospiro[3*H*-2,1-benzoxanthiole-3,9′-[9*H*]xanthen]-4′-yl)hydroxymercury *S,S*-Dioxide Monosodium Salt *see* 5912
3′,6′-Dihydroxy-4′,5′-diiodospiro[isobenzofuran-1(3*H*),9′-[9*H*]xanthen]-3-one *see* 3233
3,3′-Dihydroxy-4,4′-diketo-β-carotene *see* 890
1,11-Dihydroxy-2,10-dimethoxyaporphine *see* 2616
2,6-Dihydroxy-3,5-dimethoxyaporphine *see* 1355
2,2′-Dihydroxy-4,4′-dimethoxybenzophenone *see* 1130
5,7-Dihydroxy-4′,6-dimethoxyflavone *see* 7195
6β,14-Dihydroxy-3,4-dimethoxy-N-methylmorphinan *see* 3511
7′,12′-Dihydroxy-6,6′-dimethoxy-2,2′,2′-trimethyltubocuraranium Chloride Hydrochloride *see* 9939
3,3′-Dihydroxy-5,5′-dimethyl-2,2′-bi-*p*-benzoquinone *see* 7379

2,2′-Dihydroxy-4,4′-dimethyl[bi-1,4-cyclohexadien-1-yl]-3,3′,6,6′-tetrone *see* 7379
4-(2,4-Dihydroxy-3,3-dimethylbutyramido)butyric Acid *see* 4781
2,4-Dihydroxy-3,3-dimethylbutyric Acid γ-Lactone *see* 7145
D(+)-N-(2,4-Dihydroxy-3,3-dimethylbutyryl)-β-alanine *see* 7147
α,γ-Dihydroxy-β,β-dimethylbutyryl-β-alanyl-β-aminoethanethiol *see* 7143
3,5-Dihydroxy-4,4-dimethyl-2,5-cyclohexadien-1-one *see* 4122
(11α,13*E*,15*S*,17*S*)-11,15-Dihydroxy-17,20-dimethyl-6,9-dioxoprost-13-en-1-oic Acid Methyl Ester *see* 7003
Dihydroxydimethyldiphenylmethanedisulfonic Acid Polymer *see* 6531
1,3-Dihydroxydimethyl Ketone *see* 3225
2-[[5,7-Dihydroxy-2,2-dimethyl-8-(2-methyl-1-oxopropyl)-2*H*-1-benzopyran-6-yl]methyl]-3,5-dihydroxy-4,4-dimethyl-6-(2-methyl-1-oxopropyl)-2,5-cyclohexadien-1-one *see* 9979
3,5-Dihydroxy-4,4-dimethyl-2-(2-methyl-1-oxopropyl)-6-[[2,4,6-trihydroxy-3-(3-methyl-2-butenyl)-5-(2-methyl-1-oxopropyl)phenyl]-methyl]-2,5-cyclohexadien-1-one *see* 9979
trans-1,4-Dihydroxy-2-(3,7-dimethyl-2,6-octadienyl)benzene *see* 4413
(*R*)-N-(2,4-Dihydroxy-3,3-dimethyl-1-oxobutyl)-β-alanine *see* 7147
(*R*)-4-[(2,4-Dihydroxy-3,3-dimethyl-1-oxobutyl)amino]butanoic Acid *see* 4781
3,5-Dihydroxy-4,4-dimethyl-2-(1-oxobutyl)-6-[[2,4,6-trihydroxy-3-methyl-5-(1-oxobutyl)phenyl]methyl]-2,5-cyclohexadien-1-one *see* 4130
(2*E*,11α,13*E*,15*S*,17*S*)-11,15-Dihydroxy-17,20-dimethyl-9-oxoprosta-2,13-dien-1-oic Acid *see* 5514
(2*E*,11α,13*E*,15*R*)-11,15-Dihydroxy-16,16-dimethyl-9-oxoprosta-2,13-dien-1-oic Acid Methyl Ester *see* 4393
5,7-Dihydroxy-2,2-dimethyl-6-(2,4,6-trihydroxy-3-methyl-5-acetylbenzyl)-8-cinnamoyl-1,2-chromene *see* 8429
2,2′-Dihydroxy-6,6′-dinaphthyl Disulfide *see* 2897
2,2′-Dihydroxy-1,1′-dinaphthylmethane-3,3′-dicarboxylic Acid *see* 7136
2,2′-Dihydroxy-5,5′-dinitrobenzil *see* 3338
2,5-Dihydroxy-3,6-dinitro-*p*-benzoquinone *see* 6665
2,5-Dihydroxy-3,6-dinitro-2,5-cyclohexadiene-1,4-dione *see* 6665
2,5-Dihydroxy-3,6-dinitroquinone *see* 6665
11β,17α-Dihydroxy-3,20-dioxo-21-(4-methyl-1-piperazinyl)pregna-1,4-diene *see* 5802
(*Z*)-6β,16β-Dihydroxy-3,7-dioxo-29-nor-8α,9β,13α,14β-dammara-1,17-(20),24-trien-21-oic Acid Diacetate *see* 4666
3β,21-Dihydroxy-11,20-dioxo-5α-pregnane *see* 271
(11β)-11,21-Dihydroxy-3,20-dioxopregn-4-en-18-al *see* 226
4,4′-Dihydroxy-γ,δ-diphenyl-β,δ-hexadiene *see* 3153
4,4′-Dihydroxy-γ,δ-diphenylhexane *see* 4738

4,4′-Dihydroxydiphenylmethane-3,3′-dicarboxylic Acid *see* 6142
2,2′-Dihydroxy-N-(4,5-diphenyloxazol-2-yl)diethylamine *see* 3432
threo-2,3-Dihydroxy-1,4-dithiolbutane *see* 3441
Dihydroxydurene *see* 3521
[*S*-[*R**,*S**-(*E,Z,E,Z*)]]-5,12-Dihydroxy-6,8,10,14-eicosatetraenoic Acid *see* 5482
3,17-Dihydroxyestratriene *see* 3747
3,16α-Dihydroxyestra-1,3,5(10)-trien-17-one Diacetate *see* 2554
3β,17β-Dihydroxyestr-4-ene *see* 1352
(17β)-4,17-Dihydroxyestr-4-en-3-one *see* 7034
Dihydroxyestrin *see* 3746
1,2-Dihydroxy-1,2-ethanedisulfonic Acid Disodium Salt *see* 4520
N,N-Di(hydroxyethyl)aminoacetic Acid *see* 1248
17-(1,2-Dihydroxyethyl)androstane-3,-17-diol *see* 275
17-(1,2-Dihydroxyethyl)androstane-3,-11,17-triol *see* 274
17-(1,2-Dihydroxyethyl)androsten-3-one-11,17-diol *see* 7912
17-(1,2-Dihydroxyethyl)-Δ⁴-androsten-3-on-17α-ol *see* 7914
α,β-Dihydroxyethylbenzene *see* 9029
1,2-Dihydroxyethylenedicarboxylic Acid *see* 3229
(5*Z*,13*E*)-(8*R*,9*S*,11*R*,12*R*)-9,11-Dihydroxy-15,15-ethylenedioxy-16-phenoxy-17,18,19,20-tetranorprostadien-oic Acid *see* 3916
8-(1,2-Dihydroxyethyl)-4-[4-(D-glucopyranosyloxy)phenyl]-9-hydroxy-1,-7-dioxaspiro[4.4]nonane-2,6-dione *see* 5479
Di(hydroxyethyl)glycine *see* 1248
[5*S*-[5α(*S**),8α(*R**),9β]]-8-(1,2-Dihydroxyethyl)-9-hydroxy-4-(4-hydroxyphenyl)-1,7-dioxaspiro[4.4]-nonane-2,6-dione *see* 5478
Di-(2-hydroxyethyl)nitrosamine *see* 6733
5,7-Dihydroxyflavone *see* 2316
3′,6′-Dihydroxyfluoran *see* 4194
11β,17β-Dihydroxy-9α-fluoro-17α-methyl-4-androsten-3-one *see* 4223
(3β,5α,25*R*)-3,26-Dihydroxyfurost-20-(22)-en-12-one *see* 4652
(1α,7β)-1,7-Dihydroxyginkgolide A *see* 4433
4,6-Dihydroxy-2-(β-D-glucosido)-β-(*p*-hydroxyphenyl)propiophenone *see* 7481
Dihydroxy(glycinato)aluminum *see* 3227
2,2′-Dihydroxy-3,3′,5,5′,6,6′-hexachlorodiphenylmethane *see* 4716
1,6-Dihydroxyhexane *see* 4726
5,7-Dihydroxy-3-(3-hydroxy-4,5-dimethoxyphenyl)-6-methoxy-4*H*-1-benzopyran-4-one *see* 5103
γ,4-Dihydroxy-2-(6-hydroxy-1-heptenyl)-4-cyclopentanecrotonic Acid λ-Lactone *see* 1393
3,5-Dihydroxy-4-(3-hydroxy-4-methoxyhydrocinnamoyl)phenyl-2-*O*-(6-deoxy-α-L-mannopyranosyl)-β-D-glucopyranoside *see* 6540
5,7-Dihydroxy-2-(3-hydroxy-4-methoxyphenyl)-4*H*-1-benzopyran-4-one *see* 3349
5,7-Dihydroxy-3-(3-hydroxy-4-methoxyphenyl)-4*H*-1-benzopyran-4-one *see* 7893
l-3,4-Dihydroxy-1-[1-hydroxy-2-

Name Index

Acid Compound with 2,2'-Iminobis-[ethanol] (1:1) see 5061
3,5-Diiodo-4-oxo-1(4*H*)-pyridineacetic Acid Propyl Ester see 8048
Diiodo-oxyquinoline see 5063
2,6-Diiodophenol-4-sulfonic Acid see 8881
3,5-Diiodo-α-phenylphloretic Acid see 5047
1,3-Diiodo-2-propanol see 5081
3,5-Diiodo-4(1*H*)-pyridinone see 5079
3,5-Diiodo-4-pyridone-*N*-acetic Acid Diethanolamine Salt see 5061
3,5-Diiodo-4-pyridone-*N*-acetic Acid Propyl Ester see 8048
5,7-Diiodo-8-quinolinol see 5063
Diiodoricinstearolic Acid see 7313
3,5-Diiodosalicylic Acid, 3234
3,5-Diiodothyronine, 3235
3,5-Diiodotyrosine, 3236
β-Diiodotyrosine see 1228
Diiodyl see 7313
Diiron Phosphide see 4102
Diisoamylamine, 3237
Di-*sec*-isoamylborine see 1290
Diisoamyl Ether see 5132
4,4'-Di(isoamyloxy)thiocarbanilide see 9593
Diisoamyl Phthalate see 5138
Diisobutyl Ether see 5156
Diisobutylphenoxyethoxyethyl Dimethyl Benzyl Ammonium Chloride see 1103
Diisobutyl Sodium Sulfosuccinate, 3238
Diisobutyl Sulfide see 5166
Diisobutylthiocarbamic Acid *S*-Ethyl Ester see 1581
3,4-Di-*o*-isobutyryl Epinine see 4921
Diisocarb see 1581
2,4-Diisocyanato-1-methylbenzene see 9668
2,4-Diisocyanatotoluene see 9668
4,4'-(2,3-Diisocyano-1,3-butadiene-1,-4-diyl)bis1,2-benzenediol see 10195
4,4'-(2,3-Diisocyano-1,3-butadiene-1,-4-diyl)bisphenol see 10195
Diisopromine, 3239
Diisopropenyl see 3288
(2,6-Diisopropylacetanilido)iminodiacetic Acid see 3422
Diisopropylamine, 3240
Diisopropylamine Dichloroacetate, 3241
Diisopropylamine Dichloroethanoate see 3241
2-[2-(Diisopropylamino)ethyl]-4-methyl-2-(2-pyridyl)pentanamide see 7271
N-[2-(Diisopropylamino)ethyl]-2-oxo-1-pyrrolidineacetamide see 7886
α-[2-(Diisopropylamino)ethyl]-α-phenyl-2-pyridineacetamide see 3424
β-Diisopropylaminoethyl 9-Xanthenecarboxylate Methobromide see 7989
4-(Diisopropylamino)-2-phenyl-2-(2-pyridyl)butyramide see 3424
Diisopropylammonium Dichloroacetate see 3241
5',5''-Diisopropyl-2',2''-dimethylphenolphthalein see 9542
N,*N*-Diisopropyl-3,3-diphenylpropylamine see 3239
Diisopropyl 1,3-Dithiol-2-ylidenemalonate see 5751
Diisopropyl Ether see 5231
Diisopropyl Fluorophosphate see 5192
Diisopropyl Fluorophosphonate see 5192
Diisopropylideneacetone see 7488

2,3:4,5-Di-*O*-isopropylidene-β-D-fructopyranose Sulfamate see 9686
1,2:5,6-Diisopropylidene-D-glucofuranose see 3009
1,2:5,6-Diisopropylidene-D-glucose see 3009
α-2,3:4,6-Di-*O*-isopropylidene-L-*xylo*-hexulofuranosonic Acid see 3243
Di-*O*-isopropylidene-2-keto-L-gulonic Acid see 3243
Diisopropyl *p*-Nitrophenyl phosphate see 3242
Diisopropyl Paraoxon, 3242
2,6-Diisopropylphenol see 8020
[[[(2,6-Diisopropylphenyl)carbamoyl]-methyl]imino]diacetic Acid see 3422
N,*N*'-Diisopropylphosphorodiamidic Fluoride see 6292
Diisopropylphosphorofluoridate see 5192
Diisopropylthiocarbamic Acid *S*-2,3-Dichloroallyl Ester 3014
Diisopropylthiocarbamic Acid *S*-(2,3,-3-Trichloroallyl) Ester see 9726
1,4-Diisothiocyanatobenzene see 1345
Dijozol [Trommsdorff] see 8881
Dikalium Phosphate see 7828
Dikegulac, 3243
Dikegulac-sodium see 3243
2,3-Diketoadipic Acid see 5311
2,3-Diketobutane see 3010
2,3-Diketoindoline see 5116
3,20-Diketo-11β,18-oxido-4-pregnene-18,21-diol see 226
2,5-Diketopiperazine see 7619
2,5-Diketopyrrolidine see 9040
Diketosuccinic Acid see 3231
2,5-Diketotetrahydrofuran see 9039
3,5-Diketotriazolidine see 10004
Dikol see 5047
Dilabar [Vita] see 1817
Dilabron [Sterling-Winthrop] see 5185
Diladel [Delelande] see 3247
Dilafurane see 1113
Dilan [Comm. Solvents] see 1505 and 6691
Dilanacin [Arzneimittelwerk Dresden] see 3210
Dilangil see 5789
Dilangio [Andreu] see 1060
Dilantin [Parke, Davis] see 7475
Dilar [Cassenne] see 7162
Dilasenil [Celtia] see 6580
Dilaster [Roussel-UCLAF] see 2603
Dilatal see 6830
Dilatol [Troponwerke] see 6830
Dilatrate [Block] see 5245
Dilatrend [Boehringer, Mann.] see 1924
Dilatropon [Draco] see 6830
Dilaudid [Knoll] see 4847
Dila-Vasal see 1113
Dilavase see 5259
Dilazep, 3244
Dilcit see 5009
Dilcoran-80 [Gödecke] see 7249
Dilevalol, 3245
Dilevalon [Shionogi] see 3245
Dilexpal [Winthrop] see 5009
Dilituric Acid see 6683
Dill Apiole see 776
Dillar [Sarva-Syntex] see 7162
Dill Fruit Oil see 6899
Dill Seed Oil see 6899
Diloderm [Schering] see 3099
Dilombrin [Pfizer] see 3434
Dilongifolylborane see 5597
Dilor [Savage] see 3529
Dilospan S [Roussel-UCLAF] see 7482

Dilosyn see 6034
Diloxanide, 3246
Diloxanide 2-Furoic Acid Ester see 3246
Dilpral [Medipolar] see 3247
Dilrene [Sanofi Winthrop] see 3247
Diltiazem, 3247
Dilurgen see 5913
Diluted Alcohol see 3806
Dilvasene see 7056
Dilvax [Promeco] see 4936
Dilydrin [Medichemie] see 6830
Dilzem [Gödecke] see 3247
Dilzene [Sigma-Tau] see 3247
Dimacef [Fisons] see 2032
Dimalone see 3290
Dimantine see 3525
Dimapyrin see 495
Dimate [Tutag] see 3252
Dimaten [Promeco] see 9590
Dimaval [Heyl] see 3256
Dimazine see 3296
Dimazole Dihydrochloride see 3033
Dimazon see 3004
Dimebenzcozamide see 2513
Dimeclophenone see 6350
Dimecron [Ciba] see 7493
Dimecrotic Acid, 3248
Dimedione [AB Leo] see 3762
Dimedone see 3291
Dimefline, 3249
Dimefox, 3250
Dimegan [Dexo] see 1467
Dimeglumine Iocarmate see 5029
Dimeglumine Ioglycamide see 5067
Dimelin [Shionogi] see 59
Dimelor [Lilly] see 59
Dimemorfan, 3251
Dimene see 2784
Dimenformon [Organon] see 3746
Dimenhydrinate, 3252
Dimenoxadol, 3253
Dimension [Monsanto] see 3439
Dimepheprimine see 7959
Dimepheptanol, 3254
Dimepropion see 5984
Diméprotane see 8024
Dimeray [Mallinckrodt] see 5029
Dimercaprol, 3255
(*R**,*S**)-2,3-Dimercaptobutanedioic Acid see 9034
(*R**,*R**)-1,4-Dimercapto-2,3-butanediol see 3441
1,2-Dimercapto-4-methylbenzene see 9669
1,3-Dimercaptopropane see 7984
2,3-Dimercapto-1-propanesulfonic Acid, 3256
2,3-Dimercapto-1-propanol 3255
meso-2,3-Dimercaptosuccinic Acid see 9034
2,3-Dimercaptosuccinic Acid Cyclic Ester with Antimonic Acid, Diester with 2,3-Dimercaptosuccinic Acid Hexasodium Salt see 8966
2,3-Dimercaptosuccinic Acid Cyclic Thioantimonate (III) *S*,*S*-Diester with 2,3-Dimercaptosuccinate Hexasodium Salt see 8966
Dimerin see 6216
Dimer-X [Byk-Gulden] see 5029
Dimestrol, 3257
Dimetacrine, 3258
Dimetan [Geigy], 3259
Dimetane [Robins] see 1467
Dimetfol see 2944
Dimethacrine see 3258
Dimethadione, 3260
1,4-Di(methanesulfonyloxy)butane see 1529

methylquinolinium Salt with 4,4'-Methylenebis[3-hydroxy-2-naphthalenecarboxylic Acid] (2:1) *see* 8206

6-(Dimethylamino)-2-[2-(2,5-dimethyl-1-phenyl-3-pyrryl)vinyl]-1-methylquinolinium Salt of 2,2'-Dihydroxy-1,1'-dinaphthylmethane-3,3'-dicarboxylic Acid *see* 8206

2-(Dimethylamino)-5,6-dimethyl-4-pyrimidinyl Dimethylcarbamate *see* 7651

1-Dimethylamino-2,3-dioxamethylenepropane Methiodide *see* 7056

3-Dimethylamino-1,1-diphenylbutane *see* 9843

4-Dimethylamino-2,2-diphenylbutyramide Ethyl Bromide *see* 406

2-(Dimethylamino)-1,2-diphenyldiethyl Ether Methyl Bromide *see* 1244

2-Dimethylamino-4,4-diphenyl-5-heptanol *see* 3254

6-(Dimethylamino)-4,4-diphenyl-3-heptanol *see* 3254

(—)-6-(Dimethylamino)-4,4-diphenyl-3-heptanol Acetate (Ester) *see* 5492

6-(Dimethylamino)-4,4-diphenyl-3-heptanol Acetate (Ester) *see* 6009

6-Dimethylamino-4,4-diphenyl-3-heptanone Hydrochloride *see* 6008

1-Dimethylamino-3,3-diphenyl-4-hexanone *see* 6804

6-(Dimethylamino)-4,4-diphenyl-3-hexanone *see* 6804

α-*l*-4-Dimethylamino-1,2-diphenyl-3-methyl-2-butanol Propionate *see* 5495

6-Dimethylamino-4,4-diphenyl-5-methyl-3-hexanone *see* 5201

(+)-4-Dimethylamino-1,2-diphenyl-3-methyl-2-propionyloxybutane *see* 8024

2-(Dimethylamino)-*N*-(1,3-diphenyl-1*H*-pyrazol-5-yl)propanamide *see* 3181

4-Dimethylamino-2,2-diphenylvaleramide *see* 479

3-Dimethylamino-1,1-di(2'-thienyl)-but-1-ene *see* 3309

2-(Dimethylamino)ethanol *see* 2900

2-Dimethylaminoethanol Dihydrogen Phosphate *see* 2935

β-Dimethylaminoethanol Diphenylmethyl Ether *see* 3367

1-(Dimethylaminoethoxyacetamido)-adamantane *see* 9901

N-[(2-Dimethylaminoethoxy)benzyl]-3,4,5-trimethoxybenzamide *see* 9839

1-(β-Dimethylaminoethoxy)-3-*n*-butylisoquinoline *see* 3267

5-(2-Dimethylaminoethoxy)carvacrol Acetate *see* 6374

2-(2-Dimethylaminoethoxy)-*N*,*N*-diethylpropylamine Dimethiodide *see* 7693

2-Dimethylaminoethoxy)diphenylmethane *see* 7469

α-(2-Dimethylaminoethoxy)diphenylmethane *see* 3367

β-Dimethylaminoethoxyethyl Phenothiazine-10-carboxylate *see* 3272

5-[β-(Dimethylamino)ethoxyimino]-10,11-dihydro-5*H*-dibenzo[*a*,*d*]cycloheptene *see* 6821

4-(2-Dimethylaminoethoxy)-5-isopropyl-2-methylphenyl Acetate *see* 6374

2-[α-(2-Dimethylaminoethoxy)-α-methylbenzyl]pyridine *see* 3497

4-(2-Dimethylaminoethoxy)-2-methyl-5-isopropylphenyl Acetate *see* 6374

4-[2-(Dimethylamino)ethoxy]-2-methyl-5-(1-methylethyl)phenol Acetate (Ester) *see* 6374

[1*R*-(1α,2α,4aα,4bβ,7*E*,8β,8aα,-10aβ)]-7-[2-[2-(Dimethylamino)ethoxy]-2-oxoethylidene]tetradecahydro-2-hydroxy-1,4a,8-trimethyl-9-oxo-1-phenanthrenecarboxylic Acid Methyl Ester *see* 3727

7-[2-[2-(Dimethylamino)ethoxy]-2-oxoethylidene]tetradecahydro-1,4a,-8-trimethyl-9-oxo-1-phenanthrenecarboxylic Acid Methyl Ester *see* 1939

1-*p*-β-Dimethylaminoethoxyphenyl-*trans*-1,2-diphenylbut-1-ene *see* 9216

(*E*)-α-[*p*-[2-(Dimethylamino)ethoxy]phenyl]-α'-ethyl-3-stilbenol *see* 3502

(*E*)-1-[4'-(2-Dimethylaminoethoxy)phenyl]-1-(3-hydroxyphenyl)-2-phenylbut-1-ene *see* 3502

2-Dimethylaminoethoxyphenylmethyl-2-picoline *see* 3497

(*E*)-3-[1-[4-[2-(Dimethylamino)ethoxy]phenyl]-2-phenyl-1-butenyl]phenol *see* 3502

2-[2-(Dimethylamino)ethoxy]-*N*-tricyclo[3.3.1.1³,⁷]*dec*-1-ylacetamide *see* 9901

4-(2-Dimethylaminoethoxy)-*N*-(3,4,5-trimethoxybenzoyl)benzylamine *see* 9839

β-Dimethylaminoethyl Alcohol *see* 2900

α-[1-(Dimethylamino)ethyl]benzenemethanol *see* 6145

β-Dimethylaminoethyl Benzhydryl Ether *see* 3367

β-Dimethylaminoethyl Benzhydryl Ether 1,3-Dimethyl-8-chloroxanthine *see* 3252

erythro-α-[1-(Dimethylamino)ethyl]-benzyl Alcohol *see* 6145

2-[α-(2-Dimethylaminoethyl)benzyl]pyridine *see* 7383

N,*N*-Dimethylaminoethyl α-(Bicyclo-[2.2.1]-5-heptenyl)mandelate Methyl Bromide *see* 3612

β-Dimethylaminoethyl *p*-Bromobenzhydryl Ether *see* 1439

β-Dimethylaminoethyl *p*-Bromo-α-methylbenzhydryl Ether *see* 3596

2-Dimethylaminoethyl 4-*n*-Butylaminobenzoate Hydrochloride *see* 9330

2-Dimethylaminoethyl *p*-Butylaminosalicylate *see* 4894

α-(Dimethylaminoethyl)-*o*-chlorobenzhydrol *see* 2106

N-Dimethylaminoethyl-*N*-*p*-chlorobenzyl-α-aminopyridine *see* 2214

β-Dimethylaminoethyl (*p*-Chloro-α-methylbenzhydryl) Ether *see* 2234

Dimethylaminoethyl *p*-Chlorophenoxyacetate *see* 5820

2-Dimethylaminoethyl 2'-Diethylaminoisopropyl Ether Bismethiodide *see* 7693

β-Dimethylaminoethyl β'-Diethylamino-α'-methylethyl Ether Dimethiodide *see* 7693

S-β-Dimethylaminoethyl-*O*,*O*-diethylthionophosphate Methiodide *see* 3549

7-[2-(Dimethylamino)ethyl]-3,7-dihydro-1,3-dimethyl-1*H*-purine-2,6-dione *see* 3261

(+)-*cis*-5-[2-(Dimethylamino)ethyl]-2,-3-dihydro-3-hydroxy-2-(*p*-methoxyphenyl)-1,5-benzothiazepin-4(5*H*)-one Acetate (Ester) *see* 3247

10-[2-(Dimethylamino)ethyl]-5,10-dihydro-5-methyl-11*H*-dibenzo[*b*,*e*]-[1,4]diazepin-11-one *see* 3055

5-[2-(Dimethylamino)ethyl]-2,3-dihydro-2-phenyl-1,5-benzothiazepin-4(5*H*)-one *see* 9440

2-Dimethylaminoethyl Ethoxydiphenylacetate *see* 3253

2-[(2-Dimethylaminoethyl)furfurylamino]pyridine *see* 6010

β-Dimethylaminoethyl (1-Hydroxycyclopentyl)phenylacetate *see* 2815

2-(Dimethylamino)ethyl [5-Hydroxy-4-(hydroxymethyl)-6-methyl-3-pyridyl]succinic Acid Methyl Ester *see* 8175

3-(β-Dimethylaminoethyl)-5-hydroxyindole *see* 1502

2-Dimethylaminoethyl 1-Hydroxy-α-phenylcyclopentaneacetate *see* 2815

2-[1-[2-[2-(Dimethylamino)ethyl]inden-3-yl]ethyl]pyridine *see* 3265

3-[2-(Dimethylamino)ethyl]indole *see* 3311

3-[2-(Dimethylamino)ethyl]-1*H*-indol-4-ol *see* 8110

3-[2-(Dimethylamino)ethyl]-1*H*-indol-5-ol *see* 1502

3-(2-Dimethylaminoethyl)-5-indolol *see* 1502

3-[2-(Dimethylamino)ethyl]-1*H*-indol-4-ol Dihydrogen Phosphate Ester *see* 8111

6-[[6-[2-(Dimethylamino)ethyl]-4-methoxy-1,3-benzodioxol-5-yl]acetyl]-2,3-dimethoxybenzoic Acid *see* 6508

2-[(2-Dimethylaminoethyl)(*p*-methoxybenzyl)amino]pyridine *see* 8168

2-[(2-Dimethylaminoethyl)(*p*-methoxybenzyl)amino]pyrimidine Hydrochloride *see* 9513

2-[(2-Dimethylaminoethyl)(*p*-methoxybenzyl)amino]thiazole *see* 10319

6-[[6-[2-(Dimethylamino)ethyl]-2-methoxy-3,4-(methylenedioxy)phenyl]acetyl]-*o*-veratric Acid *see* 6508

1-[[2-(Dimethylamino)ethyl]methylamino]-3-phenylindole *see* 1266

β-Dimethylaminoethyl 2-Methylbenzhydryl Ether *see* 7007

β-Dimethylaminoethyl 4-Methylbenzhydryl Ether *see* 6130

5-Dimethylamino-3-ethyl-2-methylindole *see* 5835

3-[2-(Dimethylamino)ethyl]-*N*-methyl-1*H*-indole-5-methanesulfonamide *see* 9172

5-(2-Dimethylaminoethyloximino)-5*H*-dibenzo[*a*,*d*]cyclohepta-1,4-diene *see* 6821

4-[2-(Dimethylamino)ethyl]phenol *see* 4783

10-(2-Dimethylaminoethyl)phenothiazine *see* 4013

10-(β-Dimethylaminoethyl)phenothiazine *see* 4013

2-[2-(Dimethylamino)ethyl]-4-(phenylmethyl)-1(2*H*)-phthalazinone *see* 9205

β-Dimethylaminoethyl-2-pyridylaminotoluene *see* 9868

β-Dimethylaminoethyl-2-pyridylbenzylamine *see* 9868

N-(2-Dimethylaminoethyl)-*N*-2-pyridyl-3-thenylamine *see* 9416

2-Dimethylaminoethyl Succinate Dimethochloride *see* 9044

2-[(2-Dimethylaminoethyl)-2-thenylamino]pyridine *see* 6027

Dimethylarsinic Acid *see* 1640

[(Dimethylarsino)oxy]sodium As-oxide *see* 8738

1,3-Dimethylazalumazine *see* 4111

7,12-Dimethylbenz[*a*]anthracene *see* 3285

9,10-Dimethyl-1,2-benzanthracene, 3285

ar,ar-Dimethylbenzenamine *see* 10217

N,N-Dimethylbenzenamine *see* 3284

Dimethylbenzene *see* 10214

N,N-Dimethyl-1,4-benzenediamine *see* 3303

Dimethyl 1,2-Benzenedicarboxylate *see* 3304

N,β-Dimethylbenzeneethanamine *see* 7462

(*S*)-*N,α*-Dimethylbenzeneethanamine *see* 6015

α,α-Dimethylbenzeneethanamine *see* 7415

α,4-Dimethylbenzenemethanol *see* 3287

3,3′-Dimethylbenzidine *see* 9651

5,6-Dimethylbenzimidazole, 3286

5,6-Dimethylbenzimidazolylcobamide Coenzyme *see* 2513

5,6-Dimethylbenzimidazolylcobamide 5′-Deoxyadenosine *see* 2513

5,6-Dimethylbenzimidazolyl Cyano-cobamide *see* 10152

α-(5,6-Dimethylbenzimidazolyl)hy-droxocobamide *see* 4854

N,α-Dimethyl-1,3-benzodioxole-5-eth-anamine *see* 5806

2,2-Dimethyl-1,3-benzodioxol-4-ol Methylcarbamate *see* 1063

8,8-Dimethyl-2*H*,8*H*-benzo[1,2-*b*:5,4-*b*′]dipyran-2-one *see* 10203

N,N-Dimethylbenzofuro[3,2-*c*][1]benz-oxepin-Δ⁶⁽¹²ᴴ⁾·ᵞ-propylamine *see* 7068

3,5-Dimethylbenzoic Acid 1-(1,1-Dimethylethyl)-2-(4-ethylbenzoyl)-hydrazide *see* 9254

7,8-Dimethylbenzo[*g*]pteridine-2,4-(1*H*,3*H*)-dione *see* 5626

N¹-(3,4-Dimethylbenzoyl)sulfanil-amide *see* 9102

1,4-Dimethyl-2,3-benzphenanthrene *see* 3285

p,α-Dimethylbenzyl Alcohol, 3287

p,α-Dimethylbenzyl Camphorate Diethanolamine Salt *see* 9630

N,N-Dimethyl-*N*-benzyl-*N*-chloro-*N*′-dodecylglycinamide *see* 3466

4-(2′,3′-Dimethylbenzyl)imidazole *see* 2981

N,N-Dimethyl-*N*′-benzyl-*N*′-(α-pyrid-yl)ethylenediamine *see* 9868

2-[2-(6,6-Dimethylbicyclo[3.1.1]hept-2-en-2-yl)ethoxy]-*N,N*-diethyl-ethanamine *see* 6422

1,1-Dimethylbiguanide *see* 6001

3,3′-Dimethyl-[1,1′-biphenyl]-4,4′-di-amine *see* 9651

3,3′-[(3,3′-Dimethyl[1,1′-biphenyl]-4,-4′-diyl)bis(azo)]bis[5-amino-4-hydr-oxy-2,7-naphthalenedisulfonic Acid] Tetrasodium Salt *see* 9923

6,6′-[(3,3′-Dimethyl[1,1′-biphenyl]-4,-4′-diyl)bis(azo)]bis[4-amino-5-hydr-oxy-1,3-naphthalenedisulfonic Acid] Tetrasodium Salt *see* 3952

3,3′-[(3,3′-Dimethyl[1,1′-biphenyl]-4,-4′-diyl)bis(azo)]bis[4-amino-1-naph-thalenesulfonic Acid] Disodium Salt *see* 1132

3,3′-[(3,3′-Dimethyl-4,4′-biphenylene)-bis(azo)]bis(5-amino-4-hydroxy-2,7-

naphthalenedisulfonic Acid) Tetra-sodium Salt *see* 9923

1,1′-Dimethyl-4,4′-bipyridinium *see* 7165

Dimethyl 1,3-Bis(carbomethoxy)-1-propen-2-yl Phosphate *see* 1360

β,γ-Dimethyl-α,δ-bis(3,4-dihydroxy-phenyl)butane *see* 6786

7,13-Dimethyl-2,3:9,10-bis(methylene-dioxy)-7,13a-secoberbin-13a-one *see* 2607

N,N′-Dimethyl-*N,N*′-bis[3-(3′,4′,5′-tri-methoxybenzoxy)propyl]ethyl-enediamine *see* 4743

(2*S*,2′*S*)-*N,N*′-Dimethyl-*N,N*′-bis-1-(3′,-4′,5′-trimethoxybenzoyloxy)butyl-2-ethylenediamine *see* 1560

2,3-Dimethyl-1,3-butadiene, 3288

β,γ-Dimethyl-Δ-α,γ-butadiene *see* 3288

2,3-Dimethyl-2,3-butanediol *see* 7593

(*R**,*S**)-4,4′-(2,3-Dimethyl-1,4-butane-diyl)bis[1,2-benzenediol] *see* 6786

3,3-Dimethylbutanoic Acid *see* 1573

[1*S*-[1α,3α,7β,8β(2*S**,4*S**),8aβ]]-2,2-Dimethylbutanoic Acid 1,2,3,7,8,8a-Hexahydro-3,7-dimethyl-8-[2-(tetra-hydro-4-hydroxy-6-oxo-2*H*-pyran-2-yl)ethyl]-1-naphthalenyl Ester *see* 8686

3,3-Dimethyl-2-butanone *see* 7594

(2a*E*,4*E*,5′*R*,6*R*,6′*S*,8*E*,11*R*,13*S*,15*S*,-17a*R*,20*R*,20a*R*,20b*S*)-6′-[(*E*)-1,3-Dimethyl-1-butenyl]-5′,6,6′,7,10,11,-14,15,17a,20,20a,20b-dodecahydro-20,20b-dihydroxy-5′,6,8,19-tetra-methylspiro[11,15-methano-2*H*,-13*H*,17*H*-furo[4,3,2-*pq*]benzodioxa-cyclooctadecin-13,2′-[2*H*]pyran]-4′,-17(3′*H*)-dione 4′-(*E*)-(*O*-Methyl-oxime) *see* 6373

3,3-Dimethylbutyric Acid *see* 1573

2,2-Dimethylbutyric Acid 8-Ester with (4*R*,6*R*)-6-[2-[(1*S*,2*S*,6*R*,8*S*,-8a*R*)-1,2,6,7,8,8a-Hexahydro-8-hydroxy-2,6-dimethyl-1-naphthyl]-ethyl]tetrahydro-4-hydroxy-2*H*-pyran-2-one *see* 8686

21-(3,3-Dimethylbutyryloxy)-9α-fluo-ro-11β-hydroxy-16α,17α-(isopro-pylidenedioxy)pregna-1,4-diene-3,-20-dione *see* 9730

Dimethylcadmium, 3289

Dimethylcarbamic Acid 7-Chloro-2,3-dihydro-1-methyl-2-oxo-5-phenyl-1*H*-1,4-benzodiazepin-3-yl Ester *see* 1773

Dimethylcarbamic Acid 1-[(Dimethyl-amino)carbonyl]-5-methyl-1*H*-py-razol-3-yl Ester *see* 3313

Dimethylcarbamic Acid 2-(Dimethyl-amino)-5,6-dimethyl-4-pyrimidinyl Ester *see* 7651

Dimethylcarbamic Acid 5-[2-[(1,1-Dimethylethyl)amino]-1-hydroxy-ethyl]-1,3-phenylene Ester *see* 980

Dimethylcarbamic Acid 5,5-Dimethyl-3-oxo-1-cyclohexen-1-yl Ester *see* 3259

Dimethylcarbamic Acid Ester with 3-Hydroxy-*N,N*,5-trimethylpyrazole-1-carboxamide *see* 3313

Dimethylcarbamic Acid 3-Methyl-1-(1-methylethyl)-1*H*-pyrazol-5-yl Ester *see* 5195

Dimethylcarbamic Acid 3-Methyl-1-phenyl-1*H*-pyrazol-5-yl Ester *see* 8186

(3-Dimethylcarbamoxyphenyl) Tri-methylammonium *see* 6553

N,N-Dimethylcarbamoylmethyl-*p*-(*p*-guanidinobenzoyloxy)phenylacetate *see* 1775

2-Dimethylcarbamoyl-3-methyl-5-pyrazolyl Dimethylcarbamate *see* 3313

N-(1-Dimethylcarbamoylpropyl)-*N*-ethylcrotonamide *see* 2662

N-(1-Dimethylcarbamoylpropyl)-*N*-propylcrotonamide *see* 2659

3-(Dimethylcarbamoyloxy)-1-methyl-pyridinium Bromide *see* 8161

Dimethyl Carbate, 3290

1,6-Dimethyl-3-carbethoxy-4-oxo-6,7,-8,9-tetrahydrohomopyrimidazole Methyl Sulfate *see* 8391

Dimethyl Carbinol *see* 5227

2,6-Dimethyl-3-carbomethoxy-4-(2-nitrophenyl)-5-carbisobutoxy-1,4-dihydropyridine *see* 6658

O,O-Dimethyl 1-Carbomethoxy-1-pro-pen-2-yl Phosphate *see* 6252

α,α-Dimethyl-α′-carboxydihydro-γ-pyrone Butyl Ester *see* 1564

Dimethyl Cellosolve *see* 3274

Dimethylcetyl-2-cyclohexanolammo-nium Bromide *see* 2061

1,4-Dimethyl-5-(*p*-chlorobenzoyl)pyr-role-2-acetic Acid *see* 10322

N,N-Dimethyl-*N*′-(*p*-chlorobenzyl)-*N*′-(2-pyridyl)ethylenediamine *see* 2214

O,O-Dimethyl *O*-(2-Chloro-4-nitro-phenyl)phosphorothioate *see* 3091

O,O-Dimethyl-*O*-(2-chloro-4-nitro-phenyl)thionophosphate *see* 3091

O,O-Dimethyl *O*-(3-Chloro-4-nitro-phenyl) Thionophosphate *see* 2248

α,α-Dimethyl-*p*-chlorophenethylamine *see* 2235

1,1-Dimethyl-3-(*p*-chlorophenyl)urea *see* 6342

(+)-*O,O*′-Dimethylchondrocurarine Diiodide *see* 6227

2,2-Dimethylchromenocoumarin *see* 10203

(3β)-*N,N*-Dimethylcon-5-enin-3-amine *see* 2561

2,2-Dimethyl-7-coumaranyl *N*-Meth-ylcarbamate *see* 1851

cis-2-Dimethylcrotonic Acid *see* 685

O,O-Dimethyl *O*-(4-Cyanophenyl) Phosphorothioate *see* 2321

O,O-Dimethyl *O*-(4-Cyanophenyl) Thionophosphate *see* 2321

N,N′-Dimethylcyclobuxine D *see* 2785

4,4-Dimethylcyclohepta-2,5,7-triene-carboxylic Acid *see* 9532

5,5-Dimethyl-1,3,6-cycloheptatriene-1-carboxylic Acid *see* 9532

1,1-Dimethyl-3,5-cyclohexanedione *see* 3291

5,5-Dimethyl-1,3-cyclohexanedione, 3291

N,α-Dimethylcyclohexaneethanamine *see* 8045

N,β-Dimethylcyclohexaneethanamine *see* 2776

1,1-Dimethylcyclohexane-2,4,6-trione *see* 4122

N,α-Dimethylcyclopentaneethanamine *see* 2808

N,α-Dimethylcyclopentaneethylamine *see* 2808

2,4-Dimethylcyclotetramethylene Sulfone *see* 3306

β,β-Dimethylcysteine *see* 7214

1,10-*trans*-Dimethyl-*trans*-(9)-decalol *see* 4408

6,17α-Dimethyl-6-dehydroprogester-one *see* 5836

Name Index

thiazin-10-yl)ethyl]-*N*-(β-hydroxy-ethyl)ammonium Chloride *see* 4870

N,*N*-Dimethyl-2-(*o*-methyl-α-phenyl-benzyloxy)ethylamine *see* 7007

N,*N*-Dimethyl-2-[(*p*-methyl-α-phenyl-benzyl)oxy]ethylamine *see* 6130

O,*O*-Dimethyl *O*-[1-Methyl-2-(1-phenylcarbethoxy)vinyl] Phosphate *see* 2667

[2*S*-(2α,5α,6β)]-3,3-Dimethyl-6-[[(5-methyl-3-phenyl-4-isoxazolyl)car-bonyl]amino]-7-oxo-4-thia-1-azabi-cyclo[3.2.0]heptane-2-carboxylic Acid *see* 7036

N,*N*-Dimethyl-2-[(2-methylphenyl)-phenylmethoxy]ethanamine *see* 7007

N,*N*-Dimethyl-2-[(4-methylphenyl)-phenylmethoxy]ethanamine *see* 6130

N,*N*-Dimethyl-9-[3-(4-methyl-1-piper-azinyl)propylidene]thioxanthene-2-sulfonamide *see* 9503

N,*N*-Dimethyl-10-[3-(4-methyl-1-piperazinyl)propyl]phenothiazine-2-sulfonamide *see* 9494

2,2-Dimethyl-3-(2-methyl-1-propenyl)-cyclopropanecarboxylic Acid *see* 2310

trans-(+)-2,2-Dimethyl-3-(2-methyl-propenyl)cyclopropanecarboxylic Acid (5-Benzyl-3-furyl)methyl Ester *see* 1271

[1*R*-[1α[*S**(*Z*)],3β]]-2,2-Dimethyl-3-(2-methyl-1-propenyl)cyclopropane-carboxylic Acid 3-(2-Butenyl)-2-methyl-4-oxo-2-cyclopenten-1-yl-ester *see* 2351

2,2-Dimethyl-3-(2-methyl-1-propenyl)-cyclopropanecarboxylic Acid (6-Chloro-1,3-benzodioxol-5-yl) Meth-yl Ester *see* 1032

2,2-Dimethyl-3-(2-methylpropenyl)-cyclopropanecarboxylic Acid 6-Chloropiperonyl Ester *see* 1032

2,2-Dimethyl-3-(2-methyl-1-propenyl)-cyclopropanecarboxylic Acid Cya-no(3-phenoxyphenyl)methyl Ester *see* 2837

2,2-Dimethyl-3-(2-methyl-1-propenyl)-cyclopropanecarboxylic Acid 3-(2-Cyclopenten-1-yl)-2-methyl-4-oxo-2-cyclopenten-1-yl Ester *see* 2774

2,2-Dimethyl-3-(2-methylpropenyl)-cyclopropanecarboxylic Acid Ester with *N*-(Hydroxymethyl)-1-cyclo-hexene-1,2-dicarboximide *see* 9362

2,2-Dimethyl-3-(2-methyl-1-propenyl)-cyclopropanecarboxylic Acid (1,3,-4,5,6,7-Hexahydro-1,3-dioxo-2*H*-isoindol-2-yl)methyl Ester *see* 9362

[1*R*-[1α[*S**(*Z*)],3β]]-2,2-Dimethyl-3-(2-methyl-1-propenyl)cyclopropane-carboxylic Acid 2-Methyl-4-oxo-3-(2,4-pentadienyl)-2-cyclopenten-1-yl Ester *see* 8148

[1*R*-[1α[*S**(*Z*)],3β]]-2,2-Dimethyl-3-(2-methyl-1-propenyl)cyclopropane-carboxylic Acid 2-Methyl-4-oxo-3-(2-pentenyl)-2-cyclopenten-1-yl Ester *see* 5272

2,2-Dimethyl-3-(2-methyl-1-propenyl)-cyclopropanecarboxylic Acid 2-Methyl-4-oxo-3-(2-propenyl)-2-cyclopenten-1-yl Ester *see* 257

2,2-Dimethyl-3-(2-methylpropenyl)-cyclopropanecarboxylic Acid *m*-Phenoxybenzyl Ester *see* 7405

2,2-Dimethyl-3-(2-methyl-1-propenyl)-cyclopropanecarboxylic Acid (3-Phenoxyphenyl)methyl Ester *see* 7405

(1*R*-*trans*)-2,2-Dimethyl-3-(2-methyl-1-propenyl)cyclopropanecarboxylic Acid [5-(Phenylmethyl)-3-furanyl]-methyl Ester *see* 1271

[2*S*-[2α,5α,6β(*S**)]]-3,3-Dimethyl-6-[[[[[3-(methylsulfonyl)-2-oxo-1-imid-azolidinyl]carbonyl]amino]phenyl-acetyl]amino]-7-oxo-4-thia-1-aza-bicyclo[3.2.0]heptane-2-carboxylic Acid *see* 6259

N,*N*-Dimethyl-*N*-[2-[2-[methyl-4-(1,1,-3,3-tetramethylbutyl)phenoxy]eth-oxy]ethyl]benzenemethanaminium Chloride *see* 6103

2,4-Dimethyl-*N*-(3-methyl-2(3*H*)-thiazolylidene)benzenamine *see* 2833

O,*O*-Dimethyl *O*-(4-Methylthio-3-methylphenyl) Thiophosphate *see* 4044

3,5-Dimethyl-4-(methylthio)phenol Methylcarbamate *see* 6050

O,*O*-Dimethyl *O*-[4-(Methylthio)-*m*-tolyl] Phosphorothioate *see* 4044

[1α,2α,5α(1*E*,3*E*),6β(1*E*,3*E*,5*E*)]-3,6-Dimethyl-5-[2-methyl-4-(2,6,6-tri-methyl-1-cyclohexen-1-yl)-1,3-buta-dienyl]-6-[4-methyl-6-(2,6,6-trimeth-yl-1-cyclohexen-1-yl)-1,3,5-hexatri-enyl]-3-cyclohexene-1,2-dimethanol *see* 5328

(9α,13α,14α)-3,17-Dimethylmor-phinan *see* 3251

1,3-Dimethyl-7-morpholinomethylxan-thine *see* 6363

N,*N*-Dimethyl-1-naphthalenamine *see* 3301

2,7-Dimethyl-1,4-naphthalenedione *see* 2100

2,7-Dimethyl-1,4-naphthoquinone *see* 2100

N,*N*-Dimethyl-1-naphthylamine, 3301

2,3-Dimethyl-4-nicotinamido-1-phen-yl-5-pyrazolone *see* 6619

N,*N*-Dimethyl-*N*′-(1-nitro-9-acridinyl)-1,3-propanediamine *see* 6661

1,2-Dimethyl-5-nitro-1*H*-imidazole *see* 3315

α,2-Dimethyl-5-nitro-1*H*-imidazole-1-ethanol *see* 8562

(±)-2,6-Dimethyl-4-(3-nitrophenyl)-1,-4-dihydropyridine-3,5-dicarboxylic Acid-3-(1-benzyl-3-piperidyl) Ester-5-methyl Ester *see* 1071

2,6-Dimethyl-4-(3′-nitrophenyl)-1,4-dihydropyridine-3,5-dicarboxylic Acid 3-β-Methoxyethyl Ester 5-Iso-propyl Ester *see* 6643

O,*O*-Dimethyl *O*-*p*-Nitrophenyl Phos-phorothioate *see* 6183

O,*O*-Dimethyl *O*-*p*-Nitrophenyl Thio-phosphate *see* 6183

Dimethylnitrosamine *see* 6735

N,*N*-Dimethyl-4-nitrosobenzenamine *see* 6736

N,4-Dimethyl-*N*-nitrosobenzenesulfon-amide *see* 9681

O,*O*-Dimethyl *O*-4-Nitro-*m*-tolyl Phosphorothioate *see* 4017

O,*O*-Dimethyl *O*-4-Nitro-*m*-tolyl Thio-phosphate *see* 4017

5,5-Dimethyl-3-[4-nitro-3-(trifluoro-methyl)phenyl]-2,4-imidazolidine-dione *see* 6636

(*Z*,*E*,*E*)-2-(4,8-Dimethyl-3,7-nonadien-yl)-6-methyl-2,6-octadiene-1,8-diol *see* 7692

(*all*-*E*)-2,3-Dimethyl-5-(3,7,11,15,19,-23,27,31,35-nonamethyl-2,6,10,14,-18,22,26,30,34-hexatriacontanona-

enyl)-2,5-cyclohexadiene-1,4-dione *see* 7681

N,*N*-Dimethylnorephedrine *see* 6145

2-[2-(6,6-Dimethyl-2-norpinen-2-yl)-ethoxy]triethylamine *see* 6422

17α,21-Dimethyl-19-norpregna-4,9-diene-3,20-dione *see* 7969

7α,17α-Dimethyl-19-nortestosterone *see* 6262

N,*N*-Dimethyl-1-octadecanamine *see* 3525

N,*N*-Dimethyloctadecylamine *see* 3525

3,7-Dimethyl-2,6-octadienal *see* 2383

2,6-Dimethyl-2,7-octadien-6-ol *see* 5520

3,7-Dimethyl-1,6-octadien-3-ol *see* 5520

cis-2,6-Dimethyl-2,6-octadien-8-ol *see* 6560

(*E*)-3,7-Dimethyl-2,6-octadien-1-ol *see* 4411

trans-3,7-Dimethyl-2,6-octadien-8-ol *see* 4411

(*Z*)-3,7-Dimethyl-2,6-octadien-1-ol *see* 6560

3,7-Dimethyl-1,6-octadien-3-yl Acetate *see* 5521

(*E*)-2-(3,7-Dimethyl-2,6-octadienyl)-1,-4-benzenediol *see* 4413

trans-(3,7-Dimethyl-2,6-octadienyl)-hydroquinone *see* 4413

(*E*)-6-(3,7-Dimethyl-2,6-octadienyl)-7-hydroxy-2*H*-1-benzopyran-2-one *see* 7027

(*E*)-6-(3,7-Dimethyl-2,6-octadienyl)-7-hydroxycoumarin *see* 7027

[*S*-(?,*E*)]-3-[[4-[(3,7-Dimethyl-2,6-octadienyl)oxy]phenyl]methylene]-1,4-dihydroxy-6-methyl-2,5-piper-azinedione *see* 6400

6-(3,7-Dimethyl-2,6-octadienyl)um-belliferone *see* 7027

N,*N*-Dimethyl-*N*′-(octahydro-4,7-meth-ano-1*H*-inden-5-yl)urea *see* 6787

3,7-Dimethyl-1,3,6-octatriene *see* 6837

3,7-Dimethyl-1,3,7-octatriene *see* 6837

3,7-Dimethyl-6-octenal *see* 2390

3,7-Dimethyl-7-octenal *see* 2390

2,6-Dimethyl-2-octen-8-ol *see* 2391

3,7-Dimethyl-6-octen-1-ol *see* 2391

S-(−)-3,7-Dimethyl-7-octen-1-ol *see* 8352

Dimethylolpropionic Acid, 3302

2,3-Dimethyl-7-oxabicyclo[2.2.1]hept-ane-2,3-dicarboxylic Anhydride *see* 1797

Dimethyl Oxalate *see* 6181

5,5-Dimethyl-2,4-oxazolidinedione *see* 3260

*N*¹-[(4,5-Dimethyl-2-oxazolyl)amidi-no]sulfanilamide *see* 9077

*N*¹-(4,5-Dimethyl-2-oxazolyl)sulfanil-amide *see* 9093

1-(4,5-Dimethyloxazol-2-yl)-3-sulfa-nilylguanidine *see* 9077

[*R*-(*E*,*E*)]-9-(3,3-Dimethyloxiranyl)-3,-7-dimethyl-2,6-nonadienoic Acid Methyl Ester *see* 5287

7,7-Dimethyl-2-oxobicyclo[2.2.1]hept-ane-1-methanesulfonic Acid *see* 1781

(11β,16α)-21-(3,3-Dimethyl-1-oxobut-oxy)-9-fluoro-11-hydroxy-16,17-[(1-methylethylidene)bis(oxy)]pregna-1,4-diene-3,20-dione *see* 9730

N-(1,1-Dimethyl-3-oxobutyl)acryl-amide *see* 3007

N-(1,1-Dimethyl-3-oxobutyl)-2-pro-penamide *see* 3007

3-[2-(3,5-Dimethyl-2-oxocyclohexyl)-2-hydroxyethyl]glutarimide *see* 2797

Di-*m*-nitrobenzenesulfonylethylene-
diamine *see* 3343
3,4-Dinitrobenzoic Acid, 3327
3,5-Dinitrobenzoic Acid, 3328
3,5-Dinitrobenzoyl Chloride, 3329
4,6-Dinitro-2-*sec*-butylphenyl β,β-
Dimethylacrylate *see* 1265
Dinitro-*sec*-butylphenyl Isopropyl
Carbonate *see* 3339
4,4′-Dinitrocarbanilide, 3330
4,4′-Dinitrocarbanilide Compd with
4,6-Dimethyl-2-pyrimidinol (1:1) *see*
6578
2,4-Dinitro-1-chlorobenzene *see* 2187
2,6-Dinitro-1-chlorobenzene *see* 2188
Dinitrochlorohydrin *see* 2452
Dinitrocresol, 3331
3,5-Dinitro-*o*-cresol *see* 3331
4,6-Dinitro-*o*-cresol *see* 3331
2,4-Dinitro-6-cyclohexylphenol *see*
2802
Dinitro-*o*-cyclohexylphenol *see* 2802
Dinitrodiphenylamine Sulfoxide *see*
3332
m,m′-Dinitrodiphenyl Disulfide *see*
6715
3,5-Dinitro-4-dipropylaminotoluene
see 3404
2,6-Dinitro-*N,N*-dipropylcumidine *see*
5220
2,6-Dinitro-*N,N*-dipropyl-4-methyl-
aniline *see* 3404
3,5-Dinitro-*N*⁴,*N*⁴-dipropylsulfanil-
amide *see* 7015
2,6-Dinitro-*N,N*-dipropyl-*p*-toluidine
see 3404
2,6-Dinitro-*N,N*-dipropyl-4-trifluoro-
methylaniline *see* 9815
2,6-Dinitro-*N,N*-dipropyl-4-(trifluoromethyl)benzenamine *see* 9815
2,4-Dinitro-*N*³,*N*³-dipropyl-6-(trifluoromethyl)-1,3-benzenediamine *see*
7947
2,6-Dinitro-*N*¹,*N*¹-dipropyl-4-trifluoro-
methyl-*m*-phenylenediamine *see*
7947
2,6-Dinitro-*N,N*-dipropyl-α,α,α-tri-
fluoro-*p*-toluidine *see* 9815
2,4-Dinitro-1-fluorobenzene *see* 4210
Dinitrogen Monoxide *see* 6751
Dinitrogen Pentoxide *see* 6702
3,5-Dinitro-2-hydroxytoluene *see* 3331
Dinitrol *see* 3331
3,5-Dinitro-2-methylbenzamide *see*
3321
3,7-Dinitro-5-oxophenothiazine, 3332
2,4-Dinitrophenol, 3333
2,5-Dinitrophenol, 3334
2,6-Dinitrophenol, 3335
α-Dinitrophenol *see* 3333
3,7-Dinitro-10*H*-phenothiazine 5-
Oxide *see* 3332
2,4-Dinitrophenylhydrazine, 3336
2,4-Dinitroresorcinol, 3337
5,5′-Dinitrosalicil, 3338
Dinitrosorbide *see* 5245
3,5-Dinitro-*o*-toluamide *see* 3321
Dinkum Oil *see* 6901
Dinobuton, 3339
Dinocap, 3340
Dinoleine *see* 5063
Dinoprost *see* 8065
Dinoprostone *see* 8064
Dinormon [Lusofarmaco] *see* 2986
Dinoseb, 3341
Dinoseb Methacrylate *see* 1265
Dinosterol, 3342
Dinovex [Ortho] *see* 3153
Dinsed, 3343
Dinsidon [Geigy] *see* 6985

Dinulcid [Pharmascience] *see* 7048
Diocimex [Cimex] *see* 3496
Diocodal [Norwich] *see* 6504
Dioctyl [Medo] *see* 3460
Dioctylal [Cont. Pharma] *see* 3460
Dioctyl Calcium Sulfosuccinate *see*
3459
Dioctyl Phthalate *see* 1291
Dioctyl Sodium Sulfosuccinate *see*
3460
Diocurb *see* 2996
Diocyclin [Cimex] *see* 9337
Diocyl Hydrochloride *see* 3147
Dioderm [Dermal] *see* 4828
Diodone *see* 5061
Diodoquin [Searle] *see* 5063
Diodoxylin *see* 5063
Diodrast [Winthrop] *see* 5061
Diodrast Compound Solution [Win-
throp] *see* 5061
Diodrast Concentrated Solution [Win-
throp] *see* 5061
Dioform *see* 93
Diogyn [Pfizer] *see* 3746
Diogyn E [Pfizer] *see* 3780
Diolandrone *see* 6017
Diolen [Glanzstoff] *see* 7730
Diolostene *see* 6017
Dioloxol [Reed & Carnrick] *see* 5895
Dionin [E. Merck] *see* 3876
Dionosil [Glaxo] *see* 8048
Diopal *see* 6084
Diophyllin *see* 485
Diopine [Allergan] *see* 3400
Diopterin, 3344
Dioscin, 3345
Dioscorea, 3346
Dioscorine, 3347
Diosgenin, 3348
Diosgenin Bis-α-L-rhamnopyranosyl-
(1→2 and 1→4)-β-D-Glucopyrano-
side *see* 3345
Diosmetin, 3349
Diosmetin 7-β-Rutinoside *see* 3350
Diosmil [Bellon] *see* 3350
Diosmin, 3350
Diosmol [Merck & Co.] *see* 5788
Diosphenol, 3351
Diosven [CT] *see* 3350
Diothane Hydrochloride [Merrell] *see*
3360
Diothyl *see* 8172
Diotilan [Chinoin] *see* 3460
Diovenor [Innothéra] *see* 3350
3,8-Dioxadecane-4,7-dione-1,10-bis-
[ethyldimethylammonium Bromide]
see 9184
3,8-Dioxadecane-4,7-dione-1,10-bis-
[trimethylammonium Bromide] *see*
9043
trans-1,7-Dioxadispiro[4.0.4.2]dodeca-
3,9-diene-2,8-dione *see* 681
Dioxadrol, 3352
9-(2′,5′-Dioxahexyloxyimino)erythro-
mycin *see* 8433
Dioxane, 3353
2,3-*p*-Dioxanedithiol *S,S*-Bis(*O,O*-di-
ethyl Phosphorodithioate) *see* 3355
Dioxanin *see* 3210
1,3-Dioxan-5-ol *see* 4494
10-[3-[4-[2-(1,3-Dioxan-2-yl)ethyl]-1-
piperazinyl]propyl]-2-(trifluoromethyl)-10*H*-phenothiazine *see* 7040
Dioxaphetyl Butyrate, 3354
(1,4-Dioxaspiro[4.5]*dec*-2-ylmethyl)-
guanidine *see* 4587
Dioxathion, 3355
Dioxatrine [Janssen] *see* 1104
Dioxethedrine, 3356
Dioxin *see* 9252

Dioxin (obsolete) *see* 3273
Dioxoaminopyrine *see* 3358
9,10-Dioxoanthracene *see* 726
3,3′-Dioxo-[Δ²,²′-biindoline]-5,5′-disul-
fonic Acid Disodium Salt *see* 4978
2,2′-[(1,4-Dioxo-1,4-butanediyl)bis-
(oxy)]bis[*N*-ethyl-*N,N*-dimethyl-
ethanaminium] Dibromide *see* 9184
2,2′-[(1,4-Dioxo-1,4-butanediyl)bis-
(oxy)]bis[*N,N,N*-trimethylethana-
minium] Dibromide *see* 9043
2,2′-[(1,4-Dioxo-1,4-butanediyl)bis-
(oxy)]bis[*N,N,N*-trimethylethana-
minium] Dichloride *see* 9044
[1α,2β(1′*S**,2′*R**)]-2,2′-[(1,4-Dioxo-
1,4-butanediyl)bis(oxy-3,1-propane-
diyl)]bis[1,2,3,4-tetrahydro-6,7,8-tri-
methoxy-2-methyl-1-[(3,4,5-trimeth-
oxyphenyl)methyl]isoquinolinium]
Dichloride *see* 3487
4,4′-Dioxo-β-carotene *see* 1798
N,N′-[(3,6-Dioxo-1,4-cyclohexadiene-
1,4-diyl)bis(imino-3,1-propanediyl)]-
bis[*N,N*-diethylbenzenemethanamin-
ium]dichloride *see* 1137
2,4-Dioxo-5-diazopyrimidine *see* 3050
2,4-Dioxo-3,3-diethyl-5-methylpiperi-
dine *see* 6216
2,4-Dioxo-3,3-diethyltetrahydropyr-
idine *see* 8179
2,4-Dioxo-3,3-diethyl-1,2,3,4-tetrahy-
dropyridine *see* 8179
3,3′-(1,3-Dioxo-1,3-digermoxanediyl)-
bispropanioic Acid *see* 7979
2,4-Dioxodihydro-5,6-benzo-1,3-oxa-
zine *see* 1915
5,6-Dioxo-5,6-dihydro-4,7-phenan-
throline *see* 7337
3,5-Dioxo-1,2-diphenyl-4-*n*-butylpyr-
azolidine *see* 7431
N,N′-[(1,2-Dioxo-1,2-ethanediyl)bis-
(imino-2,1-ethanediyl)]bis[2-chloro-
N,N-diethylbenzenemethanaminium]
Dichloride *see* 396
4,6-Dioxo-1-ethyl-10-propyl-4*H*,6*H*-
pyrano[3,2-*g*]quinoline-2,8-dicarb-
oxylic Acid *see* 6524
1,1-Dioxo-3-(*p*-fluorophenylmethyl)-
3,4-dihydro-6-chloro-1,2,4-benzo-
thiadiazine-7-sulfonamide *see* 7157
2,4-Dioxo-5-fluoropyrimidine *see* 4219
3,4-Dioxohexanedioic Acid *see* 5311
3,3′-[(1,6-Dioxo-1,6-hexanediyl)di-
imino]bis[2,4,6-triiodobenzoic Acid]
see 5043
3,3′-[(1,6-Dioxo-1,6-hexanediyl)di-
imino]bis[2,4,6-triiodo-5-[(methylamino)carbonyl]ben-zoic Acid] *see*
5029
(2,5-Dioxo-4-imidazolidinyl)urea *see*
255
[(2,5-Dioxo-4-imidazolidinyl)ureato]-
dihydroxyaluminum *see* 224
1,3-Dioxolane-4-methanol *see* 4494
7-(1,3-Dioxolan-2-ylmethyl)-3,7-dihy-
dro-1,3-dimethyl-1*H*-purine-2,6-
dione *see* 3494
7-(1,3-Dioxolan-2-ylmethyl)theophyl-
line *see* 3494
(1,3-Dioxolan-4-ylmethyl)trimethyl-
ammonium Iodide *see* 7056
2,6-Dioxo-4-methyl-4-ethylpiperidine
see 1054
Dioxone *see* 3155
[*R*-[*R**,*R**-(*E*)]]-2,2′-(1,8-Dioxo-4-oct-
ene-1,8-diyl)bis(oxy-3,1-propanedi-
yl)bis[1,2,3,4-tetrahydro-6,7-dimeth-
oxy-2-methyl-1-[(3,4,5-trimethoxy-
phenyl)methyl]isoquinolinium)
Dichloride *see* 6305

α-(4-Diphenylyl)butyric Acid *see* 10205
4-Diphenylylethylacetic Acid *see* 10205
Diphesatin *see* 7108
Diphetarsone, 3394
Diphexamide Methiodide *see* 1635
Diphos [Boehringer, Mann.] *see* 3908
Diphosgene, 3395
(Diphosphonomethyl)butanedioic Acid *see* 1546
(Diphosphonomethyl)succinic Acid *see* 1546
Diphosphopyridine Nucleotide *see* 6429
Diphosphoric Acid *see* 8191
Diphosphoric Acid Tetraethyl Ester *see* 9345
Diphosphoric Acid Tetrapotassium Salt *see* 7833
Diphosphoric Acid Tin(2+) Salt (1:2) *see* 8945
Diphosphoric Acid, Zinc Salt (1:2) *see* 10287
Diphosphorus Pentoxide *see* 7512
Diphosphorus Trioxide *see* 7517
Diphylline *see* 2095
Diphyllin Methyl Ether *see* 5286
Dipicrylamine, 3396
Dipidolor [Janssen] *see* 7653
Dipin, 3397
Dipipanone, 3398
2β,16β-Dipiperidino-5α-androstane-3α,17β-diol Diacetate Dimethobromide *see* 7139
2,2′,2″,2‴-[(4,8-Di-1-piperidinylpyrimido[5,4-*d*]pyrimidine-2,6-diyl)dinitrilo]tetrakisethanol *see* 3410
Dipiperon [Janssen] *see* 7608
Dipiproverine, 3399
Dipirin *see* 495
1-(3′,4′-Dipivaloyloxyphenyl)-2-methylamino-1-ethanol *see* 3400
Dipivalyl Epinephrine *see* 3400
Dipivefrin, 3400
Diploicin, 3401
Diplosal Acetate *see* 102
Dipolyoxyethylated Polypropyleneglycol Ether *see* 7721
Diponium Bromide, 3402
Dipotassium Bis[μ-[2,3-dihydroxybutanedioato(4—)-01,02:03,04]]diantimonate(2—) Trihydrate Stereoisomer *see* 741
Dipotassium Bis[ethanedioato(2—)-*O,O′*]oxotitanate(2—) *see* 7868
Dipotassium Hexakis(thiocyanato-*S*)-platinate(2—) *see* 7805
Dipotassium Hydrogen Phosphate *see* 7828
Dipotassium Pentakis(cyano-*C*)nitrosylferrate(2—) *see* 7817
Dipotassium Phosphate *see* 7828
Dipotassium Tetraiodocadmate(2—) *see* 7861
Dipotassium Tetrakis(cyano-*C*)mercurate(2—) *see* 7856
Dipotassium Tetrakis(cyano-*C*)-nickelate(2—) *see* 7857
Dipotassium Tetrakis(cyano-*C*)-platinate(2—) *see* 7858
Diprazin *see* 5219
Diprenorphine, 3403
Diprivan [ICI] *see* 8020
Diproderm [Schering] *see* 1226
Diprolene [Schering] *see* 1226
Dipropalin, 3404
Di-2-propenylamine *see* 3015
3′,5-Di-2-propenyl-[1,1′-biphenyl]-2,-4′-diol *see* 4780

5,5′-Di-2-propenyl-[1,1′-biphenyl]-2,-2′-diol *see* 5738
Di-2-propenylcyanamide *see* 3016
5,5-Di-2-propenyl-2,4,6(1*H*,3*H*,5*H*)-pyrimidinetrione *see* 261
Dipropetryn, 3405
Diprophos [Schering] *see* 1226
Diprophylline *see* 3529
3β,17β-Dipropionyloxy-4-estrene *see* 1352
Dipropylacetamide *see* 10050
Di-*n*-propylacetic Acid *see* 10049
n-Dipropylamine, 3406
4-(Dipropylamino)-3,5-dinitrobenzenesulfonamide *see* 7015
4-[2-(Dipropylamino)ethyl]-1,3-dihydro-2*H*-indol-2-one *see* 8416
4-[2-(Di-*n*-propylamino)ethyl]-2(3*H*)-indolone *see* 8416
4-[(Dipropylamino)sulfonyl]benzoic Acid *see* 7934
5-Dipropylamino-α,α,α-trifluoro-4,6-dinitro-*o*-toluidine *see* 7947
Dipropylcarbamothioic Acid *S*-Ethyl Ester *see* 3674
Dipropylcarbamothioic Acid *S*-Propyl Ester *see* 10097
Dipropyl 3,4-Dihydroxy-2,5-thiophenedicarboxylate *see* 8075
N,N-Dipropyl-2,6-dinitro-4-methylaniline *see* 3404
N,N-Dipropyl-2,6-dinitro-4-trifluoromethylaniline *see* 9815
N³,N³-Di-*n*-propyl-2,4-dinitro-6-trifluoromethyl-1,3-phenylenediamine *see* 7947
Dipropylene Glycol Monomethyl Ether, 3407
Dipropyl Ether *see* 8042
Dipropyl Ketone, 3408
5,5-Dipropyl-2,4-oxazolidinedione, 3409
p-(Dipropylsulfamoyl)benzoic Acid *see* 7934
p-(Dipropylsulfamyl)benzoic Acid *see* 7934
Dipropyl Sulfide *see* 8053
Dipropylthiocarbamic Acid *S*-Ethyl Ester *see* 3674
Dipropyl Thiocarbamic Acid *S*-Propyl Ester *see* 10097
Diprosis [Essex] *see* 1226
Diprosone [Schering] *see* 1226
Dipsan [Lederle] *see* 1703
Dipterex [Bayer] *see* 9753
Dipyridamole, 3410
Dipyridan [Hokuriku] *see* 3410
α-[3-(Di-2-pyridinylmethylene)-1,4-cyclopentadien-1-yl]-α-2-pyridinyl-2-pyridinemethanol *see* 8173
Dipyridoxolyldisulfide *see* 8180
2,2′-Dipyridyl *see* 3411
4,4′-Dipyridyl *see* 3412
α,α′-Dipyridyl, 3411
γ,γ′-Dipyridyl, 3412
Di-2-pyridyl-(6,6-di-2-pyridylfulven-2-yl)methanol *see* 8173
3,3-Dipyridyl Disulfide Dihydrochloride *see* 3438
3-(Di-2-pyridylmethylene)-α,α-di-2-pyridyl-1,4-cyclopentadiene-1-methanol *see* 8173
2,4-Di(3-pyridyl)pyridine *see* 6609
2,4-Di(β-pyridyl)pyridine *see* 6609
Dipyrine *see* 495
Dipyrithione *see* 8178
Dipyrocetyl, 3413
Dipyrone, 3414
1,4-Dipyrrolidino-2-butyne *see* 9715
(16α)-21-[4-(2,6-Di-1-pyrrolidinyl-4-

pyrimidinyl)-1-piperazinyl]-16-methylpregna-1,4,9(11)-triene-3,20-dione *see* 9604
Diquat Dibromide, 3415
Diquel *see* 3938
N,N′-Di-6-quinolinylurea *see* 3416
1,3-Di-6-quinolylurea, 3416
6,6′-Diquinolylurea *see* 3416
sym-Di-(6-quinolyl)urea *see* 3416
Dira [Kaken] *see* 8917
Diralgan [Roussel-UCLAF] *see* 4140
Dirax [Daiichi] *see* 5029
Direktan [Gerot] *see* 6612
Direma *see* 4822
Diresorcinol, 3417
Direxiode [Delalande] *see* 5063
Dirgotarl [Horita] *see* 3217
Dirhenium Heptoxide *see* 8344
Dirimal [Lilly] *see* 7015
Dirithromycin, 3418
Dirnate [Merck & Co.] *see* 1927
Dirocide [Squibb] *see* 3165
Dironyl [Schering AG] *see* 9308
Dirox [Winthrop] *see* 45
Dirythmin SA [Astra] *see* 3424
Disadine D.P. [Stuart] *see* 7880
Disalcid [3M] *see* 8491
Disalgesic [3M] *see* 8491
Disalicylic Acid *see* 8491
Disalunil *see* 4822
Disamide [BDH] *see* 3427
Disarim [Bial] *see* 2132
Discase [Travenol] *see* 2319
Dis-Cinil [Lusofarmaco] *see* 7094
Discoid [Sagitta] *see* 4331
Discotrine [3M Pharma] *see* 6704
Disdolen [Uriach] *see* 4281
Disepron [Yoshitomi] *see* 2474
Diseptal B *see* 6202
Diseptal C *see* 9100
Disgren [Uriach] *see* 9817
Disiamylborane *see* 1290
DISIDA *see* 3422
Disilane, 3419
Disilicane *see* 3419
Disilicoethane *see* 3419
Disilicon Hexahydride *see* 3419
Disilver Fluoride *see* 8675
Disintegrins *see* 5327
Disipal [Doetsch, Grether] *see* 7007
Di-Sipidin *see* 7666
Diskin *see* 7094
Dismutec [Sterling Winthrop] *see* 9177
Disocarban [Chem. Berlin] *see* 9593
Disoderm [Schering] *see* 3099
Disodium Acetarsenate *see* 850
Disodium Arsenate *see* 8720
Disodium Arsenoacetate *see* 848
Disodium Arsonoacetate *see* 850
Disodium Azodisalicylate *see* 6976
Disodium Citrate *see* 8747
Disodium Cromoglycate *see* 2658
Disodium 3,3′-Diamino-4,4′-dihydroxyarsenobenzene *N*-Dimethylenesulfonate *see* 9111
Disodium *p,p′*-Diaminodiphenylsulfone-*N,N′*-diglucose Sulfonate *see* 4471
Disodium *o*-Dianisidinediazobis(1-naphthol-4-sulfonate) *see* 1115
Disodium Dihydrogen (1-Hydroxyethylidene)bis[phosphonate] *see* 3908
Disodium Dihydrogen Hypophosphate, 3420
Disodium Dihydrogen Pyrophosphate *see* 8713
Disodium Dihydrogen Subphosphate *see* 3420

Disodium-1,2-dihydroxybenzene-3,5-disulfonate *see* 9606

N^4-(Disodium 1,3-Disulfo-3-phenylpropyl)sulfanilamide *see* 6777

Disodium 4,4'-Disulfoxydiphenyl-(2-pyridyl)methane *see* 7559

Di-sodium Edathamil *see* 3556

Disodium Edetate *see* 3556

Disodium Ethylenebis[dithiocarbamate] *see* 6426

Disodium Ethylenediaminetetraacetate *see* 3556

Disodium Formaldehydesulfoxylate-diaminodiphenylsulfone *see* 9141

Disodium Hexachloroplatinate(2—) *see* 8768

Disodium Hydrogen Citrate *see* 8747

Disodium Hydrogen Phosphate *see* 8805

Disodium 5,5'-Indigotin Disulfonate *see* 4978

Disodium *N*-Methyl-3,5-diiodo-4-pyridone-2,6-dicarboxylate *see* 8779

Disodium Monohydroxy Methylene Diphosphonate *see* 7074

Disodium Monomethanearsonate *see* 6020

Disodium Orthophosphate *see* 8805

Disodium Pentaiodobismuthate(2—) *see* 1320

Disodium Phenoltetrabromophthalein Sulfonate *see* 9127

Disodium Phenyl Phosphate, 3421

Disodium *p*-(γ-Phenylpropylamino)-benzenesulfonamide-α,γ-disulfonate *see* 6777

Disodium 1-Phenyl-3-*p*-sulfamoylanilino-1,3-propanedisulfonate *see* 6777

Disodium Phosphate *see* 8805

Disodium Prednisolone 21-Phosphate *see* 7903

Disodium Pyrocatechol-3,5-disulfonate *see* 9606

Disodium Succinate *see* 8828

Disodium[sulfonylbis(*p*-phenyleninino)]dimethanesulfinate *see* 9141

Disodium *o*-Tolidinediazobis(1-naphthylamine-4-sulfonate) *see* 1132

Diso-Duriles [Astra] *see* 3424

Disofenin, 3422

Disomer [Schering] *see* 1467

Disonate [Lannett] *see* 3460

Disophenol, 3423

Disoprivan [ICI] *see* 8020

Disoprofol *see* 8020

Disopromine *see* 3239

Disopyramide, 3424

Disoquin *see* 5063

Disorat [Boehringer, Mann.] *see* 6221

Disorlon [Nativelle] *see* 5245

Disotat [Isis] *see* 3241

Dispadol [Specia] *see* 5894

Dispan *see* 5491

Disparlure, 3425

Dispas [Ankerfarm] *see* 3211

Disphex [ICI] *see* 7880

Disposaject [Pitman-Moore] *see* 3460

Dispranol *see* 7247

Dispril *see* 1684

Disprin *see* 1684

Disprol [Reckitt & Colman] *see* 45

Dispromil [Liberman] *see* 3972

Disrupt [Hercon] *see* 3425

Dissenten [SPA] *see* 5601

Distaclor [Lilly] *see* 1962

Distakaps V-K [Dista] *see* 7230

Distamine [Dista] *see* 7214

Distamycin A *see* 8927

Distaquaine *see* 7226

Distaquaine V [Dista] *see* 7230

Distaquaine V-K [Dista] *see* 7230

Distaval *see* 9390

Distaxid [Dista] *see* 6758

Distigmine Bromide, 3426

Disto-5 [Cogla] *see* 1343

Distol 8 *see* 3557

Distraneurin [Pharma-Stern] *see* 2444

Distyryl *see* 3378

Distyryl Ketone *see* 3053

Disulfamide, 3427

1,3-Disulfamyl-4,5-dichlorobenzene *see* 3127

Disulfan *see* 9100

Disulfatozirconic Acid *see* 10315

3,3'-Disulfinyldialanine *see* 2852

Disulfiram, 3428

Disulfoton, 3429

Disulfur Dichloride *see* 9143

Disulfuric Acid *see* 8192

Disulfuric Acid Dipotassium Salt *see* 7834

Disulfur Pentoxydichloride *see* 8193

Disulfuryl Chloride *see* 8193

Disulon [Specia] *see* 9100

Disulone [Specia] *see* 2885

Disulphamide *see* 3427

Disulphine Blue [ICI] *see* 9161

Disul-sodium, 3430

Disyncran *see* 6034

Disynformon *see* 3751

Di-Syntramine *see* 3147

Disyston [Bayer] *see* 3429

Dita Bark, 3431

Ditaine *see* 3548

Ditan *see* 3387

Ditaven [Cascan] *see* 3206

Ditazol, 3432

Diteftin *see* 4333

Ditetrazolium Chloride *see* 9380

Dithane D-14 [Rohm & Haas] *see* 6426

Dithane M-22 [Rohm & Haas] *see* 5761

Dithane M-45 [Rohm & Haas] *see* 5756

Dithane Z-78 [Rohm & Haas] *see* 10300

1,4-Dithiaanthraquinone-2,3-dicarbonitrile *see* 3433

δ-[3-(1,2-Dithiacyclopentyl)]pentanoic Acid *see* 9462

Dithianone, 3433

1,2-Dithia-5,8,11,14,17-pentaazacycloeicosane Cyclic Peptide Deriv *see* 6859

2,6-Dithia-1,3,5,7-tetraazaadamantane 2,2,6,6-Tetraoxide *see* 9366

2,6-Dithia-1,3,5,7-tetraazatricyclo-[3.3.1.1^{3,7}]decane 2,2,6,6-Tetraoxide *see* 9366

Dithiazanine Iodide, 3434

α-[1-[(3,3-Di-3-thienylallyl)amino]ethyl]benzyl Alcohol *see* 9589

Di-2-thienylglycolate *see* 9598

3-(Di-2-thienylmethylene)-5-methoxy-1,1-dimethylpiperidinium Bromide *see* 9583

3-(Di-2-thienylmethylene)-1-methylpiperidine *see* 9601

3-(Di-2-thienylmethylene)-5-methyl-*trans*-quinolizidinium Bromide *see* 9602

trans-3-(Di-2-thienylmethylene)octahydro-5-methyl-2*H*-quinolizinium Bromide *see* 9602

α-[1-[(3,3-Di-3-thienyl-2-propenyl)-amino]ethyl]benzenemethanol *see* 9589

1,3-Dithietan-2-ylidenephosphoramidic Acid Diethyl Ester *see* 4284

Dithio *see* 9139

4,4'-Dithiobis[2-aminobutanoic Acid] *see* 4772

4,4'-Dithiobis[2-aminobutyric Acid] *see* 4772

[*R*-(*R**,*R**)]-3,3'-Dithiobis[2-aminopropanoic Acid] *see* 2851

2,2'-Dithiobis[benzothiazole], 3435

N,*N*'-[Dithiobis[2-[2-(benzoyloxy)ethyl]-1-methyl-2,1-ethenediyl]]bis[*N*-[(4-amino-2-methyl-5-pyrimidinyl)-methyl]formamide] *see* 1286

N,*N*'-[Dithiobis[1-[(carboxymethyl)-carbamoyl]ethylene]]diglutamine *see* 4483

2,2'-Dithiobisethanamine *see* 2846

N,*N*'-[Dithiobis[2,1-ethanediylimino(3-oxo-3,1-propanediyl)]]bis[2,4-dihydroxy-3,3-dimethylbutanamide] *see* 7144

2,2'-Dithiobis[ethylamine] *see* 2846

N,*N*'-[Dithiobis(ethyleneiminocarbonylethylene)]bis(2,4-dihydroxy-3,3-dimethylbutyramide) *see* 7144

N,*N*'-[Dithiobis[2-(2-hydroxyethyl)-1-methyl-2,1-ethenediyl]]bis[*N*-[(4-amino-2-methyl-5-pyrimidinyl)methyl]formamide] *see* 9432

N,*N*'-[Dithiobis[2-(2-hydroxyethyl)-1-methylvinylene]]bis[*N*-[(4-amino-2-methyl-5-pyrimidinyl)methyl]formamide] Dibenzoate *see* 1286

3,3'-[Dithiobis(methylene)]bis[5-hydroxy-6-methyl-4-pyridinemethanol] *see* 8180

4,4'-Dithiobis[morpholine] *see* 3437

6,6'-Dithiobis-2-naphthalenol *see* 2897

6,6'-Dithiobis(2-naphthol) *see* 2897

3,3'-Dithiobis[pyridine] Dihydrochloride *see* 3438

2,2'-Dithiobispyridine 1,1'-Dioxide *see* 8178

Dithiobis[thioformic Acid] *O*,*O*-Diethyl Ester *see* 3449

3,3'-Dithiobisvaline *see* 7216

2,4-Dithiobiuret, 3436

Dithiocarb *see* 3443

Dithiocarbamic Acid Monoammonium Salt *see* 545

Dithiocarbonic Acid Cyclic *S*,*S*-(6-Methyl-2,3-quinoxalinediyl) Ester *see* 7113

Dithiocarbonic Anhydride *see* 1859

Dithiodemeton *see* 3429

β,β'-Dithiodialanine *see* 2851

4,4'-Dithiodimorpholine, 3437

6,6'-Dithiodi-2-naphthol *see* 2897

3,3-Dithiodipyridine Dihydrochloride, 3438

3,3'-Dithiodivaline *see* 7216

Dithioethyleneglycol *see* 3770

1,2-Dithioglycerol *see* 3255

Dithioglycolyl *p*-Arsenobenzamide *see* 831

"Dithiol" *see* 9669

1,2-Dithiolane-3-pentanoic Acid *see* 9462

1,2-Dithiolane-3-valeric Acid *see* 9462

1,3-Dithiolan-2-ylidenephosphoramidic Acid Diethyl Ester *see* 7490

5-[3-(1,2-Dithiolanyl)]pentanoic Acid *see* 9462

5-(1,2-Dithiolan-3-yl)valeric Acid *see* 9462

1,3-Dithiolo[4,5-*b*]quinoxaline-2-thione *see* 9495

2,3-Dithiolpropanesulfonic Acid *see* 3256

1,3-Dithiol-2-ylidenepropanedioic Acid Bis(1-methylethyl) Ester *see* 5751
Dithion [Agrimont] *see* 2635
Dithione *see* 9139
Dithionous Acid Disodium Salt *see* 8756
Dithiooxamide *see* 8436
Dithiophos *see* 9139
Dithiophosphoric Acid *O,O'*-Dimethyl-*S*-[(5-methoxy-1,3,4-thiadiazol-2(3*H*)-one-3-yl)methyl] Ester *see* 6048
Dithiopropylthiamine *see* 8068
Dithiopyr, 3439
Dithiosalicylic Acid, 3440
Dithiosteine *see* 3684
Dithiosystox [Bayer] *see* 3429
1,4-Dithiothreitol, 3441
Dithiotrimethyleneglycol *see* 7984
Dithizone, 3442
Dithranol *see* 723
Dithrocream [Dermal] *see* 723
Diticyl *see* 9867
Ditin Diphosphate *see* 8945
Ditin Pyrophosphate *see* 8945
Ditiocarb Sodium, 3443
Dition [Agrimont] *see* 2635
1,4-Di-*p*-toluidinoanthraquinone *see* 8251
Ditolyldiazo-3,6-disulfo-β-naphthylamine-β-naphthylamine-6-sulfonic Acid Sodium Salt *see* 10149
1,1-Di-*p*-tolylethane, 3444
1,2-Di-*p*-tolylethane, 3445
asym-Di-*p*-tolylethane *see* 3444
sym-Di-*p*-tolylethane *see* 3445
α,α-Di-*p*-tolylethane *see* 3444
α,β-Di-*p*-tolylethane *see* 3445
p-Ditolylmercury, 3446
Ditrazin *see* 3165
1,4-Di(2,2,2-trichloro-1-formamidoethyl)piperazine *see* 9819
Di[tri-(2-methyl-2-phenylpropyl)tin]-oxide *see* 4004
Ditripentat [Heyl] *see* 7265
Ditropan [Marion] *see* 7089
Ditropyl Isatropate *see* 1053
Ditrosol *see* 3331
Ditubin [Schering] *see* 5203
Diucardin [Ayerst] *see* 4830
Diucardyn Sodium [Ayerst] *see* 5918
Diucen [Central Pharm.] *see* 1155
Diulo [Searle] *see* 6231
Diumax [Cusi-Norte] *see* 7647
Diu-melusin [Schwarz] *see* 4822
Diural [Apothokermes] *see* 4331
Diurapid [Boehringer, Mann.] *see* 953
Diurazine *see* 2123
Diuresal *see* 2221
Diurese [Am. Urologicals] *see* 9754
Diuretin [Knoll] *see* 9418
Diurex [Makhteshim-Agan] *see* 3447
Diurexan [E. Merck] *see* 10212
Diuril [Merck & Co.] *see* 2221
Diurilix [Théraplix] *see* 2221
Diuril Lyovac [Merck & Co.] *see* 2221
Diurite *see* 2221
Diuriwas [IFI] *see* 50
Diuron, 3447
Diurone *see* 2154
Diursal *see* 5962
Divalproex Sodium *see* 10049
Divanil *see* 2823
Divanillalcyclohexanone *see* 2823
2,6-Divanillylidenecyclohexanone *see* 2823
Divanon (obsolete) *see* 2823
Divarine [Tutag] *see* 3414
Divasil *see* 8666

Divegal [Waldheim] *see* 3217
Diverine [Courtois] *see* 3211
Divicine, 3448
Divicine 5-Glucoside *see* 10112
Divicine-β-glucoside *see* 10112
Dividend [Ciba] *see* 3182
Dividol [Zambon] *see* 10117
Diviminol *see* 10117
Divinyl *see* 1534
Divinylene Oxide *see* 4316
Divinylene Sulfide *see* 9490
Divinylenimine *see* 8198
Divinyl Ether *see* 10133
Divinyl Oxide *see* 10133
Divipan [Makhteshim-Agan] *see* 3129
Divit Urto *see* 10156
Dixanthogen, 3449
Dixarit [Boehringer, Ing.] *see* 2450
Dixeran [Lundbeck] *see* 5866
Dixiben [Benvegna] *see* 6446
Dixina *see* 3210
Dixnalate [Sana] *see* 4385
Di-*m*-xylylene *see* 5981
m-Dixylylene *see* 5981
N,N-Di-(2,4-xylyliminomethyl)methylamine *see* 510
Dixyrazine, 3450
Dizan [Elanco] *see* 3434
Dizocilpine, 3451
Dizol *see* 505
DJ-1461 *see* 1492
DJ-7041 *see* 8412
Djenkolic Acid, 3452
DK-7419 *see* 816
DKB *see* 3052
DKP *see* 7828
DL-152 *see* 1254
DL-458-IT *see* 2916
DL-473 *see* 8385
DL-717-IT *see* 5615
DL-832 *see* 4445
DL-3117 *see* 5082
DL-8280 *see* 6865
DLP [Dawbarn] *see* 7741
DLP-787 *see* 8171
DM *see* 7357
DM-9384 *see* 6528
DMAA *see* 5872
DMAC *see* 3277
DMC *see* 2136 and 7214
DMDL Protein *see* 10033
DMDP *see* 2432
DMDR *see* 4931
DMDT *see* 6070
DMDZ *see* 6784
DMF *see* 3292
DMFA *see* 3292
DMG *see* 3293
DMGG *see* 6001
DMI *see* 3298
DMMPA *see* 6528
DMN *see* 6735
DMNA *see* 6735
DMO *see* 3260
DMP *see* 3304
DMPA *see* 3453
DMPA [Rhône-Poulenc] *see* 3302
DMPO, 3454
DMPS *see* 3256
DMS *see* 3305 and 9034
DMS-70 *see* 3308
DMS-90 *see* 3308
DMSO *see* 3308
DMSO₂ *see* 3307
DMT *see* 3311 and 9306
DMTT *see* 2892
DN *see* 3331
DN-289 *see* 3341
DN-2327 *see* 7187
DNA *see* 2954

DNase I *see* 2953
DNBP *see* 3341
DNC *see* 3330 and 3331
DNFB *see* 4210
DNOCHP *see* 2802
DNOCP *see* 3340
DNP *see* 3423
DNS-55 *see* 3338
DNTP *see* 7167
DO-14 *see* 7992
Dobell's Soln *see* 8734
Dobenam [Angelini] *see* 3512
Dobendan [Merrell] *see* 2074
Doberol [Boehringer, Ing.] *see* 9653
Dobesilate Calcium, 3455
Dobesin [Pharmacia] *see* 3175
Dobren [Ravizza] *see* 9163
Doburil [Boehringer, Ing.] *see* 2822
Dobutamine, 3456
Dobutrex [Lilly] *see* 3456
Doca [Organon] *see* 2948
Docarpamine, 3457
Docelan [Chugai] *see* 4854
Docémine [Roussel] *see* 10152
Docetaxel, 3458
Docibin [Merrell] *see* 10152
Docigram [Lipha] *see* 10152
Dociton [Rhein-Pharma] *see* 8025
Docivit [Robisch] *see* 10152
Doclizid-T *see* 2123
[4a*S*-(4a*R**,5a*S**,10a*S**,11a*R**,15a*S**,-15b*R**,16*E*,18*E*,19a*R**,19b*S**)]-2,3,-4,4a,5,5a,6,7,8,9,10,10a,11,11a,12,-13,14,15,15a,15b,16,19a,19b-Docosahydro-1*H*-diindeno[2,1-*f*:1',2'-1][1,5]-diazacyclotridecine *see* 7153
6,7,8,9,10,11,12,13,14,15,16,17,24,25,-26,27,28,29,30,31,32,33-Docosahydro-35,37-dimethyl-5,34:18,23-diethenodibenzo[*b,r*][1,5,16,20]tetra-azacyclotriacontine-23,34-diium Dichloride *see* 1288
Docosanoic Acid *see* 1051
1-Docosanol *see* 8135
(*E*)-13-Docosenoic Acid *see* 1389
(*Z*)-13-Docosenoic Acid *see* 3713
Δ¹³-*cis*-Docosenoic Acid *see* 3713
Docoxyl Alcohol *see* 8135
Docusate Calcium, 3459
Docusate Potassium *see* 3460
Docusate Sodium, 3460
1,2,3,4,7,8,9,10,13,13,14,14-Dodecachloro-1,4,4a,5,6,6a,7,10,10a,11,12,-12a-dodecahydro-1,4:7,10-dimethanodibenzo[*a,e*]cyclooctene *see* 2908
1,1a,2,2,3,3a,4,5,5,5a,5b,6-Dodecachlorooctahydro-1,3,4-metheno-1*H*-cyclobuta[*cd*]pentalene *see* 6294
Dodecahedrane, 3461
(Dodecahydro-7β,10-dihydroxy-1α,-4bβ,8,8-tetramethyl-2(1*H*)-phenanthrenylidene)acetic Acid 2-(Dimethylamino)ethyl Ester *see* 1937
1,2,3,3a,3b,7,10,10a,10b,11,12,12a-Dodecahydro-1,11-dihydroxy-2,5,-10a,12a-tetramethyl-7-phenylcyclopenta[7,8]phenanthro[2,3-*c*]pyrazol-1-yl Hydroxymethyl Ketone Acetate *see* 2603
Dodecahydrodiphenylamine *see* 3145
Dodecahydro-3-hydroxy-6-(hydroxymethyl)-3,3a,6-trimethyl-1*H*-benz-[*e*]indene-7-acetic Acid δ-Lactone *see* 7054
[2*S*-(2α,7β,7aβ,14β,14aα)]-Dodecahydro-2-hydroxy-7,14-methano-2*H*,-11*H*-dipyrido[1,2-*a*:1',2'-*e*]-[1,5]diazocin-11-one *see* 4875
[3b*S*-(3bα,5aβ,7β,8β,10aα,10bβ)]-3b,-4,5,6,7,8,9,10,10a,10b,11,12-

Dorme [A.V.P.] see 7970
Dormicum [Roche] see 6270
Dormidin see 5890
Dormigen see 5890
Dormigene [Pharmacobel] see 1418
Dormigoa [Scheurich] see 6028
Dormin see 7
Dorminal see 607
Dormiphen see 5890
Dormison [Schering] see 5890
Dormodor [Roche] see 4233
Dormogen [Spofa] see 6028
Dormonal see 989
Dormonoct [Roussel] see 5604
Dormosan see 5890
Dormovit see 4326
Dormutil see 6028
Dormwell see 3097
Dormytal see 607
Dornase Alfa see 2953
Dornavac [Merck & Co.] see 2953
Dorner [Toray] see 1190
Dornwal [Maltbie] see 621
Dorsacaine [Dorsey] see 1076
Dorsallin "A.R." [Dorsey] see 7226
Dorsedin see 6028
Dorsiflex see 5896
Dorsilon see 5896
Doryl [E. Merck] see 1823
Doryx [Parke, Davis] see 3496
Dorzolamide, 3484
Dosberotec [Boehringer, Ing.] see 4022
Dostinex [Farmitalia] see 1637
Dosulepin see 3485
Dosulfin [Geigy] see 9106
Dothiepin, 3485
Double-mycin [Ileyl] see 3222
Double Thiosulfate of Gold and
  Sodium see 4537
Dovenix [Specia] see 6754
Dovonex [Leo] see 1679
Dow 1329 see 3453
Dowanol DPM [Dow] see 3407
Dowco 118 see 3453
Dowco 132 see 2671
Dowco 179 see 2242
Dowco 213 [Dow] see 2829
Dowco 214 see 2242
Dowco 217 see 4283
Dowco 233 see 9789
Dowco 290 see 2462
Dowco 356 see 9795
Dowco 433 see 4238
Dowex 1-X2-Cl see 2257
Dowfax 9N [Dow] see 6772
Dowfax 9N9 [Dow] see 6772
Dowfume W 85 see 3842
Dow General [Dow] see 3341
Dowicide 1 [Dow] see 7458
Dowicide 2 [Dow] see 9772
Dowicide 2S [Dow] see 9773
Dowicide 9 [Dow], 3486
Dowicide A [Dow] see 7458
Dowicide B [Dow] see 9772
Dowicide G [Dow] see 7242
Dowicide Q [Dow] see 2168
Dowlap see 9771
Dowmycin E [Dow] see 3726
Dowpen VK [Merrell-Dow] see 7230
Dowpon see 2869
Dow Selective [Dow] see 3341
Doxacurium Chloride, 3487
Doxans [Schiapparelli] see 3490
Doxapram, 3488
Doxapril [Erba] see 3488
Doxatet [Cox] see 3496
Doxazosin, 3489
Doxazosin Mesylate see 3489
Doxefazepam, 3490
Doxenitoin, 3491

Doxephrin see 6015
Doxepin, 3492
Doxergan see 7080
Doxicrisol [Quimifar] see 3496
Doxidan [Hoechst] see 3459
Doxifluridine, 3493
Doxinate [Hoechst] see 3460
Doxium [Carrion] see 3455
Doxofylline, 3494
Doxol [Blair] see 3460
Doxophylline see 3494
Doxorubicin, 3495
Doxychel Hyclate [Rachelle] see 3496
Doxychol see 7062
Doxycycline, 3496
Doxycycline Fosfatex see 3496
Doxycycline Hyclate see 3496
Doxylamine, 3497
Doxylar [Lagap] see 3496
Doxytem [Temmler] see 3496
Doyle [Tanabe Seiyaku] see 887
Dozic [Rosemont] see 4629
2,4-DP see 3128
DPA see 3370 and 7987
DPA Sodium see 10049
DPC see 3366
DPD see 1546
DPE see 3400
DPGME see 3407
DPM see 3407
DPN see 6429
DPNH see 6429
DPPC see 2540
DPPD see 3388
DPPH see 3389
D-Pron [Minnesota Pharm.] see 3414
DPX-1410 see 7052
DPX-3674 see 4734
DPX 4189 see 2244
DPX-5648 see 9130
DPX-F6025 see 2142
DPX-H6573 see 4240
DPX-PE350 see 8177
DPX-T6376 see 6244
DPX-Y5893-9 see 4751
DPX-Y6202 see 8269
DQ-2466 see 1924
DQ-2805 see 5104
D.Q.V.K. [Dista] see 7230
DR-3355 see 6865
Dracylic Acid see 1122
Draganon [Roche] see 700
Dragnet [FMC] see 7321
Dragon's Blood, 3498
Dragosantol [Dragoco] see 1281
Drainasept [Geistlich] see 9243
Drakeol [Penreco] see 7327
Dramamine [Searle] see 3252
Dramarin see 3252
Dramcillin-S [White] see 7369
Dramocen [Central Pharm.] see 3252
Dramyl see 3252
Drapolene [Calmic] see 1086
Drapolex see 1086
Drat [M & B] see 2204
Draza [Bayer] see 6050
Drazine [Smith & Nephew] see 7411
Drazoxolon, 3499
DRC-1201 see 5030
Drenaren see 9867
Drenison [Lilly] see 4232
Drenusil [Pfizer] see 7744
Driclor [Stiefel] see 348
Dricol [Bristol] see 418
Dridase [Pharm. Arzneimittel] see
  7089
Dri-Die [Roussel-UCLAF] see 8634
Dridol see 3505
Dried Barium Hydroxide see 1006
Dried Calcium Sulfate see 1753

Dried Cupric Sulfate see 2722
Dried Egg White see 216
Dried Ferrous Sulfate see 4105
Dried Gypsum see 1753
Dried Magnesium Sulfate see 5731
Dried Zinc Sulfate see 10293
Drierite see 1753
Drimenin, 3500
Drimyl [Cassenne] see 3921
Dri-Na see 8783
Drinalfa [Squibb] see 6015
Drinox see 4690
Drinupal see 2138
Driol [Sanofi] see 7018
Drisdol [Winthrop] see 10156
Drixin see 7100
Drocarbil see 815
Drocode see 3214
Drocort [Lilly] see 4232
Drofenine, 3501
Drogenil [Schering] see 4242
Drolban [Lilly] see 3504
Droleptan [Le Brun] see 3505
Droloxifene, 3502
Drometrizole, 3503
Dromilac see 4767
Dromisol [Merck & Co.] see 3308
Dromoran [Roche] see 5496
Dromostanolone, 3504
Dronabinol see 9349
Droncit [Bayer] see 7896
Drop Chalk see 1697
Droperidol, 3505
Dropp [Schering AG] see 9447
Droprenilamine, 3506
Droprenylamine see 3506
Dropropizine, 3507
Drosera, 3508
Drosophilin A, 3509
Drosophilin B see 7694
Drospirenone, 3510
Drostanolone see 3504
Drosteakard [Plantorgan] see 10083
Drotebanol, 3511
Droxar [Upjohn] see 3512
Droxarol [Pharmos] see 1497
Droxaryl [Continental Pharma] see
  1497
Droxicam, 3512
Droxidopa, 3513
Droxomin [Tutag] see 4854
Droxone (obsolete) [Squibb] see 239
DRP see 10033
Drumstick see 1942
Dry-Clox [Bristol] see 2480
Dry Ice see 1857
Drylin [Merckle] see 9086
Drynap [Wako] see 8783
Dryobalanops Camphor see 1366
Dryophantin see 8133
Dryptal [Berk] see 4331
DS-103-282 see 9624
DS-4823 see 8590
DSCG see 2658
DSIP, 3514
DSMA see 6020
DSP see 8805
DSS see 3460
DST see 3222
DT-327 see 2454
DTBP, 3515
DTC see 3443
DTF see 3144
DTIC see 2866
DTIC-Dome [Bayer] see 2866
DTMC see 3136
DTPA see 7266
[DTPA-D-Phe¹]-octreotide see 7267
DTPA-SMS see 7267
DTPT see 8068

D-Tracetten see 10156
DU see 3050
DU-1219 see 202
DU-21220 see 8401
DU-23000 see 4251
DU-28853 see 3592
DU-29373 see 4138
DU-112307 see 3188
DU-319722 see 4159
Duact [Burroughs Wellcome] see 129
Dual [Ciba-Geigy] see 6230
Dualar [Armour Pharm.] see 1118
Duatok [Am. Cyanamid] see 9115
Dubimax [Oberval] see 6440
Dubnium see 3580
Duboisine see 4907
"Dubos Crude Crystals" see 9972
Dufalone [Merck & Co.] see 3140
Dufaston [ISM] see 3524
Dufrenite see 4074
Dugro [Merck & Co.] see 8413
Duhnul-balsan see 975
Dulasi [Dukron] see 9160
Dulcamara, 3516
Dulcin, 3517
Dulcion [Dulcis] see 3692
Dulcite see 4350
Dulcitol see 4350
Dulcivac [Harvey] see 3460
Dulcolan see 1282
Dulcolax [Thomae] see 1282
Dulcose see 4350
Duloctil [Searle] see 9160
Duloxetine, 3518
Dumirox [Duphar] see 4251
Dumitone see 2885
Dumocyclin [Dumex] see 9337
Dumopen [Dumex] see 628
Duna [Tedec Zambeletti] see 7596
Duncaine see 5505
Duodin see 4826
Duofas [Merck & Co.] see 8280
Duogastral [ISM] see 7646
Duogastrone [Merrell] see 1839
Duolip [Merckle] see 9420
Duosan see 3632
Duoscorb [Upsher-Smith] see 867
Duosol [Interdelta] see 2549
Duotal see 4577
DUP-453 see 3756
DUP-753 see 5613
DUP-785 see 1394
Dupéran [Cassenne] see 2443
Duphacid [Duphar] see 3188
Duphacycline [Solvay-Duphar] see 7111
Duphafral D₃ 1000 see 10157
Duphalac [Duphar] see 5360
Duphaston [Roxane] see 3524
Duplamin [Bruschettini] see 7970
Duplosan KV [BASF] see 5826
Duponol see 4378
Du Pont 634 see 5459
Du Pont 753 see 5613
Du Pont Herbicide 732 see 9298
Du Pont Herbicide 326 see 5534
Du Pont Herbicide 976 see 1402
Duprene see 6550
Dura AX [Durachemie] see 617
Durabetason [Durachemie] see 1226
Durabol see 6452
Durabolin [Organon] see 6452
Durabolin-O [Organon] see 3851
Duracebrol [Durachemie] see 6580
Duracef [Bristol] see 1963
Duracide [Endura] see 7629
Duracillin [Lilly] see 7226
Duraclamid [Durachemie] see 6226
Duracreme [LRC] see 6772
Duracroman [Durachemie] see 2658

Duradoce [Atlantis] see 4854
Duradoxal [Durachemie] see 3496
Duraflex [McNeil] see 4176
Durafurid [Durachemie] see 4331
Duragel [LRC] see 6772
Duragentam [Durachemie] see 4398
Duraglucon [Durachemie] see 4486
Duralta-12 [Merck & Co.] see 4854
Duramax see 886
Durametacin [Durachemie] see 4998
Duramipress [Durachemie] see 7897
Duramucal [Durachemie] see 401
Duranest [Astra] see 3907
Duranifin [Durachemie] see 6617
Duranitrat [Durachemie] see 5245
Duranol [Elan] see 8025
Durapaediat [Durachemie] see 3720
Durapatite, 3519
Durapental [Durachemie] see 7278
Duraphat [Woelm] see 8762
Duraphyllin [Durachemie] see 9421
Durapindol [Durachemie] see 7597
Durapirenz [Durachemie] see 7646
Durapro [Wyeth-Ayerst] see 7057
Duraprost [Am. Home] see 7057
Duraquin [Warner Chilcott] see 8244
Duraset see 6188
Duraset 20W see 6188
Duraspiron [Durachemie] see 8917
Duravolten [Durachemie] see 3132
Durazanil [Durachemie] see 1406
Durazepam [Durachemie] see 7059
Durenat [Schering AG] see 9082
Durene, 3520
Duretic see 6086
Duricef [Mead Johnson] see 1963
Durohydroquinone, 3521
Durol see 3520
Durolax see 1282
Duromine [Riker] see 7415
Duromorph [L.A.B.] see 6359
Duronitrin [Andrews] see 9900
Duropenin see 7221
Duroprocin [Minerva] see 9085
Duroquinone, 3522
Durrax [Dermik] see 4897
Dursban [DowElanco] see 2242
Dusodril [Roland] see 6440
Duspatal [Duphar] see 5808
Duspatalin [Duphar] see 5808
Dutch Liquid see 3843
Duteplase see 9608
Du-Ter [Duphar] see 9875
Dutonin [Bristol-Myers Squibb] see 6527
Duvadilan [Duphar] see 5259
Duvaline [Almirall] see 8158
Duviculine see 5259
Duvoid [Norwich] see 1232
Duxima [Dukron] see 2002
DV1 see 5810
DV [Merrell-Dow] see 3153
DV-714 see 5457
DV-1006 see 2067
DVC see 2823
D₃-Vicotrat see 10157
DW-61 see 4135
DW-3418 see 2755
DWA-2114R see 6263
Dwale see 1052
Dwarf Bay see 6258
DX.C see 2982
Dya-Tron [Myers-Carter] see 3414
Dybar [Du Pont] see 4050
Dybenal see 3110
Dycholium see 2922
Dycill [Beecham] see 3134
Dyclone [Dow] see 3523
Dyclonine, 3523
Dydrogesterone, 3524

Dyer's Alkanet see 252
Dyer's Saffron see 1919
Dyflos see 5192
Dyfonate [Stauffer] see 4261
Dygratyl [Ferrosan] see 3223
Dykon see 8752
Dylate [Unimed] see 2452
Dylene [Arco] see 9028
Dylox [Bayer] see 9753
Dymadon [Wellcome] see 45
Dymanthine, 3525
Dymelor [Lilly] see 59
Dymid [Lilly] see 3364
Dymion [Pulitzer] see 9974
Dymixal [McNeil], 3526
Dynabac [Lilly] see 3418
Dynacaine see 8197
DynaCirc [Sandoz] see 5260
Dynacoryl see 6635
Dynacrine [Sandoz] see 5260
Dynafac see 2069
Dynamisan [Sandoz] see 875
Dynamite see 6704
Dynamos [Siphar] see 3210
Dynamutilin [Sintopharm] see 9559
Dynapen [Bristol] see 3134
Dynaphylline [Welcker-Lyster] see 23
Dynarsan see 49
Dynatra [Simes] see 3479
Dynel, 3527
Dynel [Union Carbide] see 6310
Dyneric [Merrell] see 2446
Dynese [Galen] see 5685
Dynorm [E. Merck] see 2332
Dynorphin, 3528
Dynorphin₁₋₁₃ see 3528
1-13-Dynorphin (Pig) see 3528
Dynothel [Henning] see 9555
Dynovas [Mayrand] see 7151
Dyodin see 5063
Dyphylline, 3529
Dypnone, 3530
Dyprin [Lincoln] see 6053
Dyren [Sejin] see 9731
Dyrene [Bayer] see 694
Dyrenium [SKB] see 9731
Dyscural see 3773
Dysect [Deosan] see 2836
Dysedon see 7080
Dysentulin see 4503
Dyskinébyl [Saunier] see 7094
Dysmalgine see 8273
Dysmenalgit [Krewel] see 6504
Dyspamet [Bridge] see 2337
Dyspas [Byk-Gulden] see 3147
Dysprosia see 3531
Dysprosium, 3531
Dystrophin, 3532
Dystrophin-related Protein see 10033
Dytac [Bridge] see 9731
Dytransin see 4924
DZL-221 see 6528
D-z-n [Ciba-Geigy] see 3043
"E" see 5806
E see 4477
E-3 see 5343
E36U31 see 9296
E-141 see 3766
E-187 see 6276
E-212 see 9698
E-250 see 8569
E-265 see 9701
E-600 see 7164
E-601 see 6183
E-605 see 7167
E-607 see 5379
E-614 see 9866
E-643 see 1512
E-646 see 3642
E-0659 see 939

E-0671 *see* 9296
E-0687 *see* 1255
E-838 *see* 4904
E-1020 *see* 5605
E-1040 *see* 1970
E-1059 *see* 2940
E-2020, 3533
E-2663 *see* 1081
E-3128 *see* 3512
E-3314 *see* 4690
E-3810 *see* 8272
E-4424 *see* 5470
EA-1299 *see* 5804
EACA *see* 451
Eagle Vine *see* 2560
EAK *see* 3810
Eaklite *see* 1749
Early Bird [Mentholatum] *see* 8139
Earthnut *see* 7190
Earthnut Oil *see* 7191
Earth Wax *see* 2033
Easter Flower *see* 8125
Eastern Poison Oak *see* 7710
East Indian Balmony *see* 2104
"East Indian Copaiba" *see* 974
East Indian Sandalwood Oil *see* 6940
Eastman No. 1361 *see* 8230
Eatan [Desitin] *see* 6667
Eau de Javelle *see* 4912
Eau de Labarraque *see* 8773
Eazaminum *see* 3157
EB-382 *see* 308
EB-1856 *see* 243
Ebalin [Allergopharma] *see* 1467
E-Base [Barr] *see* 3720
Ebastel [Almirall] *see* 3534
Ebastine, 3534
Ebert-Merz α-Acid *see* 6462
Ebert-Merz β-Acid *see* 6461
Ebesal [Hoechst] *see* 265
Ebimar [Evans], 3535
Ebivit *see* 10157
Ebrantil [Byk-Gulden] *see* 10003
Ebrodin [Ferrer] *see* 3536
Ebrotidine, 3536
Ebucin *see* 1712
Ebufac [DDSA] *see* 4925
Ebufos *see* 1670
Eburnal [Chiesi] *see* 3537
(3α,16α)-Eburnamenine-14-carboxylic
   Acid Ethyl Ester *see* 10128
(3α,16α)-Eburnamenin-14(15H)-one
   see 3537
Eburnamenin-14(15H)-one *see* 3537
(±)-Eburnamenin-14(15H)-one *see*
   3537
Eburnamonine, 3537
(+)-*cis*-Eburnamonine *see* 3537
Ebutol [Kaken] *see* 3764
E.C. 1.4.3.6 *see* 4755
E.C. 3.2.1.11 *see* 2990
E.C. 3.4.21.5 *see* 9525
E.C. 3.4.24.4 *see* 9422
E.C. 3.5.1.1 *see* 871
Ecabet, 3538
Ecadotril *see* 72
Ecarazine *see* 9640
Ecasolv [Lepetit] *see* 4685
Ecazide [Squibb] *see* 1817
Ecboline *see* 3707
Eccothal [Centerchem] *see* 9405
Ecdysone *see* 3539
α-Ecdysone *see* 3539
β-Ecdysone *see* 3539
Ecdysteroids, 3539
Ecdysterone *see* 3539
ECGF *see* 4117
Ecgonidine, 3540
Ecgonine, 3541
Ecgonine Benzoate *see* 1146

Ecgonine Cinnamate Methyl Ester *see*
   2361
Ecgonine Ethyl Ester Benzoate
   (Ester) *see* 2516
Ecgonine Methyl Ester Benzoate *see*
   2517
Echinacea, 3542
Echinacin *see* 3542
Echinenone, 3543
Echinochrome A, 3544
Echinomycin, 3545
Echinopsine, 3546
Echinuline, 3547
Echitamine, 3548
Echothiophate Iodide, 3549
Echovist [Schering] *see* 4353
Echujetin *see* 3205
Eclabron [U.S. Ethicals] *see* 4582
Eclipse Red *see* 1132
ECM [Grove] *see* 886
Ecobutazone [Empire] *see* 7431
Ecodipin [Ecosol] *see* 6617
Ecodurex [Ecosol] *see* 426
Ecofenac [Ecosol] *see* 3132
Ecolid [Ciba] *see* 2151
Ecolid Chloride [Ciba] *see* 2151
Econazole, 3550
Ecosporina [Ecobi] *see* 2032
Ecothiopate Iodide *see* 3549
Ecotrin [SK & F] *see* 886
Ecoval 70 *see* 1226
Ecovent [Ecosol] *see* 217
ECP *see* 3746
Ecrinal [Pharma Investi] *see* 4136
Ecristidine *see* 4758
"Ecstasy" *see* 5806
ECTEOLA-Cellulose, 3551
Ectiban [ICI] *see* 7321
Ectimar [Bayer] *see* 3918
Ectomin [Ciba-Geigy] *see* 2836
Ectopor [Ciba-Geigy] *see* 2836
Ectoral [Pitman-Moore] *see* 8415
Ectylurea, 3552
Ecuanil [Erba] *see* 5908
Eczecidin *see* 5052
Eczederm [Quinoderm] *see* 1675
ED *see* 3905
10-EDAAM *see* 3553
Edalene [Pharmuka] *see* 2337
10-EdAM *see* 3553
Edathamil *see* 3559
Edathamil Calcium Disodium *see* 3555
Edathamil Disodium *see* 3556
Edatrexate, 3553
EDB *see* 3842
EDC *see* 3843
EDDP *see* 3560
Edecril [Merck & Co.] *see* 3761
Edecrin [Merck & Co.] *see* 3761
Edemex [Savage] *see* 1155
Edemox [Wassermann] *see* 50
Edestin, 3554
Edetate Calcium Disodium, 3555
Edetate Dipotassium *see* 3559
Edetate Disodium, 3556
Edetate Sodium, 3557
Edetate Trisodium, 3558
Edetic Acid, 3559
Edetic Acid Calcium Disodium Salt
   see 3555
Edetic Acid Disodium Salt *see* 3556
Edetic Acid Sodium Iron Salt *see* 4076
Edetic Acid Tetrasodium Salt *see* 3557
Edetic Acid Trisodium Salt *see* 3558
EDGF-II *see* 4117
Edifenphos, 3560
Edifenphos *see* 3560
Edolan [Lepetit] *see* 3920
Edornat [Albert-Roussel] *see* 4869
Edoxudine, 3561

EDPA *see* 3909
Edpetiline *see* 4961
Edrophone Bromide *see* 3562
Edrophonium Chloride, 3562
Edrul [Bayer] *see* 6397
EDTA *see* 3559
EDTA Calcium *see* 3555
EDTA Dipotassium *see* 3559
EDTA Disodium *see* 3556
EDTA Tetrasodium *see* 3557
EDTA Trisodium *see* 3558
EDU *see* 3561
Edurid [Robugen] *see* 3561
EEDQ, 3563
E.E.S. [Abbott] *see* 3720
EFA *see* 3743
Efalexin [Maipe] *see* 2021
Efamast [Scotia] *see* 3953
Efamol [Efamol Res.] *see* 3953
Efcorlin [Glaxo] *see* 4828
Efcortelan [Glaxo] *see* 4828
Efcortelan Soluble [Glaxo] *see* 4828
Efcortesol [Glaxo] *see* 4828
Efektolol [Brenner-Efeka] *see* 8025
Effectin [Winthrop] *see* 1344
Effekton [Brenner-Efeka] *see* 3132
Effer-Syllium [J&J-Merck] *see* 7674
Effexor [Wyeth-Ayerst] *see* 10079
Effilone *see* 5984
Effluderm [Allergan] *see* 4219
Efflumidex [Allergan] *see* 4213
Effontil *see* 3911
Effortil [Boehringer, Ing.] *see* 3911
Effortilvet [Boehringer, Ing.] *see* 3911
Effusan *see* 3331
Efisol [Roland] *see* 2959
Eflornithine, 3564
Efloxate, 3565
Efodine [Fougera] *see* 7880
Efonidipine, 3566
Efortil *see* 3911
Efosite Al *see* 4278
Efpenix [Toyo Jozo] *see* 617
Efrane [Abbott] *see* 3621
Efrotomycin, 3567
Efroxine [Pennwalt] *see* 6015
Eftapan [Merckle] *see* 3666
Efudex [Roche] *see* 4219
Efudix [Roche] *see* 4219
Efuranol [Pfizer Taito] *see* 4955
Egacene *see* 4907
Egalin [Dainippon] *see* 8293
Egazil Duretter *see* 4907
EGb 761 *see* 4432
EGCG, 3568
EGDN *see* 3846
EGF *see* 3569
EGF-URO *see* 3569
EGF-Urogastrone, 3569
Egg Albumin *see* 7032
Egg Oil, 3570
Egg White *see* 216
EGIS-2062 *see* 8619
Eglen [Tatsumi] *see* 2365
Egmol *see* 7395
Egressin *see* 9550
EGYT-201 *see* 1060
EGYT 341 *see* 9642
EGYT 475 *see* 7547
EGYT-739 *see* 4588
EGYT-1050 *see* 2909
EGYT-2062 *see* 8619
EHDP [Monsanto] *see* 3908
Ehrlich 5 *see* 7082
Ehrlich 594 *see* 49
Ehrlich 606 *see* 851
Ehrlich's Reagent *see* 3280
EHS-laminin *see* 5364
EI-3911 *see* 7486
EI-47031 *see* 7490

EI-47470 *see* 5900
Eicosamethylnonasiloxane, 3571
Eicosanoic Acid *see* 802
all-cis-5,8,11,14,17-Eicosapentaenoic Acid *see* 3572
(all-Z)-5,8,11,14,17-Eicosapentaenoic Acid, 3572
(all-Z)-5,8,11,14-Eicosatetraenoic Acid *see* 803
Eikonogen *see* 475
Einalon S [Kodama] *see* 4629
EinsAlpha [Thomae] *see* 4864
Einsteinium, 3573
Ejibil *see* 3816
Eka-cesium *see* 4286
Ekamet [Sandoz] *see* 3936
Ekilan *see* 5896
Ekko [Fleming] *see* 7475
Ekomine [Hoechst] *see* 907
Ekonal [Yoshitomi] *see* 6759
Eksmin [Sumitomo] *see* 7321
Ektafos [Ciba] *see* 3137
Ektebin [Hefa] *see* 8076
Ektyl *see* 3552
Ekvacillin [Astra] *see* 2480
EL-349 *see* 8864
EL-103 *see* 9255
EL-107 *see* 5256
EL-110 *see* 1067
EL-119 *see* 7015
EL-161 *see* 3763
EL-179 *see* 5220
EL-222 *see* 3999
EL-466 *see* 7650
EL-500 *see* 4239
EL-531 *see* 672
EL-614 *see* 1410
EL-737 *see* 8275
EL-857 *see* 792
EL-857/820 *see* 792
EL-870 *see* 9580
EL-970 *see* 3974
EL-974 *see* 9567
Elaidic Acid, 3574
Elaiomycin, 3575
ELAM-1 *see* 8568
Elamol [Brocades] *see* 9641
Elan [Schwarz] *see* 5245
Elanpres *see* 6132
Elantan [Schwarz] *see* 5245
Elarzone [Alcon] *see* 7610
Elastase, 3576
Elasterin [Phoenix] *see* 4019
Elastin, 3577
Elastonon *see* 623
Elaszym [Eisai] *see* 3576
α-Elaterin *see* 2681
Elavil [Merck & Co.] *see* 511
Elayl *see* 3837
Elazor [Sigma-Tau] *see* 4158
Elbrol [Pfleger] *see* 8025
Elcatonin, 3578
Elcitonin [Toyo Jozo] *see* 3578
Elcosal *see* 3227
Elcosine *see* 9124
Eldéprine [Schering] *see* 8569
Eldepryl [Somerset] *see* 8569
Elder *see* 8499
Elderfield Pyrimidine Mustard *see* 3480
Eldisine [Lilly] *see* 10125
Eldopal [Brocades-Stheeman] *see* 5490
Eldopaque [Elder] *see* 4853
Eldopar [Weifa] *see* 5490
Eldopatec [Labatec] *see* 5490
Eldoquin [Elder] *see* 4853
Eldrin *see* 8456
Elecampane *see* 5023
Elecampane Camphor *see* 208
Elecol [Andromaco] *see* 9000

Elecor [Galenica] *see* 7919
Electrocortin [Ciba] *see* 226
Eledoisin, 3579
Element 104, 3580
Element 105, 3581
Element 106, 3582
Element 107, 3583
Element 108, 3584
Element 109, 3585
Elen [Yamanouchi] *see* 4972
Elenium [Polfa] *see* 2132
Elenolide, 3586
Eleuthera Bark *see* 1932
Elfwort *see* 5023
Elgetol *see* 3331
Elgodipine, 3587
Elics [Senju] *see* 515
Elidin [Endo] *see* 3261
Elieten [Nippon Kayaku] *see* 6226
Elimin [Sumitomo] *see* 6641
Eliminoxy *see* 2358
Elipten [Ciba] *see* 460
Elisal [Specia] *see* 9167
Elisor [Bristol-Myers Squibb] *see* 7894
Elite [Bayer] *see* 9253
Elixophyllin [Forest] *see* 9421
Elkapin [Gödecke] *see* 3934
Elkosil *see* 9124
Elkosin [Ciba] *see* 9124
Ellagic Acid, 3588
Ellatun [Basotherm] *see* 9702
Ellipticine, 3589
Elliptinium Acetate, 3590
Elliptone, 3591
Elmarin *see* 2238
Elmedal [Thiemann] *see* 7431
Elmetacin [Luitpold] *see* 4998
Elmifarma [Farmitalia] *see* 7548
Elmiron [Baker Norton] *see* 7276
Elobact [Cascan] *see* 2002
Elobromol [Chinoin] *see* 6300
Elocon [Schering-Plough] *see* 6324
Elodrine *see* 4830
Eloisin [Farmitalia] *see* 3579
Elon *see* 6097
Elrodorm *see* 4485
Elspar [Merck & Co.] *see* 871
Elsyl [Lakeside] *see* 6223
Eltanolone *see* 7910
Eltoprazine, 3592
Eltroxin [Glaxo] *see* 5497
Elvanol [Du Pont] *see* 7745
Elvaron [Bayer] *see* 3095
Elvorine [Lederle] *see* 4254
Elymoclavine, 3593
Elyzol [Dumex] *see* 6242
Elzogram [Lilly] *see* 1967
EM-923 *see* 4397
Emamectin, 3594
Emanation *see* 8278
Emanil *see* 4934
EMB *see* 3764
Embafume *see* 6108
Embanox *see* 1582
Embarin [Diabetylin] *see* 287
Embark [3M] *see* 5846
EMBAY 8440 *see* 7896
Embden Ester *see* 4298
"Embelic Acid" *see* 3595
Embelin, 3595
Embequin [M & B] *see* 5063
Embichen *see* 5815
Embichin 7 *see* 6817
Embikhin 7 *see* 6817
Embikhine *see* 5815
Embinal *see* 989
Embonic Acid *see* 7136
Embramine, 3596
Embutox *see* 2893
Emcol E-607 [Witco] *see* 5379

Emcol E-607S [Witco] *see* 5379
Emconcor [E. Merck] *see* 1336
Emcor [E. Merck] *see* 1336
Emcyt [Roche] *see* 3749
EMD-9806 *see* 7887
EMD-15700 *see* 6668
EMD-16923 *see* 5902
EMD-30087 *see* 1966
EMD-33512 *see* 1336
EMD-34946 *see* 5638
EMD-38362 *see* 8432
EMD-45609 *see* 1893
EMD-49980 *see* 8432
Emdabol [E. Merck] *see* 9596
Emdabolin [Chugai] *see* 9596
Emdisterone *see* 3504
Emedastine, 3597
Emedyl *see* 3252
Emepride [Roche] *see* 1454
Emepronium Bromide, 3598
Emerald *see* 1201
Emerald Green *see* 1398 and 2692
Emerest [Emery] *see* 7738
Emerest 2711 [Emery] *see* 7738
Emergil [Labaz] *see* 4224
Emericid *see* 5599
Emersol 132 [Emery] *see* 8959
Emes *see* 3252
Emesazine *see* 2365
Emeside [L.A.B.] *see* 3794
Emetamine, 3599
Emete-Con [Pfizer] *see* 1154
Emethibutin *see* 3875
Emetic Herb *see* 5579
Emeticon [Pfizer] *see* 1154
Emetid [Virbac] *see* 6226
Emetine, 3600
Emetine Hydrochloride *see* 3600
Emeto-Na *see* 743
Emex [Archifar] *see* 1833
Emflex [E. Merck] *see* 27
Emicholine-F [Dojin] *see* 2380
Emilace [Yamanouchi] *see* 6533
Emilene *see* 5683
Eminase [Beecham] *see* 712
Emitefur, 3601
Emivan [USV] *see* 3765
Emmolic Acid *see* 1959
Emodin, 3602
Emodin [Upjohn] *see* 4925
Emodin-*l*-rhamnoside *see* 4288
Emonapride *see* 6533
Emoren [Wassermann] *see* 7067
Emorfazone, 3603
Emorhalt [Sigurtà] *see* 9704
Emoril [Roemmers] *see* 1454
Emotival [Armstrong] *see* 5609
Emovate [Glaxo] *see* 2424
Empecid [Bayer-Takeda] *see* 2478
Emperal [Orion] *see* 6226
Empirin [Burroughs Wellcome] *see* 886
EMPP *see* 3642
Empyreumatic Oil of Juniper *see* 5285
EMQ *see* 3800
EMS *see* 3873
Emsorb 2500 [Henkel] *see* 8872
Emsorb 2515 [Henkel] *see* 8872
Emsorb 6900 [Emery] *see* 7742
Emtexate [Nordic] *see* 6065
Emtryl [Rhône-Poulenc] *see* 3315
EMU *see* 3720
Emulphogene [GAF] *see* 7737
Emulphor [GAF] *see* 7738
Emulsept (obsolete) *see* 5379
Emulsion 212 [Apex] *see* 4342
Emulsynt [Van Dyk] *see* 7738
E-Mycin [Upjohn] *see* 3720
E-Mycin E [Upjohn] *see* 3720
Emylcamate, 3604

Eraldin [ICI] see 7882
Erantin [Boehringer, Mann.] see 8024
Erasis [Orion] see 3721
Erasol see 5815
Eratrex [Bristol] see 3726
Eraverm [Asta] see 7617
Erazon [Krka] see 7661
Erbalax-N [Erba] see 4460
Erbia see 3679
Erbium, 3679
Erbium Sulfate, 3680
Erbocain [Heilit] see 4259
Erbon, 3681
Ercefurol see 6626
Ercefuryl [Robert et Carriére] see 6626
Erco-Fer see 4094
Ercoquin see 4863
Ercotina [Orion-Farmos] see 7989
Erdin, 3682
Erdmann's Salt, 3683
Erdosteine, 3684
Erebile see 2922
Eremeyevite see 339
Eremfat [Fatol] see 8382
4βH,5α-Eremophila-1(10),7(11)-dien-2-one see 10106
Eremursine see 4783
Ergadenylic Acid see 157
Ergamine [Burroughs Wellcome] see 4756
Ergamisol [Janssen] see 5486
Ergate [Vernleigh] see 3703
Ergenyl [Labaz] see 10049
Ergine see 5663
Ergobasine see 3694
Ergobel [Hormosan] see 6580
Ergocalciferol see 10156
Ergocalm [Brenner-Efeka] see 5611
Ergochrome AA(2,2') see 8561
Ergochrome CC(2,2') see 3691
Ergoclavine see 3695
Ergoclavinine see 3695
Ergocornine, 3685
Ergocorninine, 3686
Ergocristine, 3687
Ergocristinine, 3688
Ergocryptine, 3689
α-Ergocryptine see 3689
β-Ergocryptine see 3689
Ergocryptinine, 3690
Ergodesit [Graf] see 3692
Ergoflavin, 3691
Ergohydrin [Streuli] see 3692
Ergoklinine see 3694
Ergokryptine see 3689
Ergokryptinine see 3690
Ergoloid Mesylates, 3692
Ergomar [Fisons] see 3703
Ergometrine see 3694
Ergometrinine, 3693
Ergomimet [Klinge] see 3217
Ergonovine, 3694
Ergonovinine see 3693
Ergont [Desitin] see 3217
Ergoplus [Formosan] see 3692
Ergorone see 10156
Ergosine, 3695
Ergosinine see 3695
Ergostane, 3696
5α-Ergostane see 3696
5β-Ergostane see 2585
Ergostanol, 3697
(3β,5α)-Ergostan-3-ol see 3697
Ergostat [Parke, Davis] see 3703
(3β,22E)-Ergosta-5,7,9(11),22-tetraen-3-ol see 2925
(3β,5α,22E)-Ergosta-6,8,22-trien-3-ol see 4313

(3β,9β,10α,22E)-Ergosta-5,7,22-trien-3-ol see 5629
(3β,9β,22E)-9-Ergosta-5,7,22-trien-3-ol see 5238
(3β,10α,22E)-Ergosta-5,7,22-trien-3-ol see 8181
(3β,22E)-Ergosta-5,7,22-trien-3-ol see 3701
Ergosta-5:6,7:8,22:23-trien-3-ol see 3701
(3β,5α)-Ergost-7-en-3-ol see 3700
(3β,5α)-Ergost-8(14)-en-3-ol see 3698
(3β,5α)-Ergost-14-en-3-ol see 3699
(3β,24R)-Ergost-5-en-3-ol see 1776
α-Ergostenol, 3698
β-Ergostenol, 3699
γ-Ergostenol, 3700
Ergosterin see 3701
Ergosterol, 3701
9β-Ergosterol see 5238
Ergostetrine see 3694
Ergot, 3702
Ergotamine, 3703
Ergotaminine, 3704
Ergotartrat [Rosch & Handel] see 3703
Ergothioneine, 3705
L(+)-Ergothioneine see 3705
Ergotinine, 3706
Ergotocine see 3694
Ergoton-A [Azusa] see 3703
Ergotonin [Streuli] see 3217
Ergotop [Kwizda] see 6580
Ergotoxine, 3707
Ergotrate [Lilly] see 3694
Ergotrate-H [Lilly] see 3694
Ergotrate Maleate [Lilly] see 3694
Eridan see 3042
Erigeron, 3708
Eril [Asahi] see 3980
Erinitrit see 8793
Eriochrome Black T [Geigy], 3709
Eriodictin see 3710
Eriodictyol, 3710
Eriodictyon, 3711
Eriodictyonone see 4773
Eriosept [Kreussler] see 2959
Eriscel [Rachelle] see 3722
Erisimin see 4665
Erispan [Sumitomo] see 4164
Eritadenine, 3712
Eritrocina [Abbott] see 3720
Eritroger [Isnardi] see 3722
Ermetrine [Organon] see 3694
Ermysin [Britannia] see 3720
Ernodasa [ERN] see 8979
Eromycin see 3722
Erpalfa [Intes] see 2853
Errolon see 4331
Ertron see 10156
Erucic Acid, 3713
Ervasil [Gödecke] see 362
ERYC [Parke, Davis] see 3720
Erycen [Berk] see 3720
Erycin see 3720
Erycinum [Schering AG] see 3720
Erycorbin see 5141
Ery Derm [Abbott] see 3720
Eryliquid see 3720
Erymax [Parke, Davis] see 3720
Erypar [Parke, Davis] see 3726
Eryped [Abbott] see 3720
Erypo [Ortho-Cilag] see 3729
Eryprim [Scarium] see 3726
Erysan see 2157
Erysimin see 4665
Ery-Tab [Abbott] see 3720
Erythorbic Acid see 5141
Erythrene see 1534
Erythricine see 4400

Erythrite see 2495 and 3714
Erythritol, 3714
meso-Erythritol see 3714
Erythritol Anhydride, 3715
Erythritol Tetranitrate see 3716
Erythrityl Tetranitrate, 3716
Erythrocentaurin, 3717
Erythrocin [Abbott] see 3726
Erythrocin Lactobionate [Abbott] see 3724
Erythrocuprein see 9177
Erythro ES [Sanko] see 3720
Erythrogenic Acid see 5114
D-Erythrohexulose see 8109
Erythro-Holz [Pharma Holz] see 3720
α-Erythroidine, 3718
β-Erythroidine, 3719
Erythrol see 3714
Erythrol Tetranitrate see 3716
Erythromast 36 see 3720
Erythromid [Abbott] see 3720
Erythromycin, 3720
Erythromycin A see 3720
Erythromycin 2'-Acetate Octadecanoate (Salt) see 3721
Erythromycin Acistrate, 3721
Erythromycin Estolate, 3722
Erythromycin Gluceptate see 3723
Erythromycin Glucoheptonate, 3723
Erythromycin Glucoheptonic Acid Salt see 3723
Erythromycin Lactobionate, 3724
Erythromycin 9-[O-[(2-Methoxyethoxy)methyl]oxime] see 8433
Erythromycin Octadecanoate (Salt) see 3726
Erythromycin 2'-Propanoate see 3725
Erythromycin 2'-Propanoate Dodecyl Sulfate (Salt) see 3722
Erythromycin Propionate, 3725
Erythromycin Propionate Lauryl Sulfate see 3722
Erythromycin Stearate, 3726
Erythroped [Abbott] see 3720
Erythrophlamine, 3727
Erythrophleine, 3728
Erythropoiesis Stimulating Factor see 3729
Erythropoietin, 3729
Erythropterin, 3730
Erythro S [Sanko] see 3726
D-Erythrose, 3731
L-Erythrose, 3732
D-Erythrose 4-Phosphate, 3733
4-D-Erythrosephosphoric Acid see 3733
Erythrosine, 3734
Erythrosine B see 3734
Erythrosine BS see 3734
Erythrosine Extra Yellowish see 3233
Erythroxylon see 2515
L-Erythrulose, 3735
Erytrarsin see 1639
ES-132 see 1244
ES-771 see 7584
ES-902 see 7584
Esafosfina [Foscama] see 4297
Esametina see 4724
Esantene see 5009
Esaprazole, 3736
Esbatal [Burroughs Wellcome] see 1233
Esbecythrin see 2934
Esbericard [Schaper & Brümmer] see 2636
Esberidin [Schaper & Brümmer] see 10120
Esberiven [Schaper & Brümmer] see 5862

Ethanedioic Acid Barium Salt *see* 1014
Ethanedioic Acid Calcium Salt *see* 1732
Ethanedioic Acid Copper Salt *see* 2712
Ethanedioic Acid Diamide *see* 7050
Ethanedioic Acid Diammonium Salt *see* 571
Ethanedioic Acid Diethyl Ester *see* 3174
Ethanedioic Acid Dimethyl Ester *see* 6181
Ethanedioic Acid Disodium Salt *see* 8795
Ethanedioic Acid Magnesium Salt *see* 5712
Ethanedioic Acid Silver Salt *see* 8664
Ethanedioic Acid Strontium Salt *see* 9008
1,2-Ethanediol *see* 3844
1,1-Ethanediol Diacetate *see* 3857
1,2-Ethanediol Diacetate *see* 3845
1,2-Ethanediol Dinitrate *see* 3846
1,2-Ethanediol Monoacetate *see* 3847
Ethanedioyl Dichloride *see* 7046
1,2-Ethanedisulfonic Acid, 3769
Ethanedithioamide *see* 8436
1,2-Ethanedithiol, 3770
1,1'-(1,2-Ethanediyl)bisbenzene *see* 1245
N,N'-1,2-Ethanediylbis[*N*-butyl-4-morpholinecarboxamide] *see* 3317
[[1,2-Ethanediylbis[carbamodithioato]](2—)]manganese *see* 5761
[[1,2-Ethanediylbis(carbamodithioato)](2—)]manganese Mixt. with [[1,2-Ethanediylbis(carbamodithioato)](2—)]zinc *see* 5756
[[1,2-Ethanediylbis[carbamodithioato]](2—)]zinc *see* 10300
1,2-Ethanediylbiscarbamodithioic Acid Disodium Salt *see* 6426
[[*N,N'*-1,2-Ethanediylbis[*N*-(carboxymethyl)glycinato]](4—)-*N,N',O,-O',O^N,O^N*]calciate(2—)disodium *see* 3555
[[*N,N'*-Ethanediylbis[*N*-(carboxymethyl)glycinato]](4—)]-*N,N',O,O',O^N,-O^N*-ferrate(1—) Sodium *see* 4076
N,N'-1,2-Ethanediylbis[*N*-(carboxymethyl)glycine] *see* 3559
N,N'-1,2-Ethanediylbis[*N*-(carboxymethyl)glycine] Disodium Salt *see* 3556
N,N'-1,2-Ethanediylbis[*N*-(carboxymethyl)glycine] Tetrasodium Salt *see* 3557
N,N'-1,2-Ethanediylbis[*N*-(carboxymethyl)glycine] Trisodium Salt *see* 3558
1,2-Ethanediylbis[(2,6-dioxo-4,1-piperazinediyl)methylene]carbonic Acid Bis(2-methylpropyl) Ester *see* 8708
[1,2-Ethanediylbis(imino-4,1-phenylene)]bis[arsonic Acid] Disodium Salt *see* 3394
4,4'-(1,2-Ethanediyl)bis(1-isobutoxycarbonyloxymethyl)-2,6-piperazinedione *see* 8708
1,1'-(1,2-Ethanediyl)bis[4-methylbenzene] *see* 3445
(*E,E*)-2,2'-[1,2-Ethanediylbis(nitrilomethylidyne)]bisphenol *see* 8474
N,N'-1,2-Ethanediylbis[3-nitrobenzenesulfonamide] *see* 3343
2,2'-[1,2-Ethanediylbis(oxy)]bis-ethanol *see* 9802
2,2'-(1,2-Ethanediyldiimino)bis-1-butanol *see* 3764

1,1',1'',1'''-(1,2-Ethanediyldinitrilo)-tetrakis[2-propanol] *see* 3635
Ethane-1-hydroxy-1,1-diphosphonic Acid *see* 3908
Ethanenitrile *see* 68
Ethaneperoxoic Acid *see* 7293
Ethanethioamide *see* 9453
Ethanethioic Acid *see* 9454
Ethanethioic Acid *S*-[1-[2-(Acetyloxy)ethyl]-2-[[(4-amino-2-methyl-5-pyrimidinyl)methyl]formylamino]-1-propenyl] Ester *see* 51
Ethanethiol, 3771
Ethanimidamide Hydrochloride *see* 42
Ethanion *see* 1707
Ethanite *see* 9475
Ethanol *see* 3806
Ethanol Aluminum Salt *see* 350
Ethanolamine, 3772
Ethanolamine Oxiniacate *see* 7075
"Ethanolamine Plasmalogen" *see* 7677
Ethanoylaminoethanoic Acid *see* 81
Ethaphene *see* 9189
Ethavan [Monsanto] *see* 3904
Ethaverine, 3773
Ethbenzamide *see* 3776
Ethchlorvynol, 3774
Ethclorvynol *see* 3774
Ethebenecid, 3775
Ethene *see* 3837
2,2'-(1,2-Ethenediyl)bis[5-aminobenzenesulfonic Acid] *see* 636
1,1'-(1,2-Ethenediyl)bis[benzene] *see* 8972
4,4'-(1,2-Ethenediyl)bisbenzenecarboximidamide *see* 8970
2,2'-(1,2-Ethenediyl)bis[5-[[(phenylamino)carbonyl]amino]benzenesulfonic Acid] Disodium Salt *see* 1349
Ethene Homopolymer *see* 7728
Ethenesulfonic Acid Homopolymer Sodium Salt *see* 5645
Ethenetetracarbonitrile *see* 9336
Ethenol Homopolymer *see* 7745
Ethenone *see* 5309
Ethenylamidine Hydrochloride *see* 42
[1*S*-(1α,2α,4α,5β)]-2-(5-Ethenyl-1-azabicyclo[2.2.2]oct-2-yl)-1*H*-indole-3-ethanol *see* 2344
[1*S*-[1α,2α(3a*S**,8a*R**),4α,5β]]-8a-(5-Ethenyl-1-azabicyclo[2.2.2]oct-2-yl)-2,3,8,8a-tetrahydro-3a*H*-furo[2,-3-*b*]indol-3a-ol *see* 8232
Ethenylbenzene *see* 9028
[4*S*-(4*R**,7*R**,10*S**,13*S**)]-13-Ethenyl-1,3,4,5,7,8,10,11,12,13-decahydro-4-(hydroxymethyl)-8,10,13-trimethyl-7,10-bis(1-methylethyl)-6*H*-benzo-[*g*][1,4]diazonino[7,6,5-*cd*]indol-6-one *see* 9271
5-Ethenyl-3,4-dihydro-1*H*-pyrano[3,4-*c*]pyridin-1-one *see* 4400
[1*R*-(1α,4aβ,4bα,7α,10aα)]-7-Ethenyl-1,2,3,4,4a,4b,5,6,7,8,10,10a-dodecahydro-1,4a,7-trimethyl-1-phenanthrenecarboxylic Acid *see* 5217
[1*R*-(1α,4aβ,4bα,7β,10aα)]-7-Ethenyl-1,2,3,4,4a,4b,5,6,7,9,10,10a-dodecahydro-1,4a,7-trimethyl-1-phenanthrenecarboxylic Acid *see* 7582
(5*R*-*trans*)-5-Ethenyl-6-(β-D-glucopyranosyloxy)-5,6-dihydro-1*H*,3*H*-pyrano[3,4-*c*]pyran-1-one *see* 4403
[3*S*-(3α,4β,4aα)]-4-Ethenyl-3-(β-D-glucopyranosyloxy)-3,4,4a,5,6,7-hexahydro-8*H*-pyrano[3,4-*c*]pyridin-8-one *see* 972
[4a*R*-(4aα,5β,6α)]-5-Ethenyl-6-(β-D-glucopyranosyloxy)-4,4a,5,6-tetra-

hydro-4a-hydroxy-1*H*,3*H*-pyrano-[3,4-*c*]pyran-1-one *see* 9186
1,1'-Ethenylidenebis[benzene] *see* 3381
5-Ethenyl-5-(1-methylbutyl)-2,4,6(1*H*,-3*H*,5*H*)-pyrimidinetrione *see* 10131
[3a*R*-(3aα,4α,5aα,9aα,9bβ)]-5a-Ethenyloctahydro-4-hydroxy-3,9-bis(methylene)-2*H*-furo[2,3-*f*][2]-benzopyran-2,8(3*H*)-dione *see* 10098
(*S*)-5-Ethenyl-2-oxazolidinethione *see* 4525
Ethenyloxyethene *see* 10133
(3*R*-*cis*)-3-(3-Ethenyl-4-piperidinyl)-1-(6-methoxy-4-quinolinyl)-1-propanone *see* 10141
(3*R*-*cis*)-3-(3-Ethenyl-4-piperidinyl)-1-(4-quinolinyl)-1-propanone *see* 2348
1-Ethenyl-2-pyrrolidinone Homopolymer Compd with Iodine *see* 7880
1-Ethenyl-2-pyrrolidinone Polymers *see* 7879
Ethenzamide, 3776
Etheophyl [Lindopharm] *see* 9421
Ethephon, 3777
Ether *see* 3852
Ether Chloratus *see* 3829
Ether Hydrochloric *see* 3829
Ether Muriatic *see* 3829
Etherylate *see* 3759
Ethiazide, 3778
Ethidium *see* 4767
Ethimide *see* 3783
Ethinamate, 3779
Ethinazone *see* 3757
Ethine *see* 91
Ethinyl Estradiol, 3780
17-Ethinylestradiol *see* 3780
17α-Ethinylestradiol 3-Cyclopentyl Ether *see* 8239
Ethinylestrenol *see* 5659
17α-Ethinyl-17β-hydroxyestr-4-ene *see* 5659
17α-Ethinyltestosterone *see* 3787
Ethinyl Trichloride *see* 9769
Ethiodan *see* 5074
Ethiodized Oil, 3781
Ethiodol [Savage] *see* 3781
Ethiofos *see* 424
Ethion, 3782
Ethionamide *see* 3783
Ethioniamide *see* 3783
Ethionine, 3784
Ethiops Iron *see* 4087
Ethiops Mineral *see* 5944
Ethiozin, 3785
Ethirimol, 3786
Ethisterone, 3787
Ethizone *see* 9030
Ethmosine *see* 6351
Ethmozin *see* 6351
Ethmozine [Roberts] *see* 6351
Ethnine [Allen & Hanburys] *see* 7484
Ethocel *see* 3828
Ethodin [Winthrop] *see* 3760
Ethodryl [BDH] *see* 3165
Ethofat [Akzo] *see* 7738
Ethofibrate *see* 3923
Ethofumesate, 3788
Ethoheptazine *see* 3789
Ethohexadiol *see* 3790
Ethol *see* 2070
Ethopabate *see* 3791
Ethoprop, 3792
Ethopropazine, 3793
Ethoprophos *see* 3792
Ethosperse [Glyco] *see* 7737
Ethosuximide, 3794
Ethotoin, 3795
Ethotrimeprazine *see* 3938

Ethyl Chloroformate, 3831

Ethyl 2-[[[[(4-Chloro-6-methoxypyrimidin-2-yl)amino]carbonyl]amino]sulfonyl]benzoate *see* 2142

Ethyl *p*-Chlorophenoxyisobutyrate *see* 2436

Ethyl 2-(*p*-Chlorophenoxy)-2-methylpropionate *see* 2436

Ethyl (±)-2-[[α-(*p*-Chlorophenyl)-*p*-tolyl]oxy]-2-methylbutyrate *see* 1046

Ethyl Chlorophosphite *see* 3886

Ethyl α-Chloropropionate, 3832

Ethyl 2-[4-(6-Chloro-2-quinoxalinyloxy)phenoxy]propionate *see* 8269

Ethyl β-Chlorovinyl Ethynyl Carbinol *see* 3774

24α-Ethylcholestanol *see* 8968

24β-Ethyl-Δ⁵-cholesten-3β-ol *see* 8697

Ethyl Cinepazate *see* 2349

Ethyl Cinnamate *see* 2358

Ethyl Citrate *see* 2387

N-Ethyl-*o*-crotonotoluidide *see* 2661

(2-Ethylcrotonoyl)urea *see* 3552

(α-Ethylcrotonyl)carbamide *see* 3552

(α-Ethylcrotonyl)urea *see* 3552

Ethyl Cyanide *see* 8012

Ethyl Cyanoacetate, 3833

Ethyl 1-(3-Cyano-3,3-diphenylpropyl)-4-phenylisonipecotate *see* 3371

Ethyl 1-(3-Cyano-3,3-diphenylpropyl)-4-phenyl-4-piperidinecarboxylate *see* 3371

Ethyl Cyanoethanoate *see* 3833

5-Ethyl-5-cycloheptenylbarbituric Acid *see* 4689

α-Ethylcyclohexanemethanol Carbamate *see* 7945

3-Ethyl-4-cyclohexyl-4*H*-1,2,4-triazole *see* 4735

10-Ethyl-10-deazaaminopterin *see* 3553

Ethyl Decanoate *see* 3823

Ethyl 6-(*n*-Decyloxy)-7-ethoxy-4-hydroxyquinoline-3-carboxylate *see* 2910

5-Ethyl-3-(2′-deoxyribosyl)uracil *see* 3561

5-Ethyl-2′-deoxyuridine *see* 3561

Ethyl Diazoacetate *see* 3045

Ethyl Diazoethanoate *see* 3045

2-Ethyl-3-(3,5-dibromo-4-hydroxybenzoyl)benzofuran *see* 1093

2-Ethyl-3-(3,5-dibromo-4-hydroxybenzoyl)oxaindene *see* 1093

Ethyl Dibunate, 3834

Ethyl 2,7-Di-*tert*-butylnaphthalene-4-sulfonate *see* 3834

Ethyl 3,6-Di-*tert*-butyl-1-naphthalenesulfonate *see* 3834

Ethyl 4,4′-Dichlorobenzilate *see* 2174

Ethyldicoumarol Acetate *see* 3818

Ethyl *N*-[2-(Diethylamino)ethyl]-2-ethyl-2-phenylmalonamate *see* 3995

Ethyl [[3-[2-(Diethylamino)ethyl]-4-methyl-2-oxo-2*H*-1-benzopyran-7-yl]oxy]acetate *see* 2296

Ethyl Diethylmalonate, 3835

1-Ethyl-6,8-difluoro-1,4-dihydro-7-(3-methyl-1-piperazinyl)-4-oxo-3-quinolinecarboxylic Acid *see* 5592

Ethyl Digol *see* 1847

Ethyl 1,4-Dihydro-5-(acetoxycarbonyl)-2,6-dimethyl-4-(3-nitrophenyl)-3-pyridinecarboxylate *see* 6669

3-Ethyl-3,4-dihydro-6-chloro-7-sulfamyl-1,2,4-benzothiadiazine 1,1-Dioxide *see* 3778

9-Ethyl-6,9-dihydro-4,6-dioxo-10-propyl-4*H*-pyrano[3,2-*g*]quinoline-2,8-dicarboxylic Acid *see* 6524

4-(2-Ethyl-2,3-dihydro-1*H*-inden-2-yl)-1*H*-imidazole *see* 894

5-Ethyldihydro-5-(1-methylbutyl)-2-thioxo-4,6(1*H*,5*H*)-pyrimidinedione Monosodium Salt *see* 9487

1-Ethyl-1,4-dihydro-6,7-methylenedioxy-4-oxo-3-quinolinecarboxylic Acid *see* 7079

(3*S*-*cis*)-3-Ethyldihydro-4-[(1-methyl-1*H*-imidazol-5-yl)methyl]-2(3*H*)-furanone *see* 7578

trans-3-Ethyl-2,5-dihydro-4-methyl-*N*-[2-[4-[[[[(4-methylcyclohexyl)amino]carbonyl]amino]sulfonyl]phenyl]ethyl]-2-oxo-1*H*-pyrrole-1-carboxamide *see* 4449

3-Ethyl-6,7-dihydro-2-methyl-5-(morpholinomethyl)indol-4(5*H*)-one *see* 6315

1-Ethyl-1,4-dihydro-7-methyl-4-oxo-1,8-naphthyridine-3-carboxylic Acid *see* 6446

4-Ethyl-1,2-dihydro-2-(1-methyl-4-piperidinyl)-5-phenyl-3*H*-pyrazol-3-one *see* 7632

5-Ethyldihydro-5-(2-methyl-2-propenyl)-2-thioxo-4,6(1*H*,5*H*)-pyrimidinedione *see* 6011

5-Ethyldihydro-5-(1-methylpropyl)-2-thioxo-4,6(1*H*,5*H*)-pyrimidinedione *see* 9457

1-Ethyl-1,4-dihydro-4-oxo[1,3]dioxolo[4,5-*g*]cinnoline-3-carboxylic Acid *see* 2369

5-Ethyl-5,8-dihydro-8-oxo-1,3-dioxolo[4,5-*g*]quinoline-7-carboxylic Acid *see* 7079

8-Ethyl-5,8-dihydro-5-oxo-2-(1-piperazinyl)pyrido[2,3-*d*]pyrimidine-6-carboxylic Acid *see* 7613

1-Ethyl-1,4-dihydro-4-oxo-7-(4-pyridinyl)-3-quinolinecarboxylic Acid *see* 8426

N′-(1-Ethyl-1,2-dihydro-2-oxo-4-pyrimidinyl)sulfanilamide *see* 9070

8-Ethyl-5,8-dihydro-5-oxo-2-(1-pyrrolidinyl)pyrido[2,3-*d*]pyrimidine-6-carboxylic Acid *see* 7660

N-[1-[2-(4-Ethyl-4,5-dihydro-5-oxo-1*H*-tetrazol-1-yl)ethyl]-4-(methoxymethyl)-4-piperidinyl]-*N*-phenylpropanamide *see* 236

5-Ethyldihydro-5-phenyl-4,6(1*H*,5*H*)-pyrimidinedione *see* 7927

2-Ethyl-2,3-dihydro-3-[[4-[2-(1-piperidinyl)ethoxy]phenyl]amino]-1*H*-isoindol-1-one *see* 3928

Ethyl 4,4′-Dihydroxydicoumarinyl-3,3′-acetate *see* 3818

(1α,15β)-21-Ethyl-1,15-dihydroxy-4-methyl-16-methylene-7,20-cycloveatchan-12-one *see* 8865

N-Ethyl-3,4-dihydroxynorephedrine *see* 3356

2-Ethyl-3-(3′,5′-diiodo-4′-hydroxybenzoyl)benzofuran *see* 1113

2-Ethyl-3-(3,5-diiodo-4-hydroxybenzoyl)coumarone *see* 1113

2-Ethyl-3-(3′,5′-diiodo-4′-hydroxybenzoyl)oxaindene *see* 1113

Ethyl 3,5-Diiodosalicylate. *see* 3234

Ethyl 6,7-Diisobutoxy-4-hydroxyquinoline-3-carboxylate *see* 1524

S-Ethyl *N*,*N*-Diisobutylthiocarbamate *see* 1581

O-Ethyl *S*-[2-(Diisopropylamino)ethyl]methylphosphonothioate *see* 10172

3-Ethyl-6,7-dimethoxy-1-(phenylmethyl)isoquinoline *see* 6371

3-Ethyl-10-(3-dimethylamino-2-methylpropyl)phenothiazine *see* 3938

Ethyl 2-(Dimethylamino)-1-phenyl-3-cyclohexene-1-carboxylate *see* 9578

1-Ethyl-3-(3′-dimethylaminopropyl)-3-(6′-allylergoline-8′β-carbonyl)urea *see* 1637

7-Ethyl-1,4-dimethylazulene *see* 2081

Ethyl Dimethyl Carbinol *see* 7282

Ethyl *N*-[α,α-Dimethyl-β-(*p*-chlorophenyl)ethyl]carbamate *see* 2440

3-Ethyl-5,5-dimethyl-2,4-diketooxazolidine *see* 3762

N-Ethyl-*N*,1-dimethyl-3,3-di-2-thienylallylamine *see* 3875

N-Ethyl-*N*,1-dimethyl-3,3-di-2-thienyl-2-propenamine *see* 3875

Ethyl 1,1-Dimethylethyl Ether *see* 3821

N-Ethyl-*N*,*N*-dimethyl-1-hexadecanaminium Bromide *see* 2071

Ethyldimethyl(1-methyl-3,3-diphenylpropyl)ammonium Bromide *see* 3598

N-Ethyl-*N*,*N*-dimethyl-9-octadecen-1-aminium Bromide *see* 3836

Ethyldimethyl-9-octadecenylammonium Bromide, 3836

3-Ethyl-5,5-dimethyl-2,4-oxazolidinedione *see* 3762

5-Ethyl-3,5-dimethyl-2,4-oxazolidinedione *see* 7161

Ethyl *N*-Dimethylphosphoramidocyanidate *see* 9196

Ethyl 6-[2-(4,4-Dimethylthiochroman-6-yl)ethynyl] Nicotinate *see* 9249

(+)-(*R*)-α-Ethyl-*N*,*N*-dimethyl-α-[[(3,4,5-trimethoxybenzyl)oxy]methyl]benzylamine *see* 3987

(*R*)-α-Ethyl-*N*,*N*-dimethyl-α-[[(3,4,5-trimethoxyphenyl)methoxy]methyl]benzenemethanamine *see* 3987

1-Ethyl-2,5-dioxo-4-phenylimidazolidine *see* 3795

4-Ethyl-2,3-dioxopiperazine Carbonyl Ampicillin *see* 7616

7β-[D-α-(4-Ethyl-2,3-dioxo-1-piperazinecarboxamido)-β-(*S*)-hydroxybutanamido]-7α-methoxy-3-[5-(1-methyl-1,2,3,4-tetrazolyl)thiomethyl]-Δ³-cephem-4-carboxylic Acid *see* 1968

(6*R*,7*S*)-7-[(2*R*,3*S*)-2-(4-Ethyl-2,3-dioxo-1-piperazinecarboxamido)-3-hydroxybutyramido]-7-methoxy-3-[[(1-methyl-1*H*-tetrazol-5-yl)thio]methyl]-8-oxo-5-thia-1-azabicyclo[4.2.0]oct-2-ene-2-carboxylic Acid *see* 1968

7-[D-(—)-α-(4-Ethyl-2,3-dioxo-1-piperazinecarboxamido)-α-(4-hydroxyphenyl)acetamido]-3-[[(1-methyl-1*H*-tetrazol-5-yl)thio]methyl]-3-cephem-4-carboxylic Acid *see* 1981

(2*S*,5*R*,6*R*)-6-[(*R*)-2-(4-Ethyl-2,3-dioxo-1-piperazinecarboxamido)-2-phenylacetamido]-3,3-dimethyl-7-oxo-4-thia-1-azabicyclo[3.2.0]heptane-2-carboxylic Acid *see* 7616

[6*R*-[6α,7α,7(2*R**,3*S**)]]-7-[[2-[[(4-Ethyl-2,3-dioxo-1-piperazinyl)carbonyl]amino]-3-hydroxy-1-oxobutyl]amino]-7-methoxy-3-[[(1-methyl-1*H*-tetrazol-5-yl)thio]methyl]-8-oxo-5-thia-1-azabicyclo[4.2.0]oct-2-ene-2-carboxylic Acid *see* 1968

[6*R*-[6α,7β(*R**)]]-7-[[[[(4-Ethyl-2,3-dioxo-1-piperazinyl)carbonyl]amino]-(4-hydroxyphenyl)acetyl]amino]-3-[[(1-methyl-1*H*-tetrazol-5-yl)thio]-

Ethyl Malonate, 3869
Ethyl Menthane Carboxamide, 3870
N-Ethyl-4-menthane-3-carboxamide
see 3870
Ethyl Mercaptan see 3771
Ethyl(4-mercaptobenzenesulfonato-
S⁴)mercury Sodium Salt see 9450
Ethyl[2-mercaptobenzoato(2—)-O,S]-
mercurate(1—) Sodium see 9451
3-Ethylmercapto-10-(1'-methylpiper-
azinyl-4'-propyl)phenothiazine see
9449
Ethylmercuric Chloride, 3871
N-(Ethylmercuri)-p-toluenesulfonani-
lide, 3872
Ethyl Mesylate see 3873
N-Ethyl-N-methallyl-4-trifluoromethyl-2,6-dinitroaniline see 3763
Ethyl Methanesulfonate, 3873
Ethyl Methanesulfonic Acid see 3873
Ethylmethiambutene see 3875
β-Ethyl-6-methoxy-α,α-dimethyl-2-
naphthalenepropanoic Acid see 6012
4-[Ethyl(p-methoxy-α-methylphenethyl)amino]butyl 3,4-Dimethoxyben-
zoate see 5808
(E)-4-[1-Ethyl-2-(4-methoxyphenyl)-1-
butenyl]phenol see 5975
(all E)-N-Ethyl-9-(4-methoxy-2,3,6-
trimethylphenyl)-3,7-dimethyl-2,4,-
6,8-nonatetraenamide see 6367
5-Ethyl-5-(2-methyllallyl)-2-thiobar-
bituric Acid see 6011
Ethyl-β-methylallylthiobarbituric
Acid see 6011
3-Ethylmethylamino-1,1-di(2'-thien-
yl)but-1-ene see 3875
α-[1-(Ethylmethylamino)ethyl]ben-
zenemethanol see 3752
l-α-[1-(Ethylmethylamino)ethyl]benz-
yl Alcohol see 3752
α-Ethyl-β-[2-(methylamino)propyl]-β-
phenylbenzeneethanol Acetate see
6778
N-Ethyl-α-methylbenzeneethanamine
see 3809
(±)-N-Ethyl-α-methyl-1,3-benzo-
dioxole-5-ethanamine see 5805
5-Ethyl-5-(1-methyl-1-butenyl)bar-
bituric Acid Sodium Salt see 10118
5-Ethyl-5-(1-methyl-1-butenyl)-2,4,6-
(1H,3H,5H)-pyrimidinetrione
Sodium Salt see 10118
5-Ethyl-5-(1-methylbutyl)barbituric
Acid see 7272
5-Ethyl-5-(1-methylbutyl)-2,4,6(1H,-
3H,5H)-pyrimidinetrione see 7272
5-Ethyl-5-(3-methylbutyl)-2,4,6-(1H,-
3H,5H)-pyrimidinetrione see 607
5-Ethyl-5-(1-methylbutyl)-2-thiobar-
bituric Acid Sodium Salt see 9487
3-Ethyl-5-methyl-1,4-dihydro-2,6-
dimethyl-4-(3-nitrophenyl)-3,5-pyri-
dinedicarboxylate see 6669
3-Ethyl-2-methyl-5-dimethylamino-
indole see 5835
4-Ethyl-4-methyl-2,6-dioxopiperidine
see 1054
N-Ethyl-N-methyl-4,4-di-2-thienyl-3-
buten-2-amine see 3875
(17α)-13-Ethyl-11-methylene-18,19-
dinorpregn-4-en-20-yn-17-ol see
2971
N-Ethyl-3,4-methylenedioxyamphet-
amine see 5805
1-Ethyl-6,7-methylenedioxy-4(1H)-
oxocinnoline-3-carboxylic Acid see
2369
N-Ethyl-3,4-methylenedioxyphenyliso-
propylamine see 5805

1-Ethyl-6,7-methylenedioxy-4-quin-
olone-3-carboxylic Acid see 7079
Ethyl Methylene Phosphorodithioate
see 3782
(R*,S*)-4,4'-(1-Ethyl-2-methyl-1,2-
ethanediyl)bis[2-fluorophenol] see
1259
Ethyl Methyl Ether, 3874
2'-Ethyl-6'-methyl-N-(ethoxymethyl)-
2-chloroacetanilide see 58
N-Ethyl-N'-(1-methylethyl)-6-(methylthio)-1,3,5-triazine-2,4-diamine see
411
3-Ethyl-3-methylglutarimide see 1054
7-Ethyl-2-methyl-4-hendecanol Sulfate Sodium Salt see 8838
Ethyl Methyl Ketone see 6149
O-Ethyl O-[4-(Methylmercapto)phen-
yl]-S-n-propylphosphorothionothiol-
ate see 9164
(1α,12α,15β)-21-Ethyl-4-methyl-16-
methylene-7,20-cycloveatchane-1,-
12-15-triol see 6454
N-Ethyl-N-methyl-2-(1-methylethyl)-
cyclohexanecarboxamide see 3870
Ethyl 3-Methyl-4-(methylthio)phenyl
(1-Methylethyl)phosphoramidate see
3997
Ethyl-N-methylmorphinium Iodide see
3876
1-Ethyl-7-methyl-1,8-naphthyridin-4-
one-3-carboxylic Acid see 6446
Ethyl[2-(2-methyl-5-nitro-1-imidazol-
yl)ethyl]sulfone see 9588
[2R-[2α(2E,6E),3α]]-9-(3-Ethyl-3-
methyloxiranyl)-3,7-dimethyl-2,6-
nonadienoic Acid Methyl Ester see
5287
N-[4-[2-(3-Ethyl-4-methyl-2-oxo-3-
pyrroline-1-carboxamido)-ethyl]-
benzenesulfonyl]-N'-4-methylcyclo-
hexylurea see 4449
1-[4-[2-(3-Ethyl-4-methyl-2-oxo-3-
pyrroline-1-carboxamido)ethyl]-
phenylsulfonyl]-3-(4-methylcyclo-
hexyl)urea see 4449
2-Ethyl-3-methylpentanamide see
10048
(1-Ethyl-2-methylpentanoyl)urea see
1818
N-Ethyl-α-methylphenethylamine see
3809
α-Ethyl-p-[2-[(α-methylphenethyl)-
amino]ethoxy]benzyl Alcohol see
3996
5-Ethyl-1-methyl-5-phenylbarbituric
Acid see 5899
Ethyl(4-methyl-N-phenylbenzenesul-
fonamidato-N)mercury see 3872
N-Ethyl-N-(2-methylphenyl)-2-buten-
amide see 2661
α-Ethyl-4-[2-[(1-methyl-2-phenyleth-
yl)amino]ethoxy]benzenemethanol
see 3996
5-Ethyl-1-methyl-5-phenylhydantoin
see 6046
5-Ethyl-3-methyl-5-phenylhydantoin
see 5898
5-Ethyl-1-methyl-5-phenyl-2,4-imid-
azolidinedione see 6046
5-Ethyl-3-methyl-5-phenyl-2,4-imid-
azolidinedione see 5898
Ethyl 1-Methyl-4-phenylpiperidine-4-
carboxylate see 5894
5-Ethyl-1-methyl-5-phenyl-2,4,6(1H,-
3H,5H)-pyrimidinetrione see 5899
4-Ethyl-4-methyl-2,6-piperidinedione
see 1054
4'-Ethyl-2-methyl-3-piperidinopropio-
phenone see 3642

4-Ethyl-1-(1-methyl-4-piperidyl)-3-
phenyl-3-pyrazolin-5-one see 7632
N-Ethyl-N-(2-methyl-2-propenyl)-2,6-
dinitro-4-(trifluoromethyl)benzen-
amine see 3763
5-Ethyl-5-(1-methylpropyl)barbituric
Acid Sodium Salt see 1530
α-Ethyl-4-(2-methylpropyl)benzene-
acetic Acid see 1557
1-Ethyl-1-methylpropyl Carbamate see
3604
N-[3-(1-Ethyl-1-methylpropyl)-5-isox-
azolyl]-2,6-dimethoxybenzamide see
5256
5-Ethyl-5-(1-methylpropyl)-2,4,6-
(1H,3H,5H)-pyrimidinetrione Sodium Salt see 1530
5-Ethyl-5-(1-methylpropyl)-2-thiobar-
bituric Acid see 9457
1-Ethyl-1-methyl-2-propynyl Acid
Phthalate see 7530
3-Ethyl-4-methylpyridine see 3887
3-Ethyl-6-methylpyridine see 3889
4-Ethyl-2-methylpyridine see 3888
5-Ethyl-2-methylpyridine see 3889
β-Ethyl-γ-methylpyridine see 3887
3-Ethyl-3-methyl-2,5-pyrrolidinedione
see 3794
(±)-4'-Ethyl-2-methyl-3-(1-pyrrolidin-
yl)propiophenone see 4963
2-Ethyl-2-methylsuccinimide see 3794
α-Ethyl-α-methylsuccinimide see 3794
Ethylmethylthiambutene, 3875
Ethyl 3-Methyl-2-thioimidazoline-1-
carboxylate see 1844
Ethyl 4-(Methylthio)-m-tolyl Isopropylphosphoramidate see 3997
O-Ethyl O-[4-(Methylthio)phenyl]-
phosphorodithioic Acid S-Propyl
Ester see 9164
O-Ethyl O-[4-(Methylthio)phenyl] S-
Propylphosphorodithioate see 9164
N-Ethyl-α-methyl-3-(trifluoromethyl)-
benzeneethanamine see 4015
N-Ethyl-α-methyl-m-(trifluoromethyl)phenethylamine see 4015
7-Ethyl-2-methyl-4-undecanol Hydrogen Sulfate Sodium Salt see 8838
2-Ethyl-3-methylvaleramide see
10048
α-Ethyl-β-methylvaleramide see
10048
(2-Ethyl-3-methylvaleryl)urea see
1818
Ethyl Metribuzin see 3785
Ethylmorphine, 3876
Ethyl 4-Morpholino-2,2-diphenylbu-
tyrate see 3354
Ethyl 1-(2'-Morpholinoethyl)-4-
phenylpiperidine-4-carboxylate see
6357
1-Ethyl-4-[2-(4-morpholinyl)ethyl]-3,-
3-diphenyl-2-pyrrolidinone see 3488
Ethyl Mustard Oil see 3861
Ethyl Myristate see 6416
Ethyl β-Naptholate see 3799
Ethyl β-Naphthyl Ether see 3799
Ethyl N-[2,3-dihydro-2,2-dimethyl-
benzofuran-7-yloxycarbonyl(methyl)aminothio]-N-isopropyl-β-alanin-
ate see 1069
N-Ethyl-N-[4-[[4-[ethyl[(3-sulfophen-
yl)methyl]amino]phenyl](4-hydroxy-
2-sulfophenyl)methylene]-2,5-cyclo-
hexadien-1-ylidene]-3-sulfobenzene-
methanaminium Inner Salt, Disodium Salt see 3979
3-Ethylnirvanol see 5898
Ethyl Nitrite, 3877
Ethyl Nitrobenzoate, 3878

Name Index

Ethylstibamine, 3896
Ethyl Succinate *see* 9038
5-Ethyl-2-sulfanilamido-1,3,4-thia-
diazole *see* 9075
1-Ethyl-*N*-sulfanilylcytosine *see* 9070
Ethyl Sulfate, 3897
Ethyl Sulfhydrate *see* 3771
Ethyl Sulfide, 3898
Ethyl Sulfocyanate *see* 9464
4-Ethylsulfonylbenzaldehyde 3-Thio-
semicarbazone *see* 9030
p-Ethylsulfonylbenzaldehyde Thiosemicarbazone *see* 9030
2-(Ethylsulfonyl)ethanol, 3899
Ethylsulfonylethyl Alcohol *see* 3899
1-[2-(Ethylsulfonyl)ethyl]-2-methyl-5-
nitro-1*H*-imidazole *see* 9588
1-(Ethylsulfonyl)-4-fluorobenzene *see*
4197
2-[[4-(Ethylsulfonyl)phenyl]methyl-
ene]hydrazinecarbothioamide *see*
9030
2-Ethyl-5-(3-sulfophenyl)isoxazolium
Inner Salt *see* 10186
Ethyl[(*p*-sulfophenyl)thio]mercury
Sodium Salt *see* 9450
Ethylsulfuric Acid *see* 3897
Ethyl Tartrate, 3900
Ethyl Tartrate, Acid, 3901
Ethyl 1-(Tetrahydrofurfuryloxyethyl)-
4-phenylpiperidine-4-carboxylate *see*
4323
(±)-3-Ethyl-2,3,3a,4-tetrahydro-1*H*-in-
dolo[3,2,1-*de*][1,5]naphthyridine-6-
carboxylic Acid Methyl Ester *see*
10123
7-Ethyl-9,10,12,13-tetrahydro-6,9-
methano-5*H*-pyrido[1′,2′:1,2]aze-
pino[4,5-*b*]indole-6(6a*H*)-carboxylic
Acid Methyl Ester *see* 1952
3-Ethyl-1,5,6,7-tetrahydro-2-methyl-5-
(4-morpholinylmethyl)-4*H*-indol-4-
one *see* 6315
[7*R*-(7α,8β,10β)]-8-Ethyl-7,8,9,10-
tetrahydro-1,6,7,8,11-pentahydroxy-
10-[[2,3,6-trideoxy-3-(dimethylamino)-α-L-lyxo-hexopyranosyl]-
oxy]-5,12-naphthacenedione *see*
8357
2-[5-Ethyltetrahydro-5-[tetrahydro-3-
methyl-5-[tetrahydro-6-hydroxy-6-
(hydroxymethyl)-3,5-dimethyl-2*H*-
pyran-2-yl]-2-furyl]-2-furyl]-9-hydr-
oxy-β-methoxy-α,γ,2,8-tetrameth-
yl-1,6-dioxaspiro[4.5]decane-7-bu-
tyric Acid *see* 6329
[7*R*-(7α,8β,10β)]-8-Ethyl-7,8,9,10-
tetrahydro-1,6,8,11-tetrahydroxy-7,-
10-bis[[2,3,6-trideoxy-3-(dimethylamino)-α-L-lyxo-hexopyranosyl]-
oxy]-5,12-naphthacenedione *see*
8357
3′-Ethyl-5′,6′,7′,8′-tetrahydro-5′,5′,8′,-
8′-tetramethyl-2′-acetonaphthone *see*
10101
1-(3-Ethyl-5,6,7,8-tetrahydro-5,5,8,8-
tetramethyl-2-naphthalenyl)-
ethanone *see* 10101
*N*¹-(5-Ethyl-1,3,4-thiadiazol-2-yl)-
sulfanilamide *see* 9075
3-Ethylthio-4-amino-6-*tert*-butyl-1,2,4-
triazine-5-one *see* 3785
2-(Ethylthio)-4,6-bis(isopropylamino)-
s-triazine *see* 3405
6-(Ethylthio)-*N*,*N*′-bis(1-methylethyl)-
1,3,5-triazine-2,4-diamine *see* 3405
2-Ethyl-4-thiocarbamoylpyridine *see*
3783
Ethyl Thiocyanate *see* 9464
2-(Ethylthio)ethanol, 3902

2-Ethylthioisonicotinamide *see* 3783
2-(Ethylthio)-10-[3-(4-methyl-1-piper-
azinyl)propyl]phenothiazine *see*
9449
Ethyl Thiurad *see* 3428
Ethyl *p*-Toluenesulfonate, 3903
Ethyltrianol *see* 9253
Ethyl Trichloroacetate *see* 9756
Ethyl (2,2,2-Trichloro-1-hydroxyeth-
yl)carbamate *see* 1849
N-Ethyl-*N*-α,α,α-trifluoro-*N*-(2-meth-
ylallyl)-2,6-dinitro-p-toluidine *see*
3763
α-Ethyl-3-(trifluoromethyl)benzhydrol
see 4170
α-Ethyl-β-(2,4,6-triiodo-3-butyrami-
dophenyl)acrylic Acid Sodium Salt
see 1511
α-Ethyl-β-(2,4,6-triiodo-3-butyr-
amidophenyl)propionic Acid Sodium
Salt *see* 9968
α-Ethyl-β-(2,4,6-triiodo-3-hydroxy-
phenyl)propionic Acid *see* 5075
α-Ethyl-2,4,6-triiodo-3-[(1-oxobutyl)-
amino]benzenepropanoic Acid
Monosodium Salt *see* 9968
(1α,14α,16β)-20-Ethyl-1,14,16-tri-
methoxyaconitane-4,8,9-triol 4-[2-
(Acetylamino)benzoate] *see* 5378
Ethyl 4-(3,4,5-Trimethoxycinnamoyl)-
1-piperazineacetate *see* 2349
(1α,3α,6α,14α,15α,16β)-20-Ethyl-1,-
6,16-trimethoxy-4-(methoxymethyl)-
aconitane-3,8,13,14,15-pentol *see*
117
(1α,3α,6α,14α,15α,16β)-20-Ethyl-1,-
6,16-trimethoxy-4-(methoxymethyl)-
aconitane-3,8,13,14,15-pentol 8-
Acetate 14-Benzoate *see* 120
(1α,3α,6α,14α,15α,16β)-20-Ethyl-1,-
6,16-trimethoxy-4-(methoxymethyl)-
aconitane-3,8,13,14,15-pentol 8-
Acetate 14-(4-Methoxybenzoate) *see*
5278
16-Ethyl-1,6,19-trimethoxy-4-(methoxymethyl)aconitane-3,8,10,11-
tetrol 8-Acetate 10-Benzoate *see*
4964
(1α,3α,6α,14α,16β)-20-Ethyl-1,6,16-
trimethoxy-4-(methoxymethyl)-
aconitane-3,8,13,14-tetrol 8-Acetate
14-(3,4-Dimethoxybenzoate) *see*
8095
(1α,6β,14α,16β)-20-Ethyl-6,14,16-tri-
methoxy-4-(methoxymethyl)aconi-
tane-1,7,8-triol *see* 2933
(1α,6α,14α,16β)-20-Ethyl-1,6,16-
trimethoxy-4-(methoxymethyl)-
aconitane-8,13,14-triol 8-Acetate
14-(3,4-Dimethoxybenzoate) *see*
1262
Ethyl (2*E*,4*E*)-3,7,11-Trimethyldo-
deca-2,4-dienoate *see* 4850
3-Ethyl-*N*,*N*,2-trimethyl-1*H*-indol-5-
amine *see* 5835
Ethyltrimethylolmethane Trinitrate
see 7995
2-Ethyl-*N*,*N*,β-trimethyl-10*H*-pheno-
thiazine-10-propanamine *see* 3938
N-Ethyl-*N*,*N*,α-trimethyl-γ-phenylben-
zenepropanaminium Bromide *see*
3598
Ethyl 3,5,6-Tris-*O*-(phenylmethyl)-D-
glucofuranoside *see* 9738
α-Ethyltryptamine *see* 3937
Ethyl Urethan *see* 10013
Ethyl *n*-Valerate *see* 10042
Ethyl Vanillin, 3904
Ethyl Viologen *see* 10137

Ethylxanthic Acid Potassium Salt *see*
7876
Ethylxanthic Disulfide *see* 3449
Ethyl-*p*-xenylacetic Acid *see* 10205
Ethymal *see* 3794
Ethyne *see* 91
1,1′-(1,2-Ethynediyl)bisbenzene *see*
9643
Ethynodiol, 3905
17α-Ethynyl-4-androsten-17β-ol-3-
one *see* 3787
Ethynylbenzene, 3906
α-Ethynylbenzenemethanol Carbamate *see* 1886
α-Ethynylbenzyl Alcohol Carbamate
see 1886
α-Ethynylbenzyl Carbamate *see* 1886
2-Ethynyl-2-butanol *see* 5890
1-Ethynylcyclohexanol Carbamate *see*
3779
Ethynylcyclohexyl Carbamate *see*
3779
17α-Ethynyl-6α,21-dimethyltestoster-
one *see* 3268
1-Ethynyl-2,3,3a,3b,4,5,10,10a,10b,-
11,12,12a-dodecahydro-10a,12a-
dimethyl-1*H*-cyclopenta[7,8]phenan-
thro[3,2-*d*]isoxazol-1-ol *see* 2875
Ethynylestradiol *see* 3780
17α-Ethynylestradiol 3-Methyl Ether
see 5976
17α-Ethynyl-1,3,5(10)-estratriene-
3,17β-diol *see* 3780
17α-Ethynyl-4,9,11-estratrien-17β-ol-
3-one *see* 6798
17α-Ethynyl-4-estrene-3β,17β-diol *see*
3905
17α-Ethynylestr-4-en-17β-ol *see* 5659
17α-Ethynyl-13-methyl-17β-hydroxy-
4,15-gonadien-3-one *see* 4421
17α-Ethynyl-17β-hydroxy-4-andro-
steno[2,3-*d*]isoxazole *see* 2875
17α-Ethynyl-17β-hydroxy-4-androst-
en-3-one *see* 3787
17α-Ethynyl-17-hydroxy-6α,21-di-
methylandrost-4-en-3-one *see* 3268
17α-Ethynyl-17-hydroxy-5(10)-estren-
3-one *see* 6791
17α-Ethynyl-17β-hydroxy-18-methyl-
4,15-estradien-3-one *see* 4421
17α-Ethynyl-17β-hydroxy-3-oxo-4,9,-
11-estratriene *see* 6798
17α-Ethynyl-11β-methoxyestra-1,3,5-
(10)-triene-3,17β-diol *see* 6372
17α-Ethynyl-3-methoxy-1,3,5(10)-
estratrien-17β-ol *see* 5976
17α-Ethynyl-13β-methyl-Δ⁴,⁹,¹¹-gona-
triene-17β-ol-3-one *see* 6798
17α-Ethynyl-18-methyl-11-methylene-
Δ⁴-estren-17β-ol *see* 2971
17-Ethynyl-18-methyl-19-nortestos-
terone *see* 6797
17α-Ethynyl-19-norandrost-4-ene-3β,-
17β-diol *see* 3905
17α-Ethynyl-19-nortestosterone *see*
6790
17α-Ethynyltestosterone *see* 3787
Ethyol [U.S. Bioscience] *see* 424
Ethysine *see* 4013
Etianic Acid *see* 3913
Etibi [Gerot] *see* 3764
Eticol *see* 7164
Eticyclin [Ciba-Geigy] *see* 3780
Eticylol [Ciba-Geigy] *see* 3780
Etidocaine, 3907
Etidron [Gentili] *see* 3908
Etidronate Disodium *see* 3908
Etidronic Acid, 3908
Etifelmin, 3909
Etifoxine, 3910

Etiladrianol *see* 3911
Etilefrin, 3911
Etilon *see* 7167
Etioallocholane *see* 675
Etiocholane, 3912
Etiocholane-17β-carboxylic Acid *see* 3913
Etiocholanic Acid, 3913
Etiocobalamin, 3914
Etioporphyrin, 3915
Etioporphyrin III *see* 3915
Etiproston, 3916
Eti-Puren *see* 3911
Etiroxate, 3917
Etisazol, 3918
Etisine *see* 4013
Etivex [Kabi] *see* 3780
Etizolam, 3919
Etoclofene *see* 9312
Etodolac, 3920
Etodolic Acid *see* 3920
Etodroxizine, 3921
Etofen [Lusafarmaco] *see* 9312
Etofenamate, 3922
Etofibrate, 3923
Etofylline, 3924
Etofylline Clofibrate *see* 9420
Etofylline Nicotinate, 3925
Etoglucid, 3926
Etomidate, 3927
Etomide (Rescinded USAN) *see* 1846
Etomidoline, 3928
Etonitazene, 3929
Etopalin [Geigy] *see* 3959
Etoperidone, 3930
Etoposide, 3931
Etoquinol Sodium *see* 143
Etoran [Il Dong] *see* 3930
Etorphine, 3932
Etoscol [Byk-Gulden] *see* 4745
Etoval *see* 1550
Etoxadrol, 3933
Etozolin, 3934
ETP *see* 9291
Etrenol [Winthrop] *see* 4795
Etretin *see* 114
Etretinate, 3935
Etrimfos, 3936
Etroflex *see* 6060
Etrolene [Dow] *see* 8415
Etruscomicina *see* 5621
Etruscomycin *see* 5621
Etrynit [Bofors] *see* 7995
Etryptamine, 3937
ETTN *see* 7995
Ettriol Trinitrate *see* 7995
ETU *see* 3849
Etulos *see* 2014
Etumine [Wander] *see* 2476
Etydion [Nordic] *see* 3762
Etylon *see* 9340
Etymemazine, 3938
Etymide *see* 1846
Etyprenaline *see* 5185
EU-1806 *see* 6440
EU-4200 *see* 7648
EU-5306 *see* 7197
Eubasin [Nordmark] *see* 9108
Eubine [Pharkosar] *see* 7093
Eubornyl *see* 1368
β-Eucaine, 3939
Eucaine B *see* 3939
Eucalyptol, 3940
Eucalyptus, 3941
Eucalyptus Gum, 3942
Eucalyptus Kino *see* 3942
Eucardic [Boehringer, Mann.] *see* 1924
Eucardion [Dompé] *see* 8295
Eucast *see* 6577

Eucatropine, 3943
Eucazulen *see* 4581
Eucil [Farmasa] *see* 6226
Eucilat [Clin Byla] *see* 1070
Eucistin [San Carlo] *see* 6446
Euciton [Roux-Ocefa] *see* 3476
Euclidan *see* 6577
Euclorina *see* 2118
Eucodal *see* 7093
Eucodin *see* 2526
Eucoran *see* 6635
Eucortone *see* 169
Euctan [Synthelabo] *see* 9657
Eucupin [White] *see* 3950
Eucupine *see* 3950
Eucytol *see* 2954
Eudatin [Abbott] *see* 7172
Eudemine Injection [Schering] *see* 3051
Eudermol *see* 6611
Eudextran [Stholl] *see* 2989
Eudigox [Simes] *see* 3210
Eudilat [Clin Byla] *see* 1070
EUDR *see* 3561
Eudyna [Nordmark] *see* 8333
Eufibron [Berlin-Chemie] *see* 8056
Eufin *see* 3827
Euflavine *see* 125
Euflex [Schering] *see* 4242
Eugallol [Knoll] *see* 8184
Eugenic Acid *see* 3944
Eugenol, 3944
Euglucon [Roussel] *see* 4486
Euhypnos [Sigma] *see* 9285
Euipnos [Farmitalia] *see* 9285
Eukodal [E. Merck] *see* 7093
Eukraton [Nordmark] *see* 1054
Eukystol [Merckle] *see* 4629
Eulaxan [Ferring] *see* 1282
Euloxin [Schering] *see* 4242
Eulip [SIT] *see* 9556
Eulipos [Boehringer, Mann.] *see* 9555
Eu-Med [Med. Fabrik] *see* 45
Eumelanins *see* 5854
Euminex [Lacer] *see* 1456
Eumotol [Byk-Gulden] *see* 1507
Eumovate [Glaxo] *see* 2424
Eumulgin [Henkel] *see* 7737
Eumydrin [Winthrop] *see* 907
Eunades [Farmitalia] *see* 10127
Eunal [Schering] *see* 5541
Eunarcon [Cassella-Riedel] *see* 6509
Eunasin [Bracco] *see* 6223
Eunatrol *see* 6965
Eunephran [Servier] *see* 1554
Eunerpan [Nordmark] *see* 5870
Eunoctal [Houdé] *see* 607
Eunoctin [Gedeon Richter] *see* 6667
Euonymit *see* 4350
Euonymus, 3945
Euparen(e) [Bayer] *see* 3095
Euparin, 3946
Eupatal *see* 208
Eupatorin, 3947
Eupatorium, 3948
Eupavarin (formerly) *see* 6171
Eupaverin [E. Merck] *see* 6371
Eupaverina [Bracco] *see* 6371
Euphorbia, 3949
Euphthalmine [Warner-Chilcott] *see* 3943
Euphyllin [Byk Gulden] *see* 9421
Euphyllin CR [Byk-Gulden] *see* 485
Euphylong [Byk Gulden] *see* 9421
Euplit [Desitin] *see* 511
Eupnéron [Lyocentre] *see* 3670
Euporphin *see* 787
Eupractone [Baxter] *see* 3260
Eupragin [Alcon-Thilo] *see* 3722
Euprax [Baxter] *see* 218

Eupressyl [Valpan] *see* 10003
Euprocin, 3950
Euprovasin [Magis] *see* 8025
Euradal [Lacer] *see* 1336
Eurax [Geigy] *see* 2661
Euraxil [Thomae] *see* 2661
Eureceptor [Zambon] *see* 2337
EureCor [Kade] *see* 5245
Eurekene *see* 10049
Eurelix [Hoechst] *see* 7647
Euresol [Knoll] *see* 8323
Eurex [Labaz] *see* 7897
Eurodin [Takeda] *see* 3744
Eurodopa [Europharma] *see* 5490
European White Birch *see* 1236
Europia *see* 3951
Europium, 3951
Eurosan [Mepha] *see* 3042
Eusaprim [Wellcome] *see* 9086
Eusmanid *see* 1233
Eusovit [Woelfer] *see* 10160
Euspiran *see* 5236
Eutagen *see* 7093
Euteberol [Merckle] *see* 8917
Eutensin [Hoechst] *see* 4331
Eutensol *see* 4589
Euthyrox [E. Merck] *see* 5497
Eutocol [Gador] *see* 3746
Eutonyl [Abbott] *see* 7172
Eutrit [Takeda] *see* 10218
Euvaderm [Gödecke] *see* 1226
Euvasal [Synthelabo] *see* 9160
Euvernil [Heyden] *see* 9101
Euvestin *see* 3177
Euvitol [Allen & Hanburys] *see* 4006
E-VA-16 *see* 4967
Evacalm [Unimed] *see* 3042
Evaclin *see* 5194
Evacort [Evans] *see* 1828
Evadene [Ayerst] *see* 1568
Evadol [Clin-Comar-Byla] *see* 5194
Evadyne [Ayerst] *see* 1568
Evan's Blue, 3952
Evansite *see* 371
Evazol [Ravensberg] *see* 2959
"Eve" *see* 5805
Evening Primrose Oil, 3953
Eventin [Minden] *see* 2776
Everfree [Cooper] *see* 6693
Everninic Acid Methyl Ester *see* 8883
Everone [Hyrex] *see* 9322
E-Vicotrat [Heyl] *see* 10160
Evik [Ciba-Geigy] *see* 411
E-Vimin [Astra] *see* 10160
Evion [E. Merck] *see* 10160
Evipal [Winthrop] *see* 4742
Evipal Sodium [Winthrop] *see* 4742
Evipan [Bayer] *see* 4742
Evipan Sodium [Bayer] *see* 4742
Evital [Sandoz] *see* 6794
Evodiamine, 3954
Evonogenin *see* 3205
Evorel [Ortho] *see* 3746
Evoxin [Sterling-Winthrop] *see* 3476
Evramycin [Wyeth] *see* 9899
Ewer-Pick Acid *see* 6460
Ex-12-095 *see* 3758
EX-4810 *see* 405
EX-4883 *see* 8409
Exacin [Toyo Jozo] *see* 5121
Exacor [Searle] *see* 2330
Exacyl [Sanofi Winthrop] *see* 9704
Exal *see* 10119
Exalamide, 3955
Exalgin *see* 6088
Exaltolide [Firmenich], 3956
Examen *see* 5575
Exametazime, 3957
Exandron [Chantal] *see* 2372
Exangit *see* 1127

Exaprazole see 3736
Exceed [Ciba-Geigy] see 8067
Exceglan [Dainippon] see 10323
Excegram [Dainippon] see 10323
Excenel [Upjohn] see 1999
EXD see 3449
Exdol [Frosst] see 45
Exelderm [ICI] see 9062
Exhelm [Pfizer] see 6348
Exhirud [Plantorgan] see 4754
Exhirudine see 4754
Exhoran see 3428
Exifone, 3958
Exiproben, 3959
Exirel [Pfizer] see 7644
Exitelite see 752
Exlan see 2644
Exlutena [Organon] see 5659
Exluton(a) [Organon] see 5659
Exmigra see 3703
ExNa [Robins] see 1155
Exoamylases see 640
Exocin [Allergan] see 6865
Exocorpol [Green Cross] see 7722
Exoderil [Sandoz] see 6442
Exodextranases see 2990
Exodor-Grun see 2207
Exofene see 4716
Exolan see 723
(±)-2-Exo-(2-methylbenzyloxy)-1-
    methyl-4-isopropyl-7-oxabicyclo-
    [2.2.1]heptane see 2354
Exonal [Toyama] see 9267
Exopan [Ciba] see 2775
Exopon [Ciba] see 2775
Exosalt [Bayer] see 1155
Exosulfonyl [Theraplix] see 9047
Exosurf [Burroughs Wellcome], 3960
Exotherm Termil [Diamond] see 2219
EXP-105-1 see 389
EXP-126 see 8390
EXP-999 see 6233
EXP-3174 see 5613
Expansine see 7185
Experimental Insecticide No. 269 see
    3618
Exponcit N [Salutas Fahlberg-List] see
    6811
Expressed Almond Oil see 6945
Exrheudon N [Optimed] see 7431
Exsel [Herbert] see 8580
Exsiccated Alum see 373
Exsiccated Ammonium Alum see 335
Exsiccated Ferrous Sulfate see 4105
Exsiccated Sodium Phosphate see
    8805
Extacol [Nikken] see 7412
Ext. D & C Orange 3 see 6989
Ext. D & C Red No. 14 see 10222
Ext. D & C Red No. 15 see 7749
Ext. D & C Yellow No. 1 see 5989
Ext. D & C Yellow No. 7 see 6480
Ext. D & C Yellow No. 9 see 10228
Ext. D & C Yellow No. 10 see 10229
Extencilline see 7221
Exterol [Dermal] see 10007
Extracort [Basotherm] see 9728
Extramycin [Bayer] see 8695
Extranase [Rorer] see 1409
Extrinsic Factor see 10152
Extrinsic Plasminogen Activator see
    9608
Exuril see 2221
Exypaque [Nyegaard] see 5068
Eye-derived Growth Factor-II see
    4117
Eyekas [Showa Shinyaku] see 8367
E-Z-Paque [E-Z-Em Co.] see 1023
F see 7425
F-4 see 930

F-151 see 6010
F-190 see 49
F-319 see 4905
F-440 see 2879
F-1500 see 9047
F-1983 see 8194
F-1991 see 1073
F-2207 see 6281
F-2559 see 4361
F-6060 see 2789
F-6066 see 2789
F-28249α see 6532
FA see 4206
FA-402 see 4048
FA-2071 see 5093
2-FAA see 4192
Fabahistin [Bayer] see 5809
Fabianol [Ciba-Geigy] see 425
Fabrol [Geigy] see 89
F Acid see 1940
Factitious Air see 6751
Factor I see 4116
Factor II see 8074
Factor III see 9527
Factor IIa see 9525
Factor V, 3961
Factor VII, 3962
Factor VIII, 3963
Factor IX, 3964
Factor IX Complex (Human) see 3964
Factor X, 3965
Factor XI, 3966
Factor XII, 3967
Factor XIII, 3968
Factor B see 3914
Factor T see 10164
Factorate [Armour Pharm.] see 3963
Factrel [Ayerst] see 5500
FAD see 4131
Fademin [Chugai] see 4131
Fado [Errekappa] see 1964
Fadormir see 6028
Fadrozole, 3969
Faecla see 7776
Faecula see 7776
Fagarasterol see 5636
Fagarine, 3970
α-Fagarine see 264
β-Fagarine see 8701
γ-Fagarine see 3970
Fagarol see 8616
Fagine see 2259
Fairy Gloves see 3201
FAK III see 4275
Falapen [Frosst] see 7225
Falcon [Cyanamid] see 7990
Falcopen-V [Faulding] see 7231
Falicain see 8016
Falignost [Hexal] see 5069
Falimint see 6726
Falithrom [Salutas Fahlberg-List] see
    7413
Falmonox [Winthrop] see 9262
Falomesin [Chugai] see 1997
False Hellebore see 165
False Indigo see 985
False Saffron see 1919
False Unicorn see 4664
False White Cedar see 9531
False Winter's Bark see 1791
Falvin [Farmades] see 4047
Famciclovir, 3971
Famet [Calmic] see 9084
Famodil [Sigma-Tau] see 3972
Famodine [Farmasa] see 3972
Famophos see 3973
Famosan [Alkaloid] see 3972
Famosept [Spofa] see 7456
Famotidine, 3972
Famoxal [Silanes] see 3972

Famphur, 3973
Fampridine, 3974
Famvir [SKB] see 3971
Fanasil [Roche] see 9074
Fangorex [Siegfried] see 6978
Fanodormo see 2780
Fanosin [Abello] see 3972
Fansidar [Roche] see 8169
Fansimef [Roche] see 5845
Fantacin [Kyowa Hakko] see 4172
Fantofarone, 3975
Fantorin [Glaxo] see 8967
Fanzil [Roche] see 9074
2-F-araA see 4162
2-F-ara-AMP see 4162
Farecef [Lafare] see 1981
Faredina [Lafare] see 2025
Faremicin [Lafare] see 4280
Fareston [Farmos] see 9688
Faretrizin [Lafare] see 1965
Farexin [Lafare] see 2021
Far-Go [Monsanto] see 9726
Farial [Nordmark] see 4967
Farlutal [Farmitalia] see 5838
Farmicetina [Rontag] see 2120
Farmidril [Farmitalia] see 5398
Farmiglucin [Farmitalia] see 7173
Farminosidin [Farmitalia] see 7173
Farmiserina [Farmitalia] see 2820
Farmitalia 204/122 see 9078
Farmolisina [Centralvet] see 3414
Farmorubicin [Farmitalia] see 3660
Farmoxin [Erba] see 1986
Farnesene see 3976
α-Farnesene, 3976
β-Farnesene, 3977
Farnesol, 3978
Farnoquinone see 5876
Fas-Cile 200 see 5908
Fasigin [Pfizer] see 9588
Fasigyn [Pfizer] see 9588
Fasinex [Ciba-Geigy] see 9784
Fastac [Agrishell] see 2836
Fast Green see 5739
Fast Green FCF, 3979
Fastin [SKB] see 7415
Faston [Duphar] see 3252
Fast Scarlet see 1132
Fastum [Menarini] see 5316
Fast Yellow see 9906
Fasudil, 3980
Fasupond [Funcke] see 6811
Fat Ponceau R see 8538
Fatroximin [Fatro] see 8386
all-cis-Fatty Acid 20:5 Omega-3 see
    3572
Faustan [East Germany] see 3042
Faverin [Duphar] see 4251
Favistan [Asta] see 6049
Fazadinium Bromide, 3981
Fazadon [Duncan, Flockhart] see 3981
Fazol [Schering] see 5176
Fazor see 5745
FB₁ see 4311
FB/2 see 3415
FB-5097 see 2478
FBA-1420 see 7986
FBA-1464 see 4586
FBA-4503 see 8017
5-FC see 4161
FC 11 see 9770
FC-54 see 243
FC-1157a see 9688
FC-3001 see 9562
FCCs see 2192
FCE-21336 see 1637
FCE-22101 see 8400
FCE-22891 see 8400
F-Cortef [Upjohn] see 4166
FCR-1272 see 2826

FCRC-A48 *see* 4291
FCV *see* 3971
FD & C Blue No. 1 *see* 1397
FD & C Blue No. 2 *see* 4978
FD & C Green 1 *see* 4603
FD & C Green No. 2 *see* 5508
FD & C Green No. 3 *see* 3979
FD & C Orange I *see* 6989
FD & C Red No. 1 *see* 7749
FD & C Red No. 2 *see* 391
FD & C Red No. 3 *see* 3734
FD & C Red No. 4 *see* 7750
FD & C Red No. 32 *see* 10222
FD & C Red No. 40 *see* 292
FD & C Yellow No. 1 *see* 6480
FD & C Yellow No. 3 *see* 10228
FD & C Yellow No. 4 *see* 10229
FD & C Yellow No. 5 *see* 9239
FD & C Yellow No. 6 *see* 9176
FDA-20 *see* 4221
FDA-1541 *see* 3674
FDC *see* 4221
^{18}FDG *see* 4163
FDNB *see* 4210
Featherfew *see* 4112
Featherfoil *see* 4112
Febantel, 3982
Febarbamate, 3983
Febramine *see* 2065
Febrifugine, 3984
Febrilex *see* 45
Febrinina *see* 495
Febuprol, 3985
Fecapentaene 12 *see* 3465
Fecarb [Ayerst] *see* 4089
Feclemine, 3986
Fectrim [DDSA] *see* 9086
Fedacilina [Fedal] *see* 5986
Fedan [ICI] *see* 6960
Fedibaretta *see* 7368
Fedotozine, 3987
Feguanide [Bengal] *see* 7376
Feinalmin [Sanko] *see* 4955
Felacrinos *see* 2922
Felbamate, 3988
Felbamyl [Wallace] *see* 3988
Felbatol [Carter-Wallace] *see* 3988
Felbinac, 3989
Feldene [Pfizer] *see* 7661
Felicur [Asche] *see* 3816
Felinine, 3990
(−)-Felinine *see* 3990
Felison [Sigurta] *see* 4233
Felitrope *see* 3816
Fellozine [Chromalloy] *see* 7970
Felmane *see* 4233
Feloday [Ciba-Geigy] *see* 3991
Felodipine, 3991
Felosan *see* 4880
Felviten [Grünenthal] *see* 683
Felypressin, 3992
Femadol [Lepetit] *see* 8024
Femadon [Dolorgiet] *see* 4925
Femergin [Sandoz] *see* 3703
Femestrone Inj *see* 3751
Femidyn *see* 3751
Feminone [Upjohn] *see* 3780
Femodene [Schering AG] *see* 4421
Femovan [Schering AG] *see* 4421
Femoxetine, 3993
Femstat [Syntex] *see* 1561
Femulen [Searle] *see* 3905
Fenac *see* 2135
Fenactil *see* 2238
Fenadiazole, 3994
Fenadone *see* 6008
Fenalamide, 3995
Fenalcomine, 3996
Fenallymal *see* 7349
Fenamate [Meiji] *see* 413

Fenamiphos, 3997
Fenamisal *see* 7426
Fenamizol *see* 505
Fenapanil, 3998
Fenarimol, 3999
Fenarol [Winthrop] *see* 2155
Fenarsone *see* 1830
Fenasprate *see* 1074
Fenate *see* 2991
Fenatin *see* 7358
Fenazil [Sella] *see* 7970
Fenazol [Hokuriku Seiyaku] *see* 4167
Fenazolina [Polfa] *see* 718
Fenazox [Meiji] *see* 413
Fenazoxine *see* 6529
Fenbendazole, 4000
Fenbenicillin, 4001
Fenbid [SK&F] *see* 4925
Fen-Bridal *see* 5219
Fenbuconazole, 4002
Fenbufen, 4003
Fenbutamide *see* 7362
Fenbutatin Oxide, 4004
Fenbutrazate, 4005
Fencamfamine, 4006
Fencamine, 4007
Fencarbamide *see* 7363
Fenchlorphos *see* 8415
d-Fenchone, 4008
Fencibutirol, 4009
Fenclor [Caffaro] *see* 7725
Fenclozic Acid, 4010
Fendilar [SPA] *see* 4011
Fendiline, 4011
Fendosal, 4012
Fenesterin *see* 7367
Fenethanil *see* 4002
Fenethazine, 4013
Fenethylline, 4014
Fenetrazole *see* 9253
Fenfluramine, 4015
Fenformin *see* 7376
Fenhydren *see* 7381
Fenibutazona *see* 7431
Fenibutol [Atral] *see* 7431
Fenicol [Alcon] *see* 2120
Fenidrone *see* 7091
Fenigam *see* 484
Fenigama *see* 484
Fenilin *see* 7381
Fenilor [UCB] *see* 1474
Fenipentol, 4016
Fenistil [Zyma] *see* 3265
Fenitrothion, 4017
Fennel, 4018
Fenobrate [Gramon] *see* 4019
Fenofibrate, 4019
Fenolactine *see* 5355
Fenoldopam, 4020
Fenoldopam Mesylate *see* 4020
Fenolovo *see* 9875
Fenoprofen, 4021
Fenopron [Dista] *see* 4021
Fenoprop *see* 8679
Fenormin *see* 7376
Fenospen [Erba] *see* 7230
Fenostil [Zyma] *see* 3265
Fenotard [Virbac] *see* 4019
Fenoterol, 4022
Fenoverine, 4023
Fenoxaprop-ethyl, 4024
Fenoxazoline, 4025
Fenoxedil, 4026
Fenoximone *see* 3627
Fenoxycarb, 4027
Fenoxypen [Novo] *see* 7230
Fenoxypropazine *see* 7411
Fenozolone, 4028
Fenpentadiol, 4029
Fenpiclonil, 4030

Fenpidon *see* 3398
Fenpipramide Methobromide *see* 4032
Fenpiprane, 4031
Fenpiverinium Bromide, 4032
Fenpropanate *see* 4033
Fenpropathrin, 4033
Fenpropidin, 4034
Fenpropimorph, 4035
Fenproporex, 4036
Fenprostalene, 4037
Fenpyrate *see* 8151
Fenpyroximate, 4038
Fenquizone, 4039
Fenretinide, 4040
Fenspiride, 4041
Fensulfothion, 4042
Fental [Kanebo] *see* 9267
Fentanest [Sankyo] *see* 4043
Fentanyl, 4043
Fentazin [Allen & Hanburys] *see* 7323
Fenthion, 4044
Fentiazac, 4045
Fentiazin *see* 7404
Fenticlor, 4046
Fenticonazole, 4047
Fentiderm [Zyma] *see* 4047
Fentigyn [Ciba-Geigy] *see* 4047
Fentin Acetate *see* 9875
Fentin Hydroxide *see* 9875
Fentonium Bromide, 4048
Fentrinol [Mead Johnson] *see* 418
Fenugreek, 4049
Fenuron, 4050
Fenuron TCA *see* 4050
Fenvalerate, 4051
Fenylhist [Mallard] *see* 3367
Fenyramidol *see* 7474
Feojectin *see* 4073
Feosol [SK & F] *see* 4105
Feospan [SK & F] *see* 4105
Fepradinol, 4052
Feprazone, 4053
Fepron [Lilly] *see* 4021
Feprona [Lilly] *see* 4021
Feraconitine *see* 8095
Ferbam, 4054
Ferbeck *see* 4054
Fergon [Miles] *see* 4095
Fergon 500 [Alfar] *see* 2021
Fer-in-Sol [Mead Johnson] *see* 4105
Ferlucon *see* 4095
Fermate *see* 4054
Fermcozyme *see* 4468
Fermentation Amyl Alcohol *see* 5212
Fermentation Butyl Alcohol *see* 5146
Fermentation L. casei Factor *see* 8119
Fermine *see* 3304
Fermium, 4055
Fernambuco *see* 7322
Fernasan [ICI] *see* 9510
Fernex [ICI] *see* 7652
Fernos [ICI] *see* 7651
Fero-Gradumet [Abbott] *see* 4105
Ferolactan [Bioindustria] *see* 7961
Feromax *see* 4105
Feron [Toray] *see* 5017
Feroritard [Nikken] *see* 4105
Ferradow *see* 4054
Ferredoxins, 4056
Ferric Acetate, Basic, 4057
Ferric Albuminate, 4058
Ferric Alum *see* 549
Ferric and Ammonium Acetate Solution, 4059
Ferric Ammonium Chromate *see* 546
Ferric Ammonium Citrate *see* 547
Ferric Ammonium Oxalate *see* 548
Ferric Ammonium Sulfate *see* 549
Ferric Bromide, 4060
Ferric Chloride, 4061

Ferric Chromate(VI), 4062
Ferric Citrate, 4063
Ferric Dimethyldithiocarbamate *see* 4054
Ferric Enterobactin *see* 3631
Ferric Ferrocyanide, 4064
Ferric Ferrous Oxide *see* 4087
Ferric Fluoride, 4065
Ferric Formate, 4066
Ferric Gallotannate *see* 4080
Ferric Hexacyanoferrate(II) *see* 4064
Ferrichrome *see* 4067
Ferrichrome A *see* 4067
Ferrichromes, 4067
Ferrichrysin *see* 4067
Ferric Hydroxide, 4068
Ferric Hydroxide Oxide *see* 4068
Ferric Hypophosphite, 4069
Ferriclate Calcium Sodium, 4070
Ferric Monosodium Ethylenediaminetetraacetate *see* 4076
Ferric Nitrate, 4071
Ferric Oxide, 4072
Ferric Oxide, Saccharated, 4073
Ferric Persulfate *see* 4079
Ferric Phosphate, 4074
Ferric Pyrophosphate, 4075
Ferricrocin *see* 4067
Ferric Sesquioxide *see* 4072
Ferric Sesquisulfate *see* 4079
Ferric Sodium Edetate, 4076
Ferric Sodium Pyrophosphate, 4077
Ferric Subsulfate Solution, 4078
Ferric Sulfate, 4079
Ferric Sulfocyanate *see* 4081
Ferric Sulfocyanide *see* 4081
Ferric Tannate, 4080
Ferric Tersulfate *see* 4079
Ferric Thiocyanate, 4081
Ferricytichrome c *see* 2860
Ferriheme Chloride *see* 4678
Ferriheme Hydroxide *see* 4668
Ferrihemoglobin *see* 6035
Ferrimycins *see* 2914
Ferrioxamine B *see* 2914
Ferriporphyrin Chloride *see* 4678
Ferriporphyrin Hydroxide *see* 4668
Ferriprotoporphyrin Basic *see* 4668
Ferriprotoporphyrin Chloride *see* 4678
Ferrirhodin *see* 4067
Ferrirubin *see* 4067
Ferrite, 4082
Ferritin, 4083
Ferritose [Calbiochem] *see* 4295
Ferrivenin *see* 4073
Ferrocal *see* 1708
Ferrocene, 4084
Ferrocholinate, 4085
Ferrocytochrome c *see* 2860
Ferrofolin [Farmades] *see* 4083
Ferrofume [Nordic] *see* 4094
Ferroglycine Sulfate, 4086
Ferroglycine Sulfate Complex *see* 4086
Ferro-Gradumet [Abbott] *see* 4105
Ferroheme *see* 4672
Ferroin *see* 7356
Ferrol [Chemil] *see* 4083
Ferrolip [Flint] *see* 4085
Ferromyn [Calmic] *see* 4104
Ferron *see* 4872
Ferronat *see* 4094
Ferrone [Wolf] *see* 4094
Ferronicum [Sandoz] *see* 4095
Ferronord [Cooper] *see* 4086
Ferroprotoporphyrin *see* 4672
Ferro sanol [Sanol] *see* 4086
Ferrosoferric Oxide, 4087
Ferrospinel *see* 4082
Ferrosprint [Poli] *see* 4083

Ferrostar [Mediolanum] *see* 4083
Ferrostrane *see* 4076
Ferrostrene *see* 4076
Ferrotemp [Medix] *see* 4094
Ferrous Aminoacetosulfate *see* 4086
Ferrous Ammonium Sulfate *see* 552
Ferrous Aspartate *see* 875
Ferrous Bromide, 4088
Ferrous Calcium Citrate *see* 1708
Ferrous Carbonate Mass, 4089
Ferrous Carbonate Saccharated, 4090
Ferrous Chloride, 4091
Ferrous Citrate, 4092
Ferrous Fluoride, 4093
Ferrous Fumarate, 4094
Ferrous Gluconate, 4095
Ferrous Hydroxide, 4096
Ferrous Iodide, 4097
Ferrous Lactate, 4098
Ferrous Oxalate, 4099
Ferrous Oxide, 4100
Ferrous Phosphate, 4101
Ferrous Phosphide, 4102
Ferrous Selenide, 4103
Ferrous Succinate, 4104
Ferrous Sulfate, 4105
Ferrous Sulfate Glycine Complex *see* 4086
Ferrous Sulfide, 4106
Ferrous Sulfocyanate *see* 4107
Ferrous Sulfocyanide *see* 4107
Ferrous Thiocyanate, 4107
Ferrox 4099
Ferroxidase *see* 2049
Ferrum [Green Cross] *see* 4094
Fersamal [Glaxo] *see* 4094
Fertagyl [Intervet] *see* 5500
Fertiletten [Chassot] *see* 2152
Fertilysin, 4108
Fertinorm [Serono] *see* 4299
Fertiral [Hoechst] *see* 5500
Fertirelin, 4109
Fertodur [Schering AG] *see* 2789
Ferulic Acid, 4110
3-Feruloylquinic Acid *see* 2194
Ferum Hausmann *see* 4073
Fervenulin, 4111
FES *see* 10246
Fesofor [SK & F] *see* 4105
Fesotyme [Elder] *see* 4105
Fespan [SK & F] *see* 4105
Festamoxin [Shionogi] *see* 6369
Festucine *see* 5591
Fevarin [Duphar] *see* 4251
Feverall *see* 3414
Feverfew, 4112
Fevonil [Century] *see* 3414
Feximac [Nicholas] *see* 1497
Fexofenadine, 4113
FG-4963 *see* 3993
FG-5111 *see* 5870
FG-5606 *see* 619
FG-7051 *see* 7175
FGF *see* 4117
aFGF *see* 4117
aFGF-1 *see* 4117
aFGF-2 *see* 4117
bFGF *see* 4117
des 1-15 bFGF *see* 4117
FHD-3 *see* 4632
FI-106 *see* 3495
FI-1163 *see* 5621
FI-3542 *see* 3536
FI-5853 *see* 7173
FI-6341 *see* 4270
F.I. 6426 *see* 8927
FI-6654 *see* 1907
FI-6714 *see* 6580
FI-6934 *see* 2048
FI-7045 *see* 8610

FI-7056 *see* 8610
Fialuridine, 4114
Fiasone (amp) [Orfi] *see* 7901
FIAU *see* 4114
Fiber A *see* 6998
Fiberall [Ciba] *see* 7674
Fiber V [DuPont] *see* 7730
Fiber X-51 *see* 2644
Fiber X-54 *see* 2644
Fiblaferon [Rentschler] *see* 5017
Fibocil [Lilly] *see* 793
Fibonel [Pharma Investi] *see* 3972
Fiboran [Christiaens] *see* 793
Fibrase [SK & F] *see* 7276
Fibravyl [Rhovyl] *see* 7746
Fibrezym [Bene-Arzneimittel] *see* 7276
Fibrin, 4115
Fibrinase *see* 3968
Fibrin-i *see* 4115
Fibrinogen, 4116
Fibrinogenase *see* 9525
Fibrinokinase *see* 9608
Fibrinolysin *see* 7678
Fibrin-s *see* 4115
Fibrin-stabilizing Factor *see* 3968
Fibrit [Beecham] *see* 712
Fibroblast Growth Factor, 4117
Fibroblast Interferon *see* 5017
Fibrogamin [Behringwerke] *see* 3968
Fibroin, 4118
Fibrolax [Gipharmex] *see* 7674
Fibronectins, 4119
Fibrotan *see* 4805
Ficalon *see* 5575
Ficam [Fisons] *see* 1063
Fichtelite, 4120
Ficin, 4121
Ficoid [Fisons] *see* 4188
Ficortril [Pfizer] *see* 4828
Ficusin *see* 8113
Ficus Protease *see* 4121
Ficus Proteinase *see* 4121
FIF *see* 5017
Filaribits [Norden] *see* 3165
Filarsen *see* 3121
Filazine *see* 3165
Fildesin [Shionogi] *see* 10125
Filgrastim *see* 4558
Filicic Acid *see* 4124
Filicin *see* 4124
Filicinic Acid, 4122
Filicinsäure (German) *see* 4122
Filimarisin *see* 4123
Filipin, 4123
Filipin III *see* 4123
Filixic Acids, 4124
Filix Mas (B.P.) *see* 883
Filixsäure (German) *see* 4124
Filon [Berk] *see* 4005
Filoral *see* 2266
Filtrax [Biomed. Foscama] *see* 7613
Finadyne [Fisons] *see* 4182
Finalin [Yamanouchi] *see* 1055
Finaplix [Hoechst] *see* 9716
Finasteride, 4125
Finastid [Neopharmed] *see* 4125
Finaten [Finadiet] *see* 4258
Finaven [Am. Cyanamid] *see* 3185
Finimal [Mepros] *see* 45
Finlepsin *see* 1826
Finquel [Argent] *see* 9751
Finuret *see* 4830
Fiobrol *see* 2247
Fipexide, 4126
Fiproles *see* 4127
Fipronil, 4127
Firefly Luciferin, 4128
Firemaster BP-6 [Michigan Chem.] *see* 7724

Firemaster LV-T 23P [Michigan Chem.] *see* 9881
Firemaster T 23P [Michigan Chem.] *see* 9881
Firmacef [Firma] *see* 1967
Firon [Beard] *see* 4094
Fir-wood Oil *see* 6936
Fisalamine *see* 5964
Fischer's Yellow *see* 2493
Fisetin, 4129
Fish-berry *see* 2518
Fish Glue *see* 5123
Fisidenolon 1521 *see* 4129
Fisiobil [S.A.L.V.A.T.] *see* 3248
Fisiodar [Gentili] *see* 3003
Fisostina *see* 4334
Fivent [Fisopharma] *see* 2658
FK-027 *see* 1975
FK-176 *see* 10052
FK-235 *see* 6637
FK-435 *see* 3502
FK-482 *see* 1971
FK-506 *see* 9200
FK-749 *see* 2000
FK-1160 *see* 9564
FL-113 *see* 5090
FL-1039 *see* 409
FL-1060 *see* 408
FLA-731 *see* 8301
Flabelline *see* 6047
Flagecidin [Pfizer] *see* 708
Flagentyl [Rhône-Poulenc] *see* 8562
Flagyl [Searle] *see* 6242
Flagyl I.V [SCS] *see* 6242
Flake Lead *see* 1034
Flamasone [Norbrook] *see* 7901
Flamazine [Smith & Nephew] *see* 9071
Flamitajin [Teisan; Iwaki] *see* 4131
Flammazine [Duphar] *see* 9071
Flammex AP *see* 9881
Flammex T 23P *see* 9881
Flanax [Syntex] *see* 6504
Flanin F [Tokyo Tanabe] *see* 4131
Flantadin [Lepetit] *see* 2916
Flatistine [Sanofi] *see* 1898
Flavacidin *see* 658
Flavan [Pharmafarm] *see* 5477
3,3',4,4',5,7-Flavanhexol *see* 5477
3,3',4',5,7-Flavanpentol *see* 1950
Flavaspidic Acid, 4130
Flavaxin [Winthrop] *see* 8367
Flavicin *see* 658
Flavine-Adenine Dinucleotide, 4131
Flavine Yellow Shade *see* 8219
Flavin Mononucleotide *see* 8368
Flavitan *see* 4131
Flavomycin [Hoechst] *see* 979
Flavone, 4132
7-Flavone Ethyl Hydroxyacetate *see* 3565
7-Flavonoxyacetic Acid Ethyl Ester *see* 3565
Flavopereirine, 4133
Flavophospholipol *see* 979
Flavoquine [Roussel-UCLAF] *see* 609
Flavoxanthin, 4134
Flavoxate, 4135
Flavurol *see* 5914
Flaxedil [Lederle; Davis & Geck] *see* 4361
Flaxseed *see* 5532
FLC-1374 *see* 9561
Fleabane *see* 3708
Flea Seed *see* 7674
Flebopex [Profarma] *see* 3350
Flebosan [Dukron] *see* 9738
Flebosmil [Bouchara] *see* 3350
Flebosten [Bonomelli] *see* 3350
Flebotropin [Bago] *see* 3350
Flecainide, 4136

Flectadol [Maggioni] *see* 5668
Flectar *see* 10205
Flectron [Shell] *see* 2836
Fleishmilchsäure *see* 5350
Flemoxin [Paines & Byrne] *see* 617
Fleroxacin, 4137
Flesinoxan, 4138
Flexal *see* 1889
Flexartal [Clin-Comar-Byla] *see* 1889
Flexazone *see* 7431
Flexeril [Merck & Co.] *see* 2782
Flexiban [Merck & Co.] *see* 2782
Flexible Collodion *see* 2547
Flexidol [Daker-Farmasimes] *see* 4052
Flexidor [Elanco] *see* 5256
Flexilon [McNeil] *see* 10328
Flexin [McNeil] *see* 10328
Flindersine, 4139
Flindix [Lusofarmaco] *see* 5245
Flint *see* 8637
Flivas [Asahi] *see* 6443
Flixonase [Allen & Hanburys] *see* 4244
Flixotide [Glaxo] *see* 4244
Flobacin [Sigmatau] *see* 6865
Flo-Cillin Aqueous [Bristol] *see* 7226
Flociprin [IBI] *see* 2374
Floctafenine, 4140
Flodil [Astra] *see* 3991
Floganol [Zyma] *see* 2419
Flogar [UCB] *see* 7048
Flogencyl [Parke-Davis] *see* 3737
Flogene [Polifarma] *see* 4045
Floginax [Lifepharma] *see* 6504
Flogitolo *see* 7106
Flogobene [Farge] *see* 7661
Flogobron [Intersint] *see* 7078
Flogoril *see* 7106
Flogos [Gentili] *see* 9185
Flogovital [Bago] *see* 6640
Flogozen [SPA] *see* 4949
Floionic Acid *see* 7478
Flolan [Wellcome] *see* 8061
Flomax [Chiesi] *see* 6620
Flomoxef, 4141
Flonatril [Specia] *see* 2466
Flopion [Kyoritsu] *see* 4142
Flopropione, 4142
Florantyrone, 4143
Floraquin [Searle] *see* 5063
Floredil, 4144
Florel [Union Carbide] *see* 3777
Flores Martis *see* 4061
Florfenicol, 4145
Florid [Mochida] *see* 6266
Floridin *see* 4306
Floridin [Coli] *see* 2025
Florimycin *see* 10139
Florinef [Squibb] *see* 4166
Floripavine *see* 8493
Florisil *see* 5727
Flormidal [Galenika] *see* 6270
Florocid *see* 8762
Florone [Upjohn] *see* 3186
Floropipamide *see* 7608
Floropryl [Merck & Co.] *see* 5192
Flosequinan, 4146
Flosequinon *see* 4146
Flosin [Erba] *see* 4999
Flosint [Erba] *see* 4999
Flou [ELEA] *see* 8090
Flovacil [Andromaco] *see* 3190
Flovent [Glaxo] *see* 4244
Flowering Dogwood *see* 2599
Flowers of Antimony *see* 752
Flowers of Tin *see* 8933
Flowers of Zinc *see* 10279
Flowery-headed Spurge *see* 3949
Floxacillin, 4147
Floxacin [Merck & Co.] *see* 6793

Floxapen [Beecham] *see* 4147
Floxicam [Menarini] *see* 5258
Floxil [Cilag] *see* 6865
Floxin [J & J] *see* 6865
Floxuridine, 4148
Floxyfral [Solvay] *see* 4251
Fluacizine, 4149
Fluanisone, 4150
Fluanxol [Lundbeck] *see* 4224
Fluanxol Dépôt [Lundbeck] *see* 4224
Fluaton [Tubi Lux] *see* 4213
Fluatox [Inpharzam] *see* 89
Fluazacort, 4151
Fluazifop-butyl, 4152
Fluazinam, 4153
Flubendazole, 4154
Flubenisolone *see* 1226
Flubenol [Janssen] *see* 4154
Flubenzimine, 4155
Flubron [SS Pharm] *see* 4248
Flubuperone *see* 5870
Flucam [Pfizer] *see* 629
Fluchloralin, 4156
Flucinom [Schering] *see* 4242
Fluclorolone Acetonide *see* 4157
Flucloronide, 4157
Flucloxacillin *see* 4147
Fluconazole, 4158
Flucort [Syntex] *see* 4173
Fluctin [Lilly] *see* 4222
Flucycloxuron, 4159
Flucythrinate, 4160
Flucytosine, 4161
Fludara [Triton] *see* 4162
Fludarabine, 4162
Fludarene [Merck & Co.] *see* 2294
Fludemil *see* 4174
Fludeoxyglucose F[18], 4163
Fluderma [Farmitalia] *see* 4270
Fludestrin [Bristol] *see* 9321
Fludex [Biopharma] *see* 4969
Fludiazepam, 4164
Fludilat [Thiemann] *see* 1060
Fludioxonil, 4165
Fludrocortisone, 4166
Fludroxycortide *see* 4232
Fluellite *see* 351
Flufenamic Acid, 4167
Flugeral [Italfarmaco] *see* 4179
Flugestone Acetate, 4235
Fluibil [Zambon] *see* 2096
Fluibron [Chiesi] *see* 401
Fluidane *see* 1414
Fluiden [Lafare] *see* 4041
Fluidil [Adria] *see* 2822
Fluifort [Dompé] *see* 1850
Fluimucetin [Inpharzam] *see* 89
Fluimucil [Zambon] *see* 89
Fluindione, 4168
Fluindostatin *see* 4250
Fluitran [Shionogi] *see* 9754
Fluixol [Ripari-Gero] *see* 401
Flukanide [Merck & Co.] *see* 8280
Flukiver [Janssen] *see* 2473
Flumadine [Roche] *see* 8390
Flumarin [Shionogi] *see* 4141
Flumark [Dainippon] *see* 3625
Flumax [Shionogi] *see* 1969
Flumazenil, 4169
Flumazepil *see* 4169
Flumecinol, 4170
Flumedroxone Acetate, 4171
Flumen *see* 2221
Flumequine, 4172
Flumesil *see* 1064
Flumethasone, 4173
Flumethiazide, 4174
Flumetholon [Santen] *see* 4213
Flumethrin, 4175
Flumetramide, 4176

Flumetsulam, 4177
Flumiclorac, 4178
Flumidin [Kabi] see 6354
Flumoperone see 9813
Flumoxal [Janssen] see 4154
Flumoxane [Lebrun] see 4154
Flunagen [Gentili] see 4179
Flunarizine, 4179
Flunarl [Kyowa Hakko] see 4179
Flunase [Glaxo] see 4244
Fluniget [Merck & Co.] see 3190
Flunisolide, 4180
Flunitrazepam, 4181
Flunixin, 4182
Flunoxaprofen, 4183
Fluoboric Acid, 4184
Fluocinolide (obsolete) see 4186
Fluocinolide Acetate (obsolete) see 4186
Fluocinolone Acetonide, 4185
Fluocinolone Acetonide Acetate see 4186
Fluocinonide, 4186
Fluocortin Butyl, 4187
Fluocortolone, 4188
Fluocortolone 21-Caproate see 4188
Fluocortolone Trimethylacetate see 4188
Fluodonil [Pirri] see 3190
Fluodrocortisone see 4166
Fluohydric Acid see 4831
Fluohydric Acid Gas see 4837
Fluohydrisone see 4166
Fluohydrocortisone see 4166
Fluometuron, 4189
Fluon [ICI] see 7743
Fluonid [Marion] see 4185
Fluonilid [Continental Pharma] see 5589
Fluophosgene see 1868
Fluophosphoric Acid Di(dimethylamide) see 3250
Fluopromazine see 9814
Fluoracisine see 4149
Fluoracizine see 4149
Fluoracyzine see 4149
Fluorakil 100 see 4204
3',6'-Fluorandiol see 4194
Fluorandrenolone see 4232
β-(8-Fluoranthoyl)propionic Acid see 4143
Fluorapatite see 7503
9H-Fluorene, 4190
9H-Fluorene-2,7-diamine, 4191
N-2-Fluorenylacetamide, 4192
N-9H-Fluoren-2-ylacetamide see 4192
4-(9H-Fluoren-9-ylidenemethyl)benzenecarboximidamide see 7163
α-Fluoren-9-ylidene-p-toluamidine see 7163
Fluorescamine, 4193
Fluorescein, 4194
Fluorescein Paper, 4195
Fluorescent Blue see 5345
Fluorescin, 4196
Fluorescite [Alcon] see 4194
Fluorescyanine see 4930
Fluoresone, 4197
Fluorets [Chauvin] see 4194
Fluoridamid, 4198
Fluorine, 4199
Fluorine compd with Graphite see 4561
Fluorine Dioxide, 4200
Fluorine Monoxide, 4201
Fluorine Nitrate, 4202
Fluorine Oxide see 4201
Fluorine Perchlorate, 4203
Fluoristan see 8940
Fluor-i-strip [Ayerst] see 4194

Fluorite see 1709
Fluormetholon see 4213
Fluormone [Benvegna] see 2986
Fluoroacetamide, 4204
4'-Fluoro-4-(4-acetamidomethyl-4-phenylpiperido)butyrophenone see 30
Fluoroacetic Acid, 4205
Fluoroacetic Acid Amide see 4204
4'-C-Fluoroadenosine 5'-Sulfamate see 6824
3-Fluoro-D-alanine, 4206
C₃-Fluoroalcohol see 9346
p-Fluoroaniline, 4207
2-Fluoro-9-β-D-arabinofuranosyladenine see 4162
4-Fluorobenzenamine see 4207
Fluorobenzene, 4208
4-Fluorobenzeneacetic Acid see 4215
3-[2-[4-(6-Fluoro-1,2-benzisoxazol-3-yl)-1-piperidinyl]ethyl]-6,7,8,9-tetrahydro-2-methyl-4H-pyrido[1,2-a]pyrimidin-4-one see 8397
p-Fluorobenzoic Acid, 4209
5-(p-Fluorobenzoyl)-2-benzimidazolecarbamic Acid Methyl Ester see 4154
[5-(4-Fluorobenzoyl)-1H-benzimidazol-2-yl]carbamic Acid Methyl Ester see 4154
3-[2-[4-(4-Fluorobenzoyl)-1-piperidinyl]ethyl]-2,4[1H,3H]-quinazolinedione see 5307
1-[γ-(4-Fluorobenzoyl)propyl]-4-acetamidomethyl-4-phenylpiperidine see 30
1'-[3-(p-Fluorobenzoyl)propyl]-[1,4'-bipiperidine]-4'-carboxamide see 7608
1-(3-p-Fluorobenzoylpropyl)-4-p-chlorophenyl-4-hydroxypiperidine see 4629
1-(3'-p-Fluorobenzoylpropyl)-4-hydroxy-4-p-tolylpiperidine see 6343
1-(3'-p-Fluorobenzoylpropyl)-4-hydroxy-4-(3''-trifluoromethylphenyl)-piperidine see 9813
N-[1-[3-(p-Fluorobenzoyl)propyl]-4-phenylpiperidin-4-ylmethyl]acetamide see 30
N-[[1-[3-(p-Fluorobenzoyl)propyl]-4-phenyl-4-piperidyl]methyl]acetamide see 30
8-[3-(p-Fluorobenzoyl)propyl]-1-phenyl-1,3,8-triazaspiro[4.5]decan-4-one see 8903
1-[γ-(4-Fluorobenzoyl)propyl]-4-piperidinopiperidine-4-carboxamide see 7608
1-[1-[3-(p-Fluorobenzoyl)propyl]-4-piperidyl]-2-benzimidazolinone see 1077
1-[1-[3-(4-Fluorobenzoyl)propyl]-4-piperidyl]-2,3-dihydrobenzimidazole-2-thione see 9584
1-[1-[3-(4-Fluorobenzoyl)propyl]-4-piperidyl]-2-mercaptobenzimidazole see 9584
1-[3-(4-Fluorobenzoyl)propyl]-4-(2-pyridyl)piperazine see 931
1-[1-[3-(p-Fluorobenzoyl)propyl]-1,2,-3,6-tetrahydro-4-pyridyl]-2-benzimidazolinone see 3505
1-(p-Fluorobenzyl)-2-[[1-(p-methoxyphenethyl)-4-piperidyl]amino]benzimidazole see 891
2-[(p-Fluorobenzyl)methylamino]ethyl Isopropyl (±)- 1,4-Dihydro-2,6-dimethyl-4-[2,3-(methylenedioxy)-

phenyl]-3,5-pyridinedicarboxylate see 3587
2-(2-Fluoro-4-biphenylyl)propionic Acid see 4234
Fluorocarbon 11 see 9770
5-Fluoro-6-carboxyuracil see 4214
9α-Fluoro-21-chloro-11β,16α,17α-trihydroxypregn-4-ene-3,20-dione 16,-17-Acetonide see 4621
9α-Fluorocortisol see 4166
C-Fluorocurarine see 2743
5-Fluorocytosine see 4161
6α-Fluoro-1-dehydrohydrocortisone see 4229
2-[¹⁸F]Fluoro-2-deoxy-D-glucose see 4163
5-Fluoro-2'-deoxy-β-uridine see 4148
6α-Fluorodexamethasone see 4173
6α-Fluoro-9α,11β-dichloro-1,4-pregnadiene-16α,17,21-triol-3,20-dione 16,17-Acetonide see 4157
9-Fluoro-6,7-dihydro-5,8-dimethyl-1-oxo-1H,5H-benzo[ij]quinolizine-2-carboxylic Acid see 4919
(±)-9-Fluoro-6,7-dihydro-8-(4-hydroxy-1-piperidinyl)-5-methyl-1-oxo-1H,5H-benzo[ij]quinolizine-2-carboxylic Acid see 6430
(±)-9-Fluoro-2,3-dihydro-3-methyl-10-(4-methyl-1-piperazinyl)-7-oxo-7H-pyrido[1,2,3-de]-1,4-benzoxazine-6-carboxylic Acid see 6865
9-Fluoro-6,7-dihydro-5-methyl-1-oxo-1H,5H-benzo[ij]quinolizine-2-carboxylic Acid see 4172
8-Fluoro-5,6-dihydro-5-methyl-6-oxo-4H-imidazo[1,5-a][1,4]benzodiazepine-3-carboxylic Acid Ethyl Ester see 4169
9-Fluoro-2,3-dihydro-10-(4-methyl-1-piperazinyl)-7-oxo-7H-pyrido[1,2,3-de]-1,4-benzothiazine-6-carboxylic Acid see 8448
(S)-6-Fluoro-2,3-dihydrospiro[4H-1-benzopyran-4,4'-imidazolidine]-2',-5'-dione see 8871
6α-Fluoro-11β,21-dihydroxy-16α,17-isopropylidenedioxy-Δ¹,⁴-pregnadiene-3,20-dione see 4180
9α-Fluoro-11β,21-dihydroxy-16α,-17α-isopropylidenedioxy-1,4-pregnadiene-3,20-dione see 9728
9-Fluoro-11β,17α-dihydroxy-17(S)-lactoylandrosta-1,4-dien-3-one 17β-Acetate see 4225
(11β,17β)-9-Fluoro-11,17-dihydroxy-17-methylandrost-4-en-3-one see 4223
(6α,11β,16α)-6-Fluoro-11,21-dihydroxy-16,17-[(1-methylethylidene)bis(oxy)]pregna-1,4-diene-3,20-dione see 4180
(11β,16α)-9-Fluoro-11,21-dihydroxy-16,17-[1-methylethylidenebis(oxy)]pregna-1,4-diene-3,20-dione see 9728
(6α,11β,16α)-6-Fluoro-11,21-dihydroxy-16,17-[(1-methylethylidene)bis-(oxy)]pregn-4-ene-3,20-dione see 4232
(6α,11β)-9-Fluoro-11,17-dihydroxy-6-methylpregna-1,4-diene-3,20-dione see 4213
(6α,11β,16α)-6-Fluoro-11,21-dihydroxy-16-methylpregna-1,4-diene-3,-20-dione see 4188
(11β,16α)-9-Fluoro-11,21-dihydroxy-16-methylpregna-1,4-diene-3,20-dione see 2975
9-Fluoro-11β,21-dihydroxy-2'-methyl-

5'βH-pregna-1,4-dieno[17,16-d]oxazole-3,20-dione 21-Acetate *see* 4151
9-Fluoro-11β,17-dihydroxypregn-4-ene-3,20-dione-17-acetate *see* 4235
9-Fluoro-11β,17-dihydroxyprogesterone 17-Acetate *see* 4235
1-Fluoro-2,4-dinitrobenzene, 4210
Fluoroethanoic Acid *see* 4205
(*endo,syn*)-8-(2-Fluoroethyl)-3-[(hydroxydiphenylacetyl)oxy]-8-methyl-8-azoniabicyclo[3.2.1]octane Bromide *see* 4248
(8*r*)-8-(2-Fluoroethyl)-3α-hydroxy-1αH,5αH-tropanium Bromide Benzilate *see* 4248
N-β-Fluoroethylnortropine Benzilate Methobromide *see* 4248
Fluorofen [Savio] *see* 9814
6-Fluoro-2-(2'-fluoro[1,1'-biphenyl]-4-yl)-3-methyl-4-quinolinecarboxylic Acid *see* 1394
(*E*)-4-Fluoro-β-(fluoromethylene)-benzenebutanamine *see* 6313
6-Fluoro-1-(4-fluorophenyl)-1,4-dihydro-7-(4-methyl-1-piperazinyl)-4-oxo-3-quinolinecarboxylic Acid *see* 3187
6-Fluoro-1-(4-fluorophenyl)-1,4-dihydro-4-oxo-7-(1-piperazinyl)-3-quinolinecarboxylic Acid *see* 8517
1-[*o*-Fluoro-α-(*p*-fluorophenyl)-α-phenylbenzyl]imidazole *see* 4247
Fluoroform, 4211
Fluoroformylon *see* 4270
Fluorogesarol *see* 3000
5-Fluoro-*N*-hexyl-3,4-dihydro-2,4-dioxo-1(2*H*)-pyrimidinecarboxamide *see* 1892
9α-Fluorohydrocortisone *see* 4166
(±)-5-Fluoro-2-(2-hydroxy-3-*t*-butylaminopropoxy)butyrophenone *see* 1563
4'-Fluoro-4-(4-hydroxy-4-*p*-chlorophenylpiperidino)butyrophenone *see* 4629
9α-Fluoro-17-hydroxycorticosterone *see* 4166
Δ¹-9α-Fluoro-16α-hydroxyhydrocortisone *see* 9727
(+)-(*S*)-*p*-Fluoro-*N*-[2-[4-[2-(hydroxymethyl)-1,4-benzodioxan-5-yl]-1-piperazinyl]ethyl]benzamide *see* 4138
(6α,11β,16α)-6-Fluoro-11-hydroxy-16-methyl-3,20-dioxopregna-1,4-dien-21-oic Acid Butyl Ester *see* 4187
p-Fluoro-4-(4'-hydroxy-4'-*p*-methylphenylpiperidino)butyrophenone *see* 6343
9α-Fluoro-11β-hydroxy-17α-methyltestosterone *see* 4223
9α-Fluoro-16α-hydroxyprednisolone *see* 9727
9α-Fluoro-16α-hydroxyprednisolone Acetonide *see* 9728
9α-Fluoro-16α-hydroxyprednisolone 16α,17α-Acetonide 21-(β-Benzoylamino)isobutyrate *see* 9729
p-Fluoro-4-[4'-hydroxy-4'-*p*-tolylpiperidino)butyrophenone *see* 6343
p-Fluoro-4-[4'-hydroxy-4'-(3''-trifluoromethyl)phenyl]piperidinobutyrophenone *see* 9813
4'-Fluoro-4-[4-hydroxy-4-(α,α,α-trifluoro-*m*-tolyl)piperidino]butyrophenone *see* 9813
9α-Fluoro-16α,17-isopropylidenedioxyprednisolone *see* 9728
6α-Fluoro-16α,17α-isopropylidenedi-

oxy-4-pregnene-11β,21-diol-3,20-dione *see* 4232
Fluoromar [Ohio Med.] *see* 4237
Fluoromethane, 4212
Fluoromethcholone, 4213
4'-Fluoro-4-[4-(*o*-methoxyphenyl)-1-piperazinyl]butyrophenone *see* 4150
Fluoromethylbenzene *see* 4218
1-Fluoro-2-methylbenzene *see* 4218
2-Fluoro-α-methyl[1,1'-biphenyl]-4-acetic Acid *see* 4234
2-Fluoro-α-methyl-4-biphenylacetic Acid *see* 4234
9α-Fluoro-16α-methyl-Δ¹-corticosterone *see* 2975
6α-Fluoro-16α-methyl-1-dehydrocorticosterone *see* 4188
9α-Fluoro-16α-methyl-17-desoxyprednisolone *see* 2975
S-Fluoromethyl 6α,9α-Difluoro-11β-hydroxy-16α-methyl-17α-propionyloxy-3-oxoandrosta-1,4-diene-17β-carbothioate *see* 4244
(*E*)-2-(Fluoromethylene)-4-(*p*-fluorophenyl)butylamine *see* 6313
9α-Fluoro-16-methyleneprednisolone 21-Acetate *see* 4228
9α-Fluoro-16-methylene-Δ¹,⁴-pregnadiene-11β,17,21-triol-3,20-dione 21-Acetate *see* 4228
Fluoromethyl 1,1,1,3,3,3-Hexafluoro-2-propyl Ether *see* 8621
cis-5-Fluoro-2-methyl-1-[*p*-(methylsulfinyl)benzylidene]indene-3-acetic Acid *see* 9155
(*Z*)-5-Fluoro-2-methyl-1-[[4-(methylsulfinyl)phenyl]methylene]-1*H*-indene-3-acetic Acid *see* 9155
7-Fluoro-1-methyl-3-(methylsulfinyl)-4(1*H*)-quinolinone *see* 4146
4'-Fluoro-4-(4-methylpiperidino)butyrophenone *see* 5870
6α-Fluoro-16α-methylprednisolone *see* 7162
9α-Fluoro-16α-methylprednisolone *see* 2986
9α-Fluoro-16β-methylprednisolone *see* 1226
6α-Fluoro-16α-methyl-Δ¹,⁴-pregnadiene-11β,21-diol-3,20-dione *see* 4188
Fluoromethyl 2,2,2-Trifluoro-1-(trifluoromethyl)ethyl Ether *see* 8621
5-Fluoroorotate *see* 4214
5-Fluoroorotic Acid, 4214
Fluorophene [Stecker] *see* 4216
(*E*)-2-(4-Fluorophenethyl)-3-fluoroallylamine *see* 6313
p-Fluorophenylacetic Acid, 4215
5-(2-Fluorophenyl)-1,3-dihydro-1-methyl-7-nitro-2*H*-1,4-benzodiazepin-2-one *see* 4181
(β*R*,δ*R*)-2-(*p*-Fluoro-phenyl)-β,δ-dihydroxy-5-isopropyl-3-phenyl-4-(phenylcarbamoyl)pyrrole-1-heptanoic Acid *see* 897
[*R*-(*R**,*R**)]-2-(4-Fluorophenyl)-β,δ-dihydroxy-5-(1-methylethyl)-3-phenyl-4-[(phenylamino)carbonyl]-1*H*-pyrrole-1-heptanoic Acid *see* 897
p-Fluorophenyl Ethyl Sulfone *see* 4197
1-[(2-Fluorophenyl)(4-fluorophenyl)-phenylmethyl]-1*H*-imidazole *see* 4247
α-(2-Fluorophenyl)-α-(4-fluorophenyl)-1*H*-1,2,4-triazole-1-ethanol *see* 4246
3-Fluoro-4-phenylhydratropic Acid *see* 4234
1-(4-Fluorophenyl)-4-[4-hydroxy-4-(4-

methylphenyl)-1-piperidinyl]-1-butanone *see* 6343
1-(4-Fluorophenyl)-4-[4-hydroxy-4-[3-(trifluoromethyl)phenyl]-1-piperidinyl]-1-butanone *see* 9813
2-(*p*-Fluorophenyl)-1,3-indandione *see* 4168
2-(4-Fluorophenyl)-1*H*-indene-1,3-(2*H*)-dione *see* 4168
(±)-(3*R**,5*S**,6*E*)-7-[3-(*p*-Fluorophenyl)-1-isopropylindol-2-yl]-3,5-dihydroxy-6-heptenoate *see* 4250
1-(4-Fluorophenyl)-4-[4-(2-methoxyphenyl)-1-piperazinyl]-1-butanone *see* 4150
(+)-2-(4-Fluorophenyl)-α-methyl-5-benzoxazoleacetic Acid *see* 4183
(−)-*trans*-4-(*p*-Fluorophenyl)-3-[[3,4-(methylenedioxy)phenoxy]methyl]-piperidine *see* 7175
(2*R*-*trans*)-5-(4-Fluorophenyl)-2-(1-methylethyl)-*N*,4-diphenyl-1-[2-(tetrahydro-4-hydroxy-6-oxo-2*H*-pyran-2-yl)ethyl]-1*H*-pyrrole-3-carboxamide *see* 897
[*R**,*S**-(*E*)]-(±)-7-[3-(4-Fluorophenyl)-1-(1-methylethyl)-1*H*-indol-2-yl]-3,-5-dihydroxy-6-heptenoic Acid *see* 4250
1-[(4-Fluorophenyl)methyl]-*N*-[1-[2-(4-methoxyphenyl)ethyl]-4-piperidinyl]-1*H*-benzimidazol-2-amine *see* 891
4-(4-Fluorophenyl)-7-methyl-1-(1-methylethyl)-2(1*H*)-quinazolinone *see* 4230
1-(4-Fluorophenyl)-4-(4-methyl-1-piperidinyl)-1-butanone *see* 5870
1'-[4-(4-Fluorophenyl)-4-oxobutyl]-[1,-4'-bipiperidine]-4'-carboxamide *see* 7608
N-[[1-[4-(4-Fluorophenyl)-4-oxobutyl]-4-phenyl-4-piperidinyl]methyl]-acetamide *see* 30
8-[4-(4-Fluorophenyl)-4-oxobutyl]-1-phenyl-1,3,8-triazaspiro[4.5]decan-4-one *see* 8903
1-[1-[4-(*p*-Fluorophenyl)-4-oxobutyl]-piperidin-4-yl]-2-benzimidazolinone *see* 1077
1-[1-[4-(4-Fluorophenyl)-4-oxobutyl]-4-piperidinyl]-1,3-dihydro-2*H*-benzimidazol-2-one *see* 1077
1-[1-[4-(4-Fluorophenyl)-4-oxobutyl]-1,2,3,6-tetrahydro-4-pyridinyl]-1,3-dihydro-2*H*-benzimidazol-2-one *see* 3505
1-[1-[4-(*p*-Fluorophenyl)-4-oxobutyl]-1,2,3,6-tetrahydro-4-pyridyl]-2-benzimidazolinone *see* 3505
1-(*p*-Fluorophenyl)-4-(4-phenyl-4-acetamidomethylpiperidino)-1-butanone *see* 30
N-[3-[4-(*p*-Fluorophenyl)-1-piperazinyl]-1-methylpropyl]nicotinamide *see* 6576
N-[3-[4-(4-Fluorophenyl)-1-piperazinyl]-1-methylpropyl]-3-pyridinecarboxamide *see* 6576
1-(*p*-Fluorophenyl)-4-(4-piperidino-4-carbamoylpiperidino)-1-butanone *see* 7608
1-(4-Fluorophenyl)-4-[4-(2-pyridinyl)-1-piperazinyl]-1-butanone *see* 931
4'-Fluoro-4-[*N*-[4-(*N*-piperidino)-4-carbamido]piperidino]butyrophenone *see* 7608
Fluoroplex [Allergan] *see* 4219
6α-Fluoroprednisolone *see* 4229
9-Fluoroprednisolone *see* 5190

6α-Fluoro-1,4-pregnadiene-11β,17α,-21-triol-3,20-dione see 4229
4'-Fluoro-4-[4-(2-pyridyl)-1-piperazinyl]butyrophenone see 931
5-Fluoro-2,4(1H,3H)-pyrimidinedione see 4219
Fluoros [Dieckmann] see 8762
Fluorosalan, 4216
Fluorosilicic Acid see 4220
(+)-(4S)-6-Fluorospiro[chroman-4,4'-imidazolidine]-2',5'-dione see 8871
4'-Fluoro-5'-O-sulfamoyladenosine see 6824
9-(4-Fluoro-5-O-sulfamoylpentofuranosyl)adenine see 6824
Fluorosulfonic Acid, 4217
Fluorosulfuric Acid see 4217
Fluorosulfuric Acid Methyl Ester see 6151
5-Fluoro-1,2,3,6-tetrahydro-2,6-dioxo-4-pyrimidinecarboxylic Acid see 4214
5-Fluoro-1-(tetrahydro-2-furanyl)-2,4-(1H,3H)-pyrimidinedione see 9267
5-Fluoro-1-(tetrahydro-2-furyl)uracil see 9267
(11β, 16α)-9-Fluoro-11,16,17,21-tetrahydroxypregna-1,4-diene-3,20-dione see 9727
6α-Fluoro-11β,16α,17,21-tetrahydroxypregna-1,4-diene-3,20-dione Cyclic 16,17-Acetal with Acetone see 4180
9α-Fluoro-11β,16α,17,21-tetrahydroxypregna-1,4-diene-3,20-dione Cyclic 16,17-Acetal with Acetone see 9728
9-Fluoro-11β,16α,17,21-tetrahydroxypregna-1,4-diene-3,20-dione Cyclic 16,17-Acetal with Acetone, 21-(3,3-Dimethylbutyrate) see 9730
9-Fluoro-11β,16α,17,21-tetrahydroxypregna-1,4-diene-3,20-dione Cyclic 16,17-Acetal with Acetone 21-Ester with N-Benzoyl-2-methyl-β-alanine see 9729
9-Fluoro-11β,16α,17,21-tetrahydroxypregna-1,4-diene-3,20-dione Cyclic 16,17-Acetal with Cyclopentanone 21-Acetate see 407
6α-Fluoro-11β,16α,17,21-tetrahydroxypregn-4-ene-3,20-dione Cyclic 16,17-Acetal with Acetone see 4232
6α-Fluoro-11β,16α,17,21-tetrahydroxyprogesterone Cyclic 16,17-Acetal with Acetone see 4232
Fluorothiamphenicol see 4145
4-Fluoro-4-[4-(2-thioxo-1-benzimidazolinyl)piperidino]butyrophenone see 9584
Fluorotoluene, 4218
m-Fluorotoluene see 4218
o-Fluorotoluene see 4218
p-Fluorotoluene see 4218
Fluorotrichloromethane see 9770
9α-Fluoro-11β,17,21-trihydroxy-16-methylenepregna-1,4-diene-3,20-dione 21-Acetate see 4228
(6α,11β,16α)-6-Fluoro-11,17,21-trihydroxy-16-methylpregna-1,4-diene-3,20-dione see 7162
(11β,16α)-9-Fluoro-11,17,21-trihydroxy-16-methylpregna-1,4-diene-3,-20-dione see 2986
(11β,16β)-9-Fluoro-11,17,21-trihydroxy-16-methylpregna-1,4-diene-3,-20-dione see 1226
9α-Fluoro-11β,17α,21-trihydroxy-21-methylpregna-1,4-diene-3,20-dione 21-Acetate see 4225

(6α,11β)-6-Fluoro-11,17,21-trihydroxypregna-1,4-diene-3,20-dione see 4229
(11β)-9-Fluoro-11,17,21-trihydroxypregna-1,4-diene-3,20-dione see 5190
(11β)-9-Fluoro-11,17,21-trihydroxypregn-4-ene-3,20-dione see 4166
Fluorouracil, 4219
2-Fluorovidarabine see 4162
Fluorspar see 1709
Fluosilicic Acid, 4220
Fluosol DA, 4221
Fluostigmine see 5192
Fluosulfonic Acid see 4217
Fluothane [Ayerst] see 4634
Fluovitef [Italfarmaco] see 4185
Fluoxeren [Menarini] see 4222
Fluoxetine, 4222
Fluoxymesterone, 4223
Flupen [Alfa] see 4147
Flupenthixol see 4224
Flupentixol, 4224
Fluperolone Acetate, 4225
Fluphenacur see 5624
Fluphenazine, 4226
Flupirtine, 4227
Fluprednidene Acetate, 4228
Fluprednisolone, 4229
Fluprednylidene 21-Acetate see 4228
Fluproquazone, 4230
Fluprostenol, 4231
Fluprowit [Thiemann] see 89
Fluracil see 4219
Flura-Drops see 8762
Fluram [Roche] see 4193
Flurandrenolide, 4232
Flurandrenolone see 4232
Flurandrenolone Acetonide (obsolete) see 4232
Flurazepam, 4233
Flurazepam Hydrochloride see 4233
Flurbiprofen, 4234
Fluril see 4219
Flurobate [Texas Pharmacal] see 1226
Fluroblastin [Farmitalia] see 4219
Flurofen [Boots] see 4234
Flurogestone Acetate, 4235
Fluropryl see 5192
Flurothyl, 4236
Fluro Uracil see 4219
Fluroxene, 4237
Fluroxypyr, 4238
Fluroxypyr-meptyl see 4238
Flurprimidol, 4239
Flusilazole, 4240
Fluspirilene, 4241
Flussorex [Lampugnani] see 2380
Flustar [Firma] see 3190
Flutamide, 4242
Flutazolam, 4243
Fluticasone Propionate, 4244
Flutoprazepam, 4245
Flutra [Schering] see 9754
Flutriafen see 4246
Flutriafol, 4246
Flutrimazole, 4247
Flutron [Roche] see 3493
Flutropium Bromide, 4248
Fluvalinate, 4249
Fluvastatin, 4250
Fluvean [Kowa] see 4185
Fluvermal [Janssen-Le Brun] see 4154
Fluversin [Searle] see 9160
Fluvin see 4822
Fluvisco [Searle] see 9160
Fluvoxamine, 4251
Fluxarten [Beecham] see 4179
Fluxema see 1060
Fluzilazol see 4240
Fluzon [Taisho] see 4185

Flypel see 2912
FMC-1240 see 3782
FMC-9260 see 9362
FMC-30980 see 2836
FMC-33297 see 7321
FMC-45498 see 2934
FMC-54800 see 1257
FMC-57020 see 3262
FMC-67825 see 1670
FML [Allergan] see 4213
FMN see 8368
5-FOA see 4214
Fobex see 1055
Fodrin see 8891
Foipan [Ono] see 1775
Folacin see 4253
Folaren [Ist. Chim. Inter.] see 4254
Folbex [Ciba-Geigy] see 2174
Folbex VA [Ciba-Geigy] see 1456
Folcodal [Syncro] see 2365
Foldine [Specia] see 4253
Foldox [Sidus] see 9829
Folescutol, 4252
Folettes [Fawns & McAllan] see 4253
Folex [Adria] see 6065
Foliamin see 4253
Foliben [Firma] see 4254
Folic Acid, 4253
Folic Acid Sodium Salt see 8763
Folicet [Mission] see 4253
Folicur [Bayer] see 9253
Folidol [Bayer] see 7167
Folidol-M [Bayer] see 6183
Foligan [Henning] see 287
Folikrin see 3751
Folinerin see 6963
Folinic Acid, 4254
Folipac see 4253
Folipex see 3751
Folisan see 3751
Folithion see 4017
Follestrine see 3751
Follicle-stimulating Hormone see 4299
Follicular Hormone see 3751
Follicular Hormone Hydrate see 3750
Folliculin see 3751
Follicunodis see 3751
Follidrin (Tablets) see 3751
Follitropin see 4299
Follutein [Squibb] see 2273
Folosan [Uniroyal] see 8264
Folpet, 4255
Folsan [Kali-Chemie] see 4253
Folsäure see 4253
Folvite [Lederle] see 4253
Fomecin A see 4256
Fomecin B see 4256
Fomecins, 4256
Fomepizole, 4257
Fominoben, 4258
Fomocaine, 4259
Fonatol see 3177
Fonazine, 4260
Fonazine Mesylate see 4260
Fonderma [Biosedra] see 8178
Fongarex [Sanofi] see 6978
Fonlipol [Recordati] see 9556
Fonofos, 4261
Fonofos [Pulitzer] see 4280
Fontamide see 9116
Fontarsan see 7082
Fontarsol see 3121
Fontego [Polifarma] see 1508
Fontex [Lilly] see 4222
Fontilix [Diamant] see 6220
Fonurit [Chinoin] see 50
Fonzylane [Orsymonde] see 1498
Food of the Gods see 858
Food Orange 8 see 1798
Food Yellow 13 see 8256

Foradil [Ciba-Geigy] see 4272
Foralamin [Eaton] see 6010
Forane [Anaquest] see 5191
Forapin [Mack, Illert.] see 5867
Forbina [Ciba] see 9115
Force [Zeneca] see 9266
Fordiuran [Thomae] see 1508
Fordonal [Almirall] see 9443
Forene [Abbott] see 5191
Forenol [Roemmers] see 6620
Forhistal [Ciba] see 3265
Forilin [Novo] see 8433
Foriod see 5059
Forit see 7105
Formagene see 7158
Formal see 6093
Formaldehyde, 4262
Formaldehyde Acetamide see 4269
Formaldehyde Dimethyl Acetal see 6093
α,α'-Formaldehyde Glycerol see 4494
α,β-Formaldehyde Glycerol see 4494
Formaldehyde Semicarbazone see 4262
Formaldehyde Sodium Bisulfite, 4263
Formaldehyde Sodium Sulfoxylate see 8764
Formaldehyde Solution see 4262
Formaldehyde-sulfathiazole see 9115
Formaldehydesulfoxylic Acid Sodium Salt see 8764
Formalin see 4262
Formamide, 4264
Formanilide, 4265
Formanol see 6038
Formebolone, 4266
Formestane, 4267
Formic Acid, 4268
Formic Acid Ammonium Salt see 554
Formic Acid Benzyl Ester see 1169
Formic Acid Beryllium Salt see 1210
Formic Acid Ethyl Ester see 3853
Formic Acid Methyl Ester see 6153
Formic Acid Phenylmethyl Ester see 1169
Formic Acid Propyl Ester see 8043
Formic Aldehyde see 4262
Formicin, 4269
N-Formimidoylthienamycin Monohydrate see 4954
Formin see 6036
Formin Salicylate see 6041
Formison see 5890
Formistin [Formenti] see 2063
Formo-Cibazol [Ciba] see 9115
Formocortal, 4270
Formol see 4262
Formononetin, 4271
Formononetol see 4271
Formophthaloylsulfanilyl Urea see 9080
Formosa Camphor see 1779
Formosa Oil of Camphor see 6879
Formosulfathiazole see 9115
Formoterol, 4272
Formothion, 4273
Formula 1 see 3807
Formula 2B see 3807
Formula 3A see 3807
Formula 6B see 3807
Formula 12A see 3807
Formula 13A see 3807
Formula 19 see 3807
Formula 20 see 3807
Formula 23A see 3807
Formula 28 see 3807
Formula 28A see 3807
Formula 30 see 3807
Formula 32 see 3807
Formula 33 see 3807

Formula 35A see 3807
Formula 39C see 3807
Formula 44 see 3807
4'-Formylacetanilide Thiosemicarbazone see 9427
3-Formylamino-4-hydroxy-α-[N-[1-methyl-2-(p-methoxyphenyl)ethyl]-aminomethyl]benzyl Alcohol see 4272
Formylaniline see 4265
3-C-Formyl-5-deoxy-L-lyxofuranose see 8987
Formyldienolone see 4266
5-Formyl-3,4-dihydro-1H-2-benzopyran-1-one see 3717
5-Formyl-3,4-dihydroisocoumarin see 3717
3-Formyl-2,4-dihydroxy-6-methylbenzoic Acid 3-Hydroxy-4-(methoxycarbonyl)-2,5-dimethylphenyl Ester see 901
6-Formyl-2,3-dimethoxybenzoic Acid see 6983
Formyl Fluoride, 4274
Formylformic Acid see 4521
16-Formylgitoxin see 4437
Formyl Hydroperoxide see 7302
N-Formyl-N-hydroxyaminoacetic Acid see 4617
N-Formyl-N-hydroxyglycine see 4617
2-Formyl-11α-hydroxy-Δ¹-methyltestosterone see 4266
N-Formyl-L-leucine Ester with (3S,-4S)-3-Hexyl-4-[(2S)-2-hydroxytridecyl]-2-oxetanone see 6997
[2S-[2α(R*),3β]]-N-Formyl-L-leucine 1-[(3-Hexyl-4-oxo-2-oxetanyl)methyl]dodecyl Ester see 6997
Nα-Formyl Melittin see 5867
N-Formyl trans-p-Methoxystyrylamine see 9938
2-Formyl-17α-methylandrosta-1,4-diene-11α,17β-diol-3-one see 4266
S-(N-Formyl-N-methylcarbamoylmethyl) O,O-Dimethyl Phosphorodithioate see 4273
1-Formyl-4-methyl-7-isopropenylazulene see 5346
2-Formyl-1-methylpyridinium Chloride Oxime see 7884
4-Formylphenol see 4856
[S-(E)]-4-(1-Formyl-1-propenyl)-3,4-dihydro-2-oxo-2H-pyran-5-carboxylic Acid Methyl Ester see 3586
Formylpteroic Acid see 8348
2-Formylquinoxaline-1,4-dioxide Carbomethoxyhydrazone see 1825
Formylsulfamethine see 4275
N²-Formylsulfisomidine, 4275
5-Formyl-5,6,7,8-tetrahydrofolic Acid see 4254
5-Formyl-5,6,7,8-tetrahydropteroyl-L-glutamic Acid see 4254
Foromacidin A see 8904
Foromacidin B see 8904
Foromacidin C see 8904
Forpen see 7225
Forskolin see 2539
Forstan [Bayer] see 7113
Forsterite see 5727
Forsythin see 7477
Fortalgesic (tabl.) [Sterling-Winthrop] see 7261
Fortam [Glaxo] see 1995
Fortamine see 2232
Fortasec [Esteve] see 5601
Fortaz [Glaxo] see 1995
Fortecortin [E. Merck] see 2986
Forthane [Lilly] see 6159
Forticef [Sasse] see 2032

Fortigro [Pfizer] see 1825
Fortimicin A see 4276
Fortimicin B see 4276
Fortimicins, 4276
Fortodyl see 10156
Fortombrine M [Dagra] see 5850
Fortoshade M see 5850
Fortracin [A.L. Labs.] see 965
Fortral [Winthrop] see 7261
Fortralin (tabl.) [Winthrop] see 7261
Fortrel [Celanese] see 7730
Fortrol [Shell] see 2755
Fortum [Glaxo] see 1995
Forturf see 2219
Fortuss [GP] see 3214
Forum [Shell] see 3271
Forza [Zeneca] see 9266
Fosamax [Merck & Co.] see 228
Foscarnet Sodium, 4277
Foscavir [Astra] see 4277
Fosenopril see 4282
Fosetyl Al, 4278
Fosfakol see 7164
Fosfalugel [De Angeli] see 371
Fosferno [Plant Protection] see 7167
Fosfestrol, 4279
Fosfobiotic [Bergamon] see 4280
Fosfocin [Crinos] see 4280
Fosfocina [C.E.P.A.] see 4280
Fosfocolina [Afarit] see 7519
Fosfogram [Firma] see 4280
Fosfomycin, 4280
Fosfomycin Trometamol see 4280
Fosfonomycin see 4280
Fosforal [Sis-Ter] see 4280
Fosforina B₁₂ [Francia] see 7520
Fosfosal, 4281
Fosfotricina [Italfarmaco] see 4280
Foshagite see 1749
Foshallasite see 1749
Fosinopril, 4282
Fosmicin [Meiji] see 4280
Fospirate, 4283
Fossil Flour see 5003
Fossyol [Merckle] see 6242
Fosten [Serono] see 796
Fosthietan, 4284
Fostion MM [Agrimont] see 3269
Foston [Hoechst] see 9649
Fotemustine, 4285
Fouadin see 8967
Fourneau 190 see 49
Fourneau 309 see 9181
Fourneau 710 see 7680
Fourneau 933 see 7631
Fourneau 1500 see 9047
Fovane [Pfizer] see 1155
Fowler's Soln see 7769
Foxetin [Gador] see 4222
Foxglove see 3201
Fox Green see 4992
Foximin [Caber] see 4280
FOY [Ono] see 4344
FOY-305 see 1775
FPL-670 see 2658
FPL-59002 see 6524
FPL-59002KP see 6524
FPL-60278 see 3482
FPL-60278AR see 3482
FPRMeCl see 7881
FR-02A see 3567
FR-860 see 2870
FR-10123 see 1997
FR-13479 see 2000
FR-17027 see 1975
FR-34235 see 6637
FR-900506 see 9200
Frabel see 7106
Fraction P see 2915
Frademicina see 5525

Fradiomycin see 6542
Fragivil see 1091
Fragivix [Labaz; Sigma-Tau] see 1091
Fragmin [Pharmacia] see 2870
Framycetin see 6542
Framygen see 6542
Francephane see 2010
Francital [Francia] see 4280
Francium, 4286
Frandol [Yamanouchi] see 5245
Frangula, 4287
Frangula Emodin see 3602
Frangulic Acid see 3602
Frangulin, 4288
Frangulin A see 4288
Frangulin B see 4288
Franguloside see 4288
Franidipine see 5786
Frankincense see 6969
Franklinite see 10255
Franocide [Burroughs Wellcome] see 3165
Franrose [Hishiyama] see 9267
Frantin see 1187
Fraquinol see 6542
Fratol see 4205
Fraxetin, 4289
Fraxetin-8-glucoside see 4290
Fraxin, 4290
Fraxiparine [Sanofi Winthrop] see 6434
Fraxoside see 4290
Frazalon [Daiichi] see 4319
(±)-FRC-8653 see 2334
Fredericamycin A, 4291
Free Benzylpenicillin see 1177
Freedox [Upjohn] see 9604
Free Penicillin II see 1177
Free Penicillin G see 1177
Freeuril see 1155
Frekentine see 3175
Frekven [Novo] see 8025
Frénactil [Clin-Comar-Byla] see 1077
Frenactyl [Janssen] see 1077
Frenantol [Laroche Navarron] see 7176
Frenapyl [Troponwerke] see 2440
French Chalk see 9207
French Purple see 2682
French Saffron see 8466
French Verdigris see 2691
Frenohypon see 7176
Frenolicin, 4292
Frenolon [EGYT] see 6228
Frenolyse [Specia] see 9704
Frenopect [Hefa] see 401
Frenoton see 925
Frenquel [Merrell] see 925
Frentirox see 6049
Freon [Du Pont] see 2192
Freon 11 [Du Pont] see 9770
Freon 12 [Du Pont] see 3114
Freon 14 [Du Pont] see 1865
Freon 114 [Du Pont] see 2672
Freon C318 [Du Pont] see 6841
Frequentic Acid see 2389
Frequentin, 4293
Frescon [Shell] see 9808
Fresmin [Takeda] see 10152
Friar's Balsam see 978
Friar's Cowl see 118
Friedclan-3-one see 4294
Friedelin, 4294
D:A-Friedooleanan-3-one see 4294
(3β)-D-Friedoolean-14-en-3-ol see 9231
Frigen [Hoechst] see 2192
Frigen 11 [Hoechst] see 9770
Frigen 12 [Hoechst] see 3114
Frigen 114 [Hoechst] see 2672
Fringanor [Sobio] see 7365

Frisium [Hoechst] see 2417
Froben [Boots] see 4234
Frone [Serono] see 5017
Frontier [Sandoz] see 3263
Frontline [Rhone Merieux] see 4127
Frowncide [ISK] see 4153
Frucote [Elanco] see 1579
β-D-Fructofuranose 1,3,4,6-Tetra-3-pyridinecarboxylate see 6605
β-Fructofuranosidases see 5025
β-D-Fructofuranosyl O-α-D-Galactopyranosyl-(1→6)-O-α-D-galactopyranosyl(1→6)-O-α-D-galactopyranosyl-(1→6)-α-D-glucopyranoside see 10092
β-D-Fructofuranosyl-O-α-D-galactopyranosyl-(1→6)-α-D-glucopyranoside see 8279
β-D-Fructofuranosyl-α-D-glucopyranoside see 9051
β-D-Fructofuranosyl-α-D-glucopyranoside Octakis(hydrogen Sulfate) Aluminum Complex see 9049
Fructose, 4295
D-Fructose see 4295
DL-Fructose, 4296
β-D-Fructose see 4295
D-Fructose 1,6-Bis(dihydrogen Phosphate) see 4297
D-Fructose 6-(Dihydrogen Phosphate) see 4298
Fructose-1,6-diphosphate, 4297
1,6-D-Fructosediphosphoric Acid see 4297
Fructose Monophosphate see 4298
Fructose-6-phosphate, 4298
D-Fructose-6-phosphoric Acid see 4298
Fructose 1,3,4,6-Tetranicotinate see 6605
β-h-Fructosidases see 5025
Fructosteril see 4295
Frugalan [Diamant] see 4327
Fruitone-N [Rhone-Poulenc] see 6458
Fruit Sugar see 4295
Frumin AL [Sandoz] see 3429
Frusemide see 4331
Frusetic [Unimed] see 4331
Frusid [DDSA] see 4331
FSF see 3968
FSH, 4299
FSR-3 see 5203
FT-207 see 9267
Ftalazol [Orion] see 7533
Ftalicetimida see 7532
Ftalofyne see 7530
F3TDR see 9816
Ftoracizine see 4149
Ftorafur [Grünenthal] see 9267
Ftorocort [Gedeon Richter] see 9728
FTPA see 4221
5-FU see 4219
Fuadin [Winthrop] see 8967
Fua-Med [Med. Fabrik] see 6696
Fuchsine see 5686
Fuchsin(e) Acid see 109
Fucidin [Sigma-Tau] see 4340
Fucidina [Leo Pharm.] see 4340
Fucidine [Leo Pharm.; Thomae] see 4340
Fucidin Intertulle [Leo Pharm.] see 4340
Fucithalmic [Leo] see 4340
Fuclasin see 10305
Fucosamine, 4300
D-Fucose, 4301
L-Fucose, 4302
Fucostanol see 8968
Fucosterol, 4303
Fucoxanthin, 4304
Fucus, 4305

FUDR [Roche] see 4148
Fugerel [Schering] see 4242
Fugillin [Upjohn] see 4308
Fugoa [Scheurich] see 6811
Fugu Poison see 9382
Fujiglucon [Fujisawa] see 4465
Fuklasin see 10305
Fulaid [Takeda] see 9267
Fulcin [ICI] see 4571
Fuldazin [Sumitomo] see 6283
Fulfeel [Kyorin] see 9267
Fulfill [Ciba] see 8136
Ful-Glo [Sola-Barnes-Hinds] see 4194
Fulgram [ABC] see 6793
[5,6]Fullerene-C₆₀-Ih see 1482
Fullerenes see 1482
Fuller's Earth, 4306
Fuller's Herb see 8511
Fullsafe [Ohta] see 4167
"Fulminating Gold" see 4527
Fulsix [Tatsumi] see 4331
Fuluminol [Tatsumi] see 2405
Fuluvamide [MECT] see 4331
Fulvicin [Schering] see 4571
Fulvoplumierin, 4307
Fumadil B see 4308
Fumafer [Labaz] see 4094
Fumagillin, 4308
Fumar F see 4094
Fumaric Acid, 4309
Fumaric Acid Dibenzyl Ester see 1170
Fumarin see 2624
Fumarine see 8083
Fumazone see 3068
Fumidil [Abbott] see 4308
Fumigacin see 4666
Fumigatin, 4310
Fumigrain see 133
Fuming Liquid Arsenic see 841
Fuming Spirit of Libavius see 8929
Fumiron [Knoll] see 4094
Fumonisin B₁, 4311
Fundal [Schering] see 2133
Funduscein [IOLAB] see 4194
Fundyl [Bellon] see 3629
Fungacetin [Harvey] see 9721
Fungarest [Janssen] see 5313
Fungibacid [Asche] see 9595
Fungichromin, 4312
Fungicidin see 6834
Fungiderm [Janssen] see 6266
Fungifos [S & K Pharma] see 9647
Fungilin [Squibb] see 627
Fungimycin see 7310
Funginex [Shell] see 9819
Fungiplex [Hermal] see 9061
Fungisdin [Isdin] see 6266
Fungistat [Cilag] see 9303
Fungisterol, 4313
Fungistop [Essex] see 9656
Fungizone [Squibb] see 627
Fungoral [Janssen] see 5313
Funtumine, 4314
Furachel [Rachelle] see 6696
Furacin [Eaton] see 6697
Furacinetten see 6697
Furacoccid see 6697
Furacrylic Acid see 4317
Furadan [FMC] see 1851
Furadantin [Röhm] see 6696
Furadantine MC [Eaton; Norwich] see 6696
Furadoin [Eaton] see 6696
Furadroxyl [Eaton] see 6615
Furafluor [Green Cross] see 9267
2-Furalacetic Acid see 4317
Furalan [Lannett] see 6696
2-Furaldehyde see 4324
Furaltadone, 4315
Furamazone [Eaton] see 6622

Furamide [Clin-Comar-Byla] see 3246
Furamon see 4334
Furan, 4316
Furanace [Dainippon] see 6628
2-Furanacrylic Acid, 4317
2-Furanacrylonitrile, 4318
2-Furancarbinol see 4325
Furancarbonyl Chloride see 4332
2-Furancarboxaldehyde see 4324
2-Furancarboxylic Acid see 4329
2-Furancarboxylic Acid 1-(Dichloro-
    acetyl)-1,2,3,4-tetrahydro-6-quin-
    olinyl Ester see 8241
2,5-Furandione see 5744
N_1-(2'-Furanidyl)-5-fluorouracil see
    9267
2-Furanmethanol see 4325
Furanol see 4334
Furantoina [Uriach] see 6696
1-(2-Furanyl)-N,N'-bis(2-furanyl-
    methylene)methanediamine see 4832
9-(3-Furanyl)decahydro-4-hydroxy-
    4a,10a-dimethyl-1,4-etheno-3H,7H-
    benzo[1,2-c:3,4-c']dipyran-3,7-dione
    see 2558
[2aS-(2aα,5aβ,6α,7α,8aα,8bα)]-6-[2-
    (3-Furanyl)ethyl]decahydro-6-hydr-
    oxy-2a,5a,7-trimethyl-2H-naphtho-
    [1,8-bc]furan-2-one see 5794
N-(2-Furanylmethyl)-N',N'-dimethyl-
    N-2-pyridinyl-1,2-ethanediamine see
    6010
5-(2-Furanylmethyl)-5-(1-methyleth-
    yl)-2,4,6(1H,3H,5H)-pyrimidinetri-
    one see 4326
1-(3-Furanyl)-4-methyl-1-pentanone
    see 7307
N-(2-Furanylmethyl)-1H-purin-6-
    amine see 5324
[1R-(1α,4β,5α,7β,9α)]-4-(3-Furan-
    yl)octahydro-1,7-dimethyl-2H-quin-
    olizine 5-Oxide see 6826
3-[1-(2-Furanyl)-3-oxobutyl]-4-hydr-
    oxy-2H-1-benzopyran-2-one see
    2624
3-(2-Furanyl)-2-propenenitrile see
    4318
3-(2-Furanyl)-2-propenoic Acid see
    4317
Furaplast see 6697
Furaspor [Eaton] see 6075
Furazabol, 4319
Furazalon see 4319
Furazolidone, 4320
Furazolin [USSR] see 4315
Furazolium Chloride, 4321
Furazol W see 6697
Furazosin see 7897
Furbenal see 6075
Furcellaran, 4322
Furcellaria Gum see 4322
Furenapyridazin see 6629
Furenazin see 6629
Furesis [Lääkefarmos] see 4331
Furesol see 6697
Furethidine, 4323
Furfenorex see 4327
Furfural, 4324
Furfuralcohol see 4325
Furfuramide see 4832
Furfuran see 4316
"Furfurol" see 4324
N^6-Furfuryladenine see 5324
Furfuryl Alcohol, 4325
6-Furfurylaminopurine see 5324
Furfurylidene Acetic Acid see 4317
5-Furfuryl-5-isopropylbarbituric Acid,
    4326
Furfurylmethylamphetamine, 4327
Furfuryltrimethylammonium see 4334

Furidiazina see 9725
Furilazone see 4330
α-Furildioxime, 4328
Furiton see 6634
Furloe [PPG] see 2240
Furmethide [SK & F] see 4334
Furmethonol see 4315
Furnace Black see 1856
Furobactina [Esteve] see 6696
7H-Furo[3,2-g][1]benzopyran-7-one
    see 8113
Furo[3,2-g]coumarin see 8113
Furofutran [Taiyo] see 9267
2-Furoic Acid, 4329
α-Furoic Acid see 4329
2-Furoic Acid Ester with 1-(Dichloro-
    acetyl)-1,2,3,4-tetrahydro-6-quino-
    linol see 8241
Furonazide, 4330
Furophen T-Caps [Krewel] see 6696
Furo-Puren [Klinge-Nattermann] see
    4331
Furore [Hoechst] see 4024
Furosedon [Santen] see 4331
Furosemide, 4331
Furovag see 4320
Furoxane see 4320
Furoxone [Eaton] see 4320
Furoyl Chloride, 4332
2-[4-(2-Furoyl)piperazin-1-yl]-4-
    amino-6,7-dimethoxyquinazoline see
    7897
Furozin [Daiichi] see 1913
Furpirinol see 6628
Furpyrinol see 6628
Fursemide see 4331
Fursultiamine, 4333
Furtrethonium, 4334
Furtrimethonium see 4334
Furtulon [Roche] see 3493
3-[α-(2-Furyl)-β-acetylethyl]-4-hydr-
    oxycoumarin see 2624
β-2-Furylacrylic Acid see 4317
3-(2-Furyl)acrylonitrile see 4318
2-Furylcarbinol see 4325
α-Furylcarbinol see 4325
1-(3-Furyl)decahydro-11-hydroxy-4b,-
    7,7-11a,13a-pentamethyloxireno(4,-
    4a)-2-benzopyrano[6,5-g](2)benz-
    oxepin-3,5,9(3aH,4bH,6H)-trione
    Acetate see 6769
8-(3-Furyl)decahydro-2,2,4a,8a-tetra-
    methyl-11H,13H-oxireno[d]pyrano-
    [4',3':3,3a]isobenzofuro[5,4-f][2]-
    benzopyran-4,6,13(2H,5aH)-trione
    see 5519
5-[2-(3-Furyl)ethyl]decahydro-5,8-
    dihydroxy-1,4a,6-trimethyl-1-naph-
    thoic Acid γ-Lactone see 5794
β-Furyl Isoamyl Ketone see 7307
5-(2-Furylmethyl)-5-isopropylbar-
    bituric Acid see 4326
2-Furyl Methyl Ketone Isonicotinoyl-
    hydrazone see 4330
1-(3-Furyl)-4-methyl-1-pentanone see
    7307
N'-(2-Furylmethyl)-N'-(2-pyridyl)-N,N-
    dimethylethylenediamine see 6010
Fusafungine, 4335
Fusaloyos see 4335
Fusaric Acid, 4336
Fusarine see 4335
Fusariotoxin T-2 see 9933
Fusarubin, 4337
Fusca [Hoei] see 2470
Fuscin, 4338
Fused Borax see 8733
Fused Boric Acid see 1365
Fused Silver Nitrate see 8662
Fused Sodium Borate see 8733

Fusel Oil, 4339
Fusidic Acid, 4340
Fusilade [ICI] see 4152
Fussol [Sankyo] see 4204
Fustin, 4341
Fustpentane see 1840
FUT-175 see 6436
Futhan [Torii] see 6436
Futraful [Taiho] see 9267
Futrican [Astra] see 2156
FW-293 see 3136
FW-734 see 7987
FW-925 see 6695
Fybrogel [Reckitt & Coleman] see
    7674
Fyracyl see 5689
Fyrex see 576
Fyrol FR-2 [Harwick], 4342
Fyrol HB 32 [Stauffer] see 9881
Fysionorm [Nourypharma] see 5659
G see 4500
G-I see 4566
G-II see 4566
G-III see 4566
G-4 see 3120
G-11 see 4716
G-020 see 2108
G-137 see 7560
G-277 see 4926
G-605 see 9132
G-13871 see 7431
G-19258 see 3259
G-22008 see 8186
G-22355 see 4955
G-22870 see 3313
G-23350 see 29
G-23611 see 5195
G-23922 see 2174
G-24480 see 3043
G-25766 see 2468
G-27202 see 7106
G-27692 see 8681
G-27901 see 9797
G-28315 see 9121
G-29505 see 44
G-30027 see 902
G-30028 see 7996
G-30320 see 2433
G-31435 see 7972
G-32883 see 1826
G-32911 see 8683
G-33182 see 2246
G-34161 see 7973
G-34162 see 411
G-34586 see 2447
G-35020 see 2966
G-704650 see 228
GA see 9196
GABA see 450
Gabacet [Carrion] see 7641
Gabapentin, 4343
Gabbromicina [Farmitalia] see 7173
Gabbromycin [Farmitalia] see 7173
Gabbropas see 498
Gabbroral [Farmitalia] see 7173
Gabexate, 4344
Gabexate Mesylate see 4344
GABOB see 464
Gabomade see 464
Gaboxadol see 9508
Gabren(e) [Synthelabo] see 7955
G Acid see 6475
Gacilin [Andromaco] see 4036
Gadexyl see 5908
Gadodiamide, 4345
Gadolinia see 4346
Gadolinium, 4346
Gadolinium(III) 1,4,7-tris(Carboxy-
    methyl)-10-(2'-hydroxypropyl)-1,4,-
    7,10-tetraazacyclododecane see 4348

Gadolinium Diethylenetriaminepenta-
acetic Acid see 4347
Gadolinium Diethylenetriamine Pen-
taacetic Acid Bismethylamide see
4345
Gadopentetic Acid, 4347
Gadoteridol, 4348
Gaduol see 2530
Gahnite see 10255
Gainex see 7428
Galactaric Acid, 4349
Galactaric Acid Compd with (5α,6α)-
7,8-Didehydro-4,5-epoxy-17-methyl-
morphinan-3,6-diol (1:1) see 6359
Galactin see 7961
Galactitol, 4350
Galactoflavin, 4351
D-Galactomethylose see 4301
L-Galactomethylose see 4302
4-D-Galactopyranosyl-4-D-fructo-
furanose see 5360
4-O-β-D-Galactopyranosyl-D-fructose
see 5360
O-α-D-Galactopyranosyl-(1→6)-[O-α-
D-galactopyranosyl-(1→6)-]₂-O-α-D-
glucopyranosyl-(1→2)-β-D-fructo-
furanoside see 10092
4-O-β-D-Galactopyranosyl-D-glucitol
see 5352
4-O-β-D-Galactopyranosyl-D-gluconic
Acid see 5354
4-O-β-D-Galactopyranosyl-D-gluconic
Acid Compd with Erythromycin
(1:1) see 3724
1-O-(4-O-β-D-Galactopyranosyl-β-D-
glucopyranosyl)ceramide see 2863
4-O-β-D-Galactopyranosyl-D-glucose
see 5356
6-O-α-D-Galactopyranosyl-D-glucose
see 5861
[1S-[1R*(S*),2S*,3E]]-N-[1-[(β-D-
Galactopyranosyloxy)methyl]-2-hy-
droxy-3-heptadecenyl]-2-hydroxy-
tetracosanamide see 7524
Galactoquin [Mundipharma] see 8244
Galactosaccharic Acid see 4349
D-Galactosamine, 4352
D-Galactose, 4353
β-Galactoside Sorbitol see 5352
4-β-D-Galactosido-D-fructose see 5360
4-(β-D-Galactosido)-D-gluconic Acid
see 5354
4-(β-D-Galactosido)-D-glucose see 5356
6-(α-D-Galactosido)-D-glucose see 5861
4-O-β-D-Galactosyl-D-fructose see
5360
D-Galacturonic Acid, 4354
Galanga, 4355
Galangal see 4355
Galangin, 4356
Galantamine see 4357
Galanthamine, 4357
Galanthidine see 5656
Galatone see 4510
Galatur [Wyeth] see 5091
Galbas [Ciba] see 4030
Galben [Isagro] see 1056
Galcodine [Galen] see 2528
Galecron [Ciba] see 2133
Galegine, 4358
Galena see 5445
Galenphol [Galen] see 7484
Galfer [Galen] see 4094
Galβ1→4Glcβ1→1cer see 2863
Galipine, 4359
Galla see 6827
Gallacetophenone, 4360
Gallaldehyde 3,5-Dimethyl Ether see
9191
Gallamine Triethiodide, 4361

(Gallato)hydroxyiodobismuth see 1310
Gallein, 4362
Gallepronin [Taiyo] see 4142
Gallery [Elanco] see 5256
Gallic Acid, 4363
Gallic Acid Bismuth Basic Salt see
1325
Gallic Acid 5,6-Dihydroxy-3-carboxy-
phenyl Ester see 3195
Gallic Acid 3-Monogallate see 3195
Gallic Acid Propyl Ester see 8044
Gallicin see 4363
Gallimycin [Abbott] see 3726
Gallium, 4364
Gallium Arsenide, 4365
Gallium Phosphide, 4366
Gallium Trifluoride, 4367
Gallobromol see 3071
Gallocatechin see 1950
Gallochrome see 5914
Gallocyanine, 4368
Gallodesoxycholic Acid see 2096
Gallopamil, 4369
Gallotannic Acid see 9221
Gallotannin see 9221
Gallotox see 7453
Galloxon see 4636
m-Galloylgallic Acid see 3195
1-Galloyl-α-D-glucose see 4461
1-Galloyl-β-D-glucose see 4462
(Galloyloxy)hydroxyiodobismuthine
see 1310
4-Galloylpyrogallol see 3958
Galls see 6827
GalN see 4352
Galoxone see 4636
Galphol [Galen] see 7484
Galpseud [Galen] see 3645
Galuteolin see 5641
Gamabufagin see 4370
Gamabufogenin see 4370
Gamabufotalin, 4370
Gamanil [E. Merck] see 5587
Gamaquil [Siegfried] see 7412
Gamasol 90 see 3308
Gamastan [Miles] see 4958
Gamatran see 386
Gambir, 4371
Gambir Catechu see 4371
Gambit [Ciba] see 4030
Gamboge, 4372
Gambogic Acid, 4373
Gamefar see 7134
Gamene [Barnes-Hind] see 5526
Gamibetal [ISF] see 464
Gamimune [Cutter] see 4958
Gamma Benzene Hexachloride see
5526
Gammabulin [Immuno] see 4958
Gammaferon [Bioferon] see 5018
Gammagard [Travenol] see 4958
Gammagee [Merck & Co.] see 4958
Gamma Globulins see 4958
Gamma Hexachlor see 5526
Gamma-interferon see 5018
Gammajust 50 [Horita] see 7016
Gammalin [ICI] see 5526
Gammalon [Daiichi] see 450
Gamma-OH [Clintec Nutrition] see
4860
Gamma-OZ [Kanebo] see 7016
Gammaphos see 424
Gamma-pipradrol see 925
Gammar [Armour] see 4958
Gammariza [Toyo] see 7016
Gammatsul [Nippon Chemiphar] see
7016
Gamma-vinyl GABA see 10114
Gammexane [ICI] see 5526
Gamolenic Acid see 5531

Gamonil [E. Merck] see 5587
Gamophen [J & J] see 9790
Ganasag [Squibb] see 3316
Ganciclovir, 4374
Ganglefene, 4375
Gangleron [USSR] see 4375
Gangliosides, 4376
Gangliostat see 4724
Gangliovet [Fidia] see 4376
Ganidan [Specia] see 9076
Ganite [Fujisawa] see 4364
Ganja see 1794
Ganlion see 929
Ganocide [ICI] see 3499
Ganor [Thomae] see 3972
Gansil see 2118
Gantanol [Roche] see 9086
Gantaprim [Lenza] see 9086
Gantrim [Geymonat] see 9086
Gantrisin [Roche] see 9125
GAP A see 4119
Garamycin [Schering] see 4398
Garantose see 8463
Garasin [Wakamoto] see 2021
Garasol [Schering] see 4398
Gardenal [Specia] see 7386
Garden Chamomile see 2082
Gardenin A see 4377
Gardenin B see 4377
Gardenin C see 4377
Gardenin D see 4377
Gardenin E see 4377
Gardenins, 4377
Garden Lavender see 5403
Garden Rosemary see 8423
Garden Tox [Geigy] see 3043
Gardinol see 4378
Gardinol Type Detergents, 4378
Gardol®, 4379
Gardona [Shell] see 9334
Gardrin(e) [Syntex] see 3629
Garget see 7546
Gargilon see 2959
Gargon [Squibb] see 9502
Garlic, 4380
Garlon [DowElanco] see 9789
Garmian [Fuso] see 981
Garnierite see 6582
Garosamine see 4398 and 8695
Garranil [Aristegui] see 1817
Garrathion [Stauffer] see 1869
Garryine, 4381
GAs see 4427
Gas Black see 1856
GASH see 4592
Gaslon [Nippon Shinyaku] see 5113
Gasoline, 4382
Gasstenon [Tatsumi] see 4142
Gaster [Yamanouchi] see 3972
Gasteril [Ripari-Gero] see 7646
Gastomax [Brocchieri] see 2125
Gastralgin [De Angeli] see 8431
Gastramine see 1230
Gastrax [Asche] see 6758
Gastrese [Robins] see 6226
Gastrhéma see 1945
Gastridin [Merck & Co.] see 3972
Gastrins, 4383
Gastripon see 10208
Gastrobid [Napp] see 6226
Gastrodiagnost [E. Merck] see 7250
Gastrofrenal [ISF-Italseber] see 2658
Gastrografin [Squibb; Schering AG]
see 5851
Gastroloc [Astra] see 6977
Gastromax [Farmitalia] see 6226
Gastromet [Bayropharm] see 2337
Gastron [Tanabe] see 3538
Gastron [Winthrop] see 6025
Gastronerton [Dolorgiet] see 6226

Gastronorm [Janssen] see 3476
Gastropen [CEPA] see 3972
Gastropidil see 5893
Gastropin see 10208
Gastrosedan see 6025
Gastrosil [Heumann] see 6226
Gastro-Tablinen [Sanorania] see 6226
Gastrotem [Temmler] see 6226
Gastro-Timelets [Temmler] see 6226
Gastrozepin [Thomae] see 7646
Gastrurol [Gibipharma] see 7660
Gastuloric see 4477
Gas-X [Sandoz] see 3264
Gatalone [Barnes-Hind] see 4510
Gaucho [Bayer] see 4946
Gaultherin, 4384
Gaultherioside see 4384
GB see 8520
GB-1 see 1396
GB-2 see 1396
GB-94 see 6260
GBE see 4432
GC-1189 see 2131
GC-2466 see 6380
GC-3707 see 1360
GCP-23339A see 7135
G-CSF see 4558
GD see 8860
Gd-DTPA see 4347
Gd-DTPA-BMA see 4345
Gd(HP-DO3A) see 4348
GE-132 see 7979
GEA-6414 see 9650
Geangin [Gea] see 10083
Geapur [Gea] see 287
Geatrim-Boli [Gea] see 9071
Gebutox [Hoechst] see 3341
Gedamycin Methyl Ester see 5891
Gefanil [Sumitomo] see 4385
Gefarnate, 4385
Gefarnil see 4385
Gefarnyl see 4385
Gefulcer [Ohta] see 4385
Gehlenite see 1685
Geigy 867 see 9102
Geigy Rodenticide Exp. 332 see 2623
Geissoschizoline, 4386
Geissospermine, 4387
Gelatin, 4388
Gelbin see 1700
Geldene [Pfizer] see 7661
Gelée Royale (French) see 8434
Gelfoam [Upjohn] see 4388
Gellan Gum, 4389
Gelocatil [Gelos] see 45
Gelomyrtol [Pohl] see 6423
Gelose see 182
Gelosedine [Bayer] see 7363
Gelovermin see 4750
Gelrite [Merck & Co.] see 4389
Gelsemine, 4390
Gelseminic Acid see 8552
Gelsemium, 4391
Gelstaph [Beecham] see 2480
Gelthix [Kerfoot] see 9285
Gelucystine [Parke-Davis] see 2851
Gelumina [Sanitas] see 355
Gelusil [Warner-Chilcott] see 362
Gelvatol [Monsanto] see 7745
Gemalgene see 9769
Gemcitabine, 4392
Gemeprost, 4393
Gemfibrozil, 4394
Gemonil [Abbott] see 6029
Gemora see 1544
Gemzar [Lilly] see 4392
Genabol [Wyeth] see 6779
Genatropine see 908
Gencor [Sandoz] see 4850
Gendon [Organon] see 8293

Generlac [My K] see 5360
Génésérine 3 [Amido] see 3740
Genetron [AlliedSignal] see 2192
Genetron 12 [AlliedSignal] see 3114
Genicide see 10196
Geniphene see 9693
Genisis [Organon] see 2571
Genistein, 4395
Genisteine-Alkaloid, 4396
Genistein-4'-glucoside see 8867
Genistein-4'-glucosidorhamnoside see 8866
Genisteol see 4395
Genistin see 4395
Genite [Allied], 4397
Genite 883 see 7033
Genitol 923 see 4397
Genkodein see 2527
Genlip [Malesci] see 4394
Genocodeine see 2527
Genogris [Vita] see 7641
Genol see 6097
Genophyllin see 485
Genoptic [Allergan] see 4398
Genoscopolamine [Amido] see 8551
Genostrychnine see 9022
Genotropin [Pharmacia] see 8864
Genphen [Tutag] see 7970
Gentacin [Schering] see 4398
Gentak [Akorn] see 4398
Gentalline [Schering] see 4398
Gentalyn [Schering] see 4398
Gentamicin, 4398
Gentamicin A see 4398
Gentamicin C$_1$ see 4398
Gentamicin C$_{1a}$ see 4398
Gentamicin C$_2$ see 4398
Gentamicin C$_{2b}$ see 6269
Gentamicin D see 4398
Gentamycin see 4398
Gentersal [Ortho] see 6772
Gentian, 4399
Gentianic Acid see 4405
Gentianin see 4405
Gentianine, 4400
Gentian Violet, 4401
Gentiaverm see 4401
Gentibioptal [Farmila] see 4398
Genticin [Roche] see 4398
Gentiin see 4405
Gentinatre see 4404
Gentiobiose, 4402
Gentiopicrin, 4403
Gentiopicroside see 4403
Gentisic Acid, 4404
Gentisin, 4405
Gentisine U.C.B. see 4404
Gentisod see 4404
Gentisyl Alcohol, 4406
Gentocin [Schering] see 4398
Gentogram [Merck-Clévenot] see 4398
Gent-Ophtal [Winzer] see 4398
Gentran [Travenol] see 2989
Gentran 40 [Travenol] see 2989
Gentran 75 [Travenol] see 2989
Gentrasul [Bausch & Lomb] see 4398
Gentrogenin, 4407
Gentrol [Sandoz] see 4850
Genurin [Recordati] see 4135
Geocillin [Roerig] see 1888
Geodin see 3682
Geoffroyine see 9183
Geomycin [Pliva] see 7111
Geon [Goodrich] see 7746
Geopen [Roerig] see 1838
Geosmin, 4408
Geotricyn see 7235
Gepefrine, 4409
Gephyrotoxin, 4410
GER-11 see 461

Geram [Vedim] see 7641
Geranial see 2383
Geraniol, 4411
Geranium, 4412
Geranyl-1,4-benzenediol see 4413
Geranyl Farnesylacetate see 4385
Geranylgeranylacetone see 9296
Geranylhydroquinone, 4413
Geratacaca see 5755
Gerdaxyl [Gerda] see 5834
Geref [Serono] see 8605
Gerfil [G. F. Chem.] see 7741
Gerhardite see 2710
Germalgene see 9769
German Chamomile see 2082
German Chamomile Oil see 2082
Germane, 4414
German Fungus see 183
Germanic Acid see 4417
Germanin [Bayer] see 9181
Germanium, 4415
Germanium Dichloride, 4416
Germanium Dioxide, 4417
Germanium Hydride see 4414
Germanium Tetrachloride, 4418
Germanium Tetrafluoride, 4419
Germapect [Thiemann] see 1840
Germiciclin [Mendelejeff] see 6007
Germine, 4420
Germinol see 1086
Germitol see 1086
Gernebcin [Lilly] see 9628
Gero [URPAC] see 7937
Gerobit [Gerot] see 6015
Gerodorm [Gerot] see 2368
Gerolin [CT] see 2380
Gerontex H3 see 4615
Gerontine see 8894
Geroquinol see 4413
Gerostop [Merckle] see 6857
Gerot-Epilan [Gerot] see 5898
Gerovital see 4615
Gerovital H3 see 4615
Gesafloc [Ciba-Geigy] see 9797
Gesafram [Ciba-Geigy] see 7972
Gesagard [Geigy] see 7973
Gesamil [Ciba-Geigy] see 7996
Gesapax [Ciba-Geigy] see 411
Gesapon [Ciba-Geigy] see 2898
Gesaprim [Geigy] see 902
Gesarex [Ciba-Geigy] see 2898
Gesarol [Geigy] see 2898
Gesatop [Geigy] see 8681
Gestafortin [E. Merck] see 2152
Gestageno see 4886
Gestamestrol [Hermal] see 2152
Gestanin [Organon] see 299
Gestanol see 299
Gestanon [Organon] see 299
Gestanyn see 299
Gestapuran see 5838
Gestatron [AB Leo] see 3524
Gestiron [Provet] see 7956
Gestodene, 4421
Gestone [Paines & Byrne] see 7956
Gestonorone Caproate, 4422
Gestovis see 7251
Gestrinone, 4423
Gestronol Caproate see 4422
Gestyl [Organon] see 2273
Gevatran [Anphar-Rolland] see 6440
Gevilon [Warner-Lambert] see 4394
Gewacalm [Heilmittelwerke] see 3042
Gewalan [Lannacher] see 6607
Gewazol see 7283
GEWO-399 see 5203
Gexane [Pennwalt] see 5526
Geyser [Ciba] see 3182
G-Farlutal [Farmitalia] see 5838
GGA see 9296

GH see 8864
GH 3 see 4615
Ghatti Gum, 4424
GHB see 4860
Ghee see 4425
Ghi, 4425
GH-RF see 8861
GH-RH see 8861
GH-RIF see 8863
GI-87084 see 8300
GI-87084B see 8300
Giant Yam see 10225
Giardil see 4320
Giarlam see 4320
Gibbane see 4427
Gibberellane see 4427
Gibberellic Acid, 4426
Gibberellin A₃ see 4426
Gibberellins, 4427
Gibberellin X see 4426
Gibbsite see 369
Gibbs Reagent, 4428
Gibicef [Gibipharma] see 2002
Gibrel [Merck & Co.] see 4426
Gichtex [Gerot] see 287
Gifblaar Poison see 4205
Giganten [Troponwerke] see 2365
Gigantine, 4429
Gilemal [Chinoin] see 4486
Gilsonite see 9977
Giltex see 8814
Gilucor see 6704
Gilurytmal [Giulini] see 194
Gilutensin [Giulini] see 3909
Gina [Winthrop] see 7995
Ginamate see 818
Ginapect [Winthrop] see 7995
Ginarsol see 49
Gindarine see 9354
Gineflavir [Crosara] see 6242
Ginger, 4430
[6]-Gingerol, 4431
Gingilli Oil see 8614
Gingisan [Chassot] see 2140
Ginkgo, 4432
Ginkgo Biloba Extract see 4432
Ginkgoic Acid see 663
Ginkgolide A see 4433
Ginkgolide B see 4433
Ginkgolide C see 4433
Ginkgolides, 4433
Ginoden [Schering AG] see 4421
Ginorite see 1691
Ginseng, 4434
Ginsenosides see 4434
Giractide, 4435
Girard Reagent P see 4436
Girard Reagents, 4436
Girard Reagent T see 4436
Girard's Reagent D see 3294
Gismondite see 1685
Gitaligin [Schering] see 4437
Gitalin, 4437
Gitaloxin see 4437
Githagenin see 4614
Gitogenin, 4438
Gitogenin β-Lycotetraoside see 4439
Gitonin see 4439
F-Gitonin, 4439
Gitoxigenin, 4440
Gitoxin, 4441
Gittalun [Thomae] see 3497
Give-Tan [Givaudan] see 2370
Giv-Gard DXN [Givaudan] see 3273
Gix see 3000
Gjellebaekite see 1749
GL-7 see 4509
Gla see 1876
GLA see 5531
Glacial Phosphoric Acid see 7501

Gladiolic Acid, 4442
Glafenine, 4443
Glamidolo [Angelini Francesco] see 2884
Glandin [TAD] see 8065
Glandubolin see 3751
Glanil [Janssen] see 2365
Glaphenine see 4443
Glasel [Troponwerke] see 3922
Glassy Sodium Metaphosphate see 8814
Glauber's Salt see 8829
Glaucarubin, 4444
Glaucine, 4445
Glaucon [Alcon] see 3656
Glauconex [Alcon-Thilo] see 1050
Glaucostat [Merck & Co.] see 8266
Glaucotat [Chibret] see 8266
Glaucotensil [Farmila] see 3801
Glauco-Visken [Wander] see 7597
Glaudin [Sifi] see 8266
Glauline [Smith & Nephew] see 6221
Glaumeba [Merck & Co.] see 4444
Glaupax [Erco] see 50
Glauposine [Alcon] see 3656
Glaurin see 3169
Glausyn [Ciba-Geigy] see 6221
Glauvent see 4445
Glazidim [Glaxo] see 1995
Glean [Du Pont] see 2244
Glentonin [Glenwood] see 5245
Gliadin, 4446
Glianimon [Troponwerke] see 1077
Gliatilin [Italfarmaco] see 2260
Glibenclamide see 4486
Glibenese [Pfizer] see 4451
Gliben-Puren N [Klinge-Nattermann] see 4486
Glibornuride, 4447
Glicerinformal [Calipe] see 4494
Gliclazide, 4448
Glidiabet [Ferrer] see 4486
Glifan see 4443
Glifanan [Albert-Roussel] see 4443
Glimepiride, 4449
Glimicron [Dainippon] see 4448
Glimidstada [Stadapharm] see 4486
Glior see 3491
Gliorosein see 912
Glioten [Bago] see 3605
Gliotoxin, 4450
Glipasol see 4487
Glipizide, 4451
Gliporal [Grossmann] see 1500
Gliquidone, 4452
Glisoxepid, 4453
Gln see 4479
Globacillin [Astra] see 942
Globaline [Arapahoe] see 9347
Globenicol [Gist-Brocades] see 2120
Globin, 4454
Globocef [Roche] see 1974
Globularicitrin see 8456
Globulin G₁ see 5671
γ-Globulins see 4958
Glofil-131 [Abbott] see 5080
Glomycin see 7111
Glonoin see 6704
Glosso-Stérandryl [Roussel] see 6206
Gloxazone see 5310
GLQ223 [Genelabs] see 9780
Glu see 4477
Glubate [Hoechst Roussel] see 4486
Gluborid [Grünenthal] see 4447
Glucagon, 4455
Glucal see 1712
Glucametacin, 4456
Glucametacine see 4456
Glucamethacin see 4456
Glucamine, 4457

D-Glucamine see 4457
α-Glucan Branching Glycosyltransferase see 8207
α-1,6-Glucan 6-Glucanohydrolase see 2990
α-1,4-Glucan 4-Glucanohydrolases see 640
α-1,4-Glucan Maltohydrolases see 640
Glucantim [Farmitalia] see 6154
Glucantime [Specia] see 6154
D-Glucaric Acid, 4458
D-Glucaric Acid Calcium Salt see 1747
D-Glucaric Acid Di-γ-lactone Diacetate see 24
Glucaron [Chugai] see 24
Glucazide see 4510
Gluceptate Calcium see 4463
Gluceptate Sodium see 4463
Glucid see 8463
Glucidoral [Servier] see 1881
Glucinan [Lipha Santé] see 6001
Glucinium see 1201
D-Glucitol see 8873
Glucitol Iron Complex, Compd with Citric Acid see 5112
Glucoamylases see 640
D-Glucoascorbic Acid, 4459
Glucobay [Bayer] see 15
Glucobiogen see 1712
Glucochloral see 2114
α-D-Glucochloralose see 2114
Glucofrangulin, 4460
Glucofrangulin A see 4460
Glucofren [Cophar] see 1881
D-Glucofuranurono-6,3-lactone see 4476
Glucogallic Acid see 4462
α-Glucogallin, 4461
β-Glucogallin, 4462
Glucoheptonic Acid, 4463
α-Glucoheptonic Acid see 4463
Glucolin see 4467
D-Glucomethylose see 8262
Glucomonocarbonic Acid see 4463
Gluconiazide see 4510
Gluconic Acid, 4464
D-Gluconic Acid see 4464
D-Gluconic Acid Calcium Salt (2:1) see 1712
D-Gluconic Acid Cyclic 4,5-Ester with Boric Acid Calcium Salt (2:1) see 1692
D-Gluconic Acid 4-O-β-D-Galactopyranosyl Calcium Salt (2:1) Compd with Calcium Bromide (CaBr₂) (1:1) see 1694
D-Gluconic Acid δ-Lactone see 4465
Gluconic Acid Potassium Salt see 7796
Gluconic Acid Quinidine Salt see 8244
Gluconic Acid Sodium Salt see 8766
Glucono Delta Lactone see 4465
Gluconolactone, 4465
Gluconsan K [Kayaku] see 7796
Glucoperiplocymarin see 7315
Glucophage [Lipha] see 6001
Glucopostin [Boehringer, Mann.] see 7376
Glucoproscillaridin A see 8543
β-D-Glucopyranose Aerodehydrogenase see 4468
β-D-Glucopyranose 6-Benzoate see 10037
α-D-Glucopyranose 1-Dihydrogenphosphate see 4470
α-D-Glucopyranose-1-gallate see 4461
β-D-Glucopyranose-1-gallate see 4462
α-D-Glucopyranose-1-phosphate see 4470

Glucurolactone *see* 4476
Glucurone *see* 4476
D-Glucuronic Acid, 4474
D-Glucuronic Acid γ-Lactone *see* 4476
D-Glucuronic Acid γ-Lactone 1-[(4-Pyridinylcarboxyl)hydrazone] *see* 4510
β-Glucuronidase, 4475
D-Glucuronolactone, 4476
D-Glucuronolactone Isonicotinoyl-hydrazone *see* 4510
Gludiase [Bellon] *see* 4488
Glufosinate *see* 7495
Glufosinate-ammonium *see* 7495
Glukagon 4455
Glukor [Hyrex] *see* 2273
Glumal [Kyowa] *see* 25
Glumamycin *see* 625
Glumin *see* 4479
Glumorin *see* 5291
Glurenorm [Thomae] *see* 4452
Gluronazide *see* 4510
Gluside *see* 8463
Glusulase [Endo] *see* 4475
Glutacid *see* 4477
Glutacyl *see* 6337
Glutamed [Boehringer, Mann.] *see* 4483
Glutamic Acid, 4477
L-Glutamic Acid *see* 4477
Glutamic Acid 5-Amide *see* 4479
Glutamic Acid α-Amide *see* 5193
Glutamic Acid Compd with L-Arginine *see* 818
L-Glutamic Acid 5-Ethyl Ester, 4478
L-Glutamic Acid γ-Ethyl Ester *see* 4478
L-Glutamic Acid 5-[2-[4-(Hydroxy-methyl)phenyl]hydrazide] *see* 185
Glutamic Acid Lactam *see* 8185
L-Glutamic Acid L-Lysine Salt *see* 5669
Glutamic Acid Monosodium Salt Monohydrate *see* 6337
L-Glutamic Acid Polymer with L-Ala-nine, L-Lysine and L-Tyrosine *see* 2578
Glutamidin *see* 4477
Glutamine, 4479
L-Glutamine *see* 4479
2-L-Glutamine-6-L-asparaginealytesin *see* 1359
Glutaminic Acid *see* 4477
Glutaminol *see* 4477
3-γ-D-Glutamyl-D-alanylamino-3-methoxyazetidin-2-one-1-sulfonic Acid *see* 9119
(3R)-3-(γ-D-Glutamyl-D-alanylamino)-3-methoxy-2-oxoazetidine-1-sulfonic Acid *see* 9119
γ-L-Glutamyl-α-amino-β-(2-methyl-enecyclopropyl)propionic Acid Dipeptide *see* 4914
1-[N-(γ-L-Glutamyl)amino]-D-proline *see* 5524
N-(N-L-γ-Glutamyl-L-cysteinyl)glycine *see* 4483
L-γ-Glutamyl-L-cysteinyl-glycine Di-sulfide *see* 4483
β-N-[γ-L(+)-Glutamyl]-4-hydroxy-methylphenylhydrazine *see* 185
γ-L-Glutamylhypoglycine *see* 4914
(R)-D-γ-Glutamyl-N-(3-methoxy-2-oxo-1-sulfo-3-azetidinyl)-D-alanin-amide *see* 9119
N-L-γ-Glutamyl-3-(2-methylenecyclo-propyl)alanine *see* 4914
Glutan-HCl [Lederle] *see* 4477
Glutaral *see* 4480
Glutaraldehyde, 4480

Glutargin *see* 818
Glutaric Acid, 4481
Glutaric Acid Dinitrile *see* 4482
Glutaric Dialdehyde *see* 4480
Glutarol [Dermal] *see* 4480
Glutaronitrile, 4482
ε-N-(L-Glutar-2-yl)-L-lysine *see* 8464
Glutasan [Sancarlo] *see* 4483
Glutasin [McNeil] *see* 4477
Glutathin [Mochida] *see* 4483
Glutathiol *see* 4483
Glutathion [Nichiiko] *see* 4483
Glutathione, 4483
L-Glutathione *see* 4483
Glutathione-SH *see* 4483
Glutaton *see* 4477
Glutavene [Cooper] *see* 6337
Gluten, 4484
Glutestere *see* 4478
Glutethimide, 4485
Glutimic Acid *see* 8185
Glutiminic Acid *see* 8185
Glutinal [Sankyo] *see* 4483
Glutril [Roche] *see* 4447
Gly *see* 4500
[Gly⁷,Asu¹,⁶]oxytocin *see* 1887
Glybenzcyclamide *see* 4486
Glyburide, 4486
Glybuthiazol(e), 4487
Glybuthizol *see* 4487
Glybuzole, 4488
Glycamine *see* 4457
Glycarsamide, 4489
Glyceol [Chugai] *see* 4493
Glyceraldehyde, 4490
Glyceraldehyde 3-Phosphate, 4491
Glyceric Acid, 4492
Glyceric Aldehyde *see* 4490
Glycerin *see* 4493
Glycerine *see* 4493
D-Glycero-D-gulo-heptonic Acid Compd with Erythromycin (1:1) *see* 3723
Glycerol, 4493
Glycerol Diacetate *see* 3005
sym-Glycerol Dichlorohydrin *see* 3124
Glycerol α,γ-Dichlorohydrin *see* 3124
Glycerol Dimethylketal *see* 5232
Glycerol Formal, 4494
Glycerol Guaiacolate *see* 4582
Glycerol α-(2-Methoxyphenyl) Ether *see* 4582
Glycerol α-Monobutyrate *see* 6332
Glycerol α-Monochlorohydrin *see* 2197
Glycerol Mono(2-methoxyphenyl) Ether *see* 4582
Glycerol Nitric Acid Triester *see* 6704
Glycerol 1-Octanoate *see* 6335
Glycerol Tribromohydrin *see* 9745
Glycerophosphoric Acid, 4495
α-Glycerophosphoric Acid *see* 4495
β-Glycerophosphoric Acid *see* 4495
α-Glycerophosphoric Acid Calcium Salt *see* 1713
β-Glycerophosphoric Acid Calcium Salt *see* 1713
sn-Glycero-3-phosphorylcholine *see* 2260
Glycerose *see* 4490
Glyceryl p-Aminobenzoate, 4496
Glycerylaminophenaquine *see* 4443
Glyceryl 2-p-Chlorophenoxyisobuty-rate-1,3-dinicotinate *see* 1267
Glyceryl Diacetate *see* 3005
α-Glyceryl Guaiacol Ether *see* 4582
Glyceryl Guaiacyl Ether *see* 4582
Glyceryl Iodide, 4497
Glyceryl Monoacetate *see* 6325
Glyceryl 3-Monobutyrate *see* 6332

Glyceryl Monostearate, 4498
L-α-Glycerylphosphorylcholine *see* 2260
Glyceryl o-Tolyl Ether *see* 5895
Glyceryl Triacetate *see* 9721
Glyceryl Tributyrate *see* 9750
Glyceryl Trimyristate *see* 9854
Glyceryl Trinitrate *see* 6704
Glyceryl Trioleate *see* 9861
Glyceryl Tripalmitate *see* 9865
Glyceryl Tripetroselinate *see* 7330
Glyceryl Tristearate *see* 9885
Glycidol, 4499
Glycidyl Methacrylate-ethyl Acrylate Polymer *see* 2577
Glycin *see* 4885
(Glycinato-N,O)dihydroxyaluminum *see* 3227
Glycine, 4500
Glycine, Aluminum Salt *see* 3227
7-Glycine-1,6-aminosuberic Acid-oxy-tocin *see* 1887
Glycine Anhydride *see* 7619
1-Glycine-18-L-argininamide-α⁽¹⁻¹⁸⁾-corticotropin *see* 4435
Glycine Betaine *see* 1225
Glycine Copper Complex *see* 2707
Glycine-ferrous Sulfate Complex *see* 4086
Glycine Nitrile *see* 432
Glycine p-Phenetidide *see* 7388
Glycine Sulfate, 4501
Glycinin, 4502
Glycinonitrile *see* 432
Glycobiarsol, 4503
Glycocholic Acid, 4504
Glycocoll *see* 4500
Glycocoll Betaine *see* 1225
Glycocoll-copper *see* 2707
Glycocoll-p-phenetidide *see* 7388
Glycocyamine, 4505
Glycodex [Burns-Bio] *see* 4582
Glycodiazine *see* 4517
Glycogen, 4506
Glycogenic Acid *see* 4464
Glycolande [Delalande] *see* 4486
Glycol Benzylcarbamate *see* 1525
Glycol Bromohydrin *see* 3838
Glycol Chlorohydrin *see* 3839
Glycol Cyanohydrin *see* 3840
Glycol Diacetate *see* 3845
Glycol Dilaurate, 4507
Glycol Dinitrate *see* 3846
Glycoleucine *see* 6800
Glycolic Acid, 4508
Glycolic Acid [o-(2,6-Dichloroani-lino)phenyl]acetate Ester *see* 19
Glycolic Acid 8-Ester with Octahy-dro-5,8-dihydroxy-4,6,9,10-tetra-methyl-6-vinyl-3a,9-propano-3aH-cyclopentacycloocten-1(4H)-one *see* 7694
Glycol-monoacetin *see* 3847
N-Glycoloylarsanilic Acid *see* 4489
N-Glycoloylarsanilic Acid Bismuth Deriv *see* 4503
3-Glycoloyl-1,2,3,4,6,11-hexahydro-3,-5,12-trihydroxy-10-methoxy-6,11-dioxo-1-naphthacenyl-3-amino-2,3,-6-tridioxy-α-L-arabino-hexopyrano-side *see* 3660
Glycol Salicylate, 4509
Glycoluril *see* 94
p-Glycolylaminobenzenearsonic Acid *see* 4489
p-Glycolylaminophenylarsinic Acid *see* 4489
N-Glycolylarsanilic Acid *see* 4489
Glycolylurea *see* 4796
Glycomul L [Lonza] *see* 8872

Glycomul O [Lonza] see 8872
Glycomul S [Lonza] see 8872
Glyconiazide, 4510
Glyconic Acid see 4464
Glyconon [DDSA] see 9646
Glyconormal see 4517
Glycophene see 5093
Glycoprotein G see 9529
Glycoprotein-P see 7335
Glycopyrrolate, 4511
Glycopyrronium Bromide see 4511
Glycosine, 4512
Glycosthène [AB Leo] see 4500
7-D-Glycosylapigenin see 774
8-Glycosyl-4′,5,7-trihydroxy-3′-meth-
oxyflavone see 8546
Glycotauro [HW & D] see 7062
Glycyclamide see 9648
4-Glycylaminophenetol see 7388
4-Glycylaminophenol Ethyl Ether see
7388
N-Glycylglycine, 4513
Glycylglycine Lactam see 7619
N-[N-(N-Glycylglycyl)glycyl]-8-L-ly-
sinevasopressin see 9310
Nα-Glycylglycylglycylvasopressin see
9310
Glycylpressin [Ferring] see 9310
Glycyrrhetic Acid see 3628
Glycyrrhetic Acid Acetate see 76
18β-Glycyrrhetic Acid Hydrogen
Succinate see 1839
18β-Glycyrrhetinic Acid see 3628
Glycyrrhetinic Acid Glycoside see
4515
Glycyrrhiza, 4514
Glycyrrhizic Acid, 4515
Glycyrrhizin see 4515
Glycyrrhizinic Acid see 4515
Glydiazinamide see 4451
Glyferro [Hek] see 4086
Glyfyllin see 3529
Glyhexamide, 4516
Glykin see 9913
Glykocellon see 1877
Glykresin see 5895
Glyme see 3274
Glymidine, 4517
Glymol see 7327
Glyodin, 4518
Glyotol see 5895
Glyoxal, 4519
Glyoxal Compd with Sodium Bisulfite
see 4520
Glyoxaldiureine see 94
Glyoxalic Acid see 4521
Glyoxaline see 4948
Glyoxaline-5-alanine see 4758
Glyoxal-Sodium Bisulfite, 4520
Gly-Oxide [Marion] see 10007
Glyoxyldiureide see 255
Glyoxylic Acid, 4521
Glypesin [Uzara-Werk] see 4741
Glyphenarsine see 9925
Glyphosate, 4522
Glyphosine, 4523
Glyphylline see 3529
Glypinamide, 4524
Glypolix [Stroder] see 4015
Glypressin [Ferring] see 9310
Glysal see 4509
Glysennid [Dorsey] see 8600
Glytheonate [CooperVision] see 9421
Glyvenol [Ciba-Geigy] see 9738
GM-CSF see 4559
GM-CSF/IL-3 Fusion Protein see
5021
GM-DF see 4558
GMP see 4600
3′,5′-GMP see 2778

GMP-140 see 8568
Gnoscopine see 6815
Gö-560 see 3983
Gö-687 see 3934
Gö-919 see 7639
Gö-1261C see 9578
Gö-3450 see 4343
Goal [Rohm & Haas] see 7097
Goa Powder see 807
GOBAB see 463
Godalax [Pfleger] see 1282
Gödecke 3282 see 3934
GOE-3450 see 4343
Goethite see 4068
Goetsch's Vitamin see 10164
Goitrin, 4525
Gokilaht [Sumitomo] see 2837
Golarsyl see 49
Gold, 4526
Gold-bloom see 1763
Gold Colloid ¹⁹⁸Au see 4533
Gold Cyanide see 4529
Golden Antimony Sulfide see 738
Golden Apple Seed see 8237
Golden Ragwort see 8594
Golden Seal see 4808
Gold, Explosive, 4527
Goldinodox see 916
Goldinomycin see 916
Gold Monochloride, 4528
Gold Monocyanide, 4529
Gold Monoiodide, 4530
Gold Monosulfide, 4531
Gold Monoxide, 4532
Gold Orange see 6180
Gold Oxide see 4545
Gold Potassium Bromide see 7853
Gold Potassium Chloride see 7854
Gold Potassium Cyanide see 7791
Gold Potassium Iodide see 7860
Gold, Radioactive, Colloidal, 4533
Gold Selenate, 4534
Gold Selenide, 4535
Gold Sesquioxide see 4545
Gold Sodium Chloride see 8837
Gold Sodium Cyanide see 8755
Gold Sodium Thiomalate, 4536
Gold Sodium Thiosulfate, 4537
Gold Stannate, 4538
Gold Thioglucose see 917
Goldthread see 2590
Gold-tin Precipitate see 4538
Gold-tin Purple see 4538
Gold Tribromide, 4539
Gold Tribromide, Acid, 4540
Gold Trichloride, 4541
Gold Trichloride, Acid, 4542
Gold Tricyanide, 4543
Gold Trihydroxide, 4544
Gold Trioxide, 4545
Gold Trisulfide, 4546
Gold Yellow see 9905
Goltix [Bayer] see 5985
Gonacrine see 125
Gonadoliberin see 5500
Gonadorelin see 5500
Gonadorelin Acetate see 5500
Gonadotraphon L.H. [Paines & Byrne]
see 2273
Gonadotropin-Releasing Factor see
5500
Gonak [Akorn] see 4889
Gondafon [Schering] see 4517
Gonic [Hauck] see 2273
Goniosol [IOLAB] see 4889
Gonorcin [Farmas] see 6793
Gonosan see 5299
Gonyaulax Toxin see 8533
"Good" Buffers see 4687
Good-Rite Nix see 8780

Gooroo Nuts see 5331
Gopten [Knoll] see 9703
Goserelin, 4547
Gossypine see 2259
Gossypitrin see 8217
Gossyplure, 4548
Gossypol, 4549
Gossypose see 8279
Gotensin [Warner-Lambert] see 5488
Gougerotin, 4550
Gourd Curare see 2741
Govern [Rohm & Haas] see 4002
Government Rubber Styrene see 8534
Goyl see 49
GP-2-121-3 see 811
GP-26872 see 7431
GP-45840 see 3132
GP-47680 see 7063
L-α-GPC see 2260
GPIIb-IIIa see 5014
G-Proteins, 4551
GR-2/234 see 235
GR-2/925 see 2423
GR-2/1214 see 2424
GR-2/1574 see 233
GR-20263 see 1995
GR-33343G see 8489
GR-33343X see 8489
GR-38032F see 6979
GR-43175 see 9172
GR-43175C see 9172
GR 43659X see 5344
GR-109714X see 5365
GR-122311X see 8286
Gradient [Polifarma] see 4179
Gradocycline see 8411
Grafestrol see 3177
Graham's Salt see 8814
Grains of Paradise see 1883
Gramalil [Fujisawa] see 9561
Gramaxin [Boehringer, Mann.] see
1967
Gramicidin C (Soviet) see 4552
Gramicidin D (Dubos) see 4553
Gramicidin S, 4552
Gramicidin(s), 4553
Gramicidin S (Soviet) see 4552
Gramine, 4554
Graminis see 9890
Graminon [Ciba-Geigy] see 5237
Gramipan [Mayoly-Spindler] see 2959
Gram-Micina [Lagap] see 4280
Grammite see 1749
Gramoderm [Schering] see 4553
Gramoxone [ICI] see 7165
Grampenil [Argentia] see 628
Gramplus [Chiesi] see 2439
Granaticin, 4555
Granaticin B see 4555
Granatum see 7747
Grandaxin [EGYT] see 9642
Grandisol, 4556
Grandlure see 4556
Granisetron, 4557
Granocyte [Chugai] see 4558
Granosan see 3871
Granudoxy [Pierre Fabre] see 3496
α-Granule Membrane Protein 140 see
8568
Granulestin see 5452
Granulocyte Colony-Stimulating
Factor, 4558
Granulocyte-Macrophage Colony-
Stimulating Factor, 4559
Granzymes see 7301
Grape Sugar see 4467
Graphite, 4560
Graphite Fluoride, 4561
Graphite Oxide see 4562
Graphitic Acid, 4562

Graphitic Oxide *see* 4562
Graphol *see* 6097
Graslan [DowElanco] *see* 9255
Grasp [Zeneca] *see* 9699
Gratiogenin, 4563
Gratioside, 4564
Gratus Strophanthin *see* 7031
Gravitol(e), 4565
Gravol [Horner] *see* 3252
"Gray Acetate of Lime" *see* 1683
(3β,6β,14R)-Grayanotoxane-3,5,6,10,-14,16-hexol *see* 4566
(3β,6β,14R)-Grayanotoxane-3,5,6,10,-14,16-hexol 14-Acetate *see* 4566
(3β,6β,14R)-Grayanotox-10(20)-ene-3,5,6,14,16-pentol *see* 4566
Grayanotoxin I *see* 4566
Grayanotoxin II *see* 4566
Grayanotoxin III *see* 4566
Grayanotoxins, 4566
GR-C507/75 *see* 6979
Greek Hay *see* 4049
Green Broom *see* 8547
Green Cinnabar *see* 2283
Greenhartin *see* 5376
Greenheart *see* 1043
Green Hellebore *see* 10090
Green Hydroquinone *see* 8242
"Green Nickel Oxide" *see* 6593
Greenockite *see* 1666
Green Oxide of Chromium *see* 2283
Green Rouge *see* 2283
Green Verdigris *see* 2691
Green Vitriol *see* 4105
Grenade [ICI] *see* 2827
Grepafloxacin, 4567
Grey Arsenic *see* 832
GRF *see* 8861
GRF(1-29)NH₂ *see* 8605
Griffith's Zinc White *see* 5572
Grifomin *see* 485
Grifulvin [McNeil] *see* 4571
Grilon [Fibron] *see* 6832
Grindelia, 4568
Grindelic Acid, 4569
Grinsil [Argentia] *see* 617
Gripenin-O [Fujisawa] *see* 1838
Grisactin [Ayerst] *see* 4571
Griséfuline [Sanofi Winthrop] *see* 4571
Grisein, 4570
Griseofulvin, 4571
Griseoviridin, 4572
Grisovin [Herbert] *see* 4571
Gris-PEG [Herbert] *see* 4571
GR-M *see* 6550
Grocreme [Grossmann] *see* 2959
Grodurex [Grossmann] *see* 426
Grolean [Pitman-Moore] *see* 8864
Groliberin [Kabi] *see* 8605
Grorm [Serono] *see* 8864
Grossularite *see* 1685
Ground Lily *see* 9825
Groundnut *see* 7190
Groundnut Oil *see* 7191
Grovex [Groves] *see* 7629
Growth Hormone *see* 8864
Growth Hormone-Release Inhibiting Factor *see* 8863
Growth Hormone-Releasing Factor *see* 8861
Growth Hormone-Releasing Hormone *see* 8861
GR-S *see* 8534
Grysio [Ayerst] *see* 4571
GS-95 *see* 9449
GS-385 *see* 9048
GS-504 *see* 2329
GS-1339 *see* 3525
GS-2989 *see* 5818

GS-3065 *see* 3496
GS-6244 *see* 1825
GS-13005 *see* 6048
GS-13332 *see* 3313
GS-16068 *see* 3405
GS-19851 *see* 1456
GS-23654 *see* 6731
G Salt *see* 6475
GSH *see* 4483
GSSG *see* 4483
G-strophanthidin *see* 7030
G-strophanthin *see* 7031
GT-41 *see* 1529
GT-92 *see* 2788
GT-1012 *see* 7883
GTN [Farillon] *see* 6704
GTP Binding Proteins *see* 4551
G-Tril *see* 3983
Guacetisal *see* 886
Guaiac, 4573
Guaiac Alcohol *see* 4583
Guaiac-Copper Sulfate Paper, 4574
Guaiacol, 4575
Guaiacol Benzoate, 4576
Guaiacol Carbonate, 4577
Guaiacol Carbonic Acid Neutral Ester *see* 4577
Guaiacol Glyceryl Ether *see* 4582
Guaiacol Glyceryl Ether Carbamate *see* 6060
Guaiacol Phosphate, 4578
Guaiacol Valerate, 4579
Guaiacum *see* 4573
Guaiacuran *see* 4582
Guaiacyl Glyceryl Ether *see* 4582
Guaiamar *see* 4582
Guaiapate, 4580
Guaiaspir [Lampugnani] *see* 886
Guaiazulene, 4581
S-Guaiazulene *see* 4581
Guaiazulene Soluble *see* 4581
Guaifenesin, 4582
Guaiol, 4583
Guaiphenesin *see* 4582
Guaithylline *see* 4582
Guajol *see* 4583
Guamecycline, 4584
Guanabenz, 4585
Guanacline, 4586
Guanadrel, 4587
Guanamprazine *see* 426
Guanatol [Lilly] *see* 2138
Guanazodine, 4588
Guanazolo *see* 928
Guaneran [Burroughs Wellcome] *see* 9434
Guanethidine, 4589
Guanfacine, 4590
Guanicil [Cilag] *see* 9076
Guanidine, 4591
Guanidineacetic Acid *see* 4505
Guanidine Compd with Aluminum Sulfate *see* 4592
Guanidinium Aluminum Sulfate Hexahydrate, 4592
4-Guanidino-1-butanol Syringate *see* 5463
N-(2-Guanidinoethyl)-4-methyl-Δ³-piperidine *see* 4586
N-(2-Guanidinoethyl)-4-methyl-1,2,3,-6-tetrahydropyridine *see* 4586
1-(2-Guanidinoethyl)octahydroazocine *see* 4589
1-(2-Guanidinoethyl)-1,2,3,6-tetrahydro-4-picoline *see* 4586
2-Guanidinomethyl-1,4-benzodioxan *see* 4598
α-Guanidinomethylheptamethylenimine *see* 4588
Guanidoacetic Acid *see* 4505

Guanine, 4593
Guanine Nucleotide Binding Proteins *see* 4551
Guanine Riboside *see* 4596
Guanine Riboside-3-phosphoric Acid *see* 4599
Guanine Riboside-5-phosphoric Acid *see* 4600
Guanite *see* 9019
Guano, 4594
Guanochlor, 4595
Guanosine, 4596
Guanosine Cyclic 3',5'-(Hydrogen Phosphate) *see* 2778
Guanosine 3',5'-Cyclic Monophosphate *see* 2778
Guanosine 3',5'-Cyclic Phosphate *see* 2778
Guanosine 3'-Monophosphate *see* 4599
Guanosine 3',5'-Monophosphate *see* 2778
Guanosine 5'-Monophosphate *see* 4600
Guanosine 5'-Phosphate *see* 4600
Guanoxabenz, 4597
Guanoxan, 4598
Guantal *see* 3383
Guanylate Cyclase *see* 2778
9-(p-Guanylbenzal)fluorene *see* 7163
9-(p-Guanylbenzylidene)fluorene *see* 7163
Guanylguanidine *see* 1261
Guanylhydrazine *see* 461
3'-Guanylic Acid, 4599
5'-Guanylic Acid, 4600
Guanylic Acid B *see* 4599
N¹-Guanylsulfanilamide *see* 9076
Guanylurea Sulfate *see* 3143
Guaran, 4601
Guaranine *see* 1674
Guardian [Boehringer, Ing.] *see* 4160
Guarem [Rybar] *see* 4602
Guar Flour *see* 4602
Guar Gum, 4602
Guarina [Norgine] *see* 4602
Guastil [Uriach] *see* 9163
Guatambuinine *see* 6971
Guayanesin *see* 4582
Guayule *see* 8435
Gubernal [Geigy] *see* 321
Guesapon [Ciba-Geigy] *see* 2898
Guethine *see* 4589
Guidazide *see* 4510
Guinea Green B, 4603
Guinea-pig-anti-stiffness Factor *see* 8969
Gujaphenyl [DDD] *see* 4575
L-Gulitol *see* 8873
Gulliostin [Taiyo] *see* 3410
D-Gulonic Acid, 4604
L-Gulonic Acid, 4605
D-Gulose, 4606
L-Gulose, 4607
Gum Ammoniac *see* 518
Gum Arabic *see* 11
Gumbaral [Asta] *see* 155
Gum Benjamin *see* 4608
Gum Benzoin, 4608
Gum Camphor *see* 1779
Gum Copal *see* 2581
Gum Cyamopsis *see* 4602
Gum Damar *see* 2872
Gum Ghatti *see* 4424
Gum Guaiac *see* 4573
Gum Karaya *see* 5296
Gum Kino *see* 5325
Gum Opium *see* 6986
Gum Plant *see* 3711
Gum-plant (of California) *see* 4568
Gum Quince Seed *see* 8237
Gum-resin Myrrh *see* 6421

Gum Thus *see* 6969 and 9957
Gum Tragacanth, 4609
Gum Wood *see* 3941
Guntrin [Zensei] *see* 7016
Gunyl [Hogapharm] *see* 4575
G.U.-Pen [Pfizer] *see* 1888
Guronsan *see* 4476
Guru Nuts *see* 5331
Gusathion A [Bayer] *see* 944
Gusathion M [Bayer] *see* 944
Gusperimus, 4610
Guthion [Bayer] *see* 944
Gutron [Nycomed] *see* 6272
Guttalax-Fher [Boehringer, Ing., Ital.]
  see 7559
Gutta-Percha, 4611
Guvacine, 4612
GV *see* 4572
GVG *see* 10114
GX-1048 *see* 5344
Gy-bon [Ciba-Geigy] *see* 8683
GYKI-41099 *see* 2464
Gymnemic Acid, 4613
Gymnemin *see* 4613
Gynamide [Merminod] *see* 4620
Gynamousse *see* 7111
Gynäsan [Bastian] *see* 3750
Gynefollin *see* 3153
Gyne-Lotrimin [Schering] *see* 2478
Gynera [Schering AG] *see* 4421
Gynergen [Sandoz] *see* 3703
Gynesine *see* 9822
Gynestrel [Recordati] *see* 6504
Gyn-Hydralin [Lefrancq] *see* 4500
Gynipral [Chemie-Linz] *see* 4745
Gynocardia Oil *see* 2088
Gynochrome *see* 5914
Gyno-Daktarin [Janssen] *see* 6266
Gynoestryl [Roussel] *see* 3746
Gynofon [Endopancrine] *see* 430
Gynofug [Wolff] *see* 4925
Gynogen LA [Forest] *see* 3746
Gynokhellan *see* 5321
Gynol II [Ortho-Cilag] *see* 6772
Gynolett [Labopharma] *see* 3780
Gyno-Monistat [Cilag-Chemie] *see*
  6266
Gyno-Myfungar [Klinge] *see* 7071
Gynomyk [Cassenne] *see* 1561
Gyno-Pevaryl [Cilag-Chemie] *see* 3550
Gynoplix [Théraplix] *see* 49
Gynorest [Mead Johnson] *see* 3524
Gyno-Sterosan [Geigy] *see* 2243
Gyno-Terazol [Cilag] *see* 9303
Gynotherax *see* 2243
Gyno-Travogen [Schering AG] *see*
  5176
Gyno-Trosyd [Pfizer] *see* 9595
Gynovlar [Schering AG] *see* 6790
Gypsogenin, 4614
Gypsophilasapogenin *see* 4614
Gypsum *see* 1753
Gyramid [Parke-Davis] *see* 3625
Gyrolite *see* 1749
H *see* 4758
H3®, 4615
H-33 *see* 3985
H-56/28 *see* 321
(—)H-80/62 *see* 7917
H-93/26 *see* 6235
cis-H-102/09 *see* 10254
H-115 *see* 2156
H-133 *see* 3093
H-133/22 *see* 7917
H-154/82 *see* 3991
H-168/68 *see* 6977
H-321 *see* 6050
H-365 *see* 7176
H-610 *see* 4006
H-814 *see* 4014

H-990 *see* 7100
H-1032 *see* 6223
H-3292 *see* 3424
H-3452 *see* 2789
H-3625 *see* 2603
H-3749 *see* 8158
H-3774 *see* 243
H-4007 *see* 5902
H-4723 *see* 2417
HA-106 *see* 9838
HA-1077 *see* 3980
Haarlem Oil *see* 5285 and 6951
HABA-DKB *see* 809
Habekacin [Meiji Seika] *see* 809
Habitrol [Ciba-Geigy] *see* 6611
Hachimycin, 4616
H Acid *see* 4472
Hadacidin, 4617
Haelan [Dista] *see* 4232
Haemodyn *see* 7879
Haemofort *see* 4105
Haemostop Injection *see* 6481
Haertolan *see* 6495
Haflutan [Cassella-Riedel] *see* 2434
Hafnium, 4618
Hagan Phosphate *see* 8814
Hageman Factor *see* 3967
Hahnium *see* 3581 and 3584
Haiprex [Riker] *see* 6039
HAL *see* 4627
Halamid *see* 2118
Halane *see* 3115
Halarsol *see* 3121
Halazepam, 4619
Halazone, 4620
Halbmond [Much] *see* 3367
Halciderm [Squibb] *see* 4621
Halcimat [Squibb-Heyden] *see* 4621
Halcinonide, 4621
Halcion [Upjohn] *see* 9734
Haldol [McNeil] *see* 4629
Haldol Decanoate [Janssen] *see* 4629
Haldrate [Lilly] *see* 7162
Haldrone [Lilly] *see* 7162
Haletazole *see* 4622
Halethazole, 4622
Halfan [SKB] *see* 4626
Halfa Wax *see* 3742
Half-cystine *see* 2850
Halibut Liver Oil, 4623
Halidor [EGYT] *see* 1060
Halimide®, 4624
Halinone [USV] *see* 1414
Halite *see* 8742
Haloanisone *see* 4150
Haloart [Taiho] *see* 4630
Halobetasol Propionate, 4625
Halochondrine A *see* 6958
Halofantrine, 4626
Halofuginone, 4627
Halog [Squibb-Heyden] *see* 4621
Halogabide *see* 7955
Halometasone, 4628
Halomonth [Dainippon] *see* 4629
Halon *see* 3114
Haloperidol, 4629
Halopredone Acetate, 4630
Haloprogin, 4631
Halopropane, 4632
Halopyramine *see* 2214
Halospor [Ciba-Geigy] *see* 1985
Halostachine, 4633
Halosten [Shionogi] *see* 4629
Halotestin [Upjohn] *see* 4223
Halotex [Mead Johnson] *see* 4631
Halothane, 4634
Haloxazolam, 4635
Haloxon, 4636
Halquinol, 4637
Halquivet *see* 4637

Halset [Strohschein] *see* 2074
Haltran [Upjohn] *see* 4925
Hamaméliode P [Oberlin] *see* 5477
Hamamelis, 4638
Hamamelitannin, 4639
Hamamelose, 4640
Hamarin [Nicholas] *see* 287
Hamburg Blue *see* 4064
Hämovannid [Bastian] *see* 5009
Hamycin, 4641
Hamycin A *see* 4641
Hansolar [Parke, Davis] *see* 20
Haocolin [Fuso] *see* 2380
HAPA-B *see* 5121
Hapadex [Schering] *see* 6564
Haplophytine, 4642
Haptocil [Cilag-Chemie] *see* 2781
Haptocorrin *see* 10152
Haptoglobins, 4643
Harden-Young Ester *see* 4297
Hard Paraffin *see* 7155
Harmaline, 4644
Harmalol, 4645
Harmalol Methyl Ether *see* 4644
Harman, 4646
Harmar [Zemmer] *see* 8024
Harmidine *see* 4644
Harmine, 4647
Harmogen [Abbott] *see* 3751
Harmonyl [Abbott] *see* 2964
Harnal [Yamanouchi] *see* 9217
Harness [Monsanto] *see* 58
Harodasc *see* 4794
Hartshorn *see* 534
Harvade [Uniroyal] *see* 7072
Harvatrate [Tilden-Yates] *see* 907
Harzol [Hoyer] *see* 8697
Hasach *see* 1794 and 4648
Hasethrol *see* 7249
Hashish, 4648
Hasubanan *see* 4649
Hasubanonine, 4649
Hatchett's Brown *see* 2703
Hauser's Salt *see* 10315
Hausmannite *see* 5776
Havapen *see* 7208
Havidote [Haver] *see* 3559
Havlane [Diamant] *see* 5604
Haw Apple *see* 2636
Haworth Lignan *see* 5509
Haws *see* 2636
Hawthorn *see* 2636
Haynon [R. P. Drugs] *see* 2232
Hayo *see* 2515
Hazol *see* 7100
HB-419 *see* 4486
HBA *see* 3955
HBBL *see* 3633
HBF-386 *see* 1642
HBK *see* 809
HC-3 *see* 4676
4-HC *see* 7303
Hc45 [Crookes] *see* 4828
HC-58 *see* 3578
HC-064 *see* 6634
HC-20511 *see* 5319
25-HCC *see* 1677
HCFU *see* 1892
HCG *see* 2273
γ-HCH *see* 5526
HCS, 4650
HDNF *see* 6570
HDP *see* 7074
HE-69 *see* 6306
HE-781 *see* 4594
Head & Shoulders [Procter & Gamble] *see* 8178
Healon [Pharmacia] *see* 4793
Healonid [Pharmacia] *see* 4793
Heartcin [Ohta] *see* 9974

Heartgard 30 [Merck & Co.] *see* 5264
Heart Muscle Extract, 4651
Heavy Hydrogen *see* 2983
Heavy Spar *see* 1023
Heavy Water *see* 2984
Hebanil *see* 2238
Hebaral *see* 4740
Hebucol *see* 2784
Hecogenin, 4652
Hectorite, 4653
Hedeoma *see* 7236
Hederagenin, 4654
α-Hederin, 4655
Hedex [Sterling] *see* 45
Hedonal DP [Bayer] *see* 3128
Hedonal MCPP [Bayer] *see* 5826
Hedulin [Merrell] *see* 7381
Hefasolon [Hefa] *see* 7903
Heferol [Alkaloid] *see* 4094
Heitrin [Abbott] *see* 9297
Hekbilin [Hek] *see* 2096
Helenalin, 4656
Helenien *see* 10197
Helenin *see* 208
Helenynolic Acid, 4657
HelfaCat [Helvepharm] *see* 3043
HelfaDog [Helvepharm] *see* 3043
Helfergin [Promonta] *see* 5820
Helgotan *see* 9222
Helianthine B *see* 6180
Helicin, 4658
Helicocerin *see* 2047
Helicon *see* 886
Heliophan [Greef] *see* 4776
Heliosupine, 4659
Heliotropin *see* 7628
Helium, 4660
Helixin C *see* 6633
Helixin (the Saponin) *see* 4655
Hellebrigenin Glucorhamnoside *see* 4661
Hellebrin, 4661
Helmatac [SK & F] *see* 7169
Helmet Flower *see* 8557
Helmetina *see* 7404
Helmex [Pfizer] *see* 8139
Helmezine [Allen & Hanburys] *see* 7617
Helmifren *see* 7617
Helminal *see* 5289
Helminthosporal, 4662
Helminthosporol, 4663
Helmintox [Innothera] *see* 8139
Helmirone *see* 4636
Helmitol [Winthrop] *see* 6038
Helogaphen [Spitzner] *see* 2132
Helonias, 4664
Helothion [Bayer] *see* 9164
Helvamox [Helvepharm] *see* 617
Helvecyclin [Helvepharm] *see* 9337
Helvetia Blue *see* 6107
Helveticoside, 4665
Helvolic Acid, 4666
Hem *see* 4672
Hemabate [Upjohn] *see* 1871
Hemanthine *see* 1518
Hematein, 4667
Hematin, 4668
Hematin-protein *see* 2860
Hematite *see* 4072
Hematoporphyrin, 4669
Hematoporphyrin IX *see* 4669
Hematoporphyrinmercury Disodium Salt *see* 5961
Hematoxiline *see* 4670
Hematoxylin, 4670
Hematoxylon, 4671
Heme, 4672
Hemel *see* 328
Hementin, 4673

Hemerven [Ciba-Geigy] *see* 3350
Hemerythrin, 4674
Hemicelluloses, 4675
Hemicholinium, 4676
Hemicholinium-3 *see* 4676
Hemi-Dewar Biphenyl, 4677
Hemiglobin *see* 6035
Hemin, 4678
Hemineurin [Debat] *see* 2444
Heminevrin [Astra Pharm.] *see* 2444
Hemipyocyanine, 4679
Hemisulfur Mustard, 4680
Hemlock *see* 2570
Hemoantin [Lepetit] *see* 9160
Hemocaprol [Delagrange] *see* 451
Hémoclar [Sanofi Winthrop] *see* 7276
Hemocoagulase *see* 1038
Hemocuprein *see* 9177
Hemocuron [Takeda] *see* 9738
Hemocyanins, 4681
Hemodex [Central Biokhim.] *see* 2989
Hemofil [Hyland] *see* 3963
Hemoglobin, 4682
Hemolidione [Sintetico] *see* 7381
Hemometina [Cusi] *see* 3600
Hemo-Pak [J & J] *see* 7073
Hemopoietine *see* 3729
Hemotrope [Andromaco] *see* 1535
Hempa *see* 4761
Henbane *see* 4908
2-Hendecanone *see* 6178
10-Hendecenoic Acid *see* 9983
Heneicosafluorotripropylamine *see* 4221
Henna, 4683
HEOD *see* 3152
Hepacholine *see* 2261
Hepadial [Biocodex] *see* 3248
Hepadist *see* 4716
Hepagin *see* 7348
Hepalande [Delalande] *see* 5878
Hepalidine [Riker] *see* 9586
Hepalon *see* 5575
Hepa-Merz [Merz] *see* 7002
Heparamine, 4684
Hepar Calcis *see* 1754
Heparegen [Syntex] *see* 9586
Heparexine *see* 7519
Heparides *see* 4684
Heparin, 4685
β-Heparin *see* 2270
Heparinic Acid *see* 4685
Heparin Sodium *see* 4685
Hepar Sulfuris *see* 7762
Hepartest [Cassella-Riedel] *see* 9127
Hepasil *see* 4009
Hepation [Nippon Chemiphar] *see* 5751
Hepatocatalase *see* 1948
Hepatolite [Du Pont] *see* 3422
Hepatopron *see* 5575
Hepatoxane [Pharm. L'Esplanade] *see* 9630
Hepaxanthin, 4686
Hepbisul *see* 4693
HEPES, 4687
Hepin *see* 451
Hepol *see* 5575
Heporal *see* 683
HEPP *see* 7270
Heprinar [Armour] *see* 4685
Hepronicate, 4688
Hepsal [Weddel] *see* 4685
Heptabarb *see* 4689
Heptabarbital, 4689
Heptachlor, 4690
1H-1,4,5,6,7,8,8-Heptachloro-3a,4,7,-7a-tetrahydro-4,7-methanoindene *see* 4690
Heptadecanoic Acid *see* 5793

(E,E,E)-(—)-8,10,12-Heptadecatriene-4,6-diyne-1,14-diol *see* 2328
(E,E,E)-2,8,10-Heptadecatriene-4,6-diyne-1,14-diol *see* 3608
2-(8-Heptadecenyl)-4,5-dihydro-1H-imidazole-1-ethanol *see* 428
2-(8-Heptadecenyl)-2-imidazoline-1-ethanol *see* 428
2-Heptadecyl-4,5-dihydro-1H-imid-azole Monoacetate *see* 4518
5-Heptadecylene-1-carboxylic Acid *see* 7330
2-Heptadecylglyoxalidine Acetate *see* 4518
[2α,3β(1E,3E),5α,6α]-3-(1,3-Hepta-dienyl)-5,6-dihydroxy-2-(hydroxy-methyl)cyclohexanone *see* 7120
6-(1,3-Heptadienyl)-3,4-dihydroxy-2-oxocyclohexanecarboxaldehyde *see* 4293
Heptadon [Ebewe] *see* 6008
Heptadorm *see* 4689
1,1,2,2,3,3,3-Heptafluoro-N,N-bis-(heptafluoropropyl)-1-propanamine *see* 4221
Heptaldehyde *see* 4692
Heptaldehyde Sodium Bisulfite *see* 4693
Heptalgin *see* 7348
Heptalin *see* 7348
2-(1-N,N-Heptamethylenimino)ethyl-guanidine *see* 4589
(all-E)-2-(3,7,11,15,19,23,27-Hepta-methyl-2,6,10,14,18,22,26-octacosa-heptaenyl)-3-methyl-1,4-naphtha-lenedione *see* 5876
Heptaminol, 4691
Heptamul *see* 4690
Hept-a-myl [Delalande] *see* 4691
Heptanal, 4692
Heptanal Sodium Bisulfite, 4693
2-Heptanamine *see* 9934
n-Heptane, 4694
1,7-Heptanedicarboxylic Acid *see* 938
Heptanedioic Acid *see* 7585
Heptanoic Acid, 4695
Heptanoic Acid Ethyl Ester *see* 3881
1-Heptanol, 4696
2-Heptanol, 4697
Heptanon [Pliva] *see* 6008
2-Heptanone, 4698
4-Heptanone *see* 3408
17β-Heptanoyloxy-1-methyl-5α-an-drost-1-en-3-one *see* 6044
Heptazone *see* 7348
Heptedrine [Bellon] *see* 9934
D-arabino-Hept-2-enonic Acid γ-Lac-tone *see* 4459
Heptenophos, 4699
n-Heptoic Acid *see* 4695
Heptone *see* 7348
D-glycero-D-gulo-Heptonic Acid *see* 4463
Heptoxime, 4700
D-manno-Heptulose, 4701
n-Heptyl Alcohol *see* 4696
Heptylaldehyde *see* 4692
2-Heptyl-4-hydroxyquinoline N-Oxide *see* 4787
19-Heptyl-10-hydroxy-1,5,10,14-tetra-azacyclononadecan-15-one *see* 7665
n-Heptylic Acid *see* 4695
2,2′-[3-[(3-Heptyl-4-methyl-2(3H)-thiazolylidene)ethylidene]-1-prop-ene-1,3-diyl]bis(3-heptyl-4-meth-ylthiazolium) Diiodide *see* 7690
Heptylon [Delalande] *see* 4691
Heptylphysostigmine *see* 3672
2-Heptyl-4-quinolinol 1-Oxide *see* 4787

Heptylstigmine *see* 3672
Hepzide [Merck & Co.] *see* 6660
Héraclène [Mauchant] *see* 2513
Heraclin *see* 1196
Heraldium [Sanofi] *see* 5822
Herapathite *see* 8246
Herbadox [Am. Cyanamid] *see* 7211
Herban [Hercules] *see* 6787
Herbesser [Tanabe Seiyaku] *see* 3247
Herbisan [Roberts Chemicals] *see* 3449
Hercules 528 *see* 3355
Hercules 3956 *see* 9693
Hercules 7531 *see* 6787
Hercules 14503 *see* 3013
Herculon [Hercules] *see* 7741
Hercynine, 4702
Herkol *see* 3129
Hermesetas *see* 8463
Hermophényl *see* 5939
Heroin *see* 3012
Herperal [Farmitalia] *see* 8927
Herpes-Gel [Master] *see* 4934
Herplex [Allergan] *see* 4934
Herqueinone, 4703
Herzberg's Paper *see* 2562
Herzolan *see* 4651
HES *see* 4707
Hesofen *see* 5890
Hesotanol *see* 3925
Hesotin *see* 3925
Hespan [DuPont Merck] *see* 4707
Hespander [Kyorin] *see* 4707
Hesperetin, 4704
Hesperetin 7-Rhamnoglucoside *see* 4705
Hesperetin-7-rutinoside *see* 4705
Hesperidin, 4705
Hestar [Otsuka] *see* 4707
Hestat [Otsuka] *see* 4707
Hestrium Chloride *see* 4724
Hestsol [Green Cross] *see* 4707
Hetacillin, 4706
Hetacin K [Bristol] *see* 4706
Hetaphenone *see* 3753
Hetastarch, 4707
5-HETE, 4708
Heteroauxin *see* 4994
Heterocodeine *see* 6359
Heteronium Bromide, 4709
Hetol *see* 1340
Hetolin [Boehringer, Mann], 4710
Hetoxol L-9 [Heterene Chem.] *see* 7717
Hetrazan [Lederle] *see* 3165
Hetrazeen [Heterochemical] *see* 5874
Hetrum Bromide [Lilly] *see* 4709
Heulandite *see* 1685
Hevyteck [Penreco] *see* 7327
Hexaaminecobalt Trichloride, 4711
Hexaaquochromium Triacetate *see* 2275
Hexaaquochromium Tribromide *see* 2276
Hexaaquochromium Trichloride *see* 2278
1,4,7,10,13,16-Hexaazacyclotricosane Cyclic Peptide Deriv *see* 3578
Hexabendin *see* 4743
Hexabetalin *see* 8166
Hexabione Hydrochloride *see* 8166
Hexaborane(10), 4712
Hexaboron Decahydride *see* 4712
Hexabrix [Guerbet] *see* 5084
2,2',4,4',5,5'-Hexabromobiphenyl *see* 7724
Hexacarbacholine Bromide, 4713
Hexachlorobenzene, 4714
1,2,3,4,7,7-Hexachlorobicyclo[2.2.1]-

2-heptene-5,6-bisoxymethylene Sulfite *see* 3614
1,4,5,6,7,7-Hexachloro-*endo*-bicyclo-[2.2.1]hept-5-ene-2,3-dicarboxylic Anhydride *see* 2134
1,4,5,6,7,7-Hexachloro-2,3-bis(chloromethyl)bicyclo[2.2.1]hept-5-ene *see* 2127
1,2,3,4,7,7-Hexachloro-5,6-bis(chloromethyl)-2-norbornene *see* 2127
(1α,2α,3β,4α,5α,6β)-1,2,3,4,5,6-Hexachlorocyclohexane *see* 5526
Hexachloroendomethylenetetrahydrophthalic Anhydride *see* 2134
1,2,3,4,10,10-Hexachloro-6,7-epoxy-1,4,4a,5,6,7,8,8a-octahydro-*endo*,-*endo*-1,4:5,8-dimethanonaphthalene *see* 3618
1,2,3,4,10,10-Hexachloro-6,7-epoxy-1,4,4a,5,6,7,8,8a-octahydro-*endo*,-*exo*-1,4:5,8-dimethanonaphthalene *see* 3152
Hexachloroethane, 4715
1,2,3,4,10,10-Hexachloro-1,4,4a,5,8,-8a-hexahydro-1,4:5,8-dimethano-naphthalene *see* 227
6,7,8,9,10,10-Hexachloro-1,5,5a,6,9,-9a-hexahydro-6,9-methano-2,4,3-benzodioxathiepin 3-Oxide *see* 3614
1,4,5,6,7,7-Hexachloro-*endo*-5-nor-bornene-2,3-dicarboxylic Anhydride *see* 2134
1,4,5,6,7,7-Hexachloro-5-norbornene-2,3-dimethanol Cyclic Sulfite *see* 3614
(1aα,2β,2aα,3β,6β,6aα,7β,7aα)-3,4,-5,6,9,9-Hexachloro-1a,2,2a,3,6,6a,-7,7a-octahydro-2,7:3,6-dimethano-naphth[2,3-*b*]oxirene *see* 3152
(1aα,2β,2aβ,3α,6α,6aβ,7β,7aα)-3,4,-5,6,9,9-Hexachloro-1a,2,2a,3,6,6a,-7,7a-octahydro-2,7:3,6-dimethano-naphth[2,3-*b*]oxirene *see* 3618
Hexachloropentadiene Dimer *see* 6294
Hexachlorophene, 4716
Hexachloroplatinic(IV) Acid *see* 7684
4,5,6,7,8,8-Hexachloro-3a,4,7,7a-tetrahydro-4,7-methanoisobenzo-furan-1,3-dione *see* 2134
α,α,α,α',α',α'-Hexachloro-*p*-xylene *see* 1340
Hexacyanoferrate(4—) Dicopper(2+) *see* 2703
Hexacyclonate Sodium, 4717
7,11-Hexadecadien-1-ol Acetate *see* 4548
Hexadecadrol *see* 2986
Hexadecahydro-5,2,1,6,3,4-[2,3]-butanediyl[1,4]diylidenedipentaleno-[2,1,6-*cde*:2',1',6'-*gha*]pentalene *see* 3461
[3S-(3α,3aα,6β,6aα,6bβ,7α,7aβ,8α,-10aβ,11β,13aα,13bα,13cβ,14bβ)]-3,3a,4,5,6,6a,6b,7,7a,8,9,10,10a,13a,-13c,14b-Hexadecahydro-6,8-dihydr-oxy-3,6,8,11,14,15-hexamethyl-2H-7,13b-ethenopentaleno[1'',2'':6,7;5'',-4'':6',7']dicyclohepta[1,2-b:1',2'-b']-difuran-2,12(11H)-dione *see* 8
[3S-[3R*[E(1S*,3S*,4S*)],4S*,5R*,-8S*,9E,12R*,14R*,15S*,16R*,18S*,-19S*,26aR*]]-5,6,8,11,12,13,14,15,-16,17,18,19,24,25,26,26a-Hexadeca-hydro-5,19-dihydroxy-3-[2-(4-hydr-oxy-3-methoxycyclohexyl)-1-methyl-ethenyl]-14,16-dimethoxy-4,10,-12,18-tetramethyl-8-(2-propenyl)-15,19-epoxy-3H-pyrido[2,1-*c*][1,4]-oxaazacyclotricosine-1,7,20,21(4H,-23H)-tetrone *see* 9200

[6R-(6α,7α,8β,9α,10β,13β,14α,15α,-16α,17β)]-1,3',3',4',6,7,8,9,10,11,12,-13,14,15,16,20,21-Hexadecahydro-10,13-dimethylspiro[17H-dicyclo-propa[6,7:15,16]cyclopenta[*a*]phen-anthrene-17,2'(5'H)-furan]-3,5'(2H)-dione *see* 3510
3,3a,4,5,6,6a,6b,7,7a,8,9,10,10a,13a,-13c,14b-Hexadecahydro-6-hydroxy-3,6,8,11,14,15-hexamethyl-2H-8,15-epoxy-7,13b-ethanopentaleno(1'',-2'':6,7;5'',4'':6',7')dicyclohepta(1,2-b:1',2'-b')difuran-2,12(11H)-dione *see* 662
1,2,3,3a,3b,4,5,5a,6,8,10,10a,10b,11,-12,12a-Hexadecahydro-1,10a,12a-trimethylcyclopenta[7,8]phen-anthro[2,3-*c*]pyrazol-1-ol *see* 8951
Hexadeca-μ-hydroxytetracosahydr-oxy[μ₈-[1,3,4,6-tetra-O-sulfo-β-D-fructofuranosyl-α-D-glucopyrano-side Tetrakis(hydrogen Sulfato)-(8—)]]hexadecaaluminum *see* 9049
Hexadecanoic Acid *see* 7128
Hexadecanoic Acid Aluminum Salt *see* 370
Hexadecanoic Acid Ammonium Salt *see* 572
Hexadecanoic Acid Calcium Salt *see* 1734
Hexadecanoic Acid Hexadecyl Ester *see* 2073
Hexadecanoic Acid 1,2,3-Propanetriyl Ester *see* 9865
1-Hexadecanol *see* 2070
1-Hexadecanol Lactate *see* 2072
(Z)-7-Hexadecen-1-ol Acetate *see* 4723
cis-7-Hexadecenyl Acetate *see* 4723
Hexadecyl Alcohol *see* 2070
n-Hexadecylcitric Acid *see* 184
N-Hexadecyl-N,N-dimethylbenzene-methanaminium Chloride *see* 2059
Hexadecyldimethylbenzylammonium Chloride *see* 2059
α-Hexadecylglyceryl Ether *see* 2102
Hexadecyl(2-hydroxycyclohexyl)di-methylammonium Bromide *see* 2061
N-Hexadecyl-2-hydroxy-N,N-dimeth-ylcyclohexanaminium Bromide *see* 2061
Hexadecyl 3-Hydroxy-2-naphthoate, 4718
Hexadecylic Acid *see* 7128
Hexadecyl[2-[(*p*-methoxybenzyl)-2-pyrimidinylamino]ethyl]dimethyl-ammonium Bromide *see* 9512
2-[[(Hexadecyloxy)hydroxyphosphin-yl]oxy]-N,N,N-trimethylethanami-nium Inner Salt *see* 6285
(+)-3-(Hexadecyloxy)-1,2-propanediol *see* 2102
Hexadecyl Palmitate *see* 2073
Hexadecylphosphocholine *see* 6285
n-Hexadecylphosphorylcholine *see* 6285
1-Hexadecylpyridinium Chloride *see* 2074
Hexadecyl 2-(N,N,N-Trimethylamino)-ethyl Phosphate *see* 6285
Hexadecyltrimethylammonium Bro-mide *see* 2068
Hexadecyltrimethylammonium Stea-rate *see* 2069
3,4,5,6,16,17-Hexadehydro-16-(meth-oxycarbonyl)-19α-methyloxayohim-banium *see* 8609
3,4,5,6,16,17-Hexadehydro-16-(meth-oxycarbonyl)-19α-methyl-20α-oxa-yohimbanium *see* 324
Hexadenol *see* 8870

ene)]bis[3,4-dihydroxybenzyl
Alcohol] *see* 4745
Hexamethylenebis[*N*-methylcarbamic
Acid Ester Bromomethylate] *see*
3426
Hexamethylenebis[methylcarbamic
Acid] Ester of 3-Hydroxy-1-methyl-
pyridinium Bromide *see* 3426
Hexamethylenebis[*N*-methylcarbami-
noyl-1-methyl-3-hydroxypyridinium
Bromide] *see* 3426
Hexamethylenebis(trimethylammo-
nium) *see* 4724
Hexamethylenediamine *see* 4730
Hexamethylenedicarbamic Acid
Choline Bromide Diester *see* 4713
4,4'-(Hexamethylenedioxy)dibenz-
amidine *see* 4728
Hexamethylene Glycol, 4726
Hexamethylenetetramine *see* 6036
Hexamethylenetetramine Anhydro-
methylenecitrate *see* 6038
Hexamethylenetetramine Hippurate
see 6039
Hexamethylenetetramine Phenylgly-
colate *see* 6040
Hexamethylenetetramine Tetraiodide
see 6043
2-(β-Hexamethyleniminoethyl)cyclo-
hexanone-2-carboxylic Acid Benzyl
Ester *see* 416
N,N,N',N',N',N'-Hexamethyl-1,6-hex-
anediaminium *see* 4724
Hexamethylmelamine *see* 328
Hexamethylolmelamine, 4727
Hexamethylpararosaniline Chloride
see 4401
N,N,N,N',N',N'-Hexamethyl-1,5-pen-
tanediaminium Bromide *see* 7253
Hexamethylphosphoramide *see* 4761
Hexamethylphosphoric Triamide *see*
4761
d,l-Hexamethylpropyleneamine
Oxime *see* 3957
Hexamethyl-*p*-rosaniline Chloride *see*
4401
(*all-E*)-2,6,10,15,19,23-Hexamethyl-2,-
6,10,14,18,22-tetracosahexaene *see*
8924
(*all-E*)-2-(3,7,11,15,19,23-Hexameth-
yl-2,6,10,14,18,22-tetracosahexaen-
yl)-3-methyl-1,4-naphthalenedione
see 5876
2,6,10,15,19,23-Hexamethyltetracos-
ane *see* 8923
N,N,N',N',N'',N''-Hexamethyl-1,3,5-
triazine-2,4,6-triamine *see* 328
1,1,1,3,3,3-Hexamethyl-2-(trimethylsilyl)trisilane *see* 9886
Hexameton Bromide [Burroughs
Wellcome] *see* 4724
Hexameton Chloride [Burroughs
Wellcome] *see* 4724
Hexamic Acid [Abbott] *see* 2770
Hexamidine, 4728
Hexamine *see* 6036
Hexammino-cobalt Chloride *see* 4711
Hexanal *see* 1804
Hexanaphthene *see* 2792
Hexanastab *see* 4742
Hexanastab Oral *see* 4742
n-Hexane, 4729
1,6-Hexanediamine, 4730
1,6-Hexanedicarboxylic Acid *see* 9031
Hexanedioic Acid *see* 161
Hexanedioic Acid Compd with Piper-
azine (1:1) *see* 7618
1,6-Hexanediol *see* 4726
2,5-Hexanedione *see* 69
2,2'-[1,6-Hexanediylbis(iminocarbon-

yloxy)]bis[*N,N,N*-trimethylethana-
minium] Dibromide *see* 4713
4,4'-[1,6-Hexanediylbis[imino(1-hydr-
oxy-2,1-ethanediyl)]]bis-1,2-benz-
enediol *see* 4745
3,3'-[1,6-Hexanediylbis[(methylimino)carbonyl]oxy]bis[1-methylpyri-
dinium]dibromide *see* 3426
4,4'-[1,6-Hexanediylbis(oxy)]bisbenz-
enecarboximidamide *see* 4728
Hexanicit [Bastian] *see* 5009
Hexanicotinoyl Inositol *see* 5009
Hexanicotinyl *cis*-1,2,3,5-*trans*-4,6-
Cyclohexane *see* 5009
2,4,6,2',4',6'-Hexanitrodiphenylamine
see 3396
Hexanitrohexaazaisowurtzitane, 4731
2,4,6,8,10,12-Hexanitro-2,4,6,8,10,12-
hexaazatetracyclo[5.5.0.0^{5,9}.0^{3,11}]-
dodecane *see* 4731
Hexanium Bromide [Adrian-Marinier]
see 4724
Hexanoestrol *see* 4738
Hexanoic Acid *see* 1803
Hexanoic Acid Ethyl Ester *see* 3824
Hexanoic Acid Pentyl Ester *see* 646
1-Hexanol, 4732
tert-Hexanol Carbamate *see* 3604
2-Hexanone *see* 6112
Hexanoyl Chloride *see* 1806
Hexanurat [Durascan] *see* 287
Hexaoctacontaoxononaantimonate-
heneicosatungstate(19—) Hepta-
decaammonium Disodium *see* 4786
1,4,7,10,13,16-Hexaoxacyclooctadec-
ane *see* 2669
Hexaprazol(e) *see* 3736
Hexapromin [Kowa] *see* 9704
Hexapropymate, 4733
Hexarese [Durascan] *see* 426
Hexasodium *sym*-Bis(*m*-aminobenzo-
yl-*m*-amino-*p*-methylbenzoyl-1-
naphthylamino-4,6,8-trisulfonate)
Carbamide *see* 9181
Hexastat [Bellon] *see* 328
Hexathide [Allen & Hanburys] *see*
4724
Hexathonide *see* 4724
Hexatrione [Lederle] *see* 9730
Hexatron [Nippon Shinyaku] *see* 9704
Hexavibex [Parke, Davis] *see* 8166
Hexazinone, 4734
Hexazole, 4735
Hexedine, 4736
Hexemal *see* 2780
Hexemal Calcium *see* 2780
Hexenal *see* 4742
3-(5-Hexene-1,3-diynyl)-6-(1-propyn-
yl)-1,2-dithiin *see* 9439
Hexene-Ol *see* 8870
3-Hexen-1-ol, 4737
δ-Δ^{α,β}-Hexenolactone *see* 7166
2-Hexen-5,1-olide *see* 7166
D-*erythro*-Hex-2-enonic Acid γ-Lac-
tone *see* 5141
Hexermin *see* 8166
Hexestrol, 4738
Hexestrol Bis(β-diethylaminoethyl
Ether), 4739
Hexestrol 4,4'-Diphosphoric Ester *see*
4738
Hexethal Sodium, 4740
Hexetidine, 4741
Hexetone *see* 4768
Hexigel [Warner-Lambert] *see* 4741
Hexobarbital, 4742
Hexobarbital Soluble *see* 4742
Hexobarbitone *see* 4742
Hexobendine, 4743
Hexobion [E. Merck] *see* 8166

Hexobutyramide *see* 1562
Hexocil [Parke-Davis] *see* 4741
Hexocyclium Methyl Sulfate, 4744
L-*threo*-2,3-Hexodiulosonic Acid γ-
Lactone *see* 2920
Hexoestrol *see* 4738
Hexofen *see* 5890
Hexogen *see* 2805
Hexomedin [Rhone-Poulenc] *see* 4728
Hexone *see* 5226
Hexone Chloride *see* 4724
Hexopal [Winthrop] *see* 5009
Hexoprenaline, 4745
Hexopropynate *see* 4733
Hexoral [Gödecke] *see* 4741
Hexosan *see* 4716
Hexose Diphosphate *see* 4297
Hexose Monophosphate *see* 4298
Hexose Phosphate *see* 4298
Hextol [Hoechst] *see* 7997
Hextril *see* 4741
D-*lyxo*-Hexulose *see* 9202
L-*xylo*-2-Hexulosonic Acid *see* 5315
Hexydaline *see* 6040
n-Hexyl Alcohol *see* 4732
4-Hexyl-1,3-benzenediol *see* 4750
4-Hexyl-1,3-benzenediol Compd with
9-Acridinamine (1:1) *see* 128
Hexylcaine Hydrochloride, 4746
1-(*n*-Hexylcarbamoyl)-5-fluorouracil
see 1892
n-Hexyl Chloride *see* 2196
2-Hexylcyclopropanedecanoic Acid
see 5353
2-Hexyldecanoic Acid, 4747
1-Hexyl-3,7-dihydro-3,7-dimethyl-1*H*-
purine-2,6-dione *see* 7269
4-Hexyl-1,3-dihydroxybenzene *see*
4750
1-Hexyl-3,7-dimethylxanthine *see*
7269
Hexylene Glycol, 4748
2-Hexyl-5-hydroxycyclopentanehep-
tanoic Acid *see* 8419
2-Hexyl-2-(hydroxymethyl)-1,3-pro-
panediol Trinicotinate *see* 4688
Hexylmethylcarbinol *see* 6850
Hexyl Methyl Ketone, 4749
2-(Hexyloxy)benzamide *see* 3955
2-[3-(Hexyloxy)-2-hydroxypropoxy]-
benzoic Acid *see* 3959
3-[4-(Hexyloxy)-1,2,5-thiadiazol-3-yl]-
1,2,5,6-tetrahydro-1-methylpyridine
see 10190
4-Hexylresorcinol, 4750
3-Hexyl-7,8,9,10-tetrahydro-6,6,9-tri-
methyl-6*H*-dibenzo[*b,d*]pyran-1-ol
see 9190
1-Hexyltheobromine *see* 7269
Hexythiazox, 4751
Heyden 611 *see* 8967
HF *see* 3967
HF-1854 *see* 2484
HF-1927 *see* 3055
HF-2333 *see* 7320
HGF *see* 4455 and 5017
HG-factor *see* 4455
HGH *see* 8864
HGP-1 *see* 5614
hGRF *see* 8861
HH-184 *see* 8091
HH-197 *see* 1539
HHDN *see* 227
HI-6 *see* 870
Hi-Alarzin [Yamanouchi] *see* 9656
Hializan [Pharma Investi] *see* 7061
Hibanil [Specia] *see* 2238
Hiberna [Yoshitomi] *see* 7970
Hibernal [Rhône-Poulenc Rorer] *see*
2238

Name Index

Hibiclens [Stuart] *see* 2140
Hibicon [Lederle] *see* 1045
Hibident [SmithKline Beecham] *see* 2140
Hibidil [ICI] *see* 2140
Hibiscrub [ICI] *see* 2140
Hibital [ICI] *see* 2140
Hibitane [ICI] *see* 2140
Hibon [Tokyo Tanabe] *see* 8367
Hibschite *see* 1685
Hiconcil [Allard] *see* 617
Hicoseen [Klimitschek] *see* 1551
HIDA *see* 5506
Hidacian *see* 2751
Hi-Deratol *see* 10156
Hidroferol [Juventus] *see* 1677
Hidroronol *see* 4822
Hi-Enterol *see* 5052
Hiestrone *see* 3751
Hi-Fresmin [Takeda] *see* 2513
High Acyl Gellan Gum *see* 4389
High Bush Cranberry *see* 10108
High Mallow *see* 5748
Higroton [Geigy] *see* 2246
Higueroxyl Delabarre *see* 4121
Hihustan M [Maruko] *see* 8274
Hijuven [Eisai] *see* 10159
Hilactan [Kyoritsu] *see* 2365
Hillebrandite *see* 1749
Hilong [Banyu] *see* 7059
Himecol [Kissei] *see* 4903
Hinosan [Bayer] *see* 3560
Hiochic Acid *see* 6250
Hiohex Chloride [Keith-Victor] *see* 4724
Hioxyl [Quinoderm] *see* 4839
Hipberries *see* 8422
Hipercilina *see* 7225
Hipertan [Gramon] *see* 6272
Hipertil [Normal] *see* 1817
Hiphyllin *see* 3529
Hipoartel [Lasa] *see* 3605
Hipocolestina *see* 9867
Hipoftalin *see* 4800
Hippocampal-derived Neurotrophic Factor *see* 6570
Hippodin *see* 5055
Hippramine [Riker] *see* 6039
Hippuran I 131 [Mallinckrodt] *see* 5055
Hippuric Acid, 4752
Hipputope [Squibb] *see* 5055
Hiprex [Merrell] *see* 6039
Hirathiol *see* 4929
Hiropon *see* 6015
Hirsutic Acid C, 4753
Hirudex *see* 4754
Hirudin, 4754
ds-Hirudin *see* 4754
r-Hirudin *see* 4754
rec-Hirudin *see* 4754
His *see* 4758
Hismanal [Janssen] *see* 891
Hispril [SK&F] *see* 3390
Histabromamine *see* 1439
Histabutizine *see* 1484
Histabutyzine *see* 1484
Histacuran *see* 2406
Histacyl [Streuli] *see* 3367
Histadur [Cooper] *see* 2232
Histadyl (formerly) [Lilly] *see* 6027
Histalog [Lilly] *see* 1230
Histalon [Hance] *see* 8168
Histamen [Polipharma] *see* 891
Histametizine *see* 5817
Histaminase, 4755
Histamine, 4756
Histamine Deaminase *see* 4755
Histamine Oxidase *see* 4755
Histaminos [Lesvi] *see* 891

Histan [Cooper] *see* 8168
Histantin [Burroughs Wellcome] *see* 2128
Histaphen(e) [UCB] *see* 5839
Histapon [Endo] *see* 4756
Histapyran [Horton & Converse] *see* 8168
Histapyrrodine, 4757
Histaspan [USV] *see* 2232
Histatex [Medix] *see* 8168
Histazine *see* 718
Histazol [Krka] *see* 891
Histidine, 4758
L-Histidine *see* 4758
Histidine-betaine *see* 4702
Histidine Trimethylbetaine *see* 4702
Histimin *see* 1230
Histionex [Pennwalt] *see* 7469
Histocarb [Rohm & Haas] *see* 1830
Histomibal *see* 3246
Histones, 4759
Histostab [Boots] *see* 718
Histrelin, 4760
Histrionicotoxin D *see* 4410
Histryl [SK&F] *see* 3390
Histyn [Hässle] *see* 3390
Hi-Ti [Kyowa] *see* 5696
Hitocobamin-M [Hishiyama] *see* 6125
Hitodesterol *see* 8901
Hittorf's Phosphorus *see* 7503
Hive Dross *see* 8021
Hivid [Roche] *see* 10242
Hiwolfia [Bowman] *see* 8293
Hi-Z [Otsuka] *see* 7016
HK-137 *see* 4011
HK-256 *see* 7918
IIL-255 *see* 1414
HL-362 *see* 2539
HL-2186 *see* 9205
HL-5746 *see* 2238
HL-8731 *see* 7693
3-HMC *see* 3826
HMD *see* 1843
HMDP *see* 7074
HMDS *see* 4725
HME *see* 3590
HMF *see* 4879
HMG *see* 5852
HMGA *see* 5852
HMM *see* 328
HMPA, 4761
d,l-HM-PAO *see* 3957
HMPT *see* 4761
HMS [Allergan] *see* 5840
HMT *see* 6036
HMTA *see* 6036
HMX *see* 6855
HN2 *see* 5815
HN3 *see* 9777
HNIW *see* 4731
HO-2474 *see* 3000
Hoarhound *see* 4784
Hoban *see* 5575
HOE-39-893d *see* 7209
HOE-045 *see* 1920
HOE-095K *see* 35
HOE-118 *see* 7647
HOE-280 *see* 6865
HOE-285 *see* 7997
HOE-296 *see* 2325
HOE-296V *see* 8322
HOE-304 *see* 2975
HOE-433 *see* 322
HOE-440 *see* 9558
HOE-490 *see* 4449
HOE-498 *see* 8283
HOE-555-02A *see* 4038
HOE-661 *see* 7495
HOE-760 *see* 8431
HOE-766 *see* 1527

HOE-777 *see* 7899
HOE-881v *see* 4000
HOE-893d *see* 7209
HOE-984 *see* 6768
HOE-2770 *see* 9388
HOE-2784 *see* 1265
HOE-2810 *see* 5534
HOE-2873 *see* 8146
HOE-2904 *see* 3341
HOE-2960 *see* 9736
HOE-2982 *see* 4699
HOE-13764 *see* 8138
HOE-16410 *see* 5237
HOE-17411 *see* 1836
HOE-23408 *see* 3133
HOE-33171 *see* 4024
HOE-36801 *see* 3910
HOE-39866 *see* 7495
HOE-40045 *see* 1920
Hoechst 10446 *see* 4883
Hoechst 10495 *see* 6810
Hoechst 10582 *see* 6804
Hoechst 10600 *see* 7348
Hoechst 10720 *see* 5312
Hoechst 10805 *see* 3398
Hoechst 10820 *see* 6008
Hoegrass [Hoechst] *see* 3133
Hoelon [Hoechst] *see* 3133
Hoffmann's Anodyne *see* 8910
Hoffmann's Drops *see* 8909
HOG *see* 7364
Hoggar N [Stada] *see* 3497
Hogival *see* 3751
Hogpax [Ferrosan] *see* 619
Hog's Bean *see* 4908
Hogweed *see* 8547
Hokunalin [Hokuriku] *see* 9942
Holarrhenine, 4762
Holbamate *see* 5908
Holligold *see* 1763
Holly-leaved Barberry *see* 1193
Holmia *see* 4763
Holmium, 4763
Holocaine Hydrochloride [Winthrop] *see* 7342
Holodorm *see* 6028
Holomycin, 4764
Holopon [Byk-Gulden] *see* 6084
Holoxan [Lucien] *see* 4937
Homapin [Mission Pharmacal] *see* 4766
Homarine, 4765
Homatrisol [Cooper] *see* 4766
Homatropine, 4766
Homidium, 4767
Homoalanin-4-yl(methyl)phosphinic Acid *see* 7495
Homoarterenol *see* 6785
Homobatrachotoxin *see* 1037
Homocaine *see* 2516
Homocamfin, 4768
Homochelidonine, 4769
α-Homochelidonine *see* 4769
Homochlorcyclizine, 4770
Homoclomin *see* 4770
Homocodeine *see* 7484
Homocysteine, 4771
Homocysteine *S*-Ethyl Ether *see* 3784
Homocystine, 4772
Homoeriodictyol, 4773
Homofenazine, 4774
Homogentisic Acid, 4775
Homoginin [Zeria] *see* 4770
Homomenthyl Salicylate *see* 4776
Homomycin *see* 4900
Homomyrtenyl β-(Diethylamino)ethyl Ether *see* 6422
2-Homomyrtenyloxy-1-(diethyl-amino)ethane *see* 6422
Homonal [SS Pharm.] *see* 5683

Homoolan [Winthrop] see 45
D-Homo-17a-oxaandrosta-1,4-diene-3,17-dione see 9321
D-Homopantothenic Acid see 4781
Homoproline see 7611
Homorestar [Ohta] see 4770
Homosalate, 4776
m-Homosalicylic Acid see 2648
o-Homosalicylic Acid see 2649
p-Homosalicylic Acid see 2650
Homoserine, 4777
Homosulfamine see 5683
4-Homosulfanilamide see 5683
4-Homosulfanilamide Salt with 1-Sulfanilyl-2-thiourea see 9117
Homotrilobine see 9826
Homovanillic Acid, 4778
HON, 4779
Honduras Balsam see 976
Honduras Bark see 1930
Honokiol, 4780
Honvan [Asta Medica] see 4279
Honvol [Horner] see 4279
Hopantenate Calcium see 4781
Hopantenic Acid, 4781
Hopate [Tanabe] see 4781
Hopeite see 10284
Hops, 4782
Hordenine, 4783
Horehound, 4784
Horizon [Yamanouchi] see 3042
Horizon [Bayer] see 9253
Hormantoxone see 5575
Hormezon [Tobishi] see 1226
Hormocardiol see 4651
Hormodin [Merck & Co.] see 4995
Hormoestrol see 4738
Hormofemin [Medo] see 3153
Hormofollin see 3751
Hormomed [Merckle] see 3750
Hormone A see 719
Hormonisene see 2225
Hormovarine [Clin-Comar-Byla] see 3751
Hornbest [Shizuoka] see 2380
Horn Silver see 8647
Horse Balm see 2546
Horseheal see 5023
Horse Nettle see 8856
Horse-radish, 4785
Horseweed see 3708
Hostacaine [Hoechst] see 1542
Hostacortin [Hoechst] see 7904
Hostacortin H [Hoechst] see 7901
Hostacyclin [Hoechst] see 9337
Hostaginan [Hoechst] see 7919
Hostalival [Hoechst] see 6768
Hostaquick [Hoechst] see 4699
Hostathion [Hoechst] see 9736
Houpo see 5738
HP-029 see 10077
HP-129 see 4012
HP-549 see 5257
HP-749 see 1223
HPA-23, 4786
HPC see 6285 and 7091
HPEK-1 see 9385
5-HPETE see 4708
hpGRF see 8861
hpGRF(1-29)NH₂ see 8605
HPL see 4650
HPMC see 4889
HPMF see 3633
(S)-HPMPC see 2329
HPP see 287
4-HPR see 4040
hPTH 1-34 acetate see 9309
HQ-495 see 181
HQNO, 4787
HR see 9920

HR-158 see 5604
HR-221 see 1979
HR-376 see 2417
HR-756 see 1983
HR-810 see 1990
HRF [Ayerst] see 5500
HRS-16 see 3154
HS-592 see 2405
HS-902 see 9602
HSA see 8613
H₂sal₂en see 8474
HSP-2986 see 7887
HSR-770 see 4963
HSR-902 see 9602
5-HT see 8607
HTH see 1717
5-HTP see 4895
L-5HTP see 4895
HTX D see 4410
HTZ see 4751
Huanghuahaosu see 856
Humafac [Parke, Davis] see 3963
Humagel [Parke, Davis] see 7173
Human Chorionic Gonadotropin see 2273
Human Chorionic Somatomammotropin see 4650
Human EGF-URO see 3569
Human Growth Hormone see 8864
Human Growth Hormone-releasing Factor(1-29)amide see 8605
Human IgE Pentapeptide see 7270
Human Pancreatic Somatoliberin(1-29)amide see 8605
Human Placental Lactogen see 4650
Human Serum Albumin see 8613
Human Serum Growth Factor see 7683
Humatin [Parke, Davis] see 7173
Humatrope [Lilly] see 8864
Humedil [Bio-Mar] see 1152
Humegon [Organon] see 4299
Humic Acids, 4788
Huminsulin [Lilly] see 5011
Humorsol [Merck & Co.] see 2936
Humoryl [Delalande] see 9659
Humulene, 4789
α-Humulene see 4789
β-Humulene see 4789
Humulin [Lilly] see 5011
Humulina [Lilly] see 5011
Humulon, 4790
Humulone see 4790
Hungarian Chamomile see 2082
HUP see 4791
Huperzine A, 4791
Hustazol [Yoshitomi] see 2456
HWA-285 see 7997
HWG-1608 see 9253
2-HxG see 1248
HY-185 see 1849
HY-770 see 4963
Hyacid [Scanvet] see 4793
Hyacin see 2931
Hyacinthin see 7419
Hyadur [Grünenthal] see 3308
Hyaenanchin see 4898
Hyalase [C.P. Pharm.] see 4794
Hyalgan [Fidia] see 4793
Hyalidase see 4794
Hyalobiuronic Acid, 4792
Hyalovet [Fidia] see 4793
Hyalozima see 4794
Hyaluronic Acid, 4793
Hyaluronidases, 4794
Hyamate [Xttrium] see 1525
Hyamine 10X [Rohm & Haas] see 6103
Hyamine 1622 see 1103

Hyanit [Burnus; Angelopharm] see 10005
Hyasmonta see 4794
Hyason see 4794
Hyasorb see 7225
Hyazyme [Abbott] see 4794
Hybolin Decanoate [Hyrex] see 6452
Hybridoma Growth Factor see 5017
Hybridomas see 4958
Hybrin see 867
Hycamtamine see 9687
Hycanthone, 4795
Hycholin [Ayerst] see 7260
Hyclorate [Funai] see 2436
Hycobal [Eisai] see 2513
Hycozid [Takeda] see 5203
Hydac [Hoechst] see 3991
Hydagen B [Henkel] see 1281
Hydantin [Lääkefarmos] see 7475
Hydantoin, 4796
Hydantol [Fujinaga-Sankyo] see 7475
Hydeltra [Merck & Co.] see 7901
Hydeltrasol [Merck & Co.] see 7903
Hydeltra-T.B.A. [Merck & Co.] see 7901
Hydergine [Sandoz] see 3692
Hydnocarpate Sodium see 4797
Hydnocarpic Acid, 4797
Hydnocarpus Oil see 2088
Hydnocarpylacetic Acid see 2089
Hydol see 4830
Hydrabamine Penicillin G see 7224
Hydrabamine Penicillin V see 7232
Hydrabamine Phenoxymethyl Penicillin see 7232
Hydracarbazine, 4798
Hydracillin see 7226
Hydracort [Galderma] see 4828
Hydracrylic Acid, 4799
Hydracrylic Acid β-Lactone see 8005
Hydracrylonitrile see 3840
Hydralazine, 4800
Hydrallostane, 4801
Hydramethylnon, 4802
Hydramitrazine, 4803
Hydramycin [Sankyo] see 3496
Hydrangea, 4804
Hydrangin see 9981
Hydraphen see 4805
Hydrargaphen, 4805
Hydrargyrum see 5957
Hydrastine, 4806
l-β-Hydrastine see 4806
Hydrastinine, 4807
Hydrastis, 4808
Hydrated Alumina see 355
Hydrated Cupric Oxide see 2709
Hydrated Ferric Oxide see 4068
Hydrated Magnesium-aluminum-iron Silicate see 10095
Hydrazine, 4809
Hydrazine Acid Tartrate see 4812
Hydrazine Anhydrous see 4809
Hydrazine Bitartrate see 4812
Hydrazinecarbothioamide see 9499
Hydrazinecarboxamide Monohydrochloride see 8588
Hydrazinecarboximidamide see 461
Hydrazine Dihydriodide see 9824
Hydrazine Hydrate, 4810
Hydrazine Hydrogen Tartrate see 4812
Hydrazine Sulfate, 4811
Hydrazine Tartrate, 4812
Hydrazine Yellow see 9239
Hydrazinium Sulfate see 4811
Hydrazinobenzene see 7447
4-Hydrazinobenzenesulfonic Acid, 4813

2-Hydrazino-4,6-bis(diethylamino)-1,-3,5-triazine see 4803
3-Hydrazino-6-carbamoylpyridazine see 4798
S-α-Hydrazino-3,4-dihydroxy-α-methylbenzenepropanoic Acid Monohydrate see 1843
(−)-L-α-Hydrazino-3,4-dihydroxy-α-methylhydrocinnamic Acid Monohydrate see 1843
2-Hydrazinoethanol, 4814
3-Hydrazino-6-[(2-hydroxypropyl)methylamino]pyridazine see 7577
α-Hydrazino-α-methyl-β-(3,4-dihydroxyphenyl)propionic Acid Monohydrate see 1843
2-Hydrazinooctane see 6845
1-(2-Hydrazino-2-oxoethyl)pyridinium Chloride see 4436
1-Hydrazinophthalazine see 4800
3-Hydrazinopyridazine-6-carboxamide see 4798
6-Hydrazino-3-pyridazinecarboxamide see 4798
(±)-1-[(6-Hydrazino-3-pyridazinyl)methylamino]-2-propanol see 7577
2-Hydrazino-N,N,N-trimethyl-2-oxoethanaminium Chloride see 4436
Hydrazodicarbonimide see 10004
Hydrazoic Acid, 4815
Hydrazonium Sulfate see 4811
Hydrea [Squibb] see 4896
Hydrenox [Boots] see 4830
Hydridocarbonyltris(triphenylphosphine)rhodium(I) see 9887
Hydridocobalamin see 10154
Hydril see 4822
Hydrindantin, 4816
Hydrindene see 4966
Hy-Drine [Zemmer] see 1155
Hydriodic Acid, 4817
Hydrion [Robert et Carriére] see 405
Hydrionic [Upjohn] see 4477
Hydro-Adreson [Organon] see 4828
Hydro-Aquil see 4822
Hydrobenzoin, 4818
Hydrobromic Acid, 4819
Hydrobromic Ether see 3819
Hydrocal [Bioglan] see 4828
Hydrocarbostyril, 4820
Hydrocerussite see 1034
Hydrochlorbenzethylamine see 3921
Hydrochloric Acid, 4821
Hydrochloroauric Acid see 4542
Hydrochlorothiazide, 4822
Hydrocinchonidine, 4823
Hydrocinchonine, 4824
Hydrocinnamic Acid, 4825
Hydrocodin see 3214
Hydrocodone, 4826
Hydrocomp [Beiersdorf] see 426
Hydroconchinine see 4851
Hydrocort [Pharmagalen] see 4828
Hydrocortamate, 4827
Hydrocortisone, 4828
Δ¹-Hydrocortisone see 7901
Hydrocortisone 21-Diethylaminoacetate see 4827
Hydrocortisone Hemisuccinate Sodium Salt see 4828
Hydrocortisone Sodium Phosphate see 4828
Hydrocortisone Valerate see 4828
Hydrocortistab [Boots] see 4828
Hydrocortisyl [Roussel] see 4828
Hydrocortone [Merck & Co.] see 4828
Hydrocortone Acetate [Merck & Co.] see 4828
Hydrocortone Phosphate [Merck & Co.] see 4828

Hydrocotarnine, 4829
Hydrocupreine see 2689
Hydrocupreine Ethyl Ether see 3855
Hydrocupreine Isopentyl Ether see 3950
Hydrocyanic Acid see 4836
Hydrocyanite see 2722
Hydrodeltalone see 7901
Hydro-Diuril [Merck & Co.] see 4822
Hydrofluoboric Acid see 4184
Hydrofluoric Acid, 4831
Hydrofluoric Acid Gas see 4837
Hydrofluosilicic Acid see 4220
α-Hydroformamine Cyanide see 6134
Hydrofuramide, 4832
Hydrogen, 4833
Hydrogen Arsenide see 849
Hydrogen Azide see 4815
Hydrogen Bis(benzenamine-N)tetrakis(thiocyanato-N)chromate(1−) see 8350
Hydrogen Bromide, 4834
Hydrogen Chloride, 4835
[Hydrogen Citrato(3−)]triaquoiron see 4085
Hydrogen Cyanamide see 2754
Hydrogen Cyanate see 2756
Hydrogen Cyanide, 4836
Hydrogen Dioxide see 4839
Hydrogen Dioxide Soln see 4840
Hydrogen Fluoride, 4837
Hydrogen Hexachloroplatinate(IV) see 7684
Hydrogen Hexafluorosilicate see 4220
Hydrogen Iodide, 4838
Hydrogen Oxide see 10175
Hydrogen Peroxide, 4839
Hydrogen Peroxide Carbamide see 10007
Hydrogen Peroxide Solution 3%, 4840
Hydrogen Peroxide Solution 30%, 4841
Hydrogen Selenide, 4842
Hydrogen Sulfide, 4843
Hydrogen Telluride, 4844
Hydrogen Tetrabromoaurate(1−) see 4540
Hydrogen Tetracarbonylferrate(II), 4845
Hydrogen Tetrachloroaurate(1−) see 4542
Hydrogen Tetrafluoroborate see 4184
Hydrogen Thiocyanate see 9464
Hydrogen Tribromide Compd with Pyridine (1:1) see 8156
Hydrogen Trioxooxalatomolybdate-(VI) see 7045
Hydro-Giene [Water Science] see 2732
Hydrohydrastinine, 4846
α-Hydro-ω-hydroxypoly(oxy-1,2-ethanediyl) see 7729
α-Hydro-ω-hydroxypoly(oxyethylene)poly(oxypropylene)poly(oxyethylene) Block Copolymers see 7722
Hydrokon see 4826
Hydro-long [Sanorania] see 2246
Hydrolose [Upjohn] see 6120
Hydro-Magma [Merck & Co.] see 5706
Hydromedin [Merck & Co.] see 3761
Hydromorphone, 4847
Hydromox [Lederle] see 8240
Hydrone [Du Pont], 4848
Hydronitric Acid see 4815
Hydronol [Stuart] see 5244
Hydrosan see 4510
Hydroorotic Acid, 4849

Hydroperoxide see 4839
cis-4-Hydroperoxycyclophosphamide see 7303
5-Hydroperoxy-6,8,11,14-eicosatetraenoic Acid see 4708
Hydroprene, 4850
Hydropyrin see 5545
Hydroquinidine, 4851
Hydroquinine, 4852
Hydroquinol see 4853
Hydroquinone, 4853
Hydroquinone Benzyl Ether see 6331
Hydroquinone Calcium Sulfonate see 3455
Hydroquinone-β-D-glucopyranoside see 812
Hydroquinone Glucose see 812
Hydroquinone Monobenzyl Ether see 6331
Hydro-rapid [Sanorania] see 4331
Hydroresorcinol see 3221
Hydroretrocortine see 7901
Hydrorhombinine see 5635
Hydrosaluric [Merck & Co.] see 4822
Hydrosarpan [Eutherapie] see 8291
Hydrosilicofluoric Acid see 4220
"Hydrosulfuric Acid" see 4843
Hydrotalcite see 5696
Hydrothide see 4822
Hydrotrichlorothiazide see 9754
Hydrotricine see 9972
Hydrourushiol see 7244
Hydroxamethocaine see 4894
Hydroxocobalamin, 4854
Hydroxocobemine see 4854
N-Hydroxyacetamide see 60
4′-Hydroxyacetanilide see 45
p-Hydroxyacetanilide see 45
Hydroxyacetic Acid see 4508
[3aS-(3aα,4β,5α,6α,8β,9α,9aβ,-10S*)]-Hydroxyacetic Acid 6-Ethenyldecahydro-5-hydroxy-4,6,9,10-tetramethyl-1-oxo-3a,9-propano-3aH-cyclopentacycloocten-8-yl Ester see 7694
p-Hydroxyacetophenone-D-glucoside see 7549
3-Hydroxy-21-acetoxy-5-pregnen-20-one see 77
[[4-[(Hydroxyacetyl)amino]phenyl]arsonato(1−)]oxobismuth see 4503
[4-[(Hydroxyacetyl)amino]phenyl]arsonic Acid see 4489
4-Hydroxy-N-acetylproline see 7035
(2α,3β)-3-Hydroxy-A(1),28-dinorlup-20(29)-ene-2,17-dicarboxylic Acid see 1959
4-Hydroxyaflatoxin B₁ see 180
4-Hydroxyaflatoxin B₂ see 180
β-Hydroxyalanine see 8604
2-Hydroxy-6-alkylbenzoic Acids see 663
17-Hydroxy-17α-allyl-4-estrene see 299
l-3-Hydroxy-N-allylmorphinan see 5485
α-Hydroxy-δ-aminobutylidenediphosphonic Acid see 228
γ-Hydroxy-β-aminobutyric Acid see 463
m-Hydroxy-α-(1-aminoethyl)benzyl Alcohol see 5993
5-Hydroxy-3-(β-aminoethyl)indole see 1959
[4-(N-Hydroxyamino)-2R-isobutyl-3S-(2-thienylthiomethyl)succinyl]-L-phenylalanine-N-methylamide see 1036
3-Hydroxy-5-aminomethylisoxazole see 6391

6-(*p*-Hydroxy-α-aminophenylacet-
amido)penicillanic Acid *see* 617
α-Hydroxy-β-aminopropylbenzene
Hydrochloride *see* 7461
Hydroxyamphetamine, 4855
p-Hydroxyampicillin *see* 617
(17β)-17-Hydroxyandrosta-1,4-dien-3-
one *see* 1354
3α-Hydroxy-17-androstanone *see* 680
(3α,5α)-3-Hydroxyandrostan-17-one
see 680
3β-Hydroxy-17-androstanone *see*
3646
(3β,5α)-3-Hydroxyandrostan-17-one
see 3646
(5α,17β)-17-Hydroxyandrostan-3-one
see 8950
17β-Hydroxy-3-androstanone *see*
8950
3α-Hydroxy-5α-androst-16-ene *see*
679
4-Hydroxyandrost-4-ene-3,17-dione
see 4267
(3β)-3-Hydroxyandrost-5-en-17-one
see 7892
(17β)-17-Hydroxyandrost-4-en-3-one
see 9322
2-Hydroxyaniline *see* 481
3-Hydroxyaniline *see* 480
p-Hydroxyaniline *see* 482
p-Hydroxyanilinoacetic Acid *see* 4885
o-Hydroxyanisole *see* 4575
9-Hydroxyanthracene *see* 725
1-Hydroxy-2-anthraquinonyl 6-*O*-β-D-
Xylopyranosyl-β-D-glucopyranoside
see 8437
Hydroxyapatite *see* 3519
3-Hydroxy-8'-apo-β,ψ-carotenal *see*
2385
2-Hydroxybenzaldehyde *see* 8478
p-Hydroxybenzaldehyde, 4856
2-Hydroxybenzaldehyde Oxime *see*
8479
o-Hydroxybenzal Isonicotinylhydrazone *see* 8487
2-Hydroxybenzamide *see* 8480
N-Hydroxybenzenamine *see* 7448
Hydroxybenzene *see* 7390
α-Hydroxybenzeneacetic Acid *see*
5757
endo-(±)-α-Hydroxybenzeneacetic
Acid 8-Methyl-8-azabicyclo[3.2.1]-
oct-3-yl Ester *see* 4766
α-Hydroxybenzeneacetic Acid 3-
Methylbutyl Ester *see* 5758
α-Hydroxybenzeneacetic Acid Monoammonium Salt *see* 563
α-Hydroxybenzeneacetic Acid 1,2,2,6-
Tetramethyl-4-piperidinyl Ester *see*
3943
α-Hydroxybenzeneacetic Acid 3,3,5-
Trimethylcyclohexyl Ester *see* 2771
α-Hydroxybenzeneacetonitrile *see*
5759
4-*p*-Hydroxybenzeneazo-1-*p*-chloro-*o*-
sulfophenyl-3-methyl-5-hydroxypyr-
azole Toluene-*p*-sulfonyl Ester
Sodium Salt *see* 7713
2-Hydroxybenzenecarbodithioic Acid
see 3440
4-Hydroxy-1,3-benzenedicarboxylic
Acid *see* 4873
4-Hydroxy-1,3-benzenedisulfonic
Acid *see* 7391
4-Hydroxy-*m*-benzenedisulfonic Acid
see 7391
2-Hydroxybenzenemethanol *see* 8477
4-Hydroxybenzenesulfonic Acid *see*
7396

p-Hydroxybenzenesulfonic Acid
Calcium Salt *see* 1737
p-Hydroxybenzenesulfonic Acid Copper Salt *see* 2715
Hydroxybenzenesulfonic Acid Potassium Salt *see* 7826
Hydroxybenzenesulfonic Acid Sodium
Salt *see* 8803
p-Hydroxybenzenesulfonic Acid Zinc
Salt *see* 10283
2-Hydroxybenzhydroxamic Acid *see*
8483
(2-Hydroxybenzoato-*O*¹)oxobismuth
see 1327
6-Hydroxy-5-benzofuranacrylic Acid
δ-Lactone *see* 8113
2-Hydroxybenzoic Acid *see* 8484
p-Hydroxybenzoic Acid, 4857
2-Hydroxybenzoic Acid Acetate 2-
Carboxyphenyl Ester *see* 102
2-Hydroxybenzoic Acid 4-(Acetylamino)phenyl Ester *see* 46
2-Hydroxybenzoic Acid Bismuth (3+)
Salt Basic *see* 1327
4-Hydroxybenzoic Acid Butyl Ester
see 1619
2-Hydroxybenzoic Acid 2-Carboxy-
phenyl Ester *see* 8491
2-Hydroxybenzoic Acid Compd with
1,2-Dihydro-1,5-dimethyl-2-phenyl-
3*H*-pyrazol-3-one (1:1) *see* 758
2-Hydroxybenzoic Acid Compd with
1*H*-Imidazole (1:1) *see* 4949
2-Hydroxybenzoic Acid Copper Salt
see 2717
2-Hydroxybenzoic Acid 4-(1,1-Di-
methylethyl)phenyl Ester *see* 1621
2-Hydroxybenzoic Acid Ethyl Ester
see 3894
4-Hydroxybenzoic Acid Ethyl Ester
see 3883
p-Hydroxybenzoic Acid Ethyl Ester
6-Guanidinohexanoate *see* 4344
2-Hydroxybenzoic Acid 2-Hydroxy-
ethyl Ester *see* 4509
2-Hydroxybenzoic Acid Magnesium
Salt *see* 5723
2-Hydroxybenzoic Acid Methyl Ester
see 6200
4-Hydroxybenzoic Acid Methyl Ester
see 6182
2-Hydroxybenzoic Acid 5-Methyl-2-
(1-methylethyl)cyclohexyl Ester *see*
5886
2-Hydroxybenzoic Acid Monoammonium Salt *see* 584
2-Hydroxybenzoic Acid Monosodium
Salt *see* 8819
2-Hydroxybenzoic Acid Monosodium
Salt Compd with 2-(1,2,3,6-Tetrahy-
dro-3-hydroxy-1-methyl-6-oxo-5*H*-
indol-5-ylidene)hydrazinecarbox-
amide (1:1) *see* 1832
2-Hydroxybenzoic Acid 1-Naphtha-
lenyl Ester *see* 6501
2-Hydroxybenzoic Acid 2-Naphtha-
lenyl Ester *see* 6502
4-Hydroxybenzoic Acid [(5-Nitro-2-
furanyl)methylene]hydrazide *see*
6626
p-Hydroxybenzoic Acid (5-Nitrofur-
furylidene)hydrazide *see* 6626
2-Hydroxybenzoic Acid Phenyl Ester
see 7464
2-Hydroxybenzoic Acid Phenylmethyl
Ester *see* 1181
4-Hydroxybenzoic Acid Propyl Ester
see 8051
2-Hydroxybenzoic Acid 1,7,7-Tri-

methylbicyclo[2.2.1]hept-2-yl Ester
see 1371
2-Hydroxybenzoic Acid 3,3,5-Tri-
methylcyclohexyl Ester *see* 4776
2-Hydroxybenzoic Acid Zinc Salt *see*
10288
2-Hydroxybenzoic-5-sulfonic Acid *see*
9138
7-Hydroxy-2*H*-1-benzopyran-2-one *see*
9981
Hydroxybenzopyridine *see* 4890
2-(6-Hydroxybenzothiazol-2-yl)-2-
thiazoline-4-carboxylic Acid *see*
4128
(±)-*N*-Hydroxy-*N*-(1-benzo[*b*]thien-2-
ylethyl)urea *see* 10253
6-Hydroxy-1,3-benzoxathiol-2-one *see*
9600
4-(2-Hydroxybenzoyl)morpholine *see*
8485
1-(*p*-Hydroxybenzoyl)-2-(5-nitrofur-
furylidene)hydrazine *see* 6626
(*S*)-2-[(2-Hydroxybenzoyl)oxy]ethyl
6,8-Dideoxy-7-*O*-methyl-6-[[(1-
methyl-2-pyrrolidinyl)carbonyl]-
amino]-1-thio-D-*erythro*-α-D-*galacto*-
octopyranoside *see* 2005
o-Hydroxybenzyl Alcohol *see* 8477
N'-*o*-Hydroxybenzylidenepyridine-4-
carboxyhydrazide *see* 8487
1-(*p*-Hydroxybenzyl)-6-methoxy-7-
hydroxy-1,2,3,4-tetrahydroisoquino-
line *see* 2521
α-Hydroxybenzylphosphinic Acid,
4858
(*p*-Hydroxybenzyl)tartaric Acid *see*
7664
1-Hydroxy[1,1'-bicyclohexyl]-2-carb-
oxylic Acid 2-(Diethylamino)-1-
methylethyl Ester *see* 8406
Hydroxybis(acetato)aluminum *see* 349
Hydroxybis(acetylsalicylato)alumi-
num *see* 338
α-Hydroxy-α,α-bis(5-hydroxycarva-
cryl)-*o*-toluenesulfonic Acid γ-Sultone *see* 9541
1-Hydroxy-2,2-bis(hydroxymethyl)-
butane Trinitrate *see* 7995
N-[2-Hydroxy-1,1-bis(hydroxymethyl)ethyl]glycine *see* 9783
α-Hydroxy-α,α-bis(*p*-hydroxyphen-
yl)-*o*-toluenesulfonic Acid γ-Sultone
see 7397
α-Hydroxy-α,α-bis(4-hydroxy-*m*-
tolyl)-*o*-toluenesulfonic Acid γ-
Sultone *see* 2647
5-Hydroxy-3,4-bis(3,4-methylenedi-
oxybenzyl)tetrahydrofuran *see* 2680
(5'α)-12'-Hydroxy-2',5'-bis(1-methyl-
ethyl)ergotaman-3',6',18-trione *see*
3685
12'-Hydroxy-2',5'α-bis(1-methyleth-
yl)-8α-ergotaman-3',6',18-trione *see*
3686
4-Hydroxy-6,7-bis(2-methylpropoxy)-
3-quinolinecarboxylic Acid Ethyl
Ester *see* 1524
9-Hydroxy-19,20-bisnorprostanoic
Acid *see* 8419
(1*R*,2*R*,3*S*,4*S*)-1-(2-Hydroxy-3-born-
yl)-3-(*p*-tolylsulfonyl)urea *see* 4447
Hydroxybrasilein *see* 4667
Hydroxybrasilin *see* 4670
Hydroxybrazilein *see* 4667
Hydroxybrazilin *see* 4670
3-Hydroxybutanal *see* 225
2-Hydroxybutane *see* 1576
Hydroxybutanedioic Acid *see* 5747
Hydroxybutanedioic Acid Monoammonium Salt *see* 524

Name Index

oic Acid *O,O*-Dimethyl Ester *see* 3973

7-Hydroxy-2,2-dimethyl-2*H*-1-benzopyran-6-acrylic Acid δ-Lactone *see* 10203

α-Hydroxy-β,β-dimethyl-γ-butyrolactone *see* 7145

(7α,17β)-17-Hydroxy-7,17-dimethylestr-4-en-3-one *see* 6262

(±)-α-[[(2-Hydroxy-1,1-dimethylethyl)amino]methyl]benzenemethanol *see* 4052

3-Hydroxy-*N*,*N*-dimethyl-*N*-ethylanilinium Chloride *see* 3562

(*Z*)-7-[(1*R*,2*R*,3*R*)-2-[(*E*)-(3*R*)-3-Hydroxy-4,4-dimethyl-1-octenyl]-3-methyl-5-oxocyclopentyl]-5-heptenoic Acid *see* 9853

3-Hydroxy-4,5-dimethylol-α-picoline Hydrochloride *see* 8166

[1*R*-[1*R**,4*Z*,8*Z*,10*S**,13*R**,15*R**-(*R**)]]-4-(15-Hydroxy-5,10-dimethyl-3-oxo-2,14-dioxabicyclo[11.3.1]heptadeca-4,8-dien-15-yl)-2-thiazolidinone *see* 5388

[1*R*-[1*R**,4*Z*,8*E*,10*Z*,12*S**,15*R**,17*R**-(*R**)]]-4-(17-Hydroxy-5,12-dimethyl-3-oxo-2,16-dioxabicyclo[13.3.1]nonadeca-4,8,10-trien-17-yl)-2-thiazolidinone *see* 5388

(11β,16α,17β)-11-Hydroxy-16,17-dimethyl-17-(1-oxopropyl)androsta-1,4-dien-3-one *see* 8392

(3*S*,4*R*)-3-Hydroxy-2,2-dimethyl-4-(2-oxo-1-pyrrolidinyl)-6-chromancarbonitrile *see* 5487

3-(2-Hydroxy-3,5-dimethylphenacyl)-glutarimide *see* 145

p-Hydroxy-*N*,*N*-dimethylphenethylamine *see* 4783

p-Hydroxy-*N*,α-dimethylphenethylamine *see* 7485

2'-Hydroxy-5,9-dimethyl-2-phenethyl-6,7-benzmorphan *see* 7360

4-[2-(2-Hydroxy-3,5-dimethylphenyl)-2-oxoethyl]-2,6-piperidinedione *see* 145

11β-Hydroxy-16α,17α-dimethyl-17-propionylandrosta-1,4-dien-3-one *see* 8392

S-(3-Hydroxy-1,1-dimethylpropyl)-L-cysteine *see* 3990

L-3-[(3-Hydroxy-1,1-dimethylpropyl)-thio]alanine *see* 3990

3-Hydroxy-1,1-dimethylpyrrolidinium Bromide α-Cyclopentylmandelate *see* 4511

3-Hydroxy-1,1-dimethylpyrrolidinium Bromide α-Phenyl-α-(2-thienyl)glycolate *see* 4709

3-Hydroxy-1,1-dimethylpyrrolidinium Bromide α-Phenyl-2-thiopheneglycolate *see* 4709

4-Hydroxy-*N*,*N*-dimethyltryptamine *see* 8110

5-Hydroxy-*N*,*N*-dimethyltryptamine *see* 1502

8-Hydroxy-5,7-dinitro-2-naphthalenesulfonic Acid Disodium Salt *see* 6480

Hydroxydione Sodium, 4868

Hydroxydione Succinate *see* 4868

3-Hydroxy-6*H*-[1,3]dioxolo[5,6]benzofuro[3,2-*c*][1]benzopyran-6-one *see* 5832

2-Hydroxydiphenyl *see* 7458

4-Hydroxydiphenyl *see* 7459

o-Hydroxydiphenyl *see* 7458

3-[(Hydroxydiphenylacetyl)oxy]-1,1-

dimethylpiperidinium Bromide *see* 5893

3-[(Hydroxydiphenylacetyl)oxy]-1-methyl-1-azoniabicyclo[2.2.2]octane Bromide *see* 2412

2-[[(Hydroxydiphenylacetyl)oxy]-methyl]-1,1-dimethylpiperidinium Methyl Sulfate (Salt) *see* 1239

2-[[(Hydroxydiphenylacetyl)oxy]-methyl]-1,1-dimethylpyrrolidinium Methyl Sulfate (Salt) *see* 7714

endo-3-[(Hydroxydiphenylacetyl)oxy]-spiro[8-azoniabicyclo[3.2.1]octane-8,1'-pyrrolidinium] Chloride *see* 9918

3-Hydroxy-4,4-diphenyl-6-dimethylaminoheptane *see* 3254

2-Hydroxy-1,2-diphenylethanone *see* 1124

2-Hydroxy-1,2-diphenylethanone Oxime *see* 1125

(2-Hydroxydiphenyl)methane *see* 1179

(4-Hydroxydiphenyl)methane *see* 1180

p-Hydroxydiphenylmethane Carbamate *see* 3365

3-Hydroxy-4,4-diphenyl-4-methyl-5-dimethylaminohexane *see* 5200

(1α,2β,4β,5α,7β)-7-[(Hydroxydi-2-thienylacetyl)oxy]-9,9-dimethyl-3-oxa-9-azoniatricyclo[3.3.1.0²,⁴]nonane Bromide *see* 9598

20-Hydroxyecdysone *see* 3539

(*S*)-5-Hydroxy-6-*trans*-8,11,14-*cis*-eicosatetraenoic Acid *see* 4708

[*S*-(*E*,*Z*,*Z*,*Z*)]-5-Hydroxy-6,8,11,14-eicosatetraenoic Acid *see* 4708

p-Hydroxyephedrine, 4869

25-Hydroxyergocalciferol *see* 10156

16α-Hydroxyestradiol *see* 3750

3-Hydroxyestra-1,3,5,7,9-pentaen-17-one *see* 3675

3-Hydroxyestra-1,3,5(10),7-tetraen-17-one *see* 3676

3-Hydroxyestra-1,3,5(10)-trien-17-one *see* 3751

(8α)-3-Hydroxyestra-1,3,5(10)-trien-17-one *see* 5184

(17β)-17-Hydroxyestra-4,9,11-trien-3-one *see* 9716

17β-Hydroxy-4-estren-3-one *see* 6452

(17β)-17-Hydroxyestr-4-en-3-one *see* 6452

16α-Hydroxyestrone Diacetate *see* 2554

2-Hydroxyethanesulfonic Acid *see* 5122

1-Hydroxyethanesulfonic Acid Sodium Salt *see* 39

2-Hydroxy-1-ethanethiol *see* 5917

Hydroxyethanoic Acid *see* 4508

1-[2-(2-Hydroxyethoxy)ethyl]-4-[2-methyl-3-(10-phenothiazinyl)propyl]piperazine *see* 3450

11-[4-[2-(2-Hydroxyethoxy)ethyl]-1-piperazinyl]dibenzo[*b*,*f*][1,4]thiazepine, Hemifumarate *see* 8606

4-(2-Hydroxyethoxy)-3-methoxycinnamic Acid *see* 2341

3-[4-(2-Hydroxyethoxy)-3-methoxyphenyl]-2-propenoic Acid *see* 2341

9-[(2-Hydroxyethoxy)methyl]guanine *see* 148

(±)-4-(1-Hydroxyethoxy)-2-methyl-*N*-2-pyridyl-2*H*-1,2-benzothiazine-3-carboxamide Ethyl Carbonate (Ester) 1,1-Dioxide *see* 629

2-(β-Hydroxyethoxy)naphthalene *see* 6500

(5*R*,6*S*)-6-[(*R*)-1-Hydroxyethyl]-2-[(*S*)-1-acetimidoylpyrrolidin-3-ylthio]-1-

carbapen-2-em-3-carboxylic Acid *see* 7141

α-(2-Hydroxyethyl)acetoacetic Acid γ-Lactone *see* 84

1-(2-Hydroxyethyl)-4-[3-(2-acetyl-10-phenothiazyl)propyl]piperazine *see* 70

2-Hydroxyethylamine *see* 3772

β-Hydroxyethylamine *see* 3772

7-[3-[*N*-(2-Hydroxyethyl)amino]-2-hydroxypropyl]theophylline Nicotinate *see* 10194

m-Hydroxy-α-(ethylaminomethyl)-benzyl Alcohol *see* 3911

17-(1-Hydroxyethyl)androstane-3,17-diol *see* 279

β-Hydroxyethylbenzene *see* 7370

2-Hydroxyethyl Benzylcarbamate *see* 1525

5-(β-Hydroxyethyl)-5-butylmalonylurea Carbamate *see* 1880

1-[[(2-Hydroxyethyl)carbamoyl]methyl]pyridinium Chloride Laurate (Ester) *see* 5379

(5*R*,6*S*,8*R*)-6α-Hydroxyethyl-2-carbamoyloxymethyl-2-penem-3-carboxylic Acid *see* 8400

1-(2-Hydroxyethyl)-4-[3-(2-chloro-10-phenothiazinyl)propyl]piperazine *see* 7323

3β-Hydroxy-24-ethyl-Δ⁵,²²-cholestadiene *see* 8969

N-(2-Hydroxyethyl)cinnamamide *see* 4935

1-Hydroxy-α-ethylcyclohexylacetic Acid *see* 2784

N-(β-Hydroxyethyl)-1-deoxynojirimycin *see* 6274

α-2-Hydroxyethyl 6,8-Dideoxy-7-*O*-methyl-6-(1-methyl-L-2-pyrrolidinecarboxamido)-1-thio-D-*erythro*-D-*galacto*-octopyranoside Monosalicylate *see* 2005

N-[4-(β-Hydroxyethyl)diethylenediamino-1-methyl]tetracycline *see* 7606

N-[4-(β-Hydroxyethyl)diethylenediaminomethyl]tetracycline Phenoxymethylpenicillinate *see* 7235

(2-Hydroxyethyl)diisopropylmethylammonium Bromide Xanthene-9-carboxylate *see* 7989

(2-Hydroxyethyl)dimethyl(1-methyl-2-phenothiazin-10-ylethyl)ammonium Chloride *see* 4870

1-(β-Hydroxyethyl)-2-(diphenylmethyl)piperidine *see* 3362

N-Hydroxyethylethylenediaminetriacetic Acid Trisodium Salt *see* 10102

1-(2-Hydroxyethyl)-2-(8-heptadecenyl)-2-imidazoline *see* 428

N-(2-Hydroxyethyl)hexadecanamide *see* 7127

2-Hydroxyethylhydrazine *see* 4814

β-Hydroxyethylhydrazine *see* 4814

N-1-Hydroxyethyl-3-hydroxyflurazepam *see* 3490

2-(1-Hydroxyethyl)-β-(hydroxymethyl)-3-methyl-5-benzofuranacrylic Acid γ-Lactone Hydrogen Succinate *see* 1070

(5*R*,6*S*)-6-[(1*R*)-1-Hydroxyethyl]-3-(hydroxymethyl)-7-oxo-4-thia-1-azabicyclo[3.2.0]hept-2-ene-2-carboxylic Acid *see* 8400

(2*R*,3*R*,4*R*,5*S*)-1-(2-Hydroxyethyl)-2-(hydroxymethyl)-3,4,5-piperidinetriol *see* 6274

5-(2-Hydroxyethyl)-3-[(4-hydroxy-2-

methyl-5-pyrimidinyl)methyl]-4-
methylthiazolium Chloride see 7112
(1-Hydroxyethylidene)bisphosphonic
Acid see 3908
(1-Hydroxyethylidene)diphosphonic
Acid see 3908
3-(2-Hydroxyethylidene)-7-oxo-4-oxa-
1-azabicyclo[3.2.0]heptane-2-car-
boxylic Acid see 2402
2,2'-[(2-Hydroxyethyl)imino]bis[N-(1,-
1-dimethyl-2-phenylethyl)-N-meth-
ylacetamide] see 7067
[5R-[3(S*),5α,6α(R*)]]-6-(1-Hydr-
oxyethyl)-3-[[1-(1-iminoethyl)-3-
pyrrolidinyl]thio]-7-oxo-1-azabicy-
clo[3.2.0]hept-2-ene-2-carboxylic
Acid see 7141
[5R-[5α,6α(R*)]]-6-(1-Hydroxyethyl)-
3-[[2-[(iminomethyl)amino]ethyl]-
thio]-7-oxo-1-azabicyclo[3.2.0]hept-
2-ene-2-carboxylic Acid Monohydrate see 4954
3-β-Hydroxyethylindole see 9930
3-ω-Hydroxyethylindole see 9930
2-Hydroxyethyl Mercaptan see 5917
8-[(2-Hydroxyethyl)methylamino]-
caffeine see 1671
(+)-(1R,2S)-1-(2'-Hydroxyethyl)-1-
methyl-2-isopropenylcyclobutane see
4556
1-(2-Hydroxyethyl)-2-methyl-5-nitro-
imidazole see 6242
N-(2-Hydroxyethyl)-3-methyl-2-quin-
oxalinecarboxamide 1,4-Dioxide see
6960
5-Hydroxyethyl-4-methylthiazole see
6208
5-(2-Hydroxyethyl)-4-methylthiazole
see 6208
N-(2-Hydroxyethyl)moranoline see
6274
β-Hydroxyethyl 2-Naphthyl Ether see
6500
N-(2-Hydroxyethyl)nicotinamide
Nitrate (Ester) see 6608
1-(2-Hydroxyethyl)-2-[(5-nitro-2-
furanyl)methylene]hydrazinecarbox-
amide see 6615
N-(2-Hydroxyethyl)-2-nitro-1H-
imidazole-1-acetamide see 3756
N-(β-Hydroxyethyl)-N-[p-(4-nitro-
phenoxy)benzyl]dichloroacetamide
see 2233
17β-Hydroxy-17α-ethyl-19-nor-4-
androstene see 3851
N-(2-Hydroxyethyl)palmitamide see
7127
N-(β-Hydroxyethyl)-N-[p-phenoxy-(4'-
nitro)benzyl]dichloroacetamide see
2233
β-Hydroxyethyl Phenyl Ether see
7410
N-(2-Hydroxyethyl)-3-phenyl-2-pro-
penamide see 4935
4-(2-Hydroxyethyl)-1-piperazine-
ethanesulfonic Acid see 4687
N-(2-Hydroxyethyl)piperazine-N'-2-
ethanesulfonic Acid see 4687
N-[(2-Hydroxyethyl)piperazinocarb-
oxymethyl]tetracycline see 772
5-[γ-[4-(β-Hydroxyethyl)piperazino]-
propyl]dibenzo[b,f]azepine see 6985
N-[3-[4-(2-Hydroxyethyl)piperazino]-
propyl]iminostilbene see 6985
N-[[4-(2-Hydroxyethyl)-1-piperazin-
yl]methyl]tetracycline see 7606
10-[3-[4-(1-Hydroxyethyl-4-piperazinyl)-
propyl]-4-azaphenothiazine see 7104
1-[10-[3-[4-(2-Hydroxyethyl)-1-piper-

azinyl]propyl]-10H-phenothiazin-2-
yl]ethanone see 70
10-[3-[4-(2-Hydroxyethyl)-1-piper-
azinyl]propyl]phenothiazin-2-yl
Methyl Ketone see 70
1-[10-[3-[4-(2-Hydroxyethyl)-1-piper-
azinyl]propyl]-10H-phenothiazin-2-
yl]-1-propanone see 1910
10-[3-[4-(2-Hydroxyethyl)-1-pipera-
zinyl]propyl]-3-propionylphenothia-
zine see 1910
10-[3-[4-(2-Hydroxyethyl)-1-pipera-
zinyl]propyl]-10H-pyrido[3,2-b]-
[1,4]benzothiazine see 7104
10-[3'-[4''-(β-Hydroxyethyl)-1''-piper-
azinyl]propyl]-3-trifluoromethyl-
phenothiazine see 4226
10-[3-[4-(2-Hydroxyethyl)piperidino]-
propyl]phenothiazin-2-yl Methyl
Ketone see 7615
10-[3-[4-(2-Hydroxyethyl)-1-piperidin-
yl]propyl]-N,N-dimethyl-10H-pheno-
thiazine-2-sulfonamide see 7636
1-[10-[3-[4-(2-Hydroxyethyl)-1-piper-
idinyl]propyl]-10H-phenothiazin-2-
yl]ethanone see 7615
N-Hydroxyethylpromethazine Chloride, 4870
O-(β-Hydroxyethyl)rutosides see 9920
2-Hydroxyethyl Salicylate see 4509
Hydroxyethyl Starch see 4707
Hydroxyethylsulfonic Acid see 5122
7-(2-Hydroxyethyl)theophylline see
3924
7-(2-Hydroxyethyl)theophylline Hydrogen Sulfate Compound with Pyridoxol see 8159
7-(2-Hydroxyethyl)theophylline
Nicotinate see 3925
4-(α-Hydroxyethyl)toluene see 3287
1-(β-Hydroxyethyl)-4-[3-(2-trifluoro-
methylphenothiazin-10-yl)propyl]-
hexahydro-1,4-diazepine see 4774
1-(2-Hydroxyethyl)-4-[3-(3-trifluoro-
methylphenothiazin-10-yl)propyl]-
homopiperazine see 4774
1-(2-Hydroxyethyl)-4-[3-(trifluoro-
methyl-10-phenothiazinyl)propyl]-
piperazine see 4226
N-(2-Hydroxyethyl)-2,4,6-triiodo-5-[2-
[2,4,6-triiodo-3-(N-methylacetamido)-5-(methylcarbamoyl)benzami-
do]acetamido]isophthalamic Acid
see 5084
(β-Hydroxyethyl)trimethylammonium
see 2259
(2-Hydroxyethyl)trimethyl Ammonium Chloride Carbamate see 1823
(2-Hydroxyethyl)trimethylammonium
Chloride Phosphate see 7519
(2-Hydroxyethyl)trimethylammonium
Citrate see 2263
(2-Hydroxyethyl)trimethylammonium
Salicylate see 2265
N-(2-Hydroxyethyl)-N,N,α-trimethyl-
10H-phenothiazine-10-ethanamini-
um Chloride see 4870
4-(2-Hydroxyethyl)-5-vinylnicotinic
Acid δ-Lactone see 4400
3-(β-Hydroxyethyl)-2-(5-vinyl-2-
quinuclidyl)indole see 2344
3α-Hydroxyetioallocholan-17-one see
680
3β-Hydroxyetioallocholan-17-one see
3646
8β-Hydroxy-4αH-eudesm-5-en-12-oic
Acid γ-Lactone see 208
Δ1-16α-Hydroxy-9α-fluorohydrocorti-
sone see 9727

16α-Hydroxy-9α-fluoroprednisolone
see 9727
N-Hydroxyformamidoacetic Acid see
4617
Hydroxyfumaric Acid see 7041
Hydroxyfumigatin see 8902
1-(3-Hydroxy-2-furanyl)ethanone see
5198
4-Hydroxy-4H-furo[3,2-c]pyran-
2(6H)-one see 7185
3-Hydroxy-2-furyl Methyl Ketone see
5198
(1β)-1-Hydroxyginkgolide A see 4433
12-Hydroxygitoxigenin see 3197
3-Hydroxyglutaconic Acid Dimethyl
Ester Dimethyl Phosphate see 1360
Hydroxyglutamic Acid, 4871
β-Hydroxyglutamic Acid see 4871
Hydroxyhemin see 4668
1-Hydroxyheptane see 4696
2-Hydroxyheptane see 4697
1-Hydroxy-1-heptanesulfonic Acid
Monosodium Salt see 4693
1-Hydroxy-2,4-hexadiene see 8870
2-Hydroxy-N,N,N,N',N',N'-hexameth-
yl-1,3-propanediaminium Diiodide
see 7965
1-Hydroxyhexane see 4732
5-Hydroxy-2-hexenoic Acid Lactone
see 7166
1-Hydroxy-3-hexyl-6,6,9-trimethyl-7,-
8,9,10-tetrahydro-6H-dibenzo[b,d]-
pyran see 9190
p-Hydroxyhydrocinnamic Acid trans-
(4-Aminomethyl)cyclohexanecar-
boxylate see 2067
16α-Hydroxy-Δ1-hydrocortisone-
16α,17α-acetonide see 2973
Hydroxyhydroquinone see 1101
[5R-(2Z,5R*,6S*,9S*,10S*,11R*)]-3-
Hydroxy-N-[10-hydroxy-5,11-
dimethyl-2-(1-methylpropylidene)-
3,7,12-trioxo-9-(3-pyridinylmethyl)-
1,4-dioxa-8-azacyclododec-6-yl]-2-
pyridinecarboxamide see 8160
[2S-[2α(S*),3β,6E]]-9-[3-Hydroxy-6-
(2-hydroxyethylidene)-2-methyl-2-
oxepanyl]-2,6-dimethyl-2-nonen-5-
one see 10318
2-Hydroxy-N-(2-hydroxyethyl)-3-
methoxy-5-(2-propenyl)benzamide
see 243
7-[2-Hydroxy-3-[(2-hydroxyethyl)-
methylamino]propyl]theophylline,
Compd with Nicotinic Acid see
10194
(2R,3R,4R)-4-Hydroxy-2-(7-hydroxy-
heptyl)-3-[(E)-(4RS)-(4-hydroxy-4-
methyl-1-octenyl)]cyclopentanone
see 8395
12-Hydroxy-13-[2-hydroxy-1-(hydroxymethyl)]podocarpa-8,11,13-
trien-16-oic Acid see 7702
(±)-erythro-8-Hydroxy-5-[1-hydroxy-
2-(isopropylamino)butyl]carbostyril
see 7939
2'-Hydroxy-5'-[1-hydroxy-2-(isopropylamino)ethyl]methanesulfonanilide
see 8877
o-[6-Hydroxy-5-(hydroxymercuri)-2,7-
diiodo-3-oxo-3H-xanthen-9-yl]benz-
enesulfonic Acid Sodium Salt see
5912
(±)-2'-Hydroxy-5'-[(RS)-1-hydroxy-2-
[[(RS)-p-methoxy-α-methylpheneth-
yl]amino]ethyl]formanilide see 4272
(S)-5-Hydroxy-1-(4-hydroxy-3-meth-
oxyphenyl)-3-decanone see 4431
5-Hydroxy-2-(3-hydroxy-4-methoxy-

phenyl)-6,7-dimethoxy-4*H*-1-benzo-pyran-4-one *see* 3947

(*E*)-3-Hydroxy-5-[2-(3-hydroxy-4-methoxyphenyl)ethenyl]phenyl β-D-Glucopyranoside *see* 8340

(*R**,*R**)-(±)-*N*-[2-Hydroxy-5-[1-hydroxy-2-[[2-(4-methoxyphenyl)-1-methylethyl]amino]ethyl]phenyl]formamide *see* 4272

5-Hydroxy-2-(3-hydroxy-4-methoxyphenyl)-7-(*O*⁶-α-L-rhamnopyranosyl-β-D-glucopyranosyloxy)chromen-4-one *see* 3350

5-Hydroxy-2-(3-hydroxy-4-methoxyphenyl)-7-β-rutinosyloxy-4*H*-chromen-4-one *see* 3350

4-Hydroxy-3-hydroxymethyl-α-[(*tert*-butylamino)methyl]benzyl Alcohol *see* 217

9-[4-Hydroxy-3-(hydroxymethyl)but-1-yl]guanine *see* 7210

1-Hydroxy-2-hydroxymethyl-6,7-dimethoxy-4-(3′,4′,5′-trimethoxyphenyl)-1,2,3,4-tetrahydronaphthalene-3-carboxylic Acid Lactone *see* 8632

N-[2-Hydroxy-5-[4-(hydroxymethyl)-1,3,2-dithiarsolan-2-yl]phenyl]acetamide *see* 852

(5α,17β)-17-Hydroxy-2-(hydroxymethylene)-17-methylandrostan-3-one *see* 7101

(*R**,*S**)-(±)-8-Hydroxy-5-[1-hydroxy-2-[(1-methylethyl)amino]butyl]-2(1*H*)-quinolinone *see* 7939

N-[2-Hydroxy-5-[1-hydroxy-2-[(1-methylethyl)amino]ethyl]phenyl]-methanesulfonamide *see* 8877

8-Hydroxy-2-hydroxymethyl-6,7-methylenedioxy-4-(4′-hydroxy-3′,5′-dimethoxyphenyl)-1,2,3,4-tetrahydronaphthalene-3-carboxylic Acid Lactone *see* 7203

1-Hydroxy-2-hydroxymethyl-6,7-methylenedioxy-4-(3′,4′,5′-trimethoxyphenyl)-1,2,3,4-tetrahydro-naphthalene-3-carboxylic Acid Lactone *see* 7704

8-Hydroxy-2-hydroxymethyl-6,7-methylenedioxy-4-(3′,4′,5′-trimethoxyphenyl)-1,2,3,4-tetrahydronaphthalene-3-carboxylic Acid Lactone *see* 7204

3-Hydroxy-5-(hydroxymethyl)-2-methylisonicotinaldehyde *see* 8162

3-Hydroxy-5-(hydroxymethyl)-2-methylisonicotinaldehyde 5-Phosphate *see* 8163

3-Hydroxy-5-(hydroxymethyl)-2-methylisonicotinic Acid *see* 8165

3-Hydroxy-2-(hydroxymethyl)-2-methylpropanoic Acid *see* 3302

3-Hydroxy-5-(hydroxymethyl)-2-methyl-4-pyridinecarboxaldehyde *see* 8162

3-Hydroxy-5-(hydroxymethyl)-2-methyl-4-pyridinecarboxylic Acid *see* 8165

5-{(*E*)-(1*S*,5*S*,6*R*,7*R*)-7-Hydroxy-6-[(*E*)-(3*S*,4*RS*)-3-hydroxy-4-methyl-1-octenyl]bicyclo[3.3.0]octan-3-ylidene}pentanoic Acid *see* 4940

(*E*,*Z*)-(1*R*,2*R*,3*R*)-7-[3-Hydroxy-2-[(3*R*)-(3-hydroxy-3-methyl-1-octen-yl)]-5-oxocyclopentyl]-5-heptenoic Acid *see* 808

3-Hydroxy-4-[(2-hydroxy-5-methyl-phenyl)azo]-1-naphthalenesulfonic Acid *see* 1766

2-Hydroxy-5-[1-hydroxy-2-[(1-meth-yl-3-phenylpropyl)amino]ethyl]benz-amide *see* 5341

[*R*-(*R**,*R**)]-2-Hydroxy-5-[1-hydroxy-2-[(1-methyl-3-phenylpropyl)amino]-ethyl]benzamide *see* 3245

5-Hydroxy-2-(hydroxymethyl)-4*H*-pyran-4-one *see* 5330

5-Hydroxy-2-(hydroxymethyl)-4-pyrone *see* 5330

3-Hydroxy-4-[(1-hydroxy-2-naphtha-lenyl)azo]-7-nitro-1-naphthalene-sulfonic Acid Monosodium Salt *see* 3709

3-Hydroxy-2-(3-hydroxy-1-octenyl)-5-oxocyclopentaneheptanoic Acid *see* 8063

7-[3-Hydroxy-2-(3-hydroxy-1-octen-yl)-5-oxocyclopentyl]-5-heptenoic Acid *see* 8064

4-Hydroxy-α-(4-hydroxy-2-oxo-2*H*-1-benzopyran-3-yl)-2-oxo-2*H*-1-benzo-pyran-3-acetic Acid Ethyl Ester *see* 3818

erythro-p-Hydroxy-α-[1-[(*p*-hydroxy-phenethyl)amino]ethyl]benzyl Alcohol *see* 8401

[1α,2β(1*E*,3*R**),3α]-7-[3-Hydroxy-2-(3-hydroxy-4-phenoxy-1-butenyl)-5-oxocyclopentyl]-4,5-heptadienoic Acid Methyl Ester *see* 3629

[1*R*-[1α(*Z*),2β(1*E*,3*R**),3α]]-7-[3-Hydroxy-2-(3-hydroxy-4-phenoxy-1-butenyl)-5-oxocyclopentyl]-*N*-(methylsulfonyl)-5-heptenamide *see* 9165

(±)-*N*-[2-[[2-Hydroxy-3-(4-hydroxy-phenoxy)propyl]amino]ethyl]-4-mor-pholinecarboxamide *see* 10189

2-Hydroxy-*N*-(4-hydroxyphenyl)benz-amide *see* 7018

7-Hydroxy-3-(4-hydroxyphenyl)-4*H*-1-benzopyran-4-one *see* 2868

[6-Hydroxy-2-(4-hydroxyphenyl)-benzo[*b*]thien-3-yl][4-[2-(1-piper-idinyl)ethoxy]phenyl]methanone *see* 8281

(*R*)-4-[1-Hydroxy-2-[[4-(4-hydroxy-phenyl)butyl]amino]ethyl]-1,2-benz-enediol *see* 811

7-Hydroxy-3-(4′-hydroxyphenyl)-chroman *see* 3677

5-Hydroxy-3-(4-hydroxyphenyl)-8,8-dimethyl-6-(3-methyl-2-butenyl)-4*H*,8*H*-benzo[1,2-*b*:3,4-*b′*]dipyran-4-one *see* 7017

(*R**,*S**)-4-Hydroxy-α-[1-[[2-(4-hydr-oxyphenyl)ethyl]amino]ethyl]-benzenemethanol *see* 8401

2-Hydroxy-6-[2-(4-hydroxyphenyl)-ethyl]benzoic Acid *see* 5634

5-Hydroxy-3-(4-hydroxyphenyl)-7-methoxy-4*H*-1-benzopyran-4-one *see* 8094

4-Hydroxy-γ-(4-hydroxyphenyl)-γ-methylbenzenebutanoic Acid *see* 3370

5-[1-Hydroxy-2-[[2-(4-hydroxyphen-yl)-1-methylethyl]amino]ethyl]-1,3-benzenediol *see* 4022

4-Hydroxy-α-[[[3-(4-hydroxyphenyl)-1-methylpropyl]amino]methyl]benz-enemethanol *see* 8275

Hydroxy(2-hydroxypropyl)mercury-¹⁹⁷Hg *see* 5959

5-Hydroxy-1*H*-imidazole-4-carbox-amide *see* 6306

5-(Hydroxyimino)barbituric Acid *see* 10138

4-[3-(Hydroxyimino)cyclohexyl]-α-methylbenzeneacetic Acid *see* 10211

2-[(Hydroxyimino)methyl]-1-methyl-pyridinium Chloride *see* 7884

1-(2-Hydroxyiminomethyl-1-pyridin-io)-3-(4-carbamoyl-1-pyridinio)-2-oxapropane Dichloride *see* 870

6-[(Hydroxyimino)phenylmethyl]-1-[(1-methylethyl)sulfonyl]-1*H*-benz-imidazol-2-amine *see* 3637

N-(2(*R*)-Hydroxy-1(*S*)-indanyl)-2(*R*)-(phenylmethyl)-4(*S*)-hydroxy-5-[1-[4-(3-pyridylmethyl)-2(*S*)-(*N-tert*-butylcarba-moyl)piperazinyl]]pentanamide *see* 4979

5-(5-Hydroxy-3-indolyl)-3-(3-oxindol-ylidene)-2-oxopyrroline *see* 10135

3-[2-(5-Hydroxyindol-3-yl)-5-oxo-2-pyrrolin-4-ylidene]-2-indolinone *see* 10135

3-[2-(5-Hydroxyindol-3-yl)-5-oxo-pyrrolin-4-ylidene]oxindole *see* 10135

4-Hydroxy-3-iodo-5-nitrobenzonitrile *see* 6754

4-Hydroxy-3-iodo-5-nitrobenzonitrile Compd with 1-Deoxy-1-(ethyl-amino)glucitol (1:1) *see* 6754

4-Hydroxy-3-iodo-5-nitrobenzonitrile Compd with D-1-Deoxy-1-(methyl-amino)glucitol (1:1) *see* 6754

4-(4-Hydroxy-3-iodophenoxy)-3,5-di-iodohydrocinnamic Acid *see* 9553

[4-(4-Hydroxy-3-iodophenoxy)-3,5-di-iodophenyl]acetic Acid *see* 9603

L-3-[4-(4-Hydroxy-3-iodophenoxy)-3,5-diiodophenyl]alanine *see* 5535

O-(4-Hydroxy-3-iodophenyl)-3,5-diiodo-L-tyrosine *see* 5535

8-Hydroxy-7-iodo-5-quinolinesulfonic Acid, 4872

2-Hydroxy-3-isobutyl-6-(1-methyl-propyl)pyrazine 1-Oxide *see* 876

α-Hydroxyisobutyronitrile *see* 65

γ-Hydroxyisoeugenol *see* 2568

(1-Hydroxy-3-isohexenyl)naphthaza-rine *see* 253

4-Hydroxyisophthalic Acid, 4873

6-Hydroxy-2-isopropenyl-5-benzofur-anyl Methyl Ketone *see* 3946

2-Hydroxyisopropylamine *see* 488

(±)-5-[1-Hydroxy-2-(isopropylamino)-ethyl]anthranilonitrile *see* 2336

4′-[1-Hydroxy-2-(isopropylamino)eth-yl]methanesulfonanilide *see* 8876

4′-[2-Hydroxy-3-(isopropylamino)-propoxy]acetanilide *see* 7882

7-[2-Hydroxy-3-(isopropylamino)-propoxy]-2-benzofuranyl Methyl Ketone *see* 1050

8-[2-Hydroxy-3-(isopropylamino)-propoxy]-3-chromanol 3-Nitrate *see* 6655

4-[2-Hydroxy-3-(isopropylamino)-propoxy]indole *see* 7597

2-[*p*-[2-Hydroxy-3-(isopropylamino)-propoxy]phenyl]acetamide *see* 892

6-Hydroxy-2-isopropyl-3-methoxy-5-benzofuranacrylic Acid δ-Lactone *see* 7332

11-Hydroxy-13-isopropylpodocarpa-7,9(11),13-triene-6,12-dione *see* 9247

(8*r*)-3α-Hydroxy-8-isopropyl-1α*H*,5α*H*-tropanium Bromide 2-Propyl-valerate *see* 8693

3α-Hydroxy-8-isopropyl-1α*H*,5α*H*-tropanium Bromide (±)-Tropate *see* 5089

(4α)-13-Hydroxykaur-16-en-18-oic Acid *see* 8964

Hydroxylamine, 4874

Name Index

[S-(E)]-N-[2-(Hydroxynitrosoamino)-3-methylbutyl]-2-butenamide *see* 3481

N-Hydroxy-N-nitrosobenzenamine Ammonium Salt *see* 2688

N-Hydroxy-N-nitroso-1-naphthalenamine Ammonium Salt *see* 6536

3-Hydroxy-4-nitroso-2,7-naphthalenedisulfonic Acid Disodium Salt *see* 6742

2-Hydroxy-1,2,3-nonadecanetricarboxylic Acid *see* 184

17β-Hydroxy-19-norandrosta-4,9,11-trien-3-one *see* 9716

17β-Hydroxy-19-nor-4-androsten-3-one *see* 6452

m-Hydroxynorephedrine *see* 5993

(17α)-17-Hydroxy-19-norpregna-4,20-dien-3-one *see* 6814

(17α)-17-Hydroxy-19-norpregna-5(10),20-dien-3-one *see* 6795

(17α)-17-Hydroxy-19-norpregna-4,9,-11-trien-20-yn-3-one *see* 6798

17-Hydroxy-19-norpregn-4-ene-3,20-dione Hexanoate *see* 4422

(17α)-17-Hydroxy-19-norpregn-4-en-3-one *see* 6789

(17α)-17-Hydroxy-19-norpregn-4-en-20-yn-3-one *see* 6790

(17α)-17-Hydroxy-19-norpregn-5(10)-en-20-yn-3-one *see* 6791

17α-Hydroxy-19-norprogesterone Caproate *see* 4422

4-Hydroxy-19-nortestosterone *see* 7034

5-Hydroxynorvaline *see* 7252

[R-(Z)]-12-Hydroxy-9-octadecenoic Acid *see* 8378

[S-(E)]-9-Hydroxy-10-octadecen-12-ynoic Acid *see* 4657

N-Hydroxyoctanamide *see* 6848

5-Hydroxy-4-octanone *see* 1631

8-Hydroxy-2,4,6-octatriynamide *see* 188

[1S-[1α,3α(1E,3R*)4β(Z),5α]]-7-[3-(3-Hydroxy-1-octenyl)-2,6-dioxabicyclo[3.1.1]hept-4-yl]-5-heptenoic Acid *see* 9530

2-Hydroxy-4-(octyloxy)benzophenone *see* 6838

[2-Hydroxy-4-(octyloxy)phenyl]phenylmethanone *see* 6838

(3β)-3-Hydroxyolean-12-en-28-oic Acid *see* 6964

d-12-Hydroxyoleic Acid *see* 8378

3-Hydroxy-4,5-oxidophenanthrene *see* 6356

6β-Hydroxy-4-oxo-10αH-ambrosa-2,-11(13)-dien-12-oic Acid γ-Lactone *see* 400

6-Hydroxy-2-oxo-1,3-benzoxathiole *see* 9600

γ-Hydroxy-β-oxobutane *see* 61

1-Hydroxy-4-oxo-2,5-cyclohexadiene-1-sulfonic Acid Compd with Diethylamine *see* 3766

3-Hydroxy-22-oxokopsan-1-carboxylic Acid Methyl Ester *see* 5332

5-Hydroxy-4-oxonorvaline *see* 4779

δ-Hydroxy-γ-oxo-L-norvaline *see* 4779

3β-Hydroxy-11-oxoolean-12-en-30-oic Acid *see* 3628

(3β,4α)-3-Hydroxy-23-oxoolean-12-en-28-oic Acid *see* 4614

(3β,20β)-3-Hydroxy-11-oxoolean-12-en-29-oic Acid *see* 3628

3β-Hydroxy-11-oxoolean-12-en-30-oic Acid Acetate *see* 76

3β-Hydroxy-11-oxoolean-12-en-30-oic Acid Hydrogen Succinate *see* 1839

8-Hydroxy-5-oxoophiobola-3,6,19-trien-25-oic Acid *see* 2024

4-Hydroxy-3-(3-oxo-1-phenylbutyl)-2H-1-benzopyran-2-one *see* 10174

[7(S)-(1α,2β,4β,5α,7β)]-7-(3-Hydroxy-1-oxo-2-phenylpropoxy)-9,9-dimethyl-3-oxa-9-azoniatricyclo-[3.3.1.0²,⁴]nonane Bromide *see* 6084

(endo,syn)-(±)-3-(3-Hydroxy-1-oxo-2-phenylpropoxy)-8-methyl-8-(1-methylethyl)-8-azoniabicyclo[3.2.1]octane Bromide *see* 5089

endo-(±)-3-(3-Hydroxy-1-oxo-2-phenylpropoxy)-8-methyl-8-(1-methylethyl)-8-azoniabicyclo[3.2.1]octane Inner Salt *see* 9170

(17α)-17-Hydroxy-3-oxopregna-4,6-diene-21-carboxylic Acid γ-Lactone *see* 1795

3α-Hydroxy-20-oxo-5α-pregnane *see* 283

3β-Hydroxy-20-oxo-5α-pregnane *see* 284

20α-Hydroxy-3-oxo-5α-pregnane *see* 285

20β-Hydroxy-3-oxo-5α-pregnane *see* 286

3-Hydroxy-4-oxo-4H-pyran-2,6-dicarboxylic Acid *see* 5824

3-Hydroxy-4-oxo-1(4H)-pyridinealanine *see* 6286

1-Hydroxy-5-oxo-5H-pyrido[3,2-a]-phenoxazine-3-carboxylic Acid *see* 7645

4-Hydroxy-2-oxo-1-pyrrolidineacetamide *see* 7076

13-Hydroxy-3-oxo-13,17-secoandrosta-1,4-dien-17-oic Acid δ-Lactone *see* 9321

[2α(2S,3S,4S,6R),3β,5α]-14-Hydroxy-19-oxo-3,2-[(tetrahydro-3,4-dihydroxy-6-methyl-2H-pyran-2,3-diyl)bis-(oxy)]card-20(22)-enolide *see* 1769

[2α(2S,3R,6R),3β,5α]-14-Hydroxy-19-oxo-2,3-[(tetrahydro-3-hydroxy-6-methyl-2H-pyran-2,3-diyl)-bis(oxy)]card-20(22)-enolide *see* 10030

6-(10-Hydroxy-6-oxo-trans-1-undecenyl)-β-resorcylic Acid Lactone *see* 10246

2'-Hydroxypelargidenolon 1522 *see* 6352

6-Hydroxypelargidenon 1465 *see* 8556

6-Hydroxypelargidenon 6,4'-dimethyl Ether 1467 *see* 7195

15-Hydroxypentadecanoic Acid ε-Lactone *see* 3956

4-Hydroxy-2,4-pentadienoic Acid γ-Lactone *see* 8078

3-Hydroxy-2-pentenedioic Acid Dimethyl Ester Dimethyl Phosphate *see* 1360

4-Hydroxy-2-pentenoic Acid γ-Lactone *see* 686

4-Hydroxy-3-pentenoic Acid γ-Lactone *see* 686

Hydroxypethidine, 4883

Hydroxyphenamate, 4884

3-Hydroxy-4,5-phenanthrylene Oxide *see* 6356

α-Hydroxyphenazine *see* 4679

4-Hydroxyphenethylamine *see* 9966

β-Hydroxyphenethylamine *see* 7441

2-(β-Hydroxyphenethylamino)pyridine *see* 7474

2'-Hydroxy-2-(N,β-phenethyl)-5,9-dimethyl-6,7-benzomorphan *see* 7360

2-[6-(β-Hydroxyphenethyl)-1-methyl-2-piperidyl]acetophenone *see* 5580

6-(p-Hydroxyphenethyl)salicylic Acid *see* 5634

(m-Hydroxyphenethyl)trimethylammonium *see* 5467

7-Hydroxy-3H-phenoxazin-3-one 10-Oxide *see* 8309

3-[4-(p-Hydroxyphenoxy)-3,5-diiodophenyl]alanine *see* 3235

1-Hydroxy-2-phenoxyethane *see* 7410

(—)-(S)-1-(p-Hydroxyphenoxy)-3-(isopropylamino)-2-propanol *see* 7917

1-(4-Hydroxyphenoxy)-3-[2-(4-morpholinocarboxamido)ethylamino]-2-propanol *see* 10189

N-(4-Hydroxyphenyl)acetamide *see* 45

α-Hydroxyphenylacetic Acid *see* 5757

α-Hydroxy-α-phenylacetophenone *see* 1124

2-Hydroxy-2-phenylacetophenone Oxime *see* 1125

[6R-[6α,7β(R*)]]-7-[(Hydroxyphenylacetyl)amino]-3-[[(1-methyl-1H-tetrazol-5-yl)thio]methyl]-8-oxo-5-thia-1-azabicyclo[4.2.0]oct-2-ene-2-carboxylic Acid *see* 1964

[6R-[6α,7β(R*)]]-7-[(Hydroxyphenylacetyl)amino]-8-oxo-3-[[[1-(sulfomethyl)-1H-tetrazol-5-yl]thio]methyl]-5-thia-1-azabicyclo[4.2.0]oct-2-ene-2-carboxylic Acid *see* 1980

3-[(Hydroxyphenylacetyl)oxy]-8-methyl-8-(2-oxo-2-phenylethyl)-8-azoniabicyclo[3.2.1]octane Chloride *see* 7346

β-[4-Hydroxyphenyl]acrylic Acid *see* 2628

3-Hydroxy-L-phenylalanine *see* 9971

β-(p-Hydroxyphenyl)alanine *see* 9970

p-Hydroxyphenylaminoacetic Acid *see* 4885

α-(4-Hydroxyphenyl)-β-aminoethane *see* 9966

1-(m-Hydroxyphenyl)-2-aminoethanol *see* 6792

1-(p-Hydroxyphenyl)-2-aminoethanol *see* 6856

1-(m-Hydroxyphenyl)-2-amino-1-propanol *see* 5993

α-(m-Hydroxyphenyl)-β-aminopropanol *see* 5993

4-Hydroxy-5-[[4-(phenylamino)-5-sulfo-1-naphthalenyl]azo]-2,7-naphthalenedisulfonic Acid Trisodium Salt *see* 669

(4-Hydroxyphenyl)arsonic Acid Polymer with Formaldehyde *see* 7723

2-Hydroxy-N-phenylbenzamide *see* 8482

α-Hydroxy-α-phenylbenzeneacetic Acid *see* 1109

α-Hydroxy-α-phenylbenzeneacetic Acid 2-(Diethylamino)ethyl Ester *see* 1055

α-Hydroxy-α-phenylbenzeneacetic Acid 2-(Dimethylamino)-2-methylpropyl Ester *see* 3180

α-Hydroxy-α-phenylbenzeneacetic Acid 2-(Ethylpropylamino)ethyl Ester *see* 1057

(3-endo, 6-exo)-α-Hydroxy-α-phenylbenzeneacetic Acid 6-Methoxy-8-methyl-8-azabicyclo[3.2.1]oct-3-yl Ester *see* 9908

endo-α-Hydroxy-α-phenylbenzeneacetic Acid 8-Methyl-8-azabicyclo-[3.2.1]oct-3-yl Ester *see* 9913

α-Hydroxy-α-phenylbenzeneacetic

3β-Hydroxy-5α-pregnan-20-one *see* 284

20α-Hydroxy-5α-pregnan-3-one *see* 285

20β-Hydroxy-5α-pregnan-3-one *see* 286

17-Hydroxypregn-4-ene-3,20-dione *see* 4886

21-Hydroxypregn-4-ene-3,20-dione *see* 2947

21-Hydroxypregn-4-ene-3,20-dione 21-Acetate *see* 2948

21-Hydroxypregn-4-ene-3,11,20-trione *see* 2923

(3β)-3-Hydroxypregn-5-en-20-one *see* 7915

17α-Hydroxypregn-4-en-20-yn-3-one *see* 3787

17α-Hydroxyprogesterone, 4886

21-Hydroxyprogesterone *see* 2947

17α-Hydroxyprogesterone 3-Cyclopentyl Enol Ether *see* 7251

17α-Hydroxyprogesterone Hexanoate *see* 4886

Hydroxyproline, 4887

(—)-(2S,4R)-4-Hydroxyproline *see* 4887

(2S,4S)-4-Hydroxyproline *see* 4887

Lₛ-Hydroxyproline *see* 4887

trans-4-Hydroxy-L-proline *see* 4887

d-4-Hydroxyproline Betaine *see* 9952

l-4-Hydroxyproline Betaine *see* 1234

m-Hydroxypropadrine *see* 5993

Hydroxypropanedioic Acid *see* 9240

5,5'-[(2-Hydroxy-1,3-propanediyl)bis-(acetylimino)]bis[N,N'-bis(2,3-dihydroxypropyl)-2,4,6-triiodo-1,3-benzenedicarboxamide] *see* 5044

5,5'-[(2-Hydroxy-1,3-propanediyl)bis-(oxy)]bis[4-oxo-4H-1-benzopyran-2-carboxylic Acid] *see* 2658

3-Hydroxypropanenitrile *see* 3840

2-Hydroxy-2-propanesulfonic Acid Sodium Salt *see* 67

2-Hydroxy-1,2,3-propanetricarboxylic Acid *see* 2387

2-Hydroxy-1,2,3-propanetricarboxylic Acid Calcium Salt (2:3) *see* 1701

2-Hydroxy-1,2,3-propanetricarboxylic Acid Copper Salt (1:2) *see* 2702

2-Hydroxy-1,2,3-propanetricarboxylic Acid Magnesium Salt (1:1) *see* 5700

2-Hydroxy-1,2,3-propanetricarboxylic Acid Magnesium Salt (2:3) *see* 5699

2-Hydroxy-1,2,3-propanetricarboxylic Acid Silver Salt *see* 8654

2-Hydroxy-1,2,3-propanetricarboxylic Acid Tributyl Ester *see* 1599

2-Hydroxy-1,2,3-propanetricarboxylic Acid Trilithium Salt *see* 5555

2-Hydroxy-1,2,3-propanetricarboxylic Acid Zinc Salt *see* 10263

2-Hydroxypropanoic Acid *see* 5349

3-Hydroxypropanoic Acid *see* 4799

(R)-2-Hydroxypropanoic Acid *see* 5348

(S)-2-Hydroxypropanoic Acid *see* 5350

2-Hydroxypropanoic Acid Calcium Salt *see* 1723

2-Hydroxypropanoic Acid 1-Carboxyethyl Ester *see* 5351

2-Hydroxypropanoic Acid Ethyl Ester *see* 3863

2-Hydroxypropanoic Acid Hexadecyl Ester *see* 2072

2-Hydroxypropanoic Acid Magnesium Salt *see* 5708

2-Hydroxypropanoic Acid Methyl Ester *see* 6167

2-Hydroxypropanoic Acid 2-Naphthalenyl Ester *see* 6499

2-Hydroxypropanoic Acid Silver Salt *see* 8660

2-Hydroxypropanoic Acid Strontium Salt *see* 9006

2-(2-Hydroxypropanoyloxy)propanoic Acid *see* 5351

4-Hydroxy-1-propenylbenzene *see* 715

(E)-4-(3-Hydroxy-1-propenyl)-2,6-dimethoxyphenyl-β-D-glucopyranoside *see* 9192

4-(3-Hydroxypropenyl)-2,6-dimethoxyphenyl-D-glucoside *see* 9192

(17β)-17-Hydroxy-17-(2-propenyl)-estra-4,9,11-trien-3-one *see* 327

4-(3-Hydroxy-1-propenyl)-2-methoxyphenol *see* 2568

4-(3-Hydroxy-1-propenyl)-2-methoxyphenyl β-D-Glucopyranoside *see* 2567

α-Hydroxypropionic Acid *see* 5349

β-Hydroxypropionic Acid *see* 4799

β-Hydroxypropionitrile *see* 3840

5-(α-Hydroxypropionylamino)-2,4,6-triiodoisophthalic Acid Di(1,3-dihydroxyisopropylamide) *see* 5071

4'-Hydroxypropiophenone *see* 7176

p-Hydroxypropiophenone *see* 7176

1-[2-[2-Hydroxy-3-(propylamino)propoxy]phenyl]-3-phenyl-1-propanone *see* 7978

2'-[2-Hydroxy-3-(propylamino)propoxy]-3-phenylpropiophenone *see* 7978

α-Hydroxypropylbenzene *see* 3816

Hydroxypropyl Cellulose, 4888

3-Hydroxypropylene Oxide *see* 4499

N-(2-Hydroxypropyl)ethylenediamine *see* 458

6-[(2-Hydroxypropyl)methylamino]-3(2H)-pyridazinone Hydrazone *see* 7577

Hydroxypropyl Methylcellulose, 4889

1-(2-Hydroxypropyl)-2-methyl-5-nitroimidazole *see* 8562

2-[N-[β-[4-(α-Hydroxypropyl)phenoxy]ethyl]amino]-1-phenylpropane *see* 3996

2-(α-Hydroxypropyl)piperidine *see* 2564

5-Hydroxy-2-propylpiperidine *see* 8099

5-(2-Hydroxypropyl)-5-(2-propenyl)-2,4,6(1H,3H,5H)-pyrimidinetrione *see* 8091

[10-(2-Hydroxypropyl)-1,4,7,10-tetraazacyclododecane-1,4,7-triacetato-(3—)-N¹,N⁴,N⁷,N¹⁰,O¹,O⁴,O⁷,O¹⁰]-gadolinium *see* 4348

10-(2-Hydroxypropyl)-1,4,7,10-tetraazacyclododecane-1,4,7-triacetic Acid Gadolinium Complex 4348

1-(2-Hydroxypropyl)theobromine *see* 8072

7-(2-Hydroxypropyl)theophylline *see* 8092

(2-Hydroxypropyl)trimethylammonium Chloride Acetate *see* 6003

(2-Hydroxypropyl)trimethylammonium Chloride Carbamate *see* 1232

4-Hydroxypyrazolo[3,4-d]pyrimidine *see* 287

1-Hydroxy-2(1H)-pyridinethione *see* 8178

2-Hydroxy-5-[[4-[(2-pyridinylamino)sulfonyl]phenyl]azo]benzoic Acid *see* 9112

[1-Hydroxy-2-(3-pyridinyl)ethylidene]bisphosphonic Acid *see* 8396

β-[N-(3-Hydroxy-4-pyridone)]-α-aminopropionic Acid *see* 6286

1-Hydroxy-5H-pyrido[3,2-a]phenoxazin-5-one-3-carboxylic Acid *see* 7645

5-(α-Hydroxy-α-2-pyridylbenzyl)-7-(α-2-pyridylbenzylidene)-5-norbornene-2,3-dicarboximide *see* 6780

4-Hydroxy-2(1H)-pyrimidinethione *see* 9504

2-Hydroxy-4(1H)-pyrimidinone *see* 9985

2-Hydroxy-4(3H)-pyrimidinone *see* 9985

4-Hydroxy-2(1H)-pyrimidinone *see* 9985

α-Hydroxypyrotartaric Acid *see* 2384

4-Hydroxy-2-pyrrolidinecarboxylic Acid *see* 4887

2-(4-Hydroxypyrrolidin-2-on-1-yl)-acetamide *see* 7076

4-Hydroxyquinaldic Acid *see* 5339

Hydroxyquinol *see* 1101

2-Hydroxyquinoline *see* 1873

8-Hydroxyquinoline, 4890

8-Hydroxyquinoline Benzoate (Ester) *see* 1143

4-Hydroxy-2-quinolinecarboxylic Acid *see* 5339

8-Hydroxy-5,7-quinolinedisulfonic Acid Copper Derivative Compound with Diethylamine *see* 2740

8-Hydroxy-5-quinolinesulfonic Acid, 4891

3-Hydroxyquinuclidine *see* 8266

14-Hydroxy-3β-(rhamnosyloxy)bufa-4,20,22-trienolide *see* 8060

5-Hydroxy-1-β-D-ribofuranosyl-1H-imidazole-4-carboxamide *see* 6306

4'-Hydroxysalicylanilide *see* 7018

5-Hydroxysalicylic Acid *see* 4404

8-Hydroxysantonin *see* 855

3-Hydroxy-16,17-secoestra-1,3,5(10)-trien-17-oic Acid *see* 3470

12-Hydroxysenecionan-11,16-dione *see* 8595

[15(20)E]-12-Hydroxysenecionan-11,-16,dione *see* 8595

3α-Hydroxyspiro[1αH,5αH-nortropane-8,1'-pyrrolidinium] Chloride Benzilate *see* 9918

(3β,5α,25R)-3-Hydroxyspirostan-12-one *see* 4652

(3β,25R)-3-Hydroxyspirost-5-en-12-one *see* 4407

d-4-Hydroxystachydrine *see* 9952

l-4-Hydroxystachydrine *see* 1234

2-Hydroxystilbamide *see* 4892

Hydroxystilbamidine, 4892

2-Hydroxy-4,4'-stilbenedicarboxamidine *see* 4892

Hydroxystreptomycin, 4893

(E)-5-(p-Hydroxystyryl)resorcinol *see* 8326

Hydroxysuccinic Acid *see* 5747

2-Hydroxy-5-sulfobenzoic Acid *see* 9138

3-Hydroxy-4-[(4-sulfo-1-naphthalenyl)azo]-2,7-naphthalenedisulfonic Acid Trisodium Salt *see* 391

5-Hydroxy-4-[(4-sulfo-1-naphthalenyl)azo]-1-(4-sulfophenyl)-1H-pyrazole-3-carboxylic Acid 3-Ethyl Ester Disodium Salt *see* 6991

6-Hydroxy-5-[(4-sulfophenyl)azo]-2-naphthalenesulfonic Acid Disodium Salt *see* 9176

5-Hydroxy-1-(p-sulfophenyl)-4-[(p-sulfophenyl)azo]pyrazole-3-car-

boxylic Acid Trisodium Salt *see*
9239
[(2-Hydroxy-3-sulfopropyl)thio]gold
Calcium Salt *see* 1689
3α-Hydroxy-8-(3-sulfopropyl)-1αH,-
5αH-tropanium Hydroxide, (±)-
Tropate, Inner Salt *see* 9170
3-Hydroxytamoxifen *see* 3502
Hydroxytetracaine, 4894
Hydroxytetracycline *see* 7111
5-Hydroxy-3-[4-(1,2,3,6-tetrahydro-4-
phenyl-1-pyridyl)butyl]indole *see*
8432
(±)-5-[4-[(6-Hydroxy-2,5,7,8-tetra-
methylchroman-2-yl)methoxy]benz-
yl]-2,4-thiazolidinedione *see* 9898
7-Hydroxy-3,4-tetramethylene-
coumarin O,O-Diethylthiophosphate
see 2635
β-4-Hydroxy-1,2,2,6-tetramethyl-
piperidine Mandelic Acid Ester *see*
3943
4-Hydroxy-N-2-thiazolylbenzene-
sulfonamide *see* 7403
4-[4-Hydroxy(thio-m-anisoyl)]mor-
pholine *see* 10071
o-Hydroxytoluene *see* 2645
α-Hydroxytoluene *see* 1159
2-Hydroxy-m-toluic Acid *see* 2649
2-Hydroxy-p-toluic Acid *see* 2648
6-Hydroxy-m-toluic Acid *see* 2650
α-Hydroxy-α-toluic Acid *see* 5757
2-(m-Hydroxy-N-p-tolylanilinomethyl)-2-imidazoline *see* 7417
ω-(4-Hydroxy-4-p-tolylpiperidino)-p-
fluorobutyrophenone *see* 6343
2-Hydroxy-5-tosyloxybenzenesulfonic
Acid *see* 9169
1-Hydroxytriacontane *see* 9722
β-Hydroxytricarballylic Acid *see* 2387
2-Hydroxytriethylamine *see* 3161
ω-[4-Hydroxy-4-(m-trifluoromethyl-
phenyl)piperidino]-p-fluorobutyro-
phenone *see* 9813
17-Hydroxy-6α-(trifluoromethyl)-
pregn-4-ene-3,20-dione Acetate *see*
4171
17-Hydroxy-6α-(trifluoromethyl)pro-
gesterone Acetate *see* 4171
7-Hydroxy-3-(3,4,5-trihydroxyphen-
yl)-4H-1-benzopyran-4-one *see* 984
1-Hydroxy-2,10,11-trimethoxyapor-
phine *see* 2612
11-Hydroxy-1,2,10-trimethoxyapor-
phine *see* 5178
(3β,16β,17α,18β,20α)-17-Hydroxy-
18-[(3,4,5-trimethoxybenzoyl)oxy]-
yohimban-16-carboxylic Acid Methyl Ester *see* 8292
3-Hydroxy-2,9,10-trimethoxyberbine
see 2615
5-Hydroxy-6,7,8-trimethoxy-2-(3,4,5-
trimethoxyphenyl)-4H-1-benzo-
pyran-4-one *see* 4377
3-Hydroxy-4-(trimethylammonio)-
butanoate *see* 1898
3-Hydroxy-N,N,N-trimethylbenzene-
ethanaminium *see* 5467
6-Hydroxy-β,2,7-trimethyl-5-benzo-
furanacrylic Acid δ-Lactone *see*
9864
2'-Hydroxy-2,5,9-trimethyl-6,7-benzo-
morphan *see* 5995
2-Hydroxy-2,6,6-trimethylbicyclo-
[3.1.1]heptan-3-one *see* 7110
3-Hydroxy-1,7,7-trimethylbicyclo-
[2.2.1]heptan-2-one *see* 4862
[1S-(endo, endo)]-N-[[(3-Hydroxy-4,7,-
7-trimethylbicyclo[2.2.1]hept-2-yl)-

amino]carbonyl]-4-methylbenzene-
sulfonamide *see* 4447
4-Hydroxy-α,α,4-trimethylcyclohex-
anemethanol *see* 9314
5-Hydroxy-α,α,4-trimethyl-3-cyclo-
hexene-1-methanol *see* 8707
5,5'-[(2-Hydroxytrimethylene)bis-
(acetylimino)]bis[N,N'-bis(2,3-di-
hydroxypropyl)-2,4,6-triiodoiso-
phthalamide] *see* 5044
(2-Hydroxytrimethylene)bis[trimeth-
ylammonium] Iodide *see* 7965
5,5'-[(2-Hydroxytrimethylene)dioxy]-
bis(4-oxo-4H-1-benzopyran-2-carb-
oxylic Acid) *see* 2658
5,5'-(2-Hydroxytrimethylenedioxy)-
bis(4-oxochromene-2-carboxylic
Acid) *see* 2658
2-Hydroxy-N,N,N-trimethylethan-
aminium *see* 2259
2-Hydroxy-N,N,N-trimethylethan-
aminium Chloride *see* 2261
2-Hydroxy-N,N,N-trimethylethan-
aminium Salt with 3,7-Dihydro-1,3-
dimethyl-1H-purine-2,6-dione *see*
2266
2-Hydroxy-N,N,N-trimethylethan-
aminium Salt with 2-Hydroxyben-
zoic Acid (1:1) *see* 2265
2-Hydroxy-N,N,N-trimethylethana-
minium (OC-6-44)-Triaqua[2-hydr-
oxy-1,2,3-propanetricarboxylato-
(4—)]ferrate(1—) *see* 4085
[S-(Z,E)]-5-(1-Hydroxy-2,6,6-trimeth-
yl-4-oxo-2-cyclohexen-1-yl)-3-meth-
yl-2,4-pentadienoic Acid *see* 7
(R)-4-Hydroxy-N,N,N-trimethyl-10-
oxo-7-[(1-oxohexadecyl)oxy]-3,5,9-
trioxa-4-phosphapentacosan-1-
aminium Hydroxide, Inner Salt, 4-
Oxide Mixture with Formaldehyde
Polymer with Oxirane and 4-(1,1,3,-
3-Tetramethylbutyl)phenol and 1-
Hexadecanol *see* 3960
(R)-4-Hydroxy-N,N,N-trimethyl-10-
oxo-7-[(1-oxohexadecyl)oxy]-3,5,9-
trioxa-4-phosphapentacosan-1-
aminium Inner Salt 4-Oxide *see*
2540
(5Z,11α,13E,15R)-15-Hydroxy-11,16,-
16-trimethyl-9-oxoprosta-5,13-dien-
1-oic Acid *see* 9853
1-(4-Hydroxy-2,3,5-trimethylphen-
oxy)-3-(isopropylamino)-2-propanol
4-Acetate *see* 6221
3-Hydroxy-4-[(2,4,5-trimethylphenyl)-
azo]-2,7-naphthalenedisulfonic Acid
Disodium Salt *see* 7749
11β-Hydroxy-16α,17α,21-trimethyl-
pregna-1,4-diene-3,20-dione *see* 8392
9-Hydroxy-2,5,11-trimethyl-6H-pyri-
do[4,3-b]carbazolium Acetate (Salt)
see 3590
Hydroxytriphenylstannane *see* 9875
Hydroxytriphenyltin *see* 9875
3β-Hydroxy-1αH,5αH-tropane-2β-
carboxylic Acid *see* 3541
3β-Hydroxy-1αH,5αH-tropane-2β-
carboxylic Acid Benzoate *see* 1146
3β-Hydroxy-1αH,5αH-tropane-2α-
carboxylic Acid Methyl Ester Benzoate *see* 8097
3β-Hydroxy-1αH,5αH-tropane-2β-
carboxylic Acid Methyl Ester Benzoate *see* 2517
5-Hydroxytryptamine *see* 8607
5-Hydroxytryptophan, 4895
3-Hydroxytyramine *see* 3479
3-Hydroxytyrosine *see* 3478
3-Hydroxy-L-tyrosine *see* 5490

Hydroxyurea, 4896
(3β)-3-Hydroxyurs-12-ene-27,28-dioic
Acid *see* 8260
(3α,4β)-3-Hydroxyurs-12-en-23-oic
Acid *see* 1382
(3β)-3-Hydroxyurs-12-en-28-oic Acid
see 10027
17-Hydroxy-17α-vinyl-4-estren-3-one
see 6814
17β-Hydroxy-17α-vinylestr-5(10)-en-
3-one *see* 6795
1α-Hydroxyvitamin D₃ *see* 4864
1β-Hydroxyvitamin D₃ *see* 4864
25-Hydroxyvitamin D₃ *see* 1677
8-Hydroxyxanthine *see* 10014
1-Hydroxy-2-[(6-O-β-D-xylopyrano-
syl-β-D-glucopyranosyl)oxy]-9,10-
anthracenedione *see* 8437
(3β,16α,17α)-17-Hydroxyyohimban-
16-carboxylic Acid Methyl Ester *see*
8107
(16α,17α)-17-Hydroxyyohimban-16-
carboxylic Acid Methyl Ester *see*
10236
(16β,17α)-17-Hydroxyyohimban-16-
carboxylic Acid Methyl Ester *see*
2614
17-Hydroxyyohimban-16-carboxylic
Acid Methyl Ester *see* 10237
17α-Hydroxy-20α-yohimban-16β-car-
boxylic Acid Methyl Ester *see*
10238
Hydroxyzine, 4897
Hydrozincite *see* 10260
Hyenanchin, 4898
Hyenancin *see* 4898
Hyflavin [Endo] *see* 6179
Hygric Acid Methylbetaine *see* 8926
Hygrine, 4899
Hygromix *see* 4901
Hygromycin, 4900
Hygromycin A *see* 4900
Hygromycin B, 4901
Hygrophylline, 4902
Hygroton [Ciba-Geigy; Rorer] *see*
2246
Hykinone [Abbott] *see* 5874
Hylenta [Ayerst] *see* 7225
Hylorel [Upjohn] *see* 4587
Hymecromone, 4903
Hymecromone O,O-Diethyl Phos-
phorothioate, 4904
Hymexazol, 4905
Hyminal [Eisai] *see* 6028
Hymorphan [Endo] *see* 4847
Hyndarin *see* 9354
Hyodeoxycholic Acid, 4906
α-Hyodeoxycholic Acid *see* 4906
Hyodesoxycholic Acid *see* 4906
Hyonate [Bayer] *see* 4793
Hyoper [Bordet] *see* 1838
Hyoscine *see* 8550
Hyoscine-N-butyl Bromide *see* 1624
Hyoscine Methyl Bromide *see* 6084
Hyoscyamine, 4907
dl-Hyoscyamine *see* 907
l-Hyoscyamine *see* 4907
Hyoscyamus, 4908
Hyp *see* 4887
Hypadil [Kowa] *see* 6655
Hypalon [Du Pont], 4909
Hypaphorine, 4910
Hypaque Cysto [Winthrop] *see* 5851
Hypaque Meglumine [Winthrop] *see*
5851
Hypaque Sodium [Winthrop] *see* 3040
Hypcol *see* 6028
Hyperan [SS Pharm.] *see* 3955
Hyperdrol [Britcair] *see* 356

Hyperglycemic-Glycogenolytic Factor see 4455
Hypericin, 4911
Hypericum Red see 4911
Hyperin see 8216
Hyperium [Biopharma] see 8388
Hyperol [Purdue Frederick] see 10007
Hyperoside see 8216
Hypersin [Zeria] see 1233
Hyperstat [Schering] see 3051
Hypertane see 3778
Hypertensin see 689
Hypertensin [Ciba] see 689
Hypertensinogen see 689
Hypertonalum [Essex] see 3051
Hy-Phos see 8814
Hyphylline see 3529
Hypnazol [Logeais] see 3994
Hypnodil [Janssen] see 6232
Hypnodin [Takeda] see 7320
Hypnogène see 989
Hypnomidate [Janssen] see 3927
Hypnon [Sumitomo] see 6641
Hypnone see 71
Hypnorex [Synthelabo] see 5552
Hypnovel [Roche] see 6270
"Hypo" see 8844
Hypoboric Acid see 3064
Hypoca [Yamanouchi] see 1031
Hypochlorous Acid, 4912
Hypochlorous Acid 1,1-Dimethylethyl Ester see 1605
Hypochlorous Anhydride see 2149
Hypochylin see 4477
Hypocrine [Tanabe Seiyaku] see 5500
Hypoglycin see 4913
Hypoglycin A see 4913
Hypoglycine A, 4913
Hypoglycine B, 4914
Hyponitrous Acid Anhydride see 6751
β-Hypophamine see 10073
Hypophosphoric Acid, 4915
Hypophosphorous Acid, 4916
Hypophthalin see 4800
Hypophyseal Growth Hormone see 8864
Hypostamine [Promedica] see 9894
Hypostat [Herbrand] see 7176
Hyposterol see 7432
Hyposulfite of Gold and Sodium see 4537
Hypothiazide see 4822
Hypovase [Pfizer] see 7897
Hypoxanthine, 4917
Hypoxanthine Riboside see 5005
Hypoxanthine Riboside-5-phosphoric Acid see 5007
Hypoxanthosine see 5005
Hyprenan [Astra] see 7917
Hyprim [Fisons] see 1401
Hypromellose see 4889
Hyproval P.A. [Tutag] see 4886
Hyptor see 6028
Hypurin [CP Pharm] see 5011
Hypusine, 4918
Hyrazin [Kowa] see 9436
Hyryl see 8368
Hyskon [Pharmacia] see 2989
Hysron [Kyowa] see 5838
Hytakerol [Winthrop] see 3223
Hyton Asa [Pharmacia] see 7206
Hytone [Dermik] see 4828
Hytracin [Dainabot-Dainippon] see 9297
Hytrin [Abbott] see 9297
Hytrinex [Abbott] see 9297
Hytrol O see 2795
Hyvar [Du Pont] see 1402
Hyvar General Weed Killer [DuPont] see 5174

Hyzaar [Du Pont] see 5613
I see 5196
I-612 see 8414
I-653 see 2965
I-2586 see 9082
I.A. 307 see 73
IAA see 4994
Ial [Fidia] see 4793
Iangene [Farmochim. Ital.] see 9160
Ibafloxacin, 4919
IBC see 1488
IBCA see 1488
IBD [IPG] see 5245
Ibenzmethyzin see 7938
Ibiamox [IBI] see 617
IBI-C83 see 8419
Ibidomide see 5341
Ibifur see 4315
Ibilex [IBI] see 2021
Ibinolo [IBI] see 892
Ibiotyzil see 1055
Ibistacin [IBI] see 8373
Ibition [Ibi] see 9455
Ibogaine, 4920
Ibopamine, 4921
Ibotenic Acid, 4922
Ibu-Attritin see 4925
Ibudilast, 4923
Ibudros [Manetti Roberts] see 4926
Ibufenac, 4924
Ibumetin [Benzon] see 4925
Ibunac [Kaken] see 4924
Ibuprocin [Nisshin] see 4925
Ibuprofen, 4925
Ibuproxam, 4926
Ibustrin [Farmitalia] see 4991
Ibutad [TAD] see 4925
Ibutid [Dumex] see 4925
Ibutilide, 4927
Ibutop [Chefaro] see 4925
Ibylcaine see 1552
Icacine [Bristol] see 3052
Icariin see 3653
"Ice" see 6015
Ice-B-Gon [Chevron] see 1725
Iceland Moss, 4928
Ice Spar see 2673
Ichden see 4929
Ichtammon see 4929
Ichthadone see 4929
Ichthalum see 4929
Ichthammol, 4929
Ichthammonium see 4929
Ichthium see 4929
Ichthosan see 4929
Ichthosauran see 4929
Ichthosulfol see 4929
Ichthymall [Mallinckrodt] see 4929
Ichthynat [Tenneco] see 4929
Ichthyocolla see 5123
Ichthyol [Stiefel] see 4929
Ichthyopon see 4929
Ichthyopterin, 4930
Ichthysalle see 4929
Ichtopur [Ichthyol] see 4929
ICI-350 see 455
ICI-29661 see 8172
ICI-32865 see 3926
ICI-33828 see 6013
ICI-35868 see 8020
ICI-38174 see 7974
ICI-45520 see 8025
ICI-45763 see 9653
ICI-46474 see 9216
ICI-47699 see 9216
ICI-48213 see 2789
ICI-50123 see 7250
ICI-50172 see 7882
ICI-54450 see 4010
ICI-55052 see 6558

ICI-55897 see 2426
ICI-58834 see 10116
ICI-59118 see 8295
ICI-66082 see 892
ICI-69653 see 769
ICI-72222 see 2060
ICI-80008 see 4231
ICI-80996 see 2461
ICI-81008 see 4231
ICI-118587 see 10189
ICI-118630 see 4547
ICI-122378 see 779
ICI-139603 see 9384
ICI-141292 see 3641
ICI-156834 see 1984
ICI-176334 see 1247
ICI-194660 see 5960
ICI-204219 see 10241
ICI-204636 see 8606
ICIA-0051 see 9063
ICIA-192 see 4153
ICI-D-1033 see 667
ICI-D-1694 see 9684
ICI-PP-333 see 7118
Icipen [ICI] see 7230
ICN-1229 see 8365
Iconyl see 4885
Icoral B see 5993
ICRF-159 see 8295
ICRF-187 see 8295
ICS-205-930 see 9914
ICSH see 5499
Ictéryl see 5878
ID-540 see 4164
ID-1937 see 4245
IDA see 4952
Idaein see 2757
Idalon [Nippon Roussel] see 4140
Idamycin [Adria] see 4931
Idarac [Diamant] see 4140
Idarubicin, 4931
Idebenone, 4932
Idein see 2757
Idexur see 4934
Idocyl [Ferrosan] see 8819
Idomethine [Kowa] see 4998
Idonor [Roger] see 7673
Idonyx see 5058
Idorese see 2119
Idose, 4933
Idoxene [Spodefell] see 4934
Idoxuridine, 4934
Idrocilamide, 4935
Idroestril see 3177
Idrolone [Maggioni] see 4039
Idro P_2 [Maggioni] see 9158
Idro P_3 [Maggioni] see 7047
IDU see 4934
Idulea see 4934
Idulian [Schering] see 934
IDUR see 4934
Iduridin [Ferring] see 4934
L-Iduronic Acid see 4474
IF see 1945
Ifada [Wassermann] see 3972
Ifenec [Italfarmaco] see 3550
Ifenprodil, 4936
Ifex [Bristol] see 4937
IFN see 5015
IFN-α see 5016
IFN-α_2 see 5016
IFN-αA see 5016
IFN-β see 5017
IFN-β_1 see 5017
IFN-β_2 see 5017
IFN-γ see 5018
IFN-beta$_{ser}$ see 5017
Ifomide [Shionogi] see 4937
Ifosfamide, 4937
IFP see 4493

Ifrasarl [Showa Shinyaku] see 2842
IgA see 4958
IgD see 4958
IgE see 4958
Igepal CA [GAF] see 6858
Igepal CA-630 [GAF] see 6858
Igepal CO [GAF] see 6772
Igepal CO-630 [GAF] see 6772
IGF-I see 8862
IGF-II see 8862
IgG see 4958
IgM see 4958
Ignatia, 4938
Ignatius Bean see 4938
Ignotine see 1899
Igralin [Zeria] see 9436
Ig's see 4958
131I HSA see 8613
III-2318 see 5892
[123I](±)-N-isopropyl-p-iodoamphet-amine see 5066
IK-2 see 8135
Ikaclomine [Ika] see 2446
Ikaran [Fabre] see 3217
IKF-1216 see 4153
Ikorel [Rhone-Poulenc Rorer] see 6608
Iktorivil [Roche] see 2449
IL-1 see 5019
IL-1ra see 5022
IL-2 see 5020
IL-3 see 5021
IL-17803A see 16
Ilcocillin P [Ciba-Geigy] see 7226
Ildamen [Asta] see 7096
Ile see 5196
α-Ile see 5196
Ilentazol [Geistlich] see 7533
Ileogastrone see 3632
Iletin II [Lilly] see 5011
Iliadin [E. Merck] see 7100
Iliren [Provet] see 9563
Ilixathin see 8456
Illcut [Nippon Zoki] see 9640
Il(l)oxan [Hoechst] see 3133
Illudin M see 4939
Illudins, 4939
Illudin S see 4939
Ilmenite see 9612
Ilomedin [Schering] see 4940
Ilopan [Warren-Teed] see 2988
Iloprost, 4940
Ilosone [Lilly] see 3722
Ilotycin [Lilly] see 3720
Ilotycin Gluceptate [Lilly] see 3723
Ilozyme [Adria] see 7138
Ilvin [E. Merck] see 1467
Imadyl [Roche] see 1912
Imadyl (obsolete) see 4756
Imagent [Alliance] see 7299
Imagon [Astra] see 2215
Imagopaque [Nycomed] see 5073
Imagotan [Sandoz] see 9137
Imakol [Rhône-Poulenc] see 7080
Imap [McNeil] see 4241
Imavate [Wyeth-Ayerst] see 4955
Imaverol [Janssen] see 3622
Imazalil see 3622
Imazamethabenz, 4941
Imazamethabenz Methyl see 4941
Imazapyr, 4942
Imazaquin, 4943
Imazethabenz see 4941
Imazethapyr, 4944
Imbaral [MSD] see 9155
Imbretil [Burroughs Wellcome] see 4713
Imbrilon [Berk] see 4998
Imbun [Merckle] see 7106

[(Im-Bzl)-D-His⁶,Pro⁹-NEt]-gonado-tropin-releasing Hormone see 4760
Imdur [Astra] see 5245
Imecromone see 4903
Imeson [Desitin] see 6667
Imex [Merz] see 9337
Imexim [Cimex] see 9086
Imferon [Lakeside] see 2991
IMI-28 see 3660
IMI-30 see 4931
Imiclopazine, 4945
Imidacloprid, 4946
Imidamine see 718
Imidan [Stauffer] see 7492
Imidapril, 4947
Imidaprilat see 4947
Imidazole, 4948
2,4-(3H,5H)-Imidazoledione see 4796
1H-Imidazole-4-ethanamine see 4756
4-Imidazoleethylamine see 4756
5-Imidazoleethylamine see 4756
Imidazole Salicylate, 4949
Imidazoletrione see 7154
2,4-Imidazolidinedione see 4796
2-Imidazolidinethione see 3849
Imidazolidinetrione see 7154
2-Imidazolidinone, 4950
D-α-[(Imidazolidin-2-on-1-yl)carbonyl-amino]benzylpenicillin see 947
2-Imidazolidone see 4950
Imidazoline-2-thiol see 3849
1-(2-Δ²-Imidazolinyl)-2,2-diphenyl-cyclopropane see 2330
N-(2-Imidazolin-2-yl)-N-(4-indanyl)-amine see 4967
β-(4-Imidazolyl)acrylcholine see 6387
3-(Imidazol-1-yl)-7-chloro-benzo-1,2,-4-triazine-1-oxide see 9737
2-(4-Imidazolyl)ethylamine see 4756
(E)-4-(Imidazol-1-ylmethyl)cinnamic Acid see 7115
(E)-3-[4-(1H-Imidazol-1-ylmethyl)-phenyl]-2-propenoic Acid see 7115
2-[[3-(1H-Imidazol-4-yl)-1-oxo-2-pro-penyl]oxy]-N,N,N-trimethylethana-minium see 6387
7H-Imidazo[4,5-d]pyrimidine see 8129
Imido [Roche] see 4756
Imidocarb, 4951
Imidodicarbonic Diamide see 1346
Imidodicarbonimidic Diamide see 1261
Imidol [Yoshitomi] see 4955
Imidole see 8198
ImIFN see 5018
Imiglucerase see 242
Imigran [Glaxo] see 9172
Imilanyle [Takata] see 4955
Iminase [Beecham] see 712
Iminazole see 4948
1,1'-Iminobis[4-amino-9,10-anthra-cenedione] see 1283
1,1'-Iminobis-9,10-anthracenedione see 729
2,2'-Iminobis[benzoic Acid] see 3376
2,2'-Iminobisethanol see 3156
α,α'-[Iminobis(methylene)]bis[6-fluo-ro-3,4-dihydro-2H-1-benzopyran-2-methanol] see 6519
3,3'-Iminobis-1-propanol Dimethane-sulfonate (Ester) see 4962
Iminodiacetic Acid, 4952
1,1'-Iminodianthraquinone see 729
4,4'-Iminodicyclohexanecarboxylic Acid, 4953
Iminodiethanoic Acid see 4952
2,2'-Iminodiethanol see 3156
α,α'-(Iminodimethylene)bis[6-fluoro-2-chromanmethanol] see 6519

3-Imino-7-(methylamino)-3H-pheno-thiazine Hydrochloride see 959
N-[Imino(phosphonoamino)methyl]-N-methylglycine see 7496
β-(5-Imino-2-pyrrolidinecarbox-amido)propamidine see 6766
(6R,7R)-7-[2-(2-Imino-4-thiazolyl)gly-oxylamido]-8-oxo-5-thia-1-azabicy-clo[4.2.0]oct-2-ene-2-carboxylic Acid 7²-(Z)-(O-Methyloxime) see 2000
Iminourea see 4591
Imipem [Neopharmed] see 4954
Imipemide see 4954
Imipenem, 4954
Imipramine, 4955
Imipramine N-Oxide, 4956
Imiprex [Dumex] see 4956
Imiprin [CP Protea] see 4955
Imiquimod, 4957
Imitrex [Glaxo] see 9172
Imizad Equine Injection [Burroughs Wellcome] see 4951
Imizin see 4955
Imizocarb see 4951
Imizol see 4951
Immenoctal [Houdé] see 8563
Immobilon [Reckitt & Colman] see 3932
Immuglobin [Savage] see 4958
Immukin [Boehringer, Ing.] see 5018
Immune Globulins see 4958
Immune IFN see 5018
Immune Protein P5 see 909
Immuneron [Biogen] see 5018
Immunoglobulins, 4958
Immunosome see 5538
Immunox [Cilag] see 9544
Immunozima [Salus] see 5671
Imnudorm see 5890
Imodium [Janssen] see 5601
Imolamine, 4959
Imosec [Janssen] see 5601
Imossel [Janssen-Solucal] see 5601
Imotryl [Cassenne] see 1157
Imovance [Rhône-Poulenc] see 10324
Imovane [Rhône-Poulenc] see 10324
Impact [ICI] see 4246
Impedex [Du Pont] see 8816
Imperacin [ICI] see 7111
Imperan [Millet] see 6226
Imperatorin, 4960
Imperialine, 4961
Importal [Zyma] see 5352
Impral [Tanabe Seiyaku] see 2320
Impromen [Janssen] see 1466
Improntal [Kabi] see 7661
Improsulfan, 4962
Impruvol see 1583
Impugan [Dumex] see 4331
Imtack [Astra] see 5245
Imunoviral [Newport] see 5006
Imuran [Burroughs Wellcome] see 935
Imurek [Burroughs Wellcome] see 935
Imurel [Burroughs Wellcome] see 935
Imuthiol see 3443
IN-60 see 5271
IN-511 see 7474
Inabrin [Upjohn] see 4925
Inacid [Merck & Co.] see 4998
Inacilin [Inibsa] see 7669
Inactin [Byk-Gulden] see 9457
Inactive Limonene see 5518
Inadine [J & J] see 7880
Inalone O [Lampugnani] see 1047
Inalone R [Lampugnani] see 1047
Inamycin [Hoechst] see 6818
Inaperisone, 4963
Inapetyl [Upjohn] see 1151
Inapsine [McNeil] see 3505

Inbestan [Maruko] see 2405
Incafolic see 4253
Incazan [USSR] see 6238
Incidal see 5809
Incital [Fabre Santé] see 5843
Incoran [ITA] see 7919
Incorporation Factor see 4493
Incortin see 2602
^{111}In-CYT-103 see 8530
Indacin [Merck & Co.] see 4998
Indaconitine, 4964
Indaflex [Lampugnani] see 4969
Indalapril see 2928
Indaliton [Geigy] see 2468
Indalone see 1564
Indalpine, 4965
Indamol [Armour Pharm.] see 4969
Indan, 4966
Indanal [Takeda] see 2411
Indanazoline, 4967
Indanthrene [General Dyestuffs Corp], 4968
Indanthrene Blue R see 4968
Indanthrene Brilliant Blue FF see 4968
1,2,3-Indantrione Monohydrate see 6646
1-(5-Indanyl) N-(2-Carboxy-3,3-dimethyl-7-oxo-4-thia-1-azabicyclo-[3.2.0]hept-6-yl)-2-phenylmalonamate see 1888
O-(5-Indanyl) m,N-Dimethylthiocarbanilate see 9652
α-(5-Indanyloxycarbonyl)benzylpenicillin see 1888
6-[2-(5-Indanyloxycarbonyl)phenylacetamido]-3,3-dimethyl-7-oxo-4-thia-1-azabicyclo[3.2.0]heptane-2-carboxylic Acid see 1888
Indapamide, 4969
Indar [Rohm & Haas] see 4002
1H-Indazole, 4970
3-Indazolealanine see 9928
Indecainide, 4971
Indeloxazine Hydrochloride, 4972
Indema see 7381
Indene, 4973
Indenolol, 4974
(±)-1-[Inden-4(or 7)-yloxy]-3-(isopropylamino)-2-propanol see 4974
1-[1H-Inden-4(or 7)-yloxy]-3-[(1-methylethyl)amino]-2-propanol see 4974
2-[(1H-Inden-7-yloxy)methyl]morpholine Hydrochloride see 4972
(±)-2-[(7-Indenyloxy)methyl]morpholine Hydrochloride see 4972
Inderal [Wyeth-Ayerst] see 8025
Inderite see 5693
Inderm [Luitpold] see 3720
Indian Aconite see 122
Indian Apple see 7705
Indian Arrow Wood see 3945
Indian Balm see 9825
Indian Balsam see 976
Indian Barberry see 1194
Indian Berry see 2518
Indian Cannabis see 1794
Indian Dogbane see 784
Indian Ginger see 862
Indian Grass Oil see 6908
Indian Gum see 4424
Indian Hemp see 1794
Indian Laburnum see 1942
Indian Melissa Oil see 6916
Indian Oil of Verbena see 6916
Indian Physic see 784
Indian Pink see 8900
Indian Poke see 10090
Indian Tobacco see 5579

Indian Tragacanth see 5296
Indian Turmeric see 4808
"Indian Yellow" see 2493
India Rubber see 8435
Indican (Metabolic Indican), 4975
Indican (Plant Indican), 4976
Indic Fluoride see 4990
Indigo, 4977
Indigo Blue see 4977
Indigo Carmine, 4978
Indigo Copper see 2724
Indigotin see 4977
Indigotine see 4978
Indigo Weed see 985
Indinavir, 4979
Indium, 4980
Indium Antimonide, 4981
Indium Arsenide, 4982
Indium Chloride see 4989
Indium Fluoride see 4990
Indium Gallium Aluminium Phosphide, 4983
Indium ^{111}In Satumomab Pendetide see 8530
Indium Oxide, 4984
Indium Phosphide, 4985
Indium Selenide, 4986
Indium Sesquioxide see 4984
Indium Sulfate, 4987
Indium Telluride, 4988
Indium Trichloride, 4989
Indium Trifluoride, 4990
Indobloc [Asta] see 8025
Indobufen, 4991
Indocid [Merck & Co.] see 4998
Indocin [Merck & Co.] see 4998
Indocyanine Green, 4992
Indocybin [Sandoz] see 8111
Indoklon [Ohio Mcd.] see 4236
Indolacin [Sumitomo] see 2353
Indole, 4993
Indoleacetic Acid, 4994
1H-Indole-3-acetic Acid see 4994
1H-Indole-3-butanoic Acid see 4995
Indolebutyric Acid, 4995
Indole-3-butyric Acid see 4995
endo-1H-Indole-3-carboxylic Acid 8-Methyl-8-azabicyclo[3.2.1]oct-3-yl Ester see 9914
1H-Indole-3-carboxylic Acid (2α,6α,-8α,9aβ)-Octahydro-3-oxo-2,6-methano-2H-quinolizin-8-yl Ester see 3471
Indole-2,3-dione see 5116
1H-Indole-3-ethanamine see 9927
3-Indoleethanol see 9930
2,3-Indolinedione see 5116
Indolmycin, 4996
1H-Indol-3-ol Hydrogen Sulfate Ester see 4975
3-Indolylacetone, 4997
l-β-3-Indolylalanine see 9929
4-(3-Indolyl)butyric Acid see 4995
2-Indolyl(3)-ethanol see 9930
2-(3-Indolyl)ethyl Alcohol see 9930
β-Indolylethyl Alcohol see 9930
2-(3-Indolyl)ethylamine see 9927
5-(1-Indol-3-ylethyl)-2-(methylamino)-2-oxazolin-4-one see 4996
[S-(R*,S*)]-5-[1-(1H-Indol-3-yl)ethyl]-2-(methylamino)-4(5H)-oxazolone see 4996
4-[2-(3-Indolyl)ethyl]piperidine see 4965
N-[1-[2-(1H-Indol-3-yl)ethyl]-4-piperidinyl]benzamide see 5000
1H-Indol-3-yl-β-D-glucopyranoside see 4976
2-C-(1H-Indol-3-ylmethyl)-β-L-lyxo-3-hexulofuranosonic Acid γ-Lactone

Mixt with 2-C-(1H-Indol-3-ylmethyl)-β-L-xylo-3-hexulofuranosonic Acid γ-Lactone see 868
1-(1H-Indol-4-yloxy)-3-[(1-methylethyl)amino]-2-propanol see 7597
Indol-3-yl Potassium Sulfate see 4975
Indol-3-yl-2-propanone see 4997
Indol-3-yl Sulfate see 4975
Indomed see 4998
Indomee [Merck & Co.] see 4998
Indomethacin, 4998
Indomethacin Glucosamide see 4456
Indomethine [Teika Seiyaku] see 4998
Indomod [Benzon] see 4998
Indon [Parke, Davis] see 7381
Indonaphthene see 4973
Indo-Phlogont [Azupharma] see 4998
Indoprofen, 4999
Indoptic [Merck & Co.] see 4998
Indoptol [Merck & Co.] see 4998
Indoramin, 5000
Indorektal [Sanorania] see 4998
Indorm [Wyeth] see 8007
Indospicine, 5001
Indo-Tablinen [Sanorania] see 4998
Indoxamic Acid see 7048
Indoxen [Sigma-Tau] see 4998
Indoxyl-β-D-glucoside see 4976
3-Indoxylsulfuric Acid see 4975
[^{111}In-DTPA-D-Phe^{1}]-octreotide see 7267
Indunox [UCB] see 3921
Indusil [Diamant; Recordati] see 2513
INF see 4330
INF-1837 see 4167
INF-3355 see 5842
INF-4668 see 5819
Infacol [Pharmax] see 3264
Infasurf, 5002
Inferno see 509
Infertine see 8332
Infiltrase [Armour Pharm.] see 4794
Inflamase [Iolab] see 7903
Inflamen [Hokuriku] see 1409
Inflanefran [Allergan] see 7901
Inflatine see 5580
Inflazon [Taisho] see 4998
Informational RNA see 8370
Infrocin [Frosst] see 4998
Infron [Whittier] see 10156
Infusorial Earth, 5003
InGaAlP see 4983
INH see 5203
INH-G see 4510
Inhibace [Roche] see 2332
Inhibins, 5004
Inhibostamin [Zyma] see 9894
Inhiston see 7383
Inicardio see 6635
Inimur [ICN] see 6623
Iniprol [Sanofi Winthrop] see 796
Initiss [Erba] see 2332
Inmetal see 607
Innohep [Leo Pharm] see 9592
Innovace [Merck & Co.] see 3605
Innovar see 3505
Innoxalon [Sanko] see 6446
Innozide [Merck & Co.] see 3605
Inocor [Sterling] see 634
Inofal [Sandoz] see 9137
Inokiton [Nippon Yakuhin] see 9974
Inolin [Tanabe Seiyaku] see 9719
Inopamil [Simes] see 4921
Inophylline [Millot] see 485
Inorganic Benzene see 1363
Inosie [Morishita] see 5005
Inosine, 5005
Inosine:dimethylaminoisopropanol Acetamidobenzoate (1:3) see 5006
Inosine Mono[4-(acetylamino)benzo-

ate] (Salt) Compd with 1-(Dimethylamino)-2-propanol (1:3) see 5006
Inosine Pranobex, 5006
Inosinic Acid, 5007
5-Inosinic Acid see 5007
5'-Inosinic Acid see 5007
Inosiplex see 5006
Inosite see 5008
Inositol, 5008
i-Inositol see 5008
meso-Inositol see 5008
myo-Inositol see 5008
myo-Inositol Hexakis(dihydrogen
Phosphate) see 7542
Inositol Hexanicotinate see 5009
meso-Inositol Hexanicotinate see 5009
Inositolhexaphosphoric Acid see 7542
myo-Inositol Hexa-3-pyridinecarbox-
ylate see 5009
Inositol Niacinate, 5009
Inostral [Syntex] see 2658
Inotrex (obsolete) [Lilly] see 3456
Inotropin [Bago] see 3479
Inovan [Kyowa] see 3479
Inoven [Janssen] see 4925
Inoxyl [Debat] see 7079
INPC see 8001
INPEA see 6618
Inpea [Selvi] see 6618
111In Pentetreotide see 7267
Inplacen [Merz] see 7672
Insane Root see 4908
Insariotoxin see 9933
Insecticide 1179 see 6062
Insecticide 3960-X14 [B.F. Goodrich]
see 8993
Insecticide ACC 4124 see 3091
Insecticide No. 497 see 3152
Insecticide No. 4049 see 5740
Insegar [Maag] see 4027
Insibrin [Byk-Liprandi] see 3692
Insidon [Geigy] see 6985
Insoluble Anhydrite see 1753
Insoluble Cyanuric Acid see 2752
Insoluble Salumin see 375
Insomin [Orion] see 6667
Insoral [Warner] see 7376
Instalac [Virbac] see 9840
Insubeta see 5026
Insulamin [Iwaki] see 1500
Insularine, 5010
Insulin, 5011
Insulin 131I see 5011
Insulinase, 5012
Insulin (emp) see 5011
Insulin-like Growth Factor I see 8862
Insulin-like Growth Factor II see 8862
Insulin (prb) see 5011
Insulton see 5898
Insumin [Kyorin] see 4233
Insuven [Berenguer] see 3350
Intal [Fisons] see 2658
Inteban SP [Sumitomo] see 4998
Integerrimine see 8595
Integrelin, 5013
Integrin (formerly) [Sterling] see 7105
Integrins, 5014
Intenkordin [Polfa] see 2296
Intensain [Abbott; Cassella-Riedel]
see 2296
Intensopan [Sandoz] see 1474
Intercept [Ortho] see 6772
Interceptor [Ciba-Geigy] see 6277
Interferon, 5015
Interferon-α, 5016
Interferon-β, 5017
Interferon-γ, 5018
Interferon Alfa-2a see 5016
Interferon Alfa-2b see 5016
Interferon Alfa-2c see 5016

Interferon Alfa-n1 see 5016
Interferon Alfa-n3 see 5016
Interferon Beta-1b see 5017
Interferon Gamma-1b see 5018
Intergravin-orales [Werfft-Chemie] see
1699
Interleukin-1, 5019
Interleukin-2, 5020
Interleukin-3, 5021
Interleukin-1 Inhibitor Protein see
5022
Interleukin-1 Receptor Antagonist,
5022
Intermedin(e) see 6377
Intermigran [Sanol] see 8025
Internal Antiseptic No. 307 see 73
Internally Compensated Tartaric Acid
see 9238
Interstitial Cell Stimulating Hormone
see 5499
Intertocine-S [Veterinaria] see 7114
Intestiazol see 7533
Intestibar [Spofa] see 1023
Intestin-Euvernil [Heyden] see 9080
Intocostrin [Squibb] see 9939
Intrabilix see 5043
Intrabutazone [Organon] see 7431
Intradex [Glaxo] see 2989
Intradine [Norbrook] see 9083
Intraglobin [Biotest] see 4958
Intraheptol see 5575
Intramycetin [Parke, Davis] see 2120
Intrapan [USV] see 2988
Intrasite [Fisch] see 2654
Intrasporin [Torlan] see 2025
Intrastigmina [Lusofarmaco] see 6553
Intrastigmina [Lusofarmaco] see 6553
Intrazone [Arnolds] see 7431
Intrinsic Factor see 1945
Introcar [Coup] see 6617
Introl [Fisons] see 2658
Intromene [Teikoku] see 9754
Introna [Schering] see 5016
Intron A [Schering] see 5016
Intybin see 5359
Inula, 5023
Inula Camphor see 208
Inulin, 5024
Invasin see 4794
Invenol [Hoechst] see 1881
Inversine [Merck & Co.] see 5814
Invertase, 5025
Invertin see 5025
Invert Sugar, 5026
Invesol see 5026
Invirase [Hoffman-LaRoche] see 8516
Inyoite see 1691
Iobac [West Chemical] see 5058
Iobenguane, 5027
Iobenzamic Acid, 5028
Iocarmic Acid, 5029
Iocetamic Acid, 5030
Iodafilina [B.O.I.] see 3754
Iodamide, 5031
Iodamide Meglumine see 5031
Iodaphyline [Millot] see 3754
Iodeikon [Mallinckrodt] see 5059
Iodic Acid, 5032
Iodic Anhydride see 5040
Iodinated Glycerol, 5033
Iodinated (131I) Human Serum
Albumin see 8613
Iodinated p-Toluidine Polyvinylpyr-
rolidone see 9661
Iodine, 5034
Iodine Colloidal, 5035
Iodine Cyanide see 2765
Iodine Heptafluoride, 5036
Iodine Monobromide, 5037

Iodine Monochloride, 5038
Iodine Pentafluoride, 5039
Iodine Pentoxide, 5040
Iodine-polyvinylpyrrolidone Complex
see 7880
Iodine Trichloride, 5041
Iodinin, 5042
Iodipamide, 5043
Iodipamide Meglumine see 5043
Iodisan see 7965
Iodixanol, 5044
Iodized Oil, 5045
Iodoacetic Acid, 5046
Iodoalphionic Acid, 5047
p-Iodoaniline, 5048
o-Iodoanisole, 5049
Iodobenzene, 5050
4-Iodobenzenesulfonyl Chloride see
7640
o-Iodobenzoic Acid, 5051
N-(2-Iodobenzoyl)glycine Mono-
sodium Salt see 5055
m-Iodobenzylguanidine see 5027
Iodobil see 5047
1-Iodobutane see 1607
2-Iodobutane see 1608
Iodochlorhydroxyquin, 5052
Iodochlorohydroxyquinoline see 5052
Iodochloroxyquinoline see 5052
5-Iodo-2'-deoxyuridine see 4934
3-Iodo-1,2-dihydroxypropane see 4497
Iododocosanoic Acid Calcium Salt see
1721
Iodoenterol see 5052
Iodoethane see 3859
2-(1-Iodoethyl)-1,3-dioxolane-4-meth-
anol see 5033
2-(2-Iodoethyl)-1,3-dioxolane-4-meth-
anol see 5033
Iodofenphos, 5053
5-Iodo-2'-fluoroarauracil see 4114
Iodoform, 5054
Iodoformine see 6043
Iodognost see 5059
Iodogorgoic Acid see 3236
o-Iodohippurate Sodium, 5055
4-(3-Iodo-4-hydroxyphenoxy)-3,5-di-
iodophenylalanine see 5535
β-[4-(3'-Iodo-4'-hydroxyphenoxy)-3,5-
diiodophenyl]propionic Acid see
9553
m-Iodo-o-hydroxyquinolineanasulfonic Acid see 4872
7-Iodo-8-hydroxyquinoline-5-sulfonic
Acid see 4872
1-(ω-p-Iodo-131I-benzylpolyethylene)-
2-pyrrolidinone see 9661
(±)-p-Iodo-123I-N-isopropyl-α-methyl-
phenethylamine see 5066
(±)-4-(Iodo-123I)-α-methyl-N-(1-meth-
ylethyl)benzeneethanamine see 5066
4-Iodo-iota-methylbenzenedecanoic
Acid Ethyl Ester see 5074
Iodol see 5062
Iodomethane see 6161
Iodomethanesulfonic Acid Sodium
Salt see 6051
1-Iodo-3-methylbutane see 5134
2-(Iodomethyl)-1,3-dioxolane-4-meth-
anol see 3473
1-Iodo-2-methylpropane see 5158
2-Iodooctadecanoic Acid Calcium Salt
see 1722
2-Iodooctane see 6863
Iodopanoic Acid see 5072
Iodopaque [Labaz] see 79
Iodophene see 5059
Iodophene Sodium see 5059
o-Iodophenol, 5056
p-Iodophenol, 5057

[(3-Iodophenyl)methyl]guanidine *see* 5027

p-Iodophenyl Sulfonyl Chloride *see* 7640

Iodophors, 5058

Iodophthalein *see* 5059

Iodophthalein Sodium, 5059

1-Iodopropane *see* 8047

2-Iodopropane *see* 5233

3-Iodo-1,2-propanediol *see* 4497

3-Iodo-1-propene *see* 302

3-Iodopropylene *see* 302

γ-Iodopropyleneglycol *see* 4497

2,3-(2-Iodopropylidenedioxy)propanol *see* 5033

2,3-(3-Iodopropylidenedioxy)propanol *see* 5033

Iodopropylidene Glycerol *see* 5033

3-Iodo-2-propynyl 2,4,5-Trichloro-phenyl Ether *see* 4631

Iodopsin, 5060

Iodopyracet, 5061

Iodopyracet Compound Solution *see* 5061

Iodopyracet Concentrated Solution *see* 5061

Iodopyracet Injection *see* 5061

Iodopyrrole, 5062

Iodoquinine Sulfate *see* 8246

Iodoquinol, 5063

Iodorayoral *see* 5059

Iodosorb [Perstorp; Stuart] *see* 1646

N-Iodosuccinimide, 5064

Iodosulfane *see* 9148

Iodothyrin *see* 9552

Iodotope [Squibb] *see* 8778

3-Iodotyrosine, 5065

Iodoxyl *see* 8779

Iodtetragnost *see* 5059

Ioduril *see* 8777

Iofetamine ¹²³I, 5066

Ioglycamic Acid, 5067

Iohexol, 5068

Iomapidol *see* 5071

Iomeglamic Acid, 5069

Ionamin [Lipha] *see* 7415

Ionol CP [Shell] *see* 1583

Ionone, 5070

Ionophore X-4537A *see* 5384

Iopamidol, 5071

Iopamiro [Bracco] *see* 5071

Iopamiron [Bracco] *see* 5071

Iopanoic Acid, 5072

Iopentol, 5073

Iophendylate, 5074

Iophenoxic Acid, 5075

Iopidine [Alcon] *see* 791

Ioprep [J & J] *see* 5058

Iopromide, 5076

Iopronic Acid, 5077

Iopropane *see* 5081

Iopydol, 5078

Iopydone, 5079

Iopyracil *see* 5061

Ioquin [Abbott] *see* 5063

Iosalide [Schering] *see* 5280

Iosan [West Chemical] *see* 5058

Iothalamic Acid, 5080

Iothion, 5081

Iotrol *see* 5082

Iotrolan, 5082

Ioversol, 5083

Ioxaglate Meglumine *see* 5084

Ioxaglate Sodium *see* 5084

Ioxaglic Acid, 5084

Ioxilan, 5085

Ioxitol *see* 5085

Ipacef [IPA] *see* 2002

Ipado *see* 2515

Ipamicina [IPA] *see* 4280

Ipamix [Gentili] *see* 4969

IPC *see* 8001

IPD-1151T *see* 9178

Ipecac, 5086

Ipecacuanha *see* 5086

Ipersed [Salus] *see* 6667

Ipertensina [Ciba] *see* 689

Ipertrofan [SPA] *see* 5891

Iphosphamid(e) *see* 4937

IPNO *see* 4956

Ipnofil *see* 6028

Ipodate, 5087

Ipoglicone [Farmitalia] *see* 9646

Ipoglon [Diana] *see* 7362

Ipolab [Von Boch] *see* 5341

Ipomea, 5088

Iporal *see* 4589

Ipotensium [Pierrel] *see* 2450

Ipotensivo *see* 5813

IPP *see* 4999

Ipradol [CL Pharma] *see* 4745

Ipratropium Bromide, 5089

Ipriflavone, 5090

Iprindole, 5091

Iproclozide, 5092

Iprodione, 5093

Ipronal [Biosedra] *see* 8091

Iproniazid, 5094

Ipronidazole, 5095

Ipropran [Roche] *see* 5095

Iprosten [Takeda] *see* 5090

Iproveratril *see* 10083

Ipsapirone, 5096

Ipsilon *see* 451

I.P.U. *see* 5237

IQB-875 *see* 3587

Iramil [Knoll] *see* 4955

IRAP *see* 5022

Irbesartan, 5097

Ircon [Lakeside] *see* 4094

Irenat [Troponwerke] *see* 8798

Irescein [Americal] *see* 4194

Iretin [Torii] *see* 2853

Irgafen *see* 9102

Irgalax [Geigy] *see* 4460

Irgalon *see* 3557

Irgamide [Dispersa] *see* 9072

Irgasan BS200 [Geigy] *see* 9333

Irgasan CF3 *see* 2438

Irgasan DP 300 [Ciba-Geigy] *see* 9790

Irgosan CF3 [Geigy] *see* 2438

Iridin *see* 5103

Iridina Due [Montefarmaco] *see* 6455

Iridine *see* 5103

Iridium, 5098

Iridium Hexafluoride, 5099

Iridium Sesquioxide, 5100

Iridium Trichloride, 5101

Iridocin [Bayer] *see* 3783

Iridomirmecina *see* 5102

Iridomyrmecin, 5102

Iridomyrmexin *see* 5102

Irifan *see* 2780

Irigenin, 5103

Irinotecan, 5104

Iris Blue B *see* 5345

Irish Broom *see* 8547

Irisolone, 5105

Irisone [Givaudan] *see* 5070

Irium *see* 8782

Iromon [Gador] *see* 4095

Iron, 5106

Iron Alum *see* 549

Ironate [Wyeth] *see* 4105

Iron Carbonyl *see* 5111

Iron Choline Citrate *see* 4085

Iron-dextran Complex *see* 2991

Irone, 5107

α-Irone, 5108

β-Irone, 5109

γ-Irone, 5110

Iron Hemiphosphide *see* 4102

Iron Hydrocarbonyl *see* 4845

Ironorm [Wallace] *see* 2991

Iron Pentacarbonyl, 5111

Iron Peptonized *see* 7292

Iron Sorbitex, 5112

Iron Sorbitol *see* 5112

Iron Sorbitol Citrate *see* 5112

Iron Sugar *see* 4073

Iron Sulfate-glycine Complex *see* 4086

Iron Sulfuret *see* 4106

Iron Tetracarbonyl *see* 5111

Iron Tetracarbonyl Dihydride *see* 4845

Iron Vitriol *see* 4105

Iron Yttrium Oxide *see* 10231

Irox *see* 4095

Irrigor [Aron] *see* 4959

Irritren [Byk-Gulden] *see* 5596

Irrodan [Biomed. Foscama] *see* 1498

Irrorin *see* 7919

Irsogladine, 5113

Irtan [Fisopharma] *see* 2658

Irtonin [Takeda] *see* 9720

Iruxol [Knoll] *see* 2544

IS-362 *see* 9184

IS-370 *see* 9043

IS-401 *see* 3241

IS-499 *see* 7714

Isacen *see* 7108

Isactid *see* 136

Isalon [Grelan] *see* 224

Isanic Acid, 5114

Isano Oil *see* 1358

Isaphen *see* 7108

Isapirone *see* 5096

Isarol V [Ciba-Geigy] *see* 9105

Isatide, 5115

Isatidine *see* 8335

Isatin, 5116

γ-Isatropaic Acid *see* 9922

ε-Isatropaic Acid *see* 9922

Isatropic Acid, 5117

trans-Isatropic Acid *see* 5117

α-Isatropic Acid *see* 5117

Isatropyldιtropeine *see* 1053

Isatyde *see* 5115

Isaxonine, 5118

Isazofos, 5119

Isazophos *see* 5119

Isbogrel, 5120

Iscotin [Daiichi] *see* 5203

Isdin [Medice] *see* 5245

ISDN [Stadapharm] *see* 5245

Isepacin [Schering] *see* 5121

Isepamicin, 5121

Isethion [MECT] *see* 4483

Isethionic Acid, 5122

ISF-2001 *see* 9889

ISF-2123 *see* 7577

ISF-2469 *see* 1669

ISF-2522 *see* 7076

ISF-09334 *see* 8448

Isicom [Isis] *see* 1843

Isindone [Erba] *see* 4999

Isinglass, 5123

Isipen [Krka] *see* 7616

Ismelin [Ciba] *see* 4589

Ismo [Wyeth-Ayerst] *see* 5245

Ismotic [Alcon] *see* 5244

Isoacetophorone *see* 5213

Isoadanone *see* 5201

Isoadrenaline *see* 6785

Isoalloxazine-adenine Dinucleotide *see* 4131

Isoamidone *see* 5201

Isoamidone I *see* 6804

Isoaminile, 5124

Isoamyl Acetate, 5125

Isoamyl Alcohol see 5212
dl-sec-Isoamyl Alcohol see 6110
Isoamylamine, 5126
2-Isoamylamino-6-methylheptane see 6846
Isoamyl Benzoate, 5127
Isoamyl Bromide, 5128
Isoamyl Butyrate, 5129
Isoamylcarbamic Acid Thymyl Ester see 9550
Isoamyl Chloride, 5130
Isoamyl Cyanide, 5131
Isoamyl N-(β-Diethylaminoethyl)-α-aminophenylacetate see 1784
Isoamyl α-[N-(β-Diethylaminoethyl)-amino]phenylacetate see 1784
N-Isoamyl-1,5-dimethylhexylamine see 6846
β-Isoamylene see 649
Isoamyleneguanidine see 4358
8-Isoamylenoxypsoralen see 4960
4-(β-Isoamylenyl)-1,2-diphenyl-3,5-pyrazolidinedione see 4053
Isoamyl Ether, 5132
5-Isoamyl-5-ethylbarbituric Acid see 607
Isoamyl Formate, 5133
Isoamylhydrocupreine see 3950
Isoamyl Iodide, 5134
Isoamyl Isovalerate, 5135
Isoamyl Mandelate see 5758
Isoamyl Nitrate, 5136
Isoamyl Nitrite, 5137
Isoamyl Oxide see 5132
4-[3-(Isoamyloxy)-2-(3,4,5-trimethoxybenzoyloxy)propyl]tetrahydro-1,4-oxazine see 611
Isoamyl Phenylaminoacetate Hydrochloride see 7350
Isoamyl Phthalate, 5138
Isoamyl Salicylate, 5139
Isoamyl Valerianate see 5135
Isoamyne see 623
Isoandrosterone see 3646
Isoapo-β-erythroidine, 5140
Isoarteril [Isola-Ibi] see 8291
Isoascorbic Acid, 5141
Isobamate see 1889
Isobarin see 4589
Isobatrachotoxin see 1037
Isobebeerine see 5173
Iso-Benzacyl [Wander] see 1148
Isobenzamic Acid see 5028
Isobenzan, 5142
1,3-Isobenzofurandione see 7528
Isobicina [Maggioni] see 5203
Iso-Bid [Geriatric Pharm.] see 5245
Isobide [Nikken] see 5244
Isobixin see 1347
Isoborneol, 5143
6-Isobornyl-3,4-dimethylphenol see 10210
Isobornyl Thiocyanoacetate, 5144
6-Isobornyl-3,4-xylenol see 10210
6-(2-Isobornyl)-3,4-xylen-1-ol see 10210
Isobromyl [Clin-Comar-Byla] see 1418
Isobutene see 5155
p-Isobutenylphenol Acetate see 6190
Isobutol see 3764
p-Isobutoxybenzoic Acid 3-(Diethylamino)-1,2-dimethylpropyl Ester see 4375
p-Isobutoxybenzoic Acid α,β-Dimethyl-γ-diethylaminopropyl Ester see 4375
1-Isobutoxy-2-pyrrolidino-3-N-benzylanilinopropane see 1188
3-Isobutoxy-2-pyrrolidino-N-phenyl-N-benzylpropylamine see 1188

Isobutyl Acetate, 5145
Isobutyl Alcohol, 5146
Isobutylaldehyde see 5169
5-Isobutyl-5-allylbarbituric Acid see 1536
Isobutylamine, 5147
Isobutyl p-Aminobenzoate, 5148
2-(Isobutylamino)ethanol p-Aminobenzoate (Ester) see 1552
2-(Isobutylamino)ethyl p-Aminobenzoate see 1552
Isobutylbenzene, 5149
Isobutyl Bromide, 5150
3-Isobutyl-6-sec-butyl-2-hydroxypyrazine 1-Oxide see 876
Isobutyl n-Butyrate, 5151
Isobutyl Carbamate, 5152
Isobutyl Carbinol see 5212
Isobutylcarbylamine see 5126
Isobutyl Chloride, 5153
Isobutyl Chlorocarbonate, 5154
Isobutyl Chloroformate see 5154
Isobutyl Cyanoacrylate see 1488
E,E-N-Isobutyl-2,4-decadienamide see 7201
(E,E,Z)-N-Isobutyl-2,6,8-decatrienamide see 177
N-Isobutyldeca-trans-2-cis-6-trans-8-trienamide see 177
Isobutyl 1,4-Dihydro-5-methoxycarbonyl-2,6-dimethyl-4-(2-nitrophenyl)-3-pyridinecarboxylate see 6658
Isobutylene, 5155
Isobutyl Ether, 5156
Isobutyl Formate, 5157
p-Isobutylhydratropic Acid see 4925
Isobutylhydrochlorothiazide see 1554
Isobutyl Iodide, 5158
Isobutyl Isobutyrate, 5159
Isobutyl Isovalerate, 5160
Isobutyl Kelo-form see 5148
Isobutyl Mercaptan, 5161
Isobutyl Methyl 1,4-Dihydro-2,6-dimethyl-4-(o-nitrophenyl)-3,5-pyridinedicarboxylate see 6658
Isobutyl Nitrate, 5162
Isobutyl Nitrite, 5163
(p-Isobutylphenyl)acetic Acid see 4924
2-(4-Isobutylphenyl)butyric Acid see 1557
(±)-2-(4-Isobutylphenyl)propionic Acid see 4925
dl-2-(4-Isobutylphenyl)propionohydroxamic Acid see 4926
Isobutyl Propionate, 5164
4-N-Isobutylspiropiperidylrifamycin S see 8380
Isobutyl Stearate, 5165
Isobutyl Sulfide, 5166
Isobutyl Sulfocyanate see 5167
Isobutyl Thiocyanate, 5167
p-Isobutyl-α-toluic Acid see 4924
Isobutyltrimethylmethane see 5210
Isobutyl Urethane, 5168
Isobutyl Valerate see 5160
Isobutyraldehyde, 5169
Isobutyric Acid, 5170
Isobutyric Aldehyde see 5169
Isobutyronitrile, 5171
3-Isobutyryl-2-isopropylpyrazolo-[1,5-a]pyridine see 4923
O-Isobutyrylthiamine Disulfide see 9432
Isocaine [Columbus] see 5148
Isocalm [Kaken] see 9660
Isocapronitrile see 5131
Isocaramidine see 2901
Isocarboxazid, 5172
Isocard [Schwarz] see 5245
Isocaryophyllene see 1926

Isocef [Schering-Plough] see 1998
Isochavicine see 7625
Isochavicinic Acid see 7620
Isochinol [Pharmakos] see 3267
Isochlorothion see 3091
Isochondrodendrine, 5173
Isochrysene see 9870
Isocid [CID] see 5203
Isocil, 5174
Isocillin [Hoechst] see 7230
Isocinchomeronic Acid, 5175
Isococaine see 8097
Isocolin [Isola-Ibi] see 7519
Isocolumbin see 2558
Isoconazole, 5176
Iso-Cornox [FBC] see 5826
Isocorybulbine, 5177
Isocorydine, 5178
Isocorypalmine, 5179
Isocothane see 3340
Isocrin [Warner-Chilcott] see 7108
Isocrotonic Acid, 5180
Isocrotyl Chloride see 2199
Isocyanatobenzene see 7449
4-Isocyanato-1,1'-biphenyl see 10207
1-Isocyanatobutane see 1609
Isocyanatomethane see 6163
1-Isocyanatonaphthalene see 6497
Isocyanic Acid, 5181
Isocyanic Acid 4-Biphenylyl Ester see 10207
Isocyanic Acid Butyl Ester see 1609
Isocyanic Acid Methyl Ester see 6163
Isocyanuric Acid see 2767
Isodiamylamine see 3237
Isodianisylethanolamine see 5194
Isodin [Tosi] see 7059
Isodine [Blair] see 7880
Isodiprene see 1885
Isodormid see 795
Isodrin see 227
Isodulcit see 8338
Isodurene, 5182
Isoendoxan see 4937
d-Isoephedrine see 3645
Isoestradiol, 5183
8-Isoestradiol-17β see 5183
8-Isoestrone, 5184
Isoetam [Ferrer] see 3764
Isoetarine see 5185
Isoetharine, 5185
Isoeugenol, 5186
Isofenphos, 5187
Isofezolac, 5188
4',7-Isoflavandiol see 3677
Isoflavone, 5189
Isofluorphate see 5192
Isoflupredone, 5190
Isoflurane, 5191
Isoflurophate, 5192
Isoforon see 5213
Isogel [Charwell] see 7674
Isoglaucon [Boehringer, Ing.] see 2450
L-Isoglutamine, 5193
Isoindazole see 5181
1H-Isoindole-1,3(2H)-dione see 7529
Isoinokosterone see 3539
Iso-Iodeikon see 7416
Iso-K [San Carlo] see 5316
Isoket [Sanol] see 5245
Isoladol, 5194
Isolait [Elder] see 5259
Isolan [Ciba-Geigy], 5195
Isoleucine, 5196
L-Isoleucine see 5196
Isolevin see 5236
Isolobelanine see 6802
α-Isolupanine see 5635
Isolyn [Abbott] see 5203
Isolysergic Acid, 5197

IsoMack [Mack, Illert.] see 5245
Isomaltol, 5198
Isomenyl [Kaken] see 5236
Isomeride [Ardix] see 4015
Isomerine see 2232
Isometamidium Chloride, 5199
Isomethadol, 5200
Isomethadone, 5201
Isomethazine see 5219
Isomeptene, 5202
Isomist see 5236
Isomonat [Boehringer, Mann.] see 5245
Isomycin [Werfft] see 8078
Isomycomycin see 6407
Isomyn see 623
Isomytal see 607
Isonal [Roussel-UCLAF] see 5899
Isonaphthoic Acid see 6469
Isonaphthol see 6471
Isonex [Dumex] see 5203
Isoniazid, 5203
Isonicotinic Acid, 5204
Isonicotinic Acid 2-[2-(Benzylcarbamoyl)ethyl]hydrazide see 6575
Isonicotinic Acid Diethylamide, 5205
Isonicotinic Acid Hydrazide see 5203
Isonicotinic Acid Hydrazide Hydrazone with Glucuronic Acid Lactone see 4510
Isonicotinic Acid Hydrazide, Hydrazone with Streptomycin, Sulfate (Salt) (2:3) see 8985
Isonicotinic Acid 2-Isopropylhydrazide see 5094
Isonicotinic Acid α-Methylfurfurylidenehydrazide see 4330
Isonicotinic Acid Salicylidenehydrazide see 8487
Isonicotinic Acid m-Sulfobenzylidene Hydrazide see 9132
Isonicotinic Acid Veratrylidenehydrazide see 10091
N-Isonicotinoyl-N'-[β-(N-benzylcarboxamido)ethyl]hydrazide see 6575
Isonicotinoylhydrazine see 5203
Isonicotinoylhydrazone of D-Glucuronic Acid Lactone see 4510
1-Isonicotinoyl-2-isopropylhydrazine see 5094
1-Isonicotinoyl-2-salicylidenehydrazine see 8487
1-Isonicotinoyl-2-veratrylidenehydrazine see 10091
Isonicotinylhydrazine see 5203
Isonicotinyl Hydrazonotoluene-m-sulfonic Acid see 9132
1-Isonicotinyl-2-isopropylhydrazine see 5094
Isonipecaine see 5894
Isonipecotic Acid, 5206
Isonitrosoacetone, 5207
Isonitrosoacetophenone, 5208
5-Isonitrosobarbituric Acid see 10138
β-Isonitrosopropane see 75
Isonixin, 5209
Isonizida [Bial] see 5203
Isonootkatone see 10106
Isonorin [SM & P] see 5236
Isonovobiocin see 6818
Isooctane, 5210
Isooctyl Alcohol, 5211
p-Isooctylpolyoxyethylenephenol Formaldehyde Polymer see 9964
Isoolean-14-en-3β-ol see 9231
Iso-oosporein see 6982
Isopaque [Winthrop] see 6241
Isopar 2582
Isopelletierine see 7200

4-(2-Isopentenyl)-1,2-diphenyl-3,5-pyrazolidinedione see 4053
Isopentyl Alcohol, 5212
sec-Isopentyl Alcohol see 6110
Isopentylamine see 5126
Isopentyl 5,6-Dihydro-7,8-dimethyl-4,5-dioxo-4H-pyrano[3,2-c]quinoline-2-carboxylate see 8305
N-Isopentyl-1,5-dimethylhexylamine see 6846
Isopentylhydrocupreine see 3950
Isopentyl Nitrite see 5137
α-[(Isopentyloxy)methyl]-4-morpholineethanol 3,4,5-Trimethoxybenzoate see 611
1-[β-(Isopentyloxy)phenethyl]pyrrolidine see 514
Isopentyl 2-Phenylglycinate Hydrochloride see 7350
Isopestox [Fisons] see 6292
Isophane Insulin see 5011
Isophen [Knoll] see 6015
Isophenergan see 5219
Isophenethanol see 6618
Isophenphos see 5187
Isophorone, 5213
α-Isophorone see 5213
Isophosphamide see 4937
Isophrin [Riker] see 7440
Isophthalic Acid, 5214
Isophytol, 5215
Isopilocarpine see 7578
Isopilosine, 5216
Isopimaric Acid, 5217
Isopiperine see 7625
Isopiperinic Acid see 7620
IsoPPC see 8001
10α-Isopregnenone see 3524
Isoprenaline see 5236
Isoprene, 5218
Isoprinosin [Newport] see 5006
Isoprinosina [Newport] see 5006
Isoprinosine [Newport] see 5006
Isoprochin P [Merckle] see 8056
Isoprofenamine see 2470
Isopromedol see 7968
Isopromethazine, 5219
Isopropalin, 5220
Isopropamide Iodide, 5221
Isopropanol see 5227
Isopropene Cyanide see 6006
Isopropenyl Acetate, 5222
4-Isopropenyl-1-cyclohexene-1-carboxaldehyde see 7308
2-Isopropenyl-2,3-dihydro-5-acetylbenzofuran see 9714
2-Isopropenyl-6-methoxy-5-benzofuranyl Methyl Ketone see 3946
7-Isopropenyl-4-methyl-1-azulenecarboxaldehyde see 5346
cis-(+)-2-Isopropenyl-1-methylcyclobutaneethanol see 4556
4-(5-Isopropenyl-2-methyl-1-cyclopenten-1-yl)-2-butanone, 5223
Isopropenylnitrile see 6006
Isoprophenamine see 2470
N1-(4-Isopropoxybenzoyl)-p-aminobenzenesulfonamide see 9106
N1-(p-Isopropoxybenzoyl)sulfanilamide see 9106
5-Isopropoxycarbonylaminothiabendazole see 1774
5-Isopropoxycarbonylamino-2-(4-thiazolyl)benzimidazole see 1774
(E)-O-2-Isopropoxycarbonyl-1-methylvinyl O-Methyl Ethylphosphoramidothioate see 8000
1-(Isopropoxycarbonyloxy)ethyl (6R,-7R)-7-[2-(2-Amino-4-thiazolyl)-(Z)-

2-(methoxyimino)acetamido]-3-methoxymethyl-3-cephem-4-carboxylate see 1991
(±)-1-[p-(2-Isopropoxyethoxymethyl)-phenoxy]-3-(isopropylamino)-2-propanol see 1336
(±)-1-[[α-(2-Isopropoxyethoxy)-p-tolyl]oxy]-3-(isopropylamino)-2-propanol see 1336
7-Isopropoxyisoflavone see 5090
Isopropoxymethylphosphoryl Fluoride see 8520
7-Isopropoxy-3-phenyl-4H-1-benzopyran-4-one see 5090
7-Isopropoxy-3-phenylchromone see 5090
o-Isopropoxyphenyl N-Methylcarbamate see 8022
2-Isopropoxypropane see 5231
p-Isopropoxy-N-sulfanilylbenzamide see 9106
Isopropydrin see 5236
Isopropyl Acetate, 5224
Isopropylacetic Acid see 5251
Isopropyl Acetoacetate, 5225
Isopropylacetone, 5226
Isopropyl Alcohol, 5227
Isopropylamine, 5228
4-Isopropylaminoantipyrine see 8282
(±)-1-(Isopropylamino)-3-[p-(cyclopropylmethoxyethyl)phenoxy]-2-propanol see 1229
4-Isopropylamino-2,3-dimethyl-1-phenyl-3-pyrazolin-5-one see 8282
1-(Isopropylamino)-2-hydroxy-3-[o-(allyloxy)phenoxy]propane see 7086
(±)-1-(Isopropylamino)-3-[p-(β-methoxyethyl)phenoxy]-2-propanol see 6235
1-(Isopropylamino)-3-(o-methoxyphenoxy)-2-propanol see 6345
Isopropylaminomethyl-(3,4-dihydroxyphenyl)carbinol see 5236
2-Isopropylamino-6-methylheptane see 6243
1-(Isopropylamino)-3-[(2-methylindol-4-yl)oxy]-2-propanol see 5901
α-[(Isopropylamino)methyl]-2-naphthalenemethanol see 7974
α-[(Isopropylamino)methyl]-p-nitrobenzyl Alcohol see 6618
α-(Isopropylaminomethyl)protocatechuyl Alcohol see 5236
2-Isopropylamino-1-(2-naphthyl)-ethanol see 7974
1-(Isopropylamino)-3-(1-naphthyloxy)-2-propanol see 8025
Isopropylaminophenazone see 8282
9-[3-(Isopropylamino)propyl]-9-(aminocarboyl)fluorene see 4971
α-(1-Isopropylaminopropyl)protocatechuyl Alcohol see 5185
1-[3-(Isopropylamino)-2-pyridyl]-4-[(5-methanesulfonamidoindol-2-yl)-carbonyl]piperazine see 2929
2-(Isopropylamino)pyrimidine see 5118
1-(Isopropylamino)-3-(m-tolyloxy)-2-propanol see 9653
p-Isopropylaniline see 2686
4-Isopropylantipyrine see 8056
Isopropylarterenol see 5236
p-Isopropylbenzaldehyde see 2687
Isopropylbenzene see 2683
p-Isopropylbenzoic Acid see 2684
3-Isopropyl-1H-2,1,3-benzothiadiazin-4(3H)-one 2,2-Dioxide see 1080
Isopropyl 4-(2,1,3-Benzoxadiazol-4-yl)-1,4-dihydro-5-methoxycarbonyl-

Isothazine *see* 3793
Isothebaine, 5247
Isothiazine *see* 3793
Isothiocyanatobenzene *see* 7450
Isothiocyanatoethane *see* 3861
Isothiocyanatomethane *see* 6165
(*R*)-1-Isothiocyanato-4-(methylsulfin-yl)butane *see* 9135
4-Isothiocyanato-1-(methylsulfinyl)-1-butene *see* 9136
1-Isothiocyanato-3-(methylsulfonyl)-propane *see* 2092
1-Isothiocyanatonaphthalene *see* 6498
4-Isothiocyanato-4′-nitrodiphenyl-amine *see* 613
1-Isothiocyanato-4-(4-nitrophenoxy)-benzene *see* 6731
4-Isothiocyanato-*N*-(4-nitrophenyl)-benzeneamine *see* 613
1-(4-Isothiocyanatophenoxy)-4-nitro-benzene *see* 6731
3-Isothiocyanato-1-propene *see* 303
Isothiocyanic Acid *see* 9464
Isothiocyanic Acid Allyl Ester *see* 303
Isothiocyanic Acid 4-(Methylsulfinyl)-3-butenyl Ester *see* 9136
Isothiocyanic Acid 3-(Methylsulfon-yl)propyl Ester *see* 2092
Isothiocyanic Acid *p*-(*p*-Nitroanilino)-phenyl Ester *see* 613
Isothiocyanic Acid *p*-(*p*-Nitrophen-oxy)phenyl Ester *see* 6731
Isothiocyanic Acid *p*-Phenylene Ester *see* 1345
Isothiocyanic Acid Phenyl Ester *see* 7450
γ-Isothiocyanopropyl Methyl Sulfone *see* 2092
Isothipendyl, 5248
d-Isothujone *see* 9533
Isothymol *see* 1923
Isotrate [Hauck] *see* 5245
Isotretinoin *see* 8333
Isotrex [Stiefel] *see* 8333
Isotrifoliin *see* 5240
Isotrilobine *see* 9826
Isotron 2 *see* 3114
Isotyl *see* 6190
Isovaleral *see* 5249
Isovaleraldehyde, 5249
Isovaleramide, 5250
Isovalerianic Acid *see* 5251
Isovaleric Acid, 5251
Isovaleric Acid *tert*-Amyl Ester *see* 652
Isovaleric Acid *d*-Bornyl Ester *see* 1370
Isovaleric Acid Diester with 8,9-Dihy-dro-9-hydroxy-8-(1-hydroxy-1-methylethyl)-2*H*-furo[2,3-*h*]-1-benzopyran-2-one *see* 893
Isovaleric Acid 8-Ester with *N*-(7-Butyl-8-hydroxy-4,9-dimethyl-2,6-dioxo-1,5-dioxonan-3-yl)-3-form-amidosalicylamide *see* 756
Isovaleric Acid 8-Ester with 3-Form-amido-*N*-(7-hexyl-8-hydroxy-4,9-dimethyl-2,6-dioxo-1,5-dioxonan-3-yl)salicylamide *see* 755
Isovaleric Acid *p*-Menth-3-yl Ester *see* 5887
Isovaleric Acid *tert*-Pentyl Ester *see* 652
Isovaleric Aldehyde *see* 5249
Isovaleryl Chloride, 5252
Isovaleryl Diethylamide, 5253
2-Isovalerylindane-1,3-dione, 5254
N-Isovaleryl-L-valyl-L-valyl-3-hydr-oxy-6-methyl-γ-aminoheptanoyl-L-

alanyl-3-hydroxy-6-methyl-γ-aminoheptanoic Acid *see* 7290
Isovaline, 5255
Isovex *see* 3773
Isoviral [Lenza] *see* 5006
Isovist [Schering AG] *see* 5082
Isovitamin C *see* 5141
Isovon *see* 5236
Isovue [Squibb] *see* 5071
Isovyl [Rhovyl] *see* 7746
Isoxaben, 5256
Isoxal [Shionogi] *see* 7318
Isoxepac, 5257
Isoxicam, 5258
Isoxsuprine, 5259
Isoxyl [Cont. Pharma] *see* 9593
Isoyohimbine *see* 10238
Isozid [Fatol] *see* 5203
Ispaghula Husk *see* 7674
Ispenoral [Chephasaar] *see* 7230
Isphamycin *see* 2245
Isradipine, 5260
Isrodipine *see* 5260
Issium [Lifepharma] *see* 4179
Isteropac E.R. [Bracco] *see* 5031
Istin [Pfizer] *see* 516
Istizin [Winthrop] *see* 2878
Istonil [Nippon Chemiphar/Siegfried] *see* 3258
Isuprel [Winthrop] *see* 5236
Isupren *see* 5236
IT-931 *see* 3433
ITA-104 *see* 7673
Itaconic Acid, 5261
Italprid [Teofarma] *see* 9561
Itamidone [Farmitalia] *see* 495
ITF-182 *see* 4949
Itobarbital *see* 1536
Itraconazole, 5262
Itramin Tosylate, 5263
Itridal *see* 2780
Itrin [Abbott] *see* 9297
Itrizole [Janssen] *see* 5262
Itrol *see* 8654
Itrop [Boehringer, Ing.] *see* 5089
Ituran [Promonta] *see* 6696
IUDR *see* 4934
Iuvacor [Inverni] *see* 9974
Ivadal [Marion Merrell Dow] *see* 10321
Ivaugan [Voigt] *see* 4822
Ivermectin, 5264
Iversal [Bayer] *see* 395
Iviron *see* 4073
Ivomec [Merck & Co.] *see* 5264
Ivosit [Hoechst] *see* 3341
Iwalexin [Iwaki] *see* 2021
Ixbut, 5265
Ixertol [Rottendorf] *see* 2365
Ixodin, 5266
Ixoten [Asta] *see* 9897
Ixprim [Siegfried] *see* 1266
Izokappa [Izo] *see* 5874
Izopiridina [Izo] *see* 9108
J-38 *see* 4273
Jaborandi *see* 7579
Jaclacin [Lundbeck] *see* 115
Jacodine *see* 8596
Jacutin [Hermal] *see* 5526
Jadit [Hoechst] *see* 1485
Jaguar *see* 4602
Jaikin [Basotherm] *see* 8733
Jalap, 5267
Jalovis *see* 4794
Jamaica Pepper *see* 7586
Jamboo *see* 5268
Jambul, 5268
Jamestown Weed *see* 8975
Jamylène [Pharmuka] *see* 3460
Janimine [Abbott] *see* 4955

Janus Green B, 5269
Japan Agar *see* 182
Japan Camphor *see* 1779
Japanese Oil of Camphor *see* 6879
Japan Isinglass *see* 182
Japan Tallow *see* 5270
Japan Wax, 5270
Japonilure, 5271
Jasmolin I *see* 5272
Jasmolin II *see* 5272
Jasmolins, 5272
Jasmone, 5273
Jateorrhizine *see* 5274
Jatroneural [Procter & Gamble] *see* 9811
Jatropur [Procter & Gamble] *see* 9731
Jatrorrhizine, 5274
Jaundice Berry *see* 988
Jaune Brilliant *see* 1666
Javanicin, 5275
Java Pepper *see* 2679
Java Plum *see* 5268
Javelle Water *see* 4912
JB-11 *see* 9851
JB-251 *see* 8082
JB-305 *see* 7623
JB-323 *see* 7614
JB-516 *see* 7382
JB-8181 *see* 2966
JD-91 *see* 3604
JDL-464 *see* 9690
Jectofer [Astra] *see* 5112
Jecto-Sal [Mallard] *see* 9498
Jee Cocktail *see* 6033
Jeff *see* 6033
Jefron [Marion Merrell Dow] *see* 7731
Jellin [Grünenthal] *see* 4185
Jenacaine *see* 7937
Jenacyclin [Jenapharm] *see* 3496
Jequirity *see* 5
Jeremejevite *see* 339
Jerusalem Artichoke, 5276
Jerusalem Tea *see* 6888
Jerva Acid *see* 2094
Jervasic Acid *see* 2094
Jervine, 5277
Jesaconitine *see* 5278
Jestmin [Kyowa] *see* 7018
Jesuit's Balsam *see* 2580
Jesuit's Bark *see* 2343
Jesuit's Tea *see* 5796
Jetrium [Hek] *see* 2997
Jeweler's Rouge *see* 4072
Jexin [Duncan, Flockhart] *see* 9939
JF-6064 *see* 9266
JH *see* 5287
JH 0 *see* 5287
Iso-JH 0 *see* 5287
JH III Bisepoxide *see* 5287
Jilkon *see* 4357
Jimpson Weed *see* 8975
Jimson Weed *see* 8975
Jinofloxacin *see* 6430
JL-991 *see* 9324
JL-1078 *see* 3211
JM-8 *see* 1870
JO-1016 *see* 9623
JO-1196 *see* 3987
Jodairol *see* 5055
Jodid [E. Merck] *see* 7809
Jodobil *see* 5047
Jodo-Metil-Fillina [Malesci] *see* 3754
Jodomiron [Bracco] *see* 5031
Jodozoat Meglumin [Dagra] *see* 5850
Joghurt *see* 10235
Johannisbrotmehl *see* 5584
Jojoba Oil, 5279
Joliotium *see* 3581
Jomezol *see* 6049
Jomybel [Sarva] *see* 5280

Jonctum [Merrell] see 7035
Jonit [Hoechst] see 1345
Jopamiro [Bracco] see 5071
Jordan Almond see 311
Joristen see 7313
Josacine [Bellon] see 5280
Josamina [Novag] see 5280
Josamy [Yamanouchi] see 5280
Josamycin, 5280
Josaxin [Boots] see 5280
JP-992 see 1066
Jubalon see 1846
Judean Pitch see 880
Judolor [Woelm] see 4333
Juglans, 5281
Juglone, 5282
Julin's Carbon Chloride see 4714
Julocrotine, 5283
Julodin [Takeda] see 3744
Jumbul see 5268
Jumex [Chiesi] see 8569
Junipene see 5597
Juniper Berries, 5284
Juniperberry Oil see 6912
Juniper Tar, 5285
Justamil [Anphar-Rolland] see 9093
Justar [Intersan] see 2323
Justicidin A see 5286
Justicidin B see 5286
Justicidins, 5286
Justor [Lederle] see 2332
Juvastigmin (amp.) [Dolder] see 6553
Juvastigmin (tabl.) [Dolder] see 6553
Juvela [Eisai] see 10160
Juvela Nicotinate [Eisai] see 10159
Juvenile Hormones, 5287
Juvenimicin A₃ see 8420
K see 5667
K III see 3331
K IV see 3331
K-17 see 9390
K-31 [Kyorin] see 6606
K-33 see 9261
K-351 see 6655
K-386 see 9648
K-1875 see 1287
K-1900 see 6644
K-2680 see 3928
K-3712 see 6350
K-3917 see 6350
K-3920 see 4991
K-4024 see 4451
K-4277 see 4999
K-6451 see 7033
K-9321 see 113
K-10033 see 3985
K-11941 see 234
K-21060E see 3502
K-22023 see 3453
Ka-2547 see 5980
Kabi 925 see 3604
Kabi 2165 see 2870
KABI 2213 see 7910
Kabiglobulin [Pharmacia] see 4958
Kabikinase [Pharmacia] see 8981
Kabipenin see 7226
Kadaya see 5296
Kaemferitrin see 5288
Kaempferol, 5288
Kaempferol 3-Robinoside 7-Rhamno-
  side see 8404
Kafocin [Lilly] see 2023
Kafylox [Bristol] see 7770
Kainic Acid, 5289
L_S-xylo-kainic Acid see 5289
α-Kainic Acid see 5289
Kaleorid [Leo Pharm.] see 7783
Kalex see 3557
Kalgut [Tanabe Seiyaku] see 2943
Kalignost see 8839

Kalimozan [Nikken] see 7796
Kalinite see 373
Kalirechin [Toho] see 5291
Kalitabs [Kabi] see 7783
Kalium see 7763
Kalium-Duriles [Astra] see 7783
Kalle's Acid see 6487
Kallidin, 5290
Kallidin I see 1386
Kallidin II see 5290
Kallidin-9 see 1386
Kallidin-10 see 5290
Kallidinogenase see 5291
Kallikrein, 5291
Kallinin see 5364
Kalma [Fresenius] see 9929
Kalmegh see 673
Kalmopyrin see 1684
Kalodil see 3241
Kalsetal see 1684
Kalutein [Tatsumi] see 2470
Kalymin [Arzneimittelwerk Dresden]
  see 8161
Kamala, 5292
Kamassin see 8212
Kameela see 5292
Kamila see 5292
Kamillosan [ASTA Pharma] see 1281
Kaminax [Ital. Res.] see 425
Kammerer's Porphyrin see 8084
Kamoran [Lilly] see 134
Kampfstoff "Lost" see 6395
Kamycin see 5293
Kamynex see 5293
Kanabristol [Bristol] see 5293
Kanacedin see 5293
Kanamycin, 5293
Kanamycin A see 5293
Kanamycin B see 5293
Kanamycin C see 5293
Kanamytrex [Bristol] see 5293
Kanaqua [Andromaco] see 5293
Kanasig see 5293
Kanatrol [Lusofarmaco] see 5293
Kanechlor [Kanegafuchi] see 7725
Kanendomycin [Meiji] see 5293
Kanendos [Crinos] see 5293
Kanescin [Torlan] see 5293
Kanicin see 5293
Kankohso 101 see 7690
Kannasyn see 5293
Kannit [MECT] see 10218
Kano [Pierrel] see 5293
Kanochol [Taiyo] see 7018
Kanone [Beecham] see 5874
Kanrenol [SPA] see 1795
Kantec [Daiichi] see 5751
Kantrex [Bristol] see 5293
Kantrexil [Bristol] see 5293
Kantrox [Bristol] see 5293
Kaolin, 5294
Kaolinite see 377 and 4306
Kaon [Warren-Teed] see 7796
Kaon-Cl [Adria] see 7783
Kapanol [Glaxo] see 6361
Kaparlem see 5285
Kapilin [Glaxo] see 5873
Kapilon Injectable see 5875
Kappabi [Erbamont] see 3052
Kappadione [Lilly] see 5873
Kappaxin [Sterling Winthrop] see 5874
Kapron [Klin] see 6832
Karanjin, 5295
Karanum [E. Merck] see 5873
Karathane [Rohm & Haas] see 3340
Karaya Gum, 5296
Karbam Black see 4054
Karbam White see 10305
Karbinon [Badrial] see 6481
Kardiamed [Medice] see 3210

Karidium see 8762
Karil [Sandoz] see 1680
Karion [Gedeon Richter] see 7583
Karmex see 3447
Karmex Monuron Herbicide see 6342
Karminomycin see 1921
Karsil, 5297
Karsivan [Provet] see 7997
Karstenite see 1753
Karsulphan see 9111
Kaskay [Ayerst] see 7783
"Kastle-Meyer Reagent" see 7394
KAT-256 see 2425
Katadolon [Asta] see 4227
Kata-Lipid [IBI] see 1083
Katapyrin [Endopharm] see 471
Katchung Oil see 7191
Katen [Spofa] see 6257
Kathon [Rohm & Haas] see 6853
Katilo see 5296
Katine see 6811
Katlex [Iwaki] see 4331
Katonil [Kali-Chemie] see 2154
Katorin see 7796
Katoseran [Hishiyama] see 2365
Katovit [Thomae] see 7964
Kaurie see 2581
Kautschin see 5518
Kava, 5298
Kavahin see 6218
Kavain see 5299
Kava-kava see 5298
Kavatin see 6218
Kawa see 5298
Kawain, 5299
Kayback [Ayerst] see 7783
Kay-Cee-L [Geistlich] see 7783
Kaydol [Sonneborn] see 7327
Kayexalate [Sanofi-Winthrop] see
  8815
Kayhydrin see 7536
Kayklot [Raymer] see 5874
Kayquinone [Abbott] see 5874
Kaytwo [Eisai] see 5876
Kayvisyn see 10162
KB-95 see 1152
KB-227 see 9702
KB-509 see 4245
KB-1585 see 5460
KB-2413 see 3597
KB-2796 see 5593
KBT-1585 see 5460
KC-404 see 4923
KC-2547 see 5980
KC-9147 see 9647
K-Contin [Napp] see 7783
KD-136 see 4629
KE see 2602
Keal [Sinbio] see 9049
Kebilis [Roussel] see 2096
Kebuzone, 5300
Kedacillina [Bracco] see 9059
Kefadim [Lilly] see 1995
Kefadol [Lilly] see 1964
Kefamin [Lilly] see 1995
Kefandol [Lilly] see 1964
Kefazim [Lilly] see 1995
Kefazon [Esseti] see 1981
Kefenid [SIT] see 5316
Kefglycin [Lilly] see 2023
Kefir Fungi, 5301
Kefir Grains see 5301
Kefir Seeds see 5301
Keflet [Lilly] see 2021
Keflex [Lilly] see 2021
Keflin [Lilly] see 2028
Keflodin [Lilly] see 2025
Keflordin (obsolete) [Lilly] see 2025
Keforal [Lilly] see 2021
Kefoxina [CT] see 1965

Kefroxil [Wharton] *see* 1963
Kefspor [Lilly] *see* 2025
Keftab [Lilly] *see* 2021
Kefurox [Lilly] *see* 2002
Keiperazon [Kaken] *see* 1968
Keitin [Chugai] *see* 1990
Kelamin *see* 5321
Kelcogel [Merck & Co.] *see* 4389
Kelecin *see* 5452
Kelene *see* 3829
Kelfer *see* 4070
Kelferon [Boehringer, Mann.] *see* 4086
Kelfiprim [Farmitalia] *see* 9078
Kelfizina [Farmitalia] *see* 9078
Kelfizine W [Farmitalia] *see* 9078
Kelgin *see* 240
Kelicor *see* 5321
Kelicorin *see* 5321
Kellin *see* 5321
Kelnac [Sankyo] *see* 7692
Kelocyanor [Laroche Navarron] *see* 3559
Keloid *see* 5321
Kelox [Elder] *see* 9385
Kelp [Omni] *see* 891
Kelpware *see* 4305
Kelthane [Rohm & Haas] *see* 3136
Keltrol F [Kelco] *see* 10191
Kelzan [Kelco] *see* 10191
Kemadrin [Burroughs Wellcome] *see* 7944
Kemicetine [Farmitalia] *see* 2120
Kemi S [Otsuka] *see* 8025
Kemithal [ICI] *see* 9428
Kemsol [Horner] *see* 3308
Kenacort [Bristol-Myers Squibb] *see* 9727
Kenacort-A [Bristol-Myers Squibb] *see* 9728
Kenalog [Apothecon] *see* 9728
Kendall's Compound A *see* 2923
Kendall's Compound B *see* 2601
Kendall's Compound C *see* 276
Kendall's Compound D *see* 274
Kendall's Compound E *see* 2602
Kendall's Compound F *see* 4828
Kendall's Compound G *see* 280
Kendall's Compound H *see* 271
Kendall's Desoxy Compound B *see* 2947
Kenicef [Taiho] *see* 1979
Keoxifene *see* 8281
Kephalins *see* 2022
Kephrine *see* 170
Kepinol [Pfleger] *see* 9086
Kepone [Allied Chemical] *see* 2131
Keracyanin *see* 2757
Keralyt [Bristol-Myers] *see* 8484
Keramik *see* 3175
Keramin *see* 3175
Keraphen *see* 5059
Keratin, 5302
Keratinamin [Kowa] *see* 10005
Keratinase, 5303
Keratol [Streuli] *see* 10293
Kerb [Rohm & Haas] *see* 8058
Kerecid *see* 4934
Kerlone [Synthelabo] *see* 1229
Kermesic Acid, 5304
Kernechtrot *see* 6571
Kerosene, 5305
Kerosine *see* 5305
Kertasin *see* 3911
Keselan [Sumitomo] *see* 4629
Kesint [Proter] *see* 2002
Kessar [Farmitalia] *see* 9216
Kessazulen *see* 4581
Kesso-Bamate [McKesson] *see* 5908
Kestrone [Cooper] *see* 3751

Ket [Irbi] *see* 5307
Ketaject [Bristol] *see* 5306
Ketalar [Parke, Davis] *see* 5306
Ketalgin [Amino] *see* 6008
Ketamine, 5306
Ketanarkon [Streuli] *see* 5306
Ketanest [Parke, Davis] *see* 5306
Ketanrift [Ohta-Ono] *see* 287
Ketanserin, 5307
Ketas [Kyorin] *see* 4923
Ketaset [Am. Home] *see* 5306
Ketason *see* 5300
Ketavet [Gellini] *see* 5306
Ketazolam, 5308
Ketazone [Beytout] *see* 5300
Ketene, 5309
Ketensin [Janssen] *see* 5307
Kethamed [Medo-Chem] *see* 7206
Kethoxal, 5310
Ketipic Acid, 5311
Ketobemidone, 5312
Ketobun-A [Isei] *see* 287
β-Ketobutyranilide *see* 55
2-Ketobutyric Acid *see* 56
3-Ketobutyric Acid *see* 56
4-Keto-β-carotene *see* 3543
Ketochromin *see* 3225
Ketoconazole, 5313
Ketocycloheptane *see* 2791
Ketocyclopentane *see* 2811
Ketoderm [Janssen] *see* 5313
Ketodestrin *see* 3751
(dl)-9-Keto-11α,15α-dihydroxy-16-phenoxy-17,18,19,20-tetranorprosta-4,5,13-*trans*-trienoic Acid Methyl Ester *see* 3629
17-(1-Ketoethyl)androstane-3,17-diol *see* 272
17β-(1-Ketoethyl)-Δ⁵-androsten-3β-ol *see* 7915
Ketogan [Lundbeck] *see* 5312
Ketogestin [Upjohn] *see* 5317
Ketogin [Lundbeck] *see* 5312
3-Keto-D-glucoheptonofuranolactone *see* 4459
α-Ketoglutaric Acid, 5314
β-Ketoglutaric Acid *see* 66
2-Keto-L-gulonic Acid, 5315
Ketoheptamethylene *see* 2791
D-*manno*-Ketoheptose *see* 4701
Ketohexamethylene *see* 2795
2-Ketohexamethylenimine *see* 1805
D-*erythro*-3-Ketohexonic Acid Lactone *see* 5141
D-*ribo*-2-Ketohexose *see* 8109
Ketohydroxyestrin *see* 3751
17-(1-Keto-2-hydroxyethyl)androstane-3,17-diol-11-one *see* 280
17-(1-Keto-2-hydroxyethyl)androstane-3,11,17-triol *see* 276
17-(1-Keto-2-hydroxyethyl)-Δ⁴-androsten-3,11-dione *see* 2923
17-(1-Keto-2-hydroxyethyl)-4-androsten-17α-ol-3-one *see* 2977
3-[β-Keto-γ-(3-hydroxy-2-piperidyl)-propyl]-4-quinazolone *see* 3984
Ketoisdin [Isdin] *see* 5313
Ketomalonic Acid *see* 5971
Ketomalonic Acid Calcium Salt *see* 1726
γ-Keto-β-methoxy-δ-methylene-Δᵅ-hexenoic Acid *see* 7218
1-Keto-6-methoxy-1,2,3,4-tetrahydronaphthalene *see* 6081
5-Keto-2-methoxy-5,6,7,8-tetrahydronaphthalene *see* 6081
Ketopentamethylene *see* 2811
D-*erythro*-2-Ketopentose *see* 8374
Ketophenylbutazone *see* 5300
Ketoprofen, 5316

11-Ketoprogesterone, 5317
Ketopron [Alcon] *see* 5316
β-Ketopropane *see* 64
2-Ketopropionaldehyde *see* 8203
α-Ketopropionic Acid *see* 8205
2-Ketopyrrolidine *see* 8200
2-Ketopyrrolidine-1-ylacetamide *see* 7641
Ketorolac, 5318
Ketoscilium [Zambon] *see* 4048
Ketosuccinic Acid *see* 7041
1-Keto-3-(3'-sulfamyl-4'-chlorophenyl)-3-hydroxyisoindoline *see* 2246
L-3-Ketothreohexuronic Acid Lactone *see* 867
Ketotifen, 5319
17-Ketotrilostane *see* 9827
2-Keto-1,7,7-trimethylnorcamphane *see* 1779
Keuten *see* 8753
Kevopril [Rhone-Poulenc] *see* 8267
Kew Tree *see* 4432
Keyhole-limpet Hemocyanin *see* 4681
Keypyrone [Key] *see* 3414
KF-868 *see* 5674
KH3 [Schwarzhaupt] *see* 4615
Kharophen [Burroughs Wellcome] *see* 49
Kharsivan *see* 851
Khas Khas Oil *see* 6953
Khat, 5320
Khelfren *see* 5321
Khellin, 5321
Khellinin *see* 5322
Khellol Glucoside, 5322
Khinocyde *see* 8252
Khloratsizin *see* 2108
Khus Oil *see* 6953
K-IAO *see* 7796
Kiditard [Belga] *see* 8244
Kidon [Ono] *see* 4536
Kidrolase [Rhone Poulenc Rorer] *see* 871
Kieselguhr *see* 5003
Kieserite *see* 5731
Kif *see* 1794 and 4648
KIH-2031 *see* 8177
KIH-8921 *see* 8177
Kiku Oil, 5323
Killax *see* 9345
Kilmicen [Farmitalia] *see* 9647
Kilmite 40 *see* 9345
Kimopsin [Eisai] *see* 2320
Kimoral [Nippon Chemiphar] *see* 2320
Kinaden *see* 4794
Kinavosyl [Riker] *see* 5895
Kinecid [Schering AG] *see* 9082
Kinedak [Ono] *see* 3640
Kineorl [Showa Shinyaku] *see* 9660
Kinetin, 5324
Kinetin [Schering] *see* 4794
Kinevac [Squibb] *see* 8689
King's Yellow *see* 846 and 5422
Kinic Acid *see* 8243
Kinichron [Biochimica] *see* 8244
Kinidin Durules [Astra] *see* 8244
Kino, 5325
Kinoprene, 5326
Kinotomin [Toa Eiyo] *see* 2405
Kino-yellow *see* 5676
Kinupril [Bellon] *see* 8267
Kirocid [Schering AG] *see* 9082
Kiron [Schering AG] *see* 9082
α-Kirondrin *see* 4444
Kir Richter [Lepetit] *see* 796
Kirromycin *see* 6308
Kisselo-mleko *see* 10235
Kistrin, 5327
Kistrin (*Agkistrodon rhodostoma* Reduced) *see* 5327

Kitasamycin *see* 5480
Kitol, 5328
KL-255 *see* 1521
KL-373 *see* 1274
Klacid [Abbott] *see* 2400
Klamar [Maggioni] *see* 4580
Klaricid [Abbott] *see* 2400
Klavikordal [Baer] *see* 6704
Klavocin [Pliva] *see* 2402
Klebcil [Beecham] *see* 5293
Kleenodyne *see* 5058
Klemidox *see* 2406
Klexane [Rhone-Poulenc Rorer] *see* 3626
KLH *see* 4681
Klimax E [Fink] *see* 3750
Klimicin [Lek] *see* 2414
Klimoral [AB Leo] *see* 3750
Klinit [Eisai] *see* 10218
Klinium [Janssen] *see* 5507
Klinomycin [Am. Cyanamid] *see* 6289
Klion [Gedeon Richter] *see* 6242
Klismacort [Benechemie] *see* 7901
Kloben Neburon [Du Pont] *see* 6523
Klockmannite *see* 2719
Klonopin [Roche] *see* 2449
Klor-Con [Upsher-Smith] *see* 7783
Klorpromex [Dumex] *see* 2238
Klort [Lemmon] *see* 5908
Klot [Warren-Teed] *see* 9658
Klotogen [Abbott] *see* 5874
Klottone [MET] *see* 5874
Klucel [Hercules] *see* 4888
K-Lyte [Mead Johnson] *see* 7770
Knight's Spur *see* 5383
Knob Root *see* 2546
Knoll H$_{75}$ *see* 7485
K-Norm [Pennwalt] *see* 7783
Ko-592 *see* 9653
Ko-1173 *see* 6257
Ko-1366 *see* 1514
Koate-HP [Miles] *see* 3963
Kobaton *see* 3309
Kodel [Eastman-Kodak], 5329
Kodocytochalasin-1 *see* 2859
Koettigite *see* 10277
Kofler's Quinone *see* 7681
Kogenate [Miles] *see* 3963
Koglucoid [Panray] *see* 8293
Koji *see* 9203
Kojic Acid, 5330
Kola, 5331
Kolikodal *see* 4826
Kolklot [Cole] *see* 5874
Kollateral [Ursapharm] *see* 6371
Kollerdormfix *see* 2780
Kollidon *see* 7879
Kolpolyn [Organon] *see* 3780
Kolpon [Organon] *see* 3751
Kolton (jelly) [Promonta] *see* 3390
Kolton (tabl.) [Promonta] *see* 3390
Kombetin [Boehringer, Mann.] *see* 9016
Kombiquens [Novo] *see* 5849
Komet RP [Zeneca] *see* 9266
Kompensan [Roerig] *see* 3228
Komplexon *see* 3557
Konakion [Roche] *see* 7536
Kondurangin *see* 2560
Konesta [Akzo] *see* 9756
Koninckite *see* 4074
Konlax [Nippon Shinyaku] *see* 7922
Kon Oil *see* 5675
Kontexin [AB Leo] *see* 7461
Konyne [Miles] *see* 3964
Kopsine, 5332
Korazol *see* 7283
Korbutone [Nippon Glaxo] *see* 1047
Kordafen [Polfa] *see* 6617
Korec [Sanofi Winthrop] *see* 8233

Koretic [Sanofi Winthrop] *see* 8233
Korglykon *see* 2575
Korlan [Dow] *see* 8415
Koromite [Sankyo] *see* 6276
Koroseal [Goodrich] *see* 7746
Korum [Geneva] *see* 45
Kosin, 5333
Koso *see* 1391
Kosso *see* 1391
K-Othrine [Roussel] *see* 2934
Kotoite *see* 5693
Kouso *see* 1391
Kousso *see* 1391
KP-363 *see* 1547
KPABA *see* 7766
KPB *see* 5300
31K-Precursor *see* 7975
Kraton [Shell] *see* 8534
Krebiozen, 5334
Krebon *see* 1500
Kredex [SKB] *see* 1924
Kremol [Sonneborn] *see* 7327
Kren *see* 4785
Kresatin *see* 2651
KRESTIN [Kureha] *see* 8112
Kriplex [Alfa] *see* 3132
Kristallose *see* 8463
Krizanol *see* 1689
KRN-8601 *see* 4558
Kronitex TCP [FMC] *see* 9892
Kryogenin *see* 7378
Kryolith *see* 2673
Kryptocur [Hoechst] *see* 5500
Krypton, 5335
Krypton Difluoride *see* 5335
Kryptosterol *see* 5372
Krysid *see* 764
KS4 *see* 142
K-Strophanthin *see* 9016
K-Strophanthin-α *see* 2830
K-Strophanthoside *see* 9016
KT-611 *see* 6443
KT-3777 *see* 5606
K-Tab [Abbott] *see* 7783
K-Thrombin [Fawns & McAllan] *see* 5874
KU-54 *see* 9921
Kubacron [Kayaku] *see* 9754
KUE 13032c *see* 3095
Kujimycin B *see* 5369
Kukoline *see* 8692
Kullo *see* 5296
Kupferkies *see* 2079
Kurchatovium *see* 3580
Kurchessine, 5336
Kurchiline *see* 5336
Kurchiphyllamine *see* 5336
Kurchiphylline *see* 5336
Kurcholessine, 5337
Kurnakovite *see* 5693
Kuromanin *see* 2757
Kuromatsuene *see* 5597
Kuron [Dow] *see* 8679
Kusso *see* 1391
Kusum Oil *see* 5675
Kuteera *see* 5296
Kutkasin [Lääkefarmos] *see* 9123
Ku-Zyme HP [Kremers-Urban] *see* 7138
K-Vitrat *see* 10162
KW-110 *see* 25
KW-1062 *see* 6269
KW-2228 *see* 4558
KW-2307 *see* 10127
KW-3049 *see* 1071
Kwai [Lichtwer Pharma] *see* 4380
KWD-2183 *see* 980
Kwell [Reed & Carnrick] *see* 5526
Kyamepromazine *see* 2753
Kyanmethin, 5338

Kyanol *see* 696
Kylar [Uniroyal] *see* 2874
Kylit [Taiho] *see* 10218
Kymo-trypure *see* 2320
Kynex [Lederle] *see* 9087
Kynurenic Acid, 5339
Kynurenine, 5340
Kyocristine [Kyorin] *see* 10124
Kyorin AP2 *see* 8682
Kytril [SKB] *see* 4557
Kyurinett [Zensei] *see* 9835
L *see* 5475
L-67 *see* 7924
L-104 *see* 2453
L-105 *see* 8386
L-105 (Lederle) *see* 2003
L-105SV *see* 8386
L-141 *see* 3989
L-300 *see* 9639
L-554 *see* 9894
L-627 *see* 1243
L-1102 *see* 8485
L-1573 *see* 2848
L-1633 *see* 8753
L-1913 *see* 7299
L-2103 *see* 4733
L-2197 *see* 1091
L-2214 *see* 1093
L-2329 *see* 1113
L-3428 *see* 504
L-5458 *see* 2916
L-6257 *see* 7068
L-6400 *see* 4151
L-12717 *see* 5615
L-34314 *see* 3364
L-35355 *see* 3404
L-36352 *see* 9815
LA-1 *see* 6667
LA-956 *see* 7206
LA-1211 *see* 4959
LA-1221 *see* 1535
LA-6023 *see* 6001
LA-III *see* 3042
LAAM *see* 5492
LAAO *see* 435
Lab *see* 8303
Labazene [Labaz] *see* 10049
Labelol [Elea] *see* 5341
Labetalol, 5341
(R,R)-Labetalol *see* 3245
Lab Ferment (German) *see* 8303
Labican [Boniscontro] *see* 2132
LaBID [Procter & Gamble] *see* 9421
Labile Factor *see* 3961
Labitan [Labinca] *see* 3244
Labosept [L.A.B.] *see* 2959
Labrocol [Lagap] *see* 5341
Labroda [Specia] *see* 4142
Labrodax [Specia] *see* 4142
Labyrin *see* 2365
Lac *see* 8623
LAC-43 *see* 1520
Lacalmin [Tatsumi] *see* 8917
Lacca *see* 8623
Lacca Coerulea *see* 5574
Laccaic Acid, 5342
Laccaic Acid A *see* 5342
Laccaic Acid A$_1$ *see* 5342
Laccaic Acid B *see* 5342
Laccaic Acid C *see* 5342
Laccaic Acid D *see* 5342
Lacca Musica *see* 5574
Lacdene [Tsuruhara] *see* 8917
Lac Dye *see* 5342
Lacflavin [Towa Yakuhin] *see* 8367
Lachesine, 5343
Lacidipine, 5344
Lacipil [Glaxo] *see* 5344
Lacirex [Guidotti] *see* 5344
Lacmoid, 5345

Lacmus see 5574
Lacolin see 8781
Lacretin [Tokyo Tanabe] see 2405
Lacril [Allergan] see 4889
Lacrimin [Santen] see 1076
Lacrisert [Merck & Co.] see 4888
Lacrypos [Alcon] see 2270
Lactacidogen see 4298
β-Lactamase see 7219
Lactamin [Teisan] see 7919
Lactaminic Acid see 8628
Lactaroviolin, 5346
Lactate Dehydrogenase, 5347
d-Lactic Acid see 5350
D-Lactic Acid, 5348
D(−)-Lactic Acid see 5348
DL-Lactic Acid, 5349
l-Lactic Acid see 5348
L-Lactic Acid, 5350
L(+)-Lactic Acid see 5350
DL-Lactic Acid Ammonium Salt see 561
Lactic Acid Cetyl Ester see 2072
Lactic Acid Hexadecyl Ester see 2072
Lactic Acid Lactate, 5351
Lactic Acid β-Naphthyl Ester see 6499
Lacticare-HC [Steifel] see 4828
Lactic Dehydrogenase see 5347
Lactinium [Roland] see 7004
Lactit see 5352
Lactite see 5352
Lactit M see 5352
Lactitol, 5352
Lactobacillamide see 5353
Lactobacillic Acid, 5353
Lactobacillus bifidus Factor see 1258
Lactobacillus bifidus Growth Factor see 1258
Lactobacillus bulgaricus Factor see 5409
Lactobacillus lactis Dorner Factor see 10152
Lactobionic Acid, 5354
Lactobionic Acid Calcium Salt Compd with Calcium Bromide see 1694
Lactobionic δ-Lactone see 5354
Lactobiosit see 5352
Lactoferrin see 9706
Lactoflavine see 8367
Lactogen see 7961
Lactogen (Human Placental) see 4650
Lactogenic Hormone (Placental Human) see 4650
Lactol see 6499
Lactonaphthol see 6499
p-Lactophenetide, 5355
p-Lactophenetidide see 5355
Lactophenin see 5355
Lactose, 5356
Lactosit see 5352
Lactositol see 5352
Lactosylceramide see 2863
Lactotransferrin see 9706
Lactovagan [Schwartzhaupt] see 5349
2-(Lactoyloxy)propanoic Acid see 5351
Lactucarium—"French", 5357
Lactucarium—"German", 5358
Lactucerin see 9229
α-Lactucerol see 9229
Lactucin, 5359
Lactucon see 9229
Lactucopicrin see 5359
Lactuflor [Chephesar] see 5360
Lactulose, 5360
Lacty [CCA Biochem] see 5352
Lactyl-p-phenetidin-N-(p-ethoxyphenyl)lactamide see 5355
Lacumin see 5892

Ladakamycin see 923
Ladogal [Winthrop] see 2875
Ladropen [Berk] see 4147
Lady's Slipper see 2839
Laetrile see 638
Laevilac [Wander] see 5360
Laevoral [Laevosan] see 4295
Laevosan [Laevosan] see 4295
Laevoxin see 5497
Laevulinic Acid see 5498
LAF see 5019
Lagistase [Labaz] see 3588
Lagosin see 4312
Laguncurin see 5676
Laidlomycin, 5361
Laki-Lorand Factor see 3968
LAM-1 see 8568
Lamar [Tokyo Tanabe] see 9267
Lamasine [Bristol-Myers] see 635
Lambdamycin see 2086
Lambeth [Lambeth Rope] see 7741
Lamb Mint see 7288
Lambral [Maggioni] see 9645
Lambratene (formerly) [Cilag Italiano] see 2848
Lamictal [Wellcome] see 5367
Lamidon [Kowa] see 4925
Lamifiban, 5362
Laminaran, 5363
Laminaran Hydrogen Sulfate see 5363
Laminarin see 5363
Laminin, 5364
Laminin M see 5364
Lamisil [Wander] see 9299
Lamivudine, 5365
Lamoparan, 5366
Lamoryl [Leo Pharm.] see 4571
Lamotrigine, 5367
Lamoxactam see 6369
Lamp Black see 1856
Lampit [Bayer] see 6630
Lampocef [Lampugnani] see 1967
Lampomandol [Von Boch] see 1964
Lamprecid [FHoechst] see 9388
Lampren(e) [Geigy] see 2433
Lampsporin [Von Boch] see 2002
Lampterol see 4939
Lamra [Merckle] see 3042
Lamuran [Boehringer, Mann.] see 8291
Lanacordin see 3210
Lanacort [Combe] see 4828
Lanadigenin see 3209
Lanain see 5371
Lanalin see 5371
Lanarkite see 5444
Lanatilin [Heilmittelwerk] see 3210
Lanatosides, 5368
Lanatoxin [Beiersdorf] see 3206
Lancetina [Aandersen] see 4280
Landamycine [Delalande] see 8373
Landel [Zeria] see 3566
Landomycin see 6962
Landrax see 2860
Landruma [Landerln] see 6620
Landsen [Sumitomo] see 2449
Lanesin see 5371
Laneth see 7737
Lanex [Nor-Am] see 4189
Lanexat [Roche] see 4169
Langite see 2723
Langoran [Merrell] see 5245
Laniazid [Lannett] see 5203
Lanichol see 5371
Lanicor [Boehringer, Mann.] see 3210
Lanimerck (Ampuls) [E. Merck] see 2967
Lanimerck (Suppositories) [E. Merck] see 5368
Laniol see 5371

Lanirapid [Yamanouchi] see 3210
Lanitop [Boehringer, Mann.] see 3210
Lankacidin A see 8566
Lankacidin C 14-Acetate see 8566
Lankamycin, 5369
Lankavose see 2080
Lannate [Du Pont] see 6062
Lanoconazole, 5370
Lanolin, 5371
(3β)-Lanosta-8,24-dien-3-ol see 5372
Lanosterol, 5372
Lanoxin [Burroughs Wellcome] see 3210
Lansfordite see 5696
Lansoprazole, 5373
Lantadin [Lepetit] see 2916
Lantanon [Ravasini] see 6260
Lanthana see 5374
Lanthanum, 5374
Lanthanum Sesquioxide see 5374
Lanthanum Trioxide see 5374
Lanthionine, 5375
L-Lanthionine see 5375
Lanum [Merck & Co.] see 5371
Lanvis [Burroughs Wellcome] see 9473
Lanzac [Lilly] see 6959
Lanzor [Houdé] see 5373
Lanzyme [Nissui] see 5671
Lapachic Acid see 5376
Lapachol, 5376
Lapis Lazuli see 5408
Lappa, 5377
Lappaconitine, 5378
Lapudrine [ICI] see 2237
Lapyrium Chloride, 5379
Laracor [Lagap] see 7086
Laractone [Lagap] see 8917
Laradopa [Roche] see 5490
Laraflex [Lagap] see 6504
Laragon [Roemmers] see 8680
Larapam [Lagap] see 7661
Laratrim [Lagap] see 9086
Larch Agaric see 183
Larch Turpentine see 10078
Lard, 5380
Lard, Benzoinated, 5381
Lard-factor see 10150
Lard Oil, 5382
Larex [Geistlich] see 7735
Largactil [SK & F] see 2238
Largaktyl see 2238
Large Fennel see 4018
Large-flowered Cereus see 1643
Largon [Wyeth] see 8007
Lariam [Roche] see 5845
Laricic Acid see 184
Laricin see 2567
Laridal [Cusi] see 891
Laristine [Roche] see 4758
Larixin [Toyama] see 2021
Larixinic Acid see 5752
Lark's-claw see 5383
Lark's-heel see 5383
Larkspur, 5383
Larnite see 1749
Larocaine [Roche] see 3270
Larocin (obsolete) [Roche] see 617
Larodopa [Roche] see 5490
Laroscorbine [Roche] see 867
Larostidin [Roche] see 4758
Larotid [Roche] see 617
Laroxyl [Roche] see 511
Larvacide 100 see 2208
Larvadex [Ciba-Geigy] see 2845
Larvin [Rhône-Poulenc] see 9465
LAS see 5668
LAS-11871 see 9443
LAS-17177 see 2352
Lasalocid A, 5384
Laser [Tosi] see 6504

Laserdil [Laser] see 5245
Laserpitin, 5385
Lasilix [Hoechst] see 4331
Lasiocarpine, 5386
Lasix [Hoechst] see 4331
Lasma [Pharmax] see 9421
Lasso [Monsanto] see 203
Lastet [Nippon Kayaku] see 3931
Lasurite see 5408
LAS-W-090 see 3534
LAT-1717 see 7595
LAT-A see 5388
Latamoxef see 6369
Latanoprost, 5387
LAT-B see 5388
Latibon [Bayer] see 3157
Laticort [Medphano] see 4828
Latocillin [Chassot] see 2480
Latoconazole see 5370
Latrunculin A see 5388
Latrunculin B see 5388
Latrunculins, 5388
Latumcidin see 4
Laubanite see 1685
Laudanidine, 5389
dl-Laudanidine see 5390
Laudanine, 5390
l-Laudanine see 5389
Laudanosine, 5391
Laudexium Methyl Sulfate, 5392
Laudicon see 4847
Laudissine see 5392
Laudolissin see 5392
Laughing Gas see 6751
Laumontite see 1685
Laurabolin [Veterinaria] see 6452
Lauracycline see 9337
Laurel Berry Oil see 5394
Laurel Camphor see 1779
Laureline, 5393
Laurel Oil, 5394
Laurent's Acid see 6491
Laurepukine, 5395
Laureth see 7737
Laureth 9 see 7717
Lauric Acid, 5396
Lauric Acid Ester with 2-Hydroxy-
  ethyl Thiocyanate see 5471
Lauric Acid Ester with Pyridoxol
  Diacetate (Ester) see 5401
Lauric Acid 2-Thiocyanatoethyl Ester
  see 5471
Laurocapram, 5397
Laurodin see 5399
Lauroguadine, 5398
Laurolinium Acetate, 5399
Lauromicina [Dukron] see 3722
Lauron [Endo] see 918
Laurostearic Acid see 5396
Laurotetanine, 5400
N-(Lauroylcolaminoformylmethyl)pyr-
  idinium Chloride see 5379
5-Lauroyloxy-6-methyl-3,4-pyridine-
  dimethanol Diacetate see 5401
3-Lauroyloxy-2-picoline-4,5-dimetha-
  nol Diacetate see 5401
3-O-Lauroylpyridoxol Diacetate, 5401
N-Lauroylsarcosine Sodium Salt see
  4379
Lauryl Alcohol see 3464
Lauryl Bromide, 5402
Lauryl Sulfate Salt of the Propionic
  Acid Ester of Erythromycin see
  3722
Lausit [Showa Yakuhin] see 4998
Lautarite see 1719
Lauth's Violet see 9483
Lavender, 5403
Laverin [Lemmon] see 3773
Lawesson's Reagent, 5404

Lawrencite see 4091
Lawrencium, 5405
Lawsone, 5406
Lawsonite see 1685
Laxadin see 1282
Laxanin N see 1282
Laxitex [Andreu] see 9156
Laxoberal [Boehringer, Ing.] see 7559
Laxoberon [Ferrosan] see 7559
Laxo-Isatin see 7108
Laxonalin see 1337
Laxorex see 1282
Layor Carang see 182
Lazabemide, 5407
Lazaroids see 9604
Lazurite, 5408
LB-46 see 7597
LB-502 see 4331
LBF, 5409
LC-2 see 5194
LC-33 see 5462
LC-44 see 4224
LCB-29 see 4935
LCR see 10124
LD-935 see 3399
LD-2988 see 4252
LD-3394 see 4028
LD-3612 see 7157
LD-4610 see 7060
LD-4644 see 7610
LDA see 3240
LEA-103 see 8417
Lead, 5410
Lead Acetate, 5411
Lead Antimonate(V), 5412
Lead Arsenate, 5413
Lead Arsenite, 5414
Lead Azide, 5415
Lead Benzoate see 1122
Lead Borate, 5416
Lead Bromate, 5417
Lead Bromide, 5418
Lead Butyrate, 5419
Lead Chlorate, 5420
Lead Chloride, 5421
Lead Chromate(VI), 5422
Lead Chromate(VI) Oxide, 5423
Lead Difluoride, 5425
Lead Dioxide, 5424
Lead Fluoride, 5425
Lead Fluosilicate see 5427
Lead Formate, 5426
Lead Hexafluorosilicate, 5427
Lead Hydroxide, 5428
Lead Hypophosphite, 5429
Lead Iodide, 5430
Lead Lactate, 5431
Lead Metavanadate see 5451
Lead Molybdate(VI), 5432
Lead Monosubacetate see 5443
Lead Monoxide, 5433
Lead Nitrate, 5434
Lead Oleate, 5435
Lead Orthoplumbate see 5449
Lead Oxalate, 5436
Lead Oxide Brown see 5424
Lead Oxide Hydrate see 5428
Lead Oxide Red see 5449
Lead Oxide Yellow see 5433
Lead Peroxide see 5424
Lead Phosphate, 5437
Lead Protoxide see 5433
Lead Selenate, 5438
Lead Selenite, 5439
Lead Sesquioxide, 5440
Lead Silicofluoride see 5427
Lead-sodium Alloy see 8783
Lead Sodium Hyposulfite see 5441
Lead Sodium Thiosulfate, 5441
Lead Stearate, 5442

Lead Subacetate, 5443
Lead Subcarbonate see 1034
Lead Sulfate, 5444
Lead Sulfide, 5445
Lead Sulfocyanate see 5450
Lead Superoxide see 5424
Lead Telluride, 5446
Lead Tetraacetate, 5447
Lead Tetraethyl see 9343
Lead Tetrafluoride, 5448
Lead Tetroxide, 5449
Lead Thiocyanate, 5450
Lead Trioxide see 5440
Lead Vanadate(V), 5451
Leaf Alcohol see 4737
Leaf Green see 2283
Lealgin [Janssen] see 7400
Leandin see 2751
Leanol [Yoshitomi] see 4745
Leanstar [Pitman-Moore] see 8864
Lebaycid see 4044
Leben Raib see 10235
Leblon [De Angeli] see 7646
Lebrufen [Lederle] see 4925
LECAM-1 see 8568
Lecasol [Kaken] see 2405
Lecedil [Zdravlje] see 3972
Lecibis [Andromaco] see 1187
Lecibral [Nezel] see 6579
Lecithin, 5452
Lecithol see 5452
Lectins, 5453
Lectopam [Roche] see 1406
Ledakrin [Polfa] see 6661
Ledclair [Sinclair] see 3555
Ledercillin [Lederle] see 7226
Ledercillin VK [Lederle] see 7230
Ledercort Cream [Lederle] see 9728
Ledercort (tabl.) [Lederle] see 9727
Lederdopa [Lederle] see 6132
Lederfen [Am. Cyanamid] see 4003
Lederfolat [Lederle] see 4254
Lederfolin [Lederle] see 4254
Lederglib [Lederle] see 4486
Lederkyn [Lederle] see 9087
Lederlon [Lederle] see 9730
Ledermycin [Lederle] see 2937
Lederpam [Lederle] see 7059
Lederplatin [Lederle] see 2378
Lederspan [Lederle] see 9730
Ledertrexate [Lederle] see 6065
Ledol, 5454
Ledopa [Lepetit] see 5490
Ledopur [Lederle] see 287
Ledosten see 3155
"Ledum Camphor" see 5454
Lefax [Asche] see 3264
Lefcar [Glaxo] see 1898
Lefetamine, 5455
Leftose [Nippon Shinyaku] see 5671
Legalon [Madaus] see 8680
Legederm [Schering] see 221
Legential see 4404
Leghemoglobin, 5456
Legoglobin see 5456
Legumex D see 2893
Legumin see 1934
Lehydan [Abigo] see 7475
LeIF see 5016
Leioplegil [Diathera] see 5457
Leiopyrrole, 5457
Leiormone see 10073
Lemakalim see 5487
Lemascorb [Ascher] see 867
Lemazide [Lemmon] see 1155
Lembrol see 3042
Lemoflur see 8762
Lemonol see 4411
Lemon Peel, 5458
Lemon Walnut see 5281

Lemon Yellow see 999
Lemoran see 5496
Lenacil, 5459
Lenampicillin, 5460
Lenamycin see 2009
Lendorm [Boehringer, Ing.] see 1472
Lendormin [Boehringer, Ing.] see 1472
Lenetran [Lakeside] see 5896
Lenicet see 349
Lenidolor [Menarini] see 5819
Lenigallol [Knoll] see 8184
Lenirit [Bonomelli] see 4828
Lenisarin see 3449
Lenitral [Besins-Iscovesco] see 6704
Lenograstim see 4558
Lenopect [Draco] see 7609
LenoxiCaps [Wellcome] see 3210
Lenoxin [Pharmos] see 5614
Lenoxin [Deutsche Wellcome] see
   3210
Lensan A [Ciba Vision] see 4839
Lentagran [Chemie Linz] see 8151
Lentaron [Ciba-Geigy] see 4267
Lenthionine, 5461
Lenticillin see 7226
Lentin [Dow] see 1823
Lentinacin see 3712
Lentinan, 5462
Lentisk see 5795
Lentizol [Warner] see 511
Lentobetic [Farmopatria] see 7376
Lentogest [Samil] see 4886
Lento-Kalium [Robin] see 7783
Lentonitrina [Pierrel] see 6704
Lentopenil see 7221
Lentotran [Farmopatria] see 2132
Lentrat see 7249
Lentysine see 3712
Lenzacef [Lenza] see 2032
Leo 640 see 5587
Leo-1031 see 7900
Leocillin [Leo Pharm.] see 7212
Leodrine [Leo Pharm.] see 4830
Leonurine, 5463
Leopard's Bane see 825
Leostesin [Leo Pharm.] see 5505
Leparan [Italfarmaco] see 4685
Lepargylic Acid see 938
Lepasen see 498
Lepetan [Otsuka] see 1522
Lepidine, 5464
Lepidocrocite see 4068
Lepidolite see 5543
Lepidopterans see 1960
Lepitoin see 7475
Leponex [Wander] see 2484
Leptanal see 4043
Leptandra, 5465
Leptilan [Ciba-Geigy] see 10049
Leptin, 5466
Leptodactyline, 5467
Leptophos, 5468
Leptryl [Bellon] see 7309
Lercanidipine, 5469
Lergefin [Larma] see 1845
Lergigan [Kabi] see 7970
Lergine [Burroughs Wellcome] see
   7944
Lergoban [Riker] see 3390
Lergobine [3M] see 3390
Lergopenin see 2406
Leritine [Merck & Co.] see 695
Leron [Bayer] see 4586
Lertus [Exa] see 5316
Lescodil [Zdravlje] see 6579
Lescol [Sandoz] see 4250
Lesidrin see 6225
Lesopitron, 5470
Lespedin see 5288
Lesser Centaury see 2016

Lestid [Upjohn] see 2538
Lethalaire G-58 see 7033
Lethane 60 [Rohm & Haas], 5471
Lethane 384 [Rohm & Haas], 5472
Lethelmin see 7404
Lethidrone [Burroughs Wellcome] see
   6448
Letosteine, 5473
Letrozole, 5474
LETS see 4119
Letter [Armour Pharm.] see 5497
"Lettuce Opium" see 5358
Letusin see 5495
Leu see 5475
Leu-8 see 8568
Leucaenine see 6286
Leucaenol see 6286
Leu-Cam Proteins see 5014
Leucarsone see 1830
Leucenine see 6286
Leucenol see 6286
Leuchtenbergite see 362
Leucine, 5475
L-Leucine see 5475
23-L-Leucinecolony-Stimulating Fac-
   tor 2 (Human Clone pHG25 Protein
   Moiety) see 4559
Leucine⁵-enkephalin see 3613
6-D-Leucine-9-(N-ethyl-L-prolina-
   mide)-10-deglycinamideluteinizing
   Hormone-releasing Factor (Pig) see
   5484
6-L-Leucine Sporidesmolide I see 8921
Leucinocaine Mesylate, 5476
Leuco-4 see 150
Leucoalizarin see 727
Leucocyanidin, 5477
Leucocyanidol see 5477
Leucodinine see 6420
Leucodrin, 5478
Leucoglycodrin, 5479
Leucoharmine see 4647
Leucokinin see 9941
Leucoline see 8253
Leucomax [Schering-Plough] see 4559
Leucomycin A₁ see 5480
Leucomycin A₃ see 5280 and 5480
Leucomycins, 5480
Leucomycin V 3-Acetate 4B-(3-meth-
   ylbutanoate) see 5280
Leucomycin V 4B-Butanoate 3B-Pro-
   panoate see 8408
Leucomycin V 3B,9-Diacetate 3,4B-
   Dipropanoate see 6291
Leucomycin V 3,4B-Dipropanoate see
   6271
Leucophor R see 1349
Leucopterin, 5481
Leucosar [Adria] see 4254
Leucotrofina [Ellem] see 9543
Leucovorin see 4254
Leucovorin [Lederle] see 4254
(D-Leu⁶)-des-Gly¹⁰-LH-RH-ethyl-
   amide see 5484
Leu⁵-E see 3613
Leu⁵-Enkephalin see 3613
Leuferon-2 [Roche] see 5020
Leukaemomycin C see 2890
Leukeran [Burroughs Wellcome] see
   2116
Leukerin see 5919
Leukine [Immunex] see 4559
Leukochthol see 4929
Leukocyte Adhesion Molecules see
   5014
Leukocyte-Endothelial Cell Adhesion
   Molecule-1 see 8568
Leukocyte Integrins see 5014
Leukocyte Interferon see 5016
Leukomycin [Bayer] see 2120

Leukomycin N [Bayer] see 941
Leukoprol [Morinaga] see 5678
Leukotriene A see 5482
Leukotriene A₄ see 5482
Leukotriene B see 5482
Leukotriene B₄ see 5482
Leukotriene C see 5482
Leukotriene C₁ see 5482
Leukotriene C₄ see 5482
Leukotriene D see 5482
Leukotriene D₄ see 5482
Leukotriene E see 5482
Leukotriene E₄ see 5482
Leukotrienes, 5482
Leunase [Kyowa] see 871
Leupeptin Ac-LL see 5483
Leupeptin Pr-LL see 5483
Leupeptins, 5483
Leuplin [Takeda] see 5484
Leuprolide, 5484
Leuprolide Acetate see 5484
Leuprorelin see 5484
Leurocristine see 10124
Leustatin [Ortho Biotech] see 2397
Leutrol [Abbott] see 10253
Levacecarnine Hydrochloride see 85
Levacide [Norbrook] see 5486
Levadil [Schering-Plough] see 3245
Levadin [Univet] see 5486
Levadone [Burroughs Wellcome] see
   6008
Levallorphan, 5485
Levamfetamine see 623
Levamisole, 5486
Levamphetamine see 623
Levanil [Upjohn] see 3552
Levant Wormseed see 8507
Levanxene [Montedison] see 9285
Levanxol [Montedison] see 9285
Levarterenol see 6788
Levarterenol Bitartrate see 6788
Levasole [Pitman-Moore] see 5486
Levatol [Lilly] see 7209
Levaxin see 5497
Levcromakalim, 5487
Levium [Sodelco] see 3042
Levius [Farmitalia] see 886
Levlen [Berlex] see 6797
Levobren [Knoll] see 9163
Levobunolol, 5488
Levocabastine, 5489
Levocarbinoxamine see 1845
Levocarnitine see 1898
Levocitrol [Firmenich] see 2391
Levocycline see 8411
Levodopa, 5490
Levo-Dromoran see 5496
Levodropropizine see 3507
Levofloxacin see 6865
Levoglutamina see 4479
Levoleucovorin Calcium see 4254
Levomenol see 1281
Levomepate, 5491
Levomeprazine see 6066
Levomepromazine see 6066
Levomethadyl Acetate, 5492
Levomethorphan Hydrobromide see
   8274
Levomoprolol see 6345
Levonordefrin see 6785
Levonorgestrel see 6797
Levopa [ICN] see 5490
Levophacetoperane, 5493
Levophed [Winthrop] see 6788
Levophta [Chauvin] see 5489
Levopimaric Acid, 5494
Levopraid [Ravizza] see 9163
Levoprome [Lederle] see 6066
Levopropoxyphene, 5495
Levopropoxyphene Napsylate see 5495

Levopropylcillin see 8002
Levorenin see 3656
Levorin see 1789
Levorin A₂ see 1789
Levorotatory Lactic Acid see 5348
Levorphan see 5496
Levorphanol, 5496
Lévospasme [Dausse] see 3399
Levosulpiride see 9163
Levotartaric Acid see 9235
Levotensin [Simes] see 6345
Levothroid [Armour] see 5497
Levothym [Promonta] see 4895
Levothyrox [Merck-Clévenot] see 5497
Levothyroxine Sodium, 5497
Lévotonine [Pan Medica] see 4895
Levotuss [Dompé] see 3507
Levovermax [Virbac] see 5486
Levovist [Schering] see 4353
Levoxadrol Hydrochloride see 3352
Levoxan [Cutter] see 3352
Levsin [Kremers-Urban] see 4907
Levugen see 4295
Levulinic Acid, 5498
Levulinic Acid Calcium Salt see 1724
Levulose see 4295
Levynite see 1685
Lewisite see 3112
Lexibiotico [Llano] see 2021
Lexinor [Astra] see 6793
Lexomil [Roche] see 1406
Lexone [Du Pont] see 6239
Lexotan [Roche] see 1406
Lexotanil [Roche] see 1406
LF-178 see 4019
LFA-2043 see 5093
LG-152 see 5834
LG-11457 see 3753
LG-30158 see 8406
LH, 5499
LH-RF see 5500
LH-RH, 5500
LH-RH/FSH-RH see 5500
Liatris, 5501
Libanil [APS] see 4486
Libethenite see 2716
Libexin [Chinoin] see 7918
Libigen [Savage] see 2273
Libratar [UCEPHA] see 2125
Libritabs [Roche] see 2132
Librium [Roche] see 2132
Licabile see 3986
Licaran see 3986
Licareol see 5520
Licarpin [Allergan] see 7578
Licheniformins, 5502
Lichenin, 5503
Licorice see 4514
Lidamidine, 5504
Lidanar [Sandoz] see 5970
Lidanil [Sandoz] see 5970
Lidaprim [Nycomed] see 9089
Lidarral [Rorer] see 5504
Liden see 5201
Lidepran see 5493
Lidesthesin [Ritsert] see 5505
Lidex [Syntex] see 4186
Lidocaine, 5505
Lidofenin, 5506
Lidoflazine, 5507
Lidone [Abbott] see 6315
Lidothesin see 5505
Life [SIT] see 8180
Lifène [Debat] see 7414
Life Root see 8594
Lifril [Kissei] see 9267
Light Green N see 5739
Light Green SF Yellowish, 5508
Light Oil of Camphor see 6879

Light Spar see 1753
Lignans, 5509
Lignavet [C-Vet] see 5505
Lignite Wax see 6339
Lignocaine see 5505
Lignocaine Benzyl Benzoate see 2942
Lignoceric Acid, 5511
Lignum Vitae, 5512
Ligroin, 5513
Ligustrin see 9192
Likinozym [Kaigai] see 5671
Likuden [Hoechst] see 4571
Lilacillin [Takeda] see 9059
Lilacin see 9192
Lilly 01526 see 5219
Lilly 35483 see 2822
Lilly 36352 see 9815
Lilly 53838 see 4021
Lilly 69323 see 4021
Lilly 99170 see 793
Lily of the Valley see 2574
Liman [Kali-Chemie] see 9293
Limaprost, 5514
Limarsol see 49
Limas [Taisho] see 5552
Lima Wood see 7322
Limbial see 7059
Limclair [Sinclair] see 3558
Lime see 1733
Limestone, 5515
Lime Sulfurated Solution, 5516
Limethasone [Green Cross] see 2986
Limettin, 5517
Limifen [Janssen] see 236
Limit [Monsanto] see 420
Limonene, 5518
Limonin, 5519
Limonite see 4068
Limonoic Acid 3,19:16,17-Dilactone see 5519
Limonoic Acid Di-δ-lactone see 5519
LIN-1418 see 9168
Linalol see 5520
Linalool, 5520
Linalyl Acetate, 5521
Linamarin, 5522
Linarigenin-glucoside see 5523
Linarin, 5523
Linaris [R.A.N.] see 9086
Linatine, 5524
Linaxar [Armour Pharm.] see 9027
Lincocin [Upjohn] see 5525
Lincolcina see 5525
Lincoln Bean see 8878
Lincolnensin see 5525
Lincomix see 5525
Lincomycin, 5525
Linctussal see 8753
Lindafor [Rhône-Poulenc] see 5526
Lindane, 5526
Lindatox see 5526
Linden Tree see 1035
Lindlar Catalyst, 5527
Lindol [Stauffer] see 9892
Lindoxyl [Lindopharm] see 401
Linear Gramicidins see 4553
Lineatin, 5528
4,6,6-Lineatin see 5528
Lingraine [Sterling-Winthrop] see 3703
Lingran [Sterling-Winthrop] see 3703
Linnaeite see 2488
Linodil [Sanofi-Winthrop] see 5009
Linoleic Acid, 5529
9,12-Linoleic Acid see 5529
Linolenic Acid, 5530
α-Linolenic Acid see 5530
γ-Linolenic Acid see 5531
Linolexamide see 5529

Linolic Acid see 5529
Linomide [AB Leo] see 8418
Linoral [Organon] see 3780
Linostil [Siegfried] see 3258
Linseed, 5532
Linseed Oil, 5533
Linton [Yoshitomi] see 4629
Lintrin see 6495
Linum see 5532
Linurex [Makhteshim-Agan] see 5534
Linuron, 5534
Linyl [Roussel] see 7415
Liometacen [Chiesi] see 4998
Lion's Tooth see 9228
Lioplacentyl [Maestretti] see 7672
Lioresal [Ciba] see 967
Liosol see 10205
Liothyronine, 5535
Liothyronine Sodium see 5535
Lioxone see 8154
Lipal [PVO Intl.] see 7737 and 7738
Lipal 9LA [PVO Intl.] see 7717
Lipal 400S [PVO Intl.] see 7738
Lipalt [Von Boch] see 2980
Lipamone see 3153
Lipan see 3331
Lipanor [Winthrop] see 2373
Lipanthyl [Fournier] see 4019
Lipantil [Fournier] see 4019
Liparon [Finadiet] see 2900
Lipase, 5536
Liphadione [Lipha] see 2204
Lipidil [Ibirn] see 4019
Lipidium [Rhone-Poulenc] see 6603
Lipid-mobilizing Hormone see 5539
Lipiodol [Guerbet] see 5045
Liple [Green Cross] see 8063
Lipoclar [Crinos] see 4019
Lipoclin [Sumitomo] see 2415
Lipocol [Lipo] see 7737
Lipocol C-20 [Lipo] see 7737
Lipodel [Delalande] see 7144
Lipofene [Teofarma] see 4019
Lipoglutaren [Ausonia] see 5852
Lipo-Hepin [Riker] see 4685
Lipo-Hepinette [Riker] see 4685
α-Lipoic Acid see 9462
Lipo-Lutin [Parke-Davis] see 7956
Lipolytic Hormone see 5539
Lipo-Merz [Merz] see 3923
Liponorm [Gentili] see 8686
Lipopeg [Lipo] see 7738
Lipopeg 4-S [Lipo] see 7738
Lipoprotein Lipase, 5537
Liposana [Farmaroma] see 10205
Liposit [SIT] see 4019
Liposolvin [Tosi] see 8684
Liposomes, 5538
Liposom Forte [Fidia] see 5538
Liposorb O-20 [Lipo Chem.] see 7742
Lipostat [Squibb] see 7894
Lipotril see 2261
Lipotrophin see 5539
Lipotropic Hormone, 5539
Lipotropin see 5539
β-Lipotropin (61-91) see 3613
β-Lipotropin C-fragment see 3613
Lipozid [Pierrel] see 4394
Liprinal [Bristol] see 2436
Liprodène [Farlamedia] see 7275
Lipsin [Fournier] see 4019
Liptan [Kowa] see 4925
Lipur [Parke-Davis] see 4394
Liquaemin Sodium [Organon] see 4685
Liquamar [Organon] see 7413
Liquémin [Roche] see 4685
Liquid Butter of Antimony see 749
Liquid Paraffin see 7327
Liquid Rosin see 9210
Liquid Silver see 5957

Liquifilm [Allergan] see 7745
Liquiphene [India Med] see 7453
LiquiVent [Alliance] see 7299
Liquoid [Roche] see 8813
Liquor Aluminii Aceticotartarici see 333
Liquorice see 4514
LIR-1660 see 10081
Liranol [Wyeth] see 7966
Lironox see 2865
Lisacef [Lisapharm] see 2032
Lisagal see 7108
Lisaglucon [Farmasa] see 4486
Liserdol [Farmitalia] see 5999
Lisergan see 4013
Lisidonil [Ciba] see 4803
Lisil [Lenza] see 1850
Lisino [Essex] see 5608
Lisinopril, 5540
Lisium [Brunton] see 2140
Liskantin [Desitin] see 7927
Liskonum [SKB] see 5552
Lisomucil [Synthelabo] see 1850
Lisozima [SPA] see 5671
Lispamol [Specia] see 487
Lispine [Nippon Chemiphar] see 3424
Lissamine Green SF [ICI] see 5508
Lissapol see 4378
Lissephen [Abbott] see 5895
Lissolamine V see 2068
Listenon see 9044
Listica [Armour Pharm.] see 4884
Listomin S [Lion] see 1562
Listrocol [Farmitalia] see 2835
Lisuride, 5541
Lita [Opta], 5542
Litalir [Bristol] see 4896
Litarex [Dumex] see 5555
Litec [Sandoz] see 7671
Litesse [Pfizer] see 7726
Lithamide see 5546
Lithane [Pfizer] see 5552
Litharge see 5433
Lithicarb [Fisons] see 5552
Lithiodiisopropylamine see 3240
Lithiophor [Vifor] see 5569
Lithium, 5543
Lithium Acetate, 5544
Lithium Acetylsalicylate, 5545
Lithium Alanate see 361
Lithium Aluminohydride see 361
Lithium Aluminum Hydride see 361
Lithium Amide, 5546
Lithium Antimoniothiomalate see 720
Lithium Antimony Thiomalate see 720
Lithium Benzoate, 5547
Lithium Biborate see 5549
Lithium Bichromate see 5556
Lithium Bitartrate, 5548
Lithium Borate, 5549
Lithium Borohydride, 5550
Lithium Bromide, 5551
Lithium Carbonate, 5552
Lithium Chloride, 5553
Lithium Chromate(VI), 5554
Lithium Citrate, 5555
Lithium Dichromate(VI), 5556
Lithium Diisopropylamide see 3240
Lithium-Duriles [Pharma-Stern] see 5569
Lithium Fluoride, 5557
Lithium Formate, 5558
Lithium Hydrate see 5560
Lithium Hydride, 5559
Lithium Hydroxide, 5560
Lithium Iodide, 5561
Lithium Metasilicate see 5568
Lithium Nitrate, 5562
Lithium Oxalate, 5563
Lithium Oxide, 5564

Lithium Perchlorate, 5565
Lithium Platinocyanide see 5570
Lithium Salicylate see 8484
Lithium Selenate, 5566
Lithium Selenite, 5567
Lithium Silicate, 5568
Lithium Sulfate, 5569
Lithium Tetraborate see 5549
Lithium Tetracyanoplatinate(II), 5570
Lithium Tetrahydroaluminate see 361
Lithium Tetrahydroborate see 5550
Lithobid [Ciba-Geigy] see 5552
Lithocholic Acid, 5571
Lithographic Stone see 5515
Lithol see 4929
Lithonate [Solvay] see 5552
Lithonate S [Rowell] see 5555
Lithopone, 5572
Lithostat [Uro-Research] see 60
Lithotabs [Solvay] see 5552
Liticon [Lagap] see 7261
Litlure A see 7950
Litmocidin, 5573
Litmomycin see 4555
Litmopyrine see 5545
Litmus, 5574
Litosmil [Evans] see 3350
Litsoeine see 5400
Little Water see 10169
Litursol [Alfa] see 10026
Livalfa [Mitsubishi] see 9894
Livazone [Specia] see 9427
Liver Extract, 5575
Liver *Lactobacillus casei* Factor see 4253
Liver of Lime see 1754
Liver of Sulfur see 7762
Liver Starch see 4506
Livetins, 5576
Livial [Organon] see 9566
Liviclina [Sierochimica] see 1967
Livonal see 3816
Livostin [Janssen] see 5489
Lixacol [Cusi] see 5964
Lixidol [Farmitalia] see 5318
Lixil [AB Leo] see 1508
LJ-48 see 5828
LJ-206 see 1850
LJC-10141 see 3989
LJC-10305 see 2003
LJC-10627 see 1243
LL-1530 see 6432
LL-1558 see 9556
LL-1656 see 1498
LL-AV290 see 922
LLD Factor see 10152
LLF see 3968
LL-F-28249α see 6532
Lloncefal [Castillon] see 2025
Llonexina [Castillon] see 2021
LM-001 see 3989
LM-91 see 2204
LM-123 see 4168
LM-192 see 10141
LM-208 see 8267
LM-209 see 5911
LM-280 see 5029
LM-427 see 8380
LM-637 see 1403
LM-2717 see 2417
LM-2909 see 2941
LM-5008 see 4965
LM-22102 see 5188
LMD see 2989
LMWD see 2989
LN-107 see 1471
Lobak see 2155
Lobamine [Lobica] see 6053
Lobelanidine, 5577
Lobelanine, 5578

Lobelia, 5579
Lobelidine see 5580
Lobeline, 5580
α-Lobeline see 5580
Lobenzarit, 5581
Lobeton see 5580
Lobidan see 5580
Lobron see 5580
Lobufen [Lederle] see 4925
Locabiotal [Servier] see 4335
Locacorten (formerly) [Ciba] see 4173
Localyn [Syntex] see 4185
Locapred [Fabre] see 2973
Loceryl [Roche] see 612
Lochnericine, 5582
Lochneridine, 5583
Locoid [Yamanouchi] see 4828
Locorten [Ciba] see 4173
Locron [Hoechst] see 356
Loctidon [Lepetit] see 9160
Locton [Lepetit] see 9160
Locula [East India] see 9067
Locust Bean Gum, 5584
Lodalès [Sanofi Winthrop] see 8686
Loderix [EGIS] see 8619
Lodine [Ayerst] see 3920
Lodopin [Fujisawa] see 10327
Lodosin [Merck & Co.] see 1843
Lodosyn [Merck & Co.] see 1843
Lodoxamide, 5585
Lodoxamide Ethyl see 5585
Lodoxamide Trometamol see 5585
Lodoxamide Tromethamine see 5585
Loestrin [Parke, Davis] see 6790
Lofentanil, 5586
Lofepramine, 5587
Lofetensin [Nattermann] see 5588
Lofexidine, 5588
Loflucarban, 5589
Lofoxin [Locatelli] see 4280
Loftan [Cascan] see 217
Lofton [Abbott] see 1498
Loftran [Beecham] see 5308
Loftyl [Abbott] see 1498
Logan [Ist. Chim. Int.] see 2380
Loganin, 5590
Logic [Maag] see 4027
Logical [Armstrong] see 10049
Logiparin [Novo] see 9592
Logna [Tanabe] see 2408
Logran [Ciba-Geigy] see 9732
Logwood see 4671
Logynon [Schering] see 6797
Lokalison F [Dorsch] see 2986
Lokarin see 3253
Loline, 5591
Lomadine [M & B] see 7352
Lomapect [TAD] see 7918
Lombristop [Septa] see 9426
Lomebact [Shionogi] see 5592
Lomefloxacin, 5592
Lomerizine, 5593
Lomeron [Lilly] see 10190
Lomexin [Recordati] see 4047
Lomidine [Rhône-Poulenc] see 7254
Lomir [Sandoz] see 5260
Lomotil [Searle] see 3371
Lomper [Esteve] see 5807
Lomudal [Fisons] see 2658
Lomupren [Fisons] see 2658
Lomusol [Fisons] see 2658
Lomuspray [Fisons] see 2658
Lomustine, 5594
LON-798 see 4590
Lonacol [Bayer] see 10300
Lonapalene, 5595
Lonavar [Searle] see 7054
Lonazolac, 5596
Londomin [Taiyo] see 9193
Londomycin see 6007

Luteinizing Hormone-Releasing Hormone *see* 5500
Luteoantine *see* 4299
Luteocobaltic Chloride *see* 4711
Luteogonin B [AFI] *see* 2273
Luteohormone *see* 7956
Luteolas [Serono] *see* 3905
Luteolin, 5641
Luteolin-4'-methyl Ether *see* 3349
Luteonorm [Serono] *see* 3905
Luteotropic Hormone *see* 7961
Luteotropin *see* 7961
Luteran [Cassenne] *see* 2152
Lutestral [Cassenne] *see* 2152
Lutetium, 5642
α,α'-Lutidin *see* 5643
2,6-Lutidine, 5643
"β-Lutidine" *see* 3892
Lutionex [Roussel-UCLAF] *see* 2939
Lutocyclin [Ciba] *see* 3787
Lutocyclin M [Ciba] *see* 7956
Lutocylol [Ciba] *see* 3787
Lutogyl [Roussel-UCLAF] *see* 7956
Luto-Metrodiol *see* 3905
Lutoral [Syntex] *see* 2152
Lutoral [Midy] *see* 5838
Lutrelef [Ferring] *see* 5500
Lutrepulse [Ortho] *see* 5500
Lutrexin [HW & D] *see* 5644
Lutromone [Endo] *see* 7956
Lututrin, 5644
Luvatren [Cilag-Chemie] *see* 6343
Luvistin [Boehringer, Ing.] *see* 4757
Luvox [Solvay Duphar] *see* 4251
Luxazone [Allergan] *see* 2986
Luxoben [Chinoin] *see* 9561
Luxon *see* 4636
LVD *see* 2989
LY-141B *see* 7304
LY-12735 *see* 6369
LY-031537 *see* 8275
LY-048740 *see* 921
LY-61017 *see* 3989
LY-061188 *see* 2021
LY-097964 *see* 1974
LY-099094 *see* 10125
LY-109514 *see* 6427
LY-110140 *see* 4222
LY-122772 *see* 3637
LY-127809 *see* 7304
LY-135837 *see* 4971
LY-139037 *see* 6758
LY-139481 *see* 8281
LY-156758 *see* 8281
LY-163892 Monohydrate *see* 5606
LY-170053 *see* 6959
LY-177370 *see* 9580
LY-177837 *see* 8864
LY-188011 *see* 4392
LY-188695 *see* 3597
LY-237216 *see* 3418
LY-246708 *see* 10190
LY-248686 *see* 3518
LY-253351 *see* 9217
Lyamine [Merck & Co.] *see* 5667
Lyapolate Sodium, 5645
Lycaconitine *see* 5646
Lycanol *see* 4517
Lycedan *see* 157
Lycine *see* 1225
Lycoctonine, 5646
Lycodine, 5647
Lycofawcine, 5648
Lycomarasmine, 5649
Lycopene, 5650
(all-trans)-Lycopene *see* 5650
(all-trans)-Lycopene-16,16'-diol *see* 5651
(all-trans)-Lycopen-16-ol *see* 5657
Lycopersicin *see* 9683

Lycophyll, 5651
Lycopodine, 5652
Lycopodium, 5653
Lycopodium Seed (Spores) *see* 5653
Lycopus, 5654
Lycoramine, 5655
Lycoremine *see* 4357
Lycorine, 5656
Lycoxanthin, 5657
Lycra [Du Pont] *see* 8882
Lyeton [Von Boch] *see* 10026
Lyman [Lubapharm] *see* 316
Lymecycline, 5658
Lymphazurin [Pharmascience] *see* 9161
Lymphoblastoid Interferon *see* 5016
Lymphochin *see* 697
Lymphocin *see* 697
Lymphocyte Activating Factor *see* 5019
Lymphocyte Adhesion Molecule-1 *see* 8568
Lymphocyte Mitogenic Factor *see* 5020
Lymphoquin *see* 697
Lymphotoxin *see* 9943
Lynamine *see* 5321
Lyndiol [Organon] *see* 5659
Lynestrenol, 5659
Lynoral [Organon] *see* 3780
Lynx [Bayer] *see* 9253
Lyobex [Lappe] *see* 6815
Lyogen [Promonta Lundbeck] *see* 4226
Lyogen Depot [Promonta Lundbeck] *see* 4226
Lyovac Antivenin [Merck & Co], 5660
Lyovac Diuril [Merck & Co.] *see* 2221
Lyovac Sodium Edecrin [Merck & Co.] *see* 3761
Lyphocin [LyphoMed] *see* 10066
Lypressin, 5661
Lys *see* 5667
Lysal [Gellini] *see* 5668
Lysalbinic Acid, 5662
Lysalgo [SIT] *see* 5842
Lysanxia [Warner-Lambert] *see* 7895
Lysbex *see* 1244
Lyseen [Zyma] *see* 7922
Lysenyl [Spofa] *see* 5541
Lysergamide, 5663
Lysergan *see* 4013
Lysergic Acid, 5664
Lysergic Acid Amide *see* 5663
D-Lysergic Acid (+)-Butanolamide-(2) *see* 6147
D-Lysergic Acid Diethylamide *see* 5665
d-Lysergic Acid-dl-hydroxybutyl-amide-2 *see* 6147
D-Lysergic Acid D-Propanolamide *see* 3693
D-Lysergic Acid L-2-Propanolamide *see* 3694
Lysergide, 5665
Lysergin [Toa Eiyo] *see* 3692
Lysergsäure Diethylamid *see* 5665
Lysibex *see* 1244
Lysidine, 5666
Lysine, 5667
L-Lysine *see* 5667
Lysine Acetylsalicylate, 5668
L-Lysine L-Glutamate, 5669
L-Lysine, Homopolymer *see* 7733
DL-Lysine Mono[2-(acetyloxy)benzo-ate] *see* 5668
Lysine Monosalicylate Acetate *see* 5668
8-L-Lysinevasopressin *see* 5661
N-Lysinomethyltetracycline *see* 5658

Lysivane [M & B] *see* 3793
Lysmucol [Schering] *see* 8707
Lysobex *see* 1244
Lysodren [Calbiochem] *see* 6302
Lysofon [Lafon] *see* 330
Lysomiol *see* 4651
"Lysophosphatidal Choline" *see* 7677
"Lysophosphatidal Ethanolamine" *see* 7677
Lysoplasmalogens *see* 7677
Lysortine [Adrian-Marinier] *see* 5667
Lysostaphin, 5670
Lysozyme, 5671
Lyspafen [Cilag-Chemie] *see* 3183
Lyspamin [Bracco] *see* 3382
Lysthenon [Nycomed] *see* 9044
Lysuride *see* 5541
Lysuron [Laevosan] *see* 287
N^2-L-Lysylbradykinin *see* 5290
Lytensium *see* 7253
Lytispasm [Medical] *see* 709
Lyxoflavine, 5672
L-Lyxoflavine *see* 5672
D-Lyxose, 5673
α-D-Lyxose *see* 5673
LZ-544 *see* 4710
M *see* 6053
M-285 *see* 2838
M & B 125 *see* 1182
M & B 693 *see* 9108
M & B 736 *see* 7352
M & B 744 *see* 8970
M & B 760 *see* 9115
M & B 782 *see* 7981
M & B 800 *see* 7254
M & B 2050A *see* 7274
M & B 2207 *see* 9043
M & B 2210 *see* 9184
M & B 4180 A *see* 5199
M & B 4486 *see* 7207
M & B 5062A *see* 414
M & B 8430 *see* 2466
M & B 15497 *see* 2910
M & B 17803A *see* 16
M & B 39831 *see* 9289
M-2H *see* 1562
M-14 *see* 8381
M-19-Q *see* 8408
M-71 *see* 1560
M-99 *see* 3932
M-115 *see* 9184
M-141 *see* 8890
M-144 *see* 904
M-183 *see* 3932
M-1028 *see* 4631
M-4209 *see* 1854
M-4365A2 *see* 8420
M-4888 *see* 2138
M-5050 *see* 3403
M-5943 *see* 2237
M-7555 *see* 8234
M-9834 *see* 1056
M-10580 *see* 2790
M-73101 *see* 3603
M-139603 *see* 9384
2-M-4-A *see* 44
Mabuterol, 5674
Macasirool [Hishiyama] *see* 4331
Macassar Oil, 5675
Machete [Monsanto] *see* 1533
Machiline *see* 2521
Mackreazid *see* 2751
Macladin [Guidotti] *see* 2400
Macleyine *see* 8083
Maclicine [Christiaens] *see* 3134
Maclurin, 5676
Macmiror [Poli] *see* 6623
Macocyn [Mack, Illert.] *see* 7111
Macol [Mazer] *see* 7737
Macquer's Salt *see* 7767

Macrabin [Glaxo] see 10152
Macrobid [Procter & Gamble] see 6696
Macrobin see 2475
Macrodantin [Eaton] see 6696
Macrodex [Pharmacia] see 2989
Macrodiol [Gotham] see 3746
Macrofusine see 4311
Macrogol see 7729
Macrogol Ester 400 see 7738
Macrogol Fatty Acid Esters see 7738
Macrogol Fatty Alcohol Ethers see 7737
Macrogol Nonylphenyl Ether see 6772
Macromerine, 5677
Macrophage Activating Factor see 5018
Macrophage Colony-Stimulating Factor, 5678
Macrophage-derived Growth Factor see 4117
Macrophage-targeted β-Glucocerebrosidase see 242
Macroscan-131 [Abbott] see 8613
Macrose see 2989
Maculotoxin see 9382
Macusine A see 5679
Macusine B see 5679
Macusine C see 5679
Macusines, 5679
MAD see 6017
Madar [Ravizza] see 6784
Maddrell's Salt, 5680
Madecassol [Laroche Navarron] see 869
Madelen [Finadiet] see 7000
Maderan [Nycomed] see 9089
Madlexin [Meiji] see 2021
Madopar [Roche] see 1079
Madribon [Roche] see 9073
Madrine [Langley] see 6015
MADU, 5681
Maduramicin, 5682
Maduramycin see 5682
MAF see 5018
Mafatate [Torii] see 5683
Mafenide, 5683
Magainins, 5684
Magaldrate, 5685
Magan [Warren-Teed] see 5723
Magcal see 5713
Magdelate see 5709
Magell see 6258
Magenta see 5686
Magenta I, 5686
Maggioni 1559 see 10205
Maghemite see 4072
Maghen [Savio] see 7646
Magic Acid [Cationics] see 4217
Magic Methyl [Aldrich] see 6151
Magic Tan see 3225
Magisal see 5689
Magistery of Bismuth see 1326
Maglite [Merck & Co.] see 5713
Magmilor [Poli] see 6623
Magnacort [Pfizer] see 4827
Magnamycin [Pfizer] see 1854
Magnamycin A see 1854
Magnamycin B see 1854
Magnesia see 5713
Magnesia Usta see 5713
Magnesite see 5696
Magnesium, 5687
Magnesium Acetate, 5688
Magnesium Acetylsalicylate, 5689
Magnesium Aluminate Hydrate see 5685
Magnesium Aluminum Silicate see 362
Magnesium Amide, 5690

Magnesium Ammonium Chloride see 562
Magnesium Ammonium Phosphate Hexahydrate see 9019
Magnesium Aspirin see 5689
Magnesium Benzoate, 5691
Magnesium Biphosphate see 5719
Magnesium Bisulfate, 5692
Magnesium Borate, 5693
Magnesium Bromate, 5694
Magnesium Bromide, 5695
Magnesium Bromoglutamate see 4477
Magnesium Butyrate see 1629
Magnesium Carbonate Hydroxide, 5696
Magnesium Chlorate, 5697
Magnesium Chloride, 5698
Magnesium Citrate, 5699
Magnesium Citrate, Dibasic, 5700
Magnesium Citrate Sol see 5700
Magnesium Clofibrate see 2437
Magnesium Diamide see 5690
Magnesium Diethyl see 3171
Magnesium Dioxide see 5717
Magnesium Diphenyl see 3386
Magnesium Fluoride, 5701
Magnesium Fluosilicate see 5704
Magnesium Formate, 5702
Magnesium Germanide, 5703
Magnesium Glucoheptonate see 4463
Magnesium Glucomonocarbonate see 4463
Magnesium Gluconate see 4464
Magnesium Glucosemonocarbonate see 4463
Magnesium Glutamate Hydrobromide see 4477
Magnesium Heparinate see 4685
Magnesium Hexafluorosilicate, 5704
Magnesium Hydrate see 5706
Magnesium Hydride, 5705
Magnesium Hydrogen Phosphate see 5718
Magnesium Hydrogen Sulfate see 5692
Magnesium Hydroxide, 5706
Magnesium Hyposulfite see 5734
Magnesium Iodide, 5707
Magnesium Lactate, 5708
Magnesium Mandelate, 5709
Magnesium Mesotrisilicate see 5727
Magnesium Metasilicate see 5727
Magnesium Nitrate, 5710
Magnesium Oleate, 5711
Magnesium Orthosilicate see 5727
Magnesium Oxalate, 5712
Magnesium Oxide, 5713
Magnesium Pemoline see 7206
Magnesium Perborate, 5714
Magnesium Perchlorate, 5715
Magnesium Perhydrol see 5717
Magnesium Permanganate, 5716
Magnesium Peroxide, 5717
Magnesium Phosphate, Dibasic, 5718
Magnesium Phosphate, Monobasic, 5719
Magnesium Phosphate, Tribasic, 5720
Magnesium Potassium Selenate, 5721
Magnesium Pyrophosphate, 5722
Magnesium Salicylate, 5723
Magnesium Selenate, 5724
Magnesium Selenide, 5725
Magnesium Selenite, 5726
Magnesium Silicates, 5727
Magnesium Silicide, 5728
Magnesium Silicofluoride see 5704
Magnesium Stannide, 5729
Magnesium Stearate, 5730
Magnesium Sulfate, 5731
Magnesium Sulfite, 5732

Magnesium Sulfocyanate see 5733
Magnesium Superoxol see 5717
Magnesium Thiocyanate, 5733
Magnesium Thiosulfate, 5734
Magnesium Trisilicate see 5727
Magneson, 5735
Magnespirin see 5689
Magnetic Iron Oxide see 4087
Magnetite see 4087
Magnetkies see 4106
Magnevist [Schering AG] see 4347
Magnipen [Clin-Comar-Byla] see 5986
Magnofenyl see 7091
Magnoflorine, 5736
Magnogene see 5698
Magnoline, 5737
Magnolol, 5738
Magnophenyl see 7091
Magnopyrol [Abbott] see 3414
Magnosil [Polfa] see 5727
Magnosulf [Normark] see 5734
Magrene see 3175
Magrilon [Sintyal] see 5801
Ma Huang see 3643
Maidenhair Tree see 4432
Maikohis [Nichiiko] see 2405
Maiorad [Rotta] see 9607
Maipedopa [Maipe] see 5490
Maitansine see 5800
Maizena [CPC] see 8954
Maize Oil see 2597
Majeptil [Rhône-Poulenc Rorer] see 9494
Majudin see 1196
Makarol [Mallinckrodt] see 3177
Maki [Lipha] see 1403
Makon see 6772
Malabar Nut see 158
Malachite see 2697
Malachite Green, 5739
Malachite Green G see 1398
Malagride see 49
Malamar 50 see 5740
Malaquin [Dolder] see 2215
Malaspray see 5740
Malathion, 5740
Malathon (obsolete) see 5740
Malayan Camphor see 1366
Malazide see 5745
Malazol see 315
Malcotran [Fisons] see 4766
Maleamic Acid, 5741
Maleanilic Acid, 5742
Malee see 3941
Male Fern see 883
Maleic Acid, 5743
Maleic Acid, Diethyl Ester see 3172
Maleic Acid Hydrazide see 5745
Maleic Acid Monoamide see 5741
Maleic Anhydride, 5744
Maleic Hydrazide, 5745
Male Shield-fern see 883
Maleuric Acid, 5746
Malexil [Ferrosan] see 3993
Maleylurea see 5746
Maliasin [Knoll] see 8045
Malic Acid, 5747
l-Malic Acid Monoammonium Salt see 524
Malidone [Schering] see 315
Malilum see 261
Malimyxin [Chassot] see 2542
Maliner Kren see 4785
Malipuran [Heumann] see 1497
Malix see 3614
Malix [Lagap] see 4486
Mallebrin [Krewel] see 347
Mallermin-F [Taiyo] see 2405
Mallophene [Mallinckrodt] see 7361
Mallorol [Sandoz] see 9497

Mallotoxin *see* 8429
Mallow, 5748
Malocide [Rhône Poulenc Rorer] *see* 8169
Malol *see* 10027
Malonal *see* 989
Malonamide Nitrile *see* 2758
Malonic Acid, 5749
Malonic Acid Cyclic Isopropylidene Ester *see* 5858
Malonic Acid Di-*tert*-butyl Ester *see* 3085
Malonic Acid Ethyl Ester Nitrile *see* 3833
Malonic Ester *see* 3869
Malonic Mononitrile *see* 2759
Malononitrile, 5750
Malononitrile Dimer *see* 502
Malononitrile Hydrazide *see* 2751
5,5'-[Malonylbis(methylimino)]bis[*N*,-*N'*-bis[2,3-dihydroxy-1-(hydroxymethyl)propyl]-2,4,6-triiodoisophthalamide] *see* 5082
Malonylurea *see* 990
Malotilate, 5751
Maltobiose *see* 5753
Maltol, 5752
Maltonic Acid *see* 4464
Maltos [Otsuka] *see* 5753
Maltose, 5753
Malt Sugar *see* 5753
Malvidin Chloride, 5754
Malvin *see* 5754
Malvoside *see* 5754
Malysol *see* 1054
Mamaloxin [Showa Yakuhin] *see* 2021
Mamallet-A [Showa] *see* 495
Mamiesan [Kyowa] *see* 3147
Mammacillin [Stricker] *see* 7226
Mammex *see* 6697
Mammotropin *see* 7961
Manaca, 5755
Manacan *see* 5755
Manchurian Bean *see* 8878
Mancona Bark *see* 8529
Mancozeb, 5756
Mandacon [Conal] *see* 6040
Mandamina *see* 6040
Mandarin G *see* 6990
Mandaverm [Asta] *see* 5758
7-D-Mandelamido-3-(1-methyl-1,2,3,4-tetrazole-5-thiomethyl)-Δ³-cephem-4-carboxylic Acid *see* 1964
7-D-Mandelamido-3-[[(1-methyl-1*H*-tetrazol-5-yl)thio]methyl]-3-cephem-4-carboxylic Acid *see* 1964
7-Mandelamido-3-[[(1-methyl-1*H*-tetrazol-5-yl)thio]methyl]-8-oxo-5-thia-1-azabicyclo[4.2.0]oct-2-ene-2-carboxylic Acid *see* 1964
(6*R*,7*R*)-7-[(*R*)-Mandelamido]-8-oxo-3-[[[1-(sulfomethyl)-1*H*-tetrazol-5-yl]thio]methyl]-5-thia-1-azabicyclo-[4.2.0]oct-2-ene-2-carboxylic Acid *see* 1980
Mandelamine [Warner-Chilcott] *see* 6040
Mandelic Acid, 5757
dl-Mandelic Acid *see* 5757
Mandelic Acid Ammonium Salt *see* 563
Mandelic Acid Isoamyl Ester, 5758
Mandelic Acid Isopentyl Ester *see* 5758
Mandelic Acid Nitrile *see* 5759
Mandelic Acid β-1,2,2,6-Tetramethyl-4-piperidyl Ester *see* 3943
Mandelic Acid 3,3,5-Trimethylcyclohexyl Ester *see* 2771
Mandelonitrile, 5759

Mandelonitrile-β-gentiobioside *see* 638
Mandelonitrile Glucoside, 5760
D-Mandelonitrile-β-D-glucosido-6-β-D-glucoside *see* 638
Mandelonitrile β-Glucuronide *see* 638
Mandelonitrile Vicianoside *see* 10110
β-4-Mandeloyloxy-1,2,2,6-tetramethylpiperidine *see* 3943
Mandelyltropeine *see* 4766
Mandenol *see* 3866
Mandokef [Lilly] *see* 1964
Mandol [Lilly] *see* 1964
Mandolsan [San Carlo] *see* 1964
Mandoz *see* 6040
Mandrake Root *see* 7705
Mandrozep [Henk] *see* 3042
Mandurin *see* 6040
Maneb, 5761
Maneon [Poli] *see* 429
Manerix [Roche] *see* 6309
Manex [Crystal Chem] *see* 5761
Manexin *see* 5789
Manganblende *see* 5783
Manganese, 5762
Manganese Acetate, 5763
Manganese Benzoate *see* 1122
Manganese Binoxide *see* 5770
Manganese Borate, 5764
Manganese Bromide, 5765
Manganese Carbonate, 5766
Manganese Carbonyl, 5767
Manganese Chloride, 5768
Manganese Dibromide *see* 5765
Manganese Dichloride *see* 5768
Manganese Difluoride, 5769
Manganese Diiodide *see* 5772
Manganese Dioxide, 5770
Manganese Ethylenebis(dithiocarbamate) (Polymeric) Complex with Zinc Salt *see* 5756
Manganese Fluoride *see* 5769
Manganese Green *see* 1010
Manganese Hypophosphite, 5771
Manganese Iodide, 5772
Manganese Monosulfide *see* 5783
Manganese Nitrate, 5773
Manganese Oleate, 5774
Manganese Oxalate, 5775
Manganese Oxide, 5776
Manganese Peroxide *see* 5770
Manganese Phosphate, Dibasic, 5777
Manganese Pyrophosphate, 5778
Manganese Selenide, 5779
Manganese Sesquioxide, 5780
Manganese Silicate, 5781
Manganese Sulfate, 5782
Manganese Sulfide, 5783
Manganese Superoxide *see* 5770
Manganese Trifluoride, 5784
Manganic Fluoride *see* 5784
Manganite *see* 5780
Manganjustite *see* 5781
Manganomanganic Oxide *see* 5776
Manganosite *see* 5762
Manganous Chloride *see* 5768
Manganous Ethylenebis[dithiocarbamate] *see* 5761
Manganous Fluoride *see* 5769
Mangostin, 5785
Manicol *see* 5788
Manidipine, 5786
Maninil [Berlin-Chemie] *see* 4486
Maniol [Morishita] *see* 3369
Manna, 5787
Manna Sugar *see* 5788
Mannidex *see* 5788
Mannite *see* 5788
Mannitol, 5788
D-Mannitol *see* 5788

Mannitol Hexanitrate, 5789
Mannitol Nitrate *see* 5789
Mannitol Nitrogen Mustard *see* 5790
Mannitrin *see* 5789
L-Mannomethylose *see* 8338
Mannomustine, 5790
D-Mannose, 5791
Mannose-terminated Human Placental Glucocerebrosidase *see* 242
Mannosidostreptomycin *see* 8984
Mannosylstreptomycin *see* 8984
Manoplax [Boots] *see* 4146
Mansil [Pfizer] *see* 7051
Mantadan [De Angeli] *see* 389
Mantadine [DuPont] *see* 389
Mantadix [Théraplix] *see* 389
Man-Tan *see* 3225
Mantomide [Winthrop] *see* 2126
Mantropine *see* 6040
Manvene *see* 6424
Manzate [Du Pont] *see* 5756
Manzeb *see* 5756
Manzin [Crystal Chem.] *see* 5756
MAO *see* 6330
MAOA *see* 6028
Maolate [Upjohn] *see* 2231
Maon [Takeda] *see* 8918
MAP *see* 5838
Mapeg [Mazer] *see* 7738
Mapeg 400 MS [Mazer] *see* 7738
Mapharsal *see* 7082
Mapharsen [Parke, Davis] *see* 7082
Mapharside *see* 7082
Maphenide *see* 5683
MAPO *see* 5998
Maposol [Procida] *see* 6016
Mappine *see* 1502
Maprofix *see* 4378
Maprotiline, 5792
Maratan [Ravizza] *see* 1337
Marathon [Velsicol] *see* 7947
Marazine [Geneva] *see* 2238
Marbadal [Hartmann] *see* 9117
Mar-Bate [Mardale] *see* 5908
Marboran [Burroughs Wellcome] *see* 6057
Marcain [BDH] *see* 1520
Marcaina [Pierrel] *see* 1520
Marcaine [Winthrop] *see* 1520
Marcoeritrex *see* 3722
Marcoumar [Roche] *see* 7413
Marcumar [Roche] *see* 7413
Maretin [Bayer] *see* 6441
Marevan [Duncan, Flockhart] *see* 10174
Marezine [Burroughs Wellcome] *see* 2779
Marfanil *see* 5683
Margaric Acid, 5793
Margarite *see* 1685
Margosan-O [Vickwood] *see* 6525
Margosa Oil *see* 6526
Maricolene *see* 2784
Marignac's Salt *see* 7843
Marigold *see* 1763
Marihuana *see* 1794
Marijuana *see* 1794
Marinco C [Merck & Co.] *see* 5696
Marinco H [Merck & Co.] *see* 5706
Marindinin *see* 5299
Marinol [Roxane] *see* 9349
Markweed *see* 7709
Marlate [Du Pont] *see* 6070
Marmelosin *see* 4960
Marme's Reagent *see* 7861
Marogen [Chugai-Upjohn] *see* 3729
Marograstim *see* 4558
Marplan [Roche] *see* 5172
Marrubiin, 5794
Marsh Gas *see* 6019

Marshite *see* 2732
Marshmallow *see* 325
Marsh Trefoil *see* 5888
Marsilid [Roche] *see* 5094
Marsin [Ikapharm] *see* 7385
Marsthine [Towa Yakuhin] *see* 2405
Martos-10 [Otsuka] *see* 5753
Martricin *see* 9972
Marucotol [Maruko] *see* 5820
Marvelon 150/30 [Akzo] *see* 2971
Marvinol [Naugatuck] *see* 7746
Mary-bud *see* 1763
Maryland Pink *see* 8900
Marzine [Burroughs Wellcome] *see* 2779
Masacin [Boehringer, Mann.] *see* 7939
Mascagnite *see* 590
Maschitt [Showa Shinyaku] *see* 4822
Masdil [Esteve] *see* 3247
Masdiol *see* 6017
Masletine [Sioe] *see* 2405
Masmoran [Pfizer] *see* 4897
Masnidipine *see* 5469
Masoprocol *see* 6786
Masor [Formenti] *see* 8963
Maspiron [Sawai] *see* 7016
Massicot *see* 5433
Masterfen [Dompé] *see* 9180
Masterid [Grünenthal] *see* 3504
Masteril [Syntex] *see* 3504
Masterone [Syntex] *see* 3504
Mastic, 5795
Mastiche *see* 5795
Mastimyxin [Chassot] *see* 7734
Mastisol *see* 5795
Mastix *see* 5795
Matacil [Bayer] *see* 452
Mataperro *see* 2560
Maté, 5796
Matenon [Upjohn] *see* 6262
Matico, 5797
Mator [Sandoz] *see* 4850
Matricaria *see* 2082
Matricaria Oil *see* 2082
Matricarin, 5798
Matridine *see* 5799
Matridin-15-one *see* 5799
Matrine, 5799
Matrol *see* 2152
Matromycin [Pfizer] *see* 6962
MATS *see* 6094
Matulane [Roche] *see* 7938
Maveral [Farmades] *see* 4251
Mavrik [Zoecon] *see* 4249
Maxair [3M Riker] *see* 7644
Maxaquin [Monsanto] *see* 5592
Maxeran [Nordic] *see* 6226
Maxforce [Cyanamid] *see* 4802
Maxibolin [Organon] *see* 3851
Maxicaine [Delagrange] *see* 7171
Maxicam [Parke, Davis] *see* 5258
Maxicortex [Manetti Roberts] *see* 169
Maxidex [Alcon] *see* 2986
Maxifen [Merck & Co.] *see* 7669
Maxiflor [Herbert] *see* 3186
Maxilase [Sanofi-Winthrop] *see* 640
Maxim [Ciba-Geigy] *see* 4165
Maximed [MSD] *see* 8088
Maxipen [Pfizer] *see* 7369
Maxipime [Bristol-Myers Squibb] *see* 1973
Maxitate [Pennwalt] *see* 5789
Maxiton *see* 2996
Maxivate [Westwood] *see* 1226
Maxivent *see* 3494
Maxolon [Beecham] *see* 6226
Maxulvet [Veterinaria] *see* 9073
Max-Uric [Labinca] *see* 1093
Maxus [Lilly] *see* 921
Maxutrim [Veterinaria] *see* 9073

May Apple *see* 7705
May Blossom *see* 2574
Maycor [Parke, Davis] *see* 5245
Maydol *see* 2597
Maygace [Bristol] *see* 5849
May Lily *see* 2574
May Pops *see* 7183
Maytansine, 5800
Mazicon [Roche] *see* 4169
Mazildene [Farmochim. Ital.] *see* 5801
Mazindol, 5801
Mazipredone, 5802
Mazola [CPC] *see* 2597
Mazun *see* 10235
MB-10064 *see* 1465
MB-10731 *see* 1465
MB-46030 *see* 4127
MBA *see* 5815
MBC *see* 1836
MBLA *see* 5864
MBOCA *see* 6136
MBR-8251 *see* 7300
MBR-12325 *see* 5846
MBT *see* 5916
MBTS *see* 3435
MBU *see* 6002
3-MC *see* 6124
MC-838 *see* 6368
MC-903 *see* 1679
MC-4379 *see* 1256
MCA *see* 2162
MCAA *see* 5607
MCI-186 *see* 6809
MCI-2016 *see* 1255
MCI-9038 *see* 816
M-Cillin [Misemer] *see* 7225
MCMN *see* 8413
MCMT *see* 6307
McN-485 *see* 10328
McN-742 *see* 497
McN-1025 *see* 6780
McN-1210 *see* 8173
McN-1546 *see* 4176
McN-2559 *see* 9655
McN-2559-21-98 *see* 9655
McN-4853 *see* 9686
McN-A-2833 *see* 7311
McN-A-2833-109 *see* 7311
McN-JR-4584 *see* 1077
McN-JR-7094 *see* 5507
McN-JR-8299 *see* 5486
McN-JR-15403-11 *see* 3183
McN-R-73-Z *see* 1845
McN-R-726-47 *see* 7714
McN-R-1967 *see* 4040
McN-X-181 *see* 10048
MCNU *see* 8285
MCP *see* 5803
MCPA, 5803
MCPP *see* 5826
M-CSF *see* 5678
MCT *see* 6333
MD-76 *see* 5851
MD-141 *see* 3766
MD-805 *see* 816
MD-6134 *see* 10174
MD-6753 *see* 2349
MD-67350 *see* 2350
MD-69276 *see* 9659
MDA, 5804
MDE, 5805
MDEA *see* 5805
M-Det *see* 2912
MDGF *see* 4117
5-M-dioxanol *see* 4494
MDL-473 *see* 8385
MDL-507 *see* 9269
MDL-9918 *see* 9307
MDL-14042A *see* 5588
MDL-16455 *see* 4113

MDL-17043 *see* 3627
MDL-71754 *see* 10114
MDL-72974A *see* 6313
MDL-73147 *see* 3471
MDL-73147EF *see* 3471
MDMA, 5806
MDMH *see* 4877
MDP *see* 5837 and 6385
MDP-Lys (L18) *see* 8412
MDS *see* 2993
ME-1206 *see* 1972
ME-1207 *see* 1972
ME-3202 *see* 7271
ME-3625 *see* 6601
ME-3737 *see* 9582
MEA *see* 2848
Meadow Anemone *see* 8125
Meadow Clover *see* 9818
Meadow Crocus *see* 2537
Meadow Saffron *see* 2537
MEAP *see* 7497
Measurin [Sterling-Winthrop] *see* 886
Meat Sugar *see* 5008
Meaverin [Woelm] *see* 5905
Mebadin [Glaxo] *see* 2924
Meballymal Sodium *see* 8563
Mebaral [Winthrop] *see* 5899
Mebendazole, 5807
Mebenvet [Janssen] *see* 5807
Meberyt *see* 3726
Mebeverine, 5808
Mebhydroline, 5809
Mebinol [Erba] *see* 2233
Mebiquine, 5810
MeBmt, 5811
Mebonat [Boehringer, Mann.] *see* 2432
Mebrofenin, 5812
Mebron [Daiichi] *see* 3659
Mebrophenhydramine *see* 3596
Mebryl *see* 3596
Mebubarbital *see* 7272
Mebutamate, 5813
Mebutina [Formenti] *see* 5813
3-MECA *see* 6124
Mecadox [Pfizer] *see* 1825
Mecalmin [Yoshitomi] *see* 3369
Mecamine *see* 5814
Mecamylamine, 5814
MeCbl *see* 6125
Mecca-galls *see* 6827
Mechlorethamine, 5815
Mechlorethamine Oxide Hydrochloride, 5816
Mechlorprop *see* 5826
Mechothane *see* 1232
Meciclin [Citobios] *see* 2937
Mecilex [C.E.P.A.] *see* 2021
Mecillinam *see* 408
Meclan [Ortho] *see* 5818
Meclastine *see* 2405
Meclizine, 5817
Meclocycline, 5818
Mecloderm *see* 5818
Meclodol [Parke-Davis] *see* 5819
Meclofenamic Acid, 5819
Meclofenoxane *see* 5820
Meclofenoxate, 5820
Meclomen [Parke, Davis] *see* 5819
Meclophenamic Acid *see* 5819
Meclopran [Lagap] *see* 6226
Mecloqualone, 5821
Mecloralurea, 5822
Meclosorb [Basotherm] *see* 5818
Mecloxamine, 5823
Meclozine *see* 5817
Meclutin [ABC] *see* 5818
Mecobalamin *see* 6125
Mecocyanin *see* 2757
Mecodin *see* 6008

Mecodrin see 623
Mecomec [PBI/Gordon] see 5826
Meconic Acid, 5824
Meconin, 5825
Meconinic Acid Lactone see 5825
Mecopar see 5826
Mecoprop, 5826
Mecoprop-P see 5826
Mecoral see 2139
Mecostrin Chloride see 9939
Mecrylate, 5827
Mectizan [Merck & Co.] see 5264
Mecysteine Hydrochloride, 5828
MED-15 see 637
Medaject [Neda] see 7937
Medan [Kowa] see 1224
Medapsol see 9122
Medaron see 4320
Medazepam, 5829
Medazepol [Farmasa] see 5829
Medemanol see 5789
Medemycin [Meiji] see 6271
Mederantil [Boehringer, Ing.] see 1472
Mederel [Riom] see 611
Medetomidine, 5830
(S)-Medetomidine see 5830
Medialan LL-33 see 4379
Medialan LL-99 see 4379
Mediatensyl [Inergie] see 10003
Mediator [Servier] see 1066
Mediaven [Syntex] see 6481
Mediaxal [Servier] see 1066
Medibazine, 5831
Mediben see 3090
Medicagol, 5832
Medicarpin, 5833
Medicil [Medici] see 6350
"Medicinal Chrysophanic Acid" see
2312
Medicoal [Lundbeck] see 1856
Medicort [Care] see 4828
Medifenac [Medici] see 220
Medifoxamine, 5834
Medifuran see 4315
Medigoxin see 3210
Medihaler-Epi [3M] see 3656
Medilave [Farillon] see 2074
Mediletten [Medivet] see 9337
Medilla [Omega] see 4903
Medinal [Warner-Chilcott] see 989
Medio-Contrix "38" see 5080
Medipectol see 1244
Medipren [McNeil] see 4925
Medmain, 5835
Medodorm [Medo] see 2139
Medomet [DDSA] see 6132
Medomin [Geigy] see 4689
Medopa [Kaigi/Nippon Kayaku] see
6132
Medopren [Malesci] see 6132
Medoxim [Medici] see 2002
Medphalan see 5871
Medrate [Upjohn] see 6189
Medrocort [Upjohn] see 5840
Medrogestone, 5836
Medroglutaric Acid see 5852
Medrol [Upjohn] see 6189
Medrol Stabisol [Upjohn] see 6189
Medrone [Upjohn] see 6189
Medronic Acid, 5837
Medroxyprogesterone, 5838
Medrylamine, 5839
Medrysone, 5840
Medullipin, 5841
Medullipin I see 5841
Medullipin II see 5841
Meerschaum see 5727
Meerzwiebel see 8925
Mefamide [Bayer] see 5683
Mefedina [Farmitalia] see 5894

Mefenacid [Streuli] see 5842
Mefenamic Acid, 5842
Mefenorex, 5843
Mefexadyne [Anphar-Rolland] see
5844
Mefexamide, 5844
Mefloquine Hydrochloride, 5845
Mefluidide, 5846
Mefoxin [Merck & Co.] see 1986
Mefoxitin [Merck & Co.] see 1986
Mefruside, 5847
Mega see 5813
Megabion (Japanese) see 6017
Megace [Mead Johnson] see 5849
Megacef [Beytout] see 2032
Megacert [Wakamoto] see 4344
Megacillin [Grünenthal] see 7230
Megacillin Suspension [Frosst] see
7221
Megacillin Tablets [Frosst] see 7225
Megacins, 5848
Megaclor [Pharmax] see 2448
Megakaryocyte Potentiating Factor
see 9528
Megalocin [Roche] see 4137
Megalone [Roche] see 4137
Megapen see 7226
Megaphen [Bayer] see 2238
Megasedan [DIF] see 5829
Megasul [Am. Cyanamid] see 6715
Megestat [Lappe] see 5849
Megestil [Boehringer, Mann.] see 5849
Megestrol Acetate, 5849
Megimide [Abbott] see 1054
Meglum [Bago] see 5486
Meglumine see 6154
Meglumine Acetrizoate, 5850
Meglumine Amidotrizoate see 5851
Meglumine Diatrizoate, 5851
Meglumine Iothalamate see 5080
Meglutol, 5852
Megrin [Yoshitomi] see 4688
Meiact [Meiji Seika] see 1972
Meicelin [Meiji Seika] see 1978
Meilax [Meiji Seika] see 3867
Mcionite see 1685
Meitnerium see 3585
MEK see 6149
Meksamin see 6083
Meladinine [Basotherm] see 6068
Meladrazine see 4803
Melamine, 5853
2-(4-Melamin-2-ylphenyl)-4-hydroxy-
methyl-1,3-dithia-2-arsolane see
5856
Melampyrin see 4350
Melampyrite see 4350
Melampyrum see 4350
Melaniline see 3383
Melanins, 5854
Melanocyte-stimulating Hormone see
6377
Melanocyte-stimulating Hormone Re-
lease-inhibiting Factor see 5855
Melanophore-affecting Hormone see
6377
Melanophore Dilating Hormone see
6377
Melanophore Expanding Hormone see
6377
Melanophore Hormone see 6377
Melanophore-stimulating Hormone
see 6377
Melanosome-dispersing Hormone see
6377
Melanostatin, 5855
Melanostatin I (Ox) see 5855
Melanotropic Hormone see 6377
Melanotropin see 6377
α-Melanotropin see 6377

β-Melanotropin see 6377
γ-Melanotropin see 6377
Melanotropin Inhibiting Factor see
5855
Melanterite see 4105
Melanthigenin see 4654
Melarsoprol, 5856
Melatonin, 5857
Mel B see 5856
Melbex [Lilly] see 6408
Meldane [Chemagro] see 2626
Meldrum's Acid, 5858
Melengestrol, 5859
Meletin see 8216
Melex [Sankyo] see 6254
Melezitose, 5860
Melfalan see 5871
Melfax [Apotex] see 8286
Melibiose, 5861
Melidorm [Asta] see 5823
Melilot, 5862
Melilotoside, 5863
Melin see 8456
Melinamide, 5864
Melinonine A, 5865
Melinonine G see 4133
Melipramine [EGYT] see 4955
Melissyl Alcohol see 9722
Melitase [Berk] see 2239
Melitose see 8279
Melitoxin see 3140
Melitracen, 5866
Melitriose see 8279
Melittin, 5867
Melittin I see 5867
Melixeran [Lusofarma] see 5866
Mellaril [Sandoz] see 9497
Melleretten [Wander] see 9497
Melleril [Wander] see 9497
Mellic Acid see 5868
Mellitic Acid, 5868
Mellitin see 5867
Mellitoxin see 4898
Melon Tree see 7152
Melopat [Medopharm] see 1224
Meloxicam, 5869
Meloxine [Upjohn] see 6068
Melperone, 5870
Melphalan, 5871
Melprex [Am. Cyanamid] see 3468
Melsedin [Boots] see 6028
Melsomin see 6028
Meltatox [BASF] see 3467
Meltrol [USV] see 7376
Melysin [Takeda] see 409
Memantine, 5872
Memcor [Lilly] see 10190
Memento 400 [Volpino] see 7613
Memine [Macfarlan Smith] see 7484
Memoq [Gödecke] see 6580
Memphenesin see 5895
Menacor [Menarini] see 2418
Menadiol, 5873
Menadione, 5874
Menadione Carboxymethoxime Am-
monium Salt see 5875
Menadoxime, 5875
Menamin [Chugai] see 5316
Menaphthone see 5874
Menaphthone Carboxymethoxime
Ammonium Salt see 5875
Menaquinone 4 see 5876
Menaquinone 6 see 5876
Menaquinone 7 see 5876
Menaquinones, 5876
Menatetrenone see 5876
Menazon, 5877
Menazone see 4330
Menbutone, 5878
Mencortex [Menarini] see 169

Mendelevium, 5879
Mendiaxon [Roland] see 4903
Mendon [Dainippon] see 2465
Mendrin see 3618
Menetryl see 3752
Menformon [Organon] see 3751
Menhaden Oil, 5880
Menhydrinate see 3252
Menichlopholan see 6601
Menitazine [Towa Yakuhin] see 1224
Menocil [Cilag-Chemie] see 497
Menogaril, 5881
Menogarol see 5881
Menopatol [Chemiphar] see 9660
Menophase [Syntex] see 5976
Menorest [Rhone-Poulenc Rorer] see 3746
Menotropins see 4299
Menova [E. Merck] see 2152
Mensiso [Menarini] see 8695
Menstridyl see 2152
Mentane [Hoechst-Roussel] see 10077
Mentax [Kaken] see 1547
p-Mentha-1,5-diene see 7340
p-Mentha-1(7),2-diene see 7341
p-Mentha-1,8-diene see 5518
p-Mentha-6,8-dien-2-one see 1925
(3R,4R)-2-p-Mentha-1,8-dien-3-yl-5-pentylresorcinol see 1792
p-Menthane-1,8-diol see 9314
3-p-Menthanol see 5882
(1R,4S)-(—)-p-Menthan-3-one see 5883
2-[(p-Mentha-1,3,5-trien-2-yloxy)-methyl]-2-imidazoline see 9965
1-p-Menthene-6,8-diol see 8707
p-Menth-6-ene-2,8-diol see 8707
1-p-Menthen-2-ol-3-one see 3351
m-Menth-6-en-5-one see 4768
p-Menth-1-en-3-one see 7626
R-(+)-p-Menth-4(8)-en-3-one see 8124
Menthol, 5882
l-Menthol see 5882
l-Menthone, 5883
Menthyl Acetate, 5884
Menthyl Borate, 5885
Menthyl Salicylate, 5886
Menthyl Valerate, 5887
Mentium [Guidotti] see 8175
Menyanthes, 5888
Menzol [Kabi] see 6790
Meobentine, 5889
Meothrin [Shell] see 4033
MEP see 4017
Mepacrine see 8225
Meparfynol, 5890
Mepartricin, 5891
Mepartricin A see 5891
Mepartricin B see 5891
Mepasin see 5892
Mepavlon [ICI] see 5908
Mepazine, 5892
Mepedyl [Allen & Hanburys] see 3390
Mepenicycline see 7235
Mepenzolate Bromide, 5893
Meperidine, 5894
Mepha-Butazon [Mepha] see 7431
Mephacyclin [Mepha] see 9337
Mephenamin [Boehringer, Mann.] see 7007
Mephenesin, 5895
Mephenhydramine see 6370
Mephenon see 6008
Mephenoxalone, 5896
Mephentermine, 5897
Mephenytoin, 5898
Mepherol see 5895
Mephesin see 5895
Mephine see 5897
Mephobarbital, 5899

Mephosfolan, 5900
Mephson see 5895
Mephyton [Merck & Co.] see 7536
Mepicaton [Pharmaton] see 5905
Mepiciclina see 7606
Mepicor [Corvi] see 5901
Mepicycline see 7606
Mepicycline Phenoxymethyl-penicillinate see 7235
Mepident [Parke, Davis] see 5905
Mepidium [Recordati] see 9583
Mepindolol, 5901
Mepiprazole, 5902
Mepiquat Chloride, 5903
Mepirizole see 3659
Mepirodipine see 1031
Mepirzepine see 6295
Mepitiostane, 5904
Mepivacaine, 5905
Mepivastesin [Espe] see 5905
Mepixanox, 5906
Mepixanthone see 5906
Meposed [Vernleigh] see 5908
Mepral [Bracco] see 6977
Meprane Dipropionate [Reed & Carnrick] see 6045
Mepred [Savage] see 6189
Meprednisone, 5907
Meprin [York] see 5908
Meprindon [Indo-Pharma] see 5908
Meprobamate, 5908
Meproban [Draco] see 5908
Meprocompren [Cascan] see 5908
Meprofen [Agips] see 5316
Meprol [Lokman] see 5908
Mepron [Burroughs-Wellcome] see 898
Meproscillarin see 8060
Meprospan [Wallace Labs.] see 5908
Meprotabs [Wallace Labs.] see 5908
Meproten see 5908
Meprotil [Bruner-Tillman] see 5908
Meprylcaine, 5909
Meptazinol, 5910
Meptid [Wyeth] see 5910
Meptin [Otsuka] see 7939
Meptran [Reid-Provident] see 5908
Mepyramine see 8168
Mepyrapone see 6246
Mequelon [Frosst] see 6028
Mequin [Lemmon] see 6028
Mequitazine, 5911
Mequiverine see 10141
MER-17 see 925
MER-27 see 7163
MER-29 see 9867
Meractinomycin see 2867
Meradan(e) [USSR] see 8390
Meraklon [Montecatini] see 7741
Meralein Sodium, 5912
Meralen [Merrell] see 4167
Meralluride, 5913
Meralop [ISF] see 2757
Meralops [Dulcis] see 2757
Meratonic [Merrell] see 7638
Meratran [Marion Merrell Dow] see 7638
Merbak [Schieffelin] see 63
Merbentul [Merrell] see 2225
Merbentyl [Merrell] see 3147
Merbromin, 5914
Mercaleukin see 5919
Mercamine see 2848
Mercamphamide, 5915
Mercaptamine see 2848
Mercaptoacetamide Antimony Derivative see 746
2-Mercaptoacetanilide S-Gold Derivative see 918
Mercaptoacetic Acid see 9472

Mercaptoacetic Acid Antimony Derivative Sodium Salt see 744
Mercaptoacetic Acid [3,5-Bis(1,1-dimethylethyl)-4-hydroxyphenyl]-methyl Ester see 2894
Mercaptoacetic Acid Calcium Derivative see 1758
Mercaptoacetic Acid Calcium Salt Gold Derivative see 6411
Mercaptoacetic Acid Sodium Salt see 8842
(SP-5-25)-[N-[N-[N-(Mercaptoacetyl)-glycyl]glycyl]glycinato(5—)-N,N',-N'',S]oxotechnetate(2—)-⁹⁹ᵐTc Disodium see 9258
β-Mercaptoalanine see 2850
Mercaptoarsenol see 852
2-Mercaptobenzimidazole see 1112
2-Mercaptobenzimidazole-1,3-dimethylol see 9446
2-Mercaptobenzoic Acid see 9498
2-Mercaptobenzothiazole, 5916
Mercaptobenzthiazyl Ether see 3435
Mercaptobutanedioic Acid see 9479
Mercaptobutanedioic Acid Antimony-(3+) Lithium Salt (3:1:6) see 720
Mercaptobutanedioic Acid Monogold-(1+) Sodium Salt see 4536
Mercaptodiacetic Acid see 9467
Mercaptodimethur see 6050
Mercaptoethane see 3771
2-Mercaptoethanesulfonic Acid Sodium Salt see 5969
2-Mercaptoethanol, 5917
β-Mercaptoethanol see 5917
β-Mercaptoethylamine see 2848
(2-Mercaptoethyl)trimethylammonium Iodide O,O-Diethyl Phosphorothioate see 3549
2-Mercaptohistidine see 9477
2-Mercapto-4-hydroxypyrimidine see 9504
2-Mercaptoimidazoline see 3849
N-(2-Mercaptoisobutyryl)-L-cysteine see 1481
Mercaptomerin Sodium, 5918
Mercaptomethane see 6023
N-(2-Mercapto-2-methyl-1-oxopropyl)-L-cysteine see 1481
(S)-1-(3-Mercapto-2-methyl-1-oxopropyl)-L-proline see 1817
N-(Mercaptomethyl)phthalimide S-(O,-O-Dimethyl Phosphorodithioate) see 7492
N-(2-Mercapto-2-methylpropanoyl)-L-cysteine see 1481
(2S)-1-(3-Mercapto-2-methylpropion-yl)-L-proline see 1817
N-[1-[(S)-3-Mercapto-2-methylpropionyl]-L-prolyl]-3-phenyl-L-alanine Acetate (Ester) see 202
1-Mercaptonaphthalene see 6465
2-Mercapto-N-2-naphthalenylacet-amide see 9480
2-Mercaptonapthalene see 6466
N-(2-Mercapto-1-oxopropyl)glycine see 9597
(2-Mercapto-N-phenylacetamidato-O,S)gold see 918
Mercaptophos see 2940
Mercaptophos [Bayer] see 4044
3-Mercapto-1,2-propanediol see 9471
2-Mercaptopropanoic Acid see 9476
α-Mercaptopropanoic Acid see 9476
1-(3-Mercaptopropanoic Acid)-8-D-arginine Vasopressin see 2969
1-(3-Mercaptopropanoic Acid)oxytocin see 2899
2-Mercaptopropionic Acid see 9476

N-(2-Mercaptopropionyl)glycine *see* 9597

α-Mercaptopropionylglycine *see* 9597

N-(2-Mercaptopropionyl)glycine 2-Thiophenecarboxylate (Ester) *see* 8963

6-Mercaptopurine, 5919

2-Mercaptopyridine 1-Oxide *see* 8178

2-Mercapto-4(1H)-pyrimidinone *see* 9504

2-Mercapto-4-pyrimidone *see* 9504

Mercaptosuccinic Acid *see* 9479

Mercaptosuccinic Acid Antimonate (III) Hexalithium Salt *see* 720

Mercaptosuccinic Acid S-Antimony Derivative Lithium Salt *see* 720

Mercaptothion *see* 5740

3-Mercapto-D-valine *see* 7214

Mercaptyl [Knoll] *see* 7214

Mercardan [Parke, Davis] *see* 5913

Mercate "5" *see* 5141

Mercate "20" *see* 5141

Mercazole *see* 6049

Mercazolyl *see* 6049

Mercilon [Organon] *see* 2971

Mercloran [Parke, Davis] *see* 2154

Mercodinone *see* 4826

Mercoral *see* 2154

Mercufenol Chloride, 5920

Mercuhydrin [Lakeside] *see* 5913

Mercumallylic Acid, 5921

Mercumatilin Sodium *see* 5921

Mercuramide *see* 5962

Mercuranine *see* 5914

Mercuretin *see* 5913

Mercuric Acetate, 5922

Mercuric Ammonium Chloride *see* 564

Mercuric Arsenate, 5923

Mercuric Barium Iodide *see* 1011

Mercuric Benzoate, 5924

Mercuric Bromide, 5925

Mercuric Cacodylate *see* 1640

Mercuric Chloride, 5926

Mercuric Chloride, Ammoniated, 5927

Mercuric Cuprous Iodide *see* 2733

Mercuric Cyanide, 5928

Mercuric Dichromate(VI), 5929

Mercuric Fluoride, 5930

Mercuric Imidosuccinate *see* 5942

Mercuric Iodate, 5931

Mercuric Iodide, Red, 5932

Mercuric Nitrate, 5933

Mercuric Oleate, 5934

Mercuric Oxide, Red, 5935

Mercuric Oxide, Yellow, 5936

Mercuric Oxycyanide, 5937

Mercuric Potassium Cyanide *see* 7856

Mercuric Potassium Iodide *see* 7862

Mercuric Potassium Iodide Soln *see* 7870

Mercuric Salicylate, 5938

Mercuric Silver Iodide *see* 8678

Mercuric Sodium p-Phenolsulfonate, 5939

Mercuric Stearate, 5940

Mercuric Subsulfate, 5941

Mercuric Succinimide, 5942

Mercuric Sulfate, 5943

Mercuric Sulfide, Black, 5944

Mercuric Sulfide, Red, 5945

Mercuric Sulfocyanate *see* 5946

Mercuric Sulfocyanide *see* 5946

Mercuric Thiocyanate, 5946

Mercuri-hematoporphyrin Disodium Salt *see* 5961

Mercuriphenoldisulfonate Sodium *see* 5939

Mercurius Vitae *see* 734

Mercurochrome [HW & D] *see* 5914

Mercurochrome-220 Soluble *see* 5914

Mercurocol *see* 5914

Mercurome *see* 5914

Mercurophage *see* 5914

Mercurophen, 5947

Mercurophylline *see* 5915

Mercurothiolate *see* 9451

Mercurous Acetate, 5948

Mercurous Bromide, 5949

Mercurous Chlorate, 5950

Mercurous Chloride, 5951

Mercurous Fluoride, 5952

Mercurous Iodide, 5953

Mercurous Nitrate, 5954

Mercurous Oxide, 5955

Mercurous Sulfate, 5956

Mercury, 5957

Mercury Amide Chloride *see* 5927

Mercury Ammonium Chloride *see* 5927

Mercury Bichloride *see* 5926

Mercury Bichromate *see* 5929

Mercury Biniodide *see* 5932

Mercury Bisulfate *see* 5943

Mercury Cyanide Oxide *see* 5937

Mercury Difluoride *see* 5930

Mercury, Dimethyl *see* 3299

Mercury Mass, 5958

Mercury Monochloride *see* 5951

Mercury Oxide Black *see* 5955

Mercury Oxide Sulfate *see* 5941

Mercury Oxonium Sulfate *see* 5941

Mercury Perchloride *see* 5926

Mercury Pernitrate *see* 5933

Mercury Protochloride *see* 5951

Mercury Protoiodide *see* 5953

Mercury Protonitrate *see* 5954

Mercury and Sodium Phenolsulfonate *see* 5939

Mercury Subchloride *see* 5951

Mercury Subsalicylate *see* 5938

Mercusal *see* 5962

Meregon [Malesci] *see* 1509

Mereprine [Merrell] *see* 3497

Meresa [Dolorgiet] *see* 9163

Merfen [Zyma] *see* 7456

Meridia [Knoll] *see* 8629

Merilid *see* 2154

Merinax *see* 4733

Merislon [Eisai] *see* 1224

Merisoprol Hg 197, 5959

Merit [Miles] *see* 4946

Merital [Hoechst] *see* 6768

Meritin *see* 403

Merlium [Merck & Co.] *see* 355

Merocet [Merrell Dow] *see* 2074

Merodicein [HW & D] *see* 5912

Meronem [Zeneca] *see* 5960

Meropenem, 5960

Meropenin [Kabi] *see* 7230

Merosin *see* 5364

Meroxyl [Wander] *see* 4401

Meroxylan [Wander] *see* 4401

Merpan *see* 1815

Merphalan *see* 5871

Merphenyl Nitrate *see* 7455

Merphyrin, 5961

Merprane [Squibb] *see* 5959

Merpress *see* 6610

Merrem [Zeneca] *see* 5960

Mersalin *see* 5962

Mersalyl, 5962

Mersolite [India Med] *see* 7453

Mertax *see* 5916

Mertect [Merck & Co.] *see* 9426

Mertestate [Sterling Winthrop] *see* 9322

Merthiolate [Lilly] *see* 9451

Mervaldin [Lannett] *see* 5895

Mervan [Cont. Pharma; Warner-Lambert] *see* 220

Mervastrept *see* 2949

Merxin [Merck & Co.] *see* 1986

Merzonin [Takeda] *see* 9451

Mesaconic Acid, 5963

Mesalamine, 5964

Mesalazine *see* 5964

Mesantoin [Sandoz] *see* 5898

Mesasal [SK & F] *see* 5964

Mescaline, 5965

Mescaline Oxidase *see* 6330

Mescopil *see* 6084

Mesembrine, 5966

Mesentol *see* 3794

Mesidicaine *see* 9830

Mesitylene, 5967

Mesityl Oxide, 5968

Mesityl Oxide (1-Phthalazinyl)hydrazone *see* 1492

Mesmar [Dey's] *see* 5908

Mesna, 5969

Mesnex [Asta] *see* 5969

Mesocaine *see* 9830

Mesoetioporphyrin *see* 3915

Mesoinosite *see* 5008

Mesokain *see* 9830

Mesonex *see* 5009

Mesontoin *see* 5898

Mesopin [Endo] *see* 4766

Mesoridazine, 5970

Mesoridazine Besylate *see* 5970

Mesotal [Galma] *see* 5009

Mesotartaric Acid *see* 9238

Mesothorium I *see* 8277

Mesothorium II *see* 138

Mesoxalic Acid, 5971

Mesoxalic Acid Calcium Salt *see* 1726

Mesoxalylcarbamide *see* 290

Mesoxalylurea *see* 290

Mesoxan *see* 1726

Mesoyohimbine *see* 10238

Mespafin [Merckle] *see* 3496

Mesquite Gum, 5972

Messenger RNA *see* 8370

Mestanolone, 5973

Mestenediol *see* 6017

Mesterolone, 5974

Mestilbol, 5975

Mestinon [Roche] *see* 8161

Mestoranum [Schering AG] *see* 5974

Mestranol, 5976

Mesudin *see* 5683

Mesudrin *see* 5683

Mesulfa [Triosol] *see* 9081

Mesulfen, 5977

Mesulid *see* 6640

Mesulphen *see* 5977

Mesurol [Bayer] *see* 6050

Mesuximide *see* 6085

Met *see* 6053

Metab-Auxil *see* 7245

Metabenzthiazuron *see* 6002

Metabolin [Takeda] *see* 9430

Metabutethamine Hydrochloride *see* 1552

Metabutoxycaine, 5978

Metacam [Boehringer, Ing.] *see* 5869

Metacetaldehyde *see* 5983

Metachlor *see* 203

Metachloral *see* 9755

Metachrome Yellow, 5979

Metacide [Bayer] *see* 6183

Metaclazepam, 5980

Metacortandracin *see* 7904

Metacortandralone *see* 7901

Metacresol Acetate *see* 2651

Metacresylacetate-Sulzberger *see* 2651

Metacycline *see* 6007

[2.2]Metacyclophane, 5981
Metadee see 10156
Metadelphene [Hercules] see 2912
Metadiazine see 8170
Metadomus [Med. Domus] see 6007
Metadorm see 6028
Metafar [Lafare] see 1977
Metaformaldehyde see 9863
Metahydrin [Merrell Dow] see 9754
Metalaxyl, 5982
Metalcaptase [Heyl] see 7214
β₁-Metal-combining Protein see 9706
Metaldehyde, 5983
Metalkamate see 1494
Metallibure see 6013
Metallic Arsenic see 832
Metallum Problematum see 9273
Metalutin [Parke, Davis] see 6805
Metamfepramone, 5984
Metamfepyramone see 5984
Metamide [Protea] see 6226
Metamin [Takeda] see 4224
Metamine [Pfizer] see 9900
Metamitron, 5985
Metamizol see 3414
Metampicillin, 5986
Metam Sodium see 6016
Metamucil [Proctor & Gamble] see 7674
Metandiol see 6017
Metandren [Ciba] see 6206
Metanephrine, 5987
Metanilic Acid, 5988
Metanil Yellow, 5989
N¹-Metanilyl-N²-n-butylurea see 1615
Metanite see 907
Metanyl see 6025
Metaoxedrin see 7440
Metaphanine, 5990
Metaphen [Abbott] see 6707
Metaphos see 6183
Metaphosphoric Acid see 7501
Metaphosphoric Acid Trisodium Salt see 8845
Metaphyllin see 485
Metaplexan [Rhone-Poulenc] see 5911
Metapramine, 5991
Metaprel [Dorsey] see 5992
Metaproterenol, 5992
Metaradrine see 5993
Metaraminol, 5993
Metarsenobillon see 9111
Metasclene see 9867
Metaspas [Leeming] see 3211
Metasqualene see 9867
Metastab [Boots] see 6189
Metastigmin [Star] see 6553
Meta-Systox see 6129
Metathion see 4017
Metatyrosine see 9971
Metavariscite see 371
Metaxalone, 5994
Metaxan see 6025
Metaxanin see 5982
Metazepium Iodide see 1635
Metazocine, 5995
Metazol [CT] see 1977
Metazolo see 6049
Metcaraphen, 5996
Met⁵-E see 3613
Metebanyl [Sankyo] see 3511
Metelilachlor see 6230
Meteloidine, 5997
Metenarin [Teikoku Zoki] see 6147
Metendiol see 6017
Metenix [Hoechst] see 6231
Met⁵-Enkephalin see 3613
Méténolone see 6044
Metepa, 5998
Meterazine see 7942

Meterfer [Sinclair] see 4094
Metergoline, 5999
Metescufylline, 6000
Metflorylthiazidine see 4830
Metformin, 6001
Metformin Pamoate see 6001
"Meth" see 6015
Methabenzthiazuron, 6002
Methacetin see 48
Methacetone see 3170
Methacholine Chloride, 6003
Methacide see 9667
Methacolimycin see 2542
Methacrifos, 6004
Methacrylic Acid, 6005
Methacrylonitrile, 6006
Methacycline, 6007
Methaderm [Taiho] see 2986
Methadol see 3254
Methadone Hydrochloride, 6008
Methadren(e) see 6146
Methadyl Acetate, 6009
Methafluoridamid see 5846
Methaform see 2180
Methafrone see 5321
Methafurylene, 6010
Methagon [Corvel] see 4173
Methallatal, 6011
Methallenestril, 6012
Methallibure, 6013
Methallylchloride see 2200
α-Methallyl Chloride see 2182
β-Methallyl Chloride see 2200
γ-Methallyl Chloride see 2181
1-Methallyl-3-methyl-6-aminotetrahydro-2,4-pyrimidinedione see 507
Methalutin see 6805
Methamidophos, 6014
Methaminodiazepoxide see 2132
Methampex [Lemmon] see 6015
Methamphetamine, 6015
Methampicillin see 5986
Methampyrone see 3414
Metham Sodium, 6016
Methanabol see 6017
Methanal see 4262
Methanamide see 4264
Methanamine see 6095
Methandienone see 6018
Methandiol see 6017
Methandriol, 6017
Methandrostenolone, 6018
Methane, 6019
Methanearsonic Acid, 6020
Methanebisphosphonic Acid see 5837
Methanedicarboxylic Acid see 5749
Methanediphosphonic Acid see 5837
Methanedisulfonic Acid see 6052
Methanedisulfonic Acid Calcium Salt see 1727
Methaneperoxoic Acid see 7302
1-(5-Methanesulfonamido-1H-indol-2-ylcarbonyl)-4-[3-(1-methylethylamino)pyridinyl]piperazine see 2929
1-(4-Methanesulfonamidophenoxy)-2-[N-(4-methanesulfonamidophenethyl)-N-methylamino]ethane see 3469
Methanesulfonic Acid, 6021
Methanesulfonic Acid Ethyl Ester see 3873
Methanesulfonic Acid Methyl Ester see 6170
Methanesulfonic Acid Tetramethylene Ester see 1529
Methanesulfonyl Chloride, 6022
N,N'-Methanetetraylbiscyclohexanamine see 3146
Methanethioamide see 9469
Methanethiol, 6023
Methanetriol see 7012

Methaniazide see 5203
Methanide see 6025
Methanol, 6024
O-(1,4-Methano-1,2,3,4-tetrahydro-6-naphthyl)-N-methyl-N-(m-tolyl)thiocarbamate see 9647
Methantheline Bromide, 6025
Methanthine Bromide see 6025
Methaphenilene, 6026
Methapoxide see 5998
Methapyrilene, 6027
Methaqualone, 6028
Metharbital, 6029
Methargen, 6030
Methasan see 10305
Methased see 6028
β-Methasone see 1226
Methazol [Werner] see 9084
Methazolamide, 6031
Methazolastone see 9289
Methazole, 6032
Met Hb see 6035
Methbipyranone see 6246
Methcathinone, 6033
Methdilazine, 6034
Methedrine [Burroughs Wellcome] see 6015
Methemoglobin, 6035
Methenamine, 6036
Methenamine Allyl Iodide, 6037
Methenamine Anhydromethylenecitrate, 6038
Methenamine Hippurate, 6039
Methenamine Mandelate, 6040
Methenamine Salicylate, 6041
Methenamine Sulfosalicylate, 6042
Methenamine Tetraiodide see 6043
Methenamine Tetraiodine, 6043
Methenolone, 6044
Methenolone Enanthate see 6044
N,N'-Methenyl-o-phenylenediamine see 1111
Metheph [Napp] see 6145
Methergin [Sandoz] see 6147
Methergine [Sandoz] see 6147
Methergoline see 5999
Methescufylline see 6000
Methestrol, 6045
Methetharimide see 1054
Methetoin, 6046
Methexenyl see 4742
Methexenyl Sodium see 4742
Methforylthiazidine see 4830
Met-HGH see 8864
Methiacil see 6210
Methiazic Acid see 6219
Methiazinic Acid see 6219
Methicil see 6210
Methicillin Sodium, 6047
Methidathion, 6048
Methimazole, 6049
Methiocarb, 6050
Methiocil see 6210
Methiodal Sodium, 6051
Methionamine see 97
Methionic Acid, 6052
Methionine, 6053
L-Methionine see 6053
Methionine⁵-enkephalin see 3613
Methionine Hydroxy Analog, 6054
1-(N-L-Methionyl-L-alanine)-3-L-threonine-4-L-tyrosine-5-L-arginine-17-L-serinecolony-stimulating Factor (Human Clone 1034) see 9647
N-L-Methionylcolony-stimulating Factor (Human Clone 1034) see 4558
Methionyl Human Growth Hormone see 8864
N²-L-Methionyl-1-139-interferon γ

6,7-dimethoxy-2-methyl-1-isoquino-
linyl)methyl]phenol *see* 5389
N-[[2-Methoxy-3-[(1,2,3,6-tetrahydro-
1,3-dimethyl-2,6-dioxopurin-7-yl)-
mercuri]propyl]carbamoyl]succin-
amic Acid *see* 5913
4-Methoxy-2-[tetrahydro-3,4,5-tri-
hydroxy-6-(hydroxymethyl)pyran-2-
yl]-α-resorcylic Acid δ-Lactone *see*
1197
6-Methoxy-α-tetralone, 6081
N¹-(4-Methoxy-1,2,5-thiadiazol-3-yl)-
sulfanilamide *see* 9089
N-[[6-Methoxy-5-(trifluoromethyl)-1-
naphthalenyl]thioxomethyl]-N-meth-
ylglycine *see* 9663
(E)-5-Methoxy-1-[4-(trifluoromethyl)-
phenyl]-1-pentanone O-(2-Aminoeth-
yl)oxime *see* 4251
5-Methoxy-4'-(trifluoromethyl)valero-
phenone (E)-O-(2-Aminoethyl)oxime
see 4251
2-Methoxytrimeprazine *see* 6066
(3β,16β,17α,18β,20α)-17-Methoxy-
18-[(3,4,5-trimethoxybenzoyl)oxy]-
3,20-yohimban-16-carboxylic Acid
Methyl Ester *see* 2964
(E,E)-11-Methoxy-3,7,11-trimethyl-2,-
4-dodecadienoic Acid 1-Methylethyl
Ester *see* 6063
(R)-2-Methoxy-N,N,β-trimethyl-10H-
phenothiazine-10-propanamine *see*
6066
(all-E)-9-(4-Methoxy-2,3,6-trimethyl-
phenyl)-3,7-dimethyl-2,4,6,8-nona-
tetraenoic Acid *see* 114
9-(4-Methoxy-2,3,6-trimethylphenyl)-
3,7-dimethylnona-2,4,6,8-tetraen-1-
oic Acid Ethyl Amide *see* 6367
(all-E)-9-(4-Methoxy-2,3,6-trimethyl-
phenyl)-3,7-dimethyl-2,4,6,8-nona-
tetraenoic Acid Ethyl Ester *see* 3935
1-Methoxy-3-(trimethylsilyloxy)-1,3-
butadiene, 6082
6-Methoxy-1αH,5αH-tropan-3α-ol
Benzilate *see* 9908
6-Methoxytropine Benzilate *see* 9908
5-Methoxytryptamine, 6083
3-Methoxy-L-tyrosine *see* 6133
6-Methoxyumbelliferone *see* 8552
Methoxyverapamil *see* 4369
Methral [Pfizer] *see* 4225
Methrazone [Boehringer, Ing.] *see*
4053
Methscopolamine Bromide, 6084
Methscopolamine Nitrate *see* 8550
Methsuximide, 6085
Methural *see* 7102
Methyclothiazide, 6086
Methycobal [Eisai] *see* 6125
Methyl Abietate, 6087
Methylacetaldehyde *see* 8008
Methyl 4-Acetamido-2-ethoxyben-
zoate *see* 3791
3-Methyl-4-acetaminophenylarsanilic
Acid *see* 6102
m-Methylacetanilide *see* 74
N-Methylacetanilide, 6088
Methyl Acetate, 6089
Methylacetic Acid *see* 8010
Methylacetic Anhydride *see* 8011
Methyl Acetoacetate, 6090
N-Methyl-2-acetonylpyrrolidine *see*
4899
Methylacetopyronone *see* 2919
6-α-Methyl-17α-acetoxyprogesterone
see 5838
3-Methyl-4-acetylaminophenylarso-
nate *see* 6102
Methyl Acetylsalicylate, 6091

β-Methylacrolein *see* 2663
Methyl Acrylate, 6092
α-Methylacrylic Acid *see* 6005
β-Methylacrylic Acid *see* 2664
α-Methylacrylonitrile *see* 6006
α-Methyl-1-adamantanemethylamine
see 8390
3-O-Methyladrenaline *see* 5987
N-Methyladrenaline *see* 6146
Methylal, 6093
Methylal Acetamide *see* 4269
2-Methylalanine *see* 465
3-Methyl-N,α-(β-alanyl)-L-histidine
see 717
Methyl Alcohol *see* 6024
Methyl Aldehyde *see* 4262
2-(p-Methylallylaminophenyl)pro-
pionic Acid *see* 308
α-Methylallyl Chloride *see* 2182
β-Methylallylchloride *see* 2200
γ-Methylallyl Chloride *see* 2181
13β-Methyl-17α-allyl-Δ⁴,⁹,¹¹-gonatri-
ene-17β-ol-3-one *see* 327
N-Methyl-5-allyl-5-isopropylbarbitur-
ic Acid *see* 3607
α-dl-1-Methyl-5-allyl-5-(1-methyl-2-
pentynyl)barbituric Acid Sodium
Salt *see* 6061
Methyl Allyl Trisulfide, 6094
Methylamine, 6095
N-Methylaminoacetic Acid *see* 8519
2-(Methylamino)-m-anisic Acid Methyl Ester *see* 2873
Methyl 2-Aminobenzoate *see* 6099
2-(4-Methylaminobutoxy)diphenyl-
methane *see* 1255
N-[[(Methylamino)carbonyl]oxy]-
ethanimidothioic Acid Methyl Ester
see 6062
4-Methylaminocetopyrocatechol *see*
170
4''-epi-(Methylamino)-4''-deoxyaver-
mectin B₁ₐ *see* 3594
4''-epi-(Methylamino)-4''-deoxyaver-
mectin B₁ᵦ *see* 3594
2-(Methylamino)-2-deoxy-α-L-gluco-
pyranose *see* 6155
5-(Methylamino)-2'-deoxyuridine *see*
5681
3-Methylamino-7-dimethylamino-
phenazathonium Chloride *see* 958
4-Methylamino-1,5-dimethyl-2-phen-
yl-3-pyrazolone Sodium Methanesulfonate *see* 3414
α-dl-6-(Methylamino)-4,4-diphenyl-3-
heptanol Acetate *see* 6778
N-Methylaminodithioformic Acid
Sodium Salt *see* 6016
N-Methylaminoethane Sodium Sulfonate *see* 6205
2-Methylaminoethanesulfonic Acid *see*
6205
β-Methylaminoethane-α-sulfonic Acid
see 6205
Methylaminoethanoic Acid *see* 8519
2-Methylaminoethanol, 6096
l-Methylaminoethanolcatechol *see*
3656
p-Methylaminoethanolphenol *see* 9189
m-Methylaminoethanolphenol Hydrochloride *see* 7440
p-Methylaminoethanolphenol Tartrate
see 9189
4-[2-(Methylamino)ethyl]-1,2-benz-
enediol *see* 2950
α-[1-(Methylamino)ethyl]benzene-
methanol *see* 3645
α-[1-(Methylamino)ethyl]benzyl
Alcohol *see* 3645

α-(1-Methylaminoethyl)-p-hydroxy-
benzyl Alcohol *see* 4869
N-Methylaminoethyl 2-Methylbenz-
hydryl Ether *see* 9641
4-[2-(Methylamino)ethyl]-o-phenylene
Diisobutyrate *see* 4921
2-[2-(Methylamino)ethyl]pyridine *see*
1224
4-[2-(Methylamino)ethyl]pyrocatechol
see 2950
2-Methyl-6-aminoheptane *see* 6854
2-Methyl-6-amino-2-heptanol *see* 4691
6-Methyl-2-amino-6-heptanol *see* 4691
Methyl 3-Amino-4-hydroxybenzoate
see 7011
2-Methyl-4-amino-1-hydroxynaphtha-
lene *see* 10162
β-Methylamino-α-(4-hydroxyphenyl)-
ethyl Alcohol *see* 9189
α-Methylamino-β-(3-indole)propionic
Acid *see* 6
2-Methylamino-5α-(β-indolyl)ethyl-2-
oxazolin-4-one *see* 4996
2-Methylaminoisocamphane *see* 5814
3-Methylaminoisocamphane *see* 5814
Methyl α-Amino-β-mercaptopropionate Hydrochloride *see* 5828
N-Methylaminomethanethionothiolic
Acid Sodium Salt *see* 6016
(R)-α-[(Methylamino)methyl]benzene-
methanol *see* 4633
(−)-α-(Methylaminomethyl)benzyl
Alcohol *see* 4633
1-(Methylaminomethyl)dibenzo[b,e]-
bicyclo[2.2.2]octadiene *see* 1117
9-(Methylaminomethyl)-9,10-dihydro-
9,10-ethanoanthracene *see* 1117
6-Methylamino-2-methylheptene *see*
5202
Methylaminomethyl 4-Hydroxyphen-
yl Carbinol *see* 9189
1-Methylamino-2-methyl-2-phenyleth-
ane *see* 7462
2-Methylamino-2-methyl-1-phenyl-
propane *see* 5897
α-(Methylaminomethyl)vanillyl
Alcohol *see* 5987
2-Methyl-4-amino-1-naphthol *see*
10162
N-[4-(6-Methylamino-7-nitro-2-thia-5-
aza-6-heptene-1-yl)-2-thiazolylmeth-
yl]-N,N-dimethylamine *see* 6758
p-Methylaminophenol Sulfate, 6097
2-Methylamino-1-phenylethanol *see*
4633
1-Methylamino-2-phenylpropane *see*
7462
2-Methylamino-1-phenyl-1-propanol
see 3645
DL-threo-2-(Methylamino)-1-phenyl-
propan-1-ol *see* 3645
L-erythro-2-(Methylamino)-1-phenyl-
propan-1-ol. *see* 3645
2-(Methylamino)-1-phenyl-1-pro-
panone *see* 6033
1-Methyl-4-amino-N-phenyl-N-(2-
thenyl)piperidine *see* 9413
α-Methylaminopropiophenone *see*
6033
β-(Methylamino)propylcyclopentane
see 2808
1-(3-Methylaminopropyl)dibenzo[b,e]-
bicyclo[2.2.2]octadiene *see* 5792
7-(3-Methylaminopropyl)-1,2;5,6-di-
benzocycloheptatriene *see* 8088
5-(3-Methylaminopropyl)-5H-dibenzo-
[a,d]cycloheptene *see* 8088
9-(γ-Methylaminopropyl)-9,10-dihy-
dro-9,10-ethanoanthracene *see* 5792

Methylmethioninesulfonium Bromide *see* 10165

Methylmethioninesulfonium Chloride *see* 10165

Methyl *N*-(2-Methoxyacetyl)-*N*-(2,6-xylyl)-DL-alaninate *see* 5982

1-Methyl-6-methoxy-3,4-dihydro-2-carboline *see* 6073

1-Methyl-7-methoxy-3,4-dihydro-*β*-carboline *see* 4644

1-Methyl-4-methoxy-7,8-methylene-dioxycarbostyril *see* 1936

l-*α*-2-Methyl-8-methoxy-6,7-methyl-enedioxy-1-(6,7-dimethoxy-3-phthal-idyl)-1,2,3,4-tetrahydroisoquinoline *see* 6815

(+)-*trans*-1-Methyl-3-[*p*-(methoxy)-phenoxymethyl]-4-phenylpiperidine *see* 3993

3,4-(2′-Methyl-2′-methoxy-4′-phenyl)-dihydropyranocoumarin *see* 2786

2-Methyl-2-methoxy-4-phenyl-5-oxo-dihydropyrano[3,2-*c*][1]benzopyran *see* 2786

2-(3-Methyl-5-methoxy-1-pyrazolyl)-4-methoxy-6-methylpyrimidine *see* 3659

2-Methyl-4-methoxy-6-sulfanilamido-1,3-diazine *see* 9085

2-Methyl-4-methoxy-6-sulfanilamido-pyrimidine *see* 9085

1-Methyl-6-methoxy-1,2,3,4-tetrahy-dro-2-carboline *see* 172

3-Methyl-8-methoxy 3*H*-1,2,5,6-tetra-hydropyrazino[1,2,3-*ab*]-*β*-carboline *see* 6238

1-Methyl-6-(1-methylallyl)-2,5-dithio-biurea *see* 6013

2-Methyl-6-methylamino-2-heptene *see* 5202

Methyl 2-(Methylamino)-3-methoxy-benzoate *see* 2873

4-[*N*-Methyl-4-(methylamino)-L-phen-ylalanine]virginiamycin S₁ *see* 10096

endo-1-Methyl-*N*-(9-methyl-9-azabicy-clo[3.3.1]non-3-yl)-1*H*-indazole-3-carboxamide *see* 4557

N-Methyl-2-(2-methylbenzhydryloxy)-ethylamine *see* 9641

1-Methyl-5-(4-methylbenzoyl)-1*H*-pyrrole-2-acetic Acid *see* 9655

N-[[1-Methyl-5-(4-methylbenzoyl)-1*H*-pyrrol-2-yl]acetyl]glycine 2-Meth-oxyphenyl Ester *see* 637

Methyl 1-(*α*-Methylbenzyl)imidazole-5-carboxylate *see* 6232

2-Methyl-6-(3-methylbutylamino)-heptane *see* 6846

3-Methyl-*N*-(3-methylbutyl)-1-butan-amine *see* 3237

6-Methyl-*N*-(3-methylbutyl)-2-heptanamine *see* 6846

S-Methyl *N*-[(Methylcarbamoyl)oxy]-thioacetimidate *see* 6062

(2*S*,3*R*)-5-Methyl-3-[[(*αS*)-*α*-(methyl-carbamoyl)phenethyl]carbamoyl]-2-[(2-thienylthio)methyl]hexanohy-droxamic Acid *see* 1036

Methyl *O*-(Methylcarbamoyl)thiol-acetohydroxamate *see* 6062

1-Methyl-2-[2-(*α*-methyl-*p*-chloro-benzhydryloxy)ethyl]pyrrolidine *see* 2405

1-Methyl-2-[2-(methyl-*p*-chlorodi-phenylmethyloxy)ethyl]pyrrolidine *see* 2405

6-Methyl-2-(4-methyl-3-cyclohexen-1-yl)-5-hepten-2-ol *see* 1281

α-[(*α*-Methyl-3,4-methylenedioxy-phenethylamino)methyl]protocate-chuyl Alcohol *see* 8082

1-Methyl-2-(3,4-methylenedioxyphen-yl)ethyl Octyl Sulfoxide *see* 9140

N-Methyl-3,4-methylenedioxyphenyl-isopropylamine *see* 5806

3-Methyl-6,7-methylenedioxy-1-piperonylisoquinoline *see* 6171

6-Methyl-16-methylene-17*α*-hydroxy-Δ⁶-progesterone *see* 5859

N-Methyl-*N*-methylenemethanamini-um Iodide *see* 3300

2-Methyl-6-methylene-1,7-octadiene *see* 6413

2-Methyl-6-methylene-2,7-octadiene *see* 6413

7-Methyl-3-methylene-1,6-octadiene *see* 6413

Methyl *O*-Methyl-18-epireserpate *see* 6237

1-Methylmethylergonovine *see* 6217

4-Methyl-7-(1-methylethenyl)-1-azulenecarboxaldehyde *see* 5346

(1*R*-*cis*)-1-Methyl-2-(1-methylethen-yl)cyclobutaneethanol *see* 4556

1-Methyl-4-(1-methylethenyl)cyclo-hexene *see* 5518

2-Methyl-5-(1-methylethenyl)-2-cyclo-hexene-1-one *see* 1925

2-[3-Methyl-6-(1-methylethenyl)-2-cyclohexen-1-yl]-5-pentyl-1,3-benz-enediol *see* 1792

4-[2-Methyl-5-(1-methylethenyl)-1-cyclopenten-1-yl]-2-butanone *see* 5223

endo-8-Methyl-8-*syn*-(1-methyl)ethyl-8-azoniabicyclo[3.2.1.]octane-3-(2-propyl)-pentanoate Bromide *see* 8693

Methyl(1-methylethyl)benzene *see* 2832

1-Methyl-2-(1-methylethyl)benzene *see* 2832

1-Methyl-3-(1-methylethyl)benzene *see* 2832

1-Methyl-4-(1-methylethyl)benzene *see* 2832

4-Methyl-1-(1-methylethyl)bicyclo-[3.1.0]hexan-3-one *see* 9533

2-Methyl-5-(1-methylethyl)-1,3-cyclo-hexadiene *see* 7340

(1*α*,2*β*,5*α*)-5-Methyl-2-(1-methyleth-yl)cyclohexanol *see* 5882

5-Methyl-2-(1-methylethyl)cyclo-hexanol Acetate *see* 5884

5-Methyl-2-(1-methylethyl)cyclo-hexanol Monoester with Boric Acid *see* 5885

(2*S*-*trans*)-5-Methyl-2-(1-methylethyl)-cyclohexanone *see* 5883

3-Methyl-5-(1-methylethyl)-2-cyclo-hexen-1-one *see* 4768

3-Methyl-6-(1-methylethyl)-2-cyclo-hexen-1-one *see* 7626

1-Methyl-4-(1-methylethyl)-2,3-dioxa-bicyclo[2.2.2]oct-5-ene *see* 864

6-Methyl-*N*-(1-methylethyl)-2-heptan-amine *see* 6243

(*R*)-5-Methyl-2-(1-methylethylidene)-cyclohexanone *see* 8124

exo-(±)-1-Methyl-4-(1-methylethyl)-2-[(2-methylphenyl)methoxy]-7-oxa-bicyclo[2.2.1]heptane *see* 2354

1-Methyl-2-(1-methylethyl)-5-nitro-1*H*-imidazole *see* 5095

[*S*-(*E*)]-8-Methyl-5-(1-methylethyl)-6,8-nonadien-2-one *see* 8855

2-Methyl-4-(1-methylethyl)-7-oxo-8-oxa-3-thia-2,4-diazadecanoic Acid

2,3-Dihydro-2,2-dimethyl-7-benzo-furanyl Ester *see* 1069

(*endo*,*syn*)-8-Methyl-8-(1-methyleth-yl)-3-[(1-oxo-2-propylpentyl)oxy]-8-azoniabicyclo[3.2.1]octane Bromide *see* 8693

1-Methyl-7-(1-methylethyl)phenan-threne *see* 8328

2-Methyl-5-(1-methylethyl)phenol *see* 1923

5-Methyl-2-(1-methylethyl)phenol *see* 9540

3-Methyl-5-(1-methylethyl)phenol Methylcarbamate *see* 7967

7-Methyl-1-(1-methylethyl)-4-phenyl-2(1*H*)-quinazolinone *see* 8059

1-Methyl-5-(1-methylethyl)-5-(2-prop-enyl)-2,4,6(1*H*,3*H*,5*H*)-pyrimidine-trione *see* 3607

2-Methyl-1-[2-(1-methylethyl)pyrazo-lo[1,5-*a*]pyridin-3-yl]-1-propanone *see* 4923

N-Methyl-*N*-(1-methylethyl)-*N*-[2-[(9*H*-xanthen-9-ylcarbonyl)oxy]eth-yl]-2-propanaminium Bromide *see* 7989

[1*S*-[1*α*,2*α*(*Z*),4*α*]]-2-Methyl-5-(2-methyl-3-methylenebicyclo[2.2.1]-hept-2-yl)-2-penten-1-ol *see* 8506

2-Methyl-5-(2-methyl-3-methylene-2-norbornyl)-2-penten-1-ol *see* 8506

β-[1-Methyl-4-[1-methyl-4-(1-methyl-4-formylaminopyrrole-2-carboxami-do)pyrrole-2-carboxamido]pyrrole-2-carboxamido]propionamidine *see* 8927

d-3-Methyl-*N*-methylmorphinan *see* 3251

Methyl *N*-Methylnipecotate, 6172

O-Methyl [2-(2-Methyl-5-nitroimid-azol-1-yl)ethyl]thiocarbamate *see* 1897

2*β*-Methyl-8*α*-(2-methyl-1-oxobut-oxy)mevinic Acid Lactone *see* 6251

1-Methyl-1-[1-methyl-2-oxo-2-(10*H*-phenothiazin-10-yl)ethyl]pyrrolidin-ium Bromide *see* 8057

α-*dl*-1-Methyl-5-(1-methyl-2-pentyn-yl)-5-allylbarbituric Acid Sodium Salt *see* 6061

1-Methyl-5-(1-methyl-2-pentynyl)-5-(2-propenyl)-2,4,6(1*H*,3*H*,5*H*)-pyri-midinetrione Sodium Salt *see* 6061

8-[[2-[Methyl(*α*-methylphenethyl)-amino]ethyl]amino]caffeine *see* 4007

N-Methyl-*N*-(*α*-methylphenethyl)fur-furylamine *see* 4327

(±)-*α*-Methyl-*N*-(1-methyl-2-phenoxy-ethyl)benzeneethanamine *see* 8273

dl-*threo*-*α*-Methyl-*N*-(1-methyl-2-phenoxyethyl)phenethylamine *see* 8273

2-Methyl-4-[(2-methylphenyl)azo]-benzenamine *see* 440

1-[[2-Methyl-4-[(2-methylphenyl)azo]-phenyl]azo]-2-naphthalenol *see* 8538

N-Methyl-2-[(*o*-methyl-*α*-phenylbenz-yl)oxy]ethylamine *see* 9641

Methyl (3-Methylphenyl)carbamothi-oic Acid *O*-(2,3-Dihydro-1*H*-inden-5-yl) Ester *see* 9652

Methyl(3-methylphenyl)carbamothioic Acid *O*-2-Naphthalenyl Ester *see* 9656

Methyl(3-methylphenyl)carbamothioic Acid *O*-(1,2,3,4-Tetrahydro-1,4-methanonaphthalen-6-yl) Ester *see* 9647

N-Methyl-*N*-(1-methyl-2-phenylethyl)-2-furanmethanamine *see* 4327

2-Methyl-2-phenylpropane see 1586
α-[1-[Methyl(3-phenyl-2-propenyl)-amino]ethyl]benzenemethanol see 2357
(E,E)-5-(2-Methyl-3-phenyl-2-propenylidene)-4-oxo-2-thioxo-3-thiazolidineacetic Acid see 3640
(E)-N-Methyl-N-(3-phenyl-2-propenyl)-1-naphthalenemethanamine see 6442
1-Methyl-4-phenyl-3-(2-propenyl)-4-piperidinol Propanoate see 304
dl-N-Methyl-2-phenylpropylamine see 7462
N-(2-Methylphenyl)-2-(propylamino)-propanamide see 7924
N-Methyl-N-(1-phenyl-2-propyl)furfurylamine see 4327
3-Methyl-1-phenyl-2-pyrazolin-5-one see 6809
3-Methyl-1-phenyl-5-pyrazolyl Dimethylcarbamate see 8186
N¹-(3-Methyl-1-phenylpyrazol-5-yl)-sulfanilamide see 9118
4-[2-[(4-Methyl-6-phenyl-3-pyridazinyl)amino]ethyl]morpholine see 6287
N-(4-Methyl-6-phenyl-3-pyridazinyl)-4-morpholineethanamine see 6287
trans-1-(4-Methylphenyl)-1-(2-pyridyl)-3-pyrrolidinoprop-1-ene see 9877
5-Methyl-5-phenyl-2,4,6(1H,3H,5H)-pyrimidinetrione see 7457
1-Methyl-3-phenyl-2,5-pyrrolidinedione see 7414
1-(4-Methylphenyl)-2-(1-pyrrolidinyl)-1-pentanone see 8194
(E)-2-[1-(4-Methylphenyl)-3-(1-pyrrolidinyl)-1-propenyl]pyridine see 9877
(E,E)-3-[6-[1-(4-Methylphenyl)-3-(1-pyrrolidinyl)-1-propenyl]-2-pyridinyl]-2-propenoic Acid see 129
N-Methyl-2-phenylsuccinimide see 7414
N-Methyl-α-phenylsuccinimide see 7414
3-Methyl-1-phenyl-5-(sulfanilamido)-pyrazole see 9118
Methyl-5-(phenylsulfinyl)-2-benzimidazolecarbamate see 7069
2-Methyl-4-phenylsulfonyltrifluoromethanesulfonanilide see 7300
5-Methyl-1-phenyl-1,3,4,6-tetrahydro-5H-benz[f]-2,5-oxazocine see 6529
3-Methyl-2-phenyltetrahydro-2H-1,4-oxazine see 7385
2-Methyl-9-phenyl-2,3,4,9-tetrahydro-1-pyridindene see 7380
1-Methyl-4-phenyl-1,2,3,6-tetrahydropyridine see 6376
1-Methyl-4-[phenyl-(2-thenyl)amino]-piperidine see 9413
1-Methyl-N-phenyl-N-(2-thienylmethyl)-4-piperidinamine see 9413
Methyl 5-(Phenylthio)-2-benzimidazolecarbamate see 4000
1-[1-Methyl-2-[(α-phenyl-o-tolyl)oxy]-ethyl]piperidine see 1078
(±)-N-Methyl-3-phenyl-3-[(α,α,α-trifluoro-p-tolyl)oxy]propylamine see 4222
3-Methyl-2-phenylvaleric Acid 2-Diethylaminoethyl Ester Methyl Bromide see 10044
3-Methyl-2-phenylvaleric Acid Diethyl(3-hydroxyethyl)methylammonium Bromide Ester see 10044
3-Methyl-2-phenylvaleric Acid 1-Methyl-4-piperidyl Ester see 7260

Methylphosphonofluoridic Acid 1-Methylethyl Ester see 8520
Methylphosphonofluoridic Acid 1,2,2-Trimethylpropyl Ester see 8860
Methylphosphonothioic Acid S-[2-[Bis(1-methylethyl)amino]ethyl] O-Ethyl Ester see 10172
Methylphosphoramidic Acid 4-tert-Butyl-2-chlorophenyl Methyl Ester see 2671
Methylphosphoramidic Acid 2-Chloro-4-(1,1-dimethylethyl)phenyl Methyl Ester see 2671
3'-Methylphthalanilic Acid, 6188
Methyl Phthalate see 3304
2-Methyl-2-phytyl-6-chromanol see 9631
2-Methyl-2-phytyl-6-hydroxychroman see 9631
2-Methyl-3-phytyl-1,4-naphthoquinone see 7536
N-Methylpicolinic Acid see 4765
dl-N-Methylpipecolic Acid 2,6-Dimethylanilide see 5905
1-Methyl-2',6'-pipecoloxylidide see 5905
4-Methyl-1-piperazinecarboxylic Acid 6-(7-Chloro-1,8-naphthyridin-2-yl)-2,3,6,7-tetrahydro-7-oxo-5H-1,4-dithiino[2,3-c]pyrrol-5-yl Ester see 9182
4-Methyl-1-piperazinecarboxylic Acid 6-(5-Chloro-2-pyridinyl)-6,7-dihydro-7-oxo-5H-pyrrolo[3,4-b]pyrazin-5-yl Ester see 10324
10-(γ-Methylpiperazinopropyl)phenothiazine see 7294
6-(4-Methyl-1-piperazinyl)-11H-dibenz[b,e]azepine see 7320
3-[[(4-Methyl-1-piperazinyl)imino]-methyl]rifamycin see 8382
6-(4-Methyl-1-piperazinyl)morphanthridine see 7320
N-[γ-(4'-Methyl-1'-piperazinyl)propyl]-3-butyrylphenothiazine see 1543
N-[γ-(4'-Methylpiperazinyl-1')propyl]-3-chlorophenothiazine see 7942
cis-9-[3-(4-Methyl-1-piperazinyl)-propylidene]-2-(dimethylsulfonamido)thioxanthene see 9503
10-[3-(4-Methyl-1-piperazinyl)propyl]-10H-phenothiazine see 7294
N-Methylpiperazinyl-N'-propylphenothiazine see 7294
1-[10-[3-(4-Methyl-1-piperazinyl)propyl]-10H-phenothiazin-2-yl]-1-butanone see 1543
10-[3-(4-Methyl-1-piperazinyl)propyl]-2-(trifluoromethyl)-10H-phenothiazine see 9811
1-Methyl-3-piperidinecarboxylic Acid Methyl Ester see 6172
cis-2,2'-(1-Methyl-2,6-piperidinediyl)-bis[1-phenylethanone] see 5578
2,2''-(1-Methyl-2,6-piperidinedlyl)diacetophenone see 5578
α,α'-[(1-Methyl-2,6-piperidinediyl)dimethylene]dibenzyl Alcohol see 5577
2-Methyl-1-piperidinepropanol Benzoate see 7627
N-(1-Methyl-2-piperidinoethyl)propionanilide see 7353
N-(1-Methyl-2-piperidinoethyl)-N-2-pyridylpropionamide see 8017
γ-(4-Methylpiperidino)-p-fluorobutyrophenone see 5870
(2-Methylpiperidino)propyl Benzoate see 7627

3-(2-Methylpiperidino)propyl p-Cyclohexyloxybenzoate see 2804
2-Methyl-3-piperidino-1-p-tolylpropan-1-one see 9660
10-[2-(1-Methyl-2-piperidinyl)ethyl]-2-(methylsulfinyl)-10H-phenothiazine see 5970
10-[2-(1-Methyl-2-piperidinyl)ethyl]-2-(methylsulfonyl)-10H-phenothiazine see 9137
10-[2-(1-Methyl-2-piperidinyl)ethyl]-2-(methylthio)-10H-phenothiazine see 9497
N-[1-Methyl-2-(1-piperidinyl)ethyl]-N-phenylpropanamide see 7353
N-[1-Methyl-2-(1-piperidinyl)ethyl]-N-2-pyridinylpropanamide see 8017
10-[(1-Methyl-3-piperidinyl)methyl]-10H-phenothiazine see 5892
1-Methyl-4-piperidyl Benzhydryl Ether see 3390
N-Methyl-3-piperidyl Benzilate Methyl Bromide see 5893
N-Methyl-3-piperidyl Diphenylglycolate Methobromide see 5893
1-Methyl-4-piperidyl Diphenylpropoxyacetate see 8018
(±)-2'-[2-(1-Methyl-2-piperidyl)ethyl]-p-anisanilide see 3609
5-(1-Methylpiperidylidene-4)-5H-dibenzo[a,d]cycloheptene see 2842
4-(1-Methyl-4-piperidylidene)-9,10-dihydro-4H-benzo[4,5]cyclohepta-[1,2-b]thiophene see 7671
1-Methyl-3-piperidylidenedi(2-thienyl)methane see 9601
(N-Methyl-3-piperidyl)methylphenothiazine see 5892
1-Methyl-4-piperidyl 3-Methyl-2-phenylvalerate see 7260
9-(N-Methyl-3-piperidylmethyl)thioxanthene see 6059
1-(N-Methyl-4-piperidyl)-3-phenyl-4-benzyl-5-pyrazolone see 1152
1-Methyl-3-piperidyl α-Phenylcyclohexaneglycolate see 7998
N-Methyl-3-piperidyl Phenylcyclohexylglycolate see 7998
1-(N-Methylpiperid-4-yl)-3-phenyl-4-ethylpyrazol-5-one see 7632
γ-(2-Methylpiperidyl)propyl Benzoate see 7627
2-Methyl-3-all-trans-polyprenyl-1,4-naphthoquinones see 5876
Methylprednisolone, 6189
16β-Methylprednisone see 5907
6-Methyl-Δ⁴,⁶-pregnadien-17α-ol-3,20-dione Acetate see 5849
6α-Methyl-4-pregnen-17α-ol-3,20-dione see 5838
6α-Methyl-4-pregnen-17α-ol-20-one see 664
Methylpromazine see 9834
2-Methylpropanal see 5169
2-Methyl-1-propanamine see 5147
2-Methyl-2-propanamine see 1580
2-Methylpropanenitrile see 5171
2-Methyl-1-propanethiol see 5161
2-Methyl-2-propanethiol see 1613
2-Methylpropanoic Acid see 5170
2-Methylpropanoic Acid Ethyl Ester see 3860
2-Methylpropanoic Acid 4-[2-(Methylamino)ethyl]-1,2-phenylene Ester see 4921
2-Methylpropanoic Acid Methyl Ester see 6162
2-Methylpropanoic Acid 2-Methylpropyl Ester see 5159
2-Methyl-1-propanol see 5146

Methyl Sulfate, 6203
Methyl Sulfhydrate *see* 6023
Methyl Sulfide, 6204
4-Methylsulfinylbutyl Isothiocyanate *see* 9135
Methyl Sulfocyanate *see* 6209
Methylsulfonal *see* 9131
Methyl Sulfone *see* 3307
Methylsulfonic Acid *see* 6021
4-(Methylsulfonyl)-2,6-dinitro-*N*,*N*-dipropylaniline *see* 6662
4-(Methylsulfonyl)-2,6-dinitro-*N*,*N*-dipropylbenzenamine *see* 6662
Methylsulfonylmethane *see* 3307
3-(Methylsulfonyl)-10-[2-(1-methyl-2-piperidyl)ethyl]phenothiazine *see* 9137
6*R*-[2-[3-(Methylsulfonyl)-2-oxo-1-imidazolidine Carboxamido]-2-phenylacetamido]penicillanic Acid *see* 6259
1-[3-[2-(Methylsulfonyl)phenothiazin-10-yl]propyl]isonipecotamide *see* 6233
1-[3-[2-(Methylsulfonyl)phenothiazin-10-yl]propyl]-4-piperidinecarboxamide *see* 6233
1-[3-[2-(Methylsulfonyl)-10*H*-phenothiazin-10-yl]propyl]-4-piperidinecarboxamide *see* 6233
D-(*threo*)-1-*p*-Methylsulfonylphenyl-2-amino-3-fluoro-1-propanol *see* 4145
2-[4-(Methylsulfonyl)phenyl]imidazo-[1,2-*a*]pyridine *see* 10320
Methyl Sulfoxide *see* 3308
Methylsulfuric Acid *see* 6203
Methylsympatol *see* 4869
Methyl Systox *see* 6129
N-Methyltaurine, 6205
17-Methyltestosterone, 6206
17α-Methyltestosterone 3-Cyclopentyl Enol Ether, 6207
dl-16-Methyl-18,19-tetradehydro-5,6,-7-trinor-4,8-inter-*m*-phenylene-prostaglandin I₂ *see* 1190
(±)-*cis*-4-Methyltetrafluorobenzyl 3-(2-Chloro-3,3,3-trifluoro-prop-1-en-1-yl)-2,2-dimethylcyclopropane Carboxylate *see* 9266
1-Methyl-1,4,5,6-tetrahydro-2-[2-(3-hydroxyphenyl)vinyl]pyrimidine *see* 7055
[1*R*-[1α,1(*Z*),3a*β*,5α,11*β*,14a*S**]]-2-Methyl-4-[3a,4,5,7-tetrahydro-8-hydroxy-3,3,11-trimethyl-13-(3-methyl-2-butenyl)-11-(4-methyl-3-pentenyl)-7,15-dioxo-1,5-methano-1*H*,3*H*,11*H*-furo[3,4-*g*]pyrano[3,2-*b*]xanthen-1-yl]-2-butenoic Acid *see* 4373
2-Methyl-*N*-[3-(2,3,5,6-tetrahydroimidazo[2,1-*b*]thiazol-6-yl)phenyl]-propanamide *see* 1540
(−)-2-Methyl-3′-(2,3,5,6-tetrahydroimidazo[2,1-*b*]thiazol-6-yl)propionanilide *see* 1540
[4*R*-[4α,5*β*(*E*)]]-2-Methyl-3-[1,3,4,5-tetrahydro-4-(methylamino)benz-[*cd*]indol-5-yl]-2-propen-1-ol *see* 2083
Methyl 1,2,5,6-Tetrahydro-1-methylnicotinate *see* 815
(2*R*,4*R*)-4-Methyl-1-[(*S*)-*N*²-[[(*R*,*S*)-1,2,3,4-tetrahydro-3-methyl-8-quinolinyl]sulfonyl]arginyl]pipecolic Acid *see* 816
N-Methyltetrahydropapaverine *see* 5391
1-Methyl-1,4,5,6-tetrahydro-2-pyrimi-

dylmethyl α-Cyclohexyl-α-phenylglycolate *see* 7107
(+)-Methyl-α-5-[4,5,6,7-tetrahydro[3,-2-*c*]thienopyridyl]-(2-chlorophenyl)-acetate *see* 2457
(*E*,*E*,*E*)-2-Methyl-3-(3,7,11,15-tetramethyl-2,6,10,14-hexadecatetraen-yl)-1,4-naphthalenedione *see* 5876
[*R*-[*R**,*R**-(*E*)]]-2-Methyl-3-(3,7,11,-15-tetramethyl-2-hexadecenyl)-1,4-naphthalenedione *see* 7536
N-Methyl-*N*,2,4,6-tetranitroaniline *see* 6664
N-Methyl-*N*,2,4,6-tetranitrobenzenamine *see* 6664
9-Methyl-3-(1*H*-tetrazol-5-yl)-4*H*-pyrido[1,2-*a*]pyrimidin-4-one *see* 7205
1-Methyl-4-*N*-2-thenylanilinopiperidine *see* 9413
Methyltheobromine *see* 1674
3-(5-Methyl-1,3,4-thiadiazolyl-2-mercaptomethyl)-7-(3,5-dichloro-4-pyridon-1-ylacetamido)-3-cephem-4-carboxylic Acid *see* 1966
*N*¹-(5-Methyl-1,3,4-thiadiazol-2-yl)-sulfanilamide *see* 9084
(6*R*-*trans*)-3[[(5-Methyl-1,3,4-thiadiazol-2-yl)thio]methyl]-8-oxo-7-[(1*H*-tetrazol-1-ylacetyl)amino]-5-thia-1-azabicyclo[4.2.0]oct-2-ene-2-carboxylic Acid *see* 1967
4-Methyl-2-thiazolamine *see* 470
4-Methyl-5-thiazoleethanol, 6208
N-(3-Methyl-4-thiazolin-2-ylidene)-2,-4-xylidene *see* 2833
4-Methyl-2-thiazolylamine *see* 470
N¹-(4-Methyl-2-thiazolyl)sulfanilamide *see* 9088
α-Methyl-4-(2-thienylcarbonyl)benzeneacetic Acid *see* 9180
γ-Methylthio-α-aminobutyric Acid *see* 6053
2-Methylthio-4,6-bis(isopropylamino)-*s*-triazine *see* 7973
2-Methylthio-4,6-bis(monoethylamino)-*s*-triazine *see* 8683
N-Methylthiocarbamoyl-*N′*-[(1-methylallyl)thiocarbamoyl]hydrazine *see* 6013
6-Methylthiochroman-7-sulfonamide 1,1-Dioxide *see* 6220
Methyl Thiocyanate, 6209
4-Methylthio-3,5-dimethylphenyl *N*-Methylcarbamate *see* 6050
2-Methylthio-4-ethylamino-6-isopropylamino-*s*-triazine *see* 411
O-[[*N*-[*N′*-(1-Methylthioethylideneiminooxycarbonyl)-*N′*-methylaminosulfenyl]-*N*-methylcarbamoyl]]-*S*-methylacetohydroximate *see* 9465
5-(2-Methylthioethyl)-5-(2-pentyl)-2-thiobarbituric Acid *see* 6058
5-[(Methylthio)methyl]-3-[[(5-nitro-2-furanyl)methylene]amino]-2-oxazolidinone *see* 6623
5-[(Methylthio)methyl]-3-[(5-nitrofurfurylidene)amino]-2-oxazolidinone *see* 6623
8-Methylthio-10-(4-methylpiperazino)-10,11-dihydrodibenzo[*b*,*f*]-thiepine *see* 6222
(8*β*)-8-[(Methylthio)methyl]-6-propylergoline *see* 7304
Methylthioninium Chloride *see* 6137
1-(Methylthio)propyl 1-Propenyl Disulfide *see* 858
Methylthiouracil, 6210
4-Methyl-2-thiouracil *see* 6210
6-Methyl-2-thiouracil *see* 6210

1-Methyl-3-(9*H*-thioxanthen-9-ylmethyl)piperidine *see* 6059
4-(Methylthio)-3,5-xylyl Methylcarbamate *see* 6050
DL-α-Methylthyroxine Ethyl Ester *see* 3917
5-Methyltocol *see* 9635
7-Methyltocol *see* 9638
8-Methyltocol *see* 9634
1-Methyl-5-*p*-toluoylpyrrole-2-acetamidoacetic Acid Guaicil Ester *see* 637
1-Methyl-5-*p*-toluoylpyrrole-2-acetic Acid *see* 9655
N-[(1-Methyl-5-*p*-toluoylpyrrol-2-yl)-acetyl]glycine *o*-Methoxyphenyl Ester *see* 637
N-Methyl-2-[α-(2-tolylbenzyl)oxy]-ethylamine *see* 9641
Methyl 3-(*m*-Tolylcarbamoyloxy)-phenylcarbamate *see* 7384
Methyl-*p*-tolylcarbinol *see* 3287
Methyl *p*-Tolylcarbinol Camphorate Diethanolamine Salt *see* 9630
2-Methyl-6-*p*-tolyl-2-hepten-4-one *see* 9956
2-Methyl-3-*o*-tolyl-4(3*H*)-quinazolinone *see* 6028
2-Methyl-3-*o*-tolyl-6-sulfamyl-7-chloro-1,2,3,4-tetrahydro-4-quinazolinone *see* 6231
6-Methyl-1,2,4-triazine-3,5(2*H*,4*H*)-dione *see* 936
α-Methyl-*N*-(2,2,2-trichloroethylidene)benzeneethanamine *see* 620
α-Methyl-*N*-(2,2,2-trichloroethylidene)phenethylamine *see* 620
2-Methyl-4-(2,2,2-trichloro-1-hydroxyethoxy)-2-pentanol *see* 2139
N-Methyl-*N′*-(2,2,2-trichloro-1-hydroxyethyl)urea *see* 5822
α-Methyltricyclo[3.3.1.1³,⁷]decane-1-methanamine *see* 8390
Methyltrienolone, 6211
2-[[[3-Methyl-4-(2,2,2-trifluoro-ethoxy)-2-pyridinyl]methyl]sulfinyl]-1*H*-benzimidazole *see* 5373
2-Methyl-3-(*β*,*β*,*β*-trifluoroethylthiomethyl)-6-chloro-7-sulfamyl-3,4-dihydro-1,2,4-benzothiadiazine 1,1-Dioxide *see* 7744
2-(2-Methyl-3-trifluoromethylanilino)-nicotinic Acid *see* 4182
2-[[α-Methyl-*m*-(trifluoromethyl)-phenethyl]amino]ethanol Benzoate (Ester) *see* 1066
(±)-*N*-Methyl-γ-[4-(trifluoromethyl)-phenoxy]benzenepropanamine *see* 4222
N-Methyl-3-(*p*-trifluoromethylphenoxy)-3-phenylpropylamine *see* 4222
2-[[2-Methyl-3-(trifluoromethyl)phenyl]amino]-3-pyridinecarboxylic Acid *see* 4182
2-[[1-Methyl-2-[3-(trifluoromethyl)-phenyl]ethyl]amino]ethanol Benzoate (Ester) *see* 1066
N-[4-Methyl-3-[[(trifluoromethyl)-sulfonyl]amino]phenyl]acetamide *see* 4198
1-Methyl-3-[4-[*p*-[(trifluoromethyl)-thio]phenoxy]-*m*-tolyl]-*s*-triazine-2,-4,6(1*H*,3*H*,5*H*)-trione *see* 9664
16-Methyl-1,11α,16*RS*-trihydroxy-prost-13*E*-en-9-one *see* 8395
N-Methyl-*N*-(2,4,6-triiodo-3-aminophenyl)glutaramidic Acid *see* 5069
N-Methyltrilobine *see* 9826
Methyl 3,4,5-Trimethoxycinnamoyl Reserpate *see* 8311

5-Methyl-6-[[(3,4,5-trimethoxyphenyl)amino]methyl]-2,4-quinazolinediamine see 9851
Methyl N-Trimethyl-γ-aminobutyrate Chloride see 1913
Methyl 1,1,N-Trimethyl-N-(3,3-diphenylpropyl)-2-aminoethyl 1,4-Dihydro-2,6-dimethyl-4-(3-nitrophenyl)pyridine-3,5-dicarboxylate see 5469
Methyltrimethylene Glycol see 1601
2-Methyl-2-(4,8,12-trimethyltridecyl)-6-chromanol see 9631
1-Methyl-2,4,6-trinitrobenzene see 9860
6α-Methyl-11β,17α,21-triol-1,4-pregnadiene-3,20-dione see 6189
1-Methyl-4-[3,3,3-tris(4-chlorophenyl)-1-oxopropyl]piperazine see 4710
1-Methyl-4-[3,3,3-tris(p-chlorophenyl)propionyl]piperazine see 4710
8-Methyltropinium Bromide 2-Propylvalerate see 709
N-Methyl-L-tryptophan see 6
α-Methyltyramine see 4855
α-Methyl-m-tyramine see 4409
N-Methyl-L-tyrosine see 9183
α-Methyltyrosine see 6248
α-Methyl-L-tyrosine see 6248
α-Methyl-m-tyrosine, 6212
α-Methyl-p-tyrosine see 6248
(2-O-Methyltyrosine)deamino-1-carbaoxytocin see 1841
4-Methylumbelliferone see 4903
β-Methylumbelliferone see 4903
4-Methylumbelliferone O,O-Diethyl Phosphorothioate see 4904
4-Methyluracil see 6213
5-Methyluracil see 9539
6-Methyluracil, 6213
Methylurethane see 6115
Methylustin see 6616
trans-8-Methyl-N-vanillyl-6-nonenamide see 1811
N-Methyl-β-vinyldiacetonalkamine Mandelate see 3943
Methyl Vinyl Ketone, 6214
13α-Methyl-13-vinylpodocarp-8(14)-ene-15-oic Acid see 7582
13β-Methyl-13-vinylpodocarp-7-ene-15-oic Acid see 5217
Methyl Viologen (2+) see 7165
Methyl Vitamin B₁₂ see 6125
1-Methyl-2-(2,6-xylyloxy)ethylamine see 6257
N-Methyl-N'-2,4-xylyl-N-(N-2,4-xylylformimidoyl)formamidine see 510
Methyl Yellow see 3279
Methylzineb see 8004
Methymycin, 6215
Methypranol see 6221
Methyprylon, 6216
Methyridine see 6247
Methysergide, 6217
Methysergide Maleate see 6217
Methysticin, 6218
Metiazic Acid see 6219
Metiazinic Acid, 6219
Meticlorpindol see 2458
Meticortelone [Schering] see 7901
Meticortelone Soluble [Schering] see 7901
Meticorten [Schering] see 7904
Meticrane, 6220
Metidione see 6017
Metifex [Albert Roussel] see 3760
Metiguanide [Erba] see 6001
Metilar [Syntex] see 7162
Metilcaf [Ital Suisse] see 8411
Metildigoxin see 3210

Metildiolo see 6017
Metilenbiotic [Coli] see 6007
Metilon [Daiichi] see 3414
Metina [Pharmacia] see 1898
Metione [Lobica] see 6053
Metipranolol, 6221
Metirosine see 6248
Metitepine, 6222
Metizoline, 6223
Metligine [Taisho] see 6272
Metmercapturon see 6050
Metobromuron, 6224
Metochalcone, 6225
Metoclol [Toyama] see 6226
Metoclopramide, 6226
Metocobil [Vita] see 6226
Metocryst see 6017
Metocurine Iodide, 6227
Metodik [Labinca] see 891
Metofane [Pitman-Moore] see 6072
Metofenazate, 6228
Metofoline, 6229
Metol see 6097
Metolachlor, 6230
Metolazone, 6231
Metolquizolone see 6028
Metomidate, 6232
Meton see 4724
Metopimazine, 6233
Metopirone [Ciba] see 6246
Metopon, 6234
Metoprolol, 6235
Metopyrone see 6246
Metoquinone, 6236
Metoral see 4755
Metoryl see 4755
Metoserpate, 6237
Metosomin [Maruishi] see 9660
Metosyn [ICI] see 4186
Metoxadone see 5896
Metralindole, 6238
Metramac see 509
Metramid [Nicholas] see 6226
Metranil see 7249
Metrasil see 9095
Metrazol [Knoll] see 7283
Metreton [Schering] see 7903
Metribuzin, 6239
Metrifonate see 9753
Metrisone see 6189
Metrizamide, 6240
Metrizoate Sodium see 6241
Metrizoic Acid, 6241
Metrodin [Serono] see 4299
Metrodiol [Clin-Comar-Byla] see 3905
MetroGel [Curatek] see 6242
Metrolag [Lagap] see 6242
Metrolyl [Lagap] see 6242
Metronidazole, 6242
Metron (Japanese) see 6243
Metron S, 6243
Metropin [Benzon] see 3813
Metropine [Pennwalt] see 907
Metroprione see 6246
Metrotop [Farmitalia] see 6242
Metrulen see 3905
Metsulfuron Methyl, 6244
Metubine Iodide [Lilly] see 6227
Metuclazepam see 5980
Meturedepa, 6245
Metycaine [Lilly] see 7627
Metyrapone, 6246
Metyridine, 6247
Metyrosine, 6248
Mevacor [Merck & Co.] see 5616
Mevaldic Acid, 6249
Mevalolactone see 6250
Mevalon [Guidotti] see 5852
Mevalonic Acid, 6250
Mevalonic Lactone see 6250

Mevalotin [Sankyo] see 7894
Mevasine [Merck & Co.] see 5814
Mevastatin, 6251
Mevinacor [Merck & Co.] see 5616
Mevinolin see 5616
Mevinphos, 6252
Mevlor [Merck & Co.] see 5616
Mexacarbate, 6253
Mexamine see 6083
Mexate [Bristol] see 6065
Mexazolam, 6254
Mexenone, 6255
Mexephenamide see 5844
Mexicain, 6256
Mexican Scammony (Root) see 5088
Mexican Tea see 6888
Mexiletine, 6257
Mexitil [Boehringer, Ing.] see 6257
Mexocine [Specia] see 2937
Meyerhofferite see 1691
Mezaton see 7440
Mezcaline see 5965
Mezereon see 6258
Mezereum, 6258
Mezineb see 8004
Mezlin [Miles] see 6259
Mezlocillin, 6259
Mezolin [Merck & Co.] see 4998
MF-201 see 3672
MF-934 see 8448
MFT see 1958
Mg 5-Sulfat [Artesan] see 5731
MG-02 see 58
MG-143 see 9158
MG-652 see 7047
MG-1559 see 10205
MG-2552 see 2468
MG-2555 see 1414
MG-4833 see 4009
MG-5454 see 4580
MG-5771 see 10205
MG-8926 see 3506
MG-13054 see 4039
MG-13608 see 3473
MGA [Upjohn] see 5859
MGI-2 see 4558
MGK-11 see 8271
MGK Repellent 11 [McLaughlin] see 8271
MH see 5745
M. H. see 5961
MH-532 see 7412
MHA [Monsanto] see 6054
MHIP see 9653
MHPG see 6074
MI-85 see 768
MI-85Di see 768
Miacalcic [Sandoz] see 1680
Miacalcin [Sandoz] see 1680
Miadenil [Francia] see 1680
Miadone see 6008
Mianine see 2118
Mianserin, 6260
Miaquin [Parke, Davis] see 609
Miarsenol see 6534
Miazine see 8170
Miazole see 4948
Mibefradil, 6261
MIBG see 5027
Mibolerone, 6262
Miboplatin, 6263
MIC see 6163
Micatex [Andromaco] see 9742
Micatin [J & J] see 6266
Michler's Base, 6264
Michler's Ketone, 6265
Michrome No. 24 see 219
Michrome No. 66 see 2006
Michrome No. 226 see 6571
Micinovo [Andreu] see 5986

Micoclorina [Zambon] see 2120
Micofugal [Biopharma] see 3550
Micofur [Norwich] see 6627
Micogin [Crosara] see 3550
Micol see 2068
Miconal Ecobi [Ecobi] see 6266
Miconazole, 6266
Micoren [Ciba-Geigy] see 2659
Micoserina see 2820
Micotef [LPB] see 6266
Micotil [Elanco] see 9580
Micoxolamina [Dompé] see 2325
Micranthine, 6267
Micrest [Beecham] see 3177
Micrin see 2959
Microbar [Max Ritter] see 1023
Micro-Cell see 1749
Microcide see 4468
Microcidin see 6471
Microcillin [Bayer] see 1838
Microcline see 7763
Micrococcin see 6268
Micrococcin P, 6268
Microcosmic Salt see 587
Micro-Dee [Diamond] see 10157
Microgynon [Schering] see 6797
Micro K [E. Merck] see 7783
Microlite see 9223
Microlut [Schering AG] see 6797
Micromerol see 10027
Micromet see 8814
Micromite [Uniroyal] see 3188
Micronase [Upjohn] see 4486
Micronomicin, 6269
Micronor [Ortho] see 6790
Micronovum [Cilag-Chemie] see 6790
Micronyl [Astra] see 1490
Micropaque [Nicholas] see 1023
Micropenin [Kabi] see 7036
Microtin [AB Leo] see 2028
Microtrast [Nicholas] see 1023
Microtrim [Chephasaar] see 9086
Microtubules see 9940
Microval [Wyeth] see 6797
Mictine [Searle] see 471
Mictone [Werner] see 1232
Mictonetten [Apogepha] see 8018
Mictonorm [Apogepha] see 8018
Mictrol [Pharmacia] see 9311
Micturin [Pharmacia] see 9311
Micturol [Pharmacia] see 9311
Midafenone see 6759
Midalcipran see 6281
Midalgan see 6174
Midamine [Roberts] see 6272
Midamor [Merck & Co.] see 426
Midantan see 389
Midarine [Wellcome] see 9044
Midazolam, 6270
Midecamycin A₁ see 6271
Midecamycin A₃ see 6271
Midecamycins, 6271
Midecin [Farmaka] see 6271
Midicel [Parke, Davis] see 9087
Midikel [Parke, Davis] see 9087
Midodrine, 6272
Midol PMS [Miles] see 7133
Midronal [Delalande] see 2365
Midsummer Daisy see 4112
Mielucin [Simes] see 1529
MIF see 5855
MIF-I see 5855
Mifegyne [Roussel-UCLAF] see 6273
Mifepristone, 6273
Miforon see 3246
Mifurol [Mitsui] see 1892
Miglitol, 6274
Migranal [Sandoz] see 3217
Migristène [Specia] see 4260
MIH see 7938

Mikado [Zeneca] see 9063
Mikametan [Mikasa] see 4998
Mikamycin, 6275
Mikamycin A see 10142
Mikamycin B see 6275
Mikamycin I_A see 6275
Mikavir [Salus] see 425
Mikedimide [Panray] see 1054
Mikelan [Otsuka] see 1917
Miketorin [Mitsui] see 511
Mikrotsid see 4468
Milban [Mallinckrodt] see 3467
Milbeknock [Sankyo] see 6276
Milbemax [Ciba-Geigy] see 6277
Milbemectin, 6276
Milbemycin α₁ see 6276
Milbemycin α₃ see 6276
Milbemycin β₁ see 6278
Milbemycin A₃ see 6276
Milbemycin A₄ see 6276
Milbemycin A₃ 5-Oxime see 6277
Milbemycin A₄ 5-Oxime see 6277
Milbemycin D see 6278
Milbemycin Oxime, 6277
Milbemycins, 6278
L-Milchsäure see 5350
Milchsäure (German) see 5349
D-Milchsäure (German) see 5348
Mil-Col [ICI] see 3499
Milcurb [ICI] see 3266
Milcurb Super [ICI] see 3786
Mild Aconitine see 121
Mildex see 3340
Mildiomycin, 6279
Mildison [Yamanouchi] see 4828
Mild Mercury Chloride see 5951
Mild Silver Protein see 8671
Milestrol [SM & P] see 3177
Milezin [Spofa] see 6066
Milfoil see 108
Milfoil Oil see 6957
Milgo [ICI] see 3786
Milhéparine [Millot] see 3754
Milibis [Winthrop] see 4503
Milid [Opfermann] see 7958
Milide [Beytout] see 7958
Milk, 6280
Milk Sugar see 5356
Milkweed see 866
Millafol [Parke, Davis] see 4253
Millerite see 6582
Millevit [Nordmark] see 10152
Millicaine [Corbiere] see 1235
Millicorten [Ciba] see 2986
Milligynon [Schering AG] see 6790
Millinese [Lampugnani] see 2239
Millisrol [Nippon Kayaku] see 6704
Millon's Base see 5957
Millophyline see 3754
Milnacipran, 6281
Milogard [Ciba-Geigy] see 7996
Milontin [Parke, Davis] see 7414
Milorganite [Milwaukee Met. Sewer. Dist], 6282
Miloxacin, 6283
Milrinone, 6284
Milstem [ICI] see 3786
Miltaun [Mack, Illert.] see 5908
Miltefosine, 6285
Miltex [Asta] see 6285
Miltown [Wallace Labs.] see 5908
Milvane [Schering AG] see 4421
Mimedran [Esteve] see 9169
Mimetite see 5410
Mimic [Rohm & Haas] see 9254
Mimosine, 6286
Minacalm [Tobishi] see 9660
Mina D₂ see 10156
Minalfene [Bouchara] see 308
Minalgin [Streuli] see 3414

Minaphil see 485
Minaprine, 6287
Mincard [Searle] see 471
Mindiab [Farmitalia] see 4451
Mindolic Acid see 2443
Minedil [Formenti] see 8154
Minelcin [Parke, Davis] see 1110
Minelsin [Parke, Davis] see 1110
Mineral Blue see 4064
Mineral Carbon see 4560
Mineral Oil see 7327 and 7328
Mineral Orange see 5449
Mineral Pitch see 880
Mineral Red see 5449
Mineral Spirits, 6288
Mineral Wax see 2033
Mineral White see 1753
Minetoin see 7475
Mingin see 10021
Minicid see 3228
Minidiab [Erba] see 4451
Minihep [Leo Pharm.] see 4685
Miniluteolas [Serono] see 3905
Minilyn [Organon] see 5659
Mini-Pe [Syntex] see 6790
"Mini-pill" see 6790
Miniplanor [Kodama] see 287
Minipress [Pfizer] see 7897
Minirin [Ferring] see 2969
Minisintrom [Ciba-Geigy] see 29
Minitran [3M Riker] see 6704
Minium see 5449
Minocin [Lederle] see 6289
Minocycline, 6289
Minocyn [Lederle] see 6289
Minolip [Master] see 1066
Minomycin [Lederle] see 6289
Minor Centaury see 2016
Minovlar [Schering AG] see 6790
Minoxidil, 6290
Minoximen [Menarini] see 6290
Minozinan see 6066
Minprog [Upjohn] see 8063
Minprostin E₂ [Upjohn] see 8064
Mint see 8889
Mintacol see 7164
Mintec [SKF] see 6933
Mintezol [Merck & Co.] see 9426
Mintic [Ayerst] see 6247
Minulet [Wyeth] see 4421
Minuric [R & C] see 1093
Minusin [Doetsch, Grether] see 6811
Minzil see 2221
Minzolum [MSD] see 9426
Mioblock [Sankyo] see 7139
Miocaina see 4582
Miocamycin [Meiji] see 6291
Miocard [Millot-Roux] see 504
Miochol [Cooper] see 88
Miocor [Ecobi] see 1898
Miodar [Ism] see 7474
Miodaron [Alcon] see 504
Miokamycin, 6291
Miolaxene [Lepetit] see 6060
Miolene [Lusofarmaco] see 8401
Miolisodal see 1889
Mional [Eisai] see 3642
Miorel [Lederle] see 9460
Mioril see 1889
Mio-Sed see 2155
Miostat [Alcon] see 1823
Mioticol see 3242
Miotisal A see 7164
Miotolon [Daiichi] see 4319
Mioton [Pharmakhim] see 1811
Miotonal [Caber] see 1898
MIP see 798
Mipafox, 6292
Mipax see 3304
Mirabilite see 8829

Miracil D *see* 5620
Miracle Fruit *see* 6293
Miraclid [Mochida] *see* 10021
Miracol *see* 5620
Miraculin, 6293
Miradol [Mitsui] *see* 9163
Miradon [Schering] *see* 706
Miral [Ciba-Geigy] *see* 5119
Miralin [Miralin] *see* 6293
Miranax [Syntex] *see* 6504
Mirapront [Mack, Illert.] *see* 7415
Mirasept [CIBA Vision] *see* 4839
Mirbanil [Boehringer, Ing.] *see* 9163
Mircol [Pharmuka] *see* 5911
Miretilan [Sandoz] *see* 3617
Mirex, 6294
Mirfat [Merckle] *see* 4331
Miridacin [Taiho] *see* 7957
Mirimostim *see* 5678
Mirion *see* 6043
Mirlon *see* 6832
Mirontin [Parke, Davis] *see* 7414
Miropinic Acid *see* 5217
Mirsol [Permamed] *see* 10303
Mirtazepine, 6295
Mirvan A [Cont. Pharma] *see* 220
Misch Metal, 6296
Mischmetall *see* 2037
Miso *see* 8878
Misoprostol, 6297
Mistabron [UCB] *see* 5969
Mistabronco [UCB] *see* 5969
Mistarel *see* 5236
Mistral [M & B] *see* 4035
Misulban [Techni-Pharma] *see* 1529
Misulvan [Bernabo] *see* 9163
Mitaban [Upjohn] *see* 510
Mitac [Boots] *see* 510
Mitanoline [Toyo Pharmar] *see* 7922
Mitarson [Asta] *see* 2917
Mithracin [Pfizer] *see* 7696
Mithramycin *see* 7696
Mitigal *see* 5977
Mitigan [Makhteshim-Agan] *see* 3136
Mitobronitol, 6298
Mitocin-C [Bristol] *see* 6301
Mitocor [Zambon] *see* 9974
Mitoguazone, 6299
Mitolac [Bristol-Myers] *see* 6300
Mitolactol, 6300
Mitomen *see* 5816
Mitomycin A *see* 6301
Mitomycin B *see* 6301
Mitomycin C *see* 6301
Mitomycins, 6301
Mitoquinones *see* 9974
Mitosan *see* 1529
Mitotane, 6302
Mitox *see* 2124
Mitoxana [Asta] *see* 4937
Mitoxantrone, 6303
Mitragynine, 6304
Mitramycin *see* 7696
Mitran *see* 7033
Mitrinermine *see* 8364
Mitrolan [Robins] *see* 1744
Mitronal [Searle] *see* 2365
Mivacron [Burroughs Wellcome] *see* 6305
Mivacurium Chloride, 6305
Mixed Rare Earth Metal *see* 6296
Mixobar [Byk Gulden] *see* 1023
Mizoribine, 6306
MJ-505 *see* 7474
MJ-1992 *see* 8877
MJ-1996 *see* 418
MJ-1999 *see* 8876
MJ-5190 *see* 418
MJ-9067 *see* 3609
MJ-10061 *see* 1093

MJ-13754-1 *see* 6527
MJF-9325 *see* 4937
MJF-11567-3 *see* 1963
MJF-12264 *see* 9267
MJF-12637 *see* 9160
MK 4 *see* 5876
MK 6 *see* 5876
MK-130 *see* 2782
MK-135 *see* 2257
MK-188 *see* 10251
MK-191 *see* 7669
MK-208 *see* 3972
MK-217 *see* 228
MK-223 *see* 4925
MK-231 *see* 9155
MK-240 *see* 8088
MK-244 *see* 3594
MK-264 *see* 4251
MK-302 *see* 828
MK-306 *see* 1986
MK-351 *see* 6132
MK-360 *see* 9426
MK-366 *see* 6793
MK-383 *see* 9605
MK-397 *see* 3667
MK-401 *see* 2471
MK-421 *see* 3605
MK-422 *see* 3606
MK-476 *see* 6340
MK-486 *see* 1843
MK-507 *see* 3484
MK-521 *see* 5540
MK-595 *see* 3761
MK-621 *see* 3567
MK-639 *see* 4979
MK-647 *see* 3190
MK-650 *see* 2603
MK-733 *see* 8686
MK-781 *see* 6248
MK-787 *see* 4954
MK-791 *see* 2331
MK-801 *see* 3451
MK-803 *see* 5616
MK-870 *see* 426
MK-905 *see* 1774
MK-906 *see* 4125
MK-933 *see* 5264
MK-936 *see* 1
MK-950 *see* 9585
MK-954 *see* 5613
MK-955 *see* 4280
MK-965 *see* 871
MK-990 *see* 8280
ML-236 B *see* 6251
ML-1024 *see* 9420
ML-1129 *see* 1190
ML-1229 *see* 1190
MM 4550 *see* 6972
MM 13902 *see* 6972
MM 14151 *see* 2402
MM 17880 *see* 6972
MM 22380 *see* 6972
MM 22381 *see* 6972
MM 22382 *see* 6972
MM 22383 *see* 6972
MMC *see* 6301
MMF *see* 6152
MMH *see* 6160
4-MMPD *see* 3018
4-MMPDS *see* 3018
MMS *see* 6170
MMSC *see* 10165
MMT, 6307
α-MMT *see* 6212
MN-10T *see* 9309
MN-1695 *see* 5113
MNE *see* 6580
Mnesis [Takeda] *see* 4932
MNNG *see* 6177
MNPA *see* 6504

MO [Mitsui Toatsu] *see* 2158
MO-338 *see* 2158
MO-911 *see* 7172
MOB *see* 7088
Moba [Mitsubishi Kasei] *see* 135
Mobactam [Takeda] *see* 1922
Moban [Endo] *see* 6315
Mobenol [Horner] *see* 9646
Mobidin [Ascher] *see* 5723
Mobiflex [Roche] *see* 9293
Mobilan [Galen] *see* 4998
Mobisyl [Ascher] *see* 9798
Mobutazon *see* 6312
Mobuzon *see* 6312
MOCA *see* 6136
Mocap [Mobil] *see* 3792
Mocimycin, 6308
Moclamine [Roche] *see* 6309
Moclobemide, 6309
Moctanin [Ethitek] *see* 6335
Mod [Irbi] *see* 3476
Modacin [Tanabe] *see* 1995
Modacor [Houdé] *see* 7096
Modacrylic Fibers, 6310
Modafinil, 6311
Modalim [Sanofi] *see* 2373
Modalina [Maggioni-Winthrop] *see* 9811
Modamide [Merck & Co.] *see* 426
Modane [Adria] *see* 2878
Modatrop [Nordmark] *see* 7275
Moddus [Ciba] *see* 9855
Modecate [Sanofi Winthrop] *see* 4226
Modenol *see* 1554
Moderatan [Thranol] *see* 3175
Moderil [Pfizer] *see* 8311
Modicare *see* 3474
Modicon [Ortho] *see* 6790
Modified Dakin's Soln *see* 8774
Modimmunal [Ravizza] *see* 5006
Modinal *see* 4378
Modirax *see* 4733
Moditen [Sanofi Winthrop] *see* 4226
Moditen Enanthate [Bristol-Myers Squibb] *see* 4226
Modown [Mobil; Rhône-Poulenc] *see* 1256
Modrasone [Schering] *see* 221
Modrastane [Winthrop] *see* 9827
Modrenal [Winthrop] *see* 9827
Modulan *see* 5371
Modulor *see* 3175
Modumate [Abbott] *see* 818
Moduretic [Merck & Co.] *see* 426
Moduretik [Frosst] *see* 426
Modustatina [Midy] *see* 8863
Moenocinol *see* 979
Moenomycin *see* 979
Moenomycin A *see* 979
Mofebutazone, 6312
Mofedione *see* 7060
Mofegiline, 6313
Mofenar [C.E.P.A.] *see* 1497
Mofesal [Medice] *see* 6312
Mofezolac, 6314
Mogadan [Roche] *see* 6667
Mogadon [Roche] *see* 6667
Moheptan *see* 6008
Mohr's Salt *see* 552
Molatoc *see* 3460
Molcer [Wallace] *see* 3460
Moldamin *see* 7221
Molded Silver Nitrate *see* 8662
Molecular Sieves *see* 2401
Molevac [Parke, Davis] *see* 8206
Molgramostim *see* 4559
Molindone, 6315
Molipaxin [Roussel] *see* 9712
Mol-Iron [Schering] *see* 4105
Mol-Iron (obsolete) *see* 6320

Molivate (obsolete) [Glaxo] see 2424
Mollinox see 6028
Molofac [Squibb] see 3460
Moloid [Südmedica] see 5789
Molsidolat [Hoechst] see 6316
Molsidomine, 6316
Molybdenite see 6318
Molybdenum, 6317
Molybdenum Disulfide, 6318
Molybdenum Hexafluoride, 6319
Molybdenum Sesquioxide, 6320
Molybdenum Trioxide, 6321
Molybdic(VI) Acid, 6322
Molybdic Acid, 85%, 6323
Molybdic Anhydride see 6321
Molybdophosphoric Acid see 7498
Molycor-R [Mepha] see 6792
Molyhibit 100 [Amax] see 8790
Molysite see 4061
MOM see 6291
Momentum [Much] see 45
Mometasone Furoate, 6324
Momicine [Morrith] see 6271
MON-097 see 58
MON-0573 see 4522
MON-2139 see 4522
MON-4621 see 420
MON-7200 see 3439
MON-15100 see 3439
Monacetin, 6325
Monacolin K see 5616
Monacrin [Winthrop] see 427
Monactin see 6770
Monalium Hydrate see 5685
Monapen [Fujisawa] see 9568
Monarch [SS Pharm.] see 287
Monarda, 6326
Monardaein Chloride see 6327
Monardein Chloride, 6327
Monardin see 7199
Monargan [Evans] see 49
Mon-Arsone see 3768
Monase [Upjohn] see 3937
Monasirup [Arznei Mller-Rorer] see
682
Monaspor [Ciba] see 1994
Monazan see 6312
Monazol. see 4885
Mondamin [CPC] see 8954
Mondus [Labinca] see 4179
Monellins, 6328
Monensic Acid (obsolete) see 6329
Monensin, 6329
Monetite see 1739
Monicor [Fabre] see 5245
Moniflagon [Dieetzout] see 9292
Monistat [Cilag-Chemie] see 6266
Monit [Stuart] see 5245
Monitan [Ives] see 7742
Monitor [Chevron] see 6014
Monkshood see 118
Monoacetin see 6325
Monoacetylneriifolin see 6559
N^l-Monoacetyl Sulfisoxazole see 9125
Monoamine Oxidase, 6330
Monoammonium, Monosodium Phos-
phoric Acid Salt see 587
Monobasic Lead Acetate see 5443
Mono-Baycuten see 2478
Monobenzone, 6331
Monobenzyl Hydroquinone see 6331
Monobromoacetanilide see 1420
Monobromobenzene see 1427
Monobromoethane see 3819
2-Monobromoisovalerylurea see 1418
Monobromomethane see 6108
Monobromosuccinic Acid see 1461
Monobutyl see 6312
Monobutyrin, 6332
α-Monobutyrin see 6332

Monocaine [Novocol] see 1552
Monocalcium Orthophosphate see
1740
Monocalcium Phosphate see 1740
Monocalcium Tetrasodium Bis[penta-
aquatetra-μ-hydroxy[D-gluconato-
(4—)]dioxotriferrate(3—)] see 4070
α-Monocaprylin see 6335
Mono-Cedocard [Tillotts] see 5245
Monochloracetone see 2164
Monochlorethane see 3829
Monochloroacetaldehyde see 2159
Monochloroacetic Acid see 2162
Monochloroacetic Acid Anhydride see
2163
Monochloroacetone see 2164
17-Monochloroacetylajmaline see
5607
Monochlorobenzene see 2172
Monochlorodibromomethane see 2186
8-Monochloro-3-(β-diethylaminoeth-
yl)-4-methyl-7-ethoxycarbonylmeth-
oxycoumarin see 2467
α-Monochlorohydrin see 2197
Monochloromethyl Ether see 2198
Monochlorphenamide see 2434
Monocid [SKB] see 1980
Monocidur [SKB] see 1980
Monocillin [Chassot] see 1178
Monoclair [Hennig] see 5245
Monoclate-P [Rorer] see 3963
Monoclonal Antibodies see 4958
Monoclonal Antibody B72.3-GYK-
DTPA-In-111 see 8530
Monocor [Am. Cyanamid] see 1336
Monocortin [Grünenthal] see 7162
Monocron [Makhteshim-Agan] see
6334
Monocrotaline, 6333
Monocrotophos, 6334
Monoctanoin, 6335
Monodie [Schering AG] see 4421
Mono[2-(dimethylamino)ethyl]phos-
phoric Acid Ester see 2935
Monodral Bromide [Winthrop] see
7268
5,6-Monoepoxyvitamin A see 4686
Monoethanolamine see 3772
Monoethanolamine DL-Acetylleucin-
ate see 96
Monoethanolamine Salt of α-Acet-
amidoisocaproic Acid see 96
Monoethylamine see 3808
Monoethyl Tartrate see 3901
Monoferrous Acid Citrate Monohy-
drate see 4092
Monofluoroacetamide see 4204
Mono-Gesic [Central] see 8491
Monoglycerol p-Aminobenzoate see
4496
Monoglycol Salicylate see 4509
Monoglyme see 3274
Monogynol B see 5636
Monohydrated Selenium Dioxide see
8571
Monohydroxyaluminum Bis(acetyl-
salicylate) see 338
Monohydroxyaluminum Diacetyl-
salicylate see 338
Mono(2-hydroxybenzoate)-1H-imid-
azole see 4949
Monohydroxymercuridiiodoresorcin-
sulfonphthalein Sodium Salt see
5912
Monoiodotyrosine see 5065
Mono-Kay [Abbott] see 7536
Monoket [Kremers-Urban] see 5245
Monolene see 458
Monolupine see 666
Mono Mack [Mack, Illert.] see 5245

Monomestrol see 5975
(Monomethoxypolyethylene Glycol
Succinimidyl)$_{74}$-L-asparaginase see
871
Monomethylamine see 6095
Monomethyl-p-aminophenol Sulfate
see 6097
Monomethylaniline see 6098
Monomethylarsinic Acid see 6020
Monomethyldiaminodiphenazothion-
ium Chloride see 959
o-Monomethyldiphenhydramine see
7007
N-Monomethylformamide see 6152
Monomethylhydrazine see 6160
Monomethyl Mercury see 3299
Monomethyloldimethylhydantoin see
4877
Monomethylpropion see 6033
Monomethyl Sulfate see 6203
Monomethylthionine Chloride see 959
Monomycin [Grünenthal] see 3720
Monomycin A see 7173
Mononine [Armour] see 3964
Mononitrogen Monoxide see 6674
Monooctadecyl Ether of Glycerol see
1039
Monooctanoin see 6335
Monopar [Burroughs Wellcome] see
8971
Monoparin [CP Pharm.] see 4685
Monopen see 7225
Monophenolase see 9969
Monophenoloxidase see 9969
Monophenylbutazone see 6312
Monophos see 623
Monopina [Bioindustria] see 516
Monopotassium Citrate see 7786
Monopotassium Phosphate see 7829
Monopril [Squibb] see 4282
Monopropionylerythromycin see 3725
Monorden, 6336
Monores [Valeas] see 2407
Monorhein see 8342
Monorheumetten [Benzon] see 6312
Monos [SKB] see 8448
Monosilane see 8633
Monosodium Glutamate, 6337
Monosodium L-Glutamate Monohy-
drate see 6337
Monosodium Methanearsonate see
6020
Monosodium Orthophosphate see 8806
Monosodium N-Phenylglycinamide-p-
arsonate see 9925
Monosorb [Schering] see 5245
Monospan [Santen] see 1624
Monostearin see 4498
Monosulfiram see 9123
Monotheamin [Lilly] see 9421
Monothioethyleneglycol see 5917
α-Monothioglycerol see 9471
Monotrim [Duphar] see 9840
Monotropein, 6338
Monotropitin see 4384
Monotropitoside see 4384
Monovent [Lagap] see 9302
Monoverin [E. Merck] see 7887
Monoxone [Plant Protection] see 2162
Monoxychlorosene see 7090
Monozol see 3177
Monsel's Soln see 4078
Montan Wax, 6339
Monteban [Lilly] see 6506
Montelukast, 6340
Montmorillonite, 6341
Montrel [Dow] see 2671
Montricin [SPA] see 5891
Monuril [Zambon] see 4280
Monuron, 6342

Monuron TCA *see* 6342
Monydrin [Draco] *see* 7461
Monzal [Thomae] *see* 10107
Monzaldon *see* 10107
Moogrol [Burroughs Wellcome] *see* 2089
5-MOP *see* 1196
8-MOP *see* 6068
Mopazine *see* 6080
Moperone, 6343
Mopidamol, 6344
Moplen [Montecatini] *see* 7741
Mopral [Astra] *see* 6977
Moprolol, 6345
MOPS, 6346
Moquizone, 6347
Moracizine *see* 6351
d-Moramid(e) *see* 2997
Moranoline *see* 2952
Morantel, 6348
Moranyl [Specia] *see* 9181
Morazone, 6349
Morbam *see* 5908
Morbicid *see* 4262
Morclofone, 6350
Mordant Rouge *see* 349
Mordant Violet 25 *see* 4362
Mordant Yellow 3R *see* 251
Morena [Kettelhack] *see* 3217
Morestan [Bayer] *see* 7113
Morfazinamide *see* 6355
Morial [Takeda] *see* 6316
Moricizine, 6351
Morin, 6352
Morinamide *see* 6355
Morindin, 6353
β-Morindin *see* 6353
Morindone *see* 6353
Moringine *see* 1160
Morintannic Acid *see* 5676
Moriperan [Morishita] *see* 6226
Morison's Paste *see* 5731
Moritannic Acid *see* 5676
Morkit [Bayer] *see* 726
Mornidine [Searle] *see* 7607
Morniflumate *see* 6620
Morning-glory *see* 5088
Morocide [Boots] *see* 1265
Moronal [Squibb-Heyden] *see* 6834
Moroxydine, 6354
Morpan T [A.B.M. Chem.] *see* 6419
Morphazinamide, 6355
Morphenol, 6356
Morpheridine, 6357
Morphia *see* 6359
Morphina *see* 6359
Morphinan, 6358
Morphinan-3-ol *see* 6801
Morphine, 6359
Morphine Bis(nicotinate) *see* 6607
Morphine Bis(pyridine-3-carboxylate) *see* 6607
Morphine Dinicotinate *see* 6607
Morphine Ester with Nicotinic Acid *see* 6607
Morphine Hydrochloride, 6360
Morphine Hyperduric *see* 6359
Morphine 3-Methyl Ether *see* 2525
Morphine Monomethyl Ether *see* 2525
Morphine Sulfate, 6361
Morphium *see* 6359
Morphodone *see* 7348
Morpholine, 6362
4-Morpholinecarboximidoylguanidine *see* 6354
Morpholine *N,N'*-Disulfide *see* 3437
4-Morpholinepropanesulfonic Acid *see* 6346
(—)-3-Morpholino-4-(3-*tert*-butylami-

no-2-hydroxypropoxy)-1,2,5-thiadi-azole *see* 9585
4-Morpholine-2,2-diphenylbutyric Acid Ethyl Ester *see* 3354
1-(β-*N*-Morpholinoethyl)-1-cyclo-pentenylcyclopentane-2-one *see* 2775
3-(2-Morpholinoethyl)morphine *see* 7484
N-2-Morpholinoethyl-5-nitroimidazole *see* 6644
Morpholinoethyl Norpethidine *see* 6357
1-(2-Morpholinoethyl)-4-phenyliso-nipecotic Acid Ethyl Ester *see* 6357
4-Morpholinomethylescutol *see* 4252
3-(Morpholinomethyl)-1-[(5-nitrofur-furylidene)amino]hydantoin *see* 6624
5-Morpholinomethyl-3-(5-nitrofur-furylideneamino)-2-oxazolidinone *see* 4315
N-Morpholinomethylpyrazinamide *see* 6355
7-Morpholinomethyltheophylline, 6363
3-(*N*-Morpholino)propanesulfonic Acid *see* 6346
10-(3-Morpholinopropionyl)phenothia-zine-2-carbamic Acid Ethyl Ester *see* 6351
γ-Morpholinopropyl 4-*n*-Butoxyphen-yl Ether *see* 7888
4-(Morpholinothiocarbonyl)guaiacol *see* 10071
6-(4-Morpholinyl)-4,4-diphenyl-3-heptanone *see* 7348
3-[2-(4-Morpholinyl)ethyl]morphine *see* 7484
β-Morpholinylethylmorphine *see* 7484
1-(2-*N*-Morpholinylethyl)-5-nitroimid-azole *see* 6644
1-[2-(Morpholinyl)ethyl]-4-phenyl-4-piperidinecarboxylic Acid Ethyl Ester *see* 6357
3-(4-Morpholinylmethyl)-1-[[(5-nitro-2-furanyl)methylene]amino]-2,4-imidazolidinedione *see* 6624
5-(4-Morpholinylmethyl)-3-[[(5-nitro-2-furanyl)methylene]amino]-2-oxa-zolidinone *see* 4315
2-(4-Morpholinylmethyl)-2-phenyl-1*H*-indene-1,3(2*H*)-dione *see* 7060
N-(4-Morpholinylmethyl)pyrazinecar-boxamide *see* 6355
[10-[3-(4-Morpholinyl)-1-oxopropyl]-10*H*-phenothiazin-2-yl]carbamic Acid Ethyl Ester *see* 6351
Morphothebaine, 6364
Morsydomine *see* 6316
Mortopal *see* 9345
Moryl *see* 1823
Mosaic Gold *see* 8936
Mosapramine, 6365
Mosatil *see* 3555
Moscontin [Sarget] *see* 6361
Mosegor [Sandoz-Wander] *see* 7671
Mosidal *see* 6011
Mosquito Plant *see* 7236
Mossbunker Oil *see* 5880
Moss Starch *see* 5503
Motazomin [Takeda] *see* 6316
Motens [Boehringer, Ing.] *see* 5344
Mother-of-Pearl Sulfur *see* 9142
Motiax [Neopharmed] *see* 3972
Motilex [Guidotti] *see* 2404
Motilin, 6366
Motilium [Janssen] *see* 3476
Motilyn [Abbott] *see* 2988
Motolon *see* 6028
Motosol [Boehringer, Ing.] *see* 401

Motox [Tenneco] *see* 9693
Motretinide, 6367
Motricit [Virbac] *see* 4925
Motrin [Upjohn] *see* 4925
Mountain Balm *see* 3711
Mountain Grape *see* 1193
Mountain Tobacco *see* 825
Mouse Antialopecia Factor *see* 5008
Mouse-bane *see* 118
Mouse EGF-URO *see* 3569
Movecil [Farmitalia] *see* 8158
Movellan (amps.) [Asta] *see* 9022
Movellan (tabl.) [Asta] *see* 9022
Moveltipril, 6368
Movens [Inverni] *see* 5819
Movergan [Asta] *see* 8569
Movirene *see* 3413
Movyl [Montecatini] *see* 7746
Mowiol [Hoechst] *see* 7745
Moxadil [Lederle] *see* 616
Moxal [Bellon] *see* 617
Moxalactam, 6369
Moxalactam [Lilly] *see* 6369
Moxaline [Mead Johnson] *see* 617
Moxam [Lilly] *see* 6369
Moxastine, 6370
Moxaverine, 6371
Moxestrol, 6372
Moxidec [Am. Cyanamid] *see* 6373
Moxidectin, 6373
Moxisylyte, 6374
Moxonidine, 6375
Moxyl [Fujirebio] *see* 6374
MP *see* 6197
4-MP *see* 4257
6MP *see* 5919
8-MP *see* 6068
MP-11 *see* 7320
MP-302 *see* 5084
MP-328 *see* 5083
MP-620 *see* 5030
MP-1727 *see* 7267
MPC-1304 *see* 806
MPMP *see* 5892
MPP[+] *see* 6376
α-MPT *see* 6248
MPTP, 6376
MPV-785 *see* 5830
MPV-1248 *see* 894
MPV-1440 *see* 5830
M-Pyrol [GAF] *see* 6197
MQPA *see* 816
MRD-108 *see* 7638
MRIH *see* 5855
MRL-41 *see* 2446
MS-222 *see* 9751
MS-752 *see* 5620
MS-932 *see* 135
MS-4101 *see* 4243
MSA *see* 8862
MS Contin [Chinoin] *see* 6361
MSG *see* 6337
MSH, 6377
MSH Release-inhibiting Factor I *see* 5855
MSH Release-inhibiting Hormone *see* 5855
MSMA *see* 6020
MST 10 Mundipharma [Mundi-pharma] *see* 6361
MST-16 *see* 8708
MST 30 Mundipharma [Mundi-pharma] *see* 6361
MST Continus [Napp] *see* 6361
L-α-MT *see* 6248
MT-141 *see* 1978
MTB-51 *see* 6025
MTBE *see* 6111
L-3-MTO *see* 6133
MTQ *see* 6028

MTU *see* 6210
MTX *see* 6065
4-MU *see* 4903
Mucalan *see* 5124
Mucara *see* 5296
Mucic Acid *see* 4349
Muciclar [Warner-Lambert] *see* 1850
Mucinol [Kali-Chemie] *see* 683
Mucins, 6378
Mucitux [Riom] *see* 3666
Muclox [Sigma Tau] *see* 3972
Muco-Burg [Burg] *see* 401
Mucocedyl [3M] *see* 89
Mucochloric Acid, 6379
Mucochloric Anhydride, 6380
Mucocis [Coli/Crosara] *see* 1850
Mucoclear [Mundipharma] *see* 401
Mucodil [Valeas] *see* 8963
Mucodyne [USV] *see* 1850
Mucofar [Farmakos] *see* 401
Mucofluid [UCB] *see* 5969
Mucolase [Lampugnani] *see* 1850
Mucolator [Allard] *see* 89
Mucolex [Warner] *see* 1850
Mucolitico [Maggioni] *see* 3473
Mucolysin [Proter] *see* 9597
Mucolyticum [Lappe] *see* 89
Mucomycin *see* 5658
Mucomyst [Mead Johnson] *see* 89
Muconic Acid, 6381
Muconomycin A *see* 10100
Muconomycin B *see* 10100
Mucopront [Mack, Illert.] *see* 1850
Mucosan [Boehringer, Ing.] *see* 401
Muco Sanigen [Thiemann] *see* 89
Mucosolvan [Thomae] *see* 401
Mucosolvin [Berlin-Chemie] *see* 89
Mucosta [Otsuka] *see* 8296
Mucotab [Pharmakon] *see* 1850
Mucovent [Byk-Gulden] *see* 401
Mucret [Astra] *see* 89
Muira Puama, 6382
Mukinyl [Searle] *see* 1850
Mulka *see* 6033
Mulsiferol *see* 10156
Mulsopaque [Alcon] *see* 5074
Multergan [Specia] *see* 9441
Multhiomycin *see* 6816
Multi-CSF *see* 5021
Multifungin [Knoll] *see* 1457
Multilase [Sigma-Tau] *see* 712
Multimycine *see* 2542
Multiplication Stimulating Activity *see* 8862
Multipotent Colony Stimulating Factor *see* 5021
Multum [Chephasaar] *see* 2132
Mundisal [Purdue Frederick] *see* 2265
Munobal [Hoechst] *see* 3991
Muphoran [Servier] *see* 4285
Mupirocin, 6383
Muracil [Organon] *see* 6210
Muramic Acid, 6384
Muramidase *see* 5671
Muramyl Dipeptide, 6385
Murazyme [Prospa] *see* 5671
Murel [Ayerst] *see* 10044
Murexan *see* 9987
Murexide, 6386
Murexine, 6387
Muriacite *see* 1753
Muriamic *see* 4477
Muriatic Acid *see* 4821
Murillo Bark *see* 8222
Muroctasin *see* 8412
Murrayin *see* 8553
Musaril [Sanofi Winthrop] *see* 9379
Muscalm [Nippon Kayaku] *see* 9660
Muscalure, 6388
Muscarine, 6389

Muscatox [Bayer] *see* 2626
Muscazone, 6390
Muscimol, 6391
Muscle Adenylic Acid *see* 157
Muscle Inosinic Acid *see* 5007
Muscone, 6392
Musco-Ril [Inverni] *see* 9460
Musculamine *see* 8894
Musculax [Nippon Organon] *see* 10075
Muskone *see* 6392
Musk Root *see* 9174
Musks, 6393
Mussel Poison *see* 8533
Mustard, Black, 6394
Mustard Chlorohydrin *see* 4680
Mustard Gas, 6395
"Mustard Oil" *see* 303
Mustard, White, 6396
Mustargen Hydrochloride [Merck & Co.] *see* 5815
Mustine Hydrochloride *see* 5815
Mustron *see* 5816
Mutabase [Schering] *see* 3051
Mutamycin [Bristol] *see* 6301
Mutesa [Wyeth] *see* 7067
Muthesa [Wyeth] *see* 7067
Muthmann's Liquid *see* 9328
Mutton Suet *see* 9055
Muzolimine, 6397
MV-119A *see* 3433
MV-678 *see* 7951
MY-41-6 *see* 7177
MY-301 *see* 4582
MY-5116 *see* 8305
Myacine *see* 6542
Myagen [Upjohn] *see* 1353
Myalex [ICI] *see* 4010
Myambutol [Lederle] *see* 3764
Myanesin [BDH] *see* 5895
Myanol *see* 5895
Myarsenol *see* 9111
Mybasan [Antigen] *see* 5203
MYC-8003 *see* 6308
Mycaminose, 6398
Mycanden [Schering] *see* 4631
Mycardol [Winthrop] *see* 7249
Mycarose, 6399
Mycelex-G [Miles] *see* 2478
Mycelianamide, 6400
Mycetins, 6401
Mychel [Rachelle] *see* 2120
Mycifradin [Upjohn] *see* 6542
Mycil [BDH] *see* 2230
Mycilan [Théraplix] *see* 4631
Mycinol [Horner] *see* 2120
Mycivin *see* 5525
Myclobutanil, 6402
Mycobacidin, 6403
Mycobacillin, 6404
Mycobactic Acid *see* 6405
Mycobactin P *see* 6405
Mycobactins, 6405
Mycobactyl *see* 4510
Mycobutin [Farmitalia] *see* 8380
Mycobutol [Cadila] *see* 3764
Mycofug [Hermal] *see* 2478
Mycoin C₃ *see* 7185
Mycolic Acids, 6406
Mycolutein *see* 913
(6-O-Mycolyl-α-D-glucopyranosyl)-6-O-mycolyl-α-D-glucopyranoside *see* 2592
Mycomycin, 6407
Mycophenolic Acid, 6408
Mycophyt [Mycofarm] *see* 6513
Myco-protein *see* 8270
Mycosamine, 6409
Mycose *see* 9713
Mycospor [Bayer] *see* 1260

Mycosporan [Bayer] *see* 1260
Mycosporin [Bayer] *see* 2478
Mycostatin [Squibb] *see* 6834
Mycoster [Fabre] *see* 2325
Mycotoxin T-2 *see* 9933
Mydantane *see* 389
Mydecamycin *see* 6271
Mydetone *see* 9660
Mydfrin [Alcon] *see* 7440
Mydocalm [Labatec] *see* 9660
Mydplegic [Cooper] *see* 2815
Mydriacyl [Alcon] *see* 9911
Mydrial [Winzer] *see* 9966
Mydriaticum [Merck & Co.] *see* 9911
Mydriatin *see* 7461
Mydrilate [Boehringer, Ing.] *see* 2815
Myebrol [Kyorin] *see* 6298
Myeleukon *see* 1529
Myelobromol [Enzypharm] *see* 6298
Myelographin [Squibb] *see* 5087
Myeloleukon *see* 1529
Myelosan *see* 1529
Myelotrast [Winthrop] *see* 5029
Myfungar [Siegfried] *see* 7071
Mygdalon [DDSA] *see* 6226
Myk [Cassenne] *see* 9062
Mykostin *see* 10156
Mykrox [Fisons] *see* 6231
Mylabris, 6410
Mylar [Du Pont] *see* 7730
Mylaxen [Mallinckrodt] *see* 4721
Mylepsin [ICI] *see* 7927
Myleran [Burroughs Wellcome] *see* 1529
Mylicon [J&J-Merck] *see* 3264
Mylipen *see* 7226
Mylis [Kanebo] *see* 7892
Mylodorm *see* 607
Mylofanol *see* 2347
Mylone *see* 2892
Mylosar [Upjohn] *see* 923
Mylproin [ICI] *see* 10049
Mynosedin [Toho Kogyo] *see* 4925
Myocaine *see* 5895
Myocardone *see* 4651
Myocholine *see* 1232
Myochrysine [Merck & Co.] *see* 4536
Myocord [Szabo] *see* 892
Myocrisin [M & B] *see* 4536
Myodetensine *see* 5895
Myodigin *see* 3206
Myodil *see* 5074
Myoflex [Adria] *see* 9798
Myoglycerin *see* 6704
Myohematin *see* 2860
Myolastan [Sanofi Winthrop] *see* 9379
Myolysin *see* 5895
Myonal [Eisai] *see* 3642
Myopan *see* 5895
Myopone *see* 10178
Myoral, 6411
Myordil [Winthrop] *see* 615
Myorexon [ICI] *see* 5245
Myosalvarsan *see* 9111
Myoscain *see* 4582
Myoserol *see* 5895
Myosin, 6412
Myoston [Martinet] *see* 157
Myoten [Central Pharm.] *see* 5895
Myotolon [Daiichi] *see* 4319
Myotonine Chloride [Glenwood] *see* 1232
Myoview [Mediphysics Amersham] *see* 9383
Myoxam [Menarini] *see* 6271
Myoxanthin *see* 3543
Myprozine [Am. Cyanamid] *see* 6513
α-Myrcene *see* 6413
β-Myrcene, 6413
Myreth *see* 7737

Name Index

Myrica see 1040
Myricetin, 6414
Myricitrin see 6414
Myricodine see 6420
Myricyl Alcohol see 9722
Myristica, 6415
Myristic Acid, 6416
Myristicin, 6417
Myristin see 9854
Myristyl Alcohol, 6418
Myristyl Benzylmorphine see 6420
Myristyltrimethylammonium Bromide, 6419
Myrizone [Squibb] see 9427
Myrj [ICI] see 7738
Myrj 45 [ICI] see 7738
Myrj 52 [ICI] see 7738
Myrj 52S [ICI] see 7738
Myrocodine see 6420
Myronate Potassium see 8691
Myrophine, 6420
Myrophinium see 6420
Myrrh, 6421
Myrtecaine, 6422
Myrticolorin see 8456
Myrtillin-a see 2931
Myrtle Wax see 1040
Myrtol, 6423
Mysalfon [Spofa] see 9308
Myser [Nikken] see 3194
Mysoline [Wyeth-Ayerst] see 7927
Mysuran [Winthrop] see 396
Mytab [Zeeland] see 6419
Mytatrienediol, 6424
Mytelase [Winthrop] see 396
Mytolon [Winthrop] see 1137
My-trans [Heather] see 5908
Myxin, 6425
Myxoviromycin see 419
MZ-144 see 8391
MZ-0780 see 8391
M-Zyme see 5303
N see 872
N-5' see 9705
N-021 see 9639
N-22 see 6314
N-0252 see 5397
N-399 see 10208
N-553 see 9660
N-696 see 9579
N-714 see 2241
N-746 see 2455
N-869 see 6016
N-2790 see 4261
N-7009 see 4224
Na III see 3909
NA 22 [DuPont] see 3849
NA-66 see 7583
NA-73 see 4751
NA-97 see 7139
NA-274 see 1412
NA-872 see 401
NA-8318 see 5256
NAA see 875 and 6458
N.A.B. see 6534
NAB-365 see 2407
NAB-365Cl see 2407
Nabadial see 6017
Nabam, 6426
Nabilone, 6427
Nabolin see 6018
Nabumetone, 6428
Nabuser [SKF] see 6428
NAC [Mead Johnson] see 89
Nacconate 100 [Allied] see 9668
Nacenyl [Syntex] see 6437
Nacid [Shionogi] see 5696
NaClex [Glaxo] see 4830
Nacom [Merck & Co.] see 1843
Nactate see 7714

Nacton [McNeil] see 7714
NAD, 6429
NAD⁺ see 6429
β-NAD see 6429
Nadeine see 3214
Nadex [Zyma-Galen] see 8175
NADH see 6429
Nadic Anhydride see 1842
Nadide see 6429
Nadifloxacin, 6430
Nadigest [Streuli] see 5838
Nadisan [Boehringer, Mann.] see 1881
Nadolol, 6431
Nadone see 2795
Nadoxolol, 6432
NADP, 6433
NADPH see 6433
NADPH₂ see 6433
NADPH₂ Diaphorase see 6961
Nadroparin, 6434
Naepaine, 6435
Nafamostat, 6436
Nafamostat Mesylate see 6436
Nafamstat see 6436
Nafarelin, 6437
Nafcil [Bristol] see 6438
Nafcillin Sodium, 6438
Naferon [Sclavo] see 5017
Nafiverine, 6439
Nafrine see 7100
Nafronyl, 6440
Naftalofos, 6441
Naftazone see 6481
Naftidan see 6439
Naftidrofuryl see 6440
Naftifine, 6442
Naftifungin see 6442
Naftin [Sandoz] see 6442
Naftopen [KN Gistfabriek] see 6438
Naftopidil, 6443
NAG see 6437
Naganol see 9181
Nagemid [Dexo] see 1467
Na¹³¹I see 8778
Naismeritin [Hishiyama] see 9660
Naixan [Tanabe] see 6504
Nalador [Schering AG] see 9165
Nalbuphine, 6444
Nalcrom [Fisons] see 2658
Nalcron [Fisons] see 2658
Nalde [Mead Johnson] see 418
Naled, 6445
Nalfon [Lilly] see 4021
Nalgesic [Lilly] see 4021
Nalidicron [Toyo Shinyaku] see 6446
Nalidixic Acid, 6446
Nalitucsan [Hishiyama] see 6446
D-Nal(2)⁶-LIIRH see 6437
Nalline [Merck & Co.] see 6448
Nalmefene, 6447
Nalmetrene see 6447
Nalone [Winthrop] see 6449
Nalorex [Robin] see 6450
Nalorphine, 6448
Naloxone, 6449
Nalpen [Beecham] see 942
Nalpen G see 1178
Naltrexone, 6450
N-[[[2-[[[2-[(Aminoiminomethyl)-amino]-4-thiazolyl]methyl]thio]eth-yl]amino]methylene]-4-bromoben-zenesulfonamide see 3536
Namphen [Toyo Iyaku] see 5842
Namuron see 2780
NAN see 8628
NANA see 8628
Nanbacine [Bellon] see 10210
Nancimycin see 8383
Nandinine, 6451
Nandrolin [Tutag] see 6452

Nandrolone, 6452
Nandrolone Laurate see 6452
Nandron [Morishita] see 3603
Nankor [Dow] see 8415
NANM see 6448
Nanormon [Nordisk] see 8864
Nantenine see 3472
Nantokite see 2730
NAPA [Arnar-Stone] see 17
Napacetin [Toyama] see 4925
Napageln [Lederle] see 3989
Napalm, 6453
Napelline, 6454
Napellonine see 8865
Naphazoline, 6455
Naphcon [Alcon] see 6455
Naphtha see 7329
Naphthacene, 6456
Naphthacetol see 473
Naphthalane see 2903
1-Naphthalenamine see 6485
2-Naphthalenamine see 6486
Naphthalene, 6457
1-Naphthaleneacetic Acid, 6458
α-Naphthaleneacetic Acid see 6458
1-Naphthalenecarboxylic Acid see 6468
2-Naphthalenecarboxylic Acid see 6469
1,8-Naphthalenediamine, 6459
1,8-Naphthalenedicarboxylic Acid see 6467
1,3-Naphthalenediol see 6483
1,2-Naphthalenedione see 6481
1,4-Naphthalenedione see 6482
1,5-Naphthalenedisulfonic Acid see 824
1,6-Naphthalenedisulfonic Acid, 6460
2,6-Naphthalenedisulfonic Acid, 6461
2,7-Naphthalenedisulfonic Acid, 6462
1,5-Naphthalenedisulfonic Acid Bis [2-[2-(Acetyloxy)-1-oxopropoxy]-N,-N,N-trimethylethanaminium]ion (2⁻) see 116
1-Naphthalenesulfonic Acid, 6463
2-Naphthalenesulfonic Acid, 6464
α-Naphthalenesulfonic Acid see 6463
β-Naphthalenesulfonic Acid see 6464
1-Naphthalenethiol, 6465
2-Naphthalenethiol, 6466
1-Naphthalenol see 6470
2-Naphthalenol see 6471
2-Naphthalenol Benzoate see 6495
1-Naphthalenol Methylcarbamate see 1831
6-[3-(2-Naphthalenyl)-D-alanine]lu-teinizing Hormone-releasing Factor-(pig) see 6437
2-[(1-Naphthalenylamino)carbonyl] Benzoic Acid see 6505
N-1-Naphthalenyl-1,2-ethanediamine see 6496
N-(1-Naphthalenylmethyl)-1-naphtha-lenemethanamine see 1335
N-2-Naphthalenyl-2-naphthalenamine see 3319
2-Naphthalenyloxyacetic Acid see 6484
2-(2-Naphthalenyloxy)ethanol see 6500
1-Naphthalenylthiourea see 764
Naphthalic Acid, 6467
Naphthalidine see 6485
Naphthalin see 6457
Naphthalol see 6502
Naphthalophos see 6441
Naphthane see 2903
Naphthanthracene see 1089
Naphthene see 6457

Nauheim Salts (Artificial), 6515
Naumannite *see* 8572
Nausidol [SK & F] *see* 7607
Nausilen [Inverni] *see* 246
Nautamine [Delagrange] *see* 1253
Nautazine *see* 2779
Nauzelin [Janssen] *see* 3476
Navadel [Hercules] *see* 3355
Navadyl *see* 1784
Navane [Roerig] *see* 9503
Navanide *see* 443
Navelbine [Pierre Fabre] *see* 10127
Navicalm *see* 5817
Navidrex [Ciba] *see* 2813
Navidrix [Ciba] *see* 2813
Navilox [Univet] *see* 5259
Navisin *see* 7100
Navoban [Sandoz] *see* 9914
Navolin *see* 4463
Naxcel [Upjohn] *see* 1999
Naxen [Syntex] *see* 6504
Naxidine [Lilly] *see* 6758
Naxofem [Farmitalia] *see* 6644
Naxogin [Erba] *see* 6644
Naxy [Sanofi Winthrop] *see* 2400
Nazasetron *see* 933
NBS *see* 1462
NC-14 *see* 791
NC-45 *see* 10075
NC-123 *see* 5970
NC-302 *see* 8269
NC-1667 *see* 9797
NC-1968 *see* 7310
NC-6897 *see* 1063
NC-8438 *see* 3788
NC-21314 *see* 2435
NCI-96683 *see* 8269
NCI-C55243 *see* 1438
NCI-C55254 *see* 2186
NCI-C56462 *see* 6333
NCS *see* 10302
NDEA *see* 6734
NDELA *see* 6733
2′NDG *see* 4374
NDGA *see* 6786
NDR-263 *see* 7998
NDR-304 *see* 3834
NDR-5998A *see* 4041
NE-10064 *see* 943
NE-58095 *see* 8396
Nealbarbital, 6516
Nealbarbitone *see* 6516
Neallymal *see* 6516
Neamine, 6517
Neamoxyl [Arcopharma] *see* 617
Neantine [Givaudan] *see* 7527
Neatsfoot Oil, 6518
Neazina [Centralvet] *see* 9083
Nebactam [Specia] *see* 955
Nebcin [Lilly] *see* 9628
Neberk [Morishita] *see* 9267
Nebicina [Lilly] *see* 9628
Nebilet [Janssen] *see* 6519
Nebivolol, 6519
dl-Nebivolol *see* 6519
Nebracetam, 6520
Nebralin [Gayoso Wellcome] *see* 9307
Nebramycin *see* 9628
Nebramycin Factor 2 *see* 792
Nebramycin Factor 6 *see* 9628
Nebularine, 6521
Nebulin, 6522
NebuPent [Fujisawa] *see* 7254
Neburon, 6523
Necatorina *see* 1864
Necic Acids *see* 8593
"Necine" Bases *see* 8334
Nectadon [Merck & Co.] *see* 6815
Nectocyd [Pfizer] *see* 3434
Nedius [Erba] *see* 4999

Nedocromil, 6524
Needle Antimony *see* 754
Neem, 6525
Neem Oil, 6526
Nefazodone, 6527
Nefco *see* 6697
Nefiracetam, 6528
Nefopam, 6529
Nefrix *see* 4822
Nefrolan [M & B] *see* 2466
Nefrosul [Riker] *see* 9068
Neftin *see* 4320
Nefurofan [Maruko] *see* 8917
Negamycin, 6530
Negatan [Savage] *see* 6531
Negatol [Valpan], 6531
Negaxid [Sigma-Tau] *see* 409
NegGram [Winthrop] *see* 6446
Negram [Winthrop] *see* 6446
Neguvon [Bayer] *see* 9753
Nehydrin [TAD] *see* 3692
Nektrohan [ICN] *see* 287
Nelbon [Sankyo] *see* 6667
Nema [Parke, Davis] *see* 9332
Nemacur [Baychem] *see* 3997
Nemadectin, 6532
Nemafax [Rhone Merieux] *see* 9489
Nemafos [Am. Cyanamid] *see* 9482
Nemafume *see* 3068
Nemagon *see* 3068
Nemapan *see* 9426
Nem-A-Tak [Am. Cyanamid] *see* 4284
Nematolyt *see* 7148
Nemazine *see* 7404
Nembutal [Abbott] *see* 7272
Nemestran [Hoechst Pharma] *see* 4423
Nemex *see* 1187
Nemicide [Coopers] *see* 5486
Nemispor [Isagro] *see* 5756
Nemonapride, 6533
Nemural [Winthrop] *see* 815
Nemurel [Takeda] *see* 3744
Nendrin *see* 3618
Neo-Absentol *see* 3762
Neoamyl Alcohol *see* 6546
Neo-Antergan [Poulenc] *see* 8168
Neoantimosan *see* 8967
Neo-Arsoluin *see* 6534
Neoarsphenamine, 6534
Neo-Atromid *see* 2436
Neo-Avagal *see* 6084
Neobar [E. Merck] *see* 1023
Neobenadol *see* 3367
Neo-Benodine *see* 6130
Neo-Betalin 12 [Lilly] *see* 4854
Neobiosamine B *see* 6542
Neobiosamine C *see* 6542
Neo-Bradoral [Ciba-Geigy] *see* 3474
Neobrettin [Norbrook] *see* 6542
Neocaine *see* 7937
Neocarcinostatin *see* 10302
Neo-Cardiamine *see* 9344
Neocardyl *see* 1301
Neocarzinostatin *see* 10302
Neocarzinostatin K *see* 10302
Neo-Cebicure *see* 5141
Neo-Cebitate *see* 5141
Neocefal [Gibipharma] *see* 1964
Neocembrene, 6535
Neocembrene A *see* 6535
Neochanin *see* 4271
Neocid [Geigy] *see* 2898
Neocidol [Ciba-Geigy] *see* 3043
Neoclym [Poli] *see* 2789
Neo-Cobefrin [Cook-Waite] *see* 6785
Neo-Codema *see* 4822
Neocon 1/35 [Ortho-Cilag] *see* 6790
Neo-Corovas *see* 7249
Neo-cupferron, 6536
Neocuproine, 6537

Neo-Cytamen [Duncan-Flockhart] *see* 4854
Neodaian [Nichiiko] *see* 9432
Neodalit [Hässle] *see* 3055
Neodecyllin [Penick] *see* 6543
Neo-Dema *see* 2221
Neo-Devomit *see* 2779
Neodigitalis *see* 3201
Neo-Dioxanin [Boehringer, Ing.] *see* 3210
Neo Dohyfral D$_3$ *see* 10157
Neodorm [Minden] *see* 7272
Neo-Douxan *see* 6714
Neodrenal *see* 5236
Neo-Duplamox [ISF] *see* 2402
Neodymium, 6538
Neodyne [Merrell] *see* 3834
Neo-Epinine [Burroughs Wellcome] *see* 5236
Neoergosterol, 6539
Neo-Erycinum [Schering AG] *see* 3722
Neoesserin [Ankerpharm] *see* 6553
Neofamid *see* 5683
Neo-Farmadol *see* 7106
Neofemergen *see* 3694
Neo-Ferrum *see* 4073
Neo-Fluimucil [Inpharzam] *see* 89
Neoflumen *see* 4822
Neofocin [Medici] *see* 4280
Neo-Fulcin *see* 4571
Neogama [Hormosan] *see* 9163
Neogel [Hexal] *see* 1839
Neogest [Schering] *see* 6797
Neo-Gilurytmal [Kali-Chemie] *see* 7883
Neo H2 [Bochringer, Ing.] *see* 8431
Neo-Hepatex *see* 5575
Neohesperidin DHC *see* 6540
Neohesperidin Dihydrochalcone, 6540
Neohetramine Hydrochloride [Warner-Lambert] *see* 9513
Neo-Hombreol (amp.) [Organon] *see* 9322
Neohombreol M [Organon] *see* 6206
Neohydrazid *see* 2751
Neohydrin [Lakeside] *see* 2154
Neo-Iopax [Schering] *see* 8779
Neoiscotin [Daiichi] *see* 5203
Neoisocodeine *see* 8098
Neo-Istafene *see* 5817
Neoisuprel *see* 5185
Neo Kodan [Schülke & Mayr] *see* 6852
Neolamin [Nippon Kayaku] *see* 9432
Neolate *see* 6542
Neolignans *see* 5509
Neolinarin *see* 7195
Neoloid [Lederle] *see* 1946
Neolutin Depositum *see* 239
Neomagnol [Chinoin] *see* 2117
Neomas *see* 6542
Neo-mercazole [Roche] *see* 1844
Neomestine [Taiyo] *see* 4260
Neo-Metantyl [Zambon] *see* 7989
Neo-Methiodal *see* 5061
Neomethymycin, 6541
Neomethynolide *see* 6541
Neomin *see* 6542
Neomix [Tuco] *see* 6542
Neomycin, 6542
Neomycin A *see* 6517
Neomycin B *see* 6542
Neomycin C *see* 6542
Neomycin E *see* 7173
Neomycin Undecylenate, 6543
Neomyson [Eisai] *see* 9436
Neon, 6544
Neo-Naclex [Glaxo] *see* 1064
Neonal [Abbott] *see* 1550
Neonicotine *see* 661

Neo-Nilorex [A.V.P.] see 7365
Neo-Octon see 6846
Neo-Oestranol I see 3177
Neo-Oestranol II see 3177
Neo-Oxypaat [Katwijk] see 8206
Neopax [Janssen] see 7559
Neopenil see 7212
Neopentane, 6545
Neopentanol see 6546
Neopentyl Alcohol, 6546
Neopentyl Glycol, 6547
Neopenyl see 2406
Neoperidole [Kyowa Hakko] see 4629
Neophryn [Winthrop] see 7440
Neophyl Chloride, 6548
Neopine, 6549
Neoplat [Chugai] see 6263
Neoplatin [Mead Johnson] see 2378
Neo-Ponden [Serono] see 674
Neoprene, 6550
Neopres [Boehringer, Ing.] see 2326
Neoproc see 7226
Neoprotoveratrine see 8086
Neopsicaine see 8097
Neopterin, 6551
Neo-Pynamin [Sumitomo] see 9362
Neopyrithiamine see 8176
Neoquassin, 6552
Neoral [Sandoz] see 2821
Neo-Rinoleina [Granelli] see 10219
Neoron [Ciba-Geigy] see 1456
Neosalvarsan [Hoechst] see 6534
Neosanamid II see 9100
Neoscan [Medi-Physics] see 4364
Neo-Skiodan see 5061
Neospasmina [Biofarmacoterapico] see 3211
Neospiran see 9344
Neostene [SM & P] see 6017
Neostenovasan see 3529
Neosteron see 6017
Neostibosan see 3896
Neostigmine, 6553
Neostigmin (inj.) [Ferrosan] see 6553
Neostigmin (tabl.) [Ferrosan] see 6553
Neoston [Beiersdorf] see 220
Neostreptal [Locatelli] see 9073
Neosulf see 6542
Neosulfine see 5977
Neo-Synephrine [Sanofi-Winthrop] see 7440
Neo-T see 6554
Neoteben [Bayer] see 5203
Neo-Tenebryl see 5061
Neotetrazolium Blue see 6554
Neotetrazolium Chloride, 6554
Neothesin see 7627
Neo-thyl see 6192
Neothylline [Lemmon] see 3529
Neo-Thyreostat [Herbrand] see 1844
Neotigason [Roche] see 114
Neotilina see 3529
Neo-Tizide [Farmitalia Carlo Erba] see 5203
Neotocopherol see 9632
Neoton [Schiapparelli-Searle] see 7496
Neotran see 1287
Neotropin see 3021
Neo-Uliron see 6202
Neo-Urofort see 2123
Neo-Vasophylline see 3529
Neoviridogrisein IV see 3755
Neoviridogriseins see 3755
Neovitamin A, 6555
Neozine [Rhône-Poulenc] see 6066
Neo-Zine [Neo] see 7385
"Nepal" Aconitine see 8095
Nepaline see 8095
Nepenthe [Evans] see 6359
Nepetalactone, 6556

Nepetalic Acid see 6556
Nephril [Pfizer] see 7744
Nephroflow [Medi-Physics] see 5055
Nephrotest [Cassella-Riedel] see 462
Nepresol [Ciba-Geigy] see 3212
Nepresol Inject [Ciba-Geigy] see 3212
Népressol [Ciba-Geigy] see 3212
Neprotin see 5274
Neptal [Rhône-Poulenc] see 5962
Neptall [Rhône-Poulenc] see 16
Neptazane [Lederle] see 6031
Neptunium, 6557
Nequinate, 6558
Neral see 2383
Neraval [Schering] see 6058
Neravan [Promonta] see 1530
Nerbowdine see 1518
Nerdipina [Ferrer] see 6579
Nerfactor [Ipsen] see 5118
Neribas [Schering AG] see 3189
Nericur [Schering] see 1149
Neriforte [Schering AG] see 3189
Neriifolin, 6559
Neriine see 2561
Neriodin [Nagase] see 3132
Neriolin see 6963
Nerisona [Schering AG] see 3189
Nerisone [Schering AG] see 3189
Nerobol see 6018
Nerol, 6560
Nerolidol, 6561
Nerolin "New" see 3799
Nerolin "Old" see 6076
Neroli Oil (Artificial) see 6099
Nervacton see 1055
Nervanaid B see 3557
Nerve Growth Factor, 6562
Nerve Growth Factor-2 see 6570
Nerve Root see 2839
Nervonus [Orion] see 5908
Nesacaine [Astra] see 2210
Nesdonal Sodium [Rhone-Poulenc Rorer] see 9487
Nesontil see 7059
Nethalide see 7974
Nethamine [Merrell] see 3752
Netillin [Schering] see 6563
Netilmicin, 6563
Netilyn [Schering] see 6563
Netobimin, 6564
Netrin [Geigy] see 5996
Netromicine [Schering] see 6563
Netromycin [Schering] see 6563
Netropsin, 6565
Netrosylla [Bayer] see 3918
Netsusarin see 495
Nettacin [Schering] see 6563
Neuberg Ester see 4298
Neuchlonic [Taiyo] see 6667
Neucolis [Nippon Shinyaku] see 2380
Neuer [Daiichi] see 2067
Neulactil [M & B] see 7306
Neuleptil [Farmitalia; Specia] see 7306
Neumolisina see 3855
Neupentedrin see 9189
Neupogen [Amgen] see 4558
Neuquinon [Eisai] see 9974
Neuracen [Promonta] see 1045
Neuractil see 6066
Neuractiv [Ciba] see 7076
Neuralex [Millot] see 1072
Neuraminic Acid, 6566
Neuraxin see 6060
Neuridine see 8894
Neurine, 6567
Neuriplege [Génévrier] see 2236
Neurocil [Bayer] see 6066
Neurofort see 2123
Neurogard [Merck & Co.] see 3451
Neurolene [Magis] see 6768

Neuroleptone see 1055
Neurolite [DuPont Merck] see 9257
Neurolytril [Dorsch] see 3042
Neuromet [ISF] see 7076
Neuronal see 3164
Neuronika see 886
Neuronotrophin-3 see 6570
Neurontin [Warner-Lambert] see 4343
Neuropeptide Tyrosine see 6568
Neuropeptide Y, 6568
Neuroprocin see 3552
Neurosedyn see 9390
Neurosin see 1713
Neurotensin, 6569
Neurotensin(ox) Triacetate(salt) see 6569
Neuroton [Nuovo Cons.] see 2380
Neurotoxin see 2666
Neurotrast see 5074
Neurotrophin-3 see 6570
Neurotrophins, 6570
Neuryl see 4717
Neustab [Boots] see 9427
Neutase [Sawai] see 5671
Neuthion [Senju] see 4483
Neutrafil see 3529
Neutral Acriflavine see 125
Neutral Ammonium Chromate see 538
Neutral Ammonium Fluoride see 553
Neutral Cupric Chromate see 2700
Neutralized Verdigris see 2690
Neutral Lead Acetate see 5411
"Neutral" Magnesium Phosphate see 5720
Neutral Potassium Chromate see 7784
Neutral Proflavine Sulfate see 7953
Neutral Protein Hagedorn Insulin see 5011
Neutral Red, 6571
Neutral Red Chloride see 6571
Neutral Sodium Chromate see 8745
Neutral Spirits, 6572
Neutrapen [Riker] see 7219
Neutraphylline [Dorsey] see 3529
Neutrexin [U.S. Bioscience] see 9851
Neutroflavine see 125
Neutrogin [Chugai] see 4558
Neutronyx 600'S [Onyx] see 6772
Neutronyx 605 [Onyx] see 6858
Neutronyx 611 [Onyx] see 6772
Neutrormone see 6017
Neutrosteron see 6017
Neuvitan [Fujisawa] see 6857
Neuzyme [Eisai] see 5671
Nevax see 3460
Nevental [Pharmacia] see 6516
Nevergor see 5607
Nevigramon [Chinoin] see 6446
Nevile and Winther's Acid see 6478
Nevin [Grünenthal] see 9093
Nevirapine, 6573
Nevite see 8739
Nevralgina [Climax] see 3414
Nevriton Comprimés [Mauchant] see 51
Newberyite see 5718
Newtol [Mohan] see 10218
New Victoria Green Extra I see 5739
New Victoria Green Extra II see 5739
New Victoria Green Extra O see 5739
Nexagan [Boehringer, Ing.] see 1453
Nexion [Boehringer, Ing.] see 1453
Nexus see 1958
Nezeril [Draco] see 7100
NF 6 see 9628
NF-64 see 6634
NF-67 see 6615
NF-180 see 4320
NF-246 see 6621
NF-260 see 4315

NF-963 see 4321
NFP see 6452
NGF see 6562
NGF-2 see 6570
β-NGF see 6562
(—)NGPB-21 see 5365
NHDC see 6540
Nia [Nova] see 5849
NIA-10242 see 1851
NIA-18739 see 1271
NIA-33297 see 7321
Niacin see 6612
Niacinamide see 6574
Niacinamide Ascorbate see 6610
Niacor [Upsher-Smith] see 6612
Niagara 1240 see 3782
Niagara 4512 see 8851
Niagara 4556 see 3138
Niagara 4562 see 5297
Niagara 5006 see 3093
Niagara 9044 see 1265
Niagara Blue see 9923
Niagaramite see 805
Niagaril see 1064
Niagestin [Novo] see 5849
Nialamide, 6575
Nialate see 3782
Niamid [Pfizer] see 6575
Niamidal see 6575
Niamine see 6635
Niaprazine, 6576
Niaquitil see 6575
Niazo [Schering] see 3021
Niazol see 6455
Nibiol see 6753
Nicabate [Marion Merrell Dow] see 6611
Nicacid [Dargent] see 6612
Nicalex [Merrell] see 364
Nicametate, 6577
Nicamide see 6635
Nicamindon see 6574
Nicangin [Astra] see 6612
Nicant [Piam] see 6579
Nicapress [Boehringer, Ing.] see 6579
Nicaragua Wood see 7322
Nicarb [Merck & Co.] see 6578
Nicarbazin, 6578
Nicardal [Italfarmaco] see 6579
Nicardipine, 6579
Nicarpin [San Carlo] see 6579
Nicastubin see 6610
Niccolite see 6582
Nicein see 5364
Nicel see 6120
Nicelate [Toyo Jozo] see 6446
Nicergolent [Ramon] see 6580
Nicergoline, 6580
Niccritrol, 6581
Nicetile [Sigma-Tau] see 85
Nichogenin [Nicholas] see 4398
Nicholin [Takeda] see 2380
Nicizina [Farmitalia] see 5203
Nickel, 6582
Nickel Acetate, 6583
Nickel Acetylacetonate, 6584
Nickel Ammonium Sulfate see 566
Nickel Benzoate see 1122
Nickel Bromide, 6585
Nickel Carbonate Hydroxide, 6586
Nickel Carbonyl, 6587
Nickel Chloride, 6588
Nickel Cyanide, 6589
Nickel Dibromide see 6585
Nickel Dichloride see 6588
Nickel Difluoride see 6591
Nickel Diiodide see 6594
Nickel Dimethylglyoxime, 6590
Nickel Fluoride, 6591
Nickel Formate, 6592

Nickel Hydroxide, 6593
Nickelic Oxide see 6599
Nickel Iodide, 6594
Nickel Monoxide, 6595
Nickel Nitrate, 6596
Nickelous Fluoride see 6591
Nickelous Oxide see 6595
Nickel Oxalate, 6597
Nickel Phosphate, 6598
Nickel Potassium Cyanide see 7857
Nickel Protoxide see 6595
Nickel Sesquioxide, 6599
Nickel Sulfate, 6600
Nickel Tetracarbonyl see 6587
Niclocide [Miles] see 6602
Niclofolan, 6601
Niclosamide, 6602
Nicobid [Rhone Poulenc Rorer] see 6612
Nicobion [E. Merck] see 6574
Nicoclonate, 6603
Nicodel [Mitsui] see 6579
Nicoderm [Alza; Marion Merrell Dow] see 6611
Nicofetamide see 3382
Nicofibrate, 6604
Nicofuranose, 6605
Nicolan [Elan] see 6611
Nicolane [Nicholas] see 6815
Nicolar [Rhone Poulenc Rorer] see 6612
Nicolen see 4320
Nicolin [Takeda] see 2380
Nicomol, 6606
Nicomorphine, 6607
Niconacid [Wander] see 6612
Niconyl [Parke, Davis] see 5203
Nicopatch [Fabre] see 6611
Nicopyron [Trommsdorff] see 6619
Nicor see 6635
Nicorandil, 6608
Nicorette [Lundbeck; Merrell Dow] see 6611
Nicorine see 6635
Nicorol [Lundbeck] see 4331
Nicoscorbine see 6610
Nico-Span [Key] see 6612
Nicotafuryl see 9536
Nicotamide see 6574
Nicotelline, 6609
Nicotell TTS [Ciba-Geigy] see 6611
Nicotergoline see 6580
Nicotibina [Lepetit] see 5203
Nicotilamide see 6574
Nicotinamide, 6574
Nicotinamide-adenine Dinucleotide see 6429
Nicotinamide-adenine Dinucleotide Phosphate see 6433
Nicotinamide Ascorbate, 6610
Nicotinamide Chloromethylate see 9821
Nicotinamide Methyl Chloride see 9821
Nicotinamide Mononucleotide see 6761
4-Nicotinamido-1,5-dimethyl-2-phenyl-3-pyrazolone see 6619
Nicotine, 6611
Nicotinell [Geigy] see 6611
Nicotine Neutral Sulfate see 6611
Nicotine Tartrate see 6611
Nicotinic Acid, 6612
Nicotinic Acid Aluminum Salt see 364
Nicotinic Acid Amide see 6574
Nicotinic Acid Benzyl Ester, 6613
Nicotinic Acid p-Chloro-α-isopropylbenzyl Ester see 6603
Nicotinic Acid Diethylamide see 6635

Nicotinic Acid 2-Diethylaminoethyl Ester see 6577
Nicotinic Acid Ester with 7-(2-Hydroxyethyl)theophylline see 3925
Nicotinic Acid 2-Hydroxy-1,3-cyclohexanediylidenetetrakismethylene Ester see 6606
Nicotinic Acid 2-Hydroxyethyl Ester 2-(p-Chlorophenoxy)-2-methylpropionate (Ester) see 3923
Nicotinic Acid N-Methylbetaine see 9822
Nicotinic Acid Methyl Ester see 6174
Nicotinic Acid Morphine Ester see 6607
Nicotinic Acid Neopentanetetrayl Ester see 6581
Nicotinic Acid 1-Oxide see 7075
Nicotinic Acid β-Phenylisopropylamide see 7358
Nicotinic Acid 1,1,3,3-Tetraester with 2-Hydroxy-1,1,3,3-cyclohexanetetramethanol see 6606
Nicotinic Acid Tetrahydrofurfuryl Ester see 9536
Nicotinic Acid Triester with 2-Hexyl-2-(hydroxymethyl)-1,3-propanediol see 4688
Nicotinic Alcohol see 6614
N-Nicotinoylaminoantipyrine see 6619
4-(N-Nicotinoylamino)-1-phenyl-2,3-dimethyl-5-pyrazolone see 6619
3-(Nicotinoyloxy)propyl p-Chlorophenoxyisobutyrate see 8414
Nicotinoyl-β-phenylisopropylamine see 7358
Nicotinyl Alcohol, 6614
1-Nicotinylamino-1,2-diphenylethane see 3382
Nicoumalone see 29
Nicoxin see 6578
Nicrazin [Merck & Co.] see 6578
Nidanthel [Cooper] see 6693
Nidantin [Sasse] see 7079
Nidaxin [Chassot] see 5838
Nidrafur see 6634
Nidran [Sankyo] see 6645
Nidrane [Phoenix] see 1045
Nidrel [Specia] see 6669
Nidroxyzone, 6615
Nidulin, 6616
Nifedicor [Schiapparelli-Searle] see 6617
Nifedin [Gentili] see 6617
Nifedipine, 6617
Nifelan [Elan] see 6617
Nifelat [Sidus] see 6617
Nifenalol, 6618
Nifenazone, 6619
Nifensar XL [Rhone-Poulenc Rorer] see 6617
Niflan [Yoshitomi] see 7890
Niflumic Acid, 6620
Nifluril see 6620
Nifluril (Suppositories) [UPSA] see 6620
Nifos T [Monsanto] see 9345
NIF-T see 4559
Niftolid see 4242
Nifulidone see 4320
Nifuradene, 6621
Nifuraldezone, 6622
Nifuratel, 6623
Nifurfoline, 6624
Nifuroquine, 6625
Nifuroxazide, 6626
Nifuroxime, 6627
Nifurpirinol, 6628
Nifurprazine, 6629
Nifurtimox, 6630

Nifurtoinol, 6631
Nifurzide, 6632
Nifuzon see 6697
Nigakilactone D see 8209
Nigalax see 1282
Nigelline see 2873
Nigericin, 6633
Night-blooming Cereus see 1643
Nigrin [Pfizer] see 8986
NIH-4185 see 9429
NIH-4542 see 3309
NIH-5145 see 3875
NIH-7274 see 7399
NIH-7440 see 304
NIH-7519 see 7360
NIH-7539 see 6801
NIH-7590 see 7587
NIH-7667 see 6778
NIH-8805 see 1522
Nihydrazone, 6634
Nikardin see 6635
Nikethamide, 6635
Nikion see 1064
Nikofezon see 6619
Nikoform see 4880
Nilatil [USV] see 5263
Nilergex [ICI] see 5248
Nilevar [Searle] see 6789
Nilhistin see 6026
Nilodin see 5620
Niltuvin [EGYT] see 6614
Nilutamide, 6636
Nilvadipine, 6637
Nilverm [Pitman-Moore] see 5486
Nim see 6525
Nimaol see 6845
Nimbecetin see 5288
Nimbex [Glaxo Wellcome] see 900
Nimbin, 6638
Nimbiol, 6639
Nimergoline see 6580
Nimesulide, 6640
Nimetazepam, 6641
Nimicor [Formenti] see 6579
Nimidane, 6642
Nimodipine, 6643
Nim Oil see 6526
Nimorazole, 6644
Nimotop [Bayer] see 6643
Nimrod [ICI] see 1519
Nimustine, 6645
Ninhydrin, 6646
Ninopterin, 6647
Nio-A-Let [Nion] see 10150
Niobium, 6648
Niobium Pentachloride, 6649
Niobium Pentafluoride, 6650
Niobium Pentoxide, 6651
Niobium Potassium Oxypentafluo-
ride, 6652
Nioform see 5052
Nionate [Nion] see 4095
Niopam [Bracco] see 5071
Niotal [Simes] see 10321
Nioxime [Delta], 6653
Nip [Nichay] see 7746
Nipagin A see 3883
Nipagin M see 6182
Nipantiox 1-F [Nipa] see 1582
Nipasol M [Nipa] see 8051
Nipaxon [Kabi Pharmacia] see 6815
Nipecotan see 695
Nipecotic Acid, 6654
Nipent [Lederle] see 7277
Niperyt see 7249
Nipolept [Klinge] see 10327
Nippas Calcium [Tanabe] see 498
Nipradilol, 6655
Nipradolol see 6655
Niprazina [Italfarmaco] see 6619

Nipride [Roche] see 8794
Nipruss [Schwarz] see 8794
Niran [Tamogan] see 2129
Niran [Monsanto] see 7167
Niridazole, 6656
Nirvan see 6066
Nirvanil see 10048
Nirvotin see 1886
NIS see 5064
Nische [Kodama] see 224
Nisentil [Roche] see 317
Nisidana [Geigy] see 6985
Nisin, 6657
Nisintel see 317
Nisoldipine, 6658
Nisotin see 3783
Nissorun [Nippon Soda] see 4751
Nisulid [Wyeth] see 6640
Nitalapram see 2379
Nitan [Teva] see 7206
Nitarsone, 6659
Niter see 7815
Nithiazide, 6660
Nithiocyamine see 613
Niticolin [Morishita] see 2380
Nitobanil [Kyowa Yakuhin] see 9267
Nitogenin see 3348
Nitoman [Roche] see 9325
Niton see 8278
Nitorol [Eisai] see 5245
Nitossil [Zyma] see 2456
Nitracrine, 6661
Nitradisc [Searle] see 6704
Nitrados [Berk] see 6667
Nitraldone see 4315
Nitralin, 6662
Nitramide, 6663
Nitramine, 6664
Nitran [Riker] see 6704
Nitranilic Acid, 6665
m-Nitraniline see 6679
o-Nitraniline see 6680
p-Nitraniline see 6681
Nitranitol [Merrell] see 5789
Nitraphen see 6695
Nitrapyrin, 6666
2-Nitratoethylaminotoluene-p-
sulfonate see 5263
(Nitrato-O)phenylmercury see 7455
Nitrazepam, 6667
Nitrefazole, 6668
Nitrendipine, 6669
Nitrenes, 6670
Nitrenpax see 6667
Nitretamin [Squibb] see 9900
Nitric Acid, 6671
Nitric Acid, Anhydrous, 6672
Nitric Acid Concentrated see 6671
Nitric Acid, Fuming, 6673
Nitric Acid Methyl Ester see 6175
Nitric Acid Propyl Ester see 8049
Nitric Anhydride see 6702
Nitric Oxide, 6674
Nitriderm-TTS [Geigy] see 6704
5,5'-Nitrilodibarbituric Acid Mono-
ammonium Salt see 6386
Nitrilotriacetic Acid, 6675
Nitrilotriacetic Acid Bismuth Com-
plex Sodium Salt see 1322
2,2',2''-Nitrilotrisethanol see 9798
2,2'2''-Nitrilotrisethanol Trinitrate
(Ester) Phosphate (1:2) (Salt) see
9900
Nitrimidazine see 6644
Nitrin, 6676
Nitritocobalamin see 4854
Nitroacetanilide, 6677
m-Nitroacetanilide see 6677
o-Nitroacetanilide see 6677
p-Nitroacetanilide see 6677

Nitro Acid Sulfite see 6746
Nitroacridine 3582 see 6678
Nitroakridin 3582, 6678
3-Nitroalizarin see 250
4-Nitro-2-aminophenetole see 6714
1-Nitro-3-amino-4-phenyl Ethyl Ether
see 6714
m-Nitroaniline, 6679
o-Nitroaniline, 6680
p-Nitroaniline, 6681
Nitroanisole, 6682
m-Nitroanisole see 6682
o-Nitroanisole see 6682
p-Nitroanisole see 6682
5-Nitrobarbituric Acid, 6683
Nitrobenzaldehyde, 6684
m-Nitrobenzaldehyde see 6684
o-Nitrobenzaldehyde see 6684
p-Nitrobenzaldehyde see 6684
3-Nitrobenzenamine see 6679
Nitrobenzene, 6685
4-Nitrobenzeneacetic Acid see 6719
4-Nitrobenzeneacetonitrile see 6688
p-Nitrobenzenearsonic Acid see 6659
p-Nitrobenzeneazochromotropic Acid
Sodium Salt see 2297
p-Nitrobenzeneazosalicylic Acid see
251
4-Nitro-1,2-benzenediamine see 6720
p-Nitrobenzenesulfonic Acid Methyl
Ester see 6176
Nitrobenzoic Acid, 6686
m-Nitrobenzoic Acid see 6686
o-Nitrobenzoic Acid see 6686
p-Nitrobenzoic Acid see 6686
Nitrobenzol see 6685
4-Nitrobenzoyl Chloride, 6687
p-Nitrobenzyl Cyanide, 6688
Nitro-Bid [Marion] see 6704
2-Nitro-1,1'-biphenyl see 6689
4-Nitrobiphenyl see 6690
4-Nitro-1,1'-biphenyl see 6690
o-Nitrobiphenyl, 6689
p-Nitrobiphenyl, 6690
2-Nitro-1,1-bis(p-chlorophenyl)butane
see 1505
2-Nitro-1,1-bis(p-chlorophenyl)pro-
pane, 6691
1,1'-(2-Nitrobutylidene)bis[4-chloro-
benzene] see 1505
N-Nitrocarbamide see 6749
Nitrocarbol see 6708
Nitrocellulose see 8195
Nitrochlorobenzene see 2203
m-Nitrochlorobenzene see 2203
Nitrochloroform see 2208
7-Nitro-5-(2-chlorophenyl)-3H-1,4-
benzodiazepin-2(1H)-one see 2449
p-Nitro-m-chlorophenyl Dimethyl
Thionophosphate see 2248
p-Nitro-o-chlorophenyl Dimethyl
Thionophosphate see 3091
Nitrocine [Sanol Schwarz] see 6704
m-Nitrocinnamic Acid, 6692
Nitrocobalamin see 4854
Nitrocontin [Asta] see 6704
Nitrodan, 6693
Nitroderm TTS [Ciba-Geigy] see 6704
4-Nitro-1,2-diaminobenzene see 6720
o-Nitrodiphenyl see 6689
p-Nitrodiphenyl see 6690
Nitrodisc [Searle] see 6704
Nitrodulcitol see 4350
Nitro-Dur [Key] see 6704
Nitroduran [Leo Pharm.] see 9900
Nitroerythrite see 3716
Nitroethane, 6694
5-Nitro-2-ethoxyaniline see 6714
Nitrofen, 6695
Nitrofene see 6695

Nitroform see 9859
Nitrofortin [Plantorgan] see 6704
Nitrofural see 6697
5-Nitro-2-furaldehyde Acetylhydrazone see 6634
5-Nitro-2-furaldehyde p-Hydroxybenzoylhydrazone see 6626
5-Nitro-2-furaldehyde 2-(2-Hydroxyethyl)semicarbazone see 6615
5-Nitro-2-furaldehyde Semicarbazone see 6697
5-Nitro-2-furaldehyde Semioxamazone see 6622
anti-5-Nitro-2-furaldoxime see 6627
5-Nitro-2-furancarboxaldehyde Oxime see 6627
Nitrofurantoin, 6696
6-[2-(5-Nitro-2-furanyl)ethenyl]-3-pyridazinamine see 6629
6-[2-(5-Nitro-2-furanyl)ethenyl]-2-pyridinemethanol see 6628
1-[[(5-Nitro-2-furanyl)methylene]amino]-2,4-imidazolidinedione see 6696
1-[[(5-Nitro-2-furanyl)methylene]amino]-2-imidazolidinone see 6621
3-[[(5-Nitro-2-furanyl)methylene]amino]-2-oxazolidinone see 4320
2-[(5-Nitro-2-furanyl)methylene]hydrazinecarboxamide see 6697
2-[3-(5-Nitro-2-furanyl)-1-[2-(5-nitro-2-furanyl)ethenyl]-2-propenylidene]hydrazinecarboximidamide see 6752
4-(5-Nitro-2-furanyl)-2-quinolinecarboxylic Acid 1-Oxide see 6625
Nitrofurazone, 6697
1-(5-Nitro-2-furfurylidene)-2-acetylhydrazine see 6634
1-(5-Nitrofurfurylideneamino)hydantoin see 6696
N-(5-Nitro-2-furfurylidene)-1-aminohydantoin see 6696
1-(5-Nitrofurfurylidene)amino]-3-(hydroxymethyl)hydantoin see 6631
1-[(5-Nitrofurfurylidene)amino]-2-methyltetrahydro-1,4-thiazine 4,4-Dioxide see 6630
4-[(5-Nitrofurfurylidene)amino]-3-methylthiomorpholine-1,1-dioxide see 6630
3-(5-Nitro-2-furfurylideneamino)-5-(4-morpholinomethyl)-2-oxazolidone see 4315
3-(5-Nitrofurfurylideneamino)-2-oxazolidinone see 4320
N-(5-Nitro-2-furfurylidene)-3-amino-2-oxazolidone see 4320
5-Nitro-2-furfurylmethyl Ether see 6075
Nitrofurmethone see 4315
N^1-(5-Nitro-2-furylacrylidene)-N^2-(5-nitro-2-thenoyl)hydrazine see 6632
1-(5-Nitro-2-furyl)-2-(6-amino-3-pyridazyl)ethylene see 6629
2-(5-Nitro-2-furyl)-5-amino-1,3,4-thiadiazole see 9725
[[3-(5-Nitro-2-furyl)-1-[2-(5-nitro-2-furyl)vinyl]allylidene]amino]guanidine see 6752
4-(5-Nitro-2-furyl)quinaldic Acid 1-Oxide see 6625
6-[2-(5-Nitro-2-furyl)vinyl]-2-pyridinemethanol see 6628
Nitrogard [Parke, Davis] see 6704
Nitrogen, 6695
Nitrogen Chloride, 6699
Nitrogen Dioxide, 6700
Nitrogen Fluoride, 6701
Nitrogen Lost see 9777
Nitrogen Monoxide see 6674

Nitrogen Mustard see 5815
Nitrogen Oxyfluoride see 6745
Nitrogen Pentoxide, 6702
Nitrogen Selenide, 6703
Nitrogen Tetroxide see 6700
Nitrogen Trichloride see 6699
Nitrogen Trifluoride see 6701
Nitrogen Trioxyfluoride see 4202
Nitro-Gesanit [Woelm] see 6704
Nitroglin [Thiemann] see 6704
Nitroglycerin, 6704
Nitroglycerol see 6704
Nitroglycol see 3846
Nitroglyn see 6704
Nitrogranulogen see 5815
Nitroguanidine, 6705
Nitrohydrochloric Acid, 6706
2-Nitro-1-hydroxybenzene-4-arsonic Acid see 8430
2-Nitro-2-(hydroxymethyl)-1,3-propanediol see 9883
1-Nitro-2-hydroxynaphthalene see 6712
3-Nitro-4-hydroxyphenylarsonic Acid see 8430
5-Nitro-8-hydroxyquinoline see 6753
2-Nitro-1H-imidazole see 952
4-[2-(5-Nitro-1H-imidazol-1-yl)ethyl]-morpholine see 6644
4-Nitro-4'-isothiocyanodiphenyl Ether see 6731
Nitrol [Eisai] see 5245
Nitrolan [Napp] see 6704
Nitrolande [Delalande] see 6704
Nitrolar [Luar] see 6704
Nitrolent [Medinova] see 6704
Nitrolime see 1702
Nitrolingual [Pohl] see 6704
Nitro Mack [Mack, Illert.] see 6704
Nitromagnesite see 5710
Nitromannite see 5789
Nitromannitol see 5789
Nitromersol, 6707
Nitromethane, 6708
4-Nitro-4'-[(N-methylpiperazinyl)thiocarbonylamino]diphenylamine see 608
Nitromex [Dumex] see 6704
Nitromide, 6709
Nitromidine see 930
Nitromin see 5816
Nitromuriatic Acid see 6706
Nitron, 6710
Nitronal [Pohl] see 6704
1-Nitronaphthalene, 6711
α-Nitronaphthalene see 6711
1-Nitro-2-naphthalenol see 6712
1-Nitro-2-naphthol, 6712
α-Nitro-β-naphthol see 6712
Nitrong [M & B] see 6704
N-[2-(Nitrooxy)ethyl]-3-pyridinecarboxamide see 6608
Nitropenta see 7249
Nitropentaerythritol see 7249
3-Nitropentane, 6713
Nitropenton see 7249
Nitrophen see 6695
5-Nitro-o-phenetidine, 6714
Nitrophenide, 6715
m-Nitrophenol, 6716
o-Nitrophenol, 6717
p-Nitrophenol, 6718
Nitrophenolarsonic Acid see 8430
4-Nitro-2-phenoxymethanesulfonanilide see 6640
p-(p-Nitrophenoxy)phenyl Isothiocyanate see 6731
N-(4-Nitro-2-phenoxyphenyl)methanesulfonamide see 6640
N-(3-Nitrophenyl)acetamide see 6677

p-Nitrophenylacetic Acid, 6719
p-Nitrophenylacetonitrile see 6688
3-(α-p-Nitrophenyl-β-acetylethyl)-4-hydroxycoumarin see 29
N-[4-[[(4-Nitrophenyl)amino]sulfonyl]phenyl]acetamide see 9103
4-Nitrophenylarsonic Acid see 6659
D-(−)-threo-1-p-Nitrophenyl-2-azidoacetylaminopropane-1,3-diol see 941
4-[(4-Nitrophenyl)azo]-1,3-benzenediol see 5735
4-(p-Nitrophenylazo)resorcinol see 5735
5-(p-Nitrophenylazo)salicylic Acid see 251
4-(2'-Nitrophenyl)-2,6-dimethyl-3,5-dicarbomethoxy-1,4-dihydropyridine see 6617
4-Nitro-1,2-phenylenediamine see 6720
4-Nitro-o-phenylenediamine, 6720
1-[[[5-(4-Nitrophenyl)-2-furanyl]methylene]amino]-2,4-imidazolidinedione see 2879
1-[[5-(p-Nitrophenyl)furfurylidene]amino]hydantoin see 2879
(4-Nitrophenyl)hydrazine, 6721
1-(4-Nitrophenyl)-2-(isopropylamino)ethanol see 6618
Nitrophenyl Methyl Ether see 6682
2-Nitro-N-(phenylmethyl)-1H-imidazole-1-acetamide see 1114
1-(4-Nitrophenyl)-2-methyl-4-nitroimidazole see 6668
3-(3-Nitrophenyl)-2-propenoic Acid see 6692
o-Nitrophenylpropiolic Acid, 6722
3-(2-Nitrophenyl)-2-propynoic Acid see 6722
N-(4-Nitrophenyl)-N'-(3-pyridinylmethyl)urea see 8171
4'-[(p-Nitrophenyl)sulfamoyl]acetanilide see 9103
4-Nitrophenyl-2,4,6-trichlorophenyl Ether see 2158
(4-Nitrophenyl)urea, 6723
Nitropress [Abbott] see 8794
NitroPRN [Warner-Chilcott] see 6704
1-Nitropropane, 6724
2-Nitropropane, 6725
5'-Nitro-2'-propoxyacetanilide, 6726
5-Nitro-2-propoxyaniline, 6727
5-Nitro-2-propoxybenzenamine see 6727
N-(5-Nitro-2-propoxyphenyl)acetamide see 6726
1,1'-(2-Nitropropylidene)bis[4-chlorobenzene] see 6691
5-Nitro-2,4,6(1H,3H,5H)-pyrimidinetrione see 6683
5-Nitroquinaldic Acid, 6728
5-Nitroquinaldinic Acid, 6728
5-Nitro-2-quinolinecarboxylic Acid see 6728
5-Nitro-8-quinolinol see 6753
Nitrorectal [Pohl] see 6704
Nitroretard [Dumex] see 6704
3-Nitrosalicylic Acid, 6729
5-Nitrosalicylic Acid, 6730
Nitroscanate, 6731
Nitrose [Fort Dodge] see 6746
Nitrosigma [Sigma-Tau] see 6704
2,2'-Nitrosiminodiethanol see 6733
3-Nitrosobenzamide, 6732
N-Nitrosodiethanolamine, 6733
N-Nitrosodiethylamine, 6734
N-Nitrosodimethylamine, 6735
p-Nitroso-N,N-dimethylaniline, 6736
p-Nitrosodiphenylamine, 6737
N-Nitroso-N-ethylurea see 3879

N-(2-Nitrosohydroxylamino-3-methyl-
  butyl)crotonamide see 3481
2,2'-(Nitrosoimino)bisethanol see 6733
4-Nitrosomorpholine see 6738
N-Nitrosomorpholine, 6738
1-Nitroso-2-naphthalenol see 6739
1-Nitroso-2-naphthol, 6739
Nitroso-β-naphthol see 6739
Nitrosonium Tetrafluoroborate see
  6747
4-Nitrosophenol, 6740
4-Nitroso-N-phenylbenzenamine see
  6737
N-Nitrosophenylhydroxylamine
  Ammonium Salt see 2688
1-Nitrosopyrrolidine see 6741
N-Nitrosopyrrolidine, 6741
Nitrosorbon [Pohl-Boskamp] see 5245
Nitroso-R Salt, 6742
Nitrososulfuric Acid see 6746
Nitrostat [Parke, Davis] see 6704
2-Nitro-4-sulfobenzoic Acid, 6743
Nitrosulfonic Acid see 6746
Nitrosyl Borofluoride see 6747
Nitrosyl Chloride, 6744
Nitrosyl Fluoborate see 6747
Nitrosyl Fluoride, 6745
Nitrosyl Hydrogen Sulfate see 6746
Nitrosyl Sulfate see 6746
Nitrosylsulfuric Acid, 6746
Nitrosyl Tetrafluoroborate, 6747
Nitrothiamidazol see 6656
5-Nitro-2-thiazolamine see 477
N-(5-Nitro-2-thiazolyl)acetamide see
  430
1-(5-Nitro-2-thiazolyl)-2-imidazolidin-
  one see 6656
1-(5-Nitro-2-thiazolyl)-2-oxotetra-
  hydroimidazole see 6656
N-(5-Nitro-2-thiazolyl)-2-thiophene-
  carboxamide see 9292
5-Nitro-2-thiophenecarboxylic Acid
  [3-(5-Nitro-2-furanyl)-2-propenylid-
  ene]hydrazide see 6632
5-Nitro-2-thiophenecarboxylic Acid
  [3-(5-Nitro-2-furyl)allylidene]hy-
  drazide see 6632
Nitrotoluene, 6748
m-Nitrotoluene see 6748
o-Nitrotoluene see 6748
p-Nitrotoluene see 6748
p-Nitro-α-toluic Acid see 6719
p-Nitro-α-tolunitrile see 6688
2-Nitro-3,4,6-trichlorophenol see 9771
4'-Nitro-3'-trifluoromethylisobutyr-
  anilide see 4242
4-Nitro-3-(trifluoromethyl)phenol see
  9388
2-[Nitro(2,4,6-trinitrophenyl)amino]-
  ethanol Nitrate (Ester) see 7281
Nitrourea, 6749
Nitrous Acid, 6750
Nitrous Acid Butyl Ester see 1617
Nitrous Acid tert-Butyl Ester see 1618
Nitrous Acid 1,1-Dimethyl Ethyl
  Ester see 1618
Nitrous Acid Ethyl Ester see 3877
Nitrous Acid Isopropyl Ester see 5235
Nitrous Acid 1-Methylethyl Ester see
  5235
Nitrous Acid n-Propyl Ester see 8050
Nitrous Acid Sodium Salt see 8793
Nitrous Ether see 3877
Nitrous Oxide, 6751
Nitrovin, 6752
Nitrox 80 see 6183
Nitroxanthic Acid see 7562
Nitroxoline, 6753
Nitroxy Fluoride see 4202
Nitroxyl Chloride see 6755

Nitroxylsulfuric Acid see 6746
Nitroxynil, 6754
Nitroxynil Eglumine see 6754
Nitroxynil Meglumine see 6754
Nitrozell-retard [Byk-Gulden] see
  6704
Nitrozone [Century] see 6697
Nitrumon [Simes] see 1894
Nitryl Chloride, 6755
Nitryl Fluoride, 6756
Nitryl Hypofluorite see 4202
Nitux [Inpharzam] see 6350
Nivadil [Fujisawa] see 6637
Nivadipine see 6637
Nivaldipine see 6637
Nivalenol, 6757
Nivalin [Sofia] see 4357
Nivaquine [M & B] see 2215
Nivaquine B see 2215
Nivelona see 3361
Nivemycin [Boots] see 6542
Nix [Burroughs Wellcome] see 7321
Nixyn [Organon] see 5209
Nizatidine, 6758
Nizax [Lilly] see 6758
Nizaxid [Lilly] see 6758
Nizin [Broemmel] see 9096
Nizofenone, 6759
Nizoral [Janssen] see 5313
NK-19 see 7690
NK-421 see 9973
NK-631 see 7286
NK-1006 see 5293
NKK-105 see 5751
NKT-01 see 4610
NMDA, 6760
NMF see 6152
NMN, 6761
NMOR see 6738
NMP see 6197
NNC-05-0328 see 9557
NNC-11-0232 see 10190
NND-318 see 5370
NNI-850 see 4038
No. 220 Sol see 5914
No. 356 see 2445
NO-328 see 9557
NO-05-0328 see 9557
Noah's Ark see 2839
Noan [Ravizza] see 3042
NOBA see 6732
Nobacter [Chantereau] see 9786
Nobedon see 45
Nobedorm see 6028
Nobelium, 6762
Nobfen [Toho/Funai] see 4925
Nobgen [Kanebo] see 4925
N-Oblivon see 5890
Nobrium [Roche] see 5829
Nocardamin, 6763
Nocardic Acids see 6406
Nocardicin A see 6764
Nocardicin B see 6764
Nocardicin(s), 6764
Nocardomycolic Acids see 6406
Nocertone [Labaz] see 7068
Noctal [UCB] see 7980
Noctamid [Schering AG] see 5611
Noctan see 6216
Noctazepam [Brenner-Efeka] see 7059
Noctec [Squibb] see 2113
Noctesed [Unimed] see 6667
Noctilene see 6028
Noctivane see 4742
Noctivane Sodium see 4742
Noctomin [Medichemie] see 3367
Noctone [Gea] see 8286
Noctosom see 4233
Nodakenetin Glucoside see 6765
Nodakenin, 6765

Nodapton see 4511
No-Doz [Bristol] see 1674
Noflo [Banyu; Kyorin] see 6793
Noformicin, 6766
Noformycin see 6766
Nogalamycin, 6767
L-Nogalose see 6767
Nogédal [Théraplix] see 6821
Nogexan [Farmex] see 1880
Nogos [Ciba-Geigy] see 3129
Nogram [Winthrop] see 6446
Noin [Schering] see 4972
Noleptan [Thomae] see 4258
Nolicin [Krka] see 6793
Nolipax [Foscama] see 4019
Nolotil [Boehringer, Ing.] see 3414
Noltam [Lederle] see 9216
Noludar [Roche] see 6216
Nolvadex [Zeneca] see 9216
Nolvasan [Fort Dodge] see 2140
Nomersan [Zeneca] see 9510
Nometan see 7618
Nomifensine, 6768
Nomilin, 6769
Nonactic Acid see 6770
Nonactin, 6770
Nonaethyleneglycol Monomethyl
  Ether p-n-Butylaminobenzoate see
  1127
Nonalol see 6774
Nonalupine see 5246
Nonanedioic Acid see 938
Nonanoic Acid see 7198
Nonanoic Acid Ethyl Ester see 3884
1-Nonanol see 6774
Nonanol Acetate see 6773
2-Nonenal, 6771
trans-2-Nonenaldehyde see 6771
Nonflamin [Yoshitomi] see 9590
Nonipol NO [Sankyo] see 6772
Nonisol [Ciba-Geigy] see 7738
Nonoic Acid see 7198
Nonoxinol see 6772
Nonoxynol, 6772
Nonoxynol-9 see 6772
Nonoxynol-11 see 6772
Nonulosaminic Acids see 8628
n-Nonyl Acetate, 6773
n-Nonyl Alcohol, 6774
Nonylcarbinol see 2911
n-Nonyl Ethanoate see 6773
Nonylic Acid see 7198
Nonyl Phenol, 6775
(p-Nonylphenoxy)acetic Acid, 6776
Nonylphenoxypolyethoxyethanol see
  6772
α-(4-Nonylphenyl)-ω-hydroxypoly-
  (oxy-1,2-ethanediyl) see 6772
Nonylphenyl Polyethyleneglycol
  Ether see 6772
Nootron [Biosintetica] see 7641
Nootrop [UCB] see 7641
Nootropil [UCB] see 7641
Nootropyl [UCB] see 7641
Nopalcol [Henckel] see 7738
Nopia [Daiichi] see 8412
Nopil [Mepha] see 9086
Nopinene see 7600
Nopoxamine see 6422
No-Press see 5813
Nopron [Carrion] see 6576
Noprylsulfamide, 6777
NO-PYR see 6741
18-Norabietane see 4120
Noracin [Chew] see 6793
Noracyclin [Ciba] see 5659
Noracymethadol, 6778
Noradrenaline see 6788
Noraminopyrine Methanesulfonate
  Sodium see 3414

Noranat [Labinca] see 4969
19-Norandrosta-4,9,11-trien-17β-ol-3-
one see 9716
19-Norandrost-4-ene-4,17β-diol-3-one
see 7034
Nor-5'-anhydrovinblastine see 10127
Norantipyrine see 6809
Norantoin see 6187
Noravid [Roussel-UCLAF] see 2915
Noraxin [TP] see 6793
Norbide [Norton] see 1374
Norbiline [Bellon] see 8093
Norbolethone, 6779
Norbormide, 6780
cis-5-Norbornene-2,3-dicarboxylic
Acid Dimethyl Ester see 3290
cis-endo-5-Norbornene-2,3-dicarbox-
ylic Anhydride see 1842
α-5-Norbornen-2-yl-α-phenyl-1-piper-
idinepropanol see 1274
Norcarane, 6781
Norcassamidine see 3728
Norcholanic Acid, 6782
(5β)-24-Norcholan-23-oic Acid see
6782
Norcodeine, 6783
Norcolut [Gedeon Richter] see 6790
Norcuron [Organon] see 10075
Norcycline see 8500
Nordaz [Bouchara] see 6784
Nordazepam, 6784
Nordefrin, 6785
Norden see 6856
2'-Nor-2'-deoxyguanosine see 4374
Nordette [Wyeth] see 6797
Nordialex [Pharmacodex] see 4448
Nordiazepam see 6784
Nordihydroguaiaretic Acid, 6786
Nordiocto [Nordisk] see 3963
Norditropin [Nordisk] see 8864
Nordox [Norton] see 3496
Nor-Durandron see 6452
Norea, 6787
Norephedrane see 623
Nor-ψ-ephedrine see 6811
ψ-Norephedrine see 6811
dl-Norephedrine Hydrochloride see
7461
Norepinephrine, 6788
(3β,22E)-19-Norergosta-5,7,9,22-
tetraen-3-ol see 6539
Norethandrolone, 6789
Norethindrone, 6790
19-Norethisterone see 6790
Norethynodrel, 6791
19-Nor-17α-ethynylandrosten-17β-ol-
3-one see 6790
19-Nor-17α-ethynyl-17β-hydroxy-4-
androsten-3-one see 6790
19-Nor-17α-ethynyltestosterone see
6790
Norfemac [Nordic] see 1497
Norfen [Morishita] see 6856
Norfenefrine, 6792
Norflex [Riker] see 7007
Norfloxacin, 6793
Norflurazon, 6794
Norgamen see 9586
Norgan see 4826
Norgesic [Kettelhack] see 4509
Norgesterone, 6795
Norgestimate, 6796
Norgeston [Schering] see 6797
Norgestrel, 6797
D-Norgestrel see 6797
Norgestrienone, 6798
Norgine see 241
Norglycin [Upjohn] see 9644
Norhomoepinephrine see 6785
Norhyoscyamine, 6799

Noriclan [Juste] see 3418
Noriday [Syntex] see 6790
Norimin [Syntex] see 6790
Norimipramine see 2966
Norinyl-1 [Syntex] see 6790
Norinyl 1+35 [Syntex] see 6790
Norisodrine [Abbott] see 5236
dl-Norisoephedrine see 467
Noristerat [Schering AG] see 6790
Norit [Norit N.V.] see 1856
Noritren [Lundbeck] see 6812
Norizalpinin see 4356
Norkel see 5321
Norlaudanosoline see 9355
Norlestrin [Parke, Davis] see 6790
Norleucine, 6800
Norlevorphanol, 6801
Norlobelanine, 6802
Norlongandron see 6452
Norlutate [Parke, Davis] see 6790
Norluten [SKB] see 6790
Norlutin [Parke, Davis] see 6790
Normabraïn [Cassella-Riedel] see 7641
Normal Butyl Thioalcohol see 1611
Normal Lead Acetate see 5411
Normase [Molteni] see 5360
Normastigmin (amp.) [Sigmapharm]
see 6553
Normastigmin (tabl.) [Sigmapharm]
see 6553
Normatensyl [Théraplix] see 4798
Normenon [Syntex] see 2152
Normet [Gedeon Richter] see 2436
Normetandrone see 6805
Normetanephrine, 6803
Normethadone, 6804
Normethandrolone see 6805
Normethandrone, 6805
Normetic [Abbott] see 426
Normicina [E. Merck] see 6271
Normi-Nox [Herbrand] see 6028
Normison [Wyeth] see 9285
Normix [Alfa] see 8386
Normobren [Medosan] see 85
Normoc [Merckle] see 1406
Normodyne [Schering] see 5341
Normoglucina see 7376
Normolipol [Delagrange] see 2436
Normonal [Eisai] see 9866
Normorix [Benzon] see 426
Normorphine, 6806
Normorphine 3-Methyl Ether see 6783
Normoson see 3774
Normosterol see 7248
Normothen [Fisons] see 3489
Normotiroide see 470
Normoxidil [Sigma-Tau] see 6290
Normud [Astra] see 10254
Normurat [Grünenthal] see 1093
Nornicotine, 6807
Nornidulin, 6808
Norocin [Merck & Co.] see 6793
Noroclox DC [Willows Francis] see
2480
Norodin [Endo] see 6015
Norodine [Norbrook] see 9071
Noroxin(e) [Merck & Co.] see 6793
Noroxylin see 971
Norpace [Searle] see 3424
Norphen (Ampules) [Byk-Gulden] see
6856
Norphenazone, 6809
Norphenylephrine see 6792
Norphytane see 7932
Norpipanone, 6810
Norplant [Roussel] see 6797
Norpramin [Lakeside] see 2966
(17α)-19-Norpregna-1,3,5(10)-trien-
20-yne-3,17-diol see 3780
Norpregneninolone see 6790

(17α)-19-Norpregn-4-en-17-ol see 3851
(3β,17α)-19-Norpregn-4-en-20-yne-3,-
17-diol see 3905
(17α)-19-Norpregn-4-en-20-yn-17-ol
see 5659
Nor-Progestelea see 6814
Norprolac [Sandoz] see 8226
Norpropandrolate see 1352
Norpseudoephedrine, 6811
Nor-QD [Syntex] see 6790
Norquen [Syntex] see 5976
Norquentiel [SK & F] see 6790
Norsulfazole see 9115
Norsympatol see 6856
Norsynephrine see 6856
Nortesto see 6452
19-Nortestosterone see 6452
19-Nortestosterone Cyclohexylpropionate see 6452
19-Nortestosterone Decanoate see
6452
19-Nortestosterone Furylpropionate
see 6452
19-Nortestosterone Hexahydro-
benzoate see 6452
19-Nortestosterone-3-(p-hexyloxy-
phenyl)propionate see 6452
19-Nortestosterone β-Phenylpropio-
nate see 6452
19-Nortestosterone Propionate see
6452
North American Sumach see 9171
North and South American Anti-
snakebite Serum see 761
Northern Pine see 10180
Norticon [Nihon] see 7624
Nortimil [Chiesi] see 2966
Nortran [Norden] see 9810
Nortrilen [Troponwerke] see 6812
Nortriptyline, 6812
Nortron [Lilly] see 3418
Nortron [Fisons] see 3788
1αH,5αH-Nortropan-3α-ol (−)-
Tropate see 6799
Noruron see 6787
Norval [Bencard] see 6260
Norvaline, 6813
7-L-Norvaline Cyclosporin A see 2821
Norvasc [Miles] see 6658
Norvedan [Wyeth] see 4045
Norvinisterone, 6814
Norvinodrel see 6795
Norybol-19 see 6452
Norzepine see 6812
Norzetam [Chemil] see 7641
Noscapalin see 6815
Noscapine, 6815
Nosiheptide, 6816
Nosim [Richet] see 5245
Nosophen see 5059
Nosophene Sodium see 5059
Nospan [Johnsons] see 9960
Nostal [Ames] see 3552
Nostyn [Ames] see 3552
Nosydrast see 5061
Nosylan see 5061
Notandron see 6017
Notaral [Wyeth] see 7225
Notatin see 4468
Notensil [Fisons] see 32
Notézine [Specia] see 3165
Nothiazine see 5892
Notul [Eurofarmaco] see 2337
Nourytam [Nourypharma] see 9216
Nova [Rohm & Haas] see 6402
Novaban [Sandoz] see 9914
Novacetyl see 5689
Novacid see 3414
Novacort [Syntex] see 2460

Novacrysin see 4537
Novacyl [Astra] see 7075
Novadral [Diwag] see 6792
Novafed [Dow] see 3645
Noval [Yamanouchi] see 5849
Novaldin [Winthrop] see 3414
Novalgin [Hoechst] see 3414
Novaloc [Schering AG] see 4947
Novamidon [Specia] see 495
Novamin [Bristol] see 425
Novamin [Shionogi] see 7942
Novantrone [Lederle] see 6303
Novapirina [Zyma] see 3132
Novarsan see 6534
Novarsenobenzol see 6534
Novarsenobillon see 6534
Novaruca [Bioglan] see 4480
Novasmasol [Zambeletti] see 5992
Novastan [Mitsubishi] see 816
Novastat W [Salsbury] see 9103
Novatec [Merck & Co.] see 5540
Novatrin [Ayerst] see 4766
Novatropine see 4766
Novazam [Génévrier] see 3042
Novazole [Merck & Co.] see 1774
Novedrin see 3752
Novembichin, 6817
Novemina [Lazar] see 3414
Noveril [Morishita] see 3055
Noverme [Gist Brocades] see 5807
Novesine [Wander] see 1076
Novex see 4046
Noviben [Merck & Co.] see 1774
Novicet [Schwarzhaupt] see 10120
Novicodin see 3214
Novidium see 4767
Novidroxin [Fatol] see 4854
Noviform [Dispersa] see 1246
Novil see 3414
Novismuth see 1326
Novobiocin, 6818
Novobiocyl [Francia] see 1981
Novocain [Sanofi Winthrop] see 7937
Novocamid [Hoechst] see 7936
Novocebrin [Homburg] see 9589
Novocef [Pliva] see 2002
Novocillin see 1178
Novocol see 4578
Novodigal [Beiersdorf] see 3210
Novodil [Augot] see 2771
Novodolan [Roussel-UCLAF] see 4140
Novodorm [Rubio] see 9734
Novodrin see 5236
Novoembichin see 6817
Novofluen [Engelhard KG] see 3692
Novogent N [Temmler] see 4925
Novohydrin [Lakeside] see 405
Novolaudon see 4847
Novoldiamine, 6819
Novolin [Novo] see 5011
Novol Ketone see 6819
Novomazina see 2238
Novomycetin [Neves] see 2120
Novon [SK & F] see 3681
Novonal, 6820
Novo-Nastizol A [Bago] see 891
Novonorm [Novo Nordisk] see 8304
Novophone see 2885
Novoridazine [Novopharm] see 9497
Novorin [Polfa] see 10219
Novosparol [E. Merck] see 1252
Novospasmin see 1784
Novotrone see 9141
Novotusil [Inpharzam] see 2456
Novrad [Lilly] see 5495
Novurit [Chinoin] see 5915
Novydrine see 623
Noxaben [Unifa] see 3134
Noxal see 3428

Noxibiol see 6753
Noxigram [Firma] see 2369
Noxiptilin, 6821
Noxiptyline see 6821
Noxiurotan see 7617
Noxyflex-S [Geistlich] see 6822
Noxythiolin, 6822
Noxytiolin see 6822
Noyap [Kyowa] see 4558
Nozinan [Rhône-Poulenc] see 6066
NP see 6715
NP-13 see 1054
NP-55 see 8620
NP-113 see 5006
NP-297 see 5844
NPA-3 see 6505
NPAB see 7883
NPH-1320 see 1465
NPH Insulin see 5011
NPT-10381 see 5006
NPY see 6568
NPYR see 6741
NRC-910 see 5093
NRDC-107 see 1271
NRDC-143 see 7321
NRDC-149 see 2836
NRDC-161 see 2934
NSC-185 see 2797
NSC-1026 see 2803
NSC-1771 see 9510
NSC-2101 see 8430
NSC-3051 see 6152
NSC-3053 see 2867
NSC-3364 see 4123
NSC-3590 see 4254
NSC-5159 see 2086
NSC-5366 see 6815
NSC-5547 see 3525
NSC-6470 see 6621
NSC-8806 see 5871
NSC-9706 see 9803
NSC-10023 see 7904
NSC-11905 see 5376
NSC-12198 see 3504
NSC-13875 see 328
NSC-15200 see 4364
NSC-15780 see 638
NSC-17777 see 7474
NSC-19893 see 4219
NSC-20264 see 157
NSC-21626 see 8005
NSC-23436 see 3480
NSC-23759 see 9321
NSC-25154 see 7634
NSC-25614 see 7530
NSC-25855 see 9586
NSC-26805 see 3873
NSC-27640 see 4148
NSC-28693 see 6333
NSC-30152 see 3260
NSC-30689 see 6299
NSC-31712 see 4214
NSC-32946 see 6299
NSC-34462 see 9986
NSC-35051 see 5871
NSC-37095 see 10011
NSC-37096 see 1118
NSC-39084 see 935
NSC-39415 see 3474
NSC-39470 see 1226
NSC-40725 see 5244
NSC-43193 see 8951
NSC-45383 see 8986
NSC-45388 see 2866
NSC-47439 see 4229
NSC-47774 see 7635
NSC-49171 see 8491
NSC-52644 see 8197
NSC-52947 see 7119
NSC-55975 see 3654

NSC-59729 see 8886
NSC-60584 see 9157
NSC-64375 see 1154
NSC-68982 see 4585
NSC-69856 see 10302
NSC-70731 see 5525
NSC-70845 see 6767
NSC-71047 see 9864
NSC-75520 see 9816
NSC-77518 see 3042
NSC-77625 see 9731
NSC-78502 see 5818
NSC-79037 see 5594
NSC-82116 see 5310
NSC-83653 see 389
NSC-84223 see 4860
NSC-85998 see 8991
NSC-91523 see 8025
NSC-102627 see 4962
NSC-102816 see 923
NSC-104800 see 6300
NSC-105014-F see 2397
NSC-107429 see 2773
NSC-107430 see 7261
NSC-107433 see 9262
NSC-108166 see 9899
NSC-109229 see 871
NSC-109723 see 9897
NSC-109724 see 4937
NSC-112931 see 9385
NSC-114901 see 2966
NSC-115944 see 3621
NSC-118218 see 4162
NSC-118218-H see 4162
NSC-119875 see 2378
NSC-122819 see 9291
NSC-123127 see 3495
NSC-125066 see 1351
NSC-125973 see 7117
NSC-129943 see 8295
NSC-130044 see 9449
NSC-138780 see 9933
NSC-140117 see 4962
NSC-140865 see 9974
NSC-141540 see 3931
NSC-141549 see 635
NSC-145668 see 670
NSC-148958 see 9267
NSC-153858 see 5800
NSC-156303 see 635
NSC-164011 see 10326
NSC-165563 see 1475
NSC-169780 see 8295
NSC-178248 see 2229
NSC-180024 see 1921
NSC-181815 see 7303
NSC-182986 see 3044
NSC-192965 see 8916
NSC-208734 see 115
NSC-218321 see 7277
NSC-226080 see 8288
NSC-234714 see 769
NSC-239336 see 3624
NSC-241240 see 1870
NSC-245382 see 6645
NSC-245467 see 10125
NSC-249008 see 9851
NSC-249992 see 635
NSC-256439 see 4931
NSC-264137 see 3590
NSC-266046 see 7044
NSC-269148 see 5881
NSC-0270516 see 8285
NSC-279836 see 6303
NSC-296961 see 424
NSC-301467 see 3756
NSC-301739 see 6303
NSC-305263 see 4291
NSC-312887 see 4162
NSC-325319. see 3150

Name Index

NSC-328002 *see* 4162
NSC-337766 *see* 1284
NSC-339140 *see* 5016
NSC-351521 *see* 7654
NSC-352122 *see* 9851
NSC-356894 *see* 4610
NSC-362856 *see* 9289
NSC-368390 *see* 1394
NSC-409962 *see* 1894
NSC-515776 *see* 3410
NSC-609669 *see* 9687
NSC-612049 *see* 3148
NSC-628503 *see* 3458
N-Serve [Dow] *see* 6666
NSILA-S *see* 8862
NT-3 *see* 6570
NTA *see* 6675
NTA-194 *see* 9564
NTANa₃ *see* 6675
NTD-2 *see* 1916
Nu-404 *see* 254
Nu-582 *see* 7583
Nu-1196 *see* 317
Nu-1504 *see* 7380
Nu-1779 *see* 317
Nu-2121 *see* 6614
Nu-2206 *see* 5496
Nu-2222 *see* 9837
Nuarsol *see* 8719
Nubain [Endo] *see* 6444
Nubarene *see* 5821
Nuces (Semen) Nucistae *see* 6415
Nucin *see* 5282
Nucite *see* 5008
Nuclear Fast Red *see* 6571
Nucleic Acids, 6823
Nuclein *see* 6823
Nucleocidin, 6824
Nucleoproteins *see* 6823
Nucleoside *see* 6823
Nucleotide *see* 6823
Nuctalon [Takeda; Cassenne] *see* 3744
Nudic Acid A *see* 6825
Nudic Acid B *see* 3039
Nudic Acids, 6825
Nudrin [Shell] *see* 6062
Nuelin [3M] *see* 9421
Nuital [Vaillant-Defresne] *see* 3938
Nujol [Schering] *see* 7327
Nu-K [Consolidated] *see* 7783
Nulcerin [Andromaco] *see* 3972
Nullapon *see* 3557
Nullatuss *see* 5124
Nullo Chlorophyll [DePree] *see* 2207
Nulogyl [Bristol] *see* 6644
Nulomoline *see* 5026
Numal *see* 794
Numidan [Coop. Farm.] *see* 6504
Numoquin *see* 3855
Numorphan [Endo] *see* 7103
Numotac [Riker] *see* 5185
Nuncital [Kabi] *see* 3604
Nuodex 84 *see* 5916
Nupa-Sal [Smith & Nephew] *see* 8487
Nupercaine Hydrochloride [Ciba] *see* 3081
Nupharidine, 6826
Nuprin [Upjohn] *see* 4925
Nuran [Frosst] *see* 2842
Nurison [Nourypharma] *see* 7904
Nurofen [Boots] *see* 4925
Nuromax [Burroughs Wellcome] *see* 3487
Nustar [Du Pont] *see* 4240
Nutgall, 6827
Nutinal [Boots] *see* 1055
Nutmeg *see* 6415
Nutmeg Butter *see* 6923

Nutracort [Galderma] *see* 4828
Nutraplus [Alcon] *see* 10005
Nutrase [Kyorin] *see* 9431
NutraSweet [Searle] *see* 874
Nutrin [Chemipharm] *see* 6577
Nutropin [Genentech] *see* 8864
Nuvacon [BDH] *see* 5849
Nuvacron [Ciba-Geigy] *see* 6334
Nuvan [Ciba-Geigy] *see* 3129
Nuvanol N [Ciba-Geigy] *see* 5053
Nuvapen [C.E.P.A.] *see* 628
Nuvarone *see* 6187
Nux Moschata *see* 6415
Nux Vomica, 6828
Nva²-cyclosporine *see* 2821
NVB *see* 10127
NY-198 *see* 5592
Nyazin *see* 6575
Nybomycin, 6829
Nycopren [Nycomed] *see* 6504
Nyctal *see* 1879
Nycton *see* 2113
Nydrazid [Bristol-Myers Squibb] *see* 5203
Nylidrin, 6830
Nylmerate *see* 7453
Nylon, 6831
Nylon 6, 6832
Nylon 46, 6833
Nysconitrine [BioTherabel] *see* 6704
Nystan [Squibb] *see* 6834
Nystatin, 6834
Nystatin A₁ *see* 6834
Nystavescent [Squibb] *see* 6834
Nytilax [Rohto] *see* 8600
Nyxolan *see* 4890
NZ-105 *see* 3566
Oakite *see* 8808
Oasil [Farmasimes] *see* 5908
Obedrex [Beta] *see* 4015
Obepar [Parmed] *see* 7365
Oberex [Draco] *see* 2440
Obesedrin *see* 2996
Obestat [Lemmon] *see* 7461
Obidoxime Chloride, 6835
Oblioser [Serono] *see* 6361
Oblivon *see* 5890
Oblivon C *see* 5890
OBPA *see* 7095
OB Protein *see* 5466
Obracin [Lilly] *see* 9628
OC-340 *see* 10123
Occlusal [Euroderma] *see* 8484
Occrycetin [Willows-Francis] *see* 7111
Ocelina [Roux-Ocefa] *see* 5986
Ocenol *see* 6968
Oceral [Roche] *see* 7071
Ocestrol *see* 1102
Ochratoxin A *see* 6836
Ochratoxin B *see* 6836
Ochratoxin C *see* 6836
Ochratoxins, 6836
Ocimene, 6837
O.C.M. [Gobbi-Novag] *see* 2132
Octaacetyl-aldehydro-cellobiose *see* 2008
Octaacetyl-α-cellobiose *see* 2008
Octaacetyl-β-cellobiose *see* 2008
Octabenzone, 6838
Octacaine, 6839
Octacarbonyldicobalt *see* 3135
Octachlor *see* 2129
1,3,4,5,6,7,10,10-Octachloro-4,7-endo-methylene-4,7,8,9-tetrahydro-phthalan *see* 5142
1,2,4,5,6,7,8,8-Octachloro-2,3,3a,4,7,-7a-hexahydro-4,7-methano-1*H*-indene *see* 2129
1,3,4,5,6,7,8,8-Octachloro-1,3,3a,4,7,-

7a-hexahydro-4,7-methanoisobenzo-furan *see* 5142
1,2,4,5,6,7,8,8-Octachloro-4,7-meth-ane-3a,4,7,7a-tetrahydroindane *see* 2129
1,2,4,5,6,7,8,8-Octachloro-3a,4,7,7a-tetrahydro-4,7-methanoindan *see* 2129
Octacosanol, 6840
1-Octacosanol *see* 6840
n-Octacosanol *see* 6840
Octacosyl Alcohol *see* 6840
(*Z,Z*)-9,12-Octadecadienoic Acid *see* 5529
(*Z,Z*)-9,12-Octadecadienoic Acid Ethyl Ester *see* 3866
9,12-Octadecadienoic Acid Methyl Ester *see* 6168
Octadecafluorodecahydronaphthalene *see* 4221
Octadecanoic Acid *see* 8959
Octadecanoic Acid Aluminum Salt *see* 379
Octadecanoic Acid Ammonium Salt *see* 588
Octadecanoic Acid Butyl Ester *see* 1625
Octadecanoic Acid Calcium Salt *see* 1750
Octadecanoic Acid 2-(1-Carboxyeth-oxy)-1-methyl-2-oxoethyl Ester, Calcium Salt *see* 1751
Octadecanoic Acid Compd with Erythromycin 2'-Acetate *see* 3721
Octadecanoic Acid Copper Salt *see* 2721
Octadecanoic Acid Magnesium Salt *see* 5730
Octadecanoic Acid Mercury Salt *see* 5940
Octadecanoic Acid Monoester with 1,-2,3-Propanetriol *see* 4498
Octadecanoic Acid 1,2,3-Propanetriyl Ester *see* 9885
Octadecanoic Acid Zinc Salt *see* 10292
1-Octadecanol *see* 8960
cis-6,*cis*-9,*cis*-12-Octadecatrienoic Acid *see* 5531
(*Z,Z,Z*)-6,9,12-Octadecatrienoic Acid *see* 5531
(*Z,Z,Z*)-9,12,15-Octadecatrienoic Acid *see* 5530
(*E*)-D-*erythro*-4-Octadecene-1,3-diol *see* 8899
17-Octadecene-9,11-diynoic Acid *see* 5114
cis-6-Octadecenoic Acid *see* 7330
(*Z*)-9-Octadecenoic Acid *see* 6965
(*E*)-9-Octadecenoic Acid *see* 3574
trans-11-Octadecenoic Acid *see* 10036
9-Octadecenoic Acid Aluminum Salt *see* 367
(*Z*)-9-Octadecenoic Acid Ammonium Salt *see* 569
9-Octadecenoic Acid Calcium Salt *see* 1731
9-Octadecenoic Acid Copper Salt *see* 2711
(*Z*)-9-Octadecenoic Acid Ethyl Ester *see* 6965
9-Octadecenoic Acid Magnesium Salt *see* 5711
9-Octadecenoic Acid Mercury Salt *see* 5934
9-Octadecenoic Acid 1,2,3-Propanetri-yl Ester *see* 9861
9-Octadecenoic Acid Zinc Salt *see* 10276
cis-9-Octadecen-1-ol *see* 6968

(Z)-9-Octadecen-1-ol *see* 6968
Octadec-*cis*-12,13-epoxy-*cis*-9-enoic
Acid *see* 10099
Δ⁵-Octadecylenic Acid *see* 7330
3-(Octadecyloxy)-1,2-propanediol *see*
1039
3,4,5,6,14,15,20,21-Octadehydro-
yohimbanium *see* 8591
Octafluorocyclobutane, 6841
1*H*,1*H*,5*H*-Octafluoropentanol *see*
6842
2,2,3,3,4,4,5,5-Octafluoro-1-pentanol,
6842
Octahedrite *see* 9612
[2-(Octahydro-1-azocinyl)ethyl]-
guanidine *see* 4589
[(Octahydro-2-azocinyl)methyl]-
guanidine *see* 4588
3a*β*,6a*β*,7,7a*α*,8,11a,11b*α*,11c-Octa-
hydro-2,10-dimethoxy-3,8*α*,11a*β*,-
11c*β*-tetramethylphenanthro[10,1-
bc]pyran-1,5,11(4*H*)-trione *see* 8209
D-[(4,6,6a,7,8,9,10,10a-Octahydro-4,7-
dimethyl-10a*α*-indolo[4,3-*fg*]quin-
olin-9*β*-yl)methyl]carbamic Acid
Benzyl Ester *see* 5999
3,3a,4,5,6,6a,9a,9b-Octahydro-6,9a-di-
methyl-3-methyleneazuleno[4,5-*b*]-
furan-2,9-dione *see* 400
[3a*R*-(3a*α*,5*β*,8a*β*,9a*α*)]-3a,5,6,7,8,8a,-
9,9a-Octahydro-5,8a-dimethyl-3-
methylenenaphtho[2,3-*b*]furan-
2(3*H*)-one *see* 208
[1a*R*-(1a*R**,4*E*,7a*S**,10a*S**,10b*R**)]-2,-
3,6,7,7a,8,10a,10b-Octahydro-1a,5-
dimethyl-8-methyleneoxireno[9,10]-
cyclodeca[1,2-*b*]furan-9(1a*H*)-one
see 7180
Octahydro-6,8a-dimethyl-3-(1-methyl-
ethyl)-1*H*-3a,6-epoxyazulen-7-ol *see*
2889
(+)-(1*R*,4a*S*,10a*R*)-1,2,3,4,5a,9,10,10a-
Octahydro-1,4a-dimethyl-7-(1-meth-
ylethyl)-6-sulfo-1-phenanthrenecar-
boxylic Acid 6-Sodium Salt *see* 3538
[1*R*-(1*α*,4a*β*,10a*α*)]-1,2,3,4,4a,9,10,-
10a-Octahydro-1,4a-dimethyl-7-(1-
methylethyl)-6-sulpho-1-phenan-
threnecarboxylic Acid *see* 3538
[4*S*-(4*α*,4a*α*,8a*β*)]-Octahydro-4,8a-di-
methyl-4a(2*H*)-naphthalenol *see*
4408
Octahydro-4*α*,8a*β*-dimethyl-4a*α*(2*H*)-
naphthol *see* 4408
[1*S*-[1*α*,3*β*,4*β*,4a*α*,7(1′*R**,3′*R**,4′*R**,-
4′a*R**,9′a*S**),9a*β*]]-1,1′,3,3′,4,4′,9a,-
9′a-Octahydro-4,4′,8,8′,9a,9′a-hexa-
hydroxy-3,3′-dimethyl-[7,7′-bi-1,4a-
(epoxymethano)-4a*H*-xanthene]-9,-
9′,11,11′(2*H*,2′*H*)-tetrone *see* 3691
[3*R*-[3*α*,4*β*,4a*β*,7(3′*R**,4′*S**,4′*AS**)]]-
2,2′,3,3′,4,4′,9,9′-octahydro-1,1′,4,4′,-
8,8′-hexahydroxy-3,3′-dimethyl-9,9′-
dioxo-[7,7′-bi-4a*H*-xanthene]-4a,4′a-
dicarboxylic Acid Dimethyl Ester
see 8561
Octahydro-1,3,4,7,8,10-hexanitro-5,2,-
6-(iminomethenimino)-1*H*-imidazo-
[4,5-*b*]pyrazine *see* 4731
[3a*S*-(3a*α*,4*α*,4a*β*,7a*α*,8*α*,9a*α*)]-3,3a,-
4,4a,7a,8,9,9a-Octahydro-4-hydr-
oxy-4a,8-dimethyl-3-methyleneazu-
leno[6,5-*b*]furan-2,5-dione *see* 4656
[3a*S*-(3a*α*,6*β*,6a*α*,9a*β*,9b*α*)]-3,3a,4,5,-
6,6a,9a,9b-Octahydro-6a-hydroxy-
6,9a-dimethyl-3-methyleneazuleno-
[4,5-*b*]furan-2,9-dione *see* 7179
[1*S*-(1*α*,4a*α*,10a*β*)]-1,2,3,4,4a,9,10,-
10a-Octahydro-6-hydroxy-1,4a-

dimethyl-1-phenanthrenecarboxylic
Acid *see* 7701
[1*S*-(1*α*,4a*α*,10a*β*)]-1,2,3,4,4a,9,10,-
10a-Octahydro-6-hydroxy-7-[2-
hydroxy-1-(hydroxymethyl)ethyl]-
1,4a-dimethyl-1-phenanthrenecar-
boxylic Acid *see* 7702
Octahydro-12-(hydroxymethyl)-2-
imino-5,9:7,10a-dimethano-10a*H*-
[1,3]dioxocino[6,5-*d*]pyrimidine-4,-
7,10,11,12-pentol *see* 9382
[1*S*-(1*α*,4a*α*,8a*β*)]-1,4,4a,5,6,7,8,8a-
Octahydro-1-hydroxy-5,5,8a-tri-
methyl-1,2-naphthalenedicarbox-
aldehyde *see* 10173
[1*S*-(1*α*,6*β*,7*α*,8*β*,8a*β*)]-Octahydro-
1,6,7,8-indolizinetetrol *see* 1944
2,3,4,5,6,7a,8,8a-Octahydro-3-isoprop-
yl-6,8a-dimethyl-1*H*-3a,6-epoxy-
azulen-7-ol *see* 2889
[7*R*-(7*α*,7a*β*,14*α*)]-7,7a,8,9,10,11,13,-
14-Octahydro-7,14-methano-4*H*,6*H*-
dipyrido[1,2-*a*:1′,2′-*e*][1,5]diazocin-
4-one *see* 666
4,6,6a,7,8,9,10,10a-Octahydro-10a*α*-
methoxy-4,7-dimethylindolo[4,3-*fg*]-
quinoline-9-methanol 5-Bromo-
nicotinate *see* 6580
[1′*R*-[1′*α*(*S**),4′*A*α,8′a*β*]]-4,4′*a*,5,5′,-
6′,7′,8′,8′*a*-Octahydro-2′,5,5′,5′,8′*a*-
pentamethylspiro[furan-2(3*H*),1′-
(4′*H*)-naphthalene]-5-acetic Acid *see*
4569
(1*α*,1a*α*,2a*β*,5*α*,5a*α*,7a*α*,7b*α*,7c*α*)-
(±)-1a,2,2a,5,5a,7a,7b,7c-Octahy-
dro-5-phenyl-1*H*-cyclobut[*bc*]-
acenaphthylene-1-acetic Acid *see*
3611
[1*α*(*E*),1a*α*,2a*β*,5*α*,5a*α*,7a*α*,7b*α*,7c*α*]-
(±)-4-(1a,2,2a,5,5a,7a,7b,7c-
Octahydro-5-phenyl-1*H*-cyclobut-
[*bc*]acenaphthylen-1-yl)-2-butenoic
Acid *see* 3611
[1*α*,(2*E*,4*E*),1a*α*,3*β*,3a*α*,6*β*,6*α*A,6b*α*,-
7*S**]-(±)-1,1a,2,3,3a,6,6a,6b-octahy-
dro-1-(5-phenyl-2,4-pentadienyl)-3,-
6-methanocyclobut[*cd*]indene-7-car-
boxylic Acid *see* 3611
[1*R*-*trans*]-Octahydro-2*H*-quinolizine-
1-methanol *see* 5637
[3*S*-(3*α*,5*α*,8*α*)]-1,2,3,4,5,6,7,8-Octa-
hydro-*α*,*α*,3,8-tetramethyl-5-
azulenemethanol *see* 4583
1,1a,4,4a,5,6,7,8-Octahydro-2,4a,8,8-
tetramethylcyclopropa[*d*]naphthal-
ene *see* 9534
[3*R*-(3*α*,3a*β*,6*α*,7*β*,8a*α*)]-Octahydro-
3,6,8,8-tetramethyl-1*H*-3a,7-meth-
anoazulen-6-ol *see* 1961
[1*R*-(1*α*,4*β*,4a*α*,6*β*,8a*α*)]-Octahydro-
4,8a,9,9-tetramethyl-1,6-methano-
naphthalen-1(2*H*)-ol *see* 7184
Octahydro-1,3,5,7-tetranitro-1,3,5,7-
tetrazocine *see* 6855
[1*R*-[1*α*,1(1*S**),3a*β*,4*α*,5*α*,7a*β*]]-Oc-
tahydro-*α*,3a,5-trimethyl-6,8-dioxo-
1,4-methano-1*H*-indene-1-acetic
Acid *see* 8508
[3*R*-(3*α*,5a*β*,6*β*,8a*β*,9*α*,12*β*,12a*R**)]-
Octahydro-3,6,9-trimethyl-3,12-
epoxy-12*H*-pyrano[4,3-*j*]-1,2-benzo-
dioxepin-10(3*H*)-one *see* 856
[1*R*-(1*α*,4a*β*,8a*α*)]-1,4,4a,5,6,7,8,8a-
Octahydro-5,5,8a-trimethyl-1,2-
naphthalenedicarboxaldehyde *see*
7732
[5a*S*-(5a*α*,9a*β*,9b*α*)]-5,5a,6,7,8,9,9a,-
9b-Octahydro-6,6,9a-trimethylnaph-
tho[1,2-*c*]furan-1(3*H*)-one *see* 3500
4,6,8,10,12,14,16,27-Octahydroxy-3-

(1-Hydroxyhexyl)-17,28-dimethyl-
oxacyclooctacosa-17,19,21,23,25-
pentaen-2-one *see* 4123
Octaldehyde *see* 1809
Octalene *see* 227
Octalox [Velsicol] *see* 3152
Octalupine *see* 4875
Octamet [Belphar] *see* 9160
Octamethylcyclotetrasiloxane, 6843
Octamethyldiphosphoramide *see* 8540
N,*N*′-Octamethylenebis(2,2-dichloro-
acetamide) *see* 4108
[1*R*-(1*R**,2*R**,5*R**,7*R**,10*S**,11*S**,-
14*S**,16*S**,19*R**,20*R**,23*R**,25*R**,-
28*S**,29*S**,32*S**,34*S**)]-2,5,11,14,20,-
23,29,32-Octamethyl-4,13,22,31,37,-
38,39,40-octaoxapentacyclo-
[32.2.1.1⁷,¹⁰.1¹⁶,¹⁹.1²⁵,²⁸]tetracontane-
3,12,21,30-tetrone *see* 6770
Octamethyl Pyrophosphoramide *see*
8540
Octamethyltrisiloxane, 6844
Octamoxin, 6845
Octamylamine, 6846
Octanal *see* 1809
Octane, 6847
1,8-Octanedicarboxylic Acid *see* 8558
Octanedioic Acid *see* 9031
N,*N*′-1,8-Octanediylbis[2,2-dichloro-
acetamide] *see* 4108
Octanil *see* 5202
Octanohydroxamic Acid, 6848
Octanoic Acid *see* 1808
Octanoic Acid Ammonium Salt *see*
532
Octanoic Acid 2,3-Dihydroxypropyl
Ester *see* 6335
Octanoic Acid Ethyl Ester *see* 3825
Octanoic Acid Zinc Salt *see* 10259
1-Octanol, 6849
2-Octanol, 6850
5-Octanol-4-one *see* 1631
2-Octanone *see* 4749
Octaphen *see* 7389
Octapressin [Sandoz] *see* 3992
Octatensine *see* 4589
Octatropine Methylbromide *see* 709
Octaverine, 6851
1-Octene *see* 1807
Octeniderm [Sterling] *see* 6852
Octenidine, 6852
Octenisept [Schülke & Mayr] *see* 6852
Oct HA *see* 6848
Octhilinone, 6853
Octin [Knoll] *see* 5202
Octinum D [Knoll] *see* 6846
Octisamyl *see* 6846
Octodrine, 6854
Octofène [Debat] *see* 2439
Octofollin [Schieffelin] *see* 1102
Octogen, 6855
Octoil *see* 1291
Octoil S *see* 1292
Octometine *see* 6846
Octomoxine *see* 6845
Octon *see* 5202
Octopamine, 6856
Octopirox [Hoechst] *see* 7657
Octostim [Ferring] *see* 2969
Octotiamine, 6857
Octoxynol, 6858
Octoxynol-9 *see* 6858
OctreoScan [Mallinckrodt] *see* 7267
Octreotide, 6859
Octyl Acetate, 6860
n-Octyl Bromide, 6861
sec-Octyl Bromide, 6862
[2-[2-(*p*-Octylcresoxy)ethoxy]ethyl]-
dimethylbenzylammonium Chloride
see 6103

Octylene *see* 1807
Octylene Glycol *see* 3790
sec-Octyl Iodide, 6863
2-Octyl-4-isothiazolin-3-one *see* 6853
2-Octyl-3(2*H*)-isothiazolone *see* 6853
Octyl Methoxycinnamate, 6864
β-*p-tert*-Octylphenoxyethyldiethylbenzylammonium Chloride *see* 7389
Octylphenoxy Polyethoxyethanol *see* 6858
5-[2-(Octylsulfinyl)propyl]-1,3-benzodioxole *see* 9140
n-Octylsulfoxide of Isosafrole *see* 9140
Ocufen [Allergan] *see* 4234
Oculinum *see* 1384
Ocusert Pilo [Cusi] *see* 7578
Ocu-Vinc [Alcon] *see* 10120
Ocytocin *see* 7114
ODA-914 *see* 2899
Odanon [Towa Yakuhin] *see* 1833
Odemase [Azupharma] *see* 4331
ODGF *see* 7683
Odontalg [Giovanardi] *see* 5505
Odorigenin *see* 10035
Odrik [Roussel] *see* 9703
Odylen [Winthrop] *see* 5977
Oedemex [Mepha] *see* 4331
Oekolp [Kade] *see* 3750
Oenanthal *see* 4692
Oenanthaldehyde *see* 4692
Oenanthic Acid *see* 4695
Oenanthic Ether *see* 3881
Oenanthol *see* 4692
Oenanthotoxin *see* 3608
Oenanthylic Acid *see* 4695
Oesipos *see* 5371
Oestrasid *see* 3153
Oestrin *see* 3751
Oestriol *see* 3750
Oestrodiene *see* 3153
Oestrogel [Besins-Iscovesco] *see* 3746
Oestrogeninc *see* 3177
Oestromenin *see* 3177
Oestromensyl *see* 3177
Oestromon [E. Merck] *see* 3177
Oestrone *see* 3751
Oestroperos *see* 3751
Oestroral *see* 3153
Off *see* 2912
Oflocet [Roussel] *see* 6865
Oflocin [Glaxo] *see* 6865
Ofloxacin, 6865
Ofloxacine *see* 6865
Oftanol [Bayer] *see* 5187
Ogast [Takeda] *see* 5373
Ogen [Abbott] *see* 3751
Ogostal [Lilly] *see* 1801
Ogyline [Roussel] *see* 6798
γ-OH *see* 4860
4-OHA *see* 4267
OHB₁₂ *see* 4854
1α-OH-CC *see* 4864
Ohlexin [Ohta] *see* 2021
l-OHP *see* 7044
Ohton *see* 3309
Oil of Allspice *see* 6935
Oil of Amber, Rectified, 6866
Oil of American Wormseed *see* 6888
Oil of Angelica, 6867
Oil of Anise, 6868
Oil Anise, Japanese, 6869
Oil of Anthemis *see* 2082
Oil of Arbor Vitae *see* 6954
Oil of Asarum, 6870
Oil of Balm, 6871
Oil of Basil, 6872
Oil of Bay, 6873
Oil Bergamot, 6874
Oil of Bitter Almond, 6875

Oil of Bitter Orange, 6876
Oil of Cade *see* 5285
Oil of Cajeput, 6877
Oil of Calamus, 6878
Oil of Camphor, Rectified, 6879
Oil of Canada Fleabane *see* 6905
Oil of Canada Snakeroot *see* 6870
Oil of Caraway, 6880
Oil of Cardamom, 6881
Oil of Cascarilla, 6882
Oil of Cashew Nut Shell, 6883
Oil of Cassia *see* 6890
Oil of Cedar Leaf, 6884
Oil of Cedar Wood, 6885
Oil of Celery, 6886
Oil of Champaca, 6887
Oil of Chenopodium, 6888
Oil of Cherry Laurel, 6889
Oil of Chinese Cinnamon *see* 6890
Oil of Cinnamon, 6890
Oil of Cinnamon, Ceylon, 6891
Oil of Citronella, 6892
Oil of Clove, 6893
Oil of Copaiba, 6894
Oil of Coriander, 6895
Oil of Crispmint *see* 6943
Oil of Cubeb, 6896
Oil of Cumin, 6897
Oil of Curled Mint *see* 6943
Oil of Cypress, 6898
Oil of Dill, 6899
Oil of Dwarf Pine Needles, 6900
Oil of Egg Yolk *see* 3570
Oil of Erigeron *see* 6905
Oil of Eucalyptus, 6901
Oil of Fennel, 6902
Oil of Fir, 6903
Oil of Fir—Siberian, 6904
Oil of Fleabane, 6905
"Oil Garlic" *see* 305
Oil of Garlic, 6906
Oil of Geranium, 6907
Oil of Geranium—East Indian, 6908
Oil of Ginger, 6909
Oil of Grapes *see* 3881
Oil Green *see* 2283
Oil of Hedeoma *see* 6930
Oil of Hops, 6910
Oil of Hyssop, 6911
Oil of Jojoba *see* 5279
Oil of Juniper, 6912
Oil of Juniper Tar *see* 5285
Oil of Juniper Wood, 6913
Oil of Lavender, 6914
Oil of Lemon, 6915
Oil of Lemon Balm *see* 6871
Oil of Lemon Grass, 6916
Oil of Levant Wormseed, 6917
Oil of Linaloe, 6918
Oil of Mace *see* 6923
Oil of Marjoram, 6919
Oil of Melissa Balm *see* 6871
Oil of Mirbane *see* 6685
Oil of Monarda *see* 6326
Oil of Mountain Pine *see* 6900
Oil of Mustard, Expressed, 6920
Oil of Myrcia *see* 6873
Oil of Myristica *see* 6924
Oil of Myrtle, 6921
Oil of Neroli *see* 6926
Oil of Niaouli, 6922
Oil of Niobe *see* 6104
Oil Nut *see* 5281
Oil of Nutmeg, Expressed, 6923
Oil of Nutmeg, Volatile, 6924
Oil of Orange, 6925
Oil of Orange Flowers, 6926
Oil of Origanum, 6927
Oil of Palma Christi *see* 1946
Oil of Parsley, 6928

Oil of Patchouli, 6929
Oil of Pelargonium Geranium *see* 6907
Oil of Pennyroyal—American, 6930
Oil of Pennyroyal—European, 6931
Oil of Pepper, 6932
Oil of Peppermint, 6933
Oil of Pettigrain, 6934
Oil of Pimenta, 6935
Oil of Pimento *see* 6935
Oil of Pine *see* 6904
Oil of Pine Needles, 6936
Oil of Pulegium *see* 6931
Oil of Rice Bran *see* 8375
Oil of Rose, 6937
Oil of Rose Geranium *see* 6907
Oil of Rosemary, 6938
Oil of Rue, 6939
Oil of Santal, 6940
Oil of Sassafras, 6941
Oil of Savin, 6942
Oil of Scotch Fir *see* 6936
Oil of Silver Fir *see* 6903
Oil of Silver Pine *see* 6903
Oil of Spearmint, 6943
Oil of Spike, 6944
Oil of Sweet Almond, 6945
Oil of Sweet Bay, 6946
Oil of Sweet Flag *see* 6878
Oil of Tansy, 6947
Oil of Thuja *see* 6954
Oil of Thyme, 6948
Oil of Turpentine, 6949
Oil of Turpentine, Rectified, 6950
Oil of Turpentine, Sulfurated, 6951
Oil of Valerian, 6952
Oil of Vetiver, 6953
Oil of Vitriol *see* 9147
Oil of White Cedar, 6954
Oil of Wild Marjoram *see* 6927
Oil of Wine *see* 3881
Oil of Wine, "Heavy", 6955
Oil of Wormwood, 6956
Oil of Yarrow, 6957
Oil Red XO *see* 10222
Oil Sweet Orange *see* 6925
Oil White Birch *see* 1279
Oizine *see* 8152
OK-174 *see* 1069
Okadaic Acid, 6958
Okenite, 1749
Oksafenamide *see* 7018
Oksilidin *see* 8266
Oktadin *see* 4589
Oktatensin *see* 4589
Oktatenzin *see* 4589
Oktaverine *see* 6851
OKY-046 *see* 7115
Olamin [Sigamed] *see* 2365
Olane [Avi-Sun] *see* 7741
Olanzapine, 6959
Olaquindox, 6960
Olaxin [SK & F] *see* 2441
Olbemox [Farmitalia] *see* 113
Olbetam [Farmitalia] *see* 113
Olcadil [Sandoz] *see* 2481
Olcenon [Lederle] *see* 9639
Oldren [Roemmers] *see* 6231
Old Tuberculin *see* 9937
Old Yellow Enzyme, 6961
Oleandomycin, 6962
Oleandomycin Triacetate Ester *see* 9899
Oleandrin, 6963
(3β)-Olean-12-en-3-ol *see* 660
Oleanol *see* 6964
Oleanolic Acid, 6964
Oleate of Mercury *see* 5934
Olefiant Gas *see* 3837
Oleic Acid, 6965

Oleic Acid Aluminum Salt *see* 367
Oleic Acid Ammonium Salt *see* 569
Oleic Acid Bismuth Salt *see* 1312
Oleic Acid Calcium Salt *see* 1731
Oleic Acid Lead Salt *see* 5435
Oleic Acid Potassium Salt *see* 7818
Olein *see* 9861
Oleochrysine *see* 1689
Oleocreosote *see* 2642
Oleoresin of Aspidium, 6966
Oleo Stock *see* 9211
Oleovitamin A *see* 10150
Oleovitamin D$_2$ *see* 10156
Oleovitamin D$_3$ *see* 10157
Oleptan [Bender] *see* 4258
Olestra *see* 9053
Oleth *see* 7737
Oleum *see* 9147
Oleum Abietis *see* 6904
Oleum Andropogonis Muricati *see* 6953
Oleum Rusci *see* 1279
Oleum Vitis Viniferae *see* 3881
Oleuropein, 6967
Oleyl Alcohol, 6968
Olibanum, 6969
Olicard [Giulini] *see* 5245
Oligomycin A *see* 6970
Oligomycin B *see* 6970
Oligomycin C *see* 6970
Oligomycin D *see* 6970
Oligomycins, 6970
Olimpen *see* 7235
Olimplex [Vaillant-Defresne] *see* 3211
Oliprevin [Bristol-Myers] *see* 7894
Olitref [Chemol] *see* 9815
Olivacine, 6971
Olivanic Acids, 6972
Olive Oil, 6973
Oliver [Sanzen] *see* 7016
Olive Spurge *see* 6258
Olivil, 6974
Olivin *see* 6975
Olivin [Lek] *see* 3605
Olivomycin A *see* 6975
Olivomycins, 6975
Olivoretin D *see* 9271
Olmagran [Squibb] *see* 4830
Olmelin *see* 4395
Olmifon [Lafon] *see* 168
Ololiuqui *see* 5088
Olprinone *see* 5605
Olsalazine, 6976
Olymp [Du Pont] *see* 4240
Olynth [Parke-Davis] *see* 10219
OM-518 *see* 5896
OM-805 *see* 816
OM-853 *see* 10123
Omadine [Olin] *see* 8178
Omadine Disulfide [Olin] *see* 8178
Omaflora [Olin] *see* 4814
Omaine *see* 2938
Omal *see* 9773
Ombolan [Esteve] *see* 3512
Omca [Squibb-Heyden] *see* 4226
Omcilon [Squibb] *see* 9727
OMD *see* 6133
OM-dopa *see* 6133
OMDS *see* 8178
7-OMEN *see* 5881
Omepral [Fujisawa] *see* 6977
Omeprazen [Malesci] *see* 6977
Omeprazole, 6977
Omeral [Simes] *see* 6345
Omeril [Bayer] *see* 5809
Omite [Uniroyal] *see* 7992
Omnes [Fumouze] *see* 6623
Omnibex [Bago] *see* 7415
Omnicain [Daiichi] *see* 7937
Omniflox [Abbott] *see* 9284

Omnipaque [Nyegaard; Schering AG] *see* 5068
Omnipen [Wyeth] *see* 628
Omnipen-N [Wyeth] *see* 628
Omnisan [Squibb] *see* 3651
Omniscan [Sanofi Winthrop] *see* 4345
Omnizole [Merck & Co.] *see* 9426
Omnopon [Roche] *see* 7150
Omnyl *see* 6028
Omoconazole, 6978
OMPA *see* 8540
Omperan [Taiho] *see* 9163
OMS-29 *see* 1831
OMS-45 *see* 4017
OMS-75 *see* 6445
OMS-115 *see* 3453
OMS-658 *see* 1453
OMS-659 *see* 1453
OMS-968 *see* 4273
OMS-1696 *see* 4850
OMS-1804 *see* 3188
OMS-3023 *see* 4051
OMS-3048 *see* 9700
Omtan *see* 5142
Onaka [Max Pharma] *see* 7574
ONB *see* 6689
Oncaspar [Enzon] *see* 871
Onclast [Merck & Co.] *see* 228
Oncol [Otsuka] *see* 1069
OncoScint [Cytogen] *see* 8530
Oncotrex [Warner-Lambert] *see* 9851
Oncovin [Lilly] *see* 10124
Ondansetron, 6979
Ondena [Bayer] *see* 2890
Ondogyne [Roussel-UCLAF] *see* 2789
Ondonid [Roussel-UCLAF] *see* 2789
One-Alpha [Leo Pharm] *see* 4864
One-Iron *see* 4094
Onion Oil, 6980
Oniria [Lifepharma] *see* 8211
Onkotin *see* 2989
ONO-802 *see* 4393
ONO-1078 *see* 7889
ONO-1206 *see* 5514
ONO-1308 *see* 7003
ONO-2235 *see* 3640
Onokrein P [Ono] *see* 5291
Onon [Ono] *see* 7889
Ononin *see* 4271
ONO-RS-411 *see* 7889
Onquinin [Ono] *see* 796
Onsukil [Grünenthal] *see* 7939
Ontianil, 6981
Ontosein [Diagnostic Data] *see* 9177
Onychomal [Hermal] *see* 10005
Onyxide *see* 3836
Oöcyan *see* 1264
Ooporphyrin *see* 8084
Oosporein, 6982
OP-1206 *see* 5514
Opacin *see* 5059
Opacist E.R. [Bracco] *see* 5031
Opacoron [Cilag-Chemie] *see* 79
Opal *see* 8634
Opalène [Théraplix] *see* 9850
Opalmon [Ono] *see* 5514
Opalwax *see* 1947
Oparenol *see* 5061
OPC-1085 *see* 1917
OPC-2009 *see* 7939
OPC-7251 *see* 6430
OPC-8212 *see* 10105
OPC-12759 *see* 8296
OPC-13013 *see* 2335
OPC-17116 *see* 4567
Opcon [Bausch & Lomb] *see* 6455
Opegan [Santen] *see* 4793
Operidine [Janssen] *see* 7400
Ophiobolin D *see* 2024
Ophtagram [Chauvin] *see* 4398

Ophtalmokalixan [Bristol] *see* 5293
Ophtamedine [De Bournonville] *see* 4728
Ophthaine [Squibb] *see* 7991
Ophthalamin (obsolete) *see* 10150
Ophthalgan [Wyeth-Ayerst] *see* 4493
Ophthalmadine [SAS] *see* 4934
Ophthetic [Allergan] *see* 7991
Ophthochlor [Parke, Davis] *see* 2120
Ophtocortin [Winzer] *see* 5840
Ophtorenin [Winzer] *see* 1521
Ophtosol [Winzer] *see* 1412
Opian *see* 6815
Opianic Acid, 6983
Opianine *see* 6815
Opianyl *see* 5825
Opilon [Parke, Davis] *see* 6374
Opiniazide, 6984
Opino [Bayropharma] *see* 6830
Opipramol, 6985
Opiran [Cassenne] *see* 7589
Opium, 6986
Opobalsam *see* 977
Opocarbyl *see* 1856
Oposim [Richet] *see* 8025
Opren [Lilly] *see* 1075
Opridan [Locatelli] *see* 1454
Opromazine, 6987
Opsins, 6988
α$_2$-Opsonins *see* 4119
Op-Sulfa [Riker] *see* 9067
Optal *see* 8027
Optanox *see* 10131
Optenyl [Stroschein] *see* 7151
Op-Thal-Zin [Alcon] *see* 10293
Opticlox [Norbrook] *see* 2480
Opticrom [Fisons] *see* 2658
Opticron [Fisons] *see* 2658
Optidase [Sanofi-Winthrop] *see* 1948
Optiflex [Lilly] *see* 8864
Optimax WV [E. Merck] *see* 9929
Optimil [Wallace Labs.] *see* 6028
Optimine [Schering] *see* 934
Optimycin [Biochemie; Sandoz] *see* 6007
Optinoxan [Robisch] *see* 6028
Optipect [Thiemann] *see* 2959
Optipen [C.S.L.] *see* 7369
Optipranolol [Bausch & Lomb] *see* 6221
Optipress [Wellcome] *see* 1917
Optiray [Mallinckrodt] *see* 5083
Optium [Disprovent] *see* 617
Optocain [Bayer] *see* 5905
Optojod *see* 8881
Optoquine *see* 3855
Optovit-A *see* 10150
OptoVit-E [Hermes] *see* 10160
Optraex *see* 10293
Opturem [Kade] *see* 4925
Opus [BASF] *see* 3665
Opyrin [Taisho] *see* 4167
Orabet [Berlin-Chemie] *see* 9646
Orabilex [Fougera] *see* 1511
Orabilix [Guerbet] *see* 1511
Orabolin [Organon] *see* 3851
Oracaine *see* 5909
Oracef [Lilly] *see* 2021
Oracéfal [Bristol] *see* 1963
Oracilline [Rhone-Poulenc Rorer] *see* 7230
Oracocin [Tobishi] *see* 2021
Oracon [Mead Johnson] *see* 3268
Oradexon [Organon] *see* 2986
Oradian [Chinoin] *see* 2239
Oradiol [Mallinckrodt] *see* 3780
Oradol [Ciba-Geigy] *see* 3474
Oraflex [Lilly] *see* 1075
Oragallin [Chemie Linz] *see* 945
Oragest [Teva] *see* 5838

Orageston [Organon] see 299
Oragrafin [Squibb] see 5087
Oragrafin Calcium [Squibb] see 5087
Oragulant see 3363
Oralcid see 49
Oraldene [Warner-Lambert] see 4741
Oralopen [Bayer] see 7369
Ora-Lutin [Parke, Davis] see 3787
Oramorph [Boehringer, Ing.] see 6361
Oranabol [Farmitalia] see 7099
Orange I, 6989
Orange II, 6990
Orange III see 6180
Orange IV see 9906
Orange B, 6991
Orange Crush [Cyanamid] see 1284
Orange GS see 9906
Orange N see 9906
Orange Peel, 6992
Orange Root see 4808
Oranixon [Organon] see 5895
Orap [Janssen] see 7589
Orapen V-K [Independent] see 7230
Oraprim [A.T.I.] see 9086
Orarsan [Boots] see 49
Orasone [Solvay] see 7904
Oraspor [Ciba-Geigy] see 1993
Orasthin [Hoechst] see 7114
Oratestin [Hoechst] see 4223
Ora-Testryl [Squibb] see 4223
Oratrast [Barnes-Hind] see 1023
Oratren [Bayer] see 7230
Oratrol [Alcon] see 3127
Oravue [Squibb] see 5077
Oraxim [Mcnarini] see 2002
Orazamide, 6993
Orbenin [Beecham] see 2480
Orbenin Dry Cow [Beecham] see 2480
Orbicin [Pfizer/Mack, Illert.] see 3052
Orbinamon [Pfizer] see 9503
Orbit [Ciba-Geigy] see 8003
Orcanette see 252
Orcein, 6994
Orchil see 2682
Orchisterone [Negroni] see 9322
Orchisterone-M [Merck & Co.] see 6206
Orcin see 6995
Orcinol, 6995
Orcinolcarboxylic Acid see 7009
Orciprenaline see 5992
Orcl [Nippon Shinyaku] see 135
Ordeal Bark see 8529
Ordeal Bean see 7539
Ordiflazine see 5507
Ordimel [Lilly] see 59
Ordinary Lactic Acid see 5349
Ordinary Tartaric Acid see 9237
Ordinator [Dausse] see 4028
Oregon Grape Root see 1193
Orelox [Diamant] see 1991
Orencil see 7222
Oreoselone Methyl Ether see 7332
Oresol see 4582
Oreson see 4582
Orestol see 3177
Orestralyn [McNeil] see 3780
Oretic [Abbott] see 4822
Oreton [Schering] see 9322
Oreton Methyl [Schering] see 6206
ORF-10131 see 6796
ORF-11676 see 6447
ORF-15244 see 9544
ORF-15817 see 3561
ORF-15927 see 8395
ORF-17070 see 4760
Orfenso see 6810
Orfiril [Desitin] see 10049
Org-817 see 3654
Org-2969 see 2971

Org-3770 see 6295
Org-6216 see 8392
Org-9426 see 8407
Org-10172 see 5366
Orgabolin (obsolete) [Organon] see 3851
Orgaboral see 3851
Orgadrone [Sankyo] see 2986
Orgametil see 5659
Orgametril [Organon] see 5659
Organidin [Wampole] see 5033
Organoderm [Mundipharma] see 5740
Orgasteron [Organon] see 6805
Orgastyptin [Organon] see 3750
Orgasuline [Organon] see 5011
Orgatrax [Organon] see 4897
Org-GB-94 see 6260
Org-NA-97 see 7139
Org-NC-45 see 10075
Org-OD-14 see 9566
Orgotein see 9177
Oricillin [Grünenthal] see 8002
Oriconazole see 5262
Oricur see 2154
Oriens [Asche] see 76
Oriental Berry see 2518
Orientomycin see 2820
Orifungal M [Janssen] see 5313
Orimeten [Ciba] see 460
Orimon see 7404
Orinase [Upjohn] see 9646
Oriodide [Abbott] see 8778
Oripavine, 6996
Orisul [Ciba] see 9105
Orisulf [Ciba] see 9105
Orix [Biomedica] see 6617
Orizaba Jalap Root see 5088
γ-Orizanol see 7016
Orkanet see 252
ORLAAM [BioDevelopment] see 5492
Orlipastat see 6997
Orlistat, 6997
Orlon [Du Pont], 6998
Ormeloxifene see 2018
Ormeta [Tanabe Seiyaku] see 7002
Ormetein (Rescinded) see 9177
Ormosanine see 6999
21-Ormosanin-20-yl Panamine see 6999
Ormosinine, 6999
Ornicetil [Logeais; Nordmark] see 5314
Ornid see 1395
Ornidal [Selvi] see 7000
Ornidazole, 7000
Ornidyl [Merrell Dow] see 3564
Ornipressin, 7001
Ornithine, 7002
L-Ornithine see 7002
L(+)-Ornithine α-Ketoglutarate see 5314
8-L-Ornithinevasopressin see 7001
Ornitrol [Searle] see 924
Ornoprostil, 7003
Orn(8)-vasopressin see 7001
Orobronze [Applipharm] see 1798
Orofar [Zyma] see 1144
Orofungin [SK & F] see 5891
Oroken [Pharmuka] see 1975
Oronol [Schering] see 917
Oropur see 7004
Orospray [Vetoquinol] see 2245
Orotic Acid, 7004
Orotic Acid Compd with 5(or 4)-Aminoimidazole-4(or 5)-carboxamide (1:1) see 6993
Orotidine, 7005
Orotric [SAT] see 9040
Oroturic see 7004

Orotyl see 7004
Oroxine see 5497
Oroxylin see 7006
Oroxylin A, 7006
Orparan [Chugai] see 7002
Orphenadrine, 7007
Orphol [Sanofi] see 3692
Orpidan [Heumann] see 2123
Orpiment see 846
Orpizin see 2123
Orris, 7008
Orris Root Oil see 7008
Orsanil [Orion] see 9497
Orseilles see 2682
o-Orsellinic Acid, 7009
Orsellinic Acid Methyl Ester 4-Methyl Ether see 8883
Orsile see 1064
Orsin see 7439
Orstanorm [Sandoz] see 3217
Orsudan see 6102
Ortacrone [Pharmainvesti] see 504
Ortal Sodium see 4740
Ortazol see 4320
Ortédrine see 623
Ortensan [Cimex] see 45
Orthanilic Acid, 7010
Orthene [Chevron] see 31
Orthesin see 1116
Ortho 5353 see 1494
Ortho 9006 see 6014
Ortho 12420 see 31
Orthoarsenic Acid see 833
[Orthoborato(3—)-O]phenylmercurate(2—) Dihydrogen see 7456
Orthoboric Acid see 1364
Orthocaine, 7011
Ortho-Cept [Ortho] see 2971
Orthocide-406 see 1815
Orthoclase see 7763
Orthocoll see 7798
Ortho-Creme [Ortho] see 6772
Ortho Cyclen [Ortho] see 6796
Orthodiazinc see 8152
Ortho-Dibrom [Chevron] see 6445
Orthodichlorobenzene see 3106
Orthoform [Winthrop] see 7011
Orthoformic Acid, 7012
Orthoform New see 7011
Ortho-Gynest [Cilag] see 3750
Ortho-Klor see 2129
Ortho-Mite see 805
Ortho-Novin 1/50 [Ortho-Cilag] see 6790
Ortho-Novum 1/35 [Ortho] see 6790
Ortho-Novum 1/50 [Ortho] see 6790
Ortho-Novum 7/7/7 [Ortho] see 6790
Orthophosphoric Acid see 7500
Orthophthalic Acid Didiethylamide see 9344
Ortho Spotless [Ortho] see 3320
Orthotelluric Acid see 9272
Orthotitanic Acid see 9609
Orthotran see 7033
Ortho Tri-Cyclen [Ortho] see 6796
Orthovaleric Acid, 7013
Orthoxenol see 7458
Orthoxine [Upjohn] see 6077
Orthoxycol see 4826
Ortin [Wild] see 9900
Ortisporina [Turro] see 2021
Ortizon (obsolete) [Bayer AG] see 10007
Ortonal see 6028
Ortrel [Ortho] see 6796
Ortyn Retard see 1110
Orudis [Rhône-Poulenc Rorer] see 5316
Orugesic [Rhône-Poulenc] see 5316

Oruvail [Rhône-Poulenc Rorer] *see* 5316
Orvagil [Galenika] *see* 6242
Oryvita [Ijaku] *see* 7016
Oryzaal [Sankei] *see* 7016
Oryzacidin, 7014
Oryzacidin A *see* 7014
Oryzalin, 7015
γ-Oryzanol, 7016
Oryzanol A *see* 7016
Oryzanol C *see* 7016
OS-1897 *see* 3068
OS-2046 *see* 6252
Osacyl *see* 498
Osadrin *see* 7401
Osajin, 7017
Osalmid, 7018
Osarsal *see* 49
Osarsol [Rosse] *see* 49
Osaterone, 7019
Osbil [Upjohn; M & B] *see* 5028
Oscine *see* 8554
Oscorel [SK & F] *see* 5316
O-sialic Acid *see* 8628
Osiren [Hoechst] *see* 8917
Osmarins *see* 7021
Osmaron B, 7020
Osmic Acid *see* 7024
Osmitrol [Baxter] *see* 5788
Osmium, 7021
Osmium Ammonium Chloride *see* 570
Osmium Hexafluoride, 7022
Osmium Octafluoride *see* 7022
Osmium Potassium Chloride *see* 7799
Osmium Tetrachloride, 7023
Osmium Tetroxide, 7024
Osmosal *see* 5788
Osmosin [Merck & Co.] *see* 4998
Osnervan [Burroughs Wellcome] *see* 7944
Ospamox [Biochemie] *see* 617
Ospen (syrup) [Sandoz] *see* 7231
Ospen (tabl.) [Sandoz] *see* 7230
Ospeneff [Sandoz] *see* 7230
Ospolot [Bayer] *see* 9167
Ossalin [Chemipharm] *see* 8762
Ossazone *see* 2247
Ossein *see* 2543
Ossian [Bioindustria] *see* 7079
Ossin [Sulzbach-Neuweiler] *see* 8762
Ossipirina *see* 2247
Ossiten [Robin] *see* 2432
Ossopan *see* 3519
Ostac [Boehringer, Mann.] *see* 2432
Ostelin *see* 10156
Osten [Takeda] *see* 5090
Ostensin [Wyeth] *see* 9838
Osteocalcin, 7025
Osteocalcin [Tosi] *see* 1680
Osteo-F [Colgate-Hoyt] *see* 8762
Osteofix [Chiesi] *see* 5090
Osteofluor [Merck-Clévenot] *see* 8762
Osteol [Yamanouchi] *see* 4385
Osteolite [DuPont] *see* 5837
Osteosarcoma-derived Growth Factor *see* 7683
Osteoscan-HDP [Mallinckrodt] *see* 7074
Osthole, 7026
Ostreogrycin A *see* 10142
Ostreogrycin B *see* 6275
Ostreogrycin B₁ *see* 10096
Ostreogrycin B₂ *see* 10096
Ostruthin, 7027
Ostruthol, 7028
Osvan *see* 1086
Osvarsan *see* 49
Osyritin *see* 8456
Osyritrin *see* 8456
Osyrol [Hoechst] *see* 8917

Otavite *see* 1652
OTC *see* 7084
OTCA *see* 7084
Oterben [Chinoin] *see* 9646
Otifuril *see* 4315
Otobain, 7029
Otobite *see* 7029
Otodyne *see* 10319
Otokalixin *see* 5293
Otreon [Luitpold] *see* 1991
Otrinol [Ciba-Geigy] *see* 3645
Otriven [Ciba] *see* 10219
Otrivin [Ciba] *see* 10219
Otrix [Ciba] *see* 10219
Otto of Rose *see* 6937
OU-1308 *see* 7003
Ouabagenin, 7030
Ouabain, 7031
Ourari *see* 2741
Oust [Du Pont] *see* 9130
Ovaban [Schering] *see* 5849
Ovahormon Benzoate [Teikoku Hormone] *see* 3746
Ovahormon Depot [Teikoku Hormone] *see* 3746
Ovalbumin, 7032
Ovalyse [Upjohn] *see* 4109
Ovanon [Nourypharma] *see* 5659
Ovaras [Serono] *see* 3905
Ovastol *see* 5976
Ovazyme *see* 4468
Ovcon [Mead Johnson] *see* 6790
Overal [Lusofarmaco] *see* 8433
Ovesterin *see* 3750
Ovestin [Organon] *see* 3750
Ovex, 7033
Ovicox [Hoechst] *see* 8488
Ovifollin *see* 3751
Ovin [Mead Johnson] *see* 3268
Oviol [Nourypharma] *see* 2971
Ovisot *see* 88
Ovitelmin [Cilag-Chemie] *see* 5807
Ovochlor *see* 7033
Ovocyclin [Ciba] *see* 3746
Ovocylin [Ciba] *see* 3746
Ovoester *see* 890
Ovoresta [Organon] *see* 5659
Ovotox *see* 7033
Ovotran *see* 7033
Ovotransferrin *see* 9706
Ovo-Vinces [Wolff] *see* 3750
Ovral [Wyeth] *see* 6797
Ovran [Wyeth] *see* 6797
Ovranette [Wyeth] *see* 6797
Ovrette [Wyeth] *see* 6797
O-V Statin [Squibb] *see* 6834
Ovulen *see* 3905
Ovysmen [Ortho-Cilag] *see* 6790
OX-373 *see* 2368
Oxabel [Sarva] *see* 7036
7-Oxabicyclo[2.2.1]heptane-2,3-dicarboxylic Acid *see* 3615
Oxabolone, 7034
Oxabolone Cypionate *see* 7034
Oxacarbazepine *see* 7063
Oxaceprol, 7035
2-Oxachlormadinone Acetate *see* 7019
Oxacillin, 7036
Oxacyclohexadecan-2-one *see* 3956
Oxacyclotetradecane Erythromycin Deriv *see* 8433
1,3,4-Oxadiazole, 7037
2-(1,3,4-Oxadiazol-2-yl)phenol *see* 3994
Oxadiazon, 7038
Oxaflozane, 7039
Oxaflumazine, 7040
Oxaflumine [Diamant] *see* 7040
Oxafuradene (Rescinded) *see* 6621
Oxaine M [Wyeth] *see* 7067

Oxalacetic Acid, 7041
Oxalaldehyde *see* 4519
Oxalamide *see* 7050
Oxalato (1*R*,2*R*-Cyclohexanediamine)platinum(II) *see* 7044
Oxalato (*trans*-*l*-1,2-Diaminocyclohexane)platinum (II) *see* 7044
Oxalatoplatin *see* 7044
Oxalatoplatinum *see* 7044
Oxaldin [Daiichi] *see* 6865
Oxalenediuramidoxime, 7042
Oxalic Acid, 7043
Oxalic Acid Bismuth Salt *see* 1313
Oxalic Acid Diamide *see* 7050
Oxalic Acid Diethyl Ester *see* 3174
Oxalic Acid Dinitrile *see* 2761
Oxalid [USV] *see* 7106
Oxaliplatin, 7044
Oxaloacetic Ester *see* 3882
Oxalodiacetic Acid *see* 5311
Oxalomolybdic Acid, 7045
[Oxalylbis(iminoethylene)]bis[(*o*-chlorobenzyl)diethylammonium Chloride] *see* 396
Oxalyl Chloride, 7046
Oxalylurea *see* 7154
Oxamarin, 7047
Oxametacine, 7048
Oxamic Acid, 7049
Oxamic Acid Hydrazide *see* 8589
Oxamide, 7050
Oxamidic Acid *see* 7049
Oxaminozoline *see* 8388
Oxammonium Hydrochloride *see* 4874
Oxammonium Sulfate *see* 4874
Oxamniquine, 7051
Oxamycin [Merck & Co.] *see* 2820
Oxamyl, 7052
Oxanamide, 7053
Oxandrolone, 7054
Oxantel, 7055
Oxantel Embonate *see* 7055
Oxaphenamide *see* 7018
Oxaphor *see* 4862
Oxapium Iodide *see* 2806
Oxapro [Wyeth] *see* 7057
Oxapropanium Iodide, 7056
Oxaprozin, 7057
Oxa-Puren [Klinge-Nattermann] *see* 7059
Oxarmin *see* 7078
Oxatets [Cox-Continental] *see* 7111
Oxatomide, 7058
Oxatone [Baxter] *see* 3225
Oxazacort *see* 2916
Oxazepam, 7059
Oxazidione, 7060
Oxazocilline *see* 7036
Oxazolam, 7061
Oxazolazepam *see* 7061
Ox Bile Extract, 7062
Oxcarbazepine, 7063
Oxcord [Alcon] *see* 6617
Oxedix [Labaz] *see* 7068
Oxedrine *see* 9189
Oxeladin, 7064
Oxendolone, 7065
Oxenin, 7066
Oxepinac *see* 5257
Oxetacaine *see* 7067
Oxetane *see* 9846
2-Oxetanone *see* 8005
Oxethazaine, 7067
Oxethazine *see* 7067
Oxetorone, 7068
Oxfendazole, 7069
Oxiamine [Made] *see* 5005
Oxiarsolan *see* 7082
Oxibendazole, 7070
Oxibuprokain *see* 1076

Name Index

boxylic Acid 3-Ethyl Ester Disodium Salt *see* 6991
[[2-Oxo-2-[(tetrahydro-2-oxo-3-thienyl)amino]ethyl]thio]acetic Acid *see* 3684
2-Oxo-1,2,3,4-tetrahydroquinoline *see* 4820
DL-S-[2-[N-3-(2-Oxotetrahydrothienyl)acetamido]]thioglycolic Acid *see* 3684
(6R-*trans*)-8-Oxo-7-[(1H-tetrazol-1-yl-acetyl)amino]-3-[(1,3,4-thiadiazol-2-ylthio)methyl]-5-thia-1-azabicyclo-[4.2.0]oct-2-ene-2-carboxylic Acid *see* 1997
N-[4-Oxo-2-(1H-tetrazol-5-yl)-4H-1-benzopyran-8-yl]-4-(4-phenylbutoxy)benzamide *see* 7889
Oxo[[3-(1H-tetrazol-5-yl)phenyl]amino]acetic Acid Butyl Ester *see* 9248
L-2-Oxo-4-thiazolidinecarboxylic Acid, 7084
(R)-2-Oxo-4-thiazolidinecarboxylic Acid *see* 7084
4-Oxo-2-thiazolidinehexanoic Acid *see* 6403
4-Oxo-4-[[4-[(2-thiazolylamino)sulfonyl]phenyl]amino]butanoic Acid *see* 9046
N-[1-Oxo-2-[(2-thienylcarbonyl)thio]propyl]glycine *see* 8963
4-Oxo-2-thionothiazolidine *see* 8351
9-Oxothioxanthene *see* 9507
Oxotremorine, 7085
2-[4-Oxo-3-[5-(trifluoromethyl)benzothiazol-2-ylmethyl]-3,4-dihydrophthalazin-1-yl]acetic Acid *see* 10325
4-[1-Oxo-3-(3,4,5-trimethoxyphenyl)-2-propenyl]-1-piperazineacetic Acid Ethyl Ester *see* 2349
22-Oxovincaleukoblastine *see* 10124
9-Oxoxanthene *see* 10196
Oxpentifylline *see* 7278
Oxphylline [Amido] *see* 3924
Oxprenolol, 7086
Oxsoralen [ICN] *see* 6068
Oxtriphylline *see* 2266
Oxucide *see* 7617
Oxurasin *see* 7618
Oxy-5 [USV] *see* 1149
Oxyacanthine, 7087
OxyBan *see* 4468
Oxybenzene *see* 7390
Oxybenzone, 7088
Oxybenzopyridine *see* 4890
Oxybiocycline *see* 7111
Oxybiotic [Philips Roxane] *see* 7111
β,β'-Oxybis[p-acetophenetidide] *see* 3017
1,1'-Oxybisbenzene *see* 7442
4,4'-Oxybisbenzenecarboximidamide *see* 7352
1,1'-Oxybisbutane *see* 1603
4,4'-Oxybis[2-butanol] *see* 7094
1,1'-Oxybis[2-chloroethane] *see* 3116
Oxybis[chloromethane] *see* 3119
5,5'-Oxybis[3,4-dichloro-2(5H)-furanone] *see* 6380
1,1'-Oxybisethane *see* 3852
N,N'-[Oxybis(2,1-ethanediyloxy-4,1-phenylene)]bisacetamide *see* 3017
2,2'-Oxybisethanol *see* 3168
1,1'-Oxybisethene *see* 10133
1,1'-Oxybis(2-ethoxy)ethane *see* 3167
Oxybismethane *see* 6148
1,1'-Oxybis[2-methoxyethane] *see* 3208
1,1'-Oxybis[3-methylbutane] *see* 5132

1,1'-[Oxybis(methylene)]bis[benzene] *see* 1167
1,1'-[Oxybis(methylene)]bis[4-(hydroxyimino)methyl]pyridinium Dichloride *see* 6835
3,3'-[Oxybis(methylenecarbonylimino)]bis[2,4,6-triiodobenzoic Acid] *see* 5067
1,1'-Oxybis[2-methylpropane] *see* 5156
2,2'-Oxybis[2-methylpropane] *see* 3084
3,3'-[Oxybis[(1-oxo-2,1-ethanediyl)-imino]]bis[2,4,6-triiodobenzoic Acid] *see* 5067
1,1'-Oxybispentane *see* 651
10,10'-Oxybis-10H-phenoxarsine *see* 7095
1,1'-Oxybispropane *see* 8042
2,2'-Oxybispropane *see* 5231
3,3'-Oxybis-1-propene *see* 300
1,1'-Oxybis[2,2,2-trifluoroethane] *see* 4236
Oxybuprocaine *see* 1076
Oxybutynin Chloride, 7089
Oxycamphor *see* 4862
Oxycel [Parke, Davis] *see* 7073
Oxychelidonic Acid *see* 5824
Oxychinolin *see* 4890
Oxychloroquine *see* 4863
Oxychlorosene, 7090
5-Oxychlorpromazine *see* 6987
Oxycinchophen, 7091
Oxyclipine *see* 7998
Oxyclozanide, 7092
Oxycodone, 7093
Oxycon *see* 7093
Oxycyclin [Dumex] *see* 7111
Oxydapatit *see* 1741
N,N-(Oxydiacetyl)bis[3-amino-2,4,6-triiodobenzoic Acid] *see* 5067
Oxydiazepam *see* 9285
Oxydiazol *see* 6032
4,4'-Oxydibenzamidine *see* 7352
4,4'-Oxydi-2-butanol, 7094
2,2'-Oxydiethanol *see* 3168
Oxydiformic Acid Diethyl Ester *see* 8182
1,1'-(Oxydimethylene)bis[4-formyl-pyridinium]dichloride Dioxime *see* 6835
Oxydimethylquinizine *see* 757
10,10'-Oxydiphenoxarsine, 7095
2,4:2',4'-O-(Oxydistibylidyne)bis[D-gluconic Acid] Sb,Sb'-Dioxide Trisodium Salt Nonahydrate *see* 742
Oxydol *see* 4840
Oxydon [R. P. Drugs] *see* 7111
Oxy-Dumocyclin [Dumex] *see* 7111
Oxyephedrin *see* 4869
Oxyethylated Tertiary Octylphenol Formaldehyde Polymer *see* 9964
Oxyethylated Tertiary Octylphenol-polymethylene Polymer *see* 9964
Oxyethylene Oxypropylene Polymer *see* 7721
8-(β-Oxyethyl)methylaminocaffeine *see* 1671
Oxyethyltheophylline *see* 3924
Oxyfan [Coli] *see* 4895
Oxyfedrine, 7096
Oxyflavil *see* 3565
Oxyfluorfen, 7097
Oxygen, 7098
Oxygen Fluoride *see* 4201
Oxygent [Alliance] *see* 7299
3-Oxygermylpropionic Acid Polymer *see* 7979
Oxygeron [Will-Pharma] *see* 10120
Oxyhemerythrin *see* 4674

Oxyhemocyanin *see* 4681
Oxyhemoglobin *see* 4682
Oxyjavanicin *see* 4337
Oxyject [Diamond] *see* 7111
Oxy-L [Woelm] *see* 1149
Oxylag [Lagap] *see* 7111
Oxylan [Burroughs Wellcome] *see* 3365
Oxylidine *see* 8266
Oxylone [Upjohn] *see* 4213
Oxylupanine *see* 4875
Oxymesterone, 7099
Oxymestrone *see* 7099
Oxymetazoline, 7100
Oxymethebanol *see* 3511
Oxymetholone, 7101
Oxymethurea, 7102
Oxymethylene *see* 4262
Oxymethyleneurea *see* 7735
Oxymorphone, 7103
Oxymycin [DDSA] *see* 7111
Oxyneurine *see* 1225
Oxypaat [Katwijk] *see* 7618
Oxypan [Propan-Lipworth] *see* 7111
Oxypangam [Sanorania] *see* 3241
Oxypendyl, 7104
Oxypertine, 7105
Oxypetidin *see* 4883
Oxyphedrine *see* 7096
Oxyphenbutazone, 7106
Oxyphencyclimine, 7107
Oxyphenisatin Acetate, 7108
Oxyphenonium Bromide, 7109
Oxyphyllin [Astra] *see* 9421
Oxypinocamphone, 7110
Oxypinone *see* 7110
Oxypropylated Cellulose *see* 4888
Oxyquinoline *see* 4890
Oxysept [Allergan] *see* 4839
Oxystin [Arzneimittelwerk Dresden] *see* 7114
Oxytetracid *see* 7111
Oxytetracycline, 7111
Oxytetrin [Pitman-Moore] *see* 7111
Oxytheonyl *see* 3924
Oxythiamine, 7112
Oxythioquinox, 7113
Oxytocin, 7114
Oxytrimethylline *see* 2266
Oxyzin (tabl.) *see* 7618
OYE *see* 6961
OZ *see* 7016
γ-OZ *see* 7016
Ozagrel, 7115
Ozex [Toyama] *see* 9692
Ozolinone *see* 3934
Ozone, 7116
P *see* 7963
P₄₅₀ *see* 2861
P-4 *see* 8018
P-7 *see* 5398
p28ˢⁱˢ *see* 7683
P-50 *see* 628
P53 *see* 9383
P-071 *see* 2063
P-71-0129 *see* 4012
P-113 *see* 8518
P-165 *see* 932
P-170 *see* 7335
P-253 *see* 3390
P-286 *see* 5084
P-350 *see* 2324
P-391 *see* 5892
P-607 *see* 2239
P-638 *see* 8130
P-652 *see* 4259
P-725 *see* 7294
P-1011 *see* 3134
P-1134 *see* 7592
P-1496 *see* 10251

Name Index

P-1742 *see* 4225
P-1779 *see* 326
P-2105 *see* 3661
P-2647 *see* 1154
P-3693A *see* 3492
P-4000 *see* 6727
P-4241 *see* 4945
P-4599 *see* 3492
P-5604 *see* 5614
P-7138 *see* 6628
P-720549 *see* 5257
PA-93 *see* 6818
PA-94 *see* 2820
PA-105 *see* 6962
PA-155A *see* 4996
PA-248 *see* 8002
PAA-701 *see* 1242
Paarlan [Lilly] *see* 5220
PABA *see* 443
Pabanol [Elder] *see* 443
Pabenol [Gentili] *see* 2900
Pabestrol D *see* 3177
Pacatal [Warner-Chilcott] *see* 5892
Pacatol *see* 5892
Pacemol [Gemballa] *see* 45
Pacetyn *see* 3552
Paceum [Orion] *see* 3042
Pacilan [Daltan] *see* 1695
Pacinol [Schering] *see* 4226
Pacinox [McNeil] *see* 1818
Pacitane [Cyanamid] *see* 9823
Pacitran [Grossman] *see* 3042
Pacitran (obsolete) [Squibb] *see* 6237
Pacitron [Rorer] *see* 9929
Paclitaxel, 7117
Paclobutrazol, 7118
Pactamycin, 7119
Pacyl [Warner-Lambert] *see* 5258
Padan [Takeda] *see* 1916
PADGEM *see* 8568
Padimate A *see* 3282
Padimate O *see* 3282
Padisal [Bayer] *see* 9441
Padreatin *see* 5291
Padrin [Fujisawa] *see* 7923
Padukrein [Bayropharm] *see* 5291
Padutin [Winthrop] *see* 5291
Paederine *see* 7196
ψ-Paederine *see* 7196
Paediathrocin [Abbott] *see* 3720
PAF *see* 7682
C₁₆-PAF *see* 7682
C₁₈-PAF *see* 7682
PAF-acether *see* 7682
Pagano-Cor [Helopharm] *see* 3753
Paginol [Jugoremedija] *see* 7209
Pagitane Hydrochloride [Lilly] *see* 2825
PAH *see* 462
Paidomal [Malesci] *see* 9421
Paint White *see* 1326
Paka Oil *see* 5675
Palafuge *see* 3365
Palaprin [Nicholas] *see* 316
Palatinol A [Advanced Solvents] *see* 7527
Palatinol M *see* 3304
Palatone [Dow] *see* 5752
Palavale [Otsuka] *see* 3550
Palcin [Sumitomo] *see* 766
Paldesic [R.P. Drugs] *see* 45
Paldomycin [Taiyo] *see* 3496
Pale Catechu *see* 4371
Pale Gentian *see* 4399
Palerol [Sandoz] *see* 9908
Palerol (Pelerol) [Sandoz] *see* 7632
Palestrol *see* 3177
Palfium [Purdue Frederick] *see* 2997
Palinum *see* 2780
Palitantin, 7120

Paliuroside *see* 8456
Palladium, 7121
Palladium(II) Acetate *see* 7123
Palladium Chloride, 7122
Palladium Diacetate, 7123
Palladium Monoxide *see* 7125
Palladium Nitrate, 7124
Palladium Oxide, 7125
Palladous Chloride *see* 7122
Palladous Nitrate *see* 7124
Palladous Oxide *see* 7125
Pallidin [E. Merck] *see* 9104
Palmarosa Oil *see* 6908
Palmatine, 7126
Palm Butter *see* 7129
Palmidrol, 7127
Palmita [Sanwa] *see* 7412
Palmitic Acid, 7128
Palmitic Acid Aluminum Salt *see* 370
Palmitic Acid Ammonium Salt *see* 572
Palmitic Acid Calcium Salt *see* 1734
Palmitic Acid Hexadecyl Ester *see* 2073
Palmitin *see* 9865
Palmityl Alcohol *see* 2070
Palmofen [Zambon] *see* 4280
Palm Oil, from Fruit, 7129
Palm Oil, from Seed, 7130
Palohex *see* 5009
Palosein [Diagnostic Data] *see* 9177
Palphium *see* 2997
Paludrine [Ayerst; ICI] *see* 2138
Palusil *see* 2138
Palustric Acid, 7131
Palux [Taisho] *see* 8063
Palytoxin, 7132
Palytoxin (C51-55 Hemiacetal) *see* 7132
2-PAM *see* 7884
L-PAM *see* 5871
Pamabrom, 7133
Pamacyl *see* 498
Pamaquine, 7134
Pamaquine Embonate *see* 7134
Pamaquine Naphthoate *see* 7134
2-PAM Chloride *see* 7884
Pamedon(e) *see* 3398
Pameion [Simes] *see* 7151
Pamelor [Sandoz] *see* 6812
Pamidronate Disodium *see* 7135
Pamidronic Acid, 7135
Pamine [Upjohn] *see* 6084
Pamisyl Sodium [Parke, Davis] *see* 498
Pamoic Acid, 7136
Pamovin [Frosst] *see* 8206
Panacef [Lilly] *see* 1962
Panacid [Dainippon] *see* 7660
Panacur [Hoechst] *see* 4000
Panadol [Winthrop] *see* 45
Panaldine [Daiichi] *see* 9569
Panaleve [Leo] *see* 45
Panama Bark *see* 8222
Panamicin [Gramon] *see* 3052
Panamine *see* 6999
Panaquilins *see* 4434
Panasorb [Winthrop] *see* 45
Panatus [Krka] *see* 1539
Panax *see* 4434
Panaxosides *see* 4434
Panazon [Toyama] *see* 6752
Panclar [Lematte & Boinot] *see* 2935
Pancodine *see* 7093
Pancoral [Eisai] *see* 4016
Pancorton *see* 169
Pancrease [Ortho-Cilag] *see* 7137
Pancrease [McNeil] *see* 7138
Pancreatic Basic Trypsin Inhibitor *see* 796

Pancreatic Desoxyribonuclease *see* 2953
Pancreatic Dornase *see* 2953
Pancreatic Trypsin Inhibitor (Kunitz) *see* 796
Pancreatin, 7137
Pancreatopeptidase E *see* 3576
Pancrelipase, 7138
Pancreozymin *see* 2253
Pancreozymin C-Terminal Octapeptide *see* 8689
Pancrex-Vet [Paines & Byrne] *see* 7137
Pancuronium Bromide, 7139
Pandermite *see* 1691
Pandrocine *see* 6207
Panectyl [Rhone-Poulenc Rorer] *see* 9834
Panediol [Therap. Pharm] *see* 5908
Panergon [Mack, Illert.] *see* 7151
Panets [Optrex] *see* 45
Panex [Mallard] *see* 45
Panflavin [Chinosolfabrik] *see* 125
Panformin [Shionogi] *see* 1500
Panfungol [Esteve] *see* 5313
Pangamic Acid, 7140
Pangerin *see* 3254
Pangram [Virbac] *see* 4398
Panheprin [Abbott] *see* 4685
Panimit [Nattermann] *see* 1521
Panimycin [Meiji] *see* 3052
Panipenem, 7141
Pankrin, 7142
Pankrotanon [Hausmann] *see* 7137
Panmycin [Upjohn] *see* 9337
Panmycin P [Upjohn] *see* 9337
Panodil [Sterling-Winthrop] *see* 45
Panolid [Sandoz] *see* 3802
Panophylline [Panray] *see* 9421
Panoral [Lilly] *see* 1962
Panoxolin *see* 7062
PanOxyl [Stiefel] *see* 1149
Panparnit [Geigy] *see* 1820
Panpurol [Nippon Shinyaku] *see* 7624
Panrone [Panray] *see* 9427
Pansporin [Takeda] *see* 1985
Pansporine [Cassenne-Takeda] *see* 1985
Pansporin T [Takeda] *see* 1985
Pantelmin [J & J] *see* 5807
Pantenyl [Kay] *see* 2988
Pantestone [Organon] *see* 9322
Pantetheine, 7143
Pantethine, 7144
Pantetina [Maggioni] *see* 7144
Panthecin [Sawai] *see* 7144
Pantheline [CP Protea] *see* 7989
Panthenol *see* 2988
Pantherine *see* 6391
Panthesin [Sandoz] *see* 5476
Panthoderm [USV] *see* 2988
Pantholin [Lilly] *see* 7147
Pantocaine [Hoechst] *see* 9330
Pantocid *see* 4620
Pantofenicol [Promesa] *see* 2120
Pantogam *see* 4781
Pantoic Acid γ-Lactone *see* 7145
Pantoic Lactone *see* 7145
Pantolactone, 7145
Pantomicina [Abbott] *see* 3726
Pantomin [Daiichi] *see* 7144
Pantopaque [Lafayette] *see* 5074
Pantopon [Roche] *see* 7150
Pantoprazole, 7146
Pantosediv *see* 9390
Pantosin [Daiichi] *see* 7144
Pantostrep *see* 3222
Pantothenic Acid, 7147
Pantothenol *see* 2988
Pantothenyl Alcohol *see* 2988

N-(Pantothenyl)-β-aminoethanethiol
see 7143
Pantothenylol see 2988
Pantovernil [Bristol-Myers Squibb]
see 2120
Pantoyl Lactone see 7145
N-Pantoyl-3-propanolamine see 2988
Panurin see 4822
Panwarfin [Abbott] see 10174
Panzid [Duncan] see 1995
Panzytrat [Knoll] see 7137
Papain, 7148
Papaveraldine, 7149
Papaveretum, 7150
Papaverine, 7151
Papaw see 7152
Papaya, 7152
Papayotin see 7148
Paper Red 4B [Crompton & Knowles]
see 1132
Papital T.R. [Zemmer] see 7151
Papoose Root see 1957
PAPP see 490
Papuamine, 7153
Parabanic Acid, 7154
Parabencil see 3365
Parabis [Dow] see 3120
Parabolan [Negma] see 9716
Parabolin see 3175
Parabromdylamine see 1467
(+)-Parabromdylamine see 1467
Paracarbinoxamine see 1845
Paracetaldehyde see 7160
Paracetamol see 45
Parachloramine see 5817
Parachlorometacresol see 2184
Parachlorometaxylenol see 2228
Parachlorophenol see 2206
Paracide see 3107
Paracipan see 498
Paracodin [Knoll] see 3214
Paracort [Parke, Davis] see 7904
Paracortol [Parke, Davis] see 7901
Paradiazine see 8141
Paradichlorobenzene see 3107
Paradione [Abbott] see 7161
Paradise Plant see 6258
Paradormalene see 6027
Paraffin, 7155
Paraffin Chlorinated, 7156
Paraffin Jelly see 7326
Paraffin Wax see 7155
Paraflex [McNeil] see 2249
Paraflu [Dainippon] see 4167
Paraflutizide, 7157
Paraform see 7158
Paraformaldehyde, 7158
Paraguay Tea see 5796
Paraherquamide, 7159
(—)-Paraherquamide see 7159
Parahexyl see 9190
Paral [Forest] see 7160
Paralactic Acid see 5350
Paralaudin see 3011
Paraldehyde, 7160
Paralergin [Vita] see 891
Paralest [Parachemie] see 9823
Paralgin see 3414
Paralkan see 1086
Paralytic Shellfish Poison see 8533
Paramandelic Acid see 5757
Paramelaconite see 2713
Paramenyl see 5683
Paramethadione, 7161
Paramethasone, 7162
Paramezone [Recordati] see 7162
Paramicina [Raglionieri] see 7173
Paramidin [Takeda] see 1487
Paraminol see 443
Paraminyl [Cooper] see 8168

Paramisan Sodium [Smith & Nephew]
see 498
Paramorfan see 3220
Paramorphan see 3220
Paramorphine see 9411
Paramoth [Esquire] see 3107
Paramycin see 498
Paraniazide [Hepatrol] see 5203
Paranol see 482
Paranyline, 7163
Paraoxon, 7164
Paraoxypropiophenone see 7176
Para-Pas [Glenwood] see 498
Paraphos see 7167
Paraplatin [Bristol-Myers] see 1870
Paraquat, 7165
Paraquat I see 7165
Parasal [Panray] see 498
Parasalicil see 498
Parasalindon see 498
Parasan see 1055
Parasepiolite see 5727
Parasitex [Virbac] see 3043
Parasiticol see 178
Parasorbic Acid, 7166
Paraspen [Fisons] see 45
Parasympatol see 9189
Paratartaric Acid see 9236
Paratect [Pfizer] see 6348
Paratensiol [Yoshitomi] see 8313
Parathar [Rorer] see 9309
Parathesin see 1116
Parathiazan see 9435
Parathion, 7167
Parathion-methyl see 6183
Parathorm [Hormone-Chemie] see
7168
Parathormone see 7168
Para-thor-mone [Lilly] see 7168
Parathyroid Hormone, 7168
Paravermin see 7617
Parawollastonite see 1749
Paraxenol see 7459
Paraxin [Boehringer, Mann.] see 2120
Paraxin Succinate A [Yamanouchi]
see 2120
Para-zene [Reliable Chem.] see 3107
Parazine see 7617
Parbendazole, 7169
Parbetan [Parke, Davis] see 1226
Parda [Parke, Davis] see 5490
Pardisol see 3793
Pardroyd [Parke, Davis] see 7101
Paredrine [SK & F] see 4855
Paredrinex [SK & F] see 4855
Paredrinol [SK & F] see 7485
Pareira, 7170
Pareira Brava see 7170
Parelan [Elan] see 45
Parenabol [Ciba] see 1354
Parenamine [Winthrop] see 8071
Parenogen [Miles] see 4116
Parentamin [Pierrel] see 8071
Parenzyme [Marion Merrell Dow] see
9926
Parenzymol [Horner] see 9926
Parest [Parke, Davis] see 6028
Parethoxycaine, 7171
Parfenac [Lederle] see 1497
Parfenal [Cyanamid] see 1497
Parfuran [Parke, Davis] see 6696
Pargin [Gibipharma] see 3550
Pargitan [Abigo] see 9823
Pargonyl see 7173
Pargyline, 7172
Paricina [Archifar] see 7173
Parid [Cooper] see 7617
Parietic Acid see 8342
Parietin see 3602
Parietrope [Marcofina] see 8216

Parigenin see 8526
Parinase [Upjohn] see 4524
Pariprazole see 8272
Paris Blue see 4064
Paris Green see 2692
Paris Red see 5449
Paris White see 1697
Paritane [Berk] see 7086
Parizac [Lacer] see 6977
Parkazin see 3157
Parkemed [Parke, Davis] see 5842
Parkin [Yoshitomi] see 3793
Parkinane [Lederle] see 9823
Parkinsan [Byk-Gulden] see 1491
Park Lily see 2574
Parkopan [Salutas Fahlberg List] see
9823
Parks 12 [Hommel] see 7922
Parlay [ICI] see 7118
Parlef [Parke, Davis] see 4167
Parlodel [Sandoz] see 1437
Parlodion [Mallinckrodt] see 8195
Parmal [Central] see 8168
Parmid [Lagap] see 6226
Parminal see 6028
Parmol [Rorer] see 45
Parnate [Procter & Gamble] see 9708
Parodyne see 757
Paroidin [Parke, Davis] see 7168
Paroleine see 7327
Paromomycin, 7173
Paromomycin I see 7173
Parotin, 7174
Paroven [Zyma] see 9920
Paroxetine, 7175
Paroxon see 7176
Paroxyl see 49
Paroxypropione, 7176
Parpanit [Geigy] see 1820
Parphezein see 3793
Parphezin see 3793
Parpon see 1055
Parsal [Midy] see 7177
Parsalmide, 7177
Parsidol [Warner-Chilcott; Specia] see
3793
Parsitan see 3793
Parsley Apiole see 777
Parsley Camphor see 777
Parsley Fruit see 7178
Parsley Seed, 7178
Parsley Seed Oil see 6928
Parsol 1789 [Givaudan] see 1616
Parsol A [Givaudan] see 1616
Parsol MCX [Givaudan] see 6864
Parsol MOX [Givaudan] see 6864
Parsotil see 3793
Partel [Lilly] see 3434
Parterol see 3223
Parthenicin see 7179
Parthenin, 7179
Parthenolide, 7180
Partner [Monsanto] see 58
Partocon [Ferring] see 7114
Partrex [Paramedical] see 9337
Partricin, 7181
Partricin A see 7181
Partricin B see 7181
Partricin Methyl Ester see 5891
Partusisten [Boehringer, Ing.] see 4022
Parvaquone, 7182
Parvex [Upjohn] see 7548
Parvolex [Duncan, Flockhart] see 89
Parzate see 6426
Parzate C [Du Pont] see 10300
Parzone [Mallinckrodt] see 3214
PAS see 498
Pasaden [Homburg] see 4774
Pasalin [Takeda] see 3181
Pasalon [Bayer] see 498

Pasara Sodium [Dorsey] see 498
PASER [Jacobus] see 498
Pasetocin [Kyowa] see 617
Pasid see 498
Pasiniazide see 5203
Paskalium [Glenwood] see 498
Paskate [Lilly] see 498
Pasmed Sodium see 498
Pasmicina see 498
Pasmus [Daiichi] see 4142
Pasolac see 498
Pasolind N [Stada] see 45
Paspaline-P see 2859
Paspertin [Kali-Chemie] see 6226
Pasque Flower see 8125
Passiflora, 7183
Passiflorin see 4646
Passion Flower see 7183
Passion Vine see 7183
Pastaron [Sato] see 10005
Patchouli Alcohol, 7184
Patchouli Camphor see 7184
Patchouli Oil see 6929
Patent Alum see 381
Paternoite see 5693
Pathilon [Lederle] see 9794
Pathocidin see 928
Pathocil [Wyeth] see 3134
Pathozone [Pfizer] see 1981
Patoran [Ciba] see 6224
Patricin [Kaken] see 4135
Patrol [ICI] see 4034
Patronite see 10053
Patrovina see 160
Pattonex [Makhtcshim-Agan] see
  6224
Patulin, 7185
Paucimycin see 7173
Pavabid [Marion] see 7151
Pavacap [Reid-Provident] see 7151
Pavacen [Central Pharm.] see 7151
Pavadel [Canright] see 7151
Pavagen [Darby] see 7151
Pavakey [Key] see 7151
Pavased [Mallard] see 7151
Pavecef [IBP] see 1964
Paveril [Lilly] see 3318
Paviin see 4290
Pavoninin-5, 7186
Pavoninins see 7186
Pavulon [Organon] see 7139
Paxate [Mead Johnson] see 3042
Paxel (obsolete) [Elliott-Marion] see
  3042
Paxeladine [Beaufour] see 7064
Paxidorm see 6028
Paxil [SKB] see 7175
Paxilon [Fisons] see 6032
Paxipam [Schering] see 4619
Paxistil see 4897
Paxital [Warner-Chilcott] see 5892
Paylean [Lilly] see 8275
Pay-Off [Am. Cyanamid] see 4160
Payzone [Am. Cyanamid] see 6752
Pazinaclone, 7187
Pazufloxacin, 7188
Pazufloxacin Mesylate see 7188
PB-89 see 4258
PB-868Cl see 5674
PBB see 4124
PBBs see 7724
PB-NOX [Roussel-Bio] see 7629
PBP see 4124
PBZ [Ciba-Geigy] see 9868
PC see 7496
PC 1 see 4016
PC-904 see 766
PC-1238 see 3612
PC-1421 see 7615
PCBs see 7725

PCC see 8157
PCNB see 8264
P-Component see 657
PCP see 7242 and 7364
PCV see 7210
PD-93 see 7660
PD-107779 see 3625
PD-109452-2 see 8233
PD-127391 see 2413
PD-131501 see 8884
PDB see 3107
3-PDC see 7244
PDDB see 3474
PDGA see 3344
PDGF see 7683
P-DHP see 239
P-DMEA see 2935
PDP see 6290
PDX Chloride see 7716
PEA see 7371
Peach Oil, Expressed, 7189
Peak [Ciba-Geigy] see 8067
Peanut, 7190
Peanut Oil, 7191
Pearl Ash see 7781
Pearl White see 1303
Pear Oil see 5125
Pearson's Creolin see 2639
Peau d'Espagne see 3814
PEBC see 7192
β-PEBG see 7376
Pebulate, 7192
Pecazine see 5892
Pecilocin, 7193
Pecitrol Veinogène see 1269
Pecnon [Sanken] see 5300
Pecram [Zyma] see 485
Pect [Rentschler] see 401
Pectamol [BDH] see 7064
Pectamon see 7064
Pectenine see 1896
Pectin, 7194
Pectin Sugar see 799
Pectipront see 416
Pectobloc [Siegfried] see 7597
Pectolin [Faulding] see 7484
Pectolinarigenin, 7195
Pectolinarin see 7195
Pectolitan [Kettelhack] see 2106
Pectox [Nattermann] see 1850
Pedameth [Forest] see 6053
Pederin, 7196
Pederine see 7196
ψ-Pederine see 7196
PEDG see 7376
Pediamycin [Ross] see 3720
Pediaprofen [J & J] see 4925
Pedisafe [Sagitta] see 2478
Peflacine [Bellon] see 7197
Peflox [Formenti] see 7197
Pefloxacin, 7197
Pefloxacine see 7197
Pefloxacin Mesylate see 7197
PEG see 7729
Peganine see 10072
Peganone [Abbott] see 3795
PEG-asparaginase see 871
Pegaspargase see 871
Pegasyl [SweVet] see 4839
PEG Fatty Acid Esters see 7738
PEG-400 Monostearate see 7738
Pegorgotein see 9177
Pegosperse [Lonza] see 7738
Pegosperse 400-MS [Lonza] see 7738
Pegosperse 1750-MS [Lonza] see 7738
PEG-SOD see 9177
Pegu Catechu see 1951
Peimine see 10104
Peiminoside see 10104
Pelanin Benzoate [Mochida] see 3746

Pelanin Depot [Mochida] see 3746
Pelargidanon 1602 see 6511
Pelargidenolon 1497 see 5288
Pelargidenon 1449 see 773
Pelargonic Acid, 7198
Pelargonidin, 7199
Pelargonidin-3-(6-p-coumaroyl)gluco-
  sido-5-glucoside Chloride see 6327
Pelargonin see 7199
Pelentan see 3818
Peligal see 5977
Pellagra Preventive Factor see 6612
Pelletierine, 7200
(±)-Pelletierine see 7200
Pellidol see 3004
Pellitorine, 7201
Pellotine, 7202
Pelmin see 6574
Pelosine see 1042
PELS see 3722
Pelson see 6667
α-Peltatin, 7203
β-Peltatin, 7204
β-Peltatin A see 7204
Peltophorin see 1197
Peluces [Isei] see 4629
Pemal [Benzon] see 3794
Pemilaston [Bristol-Myers Squibb] see
  7205
Pemirolast, 7205
Pemix [Prodes] see 7662
Pemoline, 7206
Pempidil [Farmigea] see 7207
Pempidine, 7207
Pempiten [I.G.I.] see 7207
Pen A [Pfizer] see 628
Pen-A-Brasive see 1178
Penadur [Wyeth] see 7221
Penagen [Genethic] see 7230
Penamecillin, 7208
Penamox [Beecham] see 617
Pen A/N [Pfizer] see 628
Penaquacaine G see 7226
Penavlon V [ICI] see 7230
Penbristol [Bristol] see 628
Penbritin [Beecham] see 628
Penbritin-S [Ayerst] see 628
Penbrock [Beecham] see 628
Penbutolol, 7209
Pencal [Pennwalt] see 1686
Pencal [Tatsumi] see 1840
Penciclovir, 7210
Pencompren [E. Merck] see 7230
Pen-Di-Ben see 7221
Pendimethalin, 7211
Pendiomid [Ciba] see 929
Penditan see 7221
Pendramine [E. Merck] see 7214
Penecort [Allergan] see 4828
Penetek see 7245
Penethamate Hydriodide, 7212
Penethecillin see 7212
Penetracyne see 7235
Penetrex [Warner-Lambert] see 3625
Pen-Fifty see 7226
Penfluridol, 7213
Pengitoxin see 4441
Penglobe [Astra] see 961
Penicals [Leo Pharm] see 7230
Penicidin see 7185
Penicillamine, 7214
Penicillamine Cysteine Disulfide,
  7215
Penicillamine Disulfide, 7216
Penicillanic Acid, 7217
Penicillanic Acid 1,1-Dioxide see 9058
Penicillanic Acid Sulfone see 9058
Penicillic Acid, 7218
Penicillin see 1178
Penicillin-152 see 7369

Pentanitrine *see* 7249
Pentanoic Acid *see* 10042
Pentanoic Acid Ammonium Salt *see* 604
Pentanoic Acid Zinc Salt *see* 10299
1-Pentanol, 7257
2-Pentanol, 7258
3-Pentanol, 7259
tert-Pentanol *see* 7282
2-Pentanone *see* 6193
3-Pentanone *see* 3170
Pentanyl *see* 4043
2,5,8,11,14-Pentaoxapentadecane *see* 9348
Pentaphen *see* 7284
Pentaphene *see* 3057
Pentapiperide, 7260
Pentapiperide Methylsulfate *see* 7260
Pentapiperium Methylsulfate *see* 7260
Pentapyrrolidinium Bitartrate *see* 7274
Pentasa [Ferring] *see* 5964
Pentasodium 4,4'-[(3-Sulfo-4,4'-biphenylene)bis(azo)]bis(3-amino-2,-7-naphthalenedisulfonate) *see* 9924
Pentasodium Triphosphate *see* 8846
1,2,3,5,6-Pentathiepane *see* 5461
Pentazocine, 7261
Pentcillin [Toyama-Sankyo] *see* 7616
Pentedrin *see* 9189
Pentek *see* 7245
1-Pentene, 7262
2-Pentene, 7263
2-Pentenylpenicillin Sodium, 7264
Δ-β,γ-Pentenylpenicillin Sodium *see* 7264
Pentetate Calcium Trisodium, 7265
Pentethylcyclanone *see* 2775
Pentetic Acid, 7266
Pentetrazol *see* 7283
Pentetreotide, 7267
Penthamil [Geigy] *see* 7265
Penthienate Bromide, 7268
Penthiobarbital Sodium *see* 9487
Penthonium *see* 7253
Penthrane [Abbott] *see* 6072
Penthrit *see* 7249
Penticainide *see* 7271
Penticort [Lederle] *see* 407
Pentid *see* 7225
Pentifylline, 7269
Pentigetide, 7270
Pentilium *see* 7274
Pentione *see* 5223
Pentisomide, 7271
Pentitrate *see* 7249
Pentlandite *see* 6582
Pentobarbital, 7272
Pentobarbital Calcium *see* 7272
Pentobarbital Sodium *see* 7272
Pentobarbitone *see* 7272
Pentofuryl [Karlspharma] *see* 6626
1-Pentol, 7273
Pentolinium Tartrate, 7274
Pentolonium Bitartrate *see* 7274
Pentone [Faulding] *see* 7272
Pentorex, 7275
Pentosalen *see* 4960
Pentosan Polysulfate, 7276
Pentostam [Burroughs Wellcome] *see* 742
Pentostatin, 7277
Pentothal Sodium [Abbott] *see* 9487
Pentovis [Parke, Davis] *see* 8238
Pentoxifylline, 7278
Pentoxiverin *see* 1840
Pentoxyl, 7279
Pentoxyverine *see* 1840
Pentoyl [Morishita] *see* 3603
Pentral 80 *see* 7249

Pentrane *see* 6072
Pentrex [Banyu] *see* 628
Pentrexyl [Bristol] *see* 628
Pentrinitrol, 7280
Pentrite *see* 7249
Pentritol [Armour Pharm.] *see* 7249
Pentryate *see* 7249
Pentryl, 7281
D-*erythro*-2-Pentulose *see* 8374
threo-Pentulose *see* 10221
Pentyde [Immunetech] *see* 7270
Pentyl Alcohol *see* 7257
tert-Pentyl Alcohol, 7282
Pentylamine *see* 639
2-(Pentylamino)ethanol 4-Aminobenzoate (Ester) *see* 6435
2-*n*-Pentylaminoethyl *p*-Aminobenzoate *see* 6435
Pentylbenzene *see* 641
Pentylcarbinol *see* 4732
6-Pentyl-*m*-cresol *see* 648
Pentylenetetrazole, 7283
Pentyl Ether *see* 651
Pentyl Hexanoate *see* 646
tert-Pentyl Isovalerate *see* 652
[2R-[2α(Z),3α]]-11-(3-Pentyloxiranyl)-9-undecenoic Acid *see* 10099
[2S-[2α(Z),3α]]-11-(3-Pentyloxiranyl)-9-undecenoic Acid *see* 10099
11-(3-Pentyloxiranyl)-9-undecenoic Acid *see* 10099
p-tert-Pentylphenol, 7284
Pentymal *see* 607
Pen-Vee [Wyeth] *see* 7231
Pen-Vee K [Wyeth] *see* 7230
Peon [Zeria] *see* 10244
Peonidin, 7285
Peonin *see* 7285
PEP *see* 7727
Pepcid [Merck & Co.] *see* 3972
Pepcidina [Merck & Co.] *see* 3972
Pepcidine [Merck & Co.] *see* 3972
Pepdine [Merck & Co.] *see* 3972
Pepdul [Merck & Co.] *see* 3972
Pepleo Injection [Nippon Kayaku] *see* 7286
Pepleomycin *see* 7286
Peplomycin, 7286
Pepo *see* 8128
Pepper, 7287
Pepperidge Bush *see* 988
Peppermint, 7288
Peppermint Camphor *see* 5882
Pepsamar [Winthrop] *see* 355
Pepsdol *see* 4477
Pepsin, 7289
Pepstatin, 7290
Pepstatin A *see* 7290
Pepstatin B *see* 7290
Pepstatin C *see* 7290
Peptan [Vianex] *see* 3972
Peptard [Riker] *see* 4907
Peptarom [Fresenius] *see* 10026
Peptavlon [Ayerst] *see* 7250
Pepticum [Andromaco] *see* 6977
Peptide T, 7291
Peptidoglycan *see* 6384
Peptilate [Dojin] *see* 224
Peptinimid [Iromedica] *see* 3794
Peptol [Horner] *see* 2337
Peptonized Iron, 7292
Per-Abrodil *see* 5061
Peracef [Pfizer] *see* 1981
Peracetic Acid, 7293
Peracon [Kali-Chemie] *see* 5124
Peragit [AFI] *see* 9823
Perandren (amp.) [Ciba-Geigy] *see* 9322
Perandren (Lozenges) [Ciba] *see* 6206
Perandren M [Ciba] *see* 9322

Peraprin [Taiyo] *see* 6226
Perazil [Burroughs Wellcome] *see* 2128
Perazine, 7294
Perazolin [Zenyaku Kogyo] *see* 8708
Perazyl *see* 2128
Perbenzoic Acid, 7295
Percaine *see* 3081
Percapyl *see* 2154
Perchloracap [Mallinckrodt] *see* 7822
Perchloric Acid, 7296
Perchloric Anhydride *see* 2147
Perchlorobenzene *see* 4714
Perchloroethane *see* 4715
Perchloroethylene *see* 9332
Perchloromethane *see* 1864
Perchloron *see* 1717
Perchloropentacyclo[5.2.1.0^{2,6}.0^{3,9}.-0^{5,8}]decane *see* 6294
Perchloryl Fluoride, 7297
Perclene *see* 9332
Percoccide [A.C.F.] *see* 9081
Percodan [Endo] *see* 7093
Percorten [Ciba] *see* 2948
Percutol [Reckitt & Colman] *see* 6704
Perderm [Schering] *see* 221
Perdiem Fiber [Rhone-Poulenc Rorer] *see* 7674
Perdilatal [Smith & Nephew] *see* 6830
Perdipina [Sandoz] *see* 6579
Perdipine [Syntex] *see* 6579
Perdolat [Knoll] *see* 7214
Perdorm [Lagap] *see* 9285
Perduren *see* 9475
Perebral [Beytout] *see* 2771
Perebron [Angelini Francesco] *see* 7078
Pereirine *see* 4386
Peremesin *see* 5817
Peremine [Chinoin] *see* 4756
Pérénan [Sanofi Winthrop] *see* 3692
Perenex [ICI] *see* 2734
Perenum [Delalande] *see* 9659
Perequil [Lepetit] *see* 5908
Perezone, 7298
Perfan [Lepetit] *see* 3627
Perfane [Merrell] *see* 3627
Perfekthion [BASF] *see* 3269
Perflan [DowElanco] *see* 9255
Perfluamine *see* 4221
Perflubron, 7299
Perfluidone, 7300
Perflunafene *see* 4221
Perfluoroacetic Acid *see* 9812
Perfluorobenzene *see* 4722
Perfluoroctyl Bromide *see* 7299
Perfluorocyclobutane *see* 6841
Perfluorodecalin *see* 4221
Perfluorotripropylamine *see* 4221
Perforin, 7301
Performic Acid, 7302
Perfosfamide, 7303
Perfusamine [Medi-Physics] *see* 5066
Pergamid [Scios-Nova] *see* 7303
Pergitral *see* 7249
Pergolide, 7304
Pergonal [Serono] *see* 4299
Pergotime *see* 2446
Perhexilene *see* 7305
Perhexiline, 7305
Perhydrit [E. Merck] *see* 10007
N-(2-Perhydroazocin-1-ylethyl)-guanidine *see* 4589
Perhydrol-Urea [E. Merck] *see* 10007
Perhydronaphthalene *see* 2903
Perhydrosqualene *see* 8923
Peri Acid *see* 6492
Periactin [Merck & Co.] *see* 2842
Periactinol [Merck & Co.] *see* 2842
Perichthol *see* 4929

Periciazine *see* 7306
Periciclina *see* 2937
Periclase *see* 5713
Periclor [Ives] *see* 7246
Peri-Colace *see* 1929
Peri-Colace [Mead Johnson] *see* 3460
Pericyazine, 7306
Peridamol [Boehringer, Ing.] *see* 3410
Peridex [Procter & Gamble] *see* 2140
Peridon [Italchimici] *see* 3476
Peridys [Pharmuka] *see* 3476
Perifadil [Sterling-Winthrop] *see* 9120
Perifunal [Aristegui] *see* 7673
Perilax [Nordex] *see* 1282
Perilla Ketone, 7307
Perillaldehyde, 7308
l-Perillaldehyde α-*anti*-Oxime *see* 7308
l-Perillaldehyde α-*syn*-Oxime *see* 7308
Perillartine *see* 7308
"Perilla Sugar" *see* 7308
Perimetazine *see* 7309
Perimethazine, 7309
Perimycin, 7310
Perimycin A *see* 7310
Perin [Endo] *see* 7617
Perindopril, 7311
Perindoprilat *see* 7311
Perindopril Erbumine *see* 7311
Periodic Acid, 7312
Periodyl, 7313
Periograf *see* 3519
Periplanone A *see* 7314
Periplanone B *see* 7314
Periplanones, 7314
Periplobiose *see* 9017
Periplocin, 7315
Periplocoside *see* 7315
Periplocymarin, 7316
Periplogenin, 7317
Periplum [Italfarmaco] *see* 6643
Peripress [Pfizer] *see* 7897
Perisalol [Chugai] *see* 6608
Perisoxal, 7318
Peristil [Recofarma] *see* 6347
Peristim [Mead Johnson] *see* 1929
Periston [Bayer] *see* 7879
Peritol [EGYT] *see* 2842
Peritrate [Warner-Chilcott] *see* 7249
Perityl *see* 7249
Perivine, 7319
Perketan [Inverni] *see* 5307
Perlacton [Chassot] *see* 7114
Perlapine, 7320
Perlatan *see* 3751
Perlinganit [Schwarz] *see* 6704
Perlon *see* 6832
Perlopal *see* 5890
Perlutex [Leo Pharm.] *see* 5838
Permanent Yellow *see* 999
Permanganic Acid Potassium Salt *see* 7824
Permapen [Pfizer] *see* 7221
Permastril [Cassenne] *see* 3504
Permax [Lilly] *see* 7304
Permeability Glycoprotein *see* 7335
Permease *see* 4794
Permethanoic Acid *see* 7302
Permethrin, 7321
Permiran [Laphal] *see* 10141
Permitil [Schering] *see* 4226
Permonid [Roche] *see* 2972
Pernaemon [Organon] *see* 5575
Pernaemyl *see* 5575
Pernambuco, 7322
Pernazene *see* 9965
Pernexin *see* 5575
Perniciosan *see* 5575
Pernovin [Chinoin] *see* 7380
Perocan [Toyo Jozo] *see* 5124

Perocef [Pulitzer] *see* 1981
Peroidin *see* 7822
Perolysen [M & B] *see* 7207
Peronine *see* 1176
Peronine Myristate *see* 6420
Perovskite *see* 9612
1,4-Peroxido-*p*-menthene-2 *see* 864
Peroxinorm [Grünenthal] *see* 9177
Peroxyacetic Acid *see* 7293
Peroxybenzoic Acid *see* 7295
Peroxyderm [Chassot] *see* 1149
Peroxydex [Virbac] *see* 1149
Peroxyformic Acid *see* 7302
Peroxymonosulfuric Acid *see* 1900
Perparin [Chemopuro] *see* 3773
Perperine *see* 3773
Perphenan *see* 7323
Perphenazine, 7323
Perphenazine 3,4,5-Trimethoxybenzoate *see* 6228
Per-Radiographol *see* 5061
Perrhenyl Chloride *see* 8346
Persadox [Texas Pharmacal] *see* 1149
Persa-gel [Ortho] *see* 1149
Persantine [Boehringer, Ing.] *see* 3410
Persedon [Roche] *see* 8179
Persian Bark *see* 1931
Persian Insect Powder *see* 8150
Persian Red *see* 5423
Persic Oil *see* 7189
Persio *see* 2682
Persisten *see* 3155
Persistol [Chevron] *see* 9803
Persolv [Lepetit] *see* 10024
Perspex *see* 6005
Persulfuric Acid *see* 1900
Pertestis [Orma] *see* 9322
Pertofran [Geigy] *see* 2966
Pertofrane [USV] *see* 2966
Pertranquil [Lepetit] *see* 5908
Pertrombon *see* 3858
Peruvian Balsam *see* 976
Peruvian Bark *see* 2343
Peruviol *see* 6561
Peruvoside, 7324
Perval [Lepetit] *see* 10120
Pervetral [Asta] *see* 7104
Pervincamine [Dausse] *see* 10120
Pervitin [Temmler] *see* 6015
Perviton *see* 7358
Pervone [Millot] *see* 10120
Perycit [Bofors] *see* 6581
Perylene, 7325
PES *see* 5645
Peson [Hoechst] *see* 5645
Pesos [Valeas] *see* 4015
Pestox II [FBC] *see* 8540
Pestox XIV (obsolete) [Fisons] *see* 3250
Pestox XV [Fisons] *see* 6292
PET *see* 7730
Petalite *see* 5543
Pet-Derm III [Beecham] *see* 2986
Peteha [Fatol] *see* 8076
Peter-Kal *see* 7783
Peterphyllin *see* 485
Pethidine *see* 5894
Petidiol *see* 3762
Petidion [Gerot] *see* 3762
Petidon *see* 9836
Petimin *see* 5727
Petinimid *see* 3794
Petinutin [Parke, Davis] *see* 6085
Petisan [Gerot] *see* 3762
Petit Muguet *see* 10185
PETN *see* 7249
Petnidan [Desitin] *see* 3794
Petrichloral *see* 7246
Petrin [Parke, Davis] *see* 7280

Petrohol *see* 5227
Petrolatum, 7326
Petrolatum, Liquid, 7327
Petrol (British) *see* 4382
Petroleum, 7328
Petroleum Benzin, 7329
Petroleum Jelly *see* 7326
Petroleum Naphtha *see* 7329
Petroleum Spirits *see* 6288
Petroselic Acid *see* 7330
Petroselinic Acid, 7330
Petrosulpho *see* 4929
Petty-morrel *see* 804
Petunidin, 7331
Petunidol *see* 7331
Petunin *see* 7331
Petzite *see* 4526
Peucedanin, 7332
Pevaryl [Cilag] *see* 3550
Pexid [Merrell] *see* 7305
Peyocactine *see* 4783
Peyonine, 7333
Peyote *see* 5965
Peyrone's Chloride *see* 2378
Peyrone's Salt *see* 2378
Pezetamid [Hefa-Frenon] *see* 8140
PF-38 *see* 4342
PF-1593 *see* 1508
P-FAD *see* 4468
Pfeiffer's Substance, 7334
PfiKlor [Pfizer] *see* 7783
Pfizer-E [Pfizer] *see* 3726
Pfizerpen VK [Pfizer] *see* 7230
PFOB *see* 7299
PFP *see* 7301
PFT Roche [Roche] *see* 1081
PG *see* 8044
PGA *see* 4253
PGE$_1$ *see* 8063
PGE$_2$ *see* 8064
PGF$_{2α}$ *see* 8065
PGG$_2$ *see* 8061 and 9530
PGH$_2$ *see* 8061
PGI$_2$ *see* 8061
P-Glycoprotein, 7335
[P(GMA-co-EA)] *see* 2577
P-gp *see* 7335
PGT/1A *see* 7574
PGX *see* 8061
PH-60-40 *see* 3188
PH-70-23 *see* 4159
PH-137 *see* 2633
PH-1882 *see* 1440
Phacetoperane *see* 5493
Phacetur *see* 7343
Phaenthine *see* 9369
Phaeomelanins *see* 5854
Phaeva [Schering AG] *see* 4421
Phalloidin, 7336
Phalloidine *see* 7336
Phaltan *see* 4255
Phanchinone *see* 7337
Phanodorm [Winthrop] *see* 2780
Phanodorn [Winthrop] *see* 2780
Phanquinone, 7337
Phanquone *see* 7337
Phanurane [Specia] *see* 1795
Pharlon *see* 4886
Pharmorubicin [Farmitalia] *see* 3660
Phasal [Lagap] *see* 5552
Phaseolin, 7338
Phaseollin *see* 7338
Phaseolunatin *see* 5522
Phaseomannite *see* 5008
Phasil [Reed & Carnrick] *see* 3264
Phasin, 7339
Phasins *see* 5453
Phazyme [Reed & Carnrick] *see* 3264
Ph BC *see* 4016
Ph-CJ-91B *see* 6976

2-chlorovinyl]cyclopropane Carboxylate *see* 4175

(±)-*m*-Phenoxyhydratropic Acid *see* 4021

N-Phenoxyisopropyl-*N*-benzyl-β-chloroethylamine *see* 7409

Phenoxymethylpenicillin *see* 7230

Phenoxymethylpenicillinic Acid *see* 7230

4-[3-[4-(Phenoxymethyl)phenyl]propyl]morpholine *see* 4259

[2-(4-Phenoxyphenoxy)ethyl]carbamic Acid Ethyl Ester *see* 4027

6-(α-Phenoxy-α-phenylacetamido)-penicillanate *see* 4001

6-(α-Phenoxyphenylacetamido)-penicillanic Acid *see* 4001

3-(Phenoxyphenyl)methyl (±)-*cis,-trans*-3-(2,2-Dichloroethenyl)-2,2-dimethylcyclopropanecarboxylate *see* 7321

α-*dl*-2-(3-Phenoxyphenyl)propionic Acid *see* 4021

4-Phenoxyphenyl *(RS)*-2-(2-Pyridyloxy)propyl Ether *see* 8174

3-Phenoxy-1,2-propanediol *see* 7444

Phenoxypropazine, 7411

6-(α-Phenoxypropionamido)penicillanic Acid Potassium Salt *see* 7369

2-(3-Phenoxy-2-propylamino)-1-(*p*-hydroxyphenyl)-1-propanol *see* 5259

α-Phenoxypropylpenicillin *see* 8002

4-Phenoxy-3-(1-pyrrolidinyl)-5-sulfamoylbenzoic Acid *see* 7647

4-[3-(α-Phenoxy-*p*-tolyl)propyl]morpholine *see* 4259

Phenozin *see* 10283

Phenpentermine *see* 7275

Phenpiazine *see* 8263

Phenprobamate, 7412

Phenprocoumon, 7413

Phenpromethamine *see* 7462

Phensuximide, 7414

Phentanyl *see* 4043

Phentermine, 7415

Phentetiothalein Sodium, 7416

Phentolamine, 7417

Phentolamine Mesylate *see* 7417

Phentonium Bromide *see* 4048

Phentydrone, 7418

Phenurone [Abbott] *see* 7343

Phenvalerate *see* 4051

Phenychol *see* 3816

Phenycholon *see* 3816

Phenygam *see* 484

Phenylacetaldehyde, 7419

N-Phenylacetamide *see* 47

α-Phenylacetamide, 7420

Phenyl Acetate, 7421

Phenylacetic Acid, 7422

Phenylacetone, 7423

Phenylacetonitrile *see* 1166

Phenylacetylene *see* 3906

3-α-Phenyl-β-acetylethyl-4-hydroxycoumarin *see* 10174

Phenyl Acetylsalicylate, 7424

(Phenylacetyl)urea *see* 7343

Phenylacrolein *see* 2356

α-Phenylacrylic Acid *see* 906

β-Phenylacrylic Acid *see* 2358

Phenylalanine, 7425

L-Phenylalanine *see* 7425

(R)-Phenylalanine *see* 7425

β-Phenylalanine *see* 7425

2-(Phenylalanine)-8-lysine Vasopressin *see* 3992

L-Phenylalanine Mustard *see* 5871

D-Phenylalanyl-*N*-[4-[(aminoiminomethyl)amino]-1-(chloroacetyl)butyl]-L-prolinamide *see* 7881

D-Phenylalanyl-L-cysteinyl-L-phenylalanyl-D-tryptophyl-L-lysyl-L-threonyl-*N*-[2-hydroxy-1-(hydroxymethyl)propyl]-L-cysteinamide Cyclic (2→7)-Disulfide *see* 6859

D-Phenylalanyl-prolyl-arginine Chloromethyl Ketone *see* 7881

γ-Phenylallyl Alcohol *see* 2362

5-Phenyl-5-allylbarbituric Acid *see* 7349

Phenylamine *see* 696

Phenylaminoacetic Acid Isoamyl Ester Hydrochloride *see* 7350

4-(Phenylamino)benzenesulfonic Acid *see* 7466

2-(Phenylamino)benzoic Acid *see* 7427

β-Phenyl-γ-aminobutyric Acid *see* 484

dl-α-Phenyl-β-aminoethanol Sulfate *see* 7441

2-Phenylamino-4-methyl-6-cyclopropylpyrimidine *see* 2841

8-(Phenylamino)-1-naphthalenesulfonic Acid *see* 699

3-[[4-(Phenylamino)phenyl]azo]benzenesulfonic Acid Monosodium Salt *see* 5989

4-[[4-(Phenylamino)phenyl]azo]benzenesulfonic Acid Monosodium Salt *see* 9906

1-Phenyl-2-aminopropane *see* 623

d-1-Phenyl-2-aminopropane Sulfate *see* 2996

1-Phenyl-1-amino-2-propanol *see* 467

1-Phenyl-2-amino-1-propanol Hydrochloride *see* 7461

Phenyl Aminosalicylate, 7425

Phenyl *p*-Aminosalicylic Acid *see* 7426

N-Phenylanthranilic Acid, 7427

Phenyl Arsenoxide *see* 7083

Phenylarsine Oxide *see* 7083

Phenylarsonic Acid *see* 1095

p-(Phenylazo)aniline *see* 439

4-(Phenylazo)benzenamine *see* 439

4-(Phenylazo)-1,3-benzenediamine Monohydrochloride *see* 2317

β-Phenylazo-α,α'-diaminopyridine Hydrochloride *see* 7361

1-Phenylazo-2-naphthalenamine *see* 10228

1-[[4-(Phenylazo)phenyl]azo]-2-naphthalenol *see* 9054

1-(*p*-Phenylazophenylazo)-2-naphthol *see* 9054

4-Phenylazo-*m*-phenylenediamine Hydrochloride *see* 2317

4-Phenylazo-*m*-phenylenediamine Hydrochloride Citrate *see* 2317

3-(Phenylazo)-2,6-pyridinediamine Monohydrochloride *see* 7361

(Phenylazo)thioformic Acid 2-Phenylhydrazide *see* 3442

N-Phenylbarbital *see* 7368

N-Phenylbenzamide *see* 1088

Phenylbenzene *see* 3372

α-Phenylbenzeneacetic Acid *see* 3374

α-Phenylbenzeneacetic Acid 2-(Diethylamino)ethyl Ester Hydrochloride *see* 160

α-Phenylbenzeneacetic Acid 1-Ethyl-3-piperidinyl Ester *see* 7623

endo-α-Phenylbenzeneacetic Acid 8-Methyl-8-azabicyclo[3.2.1]oct-3-yl Ester *see* 9903

N-Phenylbenzeneamine *see* 3375

α-Phenylbenzeneethanethioic Acid *S*-[2-(Diethylamino)ethyl]ester *see* 9509

N-Phenylbenzenemethanamine *see* 1161

α-Phenylbenzenemethanamine *see* 1105

α-Phenylbenzenemethanol *see* 1121

2-Phenyl-1*H*-benzimidazole, 7428

Phenyl Benzoate, 7429

2-Phenyl-4*H*-1-benzopyran-4-one *see* 4132

3-Phenyl-4*H*-1-benzopyran-4-one *see* 5189

2-Phenyl-1,4-benzopyrone *see* 4132

2-Phenyl-γ-benzopyrone *see* 4132

β-*p*-Phenylbenzoylpropionic Acid *see* 4003

N-Phenylbenzylamine *see* 1161

2-(*N*-Phenyl-*N*-benzylaminomethyl)-imidazoline *see* 718

N-Phenyl-*N*-benzyl-4-amino-1-methylpiperidine *see* 983

8-*p*-Phenylbenzylatropinium Bromide *see* 10208

p-(α-Phenylbenzylidene)-1,1-dimethylpiperidinium Methylsulfate *see* 3361

2-Phenylbicyclo[2.2.0]hexa-2,5-diene *see* 4677

Phenyl Biguanide, 7430

1-Phenylbiguanide *see* 7430

Phenylbis[2-chloroethylamine] *see* 697

N-[3-Phenyl-4,5-bis[(trifluoromethyl)-imino]-2-thiazolidinylidene]benzenamine *see* 4155

Phenylboric Acid *see* 1096

Phenylboron Dihydroxide *see* 1096

Phenylboronic Acid *see* 1096

Phenylboroxide *see* 1096

Phenyl Bromide *see* 1427

2-Phenylbutanamide *see* 7432

1-Phenylbutane *see* 1584

2-Phenylbutane *see* 1585

N-Phenylbutanedioic Acid Monoamide *see* 9037

2-Phenyl-1,2-butanediol 1-Carbamate *see* 4884

N-Phenylbutanimide *see* 9036

Phenylbutazone, 7431

Phenylbutazone 2-Amino-2-thiazoline Salt *see* 9443

4-Phenyl-3-buten-2-one *see* 1172

8-[4-(4-Phenylbutoxy)benzamido]-2-(tetrazol-5-yl)-4*H*-1-benzopyran-4-one *see* 7889

8-[*p*-(4-Phenylbutoxy)benzoyl]amino-2-(5-tetrazolyl)-4-oxo-4*H*-1-benzopyran *see* 7889

Phenyl-*tert*-butylamine *see* 7415

Phenylbutylcarbinol *see* 4016

Phenyl-*sec*-butyl Norsuprifen *see* 6830

α-Phenylbutyramide, 7432

2-Phenylbutyric Acid 2-[2-(Diethylamino)ethoxy]ethyl Ester *see* 1539

2-Phenylbutyric Acid 2-(Diethylamino)ethyl Ester *see* 1551

2-Phenylbutyric Acid 2-(3-Methyl-2-phenylmorpholino)ethyl Ester *see* 4005

(2-Phenylbutyryl)urea *see* 7375

N-(α-Phenylbutyryl)urea *see* 7375

Phenylcarbamic Acid *see* 1828

Phenylcarbamic Acid Ethyl Ester *see* 7473

Phenylcarbamic Acid 1-Methylethyl Ester *see* 8001

Phenylcarbamide *see* 7472

5-(*N*-Phenylcarbamoylamino)-1,2,3-thiadiazole *see* 9447

[[(Phenylcarbamoyl)methyl]thio]gold *see* 918

Phenylcarbimide *see* 7449

Phenylcarbinol *see* 1159

Phenyl Carbonate, 7433

O-(2,4,5-Trichlorophenyl)ester *see* 8415

Phosphorothioic Acid O-[2-(Ethylthio)ethyl] O,O-Dimethyl Ester Mixt with S-[2-(Ethylthio)ethyl] O,-O-Dimethyl Phosphorothioate *see* 6129

Phosphorous Acid, 7502

Phosphorous Bromide *see* 7514

Phosphorous Chloride *see* 7515

Phosphor-roesslerite *see* 5718

Phosphorsalz *see* 587

Phosphorus, 7503

Phosphorus Chloride *see* 7506

Phosphorus Hemitriselenide, 7504

Phosphorus Oxybromide, 7505

Phosphorus Oxychloride, 7506

Phosphorus Pentabromide, 7507

Phosphorus Pentachloride, 7508

Phosphorus Pentafluoride, 7509

Phosphorus Pentaselenide, 7510

Phosphorus Pentasulfide, 7511

Phosphorus Pentoxide, 7512

Phosphorus Perbromide *see* 7507

Phosphorus Perchloride *see* 7508

Phosphorus Persulfide *see* 7511

Phosphorus Sesquisulfide *see* 9375

Phosphorus Sulfochloride, 7513

Phosphorus Tribromide, 7514

Phosphorus Trichloride, 7515

Phosphorus Trifluoride, 7516

Phosphorus Trioxide, 7517

Phosphorus Triselenide, 7518

Phosphoryl Bromide *see* 7505

Phosphoryl Chloride *see* 7506

Phosphorylcholine, 7519

Phosphorylcholine Chloride *see* 7519

Phosphoryldimethylaminoethanol *see* 2935

Phosphoryldimethylcolamine *see* 2935

Phosphoryldimethylethanolamine *see* 2935

O-Phosphoryl-4-hydroxy-N,N-dimethyltryptamine *see* 8111

Phosphoryl Tribromide *see* 7505

Phosphoserine, 7520

O-Phospho-DL-serine *see* 7520

Phosphosiderite *see* 4074

Phosphothion *see* 5740

Phosphotope [Squibb] *see* 8807

Phosphotungstic Acid, 7521

Phostoxin [Degesch] *see* 372

Phosvel [Velsicol] *see* 5468

Phosvitin, 7522

Photine R *see* 1349

Photobiline *see* 5059

Photodyn *see* 4669

Photofrin II [Am. Cyanamid] *see* 7755

Photoglycine *see* 4885

Photol [Merck & Co.] *see* 6097

Photophor *see* 1742

Photopsins *see* 6988

Photo-Rex *see* 6097

Photosensitizer 101 *see* 7690

Phoxim, 7523

Phrenazol *see* 7283

Phrenosin, 7524

Phrilon [Phrixwerke] *see* 6832

Phthalamide, 7525

Phthalamodine *see* 2246

Phthalamudine *see* 2246

Phthalazine, 7526

1(2H)-Phthalazinone (1,3-Dimethyl-2-butenylidene)hydrazone *see* 1492

1(2H)-Phthalazinone Hydrazone *see* 4800

3-(1-Phthalazinyl)carbazic Acid Ethyl Ester *see* 9640

2-(1-Phthalazinyl)hydrazinecarboxylic Acid Ethyl Ester *see* 9640

Phthalic Acid, 7527

m-Phthalic Acid *see* 5214

p-Phthalic Acid *see* 9306

o-Phthalic Acid Bis[diethylamide] *see* 9344

Phthalic Acid Diamide *see* 7525

Phthalic Acid Dibutyl Ester *see* 1622

Phthalic Acid Dimethyl Ester *see* 3304

Phthalic Acid 1-Ethyl-1-methyl-2-propynyl Ester *see* 7530

Phthalic Acid Potassium Acid Salt *see* 7773

Phthalic Anhydride, 7528

Phthalidyl D-α-Aminobenzylpenicillanate *see* 9204

Phthalidyl 2-(3-Trifluoromethylanilino)nicotinate *see* 9214

Phthalidyl 2-(α,α,α-Trifluoro-m-toluidino)nicotinate *see* 9214

Phthalimide, 7529

3-Phthalimidoglutarimide *see* 9390

α-Phthalimidoglutarimide *see* 9390

Phthalin *see* 7394

(SP-4-1)-[29H,31H-Phthalocyaninato-(2—)-N²⁹,N³⁰,N³¹,N³²]copper *see* 2584

Phthalofyne, 7530

Phthalophos *see* 6441

Phthalophos (USSR) *see* 7492

Phthaloyl Chloride, 7531

N-Phthaloylglutamimide *see* 9390

Phthaloylsulfacetamide *see* 7532

Phthalthrin *see* 9362

2-(N⁴-Phthalylaminobenzenesulfonamido)thiazole *see* 7533

N-Phthalylglutamic Acid Imide *see* 9390

Phthalylsulfacetamide, 7532

Phthalylsulfacetimide *see* 7532

2-(N⁴-Phthalylsulfanilamido)thiazole *see* 7533

Phthalylsulfathiazole, 7533

Phthalylsulfonazole *see* 7533

Phthiocol, 7534

Phtorazisin [Medexport] *see* 4149

PhXA-41 *see* 5387

Phycobilin *see* 7535

Phycobiliproteins, 7535

Phycobilisomes *see* 7535

Phycocyanins *see* 7535

Phycoerythrins *see* 7535

Phygon *see* 3096

Phygon Paste *see* 3096

Phygon XL *see* 3096

Phylcardin *see* 485

Phylletten [Arznei Müller-Rorer] *see* 2959

Phyllindon *see* 485

Phyllocontin [Napp] *see* 485

Phyllocormin N *see* 3924

α-Phyllohydroquinone *see* 7536

Phylloquinone, 7536

Phyllyrin *see* 7477

Phyomone [ICI] *see* 6458

Phyone *see* 8864

Physalaemin, 7537

Physalemin *see* 7537

Physalien *see* 10248

Physalin *see* 10248

Physcione *see* 3602

Physeptone [Calmic] *see* 6008

Physex [Leo Pharm.] *see* 2273

Physiomycine *see* 6007

Physiotens [Giulini] *see* 6375

Physodalin *see* 7538

Physodic Acid, 7538

Physostigma, 7539

Physostigmine, 7540

Physostigmine Aminoxide *see* 3740

Physovenine, 7541

Physpan [Savage] *see* 9421

Phytar [Monterey] *see* 1640

Phytat D. B. [Daniel-Brunet] *see* 7542

Phytic Acid, 7542

Phytin *see* 7542

Phytoagglutinins *see* 5453

Phytoalexins *see* 8113

Phytochlorin, 7543

Phytochlorin(e) e *see* 7543

Phytochrome *see* 7535

Phytoecdysteroids *see* 3539

Phytofluene, 7544

Phytohemagglutinin *see* 5453

Phytol, 7545

Phytolacca, 7546

Phytomelin *see* 8456

Phytomenadione *see* 7536

Phytomonic Acid *see* 5353

Phytonadiol *see* 7536

Phytonadiol Sodium Diphosphate *see* 7536

Phytonadione *see* 7536

α-Phytosterol *see* 8697

3-Phytylmenadione *see* 7536

Piazofolina *see* 6355

Piazolin *see* 6355

Piberaline, 7547

Picadex, 7548

Piceid *see* 8326

Picein, 7549

Picene, 7550

Piceol *see* 7549

Piceoside *see* 7549

Picfume [Dow] *see* 2208

Picilorex, 7551

Picloram, 7552

Picloxydine, 7553

Pi-Coli [Ist. Vitamine] *see* 7613

4-Picoline *see* 7556

α-Picoline, 7554

β-Picoline, 7555

γ-Picoline, 7556

Picolinic Acid, 7557

γ-Picolinic Acid *see* 5204

Picolinic Acid N-Methylbetaine *see* 4765

7-(β-3'-Picolylaminoethyl)theophylline *see* 7584

2-Picolylidenebis(p-phenyl Sodium Sulfate) *see* 7559

4,4'-(2-Picolylidene)bis(phenylsulfuric Acid) Disodium Salt *see* 7559

N-(2-Picolyl)-N-phenyl-N-(2-piperidinoethyl)amine *see* 7558

Picoperidamine *see* 7558

Picoperine, 7558

Pico-Salax [Ferring] *see* 7559

Picosulfate Sodium, 7559

Picosulfol *see* 7559

Picotamide, 7560

Picragol [Wyeth] *see* 8670

Picramic Acid, 7561

Picrasmin *see* 5239

Picric Acid, 7562

Picric Acid Ammonium Salt *see* 581

Picric Acid Silver Derivative *see* 8670

Picrocine *see* 2974

Picrocrocin, 7563

Picrolichenic Acid, 7564

Picrolichenin *see* 7564

Picrolonic Acid, 7565

Picromycin, 7566

Picronitric Acid *see* 7562

Picropodophyllic Acid *see* 7703

Picropodophyllic Acid Hydrazide *see* 7703

Picropodophyllin, 7567

Picropodophyllinic Acid 2-Ethylhydrazide *see* 7703

Picropodophyllinic Acid Hydrazide
see 7703
Picropodophyllinic Acid Lactone see
7567
Picrorhiza, 7568
Picrotin, 7569
Picrotol see 8670
Picrotoxin, 7570
Picrotoxinin, 7571
Picryl Chloride, 7572
Picrylmethylnitramine see 6664
Picrylnitromethylamine see 6664
Pictol [Mallinckrodt] see 6097
Picumast, 7573
PID see 7381
Pidilat [Giulini] see 6617
Pidorubicin see 3660
Pidotimod, 7574
Pierami [Pierrel] see 425
Pierminox [Pierrel] see 6290
Pietil [Argentia] see 7079
Pifarnine, 7575
Pifatidine see 8431
Pifazin [Pierrel] see 7575
Pigitil [Fidia] see 7574
Piglet Pro-Gen V see 8719
Pigmentation Hormone see 6377
Pigmex see 6331
Pigweed see 392
PIH see 7382
Piketoprofen, 7576
Pikromycin see 7566
Pilagan [Allergan] see 7578
Pil-Digis [Lakeside] see 3201
Pildralazine, 7577
Piliophen see 5059
Pill-bearing Spurge see 3949
Pilo [Novopharma] see 7578
Pilocarpine, 7578
β-Pilocarpine see 7578
Pilocarpus, 7579
Pilocereine, 7580
Pilofrin [Allergan] see 7578
Pilopine HS [Alcon] see 7578
Piloral [Nippon Kayaku] see 2405
Pilosine see 5216
Pilostat [Bausch & Lomb] see 7578
Pilot [Schering] see 8269
Pilsicainide, 7581
Pilzcin [Shionogi] see 2429
Pimadin see 7587
Pimafucin [Basotherm] see 6513
Pimagedine see 461
Pimaric Acid, 7582
d-Pimaric Acid see 7582
l-Pimaric Acid see 5494
α-Pimaric Acid see 7582
β-Pimaric Acid see 5494
Pimaricin see 6513
Pimavecort [Mycofarm] see 6542
Pimeclone, 7583
Pimefylline, 7584
Pimelic Acid, 7585
Pimelic Ketone see 2795
Pimenta, 7586
Pimephylline see 7584
Pimexone [Formenti] see 5906
Piminodine, 7587
Pimobendan, 7588
Pimozide, 7589
Pimpernel see 7590
Pimpinella, 7590
Pimpinellin, 7591
Pinacidil, 7592
Pinacol, 7593
Pinacolin see 7594
Pinacolone, 7594
Pinacoloxymethylphosphoryl Fluoride
see 8860

Pinacolyl Methylphosphonofluoridate
see 8860
Pinacone see 7593
Pinacyanol Chloride see 8229
Pinakon. see 4748
Pinang see 813
Pinaverium Bromide, 7595
Pinazepam, 7596
Pinbetol [Dolorgiet] see 7597
Pincets [Merrell] see 7617
Pindac [Leo Pharm] see 7592
Pindione [Oberval] see 7381
Pindolol, 7597
Pindone, 7598
"Pineapple Oil" see 3822
Pinene see 7599
2-Pinene see 7599
α-Pinene, 7599
β-Pinene, 7600
Pinene Hydrochloride see 1369
(1R,5R)-(+)-2-Pinen-4-one see 10094
2-Pinen-7-one see 2311
Pine Oil, 7601
Pineroro [Maruko] see 3369
Pine Tar, 7602
Pine Tar Oil see 9233
Pinguinain, 7603
Pinkroot see 8900
Pinnoite see 5693
Pinol Hydrate see 8707
Pinorubin [Sanraku] see 7642
Pinosylvin, 7604
Pinozan see 7617
Pinrou [Century] see 7617
Pinsirup [Merrell] see 7617
Pinus Montana Oil see 6900
Pinus Pumilio Oil see 6900
PIO see 7206
Pioglitazone, 7605
Pioxol see 7206
Pipacycline, 7606
Pipadone [Burroughs Wellcome] see
3398
Pipadox [Dagra] see 7618
Pipamazine, 7607
Pipamperone, 7608
Pipanol [Winthrop] see 9823
Pipazethate, 7609
Pipeacid [SIT] see 7613
Pipebuzone, 7610
Pipecolic Acid, 7611
Pipecolinic Acid see 7611
Pipecurium Bromide, 7612
Pipecuronium Bromide see 7612
Pipedac [Teofarma] see 7613
Pipemid [Gentili] see 7613
Pipemidic Acid, 7613
Pipenale [Fuso] see 7624
Pipenzolate Bromide, 7614
Pipenzolate Methylbromide see 7614
Pipenzolone Bromide see 7614
Piperacetazine, 7615
Piperacid (syrup) [Universal Drug] see
7617
Piperacillin, 7616
Piperamic Acid see 7613
Piperate see 7617
Piperazate [Leeming] see 7617
Piperazidine see 7617
Piperazine, 7617
Piperazine Adipate, 7618
Piperazine-1,4-bis(N,N'-diethylene-
phosphonediamide) see 3397
1,4-Piperazinebis(ethanesulfonic Acid)
see 7633
Piperazine-N,N'-bis(2-ethanesulfonic
Acid) see 7633
Piperazine Calcium Edathamil see
7617

1-Piperazinecarbodithioic Acid see
7548
1-Piperazinecarbodithioic Acid
Betaine see 7548
1-Piperazinecarbodithioic Acid Inner
Salt see 7548
Piperazine Carbon Disulfide Complex
see 7548
1,4-Piperazinediethanesulfonic Acid
see 7633
1,4-Piperazinediethanol Bis(α-methyl-
1-naphthaleneacetate) see 6439
2,5-Piperazinedione, 7619
Piperazine Dithiocarbamate see 7548
1,4-Piperazinediylbis[bis(1-aziridinyl)-
phosphine Oxide] see 3397
1,1'-[1,4-Piperazinediylbis(imidocarbonyl)]bis[3-(p-chlorophenyl)-
guanidine] see 7553
N,N'-[1,4-Piperazinediylbis(2,2,2-tri-
chloroethylidene)]bisformamide see
9819
Piperazine Estrone Sulfate see 3751
Piperazine Salt of Bis(2,4,5-trichloro-
phenol) see 9787
Piperazine Sultosylate see 9169
Piperazine Theophylline-7-acetate see
23
Piperazine Theophylline Ethanoate
see 23
N-(1-Piperazinylacetyl)cyclohexyl-
amine see 3736
1-Piperazinyl-4-methylenecarbonyl-
cyclohexylamine see 3736
Piperic Acid, 7620
Piperidic Acid see 450
Piperidine, 7621
2-Piperidinecarboxylic Acid see 7611
3-Piperidinecarboxylic Acid see 6654
4-Piperidinecarboxylic Acid see 5206
2,2'-(2,6-Piperidinediyl)bis[1-phenyl-
ethanone] see 6802
2-Piperidineethanol 2-Aminobenzoate
(Ester) see 7649
2-Piperidineethanol Anthranilate see
7649
1-Piperidineethanol Benzilate see 7624
3-Piperidino-4'-butoxypropiophenone
see 3523
5-[3-(4-Piperidino-4-carbamoylpiperi-
dino)propyl]-10,11-dihydro-5H-di-
benz[b,f]azepine see 1911
6-Piperidino-2,4-diaminopyrimidine 3-
Oxide see 6290
2-(1-Piperidino)-4,4-diphenyl-5-
heptanone see 3398
6-Piperidino-4,4-diphenylheptan-3-one
see 3398
1-Piperidino-3,3-diphenyl-4-hexanone
see 6810
3-Piperidino-1,1-diphenyl-1-propanol
see 7922
2-(2-Piperidinoethoxy)ethyl 10H-Pyri-
do[3,2-b][1,4]benzothiadiazine-10-
carboxylate see 7609
2-(2-Piperidinoethoxy)ethyl 10-Thia-
1,9-diazaanthracene-10-carboxylate
see 7609
2-(1-Piperidino)ethyl Benzilate see
7624
β-Piperidinoethyl-4-butoxyphenyl
Ketone see 3523
2-Piperidinoethyl 3-Methylflavone-8-
carboxylate see 4135
2-Piperidinoethyl 3-Methyl-4-oxo-2-
phenyl-4H-1-benzopyran-8-carb-
oxylate see 4135
Piperidinoethyl α-Piperidinophenylacetate see 3399

β-Piperidinoethyl 4-Propoxyphenyl
Ketone *see* 8016
N-(2-Piperidinoethyl)-N-(2-pyridyl-
methyl)aniline *see* 7558
3-(2-Piperidino-1-hydroxyethyl)-5-
phenylisoxazole *see* 7318
2-Piperidinomethyl-1,4-benzodioxan
see 7631
2-(Piperidinomethyl)cyclohexanone
see 7583
1-Piperidino-2-methyl-3-(*p*-tolyl)-3-
propanone *see* 9660
3-Piperidino-1-phenyl-1-bicyclohep-
tenyl-1-propanol *see* 1274
3-Piperidino-1-phenyl-(Δ⁵-bicyclo-
[2.2.1]hepten-2-yl)-1-propanol *see*
1274
3-Piperidino-1,2-propanediol Dicarbanilate Hydrochloride *see* 3360
1-Piperidino-2-(N-propionylanilino)-
propane *see* 7353
3-Piperidino-4'-propoxypropiophenone
see 8016
2,2',2'',2'''-[(4-Piperidinopyrimido[5,4-
d]-pyrimidine-2,6-diyl)dinitrilo]-
tetraethanol *see* 6344
(*R**,*S**)-(±)-α-2-Piperidinyl-2,8-bis-
(trifluoromethyl)-4-quinolinemetha-
nol Monohydrochloride *see* 5845
3-[2-(4-Piperidinyl)ethyl]-1*H*-indole
see 4965
N-(2-Piperidinylmethyl)-2,5-bis(2,2,2-
trifluoroethoxy)benzamide *see* 4136
2-(1-Piperidinylmethyl)cyclohexanone
see 7583
3-(1-Piperidinyl)-1,2-propanediol Bis-
(phenylcarbamate) (Ester) Monohydrochloride *see* 3360
1-(2-Piperidinyl)-2-propanone *see* 7200
3-(1-Piperidinyl)-1-(4-propoxyphenyl)-
1-propanone *see* 8016
3-(2-Piperidinyl)pyridine *see* 661
6-(1-Piperidinyl)-2,4-pyrimidinedia-
mine 3-Oxide *see* 6290
2,2',2'',2'''-[[4-(1-Piperidinyl)pyrim-
ido[5,4-*d*]pyrimidine-2,6-diyl]dini-
trilo]tetrakisethanol *see* 6344
Piperidione, 7622
Piperidolate, 7623
Piperidyl Amidone *see* 3398
α-(2-Piperidyl)benzhydrol *see* 7638
α-(2-Piperidyl)benzhydrol *see* 925
DL-*erythro*-α-2-Piperidyl-2,8-bis(trifluoromethyl)-4-quinolinemethanol
Monohydrochloride *see* 5845
3-(1-Piperidyl)-1-cyclohexyl-1-phenyl-
1-propanol Hydrochloride *see* 9823
6-(1-Piperidyl)-4,4-diphenyl-3-hex-
anone *see* 6810
3-(N-Piperidyl)-1,1-diphenyl-1-pro-
panol *see* 7922
β-(2-Piperidyl)ethyl *o*-Aminobenzoate
see 7649
β-Piperidylethyl Benzilate *see* 7624
2-(1-Piperidyl)ethyl *p*-Butoxyphenyl
Ketone *see* 3523
2-(β-N-Piperidylethyl)-4,4-diphenyl-
1,3-dioxolan-5-one Hydrochloride
see 7637
2-Piperidylmethyl)-1,4-benzodioxan
see 7631
3-(1-Piperidyl)-1,2-propanediol Dicarbanilate Hydrochloride *see* 3360
Piperilate, 7624
Piperine, 7625
Piperinic Acid *see* 7620
Piperinsäure (German) *see* 7620
Piperitone, 7626
Piperocaine, 7627
Piperonal, 7628

Piperonil [Lusofarmaco] *see* 7608
Piperonylaldehyde *see* 7628
Piperonyl Butoxide, 7629
Piperonylic Acid, 7630
10-[(4-Piperonyl-1-piperazinyl)acetyl]-
phenothiazine *see* 4023
2-(4-Piperonyl-1-piperazinyl)pyrimi-
dine *see* 7648
1-Piperonyl-4-(3,7,11-trimethyl-2,6,-
10-dodecatrienyl)piperazine *see*
7575
Piperoxan, 7631
(*E,E*)-1-Piperoylpiperidine *see* 7625
(*Z,Z*)-1-Piperoylpiperidine *see* 2090
Piperverm [Wild] *see* 7617
Piperylone, 7632
PIPES, 7633
Pipethanate *see* 7624
Pipitzahoic Acid *see* 7298
Pipizan Citrate [Merck & Co.] *see*
7617
Pipobroman, 7634
Piportil [Specia] *see* 7636
Piportil L4 [Specia] *see* 7636
Piportil M2 [Specia] *see* 7636
Piposulfan, 7635
Pipothiazine *see* 7636
Pipotiazine, 7636
Pipotiazine Palmitate *see* 7636
Pipotiazine Undecylenate *see* 7636
Pipoxolan Hydrochloride *see* 7637
Pipracil [Lederle] *see* 7616
Pipradol *see* 7638
Pipradrol, 7638
Pipram [Rhône-Poulenc Rorer] *see*
7613
Pipril [Lederle] *see* 7616
Piprinhydrinate *see* 3390
Piproxen *see* 6504
Piprozolin, 7639
Pipsissewa *see* 2099
Pipsyl Chloride, 7640
Piptal [Boehringer, Mann.] *see* 7614
Pipurin [Brocchieri] *see* 7613
Piracetam, 7641
Piraldina [Bracco] *see* 8140
Piramox [Radiumpharma] *see* 617
Piranver [Usafarma] *see* 8139
Pirarubicin, 7642
Pirazolac, 7643
Pirbuterol, 7644
Pirecin [Yoshitomi] *see* 8090
Pirem [Gödecke] *see* 1882
Pirenoxine, 7645
Pirenzepine, 7646
Piretanide, 7647
Pirevan *see* 3416
Pirexyl [Pharmacia] *see* 1078
Pirfenoxone *see* 7645
Piria's Acid *see* 6490
Piribedil, 7648
Piribenzil Methyl Sulfate *see* 1239
Piricardio [Nagel] *see* 486
Piricef [CT] *see* 2030
Piridocaine, 7649
Piridol *see* 495
Piridolan [Janssen] *see* 7653
Pirifibrate, 7650
Pirilène [Marion Merrell Dow] *see*
8140
Pirimicarb, 7651
Pirimiphos-ethyl, 7652
Pirimiphos-methyl *see* 7652
Pirimor [ICI] *see* 7651
Pirinitramide *see* 7653
Pirisudanol *see* 8175
Piriton [Allen & Hanburys] *see* 2232
Piritramide, 7653
Piritrexim, 7654
Pirkam [DAK] *see* 7661

Pirlimycin, 7655
Pirmenol, 7656
Piroan [Towa Yakuhin] *see* 3410
Pirocrid [Mochida] *see* 8077
Piroctone, 7657
Pirodal [SIT] *see* 7660
Piroflex [Lagap] *see* 7661
Piroheptine, 7658
Piromen, 7659
Piromidic Acid, 7660
Piroxicam, 7661
Piroxicam Cinnamate *see* 7661
Piroxina *see* 2247
Pirozadil, 7662
Pirprofen, 7663
Pirroxil [SIT] *see* 7641
Pirsue [Upjohn] *see* 7655
Piscarol *see* 4929
Piscidic Acid, 7664
Pisciol *see* 4929
Pistachia Galls *see* 5795
Pitayine *see* 8244
Pitchblende *see* 9991
Piteranometozine *see* 10105
Pithecolobine, 7665
Pitocin [Parke, Davis] *see* 7114
Pitocin-Buccal [Parke, Davis] *see* 7114
Pitressin [Parke, Davis] *see* 10073
Pituamin *see* 7666
Pituitary Growth Hormone *see* 8864
Pituitary Lactogenic Hormone *see*
7961
Pituitary Lipotropic Hormone *see*
5539
Pituitary, Posterior, 7666
Pituitrin [Parke, Davis] *see* 7666
Pivacef [Firma] *see* 7670
Pivadorm [Specia] *see* 1418
Pival [Kilgore] *see* 7598
Pivaldione *see* 7598
Pivalephrine [Santen] *see* 3400
Pivalic Acid, 7667
Pivalic Acid 2-Benzylhydrazide *see*
7668
Pivalone [Jouveinal] *see* 9623
N¹-Pivaloyl-N²-benzylhydrazine *see*
7668
2-Pivaloyl-1,3-indandione *see* 7598
Pivaloyloxymethyl D-α-Aminobenzyl-
penicillinate *see* 7669
Pivaloyloxymethyl Ampicillinate *see*
7669
Pivaloyloxymethyl 6-[(Hexahydro-
1*H*-azepin-1-yl)methyleneamino]-
penicillanate *see* 409
Pivalylbenzhydrazine, 7668
Pivalyl Indandione *see* 7598
2-Pivalyl-1,3-indandione *see* 7598
Pivalyl Valone [Kilgore] *see* 7598
Pivalyn [Kilgore] *see* 7598
Pivamdinocillin *see* 409
Pivampicillin, 7669
Pivatil [Merck & Co.] *see* 7669
Pivcefalexin *see* 7670
Pivcephalexin *see* 7670
Pivhydrazine *see* 7668
Pivmecillinam *see* 409
Pivot [Am Cyan.] *see* 4944
Pix [BASF] *see* 5903
Pixalbol *see* 2487
PIXY321 *see* 5021
Pizotifan *see* 7671
Pizotifen *see* 7671
Pizotyline, 7671
PJ-185 *see* 5601
PK-10169 *see* 3626
PK-26124 *see* 8389
PKhNB *see* 8264
PK-Merz [Merz] *see* 389
Placenta, 7672

Placentafil [Fisons] see 7672
Placidex see 5896
Placidon see 5908
Placidyl [Abbott] see 3774
Plac Out [Bernabo] see 2140
Plactamin [Morishita] see 7919
Plactidil [Samil] see 7560
Plafibride, 7673
Planadalin see 1879
Planate [ICI] see 2461
Planavin [Shell] see 6662
Planavit C see 867
Plancol [Chugai] see 4828
Plandom [Nihon Nohyaku] see 3182
Planetol see 6097
Planocaine [M & B] see 7937
Planochrome see 5914
Planofix [Hortichem] see 6458
Planoform see 1538
Planomycin see 4111
Planovin [Novo] see 5849
Plantago Seed, 7674
Plantain Seed see 7674
Plant Amine Oxidase see 6330
Plantisul, 7675
Planum [Farmitalia] see 9285
Plaquenil Sulfate [Winthrop] see 4863
Plasdone see 7879
Plasil [Lepetit] see 6226
Plasma (In Physics), 7676
Plasmalogens, 7677
Plasmasteril [Fresenius] see 4707
Plasma Thromboplastin Antecedent
  see 3966
Plasma Thromboplastin Component
  see 3964
Plasma Trypsinogen see 7679
Plasmin, 7678
Plasminogen, 7679
Plasminokinase see 8981
Plasmochin see 7134
Plasmocid, 7680
Plasmoquine see 7134
Plasmosan see 7879
Plastenan [Sanofi-Winthrop] see 43
Plaster of Paris see 1753
Plastoquinone 9 see 7681
Plastoquinone 45 see 7681
Plastoquinone A see 7681
Plastoquinones, 7681
Platamine [Montedison] see 2378
Platelet Activating Factor, 7682
Platelet Activation-Dependent
  Granule-External Membrane Pro-
  tein see 8568
Platelet-Derived Growth Factor, 7683
Platelet Glycoprotein IIb/IIIa see
  5014
Platet [Nicholas] see 886
Plath-Lyse [Génévrier] see 3120
Platiblastin [Farmitalia] see 2378
Platifillin see 7691
Platilon [Toyo] see 7728
Platinex [Bristol] see 2378
Platinic Ammonium Chloride see 582
Platinic Chloride, 7684
Platinic Iodide, 7685
Platinic Oxide, 7686
Platinic Potassium Chloride see 7800
Platinic Potassium Thiocyanate see
  7805
Platinol [Bristol] see 2378
Platinous Ammonium Chloride see 583
Platinous Barium Cyanide see 1020
Platinous Chloride, 7687
Platinous Iodide, 7688
Platinous Lithium Cyanide see 5570
Platinous Potassium Chloride see
  7855
Platinous Potassium Cyanide see 7858

Platinous Thorium Cyanide see 9520
Platinoxan [Degussa] see 2378
Platinum, 7689
cis-Platinum II see 2378
Platinum Diiodide see 7688
Platinum Dioxide see 7686
Platistin [Farmitalia] see 2378
Platonin, 7690
Platonin J see 7690
Platosin [Pharmachemie] see 2378
Plattnerite see 5424
Platyphylline, 7691
Plaunotol, 7692
Plausitin [Erba] see 6350
Plavix [Sanofi] see 2457
Plavolex [Wyeth] see 2989
Plazolite see 1685
Ple-1053 see 953
Plebilin see 7062
Plegatil, 7693
Plegicil [Clin-Comar-Byla] see 32
Plegine [Ayerst] see 7365
Plegomazin see 2238
Plemocil [Farmitalia] see 2835
Plenastril [Protochemie] see 7101
Plendil [Astra] see 3991
Pleniron see 4086
Plenur [Lasa] see 5552
Pleocide [Merck & Co.] see 430
Plesium [Chiesi] see 1454
Plesmet [Coates & Cooper] see 4086
Plessy's Green see 2284
Pletaal [Otsuka] see 2335
Pletil [Biolab-Searle] see 9588
Pleurisy Root see 865
Pleuromutilin, 7694
Pleurotin(e), 7695
Plexamine [Biodica] see 4758
Plexan see 5575
Plexiglas see 6005
Plexol 201 see 1292
Plicamycin, 7696
Plictran [Dow] see 2829
Plitican [Delagrange] see 246
PLS see 7754
Pluchine see 1225
Plumbagin, 7697
Plumbago see 4560
Plumbic Fluoride see 5448
Plumbous Fluoride see 5425
Plumbous Oxide see 5433
Plumbous Plumbate see 5440
Plumericin, 7698
Plumieride, 7699
Pluracol E [BASF] see 7729
Plurimen [Sarget] see 8569
Pluripoietin see 4558
Pluronic F68 [BASF] see 7722
Pluronic L62LF see 7722
Pluronic L101 [BASF] see 7722
Pluronics [BASF] see 7722
Pluropon [Boehringer, Ing.] see 8680
Pluryle [Leo Pharm.] see 1064
Plusuril see 1064
Plutonium, 7700
PLV-2 see 3992
PM-185184 see 8562
PMA see 7453
PMAC see 7453
PMAS see 7453
P.M.F. see 4805
PMP see 5254
PMSG see 2273
PN see 8202
PN-200-110 see 5260
PN-205-033 see 5260
PN-205-034 see 5260
PNB see 6690
Pneumopent [Fisons] see 7254
Pneumorel [Stroder] see 4041

P.O. 12 [Biothérax] see 3628
Poast [BASF] see 8620
Pocan see 7546
Podocarpic Acid, 7701
Pododacric Acid, 7702
Podofilox see 7704
Podophyllic Acids, 7703
Podophyllin see 7707
Podophyllinic Acid see 7703
Podophyllinic Acid 2-Ethylhydrazide
  see 7703
Podophyllinic Acid Hydrazide see
  7703
Podophyllinic Acid Lactone see 7704
Podophyllotoxin, 7704
Podophyllin, 7705
Podophyllum, Indian, 7706
Podophyllum Resin, 7707
POE Alcohol Ethers see 7737
POE Fatty Acid Esters see 7738
POE (40) Monostearate see 7738
POE Sorbitan Esters see 7742
POE (20) Sorbitan Monooleate see
  7742
POE (8) Stearate see 7738
Pogy Oil see 5880
Poi, 7708
Poison Black Cherry see 1052
Poison Elder see 7711
Poison Hemlock see 2570
Poison Ivy, 7709
Poison Nut see 6828
Poison Oak, 7710
Poison Parsley see 2570
Poison Sumac, 7711
Poison Tobacco see 4908
Poison Vine see 7709
Poke Root see 7546
L-Polamidon [Hoechst] see 6008
Polamin [Schering] see 2232
Polaprezinc, 7712
Polaramine [Schering] see 2232
Polaris [Monsanto] see 4523
Polaronil [Schering] see 2232
Polar Yellow [Geigy], 7713
Poldine Methylsulfate, 7714
Poleon [Sumitomo] see 6446
Pole Reagent Paper, 7715
Polibutin [Juste] see 9829
Policydal [Eisai] see 9078
Polidexide, 7716
Polidocanol, 7717
Polik see 4631
Polimod [Poli] see 7574
Polinalin see 495
Polishing Powder see 8933
Polistin T-Caps [Trommsdorf] see
  1845
Poliuron [Lepetit] see 1064
Polival [Andromaco] see 9426
Polivasal [Coli] see 9160
Pollakisu [Kodama] see 7089
Pollucite see 2051
Polmiror [Poli] see 6623
Polonium, 7718
Polonium Dioxide, 7719
Polonium Tetrachloride, 7720
Poloxalene, 7721
Poloxalkol (obsolete) see 7722
Poloxamer 182LF see 7722
Poloxamer 188 see 7722
Poloxamer 331 see 7722
Poloxamers, 7722
Poly see 8846
Polyacrylonitrile see 6998
Poly(ADP-ribose) Polymerase see 788
Polyamide see 6831
Polyanetholesulfonic Acid Sodium
  Salt see 8813
Polyanhydroglucuronic Acid see 7073

Polybar [E-Z-EM] see 1023
Polybenzarsol, 7723
Polybrene see 4719
Polybrominated Biphenyls, 7724
Polybromobiphenyls see 7724
Poly(carbon monofluoride) see 4561
Poly-*trans*-[(2-carboxyethyl)germa-sesquioxane] see 7979
Polycell see 1877
Polychlorinated Biphenyls, 7725
Poly(2-chloro-1,3-butadiene) see 6550
Polychlorocamphene see 9693
Polychloroprene see 6550
Polychol [Croda] see 7737
Polychrome see 3739
Polycidine see 2959
Polycillin [Bristol] see 628
Polycillin-N [Bristol] see 628
Polycycline [Bristol] see 9337
Polydatin see 8326
Polydextrose, 7726
Poly(dicarbon monofluoride) see 4561
Poly[2-(diethylamino)ethyl] Polyglyc-erylene Dextran Hydrochloride see 7716
Polydimethylsiloxane see 3264
Polyestradiol Phosphate, 7727
Polyetherin A see 6633
Polyethylene, 7728
Polyethylene Glycol, 7729
Polyethylene Glycol 200 see 7729
Polyethylene Glycol 400 see 7729
Polyethylene Glycol 600 see 7729
Polyethylene Glycol 1500 see 7729
Polyethylene Glycol 4000 see 7729
Polyethylene Glycol 6000 see 7729
Polyethylene Glycol Esters of Fatty Acids see 7738
Polyethylene Glycol Fatty Alcohol Ethers see 7737
Polyethylene Glycol *p*-Isooctylphenyl Ether see 6858
Polyethylene Glycol (9) Monododecyl Ether see 7717
Polyethylene Glycol 400 Mono-stearate see 7738
Polyethyleneglycols Mono(nonylphen-yl) Ether see 6772
Polyethylenepolypropylene Glycols, Polymers see 7722
Polyethylene Sodium Sulfonate see 5645
Polyethylene Terephthalates, 7730
Polyferon [Biogen] see 5018
Polyferose, 7731
Poly-G [Olin] see 7729
Polyglucin see 2989
Poly(Glycidyl Methacrylate-co-Ethyl Acrylate) see 2577
Polyglycol E [Dow] see 7729
Polygodial, 7732
Polygris [Essex] see 4571
Poly[imino-1,4-butanediylimino(1,6-dioxo-1,6-hexanediyl)] see 6833
Poly(iminocarbonylpentamethylene) see 6832
Poly[imino(1-oxo-1,6-hexanediyl)] see 6832
Poly I . Poly (C₁₂,U) see 630
Polylysine, 7733
Polymannuronic Acid see 241
Polymeric Sodium Metaphosphate see 5680
Polymer of *N,N,N',N'*-Tetramethyl-hexamethylenediamine and Tri-methylene Bromide see 4719
Polymetatelluric Acid see 9272
Poly[methylenedi(hydroxymethyl)-urea] see 7735
Polymox [Bristol] see 617

Polymyxin, 7734
Polymyxin B see 7734
Polymyxin B₁ see 7734
Polymyxin B₂ see 7734
Polymyxin D see 7734
Polymyxin D₁ see 7734
Polymyxin D₂ see 7734
Polymyxin E see 2542
Polymyxin E₁ see 2542
Polymyxin E₂ see 2542
Polymyxin F see 7734
Polymyxin K see 7734
Polymyxin M see 7734
Polymyxin P see 7734
Polymyxin S₁ see 7734
Polymyxin T₁ see 7734
Polynoxylin, 7735
Polynoxyline see 7735
Polyoxin A see 7736
Polyoxin AL see 7736
Polyoxin B see 7736
Polyoxins, 7736
Poly[1-(2-oxo-1-pyrrolidinyl)ethylene] see 7879
Polyoxyaluminum Acetylsalicylate see 316
Polyoxyethylene Alcohols, 7737
Polyoxyethylene (20) Cetyl Ether see 7737
Polyoxyethylene Fatty Acid Esters, 7738
Polyoxyethylene Lauryl Ether see 7717
Polyoxyethylene(*n*)nonylphenyl Ether see 6772
Poly(oxyethylene)-poly(oxypropyl-ene)-poly(oxyethylene) Polymer see 7721
Polyoxyethylene Sorbitan Esters see 7742
Polyoxyethylene (20) Sorbitan Mono-oleate see 7742
Polyoxyethylene (40) Stearate see 7738
Polyoxyl 8 Stearate see 7738
Polyoxyl 40 Stearate see 7738
Polyoxymethylene see 7158
Polyoxymethyleneurea see 7735
Polypeptide β₁ 9546
Polyphenolase see 9969
Polyphenoloxidase see 9969
Polyphlogin see 2347
Polyphosphazenes, 7739
Polyphosphoric Acid, 7740
Polypodine A see 3539
Polypropylene, 7741
Poly(rI) . Poly[r(C₁₂U)ₙ] see 630
Polysaccharide B-1459 see 10191
Polysaccharide C 16 see 3535
Polysaccharide S-60 see 4389
Poly-Solv DPM [Olin] see 3407
Polysorbate 80 see 7742
Polysorbates, 7742
Polystichalbin see 210
Polystichin see 881
Polystichocitrin see 4130
Polystictine see 2355
Polystyrene see 9028
Polysulfide Rubber see 9475
Polytef see 7743
Polytergent B [Olin] see 6772
Polytergent G [Union Carbide] see 6858
Polytetrafluoroethylene, 7743
Polytetrafluoroethylene Resin see 7743
Poly(tetramethyleneadipamide) see 6833
Poly(*N,N,N',N'*-tetramethyl-*N*-trimeth-

ylenehexamethylenediammonium Dibromide) see 4719
Polythene [Du Pont] see 7728
Polythiazide, 7744
Polytrim [Wellcome] see 7734
Polytrin [Ciba-Geigy] see 2836
Polyvidone see 7879
Polyvinyl Alcohol, 7745
Polyvinyl Chloride, 7746
Polyvinylpyrrolidone see 7879
Polyvinylpyrrolidone-iodine Complex see 7880
Polyviol [Wacker] see 7745
Pomarsol [FBayer] see 9510
POMC see 7975
Pomegranate, 7747
Ponalar [Adenylchemie] see 5842
Ponalid [Sandoz] see 3802
Ponasterone A, 7748
Ponasterones see 7748
Ponceau 3R, 7749
Ponceau SX, 7750
Poncuronium Bromide (Rescinded USAN) see 7139
Poncyl-FP [Takeda] see 4571
Ponderal [Servier] see 4015
Ponderax [Selpharm] see 4015
Ponderex [Robins] see 4015
Pondex [Chinoin] see 7206
Pondimin [Robins] see 4015
Pondinil [Roche] see 5843
Pondinol [Roche] see 5843
Pondocil [Leo Pharm.] see 7669
Pondocillin [Leo Pharm.] see 7669
Pondocillina [Sigma-Tau] see 7669
Ponoxylan [Berk] see 7735
Ponsinomycin see 6291
Ponsital see 4945
Ponstan [Parke, Davis] see 5842
Ponstel [Parke, Davis] see 5842
Ponstil [Parke, Davis] see 5842
Ponstyl see 5842
Pontacyl Fast Blue R see 669
Pontal [Parke, Davis] see 5842
Pontalin see 2126
Pontamine White BR see 1349
Ponticin see 8340
Pontocaine Hydrochloride [Sanofi Winthrop] see 9330
P.O.P. see 7176
Poppy Capsules, 7751
Poppy Heads see 7751
Poppy Oil, 7752
Poppy-seed Oil see 7752
Populin, 7753
Populnetin see 5288
Populoside see 7753
Poquil [Sankyo] see 8206
POR-8 see 7001
Poractant Alfa, 7754
Porcelain Clay see 5294
Porcilene [Syntex] see 4037
Porcine Hormone Secretin see 8564
Porcine Motilin see 6366
Porcine Somatotropin see 8864
Pore-forming Protein see 7301
Porfimer Sodium, 7755
Porfiromycin, 7756
Porofor-57 see 949
Porofor BSH [Bayer], 7757
Porphin see 7758
Porphine, 7758
21*H*,23*H*-Porphine see 7758
Porphobilinogen, 7759
Porphyrillic Acid, 7760
Porphyrins, 7758
Porphyropsin, 7761
Portamycin [Upjohn] see 8982
Portolac [Zyma] see 5352
Portyn [Parke, Davis] see 1110

Potassium Stearate, 7844
Potassium Succinate see 9038
Potassium Sulfantimonate see 7864
Potassium Sulfate, 7845
Potassium Sulfhydrate see 7775
Potassium Sulfide, 7846
Potassium Sulfite, 7847
Potassium Sulfobenzoate, 7848
Potassium Sulfocarbolate see 7826
Potassium Sulfocarbonate see 7865
Potassium Sulfocyanate see 7866
Potassium Tartrate, 7849
Potassium Tartratoborate see 7778
Potassium Tellurate(IV), 7850
Potassium Tellurate(VI), 7851
Potassium Tellurite see 7850
Potassium Tetraborate, 7852
Potassium Tetrabromoaurate(III), 7853
Potassium Tetrachloroaurate(III), 7854
Potassium Tetrachloroplatinate(II), 7855
Potassium Tetracyanocadmate see 1660
Potassium Tetracyanomercurate(II), 7856
Potassium Tetracyanonickelate(II), 7857
Potassium Tetracyanoplatinate(II), 7858
Potassium Tetrafluoroberyllate see 1218
Potassium Tetrafluoroborate, 7859
Potassium Tetrahydroborate see 7777
Potassium Tetraiodoaurate(III), 7860
Potassium Tetraiodocadmate, 7861
Potassium Tetraiodomercurate(II), 7862
Potassium Tetroxalate, 7863
Potassium Thioantimonate(V), 7864
Potassium Thiocarbonate, 7865
Potassium Thiocyanate, 7866
Potassium Thiosulfate, 7867
Potassium Titanyl Oxalate, 7868
Potassium Triiodide, 7869
Potassium Triiodomercurate(II) Solution, 7870
Potassium Triiodozincate, 7871
Potassium Trioxalatochromate(III) see 2285
Potassium Trioxalatotriaquochromate(III) see 2285
Potassium Trithiocarbonate see 7865
Potassium Troclosene see 9896
Potassium Tungstate(VI), 7872
Potassium Uranate(VI), 7873
Potassium Uranyl Nitrate, 7874
Potassium Uranyl Sulfate, 7875
Potassium Xanthogenate, 7876
Potassium Zinc Iodide see 7871
Potassium Zinc Sulfate, 7877
Potassium Zirconium Fluoride see 7804
Potassium Zirconium Sulfate, 7878
Potassuril [Cochard] see 7796
Potato Branching Enzyme see 8207
Potato-spirit Oil see 5212
Potcrate see 7782
Pot Curare see 2741
Potency Wood see 6382
Pounce [FMC] see 7321
Povan [Parke, Davis] see 8206
Povanyl [Parke, Davis] see 8206
Povidone, 7879
Povidone-Iodine, 7880
Poviral [Roemmers] see 148
Powder of Algaroth see 734
Powmil [Sumitomo] see 3158
PP-009 see 4152

PP-062 see 7651
PP-148 see 7165
PP-149 see 3786
PP-175 see 5877
PP-192 see 4153
PP-211 see 7652
PP-296 see 3131
PP-333 see 7118
PP-383 see 2836
PP-450 see 4246
PP-511 see 7652
PP-557 see 7321
PP-581 see 1400
PP-588 see 1519
PP-604 see 9699
PP-675 see 3266
PP-781 see 3499
PP-910 see 7165
PP-993 see 9266
PPACK, 7881
PPD see 9937
P. P. Factor see 6612
PPL see 7233
PPN1011 see 9383
PQ-9 see 7681
PQs see 7681
PR-66 see 7632
PR-82/3 see 19
Practolol, 7882
Praecicalm [Molimin] see 7272
Praeciglucon [Pfleger] see 4486
Praenitron [Pfizer] see 9900
Praequine see 7134
Pragman Gelee [Milanfarma] see 9662
Pragmazone [UPSA] see 9712
Pragmoline [Specia] see 87
Praiden [Italchemie] see 1454
Prajmaline, 7883
Prajmaline Bitartrate see 7883
Prajmalium see 7883
Pralidoxime Chloride, 7884
Pralidoxime Iodide see 7884
Pralidoxime Mesylate see 7884
Prälumin see 2780
Pramace [Hassle] see 8283
Pramidex [Berk] see 9646
Pramiel [Teikoku Kagaku] see 6226
Pramil [Hässle] see 8563
Pramindole (obsolete) see 5091
Pramipexole, 7885
Pramiracetam, 7886
Pramitol [Ciba-Geigy] see 7972
Pramiverin, 7887
Pramiverine see 7887
Pramocaine see 7888
Pramoxine, 7888
Prandiol [Bottu] see 3410
Pranlukast, 7889
Pranone [Schering] see 3787
Pranoprofen, 7890
Prano-Puren [Klinge Nattermann] see 8025
Pranosina [Newport] see 5006
Pranosine [Newport] see 5006
Pranoxen [Napp] see 6504
Prantal [Schering] see 3361
Präparat 5968 see 4800
Präparat 9295 see 929
Praseodymium, 7891
Prasterone, 7892
Pratensein, 7893
Praticef [Zambon] see 1980
Pravachol [Squibb] see 7894
Pravacilin [Recordati] see 5986
Pravaselect [Menarini] see 7894
Pravastatin Sodium, 7894
Pravidel [Sandoz] see 1437
Praxadium [Bottu] see 6784
Praxilene [Oberval] see 6440
Praxis [Lisapharma] see 4999

Praxiten see 7059
Praxiten SP [Wyeth] see 406
Prazene [Parke, Davis] see 7895
Prazepam, 7895
Prazil [Pierzoo] see 9073
Prazine [Wyeth] see 7966
Prazinil [Specia] see 1911
Praziquantel, 7896
Prazol [Corvi] see 3736
Prazosin, 7897
Prean [Chemil] see 5813
Prebediolone Acetate see 77
Precef [Bristol] see 1982
Preceptin [Ortho] see 6858
Precipitated Bismuth see 1297
Precipitated Calcium Carbonate see 1697
Precipitated Calcium Sulfate see 1753
Precipitated Chalk see 1697
Precipitated Silica see 8634
Precipitated Sulfur see 9150
Precipité Blanc see 5951
Precocene I see 7898
Precocene II see 7898
Precocenes, 7898
Precor [Sandoz] see 6063
Precortancyl see 7901
Precortilon [Heilit] see 7901
Precortisyl [Roussel] see 7901
Precursor To ACTH-LPH-β-endorphin see 7975
Predalon [Organon] see 2273
Predalone T.B.A. [Forest] see 7901
Predef [Upjohn] see 5190
Predenema [Pharmax] see 7901
Predfoam [Pharmax] see 7901
Pred Forte [Allergan] see 7901
Pred Mild [Allergan] see 7901
Prednacinolone see 2973
Prednelan [Glaxo] see 7901
Prednesol [Glaxo] see 7903
Prednicarbate, 7899
Prednicen [Central Pharm.] see 7901
Prednilonga [Dorsch] see 7904
Prednimustine, 7900
Predniretard [Boots] see 7901
Prednisolamate see 7902
Prednisolone, 7901
Prednisolone 21-[4-[p-[Bis(2-chloroethyl)amino]phenyl]butyrate] see 7900
Prednisolone 21-Diethylaminoacetate, 7902
Prednisolone 21-N,N-Diethylglycinate see 7902
Prednisolone 17-Ethylcarbonate 21-Propionate see 7899
Prednisolone Hemisuccinate see 7901
Prednisolone Phosphate Disodium see 7903
21-Prednisolonephosphoric Acid Disodium Salt see 7903
Prednisolone 21-Pivalate see 7901
Prednisolone Sodium Metasulfobenzoate see 7901
Prednisolone Sodium Phosphate, 7903
Prednisolone Sodium Succinate see 7901
Prednisolone Steaglate see 7901
Prednisolone Tebutate see 7901
Prednisolone 17-Valerate see 7905
Prednisone, 7904
Prednitop [Hoechst] see 7899
Prednival, 7905
Prednylidene, 7906
Predonine [Shionogi] see 7901
Predsol [Glaxo] see 7903
Prefamone [Dexo] see 3175
Preferid [Astra] see 1490
Prefrin [Allergan] see 7440

Pregard [Ciba-Geigy] see 7954
Preglandin [Ono] see 4393
Pregna-1,4-diene-11β,21-diol-3,20-dione[17α,16α-d]-2'-methyloxazoline 21-Acetate see 2916
1,4-Pregnadiene-17α,21-diol-3,11,20-trione see 7904
(9β,10α)-Pregna-4,6-diene-3,20-dione see 3524
10α-Pregna-4,6-diene-3,20-dione see 3524
1,4-Pregnadiene-3,20-dione-11β,17α,-21-triol see 7901
Δ¹,⁴-Pregnadiene-16-methylene-11β,-17α,21-triol-3,20-dione see 7906
1,4-Pregnadiene-11β,17α,21-triol-3,-20-dione see 7901
(17α)-Pregna-2,4-dien-20-yno[2,3-d]-isoxazol-17-ol see 2875
Pregnancy Urine Extract see 2273
Pregnane, 7907
(5α)-Pregnane see 266
5β-Pregnane see 7907
Pregnanediol, 7908
(3α,5α,20R)-Pregnane-3,20-diol see 268
(3α,5β,20S)-Pregnane-3,20-diol see 7908
(3β,5α,20R)-Pregnane-3,20-diol see 270
5α-Pregnane-3α,20α-diol see 267
5α-Pregnane-3β,20α-diol see 269
3,20-Pregnanedione, 7909
5α-Pregnane-3,20-dione see 273
(5β)-Pregnane-3,20-dione see 7909
11,20-Pregnanedione-3α,17α,21-triol see 9350
(3α,5β,11β,20S)-Pregnane-3,11,17,-20,21-pentol see 2604
5α-Pregnane-3β,11β,17,20β,21-pentol see 274
5β-Pregnane-3α,11β,17,20α,21-pentol see 2604
5β-Pregnane-3α,11β,17,20β,21-pentol see 2604
Pregnane-3,17,20,21-tetrol see 275
3α,17α,20α,21-Pregnanetetrol-11-one see 2605
5α-Pregnane-3β,17,20α-triol see 278
5α-Pregnane-3β,17,20β-triol see 279
3α,17α,21-Pregnanetriol-11,20-dione see 9350
Pregnanolone see 7910
Pregnan-3α-ol-20-one, 7910
Pregnant Mare Serum Gonadotropin see 2273
4-Pregnene-11β,21-diol-3,20-dione see 2601
4-Pregnene-16α,17α-diol-3,20-dione see 238
4-Pregnene-17α,21-diol-3,20-dione see 2977
4-Pregnene-20,21-diol-3,11-dione, 7911
5-Pregnene-3β,21-diol-20-one 21-Acetate see 77
Δ⁴-Pregnene-17α,21-diol-3,11,20-trione see 2602
Pregn-4-ene-3,20-dione see 7956
Δ⁴-Pregnene-3,20-dione see 7956
4-Pregnene-11β,17α,20β,21-tetrol-3-one, 7912
4-Pregnene-11β,17α,21-triol-3,20-dione see 4828
4-Pregnene-17α,20β,21-triol-3,11-dione, 7913
4-Pregnene-17α,20β,21-triol-3-one, 7914
17α-Pregnenetriolone see 7914
Pregn-4-ene-3,11,20-trione see 5317

Pregneninolone see 3787
4-Pregnen-17α-ol-3,20-dione see 4886
4-Pregnen-21-ol-3,20-dione see 2947
Pregnenolone, 7915
Δ⁵-Pregnen-3β-ol-20-one see 7915
Δ⁴-Pregnen-21-ol-3,11,20-trione see 2923
Pregnesin [Serono] see 2273
Pregnyl [Organon] see 2273
Pregova [Ortho] see 4299
Prehemataminic Acid see 6566
Prelis [Ciba-Geigy] see 6235
Prelital see 1530
Prelog-Djerassi Lactone, 7916
(+)-Prelog-Djerassi Lactonic Acid see 7916
Preludin [Boehringer, Ing.] see 7385
Premarin [Ayerst] see 2571
Premarrubiin see 5794
Premerge [Dow] see 3341
Premier [Bayer] see 4946
Premocillin see 7226
Prempar [Duphar] see 8401
Premsyn PMS [Chattem] see 7133
Prenalex [Itherapia] see 9318
Prenalterol, 7917
Prenazone see 4053
Prenormine [ICI-Farma] see 892
Prenoxdiazine Hydrochloride, 7918
Prent [Bayer] see 16
Prentox [Prentiss] see 7629
Prenylamine, 7919
4-Prenyl-1,2-diphenyl-3,5-dioxopyrazolidine see 4053
4-Prenyl-1,2-diphenyl-3,5-pyrazolidinedione see 4053
Prepalin see 10150
Pre-Par [Duphar] see 8401
Preparation K see 3449
Prepared Calcium Carbonate see 1697
Prepared Chalk see 1697
Pre-Pen [Kremers-Urban] see 7233
Prephenic Acid, 7920
Prepidil [Upjohn] see 8064
PreproANF see 903
Prepulsid [J & J] see 2377
Pres [Dieckmann] see 3605
Presamine [USV] see 4955
Pre-Sate [Warner-Lambert] see 2235
Prescal [Ciba-Geigy] see 5260
Presdate [Alfa] see 5341
Presfersul see 4105
Presidal see 7243
Presidon see 8179
Presinol [Bayer] see 6132
Presmode [Sankyo] see 1669
Pressalolo [Locatelli] see 5341
Pressamina [Zambeletti] see 3314
Pressionorm [Merrell Pharma] see 4409
Pressonex [Winthrop] see 5993
Pressorol [Baxter] see 5993
Pressunic [Unipharm] see 3212
Pressural [Polifarma] see 4969
Prestociclina see 7235
PreSun [Westwood] see 443
Presuren see 4868
Prethcamide see 2659
Pretilachlor, 7921
Prétonine [Synthélabo] see 4895
Pretor [Hoechst] see 1983
Pretyrosine see 826
Prevangor see 7249
Preventol G-D see 3120
Prevex [Schering-Plough] see 3991
Previscan [Nativelle] see 4168
Prexan [LaFare] see 6504
Prexidil [Bioindustria] see 6290
PR-G 138-Cl see 2326
Priadel [Delandale] see 5552

Priamide [Janssen] see 5221
Priaxim [Ravizza] see 4183
Priceite see 1691
Pridinol, 7922
Pridinol Mesylate see 7922
Prifinium Bromide, 7923
Prilagin [Gambar] see 6266
Prilocaine, 7924
Prilon [Cassenne] see 7078
Prilosec [Astra] see 6977
Primacaine see 5978
Primacor [Sterling-Winthrop] see 6284
Primal see 395
Primalan [Pharmuka] see 5911
Primamycin see 4641
Primantron [Schering] see 2273
Primaquine, 7925
Primary Ammonium Phosphate see 577
Primary Calcium Phosphate see 1740
Primary Isoamyl Alcohol see 5212
Primary Magnesium Phosphate see 5719
Primary Sodium Phosphate see 8806
Primatene Mist [Whitehall] see 3656
Primatol [Ciba-Geigy] see 7972
Primatol A [Geigy] see 902
Primatol S [Geigy] see 8681
Primaverine see 7887
Primaxin see 4954
Primbactam [Menarini] see 955
Primeral [Master] see 6504
Primeverose, 7926
Primicid [ICI] see 7652
Primidone, 7927
Primisulfuron-methyl, 7928
Primo [Ciba] see 9855
Primobolan-Depot [Schering AG] see 6044
Primobolan Tablets [Schering AG] see 6044
Primocarcin, 7929
Primofax [Rhône-Poulenc] see 6816
Primofenac [Streuli] see 3132
Primofol [Schering AG] see 3746
Primogonyl [Schering] see 2273
Primogyn C [Schering AG] see 3780
Primolut N [Schering AG] see 6790
Primolut-Nor [Schering AG] see 6790
Primonabol see 6044
Primonabol Depot see 6044
Primosiston [Schering AG] see 6790
Primoteston [Schering AG] see 9322
Primperan [Delagrange] see 6226
Primulaverin, 7930
Primulidin see 5754
Primulin see 5754
Primycin, 7931
Prinadol [SK & F] see 7360
Prinalgin [Berk] see 220
Princep [Geigy] see 8681
Prince's Feather see 392
Prince's Pine see 2099
Princillin [Squibb] see 628
Principen [Squibb] see 628
Principen/N [Squibb] see 628
Prinil [Merck & Co.] see 5540
Prinivil [Merck & Co.] see 5540
Prinodolol see 7597
Prinzide [Merck & Co.] see 5540
Prinzone [Squibb] see 9068
Priodax [Schering] see 5047
Prioderm [Coates & Cooper] see 5740
Priomicina [San Carlo] see 4280
Priscol [Ciba] see 9645
Priscoline [Ciba] see 9645
Prisilidene see 317
Pristacin see 2074
Pristane, 7932
Pristinamycin, 7933

Pristinamycin I$_A$ see 6275
Pristinamycin I$_B$ see 10096
Pristinamycin I$_C$ see 10096
Pristinamycin II$_A$ see 10142
Privadol [Roland Marie] see 4443
Privaprol [Lepetit] see 5615
Privenal see 4742
Privine Hydrochloride [Ciba] see 6455
PRN see 7469
Pro see 7963
Proaccelerin see 3961
Pro-Actidil [Wellcome] see 9877
Pro-Air [Warner-Lambert] see 7939
Proamipide see 8296
Proaqua [Reid-Rowell] see 1155
Proasma [Radiumfarma] see 6077
Proazamine see 7970
Probamyl [Cont. Pharma] see 5908
Proban [Am. Cyanamid] see 2854
Pro-Banthine [Roberts] see 7989
Probe [Velsicol] see 6032
Probecid [Astra] see 7934
Proben [SCS] see 7934
Probenecid, 7934
Probilin [Gödecke; Warner-Lambert] see 7639
Probolin [Tennessee Pharm.] see 6017
Probon [Chinoin] see 8391
Probonal [Medipolar] see 8391
Probucol, 7935
Pro-Bumin [Lederle] see 8613
Probutylin [Rorer] see 7937
Procadil [Recordati] see 7939
Procainamide Hydrochloride, 7936
Procaine, 7937
Procaine Amide Hydrochloride see 7936
Procaine Benzylpenicillinate see 7226
Procaine Penicillin G see 7226
Procalmadiol see 5908
Procalmidol see 5908
Procamide [Sintesa] see 7936
Procanodia see 7226
Procan-SR [Parke, Davis] see 7936
Procapan [Panray] see 7936
Procaptan [Stroder] see 7311
Procarbazine, 7938
Procardia [Pfizer] see 6617
Procaterol, 7939
Procef [Bristol-Myers Squibb] see 1992
Procerin, 7940
Processine [Sanko] see 2365
Procetofen see 4019
Procetofene see 4019
Procetoken [Bernabo] see 4019
Prochloraz, 7941
Prochlorpemazine see 7942
Prochlorperazine, 7942
Prochlorperazine Mesylate see 7942
Procholon [Squibb] see 2922
Prociclide [Crinos] see 2915
Proclaim [Merck & Co.] see 3594
Proclival [Houdé] see 1495
Proclorperazine see 7942
Procodazole, 7943
Procollagen see 2543
Proconazole see 8003
Proconvertin see 3962
Procorum [Knoll] see 4369
Procrit [Ortho Biotech.] see 3729
Proctocort [Solvay] see 4828
Proctodon [Rowell] see 3360
Proctofoam [Reed & Carnrick] see 7888
Procyanidins see 1950
Procyclidine, 7944
Procyclomin see 3147
Procylin [Kaken] see 1190
Procymate, 7945

Procymidone, 7946
Procysteine [Baxter] see 7084
Procytox [Horner] see 2816
Pro-Dafalgan [UPSA] see 7976
Prodasone [Upjohn] see 5838
Prodectin [Gedeon Richter] see 8158
Prodel see 3524
Pro-Diaban [Bayer; Schering AG] see 4453
Prodiamine, 7947
Prodigiosin, 7948
Prodigiosine see 7948
α-Prodine see 317
β-Prodine see 317
Prodipine, 7949
Prodlure, 7950
Pro Dorm see 5609
Prodox [Upjohn] see 4886
Prodoxol [Warner-Chilcott] see 7079
Pro-Drone®, 7951
Producer Gas, 7952
Producil [Merck & Co.] see 3567
Proendotel [Fidia] see 2467
Pro-Entra [Wellcome-Sumitomo] see 9877
Profamina see 623
Profasi [Serono] see 2273
Profasi HP [Serono] see 2273
Profemin [Yamanouchi] see 4331
Profenamine see 3793
Profenid [Rhône-Poulenc Rorer] see 5316
Profenine see 386
Profenon [Morishita] see 4142
Profenone see 7176
Proferrin [Merck & Co.] see 4073
Profetamine Phosphate [Clark & Clark] see 623
Profibrinolysin see 7679
Profilate [Alpha Ther.] see 3963
Profilnine [Alpha Ther.] see 3964
Proflavine, 7953
Proflavine Hemisulfate see 7953
Proflax [Merck & Co.] see 9585
Proflex [Zyma] see 4925
Profluralin, 7954
Profoliol B [Schering] see 3746
Proformiphen see 7412
Proftril [Norden] see 211
Progabide, 7955
Progallin P [Nipa] see 8044
Proge [Mochida] see 4886
Progeril [Midy] see 3692
Progesic [Lilly] see 4021
Progestasert [Alza] see 7956
Progesterone, 7956
Progestin [Organon] see 7956
Progestogel [Besins-Iscovesco] see 7956
Progestol [Bruco] see 7956
Progeston [Fuji] see 7956
Progestoral [Organon] see 3787
Proglicem [Essex] see 3051
Proglumetacin, 7957
Proglumide, 7958
Proglycem [Schering] see 3051
Progon [ISM] see 2273
Prograf [Fujisawa] see 9200
Program [Ciba-Geigy] see 5624
Proguanil see 2138
Progynon [Schering] see 3746
Progynon B [Schering AG] see 3746
Progynon C [Schering AG] see 3780
Progynon Depot [Schering AG] see 3746
Progynon M [Schering AG] see 3780
Progynova [Schering AG] see 3746
ProHance [Bristol-Myers Squibb] see 4348
Proheptatriene see 2782

Proheptazine, 7959
Proinsulin, 7960
Prokarbol see 3331
Prokayvit Oral [BDH] see 5873
Proketazine [Wyeth] see 1910
Prokine [Hoechst-Roussel] see 4559
Prokrein [Tobishi] see 5291
Prolactin, 7961
Proladone [Boots] see 7093
Prolamins, 7962
Prolan [Comm. Solvents] see 6691
Prolastin [Miles] see 760
Prolate [Stauffer] see 7492
Prolax see 5895
Prolene [Industrial Rayon] see 7741
Proleukin [Chiron] see 5020
Prolidon [Carnot] see 7956
Proline, 7963
L-Proline see 7963
Prolintane, 7964
Prolixan [Siegfried; Logeais] see 768
Prolixin [Bristol-Myers Squibb] see 4226
Prolixin Decanoate [Bristol-Myers Squibb] see 4226
Prolixin Enanthate [Bristol-Myers Squibb] see 4226
Prolonium Iodide, 7965
Proloprim [Wellcome] see 9840
Proluton [Schering AG] see 7956
Proluton Depot [Schering AG] see 4886
L-Prolyl-L-leucylglycinamide see 5855
Prolysis [Wellcome] see 9608
Proma [Armour Pharm.] see 2238
Promac [Zeria] see 7712
Promacetin [Parke, Davis] see 73
Promacid [Knoll] see 2238
Promacortine see 6189
Promactil see 2238
Promanide see 4471
Promaquid [Rhône-Poulenc] see 4260
Promassol [DDR] see 395
Promassolax see 7108
Promate [Truett] see 5908
Promaxol [Esteve] see 7939
Promazil see 2238
Promazine, 7966
Promecarb, 7967
Promecon [Endopharm] see 1154
Promedol, 7968
α-Promedol see 7968
γ-Promedol see 7968
Promegestone, 7969
Promethazine, 7970
Promethazine Teoclate see 7970
Promethestrol see 6045
γ-Promethestrol see 6045
Promethium, 7971
Prometon, 7972
Prometone see 7972
Prometryn, 7973
Promid [Kaken] see 7958
Promidione see 5093
Promin [Parke, Davis] see 4471
Prominal see 5899
Promine see 8332
Promintic [Ayerst] see 6247
Promit [Knoll; Braun Melsungen] see 2989
Promizole [Parke, Davis] see 9444
Promotil [Boehringer, Ing.] see 7964
Promotin [Nippon Chemiphar] see 5291
Promwill [Will] see 7966
Pronamid see 8058
Prondol [Wyeth] see 5091
Pronestyl [Bristol-Myers Squibb] see 7936
Pronethalol, 7974

Pronilin [Akzo] see 5638
Pronison [Galenika] see 7904
Pronon [Yamanouchi] see 7978
Pronox [Reiss] see 2780
Prontalbin see 9094
Prontalgin [SPA] see 4925
Prontamid [SIT] see 9067
Prontobario [Bracco] see 1023
Prontocalcin [Dompé] see 1680
Prontoglucal see 4472
Prontoglukal see 4472
Prontosil see 9092
Prontosil Album see 9094
Prontosil Flavum see 9092
Prontosil Rubrum see 9092
Prontylin see 9094
Proopiocortin see 7975
Pro-opiomelanocortin, 7975
Propacetamol, 7976
Propachlor, 7977
Propacil [Chugai] see 8054
Propaderm [Allen & Hanburys] see 1047
1,2-Propadiene-1,3-dione see 1863
Propadrine [Merck & Co.] see 7461
Propafenone, 7978
Propagermanium, 7979
Propal see 5236
Propallylonal, 7980
Pro-Pam [Protea] see 3042
Propamidine, 7981
Propaminodiphen see 7887
Propanal see 8008
Propanamide see 8009
1-Propanamine see 8028
2-Propanamine see 5228
Propane, 7982
1-Propanearsonic Acid, 7983
1,2-Propanediamine see 8037
1,3-Propanedicarboxylic Acid see 4481
Propane-diethyl Sulfone see 9133
Propanedinitrile see 5750
Propanedioic Acid see 5749
Propanedioic Acid Bis(1,1-dimethyl-ethyl) Ester see 3085
Propanedioic Acid Diethyl Ester see 3869
Propanedioic Acid Dimethyl Ester see 6169
1,2-Propanediol see 8040
1,3-Propanediol see 9845
1,3-Propanediol Bis[α-(p-chlorophen-oxy)isobutyrate] see 8684
1,3-Propanediol Bis[2-(4-chlorophen-oxy)-2-methylpropionate] see 8684
1,2-Propanediol-3-(p-chlorophenoxy)-1-carbamate see 2231
1,3-Propanedithiol, 7984
4,4′-[1,3-Propanediylbis(oxy)]bisbenz-enecarboximidamide see 7981
4,4′-[1,3-Propanediylbis(oxy)]bis(3-bromobenzenecarboximidamide) see 3073
Propanenitrile see 8012
Propanethial S-Oxide, 7985
1,2,3-Propanetricarboxylic Acid see 9752
1,2,3-Propanetricarboxylic Acid 1,1′-[1-(12-Amino-4,9,11-trihydroxy-2-methyltridecyl)-2-(1-methylpentyl)-1,2-ethanediyl] Ester see 4311
1,2,3-Propanetriol see 4493
1,2,3-Propanetriol 1-(4-Aminoben-zoate) see 4496
1,2,3-Propanetriol Diacetate see 3005
1,2,3-Propanetriol Monoacetate see 6325
1,2,3-Propanetriol, Mono(dihydrogen

Phosphate) Calcium Salt (1:1) see 1713
1,2,3-Propanetriol Mono(dihydrogen Phosphate) Disodium Salt see 8767
1,2,3-Propanetriol Triacetate see 9721
1,2,3-Propanetriol Trinitrate see 6704
Propanidid, 7986
Propanil, 7987
Propanocaine, 7988
Propanoic Acid see 8010
Propanoic Acid Anhydride see 8011
Propanoic Acid Butyl Ester see 1623
Propanoic Acid Ethyl Ester see 3891
Propanoic Acid Methyl Ester see 6191
Propanoic Acid Propyl Ester see 8052
Propanoic Anhydride see 8011
1-Propanol see 8027
2-Propanol see 5227
2-Propanol Aluminum Salt see 359
β-Propanolamine see 488
Propanolide see 8005
2-Propanol Nitrite see 5235
2-Propanone see 64
Propanone 1-Oxime see 5207
2-Propanone Oxime see 75
Propanoyl Chloride see 8013
Propantheline Bromide, 7989
Propaphenin see 2238
Propaquizafop, 7990
Proparacaine, 7991
Propargil see 7992
Propargite, 7992
Propargyl Alcohol, 7993
Propargyl Chloride, 7994
Propargylic Acid see 8006
2-Propargyloxy-5-amino-N-butylben-zamide see 7177
Propasa [Merck & Co.] see 498
Propat [Hishiyama] see 9640
Propatyl Nitrate, 7995
Propavan [Kabi] see 8007
Propax see 7059
Propaxoline see 8090
Propazine, 7996
Propazol see 7943
Propazone see 3409
Pro-Pen [Merck & Co.] see 7226
2-Propenal see 130
2-Propenamide see 131
2-Propen-1-amine see 295
1-Propene see 8034
trans-1-Propene-1,2-dicarboxylic Acid see 5963
1-Propene Homopolymer see 7741
2-Propenenitrile see 133
Propene Oxide see 8041
2-Propene-1-sulfinothioic Acid S-2-Propenyl Ester see 258
1-Propene-1,2,3-tricarboxylic Acid see 119
2-Propenoic Acid see 132
2-Propenoic Acid Butyl Ester see 1574
2-Propenoic Acid Ethyl Ester see 3805
2-Propenoic Acid Methyl Ester see 6092
1-Propenol-3 see 294
2-Propen-1-ol see 294
1-Propen-2-ol Acetate see 5222
Propentofylline, 7997
1-Propen-2-yl Acetate see 5222
2-Propenylacrylic Acid see 8869
3-[[(2-Propenylamino)thioxomethyl]-amino]benzoic Acid Monocopper (1+) Monosodium Salt see 265
p-Propenylanisole see 682
5-(1-Propenyl)-1,3-benzodioxole see 5243
5-(2-Propenyl)-1,3-benzodioxole see 8468

(17β)-17-(2-Propenyl)estr-4-en-17-ol see 299
4-Propenylguaiacol see 5186
17-(2-Propenyl)morphinan-3-ol see 5485
4-(1-Propenyl)phenol see 715
4-(2-Propenyl)phenol see 2091
p-Propenylphenol see 715
2-(1-Propenyl)piperidine see 2565
N-2-Propenyl-2-propen-1-amine see 3015
2-Propenyl 3-(2-Propenylsulfinyl)-1-propenyl Disulfide see 195
(S)-3-(2-Propenylsulfinyl)-L-alanine see 260
(2-Propenyl)thiourea see 9500
N-2-Propenylurea see 306
Propenzolate, 7998
Properdin, 7999
Propericiazine see 7306
Propess [Roussel] see 8064
Propetamphos, 8000
Propethonum Iodide see 9794
Propham, 8001
Prophelan see 386
Prophenal see 7349
Prophenatin [Hishiyama] see 3132
Propheniramine see 7383
Prophenpyridamine see 7383
Prophyllen [Streuli] see 3529
Prophylux [Hennig] see 8025
Propicillin, 8002
Propicol see 3242
Propiconazole, 8003
Propildazine see 7577
Propine [Allergan] see 3400
Propineb, 8004
Propiocine [Roussel] see 3725
Propiocine Enfant [Roussel] see 3722
Propiodone see 8048
β-Propiolactone, 8005
Propiolic Acid, 8006
Propiomazine, 8007
Propionaldehyde, 8008
Propionamide, 8009
Propione see 3170
Propionic Acid, 8010
Propionic Acid Amide see 8009
Propionic Acid Butyl Ester see 1623
Propionic Acid Calcium Salt see 1745
Propionic Acid Sodium Salt see 8816
Propionic Anhydride, 8011
Propionitrile, 8012
β-Propionolactone see 8005
Propionopyrrothine see 914
4-Propionoxy-1,3-dimethyl-4-phenyl-hexamethylenimine see 7959
Propionylbenzene see 8015
Propionyl Chloride, 8013
3-Propionyl-10-(γ-dimethylamino-propyl)phenothiazine see 8014
Propionylerythromycin see 3725
Propionylerythromycin Lauryl Sul-fate see 3722
2-Propionyl-10-[γ-(N′-β-hydroxyeth-ylpiperazino)propyl]phenothiazine see 1910
3″-Propionylleucomycin A5 see 8408
p-Propionylphenol see 7176
N-Propionyl-2-(1-piperidinoisopropyl)-aminopyridine see 8017
Propionylpromazine, 8014
Propionylpromethazine see 8007
N-Propionyl-N-(2-pyridyl)-1-piperi-dino-2-aminopropane see 8017
Propiophenone, 8015
Propiopromazine see 8014
Propipocaine, 8016
Propiram, 8017
Propisamine see 623

Propisomide see 7271
Propitan [Eisai] see 7608
Propitocaine see 7924
Propiverine, 8018
Propizepine, 8019
Proplex T [Hyland] see 3964
Propofol, 8020
Propolis, 8021
Propolis Balsam see 8021
Propolis Resin see 8021
Propolis Wax see 8021
Proponex-Plus [Shell] see 5826
Propoxur, 8022
1-Propoxy-2-acetamino-4-nitrobenz-
ene see 6726
1-Propoxy-2-amino-4-nitrobenzene see
6727
5-Propoxy-2-benzimidazolecarbamic
Acid Methyl Ester see 7070
(5-Propoxy-1H-benzimidazol-2-yl)car-
bamic Acid Methyl Ester see 7070
(6-Propoxy-2-benzothiazolyl)carbamic
Acid Methyl Ester see 9599
Propoxycaine Hydrochloride, 8023
5-Propoxy-2-(carbomethoxyamino)-
benzimidazole see 7070
Propoxychel [Rachelle] see 8024
N-Propoxyethyl-N-chloroacetyl-2,6-
diethylaniline see 7921
Propoxyphene, 8024
l-Propoxyphene see 5495
α-d-Propoxyphene see 8024
4-Propoxyphenyl Piperidine Ethyl
Ketone see 8016
4-n-Propoxy-β-(1-piperidyl)propio-
phenone see 8016
Propranolol, 8025
Propranur [Henning] see 8025
Propulm [Chiesi] see 7939
Propulsin [Janssen] see 2377
Propycil [Kali-Chemie] see 8054
Propyl Acetate, 8026
Propylacetic Acid see 10042
N-Propylajmaline see 7883
N⁴-Propylajmalinium see 7883
n-Propyl Alcohol, 8027
Propylaldehyde see 8008
Propylamine, 8028
β-(n-Propylamino)-β,β-dimethylethyl
Benzoate see 5909
α-Propylamino-2-methylpropionanil-
ide see 7924
2-(Propylamino)-o-propionotoluidide
see 7924
3-Propylamino-α-propionylamino-2-
carbomethoxy-4-methylthiophene
see 1920
N-(α-Propylaminopropionyl)-o-tolu-
idine see 7924
n-Propylarsonic Acid see 7983
n-Propylbenzene, 8029
Propyl Bromide, 8030
S-Propyl Butylethylthiocarbamate see
7192
Propyl Butyrate, 8031
Propyl Carbinol see 1575
n-Propylcarbinyl Chloride see 1595
Propyl Chloride, 8032
N-Propyl-N'-(p-chlorobenzenesulfonyl)urea see 2239
Propyl Chlorocarbonate, 8033
Propyl Chloroformate see 8033
Propyl Cyanide see 1633
Propyldazine see 7577
Propyl [4-[(Diethylcarbamoyl)meth-
oxy]-3-methoxyphenyl]acetate see
7986
(±)-1-n-Propyl-3α-diethylsulfamoyl-
amino-6-hydroxy-1,2,3,4,4aα,5,10,-

10aβ-octahydrobenzo[g]quinoline
see 8226
Propyl 3,5-Diiodo-4-pyridone-N-
acetate see 8048
S-Propyl Dipropylthiocarbamate see
10097
Propylene, 8034
[Propylenebis(dithiocarbamato)]zinc
see 8004
Propylene Chlorohydrin, 8035
sec-Propylene Chlorohydrin, 8036
Propylenediamine, 8037
Propylene Dibromide, 8038
Propylenedicarboxylic Acid see 5261
Propylene Dichloride, 8039
(±)-4,4'-Propylenedi-2,6-piperazine-
dione see 8295
Propylene Glycol, 8040
Propylene Oxide, 8041
Propylene Polymer see 7741
Propyl Ether, 8042
Propyl Ethyl-n-butylthiolcarbamate
see 7192
Propylethylene see 7262
Propyl Formate, 8043
Propyl Gallate, 8044
n-Propyl Gallate see 8044
Propylhexedrine, 8045
Propyl Hydride see 7982
Propyl p-Hydroxybenzoate see 8051
Propylic Alcohol see 8027
Propylidene Chloride, 8046
Propyl Iodide, 8047
Propyliodone, 8048
D-6-n-Propyl-8β-methylmercapto-
methylergoline see 7304
n-Propyl Nitrate, 8049
n-Propyl Nitrite, 8050
19-Propylorvinol see 3932
Propylparaben, 8051
Propyl Parasept [Tenneco] see 8051
2-Propylpentamide see 10050
2-Propylpentanamide see 10050
2-Propylpentanoic Acid see 10049
1-(α-Propylphenethyl)pyrrolidine see
7964
l-N-n-Propylpipecolic Acid-2,6-xyli-
dide see 8417
(S)-(−)-1-Propyl-2',6'-pipecoloxyli-
dide see 8417
2-n-Propyl-Δ¹-piperidine see 2566
(S)-2-Propylpiperidine see 2569
(3S-trans)-6-Propyl-3-piperidinol see
8099
6-Propylpiperonyl Butyl Diethylene
Glycol Ether see 7629
N-Propyl-1-propanamine see 3406
Propyl Propionate, 8052
2-Propyl-4-pyridinecarbothioamide see
8076
N-Propyl-N-4-pyridinyl-1H-indol-1-
amine see 1223
1-(Propyl-4-pyridylamino)indole see
1223
6'-n-Propylspectinomycin see 9917
Propyl Sulfide, 8053
2-n-Propyl-3,4,5,6-tetrahydropyridine
see 2566
[5-(Propylthio)-1H-benzimidazol-2-
yl]carbamic Acid Methyl Ester see
211
2-Propyl-4-thiocarbamoylpyridine see
8076
5-(Propylthio)-2-carbomethoxyamino-
benzimidazole see 211
2-Propylthioisonicotinamide see 8076
Propylthiouracil, 8054
6-Propyl-2-thiouracil see 8054
Propyl-Thyracil [Frosst] see 8054
1-[N-Propyl-N-[2-(2,4,6-trichlorophen-

oxy)ethyl]carbamoyl]imidazole see
7941
N-Propyl-N-[2-(2,4,6-trichlorophen-
oxy)ethyl]-1H-imidazole-1-carbox-
amide see 7941
(E)-10-Propyl-5,9-tridecadien-1-ol
Acetate see 8055
10-Propyl-trans-5,9-tridecadienyl
Acetate see 8055
Propylure, 8055
2-Propylvaleramide see 10050
2-Propylvaleric Acid see 10049
2-Propynoic Acid see 8006
2-Propyn-1-ol see 7993
2-Propynyl (R)-2-[4-(5-Chloro-3-
fluoro-2-pyridyloxy)phenoxy]pro-
pionate see 2431
1-(2-Propynyl)cyclohexanol Car-
bamate see 4733
1-(2-Propynyl)cyclohexyl Carbamate
see 4733
2-Propynyl (E,E)-3,7,11-Trimethyl-2,-
4-dodecadienoate see 5326
Propyon [Makhteshim-Agan] see 8022
Propyphenazone, 8056
Propyromazine, 8057
Propyzamide, 8058
Proquamezine see 487
Proquazone, 8059
Prorenal [Dainippon] see 5514
Prorennin see 8303
Proresid [Sandoz] see 7703
Prorex [Hyrex] see 7970
Prosaponin B see 3737
Proscar [Merck & Co.] see 4125
Proscillan [Streuli] see 8060
Proscillaridin, 8060
Proscillaridin A see 8060
Proscillaridin-4-methyl Ether see 8060
Proscorbin see 867
Prosedar [Schering-Plough] see 8211
Proseptazine see 1182
Proserine see 6553
Proserout [Funai] see 5820
Proserum [Dow] see 8613
Proseryl [Funai] see 5820
Prosulvin [Akzo] see 5638
ProSom [Abbott] see 3744
Pro-Sonil see 2780
Prosopis Gum see 5972
Prostacyclin, 8061
Prostaglandin(s), 8062
Prostaglandin E₁, 8063
Prostaglandin E₂, 8064
Prostaglandin F₂α, 8065
Prostaglandin I₂ see 8061
Prostaglandin X see 8061
Prostal [Teikoku Zoki] see 2152
Prostalene, 8066
Prostandin [Ono] see 8063
Prostanoic Acid see 8062
Prostap [Lederle] see 5484
Prostaphlin [Bristol] see 7036
Prostaphlin-A [Galenika] see 2480
Prostarex [Biorex] see 1259
Prostarmon-E [Ono] see 8064
Prostarmon F [Ono] see 8065
Prostasal [TAD] see 8697
Prostavet [Gifavet] see 3916
Prostenoglycine see 8963
Prostetin [Takeda] see 7065
Prostianol see 5638
Prostide [Sigma-Tau] see 4125
Prostigmin [Roche] see 6553
Prostigmin (amp.) [Roche] see 6553
Prostin/15M [Upjohn] see 1871
Prostin E₂ [Upjohn] see 8064
Prostine VR [Upjohn] see 8063
Prostin F₂ Alpha [Upjohn] see 8065
Prostivas [Upjohn] see 8063

Prostosin [Iwaki] *see* 8060
Prostrumyl *see* 6210
Prosulfuron, 8067
Prosultiamine, 8068
Prosymasul [Lederle] *see* 9114
Proszine [Teisan] *see* 8060
Protabol [Draco] *see* 9596
Protacine *see* 7957
Protactinium, 8069
Protactyl [Wyeth] *see* 7966
Protagent [Alcon-Thilo] *see* 7879
Protalba *see* 8086
Protamine Zinc Insulin *see* 5011
Protamone-D [Agri-Tech] *see* 9554
Protanabol *see* 7101
Protanal *see* 240
Protandren [Ciba] *see* 6017
Protangix [Lefrancq] *see* 3410
Protara [Gensia] *see* 13
Protargin *see* 8671
Protargol [Winthrop] *see* 8671
Protasin [Bayropharm] *see* 8060
Protaxon [Opfermann] *see* 7957
α₁-Protease Inhibitor *see* 760
Protectins *see* 5453
Protecton [Yoshitomi] *see* 4962
Protein P5 *see* 909
Protein P9 *see* 1960
α₁-Proteinase Inhibitor *see* 760
Protein C, 8070
Protein Hydrolysates, 8071
Protein S *see* 8070
Proterciclina [Proter] *see* 8411
Proternol *see* 5236
Prothazine [Rhone-Poulenc Rorer] *see* 7970
Protheobromine, 8072
Prothiaden [Marion] *see* 3485
Prothil [Kali-Chemie] *see* 5836
Prothionamide *see* 8076
Prothipendyl, 8073
Prothromadin *see* 10174
Prothrombase *see* 8074
Prothrombin, 8074
Protigényl [Ronohèse] *see* 8071
Protiofate, 8075
Protionamide, 8076
Protium *see* 4833
Protizinic Acid, 8077
Protoactinium *see* 8069
Protoanemonin, 8078
Protocatechualdehyde, 8079
Protocatechualdehyde Dimethyl Ether *see* 10084
Protocatechuic Acid, 8080
Protocatechuic Acid Methylene Ether *see* 7630
Protocatechuic Aldehyde *see* 8079
Protoenstatite *see* 5727
Protoescigenin *see* 3737
Protogen A *see* 9462
Protoheme *see* 4672
Protoheme IX *see* 4672
Protokosin, 8081
Protokylol, 8082
Protolipan [Millet] *see* 4019
Protomin *see* 4471
Protopam Chloride [Wyeth-Ayerst] *see* 7884
Protophenicol [Proto] *see* 2120
Protopine, 8083
Protoporphyrin IX, 8084
Protopyrin *see* 3776
(+)-Protoquercitol *see* 8218
Protosol *see* 3225
Protostephanine, 8085
Protostib *see* 6154
Protoveratrine A *see* 8086
Protoveratrine B *see* 8086
Protoveratrines, 8086

Protoverine, 8087
Protoverine 6,7-Diacetate 15-(2-Methylbutyrate) *see* 2963
Protoverine 6,7-Diacetate 15(*l*)-2'-Methylbutyrate *see* 2963
Protoxyl *see* 8719
Protriptyline, 8088
Protropin [Genentech] *see* 8864
Pro-UK *see* 8089
Pro U-PA *see* 8089
Pro-Urokinase, 8089
Prourokinase (Enzyme-activating) *see* 8089
Prourokinase (Enzyme-activating) (Human Clone PUK4/pUK18 Protein Moiety Reduced) *see* 8089
Proustite *see* 8647
Provas [Sanol Schwarz] *see* 1498
Provatene [Microbio] *see* 1902
Provell [Lilly] *see* 8086
Pro-Vent [Wellcome] *see* 9421
Proventil [Schering] *see* 217
Provera [Upjohn] *see* 5838
Proverine *see* 386
Provest [Upjohn] *see* 5838
Provigan [Reid-Provident] *see* 7970
Proviodine [Rougier] *see* 7880
Proviron [Schering AG] *see* 5974
Provisc [Alcon] *see* 4793
Provitamin D₃ *see* 2921
Provitar [Clin-Comar-Byla] *see* 7054
Provitina [Promonta] *see* 10157
Provocholine [Roche] *see* 6003
Prower Factor *see* 3965
Prowl [Am. Cyanamid] *see* 7211
Proxagesic [Tutag] *see* 8024
Proxazocain *see* 7888
Proxazole, 8090
Proxen [Syntex] *see* 6504
Proxibarbal, 8091
Proxibarbital *see* 8091
Proxigermanium *see* 7979
Proxil [Rottapharm] *see* 7957
Proxine [Del Saz & Filippini] *see* 6504
Proxol [NOR-AM] *see* 9753
Proxymetacaine *see* 7991
Proxyphylline, 8092
Proxy-Retardoral [Artesan] *see* 8092
Prozac [Lilly] *see* 4222
Prozapine, 8093
Prozil *see* 2238
Prozinex [Makhteshim-Agan] *see* 7996
Prulaurasin *see* 5760
Prulet [Mission Pharmacal] *see* 7108
Prunasin *see* 5760
Prunetin, 8094
Prunetol *see* 4395
Prunitrin *see* 8094
Prunol *see* 10027
Prunusetin *see* 8094
Pruralgan [Pharmacia] *see* 3267
Pruralgin [Pharmacia] *see* 3267
Prussian Blue *see* 4064
Prussic Acid *see* 4836
Pryleugan [Arzneimittelwerk Dresden] *see* 4955
PS-60 *see* 4389
PS-207 *see* 9261
P-Selectin *see* 8568
Pseudoaconitine, 8095
Pseudobaptigenin, 8096
Pseudobaptisin *see* 8096
Pseudobatrachotoxin *see* 1037
Pseudobutylbenzene *see* 1586
Pseudo-butylene *see* 1549
Pseudocef [Grünenthal] *see* 1994
Pseudochelerythrine *see* 8504
Pseudocholestane *see* 2587
Pseudocinchonine *see* 4824

Pseudococaine, 8097
Pseudocodeine, 8098
Pseudoconhydrine, 8099
Pseudocubebin *see* 8616
Pseudocumene, 8100
Pseudocumol *see* 8100
Pseudodigitoxin *see* 4441
d-Pseudoephedrine *see* 3645
Pseudoergostane *see* 2585
Pseudofructose *see* 8109
Pseudohecogenin *see* 4652
Pseudohyoscyamine *see* 6799
Pseudoionone, 8101
Pseudojervine, 8102
Pseudolibethenite *see* 2716
Pseudomalachite *see* 2716
trans-Pseudomonic Acid *see* 6383
Pseudomonic Acid A *see* 6383
Pseudomonic Acid B *see* 8103
Pseudomonic Acid C *see* 8103
Pseudomonic Acid D *see* 8103
Pseudomonic Acid I *see* 8103
Pseudomonic Acids, 8103
Pseudomonil [Ciba-Geigy] *see* 1994
Pseudomorphine, 8104
Pseudomorphine (C34 Alkaloid) *see* 8104
Pseudonorephedrine *see* 6811
Pseudopaederine *see* 7196
Pseudopederin *see* 7196
Pseudopelletierine, 8105
Pseudopunicine *see* 8105
3-Pseudotropanol *see* 8106
Pseudotropine, 8106
Pseudotropine Benzoate *see* 9904
Pseudo-wollastonite *see* 1749
Pseudoyohimbine, 8107
Psicaine *see* 8097
Psicaine N *see* 8097
Psichial [Lang] *see* 2132
Psicofuranine, 8108
9β-D-Psicofuranosyladenine *see* 8108
9-β-D-Psicofuranosyl-9*H*-purin-6-amine *see* 8108
Psicopax [Vemaco] *see* 5609
Psicoperidol [Lusofarmaco] *see* 9813
D-Psicose, 8109
Psicosoma [Trommsdorf] *see* 4477
Psicosterone [ICI] *see* 7892
Psicoterina [Francia] *see* 2132
Psicronizer [Albert-Pharma] *see* 6768
Psigodal [E. Merck] *see* 5902
Psilocin, 8110
Psilocybin, 8111
Psilocyn *see* 8110
Psiquium [Sintofarma] *see* 5829
PSK [Kureha], 8112
Psoil [Basotherm] *see* 9600
Psoraderm [Goupil] *see* 1196
Psoradrate [Eaton] *see* 723
Psoralen, 8113
Psoralidin, 8114
Psoralon [Hermal] *see* 255
Psorcon [Rhone-Poulenc Rorer] *see* 3186
Psoriacide [Doak] *see* 723
Psorigel [Alcon] *see* 2487
P.S.P. *see* 7397
Psychedrine *see* 623
Psychoperidol [Ethnor] *see* 9813
Psychosan *see* 925
Psycho-Soma [Boots] *see* 4477
Psychostyl *see* 6812
Psychoton *see* 623
Psychotrine, 8115
Psychoverlan [Verla] *see* 4477
Psyllium Husk *see* 7674
Psyllium Hydrocolloid *see* 7674
Psyllium Hydrophylic Mucilloid *see* 7674

Psyllium Seed see 7674
Psyllium Seed Gum see 7674
Psymion [TAD] see 5792
Psymod [Pitman-Moore] see 7615
Psyquil [Squibb-Heyden] see 9814
PT-14 see 7649
PTA see 3966
PTC see 3964
Pteridine, 8116
2,4(1H,3H)-Pteridinedione see 5625
Pterin HB$_2$ see 1270
Pterocarpin, 8117
Pterofen see 9731
Pteroic Acid, 8118
Pterolyl-γ-glutamyl-γ-glutamyl-glutamic Acid see 8119
Pterophene see 9731
Pteropterin, 8119
Pteroyldiglutamic Acid see 3344
Pteroyldi-γ-glutamylglutamic Acid see 8119
Pteroylglutamic Acid see 4253
Pteroyl-α-glutamylglutamic Acid see 3344
N-[N-(N-Pteroyl-γ-glutamyl)-γ-glutamyl]glutamic Acid see 8119
Pteroylheptaglutamic Acid see 8120
Pteroylhexaglutamylglutamic Acid, 8120
Pteroyltriglutamic Acid see 8119
PTFE see 7743
PTGA see 8119
PTH see 7168
Ptimal [EGIS] see 9836
PTO see 8178
Pt(oxalato)(trans-l-dach) see 7044
PTX see 7132
Ptychotis Oil see 196
PU-239 see 1110
Puberulic Acid, 8121
Puberulonic Acid, 8122
Puccoon Root see 8503
Pudding Pipe see 1942
Pudding-stick see 1942
Puerzym [Firma] see 7289
PUK see 8089
Pukateine, 8123
Pularin see 4685
Pulegone, 8124
Pulmadil [Riker] see 8393
Pulmaxil N see 7212
Pulmicort [Astra] see 1490
PulmiDur [Pharma-stern] see 9421
Pulmo 500 see 7212
Pulmoclase [UCB] see 1850
Pulmo-Timelets [Temmler] see 9421
Pulmozyme [Genentech] see 2953
Pulsamin see 3911
Pulsan [Yamanouchi] see 4974
Pulsar [Medosan] see 2980
Pulsatilla, 8125
Pulsatilla Camphor see 681
Pulsoton see 4855
Pulsotyl [Chinoin] see 7485
Pulvex [Coopers Agrovet] see 7321
Puma [Hoechst] see 4024
Pumactant, 8126
Pumice, 8127
Pumpellyite see 1685
Pumpkin Seed, 8128
Punch [Du Pont] see 4240
Punicin see 7199
Punicine see 7200
Punktyl [Krewel] see 5609
Puragel see 4388
Purantix [Wander] see 2430
Purapuridine see 8857
Purapurine see 8858
Purecal [Wyandotte] see 1697
Purerin [Kowa] see 6040

Purgaceen see 7108
Purging Agaric see 183
Purging Cassia see 1942
Purified Araroba see 2312
Purified Goa Powder see 2312
Purified Oxgall see 7062
Purified Ozokerite see 2033
Purified Protein Derivative see 9937
Purified Rayon see 8294
Purim [Laphal] see 7660
1H-Purin-6-amine see 150
1H-Purine, 8129
1H-Purine-2,6-diamine see 3028
2,6(1H,3H)-Purinedione see 10193
Purine-6-thiol see 5919
Puri-Nethol [Burroughs Wellcome] see 5919
Purinethol [Burroughs Wellcome] see 5919
Purine-2,6,8-triol see 10014
Purine-2,6,8(1H,3H,9H)-trione see 10014
Purin-6(1H)-one see 4917
Purochin [Sclavo] see 10024
Purocyclina see 9337
Purodigin [Wyeth] see 3206
Puromycin, 8130
Purophyllin [Siegfried] see 8092
Purosin-TC [Tatsumi] see 8060
Purostrophan [Kali-Chemie] see 7031
Purothionin, 8131
α_1-Purothionin see 8131
α_2-Purothionin see 8131
β-Purothionin see 8131
Purothionin A I see 8131
Purothionin A II see 8131
Purple Clover see 9818
Purple Foxglove see 3201
Purple of Cassius see 4538
Purpurea Glycoside C see 2967
Purpurid [Promonta] see 3206
Purpurin, 8132
Purpurogallin, 8133
Purpurosamine see 4398
Pursennid [Sandoz] see 8600
Purshiana Bark see 1931
Pursuit [Am. Cyan.] see 4944
Pusilline see 5246
Putrescine, 8134
Putty Powder see 8933
PVA see 7745
PVC see 7746
PVK [Lilly] see 7230
P.V.P. see 7879
PVP-I see 7880
PX-917 see 9892
Pycaril see 6613
Pycazide [Smith & Nephew] see 5203
Pyctyl [Maag] see 4027
Pydrin [Shell] see 4051
Pyelectan see 8779
Pyelokon-R [Mallinckrodt] see 79
Pyelosil see 5061
Pygeum Africanum Extract, 8135
Pykaryl see 6613
Pyknolepsinum [Rhein-Pharma] see 3794
Pylapron [Nichiyaku-Kyorin] see 8025
Pylen [Toyo] see 7728
Pylorid [Glaxo] see 8286
Pylumbrin see 5061
Pymetrozine, 8136
Pynamin [Sumitomo] see 257
Pynastin [Maruko] see 7597
Pynosect [Mitchell Cotts] see 7321
Pyocefal [Cassenne-Takeda] see 1994
Pyocianil [Farmitalia] see 1838
Pyocyanine, 8137
Pyoktanin see 4401
Pyopen [Beecham] see 1838

Pyostacine [Rhône-Poulenc Rorer] see 7933
Pyoxanthose see 4679
Pyracarbolid, 8138
β-Pyracine see 8165
Pyracort D [Lemmon] see 7440
Pyradone see 495
Pyrafat [Fatol] see 8140
Pyrahexyl see 9190
Pyralene [Prodelec] see 7725
Pyralgin [Savage] see 3414
Pyralin see 2011
Pyramidon [Winthrop] see 495
Pyramidon Bicamphorate [Winthrop] see 495
Pyramidon Salicylate [Winthrop] see 495
Pyramin see 9697
Pyramine see 9697
Pyranisamine see 8168
Pyrantel, 8139
Pyrantel Embonate see 8139
Pyranton see 3008
Pyrargyrite see 8647
Pyrasanone see 7431
Pyrathyn [Flint] see 6027
Pyrazinamide, 8140
Pyrazine, 8141
Pyrazinecarboxamide see 8140
Pyrazine Carboxylamide see 8140
Pyrazinecarboxylic Acid see 8143
2,3-Pyrazinedicarboxylic Acid, 8142
Pyrazinemonocarboxylic Acid see 8143
Pyrazinobutazone see 7431
Pyrazinoic Acid, 8143
Pyrazinoic Acid Amide see 8140
Pyrazino[2,3-d]pyrimidine see 8116
N^1-2-Pyrazinylsulfanilamide see 9107
Pyrazole, 8144
1H-Pyrazole-3-ethanamine see 1230
2-Pyrazoline, 8145
1H-Pyrazolo[3,4-d]pyrimidin-4-ol see 287
Pyrazophos, 8146
Pyrcon [Meuselbach] see 8206
Pyrene, 8147
Pyresyn [Penick] see 257
Pyrethrin I see 8148
Pyrethrin II see 8148
Pyrethrins, 8148
Pyrethrosin, 8149
Pyrethrum Flowers, 8150
Pyriamid see 9108
Pyribenzamine [Ciba-Geigy] see 9868
Pyricarbate see 8158
Pyricardyl see 6635
Pyricidin [Nepera] see 5203
Pyrictal see 7368
Pyridacil [Cilag] see 7361
Pyridamal-100 [Bel-Mar] see 2232
Pyridate, 8151
Pyridazine, 8152
Pyridiate [Darby] see 7361
Pyridin [Du Pont] see 4051
2-Pyridinamine see 494
3-Pyridinamine see 494
4-Pyridinamine see 3974
2H-Pyrid[4,3-b]indole see 1853
Pyridine, 8153
3-Pyridineacetic Acid, 8154
2-Pyridine Aldoxime Methiodide see 7884
2-Pyridine Aldoxime Methyl Chloride see 7884
3-Pyridinecarboxamide see 6574
2-Pyridinecarboxylic Acid see 7557
3-Pyridinecarboxylic Acid see 6612
4-Pyridinecarboxylic Acid see 5204

Pyrinistab *see* 6027
Pyrinistol *see* 6027
Pyrinoline, 8173
Pyrinuron *see* 8171
Pyriproxyfen, 8174
Pyrisept *see* 2074
Pyrisuccideanol, 8175
Pyrithiamine, 8176
Pyrithiobac, 8177
Pyrithione, 8178
Pyrithioxin *see* 8180
Pyrithyldione, 8179
Pyritinol, 8180
Pyro *see* 9377
Pyroace [Fujisawa] *see* 8202
Pyroacetic Ether *see* 64
Pyrocalciferol, 8181
Pyrocarbonic Acid Diethyl Ester, 8182
Pyrocatechin *see* 8183
Pyrocatechol, 8183
Pyrocatechol Dimethyl Ether *see* 10089
o-Pyrocatechuic Acid Diacetate *see* 3413
Pyrochlore *see* 6648
Pyrodextrin *see* 2995
Pyrodifenium Bromide *see* 7923
Pyrogallic Acid *see* 8184
Pyrogallol, 8184
Pyrogallol 1,2,3-(Diethylaminoethyl Ether) Tris(ethyl Iodide) *see* 4361
Pyrogallolphthalein *see* 4362
Pyrogastrone [Winthrop] *see* 1839
Pyrogen-free Water *see* 10175
PyroGlu *see* 8185
L-Pyroglutamic Acid, 8185
L-Pyroglutamyl-L-histidyl-L-tryptophyl-L-seryl-L-tyrosyl-D-*N^{im}*-benzyl-histidyl-L-leucyl-L-arginyl-L-proline Ethylamide *see* 4760
3-L-Pyroglutamyl-L-thiazolidine-4-carboxylic Acid *see* 7574
Pyrojec [Maurry Biol.] *see* 3414
Pyrolan [Ciba-Geigy], 8186
Pyrolase [Chugai] *see* 9431
Pyroligneous Acid, 8187
Pyroligneous Vinegar *see* 8187
Pyrolusite *see* 5770
Pyromellitic Acid, 8188
Pyromen [Axter] *see* 7659
Pyromijin *see* 8163
Pyromorphite *see* 5410
Pyromucic Acid *see* 4329
Pyromucic Aldehyde *see* 4324
α-Pyrone-5-carboxylic Acid *see* 2625
Pyronil [Lilly] *see* 8196
Pyronine B, 8189
Pyronine G *see* 8190
Pyronine Y, 8190
Pyrope *see* 362
Pyrophosphoric Acid, 8191
Pyrophosphoric Acid Tetraethyl Ester *see* 9345
Pyroplasmin *see* 3416
Pyroracemic Acid *see* 8205
Pyrosal *see* 757
Pyrosulfuric Acid, 8192
Pyrosulfuryl Chloride, 8193
Pyrovalerone, 8194
Pyroxylin, 8195
Pyrrhotine *see* 4106
Pyrrobutamine, 8196
Pyrrocaine, 8197
Pyrrocycline-N [Banyu] *see* 8411
Pyrrodiazole *see* 9735
Pyrrolamidol *see* 2997
1*H*-Pyrrole, 8198
Pyrrolidine, 8199

1-Pyrrolidineaceto-2′,6′-xylidide *see* 8197
(*S*)-2-Pyrrolidinecarboxylic Acid *see* 7963
2,5-Pyrrolidinedione *see* 9040
1-Pyrrolidinoaceto-2,6-dimethylanilide *see* 8197
N-(Pyrrolidinomethyl)tetracycline *see* 8411
α-Pyrrolidino-*p*-methylvalerophenone *see* 8194
2-Pyrrolidinone *see* 8200
2-Pyrrolidinoneacetamide *see* 7641
2-(1-Pyrrolidinyl)-2′,6′-acetoxylidide *see* 8197
1-(1-Pyrrolidinyl)butyl *p*-Tolyl Ketone *see* 8194
1-[4-(1-Pyrrolidinyl)-2-butynyl]-2-pyrrolidinone *see* 7085
1-[(1-Pyrrolidinylcarbonyl)methyl]-4-(3,4,5-trimethoxycinnamoyl)piperazine *see* 2350
N-(1-Pyrrolidinylmethyl)tetracycline *see* 8411
10-[2-(1-Pyrrolidinyl)propionyl]phenothiazine Methobromide *see* 8057
3-(2-Pyrrolidinyl)pyridine *see* 6807
trans-2-[3-(1-Pyrrolidinyl)-1-*p*-tolylpropenyl]pyridine *see* 9877
(*E*)-6-[(*E*)-3-(1-Pyrrolidinyl)-1-*p*-tolylpropenyl]-2-pyridineacrylic Acid *see* 129
4-(1-Pyrrolidinyl)-1-(2,4,6-trimethoxyphenyl)-1-butanone *see* 1498
2-Pyrrolidone, 8200
α-Pyrrolidone *see* 8200
2-Pyrrolidoneacetamide *see* 7641
2-Pyrrolidone-5-carboxylic Acid *see* 8185
N-Pyrrolidylethyl-*N*-phenylbenzylamine *see* 4757
3-Pyrroline, 8201
Pyrrolnitrin, 8202
Pyrrolylene *see* 1534
Pyrro[*b*]monazole *see* 4948
Pyruvaldehyde, 8203
Pyruvaldehyde Bis(amidinohydrazone) *see* 6299
Pyruvaldoxime *see* 5207
Pyruvate Decarboxylase, 8204
Pyruvate Oxidation Factor *see* 9462
Pyruvic Acid, 8205
Pyruvic Carboxylase *see* 8204
Pyruvic Decarboxylase *see* 8204
Pyrvinium Chloride *see* 8206
Pyrvinium Embonate *see* 8206
Pyrvinium Pamoate, 8206
Pytetrahydroserpentine *see* 8291
PZC *see* 7323
PZI Insulin *see* 5011
Q *see* 4479
Q-254 *see* 7681
Q-275 *see* 9974
Qari [Mediolanum] *see* 8448
QCD-84924 *see* 9349
QD-10733 *see* 4226
Q-Enzyme®, 8207
Q-Gas *see* 4660
QHS *see* 856
Qinghao *see* 856
Qinghaosu *see* 856
Qing Hau Sau *see* 856
QNB *see* 8266
Q-Pam [Quantum] *see* 3042
Q.T. (Quick Tan) *see* 3225
Quaalude [Rorer] *see* 6028
Quadracyclin [Knoll] *see* 9337
Quadrafos *see* 8814
Quadrol [Wyandotte] *see* 3635
Quait [SIT] *see* 5609

Quaker Buttons *see* 6828
Quamaquil [Siegfried] *see* 7412
Quamonium *see* 2068
Quaname [Wyeth] *see* 5908
Quanil [Wyeth] *see* 5908
Quantalan [Lappe] *see* 2257
Quantril [Roerig] *see* 1154
Quarcyl [Winthrop] *see* 4489
Quark [Polifarma] *see* 8283
Quartz *see* 8637
Quarzan [Roche] *see* 2412
Quasar [Ravizza] *see* 10083
Quassia, 8208
Quassin, 8209
Quat *see* 5320
Quaternium-7 *see* 5379
Quatrachlor [Pitman-Moore] *see* 1103
Quatrimycin, 8210
Quazepam, 8211
Quazium [Schering] *see* 8211
Quebrachamine, 8212
Quebrachine *see* 10236
Quebracho *see* 884
Quebracho Colorado, 8213
Quebrachol *see* 8697
Queen Bee Jelly *see* 8434
Queensland Asthma Weed *see* 3949
Queen's Root *see* 8973
Queen Substance, 8214
Queletox *see* 4044
Quelicin [Abbott] *see* 9044
Quellada [Stafford-Miller] *see* 5526
Quemicetina [Rontag] *see* 2120
Quensyl *see* 4863
Quentan [Boehringer, Ing.] *see* 1412
Quercetagetin, 8215
Quercetagitrin *see* 8215
Quercetin, 8216
Quercetin-3-glucoside *see* 5240
Quercetin-7-D-glucoside *see* 8217
Quercetin-3-L-rhamnoside *see* 8219
Quercetin-3-rutinoside *see* 8456
Quercimelin *see* 8219
Quercimeritrin, 8217
d-Quercitol, 8218
Quercitrin, 8219
Quercitroside *see* 8219
Quercus, 8220
Querto [Byk Gulden] *see* 1924
Querton 210CL [Kemanord] *see* 3149
Quesil [EGYT] *see* 2243
Quest [Am. Cyanamid] *see* 8864
Questex *see* 3557
Questran [Mead Johnson] *see* 2257
Quiactin [Merrell] *see* 7053
Quick [Rhône-Poulenc] *see* 2204
Quick Grass *see* 9890
Quicklime *see* 1733
Quicksilver *see* 5957
Quide [Dow] *see* 7615
Quiess [Forest] *see* 4897
Quietim [Nativelle] *see* 4895
Quilan [Lilly] *see* 1067
Quilene [Ayerst] *see* 7260
Quilibrex [Isnardi] *see* 7059
Quillaic Acid, 8221
Quillaja, 8222
Quillaja Sapogenin *see* 8221
Quillaja Saponin *see* 8223
Quillay Bark *see* 8222
Quilonorm [SKB] *see* 5544
Quilonorm-retard [SKB] *see* 5552
Quilonum [SKB] *see* 5544
Quilonum-retard [SKB] *see* 5552
Quimar [Armour Pharm.] *see* 2320
Quimoral *see* 2320
Quimotrase [Farmila] *see* 2320
Quinacillin, 8224
Quinacrine, 8225
Quinaglute [Berlex] *see* 8244

Quinagolide, 8226
Quinalbarbitone Sodium *see* 8563
Quinaldic Acid, 8227
Quinaldine, 8228
Quinaldine Blue, 8229
Quinaldine Red, 8230
Quinaldinic Acid *see* 8227
Quinaldofur *see* 6625
Quinalizarin, 8231
Quinalspan [Pharmax] *see* 8563
Quinambicide *see* 5052
Quinamine, 8232
Quinamm [Merrell Dow] *see* 8245
Quinapril, 8233
Quinaprilat *see* 8233
Quinapyramine, 8234
Quinate [Rhone-Poulenc Rorer] *see* 8245
Quinazil [Malesci] *see* 8233
Quinazoline, 8235
Quinbisan [Prosana] *see* 8245
Quinbolone, 8236
Quince Seed, 8237
Quine [Rowell] *see* 8245
Quinestradiol, 8238
Quinestradol *see* 8238
Quinestrol, 8239
Quinethazone, 8240
Quinexin [Robin] *see* 1074
Quinfamide, 8241
Quinhydrone, 8242
Quinic Acid, 8243
Quinic Acid 1,5-Dicaffeic Ester *see* 2835
Quinicardine [Nativelle] *see* 8244
Quinicine *see* 10141
Quinidex Extentabs [Robins] *see* 8244
Quinidine, 8244
Quinidine Bisulfate *see* 8244
Quiniduran [Teva] *see* 8244
Quinine, 8245
β-Quinine *see* 8244
Quinine Bimuriate *see* 8245
Quinine Dichloride *see* 8245
Quinine Iodosulfate, 8246
Quinine Oleate, 8247
Quininic Acid, 8248
Quininone, 8249
Quiniofon *see* 4872
Quinisocaine *see* 3267
Quinizarin, 8250
Quinizarin Green SS, 8251
Quinocide, 8252
Quinodis [Roche] *see* 4137
Quinofen *see* 2347
Quinofop-ethyl *see* 8269
Quinol *see* 4853
Quinoline, 8253
8-Quinolineboronic Acid, 8254
2-Quinolinecarboxylic Acid *see* 8227
8-Quinolinecarboxylic Acid, 8255
Quinoline Yellow, 8256
Quinoline Yellow A *see* 8257
Quinoline Yellow Base *see* 8257
Quinoline Yellow Spirit Soluble, 8257
Quinolinic Acid, 8258
2-Quinolinol *see* 1873
8-Quinolinol *see* 4890
8-Quinolinol Benzoate (Ester) *see* 1143
2(1*H*)-Quinolinone *see* 1873
2-(2-Quinolinyl)-1*H*-indene-1,3-(2*H*)-dione *see* 8257
2(1*H*)-Quinolone *see* 1873
Quinolor [Squibb] *see* 4637
Quinomethionate *see* 7113
Quinomycin A *see* 3545
Quinone, 8259
p-Quinone *see* 8259
Quinone Monoxime *see* 6740

Quinone Oxime *see* 6740
Quinophan *see* 2347
Quinophthalone *see* 8257
Quinora [Key] *see* 8244
Quinothionate *see* 9495
Quinotoxine *see* 10141
Quinotoxol *see* 10141
Quinova-bitter *see* 8261
Quinovaic Acid *see* 8260
Quinovatine *see* 821
Quinovic Acid, 8260
Quinovic Acid β-D-Glucoside *see* 8261
Quinovic Acid β-D-Quinovoside *see* 8261
Quinovin, 8261
Quinovose, 8262
Quinovose 3-Methyl Ether *see* 9425
Quinoxaline, 8263
"Quinoxaline Antibiotic" *see* 3545
"Quinoxaline Antibiotic Complex" *see* 9862
2,3-Quinoxalinedithiol Cyclic Trithio-carbonate *see* 9495
3-(2-Quinoxalinylmethylene)carbazic Acid Methyl Ester *N,N'*-Dioxide *see* 1825
(2-Quinoxalinylmethylene)hydrazine-carboxylic Acid Methyl Ester *N,N'*-Dioxide *see* 1825
N^1-(2-Quinoxalinyl)sulfanilamide *see* 9109
N^1-(2-Quinoxalyl)sulfanilamide *see* 9109
Quinoxyl [Burroughs Wellcome] *see* 4872
Quinsan [Prosana] *see* 8245
Quintozene, 8264
Quintrate *see* 7249
Quinuclidine, 8265
3-Quinuclidinol, 8266
3-Quinuclidinyl Acetate *see* 8266
3-Quinuclidinyl Benzilate *see* 8266
3-Quinuclidinyl Benzoate *see* 8266
10-(3-Quinuclidinylmethyl)phenothia-zine *see* 5911
Quinupramine, 8267
Quinuronium Sulfate *see* 3416
Quinvet [Neopharmed] *see* 5668
Quipenyl *see* 7134
Quisqualic Acid, 8268
L-Quisqualic Acid *see* 8268
Quitaxon [Boehringer, Mann.] *see* 3492
Quival [Schering] *see* 1744
Quixalin [Squibb] *see* 4637
Quixalud *see* 4637
Quizalofop-Ethyl, 8269
Quorn [Marlow], 8270
Quotane [SK & F] *see* 3267
QZ-2 *see* 6028
R *see* 817
R-3-ZON *see* 7431
R-11, 8271
R-48 *see* 2157
R-75 *see* 5469
R-79 *see* 5221
R-100 *see* 406
R-131 *see* 7362
R-242 *see* 9162
R-381 *see* 4005
R-400 *see* 7173
R-445 *see* 6349
R-658 *see* 1635
R-738 *see* 6529
R-773 *see* 8385
R-798 *see* 8393
R-802 *see* 4172
R-805 *see* 6640
R-818 *see* 4136
R-835 *see* 4919

R-837 *see* 4957
R-875 *see* 2997
R-1132 *see* 3371
R-1303 *see* 1869
R-1504 *see* 7492
R-1513 *see* 944
R-1582 *see* 944
R-1607 *see* 10097
R-1608 *see* 3674
R-1625 *see* 4629
R-1658 *see* 6343
R-1707 *see* 4443
R-1881 *see* 6211
R-1910 *see* 1581
R-1929 *see* 931
R-2028 *see* 4150
R-2113 *see* 2975
R-2323 *see* 4423
R-2453 *see* 2939
R-2498 *see* 9813
R-2858 *see* 6372
R-3248 *see* 30
R-3345 *see* 7608
R-3365 *see* 7653
R-3746 *see* 1991
R-3763 *see* 1991
R-3959 *see* 2443
R-4263 *see* 4043
R-4318 *see* 4140
R-4584 *see* 1077
R-4749 *see* 3505
R-4845 *see* 1241
R-4929 *see* 1104
R-5020 *see* 7969
R-6128 *see* 4241
R-6238 *see* 7589
R-6700 *see* 5142
R-7315 *see* 6232
R-7465 *see* 6503
R-7904 *see* 5507
R-8299 *see* 5486
R-11333 *see* 1466
R-12563 *see* 5486
R-12564 *see* 5486
R-14827 *see* 3550
R-14889 *see* 6266
R-15403 *see* 3183
R-15454 *see* 5176
R-15889 *see* 5610
R-16341 *see* 7213
R-16470 *see* 2987
R-16659 *see* 3927
R-17147 *see* 2781
R-17635 *see* 5807
R-17889 *see* 4154
R-18553 *see* 5601
R-23979 *see* 3622
R-25061 *see* 9180
R-25831 *see* 1897
R-30730 *see* 9056
R-31520 *see* 2473
R-33800 *see* 9056
R-33812 *see* 3476
R-34995 *see* 5586
R-35443 *see* 7058
R-39209 *see* 236
R-41400 *see* 5313
R-41468 *see* 5307
R-42470 *see* 9303
R-43512 *see* 891
R-49945 *see* 5307
R-50547 *see* 5489
R-51211 *see* 5262
R-51619 *see* 2377
R-55667 *see* 8399
R-58735 *see* 8462
R-64433 *see* 3130
R-64766 *see* 8397
R-65824 *see* 6519
R-66905 *see* 8510

R-68070 *see* 8379
R-151993 *see* 9266
R-152450 *see* 4246
RA$_2$ *see* 9777
RA-8 *see* 3410
RA-233 *see* 6344
Rabalan *see* 7532
Rabeprazole, 8272
Rabon [Shell] *see* 9334
Rabond [Shell] *see* 9334
Race-acetylmethadol *see* 6009
Racefemine, 8273
Racemethionine *see* 6053
Racemethorphan, 8274
Racemic Acid *see* 9236
Racemic Desoxy-nor-ephedrine *see* 623
Racemic Ephedrine *see* 3645
Racemic Lactic Acid *see* 5349
Racemic Mandelic Acid *see* 5757
Racemic Propoxyphene *see* 8024
Racemic Tartaric Acid *see* 9236
Racemomycin A *see* 8988
Racemomycins *see* 8988
Racemorphan *see* 5496
Raceophenidol *see* 9436
Racephedrine *see* 3645
Racephedrine Hydrochloride *see* 3645
Racephedrine Sulfate *see* 3645
Racephen *see* 623
Racephenicol *see* 9436
Racepinephrine *see* 3656
Rachromate-51 [Abbott] *see* 8745
R Acid *see* 6474
Racobalamin-60 [Abbott] *see* 10153
Ractopamine, 8275
Radanil [Roche] *see* 1114
Radapon [Dow] *see* 2869
Radar [ICI] *see* 8003
Raddeamine *see* 4961
Rad-e-cate [Vineland] *see* 8738
Radecol *see* 6614
Radedorm *see* 6667
Radical Weed *see* 8856
Radicicol *see* 6336
Radicinin, 8276
Radinyl [Roberts] *see* 3756
Radioactive Colloidal Gold *see* 4533
Radioactive Cyanocobalamin *see* 10153
Radioactive Sodium Phosphate *see* 8807
Radiocaps-131 [Abbott] *see* 8778
Radiocyanocobalamin *see* 10153
Radio-gold (^{198}Au) Colloid *see* 4533
Radiographol *see* 6051
Radio-iodinated Insulin *see* 5011
Radio-iodinated (^{131}I) Tolpovidone *see* 9661
Radiopaque [Schering] *see* 1023
Radiostol *see* 10156
Radiotetrane *see* 5059
Radium, 8277
Radium Emanation *see* 8278
Radium F *see* 7718
Radon, 8278
Radsterin *see* 10156
Rafamebin *see* 5063
Raffinose, 8279
Rafoxanide, 8280
Ragadan [Hoechst] *see* 4699
Ralabol [IMC] *see* 10251
Ralgro [IMC] *see* 10251
Rally [Rohm & Haas] *see* 6402
Ralone [ICI] *see* 10251
Raloxifene, 8281
RALs *see* 10246
Ramace [Astra] *see* 8283
Rametin [Bayer] *see* 6441
Ramifenazone, 8282

Ramipril, 8283
Ramiprilat *see* 8283
Ramodar [Wyeth] *see* 3920
R/AMP *see* 8382
Rampart [Platte Chemical] *see* 7486
Ramrod [Monsanto] *see* 7977
Ramycin *see* 4340
Rancinamycin IV *see* 8079
Randa [Nippon Kayaku] *see* 2378
Randolectil [Bayer] *see* 1543
Randox *see* 259
Ranestol [Parke, Davis] *see* 1238
Ranestol (obsolete) [Parke, Davis] *see* 9787
Raney Nickel [Davison], 8284
Raney Nickel Catalyst *see* 8284
Rangasil 400 [Ciba-Geigy] *see* 7663
Raniben [Firma] *see* 8286
Ranide [Merck & Co.] *see* 8280
Ranidil [Glaxo] *see* 8286
Ranimustine, 8285
Raniplex [Fournier] *see* 8286
Ranitidine, 8286
Ranitidine Bismutrex *see* 8286
Ranolazine, 8287
Ranomustine *see* 8285
Rantudil [Troponwerke] *see* 27
Ranunculin *see* 8078
Ranvil [Gentili] *see* 6579
Raovin [Abbott] *see* 9661
RAPA *see* 8288
Rapacodin [Knoll] *see* 3214
Rapamune [Wyeth-Ayerst] *see* 8288
Rapamycin, 8288
Rapenton [Thomae] *see* 6344
Rapeseed Oil, 8289
Raphanin *see* 9136
Raphanus Rusticanus *see* 4785
Raphetamine Phosphate [Pennwalt] *see* 623
Rapicidin *see* 9967
Rapifen [Janssen] *see* 236
Rapinovet [Coopers] *see* 8020
Rapitil [Fisons] *see* 6524
Rapitux [De Angeli] *see* 3507
Rapostan [Mepha] *see* 7106
Rapynogen [Maruko] *see* 8025
Rarical [Ortho] *see* 1708
Rasapen *see* 4785
Raspberry, 8290
Rastinon [Hoechst] *see* 9646
Ratanhine *see* 9183
Rat Antispectacled Eye Factor *see* 5008
Rathimed [Pfleger] *see* 6242
Raticate *see* 6780
Ratilan [Ciba-Geigy] *see* 2623
Rattlesnake Root *see* 8597
Rattrack *see* 764
Raubasine, 8291
Raudixin [Squibb] *see* 8293
Rauhimbine *see* 2614
Raunescine, 8292
Raunormine [Penick] *see* 2964
Raunova [Zambeletti] *see* 9193
Raupine *see* 8524
Rau-sed [Squibb] *see* 8314
Rauserpol [Rhewie Pharma] *see* 8293
Rautensin [Sandoz] *see* 8293
Rauverid [OJ & F] *see* 8293
Rauwiloid [3M Pharm] *see* 8293
Rauwoldin [OJ & F] *see* 8293
Rauwolfemms [Truxton] *see* 8293
Rauwolfia Serpentina, 8293
Rauwolfine *see* 194
Rauwolscine *see* 10238
Ra-Valeas [Valeas] *see* 8293
Ravenil [Caber] *see* 8158
Raviac [Lipha] *see* 2204

Ravocaine Hydrochloride [Cook-Waite] *see* 8023
Ravyon [Makhteshim-Agan] *see* 1831
Raxil [Bayer] *see* 9253
Raylina [Robert] *see* 617
Rayon, 8294
Razoxane, 8295
Razoxin [ICI] *see* 8295
RB-1509 *see* 5594
RC-61-91 *see* 4936
RC-172 *see* 224
RC-27109 *see* 6626
RCL I *see* 8376
RCL II *see* 8376
RCL III *see* 8376
RCL IV *see* 8376
RD-292 *see* 4029
RD-1572 *see* 4767
RD-4593 *see* 5826
RD-13621 *see* 4925
RDX *see* 2805
Re-1-0185 *see* 3565
RE-4355 *see* 6445
Reacid *see* 2751
Reacthin *see* 136
Reactine [Pfizer] *see* 2063
Reactivan [Cascan] *see* 4006
Reactrol *see* 2406
Readex *see* 9495
Reagin [Baliarda] *see* 2380
Realgar *see* 834
Realphene *see* 49
Reasec [Janssen] *see* 3371
Reazide *see* 2751
Rebamipide, 8296
Rebriden [Sandoz] *see* 5260
Rec-7-0040 *see* 4135
Rec-7-0267 *see* 3249
Rec-14-0127 *see* 6347
Rec-15-0691 *see* 9565
Rec-15-1476 *see* 4047
Rec-15-2375 *see* 5469
Recanescine *see* 2964
Receptal [Hoechst] *see* 1527
Recetan [Simes] *see* 1558
Recheton *see* 5300
Recidol [Lampugnani] *see* 4925
Recipavrin [Recip] *see* 9843
Reclaim [Dow] *see* 2462
Recognan [Toyo Jozo] *see* 2380
Recolip [Benzon] *see* 2436
Recombinant Methionyl Human G-CSF *see* 4558
Recombinate [Baxter] *see* 3963
Reconin [Toyama] *see* 2405
Reconox *see* 7404
Recordil [Recordati] *see* 3565
Recormon [Boehringer, Mann.] *see* 3729
Recosen *see* 4651
Rectadione *see* 7381
Rectalad [Carter-Wallace] *see* 3460
Rectodelt [Trommsdorff] *see* 7904
Rectovalone [Jouveinal] *see* 9623
Red Arsenic Glass *see* 834
Red Arsenic Sulfide *see* 834
Red Bole *see* 1357
Red Clover *see* 9818
Red Copper Oxide *see* 2734
Redeema [Merz] *see* 10293
Redeptin [SK & F] *see* 4241
Redergin [Lek] *see* 3692
Red Gum *see* 3942
Redisol [Merck & Co.] *see* 10152
Redisol H [Merck & Co.] *see* 4854
Red Lead *see* 5449
Red Lead Chromate *see* 5423
Red Liquor *see* 349
Red Mustard *see* 6394
Red No. 2 *see* 391

Red Oil *see* 9953
Redomex [Sanofi-Labaz] *see* 511
Red Orpiment *see* 834
Redoxon [Roche] *see* 867
Red Phosphorus *see* 7503
Red Precipitate *see* 5935
Red Prussiate of Potash *see* 7792
Red Puccoon *see* 8503
Red Quebracho *see* 8213
Redroot *see* 392
Red Root *see* 8503
Red Sandalwood *see* 8531
Redskin *see* 303
Red Tetrazolium *see* 9874
Reduced Hematin *see* 4672
Reduced Nicotinamide Adenine Dinucleotide Phosphate Dehydrogenase *see* 6961
Reduced Ninhydrin *see* 4816
Reductic Acid, 8297
Reductil [Knoll] *see* 8629
Reductol *see* 112
Reducymol [Elanco] *see* 672
Redul [Schering] *see* 4517
Redupresin [Alcon-Thilo] *see* 3801
Red Uranium Oxide *see* 9997
Redusterol *see* 7432
Red-water Tree Bark *see* 8529
Redwood *see* 7322
Reelon [Sanwa] *see* 7660
Reevon [Reeves] *see* 7728
Refined Solvent Naphtha *see* 5513
Refkas [Maruko] *see* 3720
Reflux *see* 6040
Refobacin [E. Merck] *see* 4398
Refosporin [E. Merck] *see* 1966
Refugal [Bayer] *see* 2106
Refungine [A.C.F.] *see* 9061
Regacholyl *see* 4882
Regaine [Upjohn] *see* 6290
Regelan [Rhein-Pharma] *see* 2436
Regenit [Hoechst] *see* 7899
Regenon [Temmler] *see* 3175
Regianin *see* 5282
Regitine [Ciba] *see* 7417
Reglan [Robins] *see* 6226
Regletin [Teikoku Zoki] *see* 321
Reglone [ICI] *see* 3415
Regolax [Corvi] *see* 7395
Regonol [Organon] *see* 8161
Regramostim *see* 4559
Regulan [Proctor & Gamble] *see* 7674
Regulin [Young] *see* 5857
Regulipid [Heilit] *see* 2437
Regulox *see* 5745
Regulton [Nordmark; Knoll] *see* 412
Regulus of Antimony *see* 733
Regumate [Hoechst] *see* 327
Regutol [Schering-Plough] *see* 3460
Rehibin [Thames] *see* 2789
Reicaf [San Carlo] *see* 8411
Reichstein's Substance A *see* 274
Reichstein's Substance C *see* 276
Reichstein's Substance D *see* 280
Reichstein's Substance Dehydro-C *see* 289
Reichstein's Substance E *see* 7912
Reichstein's Substance Fa *see* 2602
Reichstein's Substance G *see* 174
Reichstein's Substance H *see* 2601
Reichstein's Substance J *see* 279
Reichstein's Substance K *see* 275
Reichstein's Substance L *see* 272
Reichstein's Substance M *see* 4828
Reichstein's Substance N *see* 271
Reichstein's Substance O *see* 278
Reichstein's Substance P *see* 282
Reichstein's Substance Q *see* 2947
Reichstein's Substance R *see* 281
Reichstein's Substance S *see* 2977

Reichstein's Substance T *see* 7911
Reichstein's Substance U *see* 7913
Reichstein's Substance V *see* 277
Reidamine [Reid-Rowell] *see* 3252
Reinecke Salt, 8298
Rekawan [Giulini] *see* 7783
Rela [Schering] *see* 1889
Relafen [SKB] *see* 6428
Relan Beta *see* 1064
Relane [Cutter] *see* 3352
Relanium [Polfa] *see* 3042
Relasom [RAFA] *see* 1889
Relaspium [Isei] *see* 9918
Relaxan *see* 4361
Relaxar *see* 5895
Relaxil *see* 5895
Relaxil G *see* 4582
Relaxin, 8299
Reldan [Dow] *see* 2242
Releasin (formerly) [Warner-Chilcott] *see* 8299
Relefact LH-RH [Hoechst] *see* 5500
Relestrid *see* 6060
Reliberan [Geymonat] *see* 2132
Relicor *see* 3753
Relifen [Fujisawa] *see* 6428
Relifex [Bencard] *see* 6428
Relipain [Farmitalia] *see* 6361
Reliveran [Finadiet] *see* 6226
Relvene [Pharmascience] *see* 9920
Remantadin(e) *see* 8390
Remark [Nippon Zoki] *see* 1224
Remeflin [Recordati] *see* 3249
Remeron [Organon] *see* 6295
Remestan [Wyeth] *see* 9285
Remicyclin [Schaper & Brümmer] *see* 9337
Remid [TAD] *see* 287
Remifentanil, 8300
Remifentanyl *see* 8300
Reminitrol [Schafer] *see* 6704
Reminyl [Janssen] *see* 8462
Remivox [Janssen] *see* 5610
Remnos [DDSA] *see* 6667
Removine [Brocades] *see* 3252
Remoxipride, 8301
Remsed [Du Pont Merck] *see* 7970
Remtal [AgrEvo] *see* 9797
Renacor [Merck & Co.] *see* 3605
Renafur [Eaton] *see* 6621
Renarcol *see* 5895
Renascin [Mack, Illert.] *see* 10159
Renasul [Century] *see* 9084
Rencal [Squibb] *see* 7542
Renelate *see* 6040
Renese [Pfizer] *see* 7744
Reneuron [Juste] *see* 4222
Renex 20 [ICI] *see* 7738
Renex 600'S [Atlas] *see* 6772
Rengasil [Ciba-Geigy] *see* 7663
Renilan [Mochida] *see* 1988
Renin, 8302
Renin Substrate *see* 689
Renitec [Merck & Co.] *see* 3605
Reniten [Merck & Co.] *see* 3605
Renivace [Banyu] *see* 3605
Rennase [Squibb] *see* 8303
Rennet *see* 8303
Rennin, 8303
Renografin [Squibb] *see* 5851
Reno M [Squibb] *see* 5851
Renoquid [Glenwood] *see* 9070
Renormax [Schering-Plough] *see* 8905
Ren-o-sal [Salsbury] *see* 8430
Renovist [Squibb] *see* 5851
Renovue-65 [Squibb] *see* 5031
Renovue-DIP [Squibb] *see* 5031
Renpress [Sandoz] *see* 8905
Renselin *see* 9983
Rentylin [Rentschler] *see* 7278

Renzepin [Bergamon] *see* 7646
Reocorin [Farmochimica] *see* 7919
Reomax [Bioindustria] *see* 3761
Reomucil [Astra Simes] *see* 1850
Reorganin *see* 4582
Reostral *see* 5908
Reoxyl [Byk-Gulden] *see* 4743
Repagermanium *see* 7979
Repaglinide, 8304
Repanidal [Leciva] *see* 9909
Reparil *see* 3737
Repel [Wisconsin Pharmacal] *see* 2912
Repeltin [Fabre-Medikosma] *see* 9834
Repidose [Agrovet] *see* 7069
Repirinast, 8305
Repocal [Desitin] *see* 7272
Repodral [Winthrop] *see* 8967
Repoise [Robins] *see* 1543
Repone K *see* 7783
Reposal, 8306
Reposamal *see* 8306
Reprodin [E. Merck] *see* 5638
Reproterol, 8307
Reptilase [Knoll] *see* 1038
Repulson [Mochida] *see* 796
Resacetophenone, 8308
Resantin *see* 4032
Resazoin *see* 8309
Resazurin, 8309
Resbenzophenone *see* 1138
Resbuthrin [Wellcome] *see* 1271
Rescimetol, 8310
Rescinnamine, 8311
Rescriptor [Upjohn] *see* 2929
Rescufolin [Nordic] *see* 4254
Rescula [Fujisawa] *see* 9984
Rescuvolin [Medac] *see* 4254
Resectisol [McGaw] *see* 5788
Reserpic Acid, 8312
Reserpilic Acid Dimethylaminoethyl Ester Dihydrochloride *see* 8313
Reserpiline, 8313
Reserpine, 8314
Reserpinine *see* 8311
Reserpinolic Acid *see* 8312
Reserpoid [Upjohn] *see* 8314
Resibufogenin, 8315
Resilin, 8316
Resimatil [Sanofi-Winthrop] *see* 7927
Resin Benjamin *see* 4608
Resin Benzoin *see* 4608
Resin Copal *see* 2581
Resin Damar *see* 2872
Resin Guaiac *see* 4573
Resin Ipomea, 8317
Resin Jalap, 8318
Resin Kino *see* 5325
Resin of Mexican Scammony *see* 8317
Resin Scammony, 8319
Resin Tolu *see* 977
Resistomycin, 8320
Resistomycin [Bayer] *see* 5293
Resistopen [Squibb] *see* 7036
Resitan *see* 10044
Resitox [Bayer] *see* 2626
Résivit [Oberlin] *see* 5477
Resloom M 75 [Monsanto] *see* 4727
(+)-*trans*-Resmethrin *see* 1271
Resmit [Shionogi] *see* 5829
Resochin [Bayer] *see* 2215
Resodec, 8321
Resolvable Tartaric Acid *see* 9236
Resonium A [Winthrop] *see* 8815
Résoquine *see* 2215
Resorantel, 8322
Resorcin *see* 8323
Resorcin Blue *see* 5345
Resorcinol, 8323
Resorcinolphthalein *see* 4194
Resorcinol Phthalein Sodium *see* 4194

Resorcinolphthalin *see* 4196
Resorcinol Yellow *see* 9905
β-Resorcylaldehyde, 8324
Resorcylam *see* 8322
β-Resorcylic Acid, 8325
Resorcylic Acid Lactones *see* 10246
Resotren [Bayer] *see* 4872
Resource [Valent] *see* 4178
Resovist [Schering AG] *see* 4347
Respacal [Bios-Coutelier] *see* 9942
Respaire (obsolete) [Bristol] *see* 89
Respbid [Boehringer, Ing.] *see* 9421
Respenyl *see* 4582
Respicort [Mundipharma] *see* 9728
Respifral *see* 5236
Respigon [Taisho] *see* 8315
Respilene [Lyocentre; Sigma-Tau] *see* 10303
Respirase [Gibipharma] *see* 10303
Respiride [Schiapparelli] *see* 4041
Respirot [Ciba-Geigy] *see* 2659
Resporisan [Kowa] *see* 8313
Restandol [Organon] *see* 9322
Restanolon [Isei] *see* 2470
Restar [Sumitomo] *see* 4245
Restas [Kanebo] *see* 4245
Restenacht [Kabi] *see* 3598
Restenil [Kabi] *see* 5908
Restetal [Kabi] *see* 3604
Rest-On *see* 6027
Restoril [Sandoz] *see* 9285
Restrol [Central Pharm.] *see* 3153
Restropin [Frosst] *see* 6084
Restryl *see* 6027
Resulax [Tika] *see* 8873
Resulfon [Nordmark] *see* 9076
Resveratrol, 8326
cis-Resveratrol *see* 8326
trans-Resveratrol *see* 8326
Resyl [Ciba] *see* 4582
Retabolil *see* 6452
Retalon *see* 3153
Retalon-Lingual *see* 4738
Retalon Oleosum *see* 4738
Retalon-Oral *see* 3153
Retamine, 8327
Retarcyl [Delagrange] *see* 6362
Retardon [Chassot] *see* 9104
Retcin [DDSA] *see* 3720
Retene, 8328
Retens [Wassermann] *see* 3496
Retensin *see* 4361
Reticulin, 8329
Reticulin (the Antibiotic) *see* 4893
Reticuline, 8330
Reticulogen *see* 5575
Retin-A [Ortho-McNeil] *see* 8333
Retina-derived Endothelial Cell Growth Factors *see* 4117
Retinal, 8331
Retinal 2 *see* 2926
Retinaldehyde *see* 8331
Retine, 8332
Retinene₁ *see* 8331
Retinene 2 *see* 2926
Retinoic Acid, 8333
[2*R**(4*R**,8*R**)]-(±)-Retinoic Acid 3,4-Dihydro-2,5,7,8-tetramethyl-2-(4,8,-12-trimethyltridecyl)-2*H*-1-benzo-pyran-6-yl Ester *see* 9639
Retinol *see* 10150
Retinol₂ *see* 10151
Retolen [Elmu] *see* 891
Retractyl [Rhovyl] *see* 7746
Retrangor *see* 1113
Retrocortine *see* 7904
Retroid [Roche] *see* 9717
Retrone *see* 3524
Retronecine, 8334
Retrorsine, 8335

Retrorsine *N*-Oxide *see* 8335
Retrovir [Burroughs Wellcome] *see* 10252
Reublonil *see* 1152
Reudene [ABC] *see* 7661
Reufenac [Zambon] *see* 220
Reumachlor *see* 2215
Reumagrip *see* 2247
Reumalon [Tosi] *see* 7091
Reumatox *see* 6312
Reumofene [Von Boch] *see* 4999
Reumofil [Ausonia] *see* 9155
Reumyl [Lenza] *see* 9155
Reupiron *see* 6619
Reutol [Errekappa] *see* 9655
Reuxen [Technifar] *see* 6504
REV-6000A *see* 2928
Revanil [Roche] *see* 5541
Revasc [Ciba-Geigy] *see* 4754
Reverin [Hoechst] *see* 8411
Reversil [Angelini] *see* 2884
Reversol [Organon] *see* 3562
Rev-Eyes [Angelini] *see* 2884
Reviparin Sodium, 8336
Revistel [Delagrange] *see* 319
Revivon [Reckitt & Colman] *see* 3403
Revonal [E. Merck] *see* 6028
Rexalgan [Dompé] *see* 9293
Rexan *see* 2155
Rexigen [Bago] *see* 2421
Rexitene [LPD] *see* 4585
Rexort [Cassenne] *see* 2380
Rexulfa [Medici] *see* 9104
Reychler's Acid *see* 1781
Rezifilm [Squibb] *see* 9510
Rezipas [Squibb] *see* 498
RF-46-790 *see* 4230
RF-2535 *see* 4728
RG-270 *see* 5069
RGCR *see* 242
R-gene [Kabi] *see* 817
RGH-1106 *see* 7612
RGH-3332 *see* 4170
RGH-4405 *see* 10128
RH-8 *see* 3620
RH-315 *see* 8058
RH-787 *see* 8171
RH-893 *see* 6853
RH-2161 *see* 3998
RH-2915 *see* 7097
RH-3866 *see* 6402
RH-5992 *see* 9254
RH-6201 *see* 111
RH-7592 *see* 4002
Rhabarberone *see* 313
Rhamnetin, 8337
β-Rhamnocitrin *see* 8337
Rhamnol *see* 8697
Rhamnolutein *see* 5288
Rhamnose, 8338
L-Rhamnose *see* 8338
3β-Rhamnosido-14β-hydroxy-Δ⁴,²⁰,²²-bufatrienolide *see* 8060
6-*O*-α-L-Rhamnosyl-D-glucose *see* 8457
Rhamnoxanthin *see* 4288
Rhamnus Cathartica, 8339
Rhaponticin *see* 8340
Rhapontin, 8340
rhDNase I *see* 2953
Rheadine *see* 8341
Rheic Acid *see* 8342
Rhein, 8342
Rhematan *see* 2347
Rhenic Anhydride *see* 8347
Rhenium, 8343
Rhenium Blue *see* 8347
Rhenium Chloride Oxide *see* 8346
Rhenium Heptoxide, 8344
Rhenium Hexafluoride, 8345

Rhenium Oxychloride, 8346
Rhenium Trioxide, 8347
Rhenium Trioxide Chloride *see* 8346
Rhenocain *see* 4894
Rheochrysidin *see* 3602
Rheocyclan [Grünenthal] *see* 9227
Rheomacrodex [Pharmacia] *see* 2989
Rheotran [Pharmachem] *see* 2989
Rheumacin LA [CP Pharm.] *see* 4998
Rheumatism Weed *see* 783 and 2099
Rheumatol [Tosse] *see* 1507
Rheumatrex [Am. Cyanamid] *see* 6065
Rheum Emodin *see* 3602
Rheumibis [Ibis] *see* 27
Rheumin *see* 2347
Rheumon Gel [Troponwerke; Bayer] *see* 3922
Rheumox [Robins] *see* 768
Rhex [Hobein] *see* 5895
Rhinalair [Fabre] *see* 3645
Rhinalar [Syntex] *see* 4180
Rhinanthin *see* 910
Rhinantin *see* 6455
Rhinaspray [Boehringer, Ing.] *see* 9702
Rhinathiol [Joulli] *see* 1850
Rhinocort [Astra] *see* 1490
Rhinofrenol *see* 7100
Rhinogutt [Thomae] *see* 9702
Rhinolast [Asta] *see* 939
Rhinolitan [Kettelhack] *see* 7100
Rhinoperd *see* 6455
Rhinopront [Mack, Illert.] *see* 9358
Rhinoptil [Promonta] *see* 1671
Rhinospray [Thomae] *see* 9702
Rhizopterin, 8348
Rhocya *see* 7866
Rhodalline *see* 9500
Rhodamine B, 8349
Rhodanic Acid *see* 8351
Rhodanilic Acid, 8350
Rhodanine, 8351
Rhodanwasserstoffsäure (German) *see* 9464
Rhodeose *see* 4301
Rhodialothan [Nattermann] *see* 4634
Rhodiatox *see* 7167
Rhodinal *see* 2390
Rhodine *see* 886
Rhodinol, 8352
β-Rhodinol *see* 2391
Rhodium, 8353
Rhodium Carbonyl Chloride, 8354
Rhodium Chloride, 8355
Rhodium Trichloride *see* 8355
Rhodizite *see* 2051
Rhodochrosite *see* 5766
Rhododendrin, 8356
Rhodol *see* 6097
Rhodomycin A *see* 8357
Rhodomycin B *see* 8357
β-Rhodomycin II *see* 8357
Rhodomycins, 8357
Rhodonite *see* 5781
Rhodopin, 8358
Rhodopsin, 8359
Rhodoquine *see* 7680
Rhodoquinone, 8360
Rhodoquinone-10 *see* 8360
Rhodotoxin *see* 4566
Rhodoviolascin *see* 8361
Rhodoxanthin, 8362
Rhoeadine *see* 8341
Rhombinin *see* 666
Rhomex *see* 7617
Rhothane *see* 3111
Rhovyl [Rhovyl] *see* 7746
rhSOD *see* 9177
Rhubarb, 8363
Rhubarb Yellow *see* 8342

Rhudane see 5058
rHuEpo see 3729
Rhumalgan [Lagap] see 3132
Rhynchophylline, 8364
Rhyncophylline see 8364
Rhythminal [Teikoku; Nikken] see 10209
Rhythmy [Shionogi] see 8387
Riabal [Fujisawa] see 7923
Riacen [Chiesi] see 7661
Riball [Mitsui] see 287
Ribavirin, 8365
α-Ribazole, 8366
Ribena see 867
Ribex [Formenti] see 3507
Ribitol see 166
Ribo [Tennessee Pharm.] see 8368
Ribo-Azauracil [Spofa] see 937
D-2-Ribodesose see 2955
Riboflavin, 8367
Riboflavin 5′-(Dihydrogen Phosphate) see 8368
Riboflavine 5′-Adenosine Diphosphate see 4131
Riboflavine 5′-(Trihydrogen Diphosphate) 5′→5′-Ester with Adenosine see 4131
Riboflavin Monophosphate, 8368
D-Ribofuranose-5-phosphoric Acid see 8372
9-β-D-Ribofuranosidoadenine see 152
1-β-D-Ribofuranosylcytosine see 2855
9-β-D-Ribofuranosylhypoxanthine see 5005
2-β-D-Ribofuranosylmaleimide see 8627
3-β-D-Ribofuranosylorotic Acid see 7005
9-β-D-Ribofuranosyl-9H-purin-6-amine see 152
9-β-D-Ribofuranosyl-9H-purine see 6521
9-β-D-Ribofuranosyl-9H-purine-2,6-diol see 10199
9-β-D-Ribofuranosyl-9H-purine-2,6-(1H,3H)-dione see 10199
N-(9-β-D-Ribofuranosyl-9H-purin-6-yl)butyramide Cyclic 3′,5′-(Hydrogen Phosphate) 2′-Butyrate see 1483
3-β-D-Ribofuranosyl-1H-pyrrole-2,5-dione see 8627
7-β-D-Ribofuranosyl-7H-pyrrolo[2,3-d]pyrimidin-4-amine see 9936
2-β-D-Ribofuranosyl-1,2,4-triazine-3,-5(2H,4H)-dione see 937
1-β-D-Ribofuranosyl-1H-1,2,4-triazole-3-carboxamide see 8365
1-β-D-Ribofuranosyluracil see 10016
9-β-D-Ribofuranosyl Xanthine see 10199
D-Ribohexulose see 8109
Ribolact [Taiho] see 8367
Ribomycine [Delalande] see 8373
Ribonosine [Toyo Jozo] see 5005
Ribonuclease, 8369
Ribonucleic Acid, 8370
D-Ribose, 8371
D-Ribose-5-phosphate see 8372
D-Ribose-5-phosphoric Acid, 8372
Ribosomal RNA see 8370
Ribostamin [Delalande] see 8373
Ribostamycin, 8373
Ribrain [Brocades] see 1224
D-Ribulose, 8374
RIC-272 see 6028
Ricainide see 4971
Ricamycin [Toyo Jozo] see 8408
Rice Bran Oil, 8375
Rice Oil see 8375
Richweed see 2546

Ricidine see 8377
Ricin, 8376
Ricin D see 8376
Ricinine, 8377
Ricinoleic Acid, 8378
Ricinolsulfuric Acid see 8378
Ricinstearolic Acid Diiodide see 7313
Ricinus Oil see 1946
Rickamicin see 8695
Ricketon see 10157
Ricridene [Lipha] see 6632
Rid see 2896
Ridaura [SK & F] see 911
Ridauran [SK & F] see 911
Ridazin [Taro] see 9497
Ridect Pour-On [Beecham] see 7321
Ridene [Syntex] see 6579
Ridinol see 7922
Ridogrel, 8379
Ridomil [Ciba-Geigy] see 5982
Ridzol [Merck & Co.] see 8413
Riegel's Paper see 2562
Rifa [Grünenthal] see 8382
Rifabutin, 8380
Rifacol [Wassermann] see 8386
Rifadin(e) [Marion Merrell Dow] see 8382
Rifaldazine see 8382
Rifaldin [Marion Merrell Dow] see 8382
Rifamastene [Uce] see 8384
Rifamicine SV see 8384
Rifamide, 8381
Rifampicin see 8382
Rifampin, 8382
Rifamycin AG see 8383
Rifamycin AMP see 8382
Rifamycin B see 8383
Rifamycin B Diethylamide see 8381
Rifamycin O see 8383
Rifamycin S see 8383
Rifamycin SV, 8384
Rifamycin X see 8383
Rifamycins, 8383
Rifapentine, 8385
Rifaprodin [Prodes] see 8382
Rifaxidin see 8386
Rifaximin, 8386
Rifit [Ciba-Geigy] see 7921
Rifloc Retard [Merrell] see 5245
Rifocin [Lepetit] see 8384
Rifocin M [Lepetit] see 8381
Rifoldin [Marion Merrell Dow] see 8382
Rifomycin SV see 8384
Rifomycins see 8383
Rigedal see 5245
Rigelon [Dojin] see 9436
Rikamycin see 8408
Rikavarin [Toyo Jozo] see 9704
Riker 52G see 796
Rilansyl see 2155
Rilaquil see 2155
Rilassol see 2155
Rilaten [Guidotti] see 8406
Rilmazafone, 8387
Rilmenidine, 8388
Riluzole, 8389
Rimactan(e) [Ciba-Geigy] see 8382
Rimadyl [Roche] see 1912
Rimagina [Medipolar] see 8391
Rimantadine, 8390
Rimaon see 3760
Rimatil [Santen] see 1481
Rimazole [Cheil Sugar] see 2478
Rimazolium Metilsulfate, 8391
Rimexel [Organon] see 8392
Rimexolone, 8392
Rimidin(e) [Elanco] see 3999
Rimifon [Roche] see 5203

Rimiterol, 8393
Rimitsid see 5203
Rimocidin, 8394
Rimso-50 [Res. Industries] see 3308
Rinatec [Boehringer, Ing.] see 5089
Rinderon-DP [Shionogi] see 1226
Rinder-vasopressin see 10073
Rindex [Sidus] see 6007
Rinesal [Kissei] see 2021
Rineton [Sanwa] see 9728
Ringworm Powder see 807
Rinlaxer [Taisho] see 2231
Rino-Clenil [Chiesi] see 1047
Rinogutt [Boehringer, Ing.] see 9702
Rintal [Bayer] see 3982
Riocyclin [Adcock-Ingram] see 9337
Riogen see 7453
Riol [Toa Eiyo] see 9267
Riomitsin see 7111
Riopan [Ayerst] see 5685
Rioprostil, 8395
Ripason see 5575
Ripcord [Shell] see 2836
Ripercol [Am. Cyanamid] see 5486
Ripirin [Kabi] see 3598
Ripon see 5685
Riporest [Formenti] see 6028
Riposon see 5890
Ripresil [Erba] see 2513
Riran [Yamanouchi] see 4963
rI$_n$. r(C$_{12}$U)$_n$ see 630
Ririlim see 1157
Riripen [Daiichi] see 1157
Risa-131-H [Abbott] see 8613
Risamal [Yoshitomi] see 2377
Risatarun [Ravensberg] see 2900
Rischiaril [Piam] see 2900
Rise [Yoshitomi] see 2477
Risedronate Sodium see 8396
Risedronic Acid, 8396
Riself [Gibipharma] see 5896
Risolid [Dumex] see 2132
Risordan [Théraplix] see 5245
Risperdal [Janssen] see 8397
Risperidone, 8397
Ristocetin, 8398
Ristogen [Kowa Yakuhin] see 4167
Ristomycin see 8398
Ristomycin A see 8398
Ristomycin B see 8398
Riston see 8398
Ristosamine see 134
Risumic [Dainippon] see 412
RIT-1140 see 772
Ritalin Hydrochloride [Ciba] see 6186
Ritalmex [Alkaloida] see 6257
Ritanserin, 8399
Ritipenem, 8400
Ritipenem Acoxil see 8400
Ritmocardyl [Bago] see 504
Ritmodan [Maestretti] see 3424
Ritmos [Inverni] see 194
Ritmos Elle [Inverni] see 5607
Ritmusin [Gebro] see 793
Ritodrine, 8401
Ritonavir, 8402
RIV-2093 see 319
Rivadescin [Schaper & Brummer] see 8293
Rivanol [Winthrop] see 3760
Rivasin [Giulini] see 8314
Riversideite see 1749
Rivotril [Roche] see 2449
Rixapen [RIT] see 2445
Rizaben [Kissei] see 9705
Rize [Yoshitomi] see 2477
Rizen [Puropharma] see 2477
Rizinsan K2 A2 see 4515
RL-50 see 6337
r-metHuG-CSF see 4558

RMI-17043 see 3627
RMI-71754 see 10114
RMI-71782 see 3564
RMI-83027 see 8409
RNA see 8370
mRNA see 8370
rRNA see 8370
sRNA see 8370
tRNA see 8370
RNase see 8369
Ro-1-5130 see 8161
Ro-1-5431 see 5496
Ro-1-5431/7 see 5496
Ro-1-5470 see 8274
Ro-1-5470/5 see 8274
Ro-1-5470/6 see 8274
Ro-1-6794 see 5496
Ro-1-7788 see 8274
Ro-1-7977 see 9034
Ro-1-9569 see 9325
Ro-2-0404 see 254
Ro-2-2453 see 3033
Ro-2-2985 see 5384
Ro-2-3198 see 3562
Ro-2-3308 see 8266
Ro-2-3773 see 2412
Ro-2-7113 see 304
Ro-2-9578 see 9839
Ro-2-9757 see 4219
Ro-2-9915 see 4161
Ro-2-9945 see 4214
Ro-3-4787 see 1504
Ro-3-7008 see 204
Ro-4-1544/6 see 8966
Ro-4-1575 see 511
Ro-4-1577 see 2782
Ro-4-1634 see 7668
Ro-4-1778/1 see 6229
Ro-4-3780 see 8333
Ro-4-3816 see 222
Ro-4-4393 see 9074
Ro-4-4602 see 1079
Ro-4-5282 see 5843
Ro-4-5360 see 6667
Ro-4-6467 see 7938
Ro-4-8347 see 9717
Ro-5-0810/1 see 9785
Ro-5-0831 see 5172
Ro-5-2180 see 6784
Ro-5-2807 see 3042
Ro-5-3307/1 see 2901
Ro-5-3350 see 1406
Ro-5-3438 see 4164
Ro-5-4023 see 2449
Ro-5-4200 see 4181
Ro-5-5345 see 9285
Ro-5-6901 see 4233
Ro-6-4563 see 4447
Ro-7-0207 see 7000
Ro-7-1051 see 1114
Ro-7-1554 see 5095
Ro-7-4488/1 see 6425
Ro-7-6102 see 4243
Ro-7-6145 see 3243
Ro-8-6837 see 6222
Ro-10-1670 see 114
Ro-10-5970 see 1401
Ro-10-6338 see 1508
Ro-10-9070 see 408
Ro-10-9359 see 3935
Ro-11-780 see 8391
Ro-11-1163 see 6309
Ro-11-1430 see 6367
Ro-12-0068 see 9293
Ro-13-1042 see 5610
Ro-13-5057 see 700
Ro-13-5223 see 4027
Ro-13-8996 see 7071
Ro-13-9297 see 5612
Ro-13-9904/001 see 2001

Ro-14-3169/000 see 4035
Ro-14-4767/000 see 612
Ro-14-4767/002 see 612
Ro-15-1297 see 8167
Ro-15-1788 see 4169
Ro-15-8074 see 1974
Ro-15-8075 see 1974
Ro-17-2301 see 1922
Ro-17-3664 see 7990
Ro-18-0647 see 6997
Ro-19-5247 see 1996
Ro-19-6327 see 5407
Ro-20-5720/000 see 1912
Ro-20-7234 see 4193
Ro-21-3981/001 see 6270
Ro-21-3981/003 see 6270
Ro-21-5535 see 1681
Ro-21-5998 see 5845
Ro-21-6937 see 9853
Ro-21-8837 see 3749
Ro-21-9738 see 3493
Ro-22-7796 see 2330
Ro-22-8181 see 5016
Ro-22-9000 see 234
Ro-23-6240 see 4137
Ro-31-2848 see 2332
Ro-31-3113 see 2332
Ro-31-8959 see 8516
Ro-31-8959/003 see 8516
Ro-40-5967 see 6261
Ro 44-9883 see 5362
Roaccutane [Roche] see 8333
Robac 22 [Robinson Bros.] see 3849
Robalate [Robins] see 3227
Robamol [Cenci] see 6060
Robamox [Robins] see 617
Robane [Robeco] see 8923
Robanul see 4511
Robaxin [Robins] see 6060
Robecote [Robeco] see 8622
Robenidine, 8403
Robenogatope I-131 [Bristol-Myers
    Squibb] see 8421
Robenz [Am. Cyanamid] see 8403
Robenzidene see 8403
Robigenin see 5288
Robinin, 8404
Robinul [Robins] see 4511
Robiocina [Pradel] see 6818
Robison Ester see 4469
Robitussin [Robins] see 4582
Robizone-V [Robins] see 7431
Robuoy [Robeco] see 7932
Rocain [Fuso] see 7937
Rocaltrol [Roche] see 1681
Roccal [Winthrop] see 1086
Roccellic Acid, 8405
Rocefin [Roche] see 2001
Rocephin(e) [Roche] see 2001
Rochelle Salt see 7840
Rociverine, 8406
Rock Salt see 8742
Rocornal [Rentschler; UCB] see 9709
Rocosenin see 4651
Rocuronium, 8407
Ro-Cycline [Rowell] see 9337
Rodalon see 1086
Rodameb [Specia] see 3394
Rodex [ABCO] see 10174
Rodilone see 20
Rodinal see 482
Rodipal see 3793
Rodiuran see 4830
Rody [Sumitomo] see 4033
(+)-Roemerine see 789
Roeridorm [Roerig] see 3774
Roferon-A [Roche] see 5016
Roflual [Roche] see 8390
Rogaine [Upjohn] see 6290
Rogitine [Ciba] see 7417

Rogojski's Salt see 2738
Rogor [Agrimont] see 3269
Rogue [Monsanto] see 7987
Rohrbach's Soln see 1011
Rohypnol [Roche] see 4181
Roidenin [Showa Shinyaku] see 4925
Roinin [Mohan] see 7919
Roipnol [Roche] see 4181
Rökan [Intersan] see 4432
Rokital [Formenti] see 8408
Rokitamycin, 8408
Rolicton [Searle] see 507
Rolicypram see 8409
Rolicyprine, 8409
Rolipram, 8410
Rolitetracycline, 8411
Roman Chamomile see 2082
Roman Chamomile Oil see 2082
Roman Vitriol see 2722
Romazicon [Roche] see 4169
Romensin [Elanco] see 6329
Romet [Mitsubishi] see 8305
Rometin see 5052
Romglizone see 9898
Romicil see 6962
Rompun [Bayer] see 10213
Romurtide, 8412
Ronaxan [Rhône-Merieux] see 3496
Rondar [Wyeth] see 7059
Rondase see 4794
Rondimen [Asta] see 5843
Rondomycin [Wallace] see 6007
Rongalite see 8764
Rongalite C see 8764
Roniacol [Roche] see 6614
Roniacol Tartrate [Roche] see 6614
Ronicol see 6614
Ronidase see 4794
Ronidazole, 8413
Ronifibrate, 8414
Ronilan [BASF] see 10122
Ronnel, 8415
Ronok [Ono] see 7003
Ronoprost see 7003
Ronphenil [Zeria] see 2120
Ronstar [Rhône-Poulenc] see 7038
Rontyl [Leo Pharm.] see 4830
ROP-500F see 5093
Ropinirole, 8416
Ropivacaine, 8417
Roptazol see 4320
Roquessine see 2561
Roquinimex, 8418
Rorer 148 see 6028
Roridins see 10100
Rosal [IBI] see 8419
Rosamicin see 8420
Rosamit see 5401
Rosampline [Rosa-Phytopharma] see
    628
Rosaniline Hydrochloride see 5686
Rosapin see 2962
Rosaprostol, 8419
Rosaramicin, 8420
Roscoelite see 10053
Rose Bengal, 8421
Rose Bengale B see 8421
Rose Bengal Extra see 8421
Rose Bengal Sodium I 131 see 8421
Rose Hips, 8422
Rosemary, 8423
Rosemide [Ono] see 4331
Rosenstiehl's Green see 1010
Rose-pink see 2017
Rosin, 8424
Rosin Oil, 8425
Rosinol see 8425
p-Rosolic Acid see 915
Rosoxacin, 8426
Rossitrol [Corvi] see 8433

Rostil [Ortho] see 8395
Rotacide [Roussel-UCLAF] see 7629
Rotenone, 8427
Rotersept [Roterpharma] see 2140
Rotesar [Argentina] see 981
Rotondin [Casasco] see 4015
Rotoxamine see 1845
Rotramin [Glaxo-Allen] see 8433
Rotraxate, 8428
Rottlerin, 8429
Rotundine see 9354
Rougoxin see 3210
Roulone [Roussel-UCLAF] see 6028
Round-leafed Sundew see 3508
Round-up [Monsanto] see 4522
Rouqualone see 6028
Rovamicina [Rhone-Poulenc Rorer]
    see 8904
Rovamycin [Rhone-Poulenc Rorer] see
    8904
Rovral [Rhône-Poulenc] see 5093
Rowamyelin see 8898
Rowapraxin [Rowa-Wagner] see 7637
Rowasa [Rowell] see 5964
Roxanol [Roxane] see 6361
Roxanthin Red 10 [Roche] see 1798
Roxarsone, 8430
Roxatidine Acetate, 8431
Roxenan [UPSA] see 7551
Roxiam [Astra] see 8301
Roxicam [Gramon] see 7661
Roxiden [Pulitzer] see 7661
Roxindole, 8432
Roxindole Mesylate see 8432
Roxion [Cela] see 3269
Roxit [Albert-Roussel] see 8431
Roxithromycin, 8433
Roxomicina see 3722
Royal Jelly, 8434
Royal Jelly Acid see 8434
Royline see 5646
Rozol see 2204
RP-40 see 6777
RP-46 see 1182
RP-143 see 7879
RP-217 see 9257
RP-866 see 8225
RP-2090 see 9115
RP-2168 see 6154
RP-2255 see 9116
RP-2259 see 4487
RP-2275 see 9076
RP-2512 see 7254
RP-2591 see 3121
RP-2632 see 9081
RP-2740 see 6026
RP-2786 see 8168
RP-2831 see 4691
RP-2987 see 3157
RP-3015 see 4013
RP-3203 see 5061
RP-3276 see 7966
RP-3277 see 7970
RP-3356 see 3793
RP-3359 see 2138
RP-3377 see 2215
RP-3389 see 7970
RP-3554 see 9441
RP-3668 see 8859
RP-3697 see 4361
RP-3735 see 5620
RP-3799 see 3165
RP-3828 see 487
RP-4460 see 5219
RP-4482 see 9110
RP-4560 see 2238
RP-4632 see 6080
RP-4753 see 8169
RP-4763 see 3394
RP-4909 see 2236

RP-5015 see 5203
RP-5337 see 8904
RP-6140 see 7942
RP-6171 see 626
RP-6484 see 3938
RP-6549 see 9834
RP-6847 see 7080
RP-7044 see 6066
RP-7162 see 9852
RP-7204 see 2753
RP-7293 see 7933
RP-7452 see 96
RP-7522 see 9087
RP-7746 see 9810
RP-7843 see 9494
RP-7891 see 4488
RP-8228 see 5493
RP-8595 see 3315
RP-8599 see 4260
RP-8823 see 6242
RP-8908 see 7306
RP-9159 see 7309
RP-9671 see 6816
RP-9715 see 2782
RP-9921 see 796
RP-9965 see 6233
RP-10192 see 2937
RP-11974 see 7489
RP-12222 see 6207
RP-12833 see 2466
RP-13057 see 2890
RP-13907 see 4142
RP-14539 see 8562
RP-16091 see 6219
RP-16272 see 1465
RP-17623 see 7038
RP-18429 see 9704
RP-19366 see 7636
RP-19551 see 7636
RP-19552 see 7636
RP-19560 see 5991
RP-19583 see 5316
RP-22050 see 10326
RP-22410 see 4453
RP-26019 see 5093
RP-27267 see 10324
RP-31264 see 9182
RP-54274 see 8389
RP-54563 see 3626
RP-54780 see 7044
RP-56976 see 3458
RPM see 8288
RR-32705 see 6970
RS-533 see 7141
RS-1301 see 2930
RS-1320 see 4180
RS-2252 see 4157
RS-3540 see 6504
RS-3650 see 6504
RS-3999 see 4180
RS-4691 see 2460
RS-5139 see 9287
RS-8858 see 7069
RS-9390 see 8066
RS-11988 see 5361
RS-21592 see 4374
RS-35887 see 1561
RS-37619 see 5318
RS-43179 see 5595
RS-43285 see 8287
RS-44872 see 9062
RS-69216 see 6579
RS-84043 see 4037
RS-84135 see 3629
RS-94991-298 see 6437
R Salt see 6474
RTCA see 8365
rt-PA see 9608
RTY-319 see 4905
RU-43-715 see 8059

RU-486 see 6273
RU-965 see 8433
RU-2267 see 327
RU-2323 see 4423
RU-5020 see 7969
RU-15060 see 9562
RU-15750 see 4140
RU-19110 see 4627
RU-22974 see 2934
RU-23908 see 6636
RU-24756 see 1983
RU-25474 see 9700
RU-28965 see 8433
RU-31158 see 5604
RU-38486 see 6273
RU-38702 see 127
RU-39780 see 1266
RU-44403 see 9703
RU-44570 see 9703
Rubber, 8435
Rubeanic Acid, 8436
Rubene see 6456
Ruberythric Acid, 8437
Ruberythrinic Acid see 8437
Rubesol [Central Pharm.] see 10152
Rubiadin, 8438
Rubian see 8437
Rubianic Acid see 8437
Rubiazol see 9069
Rubiazol I see 9092
Rubichloric Acid see 879
Rubidazone [Rhône-Poulenc] see
    10326
Rubidium, 8439
Rubidium Alum see 374
Rubidium Bromide, 8440
Rubidium Chloride, 8441
Rubidium Hydroxide, 8442
Rubidium Iodide, 8443
Rubidomycin see 2890
Rubigan [Elanco] see 3999
Rubigervine see 8444
Rubijervine, 8444
Rubimycin [Nikken] see 6271
Rubinorm [Ifi] see 8441
Rubixanthin, 8445
Rubozinc [Labcatal] see 4464
Rubramin PC [Bristol-Myers Squibb]
    see 10152
Rubratope-57 [Bristol-Myers Squibb]
    see 10153
Rubratope-60 [Bristol-Myers Squibb]
    see 10153
Rubriment [Nordmark] see 6613
Rubrogliocladin see 912
α-Rubromycin see 2545
Rubus, 8446
Ruby [Spencer] see 6602
Ruby Arsenic see 834
Ruby Wood see 8531
Rucaina see 5505
Rudilin [Darby] see 6830
Rudotel [OPW] see 5829
Ruelene [Dow] see 2671
Rufast [Roussel-UCLAF] see 127
Rufigallic Acid see 8447
Rufigallol, 8447
Rufloxacin, 8448
Rufol [Debat] see 9084
Rugby [FMC] see 1670
Rugulovasine A see 8449
Rugulovasine B see 8449
Rugulovasines, 8449
Rulicalcin [Roussel-UCLAF] see 1680
Rulid [Roussel] see 8433
Rumasal see 338
Rumatral [Wander] see 316
Rumensin [Lilly] see 6329
Rumex, 8450
Rusa Oil see 6908

Rusat see 1194
Ruswut see 1194
Rusyde [CP] see 4331
Rutaecarpine see 8451
Rutamycin see 6970
Rutecarpine, 8451
Rutergan see 4013
Rutgers 612 see 3790
Ruthenium, 8452
Ruthenium(VIII) Oxide see 8454
Ruthenium Oxychloride Ammoniated see 8453
Ruthenium Red, 8453
Ruthenium Tetroxide, 8454
Ruthenium Trichloride, 8455
Ruthenium Violet see 8453
Rutherfordium see 3580 and 3583
Ruticina [Bernabo] see 5986
Rutile see 9612
Rutin, 8456
Rutinose, 8457
Rutonal [M & B] see 7457
Rutoside see 8456
Ruven [Mepha] see 9920
RV-144 see 3684
RV-12424 see 4183
RWJ-17021-000 see 9686
RWJ-17070 see 4760
RX-5050M see 3403
RX-6029-M see 1522
Ryanex see 8458
Ryania, 8458
Ryanicide [Penick] see 8458
Ryanodine, 8459
Ryanodol 3-(1H-Pyrrole-2-carboxylate) see 8459
Rycarden [Syntex] see 6579
Rycopel [Rycovet] see 2836
Rydar [Cutter] see 3352
Rydene [Syntex] see 6579
Rydex [Sandoz] see 7947
Rydrin [Kodama] see 6830
Ryegonovin [Morishita] see 6147
Rynacrom [Fisons] see 2658
Rythmarone [Leurquin] see 504
Rythmatine [Burroughs Wellcome] see 5889
Rythminal [Teikoku; Nikken] see 10209
Rythmodan [Roussel-UCLAF] see 3424
Rythmodul [Albert-Roussel] see 3424
Rythmol [Knoll] see 7978
Rytmonorm [Knoll] see 7978
Ryzelan [Lilly] see 7015
S see 8604
S-2-676 see 9372
S-7 see 4046
S 8 [Chefaro] see 3367
S-23/46 see 710
S-041 see 4345
S-73-4118 see 7647
S-77-0777 see 7899
S-94 see 4226
S-165 see 3158
S-222 see 3432
S-314 see 4335
S-486 see 6603
S-596 see 827
S-640P see 1965
S-768 see 4015
S-780 see 1066
S-805 see 5617
S-940 see 6441
S-992 see 1066
S-1006 see 1969
S-1108 see 1969
S-1320 see 1490
S-1358 see 1555
S-1520 see 4969

S-1530 see 6641
S-1574 see 9560
S-1605 see 3158
S-1694 see 429
S-1702 see 4448
S-1752 see 4044
S-1844 see 4051
S-1942 see 1453
S-2225 see 1453
S-2395 see 9318
S-2539 see 7405
S-2620 see 309
S-2703 see 2837
S-2703 Forte see 2837
S-3151 see 7321
S-3206 see 4033
S-3308-10 see 3320
S-3308L see 3320
S-3341 see 8388
S-3341-3 see 8388
S-3500 see 1554
S-4084 see 2321
S-4087 see 2760
S-5602 see 4051
S-5602α see 4051
S-6059 see 6369
S-6810 see 5018
S-7131 see 7946
S-8527 see 2415
S-9318 see 8174
S-9490 see 7311
S-9490-3 see 7311
S-9780 see 7311
S-10036 see 4285
S-23031 see 4178
S-25930 see 4919
S-26308 see 4957
S-31183 see 8174
S-31252 see 7318
S-32165 see 3158
S-50022 see 4435
SA see 9974
SA-79 see 7978
SA-96 see 1481
SA-97 see 4770
SA-267 see 3402
SA-504 see 9583
Sabadilla, 8460
Sabadine, 8461
Sabalamin [Santen] see 2513
Sabari see 5817
Sabatine see 8461
Sabeluzole, 8462
Sabidal [Zyma] see 2266
Sabiden [Szabo] see 7425
Sabril [Merrell-Dow] see 10114
Sabromin [Sandoz] see 1473
sab simplex [Parke, Davis] see 3264
Saccharase see 5025
Saccharated Iron see 4073
Saccharic Acid see 4458
Saccharin, 8463
Saccharin Ammonium see 8463
Saccharin Insoluble see 8463
Saccharinol see 8463
Saccharinose see 8463
Saccharin Sodium see 8463
Saccharol see 8463
Saccharolactic Acid see 4349
Saccharolactone see 4458
L-Saccharopine, 8464
Saccharose see 9051
Saccharosonic Acid see 5141
Sacerno [EGYT] see 5898
Sacox [Hoechst] see 8488
Sacred Bark see 1931
Sadamin [Polfa] see 10194
Sadopan see 4378
Safe-Gard [Hoechst-Roussel] see 4000
Saferon see 7292

Safersan see 7548
Saffan [Glaxo] see 235
Safflor Carmine see 1918
Safflor Red see 1918
Safflower see 1919
Safflower Oil, 8465
Saffron, 8466
Saffron-bitter see 7563
Saffron Indian see 9955
Safranal, 8467
Safranin B Extra see 7402
Safrole, 8468
Safrotin [Sandoz] see 8000
Saga [Roussel-Uclaf EH] see 9700
Sagamicin [Kyowa] see 6269
Sagamicin (formerly) see 6269
Sage see 8495
Sagisal [Sagitta] see 8917
Sagittol [Sagitta] see 8025
SaH-42548 see 5801
SAICAR, 8469
Saigon Cinnamon see 2359
Saiodin see 1721
Saisan [ICI] see 3499
Saizen [Serono] see 8864
Sajodin see 1721
Sakuranetin, 8470
Sakuranin see 8470
Salacetamide, 8471
Salacetin see 886
Sal Acetosella see 7772 and 7863
Saladjidi see 4929
Salagen [MGI] see 7578
Salamid [Hamilton] see 8480
Sal Ammoniac see 537
Salanil [Sato Yakuhin] see 4385
Salatrim, 8472
Salazolon see 758
Salazopyrin [Kabi Pharmacia] see 9112
Salazosulfadimidine, 8473
Salazosulfapyridine see 9112
Sal Barnit see 10295
Salbulin [Riker] see 217
Salbumol [Glaxo] see 217
Salbutamol see 217
Salbutard [Lusofarmaco] see 217
Salbutine [Paucort] see 217
Salbuvent [Nycomed] see 217
Salcatonin see 1680
Salcetogen see 886
Salco see 2434
Salcyl see 8486
Salcylix see 8486
Saldox see 8479
Salen, 8474
Sal Enixum see 7774
Salep, 8475
Sal Ethyl see 3894
Saletin see 886
Salflex [Carnrick] see 8491
Salgain [C-Vet LP] see 8488
Salicain [Hoechst] see 4894
Salicin, 8476
Salicin Benzoate see 7753
Salicinerein see 7549
Salicoside see 8476
N-Salicoylaminophenol see 7018
Salicylacetylsalicylic Acid see 102
Salicyl Alcohol, 8477
Salicyl Alcohol Glucoside see 8476
Salicylaldehyde, 8478
Salicylaldehyde β-D-Glucoside see 4658
Salicylaldoxime, 8479
Salicylamide, 8480
Salicylamide O-Acetic Acid, 8481
Salicylamide o-Ethyl Ether see 3776
Salicylanilide, 8482
Salicylazosulfadimidine see 8473

Salicylazosulfamethazine *see* 8473
Salicylazosulfapyridine *see* 9112
Salicylhydroxamic Acid, 8483
Salicylic Acid, 8484
Salicylic Acid Acetate *see* 886
Salicylic Acid Acetate Calcium Salt *see* 1684
Salicylic Acid Acetate, Ester with 4-Hydroxyacetanilide *see* 1074
Salicylic Acid, Acid Sulfate *see* 8486
Salicylic Acid Benzyl Ester *see* 1181
Salicylic Acid Bimolecular Ester *see* 8491
Salicylic Acid Bornyl Ester *see* 1371
Salicylic Acid *p-tert*-Butylphenyl Ester *see* 1621
Salicylic Acid Choline Salt *see* 2265
Salicylic Acid Dihydrogen Phosphate *see* 4281
Salicylic Acid Ethyl Ester *see* 3894
Salicylic Acid Isopropyl Ester *O*-Ester with *O*-Ethyl Isopropyl-phosphoramidothioate *see* 5187
Salicylic Acid Monoammonium Salt *see* 584
Salicylic Acid 3,3,5-Trimethylcyclo-hexyl Ester *see* 4776
Salicylic Aldehyde *see* 8478
Salicylic Ether *see* 3894
Salicyl Morpholide *see* 8485
Salicyloxysalicylic Acid *see* 8491
4-Salicyloylmorpholine, 8485
Salicyl Phosphate *see* 4281
Salicylsalicylic Acid *see* 8491
Salicylsulfonic Acid *see* 9138
Salicylsulfuric Acid, 8486
Salicyl-Vasogen *see* 584
Salicyl Yellow *see* 5979
Saliformin *see* 6041
Saligenin *see* 8477
Saligenin-β-D-glucopyranoside *see* 8476
Saligenol *see* 8477
Salimid *see* 2813
Salimol [Kanebo] *see* 9084
Salinaphthol *see* 6502
Salinazid, 8487
Saliniazid *see* 8487
Salinidol [Doak] *see* 8482
Salinigrin *see* 7549
Salinomycin, 8488
Saliphenazon *see* 758
Salipran [Theraplix] *see* 1074
Salipurpol *see* 6511
Salipyrazolon *see* 758
Salipyrine *see* 758
Salit [Tenneco] *see* 1371
Saliuzid *see* 6984
Salizell [Byk-Gulden] *see* 8480
Salizell (amp.) [Byk-Gulden] *see* 8481
Salizolo *see* 4949
Salmetedur [Menarini] *see* 8489
Salmeterol, 8489
Salmeterol Xinafoate *see* 8489
Salmiac *see* 537
Salmine Sulfate, 8490
Salmotonin [Yamanouchi] *see* 1680
Salocoll *see* 7388
Salofalk [Falk] *see* 5964
Salol *see* 7464
Saloop *see* 8528
Salophen [Bayer] *see* 46
Salpetersäure (German) *see* 6671
Salpix [Ortho] *see* 79
Sal Polychrestum *see* 7845
Salsalate, 8491
Sal Soda *see* 8739
Salsoline, 8492
(+)-Salsoline *see* 8492
Salt *see* 8742

Salt Cake *see* 8829
"Salt of Lemon" *see* 7772
Salt of Phosphorus *see* 587
Salt of Saturn *see* 5411
Salt of Sorrel *see* 7772 and 7863
Salt of Tartar *see* 7781
Saltpeter *see* 7815
Saltron *see* 2434
Saltucin [Boehringer, Mann.] *see* 1554
Salufer *see* 8769
Saluminium Insolubile *see* 375
Salunil *see* 2221
Salures [Ferrosan] *see* 1064
Saluretil *see* 2221
Saluric [Merck & Co.] *see* 2221
Salurin [Syntex] *see* 9754
Saluron [Bristol] *see* 4830
Saluside *see* 6984
Salutaridine, 8493
Saluzid *see* 6984
Saluzide *see* 6984
Salvacard *see* 6635
Salvarsan *see* 851
Salverine, 8494
Salvia, 8495
Salvinin *see* 7199
Salvis *see* 498
Salvizol *see* 1288
Salymid [Biochemie] *see* 8480
Salyrgan [Winthrop] *see* 5962
Salyzoron [Hishiyama] *see* 1157
Salzburg Vitriol *see* 2722
Salzone [Wallace] *see* 45
Samaderin A *see* 8496
Samaderin B *see* 8496
Samaderin C *see* 8496
Samaderins, 8496
Samaderoside A *see* 8496
Samandarine, 8497
Samandarone *see* 8497
Samarium, 8498
Sambucin *see* 2757
Sambucus, 8499
Sambunigrin *see* 5760
SAMe *see* 155
Samedrin [Savoma] *see* 2032
S. Amet [Boehringer, Ing.] *see* 155
Samid [Knoll] *see* 8480
Samli *see* 4425
Samorin [M & B] *see* 5199
Samyr [Bioresearch] *see* 155
SAN-197 *see* 3936
SAN-322I *see* 8000
SAN-582H *see* 3263
SAN-619F *see* 2840
SAN-9789 *see* 6794
Sanabolicum *see* 6452
Sanamycin [FBA Pharmaceuticals] *see* 1642
Sanapert *see* 7108
Sanasthmax [Glaxo] *see* 1047
Sanasthmyl [Glaxo] *see* 1047
Sanatrichom [Gödecke] *see* 6242
Sancap [Ciba-Geigy] *see* 3405
Sanceller 22 [Sanshin] *see* 3849
Sanclomycine *see* 9337
Sancycline, 8500
Sand *see* 8637
Sandalwood Oil *see* 6940
Sandalwood, White, 8501
Sandarac, 8502
Sand-brier *see* 8856
Sandimmun(e) [Sandoz] *see* 2821
Sandoglobulin [Sandoz] *see* 4958
Sandolanid [Sandoz] *see* 3210
Sandolase [Sandoz] *see* 8089
Sandomigran [Sandoz] *see* 7671
Sandonorm [Sandoz] *see* 1362
Sandopart [Sandoz] *see* 2899
Sandopril [Sandoz] *see* 8905

Sandoptal [Sandoz] *see* 1536
Sandoscill [Sandoz] *see* 8060
Sandostatin [Sandoz] *see* 6859
Sandostene [Sandoz] *see* 9413
Sandoz 43-715 *see* 8059
Sandril [Lilly] *see* 8314
Sanecta [Holland Sweetener] *see* 874
Sanedrine *see* 3645
Sanegyt [EGYT] *see* 4588
Sanger's Reagent *see* 4210
Sangrel *see* 8608
Sanguicillin [Zdravlje] *see* 7669
Sanguinaria, 8503
Sanguinarine, 8504
Sanifer [Esseti] *see* 4083
Saniflor [Esseti] *see* 1157
Sanluol *see* 851
Sanmigran [Sandoz] *see* 7671
Sanochrysine *see* 4537
Sanocrisin [Ferrosan] *see* 2789
Sanodin *see* 1839
Sanogyl [Villette] *see* 49
Sanoma [E. Merck] *see* 1889
Sanomigran [Sandoz] *see* 7671
Sanopron *see* 2238
Sanoquin *see* 2215
Sanorex [Sandoz] *see* 5801
Sanorin *see* 6455
Sanorin-Spofa *see* 6455
Sanotensin *see* 4589
Sanoxit [Basotherm] *see* 1149
Sansdolor *see* 5895
Sansert [Sandoz] *see* 6217
α-Santalol, 8505
β-Santalol, 8506
Santavy's Substance F *see* 2938
Santemycin [Santen] *see* 6269
Santenol [Coop Farm] *see* 5455
Santobrite [Monsanto] *see* 7242
Santoflex *see* 3800
Santomerse #1 [Monsanto] *see* 8757
Santonica, 8507
Santonic Acid, 8508
l-Santonin *see* 8509
α-Santonin, 8509
Santophen 1 [Monsanto] *see* 2469
Santophen 20 [Monsanto] *see* 7242
Santoquin [Monsanto] *see* 3800
Santovar A *see* 3359
Santyl [Knoll] *see* 2544
Sanzyme [Sankyo] *see* 9203
SAP *see* 657
Sapecron [Ciba-Geigy] *see* 2137
SAPEP *see* 424
Saperconazole, 8510
Saphicol *see* 5877
Saphire [Ciba-Geigy] *see* 4165
Saphizon-DP *see* 5877
Sapidan *see* 4378
l-Sapietic Acid *see* 5494
Sapilent [EGYT] *see* 9852
Sapoderm [Reckitt & Colman] *see* 9790
Saponaretin *see* 8512
Saponaria, 8511
Saponarin, 8512
Saponins, 8513
Saponite *see* 362
Sapphire, 8514
Sapphirine *see* 362
Saprol [Shell Agrar] *see* 9819
Sapropterin, 8515
Saprosan *see* 2243
Saquinavir, 8516
Saquinavir Mesylate *see* 8516
Sar *see* 519
Sarafloxacin, 8517
Sarafotoxins S6 *see* 3616
1-Sar-8-ala-angiotensin II *see* 8518
Saralasin, 8518

Saralasin Acetate see 8518
Saran see 10134
Sarcine see 4917
Sarcoclorin [Simes] see 5871
Sarcodinine see 5336
Sarcolactic Acid see 5350
Sarcolysine see 5871
D-Sarcolysine see 5871
L-Sarcolysine see 5871
Sarcomycin see 8521
Sarcomycin A see 8521
Sarcosine, 8519
Sarelon see 7190
Sarenin [Röhm] see 8518
Sargenor [Mundipharma] see 875
Sargramostim see 4559
Sarin, 8520
Sarisol [Halsey] see 1530
Sarkin see 4917
Sarkomycin, 8521
Sarkomycin A see 8521
Sarmentogenin, 8522
Sarmentose, 8523
Sarolex see 3043
Saroten [Lundbeck; Warner] see 511
Sarotex [Lundbeck] see 511
Sarpagan-17-al see 10076
Sarpagan-10,17-diol see 8524
Sarpagine, 8524
Sarpul [Roche] see 700
Sarsaparilla, 8525
Sarsasapogenin, 8526
Sartosona [Sastre] see 2021
Saruplase see 8089
Sarverogenin, 8527
Sarvinal see 759
Saryuurin [Nippon Shoji] see 7018
SAS-643 see 3490
Sassafras, 8528
Sassolite see 1364
Sassy Bark, 8529
Sastridex [Lindopharm] see 4167
Sasulen [Andreu] see 7661
Satanolon [Tatsumi] see 3369
Satinite see 1753
Satin Spar see 1753
Satisfar [Sandoz] see 3936
Satumomab, 8530
Saturn Red see 5449
Saucy Bark see 8529
Saunders, Red, 8531
Saurol see 4929
Savac [Am. Cyanamid] see 5061
Savapyrin see 3021
Saventrine [Pharmax] see 5236
Savey [Du Pont] see 4751
Savin, 8532
Savinin see 8532
Sawacillin [Fujisawa] see 617
Saxifrax see 8528
Saxin [Burroughs Wellcome] see 8463
Saxitoxin, 8533
Saxizon [Nikken] see 4828
Saxol see 7327
Saxoline see 7326
Saxon Bark see 8529
Sayfos [Chipman] see 5877
Sazio [Scharper] see 241
SB-1 see 1615
SB-58 see 8966
SB-5833 see 1773
SB-7505 see 4921
SBA see 5453
α_2-SB Glycoproteins see 4119
SBP-1390 see 1271
SBP-1513 see 7321
SBR Rubber, 8534
SC-0051 see 9063
SC-1627 see 4013
SC-2910 see 6025
SC-6924 see 6424

SC-7031 see 3424
SC-7525 see 1352
SC-9387 see 7607
SC-9420 see 8917
SC-9880 see 4235
SC-11800 see 3905
SC-12937 see 924
SC-13957 see 3424
SC-18862 see 874
SC-29333 see 6297
SC-40230 see 1251
SC-47111 see 5592
SC-326421 see 6242
Scabiolide, 8535
Scabwort see 5023
Scammony Root, 8536
Scandia see 8537
Scandicain [Bofors] see 5905
Scandine [Zambon] see 4921
Scandium, 8537
Scaniver [Intervet] see 6731
Scarlet Berry see 3516
Scarlet Red, 8538
SCE-129 see 1994
SCE-963 see 1985
SCE-1365 see 1976
SCE-2174 see 1985
SCE-2787 see 1987
Scepter [Am. Cyanamid] see 4943
Sch-1000 see 5089
Sch-3132 see 6058
Sch-3940 see 7323
Sch-4831 see 1226
Sch-7056 see 128
Sch-10144 see 9656
Sch-10159 see 9788
Sch-10304 see 2453
Sch-10649 see 934
Sch-11460 see 1226
Sch-12041 see 4619
Sch-13475 see 8695
Sch-13521 see 4242
Sch-14714 see 4182
Sch-14947 see 8420
Sch-15719W see 5341
Sch-16134 see 8211
Sch-18020W see 1047
Sch-19927 see 3245
Sch-20569 see 6563
Sch-21420 see 5121
Sch-21480 see 9599
Sch-22219 see 221
Sch-25298 see 4145
Sch-28316Z see 4974
Sch-29851 see 5608
Sch-30500 see 5016
Sch-32088 see 6324
Sch-32481 see 6564
Sch-33844 see 8905
Sch-33861 see 8905
Sch-39300 see 4559
Sch-39720 see 1998
Schaeffer's α-Acid see 6477
Schaeffer's β-Acid see 6479
Schaeffer's Salt see 6479
Schardinger Dextrins see 2787
Scheele's Green see 2693
Scheelite see 1760
Schering 36268 see 2133
Schering 35833 see 7384
Scherisolon [Schering] see 7901
Scheroson see 2602
Scheroson F [Schering AG] see 4828
Schiff Bases, 8539
Schistomide [Specia] see 626
Schizophyllan see 8699
Schleimsäure (German) see 4349
Schlippe's Salt see 8840
Schmiedeberg's Digitalin see 3200
Schöllkopf's Acid see 6473

Schönbein-Pagenstecher's Paper see 4574
Schradan, 8540
Schultenite see 5413
Schwefelstickstoff see 9378
Schweine-Vasopressin see 5661
Schweizer's Reagent, 8541
Scifluorfen see 111
Scillabiose, 8542
Scillacrist [Draco] see 8060
Scillaren, 8543
Scillaren A see 8543
Scillaren B see 8543
Scillarenin, 8544
Scillarenin 3β-Rhamnoside see 8060
Scilliroside, 8545
Sclane see 9867
Sclerosol [Res. Med.] see 3308
Scobro [Ono] see 1624
Scobron [Mohan] see 1624
Scobutil [Rumanian] see 1624
Scoke see 7546
Scolaban [Cooper] see 1510
Scolecite see 1685
Scoline [Duncan Flockhart] see 9044
Scop [Ciba-Geigy] see 8550
Scoparin, 8546
Scoparius, 8547
Scoparone, 8548
Scoparoside see 8546
Scopine Tropate see 8550
Scopoderm TTS [Ciba-Geigy] see 8550
Scopola, 8549
Scopolamine, 8550
l-Scopolamine see 8550
Scopolamine Aminoxide see 8551
Scopolamine Bromobutylate see 1624
Scopolamine-N-butyl Bromide see 1624
Scopolamine Methobromide see 6084
Scopolamine Methyl Bromide see 6084
Scopolamine N-Oxide, 8551
Scopolammonium Bromide see 8550
Scopoletin, 8552
Scopolin, 8553
Scopoline, 8554
Scopos [Pharmuka] see 8550
Scorbacid see 867
Scorbu-C see 867
Score [Ciba] see 3182
Scorprin [Willows] see 9071
Scotch Broom see 8547
Scotine [Unimed] see 2619
Scotophobin, 8555
Scotopsins see 6988
Scourge [Roussel-UCLAF] see 7629
Scout [Roussel-Uclaf/AgrEvo] see 9700
SCP see 8740
S.C.T.Z see 2444
Scu-PA see 8089
Scuroforme see 1538
Scutellarein, 8556
Scutellaria, 8557
Scutellarin see 8556
Scutl see 7453
SD-104-19 see 8057
SD-270-31 see 7040
SD-271-12 see 2421
SD-709 see 3258
SD-1601 see 6345
SD-1750 see 3129
SD-3211 see 3590
SD-3562 see 3137
SD-4294 see 2667
SD-4402 see 5142
SD-7859 see 2137
SD-8447 see 9334
SD-9129 see 6334

SD-11831 see 6662
SD-14114 see 4004
SD-15418 see 2755
SD-41706 see 4033
SD-43775 see 4051
SD-95481 see 2354
S-Dimidine [Protea] see 9083
SDMH see 3297
SDS see 8782
Sdt-91 see 8967
SDZ-205-502 see 8226
SDZ-215-811 see 7267
SE-780 see 1066
SE-1520 see 4969
SE-2395 see 9318
Sea-Legs see 5817
Sea-oak see 4305
Sea Onion see 8925
Searlequin see 5063
Sea Salt see 8742
Sea-wrack see 4305
Sebacic Acid, 8558
Sebacil, 8559
Sebacil [Bayer] see 7523
Sebaclen [Carter-Wallace] see 1275
Sebacoin, 8560
Sebaklen [Fumouze] see 1275
Sebercim [ISF-Italseber] see 6793
Sebizon [Schering] see 9067
Secaclavine see 2083
Secalan [Zyma] see 2140
Secale Cornutum see 3702
Secalip [Fournier] see 4019
Secalonic Acid A see 8561
Secalonic Acid B see 8561
Secalonic Acid C see 8561
Secalonic Acid D see 8561
Secalonic Acid E see 8561
Secalonic Acid F see 8561
Secalonic Acid G see 8561
Secalonic Acids, 8561
Secalysat [Ysatfabrik] see 2618
Secbutobarbitone Sodium see 1530
Seccidin [Nippon Kayaku] see 7919
Secholex [Pharmacia] see 7716
Sechvitan see 8163
Seclar [Andromaco] see 1045
Secletan [Ipsen] see 2323
Seclin [Panthox & Burck] see 3211
Seclodin [Much] see 4925
Secnidazole, 8562
Secobarbital Sodium, 8563
(1α,3β,5Z,7E)-9,10-Secocholesta-5,7,-
    10(19)-triene-1,3-diol see 4864
(3β,5Z,7E)-9,10-Secocholesta-5,7,10-
    (19)-triene-3,25-diol see 1677
(1α,3β,5Z,7E)-9,10-Secocholesta-5,7,-
    10(19)-triene-1,3,25-triol see 1681
(1α,3β-5Z,7E,24R)-9,10-Secocholesta-
    5,7,10(19)-triene-1,3,24-triol see
    9197
(3β,5Z,7E)-9,10-Secocholesta-5,7,10-
    (19)-trien-3-ol see 10157
(1S,1'E,3R,5Z,7E,20R)-9,10-Seco-20-
    (3'-cyclopropyl-3'-hydroxyprop-1'-
    enyl)-1,3-dihydroxypregna-5,7,10-
    (19)-triene see 1679
(3β,5Z,7E,22E)-9,10-Secoergosta-5,7,-
    10(19),22-tetraen-3-ol see 10156
(3β,6E,22E)-9,10-Secoergosta-
    5(10),6,8,22-tetraen-3-ol see 9198
(3β,5E,7E,10α,22E)-9,10-Secoergosta-
    5,7,22-trien-3-ol see 3223
(3β,5Z,7E)-9,10-Secoergosta-5,7,10-
    (19)-trien-3-ol see 10158
Secoiridoid see 6967
Seconal Sodium [Lilly] see 8563
Secondary Ammonium Phosphate see
    576
Secondary Barium Phosphate see 1019

Secondary Calcium Phosphate see
    1739
Secondary Caprylic Alcohol see 6850
Secondary Magnesium Phosphate see
    5718
Secondary Propyl Alcohol see 5227
Secondary Sodium Phosphate see 8805
Secorvas [Squibb] see 4282
Secotil see 6987
Secrebil [Isnardi] see 7639
Secresolv [Refarmed] see 3684
Secretin, 8564
Secretine [Sinbio] see 8564
Secretin [Ox] see 8564
Secretin [Pig] see 8564
Secrosteron see 3268
Sectral [M & B] see 16
Securinan-11-one see 8565
Securinine, 8565
Securit [Pierrel] see 5609
Securon [Knoll] see 10083
Securopen [Bayer] see 947
Securpres [Poli] see 4974
Sedaform see 2180
Sedagul [Wild] see 5505
Sedalande [J & J] see 4150
Sedalin [Chassot] see 32
Sedalis see 9390
Sedamyl [Riker] see 18
Sedanoct [Woelm] see 9929
Sedantoinal [Sandoz] see 5898
Sedapain [Nichiiko] see 3673
Sedapam [Duncan, Flockhart] see
    3042
Sedaplus [Chephasaar] see 3497
Sedapran [Kowa] see 7895
Sedaquin [Therapharm] see 6028
Sedaraupin [Boehringer, Mann.] see
    8314
Sedatine see 757
Sedatival [Raffo] see 5609
Sedatromin [Takata] see 2365
Sedazin [Lagap] see 5609
Sedecamycin, 8566
Sedemesis see 2045
Sédeval see 989
Sednotic see 607
Sedobex see 1244
Sedolatan [Hoechst] see 7919
Sedoneural see 8737
Sedopretten [Schoening] see 3367
Sedor see 3097
Sedormid [Roche] see 795
Sedotussin [UCB] see 1840
Sedral [Banyu] see 1963
Sedrena [Daiichi] see 9823
Sedulon [Roche] see 7622
Sedural [Teva] see 7361
Sedutain [Key] see 8563
Seduxen [Gedeon Richter] see 3042
Sefacin [Schering] see 2025
Seffein see 1831
Seflenyl [Ciba-Geigy] see 7663
Sefona see 4872
Sefril [Squibb] see 2032
Séglor [Sanofi Winthrop] see 3217
Segontin [Hoechst] see 7919
Segosin see 6051
Seidlitz Mixture, 8567
Seignette Salt see 7840
Seki [Simes] see 2456
Sekisanine see 9250
Sekisanoline see 9250
Sekretolin [Hoechst] see 8564
Sekundal-D [Rorer] see 3367
Selacryn [SKB] see 9570
Selagine see 4791
Selagnine see 5653
Selapin [Teikoku/Nikken] see 10209
Selbex [Eisai] see 9296

Seldane [Marion Merrell Dow] see
    9307
Sel De Fordos et Gelis see 4537
Selecal [Toyama] see 9579
Selecor [Rorer] see 2007
Selectin [Bristol-Myers Squibb] see
    7894
Selectins, 8568
Selectol [Chemie Linz] see 2007
Selectomycin [Grünenthal] see 8904
Selegiline, 8569
Selemicina [Italchemi] see 4280
Selenic Acid, 8570
Selenic Acid Dilithium Salt see 5566
Seleninyl Bromide see 8577
Seleninyl Chloride see 8578
Seleninyl Fluoride see 8579
Selenious Acid, 8571
Selenious Acid Dilithium Salt see
    5567
Selenious Acid Disodium Salt see
    8822
Selenious Anhydride see 8576
Selenite see 1753
Selenium, 8572
Selenium Bromide, 8573
Selenium Chloride, 8574
Selenium Cysteine see 8584
Selenium Dioxide see 8576
Selenium Hexafluoride, 8575
Selenium Hydride see 4842
Selenium Monobromide see 8573
Selenium Monochloride see 8574
Selenium Nitride see 6703
Selenium Oxide, 8576
Selenium Oxybromide, 8577
Selenium Oxychloride, 8578
Selenium Oxyfluoride, 8579
Selenium Sulfides, 8580
Selenium Tetrabromide, 8581
Selenium Tetrachloride, 8582
Selenium Tetrafluoride, 8583
Selenkupfer see 2736
Selenocysteine, 8584
Selenocystine see 8584
Selenomethionine, 8585
Selenous Acid see 8571
3-Selenyl-L-alanine see 8584
Selenyl Bromide see 8577
Selenyl Fluoride see 8579
Selepam [Schering-Plough] see 8211
Seles Beta [Farmitalia] see 892
Selexidin [Leo Pharm.] see 408
Selexid (susp.) [Leo Pharm.] see 409
Selexid (tabl.) [Leo Pharm.] see 409
Selezen [Italfarmaco] see 4949
Seligocidin see 3634
Selinon see 3331
Selipran [Squibb] see 7894
Sellagen see 5727
Sellaite see 5701
Selobloc [Lagap] see 892
Seloken [Hässle] see 6235
Selopral [Hässle] see 6235
Selo-Zok [Hässle] see 6235
Selsun [Abbott] see 8580
Selvigon [ICN-Galenika] see 7609
Selvjgon [Rorer] see 7609
Semap [Janssen] see 7213
Sematilide, 8586
Sembrina see 6132
Semduramicin, 8587
Semen Amomi see 7586
Semen Cydonia see 8237
Semesan [Du Pont] see 4876
Semicarbazide Hydrochloride, 8588
Semicid [Am. Home] see 6772
Semikon [Beecham] see 6027
Semi-mustard Gas see 4680
Seminose see 5791

Semioxamazide, 8589
Semisulfur Mustard *see* 4680
Semotiadil, 8590
Semoxydrine [Beecham] *see* 6015
Sempervirene *see* 8591
Sempervirine, 8591
Semprex [Burroughs Wellcome] *see* 129
Semtex *see* 2805
Semtex A *see* 2805
Semtex H *see* 2805
Senarmontite *see* 752
Sencephalin [Takeda] *see* 2021
Sencor [Bayer] *see* 6239
Sencoral *see* 6239
Sendoxan [Pharmacia] *see* 2816
Seneca Snakeroot *see* 8597
Senecialdehyde, 8592
Senecic Acid, 8593
Senecio, 8594
Senecioaldehyde *see* 8592
Senecioic Acid 2-*sec*-Butyl-4,6-dinitrophenyl Ester *see* 1265
Senecionine, 8595
N^l-Senecioylsulfanilamide *see* 9072
Seneciphylline, 8596
Senega, 8597
Senega Snakeroot *see* 8597
Senegenin, 8598
Senlor [Strosohein] *see* 7206
Seniramin [Hokuriku] *see* 9193
Senna, 8599
Sennoside A *see* 8600
Sennoside B *see* 8600
Sennosides, 8600
Senociclin *see* 8411
Sensaval *see* 6812
Sensibion *see* 4669
Sensidyn [Leiras] *see* 2232
Sensit [Thiemann] *see* 4011
Sensival *see* 6812
Sentiloc [Benzon] *see* 1238
Seotalnatrium [Lilly] *see* 8563
Sepamit [Kanebo] *see* 6617
Sepan [Janssen] *see* 2365
Sepatren [Sumitomo] *see* 1989
Sepazon [Sankyo] *see* 2481
Sephadex 2-(Diethylamino)ethyl Ether *see* 7716
Sepia, 8601
Sepiolite *see* 5727
Sepiomelanin, 8602
Seponver [Ethnor] *see* 2473
Sepsinol *see* 4315
Septalan [Polfa] *see* 255
Septazine [Specia] *see* 1182
Septicid *see* 5683
Septicide [Bago] *see* 2374
Septiphene *see* 2469
Septipulmon [Pharmacia] *see* 9108
Septocillin [Leo Pharm] *see* 7230
Septopal [E. Merck] *see* 4398
Septoplix *see* 9094
Septosan [Mallard] *see* 9092
Septotan *see* 4805
Septra [Burroughs Wellcome] *see* 9086
Septrin [Burroughs Wellcome] *see* 9086
Septural [Grünenthal] *see* 7660
Sequel [Nihon] *see* 4038
Sequens *see* 2152
Sequestrene *see* 3557
Sequestrene NA 2 [Ciba-Geigy] *see* 3556
Sequestrene NA 3 *see* 3558
Ser *see* 8604
Seradix *see* 4995
Seragon [Ferring] *see* 2273
Seragonin [Ferring] *see* 2273
Seral *see* 5890

Seratrodast, 8603
Serax [Wyeth] *see* 7059
D-Ser(Bu^t)^{6}Azgly^{10}-gonadorelin *see* 4547
D-Ser(Bu^t)^{6}Azgly^{10}-luliberin *see* 4547
Serc [Unimed] *see* 1224
Serecor [Houdé] *see* 4851
Serefrex [Janssen] *see* 5307
Serenace [Searle] *see* 4629
Serenal [JWB] *see* 7059
Serenal [Sankyo] *see* 7061
Serenase [Lusofarmaco] *see* 4629
Serenesil [Abbott] *see* 3774
Serenid *see* 7059
Serenium [Squibb] *see* 3796
Serentil [Boehringer, Ing.] *see* 5970
Seren Vita [Synthelabo] *see* 2132
Serepax [Wyeth] *see* 7059
Serepress [Formenti] *see* 5307
Sereprile [Synthelabo] *see* 9561
Seresta *see* 7059
Serevent [Glaxo] *see* 8489
Serfin [Parke, Davis] *see* 8314
Serfolia [Hauck] *see* 8293
Sergetyl [Vaillant-Defresne] *see* 3938
Sericin *see* 4118
Sericosol N *see* 4494
Seriel [Biogalenique] *see* 9642
Serine, 8604
L-Serine *see* 8604
L-Serine Diazoacetate(ester) *see* 932
Serine Dihydrogen Phosphate (Ester) *see* 7520
Serine Phosphate *see* 7520
DL-Serine 2-[(2,3,4-Trihydroxyphenyl)methyl] Hydrazide *see* 1079
Sermaka [Lilly] *see* 4232
Sermion [Farmitalia] *see* 6580
Sermorelin, 8605
Sernevin [Toho Yakuhin] *see* 9163
Sernyl *see* 7364
Sernylan *see* 7364
Serocion *see* 7979
Seroden [Allen & Hanburys] *see* 9427
Seromycin [Lilly] *see* 2820
Serophene [Serono] *see* 2446
Serotinex *see* 2436
Serotone [Yoshitomi; Japan Tobacco] *see* 933
Serotonin, 8607
Serotonyl [ICT] *see* 4895
Serotropin [Teikoku Zoki] *see* 2273
Seroxat [Beecham] *see* 7175
Serozyme *see* 8074
Serpasil [Ciba] *see* 8314
Serpasol [Ciba] *see* 8314
Serpentaria, 8608
Serpentine *see* 863 and 5727
Serpentine (Alkaloid), 8609
Serpiloid [Riker] *see* 8314
Serpina [Himalaya] *see* 8293
Serpine [Pitman-Moore] *see* 8314
Serral *see* 3177
Sertaconazole, 8610
Sertan [Chinoin] *see* 7927
Sertindole, 8611
Sertraline, 8612
Serum Albumin, 8613
Serum Gonadotropin *see* 2273
Serum Lactic Dehydrogenase *see* 5347
Serum Prothrombin Conversion Accelerator *see* 3962
Serum Spreading Factor *see* 10167
Serum Transferrin *see* 9706
Serum Tryptase *see* 7678
Serutan [Menley & James] *see* 7674
Servanolol [Servipharm] *see* 8025
Servispor [Servipharm] *see* 2021
Servizepam [Servipharm] *see* 3042

L-Seryl-L-aspartyl-L-asparaginyl-L-asparaginyl-L-glutaminyl-L-glutaminylglycyl-L-lysyl-L-seryl-L-alanyl-L-glutaminyl-L-glutaminylglycyl-glycyl-L-tyrosinamide *see* 8555
N-(DL-Seryl)-N'-(2,3,4-trihydroxybenzyl) Hydrazine *see* 1079
Serzone [Bristol-Myers Squibb] *see* 6527
SES *see* 3430
Sesame Oil, 8614
Sesamex, 8615
Sesamin, 8616
Sesamodil *see* 8590
Sesamolin, 8617
Sesden [Tanabe] *see* 9583
Sesin, 8618
Sesone *see* 3430
Sesoxane [Shulton] *see* 8615
Sesastine, 8619
Setazine [Merck & Co.] *see* 1182
Sethotope [Squibb] *see* 8585
Sethoxydim, 8620
Sethyl *see* 4766
Setonil *see* 3042
Setran [Merck & Co.] *see* 5908
Setrol [Baxter] *see* 7107
Settima [IMA] *see* 7895
Seven Barks *see* 4804
Sevin [Union Carbide] *see* 1831
Sevinon *see* 9983
Sevoflurane, 8621
Sevofrane [Maruishi] *see* 8621
Sexadieno [AB Leo] *see* 2789
Sex-Ex *see* 4615
Sexocretin *see* 3177
Sexovid [Ferrosan] *see* 2789
SF-86-327 *see* 9299
SF-277 *see* 3003
SF-733 Antibiotic *see* 8373
SF-837 *see* 6271
SF-1293 *see* 7495
SFAE *see* 8872
SG-75 *see* 6608
Sgd-301-76 *see* 7071
Sgd-12878 *see* 6978
Sgd-24774 *see* 1046
Sgd-Scha-1059 *see* 1266
S-GI *see* 2952
SH-100 *see* 2806
SH-261 *see* 3816
SH-406 *see* 9308
SH-437 *see* 5082
SH-582 *see* 4422
SH-617L *see* 5087
SH-714 *see* 2844
SH-717 *see* 4517
SH-770 *see* 4188
SH-818 *see* 2430
SH-863 *see* 2430
SH-881 *see* 2844
SH-926 *see* 5031
SH-80881 *see* 2844
SHA-486300 *see* 4850
Shadocol *see* 5059
Shark Liver Oil, 8622
SHB-286 *see* 9165
SH B 331 *see* 4421
SHCH-58 *see* 2687
SH-E-222 *see* 5901
Sheep Berry *see* 10109
Shekanin *see* 9264
Shellac, 8623
Shellolic Acid, 8624
Sheridanite *see* 362
Sherpa [Rhône-Poulenc] *see* 2836
Shield [Dow] *see* 2462
Shigatox [Fawns & McAllan] *see* 9076
Shikalkin *see* 253
Shikimic Acid, 8625

Shikonin see 253
Shiomarin [Shionogi] see 6369
Shionone, 8626
Shiosol [Shionogi] see 4536
Shirlan [Zeneca] see 4153
Shisonin A see 2757
SH K 203 see 4187
SHL-451A see 4347
Shock-Ferol see 10156
Shogun [Ciba] see 7990
Showdomycin, 8627
Showersan see 5058
Showrone [Mitsui Toatsu] see 2158
Shoxin see 6780
SH U 454 see 4353
SH U 508A see 4353
Siagoside see 4376
Sialic Acids, 8628
Sibelium [Janssen] see 4179
Siberian Pine Needle Oil see 6904
Sibol see 3177
Sibutol see 1342
Sibutramine, 8629
Sicarol [Hoechst] see 8138
Siccanin, 8630
Sicoclor [Gramon] see 4007
Sicorten [Ciba-Geigy] see 4628
Sideramines see 2914
Siderin Yellow see 4062
Siderite see 5106
Sideromycins see 2914
Siderophilin see 9706
Siderophores see 6405
Sideros [Inverni] see 4083
Siderotil see 4105
Sideryl [Dandoy] see 875
Siduron, 8631
Sieromicin [Sierochimica] see 7606
Sigacalm [Siegfried] see 7059
Sigacyclat [Siegfried] see 3496
Sigadoxin [Siegfried] see 3496
Sigamopen [Siegfried] see 617
Sigapedil [Siegfried] see 3720
Sigaperidol [Siegfried] see 4629
Sigaprim [Siegfried] see 9086
Sigapurol [Siegfried] see 287
Sigmacef [Sigma Tau] see 7670
Sigmacort [Sigma] see 4828
Sigmadyn [Lipha] see 7206
Sigmafon [Fontana] see 5813
Sigmaform see 1333
Sigmart [Chugai] see 6608
Sikkimotoxin, 8632
Silain [Robins] see 3264
Silamox [Karlspharma] see 617
Silane, 8633
Silarine [Vir] see 8680
Silbephylline [Berk] see 3529
Silene see 1749
Silepar [Ibirn] see 8680
Silibinin see 8680
Silibrin [Silesia] see 2132
Silica see 8637
Silica Gel see 8634
Silicane see 8633
Siliceous Earth see 5003
Silicic Acid, 8634
Silicic Acid Tetraethyl Ester see 3895
Silicic Anhydride see 8637
Silicobromoform see 9746
Silicobutane see 9376
Silicochloroform see 9776
Silicoethane see 3419
Silicofluoric Acid see 4220
Silicon, 8635
Silicon Bromide see 8643
Silicon Carbide, 8636
Silicon Chloride see 8644
Silicon Dioxide, 8637
Silicon Disulfide, 8638

Silicone Glycol Copolymers see 8639
Silicone Glycols see 8639
Silicone Polyethers see 8639
Silicones, 8639
Silicon Fluoride see 8645
Silicon Monoxide, 8640
Silicon Nitride, 8641
Silicon Tetraacetate, 8642
Silicon Tetrabromide, 8643
Silicon Tetrachloride, 8644
Silicon Tetrafluoride, 8645
Silicon Tetrahydride see 8633
Silicopropane see 9884
Silicotungstic Acid, 8646
Silicristin see 8680
Silidianin see 8680
Silirex [Lampugnani] see 8680
Silkweed see 866
Sillimanite see 377
Silliver [Abbott] see 8680
Silmar [Von Boch] see 8680
Silomat [Thomae] see 2425
Silopentol [Ciba-Geigy] see 7064
Silosan [ICI] see 7652
Silubin see 1500
Silvacur [Bayer] see 9253
Silvadene [Marion] see 9071
Silver, 8647
Silver Acetate, 8648
Silver Benzoate see 1122
Silver Bromide, 8649
Silver Carbonate, 8650
Silver Chlorate, 8651
Silver Chloride, 8652
Silver Chromate(VI), 8653
Silver Citrate, 8654
Silver Cyanide, 8655
Silver Difluoride, 8656
Silver Dinaphthylmethane Disulfonate see 6030
Silver Fluoride, 8657
Silver Iodate, 8658
Silver Iodide, 8659
Silver Lactate, 8660
Silver Leaf see 8973
Silver Manganate(VII) see 8668
Silver Methylenebis(2-naphthyl-3-sulfonate) see 6030
Silver Monofluoride see 8657
Silver Nitrate, 8661
Silver Nitrate, Toughened, 8662
Silver Nitrite, 8663
Silver Nucleate see 8671
Silver Nucleinate see 8671
Silver Orthophosphate see 8669
Silver Oxalate, 8664
Silver Oxide, 8665
Silver(II) Oxide, 8666
Silver Perchlorate, 8667
Silver Permanganate, 8668
Silver Peroxide see 8666
Silver Phosphate, 8669
Silver Picrate, 8670
Silver Potassium Cyanide see 7839
Silver Protein, 8671
Silver Proteinate see 8671
Silver Salicylate see 8484
Silver Selenate, 8672
Silver Selenide, 8673
Silver Selenite, 8674
Silver Subfluoride, 8675
Silver Suboxide see 8666
Silver Sulfate, 8676
Silver Sulfide, 8677
Silver Tetraiodomercurate(II), 8678
Silver Trinitrophenolate see 8670
Silvex, 8679
Silvexamine see 8679
Silvisar see 8738
Silvol [Parke, Davis] see 8671

Silybin see 8680
Silybum Substance E$_6$ see 8680
Silymarin, 8680
Silymarin I see 8680
Simactil [Rorer] see 3692
Simanex [Makhteshim-Agan] see 8681
Simatin [Geistlich] see 3794
Simazine, 8681
Simeon [Toyo] see 8060
Simeskellina see 5321
Simethicone see 3264
Simetride, 8682
Simetryn, 8683
Simfibrate, 8684
Simovil [Riesel] see 8686
Simoxil [Herdel] see 617
Simpadren see 9189
Simpalon see 9189
Simpamina see 623
Simpamina-D [Recordati] see 2996
Simpatedrin see 623
Simpatoblock see 4724
Simplene [Chauvin] see 3656
Simplesse [Simplesse Co], 8685
Simplotan [Pfizer] see 9588
Simvastatin, 8686
SIN-10 see 6316
Sinalbin, 8687
Sinan see 5895
Sinanomycin see 6565
Sinapic Acid Choline Ester see 8688
Sinapine, 8688
Sinapine Glucosinalbate see 8687
Sinartrol [SPA] see 7661
Sinaxar [Armour Pharm.] see 9027
Sinbar [Du Pont] see 9298
Sincalide, 8689
Sincaline see 2259
Sincodix [Beta] see 1539
Sincomen [Schering AG] see 8917
Sindesvel see 6028
Sindiatil see 1500
Sinecod [Zyma] see 1539
Sinefungin, 8690
Sinemet [Merck & Co.] see 1843
Sineptina [Antibioticos] see 5480
Sinequan [Pfizer] see 3492
Sinerol see 7100
Sinesalin [Zeneca] see 1064
Sin-estrol see 3177
Sinetens [Erba] see 7897
Sinex [LaChartre] see 7100
Sinfibrate see 8684
Single-chain Pro-urokinase see 8089
Single-chain Urokinase-type Plasminogen Activator see 8089
Single Nickel Salts see 6600
Singoserp [Ciba] see 9193
Singulair [Merck & Co.] see 6340
Sinigrin, 8691
Sinigroside see 8691
Sinkron [Ripari-Gero] see 2380
Sinlestal [Daiichi Seiyaku] see 7935
Sinnamin [Siegfried] see 768
Sinoaculine see 8493
Sinoflurol [Kaken] see 9267
Sinogan-Debil see 6066
Sinomenine, 8692
Sinomin [Shionogi] see 9086
Sinorphan see 72
Sinorytmal [Giulini] see 9653
Sinos see 2808
Sinosid [SIF] see 7173
Sinovial [Irbi] see 7177
Sinox see 3331
Sinox W [FMC] see 3341
Sintespasmil see 1784
Sinthrome [Geigy] see 29
Sintisone [Farmitalia] see 7901
Sintoclar [Pulitzer] see 2380

Sintolexyn [ISF-Italseber] see 2021
Sintomicetina [Lepetit] see 2120
Sintomodulina [Italfarmaco] see 9544
Sintonal [Europharma] see 1472
Sintosulfa [Afi] see 9104
Sintotrat [Edmond] see 4828
Sintrom [Ciba-Geigy] see 29
Sintropium Bromide, 8693
Sinufed [Hauck] see 3645
Sinvacor [Merck & Co.] see 8686
Siogène [Geigy] see 2243
Siomine see 6043
Siosteran [Geigy] see 2243
Sipeimine see 4961
Sipiri Bark see 1043
Siplarol [Erba] see 4224
Siponic [Alcolac] see 7737
Siptazin [Isei] see 2365
Siqualine [Bristol-Myers Squibb] see 4226
S Iqualone [Bristol-Myers Squibb] see 4226
Siquil [Squibb] see 9814
SIR-8514 see 9809
Sirdalud [Sandoz] see 9624
Siringina [Toyo Jozo] see 9193
Sirolimus see 8288
Sirotol [Stada] see 4582
Siroxyl [Sopar] see 1850
Sirtal [Labaz] see 1826
Sisal, 8694
Siseptin [Schering] see 8695
Sisobiotic [Von Boch] see 8695
Sisolline [Schering-Plough] see 8695
Sisomicin, 8695
Sisomin [Max] see 8695
Sisosamine see 8695
Sistalgin [E. Merck] see 7887
Sistan [Du Pont] see 6016
Sisthane [Rohm & Haas] see 3998
Sisuril see 4830
Sito-Lande [Delalande] see 8697
β-Sitostanol see 8968
Sitosterin see 8697
α₁-Sitosterol, 8696
β-Sitosterol, 8697
γ-Sitosterol, 8698
Sivastin [Sigma-Tau] see 8686
Sivlor [Sidus] see 5616
Sizofiran, 8699
SK-7 see 7269
SK-331-A see 10194
SK-818 see 7979
Skatole, 8700
Skelaxin [Robins] see 5994
Skelid [Sanofi] see 9582
SK-Erythromycin [SK&F] see 3726
SKF-51 see 6854
SKF-trans-385 see 9708
SKF-478 see 3369
SKF-478-A see 3369
SKF-1700-A see 6830
SKF-2599 see 3491
SKF-2601-A see 2238
SKF-4657 see 7942
SKF-4965 see 1545
SKF-5137 see 2997
SKF-5883 see 9494
SKF-6574 see 7360
SKF-7988 see 10142
SKF-8318 see 10204
SKF-8542 see 9731
SKF-14336 see 2441
SKF-14336D see 2441
SKF-16214-A2 see 123
SKF-18667 see 7721
SKF-29044 see 7169
SKF-30310 see 7070
SKF-39162 see 911
SKF-40383 see 1882

SKF-41558 see 1967
SKF-60771 see 1965
SKF-62698 see 9570
SKF-62979 see 211
SKF-75073 see 1980
SKF-82526 see 4020
SKF-82526-J see 4020
SKF-83088 see 1977
SKF-88373 see 2000
SKF-92334 see 2337
SKF-96022 see 7146
SKF-96148 see 2871
SKF-101468 see 8416
SKF-101468A see 8416
SKF-102362 see 6637
SKF-102886 see 4626
SKF-104864 see 9687
SKF-104864A see 9687
SKF-105657 see 3668
SKF-108566 see 3669
SKF-108566J see 3669
Skimmetin see 9981
Skimmianine, 8701
Skimmin, 8702
Skimmiol see 9231
Skinoren [Schering AG] see 938
Skiodan [Winthrop] see 6051
Sklerolip [Mepha] see 2436
Skleromexe [L. Merckle] see 2436
Skleronorm [Grünenthal] see 3917
Sklero-Tablinen [Sanorania] see 2436
SK-Lygen [SKB] see 2132
Skopyl [Kabi] see 8550
SK-Penicillin VK [SK & F] see 7230
Skullcap see 8557
Skunk Oil, 8703
SL-73.033 see 763
SL-75.212 see 1229
SL-76.002 see 7955
SL-77.499 see 237
SL-77.499-10 see 237
SL-80.0342 see 318
SL-80.0750 see 10321
SL-80.0750-23N see 10321
Slaked Lime see 1716
S-laminin see 5364
SLD-212 see 1229
Sleepinal see 6028
Sleepwell see 6027
Slendid [Hercules], 8704
SLJ-312 see 4155
Slo-Bid [Rhone-Polenc Rorer] see 9421
Sloe-leaved Viburnum see 10109
Slonnon [Daiichi] see 816
Slo-Phyllin [Rhone-Poulenc Rorer] see 9421
Sloprolol [Generics] see 8025
Slow-Fluoride [Mission] see 8762
Slow-K [Ciba] see 7783
Slow-Pren [Norton] see 7086
Slow-reversing Endorphin see 3528
S.L.R. Factor see 8348
SM-108 see 6306
SM-224 see 854
SM-227 see 853
SM-804 see 857
SM-1213 see 506
SM-1652 see 1989
SM-3997 see 9219
SM-5688 see 3513
SM-7338 see 5960
SM-8144 see 8605
SM-A see 8862
Small Saxifrage see 7590
Smaltite see 2488
SM-C see 8862
SMDC see 6016
Smedolin [Yamanouchi] see 3928
Smilagenin, 8705

Smite [PPG] see 8724
Smithsonite see 10260
Smith's Sulfone see 9097
Smodin [Cheil Sugar] see 5504
SMS-201-995 see 6859
SMY-1500 see 3785
SN-38 see 5104
SN-46 see 2802
SN-105-843 see 6442
SN-186 see 123
SN-263 see 8718
SN-307 see 6979
SN-390 see 8225
SN-612 see 3796
SN-654 see 5891
SN-3115 see 7680
SN-3517 see 9066
SN-4395 see 8206
SN-4455 see 42
SN-5870 see 3416
SN-6771 see 1242
SN-7618 see 2215
SN-10751 see 609
SN-11841 see 635
SN-12837 see 2138
SN-13272 see 7925
SN-34615 see 7967
SN-49537 see 9447
SN-108266 see 2840
Snakeroot see 8608
Snake-weed see 3949
Snakeweed see 8608
Snapping Hazel see 4638
SND-919 see 7885
S.N.G. see 6704
SNGA see 4489
Snip [Geigy] see 3313
Sno Phenicol [Chauvin] see 2120
Sno Tears [Chauvin] see 7745
Snowball Bush see 10108
S.N.P. see 7167
Snup [Karlspharma] see 4025
Soamin [Burroughs Wellcome] see 8719
Soap, 8706
Soap Bark see 8222
Soaproot see 8511
Soapstone see 9207
Soapwort see 8511
Sobelin [Upjohn] see 2414
Sobrepin [Corvi] see 8707
Sobrerol, 8707
Sobril see 7059
Sobuzoxane, 8708
Socatil [Ciba-Geigy] see 9115
Soccerballene see 1482
Socian [Delagrange] see 508
Soclidan see 6577
SOD see 9177
Soda see 8739
Soda Alum see 378
Soda Ash see 8739
Soda Bordeaux see 1526
Soda Lime, 8709
Soda Lye see 8772
Sodamide see 8717
Soda Niter see 8792
Sodar see 6020
Sodascorbate see 8723
Sodelut G [Sodex] see 5838
α-Sodioacetic Acid Sodium Salt see 8712
Sodium, 8710
Sodium Acetate, 8711
Sodium Acetate Sodium Derivative see 8712
Sodiumacetic Acid Sodium Salt, 8712
Sodium Acetoarsinate see 850
Sodium 2-N-Acetylsulfamyl-4,4'-diaminodiphenylsulfone see 73

Sodium Tetrachloroaurate(III), 8837
Sodium Tetradecyl Sulfate, 8838
Sodium Tetrafluoroberyllate *see* 1221
Sodium Tetrafluoroborate *see* 8761
Sodium Tetrahydroborate *see* 8735
Sodium Tetraphenylborate, 8839
Sodium 2-Thenoate *see* 9491
Sodium Theobromine Acetate *see* 9419
Sodium Thioantimonate(V), 8840
Sodium Thiocarbonate, 8841
Sodium Thiocyanate *see* 9463
Sodium Thioglycolate, 8842
Sodium 2-Thiophenecarboxylate *see* 9491
Sodium Thiophosphate, 8843
Sodium Thiosulfate, 8844
Sodium Tin Oxide *see* 8826
Sodium *p*-Toluenesulfonchloramide *see* 2118
Sodium Trichloroacetate *see* 9756
Sodium L-Triiodothyronine *see* 5535
Sodium Trimetaphosphate, 8845
Sodium Triphenyl-*p*-rosanilinetrisulfate *see* 6107
Sodium Triphosphate *see* 8846
Sodium Tripolyphosphate, 8846
Sodium Tungstate(VI), 8847
Sodium Tungstophosphate *see* 8811
Sodium Tyropanoate *see* 9968
Sodium Uranate(VI), 8848
Sodium Valproate *see* 10049
Sodium Vanadate(V), 8849
Sodium Versenate [Riker] *see* 3556
Sodium Xylan Polysulfate *see* 7276
Sodiuretic *see* 1064
Sodothiol *see* 8844
Soduxin *see* 8828
Sofalcone, 8850
Sofenac [Rhône-Poulenc] *see* 5188
Sofmin [Dainippon] *see* 6066
Soframycin *see* 6542
Sofro [Alcon-Thilo] *see* 2658
Softenon *see* 9390
Softran [Stuart] *see* 1484
Soja Bean *see* 8878
Solacen [Wallace Labs.] *see* 9960
Solacthyl [Squibb] *see* 136
Solamin *see* 1103
Solan, 8851
Solancarpidine *see* 8857
Solandrine *see* 6799
(3β,12α)-Solanid-5-ene-3,12-diol *see* 8444
Solanid-5-ene-3β,18-diol *see* 5242
(3β)-Solanid-5-en-3-ol *see* 8852
Solanidine, 8852
Solanidine-S *see* 8857
Solanine, 8853
Solanine-S *see* 8858
Solanocapsin *see* 8854
Solanocapsine, 8854
Solanone, 8855
Solantal [Fujisawa] *see* 9564
Solantyl [Roussel-UCLAF] *see* 7475
Solanum, 8856
Solapsone *see* 8859
Solar NP [Swift] *see* 6772
Solarson *see* 2170
Solart [Bioindustria] *see* 27
Solaskil [Rhône-Poulenc Rorer] *see* 5486
Solasod-5-en-3β-ol *see* 8857
Solasodine, 8857
Solasonine, 8858
Solaspin [Indian Health] *see* 1684
Solasulfone, 8859
Solatene [Roche] *see* 1902
Solatubine *see* 8852
Solatunine *see* 8853
Solaxin *see* 2249

Solbrol A *see* 3883
Solbrol P [Bayer] *see* 8051
Sol Calomel *see* 1768
Soldactone [Searle] *see* 1795
Soldesam [Pharmacologico] *see* 2986
Solestril [Arzneimittelwerk Dresden] *see* 8060
Solevar *see* 6789
Solfa [Takeda] *see* 515
Solfocrisol *see* 4537
Solganal [Schering] *see* 917
Solganal B [Schering] *see* 917
Solgol [Squibb-Heyden] *see* 6431
Solian [Delagrange] *see* 508
Solicam [Sandoz] *see* 6794
Solid Crotonic Acid *see* 2664
Solid Green O *see* 5739
Solimidin [Selvi] *see* 10320
Soliphylline *see* 2266
Solis [Galen] *see* 3042
Soliwax [Concept Pharm.] *see* 3460
Solketal *see* 5232
Sol-Mycin [Roxane] *see* 3222
Solnet [Takeda] *see* 7921
Solnhofen Stone *see* 5515
Soln of Glyceryl Trinitrate *see* 8913
Soln Potassium Iodohydrargyrate *see* 7870
Solocalm [Microsules] *see* 7661
Solodelf [Cyanamid] *see* 9728
Solon [Taisho] *see* 8850
Solone [Fawns & McAllan] *see* 7901
Solosin [Cassella-Riedel] *see* 9421
Solozone *see* 8800
Sol Phenobarbital *see* 7386
Sol Phenobarbitone *see* 7386
Solprene [Phillips Petr.] *see* 8534
Solprin [Reckitt & Colman] *see* 886
Solprofen [Phoenix] *see* 4925
Solpyron [SKB] *see* 886
Solubacter [Pharmethic] *see* 9786
Solu-Biloptin [Schering AG] *see* 5087
Soluble Anhydrite *see* 1753
Soluble Aspirin *see* 1684
Soluble Barbital *see* 989
Soluble Cream of Tartar *see* 7778
Soluble Ferric Citrate *see* 547
Soluble Fluorescein *see* 4194
Soluble Glass *see* 8824
Soluble Gun Cotton *see* 8195
Soluble Indigo Blue *see* 4978
Soluble Iodophthalein *see* 5059
Soluble Pentobarbital *see* 7272
Soluble Potash Glass *see* 7838
Soluble Potash Water Glass *see* 7838
Soluble RNA *see* 8370
Soluble Saccharin *see* 8463
Soluble Sulfacetamide *see* 9067
Soluble Sulfadiazine *see* 9071
Soluble Sulfamerazine *see* 9081
Soluble Sulfapyridine *see* 9108
Soluble Sulfathiazole *see* 9115
Soluble Tartar *see* 7849
Solucin *see* 6777
Solucort [Chibret] *see* 7903
Solu-Cortef [Upjohn] *see* 4828
Soludagenan *see* 9108
Solu-Decadron [Merck & Co.] *see* 2986
Solu-Decortin-H [E. Merck] *see* 7901
Soludillar [Syntex] *see* 7162
Solufilin *see* 3529
Solufontamide *see* 9116
Solu-Forte-Cortin [E. Merck] *see* 2986
Solufyllin *see* 3529
Soluglaucit *see* 7164
Solu-Glyc [Erco] *see* 4828
Solumédine [Specia] *see* 9081
Solu-Medrol [Upjohn] *see* 6189
Solupemid [Recordati] *see* 7613

Solupred [Houdé] *see* 7901
Solu-Predalone [Forest] *see* 7903
Solupsan [UPSA] *see* 1684
Soluran *see* 2434
Solurol [Delalande] *see* 9185
Soluseptasine *see* 6777
Soluseptazine *see* 6777
Solusetazine *see* 6777
Solustibosan [Bayer] *see* 742
Solutedarol [Specia] *see* 9728
Solutrast [Bracco] *see* 5071
Solutrat [Labinca] *see* 10026
Solutricine [Bellon] *see* 9972
Solvan [Simes] *see* 3363
Solvay Soda *see* 8739
Solvazinc [Thames] *see* 10293
Solvent Blue 8 *see* 6137
Solvent Naphtha *see* 5513
Solvezink [Draco] *see* 10293
Solvirex [Sandoz] *see* 3429
Solvolan [Krka] *see* 401
Solvolip [Knoll] *see* 4036
Soma [Wallace Labs.] *see* 1889
Somacton [Ferring] *see* 8864
Somadril [Dumex] *see* 1889
Somagard [Roberts] *see* 2968
Somagerol [Brenner-Efeka] *see* 5609
Somagest [Riom] *see* 514
Somagrebove *see* 8864
Somalapor *see* 8864
Somalgen [Bago] *see* 9214
Somalgit *see* 1889
Soman, 8860
Somatocrinin *see* 8861
Somatofalk [Falk] *see* 8863
Somatoliberin, 8861
Somatomax PM *see* 4860
Somatomedin *see* 8862
Somatomedin A *see* 8862
Somatomedin C *see* 8862
Somatomedins, 8862
Somatonorm [Pharmacia] *see* 8864
Somatostatin, 8863
Somatotropic Hormone *see* 8864
Somatotropin, 8864
Somatotropin Release Inhibiting Factor *see* 8863
Somatrem *see* 8864
Somatropin *see* 8864
Somatyl [Rolland] *see* 1225
Somavubove *see* 8864
Somazina [Andromaco] *see* 2380
Somberol *see* 6028
Sombrevin [Gedeon Richter] *see* 7986
Sombucaps *see* 4742
Sombulex [Riker] *see* 4742
Somelin [Sankyo] *see* 4635
Somenopor *see* 8864
Sometribove *see* 8864
Sometripor *see* 8864
Somfasepor *see* 8864
Somidobove *see* 8864
Somilan [BDH] *see* 2111
Sominat [Merrell] *see* 3097
Somio *see* 2114
Somiproar [Mack, Illert.] *see* 3308
Somlan [Sintyal] *see* 4233
Somnafac [Cooper] *see* 6028
Somnal *see* 607
Somnalert [Warren-Teed] *see* 4742
Somnased [Duncan, Flockhart] *see* 6667
Somnatrol [Takeda] *see* 3744
Somnesin [Central Pharm.] *see* 5890
Somnibel [UCB] *see* 6667
Somnipron *see* 794
Somnite [Norgine] *see* 6667
Somnium *see* 6028
Somnomed *see* 6028

Name Index

Somnopentyl [Pitman-Moore] see 7272
Somnos [Merck & Co.] see 2113
Somnovit [Hosbon] see 5604
Somnurol [Synochem] see 1418
Somophyllin [Fisons] see 485
Somsanit [Köhler] see 4860
Sonaform see 2780
Sonal see 6028
Sonalan [Lilly] see 3763
Sonalen [Lilly] see 3763
Sonate see 8719
Sonazine [Cord] see 2238
Sonbutal see 1537
Sone [Fawns & McAllan] see 7904
Sonebon see 6667
Soneryl [M & B] see 1550
Songar [Valeas] see 9734
Songorine, 8865
Sonicated Human Serum Albumin see 8613
Sonifilan [Kaken] see 8699
Sonilyn [Mallinckrodt] see 9068
Sonin [Lipha] see 5604
Soni-Slo [Lipha] see 5245
Sonora see 5972
Sopental [Continental] see 7272
Sophocarpidine see 5799
Sophorabioside, 8866
Sophoretin see 8216
Sophoricoside, 8867
Sophorin see 8456
Sophorine see 2858
Sophorose, 8868
Sopivan [Formenti] see 10324
Soprol [Lederle] see 1336
Soprontin [Knoll] see 32
Sorbangil [Pharmacia] see 5245
Sorbic Acid, 8869
Sorbic Acid Potassium Salt see 7841
Sorbic Alcohol, 8870
Sorbichew [Stuart] see 5245
Sorbic Oil see 7166
Sorbidilat [Globopharm] see 5245
Sorbid SA [Stuart] see 5245
Sorbilande [Delalande] see 8873
Sorbilax [Upjohn] see 8873
Sorbin see 8874
Sorbinil, 8871
Sorbinose see 8874
Sorbistat [Ono] see 3640
Sorbit see 8873
Sorbitan Esters, 8872
Sorbitan Fatty Acid Esters see 8872
Sorbitan Laurate see 8872
Sorbitan Monolaurate see 8872
Sorbitan Monooleate see 8872
Sorbitan Monostearate see 8872
Sorbitan Oleate see 8872
Sorbitan Stearate see 8872
Sorbitol, 8873
D-Sorbitol see 8873
Sorbitrate [Stuart] see 5245
Sorbitur [Baxter] see 8873
Sorbo [ICI Americas] see 8873
Sorboquel [Schering] see 1744
Sorbosan see 6371
Sorbose, 8874
L-Sorbose see 8874
Sorbostyl [Synthelabo] see 8873
Sorbsan [Ross Fraser] see 241
Sordenac [Lundbeck] see 2455
Sordinol [Ayerst] see 2455
Sorensen's Phosphate see 8805
Sörensen's Potassium Phosphate see 7829
Sorensen's Sodium Phosphate see 8805
Soreton [Nippon Chemiphar] see 10244

Soriatane [Roche] see 114
Soricin [Merrell] see 8378
Soridermal [Rhône-Poulenc] see 6219
Soriflor [Basoderm] see 3186
Soripal [Specia] see 6219
Sorivudine, 8875
Sorlate [Abbott] see 7742
Sormetal see 3555
Sorot [Ravensberg] see 2959
Sorquad [Tutag] see 5245
Sorquetan [Basotherm] see 9588
"Sosan" see 5105
Sosegon [Winthrop] see 7261
Sosol [McKesson] see 9125
Sospitan [Kali-Chemie] see 8158
Sostenil [Pfizer] see 10120
Sostril [Cascan] see 8286
Sotacor [Bristol-Myers] see 8876
Sotal [Gramon] see 9601
Sotalex [Bristol] see 8876
Sotalol, 8876
Soterenol, 8877
Sotradecol [Elkins-Sinn] see 8838
Soudan Coffee see 5331
Souframine see 7404
Sour-spine see 988
Sovcaine see 3081
Soventol [Knoll] see 983
Soverin see 6028
Sowberry see 988
Sowell see 5908
Soxisol [Fort Dodge] see 9125
Soxomide [Upjohn] see 9125
Soxysympamine [Ferndale] see 6015
Soya Bean see 8878
Soybean, 8878
Soybean Agglutinin see 5453
Soybean Oil, 8879
Soy Sauce, 8880
Sozoiodole-mercury see 8881
Sozoiodole-sodium see 8881
Sozoiodole-zinc see 8881
Sozoiodolic Acid, 8881
SP see 8919 and 9032
SP-I see 7703
SP-54 [Bene-Chemie] see 7276
SP-281 see 10107
SP-732 see 7964
SP-1103 see 9362
SPA [Santen] see 5455
Spabucol [Lagap] see 9829
Spacolin [Roxane] see 386
Spaderizine [Kotobuki] see 2365
Spametrin-M [Sanzen-Yamanouchi] see 6147
Spamol see 487
Spamorin [Tanabe Seiyaku] see 4142
Span 20 [ICI] see 8872
Span 60 [ICI] see 8872
Span 80 [ICI] see 8872
Spandex, 8882
Spanidin [Nippon Kayaku] see 4610
Spanish Fly see 1796
"Spanish Fly" see 1797
Spanish Grass Wax see 3742
Spanish Saffron see 8466
Spanish Tea see 6888
Spanish White see 1326
Span-K [Fisons] see 7783
Spanon see 2133
Spanor [Biotherapie] see 3496
Spantin [Pharmacia Arz.] see 8092
Spantol [Nippon Chemiphar] see 7412
Spara [Dainippon] see 8884
Sparassol, 8883
Sparfloxacin, 8884
Sparicon [Yamanouchi] see 1624
Sparine [Wyeth] see 7966
Sparsamycin A see 9936
Sparsiflorine, 8885

Sparsomycin, 8886
(+)-Sparsomycin see 8886
Spartakon [Janssen] see 5486
Spartase [Wyeth] see 875
Sparteine, 8887
l-Sparteine see 8887
Spartocine [UCB] see 875
Spartopan [Pailluseau] see 8887
Spartrix [Janssen] see 1897
SPA-S-132 see 7181
SPA-S-160 see 5891
SPA-S-222 see 5891
SPA-S-510 see 7661
Spasfon-Lyoc [Lafon] see 7482
Spasmalex see 3211
Spasmamide [Schering AG] see 3995
Spasmaverine [Ethica] see 386
Spasmentral [Janssen] see 1104
Spasmex [Pfleger] see 9918
Spasmexan see 3986
Spasmium [Donau] see 1906
Spasmocan see 1784
Spasmocyclon [3M] see 2771
Spasmodex [Crinex] see 3211
Spasmo-Dolviran [Bayer] see 7363
Spasmol see 5758
Spasmolyn see 5895
Spasmolysin [Kade] see 8092
Spasmolytin see 160
Spasmolytol, 8888
Spasmonal [Arzneimittelwerk Dresden] see 3399
Spasmo-Nit [Stroschein] see 7151
Spasmophen see 7109
Spasmopriv [Pailluseau] see 4023
Spasmoril see 4142
Spasmostenyl see 5758
Spasmoxal(e) see 3354
Spasuret [Asche] see 4135
Spathulatine see 5246
Spatomac see 1055
Spatonin see 3165
SPC-97D see 942
SPCA see 3962
[SP-5-13-(E,E)]-Chloro[7-ethenyl-17-formyl-12-(1-hydroxy-5,9,13-trimethyl-4,8,12-tetradecatrienyl)-3,8,-13-trimethyl-21H,23H-porphine-2-18-dipropanoato(4—)-$N^{21},N^{22},N^{23},-N^{24}$]ferrate(2—) Dihydrogen see 2862
SPE see 9053
SPE-2792 see 706
Spearmint, 8889
Specifin [Bergamon] see 6446
Spectacillin [Sandoz] see 3651
Spectam [Biokema] see 8890
Spectamedryn [Allergan] see 5840
Spectamine [Medi-Physics] see 5066
Spectazole [Ortho] see 3550
Spectinomycin, 8890
Spectogard [Diamond] see 8890
Spectracide see 3043
Spectra-Sorb UV 9 [Am. Cyanamid] see 7088
Spectra-Sorb UV 24 [Am. Cyanamid] see 3357
Spectra-Sorb UV 284 [Am. Cyanamid] see 9157
Spectra-Sorb UV 531 [Am. Cyanamid] see 6838
Spectrazole [Pitman-Moore] see 2002
Spectrin, 8891
Spectrobid [Pfizer] see 961
Spectrum [Sigma-Tau] see 1995
Speda see 10131
"Speed" see 6015
Spergon see 2121
Spergualin see 4610
Spermaceti, 8892

ST-9067 see 945
STA-307 see 9596
Stabicilline [Vifor] see 7230
Stabilene [Auclair] see 3818
Stabillin V-K [Boots] see 7230
Stabinol [Horner] see 2239
Stabisol [Upjohn] see 1327
Stablon [Ardix] see 9560
Stachydrine, 8926
Stadacain [Stada] see 1566
Stadadorm see 607
Stadalax see 1282
Stadol [Bristol-Myers] see 1565
Stafac [SK & F] see 10142
Staff [Sigma-Tau] see 3930
Stafoxil [Brocades] see 4147
Stag Bush see 10109
Staggerweed see 5383
Stagid [Merck-Clévenot] see 6001
Stagural [Stada-Chemie] see 6792
Stakane [Dausse] see 763
Stalcin [Locatelli] see 1680
Stalleril [Pharmacal] see 9497
Stallimycin, 8927
Stam [Rohm & Haas] see 7987
Stamine [Tutag] see 8168
Stampede [Rohm & Haas] see 7987
Stampen [Beecham] see 3134
Stanazol see 8951
Stangen [Phys. Drug] see 8168
Stangyl (amp) [Rhone-Poulenc] see 9852
Stangyl (tabl) [Rhone-Poulenc] see 9852
Stanilo [Upjohn] see 8890
Stannic Anhydride see 8933
Stannic Bromide, 8928
Stannic Chloride, 8929
Stannic Chromate(VI), 8930
Stannic Fluoride, 8931
Stannic Iodide, 8932
Stannic Oxide, 8933
Stannic Selenide, 8934
Stannic Selenite, 8935
Stannic Sulfide, 8936
Stannochlor see 8939
Stannous Acetate, 8937
Stannous Bromide, 8938
Stannous Chloride, 8939
Stannous Fluoride, 8940
Stannous Fluozirconate(IV) see 8941
Stannous Hexafluorozirconate(IV), 8941
Stannous Iodide, 8942
Stannous Oxalate, 8943
Stannous Oxide, 8944
Stannous Pyrophosphate, 8945
Stannous Selenide, 8946
Stannous Sulfate, 8947
Stannous Sulfide, 8948
Stannous Tartrate, 8949
Stanolene [Standard Oil] see 7326
Stanolind [Standard Oil] see 7326
Stanolone, 8950
Stanozolol, 8951
Stanquinate see 5063
Stanyl [DSM] see 6833
Stapenor [Bayer] see 7036
Staphcil [Lederle] see 4147
Staphcillin [Bristol] see 6047
Staphisagria, 8952
Staphlipen [Lederle] see 4147
Staphobristol-250 [Bristol] see 2480
Staphybiotic [Delagrange] see 2480
Staphylex [Beecham] see 4147
Staphylomycin see 10142
Staphylomycine [SK & F] see 10142
Staphylomycin M₁ see 10142
Staphylomycin S see 10142
Staple [DuPont] see 8177

Staporos [Pharmuka] see 1680
Starane [Dow Elanco] see 4238
Star Anise, 8953
Starcef [Firma] see 1995
Starch, 8954
Starch Gum see 2995
Starch 2-Hydroxyethyl Ether see 4707
Starch Polymer with 2-Propenenitrile see 2654
Starch, Soluble, 8955
Star Grass see 229
Staril [Squibb] see 4282
Starwort see 229 and 4664
Starycide [Bayer] see 9809
Stas-Hustenlöser [Stada] see 401
Staticin [Westwood] see 3720
Statimo [Merck & Co.] see 1832
Statine, 8956
Statobex [Lemmon] see 7365
Statocin [Yoshitomi] see 1887
Statolon, 8957
Statran [Delagrange] see 3604
Statyl [Ayerst] see 6558
Stauffer 2061 see 7192
Stauroderm see 4233
Stavesacre see 8952
Staviban [Boots] see 4234
Staycept [Syntex] see 6772
Stazepine [Polfa] see 1826
St. Bartholomew's Tea see 5796
Steareth see 7737
Stearic Acid, 8959
Stearic Acid Aluminum Salt see 379
Stearic Acid Ammonium Salt see 588
Stearic Acid Calcium Salt see 1750
Stearic Acid Ester with Lactate of Lactic Acid Calcium Salt see 1751
Stearic Acid Lead Salt see 5442
Stearic Acid Potassium Salt see 7844
Stearic Acid Sodium Salt see 8827
Stearin see 9885
Stearodine see 1722
Stearyl Alcohol, 8960
Steatite see 9207
Steclin [Squibb-Heyden] see 9337
Stecsolin [Squibb] see 7111
Stédiril [Wyeth] see 6797
Steinbühl Yellow see 999
Stelabid [SK & F] see 5221
Steladone [Ciba-Geigy] see 2137
Stelazine [SKB] see 9811
Stellamicina see 3722
Stellar [Staley], 8961
Stellarid [SKB] see 8060
Stellerite see 1685
Stemetil [M & B] see 7942
Stemex [Syntex] see 7162
Stemphylone see 8276
Stenandiol see 677
Stenbolone, 8962
Stenediol [Organon] see 6017
Stenobolone (Rescinded USAN) see 8962
Stenol see 8960
Stenolon see 6018
Stenorol [Roussel] see 4627
Stenovasan see 485
Stephanine see 5990
Stepin [Basotherm] see 9600
Stepronin, 8963
Steranabol [Farmitalia] see 2475
Steranabol Long-acting see 7034
Steranabol Ritardo see 7034
Sterane [Pfizer] see 7901
Sterathal [Ward, Blenkinsop] see 7532
Sterax [Alcon] see 2973
Stercobilin see 10022
Stercofuge see 241
Stercorin see 2588

Sterculia see 5296
Sterecyt [Pharmaleo] see 7900
Stereocyt [Bellon] see 7900
Stereomycine [Touraude; Ayerst] see 5480
Stereon [Firestone] see 8534
Sterilon [Roter] see 2140
Sterinor [Heumann] see 9386
Sterisil [Warner-Lambert] see 4741
Steri/Sol [Warner-Lambert] see 4741
Sterisol (Domestic) [Warner-Lambert] see 4736
Sterling [Ciba] see 8136
Sterocort [Draco] see 7906
Steroderm [De Angeli] see 2973
Sterogyl [Roussel-UCLAF] see 10156
Sterolone [Rowell] see 7901
Sterosan [Geigy] see 2243
Sterox [Monsanto] see 6772
Steroxin [Geigy] see 2243
Sterretite see 371
Sterwin 904 see 231
Ster-Zac [Hough, Hoseason] see 9790
Stesolid [Dumex] see 3042
Stevacin [Intervet] see 7111
Steviol see 8964
Stevioside, 8964
Steviosin see 8964
STH see 8864
"Stibic" Anhydride see 739
2,2',2''-[Stibilidynetris(thio)]trisbutanedioic Acid Hexalithium Salt see 720
Stibine, 8965
Stibium see 733
Stibnite see 754
Stibocaptate, 8966
Stibophen, 8967
Stibosamine see 3896
Stibron [Iwaki] see 1624
Stibunal see 743
Stickstofflost see 5815
Stickstoffwasserstoffsäure (German) see 4815
Stiedex [Stiefel] see 2975
Stiemycin [Stiefel] see 3720
Stiglyn [Pitman-Moore] see 6553
(3β)-Stigmasta-5,7-dien-3-ol see 2927
(3β,5α,22E)-Stigmasta-7,22-dien-3-ol see 8901
(3β,5α,22E,24R)-Stigmasta-7,22-dien-3-ol see 2267
(3β,22E)-Stigmasta-5,22-dien-3-ol see 8969
(3β,24E)-Stigmasta-5,24(28)-dien-3-ol see 4303
Stigmastanol, 8968
(3β,5α)-Stigmastan-3-ol see 8968
(3β)-Stigmast-5-en-3-ol see 8697
(3β,24S)-Stigmast-5-en-3-ol see 8698
Δ⁵-Stigmasten-3β-ol see 8697
Stigmasterol, 8969
Stigmenene Bromide see 1153
Stigmonene Bromide [Warner-Chilcott] see 1153
St. Ignatius' Bean see 4938
Stilalgin see 5895
Stilamin [Serono] see 8863
Stilbamidine, 8970
Stilbazium Iodide, 8971
Stilbene, 8972
[Stilbene-(4,4')]bis[ω-phenylurea]-2,2'-disulfonic Acid Disodium Salt see 1349
Stilbene-α-carboxylic Acid see 7436
4,4'-Stilbenedicarboxamidine see 8970
E-3,5-Stilbenediol see 7604
3,4',5-Stilbenetriol see 8326
Stilbestrol see 3177
Stilbestrol Dimethyl Ether see 3257

Stilbestrol Diphosphate see 4279
Stilbestronate see 3177
Stilbetin [Squibb] see 3177
Stilboefral see 3177
Stilboestroform see 3177
Stilboestrol see 3177
Stilboestrol DP see 3177
Stilbofax see 3177
Stilbostatin [Taro] see 4279
Stilkap see 3177
Stillacor [Wolff] see 3210
Stilny [Will Pharma] see 6784
Stilpalmitate see 3177
Stilphostrol [Miles] see 4279
Stilronate see 3177
Stilwet L [Union Carbide] see 8639
Stimol [Biocodex] see 2392
Stimolag Fortis [Lagap] see 7638
Stimovul [Organon] see 3654
Stimsen [Lederle] see 9521
Stimul [Medinova] see 7206
Stimulexin [Robins] see 3488
Stimulin see 6635
Stimulina see 4479
Stimu-TSH [Roussel] see 9720
Stinerval [Wander] see 7366
Stinkweed see 8975
Stipolac [Burroughs Wellcome] see 5059
Stiptanon [Organon] see 3750
Stirofos see 9334
Stivane [Beaufour] see 8175
StL-1106 see 4228
Stock Guard [Cyanamid UK] see 4160
Stoddard Solvent see 6288
Stomamycin [Chassot] see 8904
Stomp [Am. Cyanamid] see 7211
Stone Oil see 4929
Stone-root see 2546
Stopetyl see 3428
Stopspot [ICI] see 7454
Storax, 8974
Storinal see 1352
Storksbill see 4412
Storocain [Eisai] see 7067
Stovaine see 656
Stovarsol [M & B; Specia] see 49
Stovarsolan see 49
Stoxil [SK & F] see 4934
Stozzon-Chlorophyll see 2207
STPP see 8846
Strabolene see 6452
Straderm [ITA] see 4186
Straminol [Cilag-Chemie] see 3466
Stramonium, 8975
Stratene [Innothera; Sigma-Tau] see 2062
Strazide [Luso] see 8985
Strengite see 4074
Strepogenin, 8976
Streptase [Hoechst] see 8981
Streptidine, 8977
Streptobiosamine, 8978
Streptobrettin [Norbrook] see 8983
Streptocid Album see 9094
Streptocide [Frosst] see 9094
Streptocide [Evans] see 9092
Streptococcal Deoxyribonuclease see 8979
Streptococcal Fibrinolysin see 8981
Streptodornase, 8979
Streptogramin, 8980
Streptogramin A see 10142
Streptogramin B see 6275
Streptohydrazid [Pfizer] see 8985
Streptokinase, 8981
Streptolin see 8988

Streptolydigin, 8982
Streptomagma [Wyeth] see 3222
Streptomycin, 8983
Streptomycin A see 8983
Streptomycin B, 8984
Streptomycin Hydrochloride see 8983
Streptomycin Hydrochloride-calcium Chloride Complex see 8983
Streptomycin Sulfate see 8983
Streptomyclidene Isonicotinyl Hydrazine Sulfate see 8985
Streptoniazide see 8985
Streptonicozid, 8985
Streptonigrin, 8986
Streptonivicin see 6818
L-Streptose, 8987
Streptothenat [Grünenthal] see 8983
Streptothricin VI see 8988
Streptothricin F see 8988
Streptothricins, 8988
Streptovaricin, 8989
Streptovaricin A see 8989
Streptovaricin B see 8989
Streptovaricin C see 8989
Streptovaricin D see 8989
Streptovaricin E see 8989
Streptovaricin F see 8989
Streptovaricin G see 8989
Streptovaricin J see 8989
Streptovarycin see 8989
Streptovirudin, 8990
Streptozocin, 8991
Streptozon see 9092
Streptozotocin see 8991
Stresam [Beaufour] see 3910
Strese & Hofmann's Hectorite see 4653
Stresnil [Janssen] see 931
Stresson [Boehringer, Ing.] see 1514
Streunex see 5526
Striadyne see 154
Striatran [Merck & Co.] see 3604
Strictylon see 6455
Strigol, 8992
Striped Alder see 4638
Strobane [Tenneco], 8993
Strobane-T [Tenneco] see 9693
Strodival [Herbert] see 7031
Stromba [Winthrop] see 8951
Strombaject [Winthrop] see 8951
Strongid [Pfizer] see 8139
Strong Silver Protein see 8671
Strontia see 9009
Strontianite see 8998
Strontium, 8994
Strontium Acetate, 8995
Strontium Bromate, 8996
Strontium Bromide, 8997
Strontium Carbonate, 8998
Strontium Chlorate, 8999
Strontium Chloride, 9000
Strontium Chromate(VI), 9001
Strontium Fluoride, 9002
Strontium Formate, 9003
Strontium Hydrate see 9004
Strontium Hydride see 8994
Strontium Hydroxide, 9004
Strontium Iodide, 9005
Strontium Lactate, 9006
Strontium Monoxide see 9009
Strontium Nitrate, 9007
Strontium Nitride see 8994
Strontium Oxalate, 9008
Strontium Oxide, 9009
Strontium Peroxide, 9010
Strontium Phosphate, Tribasic, 9011
Strontium Selenate, 9012
Strontium Sulfate, 9013
Strontium Sulfide, 9014
Strontolac [Wyeth] see 9006

Strophadogenin see 9015
Strophanthidin, 9015
Strophanthidin α-L-Rhamnoside see 2575
Strophanthin, 9016
Strophanthobiose, 9017
Strophanthus, 9018
Strophoperm [Sertürner] see 7031
Structum [SKB] see 2270
Strumacil see 6210
Strumazol see 6049
Struvite, 9019
Strychnidin-10-one see 9020
Strychnine, 9020
Strychnine Nitrate, 9021
Strychnine N⁶-Oxide, 9022
Strychnine Phosphate, 9023
Strychnine Sulfate, 9024
C-Strychnotoxine. see 1762
Stryphnonasal see 170
Stryphnone see 170
Stuart Factor see 3965
Stuart-Prower Factor see 3965
Stugeron [Janssen] see 2365
Stutgeron [Janssen] see 2365
Stutgin see 2365
STX see 8533
Stylomycin see 8130
Stylophorin see 2095
Stylopine, 9025
Styphnic Acid, 9026
Stypnone see 170
Styptic Collodion see 2547
Stypticin [Merck & Co.] see 2618
Styptol see 2618
Styquin see 1540
Styracin see 2364
Styramate, 9027
Styrax see 8974
Styrene, 9028
Styrene-butadiene Rubber see 8534
Styrene Glycol, 9029
Styrol see 9028
Styrolene see 9028
Styryl Carbinol see 2362
5-Styrylresorcinol see 7604
Su-88 see 8850
Su-3088 see 2151
Su-3118 see 9193
Su-4885 see 6246
Su-5864 see 4589
Su-6187 see 1554
Su-8341 see 2813
Su-9064 see 6237
Su-10568 see 2472
Su-18137 see 2843
Su-21524 see 7663
Suacron [Praemix] see 1822
Suanovil [Biokema] see 8904
Suavitil [Merck & Co.] see 1055
Subathizone, 9030
Sub-boric Acid see 3064
Subdue [Ciba-Geigy] see 5982
Suberic Acid, 9031
Suberone see 2791
Suberoxime see 2791
Subicard see 7249
Subitex [Hoechst] see 3341
Subitol see 4929
Sublimaze [McNeil] see 4043
Sublimed Sulfur see 9150
Subose [Squibb] see 4516
Substance P, 9032
Subtilin, 9033
Subtosan see 7879
Sucaryl [Abbott] see 8463
Sucaryl Calcium [Abbott] see 1705
Sucaryl Sodium see 2770
Succimal [Clin-Comar-Byla] see 3794
Succimer, 9034

Succinamide, 9035
Succinanil, 9036
Succinanilic Acid, 9037
Succinbromimide see 1462
Succinchlorimide see 2217
Succinic Acid, 9038
Succinic Acid Anhydride see 9039
Succinic Acid Bis[β-dimethylamino-ethyl] Ester Dimethochloride see 9044
Succinic Acid Cadmium Salt see 1664
Succinic Acid Calcium Salt see 1752
Succinic Acid Dibutyl Ester see 3087
Succinic Acid Di-tert-butyl Ester see 3088
Succinic Acid 2,2-Dimethylhydrazide see 2874
Succinic Acid Dinitrile see 9041
Succinic Acid Monoanilide see 9037
Succinic Acid Monoester with 4-Butyl-4-(hydroxymethyl)-1,2-di-phenyl-3,5-pyrazolidinedione see 9185
Succinic Acid Monoester with N-(2-Ethylhexyl)-3-hydroxybutyramide see 1562
Succinic Acid Monoester with 4-[2-(1-Hydroxyethyl)-3-methyl-5-benzo-furanyl]-2(5H)-furanone see 1070
Succinic Acid Sodium Salt see 8828
Succinic Anhydride, 9039
Succinimide, 9040
Succiniodimide see 5064
Succinite see 397
Succino-AICAR see 8469
Succinonitrile, 9041
2-[1-(Succinoyloxy)ethyl]-3-methyl-5-(2-oxo-2,5-dihydro-4-furyl)benzo[b]-furan see 1070
4-Succinylamido-4'-aminodiphenyl-sulfone see 9047
Succinyl Chloride, 9042
Succinylcholine Bromide, 9043
Succinylcholine Chloride, 9044
Succinylcholine Iodide, 9045
Succinyl Dichloride see 9042
O,O-Succinyldicholine Iodide see 9045
Succinyl Oxide see 9039
2-(N⁴-Succinylsulfanilamido)thiazole see 9046
Succinylsulfathiazole, 9046
Succisulfone, 9047
Succitimal [Katwijk] see 7414
Succosa [Astra Hässle] see 9049
Sucline [Soekami] see 8463
Suclofenide, 9048
Suconox-4 see 4859
Sucostrin [Squibb] see 9044
Sucralfate, 9049
Sucralfin [Inverni] see 9049
Sucralose, 9050
Sucrase see 5025
Sucrate [Lisapharma] see 9049
Sucrédulcor [Europ. Med.] see 8463
Sucrets [Merck & Co.] see 4750
Sucrida Berna see 1615
Sucrofer see 4073
Sucrol see 3517
Sucromat [Mayoli-Spindler] see 8463
Sucrosa see 2770
Sucrose, 9051
Sucrose Octaacetate, 9052
Sucrose Octakis(hydrogen Sulfate) Aluminum Complex see 9049
Sucrose Polyester, 9053
Sudac [Errekappa] see 9155
Sudafed [Burroughs Wellcome] see 3645
Sudan III, 9054
Sudan IV see 8538

Sudan Red BK see 9054
Sudermo see 5977
Sudil [Caber] see 9160
Sudine [Affiliated Labs.] see 9073
Suet, Prepared, 9055
Sufenta [Janssen] see 9056
Sufentanil, 9056
Sufentanyl see 9056
Sufortan [Asta] see 7214
Sufralem see 683
Sufrexal [Janssen] see 5307
Sugar see 9051
Sugar of Lead see 5411
Sugast [Alfa] see 9049
Suicalm [Janssen] see 931
Suiminth [Pfizer] see 6348
Suint, 9057
Sukor [Calif. Aromatics & Flavor] see 6540
Suladrin [Alcon] see 9125
Sulbactam, 9058
Sulbenicillin, 9059
Sulbenox, 9060
Sulbentine, 9061
Sulbutiamine see 9432
Sulcephalosporin see 1994
Sulconazole, 9062
Sulcosyn [Syntex] see 9062
Sulcotrione, 9063
Sulcrate [Nordic] see 9049
Suldixine [Bayer] see 9073
Suleo-M [International] see 5740
Sulestrex Piperazine [Abbott] see 3751
Sulf-10 [IOLAB] see 9067
Sulfabenz, 9064
Sulfabenzamide, 9065
Sulfabenzide see 9065
Sulfabenzpyrazine see 9109
Sulfabon [Vaillant] see 9073
Sulfabrom [Merck & Co.] see 9066
Sulfabromomethazine, 9066
Sulfabutin [Lentia] see 1529
Sulfacarbamide see 9101
Sulfacetamide, 9067
Sulfacetimide see 9067
Sulfachlorpyridazine, 9068
Sulfachrysoidine, 9069
Sulfactin see 3255
Sulfacytine, 9070
Sulfadiamine see 20
Sulfadiasulfone Sodium see 73
Sulfadiazine, 9071
Sulfadiazine Sodium see 9071
Sulfadicramide, 9072
Sulfadimerazine see 9083
Sulfadimethoxine, 9073
Sulfadimethyloxazole see 9093
Sulfadimethyloxazolylguanidine see 9077
Sulfadimethylpyrimidine see 9083
Sulfadimetine see 9124
Sulfadimidine see 9083
Sulfadine [Ayerst] see 9083
Sulfadoxine, 9074
Sulfaethidole, 9075
Sulfaethylthiadiazole see 9075
Sulfafurazole see 9125
Sulfaguanidine, 9076
Sulfaguanole, 9077
Sulfaisodimidine see 9124
Sulfalar [Parke, Davis] see 9125
Sulfalene, 9078
Sulfalepsin [Geistlich] see 9048
Sulfalex [DeAngeli] see 9087
Sulfallate, 9079
Sulfaloxic Acid, 9080
Sulfamerazine, 9081
Sulfameter, 9082
Sulfamethazine, 9083

Sulfamethizole, 9084
Sulfamethomidine, 9085
Sulfamethopyrazine see 9078
Sulfamethoxazole, 9086
Sulfamethoxizole see 9086
Sulfamethoxydiazine see 9082
Sulfamethoxypyrazine see 9078
Sulfamethoxypyridazine, 9087
Sulfa-5-methoxypyrimidine see 9082
Sulfamethyldiazine see 9081
Sulfamethylisoxazole see 9086
Sulfamethylphenazole see 9118
Sulfamethylthiadiazole see 9084
Sulfamethylthiazole, 9088
Sulfametin (Rescinded USAN) see 9082
Sulfametorine see 9082
Sulfametrole, 9089
Sulfamezathine see 9083
Sulfamic Acid, 9090
Sulfamic Acid Monoammonium Salt see 589
Sulfamide, 9091
Sulfamidine see 9083
p-Sulfamidoaniline see 9094
Sulfamidochrysoidine, 9092
Sulfamoxole, 9093
4-Sulfamoyl-o-arsanilic Acid see 9110
p-Sulfamoylbenzoic Acid see 1927
N-Sulfamoyl-3-[(2-guanidinothiazol-4-yl)methylthio]propionamide see 3972
3-(Sulfamoylmethyl)-1,2-benzisox-azole see 10323
D-threo-1-(p-Sulfamoylphenyl)-2-(2',2'-dichloroacetamido)-1,3-propanediol see 9387
4'-Sulfamoylsulfanilanilide see 9100
N-(5-Sulfamoyl-1,3,4-thiadiazol-2-yl)-butyramide see 1545
Sulfamul [Frosst] see 9115
p-Sulfamylacetanilide see 104
p-Sulfamylbenzoic Acid see 1927
N-(3-Sulfamyl-4-chlorobenzamido)-2-methylindoline see 4969
4'-Sulfamyl-2,4-diaminoazobenzene see 9092
Sulfamylon [Winthrop] see 5683
N-(4'-Sulfamylphenyl)-1,4-butane-sultam see 9167
N-p-Sulfamylphenyl-D-glucosylamine see 4472
Sulfanilamide, 9094
2-Sulfanilamide-5-bromo-4,6-dimethyl Pyrimidine see 9066
Sulfanilamide-d-glucoside see 4472
3-Sulfanilamido-6-chloropyridazine see 9068
2-Sulfanilamido-4,6-diethyl-1,3,5-triazine see 9114
4-Sulfanilamido-5,6-dimethoxypyrimi-dine see 9074
6-Sulfanilamido-2,4-dimethoxypyrimi-dine see 9073
6-(Sulfanilamido)-2,4-dimethyl-1,3-diazine see 9124
6-Sulfanilamido-2,4-dimethylpyrimi-dine see 9124
2-Sulfanilamido-3-methoxypyrazine see 9078
2-Sulfanilamido-5-methoxypyrimidine see 9082
5-Sulfanilamido-3-methylisothiazole see 9113
3-Sulfanilamido-5-methylisoxazole see 9086
4-Sulfanilamido-2-methyl-6-methoxy-pyrimidine see 9085
2-Sulfanilamido-4-methylpyrimidine see 9081

Sulfuric Acid Monoanhydride with Nitrous Acid *see* 6746
Sulfuric Acid Monododecyl Ester Sodium Salt *see* 8782
Sulfuric Acid Monoethyl Ester *see* 3897
Sulfuric Acid Monomethyl Ester *see* 6203
Sulfuric Anhydride *see* 9152
Sulfuric Chlorohydrin *see* 2218
Sulfuric Ether *see* 3852
Sulfur Iodide, 9148
Sulfur Monochloride *see* 9143
Sulfur Mustard *see* 6395
Sulfurous Acid, 9149
Sulfurous Acid 2-(*p-tert*-Butylphenoxy)cyclohexyl 2-Propynyl Ester *see* 7992
Sulfurous Acid 2-(*p-tert*-Butylphenoxy)-1-methylethyl 2-Chloroethyl Ester *see* 805
Sulfurous Acid 2-Chloroethyl 2-[4-(1,1-Dimethylethyl)phenoxy]-1-methylethyl Ester *see* 805
Sulfurous Acid 2-[4-(1,1-Dimethylethyl)phenoxy]cyclohexyl 2-Propynyl Ester *see* 7992
Sulfurous Anhydride *see* 9144
Sulfurous Oxide *see* 9144
Sulfurous Oxychloride *see* 9485
Sulfur, Pharmaceutical, 9150
Sulfur Subchloride *see* 9143
Sulfur Tetrafluoride, 9151
Sulfur Trioxide, 9152
Sulfur Yellow S *see* 6480
Sulfuryl Amide *see* 9091
Sulfuryl Chloride, 9153
Sulfuryl Fluoride, 9154
Sulindac, 9155
Sulinol [ICI] *see* 9155
Sulisatin, 9156
Sulisobenzone, 9157
Sulla [Robins] *see* 9082
Sulmarin, 9158
Sulmazole, 9159
Sulmet [Am. Cyanamid] *see* 9083
Sulmetozine (Rescinded INN) *see* 9889
Sulmycin [Essex] *see* 4398
Suloctidil, 9160
Sulocton [Cont. Pharma] *see* 9160
Sulodene [Alfa] *see* 9160
Sulpelin [Senju] *see* 9059
Sulperazone [Pfizer] *see* 9058
Sulphadione *see* 2885
Sulphafurazole *see* 9125
Sulphaloxic Acid *see* 9080
Sulphan Blue, 9161
Sulphaproxyline *see* 9106
Sulphasalazine *see* 9112
Sulphasolucin *see* 6777
Sulphasolutin *see* 6777
Sulphasomidine *see* 9124
Sulphasomizole *see* 9113
"Sulphation Factors" *see* 8862
Sulphenone [Stauffer], 9162
Sulphetrone [Burroughs Wellcome] *see* 8859
7-(α-Sulphophenylacetamido)-3-(4'-carbamoylpyridinium)methyl-3-cephem-4-carboxylic Acid *see* 1994
Sulphormethoxine *see* 9074
Sulphsymazine *see* 9114
Sulphur *see* 9142
Sulpiride, 9163
S(—)-Sulpiride *see* 9163
Sulpitil [Tillotts] *see* 9163
Sulprim [Polfa] *see* 9086
Sulprofos, 9164
Sulprostone, 9165

Sulprotin [Taiyo] *see* 9180
Sulpyrin *see* 3414
Sulquin [Salsbury] *see* 9109
Sulreuma [Von Boch] *see* 9155
Sulsoxin [Reid-Provident] *see* 9125
Sul-Spantab [SK & F] *see* 9075
Sultamicillin, 9166
Sultanol [Nippon Glaxo] *see* 217
Sultan Red 4B *see* 1132
Sulthiame, 9167
Sultirene [Specia] *see* 9087
Sultopride, 9168
Sultosilic Acid, 9169
Sultropan *see* 9170
Sultroponium, 9170
Sulxin [Chugai] *see* 9073
Sulzol [Yoshitomi] *see* 9115
SUM-3170 *see* 5617
Sumach, 9171
Sumach Wax *see* 5270
Sumamed [Pliva] *see* 946
Sumatra Camphor *see* 1366
Sumatriptan, 9172
Sumatrol, 9173
Sumbul, 9174
Sumetrolim [EGYT] *see* 9086
Sumi-8 [Sumitomo] *see* 3320
Sumial [IC Pharma] *see* 8025
Sumi-alpha [Sumitomo] *see* 4051
Sumicidin [Sumitomo] *see* 4051
Sumiferon [Sumitomo] *see* 5016
Sumilarv [Sumitomo] *see* 8174
Sumilex [Sumitomo] *see* 7946
Sumisclex [Sumitomo] *see* 7946
Sumithion *see* 4017
Sumithrin [Sumitomo] *see* 7405
Sumitomo S-4084 *see* 2321
Summetrin *see* 7148
Summit [Mobay] *see* 9724
Sumox [Reid-Rowell] *see* 617
Sumycin [Apothekon] *see* 9337
SUN-0588 *see* 8515
SUN-1165 *see* 7581
Sunbrella [Dorsey] *see* 443
Suncefal [Yamanouchi] *see* 1989
Suncholin [Mohan] *see* 2380
Suncide [Bayer] *see* 8022
SunDare [Texas Pharmacal] *see* 2370
Sundown *see* 3282
Sundralen [Delalande] *see* 9558
Sunette [Hoechst Celanese] *see* 35
Sunflower Seed Oil, 9175
Sunfural [Asahi Kasei] *see* 9267
Sungard [Miles] *see* 9157
Sunrabin [Asahi] *see* 3624
Sunrhythm [Suntory] *see* 7581
Sunril [Emko] *see* 7133
Sunset Yellow FCF, 9176
Supacal [Lederle] *see* 9718
Supanate [Nippon Shinyaku] *see* 4142
Supazlun [Kowa] *see* 4142
Superanabolon *see* 6452
Super-Caid [Lipha] *see* 1403
Super-Cel *see* 5003
Superinone [Winthrop] *see* 9964
Superlutin *see* 4878
Superoxide Dismutase, 9177
Superoxol [Merck & Co.] *see* 4841
Superpyrin *see* 316
Super-Rozol [Lipha] *see* 1403
Super-Slurper *see* 2654
Supirdyl *see* 7172
Suplatast Tosylate, 9178
Suplexedil [Hépatrol] *see* 4026
Supona [Shell] *see* 2137
Supotran [Winthrop] *see* 2155
Supprelin [Roberts] *see* 4760
Suprachol *see* 2922
Supracide [Ciba-Geigy] *see* 6048

Supracombin [Grünenthal] *see* 9086
Supracort [Samil] *see* 169
Supracyclin [Grünenthal] *see* 3496
Supraene [Robeco] *see* 8924
Supral [Basotherm] *see* 7193
Supralgin *see* 7348
Suprametil [Geistlich] *see* 6189
Supramid [AB Leo] *see* 9082
Supramycin [Grünenthal] *see* 9337
Suprane [Anaquest] *see* 2965
Supranol [Cilag] *see* 9180
Supra-Puren [Klinge Nattermann Puren] *see* 8917
Suprarenaline [Armour] *see* 3656
Suprarenin [Hoechst] *see* 3656
Suprasec [J & J] *see* 5601
Suprasterol II, 9179
Supratonin *see* 412
Suprax [Lederle] *see* 1975
Suprecur [Hoechst] *see* 1527
Suprefact [Behringwerke] *see* 1527
Supres [Pfizer] *see* 9828
Supressin *see* 3489
Suprifen *see* 4869
Suprilent [Galenica] *see* 5259
Suprim [Valeas] *see* 9086
Supristol [Gallier] *see* 9093
Suprocil [Cilag] *see* 9180
Suprofen, 9180
Suprol [Cilag] *see* 9180
Suprotan *see* 2155
Suractin [Lappe] *see* 628
Suracton [Rorer Kanebo] *see* 8917
Sural [Chinoin] *see* 3764
Suralgan [Poli] *see* 9562
Suramin Sodium, 9181
Surbronc [Boehringer, Ing.] *see* 401
Surcopur [Bayer] *see* 7987
Surecide [Sumitomo] *see* 2760
Surem [Ciba] *see* 1535
Surem [Galen] *see* 6667
Surestryl [Roussel] *see* 6372
Surexin [McNeil] *see* 8173
Surfacaine [Lilly] *see* 2804
Surfactal [De Angeli] *see* 401
Surfactant TA *see* 1189
Surfacten [Tokyo Tanabe] *see* 1189
Surfak [Hoechst] *see* 3459
Surfathesin *see* 2804
Surfen [Hoechst] *see* 496
Surfexo [Wellcome] *see* 3960
Surflan [Elanco] *see* 7015
Surfonic N [Jefferson] *see* 6772
Surgam [Hoechst] *see* 9562
Surgestone [Roussel] *see* 7969
Surgical Chlorinated Soda Soln *see* 8774
Surgi-Cen *see* 4716
Surheme [Aron] *see* 1535
Suriclone, 9182
Surika [Thiemann] *see* 4167
Suril [Rhône-Poulenc Rorer] *see* 9182
Surinam Balsam *see* 976
Surinamine, 9183
Surital [Parke, Davis] *see* 9437
Surlid [Roussel] *see* 8433
Surmax [Lilly] *see* 921
Surmontil (amp) [Rhone-Poulenc] *see* 9852
Surmontil (tabl) [Rhone-Poulenc] *see* 9852
Surofene *see* 4716
Surrectan *see* 176
Sursum *see* 5092
Sursumid [Labif] *see* 9163
Survanta [Ross] *see* 1189
Survector [Servier] *see* 429
Suscard [Pharmax] *see* 6704
Suscardia [Pharmax] *see* 5236
Suspen [Circle] *see* 7230

Tachigaren [Sankyo] see 4905
Tachionin [Sana] see 9754
Tachmalin see 194
Tachostyptan [Consolidated] see 9527
Tachyrol [Duphar] see 3223
Tachysterol, 9198
Tacitin [Ciba] see 1117
Tackle [Sankyo] see 8630
Tacosal [Helvepharm] see 7475
Tacrine, 9199
Tacrolimus, 9200
Tacryl®, 9201
Tadenan [Debat] see 8135
Tadeonal see 7732
Tafil [Upjohn] see 320
TAG-72 see 8530
Tagamet [SK & F] see 2337
Tagathen [Lederle] see 2220
D-Tagatose, 9202
L-Tagatose see 9202
Tag Fungicide [Chevron] see 7453
Tag HL-331 [Chevron] see 7453
Tagilite see 2716
TAI-284 see 2411
Taicelexin [Taiyo] see 2021
Taidecanone [Taiyo] see 9974
Taiguic Acid see 5376
Tailed Pepper see 2679
Takacillin [Torii] see 5460
Taka-Diastase, 9203
Takanarumin [Takata] see 287
Takaton see 3309
Takelan [Takeda] see 8566
Takepron [Takeda] see 5373
Takesulin [Takeda] see 1994
Taketiam [Takeda] see 1985
Taktic [Boots] see 510
Takus [Montedison] see 2048
Taladren [Malesci] see 3761
Talampicillin, 9204
Talastine, 9205
Talat [Polifarma] see 9204
Talatrol see 9902
Talbutal, 9206
Talc, 9207
Talcid [Bayer] see 5696
Talcum see 9207
Talecid [Schering] see 7532
Taleudron see 7533
Talidine see 7533
Talimol see 9390
Talin [Tate & Lyle] see 9408
Talinolol, 9208
Talipexole, 9209
Talis [Kali-Chemie] see 5980
Talisomycin see 9213
Talisomycin B see 9213
Talleol see 9210
Tall Oil, 9210
Tallol see 9210
Tallow, 9211
Tallow Alcohol, 9212
Tallow Shrub see 1040
Tallysomycin, 9213
Tallysomycin A see 9213
Tallysomycin B see 9213
Talniflumate, 9214
Talodex see 4044
Talofen [Pierrel] see 7966
Talon [ICI] see 1400
Talotren [Sandoz-Wander] see 9421
Taloxa [Schering-Plough] see 3988
Talpen [Beecham] see 9204
Talsis [Wyeth] see 1337
Talstar [FMC] see 1257
Talucard [Knoll] see 8060
Talusin [Knoll] see 8060
Talwin (inj.) [Sanofi-Winthrop] see 7261
Tamarind, 9215

Tamaron [Bayer] see 6014
Tamas see 1157
Tamate [Merrell] see 5908
Tambocor [3M] see 4136
TAME see 654
Tamed Iodine [West] see 5058
Tametin [ISF] see 2337
Tamex [CFPI] see 1567
Tamilan [Gador] see 7206
Tamofen [Rhone-Poulenc] see 9216
Tamoxasta [Asta] see 9216
Tamoxifen, 9216
Tamsulosin, 9217
Tamyl [Italchimici] see 1965
Tanacetin, 9218
Tanaclone [Amsa] see 3523
Tanadopa [Tanabe Seiyaku] see 3457
Tanafol see 2155
Tanakan [IPSEN] see 4432
Tanapril [Tanabe] see 4947
Tanbismuth see 1330
Tandacote [Geigy] see 7106
Tandearil [Geigy] see 7106
Tandem [Dow] see 9795
Tandix [Crinos] see 4969
Tandospirone, 9219
Tanganil [Specia] see 96
Tangantangan Oil see 1946
Tanghinigenin, 9220
Tanghinin see 9220
Tannex [Duncan, Flockhart] see 4998
Tannic Acid, 9221
Tannic Acid Bismuth Derivative see 1330
Tannigen see 107
Tannin see 9221
Tannin-formadehyde see 9222
Tannoform, 9222
Tannyl Acetate see 107
Tanorama see 3225
Tanphetamin see 623
Tanston [Parke, Davis] see 5842
Tantalic Acid Anhydride see 9226
Tantallite see 9223
Tantalum, 9223
Tantalum Pentachloride, 9224
Tantalum Pentafluoride, 9225
Tantalum Pentoxide, 9226
Tan Tone see 3225
Tantum [Angelini Francesco] see 1157
TAO [Roerig] see 9899
Taoryl [Geigy] see 1820
TAP-031 see 4109
TAP-144 see 5484
Tapar [Parke, Davis] see 45
Tapazole [Lilly] see 6049
Tapopote see 3644
Taprostene, 9227
Taquidil [Roemmers] see 9629
Taractan [Roche] see 2241
Tarasan [Roche] see 2241
Tarasyn [Syntex] see 5318
Taraxacum, 9228
Taraxast-20(30)-en-3β-ol see 9229
Taraxasterin see 9229
Taraxasterol, 9229
Taraxein, 9230
Taraxerol, 9231
Tar Camphor see 6457
Tardamide [Grünenthal] see 9093
Tardastrex [Gruntex] see 2930
Tarden [Veterinaria] see 2930
Tardigal [Beiersdorf] see 3206
Tardisal [Sigma-Tau] see 8485
Tardocillin see 7221
Taredan [FMC] see 1670
Targa [Nissan] see 8269
Targocid [Marion Merrell Dow] see 9269
Targosid [Lepetit] see 9269

Tarichatoxin see 9382
Tarivid [Hoechst] see 6865
Taroctyl see 2238
Tarodyl see 4511
Tarodyn see 4511
Tar Oil, 9232
Tar Oil, Rectified, 9233
Tarpan [Wander] see 2420
Tarquinor [Squibb] see 4637
Tarragon, 9234
Tarragon Oil see 3748
Tartar Emetic see 741
(−)-Tartaric Acid see 9235
(+)-Tartaric Acid see 9237
d-Tartaric Acid see 9237
D-Tartaric Acid, 9235
dl-Tartaric Acid see 9236
DL-Tartaric Acid, 9236
l-Tartaric Acid see 9235
L-Tartaric Acid, 9237
meso-Tartaric Acid, 9238
Tartaric Acid Bismuth Complex
  Potassium Salt see 1318
Tartaric Acid Bismuth Complex
  Sodium Salt see 1321
L-Tartaric Acid Monoammonium Salt
  see 529
Tartarized Antimony see 741
Tartarlithine see 5548
Tartarus Vitriolatus see 7845
Tartrated Antimony see 741
Tartrazine, 9239
Tartrol see 1321
Tartronic Acid, 9240
Tarugan [Ravensberg] see 6349
Taselin [Eisai] see 6848
Taseron [Janssen] see 5307
Tasis [Wyeth] see 1337
Task [Shell] see 3129
Tasmaderm [Roche] see 6367
Tasnon (elixir) [Tropon] see 7617
Tasnon (tabl.) [Tropon] see 7617
Taste-modifying Protein see 6293
TAT-3 [Takeda] see 7558
TATBA see 9730
TATD see 6857
Tathiclon [Ohta] see 4483
Tathion [Yamanouchi] see 4483
Tationil [Boehringer Biochemia] see 4483
Tauliz [Hoechst] see 7647
Taumidrine [Knoll] see 983
Taural [Roemmers] see 8286
Tauredon [Byk-Gulden] see 4536
Taurine, 9241
Tauriscite see 4105
Taurocholic Acid, 9242
Tauroflex [Geistlich] see 9243
Taurolidine, 9243
Taurolin [Geistlich; Thomae] see 9243
Tavegil [Sandoz] see 2405
Tavegyl [Sandoz] see 2405
Ta-Verm see 7617
Tavist [Sandoz] see 2405
Tavor [Wyeth] see 5609
Taxicatin, 9244
Taxicin-I see 9245
Taxicin-II see 9245
Taxicins, 9245
Taxilan see 7294
Taxine, 9246
Taxine-I see 9246
Taxine-II see 9246
Taxine A see 9246
Taxine B see 9246
Taxinine see 9245
Taxodione, 9247
Taxol [Bristol-Myers Squibb] see 7117
Taxol A see 7117

Taxotere [Rhône-Poulenc Sante] *see* 3458
Taxusin *see* 7117
Tazalest [Wakamoto] *see* 9248
Tazanol [Wakamoto] *see* 9248
Tazanolast, 9248
Tazarotene, 9249
Tazepam [Polfa] *see* 7059
Tazettine, 9250
Tazicef [SK&F] *see* 1995
Tazidime [Lilly] *see* 1995
Tazin [Grelan] *see* 1833
Tazobactam, 9251
Tazocilline [Lederle] *see* 9251
Tazocin [Lederle] *see* 9251
TB-1 *see* 9950
TB-ACA *see* 1799
TBI *see* 9729
Tb I-698 *see* 9427
TBP *see* 1343
TBS *see* 1621 and 9747
TBX *see* 7205
TBZ [Merck & Co.] *see* 9426
2,4,5-TC *see* 8679
TC-80 *see* 5090
TC-109 *see* 9292
Tc-924 (DPD) *see* 1546
^{99m}Tc Diisopropyl IDA *see* 3422
^{99m}Tc-DISIDA *see* 3422
^{99m}Tc-L,L-ECD *see* 9257
^{99m}Tc Gluceptate *see* 4463
^{99m}Tc Gluheptonate *see* 4463
^{99M}Tc Hexakis(2-methoxyisobutyliso-nitrile) *see* 9259
^{99m}Tc-HEXAMIBI *see* 9259
^{99m}Tc-HIDA *see* 5506
^{99m}Tc-MAG$_3$ *see* 9258
^{99m}Tc-MDP *see* 5837
^{99m}Tc Mercaptoacetyltriglycine *see* 9258
^{99m}Tc-MIBI *see* 9259
^{99m}Tc-PPi *see* 8945
^{99m}Tc-PYP *see* 8945
^{99m}Tc-RP-30A *see* 9259
^{99m}Tc Tetrofosmin *see* 9383
[^{99m}Tc(tetrofosmin)$_2$O$_2$]$^+$ *see* 9383
TCA *see* 1680 and 9756
TCC *see* 9786
TCDBD *see* 9252
TCDD, 9252
T-Cell Growth Factor *see* 5020
T-Cell Replacing Factor *see* 5019
TCGF *see* 5020
TCl *see* 2790
TCMn *see* 6307
TCNE *see* 9336
TCP *see* 9892
TCPP *see* 4342
TCTH *see* 2829
TCV-116 *see* 1788
cis-9,trans-11-TDDA *see* 7950
TDE *see* 3111 and 3926
T-DET-N [Thompson-Hayward] *see* 6772
TDHL *see* 9308
TDI *see* 9668
TDP *see* 9431
TDZ *see* 9447
TE-031 *see* 2400
TEAB *see* 9340
Teaberry Oil *see* 6200
T.E.A. Chloride *see* 9341
Teamsters' Tea *see* 3644
Tebamin *see* 7426
Tebanyl *see* 7426
Tebe [Simes] *see* 8072
Tebethion *see* 9427
Tebloc [Dukron] *see* 5601
Tebonin [Schwabe] *see* 4432
Tebraxin [Bracco] *see* 8448

Tebrazid [Searle] *see* 8140
Tebron *see* 4632
Tebuconazole, 9253
Tebufenozide, 9254
Tebuthiuron, 9255
Tecaf [Farmochim. Ital.] *see* 8411
Técarine *see* 9419
Teceos [Hoechst Roussel] *see* 1546
Tecesal *see* 1759
TechneScan Gluceptate [Mallin-ckrodt] *see* 4463
TechneScan HDP [Mallinckrodt] *see* 7074
TechneScan HIDA [Mallinckrodt] *see* 5506
TechneScan MAG3 [Mallinckrodt] *see* 9258
TechneScan MDP [Mallinckrodt] *see* 5837
TechneScan PYP [Mallinckrodt] *see* 8945
Technetium, 9256
Technetium ^{99m}Tc Bicisate, 9257
Technetium ^{99m}Tc Ethyl Cysteinate Dimer *see* 9257
Technetium ^{99m}Tc Mertiatide, 9258
Technetium ^{99m}Tc Oxidronate *see* 7074
Technetium ^{99m}Tc Sestamibi, 9259
Technetium ^{99m}Tc Teboroxime, 9260
Teclosan *see* 9262
Teclosine *see* 9262
Teclothiazide, 9261
Teclozan, 9262
Teclozine *see* 9262
Tecodin *see* 7093
Tecomanine, 9263
Tecomin *see* 5376
Tecomine *see* 9263
Tecquinol *see* 4853
Tecramine *see* 4167
Tecto [Merck & Co.] *see* 9426
Tectochrysin *see* 2316
Tectoridin *see* 9264
Tectorigenin, 9264
TECZA *see* 9796
Tedegliparin *see* 6434
Tedelparin *see* 2870
Tedion *see* 9339
Tedion-V18 [Solvay Duphar] *see* 9339
Tedolan [Leo Pharm] *see* 3920
TEDP *see* 9139
Teel Oil *see* 8614
Teepol *see* 4378
Tefamin *see* 485
Tefilin [Hermal] *see* 9337
Teflon [Du Pont] *see* 7743
Teflox [Abbott] *see* 9284
Teflurane, 9265
Tefluthrin, 9266
Tefsiel C [Towa Yakuhin] *see* 9267
Tegafur, 9267
Tega-Pyrone [Ortega] *see* 3414
Tegencia [Elmu] *see* 4041
Tegison [Roche] *see* 3935
Tegopen [Bristol] *see* 2480
Tegosept B *see* 1619
Tegosept M *see* 6182
Tegotin *see* 10164
Tegretal [Geigy] *see* 1826
Tegretol [Geigy] *see* 1826
TEI-5103 *see* 8428
Teichmann's Crystals *see* 4678
Teichoic Acids, 9268
Teichomycin A$_2$ *see* 9269
Teicoplanin, 9269
Teicoplanin A$_2$ *see* 9269
Teicoplanin A$_2$-1 *see* 9269
Teicoplanin A$_2$-2 *see* 9269
Teicoplanin A$_2$-3 *see* 9269
Teicoplanin A$_2$-4 *see* 9269

Teicoplanin A$_2$-5 *see* 9269
Tekodin *see* 7093
Tektamer 38 [Calgon] *see* 3069
Tektin A *see* 8891
TEL *see* 9343
Telar [Du Pont] *see* 2244
Teldane [Marion Merrell Dow] *see* 9307
Teldanex [Astra] *see* 9307
Teldrin [SK & F] *see* 2232
Telebar [Guerbet] *see* 1023
Télémid [Nativelle] *see* 9085
Telemin [Funai] *see* 1282
Telenzepine, 9270
Teleocidin A *see* 9271
Teleocidin B *see* 9271
Teleocidin B$_4$ *see* 9271
Teleocidins, 9271
Telepaque [Winthrop] *see* 5072
Telepathine *see* 4647
Teleprim [Nativelle] *see* 9086
Telesmin [Yoshitomi] *see* 1826
Telesol [Lasa] *see* 4895
Teletrast [Astra] *see* 5072
Telgin-G [Takeda] *see* 2405
Teli *see* 8529
Telluric(IV) Acid *see* 9281
Telluric(VI) Acid, 9272
Telluric Bromide *see* 9278
Telluric Chloride *see* 9279
Tellurite *see* 9276
Tellurium, 9273
Tellurium Dibromide, 9274
Tellurium Dichloride, 9275
Tellurium Dioxide, 9276
Tellurium Hexafluoride, 9277
Tellurium Tetrabromide, 9278
Tellurium Tetrachloride, 9279
Tellurium Tetraiodide, 9280
Tellurobismuthite *see* 1331
Tellurous Acid, 9281
Tellurous Bromide *see* 9274
Tellurous Chloride *see* 9275
Telmicid [Lilly] *see* 3434
Telmid [Lilly] *see* 3434
Telmin [Janssen] *see* 5807
Telodrin *see* 5142
Telomerase, 9282
Telomere Terminal Transferase *see* 9282
Telomycin, 9283
Telon [Sandoz] *see* 1152
Telone [Dow] *see* 3125
Telone II [Dow] *see* 3125
Telopar [Pfizer] *see* 7055
Telotrex [Mead Johnson] *see* 9337
Telvar [Du Pont] *see* 6342
TEM *see* 9803
Temac [Abbott] *see* 9284
Temafloxacin, 9284
Temaril [Allergan] *see* 9834
Temasept IV [Hexcel] *see* 9747
Temazepam, 9285
Tementil [Specia] *see* 7942
Temephos, 9286
Temesta [Wyeth] *see* 5609
Temetex [Sauter] *see* 3189
Temgesic [Reckitt & Colman] *see* 1522
Temik [Union Carbide] *see* 223
Temina *see* 10164
Temocapril, 9287
Temocaprilat *see* 9287
Temocillin, 9288
Temopen [SmithKline Beecham] *see* 9288
Temovate [Glaxo] *see* 2423
Temozolomide, 9289
Temposil [Lederle] *see* 1703
Tempra [Mead Johnson] *see* 45
Temserin [Merck & Co.] *see* 9585

Temur see 9368
Tenacid [Sigma-Tau] see 4954
Tenalet [Otsuka] see 1917
Tenalin [Otsuka] see 1917
Tenathan [Robins] see 1233
Tendor [Chinoin] see 2901
Tenebrimycin see 9628
Tenelid [Helsinn] see 4585
Tenelon see 5575
Tenemycin see 9628
Tenesdol [Hishiyama] see 3369
Tenex [Robins] see 4590
Teniathane [Pitman-Moore] see 3120
Teniatol [Pitman-Moore] see 3120
Tenicid see 5047
Tenidap, 9290
Teniposide, 9291
Tennecetin see 6513
Tenntuss see 4582
Teno-basan [Sagitta] see 892
Tenoblock [Leiras] see 892
Tenonitrozole, 9292
Tenopt [Sigma] see 9585
Tenorite see 2713
Tenormal [ICI] see 7207
Tenormin [ICI] see 892
Tenox BHA [Eastman-Kodak] see 1582
Tenox BHT [Eastman-Kodak] see 1583
Tenoxicam, 9293
Tenox PG [Eastman] see 8044
Tensatrin [Gedeon Richter] see 8086
Tensibar [Clin-Comar-Byla] see 1254
Tensicor see 3241
Tensilon see 3562
Tensinase D [Nippon Chemiphar] see 3909
Tensinol [Farmaroma] see 7207
Tensobon [Schwarz] see 1817
Tensofin [Bristol-Myers Squibb] see 4226
Tensoprel [Rubio] see 1817
Tensopril [Syncro] see 5540
Tensoral [Diana] see 7207
Tenso-Timelets see 2450
Tenstaten [Ipsen] see 2323
Tentone see 6080
Tenuate [Merrell] see 3175
Tenuate Dospan [Merrell] see 3175
Tenuazonic Acid, 9294
Tenyl [Nippon Kayaku] see 1224
Teofilcolina see 2266
Teokolin see 2266
Teonicon [Neopharmed] see 7584
Teoptic [Dispersa] see 1917
Teorema [Panchemie] see 4456
Teoremac [Farmades] see 4456
Teosona [Phoenix] see 9421
TEPA see 9804
Tepanil [Riker] see 3175
Tephroite see 5781
Tephrosin, 9295
Tephthol see 9306
Tepilta [Wyeth] see 7067
Tepirubicin see 7642
Tepogen see 2480
TEPP see 9345
Teprenone, 9296
Teprin [Endo] see 2422
Téralithe [Specia] see 5552
Teralutil see 4886
Teramine [Darby] see 2235
Terazol [Cilag] see 9303
Terazosin, 9297
Terbacil, 9298
Terbasmin [Farmitalia] see 9302
Terbenol [Bernabo] see 6815
Terbia see 9300
Terbinafine, 9299

Terbium, 9300
Terbuconazole see 9253
Terbufos, 9301
Terbul [Hexal] see 9302
Terbutaline, 9302
Terbutaline Bisdimethylcarbamate see 980
Terbutrazole see 9253
Tercian [Théraplix] see 2753
Terconazole, 9303
Tercospor [Cilag] see 9303
Terdin [Cheil Sugar] see 9307
Terebene, 9304
Terebic Acid, 9305
Terebinic Acid see 9305
Terenac [Sandoz] see 5801
Terenol [Hoechst] see 8322
Terephthalic Acid, 9306
Terfenadine, 9307
Terfenadine Carboxylate see 4113
Terfex [Leiras] see 9307
Terflurane see 9265
Terfluzine [Théraplix] see 9811
Tergal [Rhône-Poulenc] see 7730
Tergavon see 4378
Tergitol [Union Carbide] see 7737
Tergitol 4 see 8838
Tergitol NP [Union Carbide] see 6772
Tergitol TP-9 [Union Carbide] see 6772
Terguride, 9308
Teriam [Roussel] see 9731
Teridax [Schering] see 5075
Terion [Lusofarmaco] see 4258
Teriparatide Acetate, 9309
Terit [Hoechst] see 2325
Terital [Montecatini] see 7730
Terlenka [Kunststijde] see 7730
Terlipressin, 9310
Termil [Diamond] see 2219
Termitin see 10164
Ternadin [Cantabria] see 9307
Ternelin [Sandoz] see 9624
Terodiline, 9311
Terofenamate, 9312
Teronac [Wander] see 5801
Terondit [Hoechst] see 6981
Teropterin see 8119
Terpate [Geneva] see 7249
"Terpene" Hydrochloride see 1369
Terpene Polychlorinates see 8993
Terpenolic Acid see 9313
Terpenylic Acid, 9313
Terpin, 9314
Terpinene, 9315
α-Terpineol, 9316
Terpin Hydrate see 9314
Terpinol see 9314
Terpinyl Thiocyanoacetate see 5144
Terposen [Vir] see 8286
3,2':4',3''-Terpyridine see 6609
Terra Alba see 1753
Terrachlor see 8264
Terraclor [Uniroyal] see 8264
Terracur P [Bayer] see 4042
Terrafungine [Virbac] see 7111
Terra Japonica see 4371
Terraject [Pfizer] see 7111
Terralon-LA [Virbac] see 7111
Terramycin [Pfizer] see 7111
Terrasytam [Murphy] see 3250
Terratrex [Volpino] see 4584
Terreic Acid, 9317
Tersan [Du Pont] see 9510
Tersavid see 7668
Tersigat [Boehringer, Ing.] see 7077
Tertatolol, 9318
α-Terthienyl, 9319
2,2':5'2''-Terthiophene see 9319
Tertiary Calcium Phosphate see 1741

Tertiary Magnesium Phosphate see 5720
Tertiomycin A see 9320
Tertiomycin B see 9320
Tertiomycins, 9320
Tertran [Wyeth] see 5091
Tertroxin [Glaxo] see 5535
Terylene [ICI] see 7730
Teslac [Bristol-Myers Squibb] see 9321
Tesnol [Kowa] see 8025
Tesoprel [Organon] see 1466
Tespamin [Sumitomo] see 9805
Tessalon [Ciba-Geigy] see 1127
Testarpen [Polfa] see 7233
Testascorbic see 867
Testavol see 10150
Testex [Pasadena] see 9322
Testinon [Mochida] see 9322
Testoderm [Alza] see 9322
Testo-Enant [Geymonat] see 9322
Testolactone, 9321
Δ¹-Testolactone see 9321
Testolin [Pasadena] see 9322
Testoral see 4223
Testosterone, 9322
trans-Testosterone see 9322
Testosterone 17-Chloral Hemiacetal see 2482
Testosterone Cypionate see 9322
Testosterone 17β-Undecylate see 9322
Testoviron [Schering AG] see 9322
Testred [ICN] see 6206
Testriol see 2102
Testro AQ [Truxton] see 9322
TETA see 9796
Tetaine see 963
Tetiothalein Sodium see 5059
Tetmosol [Care] see 9123
2,3,4,6-Tetraacetyl-α-D-glucopyranos-yl Bromide see 57
Tetra-O-acetylpentaerythritol see 7248
(2,3,4,6-Tetra-O-acetyl-1-thio-β-D-glu-copyranosato-S)(triethylphosphine)-gold see 911
Tetraamminecopper Dihydroxide see 8541
Tetraamminecopper Sulfate, 9323
Tetraaquochromium Dichloride see 2301
1,3,5,7-Tetraazaadamantane see 6036
1,4,7,10-Tetraazadecane see 9796
1,3,5,8-Tetraazanaphthalene see 8116
1,3,5,7-Tetraazatricyclo[3.3.1.1³,⁷]-decane see 6036
Tetrabakat [Dorsch] see 9337
Tetrabarbital, 9324
Tetra-base see 6264
Tetrabenazine, 9325
Tetrabid [Organon] see 9337
Tetrabiguanide see 4584
Tetrablet [Makara] see 9337
Tetrabon [Pfizer] see 9337
Tetraborane(10), 9326
Tetraboron Decahydride see 9326
Tetrabrom see 9329
3,3',5,5'-Tetrabromo-(1,1'-biphenyl)-2,2'-diol Mono(dihydrogen Phos-phate) see 1440
3,3',5,5'-Tetrabromo-2,2'-biphenyldiol Mono(dihydrogen Phosphate) see 1440
4,4',6,6'-Tetrabromobiphenyl-2,2'-diol Mono(dihydrogen Phosphate) see 1440
3,4,5,6-Tetrabromo-o-cresol, 9327
3,3',5,5'-Tetrabromo-m-cresolsulfon-phthalein see 1407
2',4',5',7'-Tetrabromo-3',6'-dihydroxy-spiro[isobenzofuran-1(3H),9'-[9H]-

xanthen]-3-one Disodium Salt *see* 3639

1,1,2,2-Tetrabromoethane *see* 9328

sym-Tetrabromoethane, 9328

2',4',5',7'-Tetrabromofluorescein *see* 3639

Tetrabromofluorescein Sol *see* 3639

4,5,6,7-Tetrabromo-2-hydroxy-1,3,2-benzodioxabismole *see* 1246

3,3'-(4,5,6,7-Tetrabromo-3-oxo-1(3*H*)-isobenzofuranylidene)bis[6-hydroxybenzenesulfonic Acid] Disodium Salt *see* 9127

3',3'',5',5''-Tetrabromophenolphthalein, 9329

Tetrabromophenolphthalein Sodium *see* 9329

3,3',5,5'-Tetrabromophenolsulfonphthalein *see* 1468

Tetrabromopyrocatechol Bismuth Derivative *see* 1246

Tetracaine Hydrochloride, 9330

Tetracap *see* 9332

Tetracarbonyldi-μ-chlorodirhodium *see* 8354

Tetracarbonylhydridocobalt *see* 3135

Tetracarbonylhydrocobalt *see* 3135

Tetracemate Disodium *see* 3556

Tetracemate Tetrasodium *see* 3557

Tetracemin *see* 3557

Tetracene *see* 6456

Tetrachel [Rachelle] *see* 9337

Tetrachlorethylene *see* 9332

Tetrachlormethiazide *see* 9261

2,4,5,6-Tetrachloro-1,3-benzenedicarbonitrile *see* 2219

2,3,5,6-Tetrachloro-1,4-benzenedicarboxylic Acid Dimethyl Ester *see* 2896

2,3,5,6-Tetrachloro-*p*-benzoquinone *see* 2121

4,5,6,7-Tetrachloro-3,3-bis(4-hydroxyphenyl)-1(3*H*)-isobenzofuranone *see* 7398

2,3,5,6-Tetrachloro-2,5-cyclohexadiene-1,4-dione *see* 2121

2,3,6,7-Tetrachlorodibenzodioxin *see* 9252

2,3,7,8-Tetrachlorodibenzo[*b,e*][1,4]dioxin *see* 9252

2,3,7,8-Tetrachlorodibenzo-*p*-dioxin *see* 9252

Tetrachlorodiborane(4) *see* 3063

2,4,5,6-Tetrachloro-1,3-dicyanobenzene *see* 2219

4,5,6,7-Tetrachloro-2,3-dihydro-2-methyl-2-[2-(trimethylammonio)ethyl]-2*H*-isoindolium Dichloride *see* 2151

4,5,6,7-Tetrachloro-3',6'-dihydroxy-2',4',5',7'-tetraiodospiro[isobenzofuran-1(3*H*),9'-[9*H*]xanthen]-3-one Dipotassium Salt *see* 8421

4,5,6,7-Tetrachloro-2-(2-dimethylaminoethyl)isoindoline Dimethylchloride *see* 2151

4,5,6,7-Tetrachloro-2-(2-dimethylaminoethyl)-2-methylisoindolinium Chloride Methochloride *see* 2151

Tetrachlorodiphenylethane *see* 3111

2,4,5,4'-Tetrachlorodiphenyl Sulfone *see* 9339

Tetrachloroethane, 9331

1,1,2,2-Tetrachloroethane *see* 9331

sym-Tetrachloroethane *see* 9331

Tetrachloroethene *see* 9332

Tetrachloroethylene, 9332

N-(1,1,2,2-Tetrachloroethylmercapto)-4-cyclohexene-1,2-dicarboximide *see* 1814

N-(1,1,2,2-Tetrachloroethylsulfenyl)-*cis*-4-cyclohexene-1,2-dicarboximide *see* 1814

N-(1,1,2,2-Tetrachloroethylthio)-4-cyclohexene-1,2-dicarboximide *see* 1814

N-(1,1,2,2-Tetrachloroethylthio)-Δ⁴-tetrahydrophthalimide *see* 1814

2,4,7,9-Tetrachloro-3-hydroxy-8-methoxy-1,6-dimethyl-11*H*-dibenzo-[*b,e*][1,4]dioxepin-11-one *see* 3401

Tetrachloroisophthalonitrile *see* 2219

Tetrachloromethane *see* 1864

2,3,5,6-Tetrachloro-4-methoxyphenol *see* 3509

3,4,5,6-Tetrachlorophenolphthalein *see* 7398

m-Tetrachlorophthalodinitrile *see* 2219

α,α,α,6-Tetrachloro-2-picoline *see* 6666

Tetrachloroquinone *see* 2121

3,3',4',5-Tetrachlorosalicylanilide, 9333

2,3,5,6-Tetrachloroterephthalic Acid Dimethyl Ester *see* 2896

4,5,6,7-Tetrachloro-2',4',5',7'-tetraiodofluorescein Dipotassium Salt *see* 8421

Tetrachlorvinphos, 9334

Tetracompren [E. Merck] *see* 9337

Tetracosactide *see* 2617

Tetracosactrin *see* 2617

Tetracosamethylhendecasiloxane, 9335

Tetracosamethylundecasiloxane *see* 9335

Tetracosanoic Acid *see* 5511

1,1,2,2-Tetracyanoethene *see* 9336

Tetracyanoethylene, 9336

Tetracycline, 9337

Tetracyclinemethylene Lysine *see* 5658

(*E,Z*)-9,11-Tetradecadien-1-ol Acetate *see* 7950

(*Z,E*)-9,12-Tetradecadien-1-ol Acetate *see* 7950

cis-9,*trans*-11-Tetradecadienyl Acetate *see* 7950

Tetradecahydro-3,9-dihydroxy-4,11b-dimethyl-8,11a-methano-11a*H*-cyclohepta[*a*]naphthalene-4,9-dimethanol *see* 769

[1*S*-(1α,4aα,4bβ,7β,8aα,10aβ)]-Tetradecahydro-1,4a-dimethyl-7-(1-methylethyl)phenanthrene *see* 4120

[1*S*-(1α,4aα,4bβ,7*E*,8β,8aα,9α,-10aβ)]-Tetradecahydro-9-hydroxy-1,4a,8-trimethyl-7-[2-[2-(methylamino)ethoxy]-2-oxoethylidene]-1-phenanthrenecarboxylic Acid Methyl Ester *see* 3728

2,3,3a,3b,4,5,5a,6,10,10a,10b,11,12,-12a-Tetradecahydro-1,10a,12a-trimethyl-1*H*-cyclopenta[7,8]phenanthro[2,3-*c*]isoxazol-1-ol *see* 674

Tetradecamethylhexasiloxane, 9338

Tetradecanoic Acid *see* 6416

Tetradecanoic Acid 1-Methylethyl Ester *see* 5234

1-Tetradecanol *see* 6418

12-*O*-Tetradecanoylphorbol-13-acetate *see* 7487

[2*S*-[2α,3β(1*E*,3*E*,5*Z*,8*Z*)]]-3-(1,3,5,8-Tetradecatetraenyl)oxiranebutanoic Acid *see* 5482

[(4*R*,5*Z*)]-Tetradecen-4-olide *see* 5271

Tetradecin *see* 9337

Tetradecin Novum [Astra] *see* 9337

Tetradecyl Alcohol *see* 6418

7,8,13,13a-Tetradehydro-2,3:9,10-bismethylenedioxyberbinium *see* 2591

2,16,19-20-Tetradehydrocuran-17-oic Acid Methyl Ester *see* 199

(5'α,6'α)-7,7',8,8'-Tetradehydro-4,-5:4',5'-diepoxy-17,17'-dimethyl-[2,-2'-bimorphinan]-3,3',6,6'-tetrol *see* 8104

(3β)-1,2,6,7-Tetradehydro-14,17-dihydro-3-methoxy-16(15*H*)-oxaerythrinan-15-one *see* 3719

(3β,12β)-1,2,6,7-Tetradehydro-12,17-dihydro-3-methoxy-16(15*H*)-oxaerythrinan-15-one *see* 3718

7,8,13,13a-Tetradehydro-9,10-dimethoxy-2,3-(methylenedioxy)berbinium *see* 1192

(5α)-6,7,8,14-Tetradehydro-4,5-epoxy-3,6-dimethoxy-17-methylmorphinan *see* 9411

5,6,8,14-Tetradehydro-4-hydroxy-3,6-dimethoxy-17-methylmorphinan-7-one *see* 8493

7,8,13,13a-Tetradehydro-3-hydroxy-2,9,10-trimethoxyberbinium *see* 5274

7,8,13,13α-Tetradehydro-2-hydroxy-3,9,10-trimethoxyberbinium *see* 2557

(16*E*)-16,17,18,19-Tetradehydro-17-methoxycorynan-16-carboxylic Acid Methyl Ester *see* 2613

(*E*)-16,17,18,19-Tetradehydro-17-methoxy-17,18-secoyohimban-16-carboxylic Acid Methyl Ester *see* 2613

3,3',4,4'-Tetradehydro-1,1',2,2'-tetrahydro-1,1'-dimethoxy-ψ,ψ-carotene *see* 8361

3,3',4,4'-Tetradehydro-1,1',2,2'-tetrahydro-1,1'-dimethoxylycopene *see* 8361

7,8,13,13a-Tetradehydro-2,3,9,10-tetramethoxyberbinium *see* 7126

1',2',3',4'-Tetradehydro-6',7',10,11-tetramethoxyemetan *see* 3599

(3β,23β)-14,15,16,17-Tetradehydroveratraman-3,23-diol *see* 10085

Tetradifon, 9339

Tetradine *see* 3428

Tetradonium Bromide *see* 6419

Tetradox [Ranbaxy] *see* 3496

Tetraethylammonium Bromide, 9340

Tetraethylammonium Chloride, 9341

Tetraethylammonium Hydroxide, 9342

N,*N*,*N*',*N*'-Tetraethyl-1,2-benzenedicarboxamide *see* 9344

Tetraethylbis(phthalamide) *see* 9344

Tetraethylene Glycol Dimethyl Ether *see* 9348

Tetraethyleneimidopiperazine-*N*,*N*'-diphosphoric Acid *see* 3397

Tetraethyllead, 9343

O,*O*,*O*',*O*'-Tetraethyl *S*,*S*'-Methylenediphosphorodithioate *see* 3782

N,*N*,*N*',*N*'-Tetraethylphthalamide, 9344

Tetraethylplumbane *see* 9343

Tetraethyl Pyrophosphate, 9345

Tetraethylrhodamine *see* 8349

Tetraethyl Silicate *see* 3895

2,7,12,18-Tetraethyl-3,8,13,17-tetramethyl-21*H*,23*H*-porphine *see* 3915

Tetraethylthiodicarbonic Diamide *see* 9123

Tetraethylthioperoxydicarbonic Diamide *see* 3428

Tetraethylthiuram Disulfide *see* 3428

Name Index

Thallous Sulfate *see* 9405
Thallous Sulfide *see* 9406
THAM [Fisher Sci.] *see* 9902
Thanite [Hercules] *see* 5144
Thaumatin, 9408
R-THBP *see* 8515
Δ^1-THC *see* 9349
Δ^6-THC *see* 9349
Δ^8-THC *see* 9349
Δ^9-THC *see* 9349
THE *see* 9350
Theaflavine, 9409
Theal Ampules [Boehringer, Mann.] *see* 3529
Thean [Astra] *see* 8092
Thearubigins, 9410
Thebacon *see* 3215
Thebaine, 9411
Thebainone, 9412
Thebainone-A *see* 9412
Thecodine *see* 7093
Theelin [Parke, Davis] *see* 3751
Theelol [Parke, Davis] *see* 3750
Thefylan *see* 3529
Thein *see* 1674
Theiomycetin *see* 1200
Thekodin *see* 7093
Thelestrin [Carnrick] *see* 3751
Thelmesan *see* 3525
Thelykinin *see* 3751
Themalon [Burroughs Wellcome] *see* 9429
Themisone [Mallinckrodt] *see* 904
Thenaldine, 9413
Thenalidine *see* 9413
Thenardite *see* 8829
Thenclor *see* 2220
Thenfadil [Winthrop] *see* 9416
Thenitrazole *see* 9292
Thenium Closylate, 9414
2-Thenoic Acid *see* 9491
3-Thenoic Acid, 9415
Thenophenopiperidine *see* 9413
2-(α-Thenoylamino)-5-nitrothiazole *see* 9292
p-(2-Thenoyl)hydratropic Acid *see* 9180
2-(α-Thenoylthio)propionylglycine *see* 8963
Thenyldiamine, 9416
Thenylene [Abbott] *see* 6027
Thenylpyramine *see* 6027
Theobid [Whitby] *see* 9421
Theobroma Oil, 9417
Theobromine, 9418
1-Theobromineacetic Acid, 9419
Theocalcin [Knoll] *see* 9418
Theochron [Inwood] *see* 9421
Theocin [Winthrop] *see* 9421
Theoclear [Central Pharm.] *see* 9421
Theodrox *see* 485
Theo-Dur [Key] *see* 9421
Theofibrate, 9420
Theograd [Abbott] *see* 9421
Theolair [3M] *see* 9421
Theolamine [Merck & Co.] *see* 485
Theomin [IPG] *see* 485
Theon [Draco] *see* 8092
Theon [Klinge] *see* 9421
Theophyl [Knoll] *see* 9421
Theophyldine *see* 485
Theophyllamine *see* 485
Theophylline, 9421
7-Theophyllineacetic Acid *see* 22
7-Theophyllineacetic Acid Piperazine Salt *see* 23
Theophylline Aminoisobutanol *see* 404
Theophylline Cholinate *see* 2266
Theophylline Compd with 2-Amino-2-methyl-1-propanol *see* 404

Theophylline Compd with Ethylene-diamine *see* 485
Theophyllineethylamphetamine *see* 4014
Theophylline Ethylenediamine *see* 485
Theophyllineethyl Sulfate of Pyridox-ine *see* 8159
Theophylline Methesculetol *see* 6000
2-(7′-Theophyllinemethyl)-1,3-dioxo-lane *see* 3494
Theophylline Salt of Choline *see* 2266
1-(Theophyllin-7-yl)ethyl 2-(*p*-Chlorophenoxy)isobutyrate *see* 9420
Theostat [Laser] *see* 9421
Theovent [Schering] *see* 9421
Theoxylline *see* 2266
Thephorin [Sinclair] *see* 7380
Theprubicin [Meiji Seika] *see* 7642
Therabloat [Norden] *see* 7721
Theraderm [Bristol-Meyers] *see* 1149
Therafectin [Greenwich] *see* 506
Theralax [Beecham] *see* 1282
Theralene [Rhône-Poulenc] *see* 9834
Théraleptique *see* 3317
Theranabol *see* 7099
Therapas [Barnes-Hind] *see* 1148
Therapav [Berlex] *see* 7151
Therapin [Nattermann] *see* 10219
Theraptique *see* 3317
Therarubicin [Meiji Seika] *see* 7642
Theratuss [Squibb] *see* 7609
Theriodide-131 [Abbott] *see* 8778
Thermolysin, 9422
Thermonatrite *see* 8739
Thermorubin, 9423
Thermorubin A *see* 9423
Thermovyl [Rhovyl] *see* 7746
Thesit [Declimed] *see* 7717
Thesodate *see* 9418
Thevetigenin *see* 3205
Thevetin A, 9424
Thevetose, 9425
THF *see* 7418
THFA *see* 9353
THF Glycol *see* 9352
Thiaben *see* 9426
Thiabendazole, 9426
Thiabenzazonium Iodide *see* 9565
Thiabutazide *see* 1554
Thiacetarsamide *see* 831
Thiacetazone, 9427
Thiacetic Acid *see* 9454
Thiactin *see* 9502
Thiacyclopentane *see* 9357
1-Thia-3-cyclopentene 1,1-Dioxide *see* 9129
Thiadiazuron *see* 9447
3′-Thia-2′,3′-dideoxycytidine *see* 5365
Thialbarbital, 9428
Thialbarbitone *see* 9428
Thialbutone Sodium *see* 1553
Thialisobumalnatrium *see* 1553
Thiamazole *see* 6049
Thiambutene, 9429
Thiamcol [Morishita] *see* 9436
Thiamenidine *see* 9558
Thiamine, 9430
Thiamine Chloride Hydrochloride *see* 9430
Thiamine Dichloride *see* 9430
Thiamine Diphosphate, 9431
Thiamine Diphosphate Ester Chloride *see* 9431
Thiamine Disulfide, 9432
Thiamine 8-(Methyl 6-Acetyldihydro-thioctate) Disulfide *see* 6857
Thiamine Monochloride *see* 9430
Thiamine Monophosphate *see* 9431
Thiamine Monophosphate Chloride *see* 9431

Thiamine Monophosphate Disulfide *see* 9432
Thiamine Propyl Disulfide *see* 8068
Thiamine Pyrophosphate *see* 9431
Thiamine Tetrahydrofurfuryl Disul-fide *see* 4333
Thiamine Trihydrogen Pyrophosphate (Ester) *see* 9431
Thiamine Triphosphate, 9433
Thiamine Triphosphoric Acid Ester *see* 9433
Thiaminium Chloride *see* 9430
Thiaminogen [Ohta] *see* 7016
Thiamiprine, 9434
Thiamorpholine, 9435
Thiamphenicol, 9436
Thiamutilin *see* 9559
Thiamylal, 9437
Thianaphthene, 9438
Thianeuron *see* 51
Thiantan *see* 3157
Thiantoin(e) Sodium *see* 7476
1-Thia-4,7,10,13,16-pentaazacyclo-eicosane Cyclic Peptide Deriv *see* 1841
Thiapride *see* 9561
Thiaretic *see* 4822
Thiarubrin(e) A *see* 9439
Thiarubrin(e)B *see* 9439
Thiarubrines, 9439
Thiasine *see* 3705
Thiaton [Hokuriku Seiyaku] *see* 9602
Thiaver [Riker] *see* 3661
Thiaxanthene *see* 9506
Thiaxanthone *see* 9507
Thiazamide [May & Baker] *see* 9115
1,4-Thiazan *see* 9435
Thiazenone *see* 9440
Thiazesim, 9440
Thiazinamium Methylsulfate, 9441
2-Thiazolamine *see* 501
Thiazole, 9442
Thiazolesulfone *see* 9444
4-Thiazolidinecarboxylic Acid *see* 9586
4-Thiazolidone-2-caproic Acid *see* 6403
ε-[2-(4-Thiazolidone)]hexanoic Acid *see* 6403
2-Thiazoline-2-ammonium 4-*n*-Butyl-1,2-diphenyl-3,5-pyrazolidinedion-ate *see* 9443
Thiazolinobutazone, 9443
Thiazolsulfone, 9444
Thiazol Yellow G, 9445
2-[[[4-[6-(2-Thiazolylamino)sulfonyl]-phenyl]amino]carbonyl]benzoic Acid *see* 7533
2-(4-Thiazolyl)-1*H*-benzimidazole *see* 9426
2-(4-Thiazolyl)-5-benzimidazolecar-bamate Isopropyl Ester *see* 1774
[2-(4-Thiazolyl)-1*H*-benzimidazol-5-yl]carbamic Acid 1-Methylethyl Ester *see* 1774
N-(2-Thiazolyl)-1-phenol-4-sulfon-amide *see* 7403
4′-(2-Thiazolylsulfamoyl)succinanilic Acid *see* 9046
4′-(2-Thiazolylsulfamyl)phthalanilic Acid *see* 7533
p-2-Thiazolylsulfamylsuccinanilic Acid *see* 9046
*N*¹-2-Thiazolylsulfanilamide *see* 9115
Thiazosulfone *see* 9444
Thibenzazoline, 9446
Thibenzole [Merck & Co.] *see* 9426
Thibone *see* 9427
Thibutabarbital *see* 9457
Thidiazuron, 9447

Thidicur [A.C.F.] see 9084
Thienamycin, 9448
7-(2-Thienylacetamido)cephalosporanic Acid see 2028
N-[7-[(2-Thienyl)acetamido]ceph-3-em-3-ylmethyl]pyridinium-4-carboxylate see 2025
N-[7-(2'-Thienylacetamidoceph-3-ylmethyl)]pyridinium-2-carboxylate see 2025
1-[(7'-β-[2-(2-Thienyl)acetamido]-8'-oxo-1'-aza-5'-thiabicyclo[4.2.0]oct-2'-en-3'-yl)methyl]pyridinium-2'-carboxylate see 2025
5-(2-Thienyl)-2,2'-bithiophene see 9319
Thienylic Acid see 9570
(15-R,S)-16-(3-Thienyloxy)-ω-tetranor-PGF$_{2\alpha}$ see 9563
Thiethylperazine, 9449
Thilaven see 4929
Thimecil see 6210
Thimerfonate Sodium, 9450
Thimerosal, 9451
Thimet [Am. Cyanamid] see 7486
Thioacetaldehyde, 9452
Thioacetamide, 9453
Thioacetazone see 9427
Thioacetic Acid, 9454
Thioacetic Acid S-Ester with N-[(4-Amino-2-methyl-5-pyrimidinyl)methyl]-N-(4-hydroxy-2-mercapto-1-methyl-1-butenyl)formamide Acetate see 51
Thioallyl Ether see 305
Thioantimonic Acid Tris(2-amino-2-oxoethyl) Ester see 746
Thioarsenite see 831
Thiobarbital, 9455
Thiobenzoic Acid S-Ester with N-[(4-Amino-2-methyl-5-pyrimidinyl)methyl]-N-(4-hydroxy-2-mercapto-1-methyl-1-butenyl)formamide O-Phosphate see 1068
Thiobenzyl Alcohol, 9456
2,2'-Thiobis[acetic Acid] see 9467
1,1'-Thiobis[benzene] see 7467
1,1'-Thiobisbutane see 1626
1,1'-Thiobis[2-chloroethane] see 6395
2,2'-Thiobis[4-chlorophenol] see 4046
2,2'-Thiobis[4,6-dichlorophenol] see 1343
1,1'-Thiobisethane see 3898
Thiobismethane see 6204
1,1'-[Thiobis(methylene)]bisbenzene see 1183
N,N'-[Thiobis[(methylimino)carbonyloxy]]bisethanimidothioic Acid Dimethyl Ester see 9465
1,1'-Thiobis[2-methylpropane] see 5166
1,1'-Thiobispropane see 8053
3,3'-Thiobis[propanoic Acid] see 9468
3,3'-Thiobis-1-propene see 305
Thiobutabarbital, 9457
Thiobutarit see 1481
Thiobutyl Alcohol see 1611
Thiocarbamide see 9505
Thiocarbamisin see 9458
Thiocarbamizine, 9458
Thiocarbanil see 7450
Thiocarbanilide see 3393
Thiocarbarsone, 9459
Thiocarbazil see 9427
Thiocarbonic Acid O-Ethyl Ester, S-Ester with N-[(4-Amino-2-methyl-5-pyrimidinyl)methyl]-N-(4-hydroxy-2-mercapto-1-methyl-1-butenyl)formamide Ethyl Carbonate (Ester) see 2064
Thiocarlide see 9593

Thiochrysine see 4537
Thiocol [Roche] see 7798
Thiocolchicine, 9460
Thiocolchicoside see 9460
Thiocresol, 9461
Thioctacid [Asta] see 9462
Thioctan [Katwijk] see 9462
Thioctic Acid, 9462
6,8-Thioctic Acid see 9462
Thiocuran [Sagitta] see 9086
Thiocyanate Sodium, 9463
Thiocyanatoacetic Acid exo-1,7,7-Trimethylbicyclo[2.2.1]hept-2-yl Ester see 5144
Thiocyanic Acid, 9464
Thiocyanic Acid Ammonium Salt see 597
Thiocyanic Acid 2-(2-Butoxyethoxy)ethyl Ester see 5472
2-Thiocyanoethyl Dodecanoate see 5471
2-Thiocyanoethyl Laurate see 5471
Thiocyl see 9498
Thiocymetin [Winthrop] see 9436
Thiodan [Hoechst] see 3614
Thiodelone [Shionogi] see 5904
Thiodemeton see 3429
Thioderon [Shionogi] see 5904
3,3'-Thiodi-L-alanine see 5375
Thiodicarb, 9465
2,2'-Thiodiethanol, 9466
Thiodiethylene Glycol see 9466
Thiodiglycol see 9466
Thiodiglycolic Acid, 9467
Thiodihydracrylic Acid see 9468
2-Thio-3,5-dimethyltetrahydro-1,3,5-thiadiazine see 2892
Thiodiphenylamine see 7404
O,O'-(Thiodi-4,1-phenylene)bis(O,O'-dimethylphosphorothioate) see 9286
O,O'-(Thiodi-4,1-phenylene)phosphorothioic Acid O,O,O',O'-Tetramethyl Ester see 9286
Thiodiphosphoric Acid Tetraethyl Ester see 9139
3,3'-Thiodipropionic Acid, 9468
β,β-Thiodipropionic Acid see 9468
2-Thio-1,3-dithiolo[4,5-b]quinoxaline see 9495
Thiodril [Berk] see 1850
Thiodrol [Shionogi] see 3662
Thioethanolamine see 2848
Thioethyl Alcohol see 3771
Thioethyl Ether see 3898
Thiofeed see 9488
Thiofide see 3435
Thioformamide, 9469
Thiofuran see 9490
Thiofurfuran see 9490
Thiogenal see 6058
1-Thio-D-glucitol see 9501
(1-Thio-D-glucopyranosato-O2,S1)gold see 917
(1-Thio-β-D-glucopyranosato)(triethylphosphine)gold 2,3,4,6-Tetraacetate see 911
5-Thio-α-D-glucopyranose see 9470
1-Thio-β-D-glucopyranose 1-[N-(Sulfooxy)-3-butenimidate] Monopotassium Salt see 8691
(1-Thio-β-D-glucopyranose-2,3,4,6-tetraacetato-S)(triethylphosphine)gold see 911
5-Thio-D-glucose, 9470
Thioglycerin see 9471
Thioglycerol, 9471
Thioglycol see 5917
Thioglycolic Acid, 9472
Thioglycollic Acid see 9472

Thioglycollic-β-aminonaphthalide see 9480
Thioguanine, 9473
Thioguanosine, 9474
Thioimidodicarbonic Diamide see 3436
2-Thio-4-ketothiazolidine see 8351
Thiokol [Thiokol], 9475
Thiokol A see 9475
Thiokol FA see 9475
Thiola [Coop. Farm.] see 9597
Thiolacetic Acid see 9454
Thiolactic Acid, 9476
Thiole see 9490
2-Thiolhistidine, 9477
Thiolhistidine-betaine see 3705
Thiolin see 4929
2-Thiolpropionic Acid see 9476
Thiolutin, 9478
Thiomalic Acid, 9479
Thiomebumal Sodium see 9487
Thiomerin Sodium [Wyeth] see 5918
Thiomersalate see 9451
Thiomesterone see 9596
Thiomethyl Alcohol see 6023
Thiomorpholine see 9435
Thiomucase [Sanofi Winthrop] see 4794
Thionalide, 9480
Thionaphthene-2-carboxylic Acid, 9481
1-Thionaphthol see 6465
2-Thionaphthol see 6466
α-Thionaphthol see 6465
β-Thionaphthol see 6466
Thionazin, 9482
Thioneine see 3705
Thionembutal see 9487
Thionex [Makhteshim-Agan] see 3614
Thionicol [Mohan] see 9436
Thionine, 9483
Thiontan see 3157
Thionylan see 6027
Thionyl Bromide, 9484
Thionyl Chloride, 9485
Thionyl Difluoride see 9486
Thionyl Fluoride, 9486
2-Thio-4-oxo-6-propyl-1,3-pyrimidine see 8054
Thioparamizone [Smith & Nephew] see 9427
Thiopental Sodium, 9487
Thiopentone Sodium see 9487
Thiopeptin, 9488
Thioperazine see 9494
Thioperoxydicarbonic Acid Diethyl Ester see 3449
Thiophanate, 9489
Thiophanate-methyl see 9489
Thiophan Sulfone see 9128
Thiophene, 9490
7-(Thiophene-2-acetamido)cephalosporanic Acid see 2028
2-Thiophenecarboxylic Acid, 9491
3-Thiophenecarboxylic Acid see 9415
β-Thiophenic Acid see 9415
Thiophenicol [Clin Comar] see 9436
Thiophenol, 9492
Thiophenylpyridylamino-10-carboxylic Acid Piperidinoethoxyethyl Ester see 7609
Thiophos [Am. Cyanamid] see 7167
Thiophosphoric Acid 2-Isopropyl-4-methyl-6-pyrimidyl Diethyl Ester see 3043
Thiophosphoric Anhydride see 7511
Thiophosphoryl Chloride see 7513
Thioproline see 9586
Thiopropanal S-Oxide see 7985
Thiopropazate, 9493

Thio-2-propene-1-sulfinic Acid S-Allyl Ester see 258
Thioproperazine, 9494
Thioproperazine Mesylate see 9494
Thiopropionaldehyde S-Oxide see 7985
2-Thio-6-propyl-1,3-pyrimidin-4-one see 8054
Thiopyrophosphoric Acid Tetraethyl Ester see 9139
Thioquinox, 9495
Thioredoxin, 9496
Thioridazine, 9497
Thioridazine-2-sulfone see 9137
Thioridazine-2-sulfoxide see 5970
Thiorubber see 9475
Thiosalicylic Acid, 9498
Thiosarmine see 9111
Thioseconal see 9437
Thiosemicarbazide, 9499
Thioserine. see 2850
Thiosinamine, 9500
Thiosol [Coop. Farm.] see 9597
1-Thiosorbitol, 9501
Thiostrepton, 9502
Thiosulfil [Wyeth-Ayerst] see 9084
Thiotax see 5916
Thiotepa see 9805
Thiotepp see 9139
Thiotetrole see 9490
Thiothixene, 9503
2-Thiouracil, 9504
Thiourea, 9505
4-(Thiovanilloyl)morpholine see 10071
Thioxamyl see 7052
Thioxanthene, 9506
Thioxanthen-9-one see 9507
Thioxanthone, 9507
Thioxidrene [Bottu] see 2381
2-Thioxo-3,5-dibenzyltetrahydro-1,3,-5-thiadiazine see 9061
Thioxolone see 9600
2-Thioxo-4-thiazolidinone see 8351
4-[Thioxo(3,4,5-trimethoxyphenyl)-methyl]morpholine see 9889
Thiozin see 4929
THIP, 9508
Thiphen see 9509
Thiphenamil, 9509
Thiram, 9510
Thistle Saffron see 1919
Thiurad [Monsanto] see 9510
Thiuranide see 3428
Thiuretic [Parke, Davis] see 4822
Thixokon [Mallinckrodt] see 79
Thomas Balsam see 977
Thomas Flour see 9511
Thomas Phosphate, 9511
Thombolyse [Green Cross] see 8089
Thombran [Thomae] see 9712
Thonzide [Warner-Chilcott] see 9512
Thonzonium Bromide, 9512
Thonzylamine Hydrochloride, 9513
Thoragol see 1244
Thorazine [SKB] see 2238
Thoria see 9518
Thorium, 9514
Thorium Chloride, 9515
Thorium Dioxide see 9518
Thorium Iodide, 9516
Thorium Nitrate, 9517
Thorium Oxide, 9518
Thorium Platinocyanide see 9520
Thorium Sulfate, 9519
Thorium Tetrachloride see 9515
Thorium Tetracyanoplatinate(II), 9520
Thorium Tetraiodide see 9516
Thorium(4+) Tetrakis(cyano-C)-platinate(2—) (1:2) see 9520

Thorium X see 8277
Thorn Apple see 8975
Thoron see 8278
Thorotrast see 9518
Thoroughwort see 3948
Thoulet's Soln see 7870
Thousand-leaf see 108
Thoxan [Cutter] see 3933
Thoxidil see 5895
Thozalinone, 9521
THP see 7642 and 9355
THQ see 9385
Thr see 9522
THR-221 see 1979
Thr²-cyclosporine see 2821
Threonine, 9522
L-Threonine see 9522
N²-[1-(N²-L-Threonyl-L-lysyl)-L-prol-yl]-L-arginine see 9941
D-Threose, 9523
L-Threose, 9524
Thridace see 5357
Thrombasal see 7381
Thrombin, 9525
Thrombinar [Armour] see 9525
Thrombin-sensitive Protein see 9529
Thrombocid [Bene-Chemie] see 7276
Thrombocytin see 8607
Thrombocytopoiesis-stimulating Factor see 9528
Thrombogen see 8074
Thrombogen [J & J] see 9525
Thrombo-Hepin [Riker] see 4685
Thrombokinase see 9527
Thrombokinin see 9527
Thromboliquine see 4685
Thrombolysin [Merck & Co.] see 7678
Thrombomodulin, 9526
Thrombophob [Nordmark] see 4685
Thromboplastin, 9527
Thrombopoietin, 9528
Thrombospondin, 9529
Thrombostat [Parke, Davis] see 9525
Thrombotonin see 8607
Thromboxane A₂ see 9530
Thromboxane B₂ see 9530
Thromboxanes, 9530
THS-101 see 2986
THT see 9357
THTP see 9357
Thuja, 9531
3-Thujanone see 9533
Thujic Acid, 9532
Thujin see 8219
Thujone, 9533
α-Thujone see 9533
β-Thujone see 9533
Thujopsene, 9534
Thulia see 9535
Thulium, 9535
Thurfyl Nicotinate, 9536
Thybon [Hoechst] see 5535
Thycapsol see 6049
Thylakentrin see 4299
Thylate [Du Pont] see 9510
Thylin see 6619
Thylogen [Rorer] see 8168
Thyloquinone [Squibb] see 5874
Thyme, 9537
Thyme Camphor see 9540
Thymergix [Joulli] see 8194
Thymerin [Tokyo Tanabe] see 8285
Thymidine, 9538
Thymin (formerly) see 9545
Thymine, 9539
Thymine-2-desoxyriboside see 9538
Thyminose see 2955
Thymocyte Stimulating Factor see 5020
Thymol, 9540

m-Thymol see 9540
Thymolated Silver Sulfone see 3037
Thymol Blue, 9541
Thymolphthalein, 9542
Thymol Sulfone see 3037
Thymolsulfonephthalein see 9541
Thymomodulin, 9543
Thymopentin, 9544
Thymopoietin, 9545
Thymopoietin III see 8919
Thymopoietin Pentapeptide see 9544
Thymosin, 9546
Thymosin $α_1$ see 9546
Thymosin $β_4$ see 9546
Thymosin $β_8$ see 9546
Thymosin $β_9$ see 9546
Thymosin Fraction 5 see 9546
Thymostatin, 9547
Thymostimulin, 9548
o-Thymotic Acid, 9549
o-Thymotinic Acid see 9549
Thymoxamine see 6374
Thymus Nucleic Acid see 2954
Thymyl Acetate see 9540
Thymyl N-Isoamylcarbamate, 9550
2-(Thymyloxymethyl)glyoxalidine see 9965
2-[(Thymyloxy)methyl]-2-imidazoline see 9965
Thypinone [Abbott] see 9720
Thyradin see 9551
Thyrefact [Hoechst] see 9720
Thyreocordon [Byk-Gulden] see 9446
Thyreostat I see 6210
Thyreostat II [Herbrand] see 8054
Thyreotrophic Hormone see 9931
Thyroactive Protein see 9554
Thyroblock [Horner] see 7809
Thyrocalcitonin see 1680
Thyrocrine [Lemmon] see 9551
Thyroid, 9551
Thyroidin, 9552
Thyroid-stimulating Hormone see 9931
Thyrojod [Henning] see 7809
Thyropropic Acid, 9553
Thyroprotein, 9554
Thyrotropic Hormone see 9931
Thyrotropin see 9931
Thyrotropin-Releasing Factor see 9720
Thyrotropin Releasing Hormone see 9720
Thyroxevan see 5497
Thyroxine, 9555
L-Thyroxine Sodium Salt see 5497
Thytropar [Rhône-Poulenc Rorer] see 9931
TI-211-950 see 8905
Tiaden [Malesci] see 9556
Tiadenol, 9556
Tiagabine, 9557
Tialamide see 9564
Tiamenidine, 9558
Tiamulin, 9559
Tiamutin see 9559
Tiamutin [Leo] see 9559
Tianeptine, 9560
Tiapridal [Delagrange] see 9561
Tiapride, 9561
Tiapridex [Schuerholz] see 9561
Tiaprofenic Acid, 9562
Tiaprost, 9563
Tiaprost Trometamol see 9563
Tiaramide, 9564
Tiase [Mediolanum] see 8963
Tiaterol [Midy] see 9556
Tiazesim see 9440
Tiazolidin [UCM] see 9586
Tibatin [Dak] see 2478

Tiberal [Roche] see 7000
Tibexin [Christiaens] see 7918
Tibezonium Iodide, 9565
Tibicorten [Stiefel] see 9729
Tibinide [Ferrosan] see 5203
Tibione [Schenley] see 9427
Tibirox [Roche] see 9386
Tibolone, 9566
Tibricol [Pfizer] see 6617
Tibutol [Bracco] see 3764
Ticar [Goldshield] see 9568
Ticarbodine, 9567
Ticarcillin, 9568
Ticarda [Hoechst] see 6804
Ticarpen [SKB] see 9568
Ticillin [Beecham] see 9568
Ticinil Calcico [De Angeli] see 7431
Ticlid [Sanofi Winthrop] see 9569
Ticlobran [Siegfried] see 2436
Ticlodix [Euro-Labor] see 9569
Ticlodone [Berenguer] see 9569
Ticlopidine, 9569
Ticlosin [Maggioni-Winthrop] see 9569
Ticrynafen, 9570
Tidemol [Roche] see 1500
Tiemannite see 8572
Tiemonium Iodide, 9571
Tiempe [DDSA] see 9840
Tienam [Merck & Co.] see 4954
Tienilic Acid see 9570
Ticnor [Farmaka] see 2477
Tiesene [Isagro] see 10300
Tifatol [Ciba-Geigy] see 2833
Tifomycine [Roussel-UCLAF] see 2120
Tifosyl [Astra] see 9805
Tigan [SKB] see 9839
Tigason [Roche] see 3935
Tigemen [Squibb] see 9572
Tigemonam, 9572
Tiglic Acid, 9573
Tiglic Acid Ester with Pseudotropine see 9574
Tigloidine, 9574
3β-Tigloyloxytropane see 9574
Tiglylpseudotropeine see 9574
Tiglyssin [BDH] see 9574
Tigogenin, 9575
Tigonin, 9576
Tiguvon [Bayer] see 4044
Tiklid [Sanofi Winthrop] see 9569
Tikofuran see 4320
Tilade [Fisons] see 6524
Tilarin [Fisons] see 6524
Tilatil [Roche] see 9293
Tilcarex [Bayer] see 8264
Tilcotil [Roche] see 9293
Tildiem [Synthelabo] see 3247
Tildin see 1879
Tiliacorine, 9577
Tiliadin see 9231
Tilidine, 9578
Tilisolol, 9579
Tillam see 7192
Tillman's Reagent see 3118
Tilmapor [Ciba-Geigy] see 1994
Tilmicosin, 9580
Tilorone, 9581
Tilt [Ciba-Geigy] see 8003
Tiludronic Acid, 9582
Timacar [Merck & Co.] see 9585
Timacor [Merck & Co.] see 9585
Timaxel [Rhône-Poulenc] see 5991
Timazin [Torii] see 4219
Timbrel [DowElanco] see 9789
Timecef [Roussel-UCLAF] see 1979
Timelit [UCB] see 5587
Timentin [SKB] see 2402
Timepidium Bromide, 9583

Timiperone, 9584
Timocort [Reckitt & Colman] see 4828
Timodyne see 5844
Timolol, 9585
Timonacic, 9586
Timonil [Desitin] see 1826
Timoped [Reckitt & Colman] see 9656
Timoptic [Merck & Co.] see 9585
Timoptol [Merck & Co.] see 9585
Timostenil [Farmitalia] see 1907
Timunox [Cilag] see 9544
Tin, 9587
Tinactin [Schering] see 9656
Tinaderm [Schering] see 9656
Tinarhinin [Berlin-Chemie] see 9358
Tin Ash see 8933
"Tin Bichloride" see 8929
Tin Bronze see 8936
Tindal [Schering] see 70
Tin Dibromide see 8938
Tin Dichloride see 8939
Tin Difluoride see 8940
Tin Diiodide see 8942
Tin Dioxide see 8933
Tin Diselenide see 8934
Tin Disulfide see 8936
Tindurin [EGYT] see 8169
Tinidazole, 9588
Tin Monoselenide see 8946
Tin Monosulfide see 8948
Tin Monoxide see 8944
Tinofedrine, 9589
Tinoridine, 9590
T-inosinic Acid see 5007
Tinostat see 3089
Tin Phosphides, 9591
Tin Protochloride see 8939
Tin Protosulfide see 8948
Tin Protoxide see 8944
Tinset [Janssen] see 7058
Tin Tetrabromide see 8928
Tin Tetrachloride see 8929
Tin Tetrafluoride see 8931
Tin Tetraiodide see 8932
Tintophen X see 1349
Tintorane see 10174
Tinuvin P [Ciba-Geigy] see 3503
Tinzaparin, 9592
Tiobicina see 9427
Tiobutarit see 1481
Tiocarlide, 9593
Tioclomarol, 9594
Tioconazole, 9595
Tioctan [Fujisawa] see 9462
Tioctidasi [ISI] see 9462
Tiomesterone, 9596
Tio-Mld see 3783
Tiopronin, 9597
Tiorfan [Bioprojet] see 72
Tioten [Mediolanum] see 8963
Tiotixene see 9503
Tiotrifar see 683
Tiotropium Bromide, 9598
Tiovalon [Intersan] see 9623
Tiox [Schering-Plough] see 9599
Tioxidazole, 9599
Tioxolone, 9600
γ-TIP see 798
Tipedine see 9601
Tipepidine, 9601
Tiplagen [Burroughs Wellcome] see 9608
T.I.P.P.S. see 5059
Tiquizium Bromide, 9602
Tirade [Sumitomo] see 4051
Tirantil see 1624
Tiratricol, 9603
Tirian see 2138
Tirilazad, 9604
Tirilazad Mesylate see 9604

Tirofiban, 9605
Tiroidina see 9551
Tiron [LaMotte], 9606
Tiropramide, 9607
Tisercin [EGYT] see 6066
Tiserton [Janssen] see 8399
Tissue Factor see 9527
Tissue Plasminogen Activator, 9608
Tissue Thromboplastin see 9527
Tissulina [Hefa-Frenon] see 7672
Tissural see 7672
Titania see 9612
Titanic(IV) Acid, 9609
Titanic Hydroxide see 9609
Titanite see 9610
Titanium, 9610
Titanium Ammonium Oxalate see 599
Titanium Dichloride, 9611
Titanium Dioxide, 9612
Titanium Hydride, 9613
Titanium Isopropylate, 9614
Titanium Potassium Oxalate see 7868
Titanium Sesquisulfate, 9615
Titanium Sulfate, 9616
Titanium Tetrabromide, 9617
Titanium Tetrachloride, 9618
Titanium Tetrafluoride, 9619
Titanium Trichloride, 9620
Titanocene Dichloride, 9621
Titanous Chloride see 9620
Titanous Sulfate see 9615
Titanoxy Sulfate see 9616
Titan Yellow see 9445
Titanyl Ammonium Oxalate see 599
Titanyl Potassium Oxalate see 7868
Titanyl Sulfate see 9616
Titin, 9622
Titriplex III see 3556
Tixair [Valpan] see 89
Tixantone see 5620
Tixocortol, 9623
Tizanidine, 9624
TJN-318 see 5370
TM-723 see 116
TM-906 see 9829
T-Maz 80 [Mazur] see 7742
TMD, 9625
TMD-10 see 9340
TM-PGE₂ see 9853
TMQ see 9851
TMS-19Q see 8408
TMTD see 9510
TMU see 9368
TMV see 9627
TNAZ, 9626
TNCA see 9481
TNF see 9858 and 9943
TNF-β see 9943
TNT see 9860
TO-188 see 9248
Tobacco Mosaic Virus, 9627
Tobacco Wood see 4638
Toban [USV] see 5580
Tobanum [Gedeon Richter] see 2464
Tobermorite see 1749
Tobias Acid see 6493
Tobra [Lilly] see 9628
Tobracin [Shionogi] see 9628
Tobradistin [Dista] see 9628
Tobralex [Alcon] see 9628
Tobramaxin [Alcon-Thilo] see 9628
Tobramycin, 9628
Tobrex [Alcon] see 9628
Tocainide, 9629
Tocamphyl, 9630
Tocèn [Lepetit] see 3155
Tochlorine see 2118
Toclase [Pfizer] see 1840
Toco 500 [Pharma 2000] see 10160
Tocodilydrin [Swiss Pharma] see 6830

Tocodrin [Medichemie] *see* 6830
Tocol, 9631
Tocopherex [Squibb] *see* 10160
(+)-α-Tocopherol *see* 10159
α-Tocopherol *see* 10159
β-Tocopherol, 9632
γ-Tocopherol, 9633
δ-Tocopherol, 9634
ε-Tocopherol, 9635
ζ$_1$-Tocopherol, 9636
ζ$_2$-Tocopherol, 9637
η-Tocopherol, 9638
α-Tocopherol Acetate *see* 10160
α-Tocopheryl Acetate *see* 10160
DL-α-Tocopheryl Retinoate *see* 9639
Tocoretinate, 9639
Tocosamine [Trent] *see* 8887
ToDay [Bristol] *see* 2030
Toddaline *see* 2093
Todralazine, 9640
Tofacine [Brocades] *see* 9641
Tofenacin, 9641
Tofisopam, 9642
Tofranil [Ciba-Geigy] *see* 4955
Tofranil-PM [Geigy] *see* 4955
Tofu *see* 8878
Togamycin [Upjohn] *see* 8890
Togiren [Schwarzhaupt] *see* 3722
TOK [Rohm & Haas] *see* 6695
Tokiocillin [Isei] *see* 628
Tokiolexin [Isei] *see* 2021
Tokocin [Meiji Seika] *see* 3052
Tokokin *see* 3751
Toksobidin [Polfa] *see* 6835
Tokuderm [Nichiban] *see* 1226
Toladryl *see* 6130
Tolamine *see* 2118
Tolan, 9643
Tolanase [Upjohn] *see* 9644
Tolansin *see* 5895
Tolazamide, 9644
Tolazolamide *see* 9644
Tolazoline, 9645
Tolazul *see* 9658
Tolban [Ciba-Geigy] *see* 7954
Tolbusal [Krka] *see* 9646
Tolbutamide, 9646
Tolcasone [Rhone-Poulenc Rorer] *see* 9754
Tolciclate, 9647
Tolcil *see* 5895
Tolcyclamide, 9648
Toldimfos Sodium, 9649
Tolectin [McNeil] *see* 9655
Toleron [Mallinckrodt] *see* 4094
Tolestan [Roemmers] *see* 2481
Tolfedine [Vetoquinol] *see* 9650
Tolfenamic Acid, 9650
Tolferain [Ascher] *see* 4094
Tolfine [Vetoquinol] *see* 9650
Tolhart *see* 5895
Tolhexamide *see* 9648
o-Tolidine, 9651
Tolifer [Marion] *see* 4094
Toliman [Corvi] *see* 2365
Tolinase [Upjohn] *see* 9644
Tolindate, 9652
Toliprolol, 9653
Tolisartine [Kowa] *see* 9660
Tolit *see* 9860
Tolkan [Rhône-Poulenc] *see* 5237
Tollens Reagent, 9654
Tolmene [Sigma-Tau] *see* 9655
Tolmetin, 9655
Tolmicen [Farmitalia] *see* 9647
Tolnaftate, 9656
Tolnate [SK & F] *see* 8073
Tolonidine, 9657
Tolonium Chloride, 9658
Tolopelon [Daiichi] *see* 9584

Tolosate *see* 5895
Toloxatone, 9659
Toloxyn *see* 5895
Tolperisone, 9660
Tolpovidone ^{131}I, 9661
Tolpropamine, 9662
Tolrestat, 9663
Tolrestatin *see* 9663
Tolseram [Squibb] *see* 5895
Tolserol [Squibb] *see* 5895
Tolseron [Squibb] *see* 5895
Toltrazuril, 9664
o-Tolualdehyde, 9665
α-Tolualdehyde *see* 7419
o-Toluamide, 9666
α-Toluamide *see* 7420
Toluazotoluidine *see* 440
Toluene, 9667
Toluene 2,4-Diisocyanate, 9668
Toluene-3,4-dithiol, 9669
p-Toluenesulfinic Acid, 9670
p-α-Toluenesulfonamidobenzoic Acid *see* 1184
p-Toluenesulfonic Acid, 9671
p-Toluenesulfonic Acid Dichloramide *see* 3098
p-Toluenesulfonyl Chloride, 9672
N-(p-Toluenesulfonyl)-N'-hexamethyl-eniminourea *see* 9644
Toluenethiol *see* 9461
α-Toluenethiol *see* 9456
Toluene Trichloride *see* 1141
Toluic Acid, 9673
m-Toluic Acid *see* 9673
o-Toluic Acid *see* 9673
p-Toluic Acid *see* 9673
α-Toluic Acid *see* 7422
p-Toluic Acid 4-[2-(t-Butylamino)-1-hydroxyethyl]-o-phenylene Ester *see* 1344
α-Toluic Acid Ethyl Ester *see* 3885
α-Toluic Aldehyde *see* 7419
Toluidine, 9674
m-Toluidine *see* 9674
o-Toluidine *see* 9674
p-Toluidine *see* 9674
Toluidine Blue O *see* 9658
Tolulexin *see* 5895
Tolulox *see* 5895
o-Tolunitrile, 9675
p-Tolunitrile, 9676
α-Tolunitrile *see* 1166
Toluol *see* 9667
p-Toluoylmethylcarbinolmono-d-cam-phoric Acid Ester Diethanolamine Salt *see* 9630
5-(p-Toluoyl)-1-methylpyrrole-2-acetic Acid *see* 9655
o-Toluylaldehyde *see* 9665
2-(p-Toluyl)benzoic Acid, 9677
p-Toluyl-o-benzoic Acid *see* 9677
Toluylene Blue, 9678
Toluylene Red *see* 6571
Tolvin [Organon] *see* 6260
Tolvon [Organon] *see* 6260
Tolycaine, 9679
Tolycar [Jugoremedija] *see* 1983
m-Tolylacetamide *see* 74
m-Tolyl Acetate *see* 2651
o-Tolyl Acetate *see* 2652
5-(o-Tolylazo)-o-aminotoluene *see* 440
4''-(o-Tolylazo)-o-diacetotoluidide *see* 3004
1-o-Tolylazo-2-naphthylamine *see* 10229
4-(o-Tolylazo)-o-toluidine *see* 440
1-(4-o-Tolylazo-o-tolylazo)-2-naphthol *see* 8538
o-Tolylazo-o-tolylazo-β-naphthol *see* 8538

m-Tolylbromide. *see* 1463
2,4-Tolylene Diisocyanate *see* 9668
1-p-Tolyl-1-ethanol *see* 3287
α-(o-Tolyl)glyceryl Ether *see* 5895
Tolylhydrazine, 9680
m-Tolylhydrazine *see* 9680
o-Tolylhydrazine *see* 9680
p-Tolylhydrazine *see* 9680
2-(N'-p-Tolyl-N'-m-hydroxyphenyl-aminomethyl)-2-imidazoline *see* 7417
Tolylmercaptan *see* 9461
p-Tolylmethylcarbinol *see* 3287
p-Tolylmethylcarbinol Camphoric Acid Ester Diethanolamine Salt *see* 9630
1-(p-Tolyl)-1-oxo-2-pyrrolidino-n-pentane *see* 8194
3-(o-Tolyloxy)-1,2-propanediol *see* 5895
N-m-Tolylphthalamic Acid *see* 6188
1-(p-Tolyl)-2-pyrrolidino-1-pentanone *see* 8194
Tolylsulfonylbutylurea *see* 9646
3-(p-Tolyl-4-sulfonyl)-1-butylurea *see* 9646
1-(p-Tolylsulfonyl)-3-(2-endo-hydroxy-3-endo-D-bornyl)urea *see* 4447
p-Tolylsulfonylmethylnitrosamide, 9681
Tolyprin [Du Pont] *see* 768
Tolyspaz [Conal] *see* 5895
Tomabef [Aandersen] *see* 1981
Tomanol [Byk Gulden] *see* 8282
Tomaset *see* 6188
5α-Tomatidan-3β-ol *see* 9682
Tomatidine, 9682
Tomatine, 9683
Tombran [Boehringer, Ing.] *see* 9712
Tomieze [Yoshitomi] *see* 8089
Tomiporan [Toyama] *see* 1968
Tomiron [Toyama] *see* 1996
Tomorin [Ciba-Geigy] *see* 2623
Tomoscar [Upjohn] *see* 5881
Tomudex [Zenaca], 9684
Tonamil [Ecobi] *see* 9513
Tonarsan. *see* 848
Tonedron [Grimault] *see* 6015
Tonephin *see* 10073
Tonexol [Pharmaton] *see* 9330
Tonibral [Bouchara] *see* 2900
Tonicorine [Lematte et Boinot] *see* 1782
Tonin, 9685
Tonka Bean Camphor *see* 2630
Tonocalcin [Schiapparelli-Searle] *see* 1680
Tonocard [Astra] *see* 9629
Tonocholin B *see* 87
Tonofolin [Zyma] *see* 4254
Tonofosfan [Hoechst] *see* 9649
Tonoftal [Essex] *see* 9656
Tonolift [Teisan] *see* 6792
γ-Tonoplast Intrinsic Protein *see* 798
Tonopres [Boehringer, Ing.] *see* 3217
Tonsillosan (Lösung) [Spitzner] *see* 5349
Tonus-forte *see* 3911
Tony Red *see* 9054
Topalgic [Nippon Roussel] *see* 9180
Topamax [J & J] *see* 9686
Topaz *see* 352
Topazone *see* 4320
Topclip Parasol [Ciba-Geigy] *see* 2836
Top Form Wormer *see* 9426
Tophol *see* 2347
Tophosan *see* 2347
Topicain [Chugai] *see* 7067
Topical Thrombin *see* 9525

Topicon [Pierrel] see 4630
Topicorte [Roussel-UCLAF] see 2975
Topicycline [Procter & Gamble] see 9337
Topifug [Wolff] see 2973
Topik [Ciba-Geigy] see 2431
Topilan [Syntex] see 2209
Topilar [Syntex] see 4157
Topinambur see 5276
Topiramate, 9686
Topisolon [Hoechst] see 2975
Topitracin [Reed & Carnrick] see 965
Topocaine [Lilly] see 2804
Topotecan, 9687
Toprec [Rhône-Poulenc Rorer] see 5316
Toprek [Rhône-Poulenc Rorer] see 5316
Toprol-XL [Astra] see 6235
Topsin [Nippon Soda] see 9489
Topsin-M [Nippon Soda] see 9489
Topsym [Grünenthal] see 4186
Topsymin [Grünenthal] see 4186
Topsyn [Syntex] see 4186
Topsyne [Cassenne] see 4186
Toquilone see 6028
Toracizin see 4149
Toradiur [Boehringer, Mann.] see 9690
Toradol [Syntex] see 5318
Toraflon see 6028
Torak [Hercules] see 3013
Torantil see 4755
Torantyl see 4755
Torasemide see 9690
Torate [Bristol] see 1565
Toratex [Syntex] see 5318
Torazina see 2238
Torbugesic [C-Vet] see 1565
Torbutrol [Bristol] see 1565
Torch [Sanwa] see 10218
Tordon [Dow] see 7552
Torecan Maleate [Boehringer, Ing.] see 2449
Torelle [Dow] see 4283
Torem [Boehringer, Mann.] see 9690
Toremifene, 9688
Torental [Hoechst] see 7278
Toresten [Sandoz] see 9449
Toricelocin [Torii] see 2028
Toril Oil, 9689
Torinal see 6028
Toripamide [Upjohn] see 9866
Torlamicina see 3720
Tormosyl see 4230
Tornalate [Sterling-Winthrop] see 1344
Torque [Shell] see 4004
Torrefaction Dextrin see 2995
Torsemide, 9690
Torularhodin, 9691
Torus [Maag] see 4027
Torutilin see 10164
Toryn [SK & F] see 1820
Toscarna [Nippon Chemiphar] see 8310
Tosic Acid see 9671
Tosmilen [Chemie Linz] see 2936
Tosnone [Nippon Shoji] see 1840
Tostram see 5263
Tosufloxacin, 9692
Tosufloxacin Tosilate see 9692
Tosuxacin [Dainabot] see 9692
Tosyl Chloride see 9672
D-3-endo-p-Tosylureidoborneol see 4447
Totabin see 6564
Totacef [Bristol] see 1967
Totacillin [Beecham] see 628
Total [Hoechst] see 7495

Totalciclina [Benvegna] see 628
Totalon [Pitman-Moore] see 5486
Totapen [Bristol] see 628
Totazina see 2542
Totifen [Master] see 5319
Totomycin [Boots] see 9337
Touch Wood see 183
Tournesol see 5574
Tova [BDH] see 3268
Tovene [Kali-Chemie] see 3350
Towk [Tanabe] see 9702
Toxakil see 9693
Toxalbumin see 5
Toxaphene, 9693
Toxicarol see 9295
Toxichlor see 2129
Toxiferene [Roche] see 222
Toxiferine I, 9694
C-Toxiferine I see 9694
C-Toxiferine II see 1762
Toxiferine V see 9694
Toxiferine XI see 9694
Toxifren see 4130
Toxilic Acid see 5743
Toxilic Anhydride see 5744
Toxinal [Pharma-Selz] see 7111
Toxoflavin, 9695
Toxogonin [E. Merck] see 6835
Toxohormone, 9696
Toxopyrimidine, 9697
Toyocamycin, 9698
Toyolysom-DS [Maruko] see 5671
Toyomycin [Takeda] see 2295
Tozalinone see 9521
TP see 6554 and 9545
TP-1 see 9548
TP-5 see 9544
TP-21 see 9497
t-PA see 9608
TPD see 8068
TPN see 6433
TPN-12 see 9137
TPNH see 6433
TPO see 9528
TPP see 9431
TPS-23 see 5970
TPTZ see 9874 and 9880
TR-495 see 6028
TR-4698 see 8395
Trabest [Hoei] see 2405
Trachyl see 3876
Tracilon [Savage] see 9727
Tracix [Abello] see 4954
Tracker [DuPont] see 9700
Tracrium [Wellcome] see 900
Tractur [Damor] see 7613
Tradon [Beiersdorf] see 7206
Trafuril [Ciba-Geigy] see 9536
Tragacanth see 4609
Tragacanthin see 4609
Tral [Abbott] see 4744
Tralate [DuPont] see 9700
Tralgon [Squibb] see 45
Tralin see 4744
Tralkoxydim, 9699
Tralomethrin, 9700
Tralox [Roussel-Uclaf AH] see 9700
Tramacin [J & J] see 9728
Tramadol, 9701
Tramal [Grünenthal] see 9701
Tramat [Schering AG] see 3788
Tramazoline, 9702
Tramisol [Am. Cyanamid] see 5486
Trancalgyl see 3776
Trancolon see 5893
Trancopal [Winthrop] see 2155
Trancote [Sawai] see 2155
Trandate [Allen & Hanburys] see 5341
Trandolapril, 9703
Trandolaprilat see 9703

Tranex [Malesci] see 9704
Tranexamic Acid, 9704
Tranexamic Acid p-(2-Carboxyethyl)-phenyl Ester see 2067
Tranexan [Taiyo] see 9704
Trangorex [Labaz] see 504
Tranilast, 9705
Trankimazin [Upjohn] see 320
Trankvilan see 5908
Tranlisant see 5908
Tranpoise [Robins] see 5896
Tran-Q [Pfizer] see 4897
Tranquase [Azupharma] see 3042
Tranquilan [Bonin] see 5908
Tranquilax [Hokuriku] see 5829
Tranquiline [Intra] see 5908
Tranquillin [Teva] see 1055
Tranquinal [Profarma] see 320
Tranquit [Promonta] see 7061
Tranquizine see 4897
Tranquo-Puren [Klinge-Nattermann] see 3042
Tranquo-Tablinen [Sanorania] see 3042
Transamin [Daiichi] see 9704
Transanate [Teikoku Zoki] see 2155
Transbilix [Guerbet] see 5043
Transbronchin [Asta] see 1850
Transcobalamins see 10152
Transcop [Recordati] see 8550
Transcortin see 4828
Transcycline [Hoechst] see 8411
Transderm-Nitro [Ciba] see 6704
Transderm Scop [Ciba-Geigy] see 8550
9,10-Transdihydrolisuride see 9308
Transducin see 4551
Transene [Clin-Comar-Byla] see 2465
Transferrins, 9706
Transfer RNA see 8370
Transforming Growth Factor-β, 9707
Transiderm-Nitro [Geigy] see 6704
Transidione [Dausse] see 7060
Transithal [M & B] see 1553
Translon [Daiichi] see 6528
Transoddi [Anphar-Rolland] see 2341
Transvaalin see 8543
Trantan [Upjohn; Sumitomo] see 8890
Trantoin [McKesson] see 6696
Tranvet [Syntex] see 8014
Tranxene [Abbott] see 2465
Tranxilène [Clin-Comar-Byla] see 2465
Tranxilium [Mack, Illert.] see 2465
Tranylcypromine, 9708
Trapanal [Byk-Gulden] see 9487
Trapex see 6165
Trapidil, 9709
Trapymin see 9709
Trasacor [Ciba-Geigy] see 7086
Trasamlon [Kobayashi Seiyaku] see 9704
Trasentine-A [Ciba] see 3501
Trasentine 6-H [Ciba] see 3501
Trasentine Hydrochloride [Ciba] see 160
Traserit [RIT] see 772
Trasicor [Ciba] see 7086
Traslan [Gramon] see 2152
Trasulphane see 4929
Trasylol [Bayer] see 796
Tratul [Gador] see 2337
Traubensäure (German) see 9236
Traumacut [Brenner-Efeka] see 6060
Traumanase [Rorer] see 1409
Traumasept [SmithKline] see 7880
Traumatic Acid, 9710
Traumatociclina [Biomed. Foscama] see 5818
Traumon Gel [Troponwerke] see 3922

Triton A-20 [Rohm & Haas] *see* 9964
Triton N [Rohm & Haas] *see* 6772
Triton WR-1339 [Rohm & Haas] *see* 9964
Triton X [Rohm & Haas] *see* 6858
Triton X-100 [Rohm & Haas] *see* 6858
Tritopine *see* 5389
Tritoqualine, 9894
Tritox [Pitman-Moore] *see* 9757
Trittico [Angelini Francesco] *see* 9712
4-Tritylmorpholine *see* 9808
Triumph [Ciba-Geigy] *see* 5119
Triuret, 9895
Triurol [Lundbeck] *see* 79
Trivalbon [Ist. Vitamine] *see* 9073
Trivaline [Merck-Clévenot] *see* 389
Trivastal [Itherapia] *see* 7648
Trivazol [Parke-Davis] *see* 6242
Trivitan *see* 10157
Trizma [Sigma] *see* 9902
TRK-100 *see* 1190
Trobicin [Upjohn] *see* 8890
Trochol *see* 1237
Trocinate [Poythress] *see* 9509
Troclosene Potassium, 9896
Trodax *see* 6754
Trofan [Upsher-Smith] *see* 9929
Trofosfamide, 9897
Trofurit [Chinoin] *see* 4331
Troglitazone, 9898
Troillite *see* 4106
Trolamine *see* 9798
Troleandomycin, 9899
Trolene [Dow] *see* 8415
Trolnitrate Phosphate, 9900
Trolone [Riker] *see* 9167
Trolovol [Bayer] *see* 7214
Tromal [Burroughs Wellcome] *see* 1532
Tromantadine, 9901
Tromaril [Unichem] *see* 3620
Tromasedan [Millot] *see* 1062
Tromasin *see* 7148
Trombavar *see* 8838
Trombostop *see* 9527
Trombovar *see* 8838
Trometamol *see* 9902
Tromethamine, 9902
Tromethane *see* 9902
Tromexan [Geigy] *see* 3818
Tromexan Ethyl Acetate [Geigy] *see* 3818
Trommsdorff's Starch *see* 10270
Trona *see* 8823
Tronolane [Abbott] *see* 7888
Tronothane [Abbott] *see* 7888
Tropacaine *see* 9904
Tropacine, 9903
Tropacocaine, 9904
Tropaeolin D *see* 6180
Tropaeolin G *see* 5989
Tropaeolin O, 9905
Tropaeolin OO, 9906
Tropaeolin OOO No. 1 *see* 6989
Tropaeolin OOO No. 2 *see* 6990
Tropaeolin R *see* 9905
Tropaic Acid *see* 9910
Tropalin *see* 9867
Tropane, 9907
1αH,5αH-Tropane *see* 9907
(E)-1αH,5αH-Tropane-3α,6β,7β-triol 3-(2-Methylcrotonate) *see* 5997
3-(3,6,7-Tropanetriol) Tiglate *see* 5997
1αH,5αH-Tropan-3α-ol *see* 9912
1αH,5αH-Tropan-3β-ol *see* 8106
3β-Tropanol *see* 8106
1αH,5αH-Tropan-3α-ol Atropate *see* 780
1αH,5αH-Tropan-3α-ol Benzilate *see* 9913

1αH,5αH-Tropan-3β-ol Benzoate *see* 9904
1αH,5αH-Tropan-3α-ol Diphenylacetate *see* 9903
1αH,5αH-Tropan-3α-ol Mandelate *see* 4766
(E)-1αH,5αH-Tropan-3β-ol 2-Methylcrotonate *see* 9574
1αH,5αH-Tropan-3α-ol (−)-2-Methyl-2-phenylhydracrylate (Ester) *see* 5491
1αH,5αH-Tropan-3α-ol (±)-Tropate (Ester) *see* 907
1αH,5αH-Tropan-3α-ol(−)-tropate (Ester) *see* 4907
1αH,5αH-Tropan-3α-ol (±)-Tropate (Ester), 8-Oxide *see* 908
1αH,5αH-Tropan-3α-yl Indole-3-carboxylate *see* 9914
3α-Tropanyl-1H-indole-3-carboxylic Acid Ester *see* 9914
3α-Tropanyl (−)-2-Methyl-2-phenylhydracrylate *see* 5491
3α-Tropanyl S-(−)-Tropate *see* 4907
Tropax [Mead Johnson] *see* 7089
Tropeic Acid *see* 9910
Tropeines *see* 9912
Tropene [Degussa] *see* 3930
Tropenzile, 9908
Tropenzilium Bromide *see* 9908
Tropeolin O *see* 9905
Tropesin, 9909
Tropethydrylin *see* 3802
Trophenium [Duncan, Flockhart] *see* 7346
Trophicard *see* 875
Trophicardyl *see* 5005
Trophires *see* 9491
Trophosphamide *see* 9897
Tropic Acid, 9910
Tropic Acid Ester with Scopine *see* 8550
Tropic Acid Ester with Tropine *see* 907
l-Tropic Acid Ester with Tropine *see* 4907
1-Tropic Acid 3α-Nortropanyl Ester *see* 6799
Tropicamide, 9911
2-Tropidinecarboxylic Acid *see* 3540
Tropin *see* 907
Tropine, 9912
ψ-Tropine *see* 8106
ψ-Tropine Benzoate *see* 9904
Tropine Benzohydryl Ether Methanesulfonate *see* 1156
Tropine Benzylate, 9913
Tropine 4-Chlorobenzhydryl Ether *see* 2422
Tropine Diphenylacetate *see* 9903
Tropine Mandelate *see* 4766
Tropine (−)-α-Methyltropate *see* 5491
Tropine Tropate *see* 907
l-Tropine Tropate *see* 4907
Tropinox *see* 908
Tropisetron, 9914
Tropium [DDSA] *see* 2132
Tropomyosins, 9915
Troponin *see* 9915
Tropyl Isatropate *see* 1053
Tropylium Bromide, 9916
dl-Tropyl Tropate *see* 907
Tropyl Tropate Sulfo-betaine *see* 9170
Trosinone [Abbott] *see* 3787
Trospectomycin, 9917
Trospium Chloride, 9918
Trosyd [Pfizer] *see* 9595
Trosyl [Pfizer] *see* 9595

Trotyl *see* 9860
Trovafloxacin, 9919
Trovafloxacin Mesylate *see* 9919
Troxerutin, 9920
Troxidone *see* 9836
Troximin [Galenica] *see* 9386
Troxipide, 9921
Trozocina [Sigma-Tau] *see* 946
Trp *see* 9929
[D-Trp⁶,des-Gly¹⁰]-LH-RH Ethylamide *see* 2968
D-Trp⁶-gonadorelin *see* 9878
D-Trp⁶-LHRH *see* 9878
D-Trp⁶LRH *see* 9878
Tru [Elea] *see* 8206
True Ammonium Sulfide *see* 591
True Cinnamon *see* 2359
True Lavender *see* 5403
Trueno [Dow Elanco] *see* 4720
Trusono *see* 5890
Trusopt [Merck & Co.] *see* 3484
Truxal [Lundbeck] *see* 2241
Truxaletten [Troponwerke] *see* 2241
Truxillic Acid, 9922
η-Truxillic Acid *see* 9922
Trycite [Dow] *see* 9028
Trycol [Emery] *see* 7737
Trypaflavine *see* 125
Trypan Blue, 9923
Trypan Red, 9924
Tryparsamide, 9925
Tryparsone *see* 9925
Tryponarsyl *see* 9925
Trypothane *see* 9925
Trypoxyl *see* 8719
Trypsin, 9926
α_1-Trypsin Inhibitor *see* 760
Tryptamine, 9927
Tryptan [ICN] *see* 9929
Tryptanol [Merck & Co.] *see* 511
Tryptar [USV] *see* 9926
Tryptazan, 9928
Tryptizol [Merck & Co.] *see* 511
Tryptophan, 9929
L-Tryptophan *see* 9929
6-D-Tryptophan-9-(N-ethyl-L-prolinamide)-10-deglycinamideluteinizing Hormone-Releasing Factor (Pig) *see* 2968
6-D-Tryptophan-LH-RH *see* 9878
6-D-Tryptophanluteinizing Hormone-Releasing Factor (Pig) *see* 9878
Tryptophol, 9930
Trypure [Novo] *see* 9926
TS-110 *see* 5612
TSAA-291 *see* 7065
TSF *see* 9528
TSH, 9931
Tsiklodol *see* 9823
Tsiklometiazid *see* 2813
Tsiklomitsin *see* 9337
TSP *see* 8808 and 9529
TSPP *see* 9377
Tsudohmin [Toho] *see* 3132
Tsuduranine, 9932
TTD *see* 3428
TTFD *see* 4333
TTH *see* 9931
TTMS *see* 9123
T-2 Toxin, 9933
TTP *see* 9433
TTPG *see* 8963
TTX *see* 9382
Tuads [Vanderbilt] *see* 9510
Tuamine [Lilly] *see* 9934
Tuaminoheptane, 9934
Tuasol 100 [Merrell Dow] *see* 9747
Tuazol [Strasenburgh] *see* 6028
Tuazolone *see* 6028
Tubadil [Endo] *see* 9939

Tubarine [Burroughs Wellcome] *see* 9939
Tuba Root *see* 2961
Tubazid *see* 5203
Tube Cucare *see* 2741
Tuberactin [Toyo Jozo] *see* 3636
Tuberactinomycin, 9935
Tuberactinomycin A *see* 9935
Tuberactinomycin B *see* 10139
Tuberactinomycin N *see* 3636
Tuberactinomycin O *see* 9935
Tubercidin, 9936
Tuberculin, 9937
Tubercuprose *see* 2705
Tuberin, 9938
Tubersol [Connaught] *see* 9937
Tubex [Wyeth] *see* 136
Tubilysin [Orion] *see* 5203
Tubocurarine Chloride, 9939
d-Tubocurarine Chloride Dimethyl Ether *see* 9939
Tubocurarine Dimethyl Ether Iodide *see* 6227
d-Tubocurarine Iodide Dimethyl Ether *see* 6227
d-Tubocurine *see* 2268
Tubulin, 9940
α-Tubulin *see* 9940
β-Tubulin *see* 9940
Tuclase [Pfizer] *see* 1840
Tuduranine *see* 9932
Tuff-Lite [Indust. Plastic] *see* 7741
Tuftsin, 9941
Tugon [Bayer] *see* 9753
Tulisan *see* 6815
Tulobuterol, 9942
Tulyn *see* 4582
Tumenol *see* 4929
Tumeric *see* 9955
Tumil-K [Daniels] *see* 7796
Tumor Necrosis Factor, 9943
Tung Oil, 9944
Tungsten, 9945
Tungsten Carbides *see* 9945
Tungsten Hexafluoride, 9946
Tungsten Trioxide, 9947
Tungstic(VI) Acid, 9948
Tungstic Anhydride *see* 9947
5-Tungsto-2-antimonate *see* 4786
21-Tungsto-9-antimonate *see* 4786
Tungstophosphoric Acid *see* 7521
Tungstosilicic Acid *see* 8646
Tunicamine *see* 9949
Tunicamycin, 9949
Tunichrome B-1, 9950
Tunichromes *see* 9950
Tunol *see* 2530
Tupersan [Du Pont] *see* 8631
Turanose, 9951
Turbocalcin [ISF] *see* 3578
Turec [Zyma] *see* 1046
Turicine, 9952
L-Turicine *see* 9952
Turimycin P3 *see* 6271
Turinal [Gedeon Richter] *see* 299
Turixin [SKB] *see* 6383
Turkey Corn *see* 2611
Turkey-galls *see* 6827
Turkey-Red Oil, 9953
Turkish Geranium Oil *see* 6908
Turks Island Salt, 9954
Turlington's Balsam *see* 978
Turloc [Armour Pharm.] *see* 6245
Turmeric, 9955
Turmeric Yellow *see* 2744
ar-Turmerone, 9956
Turnsole *see* 5574
Turoptin [Dispersa] *see* 6221
Turpentine, 9957
"Turpentine Camphor" *see* 1369

Turpentine Substitutes *see* 6288
Turpeth Mineral *see* 5941
Tussafug [Robugen] *see* 1078
Tusscapine [Fisons] *see* 6815
Tusseval [Marco Viti] *see* 7622
Tussidin *see* 3272
Tussirama [Serpero] *see* 4258
Tussistop *see* 2106
Tussol [Burroughs Wellcome] *see* 6008
Tuss-Ornade [SK & F] *see* 1820
TUT-7 *see* 5881
Tutane [Elanco] *see* 1579
Tutin, 9958
Tuttomycin *see* 6542
TV-02 *see* 9197
TV-485 *see* 3922
TV-1322 *see* 27
TVX Q 7821 *see* 5096
Tween 80 [ICI] *see* 7742
Twistane, 9959
Twiston [McNeil] *see* 1845
TWSb *see* 8966
Tx 60 *see* 10172
TXA_2 *see* 9530
TXB_2 *see* 9530
Tybamate, 9960
Tybatran [Robins] *see* 9960
Tybraine [Nicholas] *see* 6145
Tyclarosol *see* 3557
Tycor [Mobay] *see* 3785
Tydantil [Polichimica Sap] *see* 6623
Tylan [Lilly] *see* 9963
Tylcalsin *see* 1684
Tylciprine [Theraplix] *see* 9708
Tylenol [McNeil] *see* 45
Tylinal [Kabi] *see* 3175
Tyllithin *see* 5545
Tylocrebrine, 9961
Tylon *see* 9963
Tylonolide *see* 9963
Tylophorine, 9962
Tylose *see* 6120
Tylose MGA *see* 1877
Tylosin, 9963
Tyloxapol, 9964
Tyloxypal *see* 9964
Tymazoline, 9965
Tymelyt [Lundbeck] *see* 5587
Tymium [Sapos] *see* 3983
Tympanol *see* 6201
Tymtran [Adria] *see* 2048
Type II Interferon *see* 5018
Tyr *see* 9970
Tyraminase *see* 6330
Tyramine, 9966
Tyreno *see* 9758
Tyri 10 [Lindopharm] *see* 9972
Tyrimide *see* 5221
Tyrocidine, 9967
Tyrocidine A *see* 9967
Tyrocidine B *see* 9967
Tyrocidine C *see* 9967
Tyroderm [Merck & Co.] *see* 9972
Tyropanoate Sodium, 9968
Tyropaque [Torii] *see* 9968
Tyrosamine *see* 9966
Tyrosinase, 9969
Tyrosine, 9970
L-Tyrosine *see* 9970
m-Tyrosine, 9971
L-Tyrosylglycylglycyl-L-phenylalanyl-L-methionine *see* 3613
L-Tyrosylglycylglycyl-L-phenylalanyl-L-leucine *see* 3613
Tyrothricin, 9972
Tyrylen *see* 1543
Tyvid [Marion Merrell Dow] *see* 5203
Tyzanol [Pfizer] *see* 9358
Tyzine [Pfizer] *see* 9358

TZB *see* 9443
TZP-4238 *see* 7019
TZU-0460 *see* 8431
U-27 *see* 7575
U-1258 *see* 5317
U-1363 *see* 3363
U-2032 *see* 5310
U-2043 *see* 9646
U-4527 *see* 2797
U-5956 *see* 4123
U-6013 *see* 5190
U-6591 *see* 6818
U-6987 *see* 1881
U-7800 *see* 4229
U-8344 *see* 9986
U-8471 *see* 5840
U-9586 *see* 8108
U-9889 *see* 8991
U-10071 *see* 9936
U-10136 *see* 8063
U-10149 *see* 5525
U-10858 *see* 6290
U-10997 *see* 6262
U-12062 *see* 8064
U-14583 *see* 8065
U-14743 *see* 7756
U-15167 *see* 6767
U-15800 *see* 7119
U-17835 *see* 9644
U-18496 *see* 923
U-19183 *see* 8886
U-19920 *see* 2853
U-21251 *see* 2414
U-22550 *see* 1771
U-24973A *see* 5866
U-26225A *see* 9701
U-26452 *see* 4486
U-26597A *see* 2538
U-27182 *see* 9214
U 28288D *see* 4587
U-28508 *see* 2414
U-28774 *see* 5308
U-31889 *see* 320
U-32070E *see* 1677
U-32921 *see* 1871
U-32921E *see* 1871
U-33030 *see* 9734
U-34865 *see* 3186
U-36059 *see* 510
U-36384 *see* 1871
U-41123 *see* 159
U-42585E *see* 5585
U-42718 *see* 5585
U-42842 *see* 808
U-52047 *see* 5881
U-53217 *see* 8061
U-53217A *see* 8061
U-57903E *see* 7655
U-63196 *see* 1988
U-63196E *see* 1988
U-63366F *see* 9917
U-64279E *see* 1999
U-67279A *see* 1999
U-69689E *see* 4109
U-70226E *see* 4927
U-72107A *see* 7605
U-74006F *see* 9604
U-76252 *see* 1991
U-76253 *see* 1991
U-76253A *see* 1991
U-90152 *see* 2929
U-90152S *see* 2929
Ub *see* 9975
Ubenimex, 9973
UBI-A1335 *see* 4159
Ubichromenol 50 *see* 9974
Ubidecarenone *see* 9974
Ubifactor [San Carlo] *see* 9974
UBI-N252 *see* 7072
UBIP *see* 9975

Ubiquasar [Brocchieri] see 9974
Ubiquinone 10 see 9974
Ubiquinone 50 see 9974
Ubiquinones, 9974
Ubiquitin, 9975
Ubiquitous Immunopoietic Polypeptide see 9975
Ubisan [Von Boch] see 9974
Ubiten [Italfarmaco] see 9974
Ubivis [Agips] see 9974
Ubretid [Nycomed] see 3426
UC-7744 see 1831
UC-21149 see 223
UC-51762 see 9465
UCB-1474 see 2125
UCB-1545 see 3986
UCB-1967 see 3507
UCB-2543 see 1840
UCB-3412 see 3450
UCB-3983 see 5969
UCB-4445 see 1484
UCB-4492 see 4897
UCB-5062 see 5817
UCB-6215 see 7641
UD-CG 115 see 7588
UD-CG 115 BS see 7588
Udekinon [Tohishi] see 9974
Udicil [Upjohn] see 2853
UDMH see 3296
Udolac [ICI] see 2885
UDP see 10017
UDPG see 10018
Ue-5908 see 8182
UF-021 see 9984
U-Gencin [Upjohn] see 4398
Uglow Black Silver, 9976
Ugurol [Bayer] see 9704
Uintahite, 9977
Uintaite see 9977
Ujothion, 9978
UK-4271 see 7051
UK-14304 see 1399
UK-14304-18 see 1399
UK-20349 see 9595
UK-33274 see 3489
UK-33274-27 see 3489
UK-48340 see 516
UK-48340-11. see 516
UK-48340-26 see 516
UK-49858 see 4158
UK-61689 see 8587
UK-61689-2 see 8587
UK-67994 see 3483
UK-68798 see 3469
Ukapen [Pharmuka] see 628
Ukidan [Serono] see 10024
Ulbreval see 1553
Ulcar [Houdé] see 9049
Ulcedin [Agips] see 2337
Ulcedine [Sanofi-Pharma] see 2337
Ulcerfen [Finadiet] see 2337
Ulcerlmin [Chugai] see 9049
Ulcerone [Brocades] see 2549
Ulcesium [Zambon] see 4048
Ulcetrax [Salvat] see 3972
Ulcex [Guidotti] see 8286
Ulcimet [Temis Lostalo] see 2337
Ulcoban [Parke, Davis] see 1110
Ulcocid see 371
Ulcofalk [Interfalk Italia] see 2337
Ulcogant [Lipha; E. Merck] see 9049
Ulcolax [Ulmer] see 1282
Ulcomedina [Von Boch] see 2337
Ulcomet [Italfarmaco] see 2337
Ulcort see 4827
Ulcosan [Dompé] see 7646
Ulcuforton [Plantorgan] see 7646
Ulcus-Tablinen [Sanorania] see 1839
Ulcyn [Pitman-Moore] see 3612
Uldumont see 6025

Ulexine see 2858
Ulfamid [Krka] see 3972
Ulfaret [Takeda] see 1994
Ulfinol [Roemmers] see 3972
UL-FS-49 see 10245
Ulgut [Shionogi] see 1065
Ulhys [Lafare] see 2337
Uliginosin A see 9979
Uliginosin B see 9979
Uliginosins, 9979
Ulinastatin see 10021
Uliron C see 9100
Ulmenide [Roche] see 2096
ULO [Riker] see 2106
Ulobetasol Propionate see 4625
Ulone [Riker] see 2106
Ulsanic [Ferrer] see 3536
Ulstar [Roche] see 9853
Ulstron [ICI] see 7741
Ultandren [Ciba] see 4223
Ultidine [Ferrosan] see 8286
Ultin [Rentschler] see 362
Ultrabion [Sabater] see 628
Ultracain [Hoechst] see 1920
Ultracarbon see 1856
Ultracef [Bristol] see 1963
Ultracid [Geigy] see 6048
Ultracillin [Grünenthal] see 2769
Ultracorten [Ciba-Geigy] see 7904
Ultracortenol [Ciba] see 7901
Ultradol [Ayerst] see 3920
Ultrafur see 4315
Ultralan Oral [Schering AG] see 4188
Ultralanum [Schering AG] see 4188
Ultram [Ortho-McNeil] see 9701
Ultramarine, 9980
Ultramarine Green see 2283
Ultramarine Yellow see 999
Ultra-Mg [Sopar] see 4464
Ultramicina [Lisapharma] see 4280
Ultraparin [Nattermann] see 3626
Ultrapen see 8002
Ultraquinine see 2689
Ultraseptyl [Chinoin] see 9088
Ultrasulfon [Streuli] see 9073
Ultrasüss see 6727
Ultra Tears [Alcon] see 4889
Ultra-Technekow [Mallinkrodt] see 8802
Ultratiazol see 7533
Ultravate [Bristol-Myers Squibb] see 4625
Ultravist [Schering AG] see 5076
Ultrax [Chemie Linz] see 9082
Ultroxim [Duncan, Flockhart] see 2002
UM-792 see 6450
UM-952 see 1522
Umatrope [Lilly] see 8864
Umbellatine see 1192
Umbelliferone, 9981
Umbelliferone Glucoside see 8702
Umbradil see 5061
UML-491 see 6217
UMP see 10020
Unacid [Pfizer] see 9058
Unacid PD oral [Pfizer] see 9166
Unacil [Firma] see 3496
Unacim (inj.) [Jouveinal] see 9058
Unacim Orale [Jouveinal] see 9166
Unagen see 3414
Unakalm [Upjohn] see 5308
Unal see 482
Unasyn [Pfizer] see 9166
Unat [Boehringer, Mann.] see 9690
2-Undecanone see 6178
10-Undecenoic Acid see 9983
Undecoylium Chloride, 9982
Undecylenic Acid, 9983
9-Undecylenic Acid see 9983

Unden [Bayer] see 8022
Unergol [Poli] see 3692
Ungernine see 9250
Unibaryt [Röhm] see 1023
Unicard [Schering-Plough] see 3245
Unicin [Rachelle] see 9337
Unidasa see 4794
Unidigin [Merrell] see 3206
Unidone [Schering] see 706
Unifer [Tosi] see 4083
Unifur see 4315
Unifyl [Mundipharma] see 9421
Unihep [AB Leo] see 4685
Unilobin see 5580
Uniloc [Benzon] see 892
Unimycin [Unigreg] see 7111
Unipen [Wyeth] see 6438
Uniphyl [Purdue Frederick] see 9421
Uniphyllin [Mundipharma] see 9421
Unipril [Zambon] see 8283
Uniprofen [Lilly] see 1075
Unipyranamide [Unichem] see 8140
Uniquin [Beta] see 5592
Unisedil [Laquifa] see 3042
Unisept [Schering] see 2140
Unisom [Pfizer] see 3497
Unisomnia [Unigreg] see 6667
Unispiran see 9344
Unistat see 6709
Unistat [Salsbury] see 9103
Unistat-3 [Salsbury] see 6709
Unitane see 9612
Unitensen Tannate [Wallace Labs.] see 2674
Unitiol see 3256
UNITOP [Roche] see 6425
Unitrim [Master] see 1401
Univer [Rorer] see 10083
Unix [Ciba-Geigy] see 2841
Unixime [Firma] see 1975
Unizole see 3315
Unnatural Tartaric Acid see 9235
Unnilennium see 3585
Unnilhexium see 3582
Unniloctium see 3584
Unnilpentium see 3581
Unnilquadium see 3580
Unnilseptium see 3583
Unoprostone, 9984
Unospaston [Siegfried] see 3402
Unresolvable Tartaric Acid see 9238
Unusual Tartaric Acid see 9235
UP-33-901 see 2330
UP-83 see 6620
UP-106 see 8019
UP-507-04 see 7551
UP-34101 see 7976
Upcyclin [Cephar] see 9337
Upixon [Bayer] see 7617
Upstene [Pharmuka] see 4965
UR-112 see 2437
UR-1501 see 9817
UR-1521 see 4281
UR-4056 see 4247
Urab [Hopkins] see 4050
Uracid [Astra] see 355
Uracil, 9985
Uracil-6-carboxylic Acid see 7004
Uracil Deoxyriboside see 2957
Uracil Mustard, 9986
Uracil Riboside see 10016
Uractyl see 9101
Uradal see 1879
Uragon [Makhteshim-Agan] see 1402
Ural see 1849
Uralenic Acid see 3628
Uralgin [Ceccarelli] see 6446
Uraline see 1849
Uralium see 1849
Uramid [Spofa] see 9101

Uramil, 9987
Uramite [Du Pont], 9988
Uramustine see 9986
Uramycin B see 9698
Uranediol, 9989
Uranic Oxide see 9997
Uranin(e) see 4194
Uranine Yellow see 4194
Uraninite see 9991
Uranium, 9990
Uranium Ammonium Carbonate see 602
Uranium Ammonium Fluoride see 603
Uranium Barium Oxide see 1030
Uranium Benzoate 1122
Uranium Dioxide, 9991
Uranium Dioxydichloride see 9999
Uranium Hexafluoride, 9992
Uranium Oxide Orange see 7873
Uranium Oxide Yellow see 8848
Uranium Peroxide, 9993
Uranium Tetrachloride, 9994
Uranium Tetrafluoride, 9995
Uranium Trichloride, 9996
Uranium Trioxide, 9997
Uranium X₂ see 8069
"Uranium Yellow" see 601
Uranium Yellow see 8848
Uranium Z see 8069
Uranous Oxide see 9991
Urantoin [DDSA] see 6696
Uranyl Acetate, 9998
Uranyl Ammonium Carbonate see 602
Uranyl Ammonium Fluoride see 603
Uranyl Benzoate see 1122
Uranyl Chloride, 9999
Uranyl Nitrate, 10000
Uranyl Phosphate, 10001
Uranyl Potassium Nitrate see 7874
Uranyl Potassium Sulfate see 7875
Uranyl Sulfate, 10002
Urao see 8823
Urapidil, 10003
Uraprene [IBI] see 10003
Urari see 2741
Urastrat [Warner-Lambert] see 10009
Urate Oxidase see 10015
Urazole, 10004
Urbac [Merck-Clévenot] see 6624
Urbadan [Roussel] see 2417
Urbanyl [Diamant] see 2417
Urbason [Hoechst] see 6189
Urbason-Solubile [Hoechst] see 6189
Urbilat [Hor-Fer-Vit] see 5908
Urbol [Heilit] see 287
Urdes [Errekappa] see 10026
Urea, 10005
Urea Amidohydrolase see 10009
Urea Calcium Acetylsalicylate see 1684
Urea Hydrochloride, 10006
Urea Hydrogen Peroxide, 10007
Urea Nitrate, 10008
Ureaphil [Abbott] see 10005
Urea Polymer with Formaldehyde see 7735
Urease, 10009
Urea Stibamine, 10010
Urecholine Chloride [Merck & Co.] see 1232
Uredepa, 10011
Uregit [EGYT] see 3761
p-Ureidobenzenearsonic Acid see 1830
(p-Ureidobenzenearsylenedithio)di-o-benzoic Acid see 9458
5-Ureidohydantoin see 255
δ-Ureidonorvaline see 2392
4-Ureido-1-phenylarsonic Acid see 1830

(p-Ureidophenylarsylenedithio)-diacetic Acid see 9459
(p-Ureidophenylarsylenedithio)di-o-benzoic Acid see 9458
Urelim see 3775
Urem [Kade] see 4925
Urena, 10012
Urenil see 9101
Ureophil [Abbott] see 10005
Urepearl [Otsuka] see 10005
Urese [Pfizer] see 1155
Urethan, 10013
Urethane see 10013
Urethan of β-Methylcholine Chloride see 1232
Urethimine see 10011
Urethylane see 6115
Uretrim [Bastian] see 9840
Urex [Riker] see 6039
Urex [Mochida] see 4331
6,6'-Ureylenebis[1,1'-dimethylquino-linium] Sulfate see 3416
6,6'-Ureylenebis(1-methylquinolin-ium)bis(methosulfate) see 3416
Urfadyn [Zambon] see 6631
Urfamycine [Inpharzam] see 9436
Urgilan [Sintesa] see 8060
Uriben [Rosemont] see 6446
Uric Acid, 10014
Uricase, 10015
Uricemil [ICT-Lodi] see 287
Uriclar [Crosara] see 6446
Urico-oxidase see 10015
Uricovac [Labaz] see 1093
Uric Oxidase see 10015
Uridine, 10016
Uridine 5'-Diphosphate, 10017
Uridine Diphosphate Glucose, 10018
Uridine-5'-diphosphoglucose see 10018
Uridine 5'-Monophosphate see 10020
Uridine 5'-Phosphoric Acid see 10020
Uridine 5'-Pyrophosphate see 10017
Uridine 5'-Pyrophosphate Glucose Ester see 10018
Uridine-5-pyrophosphoric Acid see 10017
Uridine 5'-(Tetrahydrogen Triphos-phate) see 10019
Uridine 5'-(Trihydrogen Diphosphate) see 10017
Uridine 5'-(Trihydrogen Diphosphate) Mono-α-D-glucopyranosyl Ester see 10018
Uridine 5'-Triphosphate, 10019
5'-Uridylic Acid, 10020
UriKoxidase see 10015
Urimeth [North Am. Pharm.] see 6053
Urinary Indican see 4975
Urinary Trypsin Inhibitor see 10021
Urinastatin, 10021
Urinex see 2221
Urinorm [Torii] see 1093
Urinox [Parke, Davis] see 7079
Uriodone see 5061
Urion [Zambon] see 237
Uripurinol [Azupharma] see 287
Urisal [Winthrop] see 8746
Urispas [SK & F] see 4135
Uritone [Parke, Davis] see 6036
Uritrate [Parke, Davis] see 7079
Uritrol see 6753
Urizept [RPR] see 6696
Urlea see 1064
Uroalpha [Debat] see 6374
Uro-Alvar [Gomez] see 7079
Uroanthelone see 3569
Urobenyl [Searle] see 287
d-Urobilin see 10022
d-Urobilin IXα see 10022
i-Urobilin see 10022

Urobilin IXα see 10022
Urobilins, 10022
Urocanylcholine see 6387
Uro-Carb [Hamilton] see 1232
Urocarf [SPA] see 1838
Urocaudal [Jorba] see 9731
Uro-Cedulamin [Cosmopharma] see 6040
Urochloralic Acid, 10023
Urocit-K [Mission] see 7785
Uro-Clamoxyl [Beecham] see 617
Urocoli see 6753
Urocomb [Jenapharm] see 8016
Urodiazin see 4822
Urodie [Abbott] see 9297
Urodin [Streuli] see 6696
Urodine [Kenyon] see 7361
Urodixin [Italchimici] see 6446
Uroenterone see 3569
Urofollitrophin see 4299
Urofort see 388
Urogastrone see 3569
β-Urogastrone see 3569
γ-Urogastrone see 3569
Urografic Acid Methylglucamine Salt see 5851
Urografic Acid Sodium Salt see 3040
Urografin [Schering AG] see 5851
Urokinase, 10024
Urokon Sodium [Mallinckrodt] see 79
Urolene Blue [Star] see 6137
Urolocide see 3466
Urolong [Thiemann] see 6696
Urolucosil [Parke, Davis] see 9084
Uromaline see 5757
Uroman [Toho Kogyo] see 6446
Uromandelin see 6040
Uromiro [Bracco] see 5031
Uromitexan [WB Pharm.] see 5969
Uronal see 989
Uronamin see 6040
Uronase [Mochida] see 10024
Uroneg [Ibirn] see 6446
Uronorm [Alfa] see 2369
Uropac [M & B] see 8779
Uropan [Tiber] see 6446
Uropen [Bristol] see 4706
Urophenil [Nippon Seiyaku] see 9436
Uropimid [CT] see 7613
Uropir [Sarm] see 7660
Uroplus [Shionogi] see 9086
Uropurgol see 6038
Uro-Ripirin [Kabi] see 3598
Uro-Septra [Wellcome] see 9086
Urosin [Galenus Mannheim] see 287
Urosten [Bonomelli] see 7613
Urosulfan see 9101
Uro-Tablinen [Sanorania] see 6696
Urothion, 10025
Urotractan [Klinge] see 6039
Urotrate [Warner-Lambert] see 7079
Urotropin [Warner-Chilcott] see 6036
Urotropin New [Warner-Chilcott] see 6038
Uroval [Firma] see 7613
Urovist [Schering AG] see 5851
Urox [Allied] see 6342
Uroxacin [Lazar] see 6793
Urox B [Hopkins] see 1402
Urox D [Hopkins] see 3447
Uroxin see 291
Uroxin [Von Boch] see 7079
Uroxol [Ausonia] see 7079
Ursacol [Zambon] see 10026
(3β)-Urs-12-en-3-ol see 659
(3β,18α,19α)-Urs-20(30)-en-3-ol see 9229
Ursin see 812
Ursnon [Nippon Yakuhin] see 4213
Urso [Tokyo Tanabe] see 10026

Ursobilin [Hek] see 10026
Ursochol [Inpharzam] see 10026
Ursocholanic Acid see 2251
Ursodamor [Damor] see 10026
Ursodeoxycholic Acid see 10026
Ursodiol, 10026
Ursofalk [Falk] see 10026
Ursol D see 7439
Ursolic Acid, 10027
Ursol P see 482
Ursolvan [Robert et Carrière] see 10026
Urson see 10027
Urtias [Sagitta] see 287
Urtosal [Lifepharma] see 8480
Urupan [Merckle] see 2988
Urushiol, 10028
Urusonin [Isei] see 8917
Urylon, 10029
USB-3153 see 7947
Uscharidin, 10030
Usevir [Yamasa Shoyu] see 8875
Uskan [Desitin] see 7059
Usnein see 10031
Usniacin see 10031
Usnic Acid, 10031
Usninic Acid see 10031
Uspulun see 4876
USR-604 see 3096
Ustilagic Acid, 10032
Ustimon see 4743
Ustin see 6808
Ustin Methyl Ether see 6616
Ustizeain B see 10032
Utemerin [Kissei] see 8401
Uteplex [Wyeth] see 10019
Uteracon [National] see 7114
Uteramin [Zyma] see 9966
Uterol see 4565
Uteroverdine see 1264
UTI see 10021
Utibid [Warner-Chilcott] see 7079
Uticillin [Beecham] see 1838
Uticillin VK [Upjohn] see 7230
Uticort [Parke, Davis] see 1226
Utimox [Parke, Davis] see 617
Utinor [Merck & Co.] see 6793
Utipina see 10019
Utopar [Ferrosan] see 8401
Utovlan [Syntex] see 6790
UTP see 10019
Utrogestan [Besins-Iscovesco] see 7956
Utrophin, 10033
Uval [Dorsey] see 9157
Uvaleral see 1418
Uvasol see 812
Uva Ursi, 10034
Uvesterol-D [Crinex] see 10156
Uviban [Alcon] see 143
Uvic Acid see 9236
Uvilon [Bayer] see 7617
Uvinul 400 [GAF] see 1138
Uvinul D49 [GAF] see 1130
Uvinul M-40 [GAF] see 7088
Uvinul MS-40 [GAF] see 9157
Uvistat see 6255
UX₂ see 8069
UZ see 8069
Uzarigenin see 10035
Uzarin, 10035
Uzone [Douglas] see 7431
V see 10046
V-7 see 9850
V-12 see 6011
V-1326 see 8135
V-23031 see 4178
Vabrocid see 6697
Vaccenic Acid, 10036
Vacciniin, 10037

Vacor [Rohm & Haas] see 8171
Vadebex see 6815
Vaderm [Schering] see 221
Vadilex [Robert et Carrière] see 4936
Vadosilan see 5259
Vaflol see 10150
Vagamin see 6025
Vagantin see 6025
Vagestrol [Norwich] see 3177
Vagifem [Novo] see 3746
Vagilen [Farmigea] see 6242
Vagimid [Apogepha] see 6242
Vagistat [Bristol-Myers Squibb] see 9595
Vagran [UPSA] see 8019
Vakerin see 1197
Val see 10046
VAL-480 see 8693
VAL-4000 see 8693
Valaciclovir see 10039
Valacidin, 10038
ValACV see 10039
Valacyclovir, 10039
Valadol [Squibb] see 45
Valamin [Schering AG] see 3779
Valamina [Schering] see 4226
Val⁵-angiotensin II-asp¹-β-amide see 689
Valaxona [Schaper & Brummer] see 3042
Valbazen [SK&F] see 211
Valbil [Klinge] see 3985
Valbilan [Merck & Co.] see 3985
Valcor [Maggioni] see 3506
Valcote [Abbott] see 10049
Val²-cyclosporine see 2821
Valemicina [Farmochim. Ital.] see 4280
Valentinite see 752
Valeral see 10040
n-Valeraldehyde, 10040
Valeramide-OM see 479
Valerbé see 10041
Valergen [Hyrex] see 3746
Valerian, 10041
Valerianic Acid see 10042
Valeric Acid Ammonium Salt see 604
Valeric Acid Bismuth Basic Salt see 1334
Valeric Acid, Normal, 10042
Valeric Aldehyde see 10040
Valeronitrile, 10043
Valetan [Tobishi] see 3132
Valethamate Bromide, 10044
Validacin [Takeda] see 10045
Validamycin A see 10045
Validamycins, 10045
Validol see 5887
Valimon [Takeda] see 10045
Valine, 10046
L-Valine see 10046
L-Valine 2-[(2-Amino-1,6-dihydro-6-oxo-9H-purin-9-yl)methoxy]ethyl Ester see 10039
5-Valine-angiotensin II Amide see 689
L-Valine Ester with 9-[(2-Hydroxyethoxy)methyl]guanine see 10039
Valinomycin, 10047
Valip see 9867
Valiquid [Roche] see 3042
Valisan see 1368
Valisone [Schering] see 1226
Valium [Roche] see 3042
Vallene see 5813
Vallergan [M & B] see 9834
Vallesine see 885
Vallestril [Searle] see 6012
Vallet's Mass see 4089
Valmethamide see 10048
Valmid [Lilly] see 3779

Valmidate see 3779
Valmorin [Roerig] see 2247
Valnoctamide, 10048
Valodin [Ferrer] see 3418
Valoid [Burroughs Wellcome] see 2779
Valone see 5254
Valopride [Vita] see 1454
Valoron [Gödecke] see 9578
Valpin [Endo] see 709
Valproic Acid, 10049
Valpromide, 10050
Valrelease [Roche] see 3042
Valsartan, 10051
Valsyn [Norwich] see 4315
Valtomicina [Midy] see 7606
Valtorin [Boehringer, Ing.] see 2247
Valtrex [Burroughs Wellcome] see 10039
Valyl see 5253
Valzin see 3517
Vamicamide, 10052
Vanadic Anhydride see 10056
Vanadic Oxide see 10059
Vanadinite see 10053
Vanadium, 10053
Vanadium Carbonyl, 10054
Vanadium Hexacarbonyl see 10054
Vanadium Oxydichloride see 10062
Vanadium Oxysulfate see 10063
Vanadium Oxytrichloride see 10064
Vanadium Pentafluoride, 10055
Vanadium Pentoxide, 10056
Vanadium Sesquioxide see 10059
Vanadium Sesquisulfide see 10061
Vanadium Tetrafluoride, 10057
Vanadium Trifluoride, 10058
Vanadium Trioxide, 10059
Vanadium Trisulfate, 10060
Vanadium Trisulfide, 10061
Vanadyl Dichloride, 10062
Vanadyl Sulfate, 10063
Vanadyl Trichloride, 10064
Vanaspati, 10065
Vancenase [Schering] see 1047
Vanceril [Schering] see 1047
Vancide 89 see 1815
Vancide BN [Vanderbilt] see 1343
Vancocin [Lilly] see 10066
Vancomycin, 10066
Vancor [Adria] see 10066
Vandid [Riker] see 3765
Vanilla, 10067
Vanillal see 3904
Vanilla Plant see 5501
Vanillic Acid, 10068
Vanillic Acid Diethylamide see 3765
Vanillic Aldehyde see 10069
Vanillic Diethylamide see 3765
Vanillin, 10069
Vanillinemandelic Acid see 10070
Vanillin-D-glucoside see 4473
Vanilloside see 4473
Vanillylacetone see 10301
Vanillylmandelic Acid see 10070
Vanilmandelic Acid, 10070
Vanitiolide, 10071
Vanobid [Merrell] see 1789
Vanoxin [Vanguard] see 3210
Vanquin [Parke, Davis] see 8206
Vansil [Pfizer] see 7051
Vantin [Upjohn] see 1991
Vantol [Parke-Davis] see 1238
Vantyl see 2347
Vanylglycol see 6074
Vapam see 6016
Vapona [Shell] see 3129
Vaponefrin [Fisons] see 3656
Vapo-N-Iso [USV] see 5236
Vaporpac [Menley & James] see 6854
Vapotone [Chevron] see 9345

Varacillin [Kanebo] see 5460
Varbian [Ciba-Geigy] see 7917
Vardax [Thomae] see 9159
Varemoid [Zyma] see 9920
Variamine Blue Base see 6079
Varidase [Lederle] see 8979
Varikill [Maag] see 4027
Varinon [Exa] see 3350
Variotin [Leo Pharm.] see 7193
Variscite see 371
Variton [Schering] see 3361
Varitox [M & B] see 9756
Varlane [Schering AG] see 4187
Varnish Makers' and Painters' Naphtha see 5513
Varnoline [Organon] see 2971
Varonic [Sherex] see 7738
Varsol 1 [Exxon] see 6288
Varsol 3 [Exxon] see 6288
Varunax see 486
Vasaca see 158
Vasagin [Salus] see 8158
Vasal [Tutag] see 7151
Vasangor [Winthrop] see 7995
Vasapril [Ciba] see 8158
Vascardin [Nicholas] see 5245
Vascase [Roche] see 2332
Vascor [J & J] see 1188
Vascoril [Delelande] see 2349
Vasculat [Boehringer, Ing.] see 981
Vasculit [Boehringer, Ing.] see 981
Vascunicol [Boehringer, Ing.] see 981
Vaseline [Chesebrough] see 7326
Vaseretic [Merck & Co.] see 3605
Vashegyite see 371
Vasicine, 10072
Vasiodone see 5061
Vaskulat see 981
Vasoactive Intestinal Peptide see 10140
Vasoactive Intestinal Polypeptide see 10140
Vasoc [Lindopharm] see 1091
Vasocard [Abbott] see 9297
Vasocet [Winthrop] see 2062
Vasocil [Magis] see 8158
Vasodiatol [Rowell] see 7249
Vasodilan [Mead Johnson] see 5259
Vaso-Dilatan [Agepha] see 9645
Vasodin [Alfa] see 6579
Vasodistal [Delalande] see 2350
Vasoglyn [Reid-Powell] see 6704
Vasokellina see 5321
Vasoklin [Gödecke] see 6374
Vasoliment see 7326
Vasomed [Medo-Chem.] see 9900
Vasomet [Sanko-Nikken] see 9297
Vasomotal [Philips-Duphar] see 1224
Vasonase [Syntex Latino] see 6579
Vasophysin see 10073
Vasoplex [Lappe] see 5259
Vasopressin, 10073
Vasopressin 2-(L-Phenylalanine)-8-L-lysine see 3992
Vasorbate [Euderma] see 5245
Vasorelax [Farmasa] see 1060
Vasorome [Kowa] see 7054
Vasospan [Exa] see 6580
Vasospan [Ulmer] see 7151
Vasotec [Merck & Co.] see 3605
Vasotec IV [Merck & Co.] see 3606
Vasotran [Bristol] see 5259
Vasotrate [Reid-Provident] see 5245
Vasoverin [Banyu] see 8158
Vasoxine [Burroughs Wellcome] see 6067
Vasoxyl [Burroughs Wellcome] see 6067

Vasperdil see 6605
Vaspit [Schering AG] see 4187
Vastarel F [Kyoto] see 9835
Vastcillin [Takeda] see 2769
Vasten [Specia] see 7894
Vastribil [Farmasan] see 9920
Vasurix [Guerbet] see 5850
Vasylox [Burroughs Wellcome] see 6067
Vatensol [Pfizer] see 4595
Vaterite see 1697
Vatracin [Tobishi] see 2769
Vaumigan see 4581
Vazofirin see 7278
VC9-104 see 3792
VC-13 Nemacide see 3094
V-Cil-K [Lilly] see 7230
V-Cillin [Lilly] see 7230
V-Cillin K [Lilly] see 7230
VCR see 10124
VCS-438 see 6032
VCS-506 see 5468
VD-1827 see 9166
VDS see 10125
Veatchine, 10074
Veclam [Zambon] see 2400
Vectacin [Schering] see 6563
Vectarion [Servier] see 309
Vectren [Warner-Lambert] see 5258
Vectrin [Parke, Davis] see 6289
Vecuronium Bromide, 10075
Veda-K₁ [Vedco] see 7536
Veetids [Squibb] see 7230
Vegadex [Monsanto] see 9079
Vegetable Calomel see 7705
Vegetable Casein see 1934
Vegetable Lutein see 10197
Vegetable Luteol see 10197
Vegetable Mercury see 5755
Vegetable Pepsin see 7148
Vegetable Sulfur see 5653
Vegetable Wax see 5270
Vegiben see 2115
Vegolysen [M & B] see 4724
Vegolysen-T [M & B] see 4724
Vegolysin see 4724
Vehem-Sandoz [Sandoz] see 9291
Veinamitol [Negma] see 9920
Veinartan see 6000
VEL-3973 see 5846
Velardon [Organon] see 7148
Velban [Lilly] see 10119
Velbe [Lilly] see 10119
Veldopa (formerly Weldopa) [Smith & Nephew] see 5490
Vellosimine, 10076
Velmol [Berlex] see 3460
Velmonit [Esteve] see 2374
Velnacrine, 10077
Velocef [Squibb] see 2032
Velon see 10134
Velopural [Optimed] see 4828
Velosef [Squibb] see 2032
Velosimine see 10076
Velosulin [Novo] see 5011
Velpar [Du Pont] see 4734
Velsicol 58-CS-11 see 3090
Velsicol 104 see 4690
Velsicol 1068 see 2129
Veltane [Lannett] see 1467
Veltol [Pfizer] see 5752
Vems [Searle] see 9719
Venacil (formerly) [Abbott] see 671
Venactone [Lepetit] see 1795
Venarterin [B.O.I.] see 6000
Vendarcin [Gist-Brocades] see 7111
Ven-Detrex [Zyma] see 3350
Vendex [Shell] see 4004
Venen [Tanabe] see 9877
Veneniferin see 6559

Venetlin [Sankyo] see 217
Venex [Recordati] see 9738
Vengeance [Velsicol] see 1410
Vengicide see 9698
Venice Turpentine, 10078
Venilon [Teijin] see 4958
Veniten [Wyeth] see 9920
Venlafaxine, 10079
Venlafexine see 10079
Venopirin [Green Cross] see 5668
Venoruton [Zyma] see 9920
Venosmine [Geymonat] see 3350
Ventaire [Marion] see 8082
Ventax [Robert] see 3494
Ventilat [Thomae] see 7077
Ventipulmin [Boehringer, Ing.] see 2407
Ventodisks [Glaxo] see 217
Ventolin [Allen & Hanburys] see 217
Ventox see 133
Ventramine see 6635
Ventrazol see 7283
Venturicidin A see 10080
Venturicidin B see 10080
Venturicidin B 3'-Carbamate see 10080
Venturicidins, 10080
Ventussin [Warren-Teed] see 1127
Venzar [Du Pont] see 5459
Venzonate see 1162
Venzoquimpe [Quimpe] see 5986
Vepen [Orion] see 7230
Vepesid [Bristol-Myers] see 3931
Veracillin [Ayerst] see 3134
Veracim [Cimex] see 10083
Veractil see 6066
Veracur [Typharm] see 4262
Veradol [Schering] see 6504
Veralba [Pitman-Moore] see 8086
Veralcamine see 10082
Veralipride, 10081
Veralipril [Midy] see 10081
Veralkamine, 10082
Veramex [Labaz] see 10083
Veramix see 5838
Verapamil, 10083
Veraptin [Lagap] see 10083
Veratetrine see 8086
Veratraldehyde, 10084
Veratramine, 10085
Veratran [Latema] see 2477
Veratric Acid, 10086
Veratric Acid 4-[Ethyl(p-methoxy-α-methylphencthyl)amino]butyl Ester see 5808
Veratric Aldehyde see 10084
Veratridine, 10087
Veratrine see 2075
Veratrine (Mixture), 10088
Veratrole, 10089
Veratroylaconine see 8095
3-Veratroylveracevine see 10087
Veratrum Viride, 10090
Verax [Tosi] see 1157
Verazide, 10091
Verbascose, 10092
Verbenalin, 10093
Verbenalol see 10093
d-Verbenone, 10094
Vercyte [Abbott] see 7634
Verdiana see 9867
Verdisol [Arovet] see 3129
Verecolene see 4009
Verel [Eastman Kodak] see 6310
Verelan [Elan] see 10083
Verexamil [Verex] see 10083
Vergentan [Schrholz] see 246
Vergitryl [Squibb] see 10090
Vergonil see 4830
Veriloid [Riker] see 10090

Veritab [Vista] see 5817
Veritol [Knoll] see 7485
Verladyn [Verla] see 3217
Vermicidin [Rivopharm] see 5807
Vermicompren (tabl.) [E. Merck] see 7618
Vermiculite, 10095
Vermilion see 5945
Verminum [Squibb] see 7169
Vermiparin see 5758
Vermirax [Biosintetica] see 5807
Vermitiber [Tiber] see 8206
Vermitin see 7404
Vermizym see 7148
Vermouth see 9
Vermox [Janssen] see 5807
Vernal Pheasant's Eye see 165
Vernam [Stauffer] see 10097
Vernamycin A see 10142
Vernamycin B, 10096
Vernamycin B_α see 6275
Vernamycin B_β see 10096
Vernamycin B_γ see 10096
Vernamycin B_δ see 10096
Vernine see 4596
Vernitest Reagent see 8229
Vernolate, 10097
Vernolepin, 10098
Vernolic Acid, 10099
Verol see 6097
Veroletten see 989
Veronal see 989
Veronal Sodium see 989
Verospiron [Hormosan] see 8917
Veroxil [Baldacci] see 4969
Verrucarin A see 10100
Verrucarin B see 10100
Verrucarin J see 10100
Verrucarin K see 10100
Verrucarins, 10100
Verrucarol see 10100
Verrugon [Pickles] see 8484
Versalide [Givaudan-Delawanna], 10101
Versapen [Bristol] see 4706
Versapen K [Bristol] see 4706
Versatrex [Bristol] see 4706
Versed [Roche] see 6270
Versene [Dow] see 3557
Versene-9 [Dow] see 3558
Versene Acid [Dow] see 3559
Versene CA [Dow] see 3555
Versene Disodium Salt see 3556
Versen-Ol [Dow], 10102
Versotrane see 4805
Verstran [Warner-Lambert] see 7895
Versulin see 773
Versus [Angelini Francesco] see 1061
Verticillin A see 10103
Verticillin B see 10103
Verticillin C see 10103
Verticillins, 10103
Verticine, 10104
Vertigon [SK&F] see 7942
Vertimec [Merck & Co.] see 1
Verton [Squibb] see 2152
Verucasep [Galen] see 4480
Verv-Ca see 1751
Verxite see 10095
Very Late Activation Antigens see 5014
Vesadin [Rhone Merieux] see 9083
Vésadol [Janssen] see 1635
Vesamin [Byk-Gulden] see 79
Vesdil [Astra] see 8283
Vesicating Collodion see 2548
Vesidril see 6225
Vesidryl [Clin-Comar-Byla] see 6225
Vésipaque [Warner-Lambert] see 7387
Vesipyrin see 7424

Vesistol [Inverni] see 634
Vesnarinone, 10105
Vesnaroid [Roche] see 8333
Vespéral see 989
Vesperone [UCB] see 1387
Vespral [Squibb] see 9814
Vesprin [Squibb] see 9814
Vestalin see 6795
Vesulong [Ciba-Geigy] see 9118
Vesuprim [Ciba-Geigy] see 9071
Vesuvianite see 1685
Vesuvine see 1293
Vetacortyl [Stricker] see 6189
Veta-K_1 [Sanofi] see 7536
Vetalar [Am. Home] see 5306
Vetalgina [ICC] see 5668
Vetalog [Solvay] see 9728
Vetalpha [A.T.I.] see 4864
Vetame [Squibb] see 9814
Vetamox [Am. Cyanamid] see 50
Vetarsenobillon see 6534
Vetdectin [Am. Cyanamid] see 6373
Veteusan [Veterinaria] see 2661
Vetibenzamina [Ciba-Geigy] see 9868
Veticillin [Am. Cyanamid] see 1178
Veticol [Copanos] see 2120
Vetidrex [Ciba] see 4822
Vetimast [Ciba-Geigy] see 2019
Vetiprim [Angelini] see 9073
Vetiprost [Virbac] see 3916
Vetisulid [Solvay] see 9068
Vetiver Oil see 6953
Vetiver Oil Haiti see 6953
Vetiver Oil Java see 6953
Vetiver Oil Reunion (Bourbon) see 6953
Vetivert Oil see 6953
α-Vetivone see 10106
β-Vetivone see 10106
Vetivones, 10106
Vetkelfizina [Centralvet] see 9078
Vetomazin see 6080
Vetquamycin-324 [Rachelle] see 9337
Vetrabutine, 10107
Vetranquil [Lathevet] see 32
Vetrazin [Ciba-Geigy] see 2845
Vetsin see 6337
Vetstrep [Merck & Co.] see 8983
Vetyver Oil see 6953
Vexol [Alcon] see 8392
Viaben [Schrholz] see 1454
Viacutan [Boehringer, Ing.] see 5531
Viacutan see 6030
Viadent [Vipont] see 8504
Viadil [Pharma Investi] see 1454
Viadril see 4868
Vialibran [Servier] see 5831
Vialidon [Italfarmaco] see 5842
Vi-Alpha see 10150
Vianin see 4401
Vianol see 1583
Viansin [Farmocosmici] see 2132
Viapta [Riker] see 6039
Viarespan [Servier] see 4041
Viarex [Schering] see 1047
Viarox [Schering] see 1047
Viasept see 4503
Vibalt [Roerig] see 10152
Vibazine [Pfizer] see 1484
Vibeline [Bellon] see 10147
Vibramycin [Pfizer] see 3496
Vibramycin Hyclate [Pfizer] see 3496
Vibra-Tabs [Pfizer] see 3496
Vibraveineuse [Pfizer] see 3496
Vibravenös [Pfizer] see 3496
Vibriomycin see 3222
Viburnum Opulus, 10108
Viburnum Prunifolium, 10109
Vi-Cad [Vineland] see 1653
Vicard [Abbott] see 9297

Viccillin [Meiji] see 628
Vicelat [Bayer] see 867
Viceton [Osborn] see 2120
Vicianin, 10110
Vicianose, 10111
Vicilan [ICI] see 10116
Vicin see 10112
Vicine, 10112
Vicin Neolin see 7221
Vicioside see 10112
Victan [Sanofi Winthrop] see 3867
Victoria Green B see 5739
Victoria Green WB see 5739
Vidarabine, 10113
Vidden D [Dow] see 3125
Vi-De-3-hydrosol see 10157
Videne [Riker] see 7880
Videobil [Bracco] see 5077
Videophel see 5059
Viderpen [Sintex] see 5986
Videx [Bristol-Myers Squibb] see 3148
Vidine see 2259
Vidopen [Berk] see 628
Vidora [Wyeth] see 5000
Vigabatrin, 10114
Vigantol [E. Merck; Bayer] see 10157
Vigazoo [Am. Cyanamid] see 9060
Vigil [ICI] see 3131
Vigilor [Bouchard] see 4126
Vigorsan [Albert-Roussel] see 10157
Vikane [Dow] see 9154
Vilan [Lannacher] see 6607
Villiaumite see 8762
Villikinin, 10115
Viloxazine, 10116
Vimicon [Frosst] see 2842
Viminalol see 659
Viminol, 10117
Vinarol [Gen. Dyes] see 7745
Vinbarbital Sodium, 10118
Vinblastine, 10119
Vincadar [Roussel-UCLAF] see 10120
Vincafarm [Radiumfarma] see 10120
Vincafolina [Lampugnani] see 10120
Vincafor [Clin-Comar-Byla] see 10120
Vincagil [Sarsa] see 10120
Vincalen [Firma] see 10120
Vincaleukoblastine see 10119
Vincamajoridine see 200
Vincamidol [Magis] see 10120
Vincamine, 10120
Vincaminine see 10120
Vincamone see 3537
Vincanorine see 3537
Vincapront [Mack, Illert.] see 10120
Vincasaunier [Saunier] see 10120
Vincetoxin, 10121
Vincimax [Robert et Carriére] see 10120
Vincine see 10120
Vincinine see 10120
Vinclozolin, 10122
Vinconate, 10123
Vincosid [AB Leo] see 10124
Vincrex [Bristol] see 10124
Vincristine, 10124
Vindesine, 10125
Vindolidine see 10126
Vindoline, 10126
Vindorosine see 10126
Vinegar Naphtha see 3803
Vinesthene see 10133
Vinethene see 10133
Vinetine see 7087
Vingsal [Stroschein] see 6792
Vinisil [Abbott] see 7879
Vinodrel Retard [Lepetit] see 10120
Vinol [Air Prods. Chem.] see 7745
Vinorelbine, 10127
Vinpocetine, 10128

Vintiamol, 10129
Vinyl Acetate, 10130
Vinylacetonitrile see 298
γ-Vinyl-γ-aminobutyric Acid see 10114
Vinylbenzene see 9028
Vinylbital, 10131
Vinyl Carbinol see 294
Vinyl Chloride, 10132
Vinyl Cyanide see 133
α-Vinyldiacetonalkamine Benzoate see 3939
Vinylestrenolone see 6795
17α-Vinyl-5(10)-estren-17β-ol-3-one see 6795
Vinyl Ether, 10133
Vinylethylene see 1534
Vinylformic Acid see 132
γ-Vinyl GABA see 10114
Vinylidene Chloride, 10134
Vinylidene Cyanide-vinyl Acetate Co-polymer see 2886
5-Vinyl-5-(1-methylbutyl)barbituric Acid see 10131
17α-Vinyl-19-nortestosterone see 6814
(−)-5-Vinyl-2-oxazolidinethione see 4525
1-Vinyl-2-pyrrolidinone Polymers see 7879
1-Vinyl-2-pyrrolidinone Polymers, Iodine Complex see 7880
5-Vinyl-2-thiooxazolidone see 4525
Vinyl Trichloride see 9767
Vinyzene see 7095
Viocid see 4401
Viocin [Pfizer] see 10139
Vio-D [Rowell] see 10156
Vioform [Ciba] see 5052
Vioformio [Ciba-Geigy] see 5052
Viokase [Robins] see 7138
Violacein, 10135
Violanin see 2931
Violaquercitrin see 8456
Violaxanthin, 10136
Viologen, 10137
Violuric Acid, 10138
Viomycin, 10139
Vionactan [Ciba] see 10139
Viophan see 2347
Viosterol see 10156
Viothenat [Grünenthal] see 10139
Vio-Thene [Rowell] see 7107
Viozene see 8415
VIP, 10140
Vipicil [Wyeth] see 2769
Viprynium Chloride see 8206
Viprynium Embonate see 8206
Viquidil, 10141
Vira-A [Parke, Davis] see 10113
Virac see 9982
Viraferon [Schering] see 5016
Viramid [Alfa] see 8365
Viras [Mect] see 8367
Virazene see 7403
Virazid [Brittania] see 8365
Virazole [ICN] see 8365
Virgimycin see 10142
Virginiamycin, 10142
Virginiamycin M_1 see 10142
Virginiamycin S_1 see 10142
Viridicatin, 10143
Viridin, 10144
Viridofulvin see 2960
Viridogrisein see 3755
Virilon [Star] see 9322
Virlix [Karion] see 2063
Virofral [Boehringer, Mann.] see 389
Virophta [Dulcis] see 9816
Viroptic [Burroughs Wellcome] see 9816

Virorax [Siam Bheasach] see 148
Virormone [Ferring] see 9322
Virosterone [Endo] see 9322
Virudox [Bioglan] see 4934
Viru-Merz [Merz] see 9901
Viruserol [Zyma] see 9901
Virusmin [Sumitomo] see 6354
Virustat [Delagrange] see 6354
Viruxan [Newport] see 5006
Viruzona see 6057
Visacor [ICI] see 3641
Visammin see 5321
Viscardan see 5321
Visceralgina [SIT] see 9571
Visclair [Sinclair] see 5828
β-Viscol see 5636
Viscoleo [Lundbeck] see 4224
Viscose, 10145
Viscosin, 10146
Viscotiol [Erba] see 5473
Visha see 122
Visine [Pfizer] see 9358
Vision [Ciba] see 9855
Visipaque [Nycomed] see 5044
Visiren [Jugoremedija] see 6865
Viskaldix [Sandoz] see 2454
Visken [Wander] see 7597
Visnacorin see 10148
Visnadine, 10147
Visnagalin see 5321
Visnagen see 5321
Visnagin, 10148
Visnamine [Chinoin] see 10147
Visotrast see 79
Vistagan [Allergan] see 5488
Vista-Methasone [Richard Daniel] see 1226
Vistamycin [Meiji] see 8373
Vistar [3M] see 5846
Vistaril Pamoate [Pfizer] see 4897
Vistaril Parenteral [Pfizer] see 4897
Vistide [Gilead] see 2329
Visual Purple see 8359
Visubeta see 1226
Visubutina see 7106
Vitabact [Faure] see 7553
Vitaberin [Taisho] see 9432
Vitacarn [Kendall McGaw] see 1898
Vitacee [Endo] see 867
Vitacimin see 867
Vitacin see 867
Vitagutt [Schwarzhaupt] see 10160
Vital Red, 10149
Vital Red Evans see 10149
Vitamin A, 10150
5-cis-Vitamin A see 6555
Vitamin A Acid see 8333
Vitamin A Alcohol see 10150
Vitamin A Aldehyde see 8331
Vitamin A Epoxide see 4686
Vitamin A_1 see 10150
Vitamin A_2, 10151
Vitamin A_2 Aldehyde see 2926
Vitamin B_1 see 9430
Vitamin B_1 O,S-Diacetate see 51
Vitamin B_1 Disulfide see 9432
Vitamin B_1 Propyl Disulfide see 8068
Vitamin B_2 see 8367
Vitamin B_2 Phosphate see 8368
Vitamin B_3 see 6574
Vitamin B_4 see 150
Vitamin B_5 see 7147
Vitamin B_6 Hydrochloride see 8166
Vitamin B_{12}, 10152
Vitamin B_{12}-Co(II) see 10154
Vitamin B_{12} Coenzyme see 2513
Vitamin B_{12}, Radioactive, 10153
Vitamin B_{12a} see 4854
Vitamin B_{12b} see 4854
Vitamin B_{12c} see 4854

Vitamin B_{12d} see 4854
Vitamin B_{12p} see 3914
Vitamin B_{12r}, 10154
Vitamin B_{12s} see 10154
Vitamin B_{13} see 7004
Vitamin B_{15} see 7140
Vitamin B_{17} see 638
Vitamin B_T see 1898
Vitamin B_T Acetate see 85
Vitamin B_x see 443
Vitamin Bc see 4253
Vitamin Bc Conjugate see 8120
Vitamin C see 867
Vitamin C Sodium see 8723
Vitamin D_1, 10155
Vitamin D_2, 10156
Vitamin D_3, 10157
Vitamin D_4, 10158
Vitamin E, 10159
Vitamin E Acetate, 10160
Vitamin E Acid Succinate see 10159
Vitamin F see 3743
Vitamin G see 8367
Vitamin H see 1272
Vitamin K, 10161
Vitamin K Epoxide see 10161
Vitamin K_1 see 7536
Vitamin K_1 Epoxide see 7536
Vitamin K_1 Oxide see 7536
Vitamin K_2 see 5876
Vitamin $K_{2(0)}$ see 5874
Vitamin $K_{2(20)}$ see 5876
Vitamin $K_{2(30)}$ see 5876
Vitamin $K_{2(35)}$ see 5876
Vitamin K_3 see 5874
Vitamin K_4 see 5873
Vitamin K_5, 10162
Vitamin(s) L, 10163
Vitamin L_1 see 442
Vitamin M see 4253
Vitamin P Complex see 1269
Vitamin PP see 6574
Vitamin T, 10164
Vitamin T Goetsch see 10164
Vitamin U, 10165
Vitamogen [Tanabe] see 9432
Vitaneurin [Mepha] see 9430
Vitanevril [Sanofi Winthrop] see 1068
Vitarubin [Streuli] see 10152
Vitascorbol [Specia] see 867
Vitaseptol [Faure] see 9451
Vitasprint B_{12} [Brenner-Efeka] see 7520
VitaStain [Arapahoe] see 9874
Vitas-U see 10165
Vitavax [Uniroyal] see 1874
Vitavel K [Vitamins Ltd] see 5873
Vitazechs see 8163
Vitellin see 5452
Viton [Mallinckrodt] see 5526
Vitpex see 10150
Vitride [Nat. Patent Dev. Corp], 10166
Vitronectin, 10167
Vivactil [Merck & Co.] see 8088
Vival see 3042
Vivalan [ICI] see 10116
Vivarint [ICI] see 10116
Vivatec [Merck & Co.] see 5540
Vividrin [Mann] see 2658
Vividyl [Lilly] see 6812
Vivol [Horner] see 3042
VK-55 see 9075
VLA Antigens see 5014
VLA Integrins see 5014
VLB see 10119
Vleminckx's Lotion see 5516
Vleminckx's Soln see 5516
V.M.&P. Naphtha see 5513
VM-26 see 9291

VMA see 10070
Voacamine, 10168
Voacangine see 10168
Voacanginine see 10168
Vodka, 10169
Vodol [Andromaco] see 6266
Voelicherite see 1741
Vogalene [Thraplix] see 6233
Vogan see 10150
Vogan-Neu see 10150
Vogesensäure (German) see 9236
Volatile Oil of Laurel see 6946
Volatile Oil of Mustard see 303
Volaton [Bayer] see 7523
Voldal [Zyma] see 3132
Volex [McGaw] see 4707
Volidan [BDH; Searle] see 5849
Volital [L.A.B.] see 7206
Volma [Alza] see 217
Volon [Squibb-Heyden] see 9727
Volon A [Squibb-Heyden] see 9728
Volonimat [Squibb-Heyden] see 9728
Volpo [Croda] see 7737
Voltaren [Ciba-Geigy] see 3132
Voltarol [Ciba-Geigy] see 3132
Vomex A [Brocades] see 3252
Vomicine, 10170
Vomit Nut see 6828
Vomitoxin, 10171
Vomit Wort see 5579
Vonamycin Powder V see 6542
Vondozeb [ELF Atochem] see 5756
Vonedrine see 7462
Vontil [SKB] see 9494
Vontrol [SK & F] see 3369
Vonum [Lichtenstein] see 4998
Voranil [USV] see 2472
Voren [Boehringer, Ing.] see 2986
Vorlan [Mallinckrodt] see 10122
VP-16-213 see 3931
VPM see 6016
Vraap [Inverni] see 10120
Vucine see 3760
VUFB-6453 see 6221
VUFB-6638 see 9308
Vulcamicina see 6818
Vulcamycin see 6818
Vulgarobufotoxin see 1503
Vulkacit NPV/C [Mobay] see 3849
Vulkamycin see 6818
Vulkazit see 3383
Vulklor see 2121
Vumide [Burroughs Wellcome] see
   2371
Vumon [Bristol] see 9291
VX, 10172
Vydate [Du Pont] see 7052
Vyrene [U.S. Rubber] see 8882
W see 9929
W-37 see 1500
W-45 see 486
W-50 Base see 6026
W-483 see 3793
W-491 see 2655
W-554 see 3988
W-583 see 5813
W-1191-2 see 388
W-1544a see 7366
W-1889 see 6678
W-2197 see 7280
W-2900A see 3934
W-2946M see 8307
W-2979M see 939
W-3566 see 8239
W-3699 see 7639
W-4020 see 7895
W-4565 see 7079
W-4869 see 7905
W-5219 see 7958
W-5759A see 9578

W-5975 see 1226
W-6309 see 3194
W-6421A see 5488
W-7000A see 5488
W-7320 see 220
W-8495 see 5258
W-19053 see 3907
W-36095 see 9629
WAC-104 see 1267
Wahoo see 3945
Wakaflavin L [Wakamoto] see 8367
Wakoboku see 5738
WAL-801 see 3655
WAL-801CL see 3655
Walconesin see 5895
Waldmeister (German) see 10185
Wampocap [Wallace] see 6612
Wandonorm [Wander] see 1362
Wansar [Hoei] see 3369
Waran see 10174
Warbex [Am. Cyanamid] see 3973
Warburganal, 10173
Warduzide see 2221
Warfarin, 10174
Warfarin-deanol see 10174
WARF Compound 42 see 10174
Warfilone [Frosst] see 10174
Wartec [Kabi] see 7704
Warticon [Perstorp] see 7704
Waruzol [Farmanic] see 891
Washed Sulfur see 9150
Washing Soda see 8739
Water, 10175
Water-d_2 see 2984
Water Bugle see 5654
Water Channel of Collecting Duct see
   798
Water Elder see 10108
Water Gas, 10176
Water Glass see 8824
Water Hemp see 392
Watermelon, 10177
Water Shamrock see 5888
Wavellite see 371
Waxsol [Norgine] see 3460
Waynecomycin [Upjohn] see 5525
WBA-8119 see 1400
WCH-CD see 798
WE-941-BS see 1472
WEB-1881 see 6520
WEB-1881FU see 6520
WEB-2086 see 765
Weedazol [Rhone-Poulenc] see 513
Weed Hoe [Vineland] see 6020
Weedone [Nufarm] see 2865
Weedone [Rhône-Poulenc] see 9194
Wehydryl [Hauck] see 3367
Weifacodine [Weifa] see 7484
Weinsäure (German) see 9237
dl-Weinsäure (German) see 9236
Weinstein see 819
Weinsteinsäure (German) see 9237
Weiselfuttersaft (German) see 8434
Weisspiessglanz see 752
Welfurin [GAP] see 6696
Welkstoff see 5649
Wellbatrin [Burroughs Wellcome] see
   1523
Wellbutrin [Wellcome] see 1523
Wellcome 33-A-74 see 900
Wellcome 248U see 148
Wellcome Prepn 47-83 see 2779
Wellcoprim [Wellcome] see 9840
Wellcovorin [Burroughs Wellcome]
   see 4254
Welldorm [Smith & Nephew] see 3097
Wellferon [Burroughs Wellcome] see
   5016
Wellvone [Wellcome] see 898
Welvic [European Vinyls] see 7746

Wemid [Bernabo] see 3726
Weradys [Weimer] see 5849
Wescodyne [West Chemical] see 5058
Wespuril [Spitzner] see 3120
Westamine X see 5058
Westcort [Westwood] see 4828
Western Poison Oak see 7710
Westrosol see 9769
Weymouth Pine see 10180
WG-253 see 8393
WG-537 see 4171
WG-696 see 9779
WGA see 5453
Wh-7286 see 10213
Wheat Germ Agglutinin see 5453
Wheat Germ Oil, 10178
Whey Factor see 7004
Whip [Hoechst Roussel] see 4024
Whipcide [Pitman-Moore] see 7530
Whiskey see 10179
Whisky, 10179
White Agaric see 183
White Arsenic see 844
White Beeswax see 1049
White Bole see 5294
White Cinnamon see 1791
White Dextrin see 2995
White Flag see 7008
White Lead see 1034
White Mercuric Precipitate see 5927
White Oak see 8220
White Oil of Camphor see 6879
White Phosphorus see 7503
White Pine, 10180
White Precipitate see 5927
White Saunders see 8501
White Spirits see 6288
White Tin Oxide see 8933
White Vitriol see 10293
White Walnut see 5281
White Wax see 1049
Whiting see 1697
Whitlockite see 1741
WHO-4939 see 8310
Whorehouse Tea see 3644
WHR-1142A see 5504
Widdrene see 9534
Widecillin [Meiji] see 617
Wieland-Gumlich Aldehyde, 10181
Wijs' Chloride see 5038
Wild Bergamot see 6326
Wild Black Cherry Bark see 10182
Wild Canilla see 1791
Wild Chamomile see 2082
Wild Cherry, 10182
Wild Cinnamon see 1791
Wild Cotton see 866
Wildfire Toxin, 10183
Wild Ginger see 862
Wild Indigo see 985
Wild Pepper see 6258
Wild Saffron see 2537
Wild Tobacco see 5579
Wild Woodbine see 4391
Wild Yam see 3346
Wilkinite see 1082
Wilkinson's Catalyst see 2226
Willemite see 10291
Willestrol see 3177
Wilpo [Dorsey] see 7415
Wilprafen [Mack, Illert.] see 5280
Win-771 see 4883
Win-1539 see 5312
Win-2747 see 1137
Win-2848 see 9416
Win-3046 see 5185
Win-3706 see 402
Win-4369 see 7268
Win-5047 see 2126
Win 5063-2 see 9436

Xerac BP 5 [Person & Covey] see 1149
Xerac BP 10 [Person & Covey] see 1149
Xerene [Martinet] see 5896
Xeroform [Warner-Chilcott] see 1333
Xibenolol, 10209
Xibornol, 10210
XIFAM see 10211
Ximaol see 6845
Ximoprofen, 10211
Xipamide, 10212
Xitix see 867
XK-62-2 see 6269
XL-7 see 1343
XL-90 see 4582
Xobaline [Albert-Roussel] see 2513
Xolamin [Sanko] see 2405
Xonaltite see 1749
Xonotlite see 1749
X-Otag [Tutag] see 7007
XRD-473 see 4720
"XTC" see 5806
Xtro [Xttrium] see 908
XU 62-320 see 4250
Xumbradil [Astra Pharm.] see 5061
Xuret [Galen] see 6231
Xyduril [Dorsch] see 2436
Xylamide see 7958
Xylan Hydrogen Sulfate see 7276
Xylan Polysulfate see 7276
Xylans see 4675
Xylapan [Chassot] see 10213
Xylasol [Graeub] see 10213
Xylazine, 10213
Xylene, 10214
m-Xylene see 10214
o-Xylene see 10214
p-Xylene see 10214
Xylenol, 10215
as-m-Xylenol see 10215
as-o-Xylenol see 10215
p-Xylenol see 10215
sym-m-Xylenol see 10215
vic-m-Xylenol see 10215
vic-o-Xylenol see 10215
Xylenol Blue, 10216
p-Xylenolsulfonephthalein see 10216
Xylidine, 10217
Xylite see 10218
Xylitol, 10218
Xyliton [Morishita] see 10218
L-Xyloascorbic Acid see 867
Xylocaine [Astra Pharm.] see 5505
Xylocard [Astra] see 5505
Xylocitin see 5505
Xyloidin see 8195
Xylol see 10214
Xylomed [Bio-Medical] see 10220
Xylometazoline, 10219
Xylo-Mucine see 1877
Xylonest [Astra Pharm.] see 7924
Xyloneural [Nicholas] see 5505
Xylo-Pfan [Adria] see 10220
2-[(6-O-β-D-Xylopyranosyl-β-D-gluco-pyranosyl)oxy]benzoic Acid Methyl Ester see 4384
6-O-β-D-Xylopyranosyl-D-glucose see 7926
Xylose, 10220
D-Xylose see 10220
Xylosecarboxylic Acid see 4605
6-(β-D-Xylosido)-D-glucose see 7926
Xylotocan [Astra] see 9629
o-Xylotocopherol see 9633
p-Xylotocopherol see 9632
Xylotox see 5505
Xylulose, 10221
N-(2,3-Xylyl)anthranilic Acid see 5842
1-Xylylazo-2-naphthol, 10222

1-(2,4-Xylylazo)-2-naphthol see 10222
Xylyl Bromide, 10223
m-Xylyl Bromide see 10223
o-Xylyl Bromide see 10223
p-Xylyl Bromide see 10223
[[(2,6-Xylylcarbamoyl)methyl]imino]-diacetic Acid see 5506
Xylyl Chloride, 10224
m-Xylyl Chloride see 10224
o-Xylyl Chloride see 10224
p-Xylyl Chloride see 10224
1-(2,6-Xylyloxy)-2-propylamine see 6257
Xymelin [DAK] see 10219
Xymiazole see 2833
Xynertec [Merck & Co.] see 3605
XZ-450 see 946
Y see 9970
Y-516 see 6365
Y-3642 see 9590
Y-4153 see 2427
Y-5350 see 1887
Y-6047 see 2477
Y-6124 see 1496
Y-7131 see 3919
Y-8004 see 7890
Y-9179 see 6759
Y-12141 see 9711
Y-25130 see 933
Yadulan see 2221
Yageine see 4647
Yal [Ritter] see 3460
Yamacillin [Yamanouchi] see 9204
Yamaful [Yamanouchi] see 1892
Yambolap [Chinoin] see 5090
Yam, Mexican, 10225
Yangonin, 10226
Yara Yara see 6076
Yarmor [Hercules] see 7601
Yarrow see 108
Yatren [Bayer] see 4872
Yaw Root see 8973
Yazumycin A see 8988
Yazumycins see 8988
YB-2 see 4974
YC-93 see 6579
Yeast, 10227
Yeast Adenylic Acid see 156
Yeast Nucleic Acid see 8370
Yellow AB, 10228
Yellow Arsenic Sulfide see 846
Yellow Beeswax see 1049
Yellow Cedar see 9531
Yellow Copper see 2079
Yellow Cross Liquid see 6395
Yellow Cuprocide [Rohm & Haas] see 2734
Yellow Dextrin see 2995
Yellow Dock see 8450
Yellow Gentian see 4399
Yellow Indigo see 985
Yellow Jasmine see 4391
Yellow Jessamine see 4391
Yellow Melilot see 5862
Yellow Mercury Iodide see 5953
Yellow Moccasin Flower see 2839
Yellow Mustard see 6396
Yellow OB, 10229
Yellow Phenolphthalein, 10230
Yellow Phosphorus see 7503
Yellow Precipitate see 5936
Yellow Prussiate of Potash see 7793
Yellow Prussiate of Soda see 8760
Yellow Puccoon see 4808
Yellow Resin see 8424
Yellow Root see 4808
Yellow Saunders see 8501
Yellow Sweet Clover see 5862
Yellow T see 9905
Yellow Ultramarine see 1700

Yerba Maté see 5796
Yerba Santa see 3711
Yermonil [Geigy] see 5659
YH-1 see 7206
Yig, 10231
Yinyanghuo see 3653
YL-704B1 see 6271
Ylangene, 10232
α-Ylangene see 10232
Ylang-Ylang Oil, 10233
Ylides, 10234
YM-617 see 9217
YM-08054-1 see 4972
YM-08310 see 424
YM-09151-2 see 6533
YM-09330 see 1984
YM-09538 see 614
YM-09730-5 see 1031
YM-11170 see 3972
YM-12617 see 9217
YM-12617-1 see 9217
YM-12617-2 see 9217
YM-14090 see 5016
YN-72 see 8875
Yobir [Maruko] see 321
Yochinol see 4872
Yocon [Palisades] see 10236
Yodoxin [Searle] see 5063
Yoghurt see 10235
Yogurt, 10235
Yohimbine, 10236
allo-Yohimbine, 10237
α-Yohimbine, 10238
δ-Yohimbine see 8291
Yohimex [Kramer] see 10236
Yohydrol [Riedel-Zabinka] see 10236
Yomesan [Bayer] see 6602
Yoristen [Kodama] see 4260
Yoshi 864 see 4962
Yoshicol [Yoshitomi] see 7018
Yoshimilon [Kowa Yakuhin] see 9835
Youthwort see 3508
Yperite see 6395
YTR-830 see 9251
YTR-830H see 9251
Ytterbia see 10239
Ytterbium, 10239
Yttria see 10240
Yttrium, 10240
Yttrium Iron Garnet see 10231
Yulocrotine see 5283
Yutopar [Astra] see 8401
Yxin [Pfizer] see 9358
Z-103 see 7712
Z-326 see 4048
Z-424 see 10117
Z-876 see 3184
Z-905 see 7596
Z-4828 see 9897
Z-4942 see 4937
Zabromin [Sandoz] see 1473
Zackal [SS Pharm.] see 4385
Zactane [Wyeth] see 3789
Zadine [Schering] see 934
Zadipina [Zambeletti] see 6658
Zaditen [Sandoz] see 5319
Zadorin [Mepha] see 3496
Zadstat [Lederle] see 6242
Zafirlukast, 10241
Zagam [Specia] see 8884
Zalain [Robert] see 8610
Zalcitabine, 10242
Zaldaride, 10243
Zaltoprofen, 10244
Zamanil see 7107
Zamocillin [Inpharzam] see 617
Zanchol [Searle] see 4143
Zanil see 7092
Zanizal [Italfarmaco] see 6758
Zanosar [Upjohn] see 8991

Zantac [Glaxo] see 8286
Zantic [Glaxo] see 8286
Zantrène [Lederle] see 1284
Zaratite see 6586
Zarontin [Parke, Davis] see 3794
Zaroxolyn [Fisons] see 6231
Zasten [Sandoz] see 5319
Zatebradine, 10245
Zavedos [Farmitalia] see 4931
Zaxopam [Quantum] see 7059
ZD-1033 see 667
ZD-1694 see 9684
Z-DNA see 2954
ZE-101 see 6758
Zearalanol see 10251
Zearalenone, 10246
Zeatin, 10247
trans-Zeatin see 10247
Zeaxanthin, 10248
Zeaxanthin Diepoxide see 10136
Zeaxanthol see 10248
Zebedassite see 362
Zebeta [Lederle] see 1336
Zeclar [Abbott] see 2400
Zectran see 6253
Zedolac [Maggioni-Winthrop] see 3920
Zefazone [Upjohn] see 1977
Zein, 10249
Zeldox [Zeltia] see 4751
Zelis [Proter] see 7661
Zellner's Paper see 4195
Zelmid [Astra] see 10254
Zemide [Wyeth] see 9216
Zemuron [Akzo] see 8407
Zen [Prophin] see 7661
Zenadrex [Veterinaria] see 2930
Zenoxone [Biorex] see 4828
Zentel [SK&F] see 211
Zentropil [Nordmark] see 7475
Zenusin [Mepha] see 6617
Zeolites, 10250
Zepelin [De Angeli] see 4053
Zepharovicht see 371
Zephiran Chloride [Winthrop] see 1086
Zephirol [Bayer] see 1086
Zephyr [Merck & Co.] see 1
Zepolas [Mikasa] see 4234
Zeranol, 10251
Zerit [Bristol-Myers Squibb] see 8958
Zestril [ICI] see 5540
Zesulan [Toyo Jozo] see 5911
Zetamicin [Menarini] see 6563
Zeta Protein see 4119
Zettyn see 2059
Ziac [Am. Cyanamid] see 1336
Ziavetine see 1500
Zibeth see 2395
Zibren [Duncan] see 85
Zidovudine, 10252
Zienam see 4954
Zildasac [Chugai] see 1061
Zileuton, 10253
Zimate see 10305
Zimecterin [Merck & Co.] see 5264
Zimeldine, 10254
Zimelidine see 10254
Zimovane [M & B] see 10324
Zinacef [Glaxo-Hoechst] see 2002
Zinamide [Merck & Co.] see 8140
Zinat [Glaxo] see 2002
Zinc, 10255
Zinc Acetate, 10256
Zinc Acexamate see 43
Zinc Alum see 385
Zinc Ammonium Chloride see 573
Zincaps [Protea] see 10293
Zincate see 10293

"Zincates" see 10255
Zinc Bacitracin, 10257
Zinc Bis(dimethylthiocarbamoyl) Disulfide see 10305
Zinc Blende see 10294
Zinc Bromide, 10258
Zinc Caprylate, 10259
Zinc Carbonate, 10260
Zinc Carbonate Hydroxide see 10260
Zinc L-Carnosine see 7712
Zinc Chloride, 10261
Zinc Chromate(VI) Hydroxide, 10262
Zinc Citrate, 10263
Zinc Cyanide, 10264
Zinc Diethyl see 3179
Zinc Dimethyl see 3312
Zinc Dimethyldithiocarbamate see 10305
Zinc Ethylenebis(dithiocarbamate) see 10300
Zinc Fluoride, 10265
Zinc Fluosilicate see 10267
Zinc Formate, 10266
Zinc Hexafluorosilicate, 10267
Zinc p-Hydroxybenzenesulfonate see 10283
Zinc Insulin see 5011
Zinc Iodate, 10268
Zinc Iodide, 10269
Zinc Iodide-Starch, 10270
Zincite see 10279
Zinc Lactate, 10271
Zinc Manganese Ethylenebisdithiocarbamate see 5756
Zinc Meta-arsenite, 10272
Zinc Nitrate, 10273
Zinc Nitride, 10274
Zinc Nitrite, 10275
Zinc Oleate, 10276
Zincomed [Medo] see 10293
Zinc Ortho-arsenate, 10277
Zinc Orthosilicate see 10291
Zinc Oxalate, 10278
Zinc Oxide, 10279
Zinc Perchlorate, 10280
Zinc Permanganate, 10281
Zinc Peroxide, 10282
Zinc p-Phenolsulfonate, 10283
Zinc Phosphate, 10284
Zinc Phosphide, 10285
Zinc Potassium Iodide see 7871
Zinc Potassium Sulfate see 7877
Zinc Propionate, 10286
Zinc 1,2-Propylene Bisdithiocarbamate see 8004
Zinc Pyridinethione see 8178
Zinc Pyrithione see 8178
Zinc Pyrophosphate, 10287
Zinc Salicylate, 10288
Zinc Selenate, 10289
Zinc Selenide, 10290
Zinc Silicate, 10291
Zinc Silicofluoride see 10267
Zincspar see 10260
Zinc Stearate, 10292
Zinc Subcarbonate see 10260
Zinc Sulfanilate Tetrahydrate see 9096
Zinc Sulfate, 10293
Zinc Sulfide, 10294
Zinc Sulfocarbolate see 10283
Zinc Sulfocyanate see 10298
Zinc Sulfophenate see 10283
Zinc Superoxide see 10282
Zinc Tannate, 10295
Zinc Tartrate, 10296
Zinc Telluride, 10297
Zinc Thiocyanate, 10298
Zinc Undecylenate see 9983
Zinc Valerate, 10299

Zinc Vitriol see 10293
Zinc White see 10279
Zinc Yellow see 10262
Zineb, 10300
Zinecard [Adria] see 8295
Zingerone, 10301
Zingherone see 10301
Zingiberone see 10301
Zinnat [Glaxo] see 2002
Zinophos [Am. Cyanamid] see 9482
Zinostatin, 10302
Zinviroxime see 3637
Zipeprol, 10303
Ziprasidone, 10304
Ziram, 10305
Zirberk see 10305
Zircon see 10314
Zirconia see 10313
Zirconic Anhydride see 10313
Zirconium, 10306
Zirconium Chloride, 10307
Zirconium Dioxide see 10313
Zirconium Fluoride, 10308
Zirconium Hydride, 10309
Zirconium Hydroxide, 10310
Zirconium Iodide, 10311
Zirconium Nitrate, 10312
Zirconium Orthosilicate see 10314
Zirconium Oxide, 10313
Zirconium Oxychloride see 10317
Zirconium Potassium Fluoride see 7804
Zirconium Potassium Sulfate see 7878
Zirconium Silicate, 10314
Zirconium Sulfate, 10315
Zirconium Tetrachloride see 10307
Zirconium Tetrafluoride see 10308
Zirconium Tetraiodide see 10311
Zirconyl Acetate, 10316
Zirconyl Chloride, 10317
Zirtek [Allen & Hanburys] see 2063
Zithromax [Pfizer] see 946
Zitoxil [Farmochimica] see 10303
Zitrix [Gibipharma] see 1965
Zitromax [Pfizer] see 946
Zixoryn [Gedeon Richter] see 4170
Zizanal see 6953
ZK-30595 see 3510
ZK-31224 see 9308
ZK-36374 see 4940
ZK-39482 see 5082
ZK-57671 see 9165
ZK-62711 see 8410
ZK-76604 see 7643
ZK-93035 see 4347
ZK-112119 see 2
ZL-101 see 6758
ZMA see 10272
ZN-6 see 4340
Zoalene [Dow] see 3321
Zoamix [Dow] see 3321
Zoapatanol, 10318
Zoaquin see 5063
Zocor [Merck & Co.] see 8686
Zocord [Merck & Co.] see 8686
Zofran [Glaxo] see 6979
Zoisite see 1685
Zoladex [ICI] see 4547
Zolamine, 10319
Zolicef [Lappe] see 1967
Zolimidine, 10320
Zoliridine see 10320
Zoloft [Pfizer] see 8612
Zolone [Rhône-Poulenc] see 7489
Zolpidem, 10321
Zolyse [Alcon] see 2320
Zomax [McNeil] see 10322
Zomaxin [Cilag] see 10322
Zomepirac, 10322
Zondel [Grelan] see 6792

Zongorine *see* 8865
Zoniden [Irbi] *see* 9595
Zonisamide, 10323
Zooecdysteroids *see* 3539
Zoolobelin *see* 5580
"Zoomelanoidic" Acids *see* 4929
Zophren [Glaxo] *see* 6979
Zopiclone, 10324
Zopirac [Sintyal] *see* 10322
Zopolrestat, 10325
Zorial [Sandoz] *see* 6794
Zoroxin [Merck & Co.] *see* 6793
Zorubicin, 10326
Zostrix [GenDerm] *see* 1811
Zosyn [Lederle] *see* 9251
Zotepine, 10327
Zothelone *see* 3416
Zoton [Lederle] *see* 5373

Zovirax [Burroughs Wellcome] *see* 148
Zoxamin *see* 10328
Zoxazolamine, 10328
Zoxine *see* 10328
ZPO *see* 10282
ZR-512 *see* 4850
ZR-515 *see* 6063
ZR-777 *see* 5326
ZR-3210 *see* 4249
Z Span [SKB] *see* 10293
Zuclomiphene *see* 2446
Zuclopenthixol *see* 2455
Zumaril [Abbott] *see* 220
Zumenon [Duphar] *see* 3746
Zunden [Luitpold] *see* 7661
ZY-15021 *see* 1144

ZY-15028 *see* 2419
ZY-17617B *see* 10243
Zydol [Searle] *see* 9701
Zygadenine, 10329
Zygomycin A *see* 7173
Zyklolat [Mann] *see* 2815
Zylonite *see* 2011
Zyloprim [Burroughs Wellcome] *see* 287
Zyloric [Burroughs Wellcome] *see* 287
Zymafluor [Zyma] *see* 8762
Zymofren [Specia] *see* 796
Zymoplastic Substance *see* 9527
Zymosan, 10330
Zypanar [Armour Pharm.] *see* 7137
Zyrlex [UCB] *see* 2063
Zyrtec [UCB] *see* 2063
Zytron [Dow] *see* 3453
Zyxorin [Gedeon Richter] *see* 4170

NOTES

NOTES

NOTES

NOTES

NOTES

NOTES

NOTES

NOTES

NOTES

NOTES

NOTES

NOTES

NOTES

NOTES

ATOMIC WEIGHTS

(Alphabetical Order)

Element	Symbol	Atomic number	Atomic weight	Element	Symbol	Atomic number	Atomic weight
Actinium	Ac	89	227.0278*	Neodymium	Nd	60	144.24
Aluminum	Al	13	26.981539	Neon	Ne	10	20.1797
Americium	Am	95	243.0614*	Neptunium	Np	93	237.0482*
Antimony	Sb	51	121.760	Nickel	Ni	28	58.6934
Argon	Ar	18	39.948	Niobium	Nb	41	92.90638
Arsenic	As	33	74.92159	Nitrogen	N	7	14.00674
Astatine	At	85	209.9871*	Nobelium	No	102	259.1009*
Barium	Ba	56	137.327	Osmium	Os	76	190.23
Berkelium	Bk	97	247.0703*	Oxygen	O	8	15.9994
Beryllium	Be	4	9.012182	Palladium	Pd	46	106.42
Bismuth	Bi	83	208.98037	Phosphorus	P	15	30.973762
Boron	B	5	10.811	Platinum	Pt	78	195.08
Bromine	Br	35	79.904	Plutonium	Pu	94	244.0642*
Cadmium	Cd	48	112.411	Polonium	Po	84	208.9824*
Calcium	Ca	20	40.078	Potassium	K	19	39.0983
Californium	Cf	98	251.0796*	Praseodymium	Pr	59	140.90765
Carbon	C	6	12.011	Promethium	Pm	61	144.9127*
Cerium	Ce	58	140.115	Protactinium	Pa	91	231.0388*
Cesium	Cs	55	132.90543	Radium	Ra	88	226.0254*
Chlorine	Cl	17	35.4527	Radon	Rn	86	222.0176*
Chromium	Cr	24	51.9961	Rhenium	Re	75	186.207
Cobalt	Co	27	58.93320	Rhodium	Rh	45	102.90550
Copper	Cu	29	63.546	Rubidium	Rb	37	85.4678
Curium	Cm	96	247.0703*	Ruthenium	Ru	44	101.07
Dysprosium	Dy	66	162.50	Samarium	Sm	62	150.36
Einsteinium	Es	99	252.083*	Scandium	Sc	21	44.955910
Erbium	Er	68	167.26	Selenium	Se	34	78.96
Europium	Eu	63	151.965	Silicon	Si	14	28.0855
Fermium	Fm	100	257.0951*	Silver	Ag	47	107.8682
Fluorine	F	9	18.9984032	Sodium	Na	11	22.989768
Francium	Fr	87	223.0197*	Strontium	Sr	38	87.62
Gadolinium	Gd	64	157.25	Sulfur	S	16	32.066
Gallium	Ga	31	69.723	Tantalum	Ta	73	180.9479
Germanium	Ge	32	72.61	Technetium	Tc	43	97.9072*
Gold	Au	79	196.96654	Tellurium	Te	52	127.60
Hafnium	Hf	72	178.49	Terbium	Tb	65	158.92534
Helium	He	2	4.002602	Thallium	Tl	81	204.3833
Holmium	Ho	67	164.93032	Thorium	Th	90	232.0381
Hydrogen	H	1	1.00794	Thulium	Tm	69	168.93421
Indium	In	49	114.818	Tin	Sn	50	118.710
Iodine	I	53	126.90447	Titanium	Ti	22	47.867
Iridium	Ir	77	192.217	Tungsten	W	74	183.84
Iron	Fe	26	55.845	Unnilquadium	Unq	104	261.11*
Krypton	Kr	36	83.80	Unnilpentium	Unp	105	262.114*
Lanthanum	La	57	138.9055	Unnilhexium	Unh	106	263.118*
Lawrencium	Lr	103	262.11*	Unnilseptium	Uns	107	262.12*
Lead	Pb	82	207.2	Uranium	U	92	238.0289
Lithium	Li	3	6.941	Vanadium	V	23	50.9415
Lutetium	Lu	71	174.967	Xenon	Xe	54	131.29
Magnesium	Mg	12	24.3050	Ytterbium	Yb	70	173.04
Manganese	Mn	25	54.93805	Yttrium	Y	39	88.90585
Mendelevium	Md	101	258.10*	Zinc	Zn	30	65.39
Mercury	Hg	80	200.59	Zirconium	Zr	40	91.224
Molybdenum	Mo	42	95.94				

Based on 1993 IUPAC Table of Standard Atomic Weights of the Elements.
* Relative atomic mass of the isotope of that element with the longest known half-life.